Your Place In The Cosmos
Volume IV

A layman's book of Astronomy
and
The Mythology
of the eighty-eight celestial constellations
and
Registry IV

by

James Edmund Magee

Mosele & Associates, Inc.
551 Edens Lane
Northfield, Illinois 60093
U.S.A.

Library of Congress Catalog Number 84-63015

ISBN No. 0-9614354-3-7

Printed in the United States of America

Preface

This is neither a dictionary of stars nor a road map through the heavens. In the preface to his dictionary, Samuel Johnson wrote:

> "To search was not always to be informed. . . . Dictionaries are like watches; the worst is better than none and the best cannot be expected to go quite true."

No, not a dictionary, nor a map; this book is a glance at the world of astronomy. A primer for the person who knows he ought to be more familiar with the universe in which we live. It is as much our universe as our home, our town, our country.

And what is a home? As quoted by Robert Frost.

"Home is where, when you go there, they have to let you in."

This is your home. This little speck of dust called Earth. But it is also your solar system. You are as much the rightful owner as anyone else in the universe.

For as much as we know right now, this is your galaxy. No one can come along and start making new laws about how to run it. To the best of our knowledge, no one here on Earth or out there in space has any way of altering the present laws or making new ones. The natural laws are already in force, and only the legislator of those laws of nature can revoke or suspend them.

So let's have a look around our home, our solar system, our galaxy, our universe. Might as well get to know the place because, like it nor not, you and I are stuck with it for the next 60 to 100 billion years or thereabouts. As the man said, "It may not look like much to you, but what the heck, it's home."

James E. Magee

Contents

Credits James E. Magee
Jane Johnson Nelson
Bob Ritt

Illustrations Johannes Hevelius

Introduction

This book is about common sense. It is also about astronomy, astrology, astrophysics, mythology, philosophy, and cosmology. By implication, it is also about metaphysics, which, in the truest definition, means the study of the nature of the universe. But basically it is about the universe and your place in it.

Taken one step at a time, each of the sciences used in this book has a long, scholarly history of intellectual application, spent by minds that have changed the course of human thought and experience.

Just sitting where you are, reading these words, you are comprised of minute particles which were set in motion at the beginning of time. Everything that has ever happened, in a very real sense, happened to you, for you are one of the very few alive for which it all came to pass. Perhaps you have already guessed that. Just as John Donne wrote: ". . . do not send to find for whom the bell tolls. It tolls for thee."

Since intuitive people throughout history have speculated about this thought, we also can consider the art and fantasy of astrology. Since man in his infinite egoism believes himself to be the center of the universe, he has also determined that all the stars, planets, space and even time itself existed solely for the purpose of providing a place of existence just for him. For man, as comedian George Carlin might say, the Earth is nothing more than "a place for his stuff." And the universe around him is only a place to go when he goes outside. . ."really outside!"

So while you have a place in the cosmos, you are also a part of the cosmos. You already are in the stars as you read these words. Like the space ship, which is what each star and planet is, you are moving so fast that romantic philosophers refer to it as "unfolding in time."

Before we embark on a search for an understanding of your place in the cosmos, let us locate you now, wherever you are as you read these words. To pick you out we have to aim at a moving target. In fact, as you will soon discover, the whole business is very moving.

Right now you are probably in the northern hemisphere of a tiny, egg-shaped speck of galactic dust called Earth. You, personally, are moving in a spin from west to east at 860 miles per hour: faster if you are in the south, because there is a bulge at the equator. You are also moving in orbit around the sun at 66,661 miles an hour, and the sun is carrying the earth and its sister planets outward and away from each other at about 31,550 miles an hour.

The sun moves around our galaxy at the incredible speed of 700,000 miles an hour, and the entire galaxy itself rotates at 559,450 miles per hour. And there is more. Our galaxy moves relative to all other galaxies as the entire galactic system speeds blindingly through the universe at more than one million miles per hour.

So you are dizzily moving in six separate directions at a speed in excess of two-and-one-half million miles an hour through time- . . .which in itself travels as a continuum.

As you sit there, speeding along at 2.5 million miles an hour, imagine yourself looking upward between the two farthest stars you can see. Imagine infinite space beyond that, and then ask yourself why you become impatient when the waitress brings you grits instead of hash brown potatoes with your breakfast.

The Sun and Moon are part of our home. The planets in our solar system are wanderers or *goats*. Man, in his desperate need to offset his fears about his destiny, used these planets to explain all the things over which he had no control. Thus mythology often was closely related to theology. Theology in turn was very often related to philosophy. Mythology is the study of myths. Theology is the study of the ultimate through the light of scriptures or revelation. Philosophy is the study of the ultimate through the light of human reason.

The nine bodies (once there seemed to be only seven planets) in our solar system move enough to be blamed by the ancients for that which otherwise could not be understood. They represented gods with the power to intervene in life. In the astronomical identifications established in ancient Babylon, a complete set of gods had been named and characterized in the first pantheon. Each god has attributes with a particular trait, character, adventure, or personality. For instance, even in 5000 B.C. the red and fiery planet Mars had been associated with violence and war. Jupiter was majestic, royal, authoritative. Saturn was considered remote, brooding, cool, but also very capable of losing his temper without warning and striking out in cruel, inexplicable action. Mercury was quick, wise and unpredictable.

Going back to the ice age, carvings on bone indicated man was using the stars and planets to predict changes 32,000 years ago. Relatively recently, in 4200 B.C., Egyptian star charts were plotted

to predict the future from thence forward.

Indeed, until the age of reason in the last 400 years or so, astrology and astronomy were considered two parts of the same science. But what a difference a few centuries can make. . .along with a man named Johannes Kepler. We shall hear more of him later.

It was only recently, in astrophysical terms only a split second ago, 1925 in our reckoning, that Edwin Hubble discovered that the farther away a galaxy is, the faster it is receding. We measure the receding speed of celestial objects by observing the red shifts in the spectrum, a method which works no matter how far away the galaxy. Galactic distances have turned out to be enormous, indeed, astronomical.

Our nearest galaxy is the great Magellanic Nebula, but the nearest one like our own Milky Way which we can observe with the naked eye in the northern hemisphere is Andromeda. Yet it's so far away its light by which we now see it has been traveling for 2.2 million years at more than 186,000 miles per second, the speed of light, to reach us. When we look at Andromeda we are therefore looking back 2.2 million years! On the other side of the universe we can see the constellation Hercules 500 million light years away. When you realize only one light year is more than 9 trillion miles away, 500 million light years adds up to quite a trek!*

You may ask why nature or the legislator of the laws of nature, would find it necessary to speed things up simply because they are further away. Because—get ready—that is the only way it *could* work if the universe is expanding uniformly. One sure thing we do know about the laws of nature: they *are* uniform. Even the nature of chaos probably has an element of uniformity about it, and vice versa.

Look down a football field crossed by a line of people spaced two feet apart. Imagine that you are standing in the middle of the field. The field size is then doubled. If you are in the middle, the people immediately on either side of you move four feet away from you. But the ones at the edges of the field, who originally were 50 feet away, are now 100 feet away. In effect, those farthest away from you actually moved faster because they were farthest away. The same with a galaxy. The farther away it is, the faster it moves away from you.

Another way to say it is that while you stayed in the middle of the line across the field, those furthest away receded at a faster speed. Like a loaf of bread which expands uniformly, the dough nearest the outside moves farther and faster than the dough in the center of the loaf.

How is the speed determined for an outer galaxy? To measure a change in our nearest galaxy, Andromeda, would require 500 years of photograph trace. The astronomer Slipher learned, however, that when a light source moves away from an observer the waves stretch, or are lengthened, by the motion of the receding object. These waves are seen as color by the eye; short waves are blue, long waves are seen as red.

A prism is used to measure the reddening effect in a phenomenon known as the *red shift*. This is the basic tool for providing measurement of the expanding galaxy. If you agree that everything in the universe expands uniformly, you can apply this same measure to the movement and direction in which the universe is traveling.

Like so many discoveries in science, Slipher's was accidental. At the request of Percival Lowell, director of the famous Lowell Observatory in Flagstaff, Arizona, Slipher was studying the Andromeda nebula which Lowell thought might only be a solar system like ours in birth, rather than a galaxy much like the Milky Way. Slipher began by looking for the rotation which characterizes our solar system. During his search, he discovered that the entire galaxy was moving at 700,000 miles per hour relative to Earth. Looking at other galaxies, he realized that all were moving away from Earth. He had discovered the first proof that the entire observable universe was expanding.

He reported his findings on the motion of galaxies at a meeting of the American Astronomical Society in 1914. When he was finished, the entire gathering rose and gave him a standing ovation. It was later reported that perhaps less than half of the 76 scientists in attendance understood the real significance of his report, yet everyone was overwhelmed by the fact that it seemed very important.

Largely, however, we can agree that there are no privileged observers, and that we can use a set point in time although it's not really a constant. The speed of light and the basic laws of gravity are the same everywhere, and neither the laws nor the constants are changing with time. Yet, responsible cosmologists, students of the Cosmos, agree we live in an expanding universe which changes as it ages. One paradox is an example of how one can extrapolate cosmologists' thinking into absurdity. It goes as follows.

The dark night sky provides us with an important observational

* For those who like to compute this sort of thing: Hubble's Constant = the rate at which velocity increases with distance: 75km/sec/maga par sec – 23 km/sec/per million light years.

fact about our universe: The sky is dark except where there are stars. Suppose the universe stretched into infinity. It's fair to assume that the universe is uniformly populated with stars and galaxies. Because the light intensity received from any one star or galaxy will be inversely proportional to the square of its distance from us, the faraway galaxies contribute little light on an individual basis. But the total number of visible galaxies should increase as the cube of the distance, since the volume of space surveyed increases as the cube of the distance. But when we add up the total light received from all the galaxies in an inifinite and stationary universe, we find it is infinity. The dimming of the light from remote (and faint) galaxies by the distance is exactly cancelled out by much larger numbers of galaxies in the bigger volume of space. Therefore, if the argument is right, the night sky should have a blinding intensity. That, of course, is an absurd conclusion because we can see it isn't that bright at all.

The fact that the sky is dark at night shows that the assumption is wrong. In coming to the above conclusion, our mistake was in omitting the fact that the universe is expanding. Faraway parts of it send us red-shifted light, which has lost some of its energy over and above that diminished according to the inverse square law. The dark night sky, therefore, is evidence that we live in an expanding universe.

Radio astronomy has provided us with vital clues about the nature of the universe. The information gap closes more and more every day because galaxies are at such great distances, some of them as much as 100,000,000 million light years from Earth. Their ancient light tells a unique story.

Since galaxies are the most distant objects known to us, we can study history—astronomical history—first hand. We can look right at it as it happens. So when you think of the old science fiction stories about time machines, realize they are not idle ideas. Time travel *is* within the realm of scientific possibility!

It has been found that the density of radio sources at first increases with the distance from the Milky Way. Then, at very great distances, beyond the reach of telescopes, the density declines. That tells us that the density of radio sources in the universe varies with time. In other words, there are more bodies and galaxies now than there were only 10,000 million years ago. Thus we know that as the universe expands, it becomes more heavily populated, with new stars being created every millenia or so.

Radio astronomists, have also detected a background of radiation that is uniform throughout all of the observable universe. It matches the continuous spectrum of a body with a temperature of 2.7K. This signal is believed to be the relic of the universe when it was very much younger, hotter, and denser. Some maintain it is a kind of echo from the Big Bang.

Does the universe look the same from every observation point? If we were standing on Alpha Centauri would it look the same to us as it does from here? The answer seems to be yes. Wherever we look, the density of galaxies and clusters is uniform. Background radiation is also uniform, and the distant radio sources we detect and study are uniformly distributed throughout space.

So much for observational cosmology. We now turn to the other side, the objective of theoretical cosmology is to explain why the universe looks as it does.

In pre-Einstein days it was thought that everything was in a *steady state* because things were then what they always have been and will be. Another way of saying it is that the universe is infinite, never having had a beginning and destined never to end but to continue to stretch over an infinite amount of time. Steady state cosmology proposed that new matter appeared spontaneously to fill the gaps between expanding galaxies. That the density of radio sources varies with distance on one hand, and that the universe is filled with a generally weak radiation on the other hand, was in disagreement with the whole principle of the steady state theory.

The alternative to this, of course, is the *Big Bang* theory. The *Big Bang* theorizes that the universe once was very small, extremely dense, and consequently extremely hot. This is sometimes referred to as the primeval atom. Background radiation is merely a kind of echo from that first extremely intense expansion or explosion. This expansion caused the observed recession of galaxies throughout the universe. All of this, though it seems to answer the central question of creation, leads the cosmologist to ask himself questions. What preceded the Big Bang? How long will the universe continue to expand? And, how much invisible matter such as black holes is yet to be discovered?

In a way, the astrologer and the cosmologist have something in common. Where the astrologer once based his erroneous conclusions on observance of physical evidence, the modern cosmologist observes that physical evidence and interprets his conclusions based upon it. For instance, we have physical evidence of black

holes, but what can we conclude from them? We are collecting evidence and speculating about the validity of the conclusions based on it.

In a word, the cosmologist studies the cosmos. What exactly is that? In his book, "Cosmos", Carl Sagan says, "The Cosmos is all that ever was or ever will be."

In view of this definition, imagine the kind of ego it takes to call oneself a cosmologist. What was. . .is. . .or will be? That doesn't leave out much, does it?

Maybe the cosmos is not everything, but we know it comes pretty close to being just that. The exquisite balance and infinitely intricate interrelationships which are inextricably interwoven in God's loom uncover greater mysteries every day. Observation continues to demand answers which boggle the greatest minds. It still has us making discoveries about the mind of the prime mover, the uncaused cause, the supreme being. . .God.

We learn more and more every day about how childlike we really are. Should it come to pass that we think we know it all, we have only to wait until nightfall and gaze at the stars. It is then that we realize we really don't know very much at all. That is the cosmologist's first lesson: we have yet to learn so much.

We will get back to the causes of the universe again. It is, needless to say, a rather big question. Astronomical, in fact.

Hocce Hemisphaerium Firmamenti Sobiescianum Boreale
Diversis novis Asterismis, et quam plurimis novis Stellis
exornatum atque auctum Invictissimo IOHANNI III
Polonarum Regi etc Dno Suo Clementissimo, ad immortalem Gloriam ac perpetuam Ejus Memoriam, ob Heroicas
Suae Reg. Mttis Virtutes, et ob Res sub eo Hemisphaerio
fortiter in maximum Commodum, et Incrementum
Christianae Religionis gestas, ac felicissime peractas
In lucem prodit Anno Aerae Christianae
Auctoris Sumptibus et Curâ.
Pictore verò Andr Stech et
Sculptore Carol de la Haye
Gedani
Magnitudines Stellarum
Cancer
Pollux
Castor
Taurus
Auriga
Lynx
Ursa Major
Perseus
Musca
Triangulum
Triangulum Majus
Aries
Leo
Leo Minor
Camelopardalus
Cassiopea
Cepheus
Ursa Minor
Andromeda
Coma Berenices
Asterion
Chara
Draco
Bootes
Lacerta
Cygnus
Lyra
Pegasus
Tropicus
Serpens
Vulpecula
Delphinus
Aquarius
Hercules
Cerberus
Aquila
Æquator
Equuleus
Serpentarius
Antinous
Libra
Capricornus
Sobiesc
Sagittarius

Astrological Aspects

Silently one by one, in the infinite meadows of heaven, blossomed the lovely stars, the forget-me-nots of the angels.

–Longfellow

The philosopher and one of the founders of the Astrological Association, John Addy, was convinced that the rhythm of the universe affected everything within the cosmos. Who can argue that the tides of day and night do not affect us? We will.

The history of planetary astronomy may be divided into three periods of time. The first began when man discovered that the planets were not lights, but bodies in the sky. This period lasted until the 16th century. The second period lasted from the 17th through the 19th centuries, and the third began in the 20th century and is still unfolding.

In all that time we have yet to dispense with the idea that planets have a direct and predictable effect on our lives and relationships with others. It is like spinach or tomatoes. At various times they have been considered poisonous, nutritional, or merely decorative.

For reasons which can only be described as very human, astrology persists as a quasi-serious study. Many people are also convinced that men have one more rib than women. Why do these thoughts continue for centuries when the proof is simply in the counting?

It is easy to understand how astronomy and astrology overlapped and intertwined. Unfortunately, whenever the astrologers left the reality of the present and embarked on predictions about the future they receded from the realm of inductive reasoning and crossed into the world of deductive speculation, a most unsuitable place for an astronomer, who is primarily an empirical physical scientist. But who can blame the astrologers—or the poets—for singing of stars? As Byron said, "Stars are the poetry of heaven." Or to quote Carlyle:

"When I gaze into the stars, they look down on me with pity from their serene and silent spaces, like eyes glistening with tears over the little lot of man.

Thousands of generations, all as noisy as our own, have been swallowed up by time, and there remains no record of them any more. Yet Arcturus and Orion, Sirius and Pleiades, are still shining in their courses, clear and young as when the shepherd first noted them in the Plain of Shinar."

For imperious astronomers and other empirical scientists who have decided astrology is of no importance whatsoever in the modern world, let us point out that they are wrong in at least one somewhat remote respect. Consider the career of Karl Ernst Krafft, the Swiss astrologer who became a Nazi, and the effect he had on the tides of our time. He successfully predicted the first attempt on Hitler's life. Because of this, he was accused of complicity in the plot. When he was cleared he was employed by the Gestapo to translate the prophesies of Nostrodamus at the suggestion of Frau Goebbels who thought they could be of value when used in Nazi propaganda. Many on the general staff, and later Hitler himself, read his findings avidly. Right or wrong, Krafft had a considerable effect on the course of public opinion and, some said, on the eventual course of the war.

Then there was Ludwig VonWohl, who persuaded the British government that Karl Krafft was influencing Hitler and in turn the entire German general staff. In other words, in World War II both the British and the Germans used astrology as an instrument of intelligence weaponry.

It may be said that though astrology itself has no effect on the way the world turns, it must be admitted that as long as people *think* it does, it does!

On the other side of the celestial coin was the great astronomer Johannes Kepler. A genius, Kepler was the imperial mathematician and court astronomer who succeeded the great Tycho Brahe. In 1601 he disproved his predecessor's calculations that the Earth was the center of a perfect solar system each of whose planets traveled in a perfect circle. Brahe's calculations were off by a few minutes here and a few degrees there. It was his subordinate who finally caught on that everything was also speeding up and slowing down because they were really flying in ellipses and not in perfect circular orbits as everyone since Copernicus had believed.

There are those who mark Kepler as the man who formally separated astronomy from astrology. He was an astute mathematician who required mathematical proof. He calculated new Laws of Planetary Motion regarding the nature and velocity of bodies around the Sun.

Earlier, before Brahe's death, Kepler made a little money on the side to support himself by drawing and interpreting horoscopes, a task which probably became odious to him because it took him

from his primary work as an astronomical mathematician. To this day, modern astrologers quote this statement he made while a court astrologer: "The belief in the effect of the constellations derives in the first place from experience, which is so convincing that it cannot be denied except by people who have failed to examine it."

In any case, the purpose of this book is not to further the cause of astrology at the expense of astronomy, nor vice versa, for that matter. We will not gather to refute the Flat Earth Society nor the constellationists. It is difficult to take the travels of a constellation as a true course of our future when it does not really travel as a constellation but only as its parts. It is not a real grouping of stars *per se,* but rather a grouping as seen from the Earth: that is, an unconnected tableau. Some stars in the same constellation are millions of light years deeper in space than their sister stars in the figure or object. But because they are *perceived* to be together, the ancients derived from them the characteristics of the celestial groupings we now know as constellations.

How, therefore, can we present a book about the stars, their origin, the mythology, the astronomical and even astrophysical aspects of the universe without some reference to astrology? It is, after all, a science which dates back 4,000 years. Not only that, but astrology is older than most religions and has at least as many adherents as the population of most modern countries.

It is unnecessary to deny or affirm the effects or character of the weights or movements of our solar system either astrologically or theologically. The absurdity or validity of astrology is totally evident to its believers or disbelievers so that people who prefer to believe in it are going to anyway.

However, we have gathered from highly accredited sources in astrology some of those characterizations which appear to be accepted by many who practice the discipline; knowing that even such wide generalizations will draw disagreement among some of those who read the planets, the stars, and sometimes the constellations in drawing up horoscopes.

So the following pages contain the signs of the zodiac together with a summary of the most commonly accepted aspects, both negative and positive, which astrologers generally believe characteristic of each sign of the old and new zodiacs. We say "old" zodiac because the modern school of astrological mathematicians points out that since the first formulations thousands of years ago some planets, and therefore some zodiacal signs, are so changed in their location they must be removed from further astrological calculations. And yet there are planets so altered over the centuries in their placement in the heavens that they have been replaced by signs other than traditionally used by astrologers in their analytical calculations.

In any case, what follows is a summary of comments gleaned from many expert astrologists, averaging their opinions. This exercise leads us to remember the story of the government which asked the astronomer how many different lists of stars existed. The astronomer said, "There are 27 lists." The government then said, "Well, why don't we reduce all of them to only one?" "Because," the astronomer said, "then there would be 28 of them."

So, without any feelings of guilt, what follows is a zodiac which might prove useful to those who look for a quick astrological fix on a subject.

Besides the mythology summaries you will find the positive and negative aspects, the ruling planet, and what you can most expect from a person born under that sign. Then comes a list of what to expect of that person as a child, his emotional character, his personality in terms of a plant, and the kind of career in which he might be expected to succeed. Following all this is a description of the more classical material the reader can use as a quick reference to each constellation's mythological genealogy.

Before proceeding with the myths, let us look at how the ancients divided the zodiac into 12 equal parts. The division was both simple and ingenious. As you read this, keep in mind that the author of this book is among the first to point out that if there is any similarity in personality among people born under the same sign, it must be left to antiquity itself to bear the proof. Time and billions of hours of study by millions of sincere students of astrology must carry with it some valid conclusions if there is any real validity to be found in such an exhaustively pursued study in the first place.

We now present a scenario for the establishment of the zodiac.

Having no instrument which would measure time adequately, the ancients filled a special vessel with water. They spilled the water drop by drop into another vessel set beneath the first. This process began as closely as possible to the moment when a given star rose, and continued until the star rose again the following night to complete one full cycle.

The original amount of water was then divided into 12 equal parts. Having 12 smaller vessels ready, each capable of containing one of these parts, again all water was poured into the upper vessel. When a given star in the zodiac rose, the ancients then let the water drop into one of the small vessels. As soon as that was full, the level of the water was marked. The vessel was removed and an empty one took its place. The process continued all year until the 12 vessels were marked as having been filled.

This divided the zodiac into 12 equal portions, corresponding to the 12 months of the year commencing with the vernal equinox. Each portion served as the visible representative, or sign, of the month in which it was collected.

All those stars in the zodiac which were observed to rise while the first vessel was filling were constellated and included in the first sign, called Aries in honor of the ram held in great respect by the shepherds of ancient Chaldea. The stars rising while the second vessel was filling were included in the second sign, which honored Taurus. The stars for the third vessel were called Gemini in allusion to the twin season of the flocks in Chaldea.

In a similar manner, each sign of 30 days in the zodiac received a distinctive appelation according to the superstition of the originators whose names have been forever erased from the sands of time.

In other words, the sign Aries generally included all the stars embraced in the first 30 days of the zodiac. The sign Taurus, in like manner, included all those stars embraced in the next 30 days, and so on for all the others. The constellations themselves have since left their exact locations. Now, thousands of years after the first Zodiac was charted, the heavens have changed, though the number of days (30) remains generally the same.

After centuries of scientific progress we now know that not only are the constellations which made up the original zodiac in different locations in space, but generally the mathematics used in propounding and casting horoscopes have not been proportionately altered to account for those changes. Therefore, the constellations have no relationship to our lives nor to each other.

Nevertheless, just as lovers gaze at the moon and stars, transporting themselves into the ecstasy of infinite love, the trajectory of their dreams remains largely immaterial. So, too, are man's aspirations toward being able to predict his future by means of astrological prognostication.

Man's dreams of love and success as well as the quality of his projections of the future will probably always be with us. Witness the number of newspapers which carry horoscopes every day. It's a small example of how ineffective empirical arguments are since they have virtually no effect on the importance placed on old wives' tales.

An intelligent astrologer, and doubtless they exist, ought to investigate the statistical relationships and the behavior of society in general during a full moon.

How often have you heard it argued that there is no relationship between the planets and stars and what happens to us. How then do we explain that the word lunatic refers to a person acting crazily during a full moon? Why, it has been asked, are there more crimes, births, domestic arguments, etc. during a full moon than at other times?

First, there have never been any firm, statistically controlled studies which indicate that behavior is altered in any respect where planets' moves are supposed to alter behavior, the moon notwithstanding. With respect to the tides and lunar changes, however, yes, there is an effect. Gravitational changes do occur because of the proximity of the moon. And since there are fluids in the cranial cavity, these fluids are subject to minute changes which might alter behavior. The history of behavioral changes traditionally has been related to the menstrual cycle. The root of the word menstrual is month. The root of the word month is *moonth*.

Another explanation for alterations in mental attitudes might someday prove to be the effect of air pressure, although such changes have not yet been studied. However, an aneroid barometer, an altimeter and many other instruments use the aneroid for measurement purposes. An aneroid is an empty, sealed, airtight wafer-like chamber or vessel. As air pressure increases or decreases, per square inch, the aneroid is depressed or relaxed thereby measuring the alteration in the atmosphere just as a depthometer measures water depth as it goes deeper or shallower.

Think of the cranium as a kind of aneroid. As air pressure in the atmosphere increases, it must have some effect on the cranium in which the brain floats. Thus, once we have mastered weather and atmospheric prognostication, we might someday have a predictable mood alteration study under controlled conditions.

But however minuscule the change in brain activity we can expect as a result of atmospheric change, it is infinitely more effective upon us than any planetary movement which generally is the basis

for astrological horoscopic prognostication. Astrology is based on planetary changes. The moon is not a planet. It is a satellite.

Perhaps if someone would take the time to study all the numerical progressions of all the most popular songs throughout history, it is possible that all those songs written in January might have a certain amount of similarity with each other. In any event, the lyrics of songs written in the winter might be more nearly similar to each other than to all those songs written in summer.

Let us suppose, therefore, that there is also a similarity among Capricorns compared to all those born in April. Accordingly those born under the sign of Aries might be more nearly like each other than they are like Capricorns.

What follows, then, is an accumulation of information gathered from the speculations of many astrolgers; the truth of their calculations rivals only that of the tarot card readers, crystal ball gazers, palm readers and the myths created in the names of such august figures as Leo, Libra, etc.

Aries, the Ram

March 21—April 19

Ruling planet: Mars *Trait:* Assertive, creative

Positive aspects

Creative, brave, loyal, direct, always concerned with the larger view, hating restriction, pioneering, energetic, humorous

Negative aspects

Selfish, impulsive, pugnacious, very "now" minded, impatient, overlooks important details

As a child

"Me first"

Emotional character

Strong need to express sexuality

As a plant

A rose

As a career

Artist, composer, etc.

According to fable, this is the ram of the golden fleece who flew Phrixus and his sister Helle from Thesaly to escape the persecution of their stepmother Ino. The speed of the flight made Helle dizzy, and she fell into the sea at a point afterward designated as the "Hellespont", now known as the Dardanells in Turkey, the strategic channel between the Agean Sea and the Sea of Marmara. Phrixus was carried safely to Colchis on the Black Sea in Georgian Russia, but was soon murdered by his father-in-law, Aeetes, who envied him his golden treasure, the fleece of Aries. This gave rise to the celebrated expedition in which Jason sailed in the Argo Navis to search for the famed golden fleece.

In the meantime, Naphele, queen of Thebes, had given the noble ram to her children. She was changed into a cloud as a reward for her parental solicitude, and the Greeks ever after called the clouds by her name.

Another story credits the constellation as referring to the flocks of the Chaldean shepherds, who worshipped Aries. Aries is also part of other cultural traditions.

As one of the twelve Zodiac signs in China, it is known as Heang Low, or the dog. The twelve tribes of Israel were assigned to the twelve different signs of the Zodiac. The tribe of Aries is Judah.

Aries the ram is an unimpressive constellation, but it does contain a +2 magnitude star called Hamal. It is usually positioned with a horn into Pisces and a leg into Cetus. The beta star is a +2.7 star which with the gamma binary star completes the horn goring Pisces.

Gaius Manilius, a Roman poet, in the first century wrote:

> "The Ram having passed the sea serenely shines, and leads the year, the prince of all the signs."

Around 2000 years ago the Vernal Equinox was in the constellation Aries, making Aries the first sign of the Zodiac. This, due to the processions of the Equinoxes, is no longer true. The zero point of right ascension, or Vernal Equinox, is now in Pisces. In about 4500 B.C. Orion marked the Vernal Equinox, which would indicate that the Zodiac was not fully formed or addressed in its present order or selection. There is a 26,000 year cycle, so in another 24,000 years, Aries will again be the number one sign of the Zodiac.

Taurus, the Bull

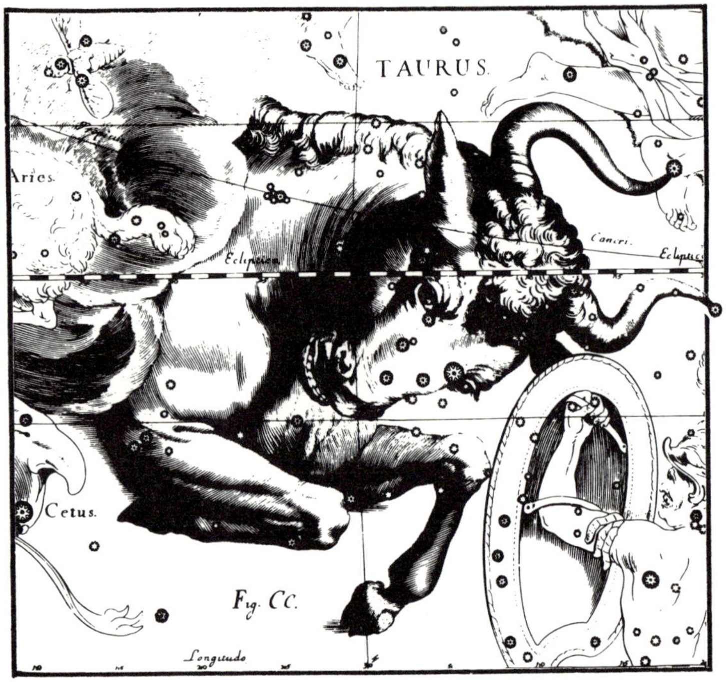

April 20—May 20

Ruling planet: Venus *Trait:* Possessive, permanent

Positive aspects

A Taurean is very much subject to two sides of the same reality. The good aspects are practical, reliable, adept in response to rules, business, moral values. Taureans love luxury, good food, are persistent, solid, determined, affectionate, strong-willed, trustworthy.

Negative aspects

Possessive, self-indulgent, capable of being a great bore, opinionated, greedy, stubborn, obsessed with routine, resentful.

As a child

Stubborn

Emotional character

"You're mine!"

As a plant

An oak tree

As a career

A banker

Represented in an attitude of rage, Taurus is always about ready to plunge into Orion, who seems to invite the battle by his antagonistic stance of attack. Only the head and shoulders of Taurus are depicted by the stars.

The constellation Taurus contains the Pleiades, a bright grouping of stars representing the seven sisters of Greek mythology, the daughters of Atlas. It also contains Hyades, a cluster of twenty-four stars, visible to the naked eye, marking the head of the bull.

The Crab Nebula, or M-1, in the northern part of Taurus is a remnant of a supernova, a star that exploded in 1054 A.D. It is interesting to astronomers because it is recent and still expanding. M-1 is also known as Taurus A, a designation for its radio emission sources.

According to Greek mythology, this is the animal which bore Europa over the seas to that country which now bears her name. Daughter of Agenor and princess of Phoenicia, Europa was so beautiful that Jupiter became enamored of her. Jupiter assumed the shape of the snow-white bull, and mingled with the herds of Agenor while Europa and her female attendants were gathering flowers in the meadows. Europa caressed the beautiful animal and at last had the courage to sit on his back. Taking advantage of the situation, Jupiter safely crossed the sea with her to Crete.

Some suppose Europa lived about 1552 years before the Christian era. It is more likely, however, that this constellation had a place in the zodiac before the Greeks began to cultivate a knowledge of the stars, and that it was an invention of the Egyptians or Chaldeans. Both the Egyptians and Persians worshipped a deity named Apis under this figure. The archeologist Belzoni is said to have found an embalmed bull in the tomb of Seti near Thebes.

Taurus also was known in China as TaLeang, the great bridge, between Hyades and Pleiades. Jesuit missionaries anglicized this to Kin Neu, the golden ox, to fit in with the western Taurus.

In the Hebrew zodiac, Taurus is ascribed to Joseph, according to some lists, but to Issachar according to the Encyclopedia Judaica.

Gemini, the Twins

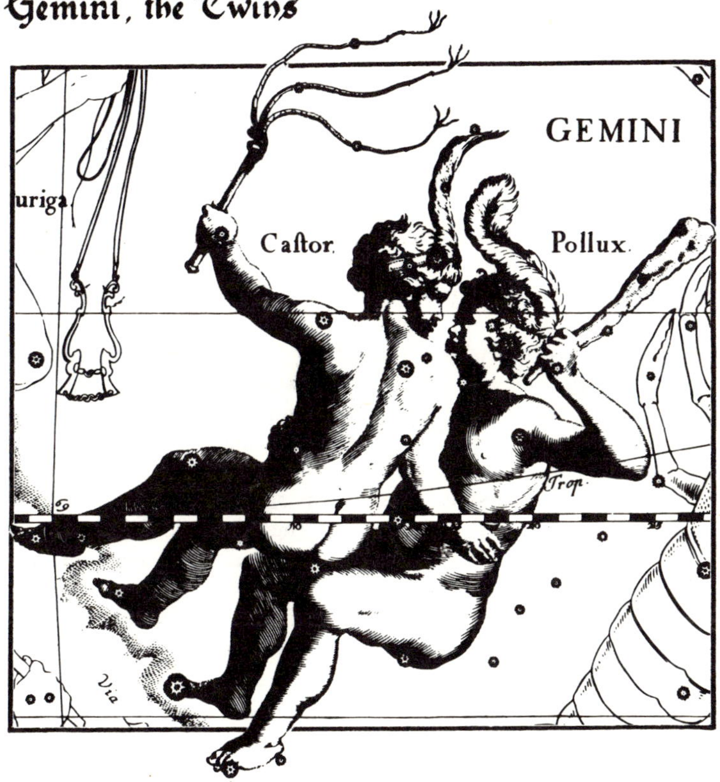

May 21—June 21

Ruling planet: Mercury | *Trait:* Communicative

Positive aspects
Witty, intellectual, talkative, adaptable, perennially youthful, busy, spontaneous flair for writing, abundant energy

Negative aspects
Changeable, restless, cunning, two-faced, nervous, superficial, gossipy, inconsistent

As a child
Needs much laughter—stimulation

As a plant
A palm tree

Emotional character
Expresses himself well in love

As a career
Communicator

The twins are seen sitting together south of Lynx, between Cancer on the east and Taurus on the west.

Castor and Pollux were twin brothers, sons of Zeus by Leda, the wife of Tyndarus, king of Sparta. The manner of their birth was singular; Castor was sired by Tyndirus, Pollux by Zeus.

They were educated at Pallena, and afterward embarked with Jason in the celebrated contest for the golden fleece. In their support of the Argonautic expedition the brothers acted with unparalled courage and bravery. Pollux distinguished himself by his achievements in arms and personal prowess, and Castor in equestrian exercises and the management of horses. This is why they're represented in Grecian temples on white horses, armed with spears, riding side by side, on their heads are crowns on whose tops glitter stars. Among the ancients, and especially among the Romans, there prevailed a superstition that Castor and Pollux often appeared at the head of their army to lead their troops to victorious battle.

"Castor alert to tame the foaming steed,
And Pollux strong to deal the manly deed."

–Martial

The brothers cleared the Hellespont and the neighboring seas from pirates after their return from Colchis. Henceforth they were regarded as the friends and protectors of navigation. It's said that during a violent storm on the Argo Navis two flames were seen to play around their heads, and immediately the tempest ceased and the sea was calm. From this sailors inferred that whenever both fires appeared in the sky there would be fair weather. When only one appeared, there would be storms.

The most commonly recounted myth is that upon Castor's death, Pollux was overwhelmed with grief and wanted to share his immortality with his twin (Pollux, being the son of Zeus, was immortal), and so Zeus reunited them in the heavens.

In the Hebrew zodiac, the constellation of the twins refers to the tribe of Benjamin in some lists, but to Zebulun according to the Encyclopedia Judaica. Sometimes they are referred to as the twin sons of Rebecca.

The twins indicate good fortune and our positive phrase *"by jimini"* comes from "by Gemini".

Castor and Pollux reach their zenith at nine o'clock P.M. in early March. Castor is a bright white *binary* and very visible. Pollux is a visible orange colored star.

Cancer, the Crab

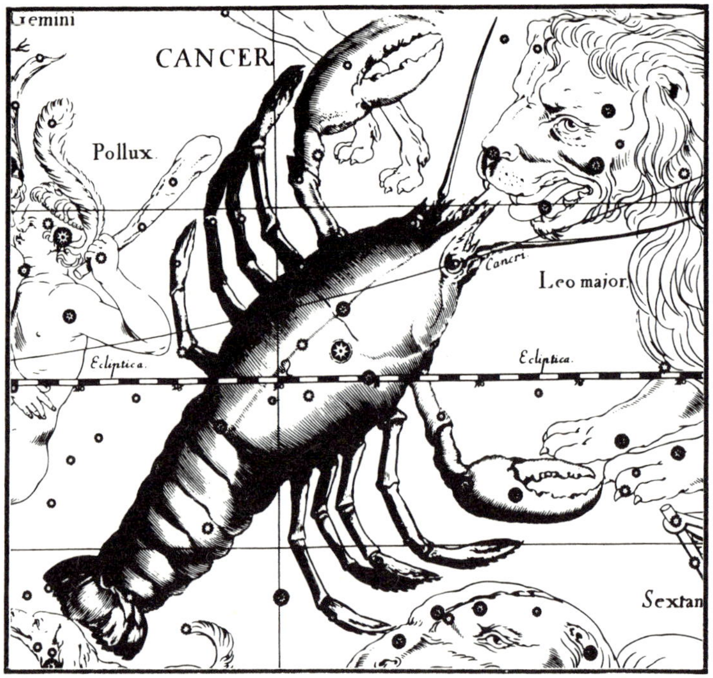

June 22—July 22

Ruling planet: Moon *Trait:* Understanding

Positive aspects

Protective, sensitive, strong maternal/paternal instinct, cautious, tenacious, resourceful, greatly imaginative

Negative aspects

Hyper-protective, mean-tempered, unforgiving, hidden personal weaknesses, easily flattered, self-pitying, petulant

As a child

Repentant, dolly-loving

Emotional character

Reticent

As a plant

Cactus

As a career

Nurse, Doctor

Cancer is represented in the heavens as a crab situated between Leo on the east and Gemini on the west.

Cancer is the faintest constellation in the entire zodiac, its brightest star is only of the fourth magnitude. M-44 or "the Bee Hive Cluster" can be seen with binoculars.

In the zodiacs of Esne and Dendera, and in most of the astrological remnants of ancient Egypt, a scarab or beetle symbolizes Cancer. In Sir William Jones's Oriental Zodiac, and in some others found in India, we meet with the figure of a crab. As the Hindus probably derived their knowledge of the stars from the Chaldeans, it is supposed that the crab sign is more ancient than the beetle.

The sign of Cancer usually is assigned to the tribe of Reuben in Jewish astrological lore.

In some eastern representations of this sign, two animals which look like jack-asses are found. Because the Chaldaic name for the ass may be translated "muddiness," it is supposed to allude to the discoloring of the Nile, since that river rose when the sun entered Cancer. In copying this sign, the Greeks also symbolized it with two jack-asses because these animals assisted Zeus in his victory over the giants.

Mythologists give different accounts of the origin of this constellation. The prevailing opinion is that while Hercules was engaged in his famous contest with the dreadful Hydra, Juno, envious of the fame of his achievements, sent a sea-crab to nip, bite and annoy the hero. Though the crab was soon crushed underfoot by Hercules, the goddess placed it among the constellations in reward for its services.

"The Scorpions's claws here clasp a wide extent,
And here the Crab's in lesser clasps are bent."

—*Unknown*

Leo, the Lion

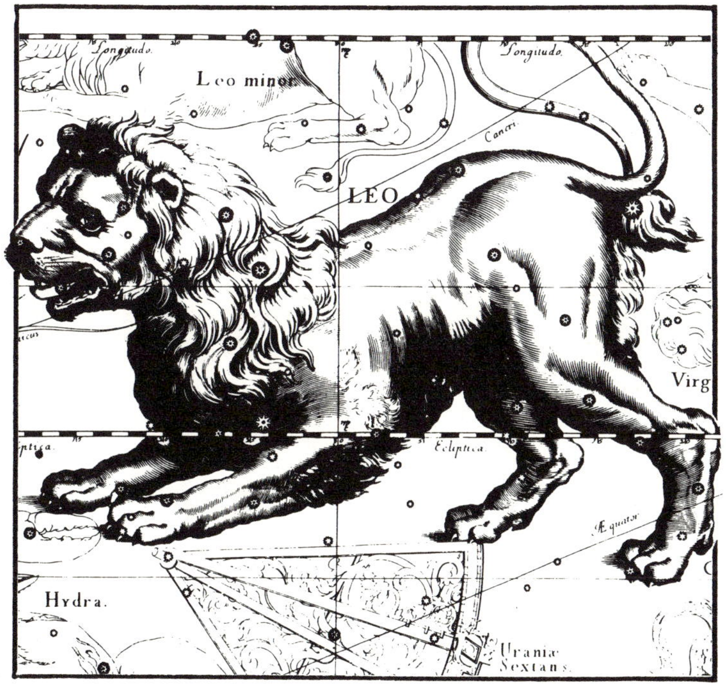

July 23—August 22

Ruling planet: Sun *Trait:* Impressive

Positive aspects
Creative, broadminded, enthusiastic, powerful, an organizer, generous, dramatic, showy

Negative aspects
A bully, pompous, indolent, conceited, dogmatic, stubborn, dramatic, liking power, a show-off

As a child
Too spirited

Emotional character
Gregarious

As a plant
Pine tree

As a career
Entrepreneur

Leo is the largest and lowest of a group of five or six bright stars which form a figure somewhat resembling a sickle in the neck and shoulder of the lion.

Leo is an easy constellation to locate. The pointer stars of the Big Dipper extended south approximately locates Regulus, the bright blue-white star in the chest of the lion. There is a meteor shower, Leonids, in mid November from the gamma star, the lion's mouth.

According to Greek fable, this lion represents the formidable animal which roamed the forests of Nemaea near Corinth, Greece. It was slain by Hercules and placed by Zeus among the stars in commemoration of the dreadful conflict. Some writers have applied the story of the 12 labors of Hercules to the progress of the sun through the 12 signs of the ecliptic; as the combat of that celebrated hero with the lion was his first labor, they have placed Leo as the first sign.

The figure of the lion, however, was on Egyptian charts long before the fable of Hercules.

In hieroglyphic writing the lion was an emblem of violence and fury. In the zodiac it signified the intense heat the sun created when it entered that part of the *ecliptic.* The Egyptians were much annoyed by lions that left the desert during the hot summer and hunted the banks of the Nile at its greatest elevation. It was, therefore, natural for their astronomers to place the lion in the zodiac.

The figure of Leo much as we know it is in all the Hindu and Egyptian zodiacs. The overflowing of the Nile, which the Egyptians regularly and anxiously expected every year, took place when the sun was in this sign. Accordingly they paid more attention to this sign than to any other.

In the Hebrew zodiac, Leo is assigned to Judah, upon whose standard a lion traditionally is painted according to several passages of Hebrew writings. On the other hand, Leo is assigned to Simon in the Encyclopedia Judaica.

Christians of the middle ages tried to replace the mythological characters of the constellation with biblical connotations. Leo was subsequently and separately notated as one of Daniel's Lions or as Doubting Thomas.

Virgo, the Virgin

August 23—September 22

Ruling planet: Mercury *Trait:* Analytical

Positive aspects
Analytical, discriminating, meticulous, prejudiced, modest, conventional, tidy, concerned, caring, assimilative

Negative aspects
Too hard a worker, hypercritical, finicky, a worrier, fussy, abnormally cautious, lacking in vision

As a child
Picky

Emotional character
Cautious

As a plant
Wheat

As a career
Secretarial, inspectorial

Virgo is the second largest constellation in the sky and is at its highest or most visible in May and June. It contains a major cluster of galaxies, among which is M-104 or the Sombrero Galaxy. The brightest quasar in the heavens, 3C 273, is in Virgo and can be seen with a small telescope. Spica, the largest star in Virgo, can be seen below the handle of Ursa Major and on a line south through the brighter star Arcturus.

Virgo, the virgin, is sometimes called Astraea. She is also known as Ceres, the goddess of the fields and agriculture in general. She is also identified with Proserpina, Ceres' daughter. Virgo was the daughter of Jupiter and Themsis, and is the Goddess of Justice as well as the goddess of the harvest. Virgo was a busy goddess. This constellation's brightest star is Spica, meaning an ear of corn or wheat. Her grains were weighted in the full moon of Libra, the Scales, which she held in her left hand. Can you believe how that Virgo gets around?

This is the seventh constellation in the zodiac and is situated east of Leo and midway between Coma Berenices on the north and Corvus on the south.

In describing the physical state of the world mythology invented a symbolic language which personified inanimate objects. The priests reduced the whole of their noblest science to fables. The people believed these as true histories representing the moral condition of mankind during the first ages of civilization.

According to the ancient poets, this constellation represents the virgin Astraea, the goddess of justice, who lived on earth during the golden age in about 3000 B.C. Offended by the wickedness and impiety of mankind during the bronze and iron ages, she returned to heaven and was placed among the constellations of the zodiac with a pair of scales in one hand and in the other, a sword.

As Virgil relates, mankind degenerated, one god after another deserted their beloved haunts. Astraea lingered till the last, but finding Earth steeped in inhumanity she also eventually flew away to the celestial regions.

Schiller ascribed Virgo to St. James the Less. The Chinese call it She Sang Nue—The frigid maiden. In most Hebrew writings, Virgo is assigned to Gad's tribe.

Libra, the Balance or Scales

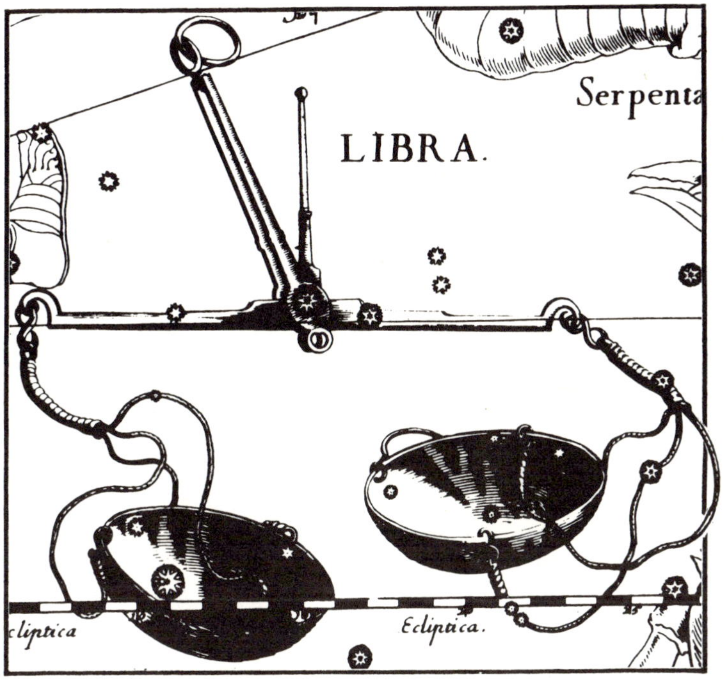

September 23—October 23

Ruling planet: Venus   *Trait:* Harmonious together

Positive aspects

Charming, pleasant to live with, a firm friend once a friend, great romantic, idealistic, capable of elegance and diplomacy

Negative aspects

Indecisive, frivolous, flirtatious, operating between two ex treme attitudes too close to each other, oscillating, cruelly humorous

As a child

Affectionate

Emotional character

In love with love

As a plant

Lily

As a career

Involved in travel

The Autumnal Equinox was signaled by Libra. The days and the nights were of equal length, or the sun and moon were in balance. The sun passes through Libra in November. Alpha Librae is a blue-white star of 3 magnitude and is also known as Zubenelgenubi or Southern Claw.

Libra is the balance. Virgo, sometimes called Astraea, was the goddess of justice. Libra was not actually a person, but the scales usually seen in Virgo's left hand, the appropriate emblem for the office of judgment.

Libra is seen on all of the Egyptian heiroglyphics. Simultaneously an argument of the great antiquity of this asterism and of the probability of having been fabricated originally by the astronomical sons of *Misraim*.

In a few very old versions of the zodiac, Astraea, the virgin who holds the balance in her hand as an emblem of equal justice, is not shown; the zodiacs of Esne and Dendera are examples. Humboldt thinks that although the Romans introduced this constellation into their zodiac in the reign of Julius Ceaser, it may have been used by the Egyptians and other nations of very remote antiquity.

It is generally supposed that the figure of the scales has been used by all nations to denote the balance of temperature between the days and nights. It has been observed that when the sun arrives at this sign, the season has a greater uniformity in temperature over all the Earth's surface, not merely between day and night.

Others say only the beam of the balance was at first placed among the stars, and that the Egyptians thus honored it as their *Nileometer* with which they measured the inundations of the Nile. The biblical prophet is thought to be alluding to this scale of measurement when he describes the Almighty as "measuring the waters in the hollow of his hand." (Isa. xi. 12)

The Greeks say the balance was placed among the stars to perpetuate the memory of Mochos, the inventor of the scale or balance. Early Christians assigned St. Philip the Apostle as representing Libra. In the book of Daniel it is identified as the balance. (verse 27.) To the Chinese it was Show Sing, sign of longevity and also Tien Ching, the celestial balance. Those who refer the constellations of the zodiac to the 12 tribes of Israel ascribe the balance to Asher. Encyclopedia Judaica, however, assigns Libra to Ephraim.

Scorpio, or Scorpius, the Scorpion

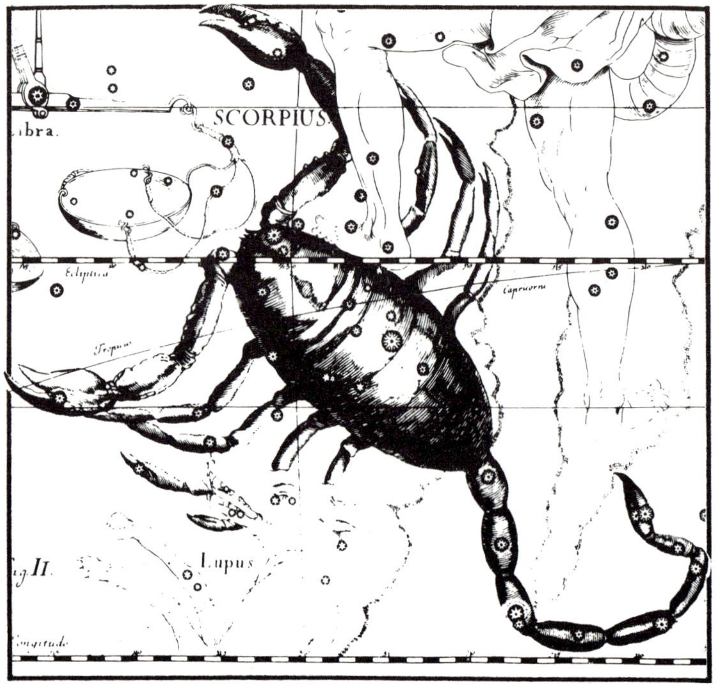

October 24—November 21

Ruling Planet: Mars *Trait:* Inventive

Positive aspects
Purposeful, discerning, subtle, persistent, very imaginative, resourceful, emotional

Negative aspects
Obstinate, suspicious, secretive, jealous, stubborn, crafty

As a child
Demanding

As a plant
Water lily

Emotional character
Sexually expressive

As a career
Policeman

Scorpio is a bright constellation in the southern skies most visible from Chicago at the end of July. Antares, the Alpha, is a 1 magnitude red star, so named because Antares was the rival of Mars, also red in color. Antares is the heart of the Scorpion. Shaula, Lambda Scorpii, is a 1.6 magnitude star and is the tail. Scorpio contains many large clusters and nebulae easily seen with binoculars.

Scorpio is situated southward and east of Libra and is on the meridian about the tenth of July, right after the Corona Borealis by about fifty minutes if anyone should happen to ask you.

One myth has it that Apollo, worried about his sister Artemis' chastity, sent the scorpion to kill Orion. Scorpio and Orion were placed as far away from each other as possible in the heavens to avoid any further battles.

This sign was represented in ancient times by various symbols, sometimes as a snake or crocodile. Most commonly it was depicted as a scorpion as found on the Mithraic monuments before 200 B.C. Evidence exists that these monuments were constructed when the vernal equinox accorded with Taurus.

The Egyptians, or Chaldeans, who first arranged the zodiac, might have placed Scorpio here to denote that when the sun enters this sign the diseases incident to the fruit season would prevail. Autumn abounded in fruit and often brought with it a great variety of diseases. It seems fitting to represent this season with the venomous scorpion which wounds with a sting of its tail.

Scorpio is an ancient constellation. Ovid mentions it in his beautiful fable of Phaeton:

"There is a place above, where Scorpio bent,
intail and arms surrounds a vast extent
In a wide circuit of the heavens he shines
And fills the place of two celestial signs."

According to Ovid, this famous scorpion sprang from the earth at the command of the imperious Juno and fatally stung Orion to punish the hero/hunter's vanity for boasting that there was no animal on earth he couldn't conquer.

It has been known as the constellation of the Apostle Bartholomew.

In China it was Ta Who, great fire and later, Tien He, the celestial Scorpion.

In the Hebrew zodiac this sign is alloted to Dan because it is written, "Dan shall be a serpent by the way, an adder in the Path."

Sagittarius, the Archer

November 22—December 21

Ruling Planet: Jupiter *Trait:* Adventurous

Positive aspects

Optimistic, jovial, freedom loving, sincere, dependable, a go-getter, explorer, philosophical

Negative aspects

Extreme, tactless, capricious, a lack of the amenities, too optimistic

As a child

A challenge

Emotional character

Sexually inventive

As a plant

Bird of paradise

As a career

Lawyer

The sun passes through Sagittarius, the ninth zodiac sign, around January first. It has many things to interest astronomers. Messier, the French astronomer, catalogued fifteen objects, the most interesting being the Lagoon Nebula and the Omega Nebula. The finest view of the Milky Way is the great Sagittarius star cloud, just north of gamma Sagittarius.

Sagittarius, the archer, was depicted as a centaur named Chiron. A centaur is a mythological creature half man and half horse. Though we have no evidence there was much of an overabundance of them, there appears to have been at least one mythological family of centaurs. The centaurs, except for Chiron, were known for their generally bad nature.

At the wedding feast of Pirithous, prince of Thessaly (Marathon, Greece), and Hippodamia, one of the centaurs drank too much and made a pass at the bride. Dreadful violence ensued, many centaurs were slain.

Chiron, unlike the other centaurs, was famous for gentleness and his knowledge of music. A prophet, a doctor of medicine, and a poet, he was also an expert archer. He taught mankind the use of plants and medicinal herbs, and instructed the greatest heroes of his age in all the polite arts. He taught Aesculapius physics, Apollo music, Hercules astronomy, and tutored Achilles, Jason and Aeneas.

According to Ovid, he was mistakenly shot by Hercules at the river Evenus for offering an indignity to his newly married bride. The arrow Hercules aimed at the centaur had been dipped in the blood of the Lernaean sea snake Hydra, rendering the wound incurable even by the father of medicine himself. Chiron (Saggitarius) begged Jupiter to deprive him of immortality if that might relieve him of his excruciating pains. Jupiter granted his request, yet still gave him a place among the constellations. Sagittarius' arrow is always aimed at the treacherous Scorpion's heart.

As this constellation appears on the ancient zodiacs of Egypt, Dendera, Esne and India, it seems conclusive that the Greeks only borrowed the figure though they invented the fable. This is true of many of the ancient constellations; hence the jargon of the conflicting accounts which have descended to us. Some Christians assign Sagittarius to St. Matthew the Apostle, and others to Ismael, son of Abraham and Hagar. The Chinese call it Seih Nuh, meaning firewood and later as Jin Ma the manhorse. In the Hebrew zodiac, Sagittarius is assigned to Benjamin.

Capricornus

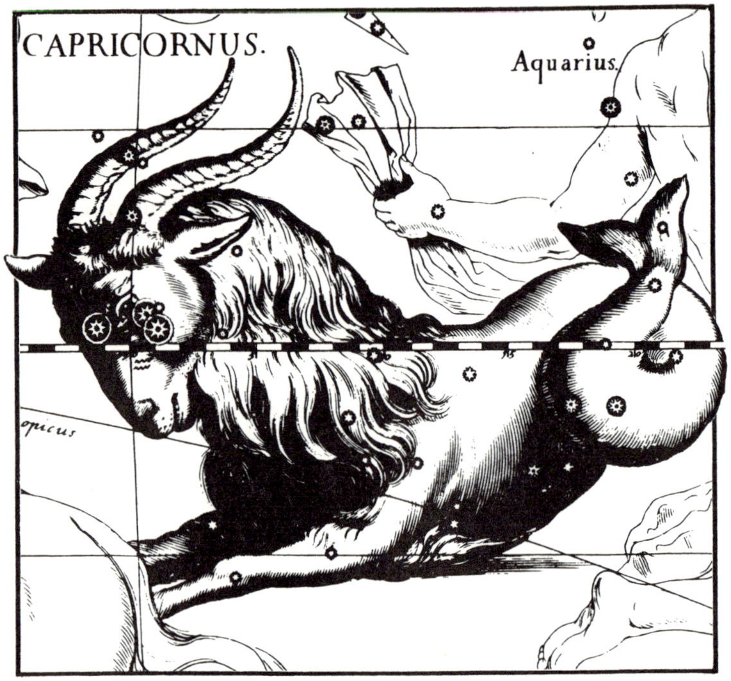

December 25—January 19

Ruling planet: Saturn *Trait:* Calculating

Positive aspects

Ambitious, reliable, loyal, patient, persevering, disciplined, headstrong

Negative aspects

Rigid outlook, miserly, conventional, exacting, merciless, mean

As a child

Dominates others

Emotional character

A loner

As a plant

Poplar tree

As a career

Civil servant, engineer

Capricorn is the tenth member of the zodiac, receiving the Sun in January and February. The Alpha star, called Algedi, meaning goat or ibex, is actually two stars each visible without magnification. Capricon is situated south of the Dolphin and next to Sagittarius.

Capricornus is said to be Pan or Bacchus. Along with some other deities, he was feasting near the banks of the Nile when suddenly the dreadful giant Typhon roared up and compelled them to assume different shapes in order to escape his fury.

Pan led the plunge into the Nile. The part of his body under water assumed the form of a fish while the other part assumed that of a goat with little horns and a goat-like beard. But before we say how his new metamorphosed shape preserved the memory of this adventure, let us look first at Pan.

Pan is derived from a Greek word signifying all things. He was often considered as the great principal of vegetable and animal life. He resided chiefly in Arcadia, in the woods and the most rugged mountains. Pan usually was terrified of the inhabitants of the adjacent country, even when they were nowhere to be seen. That kind of fear, which often seizes people in response to only an imaginary threat, was the source of the word "panic".

Some say this constellation was the goat Amalthea, who supported the infant Zeus with her milk. To reward her kindness, the father of the gods placed her among the constellations and gave one of her horns to the nymphs who had taken care of him in his infancy. This gift was ever after called Cornucopia or, the horn of plenty, since it possessed the virtue of imparting to the holder whatever he desired.

Going back, when Capricorn was born (that is, when Pan emerged in his new form as Capricorn), he surfaced to find Zeus in great trouble in Typhon's storm. Typhon was trying to tear off Zeus's arms and legs when Capricorn, still possessed of Pan's pipes, blew a shrill note. The note was so shrill, Typhon had to flee to escape the pain piercing his ears, leaving all the commotion of battle as well as Capricorn's continuous blowing of Pan's pipe. Mercury swept down from Olympus. Stopping himself with the wings on his feet, he pulled Zeus back together again. Zeus retreated and flew back home to Olympus, gathered an armload of

thunderbolts, and went after Typhon. Typhon escaped underground, from whence he came. There he remains except for the times when he comes back as a sea storm to threaten boats and islands with extinction.

In thanksgiving for his rescue, Zeus raised the new Pan into the stars where he lives in his constellation, Capricorn, between Aquarius and Sagittarius low in the southern horizon.

Known as Azazel, the scapegoat, in the Book of Leviticus, Capricorn has been assigned to Simon, the Apostle. In the Hebrew zodiac according to the Encyclopedia Judaica, Capricorn is ascribed to the tribe of Dan.

Pan

Far out there
In the immensity
So far beyond,
That two perfectly parallel lines
Converge and cross.
So far,
That if your telescope was powerful enough
You would see the back of your head
And even outside *that* radius
Beyond that edge of space
And beyond that too
All the way into infinity—far,
Measured by the world's longest number . . .
Plus one
There, beyond all heat, light and dark
Beyond cold
Beyond temperature
There. . .in the void, beyond-the-void
You might find
The face of God
And what can we know about him
Except he invented all
And since all that is
Is attracted to itself. . .by the immutable law of nature
We may surmise
We are him
And that all that is. . .is one.

–Edmund Steele

Aquarius, the Waterman

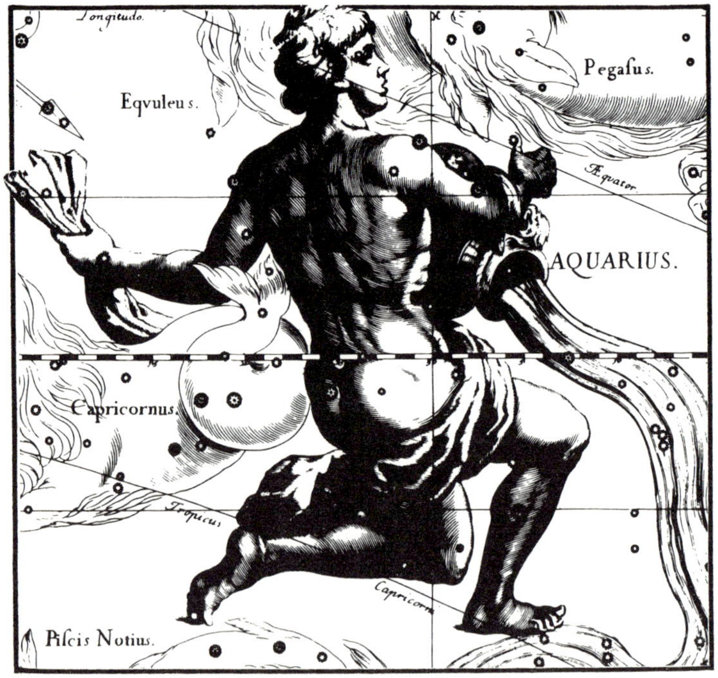

January 20—February 18

Ruling planet: Uranus *Trait:* Humane

Positive aspects

Friendly, willing, reformer, healer, idealistic, intellectual, wise, faithful

Negative aspects

Tactless, perverse, eccentric, contrary, trying to be iconoclastic, serious

As a child

Opinionated

Emotional character

Too trapped

As a plant

Ivy

As a career

Charity worker

Aquarius is the eleventh constellation of the zodiac. It is noted for its meteor showers, the strongest being Delta Aquarids on July 28. One can see about thirty-five meteors per hour. There is another on May 5, Eta Aquarids, which displays twenty each hour. Aquarius contains the Saturn Nebula and the Helix Nebula. Helix is the largest planetary nebula in the sky, but is faint and to see it binoculars are needed.

The water bearer is represented by the figure of a man carrying an urn from which water is being poured. It is situated in the zodiac immediately south of the equinoctial and bounded by the Little Horse, Pegasus and Pisces on the north, and the whale on the east.

This constellation is the famous Ganymede, a beautiful youth of Phrygia, son of Tros, king of Troy. According to Lucian, he was also the son of Dardanus. He was taken up to heaven by Jupiter as he was tending his father's flocks on Mount Ida, and became the cupbearer of the gods, replacing Hebe after she married Hercules. There are various opinions among the ancients regarding the constellation's origin. Some suppose it represents Deucalion, who was placed among the stars after the celebrated deluge of Thessaly 1500 years before the birth of Christ. Deucalion was the Noah of the Greeks. Others think it was designed to commemorate Cecrops, who came from Egypt to Greece, and founded Athens, established science and introduced the arts to polished life.

The ancient Egyptians supposed the setting or disappearance of Aquarius caused the Nile to rise when he sank his urn into the water. New Testament Christians in the seventeenth century had John the Baptist and St. Jude Thaddeus the Apostle as the two water jars. Some akin it to Moses being taken from the waters as an infant. In ancient China it was the symbol of Emperor Tchoun Hin and called Hiven Mao, commemorating a great deluge during his reign. In the zodiac of the Hebrews, Aquarius represents the tribe of Reuben. In the Encyclopedia Judaica the tribe of Asher is assigned to Aquarius.

Pisces, the Fishes

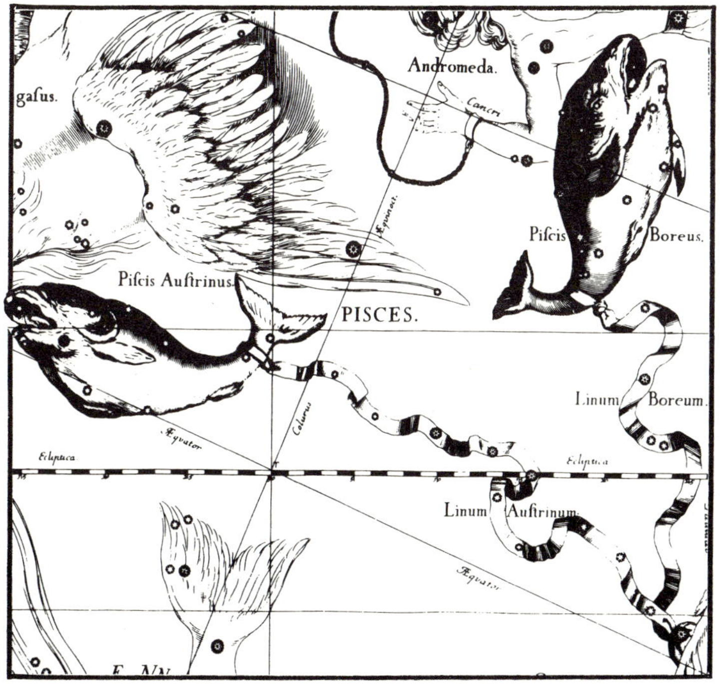

February 19—March 20

Ruling planet: Neptune *Trait:* Impressionable

Positive aspects

Compassionate, emotional, unworldly, adaptable, kind, intuitive, receptive, fun

Negative aspects

Weak-willed, vague, careless, indecisive, unable to cope, secretive

As a child

Difficult to correct

Emotional character

Easily carried away

As a plant

Lilac

As a career

Actor

Pisces is represented by two fishes tied considerably apart from each other at their extremities by a long, undulating cord or ribbon.

The sun passes through this twelfth sign in March and April. Pisces is not an easy constellation to locate. The connecting knot of the cord tying the fishes, Al Rischa (the Cord) is its brightest star.

The ancient Greeks, who have a fable to account for the origin of almost every constellation, say that Venus and her son Cupid were on the banks of the Euphrates one day when they were greatly alarmed at the appearance of the terrible giant Typhon. Throwing themselves into the river, they were changed into fishes and thus escaped danger. Minerva immortalized the event by placing two fishes among the stars.

According to Ovid, Homer and Virgil, this giant Typhon had a hundred heads like those of a serpent or dragon. Pernicious flames darted from his mouth and eyes. He was no sooner born than he made war against heaven and so frightened the gods that they fled and assumed different shapes. Jupiter became a ram, Mercury an ibis, Apollo a crow, Juno a cow, Bacchus a goat, Diana a cat, Venus a fish, etc. The father of the gods put Typhon to flight and crushed him under Mount Aetna.

The fable implies that some men's mouths are so full of cursing and bitterness, derision and violence that modest virtue sometimes is forced to disguise itself or flee from their presence.

No sign appears to have been considered of more malignant influence than Pisces. The astrological calendar describes the emblems of this constellation as indicative of violence and death. Both the Syrians and Egyptians abstained from eating fish out of dread and abhorrence of Pisces. When Egyptians heiroglyphics represented anything odious or expressed hatred, the symbol for fish was used.

In the Hebrew zodiac, Pisces is allotted to the escutcheon of Naphatali. The Chinese called Pisces, Tweu Tsze, the pig. Christians assigned it to St. Matthias, the sucessor to Judas Iscariot. Some refer to the miracle of the loaves and fishes.

The Face

To write about the stars and not include Albert Einstein is like writing about baseball and not mentioning Babe Ruth. You can do it, but you'll miss him.

Einstein was born in 1879. For years it was thought that as a child, he showed no particular signs of genius nor was he a child prodigy. This was not so. He was very extraordinary even at the age of 5. By the time he was 11 years old, he was gifted in Latin and Greek and was particularly talented on the violin. The story that he had trouble with his college entrance exams was true. They were in French and he had not yet excelled in that language as he later did. One of his teachers was supposed to have told his father, "It doesn't matter what Albert does in life. He simply will not amount to anything." New books about his life dispute that story.

After attending schools in Germany and Italy, Einstein graduated at Zurich in 1901 in engineering. Even to have attended institutions of higher learning in the 1800s was to have been elevated. After some difficulty, he landed a job at the Swiss Patent Office in Bern in 1902. He liked his job there, and someone who knew him at the time said those were probably the happiest years of his life because he could spend so much of his spare time thinking about theoretical science, a subject which always thrilled him. He was almost totally unknown and therefore largely uninterrupted in his postulations.

In 1903 he married Mileva Meric, a Serbian Physics student, with whom he had two sons. Albert continued to be happy with his work and also enjoyed his sons.

In 1905, Einstein wrote the first paper on what was to become the theory of relativity, or how things are not at all what they seem from all points of observation. It was complicated, but it went something like this: if a train is traveling at 100 miles an hour, and a man on the train drops a ball straight down, how fast is the ball going and in which direction, according to the view of an independent observer? What is the vector?

Years later, when asked about the meaning of relativity, Einstein was reported to have explained, "If you are on a park bench and enjoying yourself kissing your girlfriend, it seems it lasted only seconds. But if you sit on a hot stove, it seems like forever."

In 1913 he divorced Mileva and married his cousin, Elsa. Funny how the father of relativity married a relative.

Einstein and Elsa moved to Berlin where Einstein joined the Prussian Academy of Sciences and rose to the rank of professor. Where was that childhood teacher now—the one who predicted how he wouldn't amount to anything?

In the meantime, Einstein had been studying Hendrick Amtoon Lorentz's idea of relativity which he called the Theory of Special Relativity. So taken was Einstein with the Dutchman that he wrote, "Lorentz meant more to me personally than anybody else I have met in my lifetime."

Still, Einstein did not at all like the idea of admitting there might have been a beginning to the universe. He wrote, "The scientist is possessed by the sense of universal causation." He spoke for all scientists because generally they find it difficult to conceive of one agreeing to a beginning of the universe. It implies that the world started under conditions in which all the laws of physics are not necessarily valid. As such, it would be outside the realm of our ability to understand everything, that we'd never know the circumstances under which it all began. "What was everything before there was anything?" the scientist must ask. "A firecracker? A cosmo orgasm? A Big Bang?"

Is the universe the result of an uncaused cause? If everything is moving away from everything else and the farther away a thing is the faster it must move, where is the flash point of origination? Someplace between Andromeda and the Milky Way? Somewhere between Main St. and Elm?

It would mean that in the first moments of its existence the universe was compressed into a ball of heat and density a billion light years beyond human imagination. The shock of that microsecond must have been so great that it annihilated every particle of evidence. Or did it? In philosophical mathematics, we say nothing is ever totally annihilated. There are effects if nothing else. Echoes at the very least. But the first cause is uncaused, and no causation happens to be beyond the laws of nature as we know them. Still, the effects of a cause always exists. In this case, the effect is the existence of the universe itself.

Put another way, Einstein, like most classical scientists, couldn't cope with the history and rules of a new reality, even one he uncovered himself. He was lost in his own conjecture once he imagined beyond the moment of creation. Or, as Robert Jastrow says, "The scientist's pursuit of the past ends in the moment of creation." He was saying that different rules must have existed than the physi-

cal laws of nature as we know them now.

So the still young Einstein refused even to consider the possibility, however remote, that the whole of existence could be created by an uncaused cause. Long before, St. Augustine handled it with intellectual aristocratic sophistication by saying, "So who knows, and if he knew, he couldn't explain it anyway."

Einstein had performed a singularly astonishing achievement as far as the world of physics was concerned and particularly as far as DeSitter was concerned. Einstein was an atheist at that time. Then, with true scientific objectivity and following the accepted rules of empirical investigation, his new discovery meant that he had ripped away the ragged fabric of skepticism. And there, peeking back at him, smiling through the tatters of doubt, was the face of God.

To put everything into perspective, remember that Slipher had discovered in 1913 that a dozen galaxies he was observing were moving away from Earth at speeds of up to two million miles per hour.

He reported his findings at Northwestern University in 1914. There Edwin Hubble, one of the few people who was tuned into the most up-to-date studies of his time, picked up on what Slipher discovered and saw it as an earthshaking event.

Einstein published his equations on relativity in 1917 and DeSitter solved the unanswered questions it posed with the idea of the exploding universe. This was exactly what Slipher had observed. DeSitter, being Dutch, didn't know about what Slipher had observed because the knowledge was obscured by the war going on in Europe.

Strangely, or because he was such a lofty-thinking person, Einstein somehow failed to notice that his theory predicted the same conclusion: namely, an exploding universe. A Big Bang! It was this conclusion, the one which threatened the very language of science itself, that Einstein objected to.

The earth was trembling. Great men uncovering the solution to the greatest question man had ever posed to himself since the day he discovered he could think at all, and that was the idea that there might be a God on the other side of Creation. Yet most of the world was more concerned with World War I. The Germans, the English, the French and the Americans weren't so much concerned about the existence of a supreme being as they were about whose side He was on.

As time went by. no one had caught on to the fact that Albert had made another mistake. He had missed a solution which proved again the expanding universe part of his own theory. As it happens, a quiet little Russian mathematician named Alexander Friedman timidly noticed that his idol, the giant Einstein, had made a mistake in his calculations. Mind you, there were no computers in those days and people had to add and subtract and wade barefoot without calculators into algebra, calculus, trigonometry, etc. Still, how does a little known mathematician inform his hero that he made a small but very dumb error? One which overlooked God?

It was a simple error. Albert had divided by zero. This is not a very efficient way to get to any kind of solution. When Friedman corrected the equation, the answer was absolutely different, and correct for a change.

Einstein was upset at the upstart and vented his opinion. Friedman wrote back timidly that the maestro must be off his strudel if he thought he could fool himself into thinking he could divide zeroes into all manner of things and get away with it. Not if old Alex Friedman was still patrolling the streets of quantum mechanics. Friedman in his usual forceful manner stormed up to Einstein, demanding, in the opening of his letter, "most honored sir, please let me know if my calculations are correct."

Imagine his surprise when Einstein finally divided with a number other than zero and came up with the creation of the universe. Einstein didn't like what he discovered, but he took the blame. He wrote to the "Zeitschrift" in 1923, "My objection to Mr. Friedman's letter rested on an error in calculation. Mr. Friedman is correct."

In the meantime, everyone was buzzing with the news. DeSitter went back to watching the universe expand, wondering if it was a balloon which would burst. Everyone was finally beginning to understand the excitement generated by Slipher's report back in 1914. The man who led the British eclipse expedition which started all the Einstein flack, the famous Arthur Eddington, started picking up where DeSitter left off. So did Hubble, who laid down his famous constant and jogged around the red shift like a man gone mad.

Finally, Einstein said, "This whole thing is beginning to irritate me, this expanding universe thing." He was again complaining that if everything got any more out of hand, they were going to have to consider something like God a little more seriously.

Word was getting around that Albert was disturbed about this

idea of the expansion of a finite university. It clearly implied that the world had a beginning. In a letter to DeSitter in Lerden, Einstein wrote, "This circumstance of an expanding universe irritates me. The whole idea seems senseless." These are very strong words from a modest, imperturbable man like Einstin who ranks at the top of the list of the greatest thinkers since Kepler and Newton.

We must stop here and consider again the thought that all of science has a kind of religion in which the scientist can work. It is a religion of sorts anyway. It supposes that the rules of the faith are the laws of nature. No problem about having to go to church every Wednesday or of not eating chicken or strudel on Saturdays. The rules were the rules of nature and man could reach out, gather them in, study them and use them. In a way, a very real way, the theoretical physicist is at prayer when he is at work because he is in search of the legislator of the laws of nature. In that respect, the work of a physicist is holy work. Not all physicists will admit that.

As a matter of fact, a rabbi sent a telegram to Einstein in 1921 asking, "Do you believe in God?"

Not one to duck a question, Einstein replied, "I believe in Spinoza's God, who reveals himself in the orderly harmony of what exists." That might sound cryptic of a great man, but he was referring to the philosophy of a very gentle philosopher named Baruch Spinoza who lived in the 17th century. Spinoza was perhaps one of the truly great minds of modern times, according to Will Dirant.* Spinoza, himself a Jew and a highly accomplished Biblical scholar, wrote:

> "All scripture does not explain things by their secondary causes, but only narrates them in the order and style which has most power to move men, and especially *uneducated* men to devotion. . . It's object is not to convince the reason of man, but to attract and lay hold of his imagination."

Hence the abundant number of miracles throughout all of liturgical literature. He goes on:

> "The masses think that the power and providence of God are most clearly displayed by events that are extraordinary, and contrary to the conception which they have formed of nature. . . They suppose, indeed, that God is inactive so long as nature works in her accustomed order; and vice versa, that the power of nature and natural causes are idle so long as God is acting; that they imagine two powers distinct from one another, the power of God and the power of nature.**

* "The Story of Philosophy p. 163

** "Tractatus—Theologico—Politicus, Ch. 5

This is the basic idea of Spinoza's philosophy—that God and the processes of nature are one. Got that? He felt that nature, its laws, all matter and God were one. Or, as Durant observed:

> "Man loves to believe that God breaks the natural order of events for him."

But Spinoza didn't believe that. To him it was all one, not one a cause and one an effect. Instead, he believed in a kind of pantheism.

And here was Einstein, without any consideration of the consequences, dumping the precepts of his own and Spinoza's philosophy by leading the great minds of his time down the paths of proof that the universe, and the Being that created that universe were separate and distinct entities. How about that? Einstein, an atheist, was the man who fashioned the bit used in drilling a hole in the boat of atheism itself.

For those for whom this thought is still unclear, let it be noted here that throughout the development of civilization there have been two kinds of thought: deductive and inductive.

Throughout the history of modern philosophy men have argued that everything in existence is the effect of some cause. Man is the effect of his parents and the cause of his children. An acorn causes an oak. Conversely, an oak is the effect of an acorn. Everything is caused and is an effect of a cause. All right, then take the universe as a whole. Who caused the universe? In other words, the universe is the effect of what? The answer: a creator. Then who caused the creator?

Until recently, and speaking very generally, the physicist said the universe caused itself. Nature created nature and until someone can separate the universe from the creator they are both infinite, never had a beginning and will never end. That theory is called the Solid State Theory. That is the way the inductive thinker arrived at his conclusions. To make matters even more difficult, the inductivists, the physical scientists, are the kind of people who demand proof. They are the people who say, "Don't tell me that water will boil at 212°. Heat it a few thousand times and eventually we may agree on your proposition."

Along comes Albert Einstein, the recognized brain of all the inductive physical science thinkers, with proof that this whole business is one big cosmic orgasm separate from the Guy who threw the switch. No wonder Einstein was irritated, as he wrote to DeSitter. He was saying, "This is not at all what I had in mind."

Origins of it All

Even the most astute and confident of scholars admit that the formation of our solar system remains unclear, particularly as to the mechanism which brought it into being as we know it.

We know that all the planets revolve and that most of them do so from west to east. In our little solar system, most asteroids occur between Mars and Jupiter and may or may not be related to the solar system. They fall toward the Sun, execute an orbit about it, and return into the depths of outer space. The origin of the Sun itself, with its enormous amounts of energy exploding into the cold of space, has always inspired speculation but remains a mystery to this day.

Despite what remains unknown, we can and do arrive at certain empirical deductions about the origin of existence. For instance, in searching out the true history of our origins, we analyze the chemical content of radioactive material, rocks, accumulated decay matter, radioactive disintegration, constancy, and direction of change from every aspect.

Decay, assumed to be a constant, gives us a good idea of the age of the matter presently in our neighborhood. For instance, igneous rocks seem to have solidified about 3 to 4 billion years ago. From this and other elements we can reasonably assume that the earth probably coalesced about 4.5 to 5 billion years ago, give or take a few hundred million years.

From the rocks the astronauts brought back from the moon we learn that the moon came into being around the same time as the Earth, or about 4.5 billion years ago. In other words, everything in our solar system may have been formed at about the same time. If so we're doing better than our sister planets; after all, we have people and they don't.

Since cosmic ray particles do not penetrate our larger meteors more than a few feet, the cores of the meteors remain unaffected and provide a way to determine their general age range. Meteors are closely related to the other members of the solar system and are very probably derived from the same substances, so their age provides a framework within which we can gauge the entire solar system to have been formed.

The chemical analysis is limited because some of the decay products we measure (like helium, neon and argon) are gaseous and usually escape the rock, making accurate measurement still relatively questionable. However, by analyzing the radioactive decay we can measure the elapsed time since the primordial material solidified. Most of the Earth's matter that has been measured in this fashion is very close to the 4.8 billion year old mark.

Going further, other measurements of other solar systems in the universe appear to indicate they also came into being about 4.5 to 5 billion years ago. You'll have to admit that we are narrowing the period in which the "big bang", assuming there was one, occurred.

Since it affects us all, I have found it very amusing to peruse some of the theories propounded by the better minds of this modern era. What would be better to research than our own birth?

There are those who suggest, indeed even teach, that the Earth acquired its moon by capture. The proponents of this theory hypothesize that the sun captured the Earth and its planets the same way. They conclude that the solar system itself was captured by the galaxy, and that the galaxy in turn was captured by other solar systems. This utimately leads one to think that the Supreme Designer is a great hunter in space and that the universe is a monumental celestial zoo.

Newton proved that one body cannot capture another by means of its own gravitation. The very laws of astronomical physics show that if a body moves in closer and closer to a star or planet, it finally reaches a point of nearest approach. It must then recede in an orbit exactly like the one with which it approached.

Gravitational laws rule out spiral paths. The orbit is an ellipse, or at any rate a parabola. Celestial mechanics shows that the space-traveling body must recede to infinity if it came from infinity. Therefore, to make capture possible we should have to find a way to circumnavigate the law of gravitation.

There is a very well thought out theory of solar creation that a third body could have intervened in the course of coincidental trajectory and reacted momentarily, just long enough to have affected a capture. In any case, some form of gravitational friction would have been involved. If that sounds too far-fetched, keep in mind that a heck of a lot of coincidences can occur in a period of more than a billion years or two, and even more within five or six billion years—which is our age.

The theory goes something like this: if a large meteoroid hurled at us from outer space came near enough to glance off the gravitational pull of the Earth, it would be slowed by our atmosphere. It would then be slung back into its home orbit but with less momen-

tum. Its return orbit would be increasingly slower after each glancing blow. The body's orbit eventually would be neutralized and it would become our moon when it stabilized within the equal sphere of gravitation. However, without other forces such a meteroid actually would become continually lower until finally it plummeted to Earth or was burned up by the atmosphere.

As with all bodies revolving around the sun, the Earth possesses something called *angular momentum*. For each circular orbiting solar planet the angular momentum is the mass of the planet multiplied by its velocity and distance from the sun. Although the sun is massive, its rotational velocity is relatively low and the distance from its center is so little it contributes only a small amount of the angular momentum of the entire solar system.

Fundamental in the mathematics of planetary motion is that the total angular momentum remains constant unless acted upon by an external force. It is that constant which has given rise to so many erroneous theories of the origin of the solar system.

Marquis Pierre Laplace offered an apparently reasonable theory which was very widely accepted because of his reputation as one of the greatest astronomer/mathematicians of the 18th century.

The theory was known as the Nebular Hypothesis. Laplace figured that all matter at one time floated around in space in a huge cloud or nebula. He thought small amounts condensed by random chance; once having some cohesion, a particle began to pull other matter in by gravitation.

Laplace knew that since all celestial matter had rotation it would have to contract. And as the same people can tell you, as the clouds contracted the velocities would increase. Increased velocity would lead to flattening, and eventually the body would become first oblique and finally disc-shaped. If the rotation was unstopped, it would continue to increase until the body became unstable.

Laplace hypothesized that at this point a ring of gas would form and detach itself from the equator to temporarily relieve the instability. The compression would continue and create another instability which in turn would create another ring. Finally, the compression of rings would be so hot a star would form at its center. Each ring was then supposed to coalesce into still another body called the proto-planet. The proto-planet would then condense and increase its rate of rotation until each succeedingly shed ring eventually became a satellite. There's no record that this imaginative genius was ultimately locked up after he concluded his explanation.

Keep in mind that Laplace was considered a very great mind in his field. His theory was widely accepted for a long time by some highly respected minds of the late 18th and early 19th centuries. Those were nice centuries and people tried to get along with each other, so no one really took much of a shot at the old boy for quite some time. At least not until Phoebe came along.

An astronomer from Harvard named William H. Pickering discovered Phoebe, a satellite of Saturn in 1894. Phoebe was the first object that revolved the wrong way; that is, counter to the general motion prevalent in the solar system. This gave everyone pause about Laplace's Nebular Hypothesis because now people were saying an asteroid captured by a planet and influenced from the sun's gravitation can become a satellite that revolves in retrograde.

Furthermore, Laplace's hypothesis didn't stand up when subjected to the law of conservation of angular momentum. If his theory of rings being shed was applied to Mercury, the rotation time would have decreased to once in 475 years. That would indicate clearly that if the sun continued to expand it would swallow one planet after another. If that happened, no instabilities could occur. That in turn blows up the whole idea of the basic principle of the proto-planet nebular hypothesis, leaving Laplace no longer the great astronomer/mathematician but just another nice guy who happened to be pretty good with arithmetic.

Now then, who comes along next? The one and only Thomas C. Chamberlin, the famous geologist (1843-1928) and his cohort astronomer Forest R. Moulton (1872-1952). They popped up with the Planetesimal Theory and the ever-popular idea of Tidal Hypothesis, brought to you by none other than Sir James Jeans.

It was again at the turn of the century that the law of angular momentum came into play. You remember that one: "the total angular momentum of a system must remain constant unless acted upon by some external force." Newton had known this as the law of inertia.

Chamberlin and Moulton said the external force in this case was the close approach of passing stars that produced tidal waves in the outer layers of the sun. That in turn caused the tearing off of huge chunks of matter that were either spun into outer space or sucked along after the passing star until the star left it behind to continue to revolve around the sun independently. This matter eventually cooled and became small lumps of planetesimals. These lumps then continued to orbit the sun, gathering all sorts of loose floating

matter. Eventually they cleared out the bits and pieces of debris left after the near miss with the now disappearing star still blazing off into infinity or on a new gigantic orbit which would return in a few billion years.

This whole theory pointed to the possibility that the crust of the Earth grew as a result of planetesimals rather than from the gradual cooling of an originally gaseous core. To those doubting this theory, adherents pointed to the pockmarked surface of the moon. They suggested the craters were caused by the impact of planetesimal residue.

Geologists were ecstatic since this idea cleared up a lot of questions they couldn't otherwise answer. This hypothesis didn't really answer much at all, but it gave the questioners reason to look into the matter further, thus getting them off the backs of the geologists who figured that if you want to study rocks you should forget about peering into constellations millions of light years away and reach down and pick up a few of them.

The tidal hypothesis theory also relied on the passing star's gravitational effect. British astronomers Jeans and Sir Harold Jefferys speculated that the passing star didn't cause the wrenching away of globs of planetesimal matter. Instead, they said, the passing star pulled away a long tube of matter which stretched out between the sun and the passing star. Eventually this snake-like tube broke up into sections, cooled, and became the proto-planets of our solar system. They suggested the orbits of these bodies were long ellipticals. Due to close brushes with the gravitation of the sun, eventually they straightened out into decently behaving planets one could count on as befits proper members of a respectable solar system.

There were many variants of this passing star-planetesimal-tidal idea. One such was held by the threesome of Lyttleton, Spitzer and Hoyle. Hoyle suggested that the sun once was a double star which first exploded in half. One half then exploded again and degranulated, leaving the planets to cool individually into the solar system as we know it today. Considering that the universe is loaded with double stars, about one third of the known ones, this theory easily took hold and overshadowed the idea of interstellar collisions. Considering the forces which held that the universe in uniform balance would have to be suspended, and that such collisions are not at all common because they would have to be outside the ordinary course of natural law, such a hypothesis would be rather unique. And yet as we shall see later, celestial accidents in nature not only seem to happen, they are built into the very nature of things. They may some day prove to be predictable.

You may argue that nature breaks down as it gets older. Since the universe is approaching its six billionth birthday one might expect a breakdown now and again. Still, since breakdowns are built into the natural order of random things they occur in a natural manner and therefore are potentially predictable. Getting old, in other words, is natural. The failures which result not only are natural in themselves, but the ailments themselves are simply part of the predictable future of a cell, a body, or an entire system whether it be a galaxy or a universe. And although aberrations in nature do occur, they are relatively rare. Given that many years of experience, it's entirely possible that the sun did or will yet blow a fuse.

The eminent Dr. Donald H. Menzel, former head of the Harvard University Observatory, mentioned that, "One may briefly sum up this idea in a sentence: The solar system resulted when the sun 'blew a fuse'." (This theory) borrows significant features from most of the major earlier hypotheses, but has the merit of being a unitary hypothesis, "that is, it does not require the cooperation of a second star or outside body."

Another Harvard-Smithsonian astronomer, Fred Whipple, thought the sun might have once been surrounded by a "swarm" of meteoroids acting as an envelope. The gases encountered by those meteoroids were captured and condensed into cores which built up into planets which proceeded in orbit in the formation of our solar system.

Then there is the idea that the sun was traveling along five or six billion years ago when it ran into an enormous cloud of interstellar dust. As we have seen, unless something else intervened most of this matter would have been pulled into the sun. But as the sun got into the cloud, the ultraviolet radiation would have released electrons from the atoms and ionized them. The magnetic field of the sun, (which could have been much greater billions of years ago, not having burned off as much fuel at that point), could have deflected these charged particles into rough orbits around the sun. Once again, this could have led to the production of planets through condensations.

The other matter which makes up clouds of gas which didn't become planets could have been pulled along by the solar wind.

These clouds of hydrogen move contrary to gravity at velocities as high as hundreds of miles per second. Yet for reasons we don't as yet understand, they do not dissipate in space. Curiously, some force also keeps them from expanding. Recent motion picture studies from our space probes suggest that this force may be electric current. This current, of course, is related to magnetic fields. Such current has been shown to be more powerful than the tendency to be explosive, long enough at least to allow condensation to take place.

In any case, it's heartening to know some real meaty mysteries about the "how" of the universe are still bouncing around in the halls of higher learning. There are at least as many as there are solutions. Or, as one noted astronomer/stargazer, Howard Solomon, observed, "I don't know for sure. But it's fun looking at it if you've got a girl along."

What are the latest theories of the origins of the solar system? As we look at its farthest edges and accumulate evidence from both NASA and the Russian space programs, we have more knowledge of the extremes of space to bring to bear in our speculations. As the years go by not only do we grow closer to the origins of our own little solar system, we find ourselves closer to an understanding of the origins of our galaxy, and in turn the origin of the entire universe.

From the time man first looked up at the stars and wondered how they got there, automatically he was asking how *he* got here. That began the search to find the legislator of the laws by which we all must live. . .the laws of nature. For no matter how wise or ignorant we all are compared to each other, we all must live under the same stars, stand on the same planet and speculate from the same relative ignorance at the wonder of nature.

It seems that at this time (time being only a relative element of convenience, having no value even as a constant measurement), we've arrived at a conclusion that the universe started all at once and that it's expanding in a uniform and predictable manner.

There are laws of nature, a complex and infinite order to everything in the universe, though we have little ability to appreciate it. Wherever there is order, there is intelligence. Wherever there is intelligence, there is purpose. What can be the purpose of an infinite intelligence? It's our glory, as suggested by the medieval theologians. Glory comes only from the accolades of a higher dignity to a lower one—not from a lower to a higher one. What then is the purpose of a supreme intelligence in creating a universe?

To understand the origin of the universe let us imagine ourselves in an elevator which is descending at terminal speed. That is, we're inside a cube of space which is falling at the same rate we are. In a very real way we're unmoving compared to the box in which we fall, making us essentially weightless since we and it are falling at the same speed. While we're in that six-sided room, weightless, we must imagine what's outside the elevator. What kind of building is the elevator in? What kind of street is the building on? In what kind of town? Country? Planet? Solar system? Galaxy? And finally, what kind of universe is it all in?

Maybe we shouldn't even use the word finally. Could this be only one universe alongside many others which make up an even greater universe? One of many stacked inside each other and all of which make up one molecule, a part of which makes up all the molecules which constitute the chair on which you're sitting while you consider all of this complicated mess?

Well, we have the sun, a swarm of planets, stars which are centers of condensation, and other galaxies, but no guarantee that we will ever see the walls of the universe or even know if there are walls at all.

Currently, some popular speculation has it that the universe empties into itself rather than ending. It goes on, in other words, forever emptying while filling itself up.

Another currently popular origin theory was conceived by C. K. von Weizsacker, the German astronomer, edited by G. P. Kuiper. Like Laplace, they suggest it all began with a passing star which caused a long, distended nebula which then condensed. von Weizsacker differs from Laplace in his estimation of the frequency of the condensations in addition to that caused by the sun. He also proposed that the nebula were proto-planets which swept up nebular material after reaching orbit around the sun. This theory further proposes that when the sun was very young, the solar wind plus the pressure caused by radiation threw out part of its matter into the vast reaches of outer space. The planets were lumps left behind because they were traveling at a different velocity. They cooled. And now they are us.

Interesting theories. As we send more probes to study the comet belt, perhaps we'll discover the relationships among the matter found throughout the solar system.

We now know that time is relative as long as there are no fric-

tional forces involved; forward and backward are essentially relative only to something else. That is, if a train is moving east to west at 400 miles an hour, and the earth is moving west to east at 800 miles an hour, you're actually going 400 miles per hour the wrong way, aren't you?

At any rate, if we lived on a different planet and a space traveler brought back motion pictures of our solar system we wouldn't know which was forward and which was backward, would we? As a matter of fact, we cannot distinguish the future from the past if all we have to go on is the movement of the planets.

Just as astrology was set aside by astronomers as the basis of our fundamental knowledge of the physical universe, so chemical and geological physicists gradually were brought into greater prominence in the study of cosmology.

Let us now take a cooler look and be a little more objective about our origin. We can take all these theories and reduce them to three concrete possibilities. But first, let's look again at what we have here.

In our own Milky Way system, there are more than 100 billion stars, each a sun in its own right. Had early astronomers known that all the stars were not relatively near each other in the various constellations, the zodiac they contrived to tell time might be very different. As astrology gave way to the advent of astronomy, the astronomers at first propounded the same error as the astrologers. They thought that all the stars were roughly the same in brightness. They didn't know that the ratio between the brightness of a first magnitude star and that of the sun is about $4x10^{10}$. Therefore, neither did they know that the intensity of light varies inversely as the square root of the distance. Had they known that, they would have found that the stars seen by the naked eye are $2x10^{6}$ or 200,000 times farther away from us than the sun, and that those telescopically visible are 10 times more distant than that. As a matter of fact, our own sun is relatively lack-luster. Most of the stars one sees with the naked eye are as much as 100 times brighter than the sun.

Astronomers measure the distance of a star by the apparent motion of more distant stars as the earth moves around the sun. If that seems complicated, listen to this. The displacement which appears to result from the motion of the earth through a distance equal to the radius of a star's orbit is the parallax of the star. To put it simply, it may be said that all efforts to determine distances between stars prior to the discovery of the telescope had to fail. The measuring tool, the telescope, was missing.

In fact, it took two centuries of development for the telescope to become sufficiently refined for reliable use in measuring distance. Until photography was invented (particularly celestial photography), no one could prove that the measurement of the magnitude of the displacement would provide a determination of the parallax, and thus the distance between stars.

In addition, James Brady (1692-1762) found that the telescope does not measure the true position of the star, but the direction from which the starlight appears to be coming. This brings into the proposition that the velocity of starlight, which is called *abstraction,* (the wandering of light) is an important factor because the great distances involved make quite a difference in the time it takes for light to travel to here. All this in turn helped lead to the discovery that our own Mother Earth is itself in orbit, and provided a basis for determining the velocity of light itself.

All of this information sheds a shadow over the entire field of astrology since it means that those stars which make up a constellation actually are hardly related to each other at all except in the most contrived ways.

Going further, there has been more increased knowledge in astronomy in the last 20 years than that accumulated in the previous three centuries—since the death of Galileo. The advent of atomic engineering alone provides a periscope into the very nature of the matter of which the stars and planets are made.

As we continue to try to understand the universe in terms of our own solar system, we may take heart. We may be uncertain as to how the Earth came into being, but we can come very close to knowing how our sun, and its sole parent, the universe, came to be.

Three theories offered in the 20th century describe the birth and development of the universe. They are known as (A) the Big Bang, (B) the Oscillating Universe Theory, and (C) the Steady State Theory. Let's look at them in simple terms.

The *Big Bang Theory* holds that about 10 or 15 billion years ago there was a vast fireball of extremely hot gasses consisting of helium and hydrogen. It became so hot it exploded. As with all explosions, an expansion took place. In this case, the expansion is still going on, as evidenced by the *red shift.*

Concentration of matter developed as time went on and the gasses continued to expand. Each mass became a galaxy. Fragments

formed to become stars of the first generation. These stars emitted other elements, adding to the hydrogen and helium. The new elements were then ejected by their parent stars into interstellar gas to be used as raw material in the formation of stars.

The *Oscillating Universe Theory* suggests that expansion which began with the Big Bang sooner or later will come to a halt because of gravitational force. Then it will begin falling into itself in a giant collapse which will bring everything back into another fireball much like the one which exploded in the first place. Then the second Big Bang will occur, and the evolutionary process will commence all over again with each successive collapse, explosion and expansion becoming progressively weaker and less forceful than the previous one. After an innumerable amount of years it will all go out permanently. That is why cosmologists and astrophysicists suggest that the universe is "winding down" even though it is in the act of exploding. It *is* expanding, but in ever decreasing force until eventually it will collapse in on itself.

Finally, there is the *Steady State Theory,* which offers the idea of no beginning and no end to the universe. It has always been and always will be the way it appears now, this theory holds. The Steady State proponents insist that as galaxies get old, new galaxies are formed in the empty spaces left behind. The gasses, dust, and energy given off by the stars are the essence from which new stars are produced. Since Einstein showed that energy is mass, there is much to be said about this possibility.

As of now, the theory of the Big Bang takes scientific precedence among those who are most accredited. However, keep in mind that the majority of thinkers often have been proven very much in error in their own eras. Be warned. In no other science have there been more intelligent, responsible thinkers so wrong so often as in the study of astronomy. No matter whether or not we ever know the answer to how the universe began, to gaze at the sky is to look at once into the past, present and the future.

Mythological Stories of the Constellations

Until Victorian times, the study of classical mythological subjects denoted a truly well educated person. Latin and Greek were taught with a considerable amount of reference to mythology. In the 20th century, mythology fell into disuse as a field of academic enlightenment. It disappeared as a formal subject with the advent of trade schools which were direct, no-nonsense endeavors of practical education reminding even classical educators that there is purpose to learning, that much of that purpose is application, and that there is little purpose to which mythology can be put.

Let us now look to the constellations as catalogued by the ever popular Ptolemy. Ladies and gentlemen, let's hear it for the original Love Boat, Argo Navis.

Argo Navis
The Ship Argo and Its Parts

Carina, Puppis, Pyxis, Vela

No longer considered a single constellation, this grouping has been subdivided into four separate constellations which make up the ship: Carina, the keel; Puppis, the stern; Pyxis, the compass; and Vela, the sails.

Located entirely in the southern hemisphere, Argo Navis is the largest constellation and contains over 800 naked eye stars. Its original star designations by de Lacaille used 180 Greek letters; many repetitions lent to confusion and error. More recent astronomers have separated the original constellation of Argo Navis into four parts of the boat. Sometimes Malus, the mast, is used to further define Argo Navis, which covers almost 75 degrees of the sky. It is south of Monoceros, Hydra and Sirius (the brightest star in the sky). A few of the most northern stars of Puppis and Pyxis can be seen on the horizon in New York City in early March. Conopus, the second brightest star, is in Carina.

The Argo and its parts are of the richest fables of Greece, the quest for the golden fleece. So much has been written, story tellers often get it confused with Homer's tale of the Oddessy. Jason and Ulysseus were probably actual people performing great adventures, but they lived two or three hundred years apart. Isaac Newton says the expedition was based on a true voyage which took place about 30 years before the destruction of Troy and 43 years after the death of Solomon.

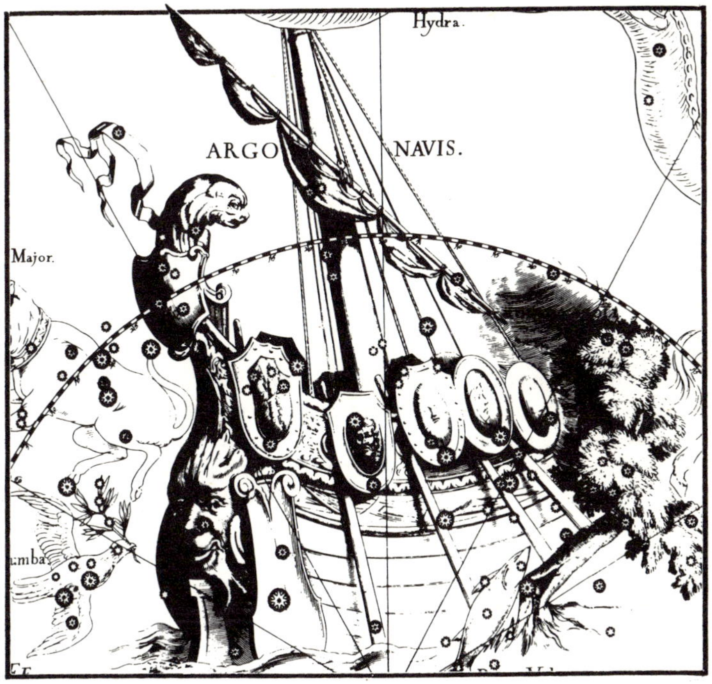

There are those more practical historians who claim that Argo Navis really comes to us as the last part of the story of Noah's Ark. Feel free to take your choice.

Our choice perpetuates the memory of the famous ship which carried Jason and his 54 companions to Colchis when they went upon the perilous journey to recover the golden fleece.

The derivation of the Argo is disputed, and has been something of a problem to scholars. Why, no one knows. Some say it's named after the man Argos, who planned the expedition originally and built the ship. Others maintain it's named after Argos, where it was built. Cicero called it Argo because it carried Greeks who com-

monly were called Argives. Diodorus called it swift, which in Diodoran or Scicilian sounded like Argos.

The ship had 50 oars and was supposed to be the biggest ship built at that time which could be carried by the crew across land passage. Until that time boats were made from hollow or dug out trees. For Argos to build a vessel like this was an engineering marvel akin to our space program. The ship was designed by Pallas Athene, who had the power to give oracle or knowledge to the Argonauts, as the crew called themselves according to other mythological scholars. The crew are called Argonauts, "Argo" the ship and "nauts" meaning sailor in Greek. From this comes astronauts, cosmonauts etc.

Of all the mythological stories from which you may discover some basically truthful elements, the story of Argo Navis is perhaps the most fascinating since its mythical fantasies may actually be based on some elements of truth. For instance, there seems to have been an actual ship and a real voyage which may have inspired the abundance of literature about it. Like Troy, at one time a city only thought to be a fanciful place, later was found actually to exist. The story of the Argo Navis appears to have had some basis in fact. Here is a brief version of this famous ship and its voyage.

Argo Navis was the ship which carried the Argonauts from Thessaly in Greece to Colchis on the Black Sea in Soviet Georgia, in Jason's search for the golden fleece.

There are so many descriptions of this ship, what it was, what it was made of, it might be best to provide an overview to give the reader a broad description from which to draw his own picture.

With Athene's help, Argos built his ship as a galley. In her prow, Athene herself set a plank from the speaking oak of Dodona, now Albania. In the foredeck, Wisdom placed the mast of truth, which spoke in Oak—a language you seldom hear these days.

It has been disputed as to how many argonauts eventually were assembled. Some stories have as many as 50 crew members. At any rate, their number at least matches the number of Divine Attributes. All of them are centered about the incarnate Self: Jason.

Considering that the ship Argos Navis was the inspiration for five new constellations plus many other mythological stories and speculations which grew from the voyage itself, there must be some basis in fact for the description of the ship and the voyage. (Remember it was from writings like these that the German archeologist, Heinrich Schliemann, deduced the fact that there really was a Troy. In digging it up, he not only found the treasures of Troy, he disguised it as a scientific effort and today Schliemann often is given the credit for establishing the science of archeology).

The enterprise undertaken by Jason and the Argonauts was a collective venture that stirred the heroic world profoundly.

Jason was a Thessalian hero. Aeson, his father, was the son of Cretheus and Tyro. (Tyro was the object of Poseidon's affections). His kingdom, Iolcus in Thessaly was stolen by his greedy half-brother Pelias, son of Tyro and Poseidon. Aeson gave his son Jason to the centaur Chiron for training and to keep safe.

When Jason reached manhood, he left Pelion, the home of giants and centaurs, and made his appearance at Iolcus wearing a panther skin, holding a lance in each hand and having no sandal on his left foot. Pelias was in the act of offering a sacrifice. When he saw Jason, he got scared, for an oracle had told him to "beware of the man who had only one sandal." When Jason walked up to the king and demanded his legal right to the kingdom, Pelias dared not openly refuse him. Instead, he delayed things a bit by asking him first to bring back the golden fleece of the ram which Phrixus and Helle had taken to Greece in earlier days. The fleece was well-reputed to be sacred to Aeetes, King of Colchis. Everyone knew that Aeetes, king of the land and son of the sea-nymph, had the fleece guarded by a fearsome dragon who never slept. Naturally, Pelias figured that this mission meant the last of the troublesome Jason.

Jason accepted the mission, then sought the advice of Argos, son of Phrixus. Prompted by Athene, Argos built the first longship, the Argo, which could take Jason and his chosen companions to Colchis on the far shore of the Black Sea. The Argo Navis was built in the port of Pagasaea in Thessaly, and was made with wood from Mount Pelion.

Jason chose 50 of the best of those volunteering to be Argonautical contenders. History has confused the complete list, but certain names crop up repeatedly: Orpheus, the musician who provided rhythm for the rowers; Tiphys, the helmsman who had been trained by Athene; the soothsayer Idmon; the two sons of Boreas the North Wind—Calais and Zetes; Castor and Pollux and their two cousins, Idas and Lynceus. Hercules is sometimes named, but he never made the return from Colchis according to other histories.

The trip started out well enough. The omens were favorable. The first stop was Lemnos, a Greek Island in the Agean, near

Turkey, where there were only women. The women had killed all their own men for being unfaithful to them after Aphrodite had put a curse on them. The Argonauts were received with great favor and stayed several months, and the subsequent population of this visit was a new race of sailors. Jason fathered twin sons with the Queen of Lemnos, Hypsipyle.

At Samothrace, another Agean Island, the Argonauts were initiated into the mysteries celebrated on that island which would protect them from shipwreck. Then they entered the Hellespont into the Sea of Marmara and the Harbor of Cyzicus. There they were favorably received by the king, who was celebrating his wedding feast. After much feasting they set sail from Cyzicus, but during the night the wind changed and blew them back to Cyzicus territory. Unrecognized because of darkness, they were taken for pirates and attacked. During the fight, the king was killed. Eventually everyone recognized everyone else and the fighting stopped. Cleite, the king's bride of one day, in sorrow hung herself. In his grief over the death of the king, Jason founded the funeral games in his honor. Later the games were said to be the original Olympic games.

The Argonauts next headed for the Mysian coast. Here young Hylas wandered away and was kidnapped by water nymphs. Hercules went to look for him and didn't come back either. Those water nymphs must have been providing something more interesting than the golden fleece.

Leaving Hercules and Hylas behind, the ship set out once more to the Island of Bebrycos, still in the Sea of Marmara. The evil King Amycus ruled this island and would challenge strangers and travelers to deadly boxing matches. Pollux, skilled in fisticuffs, took up the challenge and killed the king. According to other sources, he made the king promise to shape up his future behavior.

The next day, a storm washed the Argo up on the coast of Thrace, where Phineus ruled. He was a blind seer, a son of Poseidon, upon whom the gods had set a singular curse. Whenever he wanted to eat, the Harpies, winged demons, flew down and snatched the dishes from the table, fouling what they couldn't grab. The sons of the North Wind flew after them and made them promise by the Styx never to torment Phineus again.

A grateful Phineus prophesied the future of the Argonauts, advising them to be careful of the Symplegades (Blue Rocks). Those were reefs guarding the entrance to the Bosporus. When a ship wanted to pass, they came together to bar the way. Phineus told the Argonauts to try an experiment before going through the straits. They were to send a dove between the reefs. If it got through, the boat was to follow. If it failed, there was little point in heading for certain disaster. The Argonauts' dove caught only one of its tail feathers. So when the reefs were well apart again, the Argo sailed through with full speed ahead, losing only one plank from its stern. From this point on, the Blue Rocks remained motionless, and the way into the Bosporus has been open ever since.

After visiting Mariandyne, where the soothsayer Idmon, who had foreseen his death from the beginning, was killed by a bear while out hunting, the Argo entered the mouth of the Thermodon and reached Colchis. The helmsman, Tiphys, died shortly before this and was replaced by the hero Ancaeus.

The Argo Navis itself sailed again on a voyage to Arcadia, the traditional home of the gods, but it was lost. There have been archeologists who feel there is a relationship between the Argo and the great ship of Noah—that one fable led or gave birth to the other. That we will leave to those who would like to delve into the detective work on the origins of mythical stories. It has been considered a most classical enterprise, both intellectual and scientific if not altogether the pursuit of a romantic.

Andromeda, the Woman Chained

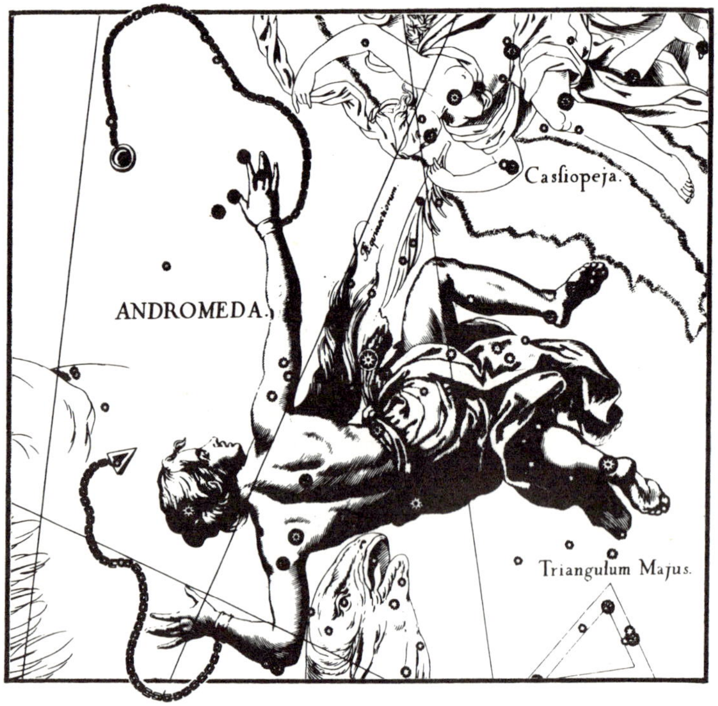

Represented by the figure of a woman with her arms extended, chained by her wrists to a rock.

The farthest object in the sky visible to the naked eye is the famous Andromeda spiral galaxy. It is very similar to our own milky way in size and structure. It usually appears as a hazy glow and is known as M-31. The head of Andromeda is shared with one of the corners of the "Great Square" of the constellation Pegasus. It is a 2.2 magnitude and named "Alpheratz" or "Horse Navel." The second brightest star in the group is Mirach, meaning waist or girdle. Mirach is a Yellow 2.3 magnitude star. Andromeda is best viewed in autumn and early winter.

The daughter of Cepheus, king of Ethiopia, and Casstopeia, Andromeda was promised in marriage to her uncle, Phineus. Meanwhile, Neptune drowned the kingdom and sent a sea monster to salvage the country to appease the resentment of his favorite nymphs against Cassiopeia. The nymphs, numbering over fifty, resented the queen because she boasted that she was fairer than Juno and the Nereides, as sea nymphs were called.

The oracle of Jupiter, Ammon, was consulted. He said nothing could pacify the anger of Neptune unless the beautiful Andromeda should be exposed to the sea monster. Accordingly, Andromeda was chained to a rock.

All of this happened near Joppa (probably now Jaffa in Syria). Anyway, the minute the monster saw her, he knew he was onto a good thing and went for her. That's when Perseus, who was returning through the air from the conquest of the Gorgons, saw her. Thinking that the monster had ideas outside the ordinary course of monsters, Perseus made his own move on her. As the eminent mythologist, Phineus Malone, put it:

"Chained to a rock she stay'd
His rapid flight, to woo the beauteous maid."

Perseus promised to deliver her and destroy the monster if Cepheus would give him her hand in marriage. Cepheus consented and Perseus changed the sea monster into a rock by showing him Medusa's head, which he held up triumphantly. The jealous, enraged Phineus, who had been betrothed to Andromeda, started a violent battle in which he, too, was turned into a rock by the petrifying influence of the Gorgon's head.

Aquila, the Eagle

Often represented as an eagle and a man, Antinous. This is a double constellation situated directly south of the fox and the swan.

Aquila is easily located by its brightest star, Altair, Arabic for eagle. Altair is a bright, white star of .9 magnitude only 16 light years away, almost a next door neighbor. Altair is part of the Summer Triangle with Deneb in Cygnus and Vega in Lyra.

Aquila was formally pictured as carrying Antinous, the favorite page of Emperor Hadrian, to heaven in its talons. This configuration was discontinued by all but the German astronomers long ago.

Mythology usually attributes the eagle to Jupiter or Zeus. One story is that Aquila carried Ganymede, the most beautiful of all mortals, to Olympus to fill the cup of Zeus and live among the gods. As a reward, the eagle was included in the constellation.

Another myth has it that this is the eagle that brought nectar to Jupiter while he was lying low in the cave on Crete, avoiding the fury of his father, Saturn.

Some of the ancient poets say this is the eagle which furnished Jupiter with weapons in his war with the giants.

"The tow'ring Eagle next doth boldly soar
As if the thunder in the claws he bore
He's worthy Jove, since he, a bird, supplies
The heaven with sacred bolts, and arms the skies."

–Manilius

The eagle is justly styled the "sovereign of birds" since he's the largest, strongest and swiftest. As Homer says, he's:

"The royal bird to whom the King of heaven
The empire of the feather's race has given."

Or, as someone else has said:

"An eagle is also not so bad a thing in golf
If you happen to be a duffer."

–Anonymous

Auriga, the Charioteer or Wagoner

Represented by the figure of a man in a declining posture, resting one foot on the horn of Taurus. A goat and her kids are in his left hand, a bridle is in his right hand.

The seventh brightest star in the sky, Capella, with a magnitude of .1, is in Auriga. Capella, the she-goat, is in the Charioteer's left shoulder, brilliant yellow in color. The Flaming Star (I.C. 405) is in Auriga and can be seen with binoculars in the winter.

The Greeks gave various accounts for this constellation. Some supposed it to be Erichthonius, the fourth king of Athens, son of Vulcan and Minerva, whose many inventions earned him his place among the constellations.

He was described as looking like a monster, which did not make him very popular, but it was Auriga who invented the chariot and the book. Maybe that is why it has been said "don't judge a book by its cover." It is written that Auriga was excellent in his management of horses. Both Virgil and Dryden wrote about him. Said Dryden of his chariot:

"Bold Erichthonius was the first who join'd
Four horses for the rapid race designed
And o'er the dusty wheels presiding sate."

Other writers say Bootes invented the chariot, and that Auriga was the son of Mercury and charioteer to Oenomaus, king of Pisa. He so excelled in charioteering that he outraced everyone in Greece. But neither of these fables account for how he came to have a goat and one of her kids as his companion. They are said to be the goats of Almathaea and her sister, Melissa, who fed Jupiter goat's milk during his infancy. As a reward for their kindness, Jupiter placed them in the arms of Auriga. Mythology has become clouded over the centuries and is at fault for this confusion.

Other mythologists insist that since it was Bootes who invented the chariot, Auriga does not appear with one. There is also the story of how Auriga's daughter, Hippodameia, arranged for Auriga to lose his chariot and his life.

Hippodameia was very beautiful and had so many suitors that her father devised a way to dispose of them. He decreed that anyone who wanted to marry her first had to win a chariot race with him. If the suitor lost he would be put to death. Naturally, that slimmed down the number of boyfriends hanging around the palace.

Then, along came Pelops. So taken was Hippodameia with this god-like lad, she spoke to her father's charioteer, Myrtilus. Flashing a promise of love for him if he would fix the race, Myrtilus pulled some of the spokes off of her father's chariot. Not only did her father lose the race, he lost his life as well. She married Pelops who took over the kingdom of Arcadia, the peninsula which we now call the Peloponnesos of Greece.

Bootes

Represented by the figure of a running huntsman grasping a club in his right hand. In his left hand, he holds the leash of his two greyhounds, Asterion and Chara, with whom he's hot on the trail of the Great Bear around the path of the heaven.

Bootes is a large constellation and easily located. To find Arcturus, the brightest northern star, just extend the handle of the Big Dipper to the big, orange-colored bright star. Arcturus is in the knee of Bootes. Izar, meaning girdle, is at the waist of Bootes. A brilliant meteor shower of 100 meteors per hour eminates from the head of Bootes between January 1st and January 4th.

Sometimes Bootes is called Arctophylax, or the bear guard. He was misunderstood by these mythmakers. He was actually *chasing* the bear, not *guarding* it.

The ancient Greeks called this constellation Lycaon, a name derived from "wolf." The Hebrews called it Caleb Anubach, the "barking dog." The Latins called it Canis, among other names.

If we go back to the time when Taurus opened the year and Virgo was the fifth zodiacal sign, we find the brilliant star, Arcturus, so remarkable for its red and firey appearance, corresponding with a period of the year just as notable for its heat.

Pythagoras, who introduced the true system of the universe into Greece, got his information on that subject from the Egyptian, Oenuphis, a priest of On. This college of the priesthood was the noblest of the east in cultivating the studies of philosophy and astronomy. Among the high honors Pharaoh bestowed upon Joseph was "a son of the priest of On." The supposed era of the book of Job, in which Arcturus is repeatedly mentioned is 1513 B.C.

Some claim Bootes is really Iacchus (Baccus), who was killed by shepherds for intoxicating them. Others say he's the same chariot inventor, Ericthonius. According to the Grecian fable and later authorities, Bootes was the son of Jupiter and Callisto and was really named Arcas.

Ovid says Juno was steamed at Jupiter for his partiality to Callisto and changed her into a bear. Her son, Arcus, who became a famous hunter, one day unexpectedly roused a bear. Not knowing the bear was his mother, he was about to kill it when Jupiter luckily snatched them both up to heaven and enshrined them as constellations.

In a work called "Pharsalia," Lucan says:

"That Brutus, on the busy times intent,
To virtuous Cato's humble dwelling went,
When bright *Callisto, with her shining son,*
Now half that circle round the pole had run."

Camelopardalis, or Camelopardus, the Giraffe

Hevelius made this constellation out of the unformed stars scattered between Perseus, Auriga, the head of Ursa Major, and the Pole Star.

It marks the northern boundary of the temperate zone, being less than one degree south of the Arctic Circle.

This strangely-named constellation took its name from the giraffe, peculiar to Ethiopia. Not surprisingly, it resembles both the camel and the leopard. Its body is spotted like that of a leopard and it has a seven-foot-long neck.

Its fore and hind legs, from the hoof to the second joint, are nearly of the same length. But from the second joint of the legs to the body, the forelegs are so long in comparison with the hind ones, no one can sit on its back without instantly sliding off the way one would from a rearing horse.

Since there are 58 small stars in the constellation, the five largest of which are only of the 4th magnitude, and its principal star is in the thigh, it seems natural that it be named after a truly strange animal.

As Edgar Guest would have said if he had known of it:

"Isn't it a marvel, Alice,
to see the Camelopardalis?"

Canes Venatici, the Hunting Dogs

The name Dogs of Set, in Egyptian gave rise to the title Canes Venatici. Set was a generic term applied to all circumpolar constellations. Canes Venatici, however, usually was accounted for in reference to the hounds of Callisto and as the hunting dogs of Bootes.

Mentioned by Hevelius, although described previously by Ptolemy, the more northern constellation is Asterion (starry) from the stars marking the body of the dog, and Chara, the one which was closest to the breast of her master.

In any case, the hounds are now firmly established and recognized by all astronomers as two greyhounds held on a leash by Bootes, ready to chase the Bear around the pole. The astronomer Proctor called them Catuli, the puppies of Bootes.

Canes Venatici contains M-51, the Whirlpool Galaxy.

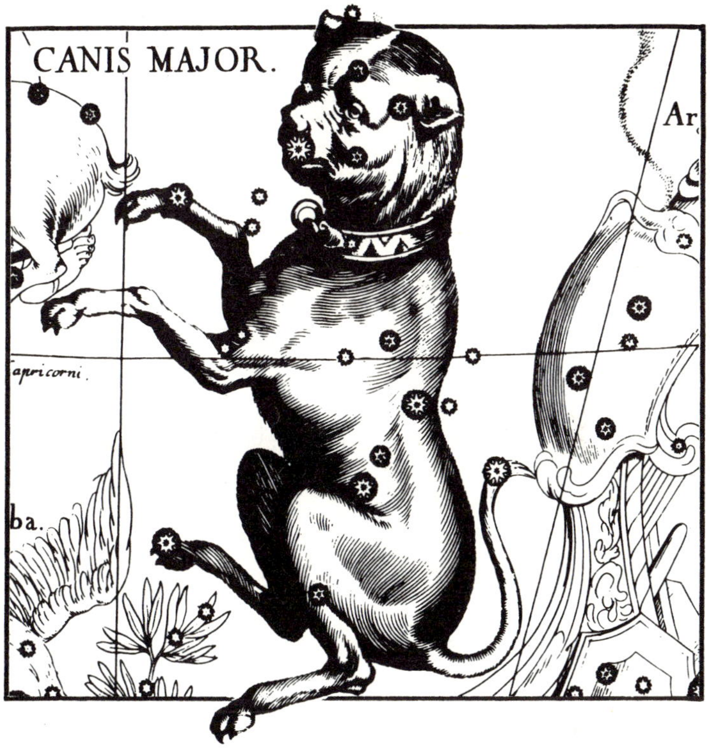

Canis Major, the Greater Dog

Represented by a dog at the feet of the twins, southeast of Orion, the celebrated huntsman.

According to some mythologists, this constellation marks one of Orion's hounds, placed alongside him to keep him company. Others say it honors the dog, Lelaps, given to Caphalus by Aurora, which was the fastest of all dogs. To prove his dog's superior speed, Cephalus pitted him against a fox, which, until then, had the reputation Caphalus was now claiming belonged to Lelaps. After the two animals had run neck and neck for a long time, Jupiter was so delighted at the speed of the dog that he immortalized him in the heavens.

The name and form of this constellation no doubt was derived from the Egyptians. They carefully watched it rise and judged the swelling of the Nile (which they called Sirius) by it. Hieroglyphically, it was depicted as a dog since it was viewed as a sentinel and the clock of the year. Clever folks that they were; the Egyptians had noticed that when Sirius became visible in the east just before dawn, the Nile immediately overflowed. Sirius thus effectively warned them like a faithful dog that they must get out of the path of inundation each time the Nile banks overflowed.

Sirius is the brightest star in the sky and the closest star to our sun. It is easily located projecting southward through the three stars in the belt of Orion to the brightest star.

Canis Minor, the Lesser Dog

Represented by a dog situated midway between Canis Major and the twins.

Procyon, a brilliant white star, is the eighth brightest star in the sky. It has a magnitude of .3 and is best seen in March.

According to Greek fable, the little dog is one of Orion's hounds. Some credit it as being the Egyptian goddess, Anubis, who was represented as a dog's head. Some say it's Diana, the goddess of hunting. Skeptics of those theories say it's the faithful dog, Maera, which belonged to Icarius. Maera showed Icarius' daughter, Erigone, where Icarius had been buried. People who don't believe any of that say it's one of Actaeon's hounds which devoured its master after Diana had transformed him into a stag to prevent his betrayal of her.

> "This said, the man began to disappear
> By slow degrees, and ended in a deer.
> Transform'd at length, he flies away in haste,
> And wonders why he flies so fast.
> But as by chance, within a neighb'ring brook,
> He saw his branching horns, and alter'd look,
> Wretched Actaeon! In a doleful tone
> He tried to speak, but only gave a groan;
> And as he went, within the watery glass,
> He saw the big round drops, with silent pace,
> Run trickling down a savage, hairy face."
>
> –*Anonymous*

It's probable that the Egyptians did invent this constellation. Because it always rises a little before the dog star (which they dreaded from season to season), it properly represented to them a watchful animal designed to give due notice that inundation time soon would be at hand.

Cassiopeia, or Cassiope

Represented in regal state seated on a throne her left hand holding a palm tree branch. Her royal head points to the Arctic Circle. She is surrounded by the royal family.

Her husband the king is on her left. Her son-in-law, Perseus, is on her right. Her daughter Andromeda is just beneath her.

All are said to be wondering when she will fix dinner.

Cassiopeia is easily recognized by its "W" shape formed by its five bright stars in the shape of her chair or throne.

Cassiopeia was the wife of Cepheus, king of Ethiopia. She was also the mother of Andromeda. Cassiopeia believed herself to be a queen of unparalleled beauty, more beautiful than Jupiter's sister Juno or the Nereides (sea nymphs).

She may indeed have deserved the reputation, but her bragging about it didn't sit too well with the ladies of the sea. They took it as a personal insult and complained about it to Neptune the brother of Jupiter. Neptune allowed that the complaints might hold water and as a start sent a sea monster to ravage the coast to punish her for insolence. Next, he demanded that she chain her daughter, Andromeda, to a rock on the beach, leaving her exposed to the fury of the sea monster. Cassiopeia did as she was told, and the monster licked his chops in delicious anticipation. That was all the taste he got, however, for Perseus stepped in just in the nick of time and saved the fair young maiden.

"The saviour youth the royal pair confess,
And with heav'd hands, their daughter's bridegroom bless."

—Eusden's Ovid

Centaurus, the Centaur

Represented by figure of a man in the body of a horse. He holds a wolf at arm's length with one hand, while he transfixes its body with a spear in the other.

Centaurus is a large, brilliant constellation in the southern hemisphere. Alpha Centauri is the nearest star to the sun and the third brightest object in the sky. It has .1 magnitude. Beta Centauri has a magnitude of .9. A line drawn north through Alpha and Beta will locate the Southern Cross for you.

Alpha Centauri, because of its close proximity to earth, has been the object of numerous space odysseys written by science fiction writers.

Mythology maintains that centaurs were fabulous half man, half-horse monsters. This is open to some interpretation. Some suppose Centaurs had the body of a shepherd and the head of a herdsman's horse. Herdsmen inhabiting the mountains of Arcadia were rich in cattle. They were probably the first cowboys.

Plutarch and Pliny believed that such monsters really existed. Not everyone believed them, others said that under the reign of King Ixion of Thessaly, a herd of bulls ran mad and ravaged the

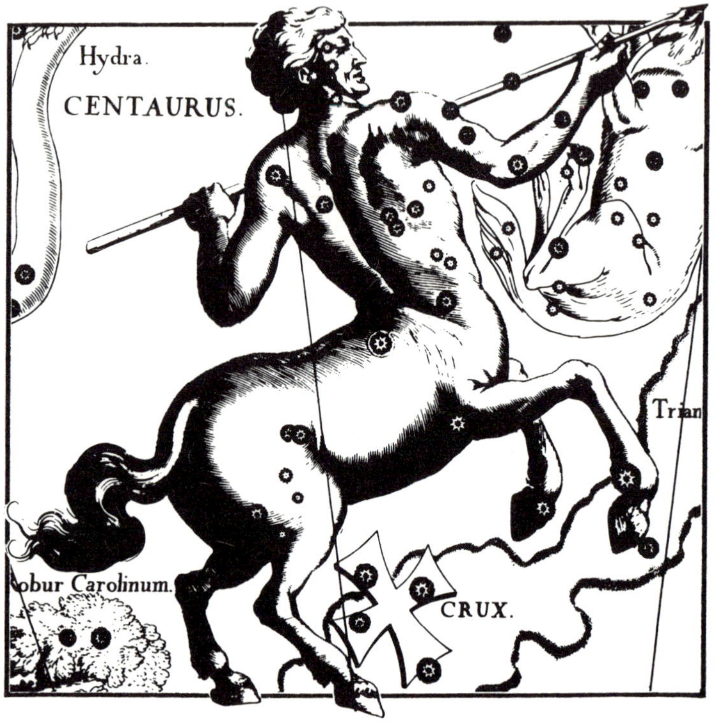

whole country, rendering the paths to the mountains inaccessible. Some of the young men who had learned to tame and mount horses decided to take the bulls by the horns, so to speak, and expel the bothersome animals. They pursued them on horseback and gained the name of Centaurs because from afar they might seem to be half man and half horse with their human heads raised above the herd.

Unfortunately, these fine young men became cocky about their success and were insolent. This insulted the people of Thessaly. When the Lapithaeans gave chase, the upstarts fled so fast they again appeared to be half horses and half men.

Horseback riders were a rarity in those days. The villagers can be forgiven for perceiving a distant man on horseback to be a single being. After all, that's how the Spanish cavalry at first appeared to the astonished Indians in the Americas, who imagined the horse and rider to be some monstrous, singularly fearsome animal.

The Centaurs actually belonged to a tribe from Lapithae near Mount Pelion, which first invented the art of breaking horses, according to Virgil:

> "The Lapithae to chariots add the state
> Of bits and bridles; taught the steed to bound;
> To turn the ring, and trace the mazy ground;
> To stop, to fly, the rules of war to know;
> To obey the rider, and to dare the foe."

Cepheus

> Represented on the map of the heavens as a king in royal robe, a crown of stars on his head. He stands like a general, his left foot over the pole, the scepter in his left hand extended toward Cassiopeia as if for a favor and in her defense.

When you speak of Cepheus, you are referring to the king against whose life almost all of mythology touches. Jupiter, Neptune, none overshadow him.

It is written that he was the son of Belus and grandson of Neptune and Libya. In the true line of mythological royalty, his great-grandfather was old Jupiter, himself.

There is not much written about Cepheus, the only king of the sky, except where it relates to his royal family.

His queen, you remember, was Cassiopeia. They were the parents of Andromeda, who was betrothed to Perseus. Cepheus was one of the famous Argonauts who accompanied Jason on his daring expedition for the Golden Fleece. That is why he was raised to live in the stars. Cepheus is closely related to the voyage of Argo Navis.

At the wedding feast of his daughter, Andromeda, and Perseus, Phineus and his gang arrived, violently protesting the marriage of his formerly betrothed, Andromeda. Cepheus, in his empirical wisdom, said, "you should have claimed her on the rock when she was the monster's intended lunch!" A fight ensued and Perseus

turned Phineus and his band into stone by exposing them to the same Gorgan's head with which he had destroyed the monster. Some versions hold that Phineus was Cepheus' brother.

Cetus, the Whale, or Sea Monster

Represented as a whale, chief monster of the deep, the largest of the aquatic race. It is fittingly a large constellation.

Most writers consider this the famous sea monster who was sent by Neptune to gobble Andromeda up after her mother, Cassiopeia, bragged about how beautiful she was. You will recall how this monster got his come-uppance from Perseus. At any rate, its origin is buried somewhere in the mists of dark antiquity.

"The winged hero now descends, now soars,
And at his pleasure the vast monster gores.
Deep in his back, swift stooping from above.
His crooked sabre to the hilt he drove."

It's been established that this constellation had a place in the heavens long before Perseus. When the equinoctial sun in Aries (right over the head of Cetus) opened the year, it was nicknamed the "Preserver" or "Deliverer" by the idolaters of the East. According to Pausanias, this is why the sun was worshipped at Eleusis under the name of the "Preserver" or "Saviour."

"With gills pulmonic breathes the enormous whale,
And spouts aquatic columns to the gale;
Sports on the shining wave at noontide hours,
And shifting rainbows crest the rising showers."

–Darwin

Coma Berenices, Berenice's Hair

The Pin Wheel Nebula and the Black Eye Galaxy are in Coma Berenices. It is a faint constellation with an amber color. Our word for varnish comes from a linguistic derivature of Berenices.

Bernice was an Egyptian princess in the third century B.C. She promised to shave her head in sacrifice to Venus if her brother Ptolemy IX, the Pharoah, would return safely from battle in Syria and Babylon. After his safe return her hair disappeared overnight before the ceremonial tonsorial exercise. The missing hair was discovered by court astronomers that evening between Bootes and Leo. Part of the Syrian/Egyptian connection is related in the Old Testament in the book of Daniel XI-6.

Corona Borealis, the Northern Crown

Represented by six principal stars placed to form a circular figure resembling a wreath. It's directly above the Serpent's head between Bootes on the west and Hercules on the east.

This beautiful little cluster of stars is said to commemorate a crown. Bacchus gave this crown to Ariadne, the daughter of Minos, second King of Crete. Her lover, King Theseus of Athens, was shut up in the celebrated labyrinth of Crete, scheduled to be the dinner of the ferocious Minotaur who lived there. The half-man, half-bull Minotaur fed upon the chosen young men and maidens which the Athenians gave as yearly tribute to the tyranny of Minos.

Daring fellow that he was, Theseus slew the monster. To keep her adored Theseus from being lost in the maze while he was slaying the creature, Ariadne gave him a thread so he could follow it back to the outside when he was through. The appreciative Theseus married Ariadne, as he'd promised and carried her away. But when he arrived at the island of Naxos, he deserted her, despite her love for him and regardless of the evidence of her endearing tenderness and obvious attachment.

Ariadne was so disconsolate at her abandonment, some say she hanged herself. Plutarch says she lived many years afterward and married Bacchus, who loved her with much tenderness. He gave her one big party after another: coming out parties, showers, bachelor parties, pajama parties. He was even getting a bad name for being a playboy. Finally, he gave her a crown of seven stars which was itself placed among the stars after her death. Only then did he finally go on the wagon, crying himself to death.

Manilius, in the first book of his "Astronomicon," says of the crown:

"Near to Bootes the bright crown is view'd
And shines with stars of different magnitude:
Or placed in front above the rest displays
A vigorous light, and darts surprising rays.
This shone, since Theseus first his faith betray'd
The monument of the forsaken maid."

Corvus

A small constellation, the "crow" is situated on the eastern part of Hydra. It contains five easily visible stars.

The crow, it is said, was not of the purest white as it had once been. Because of this, the gods argued and changed it from nondescript gray to black.

"The raven once in showy plumes was drest,
White as the whitest dove's unsullied breast,
Fair as the guardian of the capitol,
Soft as the Swan; a large and lovely fowl;
His tongue, his prating tongue, had changed him quite
To sooty blackness from the purest white."

According to Greek fable, Apollo made the crow a constellation. Apollo was jealous of Coronis, the daughter of Phtegyas and mother of Esculapius. He sent a crow to check out her behavior. Coronis, the bird perceived, was partial for Ischys the Thesalian, and immediately snitched to Apollo. The jealous god shot Coronis with an arrow, killing her instantly.

"The god was wroth; the colour left his look,
The wreath his head, the harp his hand forsook;
His silver bow and feather'd shafts he took,
And lodged an arrow in the tender breast,
that had so often to his own been prest."

As seems to have been habitual in those days, Corvus, the crow, was rewarded by being placed among the constellations.

Another legend has it that the constellation takes its name from the daughter of Coronaeus, King of Phocis, who was transformed into a crow by Minerva to rescue the maid from Neptune's hot pursuit. The following, from an eminent Latin poet of the Augustine age, is her own account of the metamorphosis as translated into English verse by Mr. Joseph Addison:

"For as my arms I lifted to the skies,
I saw black feathers from my fingers rise;
I strove to fling my garment on the ground;
My garment turned to plumes, and girt me round:
My hands to beat my naked bosom try;
Nor naked bosom now nor hands had I:
Lightly I tripp'd, nor weary as before
Sunk in the sand, but skimm'd along the shore;
Till, rising on my wings, I was preferr'd
To be the chaste Minerva's virgin bird."

Cygnus, the Swan

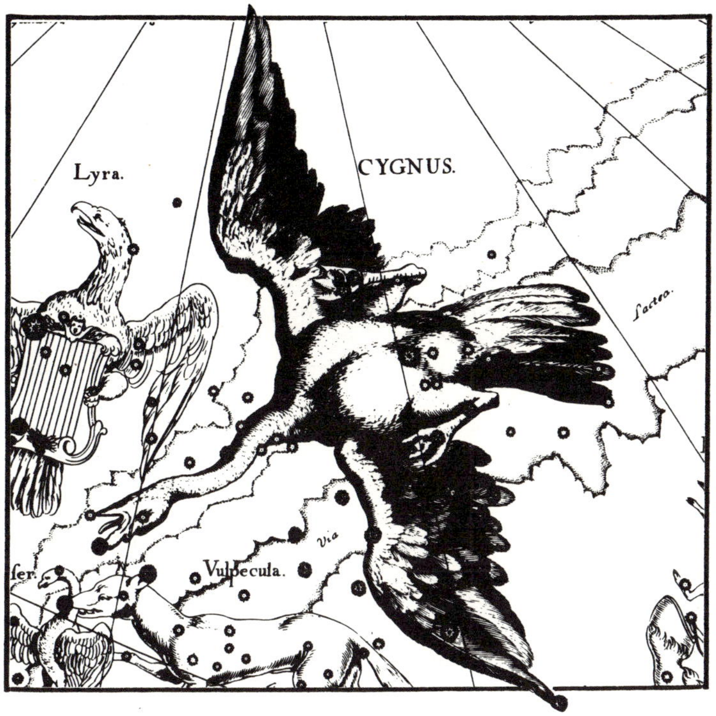

Represented as a swan flying down the Milky Way, its wings outstretched. The body and bill form a cross.

Cygnus is also known as the Northern Cross because of its configuration. It's brightest star is Deneb, which is part of the Summer Triangle with Vega and Altair. This constellation is next to Lyra and below Cepheus and is easy to locate in the summer sky.

The Veil Nebula and the North American Nebula are both visible with binoculars. A meteor shower, Cygnids, is seen from August 10th to August 20th each year.

As often is the case, mythologists have different accounts of the constellation's origin. One story says it's the celebrated musician, Orpheus, who was murdered by a cruel priestess of Bacchus. At his death, he was changed into a swan and placed near his harp in the heavens. Some refer to Cygnus as the pet swan of Queen Cassiopeia.

Another version says it is the swan into which Jupiter transformed himself while deceiving Leda, wife of the King of Sparta. The results of these liaisons were the twin, Pollux, and the lovely Helen of Troy. Another story says it was Cionus, a son of Neptune, who was so completely invulnerable that neither javelins nor arrows nor the blows of mighty Achilles in furious combat could touch him.

"Headlong he leaps from off his lofty car,
And in close fight on foot renews the war; –
But on his flesh nor wound nor blood is seen,
The Sword itself is blunted on the skin."

When Achilles saw that his darts and blows were getting him nowhere, he immediately threw Cionus on the ground and smothered him. While he was attempting to rip off his armor, Cionus was suddenly changed into a swan.

According to Ovid, the constellation took its name from Cycnus, a friend of Phaeton, who deeply lamented the untimely fate of that youth and was turned into a swan by Zeus. Phaeton's sisters mourned and wept so much that Zeus turned them into weeping willow trees to line the banks of Eridanus.

Of all the feathered race, there is probably no bird more beautiful and majestic. Almost every noted poet has so honored the swan. In virtually all ages and all countries where taste and elegance have been cultivated, the swan has been considered the emblem of poetical dignity, purity, and grace. The ancients consecrated it to Apollo and the Muses and believed that the bird foretold its own end and sang most sweetly at the approach of its death. To this day we relate sad endings to the term, "A Swan Song."

Delphinus, the Dolphin

Represented as a dolphin with four principal stars in the head. Situated northeast of the Eagle. Some know it as Job's Coffin, though the origin of the name is unknown

Delphinus is a compact constellation with five principal stars that look like a small kite with a tail. It lies below the swan, between the eagle and the flying horse, an unusual place for a dolphin to reside.

Some mythologies say the dolphin was made a constellation by Neptune because one of these beautiful creatures had persuaded the goddess, Amphitrite, who had made a vow of perpetual celibacy, to become his wife.

Others say this is the dolphin which saved the poet and musician, Arion, native of the island of Lesbos. Arion traveled to Italy with Periander, tyrant of Corinth, where he sang and played music until he became very rich. Once aboard sailors schemed to murder him and divide up his great wealth. Discovering his death was being planned, Arion requested permission to play a last tune on his lute. The melody of the instrument attracted a number of dolphins, whereupon Arion dove in among them. One of the dolphins purportedly carried the poet/musician to Taenarus, a promontory of Laconia in Peloponnesus. There Arion went to the Court of Periander, who ordered all the sailors to be crucified when they returned.

When the famous poet, Hesiod, was murdered in Naupactum, a city of Etolia in Greece, his body was thrown into the sea. Some dolphins reportedly brought the floating corpse back to shore where Hesiod's friends discovered it. Using the poet's dogs, they tracked down the assassins and drowned the murderers in the same sea.

Taras, the founder of Tarento in southern Italy, was saved from shipwreck by a dolphin. The city's inhabitants preserved the memory of this extraordinary event by placing the image of a dolphin on their coins.

Although the natural shape of a dolphin is almost straight, many artists, sculptors, and poets depict it as incurvated to accommodate a rider on its back.

The dolphin is an extremely swift fish, capable of living a long time out of the water. In fact, it seems to delight in leaping out of the water and otherwise frolicking about.

Draco, the Dragon

Represented by the head of a monster under Hercules' foot. A coil reaches down to the girdle of Cephus, then loops below Hercules to form a second coil passing down again between the heads of the Lesser and Greater Bears.

The windings of this elusive constellation are symbolic of the oblique course of the stars. Draco winds around the pole of the world, as if to indicate in the symbolic language of Egyptian astronomy the motion of the pole of the equator around the pole of

the ecliptic. The Egyptian hieroglyph for the heavens was a serpent whose scales denoted the stars. When astronomy first began to be cultivated in Chaldea, Draco was the polar constellation.

Mythologists sometimes represent Draco as the watchful dragon which guarded the golden apples in the garden of the Hesperides near Mount Atlas in Africa. The apples belonged to Juno and were the twelfth labor of Hercules. Juno took Draco up to heaven and made a constellation of him as a reward for his faithful services.

Others say that in the war with the giants, this dragon fought Minerva, who grabbed the snake and hurled its twisted body into the heavens around the axis of the world before it had time to unwind. It sleeps there to this day as Draco.

Another ancient version of Draco says that this is the dragon which Cadmus, brother of Europa, killed when Cadmus was ordered by his father to go find his sister after Jupiter had carried her away; he was told not to come back without Europa. When his search proved fruitless, he consulted the oracle of Apollo who told him to build a city where he saw a heifer stop in the grass. He was to call the city Thebes.

In good time Cadmus saw the heifer. Wishing to thank the god with a sacrifice, he sent his companions to fetch water from a neighboring grove. As luck would have it, the waters were sacred to Mars and were guarded by a fierce dragon who decided on the spot to have dinner.

When his companions didn't come back, Cadmus grew tired of waiting and went to look for them just as the dragon was licking his chops over the last of the bones. Cadmus swore either to get revenge or to die trying. He attacked the monster and, with the help of Minerva, killed him. Then he plucked out the dragon's teeth and sowed them in a plain as Athena had told him to do. Suddenly, the teeth sprang up and became armed men, startling our hero. Just as Cadmus was about to beat a hasty retreat, the men started fighting among themselves. All but five were killed. Those remaining five helped Cadmus build the promised city.

Equuleus, the Foal

> The Little Horse, or the Horse's Head, is in the head of Pegasus, and halfway between it and the Dolphin.

This constellation is supposed to be the brother of Pegasus, named Celeris. Mercury gave him to Castor, who was celebrated for his skill in the management of horses.

Another story credits him with being the famous horse which Neptune struck out of the earth with his trident when he argued with Minerva for superiority.

Only the head of Celeris is visible. It is represented in an inverted position. Equuleus is the second smallest constellation in the sky and is very faint.

Eridanus, the River

> Sometimes called the River Po part of it lies between Orion and the Whale. It resembles the shape of the northern stream And flows from Rigel in the foot of Orion to the Whale, where it makes a complete circuit and returns to its source until it disappears below the horizon for which everyone is grateful.

Eridanus is the longest constellation and the sixth largest in this celestial kaleidoscope.

Achernar, meaning "End of the River," is a blue-white star and the 9th brightest star in the sky. Achernar, being very southerly, can be seen in New Orleans at Thanksgiving time.

The Beta star, Cursa, marks the other end of the river. Cursa is a topaz-yellow star with a 2.9 magnitude. It is next to Rigel and sometimes called Orion's footstool.

Eridanus is the name of a celebrated river in Northern Italy, known today as the Po River. Virgil called it the king of rivers. The Latin poets have immortalized it in connection with the fable of Phaeton, son of Phoebus (Apollo) and the sea nymph, Clymene. Phaeton was also a favorite of Venus, who entrusted him with the care of one of her temples. Becoming vain at the honor, he asked his dad to give him an unmistakable public sign of his love so the world would be sure to know who he was. Phoebus balked at first, but then gave in and said OK.

Thinking of the grandest gesture possible, Phaeton decided to take his father's place in the sun's chariot for a day. Phoebus was most distressed at his decision and tried to convince Phaeton how

foolishly dangerous it was by describing the incredible force needed to move the stars and planets. But the impetuous youth was not to be dissuaded. If his father didn't like the idea, Phaeton concluded, it must be a worthwhile ride.

Ignoring his father's fears and advice, Phaeton leaped into the chariot and took the reins and headed off in an unpredicted direction. Properly scared, Phaeton decided this wasn't such a good idea after all. Heaven and Earth were being threatened with a universal conflagration as the fire of the sun spilled out. Jupiter saw what was happening and hurled a thunderbolt at Phaeton, propelling him overboard and into the river, Eridanus. His burned body was found by nymphs who honored him with a hero's burial. The waters of the river sometimes steam because of Phaeton's firey demise and descent. Phaeton's sisters mourned his unhappy end and were changed by Jupiter into weeping willows.

"All the long night their mournful watch they keep,
And all the day stand round the tomb and weep."

–Ovid

It is said that the tears of the sisters turned to gold with which the Phoenicians and Carthaginians carried on a lucrative secret trade. The great heat produced when the sun popped out of its usual orbit is said to have dried up the blood of the Ethiopians and turned their skins black. Worse, it is also said to have produced sterility and barrenness over the greater part of Libya.

This fable evidently alludes to some extraordinary periods of heat in a very remote time. Only this confused tradition remains today, but it may have been derived from an actual event.

Hercules

Shown with the skin of the Nemaen Lion
holding a massive club in his right hand and
the three-headed dog, Cerberus, in his left.

A large constellation betweeen Lyra and Bootes, with one foot on the head of Draco, Hercules is the fifth largest group in the sky.

It contains M-13, an impressive globular star cluster easily seen with binoculars.

Alpha, Ras Algethi, meaning "the kneeler's head" is a red-supergiant star with a blue-green companion, Beta, Korneforos, which is a pale-yellow star in the shoulder of Hercules.

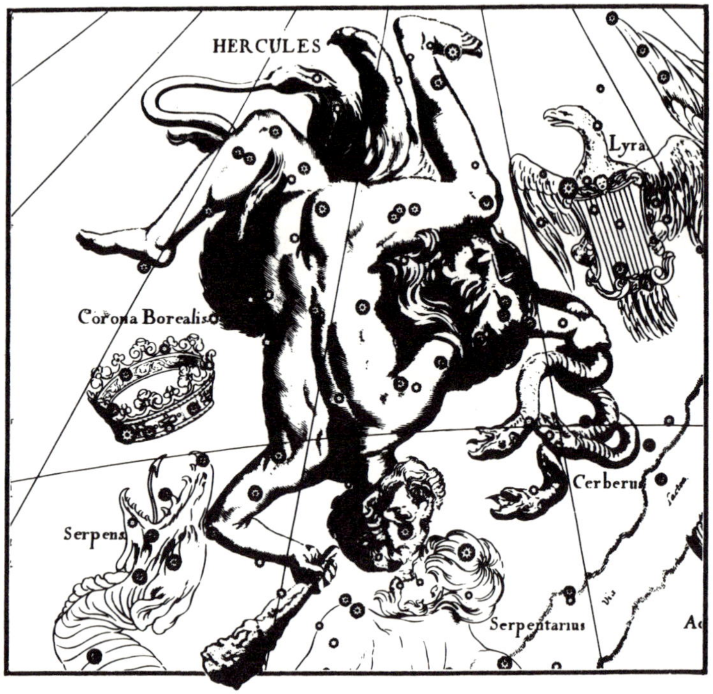

Hercules was celebrated in antiquity for his heroic valor and invincible prowess. According to the ancients, there were many people known by this name. Of these, the son of Jupiter and Alcmena is the best known, and usually gets credit for the actions of the others.

The birth of Hercules is said to have been attended with many miraculous events in Thebes. Before he was nine months old, Juno was jealous of him. She sent two snakes to devour him, but he boldly seized them and squeezed them to death. His brother, Iphicles, alerted the household with frightful shrieks. Iphicles, The Helpful, he was called!

Hercules was instructed early in the liberal arts. He soon became the pupil of the centaur, Chiron, under whom he rendered himself the most valiant and accomplished of all the heroes of antiquity. At eighteen, he started getting serious about undertaking great and glorious pursuits. He subdued the Thespian Lion of Cithaeron that

had devoured the flocks of his foster father, Amphitryon. Then he delivered his country from the annual tribute of a hundred oxen which it had been paying to Eriginus, King of Orchomenus, and the Minyan people. The Minyans were a troublesome neighbor of Thebes.

King Thespiae, a benefactor of these aforementioned deeds, rewarded Hercules by allowing him to sleep with all fifty of his daughters. After accomplishing this deed in seven nights, all fifty became pregnant, the oldest and the youngest daughters giving birth to twins. Descendants of the Thespian daughters and Hercules populated the island of Sardinia, west of Italy.

Jupiter had decreed that Hercules would be subject to the will of his cousin, Eurystheus, King of Mycenae, and must obey his every command. Eurystheus was jealous of Hercules' rising fame and power and ordered him to appear at Mycenae to perform certain tasks. Hercules refused but when he consulted the oracle of Apollo, he was told he must be subservient for twelve years because Jupiter had said that he must. Afterwards, he could straighten things out with the gods.

So, back Hercules went, determined to do the labors and get them over with while keeping his dignity intact. Upon seeing so great a man totally subject to him and also apprehensive of such a powerful enemy, Eurystheus commanded Hercules to carry out the most difficult jobs ever devised, now known as the Twelve Labors of Hercules.

Here is what he accomplished:

1. Subdued the Nemaen Lion in his den and took his hide.
2. Destroyed the Hydra, with a hundred hissing heads, and dipped his arrows in the gall of the monsters to render them venemous and incurable.
3. After a twelve-month pursuit, took alive the stag with golden horns and brazen feet famous for its incredible speed. He presented it, unharmed, to Eurystheus.
4. Took alive the Erimanthian Boar and killed the Centaurs who opposed him.
5. Cleaned the stables of Augias in which 3000 oxen had been confined for many years.
6. Killed the carnivorous birds which ravaged the country of Arcadia and fed on human flesh.
7. Took alive the wild bull of Crete which no mortal dared look upon and brought it into Peloponnesus.
8. Obtained the mares of Diomedes for Eurystheus. They fed on human flesh; Hercules fed their owner to them.
9. Obtained the girdle of the Queen of the Amazons, a formidable nation of warlike females.
10. Killed the three-headed monster, Geryon, the King of Erytheria, and brought away his man-eating flocks. He also killed his two-headed watchdog names Orthus.
11. Brought up to Earth the three-headed dog, Cerberus, guardian of the entrance to the infernal regions.
12. Obtained the golden apples from the garden of the Hesperides which had been guarded by a fearsome dragon.

This not being enough to satisfy him, Hercules set other tasks for himself. Before he delivered himself up to the King of Mycenae, he accompanied Jason and the Argonauts on Argo Navis to Colchis. He assisted the gods in their wars against the giants, and it was through him, alone, that Jupiter obtained the victory. Then he conquered and pillaged Troy and killed King Laomedon. Hercules is indeed the greatest hero in mythology.

Hydra, the Water-snake

A water serpent, probably so named because it wanders just about everywhere in a space of more than 100 degrees in length. It lies south of Cancer, Leo and Virgo, and reaches almost from Canis Minor to Libra.

Hydra has only one significant star. Then known as "Alphard" or "the solitary one", it is the heart of the sea snake. It is an orange 2 magnitude star just below the configuration of the snake's nose and head. Hydra is best viewed from February to May.

The astrologers of the east, in dividing the celestial unknown into various compartments, assigned a popular and allegorical meaning to each. Thus the sign Leo, which passes the meridian about midnight when the sun is in Pisces, was called the House of the Lions, Leo being the domicile of Sol.

Of the two serpents among the constellations, the polar one represented the oblique course of the stars, while the Hydra (Great Snake) in the southern hemisphere symbolized the moon's course. The Nodes are called the Dragon's head and tail to this day.

The Lernean hydra was a terrible monster infesting the neighborhood of Lake Lerna in the Peloponnesus. Diodorus, the first century B.C. historian who suppossedly knew about such things, claimed that it had a hundred heads. Simonides (600 B.C.) said it had 50. The more commonly agreed-upon opinion said it had nine. At any rate, as soon as one head was cut off, two immediately grew up if the wound wasn't cauterized by fire. So the stories were bound to differ depending on how many sword slices and how many fires there may have been before any witnesses got to the scene.

Destroying this cute little monster was one of Hercules' labors. Iolaus, his nephew, lent a hand by applying a burning iron to the wounds as soon as Hercules cut one of the heads off. While this was going on, Juno, jealous of the glory Hercules was accumulating, sent a sea-crab to bite off his foot. Hercules did in the sea-crab as well (no big deal after a few hydra heads), thwarting Juno's attempts to lessen his fame.

This fable of the many-headed hydra may represent nothing more than the fact that the marshes of Lerna were infested with a multitude of serpents, which seemed to multiply as fast as they were destroyed. The story of the Hydra lends more drama to the myth of Hercules.

Lacerta, the Lizard

The lizard extends from the head of Cepheus to the star at the foot of Pegasus, its northern half lying in the Milky Way. The almost inconspicuous constellation was formed by Hevelius from outlying stars between Cygnus and Andromeda.

A minor meteor shower from Beta Lacertids can be seen in August and September.

The reason for the lizard shape was simply that no other shape could fit into the available space. Hardly a romantic explanation, even though it was celebrated as an offering to Urania.

Lacerta has never been known as a pretty picture. In the past it was named a weasel, greyhound, newt and by the Chinese as the flying serpent.

Lepus, the Hare

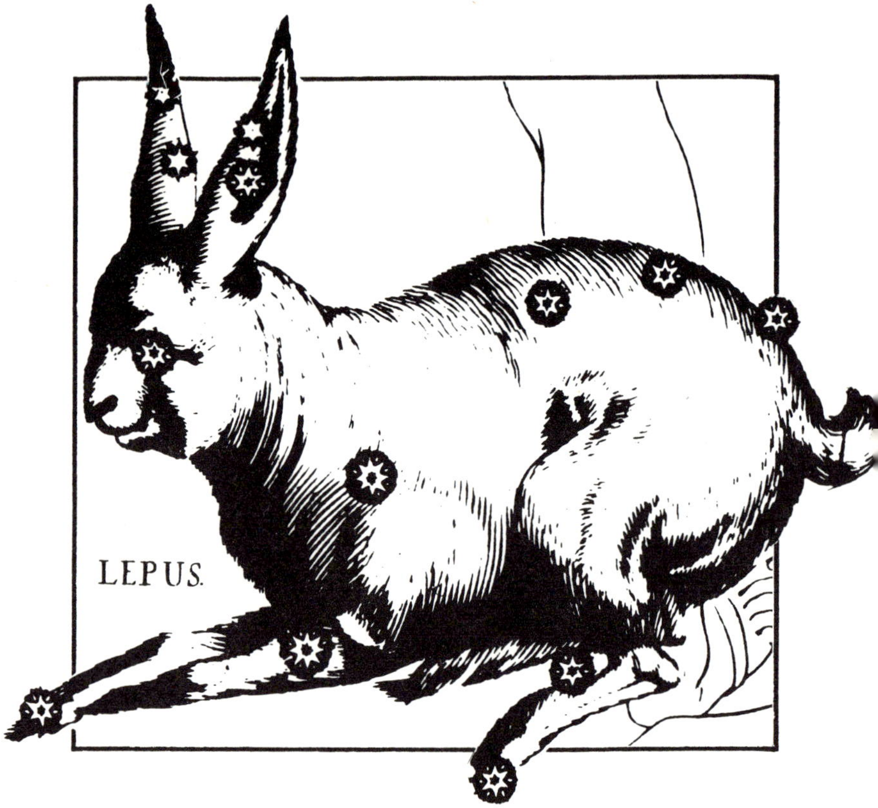

Represented as a hare, it's situated directly south of Orion.

This constellation is situated about 18 degrees west of the Great Dog. From the motion of the earth, the Great Dog seems to be pursuing the fleeing Lepus. Apparently, the dog took his clue from the Greyhounds, which pursue the Bear in a similar infinite chase scene in the sky.

It was a hare that Orion was said to have taken such delight in hunting. Because it was a favorite of his, it was put in the sky for Orion's continual pleasure. The hare probably wishes it were a bit less favored.

Leo Minor, the Lesser Lion

The "Little Lion" was one of Hevelius' creations, conjured up from unformed stars of the ancients scattered between Leo on the south and Ursa Major on the north.

Leo Minor was described as a lion because the ancients didn't know what else to name the grouping of small stars so near the great lion. Leo Minor contains 53 stars, including only one of the 3rd magnitude, and only 5 of the 4th magnitude.

Leo Minor remained undescribed for centuries. It is a relatively new constellation, having been added to our celestial maps only since the adoption of Greek notation in 1603. These are referred to by the letters of the English alphabet instead of the Greek.

"The *Smaller Lion* now succeeds; a cohort
Of fifty stars attend his steps;
And *three,* to sigh unarm'd, invisible."

This is to say that Leo Minor does not conform to the shape of a lion, but rather is simply a small constellation near the Great Lion from which it took its name.

Lupus, the Wolf

Illustrated as a wolf captured by the Centaur and south of Libra. Situated so low in the southern hemisphere that only a few stars in the group are visible to us.

According to fable, this constellation is Lycaon, King of Arcadia. He lived about 3,600 years ago and was changed into a wolf by Jupiter because he offered human victims on the altars of the god Pan

It was said that the sins of mankind had become so great, Jupiter visited the earth to punish its wickedness and impiety. He came to Arcadia where he was announced as a god. The people began to pay proper adoration to his divinity.

Lycaon, however, who used to sacrifice all strangers to his wanton cruelty, laughed at the pious prayers of his subjects. To try the divinity of the god, he served up human flesh at his dinner table. Jupiter was understandably perturbed at what he perceived as a thumbing of the nose, and immediately destroyed the house and changed the sassy Lycaon into a wolf. Lycaon, as Lupus, spends eternity in the grasp of Centaurus.

Lynx, sive Tigris, the Lynx or Tiger

Lynx is known best for both the beauty of it's white-lilac color and the profusion of double stars which number more than fifty according to the "Webbs' Celestial Objects."

There seems to be some confusion about the fact that the constellation is also sometimes called the Tiger, which is amusing. The very academic celestial etymologist, Richard Hinckley Allen, in his book, "Star Names, Their Lore and Meaning", states with a straight pen and I assume a straight face, "The alternative name, now in disuse, came from the fancied resemblance of the many stars to spots on the tiger". So academically myopic was Allen, he failed to notice that although a lynx has spots, tigers only have stripes. Such are the vagaries of celestial speculation by scientists too devoted to their school work when they should have their minds out there somewhere in the stars.

Lyra, the Lyre or Harp

The lyre, or harp, is situated directly south of the first coil of Draco between the Swan on the east and Hercules on the west.

Vega, a brilliant blue-white star of 0 magnitude is the 2nd brightest in the Northern sky. It is the brightest of the summer triangle with Deneb and Altair. Vega will replace Polaris as the pole star in about 12,000 years because of precession of the Earth's North Pole.

Lyra, because of Vega, is easy to locate. Lyra also contains the Ring Nebula of some interest to astronomers.

It's generally agreed that this is the celestial lyre which Apollo or Mercury gave to Orpheus, musician of the Argonauts. Orpheus played it so well that even the fastest rivers slowed to listen, the wild beasts of the forest forgot their ill manners, and the mountains moved to listen to his song.

Of all the nymphs who used to listen to his song, Euydice was the only one who made a deep impression on the musician. They were married, but their happiness was short-lived. Aristaeus became enamored of Eurydice. As she ran away from him, a serpent lurking in the grass bit her foot and she died of the wound.

Orpheus resolved to get her back or die trying. With his lyre in hand, he entered the infernal regions and gained an audience with Pluto. The king of hell was charmed with Orpheus' music. The wheel of Ixion stopped, the stone of Sisyphus stood still, Tantalus forgot his thirst, and even the Furies relented.

Pluto and Prosarpine were moved, and agreed to give Eurydice back, provided Orpheus agreed not to look back until he'd gotten to the farthest borders of hell. Orpheus agreed and was already in sight of the upper regions when he forgot and turned back to look at his beloved Eurydice. He saw her, but she instantly vanished from his sight. He tried to follow her, but was refused admission. Something like the biblical story of Lot's wife.

Inconsolable, Orpheus withdrew from the society of mankind. This so offended the Thracian women, they tore his body to pieces and threw his head into the Hebrus River even as he continued to call the name of Eurydice. Orpheus had been one of the Argonauts, and wrote a poetical account of that adventure which is available to us today. After his death he received divine honors, and his lyre became one of the constellations.

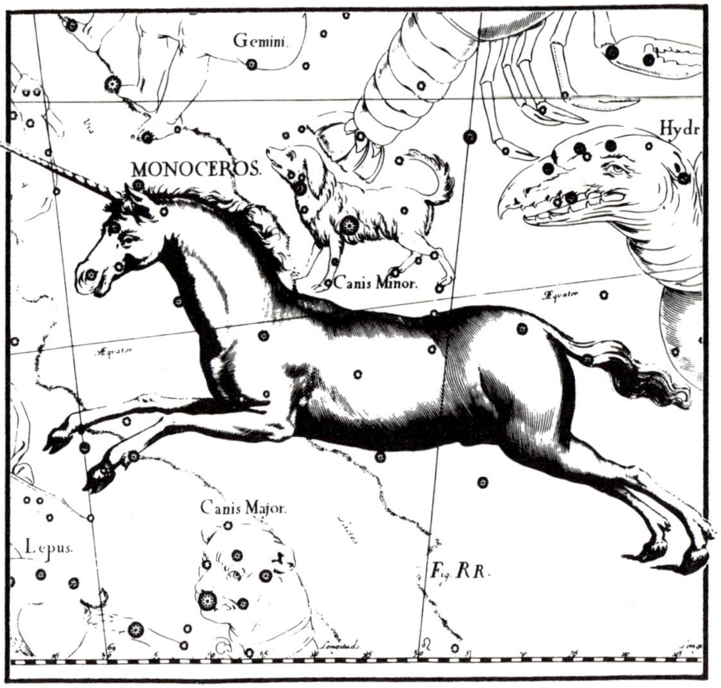

Monoceros, the Unicorn

This modern constellation is comprised of stars that are scattered over a large expanse of the heavens between Canis Major and Canis Minor. Most of the stars in this constellation are scattered broadly, making it difficult to tie them together. Its center is due south of Procyon. Monoceros is a faint constellation which contains the Rosette Nebula and the Cone Nebula.

Monoceros is the heavenly representation of the legendary *unicorn*. The unicorn was a rare and shy animal, so scarce that no one is sure what it really looked like. Some reports described the unicorn as a ferocious beast with the body of a horse, the beard of a goat, the tail of a lion and, most notably, an auger shaped horn of great strength about four feet long. This unicorn was a great swordsman with this deadly horn. He was able to vault great distances when being pursued, much like a pole vaulter.

The most common view of the unicorn is that of a pale shy horse with a beautiful horn of magical properties. This is the unicorn of which the ancient oracles counseled hunters. These oracles proclaimed that the unicorn was a lover of purity and innocence. In order to capture the unicorn a young virgin was to be used as bait. The girl was to be left in a clearing in the forest where soon the unicorn would be drawn near to lie next to the virgin and fall asleep. When the unicorn had fallen fast asleep the treacherous wench would signal the hidden hunters to capture the unsuspecting beast.

Roman mythology maintains that Jupiter so favored the unicorn that when he saw the treachery about to unfold he snatched up the unicorn to dwell forever in the heavens as Monoceros.

Ophiuchus vel Serpentarius, the Serpent-holder

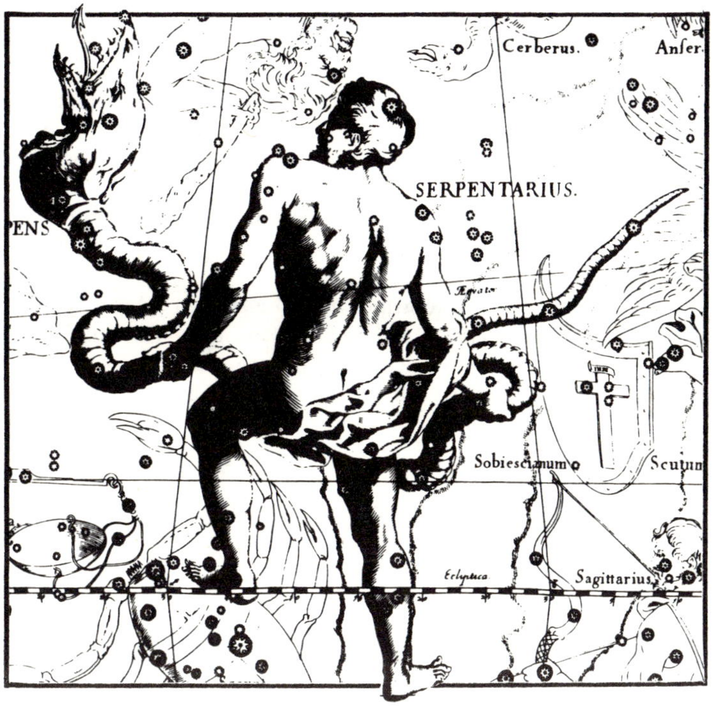

The Alpha star of Ophiuchus is "Ras Alhague", in Arabic meaning the "head of the snake charmer". Barnard's star is the second closest star to our sun, a distance of only 6 light years. It is a 9 magnitude red dwarf star.

The constellation Ophiuchus The Serpant Bearer is identified as the son of Apollo. He was the first physician of human form and gained fame by accompanying Jason and the Argonauts.

Chiron the Centaur, having been given the care of Ophiuchus by Apollo, instructed the boy in the art of healing and restoring life to the dead. One tale about Ophiuchus tells how he learned the secrets of life quite accidentally. After killing a snake by strangulation, the snake's mate appeared with an herb in its mouth. When the second snake gave some of this herb to the dead snake, life was restored. When Ophiuchus saw what had happened, he stole some of the medicinal herb and from that point on he was blessed with the ability to restore life.

The serpent in the hand of Ophiucus remains today as the symbol of medicine, called the "Caduceus".

Orion, the Giant, Hunter, and Warrior

Betelgeuse is the super red giant star in Orion's shoulder. Betelgeuse translates from Arabic "the armpit of the central one". The magnitude is .9. Rigel, "the left leg", is a blue white super giant star of .3 magnitude. Bellatrix, "the Amazon star", is Orion's other shoulder. It is yellow and is a 2 magnitude star. Delta, Epsilon and Zeta are three equally spaced bright stars forming the belt or girdle of Orion. Just below the belt of Orion, the Orion Nebula M-42 is visible with the naked eye. The spectacular Horsehead Nebula is here. Orion is best viewed from December to March.

When students, the general public or scientists are polled most reply that Orion is their favorite constellation. Perhaps that is because when a pollster asks a question, there are those who feel the need to name a favorite other than the Big and Little Dippers. Why is Orion so chosen? Maybe because it is so immediately identifiable, or so concise.

The Latin poet Manilius, who wrote five books on astronomy, said of Orion:

"First after the Twins, see great Orion rise,
His arms extended stretch o'er half the skies
His stride as large, and with a steady pace
He marches on, and measures a vast space;

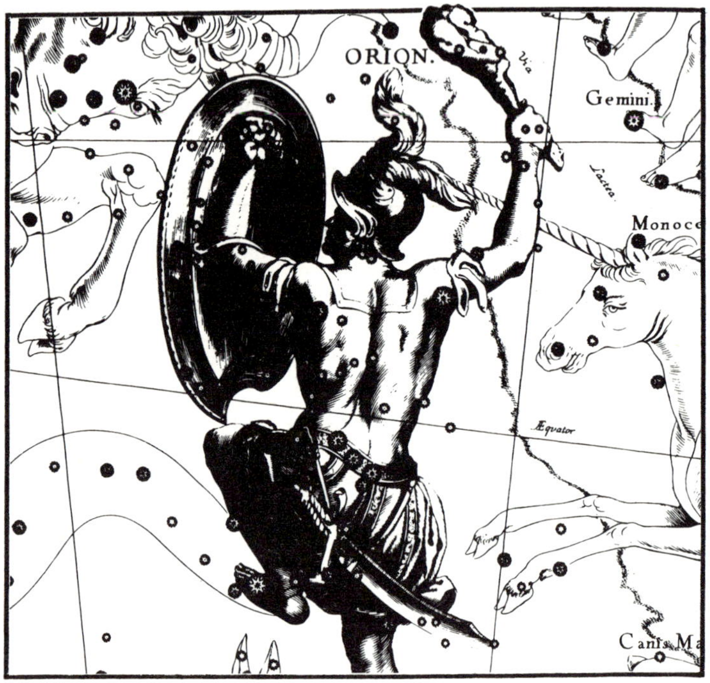

On each broad shoulder a bright star display'd.
And three obliquely grace his hanging blade.
In his vast head, immrs'd in boundless spheres,
Three stars, less bright, but yet as great, he bears,
But farther off removed, their splendour's lost;
Thus grac'd and arm'd he leads the starry host."

Whoever looks upon the constellation will never forget it. When Orion is on the meridian, it is visible to all the habitable world because the equinoctial passes along the middle of the Earth. Orion is shown on star maps as a hunter attacking a bull while bearing a sword in his belt, holding a huge club in his right hand and the skin of a lion in his left.

There's a tight triangle of three small stars in Orion's head, which forms a larger triangle with the two in his shoulders. In the middle of the constellation are three stars of the 2nd magnitude in Orion's belt, forming a straight line about 3° in length from northwest to southeast. These three stars usually are distinguished as the "Three Sisters" because no others exactly resemble them in position or brightness. Sometimes they are called the *Three Kings*. Ancient husbandmen called them *Jacob's Rod* or the *Rake*. In 1807, the University of Leipsic gave them the name *Napoleon*.

Their most common name is the *Elland Yard* from the fact that the line which unites the three stars in the belt measures just 3° in length, divided equally by the central star, like the feet on a yardstick. That makes it a standard for measuring the distance of stars from each other. A line through the three stars of the belt will point out Pleiades and Hyades on one side and Sirius, "The Dog Star," on the other.

Orion was the son of Neptune and Queen Euryale, a famous Amazonian huntress. Having the disposition of his mother, he became the greatest hunter in the world, boasting that he could conquer any animal on Earth. To punish his vanity, a scorpion sprang out of the Earth and killed him with a bite to his foot. Diana requested that he be placed among the stars opposite the Scorpion which had caused his death.

Other tales say Orion had no mother but was the gift of the gods Jupiter, Neptune, and Mercury to reward a peasant for his piety. Orion supposedly was able to walk on water and had greater strength and stature than any other mortal. A skilled blacksmith, he fabricated a subterranean palace for Vulcan. He also walled in the coasts of Sicily against the encroaching sea and built a temple to the gods there.

Orion was engaged to Merope, the daughter of Oenopion, but Oenopion decided he didn't really want to give up his beautiful little daughter. So he got Orion drunk and put out his eyes when he laid down on the seashore to sleep it off. Finding himself blind when he woke up, Orion followed familiar sounds to the nearest forge, enlisted a smithy to guide him, and headed off for the best place to see the rising sun. There he turned his face toward his father Apollo, and immediately upon recovering his sight, sped off to punish the perfidious cruelty of Oenopion. The shanghaied blacksmith, though being helpful, headed back to work at the forge, washing his hands of all kinds of gods.

The sighting of Orion is generally supposed to bring with it great rains and storms. It isn't surprising, therefore, that it has

earned the reputation in the early days of navigation of being extremely hazardous to mariners. Virgil, Ovid, and Horace all noted this fact in their writings.

The Poet Eneas even accuses Orion of causing the storm which cast him on the African coast on his way to Italy:

"To that blest shore we steer'd our destined way,
When sudden, dire Orion rous'd the sea;
All charged with tempests rose the baleful star,
and on our navy pour'd his wat'ry war."

Pegasus

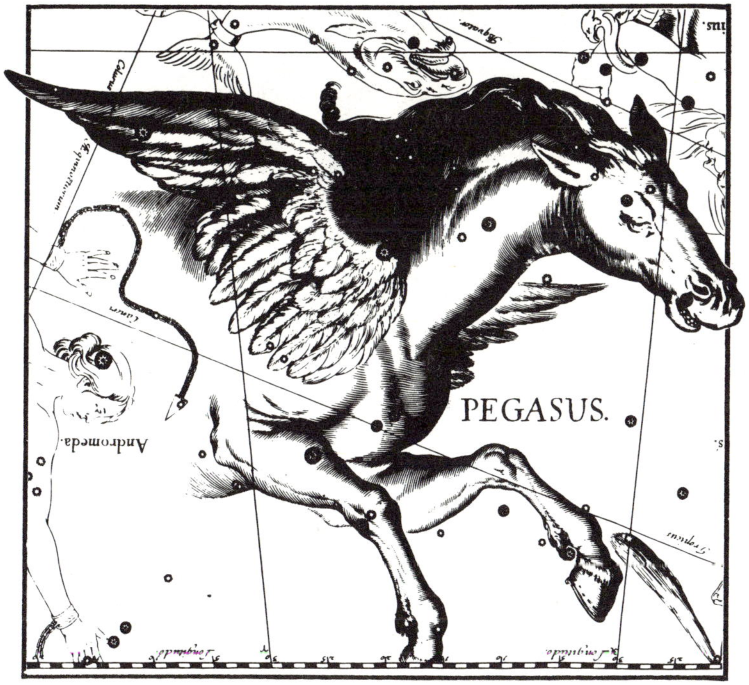

The Wings of this flying horse are shown in an inverted posture. Pegasus occupies a space between the Swan, the Dolphin and the Eagle on the west and the Northern Fish and Andromeda on the east.

Pegasus is an easy constellation to find. The center of the great square is almost devoid of any other stars and the absence is obvious.

The Alpha star of Andromeda is shared with Pegasus. Andromeda's eye is called "Alpheratz", meaning the "horse's navel" in Arabic. Alpheratz forms the upper left star of the great square and has a 2.1 magnitude.

Markab, "saddle", the lower right star of the square is a 2.5 magnitude white star. Markab is the Alpha of Pegasus.

The Beta is "Scheat," or shoulder, and is the upper right of the great square, a 2.2 magnitude and a deep yellow color.

There is a meteor shower from Pegasus on May 30. Eta Pegasides appears near the shoulder star. Remember, to find Pegasus look for the big empty square in the sky; look for nothing.

This much-talked-about flying horse, the son of Neptune and Medusa, is said to have sprung from the blood of Medusa, which spilled into the ocean after Perseus cut off her head. Hesiod, the Greek poet of the 7th Century B.C., says he got his name from being born near the sources of the ocean.

According to Ovid, he lived on Mount Helicon, where by striking the Earth with his foot he raised the fabled fountain called Hippocrene. Mt. Helicon, the home of the Muses, would be approximately 50 miles north of Athens.

The Muses were the 9 daughters of Jupiter and Mnemosyne (Memory); they were in charge of the arts and sciences. Calliope was in charge of epic poetry; Clio of history; Erato, love poetry; Euterpe, lyric; Melpomene, tragedy; Polymnia, sacred poetry; Tersichore, choral dance; Thalia, comedy; and Urania, astronomy. Pegasus at any rate became the favorite of the Muses.

The son of Glaucus and grandson of Sisyphus, founder of Corinth, was Bellerophen. He had the frightful task to slay the Chimera, a hideous monster that continually spouted flames. This monster had three heads: a lion, a goat and a dragon. The front part of its body was that of a lion, the middle that of a goat, and the back

part a dragon. It lived in Lycia, now southwest Turkey. The desolate wilderness at the top of this country was dominated by lions, the fruitful middle by goats, and the marshy bottom by serpents. Bellerophon was the first person to live there. The Chimaera was an obvious allegory to the land Belleophon conquered.

Bellerophon, after praying in the temple of Athena, was given a golden bridle to harness and break the winged horse. Mounting Pegasus, they rose into the air, found the Chimaera and quickly dispatched him with an arrow. After killing the monster, Bellerophon attempted to fly up to heaven on Pegasus. Jupiter did not appreciate this presumptuousness and sent an insect to sting the horse. The startled Pegasus thus threw his rider. Bellerophon, struck blind and lame by the fall, was left to wander the countryside and died lonely and destitute. Pegasus continued on to heaven where he was placed among the constellations.

Perseus, the Champion

Perseus holds a sword in his right hand, the head of Medusa in his left, and has wings at his feet. The head of Medusa is not a separate constellation, as is sometimes thought, but merely forms part of Perseus. It is represented as a trunkless head of a frightful Gorgon, covered with coiling snakes instead of hair, held aloft by Perseus.

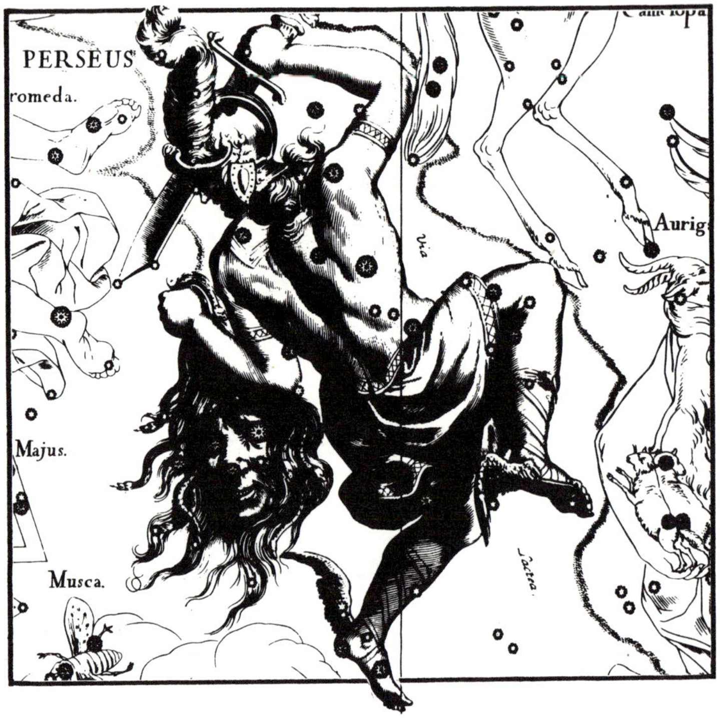

Perseus, his wife Andromeda and his in-laws, King Chepeus and Cassiopeia, have homogenous locations in the northern sky in the autumn.

Mirfak, "the elbow", is a brilliant lilac color of a 2 magnitude. A meteor shower called Perseids exhibits 60 meteors each hour and can be seen near Mirfak on August 10th to August 14th. Algol, "the demon," is a 2.3 magnitude white binary located in the eye of the Medusa.

Perseus was the son of Jupiter and Danae. No sooner was he born than he was thrown into the Aegean Sea with his mother by his grandfather Acrisius. They were swept to the coasts of one of the islands of the Cyclades called Seriphos and rescued by a fisherman. Following their rescue, they were carried to Polydectes, the king of Seriphos, who treated them with great humanity and entrusted them to the care of the priests of Minerva's Temple. Perseus' genius and manly courage soon made him a favorite of the gods.

At a great feast of Polydectes, all the nobles were expected to present the king with a superb and beautiful horse. Because Perseus felt he owed much to his benefactor, he wanted to outdo the competition. He undertook to bring his king the head of Medusa, the only one of the three Gorgons who was subject to mortality. The immortal Gorgons were Stheno and Euriale; all were snake-haired serpents with yellow wings and brazen hands. They were covered with impenetrable scales. Not a pretty sight. They had the power to turn anyone looking at them into stone.

To equip Perseus for this rather dangerous journey, Pluto, god of the Infernal regions, loaned his helmet of invisibility. Minerva, goddess of wisdom, furnished him with her mirror-like buckler (a small, round shield). Mercury gave him wings for his feet and a dagger made of diamonds.

Being suitably equipped for a hydra raid, Perseus flew through the air with Minerva's help to reach the monster. By great good fortune, Medusa was sacked out as Perseus approached. With a cunning that delighted and amazed Minerva, he maneuvered himself into position by the reflection of his shield and cut off Medusa's head with one blow. The noise awoke Medusa's companions, but because Pluto's helmet made Perseus invisible, the remaining vengeful Gorgons couldn't track him down.

Perseus then made his way through the air with Medusa's head dripping in his hand. From the blood it secreted sprang all those innumerable serpents that have ever since infested the sandy deserts of Lybia. On his way to Seriphos he rescued Andromeda, who was chained to a rock, from the sea monster Cetus and married her. Their first born son, Peres, gets credit as the first Persian.

The destruction of Medusa naturally gave Perseus lasting fame. When he died, he was immortalized as a constellation, with the head of Medusa in his hand.

Sagitta, the Arrow

Sagitta, the arrow, lies in the Milky Way directly north of Aquila and south of Cygnus, pointing eastward. It is often depicted as the arrow held in the eagle's talons because the eagle originally was the armor-bearer for Jove in some stories of mythology. It was also the arrow shot by Hercules at the Stymphalian birds, Cygnus and Aquilla.

Erostones said it was the arrow Apollo used in destroying three one-eyed monsters called Cyclopes. Early Jesuit astronomers called it the "Nail of the Crucifixion". At the very least, being the most famous arrow in history, it is the arrow Cupid always uses when he aims most true.

Scutum Sobiescianum, Sobieski's Shield

This constellation was named "Scutum Sobiescianum" by the Polish astronomer Hevelius, but shortened to Scutum by later astronomers.

The Shield, described by Hevelius, is in the Milky Way west of the feet of Amtinous, between the tail of the Serpent and the head of Sagittarius. It celebrates the Coat of Arms of John Sobieski, King of Poland, who distinguished himself during the wars with Turkey in the seventeenth century. He led twenty thousand Polish troops and defeated the Turks at Vienna, Austria on September 12, 1683, thus preserving the faith from the Turkish heathens. It is also supposed to depict a cross on his shield.

Serpens

Very ancient, the serpent has always been shown as grasped in the hands of Ophiuchus and divided by some catalogues as Serpens Caput (the head) and Serpens Cauda (the tail) on either side of the serpent holder.

This serpent, called the Caduceus, entwined around a staff is the symbol for medicine today, perhaps signifying the shedding of skin and a renewed life or well being.

The ancient biblical interpreters would suggest this constellation was another seducer of Eve. The converted Irish find a cross in the Serpens Caput and call it the cross on the miter of Saint Patrick.

Sextans Uraniae

One must understand how important the invention of the sextant was, particularly to an astronomer like Hevelius who commemorated it in 1658. Originally Sextans was comprised of twelve otherwise unclaimed stars between Leo and Hydra west of Cancer.

Sextans is a small constellation. It was known at one time as "Sudarium Verdnicae", the veil or handkerchief of Veronica of the Way of the Cross.

Triangulum

Triangulum lies just south of Andromeda on the edge of the Milky Way.

The best description was by the Greek poet-astronomer Aratos translated by a Mr. Brown. He says:

"Below Andromeda, in three measured like to Delta; equal two of them."

Transcribed by Cicero as Deltoton, naturally it was the Delta of the Nile of Egypt. Thus it was more important to the Egyptians than to the modern astronomers.

The Protestant lawyer Johann Bayer in the sixteenth century thought it astrologically represented the three known continents of the Earth: Europe, Asia and Africa. Bayer was in conflict with the Roman church, which thought it represented the Trinity or the Miter of Saint Peter.

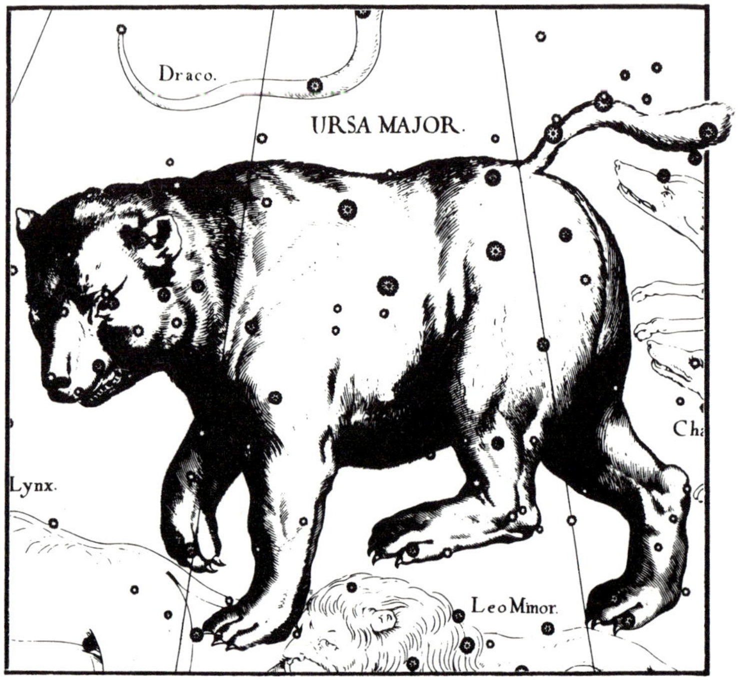

Ursa Major, the Great Bear or Big Dipper

The pointer stars are Dubhe, the back of the bear, and Merak, the loin of the bear. Dubhe is a bright yellow binary of 2 magnitude. Merak is 2.5 magnitude and greenish-white in color.

All of the stars in Ursa Major have names which relate to parts of the bear and are visible with the naked eye.

There is a meteor shower on November 10th from "Alkaid", the end of the tail or handle.

There are many remarkable stories surrounding the two tireless bears of the sky who never find rest nor water to slake their thirst.

But, like Argo Navis and Orion, there is something even more important which distinguishes the big and little dippers (or bears). That is, most laymen who want to orient themselves while gazing at the stars do so by first finding the big dipper, then drawing an imaginary line from which the dipper would spill if the liquid contained in it were spilled. That spill, if followed, leads directly to Polaris, which is the fixed North Star. Polaris is also the last star in the handle of the smaller dipper. So if you can remember that the big dipper pours on the tail of the little dipper's handle, you have oriented yourself in the heavens, located two constellations, and found the Polaris star.

A little romance music, professor, and we will recall the most famous story from celestial mythology which tells us why two bears must wander so restlessly around the sky, always above the horizon, and always looking for a bath and a drink of water.

King Lycaon of Arcadia, land of classical silvan beauty, had a beautiful daughter who, for reasons known only to mythological royalty, he named Callisto. She loved to hunt, and for that reason worshipped Diana, the beautiful lady with the hunting dog depicted on so many Greek vases and walls. So devoted to hunting was Callisto that she swore she would pledge her entire life to its pursuit. Diana, pleased by this vow, also demanded that Callisto, among other things, remain a virgin. Now if you've been into

mythology at all, you know that remainng a virgin in Arcadia is like collecting baseball cards without ever chewing the gum. It can be done, but do you know anyone who has actually done it?

You will remember the story of Phaethon, who had taken his ride in the sun chariot, took a big fall, and plunged to Earth, sending almost everything on our planet into flaming ruins. After all of this, Jupiter dropped in from Olympus to review the scene with a view to setting up shop again. The first thing Jupiter did was restore water to everything: the seas, the rivers, and the lakes. These fed the grass, and the woods and flowers bloomed again. The air drifted everywhere, bathing the lush valleys with sleepy perfume.

As he was inspecting his work, Jupiter saw someone sleeping in the bower. It was a beautiful Arcadian girl stretched out in the shade under a cool eucalyptus tree. Jupiter stopped and studied her. He was overwhelmed by the apparition of Callisto, the wench who had pledged her virginity to the Goddess of the Hunt. Realizing Callisto already had been sleeping awhile, he said, "Why not?" and with that, he changed himself into a figure which looked exactly like Diana, dog at her side and all.

"Well, my dear," Jupiter said, stroking his chin and forgetting that he no longer had a beard. "How goes the hunt?" Recognizing that Diana's voice was a full two octaves lower than the last time she had spoken, Callisto tried to free herself of Jupiter's embrace, something not even his wife, Juno, could do. Finally, she gave in to his dalliance and they made love for as long as it took her to get pregnant. After that, he handed her her bow and arrows and rose back up to Olympus where Juno stood at the front door, tapping her foot impatiently.

Feeling depressed and in shame, Callisto was approached by Diana and her followers, who asked where she had been and invited her to take a bath. When in her swollen condition she refused, Diana's giggling hunting companions undressed her. Before they pushed her into the water they saw that she was quite unmistakably with child. Not only were the nymphs aghast at how Callisto had broken her vow of chastity, they refused to let her get into the water for a much-needed bath. "Go," they ordered Callisto, "and do not pollute these nor any waters anywhere."

Not much later, the imperious Juno heard of the little boy named Arcas who had been born in the woods. She knew that Arcas was Jupiter's son, and so she cursed Callisto by having her grow fat and hairy to reconfigure her into an ugly, hairy old bear so she could never seduce any man or god again. Before Juno could do more to her, Callisto escaped into the woods.

In the years that passed, her son Arcas became a proficient hunter himself. One day he chanced upon a bear not far from where he was born. It was his mother. Forgetting her appearance and overjoyed to see him, Callisto rushed forward to give Arcas the customary bear hug. Jupiter intervened just in the nick of time before Arcas could shoot his unrecognized mother with a well-aimed arrow. Since Jupiter couldn't undo his wife's handiwork, he changed Arcas into a bear, too. Then, taking them both by the tail, he slung them into heaven with their tails stretched beyond that of any mortal bear.

When she heard that Callisto and Jupiter's bastard were elevated to the status of gods and had been given homes in the heavens, Juno went on a rampage. She contested the divinity of these two interlopers to everyone and petitioned Oceanus, god of the seas, not to let Callisto or her son ever drink or take a bath in his waters. Juno's wish was granted and to this day both bears, the great and the small, are circumpolar, never stopping in their search to dip below the horizon. They may be seen every night in summer or winter, never being allowed to rest, and forever thirstily poking around for a drink of water or a place to lie down for one night.

Ursa Minor, the Lesser Bear or Little Dipper

Polaris or Cynosura is only 1 degree from being true north in the sky. It is a yellow-white binary of 2.2 magnitude. It is easily located with the pointer stars of the Big Dipper.

The pouring edge of the Little Dipper is marked by Kochab and Pherkad. Together they are the guards or wardens of the pole. As outlined in the previous pages, the prevailing option is that Ursa Major and Ursa Minor are the hunting nymph Callisto and her son, Arcas. They were transformed into bears by the enraged and imperious Juno and afterward sent to the heavens by Jupiter so they wouldn't be destroyed by Callisto's former huntsman companions and their dogs.

The Chinese claim that the emperor Hong-ti, grandson of Noah,

first discovered the polar star and used it for navigation. It is certain in any case that it was used for this purpose in very ancient times by the Polynesians as well, a point thoroughly presented in Michener's work, *Hawaii*.

From various passages in ancient writings, it is also obvious that the Phoenecians steered by Cynosura, or the Lesser Bear. The mariners of Greece and some other nations steered by the Greater Bear, which they called Helice or Helix. The American Indians, the Chaldeans, Phoenecians, and other civilizations referred to these Ursas as bears rather than dippers. The only explanation ever offered for this unique coincidence came from the brilliant Harvard scholar, Carter Lord:

"It might be argued that since both Ursa Major and Minor are inextricably involved with the fixed Polaris star they are the most important points of navigation in the entire world. When all travelers of great distance arrived at faraway lands, they had long before agreed the dipper was a bear. Once agreed upon, it was so promulgated, American Indians notwithstanding."

Vulpecula, the Fox

Vulpecula is the home of Messiers' "M-24," the Dumbell Nebula. It has the distinct shape of a weight-lifter's dumbell and it's age is estimated at only three or four thousand years old with a maximum estimate of forty-five thousand years. The dumbell is a relative newcomer and one of the nearest such objects to the Earth.

The full title for this constellation is Vulpecula Cum Ansere, meaning "the little fox with the goose." It occupies an area between the arrow and the swan, where Via Lactea divides between two branches. Hevelius selected it because, "I wished to place a fox and a goose in the sky." Well, reason enough, I suppose. As Longfellow put it in *The Galaxy:*

"Torrent of light and river of air,
Along whose bed the glimmering stars are seen
Like gold and silver sands in some ravine
Where mountain streams have left their channels bare!"

The Other Constellations

Hocce Hemisphærium Australe
Domino Dno IACOBO
Regis Filio humillimè offert
NUMEN Suam Serenitatem
Decus atque Ornamentum
Incolumem tueatur
Gratia Benignissimè conservetur
contendit
Humillimus Devotus
Ioh Hevelius
In lucem prodiit Anno Æræ Christianæ
Auctoris Sumptibus et Curâ
Pictore vero Andr. Stech et
Sculptore Carolo de la Haye
Gedani
Magnitudines Stellarum
Serpentarius
Scorpius
Sagittarius
Corona Australis
Libra
Lupus
Ara
Indus
Pavo
Grus
Piscis
Aquarius
Capricornus
Centaurus
Crux
Musca
Triang. Aust.
Apus
Toucan
Phœnix
Chamæleon
Hydrus
Robur Carolinum
Corvus
Virgo
Piscis Volans
Dorado
Crater
Tropicus
Capricorni
Hydra
Eridanus
Sextans Uraniæ
Argo Navis
Columba
Lepus
Canis Major
Cetus
Leo
Æquator
Monoceros
Taurus
Aries
Canis Minor
Cancer
Orion
Gemini

The Other Constellations

The origins and etymologies of the major constellations come to us from many sources, both very ancient and relatively recent.

There is general agreement that the Solar Zodiac originated, as we have it in Euphratean astronomy, probably with only six signs: Taurus, Cancer, Virgo, Scorpio, Capricorn, and Pisces. These signs were later divided because of the annual occurrence of the twelve full moons. This division was thought to be the furrow of heaven, ploughed by the bull, Taurus. This idea was first recorded from about 3800 B.C. and served to define the first of the twelve. The calendar was probably taken from the stars about 2000 B.C. according to Archibald Henry Sayce of Oxford.

On the other hand, under a heading of lesser constellations are the southern ones which have minimum mythological lore or history. These more modern systems were described as late as 1687 by the Polish astronomer, Johannes Hevelius and Abbé Nicolas Louis de Lacaille (1713-1762), a French astronomer or his colleague, Joseph La Lande (1732-1807).

Antila Pneumatica, the Air Pump

Antlia Pneumatica is the water pump to most astronomers, but in Germany it is called the Luft Pumpe. Generally, in most lists of astronomy of all languages it is called more simply Antlia, since it was described in 1888 by Lacaille.

Antlia lies south of Hydra and Crater and is near Vela, the sail of Argo Navis. It has eighty-two naked eye stars and can be seen from Fort Lauderdale, Florida during spring break.

Apus, the Bird of Paradise

Apus is located south of Crater and Hydra, bordering on the Vela of Argo. It is described as The Bird of Paradise. It lies immediately below the Southern Triangle only 13° from the southern pole.

The bird of paradise for which Apus is named, was found in Papua, New Guinea. Dutch sailors named this constellation. Another translation is "Bird without Feet."

Ara, the Altar

Ara, sometimes referred to as Aitar, is an altar. It was described by the poet Aratos with the words:

"'Neath the glowing sting of that huge sign the
Scorpion, near the south, the Altar hangs."

Authors of astronomical biblical interpretations are divided. Some refer to it as the "Altar of Noah," erected after the deluge and others as the permanent golden temple of Jerusalem.

Caelum, or Scalptorium, the Burin or the Graving-tool

Caelum was formed by Lacaille, who described it first. He took it from the stars between Columbo and Eridanus. It comes to the meridian with the star Aldebaran on the 10th of January and is visible from the 40th parallel.

The English version of Caelum is "Sculptor's Tool" or "Graving Tool", not to be confused with the separate constellation, Sculptor.

Chamaeleon

This is a small constellation below the star, Carina. Octans separates it from the South Pole. Pontanus, in Chilmead's "Treatise", included it with Musca.

Circinus, the Pair of Compasses

Described by Lacaille, the two compasses lie close to the front feet of the Centaur, south from Lupus and Norma.

Columba Noae, Noah's Dove

Columba is sometimes known as Columba Noae and as Noah's Dove. The reader will remember that seven days after the rain stopped, Noah let loose a dove, but it came back. After being released a second time, it returned with an olive leaf so Noah knew that his wet days were almost over, unless you count the time he got drunk in his new vineyard a little later on.

Why it is called Columba is not clear unless it refers to the fact that Columbus also finally found a bird near the end of his first voyage, indicating to him that land was near. In any case, the name of Noah's bird is not mentioned in the Bible.

Corona Australis, the Southern Crown

Often called the Southern Crown, it is seen in the southern hemisphere close to the waist of Saggitarius on the edge of the Milky Way.

Crater, the Cup

"Wise ancient knew when Crater rose to sight,
Nile's fertile deluge had attained its height."

The cup is formed above Hydra's back just westward of Corvus and 30° south of Denebola. For that reason it often was considered part of a constellation called Hydra et Corvus et Crater. Today they are catalogued separately.

Crux, the Cross

The Southern Cross is surrounded on three sides by the Centaur. Its description is most often attributed to Frenchman Augustine Royer in about 1679, whereas Johann Bayer drew it over the hind legs of the Centaur. In any case, it is a brilliant but narrow stream three or four degrees wide, looking more like a kite than a cross. There has been much poetry and romance attached to it over the years because of the fascination it holds for mariners.

Crux is the smallest constellation in the sky. It was last seen in Jerusalem on the horizon at–31° 46′ 45″, precisely when Jesus was crucified.

Dorado, the Goldfish

Bayer described it as one of his goldfish in the southern figures. The word is from the Spanish and refers not to the little exotic cyprinoids we associate with goldfish, but rather to the large Coryphrena fish of tropical waters.

Although there was a comet then, Dorado may have been what Josephus was writing about when he described that "which for a year had hung over Jerusalem in the form of a sword." The year was 66 A.D., one of the years in which Halley's Comet appeared.

Fornax, Chemica, or Fornax Chymiae, the Chemical Furnace

Meaning "the furnace", Fornax was described by Lacaille as being within the southern bend of the river. It was named in honor of the celebrated chemist Antoine Lauraent Lavoisier and had to do with his chemical prowess.

Grus, the Crane

The crane, imagined by Caesius as the Stork of Heaven, curves gently from the southwest from the Southern Fish to become a graceful bird.

Horologium Oscillatorium, the Pendulum Clock

Horologium Oscillatorium, the pendulum clock, lies eastward of Achernan, the alpha of Eridanus, and north of Hydrus. Whitall described it in his planesphere as being between Chemica Fornax and Caela Scuptoris.

Hydrus

Still another reptile in the sky, this lesser water snake is not to be confused with its big brother, Hydra, in the north. It lies between the Clock and the Tucan and the tail almost touches Achernar of Eridanus. Schiller's biblical constellation is Raphael the Archangel.

Indus, the Indian

The Indian is one of Bayer's new constellations, south of Microscopius, between Grus and Pavo. It was named for the American Indian and drawn nude with arrows in each hand, but no bow.

The Jesuits called it the Persian. The Jews referred to this constellation as representing the Patriarch, Job.

Mensa

Lacaille named this constellation Mons Mensa. Its shape was similar to Table Mountain near Cape Town, South Africa. It lies between the southern pole and Dorado. Mensa shares a large magellanic cloud with Dorado.

Microscopium

Another of Lacaille's descriptions, Microscopium lies south of Capricornus and west of Piscis Australis. Lacaille drew this as a microscope, one of the new developing scientific instruments of the day.

The Germans thought it looked like a neper, or auger. Doctor Christian Ludwig Ideler wrote of this wood boring tool in 1804.

"It is situated at the tail of Sagittarius at Capricornis and has many stars. At the handle of the neper two and on the iron three."

Musca Australis vel Indica, the Southern, or Indian, Fly

This title, the southern fly, was substituted by Lacaille in 1752 for Bayer's Bee (Apis). Halley had catalogued it earlier in 1679 as Musca Apis. The latter prevailed.

The Chinese called this Meih Fung.

Johanne Von Schiller, in 1799, combined Apus Chamaeleon and Musca and called it "Mother of Eve." This seems rather strange since we all know that Eve was the only woman derived from man, sans mother.

Norma et Regula, the Level and Square

The ruler, level and square originally were comprised of Ara and Lupus, associated with Circinus, the compass, on the north, adjoining the two forefeet of the Centaur. It is simply called Norma by most astronomers.

Norma is north of and adjoins the Triangle. In Lacaille's description, it lies within the present limits of our Scorpio.

Octans Hadleianus

It was originally published by Lacaille in 1752 in recognition of the octant which had been invented by John Hadley in 1730. Long before, however, it has been mentioned by Ovid, Virgil and Pliny, who probably knew it as Dramasa.

The Arabs thought it had healing powers for the sick who gazed upon it. Such was astrology in the year 1000 B.C.

Octans includes the south celestial pole.

Pavo, the Peacock

The peacock lies south of Sagittarius. It has long been a symbol of immortality, said to be constantly shedding its yesteryears along with its feathers as it annually renews itself. Its starry tail was sacred to Juno, queen of all the heavens.

It was also known as Argos who became immortal after building Jason's ship. (see Argo Navis).

Phoenix

Phoenix is located between Eridanus and Grus, south of Fornax and the Sculptor, and named by Bayer. It culminates just above the horizon of New York City each November 17th.

Combined by Schiller with Grus, it was also known as Aaron the high priest of Israel. The Book of Job in the Old Testament mentions Phoenix.

In mythology the Phoenix bird was a male who's life span was always over 500 years. At the end of 500 years he would build a nest of twigs from spice-trees on which he then died by setting the nest on fire and cremating himself alive. Shortly thereafter, from the ashes of the funeral pyre, a new perfect Phoenix in the image of the father would arise. The new Phoenix would then collect what remained after the fire and fly them to Heliopolis, Egypt. Here he would sacrifice the remains on the altar of the Sun.

There is evidence that a stork-like bird called Phoenix did exist and was honored in Arabia, Assyria, Ethiopia, Egypt and India.

Pisces Australis, the Southern Fish

Known also as the Southern Fish, this constellation lies immediately south of Capricorn and Aquarius in that part of the sky earlier known as the Aratos. Oddly enough, Piscis appeared at the time best suited for the harvest of honey.

This fish is supposed to be the mother of the twin fish of the Zodiac (Pisces) because it is much brighter than its children.

Piscis Austrinus contains Fomalhaut, a bright star of the first magnitude. Fomalhaut translated from the Arabic means "Mouth of the Fish."

Pictor

The painter lies south of Columba, between Canopus and the South Pole of the ecliptic in Dorado. It is sometimes called the "Painter's Easel."

Equuleus Pictoris, introduced by Lacaille in 1750, contains little interest to astronomers.

Reticulum Rhomboidalis, the Rhomboidal Net

This is the net, the Reticulum Rhomboidalis. It was first drawn by Izaak Habrecht. It lies north of Hydrus and the Greater Cloud. Lacaille named this constellation to commemorate the reticle, used to measure star positions. A reticle is a network of fine lines, or crosshairs, placed in the focus of a telescope to assist in observations.

Sculptor

Located between Cetus and Phoenix, it was described by Lacaille not so much as the sculptor himself, but his studio or workshop. In fact, Burritt later changed it to Officina Sculptoria.

Sculptor can be seen in mid-November from New York City.

Telescopium, or Tubus Astronomicus

Observed and described by Lacaille, Telescopium Astronomicus is located between Ara and Sagittarius on the edge of the Milky Way. In his catalogue, Bailey had it restricted to the south of Scorpio, Sagittarius and Corona Australis. There is another Telescopium referred to as Telescopium Herschetti in honor of William Herschel. It was described by Abbellell in 1781 as lying between Lynx and Gemini. It is visible in August.

Triangulum Australe, the Southern Triangle

The Southern Triangle is much more noticeable than its northern (the original) cousin. It was first recognized by Bayer, who published news of its existence in 1603. Supposedly it honored the three patriarchs, Abraham, Isaac and Jacob. It lies south of Ara and Norma between the tail of pavo and the front feet of the Centaur.

Tucana, the Toucan

The toucan is mentioned by all the great astronomers including Caesius, Kepler, Ricciolo, Bayer and Burritt. Tucana lies immediately south of Phoenix, bordering on the south polar Octans, with its tail close to Achernar of Eridanus. It marks one crossing of the Equinox and Arctic circle.

Being in the southern sky it is appropriate to be named after still another tropical bird. Tucana contains the #47 Tucanae, the second largest globular cluster in the sky, and is visible with the naked eye in the Southern Hemisphere.

Volans, the Flying Fish

This is another new constellation formerly introduced by Bayer. It comprises 46 stars south of Canopus and Miaplacidus.

It is the Flying Fish or Piscis Volans, shortened to avoid confusion with all the other fish.

The Man Who Gazed at the Sun Too Long

Let us now go back for a look at existence again. Not necessarily at how it really came to pass, but how we perceived it to have occurred.

One of man's first thoughts must have been, "What are those lights in the sky?" Not long after that someone noticed that it is warmer and there is more life in the sun than in the shade. This thinker must also have watched the moon as it made its solitary circuit in the sky. This circuit became the "moonth" or "month." Eventually our thinker, or one of his students, used it as a dependable guideline for telling time.

After centuries passed, wisemen used the sun and moon as clocks to tell people when to migrate. They used those phases of the moon to forecast the floods of rivers like the Nile, the Tigris and the Ho. They used them also to predict when the valleys would turn green, meaning the tribe's survival again would be assured.

Meteors were noticed. These demanded interpretations from the elder who gleaned so much merely by staring at the sky. So it is fair to assume that the first and probably the most important scientific study was astrology, originally another name for meterology. Both studies had the gathering of food as their purpose for being. Astrology at that time had a much less important substudy known as astronomy.

So astrology, astronomy, and meteorology were really one science with the same objective: to predict where best to find and how to collect food. "Also, when it became necessary," as the Indian philosopher, Shanda-Non, said, "to blame something for his disappointments, man turned to astrology and blamed the stars."

It was a natural outgrowth to go from blaming the stars to counting on them or making predictions based on celestial calculations.

By tracing the sun, man (probably Chaldean) divided time into the ecliptic, which is the 12 signs of the *zodiac (zone of the animals).* Man recognized that the sun traveled along the zodiac and that lunar cycles had an effect on menstrual cycles and tides. This supported the cause and effect relationship between man and the heavens.

Since meteorology and astronomy were studies of the obvious, the men who calculated beyond the rise and fall of the objects in our solar system with such certainty, men who could read the stars, soon were adorned with the mantle of wise men.

It was astrology from which so many of our other sciences sprang. Mathematics was formulated in order to account for systematic celestial movements. For example, cures and herbal recipes were often prescribed during various phases of the moon. They were prepared at certain times and their ingredients were cooked at the best moments of the day or night.

Astrological recommendations were suggested in medical experiments. Chemistry and then biology gave birth to other scientific offspring as the astrologers, who played upon superstition, soon were working on alchemy to enrich themselves. It was probably the first time that someone asked the teacher, "Okay, if you're so smart, why aren't you rich?" Believing in his own ultimate wisdom, the turning of lead into gold was high on the alchemist/astrologer's list of things to do when he found the time.

There have always been those who have tried to systematize everything. The teacher or wise man of the society soon became a philosopher, speculating on uncertain answers. Finally, he achieved the idea of religion as an answer to every question. It comes down to us today as mythology, wondrously characterized as various gods cavorting in Arcadian pastures seeking and doing the crazy things which make sense only to other celestial beings. This was the fantasy of mythology.

Egyptian cosmologists saw the world as a huge box with the Earth as the floor and the sky as one big watershed from which came all water. Not a bad idea when you think about it.

Thinkers have always tried to understand the shape of the world. Xenophanes, Pythagoras and Anaxagoras all struggled with astrology which led them to the same ultimate questions concerning our origin in astronomical terms just as Plato and Eudoxus did. In the 2nd century B.C., Eudoxus finally established a theory that the planets had rotational motion. The Aristotelian conclusions of spherical bodies in space held relative validity for all astrologers whose realm seemed impenetrable from any new theories submitted by lesser men of learning such as astronomers. After all, it was the astrologer who actually interpreted the stars' movements and the meaning of the planets' frivolity. It was the astrologer's eyes, which looked into the face and heart of God as far as everyone in the tribe was concerned.

Constellation: Hercules.

Object: M13 — Star Cluster. **Distance:** 25,000 light years.

One of the finest globular star clusters in the Northern Hemisphere, over half a million stars.

By this time, portents, augers and horoscopes were cast first in importance. Oracles were contrived by the same kinds of mind as those who advised the leaders, warriors and kings. It was the astrologers, not the astronomers, who did this. The astrologers were the ones who dealt most directly with God.

Portents and horoscopes were first in importance in the world of learning until Niklas Kopernik came on the scene in the latter part of the 15th century.

Young Kopernik was allowed to read in the library of Krakow. There he discovered the works of Aristarchus, who had the temerity to postulate 17 centuries earlier, the idea that the sun was the center of the solar system. Niklas, now nicknamed Copernicus, thought this idea deserved re-evaluation, so he built a model solar system based on the Aristarchus theorem with the sun set in the center. Generally, his ideas seemed to work.

He mixed up the Epicycles of Hipparchus, Ptolemy and others at first, but finally reached agreement with a few of his predecessors' theories and discarded the others when they seemed to interfere. For instance, he concluded that the distance of the moon from Earth is about 60 times the radius of the Earth, which is pretty much on the button. He concluded that the moon had its own non-Earth-centered orbit. He also came blithely to the shattering and ignominious conclusion that The Earth was simply another planet. And where Ptolemy argued that if the Earth rotated everything would fly off into space, Copernicus posed the thought, "Not if the heavens rotated, too!"

Copernicus himself became a star in his day. He solved all kinds of mathematical problems by pointing out that the Earth wobbled, causing precession which concerns the Earth's axis in relation to the perpendicular of the Earth's orbital plane. He relocated Venus and Mercury and correctly figured the variable lengths of the retrograde motion of the nearer planets. And it was in these words that he justified his revolutionary proposals:

". . .In the midst of all, stands the sun. For who could in this most beautiful temple place this lamp in another better place than that from which it can at the same time illuminate the whole? Which some not unsuitably call the light of the world, others the soul of the ruler.

"Trismegistus calls it the visible God, the Electra of Sophocles, the all-seeing. So indeed does the sun, sitting on a royal throne, steer the revolving family of stars."

Copernicus missed a great deal of the controversy and condemnation his work caused other scholars. He died the same year his monumental work was published. The condemnation was due to the Church's view of the Earth as the center of the universe. This was truly revolutionary material and would have been the final straw that the church could have endured. Copernicus wasn't around to be tortured into recantation as others were later.

All of this leads us to the patron saint of astronomy, the Danish scholar-nobleman, Tycho Brahe, who really was as precise as anyone ever had been until then. Much more so.

He was 36 years old in 1572 when the nova blew out in the constellation Cassiopeia. It was a most brilliant display and it marked the beginning of astrophysical speculation in the modern world. It outshone Venus and every other planet or star other than the sun. It was quite visible during daylight. No one knew what it was at the time, but it certainly created a name for itself. It was the true explosion of a star which we now call a supernova.

More than one theologian thought it was the second Star of Bethlehem announcing the second coming of Christ. The great philosophers had long since concluded that the stars were immutable, invariable and eternal.

Tycho Brahe thought the event significant enough to alter the course of emerging scholastic thought. There it was: A spectacular incident of celestial expression. Neither it nor its location could be understood. It couldn't be an accident because, as St. Augustine proved, the originator of the universe and all the laws of nature couldn't have made a mistake. Perfect things like God do not make errors which include even the remote possibility of an accident. Therefore, many thought it to be a miracle. And why not?

A miracle is defined as an occurrence outside the ordinary course of nature, perceptable to the senses and traceable to the revocation of the laws of nature. The supernova seemed to fit these criteria; obviously it was an act of God.

The next logical question was why God did it. Well, no one seemed to have a ready answer for anything he did; none with any evidence to support it. Hence, the supernova's most important effect was to excite a great deal of interest in astronomy.

The most influential of the new astronomers was Tycho. What distinguished him from the others was the fact that he was rich and pretty smart to start with. He built a first-class observatory and looked upward. What should he see first off? You guessed it: an

Constellation: Lyra.

Object: M57 — Planetary nebula. **Distance:** 4,100 light years.

The famous Smoke Ring, an expanding shell of gas, blown off by its central star.

eclipse. From this phenomena he concluded that there was an error in the prediction of just when and how long they should occur. He reasoned that the error occurred because the sun was nearer to the earth and therefore seemed to move faster in winter than in summer. Could he help it if summer observations were what prognostications had been based upon?

After Tycho died, his German assistant, Kepler, used Tycho's figures, observances and calculations to present to the world a thought that cleared up all manner of mysteries and corrected innumerable errors. He reported that planets do not orbit in circles, but in *Ellipses*. He discovered also that planets travel to equal areas at equal intervals. That was to become known as *Kepler's Second Law*. Newton later used Kepler's laws in clearing up the laws of gravitation.

And then in 1564 came Galileo. He lasted until 1642. I put it that way because he's lucky he wasn't eliminated much sooner. He started with the study of medicine, but found it boring. So he went into pure science.

He was only 18 years old when he noticed that a lamp hanging from a long line in the leaning Tower of Pisa swung in equally decreasing oscillations, making a point that the duration of each oscillation depended on the length of the pendulum, not its weight. He timed the oscillation with his own pulse.

Not content to leave things alone, he then invented the pendulum clock so that doctors could measure people's pulses in their office.

Everyone probably knows the story about how Galileo dropped objects from that same tower to prove that falling bodies, regardless of their weight, fall at the same speed unless the object is light enough to be deflected by air resistance. Word around Pisa at the time was that if he didn't slow down, Galileo's experiments on the velocity of falling bodies might someday involve his own falling body. He was, as many Italians observed, "Not just another kid on the block."

Things might have been ended there except that he ran across an item about a thing called a telescope ground out by a near-sighted Dutch eyeglass blower. The invention was designed to magnify distances. Galileo ground some lenses and put them in his own telescope.

If it had been an ordinary young man, things might not have come to the attention of some of the Inquisition. But Galileo was the smart guy who measured pendulums and bounced bodies off the cobblestones at the foot of the Tower of Pisa. With his telescope he started studying the moon. First he notes the fact that it had no clouds. Then he measured the craters and mountains and watched their shadows lengthen as the sun illuminated the cold sphere.

He wrote about the fact that the moon was grossly imperfect. Remember that the Church felt why would God make an imperfect sphere? Well, the moon was one; no reason to doubt it now.

Because he was young and enthusiastic, he ignored the criticism heaped on him. Instead, he turned to other planets, stars and galaxies, not the least of which was the Milky Way. What did all this lead to?

Right.

Copernicus.

Galileo saw that Jupiter had moons of its own, which proved that the sun wasn't the sole cause of celestial motion. He wasn't far off when he announced that Jupiter was like our own system, of which the moon was a part. When he ran into doubters, he offered as proof a look through his telescope. Many critics, seeing how revolutionary his ideas were, refused to look into the thing, thinking it tainted their sense of logic. Galileo was undermining Aristotelian thought in favor of that demented heretic Copernicus. If you knew the Inquisition, you had to be out of your sense to do that.

Aristotelian conclusions about the sun didn't allow for imperfections, yet here was Galileo talking about craters on the moon and then dark spots on the surface of the sun. As he propounded his own ideas he contradicted Aristotle. Reminded about the heresies of Copernicus, the clergy brought out his work and put it under review. This study led to a posthumous popularity of Copernicus' books, which in turn resulted in still more ecclesiastical criticism. Galileo, a self-confessed thinker, was running loose around the rooftops pointing out naughty ideas, so the church put Copernicus' work on the index of what not to get caught reading on the way to work.

By now, Galileo shifted his manner of publishing his ideas and changed them into dialogues between a Copernican and an adherent to the Ptolemaic principles. This was to take the edge off his impertinence. But he wasn't fooling anyone.

Finally, all the scholars, who either didn't believe Galileo or were jealous of his genius and growing reputation, pounced on him and had him summoned to answer charges before the Court of Inquisi-

tion in 1633. Galileo admitted to many "heretical" beliefs contrary to Divine Scripture. Among other things, he admitted that he believed the sun didn't move from east to west, but that the Earth did. Furthermore, he had declared that the Earth may not be the center of the world.

Galileo was sentenced to desist speaking out about most of his beliefs, and to be confined to prison where he was to repeat the seven penitential psalms. The Pope stepped in the next day, however, and had him sent to a villa outside Rome, after which he was allowed to return to Florence. During his work there he gazed directly into the sun. As a result, one of the greatest observers in history went blind and died in 1642 on the same day Isaac Newton was born.

From Time to Time

"There are no fragments so precious as time, and none so heedlessly lost by people who cannot make a moment, and yet can waste years."

–Robert Montgomery

"You cannot kill time without injuring eternity."

–Thoreau

"Oh time. Beautifier of the dead. Adorner of the ruin; comforter and only healer when the heart hath bled.

"Time is the most undefinable yet paradoxial of things; the past is gone, the future has not come, and the present becomes the past even while we attempt to define it, and, like the flash of the lightning, at once it exists and expires."

–Colton

Our time is arrived at by the apparent motion of the sun through the sky. We divide the day into 24 equal hours, but the sun's progress is not at a steady rate. This is because the sun moves along the ecliptic, and not along the celestial equator. Also, the Earth's orbit around the sun is elliptical, not circular.

So for convenience, time is kept by an imaginary mean sun, which is moving at a regular speed and keeping *mean solar time*. Time judged by the position of the real sun is called *apparent solar time*. The difference between the apparent and mean solar times is called the *equation of time*. It ranges between plus and minus 12 minutes during the year. This means that time as measured by a sundial has to be corrected by the equation of time to give mean time.

Midday is when the sun crosses the meridian. That means that time is different for people at different longitudes on earth. As a 360° turn of the Earth takes exactly 24 hours, a difference in longitude of 15° produces a difference of one hour. That is why time zones are divided into 15°. The boundaries between time zones are not always meridians, but take into account the boundaries within towns, cities, states and nations for practicality. Astronomers use the time at longitude 0° called *universal time*.

Astronomers, of course, have an extraordinary interest in time because time is used in charting the progress of the stars through the sky. The Earth rotates once every 23 hours and 58 minutes. There is another manner of measuring time which is kept by successive meridians. This method is called *sidereal time*. The sidereal clock must run faster than the solar time clock which gains about four minutes per day.

The difference between sidereal time and an object is called the *object's hour angle*. It is called that because it gives the time which has elapsed since the object was on the meridian.

In a larger, more human sense, life to an individual is the duration of time spent between birth and death. That duration has always been considered the most precious element in existence. Yet it has been sacrificed by billions of people during wars in the name of honor, religion, love, hate and many other emotional investments. And even though it has always been considered precious, it is not unusual to hear someone say of an activity, "It's a nice way to kill time," or, on the other hand, "It's a way to pass the time."

To the man on the street, however, it was travel which first frustrated our concept of time as we had known it throughout the centuries. In fact, "travel" is inextricably tied in with time.

It was in 1522 that an earth-shaking thing happened in the civilized world. The 18 survivors of Magellan's crew arrived back in Spain on a Thursday. They had mislaid a day somewhere because in Spain it was Thursday, but to these men it was still Wednesday. They were a day older by the calendar. That meant a lot of mortal sins. If a sailor happened to get into the conventional confessional, he could drive his father confessor to distraction trying to make excuses.

What is all this about? That business about how the faster you

go, the slower time passes. As you've heard by now, time is relative and therefore a variable. That's one of the effects of relativity. The rate at which time flows depends on the speed at which the observer travels. At the speed of light, time would cease to exist and the idea of "now" would be something that would last forever. Until then, however, time slows down on the way up to the speed of light.

If we fired at the speed of light toward our nearest star, turned around and came back, the trip would have lasted about eight years according to everything on Earth such as clocks, calendars, a person's age and other devices used to measure time. But in the space ship, no one would have known the ship had taken off. It would still be "now," waiting for blast-off.

Theoretically, if a person were to look into a spaceship as it travels at even half the speed of light, everything would seem to be moving in slow motion. Everything would seem to be frozen at a standstill. To those on the spaceship everything would seem to be very normal. But imagine going at half the speed of light. The ship's instruments would indicate the ship was gaining 80,000 mph every hour. By the time the ship reached 80-85% of the speed of light, everything to the outside observer would seem to take more than half as long as expected. Two days' time would take only one day to occur. The closer the ship came to the speed of light, the closer it would get to stopping altogether—or reaching "now." Imagine everything staying at "now."

But zinging along at less than the speed of light, the ship turns around and comes back to Earth. Waiting for it is the travelers' own great-grandchildren, who aren't actually much older than the astronauts vis-a-vis Earth time.

This is no idle theory. This bit of algebra checks out to the last decimal point (unlike Newton's laws which were off a fraction). This phenomenon is called Time Dilation.

Should you need a table to use in an argument around the house or your favorite pub, here is how something can seem to begin with apparent probable fact and continue to what might seem absurd, yet still remain mathematically quite sound.

Assuming travel at an acceleration of only 1 G in a round trip, here's what you'd experience:

Duration (Out & Back) In years		**Distance Traveled In Light Years**
On the ship	**On Earth**	
1	1.0	0.06
5	6.5	1.70
10	24.0	10.00
20	270.0	137.00
50	420,000.0	208,000.00
60	5,000,000.0	2,470,000.00

Modern scientists find no natural laws which dictate against such speeds and distances being traveled. Some people believe it's even attainable in the near future. But however remote, it is fundamentally attainable.

And if it's really true that progress is its own adrenalin, then it follows that progress seems to square root itself in the most incredible manner. That being the case, who knows how long it will take to try out all these projections. After all, man broke the gravity barrier with heavier-than-air machines only at the turn of the 20th century. Now most adults in the Western world have flown in an airplane at least once. Remember, progress moves faster and faster. Beginning with Galileo who gathered more astronomical information in less time than everyone before him, so will scientists who are working today. Not only do they have all scientific data accumulated so far upon which to build, but computers make is possible for them to have this knowledge at their fingertips in *nanoseconds.*

BON VOYAGE

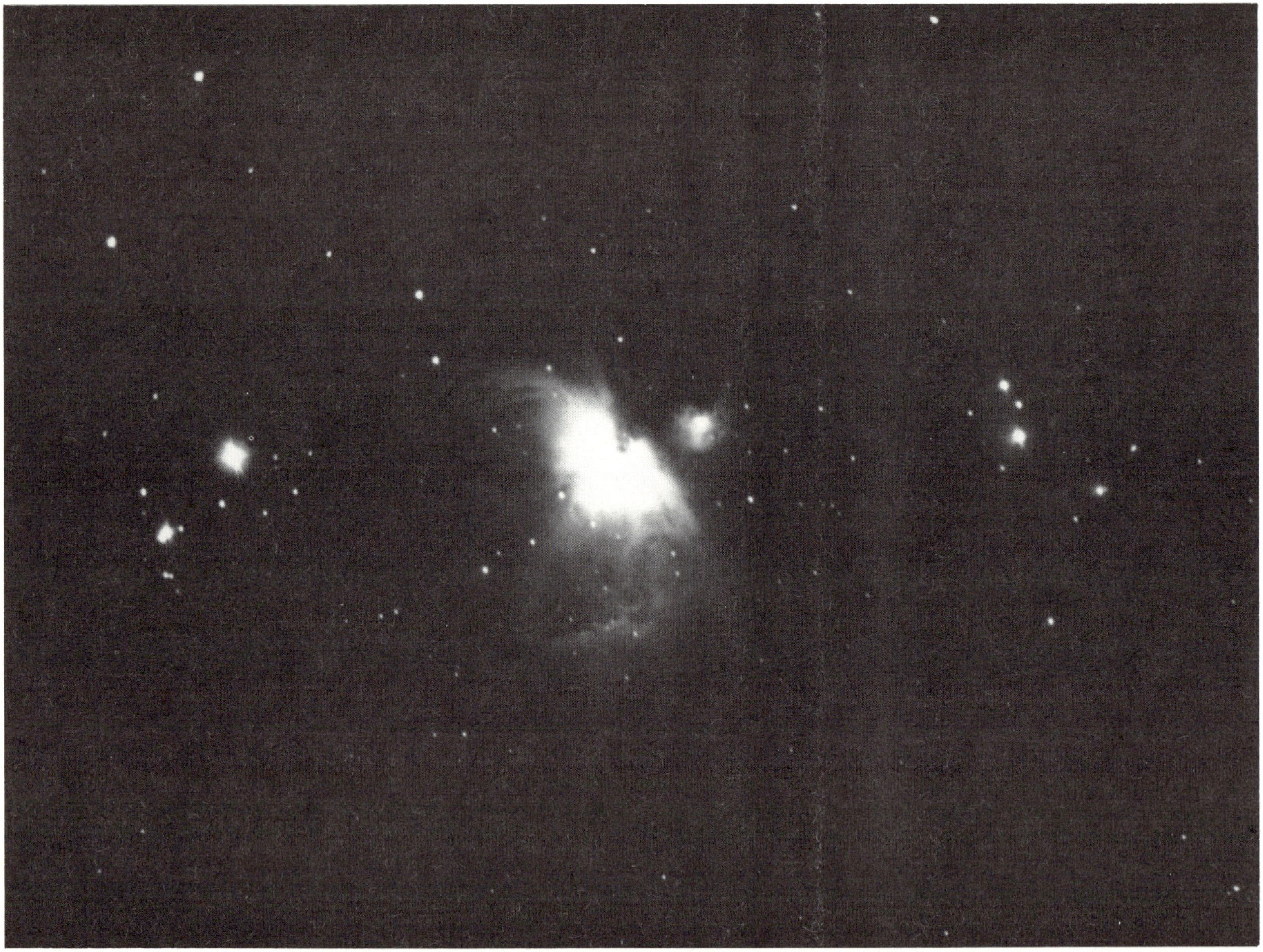

Constellation: Orion.

Object: M42 — The Great Orion Nebula. **Distance:** 1,304 light years.

A great glowing cloud of ionized-hydrogen gas in the process of forming new stars.

What on Earth ?

What about us? What about the Earth? What's inside our sphere? Oddly, while we're reaching out farther and farther into space we're curiously ignorant of our own planet. We've discussed here the birth of the universe and even the birth of our galaxy. We've even touched upon the birth of our solar system. So let's turn once more to home.

The Earth is almost a sphere with about an 8,000 mile diameter. Inside is a core of a bit more than 3,000 miles, surrounded by a thick shell that's mostly ridged. The core's essential features are that it's fluid and its density are rather higher than the rocks comprising the shell.

There is no current overall agreement as to the composition of the core. It had been generally agreed that it consists of molten metal, mostly iron. Recently, however, this idea has been challenged by a school led by W. H. Ramsey of England, who maintains that the core consists of the same rocks as the shell, though is a more complex composition. Actually, the disagreements concerning the core do center around the composition of the matter, not the matter itself. Unfortunately, this very subject is one that modern science has decided would be best left to future scientists to decide, preferring to give the weight of their pioneering attention to astronomy and marine biology rather than geology.

Isn't it amazing that we're bent on the survey of the heavens at an enormous expense even at the exclusion of what's in our own inner space, namely the study and discovery of what makes up the subfloor of the oceans and the soul of the Earth.

I suppose it's traditional to look out at what's up there because traditionally up is heavenward and down is hellbound. Curious, particularly since we have learned that there is no up or down.

Oddly, great revisions probably will have to be made concerning the true geologic nature and temperature of the Earth's center. Many current schoolbooks flatly state that the Earth's core has already proven to be molten because volcanos are nothing more than thin places in the Earth's shell. Further traditional proof concerns the fact that mines grow hotter as they grow deeper, giving rise to estimates pegging the central core of the Earth at about 3000° Centigrade because that exceeds the boiling point of iron. This would be valid if the rise in temperature in a mine were due to an outward flow of heat from Earth's center, but that just isn't true. The heat in a mine, and probably the heat which causes volcanic explosions, is almost totally due to the decay of radioactive matter which for unexplained reasons clings more to surface rocks only 20 miles deep. With this point proven, there is no evidence that the Earth is particularly hot at the core. It is warm, but not hot enough for molten iron. Many geologists say it may be compared to the heat of a wood stove.

In the future, scientists will surely answer these questions. This, however, will most assuredly raise more questions. As a wise man once said, "The more one learns, the more one realizes how much he has left to learn."

"Sometimes as I drift idly on Walden Pond,
I cease to live and begin to be me."

–Thoreau

"Know in thyself and all oneself—the same soul:
Banish the dram that sunders any part from the whole."

–Ancient Hindu Poem

"That which is necessary, does not offend me."

–Nietzsche

"Philosophy is only thinking about things in the most comprehensive possible way."

–Henry James

"If there are other intelligent beings in our universe, it is probable that they are interested in securing the same information as we are since we share the same universe and live within the same strict natural laws by which we travel in our short search to find our Maker. Indeed, we have so much in common we might very well be the same person; the alien, ourselves and God very possibly being one and the same."

–Magee

Constellation: Taurus.

Object: M45 — The Pleiades Galactic Cluster. **Distance:** 400 light years.

A galactic star cluster also known as the Seven Sisters.

Hello, Who's There?

Let us assume that among the hundreds of millions of planets which possibly could give rise to an intelligent society, at least a few might have populations which have learned to live beyond the tendency to work toward their own self-destruction. Animals and insects on Earth live that way all the time. Why not intelligent beings?

Carl Sagan and his colleague, William Newman, calculated that if a civilization which lived only 200 light years away had started a million years ago to explore other worlds, those explorers would be getting pretty close to us right now. Why might they explore? Because, at least in our experience, curiosity goes before intelligence, and it requires intelligence to survive in a hostile environment.

If they are out there, we must ask ourselves of what value these people could be to us or we to them? Surely a civilization which traveled 200 light years is already far in advance of us, and so their advanced knowledge is important to us now and in our immediate future. Perhaps they could tell us why we are condemned to such a short, finite life? There *must* be a reason our life expectancy is only 100 years. Might they know why?

But what if, just what if, we are the first species ever to arrive at this, our present state of progress. If we are first, that would certainly explain why we have not been visited thus far.

As H. G. Wells said in "The Discovery of the Future":

> "It is possible that all the past is but the twilight of the dawn. It is possible to believe that all the human mind has ever accomplished is but the dream before awakening . . .perhaps a day will come, one day in the unending succession of days, when beings, beings who are now only latent in our thoughts and hidden in our loins, shall stand upon this earth as one stands upon a footstool, and shall laugh and reach out their hands to us amidst the stars."

The thought that we on Earth are the first intelligent beings in the cosmos is anything but new. We would be derelict as intelligent people if we did not at least consider such a possibility. But then, hasn't that been the big error in the course of scientific discovery all along? Haven't we learned by now that we are not the center of the universe? Not the center of the galaxy? Not even the center of our own solar system?

Why then consider that we are first in interplanetary travel? There are two reasons. First, because we have not a shred of evidence that there exists anyone or anything that is alive, much less civilized enough to have escaped the pull of its own gravity. Secondly, not to entertain the idea that we might be the first intelligent beings in existence discredits feasible thought for no good reason.

Having said that, let us move on to the thought that we know the laws of probability dictate that there is no other life in the cosmos because of three major reasons. First, all of nature tends to support its own species' longevity. Second, intelligence is a strong factor in the preservation of a species. The longer the duration of life of that species, the greater the chance for intelligence to develop. Thirdly, there are billions of places much older than Earth capable of sustaining life. They've had enough time to develop an intelligent society capable of choosing to live rather than annihilate itself.

Thus, we can conclude that either we are the first in the cosmos to arrive as an intelligent society, or we are merely one among millions of beings who have arrived at a level of intelligent existance.

Of the two possibilities, the laws of mathematical probability dictate that not only are we not alone in this cosmos, we are comparatively primitive by implied comparison. So far, at least.

With all due respect for the purveyor of science fiction let us dispel some common irrational conclusions about the future.

One of the most blatant errors is the possibility of war between civilizations. Of one thing you may be certain; we are far more likely to be visited by an older, more intelligent society than they are likely to be visited by us. Because they are older it is only reasonable to assume they long ago learned to succeed in living with potential enemies. War, even to us, is almost primitively animalistic, being derived from the need for territorial imperative to expand undefended space. But space, real space, outer space, is endless. It's so open for conquest as to invite any and every exploration from every source.

The other reason our civilization is immune to interstellar war is that war can take place only when there is hope for eventual victory, or survival, or both. The great disparity between our capability of waging a successful war with an alien civilization is so vast, the very idea of any steadfast defense against so superior a force is

automatically an imponderable impossibility. Anyone who can travel a hundred light years or so can certainly whip us handily.

The very idea of war is an issue of comparative strengths. The smallest collateral weapons derived from the technology of beings capable of interstellar travel would be totally devastating at the very least, and incomprehensible at best. In other words, realistic consideration of a conflict with a power from outer space is so remote as to be self-contradictory. Too bad we are taking so long to evaluate the idea that we can destroy ourselves almost as easily.

Considering how poorly we tend to manage our little planet and the cavalier manner in which we address the life of our soil, our water, our atmosphere and each other's lives, it is fair to comment that we are certainly not sufficiently advanced along the road of civilization to invite another species of intelligent beings to visit us. One should straighten up the house before opening doors to important strangers. Our house doesn't even promise that it *can* be cleaned up. What we need is not a dusting, not a washing, not a paint job. What we need right now is a sand blasting and total reorganization of our priorities.

Imagine the president of the United States or the head of the Soviet Union greeting a visiting delegation from Alpha Centauri. First of all, the Office of Protocol would have no precedent for receiving an alien civilization on Earth. Just imagine the usual battalion of Marines presenting present arms. Weapons as a way of greeting, in fact, is downright threatening. We are in the habit of honoring visiting dignitaries by presenting them with a line of men holding up their guns, instruments used principally to shoot at and kill other living beings. Our formal hello is abrupt, disciplined, and threateningly in unison. Why do we "Present arms!" at all?

At dinner that night, what does one serve? Champagne? The visitors might well consider that the grapes from which it is fermented might have been their own relatives before a toast is drunk.

Be your own science fiction writer from this point forward. Might our visitors not ask the name of the steer they're being offered in the form of filet mignon? Are the grains of wheat used in baking bread really the things which make music on their planet? Or the laws? Or the wars? For all we know, *they are*.

Maybe our morals aren't really morals at all, but merely efficient rules for our behavior thus far.

We even efficiently use the dead material of our forebearers or antecedents for gasoline and oil rather than lament their passing. Should people who exhibit such behavior be trusted?

Going back once more, reflect on that time when the first artist scratched our pictures of the world around him. He was really initiating the first step on the road to culture. He may have been a prehistoric sub-human, but he was the father of art and an unwitting historian.

Since then, the artist who seeks to provide a bit of beauty to mark his passing has speculated on the duration of the work he leaves behind. How long after I die will this work live, he asks himself. Decades? Centuries? Perhaps if he is truly great, a thousand years?

What would it mean for an artist to create something about present day mankind that would last a billion years?

Those who designed the package sent up on the Voyager space vehicle did that. Taking into account cosmic dust, cosmic rays, and cold of interstellar space, there is almost nothing which can harm or mar the gold record and album jacket they sent into intergalactic travel. On it is inscribed how to play the record, where it came from, what we look like, information about the cerebral cortex and limbic system, and people saying the traditional greetings in 60 different languages. There is also the speaking of one humpback whale to another, just in case it is later learned that whales greet aliens better than we do.

Also included are photographs of humans from all over the world, showing them caring for one another, learning, creating art, using tools, and answering challenges. There is about an hour and a half of music recorded from many different cultures.

All of it expresses loneliness and the longing to meet and know another intelligent society. Everything imaginable which can express our sincere desire to meet interstellar aliens is included in the space package in the hope that when it is received it will be viewed, listened to and understood as the reach of our hand stretches across the interstellar oceans to stars. It confesses that we humans have progressed this far in the last 10 or 15 billion years of cosmic existence, admitting that we are no further along than a scared animal growling fearsomely and posturing aggressively as if everyone else wants to take from us that which we hardly cherish while we hold it in our possession.

The Habitat

As you glide toward the habitat, it will look like a miniature, a toy. That is because there is no background to which it may be compared. As your space ship orbits the city-in-space, you will see that it looks like a giant tire against a black backdrop.

This is the new frontier, the first town in outer space, populated by 15,000 pioneers, all adventurers who have passed rigorous mental and physical tests before being allowed to immigrate to this new world.

The outside is covered by a radiation shield constructed of rubble from the surface of the nearby moon. Its main purpose is to protect you and the other colonists from cosmic rays. It is about two meters thick and made from compressed rock from the moon, held together with fasteners. Long rows of mirrors extend around the surface of the habitat. These are two-way translucent windows through which only daylight minus the cosmic rays is filtered.

The entire habitat rotates within the outer protective shield at one revolution per minute around the center hub of the wheel to create normal earth gravity. It is turned by solar power, as is everything else. Solar panels unroll like giant sidewalks in space, and need be no more than 1/16 of an inch thick. Since there is virtually no dust or dirt in space, they never need to be cleaned or even swept. They have been rolled out and left without foundation, the only tether being a wire connecting the huge solar panel to the optical power cord no thicker than a pencil, manufactured in space.

Inside the city, long streets are bordered by shops, parks, fountains, homes and other buildings under which the sunshine streams in from the golden sky overhead. Although this particular city is only 100 acres in circumference, it accommodates about 15,000 people very comfortably because it terraces up the sides of the wheel.

Doors and walls are covered only for the purpose of privacy since the weather is very nearly perfectly controlled by the inhabitants. No one is ever more than a five-minute walk from an elevator which travels between floors and up the other side, turning like a gyroscope as it moves. Its motion creates its own gravity within the gravity of the wheel.

Beautiful plant life flourishes along broad avenues faced by outdoor patios stepped up to the sides of shops and offices along winding walkways. Most of the furniture is of a plastic material, wood being an expensive import from Earth. Every kitchen offers varieties of produce and meat produced on nearby wheel farms.

Like any pioneering colony, the population's average age is between 22 and 40 years. There are plenty of children because of the child-bearing ages of their parents, just as in frontier towns on Earth.

Elitist behavior will reign as a result of the screening process necessary to select the colonists. The colonists are also characterized by their willing industriousness. Improvement and innovation are a way of life because there is so much that is new upon which to work until the average age of the colonists dictates a slowdown. By then, the older inhabitants will have graduated to supervisory positions as the demand on Earth for the colony's products grows at an increasing rate. The purity of products manufactured in space will be infinitely better than those produced on Earth. The glass, steel, and ceramic products will be so pure in material and construction that everything produced on Earth will automatically be second-rate and, therefore, less expensive.

For this last reason alone, hundreds more colonies will be established. The pay rate itself will attract so many workers and artisans that successful colonization will be a foregone conclusion. Satellite power stations will have long since replaced fuel as we now know it. The sun will provide an unlimited power source via microwave direction to Earth.

We've looked at the origin of the solar system as man has seen it through history. We've traveled with learned minds into the superstitious realm of astrology until it unveiled itself and stood as an acceptable science which we now know as astronomy. We have skirted the edges of the cosmos and traced its origin back to a single instant so revolutionary as to be considered anathema by its discoverer, Einstein.

Where haven't we been?

We haven't as yet taken a good look at the very latest discoveries: Pulsars, neutron stars, black holes, and quasars. They represent the threshold of new territory.

Constellation: Vulpecula.

Object: M27 — The Dumbbell. **Distance:** 1,250 light years.

The Dumbbell Nebula is probably the most famous of the planetary nebula.

Neutron Stars

To begin, keep in mind that we cannot test our speculations about these stars; all our predictions about them are speculative. However, we've known about them since 1934 when Futz Zuicky and Walter Baade began their investigations. Additional information came from the work of J. Robert Oppenheimer and G. M. Volkoff who studied the implications of supernovae, of which the neutron star may be a remnant. Unfortunately, World War II interrupted these studies. The neutron star is correctly characterized as the only astro-physical object whose existence and properties were predicted long before they were physically discovered.

One thing we do know is that some neutron stars have only one-fifth the mass of the sun; others have two or three times more.

Pulsars

In 1968 an intriguing noise bleeped across the celestial map. It was a periodic burst of precisely spaced radio signals. Noise from outer space? Is someone out there trying to say hello?

When x-rays were found to have sources in the sky, scientists immediately began searching for the cause. It took until 1967 to find a new class of celestial objects altogether: pulsars.

Going back for a moment—about 7,000 years ago—far out in space a star exploded, blowing itself all over the universe. Very large before the explosion, it had reached a point of critical internal energy. It reacted in a violent explosion bigger than any ever recorded, and created something we now call a supernova.

Six thousand years passed as the light from this big explosion traveled to Earth from the constellation Taurus the Bull. It finally reached Earth in 1054 A.D. in the middle of the Dark Ages in Europe. In China, astronomers watched with great interest and recorded when its light reached earth and lit up the entire eastern sky. They called it the guest star and noted a date which would have been July 4 in our calendar. The brightest object in the sky, it lasted 23 days, until July 27. One could read at night by its light; it was considerably brighter than a full moon. Although it occurred during the Dark Ages, prehistoric pictographs of it were found in Indian caves, particularly in those of the Pueblo Indians in Arizona.

This entire complex, a chaotic expanding shell of gas actually contains several stars. Today we know it as the Crab Nebula. One of the central stars represents the source of all its material and is the exact star that exploded seven thousand years ago. As a neutron star, it has a surface temperature between 6 and 7 million degrees and an extremely small diameter. Today it's recognized as a pulsar.

Now then, what is a pulsar?

When a piece of iron like a horseshoe gets hot, it burns red. Continue to heat it more and it turns white hot. Continue to increase the heat and it will melt. Heat it even more and the horseshoe finally becomes a gas and the iron atoms literally explode from the high temperature mass. We know that everything organic will turn into a gaseous state if heated enough. Hence, stars are called incandescent because their sphere has surface temperatures so high the surface can't remain liquid or solid, but becomes a gas.

We now know that Nature can raise densities to such high degrees that gaseous matter can be frozen to make the gas solid again: a heat cycle. We're not talking about the kind of solid we know here on Earth. But it *is* solid nonetheless. We can't duplicate this kind of solid on Earth, at least not yet. In fact, if we did, the Earth might jump into it. But more on that later.

Moving right along, following the explosion of the supernova was such immense gravitational contraction that it shrank to a diameter of only a few miles (somewhere between 5 and 100).

BUT. . .in this little insignificant sphere is the mass of a sun. Something like 864,000 miles in diameter and one-third of a million times as massive as Earth. In fact, so massive is this star, and so dense, it turns from the incredibly blazing heated gas into a solid star.

Just imagine heating an iron poker until it turns white, melts, becomes a gas and finally becomes so hot that the thing becomes solid again.

Now imagine it 100 miles in diameter. Even to begin to extrapolate about it, we have to invoke a complex mathematical formula called the "equation of state." This takes into account temperature, pressure, rigidity, and density both individually and as they interrelate with each other.

To give you an idea of the problem here, astronomers have to become advanced mathematicians and then chemists and finally astrophysicists in order to deal with the study.

Then there is the study of electrons.

There are electrons which are not free to move in random directions. In this star they're packed so tightly the only way they can move is if the others all move to get out of the way, something like a person in a mob. The only way to move is if everyone else moves, too.

An increase in density goes on continually as the star tries harder and harder to become more and more compact. This further increases the speed of the electrons although they're already captured by the nuclei. When the electron velocities become fast enough, the neutrons become unbound to make room and move away to a point where they become a degenerate gas. When a sufficiently high density is reached, the unbound neutrons enter a state of equilibrium with the neutrons in the nucleus.

But rather than remaining stable with the neutrons and neucleus balanced, density continues to increase until the entire thing becomes incompressible, or has a resistance to compression that is 10,000 billion times that of ordinary steel. So great is this resistence that it can actually counterbalance the pull of its own gravity. A cubic inch of matter only one yard deep would weigh about 1,500 tons. You would weigh millions of tons. If you were placed on such a star you would instantly be squashed into a head-to-toe thickness less than that of a kleenex tissue.

The deeper the penetration into a neutron star (theoretically, of course), the greater the weight of the material. That cubic inch we spoke of would increase in weight to 4.5 billion tons at the center. Remember, this is only one cubic inch of material—4.5 billion tons. Now *that's* heavy!

You could not imagine the extreme heat. At the moment of implosion, which is when the business of the supernova begins, the temperature reached 10 billion degrees. You will be relieved to know that it cools at about 6 to 8 million degrees every million years or so.

Black Holes

Of all the strange new things that have come to our attention as the universe unfolds, the black hole is among the most difficult to understand of all.

Before we venture out again, let's look once more under our own feet. This puts things into perspective. The complexity of the very idea of a black hole requires that we take a breath, find our place in the cosmos, and settle back to Earth again.

As you recall, we started with the sun and our system of planets. If Earth is represented by a ball about six inches in diameter, the sun represents a town by comparison, and the nearest star is 2,000 miles away from the town. To look at the galaxy called the Milky Way, we must use the measurement of light years because, as noted earlier, no other increment of space has ever been found adequate to measure such vast distances.

We've seen that light takes several years to travel from nearby stars and that many of the stars in the Milky Way are as much as 1,000 light years away. But the Milky Way is only a small bit of a disk-shaped system of gas and stars that's turning like a tiny wheel in space. In the immense universe, our little solar galaxy is so infinitesimal, consider it as almost unchartered in its insignificance. But the diameter of the Milky Way is 60,000 light years across. From your point of view, it's so colossal that there's been only time enough for it to turn around about 20 times since its oldest stars were born about 4,000,000,000 (4 billion) years ago. And this is in spite of its blinding speed of almost a million miles per hour, the speed with which the outer parts of the disk are moving. Earth is pretty far out in the disk.

Now let's take a walk outside the galaxy again. Get out of town where the big lights interfere with our vision. What you see out there are not just a few billion stars scattered all over the sky. You can look right out of our galaxy to see other galaxies, many like our own, though in point of fact some are not very like ours. Most of the galaxies are what is called extra-galactic nebulae. Some are spherical rather than disk-like, each with something between 100,000,000 and up to 1,000,000,000,000 or so bodies.

Now look at Andromeda, the most distant object we can see with the naked eye. That's what our Milky Way looks like from the outside and it's only 2,000,000 light years away. Its light has been on its way here all that time. To the naked eye there are only a few such galaxies. To the telescope there are millions. To the radio telescope there are billions. Once we have telescopes in space, well, after billions who knows?

Now what about all those things we can't see by light and sight? What about those holes that are so dense not even light can escape their gravity?

The closest thing I can suggest to help you relate to this phenom-

enon is to ask you to hold out your hand at arm's length. Now make a fist. Now tighten it. Now hold it out there as tightly as you can and close your eyes. Next, imagine that you're holding it so tightly that everything around your fist is beginning to move toward it. . .even all the air. Then imagine that the air is emptying into your fist and that behind it everything else is following: this book, your chair, the floor, the ground beneath the floor. . .everything.

That feeling will give you an idea of what's called *gravitational collapse*. All the things emptying into that hole of your fist are emptying into a funnel called a *singularity*. The singularity is a place of space-time where infinitely intense gravitational forces deform matter beyond recognition. As the radius of a sphere shrinks to a point of zero dimension, and the volume goes to zero, matter and energy are squeezed out of our universe. Out of existence!

Most scientists agree that our understanding of the laws of nature simply isn't sufficiently progressed enough to understand the funnel called singularity. This funnel seems to pour out of the center of a black hole; through it everything seems to disappear. Either we have missed some laws, or some new laws are needed to explain this strange phenomenon.

If you're wondering, as we are often wont to do, what is left to be discovered, this is one of those things which some new young Einstein will explain to us in the future. Perhaps even in our lifetime.

The first man to suspect a black hole actually existed was John Michell in 1783. The idea was too complex as a mathematical speculation and too bizarre to be accepted except as the outpouring of the mind of a scientist who was becoming irresponsibly creative during his lifetime.

But in 1971 an x-ray observatory in Kenya detected an unusually bright emission coming from the Swan (Cygnus) constellation. It was sputtering x-rays at the rate of 1,000 times a second. They called the phenomenon Cygnus I. It was either invisible or so small it couldn't be seen. Because the signal flickered, it had to travel no faster than the speed of light across its radius. Therefore, if something was there, it was calculated to be no more than 300 kilometers in radius. [Cygnus I – 300,000 km/sec x (1/1000 sec) = 300 Km].

Now then, Cygnus I, this tiny asteroid-sized dot, was having all the effects of a super-size, full-blown star about 10 times the volume of our sun. This is not a source of a supergiant star's x-rays. That meant it was an invisible object weighing 10 times as much as our sun, with a density so powerful it collapsed into itself until the hole it left was only 300 kilometers wide.

Think of that. A thing smaller than the distance between New York and Philadelphia weighing 10 times as much as our sun. The only thing to describe it that had ever been postulated was John Michell's description of a black hole.

Since then, other dots in the sky have come to be suspected of being black holes. As an example, we know that Cassiopeia is part of a supernova explosion whose light should have been seen in the 1600s. However, no one ever reported seeing it. Why? Perhaps because not only all its matter, but the light from the super explosion and everything around it had been swallowed by a black hole.

There are those who reason that once we isolate a black hole we investigate it enough to gather evidence about the depth of the singularity. That idea is ruled out, however, because it takes an infinite amount of time to reach the gravitational radius. As the universe spans finite time, the black hole cannot have enough time, to end its singularity.

Complex? Here we go again.

What happens in space-time that could give us a graphic picture of a singularity? Picture a thin sheet of rubber stretched horizontally over a large frame. Let us assume that this rubber sheet is part of the universe. If we were to place an iron ball in the center of the stretched rubber sheet, the ball would depress the sheet by sinking into it. A heavier ball would sink lower, stretching the sheet farther. If the ball had an infinite amount of weight and the rubber sheet couldn't tear, the ball would just continue to drop into the rubber forever.

But imagine the ball creating a tear as it drops through the sheet. The ball would go through and the sheet would pop back flat with a hole in it. Space-time, like the sheet, would return, but the ball would be gone. Gone from where? Gone from our universe? To where? Is it possible it might go into another universe? To where? What would it look like there, on the other side, in another universe? What would the tear in the sheet look like?

Well, it could look like a quasar, which just might be the other end of a black hole from another universe emptying into this one. Because this is such a startling thought, much study has been poured into proving and refuting it. So far, no one has adequately

proved it to be impossible. Quite the contrary. As some astronomers look at quasars they ask themselves whether these huge energetic phenomena are black holes from some other universe suddenly flashing into being as white holes. In other words, the white hole could be the tear in the sheet now that the iron ball has popped through to the other universe at the end of the singularity.

There is a possible relationship between these two phenomena. Dr. R. M. Kjellming of the National Radio Astronomy Observatory in Greenbank, West Virginia, says:

> "It seems natural. . .to suggest that black holes, which have singularities by which matter disappears from the universe, are related in a genitive manner to white holes, defined to be singularities from which matter and energy emerge, usually in the centers of quasars and galaxies."
>
> ". . .at certain points, our universe is multipli-connected, or. . .two or more universes are connected to each other through black hole/white hole singularities. In other words, it is through these singularities that matter may be exchanged."

As the reader can imagine, speculations such as these are no longer rare among astronomers or astrophysicists. Today, such persons are not ordinarily so speculative as to suggest the idea of universes existing alongside each other, one inside another, or in antimaterial existence from one another. Yet that's where their equations lead them. Such was a part of the ultimate result of the theory of relativity. Not that everything proposed in terms of relativity is true. Correctness and truth still conform to reality regardless of how we or others view them.

J. Robert Oppenheimer and a student named H. S. Snyder were studying the end of a large mass of cold matter in 1939. Relativity predicted that once this mass began collapsing, it could not stop. That meant that eventually it would have to shrink into a black hole. That is, once a large mass like a non-rotating symmetrical star begins to collapse, it will shrink to a critical dimension called the gravitational radius. When it reaches this dimension it collapses completely, literally closing itself down.

As I said, this is not a simple speculation but the conclusion of a rigorous mathematical expression indicating that the gravitational radius of a body with the mass of an entire galaxy could be no larger than a few miles. A galaxy of more than a billion stars would have a gravitational radius large enough to extend over 2 billion miles.

When their study resumed after World War II, Oppenheimer and Snyder learned of the quasi-stellar radio sources now known as quasars. Because these entities were enormously distant from earth (billions of light years), they knew they must be emitting fantastically large amounts of energy. One explanation which came immediately to mind was that this energy came from the collapse of a giant star, or from a cluster of them. In fact, the researchers concluded that no other force except gravitation could expain such colossal outbursts.

This gave scientists the motivation to embark on the study of black holes. Here at last was the ultimate test of Einstein's laws.

Perhaps I. M. Levitt, Director Emeritus of the Fels Planetarium, put the concept of black holes most simply:

> "The very name. . .black holes. . .indicates that they represent a class of objects which cannot be seen. The gravitational field of these is so intense that if one could stand on the surface of a black hole and manipulate the most powerful searchlight (no light could be seen) even if one were no farther away from the black hole than the Earth from the sun. In fact, if we could concentrate the entire light output of the sun into this powerful searchlight we still couldn't see it, for the light could not overcome the overwhelming gravitational field to escape from the surface. That is why the surface is called the "absolute event horizon." It represents the boundary of the black hole."

Then how can we ever see a black hole? There are many answers to this, so far all of them inefficient. One answer proposed by the Soviets deals with broadband electromagnetic radiation emitted by matter that falls toward the black hole. Another deals with studying companions as they orbit prior to their plunge into apparent annihilation.

Why can't we see a black hole as it passes in front of another star, a bright one? The answer to that is that it's so shrunken from the impact of density, the diameter is too small for that kind of perception.

Yet there are stars with 25 times the mass of the sun from which we can detect behavior indicating either a neutron star or a black hole creating secondary effects. And there are other stars with 35 times the mass of our sun which behave as they might if pieces were breaking off and falling into oblivion.

In any case, as more and more people enter the field of astronomy, more and more suggestions will be made and tested until we know that the bottom of the singularity of an enormous black hole is our destiny.

The Chaos of it All

Two droplets of condensation are suspended alongside each other on the surface of a windowpane. Formed in the same room, in the same air, on the same surface, each weighs the same. Each has exactly the same volume of water. Nothing moves. Then, one runs down the pane to the sill, as if free. The other does not.

Why?

There is a faucet through which droplets are formed. The droplets fall off the lip of the faucet. The faucet never moves. But the drops form in different volumes and, for no apparent reason, fall in infrequent rhythms. Nothing about the drops is periodical. Nothing about them is predictable.

Why?

You are sitting before a waterfall. Suddenly, for no apparent reason, one splash lifts away from the rest of the flow and reaches your cheek. There is no change in the wind or the rocks or the temperature, but it doesn't happen again. Ever.

Why?

All are random, unpredictable events going on around us at all times. Why can't we predict them? The rock is not changed under the waterfall in that instant, the flow is still even, the air, friction, and general motion of the earth are still constant.

What changed?

Nothing.

It is inexplicable. Unless?

Unless it is the random chaos extant within the things themselves. That which Einstein hated to consider: the fact that God *does* indeed "play dice with the cosmos." That there is an element of variety locked in all matter. The word "luck" might really be the element which renders the roulette wheel unpredictable. Even the card counters of Las Vegas subscribe to the fact that the laws of probability are very unexact at best.

This thing, this unpredictable chaotic element, is now undergoing close scrutiny by astrophysicists because one must ask how it is that anything as chaotic as the Big Bang can generate anything as orderly as the balanced progression of the creative process and universal natural systems and laws. Yet in all this order, everything alive is profoundly influenced by chance events caused by sensitive dependence on initially unrelated incidences.

At what point do all human cells separate into uniquely different beings in the chain of heredity and does all chance come into a single predictable formula? Obviously it's when everything can be known about everything else, unless there is a shortcut, a revolutionary scientific discovery which will synthesize the entire course of what is unpredictable. How will that affect you?

For one thing, it might be invoked to warn us that a perfectly good, healthy working heart is going to go on the fritz. The reason is that in Newtonian mechanics, if you know everything there is to know about a thing, all the forces acting on it, the velocity, weight and size of its parts, the ratio of its components, etc., you can describe all its future states. You can know not only beings, but all things. You can know about all things, from weather to the study of fluid turbulence.

This study is called chaotic dynamics, and for now it is worthy of deep study. As of this publication, several groups, all rather new, are devoted to the investigation of this elusive element. It is becoming an exotic subject. As such, it attracts the promising young students and teachers who want to be at the forefront, the cutting edge of what is unfolding in the study just as Galileo and Kepler did in their time. One such person is Rob Shaw, the USCS theoretical physicist, who formed a group known as the Dynamical Systems Collective, sometimes referred to as the Chaos Cabal. Their search is for the unpredictable, strange "attractor" which will explain the nature of random results.

Consider the obvious truth that random happenstance has already interfered in our development every day since time immemorial.

The farther back one goes, the more monumental is the change we live with today. A football, a change in weather, another sunny day in a drought compared to another rainy day in a flood even before our species stood up. Our form, our internal chemistry, heredity, functional orders, our stature, our systems, laws, passions, loves, hates, fears, aggressions are more than simple accidental deviations which occurred and were then sloughed off. They stayed with us. They changed what we are, what we do, where we do it, and what, where, and how we do it now and in the future.

A solar sunspot can alter weather, which alters our genes, which alters us for all time. All this in turn alters everything else ultimately in the entire universe. So that which seems to be random

Constellation: Taurus.
Object: The Hyades Cluster.
The Face of the Bull.
The entire V-pattern of the Hyades appears in this print. The line is a Satellite streaking through the constellation.

and unimportant at the moment it happens becomes more and more important as it affects anything else on the basis of a mathematical expansion chain.

That one type of fish survived with four phalanges (bones) in its fins determined how many fingers we have in our hand. And from that we developed base ten arithmetic, according to Carl Sagan. True or not, the fact that a tree shrew, to which we once were related, had an opposite thumb certainly altered the course of our evolutionary history.

In any case, dinosaurs were rumbling around all over the face of the Earth very successfully. Standing upright with little hands quite a lot like ours, and large brains probably on their way to developing into intelligent beings, feeding on smaller animals such as our ancestors. Had the dinosaurs survived instead of man, a dinosaur could be writing or reading this book today.

No one knows positively what wiped out dinosaurs, except that it was a cosmic event which in the short calendar of life seems to have been an almost overnight catastrophe. It might well have been a supernova explosion, perhaps of the Crab Nebula. If the explosion were only fifteen or twenty light years away about fifty or sixty million years ago, it could have burned the nitrogen out of the atmosphere, which in turn would have blown away the ozone from the atmosphere. That would have allowed ultraviolet radiation inside the atmospheric envelope, thereby frying and mutating all the foods upon which the dinosaurs fed. Starvation or poisoning would have followed in very short order.

It was something random in any event, and it happened fast. We know this because there have been many dinosaur footprints discovered which otherwise would have weathered over and disappeared in the ordinary, slow evolutionary process.

This random coincidental happening opened the way for smaller mammals to survive and prosper, develop and learn, recreate and invent. In short, it looks as though once we exploded, tumbling out of the cortex of the Big Bang, once we were back on course, our species was allowed to progress into the age of the wheel and beyond. Actually, we're in a very primitive place if you take into account that if we behave ourselves we have another 50 or 60 million years ahead of us. Considering we still lived in the trees only 20 million years ago and history began only five or six thousand years ago, we are excruciatingly young.

The very randomness of our development dictates mathematically that even discovering another species exactly like ourselves is almost nil. We cannot expect to discover another species which looks like us, or even thinks like we do in what we consider logical progression. *But* it is equally apparent that if we do meet an alien species we probably will find that they look nothing like us. Their experience will have taught them to want different long- and short-term goals. You may rest assured that they will be basically intelligent simply because survival requires intelligence. Their speed of movement, defense, thoughts, and general activity will most probably be far more active than ours, having developed longer.

If there are any intelligent beings in our universe, they probably will be interested in the same information we are since we share the same universe and the same laws of nature. We should have much to learn about one another.

How will we discover and meet each other? Not to meet would be to suspend the laws of probability. Almost certainly we will eventually shake hands or noses or trunks or do whatever the equivalent greeting may require. The most obvious way they will hear from us will be through some form of radio messages via their radio telescopes or equivalent receivers. Lightning, volcanic activity, and our magnetic field in general have been sending noises for billions of years. If they are on their toes and anywhere near as capable or curious (or both) as we, they already have a fix on us as a possible place to find intelligent beings.

We are sending out stronger radio and television waves than before, sounding less and less like cosmic noise and more like deliberate messages. Sooner or later "they" will start hearing about how best to clean our toilet bowls, and have their own chance to decipher the adventures of everyone from the Lone Ranger to Archie Bunker.

Perhaps someone in Andromeda Galaxy will try to understand why we seem so unable to stabilize our species. Our alien friends beyond the solar system might eventually fall in love with Miss Piggy and Kermit the Frog, as so many humans have done, irrespective of their differences in appearance. After all, one loves his dog, cat, or horse. We're even learning to love humans from another neighborhood and color. Not that the pollution of our gene pool is a desirable goal. It is rather an inevitable marriage of these traits and experiences which makes us survive longer in a healthier lifespan.

As I pointed out in the opening of this book, the Earth travels

two million miles around the sun every day. We spin 7,000,000 miles an hour around the center of our galaxy and 500,000 miles an hour on our way toward Vega. Surely there are many, many coincidences and random happenings in store for us as super space travelers within the next million years or more. We have great excitement in our future indeed.

We are the final emergent species, able to use our opposable thumb to create machines with thousands of hands and hundreds of opposable thumbs in the trek toward our destiny. We do more than gaze at the stars like our ancestors, naming constellations and making up mythological characters to whom it is apt to pray to alter our present and future. Before us, in our path, are tens of billions of billions and billions of planets of every conceivable description to steer us into our tomorrows.

It is not an intellectual decision to pursue this course. It is an imperative, as strong in our makeup as the sex drive and the drive for knowledge. It is more than our heritage. It is our destiny to go into that good night beyond which even now we can only dream.

It is not expensive to go there. It is expensive only to hold back. Time, though only an inexact means of measurement, beckons us to experience everything we survey. After all, we are the sum of our experience. Therefore, the more we experience the sooner we will become wise and strong enough to realize our survival cannot be sacrificed on the altar of knowledge; it can only flourish there.

To Gaze at the Stars

One scans the ocean,
One surveys the land,
One peers into a microscope,
One looks at the sky.
But oh, my friend,
One "gazes" at the stars.
It is wonderment!

–James Fox

First, let us find Polaris, the North Star. If you are ever lost and can see the stars at night, you know which direction is true north merely by looking three quarters above the horizon at the brightest star.

If you can't find it that way, find the Big Dipper (Ursa Major). In England, it is called The Plough. Draw an imaginary line from the dipper's farthest lip. Extend that line directly outward from the base of the bowl straight to the brightest star. There you have the North Star: Polaris. This is the star upon which all new navigators steer their boats, because it appears to be fixed in the heavens. All the rest of the firmament moves around it. It is five times the distance from the bowl of the Big Dipper as the two pointer stars are from each other (5.5 degrees).

Polaris is not a particularly bright star, but it is the brightest in that part of the sky. It is within one degree of true celestial North Pole, utterly dependable for navigators using celestial reckoning.

Continue the line straight through Polaris and you will run into Pegasus, which looks like a great square.

If you were to draw a straight line betweeen the top star on the left side of the square of Pegasus down through the bottom star on the lower right side of the square, you will find yourself in Andromeda. Andromeda may look like a star to the naked eye, but in reality it is a whole galaxy.

Having arrived at Andromeda, draw a slight arc toward the right and you will have drawn your attention to Perseus.

If you aren't oriented by this time, go back to Polaris and begin again. It is the key to your street map of the big city of the sky.

If, on the other hand, you have familiarized yourself with Polaris, the Ursas, Pegasus, etc., let's take a stroll around the neighborhood.

This time, rather than following the pointers of the Big Dipper, follow the stars which make up the other side of the Big Dipper's bowl in the opposite direction. Draw your imaginary straight line outward and you will arrive in the constellation Leo. It is about 35° away.

If you are wondering how to compute degrees by eyesight, hold up one fist at arm's length, including the thumb. This will cover about 10° of the sky. Use it as a guide.

With a sky chart for the time of year, winter or summer, you can now plot out all the stars since you have the basics of celestial orientation.

Shortly before sunset, or just after sunrise, when the sky is barely dark, these are the best times to see the planets. Venus, the brightest, is our early evening star in the west. It is about Venus we have so often said, "Star light, star bright, first star I see tonight, I wish I may, I wish I might, have the wish I wish tonight."

As you can imagine, Venus is the wishingest star in the sky. Curiously, because it is so bright, it is also our morning star in the east.

Now then, a little recapitulation. On the front side of the Big Dipper's bowl, the side farthest from the handle, the pointers point to Polaris. By now you've discovered that Polaris is really the star which lies at the end of the handle of the Little Dipper (Ursa Minor).

Many groups of stars are visible only during certain seasons. Every night these stars rise above the eastern horizon, run above and around Polaris, and then set below the western horizon. More are visible in winter than in summer, more in fall than in springtime.

High in the sky toward Andromeda, four stars form a square. As pointed out, this is Pegasus, the Flying Horse of mythology. Just above that, if it is dark enough and the sun has been down a long time, you can probably make out the Mikly Way galaxy. It appears as a rough-edged ribbon of stars, gas and dust making ridges and furrows in black and gray bands. To look at the Milky Way galaxy is like looking at the boat on which we sail across the celestial oceans. It is actually the rest of our own Galaxy. It is our Vehicle.

We are far out on the edge of this great disk, so huge that it has only turned completely around about twenty times since the day of creation, even though it travels at millions of miles an hour. If Earth is a spaceship, the Milky Way is our launching pad, so huge we may never travel far enough to reach outside it.

Turn now to Cassiopeia, which was the brightest object in the sky in 1603, when Bayer was writing and illustrating his famous star atlas. It was exploding at the time; it was a supernova.

Perseus is the next constellation east of Cassiopeia. Opposite that is a cross located directly overhead, known as the Northern Cross. It is in the constellation Cygnus, the Swan. The star at the Swan's tail is named Deneb. To the west of the Swan is Vega, in the constellation Lyra, the Lyre.

Remember as you look at Vega that it is our target star. Vega, the third brightest star in these latitudes, is where our whole system is traveling. Learn how to find it. It is nice to know where we are going for the next few millions of years.

Continuing west, beyond Vega, you next find the constellation Hercules, named for the hero who performed the twelve labors. Next to it is a small object of light called a globular cluster. Actually, this cluster is a group of thousands of stars named M-13 by a celestial cartographer, Charles Messier, in the 18th century.

In winter, to the southeast of the Milky Way is Auriga, the Charioteer. To the south of that is the small cluster of stars which once was the constellation Pleiades, the Seven Sisters, daughters of Atlas. There are more than 100 stars in that configuration. Directly south of that is another favorite constellation, Orion, the Hunter. Actually, Orion comprises the three stars which form the belt of the Hunter. Below the belt is the sword, the Orion Nebula, where there is a hazy, gaseous area. This area is where new stars are forming.

In summer you will see Arcturus, a reddish star which is almost as bright as Vega. East of Arcturus and Vega is Cygnus, the highest point of the Milky Way. In summer you will still see the "W" which forms Cassiopeia, near the horizon.

At this point, having made the familiarization tour, it is time to avail yourself of sky and star maps and charts with which to compare your sightings before you become too firm in any error you may have made so far. For that I recommend the "Peterson Field Guide to the Stars and Planets" by Donald H. Menzel and Jay Pasachoff, two truly great authors and astronomers.

In concluding this chapter, let me point out that if you were to wake up in a strange city alone, the first thing you would do is look around the room in which you had slept. Then the house, the neighborhood would occupy your curiosity. Yet we seem to stop at the study of inner space, the oceans and the subsurface of the land. For some reason, only geologists and marine biologists take up those studies.

But the stars, the stars are there for all of us. We needn't dive nor dig. We merely need look up. Gaze. And as the poet says, "One gazes at the stars. It is wonderment."

A Summary So Far

As we move into inter-solar travel, we'll meet some other shattering revelations that rank with the final acceptance of the Copernican doctrine of planetary motion. Once we see ourselves as living on one lonely little planet swirling away totally alone in space for as far as we can see, we'll begin to realize what it means to each of us that we're all one species and therefore bound to live and let live. Something, in other words, approaching what we refer to as civilization.

From our experience, the factor which most likely will bind us as a family, separate from animals which must fight for the territorial imperative, will cease to be religion. There's no evidence that our belief in God has succeeded in assuring us a mutual appreciation of each other's privileges. Perhaps a final realization of our mutual isolation could be to unleash a new force in the thought and actions of mankind. Isn't it possible that a new cosmology might emerge to affect all of social behavior? When we *know* where we are, really get a good look and let the picture of it sink through our collective heads, chances of cooperation and understanding should be increased if for no other reason than the promulgation of humanity.

In the interest of becoming secure in whoever and wherever we are, let us pause here and draw a line from the breakthrough of human knowledge to now. This should give us a good over-all perspective, a look from the standpoint of the truly objective observer. Keep in mind that hundreds of thousands of theories of astrological as well as astronomical ideas, propounded by people of good repute in their own times, have since fallen by the wayside. Here is a chronology of ideas that brings us to our present state of knowledge.

1. Democritus in the 5th century B.C. propounded the idea that there are only four basic elements in the world: stone, water, air and fire.
2. Eratosthenes measured the diameter of the earth.
3. Archimedes established the basic laws of hydrostatics in the 3rd century B.C.
4. Ptolemy taught in the 2nd century B.C. that the sun and planets revolve (albeit around the Earth).
5. Copernicus showed that the Earth and planets revolve all right, but around the sun.
6. Galileo refuted everyone except Copernicus, and replaced the old theories with telescopic evidence as proof.
7. Johannes Kepler discovered in the 17th century that the planets need not necessarily be in perfect orbits, but that they revolve around the sun in conical sections.
8. In the 17th century, Sir Isaac Newton formulated the law of universal gravity and developed the use of the prism as an astronomical tool.
9. Dalton reinstated the ancient Greek hypotheses of atomic matter.
10. Faraday presented the basic laws of electrochemical science.
11. Dmitrii Mendeleev constructed the periodic system of chemical elements.
12. Hertz discovered the waves of electromagnetism.
13. Becquerel discovered radioactivity in the late 19th century.
14. Thompson (J. J.) discovered electrons in an electric discharge through gases in 1897.
15. Planck introduced the theory of a minimum portion of energy (a quantum), which gave Einstein a tool he needed: quantum mechanics.
16. Einstein refuted the whole idea of "world ether" in 1904.
17. In 1913 Bohr made electrons leap from orbit to orbit in the atom.
18. Rutherford discovered atomic nucleus by firing alpha particles at the atom. He was the first to transform one chemical element into another in 1919.
19. Einstein re-explained gravity through the bending of space, and relativity changed all of science forever.
20. Shipley managed to measure the galaxy in 1924.
21. Hubble proved in 1924 that the universe is expanding.
22. Hahn cracked the atom in two.
23. In 1942, Fermi produced a nuclear reaction, opening the atomic age.

For good or bad, we're on the threshold of a rather exciting few centuries because, as we said earlier, progress is its own adrenalin and it progresses geometrically.

Fortunately, we now have computers to help manage things. And there will be more minds capable of carrying us into the future than ever before in history. We also have new forms of expressing all these new ideas. As an example, just as we use the light year to express vast distances, we now have ergs, bytes and bits. Perhaps

you don't know much about bytes and bits, but you will as computers become more and more a part of your everyday experience.

And ergs? Well, an erg represents a tiny amount of energy, or the energy of motion of a mass of 2 grams (about 1/14 of an ounce) moving at 1 centimeter per second. According to the late George Gamow that's about as much energy as a mosquito exerts while flying across a room. So our energy measurements fall somewhere between one erg (which is very small) and an Egan (the measurement by which the weight of one earth is pulled by the sun to keep it in orbit). What fun!

Good Old Sol

Compared to the billions of stars in our own galaxy, the sun is just another average star. Of course, it's closer and therefore brighter and without it, we don't make it. But still, the sun is just another star in the Milky Way. Our nearest star, Alpha Centauri, is 270,000 times farther away from us than our own sun, but, coincidentally, the sun and Alpha Centauri are almost identical.

The sun is 93 million miles from Earth. It is closer to us in July than in January by about 3 million miles.

Here are some of its vital statistics:

Diameter: 864,000 miles; 109 times that of Earth
Distance from Earth:
Min: 91,500,000
Mean: 93,000,000
Max: 94,500,000
Mass: 333,400 times that of Earth
Surface gravity: 900 ft. per sec; 27.9 times Earth's
Temperature Photosphere: 5750° A
Spinning period: 24 days, 16 hours
Weight of cubic foot of sun matter: 87.4 pounds

Even though the sun is more than a hundred times larger than Earth, which means the orbit of the moon around the Earth could easily fit inside the diameter of the sun, the sun is still only an average size star. Stars run from about 4,000 miles in diameter to as much as three or four thousand times greater than our sun.

As to the volume of the sun, it is 1,250,000 times greater than that of Earth. If we were to place the sun on one side of a balance, it would equal the weight of 333,400 Earths. The sun accounts for 99.9 percent of all the matter in our solar system. Everything else comprises the other one tenth of one percent. Because it is so big, everything else revolves around it. The laws governing this solar system's motion were described by their discoverer, Johannes Kepler. They were used by Isaac Newton in describing the Universal Law of Gravity.

The sun was both the life giver in its presence and the destroyer in its absence. Primitive man, or Homo-Religious, existed even before the more mentioned Homo-Erectus. Far back in the Primordial Period, man knew that all the good and the bad that happened to him emanated from the sky. If lightning struck a tree next to him, he probably prostrated or cowered himself between some rocks and uttered the equivalent of, "Oh, my God!" If the sun was warm, all the foliage grew and provided both food and protection, protection by helping conceal him from the wild beast. If the sun was cold, food was scarce for both man and beast. The beast would forage most easily for food when the forest was bare of foliage. Man soon learned to petition the bright thing in the sky as a deity of sorts.

The first Greek god of the sun was Helios, the offspring of Hperion, the Titanic deity of light. Helios, meaning Sun in Greek, represented the sun in its daily and seasonal trek across the sky.

The history or fable of all of the Titans and their offspring was passed on to the more classic or familiar gods. Apollo, the new sun god, is depicted as driving a sun chariot across the sky. Sometimes he is called Phoebus Apollo, Phoebus meaning Bright. Apollo is always pictured as beautiful, handsome, and having a magnificent physical structure. We use this word today to describe handsomeness. As told in another story in this book, Phaeton Phoebus, Apollo's son by the nymph, Clymene, borrowed the family sun chariot for a joy ride. Fire and sparks flew all over from the wildly driven chariot, burning up most of the Earth. Zeus demolished Phaeton with a thunderbolt. That's how they dealt with mixed-up kids in the old country.

The Moon

Before we move on to the planets, let's get a look at the moon.

Our moon is not the largest in our solar system. Three moons of Jupiter, one of Neptune, and one of Saturn are physically larger than ours. Our moon is a little more than 1/4 the diameter of the Earth.

The moon orbits the Earth every 27-1/3 days and is about 1/12 of the way in our mutual orbit around the sun. The moon, of course, must orbit the Earth further to get back to the same phase, taking about 29-1/2 days. This is known as the synodic period, the interval between two successive conjunctions as seen from the Earth. These synodic months are the months referred to in lunar calendars.

If you have wondered who sees the same phase of the moon as you, the answer is that everyone on Earth who can see the moon above the horizon sees the moon in the same phase everyone else does. To us, the moon appears in different phases because we see different fractions of the lighted side. When the sun and moon are on opposite sides of the Earth, the moon is full. When the sun and moon are on the same side of the Earth we are looking past the moon at the sun.

The far side of the moon is lighted by the sun, but the side that faces us receives no light from the sun. That is when we say the moon is new. There are times, however, when we can see the moon by the sunlight that's reflected or bounced off the Earth onto the moon. This is called *Earthshine*.

There are terms describing the general phases of the moon between new and full: new moon, crescent moon, half moon, and gibbous moon (bulging just past half moon). While the moonlight's changing from new moon to full moon, the moon is *waxing*. When it's changing from full moon to new phases. the moon is *waning*.

Another interesting point is that since the moon travels in an elliptical orbit, it travels at different speeds during its passages at different locations in the orbit. That means the phases vary slightly. A new moon must rise when the sun rises and set when the sun sets. It rises an average of about 50 minutes later each night through its cycle.

Since the time the first man looked up at the moon, hundreds of thousands, perhaps millions, of years ago, we have speculated about the moon's surface. Is it water? Is it hard? Soft? Full of dust? If it *is* full of dust, would man sink into it?

Beginning in 1969 with the first Apollo landings, we finally answered the questions with 12 astronauts on six missions. We finally had first-hand knowledge.

From what we learned from 843 pounds of moon rocks, we discovered that the moon was composed primarily of basalt, common to what we have on Earth.

Analysis indicated the moon, formed by vulcanism, began to solidify about 4.6 billion years ago after being bombarded by meteorites. Radioactive elements generated such heat that lava poured out of the center, filling the larger basins. Nothing much happened to change the surface after that. The surface we see may be exactly as it has been for at least four billion years.

The Planets

The nine planets are divided into two distinct groups: Terrestrial and Jovian. Mercury, Venus, Earth, Mars and Pluto are the terrestrial planets. They are all similar in size. Jupiter, Saturn, Uranus and Neptune are Jovian. As implied by their group name, derived from Jupiter, they are much more massive than the terrestrial planets.

There is another way in which the planets can be grouped: according to distance from the sun. Mercury and Venus are known as the *inferior planets* since they're closer to the sun than Earth. The planets from Mars through Pluto are known as *superior planets* because they're farther from the sun than Earth.

Some information about the size, distances, orbits of the Planets of the Sun:

Planet	Diameter Miles	Mean Distance From the Sun (Millions of Miles)	Period of Orbit Sidereal*	Synodical Orbit Days	Rotation About Axis
Mercury	3100	36	88 Days	116	59 Days
Venus	7700	67.2	224.7 Days	584	243 Days
Earth	7927	93.0	365.25 Days		24.93 Hrs.
Mars	4200	141.6	687 Days	780	24.56 Hrs.
Jupiter	88,700	484	11.86 Yrs.	398.9	9.83 Hrs.
Saturn	74,000	887	29.46 Yrs.	378	10.24 Hrs.
Uranus	32,000	1784	84.01 Yrs.	369.66	10.82 Hrs.
Neptune	27,700	2595	164.8 Yrs.	367.49	15.8 Hrs.
Pluto	1500	3675	248.4 Yrs.	367 Days	639 Days

*A sidereal period is the time it takes the planet to complete one revolution in its orbit.

**The synodic period, which involves the motion of the Earth, is the interval between the times the sun, Earth and planet are aligned.

***Reverse to Earth's orbit (retrograde).

Mercury

Mercury is the smallest planet and is the closest to the Sun. Its surface temperature is estimated at 700 degrees and has no atmosphere.

Mercury can sometimes be seen in the morning and early evenings. Ancient Greeks thought it was 2 separate stars; they called it Apollo in the morning and Mercury or Hermes in the evening.

In Mythology, Mercury was the son of Jupiter and the father of Pan. His mother was Maia, one of Atlas' seven daughters. Maia was the eldest and most beautiful of the girls, and she was immortalized by the gods as the brightest of the stars in the Pleiades (seven sisters) of the constellation Taurus.

Mercury is pictured as a winged footed warrior with a winged hat. He has a short sword in one hand and the caduceus in the other. He was the fleet messenger of Jupiter and had many varied duties, more than any other god. Mercury was known for his good natured pranks and a buddy of Bacchus, god of the grape and the party.

Venus

Venus is the brightest of the planets and most celestial objects. It is so bright that it is visible during the day. It's bright enough at night to cast shadows after dark. It also is closest to Earth, approaching a distance of 26,000,000 miles.

The daytime brilliance of Venus is useful to sailors using celestial navigation. Venus has phases like our moon and varies in shape from crescent to full circle.

In mythology, Venus was the goddess of love and beauty. She was the daughter of Jupiter and Dione, the wife of Vulcan and appropriately the mother of Cupid (Eros). Venus wore a magic girdle which aroused passion in the gods who gazed upon her. She had many lovers including Mars and Anchises of Troy.

The story of Pygmalion is this author's favorite tale of Venus. Pygmalion, a sculptor, created a statue so beautiful that no living female could compare. He fell in love with his stone masterpiece which, of course, could not reciprocate. Venus saw how miserable he was and breathed life into the marble statue, giving her the name Galatea. Venus then performed the marriage ceremony joining Pygmalion and Galatea.

They named their first son Paphos. Paphos was the seaport on Crete most favored by Venus, and to this day has ruins of a large temple of Venus.

The musical "My Fair Lady" is based on George Bernard Shaw's play "Pygmalion". Here Galatea is replaced by a cockney girl who is trained in culture and grace by Professor Higgins (Pygmalion), who falls in love with her. It so happens that the rain in Spain falls mainly on the plain.

Earth

Of the nine planets which revolve around the sun, the earth ranks fifth in diameter and is third in distance from the sun. It is the only planet on which life exists. It is difficult to study because it has seven elementary motions: 1. Around its axis, 2. Around the sun, 3. The axis processes*, 4. The axis mutates**, 5. Through other stars, 6. The galaxy rotates and it can be pointed out that 7. The galaxy is moving as it rotates.

A common misconception is that the earth is pear-shaped. That

*Axis precess: motion of the axis

**Axis mutation: A nodding of earth due to gravitational pull of the moon

is not true. If the earth were scaled to the size of a billiard ball, the earth would be a more perfect sphere than the billiard ball. It *is* very slightly flattened at the poles, however, which means that if the earth isn't a perfect ball it would be called oblate. Thus, the true shape of the earth is one "oblate spheroid."

The most recent account we have about the why and the where of Earth's happening is front page news in the Bible. There are many similarities between mythology and the Bible occurrence.

In mythology Gaea was the personification of Earth which burst from Chaos. Chaos was the confusion of Earth, sea and air combined. Nyx, Chao's daughter, laid an egg from which hatched Earth as a singular part of Chaos. From Mother Earth (Gaea) and Father Heaven (Uranus) came the Titans.

There were six male and six female Titans. All were gigantic with tremendous strength and strange physical features and appendages. All of these nasty, erratic creatures were thunderbolted by Jupiter in good order, starting with the strongest and meanest. He then put all twelve in Haydes to stay forever.

As confusing and incoherent as these stories are, the more we learn from science, the more these preposterous tales become plausable. After Chaos (Big Bang), came gigantic monsters (dinosaurs), and then came man.

Mars

Often referred to as the "red" planet, Mars is actually a patchwork of red, gray and white. Throughout the early days of telescopic observance, astronomers often reported that they had detected canals or irrigation ditches which could have been made only by intelligent beings. For that reason, science fiction writers have written about Martians more than about beings from any other planet.

Before the Mariner 4 space probe, there was speculation that lower life forms existed on Mars. This was considered because of the visible polar ice caps which seasonally changed in size, and the appearance of green patches which varied in size as if to signify growth and dormant cycles. Mariner discovered a complete lack of water and oxygen, making growth seem impossible. The color variations, therefore, are guessed to be temperature and mineral changes.

Mars' two moons are his sons and attendants. Diemos, meaning dread, is the small inner moon. Phobos, meaning fear, is the larger outer moon.

The mythological god of war, Mars, was the son of Jupiter and Juno, and second only to his father in power and might. He was also the father of Romulus and Remus, the founders of Rome, thus making him very special to the Romans. Romans worshiped Mars as invincible, wearing a shining breast plate and helmet. He was their conquering god, one to rally the troops to outstanding bravery and even to glorious death.

Antares, the main star in Scorpio, is so named because of its red color. The Greek translation meaning "against" or "rival" to Mars.

Jupiter

Along with Saturn, Uranus and Neptune, Jupiter forms the group of major planets with large volume, large mass and low density. Of them all, Jupiter is the most voluminous. It could be considered large enough to accommodate all the others in the space it occupies. As for its mass, it would require more than 300 spheres as heavy as Earth to equal Jupiter. It is bright yellow and moves through all the constellations of the Zodiac in slightly less than 12 years.

Jupiter has 14 satellites or moons. The four largest, called Galilean moons, are Europa, Ganymede, Io and Callisto; all of whom are mentioned earlier in this text as lovers of or attendants to Jupiter,

Jupiter is also known as Jove and Zeus. He is the boss, the supreme deity. He had many marriages and many lovers. He seems to have been the father of almost all of the gods except his brothers Pluto and Neptune. Often pictured with a thunderbolt, his favorite weapon, he was the god of thunder, rain, destiny, virtue, law, and many other things—both good and bad. Most of the time it was best not to deal with Jupiter; he was quick to dispense justice at his own caprice.

Saturn

Saturn is the planet with all the rings around it, second in size only to Jupiter. It is generally a dull yellow color, very steady in light and always among the brightest in the sky.

Saturn was first thought to have ten satellites. Pioneer II, the space probe, discovered six more which have not yet been named. The first ten are named after the legendary quarrelsome Titans whom Jupiter overcame with thunderbolts and banished to the infernal regions of the underworld.

Saturn and Rhea were the parents of six gods. An oracle warned him that one of his sons would overthrow him as the main god, so Saturn ate them all, except the newborn Jupiter. Rhea hid the baby, Jupiter, and wrapped a stone in the baby's bunting. Saturn ate the stone thinking it was Jupiter. The ruse was successful and Jupiter grew up and forced Saturn to give him his throne. Saturn then vomited up Jupiter's five siblings.

Jupiter relegated Saturn to become god of the harvest. Saturn taught the Romans the art of farming and they became prosperous. This was called the return of the Golden Age of Italy. A feast honoring the harvest "Saturnalia" is still celebrated in some parts of Italy in December.

Uranus

A very slow mover, Uranus is also very hard to see because it is so far from its source of light, the sun, and so far from our observation point on Earth.

Uranus appears as a greenish disk because of the great amount of methane in its atmosphere. Other distinguishing characteristics include the fact that it rotates backward on its own axis. It was the first planet to be discovered by a telescope, quite accidentally, by William Herschel in 1781. Its equatorial plane is almost at a right angle to its orbital plane. Uranus has five satellites which revolve in a plane of the planet's equator at right angles with the orbit in which the planet moves. The satellites are Miranda, Ariel, Umbriel, Titania, and Oberon. In 1978 observers discovered nine rings of Uranus, similar to those of Saturn.

Mythology has Uranus, the personification of heaven, married to Gaea, the personification of Earth. This union produced the Titans, a race of primordial deities. This story is sometimes taken to be that of Adam and Eve and the deities to be Cain, Abel et al.

This mythology is more confusing than most because it lacks a history.

Neptune

Neptune, being so far away, is invisible to the naked eye. It has almost no variety in color due to seasonal changes. In fact, Neptune is about the dullest planet of all. The most interesting thing about Neptune is that is was discovered first by mathematical computation in 1846. Only later was it discovered by actual visual observation. It is so like Uranus that they're often referred to as twins. Neptune has two satellites; one travels east to west, the other west to east. We will be getting more information about Neptune from Voyager II in 1989.

Neptune was the god of the sea. The heavens were assigned to Jupiter and the nether world to Pluto. Neptune, as was the custom, seemed to be the father of almost half the characters in mythology. Orion was his son along with many of the lesser gods and giants.

The sign of the planet Neptune is the trident. Neptune always carried a trident, a three pronged spear, similar to a fish fork. He could perform all sorts of tricks with this trident—break rocks, call up storms or subdue them, batter coast lines and cause saints and sailors to perish.

Neptune was especially fond of animals. The first and fastest horse ever was Arion. This was the offspring of either Neptune and Ceres, or of Neptune and one of the three furies. To this day a popular name for a horse is Fury.

Pluto

The most distant planet in our solar system leaves much to be learned about. Its distance from the sun is nearly four thousand million miles. It receives only 1/1000 of the amount of heat and light from the sun that we do.

The credit for discovery of Pluto goes to Percival Lowell of Flagstaff, Arizona, in 1915 and his colleagues Pickering and Tombaugh. The discovery announcement took place on Lowell's birth-

day, March 13, 1930.

Pluto in mythology ruled the nether world or land of the dead. Pluto was considered the god of riches or wealth since gold and silver come from the earth. Our words plutocrat and plutonium reflect this belief.

Pluto was considered cold and indifferent. Men were afraid of his imperial impartiality as harvester of the dead, but as a brother of Jupiter he was not without emotion. He carried Prosperpine off to wed against the will of her mother Ceres. Jupiter intervened and to placate the mother, Pluto compromised. It was agreed that Properpine would spend the winter in Hades with him, and the summer with her mother—a little like Miami Beach and New York City.

Pluto is further immortalized by Walt Disney as Mickey Mouse's floppy-eared dog. I wonder how future historians are going to treat the mythology of the two Plutos.

Asteroids

Asteroids or minor planets which range in size from one mile or less to as much as hundreds of miles in diameter are found between the major planets. When the first asteroid was discovered, in 1801, it was thought to be a new planet and was named Ceres.

As more were discovered, they were given different names. The numbers assigned to them indicate the order in which they were discovered. Since Ceres was first, followed by Pellas, Juno, and Vesta, they were numbered in that order. Thus, 1 Ceres, 2 Pellas, 3 Juno, and 4 Vesta tells us something about them in addition to the Greek or Roman god being honored by its name.

Then it was found that there is a belt between Mars and Jupiter which came to be called the asteroid belt. We know of about 36-40 asteroids which cross the Earth's orbit. The number yet to be discovered seems limitless, at least for now. Groups such as the Apollo Group, the Aten, and the Icarus have been identified and numbered. Astronomers have catalogued and named over 1,600 asteroids so far. There is even a fat one named Albert. Some astronomers have suggested that the asteroid belt was once a planet that collided with another or its own moon. Another theory is that it is an unformed planet that never solidified into one planet.

Comets

There is a huge cloud surrounding the sun which is made up of billions and billions of small bodies far beyond all the planets in the cold of outer space. These bodies can best be compared to large, icy snowballs.

From time to time one of them is gravitationally bumped or nudged by a passing star or another object perhaps like itself, only larger. When that happens, it can be moved gravitationally toward the sun, which heats it up as it grows closer. When this happens, it gives off gas and dust to form a tail. This tail is pushed away by the outward gasses of the sun which we call the *solar wind*. The dust left behind is the gaseous tail.

The appearance of such a comet is usually unpredictable. There is one, however, which is very predictable. It is known as Halley's Comet and comes our way only once in a lifetime. That is, we can expect to see it with the naked eye once every 76 years. From historical records we know it has been seen at least 27 times since 87 B.C. Sir Edmond Halley first realized that this comet was not a series of different, infrequent visitors, but the same comet returning again and again with predictable regularity.

The closest point of Halley's Comet, its perihelion, will take place on February 8, 1986. Unfortunately, it will not be as brightly spectacular as it was in its last passage near the Earth in 1910, but it will be visible during the month before its perihelion. It will be brighter according to the latitude from which it is viewed. It will also be higher the farther south you go. For each 10° farther south you go, the comet will appear 10° higher in the sky.

Meteors

Every August you can probably see a meteor shooting across the sky. These *shooting stars,* lasting about one second or less, are considered lucky. Actually, they are solid bits, much smaller than asteroids, being burned up in the Earth's atmosphere. Some are not much larger than an atom. Others have as much as a kilometer diameter of matter.

On August 12, Earth's orbit intersects the center of a cloud of these particles, at which time we experience a phenomenon we call

the Perseid Shower. Some years these burning flashes can be seen at the rate of one hundred per minute. Known as *Leonid Shower,* this happens every 33 years. The last one occurred in 1966. When a meteor burns its way through the atmosphere and reaches Earth it is called a meteorite.

Strict meteorologists (not the weatherman, but the scientists who study meteors) categorize meteorites into many different distinctions. However, there are two basic types: nickel-iron and stone.

If you see shooting stars, look back toward the direction from which it seemed to fall. This is how you detect a known radiant from which all paths seem to converge. All meteors actually are hitting the earth in parallel lines, and so, like railroad tracks, they seem to be coming at us from the same point in the distance. When you see an unusually bright one shoot across the sky, you probably are observing a *fireball* or chip from a broken asteroid which leaves a train that hangs in the sky for a second or two after it has entered the Earth's atmosphere and burned up.

The Future of the Solar System

It is difficult to predict anything about a universe when we know so little about it. As the years, decades and centuries go by, we will learn more and more because, for some inexplicable reason, astronomy is beginning to catch the interest of more and more people.

However, from what we know right now we can hazard a fair conjecture about the natural unfolding of the universe and make a few solid predictions about our solar system. To the best of our knowledge (that is, if we let the Earth survive at the hands of man) nature itself will make some changes, the most notable being that the sun will grow old.

Right now and for the next five to six billion years, the sun gets its enormous energy from the continuing processes of thermonuclear activity. That is, we derive heat and light from the reaction of the changes which occur in alteration of hydrogen and helium.

After five billion years or more of consuming its matter and converting it to energy, the sun will begin to cool and start to become a *red giant*. It will grow bigger and redder. It might become so big as to engulf the orbiting Mercury or even Venus. It will begin to emit as much as 100 times the amount of radiation toward Earth as it does now. This will cause the oceans to heat up and boil, until all the surface water evaporates. Finally, the molecules which comprise the atmosphere which made life possible will speed up until they reach a velocity allowing them to escape our gravity. Eventually Earth will become a huge cinder with a surface heat so high life cannot be sustained.

During this period, the sun will be what we now call a red giant, remaining that way for several hundred million years. It will then become a white dwarf. This is happening to other stars all over the universe right now. Observing them, we can predict that our sun will then grow smaller until it actually becomes the size of Earth, or even smaller. It will change color from bright white-yellow to blue, then to white. It will then fade to only one-ten-thousandth of its present brightness.

Meanwhile, Earth's temperature will fall so drastically it will approach absolute zero. Nighttime will reign 24 hours a day. If we were present, our sun would appear to us as only a rather bright star. The moon would still exist, but would be invisible or very pale. From time to time a comet will pass, as happens now.

As the Earth is changing, the neighboring planets will undergo a change, too. For instance, Saturn will be sufficiently changed to allow human habitation.

If each of us could step onto that new planet-home as Neil Armstrong did onto the moon, we would have a little more to say than something about the size of our step. We could look back at Earth and realize that this Earth was the birthplace of humanity, the cradle of human knowledge, the ancestral home of us all.

As we have grown older and become wiser, we find that love is more enriching than hate, wonderment more fulfilling than greed, and concern more rewarding than fear.

As time passes, we will look not to those things about which we differ nor nourish those things which make us different. Rather, we will embrace all those things which we have in common such as each other. We will reach for and attain a new level of mobility. We will enter a richer world where we will learn to learn for knowledge's sake and use our new widsom to understand what matter, energy, light, and truth are really all about.

And then, finally, we will know why we have been invited to the banquet of life. And we will turn to our host and say, "Thank you for letting us live in your beautiful home, this magnificent, unfolding universe."

Glossary

Abberation of starlight: The tiny apparent displacement of stars resulting from the motion of the Earth through space.

Absolute magnitude: The magnitude that would be assigned to a star if it were placed at a distance of 10 parsecs (or 32.6 light years) from the observer. The sun, for example, has an absolute magnitude of 4.7.

Absolute visual magnitude: The absolute magnitude of an object measured through a special yellowish filter that approximates the visual range of the human eye.

Absorption nebula: A nebula seen in silhouette as it absorbs light from behind; also called a dark nebula.

Achromatic lens: A lens that transmits white light without dispersing it into a color spectrum. It usually consists of two component parts, cemented together to form one unit.

Albedo: Percentage of light reflected by a body, such as a planet, from the total amount of light falling on it.

Altitude: Angular distance between the horizon and a given object, measured along a vertical circle.

Analemma: The figure 8 representing the equation of time and the variation of the sun's altitude in the sky during the course of a year.

Angstrom: A unit of wavelength or distance, equivalent to 1/10,000 micrometer or 1/10,000,000,000 meter.

Annular eclipse: An eclipse of the central portion of the solar disk; an outer ring shows.

Aphelion: The point on a planet's orbit farthest from the Sun.

Apogee: Point on the moon's orbit farthest from the Earth.

Apollo: The name assigned to the U.S. project whose mission was to land men on the moon. Also the name of the vehicles used. Apollo II landed Neil Armstrong and Edwin Aldrin on the moon on July 20, 1969.

Apparent magnitude: Magnitude as seen by an observer.

Apparent solar time: Time determined by the actual position of the sun in the sky; corresponds to time on most sundials.

Asterism: A noticeable pattern of stars that makes up part of one or more constellations; not a constellation itself.

Asteroid: A minor planet, smaller than any major planet in our solar system; not one of the satellites (moons) of a major planet such as the Earth or Jupiter.

Astronomical unit: The average distance between the Earth and the sun 93 million miles or, more exactly, 92,955,700

Aurora: A diffused glow of light in the form of curtains, or bands, seen at high latitudes (70°N or 7°S). The glow is due to the interaction between the solar wind and particles in the Earth's atmosphere. The aurora in the northern hemisphere is known as the Aurora Borealis, or northern lights; in the southern hemisphere it is known as the Aurora Australis, or southern lights.

Autumnal equinox: The intersection of the ecliptic and the celestial equator that the sun passes each year on its way to southern (negative) declinations.

Baily's beads: A chain of several bright "beads" of white light, visible just before or after totality at a solar eclipse. The effect occurs when bits of photosphere shine through valleys at the moon's edge. See also diamond-ring effect.

Bayer designations: The Greek letters assigned to the stars in a constellation usually in order of brightness, by Johann Bayer in his sky atlas (1603).

Belts: Dark bands in the clouds on giant planets such as Jupiter: compare with zones.

Binary star: Two close stars held together by a gravitational force and revolving like a dumbell about a common center of gravity. The center is closer to the more massive star.

Black hole: A region of space in which mass is packed so densely that (according to Einstein's general theory of relativity) nothing, not even light, can escape.

Cassini's division: The empty space that separates the outer rings of Saturn from the bright inner rings.

Celestial equator: The imaginary great circle that lies above the Earth's equator on the celestial sphere.

Celestial longitude: Longitude measured (in degrees) along the ecliptic to the east from the vernal equinox.

Celestial poles: The points in the sky where the Earth's axis, extended into space, intersects with the celestial sphere.

Celestial sphere: An imaginary sphere of infinite radius surrounding the Earth and serving a screen against which all celestial objects are seen.

Cepheid: A star whose brightness varies periodically because of pulsations.

Chromatic aberration: (Also called "color defect.") Blurring of image due to the separation of colors by a lens. A point of white light in the object appears as a complete spectrum of colored points in the image.

Chromosphere: A layer in the sun and many other stars just above the photosphere. During eclipses, the solar chromosphere glows reddish from hydrogen emission.

Circumpolar: Refers to a star, asterism, or constellation that is close enough to the celestial pole that, from the latitude at which you are observing, never appears to set.

Collimator: (Also called "collimating lens.") A lens whose function it is to make rays of light parallel.

Colure equinoctial: (Also called "prime hour circle.") The hour circle that goes through the first point of Aris. The hour angles (same as longitude on Earth) are measured from the colure equinoctial.

Comet: A body—probably resembling a dirty snowball between 0.1 and 100 km across—which travels through the solar system in an elliptical orbit of random inclination to the ecliptic. A comet grows a tail if it comes close enough to the sun.

Conjunction: Apparent line-up of sun, Earth, and a planet. *Inferior conjuction* is when the planet is between the Earth and the sun. *Superior conjunction* is when the planet is on the opposite side of the sun.

Constellation: A group of stars apparently close together in the sky. Modern astronomy recognizes 88 such groups (e.g., Andromeda, Leo, etc.). Actually, the individual stars of a constellation may be great distances apart and moving in different directions from one another.

Contact(s): The stage(s) of an eclipse, occultation, or transit when the edges of the apparent disks of astronomical bodies seem to touch. At a solar eclipse, first contact is when the advancing edge of the sun first touches the moon; second contact is when the advancing edge of the sun touches the other side of the moon, beginning totality; third contact is when the trailing edge of the sun touches the trailing edge of the moon, ending totality; and fourth contact marks the end of the eclipse.

Copernican System: The system that assumes that the sun is at the center, and Earth and the other planets move around it.

Corona: The outermost layer of the sun and many other stars; a faint halo of extremely hot (million degrees) gas.

Crepe ring: Saturn's inner ring, also known as the C-ring, which extends inward to the planet from the brightest ring (the B ring).

Crescent: One of the phases of the moon or the inner planets (Venus and Mercury) as seen from Earth, caused by the relative angles of sunlight and the observer's viewpoint. From spacecraft, crescent phases of the Earth, Mars, Jupiter, and Saturn have also been seen.

Culmination: The position of a celestial body when it is on the meridian. A star is said to be at its "upper culmination" when it has reached its highest point for the day.

Declination: Angular distance of an object from the celestial equator, measured in degrees, minutes, and seconds. Analagous to latitude in geography.

Diamond-ring effect: An effect created as the total phase of a solar eclipse is about to begin, when the last Baily's bead—a remaining bit of photosphere—glows so intensely by contrast with the sun's faint corona that is looks like the jewel on a ring. Also refers to the equivalent phase at the end of totality.

Diffraction of light: A phenomenon exhibited by light on passing through a narrow slit or a small aperture. The light is modified to form alternate dark and bright fringes.

Discrete source: A small area in the sky—almost a point—from which very intense electromagnetic waves of radio frequency reach the Earth. These points formerly were called radio stars.

Doppler Effect: Change in frequency of light due to relative motion between observer and source of light.

Double star: A system containing two or more stars. In a true double, the stars are physically close to each other; in an optical double, they lie in approximately the same direction from the Earth and thus appear close to each other although actually they're far apart. See also *binary star.*

Earthshine: Sunlight reflected off the Earth which lights the side of the moon that does not receive direct sunlight.

Eclipse, lunar: The passage of the moon into the Earth's shadow.

Eclipse, solar: The passage of the moon's shadow across Earth.

Ecliptic: The apparent path the sun follows across the sky during the year. The same path also is followed approximately by the moon and planets.

Ejecta blanket: Chunks of rock, usually extending from one side of a crater, which were ejected during the crater's formation.

Emission lines: Extra radiation at certain specific wavelengths, allowing it to give off radiation in emission lines such as those of hydrogen. The characteristic reddish radiation of many emission nebulae is mostly from the hydrogen-alpha line.

Ephemeris: A book of tables showing computed daily positions of heavenly objects.

Equinox: One of the points of intersection between the ecliptic and the celestial equator. When the sun is at one of these two points, the length of day and night are equal everywhere on Earth. The sun is at these points every year on or about March 21 (vernal) and September 23 (autumnal).

Evening star: This is a planet, not a star, referring especially to Mercury or Venus when seen in the western sky just after sunset.

Extragalactic: Beyond our galaxy.

Faculae: Areas on the surface of the sun which appear brighter by comparison to surrounding regions.

Flyby: A research mission in which the satellite collects data while passing close to the object of research.

Fireball: An extremely bright meteor, usually with an apparent magnitude brighter than –5. Some fireballs are as bright as magnitude –20.

Flamsteed number: The number assigned to a star in a given constellation, in order of right ascension, in the 1725 catalogue of John Flamsteed.

Galactic: Pertaining to our galaxy, the Milky Way.

Galaxy: A giant collection of stars, gas, and dust. Our galaxy, the Milky Way, contains one trillion times the mass of our sun.

Gemini: The name given to the U.S. program and vehicles designed to prepare man for landing on the moon. Gemini 3 to Gemini 12 (1956-66) carried crews of two astronauts each. The program included space walks and rendezvous with other spacecraft, as well as docking techniques.

Giant: A star brighter and larger than most stars of its color and temperature. Stars become giants (normally red giants) when they use up all the hydrogen in their cores and leave the "main sequence" part of their life cycle.

Globular cluster: A spherical grouping of stars of a common origin. Globular clusters and the stars in them are very old.

Graben: A subsided long and narrow region between two faults, found on the surface of the Earth, the moon, or other planets or moons.

Half moon: The first quarter or third quarter phase, when half the visible side of the moon is illuminated.

Heliocentric parallax: The apparent motion of nearby stars seen against the background of faraway stars. The apparent motion is actually due to the revolution of the Earth around the sun.

Helmholtz contraction: The theory that the energy of the light emitted by a star is derived from the gravitational potential energy (i.e., contraction) of the star.

Hertzsprung-Russell diagram: A diagram showing a scatter distribution of stars according to luminosity and temperature. The scatter distribution is related to the various ages of the stars.

Hour angle: The sidereal time elapsed since an object was on the meridian or, if the hour angle is negative, before the object reaches the meridian. (The hour angle equals the difference between the right ascension of an object and your meridian.)

Hour circle: A line whose right ascension is constant, lying on a great circle which passes through the celestial poles and the object.

Hubble's Constant: The ratio of the velocity of recession to the distance of a galaxy. This ratio is 100 km/sec for every one million parsecs.

Hubble's Law: The relationship between the velocity and the distance of galaxies and other distant objects. It shows that the universe is expanding.

Hydrogen-alpha line: The strongest spectral line of hydrogen in the visible part of the spectrum. It falls in the red so that an emission hydrogen-alpha line is red. An absorption hydrogen-alpha line is the absence of that wavelength of red.

Inferior conjunction: The conjunction in which a planet whose orbit is inside that of the Earth passes between Earth and the sun.

Infrared radiation: Invisible radiation of wavelength slightly longer than red light.

Intrinsic brightness: The amount of energy (usually light) an object gives off. This is its true brightness, independent of the effects of distance or dimming by intervening material.

Ionosphere: Several layers of ionized air high in the atmosphere. The ionosphere plays an important part in reflecting radio waves.

Ionized hydrogen: Hydrogen which has lost its electron. Ionized hydrogen gas, commonly found in stars and nebulae, has free protons and free electrons.

Julian day: The number of days since noon on 1 January 4713 B.C. Variable-star observers and other astronomers commonly calculate the interval between dates of events by subtracting Julian days, eliminating the necessity of keeping track of leap years and other calendar details.

Libration: The turning of the visible face of the moon, which allows us to see different amounts of the lunar surface around the limb (edge).

Light-year: The distance that light travels in a year, which equals 9,460,000,000,000 km or 63,240 A.U. (astronomical units).

Limb: The edge of the apparent disk of an astronomical body, such as the sun, moon, or a planet.

Luminosity: The ratio of the total light emitted by a celestial object to the total light emitted by the sun. Also, the total energy emitted by a star, per second.

Lunar module: The term for the vehicles that carried two men in the Apollo project from the command modules to the surface of the moon and back.

Magellanic clouds: These are not clouds, they are galaxies. Two relatively nearby galaxies of irregular shape visible from the southern hemisphere, named after Magellan, the Portuguese explorer who first described them.

Magnitude: (Also called "apparent magnitude.") A number indicating the apparent brightness of a star. Bright stars are designed by small numbers (magnitude 1, say), while dim stars are designated by large numbers (e.g., magnitude 15).

Main sequence: A band in the scatter Hertzsprung-Russell diagram. It includes more than 80 percent of all stars. The energy emitted by these stars is obtained from thermonuclear reactions in the core of the stars.

Mariner: The name given to a U.S. series of space probes designed to obtain data from Venus, Mercury, and Mars. On December 14, 1962, Mariner 2 passed within 22,000 miles of Venus. Mariner 9, launched May 30, 1971, came within 900 miles of Mars. It was the first spacecraft to go into orbit around a planet other than Earth. Mariner 10, launched November 3, 1973, came within 3,600 miles of Venus and 450 miles of Mercury. This was the first probe of Mercury.

Mass-luminosity relationship: This relationship, which applies to main sequence stars, states that luminosity is proportional to M^a, where M = mass and the power a - 4½, for stars whose mass is less than half the sun; a – 3½ for stars whose mass is more than 1½ that of the sun.

Maxima: The times when a variable star reaches its maximum brightness.

Mean solar time: Time as kept by a fictitious "mean" sun that travels at a steady rate across the sky through the year.

Meridian: The great circle passing through the celestial poles and your zenith.

Messier Catalogue: The list of 103 nonstellar, deep-sky objects compiled by Charles Messier in the 1770's and subsequently expanded to 109 or 110 objects.

Meteor: A meteoroid during the time it gives off light. Also called a shooting star.

Meteorite: A meteroid which, because of its size, survived collision with the Earth's atmosphere and reached the Earth's surface. Meteorites can be seen on exhibit in many natural history museums.

Meteor shower: The appearance on many meteors during a short period of time, as the Earth passes through a comet's orbit.

Micrometeorite: A fine dust particle floating in space, too small to be seen with the unaided eye and too small to become incandescent during its passage through the atmosphere.

Milky Way: A luminous band across the sky, of which our galaxy is part. The light is due to the fact that the vast majority of stars in our (disk-shaped) galaxy are located along this narrow band on the celestial sphere.

Minima: The times of a variable star's minimum brightness.

Mira variable: A long-period variable star, like the star omicron Ceti (called "Mira").

Morning star: This is not a star, but a planet (e.g., Mercury) when seen in the eastern sky just before sunrise.

Neap tide: The lowest tide of the month.

Nebula: A vast cloud of gas, or gas and dust, in space.

NGC: The prefix used before numbers assigned to nonstellar objects in the *New General Catalogue,* published by J.L.E. Dreyer in 1888.

Neutron star: A small (20 km diameter), dense (a billion tons per cubic cm) star, resulting from the collapse of a dying star to the point where only the fact that its neutrons resist being pushed still closer together prevents further collapse.

Nova: A star which suddenly increases in brightness and later returns to its original value of brightness.

Nutation: A small nodding motion of the Earth's axis of rotation with a period of 19 years. This motion is superimposed on precession.

Oblate: A nonspherical shape formed by rotating an ellipse around its narrower axis. The equatorial diameter of an oblate body (such as Jupiter) is greater than its polar diameter.

Occultation: The hiding of one celestial body by another.

Open cluster: An irregular grouping of stars of a common and possibly recent origin. Also called a galactic cluster.

Opposition: The point in a planet's orbit at which its celestial longitude is 180° from that of the sun. A planet at opposition is visible all night long.

Parsec: The distance from which 1 A.U. appears to subtend (cover) one second of arc. One parsec equals 3.261633 . . .light-years.

Penumbra: At an eclipse, the part of the Earth's or moon's shadow from which part of the solar disk is visible. Also refers to the outer, less dark portion of a sunspot.

Perihelion: The point nearest the sun in the orbit of a planet or a comet (point A).

Photosphere: The visible surface of the sun or star. Below it is the interior of the sun; above it, the atmosphere.

Pioneer: The name given to the first series of U.S. unmanned space-probe vehicles. They were instrumental in developing, launching and guiding techniques for the Mariner, Lunar Orbiter, Ranger and Surveyor series.

Planet: One of the nine bodies revolving about the sun in almost circular orbits. Planets are made visible to us by reflected sunlight. It is reasonable to assume that many stars have planets revolving about them.

Planetary nebulae: A nebula resembling a planet in shape.

Planetoid: (Also called "asteroid.") A small, irregularly-shaped, solid body revolving about the sun. Also considered to be a minor planet.

Plasma: An ionized gas.

Polar tufts: Small spikes visible in the solar corona near the sun's poles, formed by gas following the sun's magnetic field.

Poles, celestial: Points of intersection of extensions of Earth's axis and the celestial sphere.

Precession: The slow change in direction of the Earth's axis due to the gravitational pull of the moon on the bulge at the Earth's equator. The slow change in the axis causes the westward motion of the equinoxes among the constellations.

Prominence: Gas suspended above the solar photosphere by the sun's magnetic field. Ordinarily visible at the solar limb.

Proper motion of star: The singular velocity (in seconds of angle per year) of a star in a direction perpendicular to the line of sight of a terrestrial observer.

Protostar: The portion of a nebula that is about to become a star.

Pulsar: A neutron star emitting pulsed radio signals. The first pulsar was discovered in 1967. Its pulse lasts ⅓ of a second and repeats with great regularity every 1⅓ second.

Pulsating stars: Stars that periodically vary in brightness because of periodic changes in volume.

Quadrature: An elongation of 90° east or west of the sun.

Quasar: The popular name for quasi-stellar object. These are extremely luminous objects (the most luminous known) at enormous distances (the most distant object known) which generate incredible amounts of energy. The true nature of quasars is still under study.

Radial velocity of star: The velocity (in miles or kilometers per second) in line of sight of a terrestrial observer.

Radiant: The location on the celestial sphere from which, because of perspective, meteors in a given shower appear to radiate.

Radio astronomy: The branch of astronomy which deals with the radio waves emitted by various celestial bodies, as well as the theory of their emission.

Radio star: See discrete source.

Radio telescope: An instrument used for examination of celestial objects by means of the radio waves they emit.

Radio window: The transparency of the atmosphere to radio waves ranging between .25 cm and 30 m in length.

Ranger: The name given to a series of nine U.S. lunar probe vehicles, designed to transmit photographs before crash landing on the moon. More than 17,000 photographs were obtained from the Rangers, the closest one taken .2 seconds before impact.

Red giant: A member of the giant sequence in the Hertzsprung-Russell diagram. It has a radius fifteen to thirty times larger than the sun's, and luminosity a hundred times that of the sun.

Red shift: The shift of all spectral lines toward longer wavelengths observed in all galaxies. Galactic red shift is due to the expansion of the universe. Gravitational red shift is due to the high value of the mass of the emitter.

Refracting telescope: A telescope which uses a lens in the principal stage of forming an image.

Refraction: A change in the direction of light on entering a different medium (such as glass).

Resolving power of telescope: The power to separate two close points into two distinct units.

Retrograde motion: Apparent backward (westward) motion of a planet through a starfield.

Reversing layer: The lowest of the three solar atmospheric layers. It is responsible for most of the dark lines in the solar spectrum.

Revolution: The orbiting of a planet or other object around the sun or another central body. (Compare with rotation.)

Right ascension: The angular distance from the prime meridian to a celestial body measured eastward along the celestial equator from 0° to 360° or from 0 to 24 hours. Analagous to longitude in geography.

Rocket: A tube designed to move through space, deriving its thrust by ejecting hot, expanding gasses—called a jet—which have been generated in its motor. The rocket contains within itself all the material needed for the production of the jet.

Rotation: The spinning of a planet or other object on its axis.

Saros: The interval (about 18⅓ years) between two successive lunar or solar eclipses of the same series.

Satellite: A celestial body revolving about one of the planets; e.g., the moon. Also, any small body which revolves about a larger body, man-made or otherwise.

Sedimentary rocks: Rocks formed by precipitation from water or any other solution.

Separation: The angular distance (measured in degrees, minutes, and seconds of arc) between components of a double star.

Service module: That part of the vehicle in Apollo and other programs which contains the power, supplies, and fuel.

Shadow bands: Light and dark bands which appear to sweep across the ground in the minutes before and after totality at a solar eclipse. They're caused by irregularities in the Earth's upper atmosphere.

Sidereal period: The interval of time required by a planet to make one revolution (as seen from one of the fixed stars) about the sun.

Solar constant: The quantity of radiant solar heat 1.94 calories per minute, received per square centimeter of the Earth's surface.

Solar flare: An explosive eruption on the sun reaching temperatures of millions of degrees.

Solar wind: The material, mainly protons and electrons, streaming out from the sun into space. The sun normally loses millions of tons per second of its mass via this wind.

Solstice: The point of maximum declination of the Earth on the ecliptic. The solstices are halfway between the equinoctial points. In the northern hemisphere, the summer solstice occurs when the sun is farthest north from the equator. The winter solstice occurs when the sun is farthest south.

Space probe: An unmanned vehicle sent into space to obtain scientific data.

Spectral line: A wavelength of the spectrum at which the intensity is greater than (an emission line) or less than (an absorption line) neighboring values.

Spectral type: One of several temperature classes—OBAFGKM in decreasing order of temperature—into which stars are placed, based on an analysis of their spectra.

Spectrograph: An instrument which (A) collimates (makes parallel rays of light), then (B) disperses the light (by means of a prism or grating) into a spectrum, and finally (C) produces a photograph of the spectrum.

Spectroheliograph: An instrument which photographs the sun in monochromatic (single color) light.

Spectroscopic binary: A system of two stars which can be detected only with the aid of a spectroscope.

Spectrum: (Plural: "spectra.") The radiation from an object, spread out into its component colors, wavelengths, or frequencies.

Spherical aberration: A shape defect of a lens. Light passing a spherical lens near its edge is converged more than light passing the center of the same lens, which causes the image to blur.

Spiral nebula: A galaxy of stars (not a nebula) in the form of a spiral.
Star: A large globe of intensely hot gas, shining by its own light (e.g., the sun).
Star cloud: One of several regions of the Milky Way where great numbers of stars appear.
Streamers: Large-scale structures in the sun's corona, usually near the solar equator, shaped by the sun's magnetic fields.
Sunspots: Dark (by contrast to surroundings) patches which appear from time to time on the photosphere of the sun.
Supergiant: A star even brighter and larger than giants of the same color and temperature. Only the most massive stars become supergiants, after passing through the giant stage.
Superior conjunction: The conjunction in which a planet whose orbit is inside that of the Earth passes on the far side of the sun with respect to Earth.
Supernova: A star which quite suddenly increases, perhaps a million times, in brightness. It is similar to a nova, but its increase is vastly greater. It never fully returns to its original brightness.
Supernova remnant: Gas left over from a supernova which can be seen in the sky or detected from its radio or x-ray emission. (The Crab Nebula, for example, can be detected in all three ways.)
Surface brightness: The brightness of a unit area of an object's surface. For spread-out objects such as nebulae, the surface brightness determines the amount of contrast the object has against the background sky, and whether the object's surface is bright enough to make an image on your retina. Even though the object's total brightness may be high, it still may be hard to see if it is spread out enough so its surface brightness is low.
Surveyor: The name given to a U.S. series of unmanned probes which landed on the moon to obtain data prerequisite for a manned landing.
Synodic: Related to an alignment of three bodies, often the Earth, the sun, and a third body such as the moon or a planet.
Synodic period: The interval of time required by a planet or the moon to complete one revolution as seen from the Earth.
Telescope: In astronomy, an instrument used to collect radiation from celestial objects.
Terminator: The boundary between the illuminated and dark portions of the moon or a planet.
Train: A path left in the sky by a meteor.
Transient lunar phenomena (TLPs): Changes such as emissions of gas, observed on the moon.
Transit: The motion of a small body (e.g., Mercury) across the face of a larger body (e.g., the sun).
Tropical year: The ordinary year. The year used in everyday life.
Umbra: The dark shadow cast by a planet or a satellite.
Universal time (U.T.): Solar time at the meridian of Greenwich, England.
Variable star: A star whose apparent brightness changes over time.
Velocity of escape: The velocity an object must acquire in order to escape from the gravitational pull exerted on it by another body. The velocity of escape at the Earth's surface is 7 miles per second. Any terrestrial body reaching this velocity will leave the Earth permanently.
Venera: The name given a U.S.S.R. series of space probes made to collect data about Venus.
Vernal equinox: The intersection of the ecliptic and celestial equator which the sun passes on its way to northern (positive) declinations.
White dwarf: A white star of low luminosity, small size and extremely high density.
Zenith: The point directly overhead (wherever an observer is), 90° above the horizon.
Zodiac: Traditionally, a set of 12 constellations through which the sun, moon and planets pass in the course of a year.
Zones: Bright bands in the cloud layers of the giant planets: Jupiter, Saturn, Uranus, and Neptune.

Bibliography

Allen, Richard H., *Star Names, Their Lore and Meaning,* Dover
Bulfinch, *Bulfinch's Mythology,* Crown Publishers
Bytheriver, Marylee, *A Dictionary of Astrology,* Harper and Row
Clark, Arther C., *The Promise of Space,* Harper and Row
Degani, Meir H., *Astronomy Made Simple,* Doubleday & Company
Denevi, Dane, *To the Edges of the Universe,* Celestial Arts
Dick, Thomas, *Celestial Atlas,* Huntington
Dowrick, Stephanie, *Land of Zeus,* Houghton Miffin
Durant, Will, *The Story of Philosophy,* Washington Square Press
Ganow, George, *Matter, Earth and Sky,* Prentice Hall
Gallant, Roy A., *The Constellations, How They Came to Be,* Four Winds Press
Gayley, C. M., *The Classic Myths in English Literature and In Art,* Ginn and Company
Hamilton, Edith, *Mythology,* Little, Brown and Company
Hayle, Fred, *The Nature of the Universe,* Harper Bros.
Jastrow, Robert, *God and the Astronomers,* Warner Books
Klapp, William H., *Manual of Mythology,* Tudor Publishing Company
Lovitt, I. M., *Beyond the Known Universe,* Viking Press
Martin, Martha E., *The Friendly Stars,* Van Nostrand Reinhold Company
Menzel, Donald H., *Astronomy,* Random House
Menzel, D. H. and Pasachoff, J. M., *A Field Guide to the Planets,* Houghton Mifflin Company
Milton, Jacquailine, *Astronomy Introduction,* Chas. Scribner & Sons
Neeley, Henry M., *A Primer for Star-Gazers,* Harper Bros.
Nzimov, Isaac, *Pilots of Dawn,* Doubleday
Parker, Julia and Derek, *The Complete Astrologer,* Greenwich House
Rielpath, Ian, *The Young Astronomer's Handbook,* Arco Publishing Inc.
Sagan, Carl, *Cosmos,* Random House
Smith, Benjamin E., *The Century Encyclopedia of Names,* The Century Co.
Staal, Julius, *Patterns in the Sky,* Abee Custom Graphics
Vehrenberg, Hans, *Atlas of Deep-Sky Splendors,* Sky Publishing Corporation
Winner, F. D., *Genetic Basis of Society,* Shakespeare Publishing Co.

The Roots of a Star Name

Before anyone named a constellation, someone probably named a star if for no other reason than to be able to tell others about it. No one could know that the ones which twinkled were stars and the steadily lighted ones were planets. A sparkle does not have anything to do with the radiating of the star itself, but with the atmosphere through which the light passes on its way here. The different colors come to us as through a prism of space and atmosphere, which breaks up the light.

Naming the sky patterns, or constellations, over the centuries was merely a way of dividing up the sky. As a result, we now have 88 constellations, many of which are relatively new in existence.

As a matter of record, Ptolemy listed 48 constellations in his book, *Almagest,* (which translated means "we see"). Twelve of these patterns had already been listed centuries before by wise men in the Middle East. Thousands of years before that, the constellations were used as a way of telling the time of the year and month. That calendar, of course, was the zodiac.

In 1603 The German astronomer Johann Bayer (1472-1625) published the first modern catalogue, *Uranometria*. Bayer celebrated its publication by introducing twelve new constellations of the southern sky.

Next came a Polish astronomer, Johannes Hevelius (1611-1687), whose catalogue was distributed posthumously in 1690.

The last of the great observers came from Africa. Abbé Nicolas Louis de Lacaille studied and catalogued more constellations at the Cape of Good Hope from 1751 to 1753. He was not a very imaginative man, as reflected in his uninteresting constellation configurations and the names he chose for them: Fornax (a furnace), Mensus (a table), Pictor (a painter), Telescopium (a telescope), etc.

Many other constellations have been named by various astronomers. Some remain, others have long since been in disuse. So much for constellations.

In 1930 the International Astronomical Union drew up a list of acceptable, recognizable constellations and decided to let it stand at that. These are the 88 accepted by most astronomers worldwide.

As for naming stars, again we must go back to the Chaldean shepherds who probably named the first zodiac. (The ram to a shepherd is a very noble animal.) The names used now have various origins. There are Arabic names such as Aldebaran, Rigel and Betelgeuse. There are Greek names such as Sirius and Procyon. While a stop has been put to the naming of constellations, the naming of stars goes on to this day.

As of this writing, there are in excess of twenty-six different catalogues designating stars, including their locations, magnitude of brilliance, rate of travel, etc. The Smithsonian Astrophysical Star Catalogue so far has registered not less than four volumes of stars numbered from one through 258,900 since 1950. None of these are listed with proper names. Those stars with names given to them by the International Star Registry are numbered and registered in the following compilation.

The International Star Registry provides an opportunity to those who wish to have stars registered by name.

International Star Registry

The following section of this book is a registry of new star names compiled by the International Star Registry (August 1988 to July 1992). All star names are arranged in alphabetical order and include telescope coordinates. The coordinate reference used is Epoch and Equinox 1950. It is much easier to find a star within a constellation than to decipher its Astronomical position in the sky. It was for this reason that during the Astronomical Congress of 1928 eighty-eight constellations were officially recognized. The boundaries are described in *Atlas Celeste* published at Cambridge, England in 1930. Immediately preceding the registry is a list of Abbreviations for these various constellations.

In order to locate your personalized star refer to the charts and coordinates which you received in your star package.

In the course of compiling a registry of names, it is possible that errors or ommissions may sometimes occur. If you find any such oversights please notify the International Star Registry and corrections will be included in the next printing.

Andromeda	AND	Lacerta	LAC
Antlia	ANT	Leo	LEO
Aquarius	AQR	Leo Minor	LMI
Aquila	AQL	Lepus	LEP
Ara	ARA	Libra	LIB
Aries	ARI	Lupus	LUP
Auriga	AUR	Lynx	LYN
Bootes	BOO	Lyra	LYR
Caelum	CAE	Mensa	MEN
Camelopardalis	CAM	Microscopium	MIC
Cancer	CNC	Monoceros	MON
Canes Venatici	CNV	Musca	MUS
Canis Major	CMA	Norma	NOR
Canis Minor	CMI	Octans	OCT
Capricornus	CAP	Ophiuchus	OPH
Carina	CAR	Orion	ORI
Cassiopeia	CAS	Pavo	PAV
Centaurus	CEN	Pegasus	PEG
Cepheus	CEP	Perseus	PER
Cetus	CET	Phoenix	PHO
Chamaeleon	CHA	Pictor	PIC
Circinus	CIR	Pisces	PSC
Columbia	COL	Piscis Austrinus	PSA
Coma Berenices	COM	Puppis	PUP
Corona Australis	CRA	Pyxis	PYX
Corona Borealis	CRB	Reticulum	RET
Corvus	CRV	Sagitta	SGE
Crater	CRT	Sagittarius	SGR
Crux	CRU	Scorpius	SCO
Cygnus	CYG	Scutum	SCT
Delphinus	DEL	Serpens	SER
Dorado	DOR	Sextans	SEX
Draco	DRA	Taurus	TAU
Equuleus	EQU	Telescopium	TEL
Eridanus	ERI	Triangulum	TRI
Fornax	FOR	Triangulum Australe	TRA
Gemini	GEM	Tucana	TUC
Grus	GRU	Ursa Major	UMA
Hercules	HER	Ursa Minor	UMI
Horologium	HOR	Vela	VEL
Hydra	HYA	Virgo	VIR
Hydrus	HYS	Volans	VOL
Indus	IND	Vulpecula	VUL

A

A "Keene" Star
Uma 10h46'58"60d40'
A "Ray" of love
Sge 18h58'27"19d37'
A "Spott" In Heaven
Uma 9h5'29"51d8'
A & D
Sge 19h17'43"18d42'
A & E Holiday Star
Uma 9h59'30"68d32'
A & G Nova Star
Uma 9h59'50"68d38'
A B H
Cam 11h3'35"82d14'
A Bit Of Heaven From Your Angel
Eri 3h10'16"-2d21'
A C E
Cyg 19h31'47"38d11'
A Cherri Ray
And 2h33'35"44d26'
A Christmas Star for Sue
Peg 23h3'57"22d29'
A Costar is Born
Uma 11h33'48"45d28'
A D S
Aur 5h3'0"44d27'
A D T/Forever Boo Boo
Uma 12h10'34"63d12'
A Double Flame:V & D
Vul 19h32'9"25d14'AB
A Dream Come True
Peg 21h47'13"34d27'
A E-Friends Forever
Umi 14h19'31"66d10'
A Estrela Da Tereza
Cru 12h0'60"-63d13'
A Fairy Tale
Uma 9h18'0"55d15'
A Family Star for Janice Maia & Sam
Crt 10h53'32"-8d18'
A Forget-Me Not
Aqr 22h16'8"-11d50'
A Friend Throughout Life's Journey
Vul 19h23'1"22d51'
A Grandma's Love - Forever E M S
Lyr 19h16'1"28d26'
A Grandpa's Love - Forever R W S
Lyr 19h16'18"28d25'
A Great Deal of Avoiding
Cyg 21h23'0"39d25'
A Heaven Called Rosanne
Cam 4h26'1"60d8'
A Husky's Sparkle & Shine
Ser 18h17'28"-14d20'
A J
Boo 14h10'29"54d28'
A J
Aur 6h10'13"33d28'
A J
Uma 10h46'54"55d56'
A J & Nicole,Wish On Your Star
Lyn 7h4'0"44d12'
A J O
Per 3h4'1"43d60'
A Kind of Magic R D
Her 17h25'39"42d7'
A La
Lyr 18h50'57"30d21'
A League of Our Own
Uma 12h13'15"55d55'
A Legszeb Csillag Katalin
Dra 18h23'31"68d29'
A Light Forever
Aql 19h31'40"12d5'
A Little Bit Of Heaven
Lyr 19h22'21"37d51'
A Little Bit Too Much
Ori 6h5'33"8d4'
A Little Joy in Dreamland
Vul 20h1'57"28d40'
A Little Trip To Heaven
Cnv 12h15'14"45d34'
A Lizard
Lac 22h25'52"50d36'
A Loving Mother
Cas 2h40'29"60d42'
A M C 93
Tri 1h54'55"27d56'
A M H Unforgettable
Uma 11h0'39"38d15'
A Marie,Odysee D'Un Reve
Dra 20h6'55"70d23'
A Mothers Love
Uma 8h59'1"48d27'
A New Star Of Love Is Born
Gem 7h37'19"20d53'
A P Nea
Umi 15h17'23"68d27'
A P Y O B E
Crb 16h17'22"30d16'
A Pilot's Dream
Ori 6h8'22"4d25'
A Place Called Danielle
Mon 8h2'37"-8d20'
A Place Called Annalee
Mon 6h22'1"8d59'
A Place To Meet In The End,Amy's Star
Umi 15h33'46"68d3'
A Reminder To John
Cet 3h1'17"5d0'
A Richer Moment
Uma 9h57'60"71d53'
A Shining Star In ESD-1994
Ori 5h57'1"18d5'
A Shooting Star (Mark Papa)
Leo 9h33'53"14d10'
A Silver Star
Her 17h11'0"46d36'
A Silvia,tutta ti guardano ma io solo ti
Pup 8h8'33"-28d37'
A Star Above Me
Vir 13h59'46"2d24'
A Star Called Georg
Ori 5h56'29"21d11'
A Star Called Pinguin
Peg 21h40'0"27d48'
A Star For Honey
Her 16h4'56"27d39'
A Star for Jennifer
Oph 18h17'23"8d29'
A Star for My Kimberly
Sge 19h5'18"18d4'
A Star for My Michele
Peg 21h29'12"26d41'
A Star For My Nancy
Oph 18h37'52"6d12'
A Star For Peter
Hya 9h38'29"2d23'
A Star Is Born In The Name Of B J
Aur 5h1'52"41d45'
A Star Named Saul
Boo 14h32'43"38d29'
A Star Of Shun
Sco 16h56'31"-38d59'
A Star Shines For Corey
Peg 22h36'1"10d32'
A Star to Look Up to K L
Aur 5h55'16"30d57'
A Star To Remember To Jim,Love Emma
Uma 12h4'30"56d47'
A Star-Plumarie
Boo 15h27'46"33d50'
A Study In Scarlet
Ari 2h26'6"15d23'
A S"
Aur 5h37'46"37d46'
A T & Gay Louise
Sge 20h1'44"16d47'
A Twinkle In Jody's Eye
Sge 20h1'33"16d33'
A V Steller Star Humphrys
Boo 13h41'1"19d35'
A Wedding Anniversary
Sco 16h4'30"-18d15'
A Whole New World
Boo 14h44'0"47d51'
A Wish For Carol
Lyn 7h38'1"48d2'
A Wish For Jessicca
Peg 23h44'0"31d22'
A Year & Forever
Cyg 20h15'28"40d60'
A'gou's Starlight Serenade
Uma 9h47'55"70d1'
A'IDA Star,The
Cas 0h9'11"58d44'
A'Peiron
Gru 22h12'49"-51d33'
A-B-C
Vul 20h14'44"22d35'
A-Boo & Bartress
Lyr 19h20'40"42d21'
A-GIN
Boo 13h36'38"22d7'
A-Paul-O
Ser 17h33'35"-10d59'
A-ron
Cep 23h10'44"64d29'
A1 Alex
Lyn 7h48'18"48d31'
AA-HA Computer Bersa
Ari 3h2'28"23d35'
AA-Stern fuer RENATE
Uma 9h48'29"58d16'
AABA's Lucky Star
Cas 2h39'59"61d15'
Aadland,Nathan
Aql 19h16'14"10d3'
Aagaard,Evan
Ori 6h17'17"20d13'
Aagesen,Andreas Hoier
Cep 21h36'1"70d55'
Aalderink,Sarah Elizabeth
Cyg 21h20'51"39d42'
Aalseth,Margaret
Vul 19h24'38"21d3'B
Aalseth,Oliver
Vul 19h24'38"21d3'A
Aalto-Setälä,Marjut
Peg 22h41'11"34d41'
Aamer Irfan
Cap 21h14'16"-23d46'
Aandal,The Star Of Hannele
Cas 1h58'1"60d6'
AAR+EV-24-10-93
Uma 14h2'18"52d43'
Aaraman
Her 16h41'58"8d27'
Aardema,Lt Col Roger L
Aql 20h6'42"0d44'
Aaron
Aur 7h10'57"35d37'
Aaron
Cam 6h13'1"71d13'
Aaron
Dra 19h44'21"61d34'
aaron & alycyn
Boo 14h48'24"30d23'
Aaron & Carrie
Crb 16h6'59"30d18'
Aaron & Teri Sea Dreams
Sge 19h31'34"16d48'
Aaron Bad Bear
Umi 16h29'26"71d30'
Aaron Benjamin Our Enlightened Son
Boo 14h45'40"26d15'
Aaron Daniel
Aur 6h0'0"37d52'
Aaron Frey B
Cep 22h22'13"58d9'
Aaron Jacob
Aql 19h14'32"13d43'
Aaron Joseph
Per 1h53'32"54d22'
Aaron Lynn
Lac 22h55'39"55d13'
Aaron Star
Uma 11h11'45"41d6'
Aaron Thomas
Dra 17h6'0"64d26'
Aaron's Destiny
Ori 5h57'26"-0d56'
Aaron's Eyes
Hya 8h30'30"1d44'
Aaron's Glory
Hya 8h39'54"3d58'
Aaron,Betsy
And 22h59'29"50d35'
Aaron,Janet & Bob
Eri 3h15'1"-10d37'
Aaron,Margaret & Robert F
Aur 5h4'34"40d53'
Aaron,Neil
Lmi 9h23'18"34d58'
Aaron,Ruth & Fred
Cam 5h59'1"61d12'
Aaron,Super ELF-Regis E
Uma 9h37'13"68d15'
Aaron,William E
Dra 15h13'1"61d12'
Aaron-Anne
Peg 23h22'0"26d1'
Aarons,Darryl Howard
Boo 15h7'26"11d29'
Aarons,John
Oph 17h0'29"8d45'
Aaronson,Alistair
Ori 4h55'1"0d27'
Aasha
Cmi 8h9'11"0d18'
Aashiq
Leo 9h28'22"27d31'
Abacherli,PaPa Ben
Aql 19h5'58"-0d39'
Abacus,Georgii
Gem 6h53'39"17d44'
Abalan,Magaly
Crt 11h17'18"-19d59'
Abandonato-Switzer, Gina
Cas 23h16'1"60d3'
Abarbanel,Jenine
Com 12h53'22"25d33'
Abarca,Armando
Lmi 10h45'37"25d33'
Abare,Helene Ross
Mon 6h33'16"-5d40'
Abary,Aurora Gonda
Lyr 18h24'48"37d49'
Abata,Father Russell
Aur 6h8'0"31d25'
Abate (Bebe),Rocco
Oph 16h41'7"-0d41'A
Abate (Corazon),Lucia Gargurevich
Oph 16h41'7"-0d41'B
Abate,Patricia Simpson
Cyg 19h23'29"28d42'
Abatemarco,Brooke
And 2h26'30"44d15'
Abatiell,Ethan A
Her 16h0'29"48d60'
Abaton Cerido
Cam 4h5'43"65d32'
Abaz,Debbie J
Uma 8h46'55"53d25'
Abbas,Marilyn R
Eri 3h19'1"-2d7'
Abbate,Bruce
Oph 18h41'15"8d45'
Abbate,Jillian' Michelle L
Cyg 19h25'30"33d41'
Abbate,Michael Papa
Aur 5h0'47"29d30'
Abbate,Richalyn'Nicole N
Lyn 8h1'0"39d3'
Abbate,Richard T
Her 16h39'40"34d50'
Abbaticchio,Carina
And 23h33'48"43d40'
Abbe
And 0h2'0"38d11'
Abbe,Nancy & Coleman
Uma 11h31'35"38d11'
Abbene,Anthony
Umi 16h15'59"71d48'
Abbey
Dra 17h0'57"63d22'
Abbey Marie
Equ 21h9'0"11d49'
Abbey,Cheryl Vessadini
Peg 21h24'41"10d40'
Abbey,Damion
Her 18h28'1"12d19'
Abbey,Diane
Lib 15h37'59"-22d24'
Abbey,Maureen Victoria
Cas 0h41'40"64d2'
Abbey,Paul L
Boo 18h34'0"20d11'
Abbiba,Paola Benigna
Tau 5h25'52"16d35'
Abbinanti,Dorothy
Aqr 23h5'30"-11d25'
Abboot,Julie
Cas 0h22'48"50d1'
Abbott (Nall),Doris
Ori 5h55'0"14d29'
Abbott,Bruce
Her 15h55'29"44d41'
Abbott,Chiquita
Del 20h17'0"9d24'
Abbott,Donald Arthur
Ser 15h29'14"17d48'
Abbott,Eric M
Her 18h9'33"47d39'
Abbott,Graham D
Ori 5h56'13"14d59'
Abbott,Gregg
Her 18h43'0"12d2'
Abbott,Jennifer
Mon 6h41'48"6d11'
Abbott,Margaret Svetlik
Cas 1h17'53"62d5'
Abbott,Mark W & Doreen R Gibbard
Eri 4h32'52"-10d21'
Abbott,Melissa
Aql 18h42'31"1d12'
Abbott,Meredith Lynn
Vul 19h14'54"22d26'
Abbott,Rossi & Paul
Crb 15h31'1"30d37'
Abbott,Russell
Per 2h9'0"57d12'
Abbott,The Emma
And 23h33'58"40d12'
Abbott,William J
Per 3h46'49"35d46'
Abbott,William Thomas
Mon 6h53'25"-6d24'
Abby
Lyr 18h39'12"30d40'
Abby
Cmi 7h9'11"5d44'B
Abby
And 23h32'38"43d24'
Abby & Bob
Lyr 18h49'46"32d1'
Abby J
Cnv 12h24'48"42d2'
Abby Lee
Cas 1h28'59"58d35'
Abby Lyne
Ori 5h26'35"-0d47'
Abby My Love
Cas 1h21'58"60d18'
Abby's Buddy
Ori 5h30'1"-6d51'
Abby's Dream
And 1h43'42"40d46'
Abby's Star
Cyg 20h35'54"30d43'
Abby's Star
Lyn 8h21'39"45d60'
ABC Radio Networks
Dra 17h8'33"50d54'
Abczn,Cervia
Tel 20h12'42"-46d43'
Abdallah,Nayla
Her 16h37'0"24d57'
Abdeen,Sainaja Sainul
Cyg 20h1'15"31d42'
Abdellatif
Cet 1h33'1"1d44'
Abdi-Moreno,Daniel Steven
Aur 6h27'57"38d26'
Abdon,Nonno
Gem 6h43'50"18d47'
Abdouch,Michael David
Lac 22h19'56"51d45'
Abdul,Paula
Uma 11h0'30"41d53'
Abdulaziz,Lamia
Cyg 20h39'40"45d30'
Abdulhameed,Mohammed
Her 18h7'19"18d59'
Abduljawad,Ziyad
Hya 9h10'1"1d44'
Abdulla,T
Per 1h49'37"54d9'
Abdulziz,Princess Maha M
Cam 5h55'26"67d59'
Abe & Doty
Aur 4h55'0"40d57'
Abe's Dream
Cam 12h20'1"77d54'
Abebe,Trufat
Ori 4h41'12"4d41'
Abecassis,Lauren
Ori 5h56'13"18d42'
Abecassis-Leb,Annath
And 2h12'50"42d31'
Abedini,Reza & Rouhi
Aql 19h4'41"4d17'
Abel Health Care Network,Inc
Equ 21h22'50"8d50'
Abel,Althea Laurel
Peg 23h15'53"33d11'
Abel,Francis & Mae
Cyg 19h30'11"34d40'
Abel,Griffin Joseph
Ori 6h18'41"18d60'
Abel,Jamie
Uma 10h27'35"58d22'
Abel,Peter
Equ 20h56'25"8d59'
Abela Bernard
Ori 5h56'1"20d50'
Abelah
Dra 17h56'57"68d42'
Abelar,Norma & David
Ori 5h38'0"10d54'
Abele Family,The
Cma 7h49'23"10d46'
Abele,Adam
Aql 19h55'1"1d28'
Abele,Elizabeth
Del 20h19'1"16d48'
Abelman,Jerome
Her 18h3'13"31d47'
Abeloff,Mary Sue
Hya 8h35'38"-6d15'
Abels,Rodney
Boo 15h8'38"21d37'
Abelson,Jami
Cam 7h34'53"68d30'
Abend,Jonathan
Aur 6h8'31"38d44'
Abercrombie,Annalee
Ser 15h49'13"8d14'
Abercrombie,Jason
Cet 2h54'1"5d17'
åberg,åsa
Cas 23h28'36"61d57'
Abergel,Michel
Oph 17h57'54"10d42'
Abernathy Star,The
Aur 6h33'0"38d16'
Abernathy,Dr William M
Her 16h56'39"21d20'
Abernathy,Jeter Hampton
Lyr 18h43'50"45d6'
Abernathy,Mac
Boo 13h46'45"19d11'
Abernethy,Ann L
Aur 5h54'51"38d52'
Abernethy,Emily Cathlene
Mon 8h2'12"-9d47'
Abery,Gert & Harry
Lac 22h1'25"47d27'
Abete,Luigi
Aur 05h28'11"38d0'
Abha Sudhir Ankamalika Saurabu Gupta
Aql 18h57'26"15d6'
Abhilash
Cep 21h14'30"56d15'
Abhishek Verma
Ori 6h3'30"1d49'
Abi
Pic 5h0'25"-43d4'
Abi Mohseni-Nour-Iran- 55
Umi 15h0'42"68d23'
Abid,Khalid
Her 18h3'60"14d47'
Abigail
Ori 5h12'27"-9d2'
Abigail
And 2h29'20"41d6'
Abigail C E
Eri 5h7'60"-5d51'
Abigail Li
Cam 4h56'46"60d48'
Abigail my Friend my Fantasy
Leo 9h58'21"19d35'
Abigail Robyn
And 0h18'30"36d10'
Abigail Rose
And 23h9'14"41d1'
Abigail's Jewel
Cam 3h55'24"55d59'
Abigal
Dra 18h13'48"56d34'A
Abilleira-Elwin,Manuel
Cet 2h57'47"0d16'
Abimbola Emitomo
Aql 18h57'41"-10d14'
Abir Ray
Aql 19h31'39"12d23'
Abitbol,Ralph Gordon
Dra 16h54'59"64d23'
Ablanensis,Stella Luciae
Cas 23h46'21"48d55'
Ablaze,Geraldine
Ori 5h30'22"-1d23'
Ableman,Carol Smith
Del 20h13'17"11d35'
Ableman,Stephen
Cmi 7h28'47"8d38'
Ables,Brandon
Her 16h47'1"34d33'
Ables,Stephanie
Crb 16h16'1"30d37'
Abley,Sean Michael
Cep 21h22'1"61d11'
Abner,Alan
Lac 21h56'38"36d32'
Abner,Lisa Stephenson
Aql 20h16'56"5d12'
Abney,Elizabeth Marguerite
Sct 18h29'57"-4d29'
Aboaba,Tonykarin
Aur 6h8'50"30d47'
Aboobacker
Dra 9h56'17"73d20'
Abood,Mary P
Cas 1h45'37"56d39'
Abou,Ali Ashraf
Uma 8h47'15"71d43'
Aboudiya
Ser 15h56'53"18d34'
Abouelhadi,Najette
Cyg 20h17'35"38d25'
Aboukhali,Chehade
Cap 20h6'0"-10d7'
About A Girl
Cam 5h27'59"80d19'
Abplanalp,Ron
Aql 19h27'34"13d44'
Abra
Cyg 21h26'37"28d18'
Abracadabra
Tel 18h7'1"-46d57'
Abragnani,Maria
And 2h28'28"48d15'
Abraham II,James H
Ori 5h58'30"12d10'
Abraham,Andre
Cas 23h2'11"53d2'
Abraham,Avril
Cyg 20h1'0"39d54'
Abraham,Charlie Patty Shamrock
Cyg 20h19'35"38d20'
Abraham,Eldon Leonard
Per 3h19'40"41d10'
Abraham,Elizabeth J
Uma 11h56'26"58d31'
Abraham,F Murray
Her 17h23'39"49d54'
Abraham,Gerry
Lac 22h9'0"49d55'
Abraham,Michael
Aur 5h4'44"42d12'
Abraham,Mini Mary
Aur 5h59'55"54d33'
Abraham,Nicole Hope
Mon 6h44'30"11d51'
Abraham,Nils
And 1h41'23"41d43'
Abraham,Pamela Audrey
Cyg 20h23'12"38d41'
Abraham,Seth David
Aur 5h0'41"40d34'
Abraham,Theodore O
Uma 12h6'13"56d45'
Abraham,Udo
Sge 20h14'14"17d53'
Abraham,Virginie
Mon 7h43'28"-3d25'
Abrahams's Heaven
Lmi 10h28'0"32d8'
Abrahams,J David
Her 18h10'20"50d9'
Abrahams,Jillian Kate
And 23h23'25"40d15'
Abrahamsen,Jeff
Aql 19h31'50"-6d38'
Abrahamson,Eva
Sco 16h29'13"-30d54'
Abrahamson,Gwen
Lyr 18h39'13"41d13'
Abrahamson,Katelyn Ashley
And 0h10'50"46d12'
Abrahamson,Nili
Cas 0h46'25"66d55'
Abrahamson,Sarah Nicole
And 23h3'29"45d54'
Abram,Garth E
Cet 1h39'16"-1d18'
Abram,Joyce
Crb 15h55'35"26d59'
Abram,Rosemarie T
Uma 9h54'41"56d30'
Abramovicz,Jeannette
Peg 21h54'38"33d12'
Abramowitz,Amy
And 23h19'51"47d43'
Abramowitz,Justin Scott
Her 17h38'30"23d12'
Abrams,Albert
Cnv 12h6'25"34d46'
Abrams,Anne Barnes
Cas 0h37'54"62d47'
Abrams,Cindy
Ori 6h14'59"0d33'
Abrams,Daniel Kent
Hya 8h55'54"3d10'
Abrams,Diane
Cap 20h37'16"-26d45'
Abrams,Dr Jeffrey R
Per 3h7'30"47d10'
Abrams,Hon Robert
Per 1h56'49"50d7'
Abrams,Jr,Howard E
Cam 5h23'56"80d11'
Abrams,Lorne
Cam 4h4'43"60d48'
Abrams,Marcie
Sco 16h34'53"-25d18'
Abrams,Marjorie
Cyg 19h31'30"32d8'
Abrams,Mary Ellen
Com 12h31'0"26d40'
Abrams,Michelle Lynne
Gem 7h27'26"20d21'
Abrams,Patricia
Cyg 19h32'0"37d42'
Abrams,Rhoda & Barry
Lyn 8h55'28"41d47'
Abrams,Robert Elihu
Her 17h59'59"40d40'
Abrams,Robert Joseph
Boo 15h9'17"26d58'
Abrams,Thomas Griggs
Cep 20h52'0"60d38'
Abramsky,Michele Lee
Mon 7h42'50"-5d59'
Abramson's 50 Points Of Light,Alan
Sct 18h53'1"-6d40'
Abramson's Abyss
Cyg 21h12'17"36d58'
Abramson,David A
Ori 5h59'57"16d43'
Abramson,Dr Mason
Her 18h0'17"18d46'
Abramson,Madlyn & Leonard
Crb 16h16'23"34d30'
Abramson,Sue
And 23h19'38"43d44'
Abreder,Lehla
Cas 3h30'39"69d22'
Abremski,Ken
Uma 10h3'47"62d17'
Abreu,Roberto Tadeo
Cet 2h2'19"-18d54'
Abrew,Sonya Lydia
Cnv 13h52'51"40d44'
Abrol,Lynne
Gem 6h57'45"14d30'
Abromson,Miriam
And 0h17'50"45d56'
Abruzzese,Peter A
Cep 1h59'28"55d54'
Absey,Zoë Carriveau
Uma 11h7'26"32d48'
Abshire,Alexander Wade
Her 17h14'11"18d55'
Absolut
Del 20h24'14"8d51'
Absolwentow Slaskiej Akademi I Med
Boo 14h19'47"13d17'
Abston,Cathy
Mon 6h21'40"-6d11'
Abuagla,Ali Youssif
Lyn 8h24'12"57d47'
Abundis,Jennifer Denise
Mon 6h28'0"-10d11'

Aburiziq,Toni Ann
Cas 0h7'32"60d44'
Aburto,Rosemary
Lyr 19h15'32"41d17'
Abut,Charles C
Cep 21h30'52"55d14'
Aby,Sharon Roy
Peg 22h53'14"22d29'
Academy Elan,The
Cma 6h11'45"-16d12'
Acampora
Aur 5h21'34"38d8'
Accardi,Nicholas
Her 16h19'0"25d44'
Accent
Sgr 18h6'11"-28d9'
Accetta,Mark David
Mon 8h7'43"-1d54'
Acchino,J & B
Sge 19h35'18"16d47'
Accornero,Amata
Sco 17h30'24"-30d9'
ACE
Cam 11h28'58"80d40'
ACE
Cyg 19h27'14"34d47'
Ace
Per 2h54'35"31d15'
Ace
Cep 22h14'21"55d55'
Ace,Annie & Sheila Star
Cma 6h41'12"-18d52'
Ace,The
Cep 21h9'20"70d35'
Acevedo,Jr,Frank
Her 18h19'0"12d34'
Acevedo,Juana Gena
Mon 7h5'38"-5d44'
Acevedo,Maria De Los Angeles
Eri 3h13'36"-2d30'
Acevedo,Norma
Hya 9h6'19"5d46'
ACH1
Ori 5h55'44"14d14'
Achart,Eveline
Cyg 21h38'54"40d55'
Achaume,Stephanie
Cyg 21h3'1"48d54'
Achberger,Christopher Shane
Aqr 22h27'53"0d32'
Achesineide/Alcantara
Pup 7h56'53"-24d9'
Achez-Jarrard,Diane E
Mon 7h3'55"4d46'
Achiano The Star of My Love,Pasqualina
Cas 0h21'42"61d17'
Achieve-Stella of Naoki
Tau 5h25'32"21d50'
Achileas Georgios Tzamtzis
Cep 21h40'50"61d21'
Achille
Ari 1h54'19"11d16'
Achille,Stephen Rocco
Lac 22h30'44"40d54'
Achilles
Cyg 19h37'43"38d49'
Achim
Cam 3h57'52"60d53'
ACHIM 30-04-40
Lyn 7h51'0"51d34'
Achleitner,Joachim "Sol Truxtor"
Tau 5h45'59"21d25'
Achtzehn,Angel Louise
Cas 1h43'42"67d46'
Aciköz,Fatma
Sco 17h5'51"-38d15'
Acito,Courtney Celeste
And 23h3'33"48d17'
Ack Ack
Lyn 7h59'21"52d9'
Acken,Lynn
Lyr 18h48'18"41d27'
Acken,Richard
Lyr 18h50'13"40d19'
Acker,Joyce A
Cas 1h9'41"60d17'
Acker,Ludwig Prof Dr
Vir 13h38'18"-6d48'
Ackerberg,Lisette
Ari 2h41'33"25d9'
Ackerly,Jr,Robert Nelson
Dra 17h47'56"67d53'
Ackerman
Dra 11h11'59"73d19'
Ackerman Family Star, The
Cyg 20h59'35"38d21'
Ackerman Granchildren All Stars
Her 16h38'20"41d31'
Ackerman,Alexandra Ava
Aql 18h57'21"14d31'
Ackerman,Cary
Lyr 18h35'11"40d1'
Ackerman,Dr Mona
Oph 17h50'38"13d33'
Ackerman,Elizabeth Ryan
Cas 23h32'47"61d52'
Ackerman,Frederick William
Cep 21h57'34"55d4'
Ackerman,Gary Steven
Aqr 21h33'0"-5d59'
Ackerman,Glenn R
Ori 5h26'40"-1d52'
Ackerman,Gloria Elizabeth
Cas 1h1'31"58d16'
Ackerman,Karl Richard
Ind 20h54'33"-56d7'
Ackerman,Phil
Sct 18h22'31"-14d6'
Ackerman,Robert Allan
Aur 5h9'22"41d54'
Ackerman,Roger Williams
Ori 5h49'60"19d14'
Ackermann,Frank
Cnc 8h26'14"33d5'
Ackermann,Klaus
Lyr 19h4'49"28d17'
Ackermann,Sandrine
Cam 40h48'0"60d42'
Ackerson,Richie E
Her 18h6'57"31d40'
Ackland,Esther E Krug
Aql 19h23'24"10d27'
Ackley,Samuel S
Boo 14h54'56"41d60'
Acklin
Dra 16h22'0"63d22'
Acklin,Cole Wilson
Eri 4h6'21"-8d48'A
Acklin,Scarlett-Marie
Eri 4h6'21"-8d48'B
Ackner,Larry I
Per 1h51'25"56d58'
Ackroyd,Beth
And 1h5'15"37d30'
Ackroyd,William
Cas 0h39'47"74d2'
Acksel,John
Cmi 7h37'57"4d55'
Acleve
Cas 0h36'20"68d37'
Aco & Andrea
Dra 15h11'46"63d57'
Acocella,John & Sandra
Cyg 20h7'0"39d34'
Acocella,Patsy (The Greatest W-T Power)
And 0h24'51"43d10'
Acord's Fantasy
Vul 19h46'25"22d52'
Acord,Marcia
Cyg 19h30'27"37d47'
Acosta,Alejandra Gabriela
Vul 19h47'36"28d16'
Acosta,Claro
Mon 6h55'31"10d21'
Acosta,John "Honeyman"
Cnc 9h0'55"8d32'
Acosta,Maria F
Lyn 7h25'20"50d37'
Acosta,Mark Andrew
Ori 5h32'15"-1d47'
Acosta,Teresa
Peg 21h54'1"2d10'
Acosta,Tonna Lee
Cep 22h44'0"65d37'
Acosta-King,Samuel Eloy
Aur 6h30'49"38d57'
Acquafreeda,JoAnne
Lyn 8h24'60"47d17'
Acquaviva,Bruno
Cep 22h17'22"58d37'
Acquaviva,Federica
Cnv 13h6'46"51d42'
Acquaviva,Simone
Cep 22h17'33"55d39'
Acquisitions,Inc
And 0h11'57"47d27'
Acreman,Jonathan
Aql 20h6'0"7d15'
Acres,Katie
Cam 3h58'27"61d21'
Acropole
Equ 21h1'22"7d53'
Acton,M Kate
Gem 7h55'54"32d58'
Acuff,Capt Shelley
Cyg 21h29'15"28d60'
Acuna Matata
Peg 22h40'12"3d18'
Acunto,Claudia Palmira
Lyn 8h3'0"57d39'
Acunto,Stephen Henry Richard
Boo 15h18'25"47d43'
Aczel,Georg
Cmi 7h21'48"7d39'
AD INFINITUM
Tri 1h47'58"27d52'
Ada's Light
And 2h15'49"45d58'
Ada-Diego-Renzo
Cmi 8h7'19"0d45'
ADAC
Ser 15h14'1"0d46'
Adaim,Seth
Cnv 12h44'36"42d13'
Adair
Hya 9h8'13"5d1'
Adair"Heavenly Body", Christine
Boo 14h9'26"40d27'
Adair,Andrea
Lyr 19h22'1"38d43'
Adair,Jeanne
Eri 3h44'26"-6d55'
Adair,Jeff
Ori 5h54'38"6d27'
Adair,Margaret M
Com 12h3'58"19d22'
Adair,Perry Christofer
Aql 19h7'36"3d41'
Adair,Suzanne Kathleen
Sge 19h39'26"16d14'
Adair,Tricia
Cam 7h59'0"61d22'
Adam
Aql 20h19'46"1d44'
Adam
Cep 21h58'57"55d56'
Adam
Lib 15h12'42"-8d23'
Adam
Sge 20h0'42"20d9'
Adam
Ori 4h41'20"10d32'
Adam
Her 16h14'43"23d53'
Adam
Hya 9h15'0"2d55'
Adam
Dra 16h42'0"61d17'
Adam
Cep 22h36'28"61d26'
Adam
Her 16h39'52"50d34'
Adam
Ori 5h57'30"17d45'
Adam
Ori 4h43'1"0d45'
Adam & Carrie
Sge 20h13'16"17d34'
Adam & Lucille Together Forever
Equ 21h22'44"11d20'
Adam & Tina- More Than Words
Peg 22h19'40"10d47'
Adam Charles
Ori 5h40'57"11d58'
Adam Henry
Cnc 8h35'42"30d32'
Adam Jon
Aur 4h46'25"50d15'
Adam K
Cet 2h48'30"2d9'
Adam Lawrence
Ori 5h30'45"0d21'
Adam Loves Ellie
Lac 22h1'0"49d40'
Adam Micheal
Umi 13h19'24"72d57'
Adam Ray
Sct 18h53'42"-6d7'
Adam Star
Her 16h37'1"39d32'
Adam Star Seven
Tau 3h41'43"28d13'
Adam T
Tri 2h44'50"31d9'
Adam The Great
Psc 23h20'38"5d3'
Adam's Adventure
Lyr 18h50'59"40d15'
Adam's Asteroid
Cep 22h34'16"78d25'
Adam's Piece of Heaven
Ori 4h55'36"5d19'
Adam's Silver
Gem 6h47'6"19d0'
Adam's Star
Her 17h57'34"31d42'
Adam,Anya Louise
Cas 2h11'33"61d9'
Adam,Arthur
Ori 6h7'11"7d40'
Adam,Christopher
Uma 10h2'46"47d51'
Adam,Eunice Alice Rose
Eri 3h57'36"-10d27'
Adam,Helmut
And 23h13'15"42d41'
Adam,Helmut-Oswald
Peg 23h36'34"13d45'
Adam,Jane Lange
Cas 23h21'0"63d13'
Adam,John
Per 2h43'29"40d57'
Adam,Mona & Justin's Star
Uma 8h38'19"68d39'
Adam,Petrina
Col 5h0'23"-36d23'
AdamBenJosh
Aur 4h42'28"30d50'
Adamby Badam
Cyg 21h50'50"42d6'
Adamchik
Cam 3h56'41"57d15'
Adamcik,Joan & Ted
Cyg 21h34'56"38d24'
Adamczyk,Richard Scott
Aur 5h12'23"42d40'
Adamczyk,Stella
Aur 5h17'1"41d52'
Adamdanash
Sct 18h53'13"-7d50'
Adame,Consuelo & Anthony
Aql 19h31'0"10d58'
Adame,Sarah
Eri 2h46'47"-6d54'
Adamek,Rainer
Ari 2h2'37"21d33'
Adamelle
Cam 5h5'50"68d35'
Adamidis,Kostas
Cep 3h24'50"86d43'
Adamlai
Cnv 12h43'43"40d13'
Adamo,Emilio
Tau 5h24'45"18d53'
Adamo,Stuart
Per 4h28'56"50d30'
Adamopoulos,Ava Jessie
Aqr 22h4'28"0d1'
Adams "Caledonian Legend" Star,Evelyn
Cam 4h32'33"67d32'
Adams 8-18-94,Chad Nelson
Ori 5h52'21"16d18'
Adams Allstar
Lac 22h45'1"52d43'
Adams Colway & Associates
Cam 4h51'20"69d38'
Adams III
Dra 17h23'38"60d16'
Adams III,James B
Boo 15h31'43"40d40'
Adams of Tooele,The John Henry
Ori 5h55'0"7d30'
Adams of Tucson, Shirley & John
Crb 16h20'34"31d45'
Adams Shining Forever, Fran
Peg 21h25'16"23d53'
Adams Star Fascination Kelly Megan
Mon 6h31'43"-5d57'
Adams Star of Discovery,Joey Lauren
Cma 7h6'1"-28d12'
Adams Star System,The
Sct 18h21'46"-5d44'
Adams' Star,The
Cet 0h33'24"-1d30'
Adams's Citrus Star, Bill
Hya 8h28'0"0d14'
Adams,Alan Stuart
Her 17h14'33"22d18'
Adams,Alexander Berkley
Per 4h34'16"34d23'
Adams,Alexis Morgan
Cas 2h4'15"75d9'
Adams,Allison J
Tau 4h39'43"15d42'
Adams,Ambrose Starr
Vul 20h21'38"23d37'
Adams,Amy Camber
And 23h43'51"38d36'
Adams,Aries'Dad H J
Mon 7h28'40"-6d30'
Adams,Arthur Dale
Cet 2h55'41"2d21'
Adams,Arthur Snyder
Dra 19h14'0"70d6'
Adams,Ashleigh Paige
Vul 19h38'60"20d1'
Adams,Barbara L
Peg 23h43'39"31d31'
Adams,Barrie & Sandra Alreo(WICCA)
Mon 7h51'41"-7d19'
Adams,Ben
Eri 2h59'14"-15d32'
Adams,Bernadette J
And 2h8'1"38d55'
Adams,Betty J
Lyr 18h27'50"45d25'
Adams,Bradley Daniel
Boo 14h4'46"29d27'
Adams,Bryan
Uma 9h45'26"70d14'
Adams,Bryan Mark
Ari 2h33'16"21d50'
Adams,Bryant Keith
Aur 6h20'46"37d34'
Adams,Bryon Robert
Boo 13h53'0"22d24'
Adams,Carmel
And 2h35'45"38d32'
Adams,Caroline
Lyr 19h22'34"31d29'
Adams,Cecilia Capparelli
Boo 15h46'34"44d37'
Adams,Cheri Rene
And 23h37'33"47d55'
Adams,Chris
Lac 22h29'1"52d44'
Adams,Christopher
Hya 9h34'40"5d37'
Adams,Cody Duane
Uma 11h56'40"31d18'
Adams,Crystal C
Mon 6h51'41"-5d47'A
Adams,Dale Ernest
Lac 22h53'53"55d50'
Adams,Darrell
Mon 7h55'57"-6d9'
Adams,David Albert
Hya 9h18'33"-15d24'A
Adams,David Edward John
Dra 16h50'0"70d12'
Adams,Deborah,Antonia
Eri 3h41'52"-5d8'
Adams,Diane
Cas 23h40'31"61d35'
Adams,Douglas W
Aur 6h31'44"32d35'
Adams,Dr James
Oph 17h31'45"8d30'
Adams,Dr Robert J
Aql 19h40'25"13d50'
Adams,Dustin Robert
Her 17h55'28"50d1'
Adams,E Cecile Murphy
Aql 19h28'34"-6d58'
Adams,Edward J
Aur 6h8'31"33d54'
Adams,Eric Coy
Ori 6h4'14"-0d33'
Adams,Erika Mae
And 23h22'35"41d25'
Adams,Evelyn J
Vir 12h33'28"2d22'
Adams,Fran
Tau 4h38'58"15d3'
Adams,Frank
Sco 17h53'41"-30d31'
Adams,Frank & Joyce
Cyg 21h50'58"37d33'
Adams,Gabriel
Aur 6h2'10"45d13'
Adams,George
Aur 6h22'47"31d58'
Adams,George
Per 2h23'48"54d30'
Adams,Ida & Harold
Crb 16h13'53"34d14'
Adams,Iva
And 0h50'39"33d36'
Adams,Jackquelyn
Cas 1h43'36"58d48'
Adams,Jake W
Her 17h16'0"20d5'
Adams,James Bryan
Ori 6h0'27"5d32'
Adams,James Grigsby
Crt 11h17'44"-16d15'
Adams,Jason
Ori 5h56'11"16d45'
Adams,Jayne Marie
Umi 15h7'51"66d35'
Adams,Jeffrey W
Dra 18h49'0"70d31'
Adams,Jennifer Anne Onofrio
Vul 20h15'54"23d5'
Adams,Jewel Wasson
Hya 8h58'10"-7d47'
Adams,Jim & Phyllis
Cet 2h48'13"3d48'
Adams,Joan
And 2h10'0"39d37'
Adams,Joanne
And 0h40'52"38d45'
Adams,John B
Her 16h37'51"34d42'
Adams,John Quincy
Lmi 10h2'39"32d25'
Adams,Johnny Lee
Cet 1h2'55"0d19'
Adams,Jordan Suzanne
Dra 14h11'50"64d42'
Adams,Jr,James Gerard
Vul 19h47'30"27d23'
Adams,Jr,K S"Bud"
Cet 1h48'39"-6d41'
Adams,Julie
Vul 19h52'0"20d26'
Adams,Julie & Joe
Aur 6h31'15"32d11'
Adams,Karen Maria
Uma 11h24'33"48d57'
Adams,Karin Mary
Hya 9h18'33"-15d24'B
Adams,Kate Janeane
Lyr 19h21'54"38d58'
Adams,Katie Louise
Cas 1h25'0"73d9'
Adams,Kelsea Elisbeth
And 23h22'18"51d58'
Adams,Kenneth L
Her 17h6'32"43d11'
Adams,Kerrie Lynn
Mon 6h19'56"7d49'
Adams,Kimberly A
Del 20h14'18"11d6'
Adams,Kyle Thomas
Her 18h18'42"14d54'
Adams,L B & Peggy
Dra 16h50'53"51d37'
Adams,Larry Michael
Boo 14h52'53"39d38'
Adams,Laurence
Boo 14h10'33"43d20'
Adams,Linda Lou
Lyr 18h32'24"38d39'
Adams,Loretta & Henry
Aql 18h57'55"10d8'
Adams,Lydia Joy
Cet 2h50'31"0d23'
Adams,Marion Alice
Vul 19h18'41"20d36'
Adams,Mark
Dra 16h13'1"67d38'
Adams,Mark 1
Cap 21h17'17"-22d33'
Adams,Mary
Vul 19h50'30"20d20'
Adams,Mary Elizabeth
Peg 22h2'53"4d49'
Adams,Melanie Jane
Cas 0h33'49"61d31'
Adams,Melba
Aql 19h9'35"4d19'
Adams,Michael J
Cep 23h31'15"68d4'
Adams,Michael S
Aql 19h30'44"10d35'
Adams,Michelle
Lyr 18h18'44"40d23'
Adams,Mike & Jennifer
Psc 22h53'47"6d11'
Adams,Milton W
Aur 5h27'44"30d57'
Adams,Myra Jane
Vul 19h46'23"25d9'
Adams,Nancy Jane Lawrence
Boo 15h8'0"31d32'
Adams,Nicholas Jay
Hya 8h10'34"0d2'
Adams,Patricia Ann Lynch
Del 20h13'16"15d22'
Adams,Peter
Uma 9h6'47"68d49'
Adams,Randal
Dra 15h44'28"61d54'
Adams,Ron
Dra 16h41'15"68d16'
Adams,Ronald
Aur 6h3'27"31d22'
Adams,Samuel J
Her 17h6'30"20d43'
Adams,Samuel K
Dra 17h44'49"64d49'
Adams,Shaelynne Morgein
Aql 19h53'26"15d32'
Adams,Shannon D J
Cas 2h4'32"68d45'
Adams,Stephen E
Ser 17h55'35"-13d20'
Adams,Stevie
Uma 12h52'15"60d46'
Adams,Susan Hope
Sex 9h44'51"3d24'
Adams,Tammi
Vul 19h21'30"27d4'
Adams,Taylor Kohl
Per 3h11'52"45d35'
Adams,Tia
Cnv 12h35'12"34d13'
Adams,Todd Russell
Lac 22h21'1"37d42'
Adams,Tom
Cet 2h53'1"7d12'
Adams,Tracey
Mon 6h44'13"10d56'
Adams,Trisha S
Mon 6h51'41"-5d47'B
Adams,Vanessa Fay
Cas 0h52'19"73d53'
Adams,Walter L
Per 3h11'59"45d50'
Adams,Warren Joseph
Per 3h5'1"56d41'
Adams,Wendy R
Eri 2h43'50"-5d25'
Adams,Wil
Her 17h54'48"14d25'
Adams,William E
Her 17h54'51"40d20'
Adams-"Mr Personality" Charles A
Her 18h30'0"24d42'
Adams-Bleth,Barbara K
And 0h2'1"34d17'
Adams-Bowers,Kari
Ori 5h58'30"17d56'
Adams-Hancock
Cyg 19h31'37"33d5'
Adamski,Kathleen Vallerie
And 0h28'58"38d9'
Adamson,Bridget Noel
Cas 0h55'52"73d31'
Adamson,Debra
Cyg 21h51'24"40d4'
Adamson,Luke James
Tau 5h50'12"28d46'
Adamson,Todd Edwin
Cet 1h49'40"-4d24'
Adamus,Luanne
Lyr 18h40'24"26d30'
Adani,Mark Anthony
Aur 7h1'18"41d37'
Adare
Mon 7h10'41"-6d29'
Adase #50,Joseph & Marie
Cyg 21h29'33"30d60'
Adasko,Herbert Hardy
Per 3h22'57"40d12'
Adasko,Margaret Ellen
Uma 11h17'51"70d53'
Adasko,Rochelle (Shelly) Waldman
Uma 11h2'39"67d49'
Adawi,Fatema
Cyg 19h53'1"45d11'
Adcock,Clifton C
Lyr 19h26'1"37d41'
Adcock,Jacqueline Kay
Aql 20h0'12"-0d33'
Adcock,M S
Cam 3h47'48"70d31'
Adcock,Misty Dawn
Lyr 18h56'39"31d50'
Adcock,My Love Gary A
Aur 7h23'52"40d42'
Adcock,Richard S
Umi 15h39'54"76d55'
Adcock,Ronald E
Cep 22h15'26"61d2'
Adcock,Sandy
Cet 1h57'50"0d43'
Adcox,Conal
Lyn 7h2'0"51d52'
Adcox,Orin
Boo 14h34'24"50d40'
Addante,Gianni
Aur 6h1'48"45d39'
Added,Patricia
Sge 19h58'33"20d14'
Addeo Family Star,The Frank
Uma 11h9'23"47d31'
Addeo Forever,Michael & Ronnie
Cyg 21h51'55"37d36'
Addeo Happy Valentine, Christine
And 1h35'29"39d54'
Addeo,"Special"John Anthony
Uma 11h20'0"41d5'
Addeo,Daniela
Her 18h10'37"31d6'
Addeo,Laurence T
Cam 4h3'11"61d43'
Addeo,Nancy A
Dra 19h34'13"70d12'
Addeo-Before,Lori Ann
Eri 4h17'56"-1d26'A
Addesso,Diane Margaret
Cas 0h16'28"58d37'
Addi & Bob
Per 2h39'39"48d3'AB
Addicks,Linda Joy
Aql 20h11'15"1d36'
Addicks,Lyle
Hya 9h10'44"-14d28'
Addicks,Victoria Lynn
Hya 8h54'32"3d4'
Addie,The Brightest Star
Lyr 18h27'54"32d39'
Addington,Brett Chance
Eri 4h5'14"-19d23'
Addington,Darrin Douglas
Her 17h31'1"40d14'
Addington,Logan Albert
Lac 22h4'23"40d30'
Addington,Lovella Ruth
Cas 0h28'39"60d21'
Addington,Ryan Edward
Lac 22h4'55"40d24'
Addis 60th Anniversary Laird & Ersel
Cyg 21h26'56"31d11'
Addis,Don - Heavenly Free Fall
Ori 5h56'1"10d53'
Addis,Matthew Raymond
Cmi 7h43'1"3d54'
Addis,Michael Richard
Uma 9h58'0"68d13'
Addis,Raymond O
Aur 5h4'35"40d3'
Addis,Richard B
Ser 15h58'22"4d59'
Addison,Alexander James
Umi 13h10'22"71d30'
Addison,Gini & Dick
Cet 2h53'58"4d29'
Addison,Larissa
Lyn 8h58'0"43d46'

Addison,Nicholas Edmund Umi 15h6'32"68d8'
Addison,Rebecca Kaye & John Patrick Cyg 20h44'54"38d3'
Addison,Willie Aur 4h48'59"41d5'
Addisson,Markus Ivaen Ori 4h57'39"5d23'
Addoul,Abdelkrim Cam 5h51'20"68d28'
Ade Aql 20h7'44"1d7'
Ade,Jerome C Dra 17h8'1"62d40'
Adebowale,Emmanuel Olusegun Per 1h49'29"54d6'
Adedeji,Katherine Umi 13h28'51"72d14'
Adela Llorella Boo 14h1'11"12d5'
Adelaide G And 2h28'25"44d21'
Adele Lyn 8h24'24"45d24'
Adele Celeste Cas 3h8'29"75d10'
Adelfia Pup 8h24'3"-25d12'
Adeli,Babak Lyn 8h6'35"51d7'
Adelia Crb 15h16'35"30d49'
Adelina Tau 5h45'35"27d35'
Adelina Mon 7h5'53"-6d15'
Adeline Peg 22h16'10"31d35'
Adeline,Shirley Doris Aql 20h14'49"3d58'
Adelman,Betsy & David Cam 6h11'32"70d16'
Adelman,Steven Cary Her 18h3'0"30d53'
Adelsberger,Pat Dra 9h44'38"74d2'
Adelson,Amy Mon 7h21'44"-6d16'
Adelson,Lexi Rose Del 20h13'0"12d17'
Adelson,Merv Mon 6h22'23"2d52'
Adelson,Thea Cet 0h54'1"-3d4'
Ademeit,Elke Sgr 18h50'44"-26d41'
Aden,Laura Leannette Cas 0h59'15"60d28'
Aden,Margaret Ione Psc 1h0'54"30d4'
Ader,Jürgen And 2h23'56"45d54'
Aderholt,Harry Clay "Heinie" Aur 4h55'22"51d8'
Adessa,Caterina Giovanna Freda Cam 3h11'46"60d33'
Adges,Ellyn Lyn 9h8'23"46d38'
Adherents Groupe Gram Per 1h31'15"52d51'
Adi & Rainer Cas 0h4'0"60d7'
Adiar,Harold E Boo 15h0'33"48d35'
Adidane Ori 6h5'16"5d49'
Adidas 4689 Per 3h10'44"46d11'
Adina April Peg 22h19'47"10d56'
Aditi Sge 20h4'53"20d13'
Adjan Christine Cyg 20h24'10"39d51'
Adkins,Anne Marie Lyr 19h2'25"37d32'
Adkins,Bradley Lyn 9h11'33"35d38'
Adkins,Charlie "Bud" Aur 5h0'13"44d28'
Adkins,David Her 17h8'12"38d13'
Adkins,Dorene Kaye And 0h20'41"32d58'
Adkins,Dottie Del 20h14'56"11d48'
Adkins,Guy Oph 18h25'14"8d32'
Adkins,Ivan Wayne Aur 6h2'23"38d14'
Adkins,Jean Anderson Uma 11h5'56"52d60'
Adkins,Jeffrey & Nicole Cyg 20h44'35"42d16'
Adkins,Joyce Ellen Mon 6h20'1"-6d13'
Adkins,Lisa Maria And 1h17'19"34d23'
Adkins,Marian E Lyn 7h45'34"44d33'
Adkins,Mary Ann Cyg 19h29'51"34d5'
Adkins,Stefan John Sex 9h55'59"4d19'
Adkins,Steven Benjamin Aur 4h43'40"34d52'
Adkins,William Brian Aur 6h28'21"33d33'
Adlam,Jr,Arthur Herbert Lac 22h41'20"53d24'
Adlen,Gary John Per 3h6'12"41d52'
Adler,Audrey Heather Cyg 19h56'16"37d52'
Adler,Beloved,Barry Deaton Her 16h53'1"32d37'
Adler,Bob Lyn 6h33'55"58d59'
Adler,David Per 2h32'16"56d39'
Adler,Dena Cas 0h54'54"58d34'
Adler,Haley Elise Peg 23h37'50"25d47'
Adler,Harry Dra 19h39'18"65d28'
Adler,Herbert Her 16h59'22"22d4'
Adler,Hermann Leo 9h58'1"11d14'
Adler,Howard Sgr 18h59'44"-29d12'
Adler,Kevin Her 17h28'19"40d9'
Adler,Marion S Uma 11h17'41"32d28'
Adler,Marlene Vul 19h35'27"27d31'
Adler,MD,Charles S Oph 18h18'33"7d55'
Adler,Patrick Lib 15h18'15"-25d36'
Adler,Robert Cep 22h30'30"65d31'
Adler,Sam Eric Her 16h43'51"27d51'
Adler,Sydney Ser 15h57'52"2d45'
Adlercreutz,Anne Catherine Thamazine Cas 0h25'1"62d42'
Admail 4 And 23h1'0"51d10'
Adman,Isobel Peg 21h54'13"33d4'
Admirable Christian Uma 8h7'40"70d50'
Adnett's Hope,Sireena Cyg 20h43'18"38d49'
Adolf,Hans Dieter Lyn 9h0'36"42d0'
Adolf,Kellen Cataldo Cam 7h50'1"61d45'
Adolfsen,Louis G Lyn 8h29'1"41d24'
Adolfsson,Marianne Sco 17h5'49"-31d49'
Adolphi,Florence Cnv 13h21'1"50d38'
Adolphson,Eric Stanley Sex 10h40'26"-6d29'
Adonis,Adrian Cep 22h12'54"60d34'
Adora & Fraser Uma 9h6'29"49d34'
Adorable Dor Vul 19h48'25"28d53'
Adore-e II,Patrica Mon 6h30'12"0d49'
Adored Forever,Jeff Aql 19h47'33"11d17'
Adorf,Dieter Sge 19h11'34"18d54'
Adorjan,Charles R Dra 16h0'10"63d36'
Adornetto,Dana And 23h29'14"47d16'
Adornetto,Joseph John Per 2h49'0"45d6'
Adorno 21,Joseph Luis Cyg 21h6'1"30d50'
Adrean The Pathmaker Vul 21h3'17"20d36'
Adreon,An Peg 22h0'0"25d47'
Adrian Cet 2h16'41"9d25'
Adrian Peg 22h0'34"29d59'
Adrian Uma 11h20'35"70d40'
Adrian Peg 21h58'0"11d4'
Adrian Ari 3h21'18"28d47'
Adrian Cas 23h4'18"53d11'
Adrian "Age" Cam 4h13'26"60d31'
Adrian & China Lyr 19h17'15"41d31'
Adrian (AO) Ori 5h26'0"1d6'
Adrian Matthew Her 18h18'36"12d31'
Adrian the Dragon Dra 17h45'1"57d36'
Adrian Troy Star Cam 3h32'0"63d33'
Adrian's Cenicienta Uma 11h14'18"71d26'
Adrian's Starr Cep 0h5'29"69d23'
Adriana Cma 6h56'44"-15d32'
Adriana Aql 19h0'12"-6d21'
Adriana Cet 2h7'18"2d21'
Adriana Col 6h25'18"-38d22'
Adriana Nor 16h19'48"-52d39'
Adriana Crb 15h27'30"32d21"
Adriana Aql 19h30'14"-0d3'
Adriana 30 Equ 21h21'13"12d13'
Adriana Chris Cas 1h44'36"58d46'
AdrianMarina Eri 3h8'1"-2d36'
Adrianna Cyg 21h19'55"35d54'
Adrianne Cas 0h20'52"65d27'
Adrianne & William Eri 3h53'10"-5d59'
Adrianne Marie Peg 22h12'49"33d58'
Adriano e Lory Lac 22h14'30"55d22'
Adriano J Ori 4h42'56"15d4'
Adrianus Ori 5h39'31"12d22'
Adrien & Brenda Forever Eternal Eri 3h13'60"-16d20'
Adrien's Love Lyn 8h15'31"40d39'
Adriene Lyn 8h1'49"41d7'
Adrienne Lyr 19h14'34"38d44'
Adrienne And 23h44'57"40d31'
Adrienne Mon 6h19'56"8d54'
Adrienne Lyn 7h10'51"52d37'A
Adrienne Cyg 19h22'25"28d31'
Adrienne Christina Com 13h0'35"22d20'
Adrienne In The Heavens Lyr 19h22'37"35d6'
Adrienne Lisa Cam 9h39'0"82d6'
Adrienne Nichole Lyn 8h39'14"40d16'
Adrienne Sybil Cet 2h35'12"5d4'
Adrienne's Star And 23h0'0"49d47'
Adrimateau Boo 13h57'27"15d24'
Adshead,Happy Birthday Heidi Cyg 20h18'54"39d8'
Adsitt,Matt Lac 22h46'56"54d57'
Adu Aubad Ori 5h21'28"15d27'
Adubato,Frank Dra 16h26'39"69d25'
Adunia Sgr 18h53'57"-28d10'
Advantage Computer & Leasing Aur 7h23'41"40d22'
Adversi Riccardo Per 3h29'12"51d35'
Advey,Bob & Ruth Crb 16h16'30"27d16'
Advisa Uma 12h18'23"62d40'
Advocates Cyg 19h24'56"30d57'
Adwell Communications Cnv 13h1'43"32d44'
Adé Jacky et Carine Sgr 19h29'48"-34d30'
AEA-4 Eri 2h50'49"-16d12'
Aebel,Sandra J Cyg 21h19'17"38d6'
Aegir Pukis Jupiter Thor Ori 5h10'33"-8d8'
AEL And 2h3'0"39d23'
Aelvedon Aur 5h55'22"30d43'
Aeolus/JA 173 Sct 18h42'15"-5d26'
Aeppler,Susan Cas 0h39'1"73d56'
Aeppli,Ronald J Mon 7h43'13"-3d26'
Aere Boo 15h11'1"32d28'
Aerin Elyse And 0h59'22"36d47'
Aeronotex Aql 20h1'30"10d2'
Aerts,Gérard Gabriel Cyg 20h5'49"31d22'
Aeryn Lac 22h53'30"56d13'
Aeternam,Paulus Del 20h20'11"10d32'
Aetna Relo Team Uma 11h39'0"44d48'
Aetérnam,Tua Stella Crb 16h4'23"30d55'
Afanador,Jr,Alejandro Pagan' Cam 13h17'0"77d43'
Afelt Edoudard Cnv 13h58'17"46d34'
Affa,Brittany Leigh Lyn 7h4'47"50d56'
Affection Ari 2h15'17"14d14'
Affeld,Sylvia Sgr 19h50'8"40d51'
Affeldt,Susan Cas 23h23'53"61d49'
Affenita,Alfred Paul Lac 22h20'1"38d42'
Affiez Ori 6h3'28"9d14'
Affinity Pic 5h0'36"-45d40'
Afflelou,Joelle Cas 1h0'17"58d35'
Afford,Leslie Ori 5h56'59"19d26'
Affsa,Christopher Daniel Lac 22h13'58"49d49'
Affuso,Erica Noelle Lyn 8h2'33"38d55'
Afi Star Boo 14h51'48"39d20'
Afra,Fadia Lac 22h39'23"40d26'
Afromsky,Neil I Leo 11h3'23"-2d25'
Afshari,Gazelle Aqr 22h38'19"2d5'
Aftanas,Ashley Nicole Peg 23h31'49"20d26'
After Umi 13h11'0"75d21'
Afternoon Shift,The Ori 5h55'0"12d14'
Agaaa Ultimately Irresistible Lyr 18h56'29"41d53'
Agabriel,Claire Tri 1h59'32"34d45'
Agacan,Mona Y Tau 3h56'29"2d3'
Agadir-Einstein Hor 2h49'15"-49d6'
Agamata,Vincent Joseph Cet 2h15'22"9d12'
Agan,Becky And 0h55'54"34d34'
Agans,Jacqueline & Justin Cyg 19h58'29"40d28'
Agape Lac 22h25'48"40d25'
Agape Ser 15h1'38"0d28'
Agape Trail Sco 17h53'51"-30d18'
Agape' Cam 4h9'45"80d12'
Agapetos Diane Cam 5h34'52"68d20'
Agapimon,Ronbine Cyg 21h20'0"53d40'
Agapé Eri 3h23'0"-10d58'
Agard-Jones,Vanessa Cas 0h31'58"60d10'
Agare,Sandra Elizabeth Lyr 18h16'46"30d5'
Agassi,Andre Lyr 19h4'39"41d20'A
Agassi,Andre Del 20h15'42"14d4'
Agata Eri 2h45'35"-6d2'
Agata,Abby Fran And 0h10'34"35d13'
Agathe,Claire Crb 16h8'16"28d51'
Agazzi,Michelle Emigh Eri 3h52'54"-6d33'
Agba Ali Tari Daryoush Eri 3h34'54"-2d25'
Age & Devan's Light Mother & Son Peg 23h29'31"32d56'
Agee,Jr,John Masters Hya 8h31'32"-8d16'
Agee,Phillip L Aur 7h9'24"43d32'
Agee,Robert Vernon Lyn 8h11'54"48d35'
Agee,Taylor Baley Boo 14h36'0"48d60'
Ageliky Lib 15h13'42"-22d42'
Agenter,Mike Uma 11h14'58"51d52'
Ager,Brook Atkins Aql 19h26'19"14d26'
Agerer,Karl Michael Cnv 12h40'1"36d7'
Ages Gem 7h14'38"17d29'
Agey,Mike & Sandy Cyg 21h17'1"38d29'
AGF (12/4/43-6/6/96) Her 17h22'27"27d9'
Aggie & Menasha Crt 11h21'20"-11d40'
Agi,Elizabeth Charlene Vir 13h21'0"-6d54'
Agin,Ramona Sct 18h53'53"-4d40'
Agins,Teri And 1h12'58"40d40'
Aglaia e Simone Lup 15h13'1"-38d21'
Aglalia Uma 10h38'30"51d17'
Agli,Katie Elizabeth Cam 8h39'0"81d42'
Aglialoro,John Per 1h39'18"53d28'
Aglialoro,John Sgr 19h42'35"-41d2'
Agnel,Alexandra Cep 22h53'39"58d5'
Agnello Tri 1h32'57"30d16'
Agnello,Barbara A Cam 5h55'58"70d22'
Agnello,The Marjorie Peg 21h53'42"21d56'
Agner-The Auto Suture Shooting Star,Adi Lyr 18h26'30"46d24'
Agnes Cam 4h58'39"58d29'
Agnes Lac 22h28'57"40d3'
Agnes & Tad Cyg 21h10'34"35d51'
Agnes et Michel Lmi 10h49'25"33d21'
Agnes Mon Petit bon heur Cyg 21h34'53"40d55'
Agnes Sherry,The Lyr 18h39'0"38d58'
Agneta,Jeffrey P C Ori 6h15'1"-0d2'
Agneta,Stjärnan Cas 1h29'13"73d46'
Agnetta Del 20h17'21"15d54'
Agnew,Alexandra Ori 5h56'33"16d4'
Agnew,Mary Victoria And 2h23'58"46d51'
Agnew,Susan Ori 5h57'1"15d50'
Agnew,Thomas C Per 3h54'23"37d38'
Agnès Umi 17h17'31"75d38'
Agony Acres Uma 11h19'26"37d35'
Agopian,Raffi Cyg 19h52'32"47d59'
Agopton Leo 9h45'55"16d47'
Agopton 15 Leo 9h55'22"10d6'
Agor,Weston Harris Aql 20h2'30"0d31'
Agostine,John Vincent Boo 14h57'13"36d1'
Agostinelli Sparkling Angel,Carly Anne Psc 1h19'16"20d21'
Agostinelli,Jr #1 Star,Robert Lehman Psc 1h22'49"20d12'
Agostini,Andrea Boo 14h33'0"10d31'
Agostino,Jennifer Marie Lyn 7h34'39"42d13'
Agostino,Joseph Patrick Lyn 8h25'55"43d9'
Agovino,Toni Her 16h26'47"28d48'
Agrama,Frank Uma 11h44'17"38d41'
Agranat,June Cas 3h24'23"71d8'
Agrawal,Rita Ori 5h18'0"-9d21'
Agree,Amy Lynn And 1h21'12"40d41'
Agres Ant 10h31'39"-36d53'
Agresta,Nancè Uma 11h43'19"47d23'
Agricola,Keith Allen Boo 13h43'15"14d7'
Agruso,Susan Cep 22h59'0"70d3'
Agua Azul Ant 10h31'27"-38d43'
Aguayo,Andy "Bullets" Her 17h52'16"38d53'
Aguerrevere,Antonio Oph 18h41'21"8d56'
Agui Sex 10h47'15"-5d22'
Aguiar,Antonio D Sct 18h54'41"-4d28'
Aguilar,Briannah Monique Lyr 18h46'13"35d14'
Aguilar,Derrick Mark Hya 8h19'1"6d38'
Aguilar,Frank H Aql 20h18'37"7d54'
Aguilar,John Gabriel Aql 19h6'12"-5d48'
Aguilar,John Michael Aql 19h26'46"-8d4'
Aguilar,Nicholas & Theresa Mon 6h36'27"-5d55'
Aguilera,Betty Oph 18h4'47"13d24'
Aguilera,Sr,Joe Dra 18h41'18"70d17'
Aguirre,Alba Karina Cma 7h16'51"-15d4'
Aguirre,Jesse Lac 22h27'34"54d4'
Aguirre,Paul Dra 11h49'45"67d20'
Agurs,John Cmi 7h33'0"0d13'
Agustina Lac 22h32'38"38d30'
Agustina Cmi 7h33'37"1d46'
Agápomi Cam 6h12'28"68d8'
Ah!! Ind 21h9'5"-52d7'
Ahab Cet 2h32'1"1d10'
Ahalya Cas 3h0'56"57d31'
Ahamad Family Sgr 19h45'58"-45d20'
Aharon & Tara Sge 20h0'51"19d39'
Ahava Her 16h20'41"41d9'
Ahearn III,Daniel William Vul 20h14'28"22d52'
Ahearn,Don Aur 5h4'28"41d32'
Ahearn,Evelyn Cas 0h55'12"59d19'
Ahearn,James J Dra 16h44'55"60d14'
Ahearn,The Able Dr William T Her 17h37'44"27d53'
Ahern,Jessica M Cas 0h25'1"60d48'
Ahern,John F Aql 20h15'59"0d29'
Ahern,Nancy Millicent Cen 11h32'27"47d48'
Ahern,Patric John Gru 22h34'17"52d10'
Ahern,Stephanie Mon 7h56'3"-2d7'
Aherne,Michael Ori 6h14'1"0d20'
Ahkya-Pau-Kai-pi Uma 9h20'1"52d10'
Ahlalook Family,The Vul 20h15'49"25d28'
Ahlborn,Manfred Lyr 19h19'52"31d52'
Ahlbrandt,Julia Walbridge Cam 5h58'25"68d32'
Ahle,John E Aur 6h7'23"35d0'
AHLea Mon 7h39'33"-4d20'
Ahlenius,Timothy David Dra 14h17'24"64d38'
Ahlers,Irene W Uma 12h4'31"34d34'
Ahlf,William B Her 16h53'1"28d20'
Ahlgren,Denise M Sge 20h5'42"16d31'
Ahlostrom,Christopher Dra 16h9'29"66d33'
Ahlstrand,Kodi Bear- Holly Cmi 7h44'24"8d54'
Ahmad,Hisham Cam 3h28'0"61d24'
Ahmed,Abdul Nabi Ali Lyn 8h24'25"52d6'
Ahmed,Ayman Ali Her 18h29'10"12d20'
Ahmed,Ismat Her 18h26'1"12d7'
Ahmed,Seth Khaileh Oph 17h18'21"10d1'
Ahoj Na Sobotu-Hello For Saturday Aur 5h6'44"38d49'
Ahola,Patricia Ann Cet 0h52'59"-6d1'
Ahola,Stephen Cep 23h11'34"70d35'
Ahrebinejad,Nadine Niiku Cam 7h54'29"60d19'
Ahrebinejad,Sonia Marie And 1h59'53"40d46'
Ahren Tri 2h46'25"32d8'
Ahrendt,Delilah Anna Cam 5h46'44"60d50'
Ahrens,August-Wilhelm Ori 5h57'31"19d48'
Ahrens,Fritz Tau 5h35'46"26d14'
Ahrens,Gerlinde und Karl-Heinz And 1h41'44"41d12'
Ahrens,Joachim Leo 9h20'1"8d27'
Ahrens,Margaret Theresa Wren Vir 12h37'1"1d22'
Ahrens,Nathaniel Ser 15h53'16"19d15'
Ahrens,Patsy "Shiloh" Lyr 18h56'57"38d10'
Ahrens,Paul Ori 4h58'46"-1d39'
Ahrens,William James Vir 14h56'26"7d20'
Ahsweetow Cet 0h50'24"-1d6'
Ahtee,Asko Uma 12h12'39"55d48'
Ahuvit Tau 5h37'53"26d53'
Ahwoowoo Uma 13h0'46"58d26'
Aibel,Anthony Mark Vir 13h31'12"11d36'
Aiblinger,Hans & Edith Ser 17h35'50"-13d55'
Aichinger,Klaus 16-04-41 Lyr 19h17'43"31d30'
Aida Kasum ouc & Vuk Pasic Cyg 19h29'1"38d40'
Aidan Lmi 10h11'32"39d13'
Aidé & Aldo Uma 11h22'0"58d50'
Aiei Philoi Cep 22h5'29"62d30'
Aiello Family Star, The Perry John Eri 2h50'19"-5d55'
Aiello,Ashley Paige Cet 2h7'30"-6d22'
Aiello,Jerry Lac 22h27'30'50d15'
Aiello,Lindsay Erin Sct 18h19'34"-4d43'
Aiello,Luigi Mario Aur 5h18'22"43d56'
Aiello,Mark J Cep 21h54'46"55d2'
Aiello,Peter Dra 11h11'1"74d59'
Aiello,Peter Her 17h55'15"20d6'
Aigner Crt 11h39'31"-10d20'
Aigner,Georg K Aur 5h59'1"29d23'
Aigner,Max Cyg 20h24'46"30d58"
AIJA Lmi 11h4'36"31d28'
Aiken Cma 6h29'48"-26d1'
Aiken,Amy Sgr 18h59'6"-20d25'
Aiken,Dr Simon Piers Cep 22h35'51"68d21'

Aiken,Jonathan Neal
Cnv 12h16'34"41d60'
Aiken,Tim K
Uma 9h57'0"44d46'
Aikens
Aur 6h8'42"38d9'
Aikens White Light
Lyr 19h26'54"38d41'
Aikin,Glenn & Jeannie
Dra 16h34'40"62d55'
Aikin,Jr,Arthur L
Equ 21h6'37"2d49'
Aikins,Russ
Boo 14h35'11"19d29'
Aikman,Troy
Cnv 12h19'0"41d29'
Ailee's Aspiration
Lyn 7h40'56"50d33'
Aileen
Cyg 19h51'18"38d41'
Aileen
Lyr 19h0'0"38d39'
Aileen
Mon 6h54'22"11d8'
Aileen's Eye
Lyn 6h36'54"58d43'
Ailes,Roger E
Dra 11h22'59"73d23'
Ailey,Marcia L
Her 18h50'1"14d28'B
Ailleres,Vincent
Ser 15h18'0"8d19'
Ailsa Faye
Aql 19h56'27"10d4'
Ailya Rose
Boo 14h12'26"38d24'
Aim,Gérard
Sgr 19h30'49"-30d4'
Aimable
Cep 22h53'34"57d46'
Aime
And 0h35'0"40d55'
Aimee
Cnc 8h50'51"31d59'
Aimee
Sco 17h50'0"-38d40'
Aimee
Ori 6h9'57"7d0'
Aimee & Jamie
Lyr 18h56'0"34d43'
Aimee Louise
Lyr 19h1'25"28d18'
Aimee Nicole
And 23h2'0"50d12'
Aimee Renae
Pho 23h48'35"48d57'
Aimee Rose
Cas 1h24'36"68d8'
Aimee's Brightness
Psc 1h0'0"31d51'
Aimee's Star
Lyr 19h16'0"42d27'
Aimer
Cam 3h27'1"53d1'
Aimo,Pikku
Umi 14h34'11"65d30'
Ainasoja,Pirjo Margareta
Aur 7h7'44"40d54'
Aine Babes
Leo 10h43'14"12d3'
Aino (Luik)
Mon 6h20'38"-0d20'
Ainslee,Diaan
Lyn 8h15'0"40d12'
Ainsley's Little Star
Cam 12h54'0"82d7'
Ainsley,Astra
Lyr 19h21'58"31d15'
Ainsworth,Donald
Aur 5h30'43"38d56'
Ainsworth,Donna Marie
Cyg 19h27'0"34d19'
Ainsworth,Jamie Alexander
Umi 17h12'54"86d10'
Ainsworth,Leslie & Beatrice
Cyg 19h33'58"35d44'
Aioffi,Andrea
Lyn 7h58'55"40d52'
Air Jennifer
Aql 20h11'12"1d45'
Airborne Stach
Uma 9h15'0"52d53'
Aird,Caroline
Cyg 20h48'31"30d43'B
Aird,Kenneth
Dra 13h31'58"67d44'
Airhart,Josie
Cas 0h50'25"62d14'
Airisto,Marjuana
Lyn 8h1'53"50d48'
Aisenberg,Bernard
Lmi 10h49'47"30d52'
Aisha Sian - Michelle James
Crb 16h15'18"32d3'
Aisle of You,Marvin Love
Jaenelle
Her 17h12'58"44d4'
Aislestand Bayou II
Cma 6h26'13"-15d2'
Aislingeach
Uma 9h52'0"59d49'
Aiss,Jeremy Alexander
Her 18h3'17"41d14'
Aita,Imad Subhi Mahmood
Per 1h48'41"56d52'
Aitch Erp
Per 3h10'57"41d46'
Aithammov,Michel
Per 3h58'43"39d18'
Aitken,"Darling" William
James
Dra 11h38'57"70d34'
Aitken,Fiona Claire
Cas 0h24'1"63d27'
Aitken,Hillary
And 2h26'23"38d58'
Aitken,Naomi
Cas 0h56'29"67d17'
Aitken,Rory
Aql 20h7'19"0d23'
Aitken,Sr,A Richard
Aur 6h58'46"36d51'
Aitken,William "PaPa"
Dra 11h38'35"72d19'
Aix,Courbisie Franáoise
Per 3h32'51"35d51'
Aizstrauts,Irene
Cas 0h6'13"60d18'
AJ
Aur 4h34'1"31d51'
AJ's Wish
Vul 19h21'1"26d27'
Aja
Peg 23h24'1"28d20'
Aja Khan
Ori 5h59'23"10d8'
Ajaap
Hya 8h52'40"-1d1'
Ajanwachuku's Guardian
Angel,Dr
Oph 17h56'0"11d47'
Ajaz
Uma 8h45'58"60d27'
Ajeeth
Her 17h8'39"47d36'
Ajjaraca
Aur 6h2'0"30d1'
Ajolo,Wilson B "Jo Jo"
Vul 20h2'14"22d35'
AJS-4 (Art Jos Sonnick IV)
Her 18h10'1"38d0'
AJT-7766
Crb 16h8'58"30d35'
Ak-Akr
Cam 14h15'1"81d32'
Akachin,Keiko
Vir 14h39'0"5d16'
Akad,Deniz
Per 3h15'13"56d52'
Akane
Sgr 18h57'54"-24d43'
Akasha,Alexandra
Aql 20h7'27"1d9'
Akasiadis,Georgia "Bunny"
Uma 9h38'27"51d45'
Akaslompolo,Pentti
Uma 15h38'51"68d59'
Akat,Joyce
Cyg 21h28'49"41d7'
Akbarian,Katayun
Cyg 20h11'23"37d59'
Akely,Philip
Cnc 8h2'0"8d23'
Akenhead,Eric J
Cep 3h10'11"77d54'
Akerlund,Dale
Aur 6h32'55"38d59'
Akerman,Hilbert
Umi 17h21'12"75d44'
Akerman,Lisa
And 1h38'41"38d36'
Akers
Her 17h11'55"27d53'
Akers "My Love",Mary Ielene
And 23h47'0"38d19'
Akers,Carolyn G
Cet 0h36'0"-3d31'
Akers,Kay
Cas 3h1'11"58d45'
Akers,Linda Sue
And 2h21'17"40d31'
Akers,Megan M
Cas 2h1'34"61d42'
Akers,Patsy B
Cet 3h10'20"1d27'
Akers,Rhonda
Cnc 8h53'41"17d48'
Akers,Yvonne
Lmi 10h40'27"31d39'
Akerstromic
Sex 9h41'28"-8d21'
Akey,Clare Duane
Lmi 10h46'22"31d8'
Akey,Cynthia A
Tri 2h46'37"31d21'
Akey,Olivia Mary Greenlee
Tri 2h14'11"32d41'
Akheela,M K
Aql 18h57'25"-6d18'
Akhtar,Tufail
Her 18h7'26"14d44'
Akhtar,Zohaib
Her 18h6'43"18d60'
Aki Shuto
Cam 8h15'43"74d54'
Aki Star,The
Hya 9h18'42"-14d22'
Akie
Aqr 22h5'17"-5d45'
Akiike,Jun
Boo 13h58'30"14d9'
Akin,Danielle Ann
Boo 15h4'28"28d26'
Akin,William Josiah
Boo 14h10'32"32d4'
Akinadia
Ori 5h40'0"1d13'
Akins Family Star
Del 21h2'57"12d59'
Akins Memory Star,The
Timmy
Vul 19h2'25"25d4'
Akins,Amy D
Mon 6h27'44"11d20'
Akins,Dale E
Cet 2h36'0"3d13'B
Akins,Karyn & Prince
Aql 19h59'21"-6d12'
Akins,Mike
Per 1h27'10"53d8'
Akins,Sally Brown
Cet 2h36'0"3d13'A
Akins,Vera D
Eri 3h31'46"-3d29'
Akira
Lep 4h59'11"-11d49'
Akira & Rumi
Cap 20h55'45"-19d20'
Akkasha
Peg 22h25'38"11d59'A
Aknin,Gabriel
And 23h22'0"50d26'
Ako's Star
Sgr 19h37'23"-38d55'
Akradina
Cep 2h12'0"78d28'
Akre,Barbara
Cas 0h1'55"54d27'
Akre,David James
Cep 22h29'0"60d47'
Akre,John Christian
Lac 22h16'41"46d7'
Akridge,Harold
Ser 15h35'1"1d35'
Akstull,Michael
Her 16h29'12"39d7'
Aktar,Shahid
Crb 16h12'0"37d42'
Aktiva Finanzplanung
25/05/1982
Tau 5h53'56"23d37'
Akuna,Ginger & J T
Satterwhite
Del 20h28'47"15d38"AB
Akusizu
Ser 15h26'59"1d41'
Al
Lac 21h55'0"40d36'
Al
Tau 5h0'14"16d7'
Al
Aur 5h0'29"44d50'
Al
Cet 3h16'24"1d30'
Al
Her 16h57'57"32d20'
AL
Boo 14h17'22"50d0'
Al & Edwina's Lucky Star
Crb 16h6'47"37d41'
Al Abjar
Oph 18h7'38"8d51'
Al Jr
Hya 9h30'31"-13d17'A
Al Lee Bubbah
Her 16h59'21"40d6'
Al Love
Lac 22h42'18"53d49'
Al Mandoub
Gem 7h6'22"28d42'
Al mio Chicco
Boo 14h33'0"18d22'
Al'Busaadi,Zaina
Lyr 18h26'36"42d28'
Al's Rising Star
Her 15h49'17"45d59'
Al's Star
Boo 14h50'57"51d15'
Al's Star
Cep 22h24'58"58d52'
Al's Stella
Cet 1h1'1"0d56'
Al's Weaver Lantern
Dra 15h49'12"62d37'
Al,Forever & A Day
Sge 19h5'47"19d42'
Al-Essa,Tahanni Meshart
Cep 2h4'3"81d14'B
Al-Faisal,Prince Mischal bin
Bandar
Peg 22h38'39"27d33'
Al-Fresko
Cam 3h55'60"62d13'
Al-Khalaf,Bader
Ori 6h5'41"5d40'
Al-Saoud,Princess Hala M
Cas 0h9'39"61d12'
Al-Schamma,Elizabeth Chow
Peg 23h7'41"10d57'
Al-Shaer,Alia Marie Mustafa
Peg 23h29'0"8d60'
Al-Shaer,Ebrahim Mustafa
Sge 19h53'17"16d33'
Al-Shaer,Mustafa Ebrahim
Cet 2h51'60"6d43'
AL/DI
Cyg 20h35'51"42d32'
AL/RO
Ori 5h57'18"1d39'
Aladdin-MTKG
Oph 18h42'24"10d34'
Alaemas Franáoise Jeanine
Ser 15h38'49"9d2'
Alafita,Sada Jean
Psc 1h26'59"20d1'
Alai T'Kint De Roodenbeke
Cyg 20h54'54"50d4'
Alaikia Marielle
And 1h28'35"34d13'
Alaimo,Danny
Her 18h18'24"12d9'
Alain
Peg 21h51'54"19d24'A
Alain et Domi
Cam 7h32'0"68d37'
Alain Violet
Lmi 10h27'14"32d11'
Alain,Marie-Anouk
Lyn 8h35'12"44d36'
Alaina
Her 15h51'28"50d59'
Alaina Marie
Lyn 8h53'12"33d57'
Alaina's Wishes
Uma 12h3'22"57d19'
Alakoski "Nakke",Kari
Cyg 19h27'13"33d17'
Alambeigi,Hadi
Lib 14h21'55"-23d39'
Alamo Tamale Star,The
Cet 2h36'36"-11d13'
Alamzao,Mitra
Cyg 20h22'1"30d24'
Alan
Psc 23h24'38"0d38'
Alan
Ori 6h8'12"9d4'
Alan
Cmi 7h33'40"10d35'
Alan
Her 16h16'47"10d57'
Alan & Connie "Forever"
Cyg 21h6'11"31d43'
Alan & Jenni
Aql 19h5'35"2d49'
Alan & Jessica
Umi 16h12'26"79d58'
Alan & Marcia
Cyg 21h34'23"50d16'AB
Alan & Marla
Cep 3h13'53"77d33'AB
Alan & Maureen
Lyr 18h37'1"41d42'
Alan & Molly
Cyg 20h54'41"31d34'
Alan & Morna's Lucky Star
Cyg 20h24'18"39d21'
Alan & Susan
Cnv 13h51'28"38d43'
Alan B (VIP)
Uma 8h59'0"61d42'
Alan Frank
Aql 20h12'33"10d7'
Alan June
Umi 15h34'1"68d53'
Alan Star
Ser 15h55'1"2d12'
Alan Stephen
Uma 9h54'26"54d13'
Alan's "35th"
Her 16h1'54"41d45'
Alan's Community
Her 17h13'0"27d11'
Alan's Dunn Deal
Eri 4h35'0"-12d19'
Alan's Golden Star
Aql 19h9'15"13d51'
Alan-Blanche
Cyg 21h2'40"36d21'
Alan-Tony
Her 18h6'1"14d41'
Alana
Tri 1h48'31"27d12'
Alana
Cas 0h57'21"75d49'
Alana & James
Cyg 20h56'24"38d47'
Alana Jean
Lyn 9h38'1"40d22'
Alana Lorann
And 0h16'33"33d9'
Alana-21
Uma 10h54'42"38d58'
AlanDonnaJoe
Her 17h2'35"45d2'
Alandrea
Peg 22h59'20"32d41'
Alani
Cyg 21h35'21"41d30'
Alanis,Gabriel Artemio
Del 20h18'32"10d39'
Alanki
Lyr 18h22'37"38d32'
Alanna
Lyn 8h26'22"47d34'
Alanna Eve
And 1h15'34"38d4'
Alannah
Peg 23h42'56"31d52'
Alansgem
Her 16h10'0"40d27'
Alanviv
Cyg 19h26'42"35d9'
Alape,Phil
Aur 4h59'36"48d59'
Alar,Jure
Del 20h18'41"10d50'
Alarcon,Mark
Aql 19h5'39"1d20'
Alarie,Brendan Slade
Cma 6h25'25"-18d50'
Alarie,Maxwell
Uma 11h40'60"42d55'
Alarie,Megan Elizabeth
Mon 6h41'28"1d41'
Alario,MD,Frank Charles
Cep 21h35'44"70d24'
Alarowe,Franáois
Dra 20h9'58"70d51'
Alasdair,Rory
Cyg 19h33'16"39d19'
Alaskan Romance
Umi 16h53'31"78d50'
Alastair
Cet 1h19'47"-12d30'
Alati,Marc Alessandro
Lac 22h18'33"49d33'
Alaux,Bornard
Ori 5h5'53"11d12'
Alawi,Ashwag Al
Oph 18h16'59"11d53'
Alayna Elizabeth Ann
Lyn 7h56'33"58d56'
Alazraki,Roberto M
Gem 6h47'13"17d24'
ALB
Mon 8h7'48"-8d5'
Alb-Brown,Michele
Lyr 19h22'0"41d41'
Alba
Ori 6h10'29"-1d7'
Alba,Rose et Albert Emery
Cas 22h57'36"54d23'
Albagenesis
Tri 1h59'25"28d15'
Albanese,Eda
Cas 2h35'49"70d14'
Albanese,Georgina
Uma 11h4'37"49d21'
Albanese,Georgina
Cyg 21h37'54"42d12'
Albanese,Maria E
And 2h3'55"42d28'
Albanese,Michael James
Aql 20h1'57"8d55'
Albano,Joan E
Cas 0h51'60"64d0'
Albano,John
Del 20h31'19"20d35'
Albano,Leona M
Lyn 6h29'27"54d31'
Albany,Janis
Equ 21h2'60"11d12'
Albarano,Jr,Joseph
Cyg 19h32'29"33d28'
Albarella,Rosemary & Gina
Marie
Psc 0h56'18"31d22'
Albarella,Tony & Cindy
Cep 22h46'48"70d57'
Albarron,Donna
And 23h18'49"51d51'
Albatross
Cyg 19h27'53"34d27'
Albaum,Karen
Tri 2h16'55"28d19'
Albee,Kayla Lee
Cas 1h45'51"54d16'
Albee,Tracy
Aur 5h29'25"37d57'
Albensi,Alexis Renee
Lyn 8h7'45"46d20'
Alberghetti
Her 16h53'31"28d11'
Albergo,Pat
Dra 20h0'33"63d39'
Alberico,Anthony Joseph
Cnc 9h3'33"17d37'
Alberina
Cam 5h25'36"68d37'
Albers,Clark-Alyssa- Madelyn
And 23h48'41"46d32'
Albers,Constant
Ori 5h55'32"15d14'
Albers,Danielle Leigh
Cas 0h23'44"60d49'
Albers,Dianne
Cam 4h56'42"60d40'
Albers,Jr,Ron
Boo 14h46'53"33d43'
Albers,Micayla Lee
And 2h10'57"42d25'
Albersheim,Peter
Sex 10h34'1"4d51'
Alberson,Daisy Anne
Lyn 7h57'16"45d25'
Albert
Vir 13h58'20"-7d45'
Albert
Ori 5h59'21"12d7'
Albert & Donna M
Cam 4h24'54"61d10'
Albert & Janet
Lyn 7h52'22"33d45'
Albert & Luciana
Crt 11h16'43"-20d42'
Albert & Melanie
Cyg 21h38'47"40d51'
Albert Bert
Lyr 19h21'1"30d12'
Albert Fortell Duell Team
Uma 11h53'24"50d7'
Albert,Annie
Dra 15h45'22"60d48'
Albert,Bruce Evan
Dra 16h21'59"64d33'
Albert,Carolyn Lois
Lyn 8h14'57"50d9'
Albert,Deacon Henry
Dra 16h49'25"68d17'
Albert,E Michael
Cap 20h34'45"08d53'
Albert,Gottfried
Vir 13h27'43"-12d22'
Albert,John Paul
Aur 5h2'23"47d25'
Albert,Jonathon Robert
Her 17h0'18"24d36'
Albert,Josian
Lib 15h41'11"-21d11'
Albert,Kristopher Jamon
Leo 10h36'39"12d36'
Albert,Milton L
Aql 19h56'34"1d21'
Albert,Renee
Cam 3h49'18"58d58'
Albert,Roland Thomas
Ser 15h13'31"20d54'
Albert,Rose-Marie
Cam 10h58'56"80d29
Albert,Samuel Powers
Boo 14h37'40"7d52'
Albert,Silvestre Estaban
Ori 5h5'39"8d36'
Alberta
Pup 7h56'23"-21d21'
Alberta & Donald Anniversary
Star,The
Boo 15h6'38"51d51'
Alberta T
Aql 19h40'55"14d7'
Albertazzi,Sue M
And 2h21'54"46d6'
Albertelli,Libera
Col 6h34'30"-35d59'
Alberti,Denis
Per 3h2'12"47d20'
Alberti,Lisa Marie
And 0h49'49"39d18'
Alberto
Pyx 8h40'58"-29d20'
Alberto
Del 20h14'22"11d60'
Alberto
Ant 10h45'8"-37d15'
Alberto B
Ind 21h2'35"-55d25'
Alberto's Star
Ori 5h55'34"16d1'
Alberto,Abiel
Aql 19h29'36"-0d2'
Alberts,Chari Lynn
Peg 22h0'22"11d7'
Alberts,David & Heather
Aqr 22h30'35"-17d34'
Alberts,Mercedes
Eri 3h23'38"-12d49'
Albertsen,Mary Ellen Nielsen
Vul 19h43'0"23d26'
Albertson,Bill "Stargazer"
Aql 20h30'51"-0d27'
Albertson,Greg Richard
Psc 0h1'28"0d18'
Albertson,Maynard Glenn
Ori 5h46'17"-6d39'
Albertson,Sue
Cas 0h42'41"61d10'
Albertson,The King Star
James
Aur 6h4'53"31d20'
Albertus,Fredrick
Boo 13h58'29"19d41'
Albethrita
Boo 15h22'30"34d11'
Albetski,Bernard
Aur 5h11'49"44d28'
Albetta,Claudia
Cas 0h33'59"68d31'
Albie
Cru 12h36'24"-57d2'
Albin "Forever Jeff", Jeffrey
Jon
Cnv 12h54'34"51d55'
Albina
Cyg 21h58'0"48d54'
Albini,Anthony & Michelle
Aql 20h3'45"0d7'
Albini,Dina Marie
Mon 6h36'59"-0d26'
Albini,F-F Helg-A Piva
And 0h56'31"34d17'
Albini,Flora & Mario
Crb 15h34'47"27d40'
Albion
Vul 19h21'22"25d7'
Albion
Cnv 12h34'11"38d1'
Albizo,Noah Casey
Dra 20h15'0"63d31'
Albo
Boo 14h45'27"26d13'
Albon,Scott Joseph
Her 16h15'26"25d22'
Alborada From Jazz
Psa 22h2'11"-26d51'
Alboreto,Michele
Peg 21h54'43"31d49'
Albouy,Marc
Oph 17h57'48"10d39'
Albrech,Klaus
Cyg 20h42'3"46d3'
Albrecht & Eva
Uma 11h23'33"38d56'
Albrecht,Alexander Arthur
Lac 22h23'36"48d51'
Albrecht,Cliff
Cmi 7h28'52"8d14'
Albrecht,Colin Edward
Per 2h52'54"40d24'
Albrecht,Corinne Marie
Lyn 8h29'0"51d10'
Albrecht,John
Mon 7h46'54"-6d47'
Albrecht,Mark & Teresa
Crb 16h7'0"33d19'
Albrecht,Meredith Ann
Lyn 8h5'39"35d22'
Albrecht,Paul et Eliane
Ori 5h56'0"20d43'
Albrecht,Rosi
Ori 5h20'42"1d4'
Albrecht,Sylvia
Mon 6h50'1"10d40'
Albrecht,William Carl
Per 2h57'58"43d22'
Albrecht-Meister, Simone
Aqr 22h16'31"-1d13'
Albrent,Mari "Melfy"
Peg 0h5'31"21d40'
Albrent,The Star Of Hope By
Jessica
Peg 22h16'1"33d40'
Albright,Alvenia Rhea
Uma 11h36'0"40d46'
Albright,John David
Vir 11h39'28"2d2'
Albright,Michael William
Oph 18h17'31"12d14'
Albright,Tracey A
Cam 9h34'59"84d25'
Albright-Pelon,Lonna Rae
Cas 0h56'35"54d32'
Albring-Brosda,Hella
Cep 21h36'20"65d22'
Alby
Lac 22h39'13"54d22'
Alhyn,Jr,Thomas Everett
Aur 5h9'48"38d27'
Alcala,Kimberley A
And 0h52'19"41d13'
Alcantara,Jesse
Eri 2h44'33"-2d55'
Alcaraz,Roberto
Leo 11h6'47"-2d19'
Alcaraz,Sergio Antonio
Lac 22h27'1"50d4'
Alcaro,Joan
Lyn 8h29'53"58d40'
Alcaro,Nicklaus
Per 2h36'31"57d12'
Alcina,Marlee
And 1h15'32"34d36'

Alcock,Gloria Hunter
Cas 23h28'53"54d9'
Alcock,Mary Patricia
Cyg 20h23'57"38d16'
Alcone,Matthew R
Peg 22h45'59"29d2'
Alcorn 16031948, Glynda Kathleen
Lyr 19h19'21"38d48'
Alcorn,Bryan Michael
Aur 6h33'43"31d41'
Alcorn,James Munrex
Aql 20h6'0"0d40'
Alcover,Julia
Aqr 20h57'33"032'
Alda
Aql 19h55'42"12d41'
Alda M M G
Lyr 19h2'45"28d20'
Aldag,Jennifer Elizabeth
Uma 11h31'16"37d44'
Aldag,John Maxim
Lib 15h19'20"-28d29'
Alday,Ambrose L M
Aql 20h0'45"0d24'
Alday,Beatriz
Aql 18h57'26"15d18'
Aldaya,Jolen & Alaina
Ori 4h41'1"14d20'
Alde,Frank
Sgr 19h2'13"-29d17'
Aldebaran
Vel 9h12'53"-40d51'
Alden III, John Thomas
Her 17h1'0"21d57'
Alden,Ashley Louise
Tri 2h17'12"32d15'
Alden,David M
Cet 2h14'16"4d52'
Alden,Enid George
Boo 14h55'31"53d24'
Alden,Jane Malick
Cyg 21h56'38"44d29'
Alden,Johanna
Lac 22h43'1"56d5'
Alden,Marc
Aur 5h24'36"40d40'
Alder,David A
Cep 2h12'26"80d30'
Alder,Jesse D
Gem 6h41'57"30d52'AB
Alder,John Alan
Gem 7h6'35"21d27'
Alder,John Benjamin
Lib 14h43'20"-20d57'AB
Alder,Timothy J
Gem 7h29'50"22d59'AB
Aldercine
Cyg 20h22'37"39d22'
Alderman,Alexandra Rose
Cyg 19h41'1"40d58'
Alderman,Dr Tracy
Lyn 7h7'53"44d9'
Alderman,John
Per 4h25'45"51d1'
Alderman,John Allen
Vir 14h56'0"7d26'
Alderman,Jr,Edward Marshall
Cep 21h10'23"58d44'
Alderman,Judi
Lyn 7h24'11"44d47'
Alderman,Lisa
Uma 8h41'45"71d35'
Alderman,Misty Dawn
Del 20h25'22"20d9'
Alderman,Sherri Robin
Cnc 8h14'31"30d8'
Aldermeshian,Laura
Lyr 18h58'20"46d54'
Aldermeshian,Peter
Aur 6h7'32"31d45'
Alderney,Nancy
Cas 2h22'44"61d36'
Aldi
Umi 16h2'36"79d31'

Aldieri,Anthony
Tri 1h55'25"26d26'
Aldina
Uma 11h58'60"33d13'
Aldinger,Joseph H
Peg 0h0'35"31d18'
Aldo
Sgr 19h31'47"-30d14'
Aldo
Lac 22h21'40"54d13'
Aldo e Elena
Pyx 8h51'40"-20d14'
Aldous,Alastair Robbie
Ori 6h9'45"3d39'
Aldous,Alicia Grace
Ori 6h9'43"3d36'
Aldred,Helen
And 0h43'33"41d9'
Aldrich,Andrea Joy
Cam 9h6'58"78d60'
Aldrich,Christopher
Aur 6h24'1"37d6'
Aldrich,Ken Howard
Boo 14h30'40"47d59'
Aldrich,Larry
Cam 6h20'10"82d53'
Aldrich,Lucile W
Cas 1h11'0"61d14'
Aldrich,Lynne Merrill
Lyr 18h59'12"28d22'
Aldrich,Patricia Jane Fitzgerald
Tri 2h1'1"32d28'
Aldrich,Perpetum
Sge 19h59'56"16d36'
Aldridge
Lyr 19h24'47"40d16'
Aldridge Lucky Star
Cyg 20h56'17"30d31'
Aldridge,Cleda
Mon 7h5'27"-0d20'
Aldridge,Deborah Sharon
Peg 21h58'18"2d49'
Aldridge,E W
Uma 11h50'45"30d25'
Aldridge,Hallie
Aql 19h6'53"-0d53'
Aldridge,Jenny Lea
Vul 20h57'59"28d47'
Aldridge,Trey D
Dra 15h52'1"60d54'
Aldwinckle,Stephen J
Her 16h58'26"50d36'
Ale
Sco 17h30'29"-30d14'
Ale
Lyn 7h52'34"41d29'
Ale
Cam 8h2'12"81d47'
Ale Solitario
Cam 6h39'12"68d40'
Alea
Cet 1h17'30"-16d4'A
Alebon,Christopher
Her 18h30'22"20d15'
Alebon,Colin
Ori 6h5'0"0d27'
Alebon,Ian
Peg 22h0'18"28d19'
Alebon,Kevin
Per 2h57'52"50d16'
Alec A
Uma 12h6'59"45d49'
Alec's Star
Cep 21h52'12"58d29'
Alefs,Jacob
Tau 4h45'50"18d57'
Alejandro's Deseo
Ori 5h54'13"6d13'
Alejandro,Anthony C
Peg 21h59'49"28d20'
Alejandro-Federico- Lara-Oliver
Aql 19h59'39"-1d28'

Alek The Great
Per 1h34'16"53d28'
Aleka
Aqr 20h54'39"0d45'
Aleksander
Aur 6h58'54"38d50'
Aleksandr Matt
Dra 15h38'26"58d9'
Aleksandra
And 23h41'18"41d6'
Aleli Jo
Crb 15h59'16"38d44'
Alelyunas,Carl Harry
Her 18h1'41"28d50'
Alelyunas,Stephanie Lee
Del 20h15'17"14d1'
Aleman,Alejandro
Hya 9h18'36"1d52'
Alencewicz,Helen
Lyr 18h43'16"33d21'
Alene,Esq
Mon 6h28'28"-5d45'
Alenooshmarkusstellusa morus
Cam 4h19'0"67d48'
Aleong,Annette LiFun
Lyr 18h35'45"41d41'
Alera El An Ra
Cyg 20h4'12"30d49'
Alert,Jason
Her 16h58'21"18d48'
Ales,Frank
Her 17h15'1"22d15'
Aleshin,Carol Ann
Aql 19h7'25"0d1'
Aleshire,Dorothy L
Lyr 18h40'40"45d42'
Aleshire,Leonard
Aur 5h7'1"40d44'
Aleshire,Victoria Lynn
Peg 23h32'48"23d46'
Alesi,Michelle Nicole
And 1h37'29"48d52'
Alesia,Sharon M
Tau 4h22'32"28d33'
Alessadria Jean
Ori 6h1'11"20d26'
Alessandra
Mon 7h50'51"-4d5'
Alessandra
And 0h46'35"45d51'
Alessandra
Oph 17h22'50"-20d23'
Alessandra
Cyg 19h24'38"44d47'
Alessandra
Ant 10h45'17"-35d51'
Alessandra
Cae 4h47'28"-31d54'
Alessandra
Lep 5h22'1"-24d27'
Alessandra
Crb 15h27'43"30d14'
Alessandra
Lac 22h10'28"51d10'
Alessandra
Lep 4h58'50"-18d51'
Alessandra
Per 3h39'12"51d35'
Alessandra
Del 20h14'37"14d58'
Alessandra
Cnv 13h38'12"32d9'
Alessandra '59
Cas 2h57'21"60d32'
Alessandra Barbi
Cep 23h31'32"64d14'
Alessandra Cera
Hor 2h54'44"-49d31'
Alessandra e Marcello
Cae 4h52'1"-33d24'
Alessandra P '60
Ant 10h37'49"-34d54'
Alessandra-Maria
Pyx 8h41'24"-24d20'

Alessandria Nicole
Ori 6h1'0"20d47'
Alessandro
Hya 9h39'0"5d22'
Alessandro
Dra 15h23'31"53d38"
Alessandro
Her 16h44'43"48d25'
Alessandro
Lyn 7h28'42"40d8'
Alessandro
Psc 23h25'25"6d35'
Alessandro,Elizabeth & Ralph
Uma 10h54'56"37d19'
Alessandro,Michael
Boo 14h5'32"25d33'
Alessi,Gasper
Lyr 18h59'1"46d19'
Alessi,Lillian Mary
Mon 7h38'10"-6d43'
Alessi,Mark
Cmi 7h55'32"5d28'
Alessia
Pup 7h56'20"-28d59'
Alessia
Her 16h45'0"18d55'
Alessia
Ori 5h53'0"0d47'
Alessia Ale
Per 3h36'58"51d13'
Alessia L
Hor 3h17'57"-46d22'
Alesso,Cassidy Faye
And 1h44'50"36d2'
Aleta
Lyn 9h2'12"35d43'
Aleta
Cyg 21h3'17"41d4'
Aleta Rae
Lyn 8h21'23"47d29'
Aletha
Crb 16h8'1"28d11'
Alethea
Del 20h15'46"12d55'
Aletia
And 0h0'19"47d4'
Alevras,Aris
Aql 20h1'1"4d26'
Alevropoulos,Tanya Marie
Cas 2h37'0"65d35'
Alex
Lyn 8h30'37"40d38'
Alex
Cmi 7h54'46"1d50'
Alex
Umi 16h59'1"75d54'
Alex
Mon 8h0'48"-6d2'
Alex "Mickey Mouse", Michael Ryan
Vir 11h44'22"0d45'
Alex & Boris
Uma 13h40'1"50d33'
Alex & Juliette
Cam 5h35'11"65d4'
Alex & Laura's Eternal Love Star
Eri 3h21'29"-6d51'
Alex & Oliver
Uma 10h37'22"41d18'
Alex & Roby
Pup 8h8'26"-29d16'
Alex & Steve-Always & Forever
Cyg 20h48'24"37d42'
Alex & Vicky
Lyr 18h30'16"31d6'
Alex 1 (uno)
Lyr 18h38'57"29d35'
Alex Daniel
Her 18h4'17"38d27'
Alex et Musetto
Sex 10h12'40"-2d48'
Alex from Blur
Lyr 18h24'1"37d39'

Alex Marck
Cet 3h7'47"0d58'
Alex Marie
Cam 4h3'17"58d18'
Alex mon petit coeur
Tri 2h0'8"33d39'
Alex Nancy
Her 17h3'48"40d35'
Alex Ryan
Dra 11h45'53"71d39'
Alex Starlight
Uma 11h19'11"38d39'
Alex The Great
Cap 21h16'25"-22d17'
Alex's Star
Cet 0h50'50"-2d21'
Alex's Sunshine
Ori 6h3'41"8d16'
Alex,Max
Ser 15h14'1"20d34'
Alex,Mike
Sct 18h56'37"-6d33'
Alex,Peter
Lac 22h3'16"51d56'
Alex,Richard
Aur 7h10'22"36d39'
Alex-Marco-Münchow
Cmi 7h20'50"4d22'
Alex/Carole Ann
Lyr 18h56'44"42d34'
Alex/Pauline
Cyg 21h32'58"38d55'
Alexa
Cyg 20h59'31"40d31'
Alexa
Oph 18h42'1"7d28'
Alexa
Leo 11h2'54"0d6'
Alexa
Gem 6h35'13"20d28'
Alexa
Mon 6h34'1"0d11'
Alexa
And 23h29'43"40d57'
Alexa
And 23h36'1"46d24'
Alexa 1112
Cep 2h19'57"78d23'
Alexa Adele
Lyn 6h27'17"58d23'
Alexa III,Andrew John
Aql 20h3'11"0d37'
Alexa Lauren
Lyn 7h57'47"38d18'
Alexa Leigh
Peg 22h4'57"27d54'
Alexa Margaret Daddy's Shining Star
Eri 3h13'34"-15d4'
Alexa Nicole
Uma 9h17'35"68d32'
Alexa Star,The
And 23h49'11"41d7'
Alexa-Rái
Gem 6h59'47"13d58'
Alexander
Cet 3h5'43"4d48'
Alexander
Oph 17h4'9"-23d16'
Alexander
Boo 14h11'25"30d3'
Alexander
Aur 7h8'1"38d2'
Alexander
And 23h3'33"50d5'
Alexander
Mon 6h55'18"11d18'
Alexander
Dra 16h55'36"68d4'
Alexander
Umi 13h9'11"76d29'
Alexander
Ori 4h56'28"-0d49'
Alexander
Aur 5h0'1"37d56'

Alexander
Tri 1h59'39"27d48'
Alexander
Boo 14h56'17"48d22'
Alexander
Dra 19h49'15"60d20'
Alexander
Uma 11h21'18"40d25'
Alexander
Cnc 9h17'51"31d21'
Alexander
Aql 20h1'11"10d32'
Alexander
Dra 14h26'38"61d5'
Alexander & Isabella
Cyg 19h55'34"44d36'
Alexander 17-07-1994
Lyr 19h17'26"31d19'
Alexander Ian
Lac 22h7'40"51d12'
Alexander Jacob
Ori 6h0'53"0d56'
Alexander James
Her 16h21'28"48d15'
Alexander Keegan
Aur 7h6'33"41d4'
Alexander Mark
Her 17h34'31"14d20'
Alexander Nature Center-M N,Harriet
Leo 10h29'20"12d3'
Alexander Otto
Cep 22h39'1"56d56'
Alexander R
Tau 3h52'0"1d48'
Alexander Ryan:100595
Ori 6h0'29"8d57'
Alexander Thomas
Psc 1h2'26"21d42'
Alexander's Inspiration
Ori 5h51'50"20d15'
Alexander's Abandon
Cyg 20h59'31"40d31'
Alexander's Dream
Ori 5h57'15"15d39'
Alexander's Gift
Lac 22h8'20"40d27'
Alexander's Halo
Cep 20h41'26"76d40'
Alexander's Light,S
Ara 17h49'31"-51d43'
Alexander's Star
Per 3h39'51"37d57'
Alexander's Way
Cep 23h1'36"64d15'
Alexander(Beans),Sol
Cet 1h55'0"-18d51'
Alexander,Alexis Whitney
Ori 4h56'8"14d28'C
Alexander,Alvin A
Ori 4h59'29"4d34'
Alexander,Amy
And 1h23'0"39d49'
Alexander,Arlin Ann
Eri 4h12'47"-16d54'
Alexander,B E Barbara
Peg 22h5'50"20d17'
Alexander,Barry
Cnc 8h51'20"11d22'
Alexander,Beth
Cas 0h29'12"61d2'
Alexander,Brian G
Aql 19h54'36"-0d43'
Alexander,Brian G
Dra 17h28'33"75d53'
Alexander,Brightest Star in Heavens,Karen
Per 4h4'1"40d20'
Alexander,Bryan Scott
Dra 9h49'12"73d18'
Alexander,Carrie C
Peg 22h26'34"30d29'
Alexander,Cary
Cam 5h44'18"60d53'

Alexander,Charles Meredith
Uma 9h57'52"58d15'
Alexander,Charles Malcolm Gilbert
Cra 18h20'13"-40d14'
Alexander,Debbie & Gilbert L
Boo 14h30'55"51d18'
Alexander,Denise
Peg 22h0'31"32d32'
Alexander,Dr
Vul 19h48'47"28d20'
Alexander,Erica Lynn
Del 20h20'29"10d55'
Alexander,Gary Ryan
Lac 22h11'33"49d27'
Alexander,Georgia
Lyr 18h17'55"38d33'
Alexander,Helen
Cyg 21h18'58"28d50'
Alexander,Jack R
Oph 18h7'16"1d9'
Alexander,James Robert Edward
Cyg 20h21'42"39d23'
Alexander,Jason S March 3rd 1987
Dra 12h10'1"70d42'
Alexander,Jennifer
Cnv 12h27'1"38d51'
Alexander,Jessica Riley
Vul 19h48'30"27d51'
Alexander,Jill
Dra 11h34'48"71d49'
Alexander,Jodi-Lynn
Peg 23h22'0"30d8'
Alexander,Karen
Sgr 19h36'32"-34d9'
Alexander,Kayla Marie
Del 20h20'16"10d47'
Alexander,Laurie Constance
Ori 5h45'48"10d9'
Alexander,Lee
Gem 6h45'3"14d20'
Alexander,Lisa Marie
And 1h26'16"36d20'
Alexander,Lori Ann
Com 12h17'40"22d9'
Alexander,Lori's "Outlaw",George H
Lac 22h4'13"38d6'
Alexander,Louise Anne
Lyr 19h20'25"41d17'
Alexander,Mary Blackwell
Cyg 21h10'47"35d25'
Alexander,Mary
Pho 0h32'49"-41d59'
Alexander,Mary Bryn
Lyr 18h46'27"36d20'
Alexander,Michael
Her 18h45'0"12d29'
Alexander,Michael
Lac 22h8'36"49d32'
Alexander,Michael Sterling
Ori 4h56'8"14d28'B
Alexander,Peter G
Mon 7h50'41"-2d51'
Alexander,Philip
Ori 6h3'32"9d13'
Alexander,Rachel K
Peg 22h32'30"20d52'
Alexander,Richard Elmont
Her 18h17'35"28d15'
Alexander,Robert James
Eri 2h48'0"-4d7'
Alexander,Robert W & Minerva J
Cyg 20h21'55"40d36'
Alexander,Roger "Dodger"
Lyn 6h12'34"58d48'
Alexander,Ruth Lois
Eri 4h12'29"-17d56'
Alexander,Samantha
Vul 19h48'0"27d59'
Alexander,Sara Fina
Com 12h25'24"20d59'

Alexander,Seth
Boo 15h4'55"24d42'
Alexander,Shari
And 2h0'11"38d8'
Alexander,Sharon
Lyn 8h8'1"39d22'
Alexander,Stacy Leigh
Eri 4h5'14"-0d37'
Alexander,Star of
Cyg 21h52'17"52d41'
Alexander,Stephanie Brett
Tau 4h17'32"7d57'
Alexander,Tiffani
Boo 15h13'53"38d14'
Alexander,Tracy
Uma 11h42'32"50d38'
Alexander,William Benjamin
Her 17h12'0"48d0'
Alexander,William David
Tau 4h10'0"0d55'
Alexander,Zachary Morgan
Eri 4h28'37"-12d9'
Alexander-André
Cet 2h40'1"3d25'
Alexander-Katz,Susana
Cma 6h57'34"-30d52'
Alexander-MacMillan
Uma 8h30'43"54d7'
Alexander/Paul
Lib 14h19'0"-23d43'
Alexanders Kircudbright In Sky
Cam 14h25'39"81d9'
Alexanders Light
Per 2h2'38"57d36'
Alexandra
Mon 6h25'23"-10d38'
Alexandra
Lyr 18h19'0"40d25'
Alexandra
Aql 20h22'53"0d36'
Alexandra
Peg 23h38'37"10d28'
Alexandra
Pic 4h47'28"-49d58'
Alexandra
Cyg 19h30'51"32d31'
Alexandra
Uma 10h1'15"57d33'
Alexandra
And 23h46'1"42d35'
Alexandra
Uma 9h18'59"43d6'
Alexandra
Del 20h36'31"18d59'
Alexandra
Ori 5h57'32"18d39'
Alexandra
Per 4h6'20"39d37'
Alexandra
Lib 14h21'12"-23d10'
Alexandra
Ori 6h4'30"5d46'
Alexandra
Cas 1h23'28"75d34'
Alexandra
And 23h17'58"41d59'
Alexandra Ann
And 23h15'41"41d17'
Alexandra Ariel
Cas 1h0'1"63d14'
Alexandra Charlotte
Cas 0h5'57"62d25'
Alexandra Colleen
Lyr 18h33'36"46d36'
Alexandra Eve Forever
Lmi 11h1'58"30d33'
Alexandra Lilian
Lyr 18h18'59"40d1'
Alexandra Lynn
Peg 21h54'46"31d42'
Alexandra Marie
Lyr 18h16'50"34d60'
Alexandra Noel
Mon 7h42'1"-6d4'

Alexandra Rae
Lyn 7h43'0"40d30'
Alexandra Rosaline
Cyg 19h32'14"39d31'
Alexandra The Sineu's Song
Aur 4h53'29"40d9'
Alexandra Thea
Uma 11h48'47"53d10'
Alexandra Zonia
And 1h7'30"40d55'
Alexandra's Star
And 0h12'1"39d18'
Alexandra-Dora's Twinkle Star
And 0h50'25"37d31'
Alexandragay
Peg 22h30'26"30d35'
Alexandre,Penny
Mon 7h43'34"-1d48'
Alexandre,Philippe
Per 2h18'57"58d24'
Alexandria
Lyr 18h56'42"31d48'
Alexandria
Mon 6h34'0"5d37'
Alexandria
Ori 5h58'24"9d22'
Alexandria
Vul 20h44'34"20d14'
Alexandria
Cas 2h3'0"59d2'
Alexandria & Wayne
And 22h57'12"40d37'
Alexandria Alleyne
Umi 14h25'24"65d44'
Alexandria Caroline
Cas 0h29'50"63d9'
Alexandria Forever
Tau 4h1'24"28d49'
Alexandria Joy
Cas 0h57'30"50d14'
Alexandria Michele
Cyg 21h29'26"40d59'
Alexandria Nichole
Hya 8h40'57"3d28'
Alexandrian Romance
Sct 18h50'54"-9d41'
Alexandro
Ori 6h6'11"5d57'
Alexei
Peg 22h3'27"10d19'
Alexey & Genevieve
Lyr 19h16'48"26d60'
Alexia
Uma 11h39'0"45d53'
Alexia
Pup 8h24'40"-23d38'
Alexia
Crb 15h54'54"26d22'
Alexia
Cep 2h18'20"78d46'
Alexia La Rebelle
Aql 19h32'20"1d16'
Alexia Leigh
Peg 22h37'22"24d15'
Alexiou,Joanna
Del 20h54'34"9d26'
Alexis
Cas 0h46'1"61d54'
Alexis
And 23h15'0"38d49'
Alexis
Lib 15h13'41"-20d53'
Alexis
Lyn 8h22'13"46d28'
Alexis
Lac 22h3'38"47d6'
Alexis
Lyn 8h24'1"47d16'
Alexis
Aur 5h39'20"38d56'
Alexis
Hya 10h32'55"-10d18'
Alexis
Hya 8h18'57"-6d6'

Alexis
Cam 4h51'48"65d13'
Alexis
Lac 22h29'12"50d36'A
Alexis Ann"Our Star Angel
And 0h54'27"40d13'
Alexis Jordyn
Cyg 20h17'55"38d25'
Alexis Kaye
Cas 0h32'44"63d18'
Alexis Kaye
Hya 8h13'35"5d46'
Alexis Kelly
Cas 23h13'19"61d18'
Alexis Mary Marguerite
Cnc 8h34'45"32d35'
Alexis Victoria
Cyg 20h17'10"38d45'
Alexis' Light
Dra 12h1'30"71d34'
Alexis,Dennis
Aur 6h3'50"30d10'
Alexius
Cyg 21h51'12"42d16'
Alexmolly 17694
Cnv 13h33'55"51d29'
Alexsa
Cas 0h33'25"61d24'
Alexson,Jordan Rae
Cep 23h32'20"70d14'
Alf
Cep 22h44'14"65d54'
Alf
Tri 1h59'42"34d55'
Alf
Cam 3h42'57"68d19'
ALF
Per 1h31'35"53d16'
Alfaire
Aql 19h48'14"11d39'
Alfanne
Cas 1h55'16"60d6'
Alfano,Christopher
Dra 12h2'1"71d5'
Alfano,Gregg
Per 3h21'52"54d44'
Alfano,Joseph
Hya 9h34'52"-1d23'
Alfano,Michael John & Livia
Sabrina
Ori 5h52'25"15d56'
Alfano,Odessa S
Lyn 6h13'0"60d39'
Alfaro,Ludin
Lyn 8h11'41"41d43'
Alfaro,Ludin
Crb 16h2'1"28d39'
Alferman,Richard R
Boo 15h4'38"21d10'
Alfers,Cassidy Marie
Lyn 8h48'32"42d35'
Alfie
Umi 15h8'0"81d33'
Alfie 'K'
Her 16h32'37"36d45'
Alfie's Enterprise
Hya 8h13'36"1d47'
Alfieri,Mary Anne
Cas 0h47'56"62d39'
Alfini,James
Boo 14h16'22"51d52'
Alflorursajinov
Umi 13h31'28"75d19'
Alfone "I Love You", David
Anthony
Her 17h37'11"14d35'
Alfonso
Cnv 13h52'56"45d34'
Alfonso Family Star, The
Tri 1h47'1"33d49'
Alfonso,Angie
Peg 21h41'56"24d42'
Alford,Cassie Kathleen
Cas 2h21'0"68d4'

Alford,Clare
Cyg 20h21'0"39d18'
Alford,Dude
Cam 3h58'52"53d14'
Alford,Jack
Ori 6h14'28"0d40'
Alford,John Patrick
And 22h58'58"48d48'
Alford,Simon Peter
Cyg 19h30'34"38d55'
Alfred
Uma 9h45'23"48d5'
Alfred & Nicole Everlasting
Love
Hya 8h12'15"1d26'
Alfred's Astral Palace
Ori 5h58'0"16d39'
Alfred's Light
Boo 13h56'24"21d32'
Alfred,Joshua Craig & Bradley
Uma 10h4'60"52d7'
Alfred,MD,Howard
Lac 22h8'44"49d52'
Alfreda Jane
Cas 0h11'0"58d20'
Alfredo e Clara Cristoni
Genitori
Cas 2h17'20"60d20'
Alfredo Mi Tesoro Te Quiero
Aql 20h7'1"-8d56'
Algar,Whitton John
Dra 19h45'39"70d27'
Algaret, L P S
Oph 17h6'16"8d16'
Algeciras
Lac 21h55'0"38d36'
Alger,Andy
Per 3h57'10"37d39'
Alger,Maggie Leahey Carolyn
And 22h58'43"51d13'
Algernon
Lib 14h44'18"-9d47'
Algies
Dra 15h6'20"60d34'
Algonquin
Boo 15h2'24"30d12'
Algra,Johannes Joost
Cas 1h41'42"60d59'
Alhasan,Harith
Uma 10h22'44"59d14'
Alheidis
Leo 9h57'20"30d2'
Alhoutade,Leonce
Cam 6h17'38"70d46'
Alhoutade,Leonce
Uma 9h28'35"47d21'
Ali
And 0h24'0"38d34'
Ali
Aql 19h13'28"10d5'
Ali
Eri 3h38'17"-6d49'
Ali
Cet 1h21'52"-18d58'
Ali
Aqr 20h57'31"0d36'
Ali Amor
Vul 19h32'12"27d25'
Ali Michelle
Uma 11h9'33"47d30'
Ali Star
Uma 12h8'35"45d35'
Ali's 21st
Uma 8h13'57"62d24'
Ali's Diamond
Cas 0h50'45"73d55'
Ali,Moyna M
Boo 14h21'0"28d1'
Ali,Nazzar Mohammed
Equ 21h3'1"10d36'
Aliamus,James Edward
Hya 8h57'23"-10d38'
Aliapoulios,Maxwell Matthaios
Per 2h23'34"55d45'

AliBri
Lyn 8h4'10"40d46'
Alice
Cma 6h56'21"-18d10'
Alice
And 23h37'48"40d15'
Alice
Cas 1h58'30"58d33'
Alice
Eri 4h8'30"-7d51'
Alice
Ori 4h57'30"4d48'
Alice
Dra 16h59'45"63d36'
Alice & Mina
Cam 13h20'25"81d27'
Alice Angel
Umi 15h19'14"69d25'
Alice B
And 0h25'19"45d53'
Alice Claire,The Star
Eri 2h55'36"-10d8'
Alice Elizabeth
Umi 13h8'0"76d4'
Alice Kate
And 23h33'48"42d20'
Alice Katherine
Lyn 9h19'30"40d30'
Alice Loves Harry Loves
Cyg 19h31'1"33d44'
Alice Mae's Heavenly Trail
Peg 22h11'47"4d29'
Alice Marie
And 2h24'0"49d21'
Alice Marie
Ori 6h7'1"10d53'
Alice Maureen "The Star"
Eri 3h5'45"-3d15'
Alice Meu Eterno Amor
Lup 15h12'19"-40d45'
Alice Valerius
"Schildkrötchen"
Cas 0h18'1"60d12'
Alice Victoria
And 23h17'42"41d32'
Alice's Angel
Mon 6h57'37"-6d33'
Alice's Star
Mon 6h57'38"11d55'
Alice's White Diamond
Sgr 19h28'17"-42d32'
Alicea,Gilbert John
Cep 23h14'44"78d58'
Alicea,Hunter Robert
Aur 5h10'36"40d37'
Alicelynn
Lyr 19h20'31"40d10'
Alicia
Mon 6h34'1"-1d8'
Alicia
Cyg 20h15'31"30d25'
Alicia
And 0h54'1"37d22'
Alicia 2113 Amaury
Peg 22h14'10"4d8'
Alicia Jill
Aql 19h6'0"15d38'
Alicia Maria
Cyg 19h25'53"35d8'
Alicia Marie
Peg 21h53'0"36d7'
Alicia Megan
Cas 0h46'25"71d47'
Alicia Mila
Eri 4h42'30"-2d8'
Alicya Loves Roger
And 23h6'25"43d33'
Alidoosti,Faraj
Aur 6h26'0"31d54'
Alienist,The
Cep 22h56'58"65d30'
Alienor W L
Umi 16h0'44"83d15'
Aligôte
Cam 6h18'23"68d12'

Alijon
Aur 6h57'54"37d11'
Alijon
Her 16h0'0"21d43'
Alimansky,Benjamin
Boo 15h4'1"21d10'
Alimo,David W
Dra 19h53'11"60d47'
Alimorong,Leanne
Aql 20h11'0"10d51'
Alina
Cyg 20h34'32"40d6'
Alina
And 23h49'52"42d43'
Alina Karin-Eric Alan
Ori 5h56'33"14d12'
Alindiax
Dra 18h2'17"58d23'
Aline
Eri 4h35'42"-12d23'
Aline
Vul 20h15'0"22d37'
Aline
Uma 11h33'28"37d43'
Aline Aha
Cmi 7h21'0"3d47'
Aline Ouellet Alon
Cyg 19h28'36"33d6'
Alino
Her 17h54'0"38d23'
Alion,Tyrone & Kathryn
Crb 16h11'59"32d39'
Alionka
Dra 20h21'54"70d57'
Alionushka
Uma 8h50'1"48d37'
Alioto,Dean
Mon 6h20'35"8d39"
Alioto,Jennie Marie Tessa
Peg 22h6'54"4d42'
Aliotta,Ben
Cep 23h40'48"67d18'
Alipravdi Star Baby, Max
Her 16h57'48"37d46'
Aliré
Cyg 19h34'44"31d54'
Alisa
And 2h5'1"42d55'
Alisa
And 2h14'11"50d38'
Alisa
Cas 1h3'27"61d31'
Alisa Ann
Mon 7h45'60"-2d54'
Alisa Elizabeth
Cas 0h6'11"61d38'
Alisa Ruby
Cas 2h2'33"59d29'
Alisandra
Cas 1h32'25"58d32'
Aliseda,Jr,Jose Luis
Aql 19h30'1"8d36'
Alisina
And 2h21'18"42d10'
Alison
Del 20h29'30"20d17'
Alison
Cas 0h53'38"56d0'
Alison
Cas 2h1'10"61d27'
Alison
Cas 0h46'36"60d44'
Alison
Cas 0h23'23"61d46'
Alison
Peg 22h10'52"25d41'B
Alison
Ori 5h59'42"17d40'
Alison
Crb 15h54'0"26d46'
Alison & Derrick
Dra 19h9'38"71d13'
Alison & Gerard- Soulmates
for Eternity
Cyg 19h27'1"38d43'

Alison & Peter
Cyg 19h30'12"39d36'
Alison 21
Cas 2h51'1"63d54'
Alison Jane
And 2h33'1"38d39'
Alison Jane
Cas 1h8'1"76d27'
Alison Kay
Cru 12h54'40"-63d23'
Alison Lyn
Cas 0h27'42"73d5'
Alison Lynn
Lyr 18h59'37"28d10'
Alison Moss-Boss
And 23h41'57"47d4'
Alison Verna
Uma 8h33'0"55d21'
Alison's Star
Cas 0h6'32"61d44'
Alissa Audrey
And 1h12'0"34d26'
Alissa's Light
And 1h34'20"39d17'
Alistar
Cam 4h45'32"67d37'
Alister
Per 3h26'58"40d60'
Alix
Boo 13h55'0"16d31'
Alix
Cep 22h37'35"58d20'
Alix Claude Marc
Sex 10h13'35"-6d60'
Aliya's Star
Cas 0h50'16"74d7'
Aliyev,Emil
Vir 11h41'46"3d53'
Aliza,Mollie
And 22h56'31"51d10'
Alizee
Per 2h38'45"56d43'
Alizée
Cas 23h39'58"64d14'A
Aljeris Star,The
Lmi 10h14'17"39d37'
ALJI
Tri 2h11'1"31d21'
Aljo's Little Star
Boo 14h59'33"26d48'
Aljoy
Cet 0h54'15"-1d48'
Alka Klara
Vul 19h20'49"22d53'
Alkassim,Ab S
Cet 3h13'54"0d27'
Alkassim,Dina L
Oph 17h10'17"-23d5'
Alkassim,Faris E
Peg 23h41'21"31d33'
Alkassim,Samirah A
Ser 15h16'17"10d39'
Alkassim,Sharon L
Cma 7h2'52"-18d49'
Alkire,Miss Teresa P
Cas 2h30'30"50d29'
Alkon,Bert
Cyg 21h50'16"41d56'
Alkon,Celia Simons
Cyg 21h32'58"41d30'
Alkon,Jacob
Cyg 21h35'1"41d10'
Alkon,Maxim
Ori 5h32'48"-2d9'
Alkon,Norma
And 0h58'1"35d20'
Alksninis,Clark
Cep 20h46'34"60d19'
All is Well
Aql 20h15'0"7d42'
All My Love Barbara
Del 20h17'40"10d50'
All My Love Bill
Her 17h59'40"41h11'

All My Love Forever Hayden
Umi 13h15'58"71d30'
All My Love Forever- Claudia
Nicole
And 23h37'11"40d41'
All My Love-Me
Del 20h37'0"10d29'
All My Wizzard Wishes For
Per 2h8'59"57d45'
All Of My Love
Uma 8h49'16"70d26'
All-Star Mark
Her 17h29'17"20d35'
Alla
Vul 19h52'48"20d2'
Allah
Lyr 19h16'0"41d53'
Allain,Claude
Cyg 20h19'1"40d44'
Allain,Marc
Her 15h43'0"28d37'
Allain,Raymonde
Uma 11h39'41"30d8'
Allaire
Lac 22h32'42"53d42'
Allaire,Marie- Benedicte
Tri 1h57'42"31d45'
Allaire,Nathan Lamphere
Vul 19h48'44"28d23'
Allaire,Phillip Lamphere
Crb 15h27'16"31d29'
Allais,Paul Andrew
Dra 20h17'59"64d46'
Allam,Bassil
Aur 5h37'59"50d33'
Allan
Ori 6h0'44"0d53'
Allan
Per 2h25'21"54d59'
Allan & Charlotte's Star
Sge 19h37'0"18d50'
Allan & Wendy
Lyn 8h25'1"44d52'
Allan,Amy Rhianne
Cyg 20h16'20"39d23'
Allan,Bruce
Ant 9h37'51"-30d16'
Allan,David A
Per 3h58'0"50d55'
Allan,Lisa
Lyr 18h52'0"30d1'
Allana Gayle
Ori 4h56'33"0d12'
Allane,Teana
Sco 17h23'32"-38d38'
Allard,Briana H
Cet 3h15'34"5d2'
Allard,Christine A
Cas 1h8'50"75d17'
Allard,Cindy L
Cma 6h43'37"-13d38'
Allard,Eric
Ori 5h59'1"17d0'
Allard,Evan James
Ori 6h2'44"8d49'
Allard,Ross
Cyg 21h1'27"30d18'
Allard,Simone M
Lac 22h0'55"38d5'
Allard,Stephanie L
Eri 2h58'58"-3d37'
Allardyce,Julie Ann
Aql 20h6'13"7d13'
Allas,Börje
Ori 5h34'53"-2d51'
Allastair John
Her 16h14'18"25d8'
Allbeck,Grandma & Grandpa
Cyg 19h27'29"31d26'
Allbright,Scott "Lone Wolf"
Boo 15h3'33"23d19'
Allbritten,Sir Richard Todd
Sct 18h52'9"-8d47'
Allchurch,Tim James
Ori 6h8'24"0d2'

Allcock,Michael
Ori 5h20'38"10d31'
Allder,Jr,Delbert K
Boo 14h19'16"36d49'
Allderdge,Susan
Uma 12h57'52"60d12'
Alldis,George Mary
Cet 3h5'49"1d3'
Alleach,Marina
Aur 4h50'32"40d4'
Alleaume,Paula
Cra 18h33'0"45d9'
Allebrod,Andreas
Aql 20h8'31"6d51'
Allebrod,Kerstin
Aql 20h9'20"0d22'
Alleger,Patti
Crb 15h48'12"30d2'
Allegra Brooke
Lyr 18h55'18"30d44'
Allegre,B
Lac 22h20'35"55d2'
Allegre,Gregory
Crb 16h6'23"38d41'
Allegre,Nancy
Crb 16h5'1"38d20'
Allegrini,Robert V
Aur 6h35'55"32d46'
Allegro,Joseph
Her 16h43'27"27d26'
Alleia Bethany
And 23h6'50"41d12'
Allem,Briana Noelle
Cyg 19h27'15"31d15'
Alleman,Paul
Aur 6h19'33"39d52'
Alleman,Ronald Lee
Ser 15h43'46"4d5'
Allen
Dra 16h34'37"67d37'
Allen (ARNF),Jennifer Elise
Cyg 21h8'44"30d44'
Allen (Irvin)
Hya 8h48'18"-7d40'
Allen CBBSDSB,Joe & Vivian
Cyg 21h37'39"44d43'
Allen III,Clay W
Oph 17h54'23"12d56'
Allen Star Gift From Gary
Duglin,Jeffrey
Cam 3h49'34"61d26'
Allen's Wish
Her 16h8'34"41d14'
Allen,Alison
Cas 23h27'11"60d52'
Allen,Ann E
Cas 0h57'28"75d45'
Allen,Anna
Lyr 18h54'43"37d17'
Allen,Antonia M
Dra 16h19'31"63d53'
Allen,Arthur Mabron
Aur 4h37'47"30d21'
Allen,Barbara
And 1h25'0"34d18'
Allen,Barry Quinn
Cet 3h14'26"0d28'
Allen,Beau Justin
Per 3h2'1"47d8'
Allen,Benjamin Frederick
Ori 5h51'36"19d3'
Allen,Betty Jean
Cyg 21h11'56"35d5'
Allen,Beverley Elizabeth
Boo 14h46'53"33d25'
Allen,Bill
Boo 15h6'1"30d34'
Allen,Brad Wade
Boo 15h7'47"31d50'
Allen,Brian Claude
Oph 18h1'51"7d41'
Allen,Carey
And 23h23'1"43d54'
Allen,Carl E
Lac 22h30'0"50d25'

Allen,Caroline
Cyg 20h22'1"38d17'
Allen,Cecil Calvin
Ori 5h54'14"9d3'
Allen,Christine
Cas 0h14'0"62d15'
Allen,Christopher Phillip
Ser 15h35'21"20d19'
Allen,Corey Francis
Sco 17h34'22"-38d57'
Allen,Crystal Lynn
Peg 23h29'27"22d16'
Allen,Cynthia Lee
Mon 8h3'1"-1d39'
Allen,Dan & Bridget
Crb 15h44'11"27d34'
Allen,Dave
Lac 22h9'1"50d46'
Allen,David
Per 3h11'0"41d46'
Allen,David
Her 17h26'42"20d14'
Allen,David Antony
Uma 12h4'20"57d40'
Allen,David Bottome
Cet 0h49'34"0d34'
Allen,David Lawrence
Cam 3h44'49"73d13'
Allen,David Stephen
Her 16h52'1"48d3'
Allen,Dawn & Janine
Eri 2h44'14"-4d54'
Allen,Dennis R
Aur 5h1'27"50d4'
Allen,Donna J
Cas 0h56'46"58d13'
Allen,Dorie
Eri 2h53'18"-6d40'
Allen,Dr Edward B
Oph 17h41'11"12d3'
Allen,Edmund C
Equ 21h20'37"8d50'
Allen,Edmund C
Aql 20h0'11"8d8'
Allen,Ethan Edward
Oph 16h5'21"-6d31'
Allen,Fiona Lucia Genadio
Leo 10h50'30"-5d36'
Allen,Francesca
Cyg 20h1'37"40d29'
Allen,Frank
Lyr 18h17'34"30d53'
Allen,Frank E
Cet 1h19'13"-3d1'
Allen,Gaylord
Boo 14h50'33"38d20'
Allen,George W
Lac 22h31'41"50d17'
Allen,Ginger Lynn
And 2h14'1"50d16'
Allen,Harry S
Aur 6h23'34"31d8'
Allen,Hazel Anne
Uma 8h18'32"70d37'
Allen,Heather
Cet 1h23'12"-6d38'
Allen,Heather Ursula
Cas 1h2'50"60d42'
Allen,Horace B
Hya 8h59'35"-0d25'
Allen,James Hollis
Dra 12h39'17"68d57'
Allen,James"Sweet Ray"
Ori 5h57'26"-0d3'
Allen,Jane Ellen
Peg 21h47'1"34d2'
Allen,Jason Stuart
Oph 18h5'48"12d2'
Allen,Jean W
Boo 15h7'42"32d25'
Allen,Joan Driscoll
Peg 22h2'60"8d8'
Allen,John Herbert
Her 16h27'1"33d13'

Allen,John Hunter
Peg 22h51'30"25d10'
Allen,Jr,Dozier T
Boo 14h32'1"51d38'
Allen,Jr,George W
Aql 18h55'12"-0d43'
Allen,Judy
Uma 11h17'20"62d26'
Allen,Kaden Trace
Ori 4h54'1"4d47'
Allen,Kai
Dra 18h7'38"65d33'
Allen,Kara Dawn
Lyn 7h59'50"44d44'
Allen,Kathy
Ser 15h17'1"21d10'
Allen,Kevin Scott
Cnv 12h24'40"51d13'
Allen,Kimberley
Lyr 18h33'0"38d52'
Allen,Lilo
Ari 3h1'19"22d24'
Allen,Linda
And 23h19'52"50d19'
Allen,Lindsey Joletta
And 0h46'51"22d15'
Allen,Lisa Ann
Cet 2h39'22"0d51'
Allen,Lisa Dawn
Lyr 18h40'22"37d36'
Allen,Loren Paul
Ori 4h43'44"5d28'
Allen,Lori Kristine
Umi 16h19'27"71d41'
Allen,M Lou
Vul 19h47'27"22d37'
Allen,Mae Jean
Com 13h19'37"21d43'
Allen,Maggie
Peg 21h38'27"24d0'
Allen,Mark
Cep 22h16'29"62d4'
Allen,Mark Gary
Aur 5h55'18"31d27'
Allen,Mary A
Lyr 18h45'39"37d29'
Allen,Mary Ellen
Cas 0h2'16"62d36'
Allen,Mary Joan Zimmerman
Mon 6h39'25"3d10'
Allen,Mary Wyatt
Peg 21h29'1"23d34'
Allen,Max D
Aql 19h50'27"15d40'
Allen,Maxwell
Lyr 19h3'52"26d57'
Allen,Megan Ruth
Mon 7h4'34"1d7'
Allen,Michael David
Cet 0h30'25"0d36'
Allen,Michael Lewis
Vul 19h45'40"28d41'
Allen,Ms Robin
Lyn 6h28'1"59d27'
Allen,Nancy
Cam 10h37'48"84d43'
Allen,Nigel Charles
Cep 21h48'24"65d3'
Allen,Oliver
Lyr 18h36'20"38d17'
Allen,Pat
Lmi 10h45'19"27d31'
Allen,Paul Boyd
Lac 22h12'1"38d2'
Allen,Paul Leonard
Cep 20h46'16"60d14'
Allen,Peggy
Lyn 8h50'10"35d14'
Allen,Phyllis A
Cas 3h32'46"70d0'
Allen,Rebecca
Sct 18h49'48"-6d56'
Allen,Rick
Tri 1h55'33"27d59'

Allen,Rita
Cas 1h56'15"74d32'
Allen,Robert & Georgia
Crb 15h27'53"31d38'
Allen,Robert Derr
Psc 0h55'51"30d5'
Allen,Robert Keith
Boo 14h26'17"51d17'
Allen,Robin Scott
Tau 5h51'42"28d18'
Allen,Rock
Hya 8h53'23"1d9'
Allen,Roger
Boo 13h51'23"21d58'
Allen,Ron
Dra 20h7'19"63d12'
Allen,Ronald Thomas
Aur 5h6'14"38d55'
Allen,Rosalind Lee
Mon 6h54'41"0d34'
Allen,Sabyn Elizabeth
Com 13h10'0"20d36'
Allen,Shannon M
Aqr 21h6'1"-0d39'
Allen,Sharon Darlene
Peg 21h19'58"22d41'
Allen,Shirley Wesley
Mon 8h8'51"-4d55'
Allen,Sr,Bendall Worley
Cap 20h30'20"-13d8'
Allen,Steven J
Vul 19h17'35"25d33'
Allen,Susan Carol
Lyn 8h28'24"42d1'
Allen,Susan M
Cyg 20h11'38"49d2'B
Allen,Terence Leonard
Her 16h13'17"5d26'
Allen,Terry
Tri 2h20'12"28d37'
Allen,The Star Of Jennifer Michelle
Ori 4h43'28"4d23'
Allen,Thomas
Her 17h55'55"20d8'
Allen,Toby
Her 18h39'24"13d7'
Allen,Vicki Lee
Cas 0h57'13"64d0'
Allen,William Storm
Boo 14h19'23"17d32'
Allen,Zoe
Lyr 19h4'26"28d21'
Allen-Friend,Thomas G
Ori 5h10'34"-5d7'
Allenby,Ross Malcolm Stuart
Uma 10h29'0"54d44'
Allender Family, Richard T
Lac 22h4'29"38d30'
Allender,Tina
Eri 4h5'20"-17d13'
Allendorph,Grace
Cas 23h33'15"54d8'
Allens'Celestial Millennium
Uma 13h54'0"60d41'
Allens,Santa Maria
Cyg 21h13'25"37d49'
Allenspach,Jr,Thomas
Boo 14h23'53"33d57'
Allenstein,Brandon Miles
Gem 6h57'7"12d32'
Allenstein,Taylor Ross
Vir 12h30'55"2d30'
Allensworth,Brittany Marie
Peg 22h34'1"30d19'
Allentoff,Lisa Robin
And 23h39'22"40d17'
Aller,Elke
Ari 2h23'0"20d1'
Allers,Brian Timothy
Dra 18h44'26"68d10'
Allers,Christine V
Lyn 7h33'60"40d25'
Allers,Jessica Brooke
Cyg 21h28'56"38d46'

Alles Family,The
Cep 20h45'1"60d55'
Alles,Amanda
And 2h11'55"40d39'
Alles,Kimberly
Lyr 18h51'1"40d30'
Alles,Matthew
Lac 22h18'10"51d43'
Alleson
Lyn 7h50'1"39d27'
Allessio,Phillip James
Her 18h4'30"31d24'
Alleva,Carolyn Elizabeth Jervis
Psc 0h57'24"31d19'
Alleva,Ettore
Psa 22h35'55"-28d33'
Alleva,Jr,Nicholas
Lac 22h23'0"52d33'
Alleva,Maria Russo
Cas 1h38'45"60d52'
Alley's Dream
Boo 15h45'31"42d37'
Alley,Charlie
Lac 22h13'57"49d24'
Alley,Nellie
Ori 6h5'24"1d10'
Alley,Stephen
Dra 17h32'21"61d30'
Alley,Steven J
Aql 20h5'11"7d53'
Allgeyer,James Ernest
Oph 17h1'36"11d34'
Allgood,Adam Douglas
Cet 5h52'1"2d30'
Allgood,Andrea Ellen
Peg 0h0'59"13d3'
Alli's Guiding Light
Cyg 19h28'26"30d13'
Alliance Gaming
Dra 16h3'22"63d27'
Alliance International
Hya 9h34'11"-6d20'
Alliance USA
Oph 17h52'59"7d45'
Allie
Lyn 7h32'0"51d34'
Allie & Temple
Del 20h56'17"11d7'
Allie Cat
Lmi 10h41'52"30d2'
Allie Star
Lyr 18h52'36"40d59'
Allie,Barbara Lee
Cyg 20h50'26"40d24'
Allie,Keith
Per 2h58'16"32d38'
Alliebird
Cam 4h9'55"60d33'
Allieri,Raymond Lawrence
Dra 20h12'44"70d15'
Allieri,Stephanie Anne
Lyn 7h40'45"40d39'
Alligator John
Dra 18h4'34"58d24'
Alliger,Janet
Mon 6h37'47"-4d24'B
Alliger,Martin
Mon 6h37'47"-4d24'A
Allin,Marshall Robert
Oph 18h31'18"11d24'
Allington,Harry
Uma 10h1'1"47d30'
Allington,Rolland
Peg 22h39'57"31d22'
AlLiNorJay
Her 18h15'1"31d18'
Allison
Sge 20h17'27"16d45'
Allison
Psc 22h55'0"1d16'
Allison
And 23h1'33"48d13'
Allison
Cas 0h37'53"69d45'

Allison
Lyn 7h32'55"45d57'
Allison
Mon 6h27'49"11d38'
Allison
Del 20h13'44"12d7'
Allison
Del 20h13'35"15d31'
Allison
Dra 18h53'21"48d47'B
Allison
Sge 19h5'48"19d12'
Allison
Com 12h7'52"19d39'
Allison
Boo 14h1'44"12d15'
Allison
Tri 1h30'26"33d23'
Allison
Cyg 19h59'52"37d39'
Allison
Per 1h33'0"54d4'
Allison
And 2h26'28"40d18'
Allison & Mark
Cyg 21h25'52"39d43'
Allison & Michael
Vul 21h18'51"24'8'
Allison & Michael's Shining Star
Uma 10h10'46"68d12'
Allison 1968-1991
And 0h12'59"46d41'
Allison Angel
Peg 21h43'43"28d1'
Allison Ann
Mon 6h30'24"-0d3'
Allison Crystal
Cyg 20h25'49"40d52'
Allison D
Cas 0h11'44"60d35'
Allison Jodie Sandra 9-2-90 to Forever
Peg 22h35'55"10d2'
Allison Mary
Cas 1h47'40"73d28'
Allison Rose
Lyr 18h22'59"44d49'
Allison's Bat Mitzvah
And 23h18'29"46d57'
Allison's Birthday Star
Vir 11h39'50"3d49'
Allison's Sweet Angel Star
Cyg 20h34'30"30d34'
Allison's World
And 23h0'37"45d50'
Allison, Forever!
Lyr 18h33'15"34d2'
Allison,Ashley Lynnette
Lyn 6h28'0"60d0'
Allison,Chris
Mon 7h54'0"-5d42'
Allison,Curtis
Aur 6h16'39"30d58'
Allison,Davey
Uma 10h20'28"48d56'
Allison,Davey
Lac 22h5'14"51d12'
Allison,Doug
Mon 7h2'38"5d13'
Allison,Forrest Ian
Mon 8h1'34"-8d13'
Allison,James
Ori 6h16'26"-2d56'
Allison,Karl
Cyg 19h34'46"37d24'
Allison,Margaret
Sge 20h17'58"20d35'
Allison,Michelle & Jeffrey
Aql 18h50'53"11d40'
Allison,Michelle M
Uma 8h48'49"72d1'
Allison,Pamela & Thomas James Gebbie
Cyg 21h29'55"48d49'

Allison,Patton Manuel
Oph 18h35'16"10d28'
Allison,Robert
Sex 10h33'59"0d5'
Allison,Robert & Virginia
Cyg 21h3'25"31d43'
Allison,Robert Kent
Vir 13h7'53"-1d3'
Allison,Sally
Lyn 8h8'21"58d35'
Allison,Stephen
Ori 6h6'51"1d32'
Allison-Lee-JSD-Gers 101493-Prime
Lib 15h19'12"-21d1'
Allisyn
Aql 18h58'43"-6d4'
Allman's Arcadia,Dave
Per 3h1'0"46d23'
Allman,Brian
Aur 5h18'55"41d7'
Allman,My Shining Star Jamie
Uma 9h20'55"55d6'
Allmark,Paul William
Uma 11h50'14"38d13'
Allmer,Michael P
Lib 14h57'59"-8d51'
Allo,Alexis Thom
Mon 6h18'53"7d24'
Allocca,Antonio
Cap 21h50'40"-8d17'
Allonby,Talena Marie
Peg 22h43'11"10d14'
Allor,M Jeanne
Cas 2h19'0"70d39'
Alloy,Dean
Aur 6h15'60"33d25'
Allport,James William
Ori 6h4'37"18d56'
Allred's Sirenity
Her 18h11'34"37d31'
Allred,Daniel Stewart
Aqr 22h1'27"-5d31'
Allred,Jean
Cas 1h20'18"72d35'A
Allred,Margaret Louise Hall
Leo 10h57'59"11d26'
Allred,Thurman W
Aur 6h11'23"32d45'
Alls,Gale
And 2h12'44"43d1'
Allshouse,Casandra Lynn
Leo 10h0'0"15d23'
Allsopp,Sarah
Cas 0h17'15"62d53'
Allsopp,Sarah
Cas 0h34'30"62d54'
Allstar '95
Gem 6h50'50"14d24'
AllStars' AllStar Staff Star
Aql 19h6'14"15d32'
Allsup III,Guy L
Ori 5h58'56"1d23'
Alltid Anna
Uma 10h57'46"47d33'
Alluaume
Per 3h32'17"39d59'
Allum,Lil & Bill
Peg 22h29'0"28d27'
Allums,Aaron P
Ser 15h43'1"4d41'
Allura,Kelly
Cet 2h9'40"5d57'
Allure
Aur 5h4'46"44d49'
Allure,John Gow Georgian
Her 16h5'17"18d46'
Alluring Alana
Crb 16h9'55"36d5'
Allworth,Adrian Stuart
Sct 18h43'55"-6d25'
Allworth,Hilary Dean
Peg 22h2'41"20d24'
Ally
Cyg 21h20'45"31d50'

Ally's Star
Ori 6h5'1"3d31'
Allyma & Majjah's Star
Cyg 19h15'26"44d3'
Allyn 2
Lib 15h32'27"-10d24'
Allyn Leigh
Aql 18h55'11"11d8'
Allyson
Cyg 20h50'16"38d14'
Alm Wedding Star, Debbie & Greg
Cyg 20h26'33"42d15'
Alm,Birgitta
Dra 18h48'21"71d15'
Alma
Vul 21h1'43"26d49'
Alma
Lyr 19h22'1"38d9'
Alma
Cas 0h53'41"63d14'
Alma Rose
Oph 17h40'54"10d45'
Alman,Trekkie Dianna
Peg 22h1'47"2d27'
Almax
Mon 7h6'55"0d47'
Almberg,Stephen
Ori 5h34'18"-1d60'
Almeida,Jennifer Lynn
Sgr 18h52'11"-28d36'
Almeida,Steve
Per 2h54'48"32d49'
Almeida,Toze & Carmen
Cyg 21h0'24"38d8'
Almett,John
Cet 2h10'19"3d34'
Almich,Timothy Dean
Hya 8h47'44"5d43'
Almine
Per 2h59'24"32d46'A
Almoaibed,Abdulla
Hya 8h46'21"1d38'
Almoda 8h46'21"1d38'
Lmi 9h57'19"38d58'
Almodi 9h57'19"38d58'
Lmi 9h58'36"38d43'
Almodi 9h58'36"38d43'
Lmi 10h53'41"30d38'
Almog
Lac 22h14'39"48d31'
Almon,Baylee
Uma 9h54'17"60d38'
Almond Eyes
Cam 13h36'57"78d3'
Almond,Antoinnette
Cyg 20h58'13"38d9'
Almond,Maurice René
Ori 5h25'40"1d38'
Almond,Todd Jeffrey
Her 16h8'20"51d20'
Almoni
Lep 5h5'58"-12d19'
Almonte 143
Aur 7h19'20"35d43'
Almorin
Sge 19h11'35"19d53'
ALMOST ANGEL
Umi 17h34'31"80d29'
Almost Paradise
Lyn 8h59'37"38d41'
Almquist,Douglas
Aql 19h30'0"13d28'
Almquist,Dr Glen
Her 15h58'46"41d41'
Almudena
Uma 12h4'12"47d30'
Alnas,Cynthia Valenzuela
Peg 22h16'14"31d10'
Aloha Always
Sex 10h35'1"0d22'
Aloha Nui Loa
Oph 17h5'38"11d38'
Alois
Lib 15h42'22"-20d21'

Aloise,Joseph Frank
Ori 6h7'0"10d36'
Aloisi,Ann
Lyr 18h48'31"30d38'
Aloisoyus
Cam 5h44'17"61d8'
Alomi
Lyn 9h19'15"33d59'
Alonachka
Peg 22h44'39"25d45'
Alondo's "Nightlite"
Umi 16h10'19"80d32'
Along 51832
Uma 10h16'0"55d26'
Alongi,Amanda
Lmi 10h42'24"26d19'
Alonna
Boo 15h9'59"51d56'
Alonso,Jeremy Joseph
Cep 21h52'23"60d35'
Alonzo,Arthur J
Cet 2h1'35"0d55'
Alonzo,Dr Jim
Lac 22h37'39"53d4'
Alonzo,Frank
Equ 21h20'23"8d46'
Aloupis,Jacqueline Marie
Cas 1h5'35"63d53'
Aloys
Peg 23h30'27"18d45'
Aloysius
Tri 2h34'1"31d27'
Alpa's Light Fantastic
Cyg 19h56'16"44d20'
Alpaca Express
Leo 9h57'26"19d1'
Alper,Benjamin & Mildred
Lyr 18h13'11"40d21'
Alper,Carey B
And 1h38'29"41d13'
Alper,David Matthew
Boo 14h28'36"20d24'
Alper,Gabrielle
Cam 4h3'16"58d9'
Alper,Thomas Joseph
Vul 19h15'36"25d12'
Alpha Alpha Together
Cyg 21h2'36"41d4'
Alpha Beth
Boo 14h59'31"23d7'
Alpha Bezold
Cam 12h13'49"81d28'
Alpha Cappella Gloria
Tri 1h54'26"25d40'
Alpha Carol
Ori 5h58'25"1d12'
Alpha Cen-Laurie
Aql 19h59'20"7d60'
Alpha Christophae
Her 16h40'33"32d17'
Alpha David
Aur 6h34'34"31d47'
Alpha I Hans und Ingrid Reimann
Sgr 18h57'25'-27d42'
Alpha Kappa State
Mon 6h44'59"0d57'
Alpha Leron
Lmi 10h37'16"26d36'
Alpha Meeta
Cet 2h11'22"1d6'
Alpha Michael Robert
Cyg 20h12'38"41d57'B
Alpha Nancy Lee
Eri 4h8'36"-7d60'
Alpha Natasha 22
Umi 13h52'54"76d10'
Alpha Rick
Dra 16h59'26"67d51'
Alpha Rosemary
Boo 14h30'60"50d9'
Alpha Sameen
Mon 6h49'1"11d38'
Alpha Sigma Alpha Gamma XI Chapter
Tri 2h41'0"31d13'

Alpha Stephen Gregory
Cyg 20h12'38"41d57'C
Alpha Tidburyanis
Boo 15h18'25"34d60'
Alpha Vonnabeth Gail
Cyg 20h12'38"41d57'A
Alpha Walsh-Beecher
Aql 20h1'22"1d49'
Alpha-Sam
Boo 14h21'26"31d15'
Alpigo,Claudine
Dra 9h28'44"81d14'
Alpin
Per 3h15'30"41d17'
Alquimia
Peg 22h35'48"28d10'
Alquisto,Dia Maria Regina Christina
And 22h57'54"51d16'
Alred,Mary Todd
Cam 4h57'51"70d42'
Alrizzo
Ori 6h15'42"7d51'
Alsatia
Gem 7h31'24"31d3'
Alsayeghe,Ludmila Z
Aql 18h55'21"-0d54'
Alsayeghe,Tatiana Z
Mon 6h27'20"8d0'
Alscher,Murray Kichel
Dra 19h3'55"48d1'
Alsobrook,Charles Nicholas
Hya 9h11'23"1d26'
Alsop,Caren Lorane
And 2h3'52"40d26'
Alsop,Derek
Uma 11h29'49"32d0'
Alspaugh,Frances Gwizdz
Boo 14h27'1"47d20'
Alspaugh,Richard N
Cyg 19h31'54"33d22'
Alston,Charlie Dunn
Equ 21h7'17"11d31'
Alston,Kristy
Peg 22h0'1"31d9'
Alston,Mary-Gwen
Uma 11h32'45"42d27'
Alston,Sue Dunn
Uma 11h46'44"52d50'
Alston,Vincent de Angelis
Aql 18h42'0"0d3'
Alstrup,Lynn
Ser 18h7'52"-9d34'
Alt,Dr Fredrick Wayne
Aur 7h6'29"39d55'
Alt,Richard "Dick"
Her 16h39'0"35d17'
Alta
Cyg 19h54'54"44d48'
Alta
Boo 14h57'35"30d37'
Alta Marie
Mon 6h21'56"3d26'
Alta Mira-9
Cmi 7h56'13"7d51'
Alta Quota
Cnv 13h4'34"33d15'
ALTA Space I
Tri 1h46'47"28d23'
Altarac,Miles Franklin
Cep 22h12'52"60d30'
Altavilla 1956 stella Normanna,Luigi
Aur 4h59'35"40d11'
Altazan,Elaine,P
Cas 0h40'1"61d22'
Altazan,Louis R
Oph 16h55'0"-26d14'
Altai
Cyg 21h37'17"40d19'
Altbaum,Joseph
Boo 14h8'47"50d33'
Alteirac,Laurent
Umi 13h14'1"72d24'

Altemier,Kristianna Rae
Peg 23h37'50"25d33'
Altemus,Jr,Mark Emmerich
Uma 11h23'22"41d18'
Alten,Gene Jay
Tau 4h6'19"20d10'
Altenburg,Helga
Aqr 20h54'48"0d37'
Alter,Jochen
Lac 22h1'22"51d43'
Alter,Morry
Per 1h52'25"52d40'
Alterman,Alissa
Del 20h31'14"20d10'
Alterman,Allison
Ori 6h5'43"8d7'
Alternative Learning Center (Hagerstown,MD)
Uma 8h30'59"68d8'
Althoff,Blake Joseph
Her 17h38'24"20d4'
Althoff,Carstern
Sco 17h6'16"-38d24'
Althoff,Christopher Robert
Dra 19h15'53"70d40'
Althoff,Helen Louise Campbell
Cyg 20h21'43"31d13'
Althoff,Jeffrey James
Cam 6h13'39"67d51'
Altier,William
Sex 10h29'43"-5d38'
Altieri,Claudia
Cas 2h27'39"61d7'
Altieri,Joseph Paul
Cep 23h5'47"64d0'
Altilia,Daniela
Cyg 19h48'47"29d48'
Altini,Michael
Per 3h7'15"41d30'
Altizer,Carrie Ann
Aql 19h23'35"0d27'
Altland,Jane
And 23h39'42"38d20'
Altman,Angela Rose
Cyg 21h28'41"41d12'
Altman,Crisey Elizabeth
And 23h39'53"38d28'
Altman,Emily
Dra 15h10'50"62d2'B
Altman,Leonard
Eri 3h21'53"-6d9'
Altman,Linda
Del 20h16'0"11d6'
Altman,Nancy Addison
Cam 7h53'32"67d40'
Altman,Paula
Lyn 7h11'23"58d41'
Altman,Phoenixville, PA,Elizabeth W
Cas 2h2'45"59d45'
Altman,Scott Andrew
Ori 5h57'23"15d17'
Altman,Shari I
Cas 23h32'0"60d45'
Altmann,Christian
Leo 10h52'41"20d13'
Altmann,Dereck
Ori 5h26'49"-1d17'
Altmann,Dolores
Vul 19h23'15"23d45'
Altmann,Guenther
Lyn 8h1'33"48d34'
Altmann,Michele Lynn
Mon 8h1'58"-8d51'
Altmann,Rudi
Her 15h49'49"44d28'
Altmore,Carrieann
Tri 2h12'59"32d1'
Altneu,Elizabeth
And 22h56'12"51d47'
Altomara,Ginevra Marie
And 0h53'53"40d24'
Altomare,Brent Andrew
Hya 8h26'58"-8d45'

Altomare,Dr James B
Oph 17h36'18"11d25'
Altomare,Philip V
Cep 21h43'49"58d49'
Altrimenti
Cae 4h59'47"-32d5'
Altrina
Dra 16h7'1"60d25'
Altschuler,Laurel Ellen
Lmi 10h48'27"32d7'
Altshuler Star,The Randy Neal
Sco 16h21'12"-40d30'
Altshuler,Dr Kenneth & Ruth
Crb 15h58'57"28d31'
Altvater,Christopher
Ori 5h26'1"0d41'
Altvater,Virginia
Ori 5h29'42"0d25'
Aluatam
Aql 19h58'23"0d58'
Aluco 50
Uma 9h44'56"48d52'
Aluna
And 23h43'1"41d46'
Alva's Guide
And 23'39'11"40d41'
Alvarado,Raymond Bryan
Uma 12h21'13"58d19'
Alvarado,Richard Nick
Aql 19h4'40"0d3'
Alvarado,Robert Edward
Boo 14h55'39"23d51'
Alvarez,Angelo Auli
Dra 18h44'51"67d33'
Alvarez,Arlene
Cyg 20h3'0"39d58'
Alvarez,Carl
Uma 11h57'15"58d2'
Alvarez,Eric
Hya 9h0'14"5d0'
Alvarez,Ernestina
Oph 17h34'38"11d21'
Alvarez,Jazmyn
Boo 14h34'11"20d1'
Alvarez,Jr,Martin
Ori 5h56'57"7d59'
Alvarez,Jr,Ruben Lee
Aur 6h31'14"32d26'
Alvarez,Mary Lou
Cas 1h10'0"58d39'
Alvarez,Mr & Mrs Antonio
Aur 6h18'10"45d32'
Alvarez,Refugio
Her 16h17'1"10d34'
Alvarez,Sofia Bense34'
Ori 5h56'35"20d7'
Alvarez,Tony
Aur 5h38'44"40d25'
Alvarez,William
Ari 2h59'26"30d46'
Alvarez-Serrano, Alejandro
Sct 18h44'14"-6d57'
Alvarez-Serrano,Stella Desirea
Crt 11h35'59"-10d50'
Alvaro,Gabriel Thomas
Ori 6h8'22"8d31'
Alvart,R
Cam 5h54'23"60d45'
Alveriz,Mary
Ori 5h27'32"-0d22'
Alverson,Jeffrey Allan
Vir 11h38'53"8d20'
Alverson,Joshua Eli
Cet 2h2'12"-0d8'
Alves,Lana G
Uma 14h26'1"61d57'
Alvey,Harry & Penny
Cyg 21h33'1"38d10'
Alvi,Rebekah
Mon 7h7'13"-6d59'
Alviggi,Maria
Lac 22h51'46"38d28'
Alvin
Hya 8h54'1"3d25'

Alvin
Oph 16h50'53"10d7'
Alvin
Lac 22h10'31"51d1'
Alvin Paul
Dra 20h26'47"67d27'
Alvy,Margaret
Crt 11h35'14"-20d50'
Alwara
Uma 8h41'50"51d11'
Alward,Lee C
Cam 8h49'34"78d47'
Always
Cyg 20h56'44"40d11'
Always
And 23h43'43"37d31'
Always
Boo 15h19'33"52d1'
Always
Mon 7h15'57"-6d58'
Always
And 2h10'50"41d9'
Always
Uma 10h2'42"68d26'
Always
Cyg 21h45'60"38d57'
Always & Forever
Lyn 9h13'1"36d55'
Always & Forever (10-1-90)
Cep 21h38'0"68d58'
Always & Forever Pete
Per 3h13'54"56d7'
Always & Forever Francine
Tau 4h31'0"20d7'
Always & Forever
Cet 1h34'40"-4d35'
Always & Forever
Uma 11h58'1"46d41'
Always & Forever
Lyr 19h18'47"41d44'
Always & Forever Brad
Her 15h48'59"46d41'
Always & Forever CSQ
Vul 19h23'37"25d17'
Always & Forever Dean
Cep 22h12'34"68d46'
Always & Forever Emby
Lmi 10h40'28"25d36'
Always & Forever Jim
Lmi 9h42'27"40d22'
Always & Forever Jim
Hya 8h31'22"-0d9'
Always & Forever Kevin
Her 18h8'11"40d1'
Always & Forever Sam
Tri 2h28'15"28d24'
Always & Forever Steve
Sgr 18h55'10"-26d27'
Always & Forever,Lance & Gina
Lyr 18h58'42"31d19'
Always & Forever Lia
Ind 20h54'18"48d15'
Always & Forever,PG & KM
Cyg 20h8'24"38d14'
Always & Forevermore Amie
Del 20h18'48"14d43'
Always Alan
Gem 6h27'18"13d19'
Always Alan
Her 18h20'41"12d27'
Always Alice
Lyr 19h16'56"28d34'
Always Allan
Per 3h11'49"50d18'
Always Amber
Aur 5h18'26"46d13'
Always Andrew
Ori 5h24'1"1d49'
Always Andrew
Sge 19h58'13"20d4'
Always Andy
Lyn 7h0'1"52d4'
Always Anna
Lmi 10h12'44"40d29'
Always Anthony
Aur 6h44'18"37d60'
Always B R Star
Cyg 21h50'32"37d53'
Always Bright for Bill & Debbi
Cyg 19h12'47"48d27'
Always Burrus
Her 17h30'19"40d51'
Always Chris
And 23h3'50"41d31'B
Always Denis
Uma 8h47'55"62d3'
Always Drew & Paula
Aur 6h19'58"45d4'
Always Ed
Peg 21h26'28"8d32'
Always Ethan
Her 17h38'31"14d36'
Always For 15-88
Aur 5h9'10"43d1'
Always For Alex
Peg 23h40'17"30d5'
Always Forever JDJ LRM
Cnv 13h37'1"40d31'
Always Gam
Sco 16h33'20"-43d31'
Always Grandma Clara
Peg 23h2'49"21d21'
Always John
Her 17h26'37"38d34'
Always Just S
Cyg 20h28'0"50d11'
Always Kelli
Cas 2h31'20"57d34'
Always Lora
Pup 7h28'19"22d51'
Always Marowezzi
Uma 10h17'34"47d47'
Always Mary
Uma 10h41'59"41d11'
Always Matthew
Lac 22h21'51"55d28'
Always Matthew
Cep 22h24'57"58d9'
Always Maureen
And 23h3'50"41d31'A
Always Michael
Lac 22h17'43"38d46'
Always My Lewis,Your Gracie
Aur 4h36'1"30d24'
Always My Love
Uma 9h44'39"47d24'
Always Now & A Lot
Lmi 10h35'46"28d9'
Always Olario
Per 3h16'30"40d39'
Always Pete
Hya 9h26'31"-2d25'
Always Pooh
Per 1h50'19"53d54'
Always Reggie & Louise
Cyg 21h54'58"38d38'
Always Rich
Cyg 21h33'36"36d46'
Always Rob
Her 15h52'25"41d41'
Always Robert
Per 1h47'0"56d49'
Always Sara & Irwin
Uma 9h11'24"62d5'
Always Seymore
Uma 12h3'55"42d2'
Always Sharon
Cas 1h24'36"58d13'
Always Think of Me.... and Smile
Tri 1h49'43"28d43'
Always Tl
Uma 10h21'50"41d23'
Always Together
Peg 21h38'31"27d49'
Always Whiteley-Amoss 124
Del 20h50'39"7d60'
Always Zsuzsa
Uma 14h1'55"60d34'
Always,7/25/93
Cyg 19h14'33"48d35'
Always-Dev
Lac 22h14'28"54d48'
Alwyn & Dorothy
Lyn 8h46'28"43d1'
Aly's-Sun
Cas 1h14'48"66d58'
Alyce,J.
Mon 8h5'20"-1d23'
Alycia
Mon 7h0'42"-6d11'
Alycia,Silber
Peg 22h43'41"32d48'
Alyse Siobhan
Aur 6h33'0"38d16'
Alysia G
Vul 20h19'49"22d57'
Alysia Marie
Aqr 22h52'35"-5d15'AB
Alyson & David Star, The
And 0h10'48"47d47'
Alyssa
Lyn 7h40'35"44d49'
Alyssa
Cas 0h17'59"65d32'
Alyssa
Lyn 7h39'43"51d9'
Alyssa Christine
Peg 0h11'57"18d9'
Alyssa Reneé
Cas 0h22'38"62d15'
Alyssa Zohra Moira Margaret Touati
Uma 9h4'36"57d22'
Alzayer,Zachary
Ori 5h52'0"16d32'
Alzon,Mireille
Aur 5h2'31"38d38'
Al"
Peg 23h33'0"10d35'
Am
Her 17h33'0"25d39'
Am/Can Ch Lisara's Morning After
Cas 0h47'0"62d14'
Amabile,Christina
Cas 3h10'39"61d41'
Amabile,Kerona Venera
Pic 4h58'27"-43d22'
Amadei "Tato", Cristiano
Her 18h9'27"38d39'
Amadeus
Sgr 19h37'1"-42d35'
Amadio,Lou
Aur 6h27'0"30d16'
Amado "Beloved"
Hya 10h41'1"-17d31'
Amador San Antonio (Vasa),Vicente
Aql 19h32'14"-0d28'
Amador,Hector
Sct 18h53'32"-6d47'
Amador,Raymond Howell
Ser 15h31'0"24d3'
Amaducci,Robert
Cep 20h22'1"75d14'
Amal
Aur 7h12'29"41d12'
Amal (In Andromeda) Bird of Paradise
And 1h20'28"36d56'
Amala Loves Jim Eternally
Lac 22h5'23d51"45
Amalgame le PAN
Uma 8h58'11"47d42'
Amalia
Lyr 18h16'25"44d56'
Amalie
Peg 21h29'1"20d5'
Amamus,Jacquie
Aql 19h45'34"14d4'
Aman,Thomas N
Vul 19h33'0"27d22'
Amanda
Cas 1h11'44"58d41'
Amanda
Mon 7h0'54"8d42'
Amanda
Mon 6h27'35"-10d45'
Amanda
Cyg 20h19'18"38d44'
Amanda
Lyn 9h13'18"37d41'
Amanda
And 22h56'35"41d12'
Amanda
Lyr 19h18'49"26d16'
Amanda
Lyr 18h56'48"41d36'
Amanda
Lyr 18h55'55"40d25'
Amanda
Crb 15h32'48"26d3'
Amanda
Cas 0h55'43"66d40'
Amanda
Cas 0h2'21"66d7'
Amanda
Cyg 21h40'50"31d18'
Amanda
Cyg 20h21'56"41d10'
Amanda
Cas 0h1'28"64d29'
Amanda
And 1h56'41"37d60'
Amanda
And 23h32'30"46d1'
Amanda
And 0h48'15"37d5'
Amanda
Cam 3h53'58"62d32'
Amanda "Beloved"
Crb 15h55'1"38d53'
Amanda & Pete
Cyg 21h33'54"40d2'
Amanda B
Cas 1h52'43"76d48'
Amanda Bee
Lyr 18h54'36"41d54'
Amanda Brooke
Peg 22h44'35"26d7'
Amanda D
Aur 6h4'56"37d34'
Amanda Elizabeth
Cam 5h52'1"68d54'
Amanda Grace
Del 20h37'22"11d1'
Amanda l'Étincelante
Cas 0h40'0"70d30'
Amanda Jane
Cyg 20h37'28"45d31'
Amanda Jane
And 1h1'32"37d32'
Amanda Jane
Pho 0h48'16"-47d24'
Amanda Jayne
Cas 2h53'29"57d49'
Amanda Jo's Star
And 2h20'24"48d47'
Amanda Jordan
Lyn 7h51'23"43d49'
Amanda Kate
And 23h24'12"38d31'
Amanda Lauren
Boo 14h30'26"11d56'
Amanda Lea
And 2h30'56"47d34'
Amanda Lea,The
Lyr 18h49'1"37d45'
Amanda Lorraine
Ori 4h42'18"0d58'
Amanda Lynn
Lib 15h31'0"-28d24'
Amanda M K
Cyg 21h33'42"41d52'
Amanda Marie
And 0h43'57"40d28'
Amanda Marie
And 0h1'1"35d17'
Amanda Marisa
Equ 21h11'1"10d11'
Amanda Michelle
Aql 19h57'14"14d45'
Amanda Michelle Ann
Peg 21h54'49"20d10'
Amanda Michelle Grand- ma's 3rd JoyfulSparkle
Vul 19h47'39"28d16'
Amanda Nicole
Aql 20h6'14"3d59'
Amanda R
Ori 1h18'37"37d43'
Amanda Rae
And 0h54'55"39d53'
Amanda Rae
Equ 21h2'52"7d53'
Amanda Rose D G
Vul 19h58'41"28d28'
Amanda Shannon
Per 3h18'44"54d38'
Amanda Tracy
Lmi 10h26'52"38d9'
Amanda's Aurora
And 23h5'16"38d18'
Amanda's Cella System
Ori 6h12'45"18d18'A
Amanda's Guiding Light
Tri 2h7'41"31d3'
Amanda's Jewel
Cas 1h20'20"60d1'
Amanda's Light
Cnv 13h58'42"28d15'
Amanda's Smile
Peg 21h36'50"20d2'
Amanda's Star
Lib 15h41'21"-23d49'
Amanda's Star
Peg 23h33'17"20d57'
Amanda's Wish Star
And 0h39'25"40d50'
Amanda's Wonder
Peg 22h2'23"32d55'
Amanda,Jack,Sawyer, Luke, & Taylor
Umi 14h39'11"67d34'
Amanda-Panda-1983
Uma 10h1'55"68d2'
Amandine
Peg 22h29'54"20d5'
Amann,Scott Edward
Boo 14h17'41"39d26'
Amanta,(AgÓpe) Filia Storgè
Hor 3h31'37"-49d4'
Amante,Sharon
Peg 22h13'0"10d21'
Amante,Sogno
Peg 23h15'59"31d4'
Amantes Destino Dee & G 3/8/94
Sex 9h39'46"-6d4'
Amantes Sensualez
Lac 22h8'35"50d51"
Amantiane
Vir 12h8'34"-5d45'
Amantine,Marianna
And 0h51'0"36d56'
Amanuma,Akira
Sge 20h17'54"17d37'
Amar,Barbara
Cet 2h54'37"-0d46'B
Amar,Jack
Cet 2h54'37"-0d46'A
Amar,Jason
Cet 2h59'45"0d38'
Amar,Paula & Melvin Schwartz
Lyr 18h41'19"38d46'
Amara
Cet 1h20'0"-2d40'
Amara-Prema
Vir 13h20'37"-2d17'
Amaral,Jeffrey & John R Estheimer
Cyg 21h50'27"41d55'
Amaral,Lisa Marie
And 23h46'11"43d31'
Amarando,Kelly
And 0h55'1"37d36'
Amarante
Tri 2h3'20"31d29'
Amarante
Ser 15h59'54"1d16'
Amaranth,Danielle
Cyg 20h20'0"40d16'
Amaranthine
Mon 6h31'0"-1d23'
Amarantos 78ASN
And 23h49'10"41d24'
Amare Inaeternum, Grandma Marilyn
Cmi 7h26'28"0d58'
Amare,Chantal
Cyg 19h42'53"38d52'
Amarilis
Oph 17h31'17"-20d1'
Amaro III,Joe
Umi 15h10'39"66d60'
Amarylla
Ori 6h2'32"5d25'
Amaryllis
And 23h43'15"46d27'
Amat:Alexander Mark Always Together
Ori 5h52'16"15d2'
Amatetti,John Francis
Per 4h5'11"38d5'
Amato Tu
And 23h2'33"52d48'
Amato,Ashley
Lyr 19h16'49"28d47'
Amato,Cathy
Aur 5h8'59"40d43'
Amato,Dana Marie
Cas 0h21'20"65d4'
Amato,Karen Ann Marie
Lyr 19h0'16"40d37'
Amato,Kelli & John
Cyg 21h4'37"39d17'
Amato,Michael Paul
Tau 4h11'47"20d3'
Amato,Michele
Cas 1h33'14"75d60'
Amato,Pam
Cam 5h57'51"68d53'
Amato,Regina Ann
And 1h23'41"33d43'
Amato,Vincent A
Cep 22h27'0"60d14'
Amaturo,Sr,Michael
Her 17h16'59"49d12'
Amaursra
Sct 18h55'46"-7d45'
Amaury & Carol's Star
Cyg 20h51'44"50d28'
Amaya Mesequer
And 23h15'13"50d8'
Amaya,Reynaldo
Lmi 9h44'33"38d32'
Amaya,Tim
Cep 21h4'1"65d18'
Amaya Star '69,Cilda Mercedes
Peg 22h43'1"24d29'
Amaz
Cam 5h4'33"67d42'
Amaz'n-Gra
Mon 7h46'5"-2d7'
Amazing
Aql 19h22'57"-1d44'
Amazing Grace
Gem 6h29'12"14d52'
Amazing Grace
Aql 18h41'14"-2d48'
Amazing Grace
Cas 0h28'37"68d52'
Amazing Grace
Uma 11h19'18"44d59'
Amazon Bar & Grill- Love Jackie
Tau 5h48'45"27d0'
Ambar
Cyg 20h0'31"30d16'
Ambassador Translating Inc
Cam 7h3'49"83d58'
Amber
Cet 2h31'6"-1d20'
Amber
And 23h16'40"41d50'
Amber
Lyr 19h20'33"40d2'
Amber
Mon 6h32'29"5d16'
Amber
Equ 21h4'26"11d43'
Amber Dawn
Mon 7h38'55"-1d41'
Amber Deanne
Cmi 7h22'42"7d31'
Amber Faye
Cyg 20h41'54"38d18'
Amber Grace's Eternal Light
Cyg 21h31'1"40d5'
Amber Louise
Cyg 20h41'54"45d37'
Amber Marie
Lyn 7h5'32"53d27'
Amber Rae
Mon 8h2'20"6d26'
Amber Ura
Cam 3h51'10"68d46'
Amber's Destiny
Mon 6h35'49"-1d13'
Amber's Honor
Psc 1h23'14"18d35'
Amber's Special Star
Lyr 19h4'0"37d40'
Amber's Star
Lyn 7h44'0"43d10'
Amber's Star Spangle
And 2h33'13"42d48'
Amber's Wishes, Hopes & Dreams
Tau 5h50'44"28d12'
Amber-N-Peter "Forever"
Mon 6h54'46"-6d18'
Amber-Pebbles
Mon 6h57'32"-8d7'
Amberg,J
Oph 17h9'1"11d27'
Amberg,Karen Marie
Mon 6h57'36"-1d23'
Amberleen
Lyr 18h47'47"33d25'
Amberly
Umi 13h27'43"70d39'
AMBI
Tri 2h6'35"32d34'
Ambielli,John S
Lac 22h36'42"50d30'
Amblad,Erik E
Per 2h25'0"58d30'
Amboise J Ph
And 0h27'14"40d24'
Ambra
Ant 10h45'29"-39d28'
Ambre,Anita
And 23h30'34"43d23'
Ambre,Cecilia
Lyr 18h49'16"32d34'
Ambreen
Lyr 18h15'24"38d46'
Ambrico,William
Lmi 9h49'48"34d27'
Ambroch,Fritz J
Boo 14h18'29"34d57'
Ambrogi III,Leo John Louis
Psc 22h56'1"2d9'
Ambrose
Her 18h16'24"14d54'
Ambrose,Brian
Boo 15h12'0"31d53'
Ambrose,Christine
Cyg 21h38'28"38d26'
Ambrose,Dorothy
Cas 1h27'1"58d55'
Ambrose,Dr & Grace McAlevy
Per 3h16'55"41d10'
Ambrose,Jeffrey Stuart
Mon 7h12'45"-0d4'B
Ambrose,John Vincent
Tau 4h34'20"30d50'
Ambrose,Justin Tyler
Boo 14h6'54"34d25'
Ambrose,Laura Pankratz
Mon 7h12'45"-0d4'A
Ambrose,Mary
Peg 23h39'60"31d58'
Ambrose,Maureen B
Uma 9h4'31"70d13'
Ambrose,Steven
Cep 20h54'33"55d59'
Ambrose,Thomas Gordon
Cep 21h38'50"58d29'
Ambroselli,Andrea
Cas 1h46'52"58d6'
Ambrosia
Mon 6h57'24"-10d5'
Ambrosia
Ori 5h4'32"8d4'
Ambrosini,Michael
Her 18h2'35"47d1'
Ambrosino,Dr,Salvatore V
Per 2h38'32"40d49'
Ambrosino,Joe
Boo 15h8'24"50d24'
Ambrosino,Jr,Victor
Cyg 20h19'19"38d44'
Ambrosino,Nunzio
Cyg 19h27'46"40d53'
Ambrosio,Louis & Anna
Tri 2h16'14"30d23'
Ambrosio,Zachary John
Lac 22h36'34"54d13'
Ambrozaitis IV,John Peter
Lac 22h26'18"56d24'
Ambruster,Robin A
Lyr 18h20'19"46d23'
Ambury,Karyn
Mon 8h6'19"-1d15'
Amcob
Tau 4h20'44"15d16'
Amdahl,Carlton G
Lyr 19h8'16"38d49'
Amdurer,Harry
Aur 4h57'46"51d49'
Amedeo e Liliana
And 0h10'15"38d20'
Amee's Passion
And 0h8'44"40d51'
Amelda
Uma 9h33'31"56d7'
Amelia
Lyr 18h30'50"40d17'
Amelia & Neil
Uma 10h50'19"48d52'
Amelia Hannah
Lyr 19h0'1"26d13'
Amelia KM
Lyn 8h42'46"39d54'
Amelink-Berg,Anna Maria
Vir 13h15'59"-0d20'
Amelung,Klaus
Aur 5h13'1"44d24'
Amen,Ellen
Aql 20h4'32"0d54'
Amend,Donald F
Her 16h46'35"38d31'
Amend,Loretta F
Lyn 8h50'20"34d60'
Amend,Richard
Her 16h44'12"28d12'
Amende,Karl-Heinz
Lib 15h7'54"-21d40'
Amendolaro,The Red Rose of Raquel
Lyn 7h29'1"44d44'
Amenta,Arienne
Mon 7h48'46"-6d50'
Amer,Barbara & Hisham
Lyn 8h54'53"40d41'
Amer,Barbara & Hisham
Cyg 19h27'21"31d8'
Ameri,Shadi
Uma 12h41'16"58d28'
America's Favorite Colleen
And 0h20'40"31d59'
American Beauty
Cyg 19h34'24"39d33'
American Express "Personal Rewards"
Boo 15h45'20"47d39'
American Pest Control
Cnv 12h32'1"40d13'
American Pie
Boo 14h18'1"32d59'
Americo L Traci
Aur 6h25'0"31d49'
Ameritech EBS
Her 16h43'1"38d46'
Amerkan,Capt Jo Jo
Aql 20h11'56"14d13'
Amerotti,Grazia
Psc 22h57'38"1d8'
Amerson,W A
Boo 14h31'1"21d33'
Amerson-Weible,Cynthia L
Lyr 18h27'10"35d34'
Ames
Uma 10h41'33"40d9'
Ames Technology Commercialization Cntr
Umi 14h8'0"74d30'
Ames,Alexandra Katherine
Vul 19h19'12"25d57'
Ames,Gay
Sex 10h11'28"-2d29'
Ames,Karyn Suzanne
Lyr 19h14'36"41d3'
Ames,Kenny
Leo 11h8'14"-1d49'
Ames,Kimberly Marie
Cmi 7h45'56"1d19'
Ames,Lauren Alyssa
Cyg 19h50'34"40d6'
Ames,Rebecca
Sgr 18h56'20"-26d33'
Ames,Tasha
Mon 6h38'19"11d55'
Ames,The
Ori 5h32'9"-2d9'
Ametista
Psa 22h22'51"-27d35'
Amez,Mary K
Peg 23h21'16"15d49'
Amgar,Ilan Joseph
Psc 22h51'22"5d2'
Ami
Aql 19h7'24"3d22'
Ami & Brian
Her 18h22'10"12d46'
Ami & Matt
Cyg 19h32'42"30d38'
Ami & Warren 3-4-95
Her 18h45'16"12d23'
AMI 71472
Mon 8h4'54"-6d17'
Ami Tara
Cnv 12h47'1"33d22'
Ami's Star
Cma 7h12'1"-13d12'
Ami-Amish
Com 12h5'45"24d55'
Ami-KK-Scamp
Lyn 8h29'19"46d7'
Amiard/Thylis,Isabelle
Cam 4h35'44"58d58'
Amicangioli,Linda
Lyr 18h23'23"44d55'

Amici miei
Cam 6h29'1"68d54'
Amicitia Aeternus
Lyr 19h3'57"25d37'
Amicitia Semperterna
Cam 6h7'18"60d57'
Amick,Sarah Catherine
Cas 0h26'58"63d29'
Amico,Andrea
Equ 20h55'33"3d27'
Amico,Kristopher W
Vul 20h28'58"28d52'
Amicus
Cnv 13h46'58"35d2'
Amicus Entorum
Lmi 10h36'14"30d7'
Amiee
And 0h20'24"32d14'
Amigron,Michael Christopher
Per 4h2'12"52d16'
Amil,Nicole
Ori 5h57'39"8d5'
Amilcare
Cas 23h4'18"53d24'
Amina
Lyr 19h11'43"47d17'
Aminifu,Taminifu T
Boo 14h37'50"22d28'
Amipa
And 0h4'1"40d11'
Amira,Abe
Cas 0h19'42"65d2'
Amiram
Aql 18h58'36"16d44'
Amirault,Christopher Edward
Dra 16h6'45"67d15'
Amisano,Philip L
Tri 1h57'19"34d52'
Amisisisismo,Steph Ti
Lac 22h27'27"40d21'
Amit
Uma 10h47'44"72d56'
Amit
Uma 11h18'46"38d10'
Amjadi,Saied
Vul 19h47'47"29d7'
Amjadi,Saied
Dra 18h22'35"65d37'
AML112986TJR
Tri 2h34'51"31d31'
Amler,Elizabeth Brittany
Cam 4h14'43"68d29'
Ammann,Cornelius
Umi 13h40'51"76d39'
Ammann,Klaus B B
Gem 6h41'30"30d45'
Ammann,Rudolph
Per 2h54'59"37d18'
Ammering,Shane Patrick
Per 2h24'15"55d30'
Ammerman,Dorothy June Davis
Psc 0h47'0"32d44'
Ammerschlaeger,Alois
Leo 10h48'32"1d20'
Ammie Marlene
Crb 16h10'23"26d55'
Ammirati,Frank A
Her 17h59'31"14d50'
Ammons,David
Per 3h29'50"51d31'
Ammons,Forrest Jernigan
Sco 16h29'53"-30d1'
Ammouri,Susan
Oph 18h0'53"12d23'
Amnoe
Aur 6h26'47"37d2'
Amo Paulus
Cap 21h50'25"-10d54'
Amo,Jennifer
Cas 23h5'36"58d41'
Amo,Lisa Semper
Peg 0h4'20"14d25'
Amodeo,Christopher
Cep 22h26'24"60d15'
Amodeo,Deno
Her 16h39'49"47d38'
Amodeo,Dorothy
Cas 0h33'1"64d59'
Amodio,Robert M
Aur 4h53'53"40d44'
Amoia,Loraine "Sis"
Lyn 7h26'37"50d21'
Amon-Sarmiento
Hya 8h40'28"2d55'
Amor
Aur 5h50'1"50d26'
Amor de Elena
Lyr 18h18'25"40d40'
Amor Eterno:Valentin y Maria
Sge 18h55'44"19d57'
Amor Felicitasque
Boo 15h3'15"32d3'
Amora,Miryam Michele
Per 3h30'27"51d17'
Amorae Aeturnae
Aur 6h25'21"38d38'
Amore
Lyr 19h24'15"41d37'
Amore
Pho 0h38'28"-48d6'
Amore Astro
Sge 20h4'55"16d20'
Amore,Christopher Joseph
Dra 16h33'39"64d52'
Amore,Magen
Sco 17h26'26"-31d29'
Amore,Sr,Anthony
Lmi 10h48'43"30d19'
Amorek
Per 2h54'52"31d10'
Amoria,Sunshine
Her 16h12'0"10d36'
Amoroso Bruce
Boo 14h52'48"30d28'
Amoroso,Joseph
Cnv 12h20'50"40d0'
Amoroso,Larry
Ser 15h41'27"-2d29'
Amoroso,MD,Thomas A
Ori 6h11'1"8d52'
Amorous
Cam 5h7'0"70d60'
Amorous Angel
Del 20h14'37"10d13'
Amos
Cam 5h22'15"68d2'
Amos,Claire
And 23h1'14"51d35'
Amos,Sgt Major George J
Cet 3h8'14"2d50'
Amos,Theresa Ann
Mon 5h58'39"-5d48'
Amos,Tori
And 0h57'54"33d50'
Amoul,Marie
Hya 8h43'52"5d6'
Amour
Lac 22h3'37"37d41'
Amour
Her 17h24'32"40d39'
Amour
Cas 0h31'50"68d32'
Amour
Lyn 8h44'55"40d43'
Amour Ca-ro
Cyg 21h51'60"36d60'
Amour Cathy
Mon 7h52'60"-5d45'
Amour infini
Cyg 21h4'24"38d5'
Amour J & M
Del 20h56'14"18d47'
Amour Toujours
Lyr 18h42'53"42d13'
Amourdiac
Uma 11h31'35"38d11'
AMP III
Uma 9h29'1"50d44'
Amparano,Robert
Peg 22h1'47"2d27'
Amparán-The Master, Gregg
Her 17h36'60"20d57'
Amra
And 0h24'45"32d26'
Amrhein,Al
Lac 22h9'53"50d56'
Amrita
Dra 19h34'29"67d46'
Amroian,Steven William
Dra 14h52'49"57d13'
Amrol-Davis,Phoenix Myles
Peg 21h11'23"18d55'
AMS Thunder Road TWK
Cep 23h3'11"70d56'
Amsellen,Beatrice
Dra 20h58'0"80d15'
Amsler III,John J
Aur 4h59'33"37d59'
Amstutz,Daniel Coordon
Cep 22h54'15"58d30'
Amtsberg,Gesa
Vir 12h2'19"-5d50'
Amundsen,Marion
Lyr 18h35'1"42d7'
Amundsent Family Star, Katie,Mitchell
Ori 5h5'0"11d9'
Amundson 64
Umi 15h5'40"77d46'
Amundson,Amy E R
Uma 11h22'47"60d23'
Amundson,Arlen
And 0h16'23"34d30'
Amundson,Robert
Cep 20h37'35"76d38'
Amundson,Steven Gordon & Sharon
Dra 13h49'15"64d23'
Amunson,Sean
Cmi 7h5'57"3d53'
Amy
Eri 3h20'1"-1d53'
Amy
And 2h15'57"37d16'
Amy
Lyr 19h7'46"38d35'
Amy
Lyr 18h16'17"32d46'
Amy
Lyr 18h20'49"38d10'
Amy
Cas 1h16'1"62d5'
Amy
Lyn 8h14'18"35d6'
Amy
Uma 8h53'27"53d41'
Amy
And 0h10'57"38d13'
Amy
Cam 8h39'55"73d47'
Amy
Cas 0h22'34"59d55'
Amy
Vul 19h3'25"21d17'
Amy
Cas 0h53'50"61d29'
Amy
Uma 10h52'30"59d6'
Amy
Cyg 20h6'19"42d14'A
Amy
Cas 1h0'40"48d59'
Amy
Lib 14h29'35"-22d44'
Amy
Uma 11h5'40"46d2'
Amy
And 22h55'26"36d34'
Amy
Peg 22h57'11"18d24'
Amy
Lib 14h18'27"-8d4'
Amy & Anthony
Aql 19h5'48"3d53'
Amy & Julian
Sge 20h4'0"20d36'
Amy & Leland's "Star Shines On"
Cet 2h41'41"2d9'
Amy & Morgan
Cyg 21h6'26"30d23'
Amy & Nick 6-14-96
And 0h8'37"35d44'
Amy & Steven
Vul 20h0'12"28d43'
Amy 143
And 23h35'1"40d58'
Amy Amanda
Mon 6h39'14"7d19'
Amy Angelica
Del 20h12'22"15d1'
Amy Ann
Cas 0h10'1"56d2'
Amy Avonne
Cnv 12h46'11"36d39'
Amy Beth
Eri 3h53'49"-0d3'
Amy Beth
Uma 10h33'40"58d17'
Amy Beth
Cam 5h38'21"65d4'
Amy Christine
Peg 23h21'22"32d34'
Amy Claire
Mon 8h1'25"-4d38'
Amy Dee
Lyr 18h58'0"41d40'
Amy Diane
Umi 13h58'0"70d36'
Amy Diane
Aql 18h56'0"16d21'
Amy Elizabeth
Peg 22h59'31"33d55'
Amy Elizabeth
Aql 18h59'11"-6d55'
Amy Elizabeth
Cas 23h2'41"53d49'
Amy I
And 2h4'27"37d13'
Amy Jane
Lyr 18h56'38"40d37'
Amy Jean
Cam 7h59'0"68d46'
Amy Jean
Cam 7h51'53"60d12'
Amy Katherine
Peg 22h42'15"22d49'
Amy Kathryn
Mon 6h27'45"-1d22'
Amy L
Cam 5h46'34"73d32'
Amy L
Com 12h22'32"20d5'
Amy Lee
Umi 16h45'49"84d29'
Amy Lilly
Cyg 21h26'13"37d48'
Amy Lin
And 0h47'15"40d13'
Amy Louise
Cas 0h36'46"60d54'
Amy Louise
Umi 13h35'0"71d37'
Amy Loves Cam Forever
Oph 17h34'25"11d33'
Amy Loves You
Mon 7h49'8"-5d15'
Amy Lyn
Lyn 6h53'48"52d59'
Amy Lynn
Vul 19h48'34"27d9'
Amy Lynn
Mon 7h22'1"-8d39'
Amy Lynn
And 0h53'0"33d36'
Amy Marie
Sco 17h31'53"-30d31'
Amy Marie
Mon 7h12'3"-9d31'
Amy My Girl
Crb 15h38'44"26d2'
Amy Noel
And 1h17'57"33d42'
Amy Pearl
Peg 21h55'0"29d34'
Amy Ray
Cmi 7h32'32"10d42'
Amy Renee
Peg 23h3'41"22d1'
Amy Renee
Uma 12h0'0"32d57'
Amy S
Aql 19h58'11"1d50'
Amy Sue
Hya 8h32'26"-0d57'
Amy Sweet Mother of Célia & Loriane
Cam 4h57'0"60d29'
Amy's Ardor
Peg 22h38'18"24d45'
Amy's Big Caboose
And 2h16'22"37d11'
Amy's Bright Eye
Peg 22h2'40"30d52'
Amy's Celestial Diamond
Sct 18h54'1"-4d30'
Amy's Hideaway
Mon 7h27'32"-6d29'
Amy's Hope
Peg 0h6'26"27d51'
Amy's Love
Ori 5h53'52"8d32'
Amy's Sparkle
Cas 23h5'18"53d44'
Amy's Star
And 23h33'27"37dl'
Amy's Star
Lyr 18h33'15"40d10'
Amy's Star
And 23h35'0"48d12'
Amy's World
And 0h54'37"21d36'
Amy,I Love You
Cas 0h7'1"63d22'
Amy,James D
Aql 19h31'0"13d15'
Amy-Renee
And 1h0'11"36d56'
Amyable,Immortality In The Heavens
Crb 16h2'53"32d27'
Amylou
Peg 23h23'21"15d19'
Amyotte,Rebecca Meghan
And 22h59'39"37d48'
Amélie
Ori 5h53'38"8d6'
Amélielydia
Per 2h53'1"45d14'
An
Lyn 7h9'32"58d41'
An ce gio se re ga
Lep 4h58'27"-11d27'
An Empyrean Looking-Glass;GH&JA
Uma 10h12'43"60d21'
An Eternal Embrace
Sge 19h52'22"16d27'
An Eternal Jewel From Me to You
Uma 11h10'47"47d41'
An Shan Iron & Steel (Group) Company
Her 16h58'22"23d45'
An-Lorraine
Uma 12h14'53"58d10'
Ana 12-08-1990
Lyr 19h17'21"31d23'
Ana Arie
Tau 4h13'26"22d6'
Ana Claudia
Eri 2h51'59"-5d16'
Ana Elizabeth
Cas 1h3'52"63d34'
ANA Hallo Tour
Cru 12h12'15"-60d40'
Ana Marina
Vul 19h19'40"26d9'
Ana Rosa-Pepe
Cyg 21h31'30"52d45'
Ana Tarracona i Felip
Ori 5h39'0"7d46'
Ana's,Star
Eri 3h8'0"-6d47'
Ana(Amy & Aaron)
Ori 6h13'42"7d31'
Ana-Isabel Rios Huercano
Per 3h59'1"51d41'
Ana-Maria
Cas 0h43'43"70d28'
Ana-Rosa
Tau 4h4'16"1d51'
Anabel
Equ 21h3'12"11d20'
Anabel
Cyg 19h32'47"36d16'
Anabel
Per 3h27'18"41d4'
Anaedante
Vel 9h23'47"-40d53'
Anais
Dra 16h48'20"62d3'
Analeesa-Ross
Aur 5h2'51"48d30'
Analytic Andria
Cas 0h26'29"70d55'
Anand,MD,Azad K
Aur 5h10'54"41d49'
Anand,Ravi & Priska
Cep 21h1'19"65d23'
Anand,Rocky
Her 17h22'27"27d24'
Ananda Zoe
And 1h12'21'47d33'
Anandappa,Ann Marie Estelle Deepthie
And 0h19'58"35d51'
Ananew,George J
Aur 5h19'27"41d34'
Ananian,Sharon
Eri 4h5'39"-12d60'
Ananth
Psc 1h3'13"20d6'
Anapolsky,Julian David
Per 2h53'25"38d6'
Anashua
Crb 16h21'0"28d5'
Anastacia
Cas 2h20'43"63d55'
Anastasia
Mon 6h28'52"1d52'
Anastasia
Cnv 14h1'17"39d21'
Anastasia Capri
Peg 22h57'44"34d45'
Anastasia's Star
And 23h38'22"37d47'
Anastasios
Cru 12h40'1"-57d31'
Anastasis
Mon 7h19'4"-5d31'
Anastos,C G
Tau 4h5'53"21d17'
Anastos,George
Dra 15h36'55"53d10'
Anat
And 0h8'25"28d33'
Anathan Star,The Jim & Peggy
Oph 17h48'9"7d14'AB
Anathirolef
Aur 5h8'0"41d23'
Anais,Ronard
Mon 7h50'11"-4d11'
Anberson,Erika Gwen
Del 21h5'39"12d49'
Ancelet,Jolene Rachell
Uma 9h57'15"58d39'
Ancellotti,Lisa M
Mon 6h20'17"7d52'
Anchondo,Mayra Lizbeth V
Mon 6h24'55"5d20'
Ancillotti,Barbara
Cnv 12h12'0"51d13'
Anckaert,Jean
Cam 7h34'0"80d3'
Ancker,Elizabeth
Umi 14h31'0"68d4'
Ancona,Carmelo J
Her 18h2'60"28d21'
Ancona,Katherine Elizabeth
Cet 0h36'46"1d53'
And En Al
Aur 5h25'37"37d32'
And Everything
Peg 23h2'57"30d40'
And Worlds Collide
Kelly,Anjum & Phil
Aql 19h52'30"-6d29'
Andala & Petal
Lmi 10h24'53"31d38'
Andaloro,Lisa
Lyr 19h19'10"33d57'
Andante
Cam 7h36'45"60d11'
Andee's Magic
Aur 6h24'22"30d43'
Andel De las Estrellas
And 23h3'33"41d21'
Andel,Joan Valenta
Mon 6h2'40"-8d35'
Andelbradt,Michael Murphy
Ori 5h57'54"14d40'
Andelian Friendship
Dra 19h55'22"67d52'
Andelman,Bennett Michael
Aur 7h9'21"40d20'
Andera Nicole
Cas 2h36'0"57d40'
Anders,Anna Beeka
Lib 14h40'47"-0d32'
Anders,Ian
Boo 15h3'54"20d41'
Anders,Ralph E
Aql 20h3'0"0d4'
Anders,Shirley
Cyg 21h58'27"52d47'
Andersen,Anne B
Cam 3h31'0"60d10'
Andersen,Christine
Lyr 19h0'12"35d24'
Andersen,Daniel John
Vul 19h47'12"28d21'
Andersen,David Aaron
Her 18h55'39"12d45'
Andersen,Don
Oph 18h0'49"12d27'
Andersen,Inge-List
Ori 6h2'38"8d21'
Andersen,Janet Kelly Mary
Lyn 7h31'18"44d31'
Andersen,Joyce Elizebeth
Cas 0h32'0"60d21'
Andersen,Kenneth C
Aur 5h11'24"40d36'
Andersen,Linda J
Peg 23h4'51"31d26'
Andersen,Lynne
Uma 9h59'39"57d59'
Andersen,Marilyn Kay Nelson
Crb 16h2'12"28d10'
Andersen,Melody
And 0h53'30"36d27'
Andersen,Rhea Fortuna
Lac 22h8'11"37d50'
Andersen,Robert Arthur
Leo 9h58'18"15d23'
Andersen,Ruth
Cas 23h24'40"61d38'
Andersen,Tina
Cas 0h35'31"67d36'
Anderson
Aur 6h51'1"37d57'
Anderson "Chipmunk", Julie
Lyn 8h36'40"45d13'
Anderson "Olivia", Olivia
Lyn 8h41'57"43d19'
Anderson & Children, John & Lauren
Cyg 21h50'50"42d32'
Anderson,Jason Andrew
Pho 23h47'54"41d38'
Anderson 7-18-83, Charles R
Aur 7h11'0"39d48'
Anderson 8-18-78, "Sport" Mark John
Leo 10h36'19"14d36'
Anderson II,Barb & Ray
Peg 23h44'37"31d40'
Anderson III,Clifford
Dra 14h48'59"59d17'
Anderson Light,The
Peg 22h43'15"4d28'
Anderson MCMXLV
Eri 3h22'27"-3d4'
Anderson Of Spokane, Robert Edward
Dra 14h32'1"61d14'
Anderson PGM Eastern Star NJ,Jackie
Cam 9h53'28"82d25'
Anderson Star,The
Boo 14h33'0"30d57'
Anderson Star,The Walter
Cyg 20h35'1"30d25'
Anderson,Abby Nichole
Mon 6h47'19"11d32'
Anderson,Alan Edward
Ori 5h33'52"-6d55'
Anderson,Alana & Runo
Eri 4h4'15"-10d2'
Anderson,Aleia
Cyg 21h46'1"38d32'
Anderson,Alexander Gregory
Ari 1h46'44"11d55'
Anderson,Althea R
Lyn 7h54'14"43d43'
Anderson,Amanda Elizabeth
Peg 21h45'34"28d8'
Anderson,Andrea
Cas 0h25'20"62d36'
Anderson,Anne & Tore
Dra 16h1'45"65d38'
Anderson,Anne B
Cam 3h31'0"60d10'
Anderson,Annlou Devine
Lyr 18h54'11"32d26'
Anderson,April Dawn
Cyg 20h16'18"31d29'
Anderson,Arnold Eugene
Dra 16h32'13"69d36'
Anderson,Aryn Elizabeth
Vul 19h0'23"24d34'
Anderson,Ashley
Cnc 8h39'24"18d45'
Anderson,Ashley Kay
Eri 4h13'32"-14d33'
Anderson,Audra Lynn
Cas 1h49'48"61d29'
Anderson,Avalon Boone
Peg 21h43'12"28d0'
Anderson,Baby Angel
Uma 12h33'21"62d50'
Anderson,Benjamin Monroe
Tri 1h54'44"27d17'
Anderson,Benjamin Zane
Uma 8h27'33"62d26'
Anderson,Bethany Jean
Peg 22h13'13"7d33'
Anderson,Bill
Ori 5h7'24"0d31'
Anderson,Boone Carson
Peg 22h2'58"29d51'
Anderson,Bradley Keene
Leo 11h7'45"-0d36'
Anderson,Brandon Christopher
Her 17h35'29"25d54'
Anderson,Brenda
Leo 9h40'20"10d44'
Anderson,Brian
Aur 6h31'33"32d57'
Anderson,Brian J
Her 17h17'40"21d26'
Anderson,Briana Jeannette
Peg 22h21'0"31d26'
Anderson,Cari Lynn
Cas 0h58'48"63d53'
Anderson,Carl
Her 17h14'40"45d40'
Anderson,Carlos D
Aur 5h6'0"42d5'
Anderson,Catherine Balint
Lyr 18h25'17"47d40'
Anderson,Charles Lee
Aur 5h25'0"31d33'
Anderson,Cheryl L
Cas 1h10'35"60d30'
Anderson,Chester Bernard
Ser 15h54'50"20d35'
Anderson,Chloe Betts
Vul 19h1'35"20d17'
Anderson,Chris
Del 20h13'50"12d19'
Anderson,Christine Hanjiao
Uma 8h17'31"62d21'
Anderson,Christine R
And 1h38'51"40d41'
Anderson,Christopher
Tri 2h6'42"32d30'
Anderson,Connie Sue
And 23h44'36"32d51'
Anderson,Courtney Elizabeth
And 2h3'21"46d13'
Anderson,Damion E
Ser 15h35'10"20d13'
Anderson,Danielle
Ori 4h54'57"-2d16'
Anderson,Darrick LaMont
Boo 14h14'53"20d17'
Anderson,David & Kathryn
Eri 3h43'1"-2d57'
Anderson,David Alistar
Cyg 19h27'53"36d48'
Anderson,David Louis
Aql 19h59'12"10d8'
Anderson,Dawn
And 23h20'0"45d58'
Anderson,Deborah
Peg 22h58'22"32d6'
Anderson,Delores
And 2h35'14"50d24'
Anderson,Dennis D A
Dra 20h14'29"68d9'
Anderson,Diane Ruth
Mon 6h21'29"8d53'
Anderson,Don
Cas 1h6'23"58d16'
Anderson,Doris Anne Bare
Cyg 20h30'53"48d57'
Anderson,Dorsey (Snake)
Cam 8h54'58"82d17'
Anderson,Dr Daniel E "Andy"
Aur 4h47'35"50d30'
Anderson,Drew
Per 3h39'34"36d60'
Anderson,E M (Tex)
Ser 18h4'24"-6d3'
Anderson,Eden Ann
Lib 15h41'0"-20d25'
Anderson,Edward Herbert
Vir 11h59'1"1d37'
Anderson,Edwin P
Sct 18h21'40"-14d38'
Anderson,Elizabeth
Cmi 7h26'33"1d16'
Anderson,Elizabeth Denise
Sex 9h50'26"1d35'
Anderson,Elizabeth Ann
Cas 0h33'14"67d9'
Anderson,Elyssa Marie
Lyr 18h57'52"30d23'

Anderson,Eric P
Hya 8h59'41"-1d5'
Anderson,Erling P
Uma 9h18'29"52d17'
Anderson,Evelyn G
Lyn 8h9'22"51d52'
Anderson,Faye
Cas 2h23'13"65d34'
Anderson,George B
Sct 18h42'42"-7d31'
Anderson,George M
Cyg 21h18'21"52d45'A
Anderson,Gerald Duane
Eri 3h2'21"-10d26'
Anderson,Grayson
Cet 0h27'18"-12d4'
Anderson,Gregg & Sue
Leo 10h1'0"18d24'
Anderson,Gregg Robert
Ser 15h14'18"20d34'
Anderson,Gregory David
Ori 5h55'57"21d7d
Anderson,H Michael
Aql 19h14'21"-0d23'
Anderson,Hallibe
Peg 21h29'17"27d57'
Anderson,Hanna
Peg 22h3'45"20d32'
Anderson,Helen
Cam 3h39'38"60d29'
Anderson,Howard Guy
Aql 19h0'55"13d6'
Anderson,Ian Kruer
Aql 20h32'10"0d12'
Anderson,Irma Speight
Williamson
Mon 7h57'12"-0d59'
Anderson,Jacqueline Jennifer
Cas 0h35'56"60d26'
Anderson,Jaime Elizabeth
Cyg 20h27'59"37d35'
Anderson,James Arthur
Aql 18h59'49"-6d13'
Anderson,James Ian
Tri 2h13'53"31d33'
Anderson,James L
Cma 7h16'28"-15d24'
Anderson,Jane Srah
And 0h44'12"38d26'
Anderson,Jarod M
Her 16h27'33"28d29'
Anderson,Jeanie Deland
Crb 15h58'28"31d12'
Anderson,Jeff & Karen
Umi 13h15'23"70d56'
Anderson,Jennifer R
And 1h25'28"35d9'
Anderson,Jimmy R
Crt 11h34'0"-18d47'
Anderson,John G
Aql 19h42'48"13d42'
Anderson,John H & Cindy S
Sex 09h58'20"-6d36'
Anderson,John Paul
Lac 22h11'25"51d34'
Anderson,John William
Vul 19h18'24"26d17'
Anderson,Jonathan Yves
Vul 19h21'38"27d28'
Anderson,Josa
Sex 9h55'50"5d47'
Anderson,Joshua Phillip
Cnv 12h48'42"38d34'
Anderson,Joyce Dulany
Mon 6h26'11"7d40'
Anderson,Jr,Clyde Clifton
Psc 23h29'46"0d59'
Anderson,Jr,Edwin Tarker
Boo 14h33'37"20d6'
Anderson,Jr,Jesse Earl Joan
Anderson
Oph 17h57'48"8d51'AB
Anderson,Jr,Kenny W
Oph 18h7'53"12d52'
Anderson,Jr,Richard
Boo 15h17'42"52d17'
Anderson,Jr,Rudolph
Valentino
Boo 14h13'37"31d9'
Anderson,Judy
Mon 6h38'2"-0d11'
Anderson,Julian Wayne
Eri 4h54'50"-5d49'
Anderson,Julie
Cyg 21h3'51"30d37'
Anderson,Julie
Peg 23h3'18"24d24'
Anderson,Karyn Christine
Cas 22h58'1"54d40'
Anderson,Kathleen "Daute"
Cyg 21h18'21"52d45'B
Anderson,Kathryn Anne
And 0h54'16"35d58'
Anderson,Keegan Nicole
Peg 22h29'33"33d55'
Anderson,Keith Wayne
Dra 16h58'16"62d55'
Anderson,Kendall Joseph
Isherwood
Lyr 19h13'14"40d46'
Anderson,Kevin
Cep 20h32'56"61d39'
Anderson,Kim
Ori 5h48'1"11d49'
Anderson,Kimberly & John
Lyn 7h29'33"58d54'
Anderson,Kip Alan
Ori 5h55'47"6d11'
Anderson,Kirsten Marie
Aql 19h50'39"13d58'
Anderson,Kursten
Crb 15h47'49"27d2'
Anderson,Laura Susan
Aql 19h23'31"1d25'
Anderson,Lauren R
Cyg 19h31'30"36d1'
Anderson,Lavina J
Crt 10h56'58"-18d13'
Anderson,Lawrence E.
Boo 13h42'33"23d30'
Anderson,Lea
Cyg 21h2'21"28d19'
Anderson,Lea
Vul 20h28'56"28d11'
Anderson,Leah
Peg 0h3'1"30d2'
Anderson,Lee
Vul 19h22'58"26d53'
Anderson,Lee Eric
Aql 18h54'51"7d33'
Anderson,Leif & Judy
Uma 12h24'0"54d58'
Anderson,Leigh Carlson
Lyn 7h27'0"42d22'
Anderson,Lincoln
Cet 1h29'23"-4d8'
Anderson,Logan Michelle
Sge 19h54'0"19d27'
Anderson,Lorene Crossland
And 0h12'27"46d17'
Anderson,Lorie
Cam 6h59'26"60d16'
Anderson,Lorie Joyce
Vir 13h36'35"-4d12'
Anderson,Lowry Avalon
Peg 21h55'57"28d40'
Anderson,Lynne Flaskamper
Vir 13h36'48"-7d27'
Anderson,Macey Kathryn
And 0h53'11"40d57'
Anderson,Marcus
Aur 6h1'37"41d12'
Anderson,Margaret Evelyn
Umi 15h35'1"70d4'
Anderson,Margie
And 2h19'0"39d56'
Anderson,Mari Catherine
Cyg 21h39'28"28d56'
Anderson,Marilyn Ann
Lac 22h36'19"56d16'
Anderson,Marilyn Portz
Hya 8h43'11"2d24'
Anderson,Mark & Gina
Uma 10h49'39"41d2'
Anderson,Mary Greenwood
Umi 16h23'45"71d48'
Anderson,Mary Elaine
Peg 22h55'57"22d8'
Anderson,Mary Ethel
Cyg 19h28'57"38d51'
Anderson,Mary Grace & Peter
Cam 9h12'51"77d51'
Anderson,Matthew D
Aur 7h19'43"39d30'
Anderson,Matthew J
Cet 3h20'37"7d30'
Anderson,Matthew Lee
Boo 15h12'23"30d24'
Anderson,Megan Lindora
Peg 21h53'10"30d45'
Anderson,Meghan Elizabeth
Cas 23h41'52"60d26'
Anderson,Melissa G
Aql 18h56'34"13d35'
Anderson,Michael
Per 3h12'28"48d4'
Anderson,Mike
Cep 23h2'19"64d39'
Anderson,Miss Patricia
Uma 10h4'56"61d46'
Anderson,Mr
Cnc 8h53'37"13d30'
Anderson,Ms M C Schwalbe
Nikolaisen
Cam 13h28'44"84d49'
Anderson,Nancy
Ori 5h39'29"13d6'
Anderson,Nancy Jean
Cyg 21h56'0"33d25'
Anderson,Nathan Earl
Uma 8h40'51"56d1'
Anderson,Neil
Del 20h55'12"6d21'
Anderson,Nicholas
Tau 4h13'0"22d34'
Anderson,Nicola
Sge 19h56'0"16d7'
Anderson,Noah Zane
Uma 8h40'58"62d29'
Anderson,Patricia Brady
Eri 3h59'39"-6d14'
Anderson,Paul
Her 17h3'27"18d46'
Anderson,Phyllis
Mon 8h6'54"-6d58'
Anderson,Rachel T
Mon 7h58'20"-1d51'
Anderson,Rebecca Margrethe
Lyn 8h24'31"58d0'
Anderson,Rebecca A
And 23h34'0"38d24'
Anderson,Renelle
Mon 6h55'19"11d38'
Anderson,Rhonda Marie
Lac 22h55'44"55d23'
Anderson,Richard A
Hya 8h11'17"5d29'
Anderson,Richard G & Wendy
P
Crb 15h33'44"30d17'
Anderson,Riley Danielle
Aur 4h57'47"40d36'
Anderson,Robert Alan
Lmi 10h48'27"32d7'
Anderson,Robert C
Cet 2h41'23"-1d44'
Anderson,Robert J (Skip)
Tau 5h51'22"26d32'
Anderson,Robert J "Skip"
Gem 6h49'45"19d44'
Anderson,Robert Morgan
Boo 14h32'16"45d47'
Anderson,Robin
Peg 22h29'19"27d19'
Anderson,Ron
Leo 9h39'16"10d37'
Anderson,Ronald W
Cep 20h43'26"76d30'
Anderson,Rosemary Kay
Eri 4h54'28"-8d37'
Anderson,Rudolph Valentino
Ori 5h38'11"10d27'
Anderson,Ruth Anne
And 23h21'54"41d11'
Anderson,Salena Reneé
Lyn 8h0'0"39d7'
Anderson,Samuel Robert
Lac 22h53'54"51d43'
Anderson,Sandra L
Boo 13h39'41"26d12'
Anderson,Sarah Nicole
Peg 22h1'0"31d43'
Anderson,Shaina Lea
Eri 3h1'33"-3d13'
Anderson,Shannon Marie
And 0h54'24"39d16'
Anderson,Sheriff John Wesley
Aql 19h42'10"12d25'
Anderson,Sholly Lyn
Vul 20h18'55"23d7'
Anderson,Sidney Lynn
Equ 21h21'52"11d6'
Anderson,Stephanie Lorraine
Eri 3h41'30"-12d1'
Anderson,Stephen
Her 16h59'33"38d5'
Anderson,Stephen
Per 3h15'0"41d2'
Anderson,Stephen Ernest
Hya 9h7'20"3d3'
Anderson,Steven Gary
Aql 19h23'33"13d27'
Anderson,Susan Willoughby
Cma 6h48'29"-19d27'
Anderson,Susan Adair
Aql 19h42'1"10d23'
Anderson,Susan Bennett
Lyr 18h44'26"41d46'
Anderson,Susan Burney
Mon 7h50'5"-3d7'
Anderson,Susan J
Crt 11h17'14"-13d55'
Anderson,Syd
Dra 11h31'46"70d43'
Anderson,Sylvia
Cas 0h21'24"66d46'
Anderson,Tara
Sgr 19h24'53"-43d38'
Anderson,Terri Lynne
Uma 11h43'35"37d55'
Anderson,The X Files, Gillian
Mon 6h49'34"10d24'
Anderson,Thomas Joel
Cet 2h30'30"5d4'
Anderson,Tracy
Lyr 18h43'49"45d9'
Anderson,Troy Dean
Lac 22h38'31"52d35'
Anderson,True Love Two
Albert B & Dorothy J
Cnv 12h53'22"38d14'
Anderson,Tyler William
Cet 2h5'41"6d45'
Anderson,Van Allen
Aql 18h55'16"11d38'
Anderson,Van Michael
Aql 19h15'46"10d33'
Anderson,Veronica M
Vir 13h2'33"12d56'
Anderson,Wendell R
Dra 10h45'33"74d45'
Anderson,Wes & Mary Kaye
Cyg 21h13'0"38d10'
Anderson,William Connor
Oph 17h39'2"-20d23'
Anderson,William J
Her 15h54'12"41d57'
Anderson,Wyndi Elyzabithe
And 2h29'0"42d0'
Anderson,You Are A
Star!,Kristin K
Cyg 21h21'32"38d33'
Anderson-Bond,John & Royce
Sge 19h54'47"20d28'
Anderson-Easterling's Brilliant
Union
Cyg 19h25'29"33d19'
Anderson-Huggs,Alyce Cherie
Com 12h25'11"19d58'
Anderson-Strickland
L'Mas,Lisa Marie
Vul 19h23'17"25d15'
Andersson,Christina
Cas 0h5'46"60d55'
Andersson,Gertrude Victoria
Aql 19h25'56"14d19'
Andersson,Katarina
Aql 20h6'1"1d15'
Andersson,Lars
Umi 14h35'39"66d47'
Andersson,Mark Trydal
Vul 21h12'41"20d13'
Andersson,WPJ Andy
Uma 11h47'31"49d5'
Andersson-Palme The Court
Singer,Laila
Ari 1h47'0"14d1'
Anderton,Marjorie Sue
Boo 14h29'0"12d12'
Andes,The Star Of Larry
Ori 6h2'27"4d26'
Andes,Wendolyn N
Mon 6h38'1"6d7'
Andi
Agr 20h45'14"-0d20'
Andi & Ron
Aur 4h56'42"40d43'
Andi's Gate
Peg 23h30'28"18d11'
Andiaco,Thierry
Del 20h19'46"10d13'
Andian Rose
Crb 16h22'25"34d59'
Andidebby
Tel 19h5'58"-49d15'
Andie
Cnv 12h6'30"35d6'
Andina,Harrison James
Per 2h2'37"48d48'
Andino,Edgar
Cam 6h45'47"80d6'
Andino,PhD,Miss Margie
Com 13h2'28"28d56'
Andino,Rebecco G
Lyr 19h16'40"41d46'
Andiric,Timur Reynolds
Sex 10h39'1"-0d1'
Andler
Cam 7h51'57"60d30'
Andler,Teri Susan
Cas 23h20'24"60d11'
Ando's Star
Cen 11h53'53"-48d30'
Andoe,Dawn
Peg 21h53'1"2d22'
Andonia/HPS
Aur 6h5'25'31d26'
Andorene
Vul 19h44'0"20d27'
Andou,Satumi
Umi 13h46'27"75d48'
Andra
Aql 18h41'55"-2d11'
Andrade,Donald
Aur 6h24'14"38d14'
Andrade,Gail Ann
Peg 22h51'36"26d45'
Andrade,John Edward
Lac 22h45'57"53d41'
Andrade,Jr,Edward P
Lac 22h45'28"53d5'
Andrae,Edeltraud & Fred
Uma 11h1'0"62d24'
Andranna
And 0h6'41"38d3'
Andrasko,Jessie Sophia
Cyg 20h3'41"31d7'
Andrassy,Gertrude Bingham
Coffman
Cyg 21h7'54"31d15'
Andre
Her 17h38'0"37d46'
Andre's Star of December
1993(A93)
Mon 6h54'59"-6d0'
Andre,Alfred
Dra 18h48'40"67d50'
Andre,Betty
And 1h37'0"36d51'
Andre,Evelyn & Gilbert
Cyg 20h23'12"38d31'
Andre,Neil
Cep 21h49'0"55d23'
Andre,Paula Leonie Campbell
Eri 3h36'19"-11d57'
Andre,Will
Cet 2h16'11"7d58'
Andrea
Peg 22h30'1"28d45'
Andrea
And 23h32'16"40d55'
Andrea
Lyr 18h57'33"30d45'
Andrea
Cnc 8h57'10"7d36'
Andrea
Vul 19h21'56"22d31'
Andrea
Mon 6h55'11"-10d48'
Andrea
Lyr 19h0'35"35d22'
Andrea
And 0h47'0"34d20'
Andrea
For 2h32'25"-27d36'
Andrea
Lac 22h42'32"38d9'
Andrea
Cet 0h43'53"1d30'
Andrea
Com 12h31'32"23d28'
Andrea
And 0h13'28"33d42'
Andrea
Mon 6h21'38"6d3'
Andrea
Cyg 19h28'51"34d8'
Andrea
Lyn 8h8'19"37d1'
Andrea
And 1h44'57"38d43'
Andrea
And 23h40'53"41d23'
Andrea
Peg 23h31'51"13d12'
Andrea
Lib 15h38'20"-23d7'
Andrea
Lyn 8h10'55"42d47'
Andrea
Lyn 7h47'45"40d25'
Andrea
Psa 22h35'29"-27d31'
Andrea
Com 12h33'0"26d43'
Andrea
Cas 2h2'34"59d29'
Andrea
Cas 1h25'33"54d5'
Andrea
Cas 0h29'33"58d3'A
Andrea
Aql 20h5'41"-5d54'
Andrea
Lyr 19h23'20"42d42'
Andrea
Lib 15h33'41"-10d33'
Andrea & Chris
Sge 19h17'0"16d37'
Andrea & James
Cyg 19h55'13"50d36'
Andrea & Roberta Star's
Cam 6h39'43"68d44'
Andrea & Thackery
Mon 7h19'36"-5d24'
Andrea & Tom 11-17-90
Lyr 18h22'28"40d55'
Andrea 21 A M C
Ori 5h57'33"15d39'
Andrea 6-19-1964
Mon 6h18'56"4d42'
Andrea B P
And 23h12'13"37d44'
Andrea Bright Eyes
Cyg 22h0'21"51d59'
Andrea e Patrizia
Cam 7h4'23"61d31'
Andrea Elena
And 23h38'22"42d12'
Andrea et Cri
Cas 0h2'30"59d24'
Andrea Faith
Peg 22h20'29"35d11'
Andrea I
Lyn 8h29'42"50d29'
Andrea Joy
Uma 9h40'0"44d18'
Andrea Kirsten
Lac 22h33'39"48d56'
Andrea L
Eri 3h14'24"-10d45'
Andrea L50
And 22h58'1"50d23'
Andrea Love
Lyn 8h50'11"35d15'
Andrea Luminati
Uma 13h36'31"58d15'
Andrea Lyn "Sweet Pea"
Gem 6h0'42"26d55'
Andrea Lynn
Ori 5h10'59"-4d20'
Andrea Lynne
Mon 7h0'45"-10d57'
Andrea Marie
Dra 19h34'5"71d29'A
Andrea Michelle
And 22h57'0"51d40'
Andrea My First
Cnv 12h5'0"40d53'
Andrea Nicole Meagan &
Kymry Ray
Peg 22h31'17"20d3'
Andrea Paire
And 0h22'36"44d44'
Andrea S
Cet 1h35'36"0d55'
Andrea Simona
Sex 10h12'11"-6d24'
Andrea Suzanne
Ser 15h53'50"0d29'
Andrea The Princess
And 1h22'58"39d59'
Andrea und Elmar
Lyn 8h11'30"48d39'
Andrea's & Jörg's Star
Lib 15h11'39"-22d10'
Andrea's Delight
Eri 3h1'0"-10d28'
Andrea's Light
Umi 14h58'53"68d34'
Andrea's Star
Oph 17h23'54"7d38'B
Andrea's Star
Oph 18h21'16"7d12'
Andrea's Star
Mon 6h53'36"1d11'
Andrea's Twinkling Star
Cas 23h20'13"60d10'
Andrea's Wish
And 2h15'27"37d23'
Andrea,Alyssa Michelle
Cas 0h30'24"63d44'
Andrea,Joseph F
Dra 18h29'59"80d19'
Andrea,Kellen Charles
Uma 11h13'29"48d4'
Andrea,Martin & Luise
Cyg 20h32'40"50d17'
Andrea,Rina
And 0h51'16"36d14'
Andrea-Kristen,The
And 1h17'31"34d0'
Andrea-Nicole
And 23h9'21"48d49'B
Andreaccio,D
Aql 19h25'16"12d59'
Andreadakes
Peg 21h27'51"2d46'
Andreaggi,Christian Anthony
And 0h39'54"40d16'
Andreana
Lyn 8h48'38"44d59'
Andreanna
Aql 19h31'25"13d1'
Andreano
Uma 10h22'44"40d22'
Andreas & Birgit
And 1h23'45"42d20'
Andreas,Antonio James
Uma 1h40'52"40d4'
Andreas,Sigmund
Umi 15h38'41"78d19'
Andreasen Star,The Donna
Eri 2h55'35"-5d8'
Andreasen Star,The Leo J
Eri 2h59'15"-5d15'
Andreasen's Star,Karen
Mon 7h21'45"-8d46'
Andreina
Mon 7h54'9"-4d4'
Andreotta,Jennifer
Uma 8h37'59"56d40'
Andreotti,Danielle M
Cas 2h19'1"73d14'
Andreotti,Kelli
Peg 21h59'40"34d37'
Andreotti,Michelle D
Lyr 18h31'41"43d1'
Andres,Walter
Dra 16h58'47"60d35'
Andreschek,Iris
Gem 8h4'17"30d46'
Andress,Irma Leah
Uma 10h1'28"50d37'
Andrestar
Ser 15h37'26"24d24'
Andretti;A Legend In
Racing,Mario
Per 2h23'32"54d39'
Andreu
Lyn 8h17'51"43d33'
Andreu,Antonio M
Cep 22h23'58"61d33'
Andrew
Her 17h57'0"14d30'
Andrew
Lac 22h17'60"51d58'
Andrew
Aur 5h53'32"54'33'
Andrew
Aql 20h6'56"8d59'
Andrew
Aur 6h36'21"37d39'
Andrew
Dra 14h28'25"63d18'
Andrew
Mon 6h50'14"5d57'
Andrew
Cnv 12h41'0"43d23'
Andrew
Boo 13h36'25"20d9'
Andrew
Dra 12h0'50"70d46'
Andrew
Cyg 20h6'19"42d14'B
Andrew
Uma 10h34'56"40d57'
Andrew & Alison
Cyg 21h31'42"41d48'
Andrew & Jenna
And 1h36'25"38d44'
Andrew & Karyn Tin Bears
Per 2h16'22"58d54'
Andrew & Lorraine
Sge 20h13'42"17d23'
Andrew & Melissa
Cyg 21h52'40"53d1'
Andrew & Sarah
Boo 14h38'1"8d35'
Andrew 1
Per 3h0'14"47d49'
Andrew All My Love
Lac 22h22'16"37d43'
Andrew B-612
Per 2h55'11"40d40'
Andrew Davidius-Nevada Nine
Four
Ari 2h36'59"30d57'
Andrew Family,The
Peg 22h22'40"21d44'
Andrew III
Her 17h38'35"26d11'
Andrew IV
Cnc 8h10'34"31d18'
Andrew John
Aur 6h22'25"34d32'
Andrew Joseph
Aur 5h18'19"40d32'
Andrew Joseph
Per 2h50'27"41d8'
Andrew M
Vul 19h6'33"25d14'
Andrew Martin
Vir 12h58'1"-18d9'
Andrew My Love
Ser 15h24'11"21d10'
Andrew Paul's Andrew
Hya 8h44'52"-1d38'
Andrew Ray's Place In The Sky
Cep 22h7'34"70d1'
Andrew Simon
Mon 6h21'0"9d32'
Andrew Sr
Aur 6h7'35"40d18'
Andrew Steven
Her 16h17'1"4d51'
Andrew's Asset
Ori 4h43'19"22'
Andrew's Brilliance
Cep 22h17'38"56d12'
Andrew's Guardian
Cam 4h3'42"61d17'
Andrew's Home of Love &
Laughter
Cep 3h51'28"80d35'
Andrew's Limits
Ori 5h33'56"-0d15'
Andrew's Little Star
Vul 19h40'30"26d11'
Andrew's Odyssey
Her 17h23'16"38d35'
Andrew,"Andy I"John
Vul 20h23'19"28d39'
Andrew,David
Cep 21h52'21"58d39'
Andrew,Home of John &
Mildred
Umi 13h54'18"70d18'
Andrew,Rebecca
And 23h23'14"47d11'
Andrew-Elizabeth
Cyg 20h23'38"39d14'
Andrew-It's A Bit Of Heaven
Her 17h24'55"45d49'
Andrew-Ski
Per 2h27'15"58d37'
Andrews "Pops" Star, The
David M
Her 17h23'51"47d56'

Andrews III,Sumner Robinson
Lyn 6h12'14"58d55'
Andrews Knight Light, Bonnie Lee
Lyn 9h0'22"39d25'
Andrews' Star,The
Cnv 12h10'43"48d2'
Andrews,Aaron C
Mon 8h1'11"-5d38'
Andrews,Alyssa Marie
Vul 18h58'34"24d47'
Andrews,Amy
And 0h24'35"30d9'
Andrews,Annette
Cyg 21h51'11"53d58'
Andrews,Barbara
Lmi 10h53'29"26d10'
Andrews,Betsy
Del 20h19'0"9d30'
Andrews,Charles
Ser 15h55'42"20d38'
Andrews,Cynthia Lynn
Eri 4h34'44"-11d56'
Andrews,David
Per 4h31'44"34d22'
Andrews,Destinee Nicole
And 23h19'39"46d52'
Andrews,Dolph
Del 20h13'57"15d44'
Andrews,Elizabeth Patricia
Cmi 7h54'41"1d15'B
Andrews,Ella
And 22h57'47"38d4'
Andrews,Ellen
Cas 1h12'19"62d9'
Andrews,Ellen Kathleen
Cmi 7h54'41"1d15'A
Andrews,Erica McKenzie
Com 12h12'21"30d51'
Andrews,Erika Maria
Tau 5h46'15"26d31'
Andrews,Evelyn
Cas 0h49'29"66d32'
Andrews,George
Cyg 19h25'19"32d21'
Andrews,George A
Aur 6h18'1"50d16'
Andrews,Gillian Alice
Lyr 18h27'0"47d35'
Andrews,Grace
Cas 0h5'41"50d30'
Andrews,Grandma Bannen-Ellen
Cas 3h4'32"65d18'
Andrews,James
Dra 16h5'58"63d50'
Andrews,Jeff
Cma 6h54'40"-19d34'
Andrews,John & June
Aql 20h10'38"4d12'
Andrews,Josphine A
Crb 15h53'35"27d0'
Andrews,Julie A
Sge 20h16'58"20d58'
Andrews,Lynn T
Aql 20h1'59"8d53'
Andrews,Lynne Bower
Cas 1h32'43"65d19'
Andrews,Margaret Ann Fanning
Cet 2h50'41"6d51'
Andrews,Marina C
Peg 22h0'12"24d56'
Andrews,Michael
Aql 20h6'32"7d11'
Andrews,Michael Zane
Del 20h18'27"11d12'
Andrews,Patrick Joseph
Cet 1h55'50"-6d28'
Andrews,Pearl
Lyr 18h22'20"44d28'
Andrews,Poet,Michael
Dra 19h45'46"60d9'
Andrews,Rebecca Rose
Mon 8h2'0"-0d49'
Andrews,Rick
Uma 12h1'19"31d15'
Andrews,Scott T
Ser 18h16'20"-14d59'
Andrews,Scotty
Ser 15h15'59"17d44'
Andrews,Seth William
Dra 20h4'18"68d24'
Andrews,Sherry & Stan
Peg 22h44'18"21d24'
Andrews,Susan
Mon 6h29'54"10d15'
Andrews,Thomas A
Aur 5h18'0"41d6'
Andrews,William Francis
Aur 6h5'23"32d33'
Andrews,William C
Ari 1h47'0"15d34'
Andria Joy
Cas 0h50'50"62d33'
Andriam Bolo Lona Nirina
Lmi 10h48'25"37d56'
Andriani,Nicole
Cet 0h52'25"-5d7'
Andriani,Sarah & Lance
Cyg 19h22'32"44d45'
Andrias,Deidre
Umi 13h42'16"71d30'
Andriassi,Sr,Joe & Marie
Aur 5h9'32"40d24'
Andrikokus,George Christopher
Lmi 9h48'34"33d51'
Andriks,Timothy P
Lac 22h16'13"51d58'
Androff,Michael James
Lib 15h0'38"-10d40'
Androidal Cavity
Per 3h12'12"42d38'
Androka,Charlie's
Pup 7h35'31"-14d49'
Andromedia
And 1h47'34"40d36'
Andron,Kimberly Hope
Cyg 20h33'1"39d33'
Andronaco,Dominick Gregory
Cet 3h8'0"4d23'
Androniki
Peg 23h3'1"30d33'
Andros MD,Dr George
Oph 16h50'0"-26d0'
Andros,John Bradley Abbott
Cnv 14h4'27"45d23'
Androula
Ant 10h40'0"-35d49'
Androus,Russ I
Aql 19h30'59"8d57'
Androus,Russell* Charlee Puckett
Cnc 8h48'1"30d20'
Androvica,Zvaigzne
Cam 5h6'45"68d6'
Andrulis,James Anthony
Aur 6h28'16"31d27'
Andrushko,Jerry P
Dra 17h32'19"64d55'
Andruss,Beth
Lyn 6h42'18"50d35'
Andry
And 23h3'44"50d23'
Andryszewski,Susan
Aql 20h7'16"4d53'
ANDRYTE
Her 17h40'11"18d60'
Andrzejewskar,Manta
Crb 15h54'58"26d5'
André
Umi 14h56'53"68d23'
André Emmanuelle
Boo 14h0'35"12d8'
André,Jean Bonet
Mon 6h41'13"3d19'
Andréa
Sgr 19h29'6"-32d51'
Andréa Denise
Cas 0h46'1"63d51'
Andréa II
Cyg 19h34'18"33d2'
Andréa-Chiara
Cet 1h28'46"0d43'
Andréanne
Cas 1h10'1"60d15'
Andrée,Nana
Ori 5h26'16"0d9'
Andrée-Roehmhold,Wolf Ulrich
Com 13h8'27"28d46'
Andrès,Juliette
Cyg 20h17'0"39d8'
Andrès,Luis
Hya 8h9'53"2d23'
Andtra-X
Gem 6h54'21"13d55'
Andy
Dra 18h24'49"80d22'
Andy
Cyg 19h28'31"36d32'
Andy
Cnc 8h25'59"30d16'
Andy
Ser 16h10'20"2d50'
Andy & Debbie's Star
Cyg 20h8'36"40d34'
Andy & Jen
Uma 9h8'38"52d56'
Andy & Maggie
Lyr 18h57'1"31d40'
Andy & Phyllis
Dra 16h30'22"62d19'
Andy - I Love You
Uma 12h2'39"31d33'
Andy D & Helen C
Cru 12h8'50"-56d30'
Andy F
Aql 20h5'44"4d51'
Andy N' Chelle
Cam 3h43'0"60d22'
Andy Star,The
Aur 6h27'1"31d5'
Andy's Alpha Orinis
Ori 5h37'36"-0d50'
Andy's Dream
Ori 4h47'27"4d29'
Andy's Hope
Boo 15h44'28"41d7'
Andy's My Shining Star Love Nancy
Uma 10h4'59"47d36'
Andy's Star
Cep 22h19'25"67d53'
Andy,Annie & Emma
Scu 18h54'51"-4d22'
Andylovin
Cyg 21h23'0"28d55'
Aneesa Din
Mon 6h18'45"8d56'
Anella,Albert Anthony
Sex 9h40'0"4d31'
Anella,Frances Angelina Russo
Cyg 21h15'13"34d58'
Anemone,Jr,Bill
Lac 22h34'56"53d8'
Aner - You are the Lightness of Being
Per 2h28'41"51d36'
Aneshansley,Gene I
Oph 17h13'34"-10d16'
Aneshansley,Mark Russell
Dra 15h15'19"64d12'
Aneth,Gyda
Dra 10h44'29"74d9'
Anette
Del 20h14'0"10d4'
Anette Z
Cas 2h5'29"59d3'
Anettia
And 2h25'31"48d16'
Ang
Cas 0h29'25"50d31'
Ang
And 0h12'0"38d26'
Ang,Lennor G
Uma 9h0'45"50d3'
Ang-ie Arrowood
Peg 23h19'11"10d58'
Angard "Stellar Destiny",Joe
Ari 3h1'30"28d51'
Ange Et Ambre
Aql 20h1'0"9d47'
Ange Idéal
And 1h8'27"47d26'
Ange's Heart Of Gold
And 23h13'1"41d27'
Angeias
Lyr 18h45'51"40d2'
Angel
Ori 4h41'12"13d4'
Angel
Mon 7h40'14"-10d55'
Angel
Cas 23h27'27"62d34'
Angel
And 2h24'46"45d48'
Angel
Lyr 19h25'27"37d60'
Angel
Eri 3h53'52"-0d11'
Angel
Lyr 19h22'25"31d42'
Angel
Nor 16h20'4"-52d10'
Angel
Vul 19h4'49"25d17'
Angel
Uma 9h55'16"44d34'
Angel
Uma 8h54'28"60d38'
Angel
Cas 0h22'1"74d10'
Angel
Uma 9h29'13"48d15'
Angel
Her 16h46'53"36d0'B
Angel
Cap 20h26'37"-26d29'
Angel
Lyr 18h52'29"40d15'
Angel
Lyn 7h14'59"58d22'
Angel
Sex 9h54'27"-6d57'
Angel & Forest
And 2h21'11"48d19'
Angel (To My Daughter Vanessa)
Eri 3h6'45"-4d40'
Angel Adrienne
Peg 21h40'24"22d47'
Angel Anne
Aql 19h45'37"12d17'
Angel Catherine "Smirkey"
And 0h39'0"45d57'
Angel Cheri
And 23h4'32"41d49'
Angel Cove Light
Cam 7h51'1"61d33'
Angel David
Dra 11h33'17"68d48'
Angel Deborah Ann
Cas 1h35'16"61d23'
Angel Elizabeth
Mon 6h32'60"-0d5'
Angel Ernest
Lyn 8h41'20"45d0'
Angel Eyes
Lyr 18h37'13"42d13'
Angel Eyes
Uma 9h47'37"58d48'
Angel Eyes
Mon 6h20'41"4d5'
Angel Eyes
Sge 19h2'49"19d22'
Angel Eyes
And 23h0'34"51d51'
Angel Eyes
Cam 3h31'43"63d7'AB
Angel Face
Cam 13h39'32"80d6'
Angel Gabriel
Aql 19h3'30"15d9'
Angel Girl Caci Shea
And 0h26'39"27d37'
Angel in the Sky "Angel en el Cielo"
Lyr 19h0'12"35d8'
Angel IZ
Crb 15h27'11"30d56'
Angel J
Eri 3h11'0"-2d26'
Angel Kiss
Per 3h2'48"45d34'B
Angel Love
Cas 1h0'41"61d15'
Angel M
Vul 20h39'23"23d32'
Angel Michelle
Peg 23h29'19"30d25'
Angel O D
Cyg 19h29'13"30d14'
Angel of Hope
Lyr 19h23'44"31d30'
Angel of Light
Peg 23h25'56"18d20'
Angel of Love
Uma 11h9'59"48d16'
Angel of Memory
Eri 3h26'47"-2d0'
Angel of the Morning Sun
Tau 5h23'52"18d50'
Angel Star,The
Mon 8h6'43"-8d4'
Angel Star-Mama Lu
Uma 12h16'26"60d1'
Angel's "ChrisStar"
Boo 14h56'19"47d35'
Angel's Attic
Cyg 21h28'55"38d28'
Angel's Dream
Lyn 7h0'42"50d7'
Angel's Flight
Cas 23h1'57"58d33'
Angel's Kiss
Peg 23h35'40"12d55'
Angel's Light
Peg 22h23'19"25d48'
Angel's Prince Eternal
Her 18h18'1"12d58'
Angel's Star
And 23h4'22"40d23'
Angel,David
Gem 7h35'29"20d13'
Angel,Heather Kim
Cyg 19h30'0"33d54'
Angel,Isobel Victoria
Peg 22h45'0"31d2'
Angel,Jeanne Marie
Cam 4h54'42"61d5'
Angel,Judd
Sex 10h29'30"5d12'
Angel,Matthew
Uma 11h6'36"58d39'
Angel,My Heavenly Body
And 23h48'40"44d57'
Angel,The
Cam 4h57'10"67d30'
Angel,Wesley Evan
Crb 15h53'31"27d22'
Angel-Donna
Cyg 20h22'45"38d40'
Angel-Gift of God
Peg 23h19'24"30d51'
Angela
Col 6h14'38"-41d37'
Angela
Lyr 19h20'0"38d54'
Angela
Per 3h33'19"50d17'
Angela
Cam 6h40'17"67d35'
Angela
Peg 23h41'1"30d24'
Angela
And 2h4'56"40d35'
Angela
Cas 0h36'40"68d35'
Angela
And 2h25'42"39d54'
Angela
And 2h25'0"45d21'
Angela
Lib 15h20'42"-25d25'
Angela
Aql 18h57'54"-0d29'
Angela
Crb 15h29'1"30d26'
Angela
Lyr 18h32'20"43d53'
Angela
Pic 4h41'11"-47d44'
Angela
Lac 22h9'50"51d5'
Angela
Lup 15h17'29"-44d26'
Angela
Peg 22h53'0"29d23'
Angela
Psc 1h26'36"31d26'
Angela
Lyn 9h14'0"46d33'
Angela
Boo 14h25'56"14d12'
Angela
Her 17h22'38"42d52'
Angela
Del 20h12'54"14d49'
Angela
For 2h13'26"-28d41'
Angela
Pyx 8h45'26"-29d29'
Angela
Per 3h13'54"40d18'B
Angela
And 1h1'41"37d54'
Angela & Giusy
Scl 23h12'37"-32d6'
Angela 7-11
Cas 1h42'52"60d54'
Angela 8372
Cas 2h54'59"70d15'
Angela Alison
And 0h7'16"31d43'
Angela Ann
Aql 19h14'52"15d39'
Angela Christine
Sgr 19h1'29"-28d10'
Angela Dawn
Lyn 7h56'15"45d32'
Angela G
And 0h27'54"30d22'
Angela Grand Bethel Marshal PHQ80
Cet 0h42'22"0d48'
Angela Jean
Cas 0h48'0"60d17'
Angela Joy
Mon 7h48'9"-2d37'
Angela Lucille
Lyr 19h2'43"43d48'B
Angela Lynn's Sparkle
Peg 22h7'1"4d21'
Angela Marciel
Mon 7h43'10"-2d47'
Angela Marie
Cas 2h58'1"57d32'
Angela Marie
Cas 22h59'24"54d24'
Angela Marie
Mon 6h32'55"10d8'
Angela Marie's Christmas Star
Lyn 8h46'12"39d28'
Angela Mary Madeline
Gem 6h53'60"18d3'
Angela Michelle
Boo 13h38'54"22d32'
Angela Michelle 1
Eri 2h53'54"-5d46'
Angela Noel "Windancer"
Peg 21h59'18"34d31'
Angela P '53
And 1h45'45"40d39'
Angela Ré
Aql 19h1'47"0d42'
Angela's Star
Lyr 18h51'56"38d32'
Angela's Star
Ari 1h55'58"19d46'
Angela's Star On Her Wedding Day
Cyg 19h45'38"29d29'
Angela-C
Cyg 19h50'34"40d14'
Angela-Mitchell
Sge 19h32'25"16d34'
Angela-my little lump of chewing gum
Umi 15h12'52"66d39'
Angelakos,Stavros
Her 18h4'1"30d54'
Angelaray
And 0h14'53"38d40'
Angelbeck,Robert Andrew
Aur 4h46'25"51d27'
AngelDaniel
Uma 9h35'17"67d41'
Angeldust
Peg 22h37'40"10d60'
Angelee
Lmi 10h2'19"40d32'
Angelene
Com 12h52'42"24d18'
Angeli,Famiglia
For 2h1'27"-24d58'
Angeli-Reiter,Petra & Michael
Uma 13h39'29"50d4'
Angelia
Cyg 19h23'37"28d7'
Angelia Cheri
Sct 18h54'16"-14d15'
Angelian
Cmi 7h30'1"0d8'
Angelic Aura Of Sherry Lynn,The
Psc 1h38'21"21d3'
Angelica
Scl 23h10'35"-31d47'
Angelica
Umi 15h28'22"67d40'
Angelica
Gem 7h0'0"31d45'
Angelica & Russ
Col 6h0'44"-33d39'
Angelica Patricia
Oph 17h25'7"-23d11'
Angelica's Shining Star
Aqr 21h19'19"-13d13'
Angelica's Star
Per 2h56'47"45d56'
Angelica,Karen-Valerie
And 2h23'37"44d40'
Angelicola,Lianne
And 1h56'17"37d29'
Angelika
Boo 14h25'29"14d59'
Angelika
Lib 14h23'25"-20d5'
Angelika
Aqr 20h55'25"0d13'
Angelika & Dieter 1993
Gem 6h42'54"12d17'
Angelika's Sceptre
Ari 2h37'14"21d55'
Angelillo,Forrest Walker
Her 18h21'38"12d29'
Angelillo,James & Kara
Uma 11h57'18"47d46'
Angelillo,Shane Phelan
Dra 17h58'28"61d19'
Angelin,Mauro
Cam 7h4'46"61d12'
Angelina
Cas 2h29'24"59d0'
Angelina
Peg 22h16'24"8d6'
Angelina
Del 20h33'46"10d54'
Angelina Celeste
And 23h44'39"38d16'
Angelina Theresa
Cyg 21h28'1"48d47'
Angelini,Marcello
And 1h2'22"47d51'
Angelique
Cyg 20h38'28"38d17'
Angelique
Lac 22h23'1"56d37'
Angelique
Vul 20h16'34"28d19'
Angelique
Ser 15h53'51"1d29'
Angelique,My Ribbon in the Sky!
And 23h33'1"49d42'
Angelisa Rene
Peg 22h42'55"12d20'
Angelita
Ant 10h43'43"-35d48'
Angelitus,Frank
Boo 13h40'0"14d15'
Angell II,Warren Sanford
Boo 14h28'21"22d19'
Angell,Kelly
Lyr 18h30'1"38d56'
Angell,Susie
Cas 2h55'1"70d31'
Angelle,Selina T
Oph 16h17'1"-6d13'
Angellica Marie
And 0h15'49"37d3'
Angellino,James Robert
Aqr 21h25'46"-1d30'
Angelo
Cam 4h23'1"69d28'
Angelo
Cyg 20h4'27"38d52'
Angelo
Psc 1h36'1"28d25'
Angelo Tommie "The Crazy Star"
Vir 11h43'24"7d47'
Angelo,Alison
Aur 4h52'0"51d21'
Angelo,Celari Ugo
Her 18h10'47"38d11'
Angelo,Chris Mastr
Per 3h33'51"36d18'
Angelo,Edith & Sandy
Cyg 20h57'45"31d50'
Angelo,James W
Cam 3h13'7"65d28'B
Angeloff,Casey
Per 3h8'0"47d41'
Angels Dream
Lyr 19h20'59"42d45'
Angels Rainbow Man
Aur 6h31'12"35d36'
Angels Touch
Cet 0h51'18"-6d56'
Angelstar
Uma 9h45'60"49d60'
Angelu
Per 3h13'24"43d9'
Angelucci,Louis Patrick
Lac 22h20'40"52d50'
Angelus,Alex James
Cet 3h5'54"4d24'
Angelus,Joan
Ser 16h6'24"1d36'
Angelus,Rosalind
Uma 8h36'47"47d46'
Angen
Lyn 7h9'59"51d57'
Anger,Clifford & Patricia
Eri 4h4'51"-8d45'
Angerer,Tracy Michelle
And 0h47'36"35d43'
Angerhofer Forever +2, Rick Lynn
Cam 8h23'46"78d52'
Angerhofer,Carissa
Lyn 8h3'11"40d55'
Angers,Valérie
Umi 15h20'10"70d39'
Angevin,Marie- Christine
Oph 17h2'1"10d10'
Angi
Her 17h7'0"42d52'
Angie
Cyg 21h18'54"38d27'
Angie
Lyr 18h51'13"40d13'
Angie
Cep 22h22'14"68d1'
Angie
And 23h18'22"40d60'
Angie
Umi 15h10'25"78d43'
Angie
Cap 21h40'0"-19d21'
Angie
Del 20h55'1"8d59'
Angie
Com 13h17'0"27d45'
Angie & Karen "Fun Is First"
Peg 22h23'58"31d35'
Angie B
Vul 20h5'33"28d9'
Angie P
Cyg 19h45'11"38d34'
Angie's
Ori 4h56'53"-2d43'
Angie's Dream
Lyr 19h2'14"35d12'
Angie's Faith
Aql 19h56'0"13d47'
Angie's Star
Lyr 18h37'0"38d20'
Angie's Star
Lib 14h39'32"-11d14'
Angileri,Nicholas John
Cnv 13h43'27"35d1'
Angioletta
Nor 16h23'25"-52d32'
Angiulo,Matthew John
Aql 19h24'49"8d15'
Anglade,Henry
Dra 17h56'44"70d9'
Angle,Steven Carter
Aql 18h44'30"8d59'
Angles,Peter
Hya 8h14'0"2d44'
Angley,Eugene & Lyne
Com 13h0'46"14d22'
Angley,Kevin Alexander
Cma 7h16'53"-15d26'
Angli
Lyn 8h33'44"45d5'
Anglim,Elizabeth Catherine
Tri 2h27'46"31d10'
Anglin,Darrell Glen
Oph 16h50'1"11d5'
Anglin,David Zephaniah
Ori 6h7'31"20d51'
Anglin,Eileen & John
Aql 19h52'24"15d13'
Anglin,Verne Ernest & Jane Nanette
Boo 15h3'47"20d34'
Angoid
Tri 2h1'40"31d50'
Angone Our Shining
Angel,Marceline May
Crb 15h52'47"27d56'
Angotzi,Stefano
Aur 5h23'54"50d32'
Angrilli,Mariagrazia
Aur 7h25'44"43d44'

Angrimson * My Friend,
Loretta Weflan
Cam 6h45'44"67d57'
Angrl,Joseph Paul
Aql 18h59'22"17d31'
Angst,Alexandria Kathleen
Lyn 7h16'43"58d32'
Angstadt,Lauren Melissa
And 1h58'20"37d53'
Angstrom,Lorraine Swetland
M Innes
Lyr 18h36'25"44d3'
Anguilla
Aql 20h32'44"0d27'
Angus,Beverly G
Cap 21h8'13"-22d54'
Angus,Ian
Aql 20h8'17"0d51'
Angus,John Wakefield
Tau 5h33'1"28d48'
Angus,Jr,Douglas E
Leo 10h49'36"8d16'
Angus,Orrel
Aur 4h51'59"37d51'
Angus,Peter Alexander Gerard
Cnc 8h36'12"7d30'
Angwin,Jr,Forrest G
Ser 15h18'1"5d42'
Angyelof,Lisa Marie Dare
Uma 9h2'14"57d5'
Angèle Loves Paul
Uma 8h56'30"68d58'
Anhuda's Spark
Tau 4h7'55"20d52'
Ani & Dalila Forever
Cyg 21h3'42"31d13'
Ani L'Dodi V'Dodi Li Randi &
Harlan
Cnv 12h15'55"44d45'
Ani L'Dodi V'Dodi Li Harlene
& Andy
Tri 1h58'58"26d9'
Anice Mojo
Cyg 19h14'33"44d38'
Aniek,Todd
Her 17h39'37"23d19'
Aniela Forever
Cyg 21h0'29"30d10'
Aniforms
Ori 5h49'24"21d37'
Anik
Cas 0h17'42"63d52'
Anika
Peg 22h16'19"5d41'
Anil
Her 16h14'20"10d56'
Anil 93' Christmas Star
Oph 16h50'29"-28d37'
Anile,Frank Joseph
Boo 14h31'15"47d5'
Animal's Star
Ind 20h26'41"-55d9'
Animal's Star
Ind 20h26'17" 57d51'
Anin,Caroline
Aur 5h3'24"37d41'
Anini
Cas 0h37'1"68d26'
Anipen,Elaine
Cep 22h14'11"72d57'B
Anipen,John
Cep 22h14'11"72d57'A
Anise
Peg 22h35'32"28d35'
Anisha Melina
Her 17h53'1"42d40'
Anissa & Joe
Cam 5h53'57"68d47'
Anita
Uma 9h19'32"50d12'
Anita
Cyg 19h55'30"40d28'
Anita
Per 3h30'26"36d24'
Anita
Cap 20h27'42"-26d42'
Anita
And 2h21'34"48d50'
Anita 143
Del 20h14'1"13d42'
Anita Claire
Peg 21h58'1"34d24'
Anita Elaine
Aql 20h11'11"10d21'
Anita Jane
Umi 15h5'19"67d32'
Anita Jane
Uma 11h37'22"32d59'
Anita M
Cma 7h0'15"-11d8'
Anita Marie
Eri 3h45'48"-2d2'
Anita Marie
Lac 22h37'1"56d40'
Anita Marie
And 23h39'55"47d27'
Anita Patricia
Lyr 18h34'22"29d31'
Anitra
Lyr 18h39'60"31d48'
Anitra's Star
Mon 6h59'12"-10d48'
Aniversary of Hideki & Emi
Aqr 22h3'15"-11d3'
Aniya Somebody Loves
You,Andrew
Aql 20h4'47"1d51'
Anja Daniela HFB
Sco 16h36'59"-44d3'
Anja E,Sister of Dyana C
Cyg 20h15'47"39d2'
Anjadaniel
Lib 14h39'56"-6d4'
Anjala
Mon 6h28'0"4d59'
Anjali
Uma 11h35'29"40d38'
Anjela
Lyr 19h16'41"35d16'
Anjelic Wendy Jean
Umi 16h52'48"75d13'
Anji Lynn
Mon 8h6'24"-6d38'
Anke
Com 12h27'14"31d20'
Anke 18/07/1964
Dra 17h59'40"83d0'
Anke's Stern von Elmar Kraft +
Dicki
Cnc 9h18'50"31d6'
Ankele
Dra 20h21'18"64d58'
Ankenman,Cheryl Ames
Lya 18h30'1"41d24'
Ankeny,Aaron Eugene
Ori 5h45'34"11d53'
Ankeny,Ashley
And 0h58'29"35d3'
Ankerman-Holt
Lyn 8h17'35"41d13'
Ankers,Norman C
Cep 23h25'21"63d28'
Ankers,"T M"
Cyg 21h0'26"37d14'
ANKH
Aur 6h52'38"40d52'
Ankilbeau,Chrystelle
Lyn 9h6'22"40d36'
Ankrom,Aaron Christopher
Cam 3h31'30"61d1'
Ankus,Joe
Oph 17h1'41"-12d36'A
Ankus,Stephanie
Oph 17h1'41"-12d36'B
Anky & Eugen Star,The
Cyg 19h47'54"30d43'
Ann
Mon 6h44'35"10d16'
Ann
Vul 19h47'34"28d56'
Ann
And 0h44'43"40d47'
Ann
Cyg 21h56'18"48d45'
Ann "E"
Cap 21h52'18"-20d44'
Ann & Harry Star
Vul 21h19'28"24d21'
Ann & Jim Our Love Forever
Del 20h26'20"10d50'
Ann & Mark's "Silver Light"
Del 20h16'36"14d51'
Ann & Paul's Star
Mon 6h35'1"10d30'
Ann Cecelia
Cas 2h35'37"61d39'
Ann Denise
Com 11h58'49"19d38'
Ann Grace
Lyr 18h30'36"40d60'
Ann I
Lyn 6h25'54"59d41'
Ann Julia
Cas 2h1'36"61d16'
Ann Lena
Lyr 18h36'35"28d10'
Ann Marie
Umi 13h6'38"71d49'
Ann Marie
Lyn 9h10'25"38d8'
Ann Marie
Crb 16h9'22"30d59'
Ann Marie
And 23h42'33"38d55'
Ann Veronica
Cyg 21h2'25"38d0'
Ann W
Uma 8h55'0"55d20'
Ann's Permanent Wish Granter
Lyn 7h9'24"58d44'
Ann's Star
Ara 17h54'21"-51d42'
Ann-Ann
Peg 22h22'1"21d26'
Ann-Carson-Chase- John's
All-Stars
Boo 14h57'46"24d32'
Ann-M-Oakes
Cam 3h57'49"60d4'
Ann-Margrit
Ori 5h56'1"7d57'
Ann-Marie My Alpha & My
Omega
Cep 21h3'44"55d5'
Anna
Eri 4h18'8"-14d49'
Anna
Cas 2h18'23"60d45'
Anna
Com 13h33'15"19d41'
Anna
Aqr 23h29'0"-18d6'
Anna
Lyr 19h17'12"42d45'
Anna
And 23h20'48"37d38'
Anna
Cam 14h16'20"82d10'
Anna
Boo 14h2'56"11d6'
Anna
Ori 6h3'34"7d30'
Anna
Lyn 6h39'30"58d15'
Anna
Vel 9h18'54"-42d52'
Anna
Mon 7h1'1"5d23'
Anna
Cet 2h42'33"3d22'
Anna
And 2h8'38"42d40'
Anna
Peg 22h6'37"25d25'
Anna
Mon 6h44'58"1d46'
Anna
Del 20h18'1"10d27'
Anna
And 23h16'50"51d17'
Anna & Lydia
Umi 11h30'19"88d45'
Anna '74
Umi 19h14'44"88d47'
Anna 121673
Cyg 20h54'20"30d7'
Anna 4554
Ori 5h30'14"-1d31'
Anna Alexis
Lyr 18h43'0"43d5'
Anna C
Cas 0h33'44"67d59'
Anna California Rose
Vul 19h17'21"21d52'
Anna Caroline
Mon 7h26'47"-10d38'
Anna Catarina
Mon 6h56'27"10d1'
Anna Christine
Cas 0h42'24"69d17'
Anna Darlene
Cyg 19h56'32"37d35'
Anna e Angelo
Cep 21h35'29"68d59'
Anna e Gianluca
Cep 21h51'48"58d16'
Anna e Mario
Her 18h10'50"38d7'
Anna Elizabeth
Mon 6h37'1"-6d19'
Anna Evelyn Bayh-Iola
Mon 7h46'57"-1d13'
Anna Gloria
Crt 11h14'43"-16d21'
Anna II
Ori 5h57'13"10d42'
Anna Jane
Lib 14h59'54"-5d53'
Anna Katharine
And 2h2'58"38d41'
Anna Kirsten
And 23h29'40"49d50'
Anna Laura & Astrid
Col 6h26'29"-34d20'
Anna Lee
Ori 5h10'20"-6d41'
Anna Lisa
Sgr 19h22'27"-41d59'
Anna Lisa Martina
Nor 16h21'22"-52d13'
Anna Lucile
And 0h5'52"46d60'
Anna Lynn
And 0h13'13"36d24'
Anna Lynn
And 23h21'1"48d11'
Anna M 13/07/75
Crb 16h8'23"34d33'
Anna Maria 95 una luce che
guida lontano
Ant 10h31'30"-34d39'
Anna Marie
Cas 23h33'37"60d34'
Anna Marie
Cas 0h18'10"62d5'
Anna Marie & Paul
Crb 15h50'16"38d20'
Anna Mary Margaret
And 0h23'20"45d54'
Anna Pearl
Aur 6h33'12"31d36'
Anna Sambiase di
Casamicciola Terme
Hor 3h13'30"-48d42'
Anna Simone
And 23h47'54"41d1'
Anna Viktoria
And 0h48'35"37d48'
Anna's Bat Mitzvah Star
And 0h10'40"37d54'
Anna's Place
Vul 19h40'57"26d48'
Anna's Rose
Del 20h13'19"14d18'
Anna's Tony
Aql 19h31'44"12d39'
Anna*ML-10
Sgr 19h21'55"-29d24'A
Anna, radiant
And 1h3'24"37d39'
Anna,Sherilyn M
Lyr 18h27'29"47d34'
Anna-Karina
And 23h1'45"50d8'
Anna-Lisa
Her 17h52'1"46d59'
Anna-Marie
Cyg 21h45'30"30d6'
Anna-Marie
Umi 16h40'11"75d3'
Anna-Raoul
Cma 6h11'46"-15d57'
Anna-Ruth
And 23h44'60"38d56'
Anna-Sophie
Aql 19h6'26"-6d23'
Annabannawannahas
Cas 1h13'41"60d22'
Annabel
Lyr 19h5'29"28d48'
Annabel
And 2h1'39"45d2'
Annabel
Lyn 7h51'45"36d20'
Annabel et Bernard
Cyg 20h23'41"39d37'
Annabel-Lee
Vul 21h15'41"22d59'
Annabella
Ori 5h58'25"18d23'
Annabelle
Lyr 18h34'59"31d0'
Annabelle
Cas 0h38'56"69d17'
Annabelle
And 2h0'14"41d35'
Annabelle Lee
Lyn 7h41'26"40d34'
Annabeth
And 0h9'1"46d33'
Annabeth
And 2h12'54"40d19'
Annabeth
Cyg 20h30'57"38d32'
Annabirthintine 5
Lyn 7h45'57"48d34'
Annaforeveron
Cas 0h53'38"65d33'
Annagail Lynes,The
Aql 19h56'34"10d6'
Annah Harriet
And 1h12'23"39d53'
Annaick Mabebe J'Adore
Lyn 7h53'58"41d39'
Annajean
Per 3h52'40"36d25'
Annala,Jack Howard
Sco 17h54'26"-30d30'
Annalaura
Pic 5h2'31"-45d16'
Annalea
Cnv 13h2'58"50d53'
Annalese
Cam 4h19'47"70d13'
Annaliese
And 0h32'27"30d53'
Annalisa
Lyr 18h14'22"30d55'
Annalisa
Tel 20h18'39"-49d36'
Annalisa
Vel 9h19'15"-49d22'
Annalisa
Lyr 19h22'50"40d32'
Annalisa
Aqr 22h20'12"0d56'
Annalisa
Ori 5h31'10"8d39'
Annalisa
Cam 8h58'58"81d6'
Annalisa
Cnc 9h2'26"18d4'
Annalisa
Uma 12h28'37"62d56'
Annalisa (light of my
life),Marzio
Lep 4h59'34"-19d42'
Annalisa-Alessandro-
Francesco
Pup 8h24'39"-29d46'
Annalise
Lyr 18h42'32"35d46'
Annalise,Sierra
Mon 8h2'0"-0d27'
Annaloro-Rechani,Sidra
Peg 21h58'43"36d2'
Annamaria
Cep 23h30'19"63d30'
Annamarie
And 2h27'21"44d52'
Annamarie's Brandy Rose
Lyn 7h59'24"35d45'
Annamie
And 2h34'21"40d3'
Annand,Robert Alexander
Hya 8h53'28"-7d20'
Annarita
Uma 11h57'11"32d3'
Annayah
Umi 14h11'34"68d4'
Anne
Aur 4h47'1"38d45'
Anne
Lmi 10h6'24"34d26'
Anne
And 1h22'37"33d38'
Anne
Uma 9h18'34"55d53'
Anne
Cet 2h34'59"4d28'
Anne
And 0h23'1"38d47'
Anne
Tri 2h27'24"28d26'
Anne
Per 1h28'55"52d43'
Anne
Lib 15h29'28"-28d37'
Anne
Cas 3h22'1"70d25'
Anne
Cas 3h29'0"73d35'
Anne & David's Star
Cet 1h31'37"-6d54'
Anne & Gene
Cap 21h21'28"-20d24'
Anne André
Cep 20h57'45"63d10'
Anne Aurélie
Dra 17h38'25"73d44'
Anne Casey
Cas 0h56'0"72d30'
Anne Elizabeth
Uma 12h40'29"61d8'
Anne Elizabeth
Sge 20h5'27"20d6'
Anne Elizabeth
Lyn 8h7'52"34d55'
Anne en Novembre
Sgr 19h30'42"-31d51'
Anne Et Frederic
Ser 15h38'31"9d33'
Anne Et Nicolas
Lac 22h7'0"38d55'
Anne Forever LVLP
Ori 5h57'26"21d43'
Anne Hanneman
Mon 6h38'24"9d30'B
Anne Lee
Aql 19h54'1"11d46'
Anne Louise
Cyg 20h38'16"31d18'
Anne Louise
Cyg 19h34'49"34d20'
Anne Marie
Lyr 18h37'26"30d49'
Anne Marie
And 1h42'39"36d28'
Anne Marie
Cam 3h16'16"61d27'
Anne Marie
Cap 21h52'56"-21d52'
Anne My Burning Love
Forever
Uma 10h27'25"40d2'
Anne P 1980
Boo 14h38'58"14d23'
Anne Patricia Margaret Little
Miss Perfect
Ori 5h58'1"19d12'
Anne Platzdasch=Ole's Schatz
Leo 11h1'49"2d9'
Anne R
Cyg 20h19'48"39d57'
Anne Renea
Cma 6h59'13"-19d32'
Anne Sophie
Aql 19h3'44"16d24'
Anne Star
Cam 3h31'48"60d4'
Anne und Hubert zur
Silberhochzeit
Psc 1h44'17"17d39'
Anne Victoria
Eri 4h2'16"-15d22'
Anne's Lucky Charm
Mon 7h4'33"1d49'
Anne's Star
Cas 0h0'56"58d36'
Anne's Wesbridge Flirtation
Uma 13h25'56"62d23'
Anne,Elaine,Erin & Christine-
1994
Vul 21h15'24"20d22'
Anne,Loved By Dave
Boo 13h54'31"15d53'
Anne,Yeshua's Child
Vir 13h39'41"-8d4'
Anne-Elise's Garden
Aql 20h5'1"4d16'
Anne-Kathrin
Uma 11h13'41"41d15'
Anne-Marie
Cas 1h48'12"60d40'
Anne-Marie
Her 18h19'38"18d47'
Anne-Marie
Lyr 19h6'24"26d13'
Anne-Marie
Ori 4h42'0"0d57'
Anne-Marie
Del 20h14'10"14d42'
Anne-Marie
Cas 1h4'53"74d27'
Anne-Marie
Cyg 19h29'41"31d4'
Anne-Sophie P
And 0h27'1"40d32'
Annebelle
And 23h47'42"44d9'
Annechino,John
Aur 6h7'14"31d48'
Anneke's Angel
Tau 4h20'23"28d44'
Annelies
Psc 23h0'45"0d48'
Annell's Star
Ori 5h56'48"17d57'
AnneMarie
Cas 0h53'11"75d44'
Annemiek
Vul 19h23'13"25d15'
Annenberg,Ambassador
Walter H
Per 3h14'29"40d28'
Annese,Patrick
Cyg 20h42'39"45d53'
Annett
Aqr 22h59'18"-8d3'
Annetta
Lyn 08h57'38"46d25'
Annette
Aur 5h14'1"41d6'
Annette
And 23h39'21"39d56'
Annette
Lyn 8h13'0"49d10'
Annette
Uma 10h28'0"60d6'
Annette
Mon 7h5'4"-5d7'
Annette
Lyr 18h53'12"37d47'
Annette & Paul's Nightlight
Crb 16h8'51"33d27'
Annette & Victor
Boo 15h4'1"51d17'
Annette 4-10-94
Del 20h22'60"10d17'
Annette Ester
Cas 2h52'44"70d40'
Annette Jacqueline
Cas 0h32'39"72d42'
Annette Jewel
And 0h12'11"46d54'
Annette Lee
Vul 19h37'1"20d37'
Annette Loves Russ Forever
Uma 9h52'50"46d13'
Annette's Hunka Burnin' Love
Cas 0h41'59"65d17'
Annettemylove
Aql 19h57'58"11d38'
Annetteorobert
Leonieannelemarianne
Gem 6h25'19"12d40'
Anni
Cmi 7h20'58"5d23'
Anniballi,Jason Sylvan
Her 17h27'16"26d37'
Annicharico,Kennedy Ashley
Mon 6h22'46"7d49'
Annick
Lac 22h18'21"51d55'
Annick et Pierre
Cyg 20h24'31"39d18'
Annick T C
Peg 21h59'16"10d21'
Annick,Chantal
Ori 4h50'16"5d21'
Annie
And 23h19'0"37d33'
Annie
Lyr 18h16'32"38d18'
Annie
And 22h56'42"50d19'
Annie
And 1h4'39"39d10'
Annie
And 2h34'15"49d18'
Annie
Uma 11h45'12"50d60'B
Annie
Cyg 21h22'42"53d16'
Annie
Lyr 19h0'31"31d25'
Annie
Lyn 7h7'0"44d22'
Annie
Peg 22h21'22"3d11'
Annie & Hatsia
Uma 9h20'45"56d47'
Annie & Petie's Star
Gem 6h55'7"17d53'
Annie & Scott
Lac 22h14'30"51d59'
Annie Battle
Peg 0h6'1"27d46'
Annie Bettie
And 23h48'49"44d38'
Annie et Stéphane
Cyg 19h28'59"33d50'
Annie Forever
Eri 2h47'48"-5d50'
Annie Frannie
Eri 4h1'0"-19d39'
Annie M
Cas 23h15'17"60d26'
Annie O'
And 23h21'17"41d11'
Annie's Hideaway
And 23h39'0"48d2'
Annie's Star
Cas 0h33'46"62d26'
Annie's Star
And 23h40'55"41d44'
Annie's Star
Cyg 20h19'38"39d28'
Annie's Star
Uma 10h1'55"68d2'
Annie's Star
Uma 12h2'32"32d28'
Annie's Star
Lyr 18h40'56"28d47'
Annie's Star
Ori 5h51'47"17d29'
Annie,David,Amigo, Homer
Hya 8h47'38"-7d50'
Anniebelle
Mon 6h38'25"6d51'
Annies
Lac 21h55'1"40d11'
Annika
Cas 1h51'50"73d21'
Annika
Cet 2h55'1"2d51'
Annin,Jeffrey Burt
Ori 6h0'0"11d0'
Annina
Cas 0h55'28"62d7'
Annis,Rob
Eri 4h46'33"-7d13'
Anniva
Col 6h30'36"-34d10'
Anniversary
Cnc 8h47'45"9d27'
Anniversary
Cap 20h56'20"-19d21'
Anniversary Star
Crb 15h53'57"30d25'
AnnMarie
And 23h27'28"38d37'
Annmarie
Lyr 18h40'57"41d7'
Annmarie
Cas 0h55'42"67d10'
Anno,Gabriel RyNell
Uma 9h33'0"48d18'
Anno,Nicholas Dean
Uma 9h15'34"50d29'
Anno,Rochelle Ann Garlin
Uma 9h57'1"48d45'
Annoula
Lyn 8h55'47"41d7'
Anntracy
Lyr 18h36'26"36d23'
Annunziata,Cara Elizabeth
Cas 1h31'55"77d0'
Annunziata,Caroline
Cam 5h42'24"78d47'
Ano'ai mai Akua maka'a'a
Ori 5h53'57"5d11'
Anohé Auohiyanne
Her 17h52'53"40d45'
Anon
And 23h46'18"32d47'

Anora
Cam 3h56'13"61d7'
Anora
Equ 20h59'0"10d12'
Anouche,Laura
Cam 4h6'1"58d49'
Anouck Benoit
Cet 0h53'30"1d14'
Anouk et Fabrice 09 07 1985
Cam 5h4'43"58d17'
Anoum
Cas 2h55'18"77d21'
Anouscka,Preziosa Celilia
Cyg 20h19'19"39d30'
Anquetil,Jocelyn Alice
Cyg 20h2'14"31d51'
Anqulot,Sr,Thomas
Per 3h16'22"50d15'
Anra,Stern der ewigen Liebe
Cam 4h9'26"67d58'
Anrafat,Bovdra
Crb 16h18'54"38d58'
Anrom
Lyr 18h33'24"34d13'
ANSA 13
Cmi 7h59'55"1d42'
Ansbaugh,Robert Samuel
Sex 9h43'1"3d15'
Anschutz,Michelle L & Stephen L McKee
Umi 13h52'19"76d11'
Ansel,Thora Ofelia
Ari 2h40'59"21d26'
Ansell,J4 Caroline
Cas 0h14'47"47d10'
Anselmo,Rene
Aql 19h49'41"10d23'
Anselowitz,Tara Lynn
Cas 0h10'1"61d1'
Anseneau,David
Her 16h54'41"38d27'
Anshelewitz,Dawn Marie
Cmb 13h31'10"20d14'
Ansley Elizabeth
Aql 19h43'49"13d39'
Ansley,Marybeth
Cas 0h51'41"60d51'
Anslow
Uma 10h31'44"42d13'
Ansok,Chase Bradley
Hya 8h12'45"5d37'
Anson,Linda & Paul
Cyg 21h30'39"53d18'
Ansoud,Jacqueline
Dra 15h14'1"63d51'
Anspach,Peter
Umi 15h49'0"70d37"
Anspot
Lmi 10h5'59"39d3'
Anstead,Gregory Ernest John
Ori 5h56'29"12d19'
Anstey,Lisa
Uma 9h44'14"50d59'
Anstey,Stephen Leslie
Per 3h15'29"42d29'
Answered Prayers
Sge 19h53'33"19d27'
Antaki,Antoine
Her 18h6'43"28d39'
Antal,Erika B
And 0h50'32"35d36'
Antal,Jaki
Vul 20h47'55"20d8'
Antalek,Jennifer
Peg 22h53'34"26d55'
Antaramian,Beverly
Equ 21h21'57"8d54'
Antares
Vul 21h19'25"24d9'
Antea
Psc 22h53'41"0d11'
Antel,Franz
Uma 11h47'52"40d2'
Antemann,Maestro Richard W
Ori 6h0'42"20d12'

Anter,Philip John
Lac 22h26'24"41d8'
Anter,Ryan Sheldon
Psc 1h1'39"23d49'
Anthea
And 22h59'32"40d48'
Anthes,Emily K
Cyg 19h30'0"33d10'
Anthi
Aur 4h52'0"51d15'
Anthis,Dolores "Per Tutti La Mia Vita"
Mon 7h6'55"-5d25'
Anthon,Veronica Janann
Leo 11h23'35"-6d4'
Anthony
Dra 19h46'36"68d55'
Anthony
Ant 10h42'35"37d0'
Anthony
Cet 1h20'35"-0d36'
Anthony
Aur 5h4'1"41d52'
Anthony
Sex 9h53'13"-1d44'
Anthony
Aql 20h11'28"12d10'
Anthony
Her 16h53'10"34d29'
Anthony
Her 16h53'59"32d9'
Anthony
Cep 21h43'45"67d57'
Anthony
Cyg 19h32'31"37d42'
Anthony & Danielle's Starlit Magic
Crb 16h9'60"37d33'
Anthony & Phil 11/85- 11/94
Aur 5h4'37"44d59'
Anthony IV
Her 16h56'27"26d37'
Anthony J The Sica Of The Universe
Aur 5h14'0"44d17'
Anthony Joseph
Ori 6h3'25"1d30'
Anthony Michael
Her 16h7'28"24d9'
Anthony Michael
Hya 8h23'39"-0d8'
Anthony My Sweet Arubanian Prince
Aur 6h59'26"37d8'
Anthony R M
Cep 23h3'43"80d4'
Anthony's
Tri 1h53'58"28d6'
Anthony's Future
Lac 22h49'24"53d56'
Anthony's Midnight Rambler
Ori 5h31'1"-0d54'
Anthony's Musical Dreams
Cnc 8h57'11"17d59'
Anthony's Star
Umi 15h27'20"70d46'
Anthony's Star
Boo 14h53'60'37d40'
Anthony's Star
Cep 21h46'43"55d27'
Anthony, My Deigo
Per 2h33'16"56d19'
Anthony,Barbara
Lac 22h30'16"56d40'
Anthony,Betty Jean
Del 20h24'12"7d41'
Anthony,Catherine Shen
Cas 0h33'42"58d28'
Anthony,Cathy Marie
Mon 7h7'0"-6d40'
Anthony,Christopher Michael
Sex 10h15'32"-5d36'
Anthony,Harrison James
Cmi 7h58'33"1d46'

Anthony,Leonard M
Cnv 13h30'57"40d17"
Anthony,Marissa
Cas 0h1'52"58d32'
Anthony,Michele
Per 2h7'57"56d22'
Anthony,Rita
Cas 0h58'24"75d59'
Anthony,Robert W
Ori 6h4'42"8d24'
Anthony,Taylor Catherine
Sct 18h31'36"-5d55'
Anthony,Thomas J
Aur 6h56'0"40d8'
Anthony,Todd Wheeler
Leo 9h37'19"7d2'
Anthony,Tonya, & Anthony Jr
Her 18h18'21"12d20'
Anthony,Victoria Noble
Cyg 19h21'11"28d57'
Anthony-1969
Cep 21h33'19"65d25'
Anthonys,Scott
Cep 22h10'59"56d7'
Antica
Boo 15h1'23"48d14'
Antico,Joelle
Lyn 8h25'43"41d44'
Antigone
Boo 14h58'0"29d25'
Antigone,Geneviève
Cep 20h52'15"66d17'
Antillon,Rick
Dra 18h49'31"70d7'
Antin,Nina
Lib 14h56'17"-8d40'
Antin,Roger
Cep 22h50'55"57d51'
Antin,Vivian D
Oph 17h4'17"-1d35'A
Antin,William B
Oph 17h4'17"-1d35'B
Antionette
Umi 13h56'0"78d59'
Antirion
Ori 5h57'0"18d5'
Antle,James Michael
Aur 6h22'27"38d43'
Antle,Jr,Jeffrey L
Her 18h0'34"45d46'
Antler II International
Umi 16h16'55"76d50'
Antnony B
Aql 20h10'43"10d29'
Anto
Col 6h33'10"-33d10'
Anto 27/9/68 M
Cyg 21h18'0"30d48'
Antobellibus
Pic 4h34'57"-46d22'
Antoine
Ser 15h21'52"8d19'
Antoine
Aur 5h1'47"38d35'
Antoine & Zachary- Father & Son
Aql 19h56'31"12d4'
Antoinette
And 23h22'29"50d43'
Antoinette
Mon 7h9'40"-5d1'
Antoinette et Gabriel
Cam 7h44'45"68d3'
Antoinette Mary
Del 20h18'45"14d57'
Antoinette,Countess Luanne
Tau 5h50'10"23d46'
Antol,Jason Alan
Boo 14h31'42"48d26'
Antolik,Karin
Aur 5h10'43"54d24'
Antolos,Brigid Kathleen
Cyg 21h34'59"41d8'
Anton
Cnc 9h16'26"33d9'

Anton
Lib 15h5'26"-3d39'
Anton,Edward
Hya 9h3'43"4d18'
Anton,Michael P
Ari 2h2'15"17d49'
Anton,Sarah Jean
And 0h41'1"40d21'
Anton,Thomas
Peg 0h1'47"28d11'
Anton-Martin,Joshua Eric
Per 2h54'49"40d59'
Antonacci,Alexandra Elizabeth
Cam 5h52'59"58d37'
Antonacci,Nicholas Brian
Aql 20h13'21"3d45'
Antonace,Anthony Richard
Dra 15h4'0"60d11'
Antonakottoyaoe, Konstantinoe B
Lyr 18h16'32"42d42'
Antoncich,Kenneth
Per 2h37'37"40d3'
Antone,Clifford Jamal
Cet 2h53'28"3d37'
Antone,Elaine Speight Williamson
Hya 8h14'0"-5d57'
Antone,Ellis
Del 20h54'13"9d44'
Antone,Jennifer
Cet 1h58'36"0d40'
Antonella
Cam 3h32'21"60d53'
Antonella
Scl 23h39'40"-28d42'
Antonella
Her 16h12'38"18d59'
Antonella
Oph 17h54'42"11d3'
Antonella 29 luglio 1973
Cap 21h0'55"-20d27'
Antonella e Lorenzo
Nor 16h23'47"-52d31'
Antonella e Marco
Cas 2h20'13"60d4'
Antonella Irene
Cam 13h48'57"81d55'
Antonella,Cesaro
Del 20h17'57"10d57'
ANTONELLAPERSEMPRE
Her 16h47'49"40d30'
Antonelli,Heather
Cyg 20h18'58"41d28'
Antonelli,James & Melissa
Cam 12h3'18"78d16'
Antonelli,Nicole M
Leo 10h27'32"10d47'
Antonellis,Jane Arone
Dra 20h4'27"64d4'
Antonellis,Jane Marie Arone
Lyn 8h10'0"43d43'
Antonette
And 1h43'30"40d28'
Antonetti,Joseph
Aur 5h18'19"43d6'
Antoni
Ser 18h42'30"4d30'
Antoni,Brigitte
Sgr 19h16'25"-20d1'
Antonia
Lyr 18h32'0"45d47'
Antonia et Jean-Louis
Umi 16h16'56"83d46'
Antonia y Sergio
Per 3h32'23"39d22'
Antonia's Heart
Cas 0h25'60"62d58'
Antonia,My Only Star
Col 6h1'37"-33d38'
Antonicelli,Claudia
Col 6h32'43"-37d6'
Antonides' Star To Wish Upon,Sonia
Cyg 20h18'30"37d35'

AntonietttaMauro
Lyn 7h28'0"40d11'
Antonini,Chip
Aur 7h0'21"40d23'
Antonini,Jean Claude
Oph 18h34'0"7d16'
Antonio
Cae 4h40'54"-33d4'
Antonio & Clelia
Hor 3h4'13"-49d42'
Antonio,Gladyner F
Aur 5h31'32"48d46'
Antonio,Lela,Melitone
Uma 8h34'0"70d46'
Antonio,Miguel
Per 1h57'30"50d28'
Antonio-de-Mio
Sct 18h43'42"-6d30'
Antonio-I Love You, Mary
Her 17h51'53"38d53'
Antonioli,Bruce & Linda
Cyg 20h0'0"40d11'
Antonios,Brentos
Ori 5h54'26"13d5'
Antonius
Ori 6h7'1"-2d37"
Antonius Fulginittus
Her 16h51'47"50d5'
Antonius,Roy
Aql 19h53'13"12d12'
Antonsen,Nanna N
Cas 2h2'23"60d9'
Antonucci 143MJ, Lawrence
Lmi 10h51'1"38d2'
Antonucci,Michael Anthony
Tri 2h11'1"33d25'
Antonucci,Robert & Dolly
Lyn 9h10'51"42d29'
Antonucci,Sr,Anthony
Per 2h59'0"46d43'
Antonucci,Thomas Anthony
Dra 11h24'42"78d6'
Antony,Ahwyliaid Robert
Dra 17h59'46"58d26'
Antor,Sandy & Macgill
Lyr 19h23'45"35d22'
Antorobe
Cyg 19h52'13"44d6'
Antosek II,Richard Brian
Tri 1h54'33"28d24'
Antosek,Sean Christiaan
Ser 15h19'45"5d13'
Antovan-Demir-Lucente
Cnv 13h36'42"32d4'
Antrosiglio,Jr,Ralph
Boo 14h28'30"30d48'
Antruan
Cas 23h33'16"61d30'
Antuna,Dino
Cet 0h33'22"0d51'
Antypas,Dr Philip & Mrs Marlene
Oph 18h16'47"12d42'
Antz,Marguerite
Per 3h4'13"38d58'
Anu Mäki
Cas 1h13'41"60d53'
Anulewicz,Nicole
Peg 23h28'41"24d55'
Anupa
Lyn 7h34'42"38d14'
Anupa K
Dra 12h6'49"71d43'
Anupama Varma
Peg 0h2'44"14d21'
Anuska
Com 12h33'31"14d53'
Anuzis,Princess Carolyn
And 23h33'48"45d10'
Anwen
Ser 16h8'1"0d50'
Any
Tau 5h46'20"26d44'
Any Slug Or A H Y
Uma 9h37'50"54d30'

Any Star
Peg 21h53'1"31d23'
Anya 8395
Leo 9h33'55"10d45'
Anya Ellen
Cam 5h35'53"61d26'
Anya,Angel
Peg 22h39'48"20d45'
Anzalone 7-18-81, Shannon Rose
And 1h52'0"47d20'
Anzenberger,Jessica Marie
Cam 3h55'1"56d16'
Anzevino,Angela Kathryn
Lyr 19h6'1"28d19'
Anzhelika
Lyn 7h31'0"36d45'
Anzie's Star
Lyn 9h8'55"38d28'
Anzo
Cnc 8h7'15"30d47'
AOI USAGI
Psc 1h23'2"2d42'
Aoki Family,Stella of
Psc 1h15'33"20d30'
Aossey,Pamela Kay
Uma 10h53'0"58d38'
Aouad,Georges Augustus Merrell
Ori 5h58'29"12d18'
Aoun
Dra 16h12'0"60d15'
Aoun Family,The
Boo 15h12'38"40d9'
Aoustin,Isabelle
Dra 12h29'29"63d34'
Ap
Uma 11h13'26"57d59'
APART
Mon 7h6'8"-7d6'
Apasha,Lee
Aql 19h31'47"-6d24'
Apaullo Laws I: MI Brillante Astro
Her 16h58'36"28d20'
APCMJJ Austria
Oph 17h18'55"-20d58'
Apfl,Johann
Gem 6h35'11"26d45'
Apgar,Jr,Douglas Robert
Aql 19h30'1"13d49'
Aphrodite
Cma 6h44'33"-13d58'
Aphrodite
Boo 14h28'14"27d22'
Aphrodite
Cyg 21h4'12"40d26'
Aphrodite
Cas 0h53'31"75d15'
Aphrodite Australis (Penny Darvall)
Tau 4h40'37"0d30'
Aphrodite Banks-Star of `94
Cam 12h13'58"78d39'
Aphrodite Mario Virgo
Uma 10h59'55"40d16'
Aphrodite's Peach
Cnv 13h4'42"50d52'
Aphrodite-Louise
Umi 13h15'37"71d32'
Apicella-Love Of My Life,Maria
And 2h25'45"46d30'
Apis
Per 4h6'12"50d21'
Apisson,Jeffrey
Lac 22h21'35"55d14'
Apitzsch,Renee L
Uma 9h32'1"51d50'
Apker,Louis
Oph 17h1'55"-23d13'
Aplamiss Retsis
Lyn 9h21'58"41d33'

APLI (Aileen Paskoff of Long Island)
Cam 5h47'55"61d31'
Apo With Love
Ser 15h35'46"-2d18'
Apocalypse For K N
Gem 7h11'9"24d6'
Apokryphos
Oph 17h1'18"-22d14'
Apolinario,Carol
Cas 3h1'19"60d34'
Apollo Jr High- Richardson, TX
Aur 5h7'53"41d43'
Apollo Linda
And 0h47'52"39d7'
Apollo,Billy
Ser 15h15'18"21d49'
Apollo,Francine Marie
Cas 2h19'58"61d29'
Apone,Gina Lisa
Lyr 19h17'1"41d32'
Aponte,Carlos A
Cmi 7h52'47"0d28'
Aponte-Blizzard,Laura Cristina
Umi 15h44'39"76d17'
Apostal,Dr Robert
Dra 19h42'54"65d6'
Apostel,Beate *030759* ***050679***
Per 2h5'30"58d25'
Apostol,Kimberly A
And 2h12'26"38d0'
Apostolatou,Eya A
Lac 22h49'22"38d11'
Apotheker,Allan
Dra 19h40'48"67d50'
Appel,Bryan Herrington
Aur 4h53'43"41d2'
Appel,Christopher John
Per 3h31'45"38d44'
Appel,David Joseph
Dra 18h27'19"50d36'
Appel,Estelle L
Cnc 9h14'26"23d51'B
Appel,James Robert
Aql 20h2'34"0d11'
Appel,Justin David
Boo 14h46'37"50d10'
Appel,Roger
Aur 6h12'35"31d37'
Appel,Sharon Christine
Cyg 21h35'1"41d8'
Appel,Sidney J
Cnc 9h14'26"23d51'A
Appol,Ute
Her 18h0'26"40d23'
Appelbaum,Susan
Lyn 7h11'1"58d32'
Appeldorn,Darcy
Cet 3h0'16"0d15'
Appeldorn,Jeffrey Scott
Her 17h0'14"20d58'
Appello,Allison Leigh
Cep 23h25'32"63d39'
Appello,Robert William
Umi 15h41'26"73d30'
Appelman,Karen
Crv 12h48'60"-11d50'
Appelmann,Joe
Ser 18h6'1"-13d39'
Appenrodt,Heidrun
Sco 16h39'1"-40d40'
Appenrodt,Jens
Sco 16h39'0"-40d27'
Appenrodt,Jochen
Sco 16h38'56"-43d21'
Apperly,Jim
Uma 11h25'35"32d16'
Apphia Page
Cet 2h29'26"0d18'
Appiarius,John
Peg 23h0'20"24d29'
Appiarius,Lucille
Lyr 18h18'30"40d56'

Appicello,Linda
Vir 13h5'41"-0d51'
Appignanesi,Maureen & Augusto
Lyr 18h58'12"34d38'
Appinger,Fred
Cyg 20h25'15"30d48'
Appio,Anthony Paul
Lac 22h7'22"40d18'
Appio,Christina Maria
Cam 3h50'54"61d51'
Appio,Regina Elizabeth
And 22h56'23"36d18'
Apple Dumpling Star
Boo 14h48'51"35d36'
Applebaum Star,The Sid
Ori 5h59'50"10d55'
Applebaum,Roslyn
Mon 6h39'29"10d21'
Appleby,Daniel Lee
Umi 13h12'12"72d18'
Appleby,Jasper Mark
Sgr 18h49'0"-30d51'
Appleby,Roger
Crb 16h16'43"37d33'
Appleby,Trevor
Cep 2h20'29"78d41'
Appleby,Valerie Suzanne
Lyr 18h17'57"32d26'
Applegate,Alice LaRosa
Peg 23h29'22"21d36'
Applegate,Alyssa Helen
Aql 19h40'55"12d16'
Applegate,Andrew Vincent
Cnc 8h33'57"30d45'
Applegate,Jane L
Uma 11h43'11"56d11'
Applegate,Katie A
And 23h11'0"41d2'
Appleladia
Eri 2h54'54"-6d22'
Appleton
Hya 8h40'1"3d24'
Appleton,Peter Lang
Cep 23h22'58"65d10'
Applewhite,Jr,Glen
Sct 18h43'47"-6d22'
Applewhite,Tracey C
Equ 21h19'40"3d39'
Appleyard,Fred
Boo 14h19'38"25d60'
Appleyard,Joan
Lyr 18h14'26"30d1'
Appleyard,Mary
Lyr 18h14'54"34d48'
Appretiatus Aeturnus
Lac 22h1'32"49d55'
Appuhn,John F
Her 16h58'0"20d21'
Apraham,Harry F & Patricia G
Lyn 9h7'16"42d21'
Aprati,Gabriel J
Her 18h12'56"38d44'
April
And 2h3'32"38d57'
April
Del 20h18'43"11d13'
April
And 23h48'51"37d36'
April
Cas 23h41'21"61d30'
April
Eri 3h47'1"-2d6'
April
Peg 22h47'27"20d7'
April & Brett
Sge 19h56'47"20d11'
April & Eunice & Bruce
Dra 17h52'48"67d50'
April & Timothy
Mon 7h16'1"-6d37'
April Ann
Cas 22h57'22"54d23'
April Christine
Umi 16h10'22"78d48'

April G
Vul 19h48'40"28d22'
April Jean
Peg 22h42'19"35d30'
April Marie
And 0h19'25"45d18'
April Michelle
Cas 0h3'32"60d15'
April My Love
And 2h3'0"37d35'
April Scumdog
Peg 0h3'15"28d55'
April Skies
Mon 6h55'0"-10d56'
April's Hopes & Dreams
Peg 22h53'1"21d46'
April's Star
Lyn 7h50'43"47d54'
April,Rabbi Samuel
Del 20h16'16"10d15'
April-Emily-Rachel
And 0h4'38"47d49'
Aprile Dee
Umi 16h18'16"70d11'
Aprille
And 0h22'23"36d35'
Apryl's Majestic Wolf
Aur 5h20'48"38d8'
Apsey,Ellen Abigail
Umi 15h36'40"70d40'
Apsey,Francesca Louise
Umi 15h39'17"70d19'
Apt,Heather Joan
Com 12h5'22"26d37'
Apt,Tammy Ann
Oph 17h55'32"7d57'
Apter,Dr Nathaniel
Cnv 12h15'1"46d58'
Apuna,Janel
And 0h20'47"35d29'
Apuzzo,Sally Ann
Lyn 9h0'51"39d42'
Aqeel
Aql 18h57'51"17d11'
Aquamarine
Psc 23h8'35"1d32'
Aquarius Meubelindustrie
Uma 10h57'52"40d15'
Aquaro,Georgeanna
Uma 14h15'43"59d28'
Aquesta
Cam 3h59'56"53d53'
Aquilante,James M
Aur 5h19'1"41d38'
Aquino,Amalia
Ori 5h52'10"17d3'
Aquino,Teresa
Vel 9h43'32"-49d34'
Ar Bòidheach Clann
Aql 19h6'58"0d4'
Ar Máthair - Slainté
Ori 5h56'22"17d59'
Araca,David Nicholi
Per 2h55'53"50d25'
Arachide
Cnv 12h9'52"44d38'
Arad,Tammy
And 23h29'53"40d22'
Arago,Rory
Cam 9h6'0"81d6'
Aragona,Martha
Mon 6h54'38"1d46'
Arai,Ryo
Cam 4h58'24"61d7'
Araiza,Capt Ricardo C
Cet 2h48'46"-0d18'
Araiza,Klye
Aur 7h5'45"40d2'
Araki,Tetsue "Rocky"
Uma 10h54'33"59d10'
Aramaki,Junji & Helen
Cep 22h16'1"68d26'
Arambel,Alexander Joesph
Aur 5h54'1"29d29'

Arambel,Andrew John Aur 5h55'32"29d9'

Arambula,Kenny Her 17h2'27"48d9'

Aramburo,Daniel Aur 5h48'48"50d32'

Arana,Alfred Her 17h9'55"38d28'

Arana,Caristine Marie Vul 21h26'1"24d12'

Arana-Gigliotti,Kathy Aql 20h10'23"11d44'

Arancio,Gerard Anthony Aur 4h54'29"51d21'

Arand,Thomas C Cma 6h53'18"-18d22'

Aranda,Andres M Cep 22h21'59"68d45'

Aranda,Lucy S Cas 0h23'55"61d36'

Arango Cma 6h32'47"-16d25'

Aransibia,Tracey Oph 18h7'0"12d43'

Arantxa And 0h49'49"35d59'

Arantxa Cam 13h18'45"78d46'

Aranyi,Millicent Jean Low Sex 10h1'57"-8d10'

Arash Lac 22h3'35"38d30'

Arashi Sco 16h26'38"-28d56'

Arata,Claire Marie Del 20h13'0"10d45'

Arata,Julie Ari 2h56'57"21d57'

Arata,Maureen Elizabeth Ser 15h31'2"18d44'

Arata,Robert David Gem 6h37'54"20d20'

Arati,Franáoise Per 3h22'58"54d23'

Araujo da Fonseca, Cristina Cep 22h4'32"65d17'

Araval's Eye Ori 4h50'42"4d54'

Aray Lac 22h17'12"46d60'

Arazi,Mary Sharon (Sherri) Keefer Peg 22h42'14"33d38'

Arbach,Rick Denis Lyn 7h39'10"41d7'

Arbeit,Mary Rita (Geggy) Soviero And 2h30'54"39d29'

Arbeitman,Sharon Cas 1h46'21"73d39'

Arbogast,Jr,James "ARBO-MAN" Her 18h18'37"12d38'

Arbogast,Michele Cas 0h53'21"58d59'

Arbogast,Ronald Aql 20h2'26"10d56'

Arbogast,Shane Mattew Lac 22h21'38"40d55'

Arbolino,Jack Cap 20h38'50"-15d25'

Arbore,Mark Alan Aur 4h49'14"51d22'

Arbortarium Boo 14h57'41"40d8'

Arbuckle,Alisa Lyr 18h17'42"32d3'

Arbuckle,Catherine Elizabeth Cam 3h47'45"58d14'

Arbuckle,Katharine Mon 7h0'55"4d23'

Arbuckle,Rosalind Del 20h57'17"11d12'

Arbury,Suzi & Robin Cam 7h7'36"61d49'

Arby Lewis Cep 23h29'0"65d27'

Arc-of-Elinoi Cyg 19h28'23"37d52'

Arcal Ori 5h58'28"18d13'

Arcarola Star,The Andrea Lib 15h42'13"-8d56'

Arce,Tony Sct 18h49'58"-8d52'

Arcede Cep 21h48'26"65d8'

Arceneaux,Dwayne David Cet 1h23'1"-13d22'

Arceneaux,Glynn Dale Aql 19h51'26"14d17'

Arch,Arnold Hya 8h30'60"-8d13'

Archacki,Jill Peg 23h3'48"33d47'

Archambault,Jennifer Jo "Archie" Cas 0h28'0"62d20'

Archambault-Knox,Alice Betty Cyg 19h36'28"28d34'

Archambeault,Claire Beaudry Cas 0h44'23"73d43'

Archambeault,Joyce And 0h52'60"40d33'

Archbald,Joseph Cep 3h42'42"78d56'

Archbold,Marbella Del 20h53'46"3d39'

Archer "My Honey", Chuck Per 4h24'27"50d17'

Archer,Alexandra Stella Mon 6h34'35"8d32'

Archer,Austin Allen Hya 8h21'48"-10d12'

Archer,Bradley P Her 17h30'12"20d9'

Archer,David Garland Aur 6h17'0"30d30'

Archer,Donna Garman Cru 11h58'27"-59d18'

Archer,Jane M Cnv 12h46'37"38d48'

Archer,Jr,Glynn Raymond Aql 20h6'32"4d23'

Archer,Kevin John And 23h18'42"49d12'

Archer,Lisa Renee Cas 1h3'42"50d25'

Archer,Mark Aur 5h50'46"50d18'

Archer,Marta Katherine Cet 0h57'38"1d31'

Archer,Nicole Boo 14h42'0"27d1'

Archer,Sandy Uma 8h39'12"51d46'

Archer,Tasmin And 0h36'0"41d14'

Archer-Perkins,Richard Charles Per 2h6'59"56d23'

Archer/Morris,Edna M And 0h2'40"44d25'

Archiapatti,Laura V Cyg 21h23'16"38d13'

Archibald,Mary Grace Cas 1h27'47"61d19'

Archibald,Ronald Aur 4h59'37"31d31'

Archibeque "PT's Star" Valorie Lyr 19h9'42"38d14'

Archie Aql 20h2'13"10d6'

Archie & Margaret Mon 6h21'10"3d50'

Archie's Aim Vul 19h19'13"26d18'

Archiere,Patricia Ann Cas 0h40'16"69d57'

Archimedes Aql 20h10'29"10d48'

Archimedes Vir 13h56'1"-6d30'

Archond,Jeanne Mon 6h30'55"8d18'

Archuleta,Nancy And 0h6'29"47d2'

Archuleta,Ruben Louis Cyg 21h1'12"41d1'

Arcieri,Christopher Joseph Per 3h41'18"51d14'

Arcieri,Graziella Col 6h31'50"-33d19'

Arciniaga,Mac & Aurora Cyg 20h50'33"40d15'

Arciuolo,Jill Cas 0h26'12"65d50'

Arciuolo,Matthew Vul 19h44'1"28d19'

Arcola Starburst Mon 8h7'33"-8d57'

Arctic Fox,The Lib 14h39'22"-23d20'

Arcturus,Pauline Cas 1h41'44"75d11'

Arcuri,Craig Allan Per 3h13'14"55d42'

Arcuri,Don & Cathleen Cet 0h57'47"0d50'

Arcuri,Jennifer Lynn Peg 22h25'51"20d32'

Arcuri,Michael J & Lila L Arcuri Del 20h23'46"10d23'

Arcuri,Scott & Meisje Boo 13h37'1"21d46'

Arcuri,Vincent James Lac 22h10'52"51d49'

Arcury,Michael-Ann Cam 5h8'15"60d25'

Ardelle,Marian Ser 15h14'42"4d10'

Arden & Rick Cas 1h0'56"58d53'

Arden,Dorothea Elisabeth Leo 9h21'60"17d22'

Arden,Keith Aql 19h31'18"1d4'

Arden,Vicki A Uma 9h38'32"70d8'

Arden-Griffith,Paul Lyr 18h23'43"47d5'

Ardeshir Per 1h27'22"53d10'

Ardessa Vir 14h35'1"7d22'

Ardiet,Roger And 0h22'0"34d9'

Ardilia,Mandy Del 20h36'20"7d57'

Ardill,Doris Annie Lyr 18h41'24"35d30'

Ardis Cyg 21h54'36"52d43'

Ardiss Marie Cyg 20h57'0"38d31'

Ardito,Vincent Gem 6h45'1"31d41'

Ardner,Kevin Christopher Boo 15h7'53"27d2'

Ardo Boo 14h53'50"27d13'

Ardon-Astros,Frank Hor 2h56'36"-49d38'

Arduini,BrieAnne Farlow Aql 20h4'51"0d30'

Arebalo,Jason Wade Cam 3h57'55"58d0'

Arebalo,Mandy Michael Ori 6h1'0"10d8'

Arechiga,Artemis aka Patty Uma 10h4'12"59d41'

Arechiga,Rudy & Laura Cnc 8h36'1"18d42'

Mon 8h0'18"-10d44'

Arehart,Bray Lynne Lainne Lac 22h1'13"49d54'

Arella,Msgr Gerard J Per 2h22'1"58d39'

Arellano,Mark Anthony Cnc 7h57'45"10d3'

Arena,Alan M Cep 21h59'11"55d57'

Arena,Alec Andrew Aql 19h47'22"11d36'

Arena,Carolyn Rose And 23h42'16"45d26'

Arena,Cathrine & Salvatore Oph 17h4'30"-20d9'

Arena,Chelsie Elizabeth And 1h6'58"40d15'

Arena,Claudia Lmi 10h32'41"31d28'

Arena,Debbie Cas 23h13'43"63d27'

Arena,Joseph Dominic Ori 5h52'41"15d7'

Arena,Melissa Blake Vul 20h14'35"25d43'

Arena,Pamela J And 1h30'13"37d34'

Arenas,Jr,Virgil Francis Umi 13h9'10"74d10'B

Arend,Rachel Elizabeth And 0h5'36"34d1'

Arend,Robert Joseph Per 2h48'58"50d9'

Arens,Doris Tau 5h0'54"15d44'

Arens,Kelly Wilde May 1,1990 Tau 5h53'0"23d20'

Arensmeyer,Bonnie Hya 8h9'18"1d15'

Arenz,Peter Boo 14h6'48"11d12'

Arenz,Werner Ser 15h14'42"4d10'

Arestad,B Thomas Gem 6h43'4"18d25'

Arete Sct 18h55'23"-6d15'

Aretz,Christel und Helmut Uma 11h46'30"42d51'

Arevalo I,Jonathan & Kelly Crt 11h39'54"-20d55'

Arevalo,Catherine Vir 13h33'26"12d38'

Arevalo,Noel Aql 20h21'11"7d56'

Arevel Mon 7h53'11"-5d26'

Arezina-Telling, Svetlana And 23h35'57"49d41'

Arfanis,Rubini S Lyr 18h41'34"26d50'

Arff,Dave Her 17h53'52"14d22'

Arfin,Miriam Rae & Robert Scott Rebitzer Uma 10h55'15"36d50'

Arfin,Samantha Blake Mon 7h31'0"-0d33'

Arft,Wayne Uma 11h14'38"52d28'

Arfwedson,Anders Per 2h18'1"58d11'

Argabright,Darren James Rui Her 17h24'1"30d7'

Argabright,In Honor of Edward Louis Cet 1h31'32"-11d46'

Argacha,Paula E Ignasi Lyr 19h23'0"38d37'

Argenta,Malmo Boo 15h18'23"34d5'

Argentine Dra 12h10'48"72d26'

Argenzio,Norma Elaine Vul 20h18'46"23d25'

Argenzio,Vincent James Boo 14h34'32"21d22'

Arghiere,Lulu-Marie Lyr 18h59'31"27d20'

Argila,Carmen Lloberes Boo 13h45'16"11d28'

Argiz,Mazine Greenspan Aur 5h34'40"37d38'

Argudo,Susana Cnv 12h41'49"33d51'

Argus,John Her 16h28'40"48d2'

Argyropoulos Aql 19h1'48"-6d47'

Argyros,Cristos & Eleni Lazarus Boo 14h11'23"12d13'AB

Argyrou,Eleni And 1h13'21"37d10'

Arhets,Michael John Oph 18h18'58"10d36'

Ari's Star Aql 20h1'47"7d13'

Ari's Star Boo 13h48'23"18d25'

Ariadna Oph 16h59'47"10d5'

Ariana Peg 21h55'32"28d47'

Ariana Lac 22h53'51"38d32'

Ariana Angela Cyg 20h24'56"39d39'

Ariana Rose And 1h18'0"37d25'

Ariana,Marissa Cet 3h17'60"1d36'

Ariane Tau 5h0'35"15d47'

Ariane & Eugene Uma 11h28'12"40d48'

Ariane de Genève Cam 4h57'35"58d47'

Ariani,Rojine Eri 2h54'1"-18d28'

Arianna And 23h20'17"51d31'

Arianna Ant 10h46'24"-37d27'

Arianna Ori 5h59'1"16d54'

Arianna's Dream Cas 1h6'37"70d54'

Arianne Dra 10h18'17"81d5'

Arias of the Magic Flute Lib 14h44'48"-4d4'

Arias,Esther Alicia "Jila" Martinez Mon 7h44'46"-2d32'

Arias,Hernando Cam 12h5'47"80d18'

Arias,Jr,Alfredo Luis Boo 14h54'11"48d30'

Arias,Mr & Mrs Hernan Aur 6h18'19"46d45'

Arias-Shining Star, Carolyne Lee Uma 10h18'36"59d42'

Ariel And 23h32'43"41d7'

Ariel Psc 1h3'28"21d49'

Ariel Cas 23h35'1"61d20'

Ariel Cyg 19h37'20"28d52'

Ariel Dawn Lyn 6h15'22"60d7'

Ariel Gabriella Aql 18h46'59"11d33'

Ariel Rose,The And 23h39'21"38d39'

Ariel's "Wedding Star" Crb 16h6'32"30d41'

Ariel's Star Eri 3h44'44"-7d22'

Ariel,Jason Boo 15h4'30"25d55'

Ariel-Rachel And 2h27'32"47d2'

Ariela 19/2/58 Cas 1h58'56"58d8'

Ariele,Mario Pepita Aur 7h24'16"43d29'

Arielle Umi 14h20'43"70d20'

Arielle And 1h55'59"40d19'

Arielle Fai Cnv 14h1'53"45d22'

Arielle,die Meerjungfrau Cep 21h50'56"62d51'A

Ariem (Marie) Uma 9h42'38"67d32'

Ariemma's Lodestar 50, Gloria & Ed Lyr 19h18'47"41d44'

Arien Boo 14h43'49"29d10'

Ariena,Facies Hya 9h35'43"2d18'

Aries Heavenly Body Ari 1h44'44"23d57'

Aries Tours Cru 12h48'44"-57d58'

Ariex Cam 6h25'31"67d49'

Arif,Sila Lyr 19h16'23"41d32'

Arife Eri 4h50'34"-8d40'

Ariffin,Amanda & Sasha Dhillon Sge 20h17'23"16d35'

Arih,Linda & Jessica Cyg 19h58'1"30d23'

Ariko,John Hya 8h40'22"4d58'

Arild's Romance Ori 6h9'30"4d3'

Arilotta,Lori (Love) A Uma 10h50'18"70d43'

Arinn Jean Cet 1h32'41"-0d13'

Arinsberg,Alex Aql 18h42'52"-1d46'

Arious Uma 11h44'54"44d23'

Aris Dart Crb 15h31'24"30d37'

Aris,Judy Lyn 7h10'0"50d39'

Aristizabal,Estella Marco A Eri 2h56'44"-11d24'

Aristophanes Cam 3h47'18"73d56'

Arita Ori 5h54'0"12d11'

Aritz,Maureen Cas 0h21'21"76d16'

Aritz,Maureen And 23h30'16"39d55'

Arizmendi,Ruth Mon 8h1'51"-6d3'

Arjan Lmi 10d44'0"24d2'

Arjan Ori 5h51'1"11d36'

Arjay Peg 21h59'1"8d8'

Arjucyjac Lac 22h27'58"36d38'

Arjune,Shalini U Lyr 19h13'1"40d24'

Arkell,Roderick John Tracy Aur 5h34'26"40d23'

Arkin,Alan Aql 18h43'26"11d50'

Arkin,Bruce Dra 18h22'52"50d28'

Arkin,Erica Lynn Cyg 21h20'59"39d46'

Arkle Aur 6h30'18"34d31'

Arkle,David James Ori 6h3'58"2d20'

Arkle,Marion Per 3h16'0"41d44'

Arkus,Larry Uma 10h40'59"51d3'

Arky,MD,Ronald A Per 2h3'17"56d49'

Arlanda Umi 15h46'13"71d16'

Arlandez,Rosa Frances Boo 14h24'25"51d13'

Arlati,Donatella Uma 11h46'13"50d9'

Arlati,Giuseppe Her 16h10'53"11d9'

Arle,Marilyn A Lyr 18h46'15"42d27'

Arledge,Jennifer Ari 1h48'26"17d7'

Arleij,Anita & Kyell Sunnerud Cyg 20h1'40"30d60'

Arlen,Tracey Mon 7h38'44"-2d11'

Arlene Vul 19h22'51"26d58'

Arlene Cet 1h55'0"-4d55'

Arlene Uma 8h33'30"52d1'

Arlene & Fred's Shining Star Dra 20h0'33"63d39'

Arlene & Richard Cyg 21h0'29"41d12'

Arlene & Tony's Christmas Star Peg 22h34'12"27d11'

Arlene A Precious Angel Lyn 8h5'50"39d55'

Arlene,Tony Cam 3h46'23"73d59'

Arlenzia's Light Peg 23h28'13"33d35'

Arlequin,Alexa Roma Sgr 19h0'39"-24d38'

Arlet Peg 23h36'40"31d34'

Arlette Cyg 20h54'38"50d52'B

Arlette Cam 12h14'30"82d26'

Arlette And 0h19'54"31d35'

Arlina,Judy Per 3h2'26"41d1'

Arlt,James F Lyn 8h55'47"42d31'

Arlyn's Luster Cas 0h29'0"66d54'

Armadillo Cnv 13h5'53"37d35'

Armadillo,The Oph 18h17'39"12d22'

Armand Umi 13h18'47"71d40'

Armand,Jacques Ori 6h4'57"10h39'

Armand,Joseph Umi 16h42'13"76d18'

Armand,Nedi Charlambos Cmi 7h43'44"8d1'

Armandina Gru 22h12'37"-52d24'

Armandine Uma 9h42'1"59d49'

Armando Cet 1h32'59"1d1'

Armanino Cmi 7h37'48"4d39'

Armarhkai Lyn 7h40'20"50d16'

Armas Cet 0h36'32"-3d9'

Armbruster,Pilgrim Dra 17h27'55"61d32'

Armbruster,Showgun Andy Lyn 8h25'39"33d42'

Armen & Heather's Star Forever Ori 6h17'19"-0d16'

Armendariz,Ana Alicia Equ 20h58'31"9d22'

Armendariz,Laura Patricia Mon 6h18'44"5d25'

Armendo,Mariana Pho 23h58'9"-45d42'

Armengol,Raul Ori 5h48'18"11d47'

Armeni,Happy Birthday John Aur 7h2'49"41d5'

Armeno,Jennifer Ann Tau 4h25'25"28d56'

Armenta,Andre Aur 6h8'0"38d24'

Armenta,Carmen And 2h1'32"41d15'

Armenta,Kenneth E Ser 17h56'17"-13d11'

Armenta,Nicole Rochelle Sct 18h53'15"-6d32'

Armenti Best Mom In The Galaxy,Judy Lyr 18h59'50"38d12'

Armentrout,Deb Mon 7h42'54"-6d57'

Armentrout,Erick Aql 20h4'1"8d8'

Armentrout,Howard & Delores Eri 4h2'0"-17d14'

Armentrout,Jane Cas 0h9'42"50d30'

Armentrout,Joshua Hya 9h10'14"3d15'

Armentrout,Ralph & Jan Oph 17h18'22"11d57'

Armentrout,Scott Sex 9h51'36"1d39'

Armes,Pangea-Michael Cet 1h51'19"0d29'

Armet*Police Officer #2786 SCPD,Ron Ori 5h46'1"20d26'

Armgard Sco 17h51'52"-30d8'

Armida Uma 10h9'31"42d26'

Armin Gem 6h37'54"20d3'

Armin Cep 22h4'13"61d25'

Armine,Natasha Equ 21h10'27"10d20'

Armington,Mackey, Cummings,Krem Mon 6h52'32"10d23'

Arminjon,Delphine Cyg 20h53'38"50d8'

Armintrout's Star,Paul Cep 22h58'59"70d60'

Armitage,Gilliam Margaret Cas 0h49'60"71d12'

Armitage,Jill Lyr 18h16'20"34d57'

Armitage,Julia M Sge 20h5'19"20d31'

Armitage,Michael J Per 2h41'52"40d50'

Armitstead,Paul D Aur 6h17'43"33d6'

Armon,Sarah Cutler And 0h0'31"47d4'

Armonia Nor 16h23'4"-51d1'

Armour,Jennifer Kristin And 1h53'51"38d44'

Armour,Lisalex Hope Mon 6h57'7"-1d33'

Armour,Steven Her 17h53'22"28d22'

Arms of Orion... Matthew & Marci,The Ori 5h28'43"-1d40'

Armstead,Joe Oph 17h13'22"-20d7'

Armster Bailey Ori 5h1'0"0d28'

Armstong,William C & Susan G Cmi 7h23'26"8d20'

Armstrong Her 17h3'14"21d37'

Armstrong Per 3h46'34"38d46'

Armstrong III,H C Umi 17h11'2"78d52'B

Armstrong III,S H Umi 17h11'2"78d52'A

Armstrong,Aloysius J Aur 4h54'26"50d53'

Armstrong,Amy K Lyn 7h49'32"35d57'

Armstrong,Anastasia Cam 3h47'38"71d41'

Armstrong,Brent Thomas Boo 15h46'1"42d45'

Armstrong,C J Uma 10h11'0"48d18'

Armstrong,Cash Patterson Her 15h55'52"41d17'

Armstrong,Charles Robertson Cet 0h48'1"-7d11'

Armstrong,Etta Mae Dale Umi 16h48'1"78d7'

Armstrong,Ferrell Vul 20h39'43"22d36'

Armstrong,Ginger Anne Cet 1h29'51"-10d55'

Armstrong,Guy & Jana Cyg 21h5'21"37d11'

Armstrong,Holly K Cas 1h36'59"65d24'

Armstrong,Jr,Charles Ralph Cmi 7h6'25"4d7'

Armstrong,Jr,James Edward Uma 12h58'0"61d26'

Armstrong,Katie Jane Ori 5h32'0"-0d39'

Armstrong,Ken Hya 8h58'19"2d22'

Armstrong,Kris Lyr 19h0'35"40d45'

Armstrong,Lolita V Mon 6h30'33"7d55'

Armstrong,Marie Celeste Brewer Eri 4h1'36"-18d31'

Armstrong,Marilyn Cet 2h46'30"4d42'

Armstrong,Melinda Marie Cma 6h54'7"-18d52'

Armstrong,Mercia Cet 1h16'50"-2d1'

Armstrong,Michael Oph 18h21'55"6d53'

Armstrong,Michael Lee Boo 15h4'27"53d50'

Armstrong,Mr & Mrs Jeff & Debbie Eri 2h53'42"-5d51'

Armstrong,Nicholas Graham Dra 17h55'13"60d56'

Armstrong,Nigyl Belonging to David A Cnv 13h38'28"41d20'

Armstrong,Richard Lee Roy Cet 3h7'1"0d59'

Armstrong,Robert Bryan Aql 19h54'60"8d22'

Armstrong,Roy A
Cep 23h17'1"65d3'
Armstrong,Russell
Cep 3h33'49"80d9'
Armstrong,Sarah
Mon 7h41'36"-8d53'
Armstrong,Sherry Lynn
Cam 4h53'48"58d44'
Armstrong,Star of Clan of
Cyg 20h26'24"42d39'
Armstrong,Starship James & Gail
Crt 11h21'44"-18d58'
Armstrong,Steven C
Oph 18h19'57"6d16'
Armstrong,Susan
Cyg 21h19'1"35d40'
Armstrong,Vincent James
Peg 23h29'41"17d38'
Armstrong,William Kelley
Dra 17h43'44"68d24'
Armstrong,Yared B C
Lmi 10h57'14"27d13'
Army,Barbara
Aql 19h40'39"11d42'
Arna
Ser 15h56'24"-2d59'
Arna's Shining Star
Lyn 9h3'56"39d41'
Arnal,Aleth
Eri 3h45'60"-1d10'
Arnas,Sara Lynn
Cas 22h57'45"55d16'
Arnau,Carmen Abella
Aur 5h15'55"43d9'
Arnaud,Christelle
And 0h21'14"43d45'
Arnaud,Guimard
Cep 22h53'55"68d38'
Arnaud,Jean-Philippe
Cyg 19h47'1"31d11'
Arndorfer,Katherine Ann
Cmi 7h41'43"1d34'
Arndt,Erik Arthur Nshuti
Mon 8h1'26"-6d47'
Arndt,Jürgen
Lyr 18h26'16"47d17'
Arndt,Ken & Andrea
Lyr 19h22'58"41d56'
Arndt,Richard J
Ori 6h17'1"20d49'
Arndts,Diane & David
Cyg 20h7'0"40d53'
Arndts,Robert
Cep 21h22'36"55d23'
Arne,Allene Elizabeth
Com 13h2'0"15d55'
Arne,Brian Alexander
Oph 17h35'33"-23d30'
Arne,Jason Michael
Hya 8h32'26"-0d57'
Arnelly
Vul 20h59'24"20d17'
Arnemann,Gertrud Baumann
Mon 8h2'50"-4d29'
Arnemann,Marius Gustav
Tau 4h1'37"21d20'
Arner,Harold Stanley
Aql 20h10'26"12d54'
Arner,Jim
Dra 15h54'15"60d40'
Arner,Sydney Ann
Eri 2h55'35"-2d20'
Arnes,Jr,Wilson
Crt 11h32'0"-11d22'
Arnesen,Ann Marie
Ori 5h50'29"14d14'
Arnett,Ann Canfield
Cyg 19h36'37"38d1'
Arnett,Debbie
Mon 6h22'1"8d59'
Arnett,Harriet
Aql 18h58'1"-5d18'
Arnett,Nancy Carol
Peg 22h32'18"25d50'
Arnett,Todd Loren
Uma 8h50'24"71d33'
Arnette III,Glen
Cmi 7h7'43"3d57'
Arney,Jennifer J
Vul 20h15'29"25d42'
Arney,Patricia Carol
Cas 0h2'0"63d29'
Arney,Paul Bryan
Cam 3h17'45"60d21'
Arnie's Meaning Of Life
Cep 21h25'0"84d55'
Arnie
Ant 10h20'1"32d9'
Arnim,H A von- Theodora Martienssen
Ara 17h54'37"-50d35'
Arnim,Haus Gerswalde von
Gru 22h9'2"-55d26'
Arno 11:11
Lac 22h33'48"52d40'
Arnold
And 23h19'34"41d47'
Arnold #21,Jennifer Lynn
Lyr 18h15'37"33d59'
Arnold Superstar, Rachel
And 23h1'31"50d26'
Arnold,Amy
Aur 7h17'39"40d50'
Arnold,Andrew J
Boo 14h41'45"38d24'
Arnold,Bernard
Her 17h58'46"14d34'
Arnold,Bobby
Mon 7h52'56"-3d39'
Arnold,Catherine Helen Agnes Lee
Crb 15h24'33"31d51'
Arnold,Christin
And 0h19'56"32d55'
Arnold,Christine D
And 2h30'42"42d34'
Arnold,Daniel Kelley
Lac 22h4'0"38d24'
Arnold,Dick
Aur 5h7'58"41d48'
Arnold,Dr Richard M
Dra 18h32'55"68d55'
Arnold,Emily Christine
Lyn 6h32'26"58d58'
Arnold,Emily Elizabeth
Aql 19h24'41"-0d3'
Arnold,Emily Sheridan
Cas 0h6'41"65d35'
Arnold,Erika
Leo 9h18'1"11d29'
Arnold,Frank F
Aur 5h58'54"30d4'
Arnold,Helen
Cyg 20h59'46"28d55'
Arnold,Hugh B
Tau 4h15'40"7d39'
Arnold,Jennifer & Stacy
Dra 19h31'51"78d9'B
Arnold,Jerry
Aur 5h35'54"40d1'
Arnold,Jessy James
Aql 18h57'26"-6d15'
Arnold,Jewel S
Lyn 9h15'0"37d21'
Arnold,Joan L
Com 12h2'54"27d8'
Arnold,Judi & Bob
And 2h27'0"48d35'
Arnold,Lee F
Cam 4h57'37"71d1'
Arnold,Linda
Lyn 7h51'41"33d26'
Arnold,Linda (Mum)
Lyr 18h30'59"31d35'
Arnold,Luella
Lyn 7h11'11"58d58'
Arnold,Manfred & Susan
Crb 15h20'30"30d22'
Arnold,Margie
Lyr 19h20'43"40d25'
Arnold Mary Patricia Riordan Humphrey
Lyn 7h4'41"59d57'
Arnold,Matthew Joseph
Ser 15h54'16"23d57'
Arnold,Michael J
Aql 20h0'32"6d55'
Arnold,Natalka C
Eri 2h52'31"-1d34'
Arnold,Nathaniel Stewart
Sex 10h20'53"-1d44'
Arnold,Norman
Aql 18h59'19"-5d10'
Arnold,Peter Werner
Dra 20h6'24"62d22'
Arnold,Randy Allison
Her 16h42'16"5d14'
Arnold,Raymond Douglas
Aur 5h17'17"45d9'
Arnold,Ryan Matthew
Cam 5h57'31"61d45'
Arnold,Shirley Ann
Peg 21h55'1"34d12'
Arnold,Sonja Michaela
Cas 0h40'14"73d40'
Arnold,Susan
Ind 21h3'1"45d47'
Arnold,Sr,Ernest M
Cet 2h5'26"-6d2'
Arnold,Stacy & Jennifer
Dra 19h31'51"78d9'A
Arnold,Stan Michael
Aql 19h2'36"15d4'
Arnold,Susan Peggy
Cyg 21h55'38"44d42'
Arnold,Tana Ann Lewis
Com 12h28'46"18d48'
Arnold,Thomas Huston
Aur 5h55'36"31d38'
Arnold,Thomas Wayne
Ori 5h25'12"1d33'
Arnold,Tom
Aql 18h58'44"15d15'
Arnold-Lead Guitar,Tim
Her 16h46'34"47d59'
Arnold-Mar Star
Uma 11h22'1"50d12'
Arnold-Peters,Andrea
Ari 1h46'1"11d32'
Arnold-The Star!!,Lee
Cnc 8h11'31"32d8'
Arnoldi,Liliane
Dra 15h0'50"64d25'
Arnolyn
Uma 10h47'11"71d20'
Arnone School,Dr William
Boo 14h53'10"48d15'
Arnot,Coree
Aql 19h50'54"10d52'
Arnott,Alene Louise
Gem 8h4'0"28d21'
Arnott,Linda
Mon 7h0'48"-8d7'
Arnoux,Laëtitia
Umi 15h52'26"73d15'
Arnowitz,Andrew
Dra 17h25'0"63d58'
Arnquist,Jacob David
Per 2h56'34"55d27'
Arns,Richard G
Uma 10h38'39"60d40'
Arnstein "75",Albert
Aur 4h46'10"50d33'
Arnstein,Nicky
Cep 2h42'28"78d25'
Arnstein,Selma
Eri 2h57'26"-4d47'
Arnstrom
Cam 8h45'1"73d11'
Aro,Raimo Kari Juhani
Cam 13h30'0"81d52'
Arocho,John Eli
Cep 0h59'16"77d12'
Arone,Traci
Vul 19h18'33"26d27'
Arongitler
Boo 14h9'33"30d31'
Aronica,Matthew L
Lyr 19h21'10"41d31'
Aroniss,Duke
Lib 15h0'0"-28d30'
Aronno's Argosy
Sge 19h54'34"20d32'
Aronoff,Kenny
Cep 3h10'49"75d35'
Aronovitz,RPh,Alan
Her 18h3'16"14d55'
Aronsen,Berit
Cas 3h4'1"58d12'
Aronsohn,Brigitta
Cas 0h23'25"64d10'
Aronsohn,David & Audrey
Mon 7h7'39"-7d11'
Aronson,Abbe Muecke
Oph 17h38'41"-20d10'
Aronson,Bernice
Per 2h51'38"45d26'
Aronson,Donald
Her 16h45'54"21d56'
Aronson,Donald Shelton
Cet 2h55'20"5d38'
Aronson,Elaine & Curtis Armstrong
Ser 18h22'25"4d25'
Aronson,Gary L
Lac 22h32'32"56d47'
Aronson,Jared Craig
Dra 17h15'47"64d59'
Aronson,Karol
And 23h20'58"46d31'
Aronson,Massimo,Joseph
Lmi 10h15'45"31d57'
Aronson,Michael David "Prince"
Her 18h1'17"30d58'
Aronson,Raphael
Uma 9h57'1"70d4'
Arouni,Monica
Sco 16h14'33"-22d6'
Arpa's Dream
Uma 9h15'32"56d39'
Arpino,Kelsi Rene'
Vul 19h46'30"28d19'
Arpoika,Miko James
Lac 22h12'16"50d25'
ARPS 2000
Uma 8h24'34"60d44'
Arps Star,Dave
Aur 4h48'50"37d48'
Arquette,Michelle Frances
And 0h24'22"30d34'
Arquilla
Aql 20h6'42"0d38'
Arquilla,Frank Salvador
Aql 19h53'1"11d34'
Arquitt,Helen "Hudd"
Cyg 21h9'16"39d41'
Arrcee
Vul 19h57'45"28d54'
Arredondo's "Arrde" Star,Tony
Cep 21h48'33"58d31'
Arrela,Annikki
Cam 3h39'59"60d53'
Arria,Thomas J
Ori 5h59'1"18d21'
Arriaga (Sheba),Silvia Rodriguez
Oph 18h41'28"8d23'
Arriaga,Maria Teresa Esparza
And 2h20'60"37d23'
Arriens,Ross
Sgr 19h39'42"-31d23'
Arrigo,Marc Allan
Her 17h19'1"21d22'
Arrigon,Terri & Bob
Cyg 19h26'1"31d18'
Arrigoni,Alessandra
Scl 23h17'37"-32d27'
Arrindell,Radford Arthur
Per 3h3'43"40d46'
Arrington
Cam 6h30'45"68d13'
Arrington Star,The Ramsay
Aql 19h46'52"14d6'
Arrington,Amy Leigh
Cas 2h22'1"60d41'
Arrington,Jessica Elise
And 1h2'1"40d2'
Arrington,Juanita C
Cyg 20h2'0"30d53'
Arrison,Sherri L
Cyg 20h39'10"53d27'
Arroleo,Maureen A
And 0h59'55"33d23'
Arroleo,Rachel Rose
Cas 1h29'45"75d50'
Arrowood,Horace
Dra 16h59'44"68d3'
Arrows 56
Sge 19h57'1"16d26'
Arroyo Vista Elementary
Peg 0h6'57"27d45'
Arroyo,Maria Christina
Cas 23h37'44"60d34'
Arrybae's Arrystae
Tri 2h5'1"33d18'
Arsement,John Anthony
Aur 5h1'0"50d21'
Arsenault,Angèle
Lyr 18h53'31"34d53'
Arsenault,Bonnie Leigh
Lyr 19h14'28"41d46'
Arsenault,Debra Ann
And 1h44'0"48d49'
Arsenault,Joseph Arthur
Aur 6h3'10"31d27'
Arsenault,Shelley
And 1h59'58"36d18'
Arseneau,Jöel-Evelyne Blais
Cyg 21h5'41"39d47'
Arseneault,Gary
Lac 22h48'0"54d53'
Arseneault,Julie
Lac 22h43'17"55d54'
Arseneault,Kim
Lac 22h43'26"55d29'
Arseneault,Marie- Blanche
Cyg 19h28'14"33d42'
Arseneault,Marie-Eve
Lac 22h43'35"55d0'
Arsham,Bryan Evan
Lac 22h32'56"37d53'
Arshansky,Brittany Morgan
Lyn 8h13'0"46d15'
Arsu Amor-Ananke
Mon 6h39'51"11d47'
Art
Oph 16h55'0"1d34'
Art
Boo 14h54'31"33d21'
Art
Ser 16h2'53"22d56'
Art
Cnc 9h9'43"31d37'
Art & Cynde "Destiny"
Mon 7h59'15"-2d42'
Art & Rita Forever
Her 16h39'20"4d42'
Art's Star
Her 16h56'1"40d60'
Artache,Sr,Miguel "Mick" Angel
Lib 14h26'39"-23d42'
Artaud,Armelle
Cyg 21h57'1"50d12'
Arteaga,The Roberto
Uma 10h32'32"45d54'B
Artemisia
Aql 19h48'45"10d37'
Arter,Helen H
Cas 23h35'54"61d43'
Arter,Janet S
Ori 5h37'21"11d17'
Artero,Annie
Col 6h1'26"-32d4'
Artesi,Terry E
Cet 3h8'52"1d41'
Artglier,Thomas
Vul 19h20'17"26d31'
Arthalice
Dra 16h10'44"62d26'
Arthapignet,Marie Eve
Per 3h29'32"39d38'
Arther,Carol
Mon 6h44'52"10d50'
Arther,Gregory
Hya 8h55'11"1d16'
Arthunia
Pyx 8h50'22"-27d26'
Arthur
Her 18h3'1"28d30'
Arthur
Uma 11h50'21"51d33'
Arthur
Cam 6h4'34"60d3'
Arthur
Cep 21h22'54"67d60'
Arthur
Uma 9h54'1"53d19'
Arthur
Lyn 7h10'51"52d37'B
Arthur & Barbara New Star,The
Dra 20h15'20"63d15'
Arthur 1
Hya 8h54'34"5d49'
Arthur Douglas
Uma 12h3'24"31d36'
Arthur Frederick
Ori 5h2'34"8d41'
Arthur Herbert
Cmi 7h22'38"1d0'
Arthur's Star
Oph 17h51'53"11d8'B
Arthur's Vision
Vel 10h7'20"55d40'
Arthur,Brandon Bearking
Uma 10h50'1"57d33'
Arthur,David P
Oph 18h17'40"1d17'
Arthur,Eleanor
Cas 0h50'1"71d2'
Arthur,Frank
Dra 14h52'49"55d5'
Arthur,Gabrielle A
Mon 7h14'15"-10d27'
Arthur,Kathryn Lynn
Lyn 7h41'23"45d29'
Arthur,Kay
Cas 1h3'24"64d23'
Arthur,Lori Ann
Peg 22h30'47"24d60'
Arthur,Pam "Nitz McBee"
Lyr 18h27'29"32d41'
Arthur,Patricia Ann Brooks
Peg 21h52'27"33d0'
Arthur,Ruth & Chris
Hya 10h46'32"-18d3'
Arthur,Sallie Maria
Cas 0h24'57"50d19'
Arthur,The Star Of My Life
Lac 22h35'41"55d29'
Arthur,Winifred C
Uma 11h42'0"52d42'
Arthur-Periodontist,Dr Harold
Oph 17h35'36"-23d32'
Arti
Cyg 19h24'47"35d34'
Artie
Oph 18h1'55"10d31'
Artiga,Guillermo Ernesto
Aql 19h26'36"-8d54'
Artiga,Raul R
Cmi 8h1'59"1d37'
Artigaut,Veronique
Ori 5h48'1"19d22'
Artis,Carolyn
Lyn 7h9'45"44d29'
Artist Formerly Prince & Mayte
Aur 6h3'36"37d35'
Artist Paquette,The
Ori 5h56'45"21d19'
Artist's Heart
Dra 17h3'46"61d45'
Arturi,Daniella Maria
Lyn 8h18'43"34d37'
Arturious Rex
Oph 17h1'33"11d5'
Arturito
Ori 5h57'0"14d28'
Arturo
Per 4h35'41"37d19'
Arturo,Don
Ori 5h26'13"-4d57'
Arty-My Fire in the Night Love Amy
Per 1h46'21"50d37'
Artz,Allen John
Aur 6h0'54"30d12'
Artz,Burton Lee & Kristina Wigle
Peg 22h1'56"13d24'AB
Artzi,David Alan
Ser 18h1'26"-14d30'
Aruban April
Cam 13h14'0"80d16'
Arun Paul-Flawless
Cet 2h10'29"2d39'
Arundel,Amy
Cas 0h57'1"50d36'
Arundel,Augie
Cas 0h57'12"54d41'
Arundel,Cali
And 0h52'59"40d57'
Arundel,Courtney
And 0h53'42"40d6'
Arundel,Jamie
Dra 11h27'12"77d51'
Arundel,Jordan
And 0h54'56"45d18'
Arundel,Molly
Cas 0h57'1"58d28'
Arundel,Samantha
And 0h53'57"40d31'
Arundel,Timmy
Lac 21h55'45"38d31'
Arundel,Tommy
Her 16h20'19"41d13'
Arvay,Vivian M
And 2h26'14"39d21'
Arvedson,Robert Nils
Sgr 19h29'17"-31d4'
Arveux,Jean-Franáois
Per 3h36'51"51d40'
Arvin Buddy Day,Henry
Crt 10h55'1"-11d20'
Arvin,"Amber" Gayle
Aur 6h1'17"38d7'
Arvinen,Hellin Adele Arrhenius
Sco 16h56'35"-43d41'
Arvis,Jean
Lyr 19h4'57"47d34'
Arweiler,Ruth
Vir 12h3'38"-0d52'
Aryae I
Psc 1h2'33"20d42'
Aryan
Ori 5h34'10"9d27'
Arye-40
Cyg 19h47'41"29d44'
Arzapalo,David
Per 2h7'40"47d38'
Arzenti 17-Apr-1994, Thomas Vincent
Ori 5h34'53"-0d11'
Arzenti 29-May-1991, Rosa Jean
Ori 5h39'46"-0d7'
Aráuz,Celina Suzanna
Oph 18h4'38"13d57'
AS
Uma 14h25'57"58d27'
As an Fhearann
Ori 6h5'26"5d47'
As Bright As Our Love
Cyg 20h53'13"38d15'
As Forever As KC & Sylvia's Love
Cam 13h9'0"76d58'
As You Wish
Dra 12h57'59"72d28'
As You Wish
Uma 9h22'36"56d16'
As You wish
Boo 13h54'16"20d21'
As You Wish Annie & Joe 93
Mon 7h32'13"-1d11'
As You Wish TAS Love LAT
Cnv 13h30'51"50d22'
AS-Forever J T
Dra 19h32'28"61d10'
ASA
Oph 18h40'39"8d15'
Asa
Umi 17h7'14"75d17'
Asa"25"I Love You,FN
Uma 12h20'59"57d59'
Asada,Kristen Chiemi
Aql 18h58'26"-6d57'
Asadorian,DMD,Arthur E
Boo 15h16'50"51d43'
Asaeda's
Lmi 9h57'49"38d53'
Asai,Michelle Mari
Tau 3h28'54"30d38'
Asami
Umi 13h34'0"70d49'
Asaph,Adam
Dra 17h2'26"63d29'
Asapurna
Cam 5h32'21"61d13'
Asaro,Jr,Mitchell David
Aur 7h23'42"40d55'
Asaro,Vincent
Ser 15h56'27"17d32'
Asboe Star,The Susan
Umi 15h39'48"68d5'
Asbrink,Eva
Ori 5h48'32"20d59'
Asbury,III,James Holley
Her 17h33'1"27d1'
Asbury,Mary Catherine
Cyg 21h5'38"40d36'
Ascalon
Ser 15h9'46"-1d15'
Ascani,Leonard
Aql 19h54'26"10d30'
Ascanio
Ori 5h57'0"14d50'
Ascanio,Alana Cari
And 0h6'59"47d45'
Ascari,Michael
Cet 1h20'35"-0d11'
Ascencio,Mark Anthony
Umi 14h54'0"66d35'
Ascenius,Janich J Bjerre
Umi 14h41'1"66d18'
Asch,Barry Stephen
Cyg 19h50'24"38d53'
Aschebrock,Yasmin
Lyr 18h28'45"31d52'
Aschenbrolch,Fabienne
Crb 15h35'2"37d6'
Ascher,Toby
Tau 4h19'0"10d59'
Ascherin,Hal & Barbara
Uma 10h38'57"56d58'
Asci,Gregory Louis
Per 1h43'0"52d33'
Asci,Jennifer Lynn
Tau 4h2'44"20d43'
Asci,Louis
Cep 3h21'1"78d13'
Ascolese,Vincent P
Dra 15h9'36"57d49'
Ascoli:The Star
Uma 9h44'32"58d34'
Asemani,Mehrdad
Ser 15h12'1"2d57'
Asen,Alois
Cyg 20h24'40"31d50'
Ash Brook-Flower
Nor 16h18'22"52d12'
Ash With Love,Regina Richards
Del 20h15'45"9d32'
Ash,Catherine Ryan
Lyr 18h30'24"44d3'
Ash,Emilie Antoinette
Lyr 19h9'56"47d44'
Ash,Gladys
And 0h41'37"38d47'
Ash,James Edward
Aur 6h32'40"31d24'
Ash,Morgan Trent
Cyg 21h52'31"41d30'
Ash,Pattie
Lyr 18h50'28"33d6'
Ash,Rita Kata
Cas 0h41'44"74d12'
Asha
Mon 6h18'0"-5d43'
Asha
Cas 2h39'30"65d33'
Ashabranner,Eva Mae
Mon 7h3'19"-0d56'B
Ashabranner,Gerard Dudley
Mon 7h3'19"-0d56'A
Asham,Adel
Sgr 18h52'55"-23d48'
Ashbaker,Lindsey Nicole Rush
Cyg 21h21'16"40d3'
Ashby's Bit of Heaven
Aqr 21h36'54"-6d31'
Ashby,Clarence Linden Garnett
Oph 17h39'9"-20d42'
Ashby,Keith
Per 3h26'1"40d45'
Ashby,Sharon
Cam 4h44'1"68d36'
Ashcliffe,Thaddeus
Cam 4h52'34"68d52'
Ashcraft,Albert Steven
Cet 3h3'52"3d52'
Ashcraft,Dream Come True Kelly
And 2h26'29"44d55'
Ashcraft,Maralyn Miller
Com 12h53'45"24d33'
Ashcraft,Melissa "Missy" Dawn
Tri 2h7'33"33d32'
Ashcroft,Adrienne & Richard
Peg 21h55'57"33d31'
Ashcroft,Lewis James
Cep 20h31'21"67d2'
Ashe,Gilda Ann Rizo
And 23h3'10"51d33'
Ashe,Nicole Elizabeth
Cas 1h5'43"63d19'
Ashenden
Uma 14h26'28"60d28'
Asher
Dra 9h55'0"77d46'
Asher
Ind 21h21'11"47d24'
Asher Prime
Uma 9h26'11"52d46'
Asher,Al
Aql 18h59'34"-6d50'
Asher,Ross
Aql 20h8'51"4d52'
Asher-Linnea
Com 13h6'12"20d56'
Ashera
Lyn 8h44'50"38d3'
Asherah
Sco 16h52'22"-41d2'

Ashford "Our Worm", Jeanna Marie
Peg 22h19'11"20d30'
Ashford,George L
Uma 11h59'30"58d33'
Ashford,Luke(Big Bear)
Uma 9h54'49"67d58'
Ashford,Matthew
Lac 22h8'11"50d28'
Ashkin,Justin Ross
Boo 14h11'39"39d39'
Ashkin,Justin Ross
Her 17h12'28"43d4'
Ashkinos,Andrew Samuel
Aql 20h20'54"0d5'
Ashkinos,Eric Franklin
Her 16h55'1"28d14'
Ashleigh Elizabeth
Cas 0h26'28"61d21'
Ashlen R (2=1)I
Per 2h36'29"43d8'
Ashley
And 2h8'13"50d33'
Ashley
Cas 0h28'47"67d3'
Ashley
Cyg 19h28'55"31d51'
Ashley
Cmi 7h27'56"0d3'
Ashley
Aql 18h58'16"-5d35'
Ashley
Per 2h57'35"45d27'
Ashley
Cyg 21h11'36"34d16'
Ashley & Alexandra
Eri 4h6'44"-11d32'
Ashley & Brock
Aql 19h51'34"12d41'
Ashley & J T
Cet 1h32'51"-11d35'
Ashley & Michael's Star
Cyg 19h29'42"31d7'
Ashley Angel
Cas 0h11'0"63d43'
Ashley Ann
Mon 7h1'27"-1d9'
Ashley E
And 23h21'44"48d46'
Ashley Elizabeth AKA Ashwood
Aql 20h8'48"4d36'
Ashley Elizabeth
Eri 2h46'59"-4d20'
Ashley Eve
Umi 14h0'50"72d10'
Ashley Jennifer
Psc 22h51'59"5d11'
Ashley Joy
Cyg 19h32'49"32d50'
Ashley Joy
Lyr 18h44'32"31d23'
Ashley Korn-Ponies & Rainbows
Uma 10h14'56"68d1'
Ashley Lynn
Peg 22h32'30"12d22'
Ashley Marie
Uma 9h15'24"51d46'
Ashley Nicole
Oph 18h17'40"13d36'
Ashley Nicole
Cnc 8h55'1"30d10'
Ashley Rose
Aqr 22h35'41"2d11'
Ashley's Dream Maker
Cas 0h38'39"68d57'
Ashley's Shooting Star
Mon 6h32'23"8d7'
Ashley's Star
Cyg 21h33'11"34d39'
Ashley,Amber Lynn
Uma 11h33'16"60d52'
Ashley,Cooper Graham
Cam 13h47'13"80d4'
Ashley,Craig
Aql 19h37'21"-6d54'
Ashley,Daniel Fitzpatrick
Eri 3h43'1"-6d1'
Ashley,Dr Tom
Oph 18h1'39"12d11'
Ashley,J Hayward
Cam 6h4'36"56d3'
Ashley,Like a Diamond in the Sky
Del 20h50'17"9d45'
Ashley,Pati Miller
Mon 8h4'38"-4d44'
Ashley,Rick
Aur 6h54'0"38d42'
Ashley,Roy David
Her 16h23'0"27d18'
Ashley-Kay
And 23h35'48"44d26'
Ashley-Pooh
Gem 6h42'5"14d54'
Ashlin
Ori 5h52'34"8d41'
Ashline,Rev Alfred J
Uma 12h54'36"62d4'
Ashling
Peg 21h53'0"31d15'
Ashling's Star
Dra 11h53'15"66d49'
Ashmead,Duke
Peg 0h4'27"28d22'
Ashmore,Clara Isabella
Cyg 20h24'50"38d7'
Ashok
Gem 7h35'27"27d47'
Ashor
Ori 5h53'58"15d17'
Ashoss,Katharine Lee
Del 20h34'35"11d52'B
Ashraf,Mohammed
Her 18h26'26"18d54'
Ashrafi,Mehran
Cam 3h52'35"61d36'
Ashtin,Justin Kaitlin
Equ 21h19'40"3d41'
Ashton Frederick
Oph 17h1'24"-24d10'
Ashton's Star,Robert Nathan
Boo 14h50'45"31d15'
Ashton,Andrea
Crb 15h31'44"37d46'
Ashton,Chloë
Cyg 21h4'0"28d35'
Ashton,Dean
Peg 22h48'38"24d56'
Ashton,Donald R
Her 17h12'56"42d10'
Ashton,Gail Carol
Peg 23h2'1"21d3'
Ashton,Haldene Stafford
Aur 5h9'52"42d30'
Ashton,Kama Gayle
Uma 11h19'0"50d34'
Ashton,Louisa Grace
Umi 15h7'29"69d27'
Ashton,Mary Lou
Lac 22h27'39"55d7'
Ashton,Michael James
Per 1h45'50"52d53'
Ashton,Rosemary J
Lyn 7h52'32"35d0'
Ashworth
And 0h28'52"28d32'
Ashworth,Charlotte Jean
Mon 7h29'52"-0d36'
Ashy,Mr Oussama Fahmy
Ser 15h57'50"11d11'
Asia
And 23h38'23"33d38'
Asil
Aql 19h43'35"14d0'
Asimakis,Zoe
Mon 6h38'33"-10d23'
Asimenia Efthimioy
And 23h3'25"50d13'
Asjes,Alexandria Anneke
Peg 23h27'20"33d57'
Asjes,Dirk Lucas
Lac 22h14'45"48d14'
ASK
Uma 9h6'35"57d27'
Askanas,Jordana
Ori 6h2'41"-2d13'
Asken,Edward William
Lep 15h29'7"41d23'
Asken,Harry Conner
Lep 15h29'9"42d50'
Askew,Joseph Ward
Lib 15h8'33"-1d1'
Askin,Judy
Cas 1h41'26"70d20'
Askins,Betty
And 23h31'17"49d54'
Aslakson,Eleanor Fern
Cas 23h14'49"63d12'
Asli
Ori 4h57'28"5d49'
Asmussen,Cornelia
Hya 9h32'31"5d28'
Asnath's Smile
Ori 5h1'48"13d28'
Asnes,Ray
Crb 15h49'42"37d33'
Aspagm
Lyn 7h10'34"51d24'
Aspatore,Janet
Uma 10h34'23"51d8'
Aspatore,Jennifer Lee
Cam 11h55'18"81d29'
Aspatore,Jonathan Reed
Uma 10h45'56"52d15'
Aspeck,Tania
Uma 8h47'51"50d46'
Aspenleiter,Hermann
Dra 16h20'23"58d41'
Asperger,Todd
Uma 10h0'53"52d0'
Aspes
Uma 11h44'29"50d36'
Aspgren,Johna B
Aur 6h43'35"39d46'
Aspinal,Joshua Ryan
Cyg 20h16'35"38d40'
Aspinwall,Dana
Aur 4h55'59"40d22'
Asplund,Rhonda Jeanette
Peg 23h44'36"30d46'
Asroff,Kenneth
Lyr 18h56'41"30d16'
Assaf's Special Star, Zackery Rion
Hya 8h12'59"1d6'
Assefa,Jared
Oph 17h17'40"-20d49'
Asselin et Chantal, Erika Yvan
Umi 14h21'54"69d50'
Asselin-"Esther"Carol Ann
Aqr 22h9'24"-6d11'
Assheton,Nicholas
Per 1h53'55"50d36'
Assilem XXIII,Kehoe
Ori 5h1'1" 1d15'
Association Orion
Ori 5h27'34"0d4'
Assumpton BVM Class Of 1996
Her 16h37'0"34d34'
Assunta
Aql 19h26'26"8d24'
Assunta `66,Maria
Gru 22h5'43"-53d0'
Assuras "Anchor Par Excellence",Thalia
And 23h1'42"40d17'
Ast,Steven Todd
Gem 6h50'2"14d51'
Asta & Drazen
Eri 5h2'0"-5d2'
Astar
Lyn 7h49'27"41d57'
Aster
Cap 20h54'21"-15d10'
Asterbadi,Zachary Nabil
Tau 5h50'21"26d45'
Asteri Aurea
And 1h43'24"38d5'
Asteria
Cam 4h47'22"68d12'
Asterino,Rosemarie
Tau 3h36'0"24d16'
Asterita,Adrienne Marie
And 0h39'32"45d5'
Asterita,Joseph John
Dra 16h43'45"51d57'
Asteroid B-612
Uma 8h16'42"70d37'
Asteroide B612 AS
Uma 12h11'46"58d15'
Asteroide BZ
And 0h28'27"41d4'
Astfalk,Dagmar Mercedes
Cep 22h1'49"60d2'
Asti Palaya
Cep 22h6'45"67d45'
Astings,Jean Pierre
Per 3h59'43"37d14'
Astley,Amy
Lyr 19h23'30"40d27'
Astolfi,Gaetano Walter
Umi 16h57'0"80d34'
ASTON
And 1h23'6"42d23'
Aston,Anna
Cas 23h19'32"60d31'
Aston,Laura
Del 20h18'45"10d1'
Astoo
Cyg 19h33'1"34d35'
Astore,Robert D
Lac 22h9'51"40d47'
Astoria
Cam 5h5'34"61d42'
Astorino's Flame
Peg 23h3'14"10d57'
Astra
Hya 8h58'23"-8d17'
Astra Catherinus (Buni) Tracium
Cam 8h14'24"73d42'
Astra Debbie
And 23h48'33"41d19'
Astra Luke
Oph 18h1'22"12d52'
Astra Remli Capricorni
Hya 9h2'46"5d40'
Astral Annette
Lyn 8h21'1"49d50'
Astral Rose
Cas 2h22'43"63d46'
Astral Sidney
Cep 22h12'50"62d42'
Astre Petit Young #12251993
Boo 14h57'47"17d31'
Astrid
Per 2h39'1"56d43'
Astrid
Uma 12h26'49"62d27'
Astrid Ursula
Leo 9h18'31"17d43'
Astrid,Heidi
Aql 19h1'53"-1d37'
Astro
Cmi 7h13'54"-13d28'
Astro
Uma 10h20'22"53d17'
Astro Ashburn
Vul 20h19'40"25d6'
Astro-Brack
Vul 19h23'0"23d42'
Astrum,Lindsey Cassiopeiae
Cas 0h26'46"61d58'
Astrum,Viridis
Aqr 22h27'25"-17d40'
Astuto,Gloria
And 0h18'16"33d53'
Asuman
Uma 10h59'0"50d0'
Asumi
Her 18h12'19"38d15'
Asvestis,Christophe
Com 13h33'20"20d15'
Atack,Alan
Her 16h47'0"4d27'
Ata Iti Zachary
Pup 7h28'49"28d38'
Atalla,Mitchell P
Cet 2h47'1"-0d36'
Atasoy,Selin
Umi 15h30'0"70d25'
ATB & AJB
Uma 8h39'53"49d2'A
Atcachunas,Michael John
Dra 20h15'43"80d18'
Atcheson,Zachary Tyler
Cep 0h17'0"73d36'
Atchison,Gabriella
Cyg 20h22'20"39d11'
Atchison,John
Cep 20h21'30"60d15'
Atchison,Nancy A
And 2h25'20"43d6'
Atchison,Todd
Aur 7h7'50"36d31'
Atchley,John Adams
Per 2h57'52"45d16'
Atchley,MD,William Ames
And 1h40'47"36d49'
Atea
Cmi 7h44'24"3d51'
Atef
And 23h1'1"51d56'
Atena
Ind 21h4'4"-50d59'
Ater,Dennis
Her 17h21'58"20d57'
Aternus,Amor
Aql 19h9'59"1d44'
Atha,Robert & Sallie
Cyg 20h59'1"41d9'
Athalie,Elizabeth
Lyr 18h59'46"30d47'
Athanas,Mariechen M
Lyr 18h33'0"36d19'
Athanasious
Lyr 18h58'44"46d23'
Athanassie,Dina
Mon 7h18'32"-7d9'
Athelynna
And 0h3'0"30d32'
Athena
Lyn 8h57'47"34d16'
Athena
Aql 18h48'32"11d31'
Athena,Suzanna Elisabeth
And 0h54'41"34d56'
Athene Frezados
Lyr 18h48'42"37d13'
Atherly,Elaine & John
Aql 19h11'36"13d16'
Atherton,Alva Nesbitt
Umi 16h19'34"71d0'
Atherton,Elizabeth
Cyg 21h3'24"33d52'
Atherton,Kevin James
Per 3h7'36"41d13'
Athey,John (Taff)
Cep 22h0'55"62d32'
Athiah,Sheph
Lyn 7h57'53"35d21'
Athina
Sge 20h0'23"20d21'
Atias,Marianne
Aql 19h33'31"0d9'
Atienza,Lucia
Peg 23h42'0"31d48'
Atilano,Guillermo
Lac 22h10'0"54d27'
Atkerson,Tein
Aql 19h55'1"0d59'
Atkin,Richard Dyer
Lib 15h15'45"-8d8'
Atkins,Alchemy Of
Ori 5h52'21"16d17'
Atkins,Anthony
Aur 6h26'10"35d52'
Atkins,Billy Joe "BJ"
Aql 19h55'1"-6d57'
Atkins,Brenda
Cet 1h32'0"-13d12'
Atkins,Craig William
Sex 10h31'30"-1d46'
Atkins,Don
Ori 4h47'0"0d19'
Atkins,Ellen
Lmi 10h9'27"34d60'
Atkins,Jeremy S
Dra 15h54'0"61d53'
Atkins,Luke Joseph
Per 2h28'24"56d38'
Atkins,Matthew Scott
Her 16h53'27"47d39'
Atkins,Maureen L
Ori 6h1'49"5d51'
Atkins,MD(RWA),Ronald W
Oph 18h39'1"8d35'
Atkins,Paula Jakhara
Lyn 7h6'43"44d44'
Atkins,Robert
Aur 6h25'40"38d54'
Atkins,Sarah Elizabeth
Com 12h30'1"22d22'
Atkins,Terry & Sue
Umi 15h37'48"76d39'
Atkins,Tommy Lee
Her 16h59'20"31d45'
Atkins,William Keith
Ori 6h1'40"8d12'
Atkinson
And 23h1'1"51d42'
Atkinson III,Charles Ormand Reilly
Cep 21h10'53"61d12'
Atkinson,Brenda
Cyg 19h16'15"48d56'
Atkinson,Denisa
Lyn 7h24'42"50d28'
Atkinson,Jr,Randall
Boo 14h21'0"52d52'
Atkinson,Katherine Olin
Com 12h53'16"20d49'
Atkinson,Mark
Lyn 7h52'29"43d41'
Atkinson,Raymond
Her 16h22'43"25d57'
Atkinson,Rochelle Jean
Lmi 10h14'0"37d35'
Atkinson,Shannon Ashley
Vul 21h13'22"20d22'
Atkinson,Star
Cnv 12h24'12"32d16'
Atkinson,Tayler Leigh
Sge 19h53'48"18d46'
Atlantique,Jules
Umi 13h44'35"75d45'
Atlantis
Cet 2h58'57"1d46'
Atlantis
Ser 16h4'42"4d21'
Atlantis By Roland Kästner
Peg 21h26'49"3d2'
Atle Alexander
Uma 10h53'31"47d12'
Atman
Lyn 9h18'50"38d38'
Atman,Robert E
Oph 18h16'20"10d58'
Atolli,Danielle Ruth
Cas 0h6'30"62d17'
Atomic Bull
Per 3h11'50"42d54'
Atomu
Sgr 19h6'41"-12d10'
Atriano,Lynda
Cyg 19h26'50"30d15'
ATSI
Her 15h56'0"41d31'
ATTACCA-JCH
Tri 2h18'25"32d34'
Attal,Jack
Boo 15h29'45"34d53'
Attanasio,Guido
Aur 4h53'44"51d16'
Attardi,Alfonse Adamo Allegrezza
Lyn 7h42'53"44d59'
Attardo's Fire
Cnv 12h40'55"38d25'
Attaway,Anna Elizabeth
Sge 20h7'19"20d18'
Attaway,Nora
Aur 5h58'44"38d1'
Atteberry,George C & Shirley J
Crb 15h50'47"28d14'
Attebury,Fred Oren
Hya 8h11'36"5d45'
Attem-SJTC
Crb 15h52'44"32d1'
Atten,Michel
Umi 15h53'30"78d48'
Atten,Nicolas
Ori 5h3'48"8d56'
Atterbury,Edward
Lyr 18h35'1"33d12'
Atterbury,John W
Aur 6h45'1"38d7'
Atterby,Peter
Uma 11h21'20"53d29'
Attile,Gary Scott
Her 17h39'37"14d15'
Attilio,Jackson William Rushing
Cam 11h48'55"82d20'
Attra,Broderick Cory
Dra 19h58'24"68d9'
Attrait,Jean-Luc
Lmi 10h48'1"31d17'
Attridge,Joyce Anne
Uma 12h57'38"60d52'
Attwater,Janice
And 2h34'30"40d39'
Attwell,Raymond Von
Dra 19h29'15"60d5'
Atwater's Rising Star
Cet 2h55'22"2d3'
Atwater,Marshall A
Aqr 22h21'39"0d13'
Atwell,Catherina
Uma 10h27'14"58d12'
Atwell,Erin
Cyg 20h4'54"40d26'
Atwell,Idus Morris
Cam 4h31'57"61d44'
Atwell,Linda
Boo 14h36'16"8d19'
Atwell,Mary Martha
Com 12h21'25"32d54'
Atwell,Tammy
And 22h57'51"36d45'
Atwell,William A
Psc 1h5'52"27d52'
Atwell-Frank "Celestial Union,The
Aur 6h7'37"32d5'
Atwood,Christopher Jamie
Lac 22h51'32"56d29'
Atwood,Crystal Rena
Ori 6h16'41"7d41'
Atwood,Laura Elizabeth
Cyg 20h16'45"37d41'
Atwood,Mark
Cep 23h27'19"65d22'
Atwood,Tonya & Jeff
Lac 22h9'15"51d4'
Atwood-Sanders,Lee Ann
Lac 22h46'26"56d28'
Au Grand Passage SA
Per 4h6'1"38d35'
Au,Betty
And 0h53'30"35d32'
Au,Donna D
Mon 7h0'0"4d54'
Au,Rebecca Sm
And 23h15'58"35d27'
Aube,Michael
Dra 20h37'27"68d45'
Auberger,Anne Christa
And 1h9'20"46d58'
Auberjonois,Judith
Cyg 20h6'19"42d14'B
Auberjonois,Rene
Cyg 19h55'22"33d26'
Auberjonois,Rene
Lmi 10h38'41"26d8'
Auberjonois,René
Cyg 20h6'19"42d14'A
Auberry,Jr,Bobby George
Mon 7h57'1"-6d31'
Aubert's Heavenly Body,LaDonna
Eri 3h49'40"-2d30'
Aubert,Alicia
Cyg 19h53'32"45d52'
Aubert,Isabelle
Com 13h10'16"21d21'
Aubert,Jean-Pierre
Lyr 19h4'19"47d35'
Aubert,Jerome
Dra 16h19'11"62d36'
Aubert,Marc
Sge 18h59'1"19d33'
Aubertin,Danielle
Cyg 21h3'49"39d49'
Aubertin,Jeremy
Cam 3h46'33"61d48'
Aubin,Alain
Cep 23h13'24"60d4'
Aubin,Ann Marie
Aql 20h11'10"0d59'
Aubin,Eugene Charles
Cet 3h3'20"2d7'
Aubin,Paul-Émile
Lmi 10h18'52"39d34'
Auble's Lucky Star, Serena
Cas 2h13'13"77d1'
Auble,Franáoise
Cas 2h59'21"60d25'
Aubrey
Cas 1h48'49"60d8'
Aubri Kathleen
Gem 7h23'19"34d19'
Aubry
Peg 23h3'25"12d16'
Aubry,Len
Aql 20h13'15"4d36'
Auburn Allison Nichole
Mon 6h43'42"-2d12'
Aubut,Danielle Godin
Umi 13h25'16"70d29'
Auböck,Doris
Cam 6h52'20"67d54'
Auchincloss,Sarah Sedgwick
Cas 22h57'33"56d48'
Auciello,Connie
Cas 0h53'1"76d32'
Aucoin,Martin A
Per 1h43'1"53d42'
Aucone,Jason T
Cep 1h19'59"77d49'
Aucremanne,Nancy M
Hya 9h31'10"-0d16'
Auctor
Cyg 21h2'14"31d10'
Aucutt,Emily Jane
And 0h21'37"45d51'
Audax,Colette
Cas 1h25'43"72d7'
Aude
Ori 5h52'36"21d22'
Audebert,Charles "Bibou"
And 23h1'35"40d23'
Auden PDQ
Aql 19h30'45"13d24'
Audenried,Mr & Mrs Ronald J
Sge 19h24'42"16d52'
Auderos Libiduru Infimike
Lac 22h43'28"55d39'
Audet et Ouellet
Uma 10h28'24"41d37'
Audet,Gaëtan
Per 3h9'12"40d20'
Audette,Maryann
Vul 20h15'10"28d42'
Audette,Tracy Lee
And 1h26'0"36d15'
Audi,Virgile Paul
Cnv 12h59'47"32d16'
Audiau,Jacques
Ori 6h16'44'8d53'
Audie-O
Dra 18h52'31"67d47'
Audier,Pierre
Lyr 18h54'48"30d4'
Audinot,Philippe
Cep 21h21'42"68d11'
Audit U
Sex 10h13'37"-2d45'
Audra
Cas 1h4'60"55d5'
Audra
Cas 2h36'35"57d37'
Audra
Del 20h31'11"20d11'
Audra
Lyn 7h54'1"35d47'
Audra Laurelle
Aur 5h6'1"40d50'
Audra Lea
And 0h21'15"31d39'
Audrain,Susana
Mon 7h21'29"-8d10'
Audree
And 0h3'16"43d23'
Audrey
Vul 19h57'31"28d32'
Audrey
Per 3h12'12"41d29'
Audrey
Psc 23h20'57"0d50'
Audrey & Steven, Forever More
Ori 5h28'46"0d29'
Audrey Ann
Mon 7h2'13"1d28'
Audrey Jean
Ori 5h55'19"16d15'
Audrey R
Peg 22h19'38"33d6'
Audrey's World
Lyr 19h22'1"38d49'
Audrey-Marie
Cam 3h55'36"52d37'
Audreys Star (The Beautiful Educator)
And 2h15'29"45d10'
Audron
Uma 9h59'59"51d16'
Audru,Camllle
Peg 22h44'34"3d39'
Audureau,Jerome
Lac 22h24'58"38d34'
Audy,Jean-Claude
Her 16h53'21"37d57'
Aue,Harry & Barbara
Cyg 21h13'1"39d45'
Auer,Friedl
Leo 11h18'36"-2d27'
Auer,Georg
Cmi 7h17'14"4d55'
Auerbach,Benjamin
Lyr 18h32'1"31d14'
Auerbach,Emanelle
Cas 1h3'35"53d45'
Auerbach,Jacob
Ori 5h55'39"15d42'

Auerbach,Lillian
Cas 1h3'30"55d51'
Auerbach,Rhoda
Del 20h49'35"8d29'
Auerbach,Samuel
Cas 1h3'28"50d19'
Auerbach,Sidney H
Lmi 9h50'54"37d30'
Auerbach,Sprockett
Cas 1h4'40"58d57'
Auernhammer,Albert
Vir 13h34'35"-12d10'
Augello,Jr,Vincent William
Per 2h53'18"35d14'
Augello-Cook,Alicia
Uma 9h21'1"61d35'
Augenstern,Zachary Aaron
Boo 14h38'53"37d23'
Auger,Jan Elaine
And 0h21'43"38d25'
Auger,Jennyfer
Lib 14h45'38"-20d24'
Auger,Manon
Cas 0h36'41"64d49'
Augerinos,Gregory
Lyr 18h41'17"42d41'
Augest,Joanne
Cas 0h41'48"66d21'
Augie
Cnv 13h13'39"32d14'
Augie's Infinity
Vul 19h45'24"28d41'
Augspurger,Betty Dee
Aql 19h39'41"14d17'
August Boj
Per 1h53'1"50d15'
Augusta
Cyg 21h2'24"48d56'
Augusta,Connie
Cyg 21h4'27"38d44'
Augusta,Madame
Cas 3h3'10"75d32'
Augusta,Peter J
Aur 6h9'10"46d41'
Augustin,Cailey Marie
Gem 6h35'51"12d21'
Augustin,Dominique Patricia
Sct 18h39'12"-4d53'
Augustine
Cas 0h9'48"64d25'
Augustine
Cma 6h56'38"-16d53'
Augustine,Michelle
Mon 6h24'56"4d31'
Augustiniak,Frank
Per 3h51'19"38d31'
Augustinski,Peter
Gem 7h24'13"35d26'
Augustin,Trent
Ser 15h52'34"-0d3'
Augustus Octavius
Uma 11h43'16"46d37'
Augustus,Cody
Vul 20h19'1"25d34'
Augustus,Kathleen Alice
Cyg 19h30'17"37d35'
Augustus-Keanu-The Beauty of Grey
Boo 14h22'21"30d5'
Augustyn,Julie Ann
And 23h1'58'51d16'
Augustyniak,Leonard J
Boo 14h56'0"43d18'
Auir,Syawla
Cyg 20h34'0"30d15'
Aukner,Kristine
Dra 13h19'35"63d58'
Aukofer,John
Aur 5h11'57"42d36'
Auld's Star,Great- Grandma
And 23h32'26"42d10'
Auldridge,Wesley
Ori 6h16'50"-0d54'

Auletta's Very Own Valentine Star,Mary
Cas 2h32'43"57d41'
Auletta,Jessica
And 0h19'1"31d36'
Auletta,Patrick V
Lac 22h45'0"52d33'
Aulicino,Julian
Boo 14h59'44"28d11'
Aull,Nicholas Stocks
Cep 23h0'57"61d2'
Ault,Michael J
Cet 2h48'32"1d33'
Aumaier Jun,Alfred
Tau 4h14'29"7d37'
Aument,Tina
Lyr 18h42'23"40d29'
Aunchman,John Michael
Cep 21h33'46"58d21'
Aungles,Glenn
Ori 5h6'32"1d39'
Aunio-Saasto,Irene
Cyg 21h21'33"40d1'
Aunt Alm
Lyr 18h18'44"46d56'
Aunt Bernice
Com 12h51'57"28d52'
Aunt Bert
Equ 20h58'40"7d45'
Aunt Betty's Star
Vul 19h22'59"26d56'
Aunt Daisy
Cas 0h36'38"73d7'
Aunt Debby & Eric's Star
Aur 6h13'55"31d41'
Aunt Evie
And 23h20'49"44d21'
Aunt Frankie's Star
And 0h47'30"38d42'
Aunt Hanna
Crb 15h20'32"30d21'
Aunt Ina
Umi 14h21'43"67d27'
Aunt Jane
And 0h23'33"33d43'
Aunt Jo Ann,The
Lyn 7h54'29"44d38'
Aunt Judy
And 1h4'1"38d16'
Aunt Kathy's Dreams
Vul 20h16'11"25d6'
Aunt Lee
Cam 5h53'57"68d47'
Aunt Lura
Com 12h47'29"21d33'
Aunt Margie
Com 13h16'44"21d35'
Aunt Marie
Eri 3h49'22"-5d60'
Aunt Marion
And 1h18'27"39d28'
Aunt Mary
Cas 23h20'33"61d5'
Aunt Nat & Uncle Moe
Aur 4h55'42"40d30'
Aunt P
Cas 0h0'30"54d57'
Aunt Pud
Aql 19h56'44"15d11'
Aunt Rose
Aur 4h56'42"51d26'
Aunt Susie
Mon 8h1'17"-1d28'
Aunta Majora
Peg 21h39'16"22d44'
Auntie Anne
Cas 23h20'0"61d59'
Auntie Bridget
Cas 23h19'0"60d20'
Auntie Chris
Psc 1h1'0"20d48'
Auntie Haze
Com 12h41'20"21d37'

Auntie Lesli
And 0h55'52"37d35'
Auntie's Little Piece of Heaven
Crb 15h49'33"38d42'
Auntiebomb
And 1h42'1"39d28'
Aunty Laura
And 23h19'47"51d53'
Aupetit,Isabelle
And 23h18'18"50d26'
Aura Of Sarah Lynn,The
Peg 22h41'12"12d15'
Auraleba,Lexa
Aur 5h30'0"38d41'
Auran,Sherry
Cam 3h52'16"72d49'
Auras
Umi 15h7'17"67d35'
Aurea June
Lyr 18h39'27"27d27'
Aurelia Nicolaï
Dra 17h38'54"75d33'
Aurelie Bon
Peg 23h5'10"20d43'
Aurelien
Lyr 18h50'36"40d24'
Aurelien De Barnier
Margaritora
Boo 15h19'16"37d43'
Aurelio
Aur 5h19'42"47d5'
Aurelio e Cristina
Scl 0h56'0"-26d21'
Auretta e Giuseppe
Lup 15h8'41"-37d40'
Aurette
Peg 21h56'27"21d36'
Auria 23/02/83
Aql 20h5'16"0d53'
Auriferous Jovan
Uma 13h34'0"53d27'
Auriga,Juergen
Ser 15h11'49"11d58'
Aurillarol,Alain
Her 18h3'1"28d8'
Aurita
Sco 16h34'15"-40d41'
Aurius,Dennis
Sct 18h41'36"-5d24'
Aurora
Oph 18h40'47"7d11'
Aurora
Cas 1h59'0"58d24'
Aurora
Col 6h25'36"-36d17'
Aurora
Cet 2h33'32"2d21'
Aurora
For 2h0'38"-25d19'
Aurora
Lyn 7h1'36"44d46'
Aurora Eugenia
Peg 22h28'45"12d29'
Aurora Patricia
Com 12h23'0"24d16'
Aurora Victoria
Ori 6h5'19"2d9'
Aurora,JoAnne
Cnc 8h53'33"30d44'
Aurora-Borealfred
Cas 1h41'46"61d20'
Aurore
Tri 2h3'51"30d37'
Aurorina
Peg 22h51'12"20d16'
Aurorous Angelica
Aur 6h59'0"37d32'
Ausgezeichnet
Psc 1h5'24"23d51'
Ausherman,Mary Godfrey
And 23h36'36"48d36'
Ausilia,Maria
Hor 3h26'25"-46d26'
Auslander "95", Jordan J
Cep 0h8'56"75d42'

Auslander,Stacey
And 1h32'32"38d38'
Ausman,LaVon C
Ori 5h52'32"19d33'
Ausmann,Joseph
Her 17h25'48"21d41'
Ausmus,Christian K
Cmb 12h30'25"14d42'
Aussedat,Marlotte
Cma 6h42'37"-15d18'
Ausserbauer,Dr Michele
Oph 17h3'3"-24d34'
Ausserbichler,Daniela
Lib 14h40'0"-23d42'
Aussieker,Laura A
Sge 20h0'32"20d21'
Austad,Commander Craig
Kermit
Cep 0h16'16"73d9'
Auster,Elizabeth Marie
Mon 7h59'19"-1d45'
Auster,Matthew Brett
Dra 12h3'0"68d37'
Austin
Ori 6h8'55"3d18'
Austin & Arielle's Forever Star
Lyr 18h55'47"40d42'
Austin Brianne
Uma 11h32'50"55d12'
Austin of Boston
Cet 1h26'53"1d18"
Austin Scott
Vul 19h18'23"27d17'
Austin,Alexandra Michelle
And 1h27'40"48d60'
Austin,Allen
Boo 14h14'50"15d9'
Austin,Andrew Wesley
Sct 18h40'17"-6d39'
Austin,Billy Gene
Oph 16h56'34"11d21'
Austin,Bonnie Grace
Cas 23h40'21"60d29'
Austin,Carolyn
Mon 7h46'1"-5d35'
Austin,Chante Marie
And 23h23'40"41d55'
Austin,Danny & Helen
Uma 10h45'1"40d39'
Austin,Dave
Ser 15h28'40"10d9'
Austin,David Shane
Uma 8h59'52"50d49'
Austin,Don
Aql 19h55'14"12d31'
Austin,Gregory K
Dra 17h31'36"61d44'
Austin,Harold
Cet 2h9'27"5d14'
Austin,Jack Hamilton
Umi 15h18'12"69d24'
Austin,Jackie Lynn
Lyr 18h46'10"30d57'
Austin,James Cameron
Boo 14h27'14"20d53'
Austin,John & Lynn
Tri 2h42'22"31d47'
Austin,Joshua Peter
Dra 17h43'51"64d57'
Austin,Katherine L
And 23h40'1"46d17'
Austin,Keith
Aql 19h2'40"16d14'
Austin,Kenny James
Uma 10h54'20"41d10'
Austin,Lorraine
Mon 6h53'36"-6d28'
Austin,Mark John
Ori 5h56'1"19d3'
Austin,Martha
Peg 22h52'54"20d23'
Austin,Melissa Allyson
Vul 20h57'15"20d36'
Austin,Myra A
Hya 9h38'16"2d6'

Austin,Paul
Cma 6h59'51"-19d20'
Austin,Paul Robert
Lyr 18h34'24"33d13'
Austin,Raquenel Garcia
Cas 0h31'25"67d10'
Austin,Stephanie Gayle
Mon 6h39'30"10d0'
Austin,Stephen J
Oph 17h27'30"10d47'
Austin,Vivian
Mon 6h43'11"11d48'
Austin,William H
Per 1h57'32"56d43"
Austins Star
Leo 11h7'22"0d13'
Australie
Per 3h9'52"41d10'
Auten,Alec Jeffery
Per 2h40'31"34d33'
Auten-McGill,Kiva Marie
Uma 11h29'14"64d55'
Autenrieth,Janice S
Lyr 19h14'56"42d49'
Autenzeller,Herbert
Boo 13h34'21"13d13'
Auterhoff,Roland
Cmi 7h19'28"4d31'
Auteri,Vincenza
Peg 23h48'0"16d52'
Autio,Narelle
Cru 12h9'32"-58d32'
Automatic Rain
Uma 9h18'60"55d8'
Autret,Victoria
Her 16h59'0"50d15'
Autrey,Bill
Cmi 7h57'11"1d44'
Autrina Belinda
Del 20h18'36"13d27'
Autry,Bryan Z
Aql 19h54'19"12d39'
Autum
Uma 14h15'58"61d44'
Autumn
Uma 11h5'58"41d7'
Autumn
Cnv 13h19'48"28d16'
Autumn
Dra 11h51'2"74d2'B
Autumn Dawn
And 0h9'56"27d56'
Autumn Dawn
Cas 2h42'59"60d47'
Autumn Dawn
Eri 3h23'0"-12d19'
Autumn Flair
Cma 6h15'18"-13d24'
Autumn Leif
Eri 4h12'47"-10d10'
Autumn Rebecca's Star
Lac 22h33'34"56d38'
Autuori,Emile
Cas 3h6'1"58d50'
Autuori,Kim
And 2h3'33"45d24'
Autz,Isabel Christina
Cap 21h37'60"-22d31'
Auvard,Marie
Crb 16h14'40"38d23'
Auwers Tom
Her 18h30'46"20d33'
Aux Copains, Aux Amis et Ö ceux que j'aime
Boo 14h32'34"47d38'
Auxer,Bland D
Cam 5h48'29"80d3'
Auxer-Moyta,Janet
Lyr 19h20'51"33d46'
Auyang,Emiliee
Mon 7h43'47"-5d37'
Auzzas,Patrizia
Cyg 21h27'14"48d45'
Avakian,Vahan
Cam 3h18'0"57d60'

Avallone,Maria
And 23h10'42"40d20'
Avalon
Leo 10h4'51"13d14'
Avalon
Boo 14h24'31"21d12'
Avalon
Cam 6h8'36"60d23'
Avalos,Alana Ciera
Peg 23h36'17"11d60'
Avani
Aql 19h26'0"15d59'
Avant,Jeremy K,Joshua M,Jason E
Tri 1h47'57"26d22'
Avant,Natisha Marie
And 0h21'49"36d28'
Avants,Bruce Eugene
Hya 8h33'12"-10d31'
Avants,Nick
Del 20h14'58"10d1'
Avarian Angel
Sge 19h5'41"16d59'
Avasthi,Ranjan Kumar
Hya 10h46'59"-17d37'
Avatar & Goldie 1001
Cyg 20h29'25"50d29'
Avdevich,Nicole Lynne
Com 12h27'24"20d0'
Avec Tous
Cyg 21h53'46"40d26'
Avegail
Hya 9h37'1"2d18'
Aveline
Cyg 21h54'48"52d47'
Avellano,Andrea
Hor 3h23'20"-48d30'
Avellano,Vania
Ant 10h45'29"-34d45'
Avello,Albert
Cep 22h31'30"80d20'
Avenall,Virginia
Cas 1h7'51"61d2'
Aveni,Brett J
Cam 4h6'53"58d28'
Avenir
Her 18h30'31"12d13'
Avera,Virgilia Welch
Mon 6h52'18"10d43'
Averbeck,Father Ronald J
Cep 21h38'19"55d17'
Averhart,Danae Elise
Del 20h24'28"10d19'
Averil
Aql 18h58'23"16d35'
Averil's "Glorious Light"
Mon 6h40'43"6d11'
Averill,Claire
Cas 1h4'0"58d23'
Averill,Nicholas J
Dra 18h10'50"70d16'
Averitt,Robert
Per 2h37'38"37d58'
Avers,Ellen
Peg 23h36'20"18d16'
Aversa,Frank
Aur 6h6'1"45d36'
Avertick,Shoshana Esther
Psc 23h26'35"1d14'
Averwater,Paul
Lyr 18h58'16"45d3'
Avery
Lac 22h12'22"47d20'
Avery Marie
Mon 7h41'46"-5d50'
Avery Nikolas
Her 16h40'27"39d18'
Avery Star,The Lynne
Cep 0h3'55"76d31'
Avery,Benjamin
Aql 18h48'30"11d31'
Avery,Caleb John
Cam 11h59'50"81d24'
Avery,Diane Susan
Boo 14h7'50"50d19'

Avery,Frank H
Aql 18h53'46"-0d15'
Avery,Jeff
Hya 8h45'40"0d12'
Avery,Jillian Frazier
Com 13h1'22"21d43'
Avery,John
Uma 11h54'15"41d0'
Avery,Julie
Umi 13h31'1"72d30'
Avery,Kathryn A
Lyr 18h15'46"35d30'
Avery,Myra Jacqueline Vestal
Peg 22h0'13"8d7'
Avery,Nathan
Sex 9h57'1"1d45'
Avery,Nyle Doré
Aql 19h22'40"10d0'
Avery,Rebecca Jill
Aur 5h20'11"38d59'
Avery,Robert Forrest
Per 2h0'0"50d13'
Avery,Sandra
Aql 20h14'28"0d33'
Avery,Timothy
Lac 22h4'54"47d14'
Avery-Skeen
Mon 7h15'58"-1d9'
Averys K Star Ranch, Karen
Mon 6h35'47"-0d13'
Aves,Suzanne
Eri 3h50'40"-4d53'
Avey,Gail
Crb 16h1'58"27d42'
Avgeris,Dr John A
Oph 18h35'1"7d2'
Avi & Jamie
Lyr 18h58'33"31d42'
Avila Family,The
Dra 17h15'46"71d17'
Avila USNR,Cdr Frank W
Aql 19h58'11"1d50'
Avila,Manny
Aql 19h6'60"3d20'
Avila,Maryanne Teresa Palma
Eri 3h26'35"-7d6'
Avila,Shelly Marie
Peg 21h37'55"25d34'
Avina,Veronica
Cas 1h55'45"75d26'
Avino,Dawn-Marie
And 0h24'22"34d36'
Avis
Cet 2h55'54"8d45'
Avis
Cet 0h9'1"-10d23'
Avitable,Christina
Cyg 20h3'15"58d10'
Avital,Sally
And 0h22'32"34d44'
Aviva Ben-Baruch
Cas 0h23'38"63d51'
Aviva Maia
Cas 0h14'25"63d55'
Avnit,Amir
Aql 18h43'0"10d57'
Avolese,Rosemary
Cas 0h42'1"64d37'
Avolese,Sebastian P
Aur 5h10'40"42d34'
Avondale
Cas 1h3'20"58d56'
Avrasin,Anna
Peg 23h48'15"13d27'
Avrick,Adam G
Cap 21h39'50"-23d34'
Avril
Umi 14h42'25"68d32'
Avril
Cyg 19h27'35"38d51'
Avril,Claude J
Dra 16h47'13"62d52'
Avril,Emmanuelle Aurélie Simone
Cyg 20h55'57"37d54'

Avril,Günter
Cnc 9h13'45"7d29'
Avril,Shirley
Lyr 18h53'17"40d53'
Avtobank
Cam 4h4'31"71d7'
Avus et Avia
Cyg 20h20'16"41d21'
Avval,Sussan
Cas 0h24'40"61d24'
Avy
And 23h1'26"51d23'
Awad,Ibrahim
Ori 5h59'30"14d13'
Awad,Nizar Ibrahim
Ori 5h54'10"11d47'
Awan,Hammad Ahmed
Dra 16h3'40"61d45'
Awan,Shaher Yar Tabish
Uma 9h19'38"59d26'
AWB White Weddings
Cru 12h42'57"-58d13'
Awe,Myron (Mike) Donald
Her 18h6'12"47d13'
Aweida,Jessie
Her 17h30'1"40d37'
Awi
Uma 13h52'25"50d23'
AWMQS
Cnv 12h14'13"48d56'
Awsome
And 2h20'38"39d11'
Axdorff,Alexander Asmat Saka
Aur 6h37'57"40d41'
Axe,Robert David
Boo 15h2'46"52d14'
Axel
Uma 13h39'56"60d42'
Axel in Freundschaft und Liebe
Boo 13h42'37"10d11'
Axelle M
Equ 21h2'17"8d51'
Axelrod,Alan S
Per 3h8'0"46d49'
Axelrod,Beth
Lib 14h55'48"-22d12'B
Axelrod,Charles D
Her 16h38'38"11d32'
Axelrod,Eliot S
Ori 5h55'37"6d29'
Axelrod,Jake Tyler
Sgr 19h21'47"-45d6'
Axelrod,John O
Dra 17h29'27"61d46'
Axelrod,Joyce M
Peg 21h58'0"28d54'
Axelsen,Erik Christian
Cyg 21h18'32"30d59'
Axelton,Darren
Aur 5h56'28"30d46'
Axisa,Jay
Peg 0h10'45"17d41'
Axla
Cam 5h47'44"68d18'
Axley,Capt Ray
Aur 5h48'49"50d8'
Axley,Margaret Peggy
Aql 19h57'21"1d28'
Axley,W Porter & Marion E
Cyg 20h54'0"31d17'
Axline,Kristin
Mon 6h54'1"1d43'
Axman,Katrina-Justin- Brandi
Aql 19h28'41"7d32'
AXS
Sgr 19h17'44"-29d14'
Axtater,William Lincoln
Aur 7h8'33"36d24'
Axtell
Aql 19h4'1"0d5'
Axtell,Douglas W
Aql 20h4'23"4d57'
Axtell,Dr Harold
Oph 17h31'60"0d47'

Axtell,Eleanor Ruth
Cam 5h9'21"70d34'
Aya
Lib 14h30'36"-19d5'
Aya Smile
Vir 14h25'27"-2d49'
Aya,Roderick H
Vir 13h4'56"-8d42'
Ayaka
Peg 22h37'55"2d28'
Ayako
Del 20h14'59"13d44'
Ayal,Michael
Uma 8h34'0"70d40'
Ayala,Anette
Peg 22h45'56"11d56'
Ayala,Cha Cha
Cet 0h53'17"-5d39'
Ayala,Danielle
And 2h34'34"50d24'
Ayala,Gabriel Chase
Cyg 20h56'12"30d43'
Ayala,Raquel Vazquez
Lyr 18h32'30"37d60'
Ayano-Boshi
Lib 14h33'27"-23d57'
Ayash,John
Sco 16h53'12"-37d46'
Aycaguer,Denise
Aur 6h0'50"38d15'
Aycock
Aql 20h18'29"7d35'
Aycrigg,George D
Sct 18h56'51"-5d5'
Aydas
Nor 16h22'35"-59d37'
Ayer,Dan
Dra 17h41'16"67d34'
Ayeroff,Zahina
Ori 4h59'46"-2d29'
Ayers,Ellen Johnson
And 0h47'49"45d30'
Ayers,Ernest W
Her 18h55'40"12d4'
Ayers,Frank Tyler
Ori 4h42'33"4d26'
Ayers,Jason
Cep 20h26'33"60d33'
Ayers,John C
Her 18h17'59"31d52'
Ayers,Joyce
Peg 0h8'38"13d53'
Ayers,Jr,Dennis
Aur 4h59'20"51d12'
Ayers,Murray
Aql 19h8'24"3d22'
Ayers,Ruthie
And 2h28'58"40d39'
Ayers,Scott David
Lac 22h13'28"48d46'
Ayers,Shirley
And 1h17'53"38d26'
Ayesha
Peg 23h19'43"32d18'
Ayim Keith Akyea/ Djamson
Cep 22h13'27"61d42'
Aykroyd,Dan & Donna Dixon
Sge 20h17'12"16d5'
Ayla's Dolphin
Uma 10h10'23"50d37'
Ayling,Anthony
Ori 6h16'28"0d35'
Ayling,Thomas Sebastian Sui Generis
Pho 23h41'15"-43d52'
Aylsworth Choir Director,Mary Wagner
Peg 22h46'53"20d31'
Aylward,William Thomas
Sex 10h43'1"-6d2'
Aymar,Albert George
Cyg 19h19'26"44d0'
Aymar,Amanda
Her 17h20'18"42d7'

Aymar,Peter Alfonso Her 17h10'23"44d56'
Ayo,Dennis Per 3h32'30"31d30'B
Ayotte The Star Of My Life,Carrie Tri 2h0'54"31d53'
Ayotte,David & Carolyn Cyg 20h21'28"40d29'
Ayotte,George,John, Christopher Uma 9h5'39"48d30'
Ayoub,Elizabeth And 2h26'49"46d0'
Ayre,Les & Bill Draney Cyg 21h39'38"42d14'
Ayres to the Starway Her 15h59'13"44d56'
Ayres,Jason's Star Jason Todd Dra 16h5'50"68d42'
Ayres,John J Leo 11h33'40"3d34'B
Ayres,Jr,Alfred C Lmi 10h16'45"38d1'
Ayres,Katheryn Anne Cmi 7h44'22"0d10'
Ayres,Marci A Leo 11h33'40"3d34'A
Ayres,Randall Anthony Her 17h0'19"20d27'
Ayres,Ronnie Cep 0h22'21"80d16'
Ayrev Cyg 19h54'14"40d42'
Aysan,Magic Mike Dra 18h36'41"68d50'
Ayscue,Allison Lane Mon 7h45'43"-1d57'
Aza Cyg 21h14'37"28d25'
Aza Dakroutioun Cyg 19h53'54"44d18'
Azad Boo 15h6'0"48d47'
Azadeh Her 17h8'53"48d53'
Azadi,Iris Peg 21h32'1"20d29'
Azadi,Lily Peg 22h40'52"26d42'
Azadi,Sara Mon 6h19'1"2d42'
Azara Ori 6h8'59"7d49'
Azark,Daniel Her 15h56'33"47d54'
Azarow,Arnold P Aur 5h18'52"42d32'
Azbell,Michael Joseph Cep 2h51'22"78d42'
Azcuy,Devon William Dra 18h58'23"80d6'
Azeem Cam 7h36'1"67d53'
Azerrad,Jessica Uma 10h25'38"67d38'
Azersky,Irving Per 3h15'17"38d16'B
Azersky,Ruth Frances Goldstein Per 3h15'17"38d16'A
Azevedo,Anthony J Peg 22h0'29"24d35'
Azghandi,Jheenus Sct 18h41'12"-4d22'
Aziman Cam 12h36'0"77d7'
Azimi,Sudi Ser 18h43'18"3d57'
Azimut Cnv 13h15'46"38d31'
Azinge,Harrold Cam 7h7'0"61d6'
Aziz Cnc 8h52'18"31d11'
Aziz Lyn 7h36'46"37d22'
Aziz,Kristian Peter Her 16h49'43"34d19'
Aziz,Omar Anwar Umi 15h7'34"77d34'
Aziza Tau 4h15'32"7d52"
Azmoudeh,Nicky Cas 0h12'59"61d44'
Aznar,Joe Eri 3h41'0"-3d24'
Azouz,Joël Cam 13h49'33"80d4'
Azra Aur 5h55'20"50d6'
Azul,Juan Ind 20h50'44"-53d46'
Azure K Umi 15h32'55"66d6'
Azzalini,Marta Flavia Eri 4h52'42"-6d49'
Azzani,Jacqueline Cas 1h22'0"73d39'
Azzaro Cet 2h20'40"1d7'
Azzaro,Denise Cyg 20h0'33"31d11'
Azzaro,Francesca Pyx 8h45'27"-25d56'
Azzena,Ti Amo Teresa Mon 6h36'22"6d25'
Azzimonti,Brigitte And 23h3'16"50d25'
Azzolino,Anthony C Cet 3h3'49"-1d6'
Azzopardi,Ann Cas 0h23'1"71d52'
Azzurro,Maré Peg 22h14'22"32d29'
A'B Cam 4h2'42"60d21'

B

B & B Lac 22h46'55"54d48'
B & C Forever Cyg 21h54'16"42d59'
B & E Roads Tau 4h5'54"22d0'
B A & C #5 Lac 22h24'43"50d13'
B A M Lac 22h5'17"47d38'
B A T Boo 14h38'52"10d16'
B B Juanjo Aur 5h17'11"43d16'
B B Wolf Lyr 18h46'52"35d37'
B C S III Boo 13h35'21"20d2'
B C's Mason Lyn 6h57'30"54d16'
B D N Hya 8h43'45"4d41'
B Dizdar Leo 17h20'20"30d21'
B E A M Vul 20h56'12"28d47'
B e C Eri 2h45'14"-6d1'
B G Aql 18h57'58"11d13'
B G Cyg 19h34'56"33d21'
B G Swing Productions Sct 18h53'1"-6d18'
B J Cyg 19h10'29"48d38'
B J Lmi 9h36'13"37d56'
B J Mon 6h22'48"8d42'
B J Uma 8h35'23"48d51'
B J-22 Dra 9h28'13"74d18'
B Josephine Cas 0h32'58"62d10'
B Joyous Cma 6h56'27"-18d55'
B K Her 18h2'60"28d21'
B K Peg 23h17'43"31d49'
B L J - Fish Psc 23h0'0"1d42'
B Lue Aur 7h13'0"40d9'
B M 123 Her 17h33'23"27d16'
B M G C Lac 22h40'26"54d22'
B M U I Cet 0h27'33"-0d43'
B P Charlie Umi 16h38'40"79d52'
B Ruth "Poet To The Stars" Uma 11h22'1"41d32'
B T I In the Sky Psc 1h24'18"18d36'
B Victoria Cet 3h14'54"6d38'
B Z Ori 5h1'25"10d6'
B&B Eternally Mon 7h40'14"-8d55'
B-42 Uma 9h11'0"56d31'
B-612 Sge 19h53'54"16d49'
B-COS Cam 7h48'14"60d1'
B-ELLI Sco 16h52'20"-40d58'
B-Man Lac 22h48'43"54d52'
B-Man Star,The Cep 21h54'53"55d14'
B-Radley Peg 23h48'58"15d51'
Ba Pacca Uma 8h58'33"57d35'
Baade,Joachim Dra 18h18'17"68d9'
Baar,Dr Ferdinand Aur 5h8'30"41d28'
Baar,Ferdinand Tri 2h16'36"31d42'
Baarendse,Bonnie Ball Lyn 7h43'10"38d39'
Bab Eri 3h5'37"-2d22'
Bab,Marion Uma 12h9'54"58d42'
Baba Ori 5h56'23"20d51'
Baba Uma 10h46'0"40d45'
Baba Umi 16h23'1"71d51'
Baba & Granddad Birthday Star Umi 14h10'54"71d52'
Baba Blimp Umi 10h0'48"86d7'
Baba,Johnny & Shanna Cet 1h17'36"-13d42'
Baba,Sathya Sai Sge 19h5'57"16d19'
Babak Lyn 7h47'46"41d37'
Babau Tri 1h59'45"34d39'
Babb's Lumen Victoriae Oph 17h51'0"13d21'
Babb,Cheri Oph 17h25'1"-20d54'
Babb,Kaaren Lyn 8h11'22"35d40'
Babbage,Paige Elizabeth Cas 0h48'50"75d32'
Babbin,Shirley Cam 3h58'33"56d19'
Babbitt,Jed Matthew Her 16h33'41"21d11'
Babbitt,Jenny Uma 9h33'0"55d45'
Babbitt,John David Boo 13h41'23"20d10'
Babbitt,Kelly Cas 2h53'28"73d12'
Babbs,Sr,"Christmas Star" William W Lac 22h10'43"38d51'
Babcock Aur 5h21'0"38d39'
Babcock Family Star Eri 3h42'31"-6d49'
Babcock,Carol Aql 19h55'32"8d56'
Babcock,Craig Penman Cyg 21h39'16"38d45'
Babcock,Eileen Marie Lyr 18h32'0"32d57'
Babcock,Evelyn Beatrice Peg 23h1'14"33d19'
Babcock,Harriet Mon 8h7'56"-8d5'
Babcock,Jesse Alexander Peg 23h27'19"23d31'
Babcock,John Avon Ori 5h50'15"10d55'
Babcock,Jon Lac 22h2'25"47d59'
Babcock,Jr,Robert Lewis Cyg 21h17'18"33d11'
Babcock,Jr,William S Aql 18h52'39"7d7'
Babcock,Judith Karen Lindholm Cyg 21h5'56"31d0'
Babcock,Mary Grace Cas 1h0'1"63d46'
Babcock,Richard Scott Boo 14h51'52"32d58'
Babe Uma 12h0'0"35d34'
Babe Aql 19h4'0"2d28'
Babe Cam 4h1'10"61d44'
Babe Cet 2h36'54"1d6'
Babe Sct 18h21'0"-13d21'
Babe Aqr 21h1'37"-13d58'
Babe Cet 1h11'48"-3d26'
Babe Vul 19h41'49"26d33'
Babe & Delores Lyr 19h8'12"38d29'
Babe & Ray Lyr 18h59'56"29d5'
Babe 73 Cnv 13h12'40"38d35'
Babe's Belle Starr Aql 20h2'15"8d50'
Babe,The Her 16h25'38"38d51'
Babe-Special Star- Special Person Lyr 18h41'10"40d30'
Babeix,Catherine Lyn 7h52'51"51d24'
Baber,Felicia Marie Lyn 8h5'14"43d17'
Babes-The Loveliest Bride Cyg 20h30'26"31d36'
Babestar-Cek-P And 23h21'59"35d28'
Babeth Peg 22h27'32"20d28'
Babeth Ori 5h4'30"10d49'
Babett Ari 2h58'1"20d26'
Babey,Jean-Paul Vul 19h19'17"22d50'
Babi & Dido Cam 4h13'10"68d49'
Babiar Cam 14h8'33"80d11'
Babic,Gary Thomas Aql 19h0'1"-6d55'
Babichev,Alexsei Nickolaievich Ser 15h20'25"6d19'
Babie Cam 5h52'43"68d2'
Babin's Star Mon 6h35'25"-1d7'
Babin,Angèle Her 18h5'51"49d42'A
Babin,Gloria Ann Latini Aur 5h42'18"47d27'
Babineau,My Love,Scott Uma 9h32'1"52d40'
Babino,Matthew John Tau 5h57'47"23d7'
Babis Lyn 9h1'49"40d3'
Babkoff,Ronald W Oph 17h34'41"-8d58'
Bablett,Jim Sex 10h13'22"-3d38'
Babo Cam 4h24'40"68d15'
Baboushki Cyg 19h54'44"48d36'
Babs Gem 6h29'52"14d2'
Babs Dra 11h40'51"70d35'
Babs And 22h59'34"40d51'
Babs & Boris Ori 5h52'34"13d37'
Babs,LB Ser 15h39'1"4d13'
Babsi Cas 0h32'21"62d6'
Babsi & Martin Uma 10h52'33"67d59'
Babu-Cheesebu Oph 17h50'12"13d54'
Babue,Peter Werner Cnc 8h24'24"31d58'
Babuke,Otto Sgr 19h54'19"-44d34'
Babul Hya 9h37'1"2d21'
Babushka Crt 11h2'53"-18d53'
Baby Cam 5h6'31"71d13'
Baby Cam 4h4'51"60d41'
Baby Lyn 9h8'47"42d29'
Baby Sge 19h57'1"18d55'
Baby Vul 20h16'18"25d54'
Baby Per 3h32'11"50d40'
Baby Ori 5h3'59"0d38'
Baby & Indy Cet 2h40'1"-11d44'
Baby Abbie Umi 15h14'0"66d10'
Baby Alec's Cradle Peg 22h24'21"20d14'
Baby Angel Lyr 18h49'0"30d53'
Baby Boy Per 2h52'48"40d7'
Baby Boy Boo 15h8'0"38d21'
Baby Boy Sean Ari 2h24'49"10d45'
Baby Byers Del 20h57'1"18d52'
Baby Claire Del 20h15'10"11d2'
Baby Columbo Cat Iacono Lyn 8h35'33"45d23'
Baby Danielle Vul 19h22'55"26d1'
Baby Doll Mon 6h58'30"-10d20'
Baby Doll Peg 23h40'11"27d43'
Baby Doll Gail Eri 3h52'11"-6d23'
Baby Doll MCM Cyg 19h19'56"45d30'
Baby Eddie Cyg 19h56'50"37d51'
Baby Emil Cyg 19h29'23"41d3'
Baby Face (Thomas Noel) Peg 23h3'10"10d58'
Baby Gauss Aur 7h0'44"41d11'
Baby Girl Vul 20h15'19"28d10'
Baby Girl Dee Peg 23h49'21"8d44'
Baby J Cyg 19h31'22"31d40'
Baby Jack's Star Ori 5h33'42"-1d12'
Baby John Boo 14h42'45"45d47'
Baby Kami Jo Mon 6h41'37"8d10'
Baby Kasey Lyr 19h14'51"33d56'
Baby Lisa Cam 12h58'0"80d7'
Baby M Sge 19h2'48"19d0'
Baby Maureen Mon 8h6'25"-1d4'
Baby Megan Lyn 7h31'41"51d11'
Baby Melissa Lyr 18h19'19"38d34'
Baby Michelle Eri 2h59'1"-2d21'
Baby Nini Lib 15h39'1"-28d22'
Baby Page And 0h56'17"38d11'
Baby Patranila And 0h10'31"36d31'
Baby Paule Ori 4h53'56"-2d58'
Baby Perez Vul 19h15'36"25d8'
Baby Roo Cyg 20h15'19"39d4'
Baby Spacey Cyg 20h18'57"39d3'
Baby Switzer Aql 19h28'48"7d41'
Baby Theodore Nathaniel Cam 6h15'15"63d13'
Baby Turtle Marsha Vul 20h43'51"20d2'
Baby,The Cyg 21h52'21"40d58'
Baby,The Cyg 20h5'21"40d36'
Baby-Foot de Gribouille Cam 5h4'47"60d20'
Baby-Head Lyn 6h55'35"54d10'
Babyak,Andrew Aur 7h8'56"36d10'
Babych,Kolton Michael Ori 5h45'54"11d37'
Babydoll Mon 6h54'43"-4d12'
Babydoll Cam 7h58'56"61d34'
Babyguard Tri 1h46'57"28d36'
Babykates Wishing Star Peg 22h43'44"21d12'
Babylove Aql 20h5'59"-1d45'
Baca,April Kay Peg 22h0'16"24d52'
Baca,Melanie Mon 7h1'59"-1d37'
Baca,Natalie Marie And 23h45'0"45d45'
Baccari,Charles Per 3h5'52"40d24'
Baccarini,Alessandro Nicolo Aur 7h15'17"40d48'
Baccaro,Thomas Cmi 7h19'31"1d47'
Baccei,Tom Dra 17h0'1"60d33'
Bacchus,Stella Abraham Cep 22h23'48"61d40'
Bacci,Maurizio Per 1h51'1"47d42'
Bace,Robert Donald Aur 5h2'41"48d47'
Bacellar,Maria Carolina Barboza Viana Lyr 18h49'30"36d52'
Bacen,Jr,Stephen Francis Del 20h25'0"8d7'
Bach,Alois Ari 2h52'28"22d22'
Bach,Ari Jason Peg 21h55'48"2d41'
Bach,Hannelore Leo 9h19'16"18d3'
Bach,Ken Cam 5h51'45"68d31'
Bachand,Arthur Per 3h18'21"50d16'
Bacharach,Jeremy Uma 13h19'29"60d45'
Bachelier,Christian Cep 20h16'0"60d41'
Bacher,Vernon A & Glennis A Lyr 18h51'39"40d49'
Bachert,Danielle A Vul 20h14'33"26d7'
Bachle,Alpha/S C Uma 10h39'49"50d25'
Bachman,Christopher Lac 21h56'44"42d47'
Bachman,Grandma Star- Mae Cas 1h2'13"58d56'
Bachman,Kurt Hya 8h19'51"3d46'
Bachman,Lois I Lyr 18h57'1"30d42'
Bachman,Mary A Lyr 19h22'34"37d50'
Bachmann,Marion Cmi 7h5'44"3d48'
Bachner,Candy Vul 20h0'1"28d45'
Bachrach,Per-Robert Ori 5h2'41"8d10'
Bachs-Larsen,Axel Cep 22h23'1"67d30'
Baci Boo 15h21'26"50d57'
Baci Dra 17h24'43"72d56'
Bacigalupi,Amber Marie Cas 0h32'53"60d22'
Bacigalupo,James Patrick Cep 0h3'26"68d18'
Bacik (Councilman), Terrie Aql 19h56'44"15d29'
Bacil,Angela Marie Vul 19h21'30"26d59'
Bacile,Nick J Her 16h12'16"41d15'
Bacio,Ciao Femmina Per 3h40'58"51d34'
Back,Eric Per 3h15'21"41d21'
Back,Katherine O Cyg 19h55'44"38d3'
Backes,Gerhard Dra 16h44'55"73d54'
Backes,Jürgen Aqr 22h41'14"-0d41'
Backes,Michelle Diane And 0h21'48"36d44'
Backes,Wolfgang Ori 5h1'39"13d32'
Backhamre,Ingvar Lac 22h42'23"53d4'
Backhaus,David Joseph Aql 18h58'35"16d8'
Backhaus,Wolfgang Uma 9h27'44"47d44'
Backhus,Walter & Norma Cyg 20h15'15"31d14'
Backlund,Cathrin Crb 15h30'0"26d1'
Backman,Dr Barbara Cas 23h5'0"58d14'
Backman,Drake Aubrey Sct 18h54'15"-8d55'
Backman,Keith Marcus Lac 22h11'0"51d47'
Backstrokes Mon 6h34'57"-1d5'
Bacon III,John Albert Aql 19h6'44"15d44'
Bacon,Allison Aql 18h40'32"-2d15'
Bacon,Angela (Smokey) Dra 17h58'30"58d38'
Bacon,Barbara H And 1h46'36"39d36'
Bacon,Benjamin Cep 1h41'14"77d16'
Bacon,Chuck & Kelley Cet 2h49'23"2d50'
Bacon,Douglas E Aur 6h7'56"32d48'
Bacon,Holly Cas 0h32'45"63d24'
Bacon,Jack Her 16h10'0"48d29'
Bacon,Jon & Cindy Cma 7h0'26"-16d20'
Bacon,Kate Lyn 7h54'1"42d33'
Bacon,Michel Ori 5h56'12"15d2'
Bacon,Nicole R Eri 3h57'37"-3d42'
Bacon,Ronald Z Boo 14h20'0"53d58'
Bacon,Ruth Ann Cas 0h40'26"67d33'
Bacon,Sam The Silver Fox Star Cet 1h21'1"-6d20'
Bacon,Shannon M Mon 7h42'53"-5d46'
Bacon,Stephen J Aql 19h30'11"13d55'
Bacon,Steve & Stephanie Crt 11h1'53"-6d36'
Bacsik,Greg Aur 6h30'0"38d53'
Bacuzzi,Fabrizio Hor 2h54'13"-49d9'
Bad Habits Grille, Meri & Dane Peg 0h0'20"31d23'
Bad Kreuznach Uma 13h42'36"50d39'
Bad News Cet 2h29'12"8d49'
Bada,Robert J Aur 4h51'32"40d51'
Badal,Eleanor Mon 7h2'16"-5d41'
Badalamenti,Batassano Cet 3h5'1"5d5'
Badams,Jay Darin Cam 4h0'22"68d34'
Badaro,Imad Uma 9h29'20"47d33'
Badcock,Lauren Lyr 18h43'16"42d50'
Bade,Silke & Achim Ser 15h14'33"0d10'
Badeau,Roger Cnc 8h26'24"8d11'
Badeb Ily f Vul 21h26'13"27d55'
Bader,Brianna Lynn Peg 22h26'55"20d16'
Bader,Claudia Cap 20h36'40"-18d40'A
Bader,George Dra 16h1'1"60d19'
Bader,Jacey Morgan Uma 9h23'11"47d37'
Bader,Leland Charles Cep 21h35'54"60d19'
Baderian,Lisa Ayn And 2h35'0"40d36'
Badger,Kristen R Ori 5h13'27"-8d45'
Badger,Linzi Lyr 19h16'34"28d10'
Badger,Michael H Cmi 7h6'21"4d20'
Badger,Natalie Lyr 18h19'45"43d11'
Badger,William H Cet 0h39'52"-5d4'
Badgley Winter Unity Her 17h3'0"31d45'
Badiak,Dara Tri 1h42'55"30d27'
Badiani,Eleonora Lac 22h33'23"38d34'
Badillo,Paulino Masia Cnc 7h55'27"10d25'
Badini,Ray E Cet 0h57'54"-3d57'
Badioli,Alessandra Sgr 19h1'22"-29d21'
Badman,David Her 18h12'37"38d19'
Bado,Jan & Rick Crb 15h49'14"28d33'
Badon,Russell Sex 9h52'0"-1d60'
Badorek,Carol Dix Cyg 21h13'25"39d50'
Badsgard,Richard A Her 16h12'15"5d17'
Badstieber,Werner Psc 1h16'42"21d33'
Badua,Stacia A Hya 8h22'55"0d59'
Bae,Robert J Aur 6h23'55"30d31'
Baebidahg,Aja Cmi 7h36'58"0d1'
Baecker,MD,Marvin P Aur 5h33'54"50d29'
Baedors,Kalle Sgr 18h47'54"-28d49'
Baehny,Jr,Tom Cnv 12h41'47"43d43'
Baek,Moselund Cep 23h4'32"67d55'

Baensch,Bernard Robert
Cep 22h28'26"58d12'
Baer,Captain Gerd
Aur 5h19'45"41d25'
Baer,Charles F
Her 18h3'25"46d54'
Baer,Elizabeth
And 2h13'53"38d4'
Baer,Helen Kinsman
Lyr 18h25'32"38d21'
Baer,Scott
Per 2h10'12"57d34'
Baer,Thomas Alfred
Cnc 8h1'0"8d34'
Baer-A New Beginning, Bernard A
Boo 14h11'44"41d39'
Baerga,Marcia Rachel
Lyn 7h48'21"44d48'
Baeriswyl,Pierre
Uma 9h1'11"50d12'
Baersch,Elisabeth
Uma 9h36'0"45d10'
Baert,Robert
Cnv 13h18'14"28d31'
Baeta,Love Dad, Stephanie Nicole
Leo 9h37'54"7d25'
Baethge,Diplon-Kaufmann,Hermann
Ori 5h37'20"12d20'
Baettig,Silvia
Sgr 19h14'18"-23d55'
Baety,Glenda
Mon 6h58'10"11d36'
Baeuerlein,Roland
Per 4h6'55"49d11'
Baevsky,Yvonne
And 1h57'43"35d23'
Baez,Carmen
Cyg 20h22'15"40d30'
Baez,George
Ori 5h52'46"15d13'
Baez,Rose
Mon 6h27'58"-10d28'
Baez,Walter
Cam 3h47'37"61d9'
Bafalon,Helen Athena
Peg 22h41'56"20d24'
Baffa,Sr,Robert Patrick
Aur 6h32'42"32d47'
Bag Of Peaches
Cet 2h11'50"0d21'
Bagby,Dorothy Jean
Lyn 8h15'11"37d17'
Bagdon,DO,Jeffrey D
Her 17h38'50"21d43'
Bagdziunas,Astra Liucija
Lyn 7h10'59"52d45'
Bagent,James N
Lac 22h3'23"51d29'
Bager,Line
Cas 1h28'37"64d58'
Bagetis,Gregg
Cep 21h25'43"58d11'
Baggett,Susan
Vul 20h1'22"22d30'
Baggette III,James Worth
Sct 18h54'18"-4d42'
Baggette,Alison Taylor
Peg 21h57'17"20d1'
Baggette,Caleb Alan
Ser 15h36'30"9d37'
Baggins
Boo 15h25'37"38d28'
Baggins,Bilbo
Sct 18h43'18"-5d55'
Baggs,Suzie
Gem 6h2'53"26d44'
Bagheera
Lyn 8h9'28"51d45'
Bagherzadeh,Leyli Borner
Sex 10h13'50"-5d20'
Bagley III,William Atherton
Cep 21h27'37"60d36'
Bagley,Charles Charlene Cady Lyn,& BJ
Ori 5h58'20"16d31"
Bagley,Jr,John J
Aql 20h5'32"7d15'
Bagley,Marilou
Cas 0h17'32"61d51'
Bagley,Matthew
Cet 2h55'35"0d30'
Bagley,Nell
Cyg 20h17'17"30d15'
Bagly V
Ori 5h37'58"10d19'
Bagnall,Louise
And 23h37'59"40d34'
Bagnall,Lydia M
Aql 19h31'34"12d5'
Bagnall,Madge
Aql 20h34'47"-6d27'
Bagnasacco,Clelia
Lac 22h54'53"50d6'
Bagnell,Chris
Aur 7h24'15"38d14'
Bagnell,Ken
Her 16h52'15"40d53'
Bagnell,Kit
Lyn 8h21'43"51d4'
Bagot,Jean
Uma 9h39'13"49d26'
Bagott,Prudence
Lyr 18h28'43"42d4'
Bagsby,James Arthur
Cep 22h18'1"55d31'
Bagstad
Aqr 21h3'0"-1d8'
Baguley,Peter
Cep 0h9'43"67d6'
Baguley,Sharon
Boo 14h31'58"52d14'
Bagwell,Brett Charles
Dra 19h45'37"60d50'
Bagwell,John Theodore
Dra 14h28'29"63d38'
Bahamon,Jonathan
Sex 10h12'25"-0d54'
Bahar,Fatimah Rasheed Isa
Cyg 19h40'58"37d60'
Bahar,Merrick V
Cep 20h39'16"76d40'
Bahl,Dorothy
Mon 7h43'31"-1d25'
Bahnke,SS
Crb 16h13'33"28d20'
Bahnsen,Christine M
Sge 19h26'40"16d9'
Bahr,Dr Frank R
Hya 8h19'24"0d10'
Bahr,John
Cep 20h58'40"67d53'
Bahr,John Timothy(JT)
Aur 5h0'1"47d25'
Bahr,Winnie
Leo 9h59'38"10d59'
Bahrmasel,Christine E
And 2h17'37"42d43'
Bahrmasel,Rebecca D
And 2h24'29"42d57'
Bai,Riccardo
Uma 10h6'22"48d57'
Baiardy,Robert Philip
Aql 19h51'57"12d13'
Baidal-Teruel, Francisco
Uma 11h22'39"40d12'
Baier,Adolf
Aql 19h2'55"17d47'
Baier-Schoenecker,Baby
Cyg 21h32'48"41d21'
Baierlein,Sallie
Vul 19h48'49"28d18'
Baietti,Francesco
Ori 5h2'34"8d41'
Baig,Wamiq
Per 1h48'59"56d32'
Bail,Rüdiger
Lyr 19h19'20"30d30'
Bailes,Andy
Aur 6h1'52"35d19'
Bailet,Jacques
Sgr 19h26'18"-31d51'
Bailey Constellation, Charles (Bull)
Her 17h57'0"28d18'
Bailey Jade
Cet 2h45'25"-0d18'
Bailey Rose
Lmi 10h31'47"34d13'
Bailey Star,Ken
Aur 5h3'13"50d5'
Bailey Super Star, Laura
And 23h18'27"51d28'
Bailey's Light,Stuart
Gru 22h29'35"-50d30'
Bailey's Raye
Ori 5h52'59"6d39'
Bailey's Star
Her 15h54'28"42d1'
Bailey's Star,J
Ori 6h16'17"8d20'
Bailey's World,Annette
Ori 5h54'15"1d5'
Bailey,Adam W
Her 16h40'35"41d4'
Bailey,Alexandra Ruth
Cnc 8h53'32"18d16'
Bailey,Amazing Grace B
Boo 13h42'1"19d17'
Bailey,Arlie K
Cet 2h16'1"5d1'
Bailey,Barbara
Lyr 19h6'53"25d34'
Bailey,Barbara Donovan
Lmi 10h18'1"30d25'
Bailey,Brad Thomas
Sge 20h0'1"20d18'
Bailey,Brian A
Aur 6h19'54"46d46'
Bailey,Bruce E
Aql 19h36'14"-0d2'
Bailey,Charles
Cep 20h42'34"60d31'
Bailey,Charlotte
Peg 22h39'44"27d39'
Bailey,Dale Russell
Her 15h50'1"46d1'
Bailey,David "Drummer"
Lyr 18h37'47"37d14'
Bailey,David & Lindsey
Cyg 19h30'45"39d13'
Bailey,Dayna
Ari 2h53'16"30d7'
Bailey,Deborah
And 2h24'24"48d55'
Bailey,Don E
Her 18h45'11"12d37'
Bailey,Donna L
Sct 18h19'46"-14d5'
Bailey,Duane Alan
Aur 5h58'56"30d31'
Bailey,Elise Jacqueline
Mon 7h51'56"-4d26'
Bailey,Erin Rene
Cnv 13h20'15"40d38'
Bailey,Faye & Burt
Com 13h7'36"26d35'
Bailey,Frances P
Sex 9h42'0"-6d24'
Bailey,Geoffrey Wayne
Lac 22h51'55"53d6'
Bailey,Glenda
Aur 4h52'30"50d8'
Bailey,Gloria
Equ 21h22'22"10d26'
Bailey,Harold
Uma 10h29'15"40d29'
Bailey,Harry James
Aqr 22h31'11"-11d7'
Bailey,Heather Leigh
Com 12h15'18"20d35'
Bailey,J W
Lmi 10h37'35"31d52'
Bailey,Jack
Dra 15h11'32"57d59'
Bailey,Jamie
Umi 15h0'57"67d3'
Bailey,Janelle Margaret
Cas 0h11'49"54d29'
Bailey,Jean
Boo 13h51'1"20d14'
Bailey,Jeannie
Lmi 9h51'39"34d47'
Bailey,Jennifer
Cas 1h3'0"63d47'
Bailey,Jennifer,Lynn
Peg 23h1'33"31d3'
Bailey,Jr,B James
Peg 23h3'55"17d33'
Bailey,Jr,Harry T
Her 18h8'44"38d27'
Bailey,Jr,Sean King
Ori 5h1'16"12d7'
Bailey,Katharine Allyn
Cam 7h26'18"67d31'
Bailey,Mark T
Uma 10h0'38"50d59'
Bailey,Mary Doreen
Ori 4h51'0"0d32'
Bailey,Maura
Mon 6h53'11"-2d13'
Bailey,Maybelline
And 0h58'1"34d57'
Bailey,Megan Francis
Cyg 21h38'55"40d38'
Bailey,Pat
Cam 5h44'19"73d8'
Bailey,Peggy
And 2h28'30"45d27'
Bailey,R Thompson
Lac 22h32'42"48d55'
Bailey,Ralph
Cet 1h49'59"-5d11'
Bailey,Randy
Boo 14h53'0"25d13'
Bailey,Reginald M
Boo 14h9'0"44d4'
Bailey,Robert D
Ori 5h38'54"11d45'
Bailey,Robert L
Cet 2h5'20"3d42'
Bailey,Rosa-"Short Witch of the East"
Lmi 9h41'38"38d32'
Bailey,Rosalie Duilia Fabbry
Vul 19h21'41"23d31'
Bailey,Rose Mae (McKinney)
Peg 22h28'29"28d12'
Bailey,Ryan Donovan
Per 3h16'39"50d7'
Bailey,Sarah
Cam 9h13'46"78d2'
Bailey,Sean Patrick
Boo 13h58'0"17d17'
Bailey,Shante R
Ori 5h57'21"11d13'
Bailey,Sharon
Aql 20h10'29"13d46'
Bailey,Sidney John
Cep 20h8'53"60d14'
Bailey,Susan Knight
Lyn 8h10'52"52d14'
Bailey,Thomas Gordon
Her 17h23'20"37d31'
Bailey,Tolbert B
Her 16h37'20"37d41'
Bailey,Wendelin F
Cas 23h13'1"60d22'
Baillais,Pierre
Ser 16h3'26"8d30'
Baillargeon,Robert
Lyn 9h39'38"41d33'
Baille,Jean-Pierre
Ori 5h21'27"12d23'
Baillie,Christopher
Dra 17h3'16"68d10'
Baillie,Jem Jenifer EM
Lyr 18h58'57"41d2'
Baillie,Olivia Claire
Vul 19h16'48"25d18'
Baillie,Peter & Kerrie 60/40
Psc 0h20'0"0d59'
Bailly Etoile de L'Amour
Peg 22h41'46"24d4'
Bailly,Antoine
Sgr 19h29'13"-35d33'
Bailly,Karma
Cam 3h50'59"57d10'
Bailly,Louise
Col 6h1'12"-34d14'
Bailly,Sandy
Uma 12h3'29"31d58'
Bailly,Stan
Cep 20h20'32"60d23'
Bailor,Karen
Lyn 8h19'59"52d10'
Bails,Harry Emerson
Per 2h23'30"55d50'
Baily,Jr,James Edward
Cma 6h50'49"-16d47'
Baim,Sandra Lee
Peg 23h28'43"31d32'
Baima,Alan
Dra 19h13'10"68d43'
Bain III,Andrew R
Lac 22h23'31"56d23'
Bain,John Alexander
Cet 1h48'37"-2d14'
Bain,Karen
Cyg 21h5'40"38d55'
Bain,Kristin Danielle
Hya 8h59'42"2d44'
Bain,Kyle
Oph 16h56'23"-3d38'
Bain,Lydia Powell
Cas 0h55'42"67d10'
Bain,MD,Norman H
Sex 10h36'30"-6d33'
Bain,Renee Nicole
Aql 20h18'41"5d12'
Bainbridge,Benjamin
Aur 7h12'12"40d44'
Bainbridge,Jeffrey Patrick
Dra 16h1'58"66d25'
Bainbridge,Jr,Charles W
Umi 14h53'45"70d29'
Baine,Demond L
Her 16h56'27"35d7'
Baine,Frank
Ori 5h56'45"20d36'
Baines,Kristie Lynn
Cas 0h55'1"62d45'
Baines,Rosalie
Peg 22h12'21"3d34'
Baini,John
Pho 23h37'43"-43d32'
Bains,Mandy
Sco 17h52'0"-30d38'
Bains,Tawney E
Lyr 18h31'46"30d49'
Bair,Cindy
Lyn 8h7'30"40d35'
Bair,Deborah Mignon
Peg 22h59'49"21d34'
Bair,Dr
Aur 4h48'53"40d58'
Bair,EdD,Linda H
Sco 17h2'13"-30d25'
Bair,Howard Vernon
Cyg 21h36'1"40d48'
Bair,Jason
Aur 6h6'17"32d26'
Bair,Jim
Aql 19h30'55"-8d47'
Baird,Chad Eric
Uma 10h40'23"72d31'
Baird,Irene Graham
Cas 1h45'11"61d17'
Baird,Keith W
Lac 22h2'24"48d5'
Baird,Kenneth
Pho 0h52'24"-49d43'
Baird,Margaret
Cas 0h34'26"61d9'
Baird,Marion
Peg 23h48'16"27d22'
Bairstow,Christine Sylvia
Cas 0h50'1"73d42'
Baisi,Geraldine
Com 12h29'15"26d54'
Baisley Family Star— Star of Love,George
Cam 7h58'1"60d38'
Baize,Michael Taylor
Aur 6h28'15"35d26'
Baja Brian
Per 1h50'19"50d13'
Bajgert,Dazzling Debbie Lynn
And 0h6'0"47d33'
Bajic,John
Sex 10h10'48"-4d34'
Bajmaganbet uly, Toregali
Umi 14h3'24"69d52'
Bajon,Agnieszka Anna
Uma 9h8'17"52d33'
Bajoros,April
And 23h39'18"41d12'
Bajoue,Laureline
Tri 2h21'46"30d44'
Bajoue,Syllian
Lyr 19h22'1"30d12'
Baker & Aimee Harman, The Star Of Adam
Mon 6h54'47"-1d11'
Baker 1925-1979,Sylvia Louise
And 22h56'32"50d37'
Baker Farms (Fred N. Baker)
Boo 14'6'1"23d9'
Baker ILY,Joseph T
Aur 6h28'1"32d5'
Baker Queenie Mike Squared
Aql 18h58'27"17d25'
Baker Star,The Roy Thomas
Her 17h3'20"50d13'
Baker USN,Astronaut Capt Michael A
Aql 19h26'26"14d14'
Baker's Birthday Star, Gina
Aqr 20h47'11"-6d37'
Baker's Glory
Ori 5h32'13"1d17'
Baker's Stellar Acre, F E & E K
Vul 20h22'18"26d7'
Baker,"Monkey Shines" Marie & Keith
Uma 11h12'40"42d4'
Baker,Adeana Stinnett
And 22h59'38"36d31'
Baker,Alan Sellers
Her 16h33'60"50d43'
Baker,Alexander Michael
Cyg 20h4'1"40d6'
Baker,Alonda
Cyg 19h42'27"42d16'
Baker,Alyson
Ori 6h16'28"0d35'
Baker,Amy & Dan
Ori 5h10'18"-4d57'
Baker,Amy K
Lyn 8h40'41"39d47'
Baker,Andyrew
Lyn 6h36'48"54d27'
Baker,Anne
Cas 0h13'1"61d33'
Baker,Arlene Michelle
Aql 19h35'0"15d53'
Baker,Benjiman Thompson
Her 17h11'19"46d50'
Baker,Betsy
Lyn 7h35'29"44d4'
Baker,Beulah
Lyr 19h0'21"40d28'
Baker,Billy
Aql 18h59'59"16d4'
Baker,Billy Lewis
Uma 11h1'37"71d17'
Baker,Bob
Hya 8h23'11"1d5'
Baker,Bob & Baboo
Cam 12h15'56"78d45'
Baker,Brandon Dale
Aql 19h1'1"5d50'
Baker,Brandon Michael Fredrick
Mon 8h2'28"-3d35'
Baker,Brian
Ori 6h6'49"4d56'
Baker,Brielle
Cam 5h48'47"61d4'
Baker,Caleb
Cep 23h9'1"64d59'
Baker,Cameron
Per 3h7'52"47d37'
Baker,Carl
Cma 7h26'29"7d43'
Baker,Carol Sue
Mon 6h17'59"-6d43'
Baker,Carson Reid
Cep 22h12'57"68d42'
Baker,Charles H
Cyg 20h41'28"37d50'
Baker,Cheryl
Mon 6h53'1"-6d12'
Baker,Christopher Ian
Per 3h3'0"48d47'
Baker,Clare
Gem 6h44'34"12d1'
Baker,Claudia
Del 21h1'11"12d34'
Baker,Cynthia Lynn
Lyr 18h40'19"40d32'
Baker,David
Aur 5h5'40"40d30'
Baker,David
Cnv 13h51'58"40d2'
Baker,David Edwin
Her 16h58'42"23d13'
Baker,David Stanley
Aur 6h25'57"31d38'
Baker,DeAnna Lynn Marzuola
Cet 2h45'1"0d26'
Baker,Debbie
Uma 11h55'22"40d22'
Baker,Debbie (In Pro Per)
Equ 21h15'28"2d11'
Baker,Desira
Peg 21h28'55"23d48'
Baker,Dorothy
Cet 2h28'26"-18d55'
Baker,Ethel E
Cyg 20h41'59"37d35'
Baker,Floyd Rice
Dra 19h1'34"65d35'
Baker,Francis J
Cap 20h33'11"-10d43'
Baker,Gail Lauraine
Aql 19h31'35"14d59'
Baker,Helen G
Ori 5h48'45"19d11'
Baker,Holly & David
Uma 9h41'11"51d21'
Baker,Jack
Mon 6h18'36"-5d56'
Baker,Jean Vera
Cet 3h19'22"9d23'
Baker,Jeff & Tonya
Uma 8h36'44"59d41'
Baker,Jennifer Marie
Psc 0h53'50"17d53'
Baker,Joe
Cep 22h11'29"56d2'
Baker,Jr,James Linton
Ori 5h50'29"7d9'
Baker,Jr,Robert A
Her 17h26'35"28d43'
Baker,Jr,Roger Conant
Lac 22h2'21"46d1'
Baker,Julann Irene
Mon 6h24'32"4d46'
Baker,Karen
And 0h56'42"39d40'
Baker,Karen
And 0h2'24"46d11'
Baker,Karen & Ben
Cep 23h24'27"86d8'AB
Baker,Katharine M
Uma 10h34'25"54d22'
Baker,Kathleen Clark
Crb 15h43'48"28d44'
Baker,Kenneth Lawrence
Lac 22h40'0"54d48'
Baker,Kenrick Martin
Cyg 20h0'16"40d58'
Baker,Kevin
Leo 11h47'18"18d45'
Baker,Kristi S
Del 20h49'36"13d49'
Baker,Kurtis Michael
Tau 5h53'45"28d46'
Baker,Lacey Lynn
Lyn 7h59'34"50d40'
Baker,Laura Michelle
And 23h31'47"42d23'
Baker,Lawrence Sidney Naylor
Uma 9h6'39"68d2'
Baker,Lefa & Gary
Sco 16h28'57"-30d9'
Baker,Lily Elizabeth
Vul 19h20'44"26d52'
Baker,Linda
Ari 1h59'54"10d60'
Baker,Linda L
Cnc 9h3'0"27d30'
Baker,Lori
Ori 5h57'59"10d42'
Baker,Lux Aeterna:Ian
Cep 21h28'1"68d45'
Baker,Mackenzie Aimee
Cas 0h56'59"70d47'
Baker,Mandie
Del 20h23'1"20d14'
Baker,Margaret,Rose
Aql 20h9'0"8d5'
Baker,Mark
Vul 20h14'48"25d34'
Baker,Marlene "Dawn"
And 23h37'22"46d46'
Baker,Mary Ann
Cmi 7h21'36"7d23'
Baker,Mary Elizabeth
Cam 5h57'46"60d16'
Baker,Mary Louise
Eri 3h9'1"-1d56'
Baker,Maryann
Her 16h58'12"37d32'
Baker,Mathew
Ari 2h2'48"20d17'
Baker,Matthew David
Aur 7h22'33"37d59'
Baker,MD,Carl Ned
Aur 6h15'0"30d40'
Baker,Michael
Her 16h59'46"18d51'
Baker,Michael R
Her 16h27'0"24d49'
Baker,Mona W
Cas 1h56'30"70d46'
Baker,Nicola Claire
And 0h33'50"40d35'
Baker,Olivia
Aql 19h55'0"-6d25'
Baker,Paul Ernest
Aql 19h31'40"14d25'
Baker,Peggy
Sge 19h32'28"16d29'
Baker,Phyllis
Aql 19h55'0"7d32'
Baker,Ray Allen
Her 16h26'1"38d47'
Baker,Raymond
Ori 6h5'35"50d20'
Baker,Raymond Brooks
Hya 8h14'51"2d23'
Baker,Rebecca S
Aur 4h50'34"51d18'
Baker,Rex F
Her 17h12'48"43d40'
Baker,Rhiannon
Umi 15h19'59"78d51'
Baker,Richard Allen
Her 18h2'28"30d40'
Baker,Richard Anthony "Tony"
Aur 6h20'15"40d54'
Baker,RN,Irene
Cyg 20h29'36"41d5'
Baker,Rob
Aql 19h54'59"8d59'
Baker,Robert Bruce
Aur 4h55'39"40d27'
Baker,Robert Curran
Ser 15h13'25"0d26'
Baker,Robert Warren
Aur 5h11'48"44d47'
Baker,Rodney Lee
Cep 23h26'1"71d4'
Baker,Roger Clyde
Aql 19h40'0"14d30'
Baker,Ross
Cep 22h27'29"63d40'
Baker,Ruby H
Lyr 18h57'13"31d0'
Baker,Russell Don
Cma 6h12'50"-16d42'
Baker,S&D Owens T&R
Aur 4h57'27"51d7'
Baker,Samuel E
Cet 2h59'37"6d11'
Baker,Sarah Annette
Mon 7h59'26"-3d48'
Baker,Sarah Marie
Cam 4h49'13"70d25'
Baker,Scott & Sherry
Del 20h15'36"11d48'
Baker,Shields
Lyn 7h55'54"35d20'
Baker,Shirley Ann
Ori 6h5'53"8d37'
Baker,Stephen
Her 16h43'50"48d1'
Baker,Steve
Aql 20h12'23"5d33'
Baker,Stuart
Ser 18h2'22"-14d39'
Baker,Susan K
Crt 11h51'1"-11d42'
Baker,Tamara D
Cyg 21h0'21"53d53'
Baker,Teresa Lynn
Aql 19h26'59"14d6'
Baker,Terri
Sge 19h33'4"16d26'
Baker,Theresa
Vul 19h44'46"28d17'
Baker,Thomas & Rita
Cyg 19h41'33"38d57'
Baker,Timothy Edward
Boo 14h7'50"32d6'
Baker,Tina L
Cas 0h16'40"64d51'
Baker,Tom
Dra 19h11'25"68d31'
Baker,Tom
Oph 17h56'55"11d33'
Baker,Tood
Cet 0h58'20"-6d41'
Baker,Travis Wayne Forrest
Vir 11h37'16"9d51'
Baker,Tyson
Cnv 13h23'40"28d25'
Baker,Vincent
Dra 18h55'28"65d36'
Baker-Buxton:28.12.91
Cyg 19h56'16"44d48'
Baker-My Eros,David A
Oph 17h6'18"10d53'
Baker-Stauffer,Pamela Hope
And 2h33'49"50d5'
Baker-Warrior for Jesus,Michael Glenn
Her 15h55'43"41d2'

Bakewell,Shelby Lee
Cas 23h15'32"63d19'
Bakewell-Webb,Carol Lee
Cyg 20h2'1"30d23'
Bakht,Arjumand
Aur 6h28'60"37d57'
Bakin,Carolyn Marie
And 2h24'1"39d28'
Bakken,David A
Her 17h13'23"45d38'
Bakken,Maude Olivet Miller
Vul 19h22'37"27d44'
Bakken,Skyler J
Her 17h19'57"47d5'
Bakker,Donna
And 2h27'46"45d55'
Baklaan,Sandra
Cyg 19h32'26"35d57'
Bakshi,Brandon Raj
Ori 5h59'40"17d25'
Bakshi,Nicole Jasmine
And 1h19'30"39d13'
Bakshi,Taylor Louise
Peg 22h47'29"5d45'
Baksi,Umarani Mitra
Lyr 18h43'39"30d23'
Bakula,Scott
Peg 22h33'53"10d22'
Bal Bolieu
Oph 18h17'37"12d48'
Bal-Kales,Asime
Lac 22h31'18"53d54'
Balance,Zorayda T
Mon 6h20'12"5d6'
Balandiat,Julia Anne
Peg 21h29'18"24d18'
Balara,Joseph Richard
Gem 7h58'42"32d59'
Balara,Linda & John
Gem 6h57'1"16d36'
Balas,Bertrand
Crb 16h9'33"31d8'
Balas,Marla
And 0h59'34"45d46'
Balas,Maryann
Boo 14h35'33"40d21'B
Balasa,Lillian
Cas 1h6'38"58d49'
Balassone,Ross
Cnv 13h23'18"42d22'
Balassone,Samantha
Cas 1h22'37"58d54'
Balayan,Stephanie
Aqr 23h45'31"-5d15'
Balazich,Bernadette
Cas 0h6'18"64d15'
Balazinski,Jr,John Anthony
Dra 18h13'49"68d32'
Balazot,Andre
Dra 9h53'52"81d3'
Balazs,June
And 0h4'10"43d12'
Balbay,Levon
Hya 8h50'15"-0d24'
Balbo,Antoinette
And 23h36'0"44d47'
Balcerek,Ben
Cep 21h48'41"63d46'
Balch,Dr Charles
Aur 6h9'15"38d16'
Balch,Jr,Frank Samuel
Oph 17h31'33"11d17'
Balcombe,Shaney Galaxy of Love Kim
Gem 6h6'17"26d55'
Balconi,Cheryl Lynn
Tri 2h13'40"32d30'
Bald Eagle
Uma 12h40'48"62d37'
Bald,Dolores Victoria
Com 12h36'1"21d37'
Bald,Lawrence T
Ori 5h57'1"17d49'
Bald,Mary Ann Elizabeth
And 2h34'18"45d38'

Bald,The Light of my Life,Robert Bruce
Dra 16h26'24"63d16'
Baldacchini,Simona
Eri 3h41'21"-2d25'
Baldacchino
Ori 6h4'34"1d45'
Baldaccini,Mary Louisa
Lyr 18h30'49"30d59'
Baldanzi,Alexander G
Boo 15h46'25"41d26'
Baldas,Brandon C
Ser 15h18'20"7d34'
Baldasari,Jessica Mae
Aql 19h2'51"16d46'
Baldassari,Sharon
And 23h6'19"43d53'
Baldassarri,Patrick
Her 18h5'30"28d23'
Baldassi,Richard
Cnv 13h36'1"40d7'
Baldauf,Ginny
Oph 18h0'35"8d47'
Baldelli,Laura
Per 2h12'22"58d29'
Balder,Michael Edward
Boo 15h5'34"20d27'
Balderson's Clarence, Steve
Ori 6h11'31"8d59'
Balderson,Daniel Burton Paoaka`I'ini
Cep 0h6'31"79d26'A
Balderson,Tara Leigh Holliday
Cep 0h6'31"79d26'B
Baldewein,Edith
Ari 1h46'42"23d46'
Baldi,Vittoria
Cam 6h41'18"67d30'
Baldini,Nina Catherine
Lyn 7h50'41"34d41'
Baldino/Nixon 52892
Eri 2h49'39"-5d32'
Baldocchi,Janet
Cnv 13h3'45"40d53'
Baldock Star,The
Uma 8h36'51"57d13'
Baldree,Rebecca Cecile
Aql 19h58'24"14d26'
Baldrian,Mary
Uma 11h37'20"52d6'
Baldridge,Alie
Aql 19h55'1"0d2'
Baldridge,Cynthia
Cas 0h16'43"61d46'
Balducci,Cyrielle
Cet 0h54'11"1d28'
Balducci,Guiseppe
Aqr 22h42'41"-6d16'
Baldwin Star,The
Uma 11h6'42"47d48'
Baldwin,Alaia Coeurd Alene
Vul 19h46'16"28d18'
Baldwin,Andrea Michelle
Cas 0h27'35"68d30'
Baldwin,Audrey Michelle
Cas 23h39'51"60d5'
Baldwin,Deryl
Hya 8h35'12"0d6'
Baldwin,Bessie I
Peg 23h26'37"10d17'
Baldwin,Charles H & Wilma L
Cyg 21h31'53"42d1'
Baldwin,Charles P
Her 18h17'1"14d50'
Baldwin,David Seth
Del 20h24'53"10d22'
Baldwin,Dorothy M
Mon 6h34'31"0d25'
Baldwin,Elizabeth
Cyg 19h25'52"32d10'
Baldwin,Elizabeth
Ori 5h56'26"15d27'
Baldwin,Hilary J
Lyr 19h17'17"33d54'

Baldwin,Ian Hadlow
Oph 17h17'27"-22d3'
Baldwin,Jeanette Anna
Lyr 19h16'14"28d55'
Baldwin,Jeremy
Lac 22h41'35"53d4'
Baldwin,Jonathan & Jennifer Harris
Uma 11h17'22"48d16'
Baldwin,Kahlea
Uma 12h4'27"32d7'
Baldwin,Lesley Ann
Cyg 21h33'43"38d23'
Baldwin,Linda
Sgr 20h4'31"-26d26'
Baldwin,Preal A & Reta D
Aql 19h29'42"10d25'
Baldwin,Richard
Ori 6h0'37"8d21'
Baldwin,Robert O & Marie Baldwin
Cet 1h1'18"-4d25'
Baldwin,Simon
Uma 10h45'1"40d59'
Baldwin,Steve
Lyn 7h24'14"51d39'
Baldwin,Tara
Scu 18h54'1"-4d47'
Baldwin,Taran
And 23h16'0"49d8'
Baldwin,Terra Jean
Tau 5h47'54"23d15'
Baldwin,William H
Her 18h9'53"30d11'
Baldwin-Artist & Musician,Ken
Uma 10h12'54"52d10'
Bale,Matt
Per 3h11'0"41d3'
Baleer,Elsa & Helmut
And 1h42'23"41d29'
Balensiefen,Ute
Cas 0h46'38"60d6'
Balentine,C Michael (Bunko)
Her 17h32'41"38d51'
Baler,Ferdi M.
Ori 6h17'15"8d42'
Bales Star,The Sarah & Bill
Lyn 8h3'0"34d24'
Bales,Chase
Cma 6h55'48"-18d34'
Balestriere,Robert
Her 16h8'41"42d20'
Balfany,Connor Mitchell Dodds
Lac 22h9'41"47d26'
Balfour,David
Cmi 7h40'1"4d16'
Balfour,Elizabeth
And 0h23'27"32d2'
Balfour,Maria
Cas 1h25'35"58d59'
Balhiser,Bill & Harriet
Cyg 21h31'40"31d30'
Bali's Star,Rajan
Ori 5h55'0"15d14'
Balla
Cyg 19h24'0"33d29'
Baligian,Sammuel
Ori 5h36'19"15d32'
Balina B
Cap 21h54'0"-21d10'
Balinda's Love
And 0h8'42"38d25'
Balinski,Adrian
Cyg 19h51'30"40d15'
Balinski,Adrian "Ozzie"
Lmi 9h27'1"37d57'
Balinski,Mildred
Cep 22h30'13"58d54'
Balint Brothers
Cnv 12h46'27"40d54'
Balis,James Michael
Her 15h55'41"40d55'

Balistrieri,Francesco
Uma 10h5'42"47d60'
Balistrieri,Joseph P
Ori 5h28'30"-3d13'
Balkan,Jodi
Cam 5h2'43"68d59'
Balke,Christopher Edward
Lyn 8h7'50"50d38'
Balke,Marie Elisa
Mon 7h1'42"-6d26'
Ball III,Benjamin Franklin
Aql 20h7'37"1d11'
Ball,Alexander Gregory
Mon 8h7'1"-8d0'
Ball,Allen
Boo 14h38'17"15d20'
Ball,Ashleigh Nichole
Del 20h51'56"7d42'
Ball,Catherine Mary
Cyg 21h50'39"41d12'
Ball,Christabel Anne
Equ 21h21'58"11d50'
Ball,Christian
Her 16h5'44"40d0'
Ball,Christine J
Lyr 19h20'32"38d31'
Ball,Courtney Marie
And 1h24'59"33d57'
Ball,Dale
Sex 9h51'55"0d47'
Ball,Darrell W
Cet 3h0'57"2d30'
Ball,David J
Cnv 13h45'56"39d57'
Ball,Jason Gerry
Dra 16h50'0"68d5'
Ball,Jonathan Richard
Ori 4h59'36"14d35'
Ball,M Craig
Per 2h58'50"35d18'
Ball,Megan E
Cas 0h58'12"61d10'
Ball,Melissa
Cas 1h13'26"64d35'
Ball,Michael
Her 16h58'36"27d55'
Ball,Penelope Jane
Mon 8h7'5"-7d46'
Ball,Randall
Dra 12h3'45"71d9'
Ball,Rayford L
Her 15h56'1"46d22'
Ball,Ricky Adam Kenneth
Cep 22h59'36"63d30'
Ball,Ruth M
Cyg 19h34'38"36d43'
Ball,Ryan Luke Francis
Per 2h52'48"43d39'
Ball,Sidney A
Aur 6h1'28"32d8'
Ball,Stephen "Badger"
Lyr 18h39'32"37d32'
Ball,Susan L
Peg 22h9'1"21d8'
Ball,Tina
Lyr 18h33'15"40d10'
Ball,Virginia
Uma 12h20'52"61d39'
Ball,William
Cyg 19h24'1"32d53'
Ball,William Edward
Dra 19h43'54"68d28'
Ball-Jones Family Star
Ori 6h0'1"18d49'
Ball-Rizzi,Tucker Donald
Her 16h26'44"38d52'
Balla,Sylvia K
Peg 22h9'54"2d28'
Balladares,Sandy P
Mon 7h30'8"-7d59'
Ballance,Robert
Lib 15h36'35"-21d58'
Ballantyne,Gary
Hya 9h13'44"4d1'

Ballantyne,Jacqueline
Cyg 19h24'36"30d58'
Ballantyne,Wayne
Ser 15h54'55"24d1'
Ballard 1958-1992, Victor Conway
Lyr 19h42'51"42d3'
Ballard,Brian
Uma 13h47'0"51d33'
Ballard,C Alan
Boo 14h57'42"39d52'
Ballard,Catherine F
Cas 1h5'19"63d51'
Ballard,Dolores
Mon 7h58'14"-3d3'
Ballard,Gloria J
Cam 6h11'30"61d32'
Ballard,Karie
Sge 20h17'1"20d16'
Ballard,Kim
Del 20h13'1"15d14'
Ballard,Mary Jane
Cas 23h41'51"50d11'
Ballard,Melanie Ann
Del 20h58'1"18d57'
Ballard,Natalie
Lyn 7h27'0"36d5'
Ballard,Pearl
Lyn 9h10'22"33d24'
Ballard,Sarai
Cas 23h29'44"62d14'
Ballard,Theodore
Aql 19h6'30'15d47'
Ballas,Alan
Cnc 8h38'53"18d48'
Ballas,Maribeth
Mon 6h31'26"-0d42'
Ballas,Michael Jon
Ori 6h3'23"8d59'
Ballas,Peter J & Sally O
Uma 9h28'16"52d38'
Balle,Julia
Aur 5h13'48"29d20'
Ballen,Arthur
Oph 18h42'22"7d3'
Ballenzweig,Howard
Oph 17h17'29"-20d18'
Ballerino,Donna
And 2h32'42"45d41'
Ballesteros,Carla
Mon 7h49'51"-3d30'
Ballet,Spandau
Cyg 21h53'47"52d30'
Ballew "Rambo",Randall Wayne
Uma 8h47'29"50d43'
Ballew,Catherine Weber
Peg 22h53'0"20d43'
Balli,Glenn S
Dra 20h3'1"63d19'
Balliet,Leland Butch
Per 3h12'1"50d24'
Balliet,Michael
Aur 7h10'1"41d18'
Balliett,Jr,Ned E
Aur 5h4'19"42d35'
Ballin,Richard
Aur 6h12'57"45d8'
Balling,Sandra Dickhens
Mon 7h1'41"4d4'
Ballinger,Gerald
Aql 19h58'58"-0d25'
Ballmann,Edgar
Ari 2h25'13"26d32'
Balloch,Kristine Barclay
Uma 9h55'0"53d53'
Balloon,Shelly
Mon 6h54'31"-8d11'
Balls,Dr Kent
Oph 17h6'23"-23d50'
Balluff,Michael Fredrick
Her 16h57'0"47d35'
Ballweg,Dennis D
Lac 22h6'43"50d57'

Bally,Nadia
And 1h22'1"38d0'
Bally,Nazar
Uma 12h1'46"36d53'
Balmagia,Stephen
Ori 6h10'0"8d29'
Balmain-Brown Gary
Cep 23h2'17"60d7'
Balmares,Teresa
Cyg 20h26'26"50d24'
Balmitgere,Jean Jacques
Mon 6h37'30"-1d36'
Baloche,Sandrine
Cep 22h51'1"65d21'
Balodis,Gunars
Lyn 8h45'47"38d31'A
Balodis,Tina
Lyn 8h45'47"38d31'B
Balog
Cam 5h33'56"61d33'
Balog,Connor Eric Ripken
Peg 0h2'43"30d43'
Balogh's Star,Linda June
Mon 7h41'54"-5d27'
Balough,Alexandra Rose
Lyn 7h51'29"48d27'
Balough,James Christopher
Cam 6h2'0"61d43'
Balough,Jeffrey Gerard
Per 2h23'45"54d37'
Balough,Lisa Kathleen
Cyg 20h57'1"30d40'
Balows,Albert
Ori 5h52'25"14d35'
Balsai,Robyn
And 2h22'1"42d32'
Balsai,Thomas Jacob
Her 16h13'0"22d36'
Balsam,Alan Edward
Dra 16h32'57"62d14'
Balsamel,James Vincent
Uma 11h51'34"33d41'
Balsamina "Il micio", Andrea
Eri 3h44'13"-6d56'
Balsamo,Anthony J
Cet 3h11'1"2d54'
Balsdon,Ronald Joseph
Hya 9h2'60"-12d58'
Balster,Erika
Lyr 19h15'52"40d17'
Balta
And 23h2'25"51d8'
Balter,Barbara G (Lilienfield)
Lyr 19h23'25"42d5'
Balter,Jaclyn Elise
Gem 6h48'55"19d8'
Balter,Stephanie Lynn
Gem 6h50'40"18d42'
Balthaser,Harry
Aur 7h1'45"40d50'
Balthaser,Ruth
Lyr 18h49'24"31d45'
Balthis,Amy Lee
Cas 0h53'36"54d37'
Baltimore Chrissy
Lyn 7h54'33"39d44'
Baltz,Ruth & Lowell
Cyg 20h31'47"58d19'
Baluta,Marni
Sge 19h3'26"18d42'
Balutis,Joseph A
Cep 3h4'0"78d42'
Balvin,Pat & Sandy
Cet 0h51'33"-8d15'
Balyeat,Kelly Thomas
Cyg 20h38'1"30d48'
Balysz,Marek Alexander
Ori 5h27'20"1d27'
Balz,Karlheinz
Vul 20h58'59"20d19'
Balza,Mariuccia
Cyg 20h5'46"38d33'
Balzano,Adam
Cep 22h22'1"70d51'

Balzell
Cet 1h34'40"-12d40'
Balzer,Dave-Benjamin
Ari 2h52'20"20d58'
Balzer,Frank
Mon 6h42'46"10d9'
Balzer,Jule
Psc 1h20'18"17d48'
Balzer,Virginia
Mon 6h54'51"10d9'
Bam
Lyr 18h50'0"34d13'
Bam-Bam
Her 17h7'36"41d8'
Bamana Younoussa
Com 12h10'0"21d26'
Bambace,Bunny & Bob
Eri 2h48'48"-15d1'
BamBam
Dra 18h15'50"80d17'
Bamber,Claire Constance
Boo 14h30'21"48d2'
Bamber,Norman
Per 2h39'46"37d34'
Bamberger,Ron & Marcelle
Cyg 20h22'46"39d21'
Bambi
Lyr 18h33'30"29d37'
Bambi
Uma 10h0'11"52d44'
Bambi & Hazard
Cyg 20h19'1"38d6'
Bambolina
Cas 0h45'14"76d43'
Bamborough,Barbara
Cyg 19h24'50"34d20'
Bambridge Memorial Star,The June
Cyg 21h52'49"38d48'
Bambu,Betty
Cyg 21h6'32"31d33'
Bambury,Karen
Vir 11h46'14"8d34'
Bame,Always Richard K
Per 3h20'30"41d14'
Bamel,Lauren Natalie
Cyg 19h41'19"28d53'
Bamforth's Star
Uma 11h17'1"50d3'
Bammert,Aaron Joseph
Peg 22h42'18"11d8'
Bammert,Jordan Nicole
Tau 4h44'53"16d8'
Bampf
Ori 4h54'19"1d38'
Ban,Dorothy Jean
Lyn 8h4'17"38d18'
Banaag,Sonia & Eddie
Lyr 18h53'60"30d23'
Banach,Alicia Kim
Cnv 13h55'49"42d26'
Banacki,Charlotte
And 1h30'32"38d29'
Banana Birillo, Silvia Chiufchiuf!
Lac 22h27'22"49d49'
Bananna Star,The
Lyr 19h2'47"33d46'
Banasiak,JoAnn
Dra 15h24'31"65d26'
Banasky,Charles J
Boo 13h39'57"16d19'
Banat,George
Cep 0h32'0"80d32'
Banbough,Belinda Anne
Cyg 21h36'31"42d55'
Bancroft,Donald Wayne
Uma 9h16'25"51d23'
BancStar
Ori 5h29'50"-3d53'
Band,Lindsay Jane
Cas 0h23'30"61d18'
Banda,Antonio Rioja
Her 18h22'1"12d13'

Bandeen,Chad
Cnv 13h49'40"32d4'
Bandell,Sonja
Lyn 8h10'0"51d32"
Bander,Ingram
Sgr 19h37'52"-36d26'
Bandit
Aur 5h36'15"54d33'
Bandit
Vul 20h21'21"25d42'
Bandit (Altenburg)
Dra 10h58'39"74d40'
Bandle,Jürgen
Cmi 7h6'31"4d12'
Bandstra,Ronald Leonard
Cep 21h55'45"70d11'
Banducci,Roanna
Oph 17h56'59"12d3'
Bandura,Andrew Lane
Hya 8h42'47"6d8'
Bandura,Sveta
Aql 20h2'1"10d37'
Bandy,James
Aql 18h59'43"-6d58'
Bane II,Douglas A
Her 18h3'41"47d26'
Bane,Donna
Vul 21h2'42"27d24'
Bane,Gary David
Aur 7h3'23"40d40'
Bane,Max
Aur 4h58'50"40d4'
Banegas,Alvaro
Oph 17h38'55"11d12'
Banerjee,Rathin
Ind 20h28'21"-50d21'
Banerji,Subroto
Uma 10h50'0"53d28'
Banfi,Elena
And 1h49'25"39d7'
Bang,Bill
Aur 6h4'53"37d39'
Bang,Brian Andersen
Aql 18h43'0"10d11'
Bang,Richard "Evige Lys"
Ori 5h49'52"10d48'
Banga
Cam 6h0'35"61d29'
Bangas,Brittany
Del 20h52'45"4d19'
Bange,Joseph Bernard
Aur 5h2'30"30d40'
Bangert,Christian Charles
Cep 21h55'0"68d50'
Bangert/Purcell Wedding Star
Aql 20h6'1"0d56'
Bango,Patricia
Peg 23h31'59"21d8'
Bank 30,Jeffrey
Lac 22h9'40"47d56'
Bank von Ernst
Mon 7h35'39"-0d8'
Bank,Barbara
Cas 22h58'39"57d9'
Bank,Danny & Deeb Fadel
Aql 20h5'29"1d1'
Bank,Joseph
Aur 6h36'1"37d46'
Bank,Rachel
Eri 2h48'29"-17d58'
Banke,Rita
Cyg 21h50'29"40d36'
Bankert,Doris J
Cas 0h42'56"67d15'
Bankert,Sr,William E
Aur 5h56'15"29d7'
Banket,Jeanette
Cap 20h21'38"-20d10'
Bankoski 9-25-94 And Then Some
Cam 6h10'56"60d31'
Banks "50",Tom
Aql 19h25'27"7d48'
Banks Star Geezer, Graham
Cep 0h7'34"73d35'

Banks With love Kellie Weeks,Todd
Her 17h59'21"40d45'
Banks' Celestial Rendeview,Tom & Beth
Com 12h58'37"14d41'
Banks,Adriana C
Mon 7h1'1"-0d30'
Banks,Albert Bianchi
Hya 9h17'23"6d29'
Banks,Alison A-Peter C Hall
Cyg 19h33'60"36d9'
Banks,Allison Grace
Cas 0h15'1"61d43'
Banks,B J
Sct 18h55'15"-6d41'
Banks,Bonnie Alvin
Gem 7h0'33"11d34'
Banks,Brandon Paul
Boo 14h36'59"19d39'
Banks,Briana Lynn
Cas 0h8'52"54d25'
Banks,Bruce T
Her 16h36'14"50d46'
Banks,Cheryl
And 2h20'47"42d25'
Banks,Chris & Patrece
Dra 16h41'1"60d36'
Banks,Dale Susan Corinn Ashley
Lyr 19h23'1"31d19'
Banks,David Anthony
Dra 15h19'47"60d53'
Banks,Dr David B
Oph 17h51'1"12d31'
Banks,Earl & Gisela
Eri 3h43'50"-12d48'
Banks,Joan Roberta Clegg
Cas 0h33'31"61d40'
Banks,Joesph Carson
Per 2h38'46"37d25'
Banks,John Matthews
Cyg 20h20'55"41d45'
Banks,Jr,James Paul
Aql 20h2'28"8d12'
Banks,Julie
Aql 18h39'0"-2d16'
Banks,Lillian & Don
Cyg 19h54'21"38d3'
Banks,Madison McKenzie
Ori 5h56'39"13d9'
Banks,Mr & Mrs Kenneth & Catherine
Vul 20h16'14"25d7'
Banks,Nevada
Her 16h29'11"20d12'
Banks,Rebecca Lynn
And 0h19'28"30d43'
Banks,Ronald William
Boo 14h11'58"32d7'
Banks,Sallie
Lyr 18h22'0"38d16'
Banks,Vicky Marie
Cas 1h9'21"75d23'
Banks-Star,The Barbara
Cas 1h33'35"58d13'
Bankston Family Star, The
Ori 5h55'57"12d56'
Banne
Dra 14h59'38"64d3'
Banne,Jean Yves
Lyr 19h5'52"47d9'
Banner Family & Friend Star,The Peter Byam
Gem 7h0'37"11d57'
Banner,Daniel Harris
Lac 22h7'24"49d13'
Bannerman,Brent & Nicole
Crb 15h20'18"31d47'
Bannerman,Christopher McLean
Her 18h12'30"38d19'
Banning,Caroline Christine
Cas 0h43'24"60d7'

Bannister,Elizabeth T
And 23h7'41"40d39'
Bannister,Gary,Lee
Aqr 22h24'60"0d1'
Bannister,Ian
Ori 6h4'35"5d54'
Bannon's Star
Uma 10h19'51"48d47'
Bannon,Joseph
Ori 4h56'45"0d55'
Bannon,Joseph William
Ori 5h48'23"20d56'
Bannon,Nancy J
Lyr 19h23'36"35d31'
Banom
Aur 7h16'1"35d55'
Banowsky,Sarah Elizabeth
Hya 9h9'1"2d33'
Bansal,Sarita
Cyg 21h2'24"48d56'
Banscy
Del 20h12'12"15d12'
Bansky,Robert
Dra 12h25'23"68d10'
Bant,Our Shooting Star Chris & Marla
Boo 13h59'0"23d39'
Banta,Brooklynn
Sge 19h23'18"16d48'
Banta,Jason M
Dra 16h25'45"69d48'
Banta,Laurie Alma
Vul 20h21'50"25d44'
Banta,Thomas N
Cep 22h6'50"60d46'
Bantau,Gabriela Marie
Cyg 19h23'11"28d5'
Bantman,Yves
Lyn 7h45'48"48d35'
Bantry,Bryan
Per 3h57'29"40d24'
Banuchi,Marcos Rodriguez
Sgr 19h23'57"-42d22'
Banuelos,Martha
Oph 17h43'42"13d39'
Bany,Michael
Aur 6h28'19"38d59'
Banys,Felicia G
Psc 0h30'11"5d47'
Banzai
Aql 18h52'42"-0d55'
Banzali,John Robert
Lmi 9h28'27"38d14'
Banzhaf,Floyd R
Her 17h28'25"31d12'
Bao,Jennifer
Lyr 18h28'14"46d41'
Baobab
Aql 29h0'50"3d53'
Bapiste Raymond 24-01-84
Cam 5h47'20"58d40'
Baprawski,Jr,Edmund J
Aur 5h10'36"43d19'
Baptiste Poncet
Cas 0h29'1"68d22'
Baptiste,Eric
Dra 10h5'0"73d5'
Baptiste,Lorie
Lyr 19h11'32"38d36'
Baptiste,Ryan Charles
Ser 18h2'31"-0d55'
Baptiste,Sarah Ann
Hya 8h41'27"-1d47'
Bapu
Vul 20h22'49"28d20'
Baque,Monsieur Noël
Mon 7h48'52"-4d34'
Bar Mitzvah
Cep 20h56'24"61d44'
Bar,Ruth Ali
Uma 8h34'49"49d21'
Bara,Rusty
Aur 6h25'54"30d7'
Barabas,Tom
Dra 16h55'0"70d39'
Barabis-50
Dra 16h38'0"62d16'
Baraboo,Stacy
Com 12h53'39"21d38
Baraboo,Steven
Cnv 13h44'37"32d24'
Barad,Eric Christopher Timothy
Del 20h23'58"18d57'
Baraff,Laura
Lyr 19h5'43"38d5'
Baraille,Christine
Peg 21h48'11"33d57'
Baraille,Liliane
Eri 3h44'59"-2d22'
Barajas,Humberto
Aql 20h1'16"-8d2'
Baraka
Her 18h5'46"14d10'
Baraka,Zakir
Uma 8h57'13"71d44'
Barakatt,Steve
Lyr 18h31'1"38d47'
Baral,Peter
Mon 7h54'26"-2d55'
Baran,Eddie
Lac 22h37'28"48d53'
Baran,John Paul
Ari 2h36'37"30d16'
Baran,Katharine B
And 0h16'29"31d29'
Baranca Stefania
Nor 16h23'39"-50d57'
Barancek,Joseph
Her 16h48'41"16d41'
Baranda,Mercy
Ori 5h56'57"16d20'
Baranella,Nicky
Boo 14h48'49"31d1'
Barang,Alexander Sebashan
Aur 4h59'57"38d36'
Baranovsky,Dr Max
Her 17h17'30"42d41'
Baranowski,Anna Powers
Cas 0h27'46"61d36'
Baranowski,Billy
Uma 10h33'34"52d2'
Baranowski,Jim & Peggy
Sge 19h32'55"18d46'
Bararreca,Inez Paul (Mom & Dad)
Ori 5h36'40"15d18'
Barasel,Karynna A
Lyr 18h39'1"36d24'
Barat
Per 3h36'1"51d40'
Barathur,Rajendra
Peg 21h55'24"33d49'
Baratta,Annamaria Suan
Cep 23h12'0"71d6'
Baratta,Daniela
Cep 1h18'1"80d27'
Baratz,Renee
Umi 16h29'51"70d32'
Baravalle,Laura
Mon 6h57'50"-6d56'
Barazani,Ronald
Hya 9h4'31"0d49'
Barazia,Armand
Her 17h52'21"14d35'
Barb
Cas 2h29'57"68d14'
Barb
Cyg 20h35'1"39d26'
Barb
Ori 4h57'0"5d23'
Barb's Star
Ori 5h51'52"18d38'
Barb's Star Sode-zo
Umi 16h49'1"77d54'
Barb's Willis
Aur 6h11'45"31d52'
Barb-5
Lyr 18h32'28"38d8'
Barba
Lyr 19h3'43"26d22'
Barba,Connor Killey
Lac 22h10'12"50d12'
Barba,Frank P
Her 16h37'56"50d7'
Barbagallo,Pamela Noel
Cas 2h35'49"70d45'
Barbance,Odile
Oph 18h1'35"10d19'
BarbAnn
Sct 18h55'12"-13d41'
Barbara
Lib 15h36'14"-20d2'
Barbara
Ori 5h10'56"1d55'A
Barbara
Lyn 8h4'1"48d56'
Barbara
Hya 9h18'31"5d32'
Barbara
And 1h35'58"40d5'
Barbara
Cyg 19h21'57"28d41'
Barbara
Cet 1h30'56"0d54'
Barbara
Aql 19h32'44"-5d50'
Barbara
Uma 12h8'30"62d10'
Barbara
Mon 7h1'55"-6d23'
Barbara
Lac 22h29'12"50d36'B
Barbara
Leo 10h58'58"10d25'
Barbara
Cas 1h11'0"75d57'
Barbara
Ari 3h0'28"28d44'
Barbara
Lac 22h31'29"38d59'
Barbara
Cyg 20h41'14"31d15'
Barbara
Tau 4h12'4"20d15'
Barbara
Vir 12h27'42"-7d35'
Barbara
Vir 13h35'33"-10d60'
Barbara
Cas 0h42'28"62d36'
Barbara
Per 3h35'42"51d57'
Barbara
Cam 7h4'48"61d13'
Barbara
Aur 5h0'56"41d55'
Barbara
Hya 8h52'53"6d4'
Barbara
Aql 20h14'1"4d40'
Barbara
Cas 0h33'41"67d39'
Barbara
Cep 23h18'50"68d49'
Barbara
Lac 22h17'45"51d35'
Barbara
Del 20h13'29"10d47'
Barbara
Lyr 18h48'0"32d17'
Barbara
Peg 23h31'43"10d29'
Barbara
Cnv 12h42'15"37d48'
Barbara
Cet 2h40'44"9d39'
Barbara
Nor 16h18'45"-50d59'
Barbara & Anthony
Lup 15h22'4"-44d7'
Barbara & Brian
Eri 3h52'53"-1d38'
Barbara & Jan
Cam 5h37'45"61d50'
Barbara & John
Lyn 7h21'35"50d8'
Barbara & Lora Lee
And 23h38'32"48d12'
Barbara & Robert's 38th Anniversary
Cyg 21h29'43"40d55'
Barbara & Roberto
Uma 11h11'23"38d20'
Barbara & Shelley Love Star,The
Eri 4h36'52"-18d56'
Barbara & Wendy Eternity
Cen 11h53'14"-44d13'
Barbara Ann
Cyg 21h25'54"38d12'
Barbara Ann
Cep 22h30'47"63d46'
Barbara Ann
Peg 22h10'46"29d58'
Barbara Anne
And 23h33'0"45d48'
Barbara Anne
Cru 12h53'12"-61d29'
Barbara Annie
Mon 8h4'20"-8d8'
Barbara Anthea
Vel 9h43'40"54d31'
Barbara B
Ori 6h3'50"-0d20'
Barbara B
Cyg 19h30'42"38d25'
Barbara Bright
Cas 1h50'45"75d48' B
Barbara E
Cet 3h4'54"3d6'
Barbara e Lorenzo
Pup 7h56'36"-25d24'
Barbara Ellen
Cas 1h47'24"73d48'
Barbara Gene
Hya 8h13'50"1d12'
Barbara Helen
And 0h8'49"44d41'
Barbara Jean
Del 20h21'52"2d16'
Barbara Jean
Lyn 7h44'38"52d6'
Barbara Jean
Boo 14h36'0"20d48'
Barbara Jean
Ser 15h16'18"10d36'B
Barbara Jean's A Shining Star
Com 12h2'58"26d29'
Barbara Jean's Star
Lyr 16h19'57"38d50'
Barbara Jeanne
Lac 21h57'35"40d4'
Barbara Jill
Cyg 19h59'37"31d47'
Barbara Lee
Lyr 18h47'25"40d12'
Barbara Lee SNPGC
Uma 8h55'59"47d4'
Barbara Lynn
Peg 22h6'23"20d54'
Barbara Lynn
Lyn 7h56'53"41d6'
Barbara Mae & David Wulfe Eternal Love
Cyg 20h12'29"35d5'
Barbara Marie
Lyr 18h35'14"36d58'
Barbara Olivea
Cyg 19h54'43"50d13'
Barbara R
Cet 1h31'1"1d19'
Barbara Rae
Del 20h53'16"9d25'
Barbara Roseann
Lyn 7h44'42"38d11'
Barbara Shere
Cap 20h7'26"-10d46'
Barbara Stephanie Anne
Aur 5h17'46"46d4'
Barbara's & Gary's Special Light
Sge 19h43'29"16d21'
Barbara's Bedazzling Beauty
Ori 4h55'27"-0d4'
Barbara's Brilliance
Cyg 21h30'45"41d35'
Barbara's Christmas
Lyr 18h48'45"42d54'
Barbara's Destiny
Lyr 18h59'0"37d29'
Barbara's Jewel
Aqr 21h23'24"-8d20'
Barbara's Own Little Star
Peg 22h17'42"18d49'
Barbara's Rapture
And 23h46'45"45d43'
Barbara's Skypoint
Lyr 18h46'52"30d45'
Barbara's Star
Cas 1h2'26"60d43'
Barbara's Star
And 0h20'32"35d25'
Barbara's Star
And 23h33'0"45d37'
Barbara's Starship
Sgr 19h37'29"-41d30'
Barbara,Joseph Edward
Dra 10h39'58"77d36'
Barbara-P-B-Kobek & Harry
Gem 8h1'51"30d13'
Barbaraluisa
Scl 23h21'43"-26d17'
Barbaranne,Chuck & Garrett
Peg 0h6'14"20d3'
Barbarash,Mona
Lyn 7h53"57'35d47'
Barbaresco,Remi
Crb 16h10'36"38d32'
Barbaria,Bob
Hya 9h31'35"-0d28'
Barbarino's "Beam Of Love"
Sge 19h0'34"20d19'
Barbaro,Laura
Lyn 7h54'28"40d35'
Barbaro,Tracy
Lyr 18h32'1"44d7'
Barbarod
Del 20h18'0"20d7'
Barbarotta,Vera
Cam 3h44'1"74d50'
Barbato 515,John
Per 4h45'1"50d36'
Barbato,Angela Jae
Peg 24h45'12"8d14'
Barbeau,Adrienne
Cmi 7h46'52"3d20'B
Barbeau,Edward
Boo 14h36'1"38d32'
Barbeau,Gilles
Cep 22h9'27"60d22'
Barbeau,Kathryn Ann Sirica
And 2h1'54"45d50'
Barbee,Diann
Vul 19h48'0"28d47'
Barbel-Die Standhafte
Gem 8h4'0"31d7'
Barbelin,Jean Claude
Sex 10h16'53"-3d11'
Barber of Braslow & Burners,Nicholas
Ori 5h30'6"-1d30'
Barber Star,The
Cnv 12h46'20"32d45'
Barber,Brenda Leann
Lyr 18h15'21"37d42'
Barber,Cheryl
Boo 14h30'40"57d57'
Barber,Christine
Umi 15h26'50"70d32'
Barber,Christopher Jewels
Peg 23h8'43"10d58'
Barber,Cornelia Witte
Dra 18h11'34"68d27'
Barber,Frank G (Superfly)
Aql 19h52'51"13d28'
Barber,Hank
Ser 17h53'55"-15d40'A
Barber,James
Cet 2h42'43"4d30'B
Barber,Jean
Ser 17h53'55"-15d40'B
Barber,Jerry
Mon 6h53'21"-1d28'
Barber,Jillian Lee
Psc 1h26'51"31d34'
Barber,Joanne Marie
Cyg 20h24'17"38d31'
Barber,John Francis Gonzales
Ari 2h57'22"30d1'
Barber,Luke John
Lyn 9h14'43"40d35'
Barber,Marc Lee
Aqr 22h33'51"-10d5'
Barber,Ralph
Ori 6h0'0"-0d1'
Barber,Richard T
Aur 6h3'13"37d42'
Barber,Tammy
And 23h41'10"45d13'
Barber,Thomas D
Cep 23h12'18"70d51'
Barber,Tony
Cyg 21h46'12"36d24'
Barberie,Anna Rooney
Cas 23h32'1"63d15'
Barberio,Alexandra Christina
Mon 8h4'41"-5d40'
Barberio,Edward Joseph
Psc 1h15'55"22d17'
Barberis,Hercules
Her 17h12'1"40d39'
Barberman (Elvis the King)
Ori 5h58'50"14d38'
Barbero,Monique
Ori 6h7'1"18d50'
Barbero,Nicholas
Lmi 11h0'19"30d53'
Barbero,Paolo
Per 3h32'56"51d19'
Barbers,The
Lac 22h29'41"54d22'
Barbetta,Angela Francis
Lib 14h45'19"-7d45'
Barbetta,Anthony Joseph
Aqr 22h2'1"-11d59'
Barbetta,Grace
Aqr 22h2'46"-12d13'
Barbetta,Joseph
Lib 14h45'29"-0d37'
Barbl
Uma 9h41'40"50d12'
Barbie
Lyn 7h8'45"55d53'B
Barbie
Tri 1h55'11"26d2'
Barbie
Mon 6h21'37"5d32'
Barbie
Uma 9h40'24"52d44'
Barbie & Larry Forever
Oph 18h41'59"10d32'
Barbie Doll
Boo 14h29'40"21d40'
Barbie's North Star
Eri 5h0'0"-6d4'
Barbie,I Love You, Todd!
Lyn 7h9'22"58d37'
Barbier,Rachel
Sex 10h12'23"-2d48'
Barbieri's Land, Claudia
Pyx 8h41'9"-26d33'
Barbieri,Christopher
Sgr 19h38'33"-30d43'
Barbieri,Lisa
Vir 13h24'27"-8d17'
Barbieri-Shilts
Mon 6h58'1"8d44'
Barbiero,Cosimo & Nancy Napoli
Vul 19h47'31"28d19'
Barbin,Harry C
Her 16h35'11"39d37'
Barbino,David
Dra 14h54'0"62d9'
Barbosa,Marisa Alexandra Sacadura
Lyn 6h25'55"60d23'
Barbosa,Nuno
Dra 16h9'53"62d29'
Barbosa,Robin Areas
Lib 15h42'17"-23d17'
Barbour,Anne K
Cyg 21h26'36"41d14'
Barbour,Cheryl
Cet 2h26'50"3d45'
Barbour,Daniel
Umi 14h46'28"68d12'
Barbour,Deborah Platt
And 1h15'43"38d44'
Barbour,The Brilliant
Ori 5h7'43"-2d6'
Barbour,William A
Aur 6h31'46"33d26'
Barbour/DBS,Nancy Kate
Del 20h17'27"13d36'
Barbron
Mon 5h56'17"-6d30'
Barbry,Claudine
Cep 22h26'37"57d58'
Barbuti,Monica Maria
Lyn 9h9'58"34d52'
Barby,Clara
Cam 4h20'25"60d55'
Barce,Andrea
Oph 17h6'17"8d6'
Barcelo De Castany, Sara
Aur 4h34'30"33d45'
Barchuk,Gregory
Aur 5h10'50"50d6'
Barckhoff,David Alan
Per 2h22'0"55d26'
Barclay & Sadie
Sge 20h16'41"16d43'
Barclay Star
Cas 1h58'39"76d54'
Barclay,Ailsa
Cas 23h35'11"61d15'
Barclay,Donald D
Uma 9h15'42"52d19'
Barclay,Jonathan Dean
Aur 5h5'0"41d29'
Barclay,MD,David L
Lmi 10h35'28"32d13'
Darclift,Oscar & Delpha
Cyg 19h37'1"38d21'
Barcus,Jeffrey Sebastian
Her 16h34'19"34d58'
Barczykowski,Elsa E
And 1h52'52"40d32'
Bard,Lynne Susan
Lyn 7h39'0"41d8'
Barde,Jean-Claude
Sge 19h59'0"20d31'
Bardeen,Robert Kenneth
Per 2h21'34"55d43'
Bardeleben,William
Her 16h45'21"32d16'
Barden III,James Floyd
Boo 14h51'26"33d18'
Bardes,Victoria
Mon 7h47'43"-3d51'
Bardi,Pier Luigi
Ori 6h2'48"1d7'
Bardin Family Star,The
Her 15h54'1"40d34'
Bardin,Jacques
Aur 6h8'1"30d1'
Bardin,Lucie
Oph 18h6'59"11d14'
Bardino,A J
Tri 1h38'11"35d11'
Bardino,Tony
Per 2h58'22"37d44'
Bardocz-Cowan,Barbara Ann
Per 2h16'46"58d17'
Bardon,Margot
Cam 3h29'1"60d6'
Bardovam
Cnv 12h9'0"40d11'
Bardwell,Clif & Ida
Eri 3h57'30"-6d40'
Bardwell,Kathryn G & Stanley A
Cas 1h4'1"63d54'
Bardwell,Tyler Keaton
Cam 3h19'59"60d35'
Barefoot,Jr,Reginald Gerald
Sex 10h36'42"-2d23'
Bareis,Gary Wm Jr & Tawnya Lane
Cyg 21h28'60"28d49'
Barela,Lorna Yvette
Lyn 7h4'35"60d4'
Barello,Bruce H
Uma 9h10'35"70d3'
Barens,Elione Jonn
Lyr 19h14'19"42d39'
Barentsen Candy
Aur 4h35'0"30d29'
Bares,Gordon
Her 18h14'34"48d32'
Bareslow,Mary Ann
Mon 7h38'40"-1d22'
Baretta,Carol & Lou
Uma 11h2'24"49d19'
Barfoot (Granny),Betty
Peg 22h3'12"5d57'
Bargeon,Jeffrey Joseph
Dra 14h40'0"62d2'
Barger,Juliana Maria Pennucci
And 0h46'11"34d3'
Barger,Ruth Mason
Sgr 19h40'9"-36d5'
Bargiggia,Jacopo
Lac 22h54'20"50d9'
Bargioni Family Star
Per 3h17'48"41d10'
Bargo,Pablo Daniel
Cet 2h7'53"2d10'
Barham,Joan Harting
Cas 23h22'1"58d10'
Barham,Joan Harting
Lyr 19h18'12"42d56'
Barhorst,Sr,Edwin B
Lac 22h22'57"38d21'
Bari,Ayn
Com 13h5'41"20d45'
Baria "Pablo" Paul Collins
Oph 17h17'3"-22d47'
Bariga,Karen M
Lyn 8h24'28'40d13'
Baril,Doris
Uma 8h6'41"60d46'
Barile,James Dominick
Com 12h43'1"30d8'
Barile,Paul J
Cnv 12h22'49"34d7'
Bariletti,Sergio
Del 20h21'30"10d27'
Barilleaux,Debbie Sue
And 2h35'43"37d55'
Barimaster 95
Lyn 8h23'60"40d48'
Barini,Emanuele (Lele)
Eri 3h40'53"-0d30'
Barioni,Alfred
Leo 9h59m35"10d59'
Barisch,Marcia Ruth
Mon 6h44'29"10d30'
Barkan,Barbara Ann
And 2h32'30"41d13'
Barkdoll,Jerry
Vir 13h59'25"6d55'
Barkdoll,Scott
Tau 4h22'26"28d13'
Barker II,Robert W
Aql 20h14'10"7d41'
Barker IV,K
Cnv 13h53'1"45d53'
Barker SC Hall of Fame Brother Lee
Boo 14h37'39"19d20'
Barker Star,The Cindy & Chuck
Cyg 19h32'33"28d26'
Barker's Star, Samantha Michelle
Mon 6h56'27"10d10'
Barker's Star,Allen
Del 20h13'53"10d32'
Barker's Star,Sydney Nichole
Peg 22h45'59"29d42'
Barker,Alta Steele Stroup
Lmi 10h34'45"34d15'
Barker,Angelene Gail Elizabeth Ruth
Cyg 21h57'0"53d15'
Barker,Betty Jo
Eri 3h33'26"-6d12'
Barker,Bradley S
Cma 6h51'28"-15d44'
Barker,Cameron Michelle
Lyn 7h40'47"45d22'
Barker,Charmara
Lyr 18h28'38"34d52'
Barker,Dennis Glenn
Her 16h39'0"32d48'
Barker,Dianne K
And 2h18'49"48d27'
Barker,Edward L
Oph 17h11'19"-22d30'
Barker,Emily
And 0h27'12"38d31'
Barker,Emily
And 0h22'1"45d14'
Barker,Harold
Cep 21h53'58"70d20'
Barker,James
Per 4h20'27"51d28'
Barker,Janice L
Mon 7h18'43"-6d10'
Barker,Jason Scott
Her 17h18'32"40d25'
Barker,Jeffrey Brian
Oph 17h54'23"12d56'
Barker,Jeni L
Lyn 9h8'40"38d7'
Barker,Jo Ann
And 2h0'35"45d19'
Barker,John Andrew
Ori 5h56'27"14d6'
Barker,Johnny
Cmi 7h33'56"10d5'
Barker,Lloyd James
Per 1h54'1"48d44'
Barker,Marilyn
Lyr 18h56'51"41d49'
Barker,Mary Ann
Uma 14h19'1"58d15'
Barker,Mary Louise
Mon 6h18'0"7d48'
Barker,Matthew Jordan
Aur 6h22'36"35d11'
Barker,Michael A
Sct 18h55'8"-7d44'
Barker,Peter Robert
Ori 5h55'56"15d3'
Barker,Rene King O'Brien
Ori 6h7'58"3d56'
Barker,Samuel Charles
Umi 15h42'1"77d20'
Barker,Sarah Kristen
Cyg 19h29'26"30d42'
Barker,Tom
Cep 0h1'1"70d56'
Barker,Tressie
Crt 11h2'29"-8d14'

Barket,Keith Farris
Crb 15h53'21"38d37'
Barkett,John Keith
Per 2h4'0"57d22'
Barkhau-"Papa",William David
Cnv 12h58'22"33d17'
Barki,Elia
Ori 5h46'14"-0d29'
Barking Dog,Darryl
Aql 20h8'35"0d16'
Barkley,Alcye
Aur 6h38'18"38d20'
Barkley,Charles Wade
Lac 22h22'19"37d41'
Barkley,Madison Jo
Sct 18h34'34"-6d43'
Barkley,Richard S
Uma 9h46'1"51d36'
Barkouras,Kent
Aur 5h34'0"54d37'
Barkow,Peter
Vir 11h38'22"4d21'
Barkowitz,Randy
Hya 8h54'37"1d27'
Barkowitz,Steven
Ser 16h6'31"-2d45'
Barks,Herbert Bernard
Cet 3h10'12"1d40'
Barks,The
Dra 15h18'36"60d10'
Barksdale,Sarah Marie
And 23h15'14"44d2'
Barksdale/Rice
Eri 4h2'49"-19d4'
Barlament,Annette Marie
And 23h32'1"48d15'
Barle,Gail
Eri 3h6'45"-1d36'
Barlet,Marcel
Sex 10h16'50"-1d11'
Barletta,Anthony Lawrence
Aur 6h1'38"37d54'
Barletti,Duffy
Dra 16h4'35"60d52'
Barley,Stephen Richard
Uma 11h11'53"30d50'
Barling,Neil Alan
Peg 22h33'0"28d9'
Barlotta,Arthur
Cep 20h27'1"65d12'
Barlow,Anastasia Carol Ann
Aqr 23h38'20"-5d60'
Barlow,Cathy A
Peg 23h5'0"33d12'
Barlow,Dawn Marie Hyland
Eri 4h33'15"-12d8'
Barlow,Fred W
Aur 5h0'42"30d54'
Barlow,Gary
Lyn 7h43'47"39d58'
Barlow,Gary
Lyr 19h21'58"35d24'
Barlow,Gary
Aql 20h1'50"0d4'
Barlow,James Michael
Her 18h3'37"30d42'
Barlow,James Patrick
Sex 10h15'58"-0d50'
Barlow,Jr,Ken A
Oph 16h34'34"0d21' B
Barlow,Karin
Vir 14h56'0"7d26'
Barlow,Kathleen
Uma 10h4'22"51d14'
Barlow,Mr
Her 16h59'12"41d5'
Barlow,Nellie
Cyg 21h0'28"38d27'
Barlow,Philip Thomas
Aur 5h0'53"45d6'
Barlow,Raye W
Oph 18h2'28"8d0'
Barlow,Raymond K
Cep 20h41'19"76d41'
Barlow,Stephen A
Ori 5h56'54"15d58'
Barlow,The Golden
Cnv 13h47'51"38d58'
Barmann,Mark
Ori 5h35'39"-10d7'
Barna,Terez
Cyg 21h17'2"39d32'B
Barnabei,Jean Marie
Cam 3h58'1"60d54'
Barnabus
Cam 6h7'0"60d44'
Barnaby
Per 3h58'1"40d59'
Barnabé,Michèle
Cas 0h54'0"75d40'
Barnard,Anna Jean
And 0h49'41"22d49'
Barnard,Gordon W & Louise E
Sge 19h56'40"18d48'
Barnard,James Evyn
"Beachmaster"
Hya 9h16'3"-14d46'
Barnard,Jennifer Lynn
Aqr 20h58'12"-10d34'
Barnard,Joe Allen
Dra 15h32'17"58d48'
Barnard,Julie
Lyn 7h40'18"38d41'
Barnard,Katherine O'Rear
Sge 19h33'59"18d60'
Barnard,Kathryn Kupstas
Cas 0h18'0"61d46'
Barnard,Larry
Aql 20h18'41"0d44'
Barnard,Lorriane
And 23h8'1"37d42'
Barnard,Rand
Ori 5h51'49"14d24'
Barnard,Renee Louise
Mon 6h36'31"-1d44'
Barnard,Virginia Lee
Com 12h8'16"19d37'
Barnas,Stella
Vul 19h1'25"21d34'
Barnawell,Beau
Leo 10h56'11"14d17'
Barnby,Janice
Peg 21h56'33"31d17'
Barne's Nightwing,Tom
Aql 20h1'31"12d20'
Barnebey,Ren
Hya 8h51'29"-7d0'
Barnebey,Walker H
Sct 18h41'28"-6d47'
Barner,Isabelle
Lyr 18h33'24"41d30'
Barner,Patricia A
Cam 7h28'57"68d42'
Barnes Burner,The
Crb 16h22'34"38d31'
Barnes II,Gerald A
Hya 8h13'38"4d51'
Barnes III,Bert F
Boo 14h10'37"37d37'
Barnes,Allen
Her 18h5'24"20d12'
Barnes,Amy Patricia
And 0h8'19"30d37'
Barnes,Barney
Boo 14h53'57"26d31'
Barnes,Benjamin Shaw
Aqr 22h38'55"2d7'
Barnes,Billy
Lac 22h15'20"38d17'
Barnes,C Cortlandt
Gem 6h43'0"35d17'
Barnes,Cameron William
Ori 5h55'39"8d55'
Barnes,Candace Whaley
Uma 11h25'1"30d7'
Barnes,Christine C
Lyn 6h41'55"50d5'
Barnes,Curt
Cet 2h35'45"4d2'
Barnes,Dale
Boo 14h19'31"38d23'
Barnes,David W
Lac 22h11'15"51d27'
Barnes,Donald Clifford
Aur 4h50'32"50d16'
Barnes,Emma Pauline
Ari 1h46'28"17d49'
Barnes,Fred
Peg 23h17'60"31d32'
Barnes,Galen
Aql 20h0'44"14d24'
Barnes,Gerald N
Com 13h0'20"28d38'
Barnes,James A
Boo 15h0'11"14d40'
Barnes,Jennifer Nancy
Umi 16h11'15"80d26'
Barnes,Jessica L
And 1h18'11"36d34'
Barnes,Jil Tennison
Leo 10h33'53"14d37'
Barnes,Jill Courtney
Vul 19h39'0"20d35'
Barnes,Josh
Cam 7h56'16"60d26'
Barnes,Julie & Edwin
Uma 9h59'53"49d32'
Barnes,Keith Eric
Eri 4h28'0"-11d60'
Barnes,Kelly Ann
Cas 1h31'44"61d1'
Barnes,Kenneth Alec
Vul 20h3'56"28d45'
Barnes,Laura Elizabeth
Lyr 18h32'44"34d11'
Barnes,Laurel Sue
Ori 5h32'0"-0d39'
Barnes,Lela Isabel
Vul 19h46'1"28d17'
Barnes,Maury Leigh
Tau 4h19'29"10d34'
Barnes,Michael J
Cma 7h18'44"-16d40'
Barnes,Musashi/Ricky L
Ser 16h1'17"11d34'B
Barnes,Olga
Com 12h57'1"26d58'
Barnes,Our Beacon,Love Your
Kids,Wiley
Aql 20h2'1"1d28'
Barnes,Our Brightest
Star,Polita
Ori 5h57'1"16d45'
Barnes,Our Shining
Star!,Helen
Peg 23h5'37"13d47'
Barnes,Patti A, Peter, Mark &
Mira
Per 3h59'1"51d36'
Barnes,Peter
Aql 18h55'36"-0d24'
Barnes,Reily
Aql 20h3'24"3d51'
Barnes,Robert H
Aur 5h52'54"38d53'
Barnes,Ronald Edward
Cnv 12h22'28"40d8'
Barnes,Ronald Joel
Oph 17h54'60"12d58'
Barnes,Sarah
Peg 22h4'36"29d52'
Barnes,Shannon Lynn
And 1h56'14"37d58'
Barnes,Terry
Cyg 20h42'24"49d48'
Barnes,William W
Oph 17h59'21"-3d44'
Barness,Ethan Moon
Boo 15h5'18"30d56'
Barnet,Paul
Cep 21h20'12"58d18'
Barnett
Aql 20h24'17"8d59'
Barnett Five,The Stephen
Cyg 20h53'0"38d51'
Barnett,Ali
Cet 2h32'34"5d58'
Barnett,Brenda Kay
Cet 1h53'59"-8d39'
Barnett,Bud
Dra 12h46'39"71d56'
Barnett,Dolores C
Sge 19h43'48"18d52'
Barnett,Fay
Lac 22h29'22"40d17'
Barnett,Gaynor Lindsay
Cyg 19h27'47"33d49'
Barnett,Helene Carda
Cam 3h58'22"57d29'
Barnett,Herbert
Lac 22h48'31"55d32'
Barnett,Jeffery
Lmi 11h2'20"33d7'
Barnett,Jessica Dianne
Aql 19h28'40"-8d47'
Barnett,Jim, Tink & Jimmy
Hya 9h3'13"2d41'
Barnett,Joel M
Crb 15h18'42"30d36'
Barnett,John Richard
Aql 20h22'13"0d29'
Barnett,Katherine Taylor
Cas 0h52'0"62d46'
Barnett,Keith
Aql 19h54'27"-6d27'
Barnett,Kevin
Hya 8h10'49"-0d0'
Barnett,Louis,Geri & Turbo
Del 20h36'16"10d49'
Barnett,Marvin
Lac 22h4'23"40d30'
Barnett,Maryalice
Cas 3h15'0"75d5'
Barnett,Michael & Laura
Crb 16h20'49"30d2'
Barnett,Nancy Jeanne
Lib 15h2'47"-0d34'
Barnett,Pamela "Barney"
Peg 23h2'25"32d20'
Barnett,Paul D
Her 16h35'34"37d20'
Barnett,Richard Meyer
Umi 17h10'24"76d6'
Barnett,Robert W
Peg 21h58'42"26d25'
Barnett,Samantha Anne
And 0h5'27"38d32'
Barnett,Sara Jo
Cet 2h26'54"9d20'A
Barnett,Shawn
Aur 6h2'11"38d36'
Barnett,Stephen R
Ori 5h55'14"0d59'
Barnett,Thomas Richard
Her 17h57'57"40d55'
Barnett,Wesley Gray
Cet 2h26'54"9d20'B
Barnette,Jamie
Peg 21h54'51"21d15'
Barnette,Katie
Ori 6h3'1"7d52'
Barnette,Mignonette Elissa
Lyn 7h59'1"50d45'
Barnette,Ronald Edward
Aur 5h1'56"50d3'
Barney
Boo 15h2'47"22d18'
Barney
Peg 23h17'53"30d35'
Barney
Lib 15h59'36"-10d48'
Barney,Bruce
Cet 3h4'19"3d52'
Barney,Cynthia Marie
Del 20h24'47"8d5'
Barney,Jennifer "Sweet Lips"
Aql 20h19'48"1d32'
Barney,Joshua S
Per 3h24'49"40d47'
Barney,Michael Thomas
Lac 22h6'1"49d11'
Barney,Prof Sally Anne
Cas 23h34'43"61d40'
Barney-Ritz Glow,Dr Eugene
"Jeff"
Oph 18h17'23"12d57'
Barnfather,Craig
Ori 6h3'1"8d0'
Barnfather,Jordon Jackson
Her 16h55'31"48d34'
Barnhart,Blanche
Weisenberger
Peg 23h25'1"16d32'
Barnhart,Don
Her 17h20'34"40d13'
Barnhart,Jr,William L
Per 2h57'52"31d59'
Barnhart,Lee & Georgia
Aur 6h59'51"40d55'
Barnhart,Patrick James
Dra 20h7'56"62d13'
Barnhart,R
Aql 19h55'32"13d23'
Barnhart,Ryan Michael
Ser 18h18'56"-14d12'
Barnhart,Sr,Joseph
Oph 17h34'58"-1d38'
Barnhart,The QMC
Cet 0h23'14"-11d18'
Barnhart,Tom
Dra 16h0'43"60d18'
Barnhill,April
And 1h21'20"39d49'
Barnhill,Cynthia Jean
And 22h58'1"38d11'
Barnick,Andrew John
Uma 11h36'7"45d23'A
Barnick,Scott Wilson
Uma 11h36'7"45d23'B
Barnie & Edwina's Anniversary
Star
Cyg 19h19'21"50d19'
Barns,The Sparkel Of My
Life,Patsy
Ori 5h55'53"16d28'
Barnum,Jamey
Cet 2h37'43"5d11'
Barnum,Suzi & Rob
Vul 19h18'43"26d40'
Barnumerus,Anncereus
Cnc 8h54'17"31d43'
Barnumerus,Gemarini
Gem 7h6'45"24d35'
Barnwell,David M
Per 1h54'0"56d36'
Barnwell,Josh
Ori 5h56'25"16d49'
Barnwell,Scott Hardin
Aql 19h44'12"14d18'
Barnyock,Julie Rose Amy
Cas 0h29'33"61d49'
Barofski,Denise
Boo 14h56'48"44d2'
Baron
Oph 16h56'49"-22d40'
Baron
Oph 17h52'54"12d46'
Baron Hope of Craighead
Uma 12h3'36"31d0'
Baron,Forever Allen "Loml"
Ori 5h44'33"-0d23'
Baron,Joseph & Ruth
Cnv 13h0'44"40d3'
Baron,Jr,Richard James
Joseph
Cnv 12h52'1"40d12'
Baron,Lisa
Lyn 8h27'27"47d56'
Baron,Peter
Per 4h4'11"51d28'
Baron,Peter Gabriel
Her 17h22'37"18d52'
Baron,Roger
Uma 11h34'52"37d54'
Barona XX1,Natasha
And 0h46'40"22d0'
Baroncini,James
Lac 22h10'41"50d4'
Barone,Barbara D
Uma 11h29'32"36d31'B
Barone,Benedict A
Uma 11h29'32"36d31'A
Barone,Colton Patrick
Cyg 20h56'40"40d14'
Barone,James
Boo 14h34'13"50d12'
Barone,James
Leo 11h34'52"-5d39'
Barone,John & Barbara
Cyg 20h24'1"38d57'
Barone,Joseph
Her 18h22'59"27d21'B
Barone,Leslie Lofland
Aql 19h21'24"15d33'
Barone,Matthew
Her 18h22'59"27d21'A
Barone,Nick (Super Star
Daddy)
Her 18h38'25"12d9'
Barone,Richard A
Lyn 7h8'15"58d40'
Barone,Rosalba
For 2h9'22"-28d56'
Baroni,Nichlas J
Her 16h55'24"40d47'
Baronio,Angela Marie
Tau 5h50'26"26d46'
Baroody,George T
Per 1h47'1"50d24'
Barouch,Jason William
Uma 11h12'40"48d38'
Baroux,Pascal
Crb 15h43'38"35d20'
Barozie,Jennifer
Lyr 18h21'1"42d34'
Barquero,Hes
Ser 15h15'46"7d36'
Barquinha,Pedro Almeida
Lyn 9h7'1"37d36'
Barquinha,Pedro Almeida
Sge 20h3'23"20d53'
Barr
Tau 3h39'28"23d54'
Barr 21,Joanne
Peg 22h57'35"18d29'
Barr Family Star,The Roger B
Per 1h59'29"48d16'
Barr III,John W "Trace"
Her 16h37'47"28d1'
Barr Star,The Lee
Cet 3h9'34"1d36'
Barr Welfare State Star,The
Vul 20h0'1"25d51'
Barr,Adrienne
Peg 22h49'36"21d53'
Barr,Angela
Lyn 7h36'30"51d9'
Barr,Ashley Dawn
Cas 22h56'24"54d47'
Barr,Barbara
Tau 4h21'54"18d53'
Barr,Bill Jack
Ori 5h51'1"12d17'
Barr,Caroline Miranda
Lyn 8h17'10"34d24'
Barr,Christopher R
Peg 2h51'54"38d39'
Barr,David M
Cet 0h34'19"-5d44'
Barr,David Worth
Cam 3h53'20"52d60'
Barr,Dorothy
Eri 3h40'12"-11d4'
Barr,Ed
Sex 10h15'36"-2d4'
Barr,Erin Brough
Ori 6h9'44"9d15'
Barr,F Elaine
Aql 19h45'40"11d36'
Barr,Frank & Rosa
Lyr 19h21'1"33d57'
Barr,Gary
Ori 5h57'44"15d58'
Barr,J Scott T
Lib 15h41'27"-24d1'
Barr,James
Boo 13h58'0"20d45'
Barr,James (Jim)
Her 17h15'0"44d42'
Barr,Jeremy A
Cnc 8h55'1"30d10'
Barr,John O'Banion
Oph 17h8'0"11d52'
Barr,Jr,John S
Lac 22h4'52"46d38'
Barr,Karren D
And 2h14'53"42d17'
Barr,L Rey
Cap 21h54'57"-21d35'
Barr,Lauren Laine
Tau 5h12'24"16d29'
Barr,Lisa & Donald
Peg 21h54'32"28d16'
Barr,Michael Thomas
Dra 15h45'12"61d54'
Barr,Patricia Annette Moody
Tri 1h52'54"27d38'
Barr,Ray
Psc 0h17'45"20d12'
Barr,Talia R
Aql 19h55'19"12d18'
Barr,William Stringfellow
Psc 23h1'1"0d44'
Barr-Fiora,Ian Skye
Boo 14h6'48"52d8'
Barr/The Barr Star,Lee & Pat
Uma 9h15'32"50d14'
Barra,Duane
Cam 11h41'0"78d0'
Barrack,Amy Joanna
Sco 17h54'21"-40d33'
Barrack,Samantha Albina Joy
And 23h13'38"39d44'
Barraclough,Paul Graham
Dra 17h49'1"73d12'
Barragan,Cliff
Aur 6h28'34"35d51'
Barral,Elizabeth
Cas 0h39'0"58d43'
Barranco,Brittany Ann
Cas 23h22'28"53d22'
Barranco,Jr,Big Al
Sct 18h51'51"-6d40'
Barranco,Taylor Joseph
Aur 5h0'54"31d36'
Barras,Emile Robert
Lyn 7h30'26"40d49'
Barras,Paul
Aur 4h51'25"40d47'
Barrass,Jack
Uma 12h1'34"33d11'
Barratt,David Austin
Uma 8h19'50"61d20'
Barraza,Eva
Cas 0h40'55"67d43'
Barraza,Maher Salal
Eri 3h33'13"-3d60'
Barre,Sophie
Peg 23h38'16"11d30'
Barreca,Molly Catherine
Leo 11h51'24"26d21'
Barreira,Paula
Boo 15h28'28"37d53'
Barreiro,Alberto Manuel Prieto
Ori 6h7'11"20d60'
Barreiros,Regina Soares
Mon 6h57'59"-8d47'
Barrell,Jeremy
Cep 22h19'17"70d24'
Barrenechea,Mitchell
Lmi 10h8'35"35d30'
Barrera,Jose-Javier Jimenez
Leo 9h20'41"8d10'
Barrera,Linda
Oph 17h20'26"12d57'
Barrera,Virginia Wilson
Mon 6h32'29"11d41'
Barres for Caitlin, Rick
Cnv 13h49'32"40d19'
Barrese,Louis J
Aur 5h23'45"40d35'
Barret,Christopher
Ori 5h27'14"0d42'
Barreto,Marielena
Cas 0h54'36"62d55'
Barrett III,William
Aur 5h2'1"42d57'
Barrett Married 1990, Carol &
Rob
Her 16h59'29"48d9'
Barrett,Anita
Eri 4h48'14"-9d60'
Barrett,Anna
Cet 2h13'29"1d12'
Barrett,Anne Scott
Peg 23h23'22"12d14'
Barrett,Ashley & Ryan
Mon 7h5'0"-6d54'
Barrett,Bruce Joseph
Gem 7h28'35"31d37'
Barrett,Carole
Lyr 18h16'54"42d16'
Barrett,Chase Remington
Vul 20h22'1"23d49'
Barrett,Chase Remington
Vul 19h5'28"25d26'
Barrett,Daniel J
Tau 5h18'52"16d29'
Barrett,Denise
Del 20h27'50"11d30'
Barrett,Elise
Lyn 8h5'44"38d12'
Barrett,Frank
Her 17h21'14"43d27'
Barrett,George "Mac"
Lmi 10h0'55"38d5'
Barrett,Gerald E
Mon 6h21'47"0d25'
Barrett,Jimmy
Ori 5h29'20"1d32'
Barrett,Kathryn L
Lyn 7h1'12"50d39'
Barrett,Lesli Dyan
And 1h35'19"37d47'
Barrett,Lettie Lansdowne
Ori 5h50'39"20d57'
Barrett,Lucy M
Cas 2h28'0"70d22'
Barrett,Matthew Robert
Cep 0h8'18"68d17'
Barrett,Merry Christmas 1994
Aql 20h7'27"3d48'
Barrett,Parker
Boo 15h2'21"17d22'
Barrett,Richard M
Aur 5h31'12"30d25'
Barrett,Robert F
Cam 6h9'1"70d55'
Barrett,Saint Wesley James
Cosmas
Aur 6h7'52"46d39'
Barrett,Scott Alexander
Boo 14h56'1"46d21'
Barrett,Shannon Leah
Cas 0h5'51"58d31'
Barrett,Susan
Cyg 19h30'46"38d30'
Barrett,Susan (Sammy)
Com 12h53'25"21d0'
Barrett,Tara Dawn
Lyr 18h58'54"28d8'
Barrett,Thelma Evelyn Drake
Cas 1h59'22"65d26'
Barrett,Thomas
Cep 22h21'29"55d60'
Barrett,Tom
Aur 5h9'57"41d15'
Barrett-Memorial,Kirk
Per 4h26'22"50d9'
Barrett-Roddenberry, Majel
Cnv 13h51'15"42d25'B
Barrett-Roddenberry, Majel
Peg 22h19'54"13d37'
Barrette,Walker Boone Hall
Oph 17h58'41"8d57'
Barrey,Kim
And 2h35'0"40d36'
Barriat,Ron L
Vul 19h59'51"23d23'
Barrickman,Carol Ann
Lyn 7h21'25"44d35'
Barrie
Ser 18h16'1"-14d22'
Barrie 9-22-54,Douglas Stuart
Lyn 7h55'38"45d59'
Barrie Andrew
Uma 11h23'0"56d17'
Barrie Lynn
Lyr 18h31'35"30d58'
Barrie's Star
Ari 2h27'17"22d18'
Barrie,James A
Dra 20h9'41"62d51'
Barrie,Jennifer Adams
Mon 7h2'35"-6d48'
Barriellis Star,Aurora
Cam 4h4'53"67d59'
Barrier,Jean-Luc
Dra 20h21'57"71d11'
Barrier,Luca
Boo 13h40'56"16d17'
Barrier,Tamara
Cep 21h39'22"60d38'
Barrineau,Kyle Taylor
Cet 2h45'55"3d52'
Barringer
Dra 11h20'36"78d54'
Barringer,Christine
Peg 22h21'37"27d33'
Barringer,Deborah A
And 23h39'0"47d46'
Barringer,Jody Robicon
Ori 6h2'23"-0d21'
Barringer,Katherine
Cet 0h38'17"-2d38'
Barringer,Russell Warren
Oph 18h0'15"13d40'
Barrington
Uma 8h37'17"51d36'
Barrington,Barry
Her 17h54'23"14d46'
Barrington,Douglas
Cep 21h58'39"55d40'
Barrington,Elaine
Lac 22h53'51"54d25'
Barrington,Olivia Amore
Mon 6h39'24"1d38'
Barrios,Alex C
Sct 18h54'1"-6d14'
Barrocas,Eliza Ariana
Mon 6h23'24"4d39'
Barrocas,Joshua Nolan
Hya 8h41'23"1d11'
Barron 4-19-1991,Troy R
Boo 14h20'60"36d42'
Barron 6-15-1995,Amber Lu
Cas 1h7'16"62d6'
Barron,Bruce & Mary
Aur 5h33'42"50d29'
Barron,Caroline Louise
Ori 5h51'44"5d28'
Barron,Cecilia
And 1h18'52"35d26'
Barron,David
Aur 4h37'18"30d57'
Barron,David
Cnv 13h33'45"48d49'
Barron,Edna F
Cas 23h41'46"62d25'

Barron,Gary Alan
Tau 4h6'19"23d9'
Barron,Jeanne Garrison
Mon 7h16'53"-5d18'
Barron,Mr Thomas W
Dra 16h8'29"65d53'
Barron,Nichole Lynn
Uma 9h35'50"61d16'
Barron,Rebecca J
Cyg 19h27'1"37d55'
Barron,Sandra
Cyg 20h35'1"45d29'
Barron,Ted
Cam 3h56'44"56d31'
Barros,Melissa A
And 2h18'13"42d45'
Barros,Paul
Vul 20h22'27"28d15'
Barrot,Mary
Col 6h1'11"-37d37'
Barrow,Henry Sergei Zibart
Cet 3h5'11"7d13'
Barrow,Hope Jessica
Lyr 18h34'0"28d56'
Barrow,Katherine Tatiana Zibart
Mon 6h55'31"8d36'
Barrow,Laselle Augustus
Cas 1h42'27"75d47'
Barrow,Noah Ress
Cep 22h47'27"58d57'
Barrow,Paul
Ser 15h11'0"-1d10'
Barrow,Terry
Dra 17h19'56"69d30'
Barrows,Hannah Elizabeth
Cyg 20h37'32"37d38'
Barrows,Helen & Hubert
Cet 2h37'0"-8d51'
Barrows,Suzanne
And 0h2'14"44d49'
Barrutia,Patricia D
Cam 6h13'16"56d1'
Barry
Per 4h4'55"50d51'
Barry
Peg 0h0'45"18d50'
Barry
Uma 12h4'58"31d51'
Barry
Her 16h36'59"29d4'
Barry
Boo 14h30'27"24d56'
Barry & Elizabeth
Uma 9h49'20"44d59'
Barry BSA,Susan Schlock
And 0h6'45"34d55'
Barry D Superstar
Lmi 10h41'31"25d5'
Barry Lee
Uma 8h57'29"60d57'
Barry With Love XX
Aur 7h8'60"38d48'
Barry,Alivia Lauren
Cyg 20h59'47"31d20'
Barry,Amy & Michael
Crb 16h8'31"27d40'
Barry,Barbara (Babs) A
Lyn 8h48'47"41d24'
Barry,Bill & Rose
Cyg 19h24'12"31d59'
Barry,Brad J
Boo 13h48'32"19d24'
Barry,Brian C
Aur 7h11'22"40d11'
Barry,Brian Lee
Lac 22h3'11"40d05'
Barry,Bruce E
Hya 9h5'36"5d32'
Barry,Christopher
Eri 3h32'2"-11d12'
Barry,Claude
Ori 5h56'58"20d42'
Barry,Del & Barbara
Crb 15h55'28"31d36'

Barry,Donna
Cnv 12h46'35"39d50'
Barry,Eddie
Cnv 12h26'16"40d47'
Barry,Edward Charles
Ser 15h43'45"-1d13'
Barry,George Darcy Rhett
Aur 4h34'47"34d37'
Barry,Gerald F
Per 2h46'51"40d2'
Barry,Gertrude
And 2h18'13"40d31'
Barry,J Geoffrey
Ori 6h4'59"-0d40'
Barry,James R
Aur 5h9'34"44d56'
Barry,Jessica Lynn
Cas 23h40'58"58d45'
Barry,Joe
Aur 5h44'35"29d38'A
Barry,John W
Mon 8h1'23"-3d29'
Barry,Joseph Francis
Per 2h40'48"40d6'
Barry,M Christopher
Lac 22h16'30"38d21'
Barry,Maralyn
Equ 21h7'38"10d18'
Barry,Margaret
Peg 22h58'38"30d22'
Barry,Marie
Aur 5h44'13"29d38'B
Barry,Mark & Tammy
Cyg 19h15'0"49d7'
Barry,O T
Cam 11h13'31"81d23'
Barry,Patte
Aql 18h59'50"3d9'
Barry,Renée
Del 20h14'0"10d33'
Barry,Rick
Ser 15h59'58"24d29'
Barry,Rita
Lyn 8h27'32"41d1'
Barry,Shannon Young
Peg 22h1'26"5d7'
Barry,Sidney Clarke
Leo 9h37'26"7d14'
Barry,Teresa
Cnc 8h34'26"30d36'
Barry,Teresa Luann
And 2h26'50"40d46'
Barry,Tracy Lynn Louise
Leo 11h46'12"20d23'
Barry-Lover Of Sunshine
Lac 22h45'53"52d46'
Barry-My Eternal Love
Aql 20h11'25"14d3'
Barrys,Albert Viola & Joan,The
Mon 8h2'17"-0d16'
Barsa,Joseph
Aur 6h6'52"50d30'
Barsamian,Larry
Peg 21h52'20"28d6'
Barsanti,Ronald P
Per 3h19'21"40d18'
Barsotti,David Giulio
Her 16h15'21"42d42'
Barst,Ruth Yael
Del 20h13'47"14d19'
Barstow,Richard Snunkly
Her 18h11'51"48d56'
Bart
Hya 8h59'13"4d9'
Bart
Sge 19h3'34"16d12'
Bart
Peg 22h24'1"31d45'
Bart B Dazzled
Aql 18h54'31"-0d23'
Bart Star,The
Per 2h59'12"38d41'
Bart's Star
Ori 4h57'55"4d46'

Bart,Heinz
Cep 21h15'29"63d46'
Bart,Melissa Ann
Cas 3h11'0"62d47'
Barta
Cyg 21h4'28"31d9'
Barta,Jr,Frank Michael
Boo 14h56'50"25d23'
Batalone,Annabelle
And 0h23'12"35d35'
Bartay I II III IV V, Louis Augustas
Sex 10h1'1"-11d5'
Bartay III Hall,Joan Margaret
Sco 17h53'1"-38d25'
Barteck,James Ward
Tau 5h19'0"18d57'
Bartek,Allan
Per 2h39'23"40d23'
Bartek,Sheila Hunter
Lyn 8h21'24"51d14'
Bartel,Claus Max
Her 17h28'11"43d7'
Bartels,Andrea
Leo 11h6'45"26d19'
Bartels,Audrey"Touch- The-Future"
Cnv 13h56'60"30d1'
Bartels,Brian D
Aur 6h3'57"45d56'
Bartels,Danielle Nicole
Tri 1h58'52"25d56'
Bartels,Juergen & Rachel
Aur 5h2'13"29d25'
Bartels,Michael J
Boo 14h59'15"20d56'
Bartels,Peggy
And 23h44'0"47d11'
Bartels,Phillip Conrad
Del 20h52'32"3d23'
Bartelstone,Erica Lynn
And 0h6'49"40d36'
Bartemio,Louis A
Ori 5h49'28"18d37'
Bartenbach,Constance Lynne
Ori 5h54'50"10d19'
Bartenbach,Remy Renée
Mon 7h20'54"-1d22'
Barter,David
Sex 9h48'56"0d47'
Barter,Davida D
Lac 22h16'1"46d36'
Barter,Jim
Lac 22h29'1"52d34'
Barter,Tyler
Aur 4h58'31"40d46'
Barth,Bodo
Cmi 7h17'48"8h12'
Barth,Charly
Cnc 8h27'56"18d48'
Barth,Heinz Erich Woldemar
Ari 1h58'1"25d3'
Barth,Jochen
Oph 17h0'18"1d37'
Barth,Leona
And 0h7'42"44d34'
Barth,Louis
Aur 5h12'0"41d1'
Barth,Michael
Lyn 7h43'57"48d15'
Barthe,Jacques et Nicole
Sgr 18h47'58"-32d44'
Barthe,Paul
Cam 5h4'33"67d42'
Barthel,Frederick K
Eri 4h25'45"-21d36'A
Barthel,Grandpa
Per 1h48'30"56d39'
Barthel,Ingeborg
Eri 4h50'9"-9d2'
Barthel,Nola J
Eri 4h25'45"-21d36'B
Barthel,Paul
Ser 15h15'11"-2d53'

Barthelette,N
Eri 2h53'11"-3d43'
Barthelmes,Bernd
Ser 15h12'57"10d19'
Barthelmes,Richard J
Cep 2h39'49"77d48'
Barthelmess,Ann
Tri 2h7'32"31d21'
Barthelt,Klaus
Ser 16h5'22"2d37'
Barthen,Sierra Lynden
Lyr 19h20'52"42d48'
Barthmaier,Jr,F Joseph
Tau 5h49'14"28d12'
Barthol,Jr,Robert C
Cep 21h46'20"55d6'
Bartholoma,Karen Alta Blue
Lyr 18h47'1"36d58'
Bartholomew's Brilliance
Sgr 11h37'51"-33d10'
Bartholomew,Ann
Peg 0h11'0"14d26'
Bartholomew,Ashley Marie
And 1h1'15"47d14'
Bartholomew,Dale
Aql 19h8'49"0d50'
Bartholomew,Douglas J
Cmi 7h27'54"5d21'B
Bartholomew,Jr,Gilda M
Cnv 13h46'13"31d25'
Bartholomew,Laura
Aur 5h29'49"38d1'
Bartholomew,Rebecca A
Cmi 7h5'0"3d60'
Bartholomew,Richard
Aql 18h39'1"0d7'
Bartie,Nelson
Aql 18h58'50"16d22'
Bartis,Stephen Peter
Aur 6h19'19"35d30'
Bartivic,Ruth Alexandria Sakal
Lac 22h40'49"56d32'
Bartizek,Ron & Charlotte
Eri 3h47'44"-4d8'
Bartkoski,Denise
Cyg 20h5'55"30d32'
Bartkowiak,Allison Marie
Cam 4h12'25"58d19'
Bartle,Henry
Cep 22h45'1"63d31'
Bartle,Jennifer Anne
Cas 1h47'31"75d16'
Bartlebaugh,Marc L
Aur 6h31'55"35d9'
Bartles Family,The
Eri 3h55'1"-2d0'
Bartleson,Julie
Equ 20h58'59"7d40'
Bartlett
Cas 0h16'15"60d7'
Bartlett Forever Chris,Jack Ruben
Lyn 9h11'0"37d47'
Bartlett Interlaken Ch 93,Denise
Aql 19h10'32"15d50'
Bartlett Love Mom, Heather Nicole
Lac 22h3'15"46d28'
Bartlett Love Mom, Samantha Michelle
Lac 22h7'24"46d49'
Bartlett Star,The
Lyr 18h55'22"30d36'
Bartlett,Amanda Elizabeth
Cas 1h44'35"73d9'
Bartlett,Bettie Ann
Lyr 18h40'37"27d33'
Bartlett,Christie Sue Mellinger
Aql 19h14'0"13d24'
Bartlett,Daniel
Uma 10h22'21"57d14'
Bartlett,Daniel Bay
Cmi 7h55'37"8d16'

Bartlett,Donna Lee
Eri 2h53'41"-12d23'
Bartlett,Evan Matthew
Her 16h16'0"47d58'
Bartlett,Gloria Joan & Martel Allen
And 23h45'35"32d57'
Bartlett,Jerrod
Dra 18h4'11"58d41'
Bartlett,Kimberly Michelle
And 2h30'11"49d40'
Bartlett,Melanie
Mon 7h58'25"-0d27'
Bartlett,Nancy & Andrew Keczkemethy
Cyg 21h35'47"34d58'
Bartlett,Nicholas Harvey
Cep 22h57'1"63d0'
Bartlett,Phyllis Christine
Boo 13h36'45"12d10'
Bartlett,Sussan K
Lib 15h39'55"-26d7'
Bartlett,Tanya
Del 20h13'49"10d20'
Bartlett-Sloan,Louisa Evelyn
Lyn 8h24'29"48d2'
Bartley,Jeffrey Chase
Lmi 10h56'25"26d60'
Bartley,Mae Ola
Peg 22h23'1"3d43'
Bartley,Paul
Her 16h12'22"8d28'
Bartley,Randall J
Cet 0h52'26"1d59'
Bartley,Rhiannon
Mon 6h18'33"1d32'
Bartling,Emalee Hailee
Peg 0h1'12"30d17'
Bartlow,Betty
Peg 22h0'17"28d8'
Barto,Amanda Leigh
Cam 6h0'11"68d2'
Barto,Cheryl Annette- Pray
Lyn 8h10'11"36d41'
Barto,David Bradley
Aql 19h18'1"15d57'
Bartofiewicz,Edward M
Her 16h58'17"31d37'
Bartol,Ashley Lynn Withrow
Peg 21h58'41"24d23'
Bartol,R
Lyr 18h37'24"42d53'
Bartoletti,Joseph John
Cam 4h15'55"67d31'
Bartoli,Jean-Louis
Com 12h33'12"20d14'
Bartoli,Nancy Joanne
Lyr 19h0'37"38d3'
Bartolic,"Sempiternus Amor",Nancy
Cyg 21h24'25"40d4'
Bartolomei,Maria D
Aql 19h57'41"-8d38'
Bartolomei-Lee,Annette Marie
And 2h35'20"40d14'
Bartolomeo
Vul 19h21'60"25d57'
Bartolomeo,Ralph S
Cep 0h46'31"80d8'
Bartolomeo,Vincent J
Per 2h54'45"32d51'
Bartolomé,Ana
And 23h48'41"41d58'
Bartolozzi,John M
Dra 16h32'57"62d14'
Barton
Cep 21h47'31"55d20'
Barton,Allen
Ori 5h56'55"15d44'
Barton,Bill
Aql 19h27'54"10d58'
Barton,Brittainy Nicole
Mon 7h29'35"-10d25'
Barton,David
Aur 7h9'31"38d37'

Barton,Elizabeth L
Lyr 19h4'1"40d27'
Barton,Heather
Com 13h33'51"19d38'
Barton,James
Oph 17h57'16"8d40'
Barton,Joan Irene Diell
Vul 19h22'53"26d37'
Barton,Joel
Boo 14h5'40"26d24'
Barton,Jr,Walter
Aql 20h10'37"7d47'
Barton,Keith
Cru 12h11'50"-61d37'
Barton,Linda M
Oph 16h34'34"0d21' A
Barton,Michael
Cma 6h54'35"-17d15'
Barton,Nicholas Andrew
Ori 5h29'10"-2d33'
Barton,Phyllis
Lyr 18h40'42"33d5'
Barton,Rebecca Fleur
And 0h2'51"47d30'
Barton,Ruth Louisa Feist
Mon 6h34'11"-0d52'
Barton,Sandra Sophie
Cyg 19h27'13"34d48'
Barton,Shawna
Peg 22h3'52"29d38'
Barton,W.Frank
Aur 5h23'37"38d50'
Bartone,Martha
And 2h28'28"50d7'
Bartos,Jeffrey M
Her 17h33'51"38d10'
Bartos,Myles Joseph
Aur 5h14'39"50d13'
Bartosik,Luke
Ori 6h3'23"8d40'
Bartram III,Robert
Aql 19h41'1"14d58'
Bartram,Brent Eric
Cet 1h46'1"-1d29'
Bartsch,Gwen
Peg 22h10'45"24d44'
Bartsch,Wolfgang
Psc 1h21'37"21d17'
Bartus,Anne
And 23h34'46"38d23'
Bartz,Angela Christine
Lyn 8h21'41"45d29'
Barucco,Daniele
Lac 22h8'44"38d14'
Baruch,Superstar We Love You,Elliot
Aur 5h18'26"41d22'
Baruffi,Fernando
Lac 22h20'46"50d11'
Barum,Sandy
Peg 22h58'39"18d31'
Barussell
Eri 4h4'59"-18d51'
Barve,Maddelina Monique Blixseth
Cas 0h59'29"62d35'
Barwa,Barbara Brandeau
Gem 6h49'11"19d38'
Barwell,Emma
Per 3h16'55"41d10'
Barwick,Kenneth Ray
Aql 18h58'60"-8d21'
Baryames,Paul G
Dra 10h33'58"74d15'
Baryliuk
Cet 2h50'48"0d33'
Barzach,Jonathan Peter
Cep 23h23'34"68d22'
Barzaghi,Gianni
Her 16h43'16"48d19'
Barzee,Madelyn Leigh
Cam 4h38'34"68d30'
Barzen,Andrea Lynn
Tau 4h13'16"20d58'

Barzilay,Doron
Her 15h49'50"40d56'
Barzydlo,George
Aur 7h0'12"41d43'
Bas Boy (Planet Bond)
Ori 5h43'29"10d26'
Basanese,Dominick Vincent
Tri 2h18'28"33d50'
Basara,Maria
Lmi 10h44'26"31d43'
Bascetta,Lana
Hya 9h16'40"3d24'
Bascombs,Mr Navaro
Lyr 19h6'0"40d51'
Baseman,Judith Marie
Lyn 7h31'15"45d6'
Basens,Dainis Voldemars
Boo 15h10'31"26d35'
Basevi,Manfred
Cap 21h4'46"-23d15'
Bash
Ori 6h6'18"1d28'
Basha T L B T P M D
Ori 5h28'34"-2d36'
Basha,Michelle & Michael
Lyr 18h34'25"37d19'
Basha,Patti
Cas 1h46'35"60d5'
Basham's Star,Tracy Leigh
Ori 6h14'53"7d53'
Basham,Dylan
Equ 21h22'53"8d45'
Basham,Margaret Quinn
Mon 6h59'0"0d44'
Bashert-Jim & Jeanne
Lyn 8h2'1"51d28'
Bashford,Claire
Umi 13h50'30"72d18'
Bashor,Nicole Andreá
And 23h19'39"47d6'
Bashore,Susan J
And 0h10'48"35d20'
Bashqoy,Linda
Ori 5h51'48"21d13'
Bashqoy,Mike
Ori 5h48'40"21d29'
Bashqoy,Nart
Ori 5h48'32"20d59'
Basich,Zoran
Cet 3h12'53"2d2'
Basil
Uma 10h16'19"59d30'
Basil,Collin Shea
Per 2h6'15"58d48'
Basil-Daddy All The Chocolate
Uma 8h19'44"71d20'
Basile
Peg 0h0'30"31d18'
Basile II,Richard
Her 16h14'11"48d10'
Basile,Andrea
Cap 21h28'25"-22d57'
Basile,Christine Kinkead
Com 12h31'35"20d15'
Basile,George
Her 16h16'12"22d57'
Basile,Marie Theresa
Lyn 7h45'14"41d7'
Basile,Michael Robert
Lyn 7h27'15"35d30'
Basile,Nora Ellen
Lyn 8h3'1"35d35'
Basile,Rocco F
Per 1h56'44"52d53'
Basile,Ronald Francis
Lac 22h2'20"46d10'
Basile,Vincent J
Aql 20h5'47"1d15'
Basilicata,Nicole Marie
Cas 1h12'29"60d40'
Basilico,Antonella
Pho 0h5'35"-47d34'
Basiliere's Starship, J T
Cnv 12h15'38"35d50'

Basinger,Michael Andrew
Her 17h20'10"44d36'
Baskaya,Elif Esra
Eri 3h10'0"-6d31'
Basken,Daniel James
Her 17h0'39"20d26'
Baskett,Carrie A
Cas 0h35'18"71d11'
Baskin,Dr David S
Oph 17h31'29"-22d17'
Baskin,Peter Jay
Lmi 11h1'28"28d52'
Basko,Stephen Michael
Aql 19h35'52"-5d56'
Basler,Hannah Lynn
Cyg 21h11'52"38d39'
Basly,Anne-France
Dra 16h19'17"62d36'
Basma,Maroini
Dra 20h38'58"80d36'
Bason,Charles Rogerson
Ori 5h52'1"7d42'
Basore,Sarah
Cet 2h29'57"4d8'
Basralian,Kevin Richard
Aur 6h51'57"37d59'
Bass Lease,The
Her 18h8'25"45d7'
Bass,12-6-74,Virginia Blakely
Aql 20h6'40"-6d57'
Bass,Charles
Cet 2h4'0"0d16'
Bass,Cyrus
Aur 5h56'11"38d17'
Bass,Dean & Jean
Oph 16h28'35"-7d48'AB
Bass,Donna & Michael
Dra 18h19'14"70d20'
Bass,Donna & Paul Dainty
Cra 18h24'27"-41d59'
Bass,Henry Bascom
Ser 15h37'40"5d10'
Bass,James Matthew
Aur 7h20'17"36d58'
Bass,Janice Nicole
Cmi 7h29'20"7d52'
Bass,Jordan Ashley
And 23h2'23"50d11'
Bass,Karen
And 23h44'0"45d59'
Bass,Max
Tau 5h46'0"23d48'
Bass,Melissa Ann
Cyg 19h46'15"29d29'
Bass,Patricia Ann
Com 12h13'52"26d31'
Bass,Phyllis Mae
Lyr 18h36'1"34d21'
Bass,Robert A
Cet 0h25'28"-2d53'
Bass,Steven Wayne
Sct 18h49'5"-6d15'
Bass,Terry Lee
And 23h17'14"43d25'
Bass,Thomas Andrew
Cnv 13h3'37"45d32'B
Bass,Zachary James
Sct 18h56'40"-6d28'
Bass-1993,Justin
Aql 19h35'24"-0d0'
Bassam
Uma 10h30'32"70d33'
Bassel's Shining Star
Lyr 18h48'43"41d38'
Bassen,Günther
Peg 23h34'42"16d23'
Bassermann,Julien
Per 4h45'18"51d11'
Basset,J Fred
Her 17h35'43"40d28'
Basset,Pierre
Lyr 19h18'18"40d43'
Bassett,Amy Elizabeth
Lyn 9h10'0"46d6'

Bassett,Belinda Kaye
Vul 19h38'42"20d8'
Bassett,Bruce R
Lyn 7h36'44"42d20'
Bassett,Harold F
Equ 21h23'27"2d46'
Bassett,James
Aur 6h2'51"46d20'
Bassett,Kennedy Taber
Umi 15h15'1"69d12'
Bassett,Leon the Hound
Uma 10h2'35"59d47'
Bassett,M Hunter
Umi 17h8'27"75d43'
Bassett,Martin Stanley
Cep 23h10'23"61d0'
Bassett,Sheena Jean
Cam 6h6'0"60d58'
Bassett,Stephen
Sge 20h6'1"16d15'
Bassette,Holmes & Brigitte
Aqr 19h27'50"8d23'
Bassi,Jonathan
Aur 6h17'32"38d14'
Bassi,Mrs Rita
Cas 0h0'12"63d12'
Bassi,Robert D
Peg 23h4'41"18d5'
Bassin,David & Judy
Lyr 18h31'43"42d41'
Bassinder,Madeline Paige
Com 13h18'20"28d21'
Bassinger,Leigh & Robbie
Ari 1h46'50"17d19'
Bassista,Kurt
Cmi 7h29'52"1d16'
Bassler,Katie
And 23h43'27"40d42'
Bassmaster Gary
Ser 15h24'0"9d35'
Bassner,Tony Craig
Her 18h18'55"12d32'
Bassuk,Jeffrey
Cet 2h25'0"5d45'
Bassuk,Julie R
Peg 22h59'14"27d13'
Bassy
Boo 13h45'0"11d37'
Bast,Ginger Elizabeth Katherine
Leo 11h55'0"22d33'
Bast,Nancy Jeanne
Lyn 7h56'32"42d29'
Bastecki,Savannah
And 0h3'49"40d50'
Basten,Jeremy John
Lyr 18h30'17"41d5'
Baster,Lura Swig
Lyn 7h55'26"38d26'
Bastia,Dario
Aur 7h6'44"40d16'
Bastian,Robert
Aur 5h4'59"42d47'
Bastian,Senior,John Howard
Boo 14h28'42"37d48'
Bastian-Kocic, Cristeana & Milan
Cyg 21h2'11"31d29'
Bastianoni,Fabrizio
Pup 7h56'30"-27d44'
Bastie,The
Per 2h52'46"35d37'
Bastien,Girard
Oph 18h16'47"11d38'
Bastien,Haley Rose
Cas 0h35'41"70d53'
Bastien,Jean-Pierre
Umi 15h3'19"79d18'
Bastien,Remi
Cam 3h54'59"52d48'
Bastien,Richard
Her 17h17'37"49d12'

Bastien,Tailfer
Uma 11h24'50"34d5'
Bastio,Heather Leilani
Hya 8h13'47"1d6'
Bastoky,Bruce
Gem 7h1'19"28d31'
Bastos,Maria Helena Monteiro Alves
Cyg 20h3'54"39d43'
Bastow,Pauline
Cas 0h5'28"62d34'
Basye,Genevieve Frances
Uma 11h38'34"30d4'
Baszczewski,Cathy
Cas 0h40'1"69d30'
Bataille,Paule
Boo 15h4'49"7d35'
Batchelder,Jr,Robert Lewis
Cep 21h56'29"58d50'
Batchelder,Stacia
Lyr 18h48'54"31d31'
Batcheler II,George E
Cet 2h44'1"3d55'
Batchelor,Carole
Cyg 21h42'59"38d58'
Batchelor,Diane
Lac 22h4'14"50d55'
Batchelor,Elisabeth
Cyg 19h30'52"35d48'
Batchelor,Haley Catherine
Com 12h35'33"30d40'
Batchelor,J Patrick
Sex 9h59'23"-5d25'
Batchelor,Lois V
Oph 17h19'22"-18d54'
Batchelor,Margie Inez
Peg 22h35'12"26d39'
Batchelor,Peter
Cra 18h11'0"-39d36'
Batchelor,Rom
Eri 2h51'37"-5d40'
Batcher,Dorothy Immoor
And 2h20'28"40d30'
Batchu
Aql 20h10'21"0d32'
Bate,Jane Samantha
Mon 6h18'51"8d57'
Batek-Hyde,Nancy
And 0h0'13"47d3'
Bateman,Christine
And 23h30'54"48d23'
Bateman,Cindy & Sean
Crb 16h7'19"31d7'
Bateman,Jerry & Sherry
Cep 5h0'48"80d31'
Bateman,Jr,Daniel Jon
Leo 11h50'56"22d12'
Bateman,Kate
Peg 21h9'16"18d56'
Bateman,Kristi Louise
Peg 23h0'46"11d21'
Bateman,Michael & Laurie
Cyg 21h56'33"50d14'
Bateman,Mitch
Hya 9h37'42"1d47'
Bateman,Patricia
Cam 7h34'39"61d15'
Bateman,Richard Elmer
Per 4h24'49"50d11'
Bateman,Sr,Richard M
Aql 20h1'18"10d54'
Bateman,Stephen
Leo 11h31'45"-0d54'
Bater,Julie Ann
Lyr 18h16'12"30d57'
Baterdouk,Beibers
Oph 17h51'52"13d54'
Bates,Barry
Dra 16h9'38"63d43'
Bates,Christine Susan
Cyg 20h23'12"38d40'
Bates,Darrin Lee
Uma 11h52'41"40d11'
Bates,Dawn Marie Roberts
Peg 23h45'44"8d54'
Bates,Eddie
Sex 10h20'56"1d54'
Bates,Elizabeth Carol
And 23h41'21"42d50'
Bates,Father John
Aql 20h5'51"7d39'
Bates,Faye Anne
Crb 16h12'1"34d21'
Bates,Gary
Ori 5h32'0"-1d37'
Bates,James Anderson
Oph 17h27'53"-0d20'
Bates,Jennifer
Cas 0h37'19"75d36'
Bates,Kayla Sue
Cas 23h36'57"61d46'
Bates,Little Eli
And 0h14'53"35d6'
Bates,Matt & Jennifer
Peg 23h1'11"18d26'
Bates,Mimi
Cyg 20h3'58"41d11'
Bates,Minnie
Cas 1h6'14"55d27'
Bates,Mr Shon
Cnv 12h56'46"40d14'
Bates,Najée
Ori 6h0'46"4d56'
Bates,Patrick Beryl (Bear)
Aur 6h27'51"35d42'
Bates,Sandra C
Peg 23h4'24"32d3'
Bates,Sierra Morgan
Mon 6h43'51"7d39'
Bates-My Love & My Life,Laura
Mon 6h19'45"2d34'
Bateson,Colin Richard
Her 16h50'13"41d14'
Batey,B J
Aql 19h58'12"10d21'
Batey,Jeffrey Robert
Dra 16h35'33"64d32'
Batey,Kim
Lyn 6h24'57"60d7'
Batey-August 22nd 1995,Betty
Cas 0h7'1"64d45'
Batezat,M
Cam 5h1'0"61d26'
Bath's Star,Allen Tristram
Eri 4h0'33"-12d9'
Bath,Gordon Daniel
Cep 22h13'52"60d57'
Bath,Julie
Sge 20h0'46"16d22'
Bathgate,Catriona
Mon 6h26'58"3d43'
Bathrick,Paul & Carol
Lyn 8h58'26"33d51'
Batik,Timothy
Dra 19h50'36"60d15'
Batini Cerys
Lmi 9h42'24"34d41'
Batiste,Bambi
Sgr 19h6'28'-25d34'
Batiste,G J
Leo 9h57'38"22d4'
Batiuchok,MD,John R
Ser 18h14'29"-12d4'
Batkoski 1995,J & J
Gem 6h43'60"31d3'
Batley,Maurice
Per 3h13'31"50d20'
Batlhoki,Shathiso
Cep 21h20'53"55d37'
Batman
Sct 18h54'37"-6d46'
Batman,James Robert
Cep 20h55'0"61d25'
Batner,Collin William
Aql 19h30'47"7d50'
Batson,Jodie Marie & Joseph Militello
Uma 9h50'4"65d1'AB
Battaglia,Anthony
Ori 6h0'24"8d53'
Battaglia,Cinzia
Cap 21h22'18"-23d30'
Battaglia,Michael F
Ori 5h0'51"15d13'
Battaglino,Ria
And 0h20'1"30d19'
Batteiger,Eric Dean
Dra 16h19'13"63d36'
Batten,Les & Jean
Cyg 21h56'45"50d19'
Batten,Tracie & Rikelle
Eri 3h17'1"-7d30'
Battenbough,Karen Louise
Lyn 8h20'53"48d58'
Battermann
Lac 22h26'39"54d50'
Batters,Sam
Sex 10h11'50"-5d22'
Batters,Yvonne Nicholson
Ori 5h13'45"-1d8'
Battersby,Dion
Lyr 18h36'55"27d10'
Batterton,Kimberly
Cap 21h42'51"-23d55'
Battesima,Innamoranda Nominata
Pup 7h57'51"-21d11'
Battezzata,Innamorata Amata
Pyx 8h27'6"-28d29'
Battiato,Manuela Alessandra
Pic 4h42'13"-49d59'
Battin,Allison
And 2h13'34"40d9'
Battista II,MD,Louis C
Her 17h9'54"20d48'
Battista,Christa (Munch)
And 0h5'13"47d10'
Battista,Mark & Ella
Uma 9h52'53"48d32'
Battista,Paola
Cyg 21h18'13"30d31'
Battistoni,Mark
Boo 14h13'43"46d3'
Battit,Alan Caldwell
Nor 16h27'43"-44d49'
Battle,David L
Boo 13h55'0"19d47'
Battle,James P
Her 17h20'59"47d9'
Battle,Lorenzo
Aur 6h26'0"32d5'
Battles,Lonzo & Barbara
Eri 4h12'28"-12d34'
Battson,Richard Anthony
Hya 8h19'27"3d39'
Batté,Anne M
Pup 7h56'45"-29d10'
Batuffolo Pina Macrç
Col 6h29'26"-32d14'
Batut,Astrid
Aur 5h51'57"30d41'
Batut-Giraud,Nicole
Ser 15h59'59"10d28'
Baty,Shirley
Com 12h10'55"19d17'
Baty-Tau Zeta Advisor, David
Aql 19h6'45"-0d58'
Batz Family Star
Aql 19h54'51"0d6'
Batzdorf,MD,Ulrich
Aql 18h52'50"-0d33'
Batzes,Florence & Sidney
Cam 3h58'42"57d15'
Batzloff,Kelly Christine
And 2h20'29"49d15'
Batzloff,Renee Michelle
Lyr 18h44'59"30d53'
Batç
For 2h31'1"-28d31'
Baubaro,Emmanuelle
Dra 17h38'1"73d31'
Bauch,Benita
Cas 0h27'31"61d46'
Bauchley,Louis
Her 16h40'12"5d21'
Baucom Beloved Soulmate,Robert M
Tri 2h24'59"28d17'
Baucom,Lane
Lmi 10h2'1"30d21'
Baucom,Melinda
Peg 23h21'1"32d43'
Baucum,Rosie Lee
Cyg 21h56'45"50d19'
Baucum,Sr,Earle W
Cyg 21h33'39"50d34'
Bauder,Dolores Ann
And 2h28'58"45d16'
Bauder,Peggy
Ori 5h57'27"1d16'
Bauder,Tyler James
Dra 18h26'1"70d41'
Baudet,Jean Jacques
Aur 5h2'39"31d2'
Baudier,Valerie
Lac 22h18'48"54d35'
Baudion,Christine
Cnv 12h14'14"46d13'
Baudissard,Debora
Pic 5h2'30"-47d39'
Baudouin Le Petit Prince
Uma 10h33'14"60d3'
Bauer Family Star,The Robert E
Cyg 21h19'44"34d28'
Bauer III,George Philip
Lib 14h58'39"-8d44'
Bauer Star,The Lori M
Cet 0h56'18d-1d19'
Bauer,"Buffalo Bill"
Aur 6h17'30"33d1'
Bauer,Alfred
Lyn 8h18'51"47d23'
Bauer,Anthony
Aur 6h30'22"38d12'
Bauer,Carole Maureen
Eri 4h27'45"-12d8'
Bauer,Christian Edward
Leo 10h27'0"11d1'
Bauer,Christine Ann
Mon 6h15'14"-5d47'
Bauer,Dagmar & Gerd 25-07-95
Uma 11h26'28"41d46'
Bauer,Dalene
Mon 7h4'19"-1d2'
Bauer,Ed
Dra 20h29'20"70d48'
Bauer,Eric Christopher
Aur 4h50'26"40d57'
Bauer,Ernest
Aur 6h3'50"30d10'
Bauer,Gabi
Vir 11h50'37"2d15'
Bauer,Gretchen Gerth
Cyg 19h26'24"32d46'
Bauer,J & D
Her 16h24'49"33d15'
Bauer,J R
Hya 9h54'9"-19d30'
Bauer,Janet
And 2h21'34"45d22'
Bauer,Jason Brett
Cnv 12h15'1"51d52'
Bauer,Jeffrey Alan
Boo 14h57'51"39d25'
Bauer,Jennifer
Aql 20h2'59"7d18'
Bauer,Jennifer R
Cas 0h13'48"59d45'
Bauer,Joe
Per 3h9'0"40d23'
Bauer,Josef
Peg 0h3'41"18d52'
Bauer,Joseph C
Aur 6h34'11"38d48'
Bauer,Joshua Lee
Aql 18h42'52"-2d4'
Bauer,Joshua Michael
Dra 15h49'1"66d11'
Bauer,Jr,John Trexler
Del 20h36'26"10d32'
Bauer,Julie
Lyn 7h49'12"37d7'
Bauer,Kathryn Rose
Vul 20h39'44"23d16'
Bauer,Mr & Mrs Ian
Cyg 21h35'12"38d39'
Bauer,Nicholas
Her 18h0'57"48d16'
Bauer,Nicholas Edmund
Ori 5h49'14"12d23'
Bauer,Peter
Eri 4h49'59"-9d0'
Bauer,Richard(Ridge) & Kira
Cma 6h54'55"-18d45'
Bauer,Rita Marie
Cam 4h3'37"67d36'
Bauer,Robert Clayton
Her 16h13'12"21d50'
Bauer,Sieglinde
Sco 17h1'0"-31d41'
Bauer,Sylvia
Hya 8h57'48"1d47'
Bauer,Thomas
Her 18h1'0"47d46'
Bauer,William John
Aql 20h7'49"1d18'
Bauer-Goulden,Adam Benjamin
Per 4h4'19"37d49'
Bauerlein,Peter John
Dra 19h56'38"84d47'
Bauernhuber,Raymond
Cep 2h35'50"77d11'
Bauers Northern Light
Ser 18h54'34"2d51'
Bauers,Floyd Michael
Aur 5h5'49"38d56'
Baugh,Rachel May
Vul 20h33'59"25d42'B
Baugh,Rebecca Ann
Vul 20h33'59"25d42'A
Baugher,Michelle Joy
Cyg 20h22'14"38d36'
Baughman,David Blair
Aur 5h5'54"37d58'
Baughman,George William
Uma 11h11'33"30d9'
Baughman,Helen K
Cam 13h20'25"81d27'
Baughman,James W
Per 3h5'1"41d50'
Baughman,John W
Dra 18h57'21"50d1'
Baughman,Lauren
Aql 19h59'42"11d43'
Baughman,Mr & Mrs James
Vul 20h1'19"28d33'
Baughman,Patrick Charles
Psc 0h53'36"08d44'
Baughman,Ronald J
Her 18h3'57"31d48'
Baughn,Donna Jo
Cyg 20h29'60"40d12'
Baugus,Pamela Dawn
Mon 7h15'41"-5d57'
Bauhaus
Ori 5h56'54"10d33'
Bauhofer,Scott & Mary
Uma 11h57'40"31d0'
Baum Star,Stevie- Marlene-Remy & Dylan
Cyg 19h23'1"47d48'
Baum,Elaine
Cas 2h1'23"68d29'
Baum,Irmgard
Cas 0h26'54"61d4'
Baum,Jill Amy
Ori 6h6'13"8d48'
Baum,Martin
Eri 4h4'19"-10d26'
Baum,Matthew Richard
Peg 23h1'47"13d5'
Baum,Melissa
Ori 4h55'1"1d32'
Baum,Rebecca
And 23h40'26"47d6'
Baum,Sara
Sex 10h38'0"1d30'
Baum,Tanya
And 23h32'17"41d36'
Baum,William
Aur 5h1'0"45d5'
Bauman,Joseph C
Per 2h20'36"54d42'
Bauman,Larry
Psc 1h0'40"31d25'
Bauman,Sherry
Com 13h4'0"21d2'
Bauman,Sophie Agnes
Vir 13h27'36"-9d0'
Bauman,Stewart
Sgr 19h40'20"-39d1'
Bauman,Tom & Elaine
Cma 6h50'12"-18d30'
Baumann,Don
Ori 5h37'1"0d30'
Baumann,Donald A
Oph 18h41'56"8d49'
Baumann,Dora-Maria (Mimi)
Vir 12h57'1"-10d46'
Baumann,Harry
Cma 6h11'59"-16d33'
Baumann,Karl John
Lac 22h25'17"52d36'
Baumann,Philippe
Cam 4h59'59"61d10'
Baumann,Robert
Boo 14h22'0"19d25'
Baume,Gloria
Cam 5h49'29"73d22'
Baumeister,Brandi
Sge 20h4'56"20d10'
Baumert,Lisa Lynn
Peg 22h55'11"22d29'
Baumgaertel,Herbert
Cyg 20h26'16"39d25'
Baumgardner,Harold & Naomi
Mon 7h25'46"-0d31'
Baumgardner,Leroy W
Uma 11h45'9"59d23'
Baumgardner,Tyler
Cap 20h43'12"-26d31'
Baumgart-Star of Joy, William Joseph
Crb 16h23'13"37d47'
Baumgarten,Christopher
Boo 14h21'51"14d29'
Baumgartner,CPA,Thomas
Her 16h39'1"21d42'
Baumgartner,Gerhard
Lyr 19h4'28"28d22'
Baumgartner,Heidi
Aur 5h7'30"29d35'
Baumgartner,John Jacob
Boo 15h6'18"12d15'
Baumgartner,Jr,John W
Aur 6h30'16"37d5'
Baumgartner,Steve
Aur 4h51'23"38d56'
Baumhauer,Christa
Cas 0h17'40"60d39'
Baumhower,Bob
Aql 20h18'58"8d38'
Baumler,Erik Richard Roger
Aur 5h2'1"48d42'
Baums,Georg
Sco 17h29'1" 40d12'
Baumstark,Jonathan Patrick
Her 16h59'0"32d35'
Baumwell,Kate
Cas 0h22'1"61d16'
Baumzweiger,Benjamin
Aql 19h52'38"12d5'
Baumzweiger,Sheyna
Lac 22h19'39"46d39'
Baur,Bro
Cet 2h56'35"6d52'
Baus,Jim
Tri 2h5'0"31d51'
Bausano,Jeri
Peg 22h13'16"31d39'
Bausch & Lomb
Dra 17h47'45"58d14'
Bausch,Kasey Clare
Cnc 8h53'44"20d39'
Bauschke,Christina
Lac 22h43'36"38d14'
Bausher,Shellie
And 2h29'11"39d1'
Bauske,Lori S
Cyg 21h20'44"40d41'
Bautch,Jason Eugene
Cyg 20h23'0"30d46'
Bautista,Sonny & Jannie May
Aur 4h55'15"38d58'
Bavaria An Singa,Erwin
Cas 0h16'10"60d60'
Bavaro,Thomas
Ori 5h55'44"16d37'
Bavassano,Luca
Hor 3h10'45"-46d40'
Bavassano,Sara
Lyr 18h38'33"27d35'
Bavely,Jack Jason
Cyg 21h11'17"36d2'
Bavely,Julia L
And 0h23'0"43d27'
Bavely,Michael Joseph
Cyg 21h18'58"34d16'
Bavely,Sandy Franklin
Cyg 21h19'27"39d27'
Bavely,Shirley May Bartrug
Cyg 21h16'18"38d23'
Bavely,Steven Lee
Cyg 21h17'56"37d41'
Baver,Harris Rose
Cam 7h56'21"60d16'
Baverstock,Bill
Cnv 13h17'13"32d19'
Baverstock,Rhea
Peg 22h11'24"5d52'
Bavolar,Louis M
Her 18h0'41"38d47'
Bavolar,Paul M
Aur 4h54'22"37d46'
Bavonese,Lucy Grace
Cas 0h33'13"63d10'
BaVosa,Brianne
And 0h18'19"34d27'
BaVosa,Brielle
And 0h21'44"37d46'
Bawden,Rosa
Equ 21h10'16"11d6'
Bawn,Arran
Cam 4h48'19"68d4'
Bax
Boo 15h5'26"26d20'
Baxendale,Craig
Per 2h52'52"40d54'
Baxler,Bankslee
Ori 5h56'32"15d1'
Baxmeier
Cmi 7h17'25"4d41'
Baxt,MD,Sherwood A
Her 17h55'0"40d14'
Baxter (W L),Lucy
Cyg 21h23'53"28d44'
Baxter III,James F
Cam 6h2'42"60d35'
Baxter,(Our Best Friend),Tom
Peg 3h49'38"8d22'
Baxter,Austin Thaddeus
Dra 10h20'23"73d55'
Baxter,Carol
Cam 4h10'0"60d48'
Baxter,Charles Carroll
Aur 7h13'35"35d40'
Baxter,Christine Pamela
Lyr 18h26'18"45d39'
Baxter,David
Umi 14h7'12"68d34'
Baxter,David Leonard
Per 3h1'21"38d21'
Baxter,Graham
Per 2h51'24"40d8'
Baxter,John A
Ori 5h25'10"1d16'
Baxter,John Clifford
Sct 18h38'15"-6d3'
Baxter,Judith Ann
Cnv 13h54'18"45d20'
Baxter,Judith Anne
Peg 0h6'1"13d13'
Baxter,Katie Ann
Cam 3h28'44"58d37'
Baxter,Kim
Mon 6h26'18"5d5'
Baxter,Kirk Allan
Aql 19h55'37"10d8'
Baxter,Kristina Kandi Marie
Del 20h53'44"3d58'
Baxter,Leslie Dawn
Cma 6h54'50"-15d50'
Baxter,Nicola & Paul
Cyg 20h9'24"40d58'
Baxter,Thomas C
Ser 15h54'24"12d37'A
Bay Area Showcase
Peg 23h6'16"8d58'
Bay,Christine
Cas 0h8'12"54d37'
Bay,Lukas Jerry
Uma 8h39'0"56d39'
Bayack,Jr,Ted & Elizabeth Ramos
Cep 0h28'44"86d53'
Bayan
Uma 10h14'13"56d50'
Bayer,Albert
Mon 7h54'3"-1d2'
Bayer,Bill
Per 2h46'25"36d35'
Bayer,Gustav
Boo 14h17'53"17d35'
Bayer,Joerg
Eri 4h49'51"-7d25'
Bayer,John
Boo 14h57'45"42d31'
Bayer,Michael & Suzanne
Ori 5h57'17"11d13'
Bayer,Robert & Stephanie
Crb 16h20'47"38d30'
Bayer,Stephen Robert
Ori 6h5'43"10d26'
Bayet,Monsieur Dominique
Her 16h15'13"48d30'
Bayle,Christine
Cam 3h24'59"66d9'
Bayle,Cyrtaque
Pic 4h45'4"-48d22'
Bayler,Amy Anne
Cyg 20h9'14"41d8'
Bayler,Margaret Mommy Nanny
Cyg 21h48'53"37d37'
Bayles,GF
Ori 5h52'18"17d25'
Baylocc
Sex 9h52'50"4d2'
Bayley,Cyril
Uma 10h41'26"40d31'
Bayliss"Kene",Kenneth Bruce
Aur 5h7'42"38d25'
Bayliss,Laura
And 0h20'23"30d48'
Bayliss,Roy Madison
Mon 8h1'31"-9d6'
Baylor,Kathryn Patricia Sullivan
Mon 6h43'27"3d5'
Bayly,Debbie
Mon 6h53'41"-0d35'
Bayne Amalgamated, M G
Dra 19h24'50"56d58'
Bayne,Bethany
Cyg 19h45'51"30d40'
Bayne,David Bennett
Cnc 9h12'16"30d30'
Bayne,Stephanie
Aur 6h50'33"38d58'
Baynes,Curtis D
Aur 5h48'46"38d32'A
Baynes,Jr "Will", William Lee
Cet 0h26'44"-3d36'
Baynes,RN SOF,Deborah Lynn Marie Kubinski
Peg 21h28'55"27d52'
Baynham,Michael
Cam 3h14'1"63d16'
Baynton,Barr
Aql 20h31'1"0d23'
Baynton,Bryan
Ser 16h0'12"7d56'
Baynton,Kristy
Mon 6h26'29"-1d50'
Baynton,Sally
Hya 8h44'38"4d57'
Baynum,Jr,Lynn C
Cas 0h2'20"62d34'
Bayoff,Colleen P
And 2h22'12"39d37'
Bayoff,Elizabeth C
Cas 0h39'1"67d31'
Bayoff,Samuel T
Lac 22h24'17"53d6'
Bayoff,Thomas S
Her 17h8'15"45d54'
Bayou
Aql 20h17'29"5d35'
Bayrakdar,Rula
Lyn 8h24'1"48d56'
Bayram,MD,Mehmet Oktay
Cep 20h57'1"61d49'
Bays Stephanie
Mon 7h24'22"-10d18'
Bays,Jr,Chief Raymond
Per 4h4'21"37d1'
Bays,Kenneth L
Per 3h6'33"38d53'
Baysinger,Shirley J
Aqr 21h53'15"-5d49'
Baysinger,Taylor Lauren
Mon 7h8'59"-2d2'
Bayutti
Uma 9h31'22"48d23'
Bayzick,Margaret
Cas 1h6'39"63d36'
Bazala Family Star,The
Uma 9h42'34"56d13'
Bazan,Trilina Star
Peg 0h8'33"14d34'
Bazar,Deborah
Lyn 8h6'1"37d49'
Bazarian,Dorothy
Peg 22h8'1"3d48'
Bazarov,Jacob Raphaee
Psc 23h28'37"5d46'
Bazarsky,Arlene
Peg 22h34'18"21d22'
Baze,Betty P.
Aql 19h2'20"-6d40'
Baze,Canyon Anthony
Lyn 7h0'49"52d1'
Bazeghi,Abbass
Gem 7h1'0"10d3'
Bazile,Dr Emilio
Cep 22h42'47"63d32'
Bazin,Kay Eleanor
Crb 15h54'1"28d55'
Bazinet,Henri
Cep 23h35'46"64d30'
Bazis * Sizab
Eri 4h6'13"-18d52'
Bazyli,Peter
And 2h23'46"45d40'
Bazz,The
Vul 20h17'43"23d28'
Bazzell,Carlton & Frances
Crb 15h50'1"30d54'

Bazzi,Abdelrahman
Aur 6h9'44"33d16'
Bazzini,Cheryl E
Aql 18h58'20"4d7'
Bazzinotti,Luigi
Boo 14h33'0"8d7'
Ba Boo 14h33'0"8d7'
Peg 22h16'33"4d50'
BB
Cyg 21h30'21"36d47'
BCA
Cet 1h45'43"-3d2'
Bce Da
Ori 5h55'41"15d41'
Be Fri
Peg 21h56'19"2d19'
Bea
Aur 5h5'1"37d41'
Bea & Ann
Her 16h40'1"32d25'
Bea & Ben 50th
Crb 15h45'12"29d36'
Bea's Star
Vul 21h17'1"28d14'
Bea,Mary Lynn
And 23h20'14"50d30'
Beaber,Wade Anthony "Bocci"
Aql 19h29'22"-0d2'
Beach
Aur 4h56'31"50d15'
Beach Bum,The
Cnv 12h53'0"40d9'
Beach,Annette
Cas 0h57'1"58d29'
Beach,Gary
Lyn 7h4'29"51d47'
Beach,Ironee
Sge 19h53'51"18d52'
Beach,Kimberly Mariah Toombs
Sgr 19h27'30"-44d43'
Beach,Mona Jones & Ron
Ori 5h42'53"11d56'
Beach,Robert E
Sex 10h21'56"-6d37'
Beach,Simon
Cep 0h8'25"73d51'
Beach,W Denny
Peg 22h7'1"20d18'
Beach,W Joann
Com 13h21'28"26d48'
Beachy,Kirsten
Cas 0h47'1"75d35'
Beacom,Jimmy
Lac 22h33'21"38d47'
Beacon (New Stimsonia) Grace
Lyn 7h54'58"38d17'
Beacon Melissa,The
Cas 0h22'55"61d50'
Beacon of Hope Hospice
Lyn 7h15'50"59d21'
Beacon,Palmer's Blue Jay
Aql 18h57'32"-7d45'
Beadle,Aurora Emily
Cas 0h3'47"61d53'
Beadle,Gary
Sge 20h16'27"16d6'
Beadle,Mark
Cnv 13h17'0"32d17'
Beadle,Martha & Will
Sge 18h59'17"18d39'
Beadles,Jenna Marie
Peg 23h32'53"23d6'
Beadling,"Olive Viola" Olive Viola
Cyg 20h38'48"53d12'
Beadling,Sr,"Essling"- Walter Essling
Lac 22h52'0"35d35'
Beagan,Deirdre
And 2h33'36"41d32'
Beagle
Cnv 13h51'22"38d21'
Beahm,Edward Winter
Cam 3h38'55"73d46'
Beahm,Sandra Jo
Lyn 8h40'36"40d44'
Beahm,Victoria Mae
Cyg 19h24'17"31d42'
Beaittie,Janet
Lmi 10h12'52"34d52'
Beakie
Oph 17h54'15"13d46'
Beal,Brad
Oph 17h59'35"-8d20'
Beal,Deborah Lynn
Ori 5h25'22"-4d34'
Beal,Diane & Gregory
Uma 9h5'1"72d14'
Beal,Evelyn
Eri 3h57'50"-17d4'
Beal,Gillian
Cyg 20h22'12"38d53'
Beal,Isabel Collier
Ori 6h1'24"4d21'
Beal,Kristi Leigh
Gem 6h47'24"19d38'
Beale,Christine
Cas 1h12'0"60d17'
Beale,David Michael
Per 2h53'57"37d11'
Beale,James H
Boo 14h31'55"30d3'
Beale,Nicola
Cas 0h56'37"70d43'
Beale,Stephen Joseph
Per 2h53'60"45d14'
Beales,Gina
Cas 1h9'0"70d30'
Beales,Jr,John Howard
Hya 8h23'51"5d41'
Beall Mankind's Glow, Stella Rebecca
Cet 0h56'50"-0d11'
Beall,Brittaney Naomi
Lyr 18h40'29"36d8'
Beall,Elisabeth Ann
And 23h48'59"40d53'
Beall,Jami Marie
Cnv 13h54'24"45d10'
Beall,Jean & Allein
Sge 19h1'1"19d4'
Beall,L Ruth
Del 20h14'54"11d3'
Beall-Anderson,Jamilla Dawn
Cam 9h6'0"81d6'
Bealle,Kim
Per 3h2'30"41d2'
Bealor,Lindsay Anne
Cyg 19h38'0"28d23'
Beals,Bruce Craig
Aur 7h20'41"36d0'
Beals,Catherine Stout
Equ 21h9'12"11d34'
Beals,Kendall J
Hya 8h11'31"-6d20'
Beals,MD,Carol A
And 0h20'36"40d19'
Beals,Rachel
Cyg 21h31'43"30d42'
Beaman,Dan
Her 17h30'19"50d2'
Beaman,Ethan James
Boo 15h0'49"8d36'
Beaman,Robert Dunning
Aql 19h52'60"10d27'
Beamer
Cet 3h19'1"2d6'
Beamer
Lyr 18h57'53"30d17'
Beamer,Allan
Her 17h17'52"42d3'
Beamer,Twinkle Charlie
Dra 16h22'55"68d24'
Beamers Gift
Eri 2h58'15"-3d9'
Beaming,Brilliant Bill!
Per 2h37'25"45d55'
Beamon,Jamie T
Aur 7h1'48"41d6'
Bean
Boo 15h3'12"25d40'
Bean
Uma 10h0'59"51d29'
Bean Family,The
Her 16h46'28"37d40'
Bean,Ashley Nicole
And 2h26'55"41d50'
Bean,Harold M
Ser 17h56'14"-14d29'
Bean,Jeffrey
Ser 15h57'17"24d3'
Bean,Kathleen Stacey
Cas 1h23'28"53d47'
Bean,Lorene
Cet 3h10'25"4d20'
Bean,Lulu
Peg 23h6'34"20d8'
Bean,Lynda
Dra 15h44'27"60d12'
Bean,Madison Murphy
Lib 15h33'40"-20d2'
Bean,Mae-Beth L
Cyg 19h19'0"28d5'
Bean,Meredith Lynn
Mon 6h24'14"-1d2'
Bean,Michelle Lorell
Oph 18h0'16"1d6'
Bean,Richard Brian
Aql 19h59'20"8d21'
Bean,William R
Her 17h22'23"47d44'
Bean,Zachary Daniel
Dra 13h28'54"64d23'
Beane III,Frank Eastman
Per 2h36'26"37d45'
Beane III,George Holton
Tri 1h37'32"28d47'
Beane,David
Uma 10h20'58"68d7'
Beaner
Ari 1h53'40"10d34'
Beaner-Chapman
Lyn 7h22'32"50d12'
Beanna
Lyr 19h4'39"40d25'
Beannie
Uma 9h10'51"57d52'
Bear
Umi 15h11'21"68d8'
Bear
Per 3h13'13"50d6'
Bear
Uma 11h37'1"60d44'
Bear & Sugar Bear
Uma 12h0'43"61d1'
Bear Claw Ranch Annex, The
Cet 3h14'47"9d21'
Bear Krupa
Cnv 13h54'54"37d53'
Bear of Stephenson- Burton
Umi 17h19'59"80d36'
Bear Paws
Vul 20h26'54"28d26'
Bear's Star
Com 13h6'39"28d53'
Bear,Annie
Uma 11h13'12"40d0'
Bear,Ian Edwin
Uma 11h11'0"43d51'
Bear,Jack
Uma 11h22'1"40d38'
Bear,Jeffrey
Boo 14h30'0"32d3'
Bear,Jeramy Todd
Per 4h1'12"38d36'
Bear,Jessica Shawn
And 0h29'36"40d11'
Bear,Kristy K
Del 20h19'0"10d60'
Bear,Melissa Vivienne
Uma 11h3'22"43d12'
Bear,Ted E
Cep 2h8'13"80d13'
Bear,The
Uma 10h48'24"68d23'
Bear-The Voyageur
Oph 17h55'24"13d18'
Bearak Star #90,The Charles
Her 16h26'12"28d44'
Bearak,Sandy Rose & Joe
Uma 8h20'16"68d33'
Bearchie
Per 3h4'16"47d41'
Bearcub
Ori 5h57'12"16d40'
Beard "Together Forever",Catherine
Eri 4h19'41"-4d47'A
Beard IV,Bucky's Expression,Edmund J
Aql 19h55'25"15d13'
Beard The Good Witch, Karen Makrides
Crb 16h3'56"32d42'
Beard,Anita Spaniel
Cas 1h1'10"61d12'
Beard,David Lee
Dra 15h59'54"67d34'
Beard,Dawn Renee
Cas 23h13'28"63d36'
Beard,G W
Boo 14h55'51"34d33'
Beard,George Vincent & Jean
Mon 6h29'1"2d44'
Beard,Lorene Carden
And 2h34'1"44d48'
Beard,Natalie Lynn
And 2h26'16"48d43'
Beard,Sam
Cet 1h48'53"-8d22'
Beard,Sr,"Together Forever",Gerard
Eri 4h19'41"-4d47'B
Beardmore,Dorothy Virginia Stroh
Cyg 21h10'45"35d38'
Beardslee,Bill
Aql 18h41'13"-1d58'
Beardslee,Michael
Cyg 19h31'43"34d17'
Beardsley,Elaine
Eri 2h51'59"-1d47'
Beardsley,Haden Moody
Aql 20h15'30"1d33'
Beare,Maurice Sydney
Ori 6h1'32"7d49'
Bearealis & Bill C & Maggie L
Cyg 20h8'27"38d38'
Bearheart,Jeffrey
Ori 5h55'31"8d60'
Bearmon,Lee & Barbara
Boo 15h11'34"51d51'
Bears Lair
Lmi 10h56'44"25d39'
Beas-Perez,Diego
Aur 5h58'0"54d33'
Beaser,Marjorie Eunice
And 1h12'46"40d56'
Beasley,Baileigh Elizabeth
Mon 6h42'47"1d4'
Beasley,Buzzy
Lac 22h15'11"51d40'
Beasley,Carol June
Lyr 19h12'10"41d12'
Beasley,Christie Lynne
Lyn 8h12'28"37d14'
Beasley,Geneva
Del 20h13'13"10d17'
Beasley,Kitty
Cyg 21h2'44"28d39'
Beasley,Margaret
Peg 21h23'19"23d29'
Beasley,Phyllis Hanes
Hya 8h56'37"2d32'
Beasley,Scott
Aql 19h57'0"13d18'
Beasley,Teresa A & Son Seth Rosser
Lyr 18h49'58"37d20'
Beason,Clay Brian
Ori 5h55'0"11d25'
Beason,Jerry
Tau 5h29'54"20d17'
Beast of the Field
Uma 8h39'18"61d38'
Beast,The
Dra 16h8'30"64d5'
Beaster,Leslie
Lyn 7h54'20"42d28'
Beaster,Leslie
Boo 14h9'17"52d4'
Beaster,Leslie
Eri 4h33'15"-1d22'
Beat Angels,Jon Brian Keith Kevin Michael
Her 16h15'45"20d14'
Beat Ringer
Cam 7h37'37"68d3'
Beat,Duncan Robert
Ori 5h2'1"1d36'
Beata
Cet 2h59'1"5d10'
Beata Clorinda
Cam 4h26'0"68d24'
Beata,Maria
Leo 11h21'60"-0d23'
Beate
Sgr 18h50'28"-22d58'
Beate-Ursula
Aqr 23h29'36"-12d22'
Beatiful Jennifer,Star of Virgo
Vir 13h55'41"0d40'
Beaton,Betty
Ori 4h56'0"5d55'
Beaton,Janet
Lac 22h5'14"50d36'
Beaton,Jr,Michael Lane
Cnc 9h11'13"30d25'
Beaton,Robert Campbell
Aur 5h0'0"45d19'
Beaton,Robert J
Cet 3h2'38"0d51'
Beaton,Robert J
Cet 3h16'19"2d14'
Beatrice
Umi 14h54'36"67d33'
Beatrice
Lac 22h41'32"38d9'
Beatrice
Cep 2h7'32"77d39'
Beatrice
Lyn 7h58'14"39d46'
Beatrice
Aur 4h59'47"40d26'
Beatrice
Boo 13h54'1"16d52'
Beatrice B
Tau 4h58'25"16d7'
Beatrice Josette
Uma 8h34'31"57d38'
Beatrice Laura
Cnv 18h29'0"38d52'
Beatrice's Birthday Star
Lyr 18h30'60"38d57'
Beatrice,che rende beati
Vel 9h18'44"-45d22'
Beatrice,Robert
Per 3h19'27"41d32'
Beatrix
Peg 22h41'0"32d22'
Beatrix Forever
Del 20h26'26"20d16'
Beatriz
Aql 20h18'21"0d1'
Beatriz
Lyn 8h36'16"40d53'
Beatriz
Ori 5h58'19"11d9'
Beatriz,Tonia
Uma 11h51'59"63d30'
Beatson,Patricia E
Gem 7h5'51"21d44'
Beattie,Beau Thomas
Aur 6h9'52"38d9'
Beattie,Juno Marie
Com 12h46'31"21d12'
Beattie,Kirsten
Sco 17h55'59"-40d50'
Beattie,Michael George
Aql 20h18'18"0d1'
Beatty,Andrew
Aql 18h58'58"16d20'
Beatty,Andrew
Crb 16h0'0"38d36'
Beatty,Brooke
Cmi 7h41'39"8d37'
Beatty,Geoffery Dean
Dra 11h49'18"67d14'
Beatty,Jr,Howard
Her 16h21'1"10d12'
Beatty,Kyona T
Eri 3h34'11"-2d44'
Beatty,Marc Philip
Cap 21h24'45"-8d39'
Beatus
Sco 17h52'15"-31d32'
Beatus Unus Shaun Michael
Aur 6h3'0"54d57'
Beatus Unus Sydney Anne
Cyg 21h51'42"40d17'
Beatus,Larry D
Cet 2h15'55"4d19'
Beaty & Descendents, John & Sara
Eri 2h59'45"-2d6'
Beaty,Christopher Marcus
Cyg 20h9'40"40d45'
Beaty,Claire Louise
Lyn 8h23'42"47d59'
Beaty,Ed
Sct 18h53'0"-6d47'
Beau Kismet
Cet 2h10'59"6d23'
Beau's Gazer
Per 4h19'13"50d42'
Beau's Star
Uma 10h43'1"50d50'
Beau,Lilli
Del 20h55'1"6d31'
Beauchamp,Denise Hospelhorn
And 2h31'60"48d31'
Beauchamp,Dolores
Lyn 8h57'38"41d4'
Beauchamp,Eileen
And 0h8'19"30d37'
Beauchamp,Emily Janice
Mon 6h29'44"8d16'
Beauchamp,Patrick Wade
Ori 5h13'22"15d48'
Beauchard,Roger Claude
Cyg 20h29'36"30d51'
Beauchemin,Catherine
Uma 9h54'22"61d45'
Beauchemin,Michelle
Cyg 19h46'17"29d29'
Beauchene,Fred
Peg 23h30'1"21d23'
Beaucher,Maxime
Umi 16h10'0"77d6'
Beaucousin,Albert
Lyn 9h6'18"45d22'
Beaudet,Romain
Lyr 19h20'18"42d5'
Beaudette,Norman Victor
Uma 10h29'39"70d32'
Beaudoin II,Richard Leroy
Uma 8h35'34"70d9'
Beaudoin,Linda
Umi 13h47'42"71d55'
Beaudoin,Nadine Louise
Cas 0h45'34"62d19'
Beaudoin,Richard
Del 20h21'43"11d9'
Beaudoin,Sophie
Cas 22h58'1"56d36'
Beaudoire,Corinne
Ori 6h5'22"1d12'
Beaudouin,Jean Michel
Cnv 13h28'38"38d43'
Beaudry,Carole
Peg 22h39'31"21d2'
Beaudry,Diane
Lyn 9h1'18"34d54'
Beaudry,J Douglas
Cyg 20h52'30"40d30'A
Beaudry,Lisa Ann
Cyg 20h52'30"40d30'B
Beaufays,Monique
Mon 6h56'17"-10d38'
Beaufils
Lyr 19h6'35"40d38'
Beauford,Mike-Brenda Shanee-Andre
Cep 20h54'23"61d33'
Beaulieu,Audrey
Lib 14h57'51"-22d58'
Beaulieu,Dany
Cen 11h43'3"-41d48'
Beaulieu,Hélène
Per 3h12'27"41d16'
Beaulieu,John
Cep 22h4'19"54d6'
Beaulieu,Jon C
Cep 20h54'40"61d42'
Beaulieu,Manon
Gem 6h57'18"14d29'
Beaulieu,Michelle
Vul 19h59'58"26d3'
Beaulieu,Pamela A
Crt 10h51'42"-12d24'
Beaulieu,Pamela A
Hya 8h52'50"0d54'
Beaulieu,Sylvia
Cyg 19h29'27"35d42'
Beaumier,Paul
Her 15h54'53"50d30'
Beaumont Star,The Jim & Sal
Lyn 7h11'47"58d54'
Beaumont,Alexandria Jade
And 1h7'25"37d39'
Beaumont,Michelle & Neil
Cyg 21h5'55"31d47'
Beaune,Marie Jo
Umi 16h19'17"70d54'
Beauparlant,Richard
Ori 4h50'14"5d21'
Beaupré,Marc-André
Her 16h43'35"28d29'
Beaupuy 3933
Boo 15h13'45"38d50'
Beauregard
Lyn 8h5'1"35d36'
Beauregard,Cindy
And 0h20'45"44d11'
Beauregard,Laura (Lala)
Oph 17h30'7"-20d3'
Beauregard,Patricia
And 23h46'13"47d40'
Beauregard,Sheila K
Com 13h8'32"16d32'
Beausoleil
Boo 14h42'0"36d52'
Beausoleil,Susan
Lyr 19h16'36"40d15'
Beautific Faye
Cas 0h38'22"67d30'
Beautiful
Mon 6h53'41"-6d54'
Beautiful & Precious Amrit
Ori 5h53'0"7d53'
Beautiful (RG)
And 2h30'22"40d30'
Beautiful BabyFace
Com 12h22'25"20d34'
Beautiful Barbara Gail
Cas 1h40'36"60d36'
Beautiful Beautiful Brianna Leigh
Ari 1h46'19"25d6'
Beautiful Belinda
Cru 11h53'26"-62d36'
Beautiful Beulah
Aql 19h59'0"15d19'
Beautiful Bex
Cyg 21h33'34"41d48'
Beautiful Boo Kitty's Wishing Star
Vul 19h49'51"20d28'
Beautiful Brenda Jean
Cas 0h43'48"70d40'
Beautiful Caroline
Umi 13h52'19"76d11'
Beautiful Carolyn
Cas 0h0'0"65d55'
Beautiful Christina
Cas 0h7'19"62d10'
Beautiful Christine, The
And 23h17'33"44d9'
Beautiful Cri
Cyg 19h47'32"37d41'
Beautiful Danielle
Cas 0h14'19"64d6'
Beautiful Danielle
And 0h21'28"38d51'
Beautiful Darlene
Del 20h18'14"11d14'
Beautiful Eileen
Cyg 20h2'0"38d50'
Beautiful Gail
And 23h15'28"37d53'
Beautiful Giant
Ori 4h52'47"1d7'
Beautiful In My Eyes
Crb 15h29'29"31d31'
Beautiful Kimberley
Com 13h1'39"21d26'
Beautiful Krista
Com 12h23'38"32d58'
Beautiful Maria
Sgr 19h36'16"-31d27'
Beautiful Maria of my Soul
Ori 5h51'19"16d42'
Beautiful Melissa
Com 12h7'40"24d3'
Beautiful Mother Healing Love Star
Cas 0h43'55"72d22'
Beautiful Nadja
Mon 7h10'23"-5d21'
Beautiful Nani
Mon 7h47'12"-4d20'
Beautiful Nicole Isabelle
And 23h39'46"38d11'
Beautiful Nikki
Ori 5h55'1"16d7'
Beautiful Ruth
Mon 8h1'44"-8d23'
Beautiful Stella Of Miwako
Lib 14h33'54"-23d49'
Beautiful Suzanne
Lyr 18h27'0"31d51'
Beautiful Victorias Mark
Ori 5h56'1"7d18'
Beautiful Yvonne Una
Cep 0h10'51"66d16'
Beauty
Cas 23h30'57"61d38'
Beauty
Ori 6h15'60"10d34'
Beauty Line Star
Del 20h17'40"15d45'
Beauty Moon
Psc 1h17'39"20d38'
Beauty of Aletha
Cas 1h33'10"60d36'
Beauty of Beverly's Eyes
Per 2h54'46"43d33'
Beauty Of Katie's Eyes,The
And 23h24'42"49d52'
Beauty of Mira,The
Cyg 20h22'53"38d44'
Beauty Of Nancy
Cyg 20h18'1"31d41'
Beauty of the West - KER
Lyn 8h6'15"41d32'
Beauty of visage di Gabriella
Del 20h18'1"1d53'
Beauvais Happy Anniversary,Mike
Lac 22h6'57"38d58'
Beauvais,Corin Tristan
Equ 21h2'1"2d15'
Beav,The
Dra 15h56'46"57d47'
Beaver
Psc 1h21'51"4d57'
Beaver,Kay-Ray
Aql 18h57'48"14d13'
Beaver,Patricia Lane
Cyg 21h32'18"53d45'
Beavercall
Ori 5h50'28"17d43'
Beavers,Beatrice Jury
Cam 4h54'54"61d10'
Beavers,Hailea Paige
Del 20h24'0"20d11'
Beavers,Susie
Cyg 19h31'19"32d8'
Beavin,Amanda Lee
And 22h56'43"51d22'
Beavin,Amy
Lyn 7h14'1"59d19'
Beazlie III,L Henry
Her 16h57'44"23d37'
Beb's Vincent
Cnv 12h46'48"38d55'
Beba
Lyn 8h4'47"38d17'
BeBa Star,The
Cap 20h28'49"-12d48'
Bebb,Rena
Dra 13h25'53"64d45'
Bebb,Thomas J
Boo 14h47'41"23d34'
Bebbington
Uma 10h56'45"50d6'
BeBe
Vul 20h17'44"23d57'
Bebe
Lyn 8h2'0"37d59'
Bebe
Uma 10h40'38"56d14'
Bebe Jr,Utopian Friendship
Hya 9h11'22"0d28'
Bebe Star,The
Ori 5h24'53"0d33'
Bebear,Claude
Leo 9h58'43"15d37'
Bebear,Jean-Pierre
Leo 9h20'12"10d9'
Bebel,John Warren
Com 13h7'22"20d3'
Beberniss,Shay
Mon 7h0'0"8d51'
Bebinous
Cam 8h11'28"81d46'
Becatti,Lance Norman
Cmi 7h37'53"4d51'
Because It's the "Mommy"
Umi 15h9'37"67d42'
Becca
Aql 18h58'47"8d10'
Becca
And 23h24'0"49d26'
Becca '93
And 1h37'0"39d29'
Becca V
Ori 4h55'58"4d16'
Beccaccio,Lynn
Cam 6h30'42"67d34'
BeccaZak
Hya 9h11'0"4d28'
Becerra,Jose
Ori 5h52'1"5d9'

Becerra,Stephanie M And 1h28'0"50d5'
Bechdel,Jennifer Renee Equ 21h3'15"8d42'
Becher,P Cet 1h16'48"-10d1'
Bechguenturian,Andrew Michael Hya 8h35'31"-0d19'
Bechler,Barbara Christine Cas 0h31'0"60d2'
Bechner,Michael Christopher Aur 4h50'53"51d11'
Becht,Edward Dra 16h43'57"68d42'
Bechtel,Christina Marie Lyr 18h42'40"30d7'
Bechtel,Peter Peg 23h0'45"8d45'
Bechtler,Sharon Ann And 2h23'44"39d5'
Bechtloff,Linda And 0h8'57"30d9'
Bechtloff,Richard And 0h4'3"33d22'
Bechtold,Walter Dra 18h27'48"65d15'
Bechtoldt,E Charles Aur 5h32'28"38d23'
Bechtolf,Uwe Ser 15h37'47"6d24'
Bechunas IV,Peter Lac 22h33'44"55d40'
Becich,Kayla Lynn Ari 3h0'1"21d22'
Beck's Night Light, Christopher James Aql 18h56'30"15d56'
Beck,Alexander John Mon 7h56'52"-8d8'
Beck,Andrew Graham Her 18h4'0"45d22'
Beck,Arnie Sct 18h53'17"-6d26'
Beck,Brittany Cyg 19h44'53"31d24'
Beck,C Edward Lmi 10h9'57"32d53'
Beck,Charles Glenn Boo 13h46'41"18d53'
Beck,Connor Hya 8h41'23"-18d55'
Beck,Dennis & Daniela Mon 6h41'45"7d42'
Beck,Edward Louis Aur 4h54'56"38d59'
Beck,Eileen Renee And 23h44'19"44d35'
Beck,Ellen Mon 6h35'47"6d19'
Beck,Estelle & Peter Tau 4h0'0"12d15'
Beck,Evan Miller Sco 17h52'13"-40d2'
Beck,Felix M Dra 16h9'1"60d26'
Beck,Hank & Sally Lyn 7h55'54"42d17'
Beck,Hazel Vivian Sge 19h57'1"16d17'
Beck,Hermann Erwin Gem 6h25'51"14d11'
Beck,Iris Cnc 9h16'38"31d6'
Beck,IV,William S Dra 14h46'27"63d31'
Beck,Jack Cma 6h54'28"-15d21'
Beck,Jeff Cma 7h14'35"-15d14'
Beck,Jennifer Lee Lyr 19h13'47"41d35'
Beck,Joseph and Janice Cyg 19h20'21"48d51'
Beck,Jr,Charles Hya 9h39'18"-10d13'
Beck,Karl-Heinz Cyg 20h27'46"30d36'
Beck,Katie Eri 4h40'16"-1d27'
Beck,Kelly Ann Cnv 12h14'52"42d56'
Beck,L'Ange Gardien Danny Cyg 19h29'14"31d6'
Beck,Margaret Blakeley Aql 20h8'43"0d2'
Beck,Maxwell Ward Aql 20h16'56"5d8'
Beck,Melanie Susanna Gem 6h47'10"13d14'
Beck,Mirco Mon 7h56'46"-5d57'
Beck,Norma Oph 17h53'44"10d34'
Beck,Norman Douglas Cep 23h5'20"65d32'
Beck,R S Per 2h5'58"56d59'
Beck,Rachel Per 4h2'36"38d15'
Beck,Ralph M Dra 16h53'25"62d12'
Beck,Rebecca Montgomery Peg 22h21'1"4d29'
Beck,Scott A Cas 0h2'44"58d10'
Beck,Sharon Lyn 8h13'0"42d22'
Beck,Steve Ori 6h17'51"-1d20'
Beck,Steve Ori 6h0'38"0d8'
Beck,Theresa Cap 20h38'56"-21d11'
Beck,Willi Cyg 20h18'58"47d7'
Beck,Zachary Tyler Boo 14h8'52"42d57'
Beck-Pearce,Patricia Aql 20h10'40"13d27'
Beckeleh's Bright Eye Cam 0h3'16"62d12'
Beckeleh's Bright Eye Cas 1h34'46"60d8'
Becken II,Thorwald Wilbur Per 3h25'58"53d58'
Beckenbauer,Sybille Leo 10h51'11"-5d40'
Becker 1928,67 Pete & Kitty Eri 4h4'44"-10d45'
Becker Bomber Uma 10h51'30"40d28'
Becker II,William A Cep 21h43'11"55d52'
Becker III,Raymond W "Ray" Bosen Oph 18h42'49"7d43'
Becker LIFE's Star,Pat Uma 8h57'20"57d28'
Becker,Abraham James Merlin Uma 8h42'37"68d34'
Becker,Adelbert Uma 9h29'0"46d28'
Becker,Alan Eri 3h18'47"-13d16'
Becker,Allyson Com 12h59'46"27d8'
Becker,Amy Lynn Peg 22h53'29"27d45'
Becker,Arnold H Cyg 20h21'26"38d58'
Becker,Brooke Alison Peg 22h54'30"21d9'
Becker,Bruce Frederic Cet 0h42'26"-3d12'
Becker,Charlotte Sgr 20h17'37"-37d52'
Becker,Cheryl S Crb 15h28'0"31d30'
Becker,Dana Joy Per 4h5'0"51d59'
Becker,Daniel Joseph Dra 14h15'45"64d52'
Becker,Dennis Eri 4h38'37"-0d2'
Becker,Dianne Carol Mon 7h56'2"-2d5'
Becker,Egbert Lib 15h1'18"-3d37'
Becker,Elizabeth & Eugene Cowles Mon 6h53'1"-4d32'
Becker,Ellen Peg 21h24'0"18d54'
Becker,Fern Oph 17h58'56"11d2'
Becker,Franz Josef Lac 22h31'49"37d51'
Becker,George Per 2h55'46"31d20'
Becker,Greg Hya 9h7'24"2d11'
Becker,GRM,Dorothy J Cam 12h17'39"80d47'
Becker,Heidi Peg 21h55'11"20d11'
Becker,Heinz Tau 5h32'36"28d53'
Becker,Ingeborg Lib 14h22'57"-23d12'
Becker,Jack Boo 15h11'13"51d11'
Becker,Jacqueline Claude Sge 19h53'48"18d48'
Becker,Jacqueline Mon 6h21'0"3d28'
Becker,Jenny Nicole And 0h21'51"40d16'
Becker,Jeremy Benjamin Per 4h41'35"51d12'
Becker,John & Florence Uma 11h3'33"60d5'
Becker,Julia Anne "Sparkle" Eri 3h31'13"-3d35'
Becker,Juliette Peg 21h41'33"22d53'
Becker,Kari Lynne Cyg 19h33'39"33d30'
Becker,Kenneth,Jon Her 16h37'40"50d13'
Becker,Kimberly A Cyg 20h59'18"40d2'
Becker,Lynn-Barbara Eri 4h9'0"-14d46'
Becker,Madge P Vul 20h42'35"28d53'
Becker,Martin Joseph Cyg 19h30'10"34d22'
Becker,Mary Ann Sass Lyr 18h40'0"47d25'
Becker,MD,Steven A Cep 22h16'44"68d11'
Becker,Melissa Louise Uma 12h26'45"61d22'
Becker,Michael Peg 23h38'37"22d8'
Becker,Nancy Anne And 23h31'33"44d9'
Becker,Niels Boo 14h20'55"14d21'
Becker,Poppie Richard Ori 5h1'26"15d10'
Becker,Russ Leo 9h53'19"31d30'
Becker,Russell Lac 22h18'58"51d26'
Becker,Samantha S Cas 0h32'29"66d14'
Becker,Stefan Sgr 20h3'0"-42d26'
Becker,Susan Lyr 19h1'22"25d38'
Becker,Susanna Cnc 8h34'18"30d51'
Becker,Torsten Crb 16h1'34"28d50'
Becker,William Ori 5h51'29"14d59'
Becker-Jones,Amy Cas 23h27'12"53d3'
Becker-Jones,Pamela Vul 20h27'30"28d55'
Becker-Jones,Vanessa Cas 23h27'56"53d49'
Becker-Wade,Renelda Tri 1h56'28"27d51'
Beckerich,Christophe Sgr 20h18'3"-31d1'
Beckering,Thomas E Cma 7h18'56"-15d22'
Beckerman,Cari Lyr 19h24'18"40d49'
Beckers,Franzis Bönniger Cam 14h29'28"82d25'
Beckers,Robert Cap 20h45'26"-24d4'
Beckert,Thomas H Aql 20h1'36"12d27'
Becket,Scott & Anne Aql 20h35'15"0d37'
Beckett,Larry Per 1h49'54"47d48'
Beckett,Louise Lyr 18h17'35"31d2'
Beckford,Antonio Bancroft C Vassell' Cet 1h49'1"-2d35'
Beckham (Meme),Alta Mae Edwards Cyg 20h0'50"41d9'
Beckham (Pepe),Finis Lee Mon 8h8'43"-8d52'
Beckham,C E Aur 5h0'55"50d26'
Beckham,Michelle Sandra Lyn 9h2'12"37d14'
Beckham,Steven Russell Uma 8h35'0"50d17'
Becki Lyr 18h19'0"40d26'
Becking,Vanessa Marie Mon 7h23'37"-8d39'
Beckington,Gereal Herbert L Uma 10h59'1"51d38'
Beckington,Patricia McKee Uma 10h57'51"51d11'
Beckley,Jordan James Uma 11h45'40"38d9'
Beckley,MD,Dennis J Tri 2h43'40"33d50'
Beckley,Terri Uma 8h14'24"71d56'
Becklor,Charles Alex Tristan Blanken Cep 21h48'33"58d43'
Beckman,Richard Allen Ari 3h2'26"28d56'
Beckman,Sandra I Mon 6h44'49"5d20'
Beckman,Sofie Nottoli Del 20h13'51"13d10'
Beckmann,Chase Mon 7h0'33"8d21'
Beckmann,Helga Lyn 8h16'50"40d55'
Beckmann,Leo Mon 7h53'29"-2d32'
Becknell,Linda Cyg 19h47'15"29d23'
Becks Uma 9h44'1"57d53'
Becksy Cas 0h10'13"60d33'
Beckwith 1965 Lebanon Ct USA,Brian Cep 21h36'11"67d53'
Beckwith,Alisa Marie And 1h7'18"40d30'
Beckwith,Ernest Adelbert Her 17h33'52"22d53'
Beckwith,Helen Gilleland And 0h20'55"36d6'
Beckwith,Troy Aql 19h59'59"14d49'
Becky Cyg 19h59'46"44d59'
Becky And 2h28'26"45d28'
Becky Cyg 20h54'50"37d50'
Becky Ori 6h9'0"9d34'
Becky Uma 14h25'28"61d4'
Becky Peg 22h54'21"11d34'B
Becky Cas 3h1'39"71d0'
Becky & Larry's Cosmic Lovestar Lyr 18h17'0"44d17'
Becky & Oliver Her 17h22'28"48d10'
Becky Early's "Rainbow Light" Com 12h33'14"26d29'
Becky Kate Cas 1h10'55"60d54'
Becky Lee Lyr 18h59'1"26d23'
Becky Lee Peg 23h29'16"32d22'
Becky Marie Aqr 22h40'15"-2d25'
Becky OO Lyn 9h1'49"44d53'
Becky S Mon 6h30'24"-0d3'
Becky's Beauty Com 12h29'16"21d3'
Becky's Brilliance And 22h57'10"50d3'
Becky's Brilliance And 23h27'24"40d7'
Becky's Little Stars Mon 6h32'17"7d27'
Becky's Place Cas 23h14'0"63d9'
Becky's Star And 2h30'31"48d5'
Becky's Star Aql 19h43'57"14d31'
Becky's Star Cyg 20h4'0"40d53'
Becky's Sweet Dreams Peg 21h58'20"30d23'
Becky's Wish Com 12h3'38"27d52'
Becky's Wish Cyg 21h5'59"31d40'
Becq,Vinciane Cam 6h1'31"61d4'
Becquey,Vincent Lyr 19h5'27"47d16'
Becraft,Ronnie Her 16h22'0"24d2'
Becsi II,Frank Joseph Dra 15h56'53"61d45'
Becton,Cynthia Ann Cas 0h7'44"58d25'
Becton,Wilson Prentiss Cas 0h38'54"70d18'
Becwar-World Traveler, Ray Dra 17h20'47"70d59'
Becze,Nichole Cyg 21h31'35"40d40'
Beda Mon 6h19'50"8d6'
Bedard,James Michael Aur 6h55'14"37d14'
Bedburg,Gerda und Dr Hermann Zier Aur 5h14'57"43d28'
Beddall,Jane And 0h26'0"37d36'
Beddingfield,Gary Robert Lac 21h55'16"37d14'
Beddingfield,H V Per 2h51'43"38d30'
Beddingfield,Jessie Lee Lmi 9h49'29"34d5'
Beddoe,Michael Cep 22h8'22"60d1'
Bedeau-Roopnarine, Peola Olive Sco 17h53'15"-30d27'
Bedel,Valerie Aur 4h35'12"30d53'
Bedet Sebastien Per 1h34'23"53d39'
Bedford Lyn 9h16'31"38d52'
Bedford,David J Tau 5h32'16"28d45'
Bedford,Howard E Lyr 18h29'59"38d29'
Bedford,Reginald George Cyg 19h29'10"35d15'
Bedford,Sara Lynn Lyn 8h37'29"40d24'
Bedford-Stradling, Francis Michael Dra 15h15'44"65d23'
Bedigian,Karen Peg 22h46'55"20d60'
Bedirian,Kip Ori 5h58'33"5d53'
Bedker,Robert Aql 19h39'41"14d6'
Bednar,Jill Cas 1h37'56"60d54'
Bednarczyk,Glen Her 17h11'31"47d54'
Bednarick,Michael John Boo 14h53'1"48d27'
Bednarski Da Ki Lo Ge Frank,The Oph 17h24'43"10d11'
Bednarski,Allan John Per 3h33'49"35d45'
Bednaruk,Basia Vul 19h19'51"26d30'
Bedner,Francis R Her 16h4'55"50d31'
Bedner,Michael J Ser 16h2'18"3d1'
Bednorz,James Aql 20h30'0"0d49'
Bedokaz Lofti Mashkal Egypt-Germany Cnv 12h16'19"37d37'
Bedoni,Maury Aur 4h59'27"40d7'
Bedossa,Guy Cyg 20h5'27"39d54'
Bedros Cnv 13h55'24"31d26'
Bedrosian,Aghavni Ori 5h40'14"11d28'
Bee Aur 6h23'30"33d24'
Bee & Harold Psc 1h3'42"21d8'
Bee A Star Cnc 8h34'44"30d29'
Bee Bee's Star Lyn 7h59'0"43d48'
Bee-Gee-Ess,The Ori 5h59'36"15d25'
Beebe,Christine Lyr 18h27'14"30d22'
Beebe,Eugene H & Margaret F Together Lib 15h1'0"-28d46'
Beebe,Glenn Merritt Ori 6h3'60"0d36'
Beebe,Jr,John Lyr 19h1'41"42d18'
Beebe,Keith N Dra 15h10'32"58d8'
Beebe,Kenneth Keith Ori 5h4'19"-2d17'
Beebe,Kevin Patrick Per 2h51'12"38d54'
Beebe,Kirk Edwin Ser 15h51'33"21d21'
Beebe,Martin Scott Per 1h46'56"50d17'
Beebee,David Elliot Uma 8h36'30"50d9'
Beeber,The Uma 10h56'59"70d2'
Beebs Sgr 19h59'12"44d5'
Beech,Ernest Uma 9h29'28"58d16'
Beech,Helen Eri 4h13'35"-14d43'
Beech,Pauline Lyr 18h38'51"34d21'
Beech,Tracy Donna Cyg 21h55'35"52d60'
Beecham,Sinclair Uma 11h10'22"56d57'
Beecher,Beverly Sge 19h59'1"18d57'
Beecher,Sr,W Stewart Cep 21h2'26"55d37'
Beecher-Shepard,Alpha Lmi 10h41'0"25d38'
Beeching,Cameron Philip Cet 2h52'43"4d8'
Beechum,Kevin Sct 18h43'19"-6d51'
Beechwood,Bumper Uma 11h44'36"60d52'
Beeckler,Tom Hya 8h49'55"-7d49'
Beeckman,Pierre And 0h31'41"40d6'
Beeckman-Schalen, Jeanne And 0h35'0"40d32'
Beecroft's Solitaire Ori 5h41'24"1d39'
Beecroft,Jennifer Cas 0h39'28"74d59'
Beeda,Jacqueline Peg 22h31'59"3d57'B
Beeda,Thomas Peg 22h31'59"3d57'A
Beef Her 17h4'16"48d2'
Beeg,Christina And 1h49'26"37d2'
Beegle,Anna Williams And 0h46'44"22d58'
Beegle,Kevin Her 18h10'44"38d30'
Beekar Cam 7h6'16"61d49'
Beekman,Albertus Aarnout Aur 7h1'50"35d31'
Beeler,Richard Aur 5h4'56"44d16'
Beels,Jr,Bird Aql 20h30'31"0d20'
Beem Hya 8h33'48"0d33'
Beem,Zachary Taylor Oph 17h59'35"-8d20'
Beeman,James J Sex 10h14'26"-4d14'
Beemer,Bradford Scott Cet 2h34'1"-1d14'
Beemer-Rudolphie Lup 15h42'14"48d5'
Beems,Bastiaan Pieter Cnv 12h44'0"40d23'
Been,Jon Robert Sco 17h26'45"-30d8'
Been,Lisa And 23h15'18"44d17'
Beene,Abigail Charlotte Peg 23h0'16"30d40'
Beene,Carol O Peg 22h39'48"10d54'
Beene,Roger L Per 2h53'38"34d54'
Beep's Lac 22h0'36"37d60'
Beep-Beep Cep 1h30'45"78d38'
Beer Star,The David Ralph Cnv 13h18'15"38d10'
Beer,Ashden Her 17h17'30"42d41'
Beer,Barbara E Peg 22h36'46"29d46'
Beer,George Brad Ori 5h52'48"18d42'
Beer,Paul Lmi 9h24'39"38d19'
Beer,Rebecca Peg 21h58'1"24d10'
Beer,Stephen K Ori 5h55'44"17d10'
Beers,Barbara Ann And 23h36'60"45d9'
Beers,Charlotte Cyg 19h24'11"31d14'
Beers,Diane Cam 9h9'38"78d55'
Beers,Orvas E Per 3h27'57"51d31'
Beers,Sean Franklyn Her 15h56'24"50d29'
Beerstecher,Daniel Leonard Her 16h5'57"16d22'
Beesack,Joseph Montgomery Per 3h9'27"40d4'
Beese,Joseph Her 18h15'56"31d45'
Beesley,Donna Maria Cas 0h23'41"60d14'
Beeson,Fred Charles Sex 9h55'38"3d50'
Beeson,Laura Elizabeth Ser 15h58'44"0d49'
Beeson,Mike Per 2h55'32"45d15'
Beeson,Warren Mon 7h2'18"4d31'
Beeter,Bridget Colleen Aqr 23h36'18"-10d0'
Beetz,Anneliese Leo 9h58'57"11d19'
Beezer Lyn 8h1'0"40d44'
Beezer Tri 2h16'0"31d25'
Befano,Dawn Renée And 0h22'35"36d12'
Beffa,Kevin Aql 20h3'35"0d37'
Beford,Leonard F Leo 10h56'38"17d39'
Befurt,Adolf Ori 5h59'0"7d55'
Begasse,Kenneth Boo 15h1'28"14d30'
Bege Crt 11h26'12"-17d50'
Begelman,Mark David Her 18h18'41"14d59'
Begemann,Friedrich H Sco 17h30'27"-30d55'
Beger-Bieske,Ilona Sco 17h7'52"-38d24'
Begg,Jock Cep 3h31'55"78d51'
Beggs Star,The Cyg 21h4'20"37d53'
Beggs,Randy Lac 22h30'39"37d48'
Beghin Her 18h2'24"40d46'
Begin,Jeffrey Lac 22h47'14"56d14'
Begin,Mark Richard Cet 2h42'41"8d9'
Beginnings Cyg 20h6'15"39d37'
Beginnings Cam 12h49'1"82d30'
Begley's Shining Love, Susan Cap 21h26'45"-24d15'
Begley,Bob Cet 1h54'29"0d30'
Begley,Diana Dudley Leo 9h23'0"20d33'
Bego,Jean Vul 19h16'19"25d7'
Begoin,Hubert Aur 5h52'1"38d9'
Beguier,Sylvie Boo 13h43'34"16d26'
Beguin,Emeline Sct 18h45'8"-6d47'
Beha,Thomas Lyn 8h29'56"42d10'
Beham,Danny Aur 5h1'0"46d39'
Behan,Barbara E Vir 13h7'48"-2d27'
Behan,Shirley V Nana Cyg 20h1'56"31d6'
Behanna,Clyde Dra 19h10'29"67d53'
Behar,David Durand Dra 16h36'46"63d36'
Behar,Shira Rachelle Uma 13h36'0"48d11'
Behennah Uma 9h4'1"48d6'
Behennah,Michelle Cyg 20h1'33"30d48'
Behler,Cindy Tri 1h56'0"26d31'
Behm-Gehrung,Anke & Andreas Leo 9h21'11"20d12'
Behn,Jordan Quinn Cam 4h14'24"60d17'
Behncke,Phoebe Evelyn Lyn 7h34'14"50d27'
Behnecke,Curt Cmi 7h19'52"8d55'
Behney,Chuck Cep 20h43'10"60d53'
Behnke,Mike Sct 18h50'48"-6d54'
Behnken,Tyler Jay Cam 5h54'45"68d26'
Behr,Doris Aql 19h58'46"10d47'
Behr,Glenn E Uma 8h10'54"61d3'
Behr,Leslie & Joy Cyg 19h29'46"33d22'
Behr,Steve Her 16h56'46"32d41'
Behr,Therese Sco 17h6'43"-38d57'
Behrakis,George D Aur 4h34'20"31d50'
Behre,Herr Cmi 7h17'55"5d18'
Behren,Inge Ser 15h55'12"-2d2'
Behrenbeck,Gunnar Lyn 8h10'18"43d4'
Behrendt,Edward J Boo 15h0'20"10d39'
Behrendt,Rosmarie Cap 20h36'21"-18d56'
Behrendt,Suzanne Aql 20h33'26"0d59'
Behrens,Albert Leslie Uma 11h43'0"32d42'
Behrens,Charlene Aql 20h10'52"10d34'

Behrens,Jan
Lyn 7h19'48"58d35'
Behrens,Jean Patricia
Lyn 8h8'49"57d46'
Behrens,Katie
Cas 1h4'32"55d18'
Behrens,Kristina Koenig
Dra 17h52'1"63d56'
Behrens,Peter
Oph 16h52'29"1d21'
Behrens,Robert C
Cep 22h53'47"57d60'
Behrens,Sam & Shari
Belafonte
Crb 15h55'55"37d57'
Behrens,Werner
Dra 18h5'23"68d20'
Behring,Christina
Vir 13h24'46"-0d29'
Behringer,Claudia
Cap 20h35'40"-18d49'
Behrisch,Ali
Lyn 7h54'29"41d4'
Behymer,Christopher Walter
Umi 14h51'1"77d56'
Beiboer,Kelly Edmondson
Peg 21h52'34"30d23'
Beichek,Margaret Carol
And 0h24'23"40d42'
Beichner,Josephine Cheasty
And 1h44'1"41d10'
Beichner,Marilyn Hughes
And 1h47'31"39d48'
Beier,Charlie
Cas 3h7'38"71d22'A
Beier,Walter
Lyr 19h18'47"31d49'
Beiermeister,Geoffrey
Aur 7h26'42"39d23'
Beiersdorfer,E Fred
Dra 14h7'52"63d13'
Beighlea,Charles J
Dra 11h1'55"74d39'
Beilein,Katja
Cap 21h43'45"-22d35'
Beimel,George
Boo 14h9'27"42d59'
Bein,Joyce
Peg 21h38'48"25d43'
Bein,Lindsey Russell
Aql 19h57'14"13d35'
Beinecke II,Frederick William
Per 2h51'54"38d56'
Beining,Kris A
Cyg 19h32'19"28d9'
Beinl,Dieter
Fri 4h48'56"-5d36'
Beirne,Attracta McAndrew
Equ 20h57'49"9d28'
Beiser,Dixie
Cas 3h2'44"60d12'
Beiser,Jania Dowland
Cas 1h19'18"62d37'
Beisner,Jim
Oph 18h21'1"10d41'
Beisser,My Beautiful Star,April
Lyn 8h8'21"35d17'
Beitel,Carolyn E
Cyg 20h10'34"37d34'
Beitel,Jesse J
Aur 6h29'57"31d46'
Beitler,Barry Alan
Per 2h51'7"47d56'A
Beitler,Ernest Sigmud
Dra 16h2'33"63d30'
Beitler,J Paul
Lmi 9h46'16"38d31'
Beitler,John J & Claire M
Aur 6h36'50"41d0'AB
Beitz,Roger
Uma 9h35'0"48d46'
Beitzel's Star,Julie
And 23h21'34"42d9'
Beja
Uma 8h36'22"51d9'
Bejarano,Charo del Pozo
Cyg 21h24'25"28d49'
Bejay
Uma 9h57'31"47d43'
BeJe
Lyr 19h20'23"38d51'
Bejlovec,Travis Scott Wilschke
Sgr 19h19'46"-45d7'
Bejster,Daniel Keith
Boo 14h50'38"38d5'
Bekaert,Jim
Sct 18h41'51"-6d45'
Bekash,Dalgina
Cmi 8h8'0"3d26'
Bekech,William John
Per 2h4'27"58d25'
Beker,Alison Debra
Boo 14h37'0"8d23'
Bekir & Margarida
Cyg 21h8'12"48d50'
Bekkala,John Arthur
Cep 21h50'44"58d52'
Bekker,Alfred Herbert
Lyn 7h55'14"58d22'
Bekkum,Miki S
Sge 19h57'0"16d7'
Bekoe,Martin
Leo 9h23'48"17d39'
Bekowich,Rachel Lyn
Cas 0h9'42"61d3'
Bel Amour
Lyn 8h4'43"40d12'
Bela Rolo
Mon 6h27'11"7d53'
Belalcazar,Jaime
Her 16h36'28"39d49'
Beland,Nicole Marie
And 2h23'45"44d32'
Belanger,Louie Anthony
Dra 19h11'1"70d21'
Belanger,Lynn Marie
Com 13h6'15"16d20'
Belanger,Tyler Scott
Dra 18h31'32"70d53'
Belasco,Jr,Bert L
Boo 15h3'57"38d59'
Belbo,Joyce Angela Talamone
Lyn 8h34'58"40d30'
Belcher RAF,Flying Officer
Stephen
Cyg 20h21'1"39d18'
Belcher,Kayl Wolf
Sex 10h40'59"1d36'
Belcher,Michael
Per 3h18'42"41d37'
Belcher,Ronald A
Oph 18h17'0"1d22'
Belcourt
Cam 6h9'0"68d26'
Belcove,Julie
Crb 16h5'32"26d9'
Belczak Memory,The
Everlasting
Aql 20h5'35"1d12'
Belden,Bradley Reinert
Aur 5h16'29"45d39'
Belden,Carl Elson & Evelyn
Rose
Boo 15h2'13"18d4'
Belden-Palmer,Betty
Sex 10h13'22"-3d0'
Belekevich,Steven C
Cam 4h8'53"65d4'
Belen
Mon 6h22'59"-1d33'
Belen,Anthony John
Aql 19h27'59"1d50'
Belen,Sara Nicole
Mon 8h7'27"-0d20'
Belew,Anne H
Crt 11h17'50"-14d18'
Belew,Billie
Tau 4h38'39"10d25'
Belew,Bridgette
Del 20h13'57"14d53'
Belew,Virginia
Vir 11h45'10"4d6'
Beley,Brian J
Her 17h0'23"20d38'
Belfi,Eric A
Aql 19h0'59"2d41'
Belfie,Diana
Tau 4h39'30"8d10'
Belfiore
Crb 15h21'32"31d18'
Belfiore,Maria Cleofe
Lyn 7h3'14"44d48'
Belford,Christina & Nicholas
Pryor
Crb 15h57'58"30d39'
Belgarde,Bret
Dra 12h41'33"75d33'
Belger,Lucille
Del 20h14'25"15d10'
Belgraier,Melvin
Cap 21h2'30"-18d47'
Belgraier,Michael
Aur 5h29'19"30d32'
Beliduk
Cma 7h15'0"-15d14'
Believe
Cam 7h49'1"80d38'
Believe It Or Not
Tri 1h30'60"31d5'
Believe The Children
Peg 23h15'49"30d5'
Belin-Rob
Crb 16h12'10"34d35'
Belinda
Cyg 20h42'56"44d59'
Belinda
Lyn 8h8'1"39d9'
Belinda
Peg 22h32'40"26d16'
Belinda
Uma 9h0'15"55d57'
Belinda
And 1h7'56"38d1'
Belinda
And 23h0'1"40d18'
Belinda Blue
Cyg 19h27'18"38d7'
Belinda Fay
Peg 22h31'50"24d22'
Belinda's Rose
Ind 21h21'17"49d23'
Belinda V-P
And 23h39'1"41d2'
Belinguier,Butrand
Lyr 19h5'19"47d42'
Belitski,William John
Cep 23h21'15"70d18'
Beliveau,Forever Paul
Ori 5h57'41"10d54'
Beliz
Crb 15h22'37"30d24'
Belk
Cma 6h31'14"-15d45'
Belk,Christopher Earle
Her 16h4'35"42d14'
Belk,J Blanton
Uma 8h44'29"51d48'
Belkin,Annie & Mike
Crb 16h15'17"28d16'
Belknap,Gene & Darlene
Crt 11h14'7"-14d24'
Belkys
Uma 10h41'24"68d16'
Bell Atlantic:Amnex/ Project
Saturn
Aur 7h22'25"38d8'
Bell Goddess
Cam 13h7'23"77d44'
Bell Star Of Hope,The
Lyr 19h4'55"40d56'
Bell Star,The
Aql 19h27'58"14d28'
Bell Star,The
Umi 15h11'34"68d33'
Bell's 25th Ann Star, Wayne &
Jean
Lyr 19h20'1"38d26'
Bell's Heavenly Star, Meredith
Cet 1h41'19"0d53'
Bell,Alexandra
And 2h24'33"44d8'
Bell,Amanda Marie
Peg 22h8'24"24d10'
Bell,Andrea Jean
Cas 1h44'26"75d60'
Bell,Angela
Cap 21h21'1"-24d60'
Bell,Anthony Noel
Cnv 13h30'26"48d38'
Bell,Bob
Aql 19h57'32"13d21'
Bell,Brandi Nicole
Peg 23h19'31"32d44'
Bell,Brian
Her 16h18'49"8d7'
Bell,Charles Aaron
Her 17h11'18"42d4'
Bell,Charles Nelson
Hya 8h34'35"-10d9'
Bell,Chief Robert Allen
Eri 2h56'37"-14d39'
Bell,Christina Marie
Vul 20h43'54"28d8'
Bell,Christopher Nathan
Uma 8h45'11"68d54'
Bell,Clare Joanne
Cyg 19h48'29"38d26'
Bell,Danielle Nicole
Uma 9h46'0"60d9'
Bell,Delores Jean Willis
Cet 0h8'42"-10d25'
Bell,Dominic
Cep 0h1'1"76d32'
Bell,Edward J
Cet 2h33'27"3d4'
Bell,Erin Christine
Vul 19h1'35"25d26'
Bell,Gary
Ser 15h55'52"17d36'
Bell,Heather
And 0h20'57"36d12'
Bell,James R
Dra 17h48'43"61d19'
Bell,James Ray
Cep 22h9'0"61d9'
Bell,Janis Lee
Cas 0h49'28"66d34'
Bell,JC
Vul 19h44'27"28d51'
Bell,Jeanette Moone
Ori 5h57'38"19d52'
Bell,Jordan John Mullon
Cmi 7h35'29"1d20'
Bell,Josephine G
Cyg 20h35'0"58d46'
Bell,Jr,Charles Edward
Her 16h25'48"38d33'
Bell,Jr,Chauncey F
Cnv 12h55'22"51d38'
Bell,Kayla Rae
Lib 15h46'40"-8d25'
Bell,Kenneth Douglas
Cma 7h14'15"-13d41'
Bell,Lily
Cas 0h19'18"59d33'
Bell,Liz
Cyg 20h38'27"38d22'
Bell,Lois Bernice
Aur 4h58'34"38d8'
Bell,Lucas
Umi 16h4'41"76d27'
Bell,Madison Elise
And 0h16'31"32d49'
Bell,Martin Joseph
Gem 7h3'49"21d26'
Bell,Matthew William
Aur 6h4'58"37d19'
Bell,Michelle
Lyn 7h37'24"36d32'
Bell,Myles E
Her 16h34'36"34d42'
Bell,Nicholas Patrick
Dra 20h33'18"68d34'
Bell,Nicole
Cas 0h36'1"58d9'
Bell,Oren Frank
Per 2h58'53"38d50'
Bell,Patricia Ann (Laughlin)
Cep 21h51'16"60d28'
Bell,Phillip Andrew
Ori 5h44'1"11d12'
Bell,Racheal Amarosa
Lyn 7h26'52"44d13'
Bell,Rianne
Lyr 18h43'56"47d26'
Bell,Richard T
Her 16h47'1"21d28'
Bell,Robert & Lea
Cyg 21h29'31"40d15'
Bell,Robert N
Lac 22h18'0"49d35'
Bell,Robin
Uma 11h54'0"45d19'
Bell,Rodger
Sgr 19h4'3"-24d57'
Bell,Ruth Stamey
Cas 1h48'30"67d57'
Bell,Sandra Lucille
Lib 15h16'53"-21d52'
Bell,Sarah Jacqueline
Boo 14h27'1"51d14'
Bell,Simon Christopher
Her 16h58'41"23d23'
Bell,Stephanie Dru
Mon 6h45'13"11d55'
Bell,Stephen F
Aql 19h9'36"13d9'
Bell,Suzanne
Sex 9h49'20"2d22'
Bell,Tami L
And 0h3'26"35d40'
Bell,Teresa
Cas 2h10'0"59d43'
Bell,The Rochelle Lynn
Cet 2h49'33"2d2'
Bell,The Star of David Stephen
Dra 15h7'42"60d1'
Bell,Timothy
Ser 15h14'57"8d10'
Bell,Tracy R
Cas 23h31'37"60d40'
Bell,Travis
Cnv 12h9'58"37d57'
Bell,Vincent
Aql 20h17'10"5d18'
Bell,Violet Betty Grace
Equ 21h1'25"2d22'
Bell,Wilbert Lee
Cet 2h13'10"6d26'
Bell,Willie
Cet 0h50'34"-8d29'
Bella
Cyg 20h33'40"37d47'
Bella
Cas 22h58'0"55d52'
Bella Carolina
Uma 12h27'57"62d17'
Bella Delina
Cas 0h37'12"62d5'
Bella Gloria
And 22h57'53"50d28'
Bella Lisa
And 0h22'38"45d54'
Bella Marie
Cas 1h32'1"58d17'
Bella Mira
Sct 18h52'47"-8d42'
Bella Traci
Cmi 7h28'16"0d35'
Bella Ursula Gromm
Vir 12h2'22"-5d47'
Bella-Regina
Ari 2h41'18"30d59'
Bellamy Memorial Star Donal
Stanley
Uma 9h50'30"43d22'
Bellamy,Doreen
Lyr 18h23'59"38d4'
Bellamy,Jeffrey Taylor
Ser 18h15'20"-7d5'
Bellamy,Marlene
Cet 2h13'46"0d10'
Bellan
Mon 6h54'19"-1d51'
Bellanca,Joseph A
Cep 20h56'14"58d56'
Bellante,Jr,Joseph J
Boo 15h18'54"33d40'
Bellanti,The
Aql 19h37'29"-5d46'
Bellantoni,Ernest Nathan
Vul 19h46'32"22d36'
Belle
Lyn 9h15'1"38d4'
Belle
Mon 6h21'16"-0d43'
Belle
Lyn 7h59'46"43d29'
Belle & Clyde
Lyr 18h25'21"47d34'
Belle de nuit
Cas 22h57'46"54d16'
Belle des temps
Cas 22h58'42"55d21'
Belle Lora
Cas 0h58'24"70d3'
Belle Sarah
Mon 7h4'52"-5d58'
Belle étolile Adrianne
Ori 4h54'29"0d29'
Belle,Bernard
And 23h17'1"49d54'
Belle,Ida
Crt 11h7'49"-15d52'
Belle,Ronda
Vul 19h48'33"28d12'
Belle,Tom & Chris
Uma 9h39'57"49d35'
Belle,William H
Dra 17h13'59"61d45'
Belleaux,Clayton & Koren
Cyg 20h19'20"38d11'
BelleDonna
Cyg 20h32'1"58d18'
Bellemare's Hope
Cet 2h48'11"4d22'
Bellenchia,Carolann
Lyr 19h1'24"28d31'
Beller DC, Bryan
Aur 6h22'36"38d15'
Beller,David Simon
Dra 16h45'44"72d57'
Beller,Dr Murray
Boo 14h13'0"40d16'
Beller,Dustin
Uma 9h40'35"51d13'
Beller,Melissa Masson
Cas 1h29'18"63d46'
Beller,Melody
Cas 2h31'56"61d32'
Beller,Myra
Cas 1h35'1"70d43'
Beller,Peter Copeland
Her 18h2'22"14d23'
Bellerby,John T
Cam 3h19'1"61d27'
Bellerive,Marc
Dra 19h4'29"58d26'
Bellero,Doris J
Cmi 7h55'1"8d18'
Bellesorte,DO,Joseph R
Aur 6h37'25"38d19'
Belley,Ghislain
Peg 22h44'0"31d53'
Bellezza,Incantatore
Ori 6h7'40"7d55'
Belli,Emily
Tri 2h1'34"31d52'
Bellici,John
Uma 8h56'10"49d14'
Bellig Kansas
Uma 9h51'49"57d31'
Bellile,Robert J
Hya 10h12'12"-19d42'
Bellin,Anna
Mon 6h26'47"10d15'
Bellin,Harvey J
Aur 4h48'32"41d12'
Bellingen,Emile Van
Lac 22h24'13"48d53'
Bellingrath,Charles T
Aur 5h24'49"37d37'
Bellis Family Star,The
Sct 18h53'33"-6d42'
Bellis,John
Cep 3h16'0"78d10'
Bellis,Mark A
Uma 12h2'32"32d28'
Bellisario,Anthony
Oph 17h54'1"12d11'
Bellisimo
Boo 14h19'14"54d15'
Bellitt,Gary Lee
Uma 12h9'14"62d30'
Bellitt,Kathryn Ann Marie
Vul 20h14'19"26d2'
Bellitz,Marissa Clare
Cas 2h12'28"59d38'
Belliveau,Alice
Peg 22h45'50"27d18'
Belliveau,Auréa
Lyr 18h49'49"35d1'
Belliveau,Michel
Cyg 19h28'18"33d57'
Bellm,Raymond
Aql 19h0'0"13d42'
Bellman,Eleanor "Babe" Diana
Aqr 22h20'24"0d50'
Bellman,John Anthony
Aqr 22h37'14"-1d38'
Bellman,Vernon
Aur 7h24'23"36d57'
Bellmer,Hans
Del 20h19'13"10d17'
Bello,Andy
Lib 14h57'10"-22d59'
Bello,Gaetano,Vincent
Ori 5h51'30"20d33'
Bello,JoAnn M
Uma 11h15'51"61d57'
Bello,Rosemary
Cep 22h28'46"58d9'
Bellofatto
Peg 21h7'0"13d6'
Bellomo,Dr Spartaco
Her 17h31'1"38d42'
Bellomo,Sara
Vul 20h0'58"28d14'
Bellomo,Walter J
Aur 6h11'0"37d38'
Bellomy,Glenn
Lmi 10h9'33"38d40'
Bellon,Gilbert
Boo 15h6'0"10d5'
Bellore,Michele
Vir 13h4'22"-1d58'
Bellot,Bruno
Ori 6h0'20"10d14'
Bellot,Marie-Christine
Ori 6h0'22"20d51'
Bellotti,Christopher Anthony
Aur 5h1'1"41d42'
Bellotti,Jeffrey Anthony
Per 2h53'47"37d55'
Bellovich,Rachel Grace
Cas 1h23'23"61d22'
Bellow,Jack
Dra 9h50'47"74d7'
Bells,Betty & Max = The
Golden
Uma 11h51'11"32d24'
Bellsen,Paul
Her 17h25'14"27d33'
Bellue Family,The
Cet 0h47'53"-4d15'
Belluomini Our Hero, Harry
Uma 10h18'1"47d33'
Belluomo,Elizabeth Anne
Lyn 8h3'0"37d54'
Bellybuttom
Lac 22h3'55"40d20'
Bellélaine (Elaine Brière)
Umi 14h5'36"69d18'
Belma
Ori 5h53'0"5d20'
Belmer,Louise
Peg 22h2'21"20d31'
Belmessieri,James Martin
Uma 11h25'24"71d13'
Belmont,Francis E
Aql 19h48'26"14d9'
Belmont,Mary Lou
Ori 5h42'23"3d59'C
Belmore,Leo Edward
Boo 14h56'38"53d9'
Belnak,Mary Ellen
Lyr 18h55'1"30d55'
Belonga-Mick
Hya 9h47'34"-17d55'
Belongia,Clayton B
Boo 14h38'43"41d55'
Belot,Lori & Peter
Crb 15h17'48"30d6'
Belott,Peggy
And 2h21'31"47d20'
Belova,Zina
Peg 22h1'14"33d44'
Beloved
Ori 6h6'58"0d26'
Beloved
And 2h21'53"40d46'
Beloved Aunt Arlene Star
Eri 2h46'21"-4d31'
Beloved Bear
Cet 2h54'43"0d54'
Beloved Betty
Crb 15h59'39"37d40'
Beloved David
Per 1h52'11"50d26'
Beloved Holly & Tim
Aql 19h7'23"15d26'
Beloved Iphigenia
Cas 0h52'0"73d53'
Beloved Jake
Oph 18h32'0"10d29'
Beloved Jono
Cet 1h57'33"0d24'
Beloved Martin
Aur 4h59'20"51d12'
Beloved Merroney
Lyr 18h29'36"30d48'
Beloved Mother Odessa
Cyg 21h32'25"34d22'
Beloved Nancy
Cas 0h46'42"61d44'
Beloved Rick
Cyg 20h32'46"48d54'
Beloved Sarala
Dra 17h44'57"68d60'
Beloved Sister Vivian
Eri 4h56'2"-9d12'
Beloved(22-3-1973)
Cyg 19h31'40"35d11'
Beloved, Cheryl & Dana
Cam 4h3'33"68d26'
Belpassi,Paola
Tri 2h0'32"30d26'
Belshaw,Kathy
Cmi 7h36'39"7d24'
Belsher,Sheryll J
Vul 20h20'58"23d19'
Belsito,Maddy & Jim
Tau 5h20'37"18d47'
Belski's Star,Barb
Uma 11h52'32"40d60'
Belski,Kaili Starr
Cyg 19h27'59"32d15'
Belt,David
Aur 5h1'42"45d16'
Belt,James E
Dra 16h0'46"67d30'
Belt,Martin Lesley
Tri 1h36'23"30d40'
Belt,Michael
Ser 15h37'52"5d12'
Beltershain,Hanna Von
Sgr 19h50'45"-40d18'
Belton,Jr,Robert Keith
Tau 5h19'53"28d32'
Beltramini,Ines
Leo 11h3'21"-0d56'
Beltran's Shining Light
Cet 0h33'50"0d56'
Beltrani,Al
Dra 16h31'38"67d43'
Beltz,Alvin H
Lac 22h40'57"54d57'
Beltzung,Nicole
Aur 4h52'0"40d3'
Belveal,Ashley Faith Tracy
Equ 21h5'35"10d46'
Belville,Jesse Lee
Her 16h36'33"23d32'
Belvin IV Prince of the
Cosmos,J Peter
Ori 5h47'48"18d38'
Belviso,Helen
Lyr 18h31'56"43d31'
Belvo,Jeffrey Alan
Cet 2h56'40"7d45'
Belynda Kay
Del 20h13'1"14d51'
Belynda-Eternal Light
Lyr 19h20'47"35d30'
Belyo,Katy Miller
Eri 3h44'0"-5d49'
Belz,Chris
Lac 22h9'34"37d51'
Belz,Jack
Her 17h20'57"14d41'
Belz,Jeffrey D
Aql 20h19'40"0d48'
Belz,Saul
Cet 3h16'1"4d58'
Belzer,Phillip & Sylvia
Cyg 19h27'13"31d30'
Bemah
Sct 18h43'17"-6d13'
Bemes,Russell
And 23h37'55"41d13'
Beminevermore
Lyr 18h54'33"40d7'
Bemis,Jr,Paul Richard
Cet 0h54'1"-2d56'
Ben
Cep 20h16'32"60d52'
Ben
Cep 23h12'31"64d19'
Ben
Her 16h38'56"48d17'
Ben
Cyg 20h54'49"30d59'
Ben & Agnes
Cyg 21h37'32"31d5'
Ben & Betty's Ruby Star
Cyg 21h12'56"39d44'
Ben & Jessica
Umi 14h28'47"68d50'
Ben & Mabel
Eri 3h1'51"-10d37'
Ben & Nick
Cet 0h27'26"1d24'
Ben Driss,Peter "Ben"
Leo 10h51'13"1d23'
Ben Ryan
Dra 15h53'57'68d14'
Ben Star
Aur 6h28'29"38d48'
Ben's "Birthday Star"
Aql 18h39'56"-2d35'
Ben's Forever,More
Lmi 10h10'53"32d11'

Ben's Happiness
Boo 14h29'57"27d44'
Ben-An-Anna-Decon
Cyg 19h21'13"28d30'
Ben-O-Ben
Uma 12h4'21"56d46'
Ben-Reuven,General Eyal
Her 16h59'30"48d2'
Ben-Shirl
Per 1h26'0"53d47
Bena(and the Sugar Plum Fairy),Rosine
Lib 15h3'0"-6d19'
Benabou,Salomon
Aur 4h34'44"30d23'
Benaglio,Jamie L
Leo 10h45'10"8d13'
Benain,Pierre
Per 3h53'8"35d14'
Benalil,Mohammed
Aur 6h19'0"33d31'
Benamou,Alain
Aur 4h34'30"30d13'
Benante,Martha Celeste
And 23h18'48"51d27'
Benaquisto's Star,Lisa
Lyn 9h8'1"44d8'
Benard,Yves Claude
Lup 15h12'33"-43d43'
Benardout,Kathleen A
Cmi 7h27'54"5d21'A
Benarroche,Barbara K
And 1h17'13"40d60'
Benarros,Albert
Ori 6h16'37"10d55'
Benarros,Josette
Mon 7h56'12"-5d45'
Benazet,Charlotte
Ser 15h14'0"0d0'
Benazir,Begum
Nor 16h25'45"-44d21'
Benazra,Josue
Ser 15h16'43"17d37'
Benbow,Sheila
And 23h34'38"41d37'
Benbri
Peg 21h59'18"7d56'
Bence,Brandy
Peg 22h50'48"27d32'
Bench,Steve
Aur 6h37'46"35d30'
Benchinol,Aline
Vul 19h19'38"23d21'
Benchley,Tracy P & Christopher H Turner
Lac 22h28'36"53d16'AB
Bencivenni,Justine Celeste
Cas 2h24'24"70d10'
Benda,Fritz & Christine
Lyn 7h52'28"50d45'
Benda,Fritz & Christine
Lyn 7h34'27"41d48'
Bendall,Deborah Ann
Cas 0h3'38"60d22'
Bendall,Joy
And 2h34'11"40d60'
Bendall,Sara Ann
Lyr 19h2'28"33d60'
Bende,Mary Ann
Com 12h28'55"22d4'
Bendeck,Zacarias Elias
Oph 18h3'42"0d23'
Bendel,Berne C
Uma 11h23'43"38d10'
Bendel,Wolfgang
Boo 14h26'36"11d49'
Bender January 10 93, Amy & Bruce
Sge 19h1'13"20d17'
Bender Nuptials
Cyg 21h37'14"41d35'
Bender's Lucky Star, Blake
Aql 20h33'40"0d45'
Bender,Alan R
Ori 4h53'19"-0d5'
Bender,Blake
Aql 20h11'18"5d34'
Bender,Craig
Her 17h47'47"21d44'
Bender,Doris & Merritt
Peg 21h59'59"34d28'
Bender,Jared Andrew
Gem 8h3'48"28d28'
Bender,Larry F
Cam 3h51'52"77d50'
Bender,Lisa
Mon 7h17'30"-0d6'
Bender,Lisa Joy
Aqr 22h7'17"-0d1'
Bender,Mary Anne
And 0h48'1"39d10'
Bender,Matthew Cale
Cnv 12h24'35"37d56'
Bender,Nancy Zoppa
Lyr 18h38'1"40d8'
Bender,Nuran Nafia
Dra 12h42'17"68d56'
Bender,Sarah Lynn
And 23h15'1"41d57'
Benderly,Rosie
Hya 8h36'36"1d46'
Bendes,Maida J
And 23h27'25"49d13'
Bendfeld,Hans-Juergen
Leo 10h33'40"18d0'
Bendig,Alexandra
Cap 21h2'27"-20d25'
Bendik,Holly Beth
Cyg 21h13'26"28d55'
Bending,David Paul
Leo 9h57'34"7d9'
Bendix,Barton Gage
Lac 22h29'22"50d16'
Bendix,Joseph M
Hya 8h42'53"2d27'
Bendixen,Benjamin
Gem 7h1'47"24d35'
Bendor
Cnv 14h56'33"28d16'
Bendorf,Brooke M
Cma 6h53'56"-19d15'
Bendure,Stephen W
Ori 5h52'23"14d29'
Bendzala,Judy
Cet 2h22'49"2d22'
Bendzko,Andrea
Psc 23h39'32"-1d16'
Bene & Kevin
Cep 21h39'41"65d33'
BeNea
Cma 6h53'19"-19d34'
Benecke,Diane
Aur 5h7'31"29d33'
Benecke,Roxanne Marie
Mon 7h4'39"0d32'
Benedec,Kevin
Ori 5h48'10"12d19'
Benedek,Denise
Per 3h31'30"51d47'
Benedetta
And 23h1'46"50d2'
Benedetti,Jocelyn Salva
Pho 0h32'6"-44d54'
Benedetto,Alexander Maximillian
Cnc 8h53'29"17d47'
Benedetto,Giovanni e Serena
Oph 18h5'19"10d37'
Benedetto,Sara & Sam
Cam 13h10'34"76d58'
Benedick,AFC,Baby
Equ 21h0'25"8d28'
Benedict,Andrea
Peg 22h37'46"2d26'
Benedict,Cathy P
Cas 2h41'1"68d59'
Benedict,Dorothy & Rome
Eri 4h10'28"-19d32'
Benedict,Jack A
Cet 2h34'19"5d12'
Benedict,Jeff
Per 4h20'0"50d10'
Benedict,Lois
Uma 10h52'49"40d56'
Benedict,Lorraine F L
Cyg 21h44'52"28d20'
Benedikt
Cep 21h35'37"85d54'
Benedikt
Cet 2h58'0"4d18'
Benedittini,N
Lac 22h20'10"55d22'
Benefield,David C
Aql 20h9'28"4d14'
Benefield,Paula
Crt 11h34'59"-18d36'
Benefield,Tiffany Dawn
Eri 2h54'0"-11d18'
Benel,With Love To My Mom,Helen
Vul 20h57'54"28d52'
Benell,Ted
Per 3h28'29"51d44'
Benellen
Cep 20h50'45"61d22'
Benemerito 7-13-91, Raymond & Kathleen
Her 16h56'57"38d32'
Benes,Renáta
Cmi 7h21'58"4d47'
Benesi,Kevin Matthew
Del 20h19'1"10d57'
Benet,Richie & Stacey Reiff
Her 18h5'33"41d4'
Beneth
Ori 6h6'0"9d18'
Benetti,Antonia
Cnv 13h36'23"28d34'
Benevento,Bryon
Del 20h23'26"10d54'
Benevento,Joe & Mary
Cyg 19h47'56"29d28'
Benevento,Shelly
Del 20h23'25"10d55'
Benezra,Yetta Berman
Cet 2h55'53"1d39'
Benfield,Christopher
Aur 4h58'28"38d55'
Benford,Nancy Anne
Lyr 18h37'29"35d45'
BenFredj,Leila Maria
Com 12h54'29"21d30'
Bengala
Cas 23h41'27"61d4'
Bengel,Jr,The Right Worshipful Henry
Cep 20h58'55"58d38'
Bengelsdorf,Irving
Cet 0h27'34"-12d2'
Benger,Tony
Her 17h7'1"20d25'
Bengough,Andrew & Lucinda
Eri 4h4'43"-8d50'
Bengtson,Joan M
Aql 19h31'30"0d12'
Bengtson,Nora M
Tau 4h23'53"30d37'
Bengu,Golgen
Aql 18h57'51"-7d32'
Benhalla,Fouad
Dra 10h3'13"78d33'
Benham,David
Oph 16h59'0"11d2'
Benham,John D
Ser 15h18'27"7d12'
Benham,Nicholas
Uma 11h28'13"66d33'A
Benham,The One For
Psc 1h27'57"12d26'
Benhamou,Madame Catherine
Peg 23h47'0"30d33'
Benhar 95
Lyn 8h11'46"39d33'
Benhima,Dris
Lac 22h29'1"40d19'
Benhur
Her 17h14'33"26d24'
Beniamino,The Prettiest Star Robin
Lyn 8h28'30"40d47'
Beniard,Michael Raynor
Her 16h19'21"21d59'
Benick,Hildy Weissman
Lmi 10h1'23"36d11'
Benigno,Helene
Uma 11h4'17"71d43'
Benigno,Nicholas
Her 17h22'49"44d28'
Beninati,Anthony Angelo
Aur 6h24'57"30d7'
Benincasa,Dommineque Demir
Cnc 8h54'1"30d11'
Benintends,The
Aur 5h15'56"44d50'
Benirschke,Kari
Cet 0h29'27"2d6'
Benita & Rocky
Cyg 19h30'34"36d36'
Benitez,Leonardo Bin
Tau 5h50'23"26d54'
Benitez,Louis
Boo 13h39'38"14d40'
Benitez,Lucy
Tri 2h34'30"34d50'
Benito
Cet 1h5'48"-6d42'
Benito,Isabel Diez
Cas 0h28'58"60d12'
Benito,Leihulu
Umi 14h52'22"72d15'
Benito,Lucia Cuadra
Tri 1h58'50"31d47'
Benito,Russ & George Coffman
Del 20h32'42"20d36'
Benjamin
Per 4h2'59"50d42'
Benjamin
Cep 23h12'12"61d32'
Benjamin
And 23h1'50"36d43'
Benjamin Chocolate
Oph 18h16'20"8d59'
Benjamin Luke
Cep 22h25'21"60d24'
Benjamin Michael
Cep 22h44'22"70d20'
Benjamin's Star
Sgr 18h5'37"-28d8'
Benjamin,Dashiell
Dra 20h18'34"73d57'
Benjamin,David
Cmi 7h38'44"10d58'
Benjamin,Dodie
Hya 8h14'0"0d59'
Benjamin,Howard Charles
Per 3h10'13"46d7'
Benjamin,Irene
Vul 20h39'56"20d22'
Benjamin,Jean P
And 23h43'1"45d51'
Benjamin,Jennifer Ann
Tau 4h12'47"20d9'
Benjamin,Lois Jane McQuillin
And 23h43'21"47d3'
Benjamin,Mathew Paul
Cep 22h20'45"60d3'
Benjamin,Monica
Peg 22h22'45"21d28'
Benjamin,Rich Richard
Cma 6h54'17"-19d13'
Benjamin,Robert
Cnv 12h29'38"32d30'
Benjamin,Ronald Gene
Cep 21h30'23"65d6'
Benjamin,Sheila Pauletta
Mon 6h36'42"-0d21'
Benjamin-August
Cam 5h3'17"58d9'
Benjamins,Baruh Tarum
Boo 14h49'38"47d37'
Benjaminsen,We Love U, Nana & Grandpa
Sct 18h54'54"-6d30'
Benjelloun,Farid
Oph 17h59'57"10d45'
Benji
Aur 5h50'58"50d8'
Benji
Uma 11h48'19"42d36'
Benji's Star
Per 3h18'36"41d3'
Benjie
Aur 6h54'49"37d59'
Benkan UCT Company
Uma 10h58'1"70d17'
Benkert,Jackie
Peg 22h3'57"24d25'
Benkert,Maryann
Lyr 18h41'26"42d37'
Benkoe,Stephanie Ellen
Eri 3h36'58"-15d4'
Benn,Edgar
Peg 23h4'31"18d51'
Bennaim,Gary
Uma 10h37'31"57d51'
Bennecke,Valerie
Cas 3h4'58"57d59'
Bennell's Star
Uma 11h43'16"55d16'
Benner Clover
Lyn 9h18'46"38d38'
Benner II,Louis C
Her 15h56'47"50d33'
Benner Silver Star, John & Marilyn
Uma 8h36'1"72d14'
Benner,Becky Suzanne
Aql 19h57'35"11d14'
Benner,Brandi
Peg 22h36'60"22d21'
Benner,Danielle N
Del 20h13'24"12d5'
Benner,Mark Stephan
Ser 16h7'11"5d32'
Benner,Richard Oren
Cnv 13h19'46"32d7'
Benner,Samuel Joseph
Aur 5h23'21"54d56'
Benner,Zona Marie
Cnv 13h20'50"31d29'
Bennet,Martine
Cas 2h26'21"50d60'
Bennet,Rick
Boo 14h8'48"52d53'
Bennett
Umi 16h27'31"70d18'
Bennett (Carmichael), Richard
Her 16h18'54"25d37'
Bennett,Helen
Nor 15h40'33"49d3'
Bennett (Jack),Mr John Henry
Cas 2h40'46"61d18'
Bennett 20,C B
Aur 7h25'14"38d31'
Bennett neé O'Brien (Peggy),Margaret Mary
And 1h19'55"48d49'
Bennett Superstar, Estelle
Lyr 18h14'42"45d47'
Bennett's Brillance
Sct 18h47'49"-6d19'
Bennett(Margy),Miss Margaret Mary
Com 13h16'22"26d59'
Bennett,Adam
Her 18h17'47"14d38'
Bennett,Alberta Toliver
Lac 22h15'0"47d43'
Bennett,Alison
And 2h33'60"38d44'
Bennett,Antony Paul
Cyg 20h22'1"39d58'
Bennett,Barbara L
And 2h11'47"39d51'
Bennett,Beth A M
Cnc 8h40'11"17d9'
Bennett,Brett Mansfield
Sco 17h32'0"-31d36'
Bennett,Catherine Kennedy
And 0h19'35"33d29'
Bennett,Charles W
Tri 2h34'1"35d54'
Bennett,Clinton Rob
Dra 16h13'25"62d6'
Bennett,David W
Cyg 19h53'19"38d41'
Bennett,Dean King
Dra 16h53'48"60d32'
Bennett,Delana Lorraine
Mon 7h39'26"-3d5'
Bennett,Derrick
Peg 23h43'39"30d58'
Bennett,Donald
Lac 22h20'14"40d13'
Bennett,Dorothy
And 1h46'47"37d20'
Bennett,Dustin Anthony
Oph 18h7'39"12d39'
Bennett,Edward Burton
Aql 19h31'54"10d25'
Bennett,Edward M
Cet 2h2'39"1d17'
Bennett,Elaine
Boo 15h1'47"8d16'
Bennett,Elizabeth
Mon 6h21'17"8d24'
Bennett,Emily Jane
Umi 14h26'43"67d39'
Bennett,Eric H & Paula K
Dra 16h30'37"51d43'
Bennett,Gary & Lori
Dra 17h45'1"57d36'
Bennett,Greg
Her 17h53'44"41"13'
Bennett,Hannah Grace
Lyr 18h58'13"41d12'
Bennett,Harold Harper
Per 3h11'29"44d31'
Bennett,Harry Thomas
Peg 23h0'41"14d32'
Bennett,Jackie
Sex 10h33'16"1d39'
Bennett,Jackie
Peg 0h1'1"30d45'
Bennett,James Christian
Ori 5h55'0"11d60'
Bennett,James Lawrence
Aql 20h2'12"8d8'
Bennett,Jane Ann
Lyr 19h24'38"41d33'
Bennett,Jennifer Morgan
Cas 1h1'23"58d24'
Bennett,Jill
Equ 21h16'44"3d37'
Bennett,Joan & M J Atlas "Forever"
Boo 15h3'30"32d27'
Bennett,John Russell
Per 2h55'23"38d2'
Bennett,Jonathan Harris
Her 16h12'57"23d10'
Bennett,Joseph A "Joey"
Oph 17h52'23"12d4'
Bennett,Joseph Michael
Cep 22h16'41"55d38'
Bennett,Joshua Kirk
Oph 17h1'31"11d18'
Bennett,Joyce Lynn
Eri 2h58'15"-1d34'
Bennett,Jr,Charles Harvey
Vul 20h42'40"28d55'
Bennett,Jr,Homer Lee
Cam 3h29'30"58d35'A
Bennett,Jr,James Conway
Lac 22h32'0"53d29'
Bennett,Judy Gannon
And 1h12'60"40d38'
Bennett,Julie Ann
Cas 23h42'0"61d47'
Bennett,Karen
Lyn 9h18'0"38d10'
Bennett,Kimberley Ney
Peg 22h58'28"22d57'
Bennett,Kristina
And 0h9'23"35d27'
Bennett,Linda A
Lyr 18h40'11"26d33'
Bennett,Mark Anthony
Lac 22h54'35"52d54'
Bennett,Maureen
And 23h27'43"43d27'
Bennett,MD,Jean Lester
Oph 16h45'57"2d47'
Bennett,Meryl
Per 2h44'31"40d52'
Bennett,Michele Margret
Cyg 20h22'60"31d5'
Bennett,Mollie
Mon 6h18'34"0d9'
Bennett,Molly
Peg 21h51'55"30d38'
Bennett,Nikki
Ori 5h24'60"0d24'
Bennett,Pamela
Cyg 20h18'43"40d45'
Bennett,Patricia
Oph 18h39'22"7d48'
Bennett,Paul Thomas
Her 16h10'51"23d45'
Bennett,Peter E
Ori 6h12'53"0d0'
Bennett,Rachael Victoria
Cas 23h39'13"61d11'
Bennett,Robin
Cep 23h8'18"68d20'
Bennett,Ryan "B-Luv"
Boo 13h43'18"18d13'
Bennett,Samuel
Ori 5h55'38"11d53'
Bennett,Scott Charles
Ori 5h20'1"15d45'
Bennett,Sr,James H
Del 20h53'41"2d58'
Bennett,Steve
Her 17h52'1"14d49'
Bennett,Steven Bradford
Per 1h58'31"50d4'
Bennett,Teresa Ann
Cas 0h26'14"62d17'
Bennett,Terrence R
Sct 18h53'60"-5d24'
Bennett,The King
Cep 1h9'13"78d26'
Bennett,Thomas James
Dra 16h1'57"52d27'
Bennett,Tiffany Lauren
Crt 11h13'52"-17d12'
Bennett,Tod K
Vir 11h40'16"6d57'
Bennett,Tracey Ann
Lyn 7h31'26"50d54'
Bennett,Travis
Lac 22h13'45"47d17'
Bennett,Uncle Brother
Her 17h55'20"31d8'
Bennett,Wadell "Woody" A
Cam 7h57'53"60d48'
Bennett,Ward
Cep 22h24'33"59d37'
Bennett,William & Grace
Crb 16h22'26"38d12'
Bennett,Wilma Pearl
Cam 3h29'30"58d35'B
Bennett-Rosman,Alexa
And 23h3'29"51d35'
Bennetta
Cyg 21h0'22"31d52'
Bennette
Aql 20h0'18"8d35'
Bennick,Pamela
Cas 23h36'40"61d29'
Bennington
Aur 7h11'0"35d38'
Bennington,Colin Okey
Cep 22h6'20"60d28'
Bennington,Wayne Edward
Per 1h53'23"53d8'
Bennis,I Love June
Peg 21h27'18"2d36'
Bennis,Mary (Nanny)
Cnv 12h18'0"41d8'
Bennitt,Roni A
Lyn 9h5'20"43d26'
Bennitt,Sue R
And 0h48'0"39d7'
Benno-Caris,Fan
Aql 18h57'31"15d7'
Bennorth,Judi A
And 23h12'1"40d42'
Benny
Hor 3h24'55"-47d1'
Benny
Uma 11h41'44"62d40'
Benny
Cep 22h7'58"67d46'
Benny
Ori 5h39'50"1d2'
Benny The Rat
Cma 6h12'45"-15d2'
Benny-Mae
Aql 19h50'59"15d37'
Beno
Leo 11h9'54"-0d52'
Benoit,Alan David
Aur 5h32'17"37d31'
Benoit,Annette
Ori 5h3'26"13d31'
Benoit,Bill
Boo 14h2'29"28d32'
Benoit,Bonnie
And 23h31'21"48d48'
Benoit,Brothier Guillaume
Umi 16h23'46"72d26'
Benoit,Carine
Cas 1h48'34"58d2'
Benoit,Deborah Olivia
Mon 7h21'50"-0d35'
Benoit,Jacqueline Gale
Hya 8h9'12"2d12'
Benoit,Maureen A
Uma 9h23'57"61d36'
Benoit,Pauline
Com 12h20'29"21d31'
Benoliel,The Best Uncle of All,Nick
Uma 11h38'0"31d35'
Benou
And 23h1'33"40d21'
Benoât 2-04-1952
Cam 5h57'25"56d5'
Benoât,Isabelle Brisebois
Per 3h16'0"40d49'
Bensadoun,Sylvie
Lac 22h25'54"40d1'
Bensaid,Farid
Lac 22h24'13"40d30'
Bense,David
Aur 5h0'43"48d50'
Bense,James H
Cep 22h10'34"58d17'
Bense,Kelly Ann
Cyg 20h52'58"30d35'
Bense,Scott
Per 3h46'16"36d21'
Bension,Alexander Robert
Mon 8h0'0"-8d7'
Benskin,Wesley Conner
Cep 0h17'11"73d32'
Benson "Lasting Love", William & Tammy
Uma 9h23'36"55d32'
Benson's Beacon,Benny
Lyr 18h58'36"38d54'
Benson,Alvin Keith
Boo 15h2'20"50d20'
Benson,Amy Denise
Peg 23h5'0"21d30'
Benson,Brad
Oph 18h42'23"7d57'
Benson,Brad D
Cma 7h17'27"-15d6'
Benson,Chloe
Mon 6h43'0"-10d43'
Benson,David
Her 18h19'10"12d55'
Benson,Donna
Peg 22h36'45"21d13'
Benson,Earl & Mary Lou
Mon 7h1'49"5d16'
Benson,Eric
Peg 22h41'42"12d22'
Benson,Fritz
Her 17h27'33"42d32'
Benson,George Robert
Lac 22h31'40"53d29'
Benson,Irish Love Star of Kenneth J
Cep 3h29'39"78d46'
Benson,Jeri
Umi 16h18'50"70d57'
Benson,Jill Anne
Peg 23h28'20"22d3'
Benson,Joe
Boo 14h27'23"12d19'
Benson,Jonathon
Ser 15h40'29"7d23'
Benson,Karen Ann Christy
Lyr 18h54'20"40d44'
Benson,Karen Kayserand Mason
Cyg 20h16'34"30d22'
Benson,Kenny
Dra 16h8'12"58d32'
Benson,Lisa R
And 2h24'24"42d53'
Benson,Lynsey Katelyn
Ari 1h58'30"25d22'
Benson,Melissa
Mon 6h33'59"0d1'
Benson,Susan Ann
Cet 0h29'60"-5d46'
Benson,Susan Lee
Cyg 21h16'1"28d45'
Benson,Vern & Anne
Per 3h9'43"50d22'
Bensoussan,Lorenzo
Peg 23h5'46"31d40'
Bent,Jennifer Disa
Sco 17h55'52"-38d13'
Bentaurus Bazillionaire
Cyg 19h18'22"49d3'
Bente
Cnc 8h35'18"18d53'
Bentel,Toni & Matthew
Cmi 7h32'19"10d36'
Bentley's For Crying Out Loud
Eri 2h57'34"-16d47'
Bentley,Amber Patricia
Uma 9h42'30"49d23'
Bentley,Barbara
Vir 11h51'19"8d14'
Bentley,Beryl Ann
Uma 8h21'0"61d33'
Bentley,Dawn
Lyn 6h28'35"60d38'
Bentley,Dee
Lyr 19h17'57"28d17'
Bentley,Eddie Duane
Cep 0h28'28"78d58'
Bentley,George Richard
Ser 16h8'18"14d35'
Bentley,Georgia Ewing
Cyg 19h29'19"32d19'
Bentley,Heather Elizabeth
Aql 20h34'44"-0d31'
Bentley,Jimmy
Dra 20h8'22"62d28'
Bentley,Joseph James
Aur 6h55'1"43d47'

Bentley,Linda
Peg 22h11'23"4d47'
Bentley,Lorna
Cyg 20h15'39"38d33'
Bentley,Maggie
Peg 22h11'1"3d35'
Bentley,Mari
Com 12h5'1"22d2'
Bentley,Michael Andrew
Cam 4h28'1"68d35'
Bentley,Rachelle
Sge 19h11'8"16d25'
Bentley,Robert Dewar
Boo 14h15'0"39d1'
Bentley,Roxanne Elizabeth
Vul 19h22'60"23d16'
Bentley-52086
Cnv 12h10'40"35d15'
Benton 50
Cyg 19h26'33"31d17'
Benton My Shining Star
Forever,Bob
Lmi 10h4'49"31d29'
Benton,Angelita
Cet 1h29'35"0d21'
Benton,Brian W
Per 3h13'20"41d31'
Benton,Claire
Lyr 19h11'23"40d30'
Benton,Darlene
Cas 0h35'6"67d15'
Benton,Evelyn
Aur 7h2'58"38d10'
Benton,John Mitchell
Her 16h53'37"25d35'
Benton,Jr,Joe
Boo 14h48'1"35d23'
Benton,Jude
Mon 8h5'34"-7d16'
Benton,Melva Diez
Eri 3h57'46"-10d46'
Benton,Peter Lowell
Per 4h22'59"51d47'
Benton,Sandy
Peg 0h10'21"14d5'
Benton,Steven Michael
Ori 5h49'33"18d5'
Benton,Susan Lynn
Ori 6h11'15"0d57'
Benton,Tullious
Lyn 7h8'44"58d54'
Bentsen Star,B A
Oph 17h33'0"11d24'
Bentstock,Yael
Hya 8h18'16"2d60'
Bentz,Trudy
And 23h42'19"40d55'
Bentzen,Erik
Cap 20h6'1"-10d26'
Bentzley,Tugger
Cep 21h54'0"60d0'
Bentzon,Niels Viggo
Ori 5h58'40"8d26'
Benudorion Destiny
Cap 20h58'11"-14d41'
Benuscak-Eternal Star, John F
Aur 4h59'12"50d31'
Benussi,Michele
Aur 4h59'53"40d7'
Benvenuti,Italia
Ori 5h36'34"8d38'
Benvenuti,Valerie
Lyr 18h44'52"45d11'
Benvenuti-Firenze, Chiara
Hor 3h33'39"-48d41'
Benvenuto,Remi
Boo 14h29'1"44d37'
Benvie's Eternal Love
Cyg 19h39'22"30d60'
Benyak,Sandra
Peg 22h43'19"5d47'
Benyas' "70th Birthday",Robert
Per 1h58'39"53d48'
Benyas,Bradley A
Lib 15h38'52"-23d9'
Benyas,Diane
Cas 1h5'38"58d16'
Benyomin,Malach Ben
Cep 20h51'13"68d53'
Benz,Stella Neve Forty Gary
Aql 19h56'33"15d35'
Benza,Lady Letitia
Leo 10h57'59"12d12'
Benza,Rosemary Elizabeth
Mon 6h20'35"7d31'
Benzel,Martha & Gary
Cyg 19h29'15"36d18'
Benzie,Mary
Umi 16h29'34"70d50'
Benzing,Beth
Peg 21h47'49"34d1'
Benzler,Becky
Sge 19h2'26"20d13'
Benzon,Andrija Quintrell
Cam 5h38'0"60d17'
Benèt,Jim & Ruth
Oph 17h4'0"11d19'
Beowulf
Uma 9h50'57"57d57'
Beppe & Virginia
Scl 23h39'0"-25d40'
Beppe '67
Col 6h25'37"-33d6'
Beppel,Laurette Lynn
Cas 2h16'0"65d12'
Beppel,Marian Jean
Cam 3h48'0"55d39'
Beppler,Andrew
Sgr 18h54'54"-28d55'
Beppler,Katie
Aqr 22h32'22"-10d47'
Beppler,Shana
Cnc 9h19'47"32d32'
Beppler,Shirley Mae Huber
Lyn 7h52'24"58d58'
Beppler,Steven
Per 1h55'14"48d39'
Beppo
Lyr 18h37'0"27d49'
Bera Danielle
Uma 9h28'57"50d5'
Beran,Alison & Honza
Lyn 7h30'42"36d54'
Beran,Barry Jay
Aur 6h33'58"37d40'
Beran,Herr
Cmi 7h17'29"1d20'
Berardelli,Claude
Uma 8h48'60"68d43'
Berardino,Joseph & Carol
Cyg 20h59'34"31d13'
Berarducci/01-03-51, Daniel
Aur 5h32'37"38d32'
Berarducci/08-13-80, Daniel A
Cep 22h2'48"53d2'
Berarducci/09-08-72, William
R
Dra 9h58'29"73d36'
Beratungsgesellschaft fur
Haustechnik,PB
Lyr 19h21'1"31d50'
Beraud,Janet
And 23h3'0"51d51'
Berbel
Lac 22h5'44"38d16'
Berberian,Garo
Tau 4h14'44"21d50'
Berbrier,Emilie & Danielle
Tau 3h37'49"28d32'
Berbuto,Anny
Her 16h26'0"48d22'
Berby,The Star of Tom
Per 4h5'44"37d53'
Berch,John Herbert
Lac 21h56'15"40d5'
Berch,Philip James
Lac 21h57'41"36d43'
Berch,Shannon Marie
Lac 21h59'14"40d1'
Berczi,Beatrice
Aur 4h34'27"30d55'
Berdan,Leonard & Florence
Cam 7h58'1"60d44'
Berdan,Randy
Aur 6h24'59"35d2'
Berding,Joe
Aql 19h29'40"-6d55'
Berding,Katherine Helen
Sge 20h14'27"17d3'
Berdit,Fay & Sam
Cyg 21h54'16"42d59'
Berek,Richard L
Boo 15h0'50"20d31'
Beremeneer
Cas 1h47'0"60d31'
Berend,Natalie
And 1h36'48"40d4'
Berends,DVM,John H
Per 4h44'26"51d13'
Berends,Eric Edward
Her 17h15'21"20d49'
Berends,Harald
Cam 4h26'47"58d43'
Berends,Irmhild
Leo 9h19'29"11d30'
Berends,Jeffrey & Theresa
Cam 4h50'50"70d41'
Berendsen,Lauren Dale
Cnc 8h53'27"31d9'
Berengere
And 23h7'27"43d38'
Berenholz,MD,Joseph
Oph 17h35'1"-3d39'
Berens,Catherine Jean
Lyr 18h40'15"33d23'
Berens,Janet & A James
Boo 14h47'22"30d39'
Berens,Jean
Ori 5h49'32"20d32'
Beres,Helen
Lyr 19h18'1"38d3'
Beresford,Amy
And 23h46'40"33d42'
Beresford,Robert Alexander
Uma 11h34'40"51d50'
Berets,Hilde Gimnicher
Peg 21h57'13"2d12'
Beretta,Marco
Ori 5h23'50"1d33'
Berezik,Jerry
Uma 10h58'0"53d34'
Berg Quester,Jeff
Per 3h21'21"33d21'AB
Berg,Andrew James
Aur 7h4'37"37d59'
Berg,Annette
Cnv 12h42'53"35d15'
Berg,Arlene
Umi 16h19'26"73d36'
Berg,Catherine Locke
Peg 23h20'0"28d52'
Berg,Charles
Uma 8h56'1"50d33'
Berg,Christopher
Mon 6h56'14"10d19'
Berg,Edward & Sandra
Cyg 19h41'22"38d43'
Berg,Jeff
Cet 1h26'56"-0d1'
Berg,Jr,Oswald
Boo 13h44'43"27d47'
Berg,Karen "Buddy" Lynn
Vul 20h14'54"23d60'
Berg,Katherine Elizabeth
And 1h24'40"39d54'
Berg,Kathryn Elizabeth
And 2h24'1"41d36'
Berg,Kelsey Maria
Lyr 18h40'1"45d41'
Berg,LaVonne Rommel
Crb 15h21'16"32d33'
Berg,Lawrence Herrald
Per 2h51'24"40d4'
Berg,Lee Michael
Cet 2h18'0"9d11'
Berg,Loren Christopher
Dra 15h11'31"63d12'
Berg,Pamela
And 23h2'21"50d56'
Berg,Richard
Lyr 18h17'24"30d42'
Berg,Robin
Cas 2h24'40"71d10'
Berg,Roy August Cannon
Aql 19h11'42"13d48'
Berg,Sondy Lynn
Cas 0h34'22"60d32'
Berg,Stephen
Boo 15h13'0"25d14'
Berg,Stephen Matthew
Ori 5h58'18"9d35'
Berg,Steve J
Her 17h58'29"50d27'
Berg,Tamara-Alexia
Lac 22h29'41"40d13'
Berg,Tanny
Hya 9h27'37"-0d20'
Berg,Tim R
Uma 9h29'49"42d28'
Berg-Joy,Mr & Mrs Richard
Edward
Ori 5h41'23"11d31'
Bergamini,Dr Bob's Office-Dr
Robert
Lac 22h54'52"40d40'
Bergamini,Stefania
Cas 2h11'20"60d2'
Bergantino,Antonello Et Nancy
Per 2h54'47"56d49'
Bergantino,James
Peg 22h2'21"32d41'A
Bergdoll IV,John Catieser
Her 18h4'45"38d49'
Bergdoll-Park,Jacob
Cep 21h3'1"80d13'
Berge,Ashleigh Lauren
Uma 9h18'32"51d38'
Bergel,Afton
Her 16h52'42"32d10'
Bergemann,Eric
Cnv 12h12'52"40d33'
Bergen & Family,Mae
Cas 1h4'1"55d47'
Bergen,Danielle
Lyr 18h40'36"46d54'
Bergen,Jr,Earl H
Her 16h59'26"40d39'
Bergen-Hunt Family,The
Her 18h35'0"18d48'
Bergendahl,Alexander Alfonso
Cma 6h55'59"-19d23'
Berger II,Thomas Edward
Aur 5h35'48"50d27'
Berger Shining Star, The
Stuart
Cra 18h0'16"-39d42'
Berger,Allison Rose
Peg 22h28'1"29d54'
Berger,Ann
Cmi 7h54'1"4d3'
Berger,Bev & Jeff
Cyg 19h28'36"34d51'
Berger,Christina Lynn
Lyn 8h14'18"35d55'
Berger,Christopher Russell
Boo 14h11'0"41d59'
Berger,David Matthew
Aql 18h53'42"-0d7'
Berger,Douglas Michael
Hya 8h11'51"1d25'
Berger,Dr Richard S
Cep 20h59'38"63d50'
Berger,Einar Arthur
Dra 20h33'47"70d12'
Berger,Elke
Cas 0h41'39"60d21'
Berger,Frank
Sco 17h0'1"-33d58'
Berger,Gerhard
Boo 14h19'31"10d41'
Berger,Gerry
Com 12h19'45"28d25'
Berger,Harris Joseph
Cet 2h50'21"1d2'
Berger,Ione
Lyr 18h41'60"32d10'
Berger,Jamie
Per 3h52'16"36d34'
Berger,Jeffrey E
Lac 22h51'50"53d11'
Berger,Jeremy William
Dra 19h57'27"70d2'
Berger,Jessica
Lib 15h17'28"-22d30'
Berger,Joe
Ori 5h26'1"-2d18'
Berger,Joseph
Lac 22h45'20"55d22'
Berger,Josiane
Aur 4h59'27"30d16'
Berger,Lance A
Ser 15h13'40"21d7'
Berger,Mark & Idele
Crb 16h21'55"31d17'
Berger,MD,Brent A
Dra 16h21'58"58d58'
Berger,Nina
Cmi 7h28'58"8d52'
Berger,Phillip
Lac 22h39'13"37d48'
Berger,Rachel Lee
Cyg 20h23'18"40d45'
Berger,Remi Austin
Peg 23h7'50"12d55'
Berger,Theodora Paculan
Cyg 20h27'22"40d32'
Berger,Thomas James
Her 17h11'21"48d23'
Berger,Travis Bryan
Oph 18h8'42"12d36'
Berger,William Albert
Cet 2h0'45"-1d7'
Berger,Winifred Hope
Hya 8h19'46"-0d48'
Berger-Dolphin-Kanes
Uma 11h11'1"44d45'
Bergeron,Hélène
Aql 19h54'60"13d8'
Bergeron,Jean
Umi 14h31'29"68d55'
Bergeron,Johanne
Uma 14h1'33"53d35'
Bergeron,John
Her 18h3'47"30d3'
Bergeron,John C
Her 17h4'36"47d39'
Bergeron,Matthew & Andrea
Peg 22h32'40"20d35'
Bergeron,Michelle
Cet 3h19'56"0d8'
Berges,Christina
Peg 21h59'42"23d2'
Berges,Jeremy
Oph 17h3'0"10d39'
Bergesen,Laura T
Lyr 18h55'0"37d2'
Bergeson,Alice
And 23h38'26"46d47'
Bergevin,J R "Jimbob"
Her 18h12'54"46d54'
Bergevin,Pierre
Cyg 19h27'0"33d4'
Bergez,Pioore
Ori 6h4'26"11d12'
Bergfeld,Ellen Geneva Marie
Cas 22h59'50"54d36'
Bergfield,Marisa
Cas 1h25'16"60d19'
Berggren,Maria Kristina
And 1h40'30"37d4'
Berggrun,David
Cyg 19h31'0"34d54'
Berghaus,Udo
Boo 14h7'20"11d52'
Bergheim
Oph 17h18'40"-22d33'
Berghoff,Jack & Doris
Per 4h2'26"37d42'
Berghoff,M Gail
Tau 3h59'20"2d27'
Berghold,S David Dillon
Cnv 13h32'0"51d17'
Bergin III,Richard Dana Bergin
Sct 18h51'19"-9d41'
Bergin,Thomas F
Ori 5h55'32"18d21'
Berglette Bull
Uma 10h7'0"68d30'
Berglund,Beth
Ser 15h24'13"8d60'
Berglund,Folke Waldenmar
Ori 6h4'33"11d6'
Berglund,Jon
Cmi 7h58'52"8d42'
Berglund,Rabbe Poika
Sgr 18h54'4"-25d31'
Berglund-Davenport, Vicky
Uma 11h21'51"68d44'
Bergman
Ori 5h22'45"0d32'
Bergman Star,The
Boo 14h50'32"29d58'
Bergman,Arthur Holger
Boo 15h45'24"48d57'
Bergman,Donna Ann
Peg 22h32'52"25d8'
Bergman,Grace L
And 1h21'12"36d38'
Bergman,Jake & Teddy
Cyg 20h1'37"40d38'
Bergman,Lee M
Uma 10h6'0"71d8'
Bergman,Patrick
Dra 10h41'38"80d12'
Bergmann,Claudia
Hya 9h37'39"-9d53'
Bergmann,Jesse Lee
Boo 14h10'36"38d13'
Bergmann,Nina
Gem 7h16'27"24d10'
Bergmann,Robert
Psc 22h51'0"5d9'
Bergoich-Star Drummer, Stan
Del 20h13'51"13d23'
Bergonzi,Marco
Cam 14h11'57"82d2'
Bergquist Star,Mary Elizabeth
Lyr 19h22'18"35d16'
Bergquist,Ingrid Anne
Cyg 20h0'49"30d31'
Bergren,Gail & Bob
Ori 5h53'31"12d43'
Bergson,Mitchel Myer
Dra 17h1'0"68d5'
Bergsten
Cam 4h45'51"68d6'
Bergstrom,Dori
Uma 10h15'24"52d55'
Bergstrom,Eric J
Lyn 9h15'21"35d34'A
Bergthold,Eileen
And 0h45'57"21d41'
Bergtholdt,Florine Marble
Boo 15h5'11"8d1'
Berguet,Cassy
Ser 15h58'49"10d47'
Berie MoMo
Cam 3h14'29"57d21'
Berings,Njord Cleybergh
Cep 1h28'0"85d36'
Beriot,Anne
Cyg 20h28'34"39d23'
Berisa
Her 18h8'31"40d33'
Berisa
Her 18h8'31"40d33'
Berit & Jim
Aur 6h5'32"46d5'
Berk,Bernard
Equ 21h23'1"2d33'
Berk,Brigitt & Robert
Mon 6h25'18"-10d30'
Berkana
Tri 1h57'17"30d28'
Berke,Anne
Cap 20h39'50"-20d8'
Berkebile 11-30-93, Harold
Ori 5h51'27"15d43'
Berkel,Jeffrey A
Hya 8h36'20"0d46'
Berkeley Magic Star
Boo 15h23'47"34d53'
Berkeley,Ester & Barry
Eri 3h55'23"-19d12'
Berkeley,Howard Michael
Cmi 7h42'16"4d50'
Berkemann-Wellner,Ute
Boo 14h7'23"10d43'
Berkemeier,Roswitha
Gem 5h58'30"26d50'
Berkenbrock,Henrique Philippi
Dra 16h46'37"71d39'
Berkenfield,John
Peg 21h55'27"2d45'
Berkes,John G
Cnv 13h23'1"37d41'
Berkheimer Kline Golin/Harris
Her 17h39'30"41d7'
Berkheimer,Angie Marie
Peg 23h7'21"10d35'
Berkheimer,Chase William
Aur 6h2'47"45d9'
Berkheimer,Ian Michael
Peg 23h7'39"11d26'
Berkin,Adrienne
Cas 23h20'14"60d27'
Berklacich,Ian Darcy
Peg 21h56'1"28d51'
Berkley's Beacon
Eri 3h20'46"-5d24'
Berkley,Elizabeth
Peg 22h40'0"34d48'
Berkley,Travis Jason
Aql 18h43'50"7d37'
Berkli Jean
Mon 6h23'59"-5d54'
Berklisbestar
Boo 14h48'1"33d3'
Berkman,Dilaver
Uma 10h33'1"59d29'
Berkman,Jack N
Cam 3h38'1"71d11'
Berkoff,Barbara
Cas 22h58'10"56d28'
Berkovitz,Trista
Cet 1h29'59"-3d35'
Berkowitz,Howard
Aur 5h3'34"48d46'
Berkowitz,Michael
Cep 21h32'50"60d51'
Berkowitz,Paul
Cet 3h10'56"5d56'
Berkowitz,Stacey
Lac 22h49'35"55d31'
Berkshire,Emily Susan
Cas 0h20'53"68d59'
Berkshire,Vera Claire
Cas 0h52'11"66d40'
Berkson Horus,Ras Lief
Drager
Cep 20h20'1"60d10'
Berkudos
Boo 14h47'52"34d24'
Berkule,Barry Michael
Uma 11h51'57"62d49'
Berlanga (HOB),Hector O
Aql 19h34'56"-6d40'
Berlanga,Paul
Lac 22h5'36"51d31'
Berlee,Kim
Cmi 7h35'18"11d35'
Berlekamp,Cody Randall
Vul 20h44'34"28d42'
Berliant,Susan L
Lyr 19h4'1"41d2'
Berlin & Lizy Sisters 4 Ever
Vir 13h52'50"0d47'
Berlin Super Star,The Laurent
Cma 6h57'46"-17d50'
Berlin,A Faye Eileen
And 23h28'35"46d58'
Berlin,Adam James
Dra 17h46'57"64d19'
Berlin,Amy Heather
And 1h55'0"40d12'
Berlin,Carole Merle
Ari 2h30'22"30d16'
Berlin,Charles I
Cap 20h29'41"-10d7'
Berlin,Etka
Ori 6h4'30"20d47'
Berlin,Harriet
Cap 20h27'11"-26d45'
Berlin,Jacob Alan
Aur 6h56'40"38d29'
Berlin,Joshua Alan
Cet 3h15'18"9d21'
Berlin,Marilyn
Cas 0h18'60"61d48'
Berlin,Mildred Savoy
Psc 1h40'37"28d4'
Berlin,Rainer
Ari 1h46'21"20d4'
Berlin,Richard Ian
Per 2h3'37"48d31'
Berlin,Robert Joseph Schnatz
Cep 21h57'16"55d8'
Berlin,Samantha Nicole
Cas 2h57'26"57d46'
Berliner Theater Club Otfried
Laur
Lib 15h1'45"-20d23'
Berlinger,Rhonie
Aql 19h59'39"15d16'
Berlingo,Emanuel V
Cep 21h27'16"60d13'
Berlinguette,Norman
Cep 2h37'32"78d56'
Berlon,John Paul
Ser 15h58'12"-3d33'
Berlusconi,Silvio
Boo 14h33'32"10d3'
Berluti,Matthew Christopher
Lmi 10h15'49"32d26'
Berluti,Nicholas James
Lac 22h22'17"55d55'
Berlyak,Julie
Aql 19h56'23"12d45'
Berman,Allen Marvin
Mon 7h18'3"-6d41'
Berman,Carol
Lyn 7h40'43"43d6'
Berman,Donald
Oph 18h0'33"12d38'
Berman,Gary Keith
Boo 14h11'27"42d34'
Berman,H Michael
Boo 14h34'55"51d31'
Berman,Jennifer Lynn
Tri 2h5'47"30d37'
Berman,Joanna
Mon 7h48'60"-8d56'
Berman,Jordan
Lac 22h5'28"40d18'
Berman,Leon
Equ 21h3'35"2d33'
Berman,Lucille S
And 0h3'51"38d26'
Berman,Marissa V
And 1h21'44"48d51'
Berman,Mark Daniel
Aql 19h3'1"0d10'
Berman,Meredith Lynn
Cas 0h8'42"62d25'
Berman,Natalie
And 0h25'55"44d39'
Berman,Paul"Buddy"
Dra 16h46'13"64d18'
Berman,Rachel Beth
Cyg 19h34'56"36d9'
Berman,Rachel Elexa
And 1h46'1"41d1'
Berman,Susan L
Lyn 6h59'52"58d55'
Berman,Tibor
Lac 22h20'24"53d21'
Bermellon
Lyn 7h43'13"45d19'
Bermingham,Pooh & King
Cmi 7h29'20"7d39'
Bermon,Daniel L
Per 2h56'22"43d48'
Bermont
Uma 8h37'17"51d20'
Bermosk,Greg
Cet 2h18'45"4d15'
Bermudes,Tanya M P
Mon 6h36'29"1d18'
Bern's Star
Vul 19h48'32"28d28'
Bernabeu,André
Per 3h59'50"36d37'
Bernadette
Boo 14h37'32"42d2'
Bernadette
Uma 9h46'0"58d41'
Bernadette
Lyn 9h10'21"37d60'
Bernadette
And 0h43'41"45d40'
Bernadette
Aur 5h57'49"50d27'
Bernadette,Susan
Lyn 9h19'29"40d3'
Bernadine Leora
Cam 11h13'31"81d23'
Bernal,Andrea Hernandez
Cas 23h1'20"53d43'
Bernal,Daniel A
Ser 15h12'22"9d28'
Bernal,Kristina
Lyr 18h29'0"42d44'
Bernal,Victor Eduardo
Boo 14h13'32"10d53'
Bernal-Silva,Patricia
Boo 14h39'1"45d47'
Bernanke,Jakob
Vir 13h36'10"-11d40'
Bernard
Dra 9h48'1"78d58'
Bernard
Cet 2h55'32"5d39'
Bernard
Uma 13h0'1"61d8'
Bernard & Amy
Cet 2h42'27"-10d35'
Bernard & Bernadette
Cyg 21h35'41"38d54'
Bernard 60
Cep 3h5'28"78d55'
Bernard Howard
Vul 18h55'58"24d49'
Bernard Michel
Cyg 19h38'58"38d53'
Bernard R
Dra 9h57'15"80d46'
Bernard's Stella
Del 20h14'20"14d36'
Bernard,Alain
Peg 23h35'56"10d3'
Bernard,Barry J
Sge 20h2'0"20d48'
Bernard,Barry J
Sge 20h2'0"20d48'
Bernard,Bradley J
Per 4h3'52"37d38'
Bernard,Brett Gehlmann
Cep 3h21'13"87d5'
Bernard,Cheryl
Cas 2h7'1"61d30'

Bernard,Diane
Cas 0h17'18"63d49'
Bernard,Gary Allan
Her 16h49'0"40d3'
Bernard,Michael
Hya 8h49'26"-6d36'
Bernard,Michael Arthur
Del 20h16'59"14d22'
Bernard,Mr Rouxel
Crb 16h19'33"30d39'
Bernard,Robert J
Mon 6h28'60"-5d56'
Bernard,Robert P
Mon 6h6'32"-8d33'
Bernard,Sarah
Mon 6h1'35"-6d37'
Bernard,Stewart
Aur 6h27'0"37d39'
Bernard,Sue
Mon 8h0'25"-10d5'
Bernard,The
Tau 4h5'26"18d48'
Bernard,Virginia & Robert
Uma 9h46'18"56d23'
Bernardi,Frank Michael
Aur 4h58'52"37d28'
Bernardi,Nicky
Aur 7h22'15"37d38'
Bernardi,Sara
Peg 22h52'48"24d34'
Bernardi,Tony
Dra 9h33'51"80d21'
Bernardini
Aql 19h9'23"15d39'
Bernardo,Carrie
Cyg 19h33'20"33d28'
Bernardo,Jr,Francis X
Her 16h56'30"29d14'
Bernardo,Rosa
Cyg 20h3'11"30d45'
Bernardo-Superstar, Clara
Leo 11h8'26"-1d9'
Bernatchez,Eric
Uma 8h9'54"68d42'
Bernatchez,Ginette
Uma 10h27'11"52d59'
Bernath,Merritt "Star Onysis"
Lac 22h22'15"53d41'
Bernd
Gem 7h18'17"27d44'
Bernd
Cnv 12h38'29"36d11'
Bernd
Ari 2h50'57"21d41'
Bernd Crispy
Sex 10h43'40"-11d42'
Bernd Jonas
Cnv 12h36'52"37d39'
Bernd M Michael
Vir 11h39'16"0d58'
Bernd Zimmers Stern
Lib 14h44'0"-22d31'
Bernd's Stern
Vir 14h3'59"7d23'
Bernd-Zill-50
Tau 4h25'50"30d22'
Berndt My Love Forever
Pamela Jean
Sgr 18h48'58"-23d25'
Berndt,Bill
Boo 13h40'36"26d7'
Berndt,Leon
Lyr 18h31'26"35d31'
Berndt,Sherry Ann
Uma 10h34'14"57d34'
Bernecker,Ulrich
Aql 20h9'43"3d55'
Bernel
Ori 6h5'46"-0d2'
Berner,Gerald L
Per 3h0'19"41d9'
Berner,Mark & Gayle
Cyg 21h28'29"40d26'
Bernetski,Mike
Her 16h21'47"28d27'

Bernett,Timothy Ian
Cep 22h15'0"67d44'
Bernetta,Erma
Crt 11h11'50"-13d58'
Berngen,Bunny
Uma 9h49'53"42d27'
Bernhard
Cyg 20h52'1"37d52'
Bernhard,Genore Hildagard
Cushman
Aql 19h26'39"8d29'
Bernhard,Harriett
Cet 3h17'0"-0d16'
Bernhard,Kimberly
Lyn 8h12'19"58d30'
Bernhardt,Andrew "Andy"
Cet 3h1'14"-0d51'
Bernhardt,Beatrice
Cyg 19h50'43"44d47'
Bernhardt,Christiane
Equ 21h3'15"2d14'
Bernhardt,Emily
Peg 22h43'51"11d23'
Bernhardt,Kevin Karl
Aur 5h39'40"38d17'
Bernhart,Mathias
Lmi 9h25'17"38d48'
Bernheart,Jeffrey Parker
Dra 17h29'34"71d40'
Bernica,Ann Elizabeth
And 23h11'52"38d47'
Bernice
Sgr 19h37'28"-33d31'
Bernice's "Diamond in the
Sky"
Crb 15h56'46"38d1'
Bernice,Stella Sorella
Crb 16h19'18"32d45'
Bernich,Anna "Susie"
Mon 6h29'43"-6d3'
Bernick,Dr John J
Cep 21h56'56"58d22'
Bernie
Cnc 8h48'47"30d15'
Bernie
Uma 11h1'1"70d45'
Bernie
Psc 1h31'51"28d14'
Bernie & Loretta
Cam 3h57'28"68d12'
Bernie (Grams) Ray
Sct 18h53'13"-6d32'
Bernie Bright
Aql 19h30'42"8d5'
Bernie Lee
Cnc 8h32'0"17d34'
Bernie's "Fantasy Dream"
Ser 15h53'1"-2d18'
Bernie's Golden Nugget
Cyg 19h50'38"47d48'A
Bernie's Valentine
Aur 6h35'1"37d58'
Bernie-Boop
Lyr 18h31'53"41d58'
Bernier,Anne-Sophie
Cas 0h33'0"64d30'
Bernier,Brian
Her 18h3'11"30d44'
Bernier,Carter Alan
Tau 4h17'1"18d53'
Bernier,Guy
Ori 5h59'31"13d4'
Bernier,Harold
Her 17h1'22"22d57'
Bernier,Marie
Uma 8h10'53"60d6'
Bernier,Marlaine
Cas 0h58'11"69d5'B
Bernier,Patricia
Cas 22h58'60"56d1'
Bernier,Pierre
Cyg 19h24'51"30d59'
Bernigaud,Elisabeth
Lyn 7h54'40"47d39'

Berniklau,Mary
Mon 7h5'31"1d49'
Bernin,Pierre
Sgr 19h27'26"-38d13'
Berninger,Patti
Eri 2h58'23"-11d9'
Bernini,Philip
Boo 14h12'0"40d10'
Bernius,Gudrun
Sgr 19h2'15"-23d47'
Bernjak,Konstantin
Uma 11h27'1"38d15'
Bernluis
Aur 5h59'22"29d19'
Bernocchi,Sergia
Umi 17h50'26"80d31'
Bernot,Mighty Mike
Her 17h39'1"40d4'
Bernoux,Pierre-Denis
Per 2h56'36"46d48'
Bernsen,"Crystal Princess"
Kara
Peg 23h28'37"8d44'
Bernsen,Werner
Sge 19h11'14"18d5'
Bernslaur
Cnv 13h50'1"4 1d7'
Bernstein,Alan David
Cnv 12h59'0"40d3'
Bernstein,Alexander
Aur 5h35'10"41d48'A
Bernstein,Alvin
Her 17h6'17"42d1'
Bernstein,Andrew Harris
Oph 18h40'0"8d42'
Bernstein,Brad & Marcy Gersh
Sco 17h30'25"-41d6'
Bernstein,Donald S
Her 15h55'21"43d40'
Bernstein,Elodie Deborah
Crb 16h4'12"37d55'
Bernstein,Ian Brett
Uma 11h41'54"32d15'
Bernstein,Jake David
Peg 23h46'60"31d49'
Bernstein,Joseph
Her 16h38'32"35d45'
Bernstein,Lori
And 0h4'53"40d7'
Bernstein,Rita Theiler
Cas 0h0'23"63d10'
Bernstein,Sarah Jean
Uma 10h53'45"56d5'
Bernstein,Steven Edward
Boo 14h38'10"15d28'
Bernstein,Steven Ray
Dra 16h36'50"72d51'
Bernstein,Victoria
Peg 22h37'28"29d42'
Bernstorff,Eric Johann
Peg 22h37'15"30d8'
Bernstorff,Peter Alexander
Ser 18h4'58"-1d19'
Bernt,Christian John
Sex 9h59'34"3d20'
Bernth,Nance
Sct 18h46'2" 7d34'
Berntsen,Gary J
Tau 5h49'52"28d13'
Berntsen,Robert Arnold
Cep 22h21'0"55d43'
Bero,Francis J
Lac 22h46'53"55d48'
Beronda,Donna
Mon 6h46'33"11d47'
Beronio,John & Anita
Cyg 19h12'16"48d17'
Berot,Armand
Uma 11h24'36"37d52'
Beroukhim,Asal
Peg 22h3'23"25d57'
Berquet,Jean-Pierre
Cyg 19h47'33"31d38'

Berquist Star,The Cody
Thomas
Leo 10h59'20"2d15'
Berra,Gary E
Cas 2h27'23"64d18'A
Berradi,Mehdi
Cam 5h51'45"68d31'
Berrer,Edith
Agr 22h44'36"-2d17'
Berresford,Michael P
Agr 21h19'1"-10d3'
Berri,Rodney S
Cam 6h45'46"80d31'
Berrie BA (Hons.), Julia
Lyr 19h14'11"42d26'
Berrier,Dorris Glynn
Aql 19h46'24"10d44'
Berrill,Irene Matilda
Cyg 21h51'47"38d57'
Berring-Star of Hope,
Compassion&Love,Nancy
Lyn 6h30'57"56d9'
Berriz,Armando Juan
Mon 6h35'13"-0d13'
Berroca,Ana Isabel Manzano
Com 12h18'15"21d45'
Berroll,Phil
Cep 23h13'50"60d30'
Berry III,James Morgan
Cet 3h19'0"9d59'
Berry Patch,The
Del 20h17'17"9d36'
Berry,Alison Denise
Lyr 18h16'59"40d2'
Berry,Allison Elizabeth
Uma 10h45'20"54d25'
Berry,Allison
Lyr 19h23'51"30d7'
Berry,Amy Katharine
Cnc 8h39'12"18d56'
Berry,Ann Elizabeth
And 2h32'46"39d25'
Berry,Barbara
Mon 7h4'0"-1d45'
Berry,Barbara S
Mon 6h20'17"7d26
Berry,Betty-Eric & Erica
Mon 6h34'32"1d24'
Berry,Bobby "Opie"
Boo 14h39'0"35d47'
Berry,Carole Ann
Cas 1h51'0"60d1'
Berry,Chester Bruce
Per 1h52'54"54d13'
Berry,Christine Lynn
Cas 1h54'29"65d21'
Berry,Clare Judith
Cas 1h16'45"76d26'
Berry,Cliff
Boo 14h42'54"38d8'
Berry,Daniel Patrick
Lmi 10h22'15"28d15'
Berry,Diana
Vul 20h42'27"20d19'
Berry,Doris J
Peg 22h3'39"20d33'A
Berry,Edward
Lac 22h27'50"53d23'
Berry,Elizabeth
Mon 6h42'22"10d48'
Berry,Elizabeth C
Lib 15h59'1"-8d42'
Berry,Fred
Her 17h12'56"21d31'
Berry,Gregory David
Lac 22h38'50"53d12'
Berry,Holly
Cma 6h41'0"-16d23'
Berry,I R
Ori 5h58'43"18d17'
Berry,James Findlay
Ori 4h56'39"0d1'
Berry,Jane
Mon 6h53'10"1d13'

Berry,Janie
Eri 3h13'1"-16d56'
Berry,Jennifer Brynn
And 0h46'40"22d0'
Berry,Jennifer Nicole
Mon 8h3'13"-1d21'
Berry,Jill
Lyr 18h18'37"44d23'
Berry,John M
Tau 5h52'26"28d55'
Berry,Jordan Alexandra
And 23h17'12"45d4'
Berry,Jr,CLU,ChFC,B Carroll
Cep 23h51'26"80d52'
Berry,Karla Elsa
Umi 14h2'45"69d47'
Berry,Keith Joseph
Uma 9h21'26"47d52'
Berry,Kelly
And 1h56'0"38d54'
Berry,Kelsey Lynn
Uma 10h52'0"54d52'
Berry,Louise
Com 12h28'1"22d41'
Berry,Marion
Uma 10h17'22"48d32'
Berry,Mary Elizabeth (Hull)
Mon 6h55'21"2d21'A
Berry,Michael J P
Ori 6h0'0"2d42'
Berry,Michele
Vul 19h47'0"22d45'
Berry,My Blue Angel - Martin
S
Oph 18?56"10d7'
Berry,Nan
Mon 6h53'41"-10d51'
Berry,Nichelle
Cas 0h58'1"75d58'
Berry,Patrick O'Neill
Cet 3h0'59"5d34'
Berry,Rachel René
Eri 2h50'48"-15d52'
Berry,Richard Albert
Mon 6h55'21"2d21'B
Berry,Robert Dell
Lac 22h33'16"55d33'
Berry,Robert Donald
Per 2h57'13"41d4'
Berry,Robert J
Peg 22h3'39"20d33'B
Berry,Stephen John
Her 18h28'21"24d38'
Berry,Susan
Aql 19h49'58"15d31'
Berry,Tania
Cyg 21h'25"30d58'
Berry,Theresa Kulisch
And 23h2'58"51d38'
Berry,William Frances
Her 16h56'30"23d38'
Berry,William Reid
Lib 14h55'36"-1d27'
Berryhill,Bevin Morgann
Lyn 8h8'30"36d56'
Berryhill,Keelan Rhys
Lyr 18h45'15"31d52'
Berryhill,Susan Patridge
Mon 6h21'45"8d35'
Berryman,Bob
Her 16h41'13"20d33'
Berryman,Jeny
Cas 0h24'13"71d58'
Bersch
Boo 15h1'0"53d19'
Bersch,Calvin G
Cam 7h6'0"70d40'
Bersch,Lucille L
Dra 18h26'44"58d26'
Bershaw,Lillianne Marie
Hya 9h0'51"2d17'
Berson,David Joshua
Ser 15h17'42"-0d49'
Berson,William
Lac 21h59'0"42d6'

Bert
Aur 6h39'33"38d51'
Bert
Ori 5h59'0"15d35'
Bert
Com 12h45'1"23d37'
Bert & Betty
Peg 23h21'52"10d46'
Bert 13130,Carroll
Uma 8h45'12"56d36'
Bert T
Cep 0h2'1"76d6'
Bert's Sparkle
Aur 5h15'45"40d46'
Bert's View
Cep 0h53'44"80d27'
Bert,My Love
Aql 20h7'46"4d2'
Bertagnoli,Dario
Nor 16h18'48"-50d25'
Bertain,Pamela Sue
Peg 21h52'39"28d47'
Bertannelie
Cyg 20h18'0"30d35'
Bertarelli,Elisabetta
Sco 16h59'16"-38d8'
Bertaud,Redmond E "Bert"
Ori 6h6'16"7d7'
Bertelsen,Devan
Peg 22h0'0"25d3'
Bertelsen,Tracy
Mon 6h18'10"4d7'
Bertenshaw From: Bobbi I
Love You,Bill
Her 16h53'31"38d26'
Berterame,Gianluca
Her 16h47'1"40d10'
Bertha Amelia
Cas 0h31'19"75d24'
Bertha R & Victoria Lynn
Com 12h32'55"20d44'
Bertha,Albert Frederic
Hya 8h55'33"1d44'B
Bertha,Brian Andrew
Hya 8h55'33"1d44'A
Berthelet,Louise et Yves
Her 16h54'30"23d31'
Berthelet,Pierre
Her 16h38'41"41d44'
Berthelot,Jearrnick
Per 3h56'50"37d19'
Berthiaume,Jacques
Ori 5h42'42"12d28'
Berthiaume,Mar
Per 2h54'47"35d2'
Berthold,Andreas
Cas 0h57'1"60d25'
Berthold,Melissa
Aur 5h23'48"38d17'
Bertholdi Leasing Star,Jerry
Lyn 9h0'50"36d43'
Bertholf,Theresa
And 1h47'51"40d40'
Berti,Carlo
Cyg 20h51'57"31d21'
Bertie
Vul 19h46'46"25d10'
Bertle,Alsha Leicer
Cam 5h54'26"70d26'
Bertin,Anne Marie
Per 3h29'55"38d3'
Bertin,Marie-France
Dra 12h6'10"70d23'
Bertini 1994,Cianca
Gru 22h8'22"-55d35'
Bertini,Claudia D'Antonio
Mon 6h58'44"-6d28'
Bertini,Scott & Debra
Cam 9h37'46"82d29'
Berto,Al
Cam 4h45'0"68d56'
Berto,Maria Luisa
Cnv 13h33'57"51d5'
Bertocci,Stefano
Nor 16h19'14"-51d53'

Bertoglio,Bryan Andrew
Cnv 13h52'23"39d41'
Bertola,Serenella
Del 20h12'55"14d41'
Bertolazzi,Serena
Vel 9h43'43"-45d40'
Bertoldi,Anna Maria
Aur 6h10'37"31d43'
Bertolet,Nora
Lyr 19h1'38"38d51'
Bertolini,Attilio
Aql 20h11'53"4d40'
Bertolino,Kimberly J
Lyr 18h50'33"41d41'
Bertolino,Kristen J
Umi 15h5'32"80d26'
Bertolino,Matthew L
Cep 22h46'42"68d46'
Bertollini,Stephen
Aur 6h16'33"31d42'
Bertolotti,Barbara
Cyg 21h54'0"52d53'
Berton Denis
Uma 11h26'1"31d40'
Berton,Penny
Cas 0h28'42"61d33'
Bertone,Alicia Louise
Vir 13h21'48"-4d12'
Bertone,Ann K
Cam 4h31'36"60d33'
Bertonis,James Gerard
Boo 13h34'46"20d27'
Bertovich,Martha
Cam 4h54'52"67d53'
Bertozzi,Nancy Ann
Peg 22h20'15"26d51'
Bertram,Dieter
Peg 23h32'0"16d47'
Bertram,Gabriele Heike Karin
Ari 2h1'40"20d58'
Bertram,Loris
Cam 9h45'53"81d60'
Bertram,Margaret Ann & John
Cyg 21h47'24"50d16'AB
Bertram,Matthew
Cnv 12h53'54"38d1'
Bertram,Michael Kurt
Cmi 7h59'10"8d4'
Bertram-(Rocky),Loreen
Cas 3h3'0"63d60'
Bertrand
Cep 20h57'52"63d22'
Bertrand Etoile Venard
Ori 5h34'17"8d9'
Bertrand,Angel
Cyg 21h36'40"40d40'
Bertrand,Ann M
And 0h9'50"28d42'
Bertrand,Anne Marie
Equ 21h22'26"11d11'
Bertrand,Christopher
Aur 6h9'0"37d27'
Bertrand,Jean-Pierre
Aql 19h0'11"-8d4'
Bertrand,Lucas Scott
Boo 14h3'54"27d37'
Bertrand,M Francine
Leo 11h34'51"0d37'
Bertrand,Patrice
Lyn 9h34'33"41d38'
Bertrand,Scott Richard
Ori 5h55'18"-0d5'
Bertrand,Trudi & Bill
Eri 3h54'35"-6d50'
Bertranou,Corinne
Cyg 19h42'51"30d6'
Bertreux,Anais
Aql 19h2'31"16d22'
Berts Mystic Gypsy My Love
Anna
Aql 19h13'29"14d24'
Bertsch,Cyndi
And 1h1'28"39d53'
Bertsch,Randall
Ori 6h0'31"20d21'

Bertsch,Terry Fredrich
Aql 18h53'50"-1d33'
Bertson
Aur 4h56'38"38d6'
Bertuccelli,Harry
Cyg 19h27'31"30d44'
Bertucci,Susie Lynn
Lyn 8h56'20"40d40'
Berty,Zoltan
Equ 21h22'53"3d3'
Beruffi,Virgilio
Aql 20h13'13"4d55'
Berulis,Nell
Aur 5h12'0"40d6'
Berumen,Javier
Aql 18h58'41"17d2'
Berumen,Linda
Cas 1h33'48"63d51'
Berwick,Walter M "Scott"
Del 20h54'30"9d8'
Beryer,Karl Heinz
Aur 5h9'10"40d25'
Beryl
Uma 13h39'50"51d0'
Beryl
Lyr 3h19'35"6d39'
Beryl
Aur 6h27'1"34d12'
Beryl
And 0h0'33"34d38'
Beryl Joan
Peg 22h12'27"3d23'
Beryl,Carolyn
Tri 1h57'11"25d53'
Beryl-J
Cyg 19h57'12"30d19'
Berzins,Armand John
Dra 16h6'43"61d19'
Berzins,Jenna Susan
And 1h35'50"36d31'
Besada,Joseph John
Cnv 12h54'26"32d8'
Besani,Carlo
Aur 6h9'39"31d45'
Besaw,Alyce Ronan
Cyg 19h28'12"30d15'
Besch,Robert Paul
Leo 10h37'29"12d21'
Beschoner,Heinrich
Peg 23h31'55"20d22'
Beschuetzer
Mon 7h22'20"-8d1'
Besedick,Bryan Nicholas
Boo 13h43'16"26d33'
Beseler,Carol
And 23h48'0"47d33'
Beseman,Carly Elizabeth
Cas 0h3'27"58d45'
Beseman,Heather Marie
Cyg 21h7'53"30d22'
Besemer,Thelma Porco
Oph 16h50'12"-28d44'
Beset,Wilma
Dra 10h8'38"80d49'
Beshoff,Olivia Louise
Lyr 19h22'0"38d15'
Besignano,Neil
Uma 8h46'20"51d51'
Besitos
Uma 9h16'35"71d51'
Beskin,Norma
Crt 11h11'36"-12d41'
Besley
Lmi 10h18'55"30d38'
Besnard,Stéphanie
Ori 5h55'57"1d34'
Besnier-Kuster,Loïc
Umi 13h59'41"72d28'
Beson,Robert J
Aur 6h51'44"40d21'
Besozzi,Peter J
Vul 20h15'1"28d22'
Besré,Jean
Per 3h0'19"41d9'

Bess,Amelia Elizabeth
Cas 2h15'46"65d17'
Bess,Pati Baldwin
Cas 0h7'49"64d57'
Besse,Caroline
Lyn 7h54'52"40d52'
Besse,Daniel
Her 18h12'27"31d41'
Bessede,Christine
Boo 14h29'27"40d17'
Besselievre
Ori 5h54'23"16d8'
Bessell,Dr Harold
Her 17h59'0"40d58'
Bessette,Austyn Victoria
Peg 23h28'17"15d57'
Bessette,Curt
Cam 6h4'36"58d38'
Bessette,The Navigator,Henry
Joseph
Cet 2h0'10"1d16'
Bessie M
Uma 8h11'55"70d51'
Bessinger,Bobbi
Aur 6h1'58"38d20'
Bessler,Paula Ann
Lyr 18h58'29"33d10'
Best
Lmi 10h7'33"30d52'
Best "Forever",A Elinor &
Richard C
Crb 15h16'26"31d32'
Best 1
Dra 20h14'20"62d37'
Best Daddy
Uma 11h30'1"50d35'
Best Friend
Ori 5h58'17"15d40'
Best Friends
Crt 10h55'40"-22d15'
Best Friends Forever Dean &
Judy
Ori 5h54'32"19d52'
Best Friends-In Memory of
Roxie
Cnv 13h24'30"40d19'
Best Grandpa Larry
Aur 6h15'44"37d38'
Best III,Joe A
Tau 5h48'28"23d35'
Best,Catherine
Cyg 20h5'14"40d41'
Best,Charles
Cep 0h52'25"77d57'
Best,Darrell
Oph 17h13'22"-24d55'
Best,Helen Angela
Peg 22h20'1"2d24'
Best,Howard D
Mon 6h29'54"-2d48'
Best,Janet Marie
And 23h23'31"48d23'
Best,Jr,Harry Glenn
Crt 11h31'0"-11d47'
Best,K Dale
Eri 3h35'17"-4d16'
Best,Linda Lee
Mon 7h0'59"-0d27'
Best,Michael
Psc 1h1'34"21d56'
Best,Mr & Mrs William A
Crb 15h24'56"30d14'
Best,Nancy Ellen
Scl 1h30'45"-29d20'
Best,Robert Bradley
Cnv 12h30'58"38d48'
Best,Sam
Ori 6h0'48"8d36'
Best,Sr,Harry Glenn
Aql 19h4'20"0d4'
Best,The
Cep 6h3'1"85d13'

Bestest,Freddie
Cnv 13h42'1"31d43'
Bestetti,Marianna
Boo 15h13'0"41d16'
Bestwick,Michelle Susan Dian
Com 12h38'0"21d22'
Bet-T-Zane
Peg 22h5'12"18d54'
Beta Alpha
Cet 0h34'32"-5d43'
Beta Veta Jeta & Sigma
Gamma Delta
Tri 2h0'19"28d0'
Betcha
Aql 19h50'0"15d5'
Betchley,John Charles
Aur 7h8'12"38d30'
Bete
Ori 4h52'30"0d58'
Betelgeuse Etc
Ori 6h12'23"7d31'
Beth
Peg 21h52'15"20d6'
Beth
Cas 23h42'36"60d44'
Beth
Cas 1h11'42"75d34'
Beth
Cyg 21h11'16"39d59'
Beth & Craig Forever
Cyg 21h27'17"40d36'
Beth & George's Wedding Star
Mon 6h54'24"-0d9'
Beth 35th Star
Lyn 7h27'51"50d16'
Beth Ann
Ori 6h18'0"7d33'
Beth Ann
And 2h25'56"45d22'
Beth Ann 5-22-81
Lyn 7h55'53"37d5'
Beth Anne
Cas 0h27'1"69d40'
Beth Forever
Cas 3h6'30"63d46'
Beth Gloriosa
And 1h33'31"38d9'
Beth J
Cam 6h5'24"70d57'
Beth Lee
Cas 23h22'21"61d51'
Beth Loves Dave
Boo 15h25'15"50d35'
Beth Lynn
Vul 19h43'48"23d11'
Beth My Curious Wine
Always, Sherry
Aql 19h31'35"12d2'
Beth Rhoda
Aur 6h1'46"31d34'
Beth's Bookmark
Mon 7h4'17"1d22'
Beth's Estrelleta
Peg 22h40'48"29d42'
Beth's Guardian Angel Star
Ori 6h16'38"8d58'
Beth's Love Light
And 2h12'11"38d8'
Beth's Lovelight Shining on
Miche
Tau 3h43'0"25d52'
Beth's Personal Wishing Star
And 23h6'56"44d28'
Beth's Precious Celestial Body
Aql 18h44'33"11d49'
Beth's Shining Star
Ari 1h45'38"16d29'
Beth's Star
Lyn 8h58'40"33d30'
Beth's Star
Cas 1h42'24"60d13'
Beth,I Will Always Love You
Uma 8h33'48"57d49'
Beth,Sheila
Tri 2h34'0"34d45'
Beth,So Heavenly, Forever
Uma 11h19'20"48d53'
Beth1996
Aql 19h13'1"10d3'
BethAaron
Crb 15h34'1"26d29'
Betham,Katie Marie
Lyr 19h14'47"41d12'
Bethan Rose
And 0h14'47"38d17'
Bethanie Sue
And 0h5'46"46d54'
Bethany
Cyg 22h0'1"50d23'
Bethany
Umi 13h28'24"73d23'
Bethany
Ori 6h5'22"6d48'
Bethany
Lyn 9h8'1"43d43'
Bethany
Lyr 18h25'34"40d43'
Bethany Jane
Cas 0h5'0"60d22"
Bethany Rae
And 0h4'52"34d42'
Bethany,Will You Marry Me?
Uma 10h30'59"52d29'
Bethany-Calire
Umi 13h8'43"76d16'
Bethany-Jane
And 0h14'38"39d21'
Bethards,Lou
Her 16h19'52"22d38'
Bethea,Tallie Belle
And 23h0'23"41d8'
Bethel,Carlysle Boo
Lib 15h13'38"-20d27'
Bethel,Emma Claire
Her 17h54'58"37d59'
Bethel,Frances Luckett
Cet 2h41'31"9d26'
Bethel,Meryl Anne
Ant 9h37'47"-32d5'
Bethellen
Com 12h17'51"20d3'
Betheno
Lyr 18h46'15"33d13'
Bethesda Bestest
Uma 11h33'1"50d43'
Bethesda Memorial Hospital
Ball-1995
Del 20h40'38"10d59'
Bethie
And 1h13'23"48d55'
Bethie
Lmi 10h36'41"33d39'
Bethie-Poo
Cam 5h58'25"68d32'
Bethleo
Del 20h50'25"9d28'
Bethpage Ann & The Girls
And 23h37'1"49d45'
Bethsadhim
Per 4h2'22"38d45'
Betjon
Cyg 20h28'21"41d13'
Betker Forever Darling,Emil L
Ori 6h2'49"-0d17'
Betker,Steven R
Uma 11h59'26"57d33'
Betor,Ryon Ernest
Uma 11h29'55"63d13'
Betric
Cas 0h14'49"60d53'
Betriu,Joseph
Peg 23h33'19"10d41'
Betron,Susan J
Mon 6h54'49"-10d13'
Bets
Lyn 9h12'56"43d37'
Bets
Mon 6h56'56"-6d4'
Betsabeth
Cnc 9h11'30"30d23'
Betsch Will
Her 16h9'3"47d56A
Betsey
Cas 0h45'48"62d51'
Betsy
Equ 20h54'51"2d21'
Betsy & Doug
Aur 6h32'52"32d53'
Betsy & Mark
Lac 22h20'57"51d38'
Betsy & Rick
Cyg 21h26'1"40d11'
Betsy Lou
And 22h55'44"40d9'
Betsy's Angels-Tom & Thomas
Umi 13h13'42"71d26'
Betsy's Birthday Star
Cas 1h4'25"53d8'
Betsy's Brilliance
Com 13h9'39"20d24'
Betsy-N-Bob
Crb 16h1'31"32d30'
Bett & Jose
Boo 14h50'25"38d57'
Bett,Elizbeth
Cam 3h58'54"62d50'
Bett-Nee's Little Bit
Lib 14h58'13"-8d41'
Betta
Nor 16h23'4"-50d29'
Bette
Uma 11h47'0"49d55'
Bette & Emil Some Stars
Shine Forever
Cyg 19h21'18"28d37'
Bette In the Sky
Tri 2h7'32"30d6'
Bette Jean's Blessing
Cyg 19h32'51"34d7'
Bettega,Olga Mary
Mon 6h20'17"6d38'
Betten,Richard (Rick)
Vul 20h46'11"20d22'
Bettencourt,Ashley Breanne
Mon 6h25'32"8d5'
Bettencourt,John J
Aur 7h19'0"38d42'
Bettencourt,K C
Mon 8h1'40"-10d0'
Better Val*U Store #11
Aur 6h0'36"45d15'
Betters,C J
Leo 11h6'18"-1d50'
Bettes,Jacqueline Sue
Ori 6h9'37"7d47'
Betti Irene
Lac 22h27'57"40d37'
Betti Lee
Com 12h55'1"26d16'
Bettina-290658
Cam 5h31'1"65d7'
Bettinelli,Irene
Peg 21h43'42"24d32'
Bettinghouse,Kari
Lyn 9h13'44"45d21'
Bettini,Laurent
Cam 5h54'45"68d26'
Betts,Catherine
Lac 22h35'58"41d4'
Betts,Claire
Uma 9h56'43"70d24'
Betts,Denis & Judy
Cyg 20h19'27"38d39'
Betts,Grace Elizabeth
And 0h16'0"30d11'
Betts,Jocelyn
Cyg 21h42'0"38d48'
Betts,Larry A
Hya 8h11'49"1d38'
Betts,Raymond W
Cam 7h56'1"60d1'
Betts-One Special Lady,
Madeline
Crb 15h50'42"27d17'
Betts/Sewell Rekindler The
Cyg 19h26'47"31d43'
Bettucci,Claudia
Col 5h27'10"-41d45'
Betty
Cas 1h18'11"75d14'
Betty
Cas 0h6'19"65d39'
Betty
Peg 22h0'57"30d17'
Betty
Cas 1h3'44"63d58'
Betty
Cyg 20h5'15"39d35'
Betty
Aur 5h0'40"38d8'
Betty "Spaghetti"
Uma 8h33'27"68d47'
Betty & Bluebird
And 23h37'52"46d19'
Betty & Ed Forever In Gods
Hands
Cyg 20h7'34"40d23'
Betty & Emma "Forever"
Cas 2h34'25"57d28'
Betty Ann
Cam 5h7'43"71d13'
Betty Ann
Dra 11h51'2"74d2'A
Betty C
Ori 4h59'25"10d46'
Betty Elizabeth
Cyg 21h55'56"40d6'
Betty F
Lib 15h38'50"-25d34'
Betty Jane
Aql 19h0'53"16d36'
Betty Jane
Ori 5h49'59"19d17'
Betty Jane M
Vul 19h58'36"25d7'
Betty Jean
Lyn 8h12'50"48d30'
Betty K
Cyg 19h34'20"35d38'
Betty L
Cyg 19h56'1"38d50'
Betty Lou
Uma 10h58'24"48d17'
Betty P-The Mountain Star
Peg 23h6'51"10d3'
Betty Rose
Cyg 21h10'56"39d3'
Betty's (Boop)
Mon 6h55'19"-4d53'
Betty's Beloved Susie
Uma 11h8'19"56d57'
Betty's Biggest Diamond
Vir 13h13'21"-0d49'
Betty's Birthday Star
Crb 15h58'29"30d57'
Betty's Heart
And 1h14'22"39d59'
Betty's Lumina
Cet 1h36'28"-1d13'
Betty's Star
Cyg 20h28'21"41d13'
Betty's Star
Aqr 21h1'13"-6d10'
Betty's Tomorrow Star
And 2h22'34"44d40'
BettyRon
Cyg 21h52'31"41d30'
Betz,Daniel Todd
Crb 16h12'56"30d33'
Betz,Dave & Linda
Cam 5h4'34"68d14'
Betz,Gregory William
Lac 22h43'16"52d53'
Betz,Jr,Wayne
Cam 5h44'17"61d8'
Betz,Megan Leigh
Peg 23h3'50"33d9'
Betz,Tony & Pam
Dra 19h3'28"56d37'
Betzing 100,Anna "Mom"
Uma 10h3'34"67d59'
Betzner,Charlotte
Cas 1h20'24"53d41'
Beu,Joanna Lee
Cyg 20h41'45"45d57'
Beucher,Alphonse Marie
Auguste
Mon 6h42'9"-3d51'
Beucher,Marie-Anne
Augustine
Cmi 7h22'29"9d51'
Beuda,Rita
Cmi 7h20'0"3d48'
Beuerle,Werner
And 23h11'0"40d22'
Beuger,Paul P M & Kim A
Cyg 21h31'0"52d54'
Beugin,Isabelle
Boo 14h37'0"44d19'
Beukema,Pamela Kay
And 0h11'45"47d51'
Beul,Hans
Her 17h21'14"42d49'
Beuning,Thomas J
Her 16h27'19"35d55'
Beuoy,Lora Lee
Lyr 18h48'40"39d52'
Beurer,Charles Martin
Boo 14h35'55"20d55'
Beutel,George & Antoinet
Sge 19h32'1"18d47'
Beutel,Todd Walter
Ser 17h56'42"-11d2'
Beutelspacher,Erik
Psc 23h3'18"6d53'
Beuther,Kenneth William
Sct 18h55'0"-5d50'
Bev II
Cet 2h23,15"1d48'
Bev's Star
Uma 9h33'11"57d53'
Bevacqua,Joann
Lyr 18h43'1"33d55'
Bevan's Star,Denzil
Per 1h49'15"52d59'
Bevan,Carol Anne
Mon 8h2'50"-8d48'
Bevan,Eric
Ori 5h18'59"10d53'
Bevan,Eric & Joan
Sge 20h16'1"18d50'
Bevan,Kenneth John
Uma 9h56'24"53d50'
Bevan,Sharon
Cas 0h14'12"64d3'
Bevan,Sr,James W
Her 17h32'35"38d38'
Bevel,Crystal
Peg 21h57'10"34d46'
Beveridge Star,The Andy
Steele
Ori 4h54'43"5d19'
Beveridge,Bob
Cet 0h52'0"-10d18'
Beveridge,Iain James
Dra 12h40'38"71d50'
Beveridge,Jr,David Mansfield
Aql 19h47'55"11d57'
Beveridge,Karen
Com 12h14'58"21d2'
Beveridge,T L
Cam 7h57'55"61d43'
Beverle
Lyr 19h2'1"37d51'
Beverle
Cas 1h4'21"70d13'
Beverley
Cep 20h29'1"66d42'
Beverley
Cas 0h58'0"59d53'
Beverley Anne
Lyr 18h29'51"31d48'
Beverley Jo
Uma 17h29'49"84d23'
Beverley,The
Lyr 18h31'12"41d14'
Beverley-Jane
And 1h18'0"37d41'
Beverly
Del 20h23'53"10d48'
Beverly
And 2h34'23"49d3'
Beverly
Uma 10h39'26"56d60'
Beverly
Uma 10h57'10"72d18'
Beverly
Boo 14h37'0"8d23'
Beverly
Cam 5h9'31"70d47'
Beverly
Cas 0h57'31"50d20'
Beverly & Joe Infinity 92
Mon 7h39'38"-0d56'
Beverly & Richard
Peg 22h5'1"4d39'
Beverly Alice
Cas 0h27'36"58d46'
Beverly Ann
Cas 3h4'46"67d31'
Beverly Ann
Mon 6h1'0"-5d31'
Beverly Ann
Mon 6h18'53"7d43'
Beverly Faye
And 23h19'31"45d46'
Beverly Gail
Cas 3h9'38"61d28'
Beverly Glee,The
Lyn 7h45'27"41d15'
Beverly Jean
Lyn 8h7'33"38d10'
Beverly Jo
Cas 2h38'12"60d51'
Beverly Lumpy,The
Cyg 19h25'41"31d9'
Beverly N
Mon 6h54'36"-0d11'
Beverly Teressa
Cas 0h53'44"64d46'
Beverly's Brightness
Cas 1h16'42"63d8'
Beverly's Hope
And 2h14'43"42d50'
Beverly's Star
Aql 19h6'1"-5d38'
Beverly,Therese
Cyg 21h1'12"40d1'
Beverly,Winifred Ruth
Cas 1h47'52"60d55'
Bevers,Catherine M
Cas 23h41'57"50d12'
Bevie Ann
Sex 10h15'8"-2d19'
Bevier Star,The Suella
Mon 8h6'19"-8d8'
Bevilacqua,Alessandro e
Giuseppina Milani
For 2h47'26"-27d18'
Bevilacqua,Joseph
Aur 4h49'18"51d14'
Bevinn & Rob
Dra 17h0'55"66d48'
Bevins,South C
Cet 2h56'25"0d20'
Bevis,Carol Robin
Eri 2h48'52"-2d50'
Bevis,Doyle
Cma 7h0'1"-31d34'
Bex D
Uma 9h41'27"54d32'
Bex Heart
Hya 8h29'30"-1d41'
Bey
Her 18h1'22"20d7'
Beyah,Clarshun
Vul 19h20'49"26d46'
Beydler-Maah,50,Mae
Uma 11h34'1"32d45'
Beydrichen,Armin
Cnc 8h0'12"10d15'
Beyer's,Donna
Tri 2h2'18"31d14'
Beyer,Alexa Robyn
Ari 2h41'10"25d55'
Beyer,Christopher Erich
Lib 15h3'15"-5d38'
Beyer,Chuck
Dra 16h15'0"66d20'
Beyer,Earl A
Cyg 20h8'33"43d47'A
Beyer,Ellen
Cam 5h0'50"67d47'
Beyer,Gerfried Dr
Psc 22h53'44"7d3'
Beyer,Jacqueline Alana
Eri 4h0'46"-15d9'
Beyer,Jane Dunham
And 1h14'17"48d59'
Beyer,Karin & Clemens
Cnv 12h33'15"36d19'
Beyer,Marianne- Jonathan-
Courtney
Aur 5h34'0"54d30'
Beyer,Matthew
Dra 16h42'60"63d32'
Beyer,Mr Arthur E
Cas 1h54'53"75d41'
Beyer,Mrs Gene
Cas 1h30'1"70d54'
Beyer,Sherlin
Psc 1h26'51"21d20'
Beyer,Thomas R
Per 1h58'41"50d32'
Beyer,Uwe
Uma 9h21'25"47d6'
Beyerlein,Bobbie
Mon 6h0'46"-8d56'
Beyers,Lillian R
Cas 1h5'58"63d25'
Beyl,Jessica Gabrielle
Eri 2h44'16"-16d32'
Beylin
Cap 20h24'47"-26d56'
Beylin,Anna
And 23h39'46"45d48'
Beyond Bounds
Sct 18h33'19"-6d43'
Beyond Love
Oph 16h59'27"10d52'
Beyond The Point Of No
Return
Cep 21h45'35"80d24'
Beyra,Carlus A
Cet 2h9'17"7d49'
Bez
Uma 14h19'1"60d54'
Bez,Herbert
Per 4h6'2"48d49'
Bezaire,Jeffery
Per 3h1'30"40d40'
Bezark "The Bez", Richard S
Her 18h28'1"20d22'
Bezat,Aude
Dra 15h20'0"58d30'
Bezinski,Brian M
Lmi 9h49'0"38d17'
Bezmem,William S
Cyg 21h54'17"38d52'
Bezrutczyk,Allison Suzanne
Peg 23h3'58"17d39'
Bezrutczyk,Parva Ann
And 0h9'21"35d6'
Bezy,Robert Edward
Dra 9h59'48"74d1'
Bezz
Cyg 21h25'29"40d0'
BG4E
Ori 5h52'56"21d20'
Bhakti
Cam 13h17'1"77d43'
Bhan,Madan
Per 3h53'51"41d44'B
Bhan,Suldchna
Per 3h53'51"41d44'A
Bhangoo,Kamil K
Cam 8h3'24"70d10'
Bharadwaj,Arjun Sambros
Cep 20h23'36d"60d40'
Bharat,George S
Per 4h2'13"34d8'
Bhatia,Tanya Robin
Cyg 20h37'12"30d38'
Bhatt,Chetan
Uma 9h13'12"47d25'
BHG
Dra 16h51'31"68d32'
Bhullar,Sonam
Her 16h32'1"40d15'
Bia's Star
Uma 9h43'24"47d18'
Bia-Franco,Gilda
Cam 4h59'51"61d28'
Biadac,Michael & Jennifer
Cyg 21h7'21"30d3'
Biagas,Timothy
Boo 14h44'1"27d40'
Biaggi,Francesca
Lyn 7h8'44"44d55'
Biaggioli,Elena
And 0h12'46"38d6'
Biagini,Lisa Bevinetto
Tau 3h36'40"23d14'
Bialac Family,Shelly & Stella
Mon 6h19'21"8d47'
Bialas,Dorothy T
Lyr 18h38'1"40d11'
Bialcik,Scott A
Uma 10h39'58"60d49'
Bialek,Jackson Wells
Lac 22h11'58"48d29'
Bialek,Joseph
Aur 6h27'25"31d9'
Bialick,Jerry
Aur 6h7'1"38d13'
Bialo,Mildred
Lac 22h41'35"56d18'
Bian,Roberta Rosa
Cas 1h24'35"53d38'
Bianca
Cru 12h24'48"-63d17'
Bianca
Uma 12h2'16"33d18'
Bianca
Lmi 10h3'1"31d8'
Bianca
Lmi 10h36'1"23d7'
Bianca
Lyn 8h11'0"47d4'
Bianca
For 2h8'34"-28d33'
Bianca
Boo 13h38'12"11d38'
Bianca 6-25-92
Mon 7h30'45"-8d22'
Bianca Lee
Oph 18h6'1"13d0'
Bianca,Jr,Anthony T
Dra 20h7'23"75d2'
Biancamano,Thomas Vincent
Per 2h43'47"41d12'
Bianchi,Amber Nicole
Sgr 19h4'59"-20d53'
Bianchi,Anna
Del 20h18'52"10d60'
Bianchi,Debbie
Cyg 19h6'30"54d50'
Bianchi,Jennifer
Mon 6h2'14"-5d38'
Bianchi,Massimiliano
For 2h36'1"-27d11'
Bianchi,Paolo
Sco 15h46'23"-21d21'
Bianchini,Gary
Aql 19h47'1"14d14'
Bianchini,Marcella
Cnv 13h20'49"28d12'
Bianchini,Oscar
Vel 9h18'41"-44d4'
Bianco,Alyssa Theresa
Cam 3h14'14"60d12'
Bianco,Carolyn C
Lyr 19h22'1"35d32'
Bianco,Christopher A
Cma 6h59'37"-19d29'
Bianco,Elena
Psc 22h59'23"-3d28'
Bianco,Elizabeth M
Cet 0h51'28"-1d17'
Bianconi,Fabrizio
Crt 11h40'23"-10d15'
Bianka
Gem 7h34'41"20d56'
Biardi,Albert Charles
Boo 14h20'0"23d10'
Biardi,Elinor Dougherty
Boo 14h20'40"22d34'
Biasco,Jr,Frank Joseph
Lac 22h2'50"48d57'
Biase,Christopher R
Lyr 18h58'39"26d29'
Biasi,Raffaella
Cyg 21h29'55"48d49'
Biavati,Paola
Ori 6h0'52"0d32'
Biba
Umi 15h38'58"70d24'
Bibbee,Howard Curtis
Dra 20h22'53"62d33'
Bibber,Shirley L
Cas 1h18'52"61d45'
Bibendum
Umi 15h9'32"68d57'
Bibi
Cnc 9h12'55"8d12'
Bibi & Baby
Hor 3h9'22"-46d32'
BiBi Sister
And 22h58'26"40d48'
Bibi Star
And 0h35'1"40d34'
Bibiche
And 23h3'46"40d23'
Bible,Alexandra Joanna
Lyn 8h12'1"36d6'
Bibler,Isabelle
Com 13h11'59"20d51'
Bibliowicz,Jessica M
And 0h18'16"45d28'
Bibsy
Umi 15h9'51"68d39'
Bice,Joshua David
Dra 19h1'29"54d37'
Bice,Shelby
Sct 18h32'0"6d2'
Biceman
Lyn 7h55'42"50d55'
Biche,Gregory David
Lac 22h8'38"51d34'
Bichler,Sandro Cimerlajt
Per 1h49'47"56d54'
Bicicchi,Arthur
Aqr 22h9'57"-8d15'B
Bick,Thomas
Aur 5h1'56"42d20'
Bickel Family Star,The
Ori 5h6'40"1d13'
Bickel,Katarina Joy
And 0h57'14"34d35'
Bickel,Werner
Mon 7h51'36"-2d30'
Bickel-Rees,Avril Ann
Lyr 18h39'44"33d19'
Bickenbach,Elfriede
Gem 8h3'56"31d16'
Bickerton Memorial Star,The
Stephen
Cyg 21h2'45"39d39'

Bickerton,Brian Corey
Tri 1h36'1"31d25'
Bickerton,John Henry
Cma 6h55'20"-15d50'
Bickerton,Rosemary Elisa
Sct 18h48'41"-7d37'
Bickerton,Sally Jennings
Cma 6h29'13"-11d14'
Bickford II,Frederick A
Aur 6h31'38"32d14'
Bickford,Barbara Louise
Com 13h27'0"25d41'
Bickford,Glenn & Elizabeth
Cyg 19h24'13"33d26'
Bickford,Hadley Rastad
Lac 22h18'0"54d32'
Bickford,Helen
Cas 0h42'56"68d30'
Bickford,Laura Ann
And 2h24'17"38d47'
Bickford,Robert Raymond
Lac 22h12'58"50d18'
Bickham,Charles Weston
Sex 10h10'15"-4d40'
Bickle,Diane
Lyn 7h48'57"47d38'
Bickley,Mary
Cyg 21h44'18"38d54'
Bickmore,Wm "Brad"
Boo 14h21'35"38d42'
Bicknell,Elizabeth
Com 12h9'19"20d3'
Bicknell,Fred
Ori 5h55'53"6d28'
Bicknell,Robert J
Boo 15h24'12"33d15'
Bicks,David P
Cnv 13h11'37"32d28'
Bicocchi,Christine
Her 16h58'54"16d42'C
Bicocchi,Dennis
Her 16h58'54"16d42'A
Bicocchi,Nancy
Her 16h58'54"16d42'B
Bid
Cyg 19h40'1"31d30'
Bidal,Agnes
Dra 20h3'40"70d35'
Biddelman,Miriam Bierman
Cyg 21h4'15"48d50'
Biddix,Carole Vandoren
And 1h52'36"40d32'
Biddle,Braden Byrne
Aur 6h14'15"37d51'
Biddle,Brian
Cep 22h31'22"56d60'
Biddle,Courtnee Elizabeth
Mon 6h27'11"7d25'
Biddle,Cynthia
And 2h30'53"48d8'
Biddle,Graham Stanley
Per 3h58'0"40d56'
Biddle,John
Peg 22h28'37"7d59'
Biddulph,Warren & Angela
Cyg 20h53'0"38d51'
Biden,Susanne
Ari 2h55'53"30d25'
Bidet,Louis Marie
Her 17h53'44"41d13'
Bidrawn,Micheal
Ori 5h53'26"14d5'
Bie,Carmela Diaz
Tau 4h19'52"21d21'
Biedermann,Carol
Eri 4h47'23"-7d36'
Biedermann,Franka
Sct 18h54'32" 5d38'
Biedermann,Nicky
Cma 7h50'48"10d24'
Biegel,Bryan
Vul 20h3'12"22d54'
Biegert,Mildred
Lyr 19h5'55"37d35'

Biegler,Kyle Thomas
Lyn 9h17'14"41d14'
Biehler 7-4-55,Paul Janssen
Aql 19h0'20"7d34'
Biehn,Devon
Mon 7h7'50"-5d21'
Biehn,Michael
Oph 17h34'1"-0d48'
Biehn,Taylor
Cyg 20h31'48"40d14'
Bieker,William A
Dra 19h30'43"58d60'
Biel,"Beana" Robin Lynn
Uma 11h23'15"45d47'
Biel,Edward (Big Mama)
Uma 11h53'12"30d42'
Biel,Sophia
Uma 11h52'37"32d52'
Biel,Thomas
Uma 11h53'14"38d34'
Biel,Thomas & Joellyn
Dra 16h28'35"64d31'
Biel,William
Uma 11h52'48"33d6'
Bielas,Kent & Michelle
Cyg 20h20'13"41d36'
Bieler,Denise
Psc 1h31'50"20d41'
Bielski,Francis V
Ori 5h57'40"17d16'
Biemuller MD,Martha L
Oph 17h2'13"12d7'
Biemuller,Christina L
Lyr 18h32'17"45d7'
Biemuller,Kathleen M
Cyg 21h19'39"38d42'
Bieneman,Deborah L
Cas 1h16'17"63d32'
Bienen,Henry S
Cep 2h45'0"77d39'
Bienerts Hellster
Cet 2h57'23"0d2'
Bieniewicz,Barbara
Cas 0h35'39"63d26'
Bienkowski,Witek
Uma 9h28'0"54d12'
Bienstock,Diane
Mon 6h36'45"-6d53'
Bienvenue,Edward Michael
Dra 18h25'1"50d21'
Bienz,Rolf
Lac 22h51'50"40d43'
Bierce,Amanda
And 2h20'35"49d22'
Bierce,Kaela
And 2h22'55"49d36'
Bierhaus,Kristy
Leo 10h58'12"15d36'
Bierly,Julie
Mon 7h0'41"0d12'
Bierman,Brenda D
Eri 2h54'1"-4d11'
Bierman,Jayne & Heather
Uma 9h5'38"59d15'
Bierman,Karen
And 2h21'32"37d48'
Bierman,Kristen F
And 23h48'44"47d9'
Biermanns,Gisela
Lyn 8h9'21"49d23'
Biernacki,Michael
Gem 7h33'34"20d55'
Biers,Irving
Boo 15h0'33"53d32'
Bierschenk,Karen Lee
Aqr 22h46'24"0d29'
Bierschenk,Marjory Lankenau
Peg 21h47'24"34d22'
Bierstedt,Matthew J
Boo 14h0'29"20d11'
Bierwagen,John Michael
Her 18h2'44"28d46'
Bies,Sr,(Grampy), Robert
Joseph
Dra 18h12'55"70d52'

Biesanz,John B
Lac 22h7'10"50d4'
Biese,Matt
Vul 19h2'10"22d14'
Biesemann,Hans Georg
Hya 8h55'1"-5d39'
Bieser,Raymond
Dra 19h5'26"71d10'
Bieske,Blake Gabriel
Per 2h59'0"40d59'
Biewend,Jerry
Aur 6h21'38"30d20'
Bifano,Louisa Marielle
And 0h30'0"45d32'
Biff
Her 17h10'54"42d21'
Biff & Bears Shiny Thing
Cyg 21h7'12"31d8'
Biff Collie
Cnv 12h30'14"34d13'
Big "D"
Her 17h14'1"22d50'
Big "D"
Her 16h0'45"47d52'
Big 4
Peg 22h29'0"20d15'
Big A,The
Aql 20h15'22"8d3'
Big Al
Her 18h5'25"40d2'
Big Al
Umi 15h13'23"66d52'
Big Al
Ori 5h44'0"11d28'
Big Al
Her 16h39'37"21d27'
Big Al
Dra 18h58'19"47d57'
Big Al
Hya 8h19'56"2d8'
Big Al
Her 16h32'0"38d43'
Big Al
Ori 5h57'0"17d51'
Big Al's Shining Star
Aur 6h4'13"31d37'
Big Artie
Cep 23h14'56"71d10'
Big Baldy,The
Cep 22h13'12"60d17'
Big Bang
Cam 8h23'19"74d21'
Big Bang Beanie
And 0h38'54"40d35'
Big Bear
Uma 10h10'0"59d14'
Big Bear,The
Uma 9h28'1"50d56'
Big Beaver,The
Umi 14h22'14"66d7'
Big Bee
Dra 16h50'21"62d39'
Big Ben
Uma 9h18'14"49d4'
Big Bill
Her 16h37'39"8d27'
Big Birtha
Uma 11h39'0"31d37'
Big Bob
Boo 14h26'25"18d45'
Big Boo
Hya 9h4'57"-0d26'
Big Boog,The
Aql 19h45'11"10d51'
Big Boy
Her 16h57'58"47d50'
Big Breakfast Star In Yer Face,
Mrs!,The
Uma 12h13'34"58d31'
Big Brenda
Cyg 19h43'59"34d53'A
Big Brother
Uma 12h1'30"42d42'
Big Brother
Aql 19h43'45"10d42'

Big Brother
Ser 17h33'37"-13d14'
Big Brother
Hya 9h4'52"-0d34'
Big Brother Bill
Her 16h59'38"48d16'
Big Brother Is Watching
Her 17h33'21"20d8'
Big Bubby
Cnc 8h27'34"10d49'
Big Chipper,The
Uma 10h52'0"56d9'
Big Coaxis
Peg 22h1'15"30d46'
Big D
Aur 6h31'35"33d33'
Big D
Dra 19h13'47"68d47'
Big Dad
Per 3h26'0"40d54'
Big Dad Poppy Lum,The
Cep 21h59'59"55d17'
Big Daddy
Oph 18h17'59"12d55'
Big Daddy
Aql 20h20'0"7d31'
Big Daddy Dirt Star
Ori 5h10'1"-0d17'
Big Daddy,The
Cet 2h59'41"2d14'
Big Dano
Her 17h39'21"23d8'
Big Den Star,The
Aur 6h2'56"45d8'
Big Dipper Of Masahiro
Leo 9h55'17"22d51'
Big Dream Laurent & Danielle
Dra 10h19'41"80d21'
Big Dude,The
Umi 8h44'17"88d50'
Big Ed
Per 1h45'37"53d30'
Big Ernie's Fabulous 50's
Diner
Cnv 12h29'1"33d21'
Big Foot
Boo 14h29'0"20d16'
Big G
Uma 9h46'1"70d46'
Big G
Cam 3h51'37"52d48'
Big G
Uma 10h31'33"40d10'
Big Gino
Dra 15h10'41"57d30'
Big Glick,The
Hya 8h27'26"-9d36'
Big Guy,The
Her 16h18'42"10d52'
Big Honey & Little Honey
Uma 12h1'21"39d7'
Big J Dipper,The
Uma 12h5'50"61d46'
Big Jake
Aur 6h28'45"31d27'
Big Jake
Ori 6h1'31"6d38'
Big Jer
Lac 22h52'41"38d21'
Big Jim
Per 2h10'18"57d21'
Big Jim F
Dra 16h22'57"66d18'
Big Jim's Verdict
Her 16h17'22"40d55'
Big Jimmer,The
Per 2h58'41"50d18'
Big Joe
Dra 11h21'58"78d54'
Big John
Cet 2h36'48"0d29'
Big John
Cep 23h28'0"65d50'
Big John
Umi 15h16'29"69d11'

Big K
Aur 6h3'24"35d13'
Big K B,The
Her 16h37'33"34d51'
Big Kahuna,The
Lyn 8h9'48"47d17'
Big Lar
Aur 7h21'1"40d9'
Big Large
Crb 15h23'37"32d34'
Big Leader,The
Boo 15h18'11"50d17'
Big Lin
Ori 6h9'12"0d29'
Big Ma,My Mom,Friend &
Brightest Star
Cas 23h1'49"58d10'
Big Magic Jimmy,The
Aur 7h5'41"38d53'
Big Mario
Vul 19h59'1"28d40'
Big Mel
Her 17h49'25"18d50'
Big Nab's
Uma 11h28'48"30d2'
Big Nay
Aql 20h10'49"13d41'
Big Nick
Leo 9h55'28"27d36'
Big O
Aql 20h2'33"0d43'
Big O,The
Uma 11h53'32"42d33'
Big One,The
Cam 12h15'50"84d25'
Big R
Boo 14h41'58"47d41'
Big Ray
Peg 0h5'17"13d31'
Big Red
Aur 5h46'40"31d46'B
Big Richard
Aur 6h56'43"43d20'
Big Rudy
Cam 7h8'0"60d43'
Big Schill,The
Lac 22h1'25"51d0'
Big Shot
Lmi 10h49'0"31d22'
Big Sister Erin
Uma 10h1'39"51d13'
Big Sister Rachel
Cam 5h51'45"68d31'
Big Stu,The
Per 1h56'0"50d32'
Big Sweetie
Cyg 19h19'23"28d27'
Big Sweetie,The
Ser 16h15'0"0d2'
Big T,The
Dra 17h0'0"66d36'
Big Ted
Uma 9h54'56"67d56'
Big Tuna,The
Dra 19h5'13"58d33'
Big Twinklin & Bonni
Peg 22h46'1"25d10'
Big Will,The
Peg 0h2'21"18d11'
Big Wolf,The
Lyn 8h10'46"39d45'
Big,Jessie Reed
Boo 15h44'1"40d22'
Bigarani,Christine M
Mon 7h49'2"-4d0'
Bigaud,Georgette
Ori 6h20'57"10d12'
Bigayon,Christophe
Cnv 13h16'34"38d14'
Bigbee,Michael David
Ser 16h0'23"10d14'
Bigbie,Edgar F
Cet 2h33'23"9d31'
Bigelow,Austin H.
Dra 12h24'27"64d22'

Bigelow,Marilyn Ann
Cas 2h58'43"58d47'
Bigelow,Thomas Gerald
Her 16h38'50"21d34'
Biger,Odile
Dra 11h56'35"71d32'
Bigg-Wither Star,The
Cep 0h0'33"71d15'
Biggam,Betty J
Ori 6h14'46"0d57'
Biggans,Kimberly Noel
And 1h16'0"36d40'
Bigge,Kenneth
Ori 6h7'34"8d0'
Bigge,Martin
Ori 5h56'1"21d25'
Bigge,Megan
Ori 5h55'0"20d18'
Bigge,Robert
Ori 6h11'42"7d34'
Bigger,Tali & Assaf Admoni
Cam 12h14'3"80d24'AB
Biggerstaff,Jerry Ray
Her 17h33'14"26d45'
Biggi
Cnv 12h38'1"39d35'
Biggi
Gem 7h25'35"27d27'
Biggi
Ori 6h8'11"10d14'
Biggie
Umi 16h29'27"79d55'
Biggins,Phoebe-Lee
Peg 21h53'28"33d35'
Biggio,Connor Joseph
Aur 7h16'36"40d8'
Biggs "The Skeeter", Keith
Ori 5h57'34"10d26'
Biggs,Alexander Blackman
Lac 22h11'27"46d54'
Biggs,Brooke Shelby
Peg 23h16'33"31d50'
Biggs,David
Per 3h20'36"40d32'
Biggs,Dean
Her 17h0'46"47d34'
Biggs,Larry
Tau 5h57'17"23d2'
Biggs,Malcolm John Robert
Aql 20h16'50"0d3'
Biggs,Marilyn
Ori 5h59'11"15d41'
Biggs-Nichols,Joyce Marie
Peg 22h0'31"25d25'
Biggstar,The
Uma 11h50'16"40d1'
Bigham,Michelle & Brent
Helmes
Peg 21h24'54"10d31'
Bigham,Oren
Cet 3h8'0"4d22'
Bigland (Boager),John Andrew
Dra 16h18'14"63d57'
Biglane,John & Gina
Boo 15h12'31"53d55'
Bigler,Amanda Marie
And 0h6'56"46d40'
Bigler,Diana Jean
Cas 0h43'22"61d27'
Bigler,Elizabeth
Mon 7h25'41"-5d46'
Bigler,Elyse Ann
Tri 2h4'40"31d42'
Bigler,Eric Tyson
Hya 8h24'58"5d60'
Bigler,Erin Nicole
Cas 0h45'53"64d21'
Bigler,Lisa Amelia
And 0h6'56"46d12'
Bigler,Matthew Cole
Oph 16h54'22"-28d39'
Bigler,Randy
Cet 1h10'0"-3d4'
Bigler,Richard Alan
Cet 2h22'40"-5d33'

Bigley,Christine Magdalena
Vir 11h44'33"7d32'
Bigley,Jr Star,The James M
Cep 22h20'1"60d23'
Bigley,Pat
Oph 17h34'29"11d25'
Biglin,Erin Kelly
And 1h20'51"39d35'
Biglow,F William
Lmi 10h35'21"23d22'
Bignall,Brandin Bates
Cet 0h53'28"-2d53'
Bigos,Alexander F
Her 17h8'48"48d56'
Bigos,Samuel
Uma 9h39'26"48d13'
Bigot,Maurice
Aur 4h59'44"30d25'
Bigotte,Stephanie
Cam 5h14'54"68d52'
Bigourdan,Roger
Ori 6h7'14"0d29'
Biguad,Georges
Peg 23h31'43"10d19'
Bihari Family Star
Per 4h23'11"52d20'
Bihl,Viola K
Cas 0h59'51"61d51'
Bihn,Achim
Cnc 9h1'59"18d36'
Bihrer,Andreas
Sco 17h28'57"-30d13'
Bijlani,Anita
Sct 18h53'52"-6d45'
Bijon
Cnc 8h14'1"32d55'
Bikel's-Star of
Friendship,Theodore
Uma 11h2'12"50d10'
Bikle,Bill-Sharon- Shannon-
Hope
Uma 9h26'54"42d24'
Bil Gil
Leo 10h56'33"-1d59'
Bilbe,Lisa
Cyg 19h26'46"35d43'
Bilbo
Cnv 12h17'40"41d28'
Bilden,Danae
Cep 20h12'26"60d28'
Bildersee,Adam S
Ori 5h12'28"-6d21'
Bildersee,Jennifer M
Ori 5h8'29"-6d56'
Bildersee,Lisa M
Ori 5h10'52"-6d36'
Bildman,Lars
Per 3h10'34"46d20'
Bilek,Rosann
Lac 22h20'44"51d31'
Bilello,Lenore
Peg 23h20'35"11d13'
Bilenko Star
Lmi 10h41'1"31d50'
Bilesimo,John & Elaine
Cyg 21h7'49"31d13'
Bilgray,Richard M
Cma 7h16'21"-15d48'
Bilisko's Star,Paul & Steven
Cyg 19h47'24"29d33'
Bill
Ori 5h1'40"8d39'
Bill
Aur 6h26'10"33d15'
Bill
Per 3h47'21"38d60'
Bill
Oph 17h16'54"-22d55'
Bill & Ali
Cyg 21h3'0"33d18'
Bill & Annette
Destinyontheotherside
Sex 10h27'50"0d1'
Bill & Carol
Uma 9h36'49"50d37'

Bill & Carole XX
Lac 22h8'15"38d39'
Bill & Chris
Umi 16h3'33"70d37'
Bill & Donna
Cam 7h44'24"71d5'
Bill & Erna
Ori 5h56'24"15d60'
Bill & Jane's Star
Del 20h16'39"10d23'
Bill & Jenny II Anniversary
Uma 11h14'20"41d10'
Bill & Joann-Happy
Anniversary
Cyg 21h30'21"40d9'
Bill & Jodi
Uma 10h42'14"48d35'
Bill & Kara Christmas `95
Lyr 18h31'38"30d53'
Bill & Karen's Star
Cyg 21h2'1"31d7'
Bill & Kay
Crb 15h53'15"28d41'
Bill & Phyllis
Lyr 18h58'52"31d46'
Bill & Roxie's Star
Boo 15h3'24"25d57'
Bill & Sandy
Cnv 13h58'44"40d54'
Bill & Snookie's Star
Cyg 19h29'46"41d5'
Bill & Tiffany:BFF
Lyr 18h34'26"37d3'
Bill & Valerie
Eri 3h13'20"-16d60'
Bill Aron
Tri 2h0'1"28d1'
Bill Forever
Ori 5h56'40"13d20'
Bill Jacob
Per 3h7'22"41d39'
Bill Star
Boo 13h54'25"18d20'
Bill Star,The
Per 2h36'58"50d25'
Bill The Star
Cam 3h40'39"60d29'
Bill's Amazin' Twinkly Distant
Ray O'Light
Ori 6h0'37"8d21'
Bill's Angel
Boo 14h50'40"29d21'
Bill's Dream
Cep 22h28'58"78d58'
Bill's Faith
Ser 18h14'58"-7d28'
Bill's Golden Angel
Cyg 19h50'38"47d48'B
Bill's Guiding Light "Danitra"
And 23h36'14"46d13'
Bill's Irish Star
Aur 6h4'42"46d33'
Bill's Legacy
Cet 2h51'21"3d55'
Bill's Lucky Starsky
Ser 16h0'58"24d50'
Bill's Rainbow Connection
Lac 22h17'11"51d42'
Bill's Sparkle
Ori 5h55'48"15d13'
Bill's Star
Her 17h19'59"46d28'
Bill's Star
Pro 5h35'1"-6d2'
Bill's Star
Aur 6h25'0"35d7'
Bill's Star
Ori 4h55'1"0d26'
Bill, My Everlasting Light
Cnc 8h41'18"18d48'
Bill/Ellen
Cam 13h14'20"77d3'
Billard Serge
Peg 23h34'28"20d49'

BillDee Holly
Vul 20h21'22"22d52'
Bille
Lib 15h58'35"-7d58'
Billeaud,Charlotte
Umi 14h15'1"71d1'
Billera,Joseph
Hya 8h30'0"-6d45'
Billes,I V & Eugenia
Del 20h24'19"20d30'
Billett,Bernice
Aql 18h40'1"1d5'
Billette,Christopher Alan
Ori 5h50'16"16d19'
Billi Jo
Her 17h26'21"18d54'
Billianpeg50
Cyg 21h3'41"36d6'
Billie
Ser 16h0'42"14d37'
Billie
Aql 19h1'47"-6d15'
Billie & Durf
Cyg 20h4'1"58d38'
Billie B II
Ser 15h34'0"9d29'
Billie Jo
Cas 1h1'24"61d9'
Billie's Hideway
Her 17h32'34"28d33'
Billie,Kathy
Aql 19h12'38"15d17'
Billig,Michael George
Her 17h39'20"14d48'
Billimore,Sandy
Peg 23h30'13"11d47'
Billimoria,Jeelu
Vul 19h53'12"20d17'
Billingham-Navarro
Aql 20h6'1"0d25'
Billings,Chester J
Lac 22h7'34"48d56'
Billings,Clu,ChFc, Wallace F
Cep 20h24'36"78d50'
Billings,Deborah A
Cas 1h55'33"58d55'
Billings,Jonathan Turney
Sct 18h42'39"-6d60'
Billings,Marian Janice
Peg 23h18'40"28d59'
Billings,Mary
Cas 0h20'1"64d8'
Billingsley Love & Dedication
Star,Scotty
Aql 20h5'52"6d27'
Billingsley,Barton B
Cma 6h52'15"-16d55'
Billingsley,Claire
Eri 4h30'22"-12d15'
Billingsley,Susi
Cam 5h40'48"73d12'
Billington
Cam 5h51'16"60d24'
Billington,Gwen
Lyn 8h58'51"33d27'
Billington,Mhairi
Com 12h29'0"24d17'
Billington,Natalie
Cyg 20h23'28"39d28'
Billion
Uma 13h56'1"58d39'
Billiris,Theodore
Ser 15h15'0"20d32'
Billman,James Frederick
Ori 5h3'35"0d4'
Billnton
Aql 18h57'36"14d23'
Billock,Theodore Nicholas
Vul 20h39'49"20d2'
Billon,Claude
Aur 5h2'0"31d11'
Billon,Henri
Aql 18h56'1"-6d53'
Billon,Jacques
Aur 5h2'18"30d13'

Billon,Michel
Aur 5h1'49"31d21'
Billon-Drulot, Frederique
Aur 5h1'48"38d6'
Bills,Elsie Riley
Cet 1h23'37"-10d44'
Bills,Jarrod Allen
Cma 6h40'28"-15d57'
Bills,Paige
Oph 17h0'25"-28d9'
Bills,Pamela
And 2h31'43"47d5'
Bills,Spencer Anthony
Ori 5h4'39"8d56'
Bills,Tyler Christopher
Dra 14h26'15"63d8'
Bills,Victoria Lynn
And 0h14'56"33d33'
Billson,Abigail
Cyg 21h6'59"30d52'
Billy
Her 17h23'12"43d42'
Billy
Cep 20h37'1"75d11'
Billy
Peg 22h54'21"11d34'A
Billy
Dra 16h2'13"60d24'
Billy "O"
Ori 5h52'28"12d3'
Billy & Daddy
Leo 11h21'48"0d33'
Billy & Renée
Aur 4h54'36"50d26'
Billy B
Aur 5h38'47"41d12'
Billy Baroo
Per 2h58'24"32d51'
Billy Bob
Her 16h32'40"41d26'
Billy Boy
Aur 6h26'47"33d33'
Billy D
Dra 19h17'39"70d44'
Billy G's Dream
Per 2h2'51"56d38'
Billy Jack
Boo 14h18'49"15d16'
Billy Joe
Dra 16h46'40"68d3'
Billy John
Lmi 10h42'35"33d39'
Billy Ray
Cet 0h33'35"-0d13'
Billy Scott & Hudson Lee-Stars Fanny's Eyes
Del 20h53'29"7d56'
Billy Sikorsky I Love You!
Tri 1h49'51"27d22'
Billy Star
Boo 14h32'31"46d25'
Billy The Kid
Cep 0h24'45"78d54'
Billy The Kid
Sct 18h36'24"-4d44'
Billy the Kid
Aur 6h7'14"31d48'
Billy the Kid
Cep 23h35'0"70d2'
Billy The Peanut
Ser 16h3'17"13d50'
Billy The Star I Love
Boo 14h11'57"39d52'
Billy The Writer
Uma 9h51'0"68d59'
Billy Todd
Gem 7h6'3"21d38'
Billy's Angel
Leo 10h34'0"18d36'
Billy's Dream
Lmi 10h10'39"30d54'
Billy's Forever Shining Star
Lmi 10h28'0"32d8'
Billy's Star
Hya 8h10'1"-9d30'

Billy,Christina Ariel
Cas 0h31'39"64d47'
Billy,Kathleen, Billy Jr, & Shannon
Uma 9h28'40"49d55'
Billy,Mark Anthony
Cet 2h36'23"-12d15'
Billy,Steven Roger
Uma 8h58'0"48d28'
Billy-Bob (Foulks-Christophersen)
Cnv 12h48'20"33d56'
Billyangela
Cyg 21h20'1"28d41'
Billybob
Cep 22h37'1"58d14'
BilMuzi
Del 20h18'0"18d46'
Bilobeao,Bridget E
And 23h41'28"38d50'
Bilodeau,Barbara A
And 23h41'46"38d50'
Bilodeau,Jacquelyn M
Cas 23h21'48"61d43'
Bilodeau,Muriel
Cas 0h54'27"74d52'
Bilodeau,Thomas W
Dra 19h49'48"60d25'
Bilotta,Kaitlyn Samantha
Lyr 19h3'0"25d33'
Bilotti,Tony
Sco 16h5'0"-38d25'
Bilous,Marie
Cas 2h5'18"67d50'
Bilovsdisis II
Cyg 21h2'0"28d12'
Biltoft,Kristian
Ori 5h24'24"1d42'
Bilton,Anne
Cmi 7h23'24"7d33'
Bilton,The William
Cmi 7h34'18"10d11'
Biltz,Imogene Suzette Porter
Cmi 7h26'50"0d51'
BILUS
Peg 0h6'43"13d1'
Bily,Kathryn Dana
And 0h3'21"38d17'
Bimbo
Com 12h28'17"20d46'
Bina A New Beginning, Gina
Cas 1h48'21"75d28'
Binafard,Behzad Daniel
Mon 7h49'26"-8d57'
Binasoars-Laura
Her 16h30'16"45d42'B
Binasoars-Rod
Her 16h30'16"45d42'A
Binauld,Sandra
Lac 22h28'0"40d21'
Binazeski,Michael John
Boo 14h9'48"53d21'
Binco,Daniele
Tel 20h18'59"-49d1'
Binda,Charles Ernest
Cmi 7h11'40"9d44'
Binder,Anthony Jonathan
Dra 17h52'47"68d21'
Binder,Karl Maria
Sgr 19h45'8"-44d22'
Binder,Matthew J
Tau 4h58'0"20d4'
Binder,Patricia
Per 3h58'33"39d36'
Binder,Reinhold
Leo 11h22'0"-5d56'
Binder,Robin
Aqr 23h32'43"-19d50'
Binder,Walter C
Cnv 14h0'0"38d27'
Binderstein,Pharoah
Oph 18h16'20"8d58'
Bindi,Virginia McClure
Cet 1h37'55"-10d44'

Bindon,Michael Jesse
Eri 4h53'32"-7d0'
Bine
Vir 13h0'33"-0d32'
Bine & Jörg
Uma 10h52'29"61d40'
Binet,Lorraine
Cas 0h53'20"74d37'
Binet,Michèle
Lyn 8h54'21"42d24'
Binford,Catherine Cole
Aqr 22h42'42"-5d30'
Bing
Lyn 7h37'33"52d13'
Bing,Mary Anna
Cnc 8h54'27"30d20'
Bing,Richard Taylor
Her 16h50'37"38d4'
Bingemer,R Claus
Uma 9h15'0"58d7'
Bingesser,Elsie
Cyg 19h43'0"30d36'
Binggeli,Markus
Cas 3h4'1"58d42'
Bingham,Anita & Robert
Tau 4h36'12"16d3'
Bingham,Anthony Paul
Her 16h44'23"22d52'
Bingham,Betty "Bett"
And 1h25'0"48d59'
Bingham,Bill
Cet 1h16'53"-5d48'
Bingham,Christina
Ori 5h18'17"15d53'
Bingham,Donna
Lyr 19h9'34"40d32'
Bingham,Gretchen Hennessy
And 22h57'0"51d6'
Bingham,Jeffrey J
Lac 22h16'51"46d1'
Bingham,Lesley Dawn
Cam 4h37'23"68d12'
Bingham,Michelle Jean
Cyg 21h38'35"37d55'
Bingham,William David
Oph 17h55'1"10d11'
Bingman,Jacob
Dra 17h56'1"65d10'
Bingo
Cma 7h3'37"-28d19'
Bini
Uma 9h41'30"57d45'
Binick,Jean Louis
Ser 16h5'0"8d40'
Binion,Mike
Lyn 7h26'59"58d21'
Binion,Phyllis Cope
Cyg 21h22'46"40d5'
Binkley,Edith
Peg 0h1'24"30d50'
Binko,Ann Henderson
Lyn 7h3'43"44d26'
Binks,Dottie & Fred
Aql 19h29'1"12d34'
Binks,John S
Lac 22h26'38"53d37'
Binky I
Aqr 23h6'49"-10d60'
Binky's Blinker
Cam 3h56'20"78d52'
Binnall,Benjamin James
Her 18h0'41"45d8'
Binnall,Timothy Alan
Lac 22h28'25"50d31'
Binney,Bob & Kelly
Crb 15h38'1"27d54'
Binnie
Cep 23h11'0"60d34'
Binnie-Corinne
And 1h33'0"38d46'
BinnieHerbert LoveStar
Cyg 21h40'18"38d59'
Binns,Giuliana
Crb 16h17'26"37d54'

Binns,J Leslie
Hya 9h5'19"-10d17'A
Binns,John Charles
Aur 5h2'18"45d39'
Binny Bunny
Sct 18h52'24"-9d17'
Bino
Cam 6h45'14"80d10'
Bino,Adrian
Ori 5h30'25"-6d59'
Binsami's At The End O'The Universe
Lac 22h36'22"44d24'B
Binswanger,Sylvia
Cas 1h3'19"63d6'
Bintliff,Richard
Aur 5h3'12"49d37'
Binversie,Dominic
Tri 2h36'21"33d46'
Biogio (Tony Nastasi)
Aur 4h59'32"51d19'
Biohazard
Ori 5h35'37"-6d54'
Biolchini,Joseph M
Per 2h45'48"40d38'
Biomega
Her 16h23'59"24d5'
Biondi,Frank
Cra 11h3'39"-8d30'
Biondi-Shining Forever,Michael A
Ser 15h33'44"18d46'
Biondo,Michael James
Her 18h13'18"38d28'
Biondolillo,Vincenzo
Per 3h37'55"38d54'
Biotteau Julie
Dra 16h26'48"62d10'
Bippen,Rose Frances
Umi 14h59'1"65d48'
Birch Tradition
Dra 14h48'36"63d17'
Birch' Tree
Per 2h51'11"31d23'
Birch,Charlotte Sophie-Marie Anna
Peg 22h49'32"21d46'
Birch,Jennifer Ann-Marie
And 2h33'54"45d23'
Birch,Jessica
Peg 23h46'24"29d55'
Birch,John Alexander
Uma 11h28'57"42d44'
Birch,Katherine
Cyg 19h34'54"35d49'
Birch,Mrs Everett B
Aqr 22h26'19"-1d32'
Birch,Terry
Uma 8h40'13"51d22'
Birch,Thomas
Aur 4h49'45"51d39'
Birch,William
Ori 6h5'40"5d16'
Birch-Whatever We Imagine,Paul
Gru 22h29'7"-56d11'
Bircher's Star,Dr John
Ser 15h30'47"17d47'
Birckenstock,Erna und Kurt
Gem 7h24'40"35d22'
Bird
Aql 19h3'35"-0d7'
Bird & Baboon
Aql 20h7'20"1d26'
Bird(Man),Richard Michael
Aql 19h30'59"8d25'
Bird,Anat
Uma 10h25'42"61d38'
Bird,Aston Danielle
Umi 13h34'32"73d16'
Bird,Christopher James & Kelly Kara
Lac 22h10'18"47d1'
Bird,Clifford
Boo 14h46'1"37d7'

Bird,Dale Edward
Her 18h0'27"30d2'
Bird,Dan & Connie
Sge 20h2'45"20d5'
Bird,Jame-ileen
Lyn 7h41'27"38d14'
Bird,Janet Lyons
Cam 3h44'58"71d54'
Bird,Jess Reeves
Hya 9h35'38"-1d34'
Bird,Jr,Noel Thomas
Per 4h3'37"51d31'
Bird,Katc
Boo 15h4'51"31d28'
Bird,Kevin Eric
Aur 6h3'0"45d49'
Bird,Michael John
Cep 22h20'30"58d21'
Bird,Nancy
Aql 18h41'12"-2d24'
Bird,Paula Berwick
Sgr 18h51'26"-23d47'
Bird,Peter
Cyg 21h18'32"39d44'
Bird,Sophie Rebecca
And 0h13'58"38d14'
Birdie
Lyr 19h19'25"33d53'
Birdie's Flyer
Lmi 10h21'39"28d18'
Birdsall,Mary
Cam 13h16'40"77d6'
Birdsong,Cierra Savannah
Eri 3h36'54"-17d53'
Birdt,Alyssa Robin
Cas 2h14'34"70d9'
Birdwell,Ed & Diane
Cyg 19h27'50"33d27'
Birdwell,Stephen Wayne
Vir 12h54'45"-21d19'
Birdwell,Tabatha
Aql 19h45'15"10d38'
BIRE
Dra 17h3'41"64d9'
Bireley,Heidi Lynn
Cas 2h58'33"57d32'
Birewars
Cam 10h25'1"82d7'
Birgeles,Veronica
Cyg 19h29'32"35d30'
Birgit
Cas 1h26'40"58d50'
Birgit
Lyn 7h27'1"40d42'
Birgit
And 2h15'42"45d15'
Birgit
Sco 17h55'21"-40d56'
Birgit
Cas 0h16'1"59d34'
Birgit "Locke"
Cam 7h7'41"21d53'
Birgit & Clive
Dra 20h28'47"68d14'
Birgit & Julius
Crb 16h1'35"32d26'
Birgit und Thomas
Cas 0h18'17"60d33'
Birgitta
Mon 6h6'30"-6d39'
Birk,Danielle
Lac 22h39'0"53d0'
Birk,Gloria
Cas 0h3'1"62d41'
Birk,Lisa Jahree
Equ 20h56'58"3d1'
Birk,Matthew
Cnv 13h23'46"37d49'
Birkelbach,Monika
Cep 22h6'18"60d58'
Birkelo
Dra 12h47'1"68d1'
Birkenfeld,Tinky
Del 20h24'35"10d52'

Birkett's Brilliance
Peg 22h0'28"28d21'
Birkett,Sidney
Aql 20h10'42"10d34'
Birkett,Vera & James
Cyg 20h51'41"38d34'
Birketts' Diamond Annv Star,Jim & Vina
Cyg 19h26'53"33d45'
Birkhofer-Eternal Embrace,Dave & Jewelye
Sex 10h46'24"-10d27'
Birkholm,Carl Dreiøe
Mon 6h35'14"1d15'
Birkinshaw,Jeffrey "Birk"
Lyr 19h14'52"41d50'
Birks,Joshua Steven
Peg 23h37'24"11d13'
Birkshire's Daring Dumpling,CH
Cma 6h59'47"-19d4'
Birman,Anat
Lyn 8h51'43"44d24'
Birmingham,Jeffrey Paul
Oph 17h37'40"-16d39'
Birnbaum,Brian Scott
Mon 7h9'54"-10d37'
Birnbaum,Estelle
Cas 1h46'20"73d43'
Birnbaum,Madison
Aur 6h4'12"32d30'
Birnbaum,Ruthanne Jaffe
Cas 2h33'30"57d42'
Birnbaum,Steven L
Ser 15h57'1"4d50'
Birnberg,Joshua Alan
Aur 6h13'47"38d2'
Birner,Hertha
Cas 22h57'0"55d47'
Birney,Candace
Aur 5h16'52"40d27'
Birnhak King of Hearts,Joel Robert
Boo 14h21'53"54d16'
Birnhak,Captain Bob
Cep 22h34'1"58d57'
Birnhak,Daniel Adam
Per 1h29'36"50d24'
Birnie,Katharine Charlotte
Per 4h43'1"51d31'
Birns,Daniel
Boo 13h45'26"23d18'
Birns,Joyce
Lyn 7h56'0"34d15'
Biro,Ben
Lac 22h14'56"49d2'
Biro,Courtney Marie
And 0h48'48"41d2'
Biro,Jennifer Lynn
Cyg 20h23'18"38d50'
Biro,Laszlo
Uma 8h12'36"60d56'
Biro,Paul Derek
Aql 19h47'49"10d44'
Birosel,Filomena Cuaresma
Lyn 9h18'46"34d38'
Birss,Randy
Her 16h39'43"40d34'
Birthday Of Akira
Sco 16h9'5"-26d21'
Birthe Larsen
Ori 5h36'38"8d37'
Birtwell,Lesley Margaret
Dra 16h50'0"60d26'
Bisaillon,Antonin Rossier
Umi 15h31'0"67d19'
Bisbee
Cyg 21h3'56"39d8'B
Biscaba,Beverly
And 2h20'14"39d5'
Biscay,Nora
Cet 2h24'12"8d5'
Bischler,Katie Ann
Cyg 19h31'36"34d11'

Bischof,Leonila
Sgr 18h56'3"-35d58'
Bischof,Marisa Danielle
Mon 7h27'28"-8d7'
Bischof,Stella & Richard
Cyg 19h51'23"41d12'
Bischoff,Ann Hayes
And 0h0'54"47d39'
Bischoff,Hagen
Cap 21h25'10"-23d56'
Bise,Jr,Jimmie Lee
Aur 7h0'29"36d20'
Biser's Celebration Star
Lyn 7h41'26"40d34'
Biser,Michael Christopher
Cyg 19h33'40"38d17'
Bish,Breanne
And 2h27'18"49d12'
Bish,Bruce
Boo 15h6'11"52d25'
Bish,Maranda
Lyn 9h6'0"36d9'
Bish,Raymond & Louise
Uma 11h1'54"51d50'
Bish,The
Cep 23h10'39"62d15'
Bish,Vanessa
And 2h6'13"38d2'
Bisharat,Madison Rindge
Lac 22h18'40"50d24'
Bishko,Samuel J
Aur 4h53'21"50d52'
Bishman,Janet
Mon 8h8'48"-4d1'
Bishoff,Danielle Starr
Aur 6h4'27"31d8'
Bishop
Vul 19h5'59"25d30'
Bishop & Mookie's Heavenly Love
Uma 10h44'58"58d29'
Bishop's Star
Lac 22h52'0"54d24'
Bishop's University Sesquicentennial
Uma 10h45'1"40d59'
Bishop,Anna M
Cyg 20h38'26"40d3'
Bishop,Anthony Garland
Her 18h23'43"28d9'
Bishop,Avis Marie
Eri 3h55'0"-15d0'
Bishop,Brenda
Cyg 19h29'30"38d16'
Bishop,Capt Bob
Aql 19h58'29"8d52'
Bishop,Charles "Boo"
Boo 14h52'21"33d5'
Bishop,Charles R
Ser 15h24'49"22d32'
Bishop,Christina
Peg 22h2'1"22d43'
Bishop,Chuck & Jeanette
Ori 5h34'10"-3d14'
Bishop,Donie & Don
Aql 20h3'44"1d16'
Bishop,Gabrielle
Equ 20h59'53"5d33'
Bishop,Gary John
Per 2h4'45"57d54'
Bishop,Irene & Ken
Uma 11h31'46"45d19'
Bishop,Jill Colleen
Cap 21h43'0"-20d15'
Bishop,Joseph Dean
Mon 8h8'49"-1d43'
Bishop,Joshua James
Dra 16h43'25"63d55'
Bishop,Jr USN,Captain Perry C
Ori 5h48'30"11d4'
Bishop,Julie
Lyr 18h15'32"30d4'
Bishop,Kent & Sheila
Com 13h14'13"21d50'

Bishop,Kimberly Boo Boo
Cet 2h46'11"3d35'
Bishop,Laurenne Monique
And 23h15'53"40d30'
Bishop,Linda Sue
Com 12h55'43"26d57'
Bishop,Louise
Cyg 19h18'34"44d58'
Bishop,Meghan Elizabeth
Vir 11h38'55"9d16'
Bishop,Michele
Tau 5h45'1"23d50'
Bishop,Patsy V
Peg 22h28'29"25d29'
Bishop,Robert
Uma 11h7'32"58d6'
Bishop,Sarah
Peg 23h38'21"20d19'
Bishop,Sarah Lowell
Uma 11h11'23"60d46'
Bishop,Stephanie Rose
Cas 23h23'13"63d23'
Bishop,Stephen A
Aur 7h23'47"38d42'
Bishop,Steven
Cep 3h49'44"80d18'
Bishop,Vyki
And 2h33'26"40d33'
Bishop,Wilma Beryl
Cru 12h15'2"-63d20'
Bishop-Bulan,Terre
Eri 3h21'14"-12d0'
Bishop-Sunshine & Sendy's Mom,Sandra
And 0h58'25"45d26'
Bishoprick,Stanley
Per 3h11'1"49d44'
Bisig,Nicole Maria
Cyg 21h5'44"30d29'
Bisko,Joseph
Cet 3h15'21"1d39'
Biskoff,Amy
And 0h23'53"43d54'
Biskup,Silke
Per 3h51'17"37d38'
Bisle,Michael
Cnv 13h7'1"33d1'
Bismarck v Maxquatch
Cmi 7h42'24"1d44'
Bisnett,Sr,Charles Roger
Lac 21h56'52"41d7'
Bisogni,Scott
Boo 15h7'57"16d33'
Bisounours 1
Cam 6h5'30"60d6'
Bisounours 2
Aur 4h53'1"40d4'
Bisquera,Joyce
Eri 4h29'45"-19d3'
Bissette,Bill
Lac 22h19'11"48d56'
Bissey,John
Aql 19h56'53"13d9'
Bissinger,Hans Michael
Ser 15h12'38"10d32'
Bissinger,Roger
Vir 11h46'26"8d7'
Bisson,Jennifer
And 0h32'25"45d19'
Bisson,Juliette
Uma 11h18'33"44d19'
Bisson,Timothy
Her 18h18'0"12d21'
Bissonnette,Gillian Dawn
Com 13h14'26"21d47'
Bistany,Mark D
Aql 19h40'20"8d15'A
Bistarelli,Chiara
Uma 14h19'53"60d32'
Bistarelli,Francesca
Lyn 8h57'50"45d26'
Bistarelli,Vanessa
Lmi 10h5'45"34d38'
Bister
Vul 19h59'1"28d14'

Bister,Peter
Lyn 8h4'15"40d34'
Bistline,Amy Marie
Cyg 19h54'59"40d5'
Biswell,J Patrick
Cet 0h55'46"-5d10'
Bitchette,Helen Friday
Hya 8h13'14"-5d57'
Bitcon,Christopher Richard
Lac 22h33'33"35d32'
Biter,Becky Jo
And 0h26'26"38d41'
Bitetti,Vincent
Vul 20h14'56"28d49'
Bithell,Pauline
Aur 5h19'54"49d16'
Bithray,Nick
Aur 6h14'47"36d59'
Bitman
Aql 20h2'15"7d44'
Bitner,Daniel G
Cet 0h55'57"-3d51'
Bitner,Dawn "Raven"
Peg 22h16'22"30d58'
Bito
Umi 14h54'47"67d27'
Bitonte,Eben James
Her 16h47'53"34d23'
Bitonte,Kendall Marie
And 2h18'0"46d10'
Bitonti,Angela Durazzi
Lac 22h40'59"38d49'
Bitouzé
Cas 0h30'1"68d13'
Bitre Étoile
Lyr 19h10'22"38d29'
Bitsa
Eri 4h59'44"-4d20'
Bitter,David and Julie
Uma 9h55'14"42d16'
Bitterman,Allison Lindley
Cnc 8h11'30"31d49'
Bitterman,Butch
Her 17h24'59"40d39'
Bitterman,Connie
Cas 2h40'29"67d36'B
Bitterman,Marty
Cas 2h40'29"67d36'A
Bitterman,Stephanie Rose
Gem 7h18'39"21d53'
Bitters,Ann Elizabeth
Peg 22h41'37"34d35'
Bittersuite
And 1h51'59"39d5'
Bittinger,John R
Her 17h44'27"40d16'
Bittle,Bruce Richard
Her 16h12'0"50d42'
Bittle,Noreen
Cam 3h54'0"52d34'
Bittle,Randall Paul
Aql 20h11'55"1d34'
Bittman,Laura Elizabeth
Lyn 7h47'16"38d37'
Bittmann,Patsy Peach
Lyn 8h10'33"58d32'
Bittner,Martin E
Uma 10h38'42"68d18'
Bittner,Wild Bill
Oph 18h39'53"8d43'
Bittners' Delight
Cam 4h54'51"68d43'
Bitton,Rivka
Aql 19h46'35"11d45'
Bitzas,Charlene
Aur 5h58'14"48d54'
Bitzer,Lori Lynn
Lyr 19h10'21"47d43'
Biuso,Catharine Story
Lyr 19h13'0"42d46'
Bivans-Outlaw Lisa Gwen
Cap 20h31'18"-26d54'

Bivens,Glenda
Peg 22h57'34"30d13'
Biviv,Rae
Aql 19h29'53"-10d15'
Bivona Family Star,The
Aql 19h53'17"15d24'
Bivona,Francesca Marie
Cyg 19h33'19"31d60'
Bixby,Karen
Eri 2h50'26"-15d13'
Bixby,Melissa Kathleen
Sct 18h49'40"-7d10'
Bixel,Blake Ryan
Lib 15h20'1"-25d3'
Bizar "Wedding Star"
Vul 20h15'51"26d3'
Bizarre Star,The
Cam 9h26'48"82d15'
Bizer,Herman `Tiger'
Lyn 8h9'0"50d40'
Bizot,Jean Claude
Aur 5h52'13"30d32'
Bizub,John R
Cep 22h30'42"63d59'
Bizzozero,Jackie
Cas 0h25'1"72d18'
BJ
Aur 5h20'11"38d59'
BJ
And 0h29'34"38d31'
BJ
Cam 4h44'1"68d36'
BJ
Ser 15h18'11"20d33'
BJ Loves Susan
Ori 5h57'29"15d39'
BJA11
Cam 4h48'45"71d2'
Bjelica,Vide
Ori 5h57'45"16d53'
Bjelland,Adam
Boo 14h27'23"48d48'
Bjerke,Anna Jean
Uma 11h36'30"31d11'
Bjerke,Leland
Cam 4h3'50"67d32'
Bjorklund,Heidi Lynne
Peg 22h30'1"20d10'
Bjorkman,Bo
Ori 5h30'60"0d25'
Bjorkman,Michael Kemp
Cep 22h34'17"67d55'
Bjorman,William Henry
Hya 9h1'27"4d35'
Bjorne,Jann "Loghousestar"
Cep 22h39'20"63d41'
Bjornson,Charlotte Marie Bergum
Psc 0h54'31"17d35'
Bjurling,Zorka & Jan-Olof
Uma 9h32'58"47d50'
Bjurmark,Alexander Mats William
Umi 16h22'18"78d42'
Bjørkvik,Sanne
Cam 3h20'12"60d13'
BKY 1-2-3
Sco 17h27'51"-30d8'
Blac,Thierry & Veronique
Cma 6h57'35"-16d0'
Blacet,Dianne E
Del 20h23'0"8d40'
Blacet,Essie Grigsby
Cyg 20h53'58"40d26'
Blach,Henrik
Dra 20h4'15"68d35'
Black
Del 20h14'47"12d14'
Black 10-10-91, Nicholas Ryan
Lib 15h13'55"-28d45'
Black Diamonds
Uma 8h46'51"50d29'
Black Dog
Cnv 13h40'56"39d59'
Black Family,The Chesley
Ori 5h56'27"14d33'
Black Gold
Aur 5h2'0"42d12'
Black Jack's Diamond
Per 2h54'32"38d58'
Black Lightning
Uma 12h6'0"56d44'
Black of Sawrey, Elizabeth H Née
Cas 0h31'41"61d13'
Black Sheep
Uma 8h50'54"47d56'
Black Stallion
Peg 23h16'59"30d47'
Black Star,Jacob Otto
Ari 2h0'24"20d46'
Black Swan
Cyg 21h26'19"30d52'
Black's Night Rider
Aql 19h11'34"12d32'
Black,Alexis Elizabeth
Gem 7h21'33"30d45'
Black,Andrew Karl
Mon 7h46'11"-3d25'
Black,Arlyn K
Cam 5h53'16"73d30'
Black,Art
Uma 10h53'46"56d33'
Black,Barbara
Lyn 7h29'23"50d2'
Black,Benji
Boo 14h50'54"30d55'
Black,Denise Tyson
Cas 1h13'15"75d23'
Black,Duncan
Her 16h54'12"41d9'
Black,Elaine R S
Aql 20h17'0"5d22'
Black,Frankie
Boo 15h31'36"42d28'
Black,H Kingston
Aur 6h28'21"35d17'
Black,Heath Austin
Psc 1h57'10"53d50'
Black,James Lee
Aql 19h30'18"-6d33'
Black,Janice R
Vul 19h55'21"20d05'
Black,Jennifer Leigh
Lyr 18h30'47"44d51'
Black,Jenny
Com 12h26'48"21d23'
Black,Jerome Byers
Gem 6h43'22"18d26'
Black,Joyce & Stanley
Ori 5h8'38"-5d13'
Black,Landon E
Aql 20h6'1"0d44'
Black,Larry C
Ser 17h31'45"-13d22'
Black,Lear
Cma 6h31'46"-24d45'
Black,Lisa
Cyg 19h56'56"37d47'
Black,Mairi J
Boo 14h19'54"50d53'
Black,Martha B
Cas 23h35'41"61d26'
Black,Matthew
Com 12h42'31"23d20'
Black,Maxene
Boo 14h26'1"28d8'
Black,Michael
Uma 9h12'55"68d54'
Black,Michael Allen O'Gallagher
Dra 16h29'38"61d35'
Black,Michael Dewayne
Equ 21h20'18"11d37'
Black,Mike
Cep 22h53'20"58d8'
Black,Miss Erica C
And 23h26'21"41d31'
Black,Patsie
Lmi 9h59'44"38d29'
Black,Rachel Dubbs
And 2h19'0"47d25'
Black,Robyn
Peg 23h1'54"33d45'
Black,Ron L
Cet 2h32'43"-5d28'
Black,Ronny
Uma 9h31'11"47d26'
Black,Rosella
Cas 0h21'57"66d36'
Black,S Kingston
Lac 22h35'57"53d53'
Black,Sherry E
Del 20h26'12"11d40'
Black,Shirley Palmer
Lyr 18h50'46"42d49'
Black,Tasha Nicole
Umi 17h28'33"75d53'
Black,Tina Michelle
Cas 0h34'21"68d39'
Black,Troy E
Boo 15h46'0"50d11'
Black,Victoria Leigh
Gem 8h1'0"28d10'
Black-Ingersoll,Camden Lee
Vir 12h6'0"0d8'
Black-Ingersoll,Myles Douglas
Vir 14h58'57"0d59'
Black-Mallard,Sarah & Jon
Sge 18h56'59"18d56'
Black-McLaughlin Wedding Star
Ori 5h31'33"-3d40'
Black-Milner,Karen
Uma 9h51'59"56d9'
Blackburn,Catherine Gail
Vul 20h20'1"23d29'
Blackburn,Courtney Drew
Lyr 18h57'55"30d57'
Blackburn,Easton
Boo 15h19'18"37d41'
Blackburn,Esther
And 0h28'50"45d37'
Blackburn,Jean
Cet 2h7'40"2d9'
Blackburn,Jeffery B
Dra 19h4'46"65d19'
Blackburn,JoElla
Eri 3h5'39"-1d53'
Blackburn,Juliet S
Cet 2h26'15"4d4'
Blackburn,LaVon Cram
Uma 9h16'13"61d49'
Blackburn,Nicola Jayne
Cas 23h28'0"61d37'
Blackburn,Rolande R
Her 16h11'1"20d15'
Blackburn,Russell A
Oph 17h56'44"10d18'
Blackburn,Star Lynn
Cyg 20h25'49"40d52'
Blackburn,Valorie
Lyr 19h24'1"38d47'
Blackburn-Stillwagon, Sarah
Cyg 20h25'18"38d2'
Blackenberger 8-13-94
Lyn 7h58'30"50d23'
Blacker,Dr Ann
Oph 16h53'55"-22d45'
Blackett,Lloyd
Cep 21h17'1"55d10'
Blackford,Jennifer L
Peg 23h25'36"10d48'
Blackford,Kathleen
Mon 7h55'0"-8d30'
BlackHawk
Sge 19h59'0"16d13'
Blackhurst,Carol
Aql 20h6'19"0d24'
Blackie
Cma 7h24'46"-11d15'
Blackjack,Tommy V
Lyn 7h55'21"33d21'
Blackledge,Elvira Muratdzhanovna
And 2h19'0"47d25'
Blackman
Sct 18h54'36"-6d33'
Blackman,Ariel Elizabeth
Cas 3h9'48"60d41'
Blackman,Carl
Sct 18h56'53"-6d35'
Blackman,Jay Ellsworth
Aur 4h52'15"38d43'
Blackman,Kelley Ann
Peg 22h13'17"5d38'
Blackman,Michael George
Per 3h3'25"41d40'
Blackman,Wendy Yvonne
Cyg 21h36'31"42d55'
Blackmann III,Joseph Chambers
Peg 23h8'15"10d17'
Blackmon,Dwayne
Tri 1h49'52"28d14'
Blackmon,Susan
Mon 7h57'24"-1d16'
Blackmon,Willow Heather Hoban
Lac 22h10'21"47d33'
Blackmore's Star, Jeremiah Clark
Uma 11h4'42"57d12'
Blackmore,Brian
Dra 16h30'20"61d35'
Blackney,Mary
Eri 2h43'0"-4d49'
Blacknight-DNS,J T
Lyn 7h44'51"45d39'
Blacknus,Inky
Ori 6h5'1"3d57'
Blackstock,Arley
Tri 1h59'0"27d28'
Blackstock,Reba Nell McEntire
Peg 23h5'34"8d24'
Blackstock,Sherri
Aql 19h31'51"-8d29'
Blackstone,Ted & Willene
Umi 16h20'51"78d51'
Blackwelder's Star
Equ 21h22'37"10d35'
Blackwell Family,The
Uma 12h53'19"60d56'
Blackwell's Light
Dra 17h24'23"73d21'
Blackwell,Amy N
Lyr 18h56'26"30d54'
Blackwell,Deborah Suzanne Simioni
Cam 7h58'37"71d14'
Blackwell,Jazzmonae
Lyn 8h20'22"52d12'
Blackwell,Jeff
Aql 19h53'12"0d23'
Blackwell,L Christine
Lyr 18h27'49"31d46'
Blackwell,Linda
Peg 21h38'24"26d47'
Blackwell,Linda Gail
Mon 6h57'14"8d41'
Blackwell,Lois
Peg 23h3'16"31d2'
Blackwell,Martin
Ori 5h44'0"0d16'
Blackwell,Neil
Uma 8h22'14"61d39'
Blackwell,Shahidah
Boo 14h48'1"34d3'
Blackwell,Tom
Per 2h29'41"56d19'
Blackwell,Tracey Amber
Lyn 7h37'0"40d51'
Blackwell,Trip
Aql 20h13'46"4d17'
Blackwood,Alexis Anne
Tau 5h22'48"16d49'
Blade Runner's Orb
Equ 21h7'15"11d56'
Blades,Rolen Ann
Cam 3h48'31"52d56'
Bladow,Jr,Lyle John
Per 2h23'43"58d50'
Blaemers,Katherine
Lyr 18h32'23"42d38'
Blaesing,Blaine
Dra 17h35'27"63d54'
Blagg,Bill
Ser 15h33'25"1d24'
Blaha,Jerome Arthur
Her 18h10'34"40d18'
Blaher,Jr,Peter D
Aur 5h58'35"28d8'
Blaho,Emil
Psc 1h43'1"21d31'
Blaich,Christopher
Her 17h1'49"30d15'
Blaich,Silke
Sgr 19h19'5"-20d37'
Blaikie,Donald B
Per 2h8'50"56d25'
Blailock II,Zack R
Aql 19h0'50"1d9'
Blain,Aaron Edward
Leo 10h38'1"17d40'
Blain,Daniel Eberlin
Per 4h43'1"38d16'
Blain,Marty
Per 3h56'17"40d56'
Blaine
Dra 14h59'17"62d27'
Blaine's Beam
Her 17h38'41"41d9'
Blaine,Dad's Forver Love, Star Kendall
Peg 22h13'37"33d46'
Blaine,H Terrence
Cma 6h52'2"-17d48'
Blaine,James Stanton
Aql 20h0'41"10d51'
Blaine,Margaret H
Mon 7h42'33"-5d28'
Blair "...Friendships last love..."John
Eri 4h49'60"-9d47'
Blair III,John
Lmi 10h51'57"38d42'
Blair Star,The Donald R & Margaret A
Cyg 21h33'22"50d23'
Blair's Soul
And 23h43'29"40d38'
Blair,Andrea A
Lyn 7h50'55"38d23'
Blair,Archie
Uma 8h35'56"57d12'
Blair,Bobby
Aur 5h38'1"38d14'
Blair,Brooke Aaron
Eri 3h29'39"-7d2'
Blair,Christena
And 23h29'42"46d34'
Blair,Debbie
Com 13h16'29"22d7'
Blair,Delane & Troy
Cyg 21h3'41"38d22'
Blair,Donald & Mary
Uma 11h50'1"31d12'
Blair,Elliot
Ori 5h9'51"-1d47'
Blair,Forever Jeffrey A
Per 3h12'51"48d1'
Blair,forever in my heart,Leza
Cra 18h24'39"-44d50'
Blair,Gary
Sex 10h40'56"0d31'
Blair,Gary
Per 3h25'17"40d35'
Blair,I Love You
Lyr 19h13'39"41d6'
Blair,Jason
Sex 10h41'0"2d14'
Blair,Jim
Her 16h11'59"25d30'
Blair,Julie
Peg 22h18'36"10d44'
Blair,Kathleen Lee
Peg 23h35'10"18d13'
Blair,Kathleen Mary
And 2h29'18"50d43'
Blair,Linda Stoffan
Peg 23h37'24"15d5'
Blair,Lionel
Lyn 9h34'1"40d6'
Blair,Margo H
Cas 1h41'50"58d45'
Blair,Marlene
Cet 1h4'25"-3d44'
Blair,Meghan
Cyg 20h46'44"38d43'
Blair,Morgan James
Lyr 18h43'7"41d42'
Blair,Phyliss
Oph 17h2'25"-24d37'
Blair,Robert E & Helen E
Cyg 21h22'18"31d39'
Blair,Robert Kenneth
Dra 15h1'1"55d4'
Blair,Stacy R
Oph 16h49'37"10d7'
Blair,Steve & Dana
Aql 19h6'46"5d16'
Blair,Wesley Paul
Lac 22h38'27"56d15'
Blair-Vaughn "Newtonian",The
Lyr 19h23'50"41d20'
Blais,Marilyn
Cet 1h45'32"-2d24'
Blais,Samuel Star of
Ori 5h59'34"10d17'
Blais,Sean Patrick
Cet 2h11'36"0d53'
Blaisdell,Doug
Aur 6h34'31"38d44'
Blaisdell,Jerry
Per 1h43'60"53d49'
Blaisdell,Karen & Don
Cma 6h56'2"-15d50'
Blaise
Tri 1h47'12"28d11'
Blaize & Hope
Lyr 19h20'30"38d25'
Blake
Cep 23h2'41"70d34'
Blake
Sex 9h57'31"1d59'
Blake
Cet 3h14'39"3d28'
Blake Sweet 16,Kelli C
Peg 23h2'41"8d31'
Blake's Dream
Ori 5h9'36"-6d48'
Blake's Enchanting Birthday Star
Peg 23h1'36"17d47'
Blake,"Our Star" Ken
Oph 17h26'31"-23d59'
Blake,Brendan Charles George
Ori 6h8'0"5d5'
Blake,Daria J
Lyn 7h21'12"50d10'
Blake,Dennis P
Dra 12h59'32"64d16'
Blake,Diana
And 2h15'1"49d23'
Blake,Don & Mary
Crt 11h14'57"-19d40'
Blake,Donna K
Cet 0h55'43"-2d6'
Blake,George Norris
Her 17h22'1"45d45'
Blake,Honey
Cas 0h25'32"58d43'
Blake,Irene
Cyg 19h47'43"58d42'
Blake,Kathleen O'Rourke
Cyg 21h58'14"50d24'
Blake,Kristy
Cnv 12h15'0"48d44'
Blake,Lea Ames
Cas 0h33'39"70d40'
Blake,Martha Dawson
Sct 18h55'26"-6d44'
Blake,Melynda L
And 0h54'23"40d57'
Blake,Mitchell Alan
Dra 12h8'45"71d55'
Blake,Pamela Anne
Mon 6h54'18"-10d29'
Blake,Richard
Aql 20h1'47"1d11'
Blake,Russell W.
Her 15h47m0"45d6'
Blake,Thomas Graham
Cet 0h41'33"-2d23'
Blake-Luckey
Peg 23h30'35"32d48'
Blakely
Aql 19h45'0"10d27'
Blakely,Jennifer Michelle
Cas 0h32'36"61d39'
Blakely,John & Julia
Ori 5h45'42"12d21'
Blakely,Lisa Marie
Cas 0h28'54"60d44'
Blakely,Love For Ever, Valerie Gay
Psc 23h5'40"2d13'
Blakely,Sean Ryan
Ori 5h49'46"12d25'
Blakemore,Haley Dawn
Del 20h23'53"18d59'
Blaken,Billy
Lmi 10h9'0"31d52'
Blakenbeckler,Zack
Peg 22h20'29"21d36'
Blakeney,Alice
Mon 6h4'30"-6d41'
Blakeney,Stephanie
Vul 20h18'47"23d25'
Blakeslee,Cynthia Lynn
Cyg 20h47'40"38d12'
Blakeslee,Holly Anne
Cyg 20h39'19"38d2'
Blakeslee,Karen
Aql 19h53'1"14d34'
Blakey,Greg
Dra 16h45'37"62d20'
Blakley,Glenda M
Cam 4h4'14"58d32'
Blakley,Jr,Edward Lee
Psc 0h18'1"20d39'
Blakley,Roger F
Boo 15h3'23"47d48'
Blakney,Ellen Theresa
Lmi 10h36'59"25d7'
Blalack,Marian
Aql 18h58'44"-5d24'
Blalock,Cheryl Jean
Mon 6h32'11"-0d26'
Blalock,Jr,William W
Hya 9h31'48"5d53'
Blalock,William Jackson
Boo 14h9'31"41d45'
Blamires,Todd W
Aur 7h13'23"35d29'
Blan & Faye Star
Aql 19h29'46"8d18'
Blanc's 80th Birthday, William
Her 18h1'34"31d32'
Blanc,Carmen
Peg 22h49'55"29d11'
Blanc,Jean Pierre
Cet 0h57'24"1d57'
Blanc,Jonathan
Her 16h34'1"38d47'
Blanc,Michel
Mon 7h46'29-4d27'
Blanc,Pierre
Her 16h23'17"23d20'
Blanc,Pierre
Her 16h23'17"23d20'
Blanca Alba
Cep 22h23'0"61d32'
Blanca-Roland 25
Mon 6h29'1"8d52'
Blanch,Shelley Elizabeth Hight
Tau 5h54'13"23d55'
Blanchard 1 UFP,Nexus Debra Cordes
Lyr 18h40'11"47d44'
Blanchard,Adrienne M
Cas 0h0'0"62d28'
Blanchard,Alice E
Cam 9h2'15"80d53'
Blanchard,Angele
Aql 20h31'35"0d6'
Blanchard,Carl & Rita
Mon 5h59'29"-5d55'
Blanchard,Jade LoRayne
Uma 11h22'47"42d42'
Blanchard,Jill K
Cas 0h27'51"67d50'
Blanchard,Martin Richard
Ori 6h0'26"8d14'
Blanchard,Muriel
Cyg 19h42'18"37d57'
Blanchard,Patricia
Cas 1h56'40"60d40'
Blanchard,Star Of
Cet 2h24'43"8d43'
Blancharol,Florence
Her 18h1'0"28d57'
Blanche
Sgr 18h58'59"-22d9'
Blanche Ann
Cas 2h13'41"70d37'
Blanche's Eastern Star
Mon 7h59'60"-6d36'
Blanchett,Ruby
Peg 23h37'17"17d37'
Blanchette '93
Peg 22h30'47"8d38'
Blanchette,Barbara
And 2h25'26"46d20'
Blanchette,Domenique & Daniel
Peg 22h30'23"26d54'
Blanchette,Nicholas
Peg 21h59'17"8d28'
Blanchette,Samantha A
Mon 8h7'10"-6d20'
Blanchette,Tyler Bruce
Crb 16h1'43"38d57'
Blanco,Josefina Rosa
Equ 20h58'46"8d1'
Blanco,Joseph A
Per 2h27'20"58d39'
Blanco,Joseph P
Aur 6h14'29"33d38'
Blanco,Maria
Ori 5h43'10"10d24'
Blanco,Mauricio A
Eri 4h5'1"-7d36'
Blanco,Pamela Dee
Mon 7h17'2"-5d53'
Blanco,Sonsoles
Lyr 19h16'44"42d49'
Blanco-Martinez, Antonio Luis
Aur 5h1'33"37d38'
Bland
Boo 14h24'29"31d27'
Bland II,Rickie Curtis
Eri 2h49'22"-15d33'
Bland,Ann
Equ 21h0'41"10d56'
Bland,Christina Ann
Mon 8h7'29"-10d1'
Bland,Erin A
Cam 4h43'47"67d32'
Bland,Pauline
Cas 1h28'56"65d30'
Bland,Ray & Eve
Uma 9h17'35"43d42'
Bland,Sandra (Paws)
Vul 20h2'20"28d36'
Bland,Sloan Walker
Aql 18h56'41"-7d35'
Blandford,Gwendolyn Kay
Uma 10h43'11"40d47'
Blandford,Kaitlyn Suzanne
Aur 6h59'43"38d16'
Blandina
Aql 19h58'28"11d8'
Blanding,Richard H
Boo 14h43'32"31d0'
Blando,Thomas Edward
Her 16h28'53"33d27'
Blandzinski,Vicki Jean
Cam 7h42'58"60d18'
Blang,Marcus
Dra 18h17'50"48d6'
Blangy,Mike
Dra 16h17'39"64d32'
Blank MA,MFCC,Lauren
Mon 7h1'15"1d48'
Blank,Bonnie Angel
Leo 9h58'26"11d12'
Blank,Forever Mary & Manuel
Ori 5h8'1"-6d48'
Blank,Harry
Aur 6h1'31"45d6'
Blank,Michelle Elizabeth
Com 13h9'56"21d38'
Blank,Nicholas S
Lac 22h54'38"51d42'
Blank,Patty
Crt 11h12'13"-17d33'
Blank,Russell Eugene
Hya 8h51'23"5d50'
Blank,The Zachary Evan
Her 17h33'37"24d40'
Blank-Rosenblum,The Jonathan
Her 17h32'34"25d42'
Blanke,Dietrich
Cnc 9h18'29"30d31'
Blanke,Karl-Heinz
Cmi 7h17'51"7d50'
Blanke,Ralf
Ari 2h7'38"20d50'
Blankenship,Charles V
Hya 8h44'40"4d49'
Blankenship,Edward Wade
Oph 17h27'1"10d45'
Blankenship,Kellye Lee
Lyn 7h36'47"40d46'
Blankenship,Michael Dean
Hya 9h11'41"-14d24'
Blankenship,Rett & Donna
Mon 6h19'26"0d10'
Blankenship,Sherry D
Cas 2h39'50"60d53'
Blankey Jet City,The
Nor 15h47'45"-42d57'
Blankman,Howard M
Cep 21h53'12"68d51'
Blanquier AGT
Uma 9h38'23"49d37'
Blanton,Cathleen E
Cas 1h19'22"50d34'
Blanton,Molly
Cyg 20h9'42"40d57'
Blarney Star,The
Dra 17h5'16"67d17'
Blas,Thomas G
Eri 3h25'41"-4d12'
Blaschak,Connor Louis
Dra 16h45'14"52d30'
Blaschitz,Herbert
Aur 4h47'28"40d41'
Blaser,Anke
Ori 5h54'0"21d31'
Blaser,Kim & Guy
Ori 5h53'32"5d49'
Blaser,Richard Douglas
Boo 15h7'41"53d13'

Blasey Family Star,The
Per 1h56'23"56d25'
Blashek,Robert David
Her 17h0'22"30d50'
Blasi,Leslie A
Tri 2h46'12"33d47'
Blasi,Sascha
Lac 22h44'26"52d48'
Blasi,Thomas
Per 2h41'28"40d55'
Blasi,Zachary David
Dra 19h0'23"65d20'
Blasingame,Nathan M
Oph 18h15'22"11d50'
Blass,Uwe
Gem 6h40'14"30d45'
Blatch,Barnaby
Sex 9h50'1"-5d32'
Blatchley,Alice
Peg 22h1'50"7d43'
Blatnica,Douglas P
Cyg 20h55'1"31d41'
Blatnica,Gerardine
Vir 11h59'29"4d44'
Blatt,Ashley Michelle
Vul 20h3'44"25d0'
Blatt,Emily Louise
Peg 23h29'49"32d34'
Blatt,Helga
Peg 23h32'47"18d28'
Blatt,Jeffrey Charles
Sct 18h43'37"-5d51'
Blatt,Katherine Elizabeth
Sge 19h55'50"16d15'
Blatt,Lorilynn & Robert
Crb 15h22'37"30d24'
Blatt,Marjorie T Carson
Cas 0h53'18"61d47'
Blatt,Steven & Jane- Perry
Ori 5h53'11"19d8'
Blatterman,Georgeann McBride
Cas 1h55'43"60d0'
Blattstein,Ari
Aql 19h53'18"10d53'
Blatz,Christiane
Gem 7h0'59"30d34'
Blau (Mom),Bette
Cyg 19h53'19"37d39'
Blau III,George Gafford
Aql 19h58'16"12d20'
Blau,MLB-Marshall Leigh
Aql 19h28'38"8d29'
Blauer,Henry Sporn
Cep 22h36'41"60d33'
Blaylock,Ivol Joe
Sex 9h53'55"-2d9'
Blaylock,Laura Jane
Peg 23h29'40"23d32'
Blaylock,Megan Maxie
Peg 23h29'29"23d27'
Blaylock,Michael-David
Hya 9h10'0"2d12'
Blaylock,Sr,Thomas Edwin
Aql 19h55'1"10d52'
Blaz'in Kohout's
Per 2h53'0"40d23'
Blaze
Ori 5h32'30"0d3'
Blaze
Ori 5h32'30"0d3'
Blaze,Denise & April
Lyn 8h47'18"36d7'
Blazek,Christina Marie
Cas 0h35'17"69d14'
Blazek,Ed Robert
Per 3h3'31"40d21'
Blazer,Marie
And 0h20'52"40d26'
Blazer,Regina Marie
Ori 5h5'34"10d42'
Blazer,Tanny
Vul 20h28'13"28d20'
Blazey Special,The Gerry
Dra 20h9'34"73d26'
Blazier,Dorothy Jean (Margaret Louise)
Ori 6h0'49"0d13'
Blazo,Emily E
Aur 5h39'40"41d40'A
Blazo,Joseph E
Aur 5h39'40"41d40'B
Blazovich,Heather Margret Francis
And 2h24'52"45d43'
Bleacher,Jerald S
Aql 19h1'0"16d2'
Bleakley,David Robert
Cyg 19h28'22"35d15'
Bleakley,Morgan
Cyg 19h27'58"50d12'
Bleakley,Shannon Christine
Lyr 18h28'46"31d17'
Blease,Nigel
Ori 5h33'55"0d23'
Blecha,Terrence J
Lac 22h28'0"55d15'
Blechner,Rob
Cep 23h10'1"64d28'
Blechschmidt,Angelica
Lib 15h8'8"-21d20'
Bleck,Richard D
Aur 5h8'43"40d27'
Bleckner,Vicki A
Lyr 19h19'0"38d44'
Bled,Bernard
Peg 23h38'25"10d34'
Bledsoe,Chloe Elizabeth
Del 20h36'1"18d57'
Bledsoe,Donna Mae Fitzgerald
Com 11h58'54"14d9'
Bledsoe,Travis Nicholas
Her 16h41'0"29d6'
Bledsoe/Colvin Family, The
Cet 3h5'49"1d3'
Bleeker,Michael Lawrence
Oph 16h58'54"10d20'
Bleiberg,Alain
Per 3h59'18"35d23'
Bleibinhaus,Christine
Sco 17h2'55"-33d57'
Bleich,Nancy
Cas 23h23'53"61d13'
Bleicher,Daniel Aaron
Cet 2h15'43"0d40'
Bleiel,Helmut
Ser 15h33'37"-2d17'
Bleile,Ken
Hya 8h18'44"2d43'
Bleiler,Kenneth A
Boo 14h39'25"30d1'
Bleistein,Anton
Peg 23h33'13"17d36'
Bleiweiss,Rick
Cep 21h12'29"55d1'
Blem,Diane
Vul 19h45'1"28d20'
Blence III,Frank
Dra 17h56'32"68d28'
Blencke,Jr,CJ
Per 2h2'23"48d53'
Blengino,Sirrah Annabella
Cyg 20h17'18"38d30'
Blenkinsopp,Jack
Lyr 18h17'20"43d42'
Blenman,Sergeant Mikey
Aql 19h54'24"15d19'
Blenner,Irving
Lac 22h53'18"35d3'
Blenner,Mark Alan
Dra 20h6'48"62d45'
Blennert,Cheryl C
Del 20h22'15"10d50'
Blereau,Rudy
Dra 18h55'48"70d37'
Blesius,Birgit
Leo 10h59'17"12d25'
Blesnuk,Donald J
Her 17h6'42"42d33'
Blessey,Lane Estes
Lyr 19h22'23"40d38'
Blessey,Laura Ann
Cas 23h32'16"61d58'
Blessis,John Alexander
Aur 5h30'41"48d51'
Blethen,Jean
Mon 7h4'26"1d12'
Bleumanine
Mon 6h24'48"10d11'
Blevins,Alan "Mountain Man"
Ser 15h53'30"21d35'
Blevins,Alan & Amy
Eri 2h48'56"-7d44'
Blevins,Amanda Gayle
Uma 9h22'18"67d45'
Blevins,Georgia
Mon 6h20'33"8d16'
Blevins,Larry
Boo 14h9'42"41d4'
Blevins,Maggie
Vir 11h39'15"8d27'
Blevins,Sharon
Mon 7h2'57"4d19'
Blevins,Taylor "Little Man"
Cet 2h16'16"0d21'
Blevins-Boor,Emily Katherine
Cas 2h1'21"58d45'
Blevis,Carl
Aql 19h35'50"0d38'
Blew,Gary D
Her 17h12'12"48d21
Blewitt,Michael George
Her 16h56'24"35d44'
Bley,Dr John
Her 18h17'44"38d52'
Bley,Lindsey Christine
Aql 20h7'26"0d2'
Bleyer,Lisa
Peg 23h1'28"21d44'
Blicharz,Mara E
And 0h7'1"34d5'
Blicharz,Marcia & Dennis
Cyg 20h19'13"41d54'
Bliefnick,Russell A
Uma 10h4'22"51d28'
Bligh,Patty Wenchetta
And 23h0'58"41d4'
Blin,Solene
Boo 15h1'26"20d56'
Blind Faith
Ori 4h59'25"9d55'
Blinder 8-21-83,Jason Matthew
Uma 11h55'57"62d32'
Blinder Bar Mitzvah Star,Marcus Andrew
Her 17h53'52"14d12'
Blinkhorn-I know it's round somewhere,Amanda
Tri 1h48'25"28d13'
Blinky
Sgr 20h3'50"-20d28'A
Bliss
Sge 19h33'34"17d10'
Bliss
Aql 19h6'46"3d13'
Bliss,Bob & Nancy
Sge 20h1'39"16d3'
Bliss,Debra
And 0h7'52"35d56'
Bliss,George W
Her 16h35'54"8d50'
Bliss,Jeffrey
Her 17h21'50"42d21'
Bliss,Jeretta
Mon 6h54'54"10d26'
Blitch,John Gordon
Cet 3h1'11"9d34'
Blitch,Tehani Lee
Eri 2h51'58"-18d40'
Blitz
Peg 23h33'38"21d43'
Blitz,Annick
And 0h17'28"40d48'
Blizzard,Cassandra Dawn
And 23h43'48"43d46'
Blizzard,Keila
Umi 15h14'59"77d59'
Bloated Goat
Aur 6h55'51"38d52'
Blobby-Maddox
Uma 9h56'44"55d8'
Bloc'h,Stéphane
Ori 6h16'50"8d17'
Bloch,Bruce
Dra 12h32'45"68d16'
Bloch,Lenny
Cnv 13h21'0"38d16'
Blocher,Bernard
Leo 9h37'1"14d3'
Blocherer,Pedro
Cnc 8h7'24"31d23'
Block Star,The Ed & Erika
Sge 19h10'59"16d35'
Block,Bret Lawrence
Cnc 9h11'0"31d58'
Block,Camille
And 1h52'49"39d21'
Block,Jim
Aur 6h8'45"30d7'
Block,John Jacob Seiler
Aur 5h7'43"38d42'
Block,Katelyn Ruth
Cas 23h31'41"60d48'
Block,Kenn
Uma 12h9'16"57d45'
Block,Michelle Marie Hanrahan
Cnc 9h13'40"31d2'
Block,Paul J
Oph 17h54'45"8d36'
Block,Simon Malcolm
Cap 20h36'44"-20d3'
Block,Sweet Linda Ann
Cma 6h50'45"-17d1'
Block,Tony
Boo 14h37'60"16d35'
Blocker,Alexander Weaver
Dra 10h16'13"78d32'
Blocker,Gail
Sge 19h12'19"20d17'
Blocker,Jeremy William
Dra 10h18'17"78d42'
Blocker,Megan Marjorie
Cas 1h7'39"48d52'
Blodget (The Lovey), Meredith
Col 6h1'48"-34d22'
Blodgett,Brenda J
Umi 16h24'45"71d19'
Blodgett,Noelle
And 23h36'20"46d0'
Bloecher,Dustin Michael
Ori 5h0'35"8d53'
Bloecker,Dorothy
Cyg 19h29'47"30d7'
Bloedorn,Dietrich
Tau 5h56'0"24d2'
Blom,Ellen Stewart
Cam 4h21'1"67d50'
Blom,La Donna
Aql 18h56'53"-6d35'
Blom,Perry
Del 21h4'47"12d41'
Blomberg's Blaze
Cyg 21h33'51"30d42'
Blombergs Research Star
Ori 5h39'46"1d58'
Blome,Bernhard
Del 21h1'32"12d24'
Blomeke,Phillip C
Aur 7h20'13"36d28'
Blomquist Your Special Star,Katie
Cas 0h32'42"66d11'
Blomquist,Bryan & Jodi
Cyg 20h31'0"30d12'
Blomquist,Rodney Ian
Aur 6h6'1"50d28'
Blomstrom,Christine "Pee-Wee"
Aql 20h11'25"12d54'
Blond,Pierre-Yves
Aur 5h1'33"38d48'
Blondeau,Catherine
And 0h27'1"44d15'
Blondek,Christie Marie
Lib 15h12'27"-8d21'
Blondell,Ian C
Cyg 19h31'29"35d32'
Blondie
Lyr 18h32'51"45d5'
Blondie
Del 20h17'21"10d40'
Blondie Michelle
And 0h47'16"22d15'
Blondie,Brutus & Kona
Cru 12h42'17"-56d3'
Blondy,André
Sex 10h14'59"-1d7'
Blonski,Karin Beth
Cyg 19h28'55d31d23'
Blood,Bette & Ken
Crb 15h49'32"38d19'
Bloodnut
Dra 16h52'10"67d42'
Bloodworth,Beatrice
Cet 1h47'41"0d26'
Bloodworth,Daniel
Peg 22h59'0"32d55'
Blook,Brandon Scott
Ori 5h38'59"-4d48'
Bloom 1/27,Henry H
Ori 5h54'33"15d0'
Bloom's Tale,Marty
Aur 6h27'28"35d18'
Bloom,Beranrd M
Per 2h55'59"40d49'
Bloom,Brooke Ashton
Mon 7h42'45"-1d47'
Bloom,Charles Sherman
Cam 3h32'49"52d44'
Bloom,Ellen Rose & Leonard
Crb 15h29'25"31d53'
Bloom,G Duffy
Cam 14h4'58"82d2'
Bloom,Janet Carol
And 1h30'30"41d9'
Bloom,Jay
Lib 15h36'26"-21d6'
Bloom,Jeffrey David
Boo 14h36'58"39d4'
Bloom,Lance Montgomery
Mon 7h42'31"-1d29'
Bloom,Lisa Carole
Cas 1h39'55"58d22'
Bloom,Marion
Cam 3h18'46"66d11'
Bloom,Mark S
Sct 18h40'13"-4d12'
Bloom,Marvin & Cathy
Lyr 18h45'11"31d46'
Bloom,Mary Keefer
Cap 21h42'27"-23d3'
Bloom,Melach
Per 4h1'55"51d49'
Bloom,Michelle Renae
Cas 0h54'47"54d26'
Bloom,Natalie Claire
And 23h0'0"49d26'
Bloom,PhD,A Star To Me,In Honor Of Fred
Cap 21h42'32"-24d37'
Bloom,Richard Louis
Uma 11h12'18"31d52'
Bloom,Roslyn
Lib 14h59'52"-11d11'
Bloom,Sherrie Blossom
Cas 0h11'11"65d50'
Bloom,Star Lu
Cyg 21h28'25"37d46'
Bloom,Susan (Snooze)
Uma 11h30'35"42d4'
Bloom,Susan Michelle
Sgr 20h2'51"-41d41'
Bloom,Vicki Lynn
Peg 21h57'48"26d48'
Bloom,Walter
Her 17h34'31"23d51'
Bloomberg,Gerald B
Oph 18h4'30"8d25'
Bloomberg,Roslyn
Com 12h51'26"20d24'
Bloomer,Zachary Douglas
Cep 20h59'0"65d3'
Bloomfield,Born A
Ori 6h7'12"0d30'
Bloomfield,Robert Dana
Cma 6h26'17"-15d1'
Bloomington's Bright Star The Edward H
Aql 20h0'1"8d24'
Bloomquist,Karen M
Cas 23h34'1"60d58'
Bloomquist,Lynne
Mon 6h18'1"9d41'
Bloomquist,Patrick J
Lac 22h50'23"52d46'
Bloop
Cnv 13h49'50"41d23'
Bloor,Elizabeth Anne
Lyn 7h54'53"43d21'
Bloor,Ralph Rex
Boo 15h7'45"16d12'
Blosiu,Maria C A
Eri 3h56'41"-3d42'
Bloss,Kevin
Lmi 10h0'43"40d38'
Blosser,Cheryl- Friend and Hero
Mon 6h55'0"-1d40'
Blosser,Sarah Christine
Crb 15h48'1"27d11'
Blosser,Wayne
Her 18h9'41"38d18'
Blossom
Uma 11h47'35"52d12'
Blossom
And 0h16'22"36d12'
Blossom C
Aur 6h0'41"30d54'
Blossom Phantom Star, The Howard C
Aql 19h24'37"8d58'
Blossom,Chrsitopher Frederick
Lac 22h51'24"52d37'
Blossom,Lotus
Cyg 20h32'19"37d52'
Blotcky,Pamela Mary Leczynski
Cas 0h41'32"62d45'
Blotner,Andrew Evan
Boo 15h11'49"27d10'
Blough,Robert Darrel
Per 2h7'52"57d39'
Blough,Sandy Loushawn
Peg 23h3'57"22d29'
Blouin,Christine
Cnc 8h21'22"16d22'
Blouin,David-Olivier
Cnc 8h22'56"16d29'
Blouin,Geneviève
Peg 22h45'17"31d12'
Blouin,Jerome
Cap 20h43'22"-26d23'
Blouin,Marc-Antoine
Gem 7h33'57"31d3'
Blouin,Michel
Her 17h34'45"22d13'
Blouin,Virginie
Cas 1h14'29"60d55'
Blount,Debra
And 23h49'48"40d43'
Blount,Delorie
Cas 1h15'49"64d22'
Blount,Donn
Ori 4h42'56"13d4'
Blount,Leigh Ann
Cet 1h27'28"-12d7'
Blount,Rachel
Cyg 19h34'19"34d47'
Blount,Ruth Cagle
Uma 10h58'57"48d48'
Blow,Royston
Her 17h53'1"20d22'
Blower,Ann
Lyr 18h17'12"30d21'
Blower,Jordan Ellis
Umi 15h14'54"81d2'
Blower,Margaret
Cet 0h58'0"-5d4'
Blower,Michael J
Boo 15h6'16"30d41'
Blowers,Sean
Aql 19h32'33"1d34'
Bloxham,Alan & Maureen & Carina
Cyg 21h38'47"40d51'
Bloy
Aur 5h53'44"31d39'
Bloye,Thomas R
Hya 8h29'27"5d41'
Bloze,Alice
And 23h39'31"43d16'
BLR
Cam 11h34'23"81d3'
BLT
Dra 16h18'0"63d41'
Blu,Leslie
Hya 9h15'37"1d42'
Blubaugh,David E
Dra 16h51'60"66d52'
Blue
Cma 7h22'13"4d58'
Blue
Peg 23h8'20"13d50'
Blue "Little Boy Blue" ,Stego
Cnc 8h29'57"8d25'
Blue Ann & All She Loves
Mon 7h22'48"-8d14'
Blue Baron
Hya 9h17'48"-0d40'
Blue Benzing Star
Tau 5h45'60"21d4'
Blue Butterfly Fish
Psc 1h21'23"12d9'
Blue Class of 2002, Ellen B
Tri 2h43'10"31d27'
Blue Eyed Devil
Lac 22h5'19"49d12'
Blue Eyes
Per 2h51'56"32d21'
Blue Eyes
Cet 0h30'0"-2d46'
Blue Eyes
Aql 19h54'50"12d30'
Blue Eyes
Peg 22h45'35"34d14'
Blue Eyes
Ari 2h39'55"20d27'
Blue Eyes & Jazz
Dra 19h12'30"70d18'
Blue Eyes-Baby
Tau 5h43'27"28d22'
Blue Flame
Boo 15h2'46"28d18'
Blue Flame
Cam 6h15'43"68d45'
Blue Flight Leader
Per 1h58'32"56d31'
Blue Horstmann Star
Tau 5h45'45"21d18'
Blue Kangaroo
Cet 2h40'46"3d34'
Blue Light
Gem 7h15'6"24d48'
Blue Light Fireworks
Vul 19h18'51"26d34'
Blue Marlin
Psc 1h15'36"25d48'
Blue Mini
Lyn 9h9'37"39d31'
Blue Moon
Aql 18h45'32"9d20'
Blue Moon Optical Art
Mon 6h19'11"7d21'
Blue Mumme Star
Aqr 22h43'55"-2d22'
Blue Nanee & Bapa Mike
Cma 6h14'22"-12d1'AB
Blue One Forever
Cyg 21h22'51"31d50'
Blue Pacific
Lyn 8h29'14"55d31'B
Blue Ranger,The
Aur 5h56'51"30d51'
Blue Rose (Marsha J Abell)
Cas 0h37'54"66d40'
Blue Star
Uma 11h43'15"51d32'
Blue Star
Cyg 20h6'30"40d49'
Blue Star
Com 13h10'22"30d11'
Blue Star,The
Cep 22h51'56"65d57'
Blue Thunder
Cas 22h57'29"54d21'
Blue Winged Flyer
Her 18h3'0"14d17'
Blue,Avery Lee
Cnv 13h52'19"46d42'
Blue,Bobby
Cam 3h55'29"58d59'
Blue,Lyle Andrew
Cet 2h43'22"4d58'
Blue,Martin
Ori 5h42'59"10d50'
Blue,Mary Elizabeth
Uma 10h29'54"70d25'
Blue,Matthew
Cyg 21h21'1"40d35'
Blue,Rochelle
Ori 5h9'31"-4d59'
Blue,Stephen
Dra 19h3'13"60d41'
Blue,Susan
Cas 1h59'26"61d36'
Blue-Eyed Diamond
Ori 5h59'43"20d51'
Bluebell Bill
Her 17h16'0"27d23'
Bluechel,Rudolf
Peg 23h37'35"11d51'
Bluejeanshabarankfunky butticus Star
Per 3h27'38"52d19'
Bluestein,Michael David
Aur 6h37'13"37d19'
Bluestone
Cam 7h59'37"61d51'
Bluette,Wolfie
Cnv 13h4'20"40d48'
Bluewings
Psa 22h35'29"-27d31'
Bluey
Col 6h25'30"-33d4'
Bluhm,Amy
Cas 3h6'46"73d24'
Bluhm,Ingrid
Ari 1h59'41"10d49'
Bluhm,Rolf-Dieter
Leo 10h2'37"20d25'
Bluhm,Skyler
Cep 20h48'32"60d44'
Blum 21,Wanda
Psc 0h8'1"1d45'
Blum,Alain
Per 1h32'0"53d42'
Blum,Allan Howard
Aur 6h39'35"38d29'
Blum,Beth
And 0h52'37"37d42'
Blum,Chris
Ori 6h17'21"-1d8'
Blum,Daniel
Crb 15h31'25"35d9'
Blum,Doris
Cas 23h30'29"63d1'
Blum,Elfi
Dra 10h15'51"80d27'
Blum,Elizabeth
And 2h19'38"39d40'
Blum,Jennifer,Lyne
Uma 11h23'57"32d32'
Blum,Joseph
Ori 5h40'47"11d20'
Blum,Joseph
Hya 8h13'35"5d46'
Blum,Julie Anne
Cnv 12h48'45"38d18'
Blum,Karen
Aql 19h52'43"15d5'
Blum,Karleigh Jean
Uma 9h9'44"62d15'
Blum,Kerri Lee
Uma 12h4'57"37d45'
Blum,Margaret Alyce Brewster
And 23h6'55"41d27'
Blum,Michael Christopher
Cep 7h47'27"86d34'
Blum,Monique
Crb 15h32'44"35d12'
Blum,Ralf
Lyn 9h11'42"38d30'
Blum,Robert Kenneth
Per 2h40'0"40d11'
Blum,Sheila Ann
Lmi 10h31'47"34d13'
Blum,Uwe
Lyn 8h10'57"47d7'
Blum-Freedman,Loretta Marie
Mon 6h44'37"10d2'
Blum-McGoey,Heather
Peg 22h46'6"30d50'B
Blumberg,Irvin "Uncle Irv"
Her 16h59'22"29d10'
Blume"75",Ned
Cep 21h45'39"60d45'
Blume,Brandi Jonathan Matthew
Boo 13h55'29"19d40'
Blume,John E
Dra 19h21'40"67d54'
Blumel,William August
Oph 17h54'14"12d0'
Blumenkopf,Eric Alan
Dra 16h22'18"68d56'
Blumenkopf,Howard Scott
Her 17h34'29"37d59'
Blumenstein,Lora Temple
Cnc 8h10'1"30d37'
Blumenthal "Blumer", Eric O
Cyg 19h29'12"30d33'
Blumenthal,Iris & Carl
Cyg 21h35'24"31d4'
Blumenthal,Jeannette I
Cas 23h22'50"61d51'
Blumenthal,Matthew
Sgr 18h49'18"-20d22'
Blumenthal,Max E
Del 20h14'50"11d9'
Blumenthal,Robert
Tri 2h0'39"30d41'
Blumenthal,Solomon J
Aql 20h18'46"1d23'
Blumenthal-Bywater
Oph 17h20'1"12d53'
Blumetti,Gabrielle Antonia
Mon 6h37'50"8d24'
Blumetti,Hellen E W H
And 2h22'35"42d55'
Blumetti,Michele
Uma 12h31'31"61d49'
Blumetti,Vincent James
Cet 1h34'40"-4d35'
Blumhoff,Kurt
Cnc 8h6'56"31d6'
Blumkin,"Mrs B"Rose
Cas 0h28'12"50d15'

Blumklotz,Jeff Williams & Beth
Uma 10h25'35"51d27'

Blumstein,Kyver
Cyg 21h18'51"37d0'

Blumueller,Eckard
Aur 7h3'15"43d54'

Blunk,Brandon Michael Wesley Oliver
Cet 0h48'18"-8d34'

Blunt,Ernest
Per 3h24'38"40d24'

Blunt,Judy
Cep 23h3'23"61d47'

Blunt,Tiffany
Cnv 12h18'57"47d20'

Blusch,Anette
Aqr 22h44'52"-0d29'

Blustin,Leo & Saralyn
Cyg 21h3'16"37d56'

Bluteau,Candide
Lac 22h3'53"47d59'A

Bluth,Rachel Judith
Vul 20h22'11"28d48'

Bluto Roney
Cam 6h10'1"68d1'

Bly,Kathy
Cas 0h41'35"73d17'

Blydenburgh,Louise M
And 23h2'31"51d1'

Blye,Hannah Rose
Hya 9h20'9"-17d40'B

Blye,Samuel Elliott Thomas
Hya 9h20'9"-17d40'A

Blysak,BG George J
Cnv 13h12'33"38d1'

Blyth,Myrna
Lmi 10h15'56"28d45'

Blyth,Peter & Linda
Aur 5h2'16"29d46'

Blythe Lorraine
And 1h12'45"38d53'

Blythe,Alexander
Boo 14h59'1"10d15'

Blythe,John Terry
Crb 15h55'43"26d28'

Blythe,Mary Jane
Lyn 7h38'58"38d46'

Blytte,John William
Ori 5h59'19"16d44'

Blümchenstern, Robert und Peter der
Ari 2h41'29"30d58'

BM Bella Moda
Sgr 19h19'40"-21d58'

Bo
Dra 11h49'11"71d6'

Bo
Lac 22h3'54"48d48'

Bo
Crb 16h13'46"28d54'

Bo & Diane Star Of Eternal Love
Cyg 19h21'28"28d34'

Bo Bo Bear In The Sky
Uma 10h39'27"48d28'

Bo Bo Rob
Aur 6h57'1"37d39'

Bo By Short
Cas 1h37'36"58d38'

Bo Ma Wah,Cifford P
Dra 17h12'19"67d55'

Bo Star,The
Uma 10h8'46"51d3'

Bo Youn-Stephen
Uma 8h46'0"47d34'

Bo-Orna Of Eichenberg
Eri 3h53'53"-0d45'

Boak,Sam G
Her 17h4'52"20d57'

Boakye,Irene
And 0h20'52"36d54'

Boal,Stephanie Erin
Vul 20h1'54"28d57'

Boals,Christine & David
Crb 16h5'41"37d35'

Boam,Donna & Ted
Cyg 20h22'41"40d46'

Boan,Shirley Jean
Oph 18h18'35"10d39'

Board,George Douglas
Aur 4h52'34"40d55'

Board,Kimberly Laurie
Lyr 19h17'18"28d9'

Board,Shellie
Mon 6h33'46"0d12'

Boardman,Anna Dunbar
Mon 7h7'1"-0d52'

Boardman,David Lewis
Her 18h6'31"14d52'

Boardman,Eileen A Shea
Uma 11h56'43"33d26'B

Boardman,Gordon C
Boo 15h2'45"18d10'

Boardman,Gregory
Lib 15h20'53"-22d60'

Boardman,Miles
Lib 15h19'39"-22d39'

Boardman,Paul A
Uma 11h56'43"33d26'A

Boaris Brigadier Gen of Turtle Creek
Cma 7h24'35"-13d11'

Boasberg,Chloe Avril
Cap 21h43'58"-23d39'

Boasberg,Daniella Camille
Mon 7h1'26"0d31'

Boatman,Caroline Adell
Oph 17h57'35"12d21'

Boatman,Gary Owen
Cap 20h57'56"-25d43'

Boatman,Skylar Yahola
Sco 17h35'14"-38d11'

Boatman,Tammy LouAnn
Lib 14h58'41"-22d35'

Boatman,Taylor Lauren
Cyg 19h33'41"33d16'

Boatti,Andrew Philip
Per 2h59'0"34d30'

Boatwright,Chris
Mon 6h57'52"7d59'

Boaz,Doniella Chaves
Lyn 7h15'34"58d33'

Boaz,Joyce
Cas 23h27'43"63d15'

Boaz,Ruby Jewel
Peg 22h9'54"2d28'

Bob
Cet 2h37'22"-0d40'

Bob
Cep 21h5'1"68d41'

Bob
Mon 6h22'25"3d60'

Bob & Ami
Cyg 21h40'30"38d57'

Bob & Bettie,Our Love Shines On
Cyg 19h46'35"29d55'

Bob & Binnie Christmas Star
Mon 7h42'15"-5d1'

Bob & Bonnie's Star
Uma 12h3'44"55d33'

Bob & Claudia
Crb 16h4'47"31d52'

Bob & Eleanor
Cyg 19h30'43"31d36'

Bob & Gayle
Uma 11h0'39"38d15'

Bob & Joan The Greatest Parents Ever
Cyg 20h55'17"38d22'

Bob & Marge
Cyg 19h57'0"58d56'

Bob & Marianne Forever
Tri 2h38'25"32d8'

Bob & Patt 8-90
Cyg 20h31'38"30d54'

Bob & Penny's Coffee Star
Oph 18h19'21"8d17'

Bob & Rhea
Boo 14h14'39"17d5'

Bob & Vera
Cet 2h2'44"-0d57'

Bob & Wanda
Cyg 20h15'37"30d49'

Bob Digi
Cep 23h26'18"64d7'

Bob E
Her 17h13'16"20d41'

Bob Forever
Boo 13h37'1"20d21'

Bob G
Aur 6h3'21"41d5'

Bob George
Gem 7h2'3"24d47'

Bob Lesser God
Lyr 19h13'32"40d15'

Bob P
Aur 4h56'0"52d7'

Bob Paul
Cas 0h24'11"58d21'

Bob The Cat
Lyn 7h57'37"50d55'

Bob's "Fifty"
Psc 0h59'14"32d35'

Bob's "Luminous Heavenly Body"
Leo 9h52'51"31d33'

Bob's Balloon
Cnv 13h16'35"38d37'

Bob's Beacon
Aur 6h3'27"38d1'

Bob's Bed & Breakfast
Uma 9h43'53"58d47'

Bob's Benefaction
Her 17h6'15"46d20'

Bob's Buddy
Oph 17h31'0"-0d11'

Bob's Diamond in the Sky
Tri 1h58'15"34d40'

Bob's Forever Shining Light
Dra 18h31'13"58d37'

Bob's Great Expectations
Oph 17'59"58"8d18'

Bob's Heavenly Body
Her 16h7'1"28d15'

Bob's Jewell
Uma 11h57'56"40d53'

Bob's Star
Per 3h1'26"50d6'

Bob's Star
Dra 12h11'54"68d20'

BOB'S STAR Love MJ
Cet 0h48'58"-14d2'

Bob's"Ring Ping"Star
Oph 17h14'46"-20d7'

Bob,Love Of My Life
Lyr 18h59'52"38d33'

Bob-Dog
Cmi 7h16'16"4d35'

Bobaloo
Tau 5h54'46"24d21'

Bobalou
Cyg 21h21'20"28d52'

BobAsh Star
Per 2h53'44"34d40'

Bobber,The
Cep 22h41'32"70d22'

Bobbett
Hya 8h35'22"1d15'

Bobbi
Lyn 7h14'30"50d36'

Bobbi
Oph 17h54'25"12d16'

Bobbi
Uma 9h15'33"61d13'

Bobbi & Frank's Star
Cyg 20h25'55"40d40'

Bobbi 'N' Toni
Cas 3h42'16"74d34'

Bobbi Jo
Cyg 21h19'26"37d33'

Bobbi's Wish
Mon 6h59'28"7d35'

Bobbie
Uma 13h42'1"51d10'

Bobbie
Umi 14h54'20"68d6'

Bobbie
Oph 17h8'43"10d13'

Bobbie 100332
Lib 15h39'0"-20d20'

Bobbie Jo George Jeremy Forever
Cmi 7h55'29"3d52'

Bobbie Lee 263
Ori 5h35'0"0d56'

Bobbie's Promise
Ori 5h52'23"18d20'

Bobbie's Star
Per 3h59'45"31d12'

Bobbin,Harrie
Cam 5h48'33"58d43'

Bobbinator
Per 1h53'23"53d59'

Bobbitt,Katharine Lee
Cyg 20h28'20"42d4'

Bobbitt,Rodney Dale
Aql 19h3'35"0d50'

Bobble
Peg 23h28'19"23d43'

Bobby
Per 2h36'11"45d31'

Bobby
Aur 5h6'36"42d40'

Bobby
Aur 6h8'58"30d14'

Bobby
Uma 11h47'33"33d4'

Bobby
Aql 19h8'60"1d19'

Bobby
Aql 19h57'0"15d29'

Bobby
Cep 21h17'0"67d57'

Bobby & Bambi Star,The
Aql 19h54'25"13d14'

Bobby & Rita
Vul 20h4'22"23d9'

Bobby B
Her 16h38'11"35d33'

Bobby Dean
Boo 14h28'22"25d54'

Bobby Keith's Star
Vir 13h51'22"-5d48'

Bobby O
Lac 22h9'28"40d28'

Bobby Star,The
Her 17h58'0"41d12'

Bobby's Birthday Star
Cnv 13h8'57"33d22'

Bobby's Meaning of Life
Cep 21h37'52"56d6'

Bobby's Starlight Destiny
Sct 18h41'56"-6d27'

Bobby,Jen,Forrest & Sage
Sex 10h13'24"-1d53'

Bobby-Jo
Cyg 19h28'1"36d36'

Bobby-N-Becki
Aql 18h52'39"11d13'

Bobbymeiko
Dra 15h49'17"61d32'

Bobcat
Ori 5h29'30"1d37'

Bobcat 25-Olivera
Oph 17h39'42"-21d52'

Bobchevi
Cam 11h31'48"80d1'

Bobchris
Ori 5h20'0"1d46'

Bobe,Christina
Uma 10h46'12"40d53'

Bobek,Forever,Gary Anthony
Dra 16h3'32"52d21'

Bobelilla
Lyn 8h28'51"42d9'

Bobell,Ryan Terence
Boo 15h0'46"15d37'

Bober,Stanley
Peg 0h12'1"17d36'

Bobersky,Dick
Boo 15h7'55"30d35'

Bobi
Aur 7h22'53"37d32'

Bobi J
Lyr 18h41'29"33d31'

Bobick,Thomas Gordon
Hya 9h24'59"-1d14'

Bobin,Raymond F
Aur 6h1'32"45d1'

Bobkar
Dra 17h1'0"71d33'

Bobo
Sct 18h53'44"9d26'

Bobo
Cyg 19h35'43"29d40'A

Bobo
Aql 19h53'34"15d26'

Bobo's Star
Dra 13h57'46"68d42'

Bobo,Lois J
Uma 8h38'49"67d45'

Bobolakis,Anna Marie
Cas 0h25'10"75d17'

Bobolakis,Norah-Ann
Cas 0h20'57"71d13'

Bobomisha
Aur 5h54'17"37d38'

Bobrowicz,Teresa
And 0h40'55"40d58'

Bobstar
Aur 7h24'39"40d28'

BobStar 59
Her 17h8'57"22d21'

Bobstar,The
Her 16h11'30"10d11'

Boby,Pamela E
Tri 2h6'38"32d37'

Bobzin,Donna W
Aql 19h43'23"14d55'

Bobzin,Nicola
Peg 23h7'37"13d37'

Bobe,Joseph
Boo 15h20'43"48d52'

Boc,P H
Pho 0h44'15"-42d25'

Boccardi,Antonella
Cep 23h19'1"64d27'

Bocchiaro,Ann
Lyn 7h30'11"41d6'

Bocchieri,Cody
Per 1h58'13"48d43'

Bocchino
Lac 22h22'11"56d20'

Boccia,Anna Maria
Lyr 18h39'43"38d10'

Boccia,Archie D
Boo 14h6'30"27d26'

Boccuzzi,Gloria Jane
And 0h20'57"32d36'

Bocelli,David
Gem 6h33'21"13d54'

Boch,Klaus
Aql 19h59'0"11d2'

Boch,Manfred
Peg 0h5'40"10d10'

Bochenski,Richard L
Dra 16h6'48"65d52'

Bochicchio,Anthony & Susan
Umi 14h20'59"66d50'

Bochicchio,Chucky
Dra 19h26'45"58d57'

Bochner Family Star, The
Cam 11h59'53"81d40'

Bochner,Hart
Oph 18h39'38"6d30'

Bock,Christine
Cnc 9h15'48"30d50'

Bock,Diane F
Cas 23h32'1"60d16'

Bock,Dieter
Peg 23h33'0"20d59'

Bock,Erich
Ori 5h35'48"7d54'

Bock,Erik
Leo 11h31'28"-1d35'

Bock,Marie L
And 23h7'21"41d42'

Bock,Stephanie Anna
Mon 8h1'6"-9d7'

Bock,Steve
Aur 7h2'45"38d38'

Bock,Theresa S
Vul 21h2'1"20d5'

Bockman,Shelby Lee
And 0h29'54"44d7'

Bockmann,Etta
Vul 20h22'1"22d51'

Bockmayer,Theo
Vir 13h26'37"-13d41'

Bockstahler,Helmut
Cyg 19h52'35"47d45'

Bockstein,Stanley Merrill
Per 2h35'30"57d4'

Boczek,Veronika
Uma 14h1'1"48d56'

Bod,Jasmin
And 23h24'6"45d4'

Boda,Eleanor
Cas 0h31'1"72d59'

Boda,Francis
Umi 14h37'41"67d45'

Bodacious
Cet 1h52'0"-2d38'

Bodanski,Jr,Leo Joseph
Aur 6h48'38"39d35'

Boday,Kathleen
Cyg 21h7'0"30d22'

Bodden,Ellon "Hank"
Oph 17h31'48"-20d18'

Bodden,Evely
Lac 22h4'57"51d15'

Boddenberg,Georg
Cam 7h56'37"73d0'

Boddie,Carl Edward
Equ 21h0'25"2d36'

Boddington,Sharon
And 23h16'38"40d56'

Boddy,Jeffrey W
Ori 5h57'1"14d9'

Bode,Karen Buckland
Psc 1h25'37"25d1'B

Bodeborealis
Aur 6h40'0"38d36'

Bodell,David C
Boo 14h56'13"26d45'

Bodemer,Brett
Lac 22h46'46"52d47'

Bodemer,Jane
Lyr 19h2'49"40d50'

Bodene,Thomas W
Gem 6h41'32"18d29'

Bodenheimer,Sarah
Crb 12h29'54"14d4'

Bodenrader-Durning
Cyg 20h14'38"38d27'

Bodenstein,Bill
Lac 22h34'16"50d22'

Bodgie
Aql 20h18'14"7d38'

Bodie,Janet Maryan
Cyg 21h13'54"35d37'

Bodil
Dra 20h2'42"76d46'

Bodin,Jacqueline K
Aql 19h21'0"12d56'

Bodin,Paul R & Samuel M Bodin
Hya 10h41'1"-17d31'

Bodine,Judy
Cam 5h34'57"60d41'

Bodington,Hazel Ann
And 23h1'1"50d48'

Bodisch,Laura J
And 23h28'28"44d28'

Bodisch,Paul E
Aur 6h3'37"35d57'

Bodkin Elementary School
Peg 23h1'48"30d9'

Bodkin,Renee Leann
Cyg 21h1'34"28d57'

Bodnar,Heather
Aql 18h40'31"-0d11'

Bodnar,Holly
Cyg 19h14'38"44d21'

Bodnar,The Adrienne
Psc 23h35'50"0d27'

Bodnariuc,Catherine
Lyn 9h15'14"35d58'

Bodner,Andrew Michael
Aur 6h0'1"34d3'

Bodner,Bernard B
Ori 4h52'34"-2d57'

Bodner,Donna G
Peg 21h57'31"22d39'

Bodney Beloved
Aql 20h1'23"9d31'

Bodney,Philip Gerard
Dra 16h55'10"71d36'

Bodo
Boo 13h49'37"20d20'

Bodo Clifford
Aql 19h52'18"14d4'

Bodony,Danielle
Oph 17h42'56"-3d29'B

Bodony,Edward S
Oph 17h42'56"-3d29'A

Boduch,Casey Elizabeth
Cam 7h26'22"80d14'

Bodywise
Cet 2h26'50"3d45'

Bodzin,Sidney
Peg 21h51'28"35d44'

Boe,Emilee Victoria
Cyg 21h50'19"41d53'

Boebbis,Peter
Cyg 19h42'14"42d9'

Boeck,Barbara
Cam 7h2'1"83d15'

Boecker,Lilli
Sco 16h48'49"-25d4'

Boeckmann,Winfried
Peg 23h31'60"21d36'

Boege,Roger
Hya 9h6'14"1d56'

Boege,Walter
Uma 9h24'38"45d58'

Boegeman,David
Cnv 12h8'12"37d46'

Boeger,To My Love Mary
And 2h33'14"39d6'

Boehle,"Daniel" For Daniel
Ori 6h6'45"8d60'

Boehlke,Candace Lynn
Cam 5h2'1"60d51'

Boehm,Charlotte
Hya 9h17'19"2d11'

Boehm,Frank
Sct 18h20'51"-8d53'

Boehm,Luke Joseph
Her 17h29'31"20d53'

Boehm,Marie & Harry
Cmi 7h5'32"1d36'

Boehmer,Jennifer
Hya 9h8'23"4d34'

Boehmer,Paul E
Lyn 7h53'37"41d5'

Boehner,David
Ser 18h54'34"2d39'

Boehringer,Eric Joseph
Boo 14h56'27"25d4'

Boeke,Keith Allen
Lib 14h59'56"-23d48'

Boeke,Timothy Scott
Lib 14h19'12"-18d47'

Boekenfeld,Gerhard
Cmi 7h31'45"7d57'

Boekhoud,John
Her 17h13'10"20d49'

Boekhout,Kenneth
Cmi 7h5'46"4d24'

Boelcke,Enrique Carlos
Per 2h56'12"31d22'

Boele,Cornelia
Uma 10h52'56"40d32'

Boelke,Adam
Lib 14h56'32"-11d5'

Boelkens,Gordon
Cep 1h24'36"86d57'

Boelkens,Wesley
Aur 6h29'1"40d5'

Boenker,Diane
Tri 2h0'1"31d6'

Boer,Edith Grace
Cas 0h55'21"73d17'

Boere,Yvonne
Boo 14h50'0"47d42'

Boerger,Jill
And 0h4'33"30d56'

Boergesson,Cheryl
Com 12h29'18"30d20'

Boerman,Karrel & Wilma
Aur 5h18'33"40d22'

Boerstler,Ulrich L
Uma 11h44'12"31d8'

Boesch,Wendy M
Mon 6h42'45"1d32'

Boessner,Kurt
Leo 9h18'51"17d56'

Boetger,Royce
Cmi 7h41'29"3d50'

Boettchek,Mike
Sge 19h5'42"16d25'

Boettcher,Great Grandma
Cas 1h2'40"53d18'

Boetti,Anna Francesca
Pho 0h33'47"-41d33'

Boetti,Patricia
Peg 23h38'1"13d29'

Boffi Star,The Stephen Russell
Ori 4h59'27"1d2'

Bofshever,Jessica Leigh
Lyn 7h45'48"44d35'

Bofshever,Kathyrn Sue
Lyr 18h45'18"42d29'

Bogacz,Ed G
Ori 5h32'60"-1d36'

Bogan,Anje Marie
Uma 11h33'56"38d47'

Bogan,Brenda
And 0h17'31"36d53'

Bogard,Kevin Michael
Lac 22h20'53"51d23'

Bogart,Edward
Peg 23h10'55"10d47'B

Bogart,Laura
Peg 23h10'55"10d47'A

Bogaue,Bernard
Cam 5h58'35"73d51'

Bogda,Marcela
Ori 6h6'25"-0d43'

Bogdan,John & Helen
Cyg 19h30'35"34d40'

Bogdan,Oliver John
Per 3h7'31"50d25'

Bogdanoff,Shirley
Com 12h27'24"20d0'

Bogdanoff,William Luchi
Boo 14h57'13"52d40'

Bogdanovich,Don
Cma 6h59'42"-26d6'

Bogdanovich,Marilyn
Mon 7h36'50"-1d40'

Boge,Marianne
And 23h12'30"43d34'

Boge,Millie
Lyn 8h13'42"36d17'

Bogen,Alexandra Michelle
And 0h54'57"38d42'

Bogen,Helen
Cap 20h30'30"-26d59'

Bogen,Noah John
Cep 22h10'15"60d16'

Bogena,Erich
Lyr 18h34'31"26d4'

Bogenschutz,Joyce Estella
Cas 1h4'18"58d21'

Bogers,Todd Michael
Ser 15h24'30"8d24'

Bogert Our Love Shines Eternal,Don & Karen
Aur 5h53'57"29d23'

Bogey
Uma 11h39'45"30d50'

Boggero,Claire Casanova
Cas 2h3'48"59d4'

Boggess,Jr,Jackie Eugene
Vul 20h5'1"22d39'

Boggess,William C
Boo 14h19'58"33d55'

Boggins,Jamie
Her 17h1'46"12d48'

Boggio,Kristin
Sex 10h43'1"-1d38'

Boggione,Carlo
Cam 14h0'53"81d13'

Boggs,Amber'son Bruce
Crb 15h16'20"31d6'

Boggs,Christella Marie Isham
Aql 19h31'52"13d32'

Boggs,Creighton Whitney
Equ 21h9'26"10d51'

Boggs,Crystal S
Lyr 18h30'37"35d32'

Boggs,Deborah Jean
Cyg 21h5'0"40d6'

Boggs,Elena Chastine
Cyg 20h27'22"40d32'

Boggs,Jacob Wynn
Peg 23h37'54"18d15'

Boggs,Michael M.
Dra 11h9'1"74d43'

Boggs,Rosemary
Cas 0h27'12"70d43'

Boggs,Wesley K
Oph 17h3'6"-21d30'

Bogie,Christine Karchunes
Lyn 8h58'0"41d37'

Bogle Astral Body,The Peter W
Her 17h55'18"40d10'

Bogle,Isabella
Uma 9h25'31"51d33'

Bogle,Peter
Sco 16h58'17"-44d9'

Bognar,Nadine
And 0h19'51"31d29'

Bogner,Ann
Lyn 7h48'0"50d14'

Bogner,Paul
Cet 2h55'33"1d7'

Bogoch-(Annie B),Anne Elenore
Tau 4h8'5"21d5'

Bogue,Elaine Marie
Ori 5h48'44"20d53'

Bogues,Tyrone "Mugsy"
Aql 18h43'59"0d27'

Bogus,Jessica Lynn
Cyg 20h47'21"38d52'

Boguski,Michael M
Aur 5h54'36"29d40'

Boguski,Rosemary N
Boo 15h7'15"18d28'

Boguth,Frank "Sweet Cheeks"
Lyn 8h4'54"50d10'

Bohacek,Dennie Arthur
Her 16h55'31"41d6'

Bohag
Umi 14h45'25"78d53'

Boham,Clair & Doris
Uma 11h14'26"45d21'

Bohannan,Lisa
And 23h1'53"48d26'

Bohannon,Laura
Lyr 19h5'47"25d43'

Bohannon,William
Cas 1h10'14"64d25'

Bohannon,Laura
Crb 15h57'1"38d36'

Bohannon,Laura
Aur 4h59'50"50d49'
Bohannon,Roosevelt
Del 20h18'8"14d50'
Bohaty,Mary
Cyg 21h33'55"30d36'
Bohe,Jack
Boo 13h35'0"30d37'
Bohl,Chuck
Dra 17h58'38"76d33'
Bohlen,Donna Kay
Lyr 18h52'52"30d7'
Bohlig,Irmgard
And 2h27'18"39d50'
Bohlinger,Daniel & Courtney
Cyg 21h54'28"53d44'
Bohm,Angela Christine
Equ 21h0'40"11d3'
Bohm,Mark R
Aql 20h0'18"11d10'
Bohm,Nathan M
Cmi 8h6'51"0d13'
Bohn
Dra 20h25'51"64d27'
Bohn Star,The
Lyn 9h16'33"36d20'
Bohn,Erich
Per 4h34'19"52d22'
Bohn,Gundi & Juergen
Aqr 22h41'0"-0d15'
Bohn,Hermann
Peg 23h31'57"11d34'
Bohn,Jesse
Lac 22h54'0"51d26'
Bohn,Krista
Uma 11h8'35"68d24'
Bohn,Lisa A
Peg 22h24'12"30d35'
Bohn,Peter Alexander
Boo 14h8'1"31d49'
Bohn,Theresa
Mon 6h20'28"5d44'
Bohnenberger,Susanne-Mamu
Leo 9h18'15"12d20'
Bohnert,Henry
Lac 22h4'1"51d56'
Bohnert,Malorie Kristine
Vul 19h57'55"29d16'
Bohnhof,Horst
Hya 9h17'39"-7d29'
Bohonnon,Jim
Boo 15h3'30"17d42'
Bohons Burning Bright
Vul 19h58'37"23d35'
Bohr,Andreas
Vir 13h6'16"-21d44'
Bohrer(Nee Mueller), Lillie-Marie
Lyr 19h22'13"30d54'
Bohrer,Robert
Lib 15h3'43"-0d59'
Bohrer,Tyson Lee
Mon 7h45'15"-2d19'
Bohy,Francesca/ Nathaniel
And 2h0'11"39d41'
Boiani,Joy
Lyn 7h30'35"50d41'
Boice,Johnny "JB"
Per 1h53'22"50d20'
Boikin,Andrea E
And 2h15'35"42d60'
Boilermaker Point of Light,The
Per 2h57'13"40d32'
Boiron,Marie-Christine
Mon 6h22'57"-0d17'
Bois,Fernand
Cyg 21h1'50"39d58'
Bois,Marie-Christine
Aql 19h24'30"-10d46'
Boisclair,Nadine Paris
Cas 0h36'33"63d44'
Boise,Doris Mae
Lyr 18h54'46"33d22'
Boisits (Uncle Steve), Stephen
Per 2h51'0"46d49'
Boisjoli,Gabrielle
Cas 22h59'30"55d19'
BoisJoli-Barr,Stephen Anthony
Lyr 18h59'60"45d16'
Boisrivaud,Juliette
And 23h16'48"50d29'
Boissart,Laurent
Lyn 8h25'22"33d41'
Boissat,Romain
Aql 18h56'37"-6d32'
Boisseau,Joan
Cam 14h10'40"81d2'
Boissiere,Marion & Derrick
Peg 23h1'54"33d45'
Boissinot,Franáois
Cam 5h41'57"67d54'
Boisson,Mèlanie
Peg 21h57'11"33d20'
Boisvert,Anne
Leo 10h54'1"11d52'
Boisvert,Deborah
Cas 1h2'0"58d54'
Boisvert,Gaston
Umi 13h27'26"75d44'
Boisvert,Manuel
Uma 12h9'58"60d24'
Boisvert,Marcel
Per 3h15'21"41d31'
Boitard,Jacqueline
Aql 19h57'40"13d25'
Boiteau,Brigitte
Peg 22h55'40"24d26'
Boiteau,Raymond
Peg 22h59'25"24d47'
Boiteau,Sylvain
Peg 22h59'14"24d2'
Boiteau,Sylvie
Peg 22h52'37"24d57'
Boitelle,Isabelle Genel
Cam 4h34'30"58d52'
Boitos,Jenna Lynn
Psc 22h50'17"2d14'
Boivin,Eric Joseph
Lac 22h1'31"51d26'
Boivin,Matthew Allen
Her 16h16'36"26d38'
Boizet,Anne Sophie
Uma 11h33'0"38d11'
Boizet,Anne Sophie
Umi 16h18'12"70d45'
Bokalo,Taissa
Cam 3h53'45"79d27'
Bokert Dynasty,The
Dra 9h47'16"78d37'
Bokhour,Ehsan
Cam 5h2'43"68d59'
Boklan,Bill,Phyllis & Shannon
Aql 19h25'22"8d41'
Bokobza,Marc
Sge 19h57'44"20d16'
Bokovoy,Peter Allen
Leo 9h54'60"12d6'
Bola,Pip
Aur 5h1'0"50d25'
Bola,Roberto
Cyg 21h54'54"52d44'
Boland's Love Star, Helene & Jim
Cyg 21h31'37"30d41'
Boland,Andrew Morley
Vul 21h26'45"28d14'
Boland,Cody Paul
Ori 4h54'0"1d2'
Boland,Colleen Dawn
Cas 0h7'32"64d2'
Boland,Donald
Del 20h13'23"14d28'
Boland,Jo Ann
And 23h22'0"40d58'
Boland,Johnny
Aur 6h21'43"34d51'
Bolash,Vieira
Cam 4h3'52"58d46'
Bolcek,Ivan
Vir 12h34'0"-7d56'
Bold,Cecilia Maria
Sco 17h27'34"-38d53'
Boldan,Paul G
Aql 18h55'25"7d14'
Boldireff,Vallery
Lyn 9h4'26"45d3'
Boldman,Grant Sawyer
Aur 4h57'46"30d14'
Boldosser,Daniel
Aur 5h24'52"38d52'
Boldrini,Monica
Cet 2h59'41"2d14'
Boldt,James C
Hya 9h11'48"5d44'
Boldt,John E
Eri 3h12'39"-4d23'
Boldt,Peter Wellington
Hya 8h16'27"5d37'
Boldt,Winston Edward
Vir 15h7'23"6d54'
Bolduan,Peter
Lyr 19h18'58"31d42'
Bolduc,Donald Gaston
Cyg 19h28'45"33d25'
Bolduc,Franáois
Per 3h19'21"40d18'
Bolduc,Iolanda
Cyg 19h28'14"33d16'
Bolduc,Jérime
Per 3h19'37"41d25'
Bolduc,Mr & Mrs JP
Eri 4h5'33"-17d19'
Bolduc,Nathalie
Cep 21h16'37"78d23'A
Bole,Brenda
Equ 21h20'34"3d14'
Bole,Cliff
Peg 23h17'55"32d48'
Bole,Megan Christine
Uma 9h22'1"62d7'
Bolema,Codi Lynne
Lyn 7h49'10"39d32'
Bolen II,Laurence
Aql 18h56'32"4d54'
Bolen,Christa Elise
Cas 0h8m39"60d41'
Bolen,Michael Shane
Aur 7h23'21"39d42'
Bolender,Devon Elizabeth
And 23h36'1"49d33'
Boles,Bruce
Aur 6h6'4"31d29'
Boles,Ellen
Aql 19h6'24"-1d39'
Boles,Jan
Lyr 19h13'50"40d41'
Boles,Jennifer Marie
Lyr 18h22'29"44d56'
Boles,Kathryn Ann
Lyr 19h5'30"28d9'
Boles,Robert Gerald
Hya 8h44'47"3d47'
Boles,Stella Cordelia Amanda
Dra 15h32'50"58d21'
Boles,William E
Boo 15h42'14"40d45'
Boley,Travis
Ori 5h40'14"1d5'
Bolfo,Antonio Alexander
Aur 7h3'39"39d55'
Bolfo,Francesca Alesandra
And 23h37'1"43d2'
Bolger,Jack
Her 18h3'31"50d34'
Boli Star
Lyr 18h53'31"34d54'
Bolick,Deaneann
And 2h0'43"40d17'
Bolin,Kari L
Lyr 19h19'16"42d6'
Bolind,Bruce
Her 17h26'35"40d18'
Boling,Sr,Joseph LaVance
Boo 15h39'0"40d50'
Bolinger,Caren Irene
Mon 7h3'22"-6d36'
Bolkin,Stuart Michael
Oph 17h40'25"-18d55'
Boll,Jr,David Edward
Vir 13h29'26"-9d4'
Bolla,Luisella
Vir 13h4'57"11d6'
Bolland,Kate
Cra 18h18'25"-37d20'
Bollard Family,The
Lyn 6h56'36"59d25'
Bolle,Roger
Cyg 19h47'0"38d14'
Boller III,Arthur Andrew
Peg 0h3'19"31d47'
Boller,Marlies
Leo 10h59'44"12d21'
Bolli,Susi
Psc 22h56'27"1d15'
Bolling,April
Ari 2h1'1"22d28'
Bolling,Rick
Cep 0h3'27"70d45'
Bollinger's Treasure, Kim
Cas 23h17'0"61d53'
Bollinger,Brennon Paul
Cyg 19h33'23"33d12'
Bollinger,Judith Cynthia
Lyr 19h2'54"31d35'
Bollinger,Mark
And 1h24'19"34d37'
Bollinger,Robert Dean
Aur 6h28'30"38d56'
Bollinger,Roy J
Aur 4h49'35"40d44'
Bollman,Cherri Ann
Aql 19h54'24"12d3'
Bollmeier,Josephine
Lyn 8h27'0"41d18'
Bolobanic,Christopher
Boo 15h19'10"40d48'
Bolobanic,Kristine Shone Tanya
Aur 4h54'60"40d0'
Bolognesi,Bobo ed Eleonora
And 23h0'16"52d31'
Bolognia's Star Shines Forever,Louis
Her 16h32'1"38d48'
Bolshaw,Sally Ann
Ori 5h56'33"16d28'
Bolson,Sarah "Zellidda"
Cas 1h7'0"50d12'
Bolt,Calvin Taylor
Hya 8h25'1"5d57'
Bolt,Cynthia
Aqr 22h28'51"-0d47'
Bolt,Evan Corbett
Cet 2h23'22"-0d32'
Bolt,Herman
Uma 9h19'35"51d7'
Bolt,Melanie Lynn
And 0h12'17"46d40'
Bolt,Walter Thomas
Ser 16h3'52"14d36'
Boltas,Joseph J
Cet 3h17'58"1d32'
Bolte,Jane
Mon 6h20'33"4d44'
Bolte,Maren
Tau 4h6'50"1d31'
Bolte,Richard
Aql 19h53'28"11d19'
Bolter,David Mon BÉtre D'Argent
Hya 9h52'34"-19d37'
Bolton,Cheryl & John
Peg 21h37'45"24d21'
Bolton,Dr Craig A
Oph 17h0'32"8d48'
Bolton,Edward
Cyg 19h43'11"38d21'
Bolton,Everyone's Grandma-Great,Lota
Cas 0h16'33"63d43'
Bolton,Keith A
Boo 15h9'35"28d8'
Bolton,Linda Laird
Cas 1h19'54"60d17'
Bolton,Mark E
Peg 22h55'11"26d12'
Bolton,Melissa Michele
Peg 23h39'25"32d20'
Bolton,Mindy
Peg 21h49'17"33d44'
Bolton,Steven Lanier
Umi 14h30'38"84d56'
Bolton,Thomas Frederick 14/1/1925
Cap 21h21'14"-22d14'
Bolton-A Taste of Heaven!,Michael
Uma 10h47'21"48d56'
Boltz,Brian
Dra 10h37'39"77d56'
Bolzan,Dorella- Vincenzo Franciulli
Aur 7h23'16"43d23'
Bolze,Bruce
Oph 17h29'24"-22d27'
Bolze,John
Boo 14h28'58"25d47'
Bolze,Kevin
Ser 16h1'19"2d34'
Bolze,Scott
Her 17h3'26"42d7'
BoMar
Aql 19h47'48"14d25'
Bomback,Harry Phillip
Aur 6h28'56"35d4'
Bomber,The
Dra 16h21'1"60d20'
Bombola,Brian
Peg 22h58'39"8d30'
Bomel,Benoit
Cnv 12h14'1"40d12'
Bomer,Brandon James
Aql 19h4'34"5d17'
Bommarito,Sarah Grace
Cas 23h15'41"62d20'
Bommy
Com 13h32'16"22d28'
Bomot,Charles
Dra 12h20'51"63d57'
Bomps
Dra 17h09'01"64d36'
Bon
Aql 19h6'1"-0d22'
Bon Bon
Mon 6h39'32"7d34'
Bon Point Au George
Aql 19h56'54"12d46'
Bon,Mary Jackopiii
Cyg 19h31'59"34d53'
Bon-O-Bon
Aql 18h44'21"-2d31'
Bona-EVKF,Vin Di
Eri 2h53'0"-6d22'
Bonadio,Angela Nicole
Peg 0h1'36"11d33'
Bonadio,Brittany Diane
Mon 8h1'36"-7d12'
Bonadio,Shaina
And 23h21'53"40d58'
Bonaduce,Countess Isabella Michaela
Uma 9h54'57"56d38'
Bonaduce,Dante
Per 3h33'33"37d54'
Bonalumi,Maura
Uma 12h8'0"60d30'
Bonaminio,Bennie
Cep 0h16'33"68d42'
Bonamo,Tony
Cam 4h12'36"65d10'
Bonanno,Anthony
Cnv 13h22'48"28d0'
Bonanno,Fiammetta
Eri 3h43'24"-0d49'
Bonanno,Mary A
Cas 1h7'51"58d38'
Bonanno,Sally
Lyr 19h17'0"28d39'
Bonano,Laurie Elizabeth
Ari 2h4'25"22d26'
Bonansea,Francesca
Pic 5h3'18"-44d14'
Bonansinga,Jon
Per 2h41'1"40d27'
Bonanza Jo
Vir 14h34'24"7d13'
Bonanzino,Stephanie
Cas 0h18'21"63d49'
Bonardi Light Of My Life,Georgiane
Cas 23h40'45"64d0'
Bonaro,Joseph
Cet 1h58'59"0d43'
Bonas,Siegesmund
Hya 9h39'28"1d36'
Bonasera,Sabrina
Cas 1h19'37"63d35'
Bonat,Carolyn
Uma 11h51'1"40d17'
Bonath,Christine
Aqr 21h3'38"-13d43'
Bonaventura,Adam P
Aur 6h26'35"40d28'
Bonaventura,Brandon J
Cep 1h3'58"86d40'
Bonaventura,Frances
Cas 0h10'29"58d13'
Bonaventura,Michael J
Cep 2h11'0"55d48"
Bonaventure,Joseph Frances
Uma 10h26;50"62d5'
Bonaventuro
Ori 5h12'54"12d49'A
Bonazza,Giovanna
Cma 6h24'1"-15d49'
Bond & Wai
Lyr 19h19'33"38d39'
Bond (MARB),Aurora Blue-Margaret
Cas 2h5'59"61d25'
Bond,Alec Michael
Mon 6h32'43"8d49'
Bond,Alfred H
Ori 5h33'17"0d15'
Bond,Bob
Hya 8h15'56"4d47'
Bond,Bret
Sct 18h51'44"-6d59'
Bond,Clinton Andrew
Her 15h58'34"40d4'
Bond,Dale
Her 17h59'56"28d41'
Bond,Denny & Jeanie
Uma 11h51'0"62d52'
Bond,Doug
Lac 22h15'43"49d41'
Bond,Eugene
Uma 11h55'1"33d22'
Bond,Geoff
Uma 8h36'59"57d37'
Bond,Judy Kay(Akridge)
And 0h2'43"43d57'
Bond,Kelsie Leigh
Com 12h6'17"31d32'
Bond,Kevin Ward
Uma 11h58'20"33d21'
Bond,Leonard John
Per 3h12'0"56d12'
Bond,Mark Andrew
Tau 4h11'32"1d10'
Bond,Michael Robert
Uma 11h40'56"33d14'
Bond,Paden
Cma 6h32'27"-24d48'
Bond,Peter
Lac 22h7'28"51d31'
Bond,Robert Edward
Uma 11h33'16"33d36'
Bond,Robin F
Uma 14h0'47"48d55'
Bond,Ruth & Waldo
Uma 11h31'54"31d3'
Bond,Sue
And 0h32'48"27d41'
Bond,William A
Her 16h9'27"11d30'
Bond,William Partin
Aql 20h36'0"0d1/'
Bond-Sexler,Jasmine Louise
Cam 4h37'23"68d12'
Bonda,Rich Philea
Aur 5h5'20"40d38'
Bondanini,Magy et Albert
Per 4h4'3"35d17'
Bondavi
Boo 13h39'53"22d17'
Bondch,Angela Marie
And 22h58'47"50d16'
Bondi,Barney
Cep 9h58'29"80d36'
Bonds,Mary Katherine
And 0h12'14"36d11'
Bonds:Cheri,Steve, Blake & Brett,The
Uma 9h50'44"56d41'
Bondy,Kathy
Her 17h53'48"40d34'
Bone,Elizabeth Margaret
Lyr 18h30'53"30d32'
Bone,Sarah Jane
Vul 20h15'59"25d19'
Bone,Thomas Andrew
Ser 15h17'21"12d26'
Bonelli,James
Aur 6h47'19"39d43'
Bonello,James Joseph
Vir 13h18'42"-8d13'
Bonenfant,Gilbert William
Vul 19h5'29"24d58'
Boner,Irene Magdelene
Cas 0h56'56"54d24'
Bones,Cynthia Fulmer
And 2h29'30"38d36'
Bones-Lloyd
Crb 15h27'44"31d49'
Boness,Charmaine
And 23h42'0"43d5'
Bonesz,Willard G
Sex 10h46'59"2d1'
Bonet,Angel Maurell
Lac 22h2'0"51d31'
Bonet,Lynne
Lib 15h16'1"-21d22'
Bonetti,Beautiful Beth
Cyg 20h1'45"40d12'
Bonetti,Elena
Ant 10h44'28"-39d11'
Boney M The Baby Seal
Crb 16h13'31"32d42'
Boneyre,Chantal
Cru 12h14'36"-56d51'
Bonfanti,Joseph Anthony
Her 17h37'11"40d49'
Bonferroni,Benedetta
Vel 10h12'37"-42d26'
Bonfield,George
Ori 6h4'15"8d43'
Bonfiglio,Bonnie Harrington
Peg 21h42'46"27d41'
Bonfiglio,Patrick
Cas 0h38'49"61d7'
Bonfire Of The Butterfly
Sct 18h48'8"-6d13'
Bonforte,MD,Richard J
Cam 3h53'15"58d39'
Bonforti,Angelo Emanuel
Ori 6h13'58"6d19'
Bong,Baby Joseph
Boo 14h37'34"38d28'
Bongartz,Wolfgang
Lyr 19h18'31"31d27'
Bongiorno,Matteo
Cep 22h46'13"70d59'
Bongiovanni,Ricci J
Vir 13h54'26"-21d59'
Bongiovi,John
Per 2h30'60"57d43'
Bonham,Delores McDonald
Tau 3h58'49"20d30'
Bonham,Forever with you-Jason
Ori 4h59'45"10d53'
Bonham,Greg
Aqr 22h30'12"0d50'
Bonham,LeRoy
Ori 6h5'51"7d32'
Bonham,Lewis Luther
Leo 11h17'49"-1d22'A
Bonham,Vivian Leona Armentrout
Leo 11h17'49"-1d22'B
Bonhomme Dube
Lmi 10h55'52"31d30'
Boni,Leo & Katherine
Cam 6h2'40"68d52'
Bonica,Marta
Ser 16h17'48"0d12'
Boniface,Irene Patricia
Mon 6h55'26"7d40'
Bonifanti
Cnv 13h11'46"38d8'
Bonifazi,Nicholas Adam
Uma 9h25'1"56d26'
Bonilla,Cullan Phillips
Lyn 8h37'28"40d1'
Bonilla,Ruben
Sct 18h55'11"-7d7'
Bonin,Donald J
Uma 10h34'23"41d0'
Bonin,Serge
Lyr 18h56'55"45d5'
Bonina,Mark Francis
Cas 2h34'1"70d47'
Bonine,Mark
Dra 16h53'42"62d48'
Bonini,Dixie
Ori 5h46'50"18d16'
Bonini,Theresa L
Lyr 18h29'49"35d10'
Bonino,Michael Joseph
Tri 2h5'23"30d58'
Bonior,Dennis J
Ser 17h31'13"-13d55'
Bonita
Cyg 19h21'51"44d27'
Bonita Trista Michelle
Vul 20h15'41"25d49'
Bonito,Thomas
Aur 5h2'54"40d39'
Bonlarron,Roger
Aur 5h52'13"30d10'
Bonn,Heather
Cyg 19h39'44"28d26'
Bonn,Jackie L
Peg 23h18'11"30d14'
Bonn,Jacqueline
And 2h17'0"50d14'
Bonnamy,Gillian Fiona
Crb 16h19'0"28d44'
Bonnaud,Jean-Jacques
Vul 19h32'59"27d8'
Bonne Esperance
Lyn 7h25'1"48d1'AB
Bonne Étoile Pour Angela & Jean Paul
Cam 5h56'44"67d52'
Bonnefoy,Herve
Boo 14h24'47"50d10'
Bonnell
Cam 12h14'16"76d57'
Bonnell
Vir 13h21'45"1d39'A
Bonner,Beth & Necco
Mon 7h3'54"1d49'
Bonner,Carolyn Bates
Lyn 7h28'58"41d22'
Bonner,Laura N
And 0h4'13"46d38'
Bonner,Lily
Vul 19h15'41"25d13'
Bonner,Natalie Danielle
Lyr 18h30'1"30d45'
Bonner,TJ
Aur 6h1'26"31d5'
Bonner,Veronica Marie
Cnc 8h54'10"31d38'
Bonness,Barbara Helen
Mon 7h6'33"1d4'
Bonnet Tony
Aur 7h5'27"38d54'
Bonnet,Bernard
Mon 6h29'51"-5d53'
Bonnet,Franáoise
Boo 14h30'57"41d43'
Bonnett,Vinetta Alice
Cyg 19h55'49"38d36'
Bonneuie
Boo 14h53'60"38d23'
Bonney,Curtis Dean
Lac 22h1'26"48d47'
Bonney,Gerry Nicholson
Ori 5h54'15"12d15'
Bonney,Stanley George
Aur 5h50'35"40d52'
Bonney,William
Cep 20h52'15"60d51'
Bonnie
Aur 5h18'32"40d3'
Bonnie
Lyn 8h12'1"44d43'
Bonnie
Mon 6h22'1"4d52'
Bonnie
Aur 5h55'52"40d53'
Bonnie & Clyde
Peg 0h5'27"17d42'
Bonnie & Pam Always
Umi 16h32'1"79d47'
Bonnie Doll
Ser 18h54'59"2d5'
Bonnie Evelyn 62950
Mon 7h46'44"-1d45'
Bonnie Jean
Cas 23h3'27"53d6'
Bonnie Jean
Peg 21h19'56"22d36'
Bonnie Jeanne
Boo 14h10'0"52d46'
Bonnie Jeanne
Mon 6h21'41"8d21'
Bonnie Lee
Cam 8h6'58"78d22'
Bonnie Lee
And 2h17'35"41d39'
Bonnie Leigh(Love X Infinity)
Ori 6h5'34"-2d6'
Bonnie Louis
Mon 6h43'54"10d55'
Bonnie loves John
Peg 21h25'31"7d38'
Bonnie Lynn
Peg 23h24'38"30d29'
Bonnie Marie
Peg 21h58'13"34d46'
Bonnie TCL
Equ 21h7'14"10d47'
Bonnie's Light
Lyn 8h14'55"46d45'
Bonnie's Star
Boo 15h5'49"10d54'
Bonnie's Star (27)
Cas 0h6'28"59d3'
Bonnie,Gail & Ross
Cmi 6h56'55"-16d48'
Bonnie-"Trails End"
Cas 23h29'27"61d36'
Bonnielee
Vul 20h16'11"23d25'
BonnieWilda
Cas 0h28'26"54d35'

Bonnington,MD,Donald J
Her 17h7'24"21d16'
Bonnivier,Carl
Cas 0h49'33"68d35'A
Bonnivier,Margery
Cas 0h49'33"68d35'B
Bonniwell,René S
And 0h50'12"22d24'
Bonny "The Kid"
Lib 15h42'52"-28d12'
Bonny's Joshua
Lyn 8h22'18"49d36'
Bono
Uma 9h20'18"51d25'
Bonofiglio,John Mark
Her 18h6'50"48d43'
Bonomi,Chiara
Lyn 7h1'17"44d36'
Bonopoly
Aql 18h45'18"10d52'
Bonora,Isabel
Com 12h23'35"20d59'
Bons,Sam
Lyn 9h3'48"33d44'
Bonsai
Cyg 21h35'27"41d47'
Bonsick,Cathie
Cas 1h40'40"75d36'
Bontasch,Dr Degenhard Putzi
Aur 6h0'55"30d15'
Bontempo,Arthur R
Dra 20h23'51"62d9'
Bontemps
Vul 21h20'1"24d5'
Bonton Star
Lac 22h24'58"54d56'
BonTon,The
Dra 9h51'1"74d13'
Bontrager,Doug & Sally (Pohlman)
Uma 8h34'46"50d34'
Bontrager,Eric
Uma 8h31'26"51d51'
Bontrager,Evan
Uma 8h30'49"51d53'
Bontrager,Nicholas
Uma 8h31'0"51d30'
Bonturi,Fabrizio
Dra 16h13'28"51d15'
Bonturi,Fabrizio
Her 16h43'52"40d53'
Bonura,Janet
Aql 19h18'59"15d14'
Bonus,Nancy
Aqr 21h6'1"-0d29'
Bonvicino,Alberto Frigerio
Dra 14h57'32"61d51'A
Bonvicino,Joséphine Dard Frigerio
Dra 14h57'32"61d51'B
Bonvin,Alexis
Aur 5h23'31"40d46'
Bonvini,Paolo
Per 2h37'55"56d32'
Bonwell,Arthur
Tri 1h56'39"27d16'
Bonwit,Stuart & Elaine
Cyg 21h20'30"40d1'
Bonza,Tamara
Cyg 20h22'17"30d60'
Bonzey,Charles & Eleanor
Cyg 20h53'44"52d47'
Boo
Del 20h53'20"2d50'
Boo
Uma 10h38'32"48d4'
Boo
Boo 15h2'10"52d39'
Boo
Ori 5h57'16"18d1'
Boo
Aql 18h58'1"16d58'
Boo And Me Too
Boo 14h55'31"30d40'
Boo Bear
Uma 11h20'18"40d3'
Boo Bear Bub
Uma 11h23'13"66d8'
Boo Boo
Vul 20h19'54"25d33'
Boo Boo
Eri 4h17'53"-16d33'A
Boo Boo
Peg 23h7'54"10d58'
Boo Boo & Yogi
Uma 11h55'17"31d48'
Boo Boo Star
Aql 19h2'44"1d17'
Boo Boo's
Lac 22h48'17"38d45'
Boo King & Little Kitten 2-3-9
Boo 14h41'1"37d22'
Boo's Dreamkeeper
Cma 7h15'10"-16d50'
Boo-Boo
Cam 4h11'29"70d37'
Boo-Boo
Boo 15h8'1"52d5'
Boo-Boo
Cmi 7h25'39"7d45'
Boo-Boo
Cmi 7h55'17"4d60'
Boo-Boo
Aql 20h30'0"0d13'
Boo-Boo
Dra 17h49'1"64d41'
Boo-Boo-Bear & Knish-Knosh
Uma 11h46'11"57d10'
Booba,Natasha
Lyr 18h34'13"26d5'
Boobala
Hya 8h24'40"-1d7'
Booballa
Boo 14h8'41"43d20'
Booby
Cas 1h21'1"53d19'
Booby
Cam 13h22'27"77d37'
Booby Toops
Uma 9h2'43"68d9'
Boochever 11-15-95, Jack
Cnv 12h44'45"36d15'
Booden,Alysa Brooke
And 23h17'19"48d35'
Boodhoo,Rebecca R
Aur 7h6'36"37d48'
Boof
Her 16h13'1"10d5'
Boof & Bike/WBDB
Tau 4h45'41"20d27'
Boog,Michel
Oph 17h5'56"10d49'
Booger 'n' Herb
Sct 18h39'42"-5d46'
Boogle
Cam 6h19'41"68d18'
Boojie Kid
Cnv 12h25'26"40d24'
Book,Sandy
Aql 20'24'51"-2d16'A
Booker
And 23h49'33"38d31'
Booker,D & B
Peg 23h32'1"20d43'
Booker,G
Boo 15h5'14"41d14'
Booker,Sheila
Lyr 18h15'40"38d47'
Booker,Steve & Jill
Aql 18h46'42"11d30'
Bookheimer,Julie Bartek
Lyn 7h57'56"41d59'
Bookman,Norman
Boo 14h56'17"20d10'
Books (Tessier),Karen Anne
Cas 1h4'16"60d53'
Bookwalter,Mary Stephens
Aur 6h4'51"38d45'
Bool,Viktor
Her 17h17'50"42d3'
Boole,David
Sex 10h31'34"2d24'
Boole,Lindsay
Eri 3h49'44"-5d12'
Boom-Boom & Chi-Chi
Cyg 21h27'43"38d29'
Boomer
Aqr 22h3'60"-8d55'
Boomer
Mon 6h55'59"1d35'
Boomer
Cet 2h4'41"1d7'
Boomerang Basil
Aql 19h1'43"-0d8'
Boomershine,Jackson Patrick
Boo 14h35'20"8d13'
Boomgaard,Gordon C
Lib 15h29'51"-28d17'
Boon
Her 17h25'56"21d28'
Boon,Tim
Ori 6h8'55"3d33'
Boone Forever,H R & Neta
Ori 5h22'53"0d56'
Boone,Amy
Cas 0h55'1"62d41'
Boone,Ashley Rebecca
Mon 6h36'1"8d53'
Boone,Bob & Marci
Crb 15h30'1"31d29'
Boone,Charlotte Lilly
Cas 0h8'17"64d3'
Boone,Clifford O'Neil
Sex 10h23'37"-8d48'
Boone,Diane
Crt 10h58'48"-8d60'
Boone,Donna Leah Marie
Cam 3h57'48"52d45'
Boone,Joy Renae
Ori 5h39'40"10d35'
Boone,Kristy & Randy
Crb 15h51'46"38d8'
Boone,Mr Baby
Aqr 21h56'46"-5d47'
Boone,Ray
Cet 3h15'0"1d2'
Boone,Russ
Hya 9h39'35"-19d35'
Boone,Star of "Jo" Lynne
Eri 3h2'34"-10d55'
Boone,Star of David
Cma 6h31'13"-15d27'
Boone,Star of Steve
Cmi 7h56'37"0d56'
Boonshaft,Benje
Ori 4h57'55"15d6'
Boop
Dra 15h53'40"51d35'
Boop Be Ultra Forever 5/17/93
Uma 12h56'12"53d46'
Booras Star,The
Cam 5h53'57"68d47'
Booras,Dalton C
Boo 14h19'31"18d6'
Booras,Phyllis
And 0h8'0"31d32'
Boore,Louise Anne
Boo 14h7'34"25d51'
Boorman,Joanne
Ori 5h57'41"9d31'
Boorn,James Deuel
Ser 15h17'58"10d58'
Boorom,Bob & Margaret
Del 20h53'0"9d8'
Boos
Boo 14h18'46"32d12'
Boos,Bruno
Lyr 19h21'16"31d5'
Boos,Della M
And 2h7'59"40d46'
Boos,Julie
Vul 19h20'0"25d26'
Boos,Melissa Jean
Cas 22h59'1"54d32'
Boosaba Roemer's Dream Star
Uma 9h55'56"44d13'
Boosal
Ori 4h51'56"0d54'
Boot,Charlotte Anne Louise
Cyg 19h24'13"34d36'
Boot,Maximiliaan
Per 4h5'41"36d11'
Booth's Babydoll, Jacqueline
Cet 1h1'22"-6d53'
Booth's Buddy,Patrick
Cet 1h7'47"-5d25'
Booth's Little Man, Christopher
Cet 1h0'35"-6d52'
Booth,Adam
Ser 16h2'60"-0d35'
Booth,Amy
Crb 16h14'1"32d47'
Booth,Anthony Howard
Per 1h55'57"50d29'
Booth,Barbara
And 2h8'19"38d34'
Booth,Bradley Clifton
Cet 2h33'13"-1d6'
Booth,Brian F
Ser 16h16'23"1d41'
Booth,Carrie Anne
Cyg 21h18'35"35d48'
Booth,Christopher Brian
Mon 7h17'13"-5d48'
Booth,Claire
Cyg 19h34'17"39d46'
Booth,Dr Robert E
Her 16h15'37"23d1'
Booth,Edward C
Aur 6h14'24"31d48'
Booth,Fay F
Ser 16h16'31"2d31'
Booth,Halden
Per 1h29'46"50d14'
Booth,John E
Her 17h16'25"40d42'
Booth,Kathleen
Cyg 19h27'0"35d32'
Booth,Kevin
Her 16h50'28"37d36'
Booth,Louise Dillon
Cas 23h39'17"50d8'
Booth,Marie E
Cyg 19h25'1"30d12'
Booth,Mary Katherine
Del 20h13'43"12d9'
Booth,Michael Allen
Boo 14h45'23"29d22'
Booth,Miles
Her 16h10'1"23d38'
Booth,Rachel Clare
Cyg 19h25'34"35d11'
Booth,Russell Lee
Boo 14h54'36"39d14'
Booth,Serenity Warren
Aur 5h35'0"38d26'
Booth,Simon John
Uma 11h48'30"37d52'
Booth,Spencer Ross
Sco 16h52'48"-43d13'
Booth,Stacie M
Cyg 19h51'36"38'59'
Booth,Sue
Uma 9h14'47"50d21'
Booth,The Star Of Kerry
Peg 21h56'0"34d22'
Booth-Green 1-13-45
Cyg 21h47'19"36d25'
Boothby-Yeager I
Cnv 12h23'58"44d41'
Boothman,Olivia Rose
Cyg 20h23'41"39d24'
Boothman,Phyllis Mary
Lyr 18h49'46"34d48'
Boots Webb
Boo 14h7'33"23d34'
Boots,Janet Whitaker Marshall
Boo 13h57'1"22d28'
Bootsie
Aur 6h24'20"37d49'
Booty
Boo 14h6'47"50d9'
Bootzin,Michael
Per 1h35'59"53d38'
Bootzy
Uma 10h9'18"56d26'
Booz,Eileen
Lib 14h57'51"-20d16'
Boozer,Mary Margaret
Mon 6h33'34"3d29'
Boozer,Ryan Scott
Mon 6h33'12"7d45'
Boquiren,Armando
Lac 22h52'58"55d53'
Boraas,MD,Marcia C
And 0h54'50"37d0'
Borach,Alexandra Paige
And 22h59'0"50d11'
Borasky,Paul & Stacey
Aql 19h30'17"14d27'
Boratto,Bianca Marie
Lib 14h19'27"-8d27'
Borba,Tanya
Peg 22h32'53"27d34'
Borbely,Wolfgang
Lib 15h0'34"-6d9'
Borbolla,Martha D
Mon 8h6'29"-8d6'
Borchers
Per 2h38'41"38d46'
Borchers,Arthur F
Aur 4h50'0"48d47'
Borchers,Jane Bice
Cas 1h22'1"60d22'
Borchert,Cliff R
Per 3h37'47"37d32'
Borchert,The Nancy
Cas 23h1'27"58d19'
Borcyk,Michael J
Lac 22h8'10"46d5'
Borda,Karen
And 2h11'38"40d58'
Bordas,Michel
And 23h2'30"40d5'
Bordegoni,Gabiele e Marisa
Cet 1h1'27"0d32'
Borden,Barbara Seal
Lyn 8h3'42"58d52'
Borden,Betty B
Cyg 19h46'58"38d33'
Borden,Gail Peter
Mon 6h34'28"1d34'
Borden,Kristi
Eri 3h31'57"-1d56'
Borden,Sherrie
And 23h45'45"40d38'
Bordenstein,Fred
Lyn 7h35'27"50d17'
Borders,Becky & Paige
Cam 3h23'38"55d50'
Borders,Brianna Daneal
Aqr 22h4'32"0d19'A
Borders,Nancy I
Sex 9h59'56"3d10'
Borders,Robert Donald
Aqr 22h4'32"0d19'B
Bordignon,CS,Rev Mario
Ser 15h36'48"19d7'
Bordignon,James J
Her 17h19'43"22d29'
Bordley,Darian Rose
Lyr 18h38'0"44d50'
Bordo,Caesar F
Aur 6h26'17'31d2'
Bordonaro,Matthew Joseph
Per 2h52'34"32d34'
Borealis,Bradley
Her 16h21'31"48d39'
Borean Burner
Pho 23h42'19"-48d10'
Boreham,Sarah C
And 0h44'18"38d55'
Borehert Baby Team,Joy Lynn
Peg 23h39'15"10d0'
Borel,Alain
Cam 3h57'37"53d40'
Borella II,William
Cyg 19h18'39"28d23'
Borelli,Alysa
Crt 11h9'20"-19d13'
Borelli,Jack E
Boo 13h37'38"15d9'
Boren Anniversary Star,The
Cyg 21h8'10"31d25'
Borenstein,Katie Rose
Aqr 21h0'18"0d47'
Borer,Anneke
Cam 14h24'13"80d45'
Boretos Star,The Jim Donna Chris
Cam 7h37'39"80d25'
Boretti Blaze
Vul 19h57'14"29d4'
Borg 1
Uma 11h11'1"30d19'
Borg 2
Uma 12h2'19"32d47'
Borg,Brian David
Aql 18h50'42"11d11'
Borg,Karl Andrew
Aur 4h36'14"30d21'
Borgards,Jürgen
Aur 5h11'53"43d4'
Borgatti,Alena Catherine
Cas 2h25'55"60d49'
Borgatti,Joan Ellen
Lyn 7h33'34"42d36'
Borgeat,Franáois
Ori 5h58'1"17d50'
Borgeman,William
Lib 15h17'15"-23d2'
Borgens,Troy G
Cam 5h49'17"61d44'
Borger's Star,Stacy Lynn
Lyn 8h0'13"38d44'
Borghese,Andrew "BJ"
Ari 2h0'47"25d22'
Borghese,John
Cmi 7h38'27"11d2'
Borgman-Malicoat, Constance Ann
Cas 2h9'46"59d56'
Borgmann,Ingrid
Tri 2h23'49"34d53'
Borgna,Andy & Gigi
Uma 9h16'0"61d41'
Borgnine,Ernest
Aql 19h30'38"10d38'
Borgy
Tau 5h11'42"28d12'
Boria,M Teresa Latini
Lyn 7h57'58"40d52'
Borich,Mira
Aql 20h6'28"0d50'
Borie,Pierre
Lac 22h4'32"37d46'
Borie,Véronique
Peg 23h7'44"10d12'
Boright,George Melvin
Boo 14h39'50"50d15'
Boring's Birthday Star Layne
Aql 18h59'18"-6d12'
Boring,Betty Edith Jean & Wade Carl Gross
Boo 15h25'13"51d28'
Borinstein,Joan N
And 23h35'40"42d59'B
Borinstein,Melanie Alex
And 23h35'40"42d59'A
Borinstein,Melanie Alex
Peg 22h10'1"4d19'
Borinstein,Pearl
Equ 20h59'28"9d9'
Boris
Cnv 12h42'31"51d45'
Boris Borealis
Crb 16h4'19"38d37'
Boris,Natasha "Fly Me To The Moon"
Uma 11h24'16"55d53'
Boris,Steven Charles
Mon 7h3'31"5d12'
Borison MD,PhD,Richard L
Oph 17h57'58"11d41'
Borisova,Maja
Dra 16h30'58"63d32'
Borja,Ivetta
Vir 11h38'0"4d47'
Bork,Debbie
Cyg 20h3'39"40d50'
Borko,Lyndsay Brianne
Sgr 18h53'46"-36d49'
Borkowski,Brian "Spoon"
Lac 22h16'33"50d6'
Borkowski,Cezar
Aur 6h21'10"40d35'
Borkowski,Christian J
Gem 6h57'14"20d11'
Borkowski,Michael Scott
Lmi 10h28'33"31d56'
Borlan,Louis
Dra 13h28'20"68d34'
Borland,Alison
And 0h36'29"45d2'
Borland,Michele
Lyn 7h56'0"51d45'
Borletti,Giovanna
Cet 1h7'59"0d41'
Born,Bruce William
Ori 6h0'37"0d16'
Born,Glynn Elwyn
Cmi 7h43'45"7d50'
Born,Karl
Vir 13h39'36"-1d13'
Born,Pete
Boo 15h20'19"41d20'
Born-Barwick,Florence Roberta
Oph 17h6'10"-20d35'
Borne,Anthony John
Ori 4h55'12"5d4'
Borne,Claudine et Jean Jacques
Cas 2h59'32"60d55'
Borne,Colin Robert
Peg 22h46'43"33d37'
Bornemann,Carol C
Peg 23h6'39"10d33'
Bornfeld
Boo 13h52'47"17d59'
Bornkessel,Claus
Eri 4h54'38"-9d58'
Bornmann,John & Lucille
Oph 17h2'38"-21d53'
Bornstedt,Elizabeth
Com 12h58'49"21d3'
Bornstädr,Anke
Aur 5h17'53"43d54'
Borolla,George
Hya 8h14'54"6d14'
Boronaro,David Michael
Her 17h6'38"48d11'
Boros,Jaryd Tyler
Hya 8h44'47"-6d44'
Boross,Kim Elizabeth
Lyn 7h29'0"42d24'
Borota,Paulette & Ray
Eri 3h42'57"-10d29'
Borowitz,Morris
Cnv 12h22'0"34d29'
Borowsky,Irvin J
Lac 22h7'13"47d18'
Borrani,Claude et Stéphanie Alcaraz
Per 1h51'58"50d10'
Borredon,Christine
Eri 3h35'40"-2d6'
Borrelli,Lea
Mon 6h18'55"8d41'
Borrelli,Patricia
Cas 1h36'39"58d11'
Borresen's Heavenly Body,Darryl
Ant 9h38'6"-30d41'
Borresen,A J
Her 17h4'20"42d4'
Borrey,Julian
Uma 9h30'46"48d34'
Borris
Mon 6h18'56"7d39'
Borrmann,Carl
Aur 5h8'27"42d40'
Borruso,Kenneth W
Per 2h49'59"43d14'
Borsch,Peter
Equ 20h57'24"10d47'
Borsh,Edward (Bobo)
Boo 14h6'48"39d16'
Borsky,Frances & Valentine
Lyr 19h20'50"42d50'
Borter 10-16-43, Williams Henry
Equ 21h5'0"10d34'
Borth-1994,Cydney
Peg 23h26'25"32d5'
Bortnak,Shirley
Lmi 10h6'39"39d35'
Borto
Cam 4h49'17"69d50'
Bortolazzo,Richard
Ori 6h1'49"5d21'
Bortz,Marc Lewis
Ori 6h3'0"9d26'
Boruff,Richmond
Lac 22h18'1"50d34'
Boryma
Lmi 10h10'39"39d38'
Borysiewicz,Richard Andrew
Cet 1h40'1"-3d21'
Borza,Barbie
Uma 8h44'50"62d0'
Borzym,Thomas Joseph
Dra 16h53'25"67d57'
Bos,Cameron Robert van den
Gru 22h28'24"-53d38'
Bos,Our Shining Star, Gail
Cas 2h39'40"60d6'
Bosarow,The Jonathan
Her 16h48'12"48d30'
Bosc,Elizabeth
Aur 5h4'35"30d18'
Bosch J F B III My Son,Bosch James
Aur 6h28'22"37d28'
Bosch,Doctor Lourdes
Her 17h55'15"20d33'
Bosch,Richard E
Dra 13h58'33"68d13'
Boschelli,Marianne M
Cyg 20h16'19"38d52'
Boschen,Nathan
Lac 22h3'1"46d28'
Boschert,Brian
Cnv 12h57'50"33d18'
Boschini,Peter C
Dra 15h51'20"67d27'
Boscia,Anna
And 23h7'32"40d58'
Boscia,Tony & Nicole
Aql 18h48'19"11d48'
Bosco (Pupetta I), Paola
Col 6h34'28"-36d36'
Bosco's Nova
Boo 13h58'47"14d30'
Bosco,J & D
Aql 20h6'42"4d7'
Bosco,Joanna
Oph 17h36'0"-22d14'
Bosco,Matt
Boo 15h2'20"27d22'
Boscus,André
Cnv 12h59'14"38d45'
Bose,Catherine
And 23h32'0"43d47'
Bose,T K
Lmi 10h8'0"32d21'
Boser,Meghann
And 23h19'36"48d28'
Bosha,Jim
Sex 10h25'57"3d47'
Boshi,Junpei & Rinpei
Ari 2h17'23"17d23'
Boshi,Naoki Kanan
Sgr 18h54'40"-35d11'
Bosio,John
Aql 18h53'57"10d33'
Boski 25
Cyg 21h0'1"39d10'
Bosland,Paul
Aur 5h19'31"42d57'
Boslett,William & Dolly
Hya 9h17'46"-0d10'
Bosley,G L (Bill)
Ari 2h45'16"30d23'
Bosley,Lori
And 0h21'41"30d24'
Bosma,Peter
Cnc 8h23'52"7d19'
Bosman,Jan
Cam 3h38'23"60d51'
Bosmann,Katherine Elizabeth
Vul 21h16'39"20d19'
Bosnic,Nicholas James
Ori 5h39'36"11d4'
Bosone,Jr,M Michael
Psc 1h3'28"21d59'
Boss
Cyg 21h18'28"30d27'
Boss,Larry
Lac 22h32'37"53d27'
Boss,Rhonda Ann
And 23h22'39"40d37'
Bossard,Jean-Charles
Cam 13h54'51"80d14'
Bosscher,J A K
Cas 0h57'1"62d16'
Bossert,Anne Marie
Cyg 19h28'26"33d47'
Bossert,Carl Andre
Aur 6h23'39"38d54'
Bossert,Carl William
Aql 19h44'42"10d2'
Bossert,Helene Ann
Del 20h20'48"20d23'
Bossert,Myriam
Psc 22h49'54"6d29'
Bossert,Tara Ann
And 1h21'0"35d29'
Bosset,Jean-Pierre
Lac 22h3'0"37d41'
Bossi,Barbara Jean
Del 20h13'53"9d20'
Bossi,Marco
Umi 17h47'19"80d28'
Bossi,Umberto
Umi 15h5'60"84d33'
Bossingham,Sandra Sue
Lyr 19h20'0"35d32'
Bossung,Gerald Colbert
Aql 19h8'40"1d57'
Bossy
Per 2h25'40"57d30'
Bossy,Sandra
Vul 19h21'34"23d44'
Bosta,Todd A
Oph 18h25'1"7d56'
Bostedo,Michael J
Dra 19h18'18"70d3'
Bostelmann,Herr
Cmi 7h17'49"7d57'
Bostock (Dino),Yvonne
Leo 11h53'10"21d8'
Bostock,Christina Elise
Lyr 18h35'16"43d47'
Boston Phoenix
Aql 18h53'59"-0d32'
Boston,Agnes L
Lmi 10h6'39"32d28'

Boston,Colin James
Uma 10h42'42"40d27'
Boston,Jeff
Cnc 8h2'0"8d50'
Boston,Lucille J
Lmi 10h2'51"30d24'
Bostrup,Earl W
Cep 22h2'26"61d47'
Bostwick,Mona S
Aur 7h22'15"41d10'
Bostwick,Mona S
Cet 1h18'31"-6d41'
Bosveld,J Bryan
Cet 1h31'48"0d39'
Boswell,A V Robert
Oph 16h57'15"-25d36'
Boswell,Alan (The Boz)
Aur 5h17'0"50d8'
Boswell,Alin
Equ 21h15'17"3d5'
Boswell,Ayse Rene
Eri 4h48'38"-6d42'
Boswell,Jason & Lisa
Aql 19h51'1"12d1'
Boswell,John
Cet 1h27'23"0d29'
Boswell,Steven Harris
Aur 7h8'17"40d37'
Boswood,Mary
And 23h0'46"50d36'
Bosworth,Andrew
Lac 22h10'24"47d42'
Bosworth,Ed
Aql 19h50'32"12d8'
Bosworth,Rita
Lyn 8h24'24"43d17'
Botchway,Faustina
Mon 6h58'57"-6d11'
Botek,Jacob Joseph
Gem 7h11'7"24d46'
Botek,Jr,(Forever 39), Sam
Aqr 21h58'14"-0d39'
Boteler,Louise
Cas 2h22'49"61d3'
Botelho,Jr,Robert John
Boo 14h49'1"35d20'
Botelho,Jr,Stephen M
Her 15h50'29"47d53'
Botelho,William & Rose
Lyn 8h45'34"46d10'
Botell,Stephanie
Aql 18h52'51"-1d12'
Botero,Jack C
Ori 5h44'26"11d24'
Botes,Jean-Marc
Pic 1h44'31"-48d7'
Botfield,Terry D
Hya 9h2'20"4d2'
Both,Dirk
Sct 18h35'48"-5d27'
Both,Ernst Erland
Dra 20h16'45"64d34'
Botha-Happy 70th!, Fred I
Ori 5h57'22"16d28'
Botheroyd Family Star, The
Bernie
Uma 12h3'1"30d50'
Bothner Star,The
Umi 15h43'0"70d57'
Boticelli's Angel
Her 17h13'24"48d28'
Botková,Tereza
Uma 9h59'17"48d38'
Botoms,James William
Ser 18h6'54"-14d57'
Botsacos,Laura
Cas 0h40'49"65d22'
Botson,Janet
Mon 7h28'37"-1d39'
Bott Family Star
Cru 11h58'45"-56d18'
Bott,Daniele
Cam 11h10'33"82d16'
Bott,Robert Lee
Lac 21h58'21"40d25'

Bott,Victoria Morgan
And 1h25'35"37d7'
Botta Mario
Per 4h4'18"39d26'
Botta,Mark Andrew
Aql 19h56'28"8d18'
Bottazzi,Bruno & Ana Maria
Crb 15h49'30"28d47'
Botte Di Ferro
Pup 8h8'33"-26d9'
Botter,Allan Sherwood
Cep 21h34'32"55d28'
Botterbusch,Mary
Lyn 7h50'28"51d3'
Bottger,Tess Ann
And 23h2'15"45d49'
Botthof,Rick & Chris Valez
Cyg 21h5'31"38d41'
Bottieri,Joseph
Aql 20h4'56"4d57'
Bottigliero-PF53
Lac 22h17'37"51d31'
Bottiglione,Aimee M
And 2h18'18"42d7'
Bottiglione,Carole
And 2h21'52"45d37'
Bottiglione,Cate M
And 2h12'37"42d42'
Bottiglione,Nora
And 2h14'12"42d3'
Botting,Eric D
Cep 22h57'53"57d39'
Bottoms' Star,Bill
Cet 1h41'0"-1d47'
Bottoncino
Cnv 13h28'28"38d24'
Bottone,Fanny & Chic
Uma 9h12'30"51d1'
Botts,Linda
Cyg 19h27'22"30d12'
Bottstar,The
Mon 7h39'53"-2d47'
Botwinick,Randy
Aur 6h57'0"43d30'
Botyrius,Geri
Crb 16h4'21"27d2'
Botz,Anni
Lmi 10h0'0"39d2'
Botzenhart III,Raymond A
Dra 15h49'0"66d4'
Bou Zou
Cnv 12h18'57"40d15'
Bou,Nancy
Cas 1h25'36"77d23'
Bouachru,Ali
Lyr 19h21'12"30d14'
Bouatta,Lahlou
Uma 11h7'43"50d32'
Bouazzaoui,Edith
Aur 4h34'41"34d10'
Boucard,Lissa "Marie Claire"
Del 20h33'35"18d50'
Bouch,Jennifer
And 0h22'28"36d28'
Bouchard,Croucher
Gem 7h1'19"24d29'
Bouchard,Félix
Umi 16h2'42"79d49'
Bouchard,Jr,George F
Dra 17h6'31"64d10'
Bouchard,Marie- Corinne
Laforest
Cam 3h53'47"62d29'
Bouchard,Monica
Boo 15h36'40"43d45'A
Bouchard,Peggy
Vul 21h1'51"20d34'
Bouchard,Robert
Her 16h41'57"28d48'
Bouchard,Vincent
Her 16h43'24"23d27'
Bouchard,Xavier
Umi 16h0'1"79d38'
Bouchard,Eric
Her 16h58'25"21d1'

Boucher,Andre Plantagenet
Uma 10h37'1"47d59'
Boucher,Catherine & Yvon
Umi 13h57'58"75d58'
Boucher,Christopher A
Dra 20h16'1"70d12'
Boucher,Cécile
Umi 16h56'47"79d4'
Boucher,Elaine Rosemary
Cyg 19h27'1"33d10'
Boucher,Francois
Peg 21h45'44"27d59'
Boucher,Frank JG
Aur 5h13'1"44d22'
Boucher,Gérard
Dra 15h45'1"60d43'
Boucher,Julien
Lmi 10h54'22"31d38'
Boucher,Peter Norman
Her 17h51'25"40d26'
Boucher,Rena Marie
Hennessey
Cam 4h3'50"67d32'
Boucher-Fontaine, Justin
Ori 5h55'55"11d30'
Bouchez,L Booboo
Boo 14h54'44"43d28'
Bouckaert,Daniel
Cyg 20h5'0"30d42'
Boucq,Isabelle
Uma 9h33'49"47d47'
Boudier,Stéphane
Dra 15h5'33"63d9'
Boudjellal,Amar
Cap 20h56'21"-20d21'
Boudoux,Jean Pierre
Lac 22h28'28"38d11'
Boudreau,Daniel
Her 15h48'0"41d55'
Boudreaux,Tamyra
Ser 18h43'48"3d54'
Boudry,Denise
Per 4h4'9"37d5'
Bouf,Nathalie
Dra 10h9'19"74d21'
Bouffard,Benjamin
Per 4h5'42"51d51'
Bougaud,Dominique
Boo 15h6'48"7d31'
Boughey,John Anthony
Uma 9h31'0"47d25'
Bouillant,Elise
Peg 23h30'0"21d17'
Bouknight,James A
Dra 14h41'43"62d16'
Boulais,Wanda Jo
Aql 20h12'43"12d17'
Boulanger,Jr,Richard P
Ser 15h28'28"21d30'
Boulanger,Stephen B
Cmi 7h24'57"7d48'
Boulay,Guy
Uma 11h26'47"31d35'
Bould,Queenie
Cas 0h54'15"63d15'
Boulder Head
Boo 15h29'15"50d33'
Boule de Ràve
Del 20h29'42"11d9'
Boulevard,Faáade
Uma 9h7'58"48d46'
Boulia,Hailey Elizabeth
Lyn 9h13'35"44d54'
Boulia,Taylor Anna
Cas 0h24'41"64d51'
Bouliane,Marjolaine
Com 12h29'48"20d53'
Boulis,Carrie Lucile
Cyg 20h53'36"30d28'
Boullenger,Emmonuel
Cep 22h23'52"57d38'
Boulos,Sheikh Simon J
Uma 11h35'22"31d13'
Bouloubasis,Nikolas
Her 16h12'38"47d39'

Boult,Trevor John
Psc 22h55'0"-1d2'
Boulter,Pamela Gwen
Cyg 21h2'34"38d56'
Boulton,Anya Catherine
Ellenden
Crb 16h14'53"32d17'
Boulton,Patrick
Ori 5h29'27"-1d14'
Boulton,The Joanne Magnis
Com 12h29'34"30d5'
Bouma,Paul L
Her 17h14'12"49d28'
Bouman,Ronald A
Cet 1h19'33"-12d5'
Boumans,Bill
Cet 3h12'55"4d21'
Boun,Natacha
Cnv 12h25'41"40d6'
Bounaix,Christian
Cep 23h20'34"63d44'
Bound,Grammy
Aql 19h9'42"3d31'
Boundy,Samuel Brian
Mon 7h51'58"-8d29'
Bountiful
Aql 20h6'57"-0d2'
Bouquet,Marcy
Aur 5h14'57"29d51'
Bouquet,Maurice
Ser 15h9'59"10d18'
Bourassa,David Alan
Aur 6h19'18"31d18'
Bourassa,Gillian
Dra 16h40'0"61d53'
Bourassa,Sajo-Ovide & Aline
Umi 16h42'1"79d31'
Bourassin,Christine
Tri 1h46'45"31d46'
Bourassin,Cécile
Cas 2h45'13"77d8'
Bourbeau,Corinna
Mon 6h21'56"7d46'
Bourbon,Daniel Francis
Ori 6h9'46"7d59'
Bourdages,Franáois
Cyg 19h24'54"33d48'
Bourdeau,Robert R
Her 17h17'56"46d34'
Bourdin,Anne Dominique
Lyr 18h57'0"40d44'
Bourdon,Marie
Cep 22h25'25"57d3'
Bourey 40,James Michael
Boo 14h21'0"16d53'
Bourey,Marie & Ralph Ralph
Lib 15h0'37"-18d58'
Bourgela,Jean
Her 16h39'29"40d35'
Bourgeois 205 Tanière, Pierre
Sahi
And 0h50'43"39d49'
Bourgeois,Bryce
Ori 6h8'40"9d6'
Bourgeois,Ellen A
Lyr 18h19'30"38d46'
Bourgeois,Kiersten A
Cyg 19h41'17"28d36'
Bourgeois,Timothy
Her 17h21'16"14d11'
Bourgoin,Erika & Christine
Per 3h9'1"38d35'
Bourgouin,Julien
Cas 1h33'49"64d44'
Bourguignon,Emilie
Com 12h29'48"20d53'
Bourguignon,Gabriel
Ori 6h0'58"6d50'
Bourguignon,Mario
Per 3h18'0"42d14'
Bouriachi Je T'Aime La
Jolie,Linda
Cam 4h56'54"58d14'
Bourjaily,Sharon J
Tri 2h5'1"30d52'

Bourke,Jr,Edward John
Aur 5h6'13"38d46'
Bourke,Patrick Quillan
Lac 22h15'40"51d8'
Bourke,Sherry Louise
Uma 10h51'32"60d51'
Bourlett III,John & Lisa
Crt 10h54'58'-22d49'
BOURNE
Peg 22h19'23"5d35'
Bourne,Andy
Uma 12h6'10"45d54'
Bourne,Barney Eldred
Uma 11h1'46"40d53'
Bourne,Christopher "Tiffa"
Cyg 21h0'56"36d18'
Bourne,Emily
Cyg 21h18'13"39d33'
Bourne,Jean Leroux
Uma 11h19'11"40d56'
Bourne,John
Oph 17h0'41"11d30'
Bourne,Robert W
Cet 2h25'57"-6d59'
Bournival,Anthony E
Cep 20h47'30"67d33'
Bourque,Bernard
Cet 2h26'44"6d8'
Bourrassin,Olivier
Cyg 21h57'1"50d3'
Bourrat,Rene
Cyg 19h45'0"38d58'
Bourret 3,Réjean
Peg 22h58'49"24d16'
Bourret,Denise Maxine
Cam 5h54'43"71d4'
Bourton,Clive
Sge 20h4'0"20d55'
Bouse,Clint VanNostrand
Her 16h27'50"48d47'
Bousman,Rachael
Eri 2h49'12"-6d8'
Bouson,Michael
Equ 21h6'23"11d23'
Bousquet,Suinat
Uma 9h39'13"48d46'
Bouss,Anne Marie
Uma 11h20'26"52d22'
Boussard Christophe
Cep 22h54'23"68d34'
Boussemart,Didier
Lyr 18h58'1"47d2'
Bousson,Elizabeth Rose
Lib 15h3'45"-1d40'
Boustead,Kay
Cyg 20h21'15"38d42'
Boutel,Marta
Uma 14h1'0"53d48'
Boutie Star,The
Boo 14h49'1"28d17'
Boutillette, Christopher J
Uma 15h13'16"65d42'
Boutin March 8 1948, Neal
Steven
Psc 0h2'11"0d28'
Boutin,Corrine
Aql 18h56'28"-3d14'
Boutin,Jaclyn-Marcie
Lac 22h7'56"46d28'
Boutin,Suzie
Cam 3h52'0"62d26'
Bouton,Pierre
Lac 22h19'48"50d12'
Bouttell,Carol Christine
Cyg 19h26'50"34d12'
Boutwell,Frederick Curtis
Aql 20h2'29'10d49'
Bouvier
Vul 19h48'37"20d20'
Bouvier,Jean Pierre M
Per 3h59'58"36d15'
Bouvier,Sadie
And 23h0'21"51d33'
Bouygues,Carole
Aql 19h55'12"-10d42'

Bouzu,Lucette
Aur 5h51'23"31d21'
Bouët,Marc
Peg 23h30'43"20d7'
Bova
Cyg 20h22'56"38d8'
Bove,Carmine Charles
Per 3h5'1"50d26'
Bove,Delia
Sgr 19h25'46"-41d7'
Boverini Star,The Manlio "Bo"
Her 16h21'50"23d20'
Bovi-Fabrice
Cyg 20h49'57"38d47'
Bovina,Carlo Sergio
Oph 18h3'21"10d24'
Bow,Wayne
Eri 4h12'29"-18d4'
Bowblis,Beverly
Cyg 20h18'24"40d19'
Bowblis,Constantine F
Peg 23h1'0"12d28'
Bowcutt,Gordon L
Boo 14h16'38"39d50'
Bowcutt,Michelle Lynn
And 23h46'38"43d21'
Bowd,Alan
Uma 11h58'34"42d44'
Bowden,Adeline
Uma 11h4'48"35d36'
Bowden,Amanda Elaine
And 22h57'1"50d0'
Bowden,Grace Catherine
Com 12h16'47"22d4'
Bowden,Kate Lauren
Cam 11h28'42"81d8'
Bowden,Luke Stephen
Lib 14h53'40"-1d8'
Bowden,Mandan
Cam 3h49'47"77d14'
Bowden,Troy Collin
Her 16h31'27"32d9'
Bowe,Graeme David 2/10
/1945
Pho 23h26'13"-41d19'
Bowen Family
Cyg 21h27'50"28d28'
Bowen's Reach
Lmi 10h42'54"26d33'
Bowen,Cheryl Ann
Del 20h18'18"11d9'
Bowen,Don
Aur 7h4'37"37d59'
Bowen,Doug-Aimée & Laura
Ori 5h58'34"11d1'
Bowen,Gary
Cyg 19h29'55"30d26'
Bowen,James R
Cnv 12h30'0"34d35'
Bowen,Jeanette
Com 12h34'17"23d54'
Bowen,Jim
Per 4h32'42"31d6'
Bowen,Joshua Sion
Cam 3h32'32"62d36'
Bowen,Judith Ann
Mon 6h55'1"-10d34'
Bowen,Lari
Lyr 18h33'47"31d30'
Bowen,Larry & Cathy
Cyg 19h41'11"30d55'
Bowen,Lyn Ann
Mon 6h55'15"-2d44'
Bowen,Michael
Dra 12h58'11"72d24'
Bowen,Michael David
Cet 1h50'50"-6d43'
Bowen,Peter
Per 2h55'40"34d9'
Bowen,Russell Anthony
Dra 11h52'1"70d14'
Bowen,Scott Janette Jennifer
James
Mon 7h58'56"-8d41'

Bowen,Susan Irene
Cam 6h1'0"60d11'
Bower Country Star, John
Cyg 21h49'42"53d42'
Bower Star,The
And 2h28'22"49d9'
Bower's Family Star, The
Uma 8h5'58"62d5'
Bower,Christine
And 0h2'1"40d53'
Bower,Doug & Dolores
Boo 14h37'0"19d17'
Bower,E B
Boo 13h59'49"21d38'
Bower,Edward I
Cep 3h18'39"78d26'
Bower,Elizabeth
Cas 0h24'48"60d40'
Bower,Henry A
Aur 7h19'14"35d58'
Bower,Jonathan
Her 17h28'53"38d31'
Bower,Sherry A
And 22h59'31"51d29'
Bowerman,Anna Louise
Peg 21h57'28"32d26'
Bowerman,Frank R
Cet 0h48'56"-5d53'
Bowerman,Leslie Jean
And 1h31'38"39d42'
Bowers' Destiny Keeper Doug
Cep 21h55'40"63d52'
Bowers,Alanna Christine
Mon 6h56'15"10d50'
Bowers,Andrew James
Her 16h17'55"5d4'
Bowers,Bruce
Oph 17h55'48"0d59'
Bowers,Catherine Elizabeth
Peg 0h2'40"13d41'
Bowers,Iris
Lyr 18h18'29"44d4'
Bowers,James Wesley
Oph 18h16'14"11d27'
Bowers,Johnny "Rocket"
Sex 10h3'21"-10d36'
Bowers,Jr,Hal Daniel
Lmi 9h45'55"38d18'
Bowers,Kara Jeanne
Cas 0h32'34"68d40'
Bowers,Krystal & Brian
Peg 21h50'42"33d57'
Bowers,Lauren Taylor
Peg 21h49'34"34d53'
Bowers,Myra
Cas 2h19'21"68d35'
Bowers,Rayna Sky
And 2h25'33"50d8'
Bowers,Reveta
Mon 7h19'29"-6d58'
Bowers,Stephen
Ori 5h6'1"14d36'
Bowers,Tonya J
Cet 1h29'48"0d17'
Bowers,Vera Jean
And 2h18'22"39d33'
Bowers,William E
Umi 15h44'26"71d51'
Bowersock,Daman
Hya 8h13'56"-3d4'B
Bowersock,Timmie
Hya 8h13'56"-3d4'A
Bowerson,Warren L
Oph 17h38'33"-23d18'
Bowes II,Wiz
Cmi 8h8'27"2d34'
Bowes,Christopher T
Gem 6h27'34"12d10'
Bowes,Robert Roy
Cnv 13h30'24"48d20'
Bowey,Peter
Aur 6h32'35"37d50'
Bowhunter's Boreal Nova,The
Hya 8h53'14"-0d43'

Bowi Gonnot
Cet 0h54'1"0d39'
Bowie,Amber L
Cyg 19h29'32"30d19'
Bowie,Crystal D
Com 13h12'15"21d0'
Bowie,David
Lyn 9h12'59"36d43'
Bowie,Highlander Allen
Cep 21h6'35"61d51'
Bowie,Joan Carroll
Cep 21h53'57"86d7'
Bowie,Jr,Donnell
Cet 1h6'0"-1d29'
Bowker,Martha L
Peg 23h27'51"33d0'
Bowker,Ruby
Lyn 7h49'17"58d50'
Bowlas One
Peg 22h1'43"27d38'
Bowler,David Allsop
Ori 5h59'1"8d34'
Bowler,Paul
Her 18h1'1"48d43'
Bowles,Calla Johnson
And 2h22'52"40d16'
Bowles,Colin Leonard
Cyg 20h18'23"39d1'
Bowles,Connie
Cyg 19h59'29"40d30'
Bowles,Jeffrey E H
Boo 14h39'11"46d51'
Bowles,Joyce P
Cyg 19h40'38"31d41'
Bowles,Kathy Johnson
Cas 2h4'1"73d38'
Bowles,Mathew Alexander
Sge 19h15'20"16d23'
Bowles,Robert
Her 17h37'58"24d6'
Bowles,Robert L
Lac 22h54'34"50d11'
Bowles,Sandra Kay
Cam 5h58'50"73d17'
Bowles,Susan
Per 2h13'18"57d22'
Bowles,Tabitha Annette
Lyn 7h27'25"58d18'
Bowley,John Richard
Ser 18h2'41"-1d46'
Bowlorama
Crb 15h15'16"32d48'
Bowman's "Precious
Dream",Donna
Cam 7h23'1"83d3'
Bowman's Arrow Of Light
Boo 14h54'11"48d6'
Bowman,A Star Just For
You,Sharon
Boo 14h48'57"31d14'
Bowman,Alisa Michele
Del 20h20'0"10d58'
Bowman,Amiee Vanessa
Lmi 10h17'56"30d39'
Bowman,Andrew Taylor
Cet 0h50'42"-0d12'
Bowman,Becki
Peg 22h57'17"22d43'
Bowman,Britani
Lyn 8h13'39"37d10'
Bowman,Catherine Jayne
Peg 22h32'18"12d2'
Bowman,Charles Stephan
Sex 10h22'55"-0d41'
Bowman,Clyde Dennis
Cep 21h25'25"60d45'
Bowman,Dan
Per 4h0'41"50d21'
Bowman,Dorothy
Cet 2h46'27"6d45'
Bowman,Dorothy Kroll
Cyg 19h43'42"38d1'
Bowman,Evelyn Agnes Carey
Uma 9h28'56"50d5'

Bowman,Jason J W
Lac 22h7'48"51d57'
Bowman,Jeremy R
Lyn 7h55'0"44d15'
Bowman,John Jason
Ori 5h2'27"15d34'
Bowman,Jr,"Butch" Johnny
Robert
Per 4h0'17"51d56'
Bowman,Julia Shelley
Peg 22h43'45"4d14'
Bowman,Karen Knight
Mon 7h7'0"-6d12'
Bowman,Lana Joyce
Cmi 7h57'32"0d38'
Bowman,Laura
Oph 18h2'39"13d26'
Bowman,Laurel Anne
Mon 7h24'30"-10d24'
Bowman,Marlene
And 0h57'1"37d35'
Bowman,Mary Louise
Del 20h33'14"10d29'
Bowman,Matt
Her 18h32'49"24d52'
Bowman,Michael
Dra 10h15'42"73d10'
Bowman,Papa
Cyg 21h30'44"44d36'
Bowman,Peyton
Dra 16h36'37"61d16'
Bowman,Reid Hammond
Sct 18h30'54"-5d50'
Bowman,Richard
Dra 15h25'34"65d44'
Bowman,Ronald Parker
Peg 22h12'43"4d59'
Bowman,Russell Alan
Cet 1h3'40"0d54'
Bowman,Tara Morgan
Aql 19h54'23"14d37'
Bowman,Vida
Cas 1h0'49"66d12'
Bowman,Wayne & Mary
Uma 11h54'37"45d34'
Bowman,Zachary Boyle
Her 18h3'33"38d22'
Bowman-Loved By
Honeybunch,Denise
Cyg 19h39'33"30d17'
Bowman-Trujillo, Anthony
Timothy
Boo 14h29'11"40d9'
Bowmaster,Kevin Andrew
Cep 23h0'46"64d30'
Bown,James Hugh Francis
Per 2h57'0"56d33'
Bown,Mark Bevil
Aql 19h2'1"16d60'
Bown,Nicholas Lee
Cep 21h44'15"68d40'
Bowring,Jane Louise
Peg 23h7'43"11d29'
Bows,Kelsey Wrain
Cam 7h13'26"68d44'
Bows,Kiri Wrain
Boo 14h33'37"48d47'
Bowser,Aaron Michael
Dra 15h58'19"61d11'
Bowser,Blyphe E
Peg 22h24'36"27d33'
Bowser,Dorothy
Lyn 8h43'45"39d51'
Bowser,General Al & Betty
Eri 2h44'42"-2d56'
Bowser,Jim & Susie
Crb 15h31'1"26d33'
Bowser,Star of David
Per 3h20'52"38d48'
Bowser,Taylor A
Eri 2h56'44"-6d39'
Bowyer,Harry Martin Mitford
Eri 5h8'12"-5d47'
Bowyer,Popsitude Mr BA
Cam 3h53'10"62d34'

Box,Cory Shane Dra 19h49'15"71d7'
Box,Fred William Ori 5h13'26"-0d49'
Box,Jane Lmi 10h38'11"33d11'
Box,John Ori 5h59'49"14d24'
Box,Jonathan Christian Uma 9h50'18"57d39'
Boxall,Richard Aql 20h6'10"6d2'
Boxall,Wilfred R Her 18h29'1"24d50'
Boxleitner,Bruce Tau 5h48'14"27d11'
Boy & Girl Cma 6h55'40"-18d29'
Boy Toy & Fuzzy Wuzzy Cep 3h8'16"78d30'
Boy,Barney Bedouin Cmi 7h31'1"8d51'
Boyce "My Eternal Love Star",Tommy Her 17h38'14"14d45'
Boyce's Brillance Dra 16h47'25"62d49'
Boyce,Alvy Dra 18h19'35"48d51'
Boyce,Amy Christina Hya 8h31'14"-0d58'
Boyce,Brian T Her 16h50'19"38d24'
Boyce,Cheryl And 1h18'31"38d7'
Boyce,Christian Aql 20h0'18"12d56'
Boyce,Christopher Sex 9h49'55"-5d7'
Boyce,Donna Jean And 23h32'54"48d38'
Boyce,Grant Bruce Pho 23h54'53"-45d53'
Boyce,Jill (Missouri Maude) And 23h47'1"47d58'
Boyce,John Cep 20h41'27"61d46'
Boyce,John Patrick & Jeanne Louise Vul 19h22'22"23d59'
Boyce,Robert Aur 5h2'41"46d19'
Boyce,Ruth Cas 23h21'15"53d23'
Boyce,Stephanie Mon 6h32'33"1d4'
Boyce,Susan Jennings Eri 3h51'17"-6d21'
Boyce,Thomas A Vul 19h20'51"27d18'
Boychuk,Patricia Cas 1h55'52"60d20'
Boyd Per 1h36'21"53d48'
Boyd 270634,Hugh Her 16h10'0"40d29
Boyd Beamswift GJ3 Lambda Boo 14h59'55"34d35'
Boyd Dan 12:3,Jessica Taylor Ori 5h34'36"-2d7'
Boyd Forever,Kent & Dorothy Lyn 8h1'56"50d35'
Boyd Star,The Per 4h1'27"48d46'
Boyd,Adam H B B E Lib 14h55'56"-23d8'
Boyd,Benjamin & Gloria Oph 17h6'20"11d51'
Boyd,Beulah Bell Eri 3h37'28"-2d11'
Boyd,Caitlin Dineen Lyn 7h34'58"38d25'
Boyd,Connie Sge 20h3'8"16d39'B
Boyd,Cynthia Lynne Com 13h32'58"21d27'
Boyd,Denise J Uma 12h28'36"60d45'
Boyd,Earlene Mon 7h43'36"-2d5'
Boyd,Edward Ray Cep 0h1'43"69d28'
Boyd,Elizabeth Eleanor Cam 4h143'29"70d18'
Boyd,Eric Aur 4h36'26"34d51'
Boyd,Erma Elizabeth Cas 1h0'17"60d32'
Boyd,Fern Billings Cam 3h53'49"56d15'
Boyd,Glenn Gray Aur 6h25'29"38d14'
Boyd,Hayden Andrew Gem 7h1'38"24d54'
Boyd,James Ser 18h43'1"3d31'
Boyd,James David Aur 6h31'51"38d49'
Boyd,James F F Ser 16h1'1"2d36'
Boyd,James M & Annie C Edgar Cnc 8h51'28"31d49'
Boyd,James Robert Leo 10h59'15"14d17'
Boyd,Jaymes Taylor Cep 23h13'1"61d42'
Boyd,JoEllen And 2h19'0"41d59'
Boyd,John Uma 8h6'20"62d9'
Boyd,Jonathan Boo 13h59'54"20d18'
Boyd,Joseph Lee Cep 22h19'56"55d44'
Boyd,Jr 1911,Thomas F Mon 7h26'25"-10d27'
Boyd,Jr,Mordecai James Tau 3h31'44"30d41'
Boyd,Julia Ann Boo 15h2'12"18d39'
Boyd,Julia Anna Mon 6h26'28"-10d48'
Boyd,Julia Megan Uma 11h56'1"52d20'
Boyd,Kathleen Cas 3h25'57"75d59'
Boyd,Kathy Cas 1h37'29"60d28'
Boyd,Kyle M Ori 5h43'1"11d8'
Boyd,Maegan Starr Equ 20h57'51"7d1'
Boyd,Makensie Lyne Peg 22h28'11"24d44'
Boyd,Mary Julia Mon 7h48'34"-2d53'
Boyd,Meghan McKinzie Peg 22h15'56"10d24'
Boyd,Merle Crt 11h44'45"-11d28'
Boyd,Michael Cru 18h24'54"-44d11'
Boyd,Steven Boo 14h46'0"24d50'
Boyd,Susan And 2h10'38"43d2'
Boyd,Suzanne Dra 17h26'41"68d50'
Boyd,Thomas William Her 16h8'20"16d21'
Boyd,William Michael Per 1h49'45"53d28'
Boyd,William Pinckney Sct 18h50'0"-7d56'
Boydd,Robert Cet 1h25'39"-17d31'B
Boydd,Victoria Cet 1h25'39"-17d31'A
Boyden,Christopher Jon Peg 23h2'58"12d25'
Boyden,Dick & Jean Valliere Uma 13h29'0"61d34'
Boydstun-Youngs,Reon Aql 18h43'24"-0d34'
Boyens,Hans-Günther Dra 18h28'1"65d14'
Boyer,Eric Ori 5h55'23"8d52'
Boyer,Gloria Cas 1h35'1"68d12'
Boyer,Grace Cas 0h9m21"61d25'
Boyer,Grayson Sex 10h24'1"-6d34'
Boyer,Helen Ori 5h20'6"14d18'A
Boyer,Helen And 1h20'56"33d60'
Boyer,Jill Lyn 7h31'27"41d1'
Boyer,Kenneth Edward Aql 20h10'1"0d42'B
Boyer,Laureut Cam 4h37'0"58d20'
Boyer,Lynne Ori 6h5'54"0d36'
Boyer,Marcel Lyr 18h57'59"45d18'
Boyer,Olen Ori 5h20'6"14d18'B
Boyer,Raymond H "Pop-Pop" Aur 7h16'38"35d54'
Boyer,Shirley Jane Lyn 8h6'0"50d53'
Boyer,Timothy J Aur 7h4'16"41d45'
Boyer,Windy-Lee Cyg 19h31'17"34d4'
Boyer-White,Prof René And 0h9'59"35d43'
Boyes,Hannah J Peg 23h40'50"11d31'
Boyes,Janet Lynn And 23h22'53"45d5'
Boyett,Stoncil Monroe Peg 22h37'0"31d25'
Boyette,Jr,Stoncil Monroe Cnv 12h39'55"40d25'
Boyhan,Alice Joy Mon 6h43'0"-2d24'
Boyk,David T Per 1h45'12"53d4'
Boykin,Alan J Aql 18h52'47"11d27'
Boykin,Robin Leigh Aql 20h1'44"13d22'
Boylan,Jack Aql 20h15'0"5d12'
Boylan,William Delsanter Ori 5h56'30"15d23'
Boyland,Audrey Jane & Jack Irwin Lyn 7h47'0"50d45'
Boyland,Francis S C & Susan C Aql 19h2'49"-6d43'
Boyland,Jack Ryan F & Jordan Martin Q Aur 4h58'0"40d43'
Boyland,Steve-Pat & Conor Vul 19h58'46"25d7'
Boyland,Wendy Jane Peg 23h0'0"22d26'
Boyle "JB",John (Johnny) T Uma 13h42'26"50d32'
Boyle III,Edward G Aur 5h9'35"38d10'
Boyle IV,James Joseph Her 15h49'17"50d37'
Boyle Star,The James L Her 17h22'35"42d6'
Boyle,Ann Hyde Uma 10h5'0"56d20'
Boyle,Barry A Cam 8h2'40"71d59'
Boyle,Brira Del 20h13'33"10d14'
Boyle,Carly Marie And 0h45'1"36d44'
Boyle,Christopher & Caroll (Kinion) Cam 6h58'47"70d47'
Boyle,Duane Boo 13h41'10"17d14'
Boyle,Eamand Garland Cyg 21h4'26"30d26'
Boyle,Edward J & Phyllis V Ser 16h6'12"10d11'
Boyle,Edward P Sgr 19h7'59"-22d38'
Boyle,Gerard Per 3h13'39"50d26'
Boyle,Jeremy M Per 1h25'11"50d31'
Boyle,Joe Aur 5h17'39"41d50'
Boyle,Lauren Helena Aql 19h25'23d10"53'
Boyle,Loren Jayne Cas 0h53'53"59d58'
Boyle,Marion Tushingham and Sandy Cyg 19h12'1"48d17'
Boyle,Matthew Thomas Dra 15h52'37"66d52'
Boyle,Mayor Ed Per 3h56'23"40d49'
Boyle,Melissa Adele And 23h29'39"42d1'
Boyle,Michael Oliver Per 1h55'0"54d0'
Boyle,Missie & Jason Arcillas Cyg 20h38'40"38d33'
Boyle,Paul Her 16h9'34"48d60'
Boyle,Paul Edmund Oph 18h31'54"10d57'
Boyle,Peter Uma 11h55'13"43d46'
Boyle,Shana Eve Lyr 19h22'44"35d33'
Boyle,Shirley J Peg 23h4'25"30d30'
Boyle,Teresa M And 0h23'34"35d35'
Boyle,Tess Connor Lyn 7h26'55"48d53'
Boyle,Tom Lyn 7h17'33"59d50'
Boyle-I Love You Sandy ,Jeff Her 18h5'15"31d44'
Boyles,Mary Lac 22h40'28"50d24'
Boyles,Suzanne M Cas 1h6'27"53d12'
Boymann,Melissa And 23h20'22"51d45'
Boyne,Douglas W Lac 22h4'22"40d30
Boyne,Tom Tau 5h17'0"18d59'
Boyns,Ric & Kelli Crb 16h21'58"30d12'
Boynton,Charlie Her 16h58'31"23d51'
Boynton,Pauline Lyr 18h57'25"37d35'
Boys Star,The Peg 23h1'0"28d29'
Boysen,Atle Cas 1h52'26"60d50'
Boysen,Rolf Ori 5h57'30"19d35'
Boyt,Edward A Cep 0h14'1"68d10'
Boyte,Benjamin Hamilton Cet 0h53'36'-0d3'
Boyz Her 18h54'0"12d18'
Boz Uma 9h16'17"51d35'
Bozanic,(Dusty Star) Florence Jean Cas 1h11'0"63d60'
Bozanic,Anthony T Dra 18h58'15"80d3'
Bozanic,Lisa Beth Ser 16h1'1"24d52'
Bozarth,Adam Edgar Dra 20h32'16"68d26'
Bozarth,Russ & Betty Crb 16h0'10"38d59'
Bozeman,Darla J Cam 6h4'24"58d52'
Bozeman,Matthew Cep 22h54'29"59d34'
Bozeman,Michael Per 2h21'34"58d58'
Bozena Lib 14h58'48"-8d43'
Bozic,Janet Renee Cam 7h7'25"60d18'
Bozicevic,Boze Lyn 6h59'37"44d16'
Bozika,Princess Dimitra Cyg 20h7'37"39d42'
Boziwick,Dana Vul 19h22'17"25d42'
Bozman,Bront Boysen Cet 0h47'36"-7d13'
Bozman,Susanlee Del 20h26'20"10d30'
Bozorgmehri,Ellie Equ 21h6'25"10d36'
Bozsahin,Hakan Lib 14h57'60"-22d52'
Bozza,Donna & Mario Aur 5h40'36"54d41'
Bozza,Peter Joseph Dra 17h30'40"78d56'
Bozzelli,Jacqueline And 0h10'28"36d7'
Bozzi,Domenica Aur 5h32'45"31d16'
Bozzo,Angel Peg 21h24'22"23d51'
Bozzo,Frank Joseph Aur 6h13'60"46d49'
Bozzo,Pat Oph 18h30'1"8d17'
Braam-Alberts,Diny Cet 0h38'19"2d6'
Braaten,Don Orion Consulting,Inc Ori 5h55'15"8d53'
Braatz,David & Joan Cyg 21h8'56"30d51'
Brabant,Cynthia Tri 1h43'57"30d49'
Brabant,Evelyn Lmi 9h49'58"40d56'
Brabant,Jennifer L Aql 20h11'50"13d56'
Brabant,Kenneth J Per 2h53'34"32d55'
Brabban,Anthony Ronald Vel 9h43'44"54d4'
Braboy,Penny & Richard Erbacher Ari 1h59'42"19d53'
Bracamonte Family Star,The Cyg 21h1'38"40d13'
Bracchini,Juan Cas 0h32'18"66d36'
Bracchini,Miguel Cas 0h31'24"68d45'
Bracchy,Darlene Frances Lmi 10h56'30"26d33'
Braccia,Connie Ori 5h26'31"15d10'
Bracciale,Sonia Ant 10h45'2"-33d45'
Brace,Gareth Ori 5h31'28"8d32'
Brace,James A Lyn 8h5'15"42d27'
Brace,Martin & Trisha Cyg 21h51'46"53d7'
Braceland,Chris/Kris Luetkemeyer Mon 6h56'13"1d46'
Bracero,Kathleen Lyr 18h38'59"41d21'
Bracey,Mark Leonard Per 2h28'17"51d25'
Bracey,Perry Uma 13h57'47"51d53'
Brach,Ivy Marie Vul 19h44'38"23d31'
Brach,Matthew Joseph Ori 4h44'13"0d9'
Brach,Teresa And 1h30'59"48d48'
Brache,Richard B Per 3h57'42"50d24'
Bracher,Chad Richard Her 16h37'39"37d41'
Brachmann You Are A Star,Kathy Lee Cyg 21h37'11"42d45'
Bracht,Cindy Crb 15h49'16"32d14'
Bracht,James Bernard Aur 5h5'19"30d13'
Brack,Paul Muir Uma 8h53'26"56d17'
Brackeen,L G Hya 8h27'34"-0d3'
Bracken,Jack Chevrolet Psa 22h33'35"-27d58'
Bracken,Jeanette Vul 20h18'0"22d44'
Brackenier,Yves Cas 1h45'0"58d12'
Brackett,David Bruce Dra 14h57'1"58d3'
Brackett,Kyle Eri 5h6'26"-8d54'
Brackfield Star Aur 5h18'51"49d5'
Brackney,Melissa Com 12h21'54"31d5'
Brackstone,Robert Neil & Doris Anne Umi 14h12'19"77d45'
Braconi,Mark J Sct 18h55'38"-6d46'
Brad Aur 5h25'0"38d46'
Brad Cep 21h52'19"60d11'
Brad & Jill Forever Lyr 18h51'37"31d8'
Brad & Polly's Star Peg 23h4'33"8d29'
Brad & Sara Uma 12h41'40"61d17'
Brad & Tressa Cyg 20h39'28"40d36'
Brad S Cet 2h10'0"7d47'
Brad Star,The Per 2h44'1"43d25'
Brad The "N Diamond" Star Her 17h29'30"40d54'
Brad's Forever Light Dra 17h20'0"64d13'
Brad's Tidewater In The Cosmos Oph 17h2'42"12d42'
Bradana Cyg 19h21'22"28d55'
Bradberry,Megan Cate Uma 12h57'50"61d39'
Bradburn,Gordon Ferris Aql 20h7'0"6d12'
Bradbury,Allan G Her 18h9'58"50d22'
Bradbury,Yvonne Redman Cap 21h55'25"-17d30'
Braddick,David Peg 22h42'0"33d46'
Braddock,Autumn Com 12h36'56"23d28'
Braddock,Loretta Lynn Uma 11h31'37"62d3'
Braddock,Vanessa L Cas 1h33'54"60d51'
Brade,Roger Cep 21h16'18"58d26'
Bradechard,Louison Umi 13h52'1"77d49'
Brademas,Lorri J Cas 1h5'29"63d56'
Braden,Frances L Lyn 7h29'50"40d4'
Braden,Sandy Cas 0h33'32"60d44'
Brader,Holly Anne Ori 5h0'56"8d51'
Brader,Susan Leslie Cas 0h47'59"67d52'
Bradfield "Vickita Lynbabwe",Victoria Cas 0h25'23"70d41'
Bradfield's,Jim Hya 8h11'35"2d23'
Bradfield,Geoffrey N Vir 12h57'36"12d5'
Bradford Cep 22h1'20"61d24'
Bradford's Star Ari 2h34'53"21d56'
Bradford,Amy Catherine Ori 5h31'49"-0d23'
Bradford,Brad Her 17h53'0"48d55'
Bradford,Christopher Her 17h4'38"21d42'
Bradford,Dane A Cep 22h5'15"61d21'
Bradford,Derk J Per 1h28'53"52d46'
Bradford,Gerald Edward Dra 18h52'45"70d19'
Bradford,Ivanona Aql 20h5'1"6d9'
Bradford,James Michael Cep 21h46'22"80d5'
Bradford,Kathy Cas 0h28'17"71d12'
Bradford,John Michael Dra 11h48'12"71d25'
Bradford,Trevor John And 1h3'25"40d57'
Bradie,Ilanah Lyr 18h34'16"28d43'
Bradlex & Angel's Eternal Flame Eri 4h12'58"-13d35'
Bradley Peg 22h10'52"25d41'A
Bradley Cnc 8h32'55"18d33'
Bradley Hya 9h32'17"5d56'
Bradley 21 Dra 9h53'37"81d30'
Bradley CDP,Sister Roberta And 0h12'35"34d16'
Bradley Clark Lib 14h42'1"-23d12'
Bradley Memoriam,The Cnv 13h25'51"41d60'
Bradley Star,The Michael & Victoria Dra 17h35'22"60d43'
Bradley"First Lady", Ethel A Mon 7h58'45"-1d29'
Bradley's "Wedding Star",Hilary & Rob Cyg 20h58'31"40d32'
Bradley's Beamer,James Henry Gem 6h42'35"33d51'
Bradley's Love Psc 0h51'0"28d25'
Bradley's Tee Uma 9h54'29"48d9'
Bradley's,The Lyn 7h8'35"50d37'
Bradley,Anne And 23h3'17"42d50'
Bradley,Bruce J Her 16h32'49"34d41'
Bradley,Daniel D Oph 17h25'31"11d25'B
Bradley,David Michael Aql 20h5'16"0d53'
Bradley,Dennis M Uma 11h0'15"68d17'
Bradley,Devon Umi 15h33'30"78d12'
Bradley,Elizabeth Umi 15h14'55"67d23'
Bradley,Gary William Her 18h0'32"48d42'
Bradley,Ginette Cyg 19h29'1"35d38'
Bradley,Gladys B Lyr 18h30'0"32d58'
Bradley,Gray Per 2h51'1"40d17'
Bradley,Greg Aur 5h7'25"41d47'
Bradley,Heather Hya 9h16'52"3d13'
Bradley,Heidi & John Ori 6h1'57"0d33'
Bradley,Jay Ser 15h29'14"24d1'
Bradley,Jerry Sex 10h47'30"-8d6'
Bradley,John Martin Uma 8h57'1"68d40'
Bradley,Jonathan D Cet 2h57'29"2d15'
Bradley,Jordan Taylor Dra 10h41'1"73d22'
Bradley,Lisa Cas 0h44'28"67d29'
Bradley,Lisa & Keith Peg 0h1'50"27d39'
Bradley,Margie Marie Reliford Sge 19h10'12"17d34'
Bradley,Marion & Gordon Cyg 21h1'16"28d40'
Bradley,Mary Boo 13h34'1"17d34'
Bradley,Misty Peg 21h53'1"31d27'
Bradley,Peachlyn Uma 9h1'35"61d38'
Bradley,Phyllis Uma 11h34'10"31d33'
Bradley,Richard Lyn 9h6'34"37d36'
Bradley,Rodney James Ori 5h47'53"21d13'
Bradley,Rosslyn Lyr 18h21'11"47d6'
Bradley,Sandra Reaves Peg 21h55'44"31d43'
Bradley,Scott Keith Del 21h2'50"18d53'
Bradley,Sean Boo 15h3'43"27d50'
Bradley,Sheila Anne Sge 20h7'14"20d41'
Bradley,Sheila Marie Sge 20h1'0"16d12'
Bradley,Steven Boo 14h42'26"25d13'
Bradley,Susan Lyr 18h30'27"30d1'
Bradley,Timothy A Aur 5h3'59"37d56'
Bradley,Trevor Lee Leo 10h49'0"-0d58'
Bradley,Tricia L Lyn 8h48'1"37d16'
Bradley,William Warren Her 15h51'16"49d27'
Bradley-Sailer Star, The Theresa Lyn 7h27'0"41d57'
Bradmark Ori 5h56'53"12d55'
Bradner,Carol D Vir 13h15'44"-0d50'
Bradshaw,Audrey Elizabeth Mon 6h51'38"-2d50'
Bradshaw,Christine Mon 7h23'19"-8d33'
Bradshaw,Clifford Dra 16h28'24"61d25'
Bradshaw,Curt A Boo 14h58'29"35d3'
Bradshaw,Daniel Her 18h28'56"20d31'
Bradshaw,David N Cas 1h19'1"61d50'
Bradshaw,Don Per 4h30'0"31d16'
Bradshaw,Edna May Del 20h56'19"16d25'
Bradshaw,Jean C Cas 1h21'54"61d6'
Bradshaw,Jean Marie Lyr 18h52'50"30d11'
Bradshaw,John & Edee Per 2h57'33"34d2'
Bradshaw,Kerry Lyr 18h18'32"40d51'
Bradshaw,Larry Boo 14h33'17"41d39'
Bradshaw,Lesley K Lyr 18h32'39"39d41'
Bradshaw,Love Always Warren Leo 10h53'20"21d9'
Bradshaw,Maggi Cyg 19h14'27"45d36'
Bradshaw,Michael David Boo 14h25'12"39d5'
Bradshaw,Michelle Marie And 2h10'0"39d27'
Bradshaw,Nora Lynn Cap 20h35'11"-12d26'
Bradshaw,Robert Zachary Harrison Ari 1h55'21"22d33'
Bradshaw,Robin Lyr 18h35'0"28d48'
Bradshaw,Ronald Earl Boo 14h44'0"25d21'
Bradshaw,Sandy Oph 18h2'44"8d13'
Bradstar,The Cam 8h8'23"80d40'
Brady Her 16h18'55"10d8'
Brady's Sex 10h32'1"2d31'
Brady's Brilliance Leo 10h43'1"15d23'
Brady(April 24th 65), Fred And 2h10'0"40d24'
Brady,"Mick" Aur 7h21'46"40d23'
Brady,Alexandra Violet Lyr 18h59'1"42d21'
Brady,Antoinette Stone Lyn 7h15'45"56d43'
Brady,Audrey Cas 0h38'20"60d51'

Brady,Bea
Mon 6h27'56"10d27'
Brady,Bridget
Lyr 18h42'29"40d1'
Brady,Bryan
Mon 7h7'19"-5d58'
Brady,Candace Arielle
Cyg 21h18'34"37d46'
Brady,Christopher Robert
Aql 19h12'38"12d40'
Brady,Deirdre Elena
Lyn 7h50'1"40d52'
Brady,Diana
Aql 19h18'1"13d21'
Brady,Don
Cnv 12h5'1"37d32'
Brady,Dr Thomas V
Oph 18h4'52"11d47'
Brady,Ernest
Her 16h43'20"50d30'
Brady,John Robert
Mon 7h50'57"-2d26'
Brady,Jr,Francis Martin
Boo 15h0'56"15d42'
Brady,Jr,Robert J
Lac 22h9'14"51d43'
Brady,Judy
Peg 22h6'23"29d31'
Brady,Julie
Aql 19h29'0"7d57'
Brady,Kimberly
Aqr 20h59'11"-10d6'
Brady,Kristal Ann
Cas 2h35'31"63d58'
Brady,Mark
Leo 9h38'33"33d12'
Brady,Mary Booth
Leo 11h6'1"1d41'
Brady,Mary Patricia Dietzel
Mon 7h1'1"8d38'
Brady,Michael J
Cep 22h19'55"58d59'
Brady,Pat Bob
Aql 19h28'59"13d58'
Brady,Raymond E & Patrice A
Oph 18h6'39"0d25'
Brady,Regis
Mon 6h54'57"-0d0'
Brady,Stephen
Aur 7h7'1"39d27'
Brady,Thomas
Oph 17h53'59"12d41'
Brady,Thomas
Aur 7h12'0"37d16'
Brady,William
Hya 9h36'54"-6d32'
Braeden
Aur 5h18'35"41d28'
Braedon,Aaron
Dra 17h52'48"61d44'
Braemer,Richard
Aur 6h3'27"36d30'
Braeuer,Michael
Aql 19h35'44"1d44'
Braff,Samuel B
Peg 22h50'55"8d40'
Bragdon,Paul Bowman
Ser 18h53'40"2d40'
Brage,Eurydice Gladie
Aql 20h16'34"0d6'
Brager,Alicia S
Com 12h30'1"30d44'
Brager,Betty Ann
Aqr 20h37'0"-0d4'
Brager,James & Lillian
Cyg 20h6'30"40d49'
Bragg,Audrey F
Lyn 8h4'25"51d12'
Bragg,Marjoria
Lyr 19h12'38"38d9'
Bragg,Michael
Aur 6h10'35"46d50'

Bragg-"The People's Star",Jack
Uma 11h19'53"38d19'
Braghiroli,Caria Fufola
Ori 6h5'46"0d22'
Braghiroli,Luca
Oph 18h5'55"10d33'
Braglia,Ludovica
Tel 19h8'15"-49d24'
Braglia,Silvia Pini
Tel 19h8'16"-46d13'
Brahinsky,Ben M
Sex 9h52'0"2d18'
Brahlek,James Allen
Tau 4h2'49"21d37'
Brahler,Richard V
Aql 19h55'57"11d53'
Brahley,Sandra
Mon 7h43'59"-6d8'
Brahlick,Helen Mary
And 2h29'0"49d31'
Brahm,(Dad),Robert T
Aql 19h15'56"12d49'
Braibish,Stanley M
Per 1h37'46"52d49'
Braid,Gary William
Her 16h23'20"40d54'
Braid,Grant William
Her 17h28'0"21d52'
Braid,Jessica Marie
Crb 16h12'0"28d10'
Braid,Shirley Anne
Cas 0h25'38"64d14'
Braig,Sarah Teresa Reichert
Cnc 8h11'24"32d52'
Brail,Sasha Levien
Ori 4h59'46"-2d29'
Brailsford,Mary Kathryn
Lyn 8h56'11"37d18'
Brain's Light
Uma 13h45'40"60d2'
Brain,Christopher J
Per 4h34'0"31d10'
Brain,Tony Ellis
Per 2h56'49"37d39'
Brainard,Bruce
And 0h31'14"34d21'
Brainerd,Darion Wayne 'Boomer'
Lac 22h26'59"52d46'
Brakeley,Diane D
Aqr 23h6'24"-6d50'
Brakey,Walter G
Lac 22h47'1"55d44'
Brakhage,Briana Fischer
Cam 5h1'13"60d46'
Brakhage,Desirae Fischer
Cam 5h1'40"61d42'
Brakhoff,Hartmut Falk
Sco 17h27'44"-38d18'
Braley,"Cara Mai" Megan
Lib 15h19'45"-22d31'
Brallier,Sara Anne
Cas 1h45'40"56d48'
Bram,Eric & Theodore
Aql 20h21'38"5d1'
Bramac
Cas 0h17'42"60d59'
Brambila,Paula
Vul 19h46'1"28d48'
Brambilla,Letizia
Cas 2h10'25"60d17'
Bramble,Ryan
Per 1h48'30"53d49'
Bramblett,Virginia
Lac 22h7'59"46d34'
Brame Golden Annv, Ernie & LaDonna
Cyg 20h44'54"37d45'
Brami,Chantal
Ori 5h21'0"12d31'
Bramlett III,Charles William
Cmi 7h58'10"8d30'

Bramlett's,The
Mon 7h20'35"-8d17'
Bramley,Henrietta
And 0h37'45"40d47'
Bramley,Timothy Earl
Boo 13h52'26"20d14'
Brammer,Carmen Jenny
Com 13h2'38"18d14'
Brammer,Jennifer Lynn
Cyg 21h31'51"30d26'
Brammer,Luisa Antonie
Vir 13h35'60"-8d35'
Brammer,Luke William
Ori 5h2'57"7d52'
Brammer,Stefanie
Ari 2h0'48"21d14'
Bramos II,Daniel Dennis
Mon 7h57'27"-5d54'
Bramos,Francis C
Ser 18h0'35"-13d10'
Bramucci,Ludovica
Lyr 19h10'35"38d3'
Bramwell,Vickie
And 23h20'1"51d51'
Brana,Astrid
Dra 17h55'16"73d23'
Branam,Ericka Nicole
Peg 22h59'59"11d46'
Branam,Winnie
Eri 4h33'26"-12d28'
Branaman,Christopher David
Boo 13h44'0"26d6'
Branan,Penny Michelle
Oph 17h2'46"-20d14'
Brancaccio,Rosalie Bambi
And 23h25'54"49d21'
Brancale,Grace
Cas 1h2'36"62d4'
Brancato,Annalisa
Lac 22h42'34"38d5'
Brancato,Charles D
Boo 15h17'36"41d17'
Brancato,Jeff P
Dra 19h55'56"68d45'
Brancato,Susan P
Cyg 19h27'42"35d0'
Branch,Aunt Mary
Cas 23h39'51"50d57'
Branch,Claire Julia
Vir 12h59'42"-12d25'
Branch,Girard Leon
Lac 22h40'18"55d8'
Branch,Jonathon David
Cyg 17h39'29"61d41'
Branchu,Celine
Crb 16h18'54"38d58'
Branchut,Jean Pierre
Dra 9h38'1"80d38'
Branciaroli,Bonnie Sue
Sge 19h29'56"16d44'
Branco,Dolores Rita
Lyr 19h15'40"41d30'
Branco,Lisa Braz Elizabeth
And 23h24'31"45d7'
Branco,Michael A
Aur 6h18'31"37d39'
Branco,Paula Cristina
Cyg 21h50'1"38d38'
Brancoli,"Brilliance"- Joseph B.
Cnc 8h53'35"31d35'
Brancolini,Nicoletta
Aur 5h1'58"30d57'
Brand,Anja
Leo 9h54'45"11d6'
Brand,Deree
Ori 5h51'46"12d52'
Brand,James F
Aur 6h0'21"30d34'
Brand,Joyce
Lyr 18h33'53"31d5'
Brand,Karl
Ari 2h42'17"30d2'
Brand,Mary
Aql 19h41'1"13d54'

Brandeberry,Sarah Kate
Lyr 18h57'29"30d47'
Brandee
Lyn 8h43'40"45d11'
Brandenburg's Quest
Lmi 10h9'51"36d15'
Brandenburg,Julia
Cnv 12h16'55"37d11'
Brandenburg,William F
Mon 7h16'19"-6d34'
Brander,Danielle Louise
Lyn 7h53'56"58d9'
Brander,Jr,Delbert C
Lyn 7h45'55"38d45'
Brander,Julie Anne
Cnv 14h3'40"37d44'
Brandes Star
Cas 1h14'48"61d15'
Brandes,Karen Lynne
Vul 19h44'58"22d56'
Brandes,Ryan John
Her 16h28'1"34d49'
Brandewie,Elena
Aql 19h40'37"11d59'
Brandewie,William
Aur 6h2'29"38d47'
Brandi
Tri 1h44'22"30d0'
Brandi
Peg 21h20'49"23d49'
Brandi & Rick's Star
Aql 19h44'21"14d12'
Brandi 11
Lyn 7h4'12"44d34'
Brandi Christine
Mon 6h29'41"8d33'
Brandi Nichole
Cmi 7h25'23"8d51'
Brandie
Gem 6h54'53"18d55'
Brandie
Cet 3h10'18"0d55'
Brandie
Mon 6h21'39"7d16'
Brandie Ann
Hya 8h32'59"0d23'
Brandies Keeper
Uma 11h53'1"46d42'
Brandis,Jill
Vul 19h20'47"25d33'
Brandlein,Shirl
Uma 13h21'20"58d14'
Brandlen,Peter
Ori 5h25'1"0d12'
Brandman Eva "Forever And A Day"
Cet 3h19'18"1d5'
Brando
Tel 20h14'33"-48d59'
Brandolf,Joshua
Cep 22h56'38"65d2'
Brandon
Aur 6h23'12"33d16'
Brandon
Cet 2h34'55"-5d11'
Brandon & Jennifer- Lasting Love
Her 17h42'47"14d58'
Brandon & Kristine's Star
Ori 6h2'44"8d52'
Brandon Alexander
Her 18h11'33"47d23'
Brandon Jay
Her 17h35'1"20d43'
Brandon Lee
Cep 20h55'30"65d21'
Brandon Matthew Marvelous
Cep 21h42'28"63d51'
Brandon Quoc-Vu Tran
Her 18h3'55"38d29'
Brandon's Dream
Her 17h32'40"22d37'
Brandon,Cassie
Sge 20h17'41"16d55'

Brandon,Gordon M
Cet 2h49'40"-0d4'
Brandon,Jewell
Equ 21h6'0"10d40'
Brandon,Linda Kilpatrick
Gem 6h38'55"34d18'
Brandon,Marion H Newlin
Com 12h24'17"30d47'
Brandon,My Grasshopper
Sct 18h42'41"-5d42'
Brandon,Nancy
Eri 3h42'23"-6d51'
Brandon,Ruby Rodney
And 23h24'0"43d23'
Brandonalis,Lodi Rae
Lyn 9h13'54"41d43'
Brandoni The Son
Her 16h41'45"5d0'
Brandow,Debbie Louise
Cam 7h4'22"83d52'
Brandreth,Annette
And 0h6'18"30d35'
Brandstetter,Robert
Aur 5h17'0"45d13'
Brandstrader,Stephen J
Cma 7h16'13"-16d6'
Brandt 1
Ori 5h53'20"14d22'
Brandt Memorial Star, The Mildred
Uma 13h39'1"50d55'
Brandt,Beverly
Sge 19h57'1"16d46'
Brandt,Blazing
Uma 10h47'46"48d53'
Brandt,Chester Russel
Per 2h57'26"37d36'
Brandt,Christopher Bryon
Boo 15h5'21"25d19'
Brandt,Deborah
Peg 22h36'14"26d35'
Brandt,Diana Gail
Sco 16h59'56"-40d20'
Brandt,Erich
Aqr 22h8'1"-6d41'
Brandt,Geraldine F
Cap 20h35'1"-20d9'
Brandt,Henrik
Dra 14h10'37"63d20'
Brandt,Jeanie
Tri 1h47'38"28d11'
Brandt,Jr,George H
Aur 6h15'20"38d20'
Brandt,Kelly L
Mon 6h36'23"0d2'
Brandt,Kyle Benjamin
Gem 8h4'58"33d1'
Brandt,Marianne und Herbert
Uma 10h38'15"41d2'
Brandt,Nicholas James
Cet 2h46'13"2d22'
Brandt,Paige
Peg 22h13'1"34d47'
Brandt,Renate
Cap 21h19'8"-23d34'
Brandt,Rick & Kristie
Uma 9h10'40"50d23'
Brandt,Susanne
And 23h15'52"50d7'
Brandt,Wolfgang
Del 20h55'44"8d34'
Brandum,William Alexander
Ori 6h6'24"5d40'
Brandwein,Margaget
Lyn 6h32'1"39d22'
Brandwijk,Lenita
Lyr 18h54'26"31d35'
Brandwood,Marian
Ori 6h4'11"5d34'
Brandy
Lyr 19h8'36"38d7'
Brandy
Cas 0h29'36"72d55'
Brandy
And 2h20'1"41d56'

Brandy & Doug Star,The
Oph 17h28'37"-22d14'
Brandy Lane Dog Training
Cnv 12h8'13"34d43'
Brandy Leigh
And 2h23'18"44d15'
Brandy Marie
And 0h22'34"40d38'
Brandy Nicole "A Star In My Life"
Psc 1h0'0"18d8'
Brandy Rhea
Peg 23h20'40"32d56'
Brandy's Love
Cet 1h31'27"-10d37'
Brandy's Lucky Star
Peg 22h27'11"31d28'
Brandy,Julia
Cas 0h38'20"64d18'
Brandy-Belle
Cmi 7h56'23"8d55'
Brandys,Svatloslav
Lyr 19h17'21"30d38'
Branham,Susan Elizabeth
Equ 21h11'25"10d23'
Braniecki-The One In A Billion,Stacy E
Cas 0h22'52"61d21'
Branka,Leonard
Boo 15h5'49"14d40'
Branker,Parris Jolean
And 0h15'0"37d57'
Brann,Bonnie
Peg 21h55'20"28d50'
Brann,Marc
Uma 11h4'44"47d38'
Brannan,Thomas F
Aql 18h57'0"-8d57'
Brannen,John Michael
Aql 19h16'39"14d29'
Brannen,Wilbur
Sex 10h44'19"2d16'
Branney,Glenn
Her 17h18'1"40d6'
Brannick,Margaret
Crb 16h7'54"27d56'
Brannigan
Aql 18h55'51"10d41'
Brannigan,James Richard
Per 4h26'1"50d1'
Brannock "Val's 18", Valerie Ann
Del 20h19'12"10d34'
Brannon,Beverly
Aql 18h53'31"-2d25'
Brannon,Jay L
Uma 11h26'39"30d42'A
Brannon,Retha
And 0h29'12"40d9'
Brannon,Sean David
Per 3h14'43"40d48'
Brannon,Skylar Macall
Lib 14h50'34"-1d52'
Brannon,Terry L
Uma 11h26'39"30d42'B
Branscombe-Conkling, Shirley Jeanne
And 1h13'12"34d21'
Branson,Ashley Ann
Uma 9h51'41"48d55'
Branson,C Robert
Per 2h52'33"35d10'
Branson,Linda Marie
Peg 22h15'53"32d51'
Branson,Louise Helen Patricia Geary
Cam 3h58'20"58d26'
Branson,Nancy
Sge 19h31'38"18d46'
Branstetter,Henry Eugene
Oph 17h58'13"12d0'
Branstetter,Katherine
Mon 6h21'33"5d54'
Branstetter,Olin DeWitt
Aql 19h55'44"13d46'

Branstetter,Paul Richard
Cet 2h37'0"-12d18'
Brant,Donald G
Mon 6h12'0"-10d12'
Brant,Jenna
Uma 9h16'57"55d47'
Brant,Wanda Kay
And 1h54'32"40d40'
Brantley ILYTTSAB, Sarah-Sasha
Ori 4h43'47"4d57'
Brantley,Jr,David Nolan
Oph 17h31'60"-0d42'
Brantley,Jr,McKinzie
Dra 20h13'12"63d24'
Brantley,Wilton & Jeffie
Lmi 10h43'17"28d15'
Bras,Carol Lee Rice
Cam 3h36'50"70d47'
Brasch,Adrian W
Her 15h55'58"40d39'
Brasch,Olivia August
Leo 10h52'1"-0d34'
Braschler,George
Cam 3h53'26"61d13'
Brasesco,Luis
Boo 13h40'46"17d49'
Brasewell,Harrison Senna
Dra 14h39'1"64d47'
Brasfield,Auntie Fern- Fern
Com 12h27'20"21d60'
Brasfield,Joe
Leo 10h50'10"15d59'
Brashear II,William "Bernie"
Cyg 21h5'20"37d42'
Brasher,Pollyanna & Johnathan
Crb 16h5'19"26d45'
Braski Argentum 25
Mon 6h35'14"1d15'
Brass Ring
Per 2h36'21"50d34'
Brass,Abbie-Lee
Lyr 18h19'38"44d19'
Brassel,Catherine Butterfield
Cam 5h50'50"71d22'
Brassfield,Fannie
Lyn 6h45'44"60d47'
Braswell III,Will V
Cnv 12h21'59"37d23'
Braswell,Alex Gregory
Aur 4h52'18"40d18'
Braswell,Kym T
Lyr 18h46'1"30d41'
Braswell,Starr
Aql 18h59'24"12d53'
Brat
Aur 7h1'41"38d22'
Brat(E)
Vir 12h3'45"1d42'
Bratcher,Ronald Paul
Boo 15h19'0"40d16'
Brathburry,Beatrice Bee
Mon 6h55'26"11d21'
Brathovd,Brian Christopher
Oph 18h2'41"13d38'
Brathovd,Leonard Earl
Oph 18h31'11"12d16'
Bratina,Steven
Cam 6h30'45"68d45'
Bratis,Jackie
Mon 7h48'41"-3d57'
Bratli,Elisabeth
Ori 5h33'21"-0d7'
Bratovic,Sonja
Mon 8h1'29"-6d6'
Bratt,Lisa
And 0h47'1"38d0'
Brattain,Sr,Charles Michael
Per 1h35'1"52d43'
Brattoli,Corrado
Cet 0h41'46"1d11'
Bratton,Alexandra Kent
Peg 23h3'1"10d6'

Bratton,John
Hya 8h12'52"0d56'
Bratton,Miles Ellis
Mon 7h54'17"-6d39'
Braude,Audrey Lee
Cas 1h7'28"63d37'
Braude,Jaqueline Hope
Aql 19h50'29"15d11'
Braude,Jennifer P
And 1h19'33"39d39'
Braude,MD,James Steven
Oph 17h1'49"12d8'
Brauer,Barbara
Mon 6h58'0"8d57'
Brauer,Cordula
Cep 22h24'51"60d36'
Brauer,Shawn Michael
Cam 6h44'36"80d17'
Brault,Grandma & Grandpa
Cam 4h16'34"70d42'
Brault,Pascal
Dra 16h8'21"60d15'
Braumbeck,Jim & Fran
Ser 18h43'48"3d54'
Braun Colleen
Uma 8h56'31"58d27'
Braun Make A Wish, Robert
Crt 10h55'13"-21d45'
Braun's Star,Henry Thomas
Del 20h49'31"9d36'
Braun,Audra
And 0h29'10"44d57'
Braun,Carl Alexander
Umi 15h15'47"66d46'
Braun,Chad Elliot
Oph 18h1'16"11d60"
Braun,Claudia
Peg 23h33'42"12d22'
Braun,David Joseph
Ori 5h38'23"-0d7'
Braun,Dorothea
Leo 11h6'28"-0d17'
Braun,Erna
Cmi 7h22'32"4d44'
Braun,Ervin
Aur 7h1'46"43d20'
Braun,Erwin Latein-LK 92-92 Stern
Tri 2h4'30"35d38'
Braun,Ginie Mary
Oph 17h25'20"8d28'B
Braun,Henry August
Oph 17h25'20"8d28'A
Braun,Jacob Michael
Aql 20h22'4"0d54'A
Braun,Jane
Hya 9h8'50"2d39'
Braun,Janett Marie
Del 20h16'59"10d12'
Braun,Jill
Cas 23h31'16"58d19'
Braun,Jill L
Mon 7h22'0"-8d24'
Braun,McKayla Michelle
Cas 3h0'12"61d39'
Braun,MD,Sheldon R
Cnv 14h4'47"32d12'
Braun,Norbert
Lyn 8h7'37"40d26'
Braun,Pat
Cam 7h37'46"60d3'
Braun,Pius
Lyr 18h58'41"34d59'
Braun,Regina
Lyn 8h7'45"49d58'
Braun,Rolf-Mainz Bleibt Mainz
Ari 2h29'34"26d40'
Braun,Rudolf
Cmi 7h18'0"1d49'
Braun,Sharon Katchmer
Mon 6h23'48"8d60'
Braun,Stefan
Aur 5h4'15"43d12'
Braun,Theresa Elizabeth
Aql 20h22'4"0d54'B

Braun-Hartman,Fuzzy
Cmi 7h54'48"7d19'
Braun-The Great Mother,Josephine
Cas 0h34'19"63d4'
Brauneiss,Ruth
And 23h25'27"47d33'
Brauns,Henry
Her 17h19'29"18d53'
Braunschmidt,Juergen
Del 20h19'11"12d48"
Braunschweig Belle
Aql 19h3'18"0d47'
Braunstein,Antoine Ferdinand
Lyn 8h11'17"39d42'
Braunstein,Edward Charles
Dra 13h34'25"67d39'
Braunstein,Mark
Per 2h36'44"46d17'
Braunstein,May
Del 20h17'0"9d19'
Braunstein,MD,Edward Abraham
Cep 20h55'36"58d45'
Braunstein,Michael Abraham
Lac 22h29'38"50d30'
Braunwart,Roger
Aur 6h3'37"37d7'
Bravard,Wyman Nash
Dra 13h20'29"64d0'
Brave Rose
Aql 20h21'48"3d17'
Brave,Molly
Cnc 9h1'27"30d38'
Braverman,Matthew
Her 16h13'47"48d15'
Braverman,Mickey
Lyr 18h44'24"30d26'
Bravin,Mr & Mrs Michael
Crb 15h31'36"31d21'
Bravissimo
Cam 3h43'34"68d58'
Bravo,Gavino
Lyr 19h2'11"41d9'
Bravo,J D
Boo 15h21'29"41d57'
Bravo,Rose Marie
Lyn 6h54'42"56d1'
Brawley,Jennifer
Peg 23h37'20"31d12'
Brawn-Grieves,James Stephen
Aur 6h7'53"32d52'
Brawner,Alexander Harrison
Boo 13h58'54"8d26'
Brawny
Peg 21h21'40"23d56'
Bray
Crt 11h22'54"-10d16'
Bray,David Owen
Lac 22h3'34"49d54'
Bray,Glenn Maurice
Lyn 7h21'40"44d49'
Bray,Kristine Noelle
Uma 10h11'50"51d47'
Bray,Marjorie Christine
Cyg 21h39'56"42d58'
Bray,Ronald Craig
Hya 9h7'13"2d14'
Bray,Sara Jane
Mon 7h2'0"4d25'
Bray,Sonya C
Peg 22h17'40"20d28'
Brayelle,Carl
Crb 16h17'30"30d7'
Braymen
Cam 7h56'14"60d18'
Braynin,Tamara
Sge 19h54'21"20d4'
Braystar
Peg 21h58'33"33d8'
Brayton
Cet 0h25'1"-5d20'
Brayton,Cynthia
Com 13h15'21"28d13'

Brazaski,Julie
Cet 2h22'55"0d9'
Brazeau,David
Peg 22h18'21"21d13'
Brazel,Jeanne
Uma 9h45'0"56d8'
Brazel,Katheryn Sara
Cam 6h10'11"72d38'
Brazen
Cnv 13h4'1"37d60'
Brazen,Alan
Boo 14h58'13"46d18'
Brazer,William Fletcher
Ori 6h1'34"4d16'
Brazier,Kelly
Mon 6h22'59"-0d27'
Brazil,Mark
Per 1h53'19"50d20'
Brazile,Maria
Oph 17h54'51"13d0'
Brazilian,The
Uma 11h38'23"50d1'
Bre
Mon 7h47'31"-2d37'
Bre
Lyr 19h21'47"30d47'
Brea's Quest
Ari 1h46'23"14d36'
Breakey,Scott A
Aur 6h56'53"35d38'
Breakfast Club KLCY
Hya 9h6'38"4d47'
Breakstone,Hannah Samantha
Leo 10h54'11"21d8'
Breard,Jr,Hypolite Filhiol
Aur 5h21'0"54d51'
Breasure,Brandon Lee
Lac 22h40'55"54d54'
Breathless Sarah
Cas 0h44'13"61d51'
Breault,Suzie Marie
Peg 22h19'36"3d25'
Breaux,Reuben Paul
Ser 15h21'25"4d15'
Breban,Capucine
Per 3h29'46"39d18'
Brebeuf Mothers
President,Becky Lapp
Cas 0h48'53"61d49'
Breborowicz,Joachim
Vul 20h45'44"20d11'
Breccia,Fulvio
Pic 5h15'56"-44d49'
Brecher's Seaview Star Oskar
Her 17h27'41"25d11'
Brecher,Gerald I
Aql 20h2'40"8d24'
Brechin,Ewan
Dra 16h15'42"61d33'
Brecht,Doctor Lawrence
Her 18h17'17"20d7'
Brecht,William Richard
Ser 18h55'52"3d19'
Brechtel,Nicholas Robert
Aql 19h52'0"12d56'
Breck,Alyson
And 0h5'27"38d45'
Breckenridge,Lewis Richard
Cet 1h57'38"-2d8'
Breckenridge,Marti Lynn
Com 12h0'58"25d14'
Breckler,Claude
Vul 19h33'55"27d1'
Breckler,John & Fran
Ori 5h55'38"8d20'
Breckstein,Rachel
Lyr 19h6'23"40d16'
Bredbenner,Betty E
Cas 0h38'38"67d49'
Brede,Reanna Rae
And 2h17'34"45d37'
Breden,Bridget & Matthew
Lmi 10h18'12"35d25'
Brederson,Jill- Desirée
And 2h29'27"49d5'
Brederson,Richard Paul
Her 16h52'24"32d33'
Bredeson,Duane
Ari 1h51'1"12d29'
Bredice,Anna Elizabeth
Uma 9h6'38"56d22'
Bree
Cas 22h57'47"55d56'
Bree,Tawny
Lyr 18h42'23"33d19'
Breeck,Kathie
Mon 6h21'45"-6d32'
Breed II,MD,Robert
Huntington
Her 17h22'34"21d2'
Breed,John David
Mon 8h2'37"-0d44'
Breeden,Debbra Lee
Cyg 20h40'35"31d31'
Breeden,Nancy Sue
Lyn 8h5'40"36d21'
Breeding's Shining Star,Robert
Aql 19h50'0"14d47'
Breeding,Glenn & Shirley
Cyg 21h37'34"40d46'
Breedlove,Jayne
Cas 1h1'49"48d54'
Breedlove,Nancy
Peg 23h0'42"22d6'
Breedon,Jack
Cam 8h0'1"60d40'
Breen
Cyg 20h3'28"30d22'
Breen Star,The
Oph 17h58'1"-8d49
Breen,Christine Marie
And 23h43'34"45d33'
Breen,Jonathan "Jack"
Boo 14h5'12"29d9'
Breen,Rachel
Lyr 18h36'19"45d22'
Breen,Samantha
Cas 3h0'50"60d50'
Breen,Valerie
Cas 1h30'0"58d46'
Breeze,Bob
Her 17h30'10"48d50'
Bregnard,William Edward
Her 18h4'54"38d46'
Bregoli,Mauro
Lyn 7h53'0"41d32'
Breheney,Ash
Lac 22h17'38"46d16'
Brehlyn
And 2h4'18"39d57'
Brehm,Shawn
Aur 6h14'32"45d33'
Brehmer,Steffi von
Tau 4h38'42"9d34'
Brehmer,William R
Aur 5h23'57"38d10'
Breiner,Alfred William
Sex 10h35'23"4d51'
Breining,Patti
Lyr 18h42'28"34d31'
Breinling
Aur 6h57'29"43d53'
Breit,William D
Cma 6h25'27"-18d51'
Breitenbucher,Robert B
Cep 20h49'18"67d35'
Breitenfellner,Georg
Sge 19h5'2"17d1'
Breiting,Carina
Cas 0h11'44"64d6'
Brem,Cody Mack
Psa 22h31'21"-27d54'
Bremboeck,Roswitha
Aqr 20h58'36"0d30'
Bremer,Chad
Dra 12h5'0"75d31'
Bremer,Eric
Cep 21h21'19"55d39'
Bremer,Erich V
Cep 21h1'59"60d44'
Bremer,Harry Francis Francis
Cam 4h57'25"60d8'
Bremer,Kyle
Uma 10h27'36"51d54'
Bremer,Scott Munroe
Ser 15h10'40"21d53'
Bremner,Sr,David F
Aql 20h3'1"0d46'
Brems,Sissie
Cas 23h38'1"64d18'
Bren
Cma 6h42'13"-15d54'
Brend Diane
Uma 9h58'15"44d53'
Brenda
Gem 6h53'51"18d29'
Brenda
Eri 3h56'56"-13d18'
Brenda
Lyr 18h47'30"34d23'
Brenda
Peg 22h43'43"29d0'
Brenda
Umi 14h19'45"68d46'
Brenda
Sex 10h11'22"-2d41'
Brenda
And 1h7'48"37d53'
Brenda
Sct 18h47'17"-8d53'
Brenda & Jack
Cam 11h46'12"78d37'
Brenda & Scott
Crb 15h55'1"31d18'
Brenda Anne
And 23h26'20"42d16'
Brenda Joy
Cam 3h14'53"60d0'
Brenda Kay
Lyn 8h23'34"48d29'
Brenda Kay's Light
Peg 21h19'31"21d46'
Brenda L
Cas 0h48'48"68d10'
Brenda Lea
Vul 21h0'28"20d16'
Brenda Marie
Cas 0h52'52"61d11'
Brenda Star
Cas 2h34'25"57d28'
Brenda Urmydreams
Iloveu4ever Matt
Cyg 20h23'20"40d41'
Brenda's Badawang Star
Peg 23h29'1"32d33'
Brenda's Dream
Lyr 18h46'35"42d46'
Brenda's Honaunau in the Sky
Ori 5h24'29"-0d1'
Brenda's Jewel
Cyg 21h51'43"42d47'
Brenda's Star
Cas 2h8'19"70d15'
Brenda's Super Star
Lyr 18h27'19"31d37'
Brenda-1372
Lmi 9h38'20"38d28'
Brendan
Per 2h55'0"37d48'
Brendan
Ari 2h39'0"24d56'
Brendan "Our High Flying
Hero"
Peg 22h56'24"20d25'
Brendan Joseph
Dra 16h22'0"61d49'
Brendear
Aql 20h13'18"4d10'
Brendel,Frederick
Her 16h55'38"29d6'
Brender,Dieter A
Vir 12h29'40"-10d44'
Brendie
Cas 1h39'53"61d2'
Brendie & Pesagh F
Cyg 20h35'23"50d30'
Brendler,Manfred
Cam 6h55'56"78d45'
Brendline,Debra & Michael
Dra 13h26'29"70d5'
Brendol
Lyr 18h50'24"34d28'
Brendon,Michael
Boo 14h45'33"23d34'
Brendums
Per 1h51'48"56d21'
Brener,Alfredo & Celina
Cet 1h9'48"-0d44'
Brener-Hellmund,Leon
Aql 19h57'28"12d56'
Brener-Hellmund,Nina
Aql 19h10'17"10d10'
Brenholdt,Jim
Ori 5h55'51"16d20'
Brenke,Peter
Leo 11h54'34"21d37'
Brenlite
Lac 22h29'38"56d40'
Brenn,Louis Clifford
Crt 11h11'57"-11d3'
Brenna Elizabeth
And 0h50'44"33d60'
Brenna,Loys
Eri 4h28'60"-12d13'
Brenna-F
Tau 4h3'30"20d8'
Brennan (Natu Maxima), Fiona
And 23h28'14"48d46'
Brennan 17 68
Lmi 10h38'51"27d37'
Brennan We R Together 4
Ever,Mike
Sex 10h28'58"5d43'
Brennan's (Smile)
Sex 10h15'44"-6d25'
Brennan's Bright Light
Eri 3h46'0"-4d40'
Brennan,Amelia Yackus
Lyn 8h41'47"39d6'
Brennan,Andrew P
Cep 23h11'21"62d38'
Brennan,Bill
Ser 16h1'38"10d25'
Brennan,Bob & Andrea
Umi 14h5'1"68d36'
Brennan,Brian
Cnv 12h47'54"32d8'
Brennan,Douglas Julian
Ori 5h49'18"11d33'
Brennan,Ellen
And 0h37'0"45d20'
Brennan,Emma Marie
Equ 20h59'59"8d54'
Brennan,Gerard M
Lac 22h26'27"40d54'
Brennan,Gregory T
Aql 20h1'24"11d39'
Brennan,Harrison
Her 16h44'18"38d33'
Brennan,III,James F
Boo 14h56'17"40d35'
Brennan,Jacqueline Mary
And 0h4'45"30d42'
Brennan,James Joseph
Lac 22h27'23"41d6'
Brennan,James Thomas
Per 2h31'25"57d18'
Brennan,Jessica
Dra 12h16'56"64d47'
Brennan,Jim
Sex 10h40'0"-2d22'
Brennan,Jr,Love Always -
Mom,John Joseph
Cam 4h58'46"70d43'
Brennan,Kailey Michelle
And 23h39'23"38d37'
Brennan,Kara
And 0h28'40"45d26'
Brennan,Kate Darby Miller
Aql 20h1'1"12d49'
Brennan,Katy
And 23h40'40"47d42'
Brennan,Kevin Patrick
Ori 5h58'54"8d33'
Brennan,Louise
And 23h31'37"42d52'
Brennan,Mary Lois Wade
Cas 1h48'0"68d16'
Brennan,Phyllis
Mon 7h52'6"-7d19'
Brennan,Richard
Her 16h45'18"37d35'
Brennan,Robert Jennings
Her 16h33'21"48d40'
Brennan,Sarah
Uma 11h45'45"47d48'
Brennan,Sarah Kacie
Lyn 8h20'32"48d2'
Brennan,Sean Patrick
Per 3h24'37"50d0'
Brennan,Sierra Nicole
Cas 0h26'29"70d55'
Brennan,Somer Leigh
Peg 22h0'48"35d1'
Brennan,Sr,Per Sempre,
Michael G
Hya 9h7'21"5d60'
Brennan,Susan
Lyn 8h53'29"34d8'
Brennan,Thomas
Eri 4h26'0"-1d57'
Brennan,William C
Ori 6h5'0"8d54'
Brenneisen,Louis William
Per 2h43'0"41d4'
Brenneisen,Summer John
Aur 7h4'47"40d24'
Brenneke,Frauke
Mon 7h50'54"-5d58'
Brenneman,Bonnie Lynn
And 0h52'31"38d53'
Brenneman,Conrad George
Cnc 9h1'0"30d22'
Brenneman,Steven Lynn
Leo 10h27'34"11d19'
Brenner Family,Carl A
Cet 0h53'52"-2d26'
Brenner's Shining Star
Uma 9h7'44"61d26'
Brenner,Alexander Spencer
Boo 14h39'34"20d12'
Brenner,Beth Fuchs
Lyn 7h55'18"47d53'
Brenner,Daniel S
Boo 14h51'45"33d12'
Brenner,Darrin
Del 20h19'39"10d25'
Brenner,Douglas Francis
Aql 20h3'43"7d31'
Brenner,Hans
Cnc 8h9'33"31d29'
Brenner,J Patrick Michael
Byron
Com 13h3'18"20d21'
Brenner,Pauline
Cyg 19h27'26"34d29'
Brenowitz,Scott & Andrea
Cyg 19h24'53"31d55'
Brent & Amanda's Emerald
Star
Sge 19h3'42"20d9'
Brent & Brenda Always
Hya 9h28'41"-6d27'
Brent & Mariana
Sge 18h58'30"19d49'
Brent Can Be Loved By
Jeanette
Eri 2h51'57"-6d38'
Brent's & Amy's Forever Love
Cyg 19h41'54"30d59'
Brent's Claim
Dra 10h28'40"80d42'
Brent's Passion-LuPus Oculus
Del 20h29'1"10d21'
Brent's Place
Boo 14h13'18"38d51'
Brent's Star
Cnv 13h47'27"40d29'
Brent,Anthony Jason
Hya 8h52'34"-06d23'
Brent,Major
Cet 0h57'55"0d43'
Brentlinger,David A
Ori 5h44'30"10d39'
Brenton,Emily Ann
And 0h29'1"44d35'
Brenza,John & Virginia
Uma 11h57'28"52d6'
Brer
Cet 1h11'1"-6d55'
Breschi,Carlo Maria
Aur 5h52'19"50d12'
Brescia,Catherine Larissa
Cas 0h57'40"76d17'
Bresciano,Ada
Cas 0h6'47"66d11'
Bresie,Jackie
Com 12h51'1"27d11'
Bresin,Leah Marie
And 1h48'53"39d27'
Breslau,Muriel & Harry
Ori 5h10'58"-4d1'
Breslaver,Jeffrey Steven
Per 3h42'26"51d51'
Breslin,Elizabeth Margaret
Lyn 8h55'56"38d15'
Breslin,Karen L
Lmi 10h19'46"30d36'
Breslin,M James
Cet 1h44'19"-5d45'
Breslow,Jeffrey (DT)
Dra 15h16'40"56d32'
Bresnahan,Dorothy L
Lac 22h35'1"53d42'
Bresnahan,Kevin William
Uma 8h46'1"55d7'
Bresnen,Kathleen
Peg 22h59'32"31d9'
Bresolin,Justin
Cep 0h13'27"73d15'
Bressani,Gloria
Gru 22h23'37"-54d53'
Bresser,Leo
Eri 4h49'22"-5d54'
Bresser,Rolf
Sgr 19h57'0"-44d11'
Bressler,Heide
Leo 11h51'23"23d42'
Bret Christian September
16,1991
Leo 10h50'45"-1d56'
Bret,Marie
Sex 10h7'19"-1d18'
Bret-Day,Timothy Laurence
Lyn 9h7'30"37d33'
Bretag Christine Marie
Mon 7h18'46"-6d29'
Bretag,James John Henry
Boo 14h49'19"24d12'
Bretange,Christiane
Ser 15h17'41"11d55'
Brethon,Claude
Cnv 13h28'14"38d21'
Brethous,Claude
Crb 16h6'10"37d51'
Breton,Berthe et Jean-Denis
Ori 5h56'58"16d11'
Breton,Juliette
And 0h36'1"38d58'
Breton,Yvette S
Cas 0h55'1"60d47'
Brett
Dra 10h18'57"74d60'
Brett & Ellen's Anniversary
Star
Crb 16h0'23"33d53'
Brett & Kristin
Umi 16h31'57"78d54'
Brett J
Lmi 10h15'39"31d40'
Brett Nathan & Mandy Leigh
Sge 19h17'50"18d53'
Brett Star
Peg 21h40'1"26d37'
Brett The Bear
Uma 8h57'34"52d4'
Brett's Lucky Star
Boo 13h34'18"21d16'
Brett's Star
Oph 17h2'45"10d7'
Brett's Star
Uma 10h15'54"58d44'
Brett's Star
Uma 9h1'0"53d15'
Brett,Carl & Loretta
Eri 3h23'0"-6d53'
Brett,Donna Gibbons
Lyn 7h45'59"42d47'
Brett,George C
Oph 18h1'15"11d16'
Brett,Jackson Richard
Her 16h45'23"47d35'
Brett,Stephen
Dra 16h45'26"60d43'
Brett-Jillian
Cas 0h33'52"66d17'
Brett-`My Destiny', Paul F
Dra 16h23'11"60d19'
Brettchan
Cnv 12h13'38"35d50'
Brette,Yohanna
Sgr 19h33'14"-31d26'
Bretthauer,Carol
Vul 21h13'21"20d5'
Brettle,Richard Alan
Her 16h27'54"40d33'
Bretz,John A
Hya 8h46'32"0d21'
Bretz,Robert H
Del 20h27'1"10d20'
Bretz,Sarah Elizabeth
Peg 22h26'52"20d6'
Breucha,Maiki B M
Psc 1h3'53"17d36'
Breuell,Thomas
Sgr 19h4'13"-24d12'
Breuer,Gary & Maria
Lyr 19h6'15"25d54'
Breuer,John Frederick
Oph 17h25'41"-1d16'
Breuner,Jens
Dra 16h59'0"66d58'
Breunig,Bettina
Gem 6h37'47"20d7'
Breunig,Diane
Vul 19h12'30"22d9'
Breuning,Gerhard
Lyn 7h28'43"39d59'
Breutling,Monika
Vir 12h3'57"11d51'
Brevelle,Frances V
Hya 9h8'0"4d13'
Brevig,Ryan
Aur 5h53'27"29d24'
Brevik,Jr,John Eric
Mon 7h1'56"5d21'
Brevitt,David
Ori 5h56'24"7d51'
Brevoort
Her 18h16'30"20d14'
Brewer,Acey James
Tau 5h49'30"28d47'
Brewer,Adam
Umi 15h34'27"66d17'
Brewer,Amanda Nicole
Del 20h14'41"10d35'
Brewer,Angele M
Hya 8h59'50"-6d21'
Brewer,Ashton Matthew
Sgr 18h48'51"-28d46'
Brewer,Charles Richard
Cmi 7h44'1"3d48'
Brewer,Denise
Com 12h27'26"21d50'
Brewer,Everette
Cyg 21h14'39"28d26'
Brewer,Hobie
Cet 2h29'55"3d1'
Brewer,James & Clara
Eri 3h42'31"-16d0'
Brewer,Joseph Levi
Per 1h47'1"52d52'
Brewer,Jr,Edwin Ray
Ser 18h15'32"-0d7'
Brewer,Jyl Elizabeth
And 0h30'0"40d21'
Brewer,Marjorie
Cyg 21h32'38"53d19'
Brewer,Natalie Alisa
Mon 7h44'46"-7d13'
Brewer,Nikolas Tyler
Aql 19h6'32"3d4'
Brewer,Robert C
Hya 9h32'0"5d23'
Brewer,Ruth E
Sge 20h3'0"20d17'
Brewer,Sarah Jessica
Lyn 8h7'51"37d36'
Brewer,Scott H (Dusty)
Aur 6h3'11"45d59'
Brewer,Tania Lyn
Peg 22h35'41"21d59'
Brewer,Tony
Gem 6h38'15"13d59'
Brewer,William Buddy
Aql 19h7'1"1d32'
Brewi,David
Vul 21h24'48"26d19'
Brewi,Gerard Michael George
Aur 5h10'34"43d36'
Brewis,Chris
Lyn 9h13'29"39d4'
Brewster,Alex
Aur 6h35'0"33d19'
Brewster,Corinne
Mon 7h56'36"-2d26'
Brewster,Donald Wayne
Oph 17h5'12"11d15'
Brewster,Kerri May
Cyg 19h26'18"33d19'
Brewster,Larry
Aql 18h42'1"-0d29'
Brewster,Lillian Madden
Aql 19h56'21"13d28'
Brewster,Punky
Per 3h46'33"37d25'
Brewster,Ruth I
Cas 1h49'26"60d12'
Brewster,Sarah
And 23h38'1"47d17'
Brey,Kurt Harrison
Umi 14h58'20"71d8'
Brey,Kyle Cameron
Lmi 10h33'55"28d32'
Drey,Susan Deth
Sct 18h47'37"-7d56'
Breyzen Bad Bear
Umi 16h19'59"72d25'
Breza's 40th,Michael J
Dra 16h40'10"61d11'
Brezinski,Tom & Connie
Cap 21h34'33"-21d43'
Breznican,Cyril Andrew
Cyg 19h22'19"44d41'
Brezniceanu,Dan Mihai
Cyg 20h12'1"43d13'B
Bree & John's Endless Love
Star
Cyg 19h27'27"33d18'
Bri
Her 18h2'28"28d15'
Bri' Star
Cep 22h7'23"60d39'
Bri's Farewell Tour
Sge 19h57'14"20d22'
Bri-Ang's Dream Keeper
Lyr 18h49'15"33d52'
Bri-Jes Car-Kal Chickabee
11994
Uma 9h11'31"56d10'
Bria Dawn
Cma 6h53'28"-16d19'
Bria,Anthony Francis
Lac 22h40'41"51d3'
Bria,Mary S & Pasquale L Bria
Dra 15h48'14"51d40'
Briamar XXVI
Cyg 19h56'48"45d20'
Brian
Aur 5h40'45"54d30'
Brian
Aql 18h57'15"-6d30'
Brian
Per 4h25'14"50d20'
Brian
Aur 5h5'36"40d21'
Brian
Cap 21h22'55"-14d57'
Brian
Dra 14h43'56"60d8'
Brian
Boo 14h48'22"32d32'
Brian
Boo 15h0'1"14d22'
Brian
Del 20h13'28"15d33'
Brian
Cep 22h21'36"58d22'
Brian
Her 17h15'30'18dd51'
Brian
Dra 18h56'14"58d34'
Brian
Peg 23h38'47"32d17'B
Brian
Cep 23h6'16"70d31'
Brian
Dra 14h58'29"62d13'
Brian & Amy
Peg 23h0'49"8d11'
Brian & Annie
Com 12h37'51"23d6'
Brian & Annie Forever
Cyg 21h4'1"52d55'
Brian & Christina
Per 2h50'55"37d33'
Brian & Desra's Fire & Ice Star
Peg 22h36'24"8d38'
Brian & Jennie
Her 17h21'1"27d53'
Brian & Kim Star,The
Cyg 19h31'53"33d9'
Brian & Lisa— October 1
Cyg 20h24'0"38d41'
Brian & Mara
Cyg 21h52'12"52d53'
Brian & Melissa
Ori 5h35'1"0d6'
Brian & Michele
Sge 19h2'37"18d42'
Brian & Phaedra
Boo 14h46'34"32d51'
Brian & Rebecca
Cyg 21h2'48"31d45'
Brian Allan
Per 3h5'12"40d37'
Brian Christopher
Her 16h17'52"20d7'
Brian Christopher
Per 2h58'34"40d48'
Brian Craig L
Dra 16h59'57"51d25'
Brian Daniel
Cap 20h25'51"-12d18'
Brian Forever Yours
Her 17h54'36"28d17'
Brian G
Cma 6h55'4"-19d36'
Brian H
Cyg 21h26'50"31d49'

Brian James
Mon 6h38'24"9d30'A
Brian James
Aur 4h55'15"40d2'
Brian James II
Lac 22h39'57"56d43'
Brian Jeffery
Dra 9h34'16"81d18'
Brian Joseph
Aur 6h26'12"33d38'
Brian Kristopher Michael
Aqr 22h12'48"0d8'
Brian Loves Karyn
Lyr 19h22'51"30d25'
Brian Matthew
Ori 6h0'1"1d29'
Brian Michael
Dra 19h27'51"60d60'
Brian Patrick
Boo 13h34'0"18d38'
Brian Reddin's Nitro Star
Uma 10h40'43"40d55'
Brian Zachary
Cnv 12h30'28"40d31'
Brian's
Boo 14h59'34"30d59'
Brian's Brillance
Aur 5h59'0"29d48'
Brian's Dream
Boo 15h4'50"38d46'
Brian's Enterprise
Ser 12h29'30"62d28'
Brian's Eye
Peg 22h34'53"11d19'
Brian's Fiery Moon
Sex 10h20'49"-6d19'
Brian's Paradise
Ori 4h52'53"5d7'
Brian's Twinkling Tippy
Sgr 18h55'4"-28d49'
Brian,Rachael,Adam, Laura
Peg 23h1'21"33d25'
Brian,The Light Of My Life
Sex 10h13'33"-0d37'
Brian-Lisa
Del 20h21'38"3d43'
Briana
Oph 17h54'38"10d53'
Briana
Peg 21h45'14"23d40'
Briand,Carolyn Lesley
Eri 2h47'59"-3d31'
Briand,Jr,William James
Cet 1h46'17"-5d46'
Briando Stella
Oph 18h29'32"7d53'
Brianna
Cen 11h50'16"45d47'
Brianna Angelina
And 23h31'0"44d56'
Brianna Kathleen
Del 20h52'60"3d42'
Brianna Leigh
Peg 0h1'41"10d37'
Brianna Nicole
And 1h53'18"40d28'
Brianna Renee
Lyr 18h24'49"45d12'
Brianna's Brightest
And 0h13'43"32d54'
Brianna-Shauna
Cyg 20h37'31"53d14'
Brianne
And 1h24'14"38d33'
Briano,Caroline
And 23h2'26"50d44'
Briantais,Alexandre Lloyd
Aur 5h16'30"41d29'
Briareus
Mon 6h29'59"-10d49'
Briargreen Public School 25th
Cas 0h1'11"61d9'
Briaura,Amore infinite di Dio
Peg 22h13'47"5d16'

Brice,Ali
Boo 15h0'42"42d13'
Brice,John Harold Victor
Per 2h10'0"57d50'
Brice,Sylvia Delores
Eri 4h33'22"-10d14'
Bricino
Her 17h8'33"48d12'
Briciola
For 2h32'51"-25d45'
Brick
Pup 7h56'42"-27d18'
Brickel,Shannon Leigh
Peg 23h42'19"30d29'
Bricker,Elanor Jane
Uma 10h20'60"58d60'
Bricker,Kenneth G
Umi 15h16'52"69d35'
Brickey,Hunter Gideon
Ser 17h52'45"-14d31'
Brickey,Jennifer Louise
Lep 5h52'59"-20d9'
Brickey,Marjorie Jean
And 23h38'60"40d1'
Brickey,William Christopher
Aql 20h11'22"14d27'
Brickey,William Thomas
Per 2h50'29"45d4'
Brickfield,Edmund C
Ori 5h35'30"-2d28'
Brickler,Alexander D
Oph 17h51'23"12d33'
Brickley,Abbie Victoria
Psc 0h57'27"18d30'
Brickley,David
Her 18h19'14"12d23'
Brickley,George Higgins
Gem 7h18'24"24d34'
Brickley,Hannah Rose
Aqr 22h47'34"-6d59'
Brickley,Jenna Marie
Aqr 22h47'43"0d20'
Brickley,John Gallagher
Tau 5h22'50"18d58'
Brickley,Sarah Jane
Gem 6h40'36"31d44'
Brickmeier,Kathleen "KB"
Cas 0h2'49"65d1'
Brickmeier,Paul Charles
Aql 19h16'36"14d10'
Brickner,Autumn Eve
Lyn 7h40'31"40d43'
Bridar,Ji
Uma 12h0'16"32d14'
Briddle,James Michael
Cmi 7h6'8"7d1'
Bride Leah,The
Vul 20h18'51"23d49'
Bride,William M
Umi 16h6'56"70d48'
Brideau,Jeffrey A
Aur 4h59'32"48d47'
Brideau,L'étoile du Gémeau-Renée
Gem 7h1'39"24d21'
Bridenstine,Rich & Patricia
Per 4h25'1"51d14'
Brides Magazine
Cas 2h33'41"57d26'
Bridge,Anna
Peg 22h49'21"21d32'
Bridge,Helen Theresa
And 23h36'46"42d24'
Bridge,Marilyn
Her 17h24'41"38d53'
Bridge,Ralph Edward
Per 3h8'0"47d36'
Bridge,Ron "Bear"
Her 17h5'28"37d50'
Bridgeman,David Paul
Tau 4h25'23"28d50'
Bridgeman,Gregory John
Lyr 19h9'12"37d51'
Bridgeman,Sara
Lyr 18h30'33"32d51'

Bridger,Des
Uma 10h16'34"54d60'
Bridges
Lyr 18h57'39"30d50'
Bridges Point to View, James E
Tau 4h22'31"30d1'
Bridges,Ben
Lyr 18h57'13"37d18'
Bridges,Billye Sue
Cas 0h8'16"62d11'
Bridges,Cole J
Ser 15h37'48"1d36'
Bridges,Gail Ann
Lmi 10h46'17"32d56'
Bridges,Heather
And 0h17'1"38d0'
Bridges,Joshua Luke
Her 15h50'58"43d51'
Bridges,Joy
Tau 4h3'1"11d42'
Bridges,Kenneth Allen
Sex 10h37'14"0d21'
Bridges,Susan
Uma 11h41'58"38d44'
Bridges,Tim R
Cep 22h59'29"57d59'
Bridges,Wilbern D
Per 1h46'31"53d43'
Bridget
Lyr 18h48'42"42d24'
Bridget,Johanna
Peg 23h0'1"21d41'
Bridgette
Lyn 8h0'13"50d4'
Bridgette
Cyg 19h34'23"35d21'
Bridgette "Wilson 9"
Oph 18h40'58"7d56'
Bridgewater,Kelly
Cam 11h48'21"80d34'
Bridgman,William
Dra 12h6'46"68d5'
Bridle,Kristian
And 23h36'24"41d40'
Bridport
Cyg 20h20'34"30d8'
Bridwell,Vyda & George
Umi 13h40'16"72d0'
Brieana Kay
Cas 0h33'1"61d12'
Brieanna
Lyn 9h3'0"46d1'
Brieler,Wilhelm
Uma 9h20'45"47d9'
Brielle Ami Simels
And 0h30'0"40d27'
Brien,Carolyn Louise
Aql 20h5'41"1d33'
Brien,Daniel Marc
Gru 22h10'17"-55d43'
Brien,Teresa Smith 'O'
Lyr 18h19'13"44d51'
Brienne Lee
And 23h39'29"40d15'
Brierley,John E
Ori 6h5'11"1d45'
Briers,Peter
Ori 6h7'22"8d54'
Brierton MBA MPH 1995, Robin Anne
And 2h27'1"48d2'
Brierton Star,The Karl & Claire
Crb 16h11'29"28d22'
Brigadoon
Aur 5h56'18"37d42'
Brigance,Christopher James
Peg 0h11'0"14d22'
Brigette
Del 20h54'31"8d26'
Brigette's Star
Cas 2h34'33"71d10'
Briggs,Aaron Jeffrey
Ori 5h50'25"18d51'

Briggs,Barb & Mike
Tri 1h44'34"31d16'
Briggs,Betsy
Ori 5h3'37"14d31'
Briggs,Candee
Gem 7h29'18"31d14'
Briggs,Christopher Roy
Lac 22h0'0"51d28'
Briggs,Derik Paul
Aur 5h5'41"40d7'
Briggs,Gary Philip
Dra 16h53'56"51d44'
Briggs,Helen Brewster
Sex 10h0'31"-2d10'
Briggs,Jesse
Hya 8h13'35"2d39'
Briggs,John Andrew
Boo 14h58'23"27d47'
Briggs,Jr,Henry Peter
Her 18h1'55"40d19'
Briggs,Marcus Stephen
Cep 22h6'1"53d9'
Briggs,Nina
Uma 11h42'0"58d39'
Briggs,Norman D
Ori 4h54'58"-0d12'
Briggs,Paul Francis
Dra 16h54'1"52d0'
Briggs,Paul Joseph
Cyg 20h0'29"40d58'
Briggs,Rev Dr Drucilla A
Cap 21h41'24"-22d43'
Briggs,S W "Ted"
Her 18h17'17"20d17'
Briggs,Sharron M
And 0h16'18"45d60'
Briggs,Sr,Ken
Oph 17h32'54"-1d29'
Briggs,Stephanie Kay
And 0h21'37"37d39'
Briggs,Stephen Christopher
Aur 5h2'28"51d5'
Briggs,Wendy Ann
Peg 23h19'18"33d20'
Briggs-My Kind of Music Star,Tom
Lyr 18h49'58"43d4'
Brigham Star,The
Ori 5h43'35"0d35'
Brigham,Ande Beth
Cas 0h49'47"67d44'
Brigham,Fro
Aql 19h52'1"14d38'
Brigham,The Goalie Star Kimberly H
Cep 23h33'12"70d18'
Brigham,Wingnut Aka Kelly Joy
Cas 0h45'45"68d20'
Brigham-Valedictorian 1993,Jennifer
Cam 6h41'57"68d29'
Bright 003 "Dittos"
Ori 5h0'22"1d50'
Bright Angel II
Sex 10h46'20"0d5'
Bright As Brian
Boo 14h27'1"20d31'
Bright Eyes
Lyn 7h1'52"50d31'
Bright Guardian
Lac 22h8'38"51d34'
bright nice secret
Leo 10h28'19"12d23'
Bright Star
Ori 5h47'27"11d35'
Bright Star
Ori 6h6'0"3d56'
Bright Star of Yasuhiro & Chiharu
Gem 7h2'15"15d23'
Bright Turtle
Ori 5h52'48"10d1'
Bright,Dr Frank V
Cma 6h30'35"-13d37'

Bright,John
Her 16h34'13"51d5'
Bright,Laura
Lyn 8h27'45"48d44'
Bright,Nicky
Uma 11h23'19"31d21'
Bright,Patricia
Cyg 20h37'36"53d4'
Bright,Philip Nathaniel
Sex 9h54'32"-1d17'
Bright,Sierra Heath
Uma 11h12'0"71d46"
Bright,Sr,William Charles
Boo 13h42'13"18d43'
Bright,William Shippen
Cep 0h13'26"69d36'
Brightbill III,David Ellis
Mon 6h43'22"-4d22"
Brightbill,Beth & Terry
Cyg 19h40'36"41d55'
Brightest Star Of All Beverly My Mom
Uma 11h3'0"40d32'
Brightness
Cas 23h4'33"58d41'
Brightness for Sweet Couple
Tau 5h13'44"18d18'
Brightness of Hitomi
Ari 2h19'12"15d45'
Brighton,Doris & Larry 50th Anniversary
Cyg 19h29'48"30d21'
Brighton,G Renfrew
Dra 18h13'39"80d18'
Brighton,Michelle
And 23h30'15"39d53'
Brightwell's Gold B E A D S
Uma 9h7'29"57d51'
Brightwell,Jim
Cet 1h14'19"-0d12'
Brightwell,Lori Lee
Peg 22h32'57"25d27'
Brigis,Anthony X
Aur 4h58'1"48d53'
Brigitt Mishel
Mon 8h3'37"-5d32'
Brigitta-Galactica
Boo 13h58'59"13d23'
Brigitte
Uma 9h47'47"51d23'
Brigitte
Gem 6h46'37"14d57'
Brigitte Augustine
And 23h27'44"48d46'
Brigitte et Christophe
Dra 17h57'1"70d48'
Brigitte Et Jean
Tri 1h41'37"28d30'
Brigitte et Marc
Uma 8h54'1"54d47'
Brigitte Klaudia
Lac 22h31'29"37d48'
Brigitte,Lucia
Lib 14h57'46"-20d3'
Briglands
Peg 21h40'25"27d53'
Briglia,Trudy
And 1h42'22"37d41'
Brignac,Roxanne Antionette Istre
Peg 22h0'57"25d49'
Brignac,Thomas Joseph
Cet 0h53'27"0d54'
Brignull,Ciss Olive
Com 13h29'44"26d32'
Brigstock's Star, Paula
Ori 6h6'40"0d30'
Briguglio,Anthony M
Cep 22h28'10"58d52'
BRIJON
Peg 23h5'1"20d32'
BriKel
Sct 18h46'20"-6d12'
Briles,John
Her 16h49'11"37d13'

Briley,Ann
Lac 22h19'50"51d35'
Brilhart-N-Nikita, Linda L
Lyn 9h37'25"40d11'
Brill,Ashley Rose
Cas 0h28'59"62d35'
Brill,Dorothy Y
Cas 0h54'41"58d31'
Brill,Joan & Sol
Aql 19h9'44"5d20'
Brille Dans Tout
Cmi 7h27'1"0d43'
Brillhart,11-11-71 Timothy
Per 2h57'18"40d25'
Brilliance By Burgess
Her 17h30'10"48d50'
Brilliant
Cra 18h32'17"-43d56'
Brilliant Marion
Lyr 18h56'27"31d24'
Brilliant Randy
Per 2h1'32"50d31'
Brilliant Rebecca
Uma 11h32'55"47d16'
Brilliant,Barbara
Eri 4h4'45"-2d19'
Brilliant.Geraldine (Mrs B)
Uma 9h3'35"71d4'
Brilliantly Beautiful Kerry
And 0h11'24"39d2'
Brillus,David Michael
Dra 16h17'48"63d13'
Brim,Holly Thurman
Aql 19h56'33"11d7'
Brimeister
Leo 10h54'0"18d19'
Brimley 1776,Sally Elizabeth
Ori 5h56'20"16d52'
Brimmell,Bradley
Tau 4h4'55"11d37'
Brimmer-Now & Forever, Vicki & Dave
Cyg 20h21'0"41d45'
Brin's Light
Gem 6h49'57"14d14'
Brin,Sarah
Mon 8h6'2"-7d19'
Brinckerhoff,Inger Hansen
Cas 1h32'24"60d15'
Brinckerhoff,Laurrie Jane
Cnc 8h10'47"30d48'
Brinda Kay
Com 13h3'21"18d14'
Brindamour,Laura Nicole
Cas 0h23'45"61d22'
Brindise II,Ralph S
Lmi 9h41'0"40d9'
Brindisi,Marlys Jean
Aql 19h0'52"16d32'
Brindl Psyche Willow Meadow Rainsong
Mon 7h59'53"-2d45'
Brindle,Carolyn
And 22h57'57"38d19'
Briner,Edwin Stanley
Dra 13h9'27"68d56'
Briner,Kurt
Aql 20h10'38"10d40'
Bringard,Bertrand
Per 3h58'13"35d40'
Bringham,Jay
Cet 0h49'37"1d26'
Bringsken,Jörg
Aqr 21h56'33"0d36'
Brininstool,Joey
Cep 21h18'0"55d36'
Brink,Abbey Machelle
And 23h3'0"44d34'
Brink,Henrik
Umi 13h26'29"70d5'
Brink,Holly Kristina
And 2h23'48"42d47'
Brink,Laurey
Cas 1h30'12"58d27'

Brink,Natalie Marie
And 0h50'26"34d48'
Brinkenhoff,Michael
Cet 2h54'0"1d36'
Brinker,John J & Norma C
Cyg 20h44'46"42d35'
Brinker,Kim
Aql 20h11'0"10d32'
Brinker,Kirsten
Cyg 21h34'55"40d60'
Brinker,Siegbert
Lib 14h30'31"-23d15'
Brinkerhoff,Jon Barton
Lac 22h53'59"51d23'
Brinkley,William E
Lyn 7h33'19"58d45'
Brinkman,John C
Sct 18h52'3"-6d9'
Brinkman,John C
Boo 14h17'50"18d9'
Brinkman,John Patrick
Her 17h27'57"26d54'
Brinkman,Krysta Rae
Ori 5h8'33"-5d38'
Brinkmann Star,The
Uma 10h52'45"38d48'
Brinkmann,Karl
Ori 5h49'34"20d56'
Brinley,Teenangel 1, Amanda Cathleen
Peg 22h46'51"35d27'
Brinsfield,Lauren Anna
Peg 22h41'52"34d26'
Brinsley
Lmi 10h48'41"26d59'
Brint,Casey Ray
Umi 13h7'34"72d21'
Brint,Jr,Donald Dean
Her 18h44'52"12d7'
Brinton,Howard
Oph 17h59'39"14d9'
Brintzenhofe,Richard
Aur 5h26'33"31d41'
Brinza,David E
Cyg 20h5'21"40d36'
Brion Roberto
Uma 10h28'33"48d43'
Briones,Maurice "Oso"
Ori 4h53'32"4d33'
Briony
Del 21h3'23"12d38'
Briony Helen
Lyr 18h29'22"44d25'
Briony-Denise
Boo 14h17'52"52d40'
Brior,Bob & Carolyn
Aql 20h2'33"7d59'
Brip
Lyn 7h8'45"55d53'A
Briquette
Umi 10h45'1"40d22'
Brisbane Extra
Pho 23h57'8"-47d34'
Brisbane,Thomas Milne
Cep 23h11'29"61d11'
Brisbin,Robert
Mon 6h44'26"11d2'
Brisbourne,Tina
Ari 2h27'41"22d7'
Brisby
Aql 19h30'0"7d35'
Brisco,Prudence
Uma 14h3'1"61d33'
Briscoe,Galen
Leo 10h57'1"20d59'
Briscoe,Veronica
Mon 7h43'0"-6d21'
Brisebois,Annie
Lyn 7h39'28"39d43'
Brisendine,Jacob
Aur 7h18'33"41d49'
Briseno,In Memorium of Adena
Umi 15h52'22"64d15'

Brisgel,Edward
Ser 16h2'46"4d45'
Brish,Etty
Ori 5h58'17"16d25'
Brishagen 12/21/68, Anna Cecilia
Sgr 19h29'14"-38d15'
Brisker,Sandra Kay Abercrombie
Mon 7h3'5"-7d6'
Briski,Rory G
Dra 15h54'1"65d56'
Brisley,Bill
Lib 15h3'1"-28d15'
Brisport,Beryl
Umi 14h56'16"65d54'
Brissel,Trish Bell
Lyn 6h31'1"54d18'
Brissette,Anne
Cyg 19h42'11"41d56'
Brissey,Adam Chase
Aur 4h59'25"40d1'
Brissey,Cheryl Dunlop
Cas 1h0'55"55d20'
Brissey,Edwin Jack
Per 1h40'10"52d57'
Brissey,Heather Harriet
Lyn 6h31'46"56d10'
Brissey,Ian Thomas
Per 1h40'58"52d30'
Brissey,Isabelle Briggs
Cas 0h35'1"54d32'
Brissey,Jeffrey Lee
Per 1h44'12"52d48'
Brissey,Nicholas Jack
Cep 21h6'18"58d40'
Brissey,Rachel Skipper
Cas 1h0'56"55d52'
Brissey,Thomas Edwin
Per 1h40'48"52d48'
Brister Wedding Star, The Bonnie & Bubby
Cyg 20h57'58"31d32'
Bristow,Brian
Per 2h52'16"46d28'
Bristow,Jim
Dra 18h1'34"67d31'
Bristow,Mark E
Aql 19h57'55"0d9'
Bristow,Sandra
Cyg 20h51'34"38d44'
Brita
Com 12h23'51"20d17'
Brite,Cheryl A
Com 12h17'28"32d46'
Britnee Celestial Aurora
Sge 20h1'33"17d28'
Briton,Heather
Lyr 18h18'44"40d45'
Britoria
Lyr 18h32'43"47d25'
Britt
Lyr 18h55'25"40d7'
Britt
Uma 10h13'51"71d18'A
Britt & Jan
Cmi 7h18'28"4d9'
Britt's Neon Moon
Crb 16h12'0"32d42'
Britt's Wish
Cnc 8h54'22"17d43'
Britt's Wish
Peg 23h0'34"11d21'
Britt,Beth
And 0h52'45"38d26'
Britt,Katherine
Peg 22h1'16"33d33'
Britt,Scott A
Cep 0h24'28"78d3'
Britt,Shari
Uma 12h32'26"59d24'
Britt,Stephen & Denise
Cyg 19h29'17"40d33'
Britt-Lee's Stargate
Cet 3h14'48"0d40'

Britta
Vir 11h58'17"2d1'
Britta
Eri 4h14'48"-12d2'
Britta & Sophia
Cyg 20h52'53"37d52'
Britta Andrea S
And 23h2'43"48d29'
Britta John
Per 4h42'28"41d11'
Brittain,Janice
Cam 6h59'13"67d52'
Brittain,Janice
Uma 10h10'26"62d12'
Brittain,Kristie Kay
Vul 20h19'15"23d27'
BrittAlex
Uma 9h45'17"70d26'
Brittan,Chantal Pia
Cyg 20h2'48"41d8'
Brittan,John Scott
Ser 15h54'37"4d58'
Brittany
Uma 13h8'25"62d37'
Brittany
Tri 2h22'50"35d39'
Brittany
Cyg 20h40'38"30d58'
Brittany Irin
Hya 8h35'0"-6d34'
Brittany Lee
And 0h8'23"34d42'
Brittany Nicole
Cas 2h36'17"60d44'
Brittany Rose
Aqr 23h37'41"-12d12'
Brittany Suzanne
Mon 8h8'22"-4d54'
Brittany Yvonne
Cas 0h50'39"64d1'
Britten,Hilary
Peg 22h11'28"3d38'
Brittie's Star Forever
Eri 3h10'30"-1d44'
Brittingham,Blanche
Cet 1h43'43"-2d38'
Brittingham,Cindy
Cas 2h31'28"60d58'
Brittingham,Guillermo M
Ori 5h55'12"-1d30'
Brittlyn Charlotte
And 0h51'12"21d40"
Brittman,Zachary Benjamin
Dra 15h9'23"64d48'
Brittney's Star
Peg 23h39'36"30d60'
Britton,Abby
Lyr 19h19'50"40d29'
Britton,Barbara
Com 12h56'26"14d47'
Britton,Barbara
And 0h40'52"40d4'
Britton,Christie Jo-Lynn
Eri 2h52'50"-5d47'
Britton,Daniel J
Aur 6h15'20"38d34'
Britton,Dona Lorraine
Uma 9h19'1"56d41'
Britton,Evan
Aql 18h57'38"-3d37'
Britton,Joanne
Cas 1h2'39"63d49'
Britton,Johann Jeffrey
Mon 6h34'51"4d43'
Britton,Ruth
Sge 20h6'53"16d47'
Britton,Victor Arthur
Cep 22h27'26"70d2'
Britz,Joseph E
Lyn 8h4'42"50d36'
Brix,Amanda
Peg 22h6h15"24d8'
Brixner,Veronika
Lyn 8h23'34"42d22'

Brizard,Holly Kate
Lac 22h25'34"54d8'

Brizendine,Ruth
Cas 22h59'51"54d17'

Brizland,Roy
Psc 1h28'1"31d18'

Brizoni
Uma 8h55'32"52d19'

Brkic,Adam Vlatko
Cyg 19h35'35"28d11'

Broad Star
Uma 10h28'35"42d3'

Broadbent,Walter
Her 17h19'0"14d57'

Broaddus,Kevin Paul
Uma 10h16'33"40d23'

Broadfoot,Joseph Simon Burnham
Cra 18h27'22"-44d36'

Broadfoot,Joseph Simon Burnham
Cra 18h37'4"-44d13'

Broadhurst,J D
Ser 15h34'29"-1d27'

Broadhurst,Larry
Cep 0h50'16"78d39'

Broadley,Jacqueline Ann
And 23h6'46"37d50'

Broadway Bound
Peg 21h58'39"2d4'

Broadway Plaza
Aql 19h1'25"5d5'

Broadway's Star,Tom
Boo 14h57'1"29d58'

Broadway,Lois Jean
And 1h41'45"48d57'

Broadwell's L'Avenir
Ori 6h8'15"4d39'

Broadwell,Helen
Cas 23h16'29"60d58'

Broatch,Fiona Anne
Cam 6h58'16"68d2'

Broc René
Aql 19h6'33"1d21'

Brocco,Buck Reno
Her 18h3'1"14d32'

Brocha's Star
Equ 21h1'15"3d11'

Brochet,Jean Michel
Oph 17h15'40"10d50'

Brock "Nanook", Candice Elaine
And 1h46'51"40d3'

Brock Star,The Laura & Max
Lyn 8h13'49"39d45'

Brock Star,The Sam
Aur 6h30'1"33d60'

Brock,Charles Wayne
Her 17h52'37"38d6'

Brock,Jenny
Lyr 18h46'0"33d26'

Brock,Kayla Morgan
Mon 6h34'44"10d18'

Brock,MD,William A
Boo 13h38'43"25d56'

Brock,Perri Dawn McDonald
Mon 6h20'25"-6d5'

Brock,Roy Richard
Ser 18h41'12"4d12'

Brock,Sherri Lynn
Cas 2h13'11"68d48'

Brock,Stephanie Rachel
Mon 6h27'23"10d18'

Brocke,Hans Jürgen
Ori 5h56'11"13d40'

Brocke,Julia
Aqr 22h9'37"-4d30'

Brockett,Linda
Hya 8h13'52"-5d54'

Brockett,Pauline
Cru 12h42'15"-56d32'

Brockett,The Rob Mar
Cyg 19h32'1"35d49'

Brockhoff,Alex
Mon 8h4'23"-6d25'

Brockie,Douglas `Chiefy'
Cep 23h11'29"60d51'

Brocklehurst,Eileen Ann Marie
Aqr 21h56'21"-6d27'

Brockman
Her 17h28'44"28d9'

Brockman's Future, Denise & Ed
Peg 22h0'26"23d44'

Brockman,Flo
Eri 3h55'54"-6d11'

Brockman,Mark J
Dra 15h8'49"58d39'

Brockman,Mary Ann
Del 20h54'28"9d6'

Brockmann,Reinhold
Sge 19h6'45"17d5'

Brockmeier,Lioba
Cnc 8h55'1"22d10'

Brockmeier,Marilyn Ruth
Com 13h12'36"22d20'

Brockschmidt,Bruce
Per 3h56'57"51d14'

Brockus,Tommy D
Lmi 10h40'53"26d5'

Brockway Star,The Mike
Ori 6h7'15"1d34'

Brockway,"Eros" Bruce
Aur 6h28'28"33d37'

Brockway,Elizabeth Hunt
And 2h33'53"39d40'

Brockway,James MacNeil
Cep 22h14'19"61d40'

Brockway,Leslie F
Cam 9h12'58"73d46'

Brockway,Merrill
Psc 0h58'20"21d2'

Brockway,Richard Paul
Cep 21h10'35"60d44'

Brockwell,Victoria E
Ori 6h16'1"8d1'

Brod,Jan Kirsten
Cyg 20h1'50"40d58'

Brod,Lothar
Boo 14h3'26"11d21'

Broda,Mary Patricia Talbot
And 0h0'1"47d16'

Broda,Michelle Elizabeth
And 22h57'0"51d23'

Brodak,Joseph R
Aur 6h0'1"38d30'

Brodbeck,Ernie
Per 2h19'12"58d51'

Brodbine,John James
Lac 22h29'19"52d31'

Broder Star
Com 13h5'11"16d53'

Broderick,Dr Patrick A
Her 16h58'0"23d50'

Broderick,John-Luke
Gem 5h59'1"27d51'

Broderick,Laura
Cas 1h22'57"53d28'

Broderick,Mary & John
Oph 17h55'14"12d39'

Broderick,Oonagh
And 23h0'60"50d19'

Brodersen,Marian
Ser 15h51'53"17d8'A

Brodersen,Randy
Ser 15h51'53"17d8'B

Brodeur,Andrew Douglas
Dra 18h31'25"68d56'

Brodginski,Todd M
Oph 18h6'55"8d38'

Brodhage,Steven E
Hya 8h39'33"-0d32'

Brodhecker,Casandra Jo
Aqr 20h37'59"-8d41'

Brodhecker,Julianna Rae
And 0h53'60"35d52'

Brodhurst,Jon Erik
Gem 6h37'49"12d45'

Brodie,Jacob
Cyg 19h55'13"41d3'

Brodkin,Dr Ronald H
Oph 17h32'59"7d30'

Brodowsky,Sanford
Cyg 20h42'50"47d47'

Brodsky,Anne & Danielle Wayman
Cas 0h19'49"61d44'

Brodsky,Bert E
Oph 17h55'46"7d42'

Brodsky,E Paul
Aql 20h34'56"0d32'

Brodsky,Ellen G
Peg 22h54'41"28d33'

Brodsky,Joe
Lac 22h24'28"50d20'

Brodsky,Selma
Lyr 18h38'46"35d37'

Brodsky-CMCSA,Julian A
Her 16h7'0"40d29'

Brodson,Nora J
And 0h53'15"39d24'

Brody & Pnas,JB
Dra 16h25'27"62d21'

Brody,Adam Douglas
Her 16h46'1"21d34'

Brody,Clifford G
Aur 7h24'33"37d36'

Brody,Dr Ervin C & Doris P
Cyg 20h1'0"40d48'

Brody,Hedy
Aql 18h59'1"-0d46'

Brody,Heidi Jane
Dra 11h27'47"78d52'

Brody,Jonathan
Cet 2h45'48"-0d4'

Brody,Kristin
Leo 11h34'47"8d6'

Brody,Mary Ruth
Lyr 18h49'11"37d49'

Brody,Michael Hyman
Peg 23h38'33"18d10'

Brody,Paul D
Her 17h2'56"45d40'

Brody,Reed K
Cep 22h22'14"58d41'

Brody,Thomas F
Eri 3h30'1"-6d43'

Brody-Hobbs
Lyn 7h33'0"52d1'

Broedell,Frank & Suzanne
Crb 15h18'47"30d50'

Broeder,Elaine
Cas 0h47'19"65d10'

Broehl,Steven Tyler
Boo 13h41'59"13d54'

Broekhuizen,Sr,Leo
Lmi 9h55'17"38d23'

Broennimann,Joelene M
Cam 6h14'56"67d47'

Brogan (Angell)
Peg 23h4'0"31d51'

Brogan,Jeffrey A
Her 17h0'14"40d34'

Brogan,John Terrence
Her 16h39'48"28d31'

Brogden,Brend L
Vul 19h14'52"21d59'

Brogden,Chrissa & Alex
Peg 23h8'18"14d15'

Brogden,Lauren
And 2h8'42"43d1'

Broggiato,Roberta
Uma 12h10'51"60d9'

Brogsitter-Finck,Chris & Gloria
Col 6h27'8"-39d44'

Brogère,Tessa & Jean-Luc
Her 17h18'0"42d54'

Broh-Kahn,L D K
Lmi 10h27'48"30d14'

Brohan
Cet 1h32'0"-0d34'

Brohawn,Laura D
And 2h36'34"38d20'

Broich,William J
Sex 10h9'16"-5d5'

Broj,Bethany Lynn
Cyg 19h27'45"32d55'

Brok
Dra 12h32'1"75d16'

Brokamp From Mellisa White,David J
Aql 19h3'35"2d25'

Broken Angel
Cyg 21h18'28"38d3'

Broken Arrow
Ori 6h8'45"9d2'

Broks
Uma 11h56'41"40d43'

Brola,Adam G
Lac 22h14'15"46d34'

Broll,Madeleine & Georges
Ind 20h26'12"-54d44'

Broman,Erwin
Dra 17h54'29"68d49'

Bromander,Magnus
Aur 5h6'42"40d0'

Brombach,Werner
Dra 20h28'53"68d5'

Bromberg,Liz
Aql 18h41'14"-2d48'

Bromilow (12/6/73), Sharon
And 23h16'52"35d35'

Bromilow,Lucy
And 2h34'43"40d8'

Bromley,Barbara Allen
Cyg 20h0'54"58d59'

Bromley,Devin Nicole
And 23h35'1"40d16'

Bromley,Jim
Her 16h10'0"25d35'

Bromley,Morgan Isaac
Ser 18h18'48"-14d54'

Bromley,Pat & John
Sge 20h3'1"20d6'

Bromley,Pete
Aql 20h19'39"5d30'

Bromley,Richard J
Lac 22h47'41"52d55'

Bromley,Victoria Kirschner
Mon 7h0'31"0d36'

Bromley,Walter & Karin
Cyg 19h31'25"36d32'

Bromley,Wm E (Sam)
Cep 20'58'0"60d17'

Bromm,Dorlies
Cap 20h6'26"-8d2'

Bromwich,Tammie
Cam 3h32'23"63d43'

Brondelli,Angelo
Her 16h43'0"48d24'

Bronie J
Cen 11h50'59"48d57'

Brondolin,Clement
Umi 16h21'21"72d26'

Bronfman,Jon
Uma 10h58'32"59d38'

Bronna
Mon 6h23'54"4d36'

Bronner,Irvin E
Lac 22h8'38"38d40'

Bronner,Kristopher Alan
Dra 17h21'28"73d15'

Bronner,Lorenzo- Rodolfo-Pietro
Lac 22h33'30"38d32'

Bronner,Michele- Francesco-Pio
Lac 22h33'52"38d59'

Bronntanas,Bhrendán
Cas 0h56'32"61d42'

Bronny,Almuth
Vir 13h34'44"-8d20'

Bronsema,Brand
Cnv 13h45'1"41d30'

Bronson
Cnv 12h21'32"35d56'

Bronson,Bernice
And 1h22'13"37d8'

Bronson,Dr Richard
Oph 17h55'28"10d57'

Bronson,Greer Caroline
Aqr 22h16'27"0d37'

Bronson,John F E
Boo 13h59'37"21d8'

Bronson,McCall Rose
Mon 7h46'38"-2d23'

Bronson,Stephanie R
Uma 10h44'13"41d5'

Bronstein,Laurie
Cas 3h6'18"68d52'

Bronstein,Sharon Jill
Cyg 21h3'53"36d44'

Bronston,Mia Lynn
Boo 13h54'54"20d33'

Brontë Selyn
Aql 19h6'38"3d24'

Bronwyn
Crt 11h9'50"-10d15'

Bronwyn
Cam 8h1'46"71d8'

Bronz,Ben
Ori 5h30'27"-3d7'

Bronze,Robert Michael
Cet 0h27'57"2d3'

Brook,El
Her 17h38'56"21d49'

Brook,Kellie Louise
Umi 13h46'14"70d32'

Brook,MD,Marven I
Aql 19h50'38"13d42'

Brook,Sarah Jane
Cas 0h33'1"61d20'

Brook,Stacey Lauren
Cas 0h23'1"72d24'

Brooke
Aql 18h57'17"11d7'

Brooke
Peg 21h26'42"22d57'

Brooke
Cet 1h41'35"-2d4'

Brooke
Peg 22h33'54"30d52'

Brooke "Bubba's Best Friend"
Equ 21h22'60"12d8'

Brooke Brilliant Bat Mitzvah,Melinda
Oph 17h57'10"10d1'

Brooke Jolene
Mon 6h29'11"-0d52'

Brooke Marlo
Cyg 21h38'37"38d53'

Brooke,Dawn & Hope
Lyn 8h9'15"47d8'

Brooke,Joe
Dra 17h22'11"64d6'

Brooke,Kathleen
And 0h59'22"41d8'

Brooke,Keith
Aql 20h1'29"4d7'

Brooke,Kristin
Mon 6h54'58"8d50'

Brooke,Mariel
Cas 0h18'59"60d13'

Brooke,Robert & Susan
Ser 16h1'24"8d31'

Brookelissalexisirving
Cyg 19h41'29"31d29'

Brooker,Gerald B & Susan
Her 17h35'1"20d7'

Brooker,Teresa Jeanette
Lyr 18h36'0"39d23'

Brookes,Amy
Aur 5h24'58"30d5'

Brookes,Carol Duke
Equ 21h3'38"11d4'

Brookes,David Spaulding
Mon 6h32'44"-1d40'

Brookes,Mary "Dad"
Uma 11h10'0"38d44'

Brookes-Totten,Casey
Lmi 10h50'29"32d30'

Brookfield-Dreyer
Ori 5h38'50"11d35'

Brooking,Carter
Lib 15h44'14"-24d57'

Brooking,Jayne
Cyg 21h17'35"35d38'

Brookings/Harbor
Uma 14h4'48"59d4'

Brookins,Lynn Anthony
Her 17h35'0"27d36'

Brookman,Beverly Jean
Vul 19h45'27"20d15'

Brookman,Catharine & Marc
Cnv 12h10'57"41d3'

Brookman,Corinne Marie
Aur 6h35'59"52d34'

Brookman,Michael Joseph
Aur 6h34'14"50d7'

Brookman,Ryan Michael
Cep 1h45'48"80d20'

Brooks
Cet 2h6'0"4d24'

Brooks
Uma 8h57'55"52d18'

Brooks "Big Bear", Robert T
Per 2h0'42"48d7'

Brooks "The River", Garth
Cyg 19h22'0"28d52'

Brooks & Julie Always & Forever
Cyg 19h46'53"30d48'

Brooks From Dani Girl, To Mark
Uma 9h4'1"57d58'

Brooks III,Lloyd Arthur
Uma 8h58'19"47d45'

Brooks The Big Star, Lonnie
Cep 22h28'1"58d23'

Brooks' Star,Irene
Cyg 20h24'39"37d35'

Brooks,Alma Jean
Peg 21h53'43"36d8'

Brooks,Anne M
Cet 2h49'20"6d32'

Brooks,Beatrice LeBlanc (Bebe)
Tau 5h52'54"23d25'

Brooks,Brandi Michelle
Mon 8h8'0"-5d50'

Brooks,Carol Ann
Umi 15h59'1"80d58'

Brooks,Colleen D
Vir 13h24'5"-8d58'

Brooks,Connie L
Vul 20h15'22"23d28'

Brooks,Cynthia Lynn
Sgr 19h8'25"-27d9'

Brooks,David
Ori 5h56'41"8d55'

Brooks,Deean L
Peg 22h44'30"20d58'

Brooks,Donna Carol
Aql 18h43'0"8d42'

Brooks,Donna Louise
And 0h14'31"46d3'

Brooks,Dora C
Tau 3h40'31"28d22'

Brooks,Douglas Scott
Equ 20h57'14"9d47'

Brooks,Duncan
Uma 11h24'49"56d22'

Brooks,Eternally Mark
Her 16h14'1"41d32'

Brooks,Floy Margaret
Eri 4h41'1"-8d52'

Brooks,Frederic Henry
Dra 14h20'59"63d50'

Brooks,Gary Scott
Aql 19h55'56"10d48'

Brooks,Georgia
Sgr 19h12'55"-28d2'

Brooks,Georgia
Dra 20h20'1"62d18'

Brooks,Gregory
Boo 14h14'40"15d49'

Brooks,Heather Lynn
Cma 7h15'0"-13d35'

Brooks,Hillary
Cas 23h42'21"61d54'

Brooks,James Joseph
Ser 15h30'25"10d20'

Brooks,Janine
Lyr 19h24'37"41d36'

Brooks,Jasmine Gina
Uma 10h58'58"35d22'

Brooks,Jason Nathaniel
Aql 19h56'30"10d12'

Brooks,Jennifer
Aqr 23h17'27"-6d25'

Brooks,Joanne
Eri 2h45'55"-5d38'

Brooks,John Raymond
Hya 9h36'35"-1d17'

Brooks,Kenneth Mark
Tri 2h13'46"31d44'

Brooks,Larry Gene
Cep 21h57'28"55d1'

Brooks,Lesley
Cyg 21h32'20"38d39'

Brooks,Loren
Peg 22h0'19"2d4'

Brooks,Marilyn Westphal
Tau 5h3'46"20d12'

Brooks,Marjorie Mae Morrison
Hya 10h43'56"-11d59'

Brooks,Mary Kathleen
Peg 22h6'16"28d48'

Brooks,Mary-Michaele
Uma 12h0'16"32d14'

Brooks,Matthew Seth
Uma 13h1'26"53d48'

Brooks,MD,Dr Andrew
Aqu 19h47'32"14d56'

Brooks,Naomi Ruth
Peg 23h8'28"8d54'

Brooks,Natalie
And 2h4'50"42d20'

Brooks,Olivia Kate
Cra 18h27'55"-42d7'

Brooks,Pam
Cas 1h44'29"71d12'

Brooks,Paula
Ori 5h57'24"10d56'

Brooks,Richard E
Dra 17h46'56"68d0'

Brooks,Riley Ann
Cas 2h5'27"61d31'

Brooks,Rita M
Ser 15h25'9"17d29'B

Brooks,Robert & Donna
Her 17h9'16"47d51'

Brooks,Ross
Cep 22h37'1"70d11'

Brooks,Ruth
Sge 20h0'56"16d16'

Brooks,Sarah Ann
Cnc 8h39'0"17d22'

Brooks,Sheila
And 2h25'48"42d23'

Brooks,Stephanie
Lyr 18h18'53"45d3'

Brooks,Stephanie Ann
And 1h52'0"40d1'

Brooks,Stephen
Cam 4h31'44"68d19'

Brooks,Stephen David
Peg 21h28'1"23d43'

Brooks,Susan Fran
And 0h22'23"43d41'

Brooks,Taylor Mayne Pearl
Mon 7h50'20"-1d56'

Brooks,The Star In My Sky-Arline
Cas 0h8'0"50d27'

Brooks,Thomas Charles
Sgr 19h5'11"-23d15'

Brooks,Timothy
Cam 6h9'1"60d52'

Brooks,Tommy
Aql 19h6'60"2d3'

Brooks,Vina
And 2h21'1"40d6'

Brooks-"Honey Bear", William D
Lyr 18h37'31"42d1'

Brooks-Cody-Fletcher
Uma 10h17'48"70d34'

Brookshaw,David
Per 2h59'0"41d14'

Brookshire's Brainstorm
Cnv 13h4'42"50d52'

Brookshire,Thomas Caske
Hya 9h37'55"-8d19'

Brooksie-"Our Daddy Man"
Cnv 12h56'18"38d28'

Brookstein,Barry M
Vir 13h35'32"7d10'

Broom,Angie
Cyg 20h17'45"39d8'

Brooman,Clare
Boo 15h8'24"31d29'

Broome,Jason
Cep 0h11'1"68d39'

Brooslin,Sharon
Cyg 20h59'11"30d13'

Brophy III,James Patrick
Ser 15h38'4"14d17'

Brophy,Clyde
Cmi 7h59'40"5d10'

Brophy,Mary-Beth
Lyn 7h55'54"52d27'

Brophy,Sandra Ruth
Peg 23h41'31"30d14'

Broschek,Alexis NPJ
Umi 16h46'35"77d29'

Brose,Karl
Peg 23h31'40"18d28'

Brosi,Karin & Franz
Cam 3h34'26"60d19'

Brosius,Dolores Marie
And 0h23'32"36d1'

Broskey Numen Optima, Helen Nodianos
Uma 11h20'35"33d19'

Broskin,Rose & John
Leo 11h5'46"2d22'

Brosnahan,Edwin
Oph 17h59'56"12d0'

Brosnahan,Sean
Cet 2h37'13"-11d56'

Brosnan,Patrick M
Lac 22h2'1"51d57'

Brosnan,Sheila A
Lyn 7h56'55"34d43'

Brossat,Herbert & Margarete
Cyg 19h21'47"28d28'

Brosseau,"Nanny" Mary E
Cas 1h23'1"72d24'

Brosseau,Jamison Robert Taft
Aur 5h18'29"41d34'

Brosseau,Olivier
Dra 11h59'46"72d48'

Brossia-Graham
Cam 7h28'40"83d33'

Brost,Hety
Mon 7h37'34"-1d40'

Broster,David Mark
Her 16h19'56"41d50'

Brostrom,Harry
Boo 14h17'1"46d19'

Broszath,Roswitha
Dra 17h18'1"64d49'

Brotchie,Rona Kenneth
Uma 9h16'52"53d5'

Brothagen,Ulrich
Dra 18h30'12"68d30'

Brother,The
Oph 17h25'31"11d25'A

Brotherhood of Noise,The
Aur 6h4'0"38d40'

Brotherly Affection
Lib 14h47'17"-22d21'

Brothers,Bill & Diane Chichester
Sge 19h39'37"16d9'

Brothers,David Anthony "de Ville"
Cap 20h7'23"-10d28'

Brothers,Gertrude
Cas 1h4'19"58d40'

Brothers,Ruth Ellen
Cyg 21h12'52"35d37'

Brotherton,John & Nancy
Lyn 7h14'1"56d57'

Brotherton,Susan
Com 13h14'11"21d9'

Brothwell,Eleanor
Cas 0h17'16"61d2'

Brotman's SilverStar
Ari 2h42'31"20d20'

Brott,Joe
Cet 1h24'1"-4d8'

Brotz,Joseph Daniel
Her 17h12'20"27d59'

Brotz,Kevin "Toe Boy"
Lac 22h34'1"50d26'

Brotz,Marsha Starr
Aql 19h22'24"10d60'

Brotzge,Michael
Crt 10h52'42"-17d47'

Brouard,Thierry
Dra 20h36'0"75d3'

Brough,Adam
Cep 4h0'28"80d13'

Brougham,Jim-Bob
Aur 7h24'14"38d10'

Brougham-Holder, Timothy Charles Mathew
Umi 14h26'0"68d47'

Broughman,Dorothy
Lyr 19h23'1"30d50'

Broughton,Cory Ray
Cep 20h15'0"60d48'

Broughton,Frederick B
Ori 5h59'0"16d41'

Broughton,Margaret
And 23h47'0"41d16'

Broughton,Otis Ray
Cep 23h9'1"60d27'

Broughton,Susanne Manuela Land
Cep 23h9'19"60d47'

Broughton,Tanya Lynn
Cep 20h19'14"60d12'

Brouillard,Arthur Louis
Cma 6h51'48"-17d56'

Brouillet,Pierre
Ser 15h53'30"18d29'

Brouillette,Jeané
Aql 20h10'18"0d58'

Brouillette,John Clayton
Aur 6h22'55"35d38'

Brouillette,Jr,Edward L
Her 17h29'34"28d53'

Broujet-Chaval, Elizabeth Amy
Lyr 18h29'42"31d18'

Brous,John
Her 16h38'1"48d31'

Brous,Liz
Tri 2h5'23"31d44'

Broussard Family,The
Uma 11h51'11"32d24'

Broussard,Michelle
Ori 6h3'46"6d54'

Brout,Dillon Jake
Cnv 13h19'40"51d46'

Broutin,Miguel Casafont
Hya 8h54'14"-6d49'

Brouwer,Angela Kapitan
Cam 3h17'1"61d2'

Brouwer,John C
Dra 14h56'29"56d3'

Brouwer,Paul S
Vul 19h48'16"28d42'

Brow,Elizabeth J
Del 20h39'50"10d41'

Brow,Joseph A
Ser 15h12'48"17d56'

Browder,Helga H
Peg 23h22'0"30d8'

Brower,Cliff William
Dra 18h5'1"58d27'
Brower,Elsa M
Cas 1h5'0"60d28'
Brower,Gary J
Cep 0h5'1"69d36'
Brower,Gary M
Boo 14h39'1"19d2'
Brower,Hugh W Musson
Cas 1h50'12"63d52'
Brower,John Victor
Her 17h15'21"42d43'
Brower,Lynn
Cet 2h58'1"1d22'
Brower,Roger
Her 17h55'49"18d57'
Brower,Tiffany
And 23h23'33"41d60'
Brown "Daddy",Eric
Cep 23h2'58"64d56'
Brown "The Bomber", Lisa Morgan
Equ 20h59'0"10d1'
Brown & Tex,Christine
Sex 10h47'45"0d57'
Brown (Gas),Kathleen S
And 1h19'52"40d49'
Brown 04-11-64,Charles Lee
Cnv 12h17'59"41d34'
Brown 12-20-94,Kristyn Janene
And 23h0'18"51d44'
Brown 9/22/76-9/22/94, Austin Lee
Vir 12h2'12"2d13'
Brown,Caroline
Pup 7h56'16"20d39'
Brown Christmas Star, The W F Bill
Umi 14h42'22"66d3'
Brown Clan,The
Cyg 21h33'21"38d19'
Brown Eyes
Cnv 12h10'0"42d9'
Brown Family,Travis
Lmi 10h39'51"30d54'
Brown II & III,John Wylie
Her 16h23'0"32d8'
Brown II,John A
Lib 15h33'37"-8d22'
Brown II,Nelson F
Lac 22h14'18"49d38'
Brown III,"Doubting Thomas",Clement M
Cam 7h25'51"80d9'
Brown Star,The Raymond C
Equ 20h55'38"2d47'
Brown Sugar
Cep 4h21'33"87d11'
Brown Worthy Advisor, Erin Patricia
Del 20h21'23"2d20'
Brown"Mrs H P",Carol K
Cet 2h4'42"6d41'
Brown"Pepita",Josefa
Cet 2h25'18"4d8'
Brown"The Big Iguana", Paul D
Per 2h32'45"57d8'
Brown"Wanderin Star", George
Per 2h57'52"31d55'
Brown's 50th,Don & Nina
Del 20h16'44"10d33'
Brown's All-Star,Jeff
Peg 22h39'45"12d13'
Brown's Bright Beauty
Boo 14h58'48"34d0'
Brown's Love
Cyg 20h37'0"52d43'
Brown's Star,Josh
Cmi 7h55'13"7d46'
Brown, We Love You, Timothy Brendan
Aur 5h7'44"41d6'
Brown,"Nephets" Stephen J J
Per 3h22'36"37d59'
Brown,A Heavenly Love, Jay & Kathy
Aql 18h59'41"-6d3'
Brown,Adam Nicolas
Equ 21h21'30"2d39'
Brown,Alexander Swaim
Lib 15h15'48"-18d20'
Brown,Alexander Lessels
Aur 4h54'53"50d39'
Brown,Alice
Lac 22h37'54"56d23'
Brown,Alice June
Cas 2h37'57"71d4'
Brown,Allen J
Her 16h22'31"42d23'
Brown,Allen L
Aql 18h57'19"11d52'
Brown,Allison
Mon 6h54'53"-10d50'
Brown,Allyson
Mon 6h53'31"-1d1'
Brown,Alyssa Marie
Lyr 18h57'25"40d19'
Brown,Amanda Lee
Com 12h27'52"20d36'
Brown,Amy
And 23h43'58"44d9'
Brown,An Easter Star, Dolores A
Cas 0h59'0"60d43'
Brown,Andrew Connor
Dra 16h32'0"62d37'
Brown,Andrew Watson
Dra 16h58'1"60d5'
Brown,Andy
Aur 5h10'1"42d6'
Brown,Angie
Mon 7h0'11"8d39'
Brown,Ann O
Mon 6h0'12'-8d9'
Brown,Ann Theresa
Lyr 19h23'48"37d53'
Brown,Anna
Cas 0h0'52"59d15'
Brown,Anthony
Psc 22h56'55"0d31'
Brown,Arthur Charles
Cet 2h42'17"-17d31'
Brown,Arthur E "Butch"
Her 18h0'57"45d33'
Brown,Austin Ridgway
Aql 19h19'16"10d44'
Brown,Barbara Elizabeth
Lac 22h9'41"40d10'
Brown,Barbara Joan
Eri 2h43'18"-15d7'
Brown,Bethany Claire
Tri 2h14'0"32d57'
Brown,Betty
Cas 0h12'0"61d16'
Brown,BJ
Per 1h51'32"53d19'
Brown,Brette
Oph 16h52'15"-28d36'
Brown,Brian
Dra 19h41'19"61d28'
Brown,Brittany Adair
Cyg 21h5'46"40d15'
Brown,Bruce
Aur 6h17'18"37d9'
Brown,C E
Ori 5h13'39"15d53'
Brown,Callie Ann
And 23h21'19"40d21'
Brown,Calvin & Ruth Ann
Tau 3h37'12"23d17'
Brown,Cameron James
Uma 8h8'1"67d39'
Brown,Camille
Lyr 18h46'1"33d9'
Brown,Carl
Ser 16h1'48"11d14'
Brown,Carlos & DeLores Mae
Vul 19h48'24"27d49'
Brown,Carolyn
Cam 4h16'10"70d59'
Brown,Cary
Cas 0h10'1"61d10'
Brown,Cayla Lynn
Cyg 19h55'60"40d59'
Brown,Celeste Maryn
Cas 0h56'27"60d14'
Brown,Charles H
Ori 5h53'0"15d16'
Brown,Charlie
Aql 20h15'59"1d6'
Brown,Chris
Cyg 21h24'31"37d45'
Brown,Christina
And 1h44'48"50d4'
Brown,Christopher Glenn
Sgr 19h6'48"-26d28'
Brown,Christopher Kevin
Aur 5h33'42"29d14'
Brown,Christopher A
Aql 19h9'1"3d46'
Brown,Christopher E
Cap 20h41'19"-26d58'
Brown,Cindy
Cma 6h53'30"-18d3'
Brown,Cluny
Ari 2h43'26"20d19'
Brown,Clyde Dennis
Per 3h2'1"40d3'
Brown,Col William Kavanaugh
Cyg 21h51'19"41d16'
Brown,Colin Alan
Ori 5h35'33"1d11'
Brown,Colin James
Umi 13h26'11"72d29'
Brown,Colleen Lianne
Lyr 18h37'36"40d29'
Brown,Craig Allan
Dra 17h47'56"68d55'
Brown,Craig Newton
Her 18h2'45"30d14'
Brown,Cynthia
Sco 17h29'12"-33d55'
Brown,Cynthia Renee
Lyn 8h42'15"34d51'
Brown,Daniella A
Crt 11h14'38"-19d43'
Brown,Danielle
Lyn 7h47'27"51d37'
Brown,Danielle
And 2h15'32"39d35'
Brown,David Andrew
Aql 19h49'38"13d34'
Brown,David James
Aur 6h2'40"36d1'
Brown,David M
Aur 6h12'32"38d60'
Brown,David Marcus
Aur 5h34'60"40d54'
Brown,David Shadwick
Her 18h3'35"28d20'
Brown,Davis Buchanan
Eri 3h20'0"-2d28'
Brown,Deborah
Cyg 20h37'57"31d28'
Brown,Delia
Mon 6h26'37"11d6'
Brown,Denise A
Peg 23h27'38"23d8'
Brown,Devon Michael
Dra 16h45'24"52d26'
Brown,Diane Linda
Cas 2h39'45"70d7'
Brown,Dollie
Cas 23h41'23"61d16'
Brown,Donald Edward
Aur 5h1'44"46d50'A
Brown,Donald R
Dra 16h24'39"68d30'
Brown,Donna & Ashanti
Aql 20h32'46"0d16'
Brown,Donna Teresa
Aql 19h10'34"15d17'
Brown,Donnie
Per 4h27'16"50d37'
Brown,Dorothy Alice Herrin
Oph 17h2'24"-18d45'
Brown,Doug
Cep 21h20'1"70d3'
Brown,Earl Garrett
Aql 20h8'56"0d39'
Brown,Edward & Edith
Ori 6h3'22"7d49'
Brown,Edward Paul
Aqr 20h59'31"0d20'
Brown,Edwin C
Her 17h20'26"45d56'
Brown,Elisabeth Anne
And 1h35'59"48d45'
Brown,Elizabeth
Peg 23h24'37"11d50'
Brown,Elizabeth
And 2h24'41"44d26'
Brown,Eric
Her 16h37'56"21d14'
Brown,Erin Claire
Lyr 18h58'1"31d41'
Brown,Erin Marie
Lyr 18h42'1"40d27'
Brown,Errol Antonio
Boo 14h37'38"31d46'
Brown,Ferdinand Charles
Boo 15h46'1"41d60'
Brown,Florence G & Everett R
Cyg 19h22'56"54d30'
Brown,Fraser James
Sex 10h14'23"-5d6'
Brown,Fred
Cnv 13h40'59"42'13'
Brown,Gene & Sarah
Lyr 18h55'23"31d23'
Brown,Gennifer
Cmi 8h8'10"0d45'
Brown,Geoffrey Brian Lindsey
Ori 6h6'53"5d15'
Brown,George
Ser 18h3'56"-14d60'
Brown,Georgene
And 0h20'38"44d26'
Brown,Gerald
Dra 12h6'49"71d43'
Brown,Glen Everett
Boo 14h3'40"12d52'
Brown,Glynis
Mon 8h6'51"-1d26'
Brown,Gordon Alan
Cma 6h25'1"-15d29'
Brown,Grace
Com 12h26'57"20d52'
Brown,Grandaddy
Uma 8h37'15"56d38'
Brown,Grandson Dustin T
Her 17h3'27"49d0'
Brown,Gregory A
Hya 10h20'38"-18d48'
Brown,Gregory K
Ser 15h51'19"18d28'
Brown,Gregory T
Per 2h23'43"58d23'
Brown,Harold & Lenora Jean
Oph 18h6'0"11d49'
Brown,Hazel V
And 1h9'26"40d43'
Brown,Heather
Com 12h12'39"31d19'
Brown,Helen Gurley
Cas 22h59'0"55d32'
Brown,Hope
Vul 19h4'0"25d39'
Brown,Irene N
Peg 21h50'44"30d50'
Brown,Irisha DeMel
Cap 21h55'22"-18d14'
Brown,J Harvey
Cyg 21h35'47"38d43'
Brown,James E
Aur 4h55'27"40d9'
Brown,James Michael
Ser 15h9'18"0d57'
Brown,James Sidney
Ori 5h56'0"0d5'
Brown,James W A
Ori 5h56'10"19d52'
Brown,Jamie
Lyr 18h18'55"45d7'
Brown,Jamie Lee
Peg 21h25'51"22d33'
Brown,Jamie Lynn
Cas 1h42'1"60d17'
Brown,Jane
And 23h2'37"51d43'
Brown,Janet Elizabeth
Peg 22h4'10"29d54'
Brown,Janet L
Mon 6h24'0"8d27'
Brown,Janet Teresa Stedman
Lyr 18h35'58"30d57'
Brown,Jason D Albert
Aur 6h27'36"37d13'
Brown,Jason Matthew
Cam 4h16'39"69d46'
Brown,Jean
Cam 5h52'1"68d54'
Brown,Jeffery Alan
Cet 0h52'22"-2d29'
Brown,Jeffery Todd
Her 17h23'60"21d12'
Brown,Jeffrey Allen
Oph 18h6'0"-8d3'
Brown,Jeffrey Farnham
Per 3h10'58"46d20'
Brown,Jeffrey Warren
Oph 17h0'50"11d58'
Brown,Jeffrey Willard
Lac 22h20'14"40d43'
Brown,Jeremy Michael
Aur 6h37'35"40d5'
Brown,Joan Dorgan
Umi 17h2'25"76d23'
Brown,Jody A
Lmi 10h13'16"32d26'
Brown,John
Boo 14h8'21"28d47'
Brown,John Louis
Aql 20h7'35"6d36'
Brown,John M
Crt 11h46'24"-18d51'
Brown,Jonathan
Cep 20h51'53"60d25'
Brown,Jonathan Chase
Aur 5h35'16"54d39'
Brown,Jonathan Patrick
Per 3h11'33"49d7'
Brown,Jordan Rene
Cyg 20h24'12"41d39'
Brown,Joseph Derek
Ori 5h57'43"17d30'
Brown,Joyce Stevens
Cas 0h3'52"58d17'
Brown,Jr,C Allyn
Lac 22h12'48"50d28'
Brown,Jr,Daniel Lee
Cet 0h56'51"1d10'
Brown,Jr,Dr Burnell R
Cep 21h33'56"68d25'
Brown,Jr,Fire Star,The Curtis I & Edward
Uma 9h7'28"68d4'
Brown,Jr,Karen & Walter
Her 17h13'25"40d24'
Brown,Jr,kk, "Forevermore"John R
Her 17h0'1"28d42'
Brown,Jr,Walter Edgar
Aqr 20h58'1"-13d24'
Brown,Jr,William Arthur
Cet 2h59'25"0d41'
Brown,Julia Keith
Peg 22h2'41"35d27'
Brown,Julia Marisa
Mon 7h11'58"-7d3'
Brown,Julie
Umi 16h56'46"76d2'
Brown,Justin Devin
Her 16h34'21"41d6'
Brown,Justin Will
Cyg 21h21'0"28d48'
Brown,Kalá Rae
Lyr 19h5'13"28d56'
Brown,Karen Elaine
Cyg 21h7'58"40d2'
Brown,Karen Lynne
And 2h22'43"48d4'
Brown,Karen(Cindu-May)
And 2h33'0"45d20'
Brown,Kate
Lyr 18h32'0"31d3'
Brown,Katherine & Robert
Crb 15h56'32"31d27'
Brown,Katherine A
Lyr 18h18'32"38d54'
Brown,Katie
Lyr 18h35'53"31d45'
Brown,Keith L
Her 16h31'18"38d32'
Brown,Kelly
Aql 19h40'27"14d2'
Brown,Ken
Hya 9h35'39"-1d33'
Brown,Ken(Pure of Heart)
Aql 19h55'1"-0d55'
Brown,Kenneth Lee
Cet 2h30'1'-5d56'
Brown,Kerry Frank
Crt 11h11'45"-17d37'
Brown,Kevann
Mon 6h30'57"7d41'
Brown,Kim
Peg 23h27'0"13d13'
Brown,Kirk Dewayne
Uma 12h0'21"57d31'
Brown,Kirsty
Lyr 18h32'30"34d4'
Brown,Kristin Danielle
Cam 3h58'52"53d14'
Brown,Laura Jean
Cam 6h47'15"67d57'
Brown,Laura L
Aqr 23h38'58"-10d29'
Brown,Laura S
Lyr 18h53'12"31d22'
Brown,Laurie
Crb 15h17'49"30d32'
Brown,Lee Scott
Cam 13h25'37"81d31'
Brown,Lewis Eldridge
Mon 7h16'26"-1d28'
Brown,Linda Gene
And 23h44'21"42d38'
Brown,Linda K
Lyn 8h11'1"42d27'
Brown,Lindsay Chentral
Mon 7h11'13"-10d55'
Brown,Lisa P P
Cyg 19h47'39"31d51'
Brown,Lizzie "SlotQueen"
Com 12h25'14"27d39'
Brown,Lorna
Umi 14h22'1"66d20'
Brown,Louise
Cyg 19h25'49"35d59'
Brown,Louise Sarah
Lyr 18h35'1"28d14'
Brown,Lucille
Cyg 19h18'29"44d38'
Brown,Lucy
Cas 1h36'16"60d18'
Brown,Lynne J
Cnc 9h6'18"31d17'
Brown,Madeline Ruth
Cas 1h18'35"61d1'
Brown,Maggie
Cet 1h17'27"-1d42'
Brown,Mar'Sue
Cet 2h41'43"0d28'
Brown,Marc Douglas
Sgr 19h36'28"-44d45'
Brown,Marcia A
Mon 6h34'41"-6d27'
Brown,Margaret & Elwin
Crb 15h15'0"31d43'
Brown,Margaret Estelle Thomas
Psc 1h38'43"21d21'
Brown,Marge
Ser 18h10'45"-7d10'
Brown,Mark-Carol-Dylan & Ryan
Lac 22h8'20"51d21'
Brown,Martha
Cet 1h1'32"1d10'
Brown,Martin Kenneth
Hya 9h2'21"3d26'
Brown,Martin William
Aql 20h20'49"1d42'
Brown,Mary Elizabeth
Aql 20h14'28"0d56'
Brown,Matthew & Grace
Cyg 21h4'1"37d41'
Brown,Matthew L
Cnv 12h5'51"40d38'
Brown,May Ellen
Aur 5h1'44"46d50'B
Brown,Merle Justice
Cet 2h21'17"-10d57'
Brown,Michael
Per 3h9'0"47d44'
Brown,Michael A W
Aql 20h5'44"0d55'
Brown,Michael John
Cep 21h52'40"55d55'
Brown,Mike
Cep 22h20'20"60d37'
Brown,Mike
Ori 5h24'14"1d11'
Brown,Mike,Brenda, Michelle & Mandy
Cep 20h17'38"60d4'
Brown,Monica
And 2h2'30"40d30'
Brown,Monica Lee
Ori 5h29'56"-1d59'
Brown,Myles
Cet 3h5'49"1d26'
Brown,Nancy
Dra 11h42'45"67d38'
Brown,Nathan Aaron
Her 16h55'0"38d45'
Brown,Nellie
Aql 19h47'55"14d32'
Brown,Nettie J
Crb 16h21'34"30d15'
Brown,Nicole & Kevin Burt
Lac 22h14'42"38d15'
Brown,Niles Kelly
Per 4h29'0"50d37'
Brown,Nilsa
Peg 22h58'19"22d16'
Brown,Olivia Alice Barrett
Peg 21h52'24"30d32'
Brown,Pam
Ori 5h47'18"12d19'
Brown,Pam
Aur 5h33'47"38d58'
Brown,Pam
Crb 16h1'1"26d10'
Brown,Pam
Cas 23h32'57"60d40'
Brown,Pamela T
Mon 7h56'40"-0d21'
Brown,Patrice Gaugh
Cas 0h27'25"63d25'
Brown,Patricia Jean
Peg 23h6'28"18d1'
Brown,Patrick
Ser 17h31'45"-14d18'
Brown,Patrick Santo
Ari 1h54m13"12d22'
Brown,Paul A
Cep 23h16'50"65d57'
Brown,Paul Douglas
Aql 20h4'17"1d9'
Brown,Paula Marie
Cas 23h18'27"62d42'
Brown,Pauline Ann
Cyg 20h37'12"48d51'
Brown,Peter
Dra 16h23'20"66d37'
Brown,Peter
Per 1h56'46"47d48'
Brown,Rachel Marie
Cas 1h45'57"61d17'
Brown,Randy
Aql 19h54'36"1d48'
Brown,Rea
Tri 1h51'40"28d55'
Brown,Rebecca Ann
Cas 1h45'39"68d29'
Brown,Regina Eve
Lyr 18h58'30"46d4'
Brown,Rena
Cmi 7h25'0"0d22'
Brown,Richard & Nancy
Cnv 13h6'23"40d15'
Brown,Richard T
Cam 3h23'55"61d57'
Brown,Riley Sheppard
Cyg 21h21'10"40d31'
Brown,RN,Peter J
Ori 5h55'38"-2d57'
Brown,Robert E
Aql 18h45'0"8d50'
Brown,Ron & Celeste
Vul 19h23'33"25d20'
Brown,Ronald P & Evelyn R
Hya 8h54'31"-17d14'AB
Brown,Rory
Ser 16h1'18"1d0'
Brown,Ross
Hya 8h40'47"5d36'
Brown,Ruth-Anne
Cas 0h3'45"61d0'
Brown,Ryan Grant
Oph 17h56'15"12d32'
Brown,Sabrina
Ori 4h59'36"10d31'
Brown,Sally
Uma 9h54'29"50d58'
Brown,Sam
Ari 2h34'32"21d36'
Brown,Sam & Fannie
Crt 11h36'35"-18d24'
Brown,Sandra Pauline
Del 20h13'55"15d12'
Brown,Sarah Ann
Cas 1h45'48"60d55'
Brown,Sean Patrick
Gem 6h48d31"13d3'
Brown,Shannon Daniel
Cnc 8h56'1"31d24'
Brown,Sharon Louise & Steve Alle
Tau 3h28'19"27d24'AB
Brown,Shayna
Peg 21h21'33"22d60'
Brown,Shelia
Tau 3h52'59"20d12'
Brown,Shirley A
Cas 0h48'0"65d25'
Brown,Simon Lee
Her 18h30'50"24d57'
Brown,Skip
Boo 15h26'54"40d36'
Brown,Skipper Tanner & Sarah
Dra 11h36'1"72d13'
Brown,Sonny
Eri 3h25'13"-5d27'
Brown,Sr,Frank A
Dra 20h22'20"68d57'
Brown,Sr,William Arthur
Cet 2h39'52"0d51'
Brown,Stacy Lynn
Sgr 19h38'40"-31d26'
Brown,Stephen "Tomahawk"
Aql 19h4'1"-6d33'
Brown,Steve & Patti
Eri 3h11'21"-2d58'
Brown,Steven
Dra 16h30'20"61d35'
Brown,Suzanne
Cmi 8h9'18"1d15'
Brown,T Dean
Dra 17h56'55"65d3'
Brown,Tab
Cep 0h18'53"67d49'
Brown,Tamara
Lyn 8h54'23"40d39'
Brown,Tammy
Mon 8h7'21"-1d6'
Brown,Teresa Elaine
Aql 19h13'0"10d20'
Brown,Teri
Uma 11h57'42"64d28'
Brown,Terrell Lane
Ori 5h4'33"-0d8'
Brown,Terry
Aur 6h2'13"37d39'
Brown,The King-James
Tau 5h19'0"28d37'
Brown,Thomas H
Aql 18h58'60"-8d34'
Brown,Thomas Joseph
Sct 18h46'25"-7d34'
Brown,Tiare Lakelani
Eri 3h53'32"-5d1'
Brown,Timothy D
Hya 8h15'32"2d12'
Brown,Timothy J
Lac 22h25'39"37d39'
Brown,Todd
Aur 7h23'46"41d8'
Brown,Todd R
Cet 2h42'31"9d35'
Brown,Towner Wilcox
Cep 22h31'53"67d58'
Brown,Tracie Michelle
Peg 21h55'16"34d55'
Brown,Tracy L
And 23h33'54"38d57'
Brown,Trevor Jason
Cep 21h52'25"70d44'
Brown,Valerie Patricia
Cam 3h41'43"61d37'
Brown,Vance
Per 3h25'31"51d45'
Brown,Vera Lynn
Peg 22h34'29"30d6'
Brown,Vincent Alexander
Cep 0h57'56"80d16'
Brown,Warren D
Her 17h11'42"48d52'
Brown,Wendy
And 2h23'59"44d47'
Brown,Wendy Zoa aka Wendy
Cyg 19h29'1"38d38'
Brown,Wil
Cep 0h12'0"69d44'
Brown,William Harold Thomas
Cyg 19h26'15"30d5'
Brown,William Hudson
Cas 3h34'1"71d36'
Brown,William Michael
Aur 5h21'22"38d51'
Brown,William T
Vul 19h40'40"26d7'
Brown-Cross,Jacqui
Del 20h18'18"11d9'
Brown-Fiscaletti
Lyr 18h28'0"37d37'
Brown-Jupiter,Patti J
Sct 18h46'17"-8d15'
Brown-Mon Ami,Scott
Lac 22h38'17"50d16'
Brown-Romiti,Honey "KiKu"
Cap 20h29'39"-10d1'A

Brown-Saslow
Cmi 7h56'31"8d39'
Brown-Weeks,Elsie Austin
Mon 6h22'43"8d60'
Browne III,Whitman Sinclair
Her 15h52'36"40d27'
Browne,Amanda Jayne
Boo 14h19'48"50d15'
Browne,Chandler Edwards
And 0h19'15"33d10'
Browne,Daniel Stanley
Her 17h34'1"41d9'
Browne,Dorothy Ivy
Cam 4h0'30"60d24'
Browne,Forrest Reginald
Oph 17h0'48"11d51'
Browne,James Paul
Cep 21h56'27"60d29'
Browne,Jr,Roger J
Dra 19h39'36"68d15'
Browne,Louis Robert
Aql 19h7'22"3d41'
Browne,Louis Robert
Aql 19h0'12"3d45'
Browne,Malcolm
Ari 2h25'53"11d2'
Browne,Matthew
Boo 13h43'25"20d49'
Browne,Robert
Lib 15h43'54"-24d51'
Browne,Stellar Teacher &
Friend,Anne
Lyn 8h23'40"48d20'
Browne,Tim
Aur 5h7'18"44d27'
Browneenworb Alan
Cen 13h24'13"-56d31'
Brownell,George
Lac 22h37'12"55d57'
Brownell,James E
Dra 15h51'1"62d3'
Brownell,James G
Aur 5h1'34"38d11'
Brownell,John H
Cep 20h28'24"76d49'
Browney's Sparkle
Ori 6h9'13"1d9'
Brownfeld,Megan Rae
Lyn 6h54'36"44d39'
Brownie
Aur 4h59'1"40d34'
Brownie
Eri 4h57'29"-8d37'
Brownie
Lmi 10h6'46"36d59'
Browning Love
Mon 8h5'15"-0d53'
Browning,Ardella & Phil
Sct 18h55'0"-10d14'
Browning,Claire
Peg 21h39'23"26d44'
Browning,Clifford E
Aql 20h11'55"5d1'
Browning,Isabelle Deborah
Cyg 20h2'44"41d2'
Browning,Janice J
Cam 5h55'46"60d7'
Browning,Jason Kirk
Cet 3h20'18"0d13'
Browning,Julie R
Eri 3h24'1"-7d25'
Browning,Kenneth M
Sct 18h42'29"-6d20'
Browning,Kent
Her 16h44'43"48d25'
Browning,Kimberly Dawn
Peg 22h2'54"26d4'
Browning,Margie
Cyg 19h30'1"33d44'
Browning,Nita
Mon 7h58'11"-0d48'
Browning,Peter Albert Leon
Her 18h29'1"24d45'
Browning,R Thomas
Cet 3h16'41"2d18'
Browning,Roger & Patti
Sct 18h42'0"-7d41'
Browning,Sammy Allan
Aql 19h30'27"12d24'
Browning,Sara Ann
And 1h44'12"40d3'
Browning,Sr,Jesse Robert
Boo 15h5'49"12d4'
Browning,Tyson Rodgers
Lib 15h27'46"-10d52'
Browning-Rader
Oph 18h27'52"8d47'
Brownlee,Jeff & Shana
Sex 9h52'15"2d8'
Brownlee,Jennifer Judith
And 23h13'33"35d47'
Brownlee,Michael John
Ori 6h3'44"2d40'
Brownlee,Robert Emery
Lac 23h21'0"53d50'
Brownlee,Sarah Mary
Cyg 19h30'25"33d10'
Brownlee,Thomas Gordon
Cep 20h42'23"61d19'
Brownlee,Walter
Lac 22h53'13"53d18'
Brownlees,The
Her 17h29'48"20d13'
Brownlow,Carol A
Lyn 7h56'18"38d16'
Brownrigg,Roger
Cet 2h4'20"6d16'
Brownson,Mary L
Sct 18h43'26"-4d9'
Brownstein,Harris
Per 2h51'48"50d28'
Brownstein,Kevin Wade
Boo 14h51'39"48d16'
Brownstone,Ann E Belle
Cyg 21h17'56"38d18'
Brownstone,Edwin S
Ori 5h52'42"19d34'
Brownstone,Keir
Ori 6h12'20"0d55'
Broxup,Alison Linda
Ori 5h56'19"7d34'
Broyles,Barbara & James
Cyg 20h37'40"30d4'
Broyles,Danny Alan
Ser 16h1'50'3d44'
Brozek,Joan
Uma 11h28'48"40d24'
Brozik,Brian J
Per 3h10'0"47d53'
Brozman,Erinn
Lyn 7h44'25"58d31'
Brozzo,Jennifer L
Cas 1h15'52"60d14'
Bru
Uma 10h19'33"50d57'
Bru,Claudine
Peg 23h31'36"10d52'
Brubaker,Holly
Aur 7h5'51"37d47'
Brubaker,Jr,Frank Nichols
Equ 21h7'24"10d29'
Brubaker,Sternchen Of David
Gem 8h4'22"28d48'
Brubaker,Sternchen Of Jay
Gem 8h0'1"28d47'
Brubakk,Einar
Dra 12h43'51"75d37'
Bruce
Per 2h38'0"38d42'
Bruce
Aur 5h35'47"38d27'
Bruce
Ori 5h28'59"-3d34'
Bruce
Her 17h36'38"27d58'
Bruce "Wine, Women &
Song",Scot
Hya 8h11'0"1d6'
Bruce & Angela
Her 16h59'24"32d43'
Bruce & Betsy's Bit Of Heaven
Cmi 7h56'20"0d29'
Bruce & Mary Helen
Crb 15h48'60"31d57'
Bruce & Nancy Love Eternal
1966-1994
Crb 15h37'0"27d0'
Bruce & Tristi-Forever
Boo 15h24'47"41d19'
Bruce Alexander
Cyg 19h46'51"35d11'B
Bruce Allen
Boo 15h5'1"28d53'
Bruce H
Aql 19h51'21"10d25'
Bruce II
Hya 8h42'1"3d57'
Bruce Scott/Kirk Patrick
Boo 14h59'0"11d48'
Bruce's Brilliance
Del 20h13'11"13d24'
Bruce's Far Away Star
Cet 2h33'0"-10d46'
Bruce's Little Star Adelaide
Her 16h39'16"28d3'
Bruce's Miracle
Boo 14h48'20"45d9'
Bruce's Quest
Aur 6h25'0"30d5'
Bruce's Shining Star
Cep 1h16'23"78d20'
Bruce's Star To Enlightenment
Aql 19h7'0"3d16'
Bruce's Wishing Star
Ori 5h48'51"10d4'
Bruce,Alan & Kimberly
Cma 6h51'19"-15d31'
Bruce,Alexa Marilyn
Dra 18h58'58"48d21'
Bruce,Amanda
Lyr 18h14'15"32d58'
Bruce,Amanda Anne
Mon 6h30'20"-10d9'
Bruce,Bob
Cyg 20h25'59"38d9'
Bruce,Brandy Lynn
Peg 22h22'35"31d18'
Bruce,Chichan
Per 3h32'38"52d3'
Bruce,Christina Elizabeth
Mon 7h2'37"0d41'
Bruce,Gabreon Deshawn Lee
Uma 9h33'16"60d49'
Bruce,George B
Ori 5h44'15"0d43'
Bruce,Helen Lane
Sge 19h13'53"20d33'
Bruce,Ian
Eri 4h4'48"-22d7'B
Bruce,Jackie
Peg 22h23'0"2d20'
Bruce,Jackie
And 2h30'52"50d2'
Bruce,Jillian
Umi 15h50'25"81d58'
Bruce,Joseph & Agnes
Cyg 19h58'20"45d32'
Bruco,Joyco
Tri 2h6'16"30d25'
Bruce,Kitty
Cam 8h51'26"74d57'
Bruce,Mary Jane &
James,R&D Bruce-Sons
Cnc 8h39'23"17d12'
Bruce,Neil,C
Lac 22h35'14"53d49'
Bruce,Pat
Sct 18h44'48"-6d40'
Bruce,Richard Henry
Peg 23h3'31"13d56'
Bruce,Robert Edward
Uma 11h42'49"52d58'
Bruce,Ronica J
Lyr 19h22'0"38d41'
Bruce,Stephanie Shea
Mon 6h41'29"10d23'
Bruce,Tracey Lyn
Boo 14h22'57"53d23'
Bruce,Tracy
And 0h47'14"38d12'
Bruch,Bridget Beth
Lyn 7h41'47"50d40'
Bruch,Diane
Crb 15h50'13"38d22'
Brucie de Alejo
Cet 1h7'46"-4d35'
Bruck,Jean McQuarrie
Lac 22h11'21"50d14'
Bruckenthal,Ryan Edward
Dra 14h55'28"56d22'
Brucker "Me C'28"56d22'
Cma 7h11'59"-13d42'
Brucker,Michelle Bryan &
Alyson
Lyn 8h7'38"45d46'
Bruckerl,Frank B
Aur 6h5'10"33d56'
Brucki,Christopher Robin
Cep 23h25'54"64d33'
Bruckmann,Elana
Tau 4h56'55"16d44'
Bruckmueller,Julie Ann
Cas 2h49'0"68d34'
Brucks,Jackie
Cet 2h47'24"5d21'
Bruckschwaiger, Christian
Lac 22h5'45"49d12'
Brudstar
Cnv 13h7'0"40d36'
Bruecher,Frederick Andrew
Boo 15h22'15"42d3'
Brueck,Dana Lindsey
Mon 7h15'1"-6d16'
Brueck,Julia
Gem 7h22'58"34d43'
Brueckner,Heinz
Her 16h16'0"50d39'
Brueggeman Star,The
Appreciated
Aql 19h42'12"14d17'
Bruehl,Rebecca & Geoffrey
James
Umi 16h27'1"71d30'
Bruel,Patrick
Cyg 19h21'0"28d10'
Bruenecke,Christian Arron
Cas 1h52'29"71d20'
Bruening,Heinz-Hermann
Oph 18h0'13"7d60'
Bruenning III,Leonard E (Lee)
Ori 5h57'60"16d15'
Brueser,Bert Alan
Aql 20h0'0"6d32'
Brueski,Bill
Aql 18h57'1"14d56'
Bruey,Pat & Bill
Mon 6h19'40"8d7'
Bruffy,David & Susan
Sge 20h4'15"20d5'
Brugato,Ryan Jeffrey
Lyr 18h37'48"27d12'
Bruggeman,Jeremy
Aur 5h2'15"44d31'
Brugioni,Christopher "The
Dopher"
Per 3h9'1"55d6'
Brugman,Edwin
Hya 8h40'0"2d55'
Bruhat,Anne
Cam 6h7'13"61d28'
BRUHILMAR
Lmi 10h4'40"30d33'
Bruhl,Konsul Kommerzialrat
Kurt D
Ori 6h3'39"0d15'
Bruhn,Irwin & Dorothy
Lyr 18h58'44"37d20'
Bruhn,Tammy
And 0h6'30"31d33'
Bruij,Edward
Per 3h24'59"40d58'
Bruington,Matt
Sco 16h31'50"-43d8'
BrujaeriPaul
Aql 20h9'0"6d5'
Brukcop
Hya 8h40'30"5d58'
Bruksch,Peter
Ser 16h6'42"2d19'
Brulla,Elizabeth
Com 12h0'52"27d19'
Brumbaugh,Anna Mary
Umi 18h49'31"86d13'
Brumberg,Jonathan S
Dra 19h9'13"68d45'
Brumberg-Horan,Ronnie
Aur 5h1'36"48d47'
Brumder,Christopher Cameron
Cnc 8h21'18"16d16'
Brumfield,Jon-Jon
Uma 9h53'1"53d34'
Brumfield,Merie Eugenia Ellis
Lyn 7h28'54"44d4'
Brumfield,Robert Orlan
Aql 20h2'30"11d47'
Brumfitt,David
Aql 20h1'40"10d10'
Bruml,David
Cam 3h28'13"63d14'
Brumme,Carole Sidoli
Com 13h24'58"26d50'
Brummett,Deana Annette
Mon 6h38'22"7d19'
Brummett,Jennifer Renee
Gem 7h54'37"30d4'
Brummett,John D
Boo 14h20'29"32d43'A
Brummett,Jr,John D
Cep 22h3'10"54d3'
Brummitt,Adam Cooper
Aql 18h44'0"10d24'
Brun,Corinne
Cas 2h24'47"60d24'
Brun,Michel
Uma 11h44'38"51d45'
Brun,Nathalie
Per 3h58'6"35d41'
Brun,Pierre
Per 3h58'7"36d30'
Bruna
Pho 0h4'24"-47d4'
Bruna
Lup 15h2'59"-37d49'
Bruna
Cyg 19h26'40"33d31'
Bruncucio,Penny
And 23h28'1"48d53'
Brundage,Jennifer L
Mon 6h45'46"10d26'
Brundage,Matthew D
Cmi 7h7'19"4d16'
Brundage,Michael W
Ori 4h42'40"4d41'
Brundle,Alan Keith
Per 3h6'41"40d54'
Brundrett,Cliff
Cet 1h15'0"-2d46'
Brune,Regina
And 0h54'37"21d36'
Brunead,Christian
Aur 5h50'1"38d9'
Brunell,George Lewis
Dra 16h52'26"67d49'
Brunell,In Loving Memory Of
Eva Mae
Cas 1h27'52"61d41'
Brunell,Rea Claire
And 23h28'29"43d34'
Brunell,Scott M
Uma 12h25'54"61d18'
Brunella
And 0h1'12"40d2'
Brunella
Dra 12h22'22"64d56'
Brunelle,Little Acorn
Lac 22h54'30"37d41'
Brunelle,Suzan Beth
Peg 22h2'34"32d31'
Brunelli,Cassie
Cyg 20h19'27"30d49'
Brunelli,Lynn
Lyn 6h54'37"48d58'
Brunelli,Yolande
Sgr 19h34'32"-33d37'
Brunengo,Gino & Pamela
Cyg 21h27'28"30d48'
Bruner,Damon
Cmi 7h7'1"1d32'
Bruner,Dane James
Ori 5h38'48"12d40'
Bruner,Devin Bruce
Ori 5h50'46"10d39'
Bruner,Jr,Billy
Boo 15h1'22"28d8'
Bruner,Karen
Peg 23h32'54"18d18'
Bruner,Kary Kathleen
Ori 5h26'29"15d19'
Bruner,Katy Joanne
Ori 5h53'29"12d34'
Bruner,Sheila
Eri 2h44'54"-17d40'
Bruner,Tracy
Cet 3h7'42"1d12'
Brunet,Andrée
Peg 21h24'7"19d9'A
Brunet,Claudie Anne
Sex 10h19'34"-8d47'
Brunet,Franáois
Aur 6h52'33"44d6'
Brunet,Mario
Cam 3h51'57"62d59'
Brunet,Martin
Her 17h55'58"14d56'
Brunetta-Phillips 01/10/95
Cyg 21h15'52"38d39'
Brunette,Julie Ann
Lyn 8h54'21"45d23'
Brunette,Lucas
Cyg 20h54'46"50d21'
Brunetti,Maria Paola
Dra 16h23'23"62d30'
Brunetti,Maureen
Aqr 21h6'19"0d49'
Brunetto,Damien Troy
Per 3h3'1"57d26'
Bruney,Audrey M
Aur 5h32'13"29d20'
Bruni Ever Radiant
Beauty,Angela Marie
Peg 22h45'22"11d7'
Bruni NP
Cyg 21h8'11"30d20'
Bruni,Estetica Betty
Del 20h19'58"14d2'
Bruni,Maurizio
Lyr 18h38'45"27d22'
Bruni-Bruxi
Cyg 20h40'30"31d52'
Brunick,Punkin'
Boo 14h50'26"30d59'
Brunier,Jacques
Umi 15h41'14"80d58'
Bruning,Kendra
Lib 15h8'33"-8d41'
Brunisholz,Brenda
Peg 22h0'15"27d54'
Brunjes,Hans-Dieter
Lyr 19h21'41"31d47'
Brunk,Moyer Marie Richards
Lyr 18h50'29"40d21'
Brunk,Sandra
Lmi 10h42'59"32d14'
Brunke
Mon 6h55'23"-2d28'
Brunke,Manfred
Aur 5h19'40"43d15'
Brunker Diamond,The David
Cyg 20h20'39"39d43'
Brunker,Eric Wayne
Aur 5h7'44"41d5'
Brunkhorst,Casey J
Her 17h1'42"45d10'
Brunkhorst,Desireé
And 2h21'27"44d58'
Brunkhorst,Michael W
Aur 4h54'30"41d2'
Brunkhorst,Nicholas R
Cnv 12h46'38"36d32'
Brunn,James Robert
Lac 22h12'30"51d29'
Brunnengräber,Volker
Cas 0h5'41"60d59'
Brunner,Esther
Cap 21h24'7"-23d18'
Brunner,Griffin Robert
Cet 3h11'22"5d19'
Brunner,Jennifer
Com 13h31'47"20d21'
Brunner,Jordan David
Lac 22h31'32"38d49'
Brunner,Julian James
Her 17h38'56"26d21'
Brunner,Linda Jean
Aql 18h46'33"10d13'
Brunner,Lucas Alan
Aqr 22h1'53"0d6'
Brunner,Manfred
Leo 10h1'52"18d52'
Brunner,Roy J
Aql 19h37'24"-0d12'
Brunner,Vera
Sco 17h8'0"-38d22'
Brunning,Helen Amanda
Umi 13h57'1"78d38'
Brunnock
Cep 23h11'21"62d38'
Bruno
Cep 22h40'35"63d9'
Bruno
Cep 22h15'50"55d35'
Bruno & Kathrin & Oliver geb
131291
Uma 9h49'29"58d50'
Bruno e Carla
Lyn 7h48'55"40d27'
Bruno Family Star,The
Her 16h49'43"34d9'
Bruno's Star
Her 17h6'52"48d3'
Bruno,Ali
Cas 1h18'52"61d45'
Bruno,Anna
Leo 10h49'60"4d50'
Bruno,Deborah
Cyg 21h0'14"39d27'
Bruno,Isabella
Sgr 19h24'28"-41d20'
Bruno,James
Aur 7h17'46"40d38'
Bruno,Landon Chase
Cet 0h31'47"-18d36'
Bruno,M F
Lyn 7h51'58"40d45'
Bruno,Marie & Alphonse
Sge 19h31'39"19d1'
Bruno,Maximilian P E
Cep 21h12'49"65d10'
Bruno,Mr Hedin
Com 12h11'41"21d8'
Bruno,Peter "Little Buddy"
Aul 19h56'41"-6d12'
Bruno,Rita
Cas 0h58'19"70d30'
Bruno,Robert Stanley
Hya 8h14'11"0d30'
Bruno,Rosie One
Vul 20h40'32"28d32'
Bruno,Sr,Joseph E
Dra 11h46'1"71d16'
Brunon,Charlene
Sgr 19h27'52"-35d42'
Bruns 50th,Lewis & Beverly
Cyg 19h40'39"31d44'
Bruns,Brigitte
Vir 14h38'29"-6d58'
Bruns,Bryan Austin
Lac 22h18'39"51d44'
Bruns,Janna
Peg 22h47'60"33d19'
Bruns,Leland Martin
Ser 15h29'36"19d37'
Bruns,Warren R
Peg 23h20'28"11d34'
Brunsicin,Jean Paul
Cep 20h8'52"60d40'
Brunskill,Martha
Lyr 18h39'34"44d47'
Brunson,Daniel Ian
Aql 20h8'13"4d40'
Brunson,Deborah
Mon 6h53'30"-1d41'
Brunson,Mary Von-Mohr
Cam 9h0'57"82d27'
Brunson,Sue & George
Sge 20h17'58"17d48'
Brunswick,Lucile Anne
Oph 17h6'52"1d34'
Brunswick,Nicholas Mark
Aur 7h15'0"40d42'
Brunton,Robert H
Cep 22h49'14"58d44'
Brusasco,Luna
Boo 15h4'19"47d56'
Brusca,Nancy
Cyg 20h21'38"37d59'
Bruscato,Brandon Anthony
Her 16h44'12"34d55'
Bruschini,Michael
Aur 4h47'52"40d18'
Brusewitz,Ruth Maxine
Uma 11h7'55"40d12'
Brush,Allen Sharpe
Gem 7h59'56"28d52'
Brush,Judi
Cas 0h33'18"60d13'
Brush,Selina Wenonia
Lmi 9h40'36"37d56'
Bruske,Ms Jacki
Sge 19h59'1"16d44'
Brusky,John William
Cep 22h39'16"61d27'
Brusky,Julie Wall
Lyn 8h51'33"46d41'
Bruss,Kathleen
Lyr 19h18'11"42d25'
Brusseau,Michel Mika
Tri 1h56'58"32d19'
Brusseau,Robert L (Buzz)
Equ 20h59'22"5d29'
Brusson,Thibault
Cnc 8h31'20"30d43'
Brusstarr,James
Her 17h35'17"26d7'
Brust,Winfried
Vul 19h47'28"29d4'
Brut,Jean-Louis
Aur 4h34'17"34d18'
Brutcher,June
Lib 15h17'0"-22d4'
Bruton's Bright Star, Martha
Peg 21h29'15"23d40'
Brutto,Liliana
Umi 15h31'37"67d44'
Brutus Star,The Margaret C
And 23h44'21"41d32'
Brutvan,Lori
Aur 6h40'47"40d40'A
Brutvan,Robert
Aur 6h40'47"40d40'B
Bruyand,Aurélien
Ori 6h16'0"8d41'
Bruyninckx IV,Felix Joseph
Aql 19h31'12"12d19'
Bruyntjens
Lyn 7h58'10"38d10'
Bruyère,Christiane
Lyr 18h47'0"36d9'
Bruzas,Delores Richio
And 2h25'16"49d39'
Bruzas,John
Ori 5h42'35"11d17'
Bry
Uma 10h47'33"52d4'
Bry
Dra 9h55'51"81d33'
Bry's Place
Leo 9h50'35"31d15'
Bryan
Dra 16h9'11"66d21'
Bryan
Aql 19h30'35"10d38'
Bryan & Charlie Together
Forever
Peg 23h25'0"13d60'
Bryan & Patee's Star
Dra 19h27'27"60d6'
Bryan & Shanna
Lyn 7h52'16"33d22'
Bryan Andrew
Aur 5h1'0"44d36'
Bryan Christopher
Dra 16h51'30"68d32'
Bryan Jay
Ori 5h53'19"21d40'
Bryan Lee
Dra 17h10'0"68d3'
Bryan's Blade
Ser 15h34'18"18d4'
Bryan's Dream
Ori 5h56'35"19d58'
Bryan's Dream
Aql 19h11'11"15d53'
Bryan's Imli
Aqr 22h39'26"-5d21'
Bryan's Power Star
Lyn 7h35'27"44d49'
Bryan's Star
Her 16h50'49"40d34'
Bryan's Star
Eri 3h28'41"-5d52'
Bryan,Anita
Cyg 19h28'0"37d19'
Bryan,Brenda Elizabeth
Cyg 20h32'44"39d20'B
Bryan,Derek Joe
Hya 8h59'30"-10d2'
Bryan,Judy
Ser 16h16'21"-0d31'
Bryan,Katelyn Marie
Mon 8h6'51"-1d9'
Bryan,Larry
Hya 8h38'0"-6d8'
Bryan,Rachel Adele
Boo 14h58'50"31d30'
Bryan,Raymond "Papi"
Her 18h11'12"30d7'
Bryan,Sr,Robert J
Uma 11h13'0"43d38'
Bryan,Sylvia Walker
Ori 6h7'48"5d3'
Bryan,T J
Cyg 21h7'52"30d5'
Bryan,Tanya M
Dra 15h14'37"60d16'
Bryan,Vanessa Rachael
Cyg 20h32'44"39d20'A
Bryan,William Lester
Tau 4h43'20"16d25'
Bryan-Aldous,Jillian Evelyn
Ori 6h9'53"3d1'
Bryan/5
Her 17h53'51"41d4'
Bryans,Ceri
And 0h15'1"39d41'
Bryant Worlds Best Son
Matthew Race
Ser 18h42'34"4d39'
Bryant,Amanda Virginia
And 23h41'22"40d9'
Bryant,Andrew & Lynne
Boo 15h9'50"48d30'

Bryant,Anita
Cas 0h19'40"58d21'

Bryant,Bruce
Her 16h8'0"18d49'

Bryant,Chadwick John
Dra 12h6'0"70d22'

Bryant,Colin & Austin
Uma 11h45'1"49d56'

Bryant,Darrell Ray
Ori 5h57'32"15d5'

Bryant,Dr,Ezekiel W
Oph 18h4'56"11d27'

Bryant,Eric Andrew
Her 16h31'19"32d26'

Bryant,Eric James
Ari 3h7'43"30d2'

Bryant,Flora & Ted
Cyg 19h53'33"38d40'

Bryant,Forever Mike
Sct 18h41'7"-7d40'

Bryant,Gay
Dra 17h26'53"72d29'

Bryant,Idon Weston
Lyr 18h35'0"40d18'

Bryant,James Christopher
Aur 5h1'48"40d39'

Bryant,Jessica Susan
Cyg 21h14'43"35d36'

Bryant,John
Sco 17h51'1"-37d53'

Bryant,Kyle Russell
Cep 22h6'54"53d31'

Bryant,Lindsey Meredith
Peg 22h57'16"18d9'

Bryant,Maclain Reeves
Peg 23h35'32"30d22'

Bryant,Muriel
Aql 18h58'11"-6d47'

Bryant,Nathaniel Raleigh
Aur 7h1'52"38d47'

Bryant,Nikole
Uma 10h38'34"58d47'

Bryant,Paul & Betty
Cyg 19h25'36"33d52'

Bryant,Richard & Annabelle
Lyn 9h1'21"35d29'

Bryant,Robert James
Aql 18h58'38"-1d59'

Bryant,Robert Shefford
Aur 6h21'33"38d27'

Bryant,Rosemary
And 1h22'50"38d23'

Bryant,Shane
Mon 8h6'13"-2d7'

Bryant,Shannon Rebecca
Lyn 6h36'0"60d24'

Bryant,Steven Douglas
Lac 22h49'28"53d58'

Bryant,Taylor Mackenzie
Lac 22h49'52"37d38'

Bryant,Tiffany Kristin
Peg 22h8'0"5d36'

Bryant,Tony
Cet 2h49'18"4d18'

Bryant,Virginia & Jim
Cyg 21h53'42"41d23'

Bryar,John Joseph
Boo 14h52'1"24d52'

Bryce
Aur 5h17'1"42d43'

Bryce-Buchanan-40, Carol
Sge 19h10'29"19d23'

Bryden
Uma 12h6'47"45d14'

Brydon,Charlotte Anne
Cmi 7h38'14"-0d1'

Bryerton,Matthew
Cep 22h30'36"59d8'

Brylski,Martin Andrew
Hya 9h6'58"5d44'

Brymer,Amanda Rochelle
Eri 3h45'55"-6d4'

Brymer,Sara Nicole
Mon 6h35'53"11d42'

Bryn
Cas 2h3'0"75d0'

Bryna
Mon 8h3'45"-5d41'

Brynck
Cmi 7h45'35"8d14'

Bryndon Denae
Lac 22h23'21"54d31'

Bryndzia,Tanya
And 2h19'36"41d14'

Brynildson,Marit Liv
Cyg 21h21'38"40d11'

Brynn's Star
Cas 1h3'12"63d43'

Brynnstar
Lac 22h14'45"48d36'

Bryon,Paul
Hya 9h3'50"1d46'

Bryony Elizabeth
Cas 0h27'56"60d25'

Brysch,Brendan
Her 17h37'39"25d27'

Bryslan,I Promise You Constance Sue
Cam 3h16'19"60d26'

Bryson MD,K
Oph 18h15'35"11d32'

Bryson,Trent Dana
Mon 7h1'1"7d60'

Brystowski,Bernhard
Tau 5h53'26"23d36'

Brytczuk,Walter
Cep 21h5'1"58d53'

Bryte,Judy Ann Gore
Mon 7h19'53"-8d6'

Brytin
Leo 9h55'34"13d41'

Brzenski,Bill-Joanie- Jared & Lindsay
Ori 6h2'21"8d17'

Brzostek's Midnight Dream,Colette
Vir 12h1'1"1d14'

Bräuhaus,Gisela
Lib 14h47'1"-0d22'

Bröckses,Gerhard
Cas 0h16'37"60d26'

Bröer,Jürgen
Ori 5h58'58"19d29'

Brökers,Susanne
Ari 2h21'40"20d31'

Brösler und Fichtel
Lac 22h13'49"51d32'

Brückner,Leo
Cap 20h19'37"-26d44'

Brüggemann,Hermann
Lyr 19h19'0"30d55'

Brüggemann,Klaus
Peg 23h41'59"16d56'

Brünner,Else
Tau 4h3'0"0d31'

BSL Baby Cat
Lyn 8h22'53"44d35'

BU
Gem 6h50'51"30d20'

Bu,mein Name ist RaBu
Cep 22h2'24"68d48'

Bub
Boo 14h28'48"39d10'

Bub
Cam 13h11'41"80d46'

Bub,Lois & Bill
Aur 6h0'36"45d46'

Bub-Bow
Her 6h12'38"47d39'

Buba You Are The Greatest
Dra 16h2'46"65d52'

Bubba
Ler 15h38'57"8d29'

Bubba
Ori 5h21'36"1d26'

Bubba
Sex 10h47'24"-1d50'

Bubba
Crt 11h15'10"-19d39'

Bubba
Dra 19h38'35"60d39'

Bubba
Her 16h39'49"50d30'

Bubba
Tri 1h58'32"26d16'

Bubba "Brooke's Best Friend"
Equ 21h19'10"11d20'

Bubba & Sissy
Eri 3h39'57"-10d21'

Bubba & Squeaks
Lyr 19h17'52"40d9'

Bubba's Star
Mon 6h57'44"-10d60'

Bubbaette
Umi 16h15'24"72d15'

Bubbeez & Bubbies
Dra 19h44'1"71d5'

Bubble Star
Cyg 21h46'46"36d60'

Bubbles
Peg 22h43'15"33d49'

Bubbles
Cmi 7h53'24"5d18'

Bubbles
Del 20h27'60"10d2'

Bubbles Baby Blues
Peg 23h28'37"21d51'

Bubbles in the Heavens
Cyg 21h50'1"41d49'

Bubblicious Babe
Eri 4h7'14"-3d25'

Bubbly Bill
Ori 4h59'10"-0d21'

Bubby
Cam 12h9'17"76d37'

Bubby
Per 4h4'24"50d27'

Bubby
Lac 21h56'11"42d19'

Bubeck,Jenny Lynn
Uma 11h27'54"56d43'

Bubel,William Ford
Cnc 8h50'27"30d22'

Bubenzer,Waldemar
Cyg 19h52'57"44d23'

Bubert,Jerry
Cet 1h14'1"-04d41'

Bubi
Cmi 8h6'0"1d15'

Bubier,Sally
Cas 23h2'1"53d35'

Bublitz,Emmy
Umi 15h52'28"71d54'

Bubnell,Lois Ann
Cam 6h3'39"83d30'

Bubo e Buba
Pho 0h32'33"-40d26'

Bubo's Wolke 7
Cnc 9h0'36"10d6'

Bubschnug,Bubby
Ori 5h30'55"-8d4'

Bubser,Joanne
Vir 12h27'56"1d27'

Bubsy & Pinky Always
Cnv 12h19'45"47d0'

Bubu
Lib 15h4'1"-6d37'

Bubu Chi Star
Mon 7h1'56"0d50'

Bubukittki,Karin
Cas 1h37'16"60d50'

Bucalo,Tena Roberts
Equ 20h57'27"9d32'

Buccellato,Kathleen Stingo
Cyg 21h55'38"53d34'

Buccheri,Tiffany
And 23h29'12"49d1'

Bucchiere,Alisa Helene Alexandra
And 23h1'31"51d27'

Bucchiere,III,John Joseph
Cnv 12h26'52"50d16'

Bucchiere,Jr,MD,John Joseph
Cep 21h49'57"55d28'

Bucchiere,Julia Barbara Gondek
Uma 9h44'47"71d55'

Bucchiere,Robyn Lee Julia
Cas 23h29'36"63d1'

Bucci,Adolph A
Aur 7h1'37"35d55'

Bucci,Antonella
Lyn 6h12'24"54d56'

Bucciarelli,Lisa Marie
Cam 3h54'28"58d50'

Buccigrossi,Mary
Mon 7h28'0"-8d22'

Buccola,Jennifer Kay
Tri 1h48'1"27d52'

Buccola,Nick Anthony
Ser 15h28'40"8d55'

Buccola,Salvatore
Dra 17h30'10"67d56'

Buchakjian,Alexis Nicole
Cet 0h53'24"1d34'

Buchalter,DO,Eric N
Lac 22h53'1"51d21'

Buchan,June
Cnc 8h31'49"11d10'

Buchanan's Bright Star,Harry
Her 17h28'18"30d35'

Buchanan,Barbara
Cas 1h5'35"63d53'

Buchanan,Charlotte A
Vul 19h42'46"23d5'

Buchanan,D J
Per 2h52'37"31d13'

Buchanan,Edward J
Sex 9h44'51"3d24'

Buchanan,Jessica
Uma 9h14'53"50d56'

Buchanan,Katie
Uma 9h21'47"50d5'

Buchanan,Kirsten
Uma 9h17'10"50d42'

Buchanan,Lee Erin
Eri 4h2'33"-8d34'

Buchanan,Lillian
Del 20h35'55"20d27'

Buchanan,Lin
Lyr 18h51'1"41d8'

Buchanan,Lynn
Cas 3h4'17"65d7'

Buchanan,Megan Allison
Uma 10h26'28"47d31'

Buchanan,Paul P
Dra 11h45'39"74d48'

Buchanan,Philip(Bully)
Boo 14h59'49"21d24'

Buchanan,Shane
Aur 6h16'29"38d6'

Buchanan,William Allan
Her 16h23'54"21d58'

Buchanan,William W
Aql 18h56'15"14d24'

Buchar-Roosa,Lauren
Peg 21h44'44"28d0'

Buchbinder,MD,Neil A
Dra 17h52'12"64d46'

Bucheit,Jack Ryan
Aur 5h2'49"40d55'

Buchen,Marcel
Lib 14h40'56"-20d3'

Buchen,Rainer
Psc 1h15'0"17d34'

Buchenstern
Cap 20h41'19"-26d58'

Bucher,Alfons
Umi 14h12'1"68d9'

Bucher,Brian Frederick
Hya 8h10'35"5d25'

Bucher,Dorothy Angela
Tri 2h17'16"31d11'

Bucher,Edgar
Boo 14h34'34"11d22'

Bucher,Louise
Lyn 6h47'35"60d33'

Bucher,Nicholas James
Ser 15h21'26"4d45'

Bucher,Tom
Cam 7h37'58"60d0'

Bucher,Ursula
Dra 10h33'1"81d21'

Buchert,Nancy
Aur 7h0'0"43d9'

Buchheit,Sarah Elizabeth
Cam 7h49'1"71d15'

Buchholz,Billy
Cep 21h19'29"56d6'

Buchholz,Bryan
Tri 1h51'12"26d41'

Buchholz,Egbert
Ori 5h26'1"1d48'

Buchholz,Greg "Hank"
Dra 18h32'25"58d47'

Buchholz,Lauren
Lyn 6h59'26"60d48'

Buchholz,Patricia
Uma 11h29'18"48d53'

Buchholz,Sarah
Vul 21h16'1"28d8'

Buchholz,Stacen
Lyn 18h58'33"31d42'

Buchhorn,Kerstin
Leo 9h20'51"11d15'

Buchieri,Lisa Marie
Cas 1h20'14"63d54'

Buchin,Erhard
Uma 9h13'50"48d41'

Buchko,Mary
Lyr 19h9'24"38d18'

Buchkremer,Prof Dr Gerhard
Cep 20h56'13"59d26'

Buchla,Ezra
Del 20h18'30"20d14'

Buchman,Daniel
Dra 17h11'40"72d45'

Buchman,Matthew
Boo 14h10'35"31d34'

Buchmann,E
Cmi 7h17'52"0d9'

Buchmann,Eleonore
Leo 10h44'59"21d25'

Buchmoyer,Jeanne Louise
Peg 23h19'57"33d1'

Buchner,Daniel Richard
Lmi 10h45'16"28d4'

Buchner,Elke
Sgr 19h3'49"-23d19'

Buchner,John T
Per 2h23'53"55d19'

Buchner,Sherry
Peg 21h19'32"22d49'

Bucholtz,Michael E
Per 1h58'38"56d41'

Bucholz,Darlene A
Aql 19h54'55"13d12'

Bucholz,Edythe Louise
Cet 1h55'35"0d39'

Buchsbaum,Tony & Ellen
Aql 20h12'14"1d37'

Buchschacher,Donald & Roger
Cyg 19h29'35"30d48'

Buchta Farm Outpost 1
Ori 5h53'59"15d12'

Buchta,Jessica Froome
Aql 19h45'47"12d29'

Buchta,Sharon Elaine
Peg 23h29'56"33d28'

Buchwald,Manfred
And 23h1'31"43d3'

Bucinator,Venator
Lup 15h4'29"-45d29'

Buck
Aql 20h10'28"0d6'

Buck
Eri 2h58'0"-1d44'

Buck
Dra 19h6'35"48d12'

Buck
Cma 7h14'10"-16d49'

Buck 1921-Forever,Jim
Uma 8h32'0"57d11'

Buck I & Buck II
Sge 19h54'41"19d59'

Buck's Nightlight
Aql 20h0'44"4d36'

Buck's Star
Aql 19h54'52"13d34'

Buck,Cassidy Marie
Lib 15h36'29"-28d8'

Buck,Chuck & Sally
Umi 14h38'37"78d16'

Buck,Dick & Harriet
Cyg 19h17'42"48d53'

Buck,Doris
Cyg 19h38'32"30d34'

Buck,Elizabeth Woodstock
And 23h20'1"50d31'

Buck,James William
Lib 14h20'11"-10d59'

Buck,Jason M
Cyg 21h3'13"31d37'

Buck,Jeff J
Cma 6h52'39"-16d56'

Buck,Leslie
Lyn 7h34'38"50d26'

Buck,PhD,Jacqueline N
Sco 16h57'48"-44d49'

Buck,Robert Earl
Her 18h41'52"12d12'

Buck,Sara
Uma 12h22'32"61d43'

Buck,Sheila M
Sco 17h28'25"-31d41'

Buck,Steven Eric
Aur 5h27'14"31d27'

Buck/Leber
Lyr 18h59'31"45d40'

Buckalew & Danny
Cyg 19h56'58"41d3'

Buckberry,Albert
Uma 10h37'10"47d40'

Buckel,C E "Chuck"
Umi 14h34'16"69d35'

Buckel,Darlene L
Umi 13h44'53"71d17'

Buckelew,Nikki Anina
Peg 22h59'13"17d47'

Buckholz,Lee
Per 4h3'21"51d45'

Buckingham,Albert William
Ori 6h3'27"8d30'

Buckingham,Emily Jean
Cnc 9h0'60"31d33'

Buckingham,Marian Marjorie Miller
Peg 22h4'1"26d32'

Buckingham,Susan May
Cas 0h31'46"73d3'

Buckland,Bruce
Psc 1h25'37"25d1'A

Bucklands
Lyr 18h32'42"33d3'

Buckle Star,The John
Aur 6h27'44"30d11'

Buckler Ben
Cam 4h51'44"65d13'

Buckler,Dan
Lmi 9h41'30"37d58'

Buckler,Wilma E
Uma 11h7'17"37d53'

Bucklers Landing
Peg 22h18'15"34d36'

Buckles,Michael
Per 3h31'52"38d2'

Buckles,Tim
Cmi 7h6'1"4d3'

Bucklew,John
Hya 8h21'27"1d43'

Buckley Love Eternal; MaryMatt
Vul 20h1'55"26d8'

Buckley,Angel Rose
Aql 18h57'25"6d15'

Buckley,Bonnie
Cyg 20h20'43"41d52'

Buckley,Colin Dean
Cnv 12h23'59"48d47'

Buckley,James J
Her 17h39'36"25d11'

Buckley,Jan A
Vul 19h36'46"27d17'

Buckley,Jennifer Robin
Psc 22h56'14"0d60'

Buckley,Jerry P
Boo 14h58'15"26d30'

Buckley,John
Ari 2h57'30"30d10'

Buckley,Jon Christian
Cap 20h31'2"-26d9'

Buckley,Joyce A
Cam 3h23'55"61d24'

Buckley,Joyce Marie
Peg 23h22'45"30d28'

Buckley,Jr,Raymond John
Vir 11h40'11"8d28'

Buckley,Kevin Joseph
Boo 14h7'12"42d28'

Buckley,Lisa
Cas 23h14'53"60d44'

Buckley,Lynn
Lep 5h6'0"-10d18'

Buckley,Maria & Eddie
Ori 5h56'44"16d17'

Buckley,Mark
Per 2h58'49"37d1'

Buckley,Michael F
Lyr 18h57'0"34d31'

Buckley,Michael W
Aur 5h7'28"50d14'

Buckley,Neal
Cet 0h34'59"-2d50'

Buckley,Norman
Uma 9h34'1"57d46'

Buckley,Roy & Gerry
Cyg 19h28'26"38d30'

Buckley,Ryan Jari
Dra 17h49'42"64d7'

Buckley,Tracy Michelle
Com 12h17'38"30d23'

Bucklin,Jillian Kelsey
And 23h3'30"48d41'

Bucklin-Durham,Peggy
Peg 23h28'1"15d52'

Buckman,Gary R
Crt 10h50'36"-16d38'

Buckman,James "Cody"
Crt 11h26'49"-15d60'

Buckman,Maxine
Aur 4h56'41"40d36'

Buckmaster,Holly
Lyn 8h38'21"43d6'

Buckmaster-Green, Arlene
Peg 23h18'55"30d43'

Buckner,Amy D
Peg 0h11'59"13d13'

Buckner,Debbie
Sge 20h17'55"20d27'

Buckner,Elizabeth
Mon 6h22'17"3d40'

Buckner,Gordon
Cam 3h56'47"53d34'

Buckner-Reitman,Joyce
Lib 15h16'52"-25d43'

Bucko,Eddie "Teb" 8-20-55
Uma 9h9'25"51d57'B

Bucko,Theresa "Teb" 8-20-55
Uma 9h9'25"51d57'A

Buckovetz,Martin & Monika
Mon 6h38'12"-0d7'

Bucks,Kathy Ann(Kat)
Vul 19h46'59"28d54'

Bucksbaum,Martin
Aur 6h3'56"36d51'

Bucksbaum,Matthew
Psc 23h35'46"2d4'

Buckstone,Merri
Aql 19h11'51"13d47'

Buckus,Bryan Edward
Per 2h52'23"31d27'

Bucky & Pat's Star
Lyr 18h54'47"31d49'

Bucky's Run
Ori 5h52'21"14d16'

Bucolo,Helen & Anthony
Del 20h39'49"10d8'

Bucossi,Linda
Cyg 21h23'56"40d1'

Bucove,Michael A
Cep 2h28'51"77d46'

Bud
Aur 4h47'28"41d13'

Bud
Boo 14h6'49"45d12'

Bud
Cyg 21h44'0"37d56'

Bud
Cmi 7h46'1"7d36'

Bud
Cep 23h4'32"64d44'

Bud & Dona Got Married
Aql 19h30'17"10d38'

Bud & Phyllis
Ori 5h58'56"17d26'

Bud & Shirley Together Forever
Mon 6h14'17"-10d0'

Bud Analytical Star
Oph 17h3'32"10d40'

Bud's Light
Ori 5h57'16"8d28'

Buda,Dan V
Her 17h10'50"27d54'

Budavom
Eri 3h45'26"-3d20'

Budd,Christine
And 2h21'27"45d11'

Budd,Florence
Lyn 9h6'32"38d20'

Budd,Gilleca
And 0h23'16"40d35'

Budd,Hildreth
Peg 23h22'60"30d28'

Budd,Kali Morgan
Peg 21h56'46"2d55'

Budd,Tracy
Cas 0h33'1"65d4'

Budde,Rolf
Ori 5h36'46"7d40'

Buddenbrock,Volker
Vir 13h36'47"-8d3'

Buddha Phil
Vir 13h37'3"-7d56'

Buddie-Buddie-Buddie
Ori 4h52'35"1d14'

Buddy
Oph 17h25'35"-20d15'

Buddy
Boo 14h19'18"36d4'

Buddy
Dra 17h7'3"50d54'A

Buddy
Boo 14h19'1"16d12'

Buddy
Cmi 7h38'52"5d14'

Buddy & Peggy
Lyn 7h45'16"58d39'

Buddy Buddy Little Star
Her 17h10'34"27d18'

Buddy from Buddy's Carpet Barn
Her 16h4'29"50d47'

Buddy M
Per 1h48'1"56d26'

Buddy November 1st 1980
Lyr 18h40'28"31d30'

Buddy Ro
Cep 22h8'52"54d21'

Buddy Rushton
Cmi 7h54'34"8d6'

Buddy's Flame
Oph 18h19'54"6d47'

Buddyville
Per 2h50'35"43d13'

Budgell,Bonnie Danielle
And 1h52'37"41d5'

Budgen,Alexandra Clare
Cyg 21h43'11"38d16'

Budgen-'Silverdale', Nigel
Per 3h15'58"42d34'

Budimann,Richardos
Her 17h52'24"14d53'

Budinas,Kathleen
Cmi 7h43'50"5d21'

Budlong,Willis Arthur
Ori 5h0'34"1d51'

Budner,Susan Ellen
Com 12h54'37"24d45'

Budnovich,Katherine
Eri 3h31'54"-1d38'

Budrow,Fern
Eri 4h53'42"-9d52'

Budrow,Perry
Eri 4h2'17"-10d13'

Budy,Helen
Cas 0h7'32"64d2'

Budz,Tad
Lac 22h31'50"52d52'

Budzi
Cam 6h19'0"67d31'

Budzik,Jonalyn Marie
Cnv 12h21'19"38d37'

Budzinski,Elizabeth
Cas 0h24'23"61d23'

Budzinski,Pete
Sex 9h57'0"-6d27'

Buechel,Mary Agnes
And 23h18'28"48d55'

Buechel,William J
Cep 22h37'44"60d22'

Buechele,Eugene Walter
Cep 2h34'43"87d25'

Buecherl,Prof Dr E S
Uma 9h37'25"50d60'

Buechner,Michael Edward
Lac 22h4'31"38d34'

Buegler,Klaus
Cyg 20h19'39"47d20'

Buehler,Arthur
Per 4h3'22"39d35'

Buehler,Dell Morgan
Cnc 8h58'42"17d12'

Buehler,Jr,John "Jack" Henry
Per 4h37'21"37d18'

Buehler,Lanette
Lib 14h31'57"-20d9'

Buehler,Superdoc
Cet 1h26'23"-0d52'

Buel's Celestial Trophy,Roger
Hya 9h13'1"-14d4'

Buell,Cayla
Cas 1h5'46"63d57'

Buell,Doreen Marie
And 1h31'43"39d29'

Buell,Doug
Her 18h1'22"28d43'

Buell,Mark
Cap 20h4'38"-10d34'

Buell,Richard
Boo 15h2'0"31d42'

Buell,Shawnee Lee
Com 12h58'25"27d39'

Buelow,Diane
Dra 16h47'43"70d51'

Buelow,Kenneth Martin
Dra 16h32'47"69d39'

Buelter,Nicholas Kyle
Aql 19h6'1"-5d49'

Buemi,Sally
Lyr 19h13'47"37d38'

Buenger,Mel
Cam 8h1'24"77d55'

Buentello,Salvador (Sam)
Aql 19h5'0"15d12'

Buer,Shane
Uma 10h2'0"48d15'

Buergel,Tuffer
Aur 5h19'19"42d22'

Buergermann,Horst
Lac 22h4'21"50d49'
Buescher,Helen (Wadie)
Sge 19h29'57"16d19'
Buescher,Hermann-Josef
Cmi 7h31'21"0d6'
Buescher,J Kenneth
Lac 22h50'25"53d59'
Buescher,Robert & Tammy
Crb 16h12'33"34d17'
Buesener,Frank "D"
Ari 2h43'21"30d16'
Buettner,Brea Alisabeth
And 2h33'58"44d8'
Buettner,Kimm
Leo 10h59'1"15d46'
Bufano,Alexa Rizai
Cyg 20h3'37"31d32'
Bufano,Aprile Rizai
Cas 22h58'15"53d31'
Buff & Bird
Oph 18h42'22"6d51'
Buffalo Museum Of Sci Search For A Star Team
Dra 18h43'0"68d29'
Buffalo Soldier
Aur 5h33'47"38d58'
Buffalo,Jr,Joseph L
Aur 4h34'1"34d16'
Buffam,A B
Mon 7h0'11"-6d48'
Buffard,Tom
Per 3h59'53"35d16'
Buffet,Michel
Cas 1h40'27"64d41'
Buffett,Jennifer Ann
Gem 7h4'5"34d5'A
Buffett,Peter Andrew
Gem 7h4'5"34d5'B
Buffierè,Didier
Per 3h28'35"50d16'
Buffington,Kathleen Anne
And 2h22'0"42d26'
Buffinton,Meggan & Sean
Sge 19h26'25"16d44'
Buffler,Dale & Rosemary
Cas 0h16'46"64d24'
Buffolino,A J
Dra 18h57'53"58d23'
Buffone,"Bijou" Katja
Dra 15h39'41"63d11'
Bufford,Rick
Ori 6h0'30"4d44'
Buffy
Lib 15h16'14"-23d42'
Buffy
Cet 2h18'42"2d41'
Buffy
Cam 3h22'22"66d40'
Buffy,Maxwell
Peg 22h57'59"20d49'
Bufka
Per 3h44'36"50d26'
Bufkin,Brian
Ori 6h7'0"-2d33'
Bufo
Cyg 19h58'29"44d58'
Bufo Bufo
Cep 3h12'41"80d22'
Buford & Asherah's Attitude
Cet 2h13'20"3d4'
Buford IV,Robert Pegram
Vir 11h47'54"3d26'
Buford,Anne
Lib 15h38'12"-22d31'
Bug
Umi 16h25'16"79d34'
Bugarin,Haile
Hya 8h23'0"-0d1'
Bugas,David Anthony
Cam 6h31'31"82d42'
Bugbee,Hazel
Lac 22h43'0"56d47'
Bugdahl,Nicole
Lyn 7h52'0"58d21'

Bugden,Henry Gordon
Aur 4h56'26"50d54'
Bugg,Brandi
Aql 20h13'37"1d20'
Bugg,Leon Carter
Cnv 12h42'38"37d23'
Bugs
Cet 3h7'15"0d54'
Bugsy
Peg 21h29'47"20d36'
Bugsy
Her 16h58'38"30d40'
Buguet,Christian
Per 3h5'50"31d16'
Buguloo
Her 16h4'25"50d26'
Buhl,Norman
Aql 19h49'19"14d11'
Buhl,William Reid
Cet 0h40'35"-4d10'
Buhler,Brian
Her 17h51'14"38d49'
Buhler,Marcel
Aur 5h57'11"50d12'
Buholzer,Melissa
And 23h43'21"44d43'
Buhr,Karin
Peg 23h32'15"12d58'
Buhrfeind,Susanne
Boo 14h48'27"47d39'
Buhrke,Günter
Uma 9h13'39"59d8'
Buhrman,Terry
Mon 6h5'41"-8d1'
Buhrmaster,Sarah Christine
Lyr 19h17'52"40d57'
Buhrow,Sam Edward
Cma 7h14'45"-13d55'
Buhse,Wil
Per 1h29'24"53d10'
Buhtz,Corina
Aqr 22h3'59"2d7'
Buick,Lillian
Lyr 18h32'12"33d4'
Buico,Paul
Lyn 6h15'23"58d12'
Buikema,David Henry
Boo 13h36'30"22d1'
Buikema,Nathan
Aql 18h56'24"-2d29'
Buirge,Daelynn
Lyr 18h46'56"30d60"
Buist,Edward R
Aur 5h23'37"37d31'
(Bujara)Lovely Irene
Cas 2h36'22"57d40'
Buker,General
Sex 10h30'37"0d14'
Bukhanovsky,Alexander Olimpievich
Per 3h39'40"38d32'
Bukhanovsky,Inna Borisovna
Cas 0h24'38"58d21'
Bukhanovsky,Olga Alexandrovna
Cyg 19h50'11"39d24'
Bukis,Samantha Lee
Cas 0h10'11"63d7'
Bukovsky,Jennifer Lynn
Lyr 18h18'34"38d53'
Bukowski
Uma 11h0'37"51d57'
Bukowski,Bobby
Per 2h40'29"35d3'
Bukowski,Buck
Aur 7h16'12"36d48'
Bukowski,Charles
Leo 10h42'59"14d9'
Bukowski,Linda Lee
Leo 10h39'47"14d24'
Bukowski,Uta
Cas 23h34'19"61d22'
Bulb in my Lamp,The
Ori 6h3'52"1d7'

Bulboaca,Elly
Cet 0h26'53"-5d31'
Bulck,Dennis
Cep 22h10'0"70d29'
Buldhaupt,Brent
Cep 23h16'20"70d5'
Buletta,Samuel Michael
Cma 6h27'37"-16d28'
Buleza,Karen M
Lyr 19h4'59"25d58'
Bulfinch,Susan
Vul 19h18'28"25d44'
Bulgarelli,Maria Vittoria
Pic 5h1'57"-49d33'
Bulgari
Dra 16h58'39"61d20'
Bulger,John B
Boo 15h9'55"30d17'
Bulifant,Alban & Carol
Aur 5h3'1"40d13'
Bull's Boomer
Dra 17h30'25"61d19'
Bull's Eye
Lyr 19h21'53"31d20'
Bull,Jim
Cep 2h5'23"86d58'
Bull,Nell & Julie
Peg 23h40'56"31d11'
Bull,Terry
Ori 6h7'44"5d45'
Bulla,Mickey
Ori 5h53'26"15d9'
Bullarby
Uma 12h27'26"56d55'
Bullard III,William (Bill) Jordan
Her 16h18'35"8d45'
Bullard,Ed
Per 3h34'24"40d15'
Bullard,Jr,John W
Aql 20h0'19"11d20'
Bullard,Ronnie
Dra 15h12'45"63d41'
Bullen,Kimberly
Cas 0h59'0"61d6'
Bullen,Robin Kearsley
Ori 6h8'14"2d27'
Bullet
Lyn 9h9'39"39d12'
Bulley,Allan E
Boo 14h18'53"15d34'
Bullington,Nancy Ann
Eri 4h14'5"-2d5'
Bullion,Greg G
Aql 18h58'40"-6d58'
Bullman,Carolyn Mae
Cyg 20h41'0"42d16'
Bulloch,Avril Patricia
Lyr 18h29'18"44d35'
Bullock's Sherman Oaks
Mon 6h22'41"8d36'
Bullock,Bruce
Hya 8h12'21"0d3'
Bullock,Caga December 25,1993
Uma 9h2'33"47d43'
Bullock,Douglas Lee
Her 17h6'23"38d53'
Bullock,Douglas Lee
Her 17h8'44"37d34'
Bullock,Jr,William O
Aql 19h58'1"10d58'
Bullock,Lem
Eri 3h10'1"-1d46'
Bullock,Mike & Amy
Umi 15h40'50"75d39'
Bullock,Nicola
And 23h0'45"37d9'
Bullock,Theodore Kerry
Umi 15h11'12"70d41'
Bullock,W Carl
Dra 19h5'25"78d50'
Bully for Montezuma's Revenge
Cma 7h17'15"-16d39'

Bully for Murdocks Bon Voyage
Cma 6h59'39"-19d29'
Bulmer,Katie
Lyr 18h35'21"30d19'
Bulmer,William
Aql 19h29'56"12d5'
Bulycz,Glenn Jean
Aql 19h31'0"12d23'
Bumbagiuppygigia
Col 6h29'6"-37d2'
Bumbaugh,Susan
Cyg 21h7'0"30d26'
Bumberger,Claudia Theresa
Lyn 6h28'50"61d10'
Bumbles
Uma 12h6'25"61d46'
Bumbly
Umi 13h6'37"70d57'
Bumerts,Carol
Mon 6h32'42"11d41'
Bumgardner,Gretchen
And 0h13'39"38d46'
Bumgardner,Julia Rose
Lyr 19h6'0"28d20'
Bumgarner,Marty
Per 1h30'1"53d59'
Bump Lisk
Lac 22h38'51"52d33'
Bumpas,Jr,Guy Hartwell
Her 15h51'37"40d50'
Bumper's Sunshine
Lac 22h25'38"50d2'
Bumpers,Clay
Aql 19h30'59"8d25'
Bumpers,John & Pat
Mon 8h8'53"-6d25'
Bumphrey,Stephen
Oph 18h16'46"11d45'
Bumphysis,Kohlman
Hya 8h32'46"5d59'
Bumpus,Becky Ann
Cam 4h1'50"58d34'
Bun & Iggy
Eri 3h56'43"-10d26'
Bunbo
Cet 2h29'37"-8d50'
Bunce,Donna
Sct 18h48'44"-7d33'
Bunce,Mary
Boo 14h39'33"48d16'
Bunce,Stewart
Aur 6h15'18"31d33'
Bunch,Ben
Oph 17h59'0"0d14'
Bunch,Brian
Uma 10h28'55"70d58'
Bunch,Charlotte
Cas 1h33'39"75d56'
Bunch,Ida
Uma 11h40'57"37d59'
Bunch,Jason
Aql 19h47'59"11d43'
Bunch,Joseph D
Cet 0h33'39"0d49'
Bunch,Kelly
Peg 22h41'25"21d51'
Bunch,Linda Jean
Cas 1h19'13"61d33'
Bunch,Ronnie & Ryan
Cam 5h35'13"67d51'
Buncich,Timothy K
Ser 16h7'0"5d50'
Bundalo,Milan R
Uma 11h43'29"45d44'
Bundels
Eri 3h18'48"-4d5'
Bundrick,Kallan Kathryn Parker
Lyn 8h1'48"39d27'
Bundy Star,The Thomas Michael
Her 18h12'48"31d47'
Bundy's Misty,Harvey
Dra 15h52'1"60d54'

Bundy,Audrey
Tau 4h36'33"28d17'
Bundy,John
Sct 18h56'32"-5d10'
Bundy,John Kevin
Per 2h5'30"53d23'
Bundy,Keith & Tracy
Cyg 20h26'49"48d52'
Bundy,Paul Eugene
Hya 8h16'43"1d13'
Bundy,Sarah Jo
And 23h31'0"45d42'
Bundy-Venard,Elisabeth
Dra 10h1'13"74d54'
Bunger,Chris & Ruth
Ori 5h9'59"-5d17'
Bungert,Anja Plüschtiger
Cyg 19h25'53"35d8'
Buniva,Wing & A Prayer
Aql 18h54'1"-0d42'
Bunke,Rainer Wolfgang
Aqr 21h26'0"-0d35'
Bunker,Charlotte
Mon 6h7'15"-5d57'
Bunker,Daphne
Lyr 18h16'55"38d0'
Bunker,Dynamic Dave Christopher
Per 3h15'16"40d39'
Bunker,James Henry Michael
Cma 6h51'38"-16d60'
Bunker,Julie A
Cas 0h34'17"60d4'
Bunker,Norman "Marty"
Aur 7h4'39"38d21'
Bunker,Oliver
Mon 6h19'40"-5d54'
Bunker,Tracy Michelle
Mon 7h6'21"-0d20'
Bunker,William & Shirley
Her 17h13'16"41d36'
Bunkey
Ori 5h43'18"1d19'
Bunkie Doodle
Lyn 8h15'49"38d15'
Bunky
Dra 16h9'53"62d29'
Bunn,Harvey W
Her 16h38'1"32d34'
Bunn,Henry C
Her 16h58'50"35d54'
Bunn,Richard
Per 4h7'30"51d56'
Bunnell,Richard Steven
Per 2h53'16"34d14'
Bunnell,Shelli L
Eri 3h30'42"-1d56'
Bunnell,Toni Marie
Eri 2h58'22"-11d21'
Bunner 3-31-95
Uma 10h41'0"57d35'
Bunnie
Lac 22h21'27"45d5'A
Bunny
Cma 6h53'3"-18d45'
Bunny
Del 20h13'48"14d30'
Bunny
Dra 16h30'58"62d32'
Bunny
Peg 22h12'0"33d54'
Bunny
Ori 5h52'26"5d13'
Bunny Wipe
Vul 19h48'46"25d12'
Bunnykins
And 23h12'0"41d4'
Bunola
Lyn 8h40'44"44d44'
Bunshor,Brian
Aql 20h33'1"-0d5'
Bunte,Stefan
Per 3h51'20"38d22'
Bunting,Rachael
Peg 21h54'53"29d59'

Bunting,Samuel Evan
Aur 4h57'29"36d54'
Bunty
Cyg 19h27'49"34d57'
Bunyan,Alan Donald
Dra 20h17'27"68d8'
Bunz,Dietmar
Uma 13h53'34"50d10'
Bunzel,Bernd
Cmi 7h21'25"5d3'
Buon Natale
Lyn 7h44'54"50d9'
Buona Fortuna
Aqr 22h18'2'-14d16'
Buonaiuto,Nicholas
Aur 4h54'57"51d24'
Buonamici,Sandra
And 1h2'50"47d38'
Buoncompleanno
Lep 5h6'1"-11d13'
Buonfiglio,Tony
Cep 3h3'26"80d32'
Buono,Juliette
Lyr 18h58'52"38d21'
Buono,Melissa
Dra 19h46'10"59d0'
Buono,Miriam Leon
Crb 16h0'1"37d46'
Buono,Paul A
Cma 6h33'13"-28d59'
Buono,Steve
Aur 6h28'26"35d7'
Buonocore,Carly Ann
And 0h8'46"47d58'
Buonocore-Zamacv
Cyg 21h54'20sd37d5'
Buonomo,Robert
Aur 4h35'56"31d0'
Buonvino,Clelia
Tel 20h9'24"-47d39'
Buote,Jeana
Cas 3h3'47"58d26'
Buquet III,James Joseph
Sct 18h44'48"-6d44'
Burak,Cory Dale
Aur 7h24'37"38d19'
Burak,Wasyl & Elizabeth
Cyg 21h4'46"38d27'
Burbank Bes Dad,Robert Michael
Ori 5h59'14"20d36'
Burbank,Daniel
Her 17h36'0"25d42'
Burbank,Emily
Peg 22h6'17"21d24'
Burbank,James
Lac 22h5'38"48d59'
Burbvank-Gosnell,Tracy Lynn
Peg 23h37'13"30d53'
Burch
Aur 6h21'21"38d23'
Burch,Carl Edward
Cma 7h49'33"11d28'
Burch,Donna Lynn
Lyn 8h7'38"58d53'
Burch,Greg
Cmi 7h15'25"5d12'
Burch,Jamie Lee
Leo 9h20'0"17d12'
Burch,Jennifer Leigh
Peg 22h55'52"27d39'
Burch,Jessica
Lyr 18h40'1"47d20'
Burch,Katrina
And 2h28'28"40d15'
Burch,Kristin
Lmi 10h38'10"24d27'
Burch,Richard McKenzie
Lac 22h17'20"37d44'
Burch,Stephen Matthew
Uma 9h54'1"58d36'
Burchell,A Star With God,David
Cep 22h25'59"61d1'

Burchenal,Adam B
Aql 18h56'23"-0d47'
Burchett,Jack
Aur 6h53'53"41d0'
Burchett,Jake
Cet 0h8'26"-10d27'
Burchett,Kara Renee
Del 20h13'15"10d8'
Burchett,Karson
Del 20h55'19"9d13'
Burchett,Lucas
Cep 23h3'16"62d52'
Burchett,Noelle
Boo 14h28'35"20d24'
Burchett,Tierney
Mon 8h1'25"-1d33'
Burchett,Tori
And 0h54'40"45d11'
Burchett,William
Aql 20h14'36"0d10'
Burchette,Frank
Dra 17h36'1"60d49'
Burchfield,John Wayne
Lac 22h17'45"54d31'
Burchfield,Russell
Dra 16h35'31"69d52'
Burchielli,Emilio
Oph 17h54'23"11d41'
Burchill,Kevin Joseph
Aur 5h1'0"41d9'
Burchill,Matthew Robert
Boo 15h18'42"41d47'
Burd,Paul Alan
Her 18h4'57"28d33'
Burdell,Eloise
Ori 5h4'31"8d58'
Burden,Amanda
Uma 8h41'22"60d7'
Burden,Janelle Ann
And 2h19'52"48d24'
Burden,Janice Marie
Peg 23h29'48"21d53'
Burden,Kathi
Cas 0h16'0"58d19'
Burden,Mary "The Dutchess"
Cam 9h7'29"80d19'
Burdet,Andre
Cnv 14h3'37"37d54'
Burdett,Nicola
Cyg 19h27'47"36d51'
Burdette,PhD,MD, Walter J
Oph 16h52'26"10d44'
Burdge,Betty Jo
And 22h57'1"36d41'
Burdge,Samuel T
Lac 22h33'53"55d56'
Burdick,Ginny
Vul 20h44'39"20d10'
Burdick,Richard Louis
Ori 5h38'30"-6d4'
Burdin,Thierry
Cas 3h6'54"58d36'
Burdon,Chris
Umi 15h25'20"68d13'
Burdyshaw,Rick
Boo 14h58'0"44d13'
Burdzilauskas,Laura Ruth
Cas 23h33'0"58d11'
Bureau,Florence
Boo 14h30'18"40d28'
Burel,David
Aur 6h34'31"38d44'
Burelison,Sheila Ann Kuczynski
Cnc 7h55'1"10d37'
Burelle,Lynne
Umi 15h21'1"66d50'
Buren,Jaques
Cyg 19h47'48"31d48'
Buresh,Christopher John
Cep 21h54'33"60d43'
Burey Eternal,Andrew
Aql 19h24'33"13d46'
Burfield,Christopher
Per 3h0'46"46d7'

Burfitt,Elaine
And 2h22'0"42d26'
Burford,Raymond Thomas
Cep 20h7'24"60d4'
Burford,Stan
Boo 15h28'19"41d56'
Burg,Deborah Therese
Lyr 18h33'24"35d17'
Burg,Michael"Hunny Bunny"
Her 17h32'27"40d2'
Burg,Nicole Marie
And 23h43'41"42d51'
Burgalet,mi Nana, Isabel Navaro
Ser 15h19'46"4d11'
Burgamy,Sherry Lee
Tau 3h51'52"22d17'
Burgandy's Moon
Del 20h17'50"13d23'
Burgard,Linda & Duane
Lyr 19h24'58"38d33'
Burgdorf,Sabine
Lyn 8h13'50"47d22'
Burge,Linda G
Cap 20h45'28"-26d19'
Burgener,Ruthe B
Mon 7h54'32"-1d52'
Burger,Alex
Cmi 7h38'34"5d2'
Burger,Billy
Ori 5h52'14"10d51'
Burger,David & Sally Guard
Sge 19h24'27"16d12'
Burger,Jason
Ori 4h47'51"4d59'
Burger,Jeffrey Allan
Aqr 22h3'45"-12d5'
Burger,Jeffrey Keith Scobey
Aql 19h31'18"12d1'
Burger,Jr,Dwight G
Cep 0h18'57"66d60'
Burger,Jr,Thomas Charles
Ori 4h52'29"4d49'
Burger,Jr,William Eugene
Aur 5h2'43"37d48'
Burger,Karolyn Jean
Gem 6h54'6"16d7'
Burger,Ken
Lac 22h36'25"55d16'
Burger,Mickey Weiss
Cmi 7h56'55"8d56'
Burger,Misha
Cyg 20h14'49"37d46'
Burger,Nancy
Lib 15h32'36"-28d15'
Burger,Sarah A
Cas 2h51'16"75d37'
Burgeson,Mary Ann
Crb 16h2'13"31d17'
Burgess
Leo 10h39'39"21d14'
Burgess Star,The Jim "Tequila"
Cet 2h13'21"4d21'
Burgess,Betsy A
Hya 8h10'18"0d8'
Burgess,Bobby
Umi 15h1'0"68d43'
Burgess,Bussy & Tina
Uma 10h9'27"58d39'
Burgess,Carmel S
Peg 22h38'46"25d36'
Burgess,Charles & Gladys
Dra 17h25'1"70d22'
Burgess,Craig
Dra 12h39'20"72d22'
Burgess,David P
Dra 16h40'46"63d60'
Burgess,Deborah L
Aur 6h30'44"31d15'
Burgess,Deborah L
Dra 17h53'22"75d5'
Burgess,Denise
Aql 19h42'13"14d1'

Burgess,Dianne
Lyr 19h15'59"26d56'
Burgess,Emma
Lyr 19h24'59"42d2'
Burgess,Garry
Cyg 19h49'17"38d54'
Burgess,Gary
Ori 5h0'0"10d23'
Burgess,Gelsomina
Ori 5h56'44"8d57'
Burgess,Jessica Kate
Cas 0h16'35"60d37'
Burgess,Jim
Cep 21h48'38"55d40'
Burgess,Julie A
And 23h37'53"49d47'
Burgess,Kathryn Thomas
Mon 6h44'0"11d26'
Burgess,Mary
Ori 5h57'0"10d53'
Burgess,Mary
Aql 18h58'0"-6d52'
Burgess,Melissa S
Aql 19h7'1"1d24'
Burgess,My Only Star. Love 69,Bree
Cyg 21h36'30"42d28'
Burgess,Rebecca Lucy
Peg 21h54'1"33d7'
Burgess,Richard
Aur 7h1'17"36d48'
Burgess,Scott R
Per 4h6'19"50d23'
Burgess,Vince
Tri 2h20'30"30d14'
Burgess,William G
Oph 18h17'53"8d52'
Burget,Undine Ehrman
Sge 19h38'1"16d16'
Burghardt,Kendra
Ori 4h43'21"-1d4'
Burghardt,Paul
Uma 9h35'0"43d10'
Burgin,Colleen Elizabeth
Crt 11h9'26"-11d2'
Burgin,Sherry Lynne
Vul 19h18'0"22d55'
Burgin,Simon & Katie
Cyg 20h23'0"40d54'
Burgio,David L
Sco 16h37'43"-33d57'
Burgio,Dr Maria R
Cas 0h0'57"64d56'
Burgio,Luann
Del 20h13'17"15d30'
Burgio,Robert & Elizabeth
Dra 20h5'19"64d42'
Burgio,Teresa
Pic 5h0'13"-45d43'
Burgmans,Anthony
Her 18h11'1"37d59'
Burgmeier,Franz
Per 1h48'50"54d21'
Burgoine,Steven
Lyn 8h24'33"40d5'
Burgold,Lothar
Cyg 20h25'39"30d22'
Burgoon,Danielle Lynn
Cet 1h58'0"-0d25'
Burgos Humanista, Julian
Aql 19h57'27"14d27'
Burgos"Mystique",Linda Denise
Cas 1h13'0"63d54'
Burgoyne,II,John
Aur 6h17'25"37d32'
Burguet,Ramon Sanllehi
Uma 9h32'26"47d45'
Buri,Aliana B
Com 12h23'26"32d12'
Buricand,Jean Louis
Dra 15h46'11"62d51'
Buriez,Laurence
Dra 19h5'32"80d6'

Burigo,Lorraine
And 23h21'0"45d10'
Burink,Debbie
And 0h15'1""41d1'
Burish,M A
Cam 6h7'42"60d9'
Burk,APOSTLE William Elliott
Cep 21h4'36"61d36'
Burk,Stella
Uma 11h42'0"42d12'
Burkardt,Günter 18-8- 1928
Star 73848786
Uma 12h46'11"57d40'
Burkart,Kyle William
Aqr 21h3'1"-13d55'
Burkart,Marion
Cas 0h22'58"69d47'
Burkart,Michael
Lib 15h30'34"-10d48'
Burkart-Paulson,Amy J
And 0h21'29"30d18'
Burkart-Paulson,Kate
Cas 0h33'43"72d56'
Burke
Per 2h29'52"57d44'
Burke I,Shanon Marie
Cas 0h32'18"71d27'
Burke III -Trey-,James Michael
Cet 3h16'48"2d19'
Burke Star,The Alan
Cep 23h8'35"61d14'
Burke The"Spellcaster" David
B
Aql 19h48'18"13d29'
Burke's Rest Stop, Ken
Vul 19h19'22"26d46'
Burke,Allyson
Lyn 9h15'28"37d36'
Burke,Angela Maria
And 0h34'48"45d3'
Burke,Ann Cathrynn
And 1h33'57"48d47'
Burke,Audrey Dillon
Aql 20h12'35"14d28'
Burke,Bette
Cas 1h27'13"67d58'
Burke,Brendan E
Lmi 9h40'54"40d8'
Burke,Carol
Hya 9h9'39"0d13'
Burke,Carol Ann
And 0h7'33"46d52'
Burke,Caroline
Cnc 7h56'1"10d38'
Burke,Charles W, Margaret &
Kelly
Cyg 19h28'59"38d23'
Burke,Craig
Mon 8h2'23"-7d14'
Burke,Crystal
Aql 19h40'39"10d32'
Burke,David L
Cep 22h36'28"61d26'
Burke,Edward
Dra 17h35'49"61d2'
Burke,Elizabeth
Peg 21h57'41"21d31'
Burke,Erika Faye
Lyn 8h11'47"36d10'
Burke,Frank J
Cep 21h54'40"60d55'
Burke,G B,Jenna,Kelly &
Kristen
Cyg 19h29'39"33d47'
Burke,Gail
Aql 20h7'41"0d15'
Burke,Gerd
Aur 5h8'16"43d47'
Burke,J Michael
Boo 15h6'29"10d39'
Burke,Jack
Cam 12h9'31"78d5'
Burke,Jaison Abram
Her 18h14'11"38d39'
Burke,James "Bud"
Sex 10h27'35"-5d22'
Burke,Jim
Cep 0h23'59"80d3'
Burke,Jim
Per 2h5'28"57d21'
Burke,Joan Marie
And 23h35'22"44d50'
Burke,Joel & Sandy
Cam 3h15'58"66d30'
Burke,Joey
Sct 18h32'0"-4d8'
Burke,John & Grace
Vul 21h27'30"28d9'
Burke,John & Karyn
Aql 19h45'11"12d24'
Burke,John Francis
Lac 22h33'32"53d12'
Burke,John Patrick
Cam 3h30'47"60d51'
Burke,Jonathon
Mon 7h4'1"5d2'
Burke,Joshua
Ori 5h38'53"12d54'
Burke,Jr,James P
Cep 23h17'55"64d49'
Burke,Jr.-August 13, 1943
James J
Leo 9h22'33"19d34'
Burke,Julie A
Peg 22h35'35"21d36'
Burke,Katherine M
Cam 6h53'35"70d19'
Burke,Kerry
Ori 5h49'35"19d32'
Burke,Kevin
Mon 6h54'54"0d8'
Burke,Kimberly Ann
Mon 7h19'1"-8d17'
Burke,Lauren Elizabeth
And 1h43'52"39d58'
Burke,Lieutenant Commander
Joseph S
Lmi 10h27'50"28d21'
Burke,Linda "Angel"
Cas 2h33'28"58d36'
Burke,Lisa
Lyr 19h1'0"25d56'
Burke,Mae Ellen
Cyg 21h29'0"53d11'
Burke,Margaret D
Cam 4h4'25"78d57'
Burke,Margaret L
Uma 11h55'0"52d4'
Burke,Marie
Cet 1h29'53"-14d13'
Burke,Marie Teresa
Ori 6h16'36"8d37'
Burke,Merle H
Aur 6h10'43"33d44'
Burke,Mom & Dad
Dra 10h53'54"78d3'
Burke,Patricia
Cas 2h55'49"58d36'
Burke,Patricia A
Lyn 9h9'12"40d43'
Burke,Patrick
Uma 10h28'56"56d31'
Burke,Patrick J
Uma 9h27'0"51d21'
Burke,Paul William
Lyn 7h55'33"39d36'
Burke,Reynold Brooks
Her 18h4'41"45d8'
Burke,Rita
Lyr 19h21'14"42d53'
Burke,Robert F
Cma 6h43'24"-18d5'
Burke,Ronnie
Aql 20h1'20"1d3'
Burke,Rosalyn
Crb 16h0'50"30d60'
Burke,Ryan Patrick
Boo 13h37'50"22d3'
Burke,Scott
Boo 15h4'24"30d7'
Burke,Susan
Mic 21h1'22"-27d43'
Burke,Susan
Cas 0h41'17"62d28'
Burke,Suzanne
Cas 1h2'27"62d2'
Burke,T R
Lac 21h56'37"40d15'
Burke,The Great Pumpkin aka
Allan
Her 15h56'1"42d15'
Burke,Theodosia
And 23h27'0"48d23'
Burke,Thomas J
Per 2h2'1"48d15'
Burke,Whitney Kennedy
Uma 10h57'1"71d48'
Burke,William L
Uma 10h53'0"53d2'
Burke,William Oliver
Dra 15h46'28"60d30'
Burke-Donnellan Don & Judy
Vul 19h48'50"20d20'
Burke-Gomez,Chelsea Jean
Mon 6h32'32"10d23'
Burke-Rosenberg,Collin
Hya 8h43'38"1d12'
Burke-Winger,Elaine
Cyg 21h31'25"41d32'
Burkes,Michael Edward
Uma 11h3'41"61d11'
Burkes,Paula Lynn
Cas 0h48'39"65d50'
Burket,Jr,John Zimmerman
Gem 6h59'53"16d38'
Burkett "Eagle Scout", Thomas
T
Aql 19h57'10"15d13'
Burkett,Jennifer Lynn
Lib 15h18'27"-22d59'
Burkett,Karen Michele
Cas 23h42'11"60d58'
Burkett,Raymond
Cep 22h30'0"58d25'
Burkett,Virginia Shannon
Ari 1h49'22"23d38'
Burkey,Ashlie Ann
Cyg 19h35'54"28d23'
Burkhalter,Dorothy
Cyg 21h57'14"50d17'
Burkhalter,Melody J
Peg 21h57'59"21d5'
Burkhard und Adelheid
Her 17h26'18"42d27'
Burkhard,Brooke Caroline
Psc 1h27'41"31d15'
Burkhard,Jamie Frances
Lac 22h4'28"49d7'
Burkhard,Jenna Nicole
Cas 1h12'0"62d41'
Burkhard,Joseph Clarke
Boo 15h36'32"48d17'
Burkhardt,Doreen
Sco 17h9'0"-38d18'
Burkhardt,Dorothy J
And 0h48'37"35d45'
Burkhardt,William Frederick
Cet 2h26'15"4d4'
Burkhart,Elke & Jürgen
Boo 13h34'40"11d47'
Burkhart,Fred
Tri 2h18'15"30d26'
Burkhart,Jessica Taylor
Lyn 7h35'29"36d16'
Burkhart,JM
Aql 19h29'42"1d17'
Burkhart,Walter
Her 16h10'1"24d30'
Burkhart,William C
Dra 18h59'32"68d34'
Burkher,Jeff & Denise
Aql 19h53'30"1d0'
Burkholder,Carol Ann (Hess)
And 22h58'26"50d16'
Burkle,Ron
Oph 17h15'28"-20d12'
Burkle,Sandra Lee
And 1h24'1"38d52'
Burkley,Annalee Rae
Lyr 18h51'54"31d55'
Burkley,Kathleen
And 2h25'31"42d18'
Burko Magnifico,Robert L
Lmi 10h15'35"36d57'
Burlap,Jr,Christopher John
Ori 5h37'12"-0d23'
Burle,John
Cet 3h12'1"1d54'
Burle,Romain et Corine
And 1h8'39"47d27'
Burleigh,Annaliese Ruth
Lup 15h23'56"-45d17'
Burleigh,Chase
Aur 6h44'55"38d21'
Burleigh,Emmalienne Patricia
Lup 15h27'46"-41d54'
Burleigh,Flossie
And 2h20'28"41d52'
Burleigh,Matthew Thomas
Lup 15h27'49"-44d52'
Burles-Montoya,Joseph
Cet 3h0'0"0d21'
Burleson,Sr,Dana Lee
And 23h38'41"42d27'
Burley,Arthur Edwin
Dra 20h25'22"63d22'
Burley,Michael William
Cep 22h8'34"68d26'
Burling,Thomas Robert
Anthony
Uma 8h56'28"60d57'
Burlingame,Alan
Cep 22h54'52"58d2'
Burlington,Abigail Parker
Lyn 8h32'34"43d55'
Burman,Bryan A
Ori 6h0'58"1d11'
Burman,Dr Richard
Dra 12h0'30"71d38'
Burman,Jacob Reed
Aur 7h22'1"38d52'
Burmer,Jeffrey Keith
Boo 14h35'1"18d54'
Burn's Birthday Star, Bridget
Lac 22h3'50"49d15'
Burn,Geraldine Jane
Lyr 19h4'21"26d17'
Burnand,Raechel
Cet 2h29'0"1d32'
Burner
Uma 12h35'10"56d28'
Burner,Jr,Ronald Anthony
Her 16h28'16"39d10'
Burnett "For Keeps", Michael
Alan
Aur 6h36'0"40d23'
Burnett,Alexander T
Peg 23h3'16"20d55'
Burnett,David William
Crt 11h14'56"-17d34'
Burnett,Deborah
And 2h23'15"39d33'
Burnett,Derrick Edward
Cap 21h24'25"-26d24'
Burnett,Frank
Cnv 13h19'32"32d13'
Burnett,Janet
Peg 22h8'15"26d57'
Burnett,Janet Jordahl
Uma 11h24'0"38d42'
Burnett,Jeremy O
Peg 23h0'48"30d9'
Burnett,Keeper Of My
Heart,Philip
Aql 20h35'33"0d44'
Burnett,Laurel
Aql 19h22'53"0d59'
Burnett,Linda
Mon 6h35'48"0d9'
Burnett,Margena
Eri 3h3'22"-11d55'
Burnett,Peter G
Her 16h1'47"40d8'
Burnett,Richard & Sharon
Her 17h13'55"48d43'
Burnett,Robert H
Cam 6h4'0"61d21'
Burnett,Russell
Sct 18h53'1"-5d41'
Burnett,Susan D
Cas 1h9'39"60d13'
Burnett,William E
Cnv 12h19'50"35d10'
Burnett,Yumiko Otaguro
Gem 6h44'11"18d48'
Burnett-MBP FX,Michael
Her 16h22'11"23d11'
Burnette,Charles Wilson
Boo 14h13'41"38d46'
Burnette,John B
Cet 0h55'46"-5d10'
Burnette,Karen Roberson
Com 12h17'51"20d3'
Burnette,Robbie
Her 17h30'35"26d59'
Burney,Cheri
Peg 23h0'11"30d5'
Burnham I,Thomas Paul
Umi 14h37'21"77d58'
Burnham's Star
Cam 6h9'11"80d3'
Burnham,Dan
Dra 18h59'39"80d11'
Burnham,Daniel Mercer
Boo 13h55'19"17d58'
Burnham,Darryl Anne
Per 2h55'45"43d15'
Burnham,G Craig
Cyg 21h4'55"36d56'
Burnham,John Hickey
Cyg 21h4'26"39d5'
Burnham,Marie Anderson
Cyg 21h4'41"38d23'
Burnham,Michael Scott
Per 2h55'22"43d19'
Burnham,Rose M
And 23h20'18"41d22'
Burnham-Kirkham
Ori 5h59'54"17d18'
Burning Bush RLM,The
And 0h55'49"34d14'
Burning Light Of Dawn
And 1h2'1"40d51'
Burns "Electroman", Daniel
Lee
Sex 10h42'41"1d10'
Burns (Betty),Gloria Violet
Carlton Healy
Lyn 7h19'13"51d35'
Burns 160246,Peter
Ori 5h38'58"0d1'
Burns Family Star,The
Lyr 18h47'38"42d18'
Burns Forever Loved, Ted &
Flossie
Uma 11h18'21"68d6'
Burns PJB AJ
Lac 22h43'48"53d13'
Burns Star,The Robert F
Her 18h0'25"14d54'
Burns Star,The Elma
Mon 6h28'39"4d12'
Burns The Bright, Laurie
Cas 0h51'27"60d53'
Burns,"Burnsey" John F
Dra 19h18'1"70d37'
Burns,Aliza Emily
Cyg 19h49'18"37d54'
Burns,Amanda
Lyn 8h50'1"45d1'
Burns,Amber
Lyr 19h16'19"41d52'
Burns,Anne Bosworth
Lyn 8h53'19"45d19'
Burns,Bonnie Gayle
Mon 6h45'33"10d42'
Burns,Bradley Scott
Cep 0h19'31"75d7'
Burns,Caitlin A
And 23h5'0"41d15'
Burns,Caitlin Rose
Peg 23h37'51"31d5'
Burns,Campbell Foster
Aur 5h56'27"50d1'
Burns,Carrie Michelle
Boo 14h56'32"51d49'
Burns,Cassandra RFCI
And 23h26'30"48d41'
Burns,Catherine M
Peg 21h58'23"23d43'
Burns,Charlotte
Cas 1h23'57"50d29'
Burns,Chris A
Lac 22h6'58"51d36'
Burns,Christian
Aur 5h11'50"44d16'
Burns,Christopher David
Lmi 9h22'55"34d52'
Burns,Christopher Alan
Aur 6h5'45"31d2'
Burns,Cindy
Lyn 8h3'0"35d58'
Burns,Claire
And 23h45'40"47d38'
Burns,Daniel T
Lmi 10h4'55"33d0'
Burns,Donna
Boo 14h59'32"52d8'
Burns,Elaine Mae
Mon 7h2'5"-5d49'
Burns,Forever,Jeff
Per 3h51'49"37d54'
Burns,Gene
Her 17h2'1"44d45'
Burns,Gerald
Cmi 7h43'1"8d44'
Burns,Helen Plue
Peg 21h39'51"26d16'
Burns,Isaac William
Aql 18h59'1"-0d46'
Burns,Jack L
Lac 22h2'41"51d26'
Burns,Jacob John
Umi 16h34'11"75d47'
Burns,James K
Her 16h10'34"25d16'
Burns,Jan
Lyn 7h59'28"36d51'
Burns,Jennifer N
And 23h3'14"51d48'
Burns,Jennifer Parr
Aqr 22h24'19"0d13'
Burns,Jenny Amanda
Lyr 18h35'47"40d36'
Burns,John Charles
Dra 18h23'22"68d36'
Burns,John Michael
Pup 8h0'41"-26d24'
Burns,Joseph Edward Peter
Hya 8h33'1"0d43'
Burns,Joshua M
Her 17h20'48"21d45'
Burns,Jr,William Frederick
Per 2h56'59"35d29'
Burns,Kenneth
Cmi 7h7'21"4d40'
Burns,Lara Ann
Cas 0h5'55"54d42'
Burns,Laurie Lynn
Cas 0h47'1"70d48'
Burns,Lori Lynn Szemly
Mitchum
Com 13h2'52"22d17'
Burns,Lovette
Lyn 8h53'1"40d21'
Burns,Luke & Dot
Peg 21h43'0"28d14'
Burns,Mackensie Ladd
Dra 17h21'50"60d26'
Burns,Mark A
Aur 6h29'36"38d4'
Burns,Marsha Jean
Lyr 18h59'0"29d6'
Burns,Master Ryan Patrick
Aur 5h10'52"41d24'
Burns,Matthew
Boo 14h31'32"42d47'
Burns,Michelle
Lyr 19h16'1"26d11'
Burns,Moog Ryan Nee
Margaret Ann
And 23h33'27"42d53'
Burns,Ms Janet Louise
Elizabeth
Aqr 20h58'58"-5d50'
Burns,Natalie Kim
Cas 0h27'18"58d10'
Burns,Presley
Ser 15h37'43"4d16'
Burns,Renee Sharon
Vul 19h19'35"26d36'
Burns,Robert "Bob"
Lib 15h31'20"-8d20'
Burns,Robert J
Cyg 20h36'12"33d10'A
Burns,Roy
Her 18h3'49"28d34'
Burns,Sandy Squeeze
Cas 0h29'35"67d17'
Burns,Shannon
Cam 13h34'44"80d22'
Burns,Shelia
And 0h19'25"33d59'
Burns,Sherry
Mon 6h46'0"10d47'
Burns,Steve
Per 4h4'55"50d34'
Burns,Steven Andrew
Her 16h38'11"34d23'
Burns,Teresa Kay
Eri 3h12'1"-2d32'
Burns,Thomas D
Lac 22h35'53"55d53'
Burns,Thomas G
Aur 6h24'38"37d45'
Burns,Valerie
Cet 3h8'11"6d2'
Burns-Fulkerson,Aidan Rae
Cma 6h55'37"-18d45'
Burns-Fulkerson,Keegan
Joseph
Hya 9h38'47"5d27'
Burns-Our Christmas
Angel,John P
Cyg 20h20'10"41d6'
Burns: Unforgetable, Edward
Per 2h6'45"47d45'
Burnside,Graham Craig
Gem 6h54'44"17d51'
Burow
Cam 6h54'11"80d29'
Burpee,Mary
Cnv 13h0'23"40d36'
Burpee,Mary
Lmi 10h37'46"26d15'
Burpee,Mary
Cas 0h58'28"76d3'
Burpitt,Kenneth
Per 2h7'0"57d27'
Burr Someone Special, I Allen
Eri 3h7'1"-1d56'
Burr,Dorothy Louise Tindall
Cap 20h28'45"-26d38'
Burr,Michael Devere
Aql 20h8'44"1d32'
Burr,Morgan Dawn
And 1h25'23"35d4'
Burr,Robert John
Del 20h51'55"19d35'B
Burr,Susan Ernestine
Del 20h51'55"19d35'A
Burrage,Elizabeth
Pyx 8h47'10"-23d37'
Burregi,Robyn
Cam 13h0'54"84d30'
Burrell,Christopher James
Ori 6h3'30"-2d2'
Burrell,David Charles
Per 3h39'16"38d16'
Burrell,Doris
Del 20h23'14"8d6'
Burrell,Jennifer Rebecca
Hya 9h34'20"-1d19'
Burrell,Jennifer "Jif" Ruth
Com 12h50'25"27d23'
Burrell,Jennifer Lynn
Uma 8h55'38"58d46'
Burrell,Kathleen Marie
Vul 19h6'27"25d25'
Burrell,Mark
Cep 21h15'1"58d47'
Burress,Harry Evan
Boo 14h58'55"38d35'
Burress,Muriel Marie
Vul 21h0'0"20d24'
Burridge,Daniel Luke David
Ori 5h56'38"9d1'
Burridge,James L
Lib 15h32'19"-8d8'
Burridge,Jean Elizabeth
Cyg 19h41'52"41d47'
Burridge,John
Mon 7h50'36"-2d35'
Burridge,Suzanne
Cyg 21h7'33"30d44'
Burrill,James Thomas
Aql 18h57'28"-8d34'
Burris,Kimberly Robin
Cyg 21h3'15"33d41'
Burris,Linda
And 0h57'25"35d12'
Burris,Ray
Boo 14h28'42"38d2'
Burris,Ray
Aql 19h55'19"8d46'
Burrise,Julie
Com 12h32'16"23d27'
Burriss,Ryan
Aur 6h16'42"38d40'
Burrough,Simon
Aur 5h50'48"54d46'
Burroughs,Amanda
Cas 1h8'33"70d18'
Burroughs,Benita
Crb 16h6'40"30d6'
Burroughs,Jennie Mae
Cam 4h12'14"69d38'
Burrow,Christy Belinda
Aql 19h47'36"13d29'
Burrow,Grady & Sharon
Peg 23h39'42"30d54'
Burrow,Kevin J
Ser 15h46'19"21d58'
Burroway,Gary & Marianne
Cae 4h26'32"-45d36'
Burrows
Ori 5h20'52"10d16'
Burrows,Ann Tunnell
Cyg 19h43'22"37d44'
Burrows,Christine Frances
Cep 20h27'59"60d6'
Burrows,Edmund S
Aql 19h57'33"8d24'
Burrows,Elaine Anne
Cas 2h49'1"67d38'
Burrows,Elizabeth
Cas 2h55'1"61d47'
Burrows,Frank J
Uma 11h48'0"32d17'
Burrows,Gary & Ramona
Uma 10h48'17"51d22'
Burrows,Jason
Per 1h50'23"54d11'
Burrows,John
Her 17h39'45"41d12'
Burrows,Kathleen Elaine
Del 21h4'1"12d27'
Burrows,Pam Bunny
Lyr 18h51'51"37d15'
Burrows,PhD,KeriLyn
Christine
Lmi 11h3'48"30d35'
Burrows,Rebecca A
Lyn 8h42'0"38d46'
Burrud,John
Hya 8h58'0"4d20'
Burrus' Shining Star, David W
Ori 5h4'46"1d26'
Burrus,Goodwin Douglas
Crt 10h49'49"-17d59'
Burrwell
Aur 7h10'40"40d58'
Bursee Love Star,Larry Star
Dra 16h56'14"68d37'
Bursey,The
Aql 19h47'46"12d4'
Bursk III,Edward C
Per 2h52'15"37d2'
Burson,Ronald P
Dra 14h55'31"64d8'
Burstein,Edward
Her 16h57'34"31d26'
Bursé,Benjamin
Cnc 8h30'43"8d40'
Bursé,Katharina
Aqr 22h32'10"0d22'
Burt,C Leonard
Cet 0h25'0"-6d37'
Burt,Elizabeth Anne
Hya 9h15'20"1d42'
Burt,Little Angel Dorothy
Mon 6h38'53"7d47'
Burt,Mary
Cam 10h34'46"84d29'
Burt,Noah
Lyn 7h1'47"53d46'
Burt,Richard
Tri 2h41'19"34d11'
Burt,Sean Patrick
Cmi 7h56'1"1d39'
Burt,Tyler John
Aql 20h31'51"-1d46'
Burt,Tyler John
Dra 19h24'21"58d30'
Burtin,Jacques
Her 18h2'46"14d28'
Burtness,Ken
Crt 10h50'24"-11d11'
Burtnett,Ken (Byrd)
Boo 15h22'0"40d36'
Burton & Bonnie's Brightness
Aql 19h58'43"15d14'
Burton I Love
Lyn 8h36'35"40d31'
Burton IV,James H
Her 17h2'34"21d38'
Burton Tower
Boo 14h0'0"17d60'
Burton Wedding Star, The
Cyg 20h19'0"40d37'
Burton's Class of 2000
Uma 8h35'56"52d19'
Burton's Wedding Star, Bill &
Claudia
Cyg 21h33'34"40d59'
Burton,Adrian
Aur 5h11'36"44d29'
Burton,Amy Elizabeth
And 23h25'14"41d14'
Burton,Brian Daerell
Cep 0h19'51"70d32'
Burton,Brieanna M Z
Lyr 18h44'11"32d33'
Burton,Celestine
Aql 20h2'11"12d4'
Burton,Charles Warren
Boo 15h31'31"40d42'
Burton,Christine Cierra
Peg 23h24'1"12d39'

Burton,Claudia Zeitlin
Mon 6h39'45"6d1'
Burton,Debbie & Charlie
Cyg 21h9'15"38d13'
Burton,Dorothy Thorpe
Lyr 19h23'35"35d30'
Burton,Dylan John
Boo 13h43'48"18d47'
Burton,George D
Cep 3h40'1"79d5'
Burton,Harry L
Her 17h13'1"48d59'
Burton,Jack
Uma 12h3'58"30d27'
Burton,Jack Alden
Per 3h7'10"40d59'
Burton,James
Sex 10h22'24"1d12'
Burton,Jamie Janean
Del 20h35'23"20d22'
Burton,Jeffrey Lee
Aur 6h26'0"35d26'
Burton,Jennie
Mon 6h15'1"-6d33'
Burton,Jennifer
Cyg 20h20'1"39d33'
Burton,John
Boo 14h8'1"36d60'
Burton,John F
Per 2h54'57"50d3'
Burton,Johnathon
Oph 18h1'45"12d56'
Burton,Jr,Theodore J
Cet 3h19'28"-0d9'
Burton,Maureen
Com 13h9'33"21d30'
Burton,Monique Suzanne
Eri 2h43'14"-8d8'
Burton,MSG Charles B
Her 18h12'56"31d17'
Burton,Nina
Gem 6h41'14"35d24'
Burton,Pop Pop
Aur 6h12'50"30d44'
Burton,Sharon Victoria
Com 13h5'40"21d53'
Burton,Simon Andrew
Per 3h4'11"31d49'
Burton,Tony & April
Mon 7h12'53"-7d11'
Burton,Trevor Alan
Her 17h24'47"21d43'
Burton,Wyatt Charles
Hya 8h55'54"0d20'
Burton-Dines
Cam 4h56'48"68d29'
Burton-Jim's Valentine Donna
Lyr 18h57'22"31d51'
Burts,Sarah Elizabeth Smith
Mon 7h54'40"-6d25'
Burtso
Uma 8h55'23"48d25'
Burtt,Chet & Nancy
Lyr 18h48'28"41d43'
Burum,Bethany Anne
Lyn 9h15'0"39d9
Burus,Edouard
Ori 5h58'0"10d4'
Burwell,Mae
Mon 6h23'40"8d45'
Burwell,Wilton Edward
Hya 8h44'31"3d26'
Burwinkel,Bernard
Aqr 22h43'55"-2d22'
Bury,Ben L
Dra 9h21'27"74d35'
Bury,Craig David
Aur 4h59'1"40d7'
Burya My Love,Joseph F
Boo 14h20'59"12d29'
Burzynski Love Frank, Angie
Lyn 7h38'11"38d24'
Busacca-Poppi,Gary
Lyn 9h10'48"38d11'

Busatto,Brittnee Danielle
Cam 7h57'0"70d11'
Busbin(Nikki),Ashlee Nicole
Peg 22h42'46"29d47'
Busboom,Amy Marie
And 1h54'41"39d56'
Busby,Brooklynn Kay
Mon 8h0'20"-10d4'
Busby,James
Leo 10h32'54"17d37'
Busby,Justin Brett
Ari 3h12'51"26d11'
Busby,Martin L
Her 16h58'0"38d45'
Busby,Mary Sue
Psc 23h20'30"1d55'
Buscat 23h20'30"1d55'
Boo 13h41'48"11d13'
Buscemi,Anthony P
Ser 15h27'1"21d19'
Buscemi,Constance Mary
Aql 19h2'0"16d50'
Buscemi,Jennifer Cecelia
Cyg 20h40'34"42d29'
Buscemi,Mary Rose
Aql 20h1'57"14d27'
Busch
Her 18h5'16"18d57'
Busch,Andrew F & Ruth
Schilling
Lmi 10h8'17"31d56'
Busch,Dee
Ser 15h26'29"1d30'
Busch,H William
Aur 7h23'18"40d9'
Busch,Hans Müller & Twin
Will
Lib 15h35'45"-18d60'
Busch,Jennifer L
Cyg 19h24'36"30d39'
Busch,Jr,Col Howard R
Aql 20h19'51"0d22'
Busch,Michelle
And 1h4'16"37d53'
Busch,Natasha
Eri 3h54'59"-7d7'
Busch,Zachary Jordon
Cam 3h19'27"60d25'
Busche,Ted & Mary
Psc 1h28'32"31d5'
Buschell,Ryan
Ori 4h41'25"5d8'
Buschelman,Geralyn
Cma 7h20'20"-16d32'
Buschhorn,Gary
Boo 14h40'32"48d49'
Busching,Alice
Eri 2h58'52"-6d15'
Buschmann,Hans
Cnc 8h56'1"22d9'
Buschmann,Manfred
Aur 5h14'46"42d15'
Buschner,Anja,Tom
Psc 1h25'54"20d29'
Buscombe,Martin
Tri 1h50'10"34d50'
Busdicker,Howard Luther
Boo 15h27'39"40d52'
Buse,Allison
Ori 5h51'1"9d32'
Buselli,Paolo
Cam 6h9'0"68d26'
Bush
Aur 4h50'52"40d14'
Bush "The Kub Star", Kelly
Ann
Lyn 8h0'0"35d27'
Bush,Andrew J
Sex 10h32'58"-6d12'
Bush,Aron Trey
Sct 18h43'13"-5d2'
Bush,Becki Rae
Lyr 19h1'34"40d31'
Bush,Bradley Barrett
Tri 2h25'27"28d48'

Bush,Chet M
Aur 6h4'3"36d16'A
Bush,Connie
And 1h19'12"38d35'
Bush,Elizabeth M
Aur 6h4'3"36d16'B
Bush,Gary Ellis
Dra 17h42'41"60d53'
Bush,Harry
Crt 10h50'18"-12d7'
Bush,Jacob Raymond
Aur 5h4'42"42d38'
Bush,Jefferson Fielding
Hya 8h11'0"0d41'
Bush,Jonathan Charles
Vul 19h19'19"26d45'
Bush,Jr,Carl
And 1h5'14"38d13'
Bush,Kathy
Crt 10h58'19"-17d31'
Bush,Lacy J
Lac 22h0'44"51d23'
Bush,Leah Nicole
Vul 20h58'33"28d37'
Bush,Miss Kelly
And 23h47'0"45d21'
Bush,Norma A Herrgott
And 1h40'52"40d8'
Bush,Patricia Ann
Com 13h25'0"26d57'
Bush,Phyllis
Cam 3h16'0"61d5'
Bush,Suzanne
Cas 23h2'54"58d45'
Bush,Tiffany P
Cet 3h11'35"1d34'
Bushakra,Alex
Aql 20h0'18"11d25'
Bushee,Joseph Andrew
Oph 17h29'38"8d16'
Bushell,Martin
Dra 17h28'1"61d36'
Busher,Mariah
Lyn 8h29'57"42d6'
Busher,Robin
Lyr 19h0'48"37d54'
Bushey,Doris Jean
Peg 22h57'27"22d8'
Bushey,Georges
Cnv 12h8'47"36d5'
Bushkin,Bernie
Com 12h47'10"30d15'
Bushman,Christina Marie
Boo 14h55'49"44d28'
Bushman,Dolores J
Aql 19h59'47"14d28'
Bushman,Jennifer Jené
Mon 8h7'60"-6d34'
Bushman,Matthew James
Aur 5h54'51"30d35'
Bushmiaer,Carrie Frances
Cet 3h19'17"8d15'
Bushnell
Aur 5h53'1"29d40'
Bushnell,Diana
Cyg 20h53'45"37d39'
Bushong,Colonel James T
Aql 19h6'13"14d42'
Bushong,Marva LaRae Youngs
Lyn 7h54'16"40d40'
Bushong,The Joseph F &
Mary C
Cam 5h52'43"68d2'
Bushrow,Jay Michael
Aql 20h14'11"0d52'
Bushwackers
Vir 12h59'40"-12d8'
Bushá,Austin James(AJ) (AJ)
Vul 19h48'30"28d13'
Busia & Alena
Dra 16h42'22"63d1'
Busicchia,Antonietta & Alfred
Uma 11h13'45"42d40'
Busick,Christopher
Leo 11h45'33"20d11'

Busick,Donald J
Cnv 13h59'18"41d58'
Busick,Nicholas
Gem 5h58'57"27d3'
Busija,Nickolas
Cep 21h9'31"58d44'
Busing,Pamala Kay
Mon 6h22'50"1d5'
Busk,Christina
Cas 1h20'1"52d53'
Buske,Horst
And 23h9'33"40d35'
Buskirk
Aql 19h3'24"1d40'
Buskirk,Donald & Virginia
Crb 16h19'19"33d55'
Busko,Teresa Anna
Cam 7h57'26"61d32'
Buslipp,Helen & Charles
Crb 16h15'0"38d39'
Buss,Ben
Equ 21h16'15"2d40'
Buss,Brady Ericson
Cma 6h43'44"-16d4'
Buss,Sara Davis
Cas 0h35'58"72d45'
Busscher,Robert Scott
Aur 6h24'27"32d50'
Busse,Diana & Paul
Cyg 19h24'0"31d1'
Busse,Michelle Renee
Cet 1h53'42"-0d32'
Busse,Ronald F & Faye M
Cet 1h3'19"-2d13'
Bussell,Sarah P
Uma 8h38'11"62d24'
Bussemas,Christian
Lyn 8h12'35"48d0'
Busser
Ori 6h4'0"18d47'
Bussert,S Mark "Sparky"
Aql 19h29'33"13d58'
Bussey,Alex W
Uma 10h2'46"47d51'
Bussi
Ant 10h43'4"-34d26'
Bussing,Charles
Lac 22h25'39"52d40'
Bussler,The Baku
Ori 6h7'36"3d14'
Bussmann,Franz
Lyr 19h20'45"31d36'
Bussmann,Werner
Cmi 7h32'0"1d46'
Bussolini,Bruno
And 23h34'13"36d34'
Busson,Isabelle
Cma 6h57'3"-16d15'
Bussure,Sandra Jane
Mon 7h44'20"-1d41'
Bustamante,Andrea Marie
Aql 20h33'29"-7d52'
Bustamante,Gabriella Denise
Mon 8h2'35"-6d18'
Buster
Gem 6h39'26"30d44'
Buster (110144/ 233706060)
Aql 20h0'1"12d1'
Buster H
Dra 20h23'0"68d11'
Buster,MD,Robert William
Ser 15h56'40"24d37'
Busterpher
Aur 6h30'29"38d21'
Bustillos,Jaime Trujillo
Aql 19h7'19"0d32'
Busty O
Lyn 7h47'41"48d58'
Buswell,Rob
Aql 18h59'12"17d49'
Butch
Aur 5h34'31"54d53'
Butch
Her 17h17'37"42d17'

Butch
Her 16h36'56"36d58'
Butch
Per 2h36'56"48d50'
Butch & Brutus
Cnv 12h57'50"40d9'
Butch & Harriet
Crb 15h35'43"30d52'
Butcher,Alan Christopher
Ori 6h5'56"0d29'
Butcher,Alison
Lyr 18h52'11"38d27'
Butcher,Bruce
Aql 18h41'16"-1d55'
Butcher,Carolyn
Uma 10h55'15"48d32'
Butcher,Darryl Lloyd
Lac 22h15'0"54d49'
Butcher,Daryl T
Uma 10h29'12"40d35'
Butcher,Emily R
Equ 21h19'58"12d16'
Butcher,George E
Lac 22h16'47"38d10'
Butcher,Guy Andrew
Ori 6h15'24"8d1'
Butcher,Jonathon
Cet 2h36'42"-11d45'
Butcher,Kristine & Stephen
Didziulis
Crb 15h23'40"30d50'
Butcher,Mary Lou
Tau 5h43'27"28d48'
Butcher,Pamela Doreen
Lyn 7h48'23"47d46'
Butcher,Susan
Peg 21h19'22"22d46'
Butcher,Thomas E
Aql 19h39'42"11d4'
Butchie,Saint
Peg 23h40'18"30d19'
Bute,Anna Jelden
And 0h37'0"40d18'
Buteau,Aleta S
Hya 9h12'47"2d55'
Buteau,Richard N
Ori 5h51'21"14d16'
Butenhoff,Kyle Patrick
Peg 22h29'13"20d28'
Butera,Concetta Anne
Lac 22h51'17"35d23'
Butera,Hillary Ann
Cyg 19h43'20"30d16'
Butera,Holly Ann
Cas 0h7'32"62d20'
Buth,Hans
Dra 20h21'17"68d14'
Buting,Joe
Cyg 19h17'29"45d37'
Butkis,Winifred
Boo 15h0'50"20d31'
Butler Starbright,The Louise
Gru 22h29'1"-55d52'
Butler's Star,Jeff
Vul 20h14'19"28d40'
Butler,A R
Boo 15h47'46"43d54'
Butler,Alun R
Ori 5h59'31"20d13'
Butler,Amber Desiree
Lyn 18h58'40"28d15'
Butler,Ann Marie
Cyg 20h2'44"38d45'
Butler,Ashley Nicole
Cas 0h37'52"58d48'
Butler,Bob & Harriett
Ori 5h39'45"0d35'
Butler,Brad Austin
Sge 20h17'25"16d57'
Butler,Carol & Ricky
Cyg 21h46'45"37d11'
Butler,Cheryl Ann
And 1h9'31"41d11'
Butler,Cody Patrick
Lac 22h18'42"49d10'

Butler,Connor Michael
Ori 6h1'1"20d45'
Butler,Danielle Lorna
Umi 15h52'32"78d55'
Butler,David
Her 17h10'14"45d21'
Butler,David Lawrence
Her 18h1'59"38d30'
Butler,Diana
Cyg 21h9'44"39d60'
Butler,Dorothy Leona
Aur 5h1'55"44d14'
Butler,Elliott & Butler
Lac 22h21'20"50d3'
Butler,Francile & Nolan
Crb 16h7'32"26d1'
Butler,Gray Felton
Oph 17h34'60"-25d27'
Butler,Jeffery L
Boo 13h33'20"20d50'
Butler,Jennifer Elaine
And 0h59'48"45d11'
Butler,Joan
Aur 5h21'1"37d58'
Butler,Joan G
Cyg 21h3'34"39d33'
Butler,John Christopher
Per 1h46'23"48d55'
Butler,Johnnie R
Aur 6h36'36"30d29'
Butler,Johnny
Sex 9h55'47"4d50'
Butler,Johnny "Pony"
Leo 11h29'0"-6d13'
Butler,Joshua Andrew
Oph 17h26'54"-22d59'
Butler,Joshua James
Ori 5h56'24"0d13'
Butler,Karl
Hya 8h33'52"-0d57'
Butler,Katherine Helen
Mon 6h28'27"8d8'
Butler,Keith
Her 18h16'26"14d51'
Butler,Kristen L
Cnc 8h37'48"18d41'
Butler,Lawrence Anthony
Lyr 19h15'39"40d36'
Butler,Lindsay
Mon 6h21'60"8d42'
Butler,Lorraine
Umi 14h50'21"68d24'
Butler,Mack J
Boo 14h38'40"31d12'
Butler,Marjory Gayle
Peg 22h59'0"26d14'
Butler,Mary E
Cyg 19h43'0"29d11'
Butler,Mary Francis
Peg 22h3'1"29d12'
Butler,Meghann Ilene
Mon 6h24'13"8d53'
Butler,Michael Scott
Cep 23h22'28"68d11'
Butler,Nathan
Aur 5h4'54"40d38'
Butler,Patricia Mary Scot
Mon 6h30'24"-10d32'
Butler,Rachel Barbera
Cas 2h2'25"61d21'
Butler,Richard Herman
Oph 18h15'30"11d53'
Butler,Ryan
Lac 22h14'19"49d29'
Butler,Samantha
Cas 0h34'19"68d8'
Butler,Sián Elizabeth
And 23h42'54"41d41'
Butler,Stephen
Uma 10h52'11"70d27'
Butler,Stephen
Cep 23h9'0"65d2'
Butler,Wayne Leslie
Cep 22h17'20"65d19'

Butler,William
Cep 20h7'50"60d30'
Butler,William Pearce
Ori 5h11'1"-06d13'
Butnik,Nathan John
Aur 6h11'34"45d32'
Butorac,Terry & Elaine
Cyg 21h46'1"34d55'
Butrico,Adam Joseph
Ori 5h56'37"8d27'
Butrovich,Matthew Erich
Lac 21h59'60"42d22'
Butryn,Walter
Cyg 20h38'59"40d33'
Butscher,Valentin
Dra 9h54'41"73d25'
Butt N' Ben
Uma 9h45'19"54d4'
Butt,Barbara
Lep 5h59'56"-25d9'
Butt,Marcus
Cyg 20h19'46"38d52'
Butt,Martina
Cas 1h5'24"50d13'
Butta,Amy Michelle
Lyn 7h10'20"50d46'
Buttabean
Ori 5h29'53"-3d19'
Buttacavoli,Josephine A
Oph 17h54'31"12d2'
Buttaci,Norman V
Dra 18h42'14"71d11'
Buttel
Cnc 8h35'11"16d27'
Butter,Stephen H
Eri 4h55'33"-7d48'
Buttercup
Mon 7h0'60"-1d31'
Butterfield-The Teacher,Ralph
Lyn 7h29'18"40d29'
Butterflies & Eskimoes
Cep 20h26'28"63d27'
Butterfly
Boo 15h0'1"22d45'
Butterfly Star,The
Aql 20h11'35"10d33'
Butterfly Wing
Mon 6h24'52"5d37'
Butterman's Mitzvah, Irene &
Daniel
Lyr 18h58'41"37d11'
Buttermore,Rett
Sco 16h54'17"-44d28'
Butterton,Hayley Louise
Cas 0h25'27"62d43'
Butterworth,John L
Aur 6h8'44"50d18'
Butterworth,Louis Michael
Aur 6h15'58"35d13'
Buttita,Galen
Hya 9h33'20"-9d46'
Buttitta,Carol
Lyn 8h19'59"58d43'
Buttitta,Emily Victoria
Lyr 18h28'28"30d34'
Buttitta,Joy Michelle
Peg 22h0'59"27d2'
Buttitta,Shannon L
Dra 17h52'49"68d13'
Buttner's Beauty
Cep 21h25'58"85d12'
Buttner,William John
Her 16h17'24"23d13'
Buttney,John Peter James
Sge 20h4'42"20d32'
Butto,Ron & Beth
Vul 19h47'21"25d42'
Button
Her 16h30'51"40d13A'
Button,Dan
Ori 6h17'1"20d43'
Button,Debbie Joanne
Cas 2h54'52"65d2'
Button-Dunn,Carrie
And 23h37'34"49d35'

Buttons
Ori 4h59'48"5d55'
Buttons (Debra Joyce)
Mon 7h54'51"-3d39'
Buttons Serra God's Little
Treasure
Umi 19h37'43"86d1'
Buttons,Angela
Uma 11h57'36"45d44'
Buttrick,Sid & Elaine
Vul 19h23'16"26d18'
Butts,James Robert
Oph 17h0'1"11d1'
Butz III,Maurice F
Ser 16h2'35"9d56'
Butzen,Carol
And 22h57'36"50d8'
Butzlaff,Sr,William & Susan
Cyg 20h8'55"40d46'
Buus,Les
Cep 21h56'43"55d52'
Buvic
Lac 22h48'28"53d41'
Buwen,John Bernard &
Dolores Lee
Aur 4h47'50"50d25'
Buxbaum,Art
Oph 17h12'29"-14d31'B
Buxbaum,Erich H
Per 1h31'15"52d51'
Buxbaum,Marilyn
Oph 17h12'29"-14d31'A
Buxbaum,Philipp
Lyn 8h10'45"41d48'
Buxton,Dixie May
Eri 4h1'54"-11d24'
Buxton,Harry Alexander
Uma 10h45'31"40d54'
Buxton,Lloyd Elliot
Umi 15h9'14"66d55'
Buxton,Physician, Martin
Norman
Oph 17h19'20"12d54'
Buxton,Rebecca Francine
Lyn 8h54'0"42d47'
Buxtorf,Jean-Jacques
Cas 23h39'35"61d30'
Buyers,John William
Amerman
Boo 15h10'0"30d59'
Buyoung
Boo 14h56'0"26d58'
Buywick,Liz
Cas 0h36'45"64d47'
Buzby 6/10/77,Don & Beth
Sge 19h3'16"19d15'
Buzek,Jr,Robert J
Leo 11h52'32"23d7'
Buzite
Cas 2h51'58"65d7'
Buzizi,Mohamed J
Per 2h59'30"31d41'
Buzogany,Jalyn Rupnow
Del 20h21'33"10d19'
Buzon,Mireille B
Cam 5h52'43"68d2'
Buzsan
Peg 21h38'27"27d36'
Buzz & Donna
Her 17h17'55"22d57'
Buzz's,Star
Aur 6h28'0"34d36'
Buzzallino,Joan
And 0h48'53"21d49'
Buzzanetti Earth Angel Paul
Uma 9h47'42"57d47'
Buzzard & Hart
Peg 0h9'1"22d3'
Buzzell,F Scott
Sct 18h41'42"-7d21'
Buzzell,Jeremy
And 0h27'55"28d1'
Buzzoli,Walter
Boo 14h29'53"48d10'

Buzzy
Aur 6h56'31"37d35'
Buzzy Smith
Lyr 18h41'16"28d13'
Bwall
Aur 6h39'28"38d37'
Bwana
Tri 1h56'1"27d34'
BWC Beautiful Women of
Color
And 0h53'46"39d32'
By
Boo 14h57'41"40d8'
By Knights
Cyg 20h40'11"44d56'
By the Grace of God Penny
Kirk
Cyg 19h33'21"31d36'
By The Way
Uma 9h38'32"68d8'
Byars,Helen Mactier
Cyg 20h31'14"42d23'
Byars,Michael
Dra 16h26'0"69d26'
Byart,Violet Rosina
And 23h6'28"42d23'
Byassee,Becky
Peg 22h37'1"26d21'
Byatt,Anne Shirley
Cas 0h3'39"59d59'
Byatt,Laura
Cas 0h53'29"73d20'
Byatt,Noreen
Lyr 18h28'13"47d33'
Byck,David
Cep 0h3'15"70d42'
Byczkowski,Angelika
Ori 6h18'0"8d1'
Byde,Captain
Her 16h25'40"48d41'
Bye,MD,Michael
Per 1h51'24"52d41'
Bye-A Born Star!,Craig
Mon 8h6'58"-2d13'
Bye-Bye Doggieland
Cmi 7h25'47"7d39'
Byer,Joan M
Oph 17h18'0"12d17'
Byers
Del 20h16'52"13d12'
Byers,Brook H
Sge 20h1'0"20d28'
Byers,JLB 20 Jennifer Laurel
Cas 0h58'0"63d55'
Byers,Robert Dean
Aur 5h6'20"44d21'
Byers,Ronald & Donnie
Jordan
Cet 2h54'13"1d17'
Byers,Stanley Robert
Peg 23h34'0"10d40'
Byers,Winifred
Peg 23h35'47"11d1'
Byford,Ruth Jane
Cas 0h15'22"47d45'
Byford,Sheila
Cas 0h54'26"63d33'
Bygrave,Emily
Cas 1h54'0"74d21'
Byheden,Anna Jennie
Lyn 8h1'35"48d5'
Byian's CLAN
Aur 6h16'33"45d21'
Byman,Mary Jane Cook
Mon 8h3'45"-8d38'
Byng,David
Per 4h31'30"38d17'
Bynum,Sr,John Haynes
Her 16h24'39"38d45'
Bynum-Copeland,Terry D
Psc 1h37'37"27d46'
Bynum-Together 4Ever, John
Christian
Sex 10h11'47"-1d39'

Byors,Donald R
Cep 20h22'36"60d20'
Byram's Guiding Star, Lord Eric P W
Dra 15h56'38"66d58'
Byrd Family Star,The Dr Marc Jeston
Oph 17h5'55"11d28'
Byrd Family,The
Aql 19h30'59"0d51'
Byrd II,James W
Cam 3h38'47"73d21'
Byrd III,Benjamin Calvin
Sgr 17h55'32"-36d51'A
Byrd(Westa),Daniel K
Aur 6h59'18"43d35'
Byrd,Ashley Nicole
Cas 1h19'23"60d58'
Byrd,F G
Ori 6h6'49"11d10'
Byrd,Holly
Eri 3h6'0"-5d2'
Byrd,Janet:The Star of My Life
Cep 20h30'48"75d35'
Byrd,Meghan Patricia
Uma 9h48'0"61d50'
Byrd,Rita
Lyn 7h45'34"44d33'
Byrd,Ruby Lee
Cyg 19h40'38"31d25'
Byrd,Tracy
Peg 21h36'49"20d17'
Byrne Brilliant
Aql 19h49'0"10d58'
Byrne Light,The Connor
Aur 5h7'0"41d18'
Byrne Prized Beyond Ruby,Jane & Gerard
Cyg 19h41'59"42d0'
Byrne,"Lady Di"Diane
And 0h5'39"46d28'
Byrne,Carol
And 23h2'23"50d11'
Byrne,Carol Marie
Cas 0h28'38"72d22'
Byrne,Colleen Simmons
And 02h02'11"47d0'
Byrne,Conor John
Her 18h1'0"46d57'
Byrne,David
Her 16h10'35"41d39'
Byrne,Eleanor Daisy
Cas 0h28'25"64d9'
Byrne,Jeremy J
Vir 13h34'14"2d16'
Byrne,John J
Cet 0h43'39"-6d49'
Byrne,Margot Helena
Lyr 18h47'56"31d40'
Byrne,Melvyn
Sge 20h6'16"16d5'
Byrne,Mr Thomas J
Boo 14h11'25"36d33'
Byrne,Mrs Margaret K
Lyr 18h38'45"42d44'
Byrne,Patricia
Cyg 19h54'0"38d11'
Byrne,Sarene
Cas 0h53'49"61d15'
Byrne,Sr,Donn Holland
Aur 6h45'0"48d45'
Byrne,Susan
Umi 14h46'11"67d35'
Byrne,Thomas
Equ 21h3'40"10d22'
Byrne,Tom
Oph 17h55'1"13d60'
Byrne,Veronica M
Mon 6h44'27"10d12'
Byrne,Victor
Aur 5h5'41"41d31'
Byrnes Starshine,W Gregg
Aur 5h1'0"49d44'
Byrnes,Bob & Marie
Crb 15h19'33"31d3'
Byrnes,Diane
Lyn 7h27'0"44d33'
Byrnes,Francis Xavier
Hya 9h10'0"3d26'
Byrnes,Jeanne
Cyg 20h22'20"38d37'
Byrnes,Mary Jane
Cam 7h32'0"67d46'
Byron
Pho 0h4'47"-40d54'
Byron
Cyg 21h3'56"39d8'A
Byron & Susan Star
Ori 4h46'23"4d55'
Byron,Dennis
Del 20h32'31"20d5'
Byron,Elizabeth H
And 0h35'58"31d16'
Byron,I Will Love You Always
Boo 14h46'42"34d36'
Byron,M Dona
Cam 3h14'51"58d18'
Bysouth,David
Cyg 21h0'42"36d44'
Bystriansky,Jan
Cas 0h37'28"64d35'
Bytic
Oph 17h4'36"-22d46'
Bywater,David James
Her 16h56'38"22d8'
Bywater,Grace Caroline
Peg 22h0'27"28d28'
Bywater,Grand Conductress,OES,Bettie
Uma 11h58'35"48d43'
Bywater,Richard Lamar
Her 17h54'0"14d55'
Bëatrice
Peg 22h27'46"30d48'
Bär,Birgit
Sgr 19h38'12"-45d0'
Bärbel
Aur 5h19'48"43d51'
Bärbel "Rübennase"
Psc 1h25'54"10d53'
Bäth,Stephan
Cmi 7h30'34"0d47'
Béa
Cam 4h31'57"58d45'
Béal,André
Per 3h30'20"36d27'
Béatrice
Per 3h29'35"38d16'
Béchard,Pierre
Per 3h0'48"40d47'
Béchette,Stéphane
Tri 1h29'1"33d59'
Bédard,Annie
Peg 22h39'43"21d34'
Bégin,Patrick et Christine Lacourse
Per 3h9'12"40d20'
Béhar,Henri
Lmi 9h58'39"31d16'
Bélanger,André Walter
Per 3h7'0"46d32'
Bélanger,Marilie
Uma 11h32'21"31d18'
Bélanger,Sylvie
Cas 0h1'14"61d27'
Bérubé,Annie
Cyg 19h29'34"35d30'
Bérubé,Jean-Pierre
Tau 4h44'59"16d5'
Bétymilie
Lyr 18h50'42"42d12'
Böck,Dipl-Ing Fredrich
Her 17h3'40"41d31'
Böcke,Volker
Aqr 22h42'54"-6d13'
Böckle,Ingrid
Vul 19h19'33"23d15'
Böhme Heike
Lac 22h2'25"51d58'
Böhmer,Gunnar
Vir 13h3'21"-8d18'
Böhnke,Frank & Sonja
Uma 9h43'20"50d39'
Bölli
Tau 5h30'45"25d7'
Börner,Karl-Heinz
Ori 5h55'34"13d49'
Börries,Jörg
Sge 20h13'24"17d38'
Börtz,Birgit
Cas 23h39'47"50d27'
Böttcher,Jens
Ari 2h30'36"26d8'
Böttcher,Veronika Karla geb 28/9/47
Lib 14h18'53"-20d1'
Büchner,Uwe
Sco 17h32'57"-30d55'
Büermann,Susanne
Cep 21h36'35"63d54'
Bühne,Erich
Aqr 22h49'29"-3d38'
Bürgers-Nottelmann, Maximilian
Sgr 19h4'2"-20d3'
Bürki,Willi
Ori 5h56'10"18d24'
Büsgen,Peter
Sgr 19h57'37"-42d0'
Büsing,Heiko
Ari 2h51'38"21d30"
Büttner,Hans-Hugo
Cnc 9h14'0"30d14'

C

C⁻
Cet 1h31'38"0d46'
C⁻
Uma 8h59'0"48d35'
C & P "Upon A Star"
Boo 14h53'44"40d47'
C A L L
Cam 6h16'21"83d55'
C Amanda C
Mon 6h30'0"11d9'
C Ann C III
Lyn 8h18'57"39d44'
C B
Cep 22h6'11"54d21'
C D H I J
Mon 6h44'18"10d28'
C Diane-50
Cas 0h59'13"62d21'
C Dreams
Uma 8h52'35"52d35'
C E
Col 6h25'18"-33d23'
C E M
Mon 6h32'39"-6d32'
C Fred's Star
Tau 4h42'21"28d31'
C H O C
Uma 9h26'23"67d39'
C III
Oph 17h5'40"11d14'
C J
Hya 8h18'16"6d10'
C J
Aql 19h21'36"10d13'
C J
Oph 17h4'19"10d30'
C J
And 2h6'0"38d5'
C J Barbara
Cas 1h20'25"58d34'
C J D II
Ori 4h43'51"4d9'
C J Forever
Sge 19h52'24"16d40'
C J II
Cam 3h44'0"68dd4'
C J's Star
Tau 3h38'0"23d39'
C J-101113
Aql 19h58'58"15d12'
C J-I Will Always Love You
Peg 21h43'41"23d14'
C Joc & Cook Family Star
Lyr 18h41'49"27d25'
C M Brussock
Lyn 8h2'35"40d13'
C M C Sunshine
Cet 1h40'15"0d10'
C M J III,MD
Oph 17h58'24"7d30'
C Michael-Y
Aql 20h4'1"0d50'
C O D
Her 17h7'58"46d1'
C o e a St V Elisabethina
Cap 20h32'0"-10d2'
C Oliver
Aqr 21h57'1"-18d43'
C P 95
Her 18h18'23"12d44'
C S N & C A M We Will Find A Way
Mon 6h59'10"-0d27'
C S S 4
Cmi 7h15'47"4d35'
C T O W
Hya 8h18'32"2d35'
C V B
Ori 6h1'31"8d51'
C V T III
Ori 5h30'1"-6d36'
C W 's Star
Mon 6h57'59"-8d47'
C W Doooq
Lac 22h30'59"55d43'
C W J
Per 2h1'47"48d21'
C-"Carlchen"-M 49RBLS
Leo 9h19'10"8d29'
C-N-T
Ori 5h47'0"11d41'
Ca & Ter Forever
Lac 22h10'57"49d31'
Caba,Jose Ramon
Per 3h46'38"39d28'
Cabal
Cnv 13h6'46"51d42'
Caballe,Bernabe Marri
And 0h19'30"38d40'
Caballero,Merry
Peg 22h23'26"31d1'
Caballero,Mr Armondo
Sex 10h34'58"3d49'
Caban,David
Aql 18h54'49"8d19'
Cabana,Jeffrey W
Dra 17h6'33"64d9'
Cabanial,Corinne
Umi 15h52'48"71d10'
Cabaniss,Katherine Ann
Cyg 20h16'14"30d16'
Cabare,Jean Pierre
Cnv 13h26'15"50d9'
Cabbage
Cep 23h11'35"68d14'
Cabe The Babe
Cnv 13h7'1"32d55'
Cabedanen,Lou
Tri 1h57'50"34d47'
Cabell,Sterrett
Lac 22h12'36"50d1'
Caberson,Leona E
Crb 16h11'47"37d55'
Caberto,Genalyn
Uma 9h21'0"68d22'
Cabezasculo Magee
Tau 4h4'51"5d1'AB
Cable,Norma Jean "Torri"
Aql 20h8'57"6d30'
Cabo Loves Marko
Eri 5h6'56"-6d49'
Cabot,Dianne
Psc 22h56'18"1d17'
Cabot,Lindy
Cet 3h19'27"0d52'
Cabral,Dawn Maria
Cas 1h10'37"63d11'
Cabral,Jaime
Cam 6h40'24"80d36'
Cabral,Jessica
Dra 17h45'59"65d13'
Cabral,Kristen Lee
Cas 23h3'27"58d10'
Cabrales,Damien
Gem 6h57'28"20d0'
Cabrera,Jose Perez
Her 16h3'58"40d2'
Cabrin,Bernard
Ori 6h0'13"10d31'
Cabriola
Umi 16h41'14"76d35'
Cacace,Bianca
Peg 23h19'24"30d51'
Cacace,Victoria
Cas 0h56'22"61d20'
Cacamese,Steven
Lac 22h46'33"54d14'
Cacamis,Jack
Her 17h26'44"30d42'
Caccavo,Dolores Marie Hughes
And 2h34'34"40d36'
Caccese,Mary Elizabeth
And 0h18'0"34d45'
Caccese,William
Aur 4h38'13"31d44'
Caccia,Joel R
Aur 5h14'31"43d29'
Caccialino,Lisa J
Cyg 20h18'25"38d42'
Cacciato,Michele
Cyg 20h17'0"40d24'
Cacciato,Michele
And 23h38'45"37d27'
Cacciato,Michele
Cyg 20h51'40"48d55'
Cacciato,Michele
Crb 16h1'44"50d39'
Cacciatore,Amber Shea
Mon 6h15'30"-4d49'
Cacciatore,Aubrey Kate
Cet 0h26'0"-6d33'
Cacciatore,Stéphane
Uma 11h20'57"37d33'
Cacciotti May 23,93, Joseph Alexander
Umi 17h18'23"75d29'
Cacheux,Georges
Dra 10h2'15"74d55'
Cachia,Salvo
Uma 11h57'32"32d20'
Cacossa,Kenneth
Cam 3h50'0"68d9'
Cactus & Rose
Cas 1h59'17"58d5'
Cactus Juice Café
Dra 16h58'37"68d21'
Cadden,Paul `Doodles'
Eri 4h31'27"-0d20'
Caddis,James
Cyg 20h59'33"28d21'
Caddy,Eric Jon
Cma 7h12'39"-13d21'
Caddy,K George
Cma 7h20'0"-16d1'
Caddy,Peter
Cyg 19h37'52"33d51'A
Caddy,Sarah Ann
And 2h23'15"42d10'
Cade,Jeffrey
Per 1h43'10"53d6'
Cade,Kathryn H
Lmi 10h10'14"38d49'
Cadelina,Violi
Del 20h15'36"15d55'
Caden,Dr Jesse
Uma 8h24'14"70d13'
Cadenhead 4-21-46, Harley & Marjorie
Lyr 19h21'1"30d43'
Cadieux,Adrien
Her 18h6'1"38d24'
Cadieux,James Richard
Oph 17h31'37"14d0'
Cadman,Lucienne
Ori 5h54'36"9d59'
Cadmus
Uma 9h23'21"51d58'
Cado,Roger
Cam 5h50'30"65d32'
Cadogan,Ann
Cyg 19h57'35"38d42'
Cadorette 25-06-58, Louise
Peg 22h57'23"24d24'
Cadot,Charlene
Lyr 18h55'58"31d54'
Cadoux,Christophe
Cnv 13h20'37"30d33'
Caduceus
Per 3h18'30"42d42'
Cady,Christina Maria
Equ 21h21'0"10d35'
Cady,Joanne
And 0h2'45"44d33'
Cady,Katherine Maria
Uma 10h34'17"50d16'
Cady,Steven Paul
Dra 16h3'51"63d4'
Caelestis Astrum Custos ex Albeo Ager
Cam 7h52'57"70d19'
Caen,Herbert E
Ari 1h46'1"17d38'
Caesar,Daniel Vincent
Lac 22h22'13"54d33'
Caetano,Georgiro
Cyg 19h51'31"44d2'
Cafarelle-Brockton, Mass.,Jennifer
Com 12h19'1"19d46'
Cafarelli,Mary
Cas 2h52'13"75d8'
Cafazzo,Genevieve Marie Guenther
Cas 23h22'10"53d50'
Cafe L'Europe
Sct 18h46'56"-6d24'
Caffera,Anne Hagerman
And 1h12'41"36d26'
Cafferello,Jacqueline Teresa
Cyg 19h28'0"37d50'
Cafferty,Mary Fabian
Lyn 8h12'16"37d36'
Caffiero,Alexa Ashlee
Gem 8h3'53"33d17'
Caffiero,Teresa & Lance
Lyn 7h59'40"45d4'
Caffro,Tamara Nicole
Mon 6h31'57"8d20'
Cafiero,#1 Star of our Family,Edith
Vul 19h22'52"23d17'
Caforio,Valeria
Lyr 18h20'0"47d9'
Cafritz,William N
Her 17h53'19"20d22'
Café-Eethuis "Borre"
Cnv 13h38'58"47d42'
Cagahama
Sgr 19h8'37"-24d47'
Cage Nebula (Constellation EmCare)
Uma 8h58'43"47d12'
Cage,John
Vir 13h27'5"-6d47'
Cage,Lynne & Michael
Mon 7h44'52"-3d31'
Cagead,HRH David
Cep 22h32'26"78d46'
Caglan,Arzu
Cyg 20h17'19"39d4'
Cagle,Marcus
Lmi 10h14'48"33d28'
Cagle,Victoria L
Vir 13h26'2"-4d23'
Cagley,David Charles
Her 17h7'1"48d22'
Cagnetto,Anne-Marie
Lyn 8h27'0"49d54'
Cagney,Jane
Cas 0h39'44"70d45'
Cagney-DMC,Marilyn
Cet 3h16'55"0d21'
Cahalane,Kevin James
Her 17h50'18"50d9'
Cahalane,Sean
Aur 4h37'23"30d31'
Cahalane,Thomas J
Cam 3h40'57"73d23'
Cahard,Jean Pierre
Per 3h57'12"35d0'
Cahill "Sugarbear", Bernice
Lyr 18h21'37"45d9'
Cahill,Brutus
Peg 22h41'24"11d6'
Cahill,David Starr
Per 2h9'55"58d51'
Cahill,Jeremy
Cep 21h38'43"55d47'
Cahill,Mary Ann
Ari 2h23'1"10d53"
Cahill,Michael & Peggy
Cyg 19h24'27"31d35'
Cahill,Olga Ann
Ari 2h23'15"10d59'
Cahill,Patricia A/ Joseph P Zeisler
Sge 19h38'1"16d19'
Cahill,Timothy W
Lac 22h32'34"52d43'
Cahn Star One,Cedric
And 23h22'37"50d37'
Cahn,Robert A
Eri 3h26'18"-4d44'
Cahn,Tessie
Tri 2h3'58"32d17'
Caiazza,Anthony Richard
Per 4h3'4"37d57'
Caicedo,Claudia
Lyn 7h26'0"50d20'
Caidin,Dee Dee M
Mon 7h6'19"-0d39'
Caila Nicole
Mon 6h42'53"10d3'
Cailin Rose
Lyr 18h23'32"37d43'
Cailin's Magic
Peg 22h0'34"20d56'
Caillaud,Luc
Uma 13h25'56"62d23'
Caillot,Norbert
Ori 6h1'56"10d58'
Caillouet,Gilles
Uma 8h50'1"54d32'
Cailotin
Peg 22h5'51"20d25'
Cain II,Robert Micheal
Lac 21h59'53"40d57'
Cain,Arthur
Cep 22h54'17"58d4'
Cain,Bill P
Aur 5h3'1"50d59'
Cain,Casey
Peg 22h59'45"29d40'
Cain,Clare
Cas 23h42'27"63d52'
Cain,Daniel Joseph
Boo 15h3'49"23d49'
Cain,Elsie M
And 0h58'32"45d17'
Cain,Joan Louise
Aql 19h49'20"14d8'
Cain,John Eifion
Uma 12h1'54"46d13'
Cain,Jorie Elizabeth
Lyn 9h11'20"34d28'
Cain,Michael
Ser 18h44'20"3d27'
Cain,Michael Eric
Aur 6h50'10"35d24'
Cain,Naomi Mae
Aur 6h5'31"38d14'
Cain,Omega Mae
Lyr 18h44'29"31d16'
Cain,Poe & Charlie
Lyr 19h15'59"40d5'
Cain,Susan
Lyr 18h50'1"30d30'
Cain,Victoria
Peg 21h59'1"31d48'
Caine,Jeremiah
Her 18h14'13"48d8'
Caine,Kathy
And 23h13'13"40d58'
Caine,Stephane
Uma 13h11'49"61d22'
Caiola III,Ben
Cmi 7h21'43"8d26'
Caiopoulos,James P
Aql 19h58'55"10d2'
Caire,Gil
Dra 9h23'38"77d57'
Cairns,Joshua
Ori 4h42'23"0d19'
Cairoli,Oscar Martin
Cnv 12h24'42"46d43'
Caisso,Pauline
Her 16h12'1"50d34'
Caitlin
Mon 6h30'18"8d1'
Caitlin
Psc 0h6'28"1d55'
Caitlin 19
Uma 8h57'57"60d37'
Caitlin Dian
Cmi 7h35'48"10d31'
Caitlin Patricia
Lyr 19h21'0"42d33'
Caitlin's Wish
Umi 16h0'22"72d40'
Caitlin,Noel
Cam 3h25'35"61d11'
Caitlyn
And 1h20'23"38d28'
Caito,Alan & Midge
Del 20h57'31"10d55'
Caius
Vul 19h48'42"27d32'
CAJ M 4
Sct 18h54'43"-6d49'
CAJAL IV
Ser 15h58'1"4d39'
Cajio,Jack
Cep 21h8'45"68d31'
Cajka,Curtis
Ori 5h52'36"14d33'
Cajun Queen
And 0h24'36"30d14'
Cakbub
Lyr 18h45'22"31d50'
Cake,Eric Lee
Aur 6h10'38"36d3'
Cakebread,Mike & Kelley
Cyg 21h56'27"53d11'
CAKKKKKCL 143
And 23h3'28"44d50'
Cal,Julio
Cas 2h13'54"50d53'
Cal,Nancy Amy Hoya
Lyr 19h21'59"30d47'
Cala Gogo 78
Per 3h52'36"35d10'
Calabrese Super-Star, Jeannie
Vul 20h5'20"23d40'
Calabrese,Aurora
Ori 5h51'39"16d14'
Calabrese,Carrie Leigh
Uma 8h46'25"67d30'
Calabrese,Chud
Aur 5h1'15"41d33'
Calabrese,Giovanna
Lyn 7h1'58"60d10'
Calabrese,Joan L
Lyr 19h6'17"28d48'
Calabrese,Karen
And 23h24'30"47d56'
Calabrese,Kimberly Joy
And 1h26'11"33d41'
Calabrese,Maria
Tri 2h38'47"34d2'
Calabrese,Mrs Christine
Cas 1h4'0"60d32'
Calabrese,Patrick John
Tau 3h58'1"7d46'
Calabrese,Peter
Her 16h25'55"28d60'
Calabrese,Rosalye Dicrosta
And 23h20'41"45d6'
Calabrese,Theodore
Lac 22h12'55"49d9'
Calabrese,Toni
Psc 22h51'27"5d30'
Calabrese,Vincent
Ori 6h0'1"8d19'
Calabria,Angela
Cas 2h0'35"60d13'
Calabria,Gina Marie
And 2h21'50"42d3'
Calabro 8
Aur 5h0'13"50d51'
Calabro,Jack
Cep 0h2'34"68d11'
Calabro,Judy
Sgr 18h17'31"-28d32'
Calabro,Michael P
Per 2h52'0"32d12'
Calabrotour
Vel 9h43'8"-40d36'
Calafiore,"Sammy"
Aur 4h47'21"38d14'
Calahan,Loyce
Ori 5h55'30"15d22'
Calam & Bill
Cyg 20h22'32"39d45'
Calamari,Christopher Ray
Uma 9h43'36"67d45'
Calamari,Jacqueline Grace
Uma 9h33'16"68d9'
Calamas,Nick & Beth
Cmi 7h41'24"1d49'
Calame,Bart
Ori 5h46'38"11d29'
Calamel,Valentin
Aur 6h10'53"31d12'
Calamera,Joseph
Cap 21h4'37"-22d29'
Calamia,Jr,Robert A
Her 16h57'0"35d36'
Calamigos Star C Ranch
Boo 15h3'47"20d34'
Calamity Scott
Cnv 13h59'59"39d21'
Calamusa,Giovanna
Eri 3h49'51"-1d7'
Calamusa,Stephanie Ann
Del 20h13'35"15d28'
Calandra,Joan Marie
Cyg 21h37'1"42d3'
Calandrea
Lac 22h3'52"51d3'
Calandri,Robert Theodore
Aql 20h13'1"5d24'
Calandria
Cep 20h58'1"63d17'
Calangelo,Richard L
Cam 4h23'17"68d9'
Calarco,Anthony
Lac 21h59'58"40d7'
Calatayud,Rosa
Lyn 8h13'21"48d2'

Calautti Star,The
Cyg 21h39'33"41d44'
Calaway,Mark
Ari 2h41'51"20d35'
Calbo,Mary Elena
Cas 1h6'1"53d35'
Calcao,Antonio Bernardino
Lac 22h51'18"38d43'
Calcaterra,Charles W
Aur 6h3'30"31d33'
Calcaterra,Terry
Del 20h16'42"9d24'
Calcavecchio,Angelo
Vir 11h50'27"1d0'
Caldas,Vincent James
Per 3h59'43"31d43'
Caldeira,Anthony
Ori 5h54'55"12d45'
Calder,Brendan
And 0h21'16"40d3'
Calder,David Taylor
Uma 8h52'27"56d17'
Calder,Eternally Melissa
Ant 9h38'3"-34d4'
Calder,Matthew Tyler
Boo 13h40'41"21d51'
Calder,Penny
And 1h3'12"39d43'
Calder,Russell
Ser 16h6'24"1d35'
Calderhead,Asia Tanaka
Lyr 19h3'27"40d32'
Calderon,Celina Marie
Cas 1h1'13"53d41'
Calderon,Doreen
Mon 5h57'28"-6d23'
Calderon,Garrett
Com 12h29'49"22d22'
Calderon,Leonarda S
Lmi 11h3'0"31d13'
Calderon,Richard David
Boo 15h6'59"13d56'
Calderone,Christopher
Sgr 20h14'31"-29d32'
Calderone,Philip
Aql 19h31'39"10d26'
Calderwood,Edith Mary
Cyg 20h36'22"31d26'
Calderwood,Hallie Katherine
Leo 10h51'46"18d4'
Calderwood,Joy Anne
Peg 23h0'41"10d24'
Caldicott,Eileen
Peg 23h31'0"11d8'
Caldicott,Stephen Lee
Cep 22h28'29"70d24'
Caldiroli,Michela
Cas 23h21'1"53d18'
Caldwell 1920-1993, William Fenton
Cep 22h46'51"63d21'
Caldwell's Guardian Star,Keith
Ori 6h6'0"20d57'
Caldwell,Alma
And 23h39'24"37d45'
Caldwell,Annalee Grace
Equ 20h57'30"7d17'
Caldwell,Bill
Her 16h47'1"4d46'
Caldwell,Brent Allan
Del 20h18'57"20d14'
Caldwell,Brittany Leigh
And 23h39'1"41d12'
Caldwell,Daniel Dale
Ori 5h5'52"-2d49'
Caldwell,Dawn M
Peg 22h23'1"4d4'
Caldwell,Estelle Peggy
Umi 15h15'31"67d48'
Caldwell,Floyd James
Cet 3h12'11"0d20'
Caldwell,Gail & Jim
Cam 12h47'0"77d16'
Caldwell,Glenn
Ser 15h59'32"9d39'
Caldwell,Jeremy Scott
Ori 5h56'34"11d2'
Caldwell,Joseph
Her 18h16'53"28d25'
Caldwell,Matthew & Meredith
Cyg 19h31'15"33d10'
Caldwell,Sandy K C
Eri 2h46'48"4d46'
Caldwell,Warren C
Aur 6h10'21"50d34'
Caldwell-Knight, McKenzie Emily
Lib 15h18'0"-28d12'
Caldy MS
And 2h17'30"40d23'
Cale,Christine
Lyr 19h1'29"37d33'
Cale,Robert
Boo 14h57'14"47d32'
Caleb
Mon 7h2'42"1d10'
Caleb
Uma 11h12'30"47d44'
Caledonia
And 22h57'18"38d33'
Calegari,Allison Ashley
Umi 13h25'47"74d30'
Calegari,Francesca R.
Umi 13h11'27"72d26'
Calehr,Hallym Prof Dr Med
Leo 11h24'26"11d21'
Calendarius
Col 6h30'20"-38d49'
Caley,Patricia
And 2h23'21"44d16'
Calfee,Carl
Lac 22h10'43"49d18'
Calfee,S David
Aur 4h49'0"40d53'
Calhoun,Arthur W
Ari 2h2'37"21d33'
Calhoun,Christopher Aaron
Psc 0h44'1"20d11'
Calhoun,Chuck's Forever Friend
Ori 5h53'16"15d1'
Calhoun,Jay Thomas
Aql 20h11'1"4d39'
Calhoun,Julia
And 0h9'52"37d34'
Calhoun,Thomas Bartling
Lac 22h1'23"48d52'
Calhoun,Timothy Samuel
Lib 14h58'0"-0d35'
Cali
Peg 23h33'41"30d42'
Cali
Peg 21h59'31"24d37'
Cali,Dr Saro M
Oph 17h54'45"12d58'
Cali,Richard
Her 16h41'37"11d40'
Calicdan,Calvin L
Cet 3h4'43"0d3'
Calidalla
Uma 11h26'11"34d60'
Caliendo,Jennifer
Hya 9h30'11"-0d48'
Calies,Erny (Leo)
Leo 10h1'24"10d4'
Calies,Helen Krynski
Leo 10h1'33"11d8'
Calies,Nathan A
Leo 10h2'12"11d1'
Caliesse
Cep 22h21'37"56d4'
Califano,Angela
Cas 2h11'56"59d26'
Califano,Olivia Rose
Sco 16h53'30"-41d2'
California Rose
Cyg 21h48'29"36d49'
California Snowman,The
Ser 17h36'0"-14d43'
California State Society N S D A R
Oph 16h41'27"1d35'
Calihan,Michael P
Vir 13h54'57"7d7'
Calinda
Pyx 8h46'20"-23d21'
Caliph
Ser 15h53'1"-2d18'
Calise,Mary Ann
Tri 1h50'55"26d24'
Calivert,Lidia Stefanov Sava
Lyr 18h42'29"47d37'
Calivertvert,Ronald Shotwell
Aur 5h44'32"50d1'
Caljouw,Mark
Per 3h32'48"50d21'
Calkins,Anna Phyllis
Cet 2h49'42"1d52'
Calkins,Leland
Cet 2h35'36"0d13'
Calkins,Sr,William T
Cyg 19h53'57"37d45'
Calkwood,Jonathan
Psc 1h36'1"27d35'
Call,Cash
Ori 5h37'47"8d22'
Call,Karen
Cas 1h34'1"60d13'
Call,Katherine Dianne
Cas 0h35'13"62d46'
Call,Mark
Ori 5h54'53"0d43'
Call,Nylora Joy
And 23h47'35"46d24'
Calla Fia Allison
Eri 3h16'0"-11d22'
Callaghan
Cyg 19h27'52"35d30'
Callaghan
Aql 19h46'45"11d13'
Callaghan,Christian Connolly
Lac 22h55'1"52d58'
Callaghan,John Russell
Per 4h33'57"38d14'
Callaghan,Lindy Ann
Lyn 8h10'26"47d48'
Callaghan,Lois
Uma 12h25'27"57d38'
Callaghan,Richard George
Ori 5h57'0"10d51'
Callahan & Family, Mike & Tammy
Lyr 18h40'21"33d22'
Callahan,A J
Aql 20h12'56"12d54'
Callahan,Father Kevin
Aql 20h11'44"14d14'
Callahan,G P
Lac 22h29'18"54d4'
Callahan,Kathleen Marie
Peg 22h43'33"35d14'
Callahan,Kelly
Uma 9h6'33"52d42'
Callahan,Kerry Frances
Crb 16h4'53"32d26'
Callahan,Kevin Michael
Lac 22h31'19"56d35'
Callahan,Linda
Cas 23h21'39"60d39'
Callahan,Marilyn
Vul 20h15'0"22d32'
Callahan,Pat & Jeanne
Cyg 20h31'52"37d40'
Callahan,Patrick
Her 18h19'1"28d45'
Callahan,Sr,Arthur
Oph 17h55'34"-6d59'
Callahan,Susan M
And 23h29'52"48d39'
Callahan,Susan Marie
Ori 5h50'19"15d12'
Callahan,Thomas Michael
Her 17h13'49"45d7'
Callahan,Viola Decker
Cas 23h36'1"60d8'
Callahan-Grant
Eri 4h11'0"-19d43'
Callamaras,Paul
Cyg 21h2'35"40d30'
Callan,Thomas Lyons
Lac 22h9'37"50d56'
Callanan,Michael J
Aur 5h56'54"31d36'
Callas
Lmi 11h2'35"32d48'
Callaway,Hayden
Aur 6h2'12"31d26'
Callegari,Giuseppe
Cmi 8h8'40"0d19'
Callen 8-22-76, Jennifer Lynn
Com 12h17'50"32d2'
Callen,Jeanine Marie
Cam 13h15'0"80d51'
Callender,Brandon Marcus Theodore
Dra 19h34'49"68d59'
Callender,Elisabeth Darlene
Vul 19h46'50"28d23'
Callender,John Michael
Boo 13h40'15"18d55'
Callender,Mark-Lynn Cody-Tiffany
Lyr 18h31'0"40d18'
Callender,Nancy
Lyr 18h47'20"41d53'
Callens,Paul
Aur 6h33'38"40d25'
Callens,Paul & Hilde Van Nieuwenmuyze
Cam 4h47'21"68d0'
Caller,Roy
Her 16h36'36"36d20'
Callery 4
Cam 6h7'55"58d29'
Callery MD,Charlie
Boo 13h39'20"19d24'
Callery,Charlie
Ser 18h42'52"3d50'
Calley
Cyg 19h41'0"28d38'
Callicrate,Tyler John
Oph 18h0'13"11d3'
Callie
Lyn 6h19'32"60d2'
Callie Jo
And 0h59'39"45d33'
Callie's Star
Psc 0h20'48"18d40'
Callie's Star
Cyg 20h16'1"41d39'
Callietwinkles
Cas 0h44'37"66d38'
Calligaris,Ronald
Dra 10h42'20"77d40'
Callihan,Blair"Bear"
Lyn 8h0'38"50d53'
Callihan,Dean Edward
Equ 21h5'35"10d50'
Callin,Veronika Stella
Cyg 21h56'52"53d13'
Callis,Belinda K
Aql 20h6'48"0d46'
Callis,Greg Allan
Aur 4h17'0"51d26'
Callison,Brian & Suella
Vul 20h21'59"26d2'
Callon
Dra 10h24'0"81d41'
Callow,Jeffrey E
Uma 9h4'0"53d57'
Callum Bailey
Cru 12h6'36" 64d30'
Callum John
Ori 6h7'43"3d4'
Cally
Ori 5h18'24"10d34'
Callyney
Cep 22h18'50"58d8'
Calma,Eleanor
Peg 22h31'26"26d25'
Calmac I
Per 2h38'30"37d31'
Calmar
Cas 1h13'6"60d1'
Calmon,Remi
Her 18h23'59"12d48'
Calmund,Reiner
Cmi 7h19'0"5d26'
Caloni,Helene
Aur 4h54'44"52d15'
Calonia "Heaven Sent"
Cas 23h39'30"63d45'
Calore,Anne Marie
Sge 19h52'59"16d48'
Calore,Jonathan Whitfield
Aql 19h26'24"15d33'
Calsius,Luke
Cma 6h56'45"-18d53'
Calssacy,Audrey
Cam 3h35'14"60d35'
Caltabiano,Sara
Cyg 21h24'1"48d51'
Calton,Robin & Timothy
Per 3h4'49"50d25'
Calvano,Anthony Joseph
Dra 16h8'20"61d45'
Calvano,Christopher
Leo 10h52'1"20d10'
Calvanus,Daniel
Cep 22h13'0"58d19'
Calvar,Stéphane
Cam 7h49'49"68d13'
Calvaresi,II, Alessandro "Alex"
Boo 15h28'21"37d55'
Calverley,Del "Mimi"
Cyg 19h22'60"44d46'
Calverley,Joe
Dra 15h50'51"63d53'
Calvert,Ashley Susan
And 1h6'0"38d35'
Calvert,C Michael
Mon 8h3'32"-4d18'
Calvert,David
Cep 20h30'46"60d36'
Calvert,Jonathan
Uma 9h6'17"57d32'
Calvert,Juanita Radean
Cam 3h37'52"60d26'
Calvert,Sarah Jayne
Cyg 19h33'39"39d31'
Calvert,Terry B
Hya 8h37'12"5d40'
Calvert,Valerie
Cas 23h36'60"60d17'
Calvi,Laura
Lac 22h21'10"50d12'
Calvillo,Adolfo
Cmi 7h19'42"1d32'
Calvin
Dra 20h25'30"74d32'
Calvin & Hobbes
Hya 8h23'1"-0d50'
Calvin,Frank
Sco 17h2'47"-30d5'
Calvin,Sara
Peg 22h10'24"29d30'
Calvin,Suzanne Schroeder
Peg 22h20'18"30d42'
Calvino,Alexander J
Dra 18h52'58"70d49'
Calvino,Larry
Aql 19h6'37"3d47'
Calvo,Ricardo"El Magnifico"
Ori 5h58'11"16d18'
Calwell's Reach,RB
Per 3h29'45"50d1'
Calyx
Peg 23h36'52"20d2'
Calzaretta,Lori Marie
Tri 1h51'44"28d5'
Calzolaio,Edward
Per 2h55'26"54d34'
Calzone IV,Joseph A
Aur 7h15'0"39d37'
Calzone,Alexander C
Dra 16h14'56"64d22'
Calzone,Anna Marie
And 23h37'56"43d18'
Calzone,Nicholas D
Lac 22h14'13"46d11'
Calzoni,Sara Maurizi
Umi 7h53'15"88d27'
Cam
Her 17h36'20"25d4'
CAM
Cyg 21h53'12"42d56'
Cam & Mark
Cyg 20h31'0"40d50'
Cam Le Fou Gravelle
Lmi 9h24'39"38d19'
Cam's Awakening
Cas 1h53'0"60d2'
Camacho,Elizabeth F
Crb 15h56'15"29d22'
Camacho,Juan Alberto
Her 16h25'47"26d58'
Camacho,Omar
Aql 19h47'23"10d28'
Camaggi,Elisa
Lyn 6h12'11"54d18'
Camalotto,Peter
Aql 19h30'51"7d30'
CAMALU
Vir 12h51'21"1d26'
Camara Christmas 1993, Hildegar
Cas 1h40'54"60d58'
Camara,Francisco Lopez
Cet 3h15'0"5d8'
Camara,Mathew
Aql 19h26'23"8d2'
Camarata,Chelsea E
And 2h35'55"37d57'
Camargo,Juan
Ser 15h57'57"21d22'
Camarra,Domenico
Umi 13h54'1"76d47'
Camaïeu
Dra 9h57'12"78d58'
Cambareri's Star
Dra 16h5'49"66d51'
Cambell,Mary Christine
Cyg 20h59'31"30d19'
Camberg,Leanne
Cmi 7h24'28"1d24'
Cambern,Andrea
Cam 7h43'33"70d37'
Cambor,Stephen Paul
Her 16h32'17"38d47'
Cambra,Pilar Hoskins
And 1h14'55"37d46'
Cambray,Bill- Rock Star
Aur 4h55'58"41d13'
Cambray,Doreen - Java Star
Boo 14h58'1"20d19'
Cambre,Neil Anthony
Oph 18h32'21"10d32'
Cambria Erin
Cas 0h25'26"61d6'
Cambria,Diane
Mon 6h51'37"10d45'
Cambria,Jake Nicholas
Lac 22h43'35"52d33'
Cambria,Mike
Dra 19h19'12"80d24'
Cambridge College
Aqr 22h46'49"0d29'
Cambridge,Robert H. "Bobby"
Tri 1h51'52"26d54'
Cambridge-Hall,Norma Lee
Aur 5h19'31"42d56'
Cambron,Nicole
Lyn 8h54'51"46d36'
Camden
Cep 21h59'25"58d50'
Camden,Diane Marie
Aql 19h55'37"8d26'
Camden-Britton,Heidi Christine
And 23h38'34"46d36'
Came,Hazel Jane Ward
Cam 3h57'38"61d32'
Came,Michael & Lynne
Cyg 21h34'52"30d16'
Camel I,J
Aql 19h22'49"-1d2'
Camelot
Lac 22h18'43"54d41'
Camera,David
Boo 15h29'22"48d24'
Camera,Lauren
Cas 23h24'42"51d30'
Cameron
Peg 22h0'25"10d33'
Cameron
Per 3h22'18"54d58'
Cameron
Per 4h23'26"50d2'
Cameron
Cap 21h26'34"-23d52'
Cameron
Lyr 18h41'40"30d33'
Cameron & Jeb's Star
Uma 12h3'53"36d55'
Cameron 3-15-65, Phillip M
Her 15h57'46"50d57'
Cameron By The Sea
Hya 9h9'44"1d19'
Cameron C
Lac 22h1'38"48d37'
Cameron David
Cep 22h22'54"61d40'
Cameron Dylan
Cmi 7h53'37"5d31'
Cameron III,John H
Boo 14h41'42"33d14'
Cameron L
Aur 4h50'27"50d43'
Cameron's Coach House
Uma 9h29'36"47d58'
Cameron's Five
Cep 22h36'0"63d52'
Cameron,Andrew
Ori 5h42'49"11d2'
Cameron,Barbara
Lyn 7h0'29"58d37'
Cameron,Barbara Joy
Peg 21h28'12"2d32'
Cameron,Bruce
Gem 7h25'41"28d38'
Cameron,Christopher "Chris" Samuel
Aql 19h3'17"0d55'
Cameron,Courtney Reed
Mon 6h19'52"8d45'
Cameron,Dr George L
Her 16h27'0"39d53'
Cameron,Eve
Cam 14h23'39"80d7'
Cameron,Gregory Daniel
Dra 16h0'53"62d40'
Cameron,Hayden
Cep 22h14'32"61d47'
Cameron,Iain Judson
Ori 5h2'11"12d36'
Cameron,James
Leo 9h34'43"8d23'
Cameron,James & Carolyn
Per 2h46'37"40d23'
Cameron,Janice
And 1h58'1"37d40'
Cameron,Jim
Aur 6h1'15"36d3'
Cameron,Joanne
Cyg 21h1'24"28d6'
Cameron,John
Gem 7h25'17"30d45'
Cameron,John Angus Frank
Ori 4h47'51"15d9'
Cameron,Julie
Cet 3h8'50"1d6'
Cameron,Justin
Boo 15h4'51"40d20'
Cameron,Kari
Eri 3h59'25"-11d20'
Cameron,Kimberly G
And 23h2'36"41d13'
Cameron,Marilyn Ann
Lyr 18h41'26"28d1'
Cameron,Martha & Johnnie
Lyr 18h20'52"47d37'
Cameron,Natasha
Del 20h56'31"10d10'
Cameron,Rachelle
Eri 3h17'37"-4d50'
Cameron,Ryan
Per 3h12'1"40d32'
Cameron,Scott
Dra 17h49'1"65d3'
Cameron,Sheila "Moppy"
Cas 23h1'45"58d48'
Cameron,Tyler
Peg 22h4'34"10d43'
Cameron-Olski
Cep 21h33'13"58d27'
Cameroni,Lidia
Lac 22h52'34"50d33'
Camesasca,Nando e Luisella
And 23h1'45"41d50'
Camhy,Edith
Cas 0h30'10"60d45'
Cami
Mon 8h4'6"-4d46'
Cami Leigh,The
Cam 9h14'55"84d25'
Camila
Ori 5h37'18"-0d29'
Camilla
Cyg 21h53'29"42d43'
Camilla
Cas 2h56'58"61d44'
Camilla
Mon 6h53'0"-1d25'
Camilla
Leo 10h51'52"21d16'
Camilla
Sge 20h17'24"20d36'
Camilla
Lyn 6h39'28"58d12'
Camilla
Boo 15h1'15"42d3'
Camilla
Cas 2h24'36"61d41'
Camilla
Boo 15h6'10"47d50'
Camilla
Del 20h14'44"15d13'
Camilla
Lyn 7h50'1"40d52'
Camilla 4 Martin
Cyg 21h33'25"42d20'
Camille
Peg 22h10'44"26d44'
Camille
Ori 5h12'21"-8d37'
Camille
Cam 3h46'34"67d39'
Camille
Lyn 6h32'51"54d9'
Camille
Cam 6h15'13"68d55'
Camille & Arthur
Cam 5h48'24"68d44'
Camille Denise
Leo 9h40'27"7d35'
Camille Julitta
Lyr 18h47'13"30d12'
Camille Marie
And 23h36'46"47d14'
Camille Michele
Lyn 8h23'13"40d13'
Camille's "Purple Haze"
And 23h23'1"43d7'
Camille,Ann H
Cas 0h35'14"66d30'
Camille,John Andrew N
Cep 22h19'23"58d12'
Camille,Norm
Her 16h51'57"35d46'
Camilleir,Alain
Aur 6h17'51"31d40'
Camilleri,Antoinette
And 1h9'38"37d42'
Camilleri,Catherine
Cas 0h44'49"73d43'
Camilleri,Ed
Cyg 20h38'26"40d3'
Camilli,Daniel
Peg 21h58'16"21d31'
Camilo & Fiz
Cyg 20h22'16"41d17'
Camilo's Amore
Aql 19h16'24"19d51'
Camina,Jerry
Hya 9h14'5"0d56'B
Camina,Phyllis
Hya 9h14'5"0d56'A
Caminiti,Kathryn Grace
Psc 1h3'35"28d1'
Caminker
Sex 10h46'53"0d58'
Camiré,Stéphanie
Cyg 20h22'53"39d11'
Camma,Albert J
Per 2h1'34"48d58'
Cammack,Christopher & Susan
Ori 6h10'43"0d26'
Cammack,Terridawn & Brittany
Cyg 19h45'14"31d38'
Cammarano,Penny
Lmi 10h36'0"28d48'
Cammarano,Penny
Peg 0h1'55"30d54'
Cammarata,Joseph
Lac 22h8'50"48d48'
Cammarata,Joseppina
Uma 8h13'57"72d14'
Cammarata,Robin L
Lib 14h19'21"-20d32'
Cammareri,Nicholas
Boo 15h0'57"50d3'
Cammareri,Nicholas
Uma 9h33'58"50d50'
Cammareri,Silvia
Cas 23h15'56"61d38'
Cammett,Ward
Aur 6h43'48"37d28'
Camou,Jonathan Cody
Boo 14h3'22"26d15'
Camozzi,Mary Elizabeth
Mon 7h0'12"5d35'
Camp Brookwood
Per 2h29'43"57d46'
Camp Laurel
Eri 3h2'26"-10d31'
Camp Nawakwa
Cnv 13h30'18"50d18'
Camp,"The Good One", Brooks
Sco 17h51'13"-38d25'
Camp,Billie
Sex 9h43'21"2d45'
Camp,Catherine Renee
Lyn 7h25'38"44d12'
Camp,Catherine Renee
Cet 3h3'57"3d38'
Camp,David & Lauren
Aur 5h3'58"40d23'
Camp,Doris M
Mon 7h3'50"-5d44'
Camp,Edward N
Eri 2h43'36"-16d18'
Camp,Gregory N
Cet 2h33'31"2d22'
Camp,MD,John C
Cet 0h40'33"-2d53'
Camp,Miriam M
Sge 19h40'30"16d51'

Camp,Ross O
Peg 23h47'1"10d1'
Camp,Sarah Amelia
Peg 22h36'14"21d53'
Camp,Terrance
Ori 5h39'45"-6d33'
Camp,Timothy S
Hya 8h53'31"6d19'
Camp-Robin & Brittanys Daddy,Robin Scott
Uma 8h50'18"47d40'
Campa,Suzanne Renee
Cyg 19h58'11"50d35'
Campa-Pettennude Adriana
Del 20h24'40"20d24'
Campagna,Amanda Rae Lienau
Cam 3h25'44"55d48'
Campagna,Carl
Boo 15h1'22"11d36'
Campagna,Lina V
Peg 23h5'19"10d56'
Campagna,Matthew & Kristen
Per 2h40'1"40d15'
Campagnuolo II,Vincent J
Dra 19h34'12"84d59'
Campan,Sylvie
Cyg 21h37'31"40d8'
Campana,Joe & Sue
Uma 9h38'22"49d26'
Campana,Jr,Joseph Salvatore
Her 17h29'39"28d27'
Campanella,Anna Marie Absher
Com 12h1'56"21d35'
Campanella,Emidio
Dra 19h3'52"49d50'A
Campanella,Frank
Cep 3h1'25"78d32'
Campanella,Karen
Crb 15h21'40"32d3'
Campanella,Michael
Uma 11h57'46"50d36'
Campanella,Rose
Cas 0h43'29"60d4'
Campanelli,Andrew Robert
Oph 17h19'47"-22d34'
Campanelli,Damian Anthony
Cma 6h38'42"-17d49'
Campanelli,J Steven
Hya 8h22'51"-5d47'
Campanini,Sarah Jane
Mon 6h20'0"1d19'
Campari,Daniela
And 0h22'23"45d39'
Campbell
Aql 19h30'33"12d3'
Campbell Family Star, The
Leo 9h22'52"19d31'
Campbell Lesley Jane
Peg 22h3'48"24d14'
Campbell Star,The
Cyg 20h34'44"30d25'
Campbell's Star(Go Sonics!),Doug
Aql 19h29'58"-10d43'
Campbell,Adam
Dra 16h11'25"68d54'
Campbell,Anderson Sharkey
Vul 19h33'46"27d30'
Campbell,Andrew H
Aur 6h30'1"32d4'
Campbell,Anna Rachel
Lib 15h52'45"-5d41'
Campbell,Ashley Dare
And 0h35'38"41d0'
Campbell,Austin Daniel
Lac 22h7'0"50d58'
Campbell,B R
Hya 9h1'48"4d44'
Campbell,Bette Lyle
Lyr 18h51'33"35d26'
Campbell,Beverley Mary Vaughan
Uma 10h45'10"41d22'AB

Campbell,Billy and Carol
Eri 3h42'29"-1d51'
Campbell,Brett J
Cep 22h41'44"57d40'
Campbell,Brian David
Cnv 12h19'49"47d42'
Campbell,Bruce & Kathy (Pohlman)
Uma 8h38'49"50d35'
Campbell,Bruce Thomas Hugh
Ser 15h31'1"10d54'
Campbell,Bryan Todd
Lac 22h12'56"47d14'
Campbell,Burke D
Cas 0h24'38"61d48'
Campbell,Byron & Mary Ellen
Cep 23h5'0"60d7'
Campbell,Carrie Kathleen
Del 20h18'59"10d34'
Campbell,Casie M
Cam 13h49'33"80d4'
Campbell,Cate
Lyr 18h31'0"33d1'
Campbell,Charles
Per 2h48'20"43d22'
Campbell,Christine Elizabeth
Tri 1h30'29"33d8'
Campbell,Colin
Ori 5h57'15"14d2'
Campbell,Colleen M
Lyr 18h47'47"36d30'
Campbell,Collin Wilson
Aql 19h0'1"1d8'
Campbell,Courtney Jane
Peg 0h0'17"18d42'
Campbell,Daniel J
Oph 17h6'1"10d6'
Campbell,Danielle (Dani)
And 1h50'28"38d48'
Campbell,Danton Curtis
Lac 22h40'56"51d34'
Campbell,David
Ari 2h32'48"20d32'
Campbell,David Kondwani
Cyg 21h43'0"30d6'
Campbell,David Wayne
Ori 5h54'18"8d14'
Campbell,Debbie
Cas 2h59'22"61d18'
Campbell,Devin Patrick
Eri 3h25'19"-4d24'
Campbell,Diane
Peg 0h0'40"28d55'
Campbell,Donald Lawrence
Uma 11h23'34"38d18'
Campbell,Donald
Dra 15h17'37"60d44'
Campbell,Doris C
Sgr 19h6'47"-26d43'
Campbell,E Verne
Uma 12h56'1"60d37'
Campbell,Elizabeth Mary
Lyn 7h2'30"52d60'
Campbell,Esquire,Sir Charles G
Per 2h44'47"43d21'
Campbell,Forever Your Eyes Allison
Lmi 10h17'42"30d54'
Campbell,Fr Mike
Cmi 7h22'35"1d14'
Campbell,Frank T
Dra 17h26'56"60d54'
Campbell,Freida Buchanan
Cas 0h31'38"70d30'
Campbell,Glenn F
Lac 22h34'57"52d55'
Campbell,Grace
Cas 0h31'29"64d59'
Campbell,H E
Umi 15h14'50"68d56'
Campbell,Hannah Lindley
Vul 19h47'59"28d26'
Campbell,Helen Slater
Ori 6h2'0"8d36'

Campbell,Hester Josephine
Lyr 18h29'33"46d16'
Campbell,Ian Nicholas
Aur 4h51'38"50d28'
Campbell,Isabel Avery
Cyg 20h56'39"38d3'
Campbell,Isabella Colin
Cyg 20h35'28"31d30'
Campbell,Jack
Cet 0h28'0"-0d3'
Campbell,Jacquelyn Kay
And 2h25'1"39d53'
Campbell,James P & Elizabeth M
Hya 9h7'21"5d60'
Campbell,Jeffrey D
Uma 9h14'16"49d42'
Campbell,Jeffrey Lynn
Aur 5h54'51"50d10'
Campbell,Jesse & Darlene
Aql 20h30'23"0d4'
Campbell,Jessica Dawn
Sct 18h44'15"-7d29'
Campbell,Jim
Cmi 7h37'22"1d6'
Campbell,Joanne
Cas 0h48'0"71d7'
Campbell,Jody
Cas 3h11'0"66d6'
Campbell,John
Lyr 19h0'31"28d10'
Campbell,John
Ser 18h20'1"0d48'
Campbell,John
Ori 5h33'38"-1d35'
Campbell,John Michael
Per 2h24'23"54d56'
Campbell,John N
Boo 14h51'0"38d48'
Campbell,John N
Sgr 19h6'47"-24d51'
Campbell,Judy
Tri 1h46'43"30d43'
Campbell,Julie
Ori 5h50'46"21d16'
Campbell,Karen Louise
Ori 5h59'31"20d44'
Campbell,Kevin Blaine
Boo 14h57'19"21d10'
Campbell,Kevin John
Uma 10h54'39"38d26'
Campbell,Kristy Dana
Cas 1h5'32"66d1'
Campbell,Lacy Elizabeth
Vul 19h43'29"20d35'
Campbell,Lauren Adele
Cas 1h16'0"62d30'
Campbell,Lee
Crt 11h2'47"-18d46'
Campbell,Lori Ann
Lyr 18h29'48"42d32'
Campbell,Louise & Doug
Uma 9h4'47"59d28'
Campbell,Lynne
Uma 9h50'53"56d39'
Campbell,Mabel Louise
Vul 21h16'46"20d2'
Campbell,Marjorie Ann
Lac 22h42'17"55d39'
Campbell,Marjorie D
Cep 22h20'45"60d3'
Campbell,Mary
Lyr 18h19'1"47d18'
Campbell,MD,Joe A
Oph 17h0'51"11d56'
Campbell,Meghan Alexandra
Peg 22h5'1"5d60'
Campbell,Melanie Jo
Com 13h12'18"26d52'
Campbell,Miss Fiona
Com 12h12'44"21d41'
Campbell,Mitchell James
Cep 22h29'0"61d29'
Campbell,Morag Wallace
Mon 7h3'11"-7d9'

Campbell,Nicholas
Per 3h10'16"41d12'
Campbell,Parker Blade
Her 18h45'11"12d55'
Campbell,Paul
Oph 16h49'53"11d32'
Campbell,Perry Scott
Her 16h35'48"40d33'
Campbell,Phil
Dra 17h57'59"60d38'
Campbell,Rachelle G
Cyg 20h38'56"42d34'
Campbell,Ralph D
Ser 15h12'47"21d52'
Campbell,Rebecca Dawn
Lyn 6h19'12"60d22'
Campbell,Rebecca L
Crb 16h12'0"37d42'
Campbell,Remington Alexander
Aql 18h54'31"10d40'
Campbell,Richard
Hya 8h54'0"5d48'
Campbell,Richard Matthew
Cep 23h11'39"64d47'
Campbell,Rory Adam
Dra 16h51'41"61d50'
Campbell,Rosalie Kathryn
Cas 23h37'57"61d47'
Campbell,Samuel C C
Her 18h9'0"48d11'
Campbell,Sara
Mon 6h17'15"-6d19'
Campbell,Sean
Per 3h6'52"41d52'
Campbell,Shannon
Del 20h57'28"11d7'
Campbell,Sharon Kay
Peg 22h43'53"25d2'
Campbell,Stephen
Cep 22h8'1"70d27'
Campbell,Stephen Glen
Umi 14h43'54"68d51'
Campbell,Stormy & Bob
Crb 15h26'14"30d14'
Campbell,Teena Grace
Cyg 20h22'45"38d40'
Campbell,Tom & Janelle
Cyg 21h56'53"52d54'
Campbell,Vernetta
Lyr 18h38'27"46d1'
Campbell,Victor
Boo 13h58'28"23d53'
Campbell,W H
Aql 19h45'23"11d51'
Campbell,Wayne
Lac 22h33'0"54d16'
Campbell-Judge,Mentor, Friend,Robert
Lac 22h34'28"53d17'
Campbell-One Singular Sensation,Gary M
Dra 16h25'35"64d24'
Campbell/Kronfeld Wish Star
Aql 18h47'19"11d24'
Campeau's Ride
Per 3h0'13"50d13'
Campellone,TJ "Iceman"
Per 2h35'43"56d44'
Camper,Rob
Cma 7h14'48"-18d55'
Campfield,Danielle
Lyn 7h48'56"44d13'
Campion,Camille Dana
And 23h1'1"51d21'
Campione,Jane Ann
Peg 22h48'13"11d6'
Campione,Joseph
Dra 16h58'20"64d40'
Campitiello,Sandra
And 0h15'0"37d42'
Campling,Matthew Jacob
Cep 22h40'1"65d45'
Campo Leo
Vel 9h23'41"-40d5'

Campo,Ann Elise
And 23h3'25"50d23'
Campo,Emily Ann
Mon 6h18'0"3d4'
Campo,Josie & Ray
Sge 19h59'24"16d47'
Campo,Jr,Robert P
Cep 21h55'29"55d30'
Campo,Katelyn Joan
Sgr 19h1'20"-20d13'
Campo,Ronnie S
Cas 3h9'37"73d27'
Campobasso,Susan L
Cyg 19h25'1"30d1'
Campochiaro,Joseph & Caroline
Cyg 19h44'24"38d8'
Campochiaro,Susan
Dra 11h56'27"71d0'
Campodonico,Jose Carlos
Boo 14h9'26"41d39'
Campora,Stefano
Cet 2h34'0"1d44'
Campora,Thomas D
Her 16h39'16"35d1'
Campos Neutros
Tau 5h27'12"21d18'
Campos,Patricia Ann
Mon 6h24'32"4d56'
Campoverde,Jean Baptiste
Aql 19h27'1"7d51'
Campoy,Alexandre
Cas 0h30'1"68d36'
Campregher,Claire Elizabeth
Aql 20h10'56"13d28'
Camps,Laura
Sge 20h0'40"16d43'
Campuzano,Claudine
Uma 9h15'28"48d6'
Campwala,Shamima"Our Star"
And 0h53'0"35d28'
Camélia
Peg 23h29'42"20d3'
Canaan
Ori 5h41'49"10d16'
Canaday,Andrea
Eri 4h14'20"-12d17'
Canaday,April LaRoe
And 23h28'41"49d50'
Canaday,Conni
And 23h26'46"44d21'
Canaille 1
Her 18h2'15"14d12'
Canaille 2
Peg 23h37'52"15d58'
Canal,Gabriella
Lyn 7h56'22"43d28'
Canal-Rufat
Cyg 21h40'22"28d21'
Canals,Jose Manuel
Ori 6h6'34"2d51'
CANAM
Peg 23h19'41"30d44'
Canarias,Naturaleza Cálida
Ori 5h47'59"20d59'
Canavaggio,Pierre
Cru 11h59'1"-58d12'
Canavan's Angry,Paul
Oph 17h56'20"10d17'
Canavan,Sally
And 23h20'18"46d56'
Canavan,Suzanne
Dra 18h4'0"51d1'
Cancela,Raúl
Boo 15h6'14"14d60'
Cancilla,Lisa
And 23h30'1"41d29'
Canco,Anthony & Carol
Per 1h36'49"53d58'
Candace
Cas 23h42'42"61d25'
Candace
Dra 19h32'26"61d20'
Candace
Peg 0h7'24"21d59'

Candace
And 2h30'41"44d32'
Candace
Cyg 19h32'37"32d2'
Candace
Cas 1h8'38"62d50'
Candace & David Forever
Lyr 19h22'29"30d34'
Candace I
Cep 20h59'27"68d5'
Candace K
Uma 9h44'19"56d16'
Candace Lynn
Cyg 19h25'48"30d26'
Candance
And 2h19'58"49d13'
Candance Nichole
Eri 3h22'35"-2d46'
Candela,Croce Salvatore
Mon 6h53'23"0d15'
Candela,Dr Lawrence
Vul 19h3'26"25d5'
Candela,Sarah & Elizabeth Onze
Com 12h48'10"21d55'
Candelaria,Adina L
Ori 5h53'43"8d55'
Candelaria,Felix R
Lyn 7h5'32"56d41'
Candelas,Elaine
Peg 22h56'52'22d25'
Candelore,DO,Andrew
Dra 18h23'20"68d41'
Candi's Star
And 2h22'1"41d12'
Candia,Betti
Boo 14h24'11"17d47'
Candia,Valeria
Aur 7h0'0"43d39'
Candice
Uma 14h0'0"50d24'
Candice A
And 23h17'35"51d9'
Candice Grace Wonder
Cnc 9h3'16"28d23'
Candice Lee
Com 12h3'0"22d24'
Candice Lee
Ori 5h55'40"8d3'
Candide
Cas 0h31'11"68d1'
Candido
Her 17h29'50"30d41'
Candido,Albert Michael
Per 2h52'53"50d5'
Candido,MD,Kenneth D
Oph 17h4'14"11 54'
CandiMalka
Tau 4h43'19"16d16'
Candini,Tracy
Peg 22h4'16"24d13'
Candle In The Wind
Boo 14h44'59"52d5'
Candle,Cammar
Cyg 19h47'53"30d23'
Candlelight
Cyg 21h31'16"38d60'
Cands Star
Vul 20h13'60"28d12'
Candy
Cyg 21h50'47"52d40'
Candy
And 2h24'0"45d59'
Candy
Uma 10h20'21"53d48'
Candy
Cnv 12h41'44"35d4'
Candy
Cyg 20h34'14"30d27'
Candy & Jerry
Cyg 20h15'0"31d44'
Candy Cane Dreams
Aql 19h26'22"14d45'
Candy Man Ted
Ori 6h2'51"8d22'

Candy's Star
Uma 9h53'56"58d8'
Candy, My Valentine
And 23h31'24"48d48'
Candy,M I & M G Cronin
Peg 23h4'57"20d18'AB
Candy,Molly
Cet 0h25'0"-6d37'
Candy-Bryan
Mon 6h23'22"8d24'
Candyland
Uma 9h29'39"58d20'
Cane,Gabriel Harlan
Ori 5h37'37"1d42'
Canedo,Shawn Anthony
Lac 22h14'40"51d31'
Canepa,Ti-Amo! Tom
Ori 5h54'1"8d33'
Canerday,Leslie
Peg 22h33'24"24d10'
Canerine
Boo 15h16'0"34d27'
Canessa,Caroline Bebo
Lyr 18h28'20"40d37'
Canevari,Betty & Jerry
Cyg 19h30'47"36d49'
Canevari,Michael Joseph
Ori 5h51'36"18d53'
Canfield,Anna Lee
Lyr 18h32'45"45d10'
Canfield,Beckie "Sunshine"
Mon 8h2'37"-0d44'
Canfield,Daisie Ann
Vul 19h45'41"28d49'
Canfield,Larry Ray
Cep 20h42'35"61d50'
Canfield,Susan Ann
Mon 6h55'17"-6d12'
Canga,C B
Vul 20h0'40"23d30'
Cangelose,Christopher
Aur 5h0'18"37d48'
Cangelose,Courtney
Cam 5h11'15"67d33'
Cangelose,Jacqualyne
Dra 18h7'1"70d28'
Cangelose,William J
Cnv 12h49'47"48d55'
Cangi,Nancy
Sgr 19h2'40"-26d36'
Cangialosi "Uncle Bob" Robert
Aur 5h58'13"28d38'
Cangialosi,Robin Jean Tesar
Leo 10h0'1"11d10'
Cangilos,Lyse
Tau 5h12'20"20d26'
Caniglia,Adolf Willard
Uma 8h31'13"51d27'
Canis Libertas Amermanius
Ser 15h23'28"8d36'
Canizaro,Buddy
Aur 5h6'0"40d41'
Cann,Bill
Cep 21h17'19"55d28'
Cannata,Grace Gagliardo
Aqr 23h5'52"-8d54'
Cannata,Jennifer
Lmi 10h27'0"31d44'
Cannata,MD,Rosetta
Vul 19h49'23"20d28'
Cannatella,Melanie G
Cet 2h22'24"1d50'
Cannauacciulo,Florence
Lyr 19h22'41"31d18'
Cannavino,James A
Gem 6h40'18"30d60'
Canned Foods "Rainbow"
Eri 2h52'0"-11d36'
Cannedy,Clarence & Magda
Vul 20h2'14"28d58'
Cannell,John Raymond
Dra 16h33'0"63d19'

Cannell,Kiwanis PA, Governor Bob
Cap 21h2'39"-23d1'
Cannella,Morgan Violet
Pup 7h51'16"-36d23'
Cannelle
Cas 2h52'51"77d15'
Canner,Ed
Her 17h10'57"45d1'
Cannetta,Luciana Damiani
Cnc 8h11'59"32d29'
Canney,M Ellen
And 2h21'0"44d13'
Canniford Star,The Frank
Ori 6h3'53"0d52'
Canning,Kathryn Patricia
Lyn 7h57'30"51d11'
Canning,MD,Patrick
Aur 7h22'29"38d39'
Canning,Megan
Scl 23h15'58"-25d55'
Canning,Timothy Patrick
Cam 6h19'12"83d33'
Cannistraci,Andrew T
Her 18h38'25"18d47'
Cannistraro,Peggy
Lyr 18h46'1"31d37'
Cannizzo,Dorothea Adriana
Cam 6h16'32"83d57'
Cannizzo,Michael
Cam 3h46'56"77d45'
Cannold,Mitchell
Uma 10h37'26"53d44'
Cannon"The Huggy Bear Star",James
Uma 10h50'59"56d12'
Cannon"The Poohpette Star",Marie
Uma 10h50'48"56d60'
Cannon,Barbara & James
Sgr 19h23'54"-41d2'
Cannon,Brett
Crt 10h53'58"-11d51'
Cannon,Delsie
Cyg 19h32'42"34d5'
Cannon,Don
Boo 14h50'31"37d55'
Cannon,Mary Smith
Cas 3h4'58"58d25'
Cannon,Mathew D
Per 2h43'41"40d55'
Cannon,Matthew William
Aur 7h13'34"39d38'
Cannon,Neal D
Aur 6h13'36"31d46'
Cannon,Rendy Riddick
Cam 5h10'46"70d57'
Cannon,Ron
Peg 23h30'28"21d29'
Cannon,Susan Lynn
Cam 5h43'58"61d13'
Cannon,Suzy
Eri 3h38'1"-6d59'
Cannon,The
Cmi 7h42'56"0d46'
Cannon-Brookes,Stephen
Leo 10h55'15"21d23'
Cannon-Foster
Eri 3h17'50"-5d41'
Cannuli,Beverly Jill
Mon 7h50'8"-2d18'
Cano,Amelia P
Mon 6h55'18"-1d30'
Cano,Gina Marie
Cas 23h17'59"60d48'
Cano,Janira
Cet 3h0'19"7d48'
Cano,Ruth Louise
Cet 1h32'31"-3d54'
Canon I Love You,Ricky Jamie
Uma 8h55'56"47d18'
Canon,James Trent
Pyx 8h31'8"-28d42'
Canonico,Giordano
Uma 11h58'59"31d14'

Canosa,Kathleen Marie
Cyg 19h34'27"32d58'
Cant,John Sydney
Her 17h22'13"38d44'
Cantarella,Elizabeth Anne
Lyr 18h57'41"33d7'
Cantave,Rosemonde M
Cnv 12h14'30"46d34'
Canter,Stephanie
Cas 0h47'19"66d42'
Canter,Tonya Nicole
Equ 21h10'26"12d6'
Cantin,Olier- Antoinette
Lyr 19h21'41"38d45'
Cantisano,John Joseph
Aur 5h14'47"41d47'
Canto,Chelsey Morgan
Lyr 18h33'51"44d44'
Canto,Patrolman Edward D
Lac 22h50'1"53d17'
Canton,Allison Nina
And 23h22'51"51d54'
Canton,Cheryl
And 23h31'19"41d59'
Canton,Kyle Constantin
Aur 5h0'49"37d34'
Cantone,Annmarie Manceri
Cas 1h3'57"53d40'
Cantor III,William
Lmi 10h28'59"28d58'
Cantor,David C
Lac 22h5'15"38d16'
Cantor,Kim
Cas 1h59'52"58d22'
Cantor,Leonard
Her 18h8'25"38d47'
Cantor,Ronald
Boo 14h23'43"25d56'
Cantor-Snow,Diane Helen
Dra 20h23'51"62d9'
Cantore,Sharon
Cam 7h31'0"68d42'
Cantou,Jerry
Cmi 7h36'31"0d42'
Cantrell,Bill
Aur 5h2'19"38d4'
Cantu,Elizabeth Ann
Aql 18h59'12"12d46'
Cantwell,Bernie & Charlotte
Sge 20h1'19"16d42'
Cantwell,Deane & Hank
Cyg 21h51'24"40d21'
Cantwell,Lisa Marie Borda
Cas 1h11'0"77d17'
Cantwell,Mary Agnes
Cas 1h51'42"58d36'
Cantwell,R Rea & Diane M
Cyg 19h25'11"32d53'
Cantwell,Susanne D
Cet 1h45'39"0d18'
Canty,Paul Arthur
Aur 5h3'0"50d29'
Canuel,Bryan-Michel
Cep 23h3'41"64d55'
Canyn,Emi
Cas 2h40'51"67d46'
Canyon,Christy
Cas 0h55'57"65d50'
Canzian,Margot
Her 17h7'38"40d27'
Canzoneri,Richard
Uma 10h44'54"70d27'
Canéy,Jeremy Justin
Cep 22h11'47"60d11'
Caomghin
Cma 6h54'1"-18d48'
Capaccio,Joseph
Aur 6h33'14"31d52'
Capaccio,Nancy's True Love,George
Uma 11h18'42"40d30'
Capaccioli,Remington
Ori 5h53'39"-8d1'
Capalbo,Dana A
Cas 0h37'0"62d10'

Capalbo,John G
Aql 20h2'45"4d58'
Capaldo,Johna Boyd
Lyn 8h16'1"50d26'
Capazzi,John V
Cet 2h19'49"2d52'
Capdevielle,Julia E
Lyn 9h13'26"40d29'
Capdevila,Marta
And 0h0'43d47d43'
Cape's
Umi 16h19'56"71d31'
Capela,Christopher & Clayton
Aur 6h11'53"35d49'
Capell,Allegra Celena
Cyg 19h45'37"30d2'
Capell,Misha M
Lyr 18h59'24"28d45'
Capelli,Lorey Jane
Peg 21h55'39"28d11'
Capelli,Melba
Umi 15h9'37"71d2'
Capello,Attilia Brigatti
Hor 3h20'30"-46d56'
Capello,Guglielmo
Cmi 8h4'18"5d39'
Capello,Matthew
Cep 22h34'58"63d8'
Capepsytena
Scl 23h24'51"-26d1'
Caper
Cam 10h11'59"82d5'
Caper,Tommy's Divine
Del 20h50'46"9d33'
Capes,Cherles W
Dra 16h43'56"68d42'
Caphillia
Cas 0h27'13"50d9'
Capic,Rose Lena
Ori 6h15'1"-0d48'
Capilli,John
Lyr 18h35'52"36d12'
Capistrant,John Michael
Aql 20h6'44"1d22'
Capistrant,Lisa L B
Cas 0h59'37"64d11'
Capital Star,The
Cep 22h25'29"60d24'
Capitina,Mildred Winnifred
Cap 20h28'1"-12d48'
Caplan,Beverly "Nifty Fifty"
Cyg 21h54'1"36d33'
Caplan,Carole
Dra 20h27'20"67d32'
Caplan,James Arthur
Oph 17h6'24"-6d10'
Caplan,Nathan Jay
Cnv 12h17'46"50d26'
Caplan,Sally
Lyr 18h37'24"29d15'
Caplan-Auerbach
"Capper",Jacqueline
Cet 2h15'22"7d45'
Caple,Catherine
Lyr 18h16'14"35d24'
Caplin,Benjamin
Boo 15h8'60"25d30'
Caplinger,Jeremy "The Weezer"
Aur 7h13'1"38d33'
Capo "Grandma",Rose
Tri 1h59'44"26d39'
Capo,Andrea Scacca
Peg 22h2'24"5d11'
Capo,Dennis Michael
Dra 17h0'43"68d26'
Capobianco,Daniel
Boo 14h59'17"33d45'
Capobianco,Gigi
And 0h4'11"35d5'
Capobianco,Michael
Cep 0h37'34"80d27'
Capobianco,Michael T
Her 17h8'0"41d31'

Capobianco,Robert
Lib 14h59'54"-20d24'
Capocci II,Anthony
Vul 20h3'0"28d17'
Capodilupo,Jr,Silvano A
Gem 7h58'43"31d24'
Capodilupo,Thomas Howard
Sct 18h50'25"-7d20'
Capogna,Americo
Uma 10h11'45"61d57'
Capolino,Richard L
Per 3h57'25"40d57'
Capon,David
Cep 22h28'32"70d14'
Capon,David
Cep 22h35'1"70d31'
Capon,Richard Alexander
Aur 6h1'1"36d2'
Capone II,Anthony Alexander
Her 17h59'17"50d14'
Capone,Dominic Scott
Hya 9h52'8"-17d46'
Capone,Jesselle Tomasina
Cet 1h24'22"-10d46'
Capone,Karen Melissa
Mon 7h41'27"-2d42'
Capone,Kerry
Cep 22h35'59"61d46'
Capone,Susan
Cam 4h54'52"67d53'
Capone,Vincent K
Dra 19h41'0"60d39'
Caponegro III,Francis
Boo 14h18'11"38d5'
Caponegro,MD,Robert J
Cam 3h22'32"67d42'
Caponigro,Dr Carmine A
Caponi Gro
Uma 11h40'51"64d47'
Caponigro,Lydia
Uma 11h31'47"63d44"
Caporale,Maria
Tau 5h51'55"23d3'
Caporrino,Carol
Uma 11h48'46"40d35'
CAPOVECCHIA
Boo 15h7'23"24d53'
Capozzi,Donald Paul
Cas 2h48'42"70d0'
Capozzi,Michael Vincent
Her 17h22'59"21d57'
Capozzi,Peggy
Cas 2h49'14"68d50'
Capozzi,Richard
Dra 20h11'26"80d11'
Cappachione,Laura Chantel
And 2h33'1'45d45'
Cappas-Awes,Angela (Kiki)
Dra 12h42'58"76d13'
Cappelletti,Justin
Aur 5h50'29"38d57'
Cappelletty,James
Lac 22h46'31"5544'
Cappelli,Dawn
Lib 15h0'58"-8d44'
Cappelli,Michel
Ori 5h41'22"11d35'
Cappellini Star,The Irene
Lac 22h19'22"51d53'
Cappello,Jr,Frank
Sgr 19h43'55"-43d28'
Cappello,Sheila
Cas 3h8'26"71d7'
Cappello,Tony
Boo 14h46'30"26d52'
Capper,Anthony
Cma 7h16'53" 15d4'
Capper,Maxwell
Cma 6h31'20"-15d6'
Capper,Michael
Sct 18h54'0"-6d15'
Cappi,Jan
Uma 9h2'15"53d37'
Cappiella,Elizabeth Doyle
Lib 15h11'53"-3d41'

Cappiello,Maria Stunner
Lyr 19h7'23"37d49'
Capponi,Bob & Orvie
Umi 13h57'49"71d55'
Capps Wedding Star, Dee Shoup-Mike
Eri 4h58'18"-8d41'
Capps,Bill"Big Papa"
Her 15h55'49"42d3'
Capps,David M
Ori 6h3'29"5d17'
Capps,Deborah Ann Holman
Leo 9h59'11"10d4'
Capps,Dorothy Haycraft Cash
Peg 21h57'40"23d32'
Capps,Eleanor Sappho George
Cyg 19h29'1"35d10'
Capps,George & Martha
Lyn 7h32'0"45d30'
Capps,Katherine
Mon 6h4'45"-5d57'
Cappuccilli,Anthony
Aql 19h37'16"0d12'
Cappuccino
Lyr 19h0'16"37d45'
Cappuccio,Thomas C
Aur 7h24'58"41d57'
Cappy (Carmen J Cappadona) 1936-1987
Ori 6h1'46"10d48'
Capra,Anne
Lyr 18h13'1"40d6'
Capri Pearl
Lyn 7h55'27"42d47'
Capri,Raulin S
Lyr 18h58'50"37d13'
Capricorn 2
Cap 20h31'49"-26d52'
Capricorn Man
Cap 21h26'56"-19d7'
Capricorpio
Per 3h19'0"50d20'
Caprini,Patrizia
Lib 15h1'19"-8d44'
Caprio Forever,Paul
Leo 11h8'17"-6d19'
Caprio,Taylor Leigh
Aql 19h16'57"12d40'
Caprio,Vincent
Boo 14h12'58"35d46'
Capritti,Kyle
Lac 22h25'47"50d19'
Capron Stellar Mother, Jane
Ori 4h46'41"4d4'
Capron,Bruno
Per 3h26'0"50d9'
Caprow,Marty
Aql 18h42'16"-2d18'
Capslock
Lac 22h26'0"40d3'
Capsule,The
Cam 3h39'0"72d9'
Capt Navid's Log
Sgr 18h8'22"-28d54'
Captain Blueprint
Boo 14h36'28"48d49'
Captain Dave
Sex 9h59'53"-5d39'
Captain Dowse
Ori 5h1'0"7d54'
Captain Ekika
Aql 20h8'46"7d31'
Captain Jean
Cyg 21h57'30"52d34'
Captain Kells
Lmi 9h21'13"33d52'
Captain Knute
Lac 22h21'30"38d36'
Captain Kurt
Cnc 8h27'52"31d19'
Captain Logan
Her 17h8'22"46d37'
Captain Mitch
Boo 15h15'56"41d14'

Captain Of Her Heart
Vul 19h58'22"25d9'
Captain Of My Heart
Lib 14h18'1"-22d54'
Captain Pope
Ori 6h0'44"8d51'
Captain Valentin i Sidorov
Gem 6h45'49"16d37'
Captain Walt
Cet 0h28'12"-11d17'
Captain Ward
Del 20h30'23"20d19'
Captain's Star,The
Aur 6h32'28"35d4'
Captain,The
Boo 14h26'44"21d15'
Captaine Morgan
Cet 3h15'47"2d28'
CapTel One
Lyr 19h22'14"31d4'
Captn Dick
Gem 6h39'36"14d34'
Capuano Your Star of Love,Michael J
Boo 14h41'33"38d51'
Capuano,Patricia Anne
Lyn 7h1'59"44d53'
Caputi,Christian R
Cas 0h35'51"60d30'
Caputi,James P
Dra 10h16'27"80d55'
Caputi,Victoria A
Cam 3h59'20"60d48'
Caputo,Anthony
Cep 22h9'33"60d54'
Caputo,Anthony J
Ori 5h57'29"16d4'
Caputo,James "Sonny"
Boo 15h19'13"41d57'
Caputo,Jennifer Marie
Cam 5h49'51"58d41'
Caputo,Kathy
And 23h32'0"43d27'
Caputo,Neal
Dra 16h48'41"66d30'
Caputo,Sr,Anthony S
Her 16h37'45"29d52'
Caputo,Teri
Cas 23h20'0"62d52'
Capuzzi,Patricia A
Crt 11h15'16"-11d56'
Capy,Laetitia
Lac 22h6'38"37d59'
Car,Larry Stephen
Lmi 10h42'50"30d57'
Cara
Cnc 9h13'15"10d37'
Cara
Eri 3h13'40"-7d21'
Cara Mia
Lyr 18h48'14"42d60'
Cara Rose
Lyr 18h35'1"34d21'
Cara's
Cas 23h39'13"48d52'
Cara's Guiding Light
Pup 7h58'8"26d29'
Cara,Eden
Cas 0h29'40"50d12'
Cara-Mier
Tri 1h46'19"27d15'
Caraballo,Esperanza
Ori 5h54'0"21d2'
Caraballo,Omar
Sco 17h51'40"-31d3'
Carabelli,Peter A
Ori 4h54'56"4d25'
Carabello,Alberto
Lac 22h17'19"47d7'
Caracci,Lee's Star 1995
Cyg 21h37'0"37d33'
Caradori,Patric
Oph 17h36'27"-16d37'
Caraffa,Massimilino
Pic 5h8'9"-44d23'

Caraher,Suzanne
Lyr 19h16'0"40d26'
Caralex
Hya 9h10'12"1d30'
Caralyn
Cam 3h18'55"58d33'
Caramia Angela
Lac 22h21'55"50d36'
Caramiah
Cas 0h29'26"75d46'
Carasso,Maurice
Hya 8h42'0"2d20'
Caravantes,Ted
Oph 18h41'59"10d40'
Caravati,Colleen Marie
Com 12h0'16"27d14'
Caravella, ChristineMarie
Lyn 7h54'27"58d55'
Caravelli,Maria
Cyg 20h5'54"30d31'
Caravello,Maddalena
Peg 21h53'50"33d57'
Carazola,Jamie Lynn
Lyr 18h43'40"42d26'
Carb,Barry H
Aur 7h14'39"35d59'
Carbaugh,Angela Marie
Mon 7h2'41"-6d21'
Carbaugh,Brian Allen
Boo 13h44'13"18d17'
Carbaugh,Jr,John E
Cep 8h16'44"86d20'
Carbaugh,Larissa Marie
Mon 6h18'50"-6d8'
Carberry,Ethel
Lyn 8h28'48"40d45'
Carbery,Sherrie Ann
Cas 0h37'34"71d59'
Carbone,Boo aka Anthony Bones
Her 16h53'44"37d8'
Carbone,Michael
Oph 17h53'24"8d26'
Carbonell,Miguel Tomas
Ser 17h52'32"-14d11'
Carcanade,Jean-Michel
Cam 7h37'44"68d31'
Carcathlia
And 0h51'39"40d50'
Carcedo,Maïté
And 23h1'51"45d43'
Carcieri,Suzanne Owren
Cas 1h43'0"58d53'
Carciog,Mihai
Ori 4h52'46"0d41'
Card,Carilee Rose
Lyr 18h59'13"28d45'
Card,Charles Lowell
Boo 13h41'32"26d54'
Card,Richard & Doreen
Dra 16h51'12"71d30'
Cardamenis,Byron
Aql 20h2'36"3d52'
Cardamone,Will J & Karen E Sammon
Cyg 21h11'58"34d59'
Cardell The Greatest Mum in Universe,Linda
Cas 0h29'33"60d21'
Carden,George Edward
Cep 23h9'55"60d60'
Carden,John Todd
Dra 19h28'15"68d30'
Carden,Kyla Danca
Peg 22h7'0"21d9'
Cardenas,Christopher Craig
Oph 18h0'39"13d19'
Cardenas,Jose & Gloria 1993 Family of Year
Mon 6h57'25"8d51'
Cardenas,Jose A
Eri 4h53'40"-5d33'
Cardenas,Jr,Fernando
Cmi 7h38'16"4d33'

Cardiff,Donald R
Sex 9h52'0"-0d38'
Cardiff,Jo-Ann & Gerald
Eri 4h27'12"-17d35'
Cardile,Pietro
Dra 15h20'1"58d43'
Cardilis,Sarah Josephine
Cyg 19h42'27"50d20'
Cardillo,Joe & Darlene
Crb 15h49'59"32d1'
Cardin's Avalon
Ari 3h23'2"23d40'
Cardin,Helen E
Mon 6h23'27"4d23'
Cardinal Brightness Of David
Uma 11h41'53"46d56'
Cardinal,Eric
Cyg 19h26'60"33d49'
Cardinal,Ghislaine Langlois
Cyg 19h33'1"33d46'
Cardinal,Mark Raymond
Aur 6h43'54"38d13'
Cardinal,Nathalie
Cyg 19h31'17"33d54'
Cardinal,Sharon Louise
And 23h32'53"41d55'
Cardinale,Vincent S
Aql 20h13'56"5d33'
Cardinale-With Love, Always,George
Cyg 20h56'40"30d10'
Cardinalli,Nichole Danielle
Eri 2h48'24"-3d29'
Cardinalli,Stephanie Michelle
Peg 22h2'1"11d6'
Cardine,Dominic
Dra 9h54'41"73d25'
Cardinez "Miss Maddie" Madison Ashley
Mon 6h40'31"6d58'
Cardo,Connie & Joe
Peg 22h4'39"4d55'
Cardon,Patrick
Cyg 20h29'0"39d24'
Cardona,Christopher
Ser 18h53'0"2d25'
Cardona,Claudia
Aql 19h23'35"13d35'
Cardone,Judy
Lyr 18h47'58"31d29'
Cardone,Ronald Anthony
Boo 15h4'0"29d23'
Cardoner,Marga
Tau 4h30'42"15d18'
Cardoso,Ramiro S
Uma 9h57'11"61d23'
Cardoza,Donald R
Her 17h29'52"31d43'
Carducci,Ann Mamar
And 22h56'24"40d41'
Cardwell,Doug
Cma 7h13'38"-16d40'
Cardwell,Edward
Ori 5h12'1"-8d36'
Cardwell,Sierra M
Mon 7h50'23"-1d21'
Careado,Pascual et Aurora
Lyn 8h2'32"39d0'
Careen
Peg 23h32'42"11d24'
Careghi,Jacques
Sge 19h0'40"20d20'
Carel,Kyle Robert
Cam 3h17'57"60d17'
Carella,Mariateresa
Cyg 19h34'38"33d47'
Carelli,Frank Paul
Cnv 12h50'13"34d8'
Caren
Lmi 10h38'30"24d18'
Caren Anne
Aql 20h1'1"7d35'
Caren's
Cyg 21h50'51"41d14'

Carenza,Vittoria
Her 16h51'26"40d46'
Carere,Rocco
Per 4h19'23"50d23'
Carette,Donald
Uma 12h4'1"46d6'
Carew,Christopher Michael
Cnv 12h23'24"41d41'
Carew,Michelle
Cyg 21h3'42"33d43'
Carey "11-20-93", Brenda & Thomas
Vul 21h2'22"20d14'
Carey & Brend
Lyr 18h49'57"37d10'
Carey Alexander GB
Cam 5h4'1"60d1'
Carey Anniversary Star,The
Cyg 19h32'0"34d42'
Carey Star
Lyn 8h5'40"46d27'
Carey Star,The Collin & Devin
Aql 20h17'0"5d26'
Carey's "Nifty Fifty" Jim
Cep 1h18'1"80d30'
Carey's Heavenly Body, Laurence
Ori 6h8'55"3d49'
Carey,Cathy
Cas 0h49'21"62d22'
Carey,Chole Kitty
Lyr 18h28'19"34d39'
Carey,Dale
Hya 8h40'37"-0d12'
Carey,Eugenia
And 2h24'51"39d36'
Carey,Gary Paul
Cyg 20h57'0"40d0'
Carey,Ina Clarinda Myers
Mon 6h19'1"4d56'
Carey,James Johnston Hale"Jammy's Wonder"
Psc 1h22'29"20d31'
Carey,Joseph
Ori 5h55'44"19d29'
Carey,Jr,E Tom
Aur 5h58'26"50d23'
Carey,Jr,Thomas F
Her 16h57'44"32d33'
Carey,Mariah
Uma 9h28'1"56d23'
Carey,Matthew Dennis
Cep 21h31'40"61d22'
Carey,Patricia Elaine
Cas 2h35'17"65d16'
Carey,Rhianne Emma
And 0h5'11"38d24'
Carey,Scott Charles
Aur 5h2'47"46d28'
Carey,Sr,Donald Brendan
Her 18h12'57"45d5'
Carey,Thelma G
Lyn 7h37'17"35d24'
Carey,Thomas J
Dra 16h48'33"66d40'
Carey,Timothy J
Her 17h24'0"20d27'
Carey,Victor
Por 4h32'24"38d16'
Carezani,Ricardo L
Aql 19h29'31"10d45'
Carfagni,Jr,Arthur B
Hya 8h26'25"5d48'
Carfagnini,Ed
Her 15h54'0"51d12'
Carfora,Frank Sabatino
Aur 5h18'50"48d4'
Cargas
Lyn 9h10'46"38d18'
Cargiulo,Dawn
Cyg 21h50'58"44d50'
Cargo Pup
Cma 7h14'55"-13d19'
Carhart
Per 4h20'1"50d48'

Cari
Cep 23h57'29"82d52'
Cari
Cas 0h59'12"50d0'
Cari Boo
Lyr 18h22'56"47d14'
Cari Leigh
Ori 6h6'47"6d19'
Cariad Jo
Cyg 20h19'1"39d5'
Caribbean Dream
Tau 5h47'57"23d50'
Caribiner Group,The
Aur 5h27'31"37d33'
Caricaburu,Kathy
And 2h23'29"40d46'
Cariddi,Hannah Catherine
Cam 11h56'36"78d46'
Caridi,Forever Marcello P
Hya 9h25'38"-0d20'
Caridis Five-O
Lac 22h16'25"37d55'
Carie Lee
And 2h19'54"42d47'
Carie's Diamond
And 23h25'25"41d45'
Carignan,Chantal
Lib 14h56'21"-22d42'
Carignan,Chrissy
Cas 22h57'22"54d23'
Carignan,Mary K
Lyr 18h17'17"35d10'
Cariker,Angie
Aql 19h26'45"7d41'
Cariker,Jeannie
Mon 7h4'11"-0d38'
Carin
Mon 6h20'0"0d53'
Carin#1
Vul 19h59'1"25d43'
Carin-Ann
Com 12h59'36"27d58'
Carina & Jason
Peg 21h38'14"24d21'
Carina Kathryn
Cas 1h51'0"75d32'
Carina Vera
Sco 17h26'58"-30d6'
Carina's Rose
Uma 11h52'1"41d26'
Carinci,Margaret
Lyr 18h59'1"28d17'
Carini,John & Denise Brondino-Carini
Boo 14h59'1"18d14'
Carini,Michael Thomas
Her 16h58'23"30d17'
Carino,Percy
Cet 2h25'37"7d26'
Carion,Saudrine
Peg 0h6'24"13d49'
Caris,Holly
Boo 14h53'42"32d20'
Carissa D S
And 1h10'21"39d37'
Carita Star A Soul Mate Through Time
Tri 2h0'56"31d39'
Caritas
Tau 4h59'48"16d52'
Cariau 4h59'48"16d52'
Lib 15h15'1"-20d24'
Cark & Family,Brad & Susan
Peg 23h3'41"22d1'
Carl
Her 17h54'32"39d26'B
Carl & Lou Memorial Star
Cep 23h26'38"64d5'
Carl & Paige
Cas 0h6'23"62d7'
Carl's Perfect Pillar Of Perfection
Aur 5h18'37"45d5'
Carla
Aql 19h0'56"16d51'

Carla
Cas 0h31'45"50d3'
Carla
Lyn 8h47'46"36d8'
Carla
And 23h37'56"47d15'
Carla
Cas 0h46'37"71d20'
Carla
Aur 5h17'1"49d15'
Carla "Forever Yours"
Peg 22h58'54"11d54'
Carla & Robert's "Wedding Star"
Crb 15h51'22"28d39'
Carla 81645
Del 20h32'52"20d2'
Carla G
Mon 7h58'59"-2d42'
Carla J
Cas 0h53'39"58d36'
Carla Mara
Oph 17h15'11"10d40'
Carla Michele
Aql 20h1'0"13d14'
Carla Pierre Gini
Com 12h27'36"21d53'
Carla Sue
Tau 4h41'54"16d7'
Carla-Cassiopeia
Del 20h13'13"10d47'
Carlac,Maria
Dra 9h28'22"73d10'
Carlasia
Nor 16h23'35"-59d44'
Carlberg,Kenneth GEF
Boo 15h22'20"41d40'
Carlberg,Ralph Norman
Cep 22h21'0"59d10'
Carle
Cet 1h56'54"-10d22'
Carle K
Umi 15h5'54"67d6'
Carleen & Ron's Diamond in the Sky
Lyr 18h47'20"37d35'
Carleen-Wec,Craig Langworthy
Lyn 9h19'12"36d0'
Carleo,Elizabeth
Lyr 18h37'33"36d40'
Carles,Clara J
Cep 22h11'26"61d35'
Carlesimo,Aida Bambina
Aur 4h44'51"30d15'
Carless,Dorothy
Com 12h18'38"21d48'
Carleton,Charles
Cet 1h47'34"-2d21'
Carleton,Christopher Robert
Leo 10h29'54"10d34'
Carleton,David
Dra 14h50'28"60d48'
Carleton,Peter F
Aur 6h37'25"37d57'
Carleton,Steven
Aql 19h2'42"-6d35'
Carletti,Alessandra
Pyx 8h41'5" 27d23'
Carlevato,Guy
Cyg 19h45'11"38d43'
Carley,Todd Alan
Her 15h49'22"41d54'
Carley-Do
Cyg 19h7'1"48d1'
Carli
Umi 14h45'0"68d7'
Carli,James L
Ori 5h54'51"16d0'
Carli,Melinda Kirsch
Per 2h58'36"35d16'
Carlile,Lisa
Lyr 18h19'51"38d48'
Carlin Anne
Cam 4h16'53"70d49'

Carlin,George D
Cet 2h2'54"-1d52'
Carlin,Kathryn
Oph 17h16'27"-20d8'
Carling,In Memory of Harold A
Lib 15h0'32"-8d60'
Carlino,Colleen
And 2h25'45"42d53'
Carlino,Michael L
Ori 5h56'0"14d54'
Carlis
Aql 19h50'28"14d53'
Carlisle,Alison Frances
Cyg 19h29'1"38d49'
Carlisle,Dorrie A Z
Umi 14h41'58"77d33'
Carlisle,James H S
Aur 5h39'40"33d11'A
Carlisle,Jane H
Aur 5h39'40"33d11'B
Carlisle,Jensen
Hya 8h52'35"1d57'
Carlisle,Kimberly
Cas 0h34'13"58d30'
Carlisle,Richard L
Ori 5h5'37"8d55'
Carlisle,Thomas Richard
Umi 14h56'19"77d40'
Carlisle,Troy
Aql 19h24'17"-1d52'
Carlita
Lac 22h10'43"51d21'
Carlito Mi Amor
Hya 10h1'8"-16d53'
Carll,Jennifer L
Lyr 19h17'37"41d17'
Carlo
Sgr 19h31'54"-33d36'
Carlo
Pup 7h56'49"-21d6'
Carlo
Ant 10h44'37"-33d9'
Carlo Eric
Cma 6h53'43"-18d15'
Carlo '95
Hor 3h27'25"-49d59'
Carlo,Anna Rosa Castillo
Vul 19h23'26"26d45'
Carlo,Gianmaria Alberto
Her 16h36'29"48d56'
Carlock,Ella Mae
Eri 4h8'25"-17d55'
Carlock,Tyler Michael
Aur 7h8'37"40d35'
CarloIngrid
Lyr 18h49'57"39d13'
Carlon,Aimee Lynn
Peg 21h53'12"30d27'
Carlon,Scott Michael
Lmi 9h25'12"38d29'
Carlon,Stephanie Ann
Vul 19h0'25"21d42'
Carlos
Umi 14h51'28"81d11'
Carlos
Cep 0h6'52"68d51'
Carlos & Shari
Boo 15h0'34"38d18'
Carlos I Juan
Uma 10h58'19"40d40'
Carlos Integra NSX
Aqr 20h53'1"-0d33'
Carlos Julian
Cet 2h18'25"6d4'
Carlos The Cinema Guy
Aql 18h43'51"0d23'
Carlos und Frank
Tau 3h54'14"30d42'
Carlos Y Digoras
Dra 10h9'47"73d50'
Carlos y Feliz
Uma 11h41'0"42d33'
Carlos Y Mari-Carmen
Mon 6h21'50"-1d41'
Carlos,Barbara Ann
Cyg 21h12'36"28d54'
Carlos,Hermano y Amigo
Leo 9h58'43"11d1'
Carlos,John
Cep 22h46'19"57d30'
Carlos,Roberto
Her 18h4'32"40d28'
Carlota
Cnc 9h8'26"7d15'
Carlotta
Com 12h24'50"21d19'
Carlotta Patricia
Sgr 19h27'22"-44d51'
Carlotta's Birthday Star
Cas 1h40'19"61d17'
Carlove,Joseph Robert
Cnc 8h26'48"10d6'
Carlow,Dorothy
Eri 3h30'42"-6d46'
Carlozzi,Christina
Cas 1h36'0"60d51'
Carlsen,Jan
Cep 21h59'59"60d36'
Carlsen,Niels
Cnv 12h36'0"37d13'
Carlsen,PhD,Ben A
Aql 20h0'49"8d9'
Carlsen,With Love-Eternity,Virginia May
Uma 9h25'57"55d3'
Carlson
Her 16h37'41"29d20'
Carlson (Star Shine), Sandy
And 23h42'56"40d53'
Carlson He Is Risen, Burton Carl
Sex 10h45'40"-0d39'
Carlson Mom & Dad's Pooky,Ricky
Peg 21h53'21'20d31'
Carlson Mommy's Lil' Buddy,Eric S
Cyg 21h19'4"40d8'A
Carlson Special,The Richie
Lac 22h36'34"54d17'
Carlson Star,The Carol /Paul
Hya 8h12'0'0d52'
Carlson's Star, Alexander Paul
Mon 6h45'51"10d6'
Carlson's Web
Oph 17h20'24"12d58'
Carlson,"G&G Star", Glenn & Glady's
Cyg 21h4'32"39d46'
Carlson,Allie
Lyn 8h23'14"42d47'
Carlson,Andrew Daniel
Dra 19h3'15"48d3'
Carlson,Anton
Dra 19h29'42"70d3'
Carlson,Beverly A
Umi 15h3'48"65d40'
Carlson,Bob & Johnnie
Sct 18h55'0"-6d44'
Carlson,Caryn
Mon 7h54'0"-1d5'
Carlson,Christopher Samuel
Cnv 12h56'14"51d7'
Carlson,Curt & Arleen
Com 12h6'0"31d16"
Carlson,Curtis L
Lyn 7h53'54"48d24'
Carlson,David
Cas 2h33'57"61d28'
Carlson,Dean & Elaine
Umi 15h5'47"66d18'
Carlson,Douglas Scott
Ser 15h21'45"21d2'
Carlson,Eddie Duke
Her 16h17'51"23d34'
Carlson,Edgar Alan
Dra 11h0'51"73d15'
Carlson,Erica Lynn
And 0h57'49"40d0'
Carlson,Gerald Phillip
Boo 14h43'25"29d22'
Carlson,Glenn William
Cmi 8h9'26"1d32'
Carlson,Grace & Bob 60th Anniversary Star
Cyg 21h29'26"40d59'
Carlson,Harriett & Leonard
Peg 22h43'0"27d44'
Carlson,Hazel
Boo 14h57'19"51d0'
Carlson,Herbert & Lois
Del 20h54'1"7d14'
Carlson,Janine
Cet 0h37'36"0d17'
Carlson,Jean
Aur 6h29'45"38d15'
Carlson,Jim
Aur 6h28'22"35d10'
Carlson,John Edward
Cet 0h31'51"-7d16'
Carlson,Jr,Is My Star, Parry J
Her 17h0'34"24d40'
Carlson,Judi
Cas 2h49'50"71d11'
Carlson,Karen Jeanette
Cas 0h3'52"64d52'
Carlson,Karl
Aql 19h33'21"0d33"
Carlson,Katie
Mon 7h22'26"-1d40'
Carlson,Ken & Margo
Peg 22h26'37"31d7'
Carlson,Kenneth John
Lmi 10h16'52"33d41'
Carlson,Kevin James
Dra 19h22'36"56d6'
Carlson,Kimberly M
Sco 16h35'1"-28d59'
Carlson,Laura Kathryn
Uma 11h4'30"34d18'
Carlson,Lawrence Benjamin
Dra 17h48'28"67d60'
Carlson,Lynn
Vul 19h45'51"20d36'
Carlson,Matthew Philip
Aql 20h7'15"8d8'
Carlson,Michael Paul
Lyn 7h12'41"59d56'
Carlson,Mildred
Cet 1h31'11"-0d15'
Carlson,Regina
Lib 14h59'17"-23d44'
Carlson,Richard
Ori 5h38'23"-1d8'
Carlson,Richard D
Dra 11h56'1"67d16'
Carlson,Robby
Aql 18h53'29"-0d7'
Carlson,Robert Einar
Lac 22h27'34"40d39'
Carlson,Ryan
Ori 5h56'57"18d13'
Carlson,Sven Patrick
Aql 18h53'19"-0d29'
Carlson,Tom
Hya 8h53'1"0d46'
Carlson,Valhalla Star Of Richard K
Per 1h57'26"56d49'
Carlson,Warren & Jane
Ori 5h49'1"21d60'
Carlson-Hynes,Nancy
Cas 1h9'33"58d4'
Carlson...Beloved Mother,Mary
Eri 5h3'12" 6d53'
Carlsson,Marcus
Uma 11h4'40"43d38'
Carlstroem,John
Ser 15h52'16"24d41'
Carlton's (Nor) Star in Remote
Boo 14h53'18"33d54'
Carlton,Caroline Sutton
Eri 3h54'17"-5d34'
Carlton,Francis John
Ori 5h44'29"10d34'
Carlton,John Wiley
Ser 15h33'40"18d33'
Carlton,Jr,Lobel Alva
Aql 18h58'20"6d9'
Carlton,Mickie
Boo 14h36'15"19d10'
Carlton,Richard & Blossom
Cyg 19h53'59"38d52'
Carlton,Ron
Ori 6h17'26"-0d49'
Carlton-Hyde,Troy Edward & Kimberly
Lyr 19h1'0"31d16'
Carlucci,Robert J
Dra 17h43'34"65d29'
Carlviculous I
Dra 11h43'32"72d5'
Carly
Leo 11h0'1"12d10'
Carly
Cam 3h58'51"77d22'
Carly
Cyg 21h0'17"36d51'
Carly's Wishing Star
Eri 3h1'49"-6d50'
Carlyle
Mon 6h33'18"1d18'
Carlyle,Alan
Cep 21h9'39"55d22'
Carlyle,Amy LaShawn
Cas 1h39'0"75d6'
Carlyn
Lyn 7h29'0"44d48'
Carlyn
And 2h25'18"45d22'
Carlyn Kimi
Aur 7h21'45"39d42'
Carm,Sister Norah Michael O
Cyg 19h31'25"31d40'
Carmack,Karen
Lyr 18h27'1"37d58'
Carmack,Ryan Robert
Ori 5h59'11"5d51'
Carman J
Eri 4h31'59"-19d39'
Carman,Aja L
Cam 12h1'26"84d59'
Carman,Donald Robert
Oph 18h32'35"11d22'
Carman,James C
Aur 6h54'23"37d45'
Carman,Joan Norma
Aql 19h26'29"-1d31'
Carman,Robert "Bobby"
Her 17h23'53"46d1'
Carmany,Brian
Her 16h59'38"20d16'
Carmany,Christy
Cyg 21h28'10"38d22'
Carmany,Diana
Peg 21h56'18"34d50'
Carmard,Jean Louis
Del 20h57'22"16d36'
Carme Mercado
Sgr 18h53'36"-31d1'
Carmean,Carrie
And 2h15'1"41d53'
Carmean,Sterling
Cet 0h49'1"-3d27'
Carmel
Lyn 8h3'29"38d25'
Carmel Julia
Uma 11h0'11"60d23'
Carmel of County Tyrone
Lyn 7h32'49"39d38'
Carmel,Melissa
Lyr 18h48'23"40d47'
Carmel,Peter Wagner
Cep 0h0'17"67d21'
Carmel,Robert
Cyg 21h10'1"34d54'
Carmel,Sister Marie Richard
Lyr 18h47'46"39d36'
Carmel,Spencer Paige
Dra 16h51'37'67d26'
Carmela
Per 4h0'0"50d58'
Carmela,Francesco Michele
Boo 15h0'30"38d42'
Carmela,Giulia
Cnv 13h14'1"38d4'
Carmell
Uma 9h33'40"55d36'
Carmella
Cas 2h55'19"75d34'
Carmella & Frances Mary
Cas 2h35'18"57d32'
Carmelo,Gabriele
Lyn 9h18'23"38d51'
Carmen
Cas 0h37'25"60d22'
Carmen
Lep 6h4'1"-11d3'
Carmen
Aql 19h40'1"11d40'
Carmen
Boo 13h46'32"11d35'
Carmen
Dra 16h54'20"60d34'
Carmen
Aql 19h1'0"1d6'
Carmen
Lmi 10h6'33"31d29'
Carmen (Negra ILU 4E)
Per 3h19'21"50d24'
Carmen "The Lovely Stranger
Umi 15h21'24"70d56'
Carmen & Mark
Lac 22h13'0"51d55'
Carmen & Michael
Umi 14h51'58"67d31'
Carmen & Tim "A Match Made In Heaven"
Cyg 21h39'27"40d0'
Carmen C
Gem 7h6'18"21d27'
Carmen Madame Butterfly
Eri 3h17'13"-5d6'
Carmen Silvia
Uma 9h37'16"71d5'
Carmen's Celestial Crown Radiates Love
Sge 19h24'45"18d40'
Carmen's Little Piece Of Heaven
Uma 9h11'48"56d10'
Carmen's Valentine
Lyn 8h3'33"50d11'
Carmen,Katrina Rose
Peg 0h3'39"31d10'
Carmen-3
Mon 7h11'54"-0d37'
Carmen-Sylvia
Cap 20h35'42"-10d38'
Carmichael The II, Robert J
Boo 13h44'50"26d47'
Carmichael,Abby Lynn
Peg 22h42'0"34d45'
Carmichael,Amy
Lyr 19h10'0"40d40'
Carmichael,Eric Andrew
Cet 0h59'19"0d55'
Carmichael,Ian
Boo 15h0'54"20d12'
Carmichael,Ian Devin
Aql 19h24'16"13d44'
Carmichael,John Robertson
Umi 14h44'31"69d31'
Carmichael,Keva
Aql 18h56'1"16d30'
Carmichael,Lollipop
Peg 23h43'51"10d32'
Carmichael,Nevin Stuart
Lyn 7h48'33"35d23'
Carmichael,Richard Gray
Cyg 21h45'59"30d2'
Carmichael,Robin
Mon 7h26'45"-5d40'
Carmichael,Rosemarie & Gene
Peg 23h21'55"12d60'
Carmichael,Sebastian Cole
Her 16h50'0"48d13'
Carmichael,Sharen
Sco 16h37'24"-31d32'
Carmie Corn
Cma 7h23'29"-15d32'
Carmiencke,Susan
Cas 0h57'33"64d33'
Carmina
Ari 2h39'1"25d46'
Carmody,Daniel John
Cet 2h5'42"3d6'
Carmody,Emily
Cep 21h11'1"61d4'
Carmody,Kendall Genez
Cam 3h33'37"61d37'
Carmona,Miquel Sánchez
Gem 7h2'0"31d9'
Carnagey,Summer Dorothy
Crt 11h20'28"-22d20'
Carnahan,Greg
Mon 7h2'29"-5d18'
Carnduff,George & Martha
Umi 13h48'20"76d45'
Carne,Alan John
Cyg 20h24'0"38d1'
Carne,Douglas Arthur
Hya 9h15'1"5d33'
Carneff,Sam & Evelyn
Aql 19h5'39"1d50'
Carnegie,Cecilia
Cas 0h0'48"61d36'
Carnegie,Maximilian
Umi 14h53'31"69d46'
Carnegie,Merlin
Uma 8h33'25"62d3'
Carnell,The Nicholas James
Uma 9h56'0"54d30'
Carnelli,Saverio T
Cet 2h14'58"4d27'
Carner,Talia
Uma 11h48'32"48d59'
Carnes,Jim & Nancy
Lyn 7h12'25"56d37'
Carnes,Jon Michael
Hya 8h41'39"1d8'
Carnes,M Phillip
Cmi 7h31'1"8d25'
Carnes,Mark Evan
Her 16h37'32"48d24'
Carnes,Martha
Cas 0h42'13"68d58'
Carnes,Morgan Rae
And 23h15'25"48d15'
Carnevale,Lindsay Caryn
Cam 8h5'10"80d25'
Carney II,Paul Dennis
Ori 6h8'55"9d60'
Carney's Comet
Her 16h21'41"5d1'
Carney's Crystal Of Light,Paul J
Tau 4h15'15"22d41'A
Carney's Dancing In The Light,Susie S
Tau 4h15'15"22d41'B
Carney,Alice S
Vir 11h43'34"1d43'
Carney,Bridget T
Mon 6h24'20"7d34'
Carney,Bryan
Boo 15h3'32"19d24'
Carney,Carl Michael Phillips
Per 4h7'1"51d5'
Carney,Emily Elizabeth Nicole
Cas 0h29'26"61d28'
Carney,Frances J
Lyr 18h50'17"31d24'
Carney,Jenny & Larry
Lyn 7h12'47"56d26'
Carney,Joan Lee
Peg 21h39'38"25d44'
Carney,John Michael
Sex 9h59'10"-5d47'
Carney,John William
Aql 19h57'12"13d57'
Carney,Lynn Koptionak
Sgr 19h6'43"-26d37'
Carney,Mary Joan- Joseph
Cyg 21h20'38"38d19'
Carney,Michael Sean
Per 2h1'35"57d8'
Carney,Paul
Her 17h24'11"38d6'
Carney,Rayvyn Bradberry
Per 1h29'1"53d53'
Carney,Vincent P
Boo 13h45'0"18d40'
Carnicom,Catherine Claire
Lyn 7h52'46"47d33'
Carnicom,Molly Marie
And 23h18'16"47d45'
Carniel,Clara
Lmi 10h32'42"30d41'
Carnine,Kimberly (Trixie)
Cnv 12h49'58"48d39'
Carnohan,Deane
Mon 06h53'42"10d08'
CaRo
Cyg 21h16'50"38d41'
Caro Mio Leonardo
Cma 6h53'32"-16d22'
Caro's Star
Dra 9h54'39d77'32'
Caro,Gina Lee
Cas 0h26'0"60d43'
Caro,The Dominic
Uma 10h32'32"45d54'A
Caroe,Mildred Holmes Durand
Cyg 21h26'50"38d16'
Carol
Peg 23h42'13"30d50'
Carol
Dra 17h7'3"50d54'B
Carol
Aql 19h16'1"13d53'
Carol
Mon 6h39'12"11d10'
Carol
And 2h0'58"47d30'
Carol
Umi 14h12'33"67d33'
Carol
Mon 7h37'21"-6d7'B
Carol
Cma 6h51'54"-18d33'
Carol
And 2h29'36"38d11'
Carol
Com 12h23'42"32d59'
Carol
Lyn 8h19'56'50d39'
Carol
Cet 2h37'55"2d20'
Carol
Vul 19h15'52"21d19'
Carol "B"
And 2h29'33"45d52'
Carol "Louise" Florence
And 2h23'18"48d28'
Carol "Song of Joy"
And 2h33'0"39d35'
Carol & Dan
Equ 21h23'0"7d50'
Carol & Pete Together Forever
Aur 6h31'53"38d21'
Carol & Peter
Uma 8h30'34"54d31'
Carol & Ryan Star
Ori 6h1'31"-2d16'
Carol & Vladimir
Cyg 20h57'15"41d9'
Carol Ann
Boo 15h3'0"38d8'
Carol Ann
Vul 19h13'25"21d51'
Carol Ann
Cmi 7h25'21"7d39'
Carol Ann
Mon 6h53'13"-0d30'
Carol Ann
Peg 23h25'16"31d43"
Carol Ann
Vul 20h3'44"25d50'
Carol Ann & Fred
Uma 9h52'1"68d51'
Carol Ann We All Love And Miss You
Cas 0h24'41"65d20'
Carol Ann,The
Peg 23h25'55"8d3'
Carol C
Vir 14h12'1"-8d46'
Carol Denise-Our Diamond Here On Earth
And 23h8'17"42d32'
Carol H
Cas 0h29'21"54d51'
Carol H
Peg 23h5'0"21d30'
Carol Jean Marie
And 0h54'41"41d9'
Carol K
And 2h0'12"43d4'
Carol Kivi
Com 13h4'1"15d48'
Carol Leigh
Eri 4h35'0"-12d19'
Carol Louise
Cam 5h30'21"60d35'
Carol Mae
Crb 16h19'17"30d12'
Carol Martha
Ori 6h7'24"-1d59'
Carol Rae
And 2h3'21"46d24'
Carol Renee
Cam 3h46'11"67d42'
Carol Rose
Lyr 19h22'25"33d58'
Carol Ruth 50
Eri 2h59'28"-5d11'
Carol Sue
Mon 6h5'23"-10d38'
Carol Suzanne
Oph 16h48'41"10d35'
Carol Too
And 0h50'1"36d57'
Carol V
Aqr 21h4'45"-8d26'A
Carol's 60th
Peg 22h36'30"20d40'
Carol's Christmas Star
Aql 19h55'26"8d9'
Carol's Corner
Umi 15h19'16"70d5'
Carol's Door
Aqr 21h30'59"-1d50'
Carol's Phenomenal 40th
Cas 0h13'0"60d40'
Carol's Star
Lyr 18h15'19"33d54'
Carol's Star
Ser 15h38'1"24d12'
Carol's Star
Peg 21h55'29"28d7'
Carol's Star of Aquarius
Aqr 21h43'50"0d57'
Carol's Wishing Star
Ori 5h52'13"15d22'
Carol's World
Lyr 19h9'0"38d48'
Carol,James Garbiella
Boo 14h0'44"14d59'
Carol,Light of My Life
Cmi 7h24'18"0d57'
Carol-Ann
Cas 0h8'1"59d31'
Carola
Vir 13h27'51"-13d41'
Carola
Psc 0h15'59"18d13'
Carola "Puschel"
Ari 2h55'18"20d29'
Carolan Rose
Cet 0h28'41"0d39'
Carolane
Cyg 19h47'49"29d28'
Carolane
Tau 4h34'33"15d26'
Carolann Forever
Cyg 21h23'20"40d10'
Carolann's Baby
Vul 19h42'15"20d19'
Carole
Pho 0h47'41"-47d33'
Carole
Crb 16h21'45"30d21'
Carole & Buddy
Aur 6h54'44"43d26'
Carole 23
Cyg 19h33'1"32d30'
Carole Ann
Cas 0h11'19"60d24'
Carole Ann
Mon 6h19'34"-0d17'
Carole Ann
And 0h0'28"35d31'
Carole Ann & Hank Hearts Forever
Leo 9h20'41"18d43'
Carole Courtney
And 23h49'37"44d43'
Carole Diane
Mon 6h38'55"-10d24'
Carole Et Paul
Lyn 7h55'12"50d50'
Carole J
And 2h30'47"45d20'
Carole Janette
Com 12h28'56"27d27'
Carole Zee's Kiss
Lyn 8h10'1"37d54'
Carole's 50
Cas 2h27'55"70d31'
Carole's Cluster
Tau 3h56'0"28d16'
Carole's Place
Mon 8h6'25"-1d2'
Carole's Star
Dra 16h49'16"64d43'
Carolee
Cyg 21h30'0"33d59'
Caroleo
Del 20h20'11"18d49'
Caroli,Dr G
Sgr 18h47'54"-23d39'
Carolin
Cep 22h1'44"60d25'
Carolin
Lyn 8h0'0"43d25'
Carolin
Uma 13h40'58"50d10'
Carolin Jayne
Lyr 18h33'18"43d41'
Carolina
Pho 0h7'55"-42d9'
Carolina
Pho 0h27'26"-44d21'
Carolina
And 1h33'59"39d10'
Carolina
Her 17h17'48"50d15'
Carolina
Tel 18h6'5"-47d37'
Carolina
Cas 1h59'1"58d29'
Carolina R
Peg 22h6'17"27d31'
Caroline
Cyg 19h47'31"38d29'
Caroline
Sgr 18h56'28"-26d0'
Caroline
Cas 23h41'50"63d36'

Caroline And 1h21'0"33d33'
Caroline Cas 0h36'53"63d0'
Caroline Cam 6h8'4"79d18'
Caroline Cam 6h23'11"65d36'
Caroline Dra 15h1'44"62d40'
Caroline Com 12h35'58"20d7'
Caroline Cas 0h30'51"68d43'
Caroline Cas 0h54'37"70d37'
Caroline Cas 1h37'0"58d23'
Caroline Cyg 21h38'1"40d23'
Caroline Alexandra And 1h26'4133d56'
Caroline Ann Marie Uma 8h35'50"67d44'
Caroline Candi Eri 3h24'25"-12d16'
Caroline CLT Boo 14h30'27"12d14'
Caroline Et Pascal Aql 20h4'57"1d15'
Caroline Louise Cas 0h58'0"50d15'
Caroline Rachel Aql 18h58'38"17d59'
Caroline Rose With Love WCS III Cyg 20h39'16"38d22'
Caroline '21' Lyr 19h21'11"41d21'
Caroline's Delight Ori 5h16'42"-8d44'
Caroline's Shalott Sgr 18h7'50"-28d30'
Caroline's Smile Cas 1h47'30"73d40'
Caroline's Star Tau 4h20'30"11d15'B
Caroline's Superstar Cas 0h34'24"61d50'
Caroline-Michael Cyg 20h18'53"39d9'
Carolle Per 2h58'1"45d3'
Carollo,Laurie Cas 1h9'39"61d41'
Carollove Aur 6h9'32"37d31'
Carols Candice Cas 0h28'11"64d8'
Carolus,Edith Green Cas 1h16'18"67d45'
Carolus,Edwin Barlow Her 18h11'45"40d20'
Carolus,Walt Hya 8h23'47"5d50'
Carolyn Cas 0h22'57"66d13'
Carolyn Cyg 19h56'51"38d9'
Carolyn Cas 1h51'14"60d20'
Carolyn Peg 21h37'44"23d46'
Carolyn Com 12h30'49"21d5'
Carolyn "Called To His Presence" Cas 0h22'24"61d40'
Carolyn & Al's Flame of Eternity Cyg 21h32'42"50d24'
Carolyn & John Boo 14h17'25"54d23'
Carolyn & Mark Aql 20h2'58"1d42'
Carolyn & Wally's Shining Star Cep 1h16'50"78d44'
Carolyn ad Aeternus Cam 12h19'0"81d59'
Carolyn Anne Mon 6h44'38"11d47'
Carolyn Christelle Lib 14h57'35"-0d14'
Carolyn Jade And 23hh29'58"44d49'
Carolyn Kay Ori 6h5'39"7d56'
Carolyn Lee Aql 19h51'28"14d52'
Carolyn Mae And 0h12'32"34d38'
Carolyn Marie Lyn 7h35'52"44d7'
Carolyn Michele And 1h41'27"38d37'
Carolyn Sherrie Umi 15h20'52"70d28'
Carolyn Z Uma 8h57'27"61d18'
Carolyn&scot Aql 19h58'31"14d4'
Carolyn's Peace of Heaven Oph 18h39'55"8d08'
Carolyn's Spirit Aqr 22h25'53"-5d27'
Carolyn's Starbase Crt 11h32'10"-17d39'
Carolyn,My "Terry" Forever And 2h5'38'41d3'
Carolyn/Gerald Sge 19h3'1"16d27'
Carolyne Jean Leo 10h4'36"17d46'
Carolynn Cas 1h52'1"58d31'
Carolynn Alexis Cas 3h1'34"58d18'
Carolynne Paige Cyg 19h45'50"37d50'
Caron Cam 4h18'16"61d41'
Caron #24,Jamie Cet 1h28'0"-10d52'
Caron,Juliette Vir 11h55'41"8d38'
Caron,Luc Cap 21h25'58"-22d44'
Caron,Yvan Her 16h43'55"23d29'
Caron-Lafrenière, Nataniel Lyr 18h53'22"34d52'
Caronelli,Sofia Polizzy Peg 22h27'1"3d42'
Carosella,Ann Louise Mon 6h59'28"-10d48'B
Carosella,Michael A Mon 6h59'28"-10d48'A
Carosone Love Always Scott,Nikki Uma 9h22'40"59d36'
Carotenuto,Susan M Cas 2h7'38"59d22'
Carothers,Mary Beard Lyn 7h39'55"45d6'
Carothers,Miriam Cnc 8h30'0"32d26'
Carozza,Kathleen Leahy Lyn 6h55'52"61d30'
Caroën,Frederic Aur 5h15'49"29d22'
Carp's Stellar Attitude Ori 6h3'37"9d56'
Carp,Barbara Feingold Langweiler And 1h48'42"47d15'
Carp,Larraine Cyg 19h21'15"28d30'
Carpe D M Cam 3h43'17"61d52'
Carpe Diem Psa 22h6'0"-26d10'
Carpe Diem (Leroy P de Clairlieu) Lyr 18h48'32"41d20'
Carpe Diem Cum Am41d20' Uma 10h50'32"44d38'
Carpe Diem-Eternally Tom Cep 20h48'1"75d48'
Carpe Terrenom Dolce Oph 18h39'45"8d48'
Carpediem Cnv 13h25'33"38d17'
Carpency,Gail & Joseph Tri 1h39'0"31d6'
Carpenito,Julie Ann And 2h1'17"47d8'
Carpeno,Christopher Todd Lac 22h8'1"48d15'
Carpenter Family,The Uma 10h48'24"53d46'
Carpenter Personal Wishing Star,"Hap" Uma 8h31'32"50d50'
Carpenter,Angelie Mon 7h34'23"-6d25'
Carpenter,Brenda Lee Leo 11h3'47"-2d11'
Carpenter,Carol Marsh Mon 6h54'47"0d30'
Carpenter,Catherine M Del 20h26'19"20d6'
Carpenter,Charles & Frances Eri 3h3'53"-5d10'
Carpenter,Clark Aql 20h7'24"0d36'
Carpenter,Dawn Cyg 20h0'51"30d46'
Carpenter,Douglas Paul Cyg 20h59'32"39d34'
Carpenter,Drew Whitefield Lac 22h40'36"50d26'
Carpenter,Ellen Ann Ron Cam 9h19'44"74d52'
Carpenter,Ellie Tau 5h22'53"16d52'
Carpenter,Eric A Boo 14h35'25"41d10'
Carpenter,Evelyn Cyg 20h17'32"38d22'
Carpenter,George D Cet 1h56'60"-6d58'
Carpenter,Gwenda Aql 19h1'52"4d35'
Carpenter,Helen Boo 14h30'55"50d20'
Carpenter,Ida Reim Cnv 13h40'55"41d44'
Carpenter,J Robert Dra 19h43'52"71d5'
Carpenter,Jerry Aur 5h0'46"38d11'
Carpenter,John James Uma 10h45'58"40d52'
Carpenter,Justin Aur 4h56'49"50d7'
Carpenter,Katie Lyr 18h53'26"41d6'
Carpenter,Kay C Eri 4h14'27"-10d17'
Carpenter,Kenneth Graham Cet 2h46'21"-0d24'
Carpenter,Kimmy And 0h55'42"36d5'
Carpenter,Laurel Anne Eri 3h41'30"-6d33'
Carpenter,Leo Lyn 8h50'46"40d10'
Carpenter,Mary Imogene Noble Cnv 13h48'1"42d20'
Carpenter,Meagan Elena Cet 0h59'33"-5d6'
Carpenter,My Friend Carol Lyr 19h13'47"40d2'
Carpenter,Paul S Ser 15h12'11"-1d33'
Carpenter,Peggy Doyle Cas 0h22'27"61d50'
Carpenter,Rebecca Aql 19h55'50"8dd33'
Carpenter,Richard Aql 18h58'0"16d2'
Carpenter,Rusty Aur 6h30'25"38d6'
Carpenter,Sharon L Mon 6h28'1"10d12'
Carpenter,Sue And 0h11'0"47d4'
Carpenter,Timothy J Per 3h1'43"40d37'
Carpentier,Josée Per 2h52'34"46d31'
Carpentier,Sophie Et Fabien And 23h21'13"50d53'
Carpentieri, Nicholas John Dra 18h32'34"68d28'
Carper,Frank Sex 9h55'43"1d11'
Carpio,Marco Tulio Oph 18h41'27"6d58'
Carpio,Raquel Eri 4h9'24"-7d57'
Carpio,Renee Uma 11h47'11"38d44'
Carr & Grandkids, Andrew W Per 2h20'58"55d8'
Carr Deluxe Leo 9h23'1"19d59'
Carr Starr,The Cam 4h13'35"68d23'
Carr,Bob & Kathryn Hya 8h51'57"-13d29'
Carr,Brandon Howard Michael Uma 10h58'46"71d11'
Carr,Brian Ori 5h55'15"7d49'
Carr,Bruce Her 16h1'33"41d45'
Carr,Collette And 0h24'45"35d49'
Carr,Cori Eryn Equ 21h11'25"11d8'
Carr,David Wildon Dra 11h28'41"72d34'
Carr,Jane Marie Uma 10h40'56"54d37'
Carr,Jim Cap 21h52'15"-18d48'
Carr,Joshua George Bromilow Ori 6h6'30"10d7'
Carr,Jr,Albert W Aqr 23h30'48"-8d35'
Carr,Melanie Peg 22h57'1"17d48'
Carr,Michael R Lyn 7h38'21"37d32'B
Carr,Nancy Patricia Hya 10h44'59"-12d10'
Carr,Nickolas James Cmi 7h53'14"0d26'
Carr,Patrice L Lyn 7h38'21"37d32'A
Carr,Patrick John Psc 1h1'36"22d9'
Carr,Philip Schaefer Cet 2h35'24"5d35'
Carr,Richard Wendall Ori 5h57'28"15d59'
Carr,Rita Franks Cyg 19h42'0"28d54'
Carr,Robert Gerard Crt 11h11'32"-12d13'
Carr,Sarah Cas 1h24'30"55d27'
Carr,Shirley Ann Cas 0h19'32"63d51'
Carr,Stephen Aur 7h9'44"37d47'
Carr,Tomissa Jo And 2h26'32"47d15'
Carr,Vicki Mon 7h4'38"4d22'
Carr,Victoria Rachel Lib 15h34'58"-23d38'
Carr-Russell,Robin Alaine Tri 2h21'1"31d48'
Carr/Watkins; 05/17/96 Tau 4h32'42"18d55'
Carra,Michael John Oph 16h59'1"-23d22'
Carra,Raffaella Cas 0h48'0"61d11'
Carrancejie,Margo Juliana And 2h6'0"38d43'
Carrano,Phillip Joseph Boo 14h19'41"36d57'
Carranza,Eduvijes Cantu Psc 0h25'18"7d39'
Carrara,Micheline Tau 3h44'40"2d16'
Carraro,Lucia Psc 22h49'45"5d3'
Carrasco (CJ),Carol Lee Del 20h14'21"10d6'
Carrasco,Alex Joseph Aqr 23h0'46"-5d34'
Carrasco,Marina Lyr 18h32'0"40d30'
Carraway,Christine & Robert Eri 4h10'45"-13d58'
Carraway,Edward Earl Tri 1h54'21"27d41'
Carrciro,Renee Marie & Walter Joseph Ser 15h30'22"14d16'
Carreau,Mark Cet 0h48'29"1d1'
Carreira,Maria Pena Luis Lyr 19h13'15"31d7'
Carreiro,Christopher Joseph Her 16h43'1"21d54'
Carrejo,Adan Tri 2h23'52"31d18'
Carrel,S Wayne Ori 5h39'27"11d58'
Carrel,Suzanne Oph 16h53'0"-28d39'
Carrera And 23h30'1"48d59'
Carrera,Luis Felipe Ori 5h6'29"0d37'
Carrera,Martha Cas 0h18'22"61d49'
Carreras,José Cep 20h22'24"76d49'
Carrere,Jean-Jacques Aql 19h30'0"-8d54'
Carrere,Marcel Dra 17h40'30"61d1'
Carrere-Gee,Josette Cam 3h25'19"60d10'
Carret,Philip Lord Sgr 19h58'17"-43d4'
Carriad Oph 18h1'55"8d40'
Carrick Sr,Charles Wilbur Mon 6h29'31"-10d54'
Carrick,Danielle Marie Mon 6h28'41"0d41'
Carrico,Laurie S Mon 6h52'39"1d52'
Carrico,Michelle Williams Vul 20h17'17"28d17'
Carridon Hya 9h52'43"-19d40'
Carrie Tri 2h41'15"31d37'
Carrie Aql 19h31'30"10d6'
Carrie And 2h15'1"43d6'
Carrie Mon 6h25'10"10d59'
Carrie and David Crb 16h20'14"37d41'
Carrie Anita Peg 23h24'1"32d56'
Carrie Ann "Our Special Angel" And 23h1'56"51d34'
Carrie Ann W-C Cas 23h17'1"61d22'
Carrie Anne And 2h30'22"41d41'
Carrie Katherine Ori 5h53'46"11d54'
Carrie Lyn Lyn 7h46'22"44d46'
Carrie Lynne Lyr 19h16'1"28d36'
Carrie S Mon 8h1'5"-1d5'
Carrie's Dream Lyr 19h4'0"25d44'
Carrie's Key to the Sky Com 13h1'52"20d50'
Carrie's Love Lyr 18h49'10"42d23'
Carrie's Rob Lmi 11h4'29"26d5'
Carrie's Star Tri 2h26'39"28d42'
Carrie's Wish Del 20h22'53"20d17'
Carrie-"La Lumiere De Ma Vie" Sge 19h33'30"16d35'
Carrieanne Lyr 19h1'54"37d45'
CarrieJames Cyg 20h39'28"40d36'
Carrier,Ari Leo 10h29'48"10d59'
Carrier,Brian J Tau 4h14'38"7d10'
Carrier,Donald E Boo 15h3'32"15d27'
Carrier,Janet & Michael Sex 10h28'19"-5d56'
Carrier,Maxine E Lyn 6h20'0"60d51'
Carrier,Sylvie Cyg 19h24'11"30d28'
Carrier,Yvon Lmi 11h4'56"30d51'
Carriere,Adam Cep 1h50'59"86d42'
Carriere,Claudette & Paul Cam 3h36'47"71d4'
Carriere,Jim Ori 5h56'0"18d59'
Carriere,Michel Aur 6h28'26"40d18'
Carriere,Philippe Aql 18h57'0"-8d25'
Carriere,Robert Del 20h20'19"16d1'
Carrigan "The Dreamer's Star",Steve Ser 18h16'31"-14d53'
Carrigan,Cathrine Elizabeth Cas 23h31'20"54d5'
Carrigan,Edward P Per 2h49'17"40d22'
Carrigan,Gregor Boo 15h13'38"26d35'
Carrigg,James A Dra 19h4'31"80d6'
Carrillo Aql 19h24'42"8d33'
Carrillo,Andrew Aql 19h30'51"10d18'
Carrillo,Lino J Dra 16h35'16"63d24'
Carrillo,Mario Lyr 19h2'29"41d13'
Carrillo,Ray Oph 17h52'38"8d23'
Carrillo,Sebastian Mula Leo 9h32'0"8d53'
Carrington,Bradford Robin Her 16h31'1"39d45'
Carrington,Elizabeth Rose And 23h37'11"45d43'
Carrington,Natalie J Lac 22h54'49"38d2'
Carrington,Pamela Lyn 7h54'39"52d22'
Carrino,Daniel James Her 17h25'13"31d36'
Carrino,Henry E Aur 6h0'42"38d38'
Carrino,Richard Vincent Equ 21h0'12"8d39'
Carrion,Alyssa Nicole Cyg 21h0'0"30d34'
Carrion,Misty Spring Cas 23h41'36"61d34'
Carrion,Robin DeWayne Lac 22h15'13"51d47'
Carris,Katrina Cas 0h32'12"63d39'
Carrison,Harold Francis Lac 22h19'1"50d17'
Carrithers,Zachariah Aur 6h26'41"30d13'
Carrière,Ghislaine Coco Masse Cas 22h58'20"55d36'
Carrière,Marcel Dra 17h40'30"61d1'
Carro,Eileen Lyn 8h17'0"39d44'
Carro,George Aur 6h10'24"46d30'
Carroccio,James J Crb 16h22'52"30d36'
Carroccio,Joseph J Crb 16h4'0"32d51'
Carroccio,Margaret L Crb 16h17'20"34d54'
Carrole,Patricia Elizabeth And 23h0'18"51d31'
Carroll Cyg 19h53'28"40d33'
Carroll Hya 8h30'30"1d18'A
Carroll"Light of My Life",Lisa Marie Vul 21h25'13"27d9'
Carroll,Richard Stephen Pho 23h49'42"46d24'
Carroll's Darlin, William Edward Lac 22h24'46"37d59'
Carroll's Peace of Heaven,Jay Dra 19h32'59"68d17'
Carroll,Alexander Austin Sct 18h55'0"-6d17'
Carroll,Alice Eileen Mary Margo Cam 3h13'36"63d56'
Carroll,Basia Aisha Vul 19h38'17"20d28'
Carroll,Ben Lyn 7h29'23"39d27'
Carroll,Brendan Michael Dra 16h35'0"62d38'
Carroll,Brent Dra 20h15'47"68d58'
Carroll,Cathy & Jane Oliver-"Caja" Peg 22h41'37"33d46'
Carroll,Charles T Cep 22h50'51"57d24'
Carroll,Charlotte Lyr 18h18'32"44d7'
Carroll,Daniel Joseph Cmi 7h28'49"0d10'
Carroll,Dennis Cnv 12h27'1"40d19'
Carroll,Desiree Lajuan Mon 8h2'11"-4d51'
Carroll,Doc & Elaine Mon 7h35'60"-1d47'
Carroll,Dokia Pauline And 2h4'25"37d14'
Carroll,Dr Gary Her 18h25'52"20d20'
Carroll,Dwight & Grace Del 20h13'52"9d58'
Carroll,Edna Mary Cam 7h41'57"78d49'
Carroll,Eileen Mary Lyn 6h22'40"58d26'
Carroll,Father Tom Per 2h52'0"46d13'
Carroll,Goeff Aql 19h56'20"0d26'
Carroll,Gus R Lac 22h16'55"51d6'
Carroll,Jack & Pauline Aql 20h5'57"7d47'
Carroll,James Her 18h4'27"47d52'
Carroll,James B Per 3h19'37"41d25'
Carroll,James G & Heather P Carroll Mon 6h28'59"-8d60'
Carroll,Jane Ann Mon 6h49'25"10d28'
Carroll,Janet Eri 3h54'60"-4d3'
Carroll,Jeffrey Alan Boo 13h40'15"18d46'
Carroll,Joan Ori 5h56'37"14d7'
Carroll,Lee Lyr 18h28'40"32d36'B
Carroll,Lisa "Pretty Eyes" Peg 22h29'33"11d29'
Carroll,Lynn Cyg 20h55'48"31d9'
Carroll,Margaret Com 12h55'28"24d6'
Carroll,Marie Aql 19h7'0"-6d44'
Carroll,Mary Beth Mon 6h19'27"8d19'
Carroll,McKenzie Anne Dra 11h42'1"71d37'
Carroll,Mrs Mary F Peg 23h0'55"8d50'
Carroll,Nancy Aql 19h45'27"10d5'
Carroll,Patrick Matthew Per 1h59'29"56d56'
Carroll,Patrick Davis Cmi 7h56'55"3d55'
Carroll,Patrick John Aql 20h9'17"-8d52'
Carroll,Philip James Boo 15h3'38"32d27'
Carroll,Robert Hya 9h23'16"-0d44'
Carroll,Rosanne Vul 19h18'47"26d15'
Carroll,Sarah A Eri 3h1'21"-4d43'
Carroll,Sean Patrick Cet 0h45'29"-3d31'
Carroll,Shannon Alexis Cyg 19h26'56"30d12'
Carroll,Sophie Lyr 18h17'31"43d8'
Carroll,Stephen Wesley Ori 6h8'59"8d14'
Carroll,Suzanne Catrice Peg 22h11'33"3d49'
Carroll,Teresa "TII" Cyg 21h45'19"38d53'
Carroll,Tom Sex 9h44'32"2d21'
Carroll,William A Sco 17h29'39"-30d6'
Carrollton Oaks Comets Aur 5h57'59"37d47'
Carrolton,Jr,Gordon M Crb 15h29'26"30d9'
Carronlil Ori 5h25'21"-2d11'
Carrozzo,Gloria Belinsky-Bennett Cyg 21h12'1"38d28'
Carrubba,Cathryn O'Connoll Lyr 19h0'0"30d24'
Carruolo,Jazzy Jen And 2h15'26"46d54'
Carruth,Marlies Aql 18h41'18"-1d53'
Carruthers,Deborah Rosemary Cet 0h23'37"-18d8'
Carruzzo,Franàoise Dra 10h59'21"73d27'
Carseni,Derek Joseph Her 17h14'1"45d43'
Carsenia,Gail Lyr 18h34'13"46d17'
Carsey Cet 2h43'23"9d36'
Carslake,Ginger Cyg 19h42'31"28d27'
Carson's Star Cas 0h28'34"68d0'
Carson's Star Lite Star Brite Oph 16h59'11d16'
Carson,Alex & Eunhi Lyr 18h36'1"45d41'
Carson,Anna Maria Peg 21h43'30"21d21'
Carson,Barbara Cam 7h34'0"60d1'
Carson,Brenda And 1h25'25"39d38'
Carson,Carrie Lyn 6h28'56"60d46'
Carson,Catriona Com 12h31'18"25d31'
Carson,Jamie Lmi 9h59'32"40d28'
Carson,Jane And 0h2'0"46d45'
Carson,Jay Arthur Cyg 20h50'30"40d21'
Carson,John Andrew Boo 13h59'60"21d43'
Carson,Johnny Per 2h56'43"38d39'
Carson,Maurice Angelo Her 16h33'45"39d30'
Carson,Neville William Ori 5h21'53"-4d58'
Carson,Newell & Barbara Lyr 19h12'22"37d52'
Carson,Paul John Cmi 7h57'26"8d33'
Carson,Shari And 23h1'24"50d19'
Carson,Terrence A Cnv 12h26'27"51d3'
Carsten Peg 23h30'0"17d38'
Carsten,George Dra 9h33'20"77d59'
Carsten,Katharine Gonsalves Uma 12h32'29"59d12'
Carsten,Mary Cas 1h56'15"58d10'
Carstens Star,Dr Richard Per 3h11'1"46d42'
Carstens,Elizabeth "Lizzy" Lib 14h26'57"-22d59'
Carstens,Eva Marie And 0h21'24"37d24'
Carta-Nicolosi Uma 12h0'0"32d57'
Cartagena,Maritza And 0h20'34"30d29'
Carte,Laura Lyn 8h25'1"51d4'
Cartee,Allison And 1h34'0"39d3'

Cartee,Corey Shannon
Peg 23h17'27"33d16'
Cartee,Staci Diane
Cam 3h50'54"53d59'
Cartellone,Paul A
Sex 10h34'1"5d3'
Carter
Her 17h12'20"27d31'
Carter "Moravu Treasure",John A
Boo 14h58'22"41d18'
Carter & Family, Michael & Dana
Hya 8h32'55"-0d1'
Carter Alan
Oph 18h18'44"6d38'
Carter Caring Star,The Doug
Ser 15h54'11"1d7'
Carter III,Joseph Windsor
Uma 9h18'57"55d18'
Carter Lance
Lac 22h13'40"47d34'
Carter SP 33,Paul
Ori 5h7'16"1d15'
Carter"Sperry" "1968", James Loren
Aur 5h59'54"30d16'
Carter's Dream
Uma 8h49'11"62d23'
Carter's Golden Heart, Jennifer
And 2h28'19"45d59'
Carter's Star of Light
Cep 23h6'0"65d27'
Carter's Wedding Star, Greg & Lynne
Aql 19h44'38"10d24'
Carter,Alison
Cyg 19h27'42"35d32'
Carter,Ann
Aur 6h36'23"30d13'
Carter,Ariel Elizabeth
Aqr 22h25'13"-18d55'
Carter,Bandit
Uma 11h21'57"71d32'
Carter,Barbara
Cet 1h23'1"-2d38'
Carter,Betty Lou & Bob
Boo 13h45'54"24d17'
Carter,Brianna & Justine
Umi 15h45'30"71d54'
Carter,Candice
And 1h22'22"35d15'
Carter,Cara Yvonne
Oph 17h26'12"-23d12'
Carter,Cari Ann
Cas 2h26'42"60d19'
Carter,Caroline Margaret
Cas 1h10'44"61d59'
Carter,Charlotte
Eri 3h47'14"-2d58'
Carter,Chris
Aql 20h10'43"1d30'
Carter,Christopher
Lyn 7h29'12"41d9'
Carter,Christopher Lee
Her 16h43'47"21d9'
Carter,Claire
Cas 0h23'30"74d50'
Carter,Collin Patrick
Ari 1h57'0"23d31'
Carter,D
Cmi 7h31'45"7d57'
Carter,Dannielle Gloria
Umi 16h17'54"76d23'
Carter,David Carey
Eri 2h59'52"-17d30'
Carter,David Robert
Sex 10h15'29"-1d32'
Carter,David S
Oph 16h59'50"-23d4'
Carter,DD,Michael J
Cma 6h55'33"-19d26'
Carter,Deborah Emma
Cyg 20h33'17"48d52'
Carter,Derek Michael
Mon 6h53'5"-4d57'
Carter,Donald E
Dra 18h20'33"48d40'
Carter,Edwin Charles
Cap 20h29'0"-13d29'
Carter,Elissa Dorothy
Gem 6h58'5"15d51'
Carter,Elizabeth
Cam 4h25'47"61d25'
Carter,Eve LaRoche
Cam 9h23'20"80d14'
Carter,Gina Eaves
Lyn 7h21'1"44d47'
Carter,Glen
Dra 9h50'47"74d7'
Carter,Graydon
Cam 5h3'1"67d45'
Carter,Heather & Bill
Lyr 18h36'55"42d58'
Carter,James
Aur 6h21'1"38d15'
Carter,Janet
Cas 0h23'53"70d57'
Carter,Jennifer
And 0h55'38"36d11'
Carter,Jessica L
Cyg 19h42'0"30d8'
Carter,Joanne
Lyr 18h39'17"33d35'
Carter,John D
Her 17h34'29"24d40'
Carter,John Harold
Umi 16h7'22"80d24'
Carter,John Kohl
Per 4h45'1"48d57'
Carter,John Mack
Per 2h55'16"54d30'
Carter,Jonathan Boyd
Sco 17h50'34"-38d2'
Carter,Jr,John Laughlin
Her 17h21'36"45d35'
Carter,Jr,William Benjamin
Aqr 22h25'30"-18d41'
Carter,Juanita
Del 20h25'26"11d35'
Carter,Julian B
Cas 1h9'1"61d38'
Carter,Justin Spencer
Boo 14h14'20"17d24'
Carter,Karen Holly Maze
Cap 20h33'18"-20d4'
Carter,Lawrence Quentin
Her 18h16'57"14d42'
Carter,LeeAnn Marie
Lyn 7h35'50"45d38'
Carter,Lidia I
Mon 6h32'50"8d46'
Carter,Linda (Gladys)
Mon 6h34'25"11d57'
Carter,Louise Alison
Crb 15h48'50"26d53'
Carter,Mackeel
Cam 3h16'38"66d24'
Carter,Margaret
Cnv 13h18'37"38d45'
Carter,Margaret H
Cas 1h14'58"61d35'
Carter,Marianne
Cas 0h21'50"64d49'
Carter,Mark
Cma 6h58'1"-11d13'
Carter,Mark Forrest
Dra 20h12'54"84d23'
Carter,Melanie C
And 2h6'13"42d43'
Carter,Michael "Paco"
Aur 5h21'0"40d53'
Carter,Mindy
Cas 2h9'31"59d11'
Carter,Muriel B
Mon 8h1'1"-1d32'
Carter,Nancy B Kay
Aur 6h39'37"38d48'
Carter,Nicholas Charles
Ori 6h7'59"7d43'
Carter,Nicholas Brady
Oph 17h7'31"-24d17'
Carter,Nick
Hya 8h18'49"1d40'
Carter,Nicole Alexis
Cas 0h26'24"61d24'
Carter,Noah Keith
Aql 19h58'37"14d15'
Carter,Otis
Aur 6h1'59"37d40'
Carter,Perry W
Aql 19h51'18"11d43'
Carter,Richard
Uma 11h48'1"37d50'
Carter,Robert D
Hya 8h35'26"-0d46'
Carter,Robert S
Cet 1h14'19"-3d19'
Carter,Sacha Lucia
And 23h45'52"41d21'
Carter,Sara Joan
Cas 2h33'47"67d58'
Carter,Shirley
Lac 22h31'49"56d34'
Carter,Sr,George
Aur 6h37'59"37d54'
Carter,Sr,William R
Ser 17h30'32"-13d37'
Carter,Ted
Dra 18h32'36"58d25'
Carter,Terry Russell
Sgr 17h55'32"-36d51'B
Carter,The Bruce Star
Dra 15h19'19"58d37'
Carter,Veva Louise Lowe
Vul 20h1'44"22d32'
Carter,Vincent Gene
Her 18h18'14"14d56'
Carter,William M
Sct 18h55'13"-5d37'
Carter-Jennie,Janet M
Her 17h10'26"36d52'
Carter-Melaragni
Ori 5h3'41"-2d32'
Cartey,Gavin
Mon 6h35'13"5d0'B
Cartey,Gemma
Mon 6h35'13"5d0'A
Cartey,Ron
Mon 6h53'45"-10d52'
Cartey,Terri
Mon 6h54'48"-10d7'
Carthen,Jr,(C J) Christopher
Boo 14h52'1"31d9'
Cartier,Donnie
Sct 18h42'44"-6d52'
Cartier,Julien
Uma 9h30'42"48d56'
Cartledge,Spencer Muira
Aur 5h30'25"30d10'
Cartmell,David Winn Hord
Boo 14h8'49"44d15'
Cartmill,Denise
Vul 20h40'1"28d36'
Cartner,Bobby
Aur 4h36'22"30d46'
Cartolano,Jose Miguel Gaona
Per 3h25'0"37d56'
Carton,Charles
Dra 18h10'30"58d8'
Carton,Florian
Lyn 7h54'52"58d9'
Carton,Kevin
Aur 6h15'20"31d6'
Cartoon Caricature Advertising
Sge 20h15'17"17d20'
Cartright,Margaret JoAnne Blosser
Cas 0h28'1"58d34'
Cartwright's Star
Ori 5h5'0"-0d46'
Cartwright,Darrel Steven
Ser 15h27'48"8d25'
Cartwright,John
Peg 22h3'0"29d59'
Cartwright,Joseph William
Aql 19h9'15"3d35'
Cartwright,Kelly Sue
Cam 6h7'32"58d26'
Cartwright,Morris Emory
Peg 22h2'42"24d10'
Cartwright,Rhonda Sue
Cyg 21h27'36"40d17'
Cartwright,Rosalind Dymond
Peg 23h50'33"31d39'
Cartwright,Samuel H
Aql 19h31'41"12d26'
Carty DCLTL,David J
Aql 19h56'37"10d6'
Carty,Mary Symmes
And 23h47'34"41d14'
Carty,Thomas M
Lac 22h33'35"52d37'
Carty,William Christopher
Her 16h14'21"8d35'
Caruana,Anna
Cas 1h3'18"62d51'
Caruana,Martha
Cas 23h27'48"61d28'
Caruana,Renee-Claude
Sgr 19h29'47"-34d57'
Carucci,Angela
Cam 4h0'18"67d53'
Carucci,John
Gem 6h21'0"18d50'
Carufel,Carmen
Tri 1h39'20"30d7'
Carus,Benjamin Todd
Her 16h55'12"51d14'
Caruso 12-12-82,Carol & Anthony
Sgr 18h47'54"-27d34'
Caruso,Ann
Lyr 18h51'1"32d5'
Caruso,Anna Marie
Vul 19h46'58"28d30'
Caruso,Caterina Zuccarello
Pho 0h32'49"-47d26'
Caruso,David G
Her 17h13'46"20d49'
Caruso,Jaymz David
Uma 11h9'1"44d9'
Caruso,Jennifer
Lyr 18h41'15"30d2'
Caruso,Joanne
Sco 17h31'1"-30d45'
Caruso,Joanne Lucy
Lyr 18h36'26"45d38'
Caruso,Kimberly
Uma 10h2'26"72d25'
Caruso,Lauren
And 2h34'34"42d9'
Caruso,Lexie
Dra 16h9'44"68d12'
Caruso,Mary Ellen
Lyn 7h23'16"44d7'
Caruso,Nydia
Cyg 19h24'57"31d20'
Caruso,R & R
Aur 6h8'22"32d27'
Caruso,Raymond
Peg 21h40'18"21d43'
Caruso,Rita
Aql 19h54'11"0d55'
Caruso,Sandy
And 23h44'60"43d56'
Caruso,Zachary James
Lyn 7h57'21"38d32'
Carusone,Madeleine
Lyn 8h3'42"34d26'
Caruthers,Margaret Burr
Aql 20h8'1"4d35'
Carvajal,Kelly Elizabeth
Cas 1h18'36"61d12'
Carvalho,Fernanda B M
Cas 1h16'23"62d25'
Carvalho,The L & B
Cnv 14h0'51"38d31'
Carvaltio,Simon
Ori 5h28'46"1d50'
Carver Jr,Melvin T "Ted"
Dra 18h22'1"48d50'
Carver,Martin G
Tau 5h53'0"27d32'
Carver,Patrick John
Dra 16h23'23"61d18'
Carver,Russell
Aur 5h12'51"40d30'
Carver,Shona
Peg 22h36'21"11d19'
Carvers,The
Mon 7h42'1"-1d31'
Carville,Geoffrey William
Cyg 19h31'26"35d14'
Carvino,John
Lyr 18h34'1"38d14'
Cary
And 23h48'57"40d22'
Cary & Leslie Star,The
Peg 23h29'1"20d2'
Cary's 50 Golden Years,Carl & Velma
Aur 7h4'43"43d57'
Cary's Star
Cep 21h52'0"58d53'
Cary,Christine
Aql 19h54'60"11d48'
Cary,Kathy & Doug
Crb 16h22'29"32d13'
Cary,Roslyn & Jerry
Dra 17h6'13"70d34'
Cary,Viki
Aql 20h15'54"0d50'
Caryatid
Cet 1h3'14"1d3'
Caryl
Per 4h6'0"51d19'
Caryl Elizabeth
Lyr 19h21'35"35d5'
Caryn
Lyr 19h0'1"37d46'
Caryn Laurie
Vul 19h1'54"21d29'
Caryn Mae,The
Cas 0h4'41"61d1'
Caryn XTX
Cas 0h23'41"74d32'
Caréo,Michel
Cnv 13h21'1"38d16'
Cas
Cyg 19h28'1"39d21'
Casa de Lee
Mon 7h55'24"-8d59'
Casa,Roberta
Cep 2h18'41"77d58'
Casabian,Carl Charles
Del 20h22'19"10d6'
Casado,Henry
Aql 19h55'58"-1d12'
Casady,Allen W
Hya 9h3'21"2d32'
Casady,Katherine Anne
Aql 19h53'40"14d36'
Casalb
Umi 15h43'0"70d49'
Casale,Barbara J & Michael A
Dra 17h34'1"60d11'
Casale,Drew J
Her 16h49'33"40d11'
Casale,Enzo
Nor 16h19'43"-57d10'
Casale,John Anthony
Lac 22h35'11"55d2'
Casale,Michael Mario
Dra 17h7'54"51d39'
Casale,Ms Annette M
Mon 7h3'2"-5d19'
Casale,Nicole Marie
Lyr 19h0'56"31d36'
Casalino,Bobbi A
And 2h15'47"37d26'
Casandra Lea
Cas 0h55'14"69d55'
Casanova
Her 18h24'50"20d35'
Casanova,Jean-Philippe
Boo 15h28'49"38d30'
Casanova,Manuel Marin Soto Santos
Tri 2h5'23"32d46'
Casasnovas,Roberto Rizzieri
Aql 18h59'57"6d7'
Casati,Grand Papa
Uma 11h33'37"37d50'
Casay,Anthony
Cep 21h12'42"58d5'A
Casazza,Loredana
Ant 10h38'8"-31d12'
Cascadden,Karina Cloud
Ori 5h27'20"1d33'
Cascadis Graphis
Uma 10h30'20"48d25'
Cascallar,Eduardo C
Lac 22h8'34"47d40'B
Cascallar,Lara
Lac 22h46'25"54d32'
Cascallar,Marijke I
Lac 22h8'34"47d40'A
Cascella,Guy R
Per 2h57'14"55d7'
Casci
Cma 7h17'38"-15d39'
Casciato,Peter Crawford
Per 3h6'20"46d22'
Casciero,Carol
And 23h25'50"45d57'
Cascino,Reneé
Mon 6h56'43"10d10'
Cascio,Joseph Mark
Lac 22h51'44"52d46'
Cascio,Madelyn
Eri 4h10'48"-13d7'
Cascony
Cyg 20h14'21"33d41'AB
Case "Streaker", Stephen P
Cet 2h17'37"0d38'
Case In Space
Cap 20h31'16"-26d19'
Case,Aidan M
Uma 10h34'15"52d52'
Case,Elaine
Aur 6h54'57"40d22'
Case,George Nelson
Cap 20h33'16"-11d9'
Case,Hazel Marie & Hassel
Cas 23h41'45"60d38'
Case,Jeannette Marie
Ori 5h54'39"8d48'
Case,Katherine
Vul 19h43'36"25d54'
Case,Laura Ann
Del 20h20'34"11d11'
Case,Lindsay Ann
And 0h46'56"34d18'
Case,Peter T & Lucy W
Cyg 19h45'14"30d17'
Case,Robert M
Dra 16h49'47"73d45'
Case,Sherman
Her 17h55'58"28d47'
Case,Stanley D
Ori 4h54'43"0d6'
Case-Cahoon #50
Cam 7h8'1"61d26'
Casebolt,Ryan J
Uma 11h16'43"30d46'
Casei,Nedda
Vir 13h33'39"-4d4'
Casella,Craig Wm Gus
Aur 6h17'58"33d28'
Casella,Scott
Dra 20h5'13"67d32'
Caselli,Valerie Piola
Cas 0h14'25"63d5'
Caserta,Stephen
Cas 0h55'1"66d47'
Casertano,Maria G
Cas 23h0'0"54d9'
Casewave,Rene
Aur 4h36'0"31d21'
Casey
Cnv 12h46'12"36d50'
Casey
Cet 0h51'27"1d17'
Casey
Ser 15h54'40"4d21'
Casey "P"
Crb 15h29'1"30d26'
Casey Grace
Cam 5h49'17"61d46'
Casey Loves Patrick Always & Forever
Her 18h8'19"40d49'
Casey Marie
Vul 19h41'14"20d8'
Casey Rae
Boo 14h36'35"47d37'
Casey Very Own Star, Laine
Vul 20h27'0"28d27'
Casey Z
Lyn 9h6'1"39d42'
Casey's Love For Richard
Ori 5h1'1"10d4'
Casey's Star
Gem 7h0'15"10d48'
Casey's Star
Lyn 7h30'48"45d54'
Casey's Star
Per 3h29'30"50d49'
Casey,Andrew
Aur 5h38'0"54d28'
Casey,Anne
Peg 23h38'17"27d13'
Casey,Brittany
Uma 8h35'0"58d58'
Casey,Charlie Jack
Uma 9h12'48"55d54'
Casey,Dr Judy
Oph 18h1'26"8d32'
Casey,Eddie Shears
Aql 20h2'32"-1d4'
Casey,George & Oma
Cyg 20h30'34"42d59'
Casey,James Patrick
Hya 8h54'20"5d46'
Casey,Joe
Aur 6h30'0"52d32'
Casey,Jordan Carl
Psc 0h47'20"32d28'
Casey,Kandy
Mon 7h55'21"-2d50'
Casey,Kristen Ann
Cam 3h29'36"61d16'
Casey,Lisa
Cas 0h53'42"73d36'
Casey,Mary Ellen
And 23h31'57"43d5'
Casey,Mary-Rose
Vul 19h59'0"27d19'
Casey,Meagan Kathleen
Del 20h14'16"15d8'
Casey,Pam
Psc 1h29'21"10d8'
Casey,Per
Lyr 18h49'13"31d5'
Casey,Quinn
Uma 10h52'12"40d46'
Casey,Sarah
Lib 15h7'15"-1d10'
Casey,Sr,Donald Dale
Aql 18h45'1"8d58'
Casey,Stefan W
Hya 10h12'3"-18d9'
Casey,T Randall
Uma 12h4'16"63d34'
Casey,Terry R
Aur 4h51'22"40d30'
Casey,Thomas
Lyn 7h38'36"36d59'
Casey,William G
Aur 5h13'24"41d38'
Casey-Brave & Watchful
Cep 22h10'1"60d49'
Cash,Alexander Hamilton
Aur 7h18'1"37d10'
Cash,Fred Charles
Boo 15h3'26"12d10'
Cash,Helen E
Peg 21h56'50"34d20'
Cash,Matthew Edson
Aql 18h59'35"13d6'
Cash,Michael Sterling
Cep 22h15'40"60d16'
Cash,Robin Joy
Mon 6h18'50"-0d20'
Cashatt,Alene Pressley
Peg 22h40'28"25d10'
Cashdan
Her 17h11'58"27d27'
Cashen,Beverly
Peg 21h58'0"31d15'
Cashen,Sr,Walter Ray
Aur 6h17'33"37d32'
Cashin,Ashley Nicole
Leo 10h47'15"8s42'
Cashin,Christopher Anthony
Gem 6h45'8"14d49'
Cashman,Christopher Todd
Dra 10h41'38"80d12'
Cashman,James
Ser 15h19'43"0d27'
Cashman,Margaret S
Cam 5h2'0"68d57'
Cashman,Richard
Her 18h5'17"37d50'
Cashman,Richard & Joanne
Equ 21h20'45"10d40'
Cashman,Thomas J
Aur 4h55'0"40d10'
Cashmore,Patsy
Hya 9h16'38"-14d58'
Casiano,George Albert
Lac 22h17'26"49d15'
Casias,Benjamin & Elizabeth
Lac 22h4'14"47d1'
Casias,Frank
Lac 22h40'48"54d15'
Casillas,Rosendo
Uma 11h49'7"57d3'
Casimiro,Shane Michael
Lac 22h46'49"53d11'
Casini's Star,M
Cet 2h55'0"2d12'
Casini,Ornella
Cnv 13h1'24"51d33'
Casler,Jeffrey C
Per 3h51'13"39d51'
Casm'
Cep 21h38'41"63d53'
Casmaer,Verona
Peg 22h31'20"20d30'
Casnettie,Dennis
Dra 16h0'34"60d42'
Caso,Brian
Cmi 7h54'14"1d34'
Caso,Gregory
Peg 21h24'1"20d4'
Casolo,James S
Ori 6h16'59"20d24'
Cason,Betty P
Peg 23h2'0"14d35'
Cason,Gregory Todd
Oph 17h29'47"-20d54'
Casoni,Giovanni Vacchelli
Lyn 7h39'14"50d45'
Caspar,Klaus J
Tau 5h50'12"26d43'
Caspari,Mechthild
Ari 2h27'46"26d57'
Casper
Dra 19h3'41"56d56'
Casper
Cnv 13h45'45"32d21'
Casper
Lyr 18h37'47"37d1'
Casper
Uma 8h33'35"56d22'
Casper
Cyg 21h14'24"38d3'
Casper,Alfred G T
Sex 9h42'59"-1d58'
Casper,Christy Lee
Vul 20h17'14"25d22'
Casper,Cody
Aur 7h20'16"36d42'
Casper,J H
Aql 19h24'14"13d55'
Casper,Jenna Marie
And 23h1'50"51d15'
Casper,Joe
Vir 13h21'4"-6d48'
Casper,Jr,Jay Vincent
Lyn 7h54'42"36d60'
Casper,Mary
Cyg 20h19'54"38d46'
Casper,Nancy & Howard
Lyr 18h41'33"38d53'
Casper,Nicole Lee
Lyn 7h51'41"48d42'
Caspers
Cap 20h58'44"-20d19'
Caspersen II,Robert Peter
Uma 11h44'1"61d57'
Caspersen,Marguerite Ann
Uma 11h43'36"61d59'
Caspersonus,Jerryus
Del 20h49'39"7d54'
Cass & Bryce-Best Friends For Life
Umi 16h29'45"70d32'
Cass,Donald T
Cam 4h14'16"60d22'B
Cass,Geraldine Deanna
Cam 4h14'16"60d22'A
Cass,Lyndsey
Peg 22h13'20"32d18'
Cassady,Heidi Bliss
Leo 9h49'1"33d15'
Cassady,Michael
Lyn 6h14'60"59d39'
Cassady,Natalie Karen
Lyn 7h59'56"44d28'
Cassady,Opal Agnes
Peg 22h0'0"4d56'
Cassam-Chenai,Zakir & Beatrice Mellier
Ind 20h29'53"-53d40'
Cassan,Jean-Pierre
Lac 22h2'54"37d41'
Cassandra
Cas 1h10'21"71d0'
Cassandra
And 1h1'0"36d44'
Cassandra
And 2h23'46"48d1'
Cassandra
Cas 1h16'15"66d10'
Cassandra
Psa 22h3'22"-26d31'
Cassandra Ann
Lyr 18h51'53"41d49'
Cassandra Lynn
Cam 4h31'0"68d34'
Cassandra Rae
And 23h32'11"48d40'
Cassandro,Tara Livia
Cas 0h6'24"60d35'
Cassanell,Cherylanne
Ari 1h58'53"18d26'
Cassani,Colby Tanner
Peg 21h51'34"28d5'
Cassara"Babe",Frank
Lyr 19h16'51"42d8'
Cassata,Michael James
Per 4h26'41"50d41'
Casse,Jacques
Dra 15h46'27"51d28'
Cassediche
Mon 7h59'17"-5d25'
Cassel Leigh
Lyr 19h17'37"37d53'

Cassel,Arno Aur 5h4'0"38d37'
Cassel,Suzanne Cas 23h40'22"60d34'
Cassell Student Star, 1993 Uma 8h46'55"52d9'
Cassell,Ian Boo 15h46'17"40d59'
Casselli,Mark Lyr 19h23'11"31d12'
Cassens,Jeff & Kris Tri 1h56'19"25d48'
Casserd,Freda E Com 13h0'54"22d21'
Casserly,"Ren"Erin M And 0h9'28"47d10'
Cassese,Amber Faith Lyn 7h33'15"52d0'
Cassetina,Alfred D Lyr 18h46'35"40d30'
Cassett,Aliana And 0h51'22"35d49'
Cassett,Nicole Mon 7h29'18"-6d2'
Cassi Anne And 1h31'29"35d37'
Cassibba,Blase Com 13h31'10"20d14'
Cassidy Aur 5h10'59"44d19'
Cassidy Lyr 18h41'28"26d28'
Cassidy III,William John Hya 9h32'0"1d37'
Cassidy's Star,Tracy And 1h53'38"46d54'
Cassidy,Alison Theresa Mon 6h50'44"10d38'
Cassidy,Catherine T And 1h34'41"37d50'
Cassidy,George Michael (Mike) Vul 19h2'52"24d48'
Cassidy,Gerald S J Uma 9h2'20"48d9'
Cassidy,Ian Per 2h40'34"34d58'
Cassidy,Kelly Marie Uma 11h30'29"64d20'
Cassidy,Molly Jayne Crb 15h40'30"28d39'
Cassidy,Nancy Sue Uma 10h8'1"57d22'
Cassidy,Patricia Dale Cas 2h33'0"57d1'
Cassidy,Paula Jean Mon 6h12'37"-10d9'
Cassidy,Pete Cma 6h54'16"-17d55'
Cassidy,Timothy Sct 18h43'26"-5d44'
Cassidy,William Uma 9h5'10"50d31'
Cassie Cyg 20h36'48"45d29'
Cassie Ori 5h57'1"17d20'
Cassie Cas 0h35'40"67d32'
Cassie Grace Meredith And 2h16'29"48d23'
Cassie's Star Bright Cas 0h32'22"75d43'
Cassin Family Star,The Cam 4h57'37"40d18'
Cassingham,Camille Louise Leo 10h59'18"-5d24'
Cassins my Sweet Ori 6h3'14"5d51'
Cassiopia Cas 3h34'53"73d41'
Cassise,Alanna Mon 6h43'31"7d13'
Cassity Anne Vul 20h15'37"23d57'
Cassolino,Joan Cas 3h33'41"68d37'
Casson (The Dancing Star),Lucy Lyr 18h28'25"46d60'
Cassone,Rocco D Aql 19h31'0"8d7'
Casssidy,Lindsey Kathleen Crb 15h41'16"28d25'
Cast,Rudy Dean Cma 6h50'45"-19d39'
Castagna Star,The Joe & Jeanne Lyr 18h24'37"44d5'
Castaldo,Francesca Nor 16h19'31"-51d2'
Castanaro,Jennie M Cam 5h0'50"67d47'
Castaneda,Blanca Lyn 7h37'44"45d27'
Castaneda,Denise Spisak Lyr 18h57'14"30d55'
Castaneda,Jhalainna Del 21h4'36"12d35'
Castano,George Sco 16h50'38"-43d10'
Castano,Rachel Lib 15h35'17"-18d54'
Castano,Rose Marie Lyr 18h31'53"30d58'
Castany,Sara Maria Cas 22h56'33"55d26'
Castané,Amanda Pérez Psc 1h33'1"27d47'
Castac 1h33'1"27d47' Aur 6h1'42"37d5'
Casteen,Shauna Janal Peg 22h6'0"4d54'
Castegner,Victor V Lyr 18h16'31"30d52'
Castel De Oro Aql 19h25'42"-8d7'
Castelkins Cam 8h32'16"73d27'
Castellani's Ray of Hope Lmi 10h35'1"38d20'
Castellano,Giacomo Her 16h43'22"40d13'
Castellano,Victor Joseph Oph 17h30'51"-0d54'
Castellanos,Jose' L Aql 19h3'24"-6d52'
Castelli,Maria A And 23h27'56"44d25'
Castelli,Robert L Psc 0h47'22"31d15'
Castellini,Drew Brian Her 16h38'1"32d53'
Castellini,M Adele Romagnolo Dra 17h49'30"63d54'
Castellini,Richard Aur 6h28'41"31d23'
Castellini,Robert H Aur 6h0'16"35d56'
Castello,DeLayne Alana Ori 6h4'52"5d11'
Castello,Norman Tau 4h4'21"20d33'
Castells,Jean-Louis Ori 5h1'57"10d40'
Castells,Joan Cam 13h52'17"81d25'
Castellucci,Teddy & Nelle Crb 15h27'23"30d26'
Castelluccio III, Joseph A Uma 8h36'35"71d58'
Castelluccio,Ariana T Uma 13h6'40"63d15'
Castelluccio,Maryann L Uma 11h50'30"40d25'
Casten,Don - Dave Mina Sex 10h0'0"-6d53'
Caster,Donald Richard Her 16h16'49"26d3'
Castera,Pierre Cep 23h4'34"64d20'
Casterline,Helen Cnv 13h42'40"46d33'
Castiglia,Joseph Lac 22h8'28"51d40'
Castiglione,Carl Louis Ori 5h53'17"12d1'
Castiglione,Patricia Ann And 2h29'0"45d32'
Castilla,Katherine Mary Cas 0h45'18"64d24'
Castillo,Annie L. Mon 6h22'0"8d16'
Castillo,Armida Eri 4h14'0"-7d59'
Castillo,Brad Ser 15h54'40"20d24'
Castillo,Bunnie Marie And 23h24'22"50d22'
Castillo,Carla Del Oph 17h42'29"12d11'
Castillo,Gary Enrique Per 2h58'17"40d24'
Castillo,Greg Aql 19h54'17"13d45'
Castillo,John Cet 0h56'59"-6d57'
Castillo,Katherine Vul 19h46'57"28d51'
Castillo,Michael Aur 7h17'35"41d48'
Castillo,Nicholas Tyler Her 17h33'21"40d31'
Casting,Peggy "Sue" And 2h31'0"40d56'
Castino,Anthony G Aql 20h2'35"4d38'
Castle "The Hoogie", Juliet Ori 5h59'52"16d48'
Castle In The Clouds Mon 7h18'58"-5d17'
Castle In The Sky Cyg 20h32'50"30d7'
Castle,Amber Renee Peg 21h55'47"33d41'
Castle,Cole Anthony Cep 0h22'21"77d27'
Castle,Cynthia Cas 23h39'32"50d25'
Castle,David A Dra 17h22'00"60d56'
Castle,Ed Cep 22h10'39"55d19'
Castle,James Gerald Her 17h29'22"31d37'
Castle,John & Tina Cam 5h11'13"70d17'
Castle,John Krob Uma 10h32'0"40d1'
Castle,Lida Mon 6h44'50"11d25'
Castle,Major Jack Ori 4h58'1"4d19'
Castleberry,Stephen Lorn Her 17h6'0"45d59'
Castles Anne Charles John Mary Moore Aql 20h12'0"1d41'
Castlewitz,Danielle Kristy Dra 15h31'50"65d22'
Castner,Chip Lac 22h53'0"51d50'
Castner,Rev Dr Edward W Aur 6h26'47"30d26'
Casto,James Boo 14h55'43"27d31'
Casto,Kenneth Aql 18h43'50"-1d58'
Castoldi,Robin Chi-Chi Chica Cas 0h56'30"65d55'
Caston Peg 23h30'1"18d58'
Caston,K D Ori 5h50'51"16d6'
Castonguay,Yvette Cyg 21h5'19"39d21'
Castorena,Forever Aql 20h15'54"0d59'
Castorina Sco 16h24'59"-40d44'
Castorina,Frank Michael Per 2h29'1"57d36'
Castrataro,Vincent Her 18h4'16"48d46'
Castrignanò,Luigi Pup 8h8'21"-29d33'
Castrine,Chanelle Katherine And 1h59'1"39d20'
Castriotta,Francesca Boo 15h12'51"41d35'
Castro,Alejandra And 2h26'27"38d25'
Castro,Angel A Rivera And 2h18'1"47d35'
Castro,Chandler Nichole Peg 22h20'18"32d19'
Castro,Eduardo G Cet 2h59'59"1d36'
Castro,Fran & Bob Boo 14h18'57"18d39'
Castro,John C Per 1h43'40"54d1'
Castro,Laura Aur 6h53'29"37d24'
Castro,Laurene and Bobby Cep 22h14'31"80d9'
Castro,Mara & Rick And 2h27'53"48d10'
Castro,Maria Isabel Peg 22h40'0"34d30'
Castro,Raymond Lac 22h41'21"38d60'
Castro,Roni & Lou Uma 10h38'15"58d55'
Castro,Russell Sge 19h17'52"16d51'
Castro,Sonia Cet 1h3'11"-3d37'
Castro,Virginia Peg 22h58'56"12d16'
Castro,Zachary John Cam 3h39'19"70d32'
Castro-Henderson, M2' Mon 8h2'17"-3d28'
Castro:Puddy Cat's Meow,Claudia Lyn 9h13'49"46d46'
Castruita,Daniel Gem 6h44'16"16d3'
Castsro,Lori & Rob Boo 14h17'18"18d50'
Casual Cyg 20h23'59"38d20'
Casucci,SharonAnne Uma 13h0'0"52d8'
Caswell,Susan Cas 0h38'28"74d38'
Caswell,Tommy Wesley Ori 5h37'11"15d2'
Caswell,William Benson Boo 15h6'30"8d32'
Caswick Point Uma 10h9'52"51d52'
Cat & Art Equ 21h20'21"3d29'
Cat Claudia Per 3h1'28"50d12'
Cat's Phil Per 1h50'49"50d3'
Cat's Place Lyn 7h10'1"58d55'
Catalan,Silvia Fernandez Cas 1h30'44"74d23'
Catalano,Aurelio Eri 3h51'19"2d6'
Catalano,Katelyn Marie And 23h4'10"40d44'
Catalano,Nicholas Frank Her 17h21'60"45d49'
Catalano,Nichole Marie Del 20h16'35"14d3'
Catalano,Peter Christopher Oph 18h28'51"7d60'
Catalano,Sr,Frank Cyg 20h2'30"31d7'
Catalano,Valentina Peg 23h36'46"21d25'
Catalanotto "Star of Chopin",Danny Lyr 18h27'1"34d54'
Cataldi,James F Cam 4h46'46"68d50'
Cataldo,Denise, Danielle,Rudolph Del 20h13'22"10d50'
Cataleta,Mia Cyg 20h50'52"48d56'
Catalina Aqr 22h51'1"-4d31'
Catalina,Lisa Mon 6h18'56"8d41'
Catambay,George & Trudi Aql 19h59'0"14d1'
Catambay,Nichole Rose Cyg 21h2'36"30d26'
Catambay,Trinity Ann Hya 8h14'31"-5d47'
Catanese,Thomas J Dra 19h26'38"61d37'
Catania,Joel Cep 23h34'35"68d1'
Catanio,Andrea Cas 23h37'1"62d38'
Catapano,Angela A Cas 0h45'17"61d19'
Catapano,Drew Mon 6h55'14"8d18'
Catapano,Matthew Charles Aur 7h12'25"41d5'
Catapano,Salvatore Dra 12h53'44"71d12'
Catarina And 0h57'31"37d38'
Catchpole,John Michael Her 16h57'13"50d24'
Catell,Robert B Aur 6h0'15"30d8'
Catena,Gerard S Aur 6h25'1"33d5'
Catena,Rosaria Mon 7h50'1"-4d4'
Catena,Ted & Annie Del 20h36'0"20d29'
Cater,Annette Cru 12h45'1"-59d54'
Cater,Dan Aql 18h44'16"7d40'
Cater,Julian & Christy Cyg 20h23'1"39d31'
Caterino,John Per 1h52'46"56d22'
Caterina Peg 21h58'44"20d7'
Caterina Cae 4h57'55"-32d49'
Caterina Crb 16h19'51"27d28'
Caterina Per 3h36'1"50d31'
Caterinna Cyg 19h57'19"58d60'
Cates,Anthony Vul 19h22'1"26d46'
Cates,Diane Mon 7h50'2"-3d30'
Cates,Harriett Per 1h56'1"56d53'
Cates,Kristen Renee Cyg 19h28'40"30d22'
Cates,Paul Cep 0h9'32"68d22'
Cath Com 12h54'38"23d42'
Cath-Baby Dra 20h42'52"80d15'
Cathandrew Lyr 18h30'59"40d17'
Catharine Julie And 2h28'42"39d22'
Catherine Cma 7h15'36"-16d33'
Catherine Cas 0h41'28"64d21'
Catherine Cas 0h26'54"61d44'
Catherine And 23h1'18"47d57'
Catherine Lyr 18h53'23"31d43'
Catherine Equ 21h20'52"11d19'
Catherine Cyg 19h9'36"39d7'
Catherine Umi 16h21'1"72d18'
Catherine Per 2h12'48"58d57'
Catherine Cas 2h44'47"70d37'
Catherine & Stephen Crb 15h40'43"27d49'
Catherine (Perry) Peg 22h40'21"33d46'
Catherine 143 Cas 0h56'54"67d17'
Catherine Ann Aur 5h33'54"50d29'
Catherine Anna Vul 20h15'49"25d36'
Catherine Anne Lyn 7h38'50"45d3'
Catherine B Cas 0h38'48"67d52'
Catherine Bridget Cas 0h7'23"62d33'
Catherine C Lyn 8h18'40"49d36'
Catherine Constella John Cas 0h48'26"61d2'
Catherine Dale Vul 20h39'57"23d45'
Catherine Eileen Aqr 21h6'50"-1d47'
Catherine Forever Lyr 19h0'18"26d48'
Catherine Joan Sct 18h48'4"-7d40'
Catherine Kelly And 23h14'34"37d56'
Catherine Mae Tau 4h11'35"1d33'
Catherine Margueritte Dra 20h25'1"75d1'
Catherine Mary Lyn 8h52'42"33d35'
Catherine My Love Uma 9h10'0"61d54'
Catherine Of Naon Peg 23h41'0"27d7'
Catherine Star,The And 23h3'1"47d17'
Catherine Sue And 1h39'40"36d8'
Catherine The Vamp Lmi 10h8'16"32d6'
Catherine's Bear Uma 10h58'25"58d49'
Catherine's Heart Vul 19h6'14"25d31'
Catherine's Pretty Package Crb 15h57'14"26d22'
Catherine's Shining Star of Shannon And 0h12'46"38d37'
Catherine's Treasure Aql 19h40'43"10d55'
Catherine,Mary Esther Cas 0h45'56"61d15'
Catherine-2 Mon 8h1'31"-1d18'
Catherwood,Beth Com 12h23'14"18d48'
Cathey's Star Boo 14h24'14"54d55'
Cathey,Emma Caroline Vul 19h44'19"20d3'
Cathey,Kellie Danielle Sge 19h40'32"18d49'
Cathey,Sharon Elaine Gem 6h27'1"13d47"
Cathey,Tyson Moore Sex 10h4'18"5d26'
Cathi And 23h38'25"48d8'
Cathleen And 2h21'47"45d2'
Cathlyn Alyce And 0h57'59"37d54'
Catho,29-11-92 (FY) Uma 10h27'46"51d58'
Cathrin Felicitas Herbstblume Gem 7h23'43"20d54'
Cathrina Cyg 20h1'34"30d57'
Cathro,Shannon Lee Cnc 9h19'60"30d53'
Cathryn Cyg 20h51'39"38d19'
Cathryn Cas 0h58'44"64d1'
Cathryn's Star Cyg 21h50'45"38d8'
Cathy Lyr 18h40'41"38d50'
Cathy Ori 5h55'24"16d5'
Cathy Aur 6h0'51"30d14'
Cathy & Bill Cet 2h35'54"4d10'
Cathy & Kurt Forbidden Love Star Cyg 19h18'28"44d58'
Cathy A 2-13 Uma 8h16'49"71d59'
Cathy Ann's Never Land Peg 23h2'36"8d26'
Cathy Anne-The Eternal Flame of My Soul Cra 18h20'16"-42d26'
Cathy B Lyr 18h30'55"35d16'
Cathy Jean Mon 6h45'20"-10d11'
Cathy Jean Lib 15h2'44"-20d8'
Cathy Jo 150 Cam 12h30'19"84d46'
Cathy Lynn Aql 19h9'23"10d46'
Cathy Lynne Sge 19h53'51"18d52'
Cathy's Com 12h24'39"21d53'
Cathy's Forever Shining Star Lyn 9h2'1"40d60'
Cathy's Shining Light And 23h33'48"47d31'
Cathy's Spirit Cyg 20h21'12"38d35'
Cathy's Star Peg 21h48'0"30d11'
Cathy's Star-A Piece Of Eternity And 2h29'41"39d7'
Cathy/D/469 LKN Umi 15h16'50"66d15'
CathyAnn's Happy 45th Eri 3h34'13"-5d3'
Cathys' Dust-in The Heavens Del 20h19'24"18d51'
Cati Uma 9h45'31"52d12'
Catjim Equ 21h5'40"10d9'
Catledge,June I. Oph 18h40'0"8d40'
Catlett,Edna Aqr 23h28'0"-5d42'
Catleugh,Jennifer And 0h23'20"33d44'
Catlin,Jalle Berry Vul 19h18'57"22d51'
Catlin,Odessa B Cam 3h58'53"60d6'
Catlin,You're Special Lyn 8h30'48"41d22'
Catmull,La Traviata For my Richard Lyr 18h36'54"36d48'
Cato,Thomas Aql 20h10'30"10d59'
Caton My Dearest Mother,Ruthie M Mon 6h53'7"-2d20'
Caton,Leonard Edgar Ori 5h54'30"21d57'
Catou Peg 22h57'49"31d13'
Catrillo,Ashley And 1h35'1"38d56'
Catrin Cas 0h42'1"70d15'
Catrina & John Ori 5h13'38"-5d10'
Catriona Lyr 19h2'30"26d38'
Catrionadore Umi 15h50'13"80d39'
Catron,Christopher Hal Aql 20h12'53"4d53'
Catron,Dr Phil Eri 3h33'54"-2d31'
Catron,Teresa Lynn Mon 6h55'28"-10d9'
Catsimatidis,John Alexander Aqr 21h58'43"-6d21'
Catskill Love Cyg 21h38'22"42d33'
Catsouras,Christos Aql 19h29'22"-0d31'
Catsstardance And 0h10'13"33d57'
Cattano,Noel Cma 6h54'49"-19d32'
Catterson,James Michael Vul 19h58'24"28d10'
Catterton,Gladys Rosoff And 0h10'59"37d6'
Cattle,Harry Giles Cam 3h58'19"62d30'
Catto,Elsa Her 18h54'20"18d51'
Cattoi,Robert L Cmi 7h27'30"0d53'
Catts,Adelinde Meira Mon 6h29'1"8d43'
Cattuti,Francis F Boo 14h33'33"44d5'
Catullo e Sylwan Peg 21h53'35"34d56'
Catus,Lola Felis Lyn 7h41'28"42d41'
Catusco IC 10,Iris Lyn 7h47'17"50d49'
Catwar Aur 5h15'28"40d57'
Catwell,Matthew Alexander Jason Cyg 19h55'12"48d24'
Cauchon,Frédéric Guy Benoât Uma 12h7'14"47d55'
Caudart,Marie-Annick Lyr 18h54'47"31d34'
Cauderan,Christian Boo 15h37'36"42d10'
Caudle,Howard P Cep 22h17'1"78d57'
Caufield,Douglas C Boo 15h0'37"20d7'
Caufield,Mike Aql 18h41'19"-2d9'
Caufman's Love Star, Nancy & Derek Eri 2h43'15"-3d52'
Caugh,Susan Ann Aql 19h49'48"14d23'
Caulfield,John Cam 5h18'0"80d36'
Caulfield,Marian Peg 22h11'25"4d39'
Caulfield,Thomas James Uma 11h5'13"30d8'
Caunter,Harry A Boo 14h57'38"21d18'
Causby,Debbie Renee Lyr 18h41'14"31d29'
Causer,Audrey Irene Cas 1h48'41"73d12'
Causer,Beth Cam 7h32'51"67d42'
Causey,Lily Mae Peg 22h45'1"27d54'
Cautain Michel Aur 5h59'52"31d21'
Cauthen,Jr,Robert S Ori 6h15'19"8d33'
Cauthen,Michelle And 1h24'11"37d7'
Cauthorn,Candace Mon 6h21'12"4d59'
Cava,Alex Christian Her 18h13'1"31d19'
Cava,Frank Aur 4h53'12"51d17'
Cava,James M Dra 17h37'40"64d33'
Cava,Keith Vul 20h14'17"22d39'
Cavalucci,Robert Vel 9h45'49"53d22'
Cavadi,Cesare Cas 2h32'1"70d39'
Cavalier 5C 95-96 Uma 12h11'0"60d23'
Cavaliere,Nicholas J Cas 0h29'22"61d23'
Cavaliere,Saundra L & Grover C Ward Cyg 21h35'31"38d53'
Cavaliere,Sonja Gödl Eri 3h5'21d-9d8'
Cavaliere,Steven Dra 17h12'25"69d40'
Cavaliere,Sue Cyg 20h18'43"40d28'
Cavalieri-d'Oro, Marcello Her 16h21'55"26d15'
Cavallacci Dra 17h18'0"61d17'
Cavallaio,Franco Eri 2h43'28"-6d48'
Cavallari,Anna Lyn 7h0'49"60d44'
Cavallari,Laura Cyg 20h52'51"40d11'
Cavallaro,Linda C & Christopher J Cyg 20h5'29"41d7'
Cavallaro,Raymond J Boo 15h44'0"42d29'
Cavallaro,Stephen Robert Aur 6h23'1"30d47'
Cavallero,Carmen Lyn 9h19'1"34d25'
Cavalli,Devy Peg 22h9'21"24d15'
Cavalli,Pulci Mon 5h56'49"11d58'
Cavallini,Arianna Aur 5h21'51"37d56'
Cavallini,Virginia Cas 23h40'23"63d5'

Cavallo,George C
Cep 1h0'0"77d13'
Cavallo,Michael
Lac 22h44'34"53d26'
Cavaluchi,Kelli Ann
Uma 9h54'28"70d41'
Cavana,Ashley
Lyn 7h49'42"44d5'
Cavana,Montana
Cep 22h18'40"60d56'
Cavanagh,Betty
Aql 19h2'1"-1d23'
Cavanagh,James A
Boo 14h7'45"32d6'
Cavanagh,Lawrence
Dra 11h49'27"74d14'
Cavanagh,Patrick
Her 15h56'1"46d40'
Cavanagh,PHD,Eugene J
Ori 6h4'11"8d30'
Cavanagh,Sr,Joseph E
Per 4h23'0"50d25'
Cavanaugh "Our Star", Jeffrey Thomas
Dra 19h57'59"68d4'
Cavanaugh,Beverly Sue
Lyr 18h56'0"32d21'
Cavanaugh,Catherine
Vul 20h3'19"25d35'
Cavanaugh,Dennis
Hya 8h55'56"2d57'
Cavanaugh,Eric W
Dra 15h58'60"65d2'
Cavanaugh,John R
Aur 5h19'53"43d1'
Cavanaugh,Judith
Lyn 7h37'0"40d51'
Cavanaugh,Michael
Vir 12h2'11"2d10'
Cavanaugh,Natasha
Mic 21h13'15"-30d36'
Cavanaugh,Rob
Lyn 8h1'10"42d5'
Cavanaugh,Sean Oliver
Umi 15h9'1"68d47'
Cavanaugh,Sharon K
Cyg 19h59'40"30d26'
Cavanaugh,Thomas J
Boo 14h40'36"31d7'
Cavanaugh,William Michael
Oph 18h3'47"11d51'
Cavani,Sara
Uma 12h0'46"33d5'
Cavanna,Mario
Lmi 10h32'14"37d55'
Cavari,Emily Ann
Cyg 20h2'38"30d31'
Cavataio,Steve
Ori 5h50'0"14d14'
Cave,Martha Tyson
Gem 7h22'28"24d57'
Cave,Michael
Uma 10h33'33"56d17'
Cave,Ray A
Boo 14h19'0"37d28'
Cave,Rebecca Jane
Mon 6h26'11"1d27'
Cavell,Winston Wesley
Aql 20h10'18"3d59'
Cavenaugh's Guiding Light,Timothy R
Leo 10h19'46"13d44'
Caveney 43
Boo 14h32'10"20d40'
Caveney,Dennis
Her 16h39'36"21d4'
Caveney-To Alex & Drew Grandpa
Ori 5h54'16"12d18'
Caverly,Father Patrick J
Lyr 18h34'44"40d30'
Caverly,Helen Ann
And 23h44'41"45d37'
Caverly,Jacquelyn Ann
And 0h16'1"39d17'

Caviezel My Love, Samuel Richmond
Aur 6h6'56"32d59'
Cavigliano,Jacqueline M
Cyg 19h25'40"30d57'
Caville,Jacci
And 1h57'47"37d48'
Caville,Nicole Kristine
Lyn 8h54'46"45d59'
Cavin,Raymond T
Per 3h6'34"47d29'
Cavion,Bruno
Del 20h18'11"12d37'
Cavote,Michael E
Mon 6h21'33"8d7'
Cawbron,Peter & Di
Cyg 21h52'59"52d55'
Cawdery,Anthony Devereux
Aur 5h56'13"30d56'
Cawelti,Stephanie Ann
Lyr 18h45'44"37d7'
Cawkwell,Ted
Boo 15h23'0"34d18'
Cawley Love,Eternal
Cyg 19h24'33"35d0'
Cawley Star,The Liam Michael
Cep 22h33'33"61d18'
Cawley,Beverley Ann
And 23h2'32"40d17'
Cawley,Cathy Lee
Cmi 7h26'38"7d46'
Cawley,Jen
Mon 6h15'12"-5d44'
Cawley,Scott B
Aql 19h57'58"7d54'
Cawman,George Washington
Ser 17h55'26"-13d39'
Cawthorne,Erica Lowe
Sgr 18h50'58"-33d48'
Cawthra(My Love), Dean Albert
Dra 17h41'0"68d10'
Caycsandra
Cas 2h38'29"60d25'
Cayet,Alain
Cyg 20h43'49"46d40'
Cayetan,R-Beg
Ori 5h29'35"-8d45'
Cayla Liles
Uma 10h8'13"51d49'
Caylan 9
And 23h32'21"37d34'
Caylor,Charles S
Eri 2h55'20"-13d11'
Caylor,Julie
Eri 3h5'32"-7d27'
Cayne "The Nuts",Super Star James
Uma 11h2'20"45d50'
Cayo,Gaye T
Cyg 21h50'18"44d22'
Cayouette,Elise
Cyg 21h51'1"36d58'
Cayre,Christine E V
Ori 5h2'43"10d57'
Cayse Family Star,The Doctor Robert L
Oph 18h0'50"0d2'
Caytan,Lucien
Aur 5h15'37"43d59'
Cayton,Erik Mikel
Boo 13h56'58"17d24'
Caywood,Jr,David E
Cep 20h37'31"75d45'
Caz
Peg 23h4'30"31d23'
Caz B
Crb 15h54'16"26d13'
Cazal
Cap 21h0'20"-23d15'
Cazals,Alain
Lyr 18h13'27"39d34'
Cazel,Lyle
Lyr 18h41'19"41d38'

Cazier,Adrienne Beth
Cas 0h22'36"68d50'
Cazier,Max Allen
Cep 22h30'47"63d46'
Cazort,Lee
Cmi 7h55'37"8d13'
Ca Cmi 7h55'37"8d13'
Lac 22h24'33"40d46'
CB'S Evermeet
Sct 18h22'18"-5d12'
CB,Jr
Cam 13h28'54"84d33'
CB2
Aql 19h25'1"-10d35'
CBC 5784
Ari 2h56'27"30d13'
CBG Sister I, '72
Umi 17h27'17"76d11'
CDW Star,The
Lyn 8h59'56"40d11'
Cea II,Vincent Anthony
Aur 6h34'43"38d40'
Cearlock,Cheryle Darlene
Cet 2h11'25"0d39'
Ceaubaï
Lac 22h5'1"37d38'
Ceballos,Addie L
Psc 0h46'21"32d43'
Ceballos,Samantha K
Leo 11h32'59"18d49'
Cebulko I
Lyn 9h1'33"41d4'
Cec
Cam 5h55'19"72d20'
Ceca
Uma 10h8'1"59d15'
Ceccacci,Catherine
Cam 4h44'0"78d57'
Ceccarelli,Zelinda
Lup 15h17'29"-40d43'
Ceccarini,Gaia
Dra 15h19'25"58d60'
Ceccato,Ivano Luigino
Ori 6h4'31"3d15'
Ceccato,Serge
Aql 18h57'33"-6d28'
Cecchetti,Thomas
Umi 13h32'1"71d25'
Cecchi,Brunella
Cep 23h29'38"64d2'
Cecchini,Angelo
Cep 20h38'44"75d29'
Cecchini,Elena
Leo 9h21'21"10d55'
Cecchini,Isabelle
Cas 1h16'33"60d11'
Cecchini,Joseph
Cep 21h54'26"65d28'
Cece's Star
Aur 6h36'26"38d25'
Cece,John L
Per 2h3'1"56d43'
Cecelia
Cas 23h39'49"60d24'
Cecelia & Teresa
Peg 23h0'28"33d35'
Cecelia Marie Therese
And 1h12'11"41d3'
Cech,Anthony Andrew
Dra 10h27'17"78d3'
Cech,Chuck
Cep 23h7'22"71d7'
Cech,Frantisek Ringo
Cnv 12h21'44"51d47'
Ceci,Elisa
Lmi 10h25'29"32d16'
Ceci,Nicole Rossiter
And 23h41'0"43d47'
Cecibel,Sylvia
Umi 14h24'42"66d53'
Cecil
Lmi 10h50'19"33d10'
Cecil,Jennifer
Equ 21h7'18"2d54'

Cecil,Kacie Lynne
Eri 4h5'31"-2d25'
Cecil,Melissa
Oph 18h4'1"1d3'
Cecil,Terra Alexandra
Vul 20h20'15"28d10'
Cecil-Smith,George
Umi 16h24'18"76d17'
Cecile
And 0h1'23"35d50'
Cecile
Cnv 12h10'0"38d21'
Cecile 1
And 0h22'51"40d24'
Cecile Joy
Lyn 8h2'53"46d36'
Cecilia
Dra 16h23'0"62d39'
Cecilia & Clark
Tau 5h56'36"23d14'
Cecilia B
Cep 20h49'26"73d9'
Cecilia Jean
And 0h4'14"44d19'
Cecilia Marie
Mon 7h26'46"-10d33'
Cecilia Noel
Uma 10h26'0"67d48'
Cecilione,Michael
Lmi 10h17'21"32d27'
Cecio,Monica Sofia
And 0h4'60"46d36'
Ceckowski,Donald H
Aur 6h25'59"37d35'
Cecur,Klara & Mikolaus
Uma 13h31'44"60d8'
Cedar
Sgr 18h59'49"-26d25'
Cedar Star,Micheal & Mary
Sct 18h47'41"-7d39'
Cedarbloom,Brian D
Boo 15h1'48"14d56'
Cederberg,Allison Louise
And 1h49'39"39d57'
Cederholm,Warren
Aur 7h12'14"38d46'
Cedric
Leo 11h7'0"-0d21'
Cedric,Lefevre
Dra 11h33'48"68d19'
Cedrika
Cep 22h38'52"60d6'
Cedron Star,The Ana
Psc 1h28'39"32d11'
Cee Cee
Eri 2h46'48"-1d39'
Cee nee tee yo8"-1d39'
Umi 17h4'29"76d15'
Cee,Marina
Cap 21h2'14"-24d0'
Ceejaniques/Alexander
Oph 17h15'40"10d29'
CEF
Dra 19h41'0"68d13'
Cefalu',Michael
Her 17h38'34"14d49'
Cefalù,Aldo Natale
Lyn 8h18'57"42d57'
Ceglia,Joseph
Cyg 20h23'1"39d33'
Ceil
Lmi 9h51'39"40d20'
Ceilidh Na Colleen
Del 20h16'40"9d28'
Cejmer,Regina Maria
Tau 5h24'45"18d53'
Cekauskas,Ms Cynthia Dana
Ari 1h57'59"21d57'
Cekovich,Jr,Joe
Aur 6h10'53"33d46'
Celaya,Eric Theodore
Eri 2h52'24"-18d45'
Celby
And 0h52'26"34d1'

Celebration Hunstead D-I
Uma 11h12'27"30d15'
Celebration of His Blue Desire,The
Aql 18h43'19"6d11'
Celebration of Tsuya's 60th birthday
Ari 2h18'44"26d42'
Celefte
Lmi 9h47'22"33d55'
Celen,Thomas E
Lac 22h8'30"46d44'
Celena Renae
And 0h20'12"35d38'
Celentano,Alyssa Carlotta
Gem 6h35'15"24d30'A
Celentano,Michael Vincent
Gem 6h35'15"24d30'B
Celentino,Jr,Ted
Lyn 9h11'0"36d52'
Celeski,Patsy Louise
Cyg 19h26'48"30d31'
Celesnik,Dennis G
Cam 3h18'17"60d34'
Celeste
Eri 4h9'42"-7d41'
Celeste
Lac 22h40'33"56d9'
Celeste
Cam 3h11'30"60d36'
Celeste
Peg 22h6'23"24d18'
Celeste
Dra 15h45'29"52d51'
Celeste
Uma 11h57'11"48d21'
Celeste
Cyg 20h53'1"31d35'
Celeste Alane
Cyg 19h45'52"30d16'
Celeste E L
Gem 6h57'0"20d28'
Celeste Marie
Cas 0h26'19"66d55'
Celeste,Ruth Loretta
Cep 3h23'0"78d41'
Celeste/45 Years for Cis & Les
Peg 23h1'34"31d21'
Celestial Alacrity
Cyg 20h52'31"38d16'
Celestial Bomber Tom
Ori 5h23'57"1d52'
Celestial Brian
Ori 6h4'14"5d60'
Celestial Carolyn
Cas 1h49'33"73d41'
Celestial Catherine B, The
Cas 0h53'37"70d40'
Celestial Clickie
Lyn 8h54'0"34d16'
Celestial Gracc
Uma 8h58'23"53d2'
Celestial Harry
Aur 7h1'16"43d39'
Celestial lere
Gem 6h27'17"14d40'
Celestial Jerj
Cas 0h40'17"67d7'
Celestial Joy
Uma 11h31'0"61d24'
Celestial Lyons
Leo 9h50'28"30d28'
Celestial Mia
Com 12h26'57"22d58'
Celestial Peg
And 1h19'21"37d15'
Celestial Sentinel
Cnv 13h50'33"38d21'
Celestial Trouble
Cam 5h51'43"61d30'
Celestial Zoe
Crb 16h16'40"37d30'
Celestin,Gladys
Cas 0h35'1"63d14'

Celestine Aaron
Lyn 8h45'41"37d18'
Celestinne
Peg 22h41'10"24d54'
Celestyrrell
Per 3h17'30"40d45'
Celia
Cyg 20h26'1"40d38'
Celia & Bernie's "Anniversary Star"
Lyr 19h14'41"40d53'
Celia Johanna
Com 12h2'1"27d38'
Celien,James
Cet 3h20'49"2d1'
Celina
Ori 5h57'0"15d14'
Celina
Tri 2h12'32"30d55'
Celina Marie
Uma 10h21'59"40d57'
Celine
Hya 9h34'33"5d9'
Celine
Boo 15h27'33"47d53'
Celine
Tri 1h52'54"27d38'
Celine
Lyn 7h51'24"36d12'
Celine-Amandine- Marianne
Per 3h8'14"41d13'
Celiss,Tisha
Peg 22h16'55"34d21'
Cella,Holly
Peg 22h31'58"25d54'
Cella,Linda
Per 3h20'59"54d43'
Cella,Robert Frank
Lac 22h43'31"54d7'
Cellamare,Francesca
Boo 14h33'0"10d36'
Cellar Door (Audrey's Star)
Cas 0h5'50"64d25'
Celle-Povey,Monique
Cas 1h45'58"58d24'
Cellier,Steven Lee
Lac 22h14'22"51d38'
Cellino,"Luanceal" Luis Antonio
Vir 12h32'53"2d25'
Cells,Carole
Cas 0h15'33"61d52'
Cellucci,John H
Aur 6h53'1"37d48'
Celone,Rita Naomi
Cas 1h14'28"61d41'
Celotto,Angela Jean
Uma 11h25'12"43d25'
Celtic Molly Anne
Lyr 18h40'33"32d47'
Celto,John Eugene
Dra 19h47'18"68d53'
CEM GEM Semper Una
Uma 11h35'45"55d35'
CEM-12
Leo 11h0'46"-2d15'
Cembrale,Don
Boo 15h4'0"31d7'
Cendrine
Cyg 20h22'46"39d31'
Cenerella
And 23h23'45"51d2'
Cenerentola
Pho 0h28'42"-42d34'
Ceni,Umberto
Aur 6h19'17"31d32'
Cenicola,Diane
Mon 8h6'14"-8d13'
Cenicola,Ron
Cet 3h3'25"8d38'
Cenide
Cep 2h19'43"78d6'
Cenname,Alyssa Marie
Cam 5h47'55"61d31'

Cennami,Chuck
Del 20h58'55"16d0'
Cenoukontem
Peg 21h59'21"18d47'
Centanni,Marc Anthony
Her 18h4'17"31d21'
Centano,Jason Anthony
Boo 15h4'11"19d19'
Centauri,R Griffith
Ori 6h2'32"7d56'
Center School,Old Lyme CT
Ori 5h31'37"-0d3'
Centilli,Lois A
Peg 22h3'42"9d50'B
Centilli,Sidney A
Peg 22h3'42"9d50'A
Centimark
Ori 4h44'40"0d36'
Centofanti
Dra 17h50'56"67d31'
Central Region West BBG Board 1994-1995
Ori 5h57'35"11d13'
Centrella,Melissa Anne
And 23h18'54"42d57'
Centro Estetico Anna
Del 20h18'28"14d57'
Centuri,Malamed's Demi
Aql 18h40'1"-0d8'
Centurion,Michael
Cep 22h30'45"60d29'
Cepeda,Adeline
Ari 2h38'37"30d18'
Cephaf
Cep 0h7'59"66d37'
Cephas,Sidney & Myrtle
Aql 19h13'11"10d18'
Cerajewski,Courtney A
And 2h15'35"48d32'
Ceranec,Martin
Aur 6h45'0"38d6'
Ceraso,Betsy
Vul 19h20'22"26d44'
Cerat,Robert E
Hya 8h17'50"3d34'
Ceravolo,Robert D
Aql 19h7'20"3d36'
Cerbone,John
Per 2h41'50"40d46'
Cerbulis,Janis
And 2h23'35"37d51'
Cercadillo,Monica
And 0h22'36"31d48'
Cerciello,#46,Joey "Shorty"
Aur 6h2'18"37d41'
Cercle de Vie
Ori 5h52'46"15d13'
Cercone,Dott "Sandro"
And 23h20'13"47d3'
Cerda,Nolyne
Equ 21h21'15"11d59'
Cerda,Peggy
Equ 21h21'0"11d2'
Cerda,Vanessa
Equ 21h11'29"11d35'
Cerdan
Lmi 10h36'43"37d58'
Cereda,Mario
Lmi 10h11'0"32d7'
Ceremsak,Sara
Lyr 19h2'1"31d23'
Ceresa,Donatella
Cam 13h54'31"80d47'
Cerezo,Valerie Amanda
And 23h45'59"38d10'
Cerf,James Henry
Oph 17h32'11"-20d40'
Cerf,MD,Arthur Z
Ori 5h57'11"14d57'
Cerfolio & Robby, Dr & Mrs Robert
Lyr 19h13'59"27d50'C
Cerfolio,David Ralph
Lyr 19h0'0"38d33'

Cerfolio,Dr Laverne
Lyr 19h13'59"27d50'B
Cerfolio,Dr Nina
Lyr 19h13'59"27d50'A
Ceri
Cas 0h5'33"64d36'
Cericia
Aql 18h56'25"-1d32'
Cerilli,Angelo & Antonina
Cyg 20h23'49"40d5'
Cerino,Joseph & Noreen
Cyg 22h0'22"52d30'
Cerio John (YBE) & Joy
Cep 22h13'60"65d31'
Cerio's Star,Jenifer
Lyr 19h1'21"38d8'
Ceris,Sharon
Lyr 18h20'39"38d14'
Cermak-Garfunkel,Kim
Gem 7h28'30"31d45'
Cerna,Sebastian
Her 18h16'57"28d16'
Cernia ~ Slovin
Uma 8h58'38"57d13'
Cernoch,Nikki
Cyg 20h33'56"31d25'
Cerny,George (Cactus Ken) Kenneth
Aql 19h59'17"15d31'
Cerone Family Star,The Burt
Tri 2h11'21"33d48'
Cerone,Sr,Louis Joseph
Her 17h1'59"41d7'
Cerpa,Maria A Loves Rick B Skibinski
Hya 8h27'49"-9d27'
Cerquitella,John
Cet 0h47'48"-5d53'
Cerreta,Rob
Aur 4h39'14"30d13'
Cerretani II,Laurence D
Dra 16h21'31"64d46'
Cerretani,Joseph L
Aur 6h23'25"37d30'
Cerreto,Master Christopher
Lmi 10h13'32"32d18'
Cerreto,Master Darin
Cep 21h6'45"68d23'
Cerreto,Princess Suzanne
Cas 23h31'40"60d13'
Cerrigone,George & Terri
Uma 11h52'42"50d39'
Cerrone,John Charles
Cet 0h26'48"1d2'
Cerrone,Martin "Marty" Thomas
Cam 4h2'1"58d28'
Certain,Reynaldo
Aql 18h41'23"-0d10'
Certainty
Uma 9h46'33"43d10'
Cerulli,Virginia
Uma 9h7'59"57d13'
Cerussi,Charlie Lutes
Ser 15h55'41"21d19'
Cerutti,Franchino
And 23h3'1"51d57'
Cervantes,Emma Jean
Mon 6h25'11"8d37'
Cervantes,Joseph Anthony
Cet 2h38'60"3d42'
Cervantes,Raymond Carl
Hya 8h14'0"4d1'
Cervantes,Ruben Garcia
Sct 18h28'10"-4d0'
Cerveris,Michael
Aur 4h55'42"40d38'
Cervi,Jr,Robert A
Her 17h25'49"21d15'
Cervi,Savannah Jo
Lac 22h3'22"50d51'
Cervoni,Mary-Ann Helen
And 1h25'16"48d58'
Cesaire,Jacques
Dra 10h8'33"73d23'

Cesar
Cet 0h34'56"0d34'
Cesar
Cep 22h14'41"56d6'
Cesare
Cmi 7h30'47"5d34'
Cesare,Elizabeth Rockwell
And 0h10'24"31d57'
Cesare,James Frank
Cmi 6h56'23"-18d50'
Cesari,Maurilio
Boo 14h18'45"48d35'
Ceschi,Donald J
Lac 22h38'57"52d41'
Ceschi,Tory
Vul 19h40'59"26d38'
Cesena I Love You Amor Juan Carlos
Cep 23h2'51"64d45'
Ceslik,Jr,John
Cnv 12h47'44"48d5'
Cesnovar,Alison "Starr"
Cas 0h40'39"73d30'
Cesoni,Diana M
Cas 1h8'22"62d57'
Cespedes,Doris & Sandro
Boo 14h35'38"8d18'
Cespedes,Edie
Boo 13h53'14"20d10'
Cessna,Pam
Uma 12h51'0"58d59'
Cestaro,Fr Joseph A
Umi 16h24'0"70d60'
Cestaro,Sally
And 23h36'43"49d40'
Cestra,Liberatore
Ant 10h45'39"-36d27'
Cesus
Lac 22h9'1"51d12'
Cevas,Courtney Taylor
Mon 6h39'14"10d3'
CEW II
Cep 20h18'18"76d26'
CH Castellana De Terra Nova CD
Cma 7h23'30"-16d44'
Ch Crestwoods Crackerjack
Aur 6h36'0"40d48'
Ch Great Elms Prince Charming II
Cmi 7h29'0"8d43'
Ch Liberte's Notion de la Barge
Cep 23h4'52"64d22'
Ch Reach for a Star CD CDX,The
Cmi 7h29'52"1d16'
Chaben,John
Cep 4h47'39"87d8'
Chabert,Jacques
Sex 10h16'28"-2d60'
Chabert,Jean Marc
Cep 22h24'58"57d55'
Chabner,Gary
Aql 18h54'32"0d52'
Chabot,Cynthia D
Cas 23h20'1"60d5'
Chace Wayne
Leo 9h53'13"14d20'
Chace,David G
Aur 5h24'46"50d36'
Chace,Samuel
Aur 5h0'23"50d14'
Chach's Star
Ori 5h5'48"14d33'
Chachie's Amazing Grace
Cnv 13h55'1"45d36'
Chacker,Meryl
Lyn 7h36'0"38d36'
Chacko,Kuttickal George
Lmi 10h7'14"40d38'
Chacon,Leobardo Villanueba
Cmi 7h7'1"5d1'
Chacon,Sr,Arturo
Lac 22h53'57"37d52'

Chad
Lac 22h55'46"51d24'
Chad
Aur 6h10'41"38d58'
Chad & Shannon
Eri 4h42'18"-1d1'
Chad Allen
Dra 17h33'54"75d22'
Chad Anthony My Shining Light
Aur 7h10'42"37d18'
Chad Michael
Per 2h37'39"37d11'
Chad's Starfire
Cma 6h57'46"-11d4'
Chadborn,Neil Hedley
Ori 5h33'55"-0d6'
Chadder,Daniel Thomas James
Umi 14h49'41"68d31'
Chadderton,Kourtney
Cyg 19h42'45"28d36'
Chaddock,Gary
Uma 10h48'29"48d27'
Chadduck,Aidan Pelton
Dra 17h1'13"58d35'
Chadfield,Babbette Jayne
Peg 21h50'35"30d55'
Chadfield,Julie-Ann
Peg 21h53'11"31d36'
Chadfield,Rebecca Anne
Peg 21h59'43"31d31'
Chadley
Aur 6h27'20"38d46'
Chadley Ben
Her 16h2'43"21d17'
Chadwell,Nicholas Roy
Dra 9h43'32"73d39'
Chadwick's Golden Wedding,Fred & Joyce
Cyg 19h27'43"37d9'
Chadwick,Brooke Allison
And 23h22'46"41d3'
Chadwick,Christopher Graham
Her 17h18'13"46d20'
Chadwick,Crystal Dale
Mon 6h36'50"6d25'
Chadwick,D's Yvonne
Oph 17h42'55"13d22'
Chadwick,Linda Anne
Cyg 20h15'36"39d14'
Chadwick,Mark Andrew
Aql 19h4'43"2d31'
Chadwick,Shelley Elizabeth Anne
Peg 21h53'57"2d36'
Chadwick,William Lewis
Vul 19h21'1"26d42'
Chaffee,Steven William
Her 17h21'50"46d3'
Chaffin Best In The Universe,Dr Dan
Cet 3h1'51"7d45'
Chaffin,Curt
Ori 4h59'13"4d7'
Chaffin,Margaret Mercedes
Peg 22h47'10"10d2'
Chafik,Mustapha
Oph 17h59'52"10d20'
Chafin,Carla Marie
Del 20h37'40"18d58'
Chagnon Star,The Jose Sulaiman
Oph 17h11'14"-20d52'
Chagny,Jacqueline & Roger
Tri 1h44'54"28d25'
Chagolla,John David
Mon 7h1'30"4d11'
Chai,Chin-Yi
Aur 4h58'53"30d58'
Chaidez,Efren"Chief"
Hya 8h34'32"1d40'
Chaifetz "Star Son", Matthew R
Dra 13h57'45"68d20'
Chaignot,Guenevere L
Dra 18h42'56"70d28'
Chaignot,Nadege
Ind 20h50'43"-52d12'
Chaignot,Olivia A
Cas 1h28'1"50d13'
Chaika II
Sge 19h56'1"20d33'
Chaiken,Ingrid & Albert
Cyg 21h30'21"40d9'
Chaiken,Sheldon
Boo 14h31'50"43d17'
Chaim
Cnv 12h18'32"43d49'
Chaioner,John & Caroline
Cyg 20h0'0"39d38'
Chairez,Brandon
Mon 8h8'39"-2d21'
Chaisson,Lisa & Dean
Aur 5h18'53"42d36'
Chaisson,Mandy
Uma 8h54'38"54d28'
Chakour,F M
Ori 5h40'32"11d39'
Chalabiani,Pari
Uma 10h4'60"71d10'
Chalaye,Amandine
Cep 22h38'51"63d59'
Chalberg,Joshua James
Boo 15h2'56"25d15'
Chalcraft,Arlene
Uma 10h56'51"37d43'
Chalfant,Al & Irene
Equ 20h55'1"2d43'
Chalfant,Claudine Noel
Oph 18h19'36"8d15'
Chalfant,Kathleen
And 2h5'55"39d43'
Chalfont,Kimey Jo
Mon 6h24'45"4d8'
Chalifoux,Claude
Her 17h1'37"22d21'
Chalk,Kimberly Sue
And 23h10'45"41d24'
Chalk,Stacey Lee
Peg 21h28'47"20d27'
Challis,Professor Richard E
Uma 8h46'14"57d50'
Chally,Mark David
Per 3h5'34"41d16'
Chalmers,Bruce
Ori 5h24'53"0d33'
Chalmers,Irena
Cas 1h39'23"58d32'
Chalono,Sandrine-Lee
Aql 20h1'34"10d17'
Chaloux,Lauren T
And 23h1'43"50d13'
Chalstrom,B G
Aql 20h18'43"1d43'
Chalumnae,Ilaria
Pup 8h8'43"-25d18'
Chalvin,Stephane
Aur 5h9'11"38d30'
Cham,Amanda
Cap 20h31'37"-13d13'
Chama
Psc 22h58'16"0d40'
Chamathopascis-Leruez
Ser 15h13'26"1d44'
Chamay,Chantal
And 0h53'15"40d31'
Chamayou,Bernard
Lyn 8h1'33"40d38'
Chamberlain,George Richard
Ser 15h58'40"-3d29"
Chamberlain,John W
Cnv 12h23'1"42^17'
Chamberlain,Jon & Jonina
Cyg 21h20'58"38d17'
Chamberlain,Jr,Richard L
Her 16h33'32"32d19'
Chamberlain,Marisa Ann
Cas 0h21'21"63d30'
Chamberlain,Myrtle Wood
Peg 21h47'57"35d46'
Chamberlain,Nancy
Mon 6h38'48"11d56'
Chamberlain,Neil & Cheryl
Peg 22h47'0"33d28'
Chamberlain,Peter, Wendy & Shady
Cyg 21h31'59"38d46'
Chamberlain,Philip
Her 17h12'45"40d35'
Chamberlain,Richard
Cet 2h51'0"2d1'
Chamberlain-Eadie, Louis
Her 18h18'59"18d52'
Chamberlaine,Alan L
Hya 8h31'1"-6d50'
Chamberland,Rita Montpas
Lyr 18h55'0"34d2'
Chamberlin,Robert Joseph Tracy
Her 18h16'0"31d44'
Chamberlin,Robert
Her 18h0'12"47d53'
Chamberline Family,The
Lac 22h11'0"50d5'
Chambers 12-8-80 Beloved Son,Paul Wm
Sgr 19h23'27"-40d10'
Chambers 6 (Ruth 6), Ruth Megargel
Peg 22h12'45"30d35'
Chambers Works
Cam 3h34'49"67d56'
Chambers,Anne Cox
Lac 22h35'47"54d21'
Chambers,Bobby T
Mon 7h4'40"-6d57'
Chambers,Colin Anthony
Mon 8h6'13"-1d37'
Chambers,Colin Anthony
Mon 8h6'13"-1d37'
Chambers,Donald E
Lac 22h10'25"51d41'
Chambers,Earl Wendell
Cet 2h29'44"4d51'
Chambers,G Ben
Cyg 20h21'52"39d28'
Chambers,Gerry
Lac 22h2'0"48d8'
Chambers,Janet
Eri 4h38'18"-1d45'
Chambers,John
Dra 13h14'0"68d5'
Chambers,Katherine Coley
Peg 21h57'35"29d40'
Chambers,Lambert
Peg 0h7'58"27d55'
Chambers,Lee & Chris
Cyg 20h27'0"31d14'
Chambers,Lisabeth Anne
Peg 21h39'53"25d37'
Chambers,Lynne M
Cyg 19h27'0"38d42'
Chambers,Mark
Aql 19h57'50"-0d54'
Chambers,Mark
Crt 11h52'43"-8d57'
Chambers,Melissa Jeanne
Cma 6h51'42"-16d22'
Chambers,Nicole
Uma 10h20'30"51d31'
Chambers,Patricia A
Cet 2h36'33"1d50'
Chambers,Rebecca Ellen
And 23h26'55"38d20'
Chambers,Robb
Cep 21h54'57"58d49'
Chambers,Robert
Aur 6h17'46"45d29'
Chambers,Rodger Alan
Cet 1h36'46"-14d5'
Chambers,Sheila Stramat
Cas 2h28'16"70d8'
Chambers,Steven G
Lib 15h2'18"-20d14'
Chambers,Tommy Cooksey
Cet 1h33'21"-2d3'
Chambers,Vera M
Cyg 20h16'33"38d45'
Chambers,William R
Uma 8h23'0"70d25'
Chamblee,Don
Vul 19h48'30"28d56'
Chamblee,Sr,Roland Wesley
Oph 17h13'49"-20d26'
Chambless,Destarata J
Lyr 19h21'0"40d6'
Chambliss,Don
Aur 5h11'44"42d1'
Chambolinias,TnT
Aql 18h49'36"11d21'
Chambon,Cary James
Lac 22h49'0"38d28'
Chambon,Nelly
Mon 7h49'41"-4d30'
Chambost,Lise
Cas 0h58'23"72d57'
Chamburs,Austin Robert
Aql 19h49'23"12d44'
Chamburs,Christopher John
Ser 15h58'21"0d44'
Chamburs,Crystal Lena
Peg 22h5'36"21d15'
Chamburs,David Bruce
Aql 19h50'51"12d37'
Chamburs,Joan Elizabeth
Aql 19h49'1"12d18'
Chamburs,Mark Shawn
Hya 8h44'24"-18d53'
Chamburs,Patricia
Oph 17h56'16"12d28'
Chamburs,Patrick Murphy
Hya 8h56'36"-11d14'
Chamburs,Philip Michael
Sex 9h54'34"1d12'
Chamburs,Robert Elliot
Cmi 7h54'19"0d5'
Chamburs,Tami Kuroki
Cnv 12h9'48"37d23'
Chameleon,The
Ori 5h56'39"13d21'
Chametzki,Sally
Cam 5h41'1"78d59'
Chaminade,René
Cep 22h15'53"58d28'
Chamlin,Wyatt Ross
Aur 7h24'41"40d23'
Chamness,Lt Corey D
Lyr 18h35'25"38d24'
Chamorro Spirit,The
Cra 18h15'25"-39d37'
Chamoun,Abraham
Aql 18h43'35"8d50'
Champ
Her 16h56'21"28d14'
Champ,Thomas William
Her 16h40'17"48d9'
Champagne Cork Farm Star
Crb 16h9'37"30d27'
Champagne Sam
Umi 15h8'19"67d45'
Champagne,Lincoln Scott
Lib 15h38'58"-28d17'
Champagne,Marc David De Napoli
Aur 5h1'10"41d40'
Champagne,Matthew David Denapoli
Per 2h48'55"45d28'
Champagne,Nadia Isabelle
Cas 22h56'39"54d12'
Champagne,Nathalie
Sco 16h34'53"-26d11'
Champagne,Richard
Uma 9h19'48"43d57'
Champion
Per 4h34'22"37d47'
Champion Records
Cet 1h2'46"-18d14'
Champion,Sam
Vul 19h22'20"25d17'
Champion,Skip
Aql 19h7'13"5d6'
Champlin,Jill
Mon 7h43'53"-5d34'
Champniss,Wolfie
Cnv 12h25'1"51d22'
Chamson,Eugene Charles
Ser 15h12'60"8d38'
Chan (Yan Yan),Jenny
And 1h46'47"38d4'
Chan's Star,Patsy
Lyn 7h39'27"51d0'
Chan,Bosco
Aql 19h8'31"1d57'
Chan,Carol
Boo 15h16'16"48d2'
Chan,Charles
Cam 3h25'0"59d46'B
Chan,Chi Kin
Aur 5h57'0"31d13'
Chan,David Katie Eric & Kim
Uma 12h45'27"58d49'
Chan,Erin Christie
Lyn 9h16'14"35d8'
Chan,Gloria
Cma 6h28'35"-24d37'
Chan,Harpo
Lib 15h3'34"-28d25'
Chan,Inez M
Sex 10h1'11"5d48'
Chan,Jackie
Her 16h14'17"8d38'
Chan,Jenny
Cru 12h38'22"-60d2'
Chan,Jessica
And 1h15'20"33d29'
Chan,Jillian
Vul 20h18'0"28d42'
Chan,Kenny
Aur 6h5'1"30d46'
Chan,Lawrence Walter
Cep 20h17'55"75d14'
Chan,May
Crb 16h22'1"37d39'
Chan,Richard
Her 16h58'28"24d47'
Chan,Sildic
Cyg 21h42'0"28d21'
Chan,Stella of Akko
Vir 14h25'10"6d29'
Chan,Willie
Crb 15h18'47"30d50'
Chan-Pasteur,Samuel- Yexin
Dra 16h9'0"62d41'
Chanboon
Lyn 7h49'36"51d39'
Chance
Sco 16h53'49"-41d4'
Chance
Her 17h6'6"31d16'A
Chance
Aql 19h12'15"12d18'
Chance,Debbie
Lyr 18h22'37"42d44'
Chance,John Eugene
Hya 8h17'55"5d40'
Chance,Larry
Her 16h52'36"39d58'
Chancellor Academy
Boo 15h6'15"40d13'
Chancey,Commander
Per 1h50'21"54d12'
Chancey,Heather
Mon 7h24'55"-8d2'
Chancler,Ronald Thomas
Aqr 22h29'58"-6d4'
Chanda
Cmi 8h5'43"6d38'
Chandlee III,William H
Dra 12h2'31"70d25'
Chandlee,Chad
Cam 5h41'15"61d28'
Chandler
Cet 0h9'41"-12d25'
Chandler 27/12/83-23/9/93,Mark
Cyg 19h33'16"35d11'
Chandler Golden Guardian
Cam 5h8'32"60d17'
Chandler's Birthday Star
Cyg 20h53'45"30d29'
Chandler's Princess
And 0h55'57"40d50'
Chandler,Abigail Elizabeth
And 23h1'0"51d7'
Chandler,Blake
Sex 10h2'13"-1d26'
Chandler,Caryle Elizabeth
Del 20h21'39"10d47'
Chandler,Daniel Christian
Uma 9h46'44"71d59'
Chandler,Darrell B
Mon 7h33'60"-1d42'
Chandler,Dori Ray
Ori 5h57'32"15d2'
Chandler,Isabelle Bodkin
Mon 7h50'8"-3d57'
Chandler,Jennifer Julene
Lyn 7h53'51"44d19'
Chandler,Jessica Lynn
Cam 8h6'36"83d47'
Chandler,Jillian Grace
Eri 3h6'22"-3d23'
Chandler,Joshua Adam
Boo 14h57'1"33d2'
Chandler,Kimberly M
Peg 22h28'53"24d14'
Chandler,Patricia McMillian
Peg 22h44'0"29d29'
Chandler,Robyn Lorraine
Aql 19h1'51"16d4'
Chandler,Shellie Rene
Lyn 7h28'57"45d27'
Chandler,Stan
Cnv 12h54'18"38d18'
Chandler,Tallow
Aur 6h27'16"38d39'
Chandler,Victoria Lynn
Lac 22h29'59"48d53'A
Chandlers,Miles Howard
Cnv 13h11'32"40d49'
Chandra
Cas 0h6'28"58d20'
Chandra's Rising
Lmi 9h42'40"33d49'
Chandra,Steve's Shining Star
Sge 19h59'28"18d53'
Chandra-Michael
Psc 23h29'55"6d36'
Chandrima
Cas 2h34'43"68d15'
Chaneese
Cyg 20h13'32"30d20'
Chanel B
Com 13h27'1"25d39'
Chanel,Kim
Cas 0h50'43"71d9'
Chanelle
Cyg 19h26'39"33d16'
Chanen,Lauren Francine
Lyr 19h23'1"35d2'
Chaney Enterprise
Cep 23h0'14"60d8'
Chaney,C Virginia
Crb 16h17'35"34d29'
Chaney,Daniel J
Her 16h37'14"32d5'
Chaney,Danielle
Hya 9h16'37"-16d35'B
Chaney,Dave
Cep 23h10'47"70d13'
Chaney,Emily Carol
Cyg 20h53'33"30d52'
Chaney,James Edward
Aur 6h34'24"37d43'
Chaney,Jeanene
Ori 5h34'32"0d46'
Chaney,Kyle Adam
Aur 6h28'47"31d14'
Chaney,L Cecil & Bessie T
Eri 4h34'42"-11d28'
Chaney,Matthew Peter
Uma 12h8'58"53d42'B
Chaney,Michael
Hya 9h16'37"-16d35'A
Chaney,Robert Charles
Uma 12h8'58"53d42'A
Chaney,Robert G
Her 17h56'0"38d25'
Chaney,Sita
Uma 9h42'20"48d46'
Chaney,Stephen Lee
Hya 8h55'51"5d11'
Chaney-JLNS,Deahn Debs
Del 20h23'35"11d46'
Chang,Cindy
Cas 1h5'1"66d30'
Chang,Dodi
Uma 10h20'58"50d11'
Chang,Felix
Cyg 19h41'55"41d48'
Chang,Sonia Rosa
Cas 0h11'31"62d34'
Chang,Tinjin P
Lac 22h29'57"48d58'
Changabanga
Ori 5h35'1"-6d14'
Changes In Attitudes Changes In Latitudes
Lac 22h5'30"50d35'
Changing Channels
Uma 9h0'35"56d12'
Chani
Cyg 21h23'34"28d34'
Chanilla
Mon 6h19'50"8d6'
Chaniot,Stephane
And 23h41'22"38d7'
Chanis,Barbara Ann
Cas 1h37'39"65d14'
Channa
Ori 4h48'0"0d14'
Channey Star,The
Vul 20h16'17"23d2'
Channon L
Sex 10h42'58"-6d47'
Channon,Michael
Ori 6h3'33"1d31'
Chanslor,"Big John"
Cet 1h2'44"-5d50'
Chanson,Franáois
Peg 23h35'19"10d15'
Chant,Amanda Jayne
Cnc 8h32'27"11d0'
Chant,Julie
Cam 3h51'22"60d58'
Chant,Nik
Aur 6h35'14"34d4'
Chantal
Cyg 20h54'38"50d52'A
Chantal
Peg 21h41'50"21d49'
Chantal
Lep 5h21'45"-25d17'
Chantal
Dra 16h50'54"70d21'
Chantal
Cam 9h19'55"84d53'
Chantel
Cep 0h49'12"77d36'
Chantel
Sct 18h20'17"-4d4'
Chantelle's Peace Of Heaven
Lyr 18h35'57"28d53'
Chanteloup,Stéphane
Lyn 9h4'15"40d40'
Chantrakarn 25
Mon 6h57'39"1d48'
Chanut,Pierre
Lyn 9h6'28"38d11'
Chaouy
Peg 23h24'27"13d45'
Chapa,Juan
Ser 15h23'8"14d20'
Chapaitis,Lisa A
Cyg 21h34'1"34d25'
Chaparral's Own
Sct 18h43'49"-7d16'
Chapel Star,Bob
Her 16h59'1"26d23'
Chapel,Christopher T
Psc 1h24'48"18d48'
Chapel,George
Dra 17h57'44"75d41'
Chapell,Karen
Cas 1h9'31"63d10'
Chapellaubeau,Martine
Aql 19h30'27"-10d21'
Chapiewsky,Gary
Aur 6h23'1"30d34'
Chapin III,Walter R
Aur 6h14'56"37d52'
Chapin,Elizabeth Steinway
And 23h34'57"38d33'
Chapin,Robert Francis
Uma 12h13'34"56d18'
Chapinamoureuse
Cnv 12h12'25"40d1'
Chapis
Uma 8h43'53"50d17'
Chapkin,Ryan Thomas
Aur 7h22'1"36d14'
Chaplic,Morey
Aur 5h0'1"47d9'
Chaplin's Star,Carrie
Leo 9h41'43"33d10'
Chaplin,Deanna
Sco 17h55'58"-37d55'
Chaplin,Elizabeth
Aql 20h17'22"0d8'
Chaplin,Esta & George
Crb 16h19'25"32d3'
Chaplin,Ruth
Lyr 18h31'1"32d7'
Chaplin,Stephen J
Sct 18h42'40"-6d42'
Chaplow,Mary Ann
Ori 5h51'25"9d41'
Chapman "Nana", Margaret Nolan
Cas 0h41'30"73d49'
Chapman Folly
Cyg 19h33'14"37d49'
Chapman School Class of 2002
Cam 3h40'48"70d24'
Chapman's Advantage
Aql 19h0'28"10d57'
Chapman's Silver Sparkler,John Samuel
Boo 15h3'1"20d59'
Chapman's Star,Heather Lyn
Cyg 21h7'60"31d27'
Chapman,Betsy Miller
Cam 5h3'58"60d36'
Chapman,Christopher
Ori 5h56'0"8d55'
Chapman,Daniel Richard
Lac 22h20'1"40d10'
Chapman,Daril
Cyg 20h53'15"37d42'
Chapman,Dennise
And 1h59'0"47d5'
Chapman,Don
Ser 15h30'45"19d38'
Chapman,Donna Y
Cyg 20h29'51"38d51'
Chapman,EdD,James F
Boo 15h2'28"30d28'
Chapman,Emily Anne
Sgr 18h48'4"-27d58'
Chapman,Erica Nicole
Sco 16h52'25"-40d7'
Chapman,George Victor
Cam 14h12'42"80d6'
Chapman,Holly Nicole
Lyr 18h46'31"33d3'
Chapman,Howard R
Dra 15h14'50"61d12'
Chapman,J B
Ori 4h57'29"-0d25'
Chapman,Jack
Uma 8h13'1"60d33'
Chapman,Jacquelin Santin
And 0h13'59"35d14'
Chapman,Jane B
Lyn 7h43'38"43d5'
Chapman,Jay
Her 18h10'32"47d22'
Chapman,Jennifer Graham
Mon 8h6'46"-1d40'
Chapman,John L
Lyn 7h32'1"41d21'
Chapman,Joseph T
Cep 0h28'23"78d25'
Chapman,Karen
Tau 5h46'42"28d45'
Chapman,Lee
Per 1h54'50"56d49'
Chapman,Maria & Gary
Vul 19h48'16"28d34'
Chapman,Melinda Middlebrook
Cmi 7h35'17"10d8'
Chapman,Melissa Shay
Sex 10h2'14"5d4'
Chapman,Pat
Aur 5h8'0"40d28'
Chapman,Peter Frederick
Pho 0h43'20"-44d52'
Chapman,Randall Lee
Mon 8h4'26"-7d20'
Chapman,Ray
Cam 3h56'55"57d24'
Chapman,Robert Wayne
Aur 5h4'27"40d50'
Chapman,Ronald Paul
Oph 17h16'42"-22d4'
Chapman,Sadie Louise
Gem 7h36'11"33d48'
Chapman,Sarah Rachel
And 0h36'21"40d19'
Chapman,Selwyn Robert
Cep 21h49'40"61d37'
Chapman,Van B
Cet 3h14'20"8d54'
Chapman,William Davis
Cet 2h23'41"1d8'
Chapman-Damphousse
Sex 10h43'58"3d15'
Chapot,Alain
Dra 12h6'18"70d17'
Chapoton,Shirley
Cyg 20h38'52"38d6'
Chappel,Jeff
Cep 22h8'1"53d29'
Chappelet,Jeanine
Crt 11h9'60"-10d10'
Chappell,Crystal
Lyr 18h39'32"35d21'
Chappell,Dr James C
Boo 14h41'31"50d11'
Chappell,J C
Her 15h56'17"41d20'
Chappell,Lacie Elizabeth
Mon 6h19'54"2d31'
Chappell,Laura Rhiannon
Aql 19h6'15"0d44'
Chappelle,Marion
Ser 16h1'38"10d24'
Chappuis,Irene
Lyn 8h7'39"52d4'
Chappy
Dra 16h56'1"71d36'
Chaps,Pat
Her 18h34'30"22d2'B

Chapuis,Ralph Adam
Boo 15h20'45"38d30'
Chaput,Al
Aur 5h13'58"41d22'
Chaput,Nicole
Uma 12h4'28"46d15'
Char's Star
Hya 8h42'14"4d17'
Char's Sunshine Star
Peg 22h1'37"26d36'
Char,Gary Gori
Vul 19h17'21"24d52'
Char-Star
Cas 1h52'0"58d36"
Chara
Del 20h13'10"10d5'
Chara,Eileen A
Cas 0h47'10"61d28'
Charabuska,Alexis
Cam 3h29'29"61d13'
Charade,Pierre
Per 3h11'22"49d18'
Charalambous,Stelios
Aur 5h51'36"41d14'
Charama,Mena Rosa
Boo 14h11'54"50d56'
Charann
Cep 22h32'60"60d50'
Charap,Samuel Gilmore
Aur 5h53'11"38d47'
Charapata,Sharon L
And 23h6'43"43d31'
Charb
Uma 9h55'0"55d42'
Charban,Joel
Her 16h40'11"35d7'
Charbet
Ori 4h43'49"0d9'
Charbie
Her 17h13'30"27d49'
Charbonneau,Nathan Henry
Aur 5h55'26"31d21'
Charbonnier,Ellen Sue
Cyg 20h44'48"45d54'
Charbonnier,Philippe
Uma 12h57'0"59d26'
Charden Baptiste
Cam 5h44'54"67d34'
Chardon,Linda Lee
Bonsignore
Cas 1h39'0"76d41'
Chardy
Oph 18h2'53"8d30'
Charek,Monica
And 2h21'39"44d48'
Charek,Nicholas
Aur 4h58'30"38d31'
Charest,Marie-Eve
Cam 3h35'1"63d11'
Charette,Danielle
Vir 13h56'29"-5d52'
Charger,Dianne's
Mon 6h23'35"4d52'
Chari Clothilde
Lmi 10h47'36"39d28'
Chari Lynn's Star
Peg 22h10'1"20d56'
Charibren
Boo 14h22'17"50d19'
Charina CB II- Sunnywarm
Ori 6h0'18"-0d20'
Charinander
Lyr 18h29'15"31d41'
Charity's Everlasting Trinity
Cas 0h39'0"73d11'
Charko,Dennis R
Ori 4h46'22"5d12'
Charla & Paul Forever
Lyr 18h40'43"39d12'
Charla Rae
Lyr 18h13'33"38d11'
Charland,Aubrey Loreine
Del 20h14'16"15d2'
Charland,Christopher Lewis
Aur 5h0'48"50d25'
Charland,Éric
Ori 5h57'0"14d2'
Charlap-Evans, Valentina
And 1h2'37"47d47'
Charlapararious,E
Uma 9h31'35"48d17'
Charlene
Uma 8h23'38"70d4'
Charlene
Cam 5h8'48"69d46'B
Charlene
Vir 11h37'42"7d35'
Charlene
And 1h29'24"38d54'
Charlene
Peg 21h39'0"27d8'
Charlene 04/25/69
Vul 20h1'54"23d9'
Charlene B
Leo 11h32'31"11d26'
Charlene Marie
Uma 9h49'25"58d39'
Charlene's Unspeakable Star
Ori 6h8'42"6d46'
Charleroy
Cam 6h19'37"65d5'
Charles
Lmi 9h52'51"34d31'
Charles
Boo 14h38'5"22d11'B
Charles
Peg 22h49'2"26d7'B
Charles
Dra 14h55'0"62d58'
Charles
Lyn 7h46'59"48d41'
Charles
Boo 14h44'13"27d42'
Charles
Her 18h2'43"37d54'
Charles
Pic 5h6'27"-46d20'
Charles & Dora's Star
Uma 8h57'20"56d19'
Charles & Karla Forever
Cyg 20h21'15"41d57'
Charles & Susan
Cyg 21h38'37"29d53'
Charles Christian
Her 17h34'47"28d10'
Charles Elliot
Cep 22h42'38"56d57'
Charles House
Cmi 7h15'39"5d19'
Charles Lee
Aur 5h14'42"40d27'
Charles Loves Patsy
Cyg 20h11'31"58d54'
Charles Neal
Aql 20h10'26"3d46'
Charles Paul
Ser 15h56'18"0d33'
Charles Robert Christopher
Peg 22h28'45"12d16'
Charles Scott
Cep 3h23'21"80d18'
Charles William"Bill Tatum"
Ser 15h53'1"18d6'
Charles' Treat
Sct 18h42'46"-6d37'
Charles,Andrew Martin
Uma 11h21'40"55d51'
Charles,Angela B
Com 12h14'20"22d21'
Charles,Bonnie
Lyr 18h43'17"40d37'
Charles,JoAnna
Mon 6h58'20"8d57'
Charles,John
Del 20h52'41"3d26'
Charles,Kathy Zsak
And 23h45'50"45d33'
Charles,Kerry Ann
Cyg 20h0'52"40d39'
Charles,Sarah Louise
And 0h36'1"31d24'
Charles-94,Ner Tamid
Umi 15h28'14"71d8'
Charles-Fay1995
Ori 6h2'23"6d27'
Charlestein,Jordan
Per 2h56'1"50d5'
Charley,Jeremy David
Uma 9h50'16"48d28'
Charlie
Lac 22h36'55"53d40'
Charlie
Peg 23h34'32"20d4'
Charlie
Per 2h59'33"43d56'
Charlie
Cam 7h53'17"70d10'
Charlie
Her 18h15'24"48d12'
Charlie
Her 16h6'32"40d41'
Charlie
Aur 6h56'13"38d36'
Charlie
Uma 9h14'39"70d57'
Charlie
Vul 19h22'41"26d47'
Charlie
Cma 6h14'11"-16d38'
Charlie
Leo 11h4'19"-0d42'
Charlie
Hya 8h54'17"5d55'
Charlie & Cheryl's "Fatum"
Cyg 21h30'10"38d12'
Charlie & Dollink
Ori 5h57'38"11d40'
Charlie & His Angels
Per 4h43'23"38d12'
Charlie & Martelle's Glow
Forever
Crb 15h31'11"30d32'
Charlie & Mary
Crt 11h8'7"-14d35'
Charlie 'N Betty
Sge 19h56'0"20d26'
Charlie 1916
Aql 20h5'44"0d49'
Charlie Alberto
Ori 5h5'40"14d37'
Charlie B
Ori 5h58'11"10d2'
Charlie D A F Michele
Uma 9h5'47"56d51'
Charlie Darlin'
Cnv 12h46'25"50d40'
Charlie G,The
Leo 9h32'10"8d54'
Charlie L M P
Peg 0h6'24"13d49'
Charlie Maureen Moe
Cyg 20h25'1"42d32'
Charlie The Whiz
Umi 15h34'46"68d16'
Charlie's Nexus
Cnv 13h2'19"32d6'
Charlie's Quasar
Per 3h39'18"32d46'B
Charlie's Spontaneity
Sge 19h29'51"16d26'
Charlie's Star
Her 17h17'18"48d21'
Charlie's Star My Eternal Gift
Aur 4h52'30"40d52'
Charlie,Marcia,Marisa,
Monica,Anthony
Umi 15h25'48"78d53'
Charlie-O
Per 2h7'48"57d29'
Charlillo,Samantha Marie
And 23h3'28"48d20'
Charline
Vul 20h15'1"22d59'
Charline
And 2h32'53"44d47'
Charline
Dra 9h30'45"73d41'
Charlmona-Hill
Her 17h55'56"14d60'
Charlo
Cas 1h24'58"50d5'
Charlot
Ori 5h56'20"16d13'
Charlotte
Mon 7h39'37"-1d25'
Charlotte
Tau 4h27'0"28d37'
Charlotte
Cap 21h24'19"-20d23'
Charlotte
Cas 0h44'48"73d24'
Charlotte
Aur 5h17'38"47d12'
Charlotte
Cas 0h53'36"58d38'
Charlotte
Peg 22h25'12"31d51'
Charlotte
Cas 23h2'31"58d9'
Charlotte
Sgr 19h29'4"-31d10'
Charlotte
Uma 13h13'34"62d18'
Charlotte
Aqr 22h52'57"-4d8'
Charlotte
Cnc 8h32'1"31d58'
Charlotte & John
Aur 5h4'14"29d57'
Charlotte & Richard Forever
Cyg 19h34'30"34d30'
Charlotte 30
Lyr 18h43'12"40d6'
Charlotte Belle of Beckley
Cas 1h3'57"67d43'
Charlotte Catherine Analiese
Lyr 18h35'49"28d21'
Charlotte Clark Kirsten's
Wishing Star
Cas 1h45'22"60d50'
Charlotte Elissa
Umi 15h13'11"66d54'
Charlotte et Georges
Cyg 19h47'0"30d37'
Charlotte Georgia
Cas 22h57'48"55d43'
Charlotte Grace
Lyr 18h16'32"34d51'
Charlotte Isabel
Cas 0h35'33"58d29'
Charlotte Louise
And 23h10'38"38d40'
Charlotte Rose
And 0h22'22"36d57'
Charlotte Sara
Equ 21h19'21"12d35'
Charlotte Starlet
Ori 5h55'42"16d47'
Charlotte W
Cas 0h49'1"60d8'
Charlotte's Anniversary Star
Crb 15h18'41"30d45'
Charlotte's Dream
Vul 19h39'24"20d26'
Charlotte's Light
Cas 1h2'39"55d21'
Charlotte's Star
Aql 19h1'0"16d24'
Charlotte's Star
Cas 0h37'30"60d14'
Charlotte's Star
Lyr 18h32'58"33d16'
Charlotte's Tap Shoes
Com 13h25'22"25d19'
Charlotte's Valentine '96
And 1h51'0"39d24'
Charlson,Buck
Dra 11h42'29"71d21'
Charlton's Delta Romeo
Ori 6h12'40"8d8'
Charlton,Alex Joseph Reese
Hya 10h12'29"-16d3'
Charlton,Ann
Dra 17h59'12"52d51'B
Charlton,Ann
And 2h20'58"38d40'
Charlton,Miles Joseph
Ori 5h59'34"15d53'
Charlton,Scott
Dra 17h59'12"52d51'A
Charlwood,Bianca Maria
Sebesco
Cas 23h38'47"50d35'
Charlwood,Geoff
Uma 8h52'27"61d55'
Charly 2 Alpha
Aqr 23h31'30"-10d14'
Charly I
Leo 11h24'47"10d57'
Charlyle
Del 20h14'28"10d0'
Charm,Dr Robert
Oph 16h59'20"7d52'
Charmain,Tasia
Cyg 20h39'47"40d5'
Charmaine
And 23h18'35"50d1'
Charmaine
Lyn 18h20'11"13d5'
Charmaine
Uma 11h25'16"32d50'
Charmaine & Steven Forever
Lyn 7h51'44"58d40'
Charmante Cailey
Cas 2h49'16"71d1'
Charming,Pammy
Cet 2h14'12"2d43'
Charnetsky,Lory
Mon 7h4'56"-5d42'
Charney Star,The
Ser 15h27'38"0d4'
Charnick,Skippy
Cma 7h24'54"-17d45'A
Charnick,Steven
Cma 7h24'54"-17d45'B
Charnley,Gemma Lauren
And 0h12'43"38d3'
Charnley,Keith
Umi 14h19'33"67d46'
Charno,Eddie
Cep 20h37'23"65d10'
Charol
Tau 4h37'16"18d49'
Charon,Valerie
Her 16h12'46"50d25'
Charos,Juan Gomez
Her 17h24'48"21d49'
Charpentier,Laurence
And 1h4'60"41d7'
Charpentier,Phillippe
Dra 11h33'31"68d33'
Charpy,Marc
Lac 22h5'42"38d55'
Charpy,Mousieur Franáois
And 0h26'20"44d36'
Charriez,Iris M
Sgr 18h55'8"-34d54'
Charrin,Christian Jean
Her 17h15'17"44d39'
Charrois,Richard
Cep 22h33'59"80d5'
Charron,Alain
Umi 13h39'0"73d48'
Charron,Anthony
Ori 6h0'31"7d55'
Charron,Brek Allen
Ori 5h59'0"16d11'
Charron,Eric
Per 3h32'18"36d18'
Charron,Hali
Tau 4h39'39"20d7'
Chartier Valérie
Aur 5h52'18"31d7'
Chartier,Armand
Cep 21h42'18"56d14'
Chartier,Cathleen "Cat"
Cas 1h26'10"50d11'
Chartier,Laurent
Aur 5h51'44"38d38'
Chartier,Mary
Leo 10h50'16"7d33'
Chartier,Michael Arthur
Aql 18h58'16"13d35'
Chartier,Philippe
Umi 15h42'20"81d28'
Chartier,Yvette K
Peg 22h13'36"5d46'
Chartier-Samson
Ori 5h57'11"14d42'
Chartoff,Maria
Uma 8h49'22"60d56'
Chartrand,Jean R
Dra 14h40'48"61d28'B
Chartrand,Jeri
Lyr 18h33'22"35d59'
Charwenske,Heather
And 0h21'37"35d12'
Charwillyoumarryme?
Cyg 21h31'49"38d4'
Charydriselle
Cas 0h18'59"62d18'
CHAS
And 1h59'33"41d7'
Chas-Iopeia
Cas 23h39'46"65d43'
Chasam
Per 2h55'21"34d56'
Chasan,Roslyn & Fred
Vul 19h48'20"23d26'
Chase
Sex 10h21'48"-5d32'
Chase
Mon 6h27'58"-6d57'
Chase
Lac 22h44'38"53d7'
Chase Chad
Oph 18h0'11"12d28'
Chase's Mom Heather
Cyg 10h48'39"30d15'
Chase,Barry
Dra 19h5'1"61d3'
Chase,Cheryl
Crb 16h10'34"31d19'
Chase,Christopher John
Lac 22h5'52"38d3'
Chase,Courtney
Peg 22h57'0"21d48'
Chase,David Scott
Aur 6h15'11"46d51'
Chase,Debra L K
Cyg 19h28'24"38d37'
Chase,Dorothy
Uma 13h27'36"62d3'
Chase,Ev
Mon 6h54'23"-4d53'
Chase,Florence
Cma 7h48'51"10d2'
Chase,Geoff
Ori 5h25'13"1d7'
Chase,Geoff
Ori 5h25'13"1d7'
Chase,James Arthur
Mon 7h56'39"-6d47'
Chase,John Robinson
Aql 20h33'26"0d59'
Chase,John Thomas
Aur 6h4'37"38d30'
Chase,Jr,Harry E
Aur 5h49'22"54d60'
Chase,Lindsay Ann
Mon 6h44'1"7d4'
Chase,Michael
Her 17h12'18"48d21'
Chase,Raymond J
Her 17h24'23"41d6'
Chase,Sheryl
Eri 3h33'18"-6d4'
Chase,Thomas
Lac 22h13'48"46d38'
Chase,Wally
Eri 3h54'12"-1d59'
Chasen,Emily Londner
And 0h2'50"35d5'
Chasing Hawk
Aql 20h3'21"1d14'
Chasity
Psc 0h56'1"32d28'
Chaslard,Dominique
Peg 23h4'43"8d34'
Chasse,Justin
Dra 14h56'23"63d0'
Chassin,Norma Strand
Lyr 19h2'44"46d59'
Chassé,Rita
Cyg 21h35'28"42d30'
Chastain,Breann Kelly
Peg 22h36'15"8d41'
Chastain,Jr,Homer Joseph
Crt 11h11'56"-19d48'
Chastain,Mattie
Aql 18h53'49"11d9'
Chastain-Shanor,Fritzi
Lmi 10h17'29"28d30'
Chasteen II,Henry Clay
Ori 6h5'1"4d28'
Chasteen,Christopher
Uma 11h11'1"46d46'
Chatard,Karine Michele
Daniele
Peg 22h43'1"26d14'
Chateau de St Baslemont
Per 3h43'7"37d40'
Chateau Hornchurch
Dra 16h20'37"61d44'
Chateau J R
Leo 9h19'12"14d55'
Chatel
Peg 23h28'13"20d3'
Chatel,Brittney Nichole
Peg 23h1'33"33d51'
Chatel,Stephane
Sct 18h45'41"-6d7'
Chatelain,Anne
Cet 0h55'12"0d56'
Chatelain,Erin Stewart
Lyr 18h41'52"42d47'
Chatelain,Philippe
Ser 15h13'35"8d35'
Chatenoud,Céline
Uma 8h31'1"51d32'
Chatfield,Col Williams E
Her 17h52'40"50d21'
Chatfield,Ken
Per 1h31'1"54d3'
Chattell,Heather
Cyg 19h26'1"35d30'
Chatten,Calvin Robert
Her 16h40'44"7d39'
Chau Van To
Cep 20h47'40"61d48'
Chaudhuri,Piyali
Cap 20h33'16"10d53'
Chauffeur's Star
Lac 22h9'32"48d1'
Chauhan,Ajay Kumar Singh
Lyn 7h25'11"58d43'
Chaumayrac DSG
Uma 13h7'18"62d42'
Chaumette,Franáois
Aur 6h1'17"38d7'
Chaumette,Vroni
Cyg 21h20'49"41d6'
Chaump,Louise
Ori 5h7'35"-2d51'
Chauncey
Umi 14h48'24"68d53'
Chauncey,Eliza Parish
Aql 19h53'31"14d15'
Chauncey,John Michael
Aql 18h52'60"11d31'
Chauncey,Robin
Cyg 20h52'54"37d58'
Chauncey,Sam
Boo 15h19'12"50d20'
Chaussee of Montana, Wil
Aur 6h56'35"40d25'
Chaussignand,Eric
Sgr 20h17'31"-37d0'
Chausson,Roger
Dra 15h5'51"64d40'
Chauvin,Jean-Louis
Umi 15h43'27"81d52'
Chauvin,Jean-Louis
Cas 23h1'46"53d28'
Chauvot,Christian
Dra 14h24'44"64d40'
Chauvot,Roger
Ori 6h5'0"11d0'
Chava
Peg 0h9'21"17d52'
Chavan,Karen T
Lyr 18h21'48"40d33'
Chavarria,Roxanne
Mon 6h42'53"-10d30'
Chavel,Pauline Claire
Cnc 8h52'48"31d48'
Chavenello,Peter Link
Dra 19h3'0"48d38'
Chavenson,Emma Haviland
And 23h21'26"38d46'
Chaves,Jack J
Her 17h52'51"14d51'
Chaves,Kevin Kirts
Oph 17h30'52"-22d47'
Chaves,Kimberly
Cyg 20h7'43"38d47'
Chaves,Linda F
Cap 21h42'24"-23d4'
Chavez,Araceli
Peg 21h28'0"20d9'
Chavez,Carolyn
Cas 23h30'56"61d40'
Chavez,Cruz
Ser 18h15'41"-14d60'
Chavez,David A
Her 17h54'56"14d51'
Chavez,Erick
Lmi 10h39'56"25d6'
Chavez,Eva L
Sct 18h47'23"-8d15'
Chavez,Gerard Atristain
Eri 3h41'43"-18d42'
Chavez,Hugo
Ser 15h42'60"17d25'
Chavez,Jerry S
Boo 15h10'27"31d58'
Chavez,Maria
Cyg 20h39'0"39d37'
Chavez,Maria Isabel
And 22h58'26"50d32'
Chavez,Patricia Villarreal
Cap 21h12'41"-22d0'
Chavez,Scorpio 30 Frank E
Sco 16h33'18"-41d7'
Chavez,Selena Ann
And 23h31'0"42d11'
Chavez,Zackary T
Uma 9h26'40"61d47'
Chavira,Sylvia Veronica
Mon 6h21'0"7d34'
Chavis,Joe E
Aql 19h6'55"3d35'
Chavmaz,Genevieve Et Jean-
Pierre
Aur 7h7'0"38d31'
Chavoya,Dr Gary M
Aur 6h33'0"37d2'
Chaylak,Turhan
Cam 4h20'1"70d55'
Chayse
Cet 1h23'39"-12d18'
Chaz
Cet 1h20'0"-0d49'
Chaz
Uma 10h0'57"55d31'
CHAZ*GER
Uma 13h56'29"51d35'
Chazach
Uma 11h16'26"68d16'
Chazen,Laurie
Del 20h38'52"18d49'
Chazen,Simona & Jerry
Crb 16h14'53"28d50'
Chazin,Steve M
Cnv 12h10'23"36d6'
Che
Cyg 20h39'49"45d25'
Che
Lyn 9h3'18"41d51'
Cheatham,Aaron Michael
Ser 15h25'32"21d7'
Cheatham,Chuck
Hya 8h30'26"5d54'
Cheatham,Mary
Aur 7h26'0"40d48'
Cheatham,Robert
Hya 8h30'50"-0d40'
Cheatham,Sheila & Deane
Mon 7h57'36"-8d25'
Cheatley,Gary Gene
Her 16h25'1"38d57'
Cheatum,T R
Cep 23h23'34"65d19'
Checchio,David Anthony
Lac 22h4'38"38d6'
Checcucci,Danilo
Her 17h35'43"38d39'
Chechik,Lisa
Lyn 7h35'1"50d24'
Chechik,Lucille Anna
Peg 22h28'1"24d20'
Check,Benjamin Michael
Her 16h40'33"21d49'
Check,Joey
Lac 22h47'40"54d16'
Checker-Poom
Tau 3h53'34"2d15'
Checkley,Loretta Ott
And 23h47'43"41d7'
Cheda,Cody
Peg 23h23'17"15d52'
Chedekel,Evie
Umi 4h42'50"89d38'
Chee,Ronnie James
Tri 2h41'42"34d2'
Chee-Leung,Frederick Wai
Umi 16h56'21"76d59'
Cheech
Cep 22h24'54"58d57'
Cheek to Cheek
Aql 19h0'20"11d6'
Cheek to Cheek
Ser 16h4'58"8d40'
Cheek,Clifford Sammie
Aqr 22h11'22"0d57'
Cheek,Courtney Jeneé
Cyg 20h26'23"40d2'
Cheek,Eddie
Del 20h31'2"5d16'B
Cheek,Fiona Helen
Cas 0h20'0"61d32'
Cheek,Katherine Mary
Richardson
Cam 7h53'0"60d27'
Cheek,Michelle Nicole
Cyg 20h40'24"37d58'
Cheek,Nancy
Peg 0h12'30"13d15'
Cheek,Pamela Ann
Cyg 19h35'59"28d13'
Cheek,Sr,Thomas Comer
Aql 20h17'26"5d21'
Cheek,Sueleen Renee
Mon 7h47'53"-2d51'
Cheek,Virgil
Oph 17h3'48"11d11'
Cheeky Cosmic Babe
Cyg 19h55'45"45d38'
Cheer Up
Cnv 13h48'1"40d52'
Cheers
Aur 5h33'20"48d57'

Cheeseman,Gary
Aur 5h3'10"45d12'
Cheeseman-Donelly, Giselle Moss
And 0h57'20"37d41'
Cheesman,Clare
Cas 0h59'1"58d24'
Cheetham,Alexandra Claire
And 0h24'39"36d40'
Cheetham,Colin
Ori 5h54'37"20d37'
Cheetham,John Paul
Uma 8h25'30"61d42'
Cheever,Jack & Doris
Tri 1h56'14"27d5'
Chef Boy RD
Cnc 8h10'54"30d48'
Chefarzt Dr Med Leon Brumen
Tau 5h46'0"26d30'
Cheklich,Diane Carol
Gem 7h2'19"21d29'
Chela
Uma 10h31'0"50d25'
Chele
Lyn 8h57'48"36d43'
Chele-John
Aql 19h3'46"5d26'
Cheleeee 1
Lyn 9h9'52"36d2'
Chelgren,Valerie Brighton Casler
And 2h24'44"42d37'
Chell,John
Hya 8h12'42"0d58'
Chella
Sct 18h55'16"-8d52'
Chelle
Uma 9h53'37"42d13'
Chelle-Shine
Cyg 19h26'19"32d5'
Chellee
Cyg 21h28'29"40d26'
Chelliah,Aaron
Aql 19h18'57"10d10'
Chelomia Mydas Omnis Matris
Aqr 21h0'33"-0d44'
Chelpon,Ekaterina Theodora
Peg 21h53'0"34d11'
Chelpon,Mikhail Pavlos
Ori 5h59'14"14d48'
Chels-lan-Taran,The
Uma 11h56'0"30d30'
Chelsa Dawn
Eri 2h56'51"-6d28'
Chelsea
Mon 8h6'11"-8d22'
Chelsea
Uma 11h56'0"47d5'
Chelsea
Lyn 7h45'0'42d49'
Chelsea Forever
Ori 5h6'10"13d50'
Chelsea Gae
And 1h37'26"41d2'
Chelsea Lee
Peg 23h34'1"30d18'
Chelsea Lynn
And 23h46'1"46d52'
Chelsea Morningstar
And 2h30'54"42d56'
Chelsea Nicole
Peg 23h21'14"33d13'
Chelsea Rae
Lyn 7h58'0"58d31'
Chelsea Rebekah
And 1h50'17"38d13'
Chelsea René
Peg 22h29'38"29d49'
Chelsea Rose
Peg 22h14'31"4d58'
Chelsea T
Vul 20h5'18"28d35'
Chelsea,Lovely Child, Fantastic Lady To Be
And 0h34'49"40d29'
Chelsey Addie
And 23h39'52"45d6'
Chelsey's Smiling Eyes 8th May 1995
And 0h52'1"37d20'
Chelslyn's Aurora
Peg 22h32'25"21d20'
Chemakhi,Faouzi
Aur 4h53'48"52d15'
Chemereau,Marie Chantal
Peg 22h1'44"7d51'
Chemerka,Jr,Jerome
Dra 15h22'21"65d49'
Chempinski,Yvonne
Cas 23h4'26"58d40'
Chemsubitoda
Aur 7h3'31"38d50'
Chen,Allen
And 23h18'4"45d7'
Chen,Calvin Henry
Uma 12h54'6"54d22'B
Chen,Chong Whan
Cep 22h24'13"63d39'
Chen,Christine
Cyg 19h28'52"37d54'
Chen,Dr Tei Fu Chen
Her 18h12'16"30d19'
Chen,Judy Yuchi Alison
Cet 1h38'29"0d39'
Chen,Mabel & John
Com 12h53'54"27d7'
Chen,Michele Eheler
Mon 6h54'33"-3d44'
Chen,Mindy M
Aqr 21h22'1"-10d31'
Chen,Natasha
Lyn 8h2'31"50d17'
Chen,Tony
Sct 18h44'25"-6d12'
Chen,Victor
Umi 15h37'0"67d33'
Chen,Vincent
Dra 18h25'19"65d14'
Chen,Winnie Hua Ying
Peg 21h20'15"22d40'
Chen,Yun Shiu
Lyn 7h3'39"51d2'
Chen-Beaudiquez,Joelle
Lac 22h4'54"37d33'
Chenard,Stewart S
Tri 2h2'59"31d39'
Chenault,Lynna & David
Leo 11h23'50"-6d16'
Chenchick,Ryan
Aur 6h22'51"30d8'
Cheney
Uma 11h43'60"56d54'
Cheney,Jack
Cep 3h1'32"78d47'
Cheng,Audrey Min EE
Umi 17h14'0"76d17'
Cheng,Michael Siu Man
Lib 15h42'22"-28d35'
Chenkin,Hannah Elizabeth
Aqr 21h56'55"-6d3'
Chenkin,Molly Rebecca
Peg 23h49'46"8d35'
Chenko
Per 3h3'31"41d24'
Chenoweth,Phyllis Kaiser
Uma 11h21'0"40d45'
Chenu,Marina
Her 17h26'14"40d30'
Chenut,Blanche
Ori 5h57'14"16d53'
Chepeleff,Nicole
Peg 21h59'55"30d11'
Cheplowitz,Mark
Cep 21h41'46"68d52'
Chepulis,Dennis J
Her 17h25'22"38d20'
Cher's Majestic Way
Eri 4h12'32"-16d40'
Cher's Star
Cas 2h15'37"68d6'
Cher,Maggie Lim Li
Col 6h0'48"-32d0'
Chera
Ori 5h53'18"17d41'
Chera
Uma 9h54'58"57d51'
Cherania,Nasreen
Boo 14h23'0"51d13'
Cherchi,Anna Lucia
Lac 22h29'16"50d27'
Cheri
Peg 21h59'33"26d48'
Cheri
Del 20h13'57"10d57'
Cheri
Psc 22h59'36"0d24'
Cheri
Com 13h33'24"20d27'
Cheri
And 23h38'17"47d45'
Cheri Forever The Star Of My Heart
And 1h53'1"46d57'
Cheri Lee T
And 23h48'57"40d44'
Cheri Lynn
Cap 21h21'24"-24d41'
Cherici,Jean
Lyr 18h56'0"30d8'
Cherie
Cas 1h57'39"58d10'
Cherie
And 23h0'47"37d32'
Cherie
Lyr 18h58'55"31d57'
Cherie & Charlie
Aql 19h3'45"-1d5'
Cherie & Matt
Cyg 21h45'39"34d15'
Cherie Amour
Aql 19h28'12"10d45'
Cherie's Star
Cas 0h56'45"63d5'
Cherier,Isabelle
Cas 0h2'27"59d43'
Cherise
Cam 6h10'58"58d34'
Cherish
Crt 11h23'36"-18d25'
Cherished Cathe
Ori 5h51'35"16d58'
Cherished Moments Shared
Lyr 19h5'41"37d31'
Cheritier,Florence
Dra 9h57'60"73d16'
Cherly Ann
Dra 18h20'41"70d25'
Chermak,Andrew Michael
Ori 5h10'52"-6d0'
Chernau-To Our "Forever" Love, Jack
Aql 19h24'46"-8d10'
Cherney,Edward
Per 3h21'22"40d3'
Cherney,Edward
Lac 22h48'30"55d22'
Cherney,Eric Joseph
Per 2h58'24"50d14'
Cherney,Jolene & Troy
Sct 18h49'11"-7d13'
Cherney,Sr,Robert W
Her 15h57'51"40d24'
Cherniak,Bruce
Ori 5h55'26"15d50'
Chernicoff,Elaine Shatz
And 0h54'1"36d24'
Chernik,Nina
Uma 9h57'0"70d10'
Cheron,J2F6 by Jr
Uma 14h25'28"60d57'
Cherra,Agape
Aql 19h54'17"15d15'
Cherri,Michael
Cma 6h54'55"-18d25'
Cherrill
Cas 0h34'60"73d41'
Cherroff,Alexandra M
Boo 14h29'51"21d6'
Cherroff,Kathleen P
Tri 1h38'0"28d39'
Cherry Drive Elementary Exelstar
Lyn 7h49'51"41d44'
Cherry Pop
Mon 6h33'1"-0d58'
Cherry,Adéle
Cas 0h41'60"63d58'
Cherry,Grace
Aql 19h54'28"12d32'
Cherry,Norman W
Leo 11h32'18"-2d6'
Cherry,Sidney
Sge 19h1'46"21d0'
Cherry-Ann
Eri 3h16'22"-17d23'
Cherryl
Vul 19h46'36"20d20'
Cherubic Bliss Twenty- Five
Leo 11h55'11"23d1'
Cherubim Lane
Peg 23h5'36"11d35'
Cherubini,Ralph
Aql 18h57'24"3d2'
Chervenak,DO,A Douglas
Dra 17h48'35"64d25'
Chervenak,Timothy J
Cet 0h24'0"1d50'
Cheryl
Mon 6h55'48"-6d46'
Cheryl
Peg 22h43'58"21d37'
Cheryl
Uma 9h28'1"55d27'
Cheryl
Mon 6h21'12"7d11'
Cheryl
Peg 23h6'12"17d51'
Cheryl
Eri 3h37'59"-6d16'
Cheryl
And 23h22'1"50d33'
Cheryl
Cmi 8h8'50"6d22'
Cheryl
Vul 20h2'48"28d44'
Cheryl
And 1h57'45"40d24'
Cheryl
Cas 3h4'28"68d38'
Cheryl "Big Bear"
Uma 9h19'32"71d47'
Cheryl & Bill "Eternally"
Umi 15h17'53"81d33'
Cheryl & George
Tau 5h34'43"20d18'
Cheryl & Henry
Oph 18h32'35"6d58'
Cheryl & Rebecca's Christmas Star
And 23h47'46"45d45'
Cheryl & Scot
Cyg 20h1'0"40d9'
Cheryl & Tommy
Sge 20h17'21"17d32'
Cheryl and John
Crb 16h18'24"28d58'AB
Cheryl Ann
Leo 10h1'38"18d42'
Cheryl Ann
Cas 2h38'38"75d40'
Cheryl Ann
Mon 6h22'17"8d48'
Cheryl Ann
Mon 6h26'48"-8d26'
Cheryl Anne
Mon 8h2'25"-8d20'
Cheryl Diane
And 2h9'1"38d23'
Cheryl Don-Ell
And 0h21'51"36d24'
Cheryl Joy
Cas 0h35'58"60d6'
Cheryl Kay
And 23h50'11"32d51'
Cheryl Loves Randy
Cyg 20h28'50"42d60'
Cheryl Lynn
Lyr 18h21'55"46d40'
Cheryl Lynn
Cyg 20h59'49"30d48'
Cheryl Lynn
Cas 0h26'29"50d3'
Cheryl Lynn
Cas 23h22'1"61d11'
Cheryl of the Alice
Pho 0h47'33"-44d15'
Cheryl Renee
Lyn 7h52'30"51d15'
Cheryl `Isis' Star of my World
Aqr 23h15'43"-6d40'
Cheryl's "Star Dust"
And 2h23'30"44d30'
Cheryl's & John's Wishing
Cyg 20h2'51"38d24'
Cheryl's Adventure
Cyg 19h41'12"40d36'
Cheryl's Dream
And 23h22'21"40d43'
Cheryl's Everlasting Light
And 23h37'46"47d29'
Cheryl's Star
Crb 16h20'46"30d38'
Cheryl's Wish
Hya 8h15'31"-6d46'
Cheryl-N-TJ 143 Mine For Eternity
Cyg 19h29'0"31d30'
Cherylynn
Lmi 10h54'35"27d7'
Cheré Kaye Stacey
Cet 3h5'52"2d7'
Cheshier,Dale
Sct 18h55'59"-7d42'
Cheshire,Thomas R
Aql 19h8'49"3d24'
Cheslic,Dina
Peg 22h59'53"33d49'
Chesmar,Jr,George J
Per 2h59'13"55d8'
Chesney Much Loved, Robert J
Lac 22h47'35"56d18'
Chesney,Helen Marie
And 23h17'59"47d3'
Chesniak,Joseph & Joann
Crb 15h55'55"26d13'
Chesnutt,Mark
Her 17h0'38"48d11'
Chessa,Candide
Umi 15h45'46"80d35'
Chesser,John Carl "Rusty"
Cet 3h2'31"2d59'
Chest,April
Peg 23h6'52"10d3'
Chester,Dr Daniel L
Psc 1h16'28"21d12'
Chester,With all Our Love
Hya 8h10'25"1d30'
Chestnut Hill Elementary School
Ori 6h2'36"5d33'
Chestnut,Cheyenne Marie
Uma 12h2'54"55d1'
Chestnut,Dustin Thomas
Uma 12h54'54"53d31'
Chestnut,Linda
Lyn 7h56'52"43d29'
Cheswick,Madison Ann
Sge 19h53'1"18d52'
Chet
Dra 19h7'10"68d45'
Chet Fillip
Oph 18h20'10"10d47'
Chet's "Starlight 75"
Cep 23h15'0"64d33'
Chetkowski,Daphne Elizabeth
Cyg 19h56'38"45d21'
Cheu 30,Brian
Boo 15h6'0"42d11'
Cheval,Gregory J
Boo 14h29'10"26d10'
Chevalier
Her 18h18'46"12d9'
Chevalier,Aimeé K
Peg 22h17'0"21d41'
Chevalier,Joslane
Per 4h6'24"38d34'
Chevalier,Judi
Cas 0h52'1"63d60'
Chevalier,Noëlla
Cas 0h54'0"74d50'
Chevalley,Aurore
Dra 10h34'37"81d7'
Chevallier,Michel
And 23h2'26"50d47'
Chevallier,Pascal
And 2h31'1"40d0'
Chevallier,Richard
Lac 22h6'27"37d42'
Chevanel,Marie Soleil
Peg 22h39'50"21d40'
Chevat,Edith
And 1h31'0"39d48'
Chevee,Gerard
Sex 10h13'7"-7d57'
Cheverie,Jeanne Marie
And 1h20'44"36d43'
Chevlin,Leona
Equ 21h19'17"10d17'
Chevrier,Heidi
And 23h8'54"44d7'
Chew,Fredrick J
Hya 8h44'39"4d48'
Chew,Kelsey Caitlin
Mon 7h45'5"-3d53'
Chew,Kylie Shannon
Sex 10h18'10"-1d20'
Chew,Mark H
Cet 3h7'16"3d22'
Chew,Ming
Aql 19h44'38"10d9'
Chew,Sarah Elaine
Ori 5h59'21"17d37'
Chew-Holman,Gina
Cyg 21h25'27"38d30'
Chewajahrayslo
Cam 4h38'34"68d30'
Chewerda,Debi
Uma 10h9'40"61d47'
Chewning,Martha
Peg 22h19'41"8d55'
Chewning,Ronald Steve
Cet 2h44'24"7d5'
Cheyenne
Cam 3h16'24"67d30'
Cheyenne
Cep 22h38'10"65d21'
Cheyenne
Uma 9h8'43"50d42'
Cheyenne in Essence
Lyn 7h52'45"43d33'
Cheyenne Nicole
Lyr 19h15'55"25d59'
Cheyenne's Star
Gem 7h22'42"31d44'
Cheyenne,Alexis
Cas 0h46'31"72d18'
Cheyenne,All Our Love Forever & A Day
And 23h2'44"44d9'
Cheylus,Yves
Oph 18h17'56"10d12'
Cheyne,Emily Victoria
Cas 0h26'1"61d47'
Cheyne,Fiona
Cas 0h18'54"63d37'
Cheyney,Peter R
Aur 5h35'21"38d18'
Chez Louis
Vul 19h46'41"28d52'
Chez Saby
Cyg 20h20'58"39d26'
Chez Sidney
Mon 7h48'7"-6d39'
Chezalviel,Monique
Vul 19h33'51"27d22'
Chezern
Lac 22h21'29"50d1'
Chhabra,Vijay Dr
Oph 17h14'41"-22d32'
Chi-Chi
Tri 2h42'0"31d45'
Chiabai,John O
Cet 1h28'22"-6d59'
Chiabolotti,Susanna
Lyn 7h9'33"59d22'
Chiaia,John F
Per 2h27'58"56d42'
Chiaia,Michele Anne
And 23h22'16"35d32'
Chiale's Star
Cnv 12h13'54"34d22'
Chiamos,Kyle Alexander
Aql 18h59'43"8d1'
Chiancola,Rosemarie V
Vul 19h48'30"25d5'
Chiancola,Sr,Charles A
Cet 1h25'53"-10d9'
Chiandra
Lac 22h24'43"37d42'
Chiang,Dennis
Lyn 8h25'0"33d52'
Chiang,Ed
Aql 19h3'15"2d42'
Chiang,Helen
And 1h29'45"39d18'
Chiappetta,Christopher
Mon 6h21'44"5d49'
Chiappone,Camille Schmidtgoessling
Peg 21h57'26"28d20'
Chiara
Cyg 19h38'36"28d46'
Chiara
And 0h28'1"43d51'
Chiara
Tel 20h18'54"-46d36'
Chiara
And 23h0'24"51d29'
Chiara
Cnv 13h29'42"38d19'
Chiara
Pyx 8h41'23"-24d24'
Chiara
Cas 0h33'38"61d41'
Chiara
Pyx 8h46'13"-24d24'
Chiara
Cmi 8h8'13"1d26'
Chiara & Davide
Cyg 21h56'52"52d54'
Chiara,Barbara Dalla
Cas 0h32'56"61d42'
Chiara,Colette
Mon 6h19'58"9d1'
Chiara,Dolce
Umi 16h31'53"78d28'
Chiara,Margie
And 23h28'1"41d22'
Chiara,Milani
Umi 15h16'17"68d23'
Chiara,Millie
Lyr 18h16'0"37d47'
Chiara,Monsignor Peter
Psc 1h43'15"27d51'
Chiara,Sue
Peg 22h24'30"27d19'
Chiaradio,Jeanne
And 1h40'57"40d14'
Chiaramonte,Bill
Aur 6h5'14"54d26'
Chiaramonte,Lucia
Peg 23h21'23"15d30'
Chiaramonte,Monica
And 1h58'13"38d4'
Chiaranda,Monica e Viky
Del 20h18'59"11d56'
Chiarella,John "Sweets"
Dra 16h52'55"64d54'
Chiarello,Anthony
Aur 6h1'56"45d57'
Chiaretta
Gem 6h59'6"16d6'
Chiarilli,Julie
And 0h23'11"32d50'
Chiaro Julia Mariefranáoise
Oph 18h16'47"11d2'
Chiarussi
Vir 13h2'56"12d40'
Chiate-McCartan
Mon 7h3'20"4d15'
Chiau,Stephen
Cep 4h26'24"80d29'
Chiavario,Tony
Peg 22h34'32"35d29'
Chiaverini,Ellen A
Uma 9h24'12"48d8'
CHIB
Cep 20h57'26"63d33'
Chibbaro,Christopher Andrew
Del 20h37'51"3d51'
Chica
Per 2h59'1"32d40'
Chica & Chico-Forever In Love
Cyg 19h34'1"38d52'
Chicca
Aql 20h30'11"-6d21'
Chicca,Maria
Cyg 19h30'51"37d40'
Chicco
Boo 13h46'1"12d59'
Chicco
Del 20h12'33"11d52'
Chicco 907
Aur 5h2'46"31d18'
Chicco,Francesca
Uma 11h32'19"53d36'
Chiche,Maryse Et Henry
Crb 16h14'39"38d12'
Chichester,A Loring
Lac 22h12'53"48d41'
Chick,Ronald A
Per 2h41'48"34d53'
Chicken Mama Megan
Cam 12h26'36"78d16'
Chicken Pie
Aur 4h58'19"40d15'
Chickenpet Star
Uma 10h34'40"48d41'
Chickering,Beverley Wolfe
Lyr 19h2'1"30d36'
Chickering,Ruth Joy
Cas 0h5'22"50d12'
Chickering,Winona
Cas 0h0'39"56d14'
Chickini,Ronald P
Cep 22h38'0"70d35'
Chicky Sue
And 23h21'36"40d54'
Chico's 19th Hole
Cet 0h24'36"-11d50'
Chicoine,Loïc Benoât
Her 16h52'32"33d36'
Chicola,Richard
Ser 15h47'43"24d40'
Chicvak,Joseph Edward
Lyn 8h3'45"47d39'
Chidester,Floyd
Cyg 21h3'46"38d52'
Chidester,Ruth J
Uma 10h27'24"40d10'
Chidvilasananda, Gurumayi
Tau 4h1'18"22d58'
Chief
Dra 17h16'40"60d59'
Chief Matowti
Ori 5h51'56"16d22'
Chief P
Aur 6h13'19"45d27'
Chieko
Lyn 9h8'20"40d30'
Chien
Cam 6h51'46"70d51'
Chiera,Alex
Her 17h10'13"48d20'
Chiera,Donna "OOP"
Crb 15h35'43"31d58'
Chierichella,Amy Beth
Uma 12h33'1"62d39'
Chierichella,Michael Hahn
Aur 4h52'36"50d39'
Chierichella,Rebecca Anne
Uma 12h31'1"61d16'
Chierichella,Susan Roberta
Cnv 12h50'42"43d29'
Chierus,Diane
Vul 20h18'35"28d59'
Chiesa,"Ina" Benedicte
Lyn 7h1'30"60d48'
Chiesa,Davide
Her 18h10'23"31d11'
Chiesa,Victoria Regina
Uma 9h44'48"58d7'
Chih-Han
Ori 5h47'25"21d11'
Chikako
Sgr 19h15'21"-27d53'
Chikako & George
Sgr 19h12'42"-16d57'
Chikara Abe
Cet 1h50'23"0d58'
Chilambra Starlight
Col 6h26'49"-35d7'
Child,Cody Grant
Boo 14h48'58"37d31'
Child,Daniel Proby
Cyg 19h35'1"28d1'
Child,Duncan Lewis Varley
Cyg 19h39'52"30d23'
Child,Ivy
Cyg 20h40'17"44d27'
Child,Jane
Psc 22h52'58"5d0'
Child,Mary Beth
Boo 15h18'43"37d38'
Child,Monica
Cam 3h55'46"62d14'
Childers,Allison
Cam 5h24'0"67d42'
Childers,Everett Daniel
Aql 19h52'0"14d23'
Childers,Heather
Lyn 7h45'30"41d36'
Childers-Higdon,Annie Mae
Peg 0h0'43"30d49'
Childhood Memories
Vul 20h15'16"23d19'
Children Of St Rose Of Lima
Mon 7h3'47"0d15'
Children's Hour,The
Vul 19h40'31"26d42'
Childrens Dental Center
Peg 23h6'60"10d28'
Childress,Cathy Gray
And 0h37'59"38d55'
Childress,Lee
Cmi 8h6'52"1d29'
Childress,Samuel Wesly
Her 18h18'49"20d7'
Childs,Aaron Brad
Cep 22h3'17"54d3'
Childs,Anna Josephine
Lyr 18h35'12"33d57'
Childs,Dave
Aur 7h0'28"37d39'
Childs,Ernest Lee
Lac 22h40'48"53d3'

Childs,Gretchen Beth
Cam 3h12'11"58d22'
Childs,Henry A
Cyg 21h16'1"38d39'
Childs,Linda
And 2h10'35"40d37'
Childs,William
Dra 19h56'20"67d55'
Chiles,Earle Meyer
Cep 21h45'55"55d6'
Chilgren,Sean
Cyg 19h27'24"32d54'
Chili
Uma 11h23'46"44d24'
Chili Seester
Sge 19h42'25"18d58'
Chillders,RN,Mary Wynelle Henry
Mon 6h53'1"0d35'
Chille,Frankie "The Thumper"
Aur 5h6'58"44d44'
Chillemi,Anthony J
Aur 5h9'54"41d3'
Chillemi,Christopher James
Aur 6h18'26"38d59'
Chillemi,George G
Her 18h13'42"48d7'
Chillemi,Jr,Daniel Santo
Boo 14h47'31"34d55'
Chilton,Jr,James Kenneth
Vul 19h42'0"23d3'
Chilton,Norman Barry
Oph 17h55'18"-8d0'
Chilton,Sarah Mae
Cas 0h42'60"70d8'
Chilton,Traci
Oph 17h19'49"12d53'
Chilver,Steve
Ori 5h43'22"0d53'
Chilvers,Guy
Mon 6h13'1"-4d53'A
Chilvers,Madeleine Bilodeau
Mon 6h13'1"-4d53'B
Chilvers,Susanne Marie
And 23h26'18"41d10'
Chim's Star,Sigmund
Aur 4h49'28"48d52'
Chim,Ron
Hya 9h12'8"-19d47'A
Chimayo
Cap 20h37'2"-27d58'
Chimento,Alyssa Nicole
Cam 5h56'53"70d37'
Chimes
Fri 4h55'30"-4d40'
Chin,Jacy Jennifer
And 2h3'57"38d18'
Chin,John Susan-Wong Natalie Michael
Cnv 13h20'54"31d36'
Chin,Sing Ngoe
Boo 14h14'56"54d18'
China Doll
Mon 6h16'51"-6d27'
China Moon
Umi 15h28'40"81d39'
China's Infinity,Peter
Her 16h29'22"41d36'
China,Angelo
Cep 23h32'25"63d27'
Chinchar,Mark
Aql 20h13'56"4d13'
Chinchen,John
Per 3h12'25"42d9'
Ching
Uma 11h28'1"40d55'
Chinick,Steven James
Cyg 21h50'1"41d49'
Chink
Peg 23h28'20"20d28'
Chinnis,Rae
Mon 7h15'31"-7d6'
Chinook
Lyn 9h2'15"44d24'

Chioccariello, Bernadette Evavgelista
Cyg 20h27'0"38d6'
Chiodini II,Louis A
Aql 18h57'0"-6d42'
Chiodo,Erik Aage
Aur 5h2'20"40d24'
Chiodo,Frankie J & Melissa Rose
Uma 11h34'1"32d34'
Chiodo,Janice Dreshman
Lyn 7h14'14d59d30'
Chiodo,Pamela
Cyg 21h5'29"40d3'
Chiodo,Teresa
Tri 2h20'57"30d17'
Chioffi,Anthony
Dra 20h8'13"62d53'
Chiola,Elia
Lyn 7h1'0"60d11'
Chiossone,Anabella
Lmi 10h27'60"37d37'
Chip
Cet 1h57'1"0d35'
Chip & Pam's Star
Aql 19h57'42"15d6'
Chip & Pegeen's Anniversary Star
Cyg 20h45'0"38d27'
Chip Muskrat
Per 2h23'1"55d6'
Chip's Scintillation
Per 3h24'49"38d16'
Chip-My Shining Star, Joseph
Per 4h7'20"51d36'
Chipchak,Lisa
Lyr 18h54'15"32d18'
Chipman,Jessob Robert
Lmi 10h57'42"31d49'
Chipman,Nicole Yvette
And 2h6'45"39d54'
Chipp,Dorothy & Frederick
Ori 5h57'57"9d29'
Chippas,Lou Edward
Her 16h35'58"8d41'
Chippendale,Arthur
Ari 3h24'26"28d24'
Chippo
Cep 1h26'18"80d29'
Chipps,Angela
Ori 5h50'43"17d52'
Chipps,Georgianna- Summer
And 2h33'59"37d35'
Chipps,John
Sct 18h41'4"-6d50'
Chippsi
Boo 13h58'46"12d57'
Chippy
Cet 3h18'1"3d3'
Chips Centaur
Aur 5h1'57"46d4'
Chipster
Dra 12h39'47"67d59'
Chipster
Aur 4h53'38"40d49'
Chiquita
Uma 8h45'57"52d28'
Chirac mon petit rebell
Leo 11h15'11"-5d36'
Chiriaco,Joseph L
Cet 1h22'29"-6d12'A
Chiriaco,Ruth E
Cet 1h22'29"-6d12'B
Chirico,Frances
Cam 4h12'25"60d7'
Chirikas,Michael Anthony-James
Her 17h45'37"40d29'
Chirimoya-Luna
Per 2h53'18"50d30'
Chirinos,Victor Alexander
Boo 13h37'26"16d33'
Chirko,Kenneth
And 0h45'43"36d46'

Chiron
Lyn 8h21'23"49d54'
CHIRON
Cma 7h13'0"-13d21'
Chis & Jooney's Planet of Lurve
Cyg 19h28'31"37d16'
Chisholm,Adam Kyle
Peg 0h11'48"13d12'
Chisholm,Bill
Cyg 20h21'1"38d23'
Chisholm,Dr Jullian
And 0h8'0"47d49'
Chisholm,Mary
Cas 1h24'30"52d45'
Chisholm,Robert T
Lmi 10h7'50"32d8'
Chisholm,Tara
Mon 7h22'54"-1d16'
Chislett-Trim
Her 16h43'33"50d50'
Chism,Mary Jo
Com 13h17'29"20d24'
Chism,Victoria Anne
And 23h30'22"49d12'
Chismar,Dr,Connie
Oph 16h53'1"-25d22'
Chita
Uma 8h14'19"60d20'
Chitale,Kavita
Mon 6h29'44"-8d30'
Chitale,Priya
Sct 18h53'1"-6d42'
Chitharanjan,Linda
And 1h33'27"40d29'
Chitra
Peg 23h8'25"12d21'
Chitta,Janna
Lyr 18h43'30"41d35'
Chittenden,Jerald (Jerlech)
Aur 5h57'15"30d48'
Chittester,Richard & Phyllis
Ori 4h42'25"4d59'
Chiu,Serina
Sco 16h50'0"-40d8'
Chivington,Amanda Gail
Cas 0h11'44"54d26'
Chize,Jennifer Elizabeth
Sex 10h15'11"-1d20'
Chizmar,Jennifer Ann
Cyg 19h31'30"32d56'
Chizuko
Aql 19h31'41"10d20'
Chizzola,Isabella
Vir 13h27'39"-5d7'
Chladek,Gabriela Maria
Psc 23h9'1"5d46'
Chladek,Tony
Aur 5h50'56"37d39'
Chlbel
Boo 14h14'30"53d56'
Chloe
Aur 5h57'41"48d56'
Chloe
And 0h49'1"41d3'
Chloe
Lyr 19h16'1"28d19'
Chloe
Lac 22h46'50"38d40'
Chloe
Cas 2h34'13"68d55'
Chloe
And 0h11'1"31d6'
Chloe
Aql 19h2'0"0d8'
Chloe Anna
Peg 23h33'44"12d24'
Chloe Elizabeth
Peg 21h55'15"3d25'
Chloe Melinda
Boo 15h1'58"50d45'
Chloe Rose
Cma 6h56'53"-16d16'
Chloë
Lyr 19h16'1"28d19'

Chloé
cam 6h56'22"67d31'
Chlöe
Lyr 18h34'42"33d58'
Chmarney,Paul
Lac 22h49'17"38d45'
Chmielewski,Brandon
Dra 16h59'23"61d16'
Chmielewski,Michal Aleksander
Aur 4h47'14"36d11'
Cho Nam-Hoo,Loving Jennifer
Lac 22h7'34"50d7'
Cho,MD,Michael Minn Woo
Uma 11h25'31"62d21'
Cho,Rose
Cam 5h55'0"68d23'
Cho,Yong Kee & Myoung Joo Kim
Aql 19h56'35"14d23'
Choate,Cassandra Howell
Lyr 18h38'50"31d51'
Choate,Deborah
Lyn 7h50'1"33d56'
Choate,Dotti
Aql 18h45'46"11d23'
Choate,Dr Hugh
Oph 18h1'1"11d51'
Choate,Dr,Hugh
Oph 17h4'27"7d43'
Choate,Presley V "Poop"
Oph 17h3'16"7d42'
Choate,Sally McClain
And 1h25'21"37d50'
Choban,Eric Raymond Ivan
Cnv 13h27'57"51d27'
Chobanian,Oshin
Cam 5h40'59"61d1'
Chocas,Stephanie Ann
Crb 15h53'34"30d47'
Chochensky,Gretchen
And 23h47'1"40d45'
Chocho I
Ser 18h0'38"-13d21'
Chocolate Thunder
Oph 16h32'51"-6d36'
Choe,Michele O
Cas 23h45'11"50d6'
Choel,P J
Cam 14h2'30"82d0'
Choi,Arion Thomas Thomas Choi
Eri 3h15'50"-3d0'
Choi,Belinda
And 23h35'50"40d44'
Choi,David
Lac 22h35'36"50d3'
Choi,Esther
Lyr 18h17'36"37d48'
Choi,Kiri Lena
Peg 23h16'0"32d6'
Choiniere,Dick
Cep 22h5'30"61d10'
Chole
Crb 16h19'16"30d12'
Chole-4/1/96
Aql 19h2'0"0d8'
Cholet,Pierre
Cam 3h35'13"61d0'
Cholewa,Joseph
Boo 13h40'49"21d45'
Cholewa,Lucille Ann
Umi 16h27'23"71d42'
Cholnoky,John
Per 3h6'45"48d48'
Cholnoky,Robbie
Vul 21h26'0"24d2'
Cholë Gabrielle
Pup 7h59'29"-23d5'
Chomicz,Amanda
Lyn 7h49'17"36d34'
Chomicz,David
Dra 16h10'18"68d9'
Chong,Dr Sonney L
Oph 18h18'0"10d29'

Chong,Randall T
Ori 05h52'48"14d52'
Choo Choo
Lyr 19h17'0"41d30'
Chooch,Mr & Mrs
Uma 9h53'21"51d5'
Chook
Ant 10h33'32"37d47'
Chookalooka
Dra 14h59'17"62d27'
Choon,Mary Leong Chee
Peg 23h1'16"25d7'
Choop-Choop
Cap 20h5'33"-8d14'
Chopko,Stephen Martin
Ori 5h54'20"1d47'
Choppie
Cet 0h28'41"0d39'
Chopyk,MD,John
Oph 17h53'58"0d10'
Choquette,André
Tau 5h52'55"23d23'
Chorba,Cameron Jeffery
Leo 11h9'19"-1d36'
Chorba,David
Cas 0h22'1"61d16'
Chordie Bear
Mon 6h54'47"0d30'
Chorey,Linda
Tri 2h29'44"31d41'
Chorley,Brian
Dra 12h8'1"71d8'
Chorlins Family,The
Aur 4h59'22"41d2'
Chorlins,Joy & Ivan
Per 3h54'0"40d52'
Chorn,Barbara J
Mon 6h59'33"-10d44'
Chorn,Rohland J
Mon 6h59'19"-10d5'
Chornij,Lisa Ann
Aur 5h21'22"38d7'
Chorpenning,Ryan
Equ 21h0'1"2d32'
Chosen Family Love & Light Always
Peg 22h29'46"27d44'
Choseph
Cam 12h36'26"77d54'
Chosson,Catherine
Ori 6h5'59"20d60'
Chosson,Laurent
Ori 6h5'56"20d32'
Chotin,Lana Kaye
Cyg 20h55'33"30d28'
Chotiner,Barbara & Gerald
Mon 6h24'18"7d32'
Chou Sen,Georges & Francois
Cma 6h57'51"-15d41'
Choudhoury,Bikram
Ori 5h52'0"6d25'
Choupette
Vul 19h21'29"23d39'
Choupette
Boo 14h18'37"48d50'
Choupette
Per 3h52'47"35d38'
Chourice
Cam 8h40'53"77d48'
Chovie,Chovie Jean
Aql 19h8'18"15d52'
Chow,Chu Yong
Cnv 12h5'1"37d60'
Chow,Courtney
Gem 7h3'31"21d25'
Chow,Yvonne
Peg 23h0'48"24d40'
Chow,Yvonne
Lyr 19h16'14"41d11'
Chown,Carly Goldthwaite
Cyg 19h53'22"38d46'
Chowney,Anthony B
Aur 6h14'26"33d54'

Chowning,Magical Michael Stuart O'Shea
Cma 6h55'51"-16d17'
Choy,Andy
Psc 0h26'35"2d22'
Choy,B B
Boo 15h6'25"23d59'
Chozen,Michael & Pam
Uma 11h24'34"50d10'
Chretin-Moine,Isabelle
Per 3h5'29"38d11'
Chris
Ser 15h31'60"-2d22'
Chris
Per 1h53'58"56d26'
Chris
Aur 5h54'54"29d10'
Chris
Boo 13h56'0"19d24'
Chris
Boo 13h50'23"21d38'
Chris
Per 2h29'28"58d14'B
Chris
Aql 18h59'59"-5d22'
Chris
Aur 6h16'46"38d49'
Chris
And 23h40'26"44d6'
Chris
Ser 15h32'33"24d32'
Chris
Sgr 18h48'9"-35d0'
Chris "My Lucky Star"
Cet 0h58'20"0d59'
Chris #96
Cyg 19h24'16"32d40'
Chris & Aprille In Love Forever
Uma 9h57'23"61d19'
Chris & Bernie
Cyg 19h36'14"37d44'
Chris & Brian
Cyg 19h41'0"40d20'
Chris & Carla I
Aql 19h3'40"2d59'
Chris & Dan
Cyg 21h17'13"37d31'
Chris & Faye
Sgr 18h50'2"-35d57'
Chris & Jessica With Love Always
Boo 15h5'0"32d6'
Chris & Julia's Star Of Eternal Love
Cyg 20h43'43"38d46'
Chris & Julia's Star
Cyg 21h51'34"52d46'
Chris & Kathy
Com 12h6'12"27d30'
Chris & Laura- Somewhere in the Night
Eri 4h35'42"-12d23'
Chris & Lisa's Fnoppy Star
Cyg 20h19'43"31d38'
Chris & Margaret
Aql 19h57'27"13d10'
Chris & Maria's Rock In The Sky
Cyg 19h37'30"28d17'
Chris & Mike
Cyg 20h39'1"31d41'
Chris & Paul
Uma 8h40'0"71d19'
Chris & Paul
Ori 5h55'51"18d19'
Chris & Rachal
Ori 5h7'25"1d15'
Chris & Reggie
Uma 10h38'1"61d46'
Chris & Sally
Lyr 19h20'25"35d17'
Chris & Stacy
Aql 18h48'45"10d40'

Chris & Tamara Golden Forever
Peg 0h11'1"14d25'
Chris & Tammy's Star
Aql 19h12'55"10d53'
Chris & Tonya
Aur 5h13'1"44d30'
Chris Carnate-Love Eternally, Julie
Uma 11h21'31"38d51'
Chris Et Phil G
Per 2h38'0"56d20'
Chris June Shannan & Eric Star
Cas 0h32'39"65d20'
Chris Michelle
Lyn 9h8'52"45d4'
Chris Noël
Umi 15h18'38"77d50'
Chris Our Love Burns Eternal Pam
Uma 12h9'30"48d53'
Chris Rose
Vul 20h2'27"23d40'
Chris the Mountain Lion
Leo 9h27'26"27d49'
Chris to wish upon
Umi 14h34'16"69d35'
Chris' Sirius
Cma 6h30'1"-11d8'
Chris' Star
Cet 1h33'1"-0d41'
Chris' Star
Umi 14h41'22"66d24'
Chris'Illuminating NRG
Del 20h24'38"8d2'
Chris's Navigator 94
Uma 11h57'38"64d28'
Chris's Shooting Star
Ori 5h28'32"-5d5'
Chris's Star
Lyr 19h12'59"35d4'
Chris's Star
Uma 9h43'46"47d17'
Chris,Denise & Valerie
Eri 4h4'32"-12d9'
Chris/Ashley
Lac 22h47'19"56d20'
Chris/Karen
Del 20h40'14"11d9'
Chrisalex
Com 12h12'15"20d34'
Chrischun
Cam 9h19'35"84d28'
Chrisdam,Jørgen
Umi 13h25'41"70d5'
Chrisdam,Tom
Cyg 20h44'59"45d35'
Chrisdeluca
Lyn 8h20'29"46d29'
Chrisdina
Mon 7h51'15"-5d25'
Chrise
Lyr 18h15'26"30d60'
Chrisimian
Hya 8h55'52"-5d55'
Chrisman,Don & Miriam
Vul 19h47'59"28d10'
Chrisman,Nancy Brown
And 1h31'47"38d48'
Chrismer,Dave
Hya 9h7'0"4d42'
Chrison
Peg 22h57'18"18d54'
Chriss,Alex
Dra 14h34'58"60d35'
Chrissa Eve
Lyn 7h54'53"58d49'
Chrissandra
Lyn 7h47'39"39d30'
Chrissian,Keghanoush
Uma 10h6'20"42d24'
Chrissie
Uma 8h32'54"54d16'

Chrissie
Sgr 18h52'18"-20d54'
Chrissue
Cep 22h33'1"59d11'
Chrissy
Cma 6h50'27"-16d15'
Chrissy
Cas 23h29'55"61d14'
Chrissy
Cas 0h57'44"61d11"
Chrissy
And 1h44'38"38d42'
Chrissy & Lou's Wedding Star
Cyg 19h42'1"42d2'
Chrissy Anne
Cas 0h53'28"74d8'
Chrissy Marie
And 0h45'51"39d33'
Chrissy Sue Sue
Vul 19h16'38"24d31'
Chrissystar
Lyn 6h29'49"54d51'
Christ Our Redeemer, AME Church
Crb 15h55'0"30d45'
Christ,Dan
Her 16h36'14"50d41'
Christ,Jr,Gerald J
Umi 15h13'0"70d12'
Christ,our light
Lyr 18h45'23"31d34'
Christa
Tau 5h54'12"23d57'
Christa Ann Marie
And 1h25'48"34d53'
Christa Lee
Lyr 19h23'39"31d8'
Christa und Heino
Cam 5h5'33"60d29'
Christa W Prinzessin durch AA
Tau 5h30'44"20d27'
Christainson,Rebecca
Eri 3h35'47"-7d20'
Christalain
Peg 22h39'11"21d11'
Christalee
Ser 18h52'36"3d37'
ChriStar
Eri 3h44'33"-5d13'
Christastar
Cas 23h27'31"61d24'
Christe,Raymond Georges
Dra 19h29'60"58d21'
Christeai
Lyn 7h30'31"46d15'B
Christel
Cap 20h9'1"-11d8'
Christel
Uma 11h17'59"41d26'
Christel Hexlein
Gem 8h3'27"30d6'
Christel TR
Cam 5h42'0"60d21'
Christelchen
Cyg 19h30'1"35d16'
Christeler,Agnes
Eri 4h7'41"-17d48'
Christelle
Lac 22h27'48"36d41'
Christelle et Regis
Lyn 8h5'35"39d51'
Christen,Elizabeth Dixon
Mon 6h55'20"-10d15'
Christen,Julien Jean-Rene
Cam 4h37'1"58d20'
Christensen, Christopher Michael
Ori 5h56'52"10d49'
Christensen,Alan L
Boo 14h9'60"37d31'
Christensen,Anna Louise
Lyn 8h56'10"34d52'
Christensen,Blair
Tri 1h46'56"26d54'

Christensen,Bruce Carl
Psc 23h2'33"1d15'
Christensen,Caren Marie
Lyn 7h16'39"50d1'
Christensen,Carol J
And 0h54'54"38d1'
Christensen,David Scott
Hya 9h12'43"0d30'
Christensen,David Michael
Per 1h55'51"50d19'
Christensen,Debra Faye
Cyg 20h11'28"37d37'
Christensen,Don & Arda Jean
Per 1h36'0"54d5'
Christensen,Don Carl
Lib 15h14'1"-28d27'
Christensen,Douglas Alan
Tau 5h18'36"16d6'
Christensen,Finn
Umi 13h20'10"71d39'
Christensen,Gary
Lac 22h54'46"54d28'
Christensen,Gladys
Cas 2h37'46"67d51'
Christensen,Helen & Bus
Crb 15h58'24"31d46'
Christensen,Helena
Cas 1h37'40"63d57'
Christensen,Jacob
Ser 15h31'51"18d47'
Christensen,Jacob Adam
Per 3h20'1"38d52'
Christensen,James Brian
Gem 7h36'12"24d56'
Christensen,Jeffery Don
Gem 7h24'14"20d1'
Christensen,John Thomas
Ari 2h39'54"25d23'
Christensen,Kathy L
Lac 22h53'32"38d4'
Christensen,Laura
Peg 21h58'18"30d6'
Christensen,Lisa Janell
Cet 1h8'14"-3d1'
Christensen,Lucinda
Leo 11h19'31"-6d15'
Christensen,Lykke Rahbek
Boo 13h58'55"11d35'
Christensen,Mark & Cheri
Uma 11h46'26"64d28'
Christensen,MaryAnn
Cas 0h15'10"63d49'
Christensen,Nora & Eric
Per 2h55'19"43d11'
Christensen,Panela Walker
Tau 5h18'53"16d29'
Christensen,Paul B
Hya 9h37'17"-8d59'
Christensen,Sara Ann
Sco 15h58'43"-21d25'
Christensen,Stan
Dra 16h38'16"63d10'
Christensen,The Little Deb Star
Gru 22h30'12"-56d24'
Christensen,Virginia Kesling
Cnc 9h9'47"31d42'
Christensen,Wayne Lee
Hya 8h11'14"3d59'
Christenson,Nicholas
Per 1h51'52"48d46'
Christhel
Aur 4h46'43"41d4'
Christi
Cas 23h1'53"53d11'
Christi Noel
And 0h57'18"45d8'
Christian
Dra 17h45'50"68d15'
Christian
Cep 21h47'11"58d11'
Christian
Sco 16h51'38"-44d58'
Christian
Sgr 19h55'43"-41d58'

Christian Aur 6h8'34"30d50'
Christian Lyn 7h28'26"40d41'
Christian & Line "Amitié" Lmi 10h0'53"38d45'
Christian & Nadia And 23h23'35"45d9'
Christian Alexander Oph 17h2'54"8d43'
Christian Carl Cma 6h52'50"-16d34'
Christian Forever Peg 21h47'54"31d16'
Christian James Lac 22h55'33"52d34'
Christian Jay Hya 9h35'0"-8d46'
Christian Star,The Cyg 20h30'15"40d30'
Christian,Alyxandria Cet 2h48'53"6d51'
Christian,Andre Ori 5h55'51"7d45'
Christian,Dunky Hya 9h30'58"2d17'
Christian,Honorable William A Equ 21h22'1"8d40'
Christian,Jeffrey Emerich Aur 6h8'43"48d48'
Christian,Jonathon Aqr 21h43'40"0d22'
Christian,Kelyse Jordyn Aqr 23h24'20"-6d14'
Christian,Kendall Mon 6h59'11"8d49'
Christian,Mary Leita Cyg 21h16'36"37d38'
Christian,Michael Per 4h32'39"34d11'
Christian,Pauline Cyg 19h34'51"35d11'
Christian,Raydene Yvonne Com 12h20'36"30d10'
Christian,Richard Fredrick And 01h11'38"41d5'
Christian,Richard Anthony Per 3h21'0"40d1'
Christian,Si Cet 2h16'11"7d2'
Christian,Tracey Lyn 8h19'17"36d17'
Christiana Umi 15h7'1"80d55'
Christiana Cas 1h3'0"55d11'
Christiana,Margaret Sicurella Cas 1h43'0"63d58'
Christiane Lmi 10h32'38"31d8'
Christiane Lyr 18h41'60"40d7'
Christiane,Most Beloved Friend Lyr 18h30'32"30d57'
Christiani,Robert Sco 16h37'0"-44d17'
Christians,Debra And 23h8'52"41d11'
Christiansen,Emily Marie Crb 16h11'17"32d22'
Christiansen,Holly Anne Aql 20h12'12"2d23'
Christiansen,Horst Hya 9h13'41"-6d27'
Christiansen,Jeff A Per 3h42'42"50d50'
Christiansen,Jr, David A Cep 22h41'48"58d2'
Christiansen,Lianna Lyr 18h49'49"33d33'
Christiansen,Ms Carson J And 0h0'23"44d58'
Christianson,Dawn L Cnv 13h51'13"38d52'
Christie Mon 6h58'55"-1d7'
Christie Cet 1h53'59"-0d45'
Christie & Steve Uma 9h22'1"51d37'
Christie,Charles Aur 6h25'31"40d4'
Christie,Nickolas Cep 1h15'58"86d46'
Christie,Odette Cyg 19h27'1"36d35'
Christie,Sharron A And 23h38'14"39d51'
Christie-Jade Uma 10h1'18"59d59'
Christina Leo 11h10'45"-0d54'
Christina Com 13h3'54"28d56'
Christina Mon 6h30'0"8d59'
Christina Cyg 20h23'59"38d26'
Christina And 1h22'22"39d47'
Christina And 0h53'0"38d32'
Christina And 0h55'53"36d53'
Christina Cas 1h34'16"61d42'
Christina Cas 2h43'0"70d57'
Christina Lyn 6h45'17"60d36'
Christina Leo 10h26'21"10d56'
Christina Ari 2h23'18"26d16'
Christina Eri 2h45'51"-4d32'
Christina & Derek's Wishing Star Aql 19h6'0"14d22'
Christina & Gregory Cyg 21h1'35"30d18'
Christina Adriana Isabelle Eri 2h54'13"-2d40'
Christina Alane Aur 7h5'25"40d37'
Christina Alexis Mon 7h15'45"-6d16'
Christina B Mon 6h25'58"-0d29'
Christina Beth Peg 22h5'1"21d32'
Christina Dei Cas Cyg 21h18'20"30d33'
Christina Grace Cap 21h43'16"-19d23'
Christina Isabel Peg 23h33'32"21d7'
Christina Joy Mon 6h28'58"8d46'
Christina Lorraine Peg 22h55'55"25d16'
Christina Lucia Diana And 1h23'45"48d57'
Christina M And 23h31'42"45d50'
Christina Malina Palina Uma 14h18'27"61d34'
Christina Maria Eri 4h4'53"-12d40'
Christina Marie Lac 22h8'39"51d26'
Christina Marie Cas 0h22'58"75d38'
Christina Nicole G B Cet 3h17'30"3d20'
Christina Suzanne And 23h17'1"41d49'
Christina W Lyr 18h57'45"31d47'
Christina W Lyr 18h58'0"30d52'
Christina's Light Mon 7h14'26"-10d3'
Christina's Radiance Lyr 19h2'33"38d14'
Christina's Smile Cep 22h10'31"60d21'
Christina's Sunflower Cas 0h1'49"54d53'
Christina-Diane Cam 7h44'59"70d28'
Christine Ori 5h5'25"8d18'
Christine Cam 5h53'58"72d36'
Christine Aql 18h57'37"-7d47'
Christine Mon 7h46'46"-3d37'
Christine Lyn 8h3'13"38d27'
Christine And 1h50'30"46d58'
Christine Lyn 7h38'1"50d44'
Christine Uma 9h46'54"50d38'
Christine Vul 19h50'26"20d35'
Christine Cyg 20h28'11"40d24'
Christine Leo 10h49'1"-0d38'
Christine Oph 18h17'1"11d28'
Christine Leo 11h49'41"22d4'
Christine Cyg 21h34'15"42d1'
Christine And 23h16'33"48d59'
Christine And 0h29'1"27d36'
Christine Eri 2h45'30"-17d33'
Christine Lyr 18h37'32"40d35'
Christine Aql 20h10'19"4d25'
Christine Cyg 21h6'1"30d48'
Christine Umi 16h34'20"75d6'
Christine Cas 1h17'53"60d42'
Christine Aql 19h31'0"12d17'
Christine And 23h21'56"49d8'
Christine Cyg 19h32'51"39d45'
Christine Cyg 19h23'15"47d34'
Christine & Benjamin's Dreams Uma 9h46'42"61d24'
Christine & Bob Star, The Lyr 18h31'34"38d22'
Christine & John Cyg 20h24'25"41d11'
Christine & John Cyg 20h0'50"41d9'
Christine & Nelson Cyg 21h16'52"28d36'
Christine & Sara Cas 0h55'35"54d53'
Christine & Teodor's "Wedding Star" Cyg 20h6'58"41d6'
Christine 14 02 96 Mon 6h19'47"7d23'
Christine Alicia Cyg 21h1'42"40d44'
Christine Ann Cas 0h23'56"62d17'
Christine Anne Per 4h45'12"38d20'
Christine Anne Lyn 7h13'22"58d25'
Christine Anne Uma 11h36'58"49d53'
Christine April Vul 20h14'1"28d35'
Christine C C And 0h55'53"37d47'
Christine Elizabeth 053048 Cas 0h23'32"60d38'
Christine Elizabeth Cas 0h47'25"64d37'
Christine Elizabeth Lyn 7h59'20"51d13'
Christine Ellen Sex 10h0'53"-0d25'
Christine Grace Cyg 21h30'53"42d7'
Christine J S Cas 0h0'60"54d50'
Christine Jo Peg 22h33'0"25d6'
Christine K 080658 Gem 6h46'25"13d53'
Christine Lee Cyg 20h23'34"31d2'
Christine Louise Mon 7h43'56"-2d49'
Christine Louise Cam 3h20'29"55d24'
Christine Loves Craig Star,The Lyr 19h26'54"38d41'
Christine Lynn And 0h43'12"45d29'
Christine Marie Cas 0h0'27"64d30'
Christine Marie And 2h29'19"39d30'
Christine Marie And 1h55'1"39d16'
Christine Marie 1996 Cas 1h0'40"62d47'
Christine Michael's Com 12h46'56"25d21'
Christine Noel Cas 1h47'56"73d42'
Christine Wendy Joy Sge 20h5'56"20d12'
Christine's Cat's Eye Lyn 7h40'55"48d23'
Christine's Eyes Peg 23h44'0"27d53'
Christine's Glory Cas 0h36'1"74d30'
Christine's Star Cas 1h9'18"63d29'
Christine's Star Cas 22h57'23"54d21'
Christine, I Love You 11/05/94 Lyr 18h37'29"29d9'
Christine,The Cyg 19h32'56"35d46'
Christine-Andrea-Stelk - Alpha-191266 Sgr 18h53'12"-29d45'
Christine-LSO Lac 22h5'36"47d47'
Christine Todd Cyg 19h40'32"38d4'
Christinepaul Cyg 21h0'1"38d10'
Christinii,Marcus Cyg 20h22'31"39d19'
Christl's Star Lyn 7h4'49"53d33'
Christman Star,The Cyg 19h24'42"30d25'
Christman,Cynthia Leigh Cyg 19h46'41"29d24'
Christman,Del Ori 5h54'33"15d0'
Christman,Martha "Morgan" Mon 8h6'22"-0d7'
Christman,Paul Edward Aur 5h8'18"41d4'
Christmann,Lena Vir 11h53'39"-4d45'
Christmas Dawn Boo 15h21'17"40d24'
Christmas Star For Kassi And 23h21'31"42d17'
Christmas Through Your Eyes Lib 14h47'0"-20d19'
Christmas Wish Vul 20h39'53"23d42'
Christmas,Jr,Kenneth C Per 2h37'29"56d43'
Christmas-Pasquinelli Umi 16h11'47"79d1'
Christner,Patricia Eri 3h33'35"-15d34'
Christo,Di Lmi 10h35'1"32d8'
Christodal,Jenny Cas 2h58'40"57d57'
Christodoulidou T20RT21,Evagelia Cyg 19h34'13"28d6'
Christodoulou, Efthimous Lyn 8h27'53"50d9'
Christof Aql 19h31'13"-0d45'
Christofel,Mary Beth Ori 6h7'1"8d44'
Christoffersen,Peter Per 3h8'6"34d53'
Christoffersen,Sawyer King Per 1h27'31"53d19'
Christofferson,Duane Dra 16h25'1"65d51'
Christofferson,Erik Lac 22h1'55"51d15'
Christofi,Joie Ari 2h40'55"20d36'
Christone's Fifth Cep 22h8'38"61d38'
Christonica Cam 5h55'46"60d7'
Christoph Umi 16h38'15"76d31'
Christoph Uma 13h42'0"50d25'
Christophe & Shona Per 4h0'28"37d27'
Christopher Per 3h20'0"38d46'
Christopher Ori 6h9'57"0d11'
Christopher Aur 6h52'31"37d60'
Christopher Lmi 10h1'26"28d52'
Christopher Ser 15h57'0"-0d40'
Christopher Per 1h44'26"54d3'
Christopher Gem 5h59'21"27d4'
Christopher Boo 14h55'24"25d13'
Christopher Per 4h23'55"50d16'
Christopher Aur 5h6'47"44d6'
Christopher Boo 14h45'46"38d15'
Christopher Gem 6h39'42"14d36'
Christopher Cma 7h17'50"-15d37'
Christopher Aql 19h56'13"10d49'
Christopher Psc 1h42'46"22d04'
Christopher Dra 16h9'54"61d41'
Christopher & Jennifer Uma 11h58'15"60d45'
Christopher & Amber Ori 5h28'50"-4d58'
Christopher & Carol & Sam Cyg 19h16'19"47d14'
Christopher & Melissa Forever Uma 9h42'32"53d53'
Christopher 1437 34 Cap 20h45'34"-26d54'
Christopher Amy Aur 5h18'1"47d55'
Christopher Andrew Aur 7h23'49"39d36'
Christopher B Hya 8h27'31"-6d12'
Christopher Darrel Lup 14h38'12"52d56'
Christopher David Gem 6h37'7"13d46'
Christopher Dean Mon 8h2'8"-2d56'
Christopher Eugene Forever Oph 17h14'3"-24d30'
Christopher Forever Aur 4h58'44"40d43'
Christopher Ian Cep 3h9'14"77d54'
Christopher J Her 17h12'15"41d28'
Christopher James Her 16h30'34"41d35'
Christopher John Cet 2h56'48"3d31'
Christopher John Per 4h34'20"31d50'
Christopher John Cmi 7h45'51"7d40'
Christopher John Hya 8h14'20"5d41'
Christopher Lee Boo 14h9'58"33d34'
Christopher M Aql 18h44'0"7d46'
Christopher Marc Aur 5h0'1"50d27'
Christopher Mark Ori 6h9'45"6d48'
Christopher Michael Star Her 16h45'21"32d21'
Christopher Robin Boo 14h18'1"45d24'
Christopher T's Star Uma 9h6'34"48d6'
Christopher the Knight And 0h20'28"43d33'
Christopher Thomas Boo 14h54'27"42d16'
Christopher Too Cma 6h30'20"-24d32'
Christopher William Eternal Her 17h26'26"18d48'
Christopher's Constellation Her 17h38'42"40d15'
Christopher's Valentine Uma 9h29'0"48d33'
Christopher's Christmas Star Sex 9h52"46d2'20"
Christopher's Celestial Song Aur 6h29'1"30d27'
Christopher's Clan, Looking Over The Her 17h0'1"42d38'
Christopher's Coeur de Kelly Cap 20h36'47"-10d36'
Christopher's Comet Boo 15h5'19"12d12'
Christopher's Dream Del 20h53'23"9d52'
Christopher's Hope Per 2h6'54"57d57'
Christopher's Light Boo 14h33'1"50d27'
Christopher's Soul Per 2h27'1"56d34'
Christopher's Southern Crescent Aur 6h5'33"46d21'
Christopher's Star Dra 19h34'56"65d24'
Christopher's Wish Boo 14h53'23"43d55'
Christopher,Becky Cas 1h10'0"60d47'
Christopher,Brenda Tri 2h7'49"31d58'
Christopher,Brett Dra 12h24'25"68d26'
Christopher,Brigitti Ann Gem 6h48'39"12d25'
Christopher,Christine Kreuz Cas 3h21'1"70d18'
Christopher,Diana Lyn 9h8'42"37d49'
Christopher,Dorothy And 23h23'17"43d40'
Christopher,Grant Ellery Mon 6h54'58"7d33'
Christopher,Helen And 0h17'46"35d50'
Christopher,J Her 17h23'27"46d8'
Christopher,June L And 2h4'54"42d54'
Christopher,Kira Vir 11h44'25"4d6'
Christopher-Barbara Cet 1h2'30"1d41'
Christopher-Bethany Dra 14h4'17"67d39'
Christopherson II, Kent W Aql 19h54'32"13d12'
Christopherson's Star, Robb & Kari Sct 18h49'40"-7d31'
Christopherson,A J Cam 5h52'53"58d50'
Christopherson,Dale Albert Ori 5h5'55"0d27'
Christopherson,Nate Lac 22h6'0"48d8'
Christopher" Eternity Boo 15h25'59"50d26'
Christor V96 Boo 15h0'19"18d5'
Christoun,Heather Anne Cas 0h40'42"68d56'
Christy Aql 19h31'33"14d30'
Christy Eri 3h42'1"-17d45'
Christy And 0h58'44"41d5'
Christy Del 20h13'14"11d1'
Christy Cyg 19h51'20"44d41'
Christy & David Sct 18h53'0"-5d9'
Christy Carol Mon 7h0'35"3d52'
Christy Kay Sgr 18h48'7"-25d3'
Christy Loves Joe Uma 11h36'30"40d12'
Christy Lynn And 1h1'38"37d27'
Christy's "Nightlight" Peg 22h30'42"20d37'
Christy's Wish Upon Eri 4h11'27"-10d19'
Christy,Cristen Marie Com 12h7'23"20d4'
Christy,Glenn Dra 15h6'21"64d53'
Christy,Gloria Denise Cyg 21h12'40"35d10'
Christy,Jeffrey Clayton Her 16h12'15"40d15'
Christy,Julia James Vul 19h58'36"23d37'
Christy,Martha Pauline Crb 15h52'41"26d59'
Christy,Nicholas Gerard Lyr 18h32'22"36d55'
Christyne Mon 6h59'35"0d59'
Christyne Lee Cas 23h23'46"61d6'
Chrisy's Wishing Star- 3-31-83 And 1h45'17"41d9'
Chrobak,Daniela Lyn 7h5'56"58d31'
Chrobak,Stanley Dra 12h24'1"70d10'
Chrola Cyg 21h31'17"42d48'
Chromoga,Theodore Marion Cnv 12h47'38"39d36'
Chrow,Anne Sheridan Lyr 18h58'1"38d7'
Chruma,Patricia Peg 22h25'46"31d42'
Chrysanthe Eri 3h12'38"-16d49'
Chrysler,Binnie & Bill Cam 5h47'18"70d49'
Chrysler/Plymouth Stars of Chicago Dra 20h25'22"64d27'
Chrystal,J S Uma 9h48'12"55d43'
Chrystele Lac 22h43'46"38d54'
Chrystie And 0h19'22"33d41'
Chu,Casey Christopher Ari 2h42'27"24d29'
Chu,Cecilia So Ming Umi 15h31'47"83d28'
Chu,Joe Oph 17h3'53"-21d55'
Chu,Mabel S Y And 23h6'19"42d18'
Chu,Simon Per 1h40'1"54d10'
Chua,Hoon-Hoon Lyr 18h43'38"36d57'
Chua,My Enchanting Soulmate,Johanna And 2h31'51"41d58'
ChuAm Peg 22h25'36"20d47'
Chub,Christina Leigh And 23h18'47"51d0'
Chub,Thomas Henry Boo 14h55'13"27d35'
Chubay,Steven John Paul Ori 5h55'42"11d35'
CHUBB Dra 18h12'21"80d27'
Chubb's Star,Sheila Wendy Cep 21h15'58"70d3'
Chubb,Jeffrey C Vir 11h50'48"4d59'
Chubbuck,Merry Christmas Miss Lyn 6h29'33"58d8'
Chube,David Per 3h5'34"46d5'
Chuck Ori 4h41'37"10d28'
Chuck "Forever Special" Cap 21h55'11"-22d30'
Chuck Aur 7h10'1"39d40'
Chuck Uma 8h35'20"57d52'
Chuck Boo 14h25'51"22d16'
Chuck Aql 18h58'13"3d57'
Chuck & Anita's Love Shines Forever Cyg 21h38'1"40d54'
Chuck & Cindy Peg 22h4'17"29d33'AB
Chuck & Erin's First Pride & Joy Per 2h28'35"58d13'
Chuck & Flo 1945 Tau 4h32'49"30d39'
Chuck & Jean-Together Always Uma 9h29'0"48d33'
Chuck & Jenifer Sex 10h34'22"1d10'
Chuck & Paula Forever Eri 3h41'28"-12d31'
Chuck (In Perseus) Stone Lion Per 3h19'0"49d41'
Chuck's Lucky Star Uma 11h58'46"31d14'
Chuck's Star Aur 5h51'49"30d53'
Chuck's Wish Del 20h17'0"13d19'
Chuck,Barbara J Sct 18h43'60"-7d23'
Chuck,Inna And 23h47'34"40d13'
ChuckEls Her 16h26'1"50d55'
Chuckie Baby Lyr 19h25'12"41d47'
Chuckie The Geek Cnv 12h23'17"34d10'
Chuckie's Star Light Ser 15h59'1"24d52'
Chuckiris Cyg 21h10'0"39d37'
Chud Ori 5h14'33"15d19'
Chudik,Christopher Matthew Aur 5h56'29"40d13'
Chudley,Thomas Roland Dra 16h56'1"69d10'
Chudnoff,Aron Her 17h13'0"26d40'
Chufi Uma 11h22'1"38d55'
Chui,Mitzi Elizabeth Ori 5h37'26"15d19'A
Chui,Raymond Ori 5h37'26"15d19'B
Chuises Her 17h0'54"29d55'
Chuk,Nicole Sonja And 1h45'50"37d14'
Chula Cam 3h50'30"61d33'
Chumack,Ryan Andrew Uma 11h3'48"59d50'
Chun,Duane Sco 17h50'53"-40d38'
Chun,JoAnn Hisae Matsumoto Com 12h20'55"20d5'
Chun-Ming,Holly Oph 17h15'39"-18d47'
Chundrlek I Lyn 8h29'43"41d48'
Chung,Connie And 2h20'41"40d22'
Chung,Eli Cyg 21h56'49"48d53'
Chung,Emperor "Michael" Kautai Cnv 12h44'59"38d32'

Chung,Kelly Anne
Ari 2h34'53"21d6'
Chung,Patti
Cam 3h55'57"61d30'
Chung,Ryan Sebastian
Sgr 20h24'21"-28d53'
Chung,'S L'
Uma 10h3'34"70d17'
Chunglo,William F
Per 2h50'59"32d38'
Chunk SBK
Ori 5h3'24"13d5'
Chupasko,Olga
Cam 4h56'29"67d42'
Chupik,John M & Jeffrey J
Hya 8h34'60"-6d18'
Chupka,Kevin
Aur 6h29'44"34d3'
Church III,Albert Thomas
Aur 6h26'1"38d14'
Church,Alice W
Aql 19h0'35"10d53'
Church,Carol
Cas 0h24'44"68d6'
Church,Heather Dawn
Cnc 8h58'55"32d51'
Church,Jack D
Ser 15h22'1"20d1'
Church,Jennifer Susan
Vul 19h58'20"28d23'
Church,Jill
And 23h2'32"51d48'
Church,John Aral
Per 2h45'44"40d24'
Church,Jr,Arlie Earl
Oph 17h34'15"13d17'B
Church,Kathleen B
Cam 5h59'18"58d9'
Church,Ken
Leo 9h35'41"8d33'
Church,Rachel Louise
Cyg 20h51'0"40d14'
Church,Scott Allen
Ori 5h12'15"-7d7'B
Church,Shawn James
Ori 6h7'1"8d54'
Church,Sheila Renée
Oph 17h34'15"13d17'A
Church,Stacy Lynne
Crb 16h5'31"26d34'
Churchill,Ashley Devon
And 1h10'1"38d53'
Churchill,Carol Anne
Cas 1h25'40"54d1'
Churchill,Chuck P
Ori 5h51'53"11d3'
Churchill,David
Aur 5h1'52"41d10'
Churchill,Irmgard
Cyg 21h45'17"30d8'
Churchill,John Phillip
Uma 14h1'0"48d41'
Churchill,Norma Jean "Snake"
Uma 11h55'49"43d25'
Churchill,Sereta
Mon 6h33'18"-0d4'
Churchill-Ward,Leslie
Uma 8h49'32"68d24'
Churchman,Bill
Her 16h56'42"38d34'
Churchson,Marjorie S
Del 20h55'37"10d39'
Churchwell,Tommy D
Hya 10h11'55"-15d24'
Churgin,Amy Roland
And 1h20'0"38d18'
Churilla,John Thomas
Aql 20h1'24"7d46'
Churillo,Denise Blanche
Cas 2h9'52"61d9'
Churillo,Louise Viola
Lyr 18h37'46"27d3'
Churillo,Victor
Per 2h41'15"43d37'
Chusid Ascent
Ori 5h46'44"20d36'
Chusid,Rebecca
Peg 22h34'28"27d5'
Chuspi
Per 3h36'19"38d46'
Chute,Shirley Nicholson &
Howard
Cyg 20h23'24"21d17'
Chuter,Ann V
Peg 21h51'29"33d41'
Chuting Star,The
Aur 5h2'49"49d32'
Chutsch,Michael
Cep 22h3'0"60d26'
Chvany,Nicholas Michael
Lac 22h13'52"50d6'
Chvilicek,My Sunshine Katy
And 23h35'35"46d55'
Chwat,Jr,Edwin Charles
Dra 10h30'30"73d45'
Chyke
Peg 21h20'31"23d21'
Chynna
Del 20h16'28"14d13'
Chypre,Shawn "Sheep"
Aur 5h21'27"41d10'
Chytla,Hannah Gray
Lyn 7h7'15"44d43'
CI GU
Cas 1h59'56"58d49'
Ciabatti,L
Mon 8h8'43"-8d45'
Ciabattoni,Deirdre R
Peg 22h38'39"24d26'
Ciaccio,Amanda Downing
Peg 22h36'46"35d14'
Ciaio,Grace
Lmi 10h2'45"33d22'
Cialdi,Consuelo
Cas 0h14'52"66d24'
Ciallela,Antoinette
Lyn 8h36'17"40d45'
Ciambrone,John
Cyg 21h30'31"40d40'
Ciampa Family,The
Lyn 9h9'55"41d11'
Ciampa,Mary
Hya 8h19'20"5d13'
Ciancagalini,Alejandra
Vul 20h38'28"20d30'
Cianci,Christopher
Per 2h41'14"35d54'
Cianci,Codie James
Lac 22h14'57"54d53'
Cianci,Jennie
Uma 10h16'57"52d11'
Ciancimino,Anthony R
Per 1h29'51"53d21'
Cianciulli,Frank & Kathy
Cam 5h43'1"70d45'
Cianciulliu,David
Her 16h6'41"23d20'
Ciandrha
Tel 20h19'58"-49d57'
Cianfaglione,Susan Ellen
Lyr 18h42'1"45d6'
Cianfarini,Olivia
Lyr 19h16'11"28d11'
Cianfichi,Ken
Cep 22h20'22"56d9'
Ciano,Denise Agnes
Tau 4h8'27"22d39'
Ciano,Diane Marie
Tau 4h4'33"21d14'
Ciao Titi
Per 2h10'35"58d55'
Ciara Elena
Sge 20h3'45"20d7'
Ciara's Twinkle
Cas 1h29'18"68d0'
Ciarain
Dra 10h16'20"77d58'
Ciaramella Star
Tri 2h24'29"31d37'
Ciaramitaro,Kathleen
Oph 17h57'45"12d24'
Ciaran (Angell)
Lyr 19h3'40"28d29'
Ciaranfi,Francesca
And 23h42'26"40d25'
Ciaranfi,Martina
Uma 9h27'38"68d44'
Ciaravino,M D
Leo 9h24'58"7d27'
Ciardullo,Biagio
Cep 22h25'59"61d1'
Ciarkowski,Laura
Cas 1h42'1"63d53'
Ciarleglio,Regina J
And 2h24'1"48d27'
Ciarlone,David F
Aql 19h36'49"0d30'
Ciarmatori,Sonia
Cyg 20h40'51"31d15'
Ciasulli,Anne
Lyn 7h7'34"44d16'
Ciavonne,Linda June
And 0h21'17"32d34'
Ciber,Paul & Corynn Scully
Dra 19h35'44"60d2'
Cibriano,Danny & Clara
Gem 7h14'58"21d53'
Cibulsky,Bill
Ser 15h34'26"-0d6'
Cicalese,Brianna Nicole
And 0h21'44"37d4'
Cicalese,Jimmy
Her 17h2'28"44d51'
Cicalese,Thomas & Filomena
Cam 8h18'35"78d34'
Ciccarelli,Rita
Com 12h23'28"20d10'
Ciccarone,Sofia
Peg 23h29'49"21d10'
Cicchetti 7-19-91
Cam 3h54'12"56d12'
Cicciotta
Leo 10h6'56"18d17'
Cicco,Margaret
And 23h18'22"46d21'
Ciccone,Anthony Michael
Psc 0h58'36"31d49'
Ciccone,Edward Thomas Paul
For 2h34'5"37d40'
Ciccone,Giovanna
Her 16h3'39"16d3'
Ciccone,Madonna Fortin
Com 13h14'0"20d7'
Ciccone,Marsha
Cas 1h17'43"68d2'
Ciccone,Maryann & Jim
Sge 19h38'2"16d29'
Cicconi,Ennio
Dra 15h24'25"53d53'
Ciccotosto,Lorraine
Cas 1h4'0"58d23'
Cicely Rose
And 22h58'21"50d26'
Cicerello,Thomas Joseph
Her 16h28'0"48d38'
Cicero,Chic
Her 18h24'0"20d11'
Cicero,John R
Cep 21h33'1"55d54'
Cicero,Louis Joseph
Cet 2h17'59"4'45'
Cicero,Maria Donata
Cas 1h16'11"63d58'
Cicero,Matthew John
Cet 1h4'1"-1d44'
Cicero,Sandra Tabatha
Del 20h22'21"6d6'
Cicero-Czz-BobRob
Aql 20h15'39"5d27'
Cichonski,Walter
Cyg 20h0'13"37d52'
Cicinskas,Niamh Synclere
Ori 4h56'1"5d14'
Cicio,Adele,Andrea, Marina
Mario
Ant 10h30'55"-31d4'
Cicio,Jillian
Vul 19h46'17"28d9'
Cicio,Joseph
Her 18h11'43"40d3'
Cico
Sco 16h39'42"-38d57'
Cicuttin,Nevea
Oph 17h56'41"10d25'
Cid
Dra 20h8'0"71d12'
Cid's Cricket
Uma 11h51'27"50d34'
Cidela,Colleen Margaret
Cronin
Lyn 7h54'1"43d48'
Cielencki,Christopher Eric
"Worm"
Hya 8h13'46"0d6'
Cieplik,Dziadziu
Crb 15h52'1"28d22'
Ciero,Megan Elizabeth
And 2h22'46"44d9'
Cierra Rose
Vul 19h42'15"20d17'
Ciesielski,Heinc
Lyn 6h31'42"60d59'
Ciesinski,Arthur A
Aur 7h24'47"40d52'
Ciesinski,Wanda
Cas 2h36'0"57d40'
Ciesla,Jeffrey
Boo 14h55'21"35d21'
Ciesla,Maggie
Lyr 18h47'39"33d17'
Ciesla,Virginie
Cam 7h36'12"68d27'
Cieszokowski,Ludomir W
Boo 14h10'30"38d44'
Cifaloglio,Pdpal
Mon 7h9'25"-10d33'
Cifarelli,Denise
And 23h47'1"46d41'
Cifelli,Anthony
Her 18h2'54"28d34'
Cifelli,George A
Cep 23h9'59"65d35'
Cifo '65
Lyr 18h37'50"27d1'
Cifu,Eileen
Ori 5h37'58'15d17'
Ciletti,Larry
Her 17h14'43"50d21'
Ciletti,Larry
Cep 21h6'21"55d5'
Ciliberto,Bernard K
Her 17h6'33"44d21'
Ciliento,Karissa Leigh
Cas 1h4'42"55d55'
Ciliento,Salvatore
Dra 15h56'13"62d19'
Cilione,Peter Stanley
Her 17h23'59"37d56'
Cillien,Danièle
Per 2h49'44"50d6'
Cillo,Geno David
Boo 15h31'15"42d22'
Cimador 3,A C
Peg 22h43'14"30d11'A
Cimaglia,Mary Ann
Vul 19h57'31"28d19'
Cimbolic,Katie
Lyn 6h25'39"60d36'
Ciminelli,Joeseph E
Dra 14h15'36"64d16'
Ciminelli,Joseph J
Boo 15h2'0"26d44'
Ciminera 1995,Michael J
Aur 5h1'13"41d14'
Cimino,Joseph
Oph 18h5'1"10d10'
Cimino,Maria Grazia
Cyg 21h38'50"41d42'
Cimino,Pietro
Lac 22h2'55"50d35'
Cimino,Steven
Cep 22h18'56"55d49'
Cimmarrusti,Linda
Eri 2h57'59"-12d12'
CIN-ALE
Cam 5h27'44"67d57'
Cina,Salvatore
Uma 11h13'1"40d11'
Cinadr,Steven
Hya 8h27'34"-1d27'
Cinatas,Somsoc Spimp
Eripmegaerr
Cma 6h13'55"-18d54'
Cincenelli,Arnaud
Aql 19h29'25"7d44'
Cinda
And 1h24'26"40d7'
Cindee & Kevin's Star
Aur 5h19'1"40d5'
Cindee's Star
Cyg 21h50'21"40d28'
Cindel
Peg 0h1'0"20d37'
Cinders
Mon 7h28'44"-1d1'B
Cindi
And 23h37'54"49d44'
Cindie C
Dra 14h50'14"55d36'
Cindjack
Cas 22h58'34"55d51'
Cindy
Cyg 20h4'52"34d
Cindy
Cas 3h19'15"70d17'
Cindy
Cas 22h57'16"56d57'
Cindy
Cas 0h16'42"58d54'
Cindy
Cma 6h55'54"-15d17'
Cindy
Del 20h32'49"10d44'
Cindy Claire
Cas 23h26'20"61d47'
Cindy Jo
And 23h42'49"38d55'
Cindy Kay
And 2h16'18"42d58'
Cindy Lee
Peg 23h42'24"27d12'
Cindy Lou
Uma 11h22'1"63d50'
Cindy Lou
Vul 20h0'0"22d34'
Cindy Lou
Cas 3h31'23"70d26'
Cindy Lou-Forever Sunshine
Oph 16h56'30"-22d43'
Cindy Lu
Cas 2h2'55"59d39'
Cindy Lynn
Lyr 19h0'42"38d40'
Cindy M
Lib 14h57'55"-18d53'
Cindy M
Vul 19h46'17"20d22'
Cindy Sue
Aql 18h56'25"-2d36'
Cindy Sue
Vel 9h59'21"56d2'
Cindy's Dream
Cas 1h52'33"58d22'
Cindy's Heart
Lmi 9h59'0"33d60'
Cindy's Star
Vir 13h50'55"7d1'
Cindy's Star
Com 12h54'54"30d29'
Cindy's Star
Mon 6h53'54"10d26'
Cindy's Wish
Mon 7h19'22"-5d8'
Cindy's Wishing Star
Cas 0h31'0"75d45'
Cindy(Sonny)
Cyg 21h37'31"41d11'
Cindy, Life is Your Circus
And 0h15'14"33d47'
Cindy-Boo
Crt 10h56'33"-23d32'
Cindykins
Cyg 20h19'23"39d35'
Cinelle
Gem 6h28'33"13d13'
Cinelli,Linda A
Peg 22h59'18"10d4'
Cinelli-Bologna, Francesca
Hor 3h33'44"-49d34'
CineMa star
Sgr 20h4'20"-43d45'
Cingolo,Maurizio
Ori 5h31'1"8d58'
Cinguli
Cma 6h10'38"-16d28'
Cinnabar's"Little Bit"
Uma 11h2'30"56d22'
Cinnamon
Lmi 10h32'24"28d1'
Cinnamon Girl
And 0h49'1"38d16'
Cino,John Christopher
Oph 18h32'28"11d36'
Cinotti,Gina "Cleedus"
And 2h18'41"47d40'
Cinqué
Ori 5h55'0"10d45'
Cinron
Lyr 18h37'54"40d11'
Cinti,Francesca Romana
Col 6h34'45"-39d1'
Cintia
Ori 6h1'36"0d47'
Cintorino,Kelsey May
Cyg 20h3'58"41d11'
Cintron,Carlos
Cyg 21h51'57"37d13'
Cintrg 21h51'57"37d13'
Cyg 21h15'15"28d42'
Cinzia
Nor 16h19'52"-58d31'
Cinzia
Ant 10h45'24"-33d57'
Cinzia
Gem 6h49'59"17d49'
Cinzia il mio rospetto
Sco 16h58'35"-38d3'
Ciochon,Louise
Lmi 9h59'56"33d56'
Ciofani,Luisa
Ori 5h18'29"-4d49'
Cioffi 1993 My Lone
Star,Joann M
Cyg 19h54'39"38d29'
Cioffi,C
Cep 21h38'36"55d51'
Cioffi,Ryan Montana Banks
Cep 21h1'54"55d36'
Cionca,Norbert
Cas 1h40'49"60d22'
Cione,Assunta
Cam 4h11'58"61d8'
Ciotti,Tommy
Uma 9h42'54"70d2'
Ciotti,Tullio
Lac 22h17'43"37d57'
Cipie
Peg 22h1'37"20d23'
Cipolla
Peg 23h3'53"33d48'
Cipolla,Robert John
Ori 5h2'0"0d29'
Cipollini,Giovanna
Cyg 19h28'11"40d46'
Cippone ed Elena, Francesca
Pho 0h28'47"-42d9'
Cipriano,Francesco
Dra 14h52'26"60d5'
Cipriano,Rocco J
Her 17h11'36"22d40'
Cipriano,Sean
Boo 15h6'28"11d10'
Cipsis
Lib 15h44'32"-28d12'
Circle
Ori 6H22'14"10d32'B
Circle Of Family
Aqr 22h7'54"0d29'
Circle Of Harmony Circle Of
Life '94
Umi 15h13'14"78d34'
Circle Unbroken
Sge 20h2'11"20d23'
Circlet
Aql 19h44'0"14d22'
Circo,Dennis & Heather
Crb 15h17'30"31d21'
Circolone,Dr Nicholas J
Oph 17h13'19"-20d36'
Ciric,Svetislav
Uma 11h43'32"30d16'
Cirigliano,Sherri
Cas 2h12'18"60d5'
Cirilli,Brett James
Boo 14h39'0"39d22'
Cirilli,Drew Thomas
Lac 22h27'38"50d13'
Cirillo,Anna
Psc 23h24'26"5d34'
Cirillo,Elisabeth B
And 2h20'30"48d58'
Cirio,Antonella
Boo 14h32'21"11d15'
Ciro
Ser 15h20'38"5d42'
Cirone,Florence Severini
Lyr 18h56'32"40d19'
Cironi,Jeannie
And 0h5'0"46d32'
Cisneros,Diego Alejandro
Lac 22h8'55"47d30'
Cisneros,Francisco José
Aur 7h20'58"38d10'
Cisneros,Roy M
Lac 22h43'26"52d37'
Cisney,Joshua Lee
Her 17h13'25"20d30'
Cissell,Paul "Trader One"
Aql 19h0'33"16d36'
Cister
And 2h19'12"48d58'
Citarella,Paul Joseph
Her 18h54'26"18d57'
Citera,Garry D
Per 4h3'34"52d13'
Citerne,Estelle
Cnv 12h15'15"47d3'
Cito,Karen & Mick
Sge 20h0'1"20d30'
Citrin,Marilyn
And 0h45'34"31d51'
Citron,Dalia
Eri 5h7'35"-5d36'
Citron,David Robert Pastor
Cet 2h11'33"1d48'
Citron,Ronald
Her 17h29'1"31d1'
Cittadino,Susan Anderson
Cyg 20h15'29"31d25'
Citterio,Diego
Pic 5h8'18"-45d27'
Citton,Cristina
Pho 0h6'36"-47d35'
City of Napa
Sct 18h52'56"-7d44'
City Star
Uma 8h50'46"62d19'
Citybeat
Ori 5h57'58"16d57'
Ciudad-Real,Thea "Teddi"
Cep 0h51'24"77d59'
Ciuffani,Chris
Cyg 19h45'37"30d21'
Ciuiletti Star,The Rob & Tracey
Lyn 7h56'49"50d51'
Ciuppa Happy 40th, John I
Lmi 10h11'1"32d7'
Civello,Anthony
Per 3h54'0"31d51'
Civera,Sonia Angel
Cyg 20h3'24"31d37'
Civit,Karla J
And 1h18'22"35d34'
Civitano,Gerri
Mon 8h8'29"-0d55'
Civitella,Oscar
Cma 6h24'13"-15d25'
Civitella,William N
Leo 9h20'22"14d41'
Civitillo,Sr,Nick & Lee
Cnv 12h12'10"46d30'
Cizauskas' Star
Lyr 18h45'53"33d40'
CJ & Nancy
Cyg 19h26'55"32d4'
CJ's White Wolf
Crb 16h8'26"28d13'
CKV Loves NWS 3/14/94
Uma 9h18'12"58d43'
CL-OL-MA-DA-MI
Cam 5h57'28"58d41'
Claasen Cum Caelum
Undique,Annelies V
Cyg 20h26'50"38d38'
Clabaugh,Shirley Frances
Mon 7h46'25"-3d57'
Clack,Roy Douglas
Aur 5h50'25"41d8'
Claeys,Lois Ann
Cyg 21h39'17"41d42'
Claffey,Kathy Murphy
Lyn 7h48'18"51d56'
Claggett,Tom
Aql 18h59'11"0d48'
Clahar,Dorothy
Cam 5h35'31"60d26'
Clair
Crb 16h3'0"30d49'
Clair & Kathy
Com 12h47'25"21d14'
Clair De Luna
Vir 14h28'48"5d59'
Clair NSW
And 2h34'53"50d16'
Clair,Christina Marie
And 0h17'54"36d55'
Claire
Sge 20h3'58"16d1'
Claire
Lyn 7h59'34"40d40'
Claire
Her 18h19'1"12d42'
Claire
Sge 19h27'40"18d46'
Claire
Del 20h18'0"16d27'
Claire
Del 20h50'45"8d42'
Claire & Jessica
Peg 21h55'13"34d38'
Claire et Bill Etoile D'Amour
Umi 16h15'53"70d59'
Claire Jane
Peg 23h26'47"24d3'
Claire T
Lyr 18h57'13"45d23'
Claire's Candle
Uma 14h1'33"48d42'
Claire's Celestial Celebration
Aur 6h5'41"36d25'
Claire's Friend
Eri 3h56'20"-6d10'
Claire's Sapphire
Oph 18h16'39'12d23'
Claire's Star
Cas 1h54'44"73d38'
Claire,John
Cnv 12h50'0"40d44'
Claire,Martin
Dra 20h21'46"62d4'
Claire-Emily Etoile
Lyr 18h57'55"32d22'
Claire-Line
Cyg 20h24'15"39d21'
Clairon et Milou
Lyr 18h36'37"34d3'
Claiser,Gary & Vicky
Lyr 19h2'44"31d19'
Clamity Janis
Cnv 12h43'53"32d26'
Clammer,Samuel R
Aql 19h30'17"10d38'
Clancey The Bum
Mon 8h3'41"-9d16'
Clancy
Ori 5h57'34"16d42'
Clancy,Aisling
Cam 5h37'53"60d53'
Clancy,Carmel
Uma 9h36'1"58d44'
Clancy,Christine
Uma 12h28'16"61d3'
Clancy,Eileen"Mumzo"
Cas 3h33'13"70d39'
Clancy,Gail
Uma 9h59'36"51d15'
Clancy,Gerard Mary
Peg 23h4'1"31d9'
Clancy,Janet Lynn Matayosian
Ori 5h32'12"-1d58'
Clancy,Jr,Patrick Joseph
Dra 16h54'50"68d20'
Clancy,Linda
Eri 4h13'53"-12d50'
Clancy,Niall
Aur 6h31'21"38d50'
Clancy,Susan
And 23h11'16"41d36'
Clann Chaomhanach
Dra 16h16'47"67d15'
Clanton,Don
Tri 1h55'47"27d31'
Clanton,Joe
Cyg 20h17'13"37d49'
Clanton,Jr,Fred E
Boo 15h5'38"48d2'
Clanton,Shirley
Lyr 18h28'1"34d32'
Clapp,Jacqueline
Cas 1h1'21"66d32'
Clapp,Roger William
Aql 19h19'0"15d9'
Clappa,Frederic
Lyn 7h45'14"41d7'
Clapper,David Paul
Aql 19h49'49"10d36'
Clapper,David R
Cep 23h24'1"67d48'
Clappsy,Lisa
Lyn 7h59'43"44d16'
Clapton,Conor
Uma 8h34'41"67d39'
Clapton,Eric"Slowhand"
Cep 22h12'0"55d18'
Claquin,Patrick
Her 18h24'11"12d30'
Clara
Peg 23h39'19"15d47'
Clara
Cet 3h3'55"5d15'
Clara
Vel 9h43'0"-47d15'
Clara
And 1h56'19"38d22'
Clara
Mon 7h41'48"-5d35'
Clara
And 23h20'1"40d37'
Clara & Thomas Benson
Cyg 19h43'36"31d13'
Clara D's Nova
Cyg 21h9'31"39d51'

Clara Jo
Aql 19h2'54"16d26'
Clara Rose
Peg 22h14'37"33d4'
Clara Sofia
Lyr 19h20'53"30d21'
Clara T,The
Mon 7h41'22"-3d58'
Clarabelle's
Lyr 19h22'23"31d12'
Clare
Boo 14h10'36"52d13'
Clare
Cas 23h35'35"61d8'
Clare 402882271
Peg 23h29'15"23d18'
Clare Alexandra
Umi 15h53'36"70d13'
Clare Frances
Cyg 21h24'53"28d43'
Clare,James Hugh
Aql 18h57'0"4d59'
Clare Luke
Gru 23h16'57"54d59'
Clare,Vera
Cep 22h5'17"67d44'
Clare-Louise
Umi 13h42'19"73d55'
Clarence & Betty Friendship Star,The
Umi 14h40'22"80d16'
Clarese
Cam 11h38'27"78d13'
Claretta
Cnc 7h56'30"18d51'
Clarice's Christmas Star
Cam 3h19'52"55d5'
Claridy's Star
Mon 7h1'0"7d36'
Claris
Cyg 21h30'59"40d15'
Clarissa
Cap 20h4'17"-10d18'
Clarissa's Star
Cma 7h0'34"-13d22'
Clark
Aur 6h28'60"33d59'
Clark "50",Adele & Gerard
Peg 21h43'1"23d35'
Clark & Lois
Cyg 21h41'39"29d12'
Clark 1960-1992, Stephen Maynard
Sex 10h33'24"3d50'
Clark 60,The
Vul 19h17'24"25d24'
Clark 9032,Gemma
And 0h48'50"38d7'
Clark Amos & Mazzie
Ori 6h6'46"7d39'
Clark APMD Marketing Manager,Jeffery J
Uma 12h5'38"62d25'
Clark Boys,The
Her 17h8'45"44d17'
Clark III,Francis Edward
Aur 6h13'1"37d28'
Clark III,Robert Thomas
Lmi 10h2'22"40d35'
Clark Our Wishing Star,Casey Robert
Peg 23h7'38"13d14'
Clark Star,The Susan Ann
Ori 5h15'19"11d20'
Clark"The Hound",Camas C
Gem 6h27'53"13d5'
Clark's Lucky Star
Lac 22h15'10"49d51'
Clark,"Big Brother" Richard Burton
Lyn 8h15'18"58d40'
Clark,Adam
Crb 16h14'33"37d34'
Clark,Agnes Clara
Vul 19h43'14"23d17'
Clark,Alana Ann "Sunshine"
And 2h7'50"42d20'
Clark,Alfreda H
Sge 20h16'11"20d36'
Clark,Alice L
Cas 23h35'12"60d11'
Clark,Allen Gauld
Cep 22h21'20"63d53'
Clark,Amanda Lee
And 0h6'0"47d33'
Clark,Andrea
Uma 14h35'56"80d45'
Clark,Andrew Daniel
Cma 7h13'1"-13d26'
Clark,Andrew"The Earl"
Aql 19h30'35"1d30'
Clark,Ann Lindsey
Cet 0h6'53"-8d24'
Clark,Arlene
And 23h26'16"48d55'
Clark,Arthur & Mary
Hya 9h4'51"-11d31'
Clark,Ashlie Victoria
Cyg 21h11'13"38d48'
Clark,Barbara Ann
Uma 9h34'48"68d22'
Clark,Barbara DePalma
Oph 18h3'31"11d13'
Clark,Benjamin Lee
Aql 18h42'31"-2d2'
Clark,Benjamin Leroy
Cmi 8h2'20"5d41'
Clark,Betty J
And 0h51'35"34d11'
Clark,Bill Superman
Boo 14h26'56"22d19'
Clark,Boofy
Her 17h16'47"46d54'
Clark,Brian
Dra 19h5'1"70d38'
Clark,Brian
Sex 10h18'17"-3d11'
Clark,Brian W
Aql 18h41'17"-0d13'
Clark,Bruce
Ori 5h3'37"1d34'
Clark,Bunny
Cyg 20h6'27"40d58'
Clark,Carol
Aqr 23h7'14"-11d51'
Clark,Carol M
Peg 23h30'43"20d7'
Clark,Carolyn
Cas 1h23'1"61d10'
Clark,Charles
Sge 19h6'18"18d21'
Clark,Cheryl Leigh
Cas 2h37'44"70d21'
Clark,Christian Whittemore
Aql 19h48'1"12d18'
Clark,Christine
Uma 9h28'1"52d38'
Clark,Christopher
Cep 23h0'18"68d36'
Clark,Cleve
Cma 6h55'60"-19d52'
Clark,Colin
Uma 14h44'32"80d33'
Clark,Connie Joan
Sct 18h36'0"-4d49'
Clark,Cruse P
Oph 17h22'48"-24d31'
Clark,Dan
Per 2h36'57"57d7'
Clark,Danielle Louise
Lyn 8h37'35"43d6'
Clark,Dave
Crt 11h13'58"-18d3'
Clark,Dede
Mon 7h41'8"-2d0'
Clark,Derbigny Murrell
Mon 7h15'9"-5d22'
Clark,Deshon
Aql 20h15'52"0d23'
Clark,Dr Dayton Reed
Oph 17h14'41"-23d4'
Clark,Drucilla Martin
And 2h34'40"37d39'
Clark,Dwight 87
Oph 16h50'24"-28d32'
Clark,Dylan Jason
Vul 19h47'27"28d14'
Clark,Edward
Aqr 22h49'35"0d19'
Clark,Edward George
Oph 17h38'13"-16d27'
Clark,Edward Leo
Mon 7h9'60"-10d46'
Clark,Eleanor Louise
Hya 8h52'27"1d28'
Clark,Elizabeth Anna
Peg 23h21'0"31d58'
Clark,Emilia Ann Serendipity
Del 20h20'41"9d48'
Clark,Emily Anne
Cas 1h49'13"58d15'
Clark,Emmelia Grace
Lyr 18h16'29"47d24'
Clark,Erin E
Sgr 18h48'17"-35d26'
Clark,Evelyn C
Cyg 20h29'30"42d49'
Clark,Fiona
Cyg 19h31'29"38d15'
Clark,Fran
And 0h45'0"31d7'
Clark,Frank
Ser 15h55'51"24d16'
Clark,Gail
Cas 23h42'26"64d3'
Clark,Gail & Richard
Uma 12h3'8"63d12'B
Clark,Gary
Her 16h29'1"33d11'
Clark,Georgia Connie
Vul 20h16'48"25d7'
Clark,Gertrude
Cam 3h29'51"61d54'
Clark,Gigi & Poppy
Lyr 18h39'26"42d56'
Clark,Gina
And 1h25'25"36d18'
Clark,Glen F
Aur 6h12'10"31d19'
Clark,Goodwin Murrell
Mon 7h8'5"-6d36'
Clark,Haley Elaine
Lyn 7h55'18"50d31'
Clark,Harold "Sparky"
Aql 20h0'43"6d1'
Clark,Hazel Murphy
Aql 20h0'34"10d19'
Clark,Heather Elizabeth
Mon 7h19'1"-1d6'
Clark,Heather & Sean
Cyg 21h19'52"38d42'
Clark,Helen A
Cet 1h33'42"-11d13'
Clark,James & Anne Granger Deckert
Dra 18h39'33"58d42'
Clark,James Donald
Lac 21h56'50"41d49'
Clark,James E
Ori 6h0'21"5d40'
Clark,Jason
Lyn 8h10'18"44d32'
Clark,Jennifer M
Cas 0h38'51"68d58'
Clark,Jerrell (Mr "C")
Ser 17h59'0"-14d58'
Clark,Jesse A
Boo 15h28'49"38d8'
Clark,Jessica
Lyn 8h6'21"40d42'
Clark,Jimmy
Her 16h39'23"24d3'
Clark,Joan Mary
Cnc 9h9'19"31d1'
Clark,John Arthur
Sex 9h53'36"0d20'
Clark,John David
Per 2h53'21"35d15'
Clark,John Ernest
Dra 17h4'27"64d17'
Clark,John R
Cep 20h33'38"63d27'
Clark,Joseph Embry
Cep 20h45'0"70d37'
Clark,Joseph George
Her 17h18'1"47h8'
Clark,Jr,John R
Cnv 13h50'41"38d5'
Clark,Jr,Roland Trenner
Vul 19h19'52"26d48'
Clark,JudArt S
Peg 23h49'52"31d48'
Clark,Kai
Lyn 6h14'39"58d47'
Clark,Katherine Francis
And 23h38'31"40d2'
Clark,Kathryne Louise
Boo 13h37'31"19d31'
Clark,Kelly Elizabeth
Cas 23h37'24"61d48'
Clark,Kelly Lynn
Com 12h5'34"19d37'
Clark,Kenneth J
Cyg 20h3'0"40d10'
Clark,Kenny
Cep 0h8'18"71d13'
Clark,Kerry Denise
Lyr 19h5'35"28d56'
Clark,Kevin Richard
Uma 12h3'8"63d12'A
Clark,Kim R
Lac 22h13'15"49d35'
Clark,Kim S
And 2h34'12"41d33'
Clark,Kimberlee Ann
Cas 0h37'33"60d21'
Clark,Kimberly Susan
Mon 7h43'35"-1d28'
Clark,Lady Susan Marie
Cet 1h23'36"-14d11'
Clark,Laura
Lyn 7h30'32"35d28'
Clark,Laura Derbigny
Mon 7h12'19"-5d9'
Clark,Lee
Cnc 8h48'45"32d35'
Clark,Linda
Peg 21h49'33"33d18'
Clark,Lisa Marie
Vul 19h48'22"25d45'
Clark,Lola
Peg 22h18'52"4d51'
Clark,Lori Anne
And 0h54'31"34d48'
Clark,Marilyn Kaye
Aql 19h7'18"15d31'
Clark,Marion Florence Ward
Umi 16h39'55"78d18'
Clark,Mark Willis
Aur 5h3'14"40d46'
Clark,Marsha
Eri 3h32'10"-2d42'
Clark,Martin Andrew
Cep 22h24'50"63d45'
Clark,Mel
Cet 1h34'24"-6d13'
Clark,Melba Maria
Cyg 19h31'11"33d32'
Clark,Melissa
Ser 15h37'59"18d38'
Clark,Michael Anthony
Lmi 10h45'0"28d10'
Clark,Michael David
Peg 23h4'1"18d58'
Clark,Moema
Hya 8h41'28"5d42'
Clark,Natalie Claudia
Lyr 18h26'47"38d20'
Clark,Neil R
Cep 23h24'22"65d38'
Clark,Nicole & Jim
Cyg 20h10'11"38d13'
Clark,Nicole T & Theresa Kolman
Uma 9h29'30"50d24'
Clark,Olive Rider
Cas 1h52'22"73d12'
Clark,Oraine
Aql 20h6'0"6d36'
Clark,Patricia Gail
Lyr 19h4'28"28d33'
Clark,Pegi
Cas 1h3'36"62d51'
Clark,Peter Bond
Vul 19h48'29"28d57'
Clark,Peter Leslie
Dra 17h47'20"73d18'
Clark,Philip
Dra 17h4'49"60d47'
Clark,Rachel Nicole
Cam 3h48'45"61d39'
Clark,Raymond R
Cam 8h4'30"71d1'
Clark,Richard
Aur 6h39'33"38d51'
Clark,Richard James
Boo 14h33'27"21d6'
Clark,RoberLind W
Lyr 18h42'50"40d26'
Clark,Robert A
Cep 21h57'46"61d20'
Clark,Robert Brett
Lac 22h8'36"51d3'
Clark,Robert G
Aql 19h29'41"13d13'
Clark,Robert Lloyd
Per 2h11'53"56d35'
Clark,Rorrie Anne
Peg 22h0'39"25d10'
Clark,Sarah T
Mon 6h55'48"1d17'
Clark,Scott Milton
Mon 7h10'1"-5d24'
Clark,Shannon L
Aqr 22h47'23"-6d8'
Clark,Simon Gordon
Dra 16h58'17"62d42'
Clark,Star Lover, Ernesto M
Ser 17h55'59"-14d47'
Clark,Star Lover, Ernesto M
Lib 15h4'1"-2d51'
Clark,Stephanie Lynn
Cam 7h37'15"61d37'
Clark,Stephen W
Boo 14h27'28"45d22'
Clark,Susan
Peg 22h46'0"25d10'
Clark,Suzanne Michelle
Cas 0h24'19"58d59'
Clark,The Star Of Jack W
Ori 5h58'28"6d6'
Clark,Tina M
Mon 6h2'53"-5d42'
Clark,Tracey
Hya 0h31'38"6d15'
Clark,Tracy Nanning
Peg 23h7'28"14d1'
Clark,Tresa Diane
And 0h26'41"27d50'
Clark,Trudy
Aql 20h11'30"1d24'
Clark,Verna
Cam 14h3'45"82d12'
Clark,Veronica Carol
Cas 23h15'25"60d5'
Clark,Virginia F
Crb 15h34'53"31d9'
Clark,Walter
Per 1h25'40"53d46'
Clark,William Henry
Dra 18h25'1"80d28'
Clark-James,Chelsea Alana
Lyr 18h31'53"30d58'
Clark-James,Graeme William
Her 16h10'15"20d32'
Clark-Pilot,Gimper, Grandpa,Robert John
Lyn 8h12'48"47d45'
Clark-VanBrunt,Abra Hannah
Ori 6h5'57"10d24'
Clarke
Aur 6h29'17"33d32'
Clarke III,Ernest J
Aur 6h14'54"32d47'
Clarke's "Twitch Switch",Russ T
Cet 0h32'58"-1d27'
Clarke,Alice Mary
Cas 1h6'41"61d16'
Clarke,Ann Marie
Uma 10h8'16"50d55'
Clarke,Annette
Lyr 18h51'40"41d7'
Clarke,Anthony Leonard
Uma 9h24'0"53d35'
Clarke,Ariel T
Boo 14h14'13"20d21'A
Clarke,Benjamin Robert
Lac 22h32'1"50d4'
Clarke,Bernadine
Uma 13h50'37"51d36'
Clarke,Beth
And 0h13'21"47d44'
Clarke,Brittany Alexa
Mon 8h5'34"-9d4'
Clarke,Caitlin Amanda
Uma 8h30'12"49d47'
Clarke,Caitlin McNair
Lyr 19h2'39"38d19'
Clarke,Christine
Lyr 18h35'56"26d0'
Clarke,Christopher Delamere
Lyr 19h12'38"34d28'B
Clarke,Colette
Lyn 8h40'47"38d24'
Clarke,Daniel Lee
Cep 22h58'24"68d0'
Clarke,Darci
And 0h20'0"31d40'
Clarke,David
Lyn 8h5'56"45d11'
Clarke,Dennis
Aql 18h57'0"12d18'
Clarke,Eileen M
Uma 9h2'17"51d58'
Clarke,Elisabeth Avery
Psc 01h22'46"12d12'
Clarke,Elizabeth C
Cas 0h45'26"61d41'
Clarke,Gill
Cas 0h17'34"63d20'
Clarke,Grace
Lac 22h4'0"50d13'
Clarke,Harry
Oph 17h36'16"-22d38'
Clarke,James A
Her 16h44'59"50d4'
Clarke,Jane Ann Efner
Uma 10h20'32"48d21'
Clarke,Janice
Uma 9h41'18"53d50'
Clarke,JB & Laurie
Aql 18h44'47"10d54'
Clarke,Joan
Cas 0h38'1"60d31'
Clarke,Joshua Nivy
Tau 5h47'43"23d30'
Clarke,Joyce
Cas 0h3'49"61d8'
Clarke,Justin John
Uma 10h22'13"57d42'
Clarke,Kerry Ann
Cyg 19h56'19"48d22'
Clarke,Kristin Nicole
Eri 4h6'59"-12d29'
Clarke,Laura Michelle
And 0h23'56"33d41'
Clarke,Leslie
Crt 11h16'38"-16d60'
Clarke,Linda
And 23h16'44"47d7'
Clarke,Marie A
And 1h6'56"37d51'A
Clarke,Marie Geraldine
Boo 15h4'25"31d44'
Clarke,Marvin F
Per 4h32'37"52d11'
Clarke,Mary Elllen
Cyg 20h18'15"38d13'
Clarke,Meagan A
Boo 14h14'13"20d21'B
Clarke,Michael Andrew
Oph 18h6'31"10d1'
Clarke,Nina Ellen
Cas 0h24'1"61d49'
Clarke,Patrick Michael
Dra 19h29'0"68d4'
Clarke,Robert J
And 1h6'56"37d51'B
Clarke,Roberta Pauline
Cas 0h32'12"63d11'
Clarke,Sandra Lee
Hya 8h10'29"0d51'
Clarke,Sheila Lorimore
Hya 8h27'55"-8d47'
Clarke,Simon Edward Crispian
Aur 5h36'46"40d27'
Clarke,Susan
And 0h4'13"44d37'
Clarke,Wilbur Patrick
Lac 22h18'28"46d55'
Clarke,Wild Bill & Angel Jana
Uma 9h17'24"57d38'
Clarke-DSS DTR,Skip
Cma 6h55'43"-18d9'
Clarke-Pulchritudo Ego Amor,Heather
Aql 19h1'0"-7d34'
Clarken,Colleen
Del 20h40'30"10d8'
Clarkson,Anthony Thomas
Gem 6h43'34"35d17'
Clarkson,C Jack
Oph 17h5'16"8d2'
Clarkson,George & Elizabeth
Crb 15h19'0"30d46'
Clarkson,Hannah
Lyn 8h4'41"44d35'
Clarkson,Jake
Cet 0h25'55"-1d29'
Clarkson,Laura Diane
Cam 5h56'31"70d57'
Clarkson,Robert Halliday
Her 17h16'21"27d29'
Claro,Kevin Scott
Lac 22h27'27"40d21'
Clarquist,Rodney E
Dra 14h59'29"62d53'
Clas/A
Cas 22h57'44"55d43'
Clasby,Greg "Pugsley"
Cnv 13h22'1"51d22'
Clasen,Laura
Cas 0h18'24"63d11'
Clash,Shannon Elyse
Cap 20h44'27"-26d56'
Class of 1993,The
Com 12h3'0"26d42'
Class,Marisel
Aur 6h22'40"33d3'
Classic,Gusi
Cnv 12h8'0"37d7'
Clastres,Paulette
Ser 16h4'28"8d47'
Clattenburg,Connie Frisky
Cam 3h41'22"61d44'
Clattenburg,Ray & Shirley
Ori 5h57'29"16d43'
Clatterbuck,Carl Hamilton
Aur 5h31'22"50d13'
Clauberg-Weil,Judith
Tau 3h56'11"1d44'
Claud,Joshua
Hya 8h30'46"1d5'
Claud,Justin
Sct 18h54'36"-7d2'
Claude
Boo 15h3'37"52d28'
Claude
Boo 14h47'38"48d23'
Claude
Dra 10h41'40"74d16'
Claude
Her 17h51'52"40d59'
Claude
Cyg 21h15'1"38d45'
Claude
Ori 5h46'29"18d46'
Claude
Cet 2h34'20"9d40'
Claude Et Lazovito Noadja
Cam 6h56'24"64d50'
Claude,Marie
Cas 2h0'13"73d42'
Claude-André
Uma 11h34'49"30d58'
Clauder,Kirk
Lyr 19h26'46"40d23'
Claudette
Cas 0h8'1"61d6'
Claudette
Cyg 19h51'43"44d8'
Claudette
And 2h21'0"47d26'
Claudette Heather Joy
Lyr 19h22'11"31d46'
Claudia
Cnc 9h12'23"7d38'
Claudia
Aqr 21h19'60"-13d10'
Claudia
Vul 19h47'54"27d53'
Claudia
Pho 0h4'41"-42d2'
Claudia
Pup 7h56'45"-28d30'
Claudia
For 2h13'1"-27d53'
Claudia
Lyn 7h52'59"34d35'
Claudia
And 23h0'37"44d26'
Claudia
Her 18h19'17"14d13'
Claudia
Del 20h19'48"20d20'
Claudia
And 23h24'35"44d45'
Claudia
Lyr 18h43'11"33d32'
Claudia
Cyg 19h26'1"35d22'
Claudia
Peg 23h31'55"12d22'
Claudia
Peg 23h31'59"18d46'
Claudia
Cep 22h6'44"61d38'
Claudia
Cet 2h54'1"1d27'
Claudia
Lyr 18h38'43"29d1'
Claudia
Aql 19h25'24"-10d7'
Claudia
Cas 1h56'59"60d37'
Claudia
Cyg 21h28'1"48d47'
Claudia
Her 16h10'43"48d14'
Claudia "Claudel"
Cap 20h22'46"-21d6'
Claudia & Jörg
Tau 3h52'16"1d15'
Claudia & Mike
Aur 4h50'13"38d16'
Claudia Alejandra
Cas 2h7'57"59d24'
Claudia Ann
And 23h42'28"46d3'
Claudia Jean
Cep 21h55'46"61d3'A
Claudia Joan
Cmb 12h7'45"27d9'
Claudia Tee
Peg 21h25'32"22d52'
Claudia"Schatzi"-Juist
Ari 2h24'0"20d59'
Claudia's Dream
Eri 4h12'55"-15d24'
Claudia-The Love of My Life
Dra 18h55'29"67d43'
Claudia65
Peg 23h26'23"17d45'
Claudine
Cas 1h29'57"65d21'
Claudine
Lmi 10h36'19"37d53'
Claudine
And 23h20'44"35d16'
Claudio
Pho 0h7'34"-41d12'
Claudio
Ant 10h45'48"-31d45'
Claudio
Ind 21h10'18"-51d7'
Claudio
Cnc 8h14'0"30d44'
Claudio
Cet 1h54'26"-5d11'
Claudio & Cynthia Forever
Per 3h11'0"47d50'
Claus,Christine V
Cam 3h56'0"56d28'
Clauschee,Berkley Sterling
Cyg 21h10'54"39d45'
Clause,Holly Olivia
Ori 5h43'17"11d27'
Clausell,Neomi "Lil' Rico"
Aql 19h31'42"1d39'
Clausen,Andy
Aql 20h33'42"-7d41'
Clausen,Carl Christian
Aur 5h32'55"38d24'
Clausen,Desty
Crb 16h0'52"37d59'
Clausen,Emil F
Vir 12h27'1"-6d32'
Clausen,George Nils
Boo 13h36'22"14d31'
Clausen,Helene Rae
Peg 22h23'27"3d17'
Clausen,Tove
Cas 0h11'24"61d15'
Clauser,Todd
Per 2h55'60"32d55'
Clauson,Lori
Leo 10h56'1"12d23'
Clauss,Nadine
And 0h56'59"38d33'
Claussen,Standley & Janice
Aql 19h56'48"15d49'
Clauzel,Pauline
Umi 15h59'21"73d23'
Clavelou,Dominique
Aur 6h0'0"38d58'
Clavenna,Carlo F
Per 1h30'16"53d44'
Clavet,Michel
Uma 10h3'0"53d8'
Clavette,Sollemne Durette
Cep 22h26'0"63d1'
Clave,Nathalie
Sgr 19h26'37"-38d40'
Clawson's Eastern Star Tammy
Mon 6h42'48"6d32'
Claxton,Kasey Marie
Cas 2h20'27"61d10'

Claxton,M Isabelle
And 0h12'53"32d27'
Clay
Aql 19h56'41"8d4'
Clay 50
Lac 22h48'18"52d59'
Clay Star,The Lisa Marie
Ori 6h9'21"9d13'
Clay's "Celestial Song"
Dra 11h31'11"73d27'
Clay,Caroline Talitha
Uma 13h31'39"60d35'
Clay,Carrlyn G
Leo 10h31'2"23d36'B
Clay,Elaine M
And 2h21'59"39d39'
Clay,Gary R
Leo 10h31'2"23d36'A
Clay,Joe M
Cet 0h55'39"0d55'
Clay,Theresa Kay
Cyg 19h40'1"41d39'
Clayes,Rose
Lyn 7h46'1"48d29'
Clayleen
Uma 8h35'33"60d47'
Clayton
Cyg 19h28'12"31d52'
Clayton
Vul 20h17'28"22d57'
Clayton John & Kristin Dawn Star,The
Ori 5h52'38"14d57'
Clayton Valley H S Grad Nite '96
Aql 19h29'29"12d4'
Clayton,Adam
Aur 4h54'0"51d1'
Clayton,Adam Planera
Cep 22h36'0"63d52'
Clayton,Anne
Peg 22h1'0"21d39'
Clayton,Caroline
Lyr 18h41'22"40d39'
Clayton,Caroline V
Lyr 18h25'39"45d17'
Clayton,Cheryl A
And 0h20'57"43d17'
Clayton,CLU,CFP,Randy J
Cep 3h45'43"78d47'
Clayton,Elizabeth D
Psc 1h39'59"20d44'
Clayton,Henry
Aql 19h9'39"13d14'
Clayton,Janice
Lyn 7h40'16"58d18'
Clayton,Jason Scott
Cnc 8h52'42"7d3'
Clayton,Jeffrey
Cep 20h39'43"75d40'
Clayton,Laura & Norma
Leo 11h21'58"-2d21'
Clayton,Lee Arthur
Aql 19h51'18"11d40'
Clayton,Martyn
Pho 0h30'58"-41d0'
Clayton,Paul Morris
Aur 7h14'20"40d15'
Clayton,Randy L
Per 1h54'55"52d58'
Clayton,Robert
Ori 5h43'44"10d8'
Clayton,Vicki Diane
And 0h10'32"47d11'
Clayton,William
Aur 5h25'0"30d53'
Claytor,Kevin L
Lac 22h46'13"56d2'
Claytor,William Scott
Her 16h9'1"5d36'
Clazone,Caylene
Lyn 9h5'34"44d13'
Cleal,Luke Daniel Raymond
Cru 12h45'54"-57d35'

CLEANDRE
Aql 19h56'0"13d44'
Clear,Yvonne E
Eri 4h36'1"-18d44'
Clearmont,Penny
Mon 8h4'52"-8d59'
Cleary,Brian & Anne
Ori 6h1'21"0d44'
Cleary,Cedric P R
Uma 9h6'44"48d13'
Cleary,Claire Elizabeth
Cnv 13h32'53"51d46'
Cleary,Eric Clapton
Her 15h59'18"50d29'
Cleary,Heather Lyn
Dra 18h1'49"58d19'
Cleary,Jennifer Ryan
Aqr 22h41'19"-0d54'
Cleary,Jim
Umi 15h20'50"71d7'
Cleary,Jr,Eugene D
Per 2h22'33"55d14'
Cleary,Judith Lackey
Mon 6h22'1"3d42'
Cleary,Linda & Jeremiah
Uma 9h46'36"42d23'
Cleary,Lorraine "Sparky"
Mon 7h56'54"-8d56'
Cleary,Mary Ann
Cap 21h35'55"-21d48'
Cleary,Patricia
Lyn 7h28'12"39d45'
Cleary,Shannon Catherine
Lyn 7h3'25"50d5'
Cleary,SJ,Reverend Herbert J
Per 3h52'0"38d54'
Cleary-Lamoureux,Kevin James
Per 2h16'33"58d58'
Clease,Bobby
Crb 16h13'24"28d17'
Cleaveland,John "Honey Bun"
Lyn 7h10'7"48d35'A
Cleaveland,Scott & Shannan
Aql 19h3'0"5d53'
Cleaver,E J "80"
Lac 22h16'1"46d7'
Cleaver,Joseph M
Sex 10h45'45"-5d16'
Cleaver,Martin
Oph 18h42'19"7d16'
Cled
Cyg 19h57'1"38d55'
Cleeland,Helen Martha Bacevice
Mon 8h5'51"-1d40'
Cleeland,Raymond T
Lac 22h44'20"56d32'
Cleenput,Denise Marie
And 2h13'52"42d52'
Clegg,Erin H
Cnv 13h41'26"35d56'
Clegg,Jordan Jennifer
And 1h44'45"38d2'
Clegg,Joseph
Dra 10h38'24"78d19'
Clegg,Jr,Peter
Ori 5h0'35"8d53'
Clegg,Leslie
Aur 6h31'40"38d50'
Clegg,Polly
Mon 6h21'17"8d41'
Cleggie
Boo 14h19'50"29d28'
Cleland,Kara L
Lyr 18h40'19"42d40'
Cleland,Kresta L
Mon 7h1'37"4d39'
Cleland,Luke T
Lac 22h17'25"51d1'
Cleland,Teresa
Mon 7h13'53"-6d57'
Cleland,Virginia K
Lmi 10h18'38"33d33'

Clelia,Maria
Lac 22h14'15"55d19'
Clem & Sophie
Cyg 21h37'16"40d36'
Clem Roy
Lac 22h7'41"49d39'
Clem,Brian Alan
Cet 2h27'55"5d27'
Clem,Clyde & Milly
Cyg 19h29'17"48d57'
Clem,Richard Joseph
Her 17h19'60"20d25'
Clem-Jed
Boo 14h56'0"41d22'
Clema
And 0h14'22"39d11'
Clemans,Barry David
Cam 8h2'36"61d8'
Clemen,Dorothy
Com 12h20'23"20d10'
Clemen,Shawn
Per 4h25'33"51d23'
Clemence
Equ 21h22'46"11d42'
Clemenko,Buffi
Cmi 7h44'1"1d58'A
Clemenko,David
Cmi 7h48'9"3d24'A
Clemenko,Maria
Cmi 7h48'9"3d24'B
Clemenko,Randi
Cmi 7h33'39"1d10'
Clemens,Geoffery John
Ori 5h55'34"19d27'
Clemens,Jackson Bryce
Tri 2h16'16"30d0'
Clemens,Judge
Her 15h59'32"47d22'
Clemens,Karen
Cas 1h35'1"61d38'
Clemens,Libby Scopino
And 22h58'18"50d18'
Clemens,Michael William
Aqr 21h36'60"-6d17'
Clemens,Sarah Rae
Cas 0h47'55"61d6'
Clemens,Virginia
Mon 6h20'0"0d35'
Clemens-Varwig,Debbie
Cyg 21h30'59"40d15'
Clemenson,Susan Beth
Vul 19h50'36"20d19'
Clement
Per 2h31'11"51d27'
Clement,Anne Lisbeth
Lyr 18h29'59"40d58'
Clement,George W
Aur 7h1'0"35d34'
Clement,Janet Mae
Mon 7h38'4"-4d8'
Clement,Jordan Austin
Mon 6h32'29"5d14'
Clement,Michael Warren
Cyg 19h25'28"30d26'
Clement,Phoebe Heather
Lyr 18h17'58"38d20'
Clement,Star of Lisa Jo
And 23h24'52"42d56'
Clement,William Kerr
Uma 9h15'0"54d12'
Clement-Thieme
Cyg 19h26'58"33d1'
Clemente,Brian John
Ori 6h3'15"1d18'
Clemente,Emil L
Aur 6h9'31"33d15'
Clemente,I Love You Anthony
Cet 1h19'57"-11d38'
Clemente,Kyle Anthony
Lac 22h36'55"55d3'
Clemente,Melissa
Aql 18h58'19"0d50'
Clemente,Ramsey A
Cam 5h57'44"73d35'

Clemente,Sr,Kurt J
Lac 22h27'42"54d16'
Clementina
Lup 15h29'17"-45d45'
Clementine
Per 3h38'1"51d18'
Clements in Heaven Dad,Billie Jean
Fri 2h46'41"-1d37'
Clements Ret'd CMsgt WFTX,Billy J
Cma 7h3'43"-15d21'
Clements Star,The
Sge 20h1'41"19d4'
Clements,"Taylor" Dawn Gloria
Mon 8h6'45"-8d31'
Clements,Billy & Janine
Dra 20h16'37"63d16'
Clements,Bryan Patrick
Oph 18h17'51"12d35'
Clements,Carla Anne
Eri 3h43'47"-6d29'
Clements,Carol Louise
Lyr 18h18'0"42d41'
Clements,Debbie
Ori 5h52'12"19d18'
Clements,Edward "Butsie"
Boo 14h6'17"25d56'
Clements,Ella Kathleen
Mon 6h35'55"-0d12'
Clements,Ginger
Peg 22h20'29"30d34'
Clements,Jason Robert
Dra 9h38'47"78d4'
Clements,Kirstie
Lyr 19h16'58"42d42'
Clements,Kyle Eric
Aur 6h2'32"31d35'
Clements,Muriel & John
Uma 11h24'24"42d59'
Clements,Nathan Tillman
Uma 9h1'0"47d49'
Clements,Oliver Stephen
Eri 4h44'1"-6d5'
Clements,Tiffany
And 2h23'18"44d56'
Clementus
Oph 18h5'0"11d10'
Clemins,Sr,Daniel R
Hya 8h40'0"5d25'
Clemmons,Christine M
And 0h57'23"34d21'
Clemmons,Colin Michael
Her 16h28'48"35d42'
Clemons
Sct 18h32'23"-6d23'
Clemons,Karen Elaine
Aql 18h44'42"8d25'
Clemons,Verna
Cas 1h41'1"58d19'
Clendenin,Ebony Marie
Cas 1h14'1"63d51'
Clendenin,Sandra Faye
Mon 6h29'15"-6d13'
Cleo
Crb 16h12'26"32d42'
Cleo
Aql 19h47'24"1d20'
Cleo Lay
And 23h0'54"50d39'
Cleona Louise
Mon 6h39'47"10d30'
Cleopatra
Oph 17h1'41"1d38'
Cleopatre
Dra 17h17'13"64d44'
Cleotelis 12/30/63, Mark Alexander
Hya 8h12'36"2d29'
Clerget,Chantal
Per 3h58'19"35d54'
Clerk,Donald & Doreen
Cyg 21h31'25"41d32'

Clerk,Peggy
Cap 20h44'39"-20d32'
Clerkin,Andrew Michael
Cnv 13h14'56"50d29'
Clermont,Audrey-Ann SE Bissonnette
Cas 0h17'11"63d53'
Clermont,Isabelle
Cep 20h31'1"76d10'
Clermont,Stéphanie Gagne
Ori 5h56'44"16d17'
Clerou,Dr Romain
Ser 15h16'34"24d34'
Clery,Constance & Howard
Lmi 10h41'38"32d56'
Clesson,Nana
Cyg 20h5'14"40d41'
Cleta
Aqr 22h7'34"0d55'
Clete Francis
And 0h50'0"34d12'
Cleve,Maximilian Josef
Umi 14h56'0"65d46'
Cleve,Peggy
Del 20h33'31"11d4'
Cleveland,David A
Cep 23h38'47"64d20'
Cleveland,Gail Parsons
Cet 1h37'47"-12d27'
Cleveland,Michelle
Lyn 6h14'22"58d8'
Cleveland,Shawn Gregory
Dra 16h11'57"68d54'
Cleveland,Stefan Allan
Her 16h55'35"27d23'
Clevenger,David Michael
Cma 6h53'19"-16d49'
Clevenger,Geri Lee
Aur 4h57'36"50d36'
Clevenger,Jacob Tyler
Her 17h32'30"38d53'
Clevenger,Jeffrey A
Boo 13h35'33"21d52'
Cleveringa,Rick "Sir John"
Lac 22h15'12"46d1'
Clibourn Memorial, Robert Morgan
Mon 8h2'11"-6d34'
Click,Estelle Withers
And 0h15'27"45d53'
Cliff "The Year Of The Rooster"
Boo 14h5'52"40d59'
Cliff Who's Light Has Touched Many
Aur 5h54'1"40d42'
Cliff,Jim
Aql 20h22'25"0d60'
Clifford
Ser 15h40'41"20d40'
Clifford,Alan Frank
Boo 14h26'28"22d28'
Clifford,Brian Paul
Cet 3h16'11"9d2'
Clifford,Dianne Lynn
Her 17h21'45"20d42'
Clifford,Elizabeth Kirby
Umi 16h21'18"70d56'
Clifford,Jeffrey Alan
Per 4h41'38"38d54'
Clifford,Joshua T
Dra 17h11'57"69d14'
Clifford,M L
Aql 19h59'19"0d44'
Clifford,Margaret E
And 0h48'29"36d33'
Clifford,Mary
And 23h36'0"38d58'
Clifford,Paul C
Sgr 19h21'31"-42d37'
Clifford,Raina
Uma 11h29'0"32d38'
Clifford,Richard A
Cet 1h14'39"-1d1'

Clifford,Sandy
Lyn 7h7'49"44d47'
Clifford,Sean
Sgr 19h30'56"-37d38'
Clifford,Shirley Mittleman
And 0h21'29"22d6'
Clifford,Stuart
Cet 1h46'50"-4d58'
Clifford,Victoria Jane
And 0h27'15"31d31'
Clifford,Virginia Petty
Cyg 21h31'14"40d58'
Cliffords Folly,Eric
Boo 14h11'33"40d54'
Clift,Kevin "KC"
Per 1h34'0"53d58'
Cliflin
Sco 16h31'40"-43d42'
Clift,Dorothy Ann
Lyr 18h43'1"47d9'
Clift,Kathleen Roderick
And 23h3'39"51d16'
Clift,Polly/Rachel
Lyn 7h24'51"44d45'
Clift,Richard
Her 16h50'0"50d58'
Clifton,Aaron S
Cam 5h40'28"65d26'
Clifton,Jr,Clifford W
Cam 3h46'56"74d29'
Clifton,Ryan Michael
Oph 17h39'57"-24d12'
Clifton-Stancliffe, Marie
Mon 7h57'53"-1d31'
Clifton-Wright
Cyg 20h18'18"38d14'
Climent,Jaime
Umi 15h54'51"81d9'
Climer,James Randall
Aql 19h58'18"12d41'
Climer,Michael Randolph
Leo 9h37'1"8d23'
Climo,"Ali"Alison Heather
Lyr 19h18'11"42d25'
Clinard,Edward Noel
Cet 1h16'1"0d47'
Clinard,Elizabeth Kercheval
Aql 19h30'0"12d49'
Clinard,Kathryn Moir
Vul 20h16'14"22d58'
Clinard,Margaret Hawthorne
Uma 10h16'27"51d38'
Clinard,Margaret Graham Robinson
Ori 6h7'47"9d23'
Clinard,Robert Noel
Tau 4h43'48"20d19'
Cline Family Star, The Michael W
Mon 7h7'29"-6d39'
cline ti amo You are my Starlight,jeff
Ori 6h8'54"6d54'
Cline,Delmas & Dorothy
Lac 22h30'43"56d30'
Cline,Jan
Mon 7h32'1"-6d10'
Cline,Robert Andrew
Boo 15h0'1"13d29'
Cline,Shaine Ann
Lyr 19h0'22"37d53'
Clines,Terence Francis
Boo 13h57'56"18d26'
Clinger,Alan
Cep 2h39'49"77d48'
Clinger,M & T
Hya 8h57'10"1d19'
Clingman,Laura Suzanne
And 1h13'11"40d47'
Clint
Ori 5h52'19"16d55'
Clint
Aur 6h32'55"38d28'
Clint Howard
Ori 5h43'47"8d37'

Clint's Cheyene Autumn
Her 16h14'19"20d25'
Clinton
Ori 5h27'58"0d51'
Clinton
Per 2h26'13"56d37'
Clinton
Cep 22h50'51"59d2'
Clinton Family Star, The
Pho 0h35'5"-44d40'
Clinton Star
Eri 3h1'15"-7d31'
Clinton,David Ray
Eri 04h7'1"-8d45'
Clinton,Hillary Rodham
Cas 1h6'12"53d35'
Clinton,John Donald
Leo 9h51'0"33d16'
Clinton,Kathleen Ann
Uma 9h36'0"56d16'
Clinton,Maria Sharon Kolter
Cet 1h53'21"-10d45'
Clinton,Melba L
Lyr 19h22'20"40d39'
Clinton,Michael
Dra 17h27'49"70d34'
Clinton,US President William
Umi 14h41'52"74d8'
Clinton,William Jefferson
Her 17h35'39"40d52'
Clintonia Emma of Sydney
Col 6h0'7"-35d35'
Clipperton,Rob
Her 15h50'22"47d12'
Clippinger,Kristen Ann
Lyn 8h28'36"45d29'
Clippinger,Lenore & Walter
Crb 15h50'13"38d32'
Clipsham,Betti
Uma 11h15'19"60d53'
Clive Ross
Lyr 19h1'5"42d17'
CLJMP
Lyr 19h24'0"38d29'
Clmuransky 1992
Lmi 9h54'58"37d50'
Clo-Clo
Peg 23h34'30"13d0'
Clockwork Studios
Aql 19h31'52"0d36'
Clod & Vale
Pup 8h8'28"-21d15'
Clodfelter,Daniel & Kathleen
Cyg 21h37'37"41d35'
Cloeren,Alexandra Jo
And 1h24'43"38d7'
Cloggy-AZ24
Boo 14h28'54"21d6'
Clohessy,John
Cep 21h18'48"70d8'
Cloke,Crystal Michelle
Lyn 7h28'55"41d23'
Cloke,Patricia Avon
Cas 0h30'53"66d36'
Cloke-Vansen
Oph 17h57'21"11d59'
Clokey,Dorothy
Cet 2h17'39"2d55'
Clolus,Michel
Ori 6h1'46"8d14'
Cloobeck,Monica
Boo 15h2'33"27d44'
Clor,Christian
Ori 6h2'48"1d7'
Close's Cosmic Beauty
Aur 5h1'21"47d30'
Close,Dillan Wesler
Psc 1h28'54"10d25'
Close,James Christopher
Cmi 7h16'43"4d45'
Close,Janet
Aql 18h43'43"8d50'
Close,Joyce
And 2h32'23"38d18'

Close,Richard Emerson
Lac 22h8'23"51d49'
Close,Samantha Edan
Gem 7h15'6"21d30'
Close,Samuel David Lindsay
Ori 6h7'1"5d47'
Closs,Larry
Cep 21h6'1"55d1'
Closson,Beverly J
Mon 7h58'9"-7d15'
Closson,Dayton E
Mon 7h46'47"-7d14'
Clot,Robert Peter
Her 16h58'16"31d17'
Clota Yesbek
Cyg 20h53'10"30d16'
Cloteaux,Patty
Com 12h30'45"27d27'
Cloud Cabrini
Uma 8h51'56"49d40'
Cloud,Annette Marie
Lyn 8h18'27"37d23'
Cloud,Bryan C
Ori 6h3'46"8d18'
Cloud,Chelsea Lynne
Vul 19h22'27"23d55'
Cloud,Kimberly Chilcutt
Psc 1h35'1"22d1'
Cloud,Nina Rosann Elizabeth
Cas 0h39'31"61d6'
Clough
Lmi 10h3'20"30d27'
Clough's Wishes & Dreams,Christopher
Ori 5h31'1"0d41'
Clough,Adele C
Ori 4h57'29"-0d33'
Clough,Denny
Cet 2h16'34"0d36'
Clouser,Charlie
Oph 16h55'0"11d33'
Clouser-Bayer
Lac 22h11'31"51d14'
Clouston,Allison J
Cas 1h7'0"63d50'
Clouston,Brendan R
Cam 3h26'45"53d19'
Cloutier,Carl
Her 17h17'0"13d5'
Cloutier,Eric Zachery
Her 17h57'55"31d25'
Cloutier,Fernand
Cep 21h50'56"65d0'
Cloutier,Geoffrey
Cep 22h30'55"63d34'
Cloutier,Guy
Cep 23h31'17"64d42'
Cloutier,Michel
Cep 21h42'26"60d7'
Cloutier,Raymond
Ari 3h14'53"19d18'
Cloutier,Simon
Ori 5h57'52"14d44'
Cloutier,Wendy Kristi
Equ 21h0'0"7d37'
Cloutman,William
Cam 3h18'49"61d21'
Clow,Elizabeth
Lyr 18h37'22"33d50'
Clow,Louise Newcomet
Del 20h15'33"12d33'
Clowdus "Susie",Karen
Mon 7h6'58"-6d35'
Clower's Star,Susan
Peg 23h26'29"12d43'
Clower,Bill
Dra 15h54'48"60d57'
Clowes Lucky Star
Peg 22h57'0"32d22'
Clowes,Max Clayton
Gem 7h9'24"21d5'
Clowes,Thomas
Aql 20h1'38"3d50'
Cloya
Cyg 20h1'31"30d31'

Clozier,Patrice
Ind 20h26'32"-51d55'
CLP & MRK-High Rise & Freefall
Uma 9h24'11"57d0'
Club Murf
Cet 2h26'14"-11d7'
Club Paradise Bart, Nick & Strat
Uma 8h45'14"56d26'
Clubb,Toby Lee
Aur 7h25'0"37d57'
Cluck,My Grandma- Esther
Sco 17h30'0"-31d21'
Clucker
Cyg 19h21'11"28d57'
Cluff,Tim
Vul 19h47'22"28d14'
Cluff,Tony
Aql 20h5'49"1d26'
Clugston,Jake Sung Hyun
Lac 22h2'47"47d26'
Clune,Joseph Michael
Per 3h1'42"41d19'
Clutter,Abrey
Cam 4h57'41"65d2'
Clutter,Alexis Gabrielle
Cas 1h35'19"61d13'
Clutter,Khara
Cyg 19h31'41"37d58'
Clutter,Linda & Dave
Aur 4h51'29"51d25'
Clutts,Doug
Lac 22h14'42"47d2'
Clyburn,Cynthia Satute
Hya 8h59'59"1d18'
Clyde
Cma 6h10'21"-11d6'
Clyde
Aql 20h3'21"1d14'
Clyde
Per 3h1'54"40d45'
Clyde "Golden" Holly
Aqr 22h2'36"-22d0'
Clyde & Diane
Lyn 7h29'58"43d22'
Clyde Lindsey
Dra 19h20'11"78d56'
Clyde's Star
Cep 21h58'0"55d34'
Clyde,Catherine Mary
Cas 0h31'40"73d19'
Clyde,Lisa
Del 20h13'25"10d17'
Clyde,Tekla & Ashley
Crb 15h20'35"30d6'
Clyde,Thomas J
Aql 19h30'22"12d12'
Clymer,Lauren Elizabeth
Cas 2h10'1"73d8'
Clyne,Marian Elizabeth
Lyn 6h53'21"48d47'
Clément,Marie Anne
Ori 5h56'1"7d59'
Clément,Norbert
Uma 10h1'14"53d23'
CMD Petit Bout de Choux 240493
Cas 0h1'34"61d14'
CMFFCM-20896
Sex 10h28'37"4d26'
Cnokaert,Marjorie
Cas 1h0'19"50d31'
Co Patist
Dra 17h18'42"64d15'
CO The Scorpio
Sco 16h24'14"-40d44'
Co-Lydia 5694
Equ 21h3'51"2d25'
Coach I
Boo 15h0'33"37d42'
Coach Link
Lib 15h29'0"-10d40'
Coach Sure-Shot
Per 2h12'52"56d50'

Coady,Jimmy & Susan
Cyg 20h5'0"40d39'
Coady,Julia
Peg 22h12'35"5d39'
Coakley,Janet J
Peg 22h1'1"33d19'
Coakley,Meghan Elizabeth
Aql 19h55'1"-6d5'
Coale My Shining Star, Daniel A
Aur 6h32'43"31d47'
Coan,Mo
Cnv 12h51'23"42d22'
Coari,Kyle Jansen
Boo 15h45'47"42d24'
Coartney,Mary Jane
Ori 6h3'56"20d32'
Coash,Meri
Peg 23h31'33"32d10'
Coates El-Unico, Jonathan Randolph
Ori 6h6'19"-2d51'
Coates,Alexander Robert Fraser
Ori 5h3'1"8d34'
Coates,Andrew John Fraser
Lac 22h4'1"51d8'
Coates,Ashley
Sge 19h41'51"16d48'
Coates,Christopher T
Sex 9h52'17"2d25'
Coates,Colin Philip
Eri 2h44'41"-4d58'
Coates,Dr Victor B
Oph 18h42'59"7d59'
Coates,Father John T
Aur 5h30'28"38d54'
Coates,Ian W
Hya 8h12'57"3d55'
Coates,John Sidney
Ser 17h53'33"-14d26'
Coates,Karen Ann
Per 1h36'57"54d10'
Coates,Luke
Umi 14h55'52"68d48'
Coates,Margaret Marie
And 1h6'51"40d49'
Coates,Maria
And 23h7'53"42d19'
Coates,Sharon Margaret
Ori 6h7'21"8d48'
Coates,Tucker W
Boo 15h5'54"50d7'
Coates,Tyler D
Aur 4h56'48"38d49'
Coates-"TJ",Thomas Jordan
Leo 9h54'40"11d14'
Coatney,Ryan Alexander
Her 16h34'10"48d20'
Coats,Laura
Gem 7h52'39"31d22'
Coats,Simon
Cru 12h2'1"-58d36'
Coatsworth,Kenneth Mark
Lyr 18h42'40"35d46'
Coatts,Cynthia
Eri 2h48'42"-5d45'
Coban,J C
Oph 17h10'54"-10d33'
Cobb "A Rising Star", Cindy
Lyr 18h48'42"32d19'
COBB'S GLOW,JAB and DANIEL
Cyg 19h36'51"37d50'
Cobb,Adam Harris
Hya 8h13'34"-5d54'
Cobb,Christopher-No Middlename-
Gru 22h29'1"-55d22'
Cobb,Davina Claire
Cas 0h22'40"72d59'
Cobb,Josephine Powell
Tau 4h6'59"20d56'
Cobb,Judith C
Lyn 7h55'53"43d47'
Cobb,Kristen
Uma 13h32'27"60d57'
Cobb,Luke Crosland
Cet 0h21'49"-18d22'
Cobb,Matthew Hammond
Oph 18h40'56"6d40'
Cobb,Michale D
Mon 8h4'1"-8d17'
Cobb,Peter Samuel
Peg 0h12'35"13d17'
Cobb,Robin Kay
Mon 6h46'0"11d23'
Cobb,Sandy
Com 12h54'0"25d25'
Cobb,Suzanne Michaels
Cyg 20h30'24"40d28'
Cobb,Thomas
Cnv 13h56'31"37d50'
Cobb,Walter Ty
Aur 6h27'59"31d59'
Cobb-My Hero,John
Her 16h43'34"26d48'
Cobbs,Harry Lee Elizabeth Martin
Aql 20h34'14"0d42'
Cobbstar
Cep 21h8'15"67d46'
Cobelli Monica
Lyn 7h27'44"40d2'
Coben,Lance Mitchell
Aur 5h56'18"30d38'
Cobi
Lyr 18h32'51"37d43'
Cobian,Jose
Oph 16h55'13"-6d39'
Coble,John & Carla
Cas 2h4'17"67d37'
Coble,Timothy Dane
Her 17h23'1"48d26"
Coblentz Star,The Deb
Del 20h15'0"10d14'
Cobley,Christina
And 23h15'1"47d47'
Cobo Wabo
Cam 4h10'50"70d17'
Cobo,Josephine Ann
Cas 0h20'42"61d45'
Cobo,Joshua Adam
Lac 22h29'0"53d60'
Cobo,Juan
Cep 22h43'20"63d22'
Coburn,Fionnuala Anne
Vul 19h34'31"24d31'
Coby
Mon 7h9'15"-7d11'
Coccinelle
Lmi 10h29'29"32d5'
Cocciolone,Maria Lynn
Ori 5h28'13"-1d31'
Cocco,Peg & Jene
Uma 10h34'28"52d2'
Cocheta
And 23h38'53"48d25'
Cochin,Joël
Cam 4h38'40"58d39'
Cochran,Brett Hale
Lmi 10h11'31"31d41'
Cochran,Cynthia
Mon 7h6'0"-5d3'
Cochran,Dawn Marie
And 2h22'16"41d42'
Cochran,Glenda Rose
Peg 21h55'23"34d59'
Cochran,Gloria D
Lyr 18h59'47"46d27'
Cochran,Jackson Arroyo
Dra 15h43'12"62d58'
Cochran,James Michael
Ori 6h0'48"18d48'
Cochran,Jennifer
Ori 5h56'36"8d38'
Cochran,Jim
Lac 22h46'25"52d54'
Cochran,Jr,Marion
Aql 18h44'19"-2d33'
Cochran,Mary
Ari 2h58'0"30d37'
Cochran,Matio G
Ori 5h54'52"8d1'
Cochran,Miss Julie
Cnc 8h55'17"18d53'
Cochran,Robert S
Lac 22h40'36"51d12'
Cochran,Wayne
Lac 22h39'47"52d44'
Cochrane 9-26-44, Elaine
Cas 0h57'53"64d44'
Cochrane,Adam M
Ori 5h54'57"8d23'
Cockayne,Claire
And 0h40'14"45d11'
Cockburn,Michelle
Cas 1h46'0"76d9'
Cocking,Ryan Scott
Hya 9h52'20"-16d20'
Cockinos,Nicholas Dimitrios
Her 17h7'35"44d44'
Cockrell,Benjamin Geoffrey William
Ori 5h43'1"10d37'
Cockroft,Shiloh-Marcy & Jack
Com 12h13'45"32d39'
Coco
Aql 18h41'0"1d47'
Coco,Arthur Orlando
Per 2h52'40"32d43'
Cocoa "65"
Mon 7h24'11"-6d2'
Cocoma,Jennifer
And 23h3'43"50d39'
Cocoon
Uma 11h12'31"43d39'
Cocopina
Aql 19h13'30"12d38'
Cocozza,Alessia
And 1h45'19"40d19'
Cocozza,Diddabie
Vel 9h43'13"-49d15'
Cocuzzi,Sandy
Uma 12h55'1"53d57'
CocUma 12h55'1"53d57'
Aql 18h56'20"-5d41'
CocAql 18h56'20"-5d41'
Cam 6h14'0"64d59'
Coda,Amanda & Brittany
Cyg 20h19'11"41d13'
Codaco
Cep 22h52'47"67d45'
Codas,Rebecca Amanda
Cas 22h57'33"54d22'
Coddington,Terri
Cas 1h1'32"60d10'
Codds' Love
Mon 6h4'36"-8d52'
Code,Charlie Jim
Lyr 18h32'1"43d8'
Coder,Craig William
Aql 20h18'33"0d44'
Coder,Megan Anne
Cyg 20h5'1"38d49'
Codner,Steven C
Per 3h36'22"38d31'
Cody
Cmi 7h30'56"0d41'
Cody
Boo 14h13'16"50d6'
Cody
Lyn 7h28'0"41d49'
Cody
Her 16h28'24"31d12'
Cody Alan
Ori 5h54'54"12d25'
Cody's Dream
Vul 19h40'48"26d25'
Cody,Cora May Baird Boswell
Cas 0h12'53"61d15'
Cody,Dr Liz
Aur 7h4'1"40d59'
Cody,James
Cnc 8h11'42"31d21'
Cody,Loye
Aql 19h50'53"15d2'
Cody,Michelle
Lyr 18h33'41"40d47'
Cody,Rebecca
Cas 0h8'48"54d40'
Cody,Tom & Jackie
Hya 8h52'50"0d42'
Cody,William Drew
Mon 6h21'54"8d44'
Coe 296,David
Aur 6h8'35"37d26'
Coe II,Max M
Her 18h39'55"12d23'
Coe,296,David
Aur 6h8'35"37d26'
Coe,Her Honor-Penny
Peg 23h49'36"30d26'
Coe,Jade Florentine
Uma 10h20'57"42d11'
Coe,Lillie M
Cas 0h3'0"65d40'
Coe,Michele Lynn
Cep 23h19'20"65d59'
Coe,Rob
Per 1h31'18"53d47'
Coe,Robin
Aqr 21h6'14"0d16'
Coe,Russell Hawkes
Boo 14h6'28"37d50'
Coe,Shawn Damien
Oph 18h17'29"7d36'
Coe,Taylor
Cru 12h43'38"-59d55'
Coe,Vada Louise
Uma 10h34'11"48d33'
Coe,Wendy
Lyr 18h44'54"40d10'
Coelao,Paulo
Mon 7h45'8"-7d18'
Coelho,Eunice
And 23h23'0"42d44'
Coeli's Own
Lmi 10h2'30"37d48'
Coellen,Fritz
Cet 2h42'48"1d53'
Coemar,Eurie Orion
Ori 6h6'12"20d3'
Coen,Buster Jacob
Lac 22h29'29"55d30'
Coen,John Paul
Her 15h52'23"41d47'
Coenaculum E V Michaelshoven
Leo 10h31'45"18d9'
Coers,Jacques
Lac 22h4'60"38d11'
Coester,Conrad Johannes
Tau 3h32'50"30d14'
Coeur D'William,The
Her 18h7'11"28d22'
Cofer's cat 1995
Lyn 8h17'15"40d44'
Cofer,Barbara Armstrong Cleveland
And 23h2'15"50d33'
Cofer,Edward Lamar
Her 17h54'57"38d4'
Coffeen,Patricia
Lyr 19h18'57"42d3'
Coffey Super Star, The Bonnie
Lyn 7h49'15"38d3'
Coffey,Brenda
And 23h47'30"47d45'
Coffey,Caroline Elizabeth
Lyn 8h54'51"41d11'
Coffey,Darnell
Lyr 18h46'37"30d16'
Coffey,Edward Joseph
Cam 5h47'26"65d17'
Coffey,Jean C
Ori 6h5'34"7d30'
Coffey,Jerome J
Boo 14h46'15"28d2'
Coffey,Kathleen
Cyg 19h14'1"45d32'
Coffey,Sean Christopher
Lac 22h0'34"51d6'
Coffil,Darcy & Irvin "Butch" Miller
Ori 5h55'35"20d52'
Coffin III,H Stanley
Oph 18h25'22"8d3'
Coffin,Margo
Cam 12h1'0"75d49'
Coffin,Win & Eric
Cyg 21h53'22"41d50'
Coffland Eternity, Richard
Aur 6h23'29"38d19'
Coffland,Daryl Brooks
Dra 17h21'55"67d50'
Coffman PS 147:4, Robert Joseph
Aql 19h27'44"-7d36'
Coffman,Beretta
Vul 20h19'15"28d23'
Coffman,Dean
Ari 3h7'30"16d48'
Coffman,Heather Lynn
And 0h14'55"33d3'
Coffman,John Edwin
Cyg 19h40'36"40d37'
Coffman,Shirley
Cet 2h4'1"1d2'
Coffran,James F
Dra 18h32'26"70d7'
Cofield,Richard James
Sex 10h15'39"-1d4'
Cogal,Walter J
Aql 20h6'47"0d55'
Cogan's White Light, Joan
Lyr 19h14'39"38d15'
Cogan,Jr,Francis B
Her 17h47'46"40d55'
Cogar,Russell
Gem 7h39'52"27d23'
Cogbill,Mary Elizabeth
Com 13h3'18"20d15'
Cogger,Kenneth
Dra 16h45'17"61d41'
Coggins,Erin Elizabeth
Mon 6h41'36"-10d42'
Coggins,Rhonda
And 0h1'1"47d48'
Coghill,Jr,George Parham
Oph 16h3'0"-6d27'
Coghlan,James Correll
Lac 22h7'47"47d35'
Cogie
Uma 8h43'59"49d8'
Cogliandro,Antonio
Sex 9h51'38"-5d4'
Cogliati II,Robert Peter
Aur 6h24'23"31d43'
Coglitore,Sebastian F
Lyr 18h32'0"46d57'
Cognard,Jennifer Anne
Uma 10h48'50"57d0'
Cognet
Cnv 12h19'49"47d42'
Cogswell,Anne Elaine
And 0h9'12"43d36'
Cogswell,Karin A
And 0h52'1"37d2'
Cogswell,Kipling
Aql 19h47'57"14d4'
Cogswell,Stephen
Dra 19h32'56"71d8'
Cohan,Evelyn
Sge 18h55'40"18d53'
Cohan,Michael J & Megan KC
Mon 7h27'43"-6d7'
Cohan,Ryann Kathryn
Lyn 7h11'56"58d34'
Cohen "Destined for Stardom",Jennifer
And 0h9'58"40d44'
Cohen "Sunshine",Jay Scott
Boo 14h30'48"41d12'
Cohen 45-95,Gerald E
Cep 22h24'18"70d8'
Cohen LXX,Bernard
Cam 14h14'12"80d45'
Cohen,"Rudy Boy"Reuben
Boo 14h12'44"30d12'
Cohen,Aaron & Melba
Cyg 20h8'1"39d51'
Cohen,Aaron Moses
Eri 2h48'18"-17d33'
Cohen,Adam Joshua
Lac 22h43'19"37d43'
Cohen,Addie & Bob
Peg 21h55'38"23d22'
Cohen,Adrienne & Ted
Cyg 20h5'57"40d45'
Cohen,Allan
Cmi 7h41'28"0d31'
Cohen,Amy
And 23h36'17"40d60'
Cohen,Aubry Joi
Boo 15h13'1"40d25'
Cohen,Barbara
And 1h25'44"41d9'
Cohen,Barry F
Ori 4h45'41"4d25'
Cohen,Benjamin
And 2h14'0"50d7'
Cohen,Carla
Aql 20h0'1"8d24'
Cohen,Caroline M
Lyn 9h5'45"36d53'
Cohen,Cary Allen
Cyg 19h36'52"28d49'
Cohen,Celeste Presseau
Aql 18h57'0"-5d52'
Cohen,Daniel Jared
Her 16h53'31"40d18'
Cohen,Darlene Ahmann
Lyr 18h30'32"31d29'
Cohen,David
Cep 22h45'24"57d23'
Cohen,David Brian
Cam 4h2'20"60d5'
Cohen,Deanna Ruth
Eri 2h57'35"-5d21'
Cohen,Debbie & Reuven
Lyn 7h15'46"56d54'
Cohen,Deborah
Mon 6h57'36"0d42'
Cohen,Donald Stuart
Oph 17h18'19"-20d27'
Cohen,Donald T
Aql 19h58'59"14d46'
Cohen,Dr Jeffrey S
Oph 18h35'0"10d43'
Cohen,Eddie
Cnv 12h24'49"40d44'
Cohen,Edith & Gary
Crb 15h17'42"31d17'
Cohen,Elayne & Alan
Crb 16h5'16"31d26'
Cohen,Esq,Stanley
Cet 2h31'12"5d23'
Cohen,Gabe
Cas 1h17'46"60d51'
Cohen,Gabriel
Aqr 20h58'56"0d27'
Cohen,Gail
And 23h19'0"50d50'
Cohen,Gail S
Lyn 7h3'47"53d12'
Cohen,Grace
Lyr 18h56'28"32d1'
Cohen,Harry S
Oph 18h36'1"11d20'
Cohen,Irving
Her 16h42'14"38d23'
Cohen,Irving E
Sgr 19h28'5"-43d11'
Cohen,Jack & Jane
Lac 22h27'45"54d18'
Cohen,Jacob David
Aur 6h26'30"30d59'
Cohen,James Marc
Ori 5h51'50"16d43'
Cohen,Jane
Ori 5h58'0"15d28'
Cohen,Jason
Dra 16h48'18"51d54'
Cohen,Jean-Jacques
Her 18h2'36"18d60'
Cohen,Jeffrey N
Aur 4h54'1"38d39'
Cohen,Jerome Lionel
Lac 22h40'23"53d17'
Cohen,Jerry L
Boo 14h38'0"51d1'
Cohen,Jim
Cmi 7h54'15"1d36'
Cohen,Joel & Janis
Cyg 21h3'39"30d1'
Cohen,Julien Harry
Com 13h4'55"17d51'
Cohen,Julius
Her 16h42'34"22d9'
Cohen,Kenneth H
Boo 14h53'28"40d25'
Cohen,Keren Fannie
Cyg 21h31'34"50d19'
Cohen,Kierstin Jeana
Peg 23h7'39"8d38'
Cohen,Larry Marc
Mon 6h21'45"8d18'
Cohen,Leslie M
Cet 2h23'50"1d30'
Cohen,Lillian
Mon 7h57'35"-8d59'
Cohen,Lisa Ellen
Com 12h14'18"27d52'
Cohen,Lori M
Cas 2h0'38"58d43'
Cohen,Louis Allan
Crb 16h7'42"27d1'
Cohen,Lovely Poet Marion Deutsche
Uma 9h0'0"70d2'
Cohen,Madeline T
Aql 20h14'55"1d5'
Cohen,Maggie Laboz
Cas 0h28'58"69d54'
Cohen,Marc J
Cep 1h55'32"80d33'
Cohen,Marcia Susan
Psc 0h18'29"20d55'
Cohen,Margie K
Sge 19h23'16"16d14'
Cohen,Marilyn
Cas 3h4'27"70d22'
Cohen,Mark James
Aqr 21h3'56"0d36'
Cohen,Marvin
Dra 12h51'28"71d30'
Cohen,Matthew Hunter
Cet 2h46'39"4d2'
Cohen,Matthew S
Aur 5h40'40"50d22'
Cohen,Maury
Her 17h13'57"40d10'
Cohen,Michael
Aur 6h53'46"40d6'
Cohen,Michael Aaron
Cep 23h1'12"70d51'
Cohen,Michael J
Per 4h5'12"51d8'
Cohen,Michael Spencer
Aur 6h15'0"46d27'
Cohen,Michal
Per 1h25'48"53d46'
Cohen,Mickey
Cam 4h20'24"70d14'
Cohen,Mr & Mrs Jesse
Lyr 18h46'1"36d24'
Cohen,Nancy L
Tau 4h39'15"10d41'
Cohen,Patricia
And 23h32'17"47d5'
Cohen,Patti & Jeff
Cyg 20h3'44"50d16'
Cohen,Paulette Parsons
Equ 21h5'59"2d23'
Cohen,Philip
Cnv 13h52'51"38d50'
Cohen,Rafi Z
Lyn 7h33'34"38d23'
Cohen,Robert
Dra 17h47'34"65d0'
Cohen,Robin
Cas 23h18'38"62d39'
Cohen,Ronald Bruce
Boo 14h1'60"20d42'
Cohen,Roy
Lac 22h21'0"40d55'
Cohen,Samy
Hya 8h55'1"1d33'
Cohen,Sara Marielle
And 2h35'25"40d52'
Cohen,Stanley
Boo 15h5'26"11d34'
Cohen,Steven Bailey
Gem 7h1'18"28d42'
Cohen,Sue
Vir 13h27'18"-1d35'
Cohen,Terry
Cep 22h53'25"57d30'
Cohen,The Star Of David
Cep 21h21'0"61d40'
Cohen,Todd Stephen
Cep 5h44'11"85d54'
Cohen,Uncle Roger
Cam 9h39'0"82d6'
Cohen,Zachary Pearce
Sex 9h58'59"1d2'
Cohen-Mr NY Life-Mr Sugar Daddy,Herb
Cep 21h25'47"55d58'
Cohen:The Lacrosse Star,Harvey
Per 4h41'51"37d23'
Cohn,Anna
Cyg 19h24'14"33d22'
Cohn,Christine Marie
Vul 19h16'1"24d52'
Cohn,Irma A
Uma 10h34'40"51d39'
Cohn,Jonathan D
Uma 11h43'59"55d3'
Cohn,Lynn Siegel
Oph 16h48'16"10d29'
Cohn,Marty
Dra 20h7'41"76d6'
Cohn,Ruth M
Crb 15h27'11"32d28'
Cohn,Sarah
Tau 5h55'40"24d17'
Cohrs,Jennafer
And 2h32'23"40d25'
Coia,Arthur A
Boo 15h1'28"11d47'
Coiffard,Isabelle
Ant 10h39'41"-34d48'
Coigne,Catherine
Oph 18h6'37"11d7'
Coiner,Charles Bartlett
Per 3h24'47"40d23'
Coira,Jennifer C
Ser 15h13'1"9d14'
Coiro's Star,Diane & Angelo
Uma 8h8'30"60d12'
Coispine,Jean Marie
Boo 15h0'37"7d29'
Coker,Daniel Wilson
Lac 22h29'32"55d3'
Coker,James Alan
Peg 22h51'54"25d53'
Coker,Lisa Ann Koenigs
Cas 23h31'36"53d11'
Coker,Michele
Cap 20h32'29"-13d36'
Coker,R Brian
Her 17h21'1"14d10'
Cola
And 2h24'51"44d33'
Cola Davide
Cep 23h17'47"70d30'
Cola,Anne
And 2h14'10"38d45'
Colabelli,Francis Michael
Tau 5h22'50"18d58'
Colache,Matthew Joseph
Aur 6h27'22"31d5"
Colacino,Candice
Cyg 20h4'27"31d31
Colacino,Greg
Lac 22h54'19"50d5'
Colacino,Thomas G
Lac 22h4'16"40d57'
Colaggie Pearl Star
Cam 3h27'1"63d20'
Colaianni,Jeannie
Cyg 20h6'27"40d58'
Colaiuta,Vincent Peter
Her 17h7'1"44d25'
Colan,Cameron Davis
Cnv 13h32'57"36d20'
Colan,PhD,Lee Joseph
Cnv 13h54'1"32d24'
Colanduoni,Ray
Lib 14h45'22"-8d28'
Colaneri,Gillian Amanda
Boo 14h17'51"38d53'
Colangelo,Barbara
Eri 3h46'29"-6d42'
Colantoni,Marguerite Karam
Vir 12h8'25"2d20'
Colantuoni,Anthoney & Marie
Cyg 20h1'1"41d9'
Colaprico,Ludovica
Cyg 19h36'23"38d27'
Colard,Meagen
Vul 19h18'41"26d33'
Colaruotolo,John Joseph
Oph 17h18'22"-20d16'
Colas,Felicia et Benoit
Uma 13h40'39"61d57'
Colas,Marissa
Peg 21h40'35"28d4'
Colasanto,Frank
Lyn 7h53'27"41d4'
Colatrella,Anthony
Cnv 12h55'11"44d5'A
Colatrella,Nancy
Cnv 12h55'11"44d5'B
Colavito,Michael A
Her 17h54'48"14d25'
Colbath,Nina Elizabeth Fitch
And 1h34'1"40d2'
Colbeck,Donald
Lyn 8h15'17"44d36'
Colbert,Dick
Aql 19h0'31"13d38'
Colbert,Dr Harvey
Leo 9h23'15"28d12'
Colbert,William
Cet 2h32'39"0d37'
Colbree,Ward
Her 16h50'24"51d15'
Colburn,David J
Cam 5h54'23"60d45'
Colburn,Dorothy L Campbell Rhodes
Peg 23h23'43"33d43'
Colburn,Evelyn Sarah
Lmi 10h15'36"28d53'
Colburn,Marjorie May
Aur 4h50'41"40d0'
Colburn,Meredith Joyce
Mon 6h53'34"-0d16'
Colby
Peg 23h5'37"10d14'
Colchagie,Donald Peter
Lac 22h18'33"37d32'
Coldeboeul,Joelle
Per 3h29'52"38d20'

Coldwell,Wendy
Lyr 19h22'39"31d50'
Cole et Pie
Per 3h26'31"41d13'
Cole Family
Aql 18h57'20"14d42'
Cole III,Louis Celestine
Ser 18h6'33"-14d59'
Cole III,Thomas James
Aur 7h1'57"38d45'
Cole's Star
Peg 22h23'28"30d49'
Cole,Alana Gabrielle
And 1h48'59"47d1'
Cole,Allan
Boo 15h29'1"50d40'
Cole,Baby William J
Cyg 19h26'14"35d19'
Cole,Barbara G
And 23h1'28"40d18'
Cole,Benjamin Lewis
Peg 21h23'33"2d23'
Cole,Betsy
Cas 22h57'0"55d40'
Cole,Bob C
Cet 1h9'24"-4d7'
Cole,Carol
Equ 21h1'1"10d40'
Cole,Catherine
Uma 11h15'59"32d60'
Cole,Catherine Camilla
Cas 0h58'1"60d28'
Cole,Chavela
Eri 2h53'33"-6d53'
Cole,Christopher & Paige
Uma 9h16'39"51d45'
Cole,Contessa
Aur 4h52'12"37d38'
Cole,Curtis A
Sex 10h15'36"-6d29'
Cole,David
Lib 14h18'35"-10d33'
Cole,Don Robert
Boo 15h6'0"15d41'
Cole,Dorothy
Cas 0h12'45"62d33'
Cole,Dorothy & Edwin
Cyg 20h24'24"40d39'
Cole,Douglas G
Lac 21h59'0"40d20'
Cole,Elizabeth M
Leo 9h37'29"7d46'
Cole,Elle Christine
Lyr 18h39'18"33d15'
Cole,Ellen Jane
Com 12h51'51"23d32'
Cole,Erik N
Cmi 7h55'54"8d48'
Cole,Gayle Sandra
Com 12h51'21"23d20'
Cole,Ida S
Cas 0h7'47"54d35'
Cole,J Weldon
Aur 5h1'33"48d54'
Cole,James & Chun
Sge 20h3'0"20d10'
Cole,James J
Aur 7h23'15"41d18'
Cole,Janie Samantha
Lyn 8h17'42"45d32'
Cole,Jewell Carmen
Eri 3h55'19"-15d49'
Cole,Jim
Uma 11h14'42"52d14'
Cole,Joe
Lac 22h25'1"53d40'
Cole,John Lovelace
Ori 4h48'49"4d55'
Cole,Johnny M
Ori 5h56'12"16d41'
Cole,Joy A
Lyn 8h37'37"40d24'
Cole,Julie
Aql 19h45'54"12d24'
Cole,Kimberley Catherine
Psc 0h50'18"24d5'
Cole,Lauren A
Vul 21h21'58"27d27'
Cole,Lauren Wendy
And 0h53'37"36d50'
Cole,Leslie Charles
Peg 22h45'19"34d7'
Cole,Linda Cohen
Ari 1h50'51"10d30'
Cole,Lola M
Mon 7h18'53"-5d30'
Cole,Lucie Catherine
Gem 7h12'53"24d54'
Cole,Marcia
Peg 23h47'51"27d13'
Cole,Maria Betina
And 0h9'37"27d59'
Cole,Mark Quinton
Dra 16h10'30"66d5'
Cole,Michael & Lynn
Lyr 18h54'1"30d35'
Cole,Michael A
Lac 22h2'26"38d18'
Cole,Michael A
Her 17h32'22"27d53'
Cole,Michelle Deon
Cas 1h24'0"60d31'
Cole,Nathan Troy
Aql 20h3'25"4d59'
Cole,Norwood
Boo 15h6'57"22d41'
Cole,Papa
Aql 19h51'0"10d0'
Cole,Patricia C
Lmi 10h43'60"25d50'
Cole,Rob"Robis"
Per 1h30'36"53d22'
Cole,Ruth Piersen
Mon 7h2'38"3d58'
Cole,Shane
Vul 19h48'13"22d54'
Cole,Stephen H
Ser 17h15'21"-13d58'
Cole,Thomas David
Aur 5h35'46"50d17'
Cole,William
Aur 6h53'55"37d56'
Cole-Whittaker,Terry
Lyr 19h23'41"30d44'
Colebourne,Peggy
Aur 6h26'16"34d15'
Coleclough,Stephen
Cep 23h11'39"64d27'
Colee,Michael MacInnes
Aqr 22h25'55"2d29'
Coleen
Cas 23h42'0"61d43'
Coleen Alicia's Farthest Star
Mon 7h52'47"-3d11'
Colefax,Joth
Her 17h39'49"22d20'
Coleman
Cam 7h48'57"67d50'
Coleman "Sparkling
Kitty",Kathryn R
Del 20h22'28"10d14'
Coleman RN,Jackie
Lyr 18h48'33"38d58'
Coleman's Star
Forever,Jessica Leigh
And 0h59'34"45d46'
Coleman,Amber Marie
Aql 19h30'13"0d12'
Coleman,Bill
Umi 15h21'45"68d48'
Coleman,Blake
Per 2h54'45"43d40'
Coleman,Bob
Aql 20h35'0"0d36'
Coleman,Brady
Cet 0h49'53"-10d30'
Coleman,Bud & Connie
Cam 6h4'31"58d59'
Coleman,Carolyn
Lyn 8h9'27"38d23'
Coleman,Catherine Bernadette
Daney
Aqr 20h59'24"0d30'
Coleman,Christopher
Mon 7h5'48"-5d16'
Coleman,Clark "Skip"& Karen
Aqr 20h57'1"-11d9'
Coleman,Debbie
Vir 11h38'29"4d35'
Coleman,Delores Alice
Aur 6h26'60"37d55'
Coleman,Dennis Jennifer Race
Uma 9h35'51"50d17'
Coleman,Duncan
Aql 19h58'56"11d7'
Coleman,Emily
And 23h16'40"51d18'
Coleman,Emmet J
Aql 20h0'30"14d52'
Coleman,Frances
Lyn 8h6'29"35d53'
Coleman,Galey
Cyg 20h1'57"39d34'
Coleman,George
Tau 5h49'1"23d45'
Coleman,George T
Per 3h40'9"39d14'
Coleman,Hurston
Lyr 18h41'18"27d37'
Coleman,J T
Lyn 7h58'12"41d18'
Coleman,Jacquelyn H
Cmi 7h7'0"-19d54'
Coleman,James Austin
Leo 10h57'1"-1d1'
Coleman,Jeffrey David Barrett
Cet 2h52'51"2d35'
Coleman,Joseph Paul
Umi 14h22'1"68d22'
Coleman,Kathy
Cas 0h39'39"69d37'
Coleman,Kevin
Umi 16h17'35"79d5'
Coleman,Laura Mae
Sge 19h38'22"16d18'
Coleman,Lauren Dale- Louise
Lyr 18h15'1"40d38'
Coleman,Leo & Dollie
Cyg 21h1'0"31d45'
Coleman,Marshall
Aql 19h40'57"14d20'
Coleman,Mary Gleason
Cas 0h2'1"63d31'
Coleman,Masarkra
Ori 6h5'0"4d53'
Coleman,Melissa Kay
Peg 23h46'59"28d31'
Coleman,Michael J
Gem 6h50'5"17d15'
Coleman,Myrtle & Howard
Aql 18h57'31"-5d40'
Coleman,Nicki
Her 16h42'1"7d45'
Coleman,Norma Jean
And 2h18'50"38d15'
Coleman,Patrick Dode
Aql 20h10'31"10d29'
Coleman,Pauline
Aql 20h11'35"13d39'
Coleman,Robert Lee
Aur 5h17'32"42d11'
Coleman,Ron
Equ 20h59'59"4d44'
Coleman,Scott
Aur 7h26'40"40d22'
Coleman,Scott
Ser 17h18'53"-14d6'
Coleman,Stephen Charles
Lyn 8h6'24"51d26'
Coleman,Susan
And 0h7'33"46d16'
Coleman,Tracey
Cas 0h17'25"63d10'
Coleman,Valerie
Lyr 18h26'50"42d33'
Coleman,Wayne
Oph 17h52'50"-0d3'
Coleman,William Bryant
Hya 8h24'1"5d48'
Coleman-Hancock
Umi 13h28'51"72d14'
Coles,Gerald Allen
Per 1h56'24"48d46'
Coles,James Patrick
Aur 5h34'50"54d24'
Colesnic,Mircea
Sge 20h1'45"16d52'
Colesnic,Raia
Sge 20h3'31"16d34'
Colestock,Randolph Gilbert
Cmi 7h39'54"5d11'
Colesworthy,Chad
Per 1h43'1"53d14'
Colesworthy,Sandy
Ser 18h20'50"-0d21'
Coletta,David
Boo 14h12'0"41d55'
Colette
Cam 3h44'14"60d50'
Colette
Aql 20h10'17"14d35'
Colette
Tri 2h3'0"30d25'
Colette
Cas 3h32'42"68d49'
Colette 1
Crb 16h13'24"33d38'
Colette D
Lyn 6h21'30"58d21'
Colette Louise
Mon 8h3'36"-8d24'
Colette V
Dra 10h29'19"80d43'
Coletti,Dina & Chad
Natichioni
Lep 5h16'36"-18d10'AB
Coletti,Theodore
Cet 0h4'12"-10d8'
Coley
Lyn 7h3'16"52d48'
Coley,Constance Alma
Lyr 18h33'27"40d59'
Coley,Kristina M
Vul 19h47'1"28d60'
Coley-McFarlane
Per 1h53'19"50d36'
Coleyshaw,Helen
Ori 4h42'47"0d30'
Colf,Scott Lee
Aur 7h15'16"35d31'
Colgan & Muller
Aql 20h31'14"14d51'
Colgan,Jo & Diane
Lyr 18h58'44"32d50'
Colhouer,James G & Doris
Cyg 19h40'30"42d46'
Colicafratti,Jamila Enzo
Francesca
Psa 22h33'13"-26d14'
Colicchio,Helisse & Pasquale
Cyg 21h35'19"42d56'
Colichio,Rosemarie "Toot"
And 23h4'44"38d5'
Colie,Timothy B
Hya 8h9'15"-9d15'
Colin
Cep 21h28'38"61d50'
Colin & Jennifer
Her 18h18'32"22d46'AB
Colin & Mari's Star
Leo 9h34'43"8d39'
Colin (CB)
Ori 5h55'39"7d1'
Colin Alias Cool
Cnv 12h23'14"34d21'
Colin Gregory
Ori 5h12'30"15d14'
Colin John
Lyn 8h16'47"41d59'
Colin my beautiful best friend
Per 3h0'47"47d17'
Colin's Woolly Bear
Uma 9h46'38"67d33'
Colin,André
Sex 10h16'29"-2d19'
Colitti,Daniel
Aur 6h2'13"31d38'
Coll
Per 3h48'10"36d56'
Coll,Serge
Umi 6h0'42"88d14'
Collacchi,Jr,Robert M
Cet 1h19'32"-0d35'
Collander,Craig Alan
Lac 21h55'55"40d45'
Collany,Christopher Michael
Ser 16h4'24"13d14'
Collar,Hazel
Cas 0h44'18"71d57'
Collard's Star
Aql 19h56'53"12d35'
Collazo,Gigette
Tri 2h21'15"31d27'
Collazo,M D
Mon 8h3'58"-9d26'
Collea,Daniel J
Cam 13h32'32"78d45'
Colledge,Bimmer
Mon 6h55'23"-0d38'
Colleen
Lyn 7h27'45"42d0'
Colleen
And 1h25'34"33d40'
Colleen
Mon 6h51'40"11d4'
Colleen
Cas 22h56'52"55d24'
Colleen
Cyg 19h27'18"33d6'
Colleen
Peg 0h4'34"18d2'
Colleen
Cyg 19h28'31"38d16'
Colleen
Uma 11h2'43"57d32'
Colleen & John
Aur 4h55'0"48d57'
Colleen & Ryan Forever
Uma 9h34'0"55d20'
Colleen and Mark
And 23h29'59"45d09'
Colleen Anne's Light
And 23h36'41"48d0'
Colleen Antoinette
Lyn 7h48'40"44d31'
Colleen Elizabeth
Lac 22h25'1"38d16'
Colleen Frances
Vir 12h30'29"0d34'
Colleen Marie
And 2h20'48"44d18'
Colleen Patricia & Brian Liam
Lyr 18h53'21"40d30'
Colleen The Butterfly
Sgr 19h25'56"-41d2'
Colleen Tracy
And 23h30'0"49d53'
Colleen's Eyes
Lyn 7h54'15"45d13'
Colleen's Love
Crb 15h24'35"30d26'
Colleen's Rose
Lyn 6h49'54"60d27'
Colleen's Until Forever
Boo 15h7'1"31d17'
Colleen-S
Ori 5h59'49"12d20'
Collen,Sheryl
Eri 3h18'41"-4d47'
Coller,Clint Ross
Boo 15h4'54"22d35'
Colleran,Bill
Lac 22h32'38"55d3'
Collet,Anne Marie
Lyn 9h4'45"42d8'
Collet,Sandrine
Boo 14h26'43"50d23'
Collette,Dee Cee
Cas 23h13'52"60d58'
Collette,Mary M
Peg 22h38'22"29d14'
Collette,Your Dream Of
Everlasting Peace
Cyg 21h39'46"37d60'
Colletti,Vince
Uma 11h16'36"50d7'
Colley Brian
Cep 3h12'21"77d43'
Colley,Eric M
Aql 18h40'50"-1d51'
Colley-Daniel,Jerlean E
Lyr 18h58'34"37d2'
Collick,Cindy
Tau 4h29'49"30d22'
Collick,Joel Nathan
Her 16h32'48"36d9'
Collie,JoAnn
Cas 0h51'49"73d54'
Collie,Mrs Kathleen G
And 2h30'34"48d2'
Collier,Alyssa Blaire
Eri 3h27'13"-4d15'
Collier,Fran
Sge 19h12'46"20d27'
Collier,Ian
Cet 0h26'57"0d54'
Collier,Joe
Cet 1h50'49"-3d11'
Collier,Kelsy
Cet 0h26'0"0d31'
Collier,Laura Ann
Hya 8h26'58"-0d21'
Collier,Leona
Peg 23h40'34"15d23'
Collier,Leslie Claire
Sct 18h52'15"-7d33'
Collier,Lisa Kay
Umi 15h20'54"82d11'
Collier,Logan
Cet 0h26'28"0d53'
Collier,Michele
Mon 6h32'21"-6d53'
Collier,Noel (Buddy)
Cma 7h0'45"-16d7'
Collier,Peggy
Peg 22h2'21"25d49'
Collier,Richard
Aql 19h30'54"11d14'
Collier,Thomas
Her 16h10'30"48d40'
Collier,Victoria
Cyg 20h13'1"38d50'
Collier-Vaughn My Lovely
Lady,Debra J
Uma 12h2'27"64d35'
Colligan,Kip
Per 2h59'60"45d27'
Collin Raye
Oph 17h54'24"13d12'
Collin The Special
Dra 16h39'36"62d35'
Collin's Magic Wish, Liz
Mon 6h54'17"-5d59'
Collin,Christian
And 23h20'29"35d10'
Collin,Céline
Uma 8h20'1"60d38'
Collin,Gustave Jean
Dra 20h9'14"62d25'
Collin,Heather Lynn
Cam 3h47'0"58d18'
Collin,Sean Seton
Aur 6h25'36"38d57'
Collinet,Liliane
Aur 5h29'55"37d46'
Collingwood,David
Lyr 18h15'0"42d18'
Collins
Boo 15h1'27"27d54'
Collins
Her 17h34'22"20d18'
Collins "1959", Maureen
Cas 0h46'17"63d59'
Collins & Family, Daphne
Jean
Cet 1h52'49"-3d27'
Collins 18/12/59,Lynn
Cyg 20h17'19"39d4'
Collins 22,Robby
Oph 17h59'21"-3d44'
Collins 9 Family Star
Sex 10h12'43"-2d1'
Collins Family,The David
Aql 19h19'41"15d9'
Collins III,Carter Compton
Cet 3h20'59"1d36'
Collins III,William Joseph
Sgr 19h33'35"-32d47'
Collins Of MD,Bob, Elaine &
Brad
Uma 8h45'38"50d33'
Collins,Alice Marie
Oph 18h18'57"6d15'
Collins,Bethandru
Aql 19h25'1"-1d6'
Collins,Betty Ann
Uma 13h39'32"51d28'
Collins,Brenda K
Cas 0h11'13"50d34'
Collins,Brian R
Cyg 19h29'17"33d40'
Collins,Catherine L
Tri 2h1'47"30d15'
Collins,Charles L
Her 17h27'13"26d55'
Collins,Charlotte Jacklene
And 1h56'40"36d7'
Collins,Christopher- Michael
Chaz
Her 17h20'0"48d8'
Collins,Clyde
Uma 11h19'12"38d7'
Collins,Colin Matthew
Boo 15h19'14"38d21'
Collins,Craig
Uma 12h4'57"37d45'
Collins,Cynthia
Vul 19h47'0"23d51'
Collins,Danielle Lavenia
Cyg 20h23'54"30d43'
Collins,Darron Asher
Per 3h8'43"40d48'
Collins,Deborah
Cas 2h4'31"61d52'
Collins,Devin Bryce
Her 17h20'32"47d47'
Collins,Dominic James
Umi 15h38'59"81d22'
Collins,Dr Mary Lynn
Uma 8h13'13"68d56'
Collins,Edward Ewart
Aql 18h59'11"-5d3'
Collins,Forever Young, Pointe
RJP
Sex 10h26'41"-6d31'
Collins,Francis John
Dra 19h2'53"54d34'
Collins,Gary Douglas
Cru 12h44'56"-58d2'
Collins,Georgie
Boo 14h55'23"27d52'
Collins,Harold Edward
Sct 18h41'32"-6d9'
Collins,Heather M
Mon 7h12'0"-10d38'
Collins,Irene Mary
Cas 0h33'13"62d3'
Collins,J B
Aql 18h43'27"11d37'
Collins,James
Aur 5h4'24"40d24'
Collins,Jamie-Lynn
Cas 23h34'11"61d10'
Collins,Janet
Cyg 19h30'33"33d22'
Collins,Janet M
Cas 0h39'41"58d45'
Collins,Jennifer
Aql 20h6'30"4d46'
Collins,Jenny
Cyg 19h27'33"35d23'
Collins,Jessica
Peg 23h49'22"30d56'
Collins,Jim
Cet 2h35'27"5d42'
Collins,Joanne
Uma 13h51'49"51d27'
Collins,John Christopher
Her 17h22'0"45d36'
Collins,John & Margaret
Cyg 19h29'1"38d40'
Collins,John Allan
Cep 0h2'16"68d39'
Collins,Joyce Masters
Cap 21h52'1"-20d43'
Collins,Julia Madeline
And 0h32'0"30d22'
Collins,Julie Kathleen
Uma 9h22'46"52d20'
Collins,Katherine
Lyn 9h2'37"35d35'
Collins,Kathleen N
Cyg 21h38'26"38d8'
Collins,Keira
Lyr 18h36'47"30d46'
Collins,Keith E
Ori 6h3'46"-0d1'
Collins,Ken & Sandee
Cam 3h53'21"56d21'
Collins,Kenneth Guy
Wyndham
Psc 0h59'1"21d5'
Collins,Kevin David
Oph 17h57'21"13d18'
Collins,Kira Kristen
Peg 23h28'0"32d32'
Collins,Kristin
Cet 3h13'55"3d37'
Collins,Lacey Elspeth
And 23h37'23"43d31'
Collins,Mark Eugene
Cep 22h24'31"63d40'
Collins,Martha Anne Haugh
Cam 4h27'39"60d4'
Collins,Mary & Geoff
Uma 12h1'28"32d26'
Collins,Mary Jane
Uma 12h39'23"60d34'
Collins,Maureen
Cam 8h34'1"78d33'
Collins,Max
Uma 11h11'55"60d42'
Collins,Meghan Anastasia
And 0h58'15"38d46'
Collins,Michael J
Sco 16h56'27"-44d51'
Collins,Michael John
Boo 14h25'26"12d29'
Collins,Michael Ray
Cet 2h33'0"-10d21'
Collins,Michael Shane
Uma 9h28'54"52d29'
Collins,Natalie
And 23h26'19"42d10'
Collins,Nicholas S
Lac 22h10'16"47d59'
Collins,Pa Poo
Aql 20h4'1"4d52'
Collins,Patrick Alexander
Dra 14h53'18"64d60'
Collins,Patti
Mon 7h51'54"-3d42'
Collins,Raymond L
Ori 5h15'1"1d2'
Collins,Rick
Ari 3h1'10"20d25'
Collins,Robert T
Her 17h14'45"20d35'
Collins,Rose
Cas 0h43'10"75d48'
Collins,Russell
Cep 20h56'14"58d56'
Collins,Ruth Harvey
Mon 7h41'7"-1d20'
Collins,Ryan Michael
Lac 22h51'32"52d41'
Collins,Shaun
Per 2h52'24"45d25'
Collins,Sunnie
Cet 3h8'32"5d11'
Collins,Teresa "Miss Terry"
Uma 12h11'57"62d51'
Collins,Theo Reynolds
Aql 18h59'45"-5d10'
Collins,Theresa Caroline
Umi 20h1'1"88d40'
Collins,Theron Tilford
Boo 14h28'0"47d42'
Collins,Thomas,
Annemarie,Mary
Cam 3h23'31"61d35'
Collins,Timothy J
Hya 8h33'38"-6d4'
Collins,Tom
Her 16h57'12"37d33'
Collins,Wallace Whittier
Aql 20h7'54"0d31'
Collins,William A
Cep 22h55'27"56d55'
Collins,William A
Cet 0h5'28"-11d50'
Collins,William H
Umi 8h7'39"88d46'
Collins-Buz
Ori 5h26'53"-3d29'A
Collins-Sam
Ori 5h26'53"-3d29'B
Collinson,Katian
Sgr 19h29'28"-32d4'
Collinsworth,Judith Ann
Boo 14h56'29"18d59'
Collis,Kevin James
Lyr 18h42'53"37d4'
Collis,Ron
Per 3h27'60"40d58'
Collis,Tony
Crv 12h34'46"-11d27'
Collishaw,Esther B
Ori 4h41'25"11d44'
Collison,Martin
Per 2h51'20"45d25'
Collmer 91143
Lyr 18h56'44"30d45'
Colloc'h,Françoise
Leo 10h18'40"10d5'
Colloff,Emily
Cyg 19h55'35"38d24'
Collot,Frédéric
Mon 7h49'11"-4d13'
Colluci,Tom & Diane
Aur 4h55'49"40d19'
Collum,Tim
Sct 18h56'37"-6d37'
Collura-Burke, Christina J
Peg 21h21'10"22d46'
Collyer Star,The Noel
Cep 21h16'25"70d15'
Colm
Ori 5h45'12"7d57'A
Colman,Alexandra Leah
Dorothy
Equ 21h2'11"8d48'
Colman,David A
Per 2h28'24"56d38'
Colman,Paul Robert
Boo 14h9'56"47d49'
Colman,Richard
Ser 16h7'45"10d43'

Colmar,John L
Her 18h5'37"14d7'
Colnon,William Smith
Sex 10h33'41"0d59'
Colo,Marshall
Tri 1h55'32"25d56'
Colo,Terry
Eri 4h2'30"-10d54'
Colobus Tim
Ori 6h0'16"1d39'
Colomb,Dan
Cnc 8h25'32"11d12'
Colombia
Cnv 13h5'1"32d21'
Colombo Andrea
Umi 14h49'42"67d55'
Colombo,Craig
Crt 11h45'49"-18d11'
Colombo,Hervé
Del 20h15'0"14d30'
Colombo,Irene
Lyr 19h20'53"42d47'
Colombo,Marta
And 0h9'0"38d32'
Colombo,Marzia Cristina
Per 3h33'0"51d10'
Colombo,Raffaella
Boo 15h36'1"42d0'
Colombo,Roger
Del 20h18'0"16d7'
Colombo,Rosaria
Aur 5h18'53"45d7'
Colombo,Sylvain
Del 20h13'42"14d12'
Colombo,Theresa
Cnc 8h33'1"7d43'
Colon,Cathy
Cas 2h3'36"59d19'
Colon,Deborah Eileen Rivera
Mon 8h5'54"-9d47'
Colon,Nicky
Lmi 10h13'13"37d40'
Colon,Our Little Angel Monica Mary
And 2h22'14"40d40'
Colon,Patricia
Leo 10h44'58"15d59'
Colon,Raquel & Michael
Lyn 8h59'25"40d17'
Colon,Rene Leslie
Aur 4h49'28"51d27'
Colon,Roberto Louis
Cnv 12h40'49"38d42'
Colone,Judy Ann
Cas 0h38'51"64d12'
Colonel Ray
Her 16h42'1"23d0'
Coloni,Richard V
Lyr 18h57'0"31d12'
Colonna,Diane
And 1h34'40"39d17'
Colonna,John & Kristine
Crb 15h30'38"31d11'
Colonna,Richard
Aur 6h27'26"37d41'
Colonna,Roger
Lac 22h6'23"51d6'
Colorado Fondue Company
Hya 8h49'25"-6d41'
Colovas,John Antone
Aur 6h3'1"32d1'
Colquett Aslanin Yildizi
Hya 9h10'17"3d23'
Colquitt,Toni
Ori 5h51'16"8d27'
Colreavy's Comet
Cyg 19h34'41"33d32'
Colstar
And 0h30'1"45d7'
Colstrom,Laura Ann
Lyn 6h26'1"58d53'
Colt,Buddy
Cet 3h5'1"2d30'
Colt,Peter
Cet 3h2'25"5d15'
Coltabaugh,Elizabeth Marie
And 23h22'0"43d59'
Colter,Shandon Michael
Sex 10h9'0"-5d53'
Colter-Memorial Star, Jamie
Uma 11h52'24"42d52'
Colteryahn Family
Eri 4h9'47"-10d6'
Coltin,Bernard
Cep 22h25'50"57d50'
Coltune,Debbie & Scott
Cet 2h17'19"2d11'
Colubriale,Nicholas Patten
Umi 15h13'31"66d5'
Colucci,Andrew & Suzanne
Uma 11h6'0"37d31'
Colucci,Marilyn
Cas 3h4'18"71d6'
Coluccio,Maryann
Gem 6h41'35"15d54'
Columbia Family Star, The
Tri 2h17'14"30d15'
Columbia,Pam & Ken
Umi 14h7'0"70d25'
Columbian 100th Year
Boo 14h57'1"41d30'
Columbine Star/ Estrella de Columbine
Uma 8h24'40"71d60'
Columbus,David & Michelle
Lyn 7h56'44"41d4'
Colver,Michael Robert
Ser 15h13'35"8d35'
Colverson,John
Cyg 19h34'31"39d21'
Colvil,Bruce
Uma 13h0'51"53d23'
Colville,Julie
Cyg 20h20'38"39d1'
Colvin,Alyssa Joyce
Cas 0h26'35"68d27'
Colvin,Clifford
Aql 19h53'45"13d9'
Colvin,David J
Ori 4h56'23"-0d6'
Colvin,Shirley
Eri 3h20'22"-18d30'
Colvin,Sonya
And 23h28'33"49d57'
Colvin,Thomas L
Aql 19h23'44"-6d18'
Colwell,Carolyn Sue
Sct 18h55'41"-6d53'
Colwell,Gramp & Allison
Crb 16h14'37"37d38'
Colwell,Jack Peter
Her 16h30'24"39d8'
Colwell,Judy
Aql 20h10'41"13d22'
Colwell,Lori
And 0h23'59"32d51'
Colwell,Susanna Fernald
Ori 5h59'1"21d0'
Colwin,Brian
Ori 5h29'23"-1d32'
Coma Berenices
Com 12h24'59"22d43'
Comar,Valentina
Pic 4h55'19"-44d16'
Combaluzier,Maurice
Lac 22h5'32"37d42'
Combes,Meghann Elizabeth
And 0h22'37"34d14'
Combet Sonia
Cep 20h15'39"60d39'
Combey,Jean-Christophe
Cam 5h52'53"65d9'
Combier,Jacques
Lyn 7h58'19"48d1'
Combier,Sylvie
Aur 4h37'26"31d16'
Combo,Danley-Hanks
Mon 8h6'9"-4d34'
Combs Memorial Star, The Kathy
Lyr 18h50'54"42d29'
Combs,Addison Reneé
Cet 3h15'28"0d15'
Combs,Bobbye
Eri 2h53'0"-3d47'
Combs,Bridget Carolyn
Cyg 19h26'34"30d40'
Combs,Cindy S
Equ 21h5'19"10d21'
Combs,Clint
Cep 20h59'47"58d56'
Combs,Darrel
Cnc 8h56'19"20d42'
Combs,Eric
Vul 19h43'13"23d40'
Combs,Gary
Cet 1h25'16"-11d9'B
Combs,J Michael
Cet 2h3'10"-18d52'
Combs,Jamie Lynne
Dra 18h35'50"68d30'
Combs,Jennifer
Peg 22h41'12"33d43'
Combs,Jr,Sammy "Cha-Ching"
Aur 5h5'14"40d13'
Combs,Judy A
And 1h17'25"34d7'
Combs,Kimberly Ann
Lib 15h3'23"-0d30'
Combs,Larry Wayne
Cet 2h7'43"5d28'
Combs,Linda Marie
Crb 15h52'0"27d56'
Combs,Nancy Lea
Del 20h15'20"10d2'
Combs,Stephanie
Cas 0h17'26"66d22'
Combs,Tony L
Umi 16h25'50"70d52'
Combs-Waldecker
Peg 23h35'22"28d39'
Come in from the Rain
Cam 12h44'40"78d36'
Come Together Esplanade
Sgr 18h58'8"-21d54'
Comeau,Therese
Cyg 21h4'1"32d20'
Comeau,Willis
Cep 22h9'40"60d20'
Comegna,Anthony
Tri 2h43'18"33d46'
Comelisa
Nor 16h19'59"-50d30'
Comella,Josep
Dra 15h3'17"62d6'
Comellas,Wendy B
Del 20h50'48"9d40'
Comenduley,Mike
Cep 20h20'58"60d52'
Comenius
Boo 15h4'18"26d33'
Comer,Alva Townes
Uma 11h46'47"32d42'
Comer,Clarence E
Ori 5h13'40"15d9'
Comer,Elizabeth Marie
Sge 20h7'1"20d1'
Comer,Kendra S
Ori 5h19'21"12d26'
Comerford,Gene L
Her 17h36'50"26d58'
Comes the Dawn
Cam 11h19'26"80d15'
Comet 101495
Aur 6h23'32"31d31'
Comet-Mateu
Peg 21h51'56"20d9'
Comfort,Caitlin Irene
Lyr 18h32'1"30d3'
Comiskey,Jane
Lyr 18h48'0"33d29'
Comisso,Riccardo
Gru 22h18'42"-55d16'
Comito,Rick
Aur 6h21'15"33d45'
Commander Cynthia
Cas 0h40'25"76d57'
Commander,The
Per 1h47'57"48d29'
Commeau,Amy
Del 20h27'0"18d49'
Commer,Joshua Glenn
Uma 11h7'13"41d17'
Commini,Mary DeLorenzo
Cyg 19h19'1"28d44'
Commitment 13
Eri 2h50'56"-6d42'
Commonier,Cleline
Uma 11h30'0"52d52'
Como,Norman Gerard Joseph
Boo 14h57'0"51d36'
Comola,Emphemera My Star Jim Paul
Aql 20h10'44"10d29'
Comolli,William Louis
Oph 17h18'22"-20d23'
Compagnucci,Ida
Cyg 19h33'0"30d45'
Compain,Aristide
Ori 5h1'48"10d13'
Compain,Bernadette
Ori 5h1'52"14d1'
Compain,Frederic
Ori 6h1'51"11d7'
Compani,Todd J
Boo 15h46'11"50d20'
Compas,Steve & Carmen
Cam 12h56'29"77d53'
Compassion
Lyr 18h30'49"37d22'
Compassion
Lyn 7h9'26"59d53'
Compau,Jennifer Gale
Mon 6h42'54"-10d12'
Compere Loveless
Hya 8h31'26"-0d15'
Completely
Crb 15h42'46"27d42'
Compton 18,Natalie
And 0h58'26"37d24'
Compton,Douglas
Dra 16h3'19"67d38'
Compton,Elizabeth Ann
Cas 0h55'21"64d30'
Compton,Malaak
Aql 20h1'55"11d20'
Compton,Rose M
Peg 22h43'29"25d40'
Compton-Pruett Spatial Relationship
Lac 22h36'22"44d24'A
Compufix Computer Repair Depot
Aql 19h28'16"13d0'
Computer Hardware Sales
Crb 15h53'1"38d14'
Computer Source
Mon 6h20'24"1d25'
Comstock,Chad Edward
Aql 19h53'40"1d27'
Comstock,Gertrude
Peg 21h52'15"34d17'
Comstock,Joseph Andrew
Dra 20h23'47"63d34'
Comstock,Susan
Cas 1h11'22"61d50'
Comstock,The Thomas
Cet 0h41'21"-4d43'
Comtesse Agathe
Cas 0h0'53"61d48'
Comunale,Martha Anne
Equ 21h21'0"11d13'
Comète,Louis-Maurice
Lyr 18h48'42"35d5'
Con
Sgr 19h19'25"-41d47'
Con
Vul 19h44'14"28d34'
Con Amore
Per 2h52'46"38d37'
Conacher,Christina Bain
Cas 0h15'21"62d15'
Conahan,John Joseph
Her 16h1'0"20d28'
Conan
Boo 14h49'59"35d5'
Conan
Dra 18h28'26"58d38'
Conant,Darcy
And 2h21'20"47d34'
Conant,Lyn
Cet 2h56'53"7d39'
Conant,William L
Aql 20h4'41"0d37'
Conaway,Jr,Herman
Boo 14h43'27"26d21'
Conaway,The Happy
Lib 15h5'12"-0d48'
Conboye,John
Cet 2h41'0"2d52'
Conca,Sr,Kenneth S
Cnv 12h23'15"38d13'
Conceiao Tude
Cet 3h6'1"5d14'
Conceiti,Antonella
And 0h11'41"38d43'
Concepcion,Zaida
Sge 20h17'53"16d32'
Concepta
Ori 5h19'12"-6d38'
Conchita
And 0h6'55"37d34'
Concordia Emma
Crb 16h1'0"27d28'
Concotelli,Gina
Peg 0h4'37"27d59'
Concrete Parachute
Peg 23h31'19"31d30'
Concupiscence
Umi 13h29'42"70d31'
Conde Star
Peg 23h47'46"15d38'
Conde,Isabelle
Cyg 19h44'35"30d47'
Conde,Julian Gutierrez
And 0h28'1"44d21'
Condello,Sabrina Victoria
Cnv 12h17'35"38d55'
Condit,Caleigh Arlene
Leo 9h34'15"8d9'
Condoleo
Uma 15h38'0"68d45'
Condoll,Lydia
And 22h56'37"50d15'
Condoluci,John Joseph
Per 4h7'15"52d15'
Condon,Camie Christine
Cas 1h45'15"70d1'
Condon,Cary Babington
Lmi 10h11'55"30d50'
Condon,Jodi Michelle
Peg 23h40'0"16d53'
Condon,Juanita
Cyg 19h50'1"30d54'
Condon,Megan Renae
Peg 22h0'60"5d58'
Condon,Sr,Richard Phillip
Dra 11h54'15"67d3'
Condon,Stephanie
Lyn 9h0'40"38d21'
Condon-Bill's Sparkle, Julia Boyd
Aql 18h56'59"14d20'
Condor Party Centre
Vul 19h20'34"25d30'
Condos,Nick
Her 16h58'39"23d4'
Condreva,Ken-Sandie- Kathy & Joanne
Aql 19h28'12"8d4'
Condro,Michael C
Cep 0h2'23"70d37'
Condron,Mark J
Mon 7h30'16"-0d19'
Condron,Stephen & Lorraine
Umi 15h45'23"73d28'
Condy,Sherry
Lyr 18h32'1"45d23'
Cone Star,The Chris
Peg 22h17'25"8d44'
Conelly III,Thomas Middleton
Aql 19h51'42"12d6'
Conery,Wayne Nolan Christopher
Ari 2h35'26"30d7'
Conesa,Alain
Sgr 19h28'2"-31d50'
Conesa,René
Sex 10h14'50"-7d58'
Coney II,William H
Aql 20h9'1"8d55'
Confer,L Schane
Aur 5h6'1"42d60'
Conford,Ellen
Lyn 8h7'0"48d5'
Conforti,Maryann
And 0h46'50"34d17'
Congdon,George F
Cam 3h20'46"61d34'
Congdon,Margaret Karen
Lyn 8h7'37"51d60'
Conger,Franklin Lee
Cep 22h39'47"59d1'
Conger,Terry W
Aql 20h4'54"7d47'
Congiundi,Lisa Angela
Peg 23h2'37"10d16'
Congiundi,Samuel Phillip
Cma 6h12'24"-13d15'
Congolandi,Cristina
Boo 15h6'22"42d25'
Congrats,Ron & Christine 11-11-95
Aql 19h41'18"14d10'
Coni's Star
Eri 3h5'21"-6d20'
Conicella,Rosemarie T
Her 17h39'36"20d56'
Conine,Jeff
Cma 6h52'14"-18d26'
Conkey IV,Harry Dunreith
Gem 6h32'11"14d41'
Conklin,Dick
Equ 21h21'22"3d26'
Conklin,Jim & Beth
Cam 3h18'54"66d21'
Conklin,John F
Tri 2h3'20"31d11'
Conklin,Jr,Joseph P
Aur 6h26'59"33d51'
Conklin,Kathleen Bennett
Cas 0h43'53"64d3'
Conklin,Neil Alan
Aql 19h3'36"0d1'
Conklin,Opal
Vul 19h3'49"21d37'
Conklin-310
Lac 22h44'45"54d45'
Conlan,James Peter
Hya 8h57'39"2d22'
Conlan,Peter R
Cyg 19h23'55"54d58'
Conlee,Jade Adele
And 0h14'12"37d56'
Conleen
Cra 18h29'40"-40d55'
Conleth Howard
Cep 22h46'39"57d49'
Conley II,John R
Aur 6h30'12"37d37'
Conley,Arlene & Charles
Hya 9h8'55"0d29'AB
Conley,Brian
Ser 17h59'50"-14d14'
Conley,Dale Marie
Cas 0h38'26"64d56'
Conley,Don
Ser 18h15'44"-0d7'
Conley,Herbert "Frankie"
Aql 20h15'0"0d16'
Conley,James
Aql 19h30'0"1d7'
Conley,Jennifer
And 0h51'53"35d60'
Conley,Jessica R
Cas 2h31'22"58d16'
Conley,Jodi Ann
And 23h2'51"46d55'
Conley,Kevin Joseph
Her 17h59'28"38d19'
Conley,L Henry
Per 2h8'58"56d21'
Conley,Matthew & Ildik
Ori 5h54'17"20d9'
Conley,Matthew Edward
Aur 5h38'0"38d12'
Conley,Max Quella
Hya 8h28'21"-8d31'
Conley,Mindy
Vul 20h38'24"20d18'
Conley,Peter A
Ori 5h58'16"15d35'
Conley,Philip K
Cet 1h10'54"-5d49'
Conley,Sarah Anne
Mon 6h52'24"10d36'
Conley,Shanna Marie
Aql 19h57'47"1d30'
Conley,Taylor Leigh
Peg 21h51'57"30d33'
Conlin,Gabrielle
Uma 11h14'10"47d47'
Conlin,Herbert
Aql 20h1'24"0d2'
Conlin,Joseph
Aur 4h36'13"31d17'
Conlin,Jr,Robert D
Per 2h57'17"40d25'
Conlin,Richard James
Aql 19h7'1"0d1'
Conlin,Steven John
Ori 5h52'18"15d29'
Conlogue,Noelita
Equ 20h59'14"6d8'
Conlon Constellation, J & K
Crb 16h10'33"38d56'
Conlon,Ann
And 0h7'27"47d44'
Conlon,Jacquelyn
Cet 0h26'30"-18d0'
Conlon,Jayne
And 23h3'27"50d25'
Conlon,Lawrence
Aur 4h47'28"41d8'
Conlon,Mark
Her 17h29'42"31d24'
Conlon,Patrick F
Lac 21h56'50"37d46'
Conlon,Peggy
Peg 0h12'17"20d53'
Conlon,Peggy
Peg 22h29'17"25d60'
Conn,Brian David
Her 18h13'32"47d54'
Conn,Deborah
Cyg 21h10'19"35d60'
Conn,Mallory
Lmi 10h24'52"31d45'
Conn,Patricia Margaret
Sge 20h17'1"20d16'
Connah,Ethan
Sge 20h7'41"20d36'
Connally III,William P
Lac 22h40'23"55d31'
Connaughton,Claire Alex
Lyr 18h14'51"38d4'
Connaughton,Lisa Ann
Lyn 8h30'1"40d17'
Connaughton,Maryann
And 2h30'25"41d6'
Connected 5-14-93
Lyn 7h37'0"44d51'
Connell (STB), Wonderful Suzanne
Lyr 18h36'28"27d51'
Connell 1993,Father Mark J
Cep 22h5'26"70d24'
Connell Star,The George & Joanne
Cyg 21h33'25"42d20'
Connell,Adam Sean
Her 18h4'15"31d33'
Connell,Colm
Dra 15h58'19"52d1'
Connell,David Day
Per 4h4'18"51d39'
Connell,Dean W
Per 2h55'29"32d46'
Connell,Mary Elizabeth Bristow
Lyn 9h18'23"34d42'
Connell,MD,Bruce Fowler
Equ 21h18'52"2d51'
Connell,Michael
Cmi 7h54'57"4d24'
Connell,Rachel
Mon 6h42'38"11d50'
Connell,Sean Timothy
Boo 14h53'1"30d2'
Connell,Suzanne
Cma 7h4'23"-31d18'
Connella,The Star Of Dawn
Ori 5h20'57"0d48'
Connellan,Avis Murphy Alexandra
Cyg 20h51'10"30d42'
Connelley,Ellisha Verna
Uma 11h1'15"58d37'
Connelley,Ellisha Verna
And 0h8'28"27d55'
Connelly,Aaron Louis
Dra 16h25'17"68d47'
Connelly,Bill
Cep 21h46'1"58d23'
Connelly,Donna
Cam 3h44'37"60d26'
Connelly,Harry
Cet 1h16'38"-1d38'
Connelly,Harry
Dra 16h35'42"73d40'
Connelly,Jacob Oliver King
Her 17h5'29"47d29'
Connelly,Joshua Bradin
Aur 5h22'28"38d48'
Connelly,Kirsten Ann
Cam 3h37'46"60d52'
Connelly,Linda S
Cas 1h15'0"67d54'
Connelly,Mary
Lyr 19h19'24"40d7'
Connelly,Michael William
Cep 21h7'55"65d2'
Connelly,Sean
Ori 5h54'14"15d2'
Connelly,Stephanie
Tau 04h02'00"12d09'
Connelly,Vincent F
Per 1h36'49"53d18'
Conner III,"Tucson Topper" Earl
Gem 6h52'55"19d17'
Conner Star,The
Ori 5h54'52"5d12'
Conner,Douglas Coleman
Her 17h1'33"46d46'
Conner,Jay (Grizzly)
Cmi 7h44'47"8d29'
Conner,Jr,Otis
Aql 19h22'32"-0d23'
Conner,Kristin
Peg 23h3'34"21d30'
Conner,Lawson Thomas
Dra 16h47'35"67d58'
Conner,Margaret Shull
Cyg 20h4'22"30d12'
Conner,Pleze J
Uma 9h0'0"59d29'
Conner,Robert James
Cnc 8h35'37"18d37'
Conner,Ryan Dean
Sex 10h44'11"3d25'
Conner,Sarah Pauline
Hya 8h26'49"0d30'
Conner,Scott
Oph 16h28'49"-6d37'
Conner-Marie
Umi 15h55'0"70d5'
Connerr,Helyn
Cas 23h27'51"61d21'
Conners,Eric Michael
Per 1h39'16"54d11'
Conners,Fredia
Cyg 19h53'53"38d39'
Conners,Scott M
Aql 20h20'57"5d11'
Connett,Dave
Her 17h10'40"41d45'
Conni
Lyn 9h0'26"40d44'
Conni Jane
Lmi 10h35'46"27d35'
Conni-JAJ
Tri 1h51'26"26d28'
Connie
Vul 19h58'48"22d38'
Connie
Cam 7h54'53"60d37'
Connie
Lyr 18h36'25"47d17'
Connie
Tau 4h17'51"1d39'
Connie
Aql 19h1'28"16d49'
Connie
Aql 19h2'0"16d12'
Connie
Cas 0h29'21"60d32'
Connie & Family
Eri 3h42'17"-7d1'
Connie Ann
Vir 13h50'21"5d20'
Connie B
Lyn 8h16'35"51d58'
Connie C
Cas 1h6'19"60d32'
Connie Connique
Cet 3h6'14"0d49'
Connie Elizabeth
Del 20h13'46"14d33'
Connie J
Her 16h18'41"26d1'
Connie Jo's
Vul 19h14'34"21d42'
Connie Lee
Tri 2h1'59"30d33'
Connie Lee
Eri 3h20'38"-16d23'
Connie Lou "Heaven On Earth"
Cas 2h27'45"70d24'
Connie's 1993 Christmas Star
Cyg 19h53'39"38d18'
Connie's Fingers
Crb 15h22'0"30d60'
Connie's Heart
Ori 5h36'40"0d8'
Connie's Light
Tri 1h53'1"27d9'
Connie's Star
Cet 1h3'55"-1d53'
Connie's Star Gemini
Gem 7h1'21"30d5'
Connington,MJ "Jack"
Del 20h49'21"9d26'
Connolly,Anne Megan
Lib 14h41'15"-23d34'
Connolly,Ben
Ori 5h57'30"14d48'

Connolly,Benjamin &
Shannon
Lyn 9h7'28"34d32'
Connolly,Brian J T
Her 17h14'1"29d9'
Connolly,Catherine Beth
Cas 22hh58'39"54d4'
Connolly,Jessica Diane
Cam 7h56'34"61d39'
Connolly,John Patrick
Per 3h6'1"46d35'
Connolly,John R
Per 3h53'26"35d25'
Connolly,Kevin J
Cep 1h18'57"86d59'
Connolly,Lisa Ann
And 1h21'50"48d58'
Connolly,Mrs & Mrs John
Cyg 20h8'36"40d34'
Connolly,Patricia Burns
Cas 3h9'39"58d20'
Connolly,Sr,Stephen W
Per 2h53'0"40d47'
Connor
Her 17h37'38"40d40'
Connor Aaron
Cam 8h6'12"80d37'
Connor BC 1973,Robert
Michael
Aur 6h50'0"37d15'
Connor David Proud
Aur 4h56'41"40d4'
Connor III,James P
Cep 0h43'0"77d37'
Connor,Angela Marie
Lyr 19h26'0"38d3'
Connor,Anne
Peg 22h0'24"2d13'
Connor,Courtney P
Cas 0h23'24"65d9'
Connor,David Peter
Cam 4h44'24"68d23'
Connor,Forever George & Kay
Aql 18h42'51"0d48'
Connor,James R D
Per 2h48'19"31d18'
Connor,Jayne Laurie
Cas 1h43'23"75d41'
Connor,John Jennings
Umi 14h58'1"72d14'
Connor,Michael P
Cep 23h7'1"65d15'
Connor,Robert James
Cet 3h11'54"-0d10'
Connor,Sharon Elaine
Cas 0h5'21"63d41'
Connoris,Astra
Per 2h55'1"37d24'
Connors
Aur 5h0'0"40d1'
Connors "Sparkler", Sharon
And 23h21'15"47d44'
Connors III,Matthew F
Aur 5h10'47"43d5'
Connors"Granny B90",
Elisabeth G
Mon 7h4'36"-0d11'
Connors,Alice I
And 1h55'12"40d17'
Connors,Christopher Charles
James
Her 18h45'1"12d10'
Connors,Elizabeth
Cyg 19h45'54"29d46'
Connors,Grandma
Tri 1h59'33"27d24'
Connors,Heidi P J
Cas 0h43'0"65d47'
Connors,Helen Bement
Cmi 7h34'3"7d7'
Connors,Jeannette Beland
Cas 1h20'12"60d26'
Connors,John Timothy
Her 17h1'32"48d20'

Connors,Jr,Jack M
Cma 7h18'33"-15d31'
Connors,Julie
Peg 23h5'14"22d4'
Connors,Katherine Marie
Lyr 19h16'23"41d49
Connors,Schoensee Paragon
Technologies
Dra 11h6'59"74d59'
Connors,Terence
Aur 5h58'43"28d34'
Connors,Thea
Cam 3h48'24"78d46'
Connoy,Patrick J
Cep 22h56'51"68d34'
Conny
Per 2h56'19"48d56'
Conny
Uma 14h22'58"60d35'
Conny
Cnc 8h25'53"30d48'
Cono
Lyn 8h24'18"40d55'
Conoly,Weston Zuehl
Hya 10h40'31"-18d37'
Conor
Ori 5h18'45"-6d55'
Conover,Edith
Mon 6h40'28"10d17'
Conover,Mark Alan
Aql 20h10'15"11d28'
Conover,Mark Allan
Cet 2h4'1"3d54'
Conover,Marnie
Cnv 14h2'42"45d11'
Conrad
Per 1h53'11"54d2'
Conrad
Ori 6h8'52"8d2'
Conrad
Hya 8h11'35"2d35'
Conrad & Lisa
Hya 10h4'51"-18d50'
Conrad III,Coleman W
Aql 20h10'33"13d13'
Conrad III,William Henry
Cyg 19h31'58"30d32'
Conrad Star,The Lucille Davis
Sco 17h1'59"-31d6'
Conrad,Bob
Ser 15h33'55"0d2'
Conrad,Carin Chari
Cas 1h55'55"58d32'
Conrad,Catherine Portwood
Stanley
Eri 3h23'42"-13d22'
Conrad,Dale
Lac 22h3'21"38d2'
Conrad,Deborah
Cyg 19h39'11"28d38'
Conrad,Dennis & Bridget
Cyg 19h13'48"47d23'
Conrad,Edward
Mon 7h52'2"-2d39'B
Conrad,J Clay
Aur 6h16'43"30d18'
Conrad,James
Her 16h36'57"33d18'
Conrad,LaVerna
And 23h32'47"42d16'
Conrad,Leyla
Mon 7h52'2"-2d39'A
Conrad,Matthew David
Peg 22h20'18"18d53'
Conrad,Maxine
Cas 3h2'45"70d42'
Conrad,Michael Anthony
Lac 22h8'1"38d56'
Conrad,Nicole Ashley
Peg 22h30'54"29d55'
Conrad,Pat & Fred
Eri 3h23'25"-3d47'
Conrad,Phyllis
Peg 21h53'30"3d34'

Conrad,The Jennifer Margaret
Lyn 7h53'36"34d58'
Conrad,Thomas R
Psc 0h26'14"8d34'
Conradi,Gerhard
Ari 2h29'1"28d19'
Conrads,Kim Renee
Boo 14h3'55"8d21'
Conradt,Susanne
Lyn 8h21'1"42d9'
Conrey,Terry
Uma 10h47'49"70d34'
Conroy,Amanda
Ori 5h56'52"18d57'
Conroy,Danny
Sex 10h43'10"-5d12'
Conroy,James
Peg 22h20'1"34d57'
Conroy,James T
Uma 14h0'16"58d4'
Conroy,Joanne Mather
Cyg 20h30'42"42d14'
Conroy,Katie DeWit
Lmi 10h21'25"30d41'
Conroy,Kyle Joseph
Sex 10h38'0"2d21'
Conroy,Margie"Jasmine"
And 22h59'26"37d20'
Conroy,Marion & Tom
Aur 6h21'24"50d26'
Conroy,Molly Colleen
Aur 5h53'46"38d51'
Conroy,Pamila K
Aql 19h3'36"0d0'
Conroy,Paul Cletus
Cep 22h33'1"56d53'
Conroy,Thomas J
Aql 19h58'45"10d27'
Consalves,Marina Christine
Mon 6h21'27"0d7'
Consenti,Claudia
Pup 7h56'34"-24d21'
Conshick,Jacob
Cep 21h53'11"56d1'
Conshick,John
Cep 22h27'0"60d14'
Conshick,Marie
Cep 22h27'11"60d7'
Considine,Colleen
Peg 21h51'0"34d3'
Consiglia
Lyn 7h5'54"52d43'
Consiglio,Dr & Mrs William
Dra 17h52'44"60d53'
Consiglio,Evelyn Dorothy
Peg 21h9'25"12d32'
Consiglio,John
Per 3h42'42"36d49'
Consiglio,Robin
Vul 21h25'25"26d27'
Consilvio,Damiano
Leo 10h27'16"10d20'
Consolazio,Christina Marie
And 23h21'17"49d11'
Consolo,Elaine Maria
Lyn 8h54'12"41d27'
Consta,Chrisa
Lyr 18h28'44"40d54'
Constable 471
Uma 8h43'57"55d11'
Constable,David Martin
Boo 13h41'25"17d41'
Constable,Geoff
Cep 21h59'0"63d55'
Constance
Cas 0h39'34"60d2'
Constance
Cas 2h40'46"61d18'
Constance
Aql 19h52'34"13d27'
Constance
Lyn 8h9'31"33d34'
Constance
And 0h0'22"38d13'

Constance
Lyn 6h31'53"59d22'
Constance Adele
Cyg 21h4'47"40d51'
Constance Diane
Cas 1h52'44"58d12'
Constance Leslie
Cas 0h6'11"50d4'
Constance Marie
And 23h23'36"46d32'
Constance Paige
And 1h2'1"37d49'
Constance Shining
Cyg 19h26'0"30d46'
Constance Two
Peg 23h37'54"10d56'
Constance,Joseph Anthony
Peter
Boo 15h8'17"50d27'
Constancia
Uma 9h27'32"50d44'
Constanczak,Barbara
Lib 15h35'7"-22d29'
Constand,Andrea
Boo 13h41'47"19d43'
Constant,Demetrios
Aur 6h11'12"32d47'
Constant,Paul Edward
Aql 19h51'43"14d20'
Constantin,Cecile
Ari 2h26'27"10d37'
Constantin,Henri
Vir 15h3'1"6d30'
Constantina "Forever in My
Heart"
Tau 4h42'18"28d34'
Constantina Thalia, 1957
Cas 1h7'18"60d12'
Constantine
Cep 1h44'14"80d38'A
Constantine Wedding Star-
Kim & Lee
Cyg 20h23'1"39d45'
Constantino,Michael Robert
Aql 19h5'28"1d33'
Constantino,Sienna Angelina
And 0h0'14"37d38'
Constantinou, Christopher
Anthony
Cep 21h25'46"61d20'
Constantinou,Briana Photini
Peg 23h22'0"8d27'
Constantinou,Brooke Susan
Peg 23h28'30"18d46'
Constanze
Tau 5h0'49"15d27'
Constel,Robin Hairabian
Lac 22h18'51"54d59'
Consuelo
Lyr 18h33'54"30d15'
Contadino,Joe
Cep 22h6'31"68d43'
Contaldi,Jr,Frank Vincent
Cnc 9h6'0"28d52'
Contaldi,Michael P
Her 18h9'42"40d1'
Contaxis,Sr,William
Her 16h39'55"34d46'
Conte Space Odyssey 1-30-
95,John R
Uma 10h54'25"70d7'
Conte,Allison Rose
Cas 1h20'15"58d19'
Conte,Amanda
Aql 18h47'25"11d45'A
Conte,Andrea Lynne
Cas 1h5'55"56d14'
Conte,Dr Louis
Dra 14h28'33"63d2'
Conte,Joseph
Aql 18h47'25"11d45'B
Conte,Joseph C
Psc 1h28'24"11d11'
Conte,Lodovico Massimo
Lyn 7h59'16"50d16'

Conte,Roberto
Her 16h12'51"5d7'
Contensaux,Claire
Ser 16h5'11"8d6'
Contentlybeautiful
Lyr 19h6'45"40d56'
Contes et Chansons A
L'Avenir,Courtney
And 23h18'48"35d20'
Contessa
Lep 5h21'0"-24d42'
Contessa Jane
Cas 2h3'59"68d58'
Contessa Rose
Cas 1h5'43"55d49'
Conti Jean-Charles
Uma 11h26'50"37d30'
Conti's Star,Ron
Dra 11h16'43"73d7'
Conti,Catherine Marie
Vul 20h22'15"22d45'
Conti,Daniel Allen
Aur 5h56'30"38d42'
Conti,Daniela
Ind 21h2'17"-57d20'
Conti,Gaetano
Cep 22h39'13"58d58'
Conti,John
Lac 22h1'28"51d22'
Conti,Joseph F
Lac 22h54'12"51d58'
Conti,Sonia
Uma 12h18'39"57d33'
Conti,Tiziana
Uma 11h57'30"32d48'
Conti-McCombs,Anna T
Lac 22h47'54"54d26'
Continental Resources, Inc
Cam 8h24'52"78d44'
Contini,Mario & Rosaria
Ser 15h17'19"19d42'
Contois,Mary
Cet 0h9'0"-12d28'
Contomanolis,Kyra
Cam 11h34'37"81d41'
Contos,Perry G
Sct 18h30'35"-6d6'
Contrary-To-Ordinary
Uma 9h43'0"48d60'
Contratti,Loren K
Aur 7h9'19"41d20'
Contratto Always, Michael T
Per 2h57'0"32d7'
Contrebas,Rodney
Cet 0h53'17'-3d4'
Contrera,Lucky
Aur 7h22'22"38d7'
Contreras,Erin Marie
And 0h36'41"40d20'
Contreras,Lara
Cet 2h54'18"8d18'
Contreras,Michael
Cet 2h54'0"6d33'
Contreras,Nicholas Alexis
Sct 18h31'14"-5d20'
Contursi's Lucky Star, Phil
Aur 6h11'59"30d15'
Conundrum
Eri 4h0'54"-17d38'
Converso,Cristina
Vel 10h11'46"-49d31'
Convery,Clark W
Per 1h30'60"54d7'
Convery,Daniel Paul
Dra 17h4'22"60d11'
Convery,Elizabeth Mary
Uma 10h49'44"40d59'
Conway
Uma 10h40'0"50d49'
Conway,"Burts" Eyeball Finola
Cam 4h30'59"68d41'
Conway,Billy
Equ 21h3'11"7d36'
Conway,Billy
Aur 7h2'12"36d31'

Conway,Brend J
Dra 18h1'35"70d16'
Conway,Candida M
Mon 6h54'0"-10d54'
Conway,Christopher
Aur 6h10'58"38d40'
Conway,Dan
Oph 17h30'22"-22d3'
Conway,Eileen T
Vul 20h16'1"25d0'
Conway,Frank
Dra 19h27'52"73d15'B
Conway,Jack
Ser 17h53'21"-13d52'
Conway,John R
Cnv 13h20'50"31d29'
Conway,Katherine Claire
Cas 0h58'49"75d35'
Conway,Kathleen D
Cas 0h23'0"60d11'
Conway,Katie Clare
And 0h44'0"40d34'
Conway,Kellyann
Ori 5h51'43"7d24'
Conway,Kristen "Red"
Cas 0h13'18"61d21'
Conway,Leonard E
Uma 10h10'35"48d6'
Conway,Michael J
Aql 20h0'26"0d11'
Conway,Patrick
Her 16h12'40"41d9'
Conway,Phyllis
Dra 19h27'52"73d15'A
Conway,Rachael Lynn
Peg 21h53'43"20d9'
Conway,Timothy J
Dra 19h5'0"58d49'
Conway,Timothy Patrick
Aur 6h6'60"32d42'
Conway,Todd B
Aql 18h58'52"-6d57'
Conway,Tracey Ann
Cas 0h57'49"59d60'
Conway,William
Ori 5h56'18"8d41'
Conway-Moule,Molimus
Maximus David
Aql 20h5'49"1d26'
Conwell-Hollingsworth,
Evelyn
Eri 3h54'56"-5d52'
Cony
Tau 4h31'52"30d6'
Conyel,Justice
Vir 12h30'42"-8d2'
Conyers,Katherine
Eri 4h2'1"-15d39'
Cooch-Nahai
Oph 17h9'59"-20d2'
Coody,Robert
Cam 4h4'52"68d17'
Coogan,A P
Cyg 19h21'19"28d21'
Coogan,David Joseph
Cmi 7h8'15"7d13'
Coogan,David M
Ser 15h55'0"20d27'
Coogan,Kathryn Sheridan
Cas 0h17'34"60d47'
Coogan,Kenneth J
Oph 17h2'45"0d55'
Coogan,Peter J
Lac 22h3'55"50d28'
Cook Bright Star, Kinsey Lyn
And 1h3'48"40d59'
Cook Family Star,The
Aql 19h44'47"14d15'
Cook III,Charles Glen
Cma 6h42'40"-13d23'
Cook III,George
Cam 3h56'49"68d33'
Cook Sober & Perfect
1993,Michael
Uma 11h34'16"31d20'

Cook Star,The Clive
Per 2h5'11"57d47'
Cook's Star of Hope 1-13-
74,Tricia
And 23h28'35"47d37'
Cook's Star,Denise
Tri 2h7'1"32d27'
Cook's Star,Ron
Aur 5h26'51"30d12'
Cook's Star,The
Fri 2h57'1"-2d25'
Cook(Zook's Star), Nancy
Com 12h54'19"27d4'
Cook,Adeline Julie
Ori 6h4'0"-0d29'
Cook,Alfred & Elizabeth
Cnv 13h4'33"52d24'
Cook,Allen James
Lac 22h44'1"54d37'
Cook,Andrew
Umi 14h45'21"68d15'
Cook,Andrew John
Aur 6h28'37"38d9'
Cook,Anthony John
Per 3h0'39"47d28'
Cook,April Michelle
Cap 21h0'35"-15d58'
Cook,Audrey
Sct 18h55'46"-13d59'
Cook,Barbara A
Leo 9h46'26"18d20'
Cook,Bernice M
Sge 19h54'39"16d18'
Cook,Betty Sue Hagood
Mon 7h23'48"-8d13'
Cook,Bill & Gayle
Sge 19h20'0"16d18'
Cook,Brenda Starr
Eri 4h12'29"-11d10'
Cook,Bruce
Boo 14h31'0"21d40'
Cook,Castle
Cnv 13h59'1"41d45'
Cook,Catherine May Welgs
Aql 18h44'33"11d26'
Cook,Charles Grady
Oph 17h58'57"-6d46'
Cook,Christopher Ty
Her 17h53'49"38d1'
Cook,Colleen G
And 22h58'52"50d15'
Cook,Courtney Christine
Cet 3h16'1"9d16'
Cook,Cynthia L
Aql 20h11'18"4d32'
Cook,Daniel Wesley
Mon 7h3'39"-8d13'
Cook,David
Dra 18h1'1"84d33'
Cook,David Haws
Cam 6h6'48"80d31'
Cook,David Johnston
Aur 5h0'57"46d11'
Cook,Debbie
Crt 11h21'54"-11d34'
Cook,Diana Lynn Daniels
Lyn 7h50'38"48d45'
Cook,Diane B
Aql 19h25'40"10d41'
Cook,Donald G
Aql 18h58'45"14d5'
Cook,Dr Kimberly Joyce
Tau 4h11'0"1d31'
Cook,Elizabeth Nicole
Peg 22h28'46"18d56'
Cook,Eric
Cam 6h31'24"80d30'
Cook,Eric & Jason
Eri 4h17'60"-16d42'
Cook,Erin-Bruce Tolcharian
Sct 18h47'12"-8d49'
Cook,Fred
Hya 8h18'58"5d53'
Cook,Gail
And 23h29'46"40d55'

Cook,Gregory E
Her 18h18'10"14d1'
Cook,Heather M
And 23h25'16"48d3'
Cook,III,Erwin B
Boo 15h13'48"37d44'
Cook,Jacob Anthony Stevens
Crb 15h20'11"32d44'
Cook,James
Uma 10h1'59"48d53'
Cook,James Alton
Boo 13h36'20"14d14'
Cook,James S
Her 17h14'48"47d5'
Cook,Jennifer A
Lyn 8h19'46"36d22'
Cook,Jennifer Leigh
Cas 0h36'39"54d23'
Cook,Jenny
And 23h2'48"51d42'
Cook,Jerry A
Boo 14h49'27"36d53'
Cook,Jerry Einar
Aur 5h16'0"48d26'
Cook,Jill
Peg 22h13'25"30d18'
Cook,Jr,John
Per 2h0'57"57d57'
Cook,Judy
Del 20h58'0"10d8'
Cook,Julia Shipton
Aur 5h31'48"31d13'
Cook,Kaitlyn Margaret
Lyr 19h6'25"28d47'
Cook,Kathleen Ann
Cnv 12h14'30"45d2'
Cook,Kayla Nicole
Aql 18h44'35"10d14'
Cook,Keith
Her 18h20'0"12d12'
Cook,Kevin J
Per 4h28'40"51d0'
Cook,Kristen L
Hya 8h36'55"0d10'
Cook,Larry Lee
Peg 23h39'42"31d40'
Cook,Lawrence
Ser 15h22'35"24d54'
Cook,Lawrence P
Boo 14h13'58"32d33'
Cook,Loretta D
Peg 23h1'39"18d14'
Cook,Lori J
And 23h40'38"43d9'
Cook,Louise Rose
Tau 4h40'41"15d2'
Cook,Lt Cmdr Douglas
Aur 7h3'18"40d54'
Cook,Mable
Boo 14h55'14"39d51'
Cook,Marvin Eugene
Lac 22h25'43"50d1'
Cook,Michael Allen
Eri 3h33'57"-5d31'
Cook,Michael Frederick
Ori 5h52'50"1d24'
Cook,Michelle & Harry
Aql 20h19'42"1d38'
Cook,Nadine L
Eri 3h41'21"-2d25'
Cook,Pamela Jo
Per 2h25'0"56d57'
Cook,Patricia
And 0h8'1"31d23'
Cook,Petronelle
Cyg 19h51'45"37d40'
Cook,Philip Michael
Dra 18h37'48"67d47'
Cook,Rebecca Sue Dierks
Peg 22h38'35"24d3'
Cook,Robert Fred
Tri 1h41'23"30d50'
Cook,Robert L
Cet 3h8'56"0d9'

Cook,Robert Pettitt
Sct 18h52'46"-6d26'
Cook,Robin Christine
Mon 7h1'41"-5d58'
Cook,Ronald Cameron
Her 16h44'21"32d27'
Cook,Sam B
Ori 6h17'0"-0d35'
Cook,Susan S
Cyg 19h27'48"30d11'
Cook,The Cooker, Michael E
Her 17h35'51"20d36'
Cook,Timothy J
Lmi 10h43'33"30d22'
Cook,Tony
Her 16h28'1"40d10'
Cook,Tony
Ori 5h36'0"-6d50'
Cook,Tony
Ori 5h36'0"-6d50'
Cook,Wiliam Richard
Aur 4h38'24"30d8'
Cook,William Arthur
Sex 10h11'40"-1d59'
Cook,William Clyde
Cep 22h44'15"65d8'
Cook-Crist,Julie Deann
Cep 23h37'0"75d0'A
Cook-Hennessey
Dra 16h12'22"61d58'
Cooke Family,The William
Christopher
Dra 16h58'20"62d22'
Cooke OBE,Raymond Edgar
Cyg 20h25'1"46d32'
Cooke,Bradley
Peg 22h12'21"3d46'
Cooke,Bradley N
Aql 20h4'42"4d15'
Cooke,Cara Lynne
Tri 1h33'57"30d6'
Cooke,Chauncey Register
Cyg 20h24'35"38d54'
Cooke,Christopher Morgan
Crb 15h56'27"37d31'
Cooke,Donna Lee
Cas 0h14'58"58d37'
Cooke,Francine Emmelick
Peg 23h30'54"18d45'
Cooke,Gerald Edward
Oph 18h42'42"8d46'
Cooke,Laurin Elizabeth
Cyg 20h16'59"40d42'
Cooke,Mark Benjamin
Ser 15h38'0"1d13'
Cooke,Michelle
Cyg 20h23'43"39d23'
Cooke,Mitzi
Aql 20h17'27"0d4'
Cooke,Nicholas Levi
Dra 9h56'18"73d20'
Cooke,Pat & Tom
Aql 19h44'46"14d39'
Cooke,Patricia "Pat"
Cas 0h51'27"64d19'
Cookie
Cmi 7h41'10"0d46'
Cookie
Uma 9h36'47"59d10'
Cookie
And 1h38'1"36d37'
Cookie's Star
Uma 12h8'38"63d25'
Cooklin,Alice L
And 2h17'23"41d9'
Cooksey,Jana B
Hya 8h43'0"6d19'
Cooksey,Mariel Hope
And 0h26'0"41d3'
Cooksey-Gay
Cyg 20h23'52"38d45'
Cookson,Charles Thomas
"Chase"
Boo 14h25'47"25d8'

Cookson,Grace E
Mon 6h19'18"2d45'
Cookston "Maverick", Brett L
Uma 8h14'19"60d56'
Cool Bob
Her 17h23'52"41d6'
Cool Breeze
Peg 22h7'50"24d30'
Cool Erin
Cas 0h4'42"56d5'
Cool Touch
Cmi 7h19'46"5d42'
Cool,Alden
Ori 5h52'50"14d39'
Cool,Alex
Ori 5h52'56"14d51'
Cool,Ashley
Ori 5h50'26"10d26'
Cool,Brent
Ori 5h54'0"11d32'
Cool,Ellen Gitlin
Leo 10h51'59"18d43'
Cool,Gayle
Ori 5h54'22"15d58'
Cool,Tami
And 23h24'13"47d51'
Cool,The Golden Star, Janice
Com 13h13'21"26d26'
Cool,Timothy
Ori 5h52'16"14d51'
Coole,Amy Marie
Lyr 18h36'32"40d55'
Cooler,Alice Irene
Cam 3h50'26"61d2'
Cooler,Rondi
Cam 5h3'13"68d33'
Cooley Star,The Martin G
Her 16h50'59"47d40'
Cooley,Chase Austin
Hya 8h42'32"2d22'
Cooley,Christopher Ryan
Cet 2h19'53"2d57'
Cooley,Denton A
Tri 1h57'26"30d24'
Cooley,Gwen
Lyn 6h55'58"59d23'
Cooley,Katheryne
Umi 15h10'59"69d13'
Cooley,Ken&Kathy
Gem 7h21'33"30d55'
Coolican (Geetock), Patricia Marshall
And 23h32'17"47d57'
Coolidge,Charles"Skip"
Dra 17h5'30"61d29'
Coolidge,Leigh Cheryl Todd Tracy
Cas 0h51'49"64d42'
Coolman,Dianne A
Cas 3h34'16"72d56'
Coolon-DeLancett-Ward 789
Lac 22h52'31"51d8'
Coombes,Julie
And 0h59'33"39d19'
Coombs,Dylan Reed
Per 2h46'42"43d58'
Coombs,Harry F
Cyg 20h24'0"41d14'
Coombs,Joe & Dorothy
Cyg 19h59'33"38d18'
Coombs,Joshua W
Aql 18h51'25"11d55'
Coombs,Steven Cody
Cep 23h29'59"65d44'
Coomes II,C Lowell
Eri 3h55'38"-11d6'
Coomo,Coleman Anthony Hart
Del 20h14'29"11d11'
Coon,Beverly Jo VanAntwerp
Vir 11h46'30"6d49'
Coon,Diane Marie
Lyr 18h17'23"30d49'
Coon,Don Robert
Mon 7h39'17"-1d43'

Coon,Forever Dennis Robert
Sco 16h25'0"-40d28'
Coon,Gayra K
Com 13h11'16"20d8'
Coon,Timothy
Per 2h57'1"38d46'
Coon,Tonia
Peg 21h53'0"34d15'
Coonan,Patrick Michael
Lac 22h1'10"50d33'
Coonce Family MJBJJEL Dad,The
Lac 22h40'30"53d10'
Cooney,Donna Marie
And 0h20'59"36d4'
Cooney,Gerald A
Cyg 20h11'38"49d2'A
Cooney,James Michael
Aur 4h54'30"40d0'
Cooney,John Edward
Gem 8h4'1"28d42'
Cooney,Lynsey Rose
Uma 10h50'59"72d26'
Cooney,Paul A
Cas 0h36'58"70d21'
Cooney,Robert
Cma 7h6'1"-28d19'
Cooney,Tom
Uma 9h48'1"43d21'
Coons,Michael J
Aur 6h24'29"30d0'
Coons,Richard E
Aur 6h47'29"39d42'
Coons,Valerie
And 0h36'48"40d41'
Coons/Medley Wedding Star
Eri 4h14'56"-11d5'
Coop
Cam 5h33'59"79d38'
Cooper
Her 17h1'1"42d51'
Cooper (Laurie), Laurence
Ori 5h26'58"1d39'
Cooper B A (Hons), Colleen
Cyg 19h43'24"50d21'
Cooper Comet
Cam 13h17'29"77d18'
Cooper II,Anthony Christopher
Uma 9h5'40"61d40'
Cooper Nova
Her 17h41'11"40d28'
Cooper R G N,Alison
Cyg 19h43'49"50d34'
Cooper's 19th Hole
Peg 22h30'54"8d39'
Cooper's Comet,Rob
Ori 5h40'52"-0d5'
Cooper,Allen G
Peg 23h30'1"20d59'
Cooper,Andrew Jeffrey
Cep 22h38'0"58d30'
Cooper,Anna Newman
Cas 0h32'29"64d17'
Cooper,Antonio Ansara
Uma 9h56'44"62d7'
Cooper,Arnie
Boo 14h47'37"46d17'
Cooper,Barbara
Uma 11h33'21"41d17'
Cooper,Bernard
Vir 11h40'27"5d38'
Cooper,Billy
Cnv 12h49'56"39d35'
Cooper,Bradford Scott
Leo 10h39'59"13d12'
Cooper,Brian John
Cep 21h19'57"58d41'
Cooper,Carol
Lyn 7h30'54"38d49'
Cooper,Case Sullivan
Mon 6h44'28"8d39'
Cooper,Catherine
Cet 0h34'43"-5d50'
Cooper,Cookie Elyce
Cas 0h42'13"68d58'

Cooper,Daniel
Aur 4h52'39"41d0'
Cooper,Dave
Hya 9h31'32"-2d29'
Cooper,David Barrett
Mon 6h53'22"0d24'
Cooper,David Meade
Vul 19h43'1"23d28'
Cooper,DDS,Bret E
Oph 17h31'45"8d35'
Cooper,Derek Alan
Ori 5h35'58"-0d19'
Cooper,Dona Hanks
Mon 6h3'1"-8d19'
Cooper,Douglas & Josephine
Lyn 7h51'37"33d55'
Cooper,Elizabeth Kathleen Lennox
Lyr 18h31'42"34d22'
Cooper,Elizabeth Eddy
Umi 13h28'46"75d51'
Cooper,Ella May
Cam 3h16'1"60d2'
Cooper,Emily Catherine
Cyg 21h11'41"38d17'
Cooper,Gillian Ashley
Lyn 7h43'18"52d19'
Cooper,Gwen
Leo 11h13'56"7d53'
Cooper,Hannah Leigh
Cam 10h36'1"84d29'
Cooper,Ivan
Ori 6h4'0"5d6'
Cooper,Jack E
Ser 15h36'40"17d55'
Cooper,James E
Dra 10h24'56"80d4'
Cooper,Janet
Uma 8h41'53"49d54'
Cooper,Janice Lee Blackwell
Eri 2h45'36"-10d45'
Cooper,Jasmine
Vul 19h3'54"21d14'
Cooper,Jayne
Cam 8h0'43"61d7'
Cooper,Jeff C
Oph 17h55'1"12d58'
Cooper,Jennifer
And 0h20'13"43d29'
Cooper,John
Lac 22h07'12"51d52'
Cooper,John
Per 3h14'1"41d47'
Cooper,John
Lac 22h4'49"47d48'
Cooper,John
Ori 5h56'1"14d47'
Cooper,John Howard
Cep 23h7'31"60d52'
Cooper,John Murray
Mon 7h0'52"5d32'
Cooper,Jr,Lawrence Andrew
Sex 10h1'22"-6d31'
Cooper,Jr,William Alan
Aur 6h31'0"37d4'
Cooper,Juanita Cline
Lmi 10h6'23"33d35'
Cooper,Julie
Eri 3h13'28"-2d15'
Cooper,June
Lyn 7h16'42"59d26'
Cooper,Kathy
Lyn 7h37'36"40d13'
Cooper,Kerri Elizabeth
Lac 22h20'50"50d18'
Cooper,Lee & Lauren
Lyn 8h3'15"44d5'
Cooper,Lindsay Michelle
Lyr 18h58'0"30d58'
Cooper,Mair & James
Cyg 21h36'0"38d42'
Cooper,Marc Gary
Per 1h36'11"53d50'
Cooper,Margaret
Peg 21h57'53"3d11'

Cooper,Mark & Leslie
Uma 11h12'54"57d40'
Cooper,Marvin
Cnv 12h42'17"38d56'
Cooper,Mary B
Oph 16h50'13"11d55'
Cooper,Mathew
Ser 17h54'29"-14d27'
Cooper,Melissa Nicole
Uma 11h42'0"55d14'
Cooper,Merla E Spurlock
Vul 20h15'22"26d13'
Cooper,Michael W
Aur 6h56'46"44d14'
Cooper,Paul
Boo 15h2'55"21d44'
Cooper,Peter
Uma 10h1'12"12d59'
Cooper,Prof Carol
Lyr 19h8'25"47d44'
Cooper,Rachel
Tri 2h28'37"28d34'
Cooper,Ralph William
Hya 8h44'0"5d18'
Cooper,Rebecca S
Cyg 19h40'38"31d3'
Cooper,Richard"Ashley"
Vul 19h2'24"22d21'
Cooper,Robin Michelle
Mon 6h32'40"-0d22'
Cooper,Roger
Aql 18h55'0"10d32'
Cooper,Roger "Bloggo"
Col 5h54'12"-41d43'
Cooper,Rondale D
Hya 9h1'0"4d56'
Cooper,Ruth
And 23h31'30"38d46'
Cooper,Ruth Ann
Cet 2h36'18"-18d57'
Cooper,Saebyn
Aur 5h17'30"41d50'
Cooper,Sara Elizabeth Bonham
Eri 2h53'36"-11d29'
Cooper,Sara Hurst
Cnc 8h35'44"31d54'
Cooper,Scott Christian
Uma 10h2'0"48d15'
Cooper,Stan
Ori 5h16'53"0d33'
Cooper,Sydney Alexandra
Peg 23h18'44"28d26'
Cooper,Vanessa Alexia
Uma 10h4'0"50d24'
Cooper,Virginia
Boo 14h30'23"37d54'
Cooper,William Thomas
Lyn 9h8'28"37d57'
Cooper,Zoë Elizabeth
And 23h48'0"44d59'
Cooper-Smith,Neil
Psc 23h1'56"0d41'
Cooper-The Silver Star Sandra
Lyn 8h32'49"43d53'
Cooperider,Cameron
Ori 5h2'20"0d22'
Cooperman,Amanda & Richard
Mon 7h56'58"-5d43'
Cooperman,Kaleb Adam
Her 18h8'46"38d58'
Cooperman,Neil D
Cnv 12h12'46"34d59'
Cooperstein,Joshua Aaron
Boo 15h0'20"24d23'
Coopmans,Marilyn
Mon 7h38'53"-3d23'
Coopoer,Barbara
Tau 4h28'17"30d12'
Coops
Mon 8h2'48"-7d53'
Coora,Matthew Toby
Tri 2h4'51"31d9'
Cooray,Devi
Her 16h24'28"26d15'

Coots,Alexandra Marie
And 0h9'28"46d52'
Coots,DeDe
Cam 10h34'25"84d35'
Coots,Dede
Uma 10h48'30"52d27'
Coots,Ray & Lola
Uma 11h29'49"30d4'
Coover,Austin
Sex 9h50'0"1d4'
Coover,Tom
Her 17h4'18"38d33'
Copacetic
Cas 0h50'11"60d18'
Copain,Paula Christine
Lac 22h45'46"56d29'
Cope,Amy M & Jon M Gibbs
Ori 6h4'13"-0d35'
Cope,Jacqueline Mitchell
Aql 20h0'25"-8d31'
Cope,Jerry
Her 18h13'48"38d13'
Cope,Jr,Michael Paul
Cet 0h55'45"-8d37'
Cope,Neville Arthur Charles
Ori 5h38'13"10d35'
Cope,Phyllis Kay
Peg 22h46'45"21d34'
Cope,Richard David
Cep 22h20'14"70d38'
Copeland & Kids, Stanley & Melanie
Peg 21h25'0"22d60'
Copeland's Cluster
Cep 21h41'1"56d4'
Copeland,Bethany Grace
Psc 22h54'21"6d31'
Copeland,Courtney Mercer
Peg 23h20'35"28d12'
Copeland,Diane
Eri 3h18'20"-1d33'
Copeland,Jr,Paul Richard
Aur 6h26'55"32d56'
Copeland,Raymond Ellis
Cet 2h33'15"-10d23'
Copeland,Star Of My Life James
Ori 5h55'30"16d9'
Copeland,Tana Hobson
Aql 19h9'17"12d57'
Copeland,The Heavenly Body Of LG
Aur 4h55'38"40d27'
Copeland,William Jess
Aql 19h29'46"14d21'
Copelin,Robert
Oph 18h42'50"10d1'
Copelyn,Sabreena
And 1h46'36"37d52'
Copeman,Angela Grace
And 2h28'14"49d47'
Copen,Dr Bruce
Aql 19h58'50"15d31'
Copenhagen,Kenneth Wels
Ser 15h59'44"0d30'
Copenheaver,Jr,Michael David
Her 17h23'32"18d53'
Copernicus Jr's Gift
Del 20h12'57"14d39'
Copestake,Colin
Cep 0h0'24"73d41'
Copihue
Crb 16h18'37"32d9'
Coplan,Harriet Lyn
Peg 0h7'42"20d42'
Coplen,Sharon
Eri 3h34'39"-5d25'
Copley Five O
Uma 10h24'1"71d50'
Copman,Jenny
Lyr 19h21'45"38d9'
Copp IV,Belton Allyn
Umi 15h2'44"71d11'
Copp Star,Douglas
Lyr 18h26'56"46d35'

Coppa,Chiara
Vel 10h12'32"-48d1'
Coppage,R W
Mon 6h20'33"8d16'
Coppellotti,Lona
And 2h10'39"38d24'
Coppeneur,Helmut
Ori 5h39'18"7d49'
Coppens/Swanson, Jennipher Ann
Cet 3h18'57"5d43'
Copper's Star
Mon 6h21'40"7d31'
Coppersmith,Terri
Lyn 7h49'0"50d10'
Coppi,Eneide
Scl 23h25'52"-28d56'
Coppi,Marilyn
Lyr 19h0'48"37d54'
Coppina,Maureen Ann
Cas 0h9'41"63d7'
Coppinger,Michael P
Boo 14h32'1"30d54'
Coppinger,Richard
Oph 17h35'38"-24d16'
Coppinger,Shawna Terpy
Cas 1h6'21"68d21'
Copple,James
Aur 5h49'54"50d22'
Coppleson,Doctor L Warwick
Cep 23h10'18"68d40'
Coppo,Irene
Lyn 7h53'37"41d5'
Coppo,The
Lyn 8h10'24"38d37'
Coppock,Elizabeth Edwards
Mon 6h14'39"-10d36'
Coppola,John W
Aur 6h22'48"32d37'
Coppola,Louis A
Aur 6h26'27"37d44'
Coppola,Matthew Anthony
Tri 2h20'44"32d2'
Coppola,Timmy
Ari 1h57'57"20d44'
Coppoli,Maria Pia
Cas 2h57'50"60d2'
Coprivnicar,Jr,Frank J
Dra 19h5'0"61d23'
Copsey,Derreck & Michelle
Vul 20h3'34"28d58'
Copy Pro Inc
Her 17h59'31"18d56'
Copy's Regal Heir
Peg 21h57'0"34d21'
Coquillaud,ElisÖbeth
Ori 6h20'40"10d25'
Cor Anglais
Lyr 19h2'48"38d22'
Cora
Lac 22h8'25"51d31'
Cora Lou
Cas 1h9'33"61d19'
Cora Lynn
Uma 11h49'13"45d12'
Cora,Gillian D
Aql 19h30'1"10d6'
Coradini,Christine
And 0h48'58"37d32'
Corado,Viola Ramos
Cas 1h9'20"61d2'
Coral N
And 0h9'10"46d12'
Corale "Brunella Maggiori"
Cep 23h32'11"64d9'
CoraLee,The Heartlight Of Harmony
Cas 0h59'0"63d49'
Coralia & Scott's Star
Cyg 21h4'54"50d12'
Coralie
Tri 1h41'1"28d49'
Coralie GBR
Aql 20h16'43"0d48'

Coralli
Vul 19h19'37"23d39'
Coran-Chalas
Cyg 21h3'0"38d60'
Corazon,Fuego de mi
Lyn 8h27'47"45d47'
Corazon,Rey de mi
Sco 17h29'1"-30d36'
Corbalis,Thomas F (Therm)
Per 1h35'47"52d60'
Corbeau,Andre
Cyg 19h22'1"28d40'
Corbell
Lyr 18h45'13"37d7'
Corbera,Yuri Naumov
Sgr 19h16'26"-44d36'
Corbet,Dylan Patrick
Cep 21h44'22"67d35'
Corbett
Cnv 12h15'45"35d26'
Corbett IV-"Cory", Donald McClure
Del 20h13'17"10d37'
Corbett's Destiny, Christopher John
Peg 22h47'0"30d47'
Corbett's Silver
Cyg 20h36'1"60d57'
Corbett,Anna Maeve
And 2h1'0"40d55'
Corbett,Carol Ann
Cas 1h56'18"71d0'
Corbett,Corynne
Dra 17h27'40"72d31'
Corbett,Duncan W
Cep 20h27'56"63d34'
Corbett,Farrley
Ori 4h48'13"4d36'
Corbett,Jennifer Lee
Lmi 10h59'49"25d8'
Corbett,Jim
Umi 12h14'44"58d53'
Corbett,John
Uma 11h12'37"61d8'
Corbett,Margaret-Ann
Uma 8h50'32"49d47'
Corbett,Meg
Vul 19h45'20"28d53'
Corbett,Renée Marie
Del 20h12'21"9d52'
Corbett,Ruth E
Cas 0h22'42"65d30'
Corbett,Sandra L
Lyr 18h49'49"32d27'
Corbett,Tammi
Aql 20h21'34"5d30'
Corbett,William
Her 17h2'44"51d24'
Corbett-Moculski, Colette Marie
Cas 0h48'12"64d7'
Corbi,Ann-Michele
Lyn 9h16'11"33d27'
Corbi,Christine M
And 23h34'0"43d31'
Corbin
Cet 3h20'16"2d9'
Corbin & Rebecca
Boo 14h4'28"11d19'
Corbin & Traci
Uma 13h9'41"72d27'
Corbin Star,Aprylle R
Lyn 7h29'37"39d45'
Corbin,Cheryl Christine
Lyr 18h41'45"30d1'
Corbin,James J
Cep 20h58'21"68d60'
Corbin,Joy
And 0h53'49"38d44'
Corbin,Kaylie Michelle
Leo 9h28'20"27d40'
Corbin,Micah Lynn
Lib 15h38'0"-20d1'
Corbin,Pamela Ann
And 2h30'29"48d49'

Corbin,Robert C & Betty E
Mon 6h55'52"-4d49'
Corbin,Samantha
Lyr 19h5'35"38d39'
Corbin,Samantha
Peg 22h3'1"5d25'
Corbin,Shawnda
Mon 6h35'0"0d51'
Corbin,Tyler
Dra 14h55'57"62d57'
Corbino Star,Anthony
Her 16h42'0"4d24'
Corbishley,Sue
Lyr 18h28'41"32d60'
Corbisiero,Darlene
Lyn 7h49'39"38d15'
Corbitt,Audrey Maranda
Cyg 20h18'14"38d15'
Corbitt,Kathi Rogers
Del 20h18'52"10d21'
Corbitt,Virginia (Belle)
Aql 19h59'45"8d23'
Corbière,John & Eileen
Lac 22h2'43"35d22'
Corbly Star,The
Hya 8h59'57"3d34'
Corbusier,Larry
Ari 3h25'39"28d44'
Corby & Jenni Forever
Cep 22h36'11"60d28'
Corby March 5, 1949, Lanny Lance J
Her 17h2'43"45d31"
Corcoran,Angelique & Ryan
Crb 15h35'36"31d58'
Corcoran,CA Loves Eleanor Percival
Crb 15h15'33"31d29'
Corcoran,Dot & Mickey
Uma 11h45'55"40d30'
Corcoran,James
Per 2h58'31"37d59'
Corcoran,John B
Lac 22h11'35"50d5'
Corcoran,Justin Thomas
Dra 15h14'13"60d32'
Corcoran,Kelly
Vul 19h53'55"20d22'
Corcoran,Margaret Cooper
Lyr 18h28'0"42d22'
Corcoran,Michael & Kathy
Peg 22h25'37"3d43'
Corcoran,Tom/Tom
Cet 0h55'0"-6d5'
Corcoran,Tommy Ann
Uma 11h18'27"32d14'
Corcoran,William Dorothy
Peg 22h42'1"11d60'
Corcoran,Zoe Grant
Aur 5h28'58"31d52'
Cord's Family Star,The
Cyg 20h39'34"37d43'
Corda,Melissa
Aur 7h8'0"38d16'
Corda,Robert & Sian Thiessen
Sge 20h6'0"16d1'
Cordalis,Costa
Tau 5h39'21"25d56'
Cordani,Robert
Crb 15h41'2"35d0'
Cordara,Laura
Pyx 8h46'21"-25d58'
Cordeau,Marc S
Umi 13h30'34"70d48'
Cordelia
Lyr 18h32'0"34d21'
Cordelia
Oph 18h40'28"7d12'
Cordell,Cathie
Crb 15h57'1"31d43'
Cordell,Cathie
Lmi 10h37'54"25d42'
Cordell,Samuel Lucien
Aql 19h31'18"10d23'

Corder 2/19 3:20
Cep 21h40'0"80d19'
Corder,Denny
Ori 5h40'36"10d12'
Cordero,Peter Michael
Aur 5h3'0"42d19'
Cordero,Rose Ellen
Crb 16h9'32"27d18'
Cordes,Randolph Anthony
Lyr 18h34'53"40d54'
Cordesman,Shirley Myers
Cap 21h31'57"-26d50'
Cordey,Anne
Uma 16h31'11"75d57'
Cordey,Ernst
Ori 5h58'15"19d48'
Cordial,Craig Michael
Cet 2h56'53"0d20'
Cordier,Patrice
Lmi 10h35'1"31d36'
Cordileone,Carla Ann
Cet 1h30'1"-0d19'
Cordina,Sharon
Ori 5h3'34"15d32'
Cordingly,John
Dra 16h40'17"60d33'
Cordini,Anna et Guiseppe
Oph 17h53'51"11d21'
Cordivari,Jack & Toni
Cyg 19h45'1"30d6'
Cordobal
Ori 5h50'52"18d50'
Cordon,Christopher
Cas 22h59'38"56d48'
Cordonnier,Anne
Boo 14h50'46"38d14'
Cordonnier,Emmanuelle
Aur 5h51'0"54d56'
Cordonnier,Ruth M
And 23h6'46"37d38'
Cordovano,David Alan
Tau 5h49'34"23d14'
Cordray,Joyce E
And 2h22'40"39d51'
Core,Eugene Howard
Hya 9h0'22"-8d9'
Corel,Serge
Aur 4h37'40"31d7'
Corello,Nicolo
Per 3h26'50"38d16'
Corene
Ari 3h2'21"28d42'
Corene My Shining Star
Lib 15h40'20"-22d24'
Corey & Laurie
Aur 6h5'1"46d25'
Corey B,The
Vel 10h7'18"51d14'
Corey Joe
Lmi 10h50'19"33d10'
Corey Libby
And 2h31'54"42d46'
Corey Rose,The
Cru 12h34'58"-57d58'
Corey's Dream
Cyg 20h1'18"40d29'
Corey's Star Forever
Lib 14h58'35"-8d50'
Corey,Anita,Louise
Vul 20h5'1"28d52'
Corey,D "Lively," Daves!
Boo 14h10'35"31d34'
Corey,David Vernon
Her 16h49'15"48d60'
Corey,My Little Bear
Per 2h39'26"34d57'
Corey,Ramona A
Cnc 7h56'53"10d50'
Corfman,Don & Nancy
Cnv 12h5'26"38d29'
Corgiat,Edwin F
Boo 15h19'12"50d20'
Cori & Brad
Crb 15h50'32"31d43'

Cori's Smile
Mon 7h27'1"-1d22'
Coriasco,Joseph S
Lac 22h16'41"54d38'
Corie-Brian Forever
Per 2h32'0"52d7'
Coriell,Rae Ellen
Uma 10h26'30"57d47'
Corigliano,Cindy "Cinderoo"
Mon 6h18'15"5d59'
Corinna
Eri 2h43'54"-3d30'
Corinna
Pho 2h15'4"-45d56'
Corinna & Bryan's "Hearlight"
Cyg 20h8'45"58d10'
Corinnas Liebe
Tau 5h49'13"24d19'
Corinne
Per 4h42'58"50d51'
Corinne
Lyr 18h32'39"41d53'
Corinne
Tau 3h57'40"30d48'
Corinne "Sister Friendly"
Peg 21h55'38"31d50'
Corinne Et Eric
Cam 3h50'14"52d45'
Corinne's Dream
And 0h48'22"33d37'
Cork,Louise M
Lmi 10h6'47"30d47'
Cork,Philip
Cet 2h0'24"-1d33'
Corkie
Sco 16h56'57"-37d47'
Corkie
And 23h23'37"43d3'
Corkill,Dorothy
Cas 23h13'43"62d53'
Corkill,John
Lac 22h18'16"51d33'
Corkill,Tyler Joseph
Vul 20h3'56"22d45'
Corkins,Ranea
Ari 2h1'32"21d38'
Corkish,Melanie Faye
Uma 13h2'14"52d60'
Corkum,Bruce
Boo 13h42'39"21d21'
Corky & Morrie Forever
Cyg 20h54'1"31d35'
Corky & Ross Forever
Umi 16h15'36"71d14'
Corky Lee
Dra 17h1'1"66d58'
Corless,Shawn
Cas 0h56'19"65d27'
Corlett,Janet Ann Ingstad
Cyg 20h40'42"47d8'
Corley's Cosmo
Oph 17h18'52"-23d24'
Corley,"Handsome Bob" aka Don
Dra 17h41'34"68d5'
Corley,Jr,Rex O
Hya 8h30'46"1d5'
Corley,William & Grace
Sge 19h16'22"16d52'
Corley,William Gene
Aur 7h15'41"38d52'
Corlyn & Leon's South Star
Ori 5h50'58"15d7'
Cormack,Graham
Aql 19h17'45"14d22'
Cormack,Stephen Thomson
Ori 4h45'50"0d24'
Corman's Birthday Star,Bob
Aql 19h3'37"2d57'
Cormane,Betty
Uma 9h58'26"46d38'
Cormany,"Hoshua"
Her 18h22'16"28d16'
Cormany,Mary Janet
And 1h57'37"37d2'
Cormier,Dennis
Leo 11h47'28"21d6'
Cormier,Dorothy A
Cas 0h59'0"69d4'
Cormier,Nider
Lyn 6h56'12"58d14'
Cormier-11,Christopher
Uma 11h57'30"42d33'
Corn,Susan Diane
Cyg 19h30'28"38d54'
Cornacchiulo,Daniel
Dra 16h14'36"66d30'
Cornalba,Elena
And 23h2'57"52d42'
Corneille,Andrew Root
Tri 2h31'0"35d40'
Cornelia
Mon 6h54'46"-10d20'
Cornelia
Lac 22h22'38"40d9'
Cornelia
Cep 22h6'43"67d53'
Cornelison,Charles E
Aur 6h7'44"33d2'
Cornelissen,Henry Joseph
Gem 6h43'49"30d38'
Cornelius,Kari Jeanne
Cam 6h7'0"70d39'
Cornelius,Tonya Lynn
Aql 20h12'20"11d14'
Cornell,Brian Ira
Cep 21h54'54"55d11'
Cornell,Dosha Fay
Cyg 19h29'41"38d14'
Cornell,Erin J O'Brien
Lyn 8h25'36"48d5'
Cornell,Ernest
Boo 14h1'39"7d47'
Cornell,Judy
And 1h15'1"48d45'
Cornell,Kristen "KC"
Cas 0h25'0"72d14'
Cornell,Nikki
Cyg 21h23'49"37d37'
Cornell,Ruth A
And 2h21'10"40d51'
Cornell,S
Boo 15h17'0"50d23'
Cornell,Sean
Per 2h7'0"57d55'
Cornella,Sara Allison
Cyg 19h32'0"30d17'
Corneluis,April
Cas 23h40'35"62d51'
Corner Stone,The
Aur 5h2'0"40d16'
Corner,Kristi
And 0h10'60"30d49'
Corner,Sherwood Lamson
Her 16h39'1"23d29'
Cornett A Legacy Of Love,Benny A
Sct 18h32'37"-6d46'
Cornett,Chris H
Peg 23h5'29"10d38'
Cornfeld,Alexander B
Per 2h54'47"37d40'
Cornfeld,Jeffrey D
Her 16h39'0"7d51'
Cornforth,T W
Her 17h16'17"27d11'
Cornicelli,Daniel Philip
Boo 14h32'0"44d11'
Cornicelli,David Paul
Lac 22h3'13"41d5'
Cornicelli,Sr,Anthony M
Sex 10h11'21"-3d3'
Cornicello,Charlotte Marie
Peg 22h57'23"33d15'
Cornil,Madeleine
Cyg 20h53'57"50d5'
Cornille,Lauren Andreé
Lyn 8h1'59"46d24'
Cornille,Renée Nicole
Lyn 8h29'53"42d1'
Cornils,Pola
Cep 22h36'38"58d42'
Cornish,Christy Lynn
Cyg 20h53'22"30d40'
Cornish,Reginald Frank
Uma 12h3'25"47d26'
Cornish,Robert James
Boo 14h51'23"32d6'
Corns Star,Joann
Aql 20h11'17"0d7'
Cornu,Pat
Uma 11h53'42"35d43'A
Cornu,Tom
Uma 11h53'42"35d43'B
Cornudet,Jean-Michel
And 23h44'40"33d35'
Cornwall,Alicia M
Cyg 20h7'0"40d53'
Cornwall,Daniel R
Aur 7h9'42"38d46'
Cornwell,David
Lyn 7h55'0"39d51'
Cornwell,Dean Robert
Boo 14h22'53"51d58'
Cornwell,Mark Jason
Per 3h1'1"46d59'
Corogin,Peter J
Dra 14h56'23"64d41'
Corona Leah
Cas 0h24'23"50d30'
Corona,Eric Manuel
Sct 18h42'44"-6d20'
Corona,Sante A
Cam 12h55'57"78d58'
Corona,Vita Barbato
Cas 0h16'26"61d49'
Coronato,Kim Rene
And 23h46'56"44d41'
Coronato,Richard Philip Paul
Her 16h42'18"39d51'
Coronato,Richard
Boo 15h3'58"17d25'
Coronato,Virginia Joan
And 23h37'34"46d42'
Coronet Claude Champagne
Gem 6h46'35"31d45'
Corpening,Charles Conrad
Oph 18h7'1"8d41'
Corpening,Donna Nix
Vul 19h43'14"26d3'
Corporate Software,Inc
Cam 9h2'53"82d23'
Corpus-4-94
Umi 14h58'49"71d25'
Corr,Donald Robert
Sco 17h9'1"-38d57'
Corr,John Michael
Uma 9h48'14"70d25'
Corr,Margaret Jeannine Jochman
Leo 9h28'55"7d9'
Corradino,Anthony C
Her 16h36'51"42d21'
Corradino,Sara
Peg 22h46'32"35d12'
Corrado
Uma 14h5'16"48d34'
Corrado,Adam Digsby
Dra 11h9'59"74d19'
Corrado,Anna
Cas 2h10'20"60d5'
Corrado,Kimberly Ann
Cam 6h12'22"58d13'
Corrado,Regina
Vul 21h3'30"20d15'
Corrado,Simontacchi
Cep 23h29'27"64d49'
Corrado,Stefan Antonio
Hya 8h45'30"-5d57'
Corral,Alejandro Y Araceli
Aql 19h48'58"11d18'
Corrall Family Star, The Brian & Margaret
Cyg 19h26'12"35d21'
Corre,Anaëlle
Uma 12h57'25"57d40'
Correa,Carol Jean
And 0h11'42"46d42'
Correa,Catalina
Uma 14h26'38"61d5'
Correa,Justin
Umi 15h23'11"68d21'
Correena Rose
And 23h1'39"46d44'
Correia,Norman & Gail
Cyg 21h32'41"30d44'
Correia,William
Lac 22h52'10"54d40'
Correira,David Wayne
Cep 21h19'47"70d15'
Correll's Star,James & Cynthia
Equ 20h55'55"3d27'
Correll,Bob
Her 17h33'51"40d57'
Correll,Randy Bear
Ari 3h0'27"23d1'
Correll,Sandy L
And 23h6'0"43d1'
Correll,Steele L
Lyr 19h21'53"40d10'
Corrente,Nick
Sex 10h24'50"-6d33'
Corres,Alejandro
Her 16h44'12"47d47'
Correy,Patricia Rose
Cyg 20h35'12"50d35'
Corrick,Jeffrey Clinton
Dra 15h49'54"66d20'
Corridan,Judith Michele
Crb 15h27'27"32d44'
Corridan,Robert E
Boo 14h28'44"21'55'
Corriero,Mark & Linda
Lyr 18h46'59"30d18'
Corrigan's Wedding Star
Cyg 21h2'58"31d28'
Corrigan,Aoife
Ori 6h4'14"10d34'
Corrigan,Christopher Girard
Cnv 12h19'0"47d43'
Corrigan,Jackson Andrew
Oph 17h1'60"-0d57'
Corrigan,JoAnne
Ori 4h42'27"0d7'
Corrigan,John T
Her 16h32'0"32d29'
Corrigan,Matthew Thomas
Lib 15h1'30"-28d10'
Corrigan,Nicholas Andrew
Her 15h51'36"46d48'
Corrignon,Christione
Cep 22h23'52"57d53'
Corrine
Cet 2h55'20"0d9'
Corrine Ann
Aur 6h6'0"37d58'
Corrine Dee
Peg 21h38'27"27d44'
Corrine G
Aur 6h53'45"37d12'
Corriveau,Claude
Cep 23h11'13"65d10'
Corriveau,Jean-Pierre
Cas 1h17'45"60d6'
Corrivo,Nicoletta
Boo 15h7'0"47d58'
Corrn,Leslie Moscou
Sgr 19h0'38"-25d14'
Corrodus,F R
Cet 3h15'58"1d44'
Corry
Lmi 10h53'51"26d18'
Corry,Bernadette Eileen
Lyn 9h15'57"37d41'
Corry,David W
Aur 5h0'40"42d21'
Corry,John Greely
Crb 15h56'43"32d25'
Corry,Megan Marie
Del 20h14'21"11d24'
Corry,Meghan Christina
And 2h20'23"47d12'
Corrymeela,Ray Of Hope
Tri 2h10'0"31d22'
Corsano,Chris
Lyn 7h52'20"44d2'
Corsi,Anthony Joseph
Cep 21h30'20"58d60'
Corsi,Brian
Dra 14h57'53"56d21'
Corsi,Brian J
Boo 14h56'43"52d55'
Corsi,Chloe Alexandria
Cyg 19h33'0"38d55'
Corsi,Katherine Elizabeth
Cas 0h24'48"58d16'
Corsi,Michael David
Lyr 19h8'58"47d41'
Corsi,Peter
Boo 14h7'40"48d30'
Corsitto,Barbara A
Vul 19h59'0"23d27'
Corso,Kimberly Ann
And 23h39'0"47d58'
Corso,Robert Eugene
Lyn 9h14'53"34d28'
Corson III,Donald M
Tau 4h2'45"21d29'
Corson,Kate
And 0h43'0"45d10'
Corson,Richard G
Boo 15h22'42"50d38'
Corszen,Helmut
Tau 4h18'49"22d51'
Cortellino,"The Big Dipper" Nunzio
Uma 11h51'53"31d24'
Cortelyou Peace Star Jim & Shirley
Sge 19h53'17"19d17'
Corter,Allen
Cet 0h53'0"0d7'
Cortes,Daniel
Aql 19h17'52"14d22'
Cortes,Kyle Christopher
Cnc 8h25'36"7d56'
Cortes,Patrick Joseph
Lyn 7h47'59"45d58'
Cortese 1993,Christian Nicholas
Cnv 12h19'43"51d29'
Cortese,Catherine Rose
Eri 2h56'0"-11d26'
Cortese,Charles Phillip
Dra 16h46'13"64d53'
Cortese,Edward F
Cam 3h42'52"61d38'
Cortese,Joe
Ser 15h58'40"4d18'
Cortese,Judith
Lyn 6h25'15"60d6'
Cortese,Linda Biggs
Tri 1h42'53"31d33'
Cortese,Mary Alice
And 23h29'11"44d23'
Cortez,Frank B
Ori 6h16'54"1d23'
Corthell,Anthony Ray
Lac 22h16'16"55d5'
Corti,Aurelio
And 0h23'44"44d29'
Cortinas,Margaux
Lac 22h13'47"51d44'
Cortland's Dream "Reach for the Stars"
Mon 8h0'36"-0d2'
Cortner,Janet
Com 13h7'52"28d44'
Cortney
Vul 20h4'34"25d37'
Cortorillo,Salvatore F
Oph 17h38'45"-16d5'
Corts
Aql 20h11'0"11d1'
Cortzen,Jan
Ori 5h59'41"7d46'
Corvacho,Isabel
Cnc 8h2'23"10d59'
Corvinis
Lac 22h12'50"46d15'
Corwin
Lyr 18h50'0"35d13'
Corwin Bliss
Ori 5h56'1"10d53'
Corwin,Joan
Cas 0h50'24"61d42'
Corwin,Mary-Beth
Cam 5h43'21"68d60'
Cory & Gina
Crb 16h4'55"33d22'
Cory Nathan
Her 17h21'50"26d38'
Cory Nicholas Grandpa's Little Love
Vul 21h25'30"27d31'
Cory of Beauty-Courage & Hope,Roseanne
Cas 1h15'23"63d32'
Cory's Christmas
Lac 22h9'13"50d10'
Cory,Dave Keegan
Aql 20h6'15"8d33"
Cory,David Nathaniel
Aql 20h8'21"0d32'
Cory,You Light Up Our Lives!
Vir 13h7'11"-21d60'
Cory-April 11,1995, Earle & Lois
Hya 8h57'0"-8d8'
Cory-My Knight
Dra 17h48'35"64d13'
Corymax
Eri 2h53'60"-3d8'
Cos-Kay
Her 17h19'1"18d53'
Cosaboom,Marvin R
Aql 20h4'57"6d42'
Cosby,Michael
Mon 6h19'1"-0d35'
Coscia,Dante G
Boo 14h44'0"35d54'
Coscia,Marina
And 0h21'16"40d3'
Cosco,Carly R
Hya 8h29'28"-0d49'
Cosco,Carson S
Oph 17h9'14"-24d44'
Cosentino,Francesca
Eri 3h2'1"-6d3'
Cosentino,Joseph K
Lyr 18h38'48"46d26'
Cosenza Happy Birthday Star,Celeste
And 23h36'46"46d33'
Cosenza,Al
Her 17h20'34"20d23'
Cosgrave,Dick
Her 17h5'0"45d56'
Cosgrove III,Richard Douglas
Cyg 19h26'36"32d56'
Cosgrove III,William Hugh
Her 16h34'1"41d17'
Cosgrove Memory
Her 17h13'41"29d33'
Cosgrove,Alanna Mary
And 23h27'12"37d33'
Cosgrove,Angelina
Cyg 21h53'43"53d39'
Cosgrove,David George
Her 16h45'23"47d35'
Cosgrove,Debed
Tri 2h17'39"33d41'
Cosgrove,G
Aur 6h7'50"38d24'
Cosgrove,Jay Patrick
Her 16h2'30"21d55'
Cosgrove,Mark David
Lyr 19h18'0"41d59'
Cosgrove,Michael T
Aql 19h30'37"7d34'
Cosgrove,Samantha Lynn
Lyr 19h13'1"35d17'
Cosgrove-Aranchio, Sheila Ann
And 02h04'41"41d03'
Cosie
Peg 22h34'21"28d42'
Cosimina
Ari 1h15'47"30d10'
Cosimo
Lyn 7h13'42"50d45'
Cosing,Ruby
Equ 21h1'38"10d56'
Cosio,Raul
Cep 23h7'0"61d45'
Coskun & Inci
Cyg 20h1'57"31d40'
Coslow,Kenneth Lee
Cap 21h52'17"-22d0'
Cosman,Les
Dra 19h17'12"68d15'
Cosme,Eduardito Ruiz
Cyg 19h30'20"34d59'
Cosmic Bond
Cep 22h41'54"58d34'
Cosmic Charlie
Dra 11h54'48"67d7'
Cosmic Charlie
Per 3h19'18"40d17'
Cosmic Clivey
Cep 23h12'24"61d19'
Cosmic Diesel Dave
Lac 22h39'43"53d5'
Cosmic Jason
Ori 5h57'27"16d42'
Cosmic Jerry
Her 16h34'44"41d43'
Cosmic Joyce
Gem 6h50'38"31d4'
Cosmic Lyn & Lyle
Cyg 21h8'0"31d24'
Cosmic Mama T
Cet 0h52'25"-5d7'
Cosmic Von Smedlapp
Oph 16h21'59"-3d29'
Cosmica Lattea Cometa
Ind 21h2'14"-57d47'
Cosmo
Lib 15h39'38"-23d19'
Cosmos Angel,The
Peg 0h11'38"18d26'
Cosmos Star
Hya 8h46'41"-0d17'
Cosmos,Lisa Christine
Cru 12h54'57"-60d47'
Cosmos,Robert Steven
Her 16h58'55"26d37'
Cosmot,Joseph D
Her 18h13'14"40d57'
Cosner,Shellee Renee
Mon 6h54'35"-0d47'
Cosnotti,Dominic
Her 17h5'38"38d26'
Cosper,Leslie
Eri 3h34'46"-11d4'
Cosquer,Paul
Per 3h30'19"39d57'
Coss,Niki Lynn
Mon 6h57'21"-10d51'
Cossairt,Terry
Aur 4h43'1"31d0'
Cossalter,Erik
Lac 22h13'49"48d58'
Cossar,Danielle
Cap 20h18'41"-26d29'
Cossette,Jane Marie
Gem 7h24'13"35d26'
Cossey,Glenda F
Cas 23h30'50"60d39'
Cossio
Cas 0h53'13"65d36'
Cossu FLNC,Claude Gerald
Sgr 19h18'57"-40d0'
Cost,A born Star Love H,Alex Lagé
Dra 17h14'37"73d39'
Costa 50th Anniversary Andy & Grace
Peg 23h38'24"25d8'
Costa della luce
Cnv 13h5'0"33d39'
Costa Rubin
Cas 3h35'33"71d7'
Costa The Third, Anthony George
Lac 22h7'12"47d43'
Costa,"Wife & Mother", Rita
Cyg 19h31'0"34d2'
Costa,Barbara
Eri 2h50'37"-7d16'
Costa,Barry Wayne
Ori 5h55'23"15d0'
Costa,Charles & Lilliana
Ser 15h57'51"21d8'
Costa,Conxa Granes
Ser 15h14'19"8d4'
Costa,Frankie
Sex 9h50'39"-5d31'
Costa,Ingeborg
Dra 18h16'45"68d55'
Costa,Jill Bonham
Cyg 19h27'18"32d46'
Costa,John Joseph McShane
Oph 18h5'0"7d37'
Costa,Joseph
Sgr 19h4'52"-23d12'
Costa,Kaylin April
Ori 5h55'23"16d51'
Costa,Manny & Rose
Lyr 19h3'51"38d19'
Costa,Marc
Lac 22h26'27"40d54'
Costa,Renée
Eri 2h51'13"-16d23'
Costa,Richard J
Boo 14h0'40"11d51'
Costa,Robert A
Aql 19h59'46"14d50'
Costa,Robyn (Sparky)
Boo 14h25'13"20d6'
Costa,Rocco
Umi 16h23'39"70d12'
Costa,Thomas Joseph
Aur 6h32'1"40d9'
Costabile,Stephanie (Little One)
Vul 20h14'19"23d43'
Costafelix
Ind 21h4'6"-55d24'
Costalas,Nicola
Cyg 19h26'0"35d26'
Costantini,Ralph
Aur 5h15'0"49d30'
Costanza
Lac 22h9'38"51d48'
Costanza,Danny Sky
Her 16h22'1"20d26'
Costanza,Frederick A
Dra 14h54'56"64d28'
Costanza,Salvadore Eugene
Ori 5h51'1"12d23'
Costanzo,Jube-John M
Uma 10h12'1"50d19'
Costanzo,Mary Jane
Com 12h10'0"22d13'
Costanzo,Nancy
And 23h49'0"40d50'
Costar,Marie Kathleen
Psc 22h54'14"0d25'
Costarakis,Dennis A
Hya 8h36'46"-1d27'
Costarakis,Jason
Ser 15h27'40"10d33'
Costas,Emily Kathryn
Cma 7h2'10"-16d40'
Costas,Mary Michel
Leo 10h16'47"10d36'
Coste,Sylvie
Pic 4h47'30"-46d37'
Costello Above & Beyond,Ellen
Vul 20h17'47"25d33'
Costello,Anne F
Cas 0h40'56"68d16'
Costello,Ashley
Cas 0h35'52"64d29'
Costello,Barbara Stelle
Ori 5h18'49"7d48'
Costello,Brandon T
Her 18h10'29"30d36'
Costello,EdD,Jean F
Cyg 20h37'46"44d44'
Costello,Heather
Lyn 8h9'39"45d12'
Costello,Jane P
And 2h3'46"37d9'
Costello,Joan B
Cas 0h41'19"67d5'
Costello,John "Captian"
Sex 10h30'57"2d50'
Costello,Judy
Lyn 6h39'18"54d36'
Costello,Keith Niles
Per 3h23'15"40d1'
Costello,Kim Marie
Pup 7h38'11"-37d15'
Costello,Kymberly
Cas 3h26'12"75d37'
Costello,Linda Homolak
Cam 7h26'42"60d12'
Costello,Palma N
Cam 3h21'1"61d03'
Costello,Raymond
Dra 16h1'55"60d50'
Costello,Sandra Jean
And 2h15'1"47d24'
Costello,Shannon Lynn
Com 12h13'13"27d7'
Costelloe,Paul Francis
Dra 16h29'38"61d35'
Costes,John
Cep 22h45'23"59d13'
Costigan,Gail
And 0h52'49"35d34'
Costigan,Marie
And 23h18'38"43d34'
Costigan,Mary Elizabeth
Lyn 8h22'27"47d14'
Costin,Michael N
Cet 2h43'60"3d58'
Costro,Paul John
Per 3h7'55"38d56'
Cota,Danielle Elizabeth
Cyg 20h8'50"41d8'
Cote,Betty
Ori 5h28'30"-4d44"
Cote,Carol
Cnv 12h43'1"32d30'
Cote,David Warren
Lac 22h15'21"50d4'
Cote,Julia Thorpe
Cyg 19h27'44"31d19'
Cote,Lisa Jean
Cas 23h3'1"58d45'
Cotellesse,Steve
Per 2h50'26"38d27'
Cotgageorge,Haley Marie
Cas 0h27'25"61d41'
Cotham,Elaine Faye
Vul 20h40'17"28d18'
Cothern,The Connie
And 2h21'33"37d14'
Cotner,Harold William
Lyn 6h13'10"54d32'
Cotnoir,Jason L
Dra 19h3'57"50d26'
Cotrona,Rebecca Lynne
Lyn 7h46'26"42d30'
Cotroneo,Carolyn Jean
Peg 22h39'46"22d28'

Cotsen
Cma 7h18'0"-15d4'
Cotsen,Lloyd E
Ori 6h17'0"-0d35'
Cottard,Bruno
Cam 14h26'42"81d32'
Cottay
Cnv 13h23'0"31d58'
Cotten,Alexis Nicole
Cyg 19h30'32"36d23'
Cotten,George
Hya 9h13'36"5d43'
Cotter 11-22-86,Emma Halley
Cas 0h32'48"50d20'
Cotter Star,The Jules
And 1h18'0"41d12'
Cotter,Billie Rebecca Jo
Cma 6h54'28"-18d28'
Cotter,Huey
Lac 22h52'40"50d30'
Cotter,Irene R
Mon 8h5'58"-3d12'
Cotter,Jason
Cam 3h31'22"60d36'
Cotter,Maryanne Arcuri
Cam 6h46'43"68d15'
Cottereau Jennifer
Del 20h36'34"9d3'
Cotterill,Mark Anthony Keith lan
Aur 6h0'0"37d52'
Cottet,Helene
Aur 6h6'0"37d58'
Cottin,Fabienne
Umi 13h18'48"75d9'
Cottin,Gerard
Mon 6h26'59"-0d1'
Cottingham,Dave
Cyg 21h17'18"38d6'
Cottle,Connor
Vul 21h18'46"28d26'
Cottle,Harold
Cmi 8h2'35"6d29'
Cotto,Jose
Aur 6h17'0"37d41'
Cotton
Cyg 20h4'18"40d26'
Cotton,Anna Mae (Puddin)
Peg 23h47'55"18d10'
Cotton,Genevieve Tracy
Cyg 21h7'20"30d25'
Cotton,Joan M
Mon 8h2'35"-6d18'
Cotton,Roberta "Missy"
Vul 19h46'19"28d29'
Cottone,Ashley
Aql 19h54'57"14d29'
Cottone,Brooke
Hya 8h53'16"0d33'
Cottone,Tiffany
Peg 21h28'1"21d28'
Cottrell,Carol Lynn
Peg 22h22'47"5d59'
Cottrell,Gods Heart Shines Thru U Tyler
Cyg 21h49'1"38d44'
Cottrell,Joan
Aql 20h9'54"6d52'
Cottrell,Joan
Uma 9h4'54"47d50'
Cottrell,Mary Theresa
Cyg 19h22'0"30d1'
Cottrell,Maude "Mom- Mom"
Her 17h31'18"40d14'
Cottrell,Vicki Jean
Ari 2h53'13"30d46'
COTY
Cmi 7h43'41"8d17'
Coty,Colette Marie
And 23h34'1"48d37'
Coty,Mark
Uma 10h53'4"39d54'
Couailhac,Famille Alexandre
Uma 9h14'40"49d31'
Couch,Amos Paul
Psc 23h1'43"2d16'
Couch,Andrew Stephen
Aql 20h31'26"0d12'
Couch,Ann
Cyg 21h17'28"38d25'
Couch,Annabelle Johnson
Aur 6h7'53"43d10'A
Couch,Cydney Danielle
Eri 4h6'31"-18d12'
Couch,Jamir
Umi 14h57'42"65d30'
Couch,Mr W E
Dra 12h43'11"68d38'
Couch,Robert B & Amy Q
Lyr 19h0'47"26d14'
Couch,Sr,William Allen
Aur 6h7'53"43d10'B
Couche,Jean
Ori 6h7'21"20d12'
Couchois' Place,Pat
Sct 18h54'18"-7d58'
Coucon,Sylvie
Uma 12h58'35"58d56'
Coucou
Aql 20h1'21"10d20'
Coudert,Annie
And 1h7'10"47d50'
Coudreau,Alexandra
Cam 5h15'12"68d26'
Coudreau,Bertrand
Com 12h19'50"20d57'
Coudreau,Evelyne
Per 1h44'1"53d31'
Coudret,Zachary Andrew
Leo 10h42'45"14d18'
Coudurier,Patrick
Lmi 10h35'53"31d10'
Cougar
Hya 8h44'24"-1d3'
Cougar Territory
Ori 5h54'23"6d38'
Coughlin,Alexander Deacon
Boo 14h18'16"46d52'
Coughlin,Diana
Uma 10h3'40"50d8'
Coughlin,John,E
Cep 22h51'1"71d11'
Coughlin,Jr,Nicholas Charles
Oph 17h4'14"8d31'
Coughlin,Mary
Lyn 9h9'13"38d4'
Coughlin,Michael Edward
Uma 8h15'1"70d25'
Coughlin,Pat
Boo 14h30'34"32d1'
Coughlin,Ryan Alexander
Ori 5h4'47"0d35'
Coughlin,Samantha Autumn
Lib 14h48'0"-7d51'
Coughlin,Stephen Lloyd
Sgr 18h53'43"-23d19'
Coughlin,Timothy James
Cyg 21h35'39"37d53'
Coughlin,Timothy P
Dra 14h56'1"62d25'
Couhig,Nancy
And 1h37'51"37d49'
Couillault,Jaques
Boo 14h19'28"51d52'
Coulbourne,Jr,Robert Louis
Aur 7h24'42"35d57'
Coulman,Donnie
Aur 6h5'21"37d46'
Coulombe,Anne
Umi 16h12'27"73d6'
Coulombe,Louis Richard
Per 3h13'19"43d4'
Coulon,Georges
Cam 4h9'59"68d43'
Coulon,Pierre
Umi 16h19'39"71d9'
Coulouris,Kara & Jimmy
Aur 6h22'46"33d48'
Coulson,Melinda J
Eri 3h41'1"-13d43'
Coulston,Clay
Her 16h40'39"7d38'
Coulter Star,The Deborah Kay
Peg 21h59'45"34d15'
Coulter,Clara Emojene
Lyn 7h54'38"43d8'
Coulter,Matthew D
Peg 22h45'28"5d28'
Coulter,Normagwen & Christopher
Cyg 21h17'1"28d57'
Coulter,Robert Lee
Ser 16h16'16"0d23'
Coulter,Vicki
Cas 0h18'45"63d47'
Coulter,William "Motorman"
Her 17h39'0"40d26'
Coulthard,Alastair
Aql 19h46'0"11d44'
Coulthurst,Cotton
Uma 8h33'36"50d29'
Coumac
Per 2h15'13"58d10'
Counihan,Agnes
Cas 1h48'1"75d24'
Counihan,Stephen P
Dra 16h36'37"61d19'
Counsellor,Zachary James
Per 2h7'21"57d37'
Counselman,Kyle David
Lib 15h6'35"-8d15'
Countach & Chepcake
Cma 7h24'39"-13d25'
Counter,Michael Patrick
Aur 7h9'44"35d55'
Country Boy
Lyn 8h56'32"33d36'
Countryman,Ginger Lee
Vul 19h47'40"27d19'
Countryside Montessori Largo Florida
Del 20h22'0"8d35'
Counts,Donald Craig
Tri 1h58'46"30d54'
Counts,Sr,Ronald
Aql 19h56'38"10d11'
Coup de foudre
Scl 23h14'40"-31d8'
Coupe,Anne-Marie
And 1h2'40"37d7'
Coupe,Jane Marilyn
Aql 19h55'27"8d23'
Couper,Andrea Renee
Mon 6h43'16"-10d17'
Courage,Daddy George
Per 3h4'46"56d17'
Courant,Joelle
Uma 9h34'38"48d18'
Courbis,Taffy
Dra 14h57'0"64d9'
Courcelle,Etienne
Lac 22h29'0"38d31'
Courchesne,Brandon
Lac 22h30'34"40d19'
Couri Cristina, Joseph & Nicolas
Sco 17h27'1"-40d7'
Courier,Gretchen
Peg 22h5'1"10d45'
Courim,James R
Uma 11h45'11"50d6'A
Courim,Mary L
Uma 11h45'11"50d6'B
Cournia,Bob & Linda
Umi 14h30'21"79d50'
Cournoyer,Vincent
Umi 15h19'1"66d36'
Courreges,Georges
Dra 10h4'45"77d49'
Coursey,Anne Brennan
Lyr 19h16'1"28d37'
Coursey,Ryan James
Dra 20h7'51"67d57'
Coursey,Sarah Chapman
Del 20h18'53"11d4'
Courson,Coleen R
Mon 6h20'14"4d50'
Court,Anne Marie
Vul 20h1'47"23d46'
Courtelis,Alec
Her 16h10'8"42d30'A
Courtelis,Louise
Her 16h10'8"42d30'B
Courtemanche,Michael S
Ari 1h47'53"16d10'
Courter,Josh
Her 16h30'50"32d35'
Courter,Marie
Lyr 18h38'52"38d42'
Courtien,Dane Christian
Boo 14h41'0"52d21'
Courtney
And 2h7'1"48d25'
Courtney
Cam 6h16'45"65d16'
Courtney
Vul 21h0'1"20d3'
Courtney
Uma 10h49'47"59d46'
Courtney Elizabeth
Lyr 18h41'43"28d45'
Courtney I
And 2h26'37"47d6'
Courtney Lynn
And 1h46'55"41d11'
Courtney Marie
And 23h30'59"46d0'
Courtney Nicole Grandpa's Little Love
Vul 21h25'50"26d35'
Courtney's "Inspiration"
Uma 11h22'0"40d18'
Courtney's Bova
And 2h24'41"42d9'
Courtney's Phoenix Rising
Ser 15h36'16"8d59'
Courtney's Reflection
Com 12h23'47"30d27'
Courtney,Brian
Vul 18h55'59"24d38'
Courtney,John Raymond
Lib 15h43'59"-23d1'
Courtney,Mark
Cep 20h57'1"62d31'
Courtney,Mary Ann Espinosa
Peg 22h18'44"10d49'
Courtney,Matt
Cmi 7h59'40"8d31'
Courtney,MD,Mark
Ori 4h56'29"4d33'
Courtney,Tom & Cathy
Aur 5h31'26"30d42'
Courtney,Valerie
Cyg 19h50'30"40d53'
Courtney-Girl
And 1h17'59"36d28'
Courtney-Our Guiding Light
Cas 3h11'15"68d56'
Courtnie Paige Star
Leo 10h39'0"12d58'
Courtois,Armelle
Per 3h29'54"36d59'
Courtois,Raymond
Del 20h19'40"14d14'
Courtot,Hugues
Uma 9h39'12"49d15'
Courtáde,Dr Daniel
Aur 5h0'47"50d4'
Coury,David
Oph 17h30'40"-2316'
Cousens,Regina
And 1h51'52"37d38'
Coushay,Mackenzie Lynn
Equ 21h8'23"10d38'
Cousin Brucie Cruisin' Forever
Lib 15h38'15"-28d17'
Cousin Matt
Dra 16h20'14"64d9'
Cousineau,Anne M
Sge 19h7'34"19d24'
Cousineau,Kelly
Peg 22h57'18"27d51'
Cousineau,Melissa May
And 2h19'57"45d29'
Cousino,Marguerite J
Lyr 19h18'17"38d20'
Cousins' Star,Chris
Per 2h26'24"54d31'
Cousins,Nancy
Mon 8h1'47"-6d44'
Couto,Crystal Aleena
Cet 1h47'19"0d44'
Couto,Jeremy Lee
Aql 19h43'41"10d35'
Couto,Jodi Belinda
Peg 22h9'39"21d42'
Couto,Tiffany Ellen
Peg 22h40'31"20d28'
Couto,Timothy Lee
Aql 19h8'36"1d49'
Couton,Isabelle
And 23h21'31"50d27'
Couts,Jillian Carpi
And 1h54'31"36d33'
Coutts,Sharon Leanne
Cra 18h22'37"-42d54'
Coutts,Steven Nicholas Katlin
Cma 6h51'32"-18d20'
Coutu,Benoât André
Lib 15h17'29"-22d24'
Couture,Jean & Bill
Crb 15h18'0"30d20'
Couture,Little Jeanie
And 23h7'48"44d59'
Couturier,Janine et Jacques
Ori 5h58'44"17d4'
Couturier,Michel Charles
Uma 8h33'13"51d29'
Couturier-Packo,Pascal et Diane
Cyg 21h5'32"39d49'
Couturire,Doug
Aql 19h57'42"13d12'
Couvrat,Franáois
Dra 20h25'15"73d37'
Couzens,Alexander Wayne Hobart
Equ 21h20'0"3d27'
Couzzo,Rocco & Penny
Crb 15h58'47"32d40'
Covach,Miriam E
Cam 3h45'47"70d21'
Covalt,Robert B
Aql 19h26'55"1d46'
Covault,Jay Ann
Peg 23h42'0"30d30'
Cove,Candace Price Malaga
Lyr 18h54'53"40d30'
Covell,Carol
Lyr 18h55'23"32d40'
Covell,Elizabeth- Bishop
Lyn 7h6'19"51d16'
Covell,Farrell
Cet 2h47'1"-0d45'
Coven "Munchkin", Jessica Ann
And 23h35'1"44d4'
Cover,Bernard J
Aur 6h30'15"33d54'
Coverage Goddess,The (Deborah Pegg)
Cet 1h29'41"-5d54'
Coverdale,Aaron
Uma 9h36'48"60d16'
Coverdale,Bridgette Katherine
Cas 0h3'16"63d38'
Coverdale,David John
Crb 16h12'36"33d25'
Coverdale,Elijah
Uma 9h38'1"60d24'
Covered Wagon Saloon,The
Uma 11h51'27"30d12'
Covert,Darcy
Psc 1h5'52"27d52'
Covey,Bruce Sheldon
Aql 18h57'56"16d44'
Covey,Jerolyn Shae Hutchings
Cam 4h14'42"67d50'
Coviello,Alice & Leonard
Cyg 21h9'12"40d19'
Coviello,Paul John
Cyg 19h34'55"39d54'
Coville,Gerald Kenneth
Lac 22h6'14"47d30'
Covington,Anthony Benett
Cap 21h25'50"-20d27'
Covington,Cassidy Erin
Cet 3h10'0"6d42'
Covington,Dylan Rajiv
Boo 14h39'48"45d5'
Covington,Fairy
Sct 18h50'6"-7d54'
Covington,Justin
Aur 6h35'50"34d6'
Covington,Mae Helen (Moore)
Cam 3h51'1"71d1'
Covit,Marion
Her 17h12'14"40d16'
Covo,Joseph
Mon 8h3'10"-1d21'
Covos
Ori 5h13'42"15d18'
Covuoto,Andrew Paul
Peg 23h6'41"8d16'
Cowan 1996 Wed 50 Yrs, Don & Velma
Aql 19h8'39"4d30'
Cowan,Ana R
Mon 7h50'35"-8d58'
Cowan,Corina R
Cas 2h52'26"68d47'
Cowan,Diane
Cet 0h51'24"-10d25'
Cowan,Garrett Dean
Aql 18h58'40"15d18'
Cowan,Joseph A
Aur 5h34'41"37d38'
Cowan,Kathleen
Aur 5h36'53"38d29'
Cowan,Kathy
Cyg 19h36'1"28d37'
Cowan,Madison Ully
Vir 13h36'39"-6d36'
Cowan,Mary
Boo 14h51'27"31d60'
Cowan,Mary L
Cmi 7h27'19"0d22'
Cowan,Nickolas William
Her 18h7'29"40d46'
Cowan,Nicole
And 1h1'56"37d59'
Cowan,Rita
Cas 1h22'27"50d36'
Cowan,Ross
Cep 0h59'0"87d53'
Cowan,Steven Brent
Her 10h4'1"30d33'
Cowan,Thomas
Lyn 7h44'12"50d4'
Cowans,L B & Parisella L G
Cyg 21h35'0"40d37'
Coward,Crystal Gale
Del 20h14'16"14d3'
Coward,Fayth Elaine
Aql 19h52'58"11d31'
Coward,H Roberts
Lac 22h0'14"40d37'
Coward,Sally Louise
Cas 0h19'49"59d8'
Cowart,David P
Ser 18h52'47"2d18'
Cowart,Melody
Aql 19h31'15"12d53'
Cowboy & Cupcake
Her 17h14'44"47d43'
Cowboy Birthday Star, The
Uma 11h15'17"45d22'
Cowboy Hat
Boo 14h43'1"50d14'
Cowboy Staton Forever
Vir 13h38'26"1d33'
Cowboy's Angel,The (Stacey Ann)
Mon 6h52'29"11d24'
Cowboy's Starlight
Cap 20h40'33"-20d24'
Cowden,Kirk Byron & Elizabeth Ann
Lyr 18h53'18"32d18'
Cowder,Patterson Reed
Lmi 10h16'54"38d13'
Cowderoy D
Cmi 7h33'28"10d36'
Cowdrey,Sr,Robert
Cyg 19h34'50"38d22'
Cowell
Mon 8h5'39"-5d42'
Cowell,Brian Paul
Her 16h41'35"33d42'
Cowell,Robin
Cas 0h22'1"66d16'
Cowell,William Michael
Ori 6h3'47"8d15'
Cowell,Winifred
Peg 22h11'30"5d14'
Cowen,Keay Jean
Cas 0h31'38"58d37'
Cowett,Patrick
Aur 6h38'57"38d49'
Cowey,Sharon Darlene
Mon 6h30'50"0d58'
Cowgill,Howard W
Oph 18h40'17"8d25'
Cowherd,George Gilbert
Ori 6h17'21"-0d55'
Cowherd,Mildred Dever
Cam 3h44'60"61d26'
Cowie,James A N
Dra 15h51'1"66d41'
Cowie,Ronald Weldon
Hya 9h1'12"5d18'
Cowles,Francine Kulpa
Cas 23h27'23"62d9'
Cowley (HAC),Hunter Alexander
Cet 2h3'51"1d45'
Cowley,Alice Mary
And 23h44'58"33d19'
Cowley,Do3h44'58"33d19'
Aql 20h7'29"-6d19'
Cowley,Esq,Raul (Bebito)
Ser 15h26'21"0d28'
Cowley,Esq,Raul (Miko)
Cet 1h33'53"-1d47'
Cowley,Norah
Cas 0h23'18"63d32'
Cowley,Stephen
Her 17h56'0"50d28'
Cowley,Thorald Rudy
Aql 18h55'21"-1d22'
Cowlin,Andrew David
Tau 4h13'42"20d52'
Cowling,Anne-Louise
And 0h43'1"45d28'
Cowling,Kim M
Lyn 7h33'47"48d56'
Cowman,Deborah
Ori 6h16'25"-1d39'
Cowper,Sally Margaret
Cyg 19h29'18"39d46'
Cox III,Samuel Adam
Cep 21h46'40"61d25'
Cox of Denby Village Star,The Louise
Cas 1h24'27"53d9'
Cox,Alexandra
Ser 16h6'1"10d57'
Cox,Allen Marshall
Cmi 7h38'60"5d17'
Cox,Amanda
Vul 19h46'42"28d51'
Cox,Amanda Diane
Cnv 13h19'22"40d14'
Cox,Amy
And 0h37'1"40d14'
Cox,Anthony M
Cyg 19h33'11"34d7'
Cox,Argentina Recio
Oph 17h24'17"0d15'
Cox,Barbara Ruth
Mon 6h20'60"9d30'
Cox,Bernice
Com 12h50'29"21d24'
Cox,Billy R
Aql 20h11'52"4d1'
Cox,Blair Jeffrey
Uma 10h45'46"40d6'
Cox,Blake Robert
Cep 23h16'41"65d4'
Cox,Brandy
Mon 6h41'28"11d54'
Cox,Brian Michael
Ser 18h43'51"3d6'
Cox,C Russell
Cnv 13h15'49"40d11'
Cox,Caprice Christine
Cam 8h36'59"78d32'
Cox,Carolyn Hatmaker
Vul 19h21'14"27d8'
Cox,Charles Mitchell
Her 18h4'39"14d54'
Cox,Charlotte
Sge 19h15'1"16d10'
Cox,Christopher Ryan
Crt 11h48'51"-11d2'
Cox,Christopher Shawn
Ari 3h1'39"25d29'
Cox,Cory Lee
Mon 6h57'10"-1d7'
Cox,Courtney Klohe
Cas 2h12'55"59d35'
Cox,Danielle Brooke
Cam 3h28'0"56d7'
Cox,Don
Aql 20h0'44"10d24'
Cox,Earl
Uma 9h36'36"55d15'
Cox,Edward Anthony
Her 17h34'49"26d55'
Cox,Erin Leigh
Peg 21h55'20"28d54'
Cox,Ernest Melville
Boo 14h39'39"16d42'
Cox,Fluffy Midnight
Lmi 10h7'46"32d9'
Cox,Geordan David James
Cep 21h55'30"60d21'
Cox,George A
Hya 9h15'25"6d30'
Cox,George Winston
Her 17h54'39"14d56'
Cox,Grady
Lyn 7h50'40"39d33'
Cox,Helen
Cet 3h6'0"3d31'
Cox,Helen Marie
Aql 19h8'42"5d7'
Cox,Jad Evret
Dra 14h20'26"64d41'
Cox,Jane
Lyr 18h28'52"42d24'
Cox,Jason William
Cet 0h55'32"-8d43'
Cox,Jean & Tip
Lyr 19h2'32"28d19'
Cox,Jeanne McCann
Lyn 7h4'47"53d45'
Cox,Jennifer Elizabeth
And 0h21'42"40d21'
Cox,Jessica Lindsey
And 1h5'1"39d22'
Cox,Joshua David
Cyg 20h3'46"30d42'
Cox,Jr,Branch S
Peg 22h34'39"7d44'
Cox,Jr,William H D
Cet 1h50'29"-0d39'
Cox,Karen Christine
And 23h40'34"40d43'
Cox,Kathryn
Eri 4h30'12"-11d55'
Cox,Keenan Marie
And 23h49'34"38d60'
Cox,Kelly J Michelle
Eri 4h55'49"-8d32'
Cox,L Dean
And 23h9'1"37d12'
Cox,Lanele Kay
And 1h23'21"39d32'
Cox,Laura
Lyr 18h51'52"38d57'
Cox,Leland Leslie
Cmi 7h6'28"5d5'
Cox,Lenora Stamey
Peg 22h37'0"22d35'
Cox,Lescott
Cep 21h12'48"58d59'
Cox,Lindsay
Gem 7h1'57"21d27'
Cox,Lindsay Marie
Eri 4h35'0"-5d40'
Cox,Louise
Umi 15h5'19"68d30'
Cox,Marcus & Adeline
Lac 22h53'13"53d49'
Cox,Mary Ellen
Cam 4h11'0"61d13'
Cox,Meriel Ann Rachel
And 0h21'27"45d23'
Cox,Michael R
Her 18h7'39"47d55'
Cox,Michele Kay
Del 20h21'37"20d8'
Cox,Nathan James
Aur 4h50'48"38d11'
Cox,Nick
Lac 22h27'45"41d7'
Cox,Owen Alexander
Cnv 13h43'32"48d3'
Cox,Paula Leslie
Ori 5h10'1"-4d40'
Cox,Rachel Brooke
Del 20h14'32"9d40'
Cox,Ray
Ori 5h36'1"-6d24'
Cox,Rob
Oph 18h32'28"10d13'
Cox,Roberta
Lyr 19h13'25"41d6'
Cox,Roger J
Ori 5h25'42"-6d48'
Cox,Ronda Kay
Lyn 8h18'50"37d36'
Cox,Ryan
Cma 6h50'9"-19d27'
Cox,S
Mon 6h51'20"11d1'
Cox,Shane & Cherie
Sge 19h29'51"10d20'
Cox,Shannon Kimberly
Cam 3h36'28"63d17'
Cox,Sharyn Ann
Mon 6h21'11"-6d35'
Cox,Shawnele Kay
Peg 21h55'31"23d27'
Cox,Sonya
Cyg 21h19'48"35d15'
Cox,Sr,Dennis Earl Lee
Mon 6h37'25"-0d4'
Cox,Stephanie
Sge 19h32'17"16d26'
Cox,Stephen W
Oph 17h18'26"11d16'
Cox,Susan Rogers
Com 12h0'19"20d36'
Cox,Sylvia Tosland
And 1h41'0"48d47'

Cox,Terry & Gill
Cyg 21h5'30"30d33'
Cox,Thomas Wayne
Sge 19h55'32"19d33'
Cox,Vanessa
Eri 3h14'30"-18d14'
Cox,Velma
Mon 8h4'25"-10d14'
Coxe,John B
Dra 18h34'43"68d48'
Coxton,Fayrene
Mon 7h46'51"-3d58'
Coy & The Elephant,
Geraldine
And 23h21'19"51d57'
Coy,Eva Blanche Palmer
Cyg 20h0'0"31d26'
Coy,Rhonda
Mon 8h8'17"-9d4'
Coy,Wayne
Lac 22h1'0"37d32'
Coyl,Tracy
And 0h19'45"34d39'
Coyle
Dra 9h35'38"77d36'
Coyle,Aileen
Lyr 18h14'55"38d26'
Coyle,Cathleen
Cas 0h10'56"64d23'
Coyle,Christian Blaine
Boo 13h55'1"17d50'
Coyle,James Martin
Mon 6h54'40"-1d38'
Coyle,Jason Scott
Uma 10h17'56"41d53'
Coyle,Lois
Del 20h52'12"9d27'
Coyle,Lois
Cas 0h9'37"64d10'
Coyle,Mari-Elaine Theresa
Lyr 19h6'22"28d58'
Coyle,Marilyn
Cas 0h0'60"58d29'
Coyle,Mildred Parrish
Cyg 19h35'47"38d37'
Coyle,Rita Ava
Equ 21h21'23"11d3'
Coyle,Sandra/Walter Kittredge
Star
Cyg 20h18'34"40d25'
Coyman,Robert F
Aur 4h39'0"34d39'
Coyne Operated Pictures
Cnv 12h18'18"50d1'
Coyne,Aidan Longworth
Lyn 8h10'54"38d24'
Coyne,B J
Aur 6h22'35"32d10'
Coyne,Dorothy Eileen
And 0h47'59"36d57'
Coyne,Gene
Lyn 8h53'53"41d8'
Coyne,Jaci Rae
Uma 9h30'49"60d54'
Coyne,John
Per 2h48'57"48d53'
Coyne,Karen Marie
Peg 22h32'55"25d57'
Coyne,Linda
Lyr 18h14'1"38d36'
Coyne,Lisa Marie
Cas 0h34'37"61d8'
Coyne,M A
Dra 16h40'32"58d49'
Coyne,Mare
Dra 16h42'0"52d6'
Coyne,Marie
Eri 2h50'14"-6d18'
Coyne,Mary Beth
Per 2h55'46"31d13'
Coyne,Michael Stewart
Hya 9h15'1"6d16'
Coyne,Niamh Margaret
Cas 1h0'50"64d20'
Coyote
Cnv 12h14'31"46d43'
Coypu,Squidney
Cyg 20h15'1"41d18'
Cozad,Jr,John David
Aur 6h14'1"37d26'
Cozakas,Michael J
Equ 21h0'58"8d48'
Cozza,Sheila M
Lyn 8h9'37"35d22'
Cozzarelli,Jr,James J
Per 3h37'52"36d53'
Cozzarin,Eugene L
Cnc 8h9'60"6d57'
Cozzo,Joseph Alexander
Cep 20h11'39"60d39'
CP 1
Ori 5h56'1"13d23'
CPS Christmas 1993
Tri 1h59'47"25d57'
CPST-1302
Ori 5h57'40"19d1'
Crab,Keri
Cmi 7h23'47"8d53'
Crabtree,John Creswell
Aql 20h1'11"0d1'
Crabtree,Randy Lynn
Aur 6h1'54"32d33'
Crabtree,Rita Anderson
Lyn 7h54'44"43d21'
Crabtree,Robert "Bobby"
Brown
Aur 6h8'45"38d11'
Crabtree,Thomas H
Mon 6h23'1"8d54'
Cracbar
Aql 18h46'11"11d32'
Cracco,Silvio
Col 6h32'45"-35d10'
Crace,James Michael
Her 15h55'1"50d28'
Craciun,Jesse Michael
Psc 23h4'23"6d21'
Craciun,Jonathan James
Lib 14h26'0"-22d49'
Crackel "Sandsstar", Sandra K
And 0h19'25"30d16'
Crackerjack Star,The
Her 17h20'39"46d54'
Cracknell,Grandpa & Grandpa
Clough
Cep 22h8'1"67d43'
Cracolici/Rohdenburg
Lac 22h32'0"53d50'
Craddock,Billy"Crash"
Boo 14h47'14"8d11'A
Craddock,Frances Mary
Cas 0h35'34"60d35'
Craddock,Gerald I I
Lac 22h26'45"50d29'
Craddock,Mae
Boo 14h47'14"8d11 B
Craddock,Ronald
Dra 10h15'17"81d28'
Cradduck,Lyle T
Cep 21h58'28"58d49'
Cradic,M Jane
Cet 2h29'12"-11d1'
Cradle of Sunshine
Cam 5h37'24"80d20'
Crady Family Star,The
Mon 6h54'18"-10d29'
Crafa,Debra & Lou
Cas 1h14'43"60d30'
Craffey,Patrick Sean
Cep 21h27'1"60d20'
Craft Counseling
Assoc,William Dee
Cnv 13h23'38"30d27'
Craft,"Crafty" - Garett S
Uma 10h30'0"47d44'
Craft,Annie
Lyr 18h56'0"31d36'
Craft,Beverly
Mon 6h44'35"10d34'
Craft,Brian
Aur 5h28'48"30d2'
Craft,Cori
And 2h21'45"37d7'
Craft,Jr,SE
Her 15h54'31"42d3'
Craft,Madison Kelly
Cas 23h31'0"61d48'
Craft,Robert Steven
Cep 21h47'30"55d33'
Craft,Rosemary Annette
Cas 0h8'41"59d13'
Craft,Sarah Lynn
Lac 21h58'42"37d51'
Craft,Van W
Tau 4h33'48"30d40'
Craftman's Bingo
Mon 7h47'49"-5d38'
Crafty Lady
Cam 8h37'0"78d57'
Crag
Aur 5h56'48"38d26'
Cragen,Patricia Teresa Torres
Peg 22h50'50"26d49'
Cragg,Anthony L T
Aur 6h13'55"31d19'
Craggan
Ori 5h53'48"12d2'
Crago,David Michael Aaron
Boo 14h55'49"48d15'
Crago,Elizabeth Ashley Reneé
And 0h54'24"38d23'
Crago,Jeff
Leo 11h3'37"0d14'
Cragun,Kathryn
Peg 22h57'0"29d16'
Craig
Dra 19h46'59"67d33'
Craig "Eddie's Angel", Edward
Morgan
Aur 5h4'24"40d39'
Craig & Carrie
Mon 7h6'50"-7d2'
Craig & Dawn's Eternal Love
Light
Aql 19h15'39"15d37'
Craig & Debi
Cyg 20h46'11"37d40'
Craig & Gabi's Grandpa Harry
Uma 10h13'0"47d41'
Craig & Kristin
Uma 10h12'37"61d21'
Craig & Mary
Uma 11h31'0"47d46'
Craig & Melanie
Ori 5h56'54"12d39'
Craig & Melina's Star
Aql 19h5'56"2d51'
Craig & Michelle
Ant 9h38'29"-36d18'
Craig & Suzi Forever
Crb 16h19'23"37d37'
Craig 1928-1995,Robert
Andrew
Cyg 19h34'44"33d52'
Craig 61727
Cet 3h16'48"4d57'
Craig 666
Boo 14h57'43"43d37'
Craig Nicholas
Uma 9h48'42"45d14'
Craig Star
Crb 16h12'55"28d33'
Craig Stephen
Ser 18h43'0"3d49'
Craig will you marry me?
Uma 9h33'42"54d14'
Craig Wish
Ori 6h2'0"-2d37'
Craig's Dream
Ori 5h59'25"15d51'
Craig's Lucky Star
Per 2h53'35"31d21'
Craig's Piece Of Heaven
Cas 2h0'32"59d39'
Craig's Special Star
Eri 3h49'16"-4d23'
Craig's Star
Aql 19h2'35"-6d28'
Craig's Star CCFI
Uma 9h31'45"47d31'
Craig's Star-CCFI
Per 2h57'1"56d32'
Craig,Aaron
Sco 16h50'36"-40d7'
Craig,Chenoa Cheris
Lyn 7h6'55"50d55'
Craig,Claudia
Com 12h53'21"21d44'
Craig,Courtney Patricia
Her 16h50'22"34d32'
Craig,David J
Hya 9h1'40"5d23'
Craig,Diana L
Cas 0h25'24"61d23'
Craig,Erica
Vul 19h58'26"26d13'
Craig,Esther Ellen
Aql 19h46'25"11d38'
Craig,Flora
Cas 1h24'39"55d42'
Craig,Ian
Lyn 9h5'15"33d24'
Craig,IV, Robert M
Lac 22h10'1"46d36'
Craig,J
Del 20h17'40"10d16'
Craig,J J
Cam 3h24'11"60d39'
Craig,James Edward
Umi 15h22'51"68d14'
Craig,Jeffrey A
Cep 23h38'19"65d53'
Craig,Jim
Cep 4h5'14"80d29'
Craig,Jr,Augustus
Cnv 12h23'0"34d23'
Craig,Jr,Clifton M
Dra 17h42'54"68d43'
Craig,Jr,Geoff
Cam 3h32'18"63d32'
Craig,Julia Mann
Lyn 7h28'25"40d3'
Craig,Kristi
Lyr 19h1'1"35d29'
Craig,Marge J
Ori 5h52'38"14d55'
Craig,Randolph K
Cma 7h23'27"-16d31'
Craig,Raymond Allen
Aur 6h35'13"33d5'
Craig,Redge
Ori 6h17'25"20d58'
Craig,Robert
Boo 14h42'0"22d41'
Craig,Robin Joseph
Peg 22h0'13"30d55'
Craig,Rosemary
Del 20h55'40"9d23'
Craig,Shana
And 2h23'20"47d27'
Craig,Shari Lynn
Cam 4h12'1"61d37'
Craig,Steven & Annalisa
Cyg 21h5'34"34d21'
Craig,Susan
Cet 2h5'0"6d5'
Craig,Ted
Ori 6h0'35"1d19'
Craig,Victoria Margaret
Lyr 18h56'56"31d6'
Craig,William Cameron
Sct 18h43'21"-6d40'
Craighead,Emilyann Coffey
Com 12h7'0"19d35'
Craigmeister
Cep 20h37'46"75d22'
Craigus,Jonious
Cep 20h23'50"60d18'
Crail,Jr,Frank O
Cep 22h9'23"70d16'
Crail,Patricia D
Cas 0h25'53"63d41'
Crain,James Bradley
Aur 5h8'24"44d18'
Crain,Katherine Joan
Cas 2h11'16"75d22'
Crain,Krysilynn Louise
Peg 22h42'17"20d54'
Crain,Philip
Cet 1h50'57"0d25'
Crain,Rachel Nicole
Uma 9h16'47"50d36'
Crain,Sean Michael
Boo 14h4'59"29d53'
Crais,Robert Kyle
Lmi 10h16'16"38d17'
Craley,William John
Per 2h55'1"37d10'
Crall,Max Richard
Her 18h1'56"18d53'
Cram,Jr,Harkness Warren
Uma 11h17'46"47d16'
Cram,Linda
Lac 21h57'19"37d50'
Cramer,Andrew Alan
Peg 21h55'56"2d23'
Cramer,Bev
Cyg 19h45'16"31d11'
Cramer,Christine Ann &
Breton James
Cyg 19h18'34"49d42'
Cramer,Douglas Schoolfield
Oph 18h16'56"12d58'
Cramer,Frank
Ser 15h36'13"19d14'
Cramer,Harrison
Dra 19h33'34"60d3'
Cramer,Hilary Tina
And 0h6'52"35d26'
Cramer,John
Tri 2h38'31"31d17'
Cramer,Ken
Cyg 19h45'14"30d17'
Cramer,Mikayla Mary
Cas 0h41'24"76d47'
Cramer,Raymond C
Cyg 21h50'26"37d41'
Cramer,Richard Morgan
Uma 11h12'0"43d57'
Cramer,Ronnie
Crb 15h29'1"30d26'
Cramer,Toni
Dra 20h22'26"62d24'
Cramlet,Roger A
Lac 22h30'37"40d56'
Crampin & Pring Architects &
Designers
Ori 5h40'21"1d43'
Crampton,Carol
Cas 23h38'46"63d52'
Crampton,W L
Peg 23h24'50"30d32'
Cran,Doreen
Peg 22h21'17"30d58'
Cranage,Dean Eugene
Cyg 19h30'23"37d32'
Cranage,Joanne
Lyr 18h28'54"30d32'
Cranberry Jello Bear
Uma 10h7'32"70d4'
Crandall,1-14,10:26PM, Amy
Leanne
Ori 5h51'38"16d40'
Crandall,Anamae
Aql 20h5'59"-6d23'
Crandall,Jaime K
And 2h21'1"48d46'
Crandall,Jeanette Ann Dwyer
Cyg 20h31'0"42d3'
Crandall,Kimberlin B
Uma 10h34'20"52d14'
Crandall,Vivien
Cas 0h13'52"63d45'
Crandell,Alberta & Terry
Cyg 21h25'0"40d53'
Crandell,Richard
Eri 4h2'20"-18d30'
Crandle,Christopher Nicholas
Dra 16h56'14"64d26'
Crane's Jewel
Ori 6h8'15"4d51'
Crane,Amber K
Cyg 19h27'30"31d4'
Crane,Ben R
Cmi 7h6'15"3d47'
Crane,Dmitry
Uma 9h54'14"50d15'
Crane,E F "Muffie"
Hya 9h10'0"2d12'
Crane,Harry
Tau 4h38'17"7d38'
Crane,John "Diddy"
Lyn 6h28'51"60d47'
Crane,John Dorr
Sgr 18h4'51"-28d8'
Crane,Judge Ronald J
Aql 19h27'14"10d54'
Crane,Katie
Cyg 20h4'18"31d8'
Crane,Kevin & Rita
Ori 5h59'59"14d3'
Crane,Lesley Karen
Cyg 21h17'46"35d53'
Crane,Mark W
Boo 15h3'35"28d32'
Crane,Martha Eileen
Cyg 21h35'54"41d32'
Crane,Patrick Vegas
Umi 16h3'20"76d36'
Crane,Peter Daniel
Lac 22h1'49"48d52'
Crane,Peter Evan
Per 3h51'56"36d49'
Crane,Richard William
Cep 22h51'51"65d39'
Crane,Robert F
Dra 13h0'18"76d2'
Crane,Therese K
Hya 10h19'47"-12d13'
Craney,Diane Frances
Ari 2h38'43"22d9'
Cranfield,Betty
Cas 0h18'1"61d9'
Cranford,Denis Wayne
Cet 1h38'0"-5d3'
Cranford,Hugh
Her 17h27'14"26d57'
Cranford,Michael
Umi 14h16'28"71d4'
Cranford,Pat
Peg 22h34'42"2d22'
Cranford,Pat
Crb 16h1'11"26d39'
Cranford,Pat
Lyn 7h53'0"42d48'
Cranker,Emily Veronica
And 0h22'50"33d41'
Cranmer Superstar, Dr Morris
Uma 10h51'49"48d13'
CRANMER USN "MUSTANG",
CAPT JOHN M
Cas 2h30'35"60d18'
Cranmer,Steven Charles
Cep 21h29'57"65d26'
Crannell,Kenneth A
Her 16h59'47"30d8'
Crannell,Michael Thomas
Her 17h10'1"21d14'
Cranston,Daniel Brian
Her 16h47'58"39d7'
Cranston,Kim
Lmi 10h32'59"34d5'
Crapomena,Nocciola
Ildebranda
Cep 1h20'13"80d2'
Crapse,Janie
And 0h1'0"43d26'
Craske,Bryson J
Aql 18h57'48"17d1'
Crassweller,Nicole
Cnv 12h7'50"38d16'
Craton,David Brian
Cmi 7h54'30"1d20'
Cratty,Justin Ross
Oph 18h15'23"7d48'
Craven's Star,Norma Page
Hya 9h33'45"-1d45'
Craven,Cecelia O'Leary
Cas 0h52'48"61d38'
Craven,Fran
And 1h12'1"40d34'
Craven,Kevin Patrick
Her 18h39'56"18d58'
Craven,Lucille
Cyg 20h4'31"31d6'
Craven,Richard R
Sex 9h50'33"-6d18'
Cravens Alpha,Thomas F
Cep 21h0'51"68d27'
Craver,Robert F
Per 3h19'28"40d54'
Cravit,Cynthia Renee Ross
Lyr 19h22'40"31d7'
Crawford OBE,Michael
Per 1h26'12"52d44'
Crawford,Andrew Michael
Cma 6h58'38"-28d35'
Crawford,April
Cet 0h57'17"0d37'
Crawford,Brett R
Cet 0h30'59"1d23'
Crawford,Carla Mancuso
Aur 5h24'49"37d54'
Crawford,Chris R
Ori 5h53'29"6d36'
Crawford,Christopher Kevin
Uma 11h39'48"57d35'
Crawford,Cindy
And 0h17'30"39d6'
Crawford,Col Ben
Aur 7h21'16"43d47'
Crawford,Damon"KHAFRE"
Aur 6h53'40"37d53'
Crawford,David R
Aur 6h28'32"38d12'
Crawford,Don & Margie
Eri 4h48'21"-9d11'
Crawford,Elizabeth Ann
Cas 2h10'1"61d16'
Crawford,Frank Conyngham
Cam 7h50'26"61d15'
Crawford,Janica Tai
Sct 18h53'0"-5d44'
Crawford,Jay
Her 16h44'30"30d53'
Crawford,Jessica Anne
Cet 1h43'22"0d20'
Crawford,JoAnn
Cnv 12h52'30"51d14'
Crawford,Joel Stephen
Boo 13h37'1"26d25'
Crawford,Joyce S
Tri 1h56'29"26d35'
Crawford,Kay Hall
And 0h50'51"35d39'
Crawford,Kelly Lynn
Mon 6h35'16"1d33'
Crawford,Kristen & Craig
Cyg 21h52'51"52d50'
Crawford,Lacey
And 0h10'31"38d58'
Crawford,Lilian Marie
Lyr 18h22'18"44d56'
Crawford,Lindsay Rhea
Sex 10h11'9"-2d58'
Crawford,Lois
Ori 6h5'58"7d57'
Crawford,Lyndsey Kathleen
Cas 0h37'42"68d22'
Crawford,Matt-Matt
Aql 19h47'0"11d58'
Crawford,Nicholas Clifford
Ser 15h21'12"1d18'
Crawford,Nicholas W
Cet 2h27'39"7d55'
Crawford,PhD,Susan N
Lyn 8h24'59"34d52'
Crawford,River Hayes
Eri 2h53'18"-6d40'
Crawford,Rod
Eri 2h50'27"-6d36'
Crawford,Ruth
Cyg 21h3'18"37d20'
Crawford,Sarah
And 23h36'47"39d51'
Crawford,Tessa
Eri 3h43'34"-17d44'
Crawford,Tiffany
Dra 15h33'1"65d39'
Crawford,Tod
Eri 2h59'1"-6d48'
Crawford,William Sherman
Ser 15h56'1"24d37'
Crawford-1st Birthday, Kyra
Saffran
Vul 19h23'38"26d40'
Crawley,Donna Marie
Archambault
Vul 19h57'58"25d10'
Crawley,Joshua Joel
Oph 17h36'36"-23d34'
Crawley,The David &
Katherine Star
Lyr 18h24'41"38d45'
Crawshaw,Derek
Lyn 8h0'14"45d55'
Crawshaw,George Alexander
Peg 21h57'37"31d46'
Cray,Jr,Hampton R
Hya 8h41'33"-1d46'
Crayola
Ori 5h58'49"20d24'
Crazy Eddy
Lac 22h6'54"47d39'
Crazy Horse
Ori 5h57'25"17d29'
Crazy Little Eddie
Dra 18h54'51"58d34'
Crazy Miane
Aql 19h53'20"15d15'
Crazygirl
And 23h30'34"48d15'
Cready,Carol
Lyr 19h14'36"37d53'
Creager,Dave
Her 16h10'17"24d56'
Creais,Bonnie-Jean
Psc 22h54'13"6d3'
Cream Puff Planet
Per 2h31'1"57d16'
Creamer
Vul 21h20'23"28d23'
Creamer,USN,Commander
George U
Aql 20h18'54"7d39'
Creammy's Star
And 23h35'32"40d59'
Crean,Drusilla Donnelly
Boo 14h20'1"39d11'
Crean,Robert Michael
Peg 0h2'43"17d45'
Creason,Timothy A
Aql 19h52'57"14d13'
Creasy-Nunweiler, Shelia
Peg 23h41'58"31d4'
Creation Graphique Meziw
Lac 22h4'16"37d53'
Creative Alliance Inc
Uma 9h53'23"56d12'
Creative Cosmetic Robbins
Star,The
Uma 11h21'43"50d9'
Creator
Aql 20h11'33"13d37'
Creatura
Col 6h29'55"-34d27'
Creazzo,Posha
Uma 11h15'46"42d26'
Crecelius,Haley
And 0h49'38"35d56'
Crede,Andreas
Aql 19h54'56"-6d53'
Credendus
Lyn 6h25'25"61d33'
Credo Satum Nos Coegisse:
Ed & Lisa
Ori 5h36'34"-1d9'
Creech,Jackie & John
Crb 15h33'0"29d30'
Creech,Loren Alan
Aur 4h38'21"31d11'
Creech,Melissa Ann
And 0h15'43"32d30'
Creech,Tiffany E
Lib 15h16'1"-25d20'
Creed,Margaret Nyce
Mon 6h2'50"-8d0'
Creed,Samantha
And 0h29'29"31d38'
Creedon,Bernard P
Lac 22h15'1"46d56'
Creedon,Joseph Christopher
Lyr 18h45'20"35d45'
Creek-Bell
Crb 15h51'39"31d13'
Creeks,Dorothy
Mon 7h29'21"-8d57'
Creel II,Horace
Boo 15h2'16"19d3'
Creel,Cathey Meller
Lmi 9h51'37"38d10'
Creel,Karola
Peg 22h2'34"25d29'
Creel,Kristi Rose
Cet 2h45'12"6d31'
Creelman,Catherine Jean
Lyn 7h50'40"43d45'
Creelman,Geoffrey Kirk
Per 1h33'1"53d3'
Creep
Ser 16h16'52"2d58'
Cregger,Kayla Lea
Cyg 20h9'40"40d21'
Cregle,Roger
Her 16h20'35"50d40'
Crehan,Sean
Her 16h12'22"20d30'
Creighton,Gerald M
Aur 6h55'16"36d54'
Creighton,Michael E
Cep 21h58'58"56d3'
Creighton,William Edward
Her 17h13'35"40d49'
Creith,Edna
And 23h21'0"40d27'
Cremades-Eva Diaz Poz, José
Ma Lopez
Ser 15h55'0"-2d13'
Cremeans Isheshake
Always,Kerry
Peg 23h44'54"15d43'
Cremer,Karl-Heinz
Aur 5h8'28"42d38'
Cremer,Ulrich
And 23h9'0"40d48'
Cremona,Ester
Her 16h11'59"11d7'
Cremona,Sylvia
And 2h23'35"42d40'
Cremonini,Theresa &
Giovanni
Umi 16h17'39"79d19'
Crenshaw,Kathy
Cmi 7h55'25"4d33'
Creo,Mikey
Aur 6h7'1"50d5'
Creo,Ryan Thomas
Her 17h28'41"26d58'
Crepieux,Xavier
Cam 3h26'38"61d0'

Crepin,Aline
Aur 4h37'44"31d23'
Crequer,Raymonde
Oph 18h16'26"10d40'
Cresap,Guy B
Hya 9h2'43"0d51'
Crescent,The
Sct 18h54'38"-6d35'
Crescenzo,Lucille "LaLu"
Eri 4h43'38"-6d35'
CresCoUnion
Uma 9h14'60"62d7'
Crespin,Lorena Silvia
Mon 7h3'37"0d52'
Crespo,1
Boo 14h25'29"26d29'
Crespo,Ellen
Mon 7h2'39"-5d4'
Cressie's Future
Cas 2h29'50"59d51'
Cressman,Arlene Felicia Ida Puleo
Lyn 8h10'45"52d9'
Cressman,David
Her 18h13'41"38d36'
Cressman,David & Arlene
Lyn 8h27'31"57d35'
Cresta
Aql 19h9'1"3d18'
Cresto,George Andrew
Ser 15h17'52"8d48'
Creston
Ori 4h47'12"2d10'
Creswell,Phoebe Camilla
Cyg 21h2'14"38d35'
Creter,Richard
Lib 15h30'59"-21d54'
Creutz,Casey A
Tri 2h11'0"30d7'
Creutz,John J & Eileen
Dra 18h2'59"50d29'
Creveau,Guy
Ori 5h19'34"12d54'
Crevier,Sylvain
Per 3h0'23"40d9'
Crevoiserat,Beth
Cas 0h37'17"70d56'
Crew 1994-1995,The
Lac 22h27'0"53d36'
Crew,Jack
Aql 18h53'50"10d58'
Crew,Winnie Swallwell
Crt 10h56'19"-10d49'
Crewdson,John Derrick
Cep 0h6'43"66d38'
Crewe,Susan
Lyr 18h15'37"30d9'
Crews,Amy Deann
Cet 0h56'33"-1d46'
Crews,D'Anne McAdams
Eri 2h56'32"-7d3'
Crews,Fletcher A
Per 4h4'53"37d12'
Crews,Louise Mullens
Oph 17h50'16"13d30'
Crews,Maxwell Garcia
Ari 1h54'20"15d59'
Crews,Michael J
Cet 1h27'25"-1d55'
Crews,Michael S
Cyg 21h52'21"40d58'
Crews,Tessa Caitlin
And 1h28'29"41d9'
Crews,Tyler Perren
Dra 19h25'28"58d10'
Cribb Star,Karilan
Dra 10h41'29"81d1'
Cribbins,Jennifer
And 2h30'18"49d12'
Cricco,Anthony Rocco
Dra 16h37'20"62d34'
Crichlow III,Allwyn Forestor
Vul 20h0'1"28d37'
Crichlow,Joan Pamela
Cas 0h42'14"67d15'
Crichton,Dale
Cam 7h34'0"80d3'
Crichton,Liam
Peg 21h59'0"33d14'
Crichton,Merton W
Aur 4h54'0"48d47'
Crichton,Robert
Boo 15h4'20"31d2'
Crickenberger, Christina
Cyg 20h27'13"40d28'
Cricket
Cet 0h32'0"-3d23'
Crickett
Uma 10h56'46"53d34'
Crickette,Troye C- Scott E Campbell
Uma 12h10'34"59d55'
Crickman,Charles Albert
Her 16h39'24"32d53'
Crickman,Molly Claire
Cas 0h45'55"64d23'
Crickman,Sarah Margaret
Peg 22h42'42"32d50'
Crickman,William Albert
Dra 19h27'31"68d1'
Cricritrente
Cas 1h31'1"58d59'
Crider,Christie
And 2h27'1"50d7'
Crider,Ginger Christopher
Lyr 18h37'36"40d29'
Cridlin,Joan Marie
Lyn 8h12'23"35d45'
Crighton,Robert Jack
Umi 15h24'1"68d9'
Criglar,Yvette
Cas 23h41'20"61d9'
Crigler,Kristy Lee
And 1h34'11"36d59'
Crigler,Lori Raye
And 1h27'59"39d12'
Crijns,Dèsiree Marie
Peg 21h49'0"31d39'
Crilli,Giusi E
And 1h1'20"47d55'
Crim
Ori 6h5'29"1d44'
Crimmins,Andrew Ryan
Aur 6h18'58"46d48'
Crimmins,Gael A
Cyg 21h3'19"31d22'
Crinklaw,Craig
Tri 2h18'23"34d9'
Cripe,Brandi Ann
Eri 3h34'54"-2d25'
Cripe,Bryan Lee
Her 17h19'55"44d23'
Cripe,Virginia Ann
Sge 19h55'14"19d41'
Crippa,Sergio
Com 13h12'16"26d42'
Cripps Star,The Steve
Uma 9h45'22"45d11'
Cripps,Julie
Ori 4h50'46"0d41'
Criquet Ellen
Equ 21h0'23"8d58'
Cris 212
Aur 5h2'48"38d18'
Cris III
Sgr 19h28'6"-33d51'
Crisafulli
Cam 5h49'41"60d48'
Crisafulli,Anne Marie
Sgr 18h59'4"-2d43'
Crisafulli-King,Linda R & Samuel C
Crb 15h58'35"31d1'
Crisalli,Luke
Per 2h37'38"41d1'
Crisalli,Michele
And 1h2'46"38d38'
Crisalli,Patricia
Lyn 6h58'0"59d35'
Crisanti,Edward
Mon 6h20'0"-6d0'
Crisci,Nora Evelyn
Cas 23h31'16"58d19'
Criscione Sr,Anthony
Lac 22h44'1"53d55'
Criselda
Aql 19h55'1"10d56'
Crisjoe
Peg 23h27'19"23d31'
Crisman,Barbara
Lyn 8h57'52"44d5'
Crisman,Carolyn
And 23h9'16"40d36'
Crisman,Kathy L
Cmi 7h43'34"4d33'
Crismaur White Star
Col 6h29'23"-38d14'
Crisp,Big Jim
Aql 18h48'27"11d35'
Crisp,Dawn & Jerry
Peg 23h27'43"23d40'
Crisp,Jonathan W
Lmi 10h10'53"32d11'
Crispin
Peg 22h42'17"33d49'
Crispin,Charles Wade
Cma 6h57'24"-17d23'
Crispin,Marc Stephen
Aql 19h55'23"13d40'
Crispin,Yanick N
Dra 19h28'18"50d55'
Criss Dreamer,The Colley William
Cep 21h41'35"68d2'
Crissi Dog
Com 13h30'60"19d8'
Crissman,Brady Walker
Lac 22h24'0"53d44'
Crissy
Cnv 12h51'29"41d44'
Crist,Barry Alexander
Lyn 8h14'45"58d38'
Crist,Haley Nicole
Cas 2h29'30"59d12'
Crist,Mackenzie Kathleen
Lyn 8h14'46"57d47'
Crist,Madison Delia
Dra 9h41'13"73d27'
Crist,Maria
Cyg 19h58'45"30d43'
Crist,Maureen
Lyn 8h10'18"57d31'
Crist,Robert Thomas
Cep 23h37'0"75d0'B
Crist,Zebulon David
Her 17h57'19"31d33'
Cristallina
Pic 4h41'25"-47d34'
Cristallo,Dennis
Dra 14h17'18"64d33'
Cristante,Christopher A
Per 2h24'13"54d27'
Criste,John D
Oph 16h54'1"-26d14'
Cristel Dawn
And 2h10'0"50d41'
Cristelle
Uma 13h20'36"61d58'
Cristello,Jenina
Uma 11h29'1"45d18'
Cristi
Cas 2h2'32"59d16'
Cristi,Leonardo Andrea Marco Carlo
Cep 23h30'13"70d23'
Cristian
Oph 17h0'41"8d41'
Cristiana
Cma 6h24'39"-16d34'
Cristiana
Cma 8h7'0"1d39'
Cristiana Alessandra
And 0h10'37"38d58'
Cristiana B
And 23h48'18"41d36'
Cristiano,Marie Antoinette
Col 6h37'36"-35d57'
Cristina
Cyg 21h51'26"52d41'
Cristina
Aur 4h59'13"40d10'
Cristina
Ant 10h43'54"-34d31'
Cristina
Pup 7h56'25"-23d3'
Cristina
Pho 2h14'59"-42d12'
Cristina
Pho 0h36'31"-48d24'
Cristina
And 02h25'38"39d19'
Cristina
Boo 15h3'33"32d24'
Cristina
Del 20h12'36"13d5'
Cristina
Umi 15h15'47"70d38'
Cristina
Vul 19h23'39"23d51'
Cristina
Dra 11h56'0"70d54'
Cristina
Umi 14h56'0"70d3'
Cristina
Peg 22h27'48"20d12'
Cristina
Per 3h42'55"39d56'
Cristina
Lib 14h23'41"-8d57'
Cristina
And 23h22'0"52d33'
Cristina & Paul Our Guiding Light
Cyg 21h0'46"28d39'
Cristina De Ignacio San Jose
Peg 22h32'10"30d46'
Cristina e Franco
Lyn 7h8'53"44d55'
Cristina e Stefano
Col 6h25'13"-36d46'
Cristina R
Hor 3h36'19"-47d50'
Cristina, Giampaoli e Raffaele
Cnv 12h14'45"51d55'
Cristina,Lorenzo
Tau 5h49'19"23d15'
Cristina-Barbara
Del 20h17'24"11d46'
Cristofani,Perry
Oph 17h16'45"-18d55'
Cristy
Mon 6h54'27"-1d54'
Cristy
Lyr 18h48'36"34d2'
Cristy's Counsel-Weiler/Mayes II
Lyr 19h1'47"40d26'
Criswell,"Lenora Cecilia"
Mon 6h40'41"3d2'
Criswell,Anastasia
Cnv 12h24'42"43d1'
Critan (Dr J),Jeannie
Ori 5h56'37"16d29'
Critan,Thomas "Mr Lucky"
Ori 5h58'10"17d54'
Critchett,Nancy L
Cam 3h47'44"56d27'
Critchley,Jr,Thomas J
Lib 14h19'18"-10d15'
Critchlow,Roark
Dra 14h20'19"64d20'
Critelli,Jayne
Sgr 18h59'9"-2d9'
Critelli,Jeffery Stephen
Her 17h17'0"45d40'
Crites
Aql 19h1'46"3d6'
Crites,Caitlyn
Mon 6h38'58"1d12'
Crites,Chad
Tau 4h4'51"10d57'
Crites,Chad
Lyr 18h50'10"34d11'
Crites,Chloe
Mon 6h37'55"1d18'
Crites,Corinne
Aql 19h56'37"12d56'
Crites,Jodie Faye
Peg 22h54'31"21d42'
Crittenden,John
Aur 6h3'32"35d60'
Crittin,Gérard
Lyn 9h0'46"34d3'
Critz,The Shining Star of Susan
Gem 6h39'25"13d55'
Crivella,Arthur R.
Aur 7h19'49"37d10'
Crivelli,Gabrielle
Lyn 8h8'0"34d54'
Crivolio,James Martin
Oph 17h1'26"-21d55'
Croatto,Christian & Michael
Uma 11h16'41"42d14'
Croc,Fifi Et Marie
Lmi 10h47'47"24d40'
Crocco,Charles
Her 17h0'1"37d40'
Crocco,Emily Noel
Lyn 8h7'15"40d28'
Crocco,Thomas Patrick
Cnv 12h54'24"33d27'
Croce della Democrazia Cristiana
Vel 9h43'10"-46d16'
Croce,Alberta
And 23h2'48"45d20'
Croce,Joanie
And 23h3'11"50d3'
Croce,Joni Lin
Lmi 9h55'27"38d00'
Croce,Scott A
Dra 18h26'40"58d23'
Crocfer,Philip
Sct 18h46'43"-6d57'
Crock,Dana Michelle
Peg 0h7'14"21d27'
Crocker,Molly
Cet 0h58'48"1d45'
Crocker,Natalie Ann
Aur 7h4'0"40d3'
Crockett
Uma 9h27'57"49d32'
Crockett's #1 Star Brenda Lewis
Ori 6h15'0"-0d24'
Crockett's Star
Equ 21h21'37"10d11"
Crockett,Catherine
Cas 0h39'40"66d48'
Crockett,Curtis Neal
Lac 21h58'27"36d16'
Crockett,David W
Lyn 9h10'46"38d18'
Crockett,George MacArthur
Cet 0h47'23"-6d49'
Crockett,John David
Umi 14h46'25"68d40'
Crockett,Olivia Jo
Uma 11h20'0"71d18'
Crockett,Richard H
Lac 22h9'16"46d43'
Croese,Rudolph
Her 18h12'1"47d9'
Croff,Don
Sex 10h45'21"2d7'
Crofford,Shawn & John Huffstetler
Sex 9h53'28"3d40'
Crofoot,Amy
Vir 13h34'56"-6d7'
Crofoot,Andrew
Vir 13h38'21"0d29'
Crofoot,Henry Joseph
Aql 19h27'25"13d23'
Croft,Alan & Carol
Cmi 8h8'10"4d55'
Croft,George & Viola
Uma 9h40'33"50d33'
Croft,Leanne
Cas 0h54'10"67d16'
Croft,S
Uma 9h52'41"51d38'
Croft,Scott
Equ 21h19'40"3d39'
Croghan,John H
Ser 15h16'16"7d0'
Croizat,Helena
Lib 14h56'35"-10d53'
Croizile,Gerard
And 23h17'12"50d40'
Crokie,Nadine Renee
Lyr 19h1'1"38d42'
Crokus,Pamela J
Cyg 21h0'24"31d16'
Croky & Pussy
Boo 15h2'56"42d21'
Croll Star,The
Cas 0h48'0"63d48'
Crombach,Dr Paul
Dra 13h23'26"64d26'
Crombie,Bernadette O - Prosperity
Cas 1h4'10"53d13'A
Crombie,Veronica E - Peace
Cas 1h4'10"53d13'B
Crombleholme,Daniel
Dra 16h41'52"70d49'
Cromeans,Robert Heron
Ari 3h2'38"25d30'
Cromer,Donna Marie
Aql 20h12'11"10d18'
Cromer,Jeffrey Robert
Tau 4h3'6"23d51'
Cromer,Joren Parker
Vul 19h45'60"28d28'
Cromien,The Star of
Aql 19h0'52"7d45'
Cromley,Nathan
Aur 6h26'1"38d6'
Cromosini,Rachel Nicole
Vul 19h45'55"28d26'
Crompton,Betty
Cas 0h27'46"73d57'
Crompton,Emily "Bean"
Cyg 20h25'18"38d38'
Crompton,Emma Louise
Uma 8h38'18"59d46'
Crompton,John Cowley
Cas 1h0'0"75d24'
Crompton,Mike
Cep 23h10'36"61d23'
Cromwell,Karen & Damien
Cyg 20h23'1"39d31'
Cromwell,Mandara
Lyn 7h24'21"44d33'
Cron,Margaret Mary
And 0h21'16"34d2'
Cronan,Dan
Hya 9h39'12"-10d44'
Crone,Hannah Estelle
Peg 21h53'52"20d32'
Crone,Kelsey Rae
Cam 8h15'0"80d44'
Cronenberg,Catherine Ann
Ori 5h27'23"-0d20'
Cronin,Bonnie J
Cet 3h19'35"6d39'
Cronin,Chris & Hajdu, Yvonne
Lac 22h21'56"50d28'
Cronin,David M
Hya 8h48'0"5d48'
Cronin,Debbie
And 0h24'1"33d17'
Cronin,Douglas Francis
Her 15h56'0"47d34'
Cronin,Douglas Phillip
Cmi 7h37'37"5d18'
Cronin,Honey Bear
Uma 11h9'1"42d10'
Cronin,James Reed
Lac 22h25'58"55d18'
Cronin,Jason Timothy
Psc 0h17'54"9d39'
Cronin,Jason William
Cet 2h5'10"-0d0'
Cronin,Jerry
Cep 21h45'24"58d9'
Cronin,Joe
Mon 6h55'11"-6d4'
Cronin,Joseph F
Hya 9h36'36"-10d48'
Cronin,Kelly
Lyn 7h28'37"36d5'
Cronin,Kevin M
Per 2h26'52"57d15'
Cronin,Leanne Marie
Cyg 19h19'34"28d9'
Cronin,Mary Gertrude
Cas 0h47'25"72d29'
Cronin,Michael
Dra 11h59'20"67d26'
Cronin,Nicholas Harrison
Dra 14h20'0"63d52'
Cronin,Paul J
Aur 6h30'36"31d3'
Cronin,Rosemary Patricia
And 1h53'1"38d14'
Cronkhite,Barry Scott
Lyr 19h19'1"41d16'
Cronkite,Kenneth
Dra 17h54'46"61d16'
Cronkite,Walter
Uma 15h2'39"17d48'
Cronmiller's Own
Dra 19h26'37"60d53'
Cronrath,Geordis M
Uma 9h28'32"52d44'
Crook,Andrew
Per 3h14'44"50d33'
Crook,Diane
Aur 4h59'36"34d2'
Crook,James
Cam 8h36'39"73d59'
Crook,Rebekah
Cep 22h3'19"67d53'
Crook,Thelma
Cas 0h43'16"64d3'
Crook,Vanessa Jane
Lyr 18h51'0"38d52'
Crook,Vanessa Jane
Lyr 19h15'19"41d44'
Crook,Vanessa Jane
Lyr 18h51'0"38d52'
Crooked House
Cyg 20h23'23"38d33'
•Crooks,Carolyn R
Sge 19h24'38"16d16'
Crooks,Francesca Margaret
Cas 0h29'15"61d23'
Crooksie
Aql 19h27'46"8d7'
Croom USCG Ret,Capt Archie
Mon 6h30'53"-0d12'
Croom,Kevin
Aql 20h19'36"5d31'
Cropp,Alexandra Mary
Cas 2h49'25"68d47'
Cros,Antonia Bermezo
Sge 20h17'13"16d23'
Crosa-Beck,Peggy
Cet 2h40'46"-11d51'
Crosbie,The Douglas Haig
Uma 8h51'42"56d32'
Crosby,Al
Her 18h36'46"18d52'
Crosby,April
Dra 11h53'0"66d45'
Crosby,Barbara
Cas 1h45'46"60d52'
Crosby,Cathy
Cyg 20h33'14"48d49'
Crosby,Charlie
Cmi 7h56'43"8d44'
Crosby,Gerald A
Her 16h26'56"50d10'
Crosby,Hannah
Peg 22h41'55"21d3'
Crosby,Jr,Wallace Clinton
Dra 17h43'50"60d48'
Crosby,Lindsay H
Cet 3h14'50"0d50'
Crosby,Patricia & Warney
Eri 4h33'29"-8d41'
Crosby,Susan
Cas 1h19'36"60d4'
Croset,Veran
Equ 21h3'15"7d35'
Croskell,Samuel John
Lyn 7h7'59"58d58'
Croson,Troy Atwood
Ari 1h55'22"25d18'
Cross MA,MFCC,Kim
Aql 19h53'0"-5d59'
Cross,Adara Brianne
Mon 6h34'0"-1d21'
Cross,Ashley K
Uma 11h42'56"41d45'
Cross,Brandon Paul
Lyr 18h41'50"28d4'
Cross,Brenda Lynn
And 2h8'59"38d37'
Cross,Brier Collier
Dra 19h33'10"70d41'
Cross,Carol Ann & Fred Della
Lyr 19h14'58"40d23'
Cross,Emily Jane
Cyg 20h43'45"45d20'
Cross,Emily Johanna Mai
Eri 2h52'0"-3d13'
Cross,Emma Palmer
Cyg 21h58'1"53d27'
Cross,Gary
Cam 3h56'41"57d23'
Cross,Ian F
Uma 8h40'26"61d57'
Cross,Jacqueline
Cas 1h28'20"58d20'
Cross,James Lyndon
Boo 14h33'32"47d43'
Cross,Jerry Lee
Boo 15h20'44"42d13'
Cross,Joan Elizabeth Weaver
Lyn 8h10'56"48d15'
Cross,Joseph John
Boo 14h24'31"31d38'
Cross,Karren Denise
Mon 7h47'29"-1d44'
Cross,Larry G
Lac 22h9'17"46d55'
Cross,Laura Margaret
Cas 0h25'28"61d8'
Cross,Linda Kay
Cas 1h47'12"70d17'
Cross,Martha & Paris
Lyn 7h39'55"51d38'
Cross,Rae Lynn
And 1h28'56"48d47'
Cross,Skip
Boo 15h9'12"26d57'
Cross,Spencer Michael
Lyr 18h26'48"36d22'
Cross,Steven Clair
Ari 2h35'27"21d41'
Cross,Theodore L
Vul 20h43'39"28d58'
Cross,Vicki Lee
Cyg 19h47'1"29d51'
Crosser,Donald Duane
Per 2h51'33"38d14'
Crossfield,Lauren Elizabeth
Lmi 10h56'45"26d16'
Crossfield,Victoria Marie
Lmi 10h52'26"27d31'
Crossing,William
Ori 4h58'42"0d44'
Crossland,Joan Ruth Minshew
Mon 7h2'35"0d47'
Crossley,Patricia Mary
Lyr 18h19'24"43d24'
Crossley,Sean
Ori 5h25'33"15d4'
Crossman,Mr & Mrs Kevin
Eri 3h22'23"-5d52'
Crossman,Stephanie
Mon 6h21'19"5d33'
Crosson
Lib 15h20'53"-22d33'
Crosson,Gene & Sheri
Lyr 18h34'1"37d55'
Crosswhite,Kellen J
Dra 19h1'51"54d40'
Crosta,Giancarlo
Dra 15h20'53"65d18'
Croston
Lac 22h55'1"53d33'
Croston,Jack Alan
Cnc 9h0'51"17d35'
Crotchet,Prince & Princess Quaver
Lyr 19h22'55"31d41'
Crotchett,Gene
Cep 22h53'50"56d54'
Croteau,Daniel Marcel
Umi 15h9'0"72d17'
Croteau,Emilie
Cas 1h32'42"76d51'
Croteau,Lael
Umi 16h57'13"77d55'
Croteau,Scott
Uma 10h6'46"56d47'
Crothamel,Paul E
Boo 14h12'55"40d18'
Croto,Ann & Willard
Aur 5h8'26"44d57'
Crotty,John
Uma 11h44'59"48d51'
Crotwell,Lauren Lane
Dra 12h48'35"71d34'
Crotwell,Mary Rebecca
Cnv 13h7'15"40d23'
Crouch,Barbara J
Cas 22h56'46"54d13'
Crouch,Clarence (Oscar)
Per 3h21'40"54d54'
Crouch,Don
Uma 11h16'42"52d17'
Crouch,Gregory Edmond
Cet 1h56'14"-1d29'
Crouch,John Franklin
Cma 6h56'13"-19d41'
Crouch,Melissa
Mon 6h54'26"0d5'
Crouch,Sally Foster
Lyr 19h23'12"31d10'
Croughan,Kaitlin
Cmi 7h6'23"5d12'
Crouse,Alison M
And 23h5'59"40d22'
Crouse,David
Aur 7h24'15"41d7'
Crouse,Don J
Lac 22h20'13"50d26'
Crouse,Hillary
Aur 4h38'21"31d11'
Crouse,Larry
And 2h4'51"42d10'
Crouse,Martha Hutchins
Crb 15h40'52"29d36'
Crouse,Yolanda Ann
Lyn 7h50'11"43d40'
Crouse-Martin,Cindy
Del 20h23'47"10d51'
Crout,Roslyn
Boo 15h26'53"34d19'
Crouthamel,Madalene D
Mon 6h55'13"-8d55'
Crouthamel,Willard M
Her 16h27'13"33d48'

Crow & Sarge,Phil
Cet 3h16'57"9d22'
Crow,April Elizabeth
And 0h55'38"38d45'
Crow,Barbara
Crt 11h14'37"-18d26'
Crow,Dana Ewing
Peg 21h56'47"31d43'
Crow,Elizabeth
Cnv 13h8'30"38d18'
Crow,J Kelly
Cmi 7h15'57"5d26'
Crow,Janice M
Mon 7h14'42"-10d14'
Crow,Lauren
Peg 22h14'56"31d16'
Crow,Nancy Lee
Com 13h10'1"28d35'
Crow,Penny
Aql 19h57'57"1d42'
Crow,Suzanne
Lyr 19h21'31"42d11'
Crow,Thomas
Cet 0h28'24"-7d24'
Crow,W Earl
Aur 6h3'50"31d10'
Crowden,David F
Per 2h56'57"54d38'
Crowder,Elizabeth Paige
Vul 21h2'15"27d33'
Crowder,Ella Fairjean
Del 20h38'18"10d45'
Crowder,LaVaughn
Mon 7h5'11"-1d36'
Crowe,Charles "Chuck"
Aql 20h3'40"1d3'
Crowe,Jerry
Aur 5h25'12"38d58'
Crowe,Joshua James
Per 4h24'1"51d15'
Crowe,Keith
Ser 18h7'1"-14d6'
Crowe,Kelly Ann
Peg 22h38'35"11d9'
Crowe,Matthew
Leo 10h44'0"15d20'
Crowe,Noel
Lac 22h32'45"52d41'
Crowe,Phyllis Parker
And 23h23'34"44d30'
Crowe,Russell
Lac 22h51'30"52d46'
Crowe,Sarah Nicole
Aqr 21h37'25"-1d26'
Crowe,Tony R
Cep 23h13'0"64d17'
Crowe,Victoria
Peg 22h32'48"27d40'
Crowell,Bob
Aur 5h0'12"47d0'
Crowell,Don & Ellen
Eri 2h53'52"-2d49'
Crowell,Elijah Joseph
Cet 1h59'24"-10d28'
Crowell,Josh
Boo 15h15'47"41d11'
Crowell,Kimberley Marie
Cas 0h40'25"64d52'
Crowell-Amor Vincit Omnia,C A W
Gem 6h44'15"31d32'
Crowells,The
Umi 14h21'51"70d44'
Crowhurst,Evelyn J
And 1h32'22"39d43'
Crowley Star,The H Ward
Boo 13h35'0"21d35'
Crowley,Alan F & Elizabeth B
Ori 5h50'0"15d6'
Crowley,Allen C
Aur 6h6'59"32d11'
Crowley,Anna Concetta
Dra 16h14'13"68d13'
Crowley,Candy
Ori 5h55'20"15d48'
Crowley,Carrie
Crb 16h18'34"28d58'
Crowley,Daniela Nicole
And 0h55'57"37d22'
Crowley,Darrel
Her 17h4'48"43d46'
Crowley,Debra Lynn Alexandria
Mon 6h53'54"-8d56'
Crowley,Denise L
Lyr 18h59'54"47d31'
Crowley,Ferne
Sex 10h30'30"2d20'
Crowley,For Eternity Daniel J
Cep 0h1'30"70d9'
Crowley,Isabel Sally
Uma 10h6'31"48d25'
Crowley,Xiao Yan Zhu- John
Lyr 19h19'50"42d54'
Crown Properties,Ltd
Aql 18h58'45"7d35'
Crown,Kristin
Cas 0h22'51"60d16'
Crown,Tim
Her 16h41'49"23d13'
Crowned One
Crb 15h28'0"31d30'
Crownhart,Margi R
Peg 21h45'0"23d37'
Crowningshield,Keith
Her 17h0'44"28d33'
Crowninshield,Tracey
Ori 5h17'35"13d1'
Crownover,C V
Cyg 20h54'48"39d33'
Crowson,Hugh The Stellar
Hya 8h27'54"1d35'
Crowther,Elwood James
Her 15h51'29"46d49'
Crowther,Eric Edmund
Cep 23h3'47"70d25'
Croxall,Gilbert
Cep 23h11'60"64d60'
Croy,Melanie Ann
Sct 18h42'42"-7d21'
Croydon (Coeur d' Hélène),Robert
Gem 6h55'55"17d18'
Crozier,Barry
Lyn 8h16'0"41d32'
Crozier,Cassandra
Cas 1h15'0"65d36'
Crozier,Timothy
Aur 5h2'1"42d6'
Cruce,Cassy
Cyg 20h20'26"38d58'
Cruddas,USMC,Capt Steven M
Aql 19h5'46"15d6'
Crudo,O & O,Paul E
Umi 14h53'0"71d24'
Cruger,Pegg Coffey
Cma 7h52'0"10d12'
Cruickshank,Doris Jean
And 0h49'0"37d4'
Cruickshank,Hanna
Tau 4h16'36"23d2'
Cruickshank,Lenore "Lee"
Lyr 18h29'56"38d15'
Cruikshanks,Lisa
Tau 4h18'31"20d1'
Cruisin-Chris
Umi 15h48'1"78d32'
Crull,Robert Andrew
Ser 15h55'47"1d5'
Cruly,Theresa
Boo 15h6'38"8d18'
Crum III,Charles Clinton
Aur 5h0'45"50d57'
Crum,Cory
Ser 15h19'26"0d55'
Crum,Deanna L
Hya 9h19'38"2d26'
Crum,Jimmy
Boo 15h0'16"51d16'
Crum,Kirby
Aql 20h3'30"4d54'
Crumley,Chelsea
Del 20h12'25"9d59'
Crumley,Justin
Aql 19h1'28"-6d33'
Crumley,Michael
Dra 12h13'58"64d10'
Crumley,Stan
Ori 4h41'1"14d26'
Crumley,Tessa
Vul 20h25'57"28d14'
Crump,Amanda Brook
Uma 10h2'44"62d9'
Crump,Elmo
Mon 7h3'14"-1d38'
Crump,Erika Michele
And 0h47'57"38d33'
Crump,Kimberly R
Eri 4h12'17"-12d49'
Crump,Walter E
Dra 11h46'29"74d53'
Crumpton,Kalina Marie
Lyr 18h50'41"42d42'
Crunchy & Cicci
Lup 15h18'17"-42d41'
Crupain,Helen
Mon 7h6'36"0d14'
Crusader Baby's Lallabee
Uma 10h40'22"48d54'
Crusan,Jerald S
Hya 8h19'53"2d43'
Crusco III,Lewis J
Per 3h5'33"50d8'
Cruse,Debra Marie
Tri 2h7'1"33d33'
Cruse,Frances Moore
Ori 5h17'56"12d33'
Cruser-Podawiltz,Dr des Anges
Uma 10h11'1"48d15'
Crush,Michael & Mary
Cyg 20h0'1"30d4'
Crusham,Kristina Kay
And 23h17'42"40d17'
Crusius,Sara
Eri 4h10'36"-14d50'
Crutcher,Robert George
Uma 10h17'51"50d43'
Crutchfield,Charles & Laurie
Lyn 7h37'0"51d1'
Crutchfield,Eric
Cyg 21h7'35"30d40'
Crute,Gillian
And 2h8'43"43d2'
Cruver,Suzanne L
Oph 17h9'20"7d57'B
Cruz III,Luis Antonio
Aql 19h59'22"10d55'
Cruz,Addy Ismary
And 23h37'17"42d51'
Cruz,Armida Santa
Ser 17h32'58"-11d8'
Cruz,Carlos
Ori 5h55'1"8d59'
Cruz,Elena
Cam 4h29'1"68d30'
Cruz,Jerriann Cecil Juliana Susico
Aql 19h24'48"-8d29'
Cruz,Max
Lmi 10h38'1"26d28'
Cruz,Princess Cindy
And 0h16'22"39d8'
Cruz,Roberto
Ori 5h56'15"16d32'
Cruz,Rory Jay
Her 17h28'34"20d45'
Cruz,Teresa
Mon 6h9'52"-10d12'
Cruz-Cody,Marcos Ian Dela
Oph 18h1'1"12d36'
Cruz-Nicastro, Christine
Cas 0h30'1"58d32'
Cruzburger & Little Giant Star,The
Cyg 19h41'56"31d32'
Cruzdom
Boo 14h34'29"41d17'
Cruze,Amy
Cyg 19h40'1"42d53'
Cruzen,Marcelline Mary
Sge 19h53'0"18d48'
Cryan,Barbara Jean
Cyg 21h32'54"40d60'
Cryder,Bethany L
Ari 3h25'57"20d12'
Cryder,Warner Alden
Lyr 18h49'43"33d53'
Cryderman,Walter
Dra 16h42'23"67d47'
Cryer,David John
Dra 16h58'44"67d52'
Crysler,Mary Ann
Vul 19h20'35"26d46'
Crystal
Cas 1h24'40"60d20'
Crystal
Tau 4h34'26"30d10'
Crystal
And 2h22'14"42d20'
Crystal
Umi 12h50'1"53d5l'
Crystal
Com 12h1'37"25d37'
Crystal
Oph 18h7'31"12d40'
Crystal Aimee's Wishes and Dreams
Cep 20h45'27"60d58'
Crystal Ann
Peg 23h2'1"22d21'
Crystal Corinne
Lyn 8h0'0"51d24'
Crystal Dawn
Vul 21h2'38"20d10'
Crystal Hearts Of Passion-Botany 93
Del 20h19'57"10d34'
Crystal Leaf,The
Cet 2h33'36"9d44'
Crystal Lynn
Cas 1h54'1"70d55'
Crystal Lynn
Cas 23h14'0"61d23'
Crystal Marie O'C
Lyr 19h23'35"31d28'
Crystal Rainbow
Vul 19h15'54"24d46'
Crystal Starr
And 2h5'27"40d7'
Crystal Starr
Sex 10h6'31"5d27'
Crystal Symphony
And 0h43'29"40d25'
Crystal Water Works
Dra 15h14'24"63d18'
Crystal,Sean Owen
Peg 22h59'55"8d25'
Crystaleyes
Peg 22h31'42"21d40'
Crys'n Dale's Hobo
Aur 6h39'25"38d8'
CS
Sge 20h5'0"16d29'
CS111MR
Lac 22h35'13"55d13'
Csardas
Lyn 7h35'20"44d2'
Csaszar,Angie
Tri 2h23'33"35d31'
Csehan,Harald u Heike
Cmi 7h19'35"8d54'
Csellak,Jr,William Robert
Boo 14h6'28"52d51'
Csendes,Michael
Cnv 12h46'40"51d27'
CSES 93
Peg 22h29'0"31d45'
Csiki,Katherine
Com 12h59'41"22d18'
Csonka,Ruth
Mon 6h59'46"-5d58'
CSS Celestial Beacon
Cam 6h3'35"68d20'
CT & KZ
Mon 6h34'1"-6d1'
CTG,FW,LJ
Cep 21h36'54"58d44'
CTK Class of '93
Cyg 20h56'11"40d27'
CTN(Pup-Pup)
Cep 20h23'0"60d50'
CU Being Glad You Care
Uma 9h22'10"54d11'
Cuadra,Lindsay Louise
Cas 23h27'18"61d45'
Cuan,Rachel
Cyg 19h44'33"30d19'
Cuanto Mé Quieres
Aql 19h51'44"12d13'
Cub Scout Pack 257
Uma 10h59'16"41d3'
Cubbedge,Jill "Beaner"
Psc 1h18'53"22d18'
Cubbie
Aql 19h54'49"11d25'
Cubby
Uma 11h23'0"48d56'
Cubero,Stephanie
Lyr 18h51'17"31d0'
Cubik,Clement
Lac 22h23'45"56d44'
Cuboose XXV
Dra 17h10'39"52d10'
Cuccaro,Ali Craig
And 0h8'40"28d43'
Cucci,Jr,Vincent Eugene
Cam 3h30'12"60d5'
Cuccia,Tony
Dra 18h37'32"58d30'
Cuccia,Vicki
And 0h22'25"30d13'
Cucciolo
Hor 3h18'33"-48d51'
Cucciolo Di Umo Saggio
Ori 6h4'52"8d22'
Cucciolotta,Laura
Vel 9h43'29"-43d29'
Cuccuru,Daniela Francesca
Hor 3h13'23"-46d27'
Cuche,Benjamin
Uma 8h44'43"54d8'
Cucibi
Cam 5h36'1"68d52'
Cucinotta,Alyxandra
Uma 11h35'32"48d10'
Cucinotta,Helen
Uma 8h39'1"52d10'
Cuda,Rozanne
Cas 0h41'14"62d38'
Cudby,Audrey
Cyg 20h23'44"41d31'
Cudd,Jr,John Franklin
Aur 6h12'1"45d10'
Cuddeford,Vanessa
Cyg 20h1'59"30d51'
Cuddemi,Jr,Paul
Lac 22h22'1"54d54'
Cuddlebearus Mikars
Sge 19h53'1"16d47'
Cuddles~N~Snuggles
Cyg 21h26'51"37d44'
Cuddly Counsel
Her 17h1'12"40d8'
Cuddy,Mel
Hya 8h42'42"3d47'
Cude,John William
Cyg 21h18'20"37d50'
Cude,Ruth L
Cyg 21h11'49"38d15'
Cudequest,Brandon William
Cep 5h58'18"85d49'
Cudo,Rachel Lyn
Lyr 18h40'56"27d42'
Cuellar,Fred
Cnv 12h11'38"47d59'
Cuellar,Heriberto "Beto"
Aql 19h55'22"8d5'
Cuenca,Hector Felix
Psc 1h20'0"22d16'
Cuerrier,Annie Yolande
Umi 13h44'22"76d43'
Cuestas,Margaret
Peg 23h25'36"33d8'
Cueva,Lucille Santaella
Mon 7h38'58"-3d5'
Cuevas,Bertha A
Eri 3h42'24"-16d40'
Cuff,Gail
Her 16h50'15"33d23'
Cuffe,Dave
Hya 9h10'1"3d45'
Cugier
Vul 19h47'20"22d55'
Cugini,Joseph Nicholas
Cep 21h59'0"58d54'
Cugno,Albert Nicholas
Uma 12h33'20"56d37'
Cui,J Victor P
Sco 17h55'45"-40d34'
Cui,Rena R
Psc 1h42'41"27d48'
Cuirlino,Healer JoAnn
Aur 5h26'14"38d32'
Cuivienen
Lac 22h23'20"55d51'
CujoBrowne
Dra 11h49'36"68d10'
Cukier,S
Cam 5h42'55"60d24'
Cukierman,Annabelle
Lac 22h23'56"53d41'A
Culberson,Cassius
Aql 18h54'41"8d51'
Culbertson,Lori
Umi 14h10'1"66d12'
Culbertson,Peter Iten
Sex 9h54'19"-0d15'
Culbertson,Robert
Ser 15h32'9"14d18'
Culbertson,Sam
Dra 16h19'22"61d22'
Culbertson,Teren WA
Cnv 12h30'14"50d10'
Culby,Ryan
Cnv 13h57'39"38d15'
Culhane,Daniel J
Her 16h32'39"34d53'
Culhane,Rosemarie
And 0h53'32"37d28'
Culian,Andreea
And 0h17'1"35d7'
Cull,Robert (Bob)
Ori 5h33'29"-0d44'
Cullen Serendipity, Elaine A
Uma 9h40'59"52d3'
Cullen,Andrew David
Cnv 12h42'1"34d3'
Cullen,Candice
Cnv 12h17'36"44d30'
Cullen,Emily M
Lyr 18h42'16"30d14'
Cullen,Fo
And 2h33'29"40d50'
Cullen,Ian Jacob
Cep 23h17'22"68d35'
Cullen,Jannine
Peg 23h48'50"28d1'
Cullen,Kevin Patrick
Lyn 8h1'54"40d30'
Cullen,Kevin Patrick
Aur 5h19'27"42d27'
Cullen,Kimberly Patricia
Cas 1h19'18"71d11'
Cullen,Mary Catherine
And 1h43'27"40d32'
Cullen,Michael Cranmer
Peg 21h28'15"3d12'
Cullen,Oonagh
Sco 16h13'42"-22d22'
Cullen,Thomas Kenneth
Cap 21h1'1"-26d36'
Cullen,Tricia
Cnc 8h40'34"18d40'
Cullers,Suzanne Gerette
Dra 17h41'29"64d19'
Cullerton 35th Anniv, Emilie & Walter
Peg 21h50'42"31d34'
Culleton,Dr Vas
Aur 5h7'31"41d7'
Culley,Grant "Squire"
Cep 21h18'46"55d39'
Culley,Helen
Lyn 8h52'52"43d8'
Culley,Stephen James
Ori 6h4'41"1d16'
Culligan I
Aql 19h39'40"13d53'
Culligan,John
Dra 16h49'20"64d12'
Cullinan Family Star, The
Cam 8h51'60"73d41'
Cullinan,MD,Stephen Austin
Oph 18h19'56"7d20'
Cullinane,Christopher M
Cep 22h38'41"58d55'
Cullinane,Mary
Lyn 8h9'50"45d29'
Cullinane,`Physio' Liam
Per 1h48'27"56d51'
Cullings
Ant 10h39'49"-39d19'
Cullire,Dominick
Boo 13h59'24"8d26'
Cullison,Christopher
Oph 18h41'27"10d5'
Cullison,MD,James P
Ori 5h54'51"15d27'
Culliton,Phaedra Elizabeth
Cas 2h36'39"60d35'
Cullum,William
Mon 6h57'0"-6d20'
Cully
Uma 9h45'0"50d21'
Cully,Robert David
Mon 8h4'8"-6d49'
Cully,Rosemary & Paul
Aql 19h57'42"13d49'
Culme-Seymour (Stickleback),Katie
Umi 16h8'1"70d8'
Culp "Culpernicus",Roy Edward
Cep 21h47'27"65d19'
Culp,Christopher Lawrence
Ser 15h57'19"-2d16'
Culp,Dagon Jay
Dra 16h25'56"62d7'
Culp,David V
Aql 19h58'43"8d40'
Culp,Elisabeth
And 23h2'44"45d36'
Culp,Harvey Carson
Eri 4h9'25"-14d1'
Culp,Jewell R
Uma 10h55'54"34d6'
Culp,Karen Lynette
Mon 6h23'32"10d40'
Culp,Nathan Andrew
Hya 8h10'27"5d56'
Culp,Peg
Cam 13h43'27"81d33'
Culp,Robin
Leo 10h40'43"16d12'
Culpepper,David Ray
Lyr 18h50'12"40d19'
Culpepper,Dillon Sinclair
Her 18h16'1"30d14'
Culpepper,Edward James
Her 16h53'19"33d10'
Culpepper,John Calvin
Per 3h7'0"40d1'
Culpepper,Ronald Louis
Aur 5h19'42"46d11'
Culpitt,Sandy
Lyr 18h55'19"38d17'
Culpitt,Shirley
Cyg 21h58'0"52d34'
Culven,Richard Arlen
Her 16h21'0"20d35'
Culver,Ann
Peg 22h47'36"18d46'
Culver,Bobi Jo
Cyg 21h17'26"28d40'
Culver,Jay Charles
Aql 19h59'0"15d24'
Culver,Jeff L
Vul 18h56'36"24d26'
Culver,John David
Her 18h55'44"12d38'
Culver,Melissa
Lyn 8h16'14"33d54'
Culver,Michael
Aur 6h12'21"46d36'
Culver,Sr,Julian Frank
Aur 6h5'39"37d46'
Culver,Victor W & David J Oliphant
Eri 3h34'43"-6d24'
Culverhouse,Christina Marie
Aql 20h3'37"6d52'
Culverhouse,Gay
Boo 13h40'20"20d2'
Culverhouse,Sr,Hugh F
Lac 22h31'0"41d11'
Cum Laude,Jeffrey
Dra 14h0'37"68d50'
cum SCHINDLING ad astra
Ari 1h49'54"22d56'
Cumba,Vianney
Ari 2h33'57"30d13'
Cumber,Courtney
Lyr 18h16'1"32d57'
Cumberland,Adam Arthur
Aql 19h0'1"-0d10'
Cumberland,Elizabeth Valentine
Peg 22h3'1"10d45'
Cumberland,J B
Leo 11h51'0"21d50'
Cumi,Talitha
Lyn 8h18'30"40d2'
Cumins,Jorday
Aql 19h30'40"1d6'
Cumiskey,Susan Marie
And 0h4'43"47d45'
Cumming,Heather Faith
And 23h25'59"43d8'
Cumming,Jane
Peg 0h3'36"31d13'
Cumming,Judith Anne
And 23h22'20"51d13'
Cumming,Kendra Marie
Cap 20h58'1"-26d53'
Cumming,Robert Gordon
Cep 20h49'53"61d51'
Cummings
Her 17h11'59"29d23'
Cummings Our Star Forever,Lelia O
And 0h6'23"43d50'
Cummings' Hannibal Heaven,David W
Aur 6h2'26"32d4'
Cummings,"Myron" Waiolu
Sex 9h57'31"-5d1'
Cummings,A Wallace
Lac 22h33'33"50d16'
Cummings,Alexandra Beth
Gem 7h31'58"35d10'
Cummings,Amy
And 23h35'55"49d50'
Cummings,Brian Lee
Lyr 18h51'17"31d27'
Cummings,Claudia
Peg 22h26'15"31d38'
Cummings,Colleen
Lyr 19h22'13"42d21'
Cummings,Corneilous G
Lib 15h58'35"-7d58'
Cummings,David Alexander
Cep 5h8'48"86d19'
Cummings,Debbie L
Del 20h13'30"10d34'
Cummings,Gary Lee
Her 18h8'39"38d38'
Cummings,George Elliott
Boo 15h8'0"20d8'
Cummings,James
Cet 3h10'42"0d10'
Cummings,Jean
Cas 0h50'33"60d31'
Cummings,Jill D
Cas 0h52'56"62d54'
Cummings,John Andrew
Lac 22h38'41"52d49'
Cummings,Joy
Vul 20h15'43"23d52'
Cummings,Jr,Lewis Balcom
Oph 17h19'19"-22d56'
Cummings,Merrilyn N
Peg 2h28'42"21d25'
Cummings,Molly Marie
Mon 6h54'54"8d55'
Cummings,Richard & Dorothy
Oph 18h18'22"8d52'
Cummings,Sharon Lane
Aur 5h20'18"38d58'
Cummings,William C
Aur 7h3'26"38d37'
Cummins,Allison Kendall
Ori 5h15'1"15d33'
Cummins,Barbara
Peg 22h2'54"21d53'
Cummins,Donald Hugh
Uma 10h45'51"44d53'
Cummins,Elizabeth Aline
Hya 9h32'13"5d15'
Cummins,Gary Lee
Mon 7h18'21"-6d54'
Cummins,Kelly
Sge 19h55'32"16d26'
Cummins,Lynn
Aur 5h18'0"49d22'
Cumpton,Joan R
Cam 3h17'1"61d3'
Cundall,Courtney Jade
Cet 1h42'29"-4d49'
Cundary My Shining Star,Melissa
Cap 21h37'42"-22d38'
Cundey,Jr,Dr Paul E
Cet 3h5'54"2d34'
Cundiff Awilda
Lyn 8h15'54"38d24'
Cundiff,Kenneth Michael
Dra 17h19'11"67d34'
Cuneonotte
And 23h21'33"50d17'
Cunetta,John
Her 16h12'1"50d28'
Cunha,Fred
Ori 5h32'15"-8d36'
Cunha,Steve
Cet 1h52'27"0d32'
Cuni,Josep
Dra 17h17'15"60d57'
Cunic-25 Years,Jack & Marion
Cyg 19h12'26"48d50'
Cuningham,Mathieu Louis André
Aqr 21h39'21"0d53'
Cunneen,Paddy
Cas 0h48'11"60d19'
Cunning,Celine & Daniel
Mon 6h55'38"7d35'
Cunningham DWC,Dennis Walter
Cam 7h54'28"60d8'

Cunningham Star,The Felicity & Randall
Mon 7h9'10"-5d28'
Cunningham,Aimee Nicole
Cyg 20h31'48"40d14'
Cunningham,Andrew Robert
Ori 4h53'44"5d34'
Cunningham,Annette McKnight
And 1h17'27"36d41'
Cunningham,Bridget Anne
And 23h2'52"51d58'
Cunningham,Catherine Marie
And 23h36'0"48d57'
Cunningham,Cathy Loves Michael Melia
Sge 20h2'56"16d18'
Cunningham,Cindy
Equ 21h7'1"10d18'
Cunningham,Claude C
Ori 6h17'16"7d47'
Cunningham,David & Janis
Crb 16h8'0"31d53'
Cunningham,Ed
Aur 6h26'58"32d31'
Cunningham,Eileen Barrett
And 2h22'45"39d59'
Cunningham,Freda J
Cas 1h59'41"61d46'
Cunningham,Gina Zanotti
Aur 7h25'1"40d2'
Cunningham,Holly
Lyr 18h59'30"29d45'
Cunningham,Jackie
Mon 7h50'12"-3d16'
Cunningham,James Kieran
Lyr 19h21'1"30d39'
Cunningham,Jeneen
Peg 22h37'53"35d6'
Cunningham,Jill
Mon 6h26'10"-6d29'
Cunningham,JoAnne
Ori 5h57'0"10d48'
Cunningham,Joanne
Vul 19h57'23"25d11'
Cunningham,Joseph James
Umi 13h54'27"70d55'
Cunningham,Jr, Cornelius A
Cet 2h25'26"0d36'
Cunningham,Marjorie H
Aur 4h38'46"31d26'
Cunningham,Mark Dante
Lac 22h0'41"38d22'
Cunningham,Mary E
Peg 21h53'44"28d59'
Cunningham,Michael E
And 0h57'18"38d02'
Cunningham,Mrs Jean (Mike)
Eri 3h49'27"-7d19'
Cunningham,Murray
Uma 9h24'53"62d17'
Cunningham,My Guiding Light,Michael
Lac 22h20'1"51d16'
Cunningham,Patricia C
Aql 19h16'46"15d44'
Cunningham,Patrick
Cet 0h43'38"0d15'
Cunningham,Richard Anthony
Hya 9h1'21"1d20'
Cunningham,Rodney
Umi 16h13'24"80d34'
Cunningham,Roger & Nancy
Cmi 7h32'0"0d43'
Cunningham,Samantha Lee
Peg 22h42'28"33d18'
Cunningham,Sister Agnes
And 2h29'36"45d29'
Cunningham,Thomas
Aql 19h24'20"-1d24'
Cunningham,Thomas C
Her 16h15'14"8d25'
Cunningham-Reid, Michael
Cnc 8h32'15"8d19'

Cunnington,Doreen
Umi 16h15'50"88d51'
Cunnngham-Bussell,Kiki
Peg 1h55'52"30d32'
Cuntz,Christian F
Lyr 19h22'16"30d36'
Cuomo of Brooklyn,Joe
Her 18h4'48"30d25'
Cuomo,Giuseppina
Per 3h32'48"50d33'
Cuomo,Matilda Raffa
Lyr 19h12'34"41d5'
Cuomo,Stacey Taylor
Cyg 21h20'31"40d42'
Cupani,John J
Oph 18h16'25"1d34'
Cupelli,Alessandra
Cet 2h55'1"-0d11'
Cupid II
Sge 19h55'22"16d11'
Cupid's Couple
Cet 2h58'59"1d5'
Cupit,Zachary
Hya 9h9'47"0d28'
Cupoli,Christopher Sean
Cnv 12h20'30"43d20'
Cupp,Robert Justin
Dra 17h24'0"72d25'
Cupp,The Star of Sarah Jo
Uma 11h41'28"51d25'
Cupp,The Star of Austin Grant
Uma 11h31'42"52d23'
Cupper
Aur 6h0'52"50d21'
Cupplo,Sherri
Cet 3h10'23"1d5'
Curator,The
Crb 16h13'44"33d26'
Curbaille,Emanuel
Ser 15h54'37"18d13'
Curbo,Bruce
Her 16h56'40"28d34'
Curci,Erica
Lyn 7h27'11"42d42'
Curcio,Anthony
Lyn 8h32'8"35d10'
Curcio,Dr Raymond
Oph 17h11'27"-23d8'
Curcio,Jeffery Walker
Dra 20h0'14"74d34'
Curcio,Noelle Gerard
Dra 13h8'32"68d22'
Curcione,James Andrew
Her 17h22'10"40d5'
Curcuruto,James
Per 2h51'0"41d4'
Curd,Mary Sue
Mon 6h53'47"10d7'
Cure,Sonya
And 22h58'59"50d11'
Curet,Veronica
Lyn 7h38'0"37d28'
Curetti,Paula & Fabrizio
Cam 8h21'28"78d51'
Curlew
Aur 6h32'23"34d3'
Curiel,Miss Linda
Cas 1h32'18"61d29'
Curl's Lovestar
Cyg 20h17'35"38d40'
Curl,Evelyn
Cas 3h4'30"67d55'
Curl,Ryan Matthew
Lac 22h35'0"50d7'
Curler,Brian Alan "Spud"
Cep 22h26'28"70d30'
Curley Family Star
And 2h24'12"39d28'
Curley,Dwight C
Aql 19h53'43"-1d28'
Curley,James D
Per 3h24'18"40d18'
Curley,Jr,Michael John
Per 1h50'56"54d3'

Curley,Michael John
Per 1h56'38"56d34'
Curley,Susan Reardon
And 0h54'56"36d28'
Curling,Rebecca
Mon 7h3'50"5d11'
Curly
Aqr 22h58'55"-10d1'
Curmudgeon
Uma 9h1'53"70d5'
Curmudgeon
Cam 8h27'34"83d53'
Curnayn,Kevin M
Her 16h49'39"47d58'
Curr CSN,Tammy Lyn
Vul 19h0'11"22d15'
Curran,Christopher A
Lib 15h8'33"-8d41'
Curran,Connor Ferguson
Dra 11h48'27"71d13'
Curran,Ed
Aur 5h42'0"54d36'
Curran,Jack
Aur 6h56'21"37d47'
Curran,Kendra Erin
Cyg 20h3'48"40d20'
Curran,Michael Joseph
Lmi 9h29'16"38d31'
Curran,M9h29'16"38d31'
Lyr 19h15'49"26d16'
Curran,Robert Thomas
Per 1h42'52"53d38'
Curran,Sandra L
Aqr 22h45'26"-6d52'
Currano,Al
Uma 9h30'42"55d57'
Currans
Aql 19h42'28"11d56'
Currence,Anna Nicole
And 2h1'37"47d7'
Currey,Kenneth & Dorothy
Cyg 21h8'55"38d0'
Currey,Ralph B
Cet 1h19'31"-10d41'
Currid,J P
Lyn 8h2'44"47d36'
Currie,"Fire Top" Trent Donald
Cnc 7h56'47"10d16'
Currie,Dillon Monroe
Gem 7h16'24"21d27'
Currie,Dilys
Cas 2h41'39"63d55'
Currie,Hana & Dave
Cyg 21h52'0"53d11'
Currie,Kevin
Dra 11h40'33"72d26'
Currie,Le
Lyr 18h41'20"41d52'
Currie,Lloyd David
Cnv 13h59'15"40d31'
Currie,Pamela E
Cas 1h22'0"52d42'
Currie,Philip
Ori 5h57'28"16d29'
Currie,Sean Robert
Her 16h24'16"26d37'
Currie,Thomas Trimble
Cet 0h54'32"-0d23'
Currier,Jr,John
Dra 17h41'59"60d27'
Currier,Laura Ann
Eri 3h28'16"-5d29'
Currier,Philip Boynton
Leo 9h33'31"14d26'
Currier,Stephen William
Hya 9h1'15"5d9'
Currin,David L Cathlic Social Services Chm BD
Her 17h53'47"38d16'
Currin,George Spencer
Cmi 7h54'57"3d58'
Currin-Happy Birthday, Becky Lynn
Eri 4h56'0"-5d54'

Currinus 5-6-64, Susanis Rachelia
Cas 1h8'1"61d7'
Curry,Alpha Omega Alpha
Oph 16h49'1"11d16'
Curry,Bo
Peg 22h23'49"20d6'
Curry,Boykin & Betty
Cra 16h51'1"52d24'
Curry,Dave
Tri 2h2'0"30d21'
Curry,Domico
Per 2h23'49"55d17'
Curry,Doug & Rosie
Peg 23h2'1"8d30'
Curry,E Lou
Uma 13h49'34"51d21'
Curry,Faye
Lyn 6h27'30"56d11'
Curry,Frank
Per 2h20'19"55d22'
Curry,George William
Tau 5h49'24"26d48'
Curry,Janette
Crb 13h8'0"20d52'
Curry,Jim & Kelly
Uma 9h23'1"59d4'
Curry,Juliet Faith
Mon 8h3'18"-4d14'
Curry,Krista Michele
Peg 22h21'46"35d26'
Curry,Lynda Perkins
Cyg 20h17'29"41d24'
Curry,Ronald R
Per 3h0'51"41d6'
Curry,S H
Ori 5h52'23"10d49'
Curry,Virginia
Del 20h36'57"20d8'
Curry,Yannick Sutherland Marie
Aur 6h25'45"37d51'
Currys,The
Cep 23h16'26"65d24'
Cursio,Michael
Cep 22h20'39"65d35'
Curt
Aur 6h25'46"37d22'
Curt & Gail Skylight
Mon 6h55'34"-1d14'
Curt Duff's Revenge
Aur 6h42'0"38d52'
Curtes,Jeanie
Uma 11h19'33"58d36'
Curti,Annamaria
Lup 15h4'59"-38d17'
Curtin,Jason
Aur 6h1'38"38d16'
Curtin,Michael Barnett
Lmi 10h34'24"28d16'
Curtis
Her 17h54'43"40d27'
Curtis
Peg 0h8'56"13d32'
Curtis
Aql 19h51'1"13d3'
Curtis 12-18-74
Uma 11h44'18"50d32'
Curtis School
Hya 9h5'27"1d14'
Curtis Screw Co
Aur 5h1'1"48d37'
Curtis Strawboss-14
Ori 5h55'34"13d26'
Curtis,Amanda Dianne
Com 12h0'33"20d4'
Curtis,Anthony Samuel
Cep 1h28'0"80d29'
Curtis,Austin Allen
Cet 2h37'36"-6d38'
Curtis,Bruni & Craig
Cep 20h13'24"60d4'
Curtis,Dorothy & Roy
Uma 11h3'18"38d8'

Curtis,Edward V-K
Aur 6h23'47"35d49'
Curtis,Gamma Rho Tau Sigma For Fred
Cma 6h59'14"-19d18'
Curtis,Gary
Hya 8h14'35"0d19'
Curtis,Jack
Dra 15h53'49"60d52'
Curtis,Jamie Susan
Cas 23h39'1"61d47'
Curtis,Janine
Aql 18h57'15"10d24'
Curtis,Jay Chamberlin
Aur 5h6'48"38d39'
Curtis,JoAnne H
Lyn 9h5'55"42d58'
Curtis,Kelly Ann
Peg 21h39'32"28d15'
Curtis,Larry N
Cep 20h40'48"70d49'
Curtis,Louise Curtis
Mon 6h24'1"1d60'
Curtis,Mary
And 1h14'43"40d2'
Curtis,Michael Glen
Cet 0h28'59"-1d32'
Curtis,Mrs Dawn Tonetta
Lyr 18h36'10"27d54'
Curtis,Patricia Rae
Peg 22h6'51"27d38'
Curtis,Paul Hamilton
Ori 5h53'60"13d2'
Curtis,Paul Ross
Dra 16h52'10"67d43'
Curtis,Pauline K
Cas 1h16'16"65d18'
Curtis,Sarah Beth
Vul 19h58'54"25d40'
Curtis,Scheree Michelle
Cep 20h13'40"60d31'
Curtis,Sion Severin
Cam 3h35'51"62d5'
Curtis,Sue
Cyg 21h18'15"38d7'
Curtis,Thomas Luke
Her 17h10'19"22d10'
Curtis,Tony
Aqr 23h3'35"-18d55'
Curtis,Warren Alexander
Lib 14h21'13"-10d53'
Curtis,Zoë Reneé
Uma 9h15'0"43d30'
Curtiss,John
Her 16h29'1"33d13'
CurtMauro
Ser 15h54'0"18d16'
Curto Star,Robert
Ori 5h56'52"11d32'
Curto-1924,Mario Dominick
Aur 6h5'0"45d47'
Curts,Tena Marie
Eri 2h58'16"-14d33'
Curylo,Marie A
And 1h12'47"35d44'
Curylo,Ronald P
Her 16h10'12"24d15'
Cusaac,Yolanda J W
Aur 7h13'54"37d25'
Cusack Family,The Thomas
Uma 8h32'15"70d34'
Cusack,David Christopher
Dra 20h8'13"63d38'
Cushing,Charles William
Cep 2h30'0"78d32'
Cushing,Susan Lee
Lyn 6h35'12"54d12'
Cushman Light,The
Cyg 19h47'41"29d38'
Cushy
Ori 5h59'34"20d43'
Cusic,Jerome
Aur 7h7'0"40d1'
Cusic,Patricia Bender
Com 12h11'59"20d33'

Cusick III,Laurence F
Oph 18h6'14"12d42'
Cusick,Delores
Cas 23h36'28"60d46'
Cusick,Father Gordy
Aur 5h29'58"38d14'
Cusick,Robert John
Mon 7h15'49"-10d29'
Cusick,Timothy J
Her 16h1'1"20d16'
Cusimano,Dorothea
Umi 16h34'24"76d46'
Cussen,Amy Marie
Boo 15h8'35"30d30'
Cusson,Martine
Cas 0h59'43"75d5'
Custer,April Lynne
Cmi 8h7'48"0d45'
Custer,Daniel Paul
Her 18h7'36"38d38'
Custer,Daniel Paul
Dra 17h9'50"61d24'
Custer,Gregory William
Cyg 21h34'35"41d6'
Custer,Lawrence B
Aql 20h11'51"4d34'
Custer,Trena Jean
Mon 7h55'11"-6d25'
Custis,Donna
Mon 7h2'18"-6d14'
Custodimini
Aur 5h19'59"47d11'
Custodio,Catalina Teresa
Cnc 7h55'1"11d31'
Custodio,Marie Thérèse
Del 20h19'40"10d58'
Cusumano,Eva
Cyg 20h54'17"30d24'
Cusworth,Matthew
Aql 20h11'0"11d11'
Cutajar,Sally C
Lyr 18h17'51"45d34'
Cutajar,Vince
Per 3h56'40"51d31'
Cute Boy Robbie
Aur 6h14'20"30d13'
Cuteamus Puteamus
Uma 10h23'12"53d19'
Cuthbert,Leah
And 2h26'60"41d33'
Cuthbert,Midge
Aql 19h53'22"11d34'
Cuthbertson,Janice
Lyn 8h32'23"42d36'
Cuthbertson,The Only, P Albert Einstein
Lmi 9h43'1"38d53'
Cuthbertson,Tommy
Ori 5h23'15"0d53'
Cutie
Mon 7h9'17"-5d14'
Cutie
And 1h40'20"40d58'
Cutie Pie
Vir 11h59'30"-0d44'
Cutileiro,Pedro
Umi 16h18'39"71d38'
Cutina
And 23h40'1"38d56'
Cutler Star,The Eunie
Peg 23h27'49"15d48'
Cutler,Alexander J
Hya 8h53'23"-6d6'
Cutler,Amy
Lyr 19h2'15"40d9'
Cutler,Angela
Peg 23h20'37"32d1'
Cutler,Caroline N
Lyn 7h6'1"50d11'
Cutler,Joel M
Sex 10h13'32"-2d19'
Cutler,Matthew Adam
Lac 22h10'52"47d2'
Cutler,Samantha Blair
Mon 6h19'1"0d36'

Cutler,Seth Adam
Boo 14h28'16"53d43'
Cutler,Sol
Com 11h56'1"14d14'
Cutler,Yvonne Boman
Eri 3h7'55"-3d38'
Cutlip,Lorie
Mon 7h1'56"-6d16'
Cutolo,Virginia & Louis
Lyr 18h50'37"42d28'
Cutrona,Henry
Hya 8h30'37"0d7'
Cutrone,Angela
Per 3h8'0"41d22'
Cutter,"Another Star", Michael
Oph 17h37'23"-21d54'
Cutter,Alison Marie
And 0h10'42"46d35'
Cutter,Gerald Henry
Ori 6h4'30"7d42'
Cutting,Shaun Joseph
Lyn 7h57'44"36d44'
Cuttitta,3/6/56, Patrick
Per 3h8'39"41d52'
Cuty,Christophe
Uma 12h59'34"54d55'
Cuva,Helen Ann Wood
Umi 16h36'25"78d15'
Cuviello,Susan & John
Cam 5h33'49"65d12'
Cuvilliez,Nathalie Benoit Julian
And 0h27'48"43d18'
Cuz
Dra 19h9'20"71d7'
Cuzi
Cas 0h2'0"58d25'
Cvietusa,Don
Cet 1h33'10"-5d59'
Cvrk,Vilibald
Gem 6h39'51"34d56'
Cwass,Evan "Schnookums"
Hya 9h8'35"5d13'
Cwik,Paul
Tri 2h17'0"31d18'
Cy,Cinta,Kamu,Lesley
Umi 14h35'46"65d48'
Cybelle,Iana
And 1h32'51"37d8'
Cyberian Rhapsody
Umi 15h47'14"70d51'
Cybil
Ari 2h41'53"25d5'
Cybulski,Anna
Cam 5h59'22"70d27'
Cydles
Boo 15h8'23"22d21'
Cygan,Christopher Adam
Cnv 12h30'23"33d14'
Cylinder,Chelsea (Chaia) Evan
Lyn 8h52'40"41d57'
Cylkowski,Diane M
And 0h11'32"35d57'
Cymbre Le
Fri 3h22'30"-14d39'
Cymbre The Royal Fortress Meadow
Peg 21h20'58"22d33'
Cymreig Joey
Cep 22h21'20"63d53'
Cynammonn Marie
Lyr 18h47'38"34d35'
Cynara Marie
Eri 3h41'23"-2d54'
Cyndee
Peg 22h53'0"29d21'
Cyndi
Lyr 18h46'38"42d26'
Cyndi
Uma 10h40'14"59d29'
Cyndi
Lib 15h28'27"-8d14'
Cyndi
Cyg 21h30'54"37d56'

Cyndie
Leo 10h43'0"14d30'
Cynosure
Uma 10h39'0"58d5'
Cynowa,Joshua Michael
Aur 7h9'16"38d40'
Cynthia
Cas 0h42'23"66d26'
Cynthia
And 1h52'44"38d14'
Cynthia
Lac 22h8'32"50d60'
Cynthia
Vul 21h26'49"26d40'
Cynthia
Cas 0h13'49"58d36'
Cynthia
Uma 10h35'13"41d53'
Cynthia & Dennis
Cas 1h7'41"66d5'
Cynthia & Manuel's "Life Light"
Cam 13h3'0"76d48'
Cynthia & Tim's Magic Star
Aur 4h51'10"40d7'
Cynthia Ann
Vul 20h2'25"28d22'
Cynthia Ann
Com 12h26'57"24d4'
Cynthia Anne
And 23h37'14"43d17'
Cynthia Anne
Uma 11h55'18"50d44'
Cynthia Anne's Star
Cas 1h59'1"58d29'
Cynthia Dee
Vul 19h37'0"20d2'
Cynthia Janine
Lyr 19h18'23"40d33'
Cynthia Jo
Uma 10h30'34"59d32'
Cynthia Lee(Forget Me Not)
Uma 8h45'55"47d17'
Cynthia Ma-Hony
Cas 0h1'43"63d58'
Cynthia Marie
Peg 22h30'20"24d39'
Cynthia Marie
Com 12h26'14"26d45'
Cynthia Marie
Mon 6h58'34"8d11'
Cynthia Sue
Uma 9h16'15"51d33'
Cynthia Susan
Cmi 7h25'0"7d48'
Cynthia Y
Mon 7h43'17"-7d24'
Cynthia's Eternal Point of Light
Leo 9h40'43"10d47'
Cynthia's Excalibur
Mon 8h2'0"-5d38'
Cynthia's Hope
Cas 0h6'38"60d50'
Cynthia's Love
Vul 20h19'55"26d7'
Cynthia's Star
Psc 1h26'34"21d36'
Cynthiaura
And 2h25'47"42d34'
Cypher,Kayla Lynn
Lyr 18h58'40"45d48'
Cypresse Crist
Cam 6h19'19"65d2'
Cyprien,Maxwell
Lac 22h20'34"40d2'
Cyr,Denise Colette
And 2h25'40"43d2'
Cyr,Gil & Jackie
Cyg 19h31'43"33d48'
Cyr,John F
Cma 7h12'46"-16d38'
Cyr,Kelsey Lynn
And 0h25'0"41d12'

Cyr,Wilman Henry
Per 2h59'37"32d58'
Cyr,Edith et Camil
Lmi 10h54'34"28d31'
Cyr-Weber,Sebastian Galen
Leo 9h54'30"12d4'
Cyr-with Love,Mom, Lynda"Sweet Pea"
Cas 0h3'1"60d6'
Cyra
Lyn 9h12'45"40d33'
Cyrenne,Victoria
And 1h49'27"47d23'
Cyril
Sge 19h55'49"19d32'
Cyril
Dra 16h19'53"62d1'
Cyril
Umi 15h15'0"70d20'
Cyril Albert
Uma 10h3'46"47d56'
Cyrill Fabian
Lyn 7h4'43"59d2'
Cyrille
Per 2h59'24"32d46'B
Cyrille et Mumu
Uma 11h6'41"50d22'
Cyriulus
Cyg 21h59'37"50d24'
Cyrulik's Star,Jackie
Com 12h29'35"18d50'
Cyrus,Billy Ray
Her 16h1'35"20d10'
Cyrus:Dreams Come True ,Billy Ray
Her 18h14'44"37d38'
Cytacki,Alfred W
Cep 22h13'37"55d44'
Cytherea
Sge 20h1'21"20d35'
Cytulik,Michael B & Brenda L
Cyg 19h22'46"28d57'
Cyvia Marlene,The
Cas 23h6'30"58d60'
Czachor,Lance
Cet 2h36'1"1d30'
Czachur,Heather Reneé
Vul 20h15'23"23d43'
Czajkowski,Skyler Gunnar
Oph 16h49'58"11d50'
Czak,Werner
Hya 8h14'0"1d8'
Czapela,Theodor
Her 17h19'30"42d24'
Czapla,Jeanette
Uma 11h23'22"62d54'
Czapla,Joe
Her 18h1'53"37d42'
Czaplicki,Alison Patricia
Uma 9h47'41"42d54'
Czaplicki,Amanda Sonya
And 1h27'56"38d24'
Czapski,Tony & Trudy
Uma 11h54'35"56d3'
Czarnecki,S Thomas
Hya 8h15'51"2d59'
Czarnecki-"The Czar Star",Alfred
Leo 10h27'39"28d2'
Czarniak,Henry
Oph 17h7'40"10d32'
Czarniecki,Michael Paul
Cep 6h6'59"86d28'
Czarnik,Susan
And 1h30'44"38d20'
Czech North
Cep 23h1'23"70d14'
Czech,Hunter Ethan
Her 17h13'15"26d33'
Czech,Lenny
Aql 19h7'29"3d15'
Czech,Mildred Mary
Mon 6h48'24"11d38'
Czechowski,Michele
And 1h27'49"36d29'

Czeck 50,Gregory P
Ser 15h20'0"9d36'
Czegledy,Dr Ferenc Paul
Cep 2h38'12"78d50'
Czekalski,John
Hya 10h13'37"-12d22'
Czekanski,Ruth Virginia
Cas 0h31'28"64d59'
Czepkiewicz II,Michael Henry
Uma 8h17'24"71d32'
Czermak,Kimberly
And 23h30'0"49d12'
Czernec,John Eric
Aur 5h31'24"50d16'
Czernek,Stanley William
Aql 20h3'21"4d16'
Czerniak,John J
Boo 14h45'37"24d20'
Czerniak,Suzanne M
Lyr 18h30'20"43d52'
Czerniejewski,Steve
Lac 22h31'54"55d37'
Czernieus Eternieus 6-6-1970
Dra 19h30'46"65d29'
Czerwin's Big Bear, Dolores & Karol
Uma 10h43'43"52d13'
Czesia
Cep 22h33'35"58d54'
Czinkota,Ilona
Mon 7h44'23"-1d39'
Czinkota,Uschi A
Eri 3h49'23"-0d50'
Czlapinski-Fournier
Aur 5h23'0"40d52"
Czohara,Caroline
Ser 15h21'17"8d31'
Czuprynski,Elizabeth Lynn
Com 12h2'42"18d52'
Czuprynski,Theodore Charles
Umi 14h8'52"70d57'
Czylek,Anthony John
Cam 13h19'12"80d49'
Czyzewicz,Savannah Marie
Eri 3h35'16"-16d1'
Czyzewski,Patrick Jakub
Dra 20h23'14"70d25'
Czyzyk,Kenneth Michael
Boo 14h31'45"31d30'
Czzowitz,Jim
Cep 20h55'40"58d15'
CÉndido,Elizenil De Matos
Peg 22h32'42"14d21'A
Cécile
Peg 21h26'16"11d0'
Céleste
Lyr 18h52'27"39d58'
Céleste Linck Du Princesse Valérie
And 2h18'1"40d17'
Céline
Ori 5h56'11"18d16'
Céré-Gil de Maniwaki, Gilles
Uma 12h11'24"61d34'
Cite,Tom David
Oph 17h39'57"-23d49'
Cité,Alexandre
Cam 3h58'20"61d51'
Cité,Francine
Tri 1h39'31"34d27'
Cité,Ida B
Cyg 19h27'58"33d10'
Cité,Myriam
Cyg 20h19'41"40d42'
Cité,Myriam
Cyg 20h19'41"40d42'
Cité,Naiade Ô Jean- Charles
Cyg 20h19'30"40d26'
Cité,Naïade Ô Jean-Charles
Cyg 20h19'30"40d26'
Cité,Yolande
Lyr 19h3'0"28d34'
Cité-O'Hara,Jocelyne
Lyr 19h3'0"28d36'
Cáspita Clementis
Cmi 7h16'33"5d25'

D

D
Mon 7h5'1"0d55'
D & D
Cma 6h43'37"-16d8'
D & D "Lovinity"
Com 12h38'42"23d45'
D & M
Aql 19h53'49"1d19'
D & M's Sunny The Star
Aql 20h6'45"1d18'
D & T's (Daren,Tori & Trevr)
Vul 20h14'34"25d11'
D & W Starlight
Ori 4h58'60"13d41'
D A J
Dra 15h37'29"61d58'
D A M P
Sct 18h32'30"-4d50'
D A S
Psc 0h57'11"32d6'
D B
Crb 16h14'0"37d54'
D B C
Uma 11h41'15"60d42'
d B MB A95
Peg 22h25'27"35d24'
D Bug
Vul 21h16'26"20d37'
D C & Me
Cyg 21h12'1"35d49'
D C G's Fire
Aur 5h51'47"29d49'
D C's Star
peg 0h11'29"18d19'
D D
Lyn 9h10'16"37d19'
D D & K N-Love of a Lifetime
Crb 16h11'0"38d46'
D D Louis
Mon 8h1'20"-8d31'
D D's Diamond
And 23h0'0"36d49'
D F M K "Dinanabanana"
Cas 0h21'33"51d44'A
D Fukuda
Mon 7h59'59"-5d47'
D H C
Peg 23h0'46"17d36'
D J
Cyg 19h27'58"31d6'
D J M-82
Vul 20h56'23"20d21'
D K B
Boo 14h22'57"28d43'
D K B
Lmi 10h48'36"31d10'
D K's Sunny
Her 16h30'19"20d2'
D L
Tri 1h58'54"26d57'
D L W's "Reality"
Uma 8h36'35"58d40'
D Michael
Aur 6h16'1"35d30'
D N 11494
Lyr 18h39'29"44d1'
D S More
Sco 16h1'12"-18d58'
D Squared
Tri 1h40'39"33d55'
D ssa Simona
Pic 4h55'42"-43d32'
D T III
Cam 3h55'21"71d14'
D W
Psc 1h0'13"21d36'
D W V W
Cep 21d50'20"61d42'A
D Z
Tri 1h36'14"30d17'
D&J Always With You Always With Us
Eri 4h9'0"-0d31'
D'Abdon,Antonio
Aur 5h28'24"38d7'
D'Acunto,Angelo
Tau 4h46'0"20d23'
D'Addario,Marie Ann
Uma 10h35'0"47d42'
D'Agliano,Andrea E Chicca Galleani
Hor 3h36'14"-49d10'
D'Agnelli,Michael
Lac 22h3'40"49d38'
D'Agnolo,Flavio
Uma 11h35'58"31d9'
D'Agnolo,France
Uma 11h34'0"37d31'
D'Agostini,Lee
Lac 22h45'19"54d6'
D'Agostino,Alexander
Per 2h5'1"56d21'
D'Agostino,Dean
Cnv 12h31'16"50d34'
D'Agostino,Debra
Ari 2h33'1"21d36'
D'Agostino,Donna
Cnv 13h54'50"45d37'
D'Agostino,Gabriel
Cam 7h6'19"67d34'
D'Agostino,Giacinta
Ind 21h1'39"-59d20'
D'Agostino,Mari
And 0h10'12"46d35'
D'Agostino,Michael
Per 2h1'25"50d18'
D'Agostino,Paul
Uma 10h18'25"56d26'
D'Agostino,Sabrina
Per 3h8'46"47d38'
D'Agustino,Jr,Richard Allen
Cet 1h21'0"1d22'
D'Aiello,Rosita
Pyx 8h52'2"-28d51'
D'Aiuto,Ralph & Rose
Mon 6h24'42"7d47'
D'Aiuto,Ronald Joseph
Her 16h10'57"4d50'
D'Alatri,Jeffrey Chaucer
Ori 4h53'42"-1d10'
D'Alessandro,Andrew J
Dra 17h15'0"69d53'
D'Alessandro,Frank Raymond
Ser 15h54'25"4d22'
D'Alessandro,Thomas Victor
Dra 15h55'39"68d2'
D'Alessandro,Troy
Umi 15h15'50"72d51'
D'Alessio,Gina
Del 20h13'15"10d42'
D'Alessio,Gina M
Del 20h13'39"12d10'
D'Alessio,Joey
Del 20h14'0"15d36'
D'Alessio,Lisa
Del 20h14'12"14d48'
D'Alessio,Maria
Del 20h14'25"13d5'
D'Alessio,Nancy Elizabeth Cecelia
Aql 19h59'53"10d10'
D'Alessio,Tom Gaetano Angelo
Cnv 12h26'1"50d0'
D'Alessio,Walter
Oph 16h59'26"10d20'
D'Aliesio,Franca
Ant 10h45'36"-34d20'
D'Almeida,Juliette
Cam 5h39'1"61d27'
D'Aloisio,Heaven's Jewel Hazel" B
Cas 0h33'21"68d24'
D'Aluisio,Emma Susan
And 0h28'14"27d49'
D'Aluisio,Robin T
Cas 23h32'23"53d52'
D'Amato,Andrea Felice
Cas 1h50'34"68d33'
D'Amato,Baby's Star -R.J. "Bobby"
Her 15h53'55"50d41'
D'Amato,Haily Marie
Mon 8h4'38"-9d59'
D'Ambiance,Créeuse
Umi 13h43'0"75d2'
D'Ambola,Dean
Lyr 18h49'0"39d7'
D'Ambriosio,Pat & Patricia
Ori 5h49'47"18d15'
D'Ambrogi,Barbara
Lyr 19h19'21"42d34'
D'Ambrosio Star,The
Cnv 12h25'30"32d40'
D'Ambrosio,Anthony
Boo 14h51'55"24d29'
D'Ambrosio,John P
Her 16h47'29"32d47'
D'Ambrosio,Juliette Rose
Tri 2h5'39"30d32'
D'Ambrosio,Lee
Her 16h39'42"27d22'
D'Ambrosio,Pamela
Lyn 8h9'23"38d29'
D'Ambroso,André P Dino
Her 17h59'35"50d20'
D'Amico,Carrie
Lyr 18h56'24"30d26'
D'Amico,Frank
Per 2h40'36"40d15'
D'Amico,Laura Ann
Ori 5h51'0"16d14'
D'Amico,Sr,Joseph Anthony
Lac 22h52'16"40d47"
D'Amico,Vincent Paul
Dra 17h48'1"60d24'
D'Amore,Christine
Uma 11h23'20"68d2'
D'Amore,Damian Joseph
Boo 14h28'37"20d19'
D'Amore,Donna
And 0h22'22"30d10'
D'Amore,Jessica Mary
And 22h55'28"40d19'
D'Amore,Marc
Aur 6h24'41"32d4'
D'Amours,Daniel Elliot
Per 3h38'0"38d59'
D'Andrea,Jack
Cet 3h9'40"2d8'
D'Andrea,Jenni Lynn
And 0h4'54"35d24'
D'Andrea,Steven
Dra 15h51'21"51d60'
D'Andrea,Sue
Cas 2h1'59"59d46'
D'Angelo,Jason Patrick
Per 2h22'44"55d50'
D'Angelo,John Philip
Boo 14h30'1"31d19'
D'Angelo,La Stella
Cam 5h7'31"67d43'
d'Angelo,Maria
For 2h8'1"-28d38'
D'Angelo,Paulette
Uma 9h21'32"59d10'
D'Angelo,Sally Daniela
Cet 3h1'1"1d34'
D'Angelo,Tiffani Starr
Cam 5h42'34"61d49'
D'Aniello,Gerald
Her 17h38'34"42d33'
D'Anna,Jack David
Ari 2h34'13"30d10'
D'Anna,Marie Inzerella
Sgr 18h49'34"-34d37'
D'antu 18h49'34"-34d37'
And 1h22'59"34d42'
D'Apice,Antonietta
Per 3h6'48"38d25'
D'Arbys Renaissance, Terence Trent
Ori 5h56'32"15d1'
D'Arcy,Caroline
And 0h1'1"47d3'
d'Arcy,Gerry
Lib 15h49'23"-8d37'
D'Arcy,Kevin & Pamela
Cyg 19h43'39"29d57'
D'Arezzo,Marina
Peg 23h26'54"23d45'
D'Arrigo,Sue
Dra 16h35'26"64d30'
D'Ascenzio,Kimberly
Tau 5h46'18"23d57'
D'Ascenzo,Dana
Cas 2h33'18"60d32'
D'Ascenzo,James
Dra 17h8'16"61d51'
D'Ascenzo,Laura
And 2h11'1"41d23'
D'Assaro Family Star
Equ 21h10'13"11d39'
D'Attili,D Maria S
Mon 6h19'33"8d46'
D'Augusta's,The
Lac 22h14'49"51d41'
D'Auria,Paul
Ori 4h45'46"0d50'
D'Avanzo,Lori
Cyg 19h45'20"37d34'
D'Avella,F H
Lyn 7h50'42"39d58'
D'Aversa,Antonella
Cep 23h31'51"63d31'
d'Avignon,Coleen
Hya 8h45'3"-16d53'B
d'Avignon,Pierre
Hya 8h45'3"-16d53'A
D'bard,Jahn
Her 17h28'31"31d38'
D'Cruz,Robert
Cep 21h16'15"70d50'
D'Eath,John Arthur
Lyn 8h23'46"48d35'
D'Elia,Amy & Michael
Cyg 20h19'31"41d13'
D'Elia,Jr,Gerardo
Vir 13h39'35"-7d23'
D'Elia,Kristin
Cyg 20h21'37"31d18'
D'Elia,Leonida Carmello
Cas 2h3'14"60d27'
D'Elia,Peter John
Dra 19h4'47"56d58'
D'Elia,Philomena
And 23h30'47"39d33'
D'Elia-Wittkofski
Tri 2h11'37"32d28'
D'Erasmo,Rose
Cyg 20h7'37"38d59'
D'Errico,Albert P
Psc 22h53'1"5d36'
D'Errico,Etiennette et Tancredi
Gem 6h36'25"14d17'
d'Ersu,Guillaume
Uma 9h38'29"55d51'
D'Esquerre,Severine
Dra 12h34'1"64d11'
D'Hennin,Marie-France Adida
Sge 19h59'54"20d35'
D'Hooghe,Laura L
And 23h21'11"48d58'
D'Imperio,Alexandra E
Lyr 18h45'44"36d17'
D'Imperio,Samantha L
Tri 2h6'14"30d19'
d'Ingillo,Francesca
Pyx 8h41'31"-24d31'
D'Iorio,Robert Anthony
Dra 14h57'39"60d30'
D'Jaen,Miriam Daniela
Uma 8h56'22"51d1'
D'LaTorre,Jaime
Sex 9h55'34"2d48'
D'Martino,Vincent Paul
Cep 21h58'1"61d37'
D'Onofrio,Carolyn Marie
Del 20h28'27"20d24'
D'Onofrio,Craig Meach
Uma 9h8'0"54d35'
D'Onofrio,Cristina
Per 2h56'36"34d14'
D'Onofrio,Nicholas Anthony
Lib 14h52'18"-0d23'
D'Oronzo,Carriann Rachel
Lyr 18h50'11"30d48'
D'Orsaneo,Gianna Marie
Aur 5h1'40"41d2'
D'Ostroph,Dick & Sandra
Lyr 18h38'20"30d15'
D'Souza,Ansettan Maria
Cas 0h56'58"61d41'
D'Squared
Umi 15h7'17"77d47'
d'Tenebres AKA Cherry
Dante,Gene
Vir 13h23'57"-4d22'
D'Urso,Joseph
Oph 17h0'17"8d41'
d'Villase7h0'17"8d41'
Hya 8h11'35"6d37'
D'Yquem,Claudia
Vul 20h21'18"22d46'
D'zik,James Anthony
Her 18h1'43"18d47'
D-Man
Her 18h6'33"48d10'
DA
Cmi 7h56'0"8d18'
DA
Eri 3h42'1"-10d53'
da Codigoro,Giovanni Musacci
Lac 22h10'0"51d13'
Da Corte,Marianna
Hor 3h19'13"-49d25'
Da Fano,Judith
Ori 5h3'43"15d43'
da Luz,Joe "Little Joe"
Per 2h39'30"38d28'
Da Silva,Daniela Mercaldi Zacarias
Sco 17h28'0"-41d2'
Da Silva,Tina Nathaniel Christiana
Peg 22h20'54"35d27'
Da-12051
Ori 4h43'17"-0d24'
Da-Da-Cookoo
Aql 18h58'1"17d6'
Da-Ri
Sgr 19h36'36"-41d25'
Dab
Cyg 21h39'26"38d27'
Dabaja,Harry
Aur 5h5'31"42d39'
Dabbs,Norris
Sct 18h39'36"-4d36'
Dabbs,Winston
Ser 15h10'0"9d30'
Dabbs-Shnutz,Allan
Ori 4h47'40"15d25'
Dabby
Cam 7h2'51"68d26'
Dabek,Darcy P "Night Diamond"
Per 3h23'30"41d25'
Dabney,Archie W
Hya 8h19'16"5d4'
Dabossa
And 1h7'49"38d59'
Dacanay,Trisha Anne
Peg 23h3'0"10d31'
Dacar's Star
Tau 5h15'49"16d5'
Dacasojupi
Cam 5h5'48"65d19'
Dace Star
Lyn 8h10'56"38d18'
Dacey,Michael Christopher
Per 2h27'37"58d29'
Dacey,Michael C
Per 4h2'27"37d20'
Dachel & Raisy
Gem 6h59'21"14d57'
Dack,Stephen C
Dra 11h32'16"71d27'
Dacombe,William
Aql 20h0'47"4d20'
Dacong,Anthony Theodore
Aqr 20h52'37"0d30'
DaCosta,Mark Anthony
Per 3h37'10"45d51'B
DaCosta,Paul Alexander
Per 3h37'10"45d51'A
Dacus,Jim "J D"
Uma 9h58'13"51d27'
Dacyshyn,John T W
Dra 16h34'24"60d13'
Dad
Cep 21h51'54"60d20'
Dad & Julie Forever
Crb 16h7'1"37d44'
Dad & Mad Star,The
Her 17h27'1"20d14'
Dad III
Boo 14h40'50"35d1'
Dad K & K & K & K & J
Cmi 8h8'52"2d25'
Dad The Grouch
Cap 20h50'28"-26d34'A
Dad's Casey-Christmas 1994
Lyn 7h42'23"43d5'
Dad's Star From Dianne & Pippen
Cep 0h13'53"68d51'
Dad-Dad 96'
Cyg 19h22'42"44d39'
Dad-O
Uma 10h4'14"47d45'
DaDa
Cep 22h0'47"54d5'
Dada'
Del 20h18'17"14d6'
Daddino,Jeannie
Peg 22h32'60"27d56'
Daddio,Elizabeth "Lilo" W
Cyg 20h17'39"38d57'
Dadds,Richard
Eri 2h55'59"-12d0'
Daddy
Cep 0h6'21"71d9'
Daddy
Lac 22h46'33"38d31'
Daddy
Her 18h38'28"12d17'
Daddy Copstick
Uma 10h43'32"60d47'
Daddy Judd
Aql 19h18'33"12d13'
Daddy Long Leggs
Per 2h55'25"35d28'
Daddy Love Hannah Elizabeth
Uma 11h9'0"62d20'
Daddy Loves Marlee-Rae
And 0h12'1"46d56'
Daddy Star
Per 2h49'25"40d23'
Daddy Trek
Aur 5h31'21"31d41'
Daddy's Boy
Cet 3h9'34"-0d54'
Daddy's Brightest "Sun"
Aql 19h0'40"8d41'
Daddy's Little Tinkerfartz
Peg 21h59'1"22d56'
Daddy's Little Girl Bree
Cyg 21h18'29"38d35'
Daddy's Little Star
Peg 21h56'17"2d20'
Daddy's Little Girl
Cep 21h51'48"55d5'
Daddy's Little Girl, Elizabeth
And 23h41'27"38d12'
Daddy's Little Girls Love Jen & Jess
Ari 01h57'59"15d37'
Daddy's Rachael
Mon 6h54'47"8d46'
Daddy's Smiling Eyes
Uma 8h57'17"57d36'
Daddy's Star
Cep 21h59'25"55d24'
Daddy's Star
Boo 15h0'18"19d39'
Daddy's Star
Cep 23h2'51"64d28'
Daddy's Twinkle
Lac 22h26'16"53d52'
Daddy-O
Her 17h36'21"38d58'
Dadino,Deborah
And 23h3'0"48d32'
Dado
Dra 17h55'11"65d22'
Dadoo
Her 16h11'12"41d32'
Dadswell,Joanne
And 0h7'22"30d49'
Dadswell,Joanne
And 0h7'22"30d49'
Daeche-Marshall, Alexandra Jean
Lyn 7h29'59"43d2'
Daedelow,Helfried
Cam 6h6'48"80d31'
Daelrudaclo's Star
Lyn 8h25'16"33d48'
Daffara,Michel
Cnc 8h14'17"30d19'
Dafford,Douglas Blaine
Aur 6h17'59"45d60'
Daffron,Gary L
Aur 5h13'55"40d8'
Daffs Star
Lyr 18h57'0"46d55'
Dafne
Mon 6h44'14"7d57'
Dafne
Pyx 8h51'47"-29d22'
Dafné
And 2h6'41"40d30'
Dafné
Per 3h32'35"35d25'
DAG Daddy I, 11-5-36
Umi 17h23'54"76d30'
Dagach,Sergio
Cyg 19h29'52"56d20'
Dagbert,Monique Thiou
Cnv 13h4'31"50d47'
Dagenais,Carole
Uma 11h25'24"61d58'
Dagenais,Patricia
Eri 3h14'25"-2d5'
Dages,Helen Manzer
Cas 2h17'49"75d60'
Daggett,Noelle Dawn-Marie "Poohke"
And 1h40'18"39d58'
Daggett,Ronald A
Lac 22h53'59"56d44'
Daggett,Suzie
Lyn 7h7'36"44d53'
Daggy
Cet 2h55'57"5d48'
Dagmar
Sgr 19h44'49"45d6'
Dagmar & Dietrich 1967
Vir 13h22'60"-3d52'
Dagnall,Amazing Grace
Com 12h43'46"21d50'
Dagney's Sweet Sixteen
Lyn 7h2'0"60d23'
Dagnino,Enrico
Lac 22h23'54"56d19'
Dagny Linnea
Cyg 21h3'26"33d11'
Dahausse,Pascal
Aur 5h8'36"29d10'
Daher,Esmail El
Ori 6h1'58"10d26'
Daher,Ronald Jerry
Cet 3h5'45"4d7'
Dahill,Jr,John Joseph Thaddeus
Boo 14h38'1"19d39'
Dahill,Star of "Friendship",Lisa
Lyn 8h11'41"47d33'
Dahl,Aaron
Lyr 18h39'0"34d6'
Dahl,Annika
Cas 1h23'47"64d41'
Dahl,Arnold
Ori 4h47'17"5d29'
Dahl,Henry S
Her 16h54'11"25d15'
Dahl,Kristofor Lee
Aur 6h2'22"32d53'
Dahl,Solveig & Erik
Aur 6h14'19"30d55'
Dahlen,Elizabeth Marguerite
And 23h33'53"37d33'
Dahlgren,Elyse S
Cnc 8h35'30"31d4'
Dahlia
Cas 23h39'58"64d14'B
Dahlmann,Isabel Susan
Cas 0h50'52"62d33'
Dahlstrom,Arnold
Uma 9h36'14"50d17'
Dahm,Donita
Del 20h15'37"11d51'
Dahm,Donita
Lac 22h11'0"48d54'
Dahm,Heinz-Günter
Peg 23h34'48"15d57'
Dahm-The Sanctity of Space,Perry
Per 2h9'13"56d21'
Dahmer,Jr-"Buddy Taps" C Warren
Dra 18h28'29"50d16'
Dahmer-"Saint Dee", Dolores Collins
Cyg 20h6'58"40d12'
Dahms,Gary H
Vul 19h48'1"22d57'
Dahn,Jr,Wilbur James
Per 3h2'51"41d5'
Dahnke,Vickie Ann
Cyg 21h19'56"38d8'
Dahut,Jean
Cyg 20h28'58"31d31'
Dai In The Sky
Ori 5h36'15"-0d25'
Daia,Euripedes
Peg 23h6'0"17d50'
Daia,Kristopher
Lac 22h17'55"50d18'
Daibhi Mac an Ri
Sgr 19h57'38"-43d16'
Daigle,Jon
Cet 0h26'57"1d9'
Daigle,Mark Ryan
Per 1h48'36"53d21'
Daigles's "Infinite Devotion"
Boo 14h36'49"48d57'
Daignault,Richard H
Aur 6h35'44"33d26'
Daigneaolt,Taylor Rose
Cet 0h29'12"-17d50'
Dailey's Dot
Aur 4h47'33"48d54'
Dailey,Brian Randall
Aql 20h12'56"4d30'
Dailey,Heather Michelle
Mon 7h4'0"4d6'
Dailey,Leon
Cap 21h54'37"-20d19'
Dailey,Morgan & Marci
Uma 8h39'21"51d11'

Dailey,Nancy Cas 1h33'37"60d23'
Dailey,Patricia Ann Cas 0h29'12"61d38'
Dailing,Frank & Michelle Ser 15h28'16"18d25'
Dailing,Rose & Amanda Cma 6h54'38"-18d37'
Dailley,Richard Michael Her 17h7'0"40d3'
Dailley,Richard Michael Her 17h7'0"40d3'
Daily,Beverly Jean Del 20h51'27"9d15'
Daily,Diane E Ari 2h33'34"20d44'
Daily,Frank Patrick Aur 6h2'53"54d36'
Daily,Madison Peg 22h29'38"34d9'A
Daily,Megan Peg 22h29'38"34d9'B
Daines,Paul Nigel Boo 14h17'48"51d34'
Daini,Anna Umi 16h46'19"85d58'
Dainty Diane And 0h27'1"27d36'
Daisey May And 23h23'52"43d37'
Daisey,Phillip R Lyn 8h34'36"40d28'
Daisuki Aqr 22h3'29"-23d17'
Daisy Lyr 18h23'58"40d37'
Daisy Lup 15h17'33"-43d18'
Daisy Scl 23h9'1"-32d48'
Daisy Hya 8h47'45"1d18'
Daisy Uma 12h10'26"60d8'
Daisy B Lyr 19h2'26"26d54'
Daisy B 522 Gem 7h2'58"30d6'
Daisy-Annie Umi 16h36'34"75d17'
Daizo,Mr & Ms Ayumi Aql 19h36'15"-0d2'
Dak's Dream Aqr 21h2'26"-1d44'
Dakini Dra 16h21'39"61d33'
Dakis,Gene Her 17h58'45"30d14'
Dakota Ori 6h2'44"7d21'
Dakota Dolphin Uma 10h38'24"47d0'
Dakota Lynn Aur 5h1'15"46d38'
Dakota's Destiny Peg 0h8'52"14d15'
Dakota,Cory Uma 12h2'33"35d16'
Dakotah Cma 6h10'45"-16d22'
Daksha Psc 1h3'29"27d54'
DAL "Bear" Uma 10h53'14"47d53'
Dal Bon,Francesca Uma 12h0'16"32d14'
Dal Maso,Lorena Tel 20h20'14"-47d8'
Dalba,Joni Lac 22h35'30"53d16'
Dalbergue-Poggi,Elaine Aur 4h36'1"31d25'
Dalby,Christie Michelle Aql 19h54'1"12d27'
Dalby,Wendy Ori 6h5'1"2d32'
Dale Umi 14h48'29"66d21'
Dale Cet 2h53'27"1d32'
Dale "Dweller In The Valley" Per 4h22'43"50d36'
Dale & Debbie's Star Ori 4h45'53"4d42'
Dale & Jennifer Eri 3h53'0"-5d59'
Dale & Wes Cet 1h50'58"0d38'
Dale James Cep 23h49'57"75d15'B
Dale Jonathan Umi 15h17'0"69d6'
Dale N Guy Eri 3h43'15"-0d5'
Dale's Dynasty Aql 20h3'26"4d37'
Dale's Nightlight Lac 22h19'31"48d53'
Dale's Nightlight Aur 7h15'43"38d57'
Dale's Sparkler Her 16h31'16"41d7'
Dale's Star Light Cmi 7h27'42"0d46'
Dale,Charles A Aur 5h58'42"38d14'
Dale,Darleen Cas 22h57'55"54d55'
Dale,Deborah Anne Cas 0h26'29"61d21'
Dale,Emily Nicole Vul 19h48'0"26d17'
Dale,Graham Scott Ori 5h55'21"17d2'
Dale,Holly Kathryn Lyr 19h3'40"28d46'
Dale,Ian Barry Pyx 8h52'2"-28d59'
Dale,Jerron Eri 2h51'31"-15d27'
Dale,Julie Cyg 21h38'45"30d20'
Dale,Richard Hya 8h50'26"0d4'
Dale,Stephanie Cnc 8h27'58"31d44'
DaLear Del 20h13'14"10d53'
Dalecki Family Star, The Ori 6h16'50"-0d54'
Dalen,Courtney Crt 11h15'45"-12d56'
Daleo Del 20h15'1"14d44'
Dalesandro,Mary Lou Pinto And 2h5'58"40d55'
Daley's Sylvester Reflecter,John M Aql 19h51'0"14d30'
Daley,Anna Marie Cnv 12h56'12"47d21'
Daley,Donald Joseph Umi 14h44'56"69d2'
Daley,Faith Christine Lyn 7h46'22"50d46'
Daley,Gloria Teubert Umi 15h9'39"68d28'
Daley,Gregory Maxwell Aur 6h31'0"37d58'
Daley,Joseph Cet 2h47'34"4d20'
Daley,Joseph Charles Dra 10h46'37"74d19'
Daley,Patricia Lyn 7h54'1"34d2'
Daley,Wayne Her 17h55'19"20d4'
Daley-Wieland,Maureen Cet 1h32'30"0d51'
Dalgleish,Catherine Marie Cas 23h38'29"58d46'
Dalglish,Pat & Jack Lac 22h20'19"52d52'
Dalhgren,Little Miss Muffet Emily E Uma 8h34'12"51d30'
Dalida,Jill Marie And 23h49'0"44d54'
Daliglow Cam 8h1'29"68d46'
Dalinkus,Christopher Albert Lac 22h30'18"52d40'
Dalke,Ashley Michelle And 23h25'38"49d2'
Dallago,Casimer N Per 1h49'25"50d1'
Dallaire,Claude Ari 2h37'1"21d0'
Dallaire,Francine Ari 2h33'42"30d1'
Dallao,Mary S Lyn 8h45'40"37d33'
Dallas Her 15h51'54"40d7'
Dallas Ser 15h9'44"9d54'
Dallas Tri 2h11'47"33d31'
Dallas Lyr 18h33'35"33d4'
Dallas,E Uma 10h2'59"60d57'
Dallas,Eileen Crb 16h11'11"33d56'
Dallenne,Geneviève Per 3h31'41"50d31'
Dallinger,Jürgen Cnc 9h18'1"31d54'
Dally,Martin R Dra 14h8'17"64d11'
Dally,MD,Al Per 3h12'52"47d36'
Dalmatini,Vanessa Cae 4h57'0"-33d41'
Dalquest,PhD,Walter W Oph 18h17'38"1d27'
Dalquin,Noémie Ori 6h7'24"10d13'
Dalrymple,Allison Marie"Ali" Mon 7h21'1"-8d9'
Dalrymple,Lauren W Ari 1h46'50"13d29'
Dalrymple,Lewis Baxter "Brac" Sex 9h52'0"5d28'
Dalton USMMA,Michael W Peg 22h1'52"30d22'
Dalton's "Fireball" Aqr 21h29'49"-0d19'
Dalton,Abigail Lynn Lyr 19h17'59"38d18d
Dalton,Carol Cyg 20h55'41"30d18'
Dalton,Darby Lance Ser 18h5'27"-1d43'
Dalton,Doris Irene Umi 13h46'16"77d47'
Dalton,Emma Marie Cyg 21h38'1"30d29'
Dalton,Frank Ori 5h59'42"12d36'
Dalton,Helen B(Snooky) Cyg 19h7'51"47d51'
Dalton,Jack Per 4h44'54"51d12'
Dalton,James Her 17h18'13"47d7'
Dalton,Jay Equ 20h58'12"4d41'
Dalton,Jerome & Dylan Lee Aql 20h10'55"11d57'
Dalton,Joanne Michelle Cas 0h18'21"62d51'
Dalton,John Per 3h32'30"31d30'A
Dalton,John A Boo 14h22'13"33d8'
Dalton,Judy Cas 23h0'54"58d54'
Dalton,Laurie Cas 0h55'48"75d50'
Dalton,Lilla Thomason Tau 5h24'32"16d10'
Dalton,Mark Hunter Per 1h39'23"52d49'
Dalton,Patrick Joseph Umi 13h41'15"71d29'
Dalton,Penny Cmi 7h33'18"10d59'
Dalton,Timothy Her 17h36'34"26d1'
Dalton,William H(Bill) Cyg 19h15'26"44d14'
Daltonhurst,David L Sge 19h3'0"16d19'
Daly Lyn 8h55'14"45d52'
Daly Faith Sge 19h30'49"16d50'
Daly,Beatriz Eugenia And 23h41'16"47d45'
Daly,Dr Joan Lyr 19h19'59"40d30'
Daly,Ian Sigismund Lac 22h20'35"51d24'
Daly,James Thomas Sco 16h57'39"-44d50"
Daly,Jeffery Banner Lac 21h59'44"40d17'
Daly,Joan Catherine Cas 1h26'12"60d38'
Daly,John Sct 18h41'43"-5d24'
Daly,Jr,Paul Vincent Aqr 22h7'28"-8d13'
Daly,Jr,Robert E Dra 18h56'35"68d26'
Daly,Kate Cas 0h7'22"64d37'
Daly,Kevin J Dra 17h42'10"75d19'
Daly,Marianne B Lyr 18h21'16"42d53'
Daly,Moira Ori 5h53'0"19d44'
Daly,Patricia K Cam 3h55'56"53d4'
Daly,Patrick James Cam 4h16'30"70d30'
Daly,Patti Ann Peg 22h46'0"24d30'
Daly,Robert J (Bullet- Bob) Cep 22h38'42"58d58'
Daly,Ryan Patrick Lac 22h5'42"46d27'
Daly,Sally Ruth Lyr 19h22'20"38d17'
Daly,Seamus Ori 6h5'24"1d27'
Daly,Thomas Edward Lib 15h31'19"-28d8'
Daly,Tom Dra 16h38'32"70d11'
Daly,Tynan Flawn Sailor Uma 11h23'57"40d14'
Daly,William Gerald Oph 17h29'46"-1d32'
Dam,Freddy A Monnicken Cma 6h13'10"-18d51'
Dama Del 20h13'31"13d1'
Dama Del 20h15'53"15d5'
Dama,Andrew J Dra 20h21'17"68d14'
Dama,Ivan And 23h2'25"51d8'
Damais,Maryse Uma 12h3'1"62d49'
Damaklion Ori 5h55'59"8d41'
Damann,Thomas Gem 7h26'31"35d18'
Damant,Sam Paul Lyn 9h11'0"39d25'
Damarell,Zak Ori 5h55'57"16d7'
Damarey,Guillaume Mon 7h57'0"-4d33'
Damaris Gloria Aql 19h0'35"-6d30'
Damaschino,Gary Aur 5h26'12"38d50'
Damasco,Finto And 0h17'34"40d45'
Damascus Cma 6h24'12"-16d49'
Damasky Wedding Star Cyg 19h24'13"31d52'
Damassa,Barbara M Sco 16h29'25"-38d54'
Damast,Eric Aur 5h55'1"30d42'
Damast,Michael Salvador Aur 4h42'36"30d34'
Damato,Derek James Dra 16h14'58"65d57'
Dambach,John William Lyr 18h19'57"37d53'
Dambeck,John K Per 3h37'29"37d37'
Dame Lulu Per 3h48'9"35d51'
Dame Nin Lib 14h59'43"-5d55'
Dame,Kelly Aql 19h52'38"0d25'
Dame,Sarah Eri 3h39'13"-6d37'
Damebianna & Pokey-O Cam 4h27'28"68d21'
Damek Aur 5h9'56"42d31'
Dameon,Courtney Peg 0h1'45"22d7'
Damerau,Nadine Cnc 8h57'36"10d51'
Dames,Beatrice Cyg 20h38'25"42d41'
Dames,Damien Cep 22h44'18"70d29'
Damewood,Ed Aur 4h52'44"41d7'
Dami Jayne Cas 0h48'44"65d50'
Damian Ori 5h22'56"15d30'
Damian Tau 5h53'30"23d18'
Damian,Nicholas Alexander Her 16h9'50"48d50'
Damiana Cybele Sco 17h52'43"-40d12'
Damiani,Antonio & Donna Lmi 9h55'43"38d7'
Damiani,Luciano Vel 9h18'56"-42d36'
Damiani,Richard Cnv 12h49'17"38d40'
Damiano Lib 14h20'40"-23d24'
Damiano,Patrick Lyn 8h22'59"47d52'
Damiano,Vincent & Maria Cyg 21h22'30"52d31'
Damianon,George Ori 4h43'0"0d50'
Damianos,Eurydice Dra 16h1'0"63d32'
Damico,Claude Cyg 19h50'34"38d1'
Damico,Rose Cas 0h4'15"56d15'
Damien Ser 15h55'48"2d19'
Damien Pho 23h48'50"42d3'
Damien 1 Vir 13h23'22"-4d29'
Damien B Cnv 12h43'0"40d16'
Damine Spirit Totem Per 2h52'44"43d30'
Damion-Dude Aur 4h58'1"48d53'
Damirel,Thomas Lmi 11h1'31"33d22'
Damit Janet Cyg 21h1'46"40d32'
Damita Cet 2h50'54"5d41'
Damkier,Robert Tri 2h4'18"31d43'
Damm,Hans Joachim Ari 2h2'26"25d53'
Dammasch,Jutta Lac 22h3'1"51d31'
Dammertz,Hermann-Josef Cep 22h8'31"65d8'
Dammicci,Jim & Rene Sge 19h26'37"16d12'
Damon Lyr 19h26'1"40d30'
Damon 3 Lac 22h15'27"51d14'
Damon Mon Estuaire Aur 6h2'54"37d39'
Damon's Chawlk Aur 6h33'18"35d12'
Damon's Song Cyg 19h59'58"40d21'
Damon's Way Dra 19h46'46"61d17'
Dampf Ori 5h5'39"1d28'
Dampier,Carol Lee And 22h56'29"40d34'
Dampier,James(Archie) Cet 2h58'0"-0d43'
Dampier,Patsy Oph 18h6'50"0d44'
Damrill,Kyrsten Feberette And 23h34'14"46d42'
Damron,Dr Bonnie L Ori 5h56'39"5d14'
Damron,Lorien Evenstar Ori 6h4'13"-0d35'
Damrow,Bryan William Per 3h30'54"51d32'
Dan Per 1h44'10"53d57'
Dan & Carole Aql 20h13'25"5d16'
Dan & Karin's Wishing Star Cyg 20h35'28"37d49'
Dan & Kate forever Cyg 21h18'1"38d54'
Dan & Lil's Star Aqr 21h0'32"-11d2'
Dan & Mary"Wedding Star",The Crb 15h19'33"31d3'
Dan Dan Star,The Uma 8h30'0"54d41'
Dan Memory "Pumpkin Boy #2" Cep 22h38'60"70d22'
Dan Vicky Nathan Adam Ori 4h41'1"5d46'
Dan's Destiny Lac 22h34'20"50d10'
Dan's Destiny Dra 16h16'30"61d36'
Dan's Oceania Astronomica Major Del 20h19'40"20d10'
Dan's Place Hya 8h54'47"-6d28'
Dan's Roonsker Psc 0h48'1"27d37'
Dan's Silver Fox Ori 5h45'29"-0d15'
Dan,Grete,Nicholas, & Michael Del 20h18'59"20d5'
Dan-Sue Cyg 21h29'56"41d6'
Dana Lmi 9h49'27"38d1'
Dana Lac 22h33'1"53d7'
Dana Aql 20h5'20"6d31'
Dana Umi 14h44'32"80d8'
Dana Lyr 18h55'22"30d36'
Dana & Casey's Star Long May They Wish Uma 8h42'24"71d25'
Dana & Misty Eri 3h56'18"-12d45'
Dana 25 November 1963 Sgr 19h52'34"-44d8'
Dana E-A Princess And 23h35'36"48d59'
Dana K Lmi 10h17'40'30d49'
Dana Lee Lyr 19h0'28"37d34'
Dana Loves Ann Marie Lyr 18h42'1"35d15'
Dana Loves Chris Forever Lyr 18h15'52"35d9'
Dana Lynn Lyr 18h54'23"42d22'
Dana Marie And 2h31'20"48d19'
Dana Scott Cas 0h22'21"58d45'
Dana Sue Cyg 19h40'32"31d26'
Dana V H And 1h24'17"40d50'
Dana's "Dancing Dipper" Peg 22h16'41"35d3'
Dana's Dawn Cyg 20h2'13"40d15'
Dana,Daniel Aql 19h55'15"7d35'
Dana,Rebecca Anne Cas 0h19'11"64d9'
Dana,Taylor Hamilton Gem 6h32'22"13d56'
Dana,The Shining Star Peg 22h36'44"26d20'
Dana-Christin Pauls Sgr 18h56'52"-21d49'
Danadan Del 21h5'0"13d44'
Danae "The Kid" Uma 10h32'59"40d12'
Danaher,Waneta Aur 4h55'20"40d58'
Danale P Cas 2h42'1"61d17'
DanaLee,The Peg 22h22'32"5d48'
Danatron Cam 5h47'56"65d9'
Dance Dance Dance Lyn 7h2'22"53d1'
Dance,Debi Moyer Lyn 7h54'27"51d10'
Dancer Aur 7h3'38"38d42'
Dancer Ari 2h56'33"22d20'
Dancer,Kerry Lyr 18h35'37"39d49'
Dancer,Tiny Lyr 18h43'28"40d17'
Dances With Whales Cet 2h5'44"4d1'
Dancey,Elizabeth C Cyg 21h3'13"28d20'
Dancey,Joy Ori 6h17'41"20d22'
Dancho,John Sex 10h43'24"-5d41'
Dancing Lyn 8h58'51"41d24'
Dancing Nancies Ori 5h37'0"13d4'
Dancing Spirit Lyr 19h4'40"26d50'
Dancing White Raven Woman Lyn 8h45'45"39d23'
Dancing With Another Drummer Peg 22h30'50"31d23'
Dancing With Open Arms Sge 20h0'26"19d7'
Danciu-Grosso,Emily "Angel" Peg 22h17'48"21d8'
Danciu-Grosso,Noelle "Angel" Del 20h25'1"11d40'
Dancu,Eric Finlayson Aql 18h42'1"-2d37'
Dancy,Blair & Allison Cyg 19h31'25"31d34'
Dancy,Matthew & Pamela Crt 10h53'31"-8d4'
Dandaneau,Patrick M Aur 5h8'1"41d2'
Dandee Cas 0h56'13"60d17'
Dandelion Danny Uma 9h0'30"48d0'
Dandois,Ariane Aur 6h21'26"38d30'
Dandot Cyg 20h42'49"46d47'
Dandrea,Raymond Dra 10h40'25"78d6'
Dandrimont,Margaux Cas 0h41'1"61d34'
Dandy Mon 7h4'14"-5d55'
Dandy Sandy Leo 9h38'36"8d14'
Dandy,Linda L Cas 0h32'11"61d56'
Dane Aur 7h9'1"36d7'
Dane,Lisa & Allison Cyg 19h46'43"30d2'
Dane,Susanne Cam 4h51'45"61d34'
Daneka & Ryan's Star Crb 15h20'46"30d58'
Danelian-Kalemkiarian, Rosemarie Mon 8h6'57"-5d45'
Danes Aql 19h50'23"10d19'
Danese,Janet Cas 0h18'24"66d23'
Danette My Shining Star Lyn 9h6'1"33d38'
Danford,Marie Lou Dussel Del 20h13'52"10d17'
Dangerfield,Bert Boo 15h20'48"38d23'
Dangerfield,Sylvia Amanda Aqr 22h20'1"0d44'
Dangman,Christopher Cole Cep 22h14'48"65d24'
Danhausen,Ronald Lac 22h17'60"51d24'
Dani Uma 9h49'26"47d48'
Dani Uma 11h59'56"47d38'
Dani Jean Mon 6h54'13"-8d30'
Dani's Birthday Star Cap 21h43'20"-23d30'
Dani's Star Peg 21h59'48"26d48'
Dani,Fotula Per 3h51'15"37d33'
Danias,Penny And 2h21'48"49d22'
Danica Cep 21h29'54"61d41'
Danica Lynn Cyg 21h56'58"50d9'
Daniel Her 16h11'55"47d33'
Daniel Her 18h29'56"20d33'
Daniel Her 16h56'1"30d35'
Daniel Oph 18h39'0"7d9'
Daniel Per 2h52'43"41d12'
Daniel Lac 22h24'42"55d37'
Daniel Hya 9h17'25"-1d45'
Daniel Dra 19h3'22"56d18'
Daniel Cam 5h56'44"67d52'
Daniel Aur 6h48'51"38d14'
Daniel Cas 1h53'48"60d32'
Daniel Oph 17h30'0"-8d25'
Daniel Dra 20h6'34"63d36'
Daniel Her 16h41'38"25d1'
Daniel Per 2h52'52"45d59'
Daniel Lac 22h3'0"38d52'
Daniel Cep 23h7'32"65d59'
Daniel Aql 19h55'29"10d19'
Daniel Cep 22h26'31"63d2'
Daniel & Alison's Star Crb 16h2'31"32d17'
Daniel & Debra Peg 22h2'10"24d41'
Daniel & Deepti Cyg 19h57'46"30d35'
Daniel & Janet Cyg 19h23'19"44d10'
Daniel & Mandy Sge 20h2'42"16d25'
Daniel & Rene Cyg 21h19'24"37d54'
Daniel & Victoria Forever Cyg 19h26'12"56d49'
Daniel Alexander Cmi 7h55'19"1d5'
Daniel Benjamin Cam 5h49'44"58d37'
Daniel David Per 1h35'59"53d58'
Daniel Edward Boo 14h29'55"50d3'
Daniel Hugh Ori 4h54'0"4d44'
Daniel il Primo Cnc 8h14'0"31d6'
Daniel James Uma 9h19'54"54d26'
Daniel James Uma 11h22'22"37d58'
Daniel Lee Sex 10h31'35"0d29'
Daniel Markos Aaron Dra 14h32'12"64d54'

Daniel Michael
Per 4h7'13"51d10'
Daniel of the Phantom Pair
Peg 23h47'18"27d24'A
Daniel Pierre
Aur 4h53'53"40d56'
Daniel Sweet
Sge 19h3'26"18d55'
Daniel William
Cam 6h0'1"60d15'
Daniel Xavier
Ori 5h57'18"12d13'
Daniel You're Our Shining Star
Cep 2h34'51"78d10'
Daniel's Beauty
Cep 0h55'38"77d28'
Daniel's Boone
Ori 5h2'23"0d39'
Daniel's Gate
Aql 20h18'18"1d9'
Daniel's Heartlight In The Sky
Cep 22h15'11"65d9'
Daniel's Light
Aur 5h10'23"40d22'
Daniel's Light
Per 2h57'27"40d16'
Daniel's Star
Aur 5h0'17"48d33'
Daniel,Debora
Mon 7h44'1"-1d19'
Daniel,Denise Parrat
Peg 23h24'1"12d46'
Daniel,Governor Bill
Cen 11h42'11"-48d8'
Daniel,Hannah Nicole
Sag 20h23'53"-28d12'
Daniel,Hans-Ullrich
Peg 23h29'1"11d27'
Daniel,Hugh Kenton
Cma 6h57'2"-17d53'
Daniel,Ian
Psc 22h56'1"6d32'
Daniel,Jr,Frederick R
Aur 5h37'24"40d44'
Daniel,Laura
Cas 3h10'45"65d33'
Daniel,Leslie S
Aql 19h55'47"14d3'
Daniel,Suzanne
Crt 11h16'3"-12d30'
Daniel-21,"Ivory" Loren
Cet 1h31'46"-1d41'
Daniel-Dragyn of the Northern Sky
Boo 13h53'46"21d12'
Daniel-Victoria
Uma 10h9'31"56d51'
Daniel-Your Beast
Cep 22h15'12"55d57'
Daniela
Cas 1h23'27"53d5'
Daniela
Cas 0h10'18"56d7'
Daniela
Cnc 8h50'36"32d24'
Daniela
Umi 17h9'1"75d23'
Daniela
Peg 23h31'39"17d41'
Daniela
Cae 4h57'0"-33d41'
Daniela
Cmi 7h20'1"8d56'
Daniela
Pyx 8h46'8"-27d34'
Daniela
Lyn 7h48'0"40d48'
Daniela
Dra 16h37'56"60d0'
Daniela Forever
Lyr 19h17'1"25d37'
Daniela T,ut,u
And 2h15'50"40d34'
Daniele
Nor 16h19'7"-59d30'
Daniele
Boo 15h9'53"42d19'
Daniele & Jessica
Cnv 13h39'0"31d44'
Daniele,Destiny
Mon 7h23'18"-1d47'
Daniele,Susie
Peg 0h7'0"14d23'
Daniella Marie
Lib 15h38'10"-28d8'
Danielle
And 2h27'37"39d52'
Danielle
Mon 6h30'24"10d38'
Danielle
Cas 1h7'13"60d48'
Danielle
Lyr 19h1'31"38d22'
Danielle
Cas 23h25'36"60d4'
Danielle
And 2h34'24"50d0'
Danielle
Eri 4h3'1"-11d50'
Danielle
Lyn 7h30'19"50d49'
Danielle
Cam 3h57'56"57d42'
Danielle
Cas 2h4'1"61d2'
Danielle & Sean
Crb 15h19'16"32d9'
Danielle Aileen
Uma 10h36'33"51d3'
Danielle B
And 0h55'1"45d33'
Danielle Elizabeth
Cnv 12h58'57"33d8'
Danielle Elizabeth
And 1h45'16"39d33'
Danielle Marie
Lyr 18h41'41"31d3'
Danielle Marie & Kristlyn Ashley
Mon 7h27'58"-8d55'
Danielle Meghan
Lyr 19h17'51"26d45'
Danielle My Belle
Sgr 19h22'27"-45d7'
Danielle Nichole
Sgr 19h8'22"-21d12'
Danielle Nicole
Cas 2h39'0"75d45'
Danielle Paige
Vul 19h14'45"22d11'
Danielle Star
Lyn 6h27'33"58d49'
Danielle Z Superstar
Cas 0h50'36"62d36'
Danielle's Angel Star
Com 13h26'48"25d1'
Danielle's Bobert
Cas 0h31'28"64d59'
Danielle's D'Light
Com 12h27'40"22d13'
Danielle's Dream 6794
Ori 6h0'36"-1d50'
Danielle's Dynamic Delight
Lyr 19h20'0"41d10'
Danielle's Shining Star
And 1h11'28"39d9'
Danielleneel
Lyr 19h21'14"38d13'
Daniello,The Star Of Chet & Blackie
Ori 5h19'59"1d48'
Danielly,Julian
Cyg 20h53'1"30d30'
Daniels Family,The
Aur 4h51'46"40d40'
Daniels the Potter, Greg
Dra 14h57'33"63d38'
Daniels,Bart
Cep 22h21'30"55d34'
Daniels,Bernadine
Peg 21h57'49"26d32'B
Daniels,Bonnie
Cet 2h46'34"5d37'
Daniels,Brett & Teresa
Crb 16h9'1"32d28'
Daniels,Cathryn Lisa
And 2h35'15"38d50'
Daniels,Charlie
Mon 6h18'60"8d26'
Daniels,Christopher W
Dra 16h44'58"60d30'
Daniels,Clyde & Mary
Cyg 21h32'40"41d23'
Daniels,Corrine Marie
Lyr 19h6'16"25d52'
Daniels,David Gray
Her 17h3'14"46d56'
Daniels,Dr Robert S
Oph 17h4'35"-1d14'
Daniels,Erica
Cyg 21h50'60"40d33'
Daniels,Gary & Cody
Dra 17h7'20"60d49'
Daniels,Jaime Lee
Umi 14h15'46"65d42'
Daniels,Jennifer
Cas 1h24'17"67d43'
Daniels,Jessica
Cas 1h27'21"58d52'
Daniels,John
Lac 22h53'45"55d6'
Daniels,Joseph
Dra 12h58'11"72d24'
Daniels,Jr,William Hopkins
Dra 16h9'47"51d48'
Daniels,Kellie
And 23h46'24"46d37'
Daniels,Lennart
Ori 5h34'21"-6d39'
Daniels,Merle Keith
Peg 22h58'39"18d25'
Daniels,Norma & Clair
Cnc 8h10'51"31d18'
Daniels,Norma Colleen
Vul 19h15'43"22d24'
Daniels,Paula
Cas 1h3'58"62d43'
Daniels,Peter & Nichol
Cyg 21h14'32"38d16'
Daniels,Raymond J
Peg 21h57'49"26d32'A
Daniels,Rebecca J
Peg 23h19'32"11d12'
Daniels,Renee
Peg 21h30'42"20d20'
Daniels,Rob & Jo Anne
Cyg 20h1'11"30d13'
Daniels,Robin Michelle
Cyg 19h12'18"47d12'
Daniels,Ross
Mon 7h41'24"-2d42'
Daniels,Ryan Joseph
Peg 22h23'11"31d5'
Daniels,Sue Lyn
Lyr 18h45'48"38d39'
Daniels,Sylvia Jean
Lyr 18h41'34"35d41'
Daniels,Vincent
Ori 6h3'43"0d42'
Daniels-Friedman Super Star,Buddy
Equ 21h7'31"11d57'
Danielson,Daren Roger
Per 1h46'1"50d3'
Danielson,Judith L
Ori 6h6'43"5d51'
Danielson,Ronald V
Per 3h46'46"37d34'
Daniely-Woolfolk,Eliza
And 23h38'38"39d30'
Danielzona
Cnv 12h48'1"51d29'
Danijosh
Aql 20h18'25"8d11'
Danila
Lyn 7h1'21"44d56'
Danilo
Cas 1h58'50"60d10'
Danilovic,Nikola
Lyn 7h51'0"51d51'
Danilovicz,Edward S
Cep 22h30'23"61d52'
Danilyuk,Larisa
And 1h37'0"38d35'
Danina,Helen
Mon 6h21'47"9d2'
Danise Lasso the Moon
Cyg 20h4'50"40d18'
Danish Revolution
Cam 6h8'25"83d24'
Danisi,Lisa Lynn
Lyr 19h22'14"31d4'
Danisi,Philip Michael
Lac 22h18'46"49d58'
Danka
Cam 6h13'16"71d33'
Danke Dir Gott, Anneliese List 6-1-22
Cap 21h28'19"-22d42'
Danke,Alan & Linda
Cyg 21h22'25"39d57'
Dankers,Josephine Benkert
Ori 5h53'53"13d27'
Dankner,Bessie & Nathan
Ori 5h57'58"18d28'
Danko Loving Memory, John Rowland
Cam 5h54'24"58d11'
Danko,Sharon
Lyr 18h51'33"41d0'
Danlis
Ori 6h0'56"-2d15'
Danmark Inc
Peg 22h23'59"2d24'
Danna Morningstar
Cas 2h57'24"58d44'
Danna,Light Of My life
Eri 2h59'21"-15d3'
Dannaker,James Patrick
Lac 22h29'1"55d57'
Dannard
Cam 3h16'0"58d24'
Dannat,Lothar H
Cap 20h24'2"-26d24'
Dannels,Bernie & Glenna
Dra 13h38'34"68d5'
Dannenberg,Richard B & Stephen Lowey
Per 3h11'49"56d58'
Dannenmann,Jay Andrew
Cnc 8h6'56"18d49'
Danner,Michael Lee
Eri 2h45'34"-7d4'
Dannheim Darling,Joyce
And 23h17'15"51d42'
Danni's Guardian Star For Nicky
Peg 22h38'35"20d8'
Dannica Robin
And 2h0'17"40d34'
Danning,Vicki
And 0h52'36"37d16'
Danny
Eri 4h0'46"-18d9'
Danny
Lyn 7h55'33"44d21'
Danny
Cep 3h9'58"78d31'
Danny
Cep 21h46'17"63d54'
Danny
Uma 8h34'41"52d34'
Danny
Aql 19h4'34"2d39'
Danny
Cnv 13h43'0"39d46'
Danny
Lac 22h4'27"46d14'
Danny
Lac 22h33'24"55d28'
Danny
Ori 4h55'1"1d20'
Danny
Umi 14h57'32"67d17'
Danny
Cep 23h2'51"62d41'
Danny
Gem 6h46'44"13d14'
Danny & Angie Forever And A Day
Lyn 8h17'19"57d58'
Danny & Anne Forever
Cyg 20h13'29"38d15'
Danny & Julie Love Star,The
Cyg 21h8'50"37d51'
Danny & Michael's Star
Sge 18h58'57"19d53'
Danny & Sally
Cyg 20h53'37"30d25'
Danny & Sandy's Star
Uma 9h59'46"44d49'
Danny & Stacey's Forever Love Star
Mon 6h19'50"8d52'
Danny B
Uma 9h1'1"50d49'
Danny Boy
Ori 6h8'33"0d49'
Danny et Elisabeth
Boo 14h34'47"12d24'
Danny Leone
Eri 4h4'11"-18d10'
Danny Love Poet
Boo 14h33'42"21d50'
Danny Loves Shawna
Hya 9h36'35"-6d55'
Danny My Shining Star
Mon 7h10'54"-5d31'
Danny Terence
Umi 15h32'32"70d58'
Danny's Dream
Gru 22h29'16"-50d35'
Danny's Dream
Her 18h9'31"50d27'
Danny's Flame
Cnc 9h0'1"32d32'
Danny's Mailbox
Aql 19h2'35"2d55'
Danny's Rose
Aql 19h54'0"11d49'
Danny's Star
Dra 18h58'17"58d22'
Danny's Star
Ori 5h1'1"-2d30'
Danny's Star
Aur 6h53'49"35d40'
Danny's Star
Cep 23h20'54"65d13'
Danny's Star
Cet 1h52'42"-5d56'
Dannys Lucky Star
Umi 15h31'58"70d13'
Dano
Aur 4h49'47"40d38'
Danon Star,The Richard
Ori 5h53'38"8d6'
Danot
Dra 16h29'44"60d41'
Dansan,Jean Patrice
Umi 16h29'29"79d30'
Dansby,Daniel Milton
Aql 19h0'23"-8d56'
Dansby,Grant & Kelly
Aql 20h14'19"1d25'
Dansereau,Richard
Gem 6h36'42"13d24'
Danson
Mon 6h48'46"-1d34'
Danson,Alexis
Equ 21h22'48"11d6'
Danson,Emily
Boo 13h58'24"11d21'
Danson,Jeffrey
Umi 17h19'13"75d28'
Danson,Jeremy
Boo 13h58'53"12d6'
Danson,Kate
Boo 13h35'56"11d17'
Danson,Kimberly
Per 2h16'58"58d43'
Danson,Natasha
Boo 13h57'34"11d40'
Danster
Dra 17h6'48"62d44'
Dante & Megan
Ori 5h49'0"10d52'
Dante Louis
Hya 8h19'49"-1d44'
Dante,Payton y
Hya 8h30'52"-8d10'
Dantro,Christine E
Cyg 19h14'23"44d3'
Dantuono's Dipper
Cep 0h46'52"78d55'
Danuser,Terence Derek
Ori 6h3'12"10d27'
Danuta
Cnv 13h32'1"40d14'
Danuta,Basia
Uma 11h32'55"42d54'
Danuta,Heléna
Cas 0h41'53"72d26'
Danuta-Anna Beloved Mother
Vul 20h0'36"23d7'
Dany
Ori 5h58'34"14d49'
Dany
Tel 18h6'22"-45d54'
Dany & Berna
Cyg 20h48'29"37d31'
Dany,Noel
Cam 7h47'14"80d20'
Danya
Umi 15h18'0"79d46'
Danyluik,Les
Dra 15h1'20"62d7'
Danylyshyn-Adams,Rhys
Aql 20h22'13"2d34'
Danyon's Star
Cam 5h57'44"61d23'
Danysh,Natalie
Cas 22h55'14"55d11'
Danz
Tri 2h27'28"30d27'
Danz,Lauren
Vul 19h5'15"24d59'
Danza,Robert Peter
Gem 6h27'41"14d40'
Danzeisen,Summer Paige
Equ 21h20'51"12d4'
Danzig A Great Mom, Delta S
And 2h25'22"48d53'
Danzig,Jeff
Aur 6h20'14"37d31'
Danzig,Purdie Meissner
Ori 5h37'48"-0d47'
Danzig:A Simple Man
Hobbes,Stephen D
Ori 5h10'1"5d47'
Danziger
And 0h45'48"37d42'
Danzis,Jo-Ann
Com 12h55'28"27d41'
Danùta 40
Cas 0h44'28"77d13'
Daoud,Dr David Said
Aur 4h51'48"41d7'
Dapalma,John & Cyndi
Lyr 19h25'39"41d36'
Daphanie
Eri 4h33'26"-12d28'
Daphne
Aqr 21h43'0"0d28'
Daphne
Crb 16h4'1"32d54'
Daphne Faithful Friend
Cet 3h4'1"2d8'
Daphne Marcella
Aqr 22h21'16"-2d5'
Daphne Star,The
Cyg 19h59'16"30d24'
Daphntom
Cyg 20h35'25"31d25'
DaPolito,Tami
Cas 1h43'34"60d55'
daPonte,Admiral Fuzeta
Aql 19h2'55"0d36'
Dapron
Uma 10h39'44"48d13'
Daquesian,Richard
Aur 5h9'26"38d6'
Dara
Mon 7h2'1"-6d34'
Dara
Pup 7h56'49"-23d17'
Dara
Lyr 19h13'13"35d33'
Dara Evelyn
Peg 22h33'25"27d53'
Darabant,Helen & Kevin Sorbo
Uma 9h22'59"59d46'
Darah Coryn
Mon 7h6'45"-6d7'
Daras
Lac 22h22'1"50d25'
Daravi,Caroline
Peg 23h22'1"16d42'
Darbashti,Mike
Her 16h43'31"20d15'
Darby
Aur 6h25'1"40d1'
Darby,Alan Ronald
Per 2h4'1"57d14'
Darby,Alexander Garrison
Boo 15h8'40"53d18'
Darby,Erin
Cas 1h25'35"68d43'
Darby,Julian
And 2h15'46"45d58'
Darby,Karen Elizabeth
Lmi 10h9'23"32d21'
Darby,Matthew Thomas
Cep 21h16'16"58d15'
Darby,Michelle C
Cas 0h29'30"60d24'
Darby,Nan E
And 0h36'12"40d42'
Darby,Paulette
Cam 5h48'25"60d6'
Darby,Rebecca
Vul 19h47'32"20d11'
Darby,Robert Michael
Aur 5h25'58"40d57'
Darby,Ruth Gilmartin
Cyg 21h19'46"36d44'
Darby,Sarah Galbraith
Mon 6h20'54"0d31'
Darbyshire,Jack F
Ori 5h57'1"11d34'
Darche,Denis
Aur 6h20'31"38d51'
Darcheville,Thierry
Dra 10h1'21"74d13'
Darck,Gregory Z
Del 20h21'55"8d4'
Darcy
Cep 23h49'8"81d21'
Darcy
Cam 6h59'52"64d27'
Darcy Celestial Love
Mates,Kim & Mike
Cyg 21h37'1"41d17'
Darcy Lynn
And 0h50'0"39d6'
Darcy's Heavenly Body
Cyg 19h57'27"38d14'
Darcy's Personal Star
Lyn 8h56'37"44d2'
Darcy,Christopher Michael
Ori 5h8'32"-5d51'
Darcy,Marguerite
Cas 2h58'0"73d20'
Darcy,Rod
Cnv 12h12'35"50d30'
Dard,Mova
Pic 4h43'17"-47d30'
Dard,Oryus
Pho 0h41'17"-48d9'
Dardard,Joseph A (Sea Eagle)
Ori 5h40'29"8d43'
Darden
Gem 5h58'45"26d36'
Darden,Butch
Dra 17h3'1"52d8'
Darden,Christopher A
Cet 2h2'32"2d18'
Darden,Willie
Lac 22h40'30"55d20'
Dare,Angela
Cyg 19h34'22"33d28'
Dare,Miss BIlly
Lyn 8h4'22"38d49'
Dargush "Pigeon", Margaret & Howard
Umi 15h17'25"65d60'
Daria
And 1h5'28"47d32'
Daria
Cyg 21h53'45"53d14'
Daria
Vul 20h20'26"25d27'
Daria
Gem 6h39'24"14d55'
Daria Lauren
Oph 17h31'0"8d57'
Darian
Lac 22h28'38"54d21'
Darienne
Lyr 18h25'38"45d14'
DaRif's Domain
Aur 6h3'25"54d39'
Darin & Alida's Star (ABZDAB)
Cyg 21h2'37"31d6'
Darin's Nova
Lmi 10h39'25"24d35'
Dario
Oph 17h3'14"10d7'
Dario
Pho 0h7'24"-42d34'
Dario il grande
Pic 4h59'9"-46d43'
Darion Robert
Cep 1h50'24"77d37'
Dariush Shalali
And 0h11'36"40d40'
Dark Sunflower
Cas 0h58'58"71d29'
Darkfawn
Cnv 12h12'42"34d26'
Darkiki
Cnv 13h28'54"50d48'
Darl
Aur 06h12'50"50d27'
Darla Jean
Lyr 19h15'14"41d17'
Darla Jo
Dra 15h30'22"65d22'
Darlak,Craig
Cmi 7h5'47"1d18'
Darleen
Cap 21h36'37"-19d29'
Darlene
Cas 0h39'14"67d39'
Darlene
Oph 17h8'48"11d57'
Darlene
Cas 1h56'23"70d41'
Darlene Ann
Cas 0h2'0"58d25'
Darlene Jayne
And 0h16'50"37d13'
Darlene Marie
Lyn 8h4'42"38d43'
Darlene's Heart
Lyn 7h50'50"45d57'
Darlene's Heart
Lyn 7h1'41"52d50'
Darlene's Heart
Lyn 8h20'1"51d27'
Darles,Jean et Suzanne
Sgr 19h31'31"-35d34'
Darling
Her 17h54'23"31d28'
Darling Amy & John Star,The
Cyg 19h55'59"47d17'
Darling Bud
Dra 17h24'11"68d42'
Darling Charlie
Cap 21h0'1"-14d46'
Darling Charlie
Crb 16h15'1"33d55'
Darling Darlene
And 0h59'11"45d45'
Darling Dean
Aql 19h24'48"13d45'
Darling Debra
Leo 11h54'48"22d12'
Darling Derek
Cep 23h19'15"80d10'
Darling Dorothy
And 23h47'45"41d28'
Darling Erika
Lyr 19h0'31"33d56'
Darling Judy's Star
Uma 12h2'1"32d56'
Darling Kathie
Ori 5h47'24"20d25'
Darling Little Bertie
Cma 6h31'17"-16d9'
Darling Nikki's Shooting Star
Aql 19h28'49"10d36'
Darling Tommy
Her 16h46'54"13d24'
Darling,Daria
Lep 5h25'1"-20d2'
Darling,Karen
Cma 6h43'57"-15d17'
Darling,Philippa
Cyg 19h34'53"33d17'
Darling,Sr,Kenneth T
Aur 5h57'26"29d26'
Darlington's Spring "Chilly Bleak"
Aur 5h56'20"40d45'
Darlington,J R D III
Her 17h17'49"45d26'
Darlington,Meredith Elizabeth
Mon 6h25'57"-6d33'
Darlington,Michael Henry
Cet 2h23'15"-1d54'
Darmanaden,Robert
Equ 21h2'55"7d45'
Darmiento,Lyndsey Marie
Cas 23h39'0"62d4'
Darmody,Christine
Peg 21h6'43"12d8'
Darmon,Zara
Per 2h51'49"48d59'
Darnall,Michael P
Peg 22h50'59"25d11'
Darnel,Virginia
Cas 23h16'47"60d6'
Darnell,Bob
Aur 6h27'17"35d48'
Darnell,Cecil
Crt 11h9'17"-12d38'
Darnell,Jennifer Lynne
Lyn 8h15'48"40d48'
Darnell,Jerry W
Ser 17h34'45"-13d46'
Darnell,Mark
Crt 11h2'36"-12d23'
Darnell,Rebecca
Lyr 19h2'20"25d41'
Darnell,Sydner
Crt 11h3'41"-15d9'
Darnik
Lyr 19h21'32"42d31'
Darnstaedt,Gail Frances
Lac 22h36'0"55d17'
Daroca,Robert Edward
Aql 20h3'11"4d51'

Darold
Cep 21h7'36"55d13'
Darolia,Renuka
Tri 2h0'1"28d30'
Daros
Ind 21h10'23"-50d20'
Darr,Sharon
Cma 7h3'0"-15d22'
Darran
Lya 18h17'58"30d25'
Darras,Christine
Peg 23h36'52"10d1'
Darras,Peter S
Sct 18h43'1"-6d1'
Darre-Mueller
Her 17h12'1"29d42'
Darrell's "Wishing Star"
Per 2h31'20"57d34'
Darrell,Friendly Father,Bud
Her 16h55'13"27d12'
Darrell,Robert E
Cet 2h1'0"0d55'
Darren
Del 20h49'41"7d36'
Darren & Rachel June 15,1996
Crb 16h6'25"33d50'
Darren & Rona Wedding Star
Cyg 21h9'17"37d40'
Darren & Wendy Forever
Vir 11h35'54"-3d7'
Darren 062373
Tri 1h48'33"27d15'
Darren's Star
Lyn 7h52'47"40d36'
Darrenia
Cas 23h19'57"63d11'
Darrian Joseph
Aur 6h30'59"32d14'
Darrin Forever,Sherie N
Lyn 7h36'41"38d29'
Darrin,Calvin H
Cam 5h59'52"60d9'
Darrin,Drake
Cam 4h39'48"68d25'
Darrol,Angela Ilene
Lyn 8h20'20"58d46'
Darron
Hya 8h13'29"1d26'
Darrow,Dean G
Aur 6h41'36"38d47'
Darrow,Duane
Her 16h36'19"27d40'
Darrow,Russell James
Ori 5h56'20"17d28'
Darryl
Ori 4h51'38"0d44'
Darryl's Dream
Aql 19h7'53"0d42'
Darryl's Star
Aur 7h20'53"38d48'
Darryn
Ori 5h36'9"-0d53'
Darsey,Patricia Ann
Cas 0h42'11"61d11'
Darshan
Uma 8h52'36"70d43'
Darshelle,Dayna
And 22h59'45"50d14'
Darshita Bellissima Stella
Cas 1h20'17"76d34'
Darsley,Norma Bowden
Uma 11h29'0"42d34'
Darst (Psalms 139), Aeryn Celine
Cet 3h19'37"2d17'
Darst,Christy Ann
Cas 23h42'16"64d34'
Dart,Anne S
Cet 0h54'16"-7d14'
Dart,Rollin & Mary Vay
Cam 3h58'30"78d37'
Dart-Hooton,Ruby J
Mon 6h40'15"10d4'
Dartmouth College Marching Band
Boo 15h28'35"37d37'
Darty,Florence
Uma 14h24'57"60d16'
Darty,Paulette & Bernard
Sge 19h58'41"20d32'
Darwich,Ghousson Al
Ori 5h56'16"21d10'
Darwin,IX01-095, Carolyn
Hya 8h53'23"1d9'
Daryl & Sandy's Castle in the Air
Cyg 20h34'53"42d16'
Daryn & Cyndee
Cyg 21h50'29"40d36'
Darzas,Paules
Cas 22h58'57"54d49'
Darzynkiewicz
Uma 10h10'0"50d52'
DAS
Cyg 19h43'19"31d0'
Dasaro,Joseph
Dra 14h15'10"63d46'
Daschbach,Charlie
Aql 20h7'29"1d30'
Dascoulias,Steven's Star-Steven
Cnc 8h31'38"31d45'
Dash #1 Dad,Bob
Oph 17h35'0"-6d59'
Dashevsky,Iris
Com 13h26'1"25d12'
DaShiell,Celeste Angelica
Cnv 12h46'24"33d56'
Dashira
Umi 15h22'27"80d15'
DaSilva,Eric Wayne
Eri 3h13'1"-18d4'
DaSilva,Osvaldo
Cep 22h2'19"62d53'
DaSilva,Yara
Ori 5h55'45"17d15'
Dastin,Elizabeth
Mon 6h14'0"-10d17'
Dastot,Christian
Com 12h30'12"31d30'
Dastur,Mary Ellen & Homi
Lyn 7h6'32"60d59'
Dasya
Crt 11h2'0"-18d49'
Daszenski,The Star of
Lac 22h37'30"38d41'
Data Resource Group, Inc
Boo 15h5'17"51d20'
Datamarc Computer
Cnv 12h53'25"50d14'
Datasource Direct, Inc
Lmi 11h2'35"32d39'
Datené,Elvira
Cap 20h25'18"-26d29'
Datermos,Kostantien
Aql 18h57'42"17d23'
Dates,Claire
Com 12h2'24"28d22'
Datin,Dennis Dean
Eri 3h20'51"-7d58'B
Datin,Evette Ross
Eri 3h20'51"-7d58'A
Dato,Joe & Crissy
Cep 20h24'14"75d15'
Datta,Doris
Cas 0h43'1"64d55'
Dattilo,Joseph L
Cma 6h10'0"-18d55'
Daub,Richard Paul
Sex 9h56'38"-5d22'
Daubersmith,Gary H
Cet 3h13'32"-0d36'
Daubert,J W
Aql 19h58'40"13d44'
Daubert,Mary Ann
Com 13h1'51"20d39'
Dauer,Daryl
Aur 6h31'39"30d32'
Daugharty,II,Duane A
Leo 10h52'32"-0d58'
Daugherty
Cyg 21h32'54"40d60'
Daugherty,Annie
Aql 19h1'26"16d14'
Daugherty,E B "Bud"
Mon 6h37'15"1d50'
Daugherty,Jean (Playlady)
And 23h25'23"47d29'
Daugherty,Jennifer Marie
Peg 21h58'28"34d9'
Daugherty,Mandy Lynn
Lyr 18h55'1"40d15'
Daugherty,Margie Eichelberger
Eri 2h47'26"-16d37'
Daugherty,Stacey Ray
Peg 22h48'36"4d43'
Daughetees Discovery
Cyg 20h23'1"30d50'
Daughtery,Maris
And 23h49'1"40d55'
Daughtry,Dave
Aql 19h30'0"10d25'
Daughtry-Daughter, Janine Ionia
Peg 22h18'14"34d51'B
Daughtry-Mother, Bernice Winona
Peg 22h18'14"34d51'A
Daul,Shannon
Lyn 7h39'12"45d2'
Daulby,Sarah
Lyn 8h23'40"44d35'
Daum,Courtenay W
Lyn 8h56'0"45d1'
Daum,Joan & Warren
Peg 23h17'1"30d20'
Daum,Ruth Elizabeth Gilkison
And 0h45'31"28d11'
Daunas,Toujours Reunis Yves & Gislaine
Per 2h50'25"37d33'
Daunis,Amanda
And 0h58'39"45d59'
Daunis,Rosemary
Cas 2h4'14"61d50'
Daunizeau,Chantal
Aur 7h9'42"38d46'
Daunt,Andrea Katherine
Peg 22h29'17"31d8'
Dauphin,Gary
Hya 10h14'24"-11d37'
Dauphinais,Ethan Tyler
Dra 15h53'1"61d5'
Daure-Bournazeau, Estelle
Dra 12h32'10"64d28'
Dauro,MD,Albert T
Oph 17h0'28"11d11'
Daury,Jacqueline
And 1h50'1"36d21'
Dav's Infinite Wishmaker
Aur 6h25'14"33d47'
Davalle,Marlene V
Aql 19h17'1"10d22'
Davalos,Elizabeth "Liz"
Mon 6h41'21"6d7'
Davamy
Ori 6h1'39"-1d35'
Davamy
Dra 12h46'45"68d32'
Davaree
Mon 6h20'40"-1d4'
Davarian,Reza
Uma 10h12'55"42d0'
Dave
Umi 14h47'21"68d35'
Dave
Ser 15h12'16"8d5'
Dave
Per 1h57'30"56d58'
Dave
Cep 2h7'1"77d37'
Dave
Cep 21h12'37"55d19'
Dave & Betty
Crb 15h50'34"28d15'
Dave & Bob Star,The
Her 16h42'31"48d16'
Dave & Carrie's Destiny Star
Cyg 19h24'40"32d46'
Dave & Debby
Hor 3h34'39"-45d36'
Dave & Dixie
Cyg 21h12'18"38d48'
Dave & Janet
Sge 19h55'22"16d7'
Dave & Kettles
Cyg 19h31'19"35d16'
Dave & Lily
Eri 3h56'17"-12d51'
Dave & Lizzie
Lyn 7h45'33"44d38'
Dave & Trix
Ari 1h45'44"18d55'
Dave Anthony
Aur 5h24'16"41d7'
Dave Frances
Boo 14h23'42"39d51'
Dave Loves Christen
Cyg 20h17'58"38d21'
Dave Part Of The Magic Ellis
Cep 21h29'10"78d45'
Dave's "Star" Trek
Sct 18h52'18"-9d35'
Dave's Asteroid B-612
Tri 2h35'0"34d38'
Dave's Big Dipper
Uma 12h3'23"58d36'
Dave's Destiny
Her 18h3'48"40d32'
Dave's Dream
Umi 14h20'50"68d2'
Dave's Happiness 0425
Dra 13h21'37"68d2'
Dave's Infinite Dreams
Cap 21h25'34"-24d39'
Dave's Lite
Her 16h19'57"20d29'
Dave's Moon
Cmi 7h36'23"1d23'
Dave's Shining Sherlock
Per 3h3'1"48d46'
Dave's Star
Ori 5h11'25"-5d41'
Dave's Star
Aql 19h3'22"-6d47'
Dave's Star
Oph 18h7'14"13d11'
Dave,Hazel
Cyg 20h1'39"30d14'
Dave,Love Forever & Ever-Amen Tami
Lyr 18h28'21"45d54'
Davel
Lac 22h25'55"55d17'
Davella
Uma 10h22'53"68d10'
Davenport "Bobby", Robert R
Cep 21h5'17"60d20'
Davenport,Diane & Craig
Lyr 19h24'43"40d53'
Davenport,Ethel
Lyn 7h29'25"58d8'
Davenport,Frank T
Ori 5h59'23"15d26'
Davenport,Glenna
Uma 10h55'18"48d3'
Davenport,Jessica
Lyn 7h29'33"50d20'
Davenport,Lance
Cep 21h12'57"68d45'
Davenport,Larry W
Boo 14h33'0"43d13'
Davenport,Mark D
Eri 2h47'50"-2d26'
Davenport,Mary Barbara
Mon 6h1'1"-5d40'
Davenport,Muriel Rae
Del 20h18'54"11d13'
Davenport,Preston
Oph 17h55'11"13d48'
Davenport,Samantha
Mon 7h4'57"4d33'
Daverio,Charles A
Crt 10h54'49"-6d34'
Daves Keia
Aur 6h33'54"31d40'
Daves,Michael James
Cet 2h35'17"-6d15'
Daves,P J
Cam 10h51'15"82d26'
Daves,Vonnie
Cyg 20h18'29"30d45'
Davet E T
Lac 22h36'44"53d45'
Davette & Che Together Forever
Cyg 21h54'1"53d29'
Davey
Cep 23h12'19"60d50'
Davey
Ori 6h6'46"4d43'
Davey N
Aql 19h48'1"14d6'
Davey's "Heavenly Place" DCA 71393
Peg 22h34'45"30d54'
Davey's Star,Erica
Eri 4h4'45"-2d30'
Davey,Becky
And 1h17'56"41d14'
Davey,Bunny
Del 29h22'17"10d50'
Davey,James S
Uma 10h49'45"44d47'
Davey,John
Lyn 8h19'27"49d48'
Davey,Michael Jerome
Aql 19h50'54"15d26'
Davey,Payton Aldah
Aur 7h17'58"37d27'
Davey,Robert Joseph
Uma 9h53'13"47d47'
Davey,Thomas Hance
Sct 18h30'52"-6d50'
Davey-Travers,Dr Rosalie
Lyn 8h25'26"46d36'
Daviau,Sylvie
Umi 15h21'19"66d26'
David
Ori 5h10'26"-1d2'
David
Her 16h23'0"23d3'
David
Aur 4h58'24"38d24'
David
Cep 22h19'55"70d10'
David
Boo 15h13'44"27d30'
David
Aql 19h0'22"12d8'
David
Cmi 7h59'53"4d17'
David
Lac 22h5'37"49d57'
David
Eri 4h18'23"-11d58'
David
Cet 2h24'58"-0d15'
David
Her 17h34'47"20d7'
David
Sco 16h25'14"-40d40'
David
Tri 2h26'44"30d16'
David
Per 4h2'13"37d59'
David
Her 16h58'34"41d5'
David
Oph 18h15'19"0d11'
David
Ser 15h15'28"10d1'
David
Aql 18h59'57"-6d55'
David
Oph 17h30'33"8d51'
David
Cet 1h27'25"-4d46'
David
Ori 5h17'52"-4d59'
David
Del 20h20'53"10d38'
David
Cam 3h30'60"60d2'
David & AnnMarie
Cyg 21h18'38"38d29'
David & Barbara's Christmas Star
Eri 3h43'33"-0d17'
David & Bevin
Per 4h40'58"52d19'
David & Breena aka Hawk & Wolf Eyes
Del 20h21'1"10d28'
David & Cindi
Col 5h59'56"-28d20'
David & Dawn's Eternal Love
Aql 18h43'41"11d30'
David & Debbie's Sparkles
Peg 21h59'27"29d55'
David & Diane
Uma 11h56'14"32d54'
David & Diane
Mon 6h54'0"0d11'
David & Erika
Ori 4h59'34"-0d16'
David & Ginny
Uma 10h56'24"41d8'
David & Goliath
Boo 13h35'51"20d52'
David & Hazel"The Lovers Formation"
And 0h38'58"40d49'
David & His Lady
Sge 19h2'55"20d10'
David & Iris
Eri 2h56'32"-2d46'
David & Jane Got Married!
Peg 23h26'53"27d34'
David & Janet
Cyg 21h54'47"37d11'
David & Jean Marie- Bonafide
Boo 14h53'44"41d15'
David & Karen
Vul 19h42'45"23d39'
David & Kiersten
Cyg 21h10'14"37d16'
David & Kim
Crb 16h18'0"30d28'
David & Kristal's Wishing Star
Cas 23h2'39"58d23'
David & Leighanna
Cyg 20h57'44"31d50'
David & Leslie Forever
Lyr 18h55'22"32d22'
David & Marty's Home
Cot 2h44'20"4d40'
David & Pam Soulmates Forever
Lyr 18h57'47"31d6'
David & Ruth Golden Anniversary Star
Cyg 21h32'38"38d4'
David & Susan
Uma 11h9'40"38d41'
David & Terry For-Ever
Cyg 19h21'32"48d58'
David & Valerie Forever
Cyg 20h56'23"48d56'
David & Wendy
Cyg 19h31'47"33d18'
David Alan
Ori 5h36'26"-4d2'
David Allen
Uma 11h51'13"38d40'
David and Myra Forever
Cet 1h55'29"-2d3'
David Andrew George
Cma 6h14'1"-24d52'
David Anthony
And 23h49'21"37d36'B
David Anthony & Dawn Michelle 1995
Cyg 19h32'55"30d48'
David Curtis
Per 4h45'48"50d12'
David Darling
Lyr 18h41'46"37d55'
David Gregory Duck
Lib 14h19'38"-22d58'
David Grégoire
Lmi 10h0'28"39d27'
David James,Our Little Star
Per 2h57'48"32d2'
David Joseph
Lib 15h15'32"-19d59'
David Joseph Power
Aur 5h55'29"31d13'
David Lee
Sgr 19h8'0"-20d3'
David Lee "Easy Evans"
Sct 18h43'36"-5d45'
David Loves Melanie
Cyg 19h26'35"33d3'
David Mark
Tri 1h46'47"26d38'
David Michael II
Uma 10h25'54"40d50'
David Patrick
Dra 16h53'1"61d55'
David Paul
Ori 5h51'47"15d15'
David R
Lac 22h27'16"53d16'
David R
Aqr 22h1'37"-8d53'
David Robert
Her 17h38'26"26d31'
David Rules OK
Cep 23h11'39"64d44'
David Saint
Boo 14h20'33"48d43'
David Schnitzer
Uma 13h45'21"51d58'
David Scott
Boo 14h12'12"44d2'
David Scott & Jennifer Lyn
Boo 15h7'26"28d17'
David Thomas
Cam 4h50'53"70d31'
David Thomas
Her 18h6'47"47d34'
David William
Cep 22h54'45"58d53'
David X
Sge 19h3'60"19d19'
David You Are My Shining Star
Sex 10h0'43"-6d5'
David's & Katy's Daddy
Mon 8h8'33"-8d2'
David's (Grin)
Sct 18h54'60"-6d51'
David's Angel
Crb 16h13'12"38d43'
David's Cherub
Cep 22h40'15"56d58'
David's Darling
Cep 3h28'0"78d30'
David's Dazzler
Dra 16h16'1"60d36'
David's Delight
Cet 2h50'17"0d6'
David's Diamond
Aur 5h28'56"38d10'
David's Dimple
Hya 8h32'56"-6d35'
David's Elf Star
Cnv 13h22'34"37d44'
David's Girl
Lyn 8h44'28"37d60'
David's Guiding Light
Per 3h19'30"40d34'
David's Heart
Dra 11h32'30"71d55'
David's Inspiration
Sct 18h46'21"-6d7'
David's Jewell
Her 18h55'11"12d20'
David's Kitchen
Cet 2h18'40"6d24'
David's Night Watch
Her 16h41'19"25d53'
David's Star
Dra 16h10'55"68d33'
David's Star
Cam 3h51'28"61d13'
David's Star
Ori 6h3'42"3d6'
David's Star
Cma 6h42'58"-15d53'
David's Star
Boo 14h5'1"32d23'
David's Star of Gloria,Earth 1993
Aql 19h58'23"-0d51'
David's Wishing Star
Ori 5h55'36"13d40'
David, We Love you Forever
Umi 16h43'35"75d4'
David,A Peaceful Warrior
Sct 18h42'34"-6d18'
David,Carol Laura
And 0h38'36"41d15'
David,Christine
Per 3h28'30"41d3'
David,Edwin Fred
Eri 4h6'0"-19d18'
David,Eric Dominic
Dra 19h0'59"58d40'
David,Gace
Ori 5h56'58"10d37'
David,Grace
Mon 7h3'27"5d32'
David,Harry Salim
Aur 6h16'53"38d38'
David,Joan
Cas 1h34'47"76d27'
David,Karen Jean
Cyg 20h34'51"46d12'
David,Ludmila
Cep 22h54'58"65d5'
David,Marie-Therese
Aur 5h2'41"50d35'
David,MD,Sami
Oph 16h21'43"1d19'
David,Michael Jude
Hya 9h3'50"3d12'
David,Roswitha
Cam 7h52'25"73d22'
David,Shining for Infinity
Sex 9h42'0"2d51'
David,Timothy Clark
Per 2h7'39"58d35'
David,You Are My Star Love,Francine
Per 3h38'48"37d41'
David-Joshua
And 23h9'21"48d49'A
David-My Hero Forever
Ori 6h17'30"8d24'
David-My Shooting Star
Aql 19h46'58"12d45'
David-Star of Destiny
Cyg 20h59'47"50d5'
David-Thomas Star
Aql 20h20'38"5d15'
Davide
Aur 5h16'25"41d24'
Davide
Del 20h17'54"11d13'
Davide e Ilaria
Scl 23h14'33"-30d41'
Davide e Paola
Umi 17h15'25"75d53'
Davidius K
Dra 17h0'58"63d36'
Davidman,Robyn
Cas 23h16'1"60d45'
Davido,Michael Sean
Peg 23h26'47"24d3'
Davido,William Joseph
Her 16h39'25"11d45'
Davidovits,Lola
Equ 21h18'18"3d22'
Davids,Betsie
Cnv 13h56'36"32d1'
Davids,Samuel William
Cam 5h46'60"74d34'
Davids,Timothy R
Per 4h2'59"38d20'
Davidson
Com 13h17'49"18d1'A
Davidson Design International, Inc
Dra 19h37'24"68d22'
Davidson III,Harry Waldo
Per 2h10'0"57d12'
Davidson,Adam Andrew
Ser 15h39'0"0d58'
Davidson,Arlene Conner
Lyr 18h54'29"34d38'
Davidson,Bryce
Tau 5h48'28"23d39'
Davidson,Claire
Cyg 20h37'11"45d43'
Davidson,David Fulton
Cep 1h5'40"78d39'
Davidson,David Jeremy
Aur 6h4'25"31d6'
Davidson,Diane
Lmi 9h50'37"34d37'
Davidson,Emily Michelle
And 23h19'13"47d23'
Davidson,Gerry M
Oph 17h33'15"-20d12'
Davidson,Hans
Cyg 21h31'17"42d48'
Davidson,Helene
Lyn 7h44'13"40d26'
Davidson,Herb & Veronica
Cyg 20h56'1"30d49'
Davidson,James Ervin
Aql 20h30'46"0d39'
Davidson,John
Cas 0h33'55"62d3'
Davidson,Jolyn
Aql 19h11'25"10d39'
Davidson,Julie
Cas 2h2'0"63d55'
Davidson,Kelsey Mae
Cyg 19h29'46"33d8'
Davidson,Larry Thomas
Cam 3h26'10"50d1'A
Davidson,Margaret
Lyr 18h33'10"38d36'
Davidson,Mark
Ori 5h57'58"-1d50'
Davidson,Mary Edith
Cam 3h26'10"50d1'B
Davidson,Michael A
Cam 6h7'0"61d3'
Davidson,Michael Scott
Peg 21h49'1"34d1'
Davidson,Richard
Aql 19h20'13"12d55'
Davidson,Robin
Cyg 21h27'0"28d53'
Davidson,Ross
Boo 14h32'0"21d9'
Davidson,Ryan
Aur 6h24'56"38d40'
Davidson,Sharon
Com 12h52'44"30d60'
Davidson,The Great Rob
Dra 16h13'24"60d15'
Davidson,Tim & Lisa
Uma 16h12'18"80d34'
Davidson,Tracy
And 1h44'30"39d15'

Davidson,Yvonne Umi 17h28'44"80d14'
Davidson-Chusid,Jesse Rae And 0h44'28"40d41'
Davidson-Lifelong Socialist,Scott Ori 5h58'14"8d35'
Davidstern Cep 22h7'59"68d3'
Davidus Raymondus Her 18h4'60"28d38'
Davidynia Lyn 6h54'1"54d50'
Davie,Barry & Julia Cyg 20h54'58"37d55'
Davieau,Axel James Per 3h55'39"40d39'
Davies Aur 4h54'12"41d1'
Davies Born 1933, Godfrey Henry Uma 14h19'33"58d52'
Davies I Love You This Far Star,The Joanne Ori 5h55'41"17d14'
Davies Jan 4 1926, Kenneth Ernest Cep 0h4'22"68d26'
Davies,Alwyn R Aur 7h8'0"38d16'
Davies,Amy Uma 12h19'1"53d27'
Davies,Andrew Martin Her 17h4'11"44d41'
Davies,Austin James Aur 5h55'13"29d11'
Davies,Chill Will J Aur 6h4'53"38d51'
Davies,Christopher Scott Cet 2h9'0"3d5'
Davies,Dawn Lyr 18h14'14"42d49'
Davies,Dianne & Gemma Cyg 21h43'35"31d52'
Davies,Elizabeth McCall Com 12h48'29"21d17'
Davies,Elson Nichols Ori 5h52'0"16d6'
Davies,Erin Maureen Per 3h14'56"50d34'
Davies,Faye Cnv 12h47'15"38d11'
Davies,In Memory of Elaine Cyg 21h3'1"36d45'
Davies,James Edward Stewart Ori 6h4'0"4d43'
Davies,Jessica Anne Cas 0h38'18"74d6'
Davies,John Ori 5h36'25"-6d43'
Davies,June Aql 18h57'29"17d47'
Davies,Kathleen Ann Cas 0h18'31"62d10'
Davies,Lianne Cyg 21h53'46"40d26'
Davies,Lisa Com 13h16'28"27d4'
Davies,Maria Lyr 19h5'21"28d54'
Davies,Maura Patricia Cas 0h38'0"74d25'
Davies,Melissa Bethany Com 12h53'28"23d45'
Davies,Micah Boo 13h36'46"22d33'
Davies,Michael Henry Uma 9h5'19"54d42'
Davies,Michael Jeffrey Aql 19h10'44"13d28'
Davies,Michael Paul Cep 20h36'25"76d3'
Davies,Nigel Cam 3h46'20"68d6'
Davies,Owen Her 17h13'19"48d25'
Davies,Paul Jackson Umi 14h31'50"71d56'
Davies,Paula "Sweets" Lyr 18h42'31"34d38'
Davies,Rick Vir 13h43'36"5d21'B
Davies,Roy & Meichelle Aql 19h2'15"-0d8'
Davies,Scott James Lyn 7h44'18"39d59'
Davies,Shawn M Ori 6h16'52"-0d57'
Davies,Stephen Ori 6h8'1"1d22'
Davignon's Alternative Moon,Barbara Cyg 21h3'38"48d46'
Davila,Robert B Dra 19h4'31"58d30'
Davileen Boo 15h16'1"34d38'
Davilyn Cet 2h49'55"6d3'
Davin Her 17h16'52"42d45'
Davin,Anne-Marie Sge 19h59'51"20d11'
Davincent Her 18h4'16"28d8'
Davingia Cam 6h18'49"78d54'
Davino Aur 5h35'17"37d55'
Davino,Donald J Aur 7h9'51"40d4'
Davino,Hank (Henry) Cep 22h41'59"57d48'
Davino,Michael Cep 2h3'30"86d44'
Davis "Starship", Ronald Junior Aql 20h30'27"0d8'
Davis (August 19,1942) Mr & Mrs Ernest Peg 22h13'59"29d47'AB
Davis (Uncle Tim), Timothy K Aur 6h15'14"45d37'
Davis 7132,Mike & Lisa Ori 5h11'32"-5d53'
Davis 870214,Dean Carl Cep 22h48'51"57d18'
Davis Family of Crosby Texas,The Peg 0h9'15"17d40'
Davis II,Donald T Scu 18h54'34"-13d36'
Davis III,Corbett Hya 9h7'28"2d20'
Davis III,John Manley Cep 22h8'52"61d9'
Davis In Loving Memory,Herbert H Cmi 7h34'50"3d17'
Davis Link-Up,The Gary & Sandra Lyn 7h54'33"58d43'
Davis Love Hec 95, Marilyn & Horace Lac 22h17'45"49d9'
Davis Sister Plus Dad, The Four Cyg 19h32'29"34d22'
Davis Spiritual Mother Terry S Peg 22h10'32"27d14'
Davis Star,Waldo & Inez Cyg 20h34'22"30d4'
Davis Starship,The Sandra Jean Lyn 7h43'31"58d41'
Davis Y Salazar,Andrés Fulton Eri 3h59'28"-19d36'
Davis' Destiny Aur 6h27'24"35d57'
Davis's Best,Eddie Aql 20h1'0"0d16'
Davis(Precious),Gary L Cep 21h45'19"61d20'
Davis,"Buddah Bear" Robert W Hya 9h3'33"-6d6'
Davis,"Spread Your Little Wings",Jacqui Cyg 19h51'49"58d33'
Davis,"Super Star" Gary Cyg 19h21'47"28d49'
Davis,"Why Stars Shine!",Carla Mon 6h54'26"1d15'
Davis,2 Linda Peg 0h11'28"13d12'
Davis,Adrianne Z Peg 22h39'44"27d23'
Davis,Alan Aur 6h1'56"31d30'
Davis,Alana Kate Sgr 15h57'31"-22d17'
Davis,Albert H Uma 10h35'0"40d20'
Davis,Alisha Leigh Aql 20h12'33"12d55'
Davis,Alizabeth La Dia Peg 23h34'36"31d35'
Davis,Allison Renee Peg 21h58'23"29d33'
Davis,Alyssa Marie Lyr 18h32'41"41d5'
Davis,Amy Jolyn Cas 23h1'52"58d39'
Davis,Andrew Peg 23h45'38"30d21'
Davis,Andrew Gerald "Andy" Lac 22h4'16"38d41'
Davis,Andrew John Per 3h18'47"41d46'
Davis,Angeline H Peg 22h22'1"5d55'
Davis,Anna Elizabeth Cam 8h4'30"83d5'
Davis,Ariel Emily Cas 1h58'30"58d33'
Davis,Arleen Sge 19h7'38"19d55'
Davis,Artie Aur 6h30'58"37d52'
Davis,Ashlee Janaye Mon 6h18'1"1d24'
Davis,Astra Peter Ivan Aur 5h5'42"44d36'
Davis,Barbara Aql 20h3'25"1d47'
Davis,Barbara Cyg 20h35'50"48d48'
Davis,Barbara Uma 8h48'53"49d13'
Davis,Ben Hya 8h10'29"1d40'
Davis,Benny J Aql 20h3'12"1d29'
Davis,Bill Ori 5h59'44"11d24'
Davis,Bill Lac 22h33'32"52d34'
Davis,Bob "B'Daddy" Aql 19h29'15"13d16'
Davis,Brandon Dra 16h54'52"65d46'
Davis,Brandon Tyler Uma 11h22'20"38d9'
Davis,Brian Christopher Dra 19h59'31"61d34'
Davis,Brian Aur 6h27'16"36d34'
Davis,C Scott, ("Scooter") Ori 4h52'56"1d11'
Davis,Carol Del 20h14'0"12d8'
Davis,Catherine Rosenberg Lyr 18h56'33"37d22'
Davis,Catherine Amelia Ori 5h55'23"17d18'
Davis,Catherine Brady Eri 4h14'1"-16d57'
Davis,Cathleen J Dra 11h43'55"70d17'
Davis,Charles "DC" Oph 18h1'22"12d46'
Davis,Chelsea Ann Mon 6h20'21"-6d8'
Davis,Cheryl Sex 9h48'12"3d52'
Davis,Christie A And 23h39'53"46d35'
Davis,Christine Moulton Aur 7h13'1"35d50'
Davis,Christopher Her 17h12'48"22d42'
Davis,Christopher John Ori 5h50'0"18d25'
Davis,Chuck Her 16h48'15"33d14'
Davis,Chuck "Soulmate" Ser 15h29'30"19d29'
Davis,Chuck & Mary Vul 19h48'26"28d19'
Davis,Collin Geoffrey Boo 14h34'11"48d11'
Davis,Constance & Leonard Hya 8h54'21"2d11'
Davis,Daniel Lyn 7h43'48"43d40'
Davis,Daniel J Her 17h29'13"31d19'
Davis,Danielle Renee Cas 0h22'34"73d50'
Davis,Dave Cmi 7h46'1"8d44'
Davis,David Per 1h50'1"56d37'
Davis,David Michael Lac 22h23'42"52d46'
Davis,Delia Madeline Marzee Equ 21h2'54"2d32'
Davis,Delores Ann And 23h26'0"47d35'
Davis,Derrill J & Sarah T Crb 15h54'57"27d32'
Davis,DF Aql 19h30'48"1d55'
Davis,Diane Crv 12h35'12"-12d16'
Davis,Diane Boo 14h48'55"25d10'
Davis,Don & Elaine Cyg 21h4'59"38d55'
Davis,Don & Sue Aur 4h49'26"48d52'
Davis,Doriann Cas 0h29'26"58d46'
Davis,Dorothy Marie Mon 6h43'30"11d36'
Davis,Dr Gregg Cep 22h8'55"61d19'
Davis,Dr Joseph Ori 5h8'34"-5d28'
Davis,Dr Joseph A Per 3h3'40"46d56'
Davis,Dr Joseph H Oph 16h20'0"2d33'
Davis,Drew Hya 9h55'28"-18d50'
Davis,Earl Boo 14h41'36"22d25'
Davis,Edith Anne Taylor Hya 8h12'0"5d42'
Davis,Edith Mae And 23h3'14"51d48'
Davis,Eleni Susan Buflaten Eri 3h26'60"-5d10'
Davis,Elizabeth V Lyr 18h56'57"30d22'
Davis,Ellen Lyn 7h25'48"58d29'
Davis,Elmer Ray Cet 0h36'32"-6d48'
Davis,Emily Jane Lyn 7h51'33"44d60'
Davis,Evan Fisher Her 17h23'23"50d2'
Davis,Frank Aql 19h9'18"13d13'
Davis,Frank Aur 4h51'14"40d42'
Davis,Fred Clifton Dra 20h16'34"67d44'
Davis,Gabriel Conley Cam 3h29'23"58d30'
Davis,Garrett E Psc 1h0'33"32d45'
Davis,Gary Edward Dra 19h35'30"68d59'
Davis,Gary John Per 3h12'39"50d30'
Davis,Gary L Cnv 12h29'55"44d6'
Davis,George Alan Aql 18h59'32"12d14'
Davis,Georgia & Bus Lmi 10h5'30"32d32'
Davis,Grayson Lac 22h32'23"50d8'
Davis,Gwen Y Mon 8h5'53"-7d23'
Davis,Harry Ori 6h2'32"7d56'
Davis,Heather Uma 10h14'39"60d34'
Davis,Heather Nicole Cyg 20h3'19"58d34'
Davis,Helen & George Mon 6h38'55"-0d0'
Davis,Helen Marie Farley Vir 13h56'52"-5d8'
Davis,Hollie And 1h39'56"37d57'
Davis,Honour Lauretta Cas 0h45'26"66d57'
Davis,Howard Her 18h7'23"14d30'
Davis,Hudson Jerry Cet 2h5'54"3d53'
Davis,Ian McNaught Her 16h15'26"11d21'
Davis,III,John A Lac 22h5'58"41d0'
Davis,Isabelle Emerson Hya 8h11'50"-1d40'
Davis,J Q & Delores Aql 19h6'32"-0d57'
Davis,Jack L Cnv 13h27'24"41d41'
Davis,Jacqui Eri 4h11'59"-17d16'
Davis,James Tri 1h52'18"25d25'
Davis,James Edward Ori 5h50'0"15d44'
Davis,James G Ser 18h25'17"1d15'
Davis,Janet Malley Mon 6h27'12"10d10'
Davis,Janice Boo 14h28'11"25d5'
Davis,Jason Cep 20h50'45"61d24'
Davis,Jean Crowley Cmi 7h23'39"8d59'
Davis,Jeana Cas 23h15'23"61d18'
Davis,Jenny Vul 20h56'56"28d43'
Davis,Jerry Sex 10h17'44"-3d57'
Davis,Jim Sct 18h53'14"-5d9'
Davis,Joan And 1h12'21"34d28'
Davis,Joe & Gayle Cyg 21h7'1"30d1'
Davis,John Sex 10h26'60"-8d40'
Davis,John K Lac 22h16'40"46d58'
Davis,John V Crt 11h48'15"-18d3'
Davis,Jonathan Cep 21h42'59"55d25'
Davis,Jordan Lynne Dra 17h15'36"61d19'
Davis,Jr,Charles Richard Cnv 12h47'58"50d44'
Davis,Jr,Corbett Hya 9h6'13"2d23'
Davis,Jr,E William Cas 1h12'10"61d36'
Davis,Jr,Harold N Ori 5h53'35"14d44'
Davis,Jr,John D Cam 4h56'50"71d8'
Davis,Judith Anne Eri 2h49'35"-6d37'
Davis,June Wise Lyn 9h22'34"41d30'
Davis,Karen Marie Equ 21h1'0"3d36'
Davis,Karl Her 16h44'17"32d41'
Davis,Karl Paul Aur 6h16'27"31d42'
Davis,Kathleen J Lmi 10h20'17"31d16'
Davis,Kathryn Harris Ori 4h42'15"14d48'
Davis,Kathryn Sidni Del 20h36'57"20d19'
Davis,Katie Lin And 2h2'58"40d38'
Davis,Kay Ellen Gem 6h39'22"13d6'
Davis,Kelly Anne And 23h20'30"41d19'
Davis,Kennith Michael Ori 5h53'33"14d32'
Davis,Kimberly Marie Cyg 21h30'46"37d31'
Davis,Kristina Kae Lyr 19h0'14"30d4'
Davis,Kristine Leann Cyg 19h17'54"49d38'
Davis,Larry H Hya 9h31'57"2d26'
Davis,Larry Steven Aql 20h11'48"1d14'
Davis,Lauren And 0h10'48"26d42'B
Davis,Lauren Paige Eri 3h35'0"-15d59'
Davis,Lawrence E Cet 0h44'35"-2d14'
Davis,Lee Lmi 10h26'48"33d22'
Davis,Leola Cam 3h42'49"60d20'
Davis,Linda Marie (Punkin) Lyr 19h21'44"30d39'
Davis,Lindsay Nichole Mon 7h42'48"-6d59'
Davis,Lindy Vendela Buflaten Sge 19h57'35"16d3'
Davis,Lisa A Mon 8h2'0"-5d38'
Davis,Lisle & Mildred Oph 17h5'23"-20d10'
Davis,Lorayne Johnson Mon 7h39'18"-1d51'
Davis,Lu & Jim Eri 4h12'28"-11d53'
Davis,Lucien Renato Peg 22h2'28"31d41'
Davis,Luella"Mama Lu" Dra 16h39'52"67d38'
Davis,Mack Oph 18h40'17"10d31'
Davis,Margaret & Ivon Aur 7h23'36"35d55'
Davis,Margaret Ann Mon 6h44'26"11d41'
Davis,Margurite B Lyr 18h49'23"31d41'
Davis,Mark Allen Her 18h2'59"30d29'
Davis,Mark L Lac 22h14'23"46d20'
Davis,Marlene Sunny Oph 17h10'52"-9d13'A
Davis,Martha Cet 2h31'30"-3d25'
Davis,Martha Elaine Del 20h52'58"9d29'
Davis,Marty Cnc 8h5'46"7d19'
Davis,Matthew Williams Hya 8h54'0"1d41'
Davis,Megan Elaine Ari 1h54'28"13d4'
Davis,Melissa Kitchens Uma 10h2'20"70d25'
Davis,Michael & Helen Gem 7h31'10"20d15'
Davis,Michael B Ser 15h58'59"23d48'
Davis,Michael Donald Lac 21h59'1"42d57'
Davis,Michael Joseph Lac 22h2'24"38d25'
Davis,Michael N Eri 3h21'11"-6d12'
Davis,Michael R Aur 5h26'41"38d10'
Davis,Mike Aur 5h23'25"38d46'
Davis,Montanna Rae Uma 10h27'51"41d31'
Davis,Mr & Mrs Guy James Umi 16h13'54"86d2'
Davis,Mutley C Tri 2h9'39"31d49'
Davis,Nathan Oliver Peg 22h37'12"8d50'
Davis,Nicole Christine Lyn 6h28'23"61d45'
Davis,Norm Cmi 7h30'52"7d16'
Davis,Norman & Audrey Umi 13h6'38"71d49'
Davis,Odie Thelman Lac 22h22'14"54d25'
Davis,Patricia "Pleiades" Tau 3h40'16"20d30'
Davis,Patricia May Mon 8h7'35"-1d23'
Davis,Patty C Peg 21h59'14"11d11'
Davis,Paul Uma 8h32'16"51d22'
Davis,Peter James Aur 7h10'12"36d47'
Davis,Phillip W Aql 19h54'60"13d8'
Davis,Phyllis Cyg 21h10'54"39d56'
Davis,Piers A W Aql 20h1'16"0d34'
Davis,R Michael Dra 17h40'38"60d58'
Davis,Rachel M Vul 20h17'40"28d48'
Davis,Rachel Mae Per 22h43'26"34d27'
Davis,Ralph Gordan Lac 22h9'52"50d57'
Davis,Ray Oph 17h36'47"-28d11'
Davis,Rebecca Ann Sct 18h48'43"-6d25'
Davis,Renee Jean Peg 22h51'18"29d45'
Davis,Rhonda Vir 11h38'25"3d14'
Davis,Rhonda O'Neal And 23h30'33"41d26'
Davis,Richard (Billy D) Aur 6h17'48"32d1'
Davis,Richard Elden Cep 0h5'29"69d23'
Davis,Richard Eric Lac 22h16'1"54d33'
Davis,Richard Everett Cet 1h34'48"-3d27'
Davis,Richard Lees Lyn 8h22'1"48d37'
Davis,Richard Page Per 1h47'13"48d32'
Davis,Rita Cyg 20h27'55"37d41'
Davis,Rodney Newton Oph 17h19'15"-21d22'
Davis,Ros V Eri 2h51'27"-3d22'
Davis,Russ Dra 11h37'47"70d52'
Davis,Ruth Cma 6h54'53"-15d40'
Davis,Ruth Alice Ori 5h52'1"11d13'
Davis,Ruth Frances Sco 17h35'1"-38d33'
Davis,Sally & Larry Crb 16h2'53"27d4'
Davis,Sammi Aql 19h0'22"7d47'
Davis,Sandra Kay Psc 1h31'29"20d43'
Davis,Sandy Mon 6h23'12"5d35'
Davis,Sandy K Cam 13h30'52"76d39'
Davis,Sara Marie Lyr 19h19'27"42d14'
Davis,Scott Her 16h57'38"24d15'
Davis,Sharon K Aql 20h22'25"0d60'
Davis,Shirley Jean Uma 8h38'28"49d19'
Davis,Sophie Marie Lyr 19h18'50"38d38'
Davis,Sr,Harold N Ori 5h52'29"10d52'
Davis,Sr,Ridgely W Her 16h37'54"11d17'
Davis,Stacey Ann And 0h10'42"47d58'
Davis,Stan & Karen Cmi 8h1'13"0d32'
Davis,Stephanie Jill Vul 20h1'0"25d6'
Davis,Stephanie Jo Per 1h26'40"52d48'
Davis,Steven Ward Per 2h0'32"50d28'
Davis,Sunny Oph 17h6'12"11d51'
Davis,Susan & Randy Aql 19h48'17"14d53'
Davis,Susan L Cas 0h39'32"74d38'
Davis,Suzanne And 23h31'10"49d56'
Davis,Ted M Cmi 7h42'0"4d31'
Davis,Terry Catherine Eri 2h50'51"-1d47'
Davis,The Celestial Barbara Mon 6h59'1"1d49'
Davis,Thelma J Eri 3h43'0"-17d33'
Davis,They Followed The "Star",Marshall Her 17h22'38"40d51'
Davis,Timothy Rhassan Lyr 18h36'14"34d27'
Davis,Toby & Hal Cet 0h36'1"1d41'
Davis,Todd "Honey" Hya 9h14'31"0d15'
Davis,Tracy "Crusher" Dra 20h28'46"68d25'
Davis,Tristan Perryman Aql 19h5'17"4d40'
Davis,Tyler James Ori 5h38'14"0d57'
Davis,Valerie Berlant Lyr 19h25'0"38d29'
Davis,Vicki M Cet 2h9'34"5d2'
Davis,Victoria Peg 21h43'1"22d32'
Davis,Virginia M And 0h16'0"32d17'
Davis,White Knight/ Robert Her 17h12'24"22d5'
Davis,William E Lac 22h5'23"38d33'
Davis,William Jeffrey Aql 19h48'59"11d25'
Davis,Yolande R Cyg 20h17'25"39d28'
Davis,Zachary Her 17h20'57"40d25'
Davis-Bowling,Meghan Lynn Cas 2h0'31"58d30'
Davis-Dream Aimer Star Scorer,Brian Sgr 19h8'25"-22d41'
Davis-Wick,Patricia MacIver Peg 21h56'20"24d2'
Davis:Cosmic Star of Biology,Ed Lac 22h52'44"40d42'
Davison,Barbara Ann Cas 1h4'56"61d13'
Davison,Christina Jo Sge 19h6'5"19d5'
Davison,Daniel P Aur 5h12'34"44d24'
Davison,Patrick Philip & Vicki Lynn Cnv 12h22'42"50d15'
Davison,Robbie Uma 12h40'0"62d49'
Davison,Ted & Karen Ori 6h1'53"8d41'
Davison,The Star Of Aqr 23h13'41"-4d48'
Davison,Thomas Norman Per 2h57'34"38d58'
Davisson,Lee D Gem 6h2'0"26d25'
Davitt,Elnora & Harry Cyg 21h29'45"48d58'
Davlin Peg 22h12'55"30d48'
Davmar Uma 8h55'16"53d34'
Davo Dra 16h47'42"68d12'
Davo Lmi 10h34'19"28d30'
Davolt,Lindsay Renee Eri 2h49'40"-16d37'
Davy S Star Sge 20h17'42"20d26'
Davy,Diamond Jim Aql 18h57'46"-6d49'
Davy,L Nevil Cam 7h31'13"60d3'
Davy,Pauline Cyg 19h35'0"38d16'
Davy,Scott Patrick Cet 1h55'60"0d47'
Daw 94 Cam 4h3'0"68d59'
Daw,Alice Uma 12h40'51"62d13'

Daw,Brianna Michelle
And 1h19'20"40d46'
Daw,Esther S
Cas 1h9'41"58d14'
Dawan,Bahiyyah
Ori 5h57'31"18d27'
Dawe,Andrew David
Aql 19h55'39"-0d22'
Dawedeit,Gunter O
Dra 17h49'1"85d17'
Dawekids' Dream
Umi 17h14'21"85d30'
Dawes,Dominic
Ori 6h7'41"0d21'
Dawes,Jean & Michael
Lyr 18h32'27"38d48'
Dawes,Karen
Oph 17h23'25"-20d18'
Dawes,Tom
Gem 8h4'46"30d31'
Dawg,Michelle & Scotti
Cam 6h7'40"56d14'
Dawgert,Amy Kathleen
And 23h23'36"40d36'
Dawgert,Sarah
And 0h56'22"35d37'
Dawit BaBa-Beloved,
Father,With Love
Cep 20h55'14"58d15'
Dawkins,Janet
Lyr 19h16'47"42d32'
Dawkins,Marion Spangler
Vir 13h34'6"-18d39'
Dawn
Sge 19h0'30"18d52'
Dawn
And 23h7'50"40d6'
Dawn
Eri 2h59'31"-17d28'
Dawn
And 23h19'31"51d1'
Dawn
Lyr 19h18'21"35d22'
Dawn
Lyn 8h59'26"41d28'
Dawn
Peg 22h42'34"24d28'
Dawn
Lyn 9h1'0"37d55'
Dawn
Peg 22h55'26"24d9'
Dawn & AJ
Cyg 19h25'30"34d24'
Dawn & Mark
Lyn 8h47'49"41d4'
Dawn & Martin
Cyg 20h56'24"53d16'
Dawn & Mike "It's Like Magic"
And 0h9'44"35d36'
Dawn & Ron
Cnv 13h40'36"41d41'
Dawn 30
And 0h5'16"38d20'
Dawn Adele
Leo 10h39'27"18d44'
Dawn C
Crt 11h7'44"-18d54'
Dawn Flaine
Lac 22h40'22"53d19'
Dawn Estelle,The
Cmi 7h34'0"11d10'
Dawn G
Lyr 18h42'51"32d16'
Dawn In The Night Sky
Cet 1h3'35"-3d32'
Dawn LD
Cas 1h0'51"58d16'
Dawn Lee
Vul 20h45'12"20d3'
Dawn Look Up You Are Not
Alone Love
And 23h31'49"48d36'
Dawn Marie
Aql 19h7'23"15d47'
Dawn Marie
Cas 2h53'0"70d6'
Dawn Marie
Cas 1h2'43"71d6'
Dawn Marie
Aur 5h52'43"37d31'
Dawn Marie
Cam 4h45'32"67d37'
Dawn Michelle
Boo 14h23'50"52d12'
Dawn of the Rose
And 1h57'30"47d12'
Dawn Robin
And 22h55'45"50d39'
Dawn Theresa
Uma 12h27'34"56d56'
Dawn's Aurora
Uma 8h40'17"47d12'
Dawn's Delight
Cas 1h1'13"70d12'
Dawn's Early Light
Cam 8h0'15"71d23'
Dawn's First Light
Lyn 8h30'0"50d34'
Dawn's Jewel
Cap 20h41'1"-26d34'
Dawn's Light
Gem 5h59'10"26d20'
Dawn's Morning Star
And 2h28'0"49d35'
Dawn's Ray
And 2h23'28"39d24'
Dawn's Rising Star
Lyn 7h39'56"50d19'
Dawn's Silver Apple Of The
Sky
And 2h20'39"42d42'
Dawn's Star
Lyr 19h16'0"26d30'
Dawn's Wish
Cas 2h35'16"57d54'
Dawn, The One I Love
Cyg 19h49'10"37d52'
Dawn,Andrea
Del 20h22'33"11d5'
Dawn,Dakota
Peg 22h51'35"26d26'
Dawn,Rebecca
Crt 11h8'10"-15d48'
Dawn,Sherri
Cap 20h58'0"-26d45'
Dawna
Aql 19h55'31"14d52'
Dawood,Majia S
Ori 5h56'19"20d15'
Daws,Robert
Per 2h56'44"37d56'
Dawson Star,The Frank
Aur 5h20'47"50d11'
Dawson,Amy
Cas 23h29'0"60d55'
Dawson,Betty Ann
Cmi 7h23'1"8d40'
Dawson,Carole Mary
Lyr 18h16'32"31d2'
Dawson,Carré
Cas 23h13'11"62d11'
Dawson,Daniel Thayer
Aql 19h23'14"-0d48'
Dawson,Edward
Cam 3h57'31"71d33'
Dawson,Elizabeth Princess
Peg 22h59'1"26d57'
Dawson,Gillian
Cas 1h24'0"55d53'
Dawson,Jennifer Lynne
Cas 0h4'59"56d8'
Dawson,Joshua
Gem 8h5'20"28d19'
Dawson,Joshua Murray
Ind 21h4'26"-49d8'
Dawson,Jr,Roy F
Sct 18h52'5"-9d50'
Dawson,Jr,William (Billy)
Cet 2h52'28"0d24'
Dawson,Karen
Cas 2h1'52"58d48'
Dawson,Kelly Lee
Lyr 18h14'0"38d35'
Dawson,Lawrence
Ori 5h57'0"16d58'
Dawson,Lucy
Lyr 18h32'27"43d47'
Dawson,Martin Henry
Her 17h24'35"49d50'
Dawson,Nora Lee Brown
Lyr 19h14'46"40d33'
Dawson,Perry B
Lmi 9h54'15"34d25'
Dawson,Robin Lee
Equ 21h19'38"11d28'
Dawson,Rolland L
Per 3h6'1"46d54'
Dawson,Sarah
Sge 20h17'57"16d9'
Dawson,Steven Kirk
Dra 19h28'14"61d22'
Dawson,Stuart Crittenden
Boo 15h1'49"41d11'
Dawson,Thomas Jay
Dra 16h3'46"64d36'
Dawson,Tyler
Ori 5h7'30"1d32'
Dawson,Victoria
Cyg 20h2'29"31d39'
Dawson,Virginia Culshaw
Equ 21h9'0"10d57'
Dawsons,The Four
Cap 20h42'15"-18d13'
Dax
Pho 0h47'46"-46d12'
Dax
Cet 0h28'47"0d43'
Day (Flower),Sandra
Peg 22h33'27"18d53'
Day Family's Star, The
Dra 17h50'18"60d2'
Day Star,The
Dra 13h32'58"67d50'
Day Star-Anne & Larry, The
Uma 11h42'20"44d25'
Day,Allen Robert
Boo 14h18'12"17d58'
Day,Bessie May & Alf
Cyg 21h30'33"53d13'
Day,Bridgette
Cam 4h26'42"58d45'
Day,Christopher Steven
Her 16h27'21"40d4'
Day,Davey Lee
Cet 3h11'40"2d9'
Day,David
Uma 10h32'31"47d45'
Day,David Alan
Dra 19h5'16"50d22'
Day,Dinah
Per 1h26'17"50d31'
Day,Donnie M
Oph 18h32'16"10d26'
Day,Doris
And 23h19'24"40d42'
Day,Edward
Her 18h47'14"18d47'
Day,George
Dra 11h41'1"68d26'
Day,Helen B & Roy T
Cyg 20h31'43"48d56'
Day,Jack
Cep 22h36'41"60d33'
Day,Jamison Edward
Boo 14h27'34"40d15'
Day,Jansen
Lyn 7h9'14"50d11'
Day,Joanne
Umi 14h21'24"67d32'
Day,John B
Her 16h59'21"50d1'
Day,Kaitlyn Elizabeth
And 23h30'41"41d10'
Day,Kelly
Mon 6h30'31"-8d7'
Day,Kelly
Lmi 10h2'12"39d37'
Day,Kelly
And 23h35'54"49d55'
Day,Kristin Lynn
Psc 1h25'13"33d16'
Day,La Stella Paul & Linda
Lac 22h23'46"56d43'
Day,Lawrence & Eilzabeth
Aql 18h44'43"11d9'
Day,Lawrence Carlton
Eri 3h22'29"-18d33'
Day,Lee Ann
Mon 6h36'0"6d13'
Day,Leslie
Ori 5h57'17"14d52'
Day,Linda
Cmi 7h58'52"4d35'
Day,Lois
Cas 1h12'39"64d15'
Day,Marie
Lyr 18h19'45"47d20'
Day,Matthew Gordon
Ind 20h55'36"-59d3'
Day,Mike
Cep 23h24'31"70d31'
Day,Patricia & Roger
Sex 9h46'1"2d13'
Day,Patrick John
Her 16h52'32"50d59'
Day,Roy McKie
Dra 14h5'32"64d21'
Day,Sandra
Del 20h21'1"20d30'
Day,Serris & Peggy
Mon 6h58'56"8d51'
Day,Sr,Charles Thomas
Oph 16h5'35"-6d1'
Day,Tonya
Cyg 20h48'56"37d55'
Day,Tonya J
And 1h52'49"36d51'
Day,Troy Ray
Cet 2h1'19"0d35'
Day,Ventla Marie
And 0h7'33"30d42'
Day-Day
Cet 3h8'59"3d19'
Daye
Cyg 19h52'22"50d1'
Daykin,Elizabeth
Cyg 19h54'16"37d35'
Daykin,Howard
Mon 7h42'8"-6d41'
Daylen Jane
Mon 6h39'2"-0d8'
Daymut,Kelley K
Umi 15h23'17"67d9'
Dayna Fortman,Douglas
Del 20h16'29"12d2'
Dayna V
Lyr 19h12'0"35d35'
Dayna's Delight
Mon 6h51'28"11d55'
Dayna's Own Star
Lyn 0h42'32"41d52'
Dayong
Ser 15h17'46"21d55'
Dayrit,Grace
And 1h58'23"36d49'
Days Together Unlimited
Cyg 19h57'1"50d9'
Dayton,John Shelby
Hcr 17h14'45"46d55'
Daza Chacin,Monique
Francesca
Sco 17h9'25"-38d2'
Daze
Mon 7h43'49"-1d46'
Dazeley Star,The
Del 20h22'0"4d15'
Dazl
Ser 15h37'57"3d40'
Dazzle Drop
Lyr 18h59'32"36d41'
Dazzling Bridge/Tree
Lyn 8h24'52"42d3'
Dazzling Deb Star
Pup 7h27'31"28d38'
Dazzling Patricia
Lyr 18h30'26"32d58'
DB 25
Boo 15h3'30"52d38'
DB's Wishing Star
Lyn 8h6'0"48d60'
DB,My Forever Soul Mate
Del 20h18'20"13d44'
DBB's Star
Mon 6h36'55"-6d25'
DBM Universal
Lyr 19h25'42"37d50'
DCL-"Granny"
Cas 1h45'39"61d52'
DD 40
Dra 16h16'1"62d6'
DDG
Peg 21h20'18"22d48'
DDR Günther Nenning Duell
Team
Uma 13h51'41"50d39'
Ddrzewiecki,Dennis, Love
Kara
Cnv 12h20'36"43d39'
DE
Cas 0h3'0"58d52'
De 5 Uur Show
Uma 13h56'1"48d58'
de Abreu,Orlando Francisco
Vieira
Aur 5h1'35"48d53'
de Ahlers,Lori Mendez
Peg 22h34'1"25d36'
De Alba,Mario Gutierrez
Cet 2h30'33"6d46'
de Almeida,Allison Marie
Cas 0h26'30"61d30'
de Almeida,Nicole Anne
Leo 9h24'59"18d57'
de Almeida-Mirai, Denyse
Maria Vieira
Uma 10h43'52"40d48'
De Amor,Lucero
Mon 6h24'46"4d27'
De Andra
Cyg 20h0'48"41d14'
De Angelis,Chiara
Per 3h32'26"51d20'
De Angelis,Simona
Per 2h10'14"58d49'
de Angulo,Carlos Martinez
Lyn 7h53'34"39d56'
De Atley,Dallas Blake
Per 4h6'11"51d28'
De Atley,David Lawrence
Cep 23h38'40"64d25'
De Atley,Gary Gray
Lac 22h8'11"51d31'
De Atley,Virginia Noelle
Lmi 9h35'39"38d7'
De Augustine,Poto
Cep 21h58'47"61d33'
De Ballisima Bazsali
Umi 16h52'29"79d49'
De Barrientos,Fidelia M
Revollo
Aql 19h1'58"4d54'
De Bartolo,Lidia
Cam 6h7'1"60d37'
de Beaumont,Michel
Lac 22h8'17"51d27'
de Beauvois,Guerlain du
Domaine
Cam 3h28'54"61d49'
de Beek,Michael
Cnv 12h41'49"37d5'
De Bella,Andrea
And 23h38'32"45d39'
de Belleville,Nathalie
Lyr 19h17'41"25d45'
De Blasio,Jeff
Aur 4h57'13"38d38'
De Boben,Louis & Dolores
Aql 18h44'28"11d22'
De Bracchini,Marisela Willars
Cas 0h32'36"68d12'
De Bruin,Drh Anton
Cas 1h52'56"61d19'
de Bruyne,Anita
And 23h6'38"42d40'
De Capdenac,Thierry Hunal
Sex 10h42'49"5d2'
De Carlo,Haley Lynn
Uma 10h40'11"56d4'
de Castille,Gabrielle Isabelle
Cas 0h51'0"61d18'
De Castries,Hewi
Leo 10h54'22"10d31'
De Castro,Dianna Elizabeth
Peg 23h29'59"20d51'
de Castro,James E
Lac 22h17'11"51d50'
De Castro,Mauricio
Dra 11h51'32"66d1'
De Caylus,Ludmila Piogey
Duchess
And 2h29'22"43d2'
De Chavigny,Elisabeth
Cep 22h16'0"55d44'
de Cillia,Jean
Lmi 10h46'44"26d34'
De Cota,James A
Psc 1h44'42"21d38'
De Croisset,Charles
And 23h46'49"33d16'
De Dalmases Pinot, Roberto
Oph 17h58'27"10d43'
De De & Pa Pa
Dra 19h58'15"67d49'
De De Ay
Boo 14h25'24"20d18'
de Domenico,Salvatore
Aur 6h19'43"38d52'
de Donder,Daniel
Ori 6h6'1"0d3'
de Espino,Giamarie Santiago
Mon 6h19'15"8d54'
De Estrella,Nicoletta
Cet 2h32'33"1d22'
De Evoli,Lee Gabriel
Lyn 6h39'58"58d26'
De Fazio Papa Moon, Michael
Cnv 12h58'31"50d44'
de Feria,Dr Americo
Aur 6h17'15"37d18'
De Flaviis,Maria
Per 3h32'1"50d3'
De Fleurieu,Jean Rene
Per 3h1'40"41d2'
De Fluri,Neil Michael
Peg 21h59'53"10d60'
De Fraguier,Henri
And 23h35'34"33d17'
de Francisco,Diego Fernández
Her 17h50'19"43d5'
De Fresquet,Valerie
Her 18h22'0"12d28'
de Funes,Eduardo
Aql 20h7'20"1d44'
de Funiak,Andrew Quinby
Boo 15h2'60"41d20'
de Galaxia,Mauve
Boo 14h25'46"48d42'
de Garrigue,Pierre Alain
Per 3h43'1"38d57'
De Gaulle,Yves
Per 4h32'17"34d9'
de Gaxiola,Kathy Ann Flynn
Cas 0h0'49"64d15'
De George,Charles
Ori 5h57'46"20d14'
de Gier,Marco
Aql 18h55'37"11d47'
De Giovanni,Caterina
Aur 5h27'54"38d8'
De Giovanni,Chiara
Aur 5h29'48"31d39'
de Giovanni,Serena
Her 18h6'59"28d16'
De Groot,Jules & Rosa
Forceville
Sge 19h3'33"16d11'
de Guerguiniou,Gaec
Tri 1h53'0"27d54'
de Guis
Peg 22h25'55"20d10'
de Guzman,Steve
Aql 18h58'32"-6d28'
de Haro,Michel
Cam 3h42'39"61d43'
de Haven
Aql 20h8'18"0d22'
de Havenon,Alieda
And 0h8'28"30d36'
de Hoernle,Count Adolph
Hya 8h19'45"5d15'
de Hollande,Charles
Umi 13h2'58"71d17'
de Hollanole,Iris
Her 18h1'22"28d48'
de Hoog,Frans J
Boo 14h48'44"47d36'
De Island Mon
Uma 11h2'23"36d16'
De Jacquelot,Beatrix
Dra 10h14'44"74d30'
de Jerphanion,Dauphine
Cas 23h42'36"60d44'
de Jerphanion,Dauphine
Uma 11h23'12"40d20'
De Kerdel,Guy-Vincent
And 23h43'18"33d39'
de Kerguinerien,Gaec
Lyn 8h11'34"39d55'
De Kler,Roy Brian
Dra 19h7'1"48d24'
De Koning,Anne Nicole
Uma 10h9'57"50d15'
De Koning,Denys Adam
Cas 1h47'12"61d25'
de Kyllmann,Margarita
Flossbach
Eri 4h35'35"-5d52'
De L'Hermitage, Christine
Cyg 20h53'54"58d54'
De La Baume,Alain
Per 4h9'51"37d38'
de la Buiyere,Isabelle
Her 17h13'15"47d58'
de La Forest,Xavier Gauthey
Aur 4h47'0"36d29'
de la Fuente-Wang
Ser 15h11'1"24d13'
de la Luz,Leo Edmund &
Maria
Cyg 21h50'16"41d56'
De La Martiviere, Gerard
Leo 9h39'51"19d21'
De La Nove,Vincent
And 23h41'0"33d26'
de la Rionda,Robert Anthony
Cnv 12h14'33"37d53'
de la Rosa,"Big Daddy" John
T
Aql 18h59'1"6d10'
De La Rosa,Charlotte
Lyn 8h10'0"58d42'
De La Rosa,Christina Gabrielle
Crt 11h11'10"-14d41'
De La Rosa,Katrina Anne
Crb 16h16'42"32d6'
De La Rue,Nigel
Cep 0h7'23"76d26'
de La Torre,Elizabeth
Lancaster
Ori 5h55'24"21d1'
De La Valliere,James A
Cep 23h1'60"62d33'
de Lamaze,Bruno
Cep 22h25'0"57d43'
De Lamberterie, Francois
Per 4h1'41"50d30'
de Lancie Family,The John
Psc 1h25'16"11d45'
De Lange,Matthew
Cyg 19h34'18"35d12'
de Langlade,Sophie
Aur 4h58'13"30d22'
De Lanoi,Gladwin
Dra 17h23'49"58d9'
De Lerrain,Courtier Jacques
Per 2h22'27"55d10'
de Levie,Gary
Cnv 13h47'45"38d18'
De Lignac,Prince Leo
Aur 6h51'55"48d46'
de Lisser,Elliot Taylor
Her 16h48'43"40d58'
De Loach,Bee
Hya 9h2'31"-1d11'
de Lorimier,Louis Hébert
Per 3h15'15"41d8'
De Los Santos,Joe
Uma 8h59'1"47d39'
De Luca,Carmen
Her 18h31'37"18d51'
De Luca,Carmen
Her 18h31'37"18d51'
De Luca,Claudio
Cep 1h22'45"78d12'
De Luca,Prof Dott Anselmo
Cnv 13h43'0"28d17'
De Luna,Rayita
Mon 6h29'33"-1d39'
de Lyon,Dr Bruce
Oph 17h53'55"10d39'
De Malignon,Daryl
Dra 13h36'1"68d19'
de Mar,Cheryl Sue8d19'
Cas 0h13'1"64d54'
De Marco,Jr,Ronald J
Aur 6h54'44"41d0'
De Mare,Abby Valerie
Cas 22h58'32"55d15'
De Mars Our Father,In Honor
of Joffré
Cet 1h20'59"1d13'
de Martino,Domitilla
Ori 5h58'18"19d35'
De Matte "Hap E
Luckie",Eugene
Cet 0h49'29"2d1'
de Matteis,Hugo
Aql 20h22'32"1d37'
De Medeiros,Valeria Maria
Sco 16h38'14"-38d14'
de Meeüss,Jean-Claude
Per 3h15'16"40d39'
de Melker,Helena Maria
Her 17h56'27"14d47'
de Messimi,Eve Denis
Lac 22h28'60"38d60'
de Mexico,Charlotte
Ori 5h42'30"11d17'
de Mol,John
Per 2h58'0"31d29'
de Monaco,Albert
Lac 22h11'18"37d43'
de Monica,Lucerno
Cyg 21h31'30"42d47'
de Monvel,Paulette Boutet
Dra 19h51'56"70d29'
de Nercy,Alix Bussiere
Per 3h33'54"41d10'
de Obaldia,Irma
Mon 6h42'29"3d14'
De Oca,Maria Carmen Montes
Eri 5h6'31"-6d53'
De Oliveira,Christine
Cnv 13h20'50"38d11'
De Pablo,Diane
Lyn 8h8'45"39d23'
de Palma,Maria Coronel
Per 2h8'18"58d55'
De Paolis,Livia
Cas 2h20'34"60d24'
De Pardo,Estrella
Ori 4h52'49"0d8'
de Paula,Lucero
Aql 20h34'29"-0d15'
de Pearson,Carmen Chellen
Cas 0h19'1"59d0'
de Pennart,Christian
Dra 12h26'35"70d16'
De Petri,Astela
Leo 11h48'47"20d7'
de Pfyffer,André
Per 4h5'40"36d55'
De Placido,Antony
Lyn 8h9'32"50d56'
de Ponte,Ron
Hya 8h54'52"0d17'
De Poo,Josephine
Mon 6h33'1"10d10'
de Poortere,Irene
Cep 22h12'38"55d31'
de Quesada,Cristina Maria
Eri 3h9'58"-5d44'
de Quincke,Esther Areco
Cyg 20h3'0"31d11'
De Raadt,Jennifer
And 0h54'48"36d52'
de Raismes,Jr, May 9,
1987,Joseph
Per 1h31'32"54d13'
de Ramos' Family Star, D & T
Cyg 20h24'56"40d33'
De Roche,Loran
Uma 9h41'30"50d52'
De Ros,Marc
Cep 21h30'37"60d20'
De Rosa's Piece of Heaven
Lyn 7h6'0"44d29'
De Rosa,Andrew
Cep 23h9'22"67d46'
De Rose,Virginia
And 2h33'38"40d13'
de Rothschild,Noemi
Lib 15h16'22"-28d20'
de Ruiter,Kim Oenen
Cnv 13h52'27"47d58'
De Sanctis,Marzia
Her 16h12'1"10d41'
De Santis,Goffredo
Tel 20h9'40"-48d28'
De Santis,Rosa
Peg 22h27'1"4d59'
de Schank,Elisabet Llobet
Cas 0h46'60"73d25'
de Sedano,Alexandra
Sgr 19h30'44"-33d15'
de Sedano,Luc
Sgr 19h31'12"-30d3'
de Servien-Kenwood, Austin
Benoât
Lyr 19h13'47"38d59'
de Shangai,Marie
Cyg 20h6'35"30d35'
De Simone,Anthony
Lmi 10h6'40"37d49'
De Simone,John
Aql 19h2'56"-1d18'
De Simone,Michele Lee
Lyr 19h22'0"30d12'
De Sin,Diane B
Lyr 19h0'58"31d0'
De Solange,Amaueua
Lyr 19h20'18"30d39'
de Spelder,Mark Emerson
Sct 18h42'16"-6d4'
de Straschnov,Franáois
Cep 21h28'45"60d54'
de Suyrot,Madelin
Ori 5h41'27"8d16'
de Sédano,Francoise
Sgr 19h30'56"-32d32'

de Sédano,Nicolas
Sgr 19h31'41"-38d34'
De Tinguy,Charles
Oph 17h58'34"10d37'
de Tinoco,Carmen Montilla
Sex 9h57'15"1d49'
de todo mi coraz1d49'
Del 20h15'19"9d11'
de Tonnancourt,Dean
Boo 14h18'49"54d48'
de Tora,Deborah Stella
And 23h42'15"33d42'
de Torres,Jacquelyn S
Leo 11h32'13"-0d6'
de Valck,Dr E F
Cam 7h19'40"67d47'
De Varennes,Nathalie Mauger
Dra 19h17'31"80d35'
De Varona,Diana
Peg 21h43'21"24d12'
De Veer,Wim
Cam 5h46'11"61d23'
de Victoria,Alexander Lopez
Vul 19h47'0"28d48'
De Villars-Sur-Glane, Nordmann Louise
Cam 3h59'17"52d24'
De Vita,Maria
Cnc 9h2'44"7d23'
De Vito,Todd Anthony
Ser 15h15'11"21d56'
de Vore,Dlorah Llen
Mon 6h22'38"5d44'
De Vries,Christine Anne Dunbar
Gem 6h2'56"25d17'
de Vries,Mieke
Ori 5h32'30"0d30'
de Vries,Stephen
Peg 22h27'44"31d3'
De Zarn,Jacob Anthony
Uma 11h29'31"32d53'
Dea
Tel 19h4'46"-48d52'
Deac
Cyg 21h24'1"41d1'
Deacon Gene
Her 18h4'1"48d38'
Deacon,John
Her 16h1'49"40d53'
Deacon,Jon Ross
Cep 23h1'11"65d36'
Deacon,Judson
Uma 8h35'25"53d11'
DeAcutis,John A
Ari 2h33'36"30d50'
Deadrich,Chuck
Ori 5h53'26"15d9'
Deadrich,July Burke
Mon 6h26'25"-8d40'
Deahni How 42N82NOU
Lib 14h56'52"-10d6'
Deak,Jennifer
Peg 23h38'21"30d55'
Deakelsie
Uma 8h38'56"50d33'
Deakin
Boo 14h11'46"53d45'
Deakin
Boo 14h11'46"53d45'
Deakin,Alexander R
Mon 6h52'0"10d21'
Deakin,Sarah Jane
Cas 0h22'40"71d53'
Deakos,Matthew K J
Cep 20h16'0"60d41'
Deal,Michael Anthony
Cep 22h10'20"61d15'
Deal,Vincent R
Her 16h18'52"10d56'
deAlvarez,Maria Cristina Zanelli
Mon 7h49'13"-1d37'
Dealy,Christine Anne
Cap 30h33'51"-14d49'
Dealy,Michael Kenneth
Cep 23h28'39"70d45"
Dealy,Peppino
Leo 9h54'29"11d32'
Deam,Lisa R
And 2h33'0"49d45'
Deamira
Ori 6h0'42"8d25'
Dean
Per 3h6'42"46d24'
Dean
Aur 7h1'40"38d40'
Dean
Lac 22h23'1"40d21'
Dean & Sowers
Eri 5h0'1"-10d43'
Dean & Steve
Aur 5h4'53"40d38'
Dean & Wendy's Forever Love Star
Lyr 18h54'45"41d1'
Dean Anthony & Jenna Anne
Lyr 18h59'51"31d50'
Dean Born 4-3-65, Michael David
Cnc 8h50'49"31d31'
Dean III,Stephen W
Cet 2h33'39"8d31'
Dean's Light
Per 2h32'1"56d49'
Dean's Star
Cet 2h50'0"1d42'
Dean's Star
Sex 10h37'19"2d1'
Dean's Star
Aur 6h2'18"45d5'
Dean's Star
Dra 14h30'13"64d3'
Dean,Alan
Dra 15h3'0"61d2'
Dean,Anna J
Gem 6h43'46"31d19'
Dean,Bob & Joy
Aql 19h30'50"7d42'
Dean,Brian Keith
Aur 7h6'19"37d59'
Dean,Charly
Hya 8h42'41"-0d37'
Dean,Christian Jon
Lyr 19h1'23"40d11'
Dean,David A
Aql 19h13'24"12d6'
Dean,Dorothy J
Com 12h54'54"25d57'
Dean,Elizabeth
Cyg 19h30'53"33d50'
Dean,Elizabeth Anne
Cyg 21h59'46"52d39'
Dean,Haley
Cyg 20h1'56"37d54'
Dean,Harold
Her 17h28'17"21d5'
Dean,Harry
Com 12h3'33"24d54'
Dean,Helen Hart
Vul 20h3'32"28d47'
Dean,James Byron
Aql 20h5'35"4d22'
Dean,Jane-Carey
Lac 22h28'57"40d3'
Dean,Joanne W
And 22h56'28"38d25'
Dean,John Big Cat
Uma 8h13'53"71d32'
Dean,John Jeffery
Ser 15h53'58"20d30'
Dean,JR
Sct 18h54'52"-6d31'
Dean,Kim Williams
Vul 19h22'19"27d19'
Dean,King,Kelly Ray
Uma 12h5'25"59d33'
Dean,Loren
Ser 15h37'47"6d24'
Dean,Mary Sue
Equ 21h1'51"10d38'
Dean,Pamela Marie
Cyg 21h55'1"53d51'
Dean,Rachel
Mon 8h5'31"-9d58'
Dean,Ruth Edna
Cas 0h55'42"58d39'
Dean,Sam
Per 1h58'59"50d9'
Dean,Sandy
And 2h17'18"49d2'
Dean,Scott Crawford
Aql 19h7'16"3d32'
Dean,Susan Elaine
Cap 20h36'4"-14d56'A
Dean,Terrance Michael
Tau 5h50'50"26d42'
Dean,Terry E & Bobbi J
Cet 0h57'26"-6d6'
Dean,Trisha Rae
Mon 6h54'24"-10d9'
Dean,William George
Cet 2h56'0"8d45'
Dean-Rikki Star,The
Sge 20h16'47"17d55'
Deana
Cas 23h17'37"60d28'
Deana
Lac 22h50'49"54d17'
Deana Louise
Cas 1h34'11"70d59'
Deana Mae
Oph 17h7'39"-23d49'
Deane,Alexandra Lysberg
Ori 5h58'52"15d12'
Deane,Drucilla
Scl 23h19'42"-31d21'
Deane,Edie Hope
And 2h21'53"39d14'
Deane,Elsa Lillian
And 0h31'53"45d23'
Deane,Jennifer Louise
Cas 1h47'0"65d17'
Deane,Katharine
Mon 6h34'31"-1d22'
Deane,Sophie Patricia Florence
And 23h15'39"42d45'
Deane,Terry
Ori 6h7'53"8d39'
Deaner,Carl
Aql 18h43'1"6d48'
Deaney,Steven James
Lac 22h48'0"38d39'
Deangelis,Kay
Lyn 7h44'42"50d8'
DeAngelis,Mario
Mon 7h19'31" 5d55'
Deangelis,Michael
Aql 19h53'0"15d12'
DeAngelo,Alice
Mon 6h22'52"5d46'
DeAngelo,Dena
Sge 20h16'31"20d24'
deAngulo,Maria Luisa Urgoiti
Equ 21h0'15"10d4'
Deanie
Peg 22h43'46"29d40'
Deanie Dear
Del 20h35'1"20d22'
Deann
Com 12h9'0"31d56'
DeAnn
Com 12h0'18"14d11'
DeAnna
Ori 5h48'12"20d18'
DeAnna
Hya 9h10'36"0h2'
Deanna
Per 1h54'33"53d32'
Deanna
Cas 23h27'60"60d7'
Deanna
Mon 6h28'0"-10d38'
Deanna
Cas 2h58'27"68d11'
Deanna
Hya 9h7'56"2d23'
Deanna Bandana
Uma 11h32'50"56d47'
Deanna C
Cet 0h55'29"0d45'
Deanna L O Ti Amo
Vir 12h31'13"0d20'
Deanna Lee
Mon 6h27'33"11d46'
Deanna Lynn
Del 20h14'47"9d19'
Deanna Lynn
Cma 6h59'0"-24d42'
Deanna Maria
Vul 20h4'37"28d22'
Deanna Marie
Cas 23h34'40"61d10'
Deanna's Diamond
And 2h2'0"41d37'
Deanna's Eternal Light
And 23h43'23"44d26'
Deanna's Star
Peg 22h30'21"24d22'
Deanna's Star
Com 13h12'22"30d50'
DeAnna's Star
Uma 9h28'60"58d18'
Deannacresta
Lyr 18h52'18"31d30'
DeAnne
Lyr 18h34'47"42d19'
Deanne Mary
Lyr 19h25'1"41d21'
Deano
Cam 5h54'42"73d38'
Deans,Bradley John
Lyn 7h36'37"39d30'
Deans,III,Robert Barr
Aur 6h55'35"44d7'
Deans,Linda Elenor
Lyr 18h19'19"47d27'
Deanstar
Uma 8h51'0"54d37'
DeAntonio,Dr Carlo
Oph 17h4'37"12d37'
Dear My Comrade Yuka
Sgr 19h6'15"-15d16'
Dear Ole Dad
Aur 5h19'27"41d26'
Dear Toki Genki
Leo 10h14'57"13d21'
Dear,Art & Sarah
Hya 9h7'50"-16d39'AB
Dearborn's,The
Per 3h4'18"50d16'
Dearborn,Stephen
Hya 8h20'26"5d55'
Deardon,Lesley Annette
Lyn 7h8'1"58d47'
Dearest David
Lmi 10h40'0"28d22'
Dearest Elizabeth
Mon 6h24'54"10d59'
Dearest Hunter I Love You Sophia
Peg 23h43'16"31d3'
Dearest Ksara
Lyr 18h56'0"37d58'
Dearinger,Maria Estella
Tau 5h45'0"24d10'
Dearmin,Robert
Boo 14h7'56"40d36'
DeArmond,Douglas R
Lac 22h25'34"55d31'
DeArmott,Daniela
And 23h31'21"47d34'
Dearns-Corabi,Carol
Tau 3h42'47"28d33'
Dearth,Ronald V
Del 20h22'46"10d39'
Deas,Cailin
Uma 9h16'26"53d21'
Deas,Joseph Harrison
Per 1h48'23"56d56'
Deas,Samuel Lawrence
Cep 21h23'58"55d54'
Dease,Kaleigh Marie
Crb 16h1'23"38d12'
Deason,Terry P
Aur 5h44'57"50d16'
Deason-Betz
Lyr 19h9'39"38d17'
Deasy,Colleen
Cas 0h7'24"59d50'
Death Rider
Sct 18h42'46"-7d28'
Death Star,The
Cap 21h1'11"-26d56'
Deatherage Aka-Sir Valiant,Scott Lloyd
Lyr 18h57'55"37d14'
Deaton,Anthony Todd
Aql 19h51'26"15d24'
Deaton,Anthony Todd
Aql 19h56'46"12d59'
Deaton,Helen
And 0h26'13"43d25'
Deatrice
Tri 1h47'0"27d40'
Deats,Marna
Cma 6h42'26"-13d20'
Deautels,Jacob Donald
Dra 17h49'27"64d24'
Deaux,Alice "Jeanne"
Cet 2h38'1"1d2'
Deavall,Damian
Uma 10h35'17"47d32'
Deaver,Chad Carvel
Cep 22h10'21"61d36'
Deaver,Christoper Joseph
Dra 15h54'24"64d46'
Deavers,Michael John
Aql 18h57'23"12d58'
Deavoll,Julie
Lyr 19h1'39"28d15'
Deavours,Susan Kae
Lyn 07h26'53"44d13'
Deb
And 2h13'25"42d40'
Deb & Marty Forever
Ori 5h52'49"8d56'
Deb & Tori
And 23h38'0"45d0'
Deb Meister,Deb A- Romma Ding Dong,The
Peg 22h35'23"20d58'
Deb's Eye
Gem 7h18'13"28d48'
Deb's Light
Ori 5h57'33"8d23'
Deb's Star
And 2h51'1"39d34'
Deb-1
Tri 1h35'16"30d25'
DeBach,Robert (Way Cool)
Oph 17h2'27"-24d18'
DeBach,Steven
Cet 3h19'51"8d43'
Deback II,Levens Peter
Dra 14h59'0"55d37'
DeBacker,Jerome
Cep 21h52'34"61d39'
DeBacy,Judith Karen
Cas 0h45'17"64d52'
DeBaene,David
Aur 5h1'17"42d55'
Debains,Catherine
Lyr 18h18'50"39d36'
Debaisieux,Leslie
Boo 15h6'28"7d42'
DeBaldo,Robert O
Tri 1h50'0"27d7'
Debanne
Cam 3h43'0"74d25'
DeBard,Christopher John & Suzanne Marie
Peg 22h28'36"8d4'
Debaty,Philippe
Aur 5h1'32"37d43'
DeBaufre,Dr Erik K
Cep 22h24'12"80d27'
DeBaun,Denise
Her 17h54'24"28d23'
DeBaun,Star Of
Ori 4h58'1"4d52'
Debb
Mon 6h56'58"10d40'
Debbe Lynn
Cas 0h32'17"61d41'
Debbi,Fanny
Cas 22h56'26"54d46'
Debbie
And 1h25'33"33d50'
Debbie
Cet 2h55'43"6d34'
Debbie
Aql 19h56'45"14d15'
Debbie
Crb 16h10'55"28d55'
Debbie
Mon 7h52'15"-3d40'
Debbie
Lyn 6h26'39"60d37'
Debbie
Mon 8h6'0"-0d34'
Debbie
Eri 5h2'36"-5d48'
Debbie & David
Lyn 7h3'29"50d53'
Debbie & George's Everlasting Love
Peg 23h38'17"21d46'
Debbie & Jay
Cep 20h27'14"75d22'
Debbie & John
Eri 3h30'1"-6d40'
Debbie & Jordan
Cas 0h52'30"64d33'
Debbie & Wayne's Star
Ori 5h58'1"17d34'
Debbie (DLR)
Uma 9h42'43"57d33'
Debbie Alpha
Mon 8h0'0"-0d15'
Debbie Anne
Cyg 19h28'25"36d8'
Debbie Circle Face
Uma 9h49'54"51d1'
Debbie Dew Drops
Uma 12h56'31"62d48'
Debbie Doo
Eri 4h2'1"-10d29'
Debbie Doo
And 0h22'0"44d15'
Debbie Du
Vul 20h17'12"25d28'
Debbie Gwynne
Cep 21h41'45"55d20'
Debbie II
Psc 1h28'35"20d35'
Debbie J
And 23h35'37"49d13'
Debbie Kay
Mon 7h21'15"-8d27'
Debbie Little Boot
Boo 14h28'60"23d17'
Debbie Lynn
Aql 19h2'37"-0d27'
Debbie Lynn's Love Light
Sgr 18h53'42"-36d48'
Debbie Rea's "Rainbow Light"
Lac 22h13'46"54d43'
Debbie The Tweedster Glick
And 2h13'17"42d54'
Debbie's Christmas Star
And 23h42'17"44d41'
Debbie's Diamond
Cam 14h12'42"80d6'
Debbie's Diamond
Com 12h26'51"26d16'
Debbie's Place
Tau 4h38'19"28d55'
Debbie's Smile
Cas 1h13'22"60d46'
Debbie's Star
Lyr 18h49'58"31d2'
Debbie's Star
Cas 0h58'15"58d42'
Debbie's Star
Aql 19h6'15"0d44'
Debbie's Wish
Cas 23h30'16"62d48'
Debbie,IWAHLY
Vul 19h6'34"24d35'
Debbie-M
Cas 0h48'19"62d5'
Debbieann
Cas 23h41'52"63d48'
Debby
Cas 0h31'21"66d42'
Debby
Aql 19h18'50"15d9'
Debby True Love
Mon 7h1'0"1d38'
Debby's Journey
Dra 14h52'39"63d11'
Debby's Piece Of Heaven
And 0h54'55"36d58'
Debby's Star
Leo 10h50'36"-5d7'
Debbye's Fire
Cas 0h2'11"60d15'
Debe
Cas 0h29'57"68d48'
DeBeau,Daniel E
Cnv 13h21'56"38d44'
Debek,Daniel Anthony
Per 3h53'22"38d44'
Debek,Evan R
Cep 1h26'0"87d26'
Debel,Anne
Uma 8h41'50"71d35'
Debelak I Love You, Sabine
And 23h37'51"41d54'
DeBellis,Cara Danielle
Del 20h39'32"11d12'
DeBellis,Lucy M
Sex 10h6'0"0d27'
DeBellis,Silicia
Ari 3h13'0"23d56'
DeBenedetto,Josephine Anne
Cas 2h26'53"60d55'
DeBenedetto,Michelle Lynn
And 1h3'36"40d41'
DeBenedict,Nicholas
Dra 20h21'56"67d14'
DeBenedictis,Tracey
Uma 11h43'25"55d13'
Debenham,Jack
Com 12h1'35"26d56'
Debenham,Sharon & Jim
Cyg 19h40'10"37d31'
Debernardy,Pierre
Ser 18h27'27"1d4'
DeBeve,The Rosalyn
Uma 10h9'34"62d16'
Debi
Lyn 8h51'33"44d33'
DEBI
Mon 6h21'44"2d44'
Debi & Frank's Star
Mon 6h55'0"-8d21'
Debi's Wishes
Cas 1h44'17"58d51'
Debi-My Special Friend
Aql 18h56'1"-8d22'
DeBiasi,Anthony Joseph
Aur 4h37'28"34d31'
DeBiasio,Jr,Thomas
Boo 15h6'13"26d3'
deBidart,George A
Ori 5h53'39"14d14'
DeBieux,Peter
Cep 21h20'1"58d19'
Debinator
And 0h20'34"44d42'
deBlanc,Eleanor Darcy
Com 12h5'29"20d2'
deBlanc,Jr,Ralph F
Oph 18h16'1"0d24'
deBlanc,Michele E
Hya 9h17'59"-7d35'
DeBlase,Elizabeth Marie
And 2h2'12"46d56'
DeBlasi,Dayna Nicole
Mon 6h29'1"8d43'
DeBlois,John M
Cet 3h14'30"1d7'
Debnbob
Lyr 18h29'16"31d29'
Debney Superstar, Thomas
Cep 1h22'27"80d30'
Debo Star,The
Aur 7h20'1"36d26'
DeBoard,Richard Kelley
Aur 5h0'1"40d8'
DeBoer,Eric Christiaan J V
Oph 17h53'18"-0d59'
DeBoer,Jessica
Lyr 19h16'35"42d20'
DeBoissiere,Thelma Beryl
Mon 8h2'10"-4d41'
DeBok,Herbert
Ori 5h57'23"16d26"
DeBoo,Brian James
Ser 15h12'53"5d25'
Deboon,Richard
Cnv 12h50'0"40d44'
Debora
Del 20h16'60"11d31'
Deborah
Cnv 13h0'1'33d10'
Deborah
Lyr 18h42'33"36d51'
Deborah
Sgr 19h28'57"-30d32'
Deborah
Leo 10h35'17"21d49'
Deborah & Darren
Ser 15h34'48"19d20'
Deborah & Jeffrey
Cyg 21h7'0"30d45'
Deborah & John's "Wedding Star"
Cyg 21h51'27"40d49'
Deborah & Mark
Cyg 21h57'0"53d35'
Deborah 96
Vul 20h17'16"23d19'
Deborah Ann
Lyr 18h18'47"40d31'
Deborah Ann
Uma 12h53'55"62d11'
Deborah Ann
Sco 16h5'15"-40d19'
Deborah Ann
Cyg 21h5'33"31d7'
Deborah Ann
Del 29h17'19"14d5'
Deborah Anna
Lyn 8h15'36"47d41'
Deborah Anne
And 1h2'12"38d6'
Deborah Anne
Cas 1h1'18"61d0'
Deborah e Antonello
Her 16h43'1"40d41'
Deborah E,The
And 0h35'46"37d47'
Deborah Helen
Peg 23h31'47"10d40'
Deborah I
Lyr 18h24'53"42d20'
Deborah J
Vul 20h46'59"28d42'
Deborah Jane
Lib 15h20'29"-24d53'
Deborah Jane
Cas 3h21'56"75d57'
Deborah Jean
Cyg 20h18'18"38d46'
Deborah Joy
Dra 11h45'42"70d30'
Deborah K
And 0h17'58"36d56'
Deborah K 59
Eri 3h26'40"-3d21
Deborah Kay
Cas 23h39'13"60d17'
Deborah Kelly
Ori 6h13'20"0d36'
Deborah L D
Vul 19h52'0"20d26'
Deborah Lynn
And 0h43'49"38d50'
Deborah Lynn 687383762535263635683
Ori 5h23'35"-4d12'
Deborah Lynn
Aql 20h30'11"0d26'
Deborah Lynn
Oph 18h3'13"12d56'
Deborah Lynn Loves Gregory Laurence
Cas 0h31'13"61d42'
Deborah Marie
Cas 23h38'48"62d12'
Deborah Marie
Sgr 18h56'27"-23d20'
Deborah Marie
Mon 6h20'31"9d29'
Deborah Mary Thresea 40 22696
And 1h53'27"37d51'
Deborah Rae
Cyg 21h8'29"31d3'
Deborah Rebecca,The
Aql 19h51'49"12d34'
Deborah Rose
Peg 21h42'55"22d45'
Deborah Terry
Lyr 18h31'37"31d48'
Deborah"Star Among Stars"
Cyg 21h53'59"52d44'
Deborah's Dream
Peg 0h8'54"18d40'
Deborah's Dream
Aql 19h14'34"14d37'
Deborah's Emerald Brilliance Star
Lyn 7h40'10"39d32'
Deborah's Star
Cyg 20h31'0"38d25'
Deborah-Scott
Cet 1h0'11"-6d51'
DeBorde's Star
Cyg 21h1'27"38d19'
Debouck,Frederique
Cyg 19h47'35"30d4'
DeBoutiere,Herve
Umi 15h44'27"83d28'
DeBow,Albert H
Cep 21h48'31"55d37'
Debra
Mon 6h55'45"-1d45'
Debra
Cas 0h56'12"74d36'
Debra
Her 17h48'1"14d59'
Debra
Uma 11h42'53"41d22'
Debra & Melanie B/F/F
Lyr 18h59'28"45d40'
Debra Ann
And 2h32'34"42d50'
Debra Elizabeth
Mon 7h58'43"-5d20'
Debra Ellen
Cas 22h59'41"56d7'
Debra K 08/12/57 (MVP)
And 23h23'17"41d11'
Debra Lee
Mon 6h59'56"-0d21'
Debra Lee
Cas 2h8'28"61d18'

Debra Lynn
Cyg 20h4'28"40d55'
Debra Lynn
Eri 3h33'23"-4d60'
Debra Marie's Star
Cyg 19h33'55"32d32'
Debra Renee
And 0h20'42"37d35'
Debra Rose Petal
Cet 1h38'57"-1d56'
Debra's Bairn
Peg 22h50'41"8d20'
Debra's Dreams
Cas 0h45'17"68d4'
Debra's Star
Mon 6h53'43"11d9'
Debra,The
Vul 19h48'41"25d30'
Debrah
Lyn 8h51'0"34d26'
Debrik Star,The
Umi 13h49'42"78d17'
DeBrincat,David & Tracey
Uma 9h17'1"52d26'
DeBrincat,Raymond & Susan
Lmi 9h20'31"38d14'
DebRon Forever & A Day
Cyg 20h57'0"41d11'
DeBrosse,Marie
Umi 16h21'1"86d0'
Debrovner,Charles Howard
Sge 18h58'29"18d55'
Debrovner,Patricia & Dr Charles
Crb 16h3'54"31d33'
Debrune,Anne-Marie
Cyg 19h50'49"44d34'
DeBruyne's Star Design
Eri 2h54'0"-11d18'
Debry Shadow, Dewane
Aql 18h57'25"-6d18'
Debski,Mr & Mrs Richard Eric
Cam 6h32'20"65d31'
Debus,Diane
Cas 0h28'27"58d55'
Debus,Jean-Théophile
Equ 21h19'24"11d0'
Debus,Kurt
Psc 1h1'25"21d38'
Debvena
Peg 22h26'42"25d9'
Debye
Cyg 21h31'1"40d48'
DeCandia,Joseph & Clara
Uma 9h22'36"51d16'
DeCandido,Patricia Herring
Ari 3h5'46"30d45'
DeCanio,Augustine Julia
Lac 22h15'54"46d56'
DeCanio,Thomas Charles
Lac 22h17'56"51d38'
Decaprio,Noelle
Boo 14h56'51"31d23'
DeCarlo,Mark
Ori 5h47'11"11d41'
DeCarlo,Shanter
Aur 6h39'32"40d22'
Decatria FGR
Her 18h1'12"38d3'
Decavagnal,Jack
Lac 22h19'33"49d53'
DeCecco,Frank
Aur 6h59'34"38d40'
DeCecio,Christmas Star Thomas & Dorothy
Oph 18h36'30"6d28'
DeCelle,Anna Ludine
Lmi 9h30'0"37d50'
December 5th,1980
Vir 13h29'56"-20d41'
DeChaubry,Marie France
Cnv 12h44'29"47d1'
Dechellis,Pamela J
And 2h30'1"45d29'

Dechessi Jonibeth
And 23h22'37"43d49'
DeChiara,Raymond B
Cet 2h48'13"3d52'
Dechmerowski Family, The
Cet 2h27'16"8d3'
Dechnik,Steve
Lac 22h52'1"55d23'
DeChupete,Elaine
Cas 23h2'13"58d25'
DeCiantis,John S
Dra 16h37'57"63d26'
DeCiantis,June M
Cas 0h20'23"75d40'
Decibel
Gru 22h18'11"-52d30'
Decicco,Edward A
Dra 13h34'49"68d11'
DeCicco,Joseph Michael
Dra 13h18'19"68d15'
DeCillis,Robert & Lisa
Crb 15h57'35"27d36'
Decio,Rossella
Lyr 18h19'58"40d32'
Decitre,Paul
Umi 6h46'0"88d34'
Deck,Angela Mae
Lyr 19h22'26"37d39'
Deck,Austin Mark
Her 18h7'43"40d9'
Deckard,Anna Elizabeth
Eri 4h33'55"-8d20'
Deckard,Betty Roots
Mon 7h11'13"-10d43'
Deckard,Shelby Nicole
Mon 7h49'16"-1d18'
Decker's Delight
Cep 22h48'39"65d37'
Decker,"Star of Auvergne",April
Cnv 12h22'46"34d12'
Decker,Carol
Lyr 18h43'27"40d7'
Decker,Christy
Cnv 13h57'18"31d0'
Decker,Cynthia Lee
Hya 9h38'58"2d16'
Decker,Daniel
Lyr 18h50'51"41d11'
Decker,David D
Oph 16h50'21"10d15'
Decker,Dorothee
Sgr 19h0'32"-26d10'
Decker,Edie "My Sweetie"
Per 3h53'22"35d2'
Decker,Esq,Scott A
Dra 16h22'46"60d35'
Decker,Inez
Cet 2h10'60"6d19'
Decker,Jr,Charles M
Cyg 21h35'53"44d22'
Decker,Kathleen M
Sgr 19h37'58"-31d45'
Decker,Kimberly
Lyn 8h0'1"35d0'
Decker,Mackenzie Leigh
Aql 19h55'57"13d11'
Decker,Paul
Boo 14h2'16"26d7'
Decker,Peter & David Hughes
Aur 6h23'51"38d6'
Decker,Tosh
Aql 19h26'32"15d21'
Decker,Willy
Per 4h5'46"48d5'
Decky-Declan Hayden
Uma 9h16'20"53d2'
DeClark,Julia
Lyn 8h23'25"40d46'
DeClercq,Jr,Raymond R
Aur 5h2'0"41d57'
DeClue,Andrew C
Cnv 12h20'33"37d9'

DeClue-Toombs,Lady Barbara F
And 22h58'50"40d56'
DeCoite,Shanna
Eri 3h0'38"-1d56'
DeCola,Gloria
Psc 22h55'23"0d47'
DeColores,Enrique
Del 20h53'11"3d24'
DeConcini,Edwin
Aur 5h17'33"45d19'
Deconna,Jay
Dra 13h28'59"68d53'
DeConti,Jr,Peter C
Cam 3h56'26"61d13'
DeCoppel,Yolanda Luken
Sex 10h11'38"-4d30'
DeCorrevont,Joshua Andrew
Ori 5h52'36"1d50'
DeCredico,David
Vul 19h36'52"27d40'
DeCrette,Danielle
And 23h30'0"44d33'
DeCristofano,Janet E
Uma 10h3'1"50d25'
Deda
Cas 23h40'1"61d30'
Dedaj,Simon
Ori 5h54'53"14d31'
Dedear,Lee Norman
Uma 11h54'15"62d17'
DeDee
Peg 22h2'12"34d28'
Dederich,Dawn "Little Bit"
Lyn 6h19'21"58d37'
Dederichs,Beatrix Karolina
Aqr 21h19'56"-10d10'
Dederick,John Emory
Aur 6h34'33"38d48'
Dedicated to Love VOI Forever
Peg 23h40'44"26d7'
Dedman,Alice Ann May
Mon 6h56'29"-10d54'
Dedman,Amanda Lousie
Lyr 18h59'28"28d34'
Dedo's Star
Cam 3h56'24"74d9'
DeDomenico,Josephine
Del 20h32'28"20d4'
Dedousis,Lisa
Cas 1h49'0"60d58'
Dedwald,Herbert
Boo 13h38'11"10d53'
Dedybella 93
Nor 16h19'16"-52d44'
Dee
And 23h5'0"43d53'
Dee
And 2h20'44"39d59'
Dee
Cmi 7h44'57"0d17'
Dee
Lyr 18h40'30"30d45'
Dee
Vul 19h40'19"20d9'
Dee & Craig's Cosmic Communion
Mon 7h43'18"-5d50'
Dee & Roy
Cyg 20h17'35"41d25'
Dee Dee
Cet 2h53'25"0d1'
Dee Dee Sue
And 2h14'24"39d28'
Dee Dee's Luz del Sol
Uma 9h9'54"53d14'
Dee Gee's
Cam 6h11'34"60d50'
Dee Star
Lyr 19h3'46"25d50'
Dee's Dream
Com 12h20'0"27d54'
Dee's Star
Umi 161h10'0"70d14'

Dee,Dalton Samuel
Boo 13h55'1"18d36'
Dee,John J
Her 16h27'0"22d49'
Dee,Our Love Is Forever,Love Nick
Cmi 7h39'47"9d7'
Dee,Sean Power
Ser 15h54'14"18d25'
Dee,Todd L
Aql 20h34'13"-0d9'
Dee-D
Lyr 18h56'58"33d8'
Dee-lish,Dee-lite
Cyg 20h28'0"58d34'
Deeam
Cep 20h37'0"76d25'
DeeDee Hanna
Peg 21h52'25"31d33'
DeeDee's Star
Cyg 19h59'55"38d40'
Deeds,Jane
And 23h26'21"48d17'
Deedy,Alexandra Frances
Cet 2h35'19"-6d9'
Deedy,Joseph Patrick
Aql 18h55'1"-0d18'
Deegan,Daniel
Aql 19h55'56"14d45'
Deegan,Shelley
Mon 8h1'53"-10d4'
Deej-30
Ori 6h15'19"8d33'
Deeker
Cnc 9h16'19"18d2'
DeeLee
Mon 6h59'39"-6d42'
Deeley,Christine
Oph 18h6h14"11d47'
Deeley,Michelle
Cyg 20h22'20"39d11'
Deem,Lisa
Cas 0h47'50"61d39'
Deemar,Sharon
Uma 9h43'38"59d7'
Deemar,Stacy "Star"
Peg 23h2'0"31d50'
Deemer,Art & Marion
Cyg 19h54'46"41d3'
Deen,Jeremy
Cet 3h1'0"3d43'
Deener,In Memory of Ruth L
Tri 2h33'21"31d56'
Deep River CT Rotary Club
Umi 14h8'31"69d45'
Deep Space Cookie
Uma 11h9'23"47d31'
Deep Thought
Oph 17h20'46"10d45'
Deepak Chopra
Aql 18h43'23"6d37'
Deepali
Boo 14h6'39"23d12'
Deer With Horns
Lmi 9h49'22"38d32'
Deer,Puff Elizabeth
Peg 22h9'16"5d36'
Deerhake,Dean
Oph 17h32'31"-8d25'
Deering,Matthew David
Her 16h56'11"28d52'
Deeringwater,Drew
Aql 18h41'11"-2d24'
Deerr,Jeff
Her 16h10'33"13d6'
Deerwood
Lyn 6h55'41"54d58'
Deery,Susan Marie J
Aql 19h54'29"12d41'
Dees,Phil
Her 16h43'60"27d2'
Deese,Danny Ray
Aql 20h5'30"6d31'
Deese,Diana Kay
And 0h13'39"38d55'

Deese,Donald H
Aur 6h40'20"37d5'
Deeter,Brian
Lyr 18h13'18"38d59'
Deetle
And 23h37'54"48d20'
Deets,Floyd H
Aql 19h26'1"8d3'
Deets,Richard M
Cep 21h0'26"67d45'
Deets,Taryn Shea
Cnc 9h8'55"30d29'
Deevine Starr,Inc
Cmi 7h30'12"8d48'
DEF
Uma 9h16'1"56d44'
Defayette,Jeffrey M
Aur 5h2'0"41d31'
DeFazio 1st Born 1994, Tina & Karl
Boo 13h40'0"17d47'
DeFazio,Anthony
Cep 23h11'54"64d17'
Defazio,Diane
Lyr 18h55'34"31d46'
DeFazio,The Carmine Vincent
Her 16h55'0"38d45'
Defeis,Sister Marian
Uma 12h6'1"57d52'
Defendini,Felix Luis
Her 17h56'0"30d10'
DeFeo,Linda Rae
Sco 16h35'26"-43d41'
DeFeo,Robert Mario
Vir 13h27'21"-7d17'
DeFeudis,Wendy Allyson
Com 12h5'50"29d15'
Defever,Karen Lynne Iannuzzi
Com 12h13'34"21d12'
DeFilippo,Jeanne
And 2h31'55"41d12'
DeFilippo,Kathleen B
Lyn 7h39'56"51d3'
DeFillippo,Frank
Cep 22h42'41"80d14'
DeFillippo,Neil
Cap 20h23'59"-26d4'
DeFiore,Emily Grace
Lyn 7h6'41"50d13'
Deflorian,Emanuela
Lup 15h11'58"-43d57'
DeFlorio,Marc Anthony
Dra 14h43'1"62d37'
Defoe,Colleen Lynn
Cas 0h30'13"56d4'
DeFoe-Gallagher, Camille
Lyn 7h53'18"43d51'
DeFoor,Paula
Peg 22h36'51"22d10'
Deford III,Leland E
Aql 18h59'10"16d5'
DeFord,David
Cet 0h49'59"-8d6'
DeFord,Sue Newhouse
Uma 10h58'36"48d13'
DeForest,Walter
Cet 1h11'0"-3d2'
DeForge,Stan
Her 18h3'1"31d34'
Defrance,Isabelle
And 23h2'50"50d53'
DeFrancesco,Marguerite
Cyg 20h22'44"38d38'
DeFrancia,Micaela Margaret
Cep 21h41'0"60d15'
DeFranco,Connie
Lyn 8h20'29"50d41'
DeFranco,Serafino J
Cyg 21h2'50"31d31'
deFreitas,Andrea Nicole
Aqr 22h7'24"-0d49'
DeFrier,Anne
Lyr 18h31'51"36d30'
DeFrisco,Louis Michael
Boo 15h8'49"28d15'
DeFronzo,Susan Aron
Com 12h55'33"28d28'

DeFuria,Nicholas Dominick
Boo 14h58'47"41d3'
DeFusco,David J
Umi 15h15'52"66d15'
Dega,Karen
Com 13h15'58"22d12'
Degan,Amy Lynn
Lyn 7h57'59"35d4'
Degan,Charlotte F
Del 21h1'0"12d42'
Degand,Lee
Aur 5h8'0"40d2'
Degano,Melissa
Com 12h8'21"32d15'
Degaris,Beloved Soul Gary John
Cet 3h1'55"-0d56'
Degarmo,L Edward
Lac 22h25'0"56d34'
Degen,Laurie Ann
Cyg 21h31'0"38d28'
Degener,Brave & Beautiful Barbara
Vir 14h9'1"-8d55'
DeGenova,Fabiana
Crt 11h3'44"-12d6'
DeGeorge,Dominic
Oph 17h54'0"10d1'
DeGeurin,Laura Gayle
Eri 2h54'1"-4d11'
DeGilio,Timothy
Hya 8h20'19"-10d46'
Degiorgis,Marita
Lyr 19h2'41"28d10'
DeGiovanni,Roselyne
Lmi 10h12'35"39d43'
DeGirolamo,Michelle Patricia
Com 12h13'34"21d12'
DeGiulio,Daniel
Her 16h53'16"29d18'
DeGive,James
Boo 15h11'31"30d58'
Degler,Judith P.
Cet 2h7'25"1d8'
Deglmann,Rudolf
And 23h10'1"40d34'
Degnan,Laura Gerilyn
Peg 23h38'0"20d31'
Degnan,William
Lac 22h19'33"48d12'
Degnats,Richard G
Aqr 20h54'47"-0d23'
Degner,Peter
Aqr 22h44'28"-1d37'
DeGoede,Kaine Daniel
Cyg 21h3'23"39d59'
Degollado,Benito
Tri 2h43'58"31d16'
deGraaf,Willem & Mauryn
Tau 4h16'13"20d6'
DeGraff,John W
Her 15h48'52"48d1'
Degrassi,David
Cnc 9h0'49"27d60'
Degrave,Angel
And 23h37'39"45d2'
DeGraw,Richard Taylor
Her 17h23'38"49d4'
DeGrazia Infinity Star,The Phyllis A
Vir 15h6'1"1d38'
DeGrazia,Rosario A
Aur 6h17'46"30d31'
DeGree,"Nifty Fifty Teacher" Sandy
And 0h8'41"35d27'
DeGregory,Jr,Robert J
Uma 9h30'11"67d1'B
DeGregory,Randy
Dra 16h7'50"68d38'
Degreve,Robert V
Aur 5h8'17"40d28'
DeGrezia,Joy C
Com 12h21'33"19d39'

DeGritis,Amanda Joy
Lyr 18h53'0"33d10'
DeGroat,Jacque
Hya 9h12'8"-19d47'B
DeGroat,Margaret Ann
Cyg 20h24'48"40d25'
DeGroff,Shirley
And 23h29'44"47d29'
DeGroot,Ben
Per 3h27'38"48d59'
deGroot,Charles F
Ori 5h55'39"21d16'
DeGroot,Jerry
Cyg 20h3'20"38d19'A
DeGroot,Natalie Ann
Cyg 19h27'45"34d52'
DeGross,Brian Branzi
Per 1h38'57"52d42'
DeGruttola,Mackenzie Dae
Dra 16h6'59"68d14'
Degucz,Mary Jane & Chet
Boo 14h33'55"40d35'
DeGuenther,Veronica
Lyn 8h25'28"43d24'
Deguerois,Marien
Dra 16h19'30"62d49'
DeGumbia,Matthew J
Boo 14h59'37"51d13'
deGuzman,Gloria Lucena
Aql 20h35'43"-5d39'
DeHahn,Helen V
Uma 11h53'40"43d28'
Dehar,Devin Michael
Lac 22h50'30"38d39'
DeHart,Nancy
Lmi 9h52'58"34d4'
Dehaven,Glenn S
Her 16h42'18"48d20'
Dehgan,Rostam Khodadad
Boo 15h44'54"40d39'
Dehler,Kimberly
Peg 22h58'58"11d53'
Dehne
Ari 1h56'22"25d20'
Dehner-Friend & Protector
Cyg 20h3'53"40d10'
Dehtiar,Eitan
Ori 4h53'56"-2d50'
Dei Leon
Sex 9h42'0"-0d18'
Deibler,Baby
Aql 19h7'1"15d52'
Deichert,Lewton Andrew
Ser 15h13'47"8d48'
Deida,Mark
Eri 4h25'46"-1d50'
Deidameia
Uma 9h16'55"50d38'
Deidi's Lotus Bud
And 23h18'39"41d32'
Deidra
Oph 17h1'1"8d45'
Deidre
Mon 7h2'1"4d29'
Deifik,Shirley
Uma 11h31'37"31d7'
Deignan,Marilu
Lyn 7d24'44"44d33'
Deimer,Brick
Cet 1h19'28"-6d36'
Dein Liebes-Stern
Com 13h8'1"28d27'
Deinarowicz,Patricia
Cyg 21h9'29"48d47'
Deines,Dale & Joe Lex
Uma 15h38'0"68d45'
Delnesta,Nancy Rivera
Mon 6h42'27"-10d6'
Deininger,Rosa
Vir 13h15'40"-6d2'
Deirdre
And 1h23'47"34d37'
Deirdre Dawn
Cet 3h10'23"1d5'

Deischter,Philippa
Umi 15h57'51"77d48'
Deiss,Richard P
Aql 18h47'11"10d53'
Deisz,John,Elder
Cep 22h8'27"62d43'
Deitch,Betty & David
Lyr 18h21'47"46d55'
Deitch,Irwin Earl
Her 17h21'59"48d45'
Deitle,Grandpa G A
Cet 2h15'18"3d33'
Deitmar,Hermann Josef
Sge 19h7'16"17d45'
Deitrick,Bret
Aql 20h14'33"3d59'
Deitrick,John Christopher
Per 2h3'21"48d54'
Deitrick,Michael Richard
Cnc 9h4'19"28d8'
Deitrick,Zachary Joseph
Cep 21h30'12"55d35'
Deits,Lauren Williams
Ari 1h50'15"12d27'
Deits,Taylor Alexandra
Ari 1h50'17"12d2'
Deitz,Geoffrey
Aur 6h10'1"38d60'
Deitz,Michele
Lib 15h37'43"-22d38'
Deitz,Robert J
Peg 21h50'38"28d15'
Deitz,Sarah Kathryn
Aqr 22h18'36"0d42'
Deitz,Teresa Owen
Lyn 7h50'56"58d31'
Deitz,William Lee
Ser 15h52'41"0d21'
Deiulio,Anthony M
Her 16h47'29"30d58'
Delulio,Christopher R
Cep 22h8'22"68d53'
Dejean,Dominique
Peg 23h5'14"31d24'
Dejean,Jacques
Lac 22h19'47"54d33'
deJesus,Forever Charlotte & Rick
Mon 6h53'21"0d6'
DeJesus,Melesio(Nuno)
Per 2h2'19"56d56'
DeJiacomo,Shelley Marie
Eri 4h0'0"-14d40'
DeJonahe Star of Healing,Kim
Uma 9h25'30"51d7'
DeJoria Family Star, The John Paul
Del 20h16'35"8d52'
DeJoseph III,Vincent
Cep 2h40'1"78d32'
Dejour,Yvette
Cyg 20h5'35"30d28'
Dejoux,Dominique
And 0h28'13"44d41'
DeJoy,Hunter Thomas
Boo 14h44'19"46d25'
DeJulio,Frank E
Peg 23h23'25"15d34'
Dekamay
Lyn 8h59'37"36d0'
Dekaral
Lyn 8h23'34"47d35'
Deke
Cam 6h15'25"68d23'
Deke & Christian
Hya 8h55'1"0d8'
Dekelver,Debra & Jeff
Lyr 18h44'23"31d20'
DeKermandjin,Brett
Her 17h23'25"41d52'
Dekeyser,Hilaire
Aql 18h57'1"-6d39'
Dekkers,Lizzie
Lyr 19h10'0"41d8'

Deklerck,Didier
Cep 22h17'46"55d38'
DeKock,Dennis Casparus Jan Dingeman
Uma 11h56'44"32d1'
DeKruif,Jeanne & Robert
Cet 1h20'48"-11d35'
Del & Doris
Cyg 21h37'56"31d47'
Del Borrello,Sr,Peter Michael
Cep 21h57'18"55d31'
del Campo I,Sergio
Mon 6h34'1"-6d57'
del Campo,Laura
Cyg 21h37'27"41d7'
Del Campo,Susan
Lyr 18h15'45"30d31'
Del Carmen,Maria & Milton Harold
Mon 7h18'47"-5d30'
Del Castillo,April Lynch Lopez
Lyn 8h11'0"52d27'
Del Donno,Steven Robert
Her 16h44'51"34d23'
Del Fabbro,Raffaella
Lyr 18h37'1"29d17'
Del Giudice,Daniele
Umi 16h39'49"83d38'
Del Grande,Hailey Brianne
Peg 23h41'49"25d39'
Del Greco,Steven
Ser 15h37'1"18d33'
Del Monte,Dennis
Cet 1h45'43"-3d19'
Del Negro,Ronald Mario
Sge 19h34'13"16d40'
Del Negro,Susan
Crt 11h16'49"-19d25'
del Portillo,Msgr Alvaro
Aql 20h6'37"3d47'
Del Prado,Ana M
Mon 6h18'32"8d43'
Del Prete,Angela
And 23h26'11"40d34'
Del Rio,Aubrey Jean
And 0h21'23"44d21'
Del Rio,Luke Edward
Aql 20h19'46"5d23'
Del Russo,Robert
Aur 6h6'44"30d42'
Del Santo,Bernard
Sex 10h23'55"-1d28'
Del Sarto,Mario
Ser 15h15'25"-2d20'
Del Sarto,Rose
Cet 2h28'51"3d49'
Del Silenzio 1990, Geodi
Lup 14h28'49"-45d13'
Del Tondo,My Love Forever,Douglas J
Leo 9h38'19"7d36'
Del Torto,Lisa
And 0h12'1"33d30'
del Vecchio,Barbara Anne
Lyn 6h54'0"44d50'
Del Vecchio,Esther
Vul 20h22'39"28d28'
Del Vecchio,Philip J (Grandpa)
Ori 5h45'38"10d41'
Del Vecchio,Teresa M
Ori 5h51'52"17d32'
Del Vento,Maria Grazia
Peg 21h53'1"33d49'
Del's Delsey
Uma 11h37'16"48d3'
Del,Tony
Her 17h17'37"20d32'
Del-Rae Bodine
Tri 1h50'27"27d4'
DelaCerna,Halona Malia
Leo 10h56'11"15d18'
Delachi,Maria Vittoria
Ari 2h24'55"32d46'

DeLachoux,Laurent
Aur 5h48'44"54d43'
Delacressonniere
Boo 14h54'46"50d13'
Delacroix,Daniel
Uma 10h5'46"48d13'
Delacruz,Alejandro Rafael
Her 16h10'48"40d19'
DeLaCruz,Eugene
Hya 8h20'46"-6d44'
DeLaCruz,Kelly Jeanne
Mon 6h35'33"-1d19'
Delacruz,Nathaniel
Dra 14h48'44"61d28'
DeLaCruz,Rafael Marcos
Aur 5h51'24"54d55'
DeLage,Alfred B
Her 15h50'50"50d22'
Delahanty,Everett J III
Per 1h43,27"52d52'
Delahoy Family,The
Cnv 12h11'20"39d14'
Delahoz,Rafael
Cam 7h48'15"61d27'
Delahunty,Phillip Paul
Per 3h0'45"40d0'
Delaleux,Jacques
Ori 6h5'41"0d53'
Delamadeleine,Laura
Aur 4h54'39"37d32'
Delamarter,Charles Wesley
Dra 15h53'24"62d7'
Delamater,Wilson Alan
Her 17h56'59"18d52'
deLambert,Elizabeth Ann
Cas 1h31'41"60d24'
Delancey,Harry E
Eri 2h57'20"-15d25'
Delancey,Julia M
Peg 0h7'14"13d36"
DeLancie,"Q" John
Psc 0h58'39"17d32'
Delaney,"Mimi" Eve M
Vul 20h20'31"28d8'
Delaney,Alexander
Per 2h21'30"54d50'
Delaney,Beverly R
Cas 1h43'38"61d37'
Delaney,Carole Jean
Eri 2h57'47"-18d46'
Delaney,Ernestene
Eri 2h57'38"-18d20'
Delaney,Gavin Eugene
Vul 19h23'26"26d28'
Delaney,Harry
Hya 8h58'37"3d25'
Delaney,Jenny
Cas 1h33'34"75d53'
Delaney,Joseph M
Aur 4h55'38"50d57'
Delaney,Jr"MD",John F
Ori 4h53'26"4d16'
Delaney,Laurel J
Lib 14h58'37"-18d47'
Delaney,Mike
Uma 9h12'1"58d21'
Delaney,Miranda
Umi 14h5'47"69d25'
Delaney,Sean Patrick
Her 17h15'0"42d23'
Delaney,Sharon Cedrone
Cnv 13h5'44"40d35'
Delaney,Super Man & Super Star,Dennis C
Her 16h29'40"47d47'
Delaney,Thomas S
Aur 5h2'1"31d41'
Delaney,Topher
Umi 14h42'25"65d48'
Delaney-90,Minnie Derusha Owens
Aqr 23h29'13"-19d33'
DelAngelo,Gloria
Umi 16h3'0"72d38'
Delannay,Anne
Umi 16h11'28"79d27'
Delano System,The
Oph 16h53'47"10d46'
Delano,Catherine L
And 23h44'57"40d1'
DeLaO-We Love You, Marisela
Aql 20h9'45"8d32'
DeLape,Frank,Stephanic & Alexander
Uma 8h48'38"70d55'
Delaplace,Alain
Umi 16h17'36"70d59'
Delaporte,Geordie
Cam 3h55'54"55d49'
Delarbre,Pierre
Dra 17h6'1"51d37'
Delaroche,Christophe
Aql 19h59'60"13d8'
DeLarosa,Marcela
Cas 0h28'12"50d9'
Delarosa,Missy
Ori 6h3'43"-1d3'
Delas,Nicole Ashley
And 23h33'25"36d28'
Delassus,Jim & Marie
Ori 5h56'44"10d56'
Delatorre,Jesus
Her 16h48'44"34d40'
Delattre,Eric
Boo 14h30'48"54d58'
Delauder,Kelly Marie
Cyg 19h59'45"38d7'
Delauder,Peggy
Uma 9h29'1"58d26'
Delaunay,Marie-Laure
And 0h4'0"28d38'
DeLaurentis,Carissa Ann
Cas 1h7'11"60d14'
Delauter 95,Dorothy (Loving Mother)
Cas 0h28'1"70d33'
Delavault,Franáoise
Aur 6h2'47"31d48'
DeLaVega,Jorge
Lac 22h18'52"54d53'
Delay,Franáois
Sex 10h16'57"-1d49'
DelBroccolo,Anthony
Peg 22h18'20"29d43'A
DelCastillo,William Ray
Ori 5h56'57"16d20'
Delcli 5h56'57"16d20'
Her 18h19'1"14d35'
Delderfield,Ernest William
Cep 22h34'46"56d54'
DelDosso,Michael F
Aql 18h56'47"8d14'
DelDuca,Joseph Anthony
Her 16h53'60"50d29'
DelDuco,Craig S
Cep 20h16'36"60d19'
Delebarre,Aurore
Dra 17h0'24"60d50'
Delebarre,Herve
Uma 14h23'39"61d41'
Delebarre,Solone
Per 3h59'29"35d6'
Delebarre,Yvonne Etedmond
Per 3h27'56"50d7'
Delecole,Sophie
Per 3h30'22"37d16'
DeLeeuw,Julia Kathleen
Vul 19h42'23"23d25'
Delegal,Darin L
Oph 18h2'60"11d60'
Delehoy,Raymond Lee
Dra 19h25'41"58d8'
Delenclos,Josiane
Ori 6h7'13"20d50'
DeLeo
Aql 19h31'0"11d18'
DeLeo,Alfred Vito
Cet 3h0'27"0d23'
deLeon,Anna
Peg 21h24'26"3d41'
Deleon,Carol
Mon 8h3'14"-8d60'
DeLeon,Christopher
Lac 22h55'55"54d48'
DeLeon,David
Aur 5h26'32"40d13'
DeLeon,Ivan Jay
Her 18h42'31"12d51'
DeLeon,Ruben
Psc 23h20'12"6d11'
Delepine,Jean-Gilbert
Cyg 19h45'20"37d38'
Deleshaw,Leslee
Lyn 8h6'53"46d47'
DeLessio,Mimi
Boo 14h5'1"31d35'
DeLette,Hattie Lenora
Uma 11h2'53"44d15'
DeLevie,Annette
Cam 8h8'25"81d49'
DeLevie,Melvin I
Umi 13h40'0"71d59'
Delfico,Sydney Nicole
Lyr 18h37'40"31d20'
Delfin,Alicia
Cas 0h23'27"64d20'
Delfina
And 0h27'15"45d49'
Delfina,Maria
Del 20h16'11"12d35'
Delfino,Marieh Mercedes
Mon 6h22'14"-6d7'
Delgadillo,Bonnie Elizabeth
Lyn 7h11'1"58d50'
Delgadillo,Julia
Cep 22h24'25"70d7'
DelGado Star,The George
Cep 22h44'0"71d10'
Delgado,Arturo Rafael
Boo 13h34'29"14d59'
Delgado,Cecil B Bird
Sct 18h43'56"-6d57'
Delgado,Florencio Hernandez
Mon 7h5'28"-4d35'A
Delgado,George Edward
Cnv 12h30'58"32d8'
Delgado,Kessa Lynn
Oph 18h42'18"7d58'
Delgado,Maria Tagle
Mon 7h5'28"-4d35'B
Delgado,Mark Anthony
Cet 1h45'42"-2d8'
Delgado,Michael
Cet 2h10'47"6d2'
Delgado,Nancy
Aql 19h54'37"12d49'
Delgado,Nico R
Dra 19h42'35"67d60'
Delgado,Richie
Aur 6h19'28"32d2'
Delgado,Roberto Enrique
Cap 19h37'0"-18d42'B
Delgator,The
Cam 4h36'44"68d24'
Delgiorno,Bernard
Cyg 21h11'40"34d22'
Delgiorno,Roger
Mon 7h4'36"-6d22'
DelGiudice,Matthew
Per 3h25'41"40d34'
DelGrande,Angelo Anthony
Lac 22h28'0"53d53'
Delgrolice,Gary
Mon 6h21'42"4d0'
Delgrosse,Richard
Aur 6h56'0"40d51'
Delgutte,Jacques
Cam 13h11'0"81d47'
Delhommeau,Nicolas
Equ 21h22'60"7d34'
Delhommeaux,Jacqueline
Ori 6h7'31"20d31'
Delia
Lyr 19h6'48"26d23'
Delia
Cas 0h26'45"62d21'
Delia
Tau 5h52'33"28d21'
Delia,Karen Ann
Mon 6h56'59"11d35'
Del iberato,Bob
Ser 16h0'1"13d19'
Delic,Pete
Per 1h34'54"53d25'
Delicata,Vincent
Uma 10h4'56"58d31'
Delight Of Marvin
Peg 23h6'51"12d46'
Delight,Bonnie Kay
Uma 11h46'0"53d54'
Delightful
And 0h13'38"31d35'
Delightful Donna's Dipper
Mon 7h2'44"0d13'
DeLin,Tara
Mon 7h0'31"8d2'
Delina Filia Fidelis
And 2h24'44"44d51'
Delinda
Ori 5h22'53"0d56'
Deling,Evelyn
Lyn 7h25'42"50d22'
Delintt,Michele
Lyr 18h57'37"41d39'
Delisa
Mon 7h58'41"-0d52'
DeLise,Charles
Com 12h23'23"30d44'
Delise,Douglas & Sabrina Brumer
Aur 6h16'29"30d14'
DeLiSh DeJu BeRo-OD96
Lyn 7h9'1"59d49'
Delisio,Terry
Aur 4h40'12"34d14'
Delisio,Terry
Cam 14h11'44"80d28'
DeLisle,Diane
Peg 0h6'43"18d30'
Delisle,Margaret F
Cas 0h12'1"63d38'
Delisle,Marion
Cas 0h31'24"68d45'
Delisle-Blair
Lyr 19h5'1"38d40'
Deliso,Lucia
Boo 14h4'1"14d25'
Delizonna,Harry
Per 1h27'38"53d37'
Dell Direct
Aql 19h0'27'3d18'
Dell'Aquila,Joseph Anthony
Boo 14h35'18"40d48'
Dell'Aquilano,Edera
Boo 14h33'12"19d2'
Dell'Ava-06 04 1980, Camille
Per 4h5'12"37d46'
Dell'Olio,Joe
Cep 22h56'49"58d47'
Dell'Orso,James C
Aur 5h5'24"38d29'
Dell,Margaret Evelyn (Ginny)
Mon 8h6'2"-8d40'
Dell,Marjorie A
And 1h36'32"48d45'
Dell,Mary Catherine
Uma 10h26'31"55d16'
Dell,Michael
Per 2h0'39"50d31'
Della (Say)
Uma 11h15'44"41d4'
Della Ann
And 23h5'32"41d5'
Della B
Cas 1h28'1"60d33'
Della Notta
Cmi 7h23'41"1d36'
Della-Fiorentina, Catherine
And 23h32'12"40d31'
Delladio,Bill & Carol
Cyg 20h42'33"47d27'
Dellaero,Joseph E
Aur 6h0'35"45d32'
DelLago,Frank
Per 1h52'45"50d4'
Dellandrea,Dave & Carol
Cyg 19h14'30"48d39'
Dellapennatour
Ind 21h9'2"-56d43'
Dellapi,Marion
Cas 0h29'39"63d15'
DellaPolla,Craig
Lyn 7h53'56"40d44'
DellaRocca,Laura Stephanie Nicholas
Uma 13h58'14"48d46'
DellaVecchia,Gary L
Cet 0h6'59"-8d47'
DellaVecchia,Mary Jane
Cet 0h5'1"-18d28'
DellaVecchia,Ryan Alter
Cet 0h7'18"-12d8'
DellaVecchia,Tyler M
Cet 0h7'23"-17d52'
DellAversano,Rosanne
And 1h50'28"36d35'
DellaVolpe,Michael C
Cnv 12h45'20"37d26'
Delle Marisa
Mon 6h41'46"10d39'
Delle,Emily
Uma 8h56'23"49d14'
Delle-Curti,Raven Winter
Peg 22h24'59"21d59'
Deller,Samantha Dominique
Mon 7h58'14"-6d49'
Delliere,Philippe
Per 2h59'47"31d28'
Dellorusso,P
Cet 0h53'34"-3d2'
Dellus
Boo 15h20'34"48d29'
Dellwo,Allison Rose
Lyr 18h43'34"32d14'
Dellwo,Jeannine Carole Roe
Lyn 8h13'53"35d25'
Dellwo,Julia Marie
And 23h25'32"42d41'
Delma
Cyg 20h0'37"30d37'
Delma-&-Grover Forever
Cet 1h55'0"-2d4'
Delmares,Gerard
Mon 6h24'60"-0d1'
Delmas,Jean-Jacques
Her 16h11'49"50d29'
DelMastro,Joseph
Her 16h17'42"24d57'
Delmez,Brett M
Cet 2h34'31"3d44'
Delmita
Mon 6h52'52"10d26'
Delmoro,Oscar S
Aql 19h51'53"12d59'
Delmus,Nathaniel
Vul 19h22'36"25d37'
Delo,Kate
Cas 0h39'0"60d17'
DeLoach,Dr Anthony M
Oph 17h12'5"-24d18'
Deloit,Robert
And 23h14'18"42d46'
deLong,Aurélie
Mon 7h58'46"-3d34'
DeLong,Bridget Stephanie
Cam 14h8'24"80d21'
DeLor,Florence
Cam 3h29'28"61d48'
DeLorenzo,Richard (Rocket Man)
Per 2h51'40"41d6'
Delores
Leo 10h43'22"15d51'
Delorey,William
Aur 7h13'25"41d17'
Delorium
Mon 8h5'22"-6d55'
Delorme,Alain
Mon 7h44'18"-4d27'
Delorme,Christiane et Jean Claude
And 23h1'1"38d35'
Delorme,Edward
Del 20h52'1"7d33'
Delorme,Jill
Cam 4h10'53"61d14'
Delos-reyes,Dennis James Michael
Ser 16h10'55"0d33'
Delostrinos,Denise & Rod
Eri 4h0'31"-10d13'
DeLott,Steve
Boo 13h51'58"19d47'
Deloye,Katherine
Cas 0h59'43"60d36'
Delp,Jennifer Elyse
Lyr 18h59'25"47d20'
Delp,John Michael
Aur 6h0'18"45d4'
Delp,Mark Eichler
Boo 13h45'18"15d5'
Delp,Matthew B
Dra 16h51'15"67d34'
Delphia
Lyn 8h23'1"40d57'
Delphine
Peg 23h46'25"31d59'
Delphine
Per 2h38'54"56d59'
Delphine
And 0h50'50"37d24'
Delporte,Curtis Mason
Sgr 18h49'10"-36d38'
DelPozzo Godfather, James R
Ser 18h15'35"-14d1'
Delram 11 10
Peg 22h1'38"34d1'
DelRicco,Rose
Mon 6h43'26"10d52'
DelRio,Vanessa
Cas 1h32'0"68d1'
DelsDot
Aql 19h39'57"11d18'
Delsino,Lorraine Nicolette
Tau 3h51'22"0d54'
Delsol,Robert
Cet 2h50'0"1d42'
Delstar
Oph 18h2'25"10d7'
Deltagraph A/S
Uma 10h50'54"44d41'
Delthin,Renee
And 2h13'11"50d7'
Delton
Cmi 7h56'28"7d58'
DeLuca,Anthony
Lyn 9h23'0"40d43'
DeLuca,Dale Anne
And 0h3'41"35d12'
DeLuca,Deborah
Cas 2h17'33"67d57'
DeLuca,Michael
Lac 22h36'33"40d2'
DeLuca,My Shooting Star Anthony
Vul 21h12'34"28d12'
DeLuca,Nicole Lynn
And 23h41'52"38d45'
DeLuca,Peter A & Tracy O
Cep 20h22'13"78d48'
Deluca,Sandra
Cam 8h9'11"80d58'
DeLuca,Tim
Aur 6h52'0"35d55'
DeLuca,Vince P
Cep 22h16'21"62d12'
Deluca,Wendy Taryn
Mon 6h27'13"1d45'
DeLuccio,CPA,William
Cep 20h55'14"58d15'
DeLucia,Dennis
Com 12h17'13"24d13'
DeLucia,Janine
And 2h30'58"40d33'
DeLucia,Jeffrey Paul
Dra 17h12'37"64d39'
DeLucia,Joann M
And 2h26'55"45d33'
DeLucia,John
Lac 22h15'50"37d43'
DeLucia,Marie Elizabeth
And 2h19'45"42d54'
DeLucia,Rocci
Lmi 9h54'17"38d22'
DeLunas Family,The
Cam 3h22'17"55d27'
Delve,George William
Cyg 20h40'1"45d29'
DelVecchio,Carl Thomas
Boo 14h9'0"38d12'
DelVecchio,Daniel Joseph
Aur 6h1'26"36d35'
Delware,John & Nancy
And 23h39'1"46d3'
Dema,Carey Beth
Aql 18h53'18"11d33'
Demailly Emeline
Aql 20h0'32"14d43'
DeMaio,Anthony
Lac 22h7'0"48d54'
DeMaio,Dolores
And 23h12'33"36d20'
Demaio,Jr,Christopher James
Lac 22h52'30"56d50'
Demaio,Laura Aiello
And 23h22'33"41d10'
Demakes "Forever Mom", Evie
Peg 22h50'15"29d56'
Demar,Eric S
Her 17h7'1"20d20'
Demaras,Alexander
Her 16h54'20"23d55'
DeMarco,Christine N
Cas 1h28'43"73d57'
DeMarco,Dianne
Boo 13h53'52"22d4'
DeMarco,Dolores
And 0h7'13"31d15'
DeMarco,Joseph & Julie
Eri 4h46'54"-5d28'
DeMarco,Laura
Lyr 19h25'12"37d47'
DeMarco,Michael O J
Cep 10h39'45"75d17'
DeMarco,My Hero-Erin
Uma 10h29'40"50d38'
DeMarco,Pat
Uma 11h46'30"32d5'
Demaree,Robert L
Cnc 9h1'1"28d9'
DeMareo Ed D,Mark Conway
Boo 13h59'40"24d26'
DeMaria,Crystal A
And 2h31'58"45d43'
DeMaria,Danielle Christina
Lyn 7h4'42"50d11'
DeMaria,Frank
Aur 5h24'22"40d36'
DeMartin De Vivies, Cecile-Denise
Cra 18h42'46"-42d30'
Demartinecourt Michel
Lyr 18h17'44"30d37'
DeMartinis,Lily
Cyg 20h3'48"40d20'
DeMarzo,William
Cep 20h29'30"65d50'
DeMascio's Star, Barbara F
Cam 8h59'32"74d30'
Demasco,Thomas Gerard
Dra 19h5'31"50d0'
DeMasi,Daniel Dingles
Per 2h50'46"43d18'
Demasi-Jaffe,Steven Alexander
Per 1h56'0"56d46'
DeMasse,Addy Elizabeth
Cyg 19h12'0"48d43'
DeMattei,James V
Dra 14h57'26"62d1'
DeMatteo,Michael A
Boo 14h19'47"15d53'
DeMatty-Wile Marriage Star
Cyg 21h36'40"40d40'
DeMauro,Vincent John
Per 3h28'32"38d46'
Demay,Amelie
Cyg 21h20'44"40d41'
Demay,Arnaud
Lac 22h3'0"38d2'
Demay,Dale Henri
Cep 21h58'48"70d24'
Demay,John & Helen
Cyg 21h5'41"31d41'
Demay,Peter Paul
Her 15h54'0"46d58'
Dembek,Keith Anthony
Uma 10h55'19"72d49'
Dembling,Nicholas
Aql 20h1'40"11d4'
Dembowski,Victor
Dra 14h38'51"65d14'
Demchak,Kimberly Jo
Sge 19h1'1"19d14'
Demeaux,Raymond & Simon
Cma 6h57'41"-15d37'
Demelis,Artemis
Boo 14h32'32"48d3'
Demello,Alison
And 23h11'26"38d24'
Demelo,Paulo Soares
Aur 5h2'17"50d11'
DeMenna,Lauren
Cas 23h40'22"63d50'
Dement,George
Uma 10h38'55"42d30'
DeMeo,Jane Florence
And 23h40'41"46d38'
DeMeo,Kaitlyn Emily
And 0h21'22"35d33'
DeMeo,Margaret Patricia McGarrity
Lyr 19h25'22"41d49'
Demercy,Janice M
Lac 22h25'13"54d4'
Demers 1981-05-25, David
Ori 5h56'15"10d41'
Demers brillera toujours,Karyne
Cas 0h48'55"73d28'
Demers,Anne-Marie
Per 2h52'39"38d59'
Demers,Béatrice
Cas 0h36'58"61d60'
Demers,Christian
Uma 9h13'34"60d49'
Demers,Gertrude
Uma 9h11'11"60d4'
Demers,Joël Morency
Per 2h15'46"58d41'
Demers,Marie-Josée
Per 3h12'37"41d44'
Demers,Sylvie
Uma 9h19'18"60d15'
DeMers-Father,Edward Joseph
Aur 4h58'18"50d55'
DeMetra-Sean
Lac 22h37'1"55d46'
Demetri & Efi
Cyg 19h40'47"31d24'
Demetriades,George
Aur 4h46'1"38d18'
Demetrion,James
Cet 2h43'49"0d13'
Demetriou,Melissa Ann
Umi 13h58'1"71d55'
Demetris Georgious Neophytou
Aur 6h14'25"36d2'
Demetrius
Sge 19h36'18"16d49'
Demetro,Jim
Aql 19h54'48"-6d0'
Demetross,Gregg
Cep 0h15'30"80d33'
Demeyer,Karry
Cas 1h5'24"58d20'
Demg,Beban
Aql 19h54'57"15d26'
Demi
Cyg 19h29'1"31d45'
Demi
Mon 6h43'1"11d27'
DeMichele,Louis John
Cet 2h29'27"3d15'
Demierre,Diane
Lyr 19h17'1"40d21'
DeMilo,Adam
Cnv 12h41'34"41d5'
DeMilo,J W
Lyn 7h54'23"58d23'
DeMito,Vincent R.
Dra 12h30'37"68d8'
Demitras,Diane
Lyn 7h39'37"45d6'
Demitro
Lyn 8h27'50"51d44'
Demitroff,Steven
Her 16h18'53"23d47'
Demitrus,Jeffrey Allan
Cep 22h8'1"68d12'
DEMJ'S Forever
Cnv 13h0'0"50d32'
Demlow,Abigail Nicole
And 23h28'18"41d2'
Demlow,Katrina
And 2h34'25"25d42d8'
Demma,Maria DeChantal
Vir 11h48'43"6d37'
Demmer,Tara
Cnc 9h0'60"31d33'
Demmer,William
Dra 17h39'41"64d40'
Demmerle,Jean-Pierre
Aur 5h24'0"30d58'
Demming
Aur 6h10'43"50d4'
Demmons,Lisa Mary & Joshua
Peg 21h40'43"27d27'
Demmy,Saige Elizabeth
And 2h32'0"41d7'
DeMolet,Damon
Oph 16h53'50"-6d42'
Demonaco,Princesse Caroline
And 23h3'0"50d60'
Demonchaux,Colette
Umi 15h43'8"73d17'
Demonchy,Philippe
For 3h29'6"-37d46'
DeMond,Gary L
Boo 15h19'27"40d25'
Demoney,Timothy Lee
Aur 5h6'43"38d7'
Demont,Gerold
Lac 22h22'0"40d52'
DeMont,Lt Col Ralph W
Per 2h55'60"32d55'
DeMont,Shannon Glen
Mon 7h44'41"-2d55'
Demonte,Claudia A
Cas 0h52'50"69d9'
DeMooy,Lawrence J
Per 3h28'44"50d20'
Demopoulos,Jason Robert
Her 17h34'19"20d16'
Demopoulos,Karen K
Lyn 7h25'19"58d26'
Demopoulos,Mary DeLuca
Cas 0h22'24"69d49'

Demopulos,Elizabeth Ann
Uma 11h17'47"43d58'
DeMore,Janice Faye
Mon 7h19'31"-8d9'
Demore,Robert Bruce
Cet 0h33'0"1d45'
Demorias,Cristiane Beloli
Lyr 18h45'49"39d5'
DeMornay,Andy
Aql 18h53'25"-2d37'
Demory Chantal
Dra 20h21'27"71d13'
Demory,Beth
Lyr 18h45'44"31d36'
DeMoss,Caroline Eva
Mon 6h42'26"10d17'
DeMoss,James Emmett
Ori 5h56'39"15d11"
Demougin,Laurence
Cas 2h27'24"60d41'
DeMoulin,Joseph
Her 17h8'37"38d4'
Dempsay,Maricela Samantha
Cas 0h34'10"63d20'
Dempsey
Aur 5h10'28"44d24'
Dempsey Bean
Her 17h0'58"51d3'
Dempsey Business Systems
Cyg 21h5'51"40d39'
Dempsey,Carmela
Lyr 19h5'11"28d8'
Dempsey,Chris
Uma 9h31'1"58d7'
Dempsey,Christine F
Peg 22h41'42"20d34'
Dempsey,Eileen
Lyr 19h1'38"38d1'
Dempsey,Gina
And 23h8'57"40d25'
Dempsey,Jack "Dad"
Aur 5h6'46"38d1'
Dempsey,Jeff
Oph 17h7'10"-1d51'
Dempsey,Joseph James
Buttitta
Uma 9h40'53"47d50'
Dempsey,Marcia
Cmi 7h41'58"5d4'
Dempsey,Patricia G
Cas 1h57'27"75d15' B
Dempsey,Paul
Cam 12h23'17"80d10'
Dempsey,Reginald E
Per 1h31'52"53d19'
Dempsey,Rev Paul R
Per 3h12'17"41d13'
Dempsey,Robert J
Cas 1h57'27"75d15' A
Dempsey,Scott
Ser 15h19'30"6d1'
Dempsey,Shannon Lynn
Uma 9h40'0"45d32'
Dempsey,William Edward
Cnv 14h1'44"38d1'
Dempsey-Mandeja,Anne
Kathleen
And 23h47'25"43d26'
Dempster,Joan
Aql 20h6'29"0d23'
Demsey,Rachel Jane
Cas 1h8'34"61d24'
Demski,Suzanna
Eri 3h25'21"-1d59'
Demulier,Sophie
Cnv 13h22'33"50d36'
Den
Eri 2h59'1"-17d34'
Den's Vision 2000
Uma 11h39'29"64d29"
Dena Beverly
And 23h22'27"40d59'
Denae,Jackie
Mon 6h25'34"10d59'

DeNapoli,Dr Anthony J
Oph 17h18'18"10d49'
DeNardi,Dean
Hya 8h59'42"-1d37'
Denarola,David & Christine
Crb 15h50'1"28d18'
Denault,Jill
Cam 4h2'39"61d5'
Denault,Mariette Nantel Jean-
Claude
Uma 11h33'27"30d48'
Denberg,Betsy Jennifer
Peg 22h43'1"20d54'
DenBleyker,Kirk
Dra 20h26'14"67d56'
Denbleyker,Samantha Marie
Cyg 20h23'49"40d5'
Dencker,Howard R
Cet 0h32'23"-5d42'
Denecke,Frances Harper
Cep 21h52'53"55d56'
Denene Hale
Boo 14h29'54"23d2'
Denett,Jack
Her 18h20'43"12d22'
Dengler,John & Richard
Lyn 7h6'30"44d58'
Denham,Terry
Cep 22h13'29"61d9'
Denholm,W Curly Carr
Uma 11h31'35"42d52'
Denholtz,Mikenzie
Boo 15h4'59"12d5'
Denholtz,Milenda
Oph 18h3'42"11d42'
Denholtz,Richard
Tau 3h56'12"30d40'
Deni's Lucky Star
Uma 10h6'49"48d13'
DeNicola,Paula
Uma 9h32'44"58d3'
DeNicolais,Sandy
Cas 23h1'46"53d26'
Deniella Ra
And 1h53'46"47d10'
Denihan,Donald G
Aur 7h17'30"39d32'
DeNino,Roselle
And 1h25'52"33d33'
DeNiro,Robert
Her 18h10'36"40d3'
Denis
Lyn 7h47'24"39d44'
Denis & Judy-35 Years We
Love You
Uma 9h35'17"68d24'
Denis Patrick
Cep 21h11'14"68d11'
Denis,Alicia
Aql 20h1'34"11d15'
Denis,Aurore
Lac 21h57'1"42d5'
Denis,Julie & Sam
Cyg 20h53'0"38d3'
Denis,Lise
Umi 16h9'22"79d24'
Denis,Elaine
Uma 0h44'25"54d10'
Denise
Lyn 7h56'44"43d12'
Denise
Lyr 19h1'59"28d32'
Denise
Cas 0h31'19"62d13'
Denise
Lyr 19h26'1"37d36'
Denise
And 2h32'20"45d56'
Denise
Lyn 9h2'38"41d59'
Denise
Cyg 19h33'1"35d53'
Denise
Peg 22h29'52"3d35'

Denise
Vul 19h45'56"23d10'
Denise
Per 3h52'27"36d21'
Denise
Cnc 9h18'32"32d29'
Denise
And 23h17'57"40d19'
Denise "Heart Of Gold"
Cas 1h45'43"58d48'
Denise & Tim
Aql 18h54'42"-1d11'
Denise ("La Rubio")
Eri 3h30'0"-7d30'
Denise De La O
Peg 22h8'40"21d0'
Denise Lee
And 2h4'20"40d58'
Denise Lynn
Cyg 19h58'24"30d41'
Denise Rose "A Star At 30"
Ari 2h57'52"30d0'
Denise Star Of Love And
Laughter
Eri 3h2'56"-5d54'
Denise Tamara
Lyr 19h23'12"31d10'
Denise The Mouse
Cas 0h36'43"54d49'
Denise's Eternal Brightness
Ori 5h56'46"16d59'
Denise's Joy
Mon 6h55'13"-4d20'
Denise,Magistra Magnifica
Uma 8h38'57"71d13'
Denison,Oliver V B
Cnv 13h45'0"32d11'
Denison,Patricia Mooney
Cyg 19h33'40"31d56'
Denivet,Stephanie
Per 3h38'46"51d49'
Denize,Danielle Allison
Cyg 19h29'50"34d58'
Denjean,Eve
Aql 20h1'32"10d32'
Denker,Robert Kenneth
Aql 19h16'49"10d41'
Denkler,Kirk Endres
Cmi 7h27'55"7d52'
Denley
Lyn 8h8'1"44d58'
Denley-Edwards,Doris Evelyn
Cnv 13h56'0"40d40'
Denlinger,Jennifer
And 2h16'15"46d58'
Denly-Gillings, Patricia Ann
Sgr 18h54'21"-27d31'
Denman,Jr,Robert Gee
Eri 2h57'30"-14d5'
Denmar
Uma 9h59'56"51d6'
Denmark,Larry
Lib 14h25'11"-10d12'
Denmarsh T Alexander
Sex 9h52'27"-2d12'
Dennehy,Angela Lattuca
Cas 1h17'0"62d56'
Dennehy,Michael Briscoe
Her 17h34'0"26d48'
Dennehy,Patrick M
Per 1h48'41"54d15'
Dennelly,Jerry
Boo 15h4'30"27d35'
Denneny Star,The
Lyn 8h0'31"52d16'
Dennert Family Star, The Larry
J
Lac 22h51'43"53d23'
Dennert,Ruth M
Mon 7h3'15"4d54'
Dennett,Barry
Ori 6h3'1"2d54'
Dennett,Catherine
Cyg 20h23'59"38d34'

Dennett,Dave
Per 2h59'24"40d55'
Dennett,Jim
Aql 18h56'59"-0d32'
Denney,Frances
Psc 23h29'0"6d22'
Denney,Jr,Earl L
Cep 0h10'13"70d39'
Denni's Birth Star
Ori 5h59'1"14d45'
Dennie
Aur 5h31'19"54d25'
Denniger,Alice & Charles
Cyg 19h30'57"32d21'
Denniger,Suzanne Wolfie
Cas 2h53'27"61d39'
Denning,MD,Carolyn R
Peg 23h43'56"31d37'
Denning,Richard
Scl 23h16'13"-28d23'
Denninger Family,The Richard
Cep 0h13'41"70d39'
Denninger,Jr,Family, The
Charles
Per 3h3'38"47d39'
Denninger,William
Boo 15h7'50"41d1'
Dennington,Joy
Aql 19h1'19"16d3'
Dennis
Aur 5h49'49"50d3'
Dennis
Cep 22h24'39"61d33'
Dennis
Ori 5h39'14"-0d25'
Dennis
Aur 6h20'39"31d47'
Dennis
Aql 19h11'1"13d3'
Dennis & Kathy
Cyg 21h8'53"39d44'
Dennis & Lisa's Star Forever
Com 12h31'59"23d7'
Dennis & Tracy's "Dreams"
Crt 11h8'1"-18d36'
Dennis & Valerie
Uma 10h58'13"56d49'
Dennis Lee
Lib 15h33'21"-8d37'
Dennis' Star
Cnv 13h56'42"31d1'
Dennis,"Baby Leigh" Leigh
Anne
Vul 20h46'35"28d26'
Dennis,Anne
Sge 19h57'47"20d15'
Dennis,Bruce And Patricia
Cam 3h17'48"58d16'
Dennis,Cathy Jean
Cas 0h34'55"60d22'
Dennis,Daryl Eugene
Boo 15h6'59"17d20'
Dennis,Dr Gary
Tri 1h47'25"28d59'
Dennis,Earl Aubrey
Dra 20h20'25"64d33'
Dennis,Heidi
Cas 2h29'40"65d23'
Dennis,Howard E
Ser 18h0'38"-14d53'
Dennis,Katelyn M
Mon 6h27'38"8d59'
Dennis,M L
Her 17h10'1"48d36'
Dennis,Marsha K
Cas 23h28'0"60d57'
Dennis,Matthew C
Cyg 21h9'53"40d18'
Dennis,Michael B
Cma 7h13'25"-15d4'
Dennis,Michael W
Aql 19h3'57"15d7'
Dennis,My Wish Come True
Cyg 21h24'33"50d33'

Dennis,Sean M
Sct 18h41'36"-7d8'
Dennis,Taylor Asley Neale
And 23h44'48"42d33'
Dennis,Virginia Collier
Equ 21h7'14"10d43'
Dennis-JM,Senior
Ser 15h16'14"8d3'
Dennis-Shannon
Vul 19h17'58"25d31'
Dennis/Steven
Aur 6h29'20"38d36'
Dennison,Anne
Mon 6h54'40"-2d8'
Dennison,Celesta Starr
And 0h58'48"45d23'
Dennison,Daniel Bassel
Her 17h52'54"14d50'
Dennison,Denise M
Tri 2h10'1"30d38'
Dennison,Helen
Eri 5h1'21"-6d27'
Denniss,Charley Stephen
Aur 7h2'42"41d57'
Denny's Little Angel
Peg 22h3'30"25d23'
Denny's Sepia Sigh
Hya 8h53'39"3d37'
Denny's Star (My Eternal
Love)
And 1h19'49"37d23'
Denny,Aaron Christian
Per 4h44'0"38d12'
Denny,Brian & Sarah
Cyg 20h32'12"42d2'
Denny,Dr John Todd
Lac 22h39'0"56d20'
Denny,II-Bill,William F
Ori 5h54'55"13d59'
Denny,Marc
Dra 15h0'46"65d21'
Denny,Megan
Cam 5h22'47"67d57'
Denny,Tim
Ori 5h32'13"-2d34'
Denny-Lover Boy
Ser 18h14'39"-7d0'
Dennyris
Cam 6h18'27"65d34'
Denomme,Albert
Ori 4h46'54"5d21'
Denoncourt,Denis
Umi 16h11'28"79d52'
Denoncourt,Jeanne D'Arc
Umi 16h54'33"77d3'
DeNoon,Roberta Ives
Mon 7h46'25"-1d28'
DenOuden,Dr Laura B
Tau 4h20'48"28d30'
DenPed/Dilworthsmith
Cas 0h42'0"62d56'
Densen,Ryan Michael
Her 16h44'53"21d46'
Densham,Mrs Lisa
Cyg 20h23'41"39d24'
Denslow,Carl
Her 17h24'25"18d51'
Densmore,James H
Her 18h16'51"30d60'
Denson Forever,Arline
Gem 7h34'57"20d23'
Denson,Jennifer I
Sex 10h13'7"-0d14'
Denson,Robert G
Hya 8h14'54"4d1'
Dent,Barbara Fountain
Cet 1h57'26"-1d54'
Dent,Calvin
Her 16h22'11"26d42'
Dent,Debi
Cas 1h30'59"58d28'
Dent,Lucia Frances
Cas 1h54'50"76d12'
Dent,Michael John
Per 2h54'1"43d34'

Dent,Mildred Ann
Cyg 20h39'44"45d52'
Dentino,Shalom Marilynne
Cyg 19h29'44"48d49'
Denton,Arlene Grace
Marchitelli
Cas 2h9'48"68d48'
Denton,Jamie Lynn
Dra 11h32'29"70d25'
Denton,Kirsten Melissa
Umi 13h27'29"72d43'
Denton,Majorie
Mon 6h44'31"10d10'
Denton,Susan
Cyg 19h26'28"33d4'
DeNur,Jack
Tri 2h41'24"34d23'
DeNur,Pauline
Mon 7h47'50"-2d6'
Denver,Silver Lady, alias,
Anne McDonald
Ori 6h5'56"0d16'
Denver,Sir Patrick Michael
Aql 20h0'23"4d51'
Denwood,Patricia Jacqueline
And 0h8'25"40d6'
Denys,Eileen
Lyr 18h32'44"36d34'
Denyse
Cas 2h18'46"75d51'
Denzer,Rita
Sge 19h5'25"18d26'
Denzie
Eri 5h5'0"-4d41'
Denë Star
And 23h46'50"44d29'
DeOca,Jr,Omar Juda
Ori 5h3'57"-2d26'
DeOcampo,Eymard Julian
Mon 8h8'0"-1d5'
Deoma,Matthew
Sex 10h15'39"-4d8'
Deon,Sweet Georgia
Hya 8h10'1"0d9'
Deonne
Mon 6h55'0"-8d36'
DeOrio,Jr,John
Aur 7h8'39"36d55'
Deorio,MD,Keith R
Oph 18h1'1"7d40'
Deovlet,Rick
Ori 5h17'23"12d42'
Depaepe,Gus
Tri 2h35'39"31d30'
Depaepe,Jim
Aur 6h3'40"31d37'
DePalermo,Kristofer William
Boo 15h34'23"48d19'
DePalma Together For
ever,Mamie & Biagio
Lyn 7h20'46"58d17'
DePalma,Annie Richards
Lyn 6h14'48"54d24'
DePalma,Carol
Lyr 18h32'1"34d57'
DePalma,Dip & Jane
Uma 10h50'55"70d12'
DePalma,Douglas
Boo 15h32'29"41d50'
DePalma,Elizabeth
Cyg 20h4'28"40d55'
DePalma,John Anthony
Cam 3h24'1"61d31'
DePalma,Jr,Johnny Joseph
Her 18h1'25"38d32'
DePalma,Maria
Ser 16h4'0"1d12'
DePalma,Mark L
Cet 0h50'47"-4d25'
DePalma,Meredith
Cyg 21h9'21"40d23'
DePalma,Michael Angelo
Tau 4h8'52"21d41'
DePalma,Susan Lynn
Lyn 7h19'55"50d12'

DePaola,Steven Bruce
Boo 14h19'60"15d20'
DePaolo,Jesse
Uma 11h32'43"61d37'
Depaolo,Stephen Joseph
Boo 13h40'51"20d53'
dePasquale,Graham Neil
Dra 14h17'29"64d49'
Depatie,Michael Allen
Cet 3h5'53"0d57'
DePaul,James Jacek
Cam 4h26'17"68d51'
DePaul,Peter
Aur 6h1'21"38d54'
Depaulo,Caroline Ellis
And 2h7'60"42d18'
DePauw,Thomas Anthony
Cet 2h44'1"6d12'
DePersia,Sr,Rudolph T
Peg 23h35'34"28d16'
Depeter,Michael
Aql 19h48'23"14d13'
Depew,Carey Lynn
Cyg 19h42'1"31d31'
DePew,Erin May
Cap 21h40'0"-23d59'
Depew,Phyllis Emily
Aql 19h52'50"-5d53'
DePicciotto,Robert F
Cam 7h34'12"60d3'
Depierre,Philippe
Aur 5h1'28"37d32'
Depietro,Erin Francesca Clare
Cas 23h26'58"58d50'
Depietro,James Josef
Francesco
Aur 7h2'49"38d48'
Depner,Kristian
Vir 13h36'11"-8d6'
Depoian,Paul b
Cam 13h55'42"80d24'
DePollo,Adam Christopher
Boo 15h45'51"47d46'
DePollo-Garygene-
Judylorraine
Cep 20h42'47"65d12'
Depontes,Amie-Jo Dinsio
And 23h56'40"40d57'
Depoy,Penelope
Aql 19h5'46"-0d36'
Deppe,Hans,Gerd
Equ 21h1'29"3d42'
Depperschmidt,Eva Marie
Jacqueline
Mon 6h23'49"-0d40'
Deppert,Karl Dr
Ari 2h23'44"21d4'
Depre,Joseph J
Aur 6h47'58"37d32'
Deprez,Christopher Louis
Boo 13h58'38"25d52'
Depsa
Lup 15h0'42"-41d34'
Depue,Dobby Troy
Hya 8h23'23"-10d51'
Der Bedrosian Superior,Cash
Ser 18h35'21"2d27'
Der Kleine Prinz
And 23h3'0"50"1'
Der kleine Volki
Cap 20h4'0"-10d5'
Der M
Aur 5h16'0"43d28'
Der Rote Senator
Sgr 19h39'18"-40d47'
Der Sempre David
Peg 21h59'42"2d27'
Der,Johann
Aql 19h4'40"0d3'
Deragon's Birthday Star,Sue
Cep 22h11'0"58d34'
Derald-Rebecca Alpha
Vul 20h38'22"21d44'A
Deranek,Steven Gregory
Cep 21h33'52"67d36'

deRauly,Daniel Dumas
Cnv 12h19'26"34d38'
Derbedrossian,Viken
Lyn 7h4'32"44d53'
Derbique,Mallorie Marie
Cas 1h37'1"75d52'
Derbolowsky,Karina
Vir 13h29'52"-7d35'
Derby,Shannon Lee
Ori 6h7'28"1d41'
Derbyshire,Rachael
Ori 6h0'24"0d42'
Derderian,Dr Paul M
Oph 17h8'31"-21d19'
Derdevanis,Gus S
Boo 14h44'33"27d38'
Derech,Debra Ann
Cap 21h1'17"-26d21'
Dereck Scott
Equ 21h2'42"8d40'
Derek
Ori 4h44'13"0d9'
Derek
Hya 8h59'33"-8d16'
Derek
Uma 11h43'37"30d51'
Derek
Sco 17h27'59"-40d50'
Derek
Cep 22h10'32"55d25'
Derek & Drew
Per 2h59'27"32d58'
Derek & Meghan
Cyg 20h18'17"31d38'
Derek Allen:"Chudha 70"
Aur 5h59'56"31d11'
Derek Sharon
Aur 6h9'29"45d6'
Derek,MTG
Cep 23h10'1"60d54'
Derene
Cas 0h8'18"58d19'
Derenia,Edward
Ori 5h52'42"15d17'
Deret
Crb 16h10'48"38d15'
Derf
Lac 21h56'51"40d9'
Derfurt,Peter
Lac 22h4'53"50d54'
DeRibere-Larkin,Ryan
Boo 14h33'53"8d19'
Derick,Tom
Boo 13h48'34"17d19'
Derie,Carol J
And 1h48'53"36d30'
Derimanoczy,Teri
Mon 6h23'56"-10d0'B
Derissen,Franz
Sco 16h14'27"-22d45'
Derivan,Mary Collins
Cam 7h29'53"60d27'
Derivaux Donny
Uma 9h15'33"51d41'
Derk
Ori 4h53'18"0d17'
Derkinderen,Kristen
Vul 19h17'41"25d16'
Dermer,Daniel A
Per 3h42'51"36d11'
Dermer,Melissa A Horn
Cyg 21h51'24"40d21'
Dermody,John & Hazel
Uma 10h42'59"62d11'
Dermody,Joyce
Peg 21h58'51"31d45'
Dermody,Kathleen Margaret
Eri 4h39'18"-8d13'
Dermont
Aql 19h43'47"14d26'
Dern,Laura
Crt 11h16'15"-16d39'
Dernanda Rosée d'Etoile
Dra 20h16'53"84d40'

DeRocco Eternal,David &
Jamie
Lyr 18h58'31"34d44'
Deron & Cathy
Cyg 19h38'0"28d4'
DeRosa's Destiny
Lyn 7h49'27"48d47'
DeRosa,Janice
And 0h20'1"41d4'
DeRosa,Ollie V
Lac 22h22'35"53d0'
DeRosa,Paul Thomas
Her 17h20'26"50d5'
Derosa,Ralph
Her 16h47'47"50d34'
DeRose,Christopher Francis
Aur 6h23'17"40d56'
DeRose,Sr,Ralph
Psc 1h0'51"23d13'
Derossi,Damon
Aur 4h48'31"40d9'
DeRouen,Abigail Elizabeth
Cru 12h0'54"-58d18'
Derouet,Franck
Lac 22h23'53"56d41'
Derouet,Mickaël
Dra 12h5'25"63d59'
DeRouin,Reneé Eileen
Dra 11h55'31"68d52'
Derr,Hermann Peter
Sgr 19h3'48"-26d18'
Derr,Pamela "Bellofatto"
Ori 5h56'22"13d12'
Derrah,Jr,Sean Albert
Boo 14h39'37"9d50'
Derrah,Marilyn
Peg 0h2'47"30d4'
Derrah,Marilyn
Cyg 19h59'26"40d15'
Derrah,Marilyn
Eri 4h35'12"-1d38'
Derrer,Frau
Her 17h20'1"42d41'
Derrick Keith
Cam 5h38'35"73d57'B
Derrick,Ellen
Scl 23h20'31"-25d13'
Derrick,Mark
Mon 7h58'24"-8d12'
Derrickson,Lew D
Uma 11h5'52"47d52'
Derricote 10/1/15, Theodore L
Cep 24h6'41"78d1'
Derrik,Yoko Nina
Lib 14h21'17"-23d44'
Derry,Frank Wayne
Her 17h20'1"42d41'
Derry,Rowan Olivia
Tau 3h41'27"25d51'
Derryn
Cyg 19h59'26"48d35'
DerTomasian,Harry
Her 16h39'1"4d48'
DeRuaz,Anne
Lyr 18h51'19"40d36'
Derum,Stella
Cas 0h42'34"60d13'
Derusha,Katelyn Marie
Cas 1h0'46"61d24'
DeRusso,Marc A
Cep 22h45'32"57d22'
Derwae,Kristine
Cas 0h25'0"72d59'
Derwent,Gary
Cam 3h59'0"58d31'
Dery,Alexander O
Per 2h57'1"34d0'
Des (Kingsmill)
Cyg 20h40'0"45d56'
Des Hanna
Per 2h6'28"57d27'
Des Lauriers,Carol Ann
Sgr 18h3'20"-28d28'
Des Lauriers,Jean- Regis
Cam 5h52'54"65d21'

Des Plunkett
Ori 5h56'47"12d12'
Desachy,Martine
Lyr 19h12'51"31d42'
DeSade,Alex
Aur 6h29'0"30d2'
Desaix,Rose
Lac 22h4'40"37d48'
DeSalva,Karen
Lyn 9h27'30"40d28'
Desanctis,Mark
Uma 10h15'60"60d18'
Desandro,Jr,Joseph A
Cyg 19h50'0"37d38'
DeSanno,Rita
Lyn 9h35'54"41d45'
DeSanno,Valerie
And 1h23'34"37d33'
DeSantillana,Eileen C
Peg 22h7'42"5d5'
DeSantis,A
Ori 5h46'50"11d29'
DeSantis,Brigida
Leo 11h48'25"23d41'
DeSantis,Joan
Vir 13h13'51"-5d9'
DeSantis,Lucy & Joe
Cyg 21h15'1"35d16'
DeSanto,George Joseph
Francis
Cnv 12h19'40"51d35'
DeSanto,Keith
Dra 16h59'56"64d36'
Desarmeaux,Maxime
Cam 3h47'21"67d59'
Desaulnier,Benjamin T
Lac 22h21'17"55d44'
Desaulniers,Gilles
Her 16h42'1"28d3'
Desaulniers,Sylvain
Cep 21h12'0"65d23'
Desbrow,James Patrick
Tau 4h34'50"20d24'
Descarfino,Martha
Lyr 19h13'1"40d46'
DeScenna,David Dixon
Boo 14h28'23"8d53'
DesChamps,Joe
Aql 19h58'38"7d34'
DesChamps,Noel Paul
Per 2h55'37"37d47'
Deschesnes,Caroline
Ori 5h53'19"8d25'
Deschesnes,Jean
Per 3h19'15"41d37'
Deschànes,Hugo
Ori 5h57'57"12d22'
DesCites,Manon
Lyn 8h32'47"43d59'
Desdemona
Pyx 8h50'14"-23d44'
Desejo Estrela
Per 3h59'35"50d19'
DeSena,Ron & Amelia
Cyg 21h3'31"38d16'
Desens,Tom "Deer"
Lmi 9h36'40"37d34'
DeSensi,Susan Tolios
And 1h51'21"39d49'
Desenzani,Gabriella
Del 20h18'27"14d9'
Desert Rain
Mon 6h18'56"1d56'
Desert Star,The
Per 3h48'22"38d32'
DeSerto,Carmella Rose
Cas 0h40'25"64d38'
deServien-Kenwood, Margaux
Murtaugh
Lyr 19h12'31"40d46'
DesForges,Sheila S
Cyg 20h20'26"40d36'
Desguioz,Robin
Equ 21h3'1"7d35'
DeShay,Aleatha
Lmi 9h33'40"38d20'
DeShayes,Le demi-coeur
Cep 21h57'55"55d55'
DeShazo,Chez
Cyg 21h31'57"37d33'
Deshenry,Chantal
Cep 21h27'47"60d18'
Desherman,Adriana Aguirre
Lyr 19h25'1"38d43'
DeShong,Barbara
Vul 19h15'58"24d54"
Deshong,Carrie
Com 12h31'0"20d28'
Desi 5/26/71
Per 3h29'46"37d9'
Desi-Lou
Lyn 7h58'20"42d36'
Desideria
Tel 20h10'3"-45d44'
Desiderio
Ari 1h21'36"31d2'
Desiena,Debbie,Anthony &
Derek
Uma 10h41'40"67d33'
Desilets,Paul
Hya 10h11'52"-18d12'
DeSillers,Jr,Ronnie
Oph 18h42'1"10d54'
DeSilva "Cookie", Patricia
Del 20h13'50"14d31'
DeSilva,Kindel
Oph 16h48'52"11d6'
DeSilva,Lane
Mon 6h22'0"3d25'
DeSilvia,Frederick Joseph
Per 3h56'23"51d43'
Desilvio,Julia H
Cyg 20h59'21"30d7'
Desimone,Dean
Hya 9h39'0"5d27'
DeSimone,Dominick F
Cep 23h3'42"71d13'
Desimone,Maggie Elizabeth
Mon 6h43'25"7d59'
Desimone,Pasquale Luigi
Lyn 8h6'50"42d10'
DeSimone,Stephen E
Mon 7h4'49"-1d13'
DeSimone,Thomas
Cnc 8h22'56"18d49'
DeSimone-Parvulus Dei, Don
Ori 6h2'39"7d56'
DeSio, Loretta Ann
Lyn 8h14'0"34d17'
Desio,David
Uma 8h23'30"72d32'
DeSio,MD,John M
Per 2h55'30"50d37'
Desire
Sgr 18h58'30"-24d12'
Desire Greg & Vera's Lucky
Star
Cyg 21h7'51"40d29'
Desiree
And 1h57'0"47d15'
Desiree's Star
Aur 4h52'15"50d17'
Desireé
And 2h25'54"38d7'
Desiré
Cma 6h10'31"-13d39'
Desirée Heaven On Earth
Aql 19h49'34"15d7'
DesJardins' Rose
Crt 10h54'0"-21d51'
Desjardins,Franâoise
And 23h33'33"50d5'
Desjardins,Frédérique
Cam 3h26'26"60d35'
Desjardins,Gilles
Sgr 19h3'0"-29d8'
DesJardins,John Andrew
Dra 12h16'43"67d51'
DesJardins,Karen
Com 12h19'41"22d29'
Desjardins,Louisette Cholette
Uma 11h54'0"43d21'
Desjardins,Michel
Her 17h21'31"20d7'
DesJardins,Sr,Don H
Umi 15h38'17"77d8'
Deslatte,Daphne Wilbert
Mon 6h43'0"10d46'
Deslauriers,Suzanne Dirk
Dachs 1995
Cas 0h2'18"62d25'
DeslauriersJimPegJames
KristenJohn
Boo 14h46'15"25d43'
Desloges,Micheline
Cyg 19h58'29"40d17'
Desmarais,Alexandra
Cas 0h15'33"61d19'
DesMarais,Anne
Cas 0h24'1"66d55'
Desmarais,Martin
Cep 21h16'37"78d23'B
DesMarais,Mary Kobbe
Uma 10h32'19"56d24'
Desmazieres,Guy
Oph 18h17'48"10d35'
DeSmet,Lumiere de Jean
Umi 14h30'52"65d53'
DeSmidt,Hans & Mary
Crb 16h3'27"33d49'
DeSmith,Elizabeth
And 23h26'44"47d21'
Desmon,David H
Per 2h40'22"40d12'
Desmond "The Greatest Star
Of All",Norma
Cas 0h36'57"63d24'
Desmond,Dan & Dee
Ori 5h56'40"10d26'
Desmond,Jeanette
Eri 4h30'45"-10d38'
Desmond,Kalam
Lib 15h13'42"-23d49'
Desmond,Leslie
Sge 20h2'26"20d31'
Desmond,Paul Alexander
Lmi 10h56'39"32d36'
Desmond,Shannon Laurel
Sco 17h28'56"-30d4'
Desmond-Davison,Mark
Andrew
Dra 16h31'12"58d6'
Desmoulin,Nathalie
Per 2h56'29"56d36'
Desna
Aur 5h18'25"47d35'
Desnizza,Daniel
Ori 5h44'0"11d27'
DeSola,Michael J
Per 2h49'37"43d53'
DeSopo,Carmine
Cam 5h49'17"61d46'
DeSopo,Mary Angela
Cas 2h0'30"75d21'
Desormigeres,Alain
Ori 6h1'29"0d19'
DeSota,James Andrew
Peg 23h5'40"17d39'
DeSousa,George "Duke"
Aur 6h28'1"33d6'
DeSouza,Elaine d'Arruda
Mon 7h50'42"-5d25'
DeSouza,Grace
And 0h16'32"33d55'
DeSouza,Marion & Mark
Eri 2h58'21"-02d2'
Despas,Katina J
Cas 0h28'50"62d48'
Despatie,Monique
Peg 23h37'57"26d54'
Despatie,Wilfrid
Peg 23h32'11"18d3'
Desper,Joann
Cas 0h25'53"68d16'
Desperado
Aur 5h55'32"31d7'
Desperado
Per 1h41'56"53d57'
Desperado
Cap 20h39'30"-21d2'
Despigno,Victorio Alfonso
Aur 6h30'49"33d11'
Despina
Cyg 20h0'13"39d24'
Despina-Dimitris
Aur 4h55'31"38d49'
Despoinaki
Lyr 18h31'0"40d18'
Desporte,Lynn Cherie
Mon 7h31'24"-1d47'
Despretz,Mousieur
Christopher
Boo 14h19'43"50d14'
Després,Eric
Uma 9h1'22"61d33'
Desrocher,Dremi
Mon 7h9'0"-10d1'
Desrocher,Tom
Sco 17h54'11"-30d20'
Desrosiers,Alex
Ser 15h12'18"0d28'
Desrosiers,Christian
Uma 8h56'1"47d58'
Desrosiers,Normand
Ori 6h0'0"7d1'
Desrosiers,Paul-Emile
Her 17h39'54"40d17'
Dessalvi Fa,Patrizia
Cyg 19h47'56"29d56'
Dessaux,Jonas
Uma 8h47'17"67d40'
Dessen,Andrea
Mon 8h7'10"-8d26'
Dessens,Norma Eulenfeld
Mon 6h51'45"11d5'
Dessert,Brett Matthew
Sex 10h41'48"3d22'
Dessi,Frances D
Cam 7h59'53"71d9'
Dessire
Cas 0h4'12"64d18'
Desson,Patricia
Peg 0h10'17"14d5'
Dessoude,André
Ori 6h5'40"10d36'
Destacamento,Christian Pedro
Uma 9h52'46"53d21'
Destefani,Michela
Pho 0h4'16"-44d49'
DeStefano,Gabriella Marie
And 2h22'60"42d10'
DeStefano,Janet Lee Cain
Vul 18h55'55"24d48'
DeStefano,Jr,Carmine
Aur 6h50'1"41d4'
DeStefano,Margie
And 0h53'1"37d33'
Destefano,Peter T
Hya 8h14'56"0d51'
Destefano,Ronald Frank
Boo 14h12'21"40d2'
Desteph,Ginny
And 1h1'25"33d35'
DeStephano,Tony
Cyg 20h2'52"39d32'
Desthun's Bealach Na B
Ori 4h45'40"15d4'
Destineé
Cyg 20h48'31"37d48'
Destini
Cyg 20h2'1"40d15'
Destinies
Boo 14h2'42"22d46'
Destino
Uma 11h51'1"48d42'
Destino
Equ 21h23'16"7d54'
Destino
Dra 20h0'43"62d49'
Destino
Cam 6h50'23"80d4'
Destiny
Cma 7h20'33"-15d20'
Destiny
Peg 0h9'17"18d6'
Destiny
Ori 5h59'53"11d49'
Destiny
Eri 2h56'11"-15d24'
Destiny
Lyr 19h23'58"33d55'
Destiny
Dra 19h48'52"70d29'
Destiny
Gem 6h42'8"15d40'
Destiny
Crb 16h20'0"28d10'
Destiny
Boo 15h0'21"28d55'
Destiny
Umi 16h24'0"76d37'
Destiny
Uma 10h7'31"50d9'
Destiny
Uma 9h44'28"56d51'
Destiny
Lac 22h5'0"49d57'
Destiny
Uma 8h31'55"58d8'
Destiny
Umi 14h53'40"84d47'
Destiny
Umi 15h29'34"68d33'
Destiny
Dra 13h57'18"64d14'
Destiny
Dra 12h32'60"67d52'
Destiny
Vul 19h45'11"28d25'
Destiny
Dra 17h37'14"68d23'
Destiny
Uma 10h46'42"62d22'
Destiny 12:03 On 2-14-95
Lyr 18h16'26"35d2'
Destiny June
Peg 21h55'0"35d59'
Destiny M
Aur 6h27'42"38d6'
Destiny Ray
Her 17h3'31"38d37'
Destiny's Choice
Umi 15h22'1"70d5'
Destiny's Eye
Dra 15h44'52"61d15'
Destiny's Light
Mon 6h33'10"-0d16'
Destito,Samuel G
Lac 22h54'48"56d14'
DeStrange,"Nick" Nicole
Michele
And 0h26'11"43d11'
Destrempe,Johanne
Lyr 18h57'17"45d21'
Desuiew,lure Teinturier
Her 18h7'33"38d51'
Desumeur,Michel
Aur 6h13'37"31d35'
DeSure,Susan
Cet 2h38'1"3d25'
Desuzinges,Erika
And 1h6'27"47d8'
Desvignes,Brian Scott
Uma 10h18'28"51d58'
DeSzily-Kouwenhoven
Uma 12h1'45"62d4'
Det JJBITTENBINDER
Celestial Safety
Her 16h48'37"38d29'
Details At 10
Per 3h1'46"38d52'
DeTamble,Richard Neal
Cet 2h19'34"-10d46'
DeTemple,Jessica Wagstaff
And 1h24'0"40d54'
Detering,Robert & Ruth
Dra 16h49'1"73d12'
Detert,Dr Francis L
Oph 17h19'51"-20d41'
Detlef
Vir 13h58'43"5d7'
Detlef
Ari 2h22'25"26d30'
Detlef "The Mercedes of Men"
Sge 19h19'6"17d21'
Detlef,Christopher
Dra 20h8'52"64d55'
Detlefsen,Lanie Anne
Cet 0h56'0"1d4'
Detmers,Ute
Uma 11h42'45"42d51'
DeToma's,La Estrella
Her 18h4'47"30d43'
DeTour,Kimerly
Cas 1h30'24"58d51'
Detouy,Grégory
Ori 6h7'1"20d12'
Detrick,Zona Marie
Cas 01h33'0"70d55'
DeTroth,Michelle
Peg 23h39'36"31d50'
DeTrude,Dr Judy
Vul 19h43'1"27d18'
Detti,Paola
Dra 16h5'42"60d35'
Dettki,Ulrich
Ori 5h40'38"10d40'
Dettloff,Carley Kay
Cep 20h39'1"56d0'
Dettloff,Dean Christopher
Cas 0h48'25"60d48'
Dettman,Michelle Katlin
Aql 18h59'20"-6d58'
Dettmann-Easler,Shawn &
Detra
Crb 15h47'41"34d32'
Dettmer,Brantley
Ser 15h28'39"24d20'
Dettmer,Kathleen
Cmi 7h37'54"7d14'
Dettmer,Nancy
Cma 6h59'56"-18d28'
Detto,Amber
Lyn 8h57'1"46d11'
Dettorre
Aql 18h56'57"15d19'
Detuerk,Tracy S
Lyr 18h48'38"30d48'
Detwiler,Craig Sax-Cat
Lac 22h3'54"48d48'
Detwiler,Nancy Louise
Mon 6h59'50"7d60'
Detwilter,Kimberly
And 23h21'12"47d57'
Deubel,Peter
Aur 5h11'24"43d49'
Deuber,Jennie V & Jennie V
Lyr 19h13'24"40d26'
Deubert,Elizabeth Hunt
Del 20h53'38"6d48'
Deuel,Jennifer & Brent
Lyr 19h25'54"38d30'
Deuel,Rodmad Rodney &
Marsha
Hya 8h19'20"3d4'
Deuhique,J W M
Aur 7h23'0"40d50'
Deuhmig-Hayes
Cyg 20h17'11"38d40'
Deulofeu Serrat,Jordi
Dra 15h0'48"62d48'
Deumié,Gudrun Michaël
Uma 12h8'1"47d33'
Deupree,James & Elizabeth
Aql 18h59'29"13d50'
Deus Magnus Donum
Cyg 20h18'15"39d14'
Deutsch,Carol Ann
Crb 15h27'53"31d38'
Deutsch,Helen Janice
Cnv 12h55'1"42d12'
Deutsch,Ila
Cyg 21h49'24"36d27'
Deutsch,Karl
Vir 13h9'4"-5d13'
Deutsch,Lester
Crt 10h56'25"-23d35'
Deutsch,Melanie
Cyg 21h5'1"33d25'
Deutsch,Sue & Bernard
Umi 15h12'0"67d51'
Deutsch,Tad L
Aql 19h53'51"10d49'
Deutsch,Ute
Sco 17h8'1"-38d3'
Deutschmann,Dr Werner
Uma 11h24'40"41d54'
Deutschmann,Walther
Lyn 8h1'24"47d27'
Deutschmeister,Jason
Per 2h4'16"47d59'
DeVaan Klein
Umi 13h14'1"70d40'
Devaire,Viviane
Sct 18h54'45"-6d33'
Devan,Cheryl & Eric
Mon 5h59'1"-6d2'
Devane,Kevin W
Tau 5h19'29"28d25'
Devane,Ryan Alexander
Oph 17h55'24"13d18'
DeVaney,James H
Hya 8h35'11"0d5'
Devaney,Jenny & Leon
Eri 3h21'25"-5d22'
DeVanna,Doniva J
Com 13h13'31"21d12'
Devanney,Jr,Michael P
Aur 6h0'50"38d15'
Devany,Marla G
Hya 9h12'60"6d18'
Devany,Shannon K
Hya 9h12'44"6d24'
Devany,Thomas A
Ori 5h51'18"14d40'
Devany,Thomas J
Ori 5h51'0"14d34'
Devaraj Aran
Mon 6h52'47"0d57'
Devarakonda,Prabhakar
Aql 19h13'52"10d5'
Devastatingly Handsome Andy
Lac 22h3'54"48d48'
Devaud,Thibault
Cam 3h50'20"58d25'
DeVault,Bruce M
Hya 8h30'28"-8d22'
DeVault,Robert & Muriel
Oph 17h53'54"11d17'
Devaux,Gene
Cas 0h19'27"61d35'
Devca
Uma 8h24'51"68d32'
Deveau,Leonard G
Cep 22h48'13"65d18'
DeVeau,Michelle Ruth
And 23h21'55"48d48'
DeVellis,James F
Her 16h37'32"50d35'
Devenney,Garrett Paul
Cet 2h33'52"1d19'
Devenney,Matthew Thomas
Aur 4h54'0"41d1'
Devenny,Megan Jacqui
Umi 14h42'37"66d36'
Devens,Barbara L
Cas 0h35'48"66d52'
Devent,Eugenie
Cep 20h17'27"61d30'
Deveny,Rosa Baker
Del 20h15'27"9d32'
Dever III,Robert H
Ori 5h34'10"-6d58'
Dever,Amanda Megan
Aql 18h59'14"-2d28'
Dever,Christopher Paul
Uma 11h33'11"49d43'
Dever,James J
Per 3h36'0"38d39'
Dever,Jodi Ann Weber
Aql 19h8'22"15d8'
DeVere,Randolph Don
Ori 6h6'13"1d24'
Devereaux,Michael
Cet 2h46'34"0d42'
Devereaux,Michael
Her 17h26'20"38d49'
Devereaux,Shelley Tiera
Oph 18h1'1"11d27'
Deverrando
Mon 6h23'55"1d38'
Devesa Michaël
Ser 15h38'43"9d47'
Devida
And 2h31'47"47d20'
Devienne,Patricia
Pho 0h42'14"-42d52'
Devilbiss,Jorie "Bear"
Ari 2h44'36"28d57'
Devilbiss,Julie "Hooliiahh"
Tau 5h20'48"16d32'
Devillers,Prudence
Dra 9h36'17"74d11'
Devillez,Caroline Marie
Cas 0h16'37"61d13'
Devin
Her 17h20'50"40d8'
Devin & Paul's Star
Uma 8h38'1"54d38'
Devin Butt
Ser 15h24'28"20d45'
Devin Krystine
Eri 3h58'42"-3d17'
Devincentis,Eric
Aur 4h59'51"30d38'
Devine II,Joseph Charles
Leo 11h0'0"2d14'
Devine,Eamonn
Cnv 13h49'59"41d58'
Devine,Kate
Cyg 20h42'1"45d32'
Devine,Kerri Ann
Aqr 22h37'21"-0d7'
Devine,Kevin
Psc 0h55'29"21d41'
Devine,Megan Erin
Cas 1h54'23"77d21'
Devine,Michele
Del 20h17'23"11d8'
Devine,Patty
And 22h58'1"38d57'
Devine,Peggy
Del 20h14'20"10d53'
Devine,Peter
Cam 13h28'23"80d21'
Devine,Roquel Jazz Johnson
Cam 3h54'50"68d53'
Devine,Spencer
Boo 15h19'48"38d46'
DeVingo,Corinda
Uma 10h56'39"34d56'
Devita,Greg
Per 2h54'59"40d10'
DeVita,Jack Joseph
Her 17h5'52"48d54'
DeVito,Angelo
Dra 12h14'31"70d9'
DeVito,Conall Douglas
Cma 6h54'16"-19d24'
Devito,Conall Douglas
Cam 8h12'22"74d14'
DeVito,Danny
Sco 16h36'49"-35d13'
Devito,Jeanette
And 23h37'1"48d57'
Devito,Judy
Lyn 7h51'25"34d40'
DeVito,Michele Carolyn
And 0h9'47"40d51'
Devito,Steven Charles
Aur 6h35'18"37d7'
Devito,Tom
Her 17h14'0"46d48'
Devitt,Bri
Cep 0h7'1"73d56'
Devitt,Christopher
Cep 23h11'27"64d45'
Devitt,John N
Aur 4h56'22"40d39'
DeVivo,Anthony J
Dra 19h28'29"58d24'
Devivo,Joseph
Boo 15h3'33"27d8'
Devivo,Katherine & Jack
Peg 21h51'11"31d9'
DeVivo,Marion E
Cyg 19h29'50"58d5'
Devletoglou,Nicos
Her 15h58'29"40d33'
DeVlieger,Michael
Her 17h31'0"21d54'
Devlin,Dorothy Valentine
Aql 19h7'1"3d15'
Devlin,Kylook Donald
Raymond
Cet 3h0'20"1d25'
Devlin,Mark Charles
Uma 11h21'20"50d4'
Devlin,Peter
Lib 14h24'35"-10d59'
Devlin,Tom
Uma 11h15'0"52d22'
Devlin,Tom & Myra
Cyg 21h37'16"40d36'
Devo
Her 17h4'50"48d38'
Devo
Uma 10h6'15"57d51'
Devoe,Mark Anthony
Boo 15h9'53"26d28'
Devoe,Robert & Agnes
Cam 5h41'17"61d26'
DeVogel,Willem F P
Per 4h1'49"51d31'
DeVogler,R J
Dra 17h27'17"61d0'
Devon 257
Hya 8h54'0"1d41'
Devon Ruth
And 0h21'0"44d24'
Devonald,Catherine Bridget
Cas 0h11'23"60d27'
Devonastra Rajorbitus
Maximus
Cma 6h42'13"-15d5'
Devonicus,Maxastra Karlouie
Cma 6h42'52"-15d29'
Devons Star
Leo 10h56'48"10d17'
Devorah Leah
And 23h1'60"50d32'
Devorah's Star
Lyr 18h32'11"33d47'
DeVore,Elizabeth Leigh
Aur 7h2'20"38d58'
Devore,Jr,Walter C
Tri 1h58'56"31d31'
DeVore,Mark David
Aur 4h52'49"40d53'
DeVos Children's Hospital
Uma 11h38'18"53d43'
DeVos,John Karl
Her 15h55'29"42d41'
DeVos,Marya Louise
Hoeglund
Cyg 21h36'18"40d20'
Devoti,Neda
Boo 15h11'30"42d21'

Devotion
Sct 18h43'17"-9d50'
DeVough,Dan
Hyd 9h3'51"0d43'
Devri
Ori 4h59'56"4d44'
Devries,Jim
Boo 14h6'40"28d29'
DeVries,Niclas Harm
Per 2h36'32"40d2'
Devries,Sharon
And 1h5'0"40d42'
DeVries,Wayne & Nancy
Uma 11h50'1"32d20'
Devroe
Aur 6h2'27"38d30'
DEW
Aur 4h52'1"51d35'
DEW's Diamond
Sco 17h28'40"-30d11'
Dew,Dale Kim Beekman
Cyg 20h34'31"40d32'
Dew,Ralph Stahler
Ser 15h17'48"0d41'
DeWald,Eleanor W
Cyg 21h52'37"38d11'
DeWald,Pat
Mon 6h26'18"8d13'
DeWalt,Lana
Mon 6h19'18"8d47'
Dewan,Katie "Did"
And 23h0'33"51d33'
DeWan,Robert
Oph 16h50'43"10d55'
Dewar,Barbara Ann
Cas 0h35'54"61d52'
Dewar,Lesley A
Uma 8h35'1"59d57'
Dewar,Linda Barnaby
Aur 6h39'35"38d29'
Dewar,Lisa Ann
Lyr 18h29'44"30d12'
Dewart,Wesley L
Cas 0h36'21"60d51'
Dewberry,James R
Aql 20h7'13"7d31'
Dewberry,Rita Ann
Ser 16h0'42"7d43'
DeWeese,Dana
Aql 19h44'59"13d53'
Deweese,Patricia
Peg 22h30'39"24d56'
Deweese,Thomas A
Cet 0h53'17"-5d39'
Dewell,Alicia Macer
Cam 8h11'54"82d16'
Dewes,Michael
Aql 20h8'14"7d54'
DeWet,Joss
Cep 22h23'41"56d11'
Dewey
Hya 8h52'21"-6d55'
Dewey
Aur 5h18'23"43d43'
Dewey,George & Ethel
Crb 15h29'36"30d4'
Dewey,John Rittenhouse
Sco 16h7'20"-30d25'
Dewey,John W
Cet 2h13'12"9d4'
Dewey,Jr,Marcellus Frederick
Aur 5h6'11"43d15'
Dewey,Marcellus Frederick
Aur 5h0'23"42d53'
Dewey,Sr,Henry Howard
Vul 19h19'0"26d55'
Dewhirst,Tyler Matthew
Sex 9h52'15"-0d11'
Dewhurst,Thomas
Her 17h18'1"46d24'
Dewig,Robert David
Her 16h26'34"23d17'
Dewind,Tamra
Lyr 18h44'38"39d58'
DeWine,James Michael
Boo 15h34'48"40d59'
DeWit,Jennifer Ellen
Tri 2h17'26"32d46'
DeWitt
Tau 3h40'7"25d31'B
DeWitt,Bobby
Aql 19h36'17"-6d17'
DeWitt,Dr William Errol
Aqr 21h53'22"0d34'
DeWitt,Jan
Cas 0h37'25"58d34'
Dewitt,Jared
Per 4h24'1"50d53'
Dewitt,Jason M
Dra 16h9'27"61d44'
Dewitt,Katie Marie
And 1h19'40"38d42'
DeWitt,Paul Adam
Cep 20h40'0"70d18'
DeWitt,Rosalie E Spirito
Aqr 21h5'47"0d31'
DeWitz,Rosalie Alice
And 23h15'37"48d14'
Dex
Her 17h6'1"21d31'
Dex-Star
Aur 5h35'0"50d21'
Dexter
Her 16h24'44"27d41'
Dexter,Chuck
Boo 13h34'0"17d31'
Dexter,Cody
Ori 5h13'55"15d60'
Dextradeur,Henry
Dra 18h13'18"58d11'
Dey,Tanvi & Bobby
Boo 14h9'32"39d37'
Deya
Aur 6h34'0"32d59'
Deyarmin,Boyd "Tom"
Per 1h44'46"50d26'
Deyne,Leona
Umi 16h59'34"79d8'
Deyo's Guiding Light, Trudy
Aqr 20h45'29"-1d52'
Deyo,Randy
Uma 11h8'0"37d49'
Deyoe,Joshua C
Boo 14h38'21"34d52'
Deyoung Birthday Star, Lavena B
Umi 14h2'16"68d60'
DeYoung Birthday Star, The William Ira
Aur 5h1'25"44d36'
DeYoung,Amy Catherine
Cam 8h27'59"78d1'
DeYoung,Dennis A
Uma 11h24'43"31d14'
DeYoung,Paula Jean
Ori 6h1'44"6d37'
Deysson,Robert
Boo 14h59'19"50d21'
Dezago,Dora L
And 23h48'47"42d50'
Dezalia,Jr,John Manuel
Dra 16h5'1"60d37'
DeZarn,Alexis Betty Christina
Cas 0h33'34"70d6'
Dezen,Andrew Colbus
Crt 11h51'0"-8d14'
Dezen,Julia Ford
Cma 7h12'32"-13d8'
Dezen,Shelly & Jeff
Hya 9h0'12"1d0'
Deziel,Donald W & Bette J
Hya 8h59'22"2d52'AB
Dezitter,Didier
Tri 2h0'32"30d18'
DeZorzi
Her 16h41'59'35d21'
Dezra Rae
Mon 8h3'12"-9d31'
Dezso,Esther Evelyn
Cas 1h49'19"60d44'
Dezso,Rev Albert George
Oph 18h28'18"7d42'
DFA 80
Cap 20h33'17"-10d22'
Dfaalm
Aql 19h48'21"14d4'
DGC-27026
Per 3h27'27"51d0'
Dgezits,Frank K
Aql 19h54'1"12d20'
Dhaliwal,Brendan William
Ceo 16h52'58"-37d43'
Dhar,Madhav
Lac 22h10'59"49d17'
Dhara's Sweet Sixteen
And 23h32'47"37d45'
Dhariwal,Sanjivan
Mon 6h6'43"-6d57'
Dharmendara
Uma 10h32'13"40d21'
Dhillon,Gurvinder & Angela
Cyg 21h42'38"30d4'
Dhillon,Mariam
Mon 8h3'19"-1d23'
Dhingra,Mr Kuldip Singh
Ori 5h32'1"8d39'
Dho,John M
Ser 15h22'35"24d54'
Dhue,Hannah Glori
Sgr 19h26'52"-44d13'
Di
Com 12h6'59"27d48'
Di 1212
Peg 22h41'22"26d49'
di Altisifa,Perla
Ind 21h12'41"-53d37'
di Ananiel,Angelo
Ind 21h1'41"-54d22'
Di Biasio,Lee Joseph
Boo 15h6'56"48d6'
Di Bitonto,Cinthya
And 23h0'30"51d27'
Di Bona,Bobbi Jo
And 2h8'55"38d41'
di Calabria,Mafalda Ruffo
Cyg 20h41'0"31d27'
Di Clementi,Carol
Cas 1h30'47"60d6'
Di Cosmo,Anna Maria
And 23h27'25"40d15'
di Dievole,Uomini e Donne
Uma 10h56'40"68d48'
Di Dio,El Principe
Oph 18h34'58"10d52'
Di Dio,Michael Adam
Cnc 8h55'22"32d43'
Di Genua,Iolanda
Pup 7h56'33'-24d49'
Di Giuseppe,Dean & Marla
Cyg 21h54'38"42d44'
Di Maggio,Toni "Wednesday"
Eri 3h37'24"-15d51'
Di Martino,Maria
Boo 14h32'0"8d25"
Di Mauro,Serena
Cyg 20h52'14"31d28'
Di Prima,Antonio
Cae 4h53'44"-31d45'
Di Rado,Richard Dana
Ser 15h34'44"19d57'
Di Rocco,Vincent & Olga
Crb 16h2'0"38d50'
Di Salvo,Matthew David
Aur 6h35'57"33d23'
di San Marco,Charles
Oph 17h1'48"-23d15'
Di Star
Vul 19h47'53"20d26'
di Stefano,Gaia
Cas 2h59'26"60d34'
Di Stefano,Roberto
Cyg 21h53'13"42d3'
di Tosto,Pasquale
Lac 22h15'51"38d53'
di Treviso,Sandra
Peg 22h3'0"2d9'
Di Venere II,Matthew F
Ori 5h59'21"9d25'
Di-Di
Ori 5h53'57"16d20'B
Dia
Cyg 20h33'27"31d52'
Diachun,Jackie
Cas 1h20'0"67d51'
Diadan
Aql 20h1'30"9d51'
Dialynne Daydream
Cyg 20h22'20"38d18'
Diamante,Coco
Vul 20h20'17"26d8'
Diamantopoulou,Natasa
Aur 7h4'37"40d36'
Diamautaire Merry Christmas for Kaz
Gem 7h11'50"29d43"
Diambra,Elizabeth Lane
Eri 2h55'57"-6d58'
Diamond Barbara B
Cyg 21h54'25"53d34'
Diamond Boy
Per 3h9'15"47d2'
Diamond Dan
Aur 5h53'47"50d28'
Diamond Dave
Aur 6h47'49"38d14'
Diamond Dick
Hya 9h0'27"6d1'
Diamond Directory
Cyg 19h23'15"44d29'
Diamond In The Sky
Cam 4h7'11"61d50'
Diamond Jim
Cnv 16h4'0"35d8'B
Diamond K
Hya 8h28'25"0d40'
Diamond,Adelina
Uma 11h10'30"43d12'
Diamond,Alison Kristen
Lac 22h32'40"53d2'
Diamond,C R Prarie
Aur 7h25'19"35d29
Diamond,Courtney Elizabeth
Cas 0h21'59"62d21'
Diamond,Craig
Per 2h8'15"58d36'
Diamond,Geri Moroh
And 0h51'56"37d5'
Diamond,Jack
Boo 14h9'48"32d24'
Diamond,Joel Nelson
And 1h52'50"40d4'
Diamond,K G
Gru 22h28'37"-55d50'
Diamond,Kirsty
And 0h5'16"46d54'
Diamond,Lauren Michelle
Ari 1h58'47"22d16'
Diamond,Matthew David
Aur 4h47'20"41d13'
Diamond,Neil
Her 16h35'20"40d20'
Diamond,Nicholas T
Equ 21h22'33"11d2'
Diamond,Richard T
Dra 18h25'57"50d36'
Diamond,Robin Elizabeth
Peg 22h30'20"21d59'
Diamond,Seymore
Ori 5h21'23"11d59'
Diamond,Shirley
Cas 1h36'34"75d33'
Diamonte Luis Costa
Ser 15h57'53"2d57'
Diana
Aur 5h7'24"42d48'
Diana
Lyn 8h18'17"45d18'
Diana
Cas 3h9'51"61d24'
Diana
Lyr 18h43'22"41d39'
Diana
Ori 05h54'42"15d16'
Diana
Del 20h25'19"20d18'
Diana
Lyn 7h27'47"42d45'
Diana
And 23h18'11"40d6'
Diana
Cep 23h19'17"68d16'
Diana
And 1h21'47"38d24'
Diana
Eri 3h35'22"-12d46'
Diana
Lyn 8h25'34"47d6'
Diana
And 1h38'17"38d58'
Diana
Sct 18h51'38"-8d53'
Diana
Cyg 19h29'28"30d16'
Diana
Aqr 22h58'59"-10d43'
Diana & Evan
Uma 9h2'32"59d28'
Diana & Jörg
Cep 22h1'25"68d11'
Diana - Lynn
Lyn 7h11'57"58d34'
Diana 831-4
Peg 22h38'14"20d39'
Diana Cecile
Umi 16h23'27"75d19'
Diana Faye
Cet 1h23'33"0d37'
Diana Kristen
Lyr 18h45'0"30d20'
Diana Louisa
Aur 5h18'15"40d17'
Diana Michelle
And 23h47'37"41d58'
Diana Renee
Uma 12h57'20"60d60'
Diana Rose
Cas 1h16'60"67d33'
Diana Star
And 1h54'13"40d14'
Diana The Star Of My Life
Lyr 19h18'58"35d1'
Diana von den sieben Inseln
Ari 2h38'39"21d52'
Diana's Diamond
Crt 10h52'60"-6d49'
Diana's Dream
Gem 6h42'54"33d47'
Diana's Star
Tau 4h59'10"16d5'
Diana's Starlight
Ari 2h1'34"21d32'
Diana,My Love
Aql 19h56'52"10d11'
Diana-Brian Alliance
Lyr 19h17'10"25d32'
Diana/John
Per 3h3'44"41d14'
DIANAYOURASTAR
Mon 6h57'28"-0d54'
Diane
Cas 22h58'60"55d60'
Diane
And 23h2'56"50d11'
Diane
Cas 23h37'58"61d6'
Diane
Vul 20h16'22"23d41'
Diane
Mon 6h1'46"-6d28'
Diane
Lyr 19h17'45"25d56'
Diane
And 1h38'28"36d26'
Diane
And 23h37'19"46d31'
Diane
Ori 5h58'19"18d54'
Diane
Lyr 18h47'1"30d12'
Diane
Aql 19h10'12"13d51'
Diane
Eri 3h42'40"-6d23'
Diane
Cyg 19h30'1"35d54'
Diane
And 23h42'58"41d59'
Diane
Ori 6h3'14"4d57'
Diane
Tau 4h57'56"20d5'
Diane
Cyg 21h8'12"31d0'
Diane
Her 16h31'32"30d36'B
Diane
Sex 9h47'30"0d15'
Diane
Cyg 19h53'1"37d48'
Diane
Cam 10h10'52"82d17'
Diane
Lmi 10h6'44"41d43'
Diane
Tri 1h58'30"31d16'
Diane & Andrew's "Wedding Star"
Crb 15h55'11"28d42'
Diane & Bob DBA Haleakala
Mon 7h51'58"-3d50'
Diane & Bob Forever
Cyg 21h3'1"33d53'
Diane & Erik Forever
Her 17h51'13"30d30'
Diane & Frank
Lyr 19h16'60"40d10'
Diane & Joe "Eternal Lovers"
Cyg 21h28'11"38d57'
Diane & John
Dra 20h2'34"64d19'
Diane & Jon
Uma 10h35'20"41d41'
Diane & Marco
Cyg 21h29'32"38d30'
Diane & T J
Cyg 20h45'58"37d58'
Diane - 4-14-94
Lyn 8h36'45"41d35'
Diane Celeste
Com 13h31'0"20d36'
Diane Christine
Oph 17h18'39"10d20'
Diane Christine
Umi 15h18'48"72d33'
Diane Christine
And 0h8'50"34d48'
Diane Christine Marie
And 1h54'59"37d33'
Diane Elaine
Cas 2h20'44"67d58'
Diane Elise
Del 20h14'1"11d54'
Diane et Luc
Cnc 8h34'57"30d13'
Diane Kay
And 2h17'1"45d40'
Diane Marie
And 23h38'58"47d15'
Diane Marie
Cam 3h28'31"61d33'
Diane Marie
Cas 23h22'11"61d21'
Diane Marie
Com 12h5'59"20d40'
Diane Marie
Hya 9h7'0"5d36'
Diane Rene
Lyn 8h12'1"40d42'
Diane Rose
Per 4h4'51"38d39'
Diane Yvonne
And 1h4'1"40d20'
Diane's (Rumination)
Eri 4h13'45"-14d9'
Diane's Angel
Sge 19h56'44"20d17'
Diane's BA
Cas 0h2'44"50d29'
Diane's Bill
Sgr 18h23'25"-20d4'
Diane's Dad
Aql 19h41'13"11d29'
Diane's Friendship Star
Del 20h23'50"20d37'
Diane's Hopes & Dreams
Vul 20h15'30"23d27'
Diane's Light
Lyn 7h14'28"59d54'
Diane's Light
Uma 11h3'22"38d24'
Diane's Nightlight
Sge 19h59'32"16d25'
Diane's Wishing Star
And 23h20'51"45d38'
Diane,Beautiful Lady
Aql 19h3'49"-0d1'
Diane-A Truly Special Lady
Lyr 19h22'28"41d31'
Diane-Marie
Lyn 8h18'55"40d50'
Diane-Roxane
Cep 22h53'1"68d20'
Diania
And 2h32'1"50d43'
Diann
Lyr 18h37'1"42d4'
Dianna Frances
Tri 1h58'0"25d43'
Dianne
Uma 11h34'48"31d57'
Dianne
Cas 1h26'14"60d0'
Dianne
Peg 22h21'12"21d41'
Dianne
Lyn 9h8'25"33d56'
Dianne Jennifer
Ori 5h55'14"12d8'
Dianne Mary Lynn
Uma 8h9'13"70d47'
Dianne's Darling
Peg 23h40'43"25d8'
Dianne's Star
Cas 0h2'58"61d21'
Dianne's Wishaponastar
Cas 0h37'41"58d39'
Dianne-Lynda's Treasured Sister
Crt 11h15'14"-11d57'
DiAntonio,Lori Ann
And 0h32'1"37d48
Diaplago
Cap 20h58'0"-22d57'
Dias,Dandy
Dra 16h50'59"63d40'
Dias,Dennis Lanning
Uma 10h31'38"68d36'
Dias,John John
Sex 10h22'46"0d18'
Dias,Margarida
Cyg 21h54'1"41d57'
Dias,William A
Uma 11h28'25"40d7'
Dias-Mi Amore, Mi Vida Carlos Luis
Lac 22h7'32"40d12'
Diatou
Cam 14h18'12"82d12'
Diaz & Liz,Alfonso
Cyg 21h9'36"38d25'
Diaz True Love,Joselyn M
Lyn 7h44'50"41d60'
Diaz,Argelio
Ori 5h56'40"16d22'
Diaz,Bradford
Crt 10h59'57"-8d43'
Diaz,Brianna Marie
Mon 7h43'3"-5d40'
Diaz,Cierra Marie
Mon 7h57'42"-5d41'
Diaz,Dora
Lyn 8h47'59"40d57'
Diaz,Eva Maria Peláez
Lyn 8h15'43"49d8'
Diaz,Evita
Boo 14h35'16"19d2'
Diaz,Guillermo
Her 16h39'18"32d32'
Diaz,Jesse Levine
Lac 22h13'45"50d37'
Diaz,Karen Esther
Lyr 19h17'27"41d8'
Diaz,Laurinda Jeannette
Peg 23h30'0"23d8'
Diaz,Luis Felipe
Aur 7h23'0"38d53'
Diaz,Melissa Renae
Mon 7h45'7"-2d6'
Diaz,Octavio Armando
Her 17h20'26"41d48'
Diaz,Pedro
Her 16h58'53"34d3'
Diaz-Cruz,Lydia
Eri 4h53'57"-4d35'
Diaz-Zayas,Victor J
Cmi 7h34'52"1d49'
DiBari,Joan
Lyr 19h23'1"31d33'
DiBari,Lara
Lyn 6h53'38"60d50'
DiBart,Joan P
Uma 14h8'0"61d56'
DiBartolo,Michael
Cep 0h17'1"70d25'
Dibbell,Jr,El Lobo Estrella,David G
Cep 21h52'47"60d6'
Dibbens,George Fredick
Aur 5h2'1"42d53'
Dibbern,Steve
Aql 19h46'11"11d50'
Dibeler,Marilyn Kelly
And 0h5'27"31d15'
DiBella,Michael P
Crt 10h49'48"-18d24'
DiBene,Denise
And 0h50'0"34d41'
DiBernardo,Sharon
Cyg 19h52'0"37d57'
DiBetta,Frank Paul
Cru 12h15'1"-58d36'
DiBlasi,Maria Christina Karina
Aql 19h25'17"-8d39'
DiBlasi,Trish
Cas 1h56'27"58d12'
Dibner,PhD,Hal
Cmi 7h37'59"8d31'
DiBona,Carmela
Lyn 8h28'41"38d37'
DiBona,Frank
Aur 4h55'30"38d55'
DiBona,Jeana Marie
Cas 23h22'53"61d10'
DiCandilo,Deena
Lyn 7h44'54"43d49'
DiCaprio,Tammy
And 23h39'48"42d8'
DiCapua,Sr,Ronald M
Aql 20h8'54"1d19'
DiCarlo,Joseph
Cnv 13h6'14"32d21'
DiCarlo,Julie Ann
Mon 7h32'50"-1d16'
DiCarlo,Justin D
Her 16h40'13"25d27'
DiCarlo,Lisa
Cnv 12h11'26"43d34'
DiCarlo,Michele
Umi 14h53'27"78d9'
DiCarlo,Michelle
Uma 11h13'54"70d18'
DiCaro,Angelo
Oph 17h21'19"13d26'B
DiCaro,Teresa
Oph 17h21'19"13d26'A
DiCenso,Michael Phillips Andrew
Dra 10h16'53"73d42'
DiChiaro,John
Aql 19h4'37"2d4'
DiCicco,Evelyn
Cnv 12h54'0"40d28'
DiCicco,Giovanna
And 0h21'15"44d35'
DiCicco,Rob
Dra 18h0'52"84d52'
Dick & Diane The Music Never Ends
Ori 5h38'35"11d27'
Dick & Juanita "I Honestly Love You"
Del 20h38'29"10d49'
Dick & Peg
Sex 10h46'52"-1d23'
Dick & Phyllis
Crb 16h19'30"32d35'
Dick 4-8-95,Rob & Michelle
Lac 22h51'1"56d20'
Dick Aka Dad,Sheldon
Lac 22h35'16"37d48'
Dick's Star
Cet 2h4'33"1d24'
Dick's Ultimate Last Resort
Lyr 19h18'29"40d5'
Dick,Alexandra Lee
Lyr 18h55'27"30d44'
Dick,Glenna T
Mon 7h0'1"8d43'
Dick,Jr,Steven Standlee
Cet 2h19'25"5d24'
Dick,Kimberly Ann
Cas 0h59'32"58d39'
Dick,Margaret Gwendoline
Lyr 19h6'27"26d6'
Dick,Stephen
Ori 6h11'39"7d54"
Dick,Todd Gregory
Boo 15h6'48"10d53'
Dick-Mick
Cap 21h2'59"-15d33'
Dickason,Ethel Scudder Thoms
Cnv 12h32'34"51d21'
Dicke,MD,Jeffrey, Michael
Her 15h50'0"45d55'
Dickel,John Lanie
Aur 5h59'49"30d57'
Dicken,DR
Ori 6h16'43"20d46'
Dickens
Lyr 18h39'42"27d53'
Dickens
Cyg 21h42'54"30d10'
Dickens,Dana Marie Hearn
Cas 1h35'45"61d28'
Dickens,John Maxwell
Boo 14h17'59"12d44'
Dickens,Officer Roy E
Uma 9h12'36"50d11'
Dickens,Oliver Scott
Ori 5h57'36"16d41'
Dickenson
Aql 20h12'24"5d33'
Dickenson,Charlie
Sex 9h52'48"4d4'
Dickenson,Dennis
Sex 9h39'24"5d11'
Dickenson,George
Uma 8h40'49"57d44'

Dickenson,Peggy
Uma 8h35'16"55d31'
Dicker,Mavis & Joseph
Eri 4h43'17"-0d24'
Dickerson,Christine Taylor
Cas 23h30'52"61d36'
Dickerson,Dee Ann
Mon 6h39'47"-10d10'
Dickerson,Jane
And 23h36'37"46d34'
Dickerson,Jennifer
Cet 1h1'23"0d59'
Dickerson,Mark Too Tall
Dra 16h35'20"57d51'
Dickerson,Marla Renee
Lyr 19h1'1"38d23'
Dickerson,Taylor Cottrell
Mon 7h59'13"-2d24'
Dickey's Dream
Peg 23h29'24"18d11'
Dickey,Sr,Phillip Wayne
Uma 13h3'46"62d26'
Dickey,Tina "Heart & Soul"
Aqr 23h3'54"-8d25'
Dickey-Doo,The
Cep 20h57'30"55d11'
Dickhans,Megan's Grandpa Marvin L
Uma 8h45'15"51d21'
Dickhaut,Clarissa
Lyn 8h10'49"45d4'
Dickie
Boo 14h2'17"21d58'
Dickie Pickie
Aur 6h7'43"45d36'
Dickie-Doodle
Psc 1h27'15"21d9'
Dickinson,Alan
Cep 22h48'30"57d58'
Dickinson,Alma Irene
Cas 0h37'20"75d 56'
Dickinson,Angie
Ori 5h42'23"3d59'A
Dickinson,Christopher Charles
Aur 6h8'52"31d24'
Dickinson,Douglas L
Her 16h52'25"41d4'
Dickinson,George
Ori 6h15'28"8d23'
Dickinson,Jessica Kaley
Mon 6h53'25"1d31'
Dickinson,Jordan Eric
Lyr 18h21'27"47d5'
Dickinson,Louis George
Peg 23h29'37"21d16'
Dickinson,Megan
Mon 7h48'22"-1d21'
Dickinson,Richard Alan
Cep 21h37'36"56d1'
Dickinson,Robert Gregory
Aql 19h50'23"14d18'
Dickinson,Stacy
And 2h33'45"37d3'
Dickinson,Troy Edward
Boo 14h0'11"22d50'
Dicklin,Marc
Her 18h42'54"12d5'
Dickman,Danielle Lee
Aur 5h9'40"40d43'
Dickman,Edith M
Ori 5h53'27"15d9'
Dickman,Marjorie J
Ori 5h11'53"-4d29'
Dicknella
Dra 9h34'47"78d20'
Dickner,Claude
Her 17h11'56"20d51'
Dickopf-Jaehn,Brigitta
Tau 4h59'46"16d17'
Dicks,Jack Edward
Mon 6h22'16"-6d34'
Dickson,Anna Marie
Lyr 18h43'1"31d2'
Dickson,Beejee
Sex 9h54'53"-5d31'
Dickson,Bryan & Jennifer
Aql 19h41'52"12d25'
Dickson,Catherine
Uma 11h2'41"55d14'
Dickson,Clive Arthur Easton
Aur 5h13'12"44d23'
Dickson,David M
Aur 5h10'46"43d26'
Dickson,Errol M
Oph 17h0'12"10d21'
Dickson,Forever Grandma
And 2h20'12"49d58'
Dickson,Francesca
Cas 0h20'48"74d36'
Dickson,George
Aur 5h2'0"40d35'
Dickson,Gina Marie
And 1h49'47"40d34'
Dickson,Kathleen
Ari 2h54'37"22d11'
Dickson,Katrina Lynn
Peg 22h45'30"34d8'
Dickson,Mary Ann Theresa
Lyn 9h24'42"41d38'
Dickson,Michael DeLane
Aur 6h56'13"38d36'
Dickson,Robert William
Uma 10h21'23"53d3'
Dickson,Roy Wilhelm
Lac 22h45'48"53d46'
Dickson,Temple
Mon 7h40'54"-10d48'
DiConza,Rita & Peter
Lyr 18h33'55"44d16'
Dicopouli 1
Boo 13h45'36"26d13'
Dicot,Dominique
Aur 6h1'21"38d54'
Dicpinigaitis,Paul
Cep 1h30'33"87d53'
DiCrescenzo,Alice & Lou
Lyr 18h57'18"46d8'
DiCristofaro,Dan
Dra 15h43'17"62d27'
Didav
Ser 15h14'34"10d33'
Diday,Roberta Ann
And 23h32'34"42d2'
Didden,Goske
Cnc 8h29'49"7d39'
Didden,Linda
Cas 0h24'15"60d35'
Didema June
Peg 22h51'18"20d8'
DiDi
Cas 1h50'25"58d19'
Didi
Crb 16h18'10"33d15'
Didi 40
Ori 5h55'23"14d47'
Didiano,Jeanne Rita
Cam 7h23'33"83d32'
Didier & Leslie's Wishing Star
Ori 5h59'52"10d3'
Didier,Bernard
Ori 6h1'40"10d55'
Didier,Tom & Lora
Eri 2h55'41"-3d50'
Didio,Greg
Lac 22h24'31"52d46'
Didomenico,Emma Rose
Uma 15h14'1"71d12'
DiDomenico,Margaret B
Always Loved
Tri 2h23'0"31d11'
Didomenico,Paul James
Dra 16h58'52"63d29'
Didonato,Mario T
Aur 7h4'1"43d20'
DiDonna,Richard F
Cas 23h2'0"58d39'
Didosaurier,Kleiner
Ori 5h43'55"0d59'
Didzbalis,Alan
Ser 15h24'27"19d26'
Die Füchsls
Lib 14h18'47"-20d37'
Die kleine Ursula
Cet 2h55'59"4d26'
Die Maschkes
Cnv 13h6'56"37d34'
Die schöne Carmen
Lib 14h18'41"-22d39'
Die süsse Sabine Heetmann 21 J jung
Per 2h27'54"58d37'
Diebold,Tina
Equ 21h23'60"3d31'
Diederich,Kathrene Renee
Lac 22h3'39"46d50'
Diedrea
Her 17h53'16"42d15'
Diego
Lyn 7h3'32"44d56'
Diego Mi Amor
Ori 6h3'0"1d31'
Diego,Alberto
Lmi 10h49'15"30d37'
Diego,Anthony Guy
Cet 2h1'33"1d27'
Diehl,Alexander Michael
Cap 21h55'54"-10d26'
Diehl,Brandy Lorissa
Mon 8h2'35"-3d14'
Diehl,Karlheinz
Peg 23h34'30"28d43'
Diehl,Kathy
Mon 8h4'31"-8d12'
Diehl,Kelly
Cas 3h34'50"70d3'
Diehl,Nancy
Lyr 19h11'16"40d22'
Diehl,Ronald S
Boo 14h21'13"34d39'
Diehl,Sylvester James
Hya 9h8'15"5d2'
Diekert,Jutta
Peg 23h30'52"16d35'
Dielle,Taryn
Cas 0h22'17"72d4'
Diem,Tama
And 23h38'51"47d27'
Diemer,Diane
And 0h11'18"33d2'
Diemer,Jacqueline Gurganus
Uma 10h29'36"51d53'
Diemer,Michael
Dra 14h55'25"63d53'
Diener The"Star"Of Healing,Dr
Oph 18h0'16"11d7'
Diener,Holly
Lyr 18h49'49"31d23'
Diener,Horst
Lyr 19h2'40"31d46'
Diener,James
Per 3h18'1"41d6'
Diener,Jean Jacques
Lyn 7h55'1"40d3'
Diener,Kimbo
Cam 5h56'58"58d50'
Diener,Patrick
Aur 5h3'0"38d26'
Diener,Scott
Ori 6h4'20"7d31'
Diener,Stephen
Cep 0h18'15"73d13'
Dieng,Albert und Liselotte
Sgr 19h18'28"-28d19'
Diepenbrock,Alexander J
Lac 22h36'24"55d10'
Diepenhorst,Ashley Elizabeth
Uma 9h52'25"70d8'
Diepinger,Maria
Eri 4h13'0"-13d14'
Diera
Cet 1h32'0"0d37'
Dierbritta 9560
Cnc 8h33'0"7d7'
Dierdra Dee
Cyg 20h1'13"40d18'
Dierick,Edward
Per 4h6'54"52d4'
Dierick,Louis
Per 1h57'0"56d37'
Dierick,Muriel
Lyn 7h6'47"50d0'
Dierkes,Courtney Hart
Aql 19h17'0"19d20'
Dierks,Donald
Uma 10h11'32"51d8'
Dierks,Horst
Sco 17h51'26"-33d59'
Dierksmeier,Rob
Gem 7h53'14"30d13'
Dierlam,Ray
Tau 5h18'27"18d59'
Diers,Anja & Silke
Lyn 8h7'0"48d52'
Diesel Iris
Dra 16h50'49"65d58'
Diesing,Bill
Hya 8h55'1"3d19'
Diesing,Sabine & Peter
Her 17h17'52"42d3'
Dietch,Robert
Aql 20h7'35"1d30'
Dietderich,Anna Collier
And 1h18'48"33d37'
Dietderich,Sarah Jane
Cas 0h32'18"65d50'
Dieter
Dra 15h9'24"62d3'
Dieter,Jonathan David Jakob
Dra 15h5'31"62d43'
Dieterre,Rose Mortilla
Lyr 19h20'55"30d35'
Diethard
Gem 6h37'22"13d30'
Diethelm,Lauren Margaret
Ori 5h33'22"-0d20'
Diethorn,Marcy L
Cnc 8h11'32"30d38'
Dietmar
Boo 14h16'59"46d11'
Dietrich (Sr Bob's Kingdom),Robert
Her 17h53'0"38d47'
Dietrich Family Star, The
Boo 14h23'41"38d30'
Dietrich's Brightest Star
Aur 6h33'47"35d59'
Dietrich,Anita Louise
And 2h15'47"37d23'
Dietrich,Donald A
Aur 5h56'34"40d33'
Dietrich,Gerhard
Cet 0h40'1"-7d6'
Dietrich,Griffen Taylor
Vul 19h42'0"22d43'
Dietrich,Hans-Peter
Cas 0h41'11"60d52'
Dietrich,Jens
Dra 18h33'56"65d20'
Dietrich,Jessi Bernice
Cas 22h56'26"54d46'
Dietrich,My Love Forever,Sandra L
Cas 1h21'56"75d31'
Dietrich,Nicole A
And 0h21'1"36d3'
Dietrich,Peter
Sge 19h38'45"17d1'
Dietrich,Shirley Ann
Com 12h30'46"20d20'
Dietrick,Lou
Tri 1h51'52"28d18'
Dietrolagola
Aql 18h58'20"-6d51'
Dietz,Doctor
Lac 22h47'55"52d40'
Dietz,John A
Lac 22h49'51"56d37'
Dietz,Karl
Sco 16h34'21"-28d26'
Dietz,Klaus
Aql 19h53'23"0d0"
Dietz,Michael Mickey R
Aur 6h31'15"31d19'
Dietz,Sandy
Cas 0h56'44"69d53'
Dietze,Günther J
Psc 23h32'51"-3d42'
Dietziker,Gottfried
And 23h46'4"35d59'B
Dietziker,Rosemarie
And 23h46'4"35d59'A
Dietzler,Matthew C
Ser 15h54'44"22d24'
Dietzler,Tim
Uma 12h18'19"56d28'
Diez,Alvaro Fl19"56d28'
Cep 20h40'25"60d23'
Diez,Guillermo Algorri
Cep 22h9'55"63d50'
DiFabio 10-21,Louis F
Aql 20h3'57"0d46'
DiFabio,Anita Rosa
And 23h42'14"43d57'
DiFabio,John I
Dra 17h20'13"68d55'
DiFabrizio,Giuseppina
And 2h9'0"40d57'
DiFalco,Rob
Her 17h14'14"42d31'
Difazio,Deanna
Cyg 20h2'1"40d6'
DiFebo,Bobby
Cam 5h45'58"65d8'
DiFebo,Joseph & Frances
Cep 20h55'34"59d11'
Diffely,Robert & Linda
Ori 5h53'0"14d40'
DiFilippo,Catherine Sarah
And 0h56'42"45d3'
Difilippo,Cecelia Mignanelli
Crb 15h31'36"30d60'
DiFrancesco,Clara
Cyg 19h32'15"28d54'
DiFrancesco,Gladys Isabel Strong
Cyg 21h37'29"38d10'
DiFrancesco,John
Her 16h29'43"32d16'
Difrankeri
Lyn 7h48'40"44d31'
DiFranza,Joy
Cap 20h33'22"-8d6'
DiFrisco,Theresa Rose
Lyr 18h37'32"40d43'
DiFronzo,Roberta
Lyr 18h42'31"30d35'
Digati,Charlie & Betty
Mon 6h23'14"0d12'
Digby
Dra 16h28'55"69d22'
Digby For Eternity
Equ 21h0'0"10d23'
Digby,Damien J
Uma 9h33'54"57d14'
Diggi
Uma 11h52'37"61d12'
Diggins Family Star, The
Boo 15h29'23"48d26'
Diggins,Joyce
And 1h1'31"38d26'
Diggins,Jr,Thomas
Cet 2h7'58"-6d3'
Diggs,Nicole D
Cyg 21h25'20"38d2'
DiGiacomo
Umi 15h31'53"70d53'
DiGiacomo,Danielle Elizabeth
Vir 13h13'27"-5d14'
DiGiacomo,Grace
Eri 4h56'59"-11d55'
DiGiacomo,Mary Lou
Lib 14h20'0"-11d4'
DiGiaimo,Jaclyn
Cas 23h23'40"60d20'
DiGilio,Joseph N
Per 2h58'0"43d59'
DiGiocomo & Donna's Star,Scott "Chefy"
Lac 22h2'37"46d5'
DiGiorgio,Alexander Kates
Hya 9h32'15"-5d58'
DiGiorgio,Jill
Del 20h14'35"10d32'
DiGiorgio,Joel Kates
Cet 1h22'12"-2d44'
Digiorgio,Louisa Cullo
Mon 7h54'51"-3d15'
DiGiorgio,Maria
Cas 0h48'1"73d22'
DiGiovanni,Constance
Cas 23h15'41"60d35'
DiGiovanni,Jesse
Psc 1h41'13"20d54'
DiGiovanni,Lauren Adair
Aur 5h35'40"54d44'
Digitandibo
Dra 17h22'54"68d50'
DiGiulian,Ashley Anne
Aur 4h48'18"50d22'
DiGloria,Stephen Edward
Per 4h23'1"50d8'
DiGregorio,Angela Dawne
Aql 19h55'18"12d8'
DiGregorio,Anthony Jacob
Her 18h12'25"47d45'
DiGregorio,Janet
Lyn 6h32'59"61d1'
DiGregorio,Maria Bernadette
And 1h27'46"39d35'
Diguet,Georges
Mon 6h22'31"-5d57'
Dijck,Anton van
Aur 5h13'58"43d45'
Dijohn,Michael Frank
Cep 22h43'28"57d18'
Dijou & Tyg
Cnv 13h33'60"40d23'
Dikbe
Ori 5h55'60"7d48'
Diker,Charles
Boo 13h54'24"22d2'
Dikun,Paula"Beamer"
Cas 0h48'22"61d15'
Dil,Jonny Douglas Nia
Boo 14h35'27"19d14'
Dilabio,Garielia
Vel 10h5'50"50d46'
DiLapigio,Ottavio Serena
Cam 9h41'55"82d19'
DiLauri,Greggory & Daniela
Uma 11h49'45"51d3'
Dilbeck,Raymond
Hya 8h16'24"2d53'
Dilbone,Butch
Aur 6h0'0"33d45'
Dilbone,Cheryl
Cyg 20h15'51"40d47'
Dilbone,Cheryl
Com 13h3'54"28d13'
Dildine,Robert
Her 17h34'24"37d35'
DiLeo Family
Lac 22h8'24"46d42'
DiLeo,Stephanie Kalena
Mon 8h0'44"-6d60'
Dilger,Kyle
Lmi 9h48'58"38d54'
Dilger,Travis
Her 17h47'39"18d46'
Diliberto,Elaine
Cas 1h15'0"61d35'
Dilillo,Angelina Mary
And 1h6'30"37d58'
Dilione III,Lawrence Robert
Lac 22h22'44"54d21'
Dilisi,Christina
Cyg 20h0'46"30d25'
Dilisi,Lucy Ann
Peg 21h59'21"33d20'
Dilite
Tri 1h49'16"28d52'
Dilks,Ernest S
Aur 6h4'46"36d59'
Dill III,Gustaf E
Her 15h59'27"40d54'
Dill,Dr George
Cet 2h45'47"1d54'
Dill,Jack
Ser 15h56'0"23d38'
Dill,Lisa Marie
Peg 22h19'14"34d24'
Dill,Rachel Noelle
Vul 19h23'15"25d28'
Dill,Tina Sue
Peg 0h8'26"17d39'
Dill-on,Gar-rett
Her 16h37'21"28d47'
Dillaman,Kimberly Carole
Cyg 21h31'42"42d8'
Dillaman,Susan Lee
And 23h42'12"37d48'
Dillard Boone II of Mickey One
Per 2h28'0"51d17'
Dillard,Anastasia
Cas 1h4'22"58d1'
Dillard,Bill
Aql 18h39'1"-1d59'
Dillard,Carol McLain
Mon 8h7'1"-5d48'
Dillard,Janet
And 0h57'47"36d6'
Dillard,Justin-Denver Crotchbuckle
Aur 5h4'40"40d21'
Dillard,Paula
Lyr 18h24'45"45d4'
Dillard,Ronald C
Aur 5h13'16"40d54'
Dillard,Sheryl Ann
Cas 1h33'1"58d29'
Dillard,Tracie
Lmi 10h18'52"30d34'
Dillard-Arborio
Cet 1h31'0"-10d12'
Dille,Adam Nathan
Uma 11h54'46"38d13'
Dille,Travis Jacob
Uma 11h54'35"30d37'
Dillenbeck,Mark
Oph 18h7'24"10d18'
Dillenbeck,Robert L
Aql 19h19'26"15d47'
Dillender's Symphonia, Dick
Cet 1h54'52"-8d28'
Diller,Anita
Cam 13h14'20"77d3'
Diller,Michelle Dusseau
And 0h13'38"39d23'
Diller,Phyllis
Equ 21h2'50"2d38'
Dilley,Diana
Uma 9h5'1"48d38'
Dilley,Elizabeth Faye
And 0h51'1"33d49'
Dilley,Joel
Oph 18h16'11"11d48'
Dillingham,Caroline Gaylord
Peg 21h49'13"19d35'B
Dillingham,Cole Parrish
Cam 6h4'44"80d15'
Dillingham,Olivia Andrews
Peg 21h49'13"19d35'A
Dillion,Dakota Christian
Cet 2h30'30"5d14'
Dillitzer,Andrea
Aql 19h58'58"10d15'
Dillmann,Elodie & Anthony
Boo 14h24'11"19d41'
Dillmeier,Hazel & William
Umi 16h25'53"71d22'
Dillmeier,Joy Winn
Crb 15h56'36"30d48'
Dillner,Frances
Cet 2h49'40"-0d4'
Dillon,Andrew
Peg 23h0'20"24d29'
Dillon,Catherine DeLisle
Uma 10h55'0"38d24'
Dillon,Charlotte Diane
And 2h29'0"45d45'
Dillon,Christina
Cas 0h44'40"64d47'
Dillon,Dale & Ron
Cyg 20h51'36"37d36'
Dillon,Dolores C
Uma 13h24'54"62d44'
Dillon,Elizabeth
And 23h11'55"37d16'
Dillon,Gerard James
Leo 10h53'0"-5d32'
Dillon,Helen
And 0h7'22"30d49'
Dillon,Jaci
Cas 1h56'50"75d43'
Dillon,Jennifer
And 2h10'60"38d1'
Dillon,John James
Aur 5h54'37"31d49'
Dillon,John Nicholas
Lac 22h23'36"48d51'
Dillon,Jr,John A
Ori 6h1'13"5d39'
Dillon,Mary
Cmi 7h18'18"0d24'
Dillon,Michael Robert Maietta Francis
Boo 14h55'1"53d12'
Dillon,Mr & Mrs Jeffrey David
Sge 20h1'52"18d54'
Dillon,Roger J
Ori 5h23'58"-1d24'
Dillon,Sean Michael
Uma 11h12'32"46d46'
Dillon,Shaun
Aur 5h57'49"31d27'
Dillon,Wanda M
Cma 7h24'23"-18d53'
Dills,Kelly A
Cep 21h28'22"66d49'B
Dillsworth,Anthony
Dra 19h39'46"70d3'
Dillweed
Mon 7h13'20"-6d28'
Dilnes,Susan Linda
Ori 5h4'47"0d35'
Dilnes,Susan Linda
Ori 5h4'47"0d35'
DiLonardo,Brynn Taylor
And 1h42'40"41d8'
DiLonardo,Michael S
Hya 8h34'10"-1d8'
Dilorenzo,Frank
Per 2h50'11"45d18'
DiLorenzo,Ronnie
Boo 15h1'16"40d57'
Diltz,Raymond Hershell
Her 18h3'1"14d26'
DiLullo,Jim & Stacey
Eri 3h39'35"-13d47'
Dilwyn
Cam 3h58'38"62d12'
Dim-Watts 1946,Jane
Uma 15h19'0"70d51'
Dimaggio,Lynn Ann
Cas 0h54'18"65d25'
DiMaggio,Russ Dominic
Lac 22h44'41"54d25'
DiMaio,Cheryl Ann
Cyg 20h22'19"38d2'
DiMaio,Jr,Joseph R
Per 3h3'51"50d22'
DiMaio,Vincent
Cep 21h40'0"58d47'
DiManna,David
Lib 15h30'23"-8d13'
DiMarco Star,The 1994 Chicky
Aql 20h5'43"6d0'
DiMarco,Arlene S
Cas 2h12'49"61d14'
Dimarco,Maryann (Ma)
Lyn 8h12'29"47d33'
DiMaria,Frank
Umi 14h47'38"68d17'
DiMartino,Chris & Donnie
Cas 23h22'34"62d47'
Dimartino,Star Susanna
Gem 6h53'10"18d11'
DiMarzio,Phil
Aur 6h25'34"31d58'
DiMascio,Felice
Cas 22h58'45"53d24'
DiMasi,Gabriel Lawrence
Uma 8h34'45"68d14'
Dimataris,Kathy
Cyg 19h44'33"30d19'
DiMattei,Danielle Lynn
Lyr 19h23'39"31d17'
DiMatteo,Natalie Anne
Cas 0h59'41"63d15'
DiMatteo,Tony
Boo 15h30'23"50d3'
DiMauro,Lisa M
And 23h6'12"40d1'
DIMEL
Uma 10h39'29"47d45'
DiMemmo 11-10-92, Alfredo Santino
Umi 16h18'55"71d11'
DiMemmo 12-12-95, Roberto Francis
Dra 14h50'34"64d26'
Dimenna,John
Cmi 7h41'44"0d26'
Dimenstein,Beth
Sco 15h55'27"-21d2'
Diment,Myrtle
Cep 22h32'1"60d45'
Diment,Robert A
Aur 5h43'1"50d22'
Dimeo-Grant,Paula
Cam 7h34'1"67d56'
Dimeo-Grant,Paula
Cas 1h3'44"55d17'
Dimetros,Jackie Spears
Lyn 8h24'58"47d30'
DiMichele,Daniel Armand
Cep 21h14'1"55d13'
Dimilia,Sergio
Ori 5h44'52"11d31'
Dimino,Bud & Joan
Lyr 18h53'48"40d36'
Dimitri
Cep 22h53'36"57d2'
Dimitroff,Daniel
Lac 22h20'43"52d28'
Dimitroff,Kate
Tri 2h29'11"28d31'
Dimitrov,Clara Monique Alice
Cas 0h53'56"54d33'
Dimitsas,Aristotelis
Cep 22h25'36"70d38'
Dimitt,James C
Tau 4h9'27"22d17'
Dimmerman,Adam Ross
Aur 6h28'14"33d7'
Dimmick,Henry & Cindy
Cyg 21h5'51"40d39'
Dimmick,Viola M/ Shirley D Hanusik
Per 1h56'27"56d44'
Dimmitt The Eternal Light,Terri
Lyn 8h15'23"37d23'
Dimmitt,Evan Taylor
Lac 22h1'53"41d5'
Dimon,June Marie Dornbach
Aql 19h25'1"-7d32'
Dimond,Nancy
Aql 18h58'1"15d59'
DiMou,Victoria
Peg 23h39'31"15d39'
DiMurro,Ceallaigh Marie
And 1h35'1"38d12'

DiMuzio,Francine "Frankie"
Uma 11h35'11"48d15'
Din
Cnv 13h54'0"32d9'
Dina
Lyn 7h47'59"48d36'
Dina
And 23h16'42"35d28'
Dina
Lyn 8h8'40"58d57'
Dina
Ind 20h25'49"-55d17'
Dina & Andy
Cyg 21h6'46"30d44'
Dina & Craig-Forever Together
Eri 3h51'15"-4d2'
Dina Gail
And 22h56'1"50d38'
Dina Lin
And 0h10'1"46d32'
Dina Marie
Cyg 20h49'45"50d8'
Dina-You Deserve The Milky Way, Virg
Peg 23h28'19"23d58'
Dinakar-S-25
Lyr 19h20'35"42d37'
DiNapoli,Deanna Marie
And 1h55'1"39d12'
DiNapoli,Joseph
Aur 4h48'49"51d36'
DiNapoli,Karin & Troy
Lyn 7h3'38"53d49'
DiNatale,Lisa
Cas 0h42'32"63d60'
Dinauer,Katrina
Crb 13h14'34"27d59'
Dincecco,James & Patricia
Lyn 7h8'0"58d48'
Dineen
Uma 9h47'23"49d57'
Dineen 1908,P
Uma 9h8'24"54d26'
Dineen,Patrick J
Aql 20h11'1"14d25'
Dinella,Maria S
Lyr 18h48'18"31d39'
Dinelli,Carla
Del 20h18'18"15d43'
DiNenna,Joyce L
And 1h13'26"34d27'
Dines,Clár
Eri 4h38'26"-0d14'
Dines,Margaret
Boo 14h36'35"48d0'
Dingel,Valborg
Lac 22h37'27"53d22'
Dingelstedt,Rudy
Her 18h13'14"30d16'
Dingfelder,Jessica
Cyg 21h10'26"34d34'
Dingg-A-Lingg
Uma 9h26'39"68d20'
Dingley-Jones,Michael
Boo 14h7'51"25d38'
Dingman,Susan W
Leo 11h5'13"25d3'
Dingo Show Star
Uma 9h31'58"67d32'
Dingsdale,Pat
Cet 1h50'0"-5d15'
Dinice,Louis Joseph
Uma 11h3'0"40d19'
DiNicolas,John Anthony
Aql 18h45'17"11d50'
Dininno,Bennett Mark
Cnv 12h11'38"48d38'
DiNino,Cynthia M
And 0h18'48"36d21'
DiNino,Robert
Per 3h14'41"40d6'
Dink
Sct 18h42'29"-6d3'

Dinker Always,Mileth
Tri 1h31'21"30d11'
Dinklage,Frank-D
Gem 7h14'18"31d3'
Dinkuhn,Robert
Dra 11h31'45"67d32'
Dinlocker,Jessica Lynn
Ori 6h6'58"0d55'
Dinlocker,Matthew Edward
Ori 6h7'25"0d55'
Dinnage,Emma
Lyr 18h54'35"38d39'
Dinneen,Elizabeth C
Peg 23h18'55"30d43'
Dinneen,Felicia Golisz
Aql 19h28'39"10d31'
Dinnes,Westin William
Boo 15h38'41"40d45'
Dinnie,Lubin Walter
Cep 23h12'10"60d54'
Dinnien,John Joseph Patrick
Her 16h39'49"28d6'
Dinnigan Star,The
Dra 19h4'22"58d34'
Dino
Col 6h37'12"-35d27'
Dino
Aur 5h13'44"41d53'
Dino El Monawir
Cep 21h36'19"60d32'
Dino's Star
Aur 6h5'31"37d47'
Dino,Eric
Cyg 20h37'12"31d31'
DiNofia,Corey
Cnv 13h52'47"39d26'
DiNovi's A Star,Jay
Per 1h53'30"56d48'
Dinsdale
Cep 20h38'19"58d56'
Dinsdale,Geoffrey Edward
Aur 4h49'31"37d49'
Dinsdale,Michael Adam
Dra 17h47'37"60d57'
Dinsdale,Tony
Vir 12h56'48"-9d35'
Dinsmore,Sue & Jim
Cam 3h41'37"60d43'
Dinto,Paul P
Dra 16h54'39"60d5'
Dinunzio,Matthew John
Cet 0h54'11"-5d50'
Dinwiddie
Her 17h2'27"48d9'
Dinwiddie,Patty
Cas 2h32'50"58d12'
Dinwoodie,Betty K M
Cas 3h3'58"65d19'
Dio Guardi,Dean
Her 18h52'25"18d45'
Diodati,Lorrie
Del 20h18'45"11d4'
Diodato,Virgil
Tri 2h16'1"30d44'
Diogo's Love For Carol
Vul 20h41'59"20d33'
Dion
Sct 18h54'45"-6d33'
Dion,Gabrielle Virginia
Cas 0h9'30"54d59'
Dion,Jean Franáois
Cep 22h25'1"57d51'
Dion,Phil
Per 1h50'1"47d37'
Dion,Rachel
Cep 20h31'31"76d9'
Dion,Shawn Patrick
Cnv 13h20'0"28d41'
Diona Gayle
Aqr 22h5'12"0d30'
Dionis,Danielle
And 0h19'23"38d5'
Dionisopoulos,Alice & Nick
Eri 4h13'55"-12d40'

Dionne,Celine
Lib 14h45'23"-8d54'
Dionne,Denis
Ari 2h38'54"20d24'
Dionne,Gérard Paul
Leo 9h49'53"30d36'
Dionne,Michael L
Sct 18h34'0"-4d20'
Dionne,Zachary Jeffrey
Sct 18h43'56"-6d48'
Dionot,Jeffrey
Lac 22h8'0"48d55'
Dionysius
Com 12h15'16"30d49'
Dionysus
Cyg 21h4'31"40d41'
Diorio,Anthony Joseph
Lyr 18h32'31"36d15'
Diorio,Dorothy M
And 1h6'54"38d35'
Diorio,III,Robert Anthony
Her 17h47'32"41d7'
DiOrio,Jake Thomas
Lac 22h14'36"38d19'
Diorio,Marianne
Cas 0h6'22"54d34'
DiOrio,Richard
Lyn 9h5'56"35d2'
DiOrio,Robert
Boo 15h3'12"20d57'
Dios,Maria
Lib 15h30'55"-2011'
Diosa Rosie
Mon 6h38'36"-7d31'A
Diotalevi,Florie- Florence
Cas 1h36'60"61d25'
Diotte,Sharon Jean Mary
Lyn 9h3'14"39d25'
Dip,The
Cyg 21h53'36"41d7'
DiPace,Andrew & Marianne L
Crb 15h32'1"38d48'
DiPalo,Nicholas Francis
Aur 6h18'53"30d39'
DiPaola,Douglas J
Aur 6h53'29"37d44'
DiPaolo,Joseph Edward
Sco 17h3'36"-31d32'
DiPaolo,Laura Paige
And 0h57'60"38d59'
DiPasquale's Piece Of Heaven,Dina
Lyn 9h27'50"41d18'
DiPasquale,Frank & Anne
Cep 22h9'0"61d36'
DiPasquale,Marilyn
Equ 21h22'1"3d29'
DiPasquale,Vincent
Her 16h53'41"28d17'
DiPierro,M L M C
Aur 5h30'51"50d20'
Dipl Ing Ottokar & Dipl VW Günther Wenzel
Vir 13h5'32"-5d11'
Dipl-Ing Klaus Wilhelm Stuttgart
Ser 15h53'28"1d14'
Dippel,Scott,A
Hya 8h16'0"2d35'
Dipper Dan
Aqr 19h31'51"0d11'
Dippong,Jasmine Alana
Lib 14h45'1"-20d29'
DiPrima,George
Per 2h58'10"55d40'
DiPrima,James Joseph
Aur 5h17'56"42d50'
DiQuattro,John
Aur 6h28'1"38d48'
Diracles,John & Marcia
Cep 21h52'42"60d28'
Dircks 100
Vir 13h17'4"-8d14'
Dircks,Gaston Michael
Mon 8h6'11"-10d9'

Dircks,Gunnar Ray
Equ 21h21'20"7d50'
Dircks,J & J
Cnv 13h57'41"38d37'
Direnzo,Silvana
Cas 0h30'22"61d55'
Dirgni
Ari 2h38'23"21d8'
Dirheimer,James R
Hya 8h9'26"1d32'
Dirickson,Janice Ann
Mon 7h4'1"1d48'
DiRienzo,Vincent Michael
Aur 6h23'37"30d51'
Diringer,Eugene
Uma 11h18'1"60d60'
Dirk & Ida 3010199312. 5 Years Holland
Cam 5h43'29"61d29'
Dirk & Jessika
And 1h54'6"41d6'
Dirk Star,The
Ori 6h0'16"1d2'
Dirk Taams 50 Jaar
Cas 1h 57'11"61d18'
Dirkanja One
Cet 2h59'39"4d19'
Dirkers,Gregory
Her 18h20'42"13d3'
Dirkrox
Aql 18h56'0"16d22'
Dirks,Delores
Del 20h19'11"8d50'
Dirks,MD,Kenneth R
Leo 11h49'59"21d20'
Dirks,Ray
Cam 12h28'42"81d22'
Dirks,Thomas
Aur 6h40'0"35d32'
Dirksen,Alfred "Alf"
Lib 15h28'54"-24d17'
Dirmeyer,Rick
Per 2h11'0"57d23'
Diroff,Lady Lu-Luann Marie
Peg 21h28'12"27d55'
Diroff,Lady Lu-Luann
Cam 4h12'26"58d27'
DiRosa,Jr,Angelo
Her 18h43'1"18d53'
Dirty Bruce
Cep 22h15'0"67d44'
Dirty Harry
Aur 5h6'1"41d9'
DiRubba,Melissa Elizabeth
Lyr 19h14'19"37d50'
DiSalvo,Noel
Dra 12h1'12"70d13'
DiSalvo,O John
Dra 19h32'48"68d7'
Disano
Oph 18h42'14"7d39'
DiSanto,Jerry
Dra 19h48'24"61d42'
DiSanto,Raffaella
Cnc 8h41'34"18d56'
Dischor,Sharon
And 1h15'59'35d6'
Disco Lulu
Lyn 6h36'17"54d55'
Discon,James Russell
Her 16h0'51"40d17'
Discovery-Lochridge & Company
Uma 10h20'34"72d7'
Disegni,Nicole
Mon 6h41'30"0d8'
Diserens,Stephen E
Sct 18h41'21"-4d2'
Disher,Cindy M
Umi 15h3'37"66d30'
Dishkin,Daniel
Her 17h18'29"29d3'
Dishon,Maguy
Peg 21h49'0"34d59'

Dishop,Valerie Arin
Aql 19h7'1"3d58'
Disimile,Maureen Francis
Cas 0h43'46"60d38'
Diskerud,Mikkel Morgenstar
Aql 19h5'35"2d48'
Diskin,Arthur L
Aqr 20h39'55"-8d31'
Diskin,Erin
Lac 22h10'44"49d43'
Diskomp Computer Sales, Inc
Dra 17h57'38"58d49'
Dismukes,Peace Nelson
Mon 6h55'11"8d26'
Disney,Carrie
And 2h29'18"47d29'
Dispeker,Thea
Her 18h4'55"14d23'
Dispenziere,Nicole J
Lyn 7h55'0"34d4'
Dispoto,Vincent
Dra 20h20'59"73d41'
Disraeli,Ari Murray
Lib 15h2'16"-8d58'
Dissick,Dennis Harvey
Cet 0h42'54"1d18'
Disston,Dianne H
Com 13h3'55"15d42'
Distant Glint-Bill's Star
And 0h9'34"37d48'
Distant,Enid Joyce
Mon 8h1'56"-6d17'
Distefano,Anna
And 2h25'30"45d57'
DiStefano,Bean
Uma 12h54'1"58d3'
DiStefano,Frances
Lac 22h53'32"56d38'
DiStefano,Francesca & Joell
Cas 0h10'0"62d29'
DiStefano,Michael James
Cnv 13h51'30"40d53'
DiStefano,Michael "Rambo"
Aur 6h29'1"30d41'
Distler,Arthur Stuart
Cep 1h54'32"77d54'
Ditas
Cam 4h23'52"60d15'
DiTolla,John James
Eri 4h36'20"-1d48'
DiTolla,Phille
Eri 4h37'59"-1d58'
DiTommaso,Michael Anthony
Her 17h21'51"44d53'
DiTrapani,Drew J
Per 2h56'0"32d11'
Ditsch,Peter
Tau 4h38'41"0d42'
Ditschler,Michael
Her 16h50'47"39d44'
Dittemore,David Michael
Cet 2h31'0"4d7'
Dittems One
Cet 3h18'57"3d60'
Dittmar,Gerhard
Cnc 8h8'38"30d13'
Dittmer Silver Anniversary Star
Cyg 19h56'38"48d47'
Dittmer,Eric Allen
Aur 6h21'44"30d46'
Dittmer,Frederick & Danielle
Cyg 21h3'33"38d54'
Ditto
Cyg 21h34'1"31d21'
Ditto
Umi 14h20'33"66d2'
Ditto
Lyn 7h8'1"59d26'
Ditto 459'er-2-135
Psc 0h59'33"32d22'
Dittrich,Bernd & Diana
Peg 0h0'27"27d51'
Dittrich,Clarence Otto
Boo 14h33'0"8d53'

Dittrich-Davis 9øø9ø1
Lac 22h26'0"40d03'
DiUbaldo,Gina
And 23h19'55"51d18'
Diva's Passion
Vir 11h58'0"-5d39'
Divan,Janet Ann
And 23h35'22"47d48'
Dively,Angela Sophia
Sge 19h29'44"17d34'
Dively,Stephanie Nichole
Sge 19h9'4"17d0'
Dively,Thomas Michael
Lyn 7h58'0"51d35'
Diver,Annette Frances
And 0h20'30"31d14'
Divers,Mary S
Leo 10h57'41"22d40'
Divers-Hobbs,Jill
Vir 13d1'49"-17d53'
Diviaio,Guy
Her 16h46'19"20d7'
Divin,Benjamin Andrew
Oph 17h56'25"-8d45'
Divina Tutela At Timm Et Swantje
Vir 13h37'52"2d20'
DiVincenzo,Mary
And 0h24'27"31d6'
Divine Donna
Del 20h17'0"14d25'
Divine June
Ser 16h2'34"13d59'
Divine Star Of Patience,Thomas
Peg 22h2'59"5d19'
Divine Victoria
Boo 14h57'35"30d37'
Divinest Karen
Lyn 8h9'42"37d32'
Divino,Estrella
Oph 18h33'0"6d53'
Divis,Nancy & Michael Howard 6/6/81
Hya 9h3'56"5d16'
DiVito,Carl & Vincenza
Cyg 19h31'46"34d12'
Divito,Jeffrey
Lac 22h42'0"56d35'
DiVito,Stephanie
Cyg 19h25'44"30d8'
DiVittorio,Charisse Ann
Ari 2h57'20"21d27'
Diviya
Dra 16h44'27"60d55'
Divoky,Cameron William
Ori 6h2'25"5d10'
Divoky,Sean Callum
Aql 19h58'36"15d15'
Divus Theolus Zebaoth Abdullah Von Madsen II
Ori 5h23'1"1d18'
Dix,Ian
Aql 20h2'19"10d5'
Dix,Judy
Ori 5h34'55"0d20'
Dix,Justin David
Aqr 23h2'45"-6d56'
Dix,Racheal
Peg 23h17'20"33d44'
Dixie
Lyn 8h37'0"41d37'
Dixie
Oph 17h5'23"12d11'
Dixie
Cet 2h52'22"3d5'
Dixie
Lyr 19h4'52"25d39'
Dixie
Hya 8h16'25"0d9'
Dixie
Peg 0h4'47"26d10'B
Dixie Chicken
Boo 14h40'35"48d6'

Dixie Kids
Lac 22h2'33"35d18'
Dixie's Star
Uma 10h9'57"50d9'1"
Dixit & Children,Mr & Mrs Paresh
Her 16h38'51"24d11'
Dixon's Colorado Star, Don "Dillon"
Aur 6h14'25"37d31'
Dixon,"Mad Dog" Mark A
Lac 21h58'52"37d18'
Dixon,Aileen Thorstenberg
Mon 7h44'1"-1d18'
Dixon,Alice
Cyg 19h53'30"37d59'
Dixon,Alice Elizabeth
Umi 14h40'0"65d39'
Dixon,Arthur
Lyr 18h54'53"31d43'
Dixon,Ballard Fulton
Vul 19h22'49"26d51'
Dixon,Brett
Ser 15h33'33"18d5'
Dixon,Chandler Michael
Uma 9h56'1"68d41'
Dixon,Charles Wayne
Ori 5h30'15"-3d14'
Dixon,Colin Pip
Uma 11h24'44"42d29'
Dixon,David
Cep 22h33'33"61d18'
Dixon,David M
Cep 22h36'10"70d31'
Dixon,Evelyn L
Cyg 20h4'53"41d9'
Dixon,Fred L
Aur 6h26'1"37d55'
Dixon,Gina
Cas 1h49'33"75d48'
Dixon,Helen
Ori 5h57'15"16d5'
Dixon,James
Ori 4h55'30"5d3'
Dixon,Jay (GOD)
Oph 18h9'0"12d28'
Dixon,Jeane
And 2h18'32"41d29'
Dixon,John & Jennifer
Umi 15h21'32"70d34'
Dixon,John E
Ari 2h31'18"22d8'
Dixon,Jr,Harry Wilson
Aur 6h6'0"35d37'
Dixon,Margaret G
Cas 0h41'38"71d14'
Dixon,Margaret Mary Hope
And 23h19'24"51d15'
Dixon,Myra
Cas 0h34'54"62d56'
Dixon,Pat
Aql 19h54'11"1d19'
Dixon,Pat
Cet 1h25'13"-6d49'
Dixon,Paul
Aur 5h46'39"54d48'
Dixon,Pauline A
Cas 1h35'42"60d15'
Dixon,Shelba
Lyr 18h48'49"36d37'
Dixon,Sue
Cyg 19h34'42"34d11'
Dixon,Timothy Alan
Boo 14h35'34"15d44'
Dixon,Tommie E
Sct 18h49'17"-7d41'
Dixon-Balsiger Star, Nancy
Cas 0h3'11"64d24'
Dizier,Edouard
Aql 19h57'0"13d14'
Dizlove
Ori 5h59'0"21d2'
Dizon,Maria Teresa (Tess)
Peg 23h28'1"8d43'

Dizzy
Peg 23h2'38"30d56'
DJ
Cam 5h11'51"68d58'
DJ 25
Cyg 21h33'27"34d19'
DJ Fred
Tri 1h45'0"26d30'
DJ JAK
Per 2h55'1"45d60'
DJ ONE
Per 2h26'49"57d50'
Djanashvili,Yana
Ori 5h50'59"9d37'
DJC 3357
Cam 6h19'52"68d59'
Djeng "Twin",Jane
Cas 0h30'0"50d13'
Djian,Aloma Janine
Cnv 12h21'13"40d36'
DJP6292
Peg 23h35'51"11d56'
Djurdjica
Ori 6h2'0"8d37'
DKHL-My Guiding Star- RTL
Ori 5h52'27"15d37'
DKIMDC
Cam 6h40'24"80d36'
Dkta Lar E Boi
Uma 10h9'29"56d53'
DLTM 102684
Hya 9h8'13"4d28'
Dlugokenski,Helen
Cas 0h34'15"66d42'
Dlugokenski,Helen
Com 13h5'18"30d11'
Dluzniewski,Matthew Albert
Cam 12h39'35"80d8'
DM/KN Partners of Choice
Cas 0h46'43"66d43'
DMG
Uma 11h2'32"34d21'
Dmitri Dylan's Dads Star
Cep 21h44'27"61d4'
Dmitronow-Bonelli,Lucy
Cra 18h12'47"-39d44'
Dmoch,Siegfried
Sgr 19h52'34"-40d31'
Dmytro D W
Aur 5h56'56"41d8'
Dmytro,Karen
And 22h56'20"36d42'
DNJ2B4FR
Mon 6h53'21"-1d28'
DNR
Uma 8h45'37"71d4'
DNV122594 Cluster
Ori 6h7'15"-0d47'
Doak Forever
Tri 1h49'31"26d1'
Doak,Debbie Jean Culver
Mon 6h42'54"7d53'
Doak,J R
Her 17h18'1"22d43'
Doak,Michael T
Cep 23h5'43"64d48'
Doan,Linda Jean Wallace
Equ 21h2'34"10d45'
Doan,Phillip Melville
Her 17h34'45"40d0'
Dobash,Kevin E
Boo 14h39'39"30d28'
Dobbe,Breanna
Equ 21h3'49"10d28'
Dobben,Sheila
Peg 23h1'0"20d12'
Dobberstein,Ulf
Psc 0h3'39"1d35'
Dobbie's Delight
Peg 21h57'45"10d53'
Dobbins,Emily
Lyn 8h20'0"51d48'
Dobbins,Rebecca
Cyg 21h22'33"41d8'

Dobbins,Richard A
Aur 5h7'10"40d53'
Dobbins,Sean Michael
Per 2h40'35"35d58'
Dobbs,Charles L
Sex 10h28'25"2d39'
Dobbs,Claramond Estelle
Sct 18h34'0"-4d9'
Dobbs,Irene D
Peg 22h28'27"25d44'
Dobbs,Jr,Fred A
Aql 19h2'45"2d30'
Dobbs,Ken
Ser 15h55'31"20d58'
Dobbs,Kim
Sex 10h37'36"1d52'
Dobbs,Richard
Lyr 18h59'47"36d54'
Dobbs,Robert Dean
Aql 19h8'38"1d10'
Dobbs,Sr,Fred A
Cet 0h52'0"0d49'
Dobbyn,Bethany
Umi 17h17'14"85d9'
Dober,Bethany Lynne Ann
And 0h20'34"30d52'
Dobert,Randen James
Oph 17h27'1"-1d39'
Dobert,Scottee Richard
Aql 20h1'0"12d13
Dobilas,Patricia Kelly
Cas 2h57'47"70d22'
Dobin,Noah
Her 18h12'0"37d43'
Dobkowski(Inga), Cheryl
Com 12h1'56"27d47'
Doblmeier,Raimondas
Cet 3h19'38"1d44'
Dobner,Samuel
Per 1h45'41"50d3'
Dobranski,Philip Dean
Dra 11h22'13"78d44'
Dobrenz,Dana Wason
Leo 10h51'56"7d55'
Dobrescu,Doina Teodora
Lac 22h42'21"54d8'
Dobris,Bud
Cnv 12h54'52"51d13'
Dobroski,Lisa M
Uma 10h51'0"54d57'
Dobroski,Mathew Paul
Boo 14h23'48"47d53'
Dobrovich,Teresa Ann
Cas 23h14'13"60d37'
Dobrow,Blanche
Aql 19h57'15"11d55'
Dobry,Darlene
Lyr 19h6'42"25d30'
Dobrzynski "Rock Star",The
Uma 9h33'27"50d9'
Dobson,Beth
Boo 15h6'26"10d37'
Dobson,Catherine
Umi 15h57'36"78d25'
Dobson,Clark
Peg 23h3'14"31d44'
Dobson,Gabrielle Brittney
Lyr 18h41'41"33d56'
Dobson,Jordan Kate
Cet 2h50'48"0d33'
Dobson,Jr,Frank Edward
Cep 22h53'0"58d1'
Dobson,Ross
Ant 10h42'26"38d36'
Dobson,Susanne Jayne
Cyg 19h24'50"44d48'
Doby 1
Dra 16h31'52"61d50'
Dobyns,Abigail
Hya 8h32'42"-1d21'
Dobyns,Dr Richard
Ori 5h54'0"14d28'
Dobyns,Rebeccca
Aql 20h30'31"0d20'

Doc
Sco 16h31'21"-43d10'
Doc
Vir 13h38'26"-0d14'
Doc
Oph 17h6'5"-24d25'
Doc & Carmen
Lyr 18h28'48"44d36'
Doc & Helen
Cyg 21h7'36"38d56'
Doc Scott
Ori 5h48'25"18d15'
Doc Stewart
Oph 17h53'21"-8d0'
Doc's Delight
Cnc 9h9'34"33d7'
Doceti,Kathy
Lyn 8h9'0"35d53'
Docherty,Eileen
Ari 2h54'0"22d15'
Docimo,Vincent J
Per 2h36'57"45d5'
Dock Star,The
Boo 14h55'25"40d56'
Dockery,Michael E
Cnc 8h8'39"31d22'
Docking,Bridget
Cas 1h24'39"53d35'
Dockins,Shelly
Cas 0h8'29"58d39'
Docstar
Aql 19h53'23"12d54'
Doctor Bob
Her 17h37'11"40d49'
Doctor Dave
Aur 5h16'1"42d51'
Doctor De
Uma 8h55'12"59d38'
Doctor Jon
Cet 3h18'17"-0d17'
Doctor K
Dra 18h19'31"48d28'
Doctor Leo
Sex 10h35'0"1d13'
Doctor Lisa
Lmi 9h57'0"38d4'
Doctor Merle The Pearl
Oph 18h1'35"10d45'
Doctor Pappaw
Oph 18h19'50"6d8'
Doctor Scott
Sct 18h54'28"-8d14'
Doctor Sherri
Uma 10h48'28"71d42'
Doctor Tim
Oph 17h5'26"10d8'
Doctor Z
Oph 17h38'5"-23d52'
Doda Laura
Dra 9h27'42"74d10'
Dodah
Uma 9h13'1"50d8'
Dodak,Jordan Mitchell
Aur 6h16'14"30d53'
Dodd,Caitlin Elizabeth
Tri 1h29'15"30d51'
Dodd,Claire & Jarrett
Oph 17h34'20"-22d51'
Dodd,Edgar W
Lmi 9h48'36"33d25'
Dodd,Joan
Hya 8h43'45"2d27'
Dodd,Kathleen
Eri 2h59'20"-5d45'
Dodd,Keith "Legs"
Leo 10h35'12"20d17'
Dodd,Michael Solomon
Per 2h4'22"57d12'
Dodd,Victor
Vir 11h42'18"1d13'
Dodds,Chrissie
Cam 14h16'20"82d10'
Dodds,Thomas Ford
Ser 15h39'52"4d5'
Dodds-Friend Forever, Kellie
Peg 23h21'0"28d22'
Dodge,Benjamin Loren
Lac 22h12'52"54d34'
Dodge,Elizabeth
Cas 0h34'16"73d30'
Dodge,Gregory Keith
Ori 5h8'49"-4d2'
Dodge,Jennifer Kaye
Aur 6h2'24"36d40'
Dodge,Kenneth James
Boo 14h50'17"32d13'
Dodge,Marcia Marie
Peg 22h3'47"21d33'
Dodge,Richard Jam
Ori 5h52'55"0d10'
Dodge,Ronald G
Dra 14h28'14"62d24'
Dodge,Tony
Lac 22h37'14"48d57'
Dodge,William Thomas
Peg 22h31'1"20d15'
Dodi Li
Psc 22h57'1"6d13'
Dodkins,Robin
Cep 23h10'13"62d38'
DoDo
Boo 14h12'12"41d59'
Dodovich,Nadezda Hope
Aur 7h6'53"40d47'
Dodrill, Robert L & Clara Jane
Aur 4h53m57"48d55'
Dodson,Charley
Ser 16h4'44"10d60'
Dodson,Claire
Lyr 18h18'38"44d24'
Dodson,Daniel Yarnell
Her 16h40'26"24d12'
Dodson,Dayla Dawn
Lyn 8h16'25"37d26'
Dodson,Erin McKenzie
Vul 20h14'1"23d28'
Dodson,Gilberto J
Ari 1h59'58"21d37'
Dodson,James Yarnell
Her 16h59'31"26d56'
Dodson,Lori
Mon 6h59'53"-10d0'
Dodson,Loyd "Papa"
Aur 5h27'14"37d43'
Dodson,Richard Hamilton
Hya 9h3'0"-10d14'
Dodson,Steven Yarnell
Her 16h58'39"25d50'
Doe
Boo 15h18'44"52d12'
Doe,Karl Christian
Aql 20h31'60"0d11'
Doedens,Klaus-Dieter
Cam 5h46'12"70d4'
Doedtmann,Josef
Oph 18h5'47"8d20'
Doeer II,Dr Thomas P
Lmi 10h19'28"32d18'
Doehring's Light
Boo 14h10'0"32d24'
Doelger,Andrew J
Boo 14h58'0"20d21'
Doelger,Marilyn
Uma 13h37'58"50d6'
Doelling,Rolf"Windy" & Joyce
Cyg 20h5'52"40d12'
Doellman,Paulette L
Cam 3h35'46"61d18'
Doemel,Jason
Lyr 18h52'42"42d17'
Doerge,Manfred
Peg 0h6'60"28d5'
Doering,Cheryl Lynn
Cyg 20h24'59"49d54'
Doering,Jo Ellen
Mon 6h56'54"7d60'
Doering,Susan J
Sct 18h55'11"-6d49"
Doerksen,Kathryn Anne
Cyg 19h44'1"30d13'
Doerle,Timothy
Her 16h51'53"38d56'
Doernberg,Albert
Lyr 19h11'28"38d32'
Doernberg,Gert
Lac 22h13'6"51d14'AB
Doerner,Geoffrey
Gem 6h49'49"19d41'
Doernhoefer,Joan Adel
And 1h54'58"40d12'
Doerr,Bruce Eric
Aql 20h0'11"10d57'
Doerr,J C
Boo 15h3'39"20d8'
Doerr,John J
Per 2h57'0"40d54'
Doerr,Rick Arthur
Cep 0h2'18"69d51'
Doerr,Stephen J
Lac 22h12'36"51d34'
Doerrer,Craig
Lac 22h45'32"55d53'
Doerries,Walter
Oph 18h5'52"8d30'
Doetsch III,Richard Frank
Her 16h25'41"42d3'
Doetsch,Jacqueline Lee
And 23h1'39"49d11'
Doetsch,Jr,Richard L
Cep 0h2'54"80d21'
Doetsch,Willibald
Peg 0h3'41"18d52'
Dog Star
And 23h33'0"37d56'
Dog,Samantha Annie Ehrlich
And 23h44'32"44d45'
Dogey Forever,Louis & Helen
Vir 12h21'53"1d55'
Doggett,Matthew Wesley
Aql 18h44'29"11d16'
Dogramaci,Prof Ihsan
Per 1h28'1"53d23'
Doguy
Per 4h3'20"36d9'
Dohanish,Ashley Ann
Cyg 19h58'16"40d37'
Doher Anniversary Star
Cyg 19h26'38"31d28'
Doherty,Bill
Dra 13h44'19"68d42'
Doherty,Brennan Michael
Ori 6h5'25"8d16'
Doherty,Dana
Cet 3h12'44"0d58'
Doherty,John James
Ori 6h4'0"4d43'
Doherty,John W
Her 17h35'56"26d45'
Doherty,Kathy A
Mon 8h2'10"-2d55'
Doherty,Katrina
Cyg 21h32'36"38d29'
Doherty,Laura Erin
And 23h23'23"43d51'
Doherty,Madeleine
Uma 11h30'24"60d42'
Doherty,Melanie
Peg 22h11'18"4d24'
Doherty,Michael (Doc)
Aur 4h36'1"34d26'
Doherty,Rosemarie
And 2h20'1"39d27'
Doherty,Sharon
Lyr 18h48'50"42d22'
Doherty,Sr,Edward J
Aur 5h3'52"30d43'
Doherty,Tim
Hya 9h34'41"-9d55'
Doherty,Virginia Shannon
Psc 0h57'55"28d37'
Dohm,Hans-Peter
Cep 21h40'17"61d5'
Dohman,Heidrun
Lyn 8h6'14"42d13'
Dohman,In Loving Memory Of Chondra
Cyg 19h40'58"30d27'
Dohnal,Susan
Lyr 19h11'28"38d32'
Dohner,Lisa
Psc 23h1'47"5d11'
Doho
Aur 6h54'17"40d19'
Dohrman,Dancin' Deb
Boo 14h51'40"53d59'
Dohs,Henry J
Cma 7h48'13"11d27'
Doi Family
Gem 7h3'38"15d52'
Doi,Corbin
Mon 8h4'56"-2d44'
Doig,Aaron M
Aql 19h50'48"10d9'
Doig,Christopher L
Aur 5h8'57"41d39'
Doig,Damian B
Uma 8h54'49"48d6'
Doig,Damian Duquette
Cep 22h16'42"70d40'
Doig,David J
Aur 7h21'1"41d2'
Doig,Erin Rose
Vul 20h56'43"28d26'
Doig,Gregory Austin
Her 17h54'21"14d51'
Doig,Jessica Ashley
Sge 19h43'35"16d15'
Doig,Keghan Rose
Lyr 18h27'56"34d1'
Doig,Mary Beth
Lyn 8h25'51"44d52'
Doig,Peter E
Aur 5h18'36"42d10'
Doig,Robert M
Per 2h52'50"46d51'
Doig,Sarah Thompson
Lyn 8h23'14"46d0'
Doig,William L
Cep 20h37'50"65d8'
Doig,William M
Dra 16h33'47"61d41'
Doig-Felton,Ellen K
Cas 0h59'32"60d42'
Doilfa
Cam 13h17'45"81d44'
Doimi,John
Per 2h28'54"56d39'
Doisher,Bobby Gene
Cep 23h3'56"62d23'
Doisnel,Aurélie
Dra 17h38'14"70d52'
Doisnel,Guillaume
Dra 12h4'49"64d17'
Dokas,Paul B
Lac 22h42'45"37d45'
Dokken,William W
Her 16h35'0"32d57'
Doktor Pups
Gem 6h46'31"15d31'
Dolan Together Forever Ann & Joe
Umi 14h49'1'66d44'
Dolan,Carol
And 0h21'31"32d60'
Dolan,Colleen Jennifer
And 0h22'35"34d46'
Dolan,David Joseph Michael
Ori 5h32'25"8d37'
Dolan,Dick & Carmi
Mon 6h25'34"-8d49'
Dolan,Frederick J
Dra 16h54'59"64d23'
Dolan,James
Per 4h21'20"50d17'
Dolan,Jamie Elizabeth
Leo 9h42'32"16d35'
Dolan,Jonathan Peter
Dra 9h49'15"73d25'
Dolan,Jr,Michael John
Tau 5h21'54"20d21'
Dolan,Kelly Kay
Peg 21h42'1"27d7'
Dolan,Kory Michael
Uma 10h3'55"70d54'
Dolber & Family,Rick
Per 3h6'1"46d39'
Dolby (Dreams),Julie
And 0h5'17"38d20'
Dolby,John
Ori 6h4'58"1d5'
Dolby,Jr,Thomas George
Crb 15h43'1"35d49'
Dolce
And 23h20'53"51d0'
Dolce Davide
Cas 2h20'0"60d34'
Dolce Vitae
Cyg 19h59'1"45d23'
Dolce,Dylan Matthew
Per 3h32'56"52d44'
Dolcino-Ziemek
Cam 7h58'39"68d57'
Dolden,Steven
Sge 20h6'36"16d19'
Dolder,Ann
Peg 22h55'23"28d26'
Dolder,Edward S
Sex 9h40'37"-0d18'
Doldinger,Klaus
Tau 3h59'0"30d37'
Dole,Albion
Leo 9h36'0"11d33'
Dole,Amanda
Sco 17h0'51"-37d50'
Dole,Clare
Umi 13h49'31"73d57'
Dole,Margaret
Cas 1h29'54"65d36'
Dole,Wendy
Tau 5h55'27"23d29'
Dolen,Gyllian Ilyce
Boo 13h49'46"20d1'
Doles,"Wedding Star" Andrew & Patricia
Crb 16h10'48"38d15'
Dolezal,Jessica
Peg 22h57'15"33d46'
Dolezal,Melitta
Vir 14h41'39"-0d26'
Dolfo,Baron
Sct 18h55'26"-6d3'
Dolgen,Jonathan
Tau 4h38'49"15d6'
Dolginow,Jamie Blyth
Dra 17h28'47"58d33'
Doliber & Family, Richard
Lyr 18h47'30"30d18'
Doliber & Family, Chris & Maxine
Cyg 19h29'20"32d7'
Doliber,Jeffery & Jeremy
Peg 21h49'57"31d27'
Doll
Mon 6h51'19"11d29'
Doll Face
Peg 22h22'18"27d33'
Doll,Brian Cooper
Aur 6h3'38"38d15'
Doll,Charles Augustus
Sct 18h48'49"-7d22'
Doll,Dipl Ing Robert
Lmi 11h0'42"32d40'
Doll,Dr Justin Thomas
Lac 22h37'42"48d53'
Doll,Jr,Alvin N
Lac 22h15'11"38d44'
Doll,William Victor
Cep 1h48'19"87d53'
Doll-Doll
Cas 23h1'18"58d10'
Dollar's Two-Nine
Aur 6h57'31"44d2'
Dollar,Cindy
Mon 6h52'47"10d38'
Dollar,Richard G
Per 2h56'43"45d11'
Dollat,Pax
And 23h20'45"50d34'
Dollee
Aql 19h31'16"8d42'
Dolley,Thomas John
Aur 7h7'0"36d2'
Dollface
Aur 5h8'17"44d3'
Dollie's Dad
Cep 23h3'0"65d57'
Dollins,Hunter Tyler
Her 17h53'20"48d52'
Dolly
And 1h44'1"37d41'
Dolly
Eri 3h31'37"-6d43'
Dolly Dimple
Cas 23h29'27"58d26'
Dolly Dot
Sgr 19h2'59"-26d19'
Dolomar 7 18 92
Lac 22h43'1"54d36'
Dolores
Peg 21h31'28"20d4'
Dolores
Crt 10h55'23"-23d49'
Dolores Becka
Lyn 8h12'1"40d44'
Dolores Jean
Aql 18h58'22"-2d14'
Dolores Jean
Cam 7h32'17"68d5'
Dolores Soria
Mon 6h33'0"-1d17'
Dolphin
Cep 22h54'38"58d27'
Dolphin Boy Blue
Del 20h20'25"9d42'
Dolphin Ring
Ari 2h12'12"18d34'
Dolsay,Richard Almering
Boo 14h32'39"37d20'
Dolsingh,Nishi "D'Nish"
Uma 9h1'46"56d23'
Dolson,Nicholas Mancini
Boo 14h38'19"36d32'
Dolson,Robert & Lorli
Uma 11h39'35"42d39'
Dolwne
Cyg 19h28'42"38d17'
Dom
Lac 22h7'40"37d55'
Dom Dom
Uma 8h31'35"49d54'
Dom's Destiny
Uma 10h21'0"61d47'
Domagala,Bozena
Vul 20h1'27"22d35'
Domaine Chandon
Cep 1h21'33"80d18'
Domange,Nicolas
Vir 12h9'26"2d19'
Domas,Betty
Cyg 21h35'25"31d39'
Dombroski,Brian
Oph 18h18'37"7d41'
Dombroski,Jacob Charles
Uma 11h14'37"46d17'
Dombroski,John
Lac 22h17'1"37d41'
Dombrowski,Edward Joseph
Cnv 12h54'22"51d45'
Dombrowski,John J
Her 17h29'32"42d34'
Dombrowski,Lindsay Rose
Cas 0h59'31"61d43'
Domenic
Aur 4h48'58"40d18'
Domenica
Lyr 18h35'49"44d47'
Domenicano,Caroline
Sgr 19h3'9"-29d18'
Domenick Liberatore Lights our Sky
Uma 12h4'46"51d50'
Domenick,Muriel Israel
Peg 22h43'56"11d41'
Domerchie (Fairy Godmother),Ann K
Cas 0h58'53"71d28'
Domers,Amy
Lyn 7h35'1"43d47'
Domi
Peg 23h21'28"33d49'
Domi C
Aur 5h4'1"37d48'
Domi,Mon Minou
Uma 11h8'25"47d40'
Domi-Meg
And 2h24'51"45d4'
Domina-Krenzer,Dagmar
Ari 2h55'20"30d29'
Domine & Family,Robert & Cynthia
Aur 7h2'18"40d47'
Domingo
Boo 14h58'32"31d44'
Domingo,Jordi i Dolors Alegre
Cam 6h7'1"61d21'
Domingo,Plácido
Ori 6h17'10"-2d47'
Domingue,Randy "Weasel"
Aql 18h59'22"17d31'
Dominguez,Alyssa Raeanne
Oph 18h5'41"12d3'
Dominguez,Efrain Maximiliano
Peg 22h42'1"12d9'
Dominguez,Fernando L
Ori 5h4'59"-1d19'
Dominguez,Francesca Nicole
Mon 6h24'0"11d10'
Dominguez,Luis Miguel
Crb 16h16'1"38d17'
Dominguez,Marina I
Peg 21h49'18"28d56'
Dominguez,Nick & Brenda
Eri 3h53'39"-6d7'
Dominguez,Steven Kirk
Cet 0h42'26"-3d12'
Dominiack,Tracey
And 23h32'28"46d38'
Dominic & Michelle
Cyg 20h0'58"39d57'
Dominic Bruno
Her 17h7'17"40d3'
Dominic C
Ori 5h25'21"1d6'
Dominic,Christopher
Uma 11h53'42"56d11'
Dominici,Caine
Per 3h5'60"50d4'
Dominick
Uma 13h36'1"51d3'
Dominick
Lac 22h31'20"50d9'
Dominick A Shining Star
Her 16h39'22"33d26'
Dominick's Ellie
Lyn 6h30'12"60d5'
Dominick,Frank
Sco 17h4'32"-31d42'
Dominico,Lynne
And 23h0'46"45d44'
Dominik,Lygia
Uma 10h4'47"42d20'
Dominik,Susan
And 0h16'36"37d17'
Dominika Dan
Umi 16h55'1"76d12'
Dominioni,Comm Pietro
Boo 14h32'15"18d40'
Dominioni,Laura
Aql 18h57'14"-6d19'
Dominioni,Valeria
Com 13h11'51"22d4'
Dominique
Hya 8h43'23"4d23'
Dominique
Sge 20h7'29"16d42'
Dominique Aimé
Oph 18h42'0"7d23'
Dominique Dawn
Uma 13h34'31"61d16'
Dominique Julia
Cyg 19h41'51"28d49'
Dominique Linda
Cep 22h9'0"67d35'
Domino
Cmi 6h13'36"-11d12'
Domino Spirit Totem
Cma 6h56'15"-13d15'
Domino-Weidinger
Boo 13h55'54"18d52'
Domis,Christopher Joseph
Psc 1h39'1"21d8'
Domitay
Aql 20h2'26"14d44'
Domitilla
Col 6h29'30"-33d50'
Domityszyn,Angela
Lyr 18h36'1"30d58'
Dommer,Dennis
Aur 5h16'1"40d39'
Domotor,Hazel & Albert
Com 13h16'26"26d33'
Dompierre,For My Sweetheart,Albert
Per 1h52'41"47d51'
Dompierre,Violaine
Per 2h52'24"38d49'
Domres,Michael
Boo 14h11'32"50d7'
Domrose,My Mom Gloria
Lyr 18h14'58"34d35'
Domstead,D J
Oph 17h56'40"12d4'
Domstead,Del
Oph 17h56'27"12d7'
Domus
Lyn 7h1'41"52d50'AB
Domus Pooh Winnie Ille
Cnv 12h49'58"38d29'
Don
Per 4h4'55"39d45'
Don
Her 17h21'36"14d12'
Don "Eagle"
Aql 18h56'25"4d9'
Don & Jenny's Star
Tri 1h50'47"28d53'
Don & Laura
Lyr 19h1'1"28d59'
Don & Lisa's Love Star
Umi 14h26'44"65d40'
Don & Valerie
Peg 23h5'14"31d24'
Don Bass' Wandering Star
Hya 9h19'11"-0d12'
Don G W,The
Ser 15h10'49"11d29'
Don Jon "60"
Cep 20h55'36"58d45'
Don Lights Up The Sky With Mary
Uma 11h39'22"56d16'
Don Star,The
Boo 14h31'13"20d10'
Don's "Have I Told You Lately" Star
Aur 6h51'0"37d42'
Don's Aurora
Equ 21h22'46"11d42'
Don's Eternal Light
Per 3h42'41"37d20'
Don's Invention
Dra 19h4'1"47d59'
Don's Ray
Lac 22h24'41"38d24'
Don's Rose
Umi 14h29'22"65d45'
Don's Star
Per 2h22'58"54d26'
Don's Stellar Rose
Lac 22h52'50"50d6'
Don,My Special Star 60th Birthday
Cep 22h35'21"60d17'
Don,The
Vir 14h58'0"1d39'
Don,The
Dra 17h30'11"58d11'
Don-Carolis,Karen Jane
Cas 0h46'31"63d11'
Don-Diana-Laine
Cet 2h36'20"-1d42'
Don-Dor
Per 2h4'1"56d52'
Dona
Tau 4h6'1"20d28'
Dona
Gru 22h12'22"-54d28'
Dona,Gordon L
Dra 18h56'17"67d45'
Donachie,Kirsty
Cas 0h34'37"74d35'
Donadio,Denise Ann
Cas 1h10'13"61d43'
Donagh,Joseph Mac
Cas 23h2'49"53d43'
Donahey,Rebecca J
And 23h46'16"40d3'
Donahey,William Kelly
Aur 5h3'36"41d46'
Donahue
Dra 17h10'32"51d13'
Donahue,Colonel Richard I
Aql 18h43'0"6d15'
Donahue,Debra M
Uma 10h41'29"60d5'
Donahue,Derek
Dra 16h6'46"67d15'
Donahue,Donald P
Per 3h40'14"36d35'
Donahue,Donald V
Lac 22h53'51"53d2'
Donahue,Erin Marie
Sgr 19h58'10"-42d32'
Donahue,Geni Dalene
Lyr 18h26'35"38d56'
Donahue,James Joseph
Dra 10h6'59"80d20'
Donahue,M Patricia
Psc 0h57'44"17d46'
Donahue,Mark J
Tau 4h5'56"20d25'
Donahue,Mary
Cyg 19h49'15"37d45'
Donahue,Rachel Marie "My Sunshine"
Cam 10h45'16"84d31'
Donahue,Sharon
Ori 5h31'46"-0d4'
Donal B
Sct 18h44'2"-6d36'
Donalbie's Star
Uma 11h17'16"38d0'
Donald
Cet 0h32'24"1d41'
DONALD & ANNORA Our Guiding Light
Per 2h25'18"54d58'
Donald & Barbara
Umi 15h4'1"72d51'
Donald & Kim Star,The
Cam 5h25'34"80d19'
Donald & Maida
Aql 19h54'53"0d5'
Donald & Susan
Lyn 8h15'32"38d24'
Donald Duane
Cnv 13h48'33"34d14'
Donald Gilbert
Her 17h36'0"24d28'
Donald Jane
Lyn 9h3'44"45d42'

Donald Jim
Cet 1h54'32"-10d21'
Donald Mi Amor
Aql 19h3'57"-6d0'
Donald The Lionhearted
Ori 6h2'36"2d40'
Donald The Ruler
Aur 6h8'46"45d1'
Donald's Morning Star
Aur 4h55'43"50d32'
Donald,Bonnie,Donald, Bonnie,Johnathan
Dra 10h13'15"81d10'
Donald,Carolyn Helen
Cas 0h3'58"61d11'
Donald,John Robertson
Uma 9h44'59"57d30'
Donaldson
Boo 14h58'40"24d21'
Donaldson Star,The Gayle
Crb 15h56'41"27d30'
Donaldson,Alec
Lyr 19h19'12"40d38'
Donaldson,Brooke Ann
Lyn 6h29'59"58d29'
Donaldson,Charlotte Arlene
Uma 8h46'1"62d16'
Donaldson,David Scott
Aur 6h12'1"31d13'
Donaldson,Dawn M Holzhauer
And 23h26'33"49d1'
Donaldson,Jennifer Mae
Lyn 7h49'1"36d56'
Donaldson,Kindi Malea
Lyn 9h15'16"36d19'
Donaldson,Marvin Mathew
Uma 10h54'49"52d4'
Donaldson,Randall Earl
Psc 0h48'16"20d15'
Donaldson,Steve & Katie
Lyn 7h37'36"40d13'
Donamer
Aur 7h25'48"40d25'
Donansea,Susan L
Lyn 7h24'37"44d42'
Donarich
Uma 10h42'1"56d22'
Donarski,David
Lac 22h7'38"51d37'
Donata
Ori 5h53'23"-1d35'
Donatella
Pic 5h4'19"-47d51'
Donatella
Tau 4h12'4"21d60'
Donatella M-"Lauretum"
Pyx 8h41'31"-23d57'
Donatelli,Anthony P
Lyr 18h45'19"33d37'
Donatelli,Daniel
Her 16h57'30"38d28'
Donath,Joshuah
Lmi 10h40'21"25d24'
Donath,Sophie
Sco 17h31'29"-30d24'
Donato e Francesca
Her 10h11'39"10d58'
Donato,Ellen Melissa
Cas 3h4'43"67d53'
Donato,Joseph
Uma 11h18'14"71d43'
Donato,Martina
Cas 1h56'10"60d9'
Donato,Richard
Boo 14h46'25"46d13'
Donato,Sam A
Lac 22h16'44"37d31'A
Donatoni,Stephanie Gene
And 2h26'20"40d48'
Donavan
Hya 9h6'27"5d29'
Doncar
Aql 19h25'57"0d3'
Donches,Sr,Charles J
Cyg 21h0'33"37d6'
Doncourt,Robert Alfred
Her 16h57'21"29d10'
Dondero,David W
Aql 20h3'57"4d13'
Dondero,Joseph & Gloria
Cyg 19h23'18"28d57'
Dondra
Mon 7h17'40"-8d11'
Donegan Star,The Evan & Colin
Hya 8h54'37"3d55'
Donegan,Stephanie A
Cyg 20h50'29"38d11'
Donelli "Binky", Laurence
Dra 10h2'10"73d29'
Donels,Jeffrey William
Uma 11h56'43"30d49'
Donelson,Ivy Suzanne
Peg 22h7'10"4d21'
Donelson,Joanne
Del 20h18'58"9d26'
Donelson,Mark
Aql 20h4'54"8d34'
Donelson,Riley Laine
Ori 5h56'22"19d25'
Donelson,Starr
Cas 2h27'45"60d9'
Doner 12-4-45,Michele Oka
Cet 2h0'0"0d6'
Dones,Joaquin (Jack)
Leo 10h56'29"2d3'
Dones,Jr,Wesley Keith
Peg 22h21'1"2d28'
Donese Rena PS
Aql 18h58'45"-2d17'
Doney,George
Her 16h3'21"41d48'
Doney,Richard E
Ser 15h13'0"2d23'
Donfried,Paul Andrew
Cnv 12h10'0"37d1'
Dong,Jin & John
Cyg 21h2'26"33d5'
Donges,Patty & George
Lyr 18h28'56"44d38'
Dongracia
Lmi 9h49'0"34d47'
Donham Family,The
Lmi 10h23'55"32d3'
Doni,Emanuele Maria
Uma 11h30'17"53d1'
Donini,Martino
Tau 5h50'29"23d15'
Donis-Campos (Rey), Reynaldo
Cap 21h39'0"-22d55'
Donithan,Cynthia Sue
Peg 21h9'50"12d60'
Donivan,John T
Cmi 7h46'1"7d36'
DonJoel Alpha
Aur 6h52'21"40d48'
Donka
Her 17h21'17"41d40'
Donker,J
Cnv 12h43'34"40d19'
Donkin,Emma Jane
Cas 0h17'18"63d41'
Donko Star,The
Sex 10h25'31"-0d32'
Donleavy,G Quinton
Cnv 13h52'24"38d51'
Donley,Christopher J
Her 18h7'1"48d35'
Donley,Michael E
Cyg 20h57'22"30d24'
Donlon,Marie Rose
Aql 19h45'10"10d11'
Donlyn
Peg 21h52'22"28d39'
Donlyn K
Mon 6h37'1"11d29'
Donna
Aql 20h0'1"11d26'
Donna
Cyg 21h12'0"38d1'
Donna
Peg 21h33'40"20d15'
Donna
And 2h19'0"50d2'
Donna
Tri 2h5'33"32d4'
Donna
Cap 21h42'12"-22d44'
Donna
And 0h47'23"39d16'
Donna
And 1h7'45"38d13'
Donna
And 1h3'0"37d32'
Donna
Cyg 19h24'46"45d13'
Donna
Cas 0h9'57"50d31'
Donna
And 1h55'1"38d22'
Donna
Cyg 21h5'14"40d46'
Donna
Uma 8h9'15"71d28'
Donna
Vul 19h53'55"20d30'
Donna
Cas 2h14'17"70d34'
Donna & Andy
Cyg 21h54'48"36d27'
Donna & Darryl Till The End Of Time
Sgr 18h48'16"-25d15'
Donna & David's "Happy Together"
Peg 22h7'15"3d23'
Donna & Jack "Our Love Forever"
Lyr 18h41'34"45d53'
Donna & Joe
Lyr 18h50'31"30d45'
Donna Ann
And 2h18'26"40d34'
Donna Ann
Vir 11h42'58"1d48'
Donna Ban Cow Woman
Tau 4h7'41"20d30'
Donna Christine
Cas 1h5'0"60d26'
Donna Claire
Lyr 19h16'56"42d7'
Donna Elizabeth
Lac 22h42'25"56d47'
Donna Forever In My Heart
Cyg 20h19'56"39d18'
Donna L G
Cmi 7h46'1"8d32'
Donna Lee
Sgr 18h49'6"-20d8'
Donna Lee
Lyn 8h5'1"35d9'
Donna Lee
Eri 3h46'51"-6d11'
Donna Lee
And 0h30'20"28d49'
Donna Lee
Lyn 8h24'33"50d52'
Donna Lee Forever
Aur 5h54'27"40d57'
Donna Lynn
Equ 20h54'51"2d51'
Donna Mae
Lyn 9h9'39"37d54'
Donna Mae
Hya 9h25'58"-1d56'
Donna Marie
Ori 6h3'1"-1d14'B
Donna Marie
Cas 0h13'57"58d23'
Donna Marie
Boo 14h27'0"23d21'
Donna Marie
Cyg 21h22'1"46d56'B
Donna Marie
Cyg 19h34'60"32d50'
Donna Marie
Ori 5h27'15"-1d36'
Donna Marie Ann
Sct 18h47'55"-7d58'
Donna Marie Mystar
Peg 21h20'0"23d4'
Donna Marie Star,The
Leo 11h33'39"2d7'
Donna Marie's Star
Cyg 20h1'0"40d48'
Donna Matt PQ
And 0h55'33"38d0'
Donna May
Tri 1h41'49"30d22'
Donna Nicole
Aql 19h1'58"-0d30'
Donna Rose
Peg 22h1'40"26d57'
Donna Samantha
Lyn 8h49'57"39d32'
Donna Star,The
Tau 3h39'45"24d44'
Donna Suzanne
Cet 1h44'35"-1d54'
Donna's Angel
Cas 0h21'1"61d28'
Donna's Dream
Hya 8h43'40"3d32'
Donna's Heavenly Body
Mon 6h22'13"4d15'
Donna's Light
Peg 21h54'0"34d44'
Donna-Darlene
Boo 14h27'43"41d28'
Donnachie,Robert Nisbet
Cyg 20h18'36"39d56'
Donnalee
Tri 2h3'26"32d43'
DonnaRoger
Lyn 7h32'29"51d45'
Donne,Naomi
Cas 1h4'35"60d27'
Donnell,Adrienne Jones
Ori 5h54'43"14d43'
Donnell,Brian Patrick
Ori 5h50'16"14d50'
Donnella,George F
Oph 17h5'1"-23d24'
Donnelle Lou
Lyn 8h46'23"36d8'
Donnellia
Cas 23h20'0"63d4'
Donnelly"Tennis Star", B J
Aql 18h40'40"-2d18'
Donnelly's Star for Heart Donors
Sge 19h57'29"16d18'
Donnelly,A Scott
Boo 14h19'15"39d33'
Donnelly,Ann Marie
And 1h40'54"36d7'
Donnelly,Brendan
Cam 5h24'1"68d17'
Donnelly,Bridget
Peg 22h47'1"29d51'
Donnelly,Don
Lac 22h20'31"41d10'
Donnelly,Donna
Cas 23h17'56"60d3'
Donnelly,Edward & Evelyn
Cam 3h39'55"61d28'
Donnelly,Edward John
Vir 13h21'8"-4d16'
Donnelly,Erin
Cnc 8h6'0"32d21'
Donnelly,James F
And 23h21'1"44d39'
Donnelly,Karen
Lyr 18h56'40"32d26'
Donnelly,Kelly Jean
Lyn 7h46'18"44d24'
Donnelly,Kevin E
Dra 20h3'33"67d35'
Donnelly,Noreen Clare
And 0h21'29"30d18'
Donnelly,Patrick
Cru 12h42'11"-63d23'
Donnelly,Scott James
Lac 21h55'15"42d53'
Donnelly,Stephen M
Dra 18h3'17"65d22'
Donnelly-Stipa, Christiane
Eri 3h0'23"-12d16'
Donners,Patricia
Lib 15h10'40"-20d37'
Donnetta My Angel Star
Lyr 18h18'46"38d38'
Donnie
Her 16h17'11"26d55'
Donnie
Cyg 20h16'1"37d34'
Donnie K
Her 16h58'45"40d54'
Donnie Noel
Sct 18h53'2"-7d46'
Donnie's Star,Star of True Friendship...
Mon 6h35'29"-5d59'
Donnie's Strike
Vul 19h46'45"20d4'
Donny Boy
Ori 5h20'45"1d21'
Donny's Star'93
Eri 3h31'21"-5d4'
Donny,My Heaven On Earth
Gem 7h2'11"30d33'
Dono,Cynthia May
Lyr 19h8'51"40d56'
Dono,Michael R
Ori 5h58'15"12d34'
Donofrio,Margie
And 0h26'27"38d35'
Donoghue,Christopher
Aur 5h54'30"40d44'
Donoghue,Ellen
Cam 7h58'1"61d48'
Donoghue,Michael Robert
Aur 4h47'1"40d7'
Donoghue,Rose Grace
Umi 16h44'43"75d27'
Donoghue,Shannon
Cas 1h41'41"75d4'
Donohue "Grandma's Galaxy",Connor
Boo 14h11'46"30d32'
Donohue,Adrian
And 23h42'24"46d3'
Donohue,Baby Shannon Rose
Lyr 18h32'18"34d44'
Donohue,Brenda Kincaid
Aql 19h10'32"19d18'
Donohue,Christine Patricia
Lyn 8h3'29"38d35'
Donohue,Denise
Tri 2h12'28"30d45'
Donohue,Edmund Michael
Cep 21h21'26"68d5'
Donohue,Heather
Lyn 8h3'30"50d2'
Donohue,Jr,Phil
Her 15h49'47"46d25'
Donohue,Kevin
Per 3h1'43"40d37'
Donohue,Kimo "Jimmie"
Peg 23h38'0"17d50'
Donohue,Michael V
Vul 19h45'51"22d38'
Donohue,Seamus Patrick
Dra 16h57'23"62d51'
Donohue,Thomas & Lori
Cyg 19h40'40"30d10'
Donoian,Hagop
Aur 6h17'21"38d29'
Donolo,Jr,Louis
And 0h23'52"44d50'
Donot,Jacques
Cyg 20h29'24"31d14'
Donotto,Kathryn Jean
Eri 3h17'12"-3d54'
Donovan 6,Mark Wilkins
Crb 16h5'56"28d34'
Donovan III
Sct 18h53'48"-6d36'
Donovan's De-Light
Boo 14h14'27"52d27'
Donovan,Bernadine Sue
Lyn 7h54'41"34d26'
Donovan,Carrie
Peg 23h4'36"17d55'
Donovan,Dan W
Aur 5h56'0"50d29'
Donovan,Elizabeth Anne-Maree
Cam 7h39'49"71d4'
Donovan,Jeanette Elizabeth
Mon 7h42'53"-2d21'
Donovan,Jim & Jennifer
Aur 6h13'38"38d19'
Donovan,Keefe
Vul 19h23'39"23d47'
Donovan,Lindsay
And 2h23'1"45d37'
Donovan,Margaret Ann
Uma 8h36'50"71d53'
Donovan,Marjorie Jack
Gem 6h51'57"21d38'B
Donovan,Michael
Cep 22h45'0"65d37'
Donovan,Michael
Aur 7h23'34"41d47"
Donovan,Michael Patrick
Ori 6h2'32"4d46'
Donovan,Mick
Ori 4h58'40"-0d0'
Donovan,Molly Kiersten
Mon 6h56'0"10d15'
Donovan,Nicole Elizabeth
Crb 15h16'48"31d29'
Donovan,Richard Lawrence
Dra 9h48'31"73d27'
Donovan,Robert Emmett
Gem 6h51'57"21d38'A
Donovan,Thomas P
Lyn 6h12'25"58d22'
Donovan,Victor R
Aur 5h56'1"38d55'
Donselaar,Ray
Oph 17h31'35"11d47'
Donstar
Uma 10h32'56"54d33'
Donum Dei
Lac 22h9'15"49d56'
Donut
Ori 4h57'33"15d13'
Donze,Jennifer
Mon 8h3'53"-8d50'
Donzella
Aql 19h51'12"15d3'
Doo,Thomas W
Cru 12h46'30"-62d15'
Dooba J
Uma 10h14'27"59d2'
Dooher
Aur 5h55'57"30d1'
Doobie
Lmi 10h11'1"30d9'
Doobie-Lee
Sge 19h37'37"16d46'
Doodle
Peg 22h2'39"8d21'
Doodle-Bug Mel
Sex 9h53'20"0d42'
Doodle-The Light of my Life
Aql 19h6'39"3d36'
Doodlebug
Boo 15h5'41"18d51'
Doody
Uma 11h15'42"61d15'
Doofus,Janellis
Peg 22h16'36"20d31'
Dooher,Emily Jade
Umi 15h58'57"70d24'
Dooher,Mary Ellen
And 0h16'54"36d55'
Dook,Jackie
Cru 12h45'22"-62d25'
Doolan,Anne
Aur 6h4'15"42d40'A
Doolan,Frances Ward
Aur 5h58'24"38d57'
Doolan,Pat
Aur 6h4'15"42d40'B
Doole,Sarah
And 2h8'44"43d7'
Dooley
Her 16h3'38"21d20'
Dooley "What's Up Doc?",Don A
Oph 17h5'6"-24d38'
Dooley "Yabba-Dabba-Do",Kevin Scott
Cet 1h20'0"-0d19'
Dooley's Light
Aur 6h25'21"38d38'
Dooley,A M
Boo 14h58'36"30d31'
Dooley,Anne
Vul 19h37'34"20d19'
Dooley,Christine Anne
Cyg 21h23'1"40d42'
Dooley,Cobby
Boo 13h48'31"17d51'
Dooley,Dee Rowan
Cyg 21h3'28"35d23'
Dooley,DeLane Bruner
Lyn 7h56'1"34d35'
Dooley,Heavenly
Cet 1h59'0"-10d40'
Dooley,Holly Weisner
Cas 1h8'28"61d18'
Dooley,Martha Evangeline
And 2h21'13"50d11'
Dooley,Milton
Her 16h57'0"37d52'
Dooley,Patricia A Murphy
Lyr 18h49'31"39d41'
Dooley,Patricia Ann
And 1h40'1"36d31'
Dooley,Patricia Ann
And 1h44'37"36d29'
Dooley,Patti
Del 20h52'18"7d31'
Dooley,Virginia Rozell
Aql 19h26'43"12d39'
Doolin,Wally & Lori
Del 20h35'55"3d10'
Doolittle,Evan Albert
Per 2h50'53"40d58'
Doom,Bruce
Aur 5h19'33"41d35'
Doom,Rex
Cet 2h33'0"0d54'
DooMurk
Ori 6h2'49"5d46'
Doonan,Andrew & Caitlin
Crb 16h21'20"38d39'
Doone,Brian
Her 10h10'44"14d58'
Doone,The Lorna
Vul 20h20'28"28d15'
Dooney,John Frances
Cam 6h12'44"60d40'
Doorish,Annunciata
Cam 6h11'27"58d59'
Doorknob Stravrakis
Mon 6h54'13"-8d30'
Doornbosch,Annelien & Jan
Lyn 8h22'54"46d46'
Doornbosch-Philippo, Maria P
Cas 0h24'46"64d46'
Doornink,Eileen
Sco 17h2'16"-33d56'
Doose,Nicholas James
Aur 7h11'1"38d38'
Dopp,John Allen
Aqr 22h34'22"-6d35'
Dopp,Michelle Marie
Cap 21h3'10"-22d9'
Doppelt,Gabe
Lyn 6h20'1"58d49'
Doppke,Tess
Hya 8h42'43"-0d56'
Doppler,Erika
Sgr 18h47'44"-26d54'
Dor,Caroline
Uma 11h18'23"51d25'
Dora
Peg 23h42'45"28d49'
Dora
Peg 22h35'1"20d42'
Dora Te-Lun Yen
Cet 1h3'0"1d7'
Dorado,Delphinus
Scl 23h48'46"-27d33'
Dorak,Sue Ellen
Aur 7h21'1"37d56'
Dorame,Genaro
Per 1h59'25"48d26'
Doran Family,The
Cyg 20h52'15"30d1'
Doran,Bill
Aur 6h0'43"30d51'
Doran,Burns Patrick
Aur 7h8'58"36d39'
Doran,Emilie Rose
Com 12h3'14"19d39'
Doran,Joseph
Dra 13h13'51"64d26'
Doran,Karen K
Aql 18h57'38"-5d17'
Doran,Kevin "Bob"
Ori 6h5'1"10d55'
Doran,Margaret
Ori 5h52'46"21d49'
Doran,Philip J
Per 2h55'56"48d47'
Doranbil
Cnv 12h23'1"38d55'
Dorand,Charley
Cet 0h52'1"-5d12'
Dorans,James & Mary
Cyg 19h27'54"38d3'
Dorathy Claire
Mon 6h44'60"8d53'
Dorato,Dana Marie
And 0h20'51"40d49'
Dorato,Michael Alan
Sex 9h51'53"2d28'
Dorato,Richard Charles
Dra 19h57'58"67d43'
Dorazzo,Regina
Lyn 6h56'26"59d35'
Dorbin,Harold Gregory
Aql 19h25'42"-0d25'
Dorca
Cet 3h1'33"1d46'
Dorci,Julia Christine
Cam 3h53'42"60d31'
Dorcy
Uma 11h11'37"58d10'
Dordolo,Benjamin
And 23h3'43"50d14'
Dore,Pietro
Cet 1h6'40"0d27'
Dore,Sally Sugar
Hya 9h34'0"-5d43'
Doreau,Agnes
Vul 19h21'23"25d26'
Doreen
Cnc 8h0'22"11d5'
Doreen
Cnc 8h28'59"8d24'
Doreen Mary Lee
Cyg 19h25'1"35d16'
Doreen Mary Lee
Cyg 20h18'25"38d40'
Doreen's Dream
Aur 5h3'55"40d29'
Doreen's Star
Tau 4h20'30"11d15'A
Doreena & Remy's Shining Star-Luna
Dra 20h18'20"68d7'
Doremus,Gregory John
Tau 5h45'1"27d21'
Doreraduc
Cet 1h52'24"-5d55'
Dorese
Uma 11h7'11"37d33'
Doreson,Diane A
Lyr 19h6'39"28d22'
Doretti,Carol Ann
Ori 5h55'53"20d10'
Dorez,Helene
Lac 22h45'52"56d39'
Dorf,Ethel & Sanford
Cyg 21h29'0"40d39'
Dorf,Michael M
Per 3h11'20"47d49'
Dorfis III,Thomas
Dra 15h12'1"63d17'
Dorfman,Carole
Peg 23h43'31"31d12'
Dorfman,Hannah Jennifer
Peg 23h38'1"18d36'
Dorfman,Joel
Cep 22h15'21"68d41'
Dorfman,Murray Lewis
Dra 16h51'41"63d46'
Dorfman,Ronald
Cma 7h12'35"-18d51'
Dori
Oph 17h55'1"10d2'
Doria's "Star-25", Marcelle Robinson
Cnv 13h58'24"46d37'
Doria,Joseph Alfred
Aur 5h14'32"41d46'
Dorian
Peg 22h45'1"29d48'
Dorian
Aql 19h55'55"-1d37'
Doriana 7
Cra 18h0'37"-39d36'
Dorie
Lyr 18h24'48"37d44'
Dorinda
Cet 3h0'40"0d12'
Dorinda
Mon 7h49'35"-5d56'
Dorinda My Heaven On Earth
And 23h22'41"49d30'
Dorindi,Nicoletta
Sgr 19h29'1"-33d17'
Dorion 25/06/96 O l'éternité,Caroline
Cyg 21h5'21"32d31'
Dorion,Dottie
And 0h50'46"40d33'
Dorion,George
Boo 14h58'40"47d60'
Dorion,Justine
Umi 13h44'53"71d17'
Doris
Mon 6h33'0"-6d0'
Doris
Peg 23h30'17"13d13'
Doris
Cas 2h8'17"60d44'
Doris
Cyg 19h53'38"40d8'
Doris
And 0h7'1"30d51'
Doris
Aqr 22h41'11"-1d27'
Doris "Hinny"
Lyr 18h29'1"31d19'
Doris & Dick
Uma 11h22'0"31d8'
Doris & Don-50th
Crb 15h53'46"37d31'
Doris Christine
Cas 1h46'18"73d12'
Doris Pauline
Cas 0h55'54"61d8'

Doris' Star
Mon 6h27'1"11d49'
Doris,Stephanie
Peg 22h54'36"29d51'
DorisnHarryUs
Uma 9h15'34"50d29'
Dorit
Cap 21h2'35"-23d32'
Dority,Clyde Thomas
Cet 3h2'0"5d43'
Dorlack,James Charles
Uma 8h34'13"70d19'
Dorle
Dra 17h47'32"85d32'
Dorman,Brian
Peg 22h34'53"28d13'
Dorman,Dale S
Dra 12h6'0"70d22'
Dorman,Donald B
Aql 19h56'50"10d26'
Dorman,Gayle W
Lyn 8h9'35"38d38'
Dorman,Jared's Daddy- James A
Dra 17h49'37"65d9'
Dorman,Jareds Grandpa George N
Cyg 20h54'51"40d33'
Dorman,Kimberly Megan
Peg 21h48'13"28d19'
Dorman,Michelle Ruth
Tau 5h49'54"23d17'
Dorman,Ruth
Del 20h34'35"11d52'A
Dorman,Thomas E
Per 2h57'1"40d44'
Dormand,Sebastian Tague
Uma 11h40'19"37d50'
Dormer,Catherine A
And 0h45'50"28d32'
Dormody,Tom
Tri 2h3'31"30d38'
Dorn Star,The Donald D
Aql 18h56'29"12d58'
Dorn,Emily
Ori 6h3'16"20d16'
Dorn,Mr Greg
Cep 0h5'56"75d1'
Dornan's-Points of Lite,Jim & Nancy
Aql 19h9'32"2d32'A
Dornan,Heather Eric & David
Aql 19h9'32"2d32'B
Dornemann's Very Own Star,Michael
Lib 15h17'59"-8d1'
Dorner,David
Doo 15h4'24"30d44'
Dorner,Henry H
Boo 15h13'29"41d26'
Dornfeld,Smudgy
Cnv 12h9'29"37d32'
Dornole
Tri 1h30'46"30d33'
dorogati
Cap 20h35'23"-10d16'
Doronis,The
Cet 2h42'24"2d45'
Dorospy
Sco 15h53'13"-21d21'
Dorothea
For 2h38'25"-28d47'
Dorothea G
Aql 19h2'0"0d8'
Dorothy
Ori 5h47'23"20d53'
Dorothy
Aqr 22h33'23"-6d19'
Dorothy
Cnv 13h43'55"46d40'
Dorothy
Uma 10h51'53"59d51'
Dorothy
Mon 6h35'1"1d3'
Dorothy
Cyg 20h36'15"30d25'
Dorothy
Cyg 20h17'55"41d50'
Dorothy
And 1h20'24"37d34'
Dorothy
Mon 6h39'52"10d24'
Dorothy
Sgr 18h4'60"-28d56'
Dorothy
Cnv 13h42'52"31d30'
Dorothy & Wally
Uma 9h50'39"49d16'
Dorothy Ann
Tau 5h49'50"28d49'
Dorothy Anne
Mon 7h3'26"4d29'
Dorothy Anne
Com 12h8'20"24d57'
Dorothy Dazzling Star
Cas 2h14'23"67d41'
Dorothy Elizabeth
Lyr 18h43'50"38d34'
Dorothy Francis
Gru 22h42'41"51d44'
Dorothy Jane
Cas 0h24'1"62d31'
Dorothy Jane
Aql 19h57'39"15d15'
Dorothy Jeannie
Mon 7h43'15"-8d34'
Dorothy Lauren Dancing Star,The
Cyg 19h28'44"34d2'
Dorothy Lee
Eri 3h1'28"-6d33'
Dorothy Lee
And 0h2'23"46d27'
Dorothy Lorene
Cyg 20h24'10"37d52'
Dorothy Louise,The
Psc 23h7'23"6d38'
Dorothy Lucky Star- Heavenly Body
Uma 10h31'15"57d35'
Dorothy Marie
Cyg 19h46'44"30d42'
Dorothy Of The Islands
Eri 3h41'47"-13d44'
Dorothy Therese
Cam 3h13'58"60d21'
Dorothy Virginia
Lyr 19h11'57"40d53'
Dorothy's Star
Peg 23h36'43"12d6'
Dorothy's Star
Mon 6h21'46"8d10'
Dorothy-Dee
Gem 6h48'17"19d34'
Dorothée Star
Lyn 8h4'58"39d24'
Dorozinsky,John Thomas
Oph 17h7'11"-20d33'
Dorr,Ally & Michael
Eri 4h32'18"-18d45'
Dorr,James F
Mon 7h3'22"4d16'
Dorr,Jason Martin
Cep 22h11'40"61d8'
Dorr,Mary Kathryn
Tau 4h15'45"21d44'
Dorr,Nicholas Robert
Her 16h59'1"51d19'
Dorr,Paul Stephane
Cet 2h39'42"-9d41'
Dorr,Shelley Elizabeth
Mon 6h55'32"8d52'
Dorrell,Amy-Louise Daisy Bluebell
Ori 4h59'19"5d3'
Dorrell,George B
Uma 14h0'47"51d17'
Dorricoth,Marie Elizabeth
Lyr 18h36'53"29d17'
Dorris,Jr,Paul Edward
Her 16h39'30"11d33'
Dorris,Stephen Randolph
Leo 10h1'44"22d10'
Dorsan
Del 20h15'44"8d46'
Dorsch,Christoph
Her 17h41'10"14d22'
Dorset,Ted N
Her 16h16'38"10d35'
Dorsey,Anderson
Tri 1h35'60"28d52'
Dorsey,Cheryl Landis
Ori 5h16'57"-5d30'
Dorsey,Dora Madelynn Wiley
Peg 23h1'31"18d27'
Dorsey,Douglas John
Dra 13h37'52"68d54'
Dorsey,Eric B
Aur 6h32'16"31d12'
Dorsey,John Robert
Cet 2h37'59"-1d51'
Dorsey,Jr,James Hollis
Sex 10h10'24"-9d44'
Dorsey,Kara Lynn & John Fort
Aur 6h31'28"35d7'
Dorsey,Kristen,B
Hya 8h55'20"4d1'
Dorsey,Robert E
Cep 21h0'36"56d4'
Dorsey,Sr,James Hollis
Cma 6h48'12"-19d26'
Dorsey,Susanne Aultman
Mon 7h43'32"-3d9'
Dorsey,Tom
Hya 8h58'48"5d48'
Dorsey,TéKay Ramon
Aur 5h1'0"51d30'
Dorsey,Wesley Wayne
Vul 19h43'38"22d31'
Dorsi
Mon 6h26'30"7d43'
Dorsy,James Andrew
Lib 15h18'36"-8d5'
Dort,Bobby
Aur 5h0'48"46d51'
Dortch,Theodore"Ted"
Cep 20h56'50"55d11'
Dortha Rose
Uma 11h0'55"40d37'
Dorthea
Cyg 19h31'58"32d40'
Dortmunder Star,The
Cyg 21h30'31"40d40'
Doruss
Cet 1h27'32"0d40'
Dorval
Umi 16h24'53"71d34'
Dorverber
Cyg 20h28'33"42d35'
Dorvis,Willard-Sheila
Lmi 10h35'40"28d18'
Dorwling-Carter, Mireille
Pho 0h41'31"-46d9'
Dory
Com 12h28'59"27d40'
Doryce
Cas 0h7'25"62d42'
Dorys Leal
Del 20h40'14"10d23'
Doré,Frédérique
Umi 13h34'36"70d24'
Doré,Jean
Cep 22h31'35"69d39'B
Doré,Michel R
Cep 21h8'42"85d14'
Doré,Rollande
Per 2h58'11"46d34'
Dos Corazh58'11"46d34'
Boo 15h7'43"37d39'
Dos Love of My Life
Umi 15h29'35"78d45'
Dos Santos,Mégane
Uma 12h16'19"60d6'
Dosad,Linda Louisa
Lyr 18h14'34"30d13'
Dosal,Kathryn Lynne "Katie"
Peg 0h3'0"20d26'
Dosal,Katie Lynne
Peg 23h3'44"13d27'
Doscher-Bernet
Cyg 20h11'41"37d45'
Doshi,Purvi
Uma 11h5'34"44d52'
Doskocil
Aur 7h23'46"41d8'
Dospisil-Rey
Aql 20h1'45"12d10'
Doss "The World's Best Sister",Donna
Mon 6h24'48"1d24'
Doss My Shining Star, Carson M
Sex 9h59'44"-6d20'
Doss Star,The
Umi 15h24'53"71d50'
Doss,Bill
Hya 9h32'0"-5d15'
Doss,Brian
Cyg 20h17'10"38d25'
Doss,Carole Ann Marie
Mon 7h42'22"-2d53'
Doss,Christopher Robin
Ori 5h57'0"18d2'
Doss,Jodi Leann
Peg 23h38'0"27d3'
DosSantos,Bill & Mary
Eri 4h13'39"-15d55'
DosSantos,Henry & Traci
Boo 13h46'47"16d22'
Dosso,John F
Cnv 13h10'58"38d7'
Dost,Elsie L
Del 21h4'34"12d31'
Doster,Anthony H
Hya 9h39'51"-0d13'
Dosumé,Star of
Boo 15h26'35"34d57'
Dot
Psc 23h37'29"2d8'
Dot
Del 20h20'23"13d11'A
DOT
And 23h17'39"40d33'
Dot & Al's Star
Aql 19h30'47"7d50"
Dot & John
Cyg 21h24'0"48d56'
Dot & Si
Cyg 20h24'33"38d31'
Dot n' Bud
Lyr 18h32'0"38d19'
Dot,Lyn P
Lmi 9h49'24"37d45'
Doten,Heather
Ori 5h55'30"18d1'
Dotger,Garry
Gem 7h16'30"21d40'
Dotleck,Stephen
Aqr 23h4'5"-15d24'
Doto,Deanna & Jim Del Giorno 4VR
Uma 11h55'1"42d5'
Dotson,Danny Ray
Per 2h24'10"55d39'
Dotsy Darling
Mon 7h42'11"-5d2'
Dott ssa Veli Lindner
And 0h11'25"38d47'
Dott,Theresa Anita
Lyn 7h54'22"52d25'
Dotterweich,Lucille
Cas 0h49'34"66d33'
Dottie
Equ 21h1'18"10d47'
Dottie
Cam 5h10'28"68d28'
Dottie & Bill Star,The
Sge 19h58'39"18d48'
Dottie & Paul
And 0h57'37"38d49'
Dottie & Peter Star, The
Cet 2h36'23"-12d15'
Dottie S
Uma 8h38'59"48d46'
Dottie's Star
And 0h15'31"34d55'
Dotts,Linda Jean
And 1h34'1"36d54'
Dotty May,The
Lyr 18h18'42"46d5'
Doty,Meghan Elizabeth
Lyr 18h39'37"37d55'
Doty,Ronald E
Per 2h40'1"40d53'
Doty,Tani Ann
Cas 2h35'50"60d2'
Doty,Tonja Nicole
Sge 19h56'24"16d1'
Dotzel,Kimberly A
Cyg 21h8'56"40d47'
Dotzel,Kimberly A
Cyg 21h9'16"40d42'
Double 'D'
Boo 14h21'43"50d49'
Double D,The
Ori 5h57'38"18d20'
Double L H,The
Aur 4h53'33"40d56'
Double L Star
Cas 0h54'1"56d4'
Double N
Cam 5h58'32"60d52'
Double,Erla Rose
Cas 0h36'40"68d54'
Doubnov Always With Hope,Boris
Aql 19h9'18"1d54'
Douce,Devan
Gem 8h3'56"30d34'
Doucet,Courtney Jo
Mon 7h58'11"-2d52'
Doucette,Raye
Uma 10h29'47"56d46'
Doud,James R
Per 1h32'14"53d16'
Doudell,Charlotte Ann
Cyg 19h57'26"31d18'
Doudoune
Boo 15h2'0"22d51'
Douds,Andrew Robert
Ori 5h54'0"-0d45'
Doug
Aql 19h31'55"11d16'
Doug
Hya 8h19'25"4d15'
Doug & Cheryl
Boo 13h53'33"21d6'
Doug & Gail Eternally
Uma 9h14'29"55d7'
Doug & Lorrie
Uma 11h56'37"60d11'
Doug & Vicky
Mon 6h29'13"0d34'
Doug Bug
Aur 4h56'14"40d57'
Doug Scott Revisited
Cnv 12h26'46"34d10'
Doug With Love
Vul 20h19'1"25d29'
Doug's Star
Mon 7h13'5"-5d29'
Doug's Star
Oph 18h40'59"6d48'
Doug,Waite For Me! I Love You Lori
Boo 14h13'0"50d31'
Dougal
Ori 6h2'56"8d50'
Dougal,Juman Alan
Leo 11h51'48"20d59'
Dougal,Patrick J
Her 17h50'53"14d50'
Dougan
Lyr 19h2'30"40d9'
Dougan,Agnes"Nancy"
Lyr 18h36'54"45d10'
Dougan,M Jill
Cyg 21h8'10"40d1'
Douggie's Star
Cmi 7h58'0"8d39'
Doughboy
Cet 2h18'23"-10d5'
Dougherty III,E
Hya 8h42'49"-2d25'A
Dougherty's Enterprise Evan
Uma 8h31'55"50d22'
Dougherty,Claire Christian
Mon 6h31'19"-0d28'
Dougherty,Clementine Butterfly
Aql 19h31'11"10d35'
Dougherty,Connor Redmond
Aql 20h6'11"1d6'
Dougherty,Constance S
Hya 8h42'49"-2d25'B
Dougherty,Dina Marie & Paul Christopher
Lmi 10h29'30"28d45'
Dougherty,Doug "Doc"
Oph 17h31'39"-20d39'
Dougherty,Jeffery Otto
Uma 11h47'1"52d5'
Dougherty,Jennifer
Mon 8h8'57"-10d0'
Dougherty,Joanna
Cas 1h50'22"58d13'
Dougherty,John Patrick
Hya 9h2'40"0d42'
Dougherty,Kayleigh Gayle
Mon 6h55'47"-3d50'
Dougherty,Lee
Lyn 7h33'1"38d35'
Dougherty,Lucas Floyd
Dra 9h20'0"74d13'
Dougherty,Mary Ann
Aql 19h2'30"0d32'
Dougherty,Mike
Her 18h4'46"40d28'
Dougherty,Patrick
Aur 7h4'41"41d39'
Dougherty,Robert Anthony
Per 3h38'48"37d41'
Dougherty,William
Lyr 19h11'0"47d1'
Dougherty,William L
Uma 10h7'43"61d41'
Dougherty-Proctor, Colleen M
And 0h11'30"46d56'
Doughman,Von Keith
Sex 10h23'23"-1d4'
Doughtery,Mark Christopher
Cet 3h20'34"0d22'
Doughty,Charles H
Her 17h15'26"21d26'
Doughty,Doug & Erin
Cam 5h46'6"70d45'
Doughty,Gabrielle Michele
Cet 2h49'58"-0d15'
Doughty,Kenneth & Eleanor
Lyn 7h37'46"36d58'
Doughty,Rosina
Peg 23h40'54"13d56'
Doughty,Sr,Philip Kenneth
Her 17h55'52"14d42'
Doughty-Wood,Gabriel Colin
Dra 16h30'55"62d21'
Doughty-Wood,Tristan Owen
Cmi 7h44'38"5d30'
Douginny
Oph 18h6'15"11d4'
Douglas
Aql 20h35'25"0d41'
Douglas
Cnv 13h18'40"32d10'
Douglas
Ori 5h59'37"16d26'
Douglas 143
Her 18h1'39"30d36'
Douglas D M D J
Her 17h22'27"44d59'
Douglas Daniel's Destiny
Uma 9h51'49"71d27'
Douglas Edward
Per 2h30'0"57d28'
Douglas L
Aur 6h27'0"33d45'
Douglas Rae
Her 18h1'12"40d31'
Douglas,Alison Ann
Lyn 8h34'37"40d50'
Douglas,Barclay Andrew
Leo 11h4'0"-0d31'
Douglas,Christopher James
Her 17h54'48"38d23'
Douglas,Deanna Raye
Vul 19h44'59"25d8'
Douglas,Emma Louise
Lyr 18h50'34"42d57'
Douglas,Frankie
And 23h17'19"46d39'
Douglas,Jason
Cet 3h14'38"0d14'
Douglas,Kathy L
And 1h34'55"39d21'
Douglas,Kirk
Aql 19h17'23"10d16'
Douglas,Lindsey Anne
And 0h23'23"45d22'
Douglas,Mark
Her 16h6'16"20d5'
Douglas,Mark Alan
Aur 5h33'42"37d50'
Douglas,Mark Edward
Cep 22h13'33"62d30'
Douglas,Mein Liebes
Equ 21h20'52"8d25'
Douglas,Pat & Dick
Boo 15h5'25"51d33'
Douglas,Paul Laurence
Cep 20h5'51"60d39'
Douglas,PhD,Priscilla H
Cam 4h50'1"68d23'
Douglas,Richard W
Crt 11h40'58"-8d48'
Douglas,Sonia Lopez
Peg 21h59'35"22d38'
Douglas,Stephen William
Her 17h12'0"41d24'
Douglas,Stephen
Aql 19h31'29"11d21'
Douglas,Steve
Per 3h4'57"42d20'
Douglas,Susan
Lyr 18h53'53"37d41'
Douglas,Susan Anne
Uma 12h4'25"43d56'
Douglass,Ivan Paul
Umi 1h17'16"69d52'
Douglass,Jaime Rose
Aur 6h54'17"38d38'
Douglass,John G "Hot Rod"
Vir 13h31'43"11d12'
Douglass,Linda
Mon 7h49'47"-5d29'
Douglass,Mary Anne
Mon 7h30'0"-8d21'
Douigo,Ruth M
Lyr 18h54'29"31d4'
Douille,Edward H
Aql 18h57'58"12d45'
Doukas,Michael
Lac 22h5'23"50d43'
Doukhan,Yves
Lyn 8h10'42"40d42'
Doukzsa,John J
Dra 19h3'10"58d9'
Doulcet,Pierre
Pic 4h48'25"-48d23'
Doumalin,Loïc
Cyg 21h58'42"50d35'
Doumar,Trish
Cet 3h16'36"8d11'
Douris,Herve
Pho 0h40'48"-43d29'
Dousman,Peter
Per 1h59'20"56d47'
Dousset,Nadege
Lmi 10h46'26"23d25'
Doust-Kupietz,Karen
Eri 3h15'42"-15d12'
Doust,Colin & Pamela E
Nor 16h18'22"54d12'
Doutheil,Bernd
Boo 14h26'0"13d27'
Douthett,Marianne S
Lyn 7h31'12"40d57'
Douthwaite,Margaret
And 0h15'45"39d45'
Douty,George & Linda
Cam 9h4'50"82d23'
Douze
Hya 8h29'55"1d3'
Dove's Astral Cabin, Pat & Paul
Cyg 21h13'27"38d35'
Dove,Angelina
Mon 6h22'37"5d58'
Dove,Debbie
Cnv 12h32'31"33d36'
Dove,Eddy & Christine
Crt 11h17'59"-17d19'
Dove,G G
Mon 6h29'49"11d8'
Dovedari,Fadia
And 0h15'26"40d14'
Dovener,Joseph John
Equ 21h21'58"8d5'
Dovey,Matthew James
Cep 21h31'23"55d24'
Dovigo,Anthony
Lac 22h25'46"38d20'
Dovigo,Franco A
Per 3h7'36"40d15'
Dovigo,Marco A
Aur 6h28'34"38d51'
Dow,Amy Lynn
Aql 20h3'47"0d25'
Dowd III,William Carey
Her 16h39'36"5d28'
Dowd,Alan G
Her 17h4'31"21d57'
Dowd,Christine M
Com 12h54'53"24d11'
Dowd,Donald & Sarah
Cyg 20h34'38"31d10'
Dowd,Ian Henry
Dra 16h53'19"68d37'
Dowd,Peter N
Aur 6h31'31"31d28'
Dowd,Robert "Tex"
Oph 16h51'1"10d57'
DowDell,Jessica
And 2h27'22"41d2'
Dowdy III,James Robert
Sge 19h28'18"16d18'
Dowdy,Marianne Terzes
Aql 18h58'22"-6d6'
Dowdy,Nikki
Mon 6h37'19"3d2'
Doweiko,Savannah Taylor
Mon 6h22'50"4d34'
Dowell's Star,Erika Jean
Cam 6h46'33"67d49'
Dowell,Anna M
Dra 19h30'19"61d22'
Dowell,John
Mon 7h50'11"-2d12'
Dowell,Vincent
Per 1h30'0"53d59'
Dowell-Thumfart, Christina
Crb 15h54'55"28d38'
Dowen,Irene
Cyg 20h53'1"31d23'
Dower,Joshua Ross
Her 17h28'44"40d36'
Dower,Matt
Her 15h50'40"41d32'
Dowie's Shining Star, Michael & Angie's
Lac 22h10'33"47d57'
Dowie,Alan
Umi 16h43'14"77d33'
Dowler,Helen A
Uma 10h50'15"71d34'
Dowler,Margaret C
Mon 6h14'56"-10d19'
Dowling
Cyg 21h6'51"30d55'
Dowling,Edward Peter
Ori 5h52'33"8d40'
Dowling,Heather Jean
Peg 22h42'29"33d8'
Dowling,Joseph
Tri 1h47'24"27d31'
Dowling,Mardi
Leo 10h43'29"15d47'
Dowling,Ruth
Lyr 18h55'35"32d26'
Dowling,Tammy Marie
Cyg 21h33'0"53d35'
Down,Barbara & Roland
Uma 11h36'43"32d50'
Down,Janice Poland
Aql 19h48'1"10d9'
Down,Kirk Matthew
Per 2h52'0"37d11'
Downer,Courtney Nicole
And 1h38'42"35d27'
Downer,Vicki
And 23h19'41"49d19'
Downes,Bridget Therese
Umi 13h52'39"70d13'
Downes,Lisa Ellen
And 2h5'45"38d19'
Downes,Maria Helga
And 0h21'49"44d29'
Downes,Paul Anthony
Her 16h24'18"36d18'
Downes,Richard
Equ 21h3'50"8d60'
Downey's Dream
Cmi 7h56'23"8d28'
Downey,Ann Marie
Peg 21h26'27"23d24'
Downey,Dorothy Ruth Elder
Uma 11h20'24"30d8'
Downey,Evelyn Agnes
Cas 23h28'0"60d33'
Downey,Jr,John Calvin
Sct 18h41'51"-6d1'
Downey,Jr,Kevin
Her 16h27'49"22d37'
Downey,Katie Michelle
Mon 6h29'24"11d32'
Downey,Matty
Dra 17h42'56"60d10'
Downey,Ronn
Lac 22h52'34"38d59'
Downey,Wilfred Joseph
Cyg 21h14'0"28d24'
Downie,James
Per 1h48'1"56d26'
Downing,Charles Grady
Ser 15h20'22"9d42'
Downing,Claire M S
And 0h13'55"38d4'
Downing,Ned J & Mary Alice
Uma 10h15'22"48d22'
Downing,Pat & Jennie
Cyg 19h25'11"30d14'
Downing,Paul A
Ori 5h21'1"11d46'
Downing,Taylor Reneé
Vul 20h24'1"28d23'
Downing,Thomas E
Ori 5h38'28"12d57'
Downs,David Erskine
Her 18h5'47"30d11'
Downs,David Francis
Aur 6h16'43"31d26'

Downs,Kent
Cep 22h48'58"71d12'
Downs,Laura Mary Skinner
Mon 6h35'1"-6d20'
Downs,Markii Christyn
Her 17h33'22"40d48'
Downs,Mary Erny
Her 16h41'29"33d9'
Downs,Michael Patrick
Psc 0h53'12"33d14'
Downs,Patrick O
Eri 4h17'16"-19d30'
Downs,Stephen Johannes
Per 2h58'51"50d24'
Downs,Whitney
Sge 19h1'56"18d55'
Downsrapp,Lauren
Ori 5h54'51"5d17'
Downton,Donald P
Oph 18h6'23"7d49'
Dowrey-Gauntt
Her 17h23'1"48d41'
Dowse,Steven Robert
Cep 1h0'56"85d35'
Dowsing,Norman
Ori 5h27'21"-2d2'
Dowzycki,Andrea Bridget
And 0h22'13"40d37'
Dowzycki,Elise Renee
Cam 3h59'34"55d24'
Dowzycki,Hannah Leigh
Lyn 7h36'11"45d13'
Dowzycki,Meghan Elizabeth
Lyr 19h2'42"40d14'
Dowzycki,Sarah Michell
Tri 1h59'53"28d26'
Doxsey,Matthew David
Her 18h3'39"30d34'
Doxtater,Rebecca
Sct 18h52'44"-9d14'
Doye,Rob Guthrie
Oph 17h6'39"10d25'
Doyle (Bud),John A
Her 17h14'44"26d29'
Doyle Family Star,The
Sge 20h0'26"16d8'
Doyle III,John A
Peg 23h38'39"30d50'
Doyle,Benjamin David
Cet 2h29'18"4d55'
Doyle,Benjamin Salazar
Lmi 10h31'47"28d50'
Doyle,Brodie Lee
Aur 6h7'55"31d46'
Doyle,Cecilia
Sgr 18h3'59"-28d48'
Doyle,Daniel M
Peg 22h28'15"24d7'
Doyle,Darren Dutch
Boo 14h50'30"31d54'
Doyle,David
Aur 5h56'49"31d33'
Doyle,Debbie
Lib 15h0'33"-18d45'
Doyle,Denise A
And 2h23'54"41d54'
Doyle,Donald W
Cep 22h40'20"65d14'
Doyle,Emily
And 2h32'1"38d31'
Doyle,Erin Lynn
Cnc 8h35'6"19d19'
Doyle,Eugene Francis
Oph 16h45'42"0d57'
Doyle,Francis W
Peg 21h59'0"2d24'
Doyle,Harriett Boggs
And 22h59'0"50d2'
Doyle,Hayley Ann
Aql 19h43'43"14d18'
Doyle,Heather A
Aqr 21h24'32"-0d36'
Doyle,James
Cep 21h48'45"67d55'
Doyle,Joe
Cnv 13h59'14"31d53'
Doyle,Judy Mighty Mite
Aql 19h44'47"13d55'
Doyle,Justin Patrick
Lac 22h1'25"48d16'
Doyle,Karen
And 1h17'44"37d18'
Doyle,Luke Jonathan
Umi 15h6'56"66d11'
Doyle,Lyn
Cas 0h3'1"59d8'
Doyle,Madeleine Kreitzer
Cyg 21h8'35"31d28'
Doyle,Marc
Ser 18h6'22"-14d56'
Doyle,Marcy Taylor
Vul 20h18'23"23d14'
Doyle,Mike
Aur 7h0'43"37d6'
Doyle,Rebekah
Umi 13h40'11"79d22'
Doyle,Richard James
Sex 10h36'12"-6d6'
Doyle,Robert J
Per 3h7'15"37d56'
Doyle,Ruth E
Lyn 8h9'15"48d27'
Doyle,Sheila Marcelle
Cas 1h33'32"76d19'
Doyle,Susan C 9-30-42
Lib 15h38'19"-25d54'
Doyle,Thomas C
Aur 7h22'13"38d34'
Doyon,Eric
Sge 18h57'18"19d10'
Doza,Greg
Cmi 7h23'44"0d17'
Dozier,Henry H Sr & Ethel L
Cyg 20h4'40"58d12'
Dozoretz,Joshua Ronald
Vir 12h29'55"2d28'
Dozzo 2
Mon 7h6'29"-1d34'
Dozzo I
Aql 20h11'33"0d35'
Do Aql 20h11'33"0d35'
Cet 2h30'34"4d26'
Dr Alan
Aur 6h1'13"45d42'
Dr Bill
Cep 22h11'1"70d45'
Dr Bob
Oph 17h54'47"12d40'
Dr Bob
Per 1h58'23"53d49'
Dr Dan's Star
Oph 18h1'52"12d7'
Dr David
Uma 11h31'1"31d31'
Dr Diane
Lyr 19h13'44"41d51'
Dr Edward
Cep 21h8'58"65d12'
Dr Fred
Ari 2h29'46"21d41'
Dr Gene C
Aqr 23h24'23"-4d24'
Dr H D B
Oph 17h56'1"12d40'
Dr J W R
Per 2h57'42"31d2'
Dr J's Magicstar
Aql 19h31'43"10d50'
Dr Jay Forever
Oph 17h15'1"11d2'
Dr JHW's Eternal Brilliance
Oph 18h32'0"6d13'
Dr Jim Fix Teeth Jinks
Uma 10h0'37"42d26'
Dr Joe MD
Oph 18h24'51"12d2'
Dr John
Aql 19h59'42"0d11'
Dr K A B
Oph 17h27'1"-6d7'
Dr Larry "Jet Jockey Green"
Oph 16h51'19"-28d33'
Dr Lou
Her 16h36'38"29d44'
Dr Mary
Uma 10h2'50"51d46'
Dr Matt (Kwas)
Her 18h42'49"12d5'
Dr O J
Aql 19h58'0"15d30'
Dr Phil "The Best Of The Best"
Aur 5h17'1"45d20'
Dr R
Uma 11h12'23"30d17'
Dr Rebel & Dr PK
Oph 17h53'60"8d11'
Dr Rich
Oph 16h58'46"8d44'
Dr Robert
Per 2h41'32"43d16'
Dr Rod
Cep 20h18'29"60d2'
Dr Ron Is In The Power Zone
Oph 17h32'13"-22d44'
Dr Sam
Cyg 19h32'0"34d14'
Dr Silly String
Aql 20h4'42"0d48'
Dr Smiley
Cen 13h30'54"-34d48'
Dr Tony
Her 18h17'1"20d33'
Dr Who
Oph 17h32'14-23d34'
Dr Z's Light
Her 17h31'56"28d52'
Draack,Nancy
Cas 1h27'29"65d29'
Drab,George J
Her 16h13'17"41d7'
Drabos,James Rodney
Oph 17h53'43"12d36'
Drach,Amanda
Ori 5h55'56"14d39'
Drach,Coach Buck
Aur 4h50'57"50d56'
Drach,Ryan William
Ori 5h53'19"8d25'
Drach,Tiffany
Ori 5h57'11"16d16'
Drachen Köder
Cnc 9h14'26"7d44'
Drachen-Baby,ganz oder gar nicht
Tau 5h1'1"16d10'
Drachkovitch,Lasta
Aql 20h32'51"-6d15'
Drachkovitch,Rasha
Mon 8h2'3"-7d15'
Dracon,Ellen C
Dra 17h1'33"58d47'
Draeger 77,Beier
Hya 8h43'46"5d17'
Draeker
Peg 22h35'30"8d7'
Draffan,Kirk Livingston
Uma 11h51'1"32d3'
Draffan,Madison Leigh
Uma 11h52'56"31d46'
Draffan,Stacy Jo Blum
Uma 11h52'25"30d57'
Draginis,Patricia
Mon 7h1'1"8d8'
Draginis,Sharon
Eri 3h53'17"-0d29'
Drago,Daniel
Dra 12h53'29"71d57'
Drago,Joan & Gaspare F
Cnv 13h55'29"30d5'
Drago,Michael Anthony
Aur 7h9'50"40d53'
Dragon Heart
Dra 14h46'18"60d27'
Dragon Lady
Dra 16h36'17"63d58'
Dragon M
Lyn 8h40'0"45d10'
Dragon Slayer,The
Leo 9h37'31"7d4'
Dragon's Pearl
And 1h4'38"46d32'B
Dragon,Christopher John
Aql 19h6'1"0d10'
Dragone,Katherine Mae Werner
Dra 16h4'1"63d1'
Dragonfly
Aur 5h32'31"38d50'
Dragonstar LSB
Ori 5h37'56"10d43'
Dragosin 2gether 4ever,Dave & Cindy
Cyg 19h27'54"34d28'
Drahovszky,Anne
Aur 5h13'1"43d7'
Drain,Jr,Theodore R
Her 16h40'39"34d53'
Drainville,Gary Patrick
Her 16h28'37"18d58'
Drak
Uma 10h22'54"58d57'
Drake,Addison John Brennan
Dra 17h2'27"61d60'
Drake,Barbara Carroll
And 0h58'20"45d46'
Drake,Benjamin Baird
Her 17h10'25"49d8'
Drake,Charles Roger
Boo 14h44'37"52d17'
Drake,Christine M
Cyg 21h6'50"31d18'
Drake,Delores
Cas 1h47'23"60d1'
Drake,Dick
Cet 2h10'30"6d36'
Drake,Donna
Peg 22h59'33"29d46'
Drake,Dorothy M
And 0h54'12"36d48'
Drake,Dr Frank
Eri 4h13'26"-13d40'
Drake,Jessica Maelynn
Eri 4h2'51"-17d6'
Drake,John Robert"JR" Badt
Aur 6h28'32"31d14'
Drake,Joseph
Aql 19h57'41"15d37'
Drake,Justin B
Aql 19h47'60"14d26'
Drake,Lady Rainlynne
Cas 23h29'57"61d31'
Drake,Lisa Gay
Mon 8h1'2"-1d48'
Drake,Lou
Cam 3h54'49"74d43'
Drake,Maria
Lyr 19h2'23"38d30'
Drake,Nancy Lee
Vul 19h16'38"25d8'
Drake,Rhonda
Cam 4h17'28"60d55'
Drake,Ronnie Merle
Cam 5h57'0"80d14'
Drake,Sarah
And 2h19'17"48d35'
Drake,Shirley M
Peg 21h26'56"23d45'
Drake,Tanner Dale
Dra 19h0'1"48d59'
Drake,Verley
And 23h19'41"49d26'
Drake,Victoria
Peg 23h19'59"32d14'
Drake-Landers
Hya 8h56'14"3d7'
Drakie
Vul 20h20'41"28d22'
Dramard,Jules
And 23h39'1"33d43'
Dramis,Terri
Vul 20h28'1"28d19'
Dramko,Joanne Marie
Lyn 9h22'26"41d45'
Drandoff,Fran "Heckle"
Cam 7h53'39"70d39'
Drane,Ashley
Eri 3h10'1"-3d10'
Drane,Gregory Kent
Aql 19h1'33"15d30'
Drange,Britta
Gem 6h37'19"28d29'
Dranger,Brandon Anthony
Aur 5h5'60"40d27'
Dranks,Guy Anthony
Her 16h50'1"32d34'
Drapac,Thomas F
Cep 20h55'41"68d43'
Drapeau,Claire
Lyr 18h58'33"34d3'
Drapeau,Sylvain
Psc 1h34'0"22d12'
Draper,Daniel P
Vul 19h47'31"28d56'
Draper,Harrison Brooks
Aur 6h22'27"30d58'
Draper,Jr,David G
Aur 5h59'0"30d55'
Draper,Lindsey Shannon
Umi 14h55'44"80d5'
Draper,Mom & Dad
Ori 5h27'12"0d8'
Draper,Richard W
Per 2h59'43"37d57'
Draper,Robert
Boo 14h52'40"33d34'
Draper,Robert Burrell
Ori 5h13'37"-1d32'
Draper,Seasons Marie
And 0h9'46"39d12'
Draper,Sid
Cep 21h50'1"65d8'
Draper,Susan"Love Dolphine"
And 1h30'35"40d34'
Drapkin,Mrs
Cam 13h0'1"77d42'
Drapkin,R A
Cam 3h36'26"70d38'
Drapper,Anthony E
Uma 8h57'19"56d52'
Drasdo,Roland Michael
Eri 3h20'24"-10d29'
Drasga,Louis Basil
Per 1h37'14"53d26'
Drasheff,Mary
Mon 6h23'56"8d36'
Drate
Aur 6h4'1"32d8'
Dratel
Cep 0h2'12"68d11'
Draugelates Prof Dr Ing Ulrich
Cnc 9h15'27"7d40'
Draves III,Edward F
Tau 3h42'12"2d23'
Draves III,Edward F
Tau 4h7'51"22d16'
Draving,Wolfgang Forbes
Peg 22h14'18"32d6'
Drawe-Davis 87ø425
Uma 9h56'26"48d12'
Dray's,The
Lac 22h13'1"51d16'
Drayer,Carol Jane
Lyr 19h15'0"40d30'
Drayton,Caroline
Mon 6h19'53"0d35'
Drazzle
Cam 3h16'1"60d28'
Dream
Aql 18h55'25"-0d8'
Dream Catcher
Peg 22h5'51"2d24'
Dream Catcher
Eri 2h50'33"-2d1'
Dream Catcher
Cyg 19h29'15"36d37'
Dream Catcher,The
Lyr 19h18'57"41d58'
Dream Foundation,The
Peg 22h3'29"2d11'
Dream Guy & Girl, Jon & Beth
Hya 8h12'12"1d2'
Dream In
Dra 17h58'28"61d19'
Dream Interiors,Inc
Cyg 21h15'54"35d9'
Dream Lover
And 1h45'31"39d24'
Dream of Genki
Gem 7h4'20"16d54'
Dream Of Kazuya & Sayuri,The
Lib 14h25'59"-9d21'
Dream of Miki
Psc 1h15'45"18d18'
Dream Patricia
Ser 16h7'21"9d43'
Dream Road
Gem 7h0'9"15d18'
Dream Star
Cnv 13h34'16"38d54'
Dream Team 7D of'93 Coon Rapids,MN
Uma 11h35'11"48d58'
Dream Weaver
Cnv 13h53'27"40d18'
Dream Weaver
Aur 5h9'26"42d44'
Dreamellsia
And 0h39'1"38d24'
Dreamlike for Peter & Ruth
Cas 0h48'1"60d4'
Dreammaker's Star
Eri 4h2'48"-14d3'
Dreams
Leo 10h0'54"10d6'
Dreams
Peg 21h58'15"10d11'
Dreams & Wishes
Cnv 12h54'38"43d52'
Dreams Come True
Somewhere In Time
Ori 6h1'60"-0d0'
Dreams Coming True! 895-Bill Brown
Tau 4h13'15"22d13'
Dreams of Ann
And 1h18'51"37d5'
Dreams of Helen,The
Cyg 20h29'26"40d19'
Dreams Of Hiroshi Takagi
Cap 20h57'11"-19d12'
Dreams of Hiroshi Takagi
Cap 20h57'11"-19d12'
Dreams of Junko
Lib 14h29'12"-12d30'
Dreams To Come
Ori 4h53'34"-0d0'
Dreamy
And 0h19'59"38d2'
Dreamy "Stera" of Yoji & Atuko
Aqr 22h19'5"-11d30'
Drechsel,Ulli-G
Cap 20h37'23"-14d6'
Drechsler,Dorothy
Aur 5h58'16"31d21'
Drechsler,Nathan
Aur 5h17'11"41d39'
Dreebin,Sheldon
Cet 1h48'46"-1d26'
Dregich (Minnie), Miriam
Lyn 7h56'1"44d1'
Dreher,Claus & Tobias
Cet 2h55'30"4d30'
Dreher,Eduard
Ser 15h36'28"18d34'
Dreher,Traci Anne
Lyn 8h17'33"51d51'
Dreiband,Sophie
Lyn 7h35'14"35d38'
Dreibelbis,Bill
Cep 21h30'1"55d29'
Dreibelbiss,James D
Cet 3h7'30"2d13'
Dreichler,Catherine
Ori 5h56'27"15d46'
Dreifuerst,Darin
Cep 20h12'47"80d2'
Dreisiebner,Beate
Uma 11h51'18"33d30'
Dreitzer,Edward & Shirley
Dra 12h57'53"75d38'
Dreixler,John "Drex"
Aur 5h34'55"38d32'
Drekonja,Christian
Tau 3h45'29"0d38'
Dreksler,Lionel
Dra 12h2'31"70d25'
Dremann's Star,Barbara
Lyn 9h4'38"46d13'
Dremia
Lyn 7h42'0"44d59'
Dremstedt,Linda
Cyg 21h3'52"31d39'
Drennan's
Her 17h12'17"21d6'
Drennan,Robert Vincent
Boo 15h3'14"18d33'
Drenovsky,Rachael
Aur 5h38'9"29d28'A
Drenth-Allport,Jasper Benjamin
Ori 5h12'48"-8d50'
Drerup,K Matthew
Boo 14h19'40"17d51'
Drerup,Sir Lucas Charles
Dra 16h48'49"69d10'
Dresbach,William & Judith
Sge 19h54'28"16d10'
Drescher,Thomas
And 23h45'50"38d38'
Dresden
Umi 16h45'15"77d53'
Dresdner Bank DDF, Zw Wehrhann
Vir 14h8'30"-6d29'
Dresdner Bank DDF, Fil Hilden
Tau 3h47'39"0d31'
Dresdner Bank DDF, Fil Ratingen
Psc 22h49'27"6d40'
Dresdner Bank Neuss, Fil Dormagen
Ari 2h24'30"21d38'
Dresen,Jayne S
Dra 17h48'57"75d44'
Dresher,Stacey
Mon 6h30'31"-8d8'
Dreslin,Bill & Joyce
Oph 18h39'1"10d28'
Dresner,Joseph
Lib 15h16'35"-28d42'
Dresner,Milton
Lib 15h37'29"-28d36'
Dressel,Essie V
And 0h33'21"38d57'
Dressel,J Mark
Cam 12h49'1"82d30'
Dressen,Coleman Scott
Cep 21h14'33"61d44'
Dresser,Dana Elizabeth
Cas 1h12'20"75d8'
Dresser,Reid
Per 2h51'33"31d10'
Dressler,Aida & Phil
Lyr 18h41'13"26d21'
Dressler,Bertha C
Cas 0h39'31"71d5'B
Dressler,David M
Dra 19h49'43"61d23'
Dressler,Deborah Menzell
Dra 19h46'35"58d56'
Dressler,Max
Cas 0h39'31"71d5'A
Dressler,Paula K
Peg 21h37'32"20d19'
Dreveniak/Meyers
Uma 9h49'46"68d38'
Dreves,Claudia
Uma 14h21'30"60d8'
Drew
Cep 22h36'43"60d57'
Drew
Peg 23h10'49"21d48'A
Drew #60,Ronald Martin
Vir 13h29'45"-21d11'
Drew Family Love Star
Lyr 19h15'24"42d39'
Drew,"Lord" Robert Thomas
Cep 3h21'40"78d48'
Drew,Alexander Jonathon
Ori 5h56'19"16d51'
Drew,Cordell James
Umi 15h15'58"77d17'
Drew,Erin K
And 23h17'11"41d1'
Drew,Ian Robert
Aur 5h13'1"40d42'
Drew,Jerry Alan
Ser 15h20'1"-2d19'
Drew,Jerry Lee
Sct 18h46'3"-7d9'
Drew,Joshua Daniel
Cet 2h38'1"2d3'
Drew,Kathleen D
Lyn 8h21'27"44d4'
Drew,Kevin Ian
Aur 7h3'29"38d49'
Drew,Kevin Michael
Aql 18h59'56"14d35'
Drew,Loran
Oph 17h53'37"12d14'
Drew,Margie
Cet 2h30'1"4d59'
Drew,Nicholas
Her 16h27'38"32d32'
Drew,Peter R
Per 2h56'34"34d56'
Drew,Tiffany Marie
Lyr 18h29'26"34d30'
Drew,Timothy James
Dra 16h43'50"67d27'
Drewer,Valarie Ann
Cyg 19h33'52"36d1'
Drewery Star,The
Lmi 10h3'26"39d2'
Drewry,Christy
Aql 19h22'59"1d3'
Drewry,Erica J
Sco 17h5'51"-31d11'
Drews,Katherine Elizabeth
Psc 0h55'33"32d48'
Drewstar
Ori 5h55'27"15d27'
Drewy
Cep 21h50'47"61d5'
Drexel
Boo 14h28'52"42d26'
Drexler,Beatrice
Cas 2h15'53"65d26'
Drexler,Bobby
Uma 10h34'40"50d46'
Drexler,Brian
Lac 22h11'18"46d53'
Drexler,Dale
Lmi 10h2'20"32d32'
Dreydorff,Michael
Vul 20h42'18"20d16'
Dreyer,Amanda Jane
Leo 9h19'26"21d18'
Dreyer,George Charles
Cnv 12h58'48"51d23'
Dreyer,Jacqueline Ashley
Peg 0h4'31"21d36'
Dreyer,Louise B
Lib 14h21'32"-8d12'
Dreyer,Susan
Cas 23h14'49"60d39'
Dreyfus,Laura Lynn
Mon 8h6'52"-6d58'
Dreyfus,Lisa Ashley
Peg 21h52'52"34d56'
Dreyfus,Marc G
Her 17h38'37"42d8'
DRG Light Of Everlasting Love
Uma 11h55'36"40d2'
Driehaus,Richard H
Cep 21h27'1"58d54'
Drier,Frances C
Cyg 20h15'52"38d13'
Driescher,Terry & Susan
Eri 4h36'38"-8d48'
Driesse,Timothy Keith
Per 2h40'26"35d46'
Drifting Cloud
Uma 9h0'22"71d22'
Driggers,E C
Cam 5h39'11"60d20'
Drilla,Ronald Michael
Dra 17h1'15"51d41'
Drinkwater,Samuel
Dra 16h3'18"63d11'
Drinkwine,Robert L
Cyg 20h44'10"50d29'B
Drinkwine,Sharon K
Cyg 20h44'10"50d29'A
Driscoll
Cep 22h1'24"61d9'
Driscoll #1,Mom & Dad
Lya 18h32'0"42d46'
Driscoll,Bea
Per 2h46'30"40d28'
Driscoll,David P
Lac 22h4'29"49d26'
Driscoll,Donna Denise
Cap 19h37'0"-18d42'A
Driscoll,Dr James E
Her 16h11'17"42d18'
Driscoll,Jennifer Marie
Aql 19h4'11"-5d51'
Driscoll,John F
Lac 22h12'31"48d36'
Driscoll,Lisa Marie
Lyr 18h30'0"31d19'
Driscoll,Mark D
Dra 17h4'42"64d8'
Driscoll,Meghan
Peg 22h51'13"27d28'
Driscoll,Michael Hardee
Aql 19h20'48"15d38'
Driscoll,Richard
Cnc 9h15'34"31d4'
Driscoll,Robert Eric
Lac 22h7'50"49d9'
Driscoll,Roberta Esther
Cas 1h49'28"58d14'
Driskell,Jerry
Cet 0h49'0"-10d24'
Driss,Bassri
Aur 6h55'57"41d9'
Driss-McTighe Harriet
Cyg 21h38'36"40d40'
Drissi,Khalid
Boo 15h28'1"30d27'
Dritsas,Andrew S & Mary M
Uma 9h45'41"56d3'
Drivdahl,Brian C
Dra 17h24'0"71d8'
Driver,Eugene "Tink"
Mon 7h17'43"-6d11'
Driver,Josephine Gibb Bruce
Cyg 19h26'17"34d33'
Driver,Margarete K M
Vir 15h5'19"5d53'
Driver,Ruth & Dick
Cyg 20h4'21"58d9'
Drljaca,Deborah
Cam 7h31'0"61d45'

Drljaca,Shea
Ser 15h36'16"17d53'
Drobek,Aurelia geb Neumann
Cnc 8h0'40"10d49'
Drobnich,Darcy
Cas 2h57'10"57d34'
Droddy,Benjamin Willis
Lac 22h30'34"37d54'
Droege,Robert Alan
Ori 6h1'39"-1d35'
Droemer,Janie
Equ 21h1'56"7d31'
Droetto,Tiphaine
Aur 5h14'30"29d45'
Drogan's Speed Star
Lyn 8h36'55"40d2'
Drohan,Barbara M
Cyg 19h55'31"38d29'
Drohan,Bob & Teena
Mon 6h54'0"-1d24'
Drolet,Clément
Lmi 11h4'38"31d36'
Drolet,Dr Jacques
Uma 12h58'34"60d18'
Drolet,Elaine
Lac 22h9'35"40d44'
Drolet,Lucie Dubois
Cas 0h42'58"73d35'
Dromby,Martine
Uma 11h49'51"51d50'
Drommen som ble virkelighet
Cam 8h45'0"80d43'
Dromson,Florence
Cyg 20h55'0"30d6'
Drooger,Ethan John
Ser 15h30'17"18d2'
Droopy,My Shining Star
Aur 5h16'60"46d47'
Dropy,Didier
Sge 18h59'51"19d50'
Droscha,Andrea
Lyr 19h20'21"35d31'
Drosdick,David J
Aur 7h20'45"38d37'
Drosendahl,Dennis Allan
Boo 15h4'22"24d16'
Drosendahl,Devon Alice
Lyr 18h47'28"36d52'
Drosendahl,Howard Allen
Aur 6h31'54"33d13'
Drosinos,Jonathan George
Lib 14h58'46"-23d51'
Drossman-Smith,Lois
Lmi 10h17'16"33d18'
Drossos,Christopher C N
Her 17h31'41"21d50'
Droszella,Constantin
Lac 22h3'34"51d55'
Drott,Milton C
Her 17h28'36"20d6'
Droubay,Helen
Cas 1h1'16"68d49'
Drouin,Nathalie
Lyr 18h23'45"37d44'
Drounth
Dra 11h20'29"78d19'
Drouvin,Laurene
Eri 3h41'1"-2d12'
Drover,Kevin James
Her 15h54'0"44d36'
Drown,Jesse & Zackry, Deb & Gary
Ori 5h38'19"11d25'
Drowning Man
Hya 8h12'51"-6d17'
Drozda,Jr,Clifford Edward "Papere"
Uma 10h59'54"47d0'
Drozdek,Lara
Ori 5h17'37"12d29'
Druar,Nancy Jane
Cyg 20h44'41"37d37'
Druce,Laurel S
Dra 17h57'53"61d50'
Druchniak,Mark Allen
Cam 5h4'19"70d11'
Drucker,David Aaron
Hya 8h14'51"0d33'
Drucker,David M
Cep 22h27'34"68d33'
Drucker,Jonathan Harris
Cet 2h12'58"5d32'
Drucker,Lisa
Cam 4h52'23"70d50'
Drucker,Lizzie
Cas 2h1'59"58d49'
DruDan II
Eri 4h42'19"-8d38'
Drue
Lyn 8h29'32"48d7'
Druecke,Jacob Paul
Boo 15h37'52'41d35'
Druen,Kelly Renee
Peg 22h10'13"20d51'
Druen,Nathan Quinn
Cam 3h54'30"60d46'
Drufuka,Patricia
Uma 8h53'36"48d27'
Drugatz,Rachel
Peg 22h43'33"35d5'
Drugs Don't Work National Stars
Uma 12h45'25"57d30'
Drugs Don't Work! Star Supporters
Aur 5h53'27"38d5'
Druhot,Stephen P
Boo 14h56'39"25d58'
Druk,Luidmila
Cas 0h42'39"63d15'
Drulot,Marie Rose
Aql 18h56'12"-6d57'
Drum,Russell J
Tau 4h34'47"30d33'
Drumm,Dennis
Aur 4h57'35"38d15'
Drummer Boy,The
Aql 18h58'54"13d2'
Drummer D
Sct 18h47'7"-6d59'
Drummond III,Anthony F
Boo 14h46'27"25d33'
Drummond,Andrea
Aqr 23h13'0"-4d16'
Drummond,Barbara Corlette
Com 13h32'52"17d34'
Drummond,Christopher Michael
Hya 8h54'26"2d4'
Drummond,Kaitlyn Nicole
Peg 21h47'39"35d59'
Drummond,Linda
Ori 6h8'16"4d20'
Drummond,Linda Louise
Aql 19h44'13"13d45'
Drummond,Paddy
Aur 6h9'31"45d52'
Drummond,Paul
Ser 16h4'1"22d53'
Drummond,Warren
Cep 21h56'21"70d22'
Drunzer,Ida Elizabeth
Del 20h13'15"10d4'
Druon,Daniel
Cam 13h55'39"80d19'
Druon,Michel
Umi 15h17'31"70d48'
Drury,Jennifer
And 2h26'41"48d59'
Drury,Kenneth
Per 2h27'1"56d34'
Drury,Kenneth Doris
Dra 16h11'27"61d16'
Drury,Little Light Of Ashleigh J
Equ 21h0'12"2d24'
Drusilla Darlene
Cnc 8h11'24"30d1'
Drutinus,Huguette
Dra 17h32'30"75d59'
Druve,Bernd
Per 8h25'43"51d48'
Druzianich,Daiviel Keith
Her 17h53'46"28d38'
Drwal,Robert S
Her 15h56'24"50d29'
Dry Run Murphy
Dra 17h22'0"61d3'
Dry Together Forever, Rich & Linda
Lyr 18h40'1"27d59'
Dryden,Bud
Oph 16h2'38"-6d1'
Dryden,Charmaine Alice
And 0h50'24"34d48'
Dryden,Michael
Her 18h54'35"18d49'
Drye,Ronnie
Dra 17h19'31"61d38'
Dryer,Beth Ann
And 0h1'21"47d17'
Dryer,Janet Jean
Cma 6h38'60"-19d37'
Dryer,Kevin Ray
Aql 19h55'55"10d43'
Dryhurst,Cris & Nic
Cet 2h54'25"-0d19'
Dryhurst,Jodie
Cyg 19h34'40"34d7'
Drymer,Eric Tasker
Uma 9h26'58"68d14'
Drysdale,Donna Maria
And 2h5'57"42d14'
Dryzgula,Lore Jean
Crt 11h8'37"-14d7'
Drzewiecki,Kelsea Renee
Peg 22h58'1"28d9'
Dräger,Günther
Lib 14h30'1"-23d25'
Dräger,Heimo
Aqr 21h57'19"0d19'
Dröge,Johannes
Boo 13h59'12"13d18'
DR˜
Aur 6h54'31"38d36'
DSA
Boo 15h4'53"47d38'
Dschida,Hans
Oph 16h55'27"0d54'
DT 67
Her 17h12'18"48d21'
DTG Brother I, `68
Umi 17h24'35"76d3'
Du Bois Marriage Star, Max & Simone
Lyr 18h32'49"38d47'
du Mesnil,Christian
Uma 11h7'1"50d8'
du Mée,Rob
Cnv 12h35'55"39d15'
du Peloux,Cyrille
Sgr 19h26'12"-32d37'
du Preez,Lynda Raye
Mon 7h58'44"-1d13'
Du Tillet,Christian
And 23h39'45"33d32'
Du Verdier,Franáois
And 23h37'19"33d17'
Du Vieux Loup Gris (RD)
Umi 15h21'28"65d56'
Du-das Licht unseres Lebens!
Cas 0h7'40"60d19'
Du-die Kraft unserer Mitte
Cas 0h7'18"60d4'
Duae Feles Felix Inter Stellas
Cam 3h24'18"60d4'AB
Duane & Chazz
Lyn 8h55'43"34d42'
Duane & Mary
Cyg 21h31'47"40d42'
Duane & Mary
Cas 2h9'0"61d25'
Duane Lewis
Crb 16h19'33"37d11'
Duane Louise
Aql 19h5'1"1d53'
Duane's Nugget
Per 2h7'46"58d48'
Duane,Jr,Thomas W
Cyg 19h55'17"48d46'
Duanlyn
Cyg 21h41'21"28d25'
Duardo,Marc
Uma 8h35'1"50d46'
Duarte
Uma 11h2'19"61d42'
Duarte,Elizabeth Ann "Little Bit"
Peg 22h40'47"35d12'
Duarte,Faith Margaret "Jellybean"
Crt 11h20'49"-20d13'
Duarteus,Edmundis
Equ 21h20'43"8d53'
Duault,Alain
Dra 9h59'55"78d10'
Dub Forever Shines Bright,J
Lac 22h20'0"51d37'
Dubaich,Mitchell
Cam 9h17'44"84d23'
Dubanowitz,Paul Richard
Aur 5h5'18"44d38'
Dube,Alyssa Michelle
Cas 1h42'50"58d51'
Dube,Michelle
Crb 16h9'51"34d6'
Dubell,James David
Peg 0h1'11"18d8'
Dubernas,Jr,Johnny T
Aur 5h25'50"30d18'
Dubich,Tracy Jo
Lyn 6h57'1"58d12'
Dubick,Gabrielle Harris
Mon 8h8'42"-6d21'
Dubin Star,Hilly Neal
Sct 18h50'10"-7d50'
Dubin,Dorothy
Mon 7h46'56"-1d59'
Dubin,Louis M
Lyr 19h24'24"38d9'
Dubini,Giorgio
Com 12h28'40"20d35'
Dubini,Laura
Lyn 7h54'30"48d7'
Dublow,Georgia Dee
Vul 20h3'59"25d23'
Dubner,Barry Hart
Sgr 19h2'59"-27d25'
Dubner,Diana
Sex 10h9'16"-2d25'
Duboff,Diane
Cas 23h30'34"61d1'
Dubofsky Family Star
Uma 11h43'19"37d58'
Dubois,Beatrice
Sgr 19h27'18"-38d6'
Dubois,Douglas
Her 16h45'42"50d30'
DuBois,Elizabeth Rose
Lyr 18h58'40"46d9'
DuBois,Evan George
Oph 18h7'14"10d4'
Dubois,Jean-Franáois
Umi 14h50'25"69d34'
DuBois,John Peirre
Her 17h38'56"25d28'
Dubois,Joël
Her 16h18'12"50d10'
Dubois,Michelle
Eri 2h49'58"-3d38'
DuBois,Paul
Lmi 10h32'59"38d25'
Dubois,Paul Robert Oscar
Uma 9h51'58"51d55'
Dubois,Raymond G
Cyg 20h58'23"31d45'
DuBois,Ronald W M
Per 4h27'34"51d21'
Dubois,Sarah Michelle
Cyg 19h28'11"34d14'
Dubois,Terrell "Terry" Ray
Ori 4h43'0"7d42'
Duboise,Robert Scott
Aqr 22h6'56"0d5'
Dubord,Raymond
Cep 21h55'0"86d6'
Dubovi,Abigail S-Y
Lac 22h26'60"50d35'
Dubrul,D
Cam 6h13'57"61d16'
Dubrulle,Laurent
Her 18h11'54"31d17'
Dubs,Shawn
Boo 15h46'1"48d43'
Dubuc,Sébastien
Her 17h0'12"40d26'
Dubucki,Dorothy
Lyn 7h32'23"52d24'
Dubus,Zoé Gomez
Boo 15h6'48"7d31'
Duby
Ori 6h6'38"1d9'
Dubé,Barbara B
Peg 21h44'49"27d48'
Dubé,Richard
Her 17h22'12"20d45'
Dubé,Sylvain
Her 18h5'51"49d42'B
Dubé,Yves
Cyg 20h48'31"30d43'A
Duca,Massimo
Cam 3h49'33"58d32'
Duca,Mike
Her 16h44'46"37d57'
Ducaj "The 10 Yr Miracle",Beth & Dan
Cyg 19h33'42"38d20'
DuCarme,Donna
Cyg 19h59'17"38d59'
Ducat "Wedding Star", Stacey & Shawn
Cyg 20h3'36"31d6'
Ducato,Joshua Caylan
Ori 4h45'14"4d4'
Duccia-Zilio Grandi
Umi 16h1'0"83d29'
Duce,Louise
Cas 0h33'50"74d9'
Ducey,Cecille I
And 1h35'15"38d6'
Duchano,Richard Louis
Cas 1h16'14"61d46'
DuCharme,Jeff C
Her 16h24'41"30d37'
Ducharme,Monique
Cmi 7h26'11"8d40'
DuCharme,Paul & Holli
Lyn 7h52'33"58d17'
Ducharme,Yan
Per 3h9'27"40d4'
Duchess Alorene
Cas 1h57'0"58d53'
Duchett,Teresa
Umi 15h35'1"70d23'
Duchovny,The X Files, David
Hya 9h32'48"0d23'
Duck,Genevieve Mary
Lyn 8h52'52"42d50'
Duck,Kim
Eri 3h46'32"-6d45'
Duck,Walter Harold
Lyn 8h55'0"42d22'
Duckers,Alexandra Caroline
Cas 0h26'54"50d27'
Duckers,Lord Bret Neil Gavin
Ori 6h7'14"2d1'
Duckett,Den
Eri 3h17'18"-4d12'
Duckett,Marie
Com 13h2'51"26d50'
Duckham,John William
Cep 20h36'52"76d27'
Duckor's 70th Birthday,Lew
Aql 19h49'54"12d57'
Duckstein,Daniel
Cet 2h7'1"6d21'
Duckwall,Bill & Peggy
Eri 2h50'22"-17d33'
Duckworth,Jack Henry
Hya 9h32'33"5d21'
Duckworth,Shasta Morris
Eri 3h36'1"-16d16'
Duclos II,George Allen
Aql 19h52'1"11d6'
Duclos,Denis
Per 3h27'54"50d33'
Ducote,Stephen Michael
Equ 21h21'23"11d18'
Duda,Chrissy
Cas 23h14'47"61d3'
Duda,Margaret
Sge 20h16'28"17d10'
Dudajek,Valerie Ann
And 23h1'1"40d56'
Dudaklyan,Steve & Jody Ober
Cet 2h44'24"7d29'
Dudas,Gene & Maryann
Del 20h13'22"10d42'
Dudas,James & Judy
Ser 16h3'0"10d17'
Dudas,Joseph & Ceil
Cam 4h53'20"61d38'
Duddy,Erin Rose
And 23h16'0"48d28'
Duddy,Peggy
Peg 23h47'46"15d24'
Duddy,Virgil Robert
Per 2h57'39"45d55'
Dudeck S S,Mark
Aur 6h56'0"38d23'
Dudek,Daniel Robert
Cep 21h31'27"61d43'
Dudek,Michael Leonard
Leo 10h1'1"14d19'
Dudek,Phillip J
Cyg 21h7'32"30d14'
Dudenhoefer,Jr,Robert Edward
Sex 9h40'31"-8d57'
Duder,Anna
Uma 13h48'30"50d48'
Duderstadt,Iliana
Lib 15h2'39"-1d30'
Dudesek,Jean Ena
And 2h24'1"38d3'
Dudesek,Peter
Dra 16h40'56"62d7'
Dudette
Cyg 20h36'52"38d55'
Dudkaytis B
Cas 0h32'50"61d12'
Dudley
Mon 7h28'44"-1d1'A
Dudley III,Robert Whittier
Vul 19h48'38"22d30'
Dudley's Orb
Vul 19h13'17"21d50'
Dudley's Star Bright
Dra 17h6'0"65d42'
Dudley,Brooke A
Tri 2h9'23"33d7'
Dudley,Carmen G
Hya 9h9'31"3d2'
Dudley,Derek & Dorothy
Cyg 20h25'0"38d10'
Dudley,Hillary Elizabeth
Cam 4h57'19"68d49'
Dudley,Minnie Florence
Lyr 18h37'0"29d31'
Dudrewilz,Sarah "Amore Mio"
Cap 20h34'19"-10d15'
Dudrick,Theresa
Eri 4h12'53"-14d8'
Dudycha,Anne
Lyn 9h15'0"36d24'
Dudziak's Star,Ed
Cep 21h9'57"55d14'
Due
Ind 20h59'55"-55d5'
Due,Felicita
Gem 7h29'55"24d55'
Due,Leslie Guynn
Aql 20h30'42"0d41'
Duel III,Arthur B - Kathleen C Mead
Uma 10h46'0"50d3'
Duell,Dave & Bonnie
Lyr 19h20'33"35d20'
Duell,George
Lyn 7h3'16"50d56'
Duell,Matthew Scott
Uma 11h19'25"40d12'
Duemmlein,Nikki M
Peg 22h2'19"33d34'
Duenk,Hendrik Jan
Her 18h12'21"47d14'
Duenk-Minsky,Elly
Cam 14h0'18"81d6'
Duensing's Star Search,Walt
Aur 5h19'21"41d39'
Duer,David Lee
Ser 16h0'1"10d52'
Duerkes,Amanda Marie
Lyn 8h51'35"42d17'
Duerr-Wonderful Tonight,Lisa
Cas 3h4'0"57d46'
Dues,Diane
Leo 9h54'19"8d23'
Duesing,Nicholas John
Dra 18h43'28"70d28'
Duesing,Paula
Cyg 20h27'1"40d28'
Duesing,Rikki Lee
Cam 9h17'33"73d40'
Duesler,Lady Di, Diane
Mon 7h56'48"-0d21'
Duessel,Lawrence
Per 3h3'41"41d1'
Duet
Uma 9h49'0"71d33'
Duettra,Sydney Walsh
Peg 22h3'44"21d16'
Duex,Wulf
Cnc 8h56'51"29d16'
Dufault,Lucie
Umi 16h54'59"79d2'
Dufault,Paulette
Aur 4h53'39"50d9'
Dufeau,Eric
Ori 5h57'12"21d12'
Dufek,Laurie
Lyn 8h3'20"38d35'
Duff 1
Crb 15h17'16"31d42'
Duff,Alan
Ori 6h1'31"3d14'
Duff,Alfred
Cep 20h41'49"70d47'
Duff,Ann
Cas 0h53'0"66d45'
Duff,David
Boo 14h57'52"31d9'
Duff,Dibby
Cyg 19h30'11"33d0'
Duff,Earl
Oph 17h5'42"7d60'
Duff,Eric Lamar
Sex 10h0'0"-6d25'
Duff,Erika Laquinta
Umi 15h52'1"81d3'
Duff,John C
Lmi 10h37'42"25d4'
Duff,Joseph
Aur 5h44'50"50d13'
Duff,Virginia Stuart
And 2h33'33"50d26'
Duff,William Henry
Cmi 7h34'45"0d46'
Duffany,Betty
Peg 21h46'57"33d6'
Duffany,Clarence A
Per 4h22'51"50d36'
Dufferacula
Per 2h27'23"56d44'
Duffey III,John B
Vul 19h44'22"28d40'
Duffey,Ben & Suzanne
Cyg 21h6'48"39d35'
Duffey,Brenden William
Cet 2h38'44"4d12'A
Duffey,Caitlin Elizabeth
Cas 0h13'47"46d56'
Duffey,Erin Marie
Cet 2h49'4"5d24'
Duffey,James W
Cet 2h29'0"6d43'
Duffey,Samantha Grace
Cet 2h38'44"4d12'B
Duffey,Sean Johnston
Cet 2h38'44"4d12'C
Duffin,James R
Aql 18h56'57"10d29'
Duffing,John Matt
Sex 9h50'46"3d31'
Duffius Maximus Stellaris Stuppendous
Aur 7h12'31"39d48'
Duffner,Aaron
Aur 7h18'48"39d30'
Duffus,Benson Beanie Bam Boozle
Cmi 7h36'30"10d10'
Duffus,Louise
And 0h2'44"47d25'
Duffus,Ryan Jordan
Ori 6h6'34"3d22'
Duffy
Per 2h54'53"34d23'
Duffy
Lac 22h23'55"37d30'
Duffy's Star
Cam 5h36'26"60d29'
Duffy,Alexandra Marie
Lyn 8h25'1"58d56'
Duffy,Barbara & Thomas
Eri 4h31'36"-7d50'
Duffy,Chandler Winslow
Ori 6h6'13"-0d0'
Duffy,Charles A
Boo 13h48'42"15d32'
Duffy,Donna S
Eri 4h6'53"-18d21'
Duffy,Erin Lynn
Lyn 7h2'35"48d60'
Duffy,Francis Aloysius
And 0h36'39"40d38'
Duffy,Frank
Her 16h42'19"39d55'
Duffy,George Lawrence Woodford
Tri 2h0'21"33d48'
Duffy,Grandma Mary Bridgit Cooney
Dra 19h41'42"70d24'
Duffy,IHTBY Kevin
Lyr 18h58'30"30d57'
Duffy,Jack & Elaine
Vul 19h48'0"28d9'
Duffy,Jennifer
And 23h47'45"41d43'
Duffy,Jill C
Dra 19h39'29"68d55'
Duffy,Kevin T
Her 16h0'1"21d23'
Duffy,Mark F
Lac 22h19'22"51d50'
Duffy,Mary Elizabeth
Hya 8h9'10"1d9'
Duffy,Michael J
Cep 20h39'37"55d2'
Duffy,Pat
Del 20h37'38"4d37'
Duffy,Rachel Elizabeth
Cyg 19h57'29"38d45'
Duffy,Regina
Cas 1h14'58"61d35'
Duffy,Shawn Vincent Patrick
Oph 17h53'52"12d3'
Duffy,Susan
Ari 2h1'32"21d38'
Duffy,Timothy
Boo 14h38'47"15d3'
Duffy,Tommy
Vul 20h44'15"20d27'
Dufilno,Karen
Tau 4h1'0"12d10'
Duflos,Baptiste Pierre
Cam 3h21'23"61d42'
Duford,Joshua
Boo 14h43'30"26d28'
Dufort,James Robert
Aur 5h37'57"37d46'
Dufour,Barbara
Cyg 20h53'56"50d21'
Dufour,Gilles
Boo 15h19'59"50d18'
Dufour,Kathleen
Cas 0h41'24"61d58'
Dufour,Marcel
Lyn 9h38'38"40d50'
Dufour,Paul-Henri
Cep 21h21'51"68d18'
Dufour,Pierre
And 0h28'23"41d4'
Dufoyel,Paul
Uma 9h42'30"42d23'
Dufraine,Melissa
Vul 20h57'41"28d59'
Dufraine-Bierbach, Ruth A
Her 17h8'39"47d36'
Dufrene,Antonin
Sex 10h12'5"-2d45'
Dufrene,Sylvie
Cyg 19h42'54"30d9'
Dufresne,Claude
Cam 8h2'22"77d33'
DuFresne,David W
Cep 21h22'1"60d38'
Dufresne,Michel
Ori 6h0'19"6d18'
Dufresne,PH
Ind 20h26'45"-57d47'
Dufresne,Theophile
Sgr 19h34'26"-34d50'
Dufus
Aql 18h59'33"-8d17'
Duga,Brady
Aql 18h43'40"10d51'
Dugal,Christian
Uma 11h34'58"31d46'
Dugal,Kevin & Jennifer
Cyg 19h36'43"38d49'
Dugally "Grandpa",Taft Ali
Boo 14h14'25"32d24'
Dugan,Bethany Ann
And 2h22'24"42d17'
Dugan,Cynthia Jane
Uma 11h16'38"68d26'
Dugan,Daniel Patrick
Gem 6h47'43"12d23'
Dugan,Emma A
Tau 4h1'50"30d55'
Dugan,Ferdinand & Patricia
Lmi 10h4'27"34d30'
Dugan,Ferdinand & Bobbie
Ori 5h50'50"16d3'
Dugan,Jr,Leonard M
Hya 8h57'54"2d18'
Dugan,K G Dug Kelly
Aql 19h44'15"11d28'
Dugan,Michael Joseph
Tri 1h59'49"27d4'
Dugan,Noah Luke
Leo 9h57'0"10d22'
Dugan,Patrick & Anne
Umi 15h21'13"70d37'
Dugan,Samuel Louis
Cet 2h19'0"7d59'

Dugandzic,Milenko
Cet 3h6'57"0d57'
Dugandzic-We Love You, Suzana Anna
Sco 16h58'16"-40d19'
Dugas,Eric
And 0h28'0"40d33'
Duggan GNSH,PhD, Sr Mary Kathleen
Cas 0h8'32"64d15'
Duggan,Chip
Her 17h16'11"18d53'
Duggan,Christopher V "Buzzy"
Cep 20h40'47"70d2'
Duggan,Damian Matthew
Her 16h56'23"32d39'
Duggan,Jared Martin
Her 18h13'13"41d0'
Duggan,Joan H
Cas 23h39'0"50d31'
Duggan,Tara & Brian
Sct 18h54'32"-5d18'
Dugger's Golden Anniversary Star
Sge 20h1'57"18d31'
Dugger,Deji
Umi 15h29'47"78d40'
Duggin,Fred
Ori 5h53'16"14d43'
Duggleby,Daniel
Per 3h40'12"38d60'
Dugmar
Lyn 8h22'19"48d26'
Dugni,Franco
Lyn 7h54'14"40d42'
Dugo,Sarah & Mark
Cyg 19h28'50"30d32'
Dugrich Alpha
Uma 8h44'10"71d48'
Dugré,Yvonne
Dra 17h27'45"50d54'B
DuGuay,Caleb Cristopher
Dra 20h38'20"68d28'
Duguay,Denise
Ori 5h46'19"20d55'
Duguay,Nicolas
Per 3h8'30"41d29'
Duguay,Simone
And 23h43'11"32d30'
DuHadaway,Collette M
Cas 1h6'21"60d18'
Duhon,Yvonne"Bonnie" B
Lmi 9h46'29"33d23'
Duhring,Cathy Alicia
Mon 7h1'45"-0d24'
Duignan,Buffy
Cmi 7h55'1"8d49'
Duilio
Uma 8h52'54"55d47'
Duin Star,The Will
Hya 10h12'33"-16d15'
Duirk,Ken
Lac 22h11'58"37d37'
Dujardin,Philippe
Oph 17h58'1"10d29'
Dukas,Geordie
Cam 5h54'7"73d30'
Duke
Peg 0h4'0"11d5'
Duke
Uma 7h14'25"-13d26'
Duke
Aur 4h50'0"48d57'
Duke
Her 16h45'1"21d5'
Duke & Paulette
Dra 17h44'41"67d57'
Duke 12
Boo 15h4'47"21d20'
Duke Of Fairfield
Tri 1h42'22"30d36'
Duke,Bobbie Ann
Ori 6h6'43"7d3'
Duke,Bradley
Ori 6h16'29"1d40'
Duke,Charles Douglas
Oph 18h17'36"7d49'
Duke,Christine
Cas 0h15'23"61d28'
Duke,Eyvonne Reneé
Vul 19h48'27"27d38'
Duke,Gail
Cas 0h33'55"67d53'
Duke,Jennifer
Aql 19h8'51"5d42'
Duke,Joanna Domenica
Peg 22h46'35"20d7'
Duke,Jr,Albert G
Uma 10h24'39"40d16'
Duke,Leonard L
Her 16h9'45"20d26'
Duke,Terry & Bobbie
Aql 19h45'0"11d49'
Duke,The
Boo 14h2'1"26d38'
Dukes,Devyn
Sgr 19h40'38"-30d59'
Dukes,James Rylie
Uma 11h36'23"46d39'
Dukes,Rick
Cyg 19h39'49"31d16'
Dula,Carlyn Leigh
Cyg 19h25'34"32d34'
Dula,Shaun Edward
Aur 6h7'17"35d38'
Dulaney,Barbra Burch
Uma 10h26'56"67d39'
Dulaney,Phillip B
Cep 21h58'46"55d54'
Dulaney,Tom,Sandy & Tommy
Aur 6h31'54"38d56'
Dulanto,Miguel "B"
Per 3h14'54"56d17'
Dulawa MD,Dr Leopold B
Oph 16h58'29"-28d54'
Dulberg,Beatrice Saltman
Uma 11h22'0"38d55'
Dulborough,Gwendoline
Cyg 21h43'15"38d12'
Dulcenaya,Diana
Cas 0h32'10"67d19'
Dulcinée
Umi 13h41'15"71d29'
Dulemba III,Edward J
Uma 10h52'10"52d9'
Dulgerian,Chuck
Aqr 22h0'30"0d2'
Dulian,Charles Donald
Sco 17h52'1"-40d39'
Dull,Donald Ramon
Hya 8h29'56"0d9'
Dullagham,Amy Louise
Lyn 8h3'1"44d60'
Dulle,Valerie
Cas 0h43'25"66d28'
Dullnig,Kurt & Jackie
Crb 16h11'1"30d49'
Dullum,Carol J
And 23h13'41"40d4'
Dully,Lois
Crt 11h8'47"-18d49'
Dulmage,Mary Jo-Ann
Lyn 7h3'0"53d52'
DuLong,Marianne
Cas 1h56'41"73d39'
Dultz,Helen Elizabeth
Eri 3h34'59"-15d39'
Duluc,Pierre
Mon 5h54'1"-4d54'
Dum-Dum & Buffy
Cnv 13h14'19"32d18'
Dumais,Joseph
Ori 6h2'20"6d9'
Dumais,Louise
Per 2h57'18"52d9'B
Dumas,Ghislaine
Cyg 19h24'0"33d34'
Dumbauld,James E
Her 18h8'15"38d31'
Dumbleton,Andrew
Ori 6h4'1"18d50'
Dumbo Tolle
Boo 14h31'17"41d44'
Dumbo's Surprise
Cyg 20h22'15"39d5'
Dumbrell,Lyn
Uma 14h14'47"58d16'
Dumesnil,Christian
Aur 4h52'1"41d12'
Dumire,Jeffrey Steven
Uma 12h5'37"55d44'A
Dumler,Paul
Per 3h0'40"41d19'
Dumler,Stephanie Kliewer
Uma 11h52'17"31d1'
Dumois,Ivette
Del 20h24'47"10d32'
Dumoitier,Daniel
Per 3h37'34"51d35'
Dumoitier,Franáois
Aur 6h5'41"45d36'
Dumond,Claire
Uma 8h50'36"57d59'
Dumont,Adam & Jennifer
Del 20h59'1"18d47'
Dumont,Guy
Ori 5h57'10"21d28'
Dumont,Mark
Mon 7h52'44"-2d29'
Dumont,Martin Robert
Aur 5h24'36"37d42'
Dumont-Bengston,Susan
Equ 21h0'23"10d10'
Dumouchel,Kimberley A
Cas 0h37'43"58d22'
Dumouchel,Olivier
Cyg 19h28'29"31d5'
Dumouchel,Véronique
And 0h43'24"44d35'B
Dumoulin,Don
Hya 8h58'0"1d40'
Dumoulin,Linda
Cas 0h53'54"64d36'
Dumper
Ori 5h55'40"6d26'
Dunagan,Dan
Umi 15h2'19"66d42'
Dunagin,Ryan Edward
Her 17h38'15"27d14'
Dunakin,Jeffrey Thomas
Tri 1h55'0"25d41'
Dunavin,Ephraim
Aqr 21h3'48"0d41'
Dunaway,Cara Elizabeth
Peg 22h37'0"27d46'
Dunaway,Stacy Ann
Eri 3h21'59"-12d26'
Dunaway,Tyler Steele
Cam 3h55'24"74d34'
Dunbar Star
Cam 8h0'24"70d16'
Dunbar,David Craig
Cep 1h28'48"77d30'
Dunbar,Dr Edward W
Aur 6h21'54"35d26'
Dunbar,James Paul
Per 4h33'46"52d10'
Dunbar,John L
Ori 5h37'22"15d17'
Dunbar,Jr,Warren
Dra 12h9'32"67d53'
Dunbar,Laura Adabelle
And 23h48'27"45d46'
Dunbar,Sara Marie
Cam 3h52'46"61d15'
Dunbier,Scott Rolf
Sct 18h38'49"-6d21'
Duncan
Aql 20h2'22"0d40'
Duncan
Cam 3h26'0"61d36'
Duncan "DareStar"
Uma 8h12'0"72d25'
Duncan "T.T.",H Robert
Lyr 19h23'59"37d31'
Duncan Gregory McGregor I Love You!
Uma 11h56'27"50d26'
Duncan II,Gene & Marissa Bailey
Ori 5h55'52"6d51'
Duncan III
Ori 5h53'46"7d59'
Duncan III,Samuel Patrick
Her 16h45'51"21d51'
Duncan's Duncstar, Terence
Aur 7h18'39"37d9'
Duncan's Star
Cyg 21h50'29"37d50'
Duncan,Allan
Lac 21h58'13"37d60'
Duncan,Ann Stewart
Leo 9h33'40"8d39'
Duncan,Antione R
Hya 8h13'50"1d12'
Duncan,Bettina
Peg 23h32'0"18d43'
Duncan,Bob & Maggie
Cyg 19h25'27"30d52'
Duncan,Bobby J
Ser 18h4'0"-5d57'
Duncan,Carl
Cyg 21h14'35"28d38'
Duncan,Carl Michael
Psc 1h43'1"27d39'
Duncan,Casey W E
Cet 1h24'10"1d6'
Duncan,David
Eri 3h23'48"-13d49'
Duncan,Elizabeth Worthington
Leo 10h38'16"18d48'
Duncan,Estrella John
Uma 11h49'53"55d5'
Duncan,Gary
Cyg 19h33'38"28d31'
Duncan,Harry & Janet
Uma 14h12'47"59d15'
Duncan,James M
Ori 4h52'25"0d55'
Duncan,Jan Lauren
Peg 22h41'33"35d29'
Duncan,John & Trudy
Oph 17h3'39"-13d52'AB
Duncan,Jonathan David
Boo 15h25'37"50d43'
Duncan,K W
Tau 5h55'56"28d38'
Duncan,Karen
Cyg 21h50'27"41d54'
Duncan,Kari
Psc 23h5'38"1d31'
Duncan,Lisa Ann Glair
Cas 1h57'1"68d3'
Duncan,Lisa Nevins
Cas 0h29'23"60d40'
Duncan,LLoyd
Her 16h23'1"29d44'
Duncan,Mary Heather
And 23h24'55"43d55'
Duncan,Melynda
Mon 7h19,13"-1d13'
Duncan,Raymond,E
Cyg 21h41'11"37d35'
Duncan,Robert John
Dra 12h53'44"71d12'
Duncan,Robert K
Her 16h6'30"40d52'
Duncan,Roger
Cet 3h14'50"0d50'
Duncan,Rosalie
Oph 17h1'10"11d41'
Duncan,Rose
Uma 9h38'48"53d1'
Duncan,Sarah Katherine
Peg 22h16'53"34d23'
Duncan,Suezette
Sex 10h14'2"-6d11'
Duncan,Taylor Fraser
Dra 17h37'60"64d55'
Duncan,Taylor John
Cnv 12h44'35"40d22'
Duncan,Terry Dean
Cet 2h43'49"4d48'
Duncan,Trudy R
Eri 3h39'54"-5d41'
Duncan,Trudy R
Oph 16h42'1"2d46'
Duncan,Veronica
Cas 0h37'56"64d24'
Duncan,Walter D
Boo 14h13'12"30d29'
Duncan,William Bruce
Eri 4h8'22"-8d4'
Duncan,Yvonne Aurelia
Leo 9h56'46"8d45'
Duncan-John,Jan
And 0h26'11"30d54'
Duncan-Kennedy,William
Her 17h12'51"43d9'
Dunchak,Donna
Cnv 13h8'20"32d60'
Duncker,Manfred
Boo 13h48'56"16d35'
Dunckley,Penelope & James
Cyg 21h19'25"34d29'
Dundas,Peter Alexander
Cep 22h28'49"70d45'
Dunday,Phillip & Sheryl
Peg 21h45'14"26d39'
Dunderman,Amy Leigh
And 0h50'0"36d54'
Dundon,H Dwyer & Gloria M
Eri 4h4'32"-12d35'
Duneen,Delia
Eri 4h39'28"-1d9'
Dunetz,Adam
Her 18h19'48"18d49'
Dunford Jr My Prayer Love M A,Albert
Aql 19h52'60"-5d52'
Dunford,Molly
Cas 0h57'0"65d45'
Dungan's,Anniversary Star,Jim & Katherine
Ori 5h53'17"13d10'
Dungan,Delora Kay
Cnc 9h6'31"27d59'
Dungan,Heather
Cnv 12h31'50"40d54'
Dungan,John Anthony
Ser 15h21'51"6d10'
Dungey,Emily Marie
Del 20h49'32"7d59'
Dunham Star,Linda Marie Guerty
Peg 21h51'35"30d18'
Dunham,Clint
Sex 10h1'53"2d13'
Dunham,Deanna Seidel
Eql 21h18'54"2d34'
Dunham,Eric Jason
Tau 4h47'13"16d12'
Dunham,George Thomas
Her 15h55'16"41d22'
Dunham,Jeff
Boo 15h5'1"15d29'
Dunham,Jenifer Ethel
Mon 8h5'59"-8d25'
Dunham,L Elizabeth "Betty"
Peg 23h31'19"33d42'
Dunham,Mary Helen
Mon 6h53'58"0d14'
Dunham,Scott R
Ori 5h25'19"0d19'
Dunia,Lyda
Cyg 20h3'45"39d24'
Dunigan,Barbara Anne
Aql 20h19'40"0d42'
Dunigan,Francis J
Cet 2h32'59"-1d24'
Dunivin,Kay Amelia
Vul 20h40'26"28d56'
Dunja
Boo 14h2'1"12d27'
Dunk,Brandon Christopher
Lac 22h14'38"37d51'
Dunk,Kara
Cam 4h59'46"68d40'
Dunkel,Anja
Lyn 8h6'0"48d27'
Dunkel,Lynn
Cas 0h56'17"50d37'
Dunker,Aaron Lee
Per 2h43'39"40d39'
Dunker,Alexander Joseph
Her 17h38'40"21d19'
Dunkerley,Andrea Lee Paxton
Lyr 18h33'35"46d41'
Dunkie
Uma 11h55'54"45d23'
Dunkin,Jeannie
And 1h23'41"40d53'
Dunklau,Katherine Ann
Eri 3h38'50"-13d52'
Dunkley & Papa,Heather & Samantha
Peg 23h43'1"27d31'
Dunlap,Christopher McNeely
Sct 18h41'60"-7d18'
Dunlap,David Brian
Dra 17h55'39"58d9'
Dunlap,John N
Her 16h33'37"36d21'
Dunlap,Kathryn
And 23h33'48"40d34'
Dunlap,Lauren Terrell
Tri 2h20'1"30d15'
Dunlap,Lucille B
Cet 2h39'1"-6d37'
Dunlap,Mary Elizabeth
Mon 6h40'31"6d42'
Dunlap,Michael Frederick
Cep 21h54'59"60d51'
Dunlap,Paula
Cas 23h41'26"61d56'
Dunlap,Tania Marie
Lyr 19h24'30"42d59'
Dunlap,Tom
Dra 12h22'0"75d7'
Dunlap,William Martin
Aur 5h57'29"31d43'
Dunlap,William R
Lmi 10h46'37"32d11'
Dunlavy,David James
Cnv 13h23'0"40d32'
Dunlavy,Richa Dae (Trixie)
Cep 22h29'0"61d22'
Dunleavy,Phillip Howard
Hya 8h58'0"1d53'
Dunlevy Family
Lyr 18h46'16"30d47'
Dunlop,John
Cam 3h15'1"63d4'
Dunlop,Mary
And 23h28'23"48d59'
Dunlop,Mary Isabelle
Lyn 6h31'44"56d14'
Dunlop,Sam
Sgr 18h48'3"-27d46'
Dunlop,Shawn F
Cam 6h2'40"68d52'
Dunn
Aql 20h5'48"4d53'
Dunn III,James William
Aur 5h0'60"48d2'
Dunn Star,The Michelle
Cas 1h46'14"75d45'
Dunn's Golden Star Of Taurus,Carol
Tau 4h40'41"28d51'
Dunn's Luck
Lyn 8h16'23"40d2'
Dunn's Star,Sabrina
And 1h5'28"39d53'
Dunn(Forty Two),Glenn
Cep 20h54'14"71d12'
Dunn,Al
Cmi 7h11'38"7d10'
Dunn,Arthur Stephen
Per 1h54'0"52d51'
Dunn,Bethany Lynne
Mon 7h58'12"-7d20'
Dunn,Bradley Keith
Sgr 19h36'30"-33d33'
Dunn,Carolyn I
Mon 6h53'31"-1d26'
Dunn,Casey
Aql 18h56'49"14d47'
Dunn,Catherine Linette
Lyn 7h35'1"45d19'
Dunn,Claudia E
And 1h56'57"40d31'
Dunn,Daniel & Donna
Crb 15h28'1"31d36'
Dunn,DarLea L
Eri 2h48'1"-2d43'
Dunn,David Burnett
Aql 20h2'16"1d47'
Dunn,David John
Cep 22h21'28"61d4'
Dunn,Dylan
Aur 4h48'60"38d6'
Dunn,Elizabeth
Cas 3h30'32"71d12'
Dunn,Emily René
Cas 1h17'14"61d18'
Dunn,Gerald
Cep 21h2'1"58d33'
Dunn,Gerald
Cep 21h2'1"58d33'
Dunn,Helen & Mel
Cru 12h43'26"-58d51'
Dunn,Jackie & John
Cyg 21h1'46"33d4'
Dunn,James F
Dra 19h11'59"70d57'
Dunn,James Ryan
Aqr 22h1'55"-0d19'
Dunn,Jamie
Vel 10h12'49"-42d18'
Dunn,Jim
Aql 19h12'54"12d50'
Dunn,Joan Elizabeth
Lyn 7h39'1"50d51'
Dunn,Jr,Charles Funster
Boo 15h20'48"38d23'
Dunn,Jr,John Walston
Aql 20h8'48"4d1'
Dunn,Jr,Walter E
Dra 14h19'18"64d26'
Dunn,Jr,Wendell Earl
Aur 5h5'15"49d3'B
Dunn,Judy Kay
Eri 3h44'59"-2d22'
Dunn,Kenneth
Her 17h49'13"18d59'
Dunn,Kerrin Lynn
Cas 23h13'1"63d33'
Dunn,Laura Stephen
Cam 8h19'25"83d45'
Dunn,Lillian Daniels
Aur 5h5'15"49d3'A
Dunn,Lorraine Kringen
Peg 21h45'40"23d44'
Dunn,Marcia Bergeson
Tau 5h43'21"28d12'
Dunn,Martha
Lyn 7h19'39"50d25'
Dunn,Mary
Aql 19h10'59"10d27'
Dunn,Mary Virginia Riley
Mon 6h23'13"-10d1'
Dunn,Merideth
And 0h58'56"39d39'
Dunn,Michael Dickson
Cep 21h17'0"70d11'
Dunn,Pamela
Cas 0h29'37"61d20'
Dunn,Robert James
Lac 22h13'7"39d31'B
Dunn,Ruth Caroline Fry
Lac 22h13'7"39d31'A
Dunn,Sandy
Del 20h16'58"10d16'
Dunn,Sophie
And 0h14'1"38d44'
Dunn,Sr,William G
Dra 15h2'12"64d45'
Dunn,Stanley Thomas
Mon 8h4'17"-2d2'
Dunn,Stephen L
Cet 0h40'0"1d6'
Dunn,Taylor Jacob
Vul 19h57'34"28d35'
Dunn,The Love Of My Life Charlie
Aur 6h4'30"37d31'
Dunn,Timothy Patrick
Her 18h4'29"28d25'
Dunn,Unique & Susan Shafeek
Cyg 21h36'40"44d51'
Dunn-Heavenly Star, Star
Cet 2h56'12"7d27'
Dunn-Ludden,Rita
Boo 14h28'12"10d31'
Dunnachie
Cep 22h25'0"56d51'
Dunnan,Nancy
And 23h15'35"48d25'
Dunne,Carolyn Anne
And 0h21'1"34d26'
Dunne,Danny
Vul 20h39'55"28d14'
Dunne,Eric"Rabbit"
Cep 22h14'39"60d0'
Dunne,Francis Henry
Lac 22h02'1"48d52'
Dunne,Maria Eliza
Com 13h15'19"21d27'
Dunne,Nicole M
Uma 11h27'28"49d48'
Dunne,Phill
Oph 16h22'17"-5d41'
Dunne,Tara
Per 2h56'1"34d22'
Dunner,Peggy Z
Com 12h58'28"26d37'
Dunning,Christy & Scott
Aur 5h0'45"42d13'
Dunning,Donald Thomas
Cyg 20h53'0"40d24'
Dunning,Matthew Ryan
Del 20h18'37"20d5'
Dunning,Susan
Peg 22h1'1"31d37'
Dunning,Terence & Laura
Peg 0h9'39"20d49'
Dunphy's Dolphin
Del 20h21'26"11d39'
Dunscomb,Bobby
Her 16h40'14"22d44'
Dunseith,Kathleen Janet
Cyg 21h51'17"40d5'
Dunseith,Sarah Jayne
Cap 21h21'16"-14d39'
Dunsing's Infinity
Del 20h55'13"9d32'
Dunsmore,Gina Marie
Peg 23h22'37"33d42'
Dunstan's Star
Cam 4h15'0"60d4'
Dunstan,William V
Lac 22h38'0"38d12'
Dunston US Navy,Lt Deborah Gail
Peg 22h2'31"31d56'
Dunton,Bobby
Uma 9h12'39"55d12'
Dunton,Pauline
Cas 0h51'59"75d6'
Dunton,Shirley
Lyn 8h51'33"34d55'
Dunwoody,Kevin
Sex 10h4'43"-8d53'
Duonnolo,Carmen L
Vir 12h0'50"8d39'
DuPerow,Lisa & Mark
Cyg 19h24'38"31d59'
Duperrier,Eric
Peg 0h1'20"20d8'
Duphorne,Calvin F
Aql 19h52'15"11d19'
Duphorne,Virginia
Peg 21h19'18"22d56'
Duplantis,Jonathan
Boo 14h34'19"7d33'
Duplessie,Sarah & Michael
Cep 23h9'16"60d18'
Duplessis,Jr,Rudolph Joseph
Cet 2h52'21"1d16'
Dupont,Alain
Per 2h53'38"37d0'
DuPont,Antoine
Dra 20h11'0"71d2'
Dupont,Bernard
Cnv 13h19'26"31d56'
Dupont,Daniel
Sgr 19h33'10"-37d57'
Dupont,Daniele
Cep 22h14'1"58d17'
Dupont,Erwan
Uma 11h24'44"37d58'
Dupont,Franáois
Uma 11h26'18"38d13'
duPont,Kay
Peg 22h19'50"5d0'
Dupont,Manuel Mendez Marie Laure
Dra 10h43'59"73d37'
Dupont,Marie-Pierre
Per 4h5'41"38d57'
Dupont,Sandra
Her 17h26'40"40d51'
Dupont-Fauville, Benoist
And 23h49'57"32d26'
Duppler,Leila "Ma Dupps"
And 23h27'11"45d3'
Dupps,Kristin Lauren
Uma 10h5'42"62d30'
Duprez,Don Brian
Cet 2h25'41"3d44'
Dupuis,Billy
Dra 19h49'0"70d56'
Dupuis,Douglas Eli
Boo 14h13'1"41d55'
Dupuis,Richard
Oph 18h40'25"8d56'
Dupuis,Roger
Her 17h34'40"23d12'
Dupuis,Taylor Elizabeth
Ori 5h48'1"20d44'
Dupuy,Philip
Ori 6h8'18"3d57'
Dupuy-Damielle
And 2h32'56"44d13'
Duque,Kyle Clayton
Her 16h59'41"37d21'
Duquesnoy,Philippe
Cyg 19h42'42"38d54'
Duquette,Jr,Joseph Arthur
Oph 18h21'0"7d55'
Duquette,Juanita Lynn Wilson
Oph 18h25'41"8d46'
Duquin,Mary E
And 2h1'18"45d12'
DuRae,Liliane
Lib 14h56'14"-4d47'A
Duran
Aur 6h44'54"38d40'
Duran,Alexis Kathryn
Cas 0h35'27"62d52'
Duran,James Austin
Aur 7h9'15"38d41'
Duran,Jason
Sex 10h26'45"-5d5'
Duran,John K
Boo 14h4'1"28d6'
Duran,Marisa Anne
Vul 20h40'32"28d32'

Durand,CB
Uma 11h53'23"43d3'
Durand,L & A
Vul 20h15'58"25d54'
Durand,Philippe
Uma 13h6'0"60d34'
Durand,Sydney Tenero
Cma 7h1'1"-20d25'
Durand,The
Cet 3h17'58"8d24'
Durand,Véronique
Cep 22h38'59"61d50'
Durand-Perdriel,Dona
Cyg 19h51'26"44d20'
Durando
Lep 5h58'1"-11d45'
Durando,Patrick
Aur 4h55'23"41d9'
Durandy,Didier
Cap 20h16'48"-20d19'
Durant's Guardian Angel,Martin
Aur 6h27'49"31d2'
Durant,Deanielle
Lyn 6h28'53"59d14'
Durant,Jennie Mae Pate
Aql 19h6'26"-0d8'
Durante III,Dominic
Her 16h52'1"39d33'
Durante,Tara
Lyn 6h20'22"58d29'
Durantin,Jean-Cedric
Dra 10h13'48"73d35'
Duration
Sct 18h41'7"-7d33'
Durban's Dream
Per 2h2'14"47d58'
Durbano,Linda Kay
Lyr 18h59'33"27d53'
Durbin (Jim),James C
Cmi 7h23'15"7d55'
Durbin,Jessica Danielle
Vul 20h19'21"23d25'
Durbin,Philip
Dra 16h48'49"67d52'
Durbin,Piper Renae
Lyn 7h55'0"43d29'
Durden,Jane Marie
Sgr 19h9'23"-23d57'
Durdic,Stéfano
Per 3h38'37"38d12'
Duren,Riley
Ori 5h50'34"16d24'
Durette,Chantal
Aqr 23h47'41"-4d20'
Durette,Gilles
Lmi 10h10'46"38d1'
Durf
Dra 10h42'29"81d14'
Durfee,Linda Ann
Cam 5h4'24"68d17'
Durfey,John Thomas
Ori 5h25'39"-4d35'
Durflinger,Thomas J
Aur 6h42'50"37d5'
Durgin,Allen
Her 16h43'45"10d11'
Durham,Arden
Ori 6h3'13"7d49'
Durham,DMD,Stephen C
Dra 16h31'11"67d34'
Durham,Dylan
Cet 3h16'25"3d26'
Durham,Elaine
Uma 10h39'37"47d24'
Durham,Elizabeth Mary Joyce
Peg 23h45'10"11d51'
Durham,F Montie
Sex 10h41'58"-1d53'
Durham,Jeff & Jennifer
Lyn 7h53'22"58d9'
Durham,Kathleen Ann
And 0h48'57"38d12'
Durham,Kenneth
Her 15h58'55"44d48'
Durham,Lynn Payne
Lyn 7h27'0"44d15'
Durham,Michael Jennifer & Digger
Hya 8h13'17"5d21'
Durham,Star Of David
Her 17h24'23"49d31'
Durham,Stephen Andrew Noel
Aur 6h11'25"45d48'
Durham-Floyde, Elizabeth & Andrew
Crb 15h50'43"38d34'
Duringer,Marcus Edward
Dra 19h24'35"56d55'
Durity,Edward Phillip
Psc 23h38'40"1d34'
Durity,Steven Daniel
Uma 11h44'18"58d36'
Durivage,Dan & Gayle
Cyg 19h24'49"31d14'
Durkin,Chris & Joan
Cyg 19h59'19"40d20'
Durkin,Gregory S
Dra 19h21'39"68d33'
Durkin,Jennifer
Cam 12h17'19"76d33'
Durkin,Kelly Marie
Cas 2h4'30"71d6'
Durkin,Patrick
Hya 9h36'54"-5d17'
Durmann,Brenda-Marie
Uma 11h53'59"40d53'
Durnberger,Lee Thomas
Tau 4h27'23"30d4'
Durney,Lucas Nathaniel
Cnv 13h50'12"30d22'
Durocher,Richard
Umi 15h11'21"68d8'
Duros,Alexander
Cep 22h34'15"63d54'
Duros,Jack
Cep 22h34'47"60d13'
Duros,Nicholas
Cep 22h34'16"61d38'
Durozard,Sylvain
Lyn 7h52'14"51d5'
Durrant,Michael
Lyn 8h5'1"44d45'
Durrett,Catherine
Cam 12h1'46"77d59'
Durrett,Sean Carole
Aur 5h51'50"38d22'
Durrieu
Her 17h8'42"40d18'
Durry,Margaret Patrice
And 23h24'54"44d31'
Dursin,Lindsay H
Cmi 7h21'37"9d18'
Durso
Lyn 8h45'28"37d59'
Durso,Steve
Eri 4h1'41"-19d31'
Durst,Douglas Langworthy
Oph 16h54'23"10d8'
Durst,Greg
Aur 6h4'48"37d38'
Durst,Joseph
Aqr 22h52'35"-5d15'A
Durst,Madison Catherine
And 23h20'21"48d6'
Durst,Peter D
Aqr 22h2'30"-8d40'
Durst,Rose F
Aqr 22h52'35"-5d15'B
DuRussell,Ellen
Cas 1h44'28"60d40'
Durward Erwin
Her 16h39'28"5d23'
Durward,Cynthia Sawyer
Aql 20h10'43"1d30'
Duryea,Donald Bruce
Dra 11h55'48"71d28'
Duryee,Cadre C
Peg 23h27'54"32d13'
Duryee,Carin Candace
And 23h13'16"40d53'
Duryee,Patricia Gayle
Cyg 21h21'0"31d30'
Dusek Star, GeCaBuKaChiTim
Cet 0h30'60"-2d23'
Dusek,Helen Wedel
Mon 7h20'55"-6d50'
Dusek,Jr,Joseph Daniel
Cep 21h22'17"68d49'
Dusenberry,Derek
Lyr 18h30'52"38d0'
Dushek,Dennis Wayne
Ori 4h42'27"0d7'
Dusigne,Marine
Pho 0h40'59"-43d25'
Duskin,Samuel "Dusty"
Aur 5h11'1"40d54'
Dusling,Robert J
Cnv 13h52'49"40d46'
Duss,Walti
Per 2h50'41"31d8'
Dussauge,Jessica
Lyr 19h20'30"41d58'
Dussault,Andrew
Uma 10h48'16"61d28'
Dussault,Chloe Anne
Uma 10h52'33"67d59'
Dussault,DJ
Uma 10h50'47"62d11'
Dussault,Haylee Alexandria
Cas 0h15'0"60d26'
Dussault,Michael
Uma 10h50'45"68d46'
Dussel,Andrew Thomas
Per 3h10'14"46d49'
Dussel,Samantha Concetta
Sco 16h55'22"-40d56'
Dusseldorp,Denise Van
Cet 1h59'53"0d43'
Dussich,Gene
Lac 22h24'0"54d37'
Dussouchet,Georges
Dra 15h45'40"63d15'
Dust Bunny
Lyr 19h26'19"40d17'
Dustan,Theresa Maria
Cyg 20h52'24"30d30'
Dusterdeck,Amy Camille
Cas 22h59'1"54d46'
Dusterdeck,Victoria Beth
And 22h58'35"50d18'
Dustin & Russell
Crt 11h14'32"-8d47'
Dustin '95
Lac 22h9'28"47d5'
Dustin Michael FK
Cep 22h45'28"68d2'
Dustin Sage
Cet 3h0'33"0d20'
Dustin The Intrepid
Peg 21h7'36"13d5'
Dustina
Cam 13h23'11"81d59'
Dustman,Jessica Marie
Lyn 7h28'28"41d22'
Dustour,Pierre
Uma 8h54'1"58d4'
Dusty
Cam 7h6'0"71d10'
Dusty
Uma 14h3'37"61d27'
Dusty Cliff
Cap 21h2'53"-15d49'
Dutch
Cet 3h7'40"2d22'
Dutch & Katy
Cam 4h53'0"60d59'
Dutchy
Lmi 9h53'22"33d15'
Dutescu,Michael P
Lac 22h13'0"46d28'
Duthie,Nicole
Mon 6h41'19"2d59'
Duthion,Jean Pierre
Cyg 20h42'43"45d55'
Dutra,Bill T
Ori 5h54'25"12d54'
Dutra,Bill T
Hya 8h12'1"0d58'
Dutt,Sunjay
Lyn 8h52'59"44d24'
Duttlinger,Leona
Lyn 7h43'42"50d42'
Dutton Thomas
Umi 21h16'45"89d46'
Dutton,Candice Rose
Leo 10h45'23"15d4'
Dutton,Joe
Umi 15h7'34"66d51'
Dutton,John & Deanna
Cam 8h14'58"80d36'
Dutton,Justin N
Oph 18h0'58"11d19'
Dutton,Margaret Leitch
Vul 19h59'44"23d58'
Dutton,Richard S
Uma 12h0'32"31d46'
Duty,Andrea Brooke
Tau 5h55'0"28d15'
Duty,Anne VanMeter
Lyr 18h55'24"37d19'
Duva,Dalton
Uma 10h58'15"71d6'
Duval Game A Foot Memorial Star,Jim
Aur 7h9'21"41d27'
Duval,Barron
Aql 18h59'14"4d5'
Duval,James Joseph
Aur 4h59'10"48d55'
Duval,Jessika Anne
And 2h32'41"40d55'
Duval,Linda Denise
Cet 0h57'36"-5d29'
Duval,Lindsey Marie
Lyn 8h53'29"41d45'
Duval,Melodia From Grandma Duval
Lyr 18h48'35"30d45'
Duval,Pierre
And 23h50'50"32d45'
Duval,Richard E
Del 20h14'41"11d17'
Duvale,Ray
Cnv 13h23'14"32d9'
Duvall's Star,Rhett Evans
Oph 17h3'1"10d28'
Duvall,Brianne Olivia
Peg 22h29'45"20d28'
Duvall,Lane
Sco 17h29'11"-30d12'
Duvall,Patricia E
Cas 2h27'23"64d18'B
DuVivier
Ori 5h57'10"16d58'
Duxbury,Carl
Aql 19h57'40"1d35'
Duxbury,Judy
Sex 9h58'22"-4d10'
Duzmal,Walt
Aql 19h7'17"3d50'
Duzzer
Uma 11h7'18"61d43'
DVI - a stellar designer & friend
Uma 8h32'57"50d7'
Dvojack,Betty Jean
Tri 1h30'50"31d18'
Dvorak 9-27-1916, Caroline Hanson
Cas 0h15'28"56d8'
Dvorak,Ashley Lynn
Lyr 18h49'22"41d53'
Dvorak,Emily Marie
Lyr 18h36'41"40d41'
Dvorak,Hans-Joachim
Per 4h5'51"49d1'
Dvorak,John & Libby
Cyg 19h24'47"30d10'
Dvorak,Joseph Thomas
Lyn 9h5'36"34d52'
Dvorak,Stephen Edward
Dra 17h27'17"61d9'
Dvorak,William K
Cep 22h1'58"61d42'
Dvorscak,James Michael
Aur 5h16'40"46d16'
Dvorscak,Richard P
Leo 9h54'28"28d38'
Dwain-Forever Yours- Pam
Aur 7h4'37"40d36'
Dwan,James Shawn
Cet 0h26'49"-6d46'
Dwani,Abdul Rahman Al
Cam 4h39'48"68d25'
Dwar II
Per 3h0'56"31d41'
Dwayne & Maxine's "Little Cherub"
Cyg 20h54'1"38d21'
Dwelly,Brian Scott
Cam 4h27'34"65d11'
Dwight
Boo 14h27'41"32d0'B
Dwight
Cep 21h17'26"55d2'
Dwight Charles
Aur 6h37'21"38d56'
Dwight,Timothy Alexander
Boo 15h9'44"48d51'
Dwinell "Dad",Roger Raymond
Vul 20h4'1"28d31'
Dwinell Our Shining Star,Elizabeth D
Cas 0h43'43"61d46'
Dwinell USAF Star,Lt Col Clifford
Oph 16h53'35"-25d7'
Dworkin,The Broadway Star,David
Aur 4h35'0"31d47'
Dwornek,Bibbie
Com 13h14'1"22d28'
Dworschak,Daniella Margarita
Cas 2h42'1"60d34'
Dwyer III,Lawrence P
Aur 6h47'36"48d60'
Dwyer,Anthony T
Cep 21h52'13"61d45'
Dwyer,Jack & Bert
Cyg 20h5'57"30d44'
Dwyer,John
Aur 5h26'21"38d5'
Dwyer,John Patrick
Ori 5h36'57"-5d52'
Dwyer,Kathryn Marie
Cyg 19h26'19"31d37'
Dwyer,LaVerne Lindsay
And 23h3'44"40d53'
Dwyer,Michael Aaron & Kimberly Jo
Peg 22h7'41"4d8'
Dwyer,Michele Mary
Lac 22h24'24"56d36'
Dwyer,Rad
Tau 4h2'56"20d19'
Dwyer,Shannon
Cas 0h20'27"63d6'
Dwyer,Sharon V
Vir 11h42'48"5d24'
Dwyer,Susan,Marie
And 2h16'27"49d25'
Dwyerama Aurora
Lyn 7h22'44"50d45'
DXC 1995 MPK CJW KLG
Uma 9h12'32"50d39'
Dy,Julie Anne
Mon 7h58'44"-6d50'
Dyal,Caroline
And 0h20'32"37d34'
Dyal,Juliet
Eri 4h28'21"-17d36'
Dyan C, Sister of Anga E
Cyg 20h15'55"39d36'
Dyan Lynn
And 2h32'43"39d1'
Dyan,The
Cas 0h55'35"67d26'
Dyanne
Cas 0h30'50"64d39'
Dybsky,Tom & Mary
Cyg 20h44'0"45d51'
Dychton,George
Her 18h10'58"31d25'
Dyckman,Walter & Charlotte
Aur 7h4'14"43d49'
Dycus,Jeff
Boo 14h56'56"22d22'
Dye,Ashley Danyelle
And 0h44'13"40d55'
Dye,Barbara
Cyg 19h33'0"31d39'
Dye,Courtney A
Del 20h17'59"10d32'
Dye,Dee Dee
Mon 6h26'27"10d55'
Dye,Ian Leslie
Cep 23h14'10"80d21'
Dyer,Adam
Aql 20h6'36"1d24'
Dyer,Alison
Mon 7h59'50"-8d16'
Dyer,Aqueous Chimera 2029 Larry
Mon 7h3'45"0d55'
Dyer,Architect, Jeffrey Edward
Boo 14h53'39"25d5'
Dyer,Billy Gene
Ser 15h13'0"8d33'
Dyer,Charles T
Lac 22h12'26"48d20'
Dyer,Christina Toby
Del 20h35'0"10d14'
Dyer,Dave
Dra 16h5'51"67d32'
Dyer,Dr Frank E
Ori 5h50'43"18d53'
Dyer,George
Uma 11h5'18"48d41'
Dyer,Ginger B
Aqr 22h0'59"-0d12'
Dyer,James
Vir 13h26'13"-4d20'
Dyer,JoAnne Elizabeth
Peg 21h51'50"34d38'
Dyer,Johnnie Ray
Her 16h4'34"50d28'
Dyer,Julie Caroline
Cas 0h16'18"46d53'
Dyer,Lucille
Dra 11h56'1"67d16'
Dyer,Mark Thomas
Cep 22h21'28"61d4'
Dyer,Maxine Margaret
Uma 11h43'29"50d50'B
Dyer,Sharon M
Cet 3h11'31"6d16'
Dyer,Thomas
Boo 15h2'0"26d11'
Dyer,Vanita
Sct 18h38'16"-5d20'
Dyer,Victoria Claire
Peg 23h44'60"30d20'
Dyer,William Harry
Uma 11h43'29"50d50'A
Dygart,Harry
Aur 7h5'35"40d14'
Dygulski,Pat Connelly (Kochanski)
Ori 5h57'20"14d45'
Dyk,Sr,5-22-11,William J
Her 16h27'13"27d31'
Dyke,April Georgina Evelyn
Cyg 19h28'51"38d13'
Dyke,Deborah Lynn
Cet 2h59'11"8d56'
Dykeman,Jake Louis
Dra 16h20'18"66d51'
Dykhuizen,Richard Charles
Sex 10h4'20"-6d0'
Dylan
Ser 15h39'1"0d44'
Dylan Drew
Sgr 18h56'43"-23d0'
Dylan Michael
Uma 13h34'43"50d23'
Dylan Milan Billy
Boo 14h59'53"20d37'
Dylan Spencer
Hya 10h12'43"-15d33'
Dylan Star
Aur 5h58'16"29d9'
Dylan's Wish Upon A Star
Mon 7h44'32"-5d38'
Dylan,Jenna
And 23h19'47"48d57'
Dyleski,Allan
Dra 15h46'22"60d9'
Dylyn,Cara
Cas 23h41'0"61d19'
Dymesich,John & Marla
Aur 5h19'16"42d37'
Dymins,Abby
Peg 23h1'0"33d22'
Dymond,Sheila-Ashley Cory & Dean
Aql 19h55'13"12d48'
Dympna Tracey
Lyn 7h23'50"50d42'
Dympna-Tyler One
Uma 9h6'29"56d38'
Dynamic David
Dra 17h32'27"50d35'
DynaMichael
Vul 19h23'36"25d49'
Dynamite Candy & The Five C's
Peg 21h25'27"22d54'
Dynamo
Vul 19h38'47"29d17'
Dyne
Gem 6h45'51"18d15'
Dynes,Brandon Douglas Robert
Cap 21h3'20"-26d19'
Dynof,Jennifer
Lyr 18h59'11"34d34'
DyReyes,Roberto
Cep 21h57'42"55d38'
Dysan,Allan James
Ori 5h49'57"18d31'
Dysart,Melissa
Com 12h26'58"23d58'
Dysarz,Cherie L
Cyg 21h32'50"38d50'
Dysinger
Cet 0h36'56"1d11'
Dyson Family
Uma 10h6'55"48d34'
Dyson,Kay
Sgr 19h4'40"-23d24'
Dyson,Matthew Thornton
Lac 22h33'1"52d50'
Dyson,Morgan Frances
Peg 23h39'10"25d50'
Dyson,Susie M
Lyr 18h48'0"39d7'
Dyson,Victoria
Cas 0h54'0"73d20'
Dzat,Jr,Raymond J
Aur 6h1'53"30d22'
Dziadosz,Marion
Cas 0h44'32"70d47'
Dziadosz,Norman
Aur 6h19'19"46d43'
Dziadzi
Peg 23h34'36"31d35'
Dziadzio,Raymond & Eric
Sge 19h28'1"16d51'
Dziak,Jake
Aur 6h1'22"32d28'
Dziamniski,Shane Patrick
Her 15h50'1"43d36'
Dziedzic,Alex B & Sally
Cas 1h43'17"68d43'
Dzierza,Katharina
Vir 13h28'18"-11d3'
Dzieszko Forever, Cecilia
Her 17h11'41"42d31'
Dziewit,Christine M
Cyg 20h4'50"40d18'
Dziewor,Ruth Maria
Tau 5h33'47"20d15'
Dzinbala,Claudia
Leo 11h17'14"-1d59'
Dzubin,Alex
Dra 17h7'33"51d26'
Dzurick,Andrew
Cam 3h48'55"56d1'
Dzwinel,Stanley & Pauline
Cyg 21h17'52"37d32'
Dédé's Star
Lyr 18h31'49"31d11'
Démo Magic
Del 20h49'59"9d54'
Döhring,Kathi
Eri 4h50'44"-6d13'
Dölle,Günter
Sgr 19h56'35"-40d56'
Dörig,Alfred
Psc 22h55'45"5d31'
Döring,Christiane
Cap 20h56'50"-20d2'
Dörr,Gerlinde und Gerhard
Dra 15h10'38"63d32'
Dörre,Jutta
Lib 15h8'43"-20d60'
Dörsch,Kerstin
Peg 23h33'31"11d53'
Düha,Uschi von
Leo 10h30'17"18d21'
Dürr,Helmut H
Dra 15h56'23"62d27'
Dürrenfeldt
Cmi 7h21'55"8d49'
Düsterdiek,Christian
Umi 14h12'56"68d31'
Düx,Kerstin & Matthias
Her 17h7'51"42d27'

E

E
Peg 23h2'15"33d51'
e
Sex 10h31'20"-6d55'
E˝
Her 16h1'28"21d2'
E & J Energie Stern
Sco 17h5'51"-31d11'
E & J Forever
Crb 15h53'18"30d24'
E B
Lyn 7h43'30"45d22'
E B 3/24/69
Mon 7h42'1"-8d11'
E C Beauty Queen
And 0h22'0"40d13'
E Clarissa
Vir 13h35'12"-8d11'
E Deyo's Irish Eyes
Leo 10h38'29"20d29'
E J & Andrea
Vul 19h48'27"28d17'
E K R
Gem 6h28'35"13d5'
E L F I
Psc 1h3'1"5d44'
E L K 96
Aql 19h6'10"15d16'
E M M
Lyn 8h45'12"41d38'
E N
Eri 2h43'54"-1d58'
E R S
Uma 8h46'0"70d21'
E T
Boo 14h15'24"53d28'
E T H 39
Del 21h3'38"12d3'
E U
Uma 11h0'39"38d60'
E Z My Star Keeper
And 23h41'12"40d58'
E-Man Kite King
Cnv 12h25'17"47d59'
E-Rock Hans #1
Uma 11h24'27"33d27'
E-VON
Peg 22h36'13"30d59'
E-Z Mike
Her 16h26'38"34d25'
EAA-Forever Yours Forever Love
Vul 19h12'46"21d14'
Eachother
Uma 11h38'1"40d47'
Eades,Alexandra Rene
Cas 0h27'17"63d45'
Eads,Angela
Mon 6h55'54"0d14'
Eads,Michelle Lynn
Vir 13h11'7"-8d18'
Eads,Nancy A
Lyr 18h38'42"36d48'
Eagan,Denise Louise
Del 20h15'0"10d48'
Eagan,Josh "Mr Baseball"
Per 3h9'41"48d54'
Eager,Fred & Joanne
Peg 22h14'56"8d14'
Eagle Spirit(A Star For Jim Long)
Aql 18h59'30"-6d31'
Eagle's Shadow
Aql 19h7'27"0d4'
Eagles,Derek
Uma 10h24'0"71d26'
Eagles,Max
Ori 5h45'47"11d33'
Eaglesham,Angela
Eri 4h5'48"-7d12'
EagleStar
Aql 19h7'14"3d52'
Eaken,Gordon Frank
Per 2h20'41"55d0'
Eakes,Lisa
Eri 2h54'23"-3d57'
Ealey,Albert & Grace
Cyg 20h2'18"38d40'
Ealon,Maude Carmen
Lyn 6h59'0"44d43'
Eamonn,Casey
Cep 23h7'37"70d13'
Eans,Jr,Fredrick William
Per 3h1'22"46d51'
Earae Infinity
Mon 7h4'40"-7d3'
Eardley,Dave
Gem 6h50'48"13d5'
Earl's Party
Cep 23h20'23"64d9'
Earl,Indiana M
Aql 20h1'40"14d17'
Earl,John G
Cep 22h11'50"62d43'
Earl,Shannon,Scott, Jenny,Tolley
Peg 23h1'59"17d38'
Earle,Barbara J
Lyr 18h43'38"32d48'
Earle,Catherine
Tau 5h27'56"20d25'
Earle,Donna
Cmi 7h56'29"1d42'

Earlene
Cyg 19h28'17"38d11'
Earles,Mary Dora
Peg 22h16'35"21d45'
Earley,Catherine J
And 0h16'45"33d52'
Earley,T Morgan
Lyn 8h18'48"38d41'
Earls II,Timothy Matthew
Cnv 12h18'42"46d50'
Earls,Jennifer Beth
Uma 11h49'14"37d50'
Earls,Michael Raymond
Ori 6h1'40"0d20'
Earls,Rodney Clark
Leo 10h51'53"2d27'
Early,Brendan
Cyg 21h1'15"37d59'
Early,Cynthia
Peg 21h39'1"23d37'
Early,Kathie
Boo 14h56'50"20d33'
Early/Easy Dog,Edward T
Lac 22h50'33"54d4'
Earlywine,Elaine Becker
Uma 11h56'1"47d3'
Earman,John Gary & Geraldine Walker
Sge 19h19'41"17d54'
Earnest,Grace Patricia
Aql 18h40'0"-2d37'
Earnest,Theresa Gamiotea
Mon 6h59'19"-1d17'
Earney,Dorothy Marie
And 0h8'21"30d38'
Earnshaw,Eric Shane
Dra 12h0'11"70d15'
Ears Paws Tik Della Pet Power
Tri 2h17'41"31d10'
Earsley
Gem 5h59'1"27d58'
Earth Shop,The
Ori 5h19'20"-6d54'
Earthscapes,Inc
Cep 23h14'54"65d50'
Eas,Claude
Per 1h53'19"48d55'
Eash,Ezra Aaron
Boo 15h14'54"38d36'
Easley,Amy Beth
Lyr 19h14'0"33d51'
Easley,Benjamin Augustus
Cep 22h35'57"80d16'
Easley,Jerry Don
Lyr 19h4'29"25d50'
Easley,Marilyn Sue
Vir 13h21'19"-4d44'
Easlick,Helen
And 2h23'36"40d51'
Eason,Michael
Dra 14h15'0"63d16'
Eason,Penny Gale
Cam 3h39'0"61d4'
Eason,Philip Andrew
Ori 6h2'1"20d14'
East Family Star,The
Eri 3h29'25"-6d43'
East Mountain Center
Lmi 10h16'14"28d50'
East Mountain Center
Cam 13h32'16"81d59'
East of John M
Dra 14h57'0"63d46'
East River
Uma 11h8'0"60d49'
East,Marguerite Ellard
Mon 7h0'35"3d45'
Easter-Utley,Isidora
Gem 6h30'2"13d5'
Easterday,Whitney
Ori 6h15'0"-0d24'
Easterlin,Mikaela Laine
Lyn 7h58'20"58d46'
Easterly
Vul 19h59'45"28d55'
Easterly,Jack
Her 16h21'49"26d32'
Easterwood,Sandra "Fuzzy"
Cas 0h0'49"60d9'
Eastin Jan 9 1756, Stephen
Del 20h55'44"9d7'
Eastin Star,The
Ori 5h55'1"10d45'
Eastland,Robert William
Crb 16h12'0"37d13'
Eastman,Anastasia Doris
And 2h12'0"39d28'
Eastman,Jill Ann
Uma 10h26'53"56d52'
Eastman,John William
Cep 22h16'35"67d47'
Eastman,Lance
Her 18h4'30"31d16'
Eastman,Ricky
Crb 15h12'1"15d24'
Eastman,Scott
Her 18h18'12"12d52'
Easton Cowboy,The
Peg 21h58'52"33d9'
Easton's Etoile Luminez,Bil
Vir 13h28'12"-1d47'
Easton,Catherine Crowninshield
Hya 8h12'30"-0d15'
Easton,Catherine
Cam 5h51'20"68d28'
Easton,Fiver
Cet 2h17'29"4d33'
Easton,Louise Rae
Cam 13h17'45"81d44'
Easton,Michael
Aqr 12h2'47"0d30'
Easton,Scotti A
Lac 22h10'41"38d51'
Eastwood Superstar, Howard William
Cep 0h6'33"73d27'
Eastwood,Anthony
Uma 10h6'0"47d53'
Eastwood,Clint
Her 16h55'21"27d1'
Eastwood,Francesca Ruth Fisher
Leo 10h38'0"20d7'
Eastwood,Pauline
Cas 0h25'20"62d36'
Easy Rider
Dra 15h2'13"62d58'
Eaton Affirmed
Mon 7h57'20"-3d0'
Eaton,Barton
Del 20h38'16"11d10'
Eaton,Bruce
Sco 16h52'35"-38d39'
Eaton,C Terence
Mon 7h57'55"-5d56'
Eaton,Chris & Dawn
Sge 19h18'48"16d16'
Eaton,Christian
Cmi 8h7'11"1d47'
Eaton,Dawn S
And 2h16'33"46d21'
Eaton,Douglas Edward
Aur 5h56'54"38d6'
Eaton,Eric
Aur 6h24'58"30d38'
Eaton,Gary
Ser 15h57'46"-1d50'
Eaton,James Patrick
Dra 15h51'35"62d37'
Eaton,Jared
Boo 15h13'17"41d29'
Eaton,Jeffrey
Her 17h4'18"20d47'
Eaton,John M
Uma 11h38'33"50d4'
Eaton,John Marshall
Uma 12h5'37"55d44'B
Eaton,Margie
Cyg 19h59'42"38d56'
Eaton,Rebecca Jean Becky
Uma 10h52'23"52d12'
Eaton,Samantha
Cas 0h23'53"70d57'
Eaton,Stacia Jeraldine
Mon 6h39'17"2d55'A
Eaton,Tamarh S
Lac 22h6'11"51d6'
Eatwell,John
Aqr 22h31'0"-1d36'
Eavenson,Alden Truitt
Eri 3h24'25"-4d2'
Eavenson,Bradley Barr
Ser 15h36'44"-2d50'
Eaves,Dennis Ray
Ori 6h0'59"6d46'
Eaves,Patricia Ann
Lyn 7h8'27"59d42'
Eaves,Taylor
Uma 9h45'10"62d8'
Eayres,Jaime M
Eri 3h57'0"-19d16'
Eayres,Kelly E
Peg 0h3'1"20d29'
EB
Aql 19h27'26"0d22'
Eb-Udab
Cmi 8h5'0"0d4'
Ebata,Burt Hiroshi
Sct 18h36'18"-6d21'
Ebba
Umi 15h17'60"67d25'
Ebba Marie
Eri 3h36'42"-5d0"
Ebdon,Claire
Ori 5h4'24"1d47'
Ebe
Cam 8h21'1"80d44'
Ebeid,Boran M
Cep 20h22'13"78d48'
Ebel,Alfred
Sco 17h51'29"-30d42'
Ebel,Charles "Grandpa"
Ori 6h7'1"-2d55'
Ebel,Peter
Aqr 23h32'37"-17d48'
Ebeling,Jolene Renee
And 23h35'20"46d25'
Ebeling,Neil F
Uma 10h20'0"48d47'
Ebell,Louis A
Cmi 7h58'52"4d34'
Ebenhofer,Günther
Cep 22h0'55"65d26'
Ebenhöch,Petra
Cet 2h59'36"5d8'
Eberdt,Meta Jessan
Uma 10h57'1"48d37'
Eberhard Urbanus
Eri 4h50'8"-5d33'
Eberhardt,Danielle Catherine Leigh
And 2h7'18"41d41'
Eberhardt,Jordan M
Lac 22h44'21"53d11'
Eberhardt,Nicholas Allen
Cap 21h23'16"-23d49'
Eberhart(12-4-1930), Ozella
Cam 3h29'43"60d30'
Eberhart,Joseph F
Lac 22h33'41"55d0'
Eberle,Denise
Lyr 18h46'51"37d6'
Eberle,Reinhardt
Ser 15h37'1"17d49'
Eberle,Wolfram-Sylvia und Stev
Peg 23h32'0"17d37'
Eberlin,Raymond & Karen
Cyg 21h8'56"40d12'
Eberly Family Tribute, EBSTAR:A C D
Mon 6h43'1"10d28'
Eberly III,William L
Dra 15h13'53"64d11'
Ebermarin
Uma 10h18'42"51d41'
Ebersold,Julian Frederick
Uma 10h7'53"50d26'
Eberspacher,Mark Lee & Kim Marie
Lac 22h40'22"51d40'
Ebert * Bad Rehburg, Friedel
Cet 2h54'49"0d29'
Ebert III,Jack A
Cam 4h15'55"67d31'
Ebert III,Jack A
Her 15h56'38"48d32'
Ebert,David & Nancy
Ori 6h0'59"6d46'
Ebert,III,Jack A
Per 2h8'14"57d3'
Ebert,Jeffrey Reuben
Lac 22h32'38"52d42'
Ebert,Manuela
Equ 21h3'22"12d1'
Ebert,Mary Lou
Cyg 20h18'23"41d56'
Ebert,Paul
And 23h15'14"45d8'
Eberwein,Cusack
And 1h42'12"41d8'
Eberwein,Kristi Leanne
Cyg 20h54'22"53d23'
Ebling,Jakob
Peg 21h10'48"12d45'
Ebner Family Star,The
Uma 13h29'15"63d20'
Eboni
Cet 1h53'38"-3d33'
Ebony
Ori 5h56'25"16d32'
Ebony Eyes
Her 17h5'22"40d32'
Ebony's Sun
Ori 5h3'0"0d42'
Eby,Gordon & Edith
Cep 21h43'38"61d2'
Eby,Susan
Ori 5h50'30"9d31'
Ecay,William Edward
Boo 15h41'1"48d32'
Eccard,Nicole
And 1h22'11"38d38'
Eccles,Jake A
Aql 19h46'49"14d7'
Eccleston,Jeffrey
Cnv 13h18'56"28d55'
Ecclestone,Roger
Her 15h59'57"40d41'
Eccoqui
Aur 6h34'18"37d33'
Echarri,Lolita et Henri Rivier
Cas 2h3'42"60d55'
Echausse,Patrick
Cep 21h44'30"61d40'
Echement,Jessica
And 1h54'43"41d0'
Echevarria,Joaquin
Lyn 8h19'1"39d43'
Echeverria,Lauren Michelle
Uma 9h49'29"52d2'
Echlinhoey,Alan
Aur 6h35'16"34d40'
Echo
Hya 9h3'0"5d7'
Echo
Uma 9h37'31"68d51'
Echo
Cam 10h54'57"80d44'
Echo's Solace
Cnv 12h34'39"41d2'
Echols,Alexes
Lyn 7h28'55"44d50'
Echols,Cheryl "Luna"
And 23h28'1"48d27'
Echols,Ethan Joseph
Her 16h37'13"35d55'
Echols,Harold Boysen
Ser 15h31'47"17d38'
EchoStar,The
Hya 8h11'28"3d37'
Echtner,Richard
Aur 4h52'21"40d51'
Eck,In Memory of Leonard H
Lyr 19h12'17"35d26'
Eck,Linda
Cas 0h8'1"65d9'
Eck,Peter George
Ser 15h9'51"-1d31'
Eck,Thomas
Ari 2h40'14"21d8'
Eckard,Sheila Ann
Mon 6h25'48"-8d48'
Eckart,Thomas
Dra 17h51'37"82d45'
Eckblad,Barbara J
And 2h34'20"48d60'
Eckel,Marjorie S
Ori 6h6'23"-0d25'
Eckelaert,Jack & Sandra
Hya 9h30'50"5d43'
Eckels,Nancy D
Del 20h13'15"13d50'
Eckenstein,John T
Dra 16h54'56"72d41'
Ecker,Aram
Her 17h54'23"14d46'
Ecker,Sr,Edward V
Cam 3h50'46"60d28'
Eckert,Bill
Cet 3h6'17"1d44'
Eckert,Christopher J
Cep 22h19'57"55d23'
Eckert,Hans
Cam 5h0'1"70d29'
Eckert,Katharine Elizabeth
Cas 3h15'13"75d21'
Eckert,Katherine
Cyg 21h53'28"53d57'
Eckert,William H
Cet 1h11'48"-0d5'
Eckhardt,Cheryl Lynn
Lib 15h2'23"-6d23'
Eckhardt,John R
Aqr 23h34'15"-18d24'
Eckhart,Edward R
Ari 1h54'0"10d16'
Eckhart,Robert
Ser 15h47'33"20d3'
Eckhaus,Janelle
And 0h12'48"46d21'
Eckhaus,Joshua & Christine
Sge 18h59'58"19d46'
Eckler's,Star Laurie
Uma 11h3'1"68d57'
Eckles,Ashley
Uma 9h59'12"62d26'
Eckman,Brian
Dra 11h48'0"71d21'
Eckmann,Günter
Sgr 19h16'2"-42d9'
Eckols,Glen Elwood
Cam 3h56'35"60d33'
Eckrote,Lindsay Marie
And 23h45'60"40d54'
Ecks,Maria
Uma 10h31'32"40d57'
Eckstein,Chuck
Lac 22h7'15"51d59'
Eckstein,Daniel
Aur 4h46'21"38d31'
Eckstein,Helene W
Com 12h55'21"27d54'
Eckstein,Michael Charles
Umi 20h20'0"88d14'
Eckstein,Michael G
Peg 21h50'0"34d34'
Eckton,Pat
Cnv 13h55'1"38d43'
Eckton,Pat
Mon 8h6'31"-6d32'
Eckton,Pat
Peg 0h1'0"30d39'
Eclipsa,Elizabeth
Cyg 21h21'27"38d27'
ECM 1950
Uma 9h57'15"43d42'
Ecobichon,Joshua
Aur 4h57'51"50d57'
Economacos,Eleni
Mon 8h6'5"-1d60'
Economakos,Jesse Michael
Hya 9h10'35"2d52'
Economos
Lyr 19h1'40"25d58'
Ecosse
Hya 9h3'1"-0d44'
Ecto-1,KITT,Slimer Ghostbusters
Pho 0h38'25"-43d21'
Ecung,Mario
Dra 16h9'21"52d20'
Ecurb Rehtaeh
Lac 22h52'51"35d32'
Ecuyer,Esther
Aql 20h9'13"8d51'
ECYRB
Cyg 19h47'18"29d53'
Ed
Boo 14h37'25"20d9'
Ed
Per 3h1'32"41d36'
Ed & Billie's Light
Cyg 20h30'49"30d21'
Ed & Eileen 30th Anniversary Star,The
Cyg 20h16'0"39d28'
Ed & Jackie
Lib 15h2'47"-1d28'
Ed & Rose
Eri 2h52'23"-6d8'
Ed Nini
Boo 14h6'39"33d57'
Ed's 50th
Aql 19h54'1"0d34'
Ed's Caber Toss
Aur 7h3'55"38d35'
Ed's Miracle
Per 1h58'1"50d9'
Ed's Place
Cma 6h59'27"-18d50'
Ed's Shining Star
Dra 16h7'28"65d47'
Ed's Star
Boo 14h46'26"50d19'
Eda
Cas 0h58'49"62d41'
Edanicol
Del 20h13'23"12d58'
Edda
And 2h20'33"39d42'
Edda
Del 20h15'0"12d1'
Eddie
Her 17h5'54"40d38'
Eddie
Cep 0h2'19"75d3'
Eddie
Umi 16h18'59"70d6'
Eddie
Boo 14h23'21"28d36'
Eddie & Dena Forever
Uma 9h20'46"58d55'
Eddie & Edwina
Ori 4h56'0"4d51'
Eddie & Jeannie
Uma 11h14'1"50d41'
Eddie & Joanna
Sge 19h17'53"20d19'
Eddie & Lisa
Aur 5h6'1"41d10'
Eddie & Lynn Forever
Crb 15h50'38"30d54'
Eddie & Mia
Cyg 21h8'11"31d15'
Eddie & Moggy Forever
Lyr 18h22'6"47d7'
Eddie & Phyllis
Cyg 21h5'47"30d56'
Eddie Joe
Ori 6h0'23"4d15'
Eddie Mair Live
Dra 16h52'37"67d41'
Eddie Our Shining Star
Aur 4h54'1"48d59'
Eddie Scott
Lac 22h47'1"56d45'
EDDIE THE K
Her 17h0'21"20d33'
Eddie's Love
Hya 9h9'33"5d23'
Eddie's Ralta
Uma 11h4'1"34d53'
Eddie's Window
Cap 20h32'53"-20d5'
Eddie's Wonder
Cep 20h57'43"58d30'
Eddie,Maryvonne Bonniot
Per 2h19'16"58d17'
Eddings,Patricia Draper
Cet 0h59'11"0d47'
Eddins,Shirley
Eri 2h57'57"-6d45'
Eddleman,Lauren Eve
Mon 6h53'1"-1d17'
Eddy
Cas 1h45'58"58d30'
Eddy Heavenly Sweetheart,Nelson
Aql 19h5'50"0d2'
Eddy,Christina
Mon 7h12'30"-8d58'B
Eddy,Jim & Betty
Cyg 20h3'52"39d36'
Eddy,John Michael
Cet 2h44'27"1d52'
Eddy,Linda
Cyg 21h12'55"28d52'
Eddy,Nola
Cyg 20h34'57"30d20'
Eddy,Patrick
Lmi 10h0'14"33d6'
Eddy,Ray
Ori 5h6'1"0d33'
Eddy,Timothy
Mon 7h12'30"-8d58'A
Ede Star,The
Cyg 19h26'24"35d44'
EDEE 25
Psc 1h24'20"32d39'
Edeen,Karl Ryan "Niko"
Dra 20h24'21"63d8'
Edel,Thomas R
Per 2h1'17"48d37'
Edelberg,Susan Allison
Cam 3h37'21"61d41'
Edelbrock,Brian Socks
Lyr 18h30'39"41d4'
Edelen,Todd
Sct 18h53'24"-6d30'
Edeling,Rolf
Oph 17h31'45"8d30'
Edell Star,Stephen
Ori 5h40'25"-0d30'
Edelman Star,The Noah
Lmi 9h49'19"37d32'
Edelman Star,The Randy
Del 20h58'30"18d57'
Edelman,Benjamin Jay
Her 16h2'19"48d15'
Edelman,Daniel Stephan
Ori 5h30'30"-0d41'
Edelman,George "BearStar"
Per 3h12'33"40d36'
Edelman,Irving
Her 17h13'0"45d20'
Edelman,Libby & Sam
Cyg 21h12'22"35d55'
Edelman,Margaret
Lyr 19h21'51"40d48'
Edelman,Randy
Aql 18h57'24"-4d12'
Edelman,Robert R
Ori 5h51'59"12d54'
Edelson,Hannah Brook
Aqr 20h36'35"-8d48'
Edelson,Jay Richard
Aur 5h5'5"29d8'
Edelstein,Michael
Ser 16h12'21"2d32'
Edelstein,Sarah Helrich
Mon 6h25'6"-0d4'
Edelstein,Stephen
And 0h13'58"39d42'
Edelston,Martin
Boo 14h40'48"30d45'
Edelstone,Steven Gregory
Ori 6h6'19"7d47'
Edeltännchen-Juanita
Boo 13h38'34"13d0'
Eden Claire
Dra 12h24'25"64d44'
Eden,Diane W
Vul 20h19'20"23d42'
Eden,Erica Ashlee
Mon 6h45'37"11d48'
Eden,Johnny F
Her 18h15'35"14d54'
Eden,Linda Ruth
Peg 23h5'46"31d40'
Eden,Michael(Mickey)
Dra 19h23'53"56d60'
Eden,Sally & Simon Banks
Cru 12h49'20"-56d26'
Eden,Wendy
Mon 7h9'50"-5d4'
Eder,Linda
Uma 11h3'13"37d47'
Eder,Linda
Mon 6h41'47"10d25'
Eder,Lori
And 1h28'33"34d5'
Eder,Lorri Ann
Crt 11h7'59"-17d10'
Eder,Walter
Umi 16h56'1"76d38'
Eder,Wayne Ludwig
Oph 16h50'54"11d42'
Eder-Best Dad,Frank S
Lac 22h25'52"52d46'
Ederer,Eva
Cap 20h33'17"-18d58'
Ederina Inez
Cas 23h17'52"60d32'
Ederlyn
Uma 12h9'55"56d43'
Edgar Tayo & Fabiola
Sct 18h41'45"-6d15'
Edgar,Edward E & Marie Lou
Mon 6h5'49"-8d11'
Edgar,Ronald Allan
Cet 1h40'20"-2d53'
Edgcombe-Brady,Cheryl L Jenkins
And 1h44'27"39d23'
Edge III,Loy L
Aur 6h12'37"33d19'
Edge Phd Univ Of Mich, Ernest R
Boo 14h53'14"27d2'
Edge,Donna-Marie Pauline Paridise
Tri 2h41'46"31d14'
Edge,Jan
Cet 1h59'11"0d7'
Edge,Samantha Lyn
Lyr 18h47'44"33d36'
Edge,William
Ori 6h0'44"-1d50'
Edgecomb,Jane Shapleigh
Cas 23h39'13"58d16'
Edgerton,Bradley
Cet 2h17'43"9d51'
Edgerton,Diane Lee "Pretty Girl"
Oph 18h24'36"7d33'
Edgerton,Elaine & George
Umi 15h6'56"71d13'
Edgett,Jr,Dr Joseph W
Lyn 9h1'49"34d45'
Edgett,Robyn
Umi 16h58'10"76d45'
Edgington
Boo 14h9'42"51d5'
Edgington,Blair
And 22h58'0"50d7'
Edgington,Brook
Cyg 20h20'44"40d27'
Edgington,Gary
Her 17h18'0"43d5'
Edgington,Laura
And 1h1'56"37d7'
Edi
Cam 6h51'46"71d2'
Edi
Per 1h44'56"53d16'
Edie
Lyn 7h49'35"38d15'
Edie M
And 2h35'48"40d22'
Edietuckbendewey
Crb 15h24'15"30d45'
Edinger
Cep 21h3'50"55d34'
Edinger,Fred B
Ser 15h51'18"24d23'
Edinger,James R
Cmi 7h55'34"8d27'
Edington Shining Star, Patrick
Cmi 7h37'5"8d45'
Edison,David Alan
Crt 10h55'12"-10d15'
Edison,Jack H
Aql 20h2'25"0d6'
Edith
Ori 5h33'34"8d20'
Edith
Cas 0h56'25"66d3'
Edith C
Uma 11h37'36"32d2'
Edith Christophe
Boo 14h26'12"48d34'
Edith und Heino
Cam 5h23'56"80d11'
Edlefson,John Franklin
Peg 22h56'25"17d49'
Edlin,Karen Lansky
Mon 6h48'44"10d5'
Edlow,Frances & Sam
Aur 5h37'32"37d59'
Edman,Eryn
Peg 23h24'55"18d30'
Edman,Jennifer
Boo 14h17'36"38d38'
Edman,Nicholas
Cep 22h59'0"70d57'
Edman,Webster T
Aur 6h0'13"30d1'
Edmond
Com 13h12'1"28d11'
Edmond,Annette
Mon 6h20'51"4d23'
Edmond,The Great Jerilynn J
Mon 8h1'46"-9d34'
Edmond,Tyler Joseph
Her 17h36'59"40d20'
Edmonds,Becky
Mon 7h30'12"-1d5'
Edmonds,Cindy
Sex 10h34'1"3d40'
Edmonds,Glyn & Freya
Peg 22h18'28"34d37'
Edmonds,Grace
Mon 7h50'56"-1d34'
Edmonds,Sweet Amy
Sge 20h17'32"20d4'
Edmonds,Theckla Elizabeth
And 1h27'41"40d19'
Edmondson III,William Franklin
Aur 5h53'56"30d8'

Edmondson,Alexandra
And 2h21'42"50d23'
Edmondson,Kassandra Vanessa Ferrero
Mon 7h31'49"-8d37'
Edmondson,Nina
Lyn 9h17'1"39d12'
Edmondson,Punkin
Oph 17h39'39"11d32'
Edmonson,Lisa
Aql 19h25'52"8d17'
Edmund Lee's Star
Cep 22h6'17"80d16'
Edmund's Light of Eternity,Julia
Per 4h2'27'38d10'
Edmund,Donna M
Cas 0h15'38"66d20'
Edmund,Megan Louise
Cas 0h25'14"62d6'
Edmunds,Frank Lloyd
Her 17h7'50"50d3'
Edmunds,Kenneth William
Cam 4h50'39"68d53'
Edmunds,Kirsty Elizabeth
And 0h37'51"40d48'
Edmunds,Robert Emlyn
Peg 21h27'60"2d17'
Edmunds,Ryan Fancisco
Per 3h19'39"50d7'
Edmunds,Tera Nicole
Ori 5h1'35"12d50'
Edmundson, Wells
Sge 19h30'40"16d43'
Edmundson,Dirk
Eri 3h35'18"-13d5'
Edna
Ori 6h6'53"-0d7'
Edna Alice
Ori 5h52'44"14d12'
Edna G
Eri 4h48'9"-7d40'
Edna M T M
Lyn 8h57'25"46d16'
Edna Mae
Lyn 8h14'49"34d54'
Edna Marie
And 2h31'28"37d45'
Edna Mary
Umi 14h6'52"68d8'
Edna May
Cyg 20h16'34"39d11'
Edna's Eternal Brightness & Beauty
Cyg 21h6'0"48d49'
Edney,Nicola
Del 20h56'26"10d49'
Edo - 5
Cyg 21h20'52"38d39'
EDO 20 12 1992
Boo 14h7'48"40d36'
Edoardo
Her 16h4'48"50d6'
Edoardo Proposizione 1
Sge 20h5'31"20d9'
Edouard II
Umi 13h40'0"75d52'
Edouard,Franck
Cyg 19h52'31"44d52'
Edrie Opal
Cas 0h20'1"63d48'
Edron
Per 2h53'20"34d26'
Edsall,Donald
Lmi 10h11'29"31d27'
Edson "Searchlight", Edward
Aur 5h0'29"42d1'
Edson,Howard
Her 18h14'49"48d26'
Edson,Walter & Madeline
Uma 9h23'57"58d50'
Edstar
Cma 7h48'27"10d25'
Edualc
Boo 14h47'34"38d16'
Educator's Best Wendy & Kyle
Aur 6h47'47"37d32'
Edundha
Cap 20h5'12"-10d57'
Edvin,Gottfried
Crb 15h33'57"38d40'
Edvise
And 23h5'60"52d46'
Edward
Per 2h56'41"40d27'
Edward
Ori 6h15'47"1d35'
Edward & Monica
Lyr 18h18'41"38d8'
Edward Morris
Lac 22h46'29"54d49'
Edward's Brillance
Cet 2h7'40"2d42'
Edward's Magic
Cep 21h50'59"56d14'
Edward's Night Light, Alexander
Aur 5h10'16"41d50'
Edward's Wish
Lmi 10h59'36"28d55'
Edward-Little Bear, Robert L
Umi 16h29'34"80d17'
Edwards
Ori 4h59'24"1d32'C
Edwards
Ori 5h10'56"1d55'C
Edwards Family Star, The
Tri 1h31'0"28d46'
Edwards,Alicia Jane
Pho 0h50'51"-44d50'
Edwards,Allan David
Her 16h49'45"35d8'
Edwards,Andrew
Dra 16h58'11"63d51'
Edwards,Angela Loraine
Eri 3h58'0"-13d47'
Edwards,Annie
Cyg 21h1'40"37d55'
Edwards,Annmarie
And 1h17'15'35d59'
Edwards,Bernard Keith
Cet 2h24'57"8d38'
Edwards,Bunny
Aql 19h37'21"-6d54'
Edwards,Carole Lynne
Lyr 18h56'57"42d25'
Edwards,Celeste Revay
Lyr 19h18'48"42d10'
Edwards,Chris
Aql 19h25'41"8d50'
Edwards,Christen Lee
Lac 22h35'16"55d4'
Edwards,Christian Joseph
Mon 6h43'16"7d13'
Edwards,Claudine
Com 13h32'0"26d12'
Edwards,Clayre D
Cas 0h25'35"61d29'
Edwards,Colin David
Aql 19h5'58"1d4'
Edwards,Colten MacKennan
Peg 23h5'33"8d10'
Edwards,Daniel
Boo 13h49'1"18d23'
Edwards,Daniel James
Pho 0h49'14"-44d2'
Edwards,Danielle
Cam 3h30'1"60d47'
Edwards,David
Cyg 19h33'34"31d43'
Edwards,Deanna Elaine
Peg 22h12'1"30d52'
Edwards,Denise Susan
Lyr 18h16'21"38d47'
Edwards,Doris L
Com 13h26'36'25d28'
Edwards,Elgin
Uma 12h37'30"62d37'
Edwards,Enid
Cyg 21h33'26"41d7'
Edwards,Frank
Cep 20h6'43"60d45'
Edwards,Freddi
And 0h12'40"43d56'B
Edwards,Gary
Boo 14h40'49"38d8'
Edwards,Gene C
Psc 22h55'47"1d21'
Edwards,Gillian
Uma 12h45'44"54d29'
Edwards,Gregory S
Her 18h19'17"20d13'
Edwards,Hilary Jean
Cyg 21h0'54"36d30'
Edwards,Ian
Dra 17h48'1"61d21'
Edwards,Ian Joseph
Peg 21h37'42"25d13'
Edwards,James Homer
Aql 18h58'20"-6d51'
Edwards,Jandav
Aur 6h6'49"36d18'
Edwards,Jane
And 23h35'47"46d60'
Edwards,Jarco
Lac 22h7'32"40d29'
Edwards,Jasmine
Peg 22h2'34"4d14'
Edwards,Jason Michael
Oph 18h0'51"12d57'
Edwards,Jean Goodband
Cas 0h5'34"63d59'
Edwards,Jewell
And 2h0'41"45d8'
Edwards,Joan Gail
Peg 21h31'15"20d7'
Edwards,John Christopher
Ori 6h16'52"12d18'A
Edwards,John Mark
Aur 7h0'0"43d39'
Edwards,John Richard
Her 18h7'15"28d27'
Edwards,Jr,Framan Lewis
Oph 18h18'0"7d55'
Edwards,Jr,James William
Per 4h22'21"51d1'
Edwards,Jr,Kenneth J
Hya 9h6'32"3d42'
Edwards,Julia
Cas 1h55'41"61d7'
Edwards,Kate
And 0h1'21"46d56'
Edwards,Kathi
Peg 21h33'11"24d13'A
Edwards,Kathryn Noel
Peg 22h55'51"27d27'
Edwards,Kelli Kaye
Mon 6h28'14"8d44'
Edwards,Kelly
Cas 0h45'25"69d39'
Edwards,Kiara
Sco 17h29'52"-30d17'
Edwards,Kym
Gem 7h24'11"28d58'
Edwards,Lance
Dra 18h23'28"67d39'
Edwards,Lauren
And 0h47'24"39d44'
Edwards,Lauren Beth
Cyg 21h9'46"40d35'
Edwards,Lorraine Raye
Mon 6h5'0"-10d60'
Edwards,M E Eddie
Cyg 20h30'14"42d10'
Edwards,Madeline Ann
Peg 22h42'50"20d23'
Edwards,Mark E
Sex 9h56'47"2d54'
Edwards,Martelle
Peg 23h19'37"8d19'
Edwards,Miss Carole Angela
Cas 0h3'19"59d36'
Edwards,Mitchell F
Hya 9h10'37"1d25'
Edwards,Norma & Irving
Uma 12h12'20"59d10'
Edwards,Patricia
Aql 19h5'14"3d5'
Edwards,Pauline
Cyg 19h31'41"39d9'
Edwards,Pauline Grace
Cyg 19h19'0"30d4'
Edwards,PD,Jody
Oph 17h24'25"10d36'
Edwards,Rhonda L
Cas 22h56'60"54d14'
Edwards,Rita
Peg 22h7'1"3d31'
Edwards,Ronald Lew
Per 3h1'10"50d10'
Edwards,Samuel Alden
Cmi 7h58'44"8d12'
Edwards,Stephen DeWitt
Aur 6h31'32"31d6'
Edwards,Stu & Eda Bessie
Her 16h56'17"37d6'
Edwards,Summer Ann
Crt 11h19'54"-20d55'
Edwards,Ted
Per 3h36'38"51d31'
Edwards,Teddi
And 0h12'40"43d56'A
Edwards,Terry
Tri 2h45'15"31d17'
Edwards,Theresa Marie
Cyg 19h24'36"34d46'
Edwards,Thomas William
Cet 2h39'35"-9d26'
Edwards,Tim
Cet 1h41'13"-1d28'
Edwards,William Brian
Ser 15h50'25"19d60'
Edwards-My Star Bright Frank Davion
Per 3h25'59"40d26'
Edwardson,John
Aur 5h38'28"40d19'
Edwin
Hya 8h32'44"0d54'
Edwin Nikolaus
Sgr 19h44'21'-43d45'
Edwin's Dream
Per 2h51'35"46d8'
Edwin's Pearl
Crb 15h16'19"30d4'
Edwina B
Cyg 21h21'12"41d14'
Edy
Cnv 13h0'10"38d31'
Eeftink,Joost
Ori 5h10'0"-1d17'
Ebeling,Winfried
Aql 19h7'21"10d27'
Eelis Benjamin
Dra 20h30'60"70d31'
EFC-III
Aur 5h32'31"38d50'
Effamagrafabits
Aur 4h59'10"48d55'
Effi
Lup 15h14'59"-43d26'
Effremtopel,Richard
Sct 18h45'39"-7d46'
Effulgent
Aql 19h7'15"2d55'
EFJADEES
Peg 22h33'41"26d58'
Efrona & Howard
Crb 16h18'44"30d59'
Efrosman,Alexander
Uma 9h53'16"45d54'
Efthim,Michael Paul
Her 16h16'32"8d13'
Egan II,Gerard F
Her 16h55'20"33d7'
Egan III,(Oker) John F
Aql 18h57'23"12d22'
Egan,Adrienne Tara
Leo 9h32'31"8d50'
Egan,Daniel J
Aur 6h21'22"31d57'
Egan,Dominique M
Cas 2h35'52"70d49'
Egan,Erin Elizabeth
Cep 0h45'22"78d32'
Egan,Gentle John
Dra 16h39'46"62d49'
Egan,James E
Boo 14h54'47"52d4'
Egan,Jim & Carolyn
Eri 4h56'1"-6d8'
Egan,Kevin T
Dra 15h18'36"61d33'
Egan,Kyle Piontkowski
Uma 11h28'7"41d33'B
Egan,Lorraine Elizabeth
Crb 15h59'45"26d39'
Egan,Niamh
Ori 6h1'51"1d37'
Egan,Philippa
Ori 5h52'27"8d57'
Egan,Richard From The CRPG
Lac 22h2'52"41d7'
Egan,Richard & Ruth
Eri 4h36'20"-0d1'
Egan,Robert F
Aur 5h17'38"45d35'
Egan,Ryan Piontkowski
Uma 11h28'7"41d33'A
Egan,Sally
Peg 22h35'12"21d22'
Egan,Shelby Jo
Eri 3h37'41"-6d60'
Egan,Stephen J
Aur 6h34'55"33d51'
Egar,Ean
Boo 14h47'21"34d40'
Egbert
Aql 19h37'1"1d47'
Egea
Cas 1h59'41"58d40'
Egebakken,Richard
Dra 19h28'53"58d24'
Egelhof,A NY Star!, Richard Joseph
Cnv 14h2'54"31d35'
Egelson,David C
Hya 8h20'45"1d26'
Eger,Joyce F
And 23h0'28"48d57'
Egersdoerfer,Barbara
Cas 1h43'58"60d45'
Egersduerfer,Horst
Boo 14h51'1"45d35'
Egert,Harry
Oph 18h27'60"8d14'
Egert,Roger
Mon 6h3'49"-6d39'
Egerton,Max
Peg 23h1'14"12d35'
Eggebrecht,Lois
Psc 23h5'58"1d10'
Eggenton,Stuart James
Cyg 21h2'57"39d17'
Egger,Glenn Allan
Aql 19h57'51"13d42'
Egger,Michael P
Dra 17h13'26"60d24'
Eggers,James Evert Walton
Aur 4h48'12"51d4'
Eggers,Jörn
Cyg 20h42'41"46d17'
Eggers,Lee Edward
Ori 5h52'21"6d54'
Eggers,Matthias
Lib 15h5'53"-1d11'
Eggers,Scott
Uma 12h2'32"32d28'
Eggers,SiEn Cassandra
Peg 22h18'56"30d45'
Eggers,Tiffany Jade
Tri 2h7'51"33d59'
Eggert,Anmarie
Cep 20h57'30"60d3'
Eggert,Debi & Charles
Lyr 18h46'14"30d8'
Eggert,Frances Daves
Oph 17h6'29"-5d43'
Eggert,Harald
Sgr 19h16'29"-26d17'
Eggert,Melanie
Lyn 8h2'27"47d32'
Eggerton,Christopher
Her 16h55'0"32d48'
Eggertsen,Karen A
Cep 1h22'48"78d18'
Eggertsen,Lynn Ellen
Lyr 18h42'1"37d18'
Eggi
Lmi 9h30'33"38d31'
Eggink,Mark Steven
Peg 22h59'0"33d15'
Eggland,Steve
Ori 6h5'54"8d40'
Eggler,Jr,Joseph
Lyn 8h22'37"48d50'
Egglesfield,Colin
Aur 6h14'15"37d44'
Eggleston,Kay
Lyn 7h59'53"58d34'
Eggleton,James Andrew
Cep 22h37'29"70d11'
Eggleton,Kerns Ronald
Aur 6h8'42"38d9'
Egidio,Dario Pietro
Cas 0h56'17"70d49'
Eglantina
Lib 15h29'16"-10d39'
Eglantine Salingue
Mon 7h57'12"-3d30'
Egle,Alessandro E
Ori 5h6'33"0d54'
Egleston,Jody
Lyr 19h3'47"26d2'
Egley,Christine Diane
Cyg 20h32'57"50d29'
Egli,Capain Dave
Aur 5h18'39"54d34'
Eglite,Ina
Cas 1h30'55"61d19'
Egloff,Christi D'Ann
Cyg 19h24'32"33d49'
Eglon,Philip
Cnc 8h31'45"30d6'
Egnal,Michael J B
Per 2h52'25"37d29'
Egner,Audrey E
Cas 0h54'48"70d38'
Egon
Cap 20h4'42"-10d30'
Egozi,Alon
Cnv 14h0'19"30d37'
Egrise-Rey,Brigitte
Per 2h20'48"58d19'
Eguia,Shelby James
Vul 20h17'13"25d22'
Eguia,Tyler Anthony
Cnv 13h54'31"31d17'
Ehepaar Dorothea und Willi Sippel
Vir 13h39'45"-6d20'
Ehinger,William
Dra 17h1'0"68d5'
Ehl,Horst
Her 17h8'19"42d50'
Ehlen,Kate E
Peg 22h38'44"22d36'
Ehler "A C E", Alexander Charles
Cyg 20h2'13"40d15'
Ehler,Carrie
Com 13h4'53"20d57'
Ehler,Michael
Vul 19h6'25"25d14'
Ehler,Sean
Boo 14h40'34"37d39'
Ehlert,Gary W
Mon 7h25'25"-10d18'
Ehlies,William Harry
Lib 15h19'18"-22d9'
Ehmke,Thomas
Lac 22h20'0"38d57'
Ehmry,Mark & Sharon
Crb 16h10'41"26d47'AB
Ehre,Stephen Jon
Aur 5h31'22"50d13'
Ehrenheim,Manfred
Peg 23h29'52"13d30'
Ehrenreich,Joe
Cnv 13h14'24"50d41'
Ehrenreich,The Star Jonah
Her 17h55'23"18d50'
Ehrens,Charles
Boo 14h38'48"46d2'
Ehresmann,Horst
Dra 17h59'54"80d37'
Ehret,Clinton Ralph
Her 17h28'1"38d34'
Ehret,Erwin
Aur 5h8'11"42d34'
Ehrhard III,Joseph A
Her 16h48'37"48d7'
Ehrhard,Robert Anthony
Cep 21h35'32"67d48'
Ehrhardt Star,The
Ori 6h8'31"8d23'
Ehrhardt,H L "Deke" & Audrey
Aur 6h53'29"37d44'
Ehrhardt,Kathryn
And 1h35'49"36d41'
Ehrhardt,Knut
Aqr 22h35'59"-8d52'
Ehrhardt,Max
Psc 23h3'1"0d10'
Ehrhart 50th Ann, Dede & Pal
Cyg 19h18'27"28d13'
Ehrhart,Leela Gulick
Cas 0h46'25"61d39'
Ehrhart,William D
Lib 15h32'0"-28d50'
Ehrhorn,Wilbert
Dra 16h34'53"70d26'
Ehrig,John Frank
Cet 1h33'21"-2d3'
Ehritt,Christina
Peg 23h31'47"10d40'
Ehrlich,Gloria
Cnc 8h2'20"8d8'
Ehrlich,Heini
Boo 13h40'13"10d33'
Ehrlich,Walter
Cyg 20h27'43"40d34'
Ehrman,Heather Ann
Cas 1h43'28"75d12'
Ehrman,Heather Ann
And 0h21'37"34d9'
Ehrmann,John William
Her 17h27'44"30d35'
Ehrnfelt,Mayor Walter
Aur 5h39'25"37d35'
Ehrsam,Walter
Hya 8h53'57"5d4'
Ehrstine Ecstasy
Cam 6h5'1"58d52'
Eib,John Westward
Aur 7h0'53"36d22'
Eibeler,Therese
Lyn 8h2'51"37d53'
Eibhlin,Realta
Cas 23h2'54"58d40'
Eibl,Ilone
Lyn 8h19'10"48d32'
Eich,Timothy Franklin
Lmi 10h50'1"28d39'
Eichel,Joshua Leigh
Aur 6h3'38"30d44'
Eichel,Karl-Heinz
Ari 2h24'20"26d45'
Eichelman,Carolyn
Mon 7h24'40"-1d4'
Eichenwald,Mel
Aur 6h29'55"37d46'
Eicher,MD,Geary M
Oph 17h56'28"12d6'
Eichert,Thomas James
Cmi 7h42'60"4d34'
Eichholz,John T
Her 16h45'1"10d50'
Eichhorn,Dr Gabriele
Lib 15h0'23"-1d19'
Eichhorn,Jacqueline Taylor
Lyn 7h31'12"40d57'
Eichhorn,Joseph
Cma 7h19'1"-15d38'
Eichhorn,Joyce
Cyg 19h31'29"31d34'
Eichhorn,Karen Dee Annis
Lyr 19h23'43"31d35'
Eichhorn,Louella & Charles
Lac 22h55'35"55d35'
Eichhorn,Martin
Sgr 19h39'46"-44d31'
Eichhorn,Theresa Helen
And 2h28'12"45d21'
Eichler,Elizabeth Anne
Leo 11h32'34"10d54'
Eichler,Fredrick Andrew
Cnc 7h54'29"10d37'
Eichler,Masters of Display Norbert
Cmi 7h16'51"4d6'
Eichman,Heidi Marie
Cas 1h19'31"62d3'
Eichner,Bodo
Sge 19h8'33"18d48'
Eichner,Christopher Michael
Her 16h46'57"34d9'
Eichner,Jean
Com 12h27'19"30d40'
Eichner,Lindsay
Aql 19h59'19"14d11'
Eichner,Marissa Ann
Cas 0h55'24"63d59'
Eichorn-Hathaway
Peg 22h17'25"21d0'
Eichstädt,Walter
And 23h1'29"43d45'
Eichwald,Richard A
Cet 2h5'22"3d56'
Eick,Lady Beverly
Aur 5h4'41"40d24'
Eick,Spencer Christian
Dra 20h25'41"74d23'
Eickenberg
Dra 15h53'43"62d22'
Eickhoff,Stephen
Cam 3h56'25"58d17'
Eickhorst,Henry Paulo Maria
Cas 0h16'31"60d19'
Eickmeyer,Herbert
Lac 22h0'43"51d48'
Eickmeyer,Robert
Dra 12h28'1"71d58'
Eid,Paulina
Sgr 20h20'43"-28d50'
Eide,Lisa Marie
Cam 12h7'40"77d17'
Eide,Mardene
Cas 1h47'0"68d47'
Eidelheit,Daddy
Cep 21h46'47"63d48'
Eidem,Lee D
Aur 6h32'17"52d33'
Eiden,Jeremy Allen
Cam 3h49'51"60d5'
Eidenier,Carol Lynn
And 0h28'1"45d50'
Eider
Cep 22h11'58"65d32'
Eidson,Cheryl Ann
And 0h2'11"46d22'
Eidson,E C
Her 16h26'1"28d14'
Eidt,Richard
Peg 22h15'33"5d36'
Eiermann,Thomas J
Boo 14h18'55"15d28'
Eifert,Sheri
And 0h59'11"33d39'
Eiff,Dorit von
Gem 7h32'53"34d47'
Eigenmann,Carolyn Marie
Uma 10h9'13"52d20'
Eiger,The
Dra 17h41'51"73d38'
831 Always
Uma 9h23'27"48d10'
831 Claire
Lyr 18h37'45"41d15'
8ECR4
Uma 11h50'0"49d6'
#88 Always A Knight
Lmi 9h47'39"40d3'
8888
Sct 18h47'54"-6d44'
18NOV94MO23Job10JOSH40 Is31#1ChocoLn
Peg 22h19'26"3d17'
Eigner,Edwin Moss
Cet 3h4'49"-1d28'
Eiji
And 23h22'0"40d53'
Eike
Leo 11h24'22"-1d58'
Eikenaar,Suzanne
Cyg 19h28'11"40d33'
Eikner,Gay
Uma 8h30'18"68d27'
Eiko
Aql 19h42'46"13d45'
Eilberg,Brandon & Jamie
Lac 22h8'10"49d9'
Eile For Ever
Uma 9h31'27"47d44'
Eileen
And 0h20'44"45d35'
Eileen
Cas 0h27'41"61d7'
Eileen
Lyr 18h15'16"45d2'
Eileen
And 23h32'25"43d35'
Eileen
Pyx 8h31'7"-29"13'
Eileen
Cas 1h22'38"58d22'
Eileen
Cyg 19h38'0"28d48'
Eileen
Sgr 19h33'50"-44d33'
Eileen
Cyg 19h40'44"38d31'
Eileen & Greg
Crb 15h38'49"29d54'
Eileen & Mark's Star
Lyr 18h32'22"37d34'
Eileen Ann
And 2h2'0"38d59'
Eileen Claire
Cam 5h7'17"70d49'
Eileen Constance
Mon 6h53'45"-8d12'
Eileen F
Lib 15h29'0"-10d38'
Eileen Louise Sean Forever
Aur 6h5'11"32d57'
Eileen Marie
Cas 0h58'23"61d39'
Eileen Marie
Tau 4h21'16"30d2'
Eileen Marina
Cas 0h6'53"62d18'
Eileen Star Of Courage & Inspiration
Peg 23h30'51"10d14'
Eileen, Daniel's Love
Lyn 8h19'32"39d48'
Eileen-Eli
And 0h2'44"40d8'
Eilermann,Tad H E E
Aql 20h13'38"5d24'

Eilert,Michelle Marie
Mon 8h3'21"-4d0'
Eimer,Cheri
Peg 22h2'1"10d57'
Eimer,D J
Peg 21h58'56"10d5'
Eimholt,Linn A
Cas 1h39'13"64d31'
Einbund,Nate
Hya 9h52'2"-15d59'
Einiger,Kenneth
Psc 23h4'53"6d32'
Einstein,Alexander
Sex 10h34'10"-1d37'
Einstein,Jay Dwight Strine
Cnv 13h43'44"31d46'
Einziger,Scott
Lac 22h16'41"46d45'
Einödshofer Karolus, Sieglinda
Cep 22h8'54"60d8'
Eir
Aql 19h5'39"15d23'
Eira
Psc 1h22'12"10d32'
Eiring,Jay Paul
Aur 6h14'0"33d48'
Eis,Susan Lynn
Lyr 18h30'23"31d18'
Eisa,Michelle
Ori 5h4'56"14d20'
Eisele,Jeanie
And 23h39'42"44d23'
Eiseman II,Floyd Cameron
Lac 22h22'38"54d19'
Eiseman,John
Cet 0h28'1"-6d35'
Eiseman,Mitchell L
Cnv 12h29'51"41d57'
Eiseman-Conroy Helen's Kid,Baby Brat
Cep 21h48'22"68d52'
Eisemann,R J
Dra 20h17'1"64d17'
Eisen,Hilary Anne
Vul 19h5'15"25d29'
Eisenback
Hya 8h30'54"1d43'
Eisenbarth,Spencer
Cnv 13h59'39"30d57'
Eisenberg
Aur 6h31'41"40d16'
Eisenberg,MD,Dr Lee
Cep 21h15'11"55d30'
Eisenberg,Michael
Lac 22h33'35"55d19'
Eisenberg,Priscilla
Mon 6h23'38"5d29'
Eisenberg,Renate Thekla
Cyg 19h26'0"30d28'
Eisenberg,Richard
Per 2h36'0"40d32'
Eisenberg,Spencer
Ser 15h37'15"18d34'
Eisendrath,Edwin & Jennifer
Crb 15h18'0"30d33'
Eisenhardt,Sr,David Paul
Her 18h6'34"31d51'
Eisenhart,Harry David
Dra 17h37'34"63d52'
Eisenhart,Roberta Marie
Aql 19h6'30"-0d12'
Eisenhotter,Joseph
Oph 17h30'21"14d1'
Eisenmann,Brandon
Eri 4h18'25"-13d27'
Eisenmann,Christopher
Aql 18h53'42"-0d7'
Eisenmann,Katja
Cap 20h4'1"-10d37'
Eisenstark,Roslyn Karol
Tau 4h40'50"28d27'
Eisenstein,Jerry M
Cnv 13h20'38"40d12'
Eisenstock,Jordan Harris
Boo 15h46'1"41d60'
Eisenstock,Lee Michael
Aur 7h0'42"41d9'
Eisgeth,Julius
Aql 19h1'12"17d52'
Eish,Sara Abu
Cnv 13h46'1"34d29'
Eisman
Cap 21h28'22"-22d40'
Eismont,MD,Frank J
Oph 17h57'21"0'12'
Eisner,Adam Russell
Lyr 18h34'37"40d7'
Eisner,Andrew Glenn
Lyr 18h30'58"40d7'
Eisner,Carlo
Lyn 8h23'34"48d29'
Eisner,Denise
And 23h3'1"49d6'
Eisner,Elaine B
Lyn 6h59'36"59d7'
Eisner,Jay Allen
Lyr 18h31'37"41d19'
Eisner,Jr,Jeffery Whitman
Lyr 18h36'41"42d9'
Eisner,Judy
Cyg 20h1'15"40d54'
Eisner,Leslie Marie
Lyr 18h32'44"41d39'
Eisner,Richard L
Dra 16h57'36"62d33'
Eissinger,Gus
Cep 22h37'23"71d10'
Eissmann,Dorothea A
Cyg 21h11'47"38d17'
Eissmann,Walter W
Cyg 21h22'14"38d16'
Eisterer,Beatrix und Herbert
Cep 22h7'34"68d45'
Eiswerth,Mary Marguerite
Cnv 12h30'25"40d1'
Eitel,Michael
Cam 5h7'1"70d14'
Eitel,Michael
Uma 9h17'0"48d2'
Eitoob
Uma 13h38'0"50d2'
Eitsee
Ari 2h51'1"30d48'
Eivers,Monique Marie
Cas 0h31'30"58d47'
EJ
Mon 6h55'16"-10d38'
Ejay
Cma 6h55'14"-18d35'
Ejgird,Michael
Aur 6h6'32"37d12'
Ejrup,Florence
Cas 2h34'0"61d24'
Ekaterini Kurek
Dra 17h2'21"73d46'
Ekberg,Cecilia
And 0h1'21"46d56'
Ekberg,Mr & Mrs Richard J
Lyr 18h37'10"29d23'
Ekblom,Christy
Cet 3h5'32"2d13'
Eke,Ken
Oph 17h22'15"-20d25'
Ekenberg,Jeffrey Warren
Cep 23h12'56"62d3'
EKG
Boo 15h41'12"40d8'
Ekiert,Jr,Thomas
Dra 19h4'0"58d38'
Ekker
Dra 20h7'39"73d27'
Eklund,Candace Prather
Lib 15h35'57"8d37'A
Eklund,George & Joan
Umi 14h32'19"73d44'
Eklund,Lowell R
Dra 20h1'15"63d33'
Eklund,Robin
Cet 2h12'10"5d14'
Eklund,Tom
Umi 14h47'19"75d13'
Ekman,Marilyn Jean
Lyn 8h50'50"45d36'
Ekman,Thomas A
Cep 23h35'30"65d49'
EKP III
Ori 5h48'1"11d48'
Ekstatis
Cnc 9h16'16"18d15'
Ekstrom,L Ellen
And 1h40'51"36d19'
El Bobbo
Aur 7h9'44"39d50'
EL CHEFO
Dra 17h48'12"85d24'
El Cid
Aur 4h46'21"48d60'
El Cid
Ori 5h58'60"17d27'
El Dakhakni,Ayman
Cet 0h55'16"-6d20'
el diaz de diciembre
Dra 20h38'40"68d57'
El Filali,Abdelkader
Oph 17h58'54"10d18'
El Freeda Lobo
Lyr 19h17'10"41d36'
El Grandé Ruglandé,J O
Umi 14h29'37"66d56'
El Hierro
Ori 5h43'19"7d48'
El Principe Azul
Ori 5h4'47"0d35'
El Stone
Hya 9h31'42"5d33'
El Sueno de Inma y Michael
Cnv 12h17'48"37d10'
El Tigre
Ori 4h54'30"0d51'
El-Amin,Katrina Lateefah
And 0h53'47"35d53'
El-Hoshan,Rakan
Uma 9h37'40"54d16'
El-Majied,Asma
Lyr 18h39'12"33d19'
El-Sid
And 0h50'18"38d54'
Elabren
Aqr 21h53'22"-6d14'
Elaine
And 23h19'39"44d56'
Elaine
Lyn 8h18'11"37d40'
Elaine
And 1h52'24"36d26'
Elaine
Cyg 19h26'12"34d29'
Elaine
Lyn 9h1'1"41d46'
Elaine
Com 12h28'1"30d20'
Elaine
Psc 1h1'45"18d21'
Elaine "Fire N' Ice"
Gem 5h58'57"27d17'
Elaine "The Star of My Life"
Cas 0h52'49"67d45'
Elaine & Dick Forever
Lyn 9h3'16"40d14'
Elaine & P J
Cyg 20h18'46"41d2'
Elaine (John's Best Pal)
And 0h41'40"31d9'
Elaine Jean
Peg 21h59'38"31d49'
Elaine Kelly
Lyr 18h56'1"30d15'
Elaine S U
Cmi 7h55'0"4d29'
Elaine Ying-Ying Su
Lyn 7h39'0"41d8'
Elaine's
Lyn 7h41'26"58d55'
Elaine's Hoochie Coochie Star,Gary
Cyg 19h25'29"34d54'
Elaine's Illuminator
Peg 23h20'21"28d44'
Elaine's Star
Cma 6h51'18"-15d60'
Elaine,Danna
Crt 10h57'23"-12d23'
Elaine-Nina-Tanay "Little Doll"
Aql 19h55'14"12d41'
Elam,Kathryn Alene
Cam 7h47'22"68d47'
Elam,Richard J
Aur 5h16'0"45d48'
Elan,Hila
Lyr 18h31'12"41d14'
Elana
Aql 20h1'45"13d12'
Elana
And 0h23'51"40d24'
Elana's Angel
Cas 1h55'30"58d4'
Elana's Enchantment
Lyn 8h15'39"58d54'
Eland,Barbara
Cet 2h42'47"4d3'
Elander
Ori 6h4'26"5d24'
Elanor
And 23h35'14"41d33'
Elardo,Kiki
Cnv 12h11'11"38d7'
Elas-Monique
Leo 11h5'15"28d32'
Elasar
Psa 22h37'36"-25d51'
Elayne 5-21-92
Com 13h3'60"19d42'
Elayne Louise
Cyg 21h10'45"38d34'
Elba
Lyn 8h12'11"47d58'
Elba
Lmi 10h40'44"24d5'
Elbaz,Jean-Sauveur
Lyn 8h31'22"42d16'
Elberling,Charity Noel
Mon 6h40'45"7d55'
Elbert's Star,Kathy Mary
Tri 2h10'1"33d1'
Elberta Clara Elda Heavenly Star
Tau 3h42'27"20d34'
Elbrick,Alexia
Cas 1h46'50"76d34'
Elbuka,Stacey
Lyr 18h17'41"32d50'
Elcon Electric Incorporated
Oph 18h4'33"13d29'
Elcorazon
Ori 5h57'59"11d56'
Elda
Aql 20h0'27"-6d45'
Elda 7 Marzo 1930
Cyg 21h26'37"39d57'
Elder,Cory Wayne
Cnv 12h7'45"36d27'
Elder,John Charles
Per 1h45'34"50d30'
Elder,Katie Marie
Sgr 18h55'50"-29d7'
Elder,Katrina A
Equ 21h10'58"11d35'
Elder,Noah Charles
Per 1h58'42"52d54'
Elder,Randy
Cep 23h1'36"64d15'
Elder,Robert C
Ori 5h44'26"11d41'
Elders,Buddy
Aur 6h7'41"37d39'
Elders,Christian David
Cyg 20h44'59"37d54'
Elderton,Don
Ori 5h18'26"15d55'
Eldnirg,Suoegrog
Peg 23h32'57"33d51'
Eldonna
Peg 233h1'0"25d10'
Eldorado
Leo 10h17'35"7d32'
Eldorado,Estelle
Cas 3h32'37"69d55'
Eldra
Aql 19h51'29"14d29'
Eldred,Anne Madison
Vul 19h19'17"26d19'
Eldred,Terry
Lac 22h32'19"53d19'
Eldredge,Katherine Marie
Peg 21h59'58"30d6'
Eldredge,Paula
Peg 21h59'33"34d15'
Eldridge,Brooke
Uma 10h58'34"48d28'
Eldridge,Chas
Cyg 19h33'49"34d12'
Eldridge,Cynthia L
Cas 0h31'19"71d52'
Eldridge,David
Her 17h04'20"42d04'
Eldridge,Debra L
Hya 8h34'54"-1d11'
Eldridge,Dr Amy
Oph 17h59'51"8d11'
Eldridge,Marion
Cyg 19h25'54"34d33'
Eldridge,Nicola
Cyg 19h26'30"35d2'
Ele 13-11-1963
Lyr 19h16'58"31d28'
Elean
Cmi 8h6'0"1d15'
Eleana Mi Novia
And 2h19'44"45d37'
Eleanor
Ori 5h53'1"15d5"
Eleanor
Equ 21h15'47"3d39'
Eleanor
Cas 1h54'37"73d40'
Eleanor
Crb 16h4'53"30d9'
Eleanor
Ori 5h3'19"13d48'
Eleanor & Frank
Lyr 18h20'0"42d54'
Eleanor & Orville
Aur 6h29'57"52d30'AB
Eleanor Anne
Lyr 19h4'0"26d16'
Eleanor Clare
Cam 6h10'0"61d2'
Eleanor Peggy
Lyr 18h16'34"31d16'
Eleanor Rose
Cas 0h49'17"63d38'
Eleanor,Mary
And 1h37'50"38d17'
Eleanor-090569
And 0h27'37"38d7'
Eleanora
Cyg 21h10'57"48d50'
Electra 92
Cep 21h25'14"67d36'
Electrafantastic
Lac 22h11'50"48d59'
Electre
Cam 5h55'26"67d59'
Electric
Cam 6h10'20"58d13'
Electric Blue
Aql 19h31'50"10d37'
Electric Blue Sandra & Michael's Star
Eri 3h29'21"-8d59'
Electric Jim
Del 20h39'15"10d56'
Electricity
Uma 11h40'55"50d20'
Electrifying Michele, The
Cam 9h53'14"82d5'
Elefthrow,Maryanne
Cas 1h41'40"75d46'
Elen Si la Lumenn Omentilmo
Sco 16h25'1"-40d36'
Elena
Cyg 20h4'28"30d19'
Elena
Cma 6h53'33"-19d52'
Elena
Cas 2h3'21"59d41'
Elena
Aur 7h21'16"43d47'
Elena
Aur 7h8'1"38d2'
Elena
Cnv 13h37'60"28d52'
Elena
Boo 14h17'21"17d25'
Elena
Her 17h21'1"40d43'
Elena
Ori 6h3'52"10d12'
Elena
Psa 22h20'60"-25d35'
Elena
For 2h41'57"-28d33'
Elena
Umi 16h3'11"83d25'
Elena
Eri 3h50'31"-1d47'
Elena
For 2h25'49"-28d45'
Elena
Pup 7h56'51"-20d57'
Elena
Gem 6h49'0"31d11'
Elena
Gru 22h14'38"-50d34'
Elena
Pic 4h59'55"-46d5'
Elena
Pup 8h8'44"-29d20'
Elena
Del 20h18'41"14d22'
Elena
Pup 8h24'26"-29d51'
Elena
And 0h17'45"38d52'
Elena
Ori 5h25'14"0d10'
Elena
Her 17h17'55"50d9'
Elena
Cet 2h29'42"-0d41'
Elena & Joseph's Promise
Cyg 21h1'47"33d25'
Elena & Victor
Cyg 20h30'33"40d15'
Elena 1st
Peg 21h55'25"22d34'
Elena is Enough
Vul 19h16'56"21d58'
Elena Marie
Gem 7h6'55"21d33'
Elena P
Mon 7h49'34"-4d8'
Elena PD
Vel 10h11'56"-43d1'
Elena's Wish
Mon 6h20'1"3d23'
Elendil
Lyn 7h34'35"35d25'
Eleni
And 0h21'47"45d22'
Eleni
Peg 22h52'17"27d15'
Eleni
Cyg 21h1'39"38d5'
Elenian
Eri 3h49'46"-1d46'
élénonore Franceschi
Cep 20h24'13"60d28'
Elens Star,The David
Her 16h12'18"11d34'
Eleny
Cyg 20h27'26"58d49'
Eleonora
Lac 22h51'39"38d32'
Eleonora
Cae 4h48'0"-33d17'
Eleonora
Pup 8h24'34"-23d18'
Elephant You
Cam 9h1'44"73d55'
Elephants Nick
Uma 9h47'60"47d33'
Eletodoleni Together Forever
Uma 8h55'40"52d50'
Eleuthere
Dra 15h43'42"53d1'
Elevator Jack
Per 3h29'56"37d34'
Eleveld,Robert J
Lac 22h4'12"47d10'
Elewski,Jr,Chester
Cnv 12h43'15"38d57'
Elexis,Erin
Aql 18h58'0"-6d34'
Eley,Cecil John
Ori 6h1'57"0d33'
Eley,David John
Ori 5h56'0"12d39'
Eley,John & Eryl
Cyg 19h27'28"38d24'
Elfie
Com 12h27'27"22d17'
Elfin's Amber
Crb 15h54'30"27d19'
Elfman,Isaac
Leo 9h40'13"14d25'
Elford,Laura & David
Boo 15h8'25"8d38'
Elfriede
Lmi 10h2'51"30d37'
Elgar,David A
Per 4h45'1"40d0'
Elgart,Susan Suddarth
Mon 6h42'42"8d48'
Elgas,Pia
Tri 1h51'34"27d27'
Elger,Dennis & Christine
Lyr 18h48'16"40d34'
Elgin,Dick
Aur 6h7'59"38d30'
Eli
Cnv 13h6'18"38d43'
Eli & Rachel
Cam 5h27'40"68d46'
Eli James
Cam 6h39'22"80d7'
Eli Torin
Dra 15h50'48"67d8'
Eli,Charles
Her 16h20'45"20d27'
Eli/66
Cnc 8h7'23"31d39'
Elia
Cas 2h0'34"60d54'
Elia 9
Peg 23h49'19"17d55'
Elia,Claudio
And 0h17'53"32d31'
Elia,Joseph
Lyn 7h55'32"33d52'
Elian
Ori 5h57'20"19d57'
Eliane et Frédéric
Cas 1h41'25"61d26'
Elias,Atremisa Margarita
Eri 2h54'15"-8d29'
Elias,Fay Harrison
Lyr 19h17'29"38d33'
Elias,Frederick Mestman
Dra 9h48'43"74d14'
Elias,Jessica Marie
Vul 21h19'0"24d3'
Elias,Lilli Sara
Mon 6h30'32"3d6'
Elias,Mary
Aql 18h58'33"16d49'
Elias,Patricia
And 23h1'1"51d10'
Elias,Sara Marie Padma
And 23h42'31"47d1'
Élias,Sébastien
Her 17h12'54"20d38'
Elias,Vamadevi Sara Marie Padma
Cas 0h32'17"62d16'
Elias-Kay
Per 3h6'15"48d56'
Eliaskelly
Lyn 8h58'1"42d22'
Eliason,Birdell
Uma 8h41'25"53d52'
Eliassen,Delores
Cyg 20h55'16"39d23'
Elicic,Michele
Cas 23h23'55"58d29'
Elicker,Earl
Boo 14h7'1"12d49'
Elicker,Jane
Cas 1h33'36"61d12'
Elicker,Michael
Lyn 8h6'23"41d59'
Eliduccia
For 2h10'14"-26d42'
Elie,Claire
Cas 0h41'14"73d26'
Elifritz,Patrick
Her 17h2'48"20d39'
Eligia
Equ 21h19'38"11d29'
Elijah's Hope
Cap 20h32'39"-13d47'
Eline
Cam 6h5'26"72d28'
Eling,Sabine
Tau 3h55'16"22d18'
Elinoff,Yetta Hyman
Sct 18h49'55"-9d17'
Elinor
And 0h23'53"36d14'
Eliobby
Ser 15h17'24"-2d0'
Eliopoulos,Gary P
Aql 19h29'39"10d54'
Elisa
Cas 0h3'20"66d17'
Elisa
Cnv 13h1'44"38d16'
Elisa
Lac 22h3'42"51d2'
Elisa
Cas 22h58'1"55d25'
Elisa
Lyn 9h18'59"38d37'
Elisa
Uma 10h39'44"54d18'
Elisa
Eri 3h20'33"-7d16'
Elisa
Com 12h28'21"23d34'
Elisa
Lyn 8h26'13"34d26'
Elisa
Per 3h50'2"36d53'
Elisa
Lyn 7h29'19"48d48'
Elisa Joanne
Aql 18h58'1"-6d14'
Elisa Marie
Lyn 8h21'52"51d38'
Elisa Michelle
Lyr 19h21'1"35d36'
Elisa's Star
Eri 3h57'36"-14d10'
Elisabeth
Cam 5h5'50"68d35'
Elisabeth
Cet 2h58'38"9d29'
Elisabeth & Peter
Cnv 13h53'32"46d55'
Elisabeth & Rainer
Com 13h0'25"28d33'
Elisabeth F
Uma 12h26'0"62d29'
Elisabeth,Emilia,Maria
Per 3h31'38"39d12'
Elisabetta
Peg 22h59'0"20d50'
Elisabetta
For 2h7'57"-24d6'
Elisabetta Anton
And 23h1'1"41d6'
Elisabetta F Hal '71
Ori 5h53'28"-1d40'
Elise
Ori 4h44'0"1d44'
Elise
Cas 2h0'27"58d31'
Elise-Diane
Aql 18h58'38"17d59'
Élise et Dennis
Per 4h43'0"51d7'
Elisha Sophie
Umi 16h20'26"75d12'
Elisheba
Crb 16h8'49"26d40'
Elisheba,Theodora
Cas 0h43'37"74d11'
Elissa & Bob
Aql 19h48'57"11d57'
Elissa Ann
Cyg 19h25'19"44d33'
Elissa Star
Vul 20h18'38"28d33'
Elissandro
Del 20h16'46"10d37'
Elite Ray Of Light, Alexandra M
Cet 0h24'36"-8d44'
Eliza
Com 13h7'56"28d55'
Eliza Dora
Tau 5h53'1"23d44'
Eliza,Caylen
And 2h26'39"39d37'
Elizabet-Eternal Sparkle
Psc 22h52'26"6d55'
Elizabeth
Ori 6h2'33"-1d30'
Elizabeth
Lyr 19h6'1"28d43'
Elizabeth
Mon 6h36'22"6d21'
Elizabeth
Aql 19h53'49"14d19'
Elizabeth
Aql 19h56'1"15d10'
Elizabeth
Peg 21h41'38"21d39'
Elizabeth
Cas 0h45'26"70d54'
Elizabeth (The Sweetest Girl)
Ori 4h52'21"0d21'
Elizabeth
Cyg 20h31'52"50d23'
Elizabeth
Cma 6h56'30"-18d56'
Elizabeth
Lyn 7h40'45"45d29'B
Elizabeth
Del 20h24'11"7d58'
Elizabeth
Uma 10h18'59"44d9'A
Elizabeth
Cas 0h29'33"58d3' B
Elizabeth
Vul 19h58'16"28d47'
Elizabeth
Vul 19h58'48"22d50'

Elizabeth
Cam 7h31'20"84d1'
Elizabeth
Cyg 19h27'16"37d42'
Elizabeth
And 23h2'23"51d8'
Elizabeth
Cas 23h22'50"61d21'
Elizabeth "Momma-Lady" Love
Sct 18h42'57"-8d1'
Elizabeth & Kevin
Cyg 20h0'48"41d14'
Elizabeth & Richard
Crb 16h2'19"38d19'
Elizabeth 100
Crt 11h3'48"-10d48'
Elizabeth Anastasia
Vul 19h21'45"27d38'
Elizabeth Ann
Uma 10h25'40"68d48'
Elizabeth Ann
Per 4h32'17"34d1'
Elizabeth Ann
Vul 19h1'31"25d39'
Elizabeth Ann
Lyn 6h47'15"60d44'
Elizabeth Ann
Peg 23h2'29"17d36'
Elizabeth Ann
Cyg 21h2'0"30d56'
Elizabeth Ann
Hya 8h12'1"5d44'
Elizabeth Ann
And 23h34'38"47d23'
Elizabeth Ann
Mon 6h55'14"10d46'
Elizabeth Ann
Cas 1h47'27"73d58'
Elizabeth Ann
And 23h15'33"35d14'
Elizabeth Anne's Design
Crb 15h44'49"29d11'
Elizabeth Ashby "Madam X"
Her 17h56'18"20d33'
Elizabeth Celeste
Mon 6h53'40"-10d20'
Elizabeth Christine
Cyg 19h26'24"31d49'
Elizabeth Clare
Cmi 7h43'11"5d16'
Elizabeth De'Anne
Peg 23h35'1"30d43'
Elizabeth Diane
Uma 11h54'37"47d12'
Elizabeth Forever
Uma 9h2'14"48d49'
Elizabeth Gigi
Com 13h4'56"22d21'
Elizabeth Grace
And 1h30'38"36d41'
Elizabeth Jane
Cas 0h40'34"74d5'
Elizabeth Jean
Crt 11h17'20"-14d44'
Elizabeth Joy
Lyr 18h33'28"37d9'
Elizabeth Jubilaeus Illuminatus
Equ 21h22'29"11d47'
Elizabeth Katalin
Tau 4h12'9"23d44'
Elizabeth Kay
And 23h30'44"41d33'
Elizabeth Lee
Vul 18h57'41"24d26'
Elizabeth Lillian
Psc 1h3'52"20d6'
Elizabeth Lindsey
Cas 0h40'17"67d7'
Elizabeth Loves Harry
Sge 19h16'54"17d21'
Elizabeth Lynn
Cas 1h2'0"75d29'
Elizabeth Marie
Peg 22h42'49"34d19'
Elizabeth Rachel
Lmi 9h53'38"37d46'
Elizabeth Rose
Cas 2h38'53"60d31'
Elizabeth Rose
Cas 0h52'1"74d9'
Elizabeth Ruth
Cas 1h23'23"60d39'
Elizabeth Viktoria
Peg 21h42'40"24d18'
Elizabeth W Star,The
Cas 23h26'32"60d38'
Elizabeth '95
Lyn 8h45'47"35d10'
Elizabeth's Magic
Cam 7h35'55"68d35'
Elizabeth's Star
Uma 11h36'49"33d31'
Elizabeth's Star
Peg 23h0'0"8d11'
Elizabeth's Star
And 2h6'29"42d34'
Elizabeth's Wish
Mon 6h25'59"4d21'
Elizabeth,Loving Sister & Friend
Uma 11h0'30"52d1'
Elizabeth,The
Oph 17h11'23"-22d50'
Elizabeth- Meinus Kleinas Buttlechae
Cam 11h4'58"80d14'
Elizabeth-Kevin
Lyn 8h45'58"41d23'
Elizabeth-Key To The Heavens
And 22h58'51"36d17'
Elizamax
Cyg 19h44'16"30d22'
Eliziam
Mon 7h1'12"8d60"
Eljeelantius
Eri 4h36'1"-17d53'
Elk Dreamer
Sct 18h41'35"-7d21'
Elk,Gerda Elisabet Kristina
And 1h6'0"40d57'
Elkas,George "CEO"
Uma 12h32'1"63d11'
Elke
Cnc 8h32'34"8d29'
Elke
Her 17h27'1"42d42'
Elke Hohmann geb Jakob
Cas 1h30'19"74d16'
Elke My Powerstar
Sgr 18h48'28"-34d4'
Elking,Samuel C
Cet 3h14'22"8d15'
Elkington,Angela
Cyg 20h21'31"38d16'
Elkins,Andrew T
Umi 13h15'37"71d32'
Elkins,Claribel
Aql 19h31'0"10d34'
Elkins,Donald,Marge, Linda & Teri
Cep 20h18'24"60d19'
Elkins,James D
Per 1h45'31"50d9'
Elkins,James H
Her 16h17'54"26d32'
Elkins,Jerald R
Umi 14h13'17"71d26'
Elkins,Philip
Sct 18h54'1"-4d35'
Elkins,William & Sandra
Cyg 20h51'30"30d31'
Elkinson,Jay Victor
Umi 16h15'1"74d58'
Elkman,Allison
And 2h19'51"47d9'
Elkman,Bradley
Aur 6h13'24"33d44'
Ella
Lyr 18h53'30"35d8'
Ella
Cet 1h30'1"-6d60'
Ella
Mon 8h8'0"-5d50'
Ella
Lyr 19h2'12"38d40'
Ella M
Cnc 7h56'38"11d12'
Ella Sons
Ori 6h7'26"-1d40'
Ellacot,Jane A N
Lyr 18h28'57"46d56'
Ellard,Gerald
Aur 5h52'45"40d8'
Ellasaid Zoe
Cam 6h11'16"72d48'
Elle
Ori 6h1'54"7d24'
Ellebasi
Lyn 8h5'56"39d48'
Ellebracht,Tim & Lori
Mon 7h18'0"-6d8'
Elledge,Jason Nicholas
Per 4h0'16"50d45'
Elleman,Nicholas
Cyg 21h0'1"37d7'
Ellemunt,Hermann
Uma 11h55'11"32d22'
Ellen
Leo 11h25'50"-0d57'
Ellen
Cyg 20h22'44"38d38'
Ellen
Uma 9h19'0"44d7'
Ellen
Lyr 18h58'18"30d40'
Ellen
Com 12h26'1"20d38'
Ellen
Cas 22h55'48"55d26'
Ellen "Always Forever"
Cyg 21h34'0"37d40'
Ellen & Bill
Aur 5h38'0"54d28'
Ellen & Gareth's Star
Mon 7h54'54"-5d58'
Ellen (S.B.) Frieder
Sgr 18h55'50"-23d24'
Ellen Claire
Peg 23h28'21"33d36'
Ellen Corinne
Gem 6h51'48"31d49'
Ellen Kay
Cas 0h9'13"58d13'
Ellen Kaye
Cas 1h48'31"73d15'
Ellen Louise
Cam 3h16'1"61d44'
Ellen Mindy
Sgr 19h36'16"-39d51'
Ellen O de V
Cas 23h42'27"63d52'
Ellen Star
Cas 2h55'51"65d36'
Ellen"Our Beloved Mom & Nana"
Cas 0h19'31"60d2'
Ellen's Brilliance
Aqu 20h57'40"0d23'
Ellen's Light
Aql 19h31'54"8d33'
Ellen's Neal
Uma 11h44'60"49d38'
Ellen-Marie
Cap 20h24'53"-15d6'
Ellen-Zoe
Uma 9h51'43"55d39'
Ellena & Jyrki
Uma 11h43'1"49d46'
Ellenberger,Christina P
Cas 0h12'50"65d59'
Ellenberger,Diana Marie
Uma 11h24'51"31d1'
Ellender,Merle Madaline
Mon 6h20'18"5d35'
Elleni
Lyr 18h58'50"27d24'
Ellenoff,Nicholas Samuel
Cnv 13h13'1"50d7'
EllenPea
Cet 2h20'56"5d2'
Ellenthia
Crb 15h42'31"26d23'
Ellentuck,Michael Jay
Per 2h52'44"48d55'
Eller,Donna Marie
Cas 1h2'1"63d24'
Eller,Gynn
Mon 8h2'0"-4d18'
Eller,Jean Elizabeth
And 23h47'51"38d21'
Eller,Phillip & Brenda
Oph 17h3'26"-20d53'
Eller,Sean E
Cet 3h18'12"6d29'
Eller,Stan
Oph 16h42'32"2d56'
Ellerbusch,Michael Max
Her 17h22'34"47d19'
Ellerich,Stella Maris Patrick
Cmi 7h58'43"1d13'
Ellermann,Niklas Vinzenz
Psc 23h20'57"0d50'
Ellerson,Mary Thomas
Cnc 8h1'54"10d2'
Ellery,Douglas
Cen 13h23'57"-59d22'
Ellery-E B C
Dra 16h35'24"68d22'
Ellets Procrustean Brilliance,Terri
Lyr 19h23'42"31d48'
Elli
And 23h48'19"46d39'
Ellie
Cyg 20h22'36"37d60'
Ellie
Lyn 7h52'55"34d40'
Ellie
Cas 1h59'14"63d55'
Ellie
And 1h57'33"36d34'
Ellie
Tri 2h6'41"32d40'
Ellie
Per 3h19'55"40d48'
Ellie & Rig
Lyr 18h39'53"46d8'
Ellie 831 xx
Cas 23h29'1"60d15'
Ellie Belle
Cyg 21h46'0"38d57'
Ellie P's Treasure
Lyr 19h21'1"31d1'
Ellie's Little Bit Of Heaven
Mon 6h53'35"1d18'
Ellifson,Lea Anastasia Karlsson
Cas 0h14'0"61d41'
Ellin,Morton
Dra 20h10'1"63d39'
Ellin,Paula
And 1h58'29"38d47'
Elling,Jean
Aql 19h0'49"16d38'
Elling,Lindsay Rae
And 0h55'32"36d55'
Elling-Walz,Susan Haynes
Mon 7h49'18"-1d11'
Ellingham,Craig
Per 2h39'59"34d37'
Ellinghoff,Franz-Josef
Uma 10h35'44"48d16'
Ellingsberg,Daniel
Dra 16h30'29"65d29'
Ellingson,Scott David
Peg 22h7'30"3d44'
Ellington,Abby
Cep 20h38'27"75d48'
Ellington,Francis
Cep 22h22'1"61d0'
Ellington,Owen Bernardo
Ser 15h22'16"2d57'
Ellingworth,David
Uma 11h8'36"53d59'
Ellingworth,Mandy
Uma 11h13'15"55d29'
Ellinoy
Leo 11h32'21"18d59'
Elliot
Cep 21h54'21"55d20'
Elliot
Boo 15h0'53"26d56'
Elliot Star,The
Ori 4h58'1"0d19'
Elliot,David Alan
Uma 8h46'59"57d47'
Elliot,Gavin Tudor
Ser 15h11'1"21d17'
Elliot,Ian Marshall
Her 17h19'38"27d17'
Elliot,Jake Grant
Her 16h31'53"50d48'
Elliot,Jane
Mon 7h1'21"5d33'
Elliot,Joseph Warden
Uma 11h11'0"31d47'
Elliot,Morgan Kieth Ramsey
Uma 10h47'21"67d40'
Elliot,Tracy
Aur 5h1'49"48d7'
Elliott AEI,Alan Fraser
Per 4h41'54"40d20'
Elliott's Star
Aur 4h52'22"40d33'
Elliott, Forever My Love,Jay C
Her 16h11'1"40d44'
Elliott,A Byron
Tri 2h25'60"28d34'
Elliott,Barbara
Ori 5h55'23"14d47'
Elliott,Carie
Cas 0h29'12"66d30'
Elliott,Casey Nicole
And 2h19'40"39d42'
Elliott,Cathryn Marie
Mon 6h53'1"-6d12'
Elliott,Clare
Lyr 18h15'45"31d48'
Elliott,Cliff
Lac 22h18'59"46d57'
Elliott,Col E Carter
Her 18h3'10"28d31"
Elliott,Cooper Dean
Dra 17h5'30"71d27'
Elliott,David Scot
Ari 2h33'26"22d6'
Elliott,Delynn
Cet 0h58'49"0d35'
Elliott,Denise Anne
Peg 22h19'36"20d25'
Elliott,Florence M
Cas 1h22'15"55d49'
Elliott,H Burton
Her 16h8'26"51d6'
Elliott,Jamie Pauline
Cas 0h44'40"67d56'
Elliott,Jennifer
Sex 10h10'31"-9d7'
Elliott,Jim
Oph 17h37'46"-20d49'
Elliott,Jimmie
Dra 18h24'59"80d23'
Elliott,John Hamilton
Aur 5h7'44"38d19'
Elliott,John Richard
Boo 15h4'44"19d15'
Elliott,Jonathan
Cyg 20h21'40"39d5'
Elliott,Joyce Anita Dobson
Cyg 19h20'38"50d27'
Elliott,Jr,Forever Mark Steven
Her 17h31'12"41d12'
Elliott,Jr,Robert A
Dra 18h11'27"70d48'
Elliott,Julie Catherine
Uma 14h21'28"58d33'
Elliott,June
Peg 22h17'44"21d33'
Elliott,Kenneth F
Uma 11h26'16"55d25'
Elliott,Lee
Her 17h32'16"28d16'
Elliott,Linda Arlene
Uma 10h53'33"61d13'
Elliott,Lindsey Dale
Eri 2h54'47"-18d38'
Elliott,Margaret & Mark
Cyg 21h18'54"36d35'
Elliott,Meriam
Hya 9h2'59'5d54'
Elliott,Nicholas
Cep 21h11'44"68d54'
Elliott,Prescilla "Sarina"
Crb 15h26'39"31d50'
Elliott,Robert
Ori 6h9'1"1d3'
Elliott,Shawn Ryan
Aur 6h30'32"40d58'
Elliott,Sophia Renée
Mon 6h58'53"11d0'
Elliott,Stephen N
Per 3h1'57"46d16'
Elliott,Stuart Alexander
Cep 21h49'41"67d56'
Elliott,Thomas R
Aur 5h5'49"30d13'
Elliott,Tim S
Hya 8h29'52"-6d24'
Elliott,Willie Mae
Eri 4h31'13"-6d8'
Ellipsys
Sex 10h33'43"5d42'
Ellis & Marge
Lyn 8h10'1"50d22'
Ellis III,H Lee
Cet 3h15'14"8d0'
Ellis Memorial Light, Alexandra Rathburn
And 2h23'0"48d51'
Ellis,Amy Christine
Vul 19h18'1"26d31'
Ellis,Andrew L
Uma 11h30'50"57d36'
Ellis,Bethany Mary
Ori 5h57'34"17d46'
Ellis,Carolyn
Cas 1h56'45"65d37'
Ellis,Charmian Sarah
And 23h20'30"42d15'
Ellis,Daniel L
Boo 14h47'41"32d32'
Ellis,Darlena
Peg 22h38'56"27d18'
Ellis,David
Hya 8h16'45"4d57'
Ellis,David William
Uma 9h29'0"58d2'
Ellis,Dee
Lmi 10h10'22"34d57'
Ellis,Ed & Lisa
Uma 9h36'32"48d5'
Ellis,Elnor S
Sex 9h49'14"2d14'
Ellis,Elsie Marie
Vul 20h2'21"23d59'
Ellis,Estelle
Cas 23h37'51"58d36'
Ellis,Gary John
Uma 9h37'30"48d31'
Ellis,Heather Kamila
Lyr 18h57'28"47d27'
Ellis,Ilene & Alvin
Aur 4h47'18"41d0'
Ellis,Jack Robert
Her 16h43'17"30d25'
Ellis,Jaclyn Marie
Cap 21h39'41"-19d47'
Ellis,James Robert
Lac 22h17'0"38d13'
Ellis,Jinger
Cas 0h59'36"59d16'
Ellis,Jonathan Raymond
Cep 20h12'47"80d2'
Ellis,Julie Ann
Lyn 8h3'16"35d54'
Ellis,Kathy
Peg 22h21'46"4d46'
Ellis,Kelly
Umi 14h34'30"70d19'
Ellis,La-La
Lyr 18h33'55"43d40'
Ellis,Laura Catherine
Mon 6h30'28"4d6'
Ellis,Louise
Cyg 20h35'52"31d38'
Ellis,Luke
Ori 6h1'11"1d47'
Ellis,Lynn Ann
Tri 2h19'56"31d8'
Ellis,Malcolm C
Boo 15h2'42"22d29'
Ellis,Molly
And 0h2'52"44d45'
Ellis,Nancy-Jane
Lyr 18h14'49"41d3'
Ellis,Pam
Cet 0h53'1"0d60'
Ellis,Patrick Michael
Per 3h9'40"41d11'
Ellis,Peggy
Psc 0h59'13"20d0'
Ellis,Robert H
Sex 9h49'11"3d4'
Ellis,Robert Malcolm Douglas
Oph 18h0'57"12d19'
Ellis,Ronald W
Dra 14h53'57"61d41'
Ellis,Russell S
Uma 11h54'43"58d2'
Ellis,Shaun Thomas
Mon 6h32'22"-5d46'
Ellis,Sr,Kenneth Leroy
Aql 18h43'1"7d9'
Ellis,Susan Elaine
Peg 22h29'53"7d47'
Ellis,Terrance Lamonte
Ser 17h34'20"-14d31'
Ellis,The Tony
Leo 10h15'54"14d31'
Ellis,Vertite
Cas 0h9'54"59d11'
Ellis-Killian,Sandra
Lyn 7h12'34"57d16'
Ellison III,Charles
Aql 20h10'30"10d29'
Ellison,Alice & Larry
Crb 15h49'33"38d41'
Ellison,Ashley Erin
Sct 18h46'21"-7d34'
Ellison,Beth
Cam 3h20'15"61d23'
Ellison,Gail
Cam 5h56'46"60d26'
Ellison,Kerry
Sgr 18h47'54"-24d34'
Ellison,Melba Louise Webb
Vul 19h46'1"28d56'
Ellison,Stuart Robert
Ser 16h1'17"10d23'
Ellison'95,Michelle Marie
Eri 3h52'0"-3d51'
Ellman,Jennifer
Peg 23h6'0"8d27'
Ellmann,Nana
And 23h43'14"43d10'
Ellner,Harvey
Aur 6h29'24"31d17'
Ellner,Melanie Nicole
And 23h3'21"40d50'
Ellner,Rachel Meredith
And 0h20'29"32d56'
Ellora
Uma 10h38'45"55d5'
Ellores
Boo 14h20'47"22d46'
Ells & Family,Penny & Bob
Aql 19h43'0"11d55'
Ellsbery
Cma 6h56'45"-19d24'
Ellson,Earle Ivers
Cet 0h26'52"-0d57'
Ellstrom,Christopher
Boo 15h13'24"52d4'
Ellsworth,Alana
Aql 19h45'53"11d16'
Ellsworth,Rakaan Talal
Ori 5h56'52"17d34'
Ellsworth,Robert M
Peg 23h18'40"28d10'
Ellsworth,William Lane
Aql 19h30'54"12d60'
Ellsworth,William Ray
Sco 16h9'32"-35d9'
Ellton
Cyg 20h22'57"38d36'
Ellwin,Shelley Brooke Masako
Mon 7h59'0"-8d14'
Ellwyn
Uma 8h19'0"62d21'
Elly
Psc 23h27'36"1d19'
Elly
Oph 17h31'43"-23d31'
Elly's Light
Sge 20h17'58"17d39'
Ellzey,Gertrude
Mon 6h15'30"-5d44'
Ellzey,Jr,Leon J
Ori 5h47'45"21d29'
Elmaalouf,Christian M
Oph 18h20'0"8d40'
Elmaleh,Stanley
Cyg 19h43'40"30d36'
Elman's Child
Cam 8h55'52"77d43'
ElMar 4393
Uma 11h4'41"38d44'
Elmegreen,Debra Anne Meloy
Lyr 18h59'52"29d9'
Elmer Paul
Tri 1h53'0"28d51'
Elmer,James
Her 16h23'1"21d31'
Elmes,Wendy
Cnc 7h57'59"10d15'
Elmira Hélène
Lyr 19h16'1"28d33'
Elmo Gabrielle
Cyg 19h28'53"38d35'
Elmo's Cheeks
Lyr 19h16'0"28d37'
Elmore's Star,Our Dad, Robert F
Cep 21h25'31"55d43'
Elmore,Boo
Eri 3h4'34"-3d26'
Elmore,Jason William
Aql 18h56'23"-0d47'
Elmore,Shelby Leigh
Cma 7h3'43"-15d49'
Elmore,Wilma
Peg 23h35'47"31d52'
Elmquist,Sydney
Peg 23h49'13"18d31'
Elms,Joshua Paul Kenneth
Umi 15h28'31"68d48'
Elms,Karen
Mon 8h2'36"-4d16'
Elms,Scott T
Aur 7h3'25"36d30'
Elmwood School
Uma 11h15'30"71d22'
Elniski,Peter Nicholas
Lac 22h16'0"49d40'
Elodie
Cet 2h8'38"1d58'
Elodie
Oph 18h23'37"8d14'
Elodie,Miranda
Ori 5h6'0"13d43'
Eloi
Lac 22h28'26"40d27'
Eloise
Vul 20h15'19"22d33'
Eloise
Crb 15h56'59"27d55'
Eloise
Uma 9h56'40"71d28'
Eloise
Mon 6h57'51"-8d45'
Elojen
Sct 18h46'57"-7d59'
Elona
Ari 2h1'1"22d28'
Elowe,Edmond Nasir
Lac 22h5'0"47d50'
Elowe,Greg
Her 17h26'17"30d15'
Eloy José
Dra 15h57'29"62d20'
Eloise
Cas 1h30'20"73d28'
Elpert,Kerstin Kaethe
Tau 5h56'57"26d50'
Elphick,Sandra
Cyg 19h39'40"30d3'
Elphick,Stephen
And 22h58'32"48d60'
Elro
And 0h23'19"40d3'
Elrod,Demi
Lyr 18h25'29"38d3'
Els Puylaert Raymakers 70 Love Carl
Uma 10h57'17"40d13'
Elsa
Pic 5h7'53"-49d2'
Elsa 001
Cas 2h57'48"60d11'
Elsa De Paz Y de Paz
Cyg 21h1'46"40d23'
Elsa Megastar
Lyr 19h3'0"38d20'
Elsa Protector of all Pussies
Lyn 8h24'23"48d59'
Elsa"JoAnn" Rose
Cam 4h13'40"68d15'
Elsa-Alisha
Pic 4h47'52"-49d54'
Elsbeth 1952
Cas 2h18'30d67d56'
Else of Lejre
And 0h48'53"21d49'
Elser,Arnold
Cnv 14h3'27"41d7'
Elsholtz,Edith
Sco 17h26'45"-30d8'
Elsie
Umi 14h49'46"78d12'
Elsie
Mon 7h37'50"-0d55'
Elsie
Lyr 18h32'22"34d58'
Elsie
And 23h39'1"46d56'
Elsie & David's Wedding Star
Uma 11h48'33"43d31'
Elsie G
And 23h34'35"38d46'14'
Elsie Noel
Cam 6h9'59"60d22'
Elsie R
Cyg 19h42'37"30d14'
Elsierra Teachers
Tri 2h4'48"33d25'
Elsing,Bernd
Leo 9h22'28"19d33'
Elsk
And 23h22'22"50d33'

Elskaar
Cet 1h38'44"0d50'
Elsner,Ardith J
Cas 1h10'35"63d43'
Elsner,Ariane-Corinna
Peg 23h34'49"13d37'
Elsner,Byron D
Her 16h26'1"30d32'
Elsner,Dorothee
Cam 12h53'35"80d60'
Elsner,Nicole
Cam 4h10'38"67d36'
Elsroth,Sr,Walter Edward
Mon 8h4'1"-5d48'
Elster,Jacob Aaron
Ori 5h48'18"21d9'
Elster,Joshua Benjamin
Ori 6h0'26"20d4'
Elster,Wolfgang
Ori 5h51'1"20d2'
Elster,Wolfgang
Umi 13h59'42"73d7'
Elstob,Doreen
Lyr 18h19'0"46d29'
Elston,Karyn
And 2h0'56"47d22'
Elston,Steven Duane
Gem 6h26'58"13d18'
Elston-God's Little Angel,Heidi
Cas 0h27'25"54d50'
Elswick (Miss Molly), Molly Ann
Mon 6h41'31"7d6'
Elswick,Avery Cassel
Cam 5h4'24"68d17'
Elten,Rolli van
Cap 20h33'20"-10d54'
Elter,Alan & Stephanie
Mon 7h38'57"-1d53'
Elting,Paul
Peg 23h36'35"31d34'
Eltman,Patricia Josephine
And 2h31'57"50d52'
Elton "Star Gazer",In Memory of Gary
Aur 5h59'46"54d37'
Eluk,Colonel Ron & Bethany
Ori 5h49'35"18d33'
Elusia Anna
Uma 9h58'1"52d45'
Elvaelf
Cas 0h43'1"72d9'
Elvin
Dra 16h51'38"67d26'
Elvira
Lyn 7h27'23"40d38'
Elvira
Cas 0h21'27"69d38'
Elvira
Peg 22h57'20"26d49'
Elvis
Ind 21h4'6"-51d25'
Elvove
Aur 4h56'17"50d22'
Elvove,Caryn
Cas 1h0'30"61d17'
Elwell,Christina Catherine
Lyr 18h30'40"40d17'
Elwell,Michael Lewis
Cep 20h28'1"61d52'
Elwell,Raymond
Ori 5h56'56"16d43'
Elwell,Shayna Lee
Cas 3h7'0"57d34'
Elwes,Sophie
Cas 1h24'1"58d55'
Elwin
Vul 19h59'24"28d59'
Ely & Roxy
Pic 5h6'48"-49d21'
Ely,Mary Lou
Ari 2h39'36"25d8'
Ely,Sarah
Cap 20h15'36"-26d29'

Elyce Anne
Lib 15h37'48"24d25'
Elyse
Peg 21h58'42"3d22'
Elyse
Cyg 19h56'49"40d25'
Elyse Danielle
Peg 22h33'0"24d39'
Elyse Rothschild
Cas 0h36'21"68d37'
Elyse's Shining Star
Cyg 21h18'0"31d41'
Elyson
Boo 15h6'24"37d36'
Elyssa-Sparkle Of Night
Lyr 19h15'48"42d17'
Elz
Cnv 12h43'47"38d30'
Eléna et Giacomo
Cet 1h31'24"0d58'
#11 RDA Macwhiskey's Rose
Ori 5h27'0"0d44'
EM
Com 13h32'13"17d9'
E-mann,Diplom-Oecotrophologin Gaby
Lib 15h41'53"-22d48'
Em's Star
Cra 18h4'25"-39d33'
Em,Rocco James
Cep 20h16'0"60d17'
Emaho
Lep 6h3'37"-11d10'
Emana FLM,Klendalyn Starr
Uma 11h57'35"46d26'
Emanamy
Cyg 19h42'1"41d31'
Emanuel,Kristin
Lyn 8h48'33"37d54'
Emanuela
Hor 3h16'53"-47d7'
Emanuela
Eri 3h45'27"-5d21'
Emanuela
Cnv 13h28'0"51d54'
Emanuela L
Peg 23h34'58"20d11'
Emanuele
Pup 8h24'26"-27d40'
Emanuele & Francesca
Her 16h27'36"23d7'
Emanuele,Catherine
Del 20h56'12"16d23'
Emanuelita
Tel 20h8'49"-46d19'
Emanuelsons' Light
Her 16h34'0"39d14'
Emarty
Boo 15h10'0"26d4'
Emazio
Pyx 8h46'12"-28d9'
Embiricos,Alex
Cep 3h14'0"30d27'
Embrace
Lmi 9h58'26"33d37'
Embraced By Fate
Lyn 8h7'58"46d15'
Embregts,Sandra
Lyr 18h34'55"39d44'
Embroy,Robert F
Cam 9h33'31"82d9'
Embrouille-Gribouille
Dra 17h2'15"60d34'
Embry, Dorothy J
Ser 18h53'32"6d6'
Embry,James Nelson
Sct 18h34'11"-4d38'
Embuana,Adolfo Gomez
Cep 20h17'26"61d22'
Emco,Amalia
Lyn 8h6'20"36d4'
EMD I
Sco 17h24'51"-31d34'
Emeline
Boo 15h5'15"48d6'

Emerald
Uma 11h31'21"32d57'
Emerald Li
And 23h36'38"40d52'
Emerald Lim
Peg 22h21'43"28d36'
Emerald Sky
And 23h46'0"46d48'
Emerich & Diane Forever
Uma 10h31'29"48d47'
Emerling,Beverly
And 1h56'1"39d34'
Emerlyn Of James
Ori 6h5'1"-2d38'
Emerson 18th,Lee
Uma 11h21'40"43d26'
Emerson MKDCP,The Star of
Aql 19h30'12"12d22'
Emerson,Crystal Rae
Vul 21h0'16"20d18'
Emerson,David
Her 16h51'1"33d35'
Emerson,Harriett
Aql 20h3'21"1d26'
Emerson,Jack
Cep 21h32'0"61d40'
Emerson,Marie Barberio
And 0h41'43"41d5'
Emerson,Millicent
And 2h25'32"49d56'
Emerson,Norma Niehoff
Cas 1h53'14"58d20'
Emerson,Paul Edward
Boo 15h47'42"44d51'
Emerson,Rae
Uma 9h42'23"43d55'
Emerson,Reverend John
Sex 10h11'52"-6d58'
Emerson,Sally
Ori 4h56'32"5d3'
Emerson,Stanley Barry
Per 2h37'28"40d1'
Emerson,William
Ser 15h41'53"20d2'
Emertz' Star
Peg 22h43'15"4d52'
Emery,Amanda Claire
Lyr 18h57'0"30d11'
Emery,Bob
Lyn 7h12'16"51d42'
Emery,Cory Megan
And 23h17'11"48d54'
Emery,Gene & Diane
Tri 1h59'1"33d47'
Emery,Lee
Eri 2h55'1"-3d16'
Emery,Robert M
Per 3h0'0"40d44'
Emery,Sharon
And 2h3'15"40d9'
Emery,Susan
Lyr 18h34'1"43d20'
Emeryus
Crb 16h15'1"37d3'
Emi
Com 12h33'1"25d17'
Emidio
Uma 9h22'54"49d7'
Emig,Charles
Aql 20h14'36"4d22'
Emiko-Ria
Del 20h21'36"2d15'
Emili
Cap 20h36'58"-18d48'
Emilia
Cas 2h57'50"60d50'
Emilia
Uma 9h42'35"55d4'
Emilie
Peg 23h34'1"21d38'
Émile JM AL
Umi 16h26'0"72d10'
Emilie Hope
Umi 13h49'0"77d36'

Emilie-Mayens
Lac 22h27'34"40d39'
Emilina
Peg 23h38'27"18d43'
Emilio
Dra 9h35'0"73d39'
Emilio
Boo 14h11'28"34d40'
Emilio
Col 6h32'47"-35d39'
Emilio
Gru 22h6'0"-52d3'
Emilio Peter Carlo
Uma 11h42'20"63d24'
Emilio,Don
Lyr 18h27'21"42d59'
Emily
Cas 0h46'15"60d23'
Emily
Mon 6h41'26"-10d44'
Emily
And 23h0'58"51d38'
Emily
Mon 6h56'32"-10d36'
Emily
Cma 6h42'25'-13d13'
Emily
Pup 8h24'34"-23d40'
Emily
And 1h20'21"37d13'
Emily
And 1h59'34"36d41'
Emily
And 1h49'37"39d17'
Emily
Uma 10h10'15"57d49'
Emily
Cyg 19h33'27"35d35'
Emily
Com 12h3'55"28d6'
Emily
Cmi 7h59'20"18d30'B
Emily
And 0h21'15"43d12'
Emily
Cam 4h46'30"68d45'
Emily
And 23h2'47"48d52'
Emily
Cas 1h1'25"48d48'
Emily B'Mily
Cet 1h23'21"-1d59'
Emily Brooke
Mon 7h1'29"-6d35'
Emily Christine
Cyg 20h37'27"42d48'
Emily Claire
And 1h57'50"41d0'
Emily Diane
Peg 23h45'31"8d33'
Emily Grace
Tri 1h48'24"27d14'
Emily Jane
And 0h6'25"40d11'
Emily Jane
Sco 17h34'19"-43d10'
Emily Jeanette
Aql 19h14'55"14d27'
Emily K D
And 1h20'53"33d46'
Emily Kate
Lyr 19h2'50"38d34'
Emily Kate
Lyr 18h46'1"33d37'
Emily Katherine
And 23h15'35"49d5'
Emily Mae
Mon 6h30'33"10d2'
Emily Marie
Umi 15h2'54"72d6'
Emily Noel
And 2h2'58"45d41'
Emily One
Peg 22h19'53"24d12'

Emily Rachel
Aur 6h3'59"30d15'
Emily Rose
And 2h27'22"40d33'
Emily Rose
Lyn 8h4'19"39d21
Emily Rose
Lyn 9h9'45"35d47'
Emily Starr
Cyg 21h56'1"52d59'
Emily Victoria
And 23h19'47"51d53'
Emily's "Dreams"
Uma 12h2'28"58d15'
Emily's Beauty
Peg 22h45'1"25d46'
Emily's Heart
Peg 21h56'0"28d35'
Emily's Star,Jack
Dra 14h24'22"63d31'
Emily,Mary
Lyn 6h55'53"58d43'
Emina Kadric Arci Tuta
Cyg 19h28'1"39d8'
Emirali-Mesarich, Zachary
Ind 20h49'45"-59d48'
Emjay
Her 17h59'19"18d55'
Emkaran
Uma 11h50'57"40d20'
Emler,Andy
Boo 14h19'34"51d43'
Emma
Lyr 18h34'45"29d27'
Emma
Ori 5h30'57"-0d41'
Emma
Cyg 19h29'15"38d53'
Emma
Mon 6h25'39"3d44'
Emma
Cet 2h37'13"-11d56'
Emma
Cas 1h3'14"52d44'
Emma
And 2h31'21"40d30'
Emma Alice
Cas 1h16'49"60d52'
Emma Christena
Tri 1h54'25"28d44'
Emma Clair & Jonathan Mark Forever
Cyg 21h4'21"48d45'
Emma Elizabeth
And 22h57'40"51d38'
Emma Jayne
Cas 0h15'47"46d36'
Emma Jayne's Star
Cyg 19h49'32"38d2'
Emma Louise
Cyg 21h46'46"53d27'
Emma Louise
And 23h20'34"50d29'
Emma Louise
Lyr 18h16'0"34d2'
Emma O
And 23h28'59"42d34'
Emma Rose
Crb 16h17'37"33d20'
Emma Rose
Lyn 10h2'55"39d49'A
Emma S W
Umi 16h29'48"84d28'
Emma's Lucky Wish
Com 12h33'54"25d55'
Emma's Sacha
Peg 23h32'57"10d38'
Emma's Star
And 0h16'37"32d48'
Emma's Star to Wish Upon
Cnv 13h43'51"30d37'
Emma-C
Peg 22h47'32"24d59'
Emmannelle
Ori 5h59'0"14d44'

Emmans,Cindy C
And 1h42'59"40d52'
Emmanuel,Leslie Stephen
Oph 17h12'32"-20d16'
Emmanuelle
And 22h59'54"37d42'
Emmanuelle
Ori 5h46'48"19d58'
Emmanuelle,Alain
And 23h16'44"50d52'
EmmaRuth
Cas 0h28'1"61d54'
Emme Katharina
Com 12h50'23"27d44'
Emmel,John Connor
Umi 15h16'18"68d40'
Emmerechts,Myranda
Peg 22h27'58"28d24'
Emmerich,Joan Marie
And 23h21'21"48d13'
Emmerich,Ora Ann
Uma 14h0'34"60d35'
Emmerich,Toby
Lac 22h24'13"53d17'
Emmerke,Burkhard
Cmi 7h29'33"7d35'
Emmerling,John & Helen
Cam 3h27'18"60d11'
Emmerson,Olive
Cas 23h34'27"63d0'
Emmert-Raab,Ilse
Cnv 12h18'54"51d59'
Emmets,David
Cas 23h15'0"61d23'
Emmett
Oph 17h57'38"13d0'
Emmett James
Aql 19h6'50"3d45'
Emmett,Fallon Rae
Lyn 7h53'35"58d13'
Emmie
Cas 0h59'52"66d11'
Emmie
Cas 0h27'22"61d19'
Emmler,Yvonne
Psc 1h15'19"21d12'
Emmma's Treasure
Cas 2h51'37"63d52'
Emmons,Alan M
Boo 14h6'0"26d15'
Emmons,Glen
Eri 2h48'45"-7d11'
Emmons,John
Dra 18h37'1"58d29'
Emmons,John F
Vul 20h46'0"28d43'
Emmons,Judy & Mike
Cam 10h56'0"80d4'
Emmons,William H
Ser 18h15'43"-0d3'
Emmott,Jr,Robert C
Aql 20h8'1"4d34'
Emmy
Peg 22h55'60"27d47'
Emmy & Andy's Star
Cyg 20h33'56"50d4'
Emmy Lee
Lyr 18h33'44"27d54'
Emo's Star
Cep 21h30'0"58d34'
Emond,Jeannine
Cyg 19h28'0"33d10'
Emore,Doug & Carol
Lyr 18h38'28"38d42'
Empacher,Gregor F
Her 17h0'0"50d41'
Emperio,Christian Nicholas
Her 16h46'45"50d42'
Empl,Sandra
Lyn 7h43'26"38d25'
Emple,Lucy Potts
And 23h47'52"40d35'
Empress Adeline
And 23h1'1"48d36'

Empress Christine
Crb 16h3'19"31d57'
Empress V,X,&XV Jack-E
And 0h49'31"38d19'
Empyrean
Lyr 18h41'13"30d6'
Emquies,Anabella Rica
And 0h20'0"30d51'
Emrich,Carol
Cnv 13h32'1"50d46'
Emrich,Patrick David
Her 18h12'52"41d15'
Emrick,Lenny
Cep 21h26'17"55d10'
Emry,Anna Marie
Lyn 7h54'52"58d34'
Emrys,Hywel
Ori 4h51'0"5d30'
Emsell,Til Lindsey Ann
Cas 2h11'30"68d15'
Emser,Elisabeth
Sgr 19h11'6"-23d9'
Emshoff,Michael R
Mon 8h6'13"-8d58'
Emski
Cyg 19h59'0"47d49'
Emslander,Becky
Uma 11h44'0"43d40'
Emslie-Alistair's Greatest Flame,Helen
Cas 2h40'19"65d19'
Emswiler,Katheryn E
Mon 6h17'26"-5d60'
Emy
Cas 2h52'36"60d7'
Emy e Luca
Ant 10h46'7"-34d11'
Emy's Empyrean
Cen 11h31'33"-48d32'
Emy's Twinkle
Cam 3h59'28"60d14'
Emy,The
And 2h0'29"39d40'
Enchant-"Ed"
Gem 7h29'46"28d29'
Enchantement Parisienne
Ind 21h2'1"-57d3'
Enchanteresse Jeannette
Cnc 8h2'0"7d7'
Enchantingly Rich & Angela
Sge 19h18'50"18d27'
Ency
Lup 14h39'17"-42d14'
End Of The World,The
Tau 5h28'44"17d15'
Endaxi
Scl 23h11'17"-32d27'
Endeavour Elementary Mukilteo-1994
Uma 12h4'27"32d6'
Endelman,Jeffrey Bert
Aql 19h7'31"1d49'
Enderlein,Mary & Harold
Crb 15h29'21"30d47'
Enders,Fridolin Pius
Aqr 22h43'36"0d49'
Enders,Karin
Cnc 8h29'11"7d50'
Endicott,Carol Ann
Oph 18h38'20"7d42'
Endless Dream
Leo 10h0'53"13d6'
Endless Dream In Maki
Gem 7h10'57"15d56'
Endless Ecumenicism
Cyg 21h33'25"50d30'
Endless Lite of Yellow Stone
Lyn 7h21'15"58d48'
Endless Love
Lac 22h22'36"53d20'
Endless Love
Aql 19h42'12"11d32'
Endless Spirit
Cma 6h47'45"-19d29'

Endless Éléna
Peg 0h1'0"21d21'
Endlessly Allen
Per 1h56'0"56d34'
Endlessy Ermanno
Cae 4h52'1"-32d59'
Endora,Precious Princess
And 2h12'51"50d48'
Endrelunas,Richard M
Umi 14h23'21"67d10'
Endresen,Linda Marie Lehti
Cas 0h25'20"66d24'
Endriss,Richard
Boo 15h6'35"20d32'
Endsley,James E
Aur 5h10'34"40d39'
Endurance
Ori 5h52'18"19d38'
Endymion
Sct 18h40'58"-6d46'
Enedelia
Mon 6h39'1"-10d43'
Eneida Socorro
Mon 7h7'26"-5d58'
Enem-27 Ott 94
Sco 17h54'20"-30d46'
Energy
Del 20h13'27"11d2'
Enevs 1
Cet 2h58'38"3d54'
Enewaldsen Christensen Nico
Dra 15h4'22"62d58'
Eney,Carolyn A
Del 20h23'50"20d2'
Eng,Deanna Susan Jerry
Cyg 19h44'1"30d13'
Eng,Reid
Cyg 20h7'33"40d21'
Eng,Steve
Oph 17h37'56"-24d0'
Engblom,Carolyn
Cyg 20h17'56"38d44'
Engblom,Janice R
And 2h32'1"42d27'
Engblom,Shirley Ada Laura Kaeser
Cas 0h57'35"63d7'
Engbrecht,Janette A
Eri 3h52'56"-2d52'
Engdahl,Suzanne Marie
Peg 23h2'19"30d31'
Engebretsen,Thorunn
Aur 5h24'0"40d35'
Engel
Ori 5h55'22"17d43'
Engel und Liebes
Uma 11h57'37"32d17'
Engel,Alexander
Lac 22h48'27"54d55'
Engel,Dana Jill
Cam 4h23'57"61d41'
Engel,David
Lmi 10h37'0"38d39'
Engel,Dieter
Her 17h17'44"54d13'
Engel,Dieter Fritz
Ari 2h38'32"21d44'
Engel,Fred Edward
Vir 11h56'39"-5d14'
Engel,Joseph R
Cep 2h45'1"77d51'
Engel,Leonard H.
Ser 15h9'37"10d38'
Engel,Richard G
Psc 0h47'47"20d7'
Engel,Robert K
Lac 22h50'27"53d38'
Engel,Stacee J
Sgr 18h52'31"-25d5'
Engelbach,Klaus und Else
Uma 11h24'0"41d10'
Engelbart,Mark A
Crt 11h48'1"-17d57'
Engelberger René
Cas 1h46'1"61d28'

Engelbrecht,Bill
Aql 18h39'22"-2d39'
Engelchen
Lib 14h57'54"-20d18'
Engelchen-Elke
Cnc 8h57'19"8d47'
Engelen,Robert
Cep 21h42'0"63d47'
Engelhard,Erich
Aql 19h5'0"10d31'
Engelhard,Max Verser
Aur 6h29'0"33d45'
Engelhardt, Moritz Markus Alexander
Lib 14h46'32"-23d51'
Engelhardt,Felix Jan Fabian
Vir 13h30'23"-12d5'
Engelhardt,Michael W
Cma 6h51'39"-18d39'
Engelhart,Lisa & Mitchell
Sge 18h58'13"19d45'
Engelhart,Manfred
Aur 5h8'30"42d48'
Engelkamp,Eshana Jolanda M M
Cnv 13h31'51"47d48'
Engelman,Heather Sullivan
Peg 23h27'32"10d56'
Engelman,Marla Claire
Lyn 6h15'43"60d22'
Engelmann
Cet 2h58'1"5d4'
Engelmann,Betty Jean
Mon 7h36'17"-0d49'
Engels,Karl-Heinz
Cnc 8h25'46"31d41'
Engels,Peter
Sco 17h28'52"-38d20'
Engels,Rainer
Boo 13h46'44"10d50'
Engels,Renate
Sgr 19h4'12"-23d56'
Engelstad,Alice
Cam 6h36'30"68d34'
Engelstad,Deborah
Cas 23h25'34"50d29'
Enger,Walter & Charlotte
Cyg 21h53'13"41d6'
Enghaoser,Thomas L
Cet 2h34'44"0d36'
Engiles,Robert D
Dra 10h22'34"74d59'
Engine 239, Fourth Avenue Express
Vul 19h15'36"25d0'
Enginger,Anna
Tri 2h2'31"30d50'
England 50 ans,Robert
Ori 6h1'18"6d27'
England,Harriet SiEn
Cyg 21h1'48"37d24'
England,Lois & Richard
Lyr 18h41'37"38d31'
England,Lou
Cnv 13h21'1"42d11'
Englander,Gary-Laura- Tobi
Aql 19h46'43"10d9'
Englander,Karen
Cnv 12h14'43"45d10'
Englander,Kristin
Aur 4h53'27"51d26'
Engle,Forever Debbie
Lac 22h30'55"56d36'
Engle,Gerald
Ori 5h55'12"10d0'
Engle,Jason Scott
Boo 15h27'43"51d23'
Engle,Princess Alicia R
Uma 8h34'58"51d34'
Engle,Scooter Pie Lexis L
Uma 8h46'30"51d21'
Engle,Tanya Marie
Aql 19h57'22"13d31'
Engle,The Robert B
Her 17h29'46"38d4'

Englebert,Jr,Kenneth Andrew
Boo 14h26'1"26d54'
Englebrecht,Bruce
Psc 1h0'15"23d26'
Englehart,Lloyd "Daddy"
Aql 18h53'0"-2d1'
Englen,Andrew James
Aur 6h1'49"34d7'
Engler,Carla
Cas 23h22'44"60d49'
Engler,Donna A
Aql 20h35'11"0d19'
Engler,Rickey Leigh
Her 17h38'22"27d7'
Englert,Julia O & Robert G
Crb 16h23'50"38d16'
Englert,Kristy
And 0h8'18"38d47'
Engles,Bill
Dra 14h51'27"58d53'
Engles,Morris Maxwell
Sco 16h35'0"-44d23'
Englesson's Light
Aql 20h2'59"4d10'
English Plastow
Lib 14h57'0"-22d54'
English,Ashley Melanie
Peg 23h6'39"20d51'
English,Connie
Hya 9h8'21"3d45'
English,In Loving Memory Of Alan
Per 2h20'38"54d50'
English,Jamie Gail
Crb 15h56'1"28d34'
English,Jane
Peg 22h32'59"27d37'
English,Jon Clayton
Aql 20h1'44"6d15'
English,Kurt
Psc 0h48'59"31d57'
English,Leona & James
Cyg 19h56'22"38d10'
English,Leslie James
Ori 6h0'50"10d34'
English,Matthew & Lisa
Cam 3h15'33"61d4'
English,Maxine
And 1h47'46"37d24'
English,Peggy
Vul 19h58'53"26d8'
English,Sallyanne Anita
Ori 4h59'49"4d38'
Englund,Sir Scott Richard
Dra 15h48'59"62d10'
Engmann,Michael W
Ser 15h31'21"20d32'
Engstrom,Donald R
Boo 15h27'55"47d31'
Eni
Crb 15h27'46"32d24'
Enid Leslie
And 23h15'28"46d32'
Enid May-Irish Eyes Keep Smiling
Tau 5h29'13"29d27'
Enidio & Mary
Peg 22h28'52"21d40'
Enigma
Cap 20h21'0"-26d47'
Enigma
Lac 22h48'55"54d15'
Enix,Dennis Dale
Aql 20h2'0"4d57'
Enjie-Y-Ay (El)
Leo 11h32'17"10d41'
Enkhyl
Peg 22h25'38"11d59'B
Enloe MBA I994,Larry A
Oph 16h23'1"-8d28'
Enlow,Donald "Teddy"
Cma 6h51'9"-15d33'
Enlow,Lana L
Mon 7h4'0"-5d49'

Ennbee
Uma 8h38'12"50d19'
Ennenga,India
Uma 12h32'1"60d48'
Ennico,Rod
Hya 8h18'37"4d45'
Ennio,Don
Per 1h58'15"56d28'
Ennis,Bernie
Ori 6h17'39"7d31'
Ennis-Eternity,Joe
Eri 4h2'50"-8d46'
Enoara Stella Volante
Per 4h6'14"50d2'
Enoch,Beulah Taylor
Eri 3h47'0"-7d26'
Enoch,Frank C
Cet 0h41'55"-7d16'
Enoch,Martin
Ser 18h16'19"-0d2'
Enond,Nathalie
Boo 15h5'0"48d59'
Enos,David
Aql 19h43'0"10d27'
Enos,Roy
Her 16h38'49"40d50'
Enotiades,Spyros
Lyr 18h49'23"31d16'
Enric Clara
Lyn 8h57'42"42d18'
Enrica
Cma 6h30'1"-13d35'
Enrica
Cae 4h56'51"-31d16'
Enrica
Lep 6h1'20"-11d1'
Enrico
Her 16h39'27"24d51'
Enrico
Dra 12h11'18"64d13'
Enrico E Stefania
Cnv 13h37'16"28d42'
Enrico,Barbara
Dra 17h56'15"63d54'
Enright,Carol Elizabeth
Sex 9h59'21"2d28'
Enright,Danny
Per 3h0'32"41d25'
Enright,David
Aur 6h25'1"35d28'
Enrique & Michelle Friends Forever
Aql 19h30'57"8d22'
Enrique & Sergio
Per 2h50'1"32d1'
Enriquez,Marianela
Cas 3h8'1"57d31'
Enriquez,Rachele
Lyn 7h54'32"40d0'
Enros
Tel 19h4'43"-48d31'
Ensell,William G
Aur 5h42'24"54d36'
Enselman,Theodore
Aur 6h35'49"31d28'
Enser,Werner
Aur 5h8'45"42d52'
Ensey,Nicole Margaret
And 23h30'50"49d50'
Ensign,Kathie
Peg 23h36'55"22d19'
Ensign-Supreme Love, Richard J
Her 17h56'23"38d26'
Ensley,The Sportsmans Friend/Harold
Cet 3h18'29"8d27'
Ensminger,Michael
Ori 4h58'41"15d13'
Enteman,John Harrison Finley
Lyn 8h4'55"38d13'
Enticknap,Jo
Lyn 7h29'39"35d32'
Entrar Al Mistico
Cas 0h27'13"60d41'

Entrekin,Logan
Psc 1h2'15"32d37'
Entrekin,Paul
Sco 17h3'1"-31d42'
Entreprenuership Begins
Del 20h19'58"10d53'
Entrup,Jennifer L
Lyn 8h31'0"52d28'
Entzion,Matthew Stephen
Her 17h15'59"45d54'
EOE
Lac 22h9'13"50d34'A
Eosso,Jennifer A
And 23h26'0"45d25'
Eowyn
Oph 18h29'36"8d59'
Epaq
Boo 13h39'0"12d37'
Epatia
Aql 18h57'1"-5d32'
Epeteibru,Dbaro Solomon
Boo 14h59'22"24d58'
Ephrem
Ori 5h56'36"16d54'
EPI
Boo 14h19'1"13d21'
EPI Methods I
Oph 18h40'39"10d49'
Epic
Aql 20h3'1"8d52'
Epichteme,Isabelle
Per 1h44'56"53d16'
Epifano,Grace
Cyg 19h41'38"37d37'
épiphanie
Lmi 9h51'56"40d22'
Epiphany
Per 2h55'50"41d5'
Eplin,Jr,Thomas Morgan
Mon 8h5'19"-4d42'
Eppa,Dianne
Lyr 19h4'11"28d42'
Epper
Lyr 18h45'1"42d54'
Eppers,Dylan Alan
Boo 15h5'34"27d6'
Eppers,Loving Memory Kenneth (Red)
Ori 5h56'18"11d7'
Epperson,Peter Joseph
Aur 4h53'21"38d57'
Eppert,Scott Benton
Boo 14h51'28"25d6'
Eppig,Sheree L
Lyn 8h48'23"46d22'
Eppihimer,Daniel
Ari 2h7'24"20d34'
Eppihimer,Mark
Ari 2h7'26"22d19'
Eppler,Amy Elizabeth
Cas 1h4'31"61d14'
Eppler,Angela
And 23h37'22"43d59'
Eppler,Daniel Ray
Sex 9h54'27"-6d57'
Eppley,Lance
Sex 10h1'0"-5d53'
Eppright,Elizabeth
Peg 22h2'32"27d25'
Epps,Jr,Willis
Crt 11h43'0"-8d55'
Epps,Shelley
And 0h9'34"34d51'
Epps,Thomas
Mon 6h46'0"-4d49'A
Epsilon Lindy
Uma 9h17'30"68d52'
Epsilon Psi
Cam 6h50'0"68d58'
Epstein,Adam Neal
Peg 22h1'26"31d17'
Epstein,Arthur
Per 1h50'50"56d56'
Epstein,Bea
Cyg 19h22'30"44d6'

Epstein,Bea
Cyg 19h22'29"48d50'
Epstein,Brett Neal
Vul 19h42'53"26d15'
Epstein,Dr Benjamin C
Sgr 19h42'39"-42d23'
Epstein,Dr Donald
Oph 17h29'15"0d25'
Epstein,Dr Marc I
Sco 17h36'17"-38d28'
Epstein,Dr Robert S
Mon 7h45'33"-3d18'
Epstein,Dylan Jezuit
Leo 10h37'58"13d33'
Epstein,Ethan Marc
Dra 16h0'0"61d52'
Epstein,Harry H
Ori 5h41'56"-0d19'
Epstein,Ira Marc
Cam 4h57'15"65d3'
Epstein,Jacob Henning
Ori 5h56'1"13d56'
Epstein,Jill
Lyr 18h57'0"33d0'
Epstein,Joshua
Her 16h53'50"38d54'
Epstein,Joshua M
Uma 8h49'19"60d57'
Epstein,Lady-Lake
Cas 0h10'35"64d8'
Epstein,Marilyn
Vul 19h48'14"25d26'
Epstein,Nicole Joanne
Tau 4h31'1"30d19'
Epstein,Nikki
Lyr 19h23'1"38d48'
Epstein,Noah Jezuit
Per 2h25'1"55d20'
Epstein,Sam
Umi 17h9'48"80d20'
Epstein,Susan Lynn
Aur 5h26'0"40d23'
Epter,Joanne Brugger
Mon 6h59'54"-6d1'
Epting,Warren
Her 16h53'12"50d48'
Epton,Harold
Aur 6h20'55"34d44'
Equinoràv
Aql 18h57'21"-6d52'
Equinox Astrological Company
Sco 17h20'44"-31d25'
Equinoxe
Psa 22h8'14"-25d4'
Equitz,Delia Price
Leo 11h7'30"1d14'
Fr No Do Chean
Hya 10h42'19"-18d9'
Eragra 10h42'19"-18d9'
Uma 11h56'50"52d20'
Erali,Michael David
Cnv 13h57'0"40d36'
Eraso,Jimmy
Oph 16h56'52"-1d41'
Erb,Jay
Aur 6h15'59"30d0'
Erba,Edo
Cam 14h8'24"80d21'
Erbar,Dave & Molly Banahan
Peg 22h58'40"20d15'
Erbe,Bettina
Cep 22h2'33"65d25'
Erbe,Denise Ann
And 0h13'26"46d53'
Erbe,Mark
Boo 15h4'1"21d17'
Erbe,Peter
Lyn 8h1'10"40d45'
Erchick,Tim
Dra 17h38'23"61d29'
Ercol,Victoria Ruth
Ari 2h56'35"28d49'
Ercolino,Ralph J
Boo 13h37'1"19d16'

Erdelt,Ken
Her 17h59'25"30d49'
Erdem,Altan
Ori 5h3'38"0d50'
Erdman-Wright,Carolyn
Mon 6h32'34"11d22'
Erdmann,Manfred
Ser 16h2'0"2d45'
Erdmann,Siegfried Albert
Hya 9h0'14"3d11'
Erdnuss
Cnv 12h19'47"50d59'
Erdo,Elaine
Lyn 8h43'18"45d47'
Erdtsieck,Fred
Aql 19h56'46"0d21'
Erebia,Jesus Dan
Aur 5h26'11"40d5'
Eren,Dmitry Hassan
Cep 23h7'56"65d21'
Erenbergus Michaellis
Crb 15h32'21"27d31'
Eres Tu
Peg 22h36'1"10d14'
Ergottino,Mr Invincible-Louis
Lac 22h17'41"37d51'
Erhard,Andreas
Uma 11h47'25"33d37'
Erhardt,Stephanie Jewel
Peg 22h31'60"25d39'
Erhart,Hans
Hya 8h59'50"1d43'
Eric
Aur 6h36'0"40d22'
Eric
Gem 7h3'18"24d56'A
Eric
Dra 16h34'35"71d42'
Eric
Uma 10h18'59"44d9'B
Eric
Aur 5h25'33"30d0'
Eric
Tri 2h33'22"31d19'
Eric
Cas 2h51'32"60d35'
Eric
Uma 12h17'47"60d52'
Eric
Her 18h13'15"48d17'
Eric
Com 13h6'24"20d57'
Eric
Oph 17h1'57"11d21'
Eric & Brian The Ring Bears
Boo 14h32'0"21d38'
Eric & Dale
Aql 19h57'23"14d46'
Eric & Danielle
Cyg 19h29'20"34d13'
Eric & Emily
Sex 10h7'30"-1d44'
Eric & Heidi
Cyg 21h3'1"37d14'
Eric & Jennifer
Peg 21h48'1"36d5'
Eric & Jenny
Cas 3h35'0"73d7'
Eric & Katie-1995
Cyg 21h36'15"37d37'
Eric & Pamela
Aql 19h53'44"15d27'
Eric & Robin's "Lovestar"
Cyg 21h50'55"40d1'
Eric & Salley
Lyr 18h24'41"42d18'
Eric & Susan's Peace of Heaven
Boo 14h35'1"47d31'
Eric & Victoria
Cyg 19h27'54"31d55'
Eric & Zack-Stars Forever 12-28-94
Dra 20h9'59"63d56'

Eric Andrew
Dra 16h41'13"62d3'
Eric David Like An Eternal Dream
Her 17h24'59"40d39'
Eric I Love You-Happy Anniversary
Uma 9h16'33"50d6'
Eric James
Uma 10h40'38"50d3'
Eric Jay
Dra 16h32'22"61d2'
Eric Kevin
Psc 1h37'0"21d37'
Eric Paul Jesse
Equ 21h20'13"11d1'
Eric Spencer
Ori 5h53'16"10d6'
Eric The Explorer
Vel 10h5'29"-45d24'
Eric The Great
Dra 19h3'39"48d38'
Eric The Magnificent
Her 16h17'35"41d50'
Eric Thomas
Ori 6h7'0"0d57'
Eric's Bit of Heaven
Per 1h55'16"50d34'
Eric's CasaNova
Uma 8h39'12"71d21'
Eric's Dream
Her 17h42'24"40d56'
Eric's Escape
Sex 10h41'33"-5d36'
Eric's Eternal Star
Uma 11h24'55"49d0'
Eric's Forever Star
Cam 6h16'45"70d25'
Eric's Hope
Per 4h19'24"51d42'
Eric's Valentine
Uma 9h37'13"68d15'
Eric-Diane-Nan-Heather - Matt-S
Lyr 18h43'43"45d20'
Erica
Peg 23h33'43"18d26'
Erica
Aql 19h28'47"-0d48'
Erica
Lyn 7h39'18"42d25'
Erica
And 23h2'39"48d37'
Erica
Lib 15h38'19"-24d4'
Erica
Sgr 18h52'59"-35d56'
Erica
Aqr 21h53'32"0d30'
Erica Ann Lynn
And 23h47'56"44d47'
Erica Anne
Ori 6h3'50"8d59'
Erica Brooke
And 1h27'50"37d28'
Erica Elizabeth's Star
Cas 3h21'33"73d36'
Erica Everlasting
Mon 6h19'36"-6d36'
Erica Lynn
Cas 23h30'45"62d58'
Erica Lynn
Vul 19h43'1"20d17'
Erica M 10 1
Cnc 18h32'11"10d55'
Erica Tobleriana 2436466
Vir 11h38'25"2d22'
Erica's Diamond
Leo 10h48'18"2d26'
Erica's Unchained Melody
Crt 11h42'19"-8d37'
Erica-N-Brian
Uma 11h31'19"44d12'
Erich,Volker
Cas 0h7'1"60d42'

Ericka's Dreamkeeper
Peg 23h46'18"10d17'
Ericka-7
Peg 23h30'46"22d40'
Ericksen,Lisa Aimeé
Tau 3h59'49"20d21'
Ericksen,Lynn Suzanne
Del 20h37'11"18d48'
Erickson's Peace,David Allen
Aql 19h58'12"14d39'
Erickson,Amanda Rae
Cas 23h31'1"60d15'
Erickson,Amy Margaret
Crb 15h27'43"30d14'
Erickson,Barbara Christine
And 0h19'23"41d6'
Erickson,Chad E
Dra 20h21'1"62d25'
Erickson,Chelsea McCall
Eri 3h51'19"-2d6'
Erickson,Daniel James
Hya 8h47'37"-7d5'18"
Erickson,Debi & Doobie
Per 2h48'50"41d13'
Erickson,Edward A
Cep 23h15'38"64d8'
Erickson,Elizabeth M
And 23h31'59"48d21'
Erickson,Eric
Hya 9h11'36"-14d18'
Erickson,Helge J
Sge 19h54'40"16d48'
Erickson,Jack
Mon 7h58'15"-5d6'
Erickson,Jane Kristine
Cas 2h26'15"67d53'
Erickson,Jeanne M
Lyn 7h55'0"43d49'
Erickson,Jonathan D
Cep 21h54'13"56d14'
Erickson,Karen Marie
Cam 8h29'46"80d26'
Erickson,Kay "Eric"
Uma 11h56'32"38d16'
Erickson,Kiote James
Her 16h44'30"39d3'
Erickson,Kristina Lee
Mon 6h56'0"10d4'
Erickson,Laura
Cas 0h5'45"59d15'
Erickson,Loren Leslie
Dra 17h49'31"61d36'
Erickson,Marianne
Eri 3h25'42"-13d32'
Erickson,Marie E
Peg 22h29'51"21d31'
Erickson,Queen Esther
Cas 23h22'0"54d1'
Erickson,Reynold
Lac 22h32'24"53d27'
Erickson,Robert F
Mon 6h55'18"0d22'
Erickson,Robert L
Cnv 13h4'0"52d34'
Erickson,S Kimberly
And 1h58'21"38d3'
Erickson,Susan Carol
Mon 7h57'57"-6d53'
Erickson,Walt & Nancy
Umi 15h6'39"67d38'
Ericsandra
Cyg 20h36'1"40d14'
Ericson,Mom & Dad
Cam 3h29'1"53d54'
Erik
Per 1h53'58"48d35'
Erik
Tau 3h37'46"23d8'
Erik & Christine
Cam 4h16'40"61d24'
Erik & Pamela Forever
Boo 14h46'1"47d30'
Erik et Brigitte au Paradis
Lyr 18h31'40"38d54'

Erik's Everlasting Starfire
Aur 7h3'59"39d49'
Erik's Piece of Heaven
Lmi 10h55'26"25d37'
Erika
Eri 4h35'0"-12d18'
Erika
And 1h57'56"37d50'
Erika
Del 20h51'45"9d5'
Erika
Lib 14h30'0"-23d33'
Erika
Cam 6h9'11"67d58'
Erika
Hya 10h47'28"-11d54'
Erika
Com 12h46'10"20d55'
Erika
Ari 2h15'21"24d2'
Erika
And 23h30'21"49d5'
Erika & Bill"Forever In Love"
Cam 3h44'38"61d26'
Erika Diamond
Cas 23h19'0"60d20'
Erika Jo
Vul 21h14'14"20d3'
Erika Renee
Lyr 19h12'26"33d48'
Erika Star,The
Lyn 9h15'1"36d22'
Erika Tu Amo
Cas 0h1'30"58d58'
Eriko
Leo 10h16'44"9d15'
Eriko-The Sweetest Confectioner
Ari 1h48'1"13d29'
Eriksen,Jean
And 0h38'44"37d45'
Eriksen,John L (Chi-Chi)
Dra 15h57'56"67d35'
Erikson,Anne E
And 23h2'1"50d57'
Erikson,Jr,Arthur
Her 17h23'15"37d43'
Erikson,PhyllisAnn
Uma 11h7'1"40d25'
Eriksson,Jonas
Uma 9h10'32"47d47'
Erin
Cyg 19h22'12"28d12'
Erin
And 0h32'28"36d33'B
Erin
Ari 3h13'0"28d43'
Erin
Cas 22h56'30"54d0'
Erin
Lyn 7h2'40"50d33'
Erin
Lyn 6h52'1"60d30'
Erin
Hya 8h26'17"-0d2'
Erin
Boo 15h38'18"48d0'
Erin
Lmi 10h58'23"33d19'
Erin & Kurtis
Cyg 19h58'22"30d8'
Erin & Paul
Eri 3h54'26"-1d6'
Erin Ann
Lyr 18h25'0"47d29'
Erin Chocolate Star, The
Peg 22h1'13"27d27'
Erin Colleen
Cmi 7h41'0"5d19'
Erin Elizabeth
Peg 23h40'49"28d36'
Erin Elizabeth
Ori 5h55'14"17d26'
Erin Gayle
Cam 6h10'36"68d45'

Erin Leia
Peg 23h20'49"32d28'
Erin Leigh's Star
Mon 7h19'9"-6d31'
Erin Louise
Cas 23h0'41"58d50'
Erin Lynn
Ori 6h7'21"8d1'
Erin Lynn
Aql 19h26'35"10d24'
Erin M
Psc 0h8'28"7d50'
Erin MacKenzie
And 23h32'13"46d8'
Erin Marie
Psc 23h27'44"6d44'
Erin Michele
Lyn 8h58'0"46d47'
Erin Nicole
And 2h29'59"39d33'
Erin P C
Cnc 8h2'15"10d49'
Erin Rose
Dra 18h23'19"70d14'
Erin Rose
Lyr 18h58'44"27d38'
Erin Star
Cam 3h51'57"78d49'
Erin Taylor Hope
Lyn 8h56'1"35d8'
Erin Victoria's Light
And 2h36'13"38d7'
Erin's Ecstasy
Mon 7h11m25"-7d5'
Erin's Star
Cas 0h13'26"46d53'
Erin's Star
Aur 6h27'52"31d21'
Erin's Star
And 2h25'10"49d3'
Erin's Star LAF JC
Cyg 19h56'15"37d39'
Erin's Wish
Psc 1h2'50"20d29'
Erin61196
And 0h21'52"30d13'
Eringaard,Mary Jane
Cas 0h29'36"75d7'
ERIPASCHMA-C V
Lib 15h7'56"-23d56'
Eris
Cam 4h19'11"71d9'
Eris
Cep 23h17'29"68d14'
Erisman,Pamela Sue
Boo 14h8'0"53d59'
Eritros
Ind 21h9'9"-51d31'
Eriv,Carol & Dr Alfred
Crb 16h0'21"31d25'
Erjie,Raymonde
Boo 15h21'47"33d28'
Erker,Anthony James
Cet 2h43'54"5d41'
Erker,Susanne
Cas 0h11'45"60d30'
Erkfitz,Erwin
Per 2h51'60"45d59'
Erki & Werner
Ari 2h51'55"20d52'
Erkie
Cnv 12h44'42"33d53'
Erlandson,PhD,Dr Robert Franz
Oph 18h1'27"8d15'
Erlich,Linda Ilene
Sco 16h2'35"-23d41'
Erlichman,Jonathan
Oph 18h16'17"11d25'
Erlos,Marielle
Umi 15h31'47"77d58'
Erma
Cnv 13h57'16"41d35'
Erma Jean
And 2h25'26"38d44'

Ermando-Ermi
Cam 3h42'35"68d15'
Ermes
Hor 3h33'47"-48d16'
Ermillo,Vincent Joseph
Her 16h56'23"18d46'
Erminia
Ant 10h43'40"-39d32'
Erna
Lyr 18h44'42"40d14'
Erna
Ari 3h1'39"23d5'
Erna
Uma 13h35'52"48d26'
Ernest's Star,Janice
Peg 23h23'39"31d39'
Ernest,Sarah
Peg 22h23'32"26d4'
Ernest-Marie
Mon 6h54'46"-8d50'
Ernestina & Carlo
Pho 0h28'22"-41d39'
Ernestine Alpha
Mon 7h2'0"3d47'
Ernestine Linn
Lyr 18h54'28"33d8'
Ernestlee
Lac 22h27'42"40d29'
Ernesto
Lib 15h40'34"-28d41'
Erni,die Liebe Meines Lebens
Aur 5h58'11"37d35'
Ernie #1 Superstar
Per 4h31'50"31d24'
Ernie B
Oph 17h31'54"-22d27'
Ernie's Heavenly Body
Per 2h26'30"58d56'
Ernie's Lucky Star
Lac 22h33'50"48d58'
Ernie's Star
Dra 16h22'18"66d40'
Ernie's Star In Loving Memory
Eri 3h58'54"-5d50'
Ernsberger,Anne-Marie
And 23h1'40"51d31'
Ernst,Amy Marie
Peg 22h33'25"21d7'
Ernst,Frank
Sgr 19h17'20"-42d0'
Ernst,George
Oph 18h18'19"6d38'
Ernst,Helga
Tau 5h45'57"21d18'
Ernst,Jane & Jim
Umi 13h58'55"71d14'
Ernst,Liselotte
Tau 5h47'0"27d54'
Ernst,Madelyn Ketover
Com 13h14'39"28d42'
Ernst,Nicholas
Sex 10h40'28"2d35'
Ernst,Samantha
Mon 6h27'1"10d31'
Ernst-Peter
Boo 13h39'32"13d33'
Ernsteins,Jake Joseph
Uma 11h18'43"30d14'
Ernster,Denny Alan
Hya 8h29'0"-5d57'
Ernster,Donna Jean
Com 12h3'27"20d1'
Ernster,Erin Diane
Cyg 19h10'45"48d24'
Ernster,Gary Alan
Boo 14h31'13"8d31'
Ernster,Kevin Barrett
Cmi 8h3'55"5d52'
Erobreren,Jon øyvind
Crb 15h24'56"30d14'
Eroica,Stella
Oph 18h34'0"11d47'
Eros
Lac 22h40'15"53d11'
Eros e Liana
Lyn 7h53'1"41d43'
Eros Eric
Her 17h32'28"27d49'
Eros,Leontine W
Eri 2h49'31"-5d35'
Erotsieck III,Freddie
Ori 6h16'1"18d57'
Erpenbach,Nena
Aql 19h30'23"14d29'
Err...
Dra 12h42'38"67d40'
Errante,Frances Oddo
Aql 18h58'11"-6d47'
Errera,Phyllis
Cnc 9h5'34"31d45'
Errickson,Billy Dick
Uma 12h22'0"54d54'
Errigo,Catherine
Cam 5h5'21"68d31'
Errington,Jr,Kevin
Per 3h1'44"47d30'
Errol
Uma 9h16'48"53d28'
Ersbak,Jamie
Ori 5h55'55"10d4'
Erschy
Her 17h57'1"42d8'
Erskine,David P
Per 4h42'52"37d35'
Erskine,Dennis
Per 2h40'25"40d11'
Erskine,Jack
Umi 17h19'1"86d12'
Erskine,Jessica
Lyr 18h41'48"34d25'
Erskine,Peter Albert
Her 17h2'25"38d2'
Erskine,Rachel Diana Mary (Rickman)
Boo 14h54'40"44d26'
Erskine,Richard "Tigger"
Her 17h10'1"27d35'
Erskine,Tara
And 23h21'29"48d14'
Erspamer,Elite
Per 2h0'43"50d17'
Ertan,Atilla
Gem 6h36'54"12d35'
Ertas,Melek
Tau 5h30'14"28d56'
Ertegun,Selma
Cas 22h56'40"55d22'
Ertl,Thomas
Aql 19h2'54"17d23'
Ertle,George Joseph
Cas 2h27'25"60d42'
Ertmoed,Ernest Edward
Aur 5h18'34"42d25'
Ertnes,Heidi & Kjetil
Cyg 21h1'51"36d6'
Eruera
Pho 23h37'48"-42d1'
Eruysal,Emily Rose
Cas 1h36'15"61d18'
Eruysal,Taylor Evan
Cas 1h20'20"61d0'
Ervin & Anthony
Cet 1h55'57"-3d17'
Ervin's Wedding Star, Maggie & Don
Lyn 8h1'60"48d42'
Ervin,Wyeth Q & Brenda G
Cep 22h16'59"60d46'
Ervin Dinari, Abdclouafi & Jerri
Eri 4h4'50"-1d59'
Erwin III,Henry Washington
Her 17h28'17"20d18'
Erwin,Avanell
Cyg 21h6'37"30d32'
Erwin,Blane
Lyr 18h46'14"32d15'
Erwin,Dale H
Cep 22h14'52'63d52'
Erwin,Deborah
Lyn 7h15'33"59d59'
Erwin,Douglas Scott
Her 17h29'18"28d16'
Erwin,E David
Cnv 12h32'59"33d20'
Erwin,Gregory DuBois
Cam 5h2'0"54d53'
Erwin,Heather Jane Allen
Com 12h28'57"27d50'
Erwin,Iva Fern
And 0h57'50"33d48'
Erwin,Margaret
Peg 23h30'14"21d10'
Erwin,Mary Ann Millot
Com 12h28'47"27d47'
Erwin,Robert & Anita
Cet 3h0'34"0d22'
Erwin,Ronald Kenneth
Lac 22h34'26"53d20'
Erwin,Terry Lee
Umi 15h28'40"81d39'
Erwin,Terry W
Her 16h39'46"28d17'
Erwood,Carleton R
Lac 22h51'29'54d53'
Erxleben,Gene
Dra 14h57'12"62d14'
Erzen-Por Vida,Bill & Tina
Peg 22h56'33"18d48'
Es Destino
Peg 22h17'56"35d26'
Esak,Jr,Joseph Ronald
Cet 1h49'60"-3d42'
Esarey,Brenda
Mon 6h53'31"-1d26'
Escada Star
Cnv 12h37'27"39d2'
Escallon,Santiago Moreno
Cep 22h49'49"59d22'
Escape
Aur 6h14'58"32d32'
Escario,Julien
Lyn 8h27'37"42d0'
Eschbacher,Robert J
Cep 23h32'0"70d23'
Eschelbach,Peggie & Karl
Aql 19h31'26"12d25'
Eschenbach,Margaret Nagel
Cas 2h32'14"57d46'
Eschenbrenner,Bill
Boo 15h29'11"41d51'
Escher,Theodora
Cam 12h5'47"80d18'
Eschweiler III,Earl Edward
Per 2h6'40"58d49'
Eschweiler,Armin
Lyr 19h21'20"30d33'
Escobar,Carlos Alberto
Lac 22h7'55"41d2'
Escobar,Diana
Cet 0h53'52"0d2'
Escobar,John
Her 16h0'1"44d53'
Escobar,M Colleen
Cam 5h59'55"67d39'
Escobedo,Jennifer K
Cas 0h34'30"60d7'
Escobido,Lenny B
Hya 8h58'41"1d13'
Escoffier,Pierre
Oph 17h2'28"10d46'
Escos,Juan Hormaechea
Sex 10h26'21"5d36'
Escoto,Catalina
Cas 2h55'36"70d0'
Escudié,Remi Christian
Cmi 7h19'37"4d43'
Esgar,Ruthie "Rhino"
Aur 6h31'51"38d49'
Esh Gupta
Lyr 18h29'40"38d23'
Eshelman,Lynn Norah
And 23'40"41d10'
Esher,Jennifer
Peg 23h47'54"28d49'
Eshima,Dr Rachel "CG" Waldron
Del 20h23'26"10d1'
Eshrat
Ori 4h50'35"0d40'
Esiacen
Aql 18h56'25"-0d45'
Eskelson,C TomTer
Cyg 20h39'49"38d19'
Eskew,David Robert
Cet 2h6'59"0d21'
Eskew,Maude V
And 2h30'36"42d38'
Eskie,Charles John
Dra 19h52'35"60d26'
Eskimos & Butterflies
Per 2h31'38"57d14'
Eskind,Neil & Sue
Peg 22h1'48"27d47'
Eslinda's Glow
Pho 0h27'30"-40d44'
ESME
Col 6h1'2"-33d18'
ESME-I Miss You & I Love You Everyday
Sco 15h48'15"-21d25'
Esmeralda
Col 6h31'47"-33d47'
Esmond,Steven
Uma 11h41'54"33d13'
Esoomsevil In Morgan Danielle
Lyn 7h39'45"38d52'
Esp,William L
Oph 17h31'53"7d33'
Espanet,George & Iris
Hya 8h10'51"1d18'
Esparza
Lyr 18h59'28"30d46'
Esparza,Joanna
Mon 6h22'40"2d54'
Esparza,Michael Daniel
Cet 0h31'1"1d18'
Espera
Cnv 12h46'57"38d28'
Esperanza
Cnv 14h1'30"46d8'
Esperanza
Cam 6h8'48"58d53'
Esperanza
Aur 5h30'1"54d56'
Esperanza
Peg 22h0'29"34d58'
Esperanza Valcárcel Grande
Cep 22h8'17"68d23'
Esperon "Silver Star"
Eri 3h37'23"-5d54'
Espeso,Erin
Uma 8h34'38"56d23'
Espiau,Mary
Peg 21h55'37"22d32'
Espie,Daniel
Peg 23h5'50"20d49'
Espino,Alexia Roxanne
Lyi 18h35'56"45d28'
Espinosa,Dr Abner J
Oph 17h31'14"-20d31'
Espinosa,Manuel Roberto
Ser 17h54'0"-13d8'
Espinoza,Gloria
Mon 6h17'35"-5d37'
Espinoza,Mark Damon
Boo 14h18'14"18d17'
Espinoza,Victoria Shannon
Cam 7h19'19"67d47'
Espinoza,Yosselyn G
Cam 7h37'30"60d28'
Espitalier-Noel,Gerard
Cru 12h13'7"-58d8'
Espo,Andrew Michael
Cep 20h46'14"60d52'
Esposito I Love You, Andrew
Lac 22h3'49"50d42'
Esposito(Boss Man), Michael
Aur 6h22'0"31d24'
Esposito,Barbara Lee
Cam 12h29'44"80d27'
Esposito,Cory Joseph
Del 20h13'30"9d44'
Esposito,Elizabeth A
Uma 10h28'20"48d56'
Esposito,Erik Christen
Lyr 19h18'0"40d52'
Esposito,Frank
Uma 8h35'0"70d50'
Esposito,James Charles
Aur 4h59'25"51d13'
Esposito,Janet C
And 0h22'53"34d46'
Esposito,Joseph M
Dra 18h52'1"70d11'
Esposito,Louis James
Aur 5h58'54"30d4'
Esposito,Luca
Sgr 19h28'55"-33d8'
Esposito,Lucrecia Lee
And 0h6'1"34d50'
Esposito,Nathaniel J
Per 3h1'34"47d51'
Esposito,Nicholas
Dra 9h30'50"73d26'
Esposito,Nicholas W
Per 3h1'1"46d27'
Esposito,Peggy
Cas 1h5'36"60d3'
Esposito,Richard H
Lib 15h16'29"-8d24'
Esposito,Scott & Elizabeth
Lyn 7h4'43"44d60'
Esposo,Jr,Richard Peter
Vul 20h45'0"28d39'
Esprit Apparenté
Cyg 21h8'47"31d6'
Esprit de Corps
Cyg 20h23'28"39d28'
Espurvoa,Deborah Anastacia
Peg 22h3'1"11d14'
Esquivel,Christina
Cas 2h45'39"61d37'
Esquivel,Dianna Jean
Eri 2h46'35"-18d54'
Esquivel,Meredith
Cas 2h50'0"61d49'
Esreikie
Per 3h6'42"56d56'
Essa,Forever Joe & Margaret
Cyg 20h56'46"31d8'
Essa,Lisa Beth
Com 13h5'43'21d8'
Essan
Ori 5h31'47"0d15'
Essebier,Jürgen
Ser 15h56'11"-1d7'
Esselljay-J1
Umi 9h32'17"88d15'
Essen,Herman
Her 17h30'0"22d21'
Essence of Woman,The
Cyg 21h2'15"38d36'
Essencsabrine
Mon 7h6'10"-6d58'
Essenfeld,Ilena
Lyn 7h57'53"44d55'
Esser,Carol
Aql 18h57'10"-5d57'
Esser,Jake Arthur
Psc 23h7'14"5d5'
Esser,Nick
Uma 11h39'23"31d2'
Esser,William F
Cnv 12h28'50"48d37'
Essert,Shane Mitchell
Peg 21h24'40"8d4'
Essex Computers,Inc
Crb 16h19'31"31d34'
Essex,Jeffrey Scott
Aql 19h54'35"10d24'
Essex,Marie
And 1h11'22"39d39'
Essing,Heinz
Vir 13h4'34"-8d49'
Essington,Jay Matthew
Mon 6h25'49"-10d32'
Essis
Mon 6h6'24"-8d18'
Esslinger,Jane
Cas 1h46'17"75d31'
Essock,Ed
Her 18h13'49"40d12'
Esson,Grandpa James William
Uma 9h46'59"45d46'
Essy
Uma 9h44'17"57d37'
Estabrook,Pinky
Eri 4h29'58"-10d13'
Estadt,C Scott
Vul 19h47'23"28d58'
Estarus
Oph 17h59'51"7d45'
Este,Robert M
Peg 22h40'42"24d48'
Estefuzzy,Angie
Uma 10h52'42"56d42'
Estel,James
Cet 0h49'25"-4d45'
Estela Rios Portela Da Cunha Seabra
Oph 17h38'34"-20d9'
Estell,Ashley Nicole
Mon 8h2'2"-10d14'
Estell,Sr,Ralph A
Her 17h53'0"20d12'
Estella
Cyg 21h56'45"52d58'
Estella
Lyr 18h44'51"40d19'
Estelle
Uma 11h32'23"31d17'
Estelle
Per 1h51'12"50d12'
Estelle
Cap 20h56'18"-23d57'
Estelle & Mark
Cyg 19h30'55"36d32'
Estelle De Jesus-Kelly
And 1h45'32"37d31'
Estelle Lee
Eri 3h14'39"-12d26'
Estelle Star
Dra 11h28'55"66d27'
Estelle,Amanda
And 1h47'1"40d31'
Estelle-Catherine
Cnv 12h39'22"35d7'
Esten,Jonathan
Per 2h7'42"47d43'
Esten,Juanita Marie Dudley
Lyr 18h32'20"36d53'
Estena,Lurine
Ori 6h9'15"0d58'
Estenota,Carine
Aur 5h50'13"54d34'
Estep "Miracle Baby", Matthew Joseph
Boo 15h3'39"30d32'
Estep,Brock Earl
Cmi 7h53'0"1d42'
Estep,Cecil
Sex 9h39'0"5d27'
Estep,Karen Denise
Aql 20h19'24"5d20'
Estephan,James George
Boo 15h11'46"47d45'
Estes
Lac 22h27'1"41d12'
Estes "Lucky",Melinda
And 23h3'28"48d20'
Estes Forever Nov 4 1950,Terry
Tri 1h56'51"28d25'
Estes,Betty R
Peg 22h31'46"10d19'
Estes,Carla
Mon 6h36'19"1d26'
Estes,Christopher William
Lmi 10h10'0"35d20'
Estes,Conor Ian
Per 2h9'41"58d26'
Estes,David Haeger
Her 16h22'33"23d24'
Estes,Frances Louise
Mon 6h55'38"0d39'
Estes,Helen T
Vul 19h23'31"26d42'
Estes,James Keith
Per 2h28'48"58d14'
Estes,Jodi
Vul 19h58'52"29d7'
Estes,LaVerne
Ori 6h16'17"8d20'
Estes,Michael Anthony
Gem 6h41'57"16d31'
Estes,Michael Keith
Uma 11h47'43"42d39'
Estes,Mr Nicholas Haeger
Uma 14h25'32"61d24'
Estes,Shirley A
Umi 16h51'0"75d54'
Estes,Thelma (Howard)
Vul 20h0'25"23d2'
Estes,Tim & Jen
Sge 18h58'38"19d18'
Estetica Annabel
Del 20h17'1"15d10'
Estetica Rita
Del 20h12'39"15d12'
Esteves,Robereo
Aql 20h1'53"8d54'
Estevez,Anne Marie
Mon 7h14'48"-10d7'
Estevez,Anthony Joseph
Sex 9h55'43"5d46'
Estevez,Cassandra
Ori 5h26'18"8d24'B
Estevez,Henry
Aur 6h39'22"38d46'
Estevez,Michael Murphy
Eri 3h43'31"-4d40'
Estey,Darlene
Cas 0h19'0"60d25'
Esthelita
Lyn 7h51'58"43d23'
Esther
Uma 9h9'51"58d59'
Esther
Cas 1h10'1"60d43'
Esther
Sgr 19h1'8"-25d11'
Esther
Ori 5h17'52"-4d27'
Esther (Aerial)
And 2h19'30"39d41'
Esther Amelia
Peg 0h8'0"13d23'
Esther Lillian
Cas 22h57'15"54d24'
Esther Mariah
Uma 14h15'40"60d40'
Esther,Dolores
Umi 16h18'33"70d31'
Estheria,Angela
Cas 0h43'40"62d54'
Estill,Jo
Pyx 8h51'53"-28d6'
Estl,Eva & Karl
Uma 8h43'16"51d40'
Estock,Marguerite M
Cyg 19h25'21"33d23'
Estrada,Alyssa Cecilia
Cyg 19h40'32"38d50'
Estrada,Dion Joseph
Per 2h53'11"31d4'
Estrada,Max Skeeter
Her 17h1'20"44d57'
Estrada,OB David
Ori 6h13'18"1d11'B
Estrada,Philip Michael
Uma 8h48'0"67d38'
Estrada,Sara Dyanne
Aql 19h34'49"1d15'
Estralita
Lyr 18h33'21"39d35'
Estrela
Her 17h23'34"27d20'
Estrella
Lyn 7h1'15"44d44'
Estrella Cristina
Ori 5h32'30"-8d35'
Estrella d'DANA
Peg 21h25'44"3d34'
Estrella de Amor
Crb 16h3'57"28d35'
Estrella de Ariel
Cma 6h54'26"-15d23'
Estrella De Daryl
Ari 2h27'0"11d11'
Estrella De Evelyn
Lyn 8h18'18"36d53'
Estrella de Evis
Lmi 9h22'36"34d0'
Estrella De Javier
Aql 20h12'57"11d0'
Estrella De Lucia Y Cecilio
Lyn 8h27'32"34d23'
Estrella De Maria
Gem 8h2'44"28d11'
Estrella De Nuestras Amore'
Cas 2h58'32"70d41'
Estrella De Un Angelito
Peg 22h2'57"2d30'
Estrella Donna
Mon 7h13'2"-6d47'
Estrella Elena
Mon 6h30'38"11d8'
Estrella Hermoso Para Ti
Peg 22h14'37"33d46'
Estrella Jose-Olga
Lmi 10h9'55"34d4'
Estrella,Marco
Lyr 18h26'40"46d1'
Estus,Rachael
And 23h17'16"48d52'
Esty,Elinor Rae
And 23h43'16"44d9'
Esty-Beske
Umi 16h10'26"70d17'
Esway,Don
Dra 17h31'0"61d45'
Eszenyi,Bertha
Aur 6h31'57"37d45'B
Eszenyi,Joseph
Aur 6h31'57"37d45'A
Etarnal love of Mikihiko & Mika
Ari 2h27'44"20d39'
Etc Mari
Cas 0h45'53"71d32'
Etcetera I
Lmi 9h59'0"28d57'
Etchegorry,Olga
Peg 23h42'55"26d37'
Etchells,Geoff & Irene
Cyg 19h24'6"30d27'
Etcheverry,Gene & Lizette
Cyg 21h31'0"30d30'
Etchison,Eric Morris
Ser 15h38'1"2d40'
ETEMEFEO My Beloved Michelle
And 2h23'51"45d11'
Eternel
Oph 18h0'20"12d6'
Eternal
Cyg 19h33'14"34d15'
Eternal Beaute Michelle
Ori 6h13'43"7d56'
Eternal Brotherhood
Equ 21h7'59"11d47'
Eternal Dreamer Tony
Uma 13h57'10"51d34'
Eternal Flame
Lyr 19h25'1"38d22'
Eternal Flame
Sge 20h4'31"20d26'
Eternal Fleur
Cnc 8h51'50"11d2'
Eternal Friends
Cyg 21h1'12"40d1'
Eternal Happiness Hiroyuki & Yasuko
Leo 10h7'51"19d7'
Eternal Heart To Hirofumi
Leo 10h3'53"8d47'
Eternal Kirk & Jennifer's Star
Uma 10h6'17"51d27'
Eternal Lee
Cet 3h4'17"1d33'
Eternal Love
Com 12h4'59"27d23'
Eternal Love
Uma 9h29'23"60d23'
Eternal Love
Gem 7h11'41"16d41'
Eternal Love
Aur 5h6'38"42d35'
Eternal Love
Cap 20h55'45"-27d23'
Eternal Love
Cyg 20h35'1"31d43'
Eternal Love
Her 17h31'24"20d35'
Eternal Love
Ori 6h8'46"5d3'
Eternal Love
Cyg 19h29'50"35d57'
Eternal Love II
Gem 7h24'29"31d34'
Eternal Love Michelle & Tom 9-10-94
Cyg 21h7'14"31d40'
Eternal Love of Richard & Kristin
Cyg 19h40'40"40d51'
Eternal Michael
Her 15h54'40"40d12'
Eternal Mikaneam Brillance
Cam 12h53'16"78d30'
Eternal Passion
Cas 0h56'56"70d14'
Eternal Rendezvous: JRC's Cloud 9
Ori 5h56'54"7d56'
Eternal Tag
Dra 19h3'57"50d26'
Eternal Thomas
Cep 21h49'43"70d40'
Eternal Tim
Her 16h15'57"48d33'
Eternal Twinkling Star - Tomomi
Gem 7h11'59"26d5'
Eternally Aaron
Cep 22h22'0"60d26'
Eternally Alex
Psc 1h0'52"22d8'
Eternally Amorous Gushy
Cyg 21h59'31"52d48'
Eternally Andy
Dra 16h34'48"51d58'
Eternally Angie
Aur 5h0'31"47d49'
Eternally Bret
Del 20h20'11"18d52'
Eternally Chad
Peg 21h7'38"18d55'
Eternally Cheri
Cnv 13h51'41"41d16'
Eternally Dan
Cep 3h13'39"80d7'
Eternally Darren's
Cep 3h51'15"80d30'
Eternally Don
Aqr 20h54'37"0d48'
Eternally Edward
Dra 10h48'30"73d45'

Eternally Elaine
Com 12h29'17"26d16'
Eternally Eldris
Hya 10h25'0"-12d19'
Eternally Eliane & John
Aur 5h35'46"41d3'
Eternally Em
Lup 15h17'25"46d7'
Eternally Eric
Per 4h7'1"51d22'
Eternally F M
Cep 1h2'0"78d16'
Eternally Freda
Lyr 18h29'0"30d41'
Eternally Jane
Aqr 21h3'25"-10d50'
Eternally Jeff
Her 17h8'11"48d10'
Eternally Joseph
Her 15h57'24"40d33'
Eternally Linda
Lac 22h14'40"54d58'
Eternally Lindsay
And 1h31'0"35d35'
Eternally Marcus & Marsha
Tau 3h57'24"20d32'
Eternally Mark
Mon 8h4'18"-7d57'
Eternally Mark
Per 1h55'54"50d25'
Eternally My Sweet Maggie Rose
Uma 9h7'25"52d22'
Eternally Neil
Dra 16h53'51"62d27'
Eternally Noble
Ant 10h33'25"35d18'
Eternally Ours
Crb 16h2'22"31d45'
Eternally Paul's
Lac 22h52'46"40d35'
Eternally Paula
Lyr 18h15'37"30d9'
Eternally Peter
Per 2h52'1"48d59'
Eternally R A S
Her 16h57'52"20d13'
Eternally Radiant Jason
Sct 18h43'23"-7d18'
Eternally Rich
Oph 18h30'15"8d31'
Eternally Richard
Ser 16h13'48"1d30'
Eternally Rick
Lac 22h21'0"40d58'
Eternally Tepper
Ori 5h56'49"13d25'
Eternally Tom
Boo 13h41'1"21d19'
Eternally Victor Forever Cheri
Boo 14h47'57"50d47'
Eternally Yours
Tri 1h58'44"31d39'
Eternally Yours (Pamela)
Psc 1h36'30"20d25'
Eternally Yours Tom
Aur 5h53'54"30d27'
Eternidad
Aql 19h31'37"7d51'
Éternité
Umi 14h46'54"69d5'
Eternitie'
Ori 5h58'58"11d13'
Eternity
Cet 2h31'50"0d33'
Eternity
Cma 7h3'24"-20d23'
Eternity
Hor 3h54'23"48d9'
Eternity
Mon 6h30'31"-8d7'
Eternity
Lyn 7h42'53"40d59'
Eternity
Gem 6h53'6"17d12'
Eternity
Her 17h11'51"42d54'
Eternity
Leo 10h26'51"11d6'
Eternity
Uma 9h17'46"47d17'
Eternity
Cyg 19h27'33"33d13'
Eternity
Crb 16h4'49"32d33'
Eternity
Eri 3h20'25"-14d15'
Eternity
Peg 22h1'60"33d3'
Eternity
Boo 14h19'1"18d16'
Eternity
Cyg 21h2'48"37d53'
Eternity
Cyg 20h11'0"41d18'
Eternity * Wes & Doris
Cyg 20h0'57"40d10'
Eternity Love K & M
Cnc 8h47'42"11d32'
Eternity Plus One
Sct 18h56'56"-6d56'
Eternity Point
Del 20h55'44"8d34'
Eternity Ron & Bonnie
Lyn 7h56'1"42d29'
Eternitá
Ori 5h34'16"-1d53'
ETG (Extra Terrstrial Gordon)
Uma 8h48'53"57d17'
Etha
Mon 6h44'43"10d22'
Ethel Kasha Mae
Lyn 8h7'41"39d36'
Ethereal Dawn
Vel 10h7'22"50d23'
Ethereal Julie
Cnv 13h20'42"40d19'
Ethereal Sphere Of Kage
Lmi 10h20'13"30d9'
Etheredge,Erin M
Uma 11h26'32"43d17'
Etheridge Family,The
Her 16h43'51"37d47'
Etheridge,Nicky
Umi 13h9'35"72d18'
Etheridge,Sr,James Dale
Oph 17h18'47"-20d40'
Etherton,Jeff
Ser 17h32'50"-10d28'
Ethesa (Alpha Et)
Ori 5h57'1"18d4'
Éthier,Marie-Thérèse
Umi 16h3'28"77d8'
Ethier,Sol D
Aqr 21h59'38"0d21'
Ethington's Star,Kerri
Uma 9h11'36"50d18'
Ethredge,Christopher
Oph 17h31'26"-23d49'
Ethridge,Shelly
Cas 1h13'43"62d50'
Etienne
For 2h31'31"-27d23'
Etienne
Lyn 8h21'31"47d46'
Etienne Blond
Ser 15h38'18"9d27'
Etincelle
Lac 22h45'47"38d35'
Etincelle de Bonheur
Uma 9h57'24"51d12'
Etman,Larry & Marci
Vul 20h41'40"20d28'
Etoil de Tranquillite
Uma 9h41'30"61d10'
Etoile D'Amour
Cyg 19h24'35"35d45'
Etoile de Dourelas
Aql 19h30'26"12d42'
Étoile de Faguy Symax 1
Uma 9h59'44"62d22'
Étoile de Glen (Glen's Star)
Cnv 12h11'1"51d51'
Étoile de Josée
Umi 16h9'43"75d4'
E'toile de Kakushima
Leo 10h2'30"20d8'
Étoile de KAZUHIKO
Sco 16h20'20"-18d39'
Etoile de L'Amour
Umi 16h39'60"75d7'
Étoile de Masayuki
Aqr 22h0'15"-23d29'
Étoile de Noriko
Sco 16h21'21"-27d49'
Etoile de Tomas
Aur 7h5'49"38d28'
Etoile du CUL
Uma 8h24'52"68d32'
Etoile Jeanelson;Omnia Vincit Amor
Mon 6h53'1"-5d40'
Etoile Priddy
Ori 5h29'30"-2d16'
Etoile,Gerda
Cap 20h50'32"-26d46'
Etoile-Savi
Peg 22h17'58"32d38'
Etoof Nnej L'Etoile
Lmi 9h51'59"40d53'
Etta
Ser 15h11'1"11d43'
Etta R
Peg 21h48'37"36d12'
Ettefagh,Shahla
Mon 6h54'54"-0d31'
Etten,Dr Mary Jean
Sgr 18h51'4"-22d59'
Etter,Chris
Boo 13h56'31"19d22'
Etter,Irene
Eri 2h57'40"-1d60'
Etter,Lindsey
Eri 3h59'25"-19d13'
Ettinger,Betty H & Stanley
Ori 5h27'48"-1d13'
Ettinger,Mark
Per 3h4'36"50d4'
Ettorri,Antonella Plescia
Aur 7h23'1"43d10'
Ettrick (Armagh)
Cam 8h3'32"70d23'
Ettwig,Andreas
Vir 12h34'33"-8d56'
Etwaroo,Henry
Per 3h0'44"46d51'
Etzdorf,Hans-Ulrich von
Sco 17h34'50"-38d13'
Etzler,Melissa Starr
Uma 12h14'58"62d8'
Etiile,Danielle
Peg 0h1'48"10d42'
Euan Star,The
Uma 14h2'0"51d54'
Eubank,Reda & Jim
Mon 6h38'59"10d29'
Eubank,Rodford
Tri 2h33'12"31d11'
Eubank,Suzanne
Mon 6h38'56"10d10'
Eubanks,Amy Marie
Peg 22h4'27"20d24'
Eubanks,Bob
Eri 4h14'31"-16d57'
Eubanks,Christopher John
Hya 8h14'0"3d15'
Eubanks,Gregg
Lac 22h26'0"56d31'
Eubanks,Lee K
Vul 20h19'54"25d37'
Eubanks,Starmate
Forever,John Kenneth
Cet 0h47'1"1d42'
Eudaly,Lydia J
Cet 2h34'22"-8d42'
Eudeline,Nicole
Aur 4h53'1"52d27'
Eudemonia
Aur 4h55'23"38d59'
Eudoxie,Christine
And 23h13'53"38d30'
Euer
Her 17h49'26"46d56'
Euer,Rose M
Cas 0h49'14"61d5'
Eufemia
Col 6h31'30"-35d25'
Euforia
Lac 22h10'34"51d56'
Eufrasia,Baronessa
Boo 15h8'49"38d37'
Eugene My Pookie Bear
Lib 15h42'1"-28d26'
Eugene's Aramis Aurastel Love, Lily
Her 16h59'56"18d54'
Eugene,Kevin
Per 2h31'20"57d55'
Eugene,Randall
Dra 9h45'54"74d59'
Eugenia
Cas 0h17'36"58d16'
Eugenia
Lyn 9h9'21"41d6'
Eugenia
Uma 9h6'0"51d11'
Eugenia D
Dra 20h19'15"64d33'
Eugenia Heredia
Cas 0h40'53"60d11'
Eugenia-24-05-CLSSM
Aur 6h32'22"38d13'
Eugenie
Cam 8h25'52"78d48'
Eugenie Marie
Peg 22h20'56"35d27'
Eugenius
Her 16h37'1"50d27'
Eul,Peter
Lyr 19h21'1"30d29'
Eula
Boo 14h12'46"50d20'
Eulalika
Cet 2h58'1"1d7'
Eulenbruch,Renate
Gem 6h24'45"14d48'
Eunice
Eri 4h1'57"-11d60'
Eunice
Sgr 18h48'5"-24d26'
Eunice Adaline
Tri 2h16'14"33d3'
Eunice W
And 1h30'42"40d53'
Eunyce
Peg 21h56'0"32d3'
Eupene,Geoff
Lup 15h28'13"50d35'
Euphoria
Cet 2h29'51"-1d49'
Eureka
Mon 7h5'32"0d46'
Euritalia
For 2h4'35"-26d36'
Euritt,Charlet Elaine
Vir 11h39'49"4d7'
European Traveler
Uma 10h42'59"50d37'
Eurostar
Sco 16h18'24"-21d32'
Euràka
Cam 5h54'58"60d36'
Eustace,Ann Lucas
Per 1h40'27"52d38'A
Eustace,Edward William
Per 1h40'27"52d38'B
Eustace,J P
Uma 8h11'0"71d38'
Eustace,J T
Aur 5h20'41"38d31'
Eustace,James Howard
Dra 15h58'11"52d14'
Eustace,John William
Aur 4h53'26"38d22'
Eustolia
Aql 19h28'1"10d11'
Euting,Hans-Joachim
Lac 22h3'21"51d19'
Eva
Lib 14h39'31"-23d15'
Eva
Cas 1h27'15"77d19'
Eva
Cnc 8h59'53"27d31'
Eva
Lyr 19h23'42"40d34'
Eva
Cma 6h24'28"-13d13'
Eva
Psa 22h32'33"-27d36'
Eva
Del 20h21'25"20d36'
Eva
Lyn 8h11'42"36d48'
Eva
Umi 14h15'1"66d33'
Eva
Dra 15h12'46"62d35'
Eva
Her 17h9'33"42d42'
Eva
Psc 1h2'50"20d29'
Eva & Marieke
Cnv 13h39'19"47d31'
Eva & Michael
Tri 2h18'0"33d34'
Eva Alexandra
And 0h9'39"28d14'
Eva Alexandra
Del 20h15'47"11d1'
Eva Christine
Umi 15h31'0"81d43'
Eva Ester In Aeternum
Uma 11h9'32"40d53'
Eva Jane
And 23h37'19"48d12'
Eva Kate
Aql 19h30'1"13d15'
Eva Laura
Cyg 20h39'13"31d3'
Eva Maria
Tri 2h4'46"32d23'
Eva Maria
Umi 14h56'51"72d1'
Eva Marie
Peg 22h8'23"29d48'
Eva Paulinc
Cap 21h1'7"-22d23'
Eva's Love
Mon 6h59'29"0d56'
Eva's Star
Cam 5h57'0"60d51'
Eva's Star
Lmi 10h29'17"33d35'
Eva,Darren
Peg 23h3'58"30d50'
Eva,Mark
Aur 6h11'32"36d2'
Eva-Olivia
Cas 1h54'44"67d33'
Evalynn
Tri 1h30'33"31d15'
Evan
Lac 21h56'28"38d16'
Evan & Camelia
Cyg 20h51'54"30d23'
Evan Anne
Cam 5h48'33"68d57'
Evan Nicholas
Ser 17h55'13"-14d25'
Evan To Know You Is To Love You
Uma 11h26'56"44d57'
Evan's First Star He Sees At Nite
And 1h5'27"39d14'
Evan's Kho Star
Cep 22h58'50"70d40'
Evan's Mama & Papa
Cnv 13h24'17"42d29'
Evan's Piece Of Heaven
Tau 5h1'25"20d3'
Evan's Piece of Heaven
Sco 17h9'27"-38d22'
Evan's Wishing Star
Boo 13h39'28"22d39'
Evan,Robert
Uma 12h6'32"48d14'
Evan,Tristram
Uma 10h48'0"71d36'
Evangelene Noelle
Lyr 19h5'21"38d12'
Evangeline
Mon 6h44'29"11d33'
Evaniroff,Paul
Her 15h52'1"40d41'
Evanly Body
Cam 4h6'22"71d6'
Evans 143
Eri 4h5'33"-11d55'
Evans 1964,Caroline
Ori 4h46'10"0d33'
Evans III,David Francis
Boo 14h16'26"54d45'
Evans Loving Husband- Dad-Grandpa,Robert
Eri 3h26'25"-2d11'
Evans Star
Uma 10h7'50"61d28'
Evans Star,The David
Cep 20h50'45"70d29'
Evans Star,The Elvey
Ori 5h55'47"7d14'
Evans Star,The Sue
Umi 15h36'0"70d18'
Evans' Star,The Scott
Sex 10h36'20"1d14'
Evans, Charles M.
Uma 8h54'0"67d50'
Evans,A Donde Hugh
Oph 17h33'51"-6d24'
Evans,A N E One- Antony Neil
Uma 10h18'60"71d24'
Evans,Alan Michael
Ori 5h55'56"15d31'
Evans,Andrea
Cas 23h22'46"53d19'
Evans,Andrew
Ori 5h54'27"21d33'
Evans,Annabelle
Uma 11h45'45"40d13'
Evans,Anne
Umi 15h12'0"77d24'
Evans,Betty
Peg 22h31'14"29d32'
Evans,Betty M
Cyg 21h30'27"30d57'
Evans,Booker Kevin
Dra 20h38'32"70d1'
Evans,Boyd A
Aql 19h12'51"10d46'
Evans,Bruce Kingsman
Ind 20h50'47"-54d18'
Evans,Calvin G
Aql 20h0'37"8d23'
Evans,Carla Ann
Peg 22h18'47"21d29'
Evans,Ceri
Cas 0h23'15"61d33'
Evans,Charles & Alice
Cyg 19h25'31"30d52'
Evans,Charles J
Umi 15h7'0"77d4'
Evans,Christopher Nathan
Ori 5h57'10"15d17'
Evans,Claude & Cindy
Eri 3h52'54"0d0'
Evans,David J
Umi 15h37'56"68d35'
Evans,David L
Ser 18h1'0"-14d2'
Evans,David Todd
Dra 16h28'33"62d59'
Evans,Dennis Charles
Her 18h44'16"12d24'
Evans,Diana
And 23h24'19"45d26'
Evans,Dolly
Cyg 21h5'54"30d25'
Evans,Don L
Oph 18h34'37"10d6'
Evans,Doyle
Oph 18h40'19"8d41'
Evans,Dwight C
Lib 15h41'43"-22d51'
Evans,Eddie
Per 2h53'50"34d1'
Evans,Elizabeth
Ori 5h29'13"-1d35'
Evans,Elvira H
Peg 0h3'57"22d1'
Evans,Fire Officer John
Uma 10h4'1"47d49'
Evans,Frank
Uma 10h11'1"54d58'
Evans,Frank W
Umi 15h2'60"78d28'
Evans,Gay
Dra 16h1'24"67d27'
Evans,Glenn David
Per 1h53'17"56d52'
Evans,Grace A
And 1h12'15"35d58'
Evans,Hannah Imogene
Uma 11h38'1"44d48'
Evans,Hazel Cornell
Crb 15h26'39"31d50'
Evans,Ida Pamela
Lyr 18h28'38"32d52'
Evans,James
Lup 15h4'21"-49d30'
Evans,James Patrick
Aql 19h7'31"5d35'
Evans,Jane
Cyg 20h32'13"40d40'
Evans,Jean M
Peg 22h46'0"35d26'
Evans,Jean L
Mon 6h33'14"7d26'
Evans,Joan Marie
Mon 7h48'59"-1d25'
Evans,Joann D
Cam 5h4'11"67d35'
Evans,JoAnne
Cet 1h41'30"0d42'
Evans,Joe Owen
Cep 1h2'44"77d53'
Evans,John
Cep 1h10'40"80d21'
Evans,John Mark
Cep 21h2'39"61d21'
Evans,John Michael
Ori 5h56'1"0d53'
Evans,Jr,George R
Vul 19h46'31"28d48'
Evans,Julie Nan
Lmi 10h30'17"33d34'
Evans,K C
Peg 23h2'16"20d31'
Evans,Kain
Sgr 18h49'52"-24d56'
Evans,Karen Kay & Juddson Jori Plum
Boo 14h11'44"37d19'
Evans,Kerry
Uma 13h51'12"51d6'
Evans,Leslie A
Umi 15h4'1"77d3'
Evans,Linda Moran
Lac 22h35'25"38d28'
Evans,Lt Col John Roberts
Ori 4h41'1"0d36'
Evans,Lucy J
Cyg 19h52'59"38d46'
Evans,Marcia
Cas 23h15'25"61d14'
Evans,Marjorie E
Uma 10h12'15"54d40'
Evans,Martin W
Ser 15h11'30"-2d3'
Evans,Michelle
Lyr 19h21'48"35d6'
Evans,N S D,Mary Helen
Aqr 22h36'47"-12d23'
Evans,Nancy
Leo 9h22'26"18d33'
Evans,Nathan Robert Francis
Her 17h56'23"50d11'
Evans,Nicholas Vincent
Gem 6h33'2"13d33'
Evans,Nicole Leigh
Peg 22h58'1"25d7'
Evans,Odahbitai Tai Debra L
Hya 8h18'47"1d2'
Evans,Pamelia May
Uma 11h30'30"40d46'
Evans,Patty P
Peg 22h8'1"4d19'
Evans,Peter M
Aur 7h14'34"41d7'
Evans,Rich
Aur 5h9'1"41d14'
Evans,Richard Edward
Sex 10h29'29"0d33'
Evans,Richard Edward
Sex 10h29'29"0d33'
Evans,Robert The Ultimate Star
Per 3h17'20"41d39'
Evans,Robert W
Sct 18h52'43"-7d38'
Evans,Rodney "Daisy"
Lmi 9h59'0"33d16'
Evans,Ruth Ann
Hya 8h59'42"-7d11'
Evans,Samuel Kenneth
Tau 5h46'0"24d9'
Evans,Sarah
Peg 21h59'0"33d31'
Evans,Sheila
Peg 23h44'39"27d1'
Evans,Sherri
Tri 2h13'23"32d43'
Evans,Simon
Ori 4h46'11"0d39'
Evans,Sophie Nicole
Peg 22h18'34"30d37'
Evans,Sr,Ronald J
Tau 4h13'27"24d6'
Evans,Stacee
Ori 5h56'40"10d26'
Evans,Stacie
And 23h45'44"32d59'
Evans,Steven Dean
Ari 2h31'25"30d57'
Evans,Thomas Robert
Oph 17h28'45"-0d29'
Evans,Tim
Per 3h4'33"41d31'
Evans,Todd
Uma 8h46'58"53d15'
Evans,Tommy E
Boo 15h18'44"41d52'
Evans,Wendy
Cyg 21h50'51"41d14'
Evans,William Thomas
Oph 17h5'11"0d10'
Evans-711,Tony
Aur 5h58'25"30d47'
Evans-Meier,Brigid & Scott
Aql 19h6'27"2d2'
Evarella
Vul 19h48'1"27d15'
Evarts,Robert
Cep 20h48'24"68d40'
Evdokimow,David
Aur 7h10'25"41d54'
Eve
Oph 18h33'50"11d12'B
Eve
Uma 8h30'54"48d26'
Eve & George
Cyg 20h49'44"37d55'
Eve Marie
Cep 23h0'36"65d23'
Eve Of Hope
And 0h41'19"40d30'
Eve's Sentinel
Pho 0h49'20"-46d57'
Eve,Diane Faris Hall
Oph 16h59'40"11d49'
Eve,Marie
Ori 5h58'1"16d53'
Eve,Phil
Cam 5h13'34"68d17'
Eve-Lyne
Lyn 8h45'34"42d1'
Eveland,Jeanne & Chuck
Eri 3h15'48"-2d8'
Eveldave
Eri 4h32'1"-8d25'
Evelie's Symphonic Star
Leo 9h33'46"10d9'
Evelina
Cas 1h36'1"65d11'
Eveline
Sge 20h5'43"20d26'
Eveline
Dra 11h28'22"66d3'
Evelini C,The
Cap 21h56'1"-22d10'
Evelyn
Vir 12h7'19"-6d4'
Evelyn
Lyr 18h44'28"46d51'
Evelyn
Cas 23h38'21"60d38'
Evelyn
Her 17h17'14"50d19'
Evelyn
Aur 6h7'5"31d32'
Evelyn
Mon 8h2'0"-0d27'
Evelyn
Mon 7h1'16"-8d42'
Evelyn
Boo 14h58'50"45d1'
Evelyn "Sunny"
And 2h29'23"50d34'
Evelyn Clarissa
Crb 15h40'35"26d48'
Evelyn Elizabeth
Umi 17h12'51"80d6'
Evelyn Margaret
Aqr 22h6'54"-8d48'
Evelyn's Evening Star
Cas 0h10'32"58d50'
Evelyn's Star
Cma 6h55'23"-16d45'
Evelyn's Star
Uma 11h9'54"61d27'
Evelyn,Salis
Gem 6h47'5"14d30'
Evelyne
Ori 4h43'14"9d53'
Evelyne
Cas 3h0'1"58d18'
Evelyne
Ori 4h43'14"9d53'
Evelyne
Per 3h15'13"41d25'
Evelyne et Sébastien
And 1h6'0"47d32'
Evelynx-Grahlii
Gem 7h26'28"26d41'
Even's Star
Mon 7h6'13"-6d51'
Even,Martin
Dra 10h2'60"73d34'
Evens,Noline
Lyr 18h46'35"38d35'

Evenson
Boo 14h37'11"21d28'
Evenson's Star
Vul 20h19'19"25d54'
Evenson,Eternally Eric -Eric
Sco 17h32'46"-31d38'
Evenstar,Dani
Umi 16h41'52"79d49'
Ever Lasting Paradise
Gem 7h13'33"23d35'
Ever-Bright Bruce
Ori 5h7'19"0d43'
Everakes,Jules
Aur 6h3'28"37d40'
Everard,Hal
Cyg 20h25'58"38d21'
Everard,Isabel Mercedes
And 2h15'31"40d3'B
Everard,Oscar Cameron
And 2h15'31"40d3'A
Everett Family,G N B L D
Cam 8h26'29"82d10'
Everett's Star
Dra 20h11'42"71d12'
Everett,"Gage d'Amour" for Camm
Dra 19h7'22"60d42'
Everett,Elsie L
Eri 3h16'17"-16d29'
Everett,James Ronald
Dra 13h52'39"64d29'
Everett,Jenny-Jo Marie
Peg 0h2'14"31d44'
Everett,John Lauchlin
Cet 1h56'0"-0d0'
Everett,Kim Ann
Cam 5h52'1"68d54'
Everett,Lisa
Lyr 19h21'10"37d35'
Everett,Nicholas Alexander Kaulana
Cma 7h20'1"-16d18'
Everett,Rebecca Lynn
Cep 23h14'36"80d12'
Everett,Richard 10/4/67
Boo 14h22'35"32d20'
Everett,Suzanne & Brian
Cyg 20h53'20"31d3'
Everett,Todd
Aur 6h6'0"46d50'
Everett,Tommye Belle
Peg 22h36'32"26d31'
Everett,Winnie
Cas 0h3'0"54d55'
Evergreen
Uma 12h1'32"30d9'
Evergreen Steve
Lyn 8h9'19"47d52'
Everhart,Margie
Mon 7h16'1"-1d8'
Everitt,Melissa Bree
Cas 23h19'49"62d40'
Everlasting
Boo 14h50'45"24d30'
Everlasting
Cmi 7h40'48"7d43'
Everlasting
Cnv 13h49'46"41d56'
Everlasting
Uma 11h30'46"30d37'
Everlasting
Peg 22h36'24"30d14'
Everlasting
Lyn 7h40'31"40d43'
Everlasting Andrew
Sct 18h52'39"-9d9'
Everlasting Glenn
Hya 8h28'40"-0d18'
Everlasting Kathryn
And 23h37'48"46d57'
Everlasting Love Of Judy & Eddie
Crb 5h33'0"31d49'
Everlasting Love- Katsuji & Chizuru
Lib 14h44'30"-13d39'
Everlasting Love-Pat & Jennifer
Peg 0h1'27"18d47'
Everlasting Mary
Cas 1h40'45"77d6'
Everlasting Press Star The
Cyg 19h34'37"34d32'
Everlasting Stella of Maki
Sgr 19h18'15"-29d34'
Everline,Lorraine W
And 0h9'18"47d57'
Everline,Lorraine W
Cyg 21h29'53"39d29'
Everly Brothers "The Flaming Legends"
Lyr 18h58'21"31d24'
Everly,Brian
Aur 5h54'0"31d13'
Evers,Ila Fern Fethke
Peg 0h3'0"20d40'
Evers,James T
Per 3h1'0"40d31'
Evers,Jan Kuipers
Cas 1h41'16"61d21'
Eversole's Dazzlin' Daddy Bob
Boo 14h37'44"9d19'
Everson,Adam,Gabriela & MacKenzie
Dra 17h12'0"70d12'
Everson,Christine Lynn
Per 3h12'1"40d32'
Everson,Rachelle Lynn
And 0h22'36"31d51'
Evert,Ethan August William
Uma 11h31'54"63d40'
Everton,Terance Charles
Aql 20h31'0"-0d25'
Everts,Richard Nicole
Boo 14h56'28"27d55'
Every Day Is Christmas
Lyr 18h31'35"42d43'
Every Generation's Author & Angel,Og
Cyg 19h57'40"38d36'
Every-Wortman,Ella Louise
Mon 8h5'49"-8d50'
Eves,Alfred Alexander
Car 7h28'59"-57d17'
Eves,Coralie
Car 7h30'6"-57d28'
Eves,Meredith
Uma 9h32'15"50d3'
Evette Kaye
Ori 5h53'28"-8d20'
Evetts,Tarron
Aql 19h52'44"14d41'
Evi
Crb 15h26'46"31d2'
Evi
And 1h7'51"41d14'
Evidente,Josielynne
And 2h33'53"38d56'
Evidenzia
Crb 15h31'38"37d58'
Evie
Cet 2h16'44"4d9'
Evie
Lyr 18h50'42"31d23'
Evie T
Lyn 7h34'37"35d32'
Evil Clown
Cet 2h19'45"1d32'
Evin
Aur 7h19'36"38d43'
Evins III,Doyle Eugene
Per 3h2'37"46d7'
Evins,Dustin Edward
Aur 6h21'21"35d45'
Evita
Cet 3h13'14"6d58'
Evitts,Allen Blair
Psc 3h1'11"55d11'
Evitts,Pamela Jane
And 23h39'41"32d36'
Evon
Vul 20h3'45"22d49'
Evon,Kathleen Anne Kolter
Peg 23h34'1"30d51'
Evonits,"The Poohs" Barb & Joe
Tri 1h49'19"26d12'
Evridiki
Lyr 18h39'0"37d45'
Evslin,Nathaniel Aren
Vir 14h8'37"7d24'
EW The Great
Cep 1h17'42"85d39'
Ewald,Brian Herbert
Eri 3h55'17"-5d36'
Ewald,Deborah & Charles
Cam 4h9'10"61d13'
Ewald,Fabian
Vir 11h54'34"-3d24'
Ewald,Linda
Lyn 8h42'14"39d58'
Ewald,Mark
Uma 10h16'38"67d44'
Ewald,Rich
Vul 19h19'0"26d39'
Ewald,Sabrina
Cas 0h36'52"70d49'
Ewalt,Kenneth Frank
Hya 8h56'47"4d24'
Ewasick,Christine Lynn
And 0h49'27"40d41'
Ewen,Terrell Benton
Umi 17h28'38"78d53'
Ewens Family,The
Mon 6h14'56"-10d19'
Ewers,Ingrid
Hya 9h7'45"3d16'
Ewersmeyer,Elmar
Peg 23h32'48"20d20'
Ewing III,Dwight Miller
Aur 6h21'44"37d46'
Ewing,Alycia
Cam 14h24'33"81d51'
Ewing,Denise Annette
Per 3h17'45"40d25'
Ewing,Emily Elizabeth
Aql 19h54'59"13d34'
Ewing,Gregory
Lac 22h6'11"51d50'
Ewing,Jr,Morris
Lmi 10h18'23"28d32'
Ewing,Michael
Dra 17h57'1"60d29'
Ewing,Sr,Danielle Frederic
Cas 23h21'29"54d2'
Ewing-Jaggard,Nancy Renee
Cep 0h10'22"66d43'
Ewings,The
Aql 20h14'51"8d13'
Ewoldt,Roy & Marilyn Hedin
Uma 9h56'29"48d26'
Ex Libris
Cnc 9h0'0"11d25'
Exalibur
Pyx 8h46'4"-24d5'
Exarus,Faye
Lyr 18h15'53"37d46'
Excalibur-For My Dad
Her 17h3'15"49d18'
Excoffier,Christian
Leo 9h57'55"19d9'
Exel,Brian Mathew
Her 16h5'43"41d34'
Exell,Ellen Louise
Boo 14h20'0"53d58'
Exia
Lyn 8h2'17"38d47'
Exile
Umi 15h11'0"78d15'
Exit the Lemming
Ori 6h7'54"0d22'
Exner,Beate
Ari 2h38'38"20d35'
Exon,Donna Lynne
Scl 23h17'48"-33d26'
Exposto,Joao
Del 20h36'40"5d34'
Exquisite Heidi Dee, The
And 2h32'1"41d1'
Exquisitely Curtis
Aql 20h4'42"6d11'
Exton,Karen
Mon 6h34'49"-0d15'
Eydely
Dra 17h55'16"61d15'
Eye & Ear Hospital of Pittsburgh
Lyn 8h59'44"36d7'
Eye In The Sky
Boo 14h56'21"53d5'
Eye Of Love
Cyg 21h52'33"41d14'
Eye of Marguerita,The
Lyn 7h57'22"48d31'
Eye Of Melisa
Lyn 8h26'57"41d54'
Eye of Monique
And 1h21'47"39d49'
Eye of The Griffin
Aur 6h0'1"32d53'
Eyer,Jason
Cma 6h13'60"-13d13'
Eyerman,Brown Lee
Boo 14h27'47"27d12'
Eyerman-My Love,Glad Alexandra
Leo 10h58'32"12d17'
Eyers,James Howard
Dra 16h56'52"58d54'
Eyert,Wilfried
Eri 4h48'54"-5d55'
Eyhab
Uma 9h2'12"56d12'
Eynard,Viola
Ari 1h22'11"33d32'
Eyraud Nathalie "Minnie"
Lac 22h7'12"38d20'
Eyre,Robert S
Aur 6h16'39"31d10'
Eyres,Beth
Lyr 18h34'50"38d19'
Eyring The Light Of My Life,Marilee
Cas 0h37'31"70d6'
EZ2LUVU
Boo 14h25'33"53d50'
Ezell,Leo & Helen
Umi 16h21'0"84d52'
Ezer,Jonathan J & Nicole G
Eri 4h32'59"-12d1'
Ezey,Helen P
Mon 7h0'10"-10d32'
Ezhilchelvan,Nason
Umi 15h51'0"83d33'
Ezio e Giangi
Cep 23h30'25"64d18'
Ezra Sean 1995
Lac 22h19'39"46d50'
Ezra,Nora
Lyr 18h54'15"30d43'
Ezzedine,Haikel et Souad
Per 3h53'7"37d22'

F

F
Peg 21h49'14"33d33'
F B Star I
Lac 22h46'32"54d54'
F C F Forever
Uma 9h47'51"51d51'
F David
Cma 7h2'35"-18d58'
F E C "Little"
Per 3h57'1"37d58'
F G G Shibumi
Lac 22h28'1"49d48'
F I L
Lac 22h10'37"47d5'
F I O N A
Cas 0h22'23"72d15'
F Martin S
Leo 9h55'24"10d44'
F R D 50
Ori 5h3'53"13d14'
Fa Feng
Pho 0h31'21"-42d52'
Faanes,Sarah Ann
Lyn 6h28'36"59d50'
Faapmjpdje
Tri 2h1'52"31d36'
Faas,David
Lyr 19h26'29"40d6'
Faas,Monica McMarthy
Crb 15h27'30"32d21'
Faas-Jackson Families, The
Ori 5h36'15"-6d24'
Faass,Axel
Cap 21h24'36"-10d57'
Fab Et Coco
Tri 1h58'46"27d39'
Fabano,Michael Vincent
Dra 14h49'32"63d37'
Fabares,Shelley & Mike Farrell
Uma 11h42'16"47d57'
Fabbri,Leo
Uma 9h9'1"47d44'
Fabbricini/Honey Bear, Luigi
Lac 22h41'41"55d55'
Fabbro,Wilma
And 23h2'43"51d5'
Faber Family,The Michael
Cyg 21h28'35"28d46'
Faber,Bob & Robin
Cas 1h35'57"58d9'
Faber,Carol A "Toni" & Daniel K
Mon 8h2'35"-10d2'
Faber,Casey
Her 16h22'44"10d27'
Faber,Dagmar Anna
Cam 6h15'53"62d28'
Faber,Dr
Sco 16h48'35"-25d51'
Faber,Heinrich-Herbert
Aur 5h4'12"43d28'
Faber,Horst
Aql 20h8'60"8d36'
Faber,Ingeborg
Aql 20h9'56"4d1'
Faber,Raimund
Umi 13h4'14"75d22'
Faberman,Martha
Cas 0h18'49"64d37'
Fabian
Tri 2h5'50"33d39'
Fabian,Blake
Aqr 20h55'0"-6d54'
Fabian,Daniel Ward
Ori 5h21'19"11d18'
Fabiano,Michelle
Lyn 7h59'19"37d9'
Fabiano,Nancy Lee
Cas 0h55'58"65d3'
Fabien Mathieu
Com 12h11'40"21d19'
Fabien,Maille
Uma 9h39'43"48d54'
Fabienne Amour
Cet 1h33'34"0d28'
Fabietti,Thomas Robert
Uma 12h2'1"64d34'
Fabio
Her 17h30'42"27d29'
Fabio & Jeanine
Lyn 9h7'53"42d50'
Fabio e Betty
Peg 23h0'50"20d1'
Fabio L'universo
Her 17h22'39"40d25'
Fabio Lukas
Cap 20h59'36"-14d54'
Fabio Star
Lup 14h41'57"-45d12'
Fabio,RoseMarie
Vir 11h58'19"-4d31'
Fabiola
Com 12h18'50"20d9'
Fabishak,Kristina M
Uma 8h52'57"50d53'
Fabius,der Fisch
Cep 21h50'56"62d51'B
Fabie,Pierre
And 23h4'12"38d9'
Fabo,Erica Lynn
And 23h42'0"42d28'
Fabo,Jessica Ann
Lyr 18h43'0"33d18'
Fabozzi,Lisa
Eri 3h34'48"-2d24'
Fabre,Olivier
Her 16h41'21"10d57'
Fabri & Ale
Cnv 13h39'41"37d45'
Fabri,Franca
Cyg 20h59'32"31d2'
Fabrizia
Pho 0h1'1"-43d20'
Fabrizio
Cas 2h1'25"60d22'
Fabrizio
Col 6h34'32"-33d26'
Fabrizio,Anthony & Alice
Boo 14h30'28"50d13'
Fabrizio,Anthony M
Cet 3h12'57"6d51'
Fabrizio,Deanna Marie
Tau 4h19'9"20d59'
Fabrizio,Gerri-Ellen
Gem 6h42'53"18d48'
Fabry 20/6/69 M
Her 18h9'52"30d29'
Fabry,Ute 06Juni1953
Gem 6h43'37"12d46'
Fabulous
Uma 12h21'53"53d3'
Fabulous Florence
Cas 0h55'11"67d27'
Fabulous Francesco
Dra 19h32'32"80d13'
Fabulous Freda
And 0h51'51"37d51'
Fabulous Miss Jen,The
And 1h26'39"36d23'
Fabulous Sue
Cap 20h34'49"-10d30'
Facchiano,Anna
Lyr 18h16'17"45d57'
Facchin,Robert A
Aqr 21h57'1"-18d43'
Face
Per 3h0'23"46d31'
Facemyer,Kelly R
Lyr 19h7'16"37d57'
Fachamara,Irene Williams
Cyg 20h1'1"30d11'
Fachet,John
Aql 18h57'38"16d20'
Faciane,Phillip
Sex 10h8'51"2d25'
Faciano,Andrew
Lyr 19h2'33"33d47'
Facincani,Robert
Per 4h24'13"50d12'
Facio,Jorge Antonio Trejos
Ari 3h3'46"30d2'
Fackler,Dale
Dra 19h1'21"50d36'
Facon,Johnathan
Cas 1h24'9"64d46'
Facque,Noemie
Dra 20h1'43"70d16'
Facque,Valentine
Lac 22h19'25"50d11'
Factor,Joseph R & Ruth J
Uma 11h50'28"55d19'
Factor,Michelle Joy
Ori 5h36'56"-1d10'
Factor,Todd Joshua
Lmi 10h52'32"28d6'
Factor,Trisha Lee
Peg 22h8'28"3d17'
Factora Family Star
Peg 23h5'54"21d7'
Facunaus,Paulus
Her 17h24'39"20d32'
Facy,Albane
Ori 6h16'33"1d18'
Fada Morgana
Ind 20h40'51"-49d25'
Faddis,Lauren Nicole
Mon 6h39'37"10d35'
Faddis,William Gray
Aur 7h12'0"37d13'
Faden,Eric Matthew
Per 2h0'1"50d10'
Fadoul,Graziella
And 0h23'28"44d47'
Faenchen
Gem 6h49'40"18d50'
Faenger,Robert
Boo 14h22'11"22d53'
Faerber,Darlene Kay McDonald
Cmi 7h54'33"0d16'
Fafafou Ciro
Ser 15h54'44"2d40'
Fafard,Denise
Cep 21h12'42"58d5'B
Fagan,Annette
Lyn 7h53'38"58d12'
Fagan,Brian Francis
Her 17h18'14"40d58'
Fagan,Charles "Chuck"
Boo 14h59'42"20d2'
Fagan,Clifford L
Oph 17h26'28"-22d53'
Fagan,Jon Gerald
Aur 5h1'34"47d30'
Fagan,Mark
Lac 22h33'21"52d35'
Fagan,Peter J
Her 16h15'37"23d60'
Fagan,Richard
Aql 19h6'0"15d49'
Fagan,Robert Joseph
Dra 9h41'39"80d28'
Fagan,Shawn David
Cnv 13h11'30"38d44'
Fagawe
Sco 16h52'33"-43d53'
Fagel,Marvin
Dra 16h13'52"61d8'
Fagelman,Sidney
Lac 22h50'30"38d39'
Fagenholz,Lori Ellen
Peg 22h57'50"27d51'
Fagenson,Henry Corfield
Cet 2h35'0"0d11'
Fageol,Michel
Cam 5h53'16"68d24'
Fager,Annette
Ari 1h54'49"12d29'
Fager,Russel & Family Star
Ori 4h55'24"-2d13'
Fagerlie,Anne-Cecilie
Umi 16h4'38"70d30'
Fagernes,Brian R
Her 17h38'25"25d1'
Fagerström,Olle
Umi 15h27'0"81d28'
Fagg,Keith
Cmi 8h8'31"6d16'
Faggi,Valentina
Cyg 20h3'42"37d34'
Faggio Sui Generis, Peter C
Her 16h3'11"42d13'
Faggioni,Giulia
Mon 7h15'59"0d45'
Fagin With Love
Vul 19h42'1"26d38'
Fagnano,Margaret C Shea
Cas 0h47'1"61d10'
Fagnano,Sr,Raymond Francis
Cas 0h33'35"63d34'
Fagnone,Alain
Cyg 20h28'39"30d14'
Fagone,Andrew
Umi 14h43'50"78d55'
Fagone,Irene
Cas 23h29'0"60d46'
Fagot,Jean Marie
Cyg 20h42'11"45d41'
Fahey,Brian K
Cnv 13h24'34"50d15'
Fahey,Daniel
Lmi 9h58'41"28d45'
Fahey,Sharon
Ser 18h36'0"5d57'
Fahje,Gerd
Dra 20h24'26"62d11'
Fahl,Hans-Juergen
Hya 9h17'16"-6d53'
Fahlgren,H Smoot
Uma 8h56'58"52d4'
Fahlgren,Smoot H
Aql 19h31'1"10d38'
Fahlman,Campbell George
Cnc 8h30'43"8d40'
Fahlsing,Max Putnam
Per 2h53'1"45d43'
Fahn Star,The Marion
Uma 11h19'1"53d17'
Fahn,Bunny
Uma 10h6'27"67d57'
Fahnstar
Per 4h4'25"51d29'
Fahoome,Joe
Cas 22h56'55"54d17'
Fahoome,Karynann
Cas 22h56'33"55d26'
Fahrenbruch,Mark Leslie
Per 1h47'14"56d30'
Fahringer,Brenda Jill
Lib 15h1'10"-10d34'
Fahy,Katie
And 1h24'57"38d55'
Faibian,Kenya
Boo 13h55'16"22d15'
Faid,Karen
Cas 0h22'31"59d10'
Faille,Rebecca Mallery
Cas 0h55'51"76d41'
Faiman,Gregg Howard
Her 16h26'40"22d42'
Fain,Andrew P
Cep 22h26'47"58d20'
Fain,Howard
Cep 23h4'25"64d5'
Fain,Robert Michael
Ori 5h0'52"13d46'
Fainberg,Colin
Ant 9h38'25"-34d30'
Faini,Jerilyn
Crb 16h22'1"30d21'
Fair O A K S Oliver & Kallyn's Star
Cyg 21h4'1"38d52'
Fair,David Frank
Boo 13h43'1"14d10'
Fair,Dawson Van Prentice
Hya 8h42'27"-1d19'
Fair,G M
Uma 8h36'1"71d10'
Fair,Hallie Elizabeth
Tri 1h49'26"28d59'
Fair,Jane
Vul 19h31'6"20d18'B
Fair,Little Star Tracy Lynn Bowser
Lyn 8h14'38"39d32'
Fair,Lori K & Jo Ann Hidey
Hya 8h37'33"-6d12'
Fair,Ray
Vul 19h31'6"20d18'A
Fairbairn,Robin Hart
Del 20h20'34"9d30'
Fairbanks,Kawika
Mon 7h6'55"-6d10'
Fairbanks,Linda Lou
Vul 20h39'54"23d26'
Fairbanks,Michael C
Per 1h58'57"50d20'
Fairbanks,Richard
Tau 4h3'46"20d14'
Fairbrother,Jerry
Her 16h45'50"38d55'
Fairbrother,Virginia & Richard
Uma 8h44'32"49d52'
Fairchild,Diane
Cyg 20h10'0"38d25'
Fairchild,Jean
Ari 1h57'27"23d39'
Fairchild,Nancy Savino
Cyg 20h18'33"42d54'
Fairchild,Susan
And 23h21'1"45d26'
Fairchok,Jr,Frank M
Cet 2h10'37"5d15'
Faircloth,Shannon Lee
Oph 17h4'33"11d37'
Fairclough,Audrey "Thelma"
And 1h0'44"40d45'
Fairclough,Karla Christine
Lyr 18h30'28"36d28'
Fairclough,William Norman
Cyg 19h30'0"37d9'
Faires,Dano
Ori 6h2'0"1d5'
Fairfax,Dawn D
Cyg 19h40'45"28d52'
Fairfeild I
Boo 14h7'0"31d44'
Fairfield,Jr,Frederick
Per 3h15'59"50d24'
Fairley,Edward
Her 18h1'1"48d43'
Fairman,Bettina L
Vul 20h16'23"26d8'
Fairman,Gabriel J
Uma 10h58'31"53d28'
Fairman,MD,Ronald M
Uma 10h43'53"50d22'
Fairman,Patricia Ann
Leo 9h40'1"10d36'
Fairweather Constellation,The
Dra 9h43'37"74d31'
Fairweather's Dragon
Eri 4h7'20"-5d31'
Fairweather,David John
Her 17h58'28"30d31'
Fairy Queen Of Ink
Eri 3h38'46"-6d34'
Fairy Tale
Leo 10h4'6"21d15'
Faiss,Rosemary A
Cas 3h9'19"67d38'
Faistman,Dulce Mami Babi-Yocheved
And 1h45'40"36d51'
Faith
Sge 20h5'0"20d23'
Faith
Cru 12h54'43"-59d52'
Faith
Cam 4h7'39"60d31'
Faith
Cep 1h52'1"78d34'
Faith Bethany
Lyr 18h36'36"35d4'
Faith Elizabeth
Eri 2h47'51"-18d17'
Faith,Jackie
And 2h32'1"41d5'
Faith,Love,Hope
Vul 20h14'51"23d36'
Faithful Heart
Oph 16h5'21"-5d38'

FAITHOMELISS
Cnv 12h44'1"41d12'
Faivre,Florence Rigot
And 1h7'55"47d2'
Fakas,Harrison
Oph 17h15'1"11d44'
Falaschi,Roberto
Her 17h23'1"48d44'
Falbo,Edith Tassa
Cyg 21h32'40"41d23'
Falcey,Cheryl
And 23h45'48"46d45'
Falch,Karl-Heinz
Dra 16h11'19"51d57'
Falciano,Frank J
Ser 18h40'57"3d52'
Falcione,Tina
Cas 0h29'60"74d49'
Falco,Domenica Bernadette
Lyn 8h42'38"33d54'
Falco,Jo Ann
Del 20h15'45"15d24'
Falco,John Mawdsley
Aur 5h58'44"30d35'
Falcon Tour-Repubblica di San Marino
Boo 15h3'1"48d58'
Falcon,John's Ebony
Mon 6h55'55"-10d56'
Falcone,Anna Rita
Nor 16h21'27"-50d37'
Falcone,Frank Anthony Rovirosa
Peg 23h22'45"31d56'
Falcone,Iris
Aql 19h26'12"0d55'
Falcone,Kaitlyn Nicole
Cas 2h12'55"60d42'
Falcone,Perrine Vincente Iris
Oph 18h7'27"12d58'
Falcone,Rachel
Vul 19h47'15"28d59'
Falcone,Rebecca
Cyg 21h52'1"41d15'
Falcone,T W
Hya 9h34'47"-9d42'
Falconer,Jeremy Ryan
Dra 14h22'29"63d34'
Falconetti,Rose Marie
Cma 7h17'29"-15d9'
Falconnet,Anne-Marie
Aur 4h37'41"31d45'
Falcus,Kayleigh Joyce
Ori 5h25'0"-0d23'
Faleiro,Dylan
Aur 6h1'21"37d55'
Falenski,Kathy
Lyn 9h6'45"33d37'
Fales,Carl F
Dra 19h56'37"84d47'
Fales,Dan & Jerry
Peg 23h2'11"32d2'
Fales,David
Per 3h11'48"45d21'
Fales,Jill & Matt
Cyg 21h30'0"42d46'
Fales,John Richmond
Her 16h49'35"32d44'
Fales,Kevin Scott
Dra 19h29'37"58d18'
Fales,Olivia Jean
Lyn 8h9'18"36d45'
Fales,Ryan Michael
Her 16h13'14"50d27'
Fales,Sean P
Vir 13h33'15"-10d35'
Faley,Ryan Edward
Oph 16h57'19"-26d4'
Falic,Samuel Moses
Hya 8h57'39"1d49'
Falick,Angela
Tau 5h45'58"16d29'
Falick,Frank
Vir 14h55'13"7d25'
Falis,Alexandra Ann
And 23h3'29"45d56'
Falk,George Sylvester & Jane Joan
Crb 15h21'58"30d60'
Falk,Jimmy
Boo 15h1'35"15d15'
Falk,Luna Alina
Lyn 8h7'46"45d24'
Falk,Paul Frederick Peter
Dra 16h37'16"64d5'
Falk,Steven
Aur 7h14'40"38d50'
Falk,Thomas
Per 2h54'29"43d42'
Falk,Wally
Sgr 19h50'17"41d14'
Falkenberg,Adelia
Cas 23h32'23"60d5'
Falkenberg,Margaret Hunting
Cnv 12h46'0"51d24'
Falkenberg,Michael Lee
Hya 9h0'25"1d41'
Falkenburg,Frank
Sex 10h46'22"-8d7'
Falkenhagen,Ernest R
Oph 17h55'25"13d17'
Falkenhahn,Patrick
Lyn 8h37'0"42d26'
Falkenhain,Klaus
Boo 13h57'30"10d23'
Falkenhausen,Katharina von
Cnc 8h29'51"30d31'
Falkenstein,Aleksandr
Her 18h4'0"37d33'
Falkenstein,W D
Cet 2h31'59"-8d56'
Falkland,Inez & Fred
Cnv 13h20'22"41d21'
Falkowski,Paula Carla
Lyn 8h6'31"41d1'
Falkus,Christopher David
Cep 22h25'54"61d47'
Fallen Eagles,The
Aql 18h59'40"16d55'
Fallenberg,Elisabeth
Crb 16h6'11"27d29'
Faller-Parrett,Jessie
Hya 9h13'30"0d54'
Faller-Raw,Oliver Harrison
Lyr 18h32'57"33d54'
Fallgatter,Tonazory
Aql 19h56'47"11d11'
Fallick,David
Cet 2h33'15"-10d23'
Falling Star,The
Aqr 23h36'1"-11d38'
Fallon,Anne Kathleen
Lyr 18h30'14"47d25'
Fallon,Clare
Cas 0h38'39"71d12'
Fallon,James
Dra 16h29'20"64d29'
Fallon,Joe
Per 3h7'28"54d26'
Fallon,John Peter
Ori 5h36'48"-0d59'
Fallon,Jr,Daniel
Cam 3h50'19"61d56'
Fallon,Julie Allison
And 23h44'0"46d6'
Fallon,Laurence
Mon 6h20'37"8d5'
Fallon,Mary Lou
Aql 19h58'16"11d8'
Fallon,Michael J
Dra 19h40'48"68d53'
Fallon,Robert
Aur 5h7'27"44d26'
Fallon,Robert James
Her 16h21'48"41d39'
Fallon,Ruth E Mondell
And 23h25'11"49d56'
Fallon,Ryan Thomas
Boo 14h59'1"8d14'
Fallon,Shonna
Sct 18h54'18"-8d7'
Fallone (Mum),Maisie
Cas 0h15'10"63d13'
Fallos,Douglas Edward
Ser 15h35'31"7d36'
Fallowes,Audrey Gail
Ari 1h55'17"18d6'
Falls,Coman Franklin
Mon 7h16'25"-6d24'
Falls,Jennifer Lee
Lyr 19h1'54"30d0'
Falls,Kelly M
Mon 6h48'0"11d51'
Falnes,In Memory of Emma Cathrine
Cyg 21h50'35"40d3'
Falotico,Robert J
Per 1h47'53"53d31'
Falsetta,Mary Ann(Roe)
Peg 21h38'37"27d20'
Falsetto,Joseph Louis
Uma 11h50'17"37d41'
Faltas,Mouni
Lyr 19h18'42"30d43'
Falter,Robert Jude
Hya 9h36'15"-10d36'
Falu,Ivette
Uma 11h1'21"48d23'
Falzone,Marcia
Vir 13h7'46"-21d0'
Falzone,Marianne
Lyn 7h59'50"51d12'
Fam Di Sera
Scl 0h4'16"-25d55'
Fama,Jacquelyn Marie
Lyn 7h6'55"60d34'
Fama,Megan
And 23h34'1"42d3'
Famelart,Pierre
Per 3h9'52"41d10'
Familetti,Jr,Thomas John
Lac 22h38'28"48d58'
Familie Bernd Pesall
Leo 10h0'19"8d38'
Familie Schwarz Germany
Lyn 7h55'1"58d51'
Familio,Theresa
Sgr 18h56'33"-34d0'
Familles,Badez Tournour-Viollet
For 3h29'49"-35d52'
Family Feiner,The
Cnv 13h24'12"38d13'
Family Lorenz
Mon 7h45'3"-5d4'
Family Star,The
Cnv 13h15'52"40d6'
Faminu,Femi
Cas 0h32'32"67d37'
Famolare III,Charles J
Cas 1h22'60"55d27'
Famous Amos
Aql 20h2'0"0d58'
Famous Baby Mikey
Uma 10h9'30"59d8'
Fan,Bijian
Cet 2h25'23'4d37'
Fancett,Joseph Michael
Her 15h51'28"51d3'
Fancher(Sanch),Richard Henry
Per 4h36'33"52d18'
Fancher,John B
Cap 21h12'51"-26d45'
Fancher,Ronald L
Hya 9h31'25"-0d12'
Fancher,Rose
Mon 8h4'47"-7d36'
Fancher,Samantha
Uma 8h57'28"49d5'
Fanchi
Lyr 18h22'13"40d9'
Fancy Face
Cas 0h24'0"50d27'
Fancy Nancy
Cas 1h6'42"67d58'
Fandango
Tri 2h10'54"33d34'
Fanell,Ronald Eugene
Cep 22h8'42"71d7'
Fanelli's Triumphal Goal
Her 17h17'48"18d51'
Fanelli,Anthony
Aur 6h3'45"37d44'
Fanelli,Katelyn Marie
Cep 20h48'48"61d6'
Fanelli,President Sean
Lac 22h5'51"50d10'
Fanfan
Lmi 11h1'28"30d25'
Fang,Annie
Cmi 0h24'19"8d23'
Fanizzo,Dr William
Ori 5h43'28"10d38'
Fanjoy,Sarah Alison
Ori 5h56'1"18d32'
Fankhauser,Mike D
Aql 19h58'32"10d42'
Fannie's Light Of Love
Lyn 8h5'12"45d3'
Fanning,Davis
Her 16h22'0"41d51'
Fanny
Lyn 8h32'27"38d28'
Fanny
Vul 20h15'18"22d56'
Fanny
Dra 16h21'17"62d29'
Fanny
Cam 5h57'20"67d33'
Fanny
Aql 20h1'37"10d55'
Fanny 1993
Lyn 0h10'12"38d33'
Fanny Forever
Tau 4h18'60"24d9'
Fano,Caroline
Cam 7h37'19"68d55'
Fansler,Diana
Equ 20h56'17"3d1'
Fansler,Kavondrea Ostara
Cet 0h48'27"0d55'
Fansmith,John F
Her 18h43'27"12d26'
Fantastic Fireworks
Ori 5h56'0"20d43'
Fantastic Kathy
Lyn 8h48'40"39d40'
Fantasy Island-Dave & Peggy
Cnc 8h25'16"30d34'
Fantaxcelsus Abnormis 21-34
Cnv 13h26'37"38d54'
Fante
Aur 5h53'53"40d32'
Fantozzi,Robert
Cnv 13h43'45"31d56'
Fantucchio,Giovanni
Dra 16h58'16"65d49'
Fantò 5/9/1967,Carmen
Dra 16h15'44"60d22'
FAO Schwarz
Peg 21h59'1"28d52'
Faouzi et Fanny
Cam 6h6'25"68d26'
Fappani,Alberto
Per 3h3'0"48d55'
Far Away But Near
Mon 6h48'14"10d25'
Far Side
Aql 18h54'23"8d15'
Far Star,The
Mon 6h32'22"-5d46'
Farabaugh,Peter John
Lib 15h16'17"-22d22'
Farace,Gail
Cas 1h16'1"66d8'
Faraco,André
Aur 5h53'56"40d4'
Faradji,Henri
Lmi 10h49'29"37d40'
Faraglia,Kathleen Ann
Com 12h28'28"21d4'
Farago,Dr Peter
Oph 17h18'17"10d55'
Farah
Umi 13h57'54"70d60'
Farah's Gentleman Jake
Her 16h57'50"22d5'
Farah,R
Boo 14h22'29"45d24'
Farahi,John Wiseman
Lac 22h52'55"40d18'
Faraj,Nicholas Wialbut
Her 17h52'48"40d38'
Faran,Michael Maximiliano
Ori 5h28'12"-4d31'
Farano,Mary Diana Davis
Equ 21h18'39"2d56'
Faraone,Seth J
Psc 22h55'44"5d7'
Faraway Fran
Peg 23h1'44"21d23'
Faraway Loveshack Hide Away
Uma 10h25'23"62d3'
Farb,Gene
Her 16h43'45"21d16'
Farber,Monte & Amy Zerner
Cyg 20h15'19"41d57'
Farda,Dominique Annette
Aur 5h31'49"50d4'
Fardad,Farah Bazeghi
Eri 3h41'45"-12d50'
Fardella's Shining Star,Cody
Aql 20h2'33"3d54'
Fardella,Margie
Lyn 6h17'34"54d1'
Farell's Angel Star, Maureen
Lyr 18h48'1"32d7'
Farenc,Jacques
Dra 15h5'13"62d31'
Farfallina
Cam 3h35'17"60d12'
Fargeas Leslie
Cet 2h21'0"1d37'
Fargher,Ina L
Com 13h16'0"27d38'
Fargo,Melissa
Dra 19h34'24"67d44'
Fargo,Peter
Aur 5h36'42"38d55'
Fargo,Rebbecca
Cas 1h6'60"50d37'
Fargone
Cep 22h8'0"71d4'
Farquhar,Gordon R
Uma 9h23'47"48d4'
Farha,W G
Boo 15h13'51"26d1'
Farhana
And 0h26'0"45d20'
Farhi,Marie-Claire
Uma 11h6'0"55d0'
Faria,Caroline
Lyr 18h29'24"44d53'
Faria,Joseph Roman
Ser 16h5'27"14d52'
Farias,Jr,John Jimenez
Aql 19h16'19"10d6'
Farias,Mark Keoki
Tri 2h25'45"30d3'
Farichild,Steven Alan
Uma 11h11'1"30d57'
Farid
Sgr 19h18'54"-24d6'
Farida
Aqr 21h52'32"0d43'
Fariello,Luca
Cam 14h3'54"80d45'
Farina,Cindy
Cet 2h54'11"9d18'
Farina,Judy
Vul 21h22'54"27d35'
Farina,Madelyn
Cas 0h55'52"58d53'
Farina,Patrizia Pinna
Col 6h30'24"-37d25'
Farina,Tanner Blakely
Aur 5h56'48"37d50'
Farineau
Cas 1h16'22"66d16'
Farino,Michael Anthony
Her 18h8'26"38d41'
Farioli,Carla
Her 18h30'0"13d5'
Faris III,George Thomas
Aql 20h16'1"5d2'
Faris,Candida Noemi
Hya 9h35'28"-1d47'
Faris,Kasey
Cet 0h5'58"-8d50'
Faris,Thomas M
Ori 5h56'20"10d59'
Fariss,Jill D
Aql 19h5'11"-1d37'
Fariss,Robert David
Boo 14h48'38"37d43'
Fariveau,Julie
Boo 14h31'22"40d40'
Farkas
Lmi 9h48'28"34d37'
Farkas,Brian D
Aur 6h32'40"35d34'
Farkas,C Blayre
Sge 19h39'0"16d44'
Farkas,Carol & Robin
Lmi 9h55'29"37d40'
Farkas,Julie Anne
Psc 1h0'46"10d15'
Farkas,Mary Frank & Frances
Aur 6h25'48"37d51'
Farkas,Neil J
Aur 6h29'17"32d59'
Farkas,Nicole Margaret
Ari 1h45'51"18d55'
Farkash,Avi
Mon 6h14'17"-10d0'
Farla,Betty
Dra 10h55'1"73d59'
Farley III,George Thomas
Her 18h6'44"40d50'
Farley,Alexander Marcus
Pyx 8h45'26"-21d23'
Farley,Carol
Mon 6h57'35"10d22'
Farley,Geri
Aur 6h55'42"38d34'
Farley,Jessica Hope
Mon 7h12'36"-6d58'
Farley,Kara Grace
And 0h5'18"40d42'
Farley,Regina
Cas 1h48'57"65d52'
Farley,Richard Anthony
Tau 4h15'24"21d48'
Farley-Brooks,Kaye
Cas 1h59'12"68d13'
Farleyview
Cep 21h48'53"67d31'
Farlo,Justin Daniel
Ori 6h5'39"-0d9'
Farlow,Judy
And 1h56'19"37d38'
Farmakis,Jack & Betsy
Lyr 19h0'17"26d16'
Farman,Jack P
Cnc 8h6'49"30d55'
Farmar,Dr,James G
Oph 17h53'52"12d3'
Farmer "Forever Star", Jim & Aimee
Cam 14h4'50"82d27'
Farmer 040552,Jo
Del 20h13'0"13d43'
Farmer 5-11-83,Jacklyn
And 23h39'40"46d56'
Farmer EMTP,Bill C
Sct 18h43'11"-6d31'
Farmer,Ayn Marie
And 23h22'16"42d48'
Farmer,Chase Collier
Vir 13h3'35"-8d24'
Farmer,Christopher
Aur 7h23'33"41d18'
Farmer,Dempsey Streetman
Eri 4h4'28"-7d0'
Farmer,Francis M
Cet 0h4'57"-11d39'
Farmer,Jane
Com 13h3'46"28d37'
Farmer,Jennie Marietta
And 1h48'19"37d30'
Farmer,Jessica
And 23h36'1"47d47'
Farmer,Jr,Henry
Per 3h19'55"40d48'
Farmer,Madelyn
Cet 2h40'16"-8d47'
Farmer,Marcie
Vir 11h42'42"8d5'
Farmer,Margaret P
Tri 1h48'1"25d29'
Farmer,Marjorie
Cam 3h37'29"71d4'
Farmer,Marvellous Marion
And 23h37'38"41d54'
Farmer,Mary Ann
Peg 23h4'1"20d11'
Farmer,Mary Ellen
And 23h36'47"44d56'
Farmer,Morgan Ayn
Lyr 18h45'13"32d46'
Farmer,Patrick
Aql 19h50'47"10d34'
Farmer,Sarah Beth
Dra 9h41'0"78d32'
Farmer,Shirley J
Mon 6h59'49"0d49'
Farmer,Stanley Joseph
Oph 17h25'34"-23d47'
Farmer,Susan Deborah
Cyg 21h51'23"37d31'
Farmero,Hed-Ted
Ori 5h55'21"15d58'
Farnam,Rodney
Boo 14h4'20"25d20'
Farnath Diamond,The Miriam & Maurice
Cyg 21h1'20"37d9'
Farness,Robert
Dra 18h43'20"70d1'
Farney,Lisa
Peg 22h7'16"5d53'
Farnham,Jane Hubbard
Peg 23h4'42"21d40'
Farnham,Kellianne & Brian
Uma 8h31'42"53d4'
Farnham,Lisa
Eri 4h13'58"-13d15'
Farnham,Patricia Reilly
Cas 22h57'35"55d45'
Farnsworth,Gwendolyn Jane
Peg 21h54'57"24d34'
Farnsworth,Merle
Cam 3h30'46"61d37'
Farnsworth,Norman R
Cep 0h40'0"77d34'
Farnsworth,Robert Joseph
Cep 21h21'34"68d13'
Farnsworth,Stacy D
Uma 10h25'10"59d23'
Farnum,Betty
Cet 2h46'1"4d25'
Farokh
Cnv 13h57'18"41d48'
Farougi,Vincent Matthew
Aur 6h39'26"37d24'
Farouk & Ronee Forever
Sge 20h1'46"17d22'
Farpoint Station
Lac 22h7'11"47d1'
Farquhar, Bill & Lila
Aql 19h43'57"14d20'
Farquhar,Gillian
Aur 7h5'0"38d10'
Farquhar,Jack & Shelley
Aur 6h27'60"31d31'
Farquhar,Nick
Ser 18h18'1"-13d45'
Farquhar,Sue Ellen
And 0h5'57"40d59'
Farquharson,Madeline Alice
And 0h55'34"36d22'
Farquharson,Raymond L
Dra 16h30'58"68d57'
Farr
Umi 15h18'0"69d5'
Farr Star
Boo 15h10'1"30d0'
Farr,Harold Fox
Boo 15h1'45"19d9'
Farr,Kendall
Cam 5h0'1"68d29'
Farr,Kevin I
Aql 18h59'14"16d18'
Farr,Mary
Cyg 20h24'46"37d60'
Farrah Forever
Nor 15h47'30"46d16'
Farrahi,Forever
Lyn 9h9'0"41d45'
Farrall,Bruce Jay
Boo 14h57'0"33d24'
Farrar,Anson
Cam 8h44'57"78d50'
Farrar,Jeff
Cam 3h18'0"60d11'
Farrar,Scott & Rosemary
Boo 14'32'21"42d30'
Farrar,William Harold
Ori 5h39'53"11d14'
Farrel,Lisa
Sct 18h53'29'-4d38'
Farrell Shining
Mon 6h52'17"10d53'
Farrell,Angi Margaret
Umi 14h37'56"66d59'
Farrell,Barbara & Jerry
Sge 20h1'54"17d59'
Farrell,Colin M
Tau 5h33'40"20d0'
Farrell,Corrina
And 0h59'28"34d58'
Farrell,Diane Lewis
Ori 6h6'39"6d47'
Farrell,Edwin M
Cet 2h43'46"-0d18'
Farrell,Eileen
And 23h47'35"32d36'
Farrell,F Kevin
Her 16h51'42"37d8'
Farrell,Jim (Kimo)
Ori 5h18'29"-8d35'
Farrell,Jr,Thomas Gerald
Lac 22h24'50"56d33'
Farrell,Kristen
Lyn 7h8'53"44d55'
Farrell,Linda Chervansky
Per 3h1'11"40d20'
Farrell,Mark
Cnv 12h57'0"51d50'
Farrell,Mary
Lyr 18h36'45"29d34'
Farrell,Meghan
Cas 0h21'34"64d37'
Farrell,Melissa Louise
Uma 13h58'49"60d22'
Farrell,Michael Charles
Aur 5h10'40"44d55'
Farrell,Michael K
Cnc 9h0'23"8d31'
Farrell,Nicolle Elizabeth
Crb 16h21'11"32d20'
Farrell,Robert Joseph
Her 16h37'32"4d39'
Farrell,Ron
Boo 13h55'51"19d12'
Farrell,Sarah Elizabeth
And 0h37'56"30d1'
Farrell,Sarah Taylor
Eri 4h30'0"-6d50'
Farrell,Sharon Ann Etzbach
And 2h15'25"41d6'
Farrell,Sr,Robert Michael
Boo 15h1'57"40d2'
Farrell,Tara & Dan
Cyg 21h48'30"37d15'
Farrell,Timothy John
Ari 1h47'1"16d40'
Farrell,Wallace
Aql 19h55'44"10d52'
Farrely,Terry
Aql 19h34'10"-0d32'
Farren Theo-Invigilare
Tri 2h19'28"31d41'
Farren,Lisa Marvel
Eri 4h54'52"-5d57'
Farrer,Judith Ann
Cas 0h25'25"59d7'
Farrer,Marcela Dolores
Equ 21h22'31"8d35'
Farrier,Robert M
Aur 4h35'40"31d27'
Farrin,Judith
Eri 2h48'60"-1d38'
Farrington,Miss Rachel
Cas 1h30'22"74d11'
Farrington,Nancy H
Mon 6h34'11"-5d59'
Farrington,Patricia A
Cas 0h41'38"64d29'
Farrior,Booker
Lac 22h13'1"46d20'
Farrior,Mary Susan
Oph 18h17'1"13d34'
Farris Star,The Earl F
Cep 21h42'59"55d51'
Farris,Ashley Nicole
Sge 20h17'25"17d11'
Farris,Grover & Jane
Crb 16h4'37"38d14'
Farris,Marvin K
Cmi 7h56'0"7d34'
Farris,Mrs Cinda
And 0h21'53"33d38'
Farrow's Star,Becky Mott
Mon 6h20'1"-6d37'
Farrow,Craig Randolph
Boo 14h36'42"50d57'
Farrow,Don C & Donna D
Ori 5h53'1"13d19'
Farrow-Chadwick,Keith
Aur 5h33'54"40d36'
Farruggio,Rosario
Aql 19h58'59"11d38'
Farrum,Linda Diane
Del 20h13'15"8d49'
Farry's Legacy
Dra 17h53'1"68d50'
Farry,Emma
Cra 18h16'50"-44d5'
Farsad
Aur 6h11'17"46d26'
Farsai,Paul
Dra 18h20'57"68d40'
Farschon,Robyn & Mike
Aur 5h40'10"50d16'
Farshian,Alain
Cam 13h54'31"80d47'
Farshin
Her 17h31'29"47d55'B
Farson Joining My Loved Ones,Persis H
Her 18h0'35"38d55'
Farthing,Heather Marie
And 0h15'32"46d3'
Farwell,Betty J
Vul 20h21'10"25d45'
Farwell,Christopher L
Boo 14h17'39"16d38'
Farwell,Gregory B
Her 16h44'45"21d26'

Farwell,Jeffrey L
Cnv 13h18'41"51d16'

Farwell,Richard W
Lac 22h37'50"53d18'

Farzaneh
Her 17h39'0"21d44'

Fasani,Alessandro
Cam 5h52'18"61d13'

Fasching,Joseph Ryan
Boo 13h45'42"17d14'

Fasciano,Isabella
Cas 0h47'1"66d47'

Fascinating Fred
Her 15h55'1"41d38'

Fascione,Lee
Ori 5h24'13"1d34'

Faseler,Bob
Eri 4h13'53"-7d34'

Fasenmyer,Sharon Jean
Peg 22h21'18"31d45'

Fashae,Alycia H
Mon 6h58'30"-10d20'

Fasola '77,Emanuela
Pho 0h7'16"-41d52'

Fasone,Rod Alen
Aur 6h54'31"38d36'

Fass,Ronald M
Hya 8h16'0"1d18'

Fassano,Donna
And 1h27'0"41d10'

Fassano,Donna Bisbee
And 2h19'42"45d27'

Fassbender,Renate
Aqr 22h50'1"-5d34'

Fassbender,Susanne Maria
And 0h48'59"35d60'

Fassbinder,Yvonne
Tri 1h45'43"25d39'

Fassburg,Terry
Lac 22h43'37"37d50'

Fasselius,John "Phiz"
Lmi 10h9'1"31d50'

Fassett,Cheryl
Cas 1h1'56"63d46'

Fassler,Suli Ann
Cyg 21h0'43"30d42'

Fassnacht,Marilyn
Aur 6h7'25"41d8'

Fast Company
Umi 15h18'36"71d7'

Fast Eddie
Aql 20h1'51"13d13'

Fast Eddie
Cam 12h58'50"77d54'

Fast Fred
Hya 9h5'0"5d38'

Fast Freddie
Hya 8h55'30"-7d32'

Fast Times At Eastmeadow High
Uma 9h42'1"44d11'

Fast,Carolyn
Ari 2h44'46"25d4'

Fast,Nancy
Lmi 9h26'44"38d17'

Fastenau,Doris
Peg 0h2'26"30d39'

Fastenrath,Heike
Boo 13h43'1"10d45'

Fasulo,Danny
Aur 6h27'0"38d16'

Fasy,Jr,Raymond
Aur 5h38'12"37d49'

Faszl,Ingeborg
Sgr 18h54'34"-23d16'

Fat Boy Jordan
Uma 11h50'15"30d46'

Fat Cat
Aur 5h42'27"50d18'

Fata,Rosemary
Lyn 9h22'22"46d21'

Fatalbert & Gustav
Boo 14h57'26"31d0'

FataStrega,Sara
Tau 5h50'12"23d38'

Fate
Aur 5h0'32"40d39'

Fate
Uma 10h58'34"48d28'

Fate
Aql 19h54'35"15d19'

Fate
Dra 15h47'23"62d19'

Fate is Kind
Cyg 21h17'45"28d41'

Fate's Wave
Uma 9h42'26"44d41'

Fath,Ashley Elizabeth
Cas 1h11'22"61d36'

Fath,Gregory C & Isabel C
Cyg 19h18'59"28d14'

Fath,Tom
Aql 18h40'1"-2d56'

Fathead
Her 16h9'17"48d48'

Father & Son
Uma 8h39'53"49d28'

Father Duffy's Star
Boo 14h44'48"52d23'

Father Jim
Cep 22h11'55"58d9'

Father Joe
Per 1h45'43"53d17'

Father Matt
Per 3h3'26"46d50'

Father of the Bride
Aur 5h17'12"41d26'

Father's Pride
Umi 14h20'13"78d34'

Fathergill,Evangelica Victorioso Scott
Cyg 21h27'1"28d38'

Fathie,Elham
Cap 20h16'46"-26d44'

Fathieh
Crb 16h12'39"37d37'

Fatiha
Cnv 12h51'1"40d7'

Fatima
Cam 4h9'0"68d48'

Fatima 94
Dra 16h22'16"62d56'

Fatima,Nadia Piovesan
Dra 15h19'48"56d38'

Fatino,Marsha
Peg 22h35'50"31d28'

Fatum
Her 15h56'0"47d1'

Fatum,James Bryant
Cep 21h57'35"68d5'

Fatum,Sr,William Roger
Aur 6h21'24"32d50'

Fatzick,Jr,Samuel
Cep 21h45'18"55d34'

Fatzick,Mary E
Cyg 20h0'15"40d17'

Faubel,Jonathan
Tri 1h52'24"28d37'

Fauber,Mary Reed
And 0h24'39"44d53'

Faubion,Charlotte
Aql 19h26'1"31d31'

Faucettis,Ellena
Cas 0h4'42"58d13'

Faucher,Anne
Crb 15h15'0"31d43'

Faucher,Bear
Dra 14h30'23"62d16'

Faucher,Francis
Ori 5h1'39"13d26'

Faucher,Madeline Edith
Eri 4h35'41"-19d24'

Faucher,Speedy
Lyr 18h59'15"34d31'

Fauchi-Drachi
Per 4h44'59"50d4'

Fauconier,Daniel
Ser 16h7'49"6d49'

Fauconnean,Mathias
Ser 15h59'58"10d46'

Fauconniere,James
Cep 21h49'26"68d17'

Fauer,Marlena Bueser
Cyg 21h18'0"38d55'

Faught,Marjorie
Cnv 12h47'26"35d39'

Faught,Richard
Aql 18h44'21"-2d31'

Faulds,Andrew N W
Psc 0h58'38"31d36'

Faulds,Craig James
Ser 15h12'1"17d57'

Faulk,Linda
Vul 20h14'1"25d15'

Faulkenberry,Darrell Glenn
Her 15h52'1"44d26'

Faulkenberry,Rickie Jo
And 0h9'46"40d12'

Faulkner's Aloha,Diane
Equ 21h22'43"11d52'

Faulkner,Amelia
Cet 2h48'26"4d49'

Faulkner,David Foreman
Dra 16h55'35"72d16'

Faulkner,Ella Louise
Lyr 19h11'50"47d18'

Faulkner,Harris Kimberley
Umi 15h10'11"78d34'

Faulkner,Helen & Martin
Ori 5h58'49"9d7'

Faulkner,Joyce K
Aur 6h31'16"30d53'

Faulkner,Katherine Lynn
Lmi 9h44'1"37d41'

Faulkner,Kelly Ann
Mon 8h2'1"-1d7'

Faulkner,Linda
Lyr 19h6'0"28d17'

Faulkner,Lisa
Cyg 19h27'34"34d56'

Faulkner,Paul & Karen
Gem 7h53'33"30d29'

Faulkner,Peter J
Hya 9h29'23"-5d49'

Faulkner,Taylor Erwin
Cep 21h7'55"60d2'

Faull,Lesley Allan
Sgr 19h31'2"-35d48'

Faull,Thomas E
Dra 16h56'44"68d53'

Faunce,Thomas L
Uma 10h40'47"61d32'

Fauntleroy,Marian Frances Marye
And 0h49'1"40d2'

Fauquet,Louise
Cmi 7h24'30"5d39'

Fauquier Family Star
Aql 19h59'51"15d46'

Faure,Beatrice
Mon 6h27'18"-6d32'

Faure,Elise
Sgr 19h30'30"-30d25'

Faure,Nathalie
Cam 7h10'50"67d46'

Faure,Sandra
Uma 13h29'0"62d12'

Faurie,Emile
Ara 17h57'10"-53d15'

Fause,Kenneth Raymond
Tri 2h20'54"34d48'

Faussemagne
Cet 0h0'27"-18d51'

Faust,Jonathan Robert
Her 17h39'58"23d14'

Faust,Manfred
Sco 16h12'42"-24d16'

Faust,Robert Maddux
Cet 2h29'26"4d26'

Faust,Ronald C
Lac 22h26'54"48d53'

Faust,Sr,Richard C
Cep 22h20'46"55d46'

Faust,Tom
Per 4h3'0"38d29'

Faust,Zoraida Enid
Eri 3h55'27"-10d10'

Faustin,Konstantin
Ori 5h47'22"18d49'

Faustini,Amanda Lauren
And 23h19'40"51d11'

Faustini,Gina Marie
And 23h2'31"50d56'

Faustini,Paola
Sgr 18h50'13"-27d10'

Fauvre,Sylvie
Aql 19h8'51"13d52'

Faux,Christine
Cas 2h54'30"58d18'

Favata,Richard
Mon 7h16'34"-6d53'

Favereaux,Aaron
Cep 2h44'42"77d21'

Faverino,Glenn James
Her 17h22'38"42d52'

Faverio,Alessia Marie, Gass
Boo 14h34'13"50d12'

Faverio,Elena Leigh Gass
Lyr 18h40'0"28d49'

Faversham,Harry
Uma 11h47'42"51d10'

Favier,Patrick
Cnv 13h20'38"38d15'

Favini,Christina
Lib 15h30'47"-8d49'

Favorite Guy
Aql 19h58'19"14d60'

Favorite Sister
Lyn 7h27'17"42d15'

Favorito,Jennifer
Lyn 7h58'12"44d36'

Favreau,Pete
Dra 16h25'39"60d7'

Favurito
Vir 14h27'20"1d2'

Fawcet,Farah
Eri 3h47'51"-5d27'

Fawcett,Charles Frederick
Aur 7h24'39"37d33'

Fawcett,Jeff
Crt 11h51'47"-7d35'

Fawcett,Patricia
Lyn 9h15'36"33d33'

Fawcett,Paul
Ori 5h58'0"15d59'

Fawcett,Rosemary
Peg 22h43'55"11d17'

Fawcett-Ward,Diane Louise
Cas 1h50'30"58d3'

Fawcus,Robert Archbold
Cyg 21h11'25"34d29'

Fawlaw,Stanley Allen
Per 1h51'1"56d23'

Fawn's Star
Cas 0h2'1"62d24'

Fawna
Cma 6h53'43"-18d52'

Faxon,Barbara Washam
Aqr 20h37'60"-8d49'

Faxon,Joan Macgowan
Sge 19h59'56"20d25'

Faxon,Jr,Robert Endicott
Aqr 21h57'34"-11d33'

Faxon,Prudence
Sge 19h54'1"16d4'

Fay DMA, N3HPZ,James Spencer
Uma 11h3'23"45d7'

Fay VI
Lyr 19h0'44"26d36'

Fay's Christmas Star
And 0h30'1"45d36'

Fay,Amelia Elena
Del 20h15'22"9d38'

Fay,Benjamin Sonny
Per 3h0'17"46d54'

Fay,Keith
Boo 13h56'36"20d12'

Fay,Laura Anne
Lib 14h44'1"-8d40'

Fay,Margaret Furay
Vir 12h8'35"-0d27'

Fay,Michael Jude
Cep 21h36'1"56d1'

Fay,Sharon E
Tri 2h34'26"32d39'

Fay,Sharyn Liese
Aur 5h6'31"40d10'

Fay,Stephanie Ann
Vul 20h2'35"23d27'

Fay,Steven Joseph
Her 17h24'1"42d26'

Fay,Susan
Lyn 7h25'38"58d56'

Fay-Bag
Cmi 7h7'0"5d5'

Fayden,Florence
Uma 10h25'15"40d55'

Faye Lim
Cyg 19h26'56"34d57'

Faye Louise
Lac 21h58'37"40d7'

Faye,Charlie
Oph 16h20'16"0d26'

Faye,Kathleen Antionette
Sge 20h2'37"18d56'

Fayed,Dr Abdel Monein A Ismail
Ari 2h31'39"30d51'

Fayerweather,John
Per 4h41'33"50d47'

Fayet,Bernard
Cyg 20h28'47"40d21'

Fayet,Georges
Oph 18h40'46"7d6'

Fayette School, Harold D
Aur 6h56'31"37d25'

Fayette,Gordon & Naomi
Uma 9h26'28"57d52'

Fayko
Aql 20h1'0"10d11'

Fayre (Fay),Champion Alilah Vanity
Cmi 7h41'1"-0d5'

Fazackerley,Mary
Boo 13h41'0"15d56'

Fazekas,George B
Uma 9h32'49"57d45'

Fazekas,Kent Charles
Boo 14h0'26"19d1'

Fazekas,Nicholas James
Aur 5h38'29"40d4'

Fazekas,Ross Webster
Lac 22h8'14"38d6'

Fazio,Al & Rose
Cnv 12h13'15"45d59'

Fazio,Christine
Cep 22h54'49"78d57'

Fazio,David
Boo 14h18'16"30d12'

Fazio,Gayle S
Cas 23h25'56"62d21'

Fazio,Georgo W
Cnv 12h45'53"42d26'

Fazio,Joe
Aur 6h7'27"33d6'

Fazio,Michael
Ori 5h51'15" 8d9'

Fazio,Nina
Uma 11h21'1"48d7'

Fazio,Virginia
Cas 1h36'27"60d29'

Fazio,Vita
Per 3h6'32"56d20'

Fazler,Madeline
Cyg 19h51'18"40d44'

Fazlin,Christina & Fazal
Hya 9h1'11"-1d38'

Fazzalari "FLI",Robyn
Cyg 19h55'1"41d1'

Fazzari,Bradley M
Dra 16h51'34"64d48'

Fazzi,Joel "Punkin Doodle"
Sct 18h54'25"-4d44'

Fazzia,Joseph Angelo
Cep 22h46'25"57d45'

Fazzinga III,Frank Anthony
Cam 3h28'1"53d43'

Fazzini,Lidiziana
Cas 0h10'13"58d55'

Fazzolare,Anthony
Oph 18h7'33"7d37'

Feaadaas,MI LV Regina
Peg 22h6'56"5d40'

Feagin,Paula
Uma 10h9'53"67d59'

Fearen,William
Aur 5h2'30"41d20'

Fearing,Bruce & Cindy
Cam 5h46'29"60d46'

Fearn
Hya 8h33'43"1d46'

Fearnley,Jr,William R
Aql 19h31'28"12d39'

Fears,Robert
Oph 16h56'43"11d1'

Fearson,Col.,Gordan T
Cma 6h56'29"-18d40'

Feasey,Michael Thomas
Aql 18h57'1"16d57'

Feaster,Ann Marie
And 1h6'49"38d20'

Feaster,George E
Cet 2h1'30"-10d23'

Feaster,Hank
Her 17h6'1"20d46'

Feaster,Mary Alliene
Boo 13h52'32"17d12'

Feather,Deborah Marie
Lyr 19h5'47"38d51'

Featherman,Danny & Jordan
Her 16h46'60"32d34'

Featherrock,Eleanor Guthrie
Peg 22h23'40'18d52'

Featherston,Richard Bunn
Ser 18h43'0"2d28'

Featherstone,I Love You Donna
Cas 1h8'32"66d35'

Featherstonhaugh,James Duane
Aur 6h13'25"46d19'

Feazell,Captain John Clayton "Tad"
Aql 19h5'45"-0d22'

Febbraro,Pasquale
Ser 15h20'59"2d52'

Febery,Valerie Ann
And 0h52'21"33d23'

FeBland,Gabrielle Simone
And 23h36'33"32d49'

February 23rd
Uma 8h38'1"70d50'

Fecher,Patrick Cody
Lyn 8h38'29"41d45'

Fechner,Lothar
Cnv 14h2'37"33d47'

Fecht,Anthony J
Boo 14h30'28"31d23'

Fecht,Kurt Vonder
Dra 11h46'25"70d39'

Fechter,Danny
Per 1h54'0"54d1'

Fechter,Dorothy Marie
Uma 11h14'41"40d3'

Fechter,Walter John August
Uma 11h12'41"38d30'

Fechtner,Dr Jerome
Boo 14h47'43"26d39'

Fecio,Linda
Lyn 7h53'37"42d41'

Feckner,Louis E
Aql 20h3'37"1d40'

Fedak,Julie
Cas 0h31'0"62d17'

Fedchenko,Max
Per 1h50'15"56d41'

Fede
Del 20h13'58"15d35'

Fede
Cep 21h59'57"60d32'

Fede,Tonyteo
Cam 6h27'59"65d1'

Fedeferdigenia Splendes
Pyx 8h41'2"-21d48'

Fedele,Ronald Dominick
Cep 22h47'1"57d31'

Fedelich,Marie-Claude
Aur 4h48'42"40d54'

Feden,Nicasius Deborah
Lyn 8h47'27"34d60'

Feder,Dora Quat
Tau 4h22'43"15d11'

Feder,Samuel L
Aur 5h10'32"41d38'

Feder,Stanley
Cma 7h13'0"-13d25'

Federer,Silvia & Urs
Tau 4h4'12"0d35'

Federica
Lac 22h21'14"50d34'

Federica
Lac 22h29'26"36d5'

Federica
Cep 23h30'11"65d22'

Federica
Her 16h5'48"50d35'

Federica
Col 6h32'35"-33d51'

Federica
Hor 2h53'47"-49d8'

Federica
Hor 3h22'13"-48d22'

Federica
Pyx 8h31'9"-23d37'

Federica 72
Peg 21h51'52"33d48

Federica 81-92
Cam 6h40'46"68d41'

Federica La Stella Della Nostra Vita
Lyn 8h37'26"41d49'

Federici,Anthony
Per 3h37'10"38d11'

Federico
Lac 22h16'40"37d44'

Federico
Her 18h10'58"40d51'

Federico
And 0h12'38"38d42'

Federico
Gru 22h23'52"-53d13'

Federico
Pup 8h24'37"-26d44'

Federico
Pyx 8h31'33"-27d38'

Federico,Andrea Marie
And 1h3'32"38d0'

Federico,Jacob Scott
Ser 15h11'22"0d9'

Federl,Gerhard
Ari 2h42'54"21d60'

Federowicz,Gregory T
Aur 5h5'43"42d33'

Fedewa,Michelle
Vul 19h14'0"21d58'

Fedler Jr,Charles Robert
Peg 23h6'39"20d51'

Fedor Aurora-Borich- Margie-Alis
Mon 7h57'59"-3d22'

Fedorchak,John&Pat
Vul 20h1'48"28d52'

Fedorka,Dawn N
Cas 0h41'26"64d16'

Fedrigon,Jerry
Aur 6h1'26"31d13'

Fedro,Rosalyn
Peg 23h18'35"28d54'

Fedry,Familie
Ser 15h15'26"8d3'

Feduniak,Robert Ben
Aur 6h2'1"38d49'

Fee,Jade Alexandra
Equ 21h21'30"12d39'

Fee,Megan Elizabeth
And 23h3'1"41d6'

Feebin
Lyn 7h49'28"39d27'

Feehan,MaryAlexis
And 23h44'56"32d38'

Feel: "The Firefly"
Gem 6h56'41"35d9'

Feeley E Wakefield NH, Erin
Uma 8h30'0"59d4'

Feeley,Andrew James
Per 3h51'32"38d52'

Feeley,Connor Robert
Sex 10h17'34"-8d53'

Feelgoodes Fireside Star
Her 17h37'17"41d12'

Feenan,Amy
Crb 16h19'33"30d39'

Feeney,Cathy
Boo 14h33'13"44d16'

Feeney,Dr Owen
Cep 22h20'20"67d60'

Feeney,Jane
Lyn 8h20'27"49d58'

Feeney,John Robert
Her 16h43'18"25d57'

Feeney,Siobhán
Cnc 8h27'48"8d33'

Fees,Kyle Elliot
Eri 3h42'25"-5d45'

Fees,Lindsey Brooke
Oph 17h14'11"-20d7'

Fees,O S
Cmi 7h55'24"0d9'

Feezer,Mischelle "Naples"
Umi 16h45'30"77d33'

Feffer,Marc Andre
Dra 12h30'38"64d3'

Fegan,J Keith
Boo 14h48'27"23d52'

Feger,Manfred
Tau 4h7'35"1d38'

Fehlauer,Chesa
Cam 5h49'11"65d27'

Fehlman,Wayne
Tau 5h45'0"23d31'

Fehmerling,Jr,G "Duke"
Cep 21h58'38"55d40'

Fehr Family,W
Umi 13h19'54"72d28'

Fehr,Donna J
And 1h52'0"37d48'

Fehrenback,Val & Jack
Her 18h5'23"28d17'

Fehring,Beautiful Sarah
Lyr 18h51'27"31d47'

Fehse,Shirley Gibson
Sct 18h48'52"-6d26'

Fehsenfeld,Mildred Louise Cornelius
And 23h2'50"49d59'

Feibus,Jonathan Robert
Her 18h7'54"16d27'A

Feibus,Mary Levinsohn
Her 18h7'54"16d27'B

Feibus,Stephen S
Oph 17h57'13"12d10'

Feichtenschlager, Günter
Lyr 19h2'36"31d52'

Feidel Merry Christmas 1993,Donald
Aql 18h59'57"-1d29'

Feig,Clayton R
Cnv 12h30'50"33d14'

Feigenbaum's 60th,Ed
Aql 19h17'11"12d43'

Feigenbaum,Mary "Grandma"
Dra 18h29'12"50d13'

Feighery,John Edward
Boo 14h8'18"22d55'

Feigin,Andrew
Per 1h48'52"48d41'

Feilberg,Nicolai
Uma 10h40'56"44d34'

Feiler,Jane & Larry
Cyg 20h2'21"41d5'

Feimer,Flosey Dotes
Peg 21h57'38"29d17'

Feimer,Poppy Seed
Ori 4h46'23"4d17'

Fein Star,A Fine Star, The Bruce
Lac 22h50'30"38d18'

Fein,Lisa Marie Towler
Lyr 19h18'60"41d45'

Fein,The Randall
Mon 6h32'33"1d19'

Feinberg,Daddy
Her 16h41'38"21d43'

Feinberg,David Allen
Cep 21h7'11"58d26'

Feinberg,Dr E Richard
Aur 6h20'30"32d46'

Feinberg,Jay
Dra 18h36'13"67d37'

Feinberg,Lenny
Vul 20h17'11"26d14'

Feinberg,Leonard
Dra 9h52'56"80d3'

Feinberg,Melvin
Aur 7h16'0"40d47'

Feinberg,Randy
Dra 16h22'60"63d9'

Feinberg,Renne
Cnv 12h59'19"31d58'

Feiner,Brett Rachel
Aur 7h11'1"36d42'

Feiner,Sorin A
Aur 6h34'1"32d9'

Feingold,Edmund
Ori 6h4'30"10d56'

Feingold,Richard E
Aur 5h17'18"48d41'

Feinman,Judy & Irving
Cyg 19h57'12"40d14'

Feinstein,Dov
Mon 7h1'47"4d59'

Feinstein,Star of Lester
Cmi 7h56'1"1d42'

Feinstein,Stephanie Yvonne
Mon 7h15'49"-0d35'

Feirer Star
Uma 10h21'40"60d54'

Feist,Marilyn
And 23h30'29"39d54'

Feitelberg,Alan Seth
Eri 3h35'40"-2d6'

Feitelberg,Richard Charles
Sge 19h56'17"19d26'

Fekas,John
Hya 8h11'25"-6d2'

Fekete,Susan
Uma 10h45'0"40d59'

Fekete,Susan
Eri 3h59'37"-17d7'

Feland Moonchild,John Thomas
Boo 15h10'36"48d39'

Feld,Kenneth
Cep 22h2'16"53d21'

Felder,Constance P
Cam 5h57'46"60d16'

Felder,Frances
Uma 8h34'18"47d6'

Feldhaus,Laverne M
Aql 18h57'48"3d39'

Feldhousen,Rosanne Denise
Cyg 19h43'37"30d39'

Feldman,Adam
Her 17h38'39"25d31'

Feldman,Amy M
Sgr 19h39'57"-36d29'

Feldman,Audrey
Vir 13h36'47"-9d13'

Feldman,David Borel
Umi 15h7'27"68d12'

Feldman,David Edward Kline
Cep 23h12'54"70d45'
Feldman,Don
Cet 3h0'34"0d22'
Feldman,Dr Douglas A
Oph 17h6'11"1d4'
Feldman,Dr Joseph A
Oph 17h32'36"-23d22'
Feldman,Esther
Peg 22h19'51"18d46'
Feldman,F Barbara
Lyr 18h15'51"30d47'
Feldman,Florence & Abner
Aql 19h4'16"10d14'
Feldman,Francis Emmanuel
Umi 15h22'24"67d29'
Feldman,Jeffrey Andler
Lac 22h48'22"55d6'
Feldman,Jr,Henry A
Her 16h17'33"48d37'
Feldman,Marcel
Umi 15h27'33"67d56'
Feldman,Phil
Lyr 18h32'54"37d45'
Feldman,Robert Alan Kline
Aur 6h12'51"45d20'
Feldman,Ruth
Lyr 18h55'16"32d31'
Feldman,Seth Andler
Cnv 13h43'32"33d58'
Feldman,Stephanie
Ori 5h50'15"9d37'
Feldman,Stewart & Richard
Cacace
Her 18h6'51"30d55'
Feldman,William Harrison
Hya 8h43'27"3d53'
Feldman,Zachary Ethan
Cet 2h0'1"1d32'
Feldmann,Stefan
Tau 3h46'44"0d24'
Feldon,Alicia
Cyg 21h1'47"31d17'
Feldstein,Jenna
Cas 23h20'47"60d11'
Feldstein,Margaret & Milton
Cet 2h54'31"-0d16'
Felenada
Pup 8h8'53"-22d44'
Feleppa,Anselmo
Mon 6h52'49"1d22'
Felice
Cam 6h28'46"67d37'
Felice,Aurora
Scl 23h47'39"-26d40'
Felice,Fuddy & Duddy Beach
Ori 4h54'29"0d29'
Felice,Grazia
Cam 5h8'30"61d15'
Felice,Marco
Per 1h32'31"52d35'
Felice,Mimi
Umi 10h30'13"41d44'
Felice,Rosalie
Lyn 7h43'19"51d41'
Felicella Annamaria
Del 20h19'26"12d35'
Felicia
Peg 21h57'38"22d53'
Felicia Lyn
Eri 5h3'52"-5d41'
Felicia Nicole
And 23h50'51"33d40'
Feliciano-Baker, Tabitha
Uma 10h18'39"55d49'
Felicio,Richard
Aur 6h8'39"46d30'
Feliciotti,Flora
Lyr 18h37'42"29d34'
Felicissimo,John
Her 16h38'48"35d50'
Felicita
Lyn 7h33'13"45d2'
Felicita
Gem 7h16'5"24d32'
Felicita
Ori 5h56'55"15d4'
Felicitas Francesca
Uma 8h39'56"55d60'
Felicitas,Eva
Aqr 22h20'35"2d2'
Felicitate,Schwanzaris
Tau 4h6'32"1d47'
Felicity
Eri 3h42'25"-10d7'
Felicity
Cyg 20h29'60"40d12'
Felicity
Mon 8h3'57"-3d4'
Felicity
Aql 19h28'44"-0d44'
Felicity
Cas 1h7'44"60d45'
Feliks
Lib 14h57'34"-22d35'
Felina
Boo 13h43'39"11d25'
Felios,Ailse Marie
Oph 18h28'1"7d41'
Felix
Cma 6h22'2"-12d56'B
Felix
Lmi 9h44'36"40d16'
Felix
Dra 17h35'16"70d60'
Felix
Leo 9h55'39"11d26'
Felix
Cep 20h27'36"63d44'
Felix
Dra 17h43'38"61d13'
Felix
Aql 19h48'43"14d53'
Felix Dvorak Duell Team
And 2h2'0"40d18'
Felix The Cat Diner
Lyn 8h6'49"40d58'
Felix,Allison,Lauren, Ellen,&
Dave
Eri 3h36m24"-2d58'
Felix,Crystal Nehcole Mad
Ups
Crb 16h12'41"32d4'
Felix,Daniela
Tau 4h5'27"20d1'
Felix,Gustav F
Uma 9h25'49"47d37'
Felix,Laura
Lup 15h11'54"-45d41'
Felix,Phil & Lucy
Cyg 21h9'1"48d52'
Felix,Phyllis & David
Eri 3h28'33"-3d14'
Felix,William E
Lac 22h7'54"51d47'
Feliz,Guillermo
Cam 5h54'34"58d24'
Feliz,Richard
Cma 6h59'0"-16d14'
Feliza
Sct 18h52'39"-5d21'
Felker,My Forever Love-Jeffrey
Per 2h20'13"54d54'
Felker,Richard J
Aur 4h49'33"38d19'
Fell,Andrew
Her 16h34'7"19d5'
Fell,Barry
Aql 18h59'0"-6d58'
Fell,Dave & Carolyn
Cyg 20h56'40"31d37'
Fell,Jenny
Psc 0h23'28"-2d29'
Fell,Louis R
And 2h33'46"43d6'
Fell,Nicholas
Her 18h3'20"38d48'
Fellenz Family Star
Uma 9h38'25"51d22'
Fellenzer,Heinz
And 23h10'13"40d12'
Feller,Geri
Umi 16h37'21"79d2'
Feller,Rachel
Cmi 7h23'58"0d22'
Felling,Bradley C
Ser 15h57'23"-0d32'
Fellner,Andreas
And 0h5'40"37d55'
Fellner,Uschi
Cnv 13h13'1"32d1'
Fello
Sge 19h57'34"16d47'
Fello,William
Aql 18h55'49"8d3'
Fellows,Karen
Lyr 18h37'32"40d35'
Fellows,Meredith Jo
And 1h40'56"40d34'
Fellows,Moira
Eri 2h59'15"-5d16'
Fellows,Nicole Lyn
Equ 21h3'38"7d54'
Fellows,Priscilla
And 1h55'1"37d47'
Fells,Frank Michael
Per 4h31'28"38d9'
Felmet,Floyd Ray
Cma 7h0'32"-28d59'
Fels,Anita
Oph 17h58'58"7d43'
Felskie,Danny
Aur 6h22'29"30d56'
Felt,Joseph M
Hya 8h19'0"-5d50'
Felten,John Roland
Aur 6h26'15"52d38'
Felten,Suzanne Carol Ford
Cam 5h41'35"70d56'
Felter,Donald B
Per 3h1'55"50d7'
Felter,Grandma
Cas 1h38'25"58d36'
Feltham,John Oliver
Uma 11h56'19"58d26'
Feltis
Hya 8h31'53"-8d14'
Feltman,Arnold
Lac 22h12'57"54d26'
Felton's Fire
Ind 20h49'20"-50d42'
Felton,Bud & Barbara
Uma 11h53'57"55d36'
Felton,Elizabeth "Libby Gail"
Vul 19h43'22"20d8'
Felton,Mclany
Mon 6h42'43"10d47'
Felton,Shannon Kelley
Peg 0h4'32"20d4'
Felton,Star-Jay
Lac 22h4'15"38d4'
Felton,William Christopher
Mon 6h27'42"11d11'
Feltz,Jeanne Marie
Cam 3h33'21"60d59'
Feltz,My-Knight-Tom
Her 18h11'55"31d17'
Felux,Petra
Sgr 18h57'36"-26d44'
Felver,Barbara E M
And 1h4'42"39d47'
Femano,Candace
And 2h25'44"47d50'
Femi's Light
Cmi 7h24'0"0d37'
Fender II,James Patrick
Trenton
Aql 19h24'22"10d30'
Fender,Angela
Cyg 21h9'21"37d41'
Fendlay,Ellsworth
Uma 9h18'59"54d42'
Fendler,Don
Uma 15h0'57"81d35'
Fendler,Joan
Cyg 20h36'45"37d50'
Fendrich,Rainhard
Psc 22h59'36"1d22'
Fendrick,Christina Diane
Aur 5h34'1"40d14'
Fendrick,Lisa A
Vul 19h58'13"29d6'
Fendt,Magician,Frank C
Aur 4h56'47"41d8'
Fenech,CPA,Atty Marie
Ori 5h50'18"17d17'
Fenech,Jean Francois
Ori 6h5'18"1d5'
Fenella F B T D
Aql 19h20'34"15d60'
Fenerty,Jack
Uma 10h5'60"58d49'
Fenev,Sarah Louise
And 1h42'23"48d58'
Feng Bao
Aur 5h56'22"40d37'
Fenger,Svante Rasmus
Cep 23h6'13"60d16'
Feniello-Coviello
Lyr 18h23'15"42d48'
Feniger,Jerome
Gem 6h44'13"12d36'
Fenkart,Bernd
Uma 10h3'45"61d14'
Fenley,Baby
Peg 22h4'0"10d53'
Fenn,Brett Whitfield
Ser 15h13'30"9d48'
Fenn,Gordon Paul
Aur 6h18'35"34d49'
Fenne
Eri 4h2'29"-17d42'
Fennelly,Dora Alice
Com 12h18'46"32d21'
Fennelly,Ginger
Eri 3h55'20"-15d37'
Fennelly,Kalee Mae
Gem 6h48'34"12d46'
Fennema,Audrie Marie
Mon 6h53'40"0d18'
Fenner III,Edward Raymond
Cyg 19h55'16"38d55'
Fenner,Arthur
Cyg 19h30'1"33d51'
Fenner,Craig
Sge 19h53'0"18d48'
Fenner,John & Carol
Aql 18h43'15"11d33'
Fenner,Jr,Marvin
Cmi 7h44'19"4d27'
Fenner,Lorry M
Lyn 8h34'43"40d30'
Fennern,Allison Emma
And 23h39'46"39d28'
Fennes,Doris & Toni
Lib 15h28'45"-20d9'
Fennessy's Star, Annmarie
Cas 1h53'55"60d26'
Fenno-Durning,Zelda
Cas 0h3'14"63d3'
Fenoglio,Margherita
Tel 19h4'31"-46d25'
Fenollera,Isabel
Lyr 19h17'1"25d53'
Fensick,Melanie Eileen
Uma 9h39'36"45d42'
Fenske,Gail
Mon 6h42'22"6d9'
Fenske,Kay Viering
Cyg 20h47'54"38d50'
Fenske,Wilfried
Vul 19h59'37"29d11'
Fenstermacher, Jacqueline
Harriet
Eri 2h53'53"-11d41'
Fenstermacher,Lauren Grace
Lyr 19h0'38"40d18'
Fenstermacher,Lori Heiser
And 23h32'44"41d21'
Fenstermacher,Nancy
Peg 23h5'41"33d49'
Fenstermaker,Richard
Aur 4h36'15"33d57'
Fenstermaker,Scott
Cep 2h44'0"77d41'
Fent,Heather Lee
Ari 1h55'1"16d19'
Fenton,Colin
Ori 4h42'1"15d23'
Fenton,Daniel Edward
Lib 15h38'50"-23d55'
Fenton,Graham Stewart
Peg 21h59'25"30d0'
Fenton,Hannah Piper
Cas 1h5'0"60d33'
Fenton,Ilyssa Randi
And 23h43'33"42d52'
Fenton,James Ross
Dra 16h55'48"68d23'
Fenton,Sarah Elizabeth
Lyr 19h1'37"28d27'
Fentress,Mary Moore
Aql 18h43'26"10d40'
Fenty,Phillip Adriaan
Ser 18h55'19"3d24'
Fenwick,Paul Joseph
Aql 18h58'40'-6d47'
Fenwick,Rosa
And 0h37'23"40d54'
Fenwick,Tyler & Marty
Cmi 7h42'14"0d19'
Fenz,Roland
Ser 18h10'40"-5d28'
Fera,Alexander Thomas
Ori 5h52'21"15d16'
Feran,Deborah Jean
Lyr 18h53'0"30d59'
Ferber,Laurence
Cnv 12h50'25"51d9'
Ferber,Megan
And 1h56'15"37d38'
Ferber,Sandy
Lyn 9h25'23"40d6'
Ferber,Stephen C
Tri 1h53'38"27d16'
Ferbus,Caroline
Per 3h36'0"51d22'
Fercello,Marion Leach
Cep 21h8'19"61d49'
Ferden,Maestro Bruce
Aur 5h57'27"30d44'
Ferdenda
Sex 10h28'16"1d8'
Ferdinand,Theo
Peg 23h32'26"20d51'
Ferdinands
Dra 17h55'48"67d51'
Ferdon,Richard G
Ser 15h25'9"17d29'A
Ferdun,Monica
Tri 2h1'20"33d20'
Fereday,Robert
Uma 9h9'1"48d59'
Ferembach,Catherine
Lyr 18h32'48"35d42'
Feren,Patrick
Sct 18h43'46"-6d27'
Ferency (Helen),Maria
Candelaria Fajardo
Cet 1h44'22"0d53'
Feresten,Nancy Laties
And 2h26'55"48d55'
Feretti,To Christine & John
Cas 1h15'30"63d35'
Fereza,Jesse Joshua
Per 1h52'41"50d15'
Ferg,Helga
Leo 10h51'39"-5d9'
Fergeson,Richard R
Lyr 18h49'37"41d13'
Fergie
Cet 3h9'53"0d14'
Fergie's Folly
Cet 2h25'44"4d55'
Fergione,Michael Edward
Dra 15h45'53"62d30'
Fergus,Allan
Cas 0h40'29"72d51'
Fergus,Eddie
Peg 23h29'24"17d36'
Fergus,Haley Elizabeth
Eri 3h21'1"-10d14'
Fergus,Jr,James Perry
Ori 4h47'30"4d34'
Fergus,Whitney Leigh
Aql 19h44'0"14d3'
Ferguson
Cyg 20h30'12"42d24'
Ferguson Family Star, The
Daryl
Cep 23h10'36"68d13'
Ferguson II,Joseph Edwin
Boo 14h56'22"50d35'
Ferguson,Austin Neil
Her 17h12'49"44d28'
Ferguson,Brianna Irene
Lmi 10h30'47"28d24'
Ferguson,Christina Colette
And 0h53'49"33d39'
Ferguson,Christine
Peg 22h39'14"22d6'
Ferguson,Connie J
Sex 10h26'0"-2d20'
Ferguson,Constance Woodhall
Ori 6h17'58"20d43'
Ferguson,Countess Harriet
Cyg 20h17'45"38d24'
Ferguson,Debra Barker
Gem 7h27'41"31d9'
Ferguson,Dick,Marilyn Megan
& Craig
Uma 14h2'21"58d48'
Ferguson,Donald Frederick
Dra 15h9'26"64d28'
Ferguson,Dorothy
Mon 6h48'39"11d3'
Ferguson,Earl of
Lamond,Barclay
Aur 6h1'45"38d5'
Ferguson,Forrest William
Lmi 10h4'13"37d57'
Ferguson,George A
Aql 19h30'50"13d46'
Ferguson,Heather Rembert
Brooks
And 23h48'33"40d14'
Ferguson,Heather
Cet 0h50'36"-4d9'
Ferguson,Helen Pappalardo
Cas 0h13'20"61d43'
Ferguson,II,David
Per 2h30'60"50d56'
Ferguson,Jess Nathaniel
Aql 19h55'1"13d27'
Ferguson,John Henry
Aur 6h24'41"32d3'
Ferguson,Jonathan Edmund
Cap 20h9'50"-10d16'
Ferguson,Jr,Walter E
Lac 22h11'18"49d21'
Ferguson,Julie Lynn
Lyn 8h27'0"41d4'
Ferguson,June
And 23h3'23"50d36'
Ferguson,Kathryn Rachel
Aur 5h16'38"40d12'
Ferguson,Keith D
Ser 15h12'60"1d8'
Ferguson,Kenneth L
Ori 6h10'48"0d35'
Ferguson,Linda Rae
Mon 7h55'23"-1d6'
Ferguson,Margaret Ashleigh
Equ 21h0'34"3d33'
Ferguson,Mark
Hya 9h37'56"-2d5'
Ferguson,Mark
Cet 2h27'55"5d27'
Ferguson,Marty
Dra 17h51'36"60d32'
Ferguson,Michael Ford
Del 21h0'42"12d50'
Ferguson,Michele
Dra 18h56'17"47d42'
Ferguson,Miriam Pew
Uma 11h24'26"67d44'
Ferguson,Monica Monique
Uma 11h29'24"44d30'
Ferguson,Murray J
Cep 22h17'33"70d36'
Ferguson,Neil
Per 2h2'14"57d10'
Ferguson,Stephanie
Ori 5h51'53"8d52'
Ferguson,Sue
Cyg 19h13'31"48d12'
Ferguson,Terry
Lyr 18h44'1"45d39'
Ferguson,Thomas
Oph 17h31'29"11d9'
Ferguson,Virgina"MeMe"
Sct 18h55'15"-6d60'
Ferguson-30,William J
Cnc 8h12'16"32d33'
Feriae Delores
Peg 22h41'36"29d31'
Ferick,Alexa Catlin
Cas 22h58'15"54d41'
Ferida
Cet 2h31'31"-6d50'
Fering:Creative Soul, Kirsten
Per 3h1'33"46d48'
Ferkin,Michael D
Her 17h22'27"40d1'
Ferko,John D
Cep 3h3'60"77d43'
Ferland Star,Armand & Jem
Umi 14h38'1"86d19'
Ferlini,Ferlini
Pyx 8h31'8"-26d0'
Ferlitsch-Egger, Hildegard
Umi 14h23'46"68d22'
Ferlo,Julie K
Peg 22h15'57"33d45'
Fern
Ori 6h4'50"10d33"
Fern
And 23h28'23"44d19'
Fern Ann
Aur 5h31'33"31d38'
Fern Extraordinaire
Vir 11h52'44"5d40'
Fern,Joseph M
Her 17h17'24"22d50'
Fern,Ruby
Eri 3h27'17"-13d22'
Fernand
Uma 9h18'11"43d44'
Fernanda Assuncao
Lyr 18h26'47"38d43'
Fernandes,Christopher J K
Eri 4h17'6"-18d31'
Fernandes,Jacob Xavier
Sge 19h57'37"20d16'
Fernandes,Margo Lissa
Mon 7h41'11"-5d46'
Fernandes,Maria Luisa
Cas 2h33'14"57d45'
Fernandes,Ron
Uma 9h34'23"70d18'
Fernandez Birthday Star,The
Claude
Com 13h15'50"22d22'
Fernandez De Paz, Tedfilo
Cep 20h16'41"61d12'
Fernandez Star,Marylyn
Tau 5h13'59"16d8'
Fernandez"Tesora Mia", Sarina
Sge 19h42'58"16d36'
Fernandez,Carmen
Cas 0h0'14"66d9'
Fernandez,Cheryl Ann
Peg 23h2'49"21d34'
Fernandez,David "Moonstar"
Lac 22h21'30"55d50'
Fernandez,David Louis
Dra 16h11'1"61d46'
Fernandez,Erica Celine
Mon 6h29'0"11d38'
Fernandez,Jack R
Peg 23h2'0"31d50'
Fernandez,Jacob J
Eri 4h10'20"-16d50'
Fernandez,Jennifer Marie
Cam 5h51'5"73d19'
Fernandez,Jessica May
Lyn 6h32'57"54d38'
Fernandez,Jose Domingo
Gamio
Cma 6h56'56"-15d59'
Fernandez,Juan-Manuel Lojo
Her 16h39'28"17d50'
Fernandez,Lizzette
Peg 22h5'0"27d37'
Fernandez,Lucia
Sgr 18h3'1"-28d23'
Fernandez,Manuel Alfredo
Lopez
Peg 22h31'40"31d17'
Fernandez,Mark Philip
Eri 3h24'10"-13d47'
Fernandez,Martin
Uma 9h39'1"47d50'
Fernandez,Mr & Mrs Celti
Lac 22h55'57"37d48'
Fernandez,Myra Elaine
And 0h13'43"38d43'
Fernandez,Orlando
Ori 6h5'47"8d36'
Fernandez,Raquel
Umi 14h2'44"68d2'
Fernandez,Ray & Vivian
Cyg 20h38'35"46d32'
Fernandez,Robert G
Mon 8h3'12"-5d17'
Fernandez,Sofia Cristina
Umi 14h22'21"66d58'
Fernandez,Stacie Danette
Peg 23h38'48"25d43'
Fernandez,Susan
Leo 10h52'30"-6d17'
Fernandez-My Love My
Hero,Sergio
Dra 11h30'29"68d33'
Fernando alias Microbos
Hor 3h29'29"-48d18'
Fernando,David
Cnv 13h21'34"28d59'
Fernando,John NLC
Per 2h40'47"43d40'
Fernando,Rosanna
Cet 0h57'24"1d57'
Fernando,Tracy
Umi 14h38'46"77d46'
Ferney,Sandra J
Eri 4h3'1"-16d56'
Ferneyhough,Brian John Peter
Lyn 7h2'58"44d32'
Fernquist,Darrell Lee
Her 16h31'36"34d20'
Fernández Zacarías
Aur 5h12'54"43d16'
Fero,Jacob
Aur 6h45'0"35d41'
Ferracci,Donald N
Dra 15h14'34"63d12'
Ferragamo,Paul J
Boo 14h24'11"27d60'
Ferrali-A Wish Come
True,Joseph
Ori 5h38'0"8d37'
Ferrall,Jr,Victor E
Boo 15h2'1"31d42'
Ferrand,Sir Richard of
Ori 4h59'0"7d32'
Ferrando,Bill
Boo 14h17'23"15d44'
Ferrante,Frank
Lmi 10h35'1"26d42'
Ferrante,Glenn
Per 3h17'48"41d10'
Ferrante,Julie
Dra 11h28'1"68d7'
Ferrante,Linda Patas
And 0h42'53"45d51'
Ferrante,Loren Elizabeth
Aql 18h45'35"9d24'
Ferrante,Scott B
Sex 10h26'32"1d2'
Ferranti Our Shining
Light,Dina
Mon 6h58'26"11d41'
Ferrara 2/96,Kathleen Emily
Aql 18h43'1"6d8'
Ferrara 42
Boo 14h26'15"20d24'
Ferrara III,Lawrence Peter
Her 17h12'30"27d48'
Ferrara,Adam
Dra 19h41'50"67d38'
Ferrara,Andrea
Aql 20h15'42"1d40'
Ferrara,Antonina Carollo
Umi 16h23'57"80d30'
Ferrara,Basi
Boo 14h39'55"57d36'B
Ferrara,Benny Paul
Aql 20h2'30"0d18'
Ferrara,Craig
Aur 6h15'46"38d59'
Ferrara,Eileen Carol
Gem 7h29'23"30d2'
Ferrara,Giulia
Cet 2h4'1"1d2'
Ferrara,Jerry
Per 2h47'60"43d37'
Ferrara,John
Boo 14h39'55"57d36'A
Ferrara,Joseph R
Per 3h10'24"50d8'
Ferrara,Kaitlyn
Dra 11h7'21"73d36'
Ferrara,Katharine Garito
And 0h51'0"35d43'
Ferrara,Krysta Camille
Lyn 8h49'35"41d44'
Ferrara,Leonard C
Hya 8h13'16"1d48'
Ferrara,Luca
Crt 11h15'25"-8d49'
Ferrara,Marco
Hya 8h18'60"2d53'
Ferrara,Stefan V
Aql 20h7'10"6d7'
Ferrara,Stefan V
Aur 5h4'45"44d45'
Ferrara,Steven Robert
Lac 22h36'47"40d24'
Ferrara,Zachery Vincent
Aql 19h52'42"10d30'
Ferrara-Star Grandma, Rose
Cas 3h1'31"58d31'
Ferrare,Nicholas A
Aur 5h17'25"45d4'
Ferrarese,Alice
Hor 3h28'10"-47d33'
Ferrarese,Matthew Vincent
Her 16h10'16"47d45'
Ferrari"Elephant Shoe" ,Robert
D
Gem 6h57'20"20d36'
Ferrari's Star
Tri 2h1'18"31d29'
Ferrari,Cheryl
Cam 7h55'0"70d21'
Ferrari,Edward J
Aur 6h37'32"37d7'
Ferrari,Gennaro
Aur 5h1'20"50d20'
Ferrari,Jason Charles
Gem 6h37'7"12d34'

Ferrari,Marilena
Cnv 13h0'22"38d56'
Ferrari,Mikaela
Cas 23h20'15"53d22'
Ferrari,Oliver
Cep 20h31'36"65d34'
Ferrari,Robert J
Boo 14h13'0"44d40'
Ferrari,Scott A
Cet 1h21'25"-4d34'
Ferrari,Vince
Aur 5h55'56"30d60'
Ferrario,Marco
Cma 6h11'33"-13d44'
Ferraro's
Significance,Nicholas
Cet 2h6'0"2d19'
Ferraro,Anna
And 1h52'52"37d32'
Ferraro,Dominick
Cam 12h23'17"80d10'
Ferraro,Frank
Oph 18h2'42"11d40'
Ferraro,Jr,Salvatore F
Dra 14h26'33"63d3'
Ferraro,Michael Anthony
Boo 14h10'56"32d22'
Ferraro,Nancy E
Leo 10h59'16"15d50'
Ferraro,Nicolas
Umi 15h31'29"70d48'
Ferraro,Raymond
Aqr 21h2'52"-1d42'
Ferraro,Rudy
Lac 21h56'36"40d12'
Ferrarri,Thomas Alan Taisho
Peg 22h59'10"17d60'
Ferravecchio,Frank
Cnc 8h56'27"18d54'
Ferree,Barbara
Tri 2h25'43"31d35'
Ferreira,Ana Monica
Uma 9h41'27"59d56'
Ferreira,Christiana Marie
Tau 5h18'57"20d27'
Ferreira,Cristini Grace
Lyr 19h23'14"31d15'
Ferreira,Daniel
Aur 6h30'56"37d45'
Ferreira,David
Aur 6h6'45"30d18'
Ferreira,Karen Michelle
And 1h19'29"40d0'
Ferreira,Katrina Elise
Sco 17h3'47"-31d18'
Ferrell III,William Owen
Per 4h43'1"38d42'
Ferrell,Ashley Elizabeth
Peg 22h32'27"31d50'
Ferrell,Betsy Lois "Grammy"
Cyg 19h25'27"31d35'
Ferrell,Dennis Edward
Lac 22h40'17"50d35'
Ferrell,Libby
Vul 19h45'44"25d58'
Ferrell,Pat
Aur 6h28'35"30d7'
Ferrell,Rene F
And 2h32'13"39d34'
Ferrell,Susan
Cet 2h17'57"8d49'
Ferren,Donald G
Cep 20h57'14"55d37'
Ferren,Sherryl L
Mon 6h28'34"-6d12'
Ferrendi,To My Son Carmine
Aur 6h31'40"38d50'
Ferrer,Christina Santos
Vir 11h47'28"1d21'
Ferrer,Christopher Alan
Tri 2h19'50"31d43'
Ferrer,Eleanor Santos
Sco 17h53'34"-40d42'
Ferrer,Enrique Rico
Her 18h24'60"12d42'
Ferrer,Fernando & Sally
Cmi 7h6'31"4d55'
Ferrer,Monica
Del 20h15'30"10d10'
Ferrera IV,Victor Joseph
Ori 6h17'39"8d45'
Ferrere,Anny
Aql 19h31'0"-10d12'
Ferrero,Marc
Cep 20h46'22"60d52'
Ferri Esquire,Albert
Crt 10h58'29"-10d46'
Ferri,Alyssa Marie
Uma 10h7'12"50d40'
Ferri,Cesare
Ori 5h52'58"7d42'
Ferri,Elisa
Umi 16h26'1"83d11'
Ferri,Jennifer Lyn
Cas 0h12'12"54d24'
Ferri,Kathryn Ann
Cyg 19h21'13"28d38'
Ferri,Kimberley S
Uma 10h37'11"59d1'
Ferri,Mark Alan
Oph 17h58'49"12d4'
Ferri,Paola
Cnv 13h0'56"38d16'
Ferrid,Martin
Dra 14h24'49"64d17'
Ferrier & Associates Star,The
Ori 5h29'18"0d59'
Ferrier,Jr,Robert Charles
Peg 22h9'24"6d38'B
Ferrier,Sr,Robert Charles
Peg 22h9'24"6d38'A
Ferriera,Harry S
Peg 23h32'57"31d8'
Ferrill,Tervor Wesley
Cep 23h4'32"67d55'
Ferrin,Rachelle Lynn
Cet 1h51'53"-3d40'
Ferringer,Jeffrey
Aur 4h58'18"50d55'
Ferrini,Bruce
Aur 4h50'30"41d8'
Ferrini,The Ralph P & Iola D
Crb 16h19'23"28d31'
Ferris
Dra 15h55'30"62d42'
Ferris,Ann-Tom
Crb 15h25'17"30d60'
Ferris,David John
Her 17h12'45"48d27'
Ferris,Donald
Dra 17h55'55"63d56'
Ferris,Jacqueline & Peter
Nowlan
Cyg 20h40'23"45d35'
Ferris,Jenna K
Lyr 18h57'40"46d57'
Ferris,Peggy
Cet 3h12'45"3d2'
Ferris,Roy
Ori 5h27'48"0d38'
Ferris,Roy Lewis
Boo 14h39'13"33d42'
Ferris,Thomas F
Ser 15h38'40"5d3'
Ferristar SFLC33
Lac 22h10'0"30d24'
Ferriz,Sofia Embid
Peg 22h58'1"21d41'
Ferro,Joseph Manuel
Aur 5h6'47"43d35'
Ferro,Leonard J
Equ 21h7'55"11d17'
Ferro,Patti & Frank
Lac 22h37'40"52d53'
Ferro,Thomas M (Cupcake)
Her 16h18'1"7d57'
Ferrol,Amber
Cmi 7h23'16"0d29'
Ferron,Paul
Lyr 18h33'57"35d51'
Ferront,Anne-Marie
Aql 19h55'41"-10d48'
Ferrucci,Silvia
Pho 0h26'56"-47d12'
Ferruccio,Michael Anthony
Aur 5h38'0"37d58'
Ferry,Carla Patricia
Vul 19h48'42"28d32'
Ferry,Marilee
Mon 6h55'50"-6d19'
Ferry,Ryan Andrew
Per 2h25'1"54d53'
Ferry,Sam & Sheila
Sct 18h54'55"-4d8'
Ferry,Thomas
Boo 13h46'1"22d40'
Ferryman,Dolin
Uma 11h26'51"33d55'
Fersini,Jolanda
Uma 12h0'1"30d54'
Fertal,Jerome J
Uma 10h22'31"47d49'
Fertsch-Röver,Wolfgang
Ori 5h50'20"21d33'
Fescina,Laneya
Cnv 12h28'0"33d37'
Fesmire,Leigha Marie
Cas 0h13'21"62d19'
Fessenden,Chef Bob
Dra 20h12'36"63d1'
Fessenden,Guy
Dra 16h22'55"68d24'
Fessenden,Myron "Glenn"
Aur 5h19'22"48d20'
Fessler,E Anthony
Aur 5h28'38"31d46'
Fessler,Guenther
Oph 18h7'11"7d33'
Festa,Frank Alexander
Per 2h59'38"40d9'
Festa,Frank Thomas
Ori 5h56'13"15d43'
Festa,Linda
Ori 5h51'19"15d25'
Festa,Matthew Thomas
Ori 5h59'47"15d44'
Festa,Paola Concetta
And 0h11'52"34d38'
Fetchel,Michael Patrick
Her 16h59'1"34d53'
Fetherston,Dan
Cep 0h12'24"69d42'
Fetherston,Jean
Lyn 7h23'35"44d38'
Fetman 3-3-1928, Bernice
Uma 10h28'0"72d52'
Fetman,Brian
Uma 8h16'1"72d9'
Fetman,Elliot
Uma 10h59'40"72d30'
Fetman,Jay A
Uma 10h59'49"70d8'
Fetman,Jayme
Uma 10h44'47"70d36'
Fetscher,Nicholas
Her 17h6'11"44d17'
Fetsko,Jack
Boo 14h10'1"34d8'
Fett,Mark Eugene
Aql 19h57'20"0d34'
Fett,Sheryl
Cam 4h29'39"80d2'
Fetter,Carolyn Marie
Cas 0h51'46"60d50'
Fetzko,Michael Allen
Ori 5h49'49"11d33'
Feuer,Nancy Estelle Koonce
Com 13h16'0"30d16'
Feuer,Regina B
And 0h53'53"40d24'
Feuerbach,Norbert
Dra 18h23'31"68d29'
Feuerhake,Morgan Bayliss
Cyg 21h52'25"52d40'
Feuerstein,Lisa
Sco 17d52'0"-31d35'
Feuk,Hans
Dra 13h25'41"70d5'
Feustel,Mr & Mrs Vincent
Cam 7h43'30"61d5'
Feutry,Dominique
Per 2h22'26"55d35'
Fev,Hostench
Dra 16h24'31"62d34'
Fevrier,Thibaut
Dra 20h41'28"80d11'
Fewel,Raymond Percy
Cnv 12h30'23"33d14'
Fewell,Robert Martin
Aql 20h4'52"6d50'
Fey,Axel
Sco 17h21'46"-38d19'
Feyk,Michelle Cathleen
Cet 3h7'36"1d4'
Fezer,Richard
Cmi 7h6'21"5d17'
Fialo,Star of Angels Vita &
Coco
Cyg 19h25'44"32d55'
Fiammetta
Lac 22h29'19"50d15'
Fiasca,Moneisa D
Cas 23h22'25"53d30'
Fiaschetti,Laura Charlotte
Ori 5h9'7"0d27'A
Fiaschetti,Martha Alice
Ori 5h9'7"0d27'B
Fica,Dave "Mike"
Lac 22h25'44"52d55'
Ficarelli,Jr,Richard J
Aur 5h4'20"31d33'
Ficarrotta,Diana
Peg 23h2'1"18d57'
Ficca,Raymond G
Lac 22h40'26"51d33'
Fichou,Adrien
Lac 22h3'0"38d35'
Fichter,Evan Jennings
Mon 7h56'37"-8d18'
Fichter,Kylie Cummings
Peg 21h55'29"22d37'
Fichter,Walter & Mary
Mon 6h38'51"-0d19'
Fichtner,Carola
Gem 6h46'26"12d13'
Fick,MaryJo
And 2h3'35"38d26'
Fick,Megan
And 1h17'13"39d14'
Ficke,Sean Christopher
Lac 22h20'57"48d49'
Fickel,Lauren Danielle
Sco 16h55'55"-38d16'
Ficklin,Clare E
Eri 3h15'13"-1d35'
Fico,Valerie Ann
Crb 15h26'46"31d2'
Fid,Frank Gade
Peg 23h32'57"10d38'
Fidanza,Janet
Ori 5h55'58"17d29'
Fidelia Star,The Lucy Ellen
Crb 16h9'58"36d20'
Fidelis Janus
Cyg 19h21'0"28d31'
Fidelma
Lyn 9h14'1"38d44'
Fidgeon,Ruth
And 23h39'52"48d19'
Fidissimus Kylie
Cyg 19h26'39"34d9'
Fidler,Ashley Paige
Uma 14h17'58"61d50'
Fidler,Janet Audery
Cyg 21h18'19"28d53'
Fido
Lyn 8h55'1"35d39'
Fiduccia,Don
Hya 8h56'49"1d18'
FIDUCIA
Per 3h22'42"40d21'
Fiebig,Eric Andrew
Lac 22h28'53"55d1'
Fiebusch,Jane Findling
Vir 11h59'12"8d16'
Fiedler,Günter
Her 16h54'5"32d51'
Fiege,Bettina
Equ 20h55'60"2d21'
Fieke
Lyn 7h34'15"50d15'
Field,Camilla
Peg 22h59'51"21d34'
Field,Carolyn Joyce
Cyg 19h51'44"40d28'
Field,Donald
Her 16h7'19"40d45'
Field,John Bernard
Cnc 8h34'0"32d29'
Field,Karen Elizabeth
Lyr 18h59'50"30d48'
Field,Kirk
Ori 6h17'30"8d24'
Field,Kristi
Uma 10h50'23"70d24'
Field,Kristi-Bella
Aur 4h59'0"41d4'
Field,Matthew Bridge
Her 16h44'0"38d47'
Field,Mike
Uma 8h50'27"56d16'
Field,Sandra
Lyr 18h14'0"42d41'
Field,Sofia
And 23h17'19"47d53'
Field,Ted
Hya 8h56'1"3d30'
Field,Teri Dale
Mon 6h43'57"11d23'
Field,Tobe A
Ari 2h39'27"25d48'
Fielder,David
Cma 7h14'0"-13d15'
Fielder,Keith J
Leo 10h48'34"-1d48'
Fielder,Mary Jane
Cas 0h36'13"61d5'
Fielding,Bruce
Lyn 8h33'32"38d56'
Fieldings' Forever
Lmi 10h14'28"39d36'
Fields of Gold (Dreams Come
True)
Lyn 7h51'17"51d46'
Fields,Alexander Barry
Cet 3h17'11"-0d51'
Fields,Andrea
Cas 23h36'18"62d2'
Fields,Athelean
Lyn 6h18'31"59d5'
Fields,Debor-Ah Ann
Cas 23h34'24"63d15'
Fields,Donald E Shatov
Cep 23h5'50"61d13'
Fields,Dr David
Her 16h40'38"23d30'
Fields,Erin
Mic 20h45'32"-45d0'
Fields,Isaac Roy
Aur 5h2'13"46d3'
Fields,Jackie
Uma 11h35'15"31d34'
Fields,Jeffrey Chad
Aur 4h59'18"50d25'
Fields,Jim
Aur 6h39'16"38d21'
Fields,John Craig
Per 2h0'32"50d28'
Fields,Jon J
Uma 11h35'25"41d30'
Fields,Larry Michael
Cep 0h3'29"68d39'
Fields,Lenora Maddox
Cnv 12h29'39"40d49'
Fields,Michael C
Aql 19h54'27"0d30'
Fields,R L & D Jean
Sgr 18h48'32"-29d22'
Fields,Richard Alan
Boo 13h39'42"24d32'
Fields,Richard,Willa, Colin, &
Dylan
Aql 20h14'51"5d36'
Fields,Robert P
Her 17h54'21"30d33'
Fields,Russell
Per 3h3'23"55d31'
Fields,Steve
Aur 6h9'28"37d6'
Fields,Stuart Jay
Boo 15h24'55"40d13'
Fields,William H(Bill)
Boo 14h57'35"40d21'
Fierce,Erik
Dra 14h32'29"62d31'
Fierro,Anna Pliego
Mon 6h54'46"-10d39'
Fierro,Kaitlyn Miranda
Cyg 20h24'12"38d45'
Fierro,Rae
Cet 0h29'57"2d26'
Fierro,Veronica
Sgr 19h32'34"-28d43'
Fierst,Dottie
Crb 15h27'20"31d5'
Fiesemann,Doris
Vul 19h45'54"27d29'
Fiest '96 Bears
Hya 8h37'60"0d7'
Fiest,Marion
Aur 4h51'23"41d2'
Fietz,Siegfried J
Dra 16h49'36"63d2'
Fietze,Katharina
Lyr 19h6'20"41d8'
Fife Star
Aur 5h23'10"38d52'
Fife,Benjamin Stuart
Cam 3h38'1"74d34'
Fife,Brooks Katherine
Tri 2h6'14"33d20'
Fife,Trevor Ames
Boo 15h7'0"18d22'
Fifer,William Henry
Her 17h30'21"14d59'
Fiffi
Boo 13h58'0"12d43'
1509
Mon 6h52'51"-0d33'
50th-Uncle Pete & Aunt
Oda,The
Eri 4h31'54"-8d51'
5103
Sgr 19h14'24"-26d35'
5185 Forestire
Uma 8h51'14"47d32'
52392 My Love 4 U 4 Ever
Lac 22h4'50"49d40'
528 i
Cas 22h59'25"54d51'
5683
Ser 15h14'38"7d45'
58 Sarah 58 Jane 58
And 23h44'19"44d52'
Fifty Fifty
Ori 5h55'58"15d40'
Fig
Mon 6h25'15"-10d32'
Figge,Anabell
Vir 12h4'47"-2d19'
Figge,Jochen
Leo 10h52'34"0d31'
Figge,Sally
Sgr 19h19'16"-22d1'
Figgett 21,Mark Thomas
Cep 21h49'38"65d3'
Figgie III,Harry E
Lac 22h7'24"51d17'
Fight,Akira & Mayumi
Cap 21h45'12"-22d51'
Fighter
Cep 21h2'1"58d55'
Figler,Jeffrey David
Aql 19h52'16"14d16'
Figlia,Victoria Denise
Umi 15h3'45"78d29'
Figment
Dra 9h52'1"73d20'
Figment
Aql 19h31'43"10d58'
Figoni"Baseball Star," Richard
Robert
Peg 23h44'22"27d1'
Figueras,David Bru
Cnv 12h18'45"36d8'
Figuerda,Ivan Daniel
Ori 6h9'56"7d18'
Figueroa,Ana Maria
Mon 7h16'58"-6d35'
Figueroa,Ernesto G
Ari 2h39'56"30d7'
Figueroa-Melidor, Melissa
Psc 0h47'46"20d49'
Figues,Jean Claude
Per 3h57'59"35d36'
Figues,Stephane
Per 3h57'56"36d31'
Figura,Donna J
Tri 1h43'27"34d27'
Figuracion,Ethan Michael
Mariano
Aql 19h58'28"14d40'
Fiji 21
Cam 4h1'35"61d42'
Fike,Cynthia Lynn
And 1h28'46"35d27'
Fike,David & Maryjean
Cyg 21h6'36"31d25'
Fike,Forever John
Cep 22h11'10"68d20'
Fikentscher,"T J" Timothy
Joseph
Boo 15h4'35"15d10'
Fikentscher,Erin Elisabeth
Peg 23h46'59"27d55'
Fikentscher,Joshua Robert
Lac 22h1'60"47d12'
Fikis,Dominik
Leo 9h56'41"28d13'
Fikis,Philipp
Leo 10h36'23"20d27'
Fikry,Laurie & Mohamed
Boo 14h13'26"51d29'
Fiksdal,Alisa Kristina
Peg 22h57'11"34d38'
Fiksdal,Erika Kirsti
Cas 0h5'21"63d14'
Fiksdal,Krystal Marja
Sge 19h58'30"18d54'
Filali,Nadia
Crb 15h34'17"32d4'
Filanowski,Katherine
And 1h12'56"39d47'
Filante
Lac 22h48'58"38d7'
Filasky,Pamela J
Lyn 9h4'34"39d2'
Filbert,John Stephen
Cet 0h52'17"-0d22'
Filbert,Madeline
Eri 4h14'17"-18d32'
Filbrant,Patricia J
Cep 5h0'44"80d0'
Filby,Julie Dawn
Cyg 19h25'53"34d15'
Filby,Kathie Susanne
Umi 14h10'52"78d34'
Filderman,Janet
Cas 2h58'0"65d17'
Filer,Caroll J
Peg 23h15'28"31d2'
Filer,Edward Scott
Lyn 8h11'19"50d56'
Filer,Mark M
Hya 9h33'19"-5d21'
Files,Dorothy E
Vul 20h46'0"28d10'
Files,Patrick
Uma 11h51'59"63d30'
Files,Russell
Ori 6h16'21"-2d25'
Filho,Victor Brecheret
Mon 6h26'46"7d59'
Filia Caritas
Uma 9h11'59"55d53'
Filiaggi,Elisea
Her 18h54'47"18d50'
Filiatrault,Harry
Aur 6h1'32"31d20'
Filiatrault,Julie
Uma 8h30'57"54d38'
Filiatreault, Catherine Jobin
Uma 8h8'48"60d36'
Filice,Gia
Eri 2h43'52"-3d47'
Filion,Annie
Dra 17h56'57"52d13'A
Filion,Shany
Dra 17h56'57"52d13'B
Filipek,Scot & Jessica
Uma 10h20'11"61d4'
Filipody
Cas 1h0'21"53d22'
Filipos,Andrew Wahl
Boo 13h38'10"17d8'
Filipovich,John T
Dra 14h24'22"63d31'
Filippazzo,Paul Vincent
Ori 5h51'25"18d54'
Filippo
Oph 18h17'36"12d30'
Filippo Carlo Papa
Psc 22h57'59"0d24'
Filippoli,Siegried
Boo 14h30'29"27d14'
Filippone M/PM Star, The
Andrew
Aur 7h16'34"41d24'
Filis,George
Cep 0h16'44"80d32'
Filisky,Christine
Uma 11h28'0"30d14'
Fill,Jochen
Lac 22h7'28"38d27'
Fillancq,Marie Christine
Aur 5h3'60"29d26'
Fillebrown,Brian "Nudge"
Uma 9h59'15"42d7'
Filler,Douglas
Per 4h4'56"37d31'
Fillery 1/7/69,Sara Jane
Lyr 18h35'55"27d57'
Fillet,Jeffrey
Sct 18h21'46"-5d44'
Fillet,Jr,Stephen
Aql 19h26'21"-10d46'
Filliard,Andre
Crb 15h58'59"35d17'
Fillingham,Robert A
Hya 8h47'50"-0d9'
Fillingim,Matthew Owen
Oph 17h17'18"-22d37'
Fillion,Gilles
Cyg 19h29'16"35d53'
Fillipo,Donna
Tau 3h40'17"28d55'
Fillman,Richard L
Aql 19h54'59"13d34'
Fillmore,Brenda
Cas 0h29'26"69d17'
Fillmore,Glen Alan
Aql 18h42'0"1d5'
Film at 11 Phil
Cnv 12h31'38"41d7'
Filmyer,Henry Tucker
Lac 22h11'51"37d54'
Filotei,Nicholas Joseph
Aur 5h55'33"31d45'
Fils-Aimé,André
Uma 11h35'20"31d41'
Fina
Uma 9h37'55"42d4'
Fina
Hya 8h37'0"0d45'
Finamore,Will & Hope
Cam 10h51'32"80d34'
Finan,Donna M & Joseph A
Finan
Lyn 7h4'57"44d36'
Finbar
Ori 5h55'37"12d19'
Finch,Abigail Arynn
Sgr 19h28'41"-42d23'
Finch,Brandon William
Dra 16h38'23"65d19'
Finch,Holly Elizabeth Nancy
And 0h53'41"45d42'
Finch,Jack
Her 18h0'0"30d25'
Finch,Peter G
Cap 20h5'0"-10d4'
Finch,Simon T K
Cep 1h6'57"77d24'
Finch,Sr,Harry Chandler
Aur 6h37'0"37d49'
Finch,W S
Aur 5h24'1"37d43'
Fincke,Edward Michael
Oph 18h39'13"7d17'
Fincken,Jr,Frank
Cam 7h51'48"60d12'
Findlay,Jaimiee Irene Block
Cam 14h4'58"82d2'
Findlay,William Harold
Cet 0h25'38"-11d3'
Findley,Dick
Aql 20h3'12"3d56'
Fine,Laura Louise
Mon 6h19'32"-0d41'
Fine,Leona
Cet 0h38'15"-7d30'A
Fine,Lynna S
Ser 18h55'60"2d44'
Fine,Michelle
Eri 3h34'18"-6d16'
Fine,Mike
Cet 0h38'15"-7d30'B
Fine,Paul & Gloria
Sge 20h1'30"18d44'
Fine,S Richard
Per 2h50'14"32d28'
Fine,Selma
Cas 1h43'17"61d43'
Fine,Shirley Lee
Aur 6h29'57"52d30'A
Fineberg,Manuel
Her 17h36'35"25d46'
Finefield,Altha D
Crb 16h19'47"38d34'
Finegan,William
Aur 5h11'12"41d8'
Finelli,Barbara
And 0h28'50"45d37'
Fineman,Shiela
Cnv 12h55'30"42d23'
Finerty,Kelly
Uma 8h38'31"58d13'
Finfrock,Harry Lee
Aur 6h15'47"35d32'
Fingar,Jr,Floyd S
Per 4h20'17"51d23'
Finger Touch
Uma 11h40'32"49d21'
Finger,Betty Mae
Dra 14h51'27"58d53'
Finger,David Gold
Dra 12h22'56"70d10'
Finger,Florence
Mon 7h3'51"4d16'
Finger,Sammy
Cet 3h18'17"02d8'
Finger,Susan Lee
Mon 7h45'59"-1d18'

Fingerhut,Anne
Ori 6h9'1"8d51'
Fingers,Rollie
Aur 6h25'30"37d2'
Finholm,Jana Rae
Crt 11h3'12"-13d26'
Fini,Mitchell Edward
Per 1h31'18"52d47'
Finizio,Amy Marie
And 0h38'40"40d25'
Fink
Aql 19h53'50"10d42'
Fink Star,The
Cep 21h19'14"58d39'
Fink,Alexander
Her 17h4'0"50d24'
Fink,Beatrice
Aqr 20h54'38"-5d56'
Fink,Bobbie
Aur 6h26'30"31d47'
Fink,Chris
Cet 0h50'53"-0d21'
Fink,Daniel Edward
Lyn 7h35'13"41d32'
Fink,Diane
Lyn 9h15'33"38d40'
Fink,Elizabeth Hosticka
Aqr 22h14'0"0d13'
Fink,Landy Lilith
And 0h23'41"34d12'
Fink,Lillian & Albert
Equ 20h59'44"9d7'
Fink,Lorna Laney
Mon 6h42'53"3d11'
Fink,Perry Bruce
Per 3h25'26"40d32'
Fink,Zoey Rae
Ari 3h13'36"25d33'
Fink-25 Years,Donna & Stewart
Del 20h21'40"16d21'
Finke,Doris Moses
Aqr 23h2'43"-6d43'
Finkel,Andrew
Per 3h36'56"33d57'B
Finkel,Karel
Per 3h36'56"33d57'A
Finkelmeier,Maureen
Boo 14h9'15"42d30'
Finkelpearl,Tom
Lac 22h27'16"55d29'
Finkelstein,April Bari
And 0h6'47"35d4'
Finkelstein,Margaret
Cas 0h35'26"63d13'
Finkelstein,Marsha & Lou
Cyg 21h18'0"38d48'
Finkelstein,Randi
Cas 1h16'28"61d42'
Finks,Glen Michael
Aql 19h0'24"13d42'
Finlay's Fancy,Fiona
Uma 9h40'42"51d14'
Finlay's Folly, Alexander
Uma 9h38'33"51d29'
Finlay,Evelyn Louise
Cam 10h20'39"81d51'
Finlay,Jacob
Aql 18h39'25"0d18'
Finlayson,Ciaran Francis
Mon 7h0'14"-8d55'
Finley,Dylan Christopher
Lac 22h6'13"40d38'
Finley,Jon M
Per 1h25'51"53d55'
Finley,Kristen
And 0h59'42"38d33'
Finley,L Terrance Spencer
Aur 4h59'17"51d14'
Finley,Lucille Ruth Budd
Cas 0h4'32"56d11'
Finley,Mary Dougherty
And 0h20'55"35d43'
Finley,Sarah J
Cyg 19h39'0"31d44'
Finley,Tara Elizabeth
Sct 18h40'44"-6d60'
Finley,Vanessa Irene
Vul 20h19'44"28d23'
Finn
Cas 3h31'26"69d58'
Finn,"My Angel",John Michael
Del 20h18'0"10d48'
Finn,Benjamin
Per 2h51'42"45d50'
Finn,Eli
Uma 9h35'46"43d8'
Finn,Geneva
Cas 0h44'1"68d53'
Finn,Ivy Alice Ethel
And 23h37'12"41d13'
Finn,Jase
Uma 13h42'1"48d32'
Finn,Melanie Felony
Cas 3h0'51"57d38'
Finn,Susan "Fifi"
Peg 22h59'59"24d5'
Finn,Timothy Timothy & Kathryn
Uma 10h32'0"53d18'
Finn-Harald
Boo 14h48'0"48d7'
Finnegan
Lac 22h54'33"54d10'
Finnegan,Bertha (Geisbrecht)
Uma 11h9'20"71d49'
Finnegan,Denise Bethlehem
Cas 1h23'28"75d34'
Finnegan,Hannah Danielle
Aql 20h7'26"0d2'
Finnegan,Harold J
Dra 15h53'37"61d37'
Finnegan,Joseph H
Aur 7h4'16"41d37'
Finnegan,Kevin
Aql 20h11'1"13d24'
Finnegan,Marian
Cyg 20h1'37"40d29'
Finnell,Cory S
Per 2h59'1"34d15'
Finneran,James A
Cmi 7h41'1"8d34'
Finnerman,Ray
Sct 18h54'10"-5d6'
Finnerty,Brian
Sct 18h19'43"-14d40'
Finnerty,Kevin Patrick
Ori 6h9'0"8d36'
Finnerty,Thomas Joseph
Boo 15h9'40"47d57'
Finney,Andrea Briana
Peg 23h48'26"8d45'
Finney,Jr,David B
Cep 21h3'49"68d48'
Finney,Julia
Ari 2h40'15"22d2'
Finney,Mary Ellen H
Cyg 21h50'55"40d1'
Finney,Mr & Mrs John Hayes
Aql 18h47'34"11d9'
Finney,Olive
Cyg 19h45'45"29d29'
Finney,Patrick
Mon 7h18'14"-1d50'
Finnicum,David A & Helen J
Can 9h10'41"30d56'
Finnie,Stuart
Uma 8h56'60"56d26'
Finnie-Ryan Friendship Star,The
Uma 11h31'0"43d51'
Finnigsmier,Alta Kay
Aql 19h54'52"13d29'
Finnley Belle
Her 16h40'36"33d41'
Finno,Reginald P
Aur 5h2'54"41d29'
Fino,Filipa
Cnv 12h53'54"38d1'
Finocchario-Cozzone, Carol
And 23h38'54"47d26'
Finocchiaro,Graziella
Pic 5h3'52"-45d1'
Finol,Enrique Alberto
Mon 8h5'53"-8d56'
Finola
Cyg 20h27'0"31d41'
Finster,Anja
Her 17h8'30"42d32'
Finster,Lars
Her 17h9'16"47d51'
Finster,Norbert
Aur 6h57'54"43d8'
Finvicdanbern
Dra 16h19'21"66d18'
Finzi,Luca
Dra 16h28'55"58d18'
Fio Star
Lac 22h16'34"38d24'
Fiocco,Bella Karen
And 0h5'56"35d42'
Fiona
Cep 0h16'53"67d3'
Fiona
Cas 0h43'40"60d51'
Fiona
And 0h18'0"38d33'
Fiona
Oph 18h2'24"12d27'
Fiona
Boo 14h27'56"38d31'
Fiona's Folly
Cyg 20h22'53"39d23'
Fiona-Suresh
Peg 23h1'57"22d13'
Fionn
Cyg 19h30'22"35d6'
Fioravanti,Lisa A
Uma 11h54'43"41d44'
Fiordellisi,Angelina
Uma 10h40'53"70d33'
Fiore di Panna
Dra 16h40'30"67d3'
Fiore,Carl J
Cep 22h19'33"61d50'
Fiore,Michael"Fury"
Boo 14h10'28"50d21'
Fiore,Rosario
Vul 20h16'1"23d21'
Fiore,Rosario Russell
Mon 6h20'52"3d8'
Fiore,Tim
Dra 20h21'41"62d20'
Fiore,Tom
Uma 11h7'31"61d21'
Fiorella
Mon 8h2'0"-6d38'
Fiorella
Peg 21h52'14"30d31'
Fiorella e Piero 1983
Lac 22h20'25"55d12'
Fiorelli,The Don
Aur 5h6'48"40d43'
Fiorente,Pasqua
Ant 10h37'57"-32d40'
Fiorentino,Gennaro A
Lyn 3h46'13"34d17'
Fiorenza
Psc 22h57'0"-3d8'
Fiorenza
Del 20h15'42"10d40'
Fiorenza
Cyg 20h0'13"40d10'
Fiorenza,Kristen
And 23h28'1"49d56'
Fiorenzano,Bianca
Lyr 19h0'27"28d26'
Fiorenzo
Ari 2h40'14"28d34'
Fioretti,David Bastian
Ser 16h2'11"0d31'
Fiori,Alyse P
And 1h48'55"47d21'
Fiori,Grace P
Lac 22h31'47"53d57'
Fiori,Madge Marie
Lyr 18h36'57"34d20'
Fiori,R John
Cet 3h0'0"6d29'
Fiorillo,Carol T
Cas 1h41'37"60d41'
Fiorillo,Jacquelyn M
Cas 1h41'21"60d30'
Fiorio,Gianfranco Roberto
Lyn 7h3'17"59d43'
Fiorito,Eileen M Gastiger
And 23h24'26"50d10'
Fiorito,Eunice K
Cet 2h20'53"0d33'
Fiorito,Louise
Aur 5h0'41"45d23'
Fioroni,Claudio
Ari 1h47'1"17d38'
Fiparo
Lup 15h18'25"-40d19'
Fire
Aur 5h5'34"42d56'
Fire Of My Heart
Aql 18h59'58"3d15'
Firebaugh,Robert Neal
Aql 20h10'21"5d15'
Fireburst
Cet 2h3'0"1d47'
Firecracker,Stephanie R
Cma 6h56'19"-15d33'
Firefly
Cet 1h59'30"-8d1'
Firefly-+' ' '6/2/94
Lac 22h48'25"54d36'
Firehammer,Mary & Waldemar Fredrick
Uma 11h41'43"48d46'
Firely,Douglas C
Umi 17h5'21"80d0'
Firemann,Alexander K
Psc 23h21'57"6d31'
Firenze,Carmen Rose
And 2h0'26"45d20'
Firenze,Tara
Uma 8h41'50"71d35'
Firestone,Kimberly Michelle
And 1h28'42"48d59'
Firestone,Sara Adiel
And 0h21'14"37d48'
Firetag,Philip L
Tri 2h1'30"30d25'
Firetto,Michael William
Her 16h32'27"50d41'
Firgard,Janice Blush
Cmi 8h5'0"1d23'
Firgeleski,Joseph & Susan Krehley
Boo 14h59'22"31d3'
Firkins,Henry Ellis "Pete"
Hya 8h12'15"1d26'
Firma Hans Bryjak
Aur 5h13'19"43d16'
Firman,Charles Sheppard
Her 17h0'11"44d57'
Firman,Kimberly Mae
Umi 15h4'58"82d27'
Firmstone,Chrissy
Ori 5h57'0"15d50'
Firon,Michael & Zehava Caspi
Peg 21h32'53"21d10'AB
Firpo,Andrew Ethan
Hya 8h19'26"4d21'
Firshein,Dr Richard N
Aur 7h17'18"40d17'
1st Anniversary Of Brad & Valerie
Ser 15h25'26"7d34'
First Anniversary
Lup 15h42'55"47d54'
First Class Mayle
Her 16h4'40"41d18'
First Dog Training Club Of North N J
Cmi 7h43'58"8d25'
First Love Star
Aur 6h9'1"46d5'
First Love,Jordan
Vul 19h18'45"26d28'
First Star
Umi 14h4'57"66d59'
First Time
Cas 2h53'16"77d11'
First United Methodist Church
Lyr 19h10'13"38d17'
First,The
Uma 10h26'0"59d4'
Firth,Alexander Stuart
Ori 5h19'30"15d3'
Firth,Debra
Cas 1h30'15"70d0'
Firth,Kieran Patrick
Peg 23h26'12"25d54'
Firth,Rachel Anne
And 0h1'12"47d20'
Firth,Sarah Caroline
Lyr 19h1'20"28d11'
Fiscante,Rosario
Cep 3h23'37"78d54'
Fisch,Ellie
And 23h36'35"44d7'
Fisch,Sherry
And 2h26'19"44d40'
Fischbach,Friedrich
Vir 12h29'24"-10d3'
Fischbein,Mia Elana
Peg 21h59'0"28d59'
Fischbugh,Paul
Her 17h34'48"38d45'
Fischer,"My Lucky Star",Byron Charles
Hya 8h20'45"-10d55'
Fischer,Angela
Lyr 18h52'18"38d55'
Fischer,Anne,Rosenberg
Cas 0h17'21"66d3'
Fischer,Barbara
Lyn 7h47'27"47d50'
Fischer,Bob
Hya 8h46'28"5d51'
Fischer,Bryan J
Cet 2h39'20"-8d52'
Fischer,Dennis Basil
Her 15h56'30"41d13'
Fischer,Dipl-Ing oec Klaus
And 23h13'58"40d47'
Fischer,Doris
Psc 0h51'29"2d24'
Fischer,Dorothy
And 2h19'27"45d0'
Fischer,Erin Christina
Leo 11h7'23"-2d29'
Fischer,Erin McKelle
Lyr 18h41'40"32d57'
Fischer,Georg
Cyg 19h52'41"47d52'
Fischer,Georg
Cyg 20h25'42"30d23'
Fischer,Herr
Cmi 7h19'36"4d53'
Fischer,Ian Alexander
Ori 4h54'19"1d38'
Fischer,Ingrid
Vul 19h40'29"26d45'
Fischer,Jacob David
Boo 14h57'53"30d51'
Fischer,Jasmine Taylor
And 1h27'27"38d42'
Fischer,Jean Christine
Aqr 22h27'22"-6d2'
Fischer,John & Dorothy
Eri 3h1'11"-17d54'
Fischer,Kenneth A
Her 16h59'47"39d15'
Fischer,Konstantin
Psc 23h22'44"6d27'
Fischer,Lukas Robert
Aur 5h0'49"40d4'
Fischer,Martin Kumin
Ori 5h56'31"12d38'
Fischer,Max
Cyg 19h52'20"44d5'
Fischer,Morgan Kelly
Vul 19h46'19"28d34'
Fischer,Nancy D
Cas 23h32'1"60d35'
Fischer,Patrick
Boo 14h25'25"14d58'
Fischer,Peter
Cmi 7h21'0"0d8'
Fischer,Peter
Cam 6h23'13"78d50'
Fischer,Pierre
Aur 5h53'45"31d33'
Fischer,Sean Wesley
Ori 5h12'50"-4d25'
Fischer,Stefan
Cmi 7h32'50"1d40'
Fischer,Thomas
Cap 20h56'28"-20d35'
Fischer,Udo
Peg 0h4'27"28d22'
Fischer,Valentin
Boo 14h0'14"13d16'
Fischer,Wally
Cnv 12h30'0"32d27'
Fischer,Wendy Suzanne
And 23h15'33"48d52'
Fischer,Wienfried
Her 18h20'1"12d19'
Fischerbock,Marcus
Dra 18h10'19"67d44'
Fischesser,Sarah
And 0h45'50"28d53'
Fischetti,Diane
And 2h22'26"39d30'
Fischlowitz,Ruth Epstein
Peg 23h7'16"12d19'
Fischman,Sidney Mays
Lac 22h17'23"49d45'
Fischtalo
Oph 17h39'57"11d52'
Fish III,Thomas Leo
Aur 7h7'57"41d37'
Fish's Finn Forever
Ori 5h58'58"1d15'
Fish,Daniel "Danny"
Her 16h45'13"50d14'
Fish,Donald R
Uma 13h23'50"62d54'
Fish,Dorothy
Cam 6h13'15"58d20'
Fish,Eric D
Lac 22h8'28"50d31'
Fish,John & Marion "Naptown Sound"
Cyg 21h53'54"53d19'
Fish,Leslie
Psc 23h5'38"6d12'
Fish,Robert E
Her 16h59'59"35d9'
Fish,T E
Cet 3h1'18"9d5'
Fishbeck,Laura E
Cam 7h8'47"71d7'
Fishbein,Bernard W
Dra 20h26'50"68d28'
Fishbein,Gerald
Per 3h11'36"46d21'
Fishbein,Marc David
Aqr 22h4'48"0d48'
Fishburn,Dale
Umi 14h59'45"65d56'
Fishel 1912-1992, Charles Eugene
Aql 20h8'13"0d30'
Fisher
Lac 22h30'17"38d25'
Fisher Family,The Sean L
Mon 7h39'55"-4d36'
Fisher II,Robert Littell
Ser 15h50'35"16d5'B
Fisher III,Melville Wiley
Her 16h58'31"32d47'
Fisher Star,The
Aql 19h58'16"13d7'
Fisher Star,The Famous Colin
Cas 0h38'46"60d22'
Fisher Star,The Jan
Cyg 19h32'46"33d9'
Fisher's Family Star, Liz & Keith
Cam 3h21'29"61d35'
Fisher's Flicker
Cyg 19h26'17"32d48'
Fisher,Alexis Joanna
And 0h26'45"40d54'
Fisher,Alice Middlebrook
Cas 0h8'13"62d56'
Fisher,Allan Scott
Her 16h32'28"17d5'
Fisher,Andrew Shaun
Aur 6h31'32"31d23'
Fisher,Angela Carrozza
And 1h47'1"37d46'
Fisher,Ann
Oph 18h39'58"6d40'
Fisher,Barbara A
Per 3h34'55"44d38'A
Fisher,Brandon Michael
Dra 19h13'51"84d56'
Fisher,Brian Alan
Boo 14h43'59"31d16'
Fisher,Carole
Aql 19h22'0"3d38'
Fisher,Carolyn Yvonne
Cas 2h12'59"60d57'
Fisher,Carrie Ann
Psc 1h26'29"30d44'
Fisher,Christopher David
Hya 8h53'1"-0d53'
Fisher,Darlene Henson
Peg 23h6'44"8d10'
Fisher,Doctor Allen Michael
Oph 17h53'0"12d2'
Fisher,Donald M
Per 3h34'55"44d38'B
Fisher,Dorothy
Aur 7h20'45"37d37'
Fisher,Emily Seltzer
And 2h18'58"38d12'
Fisher,Emma
Umi 15h30'36"66d16'
Fisher,Forest M
Mon 8h3'8"-9d20'
Fisher,Huguette
Uma 10h3'57"50d4'
Fisher,III,William L
Cep 22h52'28"57d45'
Fisher,James
Per 1h44'58"53d20'
Fisher,James
Aql 20h0'45"0d50'
Fisher,Jay
Dra 19h19'1"71d14'
Fisher,Jean
Cmi 7h29'15"1d31'
Fisher,Jennifer
Lyn 8h2'11"35d36'
Fisher,Jennifer & Brad
Cam 13h20'11"80d16'
Fisher,Jennifer Marie
Mon 6h40'58"7d58'
Fisher,Jenny
Mon 7h55'54"-5d10'
Fisher,Jeremy
Cnv 12h23'23"35d15'
Fisher,Jeremy
Dra 15h7'29"58d7'
Fisher,Jessie Lynn
Lyn 8h14'26"37d47'
Fisher,Joanne
Lyr 18h15'56"42d15'
Fisher,Jon
Cnv 12h29'1"33d21'
Fisher,Jonathan Peter
Ser 15h49'40"18d37'
Fisher,Joseph Barnard
Aur 4h50'17"41d4'
Fisher,Jr Family, Albert B
Aur 6h23'38"38d53'
Fisher,Jr,Henry
Cet 2h37'20"4d27'
Fisher,Julia
Crb 15h55'42"26d34'
Fisher,Julia F
Com 12h22'48"21d49'
Fisher,Karen
Del 20h16'47"12d35'
Fisher,Katie
And 23h27'41"48d16'
Fisher,Kelly Dawn
Uma 10h26'45"56d28'
Fisher,Kenneth P
Lyn 7h33'22"44d34'
Fisher,Kori Jo
Lyr 18h41'44"27d5'
Fisher,Lauren Elizabeth
And 2h30'18"41d54'
Fisher,Leonard
Cam 5h49'55"67d55'
Fisher,Leslie Edward
Aur 5h0'54"46d30'
Fisher,Lucky
Aur 6h14'15"38d27'
Fisher,Lyle A
Boo 13h56'27"18d3'
Fisher,Marc Michael
Per 3h12'37"42d42'
Fisher,Matthew Robert
Boo 13h45'57"22d04'
Fisher,Michael
Ser 15h30'48"20d14'
Fisher,Michael & Kristi
Uma 11h57'45"46d26'
Fisher,Michael James
Aur 4h55'1"40d05'
Fisher,Michele
Boo 14h22'16"23d51'
Fisher,Monkey B
Uma 10h30'14"47d58'
Fisher,Morgan Alan
Lac 22h48'1"56d30'
Fisher,Paul Douglas
Per 2h22'17"54d36'
Fisher,Paula Wilson
Aql 20h11'22"4d25'
Fisher,Randall T
Dra 11h2'47"74d6'
Fisher,Rina
Uma 10h52'16"71d31'
Fisher,Ronald
Boo 15h5'31"26d3'
Fisher,Rosie
Cas 0h20'43"72d14'
Fisher,Roydie Uinston
Cet 2h52'17"1d11'
Fisher,Ruth
Cas 1h10'55"61d1'
Fisher,Sara
And 0h12'23"36d3'
Fisher,Sarah Marie
Cyg 19h44'27"31d36'
Fisher,Scott B
Boo 15h18'39"41d15'
Fisher,Sherry & Stephen
Cyg 20h23'43"38d29'
Fisher,Skee
Her 18h17'47"14d38'
Fisher,Stephen Kenneth
Lac 22h0'1"41d13'
Fisher,Stone
Oph 17h2'15"11d24'
Fisher,Teodora Doretha
Crt 11h15'44"-19d56'
Fisher,Theodore
Her 16h52'57"37d14'
Fisher,Thomas C
Cam 12h20'22"81d39'
Fisher,Thomas W
Aql 19h56'34"12d34'
Fisher,Tom & Jerry
Boo 14h44'0"22d15'
Fisher,Tristan Lindsay
Lyn 8h17'12"38d13'
Fisher,Virgil Maceo
Ori 6h9'25"18d53'
Fisher,Virgil William
Cet 1h50'21"-5d19'
Fisher,William Casey
Hya 10h44'13"-17d33'
Fisher,Zachary
Equ 20h55'60"2d21'
Fisher-79,Danielle
Mon 7h0'46"-6d38'
Fisher-Carney Family
Cam 8h26'49"73d21'
Fisher-Cunningham, Joshua Shaw
Oph 16h20'47"1d54'
Fisher-Smith,Judy Lynn
Peg 23h28'15"22d11'
Fisherkeller,Peter Joseph
Aur 6h0'22"35d23'
Fisherman's Star,The
Lac 22h14'46"54d44'
Fisherman,Kids Club Winnie Penny
Peg 23h38'36"28d30'
Fishkin,Eric & Jamie
Cnv 13h57'17"40d28'
Fishler,MD,Kenneth Orson
Cep 21h54'0"55d26'
Fishman,Cary
Lyr 19h21'0"38d41'
Fishman,Frances
Cas 1h43'34"65d18'
Fishman,Hal
Dra 16h37'18"70d6'
Fishman,Ivan
Hya 8h49'50"-6d22'
Fishman,Lawrence M
Ser 15h12'60"22d49'
Fishman,Marc
Sct 18h56'45"-4d23'
Fishman,MD,Theodore D
Cep 20h55'36"58d45'
Fishman,Morton I
Her 17h59'25"14d51'
Fishman,Stacy
And 0h53'32"40d22'
Fishpaw,Andrew J
Boo 14h34'29"7d54'
Fishpaw,Barb
Aur 5h30'27"38d57'
Fishpaw,Barb & John
Cet 1h26'33"-2d30'
Fishpaw,Barb & John
Ori 5h47'43"12d4'
Fishstein,Austin Casey
Cmi 7h40'43"8d16'
Fishstein,Travis James
Uma 11h42'0"41d55'
Fishter,Dr Michael E
Her 17h54'60"14d47'
Fishy
Del 20h20'57"16d34'
Fishy
Lac 22h22'19"55d20'
Fisia
Pyx 8h40'48"-22d3'
Fisicaro,Verna
Cet 2h47'45"0d25'
Fisk University Cultural Studies Tour
Tri 1h59'12"33d48'
Fisk,Conor Boyd
Aur 7h0'47"41d24'
Fisk,Shela Baskin
Peg 21h23'46"7d53'
Fiske III,Lincoln Bryant
Lib 15h1'29"-8d27'
Fiske's Star,Arthur Byron
Her 17h32'1"28d27'

Fiske,Patricia J
Ori 5h57'36"18d50'
Fiske,Ralph & Eleanor
Cyg 20h35'43"37d54'
Fiss,Gina
Uma 9h31'14"70d58'
Fisselberger,Vanessa Anne-Maria
And 0h1'45"46d23'
Fissell,Carl
Cap 21h5'49"-22d30'
Fister,Adele
Uma 9h45'55"57d52'
Fister,Brian Leon
Uma 11h56'16"32d14'
Fitch,Dawn L
Mon 6h25'38"-1d52'
Fitch,Doctor Margaret
Eri 3h1'14"-4d13'
Fitch,Helen
Sct 18h38'42"-5d41'
Fitch,Joya Sue
Cas 23h14'50"60d16'
Fitch,Jr,Charles
Ser 15h12'41"8d56'
Fitch,Kimberly
Gem 7h2'20"33d54'
Fitch,Margaret
Lyr 18h28'27"30d36'
Fitch,Tammy Lynn
Sex 9h45'16"2d30'
Fitch,Tierney Nicole
And 2h0'1"45d26'
Fitch,Tiffany Brooke
Com 13h17'49"21d21'
Fitchet,Jon
Cep 22h14'13"68d54'
Fitchett,Emma Maria
Umi 14h8'58"69d41'
Fite,Ma Sara
Mon 6h16'18"-6d57'
Fites-Kelley,Sheila Mae
Cma 6h42'0"-15d20'
Fitez,Frances G
Aur 4h48'47"40d22'
Fitiles,George James
Aql 19h5'56"-5d60'
Fitoussi,Bernard
Per 4h2'57"50d34'
FitrZyk
Sge 20h4'15"20d5'
Fitting,Chris & Judy
Com 13h2'25"18d32'
Fittizzi Christmas Star '94
Lyn 8h34'27"44d52'
Fitts,Sherry Lee
Vul 19h43'16"26d60'
Fittz,Michael Grant
Aql 19h30'36"10d2'
Fitz's First
Umi 13h53'49"78d51'
Fitz,Paula P
Cas 2h51'0"70d17'
Fitz-1,Dennis
Sct 18h53'15"-6d33'
Fitzbugh,Jonathan T
Cnv 13h58'15"40d35'
Fitzcrane,Meshach Clay
Ari 3h11'22"22d46'C
Fitzcrane,Walker Lee
Ari 3h11'22"22d46'A
Fitzcrane,Yatarose Anna
Ari 3h11'22"22d46'B
Fitze,Tara Cassidy
Cas 0h25'22"61d20'
Fitzell,Margarethe Luise Vesely
Umi 14h59'0"65d57'
Fitzgerald Union,The
Aur 5h17'23"45d11'
Fitzgerald,Abby Jacquelyn Short
Aqr 21h7'0"-5d38'
Fitzgerald,Allen Joseph
Aql 20h6'40"1d48'
Fitzgerald,Ann Veronica
And 0h53'27"34d5'
Fitzgerald,Bamboo-June
Vul 19h52'54"20d6'
Fitzgerald,Betty
Boo 15h18'0"34d34'
Fitzgerald,Bonnie Sue Mills
Peg 23h20'13"26d3'
Fitzgerald,Brian
Oph 16h48'33"11d50'
Fitzgerald,Charles B
Oph 17h53'39"12d54'
Fitzgerald,Clare Marie
And 23h39'45"43d46'
Fitzgerald,Cosette
Mon 6h57'59"10d35'
Fitzgerald,Deirdre
Cyg 19h32'17"33d24'
Fitzgerald,E Kendall
Tri 2h7'43"30d46'
Fitzgerald,Garrett
Mon 6h44'0"10d20'
Fitzgerald,Gary Lee
Ori 5h37'35"-5d53'
Fitzgerald,Gerald
Her 17h1'1"44d57'
Fitzgerald,Gerard
Lac 22h1'30"40d59'
Fitzgerald,James Robert
Her 18h13'46"38d47'
Fitzgerald,James
Per 2h44'25"43d18'
Fitzgerald,Jr,Dr G Joseph
Dra 17h41'30"67d35'
Fitzgerald,Jr,Richard J
Lyr 18h49'44"31d2'
Fitzgerald,Marguerite Kavanagh
Cet 1h3'20"-6d33'
Fitzgerald,Matthew
Her 17h14'16"26d3'
Fitzgerald,Michael Desmond
Cnv 13h25'20"41d18'
Fitzgerald,Patricia
Cas 0h6'1"62d29'
Fitzgerald,Patrick
Aql 18h56'58"-2d26'
Fitzgerald,Scott Patrick
Her 17h11'51"42d54'
Fitzgerald,William J
Aur 5h15'13"45d27'
Fitzgibbon,Brian
Lmi 9h51'19"34d43'
Fitzgibbon,Molly Elizabeth
Lyr 18h33'52"45d3'
Fitzgibbon,Rita & Maurice
Lyr 19h3'45"40d0'
Fitzgibbons,Erin
Aur 6h59'0"37d32'
Fitzgibbons,Jacqueline Elizabeth
Uma 10h55'45"53d51'
Fitzgibbons,John
Ori 5h32'41"-0d38'
Fitzhugh,Jonathan T
Cyg 19h50'58"38d37'
Fitzinger,Karl
Lib 14h50'11"-0d56'
Fitzkenneth
Ori 6h5'53"8d37'
Fitzmaurice,Paul
Dra 19h29'52"58d43'
Fitzpatrick,Brendan Todd
Dra 19h0'21"70d11'
Fitzpatrick,Carol
Uma 11h58'26"40d27'
Fitzpatrick,Eileen
Cam 3h48'16"55d41'
Fitzpatrick,Elizabeth
And 1h47'47"37d56'
Fitzpatrick,George Edward
Cam 7h47'31"70d55'
Fitzpatrick,Jeanne
Cas 0h1'0"61d26'
Fitzpatrick,Joseph & Theresa
Cyg 19h47'49"29d29'
Fitzpatrick,Justine
Uma 14h3'43"54d51'
Fitzpatrick,Kathleen A
Oph 17h31'48"7d59'
Fitzpatrick,Louis
Crb 15h27'16"31d29'
Fitzpatrick,Mary
Cas 23h38'51"60d54'
Fitzpatrick,Patrick J
Aur 7h21'19"35d46'
Fitzpatrick,Ralph
Boo 15h7'28"52d57'
Fitzpatrick,Richard
Dra 16h13'53"68d24'
Fitzpatrick,Ruth
And 0h33'1"40d3'
Fitzpatrick,Sean Timothy
Psc 1h20'46"20d28'
Fitzpatrick,Tara Angela
Com 13h8'1"22d27'
Fitzpatrick,Timothy James
Equ 20h58'15"10d37'
Fitzrandolph,Tammy
Mon 6h25'26"11d26'
Fitzsimmons,Casey Taylor
Cnv 13h16'38"38d18'
Fitzsimmons,Edward Joseph
Her 18h1'18"31d3'
Fitzsimmons,James & Valerie
Crb 16h8'46"30d36'
Fitzsimmons,Marie Mazzeo
Cas 2h32'59"57d29'
Fitzsimmons,Rachel Carey
Cnv 13h26'20"41d15'
Fitzsimmons,Rita Marie
Dra 11h5'43"73d15'
Fitzsimmons,Rosaleen
Lmi 10h25'17"39d39'
Fitzsimons,Kaylene Ann
Cra 18h7'38"-39d46'
Fitzwater,Harry E
Equ 21h2'49"2d47'
Fiudo,Danielle Nicole
And 2h33'41"44d37'
Fiumara,Vincent
Aur 5h48'1"50d26'
Five Live
Uma 14h1'57"50d15'
Five Points
Boo 14h55'20"53d48'
Fix,Michael Robert
Hya 8h13'38"-6d2'
Fixari,Kristine Elizabeth
Cyg 19h27'18"40d2'
Fixel,Lee Jared
Ori 5h24'37"1d48'
Fixmer,Gale
Uma 12h12'0"56d39'
Fiyod,Solomon J
Her 18h16'44"18d47'
Fizgerald's Heavenly Light,Susan
And 2h34'1"44d58'
Fizzer
Per 2h37'45"38d47'
Fjelstad,Steven Carl
Cet 1h3'55"-1d53'
FJP
Aur 6h43'45"38d24'
FKBP12-rapamycin-FRAP
Uma 9h17'0"57d6'
Flacco,Barbara Ann
And 0h19'51"32d46'
Flach,Mathias
Ari 1h45'33"18d44'
Flachenecker,Karl
Ser 15h10'46"0d8'
Flack,Haley Lynn
And 0h13'14"37d56'
Flad,Dan
Per 3h25'37"50d45'
Flade,Alexandra Marie
Mon 6h43'0"11d55'
Flader,Kirsten & Ralf
Dra 17h41'56"61d27'
Flaemig,Robert Scott
Cam 5h54'58"60d36'
Flagg,Donna
Peg 21h57'32"34d32'
Flagler,Beverly Frances
Lyr 18h46'13"43d3'
Flagstar Bank
Tri 2h3'41"32d56'
Flahart,Mitchell E
Aur 4h56'42"40d42'
Flaherty,Beverley Jane
Mon 8h0'35"-8d42'
Flaherty,Jenny
Cyg 20h28'58"58d29'
Flaherty,Jerome T
Her 16h31'37"38d42'
Flaherty,Kevin Patrick
Dra 19h4'52"56d59'
Flaherty,Mary
Cas 1h43'38"60d44'
Flaherty,Robert Edward
Tri 2h0'1"33d41'
Flahiff,Marci
Boo 14h21'31"23d8'
Flail,Gary Melvin
Cet 2h8'60"6d19'
Flair
And 2h28'44"42d28'
Flake,Malcolm
Ori 5h18'38"7d49'
Flambascus
Cnv 13h51'43"38d56'
Flambise
Boo 14h34'1"48d59'
Flame
Peg 23h43'14"18d1'
Flame of Our Love
Cyg 19h29'20"32d15'
Flamer,The Sophia
Cas 2h44'11"60d14'
Flamholtz,William Michel
Aur 6h17'50"30d25'
Flamin' Ramin'
Per 3h24'58"40d50'
Flaminia
Cnv 13h0'25"51d42'
Flamion,Alexandra Marie
Cas 23h25'15"61d18'
Flamion,Sydney Clara
And 1h50'26"36d31'
Flammia,Joseph M
Cep 23h15'31"64d36'
Flammia,Nicholas Vito
Gem 7h19'3"28d39'
Flamming,Zdenek
Cyg 20h43'56"45d57'
Flanagan II,Patrick Kipp
Aur 4h49'1"51d24'
Flanagan,Audrey Deloris
Peg 22h12'38"31d11'
Flanagan,Betty Anne O'Neill
Uma 11h15'1"68d56'
Flanagan,Dave
Aur 6h26'34"38d47'
Flanagan,David Meade
Aur 6h14'16"33d22'
Flanagan,Dick
Her 16h21'0"50d53'
Flanagan,Eileen Idgic
Tri 1h58'51"25d39'
Flanagan,Elizabeth S
Lyn 7h34'47"58d27'
Flanagan,Ellen
Aql 19h11'1"19d9'
Flanagan,Jean
Cam 12h47'0"81d8'
Flanagan,John "Jack" Maurice
Her 16h36'47"27d19'
Flanagan,John J
Dra 17h2'1"62d42'
Flanagan,Jr, A Star, David M
Her 16h40'37"28d14'
Flanagan,Jr,Michael Gerard
Vir 13h33'44"-1d25'
Flanagan,Kerry Lynn
Lib 14h59'49"-20d13'
Flanagan,Marge
Cas 2h7'20"59d27'
Flanagan,Thomas Francis
Per 2h25'0"55d30'
Flanagan,Wayne,Faith & Morjan
Her 16h6'52"16d39'
Flanagan,William T
Cap 21h3'15"-22d43'
Flanary,Barry Eric
Ori 5h53'57"-8d17'
Flanders
Cyg 19h27'21"30d12'
Flanders,Kelly Lynn
Peg 23h28'16"33d5'
Flanigan,Brenda Carol Clark
Mon 7h3'34"-6d37'
Flanigan,John Patrick & Ann Allen
Uma 11h4'15"38d32'
Flanigan,Shawn Timothy
Aql 20h3'1"6d44'
Flannagan,Cherie A
Cas 23h27'32"61d21'
Flanner,Erin Meredith
Lyr 18h45'20"36d49'
Flannery,Carolyn Ann
And 0h9'22"47d9'
Flannery,Jon
Aql 18h40'31"-2d39'
Flannery,Kevin Douglas
Gem 6h49'28"30d18'
Flannery,Mary K
Cas 23h1'38"58d43'
Flannery,Michael Patrick
Per 1h56'48"53d4'
Flannigan,Leah
Mon 6h44'1"7d47'
Flasco,CFP,Michael A
Cep 23h59'32"83d32'
Flash #7
Cam 7h0'14"68d6'
Flashback
Uma 8h27'31"68d36'
Flaski,Fabio
Aql 19h53'24"-0d28'
Flasted-Maas
Vir 11h50'55"6d16'
Flat Rock Michigan Community Star
Per 3h8'0"41d22'
Flaten,Alice
Cas 1h4'0"75d52'
Flatt,Cassandra Elizabeth
Peg 22h32'31"7d57'
Flatt,Jessica Leigh
And 23h19'53"40d18'
Flaucher,Sally & Jeff
Lyn 7h49'19"39d34'
Flaugher,David A
Sct 18h55'23"-4d44'
Flautt,Sarah Elizabeth
Sco 16h56'48"-40d49'
Flaux,Clenent
Lyr 19h22'0"31d24'
Flavia Arena
Cas 1h30'30"68d44'
Flavin,Eleanor
Uma 10h45'32"70d46'
Flavin,Jodi L
Cas 0h30'15"65d31'
Flavin,Kathryn Mary
Aur 5h5'0"41d14'
Flavin,Tim
Ser 15h49'43"18d31'
Flaxa,Christiane und Carmen
Uma 9h39'51"46d52'
Flaxman,Edward
Dra 12h55'54"75d27'
Flayderman,E Norman
Cnv 12h15'40"36d20'
Flayler,James C
Hya 9h9'24"4d52'
Flea
Cnv 12h53'0"50d45'
Flebotte,Dillon Alexandra
Mon 6h53'24"-1d13'
Flechazo
Pho 0h32'52"-44d15'
Fleck,Christopher Michael
Cet 2h27'31"5d58'
Fleck,Kitty
Cmi 7h17'38"-0d0'
Fleck,Robert
Sct 18h56'18"-6d22'
Fleckenstein,Tom
Hya 8h11'36"5d59'
Fleegle I,J G
Cnc 8h30'25"7d54'
Fleener,Daniel
Ori 4h45'58"4d3'
Fleener,Eugene
Ser 15h50'0"21d53'
Fleener,Janine
Eri 3h14'58"-12d19'
Fleener,Mark
Aql 19h53'45"12d5'
Fleenor,Barbara June
Peg 22h19'22"33d30'
Fleenor,Chuck
Hya 8h37'1"5d47'
Fleenor,Jamin J C - Chris Fleenor
Ori 6h7'22"8d54'
Fleenor,Michael Dennis
Her 18h3'16"31d24'
Fleet,Eric N
Cnv 12h10'34"37d34'
Fleetwood,Bonnie
Lyn 9h9'1"45d6'
Fleetwood,Ray & Helen
Cyg 21h26'14"38d19'
Flegel,Drake Justin Orion
Ori 5h57'22"17d0'
Flegel,Thomas
Oph 18h3'36"7d49'
Fleicia
Com 12h29'1"21d15'
Fleig,Beth Pauline
Oph 18h0'12"13d14'
Fleig,Richard & Darrell
Sex 9h50'1"5d51'
Fleischaker,Michael
Lac 22h22'16"50d7'
Fleischauer,Bill
Eri 4h54'3"-9d46'
Fleischer,Detlef
Cmi 7h18'15"1d41'
Fleischer,Helmut
Aql 18h57'15"16d60'
Fleischer,Mary
Cyg 20h21'50"40d42'
Fleischer,MD,F Jean
Aql 19h30'0"-0d55'
Fleischer,Scott David
Gem 7h29'51"22d60'A
Fleischer,Siena BC
Ari 1h50'42"11d51'
Fleischer,Steven Robert
Gem 7h29'51"22d60'B
Fleischer-My Light, Robert Samuel
Leo 10h43'0"7d26'
Fleischhacker,Marianne
Sco 17h27'1"-38d21'
Fleischhacker,Stefan
Vir 13h12'47"-5d13'
Fleischhacker,Terry
Cam 7h14'37"82d36'
Fleischhauer,Stephan
Mon 7h1'38"-5d57'
Fleischman,Harold S
Lmi 10h58'53"30d47'
Fleischman,Jordyn Anne
Cet 2h52'20"4d27'
Fleischman,Mark
Aur 7h1'30"36d12'
Fleischmann,Captain Jack
Cep 0h19'17"76d35'
Fleischmann,Dan & Mary
Cyg 21h45'24"28d25'
Fleischmann,Emma Maire
Mon 6h22'30"7d45'
Fleischmann,Fritz
Mon 7h3'32"3d53'
Fleischmann,Matthew Steven
Cma 6h56'29"-16d5'
Fleishman,Edward
Lac 22h45'28"37d51'
Fleishman,Rishona
Cam 6h15'21"83d2'
Fleissner,Kathy
Peg 22h36'54"31d23'
Fleit,Linda
Lyn 8h46'26"44d56'
Fleitas,Gilbert
Cap 19h46'50"-16d32'B
Fleming's Finest Star, Beverly
Eri 3h19'45"-2d50'
Fleming,Bob & Viv
Cyg 19h29'35"35d11'
Fleming,Carl & Katherine
Dra 17h40'17"65d29'
Fleming,Clayton Paul
Ori 6h16'54"1d23'
Fleming,Donald Glenn
Hya 9h38'19"-5d19'
Fleming,Eileen C
And 1h4'0"38d31'
Fleming,Frank & Amanda
Cyg 20h0'1"39d34'
Fleming,Guinevere Tephinzy
Cas 0h2'50"64d1'
Fleming,Jack W
Ser 15h43'54"-1d58'
Fleming,James K
Lac 22h52'41"55d57'
Fleming,James L
Ser 15h52'60"24d23'
Fleming,Joe E
Hya 8h40'46"2d45'
Fleming,Katelyn
Tau 3h57'18"30d13'
Fleming,Kenneth Raymond
Per 3h11'42"48d25'
Fleming,Martha
Cas 2h49'47"71d11'
Fleming,Melissa Lamar
Peg 22h45'29"25d35'
Fleming,Patricia Ann Morrison
Eri 4h2'48"-18d49'
Fleming,Paul
Ori 4h59'47"0d32'
Fleming,Ryan W
Her 17h32'22"21d49'
Fleming,Tim
Cep 22h42'0"68d27'
Fleming,Timothy Gerard
Umi 16h1'47"70d10'
Fleming-Davies Scientist,Arietta
Oph 18h17'45"13d49'
Flemings' Field
Lyn 8h58'36"40d6'
Flemming,Cathy
Cyg 20h52'34"31d49'
Flemming,Robyn
And 23h30'41"38d18'
Flener
Oph 17h28'58"-22d25'
Flens,Carolyn Jean
Cyg 19h43'25"30d13'
Flentje,Sabine
Lib 14h18'0"-23d34'
Flenton's "Hole-in- One"
Ori 5h54'53"0d43'
Flesch,Jordan Brittany
Cas 2h37'51"57d46'
Flesch,Lisa M
Com 12h29'43"26d35'
Flesch,Shauna Danielle
Cas 2h38'1"58d19'
Fleschlight
Her 17h50'46"42d10'
Flesdrager,Lois
Cas 1h19'29"63d58'
Flesh
Dra 18h44'48"67d46'
Fletcher
Cam 12h11'1"78d4'
Fletcher #80,The Clark
Per 2h51'52"46d7'
Fletcher ABT,Molly
Eri 3h38'41"-1d51'
Fletcher,Alexander James
Per 3h5'35"40d45'
Fletcher,Amy Ellen
Eri 3h13'24"-18d37'
Fletcher,Anita
Vul 20h16'20"23d2'
Fletcher,Anna Lou
Peg 22h45'43"29d30'
Fletcher,Anthony Stanton
Aur 4h48'39"37d3'
Fletcher,Antony Kent
Ind 20h32'33"-51d4'
Fletcher,Bernard Alport
Cam 4h28'1"68d35'
Fletcher,Danette Lynn
Cas 1h7'1"60d20'
Fletcher,David "Honey Bunny"
Aur 5h35'1"37d45'
Fletcher,Elizabeth "Biz"
Eri 4h13'52"-11d18'
Fletcher,Ewan James
Cep 20h52'38"62d35'
Fletcher,Gemma
Lyr 18h14'21"46d41'
Fletcher,George
Ori 6h0'25"20d40'
Fletcher,Irene
And 1h21'56"34d50'
Fletcher,James Ross
Per 2h57'12"32d25'
Fletcher,Jay
Crt 10h51'19"-12d7'
Fletcher,Jim
Hya 8h18'40"4d25'
Fletcher,Joe & Allena
Cyg 21h35'28"42d30'
Fletcher,John
Cnc 7h58'52"20d32'
Fletcher,Jr,Jacque Kingsley
Cas 2h44'13"61d42'
Fletcher,Kara S
Mon 8h3'48"-10d8'
Fletcher,Kathryn Jane
Cam 6h49'52"83d44'
Fletcher,Kelly Louise
Del 20h36'16"10d30'
Fletcher,Ken & Amelia
Lyr 18h21'18"40d37'
Fletcher,Kim
Ori 6h15'47"1d35'
Fletcher,Loretta D
Cyg 19h24'32"33d43'
Fletcher,Mark Lee
Oph 17h17'19"10d35'
Fletcher,Maxine
Cas 0h36'14"60d39'
Fletcher,Michael William
Sgr 19h0'57"-28d16'
Fletcher,Noelle Marie
Ori 5h14'1"-4d48'
Fletcher,Paul
Ori 6h17'54"10d53'
Fletcher,Rhonda & Chuck
Mon 7h58'15"-6d56'
Fletcher,Sam Michael John
Per 3h8'1"38d53'
Fletcher,Star
Ori 6h1'46"8d25'
Fletcher,Starr Wendy
Aql 18h46'43"10d36'
Fletcher,Wallace M
Her 17h58'1"41d0'
Fletcher,William Lester
Ori 5h26'40"-2d37'
Fleur Anna Sophia
Lyr 18h36'40"37d17'
Fleur Rodda
Ind 20h49'59"-59d52'
Fleurdelys,Carine
Cep 20h58'0"65d59'
Fleury,Catherine
Cyg 19h47'44"37d38'
Fleury,David R
Aur 5h6'47"44d48'
Fleury,Karene
Lyn 9h5'56"44d6'
Fleury,Maryionne
Per 3h32'45"35d52'
Flick e Flock
Psa 22h6'14"-25d44'
Flick's Eye
Lyn 7h29'49"36d20'
Flick,Dr Friedrich Karl
Cam 7h0'36"61d43'
Flick,Victoria Ann
And 23h26'56"49d34'
Flickering Frisky
Crb 16h8'37"31d12'
Flicop,Susan & Christopher Hill
Cam 7h56'0"83d50'
Fliegel,Alison Dawn
Lyn 7h37'29"38d39'
Fliegel,David Scott
Lac 22h53'1"51d59'
Fliegel,Jessica Erin
Tri 2h12'1"31d6'
Fliegler Vancata, Monika
Cnc 8h51'49"11d3'
Flierl,Betty
Cnv 12h17'12"34d12'
Fliesenstar Bernd Regener
Lyn 8h5'51"47d25'
Flig,Remy Michelle
Peg 21h25'0"2d42'
Flight,Jennifer Leigh
Lyn 7h6'56"51d53'
Flimm,Ursula
Aqr 21h2'32"-8d40'
Flinchbaugh,Katherine Grace
Mon 6h59'31"-10d5'
Flinkingshelt,Susan
Oph 17h56'57"10d31'
Flinn,Colleen
Gem 7h55'1"30d17'
Flinn,Louise & Timothy
Sgr 18h51'39"-25d21'
Flinn,Michael Dax
Ser 18h2'49"-1d29'
Flint
Boo 13h39'27"18d4'
Flint
Tri 2h17'25"33d11'
Flint,Janet L
Com 12h32'0"27d18'
Flint,Jr,James Howard
Aql 19h7'0"3d36'
Flint,Robert Charles
Oph 17h2'50"-20d41'
Flintoft,Jim
Uma 11h38'21"31d10'
Flip
Boo 14h57'16"23d49'
Flip
Oph 17h53'28"-0d33'
Flippen,Richard Eugene
Hya 8h52'23"-0d49'
Flippin,Mason Patrick
Dra 16h30'52"71d44'
Flippin,Vance Garrett
Dra 16h8'60"61d37'
Flippo aka Grant
Philipo,Charles G
Tau 5h57'29"23d56'

Flipse,John Ori 6h15'1"8d4'
Fliss,Aaron Clarke Per 3h51'51"36d44'
Fliss,Mitchell David Lac 22h24'0"38d6'
Flitzen Ori 6h14'47"10d6'
Flo Star I Cnv 12h27'29"38d19'
Flo's Diamonds & Pearls Lyr 18h27'44"32d11'
Flo-Wyle Cyg 20h26'49"40d17'
Flobeck,The Marian Leigh Del 20h15'26"9d35'
Flocken,Andrea Lyn 7h40'18"38d45'
Flodman,Eric Aql 19h57'1"13d24'
Flogdell,Diana Aur 5h0'47"45d59'
Flojo Cas 0h11'51"61d14'
Flom,Mark Bradley Sct 18h42'31"-6d15'
Flomel Cas 0h30'24"68d15'
Flood Anniversary Star The S & K Cam 6h14'48"65d17'
Flood Federation,The Aur 5h38'51"37d51'
Flood,Abigail Emma Oph 18h4'20"12d14'
Flood,Campbell Charles Cnv 12h58'25"41d56'
Flood,Jerome Thomas Her 18h1'45"31d1'
Flood,Thomas B Dra 19h48'58"61d29'
Flook Star Vul 20h24'28"28d14'
Floom,Bev Lyr 18h32'0"46d21'
Flor,Julien Per 4h25'0"50d15'
Flor,Leah & Heath Aql 19h30'36"10d55'
Flora Her 16h2'1"50d31'
Flora Lyr 18h15'54"33d57'
Flora Del 20h18'10"11d17'
Flora "Momma" Lee Cas 0h29'31"70d43'
Flora Mae Vul 21h2'1"20d26'
Flora Star,The Richard & Mary Cyg 19h28'59"30d44'
Flora's Joy And 2h6'0"30d33'
Flora,Dick Per 1h34'44"53d2'
Flora,Lorene A Eri 3h21'26"-4d26'
Flore DC Aql 20h1'1"11d5'
Florea,Shoshona Mariah Pilip And 0h59'15"45d10'
Florence Com 12h56'19"28d7'
Florence And 0h44'57"30d31'
Florence Aql 19h3'41"-0d49'
Florence Cam 6h5'59"60d57'
Florence Mon 6h22'55"5d12'
Florence Com 12h10'38"21d29'
Florence Ori 5h53'18"7d49'
Florence & Allen Forever And 1h16'1"38d3'
Florence & Sol Lyn 7h57'33"42d15'
Florence Alice Cas 23h13'59'61d6'
Florence's Treasure Cyg 20h31'16"38d58'
Florence,Big Bob Tau 4h2'22"23d15'
Florence,Ryan Alexander Vul 20h17'56"23d43'
Florence,The Great Jacqueline E Cas 3h11'37"58d41'
Florent,Gabriel Dra 16h38'0"63d48'
Florentine Per 3h22'46"41d10'
Florentino,Michael Scott Boo 14h27'48"12d22'
Florentino,Paul Her 18h16'30"14d51'
Florentino,Teri Lyn 8h28'20"58d43'
Flores 09111976, Roberto Aql 20h5'25"4d22'
Flores,Anthony Jo Oph 17h2'36"8d53'
Flores,Anthony P Cma 6h49'55"-18d23'
Flores,Arthur Per 3h0'0"40d44'
Flores,Carlos G Uma 9h33'41"70d19'
Flores,Chad Mon 7h56'48"-8d4'
Flores,Christine Marie Hya 8h22'1"-5d41'
Flores,Dana Rhea Cyg 19h29'40"34d10'
Flores,Diana Cam 3h34'36"61d26'
Flores,Diana R Mon 6h59'41"-0d31'
Flores,Jana Eri 4h38'49"-6d54'
Flores,Jenssie M Lac 22h22'26"53d21'
Flores,Juan Antonio Ser 15h39'13"21d58'
Flores,Luis Her 17h22'27"40d1'
Flores,Maria Mon 7h2'16"1d42'
Flores,Maria Peg 0h3'23"22d3'
Flores,Mel Vul 20h1'24"28d35'
Flores,Miriam Lyr 18h29'25"32d0'
Flores,Miriam E Cra 18h3'43"-37d28'
Flores,Myriamcita And 0h2'31"35d19'
Flores,Rafael Ori 5h59'25"14d47'
Flores,Renato Vidal Hya 8h43'51"-6d9'
Flores,Robert R Her 16h23'0"38d2'
Flores,Thomas Aur 7h9'30"36d7'
Flores,V N E Her 17h35'42"26d14'
Flores,Victor R Hya 9h21'0"-0d13'
Floresca,Marie Angeline Kumiyama Mon 6h18'51"1d8'
Floria Mother Of Lois Pat Frank & Sal,Marie Cas 23h24'49"61d38'
Florian Tri 2h3'36"31d38'
Florian Umi 16h7'13"83d1'
Florian,Bernard Per 3h58'9"35d16'
Florian,Birgit Lyr 18h26'0"45d12'
Florian,Kommerzialrat Herbert Uma 10h10'0"47d59'
Florianne Cas 2h10'11"59d14'
Florianne B Crb 16h19'44"37d44'
Florice Cas 0h22'48"63d56'
Florigan Boo 14h31'33"50d16'
Florijn,Eddy J M Lac 22h22'11"48d58'
Florin Cep 22h30'48"63d48'
Florins,Guillaune Lyr 19h22'57"31d11'
Florio,Erin Uma 8h6'12"67d58'
Florio,Frank Per 1h28'40"53d13'
Florio,Frank Gerard Sco 17h26'49"-38d2'
Florio,Jim Lac 22h55'46"54d7'
Florio,Joan Lyr 18h39'46"30d29'
Florio,Laura Ann And 2h3'55"38d45'
Florio,Lori Lyn 7h55'1"48d48'
Florio,Lynne Cas 2h2'46"61d2'
Florio,Nancy And 23h1'43"38d54'
Florio,Susan C Aur 6h48'5"46d33'B
Florio,Tom Dra 17h37'45"70d7'
Florio,Tommy Cyg 21h8'13"50d7'
Florio,William J Aur 6h48'5"46d33'A
Floris,Annamaria Cmi 8h8'13"0d58'
Florita Nae Nae Aql 19h12'29"12d45'
Florkowski,Aaron Per 1h37'53"53d45'
Flormina Crb 16h19'50"38d31'
Floros,Vicki Lyr 18h13'25"37d37'
Flory's 60th Per 2h7'42"58d26'
Flory,Jean Cam 13h1'14"81d47'
Flory,Misha Yvonnc Cam 3h51'11"53d40'
Flosnik,CPL Joseph T Hurrah Aur 6h32'15"31d51'
Floss' & Walt's Star Her 18h19'25"14d19'
Flossie & Freddie Lyn 7h42'23"45d8'
Flounoy,Sondra Cas 0h56'35"63d36'
Flournoy,Sharon Hui Peg 21h26'37"23d42'
Flouty,Henry Joseph Dra 16h12'26"61d31'
Flow,Marianna Cam 5h18'28"68d6'
Flower,Bruce Wayne Ser 15h22'15"2d30'
Flower,Frank S Cep 21h49'39"85d50'
Flowerday,Jr,Donald W Cet 1h54'20"-3d28'
Flowerday,Todd Dra 19h55'1"80d33'
Flowers #4 Ori 5h48'0"12d26'
Flowers III,Stanley Ray Aql 18h56'39"17d32'
Flowers,Anthony Vito Dra 19h1'59"48d38'
Flowers,Barbara A & Daniel M Radell Boo 14h19'0"54d26'
Flowers,Elizabeth Ann Mon 7h2'7"-6d60'
Flowers,Jeffrey Scott Cet 0h39'18"0d55'
Flowers,Jim Aqr 22h22'49"-5d50'
Flowers,Joe Mathew Vir 14h38'34"7d16'
Flowers,John Michael Per 3h23'40"38d24'
Flowers,Justin R Cet 2h17'58"5d27'
Flowers,Larry Stephen Ori 5h46'56"12d12'
Flowers,Sue & Jim Gem 5h58'0"26d50'
Floyd Central Physics Class of 1994 Cam 13h56'52"81d55'
Floyd,Clifford E Ser 18h17'56"-14d20'
Floyd,Donna And 0h1'1"47d35'
Floyd,Grant Taylor Ori 5h54'0"12d3'
Floyd,James Richard Her 17h27'56"30d32'
Floyd,Jennifer Leigh Mon 6h56'15"10d5'
Floyd,John M Boo 14h27'57"25d46'
Floyd,Joshua Logan Her 17h3'53"47d53'
Floyd,Katherine Austin Mon 7h1'50"4d20'
Floyd,Kenny Aql 20h18'37"5d30'
Floyd,Matthew Christian Sex 9h40'0"2d54'
Floyd,Robert Jean Aur 5h57'45"29d52'
Floyd,Robert Norman & Carol Gray Cyg 21h9'21"40d23'
Floyd,Roger "Bear" Aur 7h14'50"40d11'
Floyd,Sherrie B Lyr 18h55'29"33d8'
Floyd,Tammy Kay Cyg 20h3'43"40d31'
Floyd-Dad,William Francis Aur 6h8'20"45d56'
Floydman & Wifey Boo 13h38'56"21d27'
Fluckiger,Brent Lac 22h12'0"54d45'
Fluegert,Frieda Augusta Cyg 20h30'56"37d35'
Fluehr,James F Her 18h17'56"14d40'
Fluffers Cma 7h51'43"11d35'
Fluffie Clinkstar Aql 20h32'29"-1d47'
Fluffy Linstar Peg 0h11'32"13d11'
Fluffy's Star 1992 Cam 8h11'28"81d46'
Flugger,My Mentor, J Cyril Ori 5h56'55"5d45'
Flugger,Our Shining Star,Ray T Aql 19h15'26"13d41'
Fluharty,Deborah Aur 4h57'57"37d22'
Fluharty,Paula Cet 1h25'0"-4d9'
Fluitt,Jack & MaryAnn Aql 19h53'28"10d55'
Fluke,Dwayne Anthony Aqr 23h10'39"-5d5'
Flum,Nancy Rose Cyg 20h54'0"31d45'
Flusilongus,Magnus D Dra 16h5'33"63d32'
Fluter,Edith Peg 22h17'29"21d16'
FluteStar Lyr 18h37'52"40d14'
Flutie 4 MVPS CFL 91-94,20 Doug Cep 21h59'0"68d7'
Fly,DeWitt Cyg 21h18'25"28d37'A
Fly,Jr,Earl D Cyg 21h10'41"38d59'
Fly,Murry L Cyg 21h16'42"38d3'
Fly,Nancy Cyg 21h18'25"28d37'B
Flyboy's Magic Cam 7h37'18"60d8'
Flyin' Ryan-"Ryan Clay Forbes" Del 20h54'35"2d11'
Flying Dutchman,The Dra 19h32'38"61d13'
Flynn (Pere),Joseph Francis Lyn 8h14'13"40d30'
Flynn III,Robert E Aql 20h18'40"0d43'
Flynn's First & Foremost Cam 3h35'50"71d27'
Flynn,Andrew Gerald Uma 12h24'59"56d15'
Flynn,Betty & Fields Peg 23h42'40"27d20'
Flynn,Bryan J Boo 15h4'39"20d32'
Flynn,Colby Lightner Dra 17h11'0"64d12'
Flynn,Colleen Erica Peg 23h44'46"15d51'
Flynn,Daniel Edward Per 2h52'0"35d26'
Flynn,David Joseph Cep 22h19'54"68d21'
Flynn,Dillion J Umi 16h12'58"72d58'
Flynn,Elizabeth Ann Cas 0h4'24"62d32'
Flynn,Genevieve Margaret Cet 2h19'0"-18d52'
Flynn,George A Wilson Hya 8h13'18"-5d41'
Flynn,Hailey Denease Peg 22h59'24"20d12'
Flynn,Haleigh Elizabeth And 1h25'0"33d55'
Flynn,Henry Joseph Per 2h22'32"55d28'
Flynn,Jeff Her 17h38'34"21d42'
Flynn,Jennifer L Mon 7h24'20"-10d56'
Flynn,Jennifer Leigh Lyn 8h1'15"57d45'
Flynn,Jessie Boo 14h24'19"51d1'
Flynn,John Michael Cet 3h10'25"2d3'
Flynn,Joseph Carl Mon 7h48'1"-7d19'
Flynn,Keith Per 1h56'32"56d46'
Flynn,Kristan Lac 22h1'30"40d59'
Flynn,Logan Aur 5h53'37"30d15'
Flynn,Margaret E Cas 23h21'49"60d20'
Flynn,Margaret Rose Lyn 7h32'50"45d8'
Flynn,Michael Francis Lyn 7h48'31"36d39'
Flynn,Michael John Hya 9h13'33"0d26'
Flynn,Michael T Cep 22h0'59"54d6'
Flynn,Ranee Lac 22h36'21"38d41'
Flynn,Robert Cnv 12h23'31"34d22'
Flynn,Sean Cas 1h0'20"55d10'
Flynn,The Eternal Light of Lara Aql 19h57'0"8d44'
Flynn,Tom Sct 18h56'0"-5d49'
Flynn,Tom Aql 19h57'40"14d25'
Flynn,Toni Marie Jayne Cyg 21h44'27"31d40'
Flynn,William Michael Cam 5h5'25"70d34'
Flynt,Chester "Pop-Pops" Per 2h55'21"34d54'
Flynt,Diana King & Jim Flynt Aql 19h26'0"-8d35'
Fobes II,William W Hya 8h58'39"4d60'
Foca,Dakota Lyn 7h28'29"35d46'
Focareta,Mary L S Vul 19h46'45"20d21'
Focke,Hans-Heinrich And 2h23'44"45d40'
Fodar Forever,Rick Lmi 10h53'42"32d4'
Foderaro,Leeann Aur 4h40'1"33d50'
Fodermayer,Hugo Dra 20h18'47"62d8'
Fodor,Christopher A Dra 18h34'50"70d31'
Fodor,Patricia Cairns Lyr 18h46'0"38d22'
Fodrey,Ted V Aur 6h32'36"37d54'
Foeller,Kurt Sge 19h9'54"18d60'
Foeller,Melanie Ann Lyr 19h20'39"38d43'
Foerch,Richard H Dra 19h35'31"68d32'
Foerst,Jr,George J Ser 15h52'40"21d5'
Foerst,Scott Aql 19h58'28"12d49'
Foerster,Michael Dra 19h0'41"40d36'
Foerster,Zeke Ras Her 18h0'45"37d55'
Fofi Lmi 10h2'30"28d38'
Fogarty,Bob & Terrie Crt 11h14'17"-11d52'
Fogarty,Christina Marie Umi 16h0'0"70d34'
Fogarty,John Cap 21h25'41"-14d22'
Foged,Ivar Ori 5h28'58"-4d42'
Fogel,"Guy R MD-Physician,Magician!!!" Per 3h3'32"46d13'
Fogel,Alexa Cas 2h36'22"57d40'
Fogelberg,Dan Vir 13h33'23"5d29'
Fogelman,Burton D Per 3h12'0"48d10'
Fogerty & Klein Lmi 9h56'0"37d58'
Fogg,Jack Per 1h48'12"53d1'
Fogg,The CHH Superstar Jay Aql 19h32'44"-6d49'
Fogle,Dan "Joker" Ori 5h53'23"-1d35'
Foglesong,Erin Lane Sgr 18h1'0"-28d44'
Foglesong,Mary Jo Peg 22h16'30"34d22'
Foglia's Star Ori 5h56'34"20d32'
Foglio,Marian Cyg 20h54'19"31d12'
Fohler,Andrea Tau 5h50'11"27d30'
Foicik,Alain Sex 10h16'17"-2d36'
Foiles,Bryan Matthew Ser 15h10'0"2d16'
Foissey,André Ori 6h4'12"8d43'
Foister,Randall Allen Her 17h55'34"14d52'
Foisy,Maurice Joseph Jean Baptiste Her 16h39'56"35d47'
Foizey,Edna And 23h11'30"35d7'
Fojan,Jennifer A Com 12h24'45"24d58'
Foka,Sevasti Cyg 21h57'18"53d55'
Folan,Caitlan Ross Cyg 19h31'45"34d37'
Folch Cep 22h23'13"63d49'
Folden,Katie Cas 23h33'14"61d59'
Foldvary,Alex Virgil Boo 15h4'43"28d47'
Foley "Chicken", Pamela J Peg 21h43'38"21d46'
Foley III,Raymond J Cap 20h39'33"-16d50'
Foley's Star Per 3h22'33"41d6'
Foley,Anna & Truman Cyg 20h15'41"30d51'
Foley,Arlene And 2h25'59"50d11'
Foley,Audra Elizabeth Per 2h54'36"40d41'
Foley,Brennan Michael Cyg 19h29'14"33d7'
Foley,Bridget Peg 23h1'52"17d56'
Foley,Bridget Lyr 19h19'58"40d9'
Foley,Christin Michele And 23h28'45"43d18'
Foley,Dr James B Aur 5h20'52"37d54'
Foley,Elissa Kathleen Sct 18h54'38"-6d48'
Foley,Emma/Damien Cyg 20h37'50"45d31'
Foley,Frank J Aql 19h59'46"14d42'
Foley,John Michael Lac 22h54'18"54d33'
Foley,Joyce Lyn 8h15'34"40d54'
Foley,Jr,J Mark Her 18h11'58"30d43'
Foley,Keenan Patrick Cam 6h39'22"78d52'
Foley,Luke Ulvestad Per 1h57'47"50d12'
Foley,Margaret V Mon 8h6'22"-8d45'
Foley,Marie Remmes Dra 19h31'41"65d24'
Foley,Martin J Dra 19h34'59"60d14'
Foley,Paulina Cas 0h29'36"58d9'
Foley,Roy Leonard Dra 14h32'39"64d37'
Foley,Sandy Krober And 1h58'0"39d45'
Foley,Stacy Alain Cyg 19h46'59"38d35'
Foley,Thomas & Julie Werner Her 17h5'33"38d0'
Foley,Treha Leo 10h55'18"-5^51'
Folgelberg,Dan & Anastasia Cyg 19h26'38"31d37'
Folger,Nicole And 0h49'44"41d3'
Folger,Richard N (Fat Boy) Boo 14h8'31"31d17'
Folger,Sarah Anne Eri 2h52'1"-2d1'
Folies Del 20h13'19"15d16'
Folino,Deanna Com 13h6'20"26d56'
Folino,Mary Andrea Lyn 7h56'48"58d55'
Folino,Paola For 2h20'28"-27d42'
Folkers,Irma Aur 7h20'53"38d48'
Folkman,Zachary Simon Uma 13h18'53"63d20'
Folks,Jeanne C Uma 9h24'20"52d50'
Folle,Nikki Eri 3h41'47"-2d17'
Follett & Sawyer,Neil Alfred Uma 9h43'54"46d34'
Follis,Jay A Cep 21h37'40"58d28'
Follmann Ori 5h57'48"19d51'
Follmer,Clive Peg 23h5'29"20d26'
Folmer,Lottie Cet 0h26'58"-1d47'
Folschette,Tod Her 16h24'50"23d49'
Folsom,Stefani Cam 6h10'34"60d49'
Folstein,Kenneth & Dorianne Crb 16h21'54"32d29'
Folstein,Robert & Carole Joyce Cyg 19h29'20"34d12'
Foltin,Charles Edward Aql 19h30'41"10d60'
Foltz,James Dean Dra 15h10'42"55d53'
Foltz,Warren Lewis Her 16h15'38"23d36'
Folweiler,Mel Cam 3h41'0"60d52'
Folz,Pamela Jean And 1h44'57"39d51'
Fonberg (Bibi), Mitchell Cet 2h17'45"5d48'
Fonchain,Eric Com 13h12'41"21d11'
Foncu Cep 22h10'31"60d52'
Fond,Karen J Ori 5h54'12"14d25'
Fond,Lennart Cam 4h7'1"58d14'
Fonda V Aql 19h1'59"16d22'
Fonda,Philomena Cam 7h5'51"83d58'
Fondaw,Rickey Carl Ser 15h22'30"-1d22'
Fondessy,Rebecca Angel Umi 14h59'22"66d35'
Fondessy,Roland Aufait Ori 5h47'32"18d49'
Fondren,Scott Aql 18h58'37"10d12'
Fondren,Terry Lynn Peg 23h39'1"28d29'
Foner,Mandel & Kathy Umi 13h40'1"78d39'
Fong,Angela Peg 22h14'30"3d18'
Fong,Angela Siu Cyg 19h48'0"38d15'
Fong,Eilina & Benson Dra 16h40'12"71d7'
Fong,Jeffery "Jeff" Sex 10h12'0"5d42'
Fong,Joana Vul 19h35'31"27d10'
Fong,Josephine Siao Fong And 0h7'1"38d6'
Fong,Larissa Jane Hya 10h48'57"-12d6'
Fonjallaz,Rita & Andre Perruchoud Crb 16h10'22"38d2'
Fonk,Günter Lyr 19h20'1"31d41'
Fonseca,Gabriel Javier Cas 2h23'58"67d41'
Fonseca,Jazmin Araiza Psc 0h2'17"-2d8'
Fontain,Pat Cyg 19h26'19"31d37'
Fontaine,Adrianne Umi 16h5'49"73d39'
Fontaine,Amanda Lyn 7h8'17"51d31'
Fontaine,Duke Caribou Lyn 8h24'25"34d34'
Fontaine,Elena Lyr 19h11'42"40d39'
Fontaine,Guy Cep 22h13'0"60d25'
Fontaine,Marianne Eri 4h53'28"-7d22'
Fontaine,Pierre Per 3h19'22"43d25'
Fontana's Star And 23h20'52"44d35'
Fontana,Beatrice Cas 2h7'0"68d11'
Fontana,Deborah Lyn 7h31'58"38d35'
Fontana,Rosaria Cam 6h30'12"68d3'
Fontanella,Francesco Dra 17h53'48"64d32'
Fontang,Cyril Cmi 7h7'23"2d49'
Fonte Lyr 18h15'41"37d56'
Fonte II Eri 4h5'11"-19d30'
Fontello,Danielle Cas 1h20'36"63d49'
Fontes,Patricia J Cam 4h11'53"68d15'
Fontinelli,Louis Aur 6h25'1"31d35'
Fonzi,Angelo A Boo 15h0'16"30d41'
Foo,Jan And 1h18'17"38d31'
Foo,Nancie Aur 5h36'49"38d53'
Foo,Paula Uma 8h30'26"58d24'
Foodman,Kim & Jason Mon 6h32'31"-1d23'
Fool For A Lifetime Ori 5h43'0"11d36'
Foolish Pleasure Peg 23h1'14"32d46'
Foord,Miriam K Cas 0h3'1"65d5'

Foos,Harley Ann
Uma 10h50'46"48d9'
Foote,David K
Aql 19h30'29"7d44'
Foote,Erin & Andrew
Gem 5h58'12"26d22'
Foote,Janie Belle
Mon 7h58'10"-6d26'
Foote,Libbie
Cas 1h59'33"70d45'
Foote,Marie
Lac 22h40'35"55d58'
Foote,Seneca Wilber
Aql 19h15'33"19d46'
Foote,William
Her 17h19'33"22d12'
For A Special Dad
Per 3h17'40"41d28'
For All Eternity Carlos
Dra 17h35'48"64d8'
For All Eternity
Lyr 19h0'1"38d12'
For All Eternity (Shawn and Todd)
Cyg 21h51'10"42d22'
For All Eternity,Lee
Uma 8h38'35"50d27'
For Always Ken
Sct 18h48'27"-7d34'
For Angela in Love
Gem 7h11'52"24d26'
For Axel in Memory of Mauritius
Peg 23h16'59"31d15'
For Bear Love Princess
Cas 1h16'44"65d28'
For Claire From Lorena
Cas 2h34'57"65d17'
For DJC:2 BNYZ ONASTR
Dra 14h42'23"65d12'
For Dominica, Friendship For Life
Vul 20h19'41"22d48'
For Elaine All My Love Always Graham
Cyg 19h27'0"35d36'
For Emily
Cas 2h3'29"68d5'
For Eternity 1103-26
Cep 2h3'22"77d19'
For Ever
Lyr 19h22'44"30d24'
For Ever & Ever Amen
Uma 11h54'35"31d13'
For Evermore, Gail
And 1h25'24"34d16'
For Iris With Love
Cap 21h3'17"-26d55'
For Jerry,Always Amy
Boo 15h16'59"53d55'
For Kate,My Quiet Star
Cyg 19h28'24"38d37'
For Larry All My Love Estelle
Aur 5h23'1"38d9'
For Michael
Boo 15h17'59"38d18'
For My Lovey
Peg 0h10'41"13d19
For My Loving Mother, Tillie
Cas 0h43'55"67d16'
For Now,Forever, Rick & Susan
Cyg 19h54'0"50d2'
For Our Dad
Gem 8h5'15"33d12'
For Puma 1993
Mon 7h31'0"-8d48'
For Ralph My Everlasting Love
Ari 1h46'28"25d6'
For Real Not For Pretend
Peg 21h57'12"24d21'
For Special Wishes Now And Always
Uma 10h45'42"47d37'
For The All Time Record JD
Boo 14h0'29"20d56'
For The Love Of Joanne
Peg 23h15'50"31d55'
For The Love of Mike
Boo 15h17'19"51d54'
For The Love Of Molly
Cas 2h8'27"59d8'
For The Love of Terry
Aur 6h15'21"37d46'
For The Phillipsburg Sheet Gang
Lyn 7h24'1"44d49'
For Us
Vul 20h16'25"23d11'
For You And Me
Cyg 19h31'55"38d44'
For Your Daughters
And 1h37'20"35d35'
For your Eyes Only Shawn
Boo 14h49'18"34d57'
Foraker-Smith,Greg & Randi
Lac 22h8'56"51d4'
Foran,Celeste
Peg 21h27'40"22d59'
Foran,Michael F
Uma 10h51'18"40d24'
Foran,Sue
Lyn 7h4'0"59d16'
Foras,Stamati
Uma 8h48'24"57d17'
Forbell,Andy
Her 17h12'51"43d44'
Forbes,Alice Kay
Cet 3h12'51"1d27'
Forbes,Anita Irene
And 2h32'47"42d34'
Forbes,Anna
Lmi 9h23'15"38d8'
Forbes,Barbara Ann
Lyr 19h4'29"38d29'
Forbes,Elaine H
And 23h31'51"45d7'
Forbes,G James
Cnv 12h45'16"40d46'
Forbes,Glenn James
Lac 22h36'44"40d3'
Forbes,James Dean
Sct 18h54'59"-9d21'
Forbes,Kathy Lee
Peg 22h43'23"26d2'
Forbes,Keith
Boo 15h19'0"40d47'
Forbes,Mike
Dra 19h56'27"65d14'
Forbes,Rosie & Scott
Dra 15h39'15"64d32'
Forbes,Steve
Cep 0h0'31"69d56'
Forbes,Steve
Her 16h9'39"48d37'
Forbes,The Nightrider Jeffrey Ray
Aql 19h58'12"1d22'
Forbes,Thomas Edward
Ori 6h6'48"-0d3'
Forbes,Walter
Cep 4h31'29"80d8'
Forbeth
Peg 22h21'37"21d32'
Forbiddance
Cet 3h15'20"3d27'
Forbidden
Aur 6h7'52"38d54'
Forbis,Richard
Eri 4h9'49"-0d20'
Forburger,Britney Ann
Tau 4h33'1"15d21'
Forbush,Red
Uma 11h57'49"46d14'
Forbuss,Terry L
Mon 7h4'17"3d55'
Force,Edna Adams
Equ 20h57'32"7d10'
Force-/G
Mon 6h32'45"11d4'
Forcella,Phillip E
Eri 2h57'19"-10d59'
Forcellati,Mario Angelo
Per 2h26'29"58d27'
Forchner-Hoffmann, Elisabeth
Cap 20h22'53"-26d34'
Forcier,Martin
Cep 22h33'21"63d34'
Forcillo,Katy
Cas 0h54'28"74d59'
Ford 50th Birthday, Mary Anne Braun
Ori 4h55'42"4d32'
Ford City High School
Dra 16h53'14"62d7'
Ford's Star,Jason
Cep 21h45'58"60d9'
Ford,Ann & Bob
Cam 13h10'36"81d31'
Ford,Avi
Peg 22h48'0"20d27'
Ford,Barbara
Mon 7h1'34"-1d50'
Ford,Bernadette
Cas 0h6'11"58d39'
Ford,Christy
Peg 22h59'0"33d4'
Ford,Cody Lee
Peg 22h58'0"20d24'
Ford,Deborah Jayne- "Slice"
Mon 6h22'35"-0d14'
Ford,Dr Elinor Rita
Lyn 8h55'59"41d53'
Ford,Evelyn M
And 1h12'54"35d43'
Ford,Flora
Vir 11h52'0"2d18'
Ford,Gary William
Her 17h17'54"20d40'
Ford,Gloria
And 0h0'23"40d32'
Ford,Haleigh Renee
Cet 1h8'0"-1d41'
Ford,Heather
Mon 6h51'39"-1d50'
Ford,Helen Louise
Cmi 7h26'15"1d43'
Ford,Henry W
Aql 19h31'28"0d6'
Ford,Hywel Stuart
Boo 15h34'0"47d39'
Ford,Ian A
Aur 7h13'24"40d43'
Ford,James V
Aur 4h56'39"41d10'
Ford,Julie Ann
Mon 6h32'1"8d30'
Ford,Laura
Crb 16h10'58"38d23'
Ford,Lou
And 1h6'0"37d50'
Ford,Marilyn J
Tau 4h7'29"21d30'
Ford,Matthew Dillon
Ori 5h2'41"8d10'
Ford,Mrs Mary Blanche Allendar
Cas 1h54'1"73d56'
Ford,Parker Gurney
Per 1h42'0"53d41'
Ford,Patricia
Cas 0h2'53"61d39'
Ford,Peter
Ori 5h55'37"19d1'
Ford,Robert P
Lac 22h25'44"50d22'
Ford,Shawn Christian
Per 2h3'13"57d14'
Ford,Sherman
Aql 19h45'52"10d36'
Ford,Skylaar Daylan
Leo 10h44'12"15d56'
Ford,Stephen Russell
Cet 2h51'46"3d13'
Ford,Susan Michelle
Peg 23h37'58"28d3'
Ford,Tomas Jarrom
Dra 11h25'40"71d14'
Ford,Trevor Christopher
Ori 5h52'54"19d13'
Ford,William David
Aur 5h54'56"31d2'
Ford-Helzer,Jessica Rae
Peg 21h30'22"20d20'
Forde,Josephine Patricia
Lyr 18h35'50"26d2'
Fordham,Jinny K
Cas 0h22'58"61d4'
Fordham,John
Dra 17h40'24"70d21'
Fordham,Martin Robert
Cyg 21h2'1"39d23'
Fordowski,Ron
Lac 22h21'30"40d13'
Fore,Richard D
Equ 21h5'15"2d22'
Fore,Sandi
Peg 22h58'35"34d33'
Forehand,Doye Dwayne
Hya 8h16'56"5d35'
Forelli,Sharon Elizabeth Briggs
Cas 0h36'31"63d60'
Foreman For Eternity, Lyle Dean
Uma 10h37'19"50d43'
Foreman III,Taylor W
Eri 3h50'54"-2d23'
Foreman's 7th Period Class of 1997,Mr
Aur 6h12'31"31d34'
Foreman,Michael E
Ori 5h57'16"16d31'
Foreman,Olivia Mae
Cas 2h57'0"61d22'
Foreman,Tyler Joseph
Ari 1h59'14"18d57'
Foreman,Virginia L - Hugh M Thompson
Cam 3h18'38"61d0'
Foreman,Wesley Daryl
Aur 6h54'49"37d59'
Foren,Keith Michael
Aqr 21h37'0"-1d34'
Forese,Margaret Emily Mary Muller
Lyn 8h2'57"38d35'
Forest Wedding Star, Ron & Karen
Cyg 21h8'30"40d16'
Forest,Camille
Aur 6h23'12"33d53'
Forest,Derek Richard
Sex 9h41'32"0d53'
Forest,Mikaela Francesca
Sex 10h20'22"-6d58'
Forestelle,John Q
Eri 4h14'29"-10d22'
Forester,Brian
Aur 6h55'56"37d16'
Forester,Margaret Burrus
Cma 7h1'0"-13d13'
Forestieri,Tess Ann
Lmi 10h12'31"37d56'
Foretich,Frank
Hya 8h16'58"3d32'
Foretich,Michael & Brenton
Ari 1h47'21"16d56'
Foreve Joe & Joni
Cyg 20h54'22"31d52'
Forever
Vul 19h48'46"28d46'
Forever
Aur 7h9'24"43d32'
Forever
Cam 3h52'13"53d57'
Forever
Tri 1h52'53"28d53'
Forever
Sge 19h55'1"19d30'
Forever
Cam 13h20'57"84d58'
Forever
Aql 18h54'22"8d12'
Forever
Lyn 7h39'10"41d7'
Forever
Sge 20h2'44"20d40'
Forever
Cyg 19h21'26"27d50'
Forever
Lyr 18h55'11"42d34'
Forever
Cep 22h39'37"61d30'
Forever
Cyg 21h14'51"38d44'
Forever
Cas 0h23'0"63d50'
Forever
Cyg 19h19'54"30d7'
Forever
Uma 14h1'21"52d30'
Forever
Peg 22h3'45"25d16'
Forever
Aql 18h58'0"17d33'
Forever
Lyn 8h13'35"39d27'
Forever
Aql 18h44'33"10d42'
Forever
Cet 0h59'27"-5d39'
Forever
Cyg 19h24'47"34d18'
Forever "Cyndi Puddin"
Del 20h54'53"3d27'
Forever & A Day
Dra 17h14'21"61d35'
Forever & A Day
Cas 3h10'48"60d26'
Forever & a Day
Per 3h6'44"40d32'
Forever & Always Shannon
Lmi 10h21'42"31d34'
Forever & Always
Cyg 20h20'44"39d47'
Forever & Ever Amen
Ori 5h58'56"1d23'
Forever & Ever Ken's
Ser 16h15'58"2d55'
Forever 21
Uma 9h2'0"70d47'
Forever a Rainbow
Cas 1h43'1"76d52'
Forever Aaron
Hya 8h55'0"0d10'
Forever Adam
Per 3h11'19"50d1'
Forever Aimee
And 23h20'42"46d4'
Forever Al
Uma 8h50'23"68d37'
Forever Alan
Her 17h30'12"37d55'
Forever Alan
Uma 10h15'41"41d5'
Forever Alan & Debbie
Com 12h17'1"27d46'
Forever Alba & Donald
Cyg 19h25'56"50d28'
Forever Alberto
Lmi 10h38'32"38d0'
Forever Alex
Mon 6h54'22"8d36'
Forever Alex
Sco 16h33'34"-41d4'
Forever Alfred
Lmi 9h25'0"38d24'
Forever Allison
Mon 6h44'1"3d39'
Forever Allyson & Steve
Lyr 19h22'35"30d12'
Forever Alyce
Lib 15h59'0"-18d54'
Forever Amy
Cyg 21h54'23"40d55'
Forever Amy
Hya 9h5'1"-6d24'
Forever Amy
And 0h16'25"38d24'
Forever Amy Sue
And 0h9'31"38d29'
Forever and Eternity Rich
Dra 18h4'1"67d44'
Forever Andrew
Cru 12h25'31"-62d40'
Forever Andrew
Tri 1h54'19"26d27'
Forever Andy
Cep 20h40'57"76d24'
Forever Andy
Aur 5h0'38"50d25'
Forever Anita
Lib 15h43'50"-20d19'
Forever Ann-Minori
Cyg 19h31'33"32d46'
Forever Anthony
Lib 15h30'33"-10d3'
Forever Anthony
Boo 15h9'13"38d2'
Forever Aplin
Cyg 20h23'59"38d26'
Forever Arden's
Cam 4h18'1"68d60'
Forever Arleen & Elmer
Ori 5h57'35"15d57'
Forever Arlene
Cas 2h31'1"76d55'
Forever Arlene
Cas 0h25'30"75d59'
Forever Armand Frederick
Cep 22h16'23"80d17'
Forever Arnie's
Cet 0h37'58"1d16'
Forever Arstar
Cnv 12h50'23"41d7'
Forever As One
Sge 19h0'55"19d21'
Forever Ashley
Ari 1h54'52"15d12'
Forever Ashley
Boo 15h45'31"42d37'A
Forever B B Bears
Mon 6h31'11"-0d46'
Forever B J's Sparkle
Peg 23h4'20"11d59'
Forever Babe
Aur 5h53'39"48d53'
Forever Barbara
Uma 10h36'14"50d6'
Forever Barry
Del 20h53'30"2d3'
Forever Barry
Per 3h14'33"50d24'
Forever Bayne
Cyg 21h54'21"38d50'
Forever Bernadette
And 1h54'38"40d27'
Forever Beth
Peg 23h26'45"15d23'
Forever Beverly
Ori 5h56'24"13d48'
Forever Beverly Frances
Lac 22h37'15"48d54'
Forever Bill
Peg 23h45'35"31d55'
Forever Bill
Ari 2h59'0"28d20'
Forever Bill
Her 17h57'46"20d13'
Forever Bill
Del 20h24'23"8d44'
Forever Billy
Dra 19h42'39"61d14'
Forever Billy-Bob
Mon 8h8'17"-7d29'
Forever Bob
Ori 5h3'24"-0d8'
Forever Bob
Her 15h49'57"42d12'
Forever Bob
Her 17h50'30"40d48'
Forever Bob
Her 17h10'17"42d51'
Forever Bob & Marcy
Crb 15h22'42"32d47'
Forever Bobby
Aur 7h0'25"40d2'
Forever Bobby
Boo 14h37'35"50d29'
Forever Bobby NG POE #70
Lyr 18h21'1"45d36'
Forever Bond
Cyg 19h58'59"48d57'
Forever Boy & Girl
Aur 5h57'58"31d43'
Forever Brad
Mon 8h5'56"-3d39'
Forever Brady
Hya 8h56'35"-6d16'
Forever Brand
Boo 14h13'26"46d48'
Forever Brenda L
Com 12h15'43"20d41'
Forever Brent
Her 16h11'0"48d19'
Forever Brian
Per 1h48'48"56d42'
Forever Brian
Del 20h14'34"14d24'
Forever Brian
Cep 20h40'31"73d8'
Forever Brian Love Always Barbara
Boo 14h56'12"32d14'
Forever Bryan
Del 20h17'22"14d20'
Forever Bryan
Her 16h55'58"48d1'
Forever Bryce
Aql 19h0'0"-10d36'
Forever Bubalalli
Aql 19h25'39"0d51'
Forever Buddy
Per 3h38'26"38d11'B
Forever Bugsy
Per 1h43'41"52d31'
Forever Butch
Ser 15h59'31"-0d38'
Forever Byron
Her 18h9'24"50d2'
Forever C C
Oph 16h56'33"-6d8'
Forever Camp Daniel
Mon 6h12'17"-10d31'
Forever Carol
Ori 5h59'0"7d55'
Forever Casey
Del 20h54'44"7d47'
Forever CC
Uma 11h51'19"30d11'
Forever Chandi
Del 20h55'28"6d40'
Forever Chantel
Cyg 21h30'40"33d52'
Forever Charles
Del 20h23'15"10d28'
Forever Charles Pierre
Gem 6h26'49"12d10'
Forever Charlie
Cep 5h6'0"80d15'
Forever Charlie
Dra 19h26'59"70d41'
Forever Charlie O'
Hya 9h39'38"-8d33'
Forever Cheri
Cmi 7h55'29"8d11'
Forever Chris
Boo 15h20'45"38d30'
Forever Chris
Lyr 19h3'45"38d51'
Forever Chris
Oph 18h17'16"11d39'
Forever Chris
Cmi 8h0'1"5d59'
Forever Chris
Aql 19h57'17"8d36'
Forever Chris & Heather
Del 20h58'25"10d14'
Forever Chris & Moira
Cyg 20h16'1"30d43'
Forever Chris's Unit
Boo 15h5'11"21d25'
Forever Christine & Christopher
Del 20h48'55"9d33'
Forever Christine's
And 2h4'32"38d59'
Forever Christopher
Tri 1h55'35"28d38'
Forever Christopher
Peg 21h57'30"30d29'
Forever Christopher
Her 17h33'22"40d48'
Forever Chuck & Rosalie
Com 13h33'20"20d50'
Forever Cindie
Del 20h27'39"20d35'
Forever Cindy
Com 13h8'40"30d29'
Forever CJ
Her 17h34'1"38d2'
Forever Claire
Cyg 19h55'49"47d46'
Forever Clay
Aur 7h5'30"40d8'
Forever Corbin
Her 18h3'1"30d39'
Forever Corey
Lyn 8h25'10"33d51'
Forever Cowboy
Equ 21h21'0"10d18'
Forever Craig
Mon 8h1'55"-8d10'
Forever Craig
Oph 18h15'52"12d28'
Forever Cris
Lmi 10h40'52"26d16'
Forever Crystal(Booty)
Lyn 7h55'25"50d26'
Forever Cynthia
And 0h35'1"45d47'
Forever Daddy Shining, Bright
Lyn 8h55'53"46d7'
Forever Dala
Cas 0h59'60"75d24'
Forever Dale
Her 17h24'46"31d19'
Forever Dan
Per 2h53'0"50d4'
Forever Dan
Aqr 21h25'29"-0d38'
Forever Dan
Cap 20h22'54"-20d19'
Forever Dan & Anne
Cyg 19h50'18"50d14'
Forever Dana & Brad
Lyr 18h31'27"30d30'
Forever Daniel
Per 2h39'57"37d39'
Forever Dave
Per 4h2'54"37d36'
Forever Dave
Cep 20h16'41"75d31'
Forever David
Her 15h58'35"44d8'
Forever David
Peg 21h50'60"28d23'
Forever David
Ser 15h54'3"15d26'
Forever David & Kristan
Mon 7h11'37"-7d8'
Forever David & Ewa
Cyg 20h1'0"30d3'
Forever Dean
Pho 32h48'42"43d51'
Forever Deana
Aqr 21h0'45"0d19'
Forever Debbie
Lyn 8h2'42"40d18'
Forever Debbie (A/K/A O C R)
Cnc 8h35'17"7d5'
Forever Debbie's Star
Mon 7h2'34"-5d13'
Forever Deborah
Cas 23h27'40"58d24'
Forever Dennis
Her 18h11'21"45d45'
Forever Devina
Eri 4h27'50"-18d1'
Forever Diane
Cas 3h1'16"58d22'
Forever Diane & Tom
Umi 13h13'42"71d26'
Forever Dianne
Tau 4h1'29"11d45'
Forever Dianne
Cyg 21h3'56"36d44'
Forever Dineen
Mon 6h27'11"-10d7'
Forever Doc & El
Uma 9h52'53"71d44'
Forever Doc Bob
Per 1h31'24"53d60'
Forever Don
Lac 22h26'21"38d32'
Forever Donnie
Ori 5h56'23"15d43'
Forever Dooney
Dra 15h46'10"58d27'
Forever Doug
Lmi 10h38'49"23d26'
Forever Doug
Cep 21h47'14"55d54"
Forever Duane
Uma 11h20'1"40d26'
Forever DX
Uma 11h22'0"42d21'
Forever Ed
Cet 2h36'23"1d60'
Forever Ed
Her 17h38'42"40d15'
Forever Ed-My Only One
Cep 0h52'26"77d19'
Forever Eddie
Cep 22h13'0"61d25'
Forever Edward
Ari 1h58'41"18d18'
Forever Ela
Cru 12h53'26"-60d23'
Forever Eloy
Lyn 8h1'27"41d13'
Forever Eric
Aql 19h29'1"8d38'
Forever Eric
Hya 8h47'45"-5d60'
Forever Eric
Per 3h4'42"47d52'
Forever Ernest
Per 1h49'54"50d32'
Forever Evan
Com 12h58'58"30d55'
Forever FA
Boo 14h53'54"42d7'
Forever Faisal
Cap 21h8'36"-22d53'
Forever Father Francis
Cep 22h36'1"58d32'
Forever Fatima
Del 20h17'1"9d4'
Forever Faye & Ralph
Cnv 13h34'0"48d52'
Forever Felicia
Lyr 18h51'1"37d48'
Forever Five
Uma 12h14'48"60d7'
Forever Fleury
Peg 23h19'50"18d44'

Forever For Thea
Cas 2h29'0"68d30'
Forever Foster's
Crb 15h16'32"30d49'
Forever Franca
Lyr 18h27'52"31d35'
Forever Frank
Ari 3h23'27"30d22'
Forever Frank
Dra 16h20'20"61d13'
Forever Franko
Her 16h3'16"48d44'
Forever Friends
Lyn 9h6'33"42d28'
Forever Friends
Del 20h15'0"13d38'
Forever Friends
Cyg 21h31'1"33d56'
Forever Friends
Ori 5h30'35"0d39'
Forever Friends Justine
Cyg 20h1'1"40d4'
Forever Friends
Eri 3h34'39"-5d46'
Forever Friends
Vul 20h17'53"25d54'
Forever Gail
Lyr 19h4'45"28d8'
Forever Gary
Her 16h43'25'35d40'
Forever Gary
Hya 8h32'23"-1d27'
Forever Gary
Uma 11h48'23"42d17'
Forever Gene
Hya 8h34'30"6d38'
Forever Geoffrey
Aur 4h54'0"50d40'
Forever George
Lac 22h3'58"41d14'
Forever Giggles & Chuckles
Uma 9h47'23"57d53'
Forever Gina
Del 20h13'31"11d9'
Forever Ginger
Cas 0h32'26"63d25'
Forever Glen
Tri 1h52'0"27d4'
Forever GNJ
Oph 17h16'27"-20d24'
Forever Gordon-I Love You G
Dra 19h4'24"61d3'
Forever Greg
Sex 9h58'53"2d38'
Forever Greg,Our Shining Star
Sex 9h57'59"-6d0'
Forever Gregg
Por 2h10'0"56d21'
Forever Gregory
Sct 18h51'11"-8d50'
Forever Gus
Aql 19h0'34"7d39'
Forever Guy
Aql 20h20'12"0d24'
Forever Guy & Jenny
Uma 11h15'1"45d54'
Forever Hal
Cet 1h3'20"-3d49'
Forever Happy
Oph 18h39'36"8d50'
Forever Happy Birthday Carol
Cas 2h40'19"73d44'
Forever Harley
Cet 1h34'0"-0d22'
Forever Heather
Peg 23h27'43"23d40'
Forever Heather
Mon 7h1'14"8d51'
Forever Heidi
Com 13h19'52"26d58'
Forever Henry
Ori 5h24'0"-0d7'
Forever Heuer's
Ori 5h55'26"13d43'
Forever Hope
Eri 3h49'56"-5d30'
Forever Howard;Love, Your Guiding Star
Cep 22h35'1"80d29'
Forever Hugh
Boo 14h35'1"31d10'
Forever II, Brenda Cheryl
Her 17h30'56"38d36'
Forever In Heaven- Ray & Joleen
Vul 20h4'43"28d21'
Forever In Love Moon Star
Sge 19h31'1"16d42'
Forever In Love
Sge 18h58'24"19d48'
Forever In Love:David & Gladys
Peg 22h0'23"3d54'
Forever In My Heart
Per 1h58'38"50d12'
Forever In My Heart
Cyg 21h3'44"41d15'
Forever In My Heart
Per 2h58'42"55d29'
Forever In My Heart Michael
Uma 10h57'45"60d3'
Forever In My Heart
Del 20h19'60"20d26'
Forever In My Heart
Ori 5h54'1"9d39'
Forever In The Stars
Cas 0h11'44"63d54'
Forever Infies
Cyg 19h13'39"47d25'
Forever Israel
Boo 15h12'40"48d19'
Forever Ivan & Catherine
Lyr 18h59'1"36d23'
Forever J K S
Cep 21h52'46"68d16'
Forever J R D
Her 17h59'41"37d48'
Forever Jack
Boo 15h3'35"28d32'
Forever Jack
Per 3h4'22"40d18'
Forever Jack
Aur 5h19'36"43d18'
Forever Jack & Margaret
Uma 11h28'54"48d46'
Forever Jaimie
Cyg 21h23'38"52d49'
Forever Jair (JR)
Leo 10h17'43"10d56'
Forever James
Per 1h45'29"50d5'
Forever James
Per 3h25'48"52d51'
Forever Jan
And 2h34'34"40d41'
Forever Janet
Peg 22h0'34"34d33'
Forever Jason
Dra 14h35'18"62d42'
Forever Jason
Boo 14h27'54"32d16'
Forever Jason
Per 2h34'26"56d29'
Forever Jason & Veronica
Lmi 9h26'19"38d10'
Forever Jason & Dannielle
Lyr 19h7'45"37d42'
Forever Jay
Boo 15h12'34"38d25'
Forever Jay
Her 17h35'34"40d24'
Forever Jay
Aur 4h51'1"40d44'
Forever Jay
Her 16h54'24"34d43'
Forever JC
Sct 18h52'6"-9d18'
Forever Jean
Dra 14h52'24"62d8'
Forever Jefe
Dra 20h14'1"71d9'
Forever Jeff
Her 16h23'20"48d26'
Forever Jeff
Her 18h31'51"20d26'
Forever Jeff
Ori 5h56'53"13d45'
Forever Jeffrey
Per 2h51'18"31d22'
Forever Jen
Com 12h57'51"30d25'
Forever Jennifer
Cas 2h28'38"70d54'
Forever Jennifer
And 23h47'22"40d9'
Forever Jennifer & Gary
Cnv 12h28'13"50d0'
Forever Jill
Lyn 6h57'40"48d59'
Forever Jill
And 0h31'41"37d41'
Forever Jim
Cep 4h41'16"80d19'
Forever Jim
Ori 5h57'24"16d48'
Forever Jim
Peg 23h5'47"11d47'
Forever Jim
Sgr 18h56'42"-25d7'
Forever Jim
Her 16h20'48"40d28'
Forever Jim
Per 3h25'26"40d21'
Forever Jim
Her 16h59'26"28d19'
Forever Jim
Aur 4h56'46"40d42'
Forever Jim
Aur 6h14'0"32d50'
Forever Jim Bear
Aqr 23h4'48"-6d37'
Forever Jimmy
Per 2h53'0"40d34'
Forever Jimmy & Sandy
Hya 9h26'15"-2d23'
Forever Jo
Aql 19h3'33"-1d13'
Forever Joan R I
Lyr 19h24'41"38d12'
Forever Joe
Vul 19h23'0"23d23'
Forever Joe
Aur 7h4'1"43d50'
Forever Joe
Sge 19h30'7"17d58'
Forever Joe
Uma 11h54'21"43d17'
Forever Joe
Cnc 9h0'58"31d8'
Forever Joey
Dra 18h1'1"67d33'
Forever Joey
Hya 9h1'11"-6d40'
Forever John
Dra 19h27'38"70d55'
Forever John
Aql 19h59'44"-10d44'
Forever John
Ori 5h54'46"13d26'
Forever John
Vul 21h3'41"27d42'
Forever John
Lac 22h38'49"35d37'
Forever John
Per 2h37'1"50d36'
Forever John
Uma 8h55'60"50d27'
Forever John
Cep 20h43'27"75d38'
Forever John
Per 1h53'31"48d55'
Forever John & Sandy
Aql 20h21'1"3d9'
Forever Johnny
Per 2h57'18"40d47'
Forever Johnny & Jenny
Cyg 21h2'21"53d46'
Forever Jolie
Cyg 20h5'54"40d45'
Forever Jon-Paul
Hya 8h54'17"-6d60'
Forever Jonathan
Dra 11h59'50"72d17'
Forever Jonathan
Per 2h3'1"57d43'
Forever Joop
Uma 11h52'42"45d48'
Forever Joseph
Sex 9h44'36"3d24'
Forever Josh
Her 16h31'1"41d13'
Forever Joshua
Cru 12h47'58"-62d1'
Forever Judi
Cyg 21h35'18"34d48'
Forever Jules
Mon 7h19'4"-7d9'
Forever Julian
Cep 20h40'42"73d46'
Forever Julianna
Umi 16h44'45"75d60'
Forever Julie
Umi 13h56'18"71d7'
Forever Julie
Lyn 7h56'26"40d9'
Forever Justin
Uma 8h26'46"68d31'
Forever Kate
Lyr 18h57'45"30d53'
Forever Katie
Aqr 20h37'21"-0d7'
Forever Kay & George
Peg 22h0'1"30d39'
Forever Keith
Aql 19h58'31"-1d33'
Forever Keith
Hya 8h11'36"5d59'
Forever Kelli & Kayla
Mon 7h4'27"0d41'
Forever Ken
Her 15h56'43"48d6'
Forever Ken
Gem 7h8'46"28d42'
Forever Ken
Cep 5h15'38"80d17'
Forever Ken & Celeste
Cyg 20h35'57"40d31'
Forever Kenny
Lmi 10h35'56"31d0'
Forever Kenny & Annette
Dra 19h50'27"61d26'
Forever Kerry
Vul 19h48'48"23d21'
Forever Kim
Mon 6h57'24"8d46'
Forever Kim
Cma 7h15'29"-16d40'
Forever Kimberly
And 23h36'34"37d57'
Forever Kirsten
Ara 17h57'16"-51d49'
Forever Kris
And 23h33'33"38d8'
Forever Krista
Vul 19h47'42"27d50'
Forever Kristin
Ori 5h24'21"-6d54'
Forever Kristy
Peg 22h2'42"10d36'
Forever Kristy
Lib 15h39'0"-24d37'
Forever Kurt
Per 2h27'29"57d48'
Forever Lana
Del 20h16'44"10d19'
Forever Larry
Cap 20h34'25"-10d41'
Forever Larry & Sandi
Cyg 20h39'11"40d20'
Forever Larry Ray
Cep 23h3'38"60d18'
Forever Laura & Alan
Ori 5h56'12"8d45'
Forever Lauren And John
Boo 15h13'1"38d21'
Forever Lawrence
Boo 15h1'13"20d15'
Forever Layla
Lyr 18h15'32"30d4'
Forever Len
Uma 13h0'51"53d23'
Forever Lester
Hya 8h57'35"-5d41'
Forever Lewis
Aqr 20h56'13"-0d14'
Forever Linda
Her 17h39'1"40d4'
Forever Lisa
Ori 5h52'43"13d54'
Forever Lori & Scott
Crb 16h0'48"28d7'
Forever Lori's
Cam 4h3'1"61d27'
Forever Lorie
Peg 23h3'49"13d25'
Forever Lou
Per 2h53'23"40d26'
Forever Lou
Cyg 19h55'22"48d55'
Forever Louisa
Vir 13h52'43"-20d57'
Forever Louise
And 0h25'1"40d5'
Forever Louise
And 23h27'30"42d43'
Forever Love
Peg 22h0'11"32d28'
Forever Love
Cyg 19h36'12"28d47'
Forever Love
Uma 11h33'14"45d16'
Forever Love
Cyg 20h9'46"37d56'
Forever Love Chandra & Mark
Lyr 18h54'58"31d52'
Forever Love K T
Per 1h52'27"53d4'
Forever Love Santese n Ammee
Cyg 19h34'54"31d38'
Forever Love Shelby & Danielle
Cyg 19h34'30"30d7'
Forever Love To Hiromi
Cnc 8h48'17"13d11'
Forever Love,Adam & Nannette
Aql 20h11'14"14d40'
Forever Love-Jake & Kristine
Uma 8h7'49"68d60'
Forever Loved
Leo 10h31'56"11d12'
Forever Loved,R E D
Sge 19h56'34"18d48'
Forever Loving Lynette
Cyg 21h45'55"28d44'
Forever Lucille
Eri 4h5'17"-5d12'
Forever Lur
Eri 4h4'1"-8d27'
Forever Mad-Dog
Uma 11h21'37"38d27'
Forever Mandr
Uma 8h39'21"58d56'
Forever Marc
Del 20h23'1"8d19'
Forever Marc
Boo 14h33'50"48d37'
Forever Marc
Sct 18h44'15"-6d6'
Forever Marc
Peg 23h28'1"18d50'
Forever Marie
Lyn 8h52'16"43d22'
Forever Marjorie
Ori 5h58'45"16d12'
Forever Mark
Lac 22h15'60"49d51'
Forever Mark N S R F T S
Aur 5h6'40"40d31'
Forever Mark
Per 1h56'43"50d2'
Forever Mark
Del 20h32'11"11d4'
Forever Mark
Hya 8h10'59"-6d16'
Forever Mark & Michele
Uma 8h46'47"51d26'
Forever Martin
Cyg 21h1'18"31d2'
Forever Marty & Gracie
Hya 8h34'56"-6d35'
Forever Marty & Sheila
Peg 22h25'45"30d22'
Forever Marty 56104416
Aur 6h34'18"37d33'
Forever Mary Jane
And 1h39'57"37d36'
Forever Marylee
Aur 6h22'48"41d14'
Forever Matthew
Cep 22h0'30"60d53'
Forever Max
Lyr 19h0'40"41d6'
Forever Max
Dra 16h32'23"73d41'
Forever Meghan
Cas 23h29'19"61d9'
Forever Meghan & Stephanie
Lyr 18h41'16"28d43'
Forever Melissa
Peg 0h3'23'18d30'
Forever Mema & Pappy
Aur 4h56'1"40d9'
Forever Michael
Her 17h1'31"20d18'
Forever Michael
Dra 17h29'38"58d12'
Forever Michael
Per 2h55'20"50d16'
Forever Michael
Lac 22h31'19"40d37'
Forever Michael
Uma 8h58'12"50d37'
Forever Michael
Cep 21h1'54"80d16'
Forever Michael
Boo 15h19'55"51d14'
Forever Michael
Tri 2h24'52"30d55'
Forever Michael
Per 1h27'60"53d18'
Forever Michael
Dra 19h42'53"60d47'
Forever Michael
Ari 3h22'54"30d9'
Forever Michael
Her 17h5'1"48d7'
Forever Michael
Boo 15h6'28"31d9'
Forever Michael
Dra 16h26'51"69d19'
Forever Michael
Aur 7h23'28"35d35'
Forever Michael J
Her 15h47'55"45d41'
Forever Michael Ryan
Cep 22h0'43"68d45'
Forever Michel
Psc 1h22'27"22d5'
Forever Michele
Com 13h1'18"26d16'
Forever Michelle
Aql 19h8'27"4d0'
Forever Mike
Gem 7h2'1"28d46'
Forever Mike
Dra 15h51'53"51d31'
Forever Mike & Jane
Sge 19h53'42"16d11'
Forever Millie & Steve
Uma 9h14'0"47d36'
Forever Mine Scott Allan
Dra 18h29'26"50d18'
Forever MK 143
Uma 10h4'22"50d23'
Forever Mo
Per 2h31'20"57d34'
Forever Mock I
Aql 20h14'27"8d4'
Forever More Todd-30
Uma 10h22'58"41d51'
Forever Morris
Cep 22h24'1"80d16'
Forever My Chantel
Aql 19h56'34"13d44'
Forever My Friend Diana
Del 20h50'35"7d46'
Forever My Hero
Her 15h47'17"46d30'
Forever My Hero
Cep 21h40'24"56d9'
Forever My K - Millumino Dimmenso
Del 20h53'49"3d34'
Forever My Kevin
Ori 5h59'34"14d57'
Forever My Love
Cyg 21h3'57"36d19'
Forever My Love Rocco
Her 15h51'17"41d26'
Forever My Philip,With Love Eileen
Cnc 8h12'34"32d58'
Forever My Pookie
Uma 10h11'29"48d15'
Forever My Prince Thomas
Aur 6h7'48"37d60'
Forever My Ric
Her 16h26'50"20d12'
Forever My Robert
Aql 19h4'1"1d53'
Forever My Rosie
Lyr 19h21'0"33d51'
Forever My Shining Terri
Boo 15h1'0"38d26'
Forever Natalie
Eri 3h33'37"-6d36'
Forever Nathan
Aur 6h19'52"46d29'
Forever Nelson
Per 1h56'1"56d40'
Forever Nicholas
Peg 21h24'16"11d10'
Forever Nichole & Trevor
Mon 7h15'60"-5d27'
Forever Nick
Uma 13h2'0"54d18'
Forever Nicole
Cam 4h59'36"68d17'
Forever Nicole
Pho 23h39'53"46d39'
Forever Norman
Boo 14h32'45"21d16'
Forever Norman's Miracle
Ser 15h26'31"21d55'
Forever Oliverio
And 2h27'0"49d33'
Forever Oren
Cep 20h15'38"75d33'
Forever Otis
Cep 2h45'59"78d18'
Forever Ours
Ori 5h55'50"15d10'
Forever Ours
Ind 21h21'20"48d22'
Forever Ours
Ind 20h54'29"46d40'
Forever Pam
Mon 7h43'0"-8d55'
Forever Pamela's Dream
Ori 4h43'16"8d43'
Forever Parker
Cep 22h16'28"60d50'
Forever Parr
Per 2h38'55"40d7'
Forever Partick
Cep 22h16'1"68d26'
Forever Pat
Per 1h56'52"56d30'
Forever Pat
Peg 21h49'32"28d3'
Forever Pat
Uma 9h1'50"60d8'
Forever Patrick
Cyg 21h26'25"53d50'
Forever Patrish
Her 17h5'47"40d13'
Forever Patti
Lyr 18h33'1"37d44'
Forever Paul
Sex 9h59'32"2d1'
Forever Paul
Tau 4h16'56"22d2'
Forever Paul
Her 15h54'37"45d23'
Forever Paul
Her 17h25'38"37d31'
Forever Paula
Leo 10h37'29"18d31'
Forever Paula
Cep 22h14'29"55d50'
Forever Paula & Bud
Aql 18h52'45"11d10'
Forever Pauline
Hya 9h4'29"-1d29'
Forever Pete
Tau 4h10'18"20d45'
Forever Peter & Judy
Uma 9h4'27"62d29'
Forever Pixy Loves The Big Honey
Crb 16h12'37"38d2'
Forever Pookie
Cnv 12h8'36"40d23'
Forever Preet
Del 20h15'46"10d60'
Forever Preston
Leo 9h23'52"7d13'
Forever Princess Shona
And 0h19'15"45d31'
Forever Princess Your Smile Shines On
Cam 5h52'18"61d13'
Forever Puss
Lyn 7h28'52"40d34'
Forever R A W
Sct 18h42'47"-9d53'
Forever R Dale
Boo 15h28'36"41d5'
Forever Rachel
And 2h8'58"38d14'
Forever Ralph
Aur 5h29'22"48d48'
Forever Randy
Ori 5h37'13"-6d35'
Forever Randy
Aqr 21h54'32"-5d39'
Forever Randy & Kim
Peg 21h26'24"18d54'
Forever Ray & Deborah
Per 2h39'44"40d49'
Forever Raycrofts
Uma 11h43'31"40d5'
Forever Renee
Hya 8h55'1"-1d45'
Forever Renee
Uma 14h16'29"58d23'
Forever Rich
Sct 18h53'38"-7d56'
Forever Richard
Her 16h41'56"47d36'
Forever Richard
Lyn 9h2'0"41d51'
Forever Richard
Uma 8h31'23"62d19'
Forever Richie
Cep 23h5'20"80d12'
Forever Rick
Aur 7h25'54"40d6'
Forever Rick
Sct 18h50'38"-7d41'
Forever Rob
Per 1h57'11"53d11'
Forever Rob
Cep 20h22'30"75d51'
Forever Rob
Cet 3h5'0"3d26'
Forever Robby
Per 4h2'35"51d44'
Forever Robert
Lmi 9h57'13"33d43'
Forever Roger
Aql 20h19'1"8d14'
Forever Ron
Boo 13h53'17"21d45'
Forever Ron
Sex 9h48'24"-1d38'
Forever Ron
Per 2h12'0"56d58'
Forever Ron
Cep 2h24'51"77d14'
Forever Rosemary
Del 20h52'39"7d47'
Forever Roy
Cep 21h49'51"58d14'
Forever Roy
Oph 16h6'54"-6d1'
Forever Roy
Aql 19h0'44"7d49'
Forever Sal
Uma 11h26'0"40d15'
Forever Sam
Mon 6h10'44"-10d32'
Forever Samantha
Cas 0h29'10"63d34'
Forever Sandra
Pup 7h28'8"29d35'
Forever Sas
Her 17h23'38"40d40'
Forever Scott
Cnc 7h55'0"11d12'
Forever Scott
Vul 19h22'23"25d18'
Forever Scott
Per 1h53'46"56d30'
Forever Scott
Her 16h57'49"47d34'
Forever Scott
Her 17h36'47"25d44'
Forever Scott
Cam 3h13'34"63d33'
Forever Scott E
Lyn 7h58'15"35d25'
Forever Shawn
Uma 10h58'1"37d52'
Forever Sheila
Com 12h22'26"22d24'
Forever Shelley
Lyn 9h34'0"40d33'
Forever Sherry
Cet 2h4'27"6d0'
Forever Shining Carolyn
Cnc 8h26'16"7d36'
Forever Shining Sam
Cam 13h12'53"78d23'
Forever Shining,Arthur & Rosalind's Love
Sge 20h16'19"16d41'
Forever Shining,Renee
Lyn 6h27'54"59d26'
Forever Shmoo
Her 18h18'0"13d7'
Forever SMGLHF
Cam 10h43'1"82d16'
Forever Snoopy
Sct 18h48'41"-6d19'
Forever Sonny
Lyr 19h0'46"38d4'

Forever Sonny
Per 2h27'29"57d48'
Forever Sophia
Mon 7h53'55"-9d22'
Forever Sophia
Lyr 18h19'14"40d29'
Forever Soul Pirates
Mon 8h6'29"-3d50'
Forever Spencer
Com 13h0'33"20d58'
Forever Spike
Her 17h57'38"40d60'
Forever Stacey
Cas 0h34'0"62d16'
Forever Stefan & Sandy
Cyg 19h25'31"33d20'
Forever Stephanie
Crb 15h37'51"26d37'
Forever Stephen
Sct 18h53'27"-9d17'
Forever Stephen N' Lace
Ori 5h59'23"16d47'
Forever Steve
Boo 13h55'38"21d36'
Forever Steve
Ori 6h2'17"7d50'
Forever Steve
Sge 19h29'54"16h34'
Forever Steve & Stacey
Peg 22h59'24"18d23'
Forever Steve & Dana
Umi 16h12'52"79d37'
Forever Steven
Uma 12h2'51"48d38'
Forever Steven
Cep 20h36'48"75d39'
Forever Stevie
Cep 21h16'20"65d36'
Forever Stoo
Aur 5h8'23"42d19'
Forever Stormy
Lyn 7h48'50"51d4'
Forever Strong
Cyg 21h30'36"37d44'
Forever Sundance
Boo 13h56'1"21d58'
Forever Sunshine
Peg 23h2'12"20d45'
Forever Surfside
Crb 15h52'44"32d1'
Forever Suzanna
Com 12h22'60"28d19'
Forever Suzanne
Oph 16h55'18"-22d41'
Forever Sweet Tim
Dra 12h58'10"72d28'
Forever Sydney
Lac 21h59'1"40d35'
Forever Tami Hadley
Peg 22h10'25"29d3'
Forever Tanja
Eri 3h47'51"-2d9'
Forever Tara
And 2h35'0"40d34'
Forever Tara
And 1h18'30"40d2'
Forever Tash
Lyr 19h18'52"41d15
Forever Ted
Ser 18h7'18"-5d28'
Forever Teddy
Her 16h49'37"48d33'
Forever Teej
Dra 11h5'30"78d23'
Forever Terry
Her 15h48'24"41d52'
Forever Thelma & Fred
Boo 14h18'55"54d5'
Forever Thomas
Per 2h50'34"40d17'
Forever Tim
Per 2h56'0"43d28'
Forever Timmy's Girl
Vul 19h57'51"23d26'
Forever Tina
Cas 1h6'18"58d41'
Forever Todd
Lmi 10h40'13"28d18'
Forever Tokzic
Her 16h2'29"42d19'
Forever Tom
Her 16h55'48"47d39'
Forever Tom & Suzanne
Lyr 18h32'22"41d12'
Forever Tommy
Her 18h8'0"37d37'
Forever Tommy Starr
Sex 9h57'0"2d13'
Forever Tonight
Lyn 8h51'14"35d18'
Forever Tony
Per 2h55'32"40d44'
Forever Tony
Her 17h39'35"41d8'
Forever Tony & Melissa
Crb 15h18'41"30d45'
Forever Tony & Pauline
Vel 9h43'52"53d43'
Forever Topo
Com 13h14'0"28d50'
Forever Traci
Cas 1h5'23"71d15'
Forever Trey
Del 20h24'1"11d3'
Forever Trish & Rick
Cyg 19h23'0"54d52'
Forever Trudles
Boo 14h17'33"53d34'
Forever Truly Uly
Peg 22h29'15"10d22'
Forever Vince
Del 21h4'28"13d0'
Forever Vinny
Cep 22h49'1"80d13'
Forever Vinny
Dra 20h6'0"64d40'
Forever Waldo's
Per 2h0'11"50d6'
Forever Warwick
Per 2h58'1"40d24'
Forever Wayne
Cyg 20h32'30"40d60'
Forever Weezie
And 1h15'38"38d9'
Forever Wendy
And 23h31'49"37d53'
Forever Wes
Cnv 13h5'41"51d11'
Forever Wes
Ser 18h16'5"-5d18'
Forever Whipple
Mon 7h4'23"4d28'
Forever William
Hya 9h10'1"1d15'
Forever Winston
Lac 22h39'43"41d1'
Forever With My David
Aur 5h35'11"50d29'
Forever Young
Cep 2h44'33"77d29'
Forever Young
Boo 15h5'43"40d42
Forever Young
Del 20h54'54"3d42'
Forever Young With Love To Denny
Hya 8h42'1"-6d44'
Forever Your Princess
Crb 15h50'19"31d34'
Forever Yours
Uma 8h43'25"68d37'
Forever Yours
Ori 5h55'18"13d9'
Forever Yours
Cyg 20h32'54"30d57'
Forever Yours
Cam 7h59'37"61d33'
Forever Yours
Aql 19h25'35"-10d24'
Forever Yours
Lyr 18h28'42"31d23'
Forever Yours
Ori 5h32'12"1d3'
Forever Yours
Uma 8h47'47"57d50'
Forever Yours
Peg 22h2'56"33d1'
Forever Yours
Cyg 19h33'47"35d28'
Forever Yours Anita & Bob R
Mon 8h6'42"-2d18'
Forever Yours Chandra
And 1h9'23"38d22'
Forever Yours Gabe
Uma 9h35'58"48d2'
Forever Yours,Bean
Ser 18h15'0"-6d47'
Forever Yours,Tamara
Peg 22h36'29"8d3'
Forever Yu-Suk's
Cam 5h57'41"65d36'
Forever Zachary
Her 17h52'14"40d47'
Forever Zachary
Lac 22h36'26"41d4'
Forever Zig
Lac 22h12'57"37d60'
Forever,AnnMarie
Cyg 21h19'50"38d59'
Forever,Brett & Lisette
Cet 2h42'54"-12d10'
Forever,Don & Patricia In Love
Uma 11h0'37"38d32'
Forever,Joseph Lee Anthony
Her 17h39'1"40d3'
Forever,Louie
Cet 2h7'10"-6d55'
Forever-And A Day
Uma 9h49'28"67d55'
Forever-Prince Charming Lil' Angel
Sct 18h34'28"-5d52'
Forever-Scott & Kim
Dra 16h42'23"51d25'
Forever...
Eri 4h11'34"-18d6'
Foreverness
Vul 21h21'25"24d21'
Foreveron
Per 4h27'36"52d15'
Foreverstan
Ari 2h31'1"30d27'
Forgach,Diane M
Aql 19h53'31"15d46'
Forgatch,Amie
Cas 23h22'29"62d45'
Forgen,Jr "My Babe" Michael E
Dra 19h2'19"58d13'
Forget Me Not
Peg 23h3'53"33d48'
Forgey,Jim
Cet 0h52'30"0d26'
Forgione,Anthony
Per 3h3'1"40d4'
Forgiono,DC,Joseph G
Her 17h17'18"20d42'
Forgit,Philip Robert
Aur 6h32'23"33d3'
Forgotten Dreams
Ori 6h0'44"0d53'
Forgue,Ken (Taz)
Cep 21h34'53"55d32'
Forgue,Samantha Reneé
Tau 4h39'25"15d6'
Forker,Ruth A "Marr"
Mon 7h56'52"-8d6'
Forker-Collins,Jane Ruth
Sgr 20h0'34"-40d4'
Forkes,Ken
Lyn 8h46'34"39d59'
Forkin,Joseph
Lmi 10h0'37"38d19'
Forlani,Annamaria
Cas 23h21'20"53d0'
Forlenza
Gru 22h20'40"-50d41'
Forman,Don
Aur 7h5'21"35d50'
Forman,Frieda
Lyr 18h44'1"34d27'A
Forman,Martin
Boo 13h58'31"14d9'
Forman,Melanie K
And 2h1'18"38d25'
Formanek,Stephen & Dawn
Cyg 20h27'53"42d23'
Formell,Keith Jeffrey
Hya 8h40'23"-6d15'
Formenti
Cmi 7h56'0"7d35'
Formentin,Elena - Marco Avezzù
Scl 23h46'46"-26d35'
Formery,Erwin Pierre
Cnc 8h28'21"10d12'
Formes,Jr,John J
Aql 20h12'13"11d17'
Formica,Debra
Cas 3h13'0"70d3'
Formica,Peter C
Gem 6h50'9"12d3'
Formidoni,Claire
Del 20h17'54"10d21'
Formosus Azure Margee
Cam 4h57'60"68d19'
Formwalt,Rebekka Elise
And 0h47'50"45d58'
Fornalski,Stefan
Oph 18h17'53"12d49'
Forner,Sabine
Sco 16h39'18"-43d18'
Fornesori,Enzo Librano et Mario
Dra 9h40'1"80d31'
Forney,Evelyn Arquette
Crb 15h47'11"37d41'
Forney,John
Ori 6h9'15"7d22'
Forney,Joseph Kent
Crt 11h45'23"-11d28'
Forni,Suzanne
And 2h9'46"42d21'
Foroulis,Andreas
Lyn 7h53'17"51d21'
Forrer #1,Patricia F
And 23h35'0"49d44'
Forrest Rain
Ori 5h59'30"10d33'
Forrest Star,The
Aql 19h31'0"14d23'
Forrest Star,The
Mon 7h2'57"-5d38'
Forrest,Belle
Cnc 8h27'24"10d19'
Forrest,Carol
Peg 21h24'17"13d28'B
Forrest,Claudia
Cet 2h54'57"1d39'
Forrest,Faye Rae Madison
Cyg 21h5'58"33d40'
Forrest,James E
Aur 6h55'42"38d34'
Forrest,Lisa Michelle Gabrielle
Ori 5h56'27"7d33'
Forrest,Marshall W
Peg 21h24'17"13d28'A
Forrest,Zachary Lewis
Leo 10h47'12"7d23'
Forrester,Ella Victoria
Ori 5h57'27"9d28'
Forrester,Mary Sue Jarrett
Cyg 21h21'22"37d36'
Forrester,Neil Allan
Her 17h14'1"27d30'
Forrester-Smith,Inc
Cet 2h7'12"1d42'
Forristall,Amy Beth
Vul 19h2'39"22d28'
Forsberg,Claire Elliston
Mon 6h37'14"0d41'
Forsberg,Hailey Hannah
Mon 6h40'28"1d44'
Forsbergh,Lars-Erik
Cam 3h14'37"60d58'
Forsha,Jeff
Lac 22h25'55"50d30'
Forshee,Bee & Katy
Lyr 18h28'33"35d10'
Forsmith,Dawn Marie
Cma 6h54'50"-17d17'
Forst 1996,Michael Anthony
Cet 2h30'30"5d4'
Forst,1993,Sheri Lynn
And 0h54'50"21d44'
Forstell,Katelyn Jane
And 23h43'53"47d11'
Forster,Diana Marie
Cas 1h48'49"60d48'
Forster,Dr John
Oph 18h4'45"11d10'
Forster,Jamie Anne
Vul 20h0'27"22d49'
Forster,Jo
Lyr 18h15'1"43d12'
Forster,Peter H
Cyg 21h1'31"31d6'
Forster,Robert C
Aql 20h3'35"6d42'
Forster,Sue
Lyn 7h55'17"48d47'
Forster,Susan E
Cyg 21h1'46"33d7'
Forsyth V,Thomas G
Cmi 7h59'48"8d23'
Forsyth,Alastair Elliott
Cep 5h28'33"86d24'
Forsyth,David
Aur 6h0'57"46d16'
Forsyth,Russell E
Sgr 18h0'37'-28d31'
Forsyth,Stacey
Cas 0h55'34"74d45'
Forsyth,William James Anthony
Her 16h3'20"40d10'
Forsythe,Charles G
Per 1h45'22d52"55'
Forsythe,Janice
Peg 22h36'33"26d42'
Forsythe,John
Aqr 22h22'1"0d43'
Forsythe,Mark
Uma 12h7'33"61d20'
Forsythes,The
Lyr 18h49'38"41d32'
Fort,Leonard Wayne & Kathy Lynn Tangeri
Ori 5h30'26"-3d21'
Fort,William T
Her 18h3'1"28d59'
Forte,Angela Reneé
Cas 1h7'36"58d2'
Forte,Christian
Cet 1h6'29"1d35'
Forte,Dana
Cam 8h1'14"68d41'
Forte,Fabian
Cyg 20h35'32"40d19'
Forte,Harriet Jeanne
Cam 4h27'30"78d48'
Forte,Julie
Mon 7h43'38"-1d48'
Forte,Matthew Joseph
Cep 22h12'17"61d15'
Forte,Special Love-Lou & Cheryl
Crb 15h28'41"31d1'
Forte,Victoria Veronica
Cnv 12h32'17"51d23'
Fortenberry,Colonel Charles P
Lac 22h6'25"51d4'
Fortes,Agnes & John
Uma 8h56'43"52d6'
Fortes,Brian
Lac 22h45'23"55d26'
Fortes-my little love- Jumara Rocha
Aql 19h2'42"-1d46'
Fortess,Rebeka
Lyr 18h31'38"30d14'
Forth,Alice Jane
Aql 18h58'57"-6d50'
Forthill
Cet 1h34'1"-1d2'
Forthuber,Marcus John
Aqr 22h47'30"-2d21'
Fortier,Craig Joseph
Sex 10h47'30"-8d6'
Fortier,Eleanore B
Cyg 19h33'40"36d21'
Fortier,Joseph E & Evelyn W
Mon 8h6'31"6d32'
Fortier,Jr,Jeffrey William
Dra 16h44'36"60d41'
Fortier,Kim
And 2h23'40"41d1'
Fortier,Mark & Julia
Mon 8h4'46"-6d46'
Fortier,Rita
Ori 5h46'36"20d52'
Fortin,Armelle
Uma 8h41'29"67d33'
Fortin,Isabelle
Sco 16h34'53"-28d22'
Fortin,Larry & Milly
Lyn 7h35'18"41d19'
Fortin,Martine
Cas 22h57'30"54d21'
Fortino,Matthew Gregory
Cep 21h14'0"55d45'
Fortis Christoferens
Boo 14h35'0"51d0'
Fortkamp,Jean "Mom"
Lyn 7h1'34"50d42'
Fortner,Bill Joe
Aql 19h10'19"12d59'
Fortner,Courtney Ann
Cyg 20h38'49"44d57'
Fortner,Craig Adam
Cmi 7h58'57"7d38'
Fortner,Debora Scott
Peg 22h40'0"32d39'
Fortner,Kent & Laura
Cyg 21h0'54"30d52'
Fortner,Laurel Elizabeth
Peg 23h27'16"30d34'
Fortner,Nona
Lyr 19h7'57"37d43'
Fortner,Richard Greer
Cnv 12h36'0"34d25'
Fortner,Scott L
Lac 22h35'39"50d12'
Fortney 271979,Kristy Marie
Lyr 18h17'21"44d16'
Fortney,Princess Kimberly
Peg 22h4'18"25d40'
Fortson,Erian J
Lyr 18h45'10"32d32'
Fortuna
Cep 23h18'29"70d8'
Fortuna
Lep 4h56'28"-19d38'
Fortuna
Uma 9h51'48"54d24'
Fortunas
Sct 18h48'31"-9d24'
Fortunata
And 2h25'42"41d57'
Fortunate Noel
Peg 21h47'47"34d24'
Fortunate,Alan D
Aur 7h4'34"40d24'
Fortunate,Alan G
Uma 9h56'1"52d46'
Fortunate,Bridget G
And 23h20'0"43d14'
Fortunate,Derek A
Dra 16h47'35"67d58'
Fortunate,Gary R
Per 3h3'33"46d47'
Fortunate,Ray A
Lac 22h4'16"46d40'
Fortunate,Raymond A
Her 17h29'51"20d22'
Fortunate,Tania L
Vul 20h21'48"25d44'
Fortune's Star Diane
Lyr 19h1'24"37d43'
Fortune,Imogen Kate
Lyr 19h17'50"26d2'
Fortune,Iris
Cyg 20h1'19"37d35'
Fortune,Iris
Cyg 19h39'46"30d2'
Fortune-Stone,Marian
Lmi 10h54'1"27d42'
Forward,Dr Susan
Mon 7h1'1"1d23'
Fory,Jon Roger
Lac 22h45'15"55d51'
Forys,Kristin
And 23h13'17"37d6'
Forys,W Michael
Aql 20h2'0"0d9'
Foscaldi,Leah M
Del 20h56'22"14d57'
Foscante,Tess
Lac 22h16'44"37d31'B
Foschi,Barbara
Eri 3h41'46"-6d59'
Fosco,Guardian of Denise Danielle
Cas 1h54'32"58d39'
Foshag,Leland J
Per 1h50'56"50d13'
Foshinbaur,Susan
Lyr 18h45'22"35d58'
Foss,Amaryllis Star
Sex 10h29'51"1d21'
Foss,Arlene P
Cas 0h27'12"69d12'
Foss,D W & Mabel
Cam 7h8'15"61d39'
Foss,Emily Faith
Cam 7h39'25"60d51'
Foss,Giz
Lyn 7h32'52"52d6'
Foss,Jane Whitney
Her 17h59'32"18d52'
Foss,Katie
Lyr 18h42'47"41d54'
Foss,Michaela
Cas 0h22'45"63d12'
Foss,Robert Christian & Laura Lee
Mon 6h43'33"10d18'
Fossat,Dominique
Her 16h18'26"50d12'
Fosse,Francis
Aur 5h14'2"29d58'
Fossella,Dylan Michael
Boo 14h40'40"37d40'
Fossett,Chad
Per 1h46'19"53d48'
Fossey,Diane
Lyr 19h23'48"38d9'
Fosso,Michael Jerome
Her 17h15'29"48d21'
Fossum,Kendra Beth
Uma 11h34'38"43d37'
Foster IV
Sct 18h55'22"-6d38'
Foster's Fate
Peg 22h0'11"8d22'
Foster's Night Light, Stephen Paul
Cep 7h40'44"86d41'
Foster,Aaron Samuel
Cmi 7h28'29"7d47'
Foster,Alice
Aql 20h2'58"6d4'
Foster,Andy
Aur 5h19'10"49d26'
Foster,Ashley Mighell
Peg 22h46'0"30d60'
Foster,Barry Keith
Cep 22h9'57"67d52'
Foster,Bud & Kay Foster
Cyg 21h5'49"32d35'
Foster,Carrie A
And 0h33'0"28d51'
Foster,Chris
Uma 14h4'28"53d19'
Foster,Cindy Lee
Peg 22h34'32"21d34'
Foster,Cynthia"Baby" Ellen
Mon 8h2'28"-1d24'
Foster,Daniel A
Oph 17h16'30"10d14'
Foster,Danna
Lyr 18h40'0"38d46'
Foster,Darren
Per 2h52'11"43d26'
Foster,David Edward
Equ 21h20'0"10d12'
Foster,Dean Christianus
Boo 14h51'20"32d32'
Foster,Deputy Dana
Oph 18h2'15"10d60'
Foster,Doc
Uma 8h35'21"58d26'
Foster,Dorothea LeCain
Com 12h14'20"26d21'
Foster,Dr Sidney & Sandra Lee
Lac 22h4'10"38d38'
Foster,Elaine
And 23h2'42"51d51'
Foster,Eric Curtis
Lac 22h54'50"55d28'
Foster,Erik Stephen
Her 16h40'51"33d4'
Foster,Farell,Jane
And 22h56'13"40d40'
Foster,Ian Charles
Aql 18h53'12"-1d22'
Foster,James
Dra 10h42'29"81d14'
Foster,James A
Aql 20h5'35"4d6'
Foster,Jamie Elizabeth
Peg 23h1'20"13d4'
Foster,Jarvis Dewayne
Peg 22h37'36"27d25'
Foster,Jerry L
Aur 6h27'42"38d2'
Foster,John Dennis
Mon 6h23'14"8d1'
Foster,Joseph
Per 3h17'1"54d58'
Foster,Jr,James Edward
Aur 6h1'32"41d10'
Foster,Kathleen Pillow
Cyg 19h51'0"37d56'
Foster,Kayla Bethanne
And 1h2'36"36d24'
Foster,Kenneth Edward
Sct 18h20'1"-8d33'
Foster,Lucia Maria Ciullo
Mon 7h54'39"-5d54'
Foster,Marissa Dara
And 0h9'48"39d8'
Foster,Mary E
Aql 20h1'30"12d19'
Foster,Mary P
Cas 0h4'55"54d26'
Foster,Matthew Kee
Aur 6h34'33"38d48'
Foster,Meaghan Victoria
Uma 10h35'24"56d18'
Foster,Mr Stewart John
Dra 16h28'36"61d38'
Foster,Olga
Cet 0h38'42"-5d46'
Foster,Our Star Cliff & Debbie
Uma 9h39'36"59d51'
Foster,Ozias (Orion)
Ori 5h52'45"-1d28'
Foster,Patty
Cas 23h13'10"61d42'
Foster,Paul
Mon 6h45'13"0d24'A
Foster,Rand William
Vul 20h5'16"22d53'
Foster,Raymond (Ray)
Eri 4h8'1"-15d58'
Foster,Richard G
Aur 6h21'53"31d52'
Foster,Shannon Neal
Uma 11h45'33"50d56'
Foster,Sheila Theresa
Uma 9h53'32"56d35'
Foster,Sr,Robert Turnbull
Cep 21h47'60"55d56'
Foster,St Nicholas T
Tau 3h45'52"23d26'
Foster,Tara Elizabeth
Cyg 21h38'55"40d38'
Foster,Terre
Boo 14h23'54"12d17'
Foster,William (Bill)
Cmi 7h23'33"-16d6'
Foster-Burns,Bonnie
Ori 5h57'44"18d4'
Foster-Burns,Boyd
Ori 5h46'21"19d48'
Foster-Tembreull,Ann & Roger
Cyg 19h45'24"29d33'
Fostle,Nancy Lois
Lyr 18h33'44"40d32'
Foti,Carmen "Pepe"
Lyn 8h36'46"43d31'
Foti,Joe
Lyr 19h17'51"40d56'
Foti,Joseph
Her 17h16'59"42d26'
Fotini
Lyr 18h37'25"33d22'
Fotini Tsiatura
Lyr 19h18'30"40d50'
Fotis
Cap 20h39'33"-14d42'
Fotis Helious LeHeux
Uma 9h47'1"53d23'
Foto,Varvara
And 1h12'1"37d15'
Fotopoulos,Voula
Lac 22h12'0"47d20'
Fottles Star,Harold Ernest
Ori 5h55'26"15d25'
Fouad
Uma 9h18'42"59d57'
Foucault,Bernard
Uma 11h38'13"52d46'
Foucher,Laurent
Lyn 8h57'46"42d14'
Foudy's Flaming Love Star
Cam 8h7'30"81d37'
Fouks,Arthur & Ancie
Cyg 21h5'54"31d9'
Fouladi,Farah
Cam 8h8'0"83d5'
Foulger,Cecilia
Com 12h43'0"30d19'
Foulk,Cassandra Noelle
And 2h11'53"42d35'
Foulk,Corey Allen
Cyg 20h39'27"40d2'
Foulk,Kyle Lawton
Aur 5h9'19"43d35'
Foulkes,Tracey
Cyg 21h1'14"28d44'
Fountain Rock Elemntry School (Hagerstown,MD)
Uma 8h29'43"71d53'
Fountain"Skip's Love Light",Edward
Cmi 7h27'46"8d43'
Fountain,Heine
Her 16h9'0"47d35'

Fountain,John Andrew
Cas 0h13'53"59d22'
Fountain,Lauren Marie
And 23h17'25"44d41'
Fountain,Travis James
Cam 3h25'16"60d3'
Fouques,Jacques
Aur 4h48'1"40d37'
Fouquet,Dominik
Aur 6h55'19"44d8'
4 C's,The
Ori 5h56'33"18d0'
4 Cory 13
Cam 5h50'1"80d6'
4 Donna
Lyn 7h46'1"42d31'
4 Infinity
Cep 23h30'47"68d2'
4 JEB's
Cep 22h32'48"58d29'
4 Lor-Mon Autre Femme
Mon 7h26'10"-8d38'
4 MOF
Uma 14h0'57"49d36'
4 Muriel Bird Only 94
Aql 20h32'31"0d35'
4 My Anne Marie
Peg 21h41'17"24d46'
4 Ned
Equ 21h06'38"03d27'
4 Sands
Her 17h47'30"14d28'
4 U Mary
Peg 21h38'60"27d2'
4-Ever Reg
Oph 18h6'24"8d1'
4-Strahan
Aur 7h8'48"40d18'
40 Anniversary Shizuko
Sgr 19h11'23"-28d54'
40th Birthday Star For Debs
Lyr 19h14'25"41d58'
411
Uma 12h17'37"54d26'
416343
Boo 14h46'59"48d25'
436-MAW-CRAF
Cam 5h38'30"60d51'
44-94 Bonfire
Peg 22h18'36"3d58'
44-94 Lovelight
Cyg 21h15'39"28d42'
440 Mike
Peg 22h43'42"32d26'
44th Year of Che & Tony,The
Ori 5h23'7"-3d1'
4588
Boo 15h8'59"53d10'
4M-Malstrac
Boo 15h5'0"14d48'
4Stuart4Always&Always
Sex 10h30'45"-5d32'
4U4ME4EVER
Crb 16h1'22"38d53'
4UJB
Cam 4h51'58"68d50'
1413:Pip OJ Mick 1955 1975 1995
Uma 9h25'0"68d25'
1437 Sam & Sara
Cyg 19h15'15"45d25'
14U Sukeena
Cap 21h0'1"-14d53'
Four Castors
Uma 8h37'0"67d41'
Four M
Lac 22h20'21"53d8'
Fouraker,Cami & Bruce Barnes
Cyg 20h6'49"41d14'
Fourantoni's Star
Lyn 8h47'38"45d55'
Fourmeaux,Maxime
Boo 15h7'0"8d48'
Fournier,Bruno
Uma 11h25'32"32d21'
Fournier,Chantal
Lyr 18h21'27"45d6'
Fournier,Connie Lynn Bass
Com 12h42'16"24d19'
Fournier,Diane
And 2h0'23"37d3'
Fournier,Eriue
Per 2h20'21"55d59'
Fournier,Isabelle
Boo 14h54'30"38d6'
Fournier,Kelly J
Tri 2h26'15"28d59'
Fournier,Lise
Cas 1h0'53"60d7'
Fournier,Maurice
Uma 11h55'43"41d44'
Fournier,Michel
Dra 15h44'52"68d26'
Fournier,Micheline
Per 3h7'22"41d39'
Fournier,Robert
Per 3h57'35"38d44'
Fournier,Sabrina
Mon 7h53'21"-5d45'
Fournis,Martine
Dra 10h12'41"73d36'
Fournié,Alain Joseph
Ori 5h56'0"20d2'
Fourton,Jean-Rene
Leo 10h39'25"18d20'
Fourvel Laurent Maxine
Aur 5h10'59"29d9'
Fousek,Ute
Aur 5h17'35"44d44'
Foust,Betty
Cyg 21h38'60"42d44'
Foust,Sr,James D & Jean S
Cyg 20h33'29"31d9'
Fout,Billy & Mary
Ori 6h4'26"7d51'
Fout,Glenn
Lac 22h27'31"55d40'
Foutain,Ross Martin
Cet 3h2'22"0d27'
#4747 Sweet Babycakes (G&L)
Lac 21h56'30"40d1'
#4779
Cas 2h8'29"70d0'
Fouty,Ruth
Cas 1h48'13"61d18'
Fow,Constance & Richard England
Vul 19h47'39"26d8'
Fowble,Hanah Marie
Vul 20h17'15"23d41'
Fowkes,Christopher Peyton
Dra 15h0'26"58d54'
Fowler III,William Henry
Oph 18h3'26"8d43'
Fowler Shine On Always Raymond D
Per 4h28'56"50d16'
Fowler,Annie M
Mon 6h54'59"-2d12'
Fowler,Boyd W
Aql 19h58'12"-0d50'
Fowler,Captain Robert E
Hya 8h56'37"4d12'
Fowler,Charles Robert
Cnv 12h41'0"38d54'
Fowler,Christopher Andrew
Boo 13h35'0"17d4'
Fowler,Christopher Thomas
Uma 9h24'54"50d56'
Fowler,Daniel L
Ori 5h50'26"18d46'
Fowler,David B
Lac 22h8'54"49d18'
Fowler,Gwendolyn
Eri 3h46'1"-2d33'
Fowler,In Memory of Wendy Kathleen
And 1h56'30"40d4'
Fowler,James W
Cap 20h34'42"-10d54'
Fowler,Joey
Equ 21h3'23"7d3'
Fowler,John
Ari 2h48'22"30d59'
Fowler,Jonnye
Oph 17h59'28"1d16'
Fowler,Kathleen M
Aqr 20h54'20"-1d29'
Fowler,Kelly Elizabeth
Peg 22h31'14"27d39'
Fowler,Kevin Charles
Sex 9h44'33"2d11'
Fowler,Lori A
Lyn 7h25'39"50d28'
Fowler,Louise
Lac 22h3'43"50d26'
Fowler,Mark
Aql 19h1'0"13d25'
Fowler,Meredith Row
Com 12h0'11"28d16'
Fowler,Papa's Star, Charles A "Red"
Ori 5h28'15"-4d58'
Fowler,Peter Livingston
Aqr 22h0'0"-5d34'
Fowler,Robert William
Cyg 19h27'39"30d1'
Fowler,Scott Hathaway
Uma 10h19'56"50d25'
Fowler,Sharen Grabova
And 23h32'55"43d22'
Fowler,Waldo
Lyr 18h41'14"33d22'
Fowler,Wendy Kathleen
Mon 7h4'16"-5d24'
Fowlers,Way Over Yonder the Big Sky
Cam 5h57'29"68d29'
Fowles Star,The Camilla
Com 12h28'20"21d2'
Fowles,Mary
Cet 1h33'58"-10d43'
Fox & Tiger Light 917321894
Vul 20h5'1"25d26'
Fox Hollow French School
Uma 11h44'33"50d32'
Fox of Virginia, Michael J.
Cep 22h21'57"59d30'
Fox Star,The
Ori 6h2'48"0d44'
Fox Star,The Marjorie Dara
Tri 2h18'1"31d38'
Fox's Heavenly Body, Paul J
Per 3h12'16"42d43'
Fox's Star,Michelle Janine
Del 20h14'49"15d38'
Fox,"The Duke"
Vul 19h0'52"22d27'
Fox,Allison Ann
Peg 23h25'52"17d24'
Fox,Andrew David
Cep 23h0'37"80d3'
Fox,Andrew Richard
Her 18h3'13"31d33'
Fox,Arnold
Boo 14h12'23"50d21'
Fox,Ashley Simone
Psc 22h55'36"0d28'
Fox,Bernard
Her 17h54'57"40d2'
Fox,Bill & Trudy
Umi 15h30'42"68d32'
Fox,Blair Fairchild
Peg 22h29'29"31d37'
Fox,Bob
Aur 5h25'0"38d54'
Fox,Brandon Joseph
Boo 15h19'49"50d22'
Fox,Brendan Nicholas
Boo 14h55'30"45d23'
Fox,Briar
Per 3h1'28"55d2'
Fox,Caroline
Umi 16h21'54"72d7'
Fox,Carrie Mae
Cma 7h23'16"-18d54'
Fox,Cybelle Alexandra
Mon 6h52'49"-1d21'
Fox,David & Joyce
Oph 17h0'14"11d27'
Fox,Donald Frederic
Her 18h3'53"31d3'
Fox,E Anne
Lyn 8h39'11"40d5'
Fox,Edwin B Geraldine J
Cam 3h26'51"61d9'
Fox,Emily
Ori 5h42'23"3d59'A
Fox,Eugene
Sex 10h32'36"-6d51'
Fox,Evelyn Veronica Schablein
Aql 19h0'30"10d55'
Fox,Garold Beauregarde
Ori 6h6'40"8d33'
Fox,Gerald Paul
Cnv 12h14'11"46d40'
Fox,Harvey
Boo 14h8'14"54d53'
Fox,Hettie Rebecca Jane
Mon 8h8'8"-2d58'
Fox,Howard A
Boo 14h46'37"50d10'
Fox,Ivan
Vul 19h45'16"28d15'
Fox,James Adrian
Cma 7h19'1"-16d48'
Fox,James Thomas
Cnv 12h7'50"37d58'
Fox,Jamie
Cam 11h31'48"80d1'
Fox,Janis Day
And 1h32'22"35d30'
Fox,Jennifer
Vul 19h59'19"22d39'
Fox,Jeremy Allen
Lmi 10h43'38"25d46'
Fox,John Noel
Boo 15h0'59"12d50'
Fox,Jonathan Morgan
Per 3h15'10"50d23'
Fox,Josephine
Cas 0h54'34"58d53'
Fox,Jr's,Star,Leland L
Per 1h32'32"53d52'
Fox,Jr,Jerry Monte
Eri 3h42'1"-2d36'
Fox,Judy
Cas 0h17'1"62d40'
Fox,Kathi
Cas 3h31'17"71d6'
Fox,Katja S
Cas 0h35'13"67d0'
Fox,Keith
Uma 11h10'17"50d24'
Fox,Kevin J
Dra 9h23'56"78d25'
Fox,Laura May
Eri 3h2'47"-11d55'
Fox,Lindsey Nicole
Uma 11h44'27"41d47'
Fox,Lisa
Tau 4h2'12"21d25'
Fox,Lori B
Cas 1h42'48"68d53'
Fox,Louis J
Her 17h11'41"42d31'
Fox,Marc Alan/Alfonso Quagmire Petrale
Aur 6h27'16"38d12'
Fox,Mark Albert
Ser 18h15'41"-7d4'
Fox,Marty
Per 1h51'54"52d46'
Fox,Matt
Lac 22h37'0"53d44'
Fox,Maureen Kelly
And 23h47'33"41d56'
Fox,Melanie
Mon 7h1'1"4d58'
Fox,Melody
Eri 2h56'26"-7d21'
Fox,Mike
Ori 5h32'52"0d12'
Fox,Patty & Ed
Crb 15h29'39"30d56'
Fox,Paul J
Dra 19h2'1"48d29'
Fox,Paul Thaddeus
Her 17h2'15"31d17'
Fox,Peter John Vakarenkov
Her 17h37'54"27d25'
Fox,Peter M
Ori 5h56'30"16d10'
Fox,Rachel
Oph 17h19'22"11d5'
Fox,Robert Burnap
Vul 19h23'0"26d47'
Fox,Ronald E Margo E Vognild-Fox
Peg 22h1'54"24d19'
Fox,Russell
Her 16h5'0"41d23'
Fox,Russell
Ser 18h16'59"-13d20'
Fox,Sarah
Sgr 18h48'5"-26d4'
Fox,Sayde
Gem 7h57'59"31d42'
Fox,Shira Simcha
Cas 3h5'33"58d28'
Fox,Sir Paul
Per 3h5'36"40d15'
Fox,Stephen D
Her 17h53'28"28d58'
Fox,Susan
Mon 8h8'50"-6d46'
Fox,Susan Marie
Ori 6h12'59"8d21'
Fox,Teresa
And 23h15'0"47d54'
Fox,Terry & Dan
Vul 20h23'26"28d15'
Fox,Thomas
Ori 5h57'1"20d35'
Fox-Barnes,Allison
Crb 16h14'53"30d57'
Fox-Frazer,Lynda
Cas 0h3'35"61d39'
Fox-Sohner,Sandra J
Mon 8h1'44"-8d10'
Foxe,Jon A
Oph 17h1'2"-24d57'
Foxley,Catherine Burns
Sco 17h54'13"-30d13'
Foxman,Alan
Per 1h34'28"53d14'
Foxo
Peg 21h57'13"3d33'
Foxworth,Deirdre
Lac 22h41'37"55d59'
Foxy Hummie
Lyn 9h15'54"36d28'
Foxy Lady
Del 20h14'53"9d29'
Foy,Kattie Marie
And 23h2'27"48d39'
Foy,Max
Cep 20h41'37"60d32'
Foy,Susan
And 23h39'0"45d47'
Foytek,Ashlyn Kristine
And 0h33'12"45d56'
Foytek,David Michael
Lac 22h35'28"37d59'
Foytek,Michelle Theresa
And 0h33'26"45d7'
Fr Mike
Aql 19h18'30"15d11'
Fracas,Albert
Lyn 8h21'15"47d39'
Fracassi,Stacey Jean
Lyr 19h13'21"41d42'
Fracchiolla,Vito
Dra 18h21'46"70d6'
Frackiewicz,Mike
Del 20h15'29"10d58'
Fraclose,Edward & Margaret
Cam 4h25'32"60d48'
Fradette,Amanda Ann
Umi 14h42'58"69d4'
Fradette,Cassandra Lee
Umi 14h57'43"69d35'
Fradette,Donald
Aur 6h0'15"30d8'
Fradkin,Melissa Hannah
Cas 0h50'15"64d0'
Frady,Billy Fred
Oph 17h56'50"-5d55'
Fradziak,Dor Elizabeth
Ori 6h15'51"1d31'
Fraenckel,Jr,Victor Hugo
Lac 22h39'1"38d2'
Fraerman,Noah Michael
Aur 6h38'32"37d6'
Fraga,Rev Claudia
And 1h51m35"38d40'
Fragale,Brigadier General Frank M
Uma 11h43'56"63d8'
Fragapane,Margaret Pearl
And 22h58'47"51d52'
Frager,Norman
Aur 5h24'0"30d34'
Fragola
Del 20h19'23"14d56'
Fragua,Paul L
Ser 18h16'12"-14d39'
Frahler,Austin Cunningham
Boo 13h41'42"26d58'
Frahler,Damon Jennings
Aql 20h11'18"3d59'
Fraiberg,Isabella
Eri 3h3'31"-2d15'
Fraiberg,Larry
Aur 6h6'12"54d36'
Frailich,Annie
Peg 22h33'0"30d0'
Frailich,Michelle J
Mon 6h42'58"3d29'
Fraioli,Matthew Christopher
Dra 16h55'0"51d31'
Frakes
Vul 20h1'30"28d44'
Frakes,"Commander Riker" Jonathan
Leo 10h51'1"-5d41'
Frale,Michael Anthony
Her 16h39'16"38d8'
Fraley,Alice
Uma 11h40'42"48d8'
Fraley,Mark D
Cet 2h15'17"1d36'
Fraley,Rebecca Christine
Lyr 18h37'29"36d37'
Framabina
Lup 15h17'33"-44d23'
Framag
Ari 2h39'12"28d43'
Frambers,Terry
Aql 19h37'0"0d24'
Frambes"Superstar", Bob
Aql 18h58'37"-6d32'
Framboise
Uma 13h17'37"61d12'
Frame,Larry J
Eri 3h15'39"-6d42'
Frame,Terrance C & Kathryn B
Cet 0h50'40"-5d28'
Framia
Boo 15h5'29"41d7'
Framir's Love
Lup 15h29'55"-41d53'
Frampton as named by Joe Domin
Per 2h21'45"55d2'
Frampton III,Fr William
Per 3h6'41"46d45'
Frampton,Colleen
Peg 22h4'45"5d57'
Fran
Peg 22h22'1"5d9'
Fran & Bud
Ori 6h6'37"-0d2'
Fran & Roberto's Wedding Star
Cyg 21h12'31"39d24'
Fran & Walt's Treasure
Crb 15h56'37"27d17'
Fran Doll
Lyr 18h41'1"35d36'
Fran's Star
Aur 7h22'15"37d39'
Fran's Wishing Star
Cas 0h59'26"63d51'
Frana,Kelly
And 0h21'36"34d51'
Frananna
Aql 19h2'53"-0d28'
Franc
Eri 4h14'12"-10d12'A
Franca
And 23h6'26"52d43'
Franca,Mela
Ari 2h40'39"21d46'
Francastar
Tel 18h8'31"-48d36'
Francavilla,Frances Rose
And 2h30'46"45d5'
Francavilla,Frank N
Her 17h50'51"40d40'
Francavillo,Alphonse
Her 16h33'57"21d59'
Franccesca, 4-3-93
Sgr 18h49'16"-24d3'
France,Christopher Cameron
Dra 16h2'50"68d53'
France,Marie
Cas 2h24'50"77d9'
Franceour,Marlene
Com 12h31'37"28d43'
Frances
Del 20h22'33"10d9'
Frances
Uma 10h44'31"60d28'
Frances
Mon 6h39'40"6d40'
Frances & Arthur
Uma 8h43'48"50d19'
Frances & Bill
Cyg 21h14'0"38d8'
Frances B
Crt 11h13'56"-19d19'
Frances Kaye's Shining Light
Uma 10h36'46"70d49'
Frances Shines Forever
Cas 2h8'33"70d59'
Frances Thomas
Eri 3h56'11"-17d58'
Frances,Alexis Gillian
Ori 4h53'53"0d4'
Frances,Beverly
Com 12h17'42"30d42'
Frances,Janet
And 0h20'35"43d41'
Frances,Rae
And 2h11'19"40d18'
Frances: The Purple Gemini
Gem 7h14'20"21d8'
Francesca
Cas 0h28'13"60d10'
Francesca
Lyr 18h39'0"29d32'
Francesca
Pic 4h52'19"-45d8'
Francesca
Pho 0h28'5"-47d18'
Francesca
Col 6h35'48"-37d26'
Francesca
Tel 19h9'38"-49d14'
Francesca
Vel 9h43'1"-42d2'
Francesca
Boo 14h13'32"47d54'
Francesca
Lmi 10h35'48"31d59'
Francesca
Cas 1h49'31"60d45'
Francesca
Cyg 20h59'21"31d40'
Francesca
Aur 7h0'51"40d27'
Francesca
Lyn 7h28'36"40d18'
Francesca 13/9/90
Dra 17h52'27"61d33'
Francesca 2
And 1h1'35"47d47'
Francesca F
Hor 3h17'47"-47d43'
Francesca Lee
Cyg 20h35'51"60d21'
Francesca's "Rainbow Fantasy"
Aur 4h51'28"51d3'
Francesca's Dream
Cam 8h25'25"74d50'
Francesca,Anje
And 1h53'51"39d37'
Francescagiulia
Scl 23h46'56"-27d19'
Francesco
Uma 12h1'58"48d1'
Francesco
Cet 2h34'1"1d26'
Francesco
Del 20h14'17"12d52'
Francesco,Roberta e
Pyx 8h50'9"-27d34'
Francesco,Sabina
And 2h32'26"50d46'
Francescon,Maria
And 1h53'38"38d43'
Francescone,Linda
Cyg 20h24'22"40d3'
Francese,Susan D
And 2h27'27"44d27'
Franchek,William E
Lac 22h25'40"50d14'
Franchetti,Jr,Joseph Arthur
Aur 6h25'11"31d57'
Franchi,Anna Louise
Lyr 18h15'0"45d6'
Franchi,Gabriele
Leo 11h3'36"-0d1'
Franchi,Presley Terese
Cas 23h20'0"62d16'
Franchini,Dom
Ori 5h56'25"10d53'
Franci
Cyg 21h54'46"53d18'
Francia,Paola
Cet 1h8'51"0d33'
Francia,Rita
Cet 1h7'43"0d45'
Francie
And 0h46'59"39d51'
Francie
Cyg 19h58'40"31d37'
Francie Faithful & Free
Del 20h14'16"9d34'
Francine
Cyg 20h4'0"31d9'
Francine
And 1h33'39"39d1'
Francine
Mon 6h39'19"7d16'
Francine et Richard
Ori 5h40'34"8d58'
Francine's Star
Cyg 21h55'52"53d40'
Francine,Diana Marie
Vul 19h53'37"20d31'
Francine-Antonio
Cyg 21h1'11"39d27'
Francine/Moss Fly Free
Hya 8h9'28"-9d43'
Francini,Paola
And 23h7'59"41d51'
Francione,Ernest Big Ern
Uma 10h55'0"40d22'
Franciose,Carolyn E
Cas 0h13'11"47d23'
Franciose,Louis W
Her 16h43'40"50d40'
Francis
Cet 1h35'58"-14d42'
Francis
Cas 23h28'11"62d2'
Francis
Dra 19h34'40"70d24'
Francis
And 1h22'56"34d16'
Francis
Dra 17h2'29"69d48'
Francis Alias Lara Cristina Sag '70
Scl 23h43'57"-28d3'
Francis Charlie II
Uma 12h37'51"59d19'
Francis et Isabelle
Cyg 19h28'31"33d47'
Francis M B
Cyg 20h26'54"31d33'
Francis,Brian
Oph 16h57'1"-23d59'
Francis,Charles J
Eri 4h59'49"-6d29'
Francis,Courtney A
Aur 4h46'46"40d25'
Francis,Harry
Tau 3h54'56"22d17'
Francis,Jason Power
Hya 8h49'59"-7d57'
Francis,Karen J
Cam 3h22'55"56d8'
Francis,Kelly
Uma 11h41'33"53d41'
Francis,Lauren A
Peg 21h52'56"33d24'
Francis,Lauren Ann
And 0h24'40"40d32'
Francis,M C
Boo 14h30'34"36d31'
Francis,Mary
Tau 3h57'18"20d12'
Francis,Mary Ann
Eri 4h36'49"-17d33'
Francis,Mary Victoria
Lyr 19h23'17"37d40'
Francis,Matthew Winchell
Cma 7h49'60"10d2'
Francis,Michael Lewis
Dra 16h15'33"61d27'
Francis,Peter Ivor
Ori 5h35'55"-0d47'
Francis,Phyllis
Uma 11h39'18"44d6'
Francis,Timothy Patrick
Mon 7h57'59"-5d53'
Francis,Virginia
And 23h35'53"38d21'
Francisca
Cyg 21h52'26"53d38'
Francisco & Katherine
Crb 16h0'34"38d4'
Francisco Chang San Roman
Cep 22h46'26"70d11'
Francisco,Adele
Hya 9h11'0"4d7'
Francisco,Erika Katherine
Uma 8h51'15"55d29'
Francisco,Jennifer R
Crt 11h42'58"-8d20'
Francisco,Vaden B & Robin L
Umi 16h8'0"72d40'
Francisé & Robert & Hannah
Peg 22h14'13"34d11'
Franck,Alessandra- Valente
Pho 2h8'20"-40d35'

Franck,Taylor Michelle
Vul 19h44'40"28d49'
Franck-Emmanuel
Tri 2h3'21"30d58'
Francke III,Richard H
Psc 1h18'24"23d16'
Francken,Ria
Vul 19h23'13"26d20'
Francky
Tri 2h7'44"31d42'
Franco
Tel 20h15'26"-49d46'
Franco
Boo 14h45'26"51d6'
Franco e Lalla
Pho 0h4'0"-44d19'
Franco e Paola
Pic 5h1'40"-45d36'
Franco,Artemio "Artie" Morales
Aur 5h1'51"48d53'
Franco,Claudia
Equ 21h0'1"10d3'
Franco,Daniella
And 0h49'25"36d3'
Franco,Diane L
Cet 2h46'41"2d29'
Franco,Ferdinando
Sco 16h54'39"-44d50'
Franco,Ferdinando & Olive Mary Gherlone
Sgr 18h59'59"-27d12'
Franco,Georgina
Cet 0h54'1"0d39'
Franco,HR
Cas 0h7'30"58d18'
Franco,Janet
And 2h0'49"46d6'
Franco,Lillian
Mon 6h54'41"0d34'
Franco,Lorry L
Eri 3h3'16"-3d54'
Franco,Marguerite
And 0h25'48"40d53'
Franco,Olive Mary Gherlone
Com 12h23'56"21d58'
Franco,Paul
Peg 23h42'56"26d52'
Francoeur,Marc
Per 3h12'0"40d52'
Francois,Nichole
Cmi 8h3'52"5d54'
Francois,Nicole "Nicky"
Uma 10h30'23"68d25'
Francois-Sezionale-Basilacato
Aql 18h58'41"-6d49'
Francoise
Ori 5h59'32"10d48'
Francoise
Lmi 9h24'1"38d57'
Francoise,Christine
Cas 0h42'23"62d50'
Francoise,Marie & Jean Louis
Cnv 12h26'53"40d25'
Francolini,Kelly
Eri 3h21'57"-13d39'
Francomano,Anne
Lyr 19h8'0"40d48'
Francone,Michael John
Aql 19h2'37"1d6'
Francy
Lyn 9h3'45"38d57'
Francy,Gatta
Peg 22h28'0"4d22'
Frandadivimi
Col 6h32'10"-38d48'
Frandsen,Lau B
Aql 19h58'52"11d3'
Franek Playmate
Per 3h3'23"40d51'
Franetovic,Vjekoslav
Aur 7h9'0"35d47'
Franey,Nancy Ann
And 1h25'37"33d43'

Franey,Shaker Noel Feathers
Cnv 13h50'45"38d51'
Franey,Timothy
Ori 5h55'39"19d37'
Franey,William Thomas
Lyn 8h20'35"43d26'
Frangen,Jochen
Sco 16h14'17"-23d1'
Frangos,Mary
Peg 23h6'22"21d55'
Frangoul,Tamara M
And 2h13'47"39d30'
Franjo
Lib 14h21'28"-22d45'
Frank
Boo 14h11'12"36d7'
Frank
Cnv 12h40'58"36d53'
Frank & Beatrice's Golden Star
Lac 22h25'54"50d29'
Frank & Beryl's Star
Cyg 19h14'56"49d29'
Frank & Christine
Cyg 20h25'45"38d52'
Frank & Emogene
Aql 20h7'1"0d2'
Frank & Glenda
Uma 11h46'23"60d5'
Frank & Julie's Star
Cep 23h6'11"60d40'
Frank & Lori & Denise & Drew
Mon 7h35'15"-1d17'
Frank & Robby
Del 20h16'20"10d25'
Frank & Tess's Barabas Girls
And 23h38'44"42d54'
Frank Anthony
Lmi 9h48'29"33d17'
Frank FMLDSZ
Per 2h51'41"50d20'
Frank John
Per 2h51'29"34d28'
Frank My White Knight
Del 20h21'1"3d9'
Frank N' Nanny
Uma 9h56'42"59d38'
Frank O
Mon 7h1'56"-8d12'
Frank Paul "Forever In My Heart"
Per 2h24'12"54d29'
Frank Sara
Dra 20h33'54"71d12'
Frank Sr
Her 15h57'57d42d20'
Frank Star,The Arnold
Her 16h56'49"25d17'
Frank's Big Sky
Boo 14h4'35"27d29'
Frank's Forever
Tri 1h55'47"26d33'
Frank's Immortal Diamond
Lup 15h27'50"-42d58'
Frank's Place
Sco 17h4'16"-31d24'
Frank's Star
Cam 4h57'5"61d0'B
Frank's Star
Aqr 21h57'30"-5d56'
Frank's Star
Ori 6h6'43"1d44'
Frank,Abraham Mark
Her 18h19'0"12d8'
Frank,Agnes
Mon 7h1'46"-4d9'
Frank,Axel
Cep 20h30'53"66d39'
Frank,Bau-Ing VSI, Elmar G
Umi 15h32'24"77d48'
Frank,Beatrice Feil
And 0h19'53"30d20'
Frank,Carolyn
Peg 22h47'31"34d44'

Frank,Christopher Jacob
Cam 3h46'57"72d31'
Frank,Deanna & Kurt
Cyg 19h41'27"41d5'
Frank,Dennis & Cheryl
Her 17h23'40"18d56'
Frank,Eleanor
Lyr 18h39'11"33d31'
Frank,Elise Nicole
Eri 4h8'37"-17d50'
Frank,Eric Ryan
Her 16h43'47"33d50'
Frank,Ernest H
Uma 12h2'55"31d14'
Frank,George
Sct 18h45'54"-6d39'
Frank,Gregory Michael Rynold
Lac 22h42'39"53d15'
Frank,Heike
Mon 7h53'4"-1d19'
Frank,Helen P
Uma 12h2'54"32d17'
Frank,Israel
Cnv 13h6'42"40d25'
Frank,Joan Craig
Gem 6h55'57"16d30'
Frank,John P
Oph 18h17'32"11d48'
Frank,Jonathan P
Boo 14h34'42"19d38'
Frank,Joyce
Cam 10h43'1"82d16'
Frank,Judy
Cyg 19h58'1"40d59'
Frank,Julia E
Cyg 19h40'1"30d29'
Frank,Karola
Ari 2h21'11"21d35'
Frank,Linda Marie
Lyr 19h21'0"38d33'
Frank,Lindsey
Lyr 18h59'28"38d26'
Frank,Maureen Alyson
Sgr 18h55'39"-28d21'
Frank,Melissa Lynn
Leo 11h0'41"8d23'
Frank,Nina Rose
Lyn 7h1'52"58d57'
Frank,Peter
Gem 6h34'22"26d52'
Frank,Robert Langer 04-02-1966
Aqr 23h32'14"-10d23'
Frank,Roger & Sharon
Dra 10h45'14"78d30'
Frank,Sara
Uma 12h12'1"60d35'
Frank,Sharon
Cnv 12h43m54"38d5'
Frank,Sheldon & Wendy
Cyg 20h28'0"42d31'
Frank,Susan
Cas 0h10'53"61d55'
Frank,Suzy
Aur 5h31'24"48d50'
Frank,Ulrich
And 23h26'37"45d2'
Frank,Warren & Jane
Lac 22h44'37"56d25'
Frank,Warren Nathan
Her 17h51'12"48d53'
Frank,William Miller
Per 2h0'11"48d48'
Frank,Wim
Ari 1h55'25"17d19'
Franke Elor Anneliese
And 23h2'0"51d10'
Franke,Anna
Tri 2h6'29"30d2'
Franke,Brandon Alexandar
Her 18h43'59"12d26'
Franke,Catherine E
Her 18h33'25"23d33'B
Franke,Daniel Grant
Dra 17h43'27"63d57'

Franke,Dietmar
Cmi 7h31'25"7d39'
Franke,Edward P
Her 18h33'25"23d33'A
Franke,Fritz Wilhelm
Dra 18h27'34"65d17'
Franke,Josef
Del 20h21'44"2d48'
Franke,Louise Scherer
Lyr 18h20'51"45d56'
Franke,Martin
Cep 23h1'17"64d38'
Franke,Paul Vincent
Cet 1h4'1"0d32'
Franke,W Michael
Aql 20h12'17"4d16'
Frankel 40,Isaac (Ozzie)
Dra 12h33'1"75d49'
Frankel,Charles
Ori 5h51'50"7d2'
Frankel,Chet & Ida
Cyg 20h49'1"37d39'
Frankel,Herman
Her 18h54'24"12d19'
Frankel,Keith I
Ori 6h17'18"-2d15'
Frankel,Linda
Cnv 13h4'1"50d52'
Frankel,Ronnie
Peg 23h4'0"31d20'
Frankel,Timothy
Ori 5h55'54"11d46'
Franken,Maria Sropes Rosewell
Umi 16h37'57"75d35'
Frankenberg,Bobby
Aur 7h5'11"38d36'
Frankenberg,Robert
Per 3h43'26"51d4'
Frankenberger,Peter
Lyr 19h20'23"30d20'
Franki
Cam 5h48'24"68d44'
Frankie
Aqr 22h31'36"2d10'
Frankie
Her 17h22'42"40d4'
Frankie
Lyr 18h53'43"43d3'
Frankie
Lyr 19h22'18"31d31'
Frankie "Soul Mate"
Cep 22h10'23"60d55'
Frankie B From N T
Sge 19h55'24"16d16'
Frankie Jo
Cet 2h27'25"8d17'
Frankie Marie
Cyg 21h22'1"46d56'A
Frankie"D"
Ori 5h59'57"-0d19'
Frankie's Faith
Lmi 11h0'15"31d4'
Frankiewich,Chester
Lyr 18h40'0"26d51'
Frankjohndadiopapakell er
Lac 22h3'56"38d56'
Frankl,Ernst & Barbara Hengst
Lyn 8h2'24"42d3'
Frankl,Rhoda Lee
Uma 9h58'0"68d44'
Frankl,Spencer Nelson
Dra 16h34'21"52d11'
Frankland,Mark P
Cmi 7h29'49"7d56'
Franklin BDB-HV Day, David A
Sct 18h44'27"-4d12'
Franklin,Amber Marie
Mon 6h31'57"2d57'
Franklin,Ann & Chet
Cnv 13h52'22"40d36'
Franklin,April Irene
Tau 5h48'14"28d26'
Franklin,Beau
Sgr 20h0'17"-40d6'

Franklin,Caroline
Lyr 18h29'35"40d2'
Franklin,Charlotte
Umi 14h37'32"67d50'
Franklin,Devin
Mon 6h54'15"08d33'
Franklin,Emily
Cam 3h21'45"56d6'
Franklin,Eric Wayne
Per 2h41'46"35d13'
Franklin,Erik
Aql 20h10'33"13d38'
Franklin,Jill
Gem 6h59'20"13d27'
Franklin,Jimmy Don
Ori 5h59'33"10d30'
Franklin,Joe
Per 2h35'14"50d24'
Franklin,Jr,William Daniel
Lac 22h1'44"47d23'
Franklin,Katherine Anne
Aqr 22h24'15"-6d32'
Franklin,Kelly Maureen
And 2h12'16"39'45'
Franklin,Kristina Dawn
And 23h34'39"47d17'
Franklin,Laura
And 2h3'1"46d27'
Franklin,Lauren G
Tri 1h53'20"25d45'
Franklin,Martha Kay
Eri 4h53'27"-6d40'
Franklin,Michael Jamal
Her 18h19'23"14d15'
Franklin,Nicholas Thompson
Vir 13h33'30"-7d34'
Franklin,Nicholas Starr
Ori 5h11'11"-4d14'
Franklin,Noah M
Aur 6h10'49"46d35'
Franklin,Paul,Gregory, Jeffrey
Uma 10h38'27"42d8'
Franklin,Tod Griggs
Aql 19h32'15"0d55'
Franklin,Tyler
Aur 5h32'54"30d30'
Franklin,Viola L
Sco 17h3'25"-30d16'
Frankly Forever
Boo 15h4'46"53d49'
Franklyn T
Dra 16h6'15"61d3'
Franklyn,Ian Scott
Her 18h44'27"12d58'
Franko
Cet 1h55'24"-3d19'
Frankowski,Matt
Lac 22h14'21"51d32'
Franks,John
Aql 19h52'1"12d36'
Franks,Leslie Leck
Cet 0h53'53"-2d5'
Franks,Randall Scott
Aql 20h10'52"5d11'
Franks,Steven E
Aur 6h50'0"37d54'
Franky
Cep 22h40'28"65d27'
Franky
Pup 8h8'31"-20d30'
Frannie,Paul's Star
Cas 0h30'15"64d12'
Franny's Zelda
Cas 23h29'58"60d60'
Franny-Twinklc Twinkle Little Star
Mon 6h28'28"-10d32'
Fransen,Larry J
Lac 22h21'31"53d22'
Fransen,Mary
Lac 22h7'1"46d36'
Fransen,T Joseph
Peg 23h3'25"13d28'
Franson,Oscar
Umi 15h30'42"68d32'

Frantz,Daniel
Aql 20h12'39"10d47'
Frantz,Elaine Mae
Uma 8h37'39"49d10'
Frantz,Gary
Per 1h43'56"52d43'
Frantz,Grace Elizabeth
Aql 20h34'23"-0d28'
Frantz,Joeri Bas
Dra 19h2'16"71d12'
Frantz,Karl
And 23h28'35"45d35'
Frantz,Rick
Sct 18h41'1"-6d12'
Frantzen,Brenda Helena
Cam 4h5'51"58d60'
Franz
Dra 17h8'13"61d22'
Franz & Mary Ellen
Lyr 19h10'11"37d48'
Franz,Jason Alan
Her 16h42'39"32d37'
Franz,Michael
Cmi 7h18'35"4d39'
Franz,Russell Edward
Aur 6h20'16"38d33'
Franz,Walter R
Dra 18h24'46"68d26'
Franz-Wilhelm
Vir 13h2'25"-8d5'
Franzen's 50th
Cyg 20h36'34"42d31'
Franzen,June Audrey
And 1h48'55"40d3'
Franzi
Leo 11h23'30"2d43'
Franzi "Mio Amico", Fulvio
Boo 13h36'55"23d17'
Franzi,Rebecca
Hya 9h6'15"2d43'
Franziska,Ursula
Dra 15h57'58"62d19'
Franzman,Taylor
Cet 0h55'1"-8d40'
Franzmann,Ib
Cyg 19h26'1"56d33'
Franzone,Gabrielle
Aur 6h43'32"38d18'
Franzoso,Flavio
Lup 15h10'48"-41d22'
Franzoy,Debbie
Cas 1h54'32"67d50'
Franáois
Aql 20h22'38"0d9'
Franáois
Her 17h19'28"18d48'
Franáois et Véronique
Boo 14h56'57"17d58'
Franáois HG
Uma 13h39'38"60d1'
Franáois-Xavier
Ori 5h35'35"-6d22'
Franáoise
Boo 14h50'1"38d11'
Franáoise
Cep 22h30'0"63d21'
Franáoise et Philippe
Oph 18h5'17"11d16'
Franáoise,Thomas Maris
Mon 6h25'47"7d49'
Frappat,Edouard
Peg 23h29'48"21d53'
Frapé,Raymond A
Aur 6h2'45"46d6'
Frarik
Lyr 19h21'31"30d12'
Frary Stardust Memory, Ted
Umi 15h11'53"68d10'
Frasca,Joseph W
Cam 3h33'28"61d36'
Frascadore,Henry C
Aur 6h20'36"32d11'
Frascia,Franca M
Lyn 7h50'1"48d1'

Frase,Joel Alexander
Per 2h51'1"43d21'
Frase,Lisa
Peg 22h52'38"25d31'
Fraser Jr,Robert D
Lac 22h14'23"51d33'
Fraser Star,The
Cma 6h13'42"-26d13'
Fraser's-Michael D & Roberta T,The
Lyn 8h24'42"48d49'
Fraser,Cynthia
Sge 19h57'55"18d49'
Fraser,Daron
Per 2h54'22"37d7'
Fraser,Deborah A
Cas 0h26'44"54d34'
Fraser,Dennis & Candi Love Is Forever
Cas 2h57'42"73d32'
Fraser,Ellie Adams
And 23h44'49"41d16'
Fraser,Garfield
Aql 19h54'36"12d41'
Fraser,Irene
Cas 23h35'1"61d29'
Fraser,Jack
Lac 22h44'34"53d33'
Fraser,Julie Ann
Cyg 19h43'0"30d15'
Fraser,Karen & Nicole
Cyg 21h23'35"48d56'
Fraser,LeeAnn Theresa
Lyn 7h27'1"58d58'
Fraser,Lloyd
Uma 12h7'27"47d39'
Fraser,Mark Robert
Per 4h42'41"38d36'
Fratt,Mildred Anna "Nana"
Del 20h18'24"13d39'
Fratti "The Sea Otter" ,Robert
Sct 18h55'26"-4d37'
Fratto,Bean
Uma 9h46'12"43d56'
Frau Bärbel Mix
Scl 23h18'11"-32d4'
Frau des Regenbogens
Tau 3h35'13"30d49'
Frauenheim,Barb
Cnv 14h3'1"41d8'
Frauenschuh,Theresa
Aqr 20h56'43"-0d22'
Frauke B
Tau 4h2'0"30d22'
Frauke Lina
Cyg 20h35'24"31d38'
Fravili,Romina
Pyx 8h45'28"-28d11'
Frawley,Janine Ann
And 2h28'31"40d41'
Frawley,Leah-Sinead
Leo 11h50'0"20d15'
Frawley,Mary Gail & Dennis O'Dea
Sge 19h17'1"16d35'
Frayne Again I Love You,Ronnie!!! Annie
Lyn 7h6'54"50d7'
Fraysse,Philippe
Aur 6h2'16"31d29'
Frazee,David Greer
Hya 9h7'29"5d17'
Frazee-Spiker,Karen
Boo 14h57'55"19d20'
Frazer,Jared Daniel
Boo 15h7'1"37d51'
Frazer,Margaret Martina Maria
Cas 0h31'48"70d46'
Frazer,Renaud
Mon 7h49'6"-4d6'
Frazer,Rhoda
Lib 15h30'34"-28d17'
Frazer,William Gordon
Umi 17h21'17"75d52'

Frazier Star
Dra 18h58'14"58d41'
Frazier,Alexandra
Lyn 7h48'35"47d45'
Frazier,Ann
Lyn 7h23'35"44d38'
Frazier,Claudia Hodari
Eri 4h29'7"-13d45'B
Frazier,Dalton T
Uma 8h46'53"47d23'
Frazier,Daniel Lawrence
Cam 3h13'28"61d5'
Frazier,Derek L
Cet 0h58'1"-2d58'
Frazier,Duncan Alexander
Uma 11h35'25"44d5'
Frazier,Frances M
Lyn 7h25'33"50d10'
Frazier,Hayley Brinn
Lyn 8h30'15"40d29'
Frazier,Ian M
Eri 2h47'26"-17d33'
Frazier,James Keel
Oph 18h29'35"6d44'B
Frazier,Jeff
Eri 4h29'7"-13d45'A
Frazier,Joan
Cas 2h2'13"58d18'
Frazier,Joan
Lyr 18h15'19"32d59'
Frazier,Margie Hodges
Aql 18h45'17"7d45'
Frazier,Marjorie Elizabeth
Lyn 7h30'0"42d37'
Frazier,Michael
Lyn 7h54'56"58d53'
Frazier,Myletta
Cyg 19h59'54"38d15'
Frazier,Neil
Cep 21h50'1"58d22'
Frazier,Robert
Aur 5h44'34"54d30'
Frazier,Robyn
Peg 21h43'19"26d49'
Frazier,Sharon L
Mon 6h53'26"-1d13'
Frazier,Terry
Her 16h25'36"38d44'
Frazin,Emily Sherry
Vul 19h41'35"26d45'
Frazzitta 1970
Ori 5h4'12"8d1'
Frear Ascending, William
Her 17h9'1"37d52'
Frear,Joshua & Benjamin
Ori 5h59'55"17d18'
Freas,Randy Dale
Aql 20h32'0"-0d13'
Frech,Morley
Ser 18h1'29"-14d47'
Frechette,Ed
Oph 18h5'57"8d15'
Frechette,Mary Ann
Lyr 18h42'0"38d42'
Freckelton,Donald
Per 2h54'48"31d59'
Frecker,Samuel Edward
Uma 10h54'1"38d51'
Freckles
Mon 6h54'11"-10d4'
Freckles Fancy Lad Silver Such Joy MomDad
Uma 10h57'48"38d12'
Fred
Boo 15h7'0"8d48'
Fred
Aql 20h4'0"1d37'
Fred
Cyg 20h24'12"39d5'
Fred & Carmela Forever
Gem 6h37'50"31d5'
Fred & Carmen
Cyg 19h30'0"34d42'
Fred & Debbie
Cyg 19h59'58"30d32'

Fred & Ginger's Star
Uma 14h19'17"61d35'
Fred & Jill = 381
Uma 11h14'20"47d43'
Fred & Lila
Ori 5h58'0"16d1'
Fred & Molly
Dra 12h36'54"71d50'
Fred & Sandra's Star
Ori 5h29'57"-1d28'
Fred & Susan
Cas 0h57'59"50d15'
Fred The Big Daddy
Uma 11h26'46"40d34'
Fred's Fabulous Forty
Umi 15h12'1"67d51'
Fred's Flight
Her 17h10'48"27d21'
Fred's Heavenly Angel
Dra 16h50'53"67d47'
Fred's Light
Lyr 19h22'38"40d18'
Fred's Quest
Lmi 10h35'19"28d52'
Fred's Star
Her 16h5'0"18d51'
Fred's Star
Del 21h4'37"13d1'
Fred's Star At Newpoint 31780
Vul 20h42'0"20d37'
Fred,D J
Tri 1h45'0"26d30'
Freda
Mon 7h4'17"-6d27'
Freda
Peg 22h7'27"5d19'
Freda Rose
Lyn 8h43'35"42d2'
Freda's Star
Mon 7h19'20"-6d39'
Freda,Joseph
Cep 21h55'37"84d55'
Freddie
Cep 0h57'50"86d51'
Freddy
Aql 19h6'57"-1d11'
Freddy
Cmi 7h40'45'5d32'
Freddy & Cathey
Crb 15h54'35"28d24'
Freddye
Sct 18h31'16"-6d37'
Fredenburg(Dad),Donald Earl
Boo 14h56'43"24d26'
Frederic Et Pierre-Yves
Lyn 8h18'30"40d37'
Frederic Nathalie
Tri 1h59'56"31d40'
Frederic,Jay Brisbon
Hya 9h4'22"2d55'
Frederica
Tri 2h43'24"33d50'
Frederica
Cam 3h45'45"68d53'
Frederica 112
Cae 4h52'1"-33d24'
Frederica's Punto di Luce
Lyr 19h9'41"37d56'
Frederick 80,Charles J
Uma 12h8'0"46d46'
Frederick Franklin
Lyr 19h4'0"47d28'
Frederick J 10-13
Boo 15h6'50"12d1'
Frederick Sunset
Uma 13h29'35"61d57'
Frederick,Brian Andrew
Per 1h55'58"56d59'
Frederick,Charles Race
Tau 5h34'50"28d24'
Frederick,Chevrolet
Umi 14h26'43"67d39'
Frederick,Chuck
Dra 19h26'53"65d25'

Frederick,Daniel Joseph
Cet 2h59'28"0d34'
Frederick,Deborah L
Cet 1h16'0"-6d43'
Frederick,Gregory Michael
Aql 19h28'20"0d46'
Frederick,Margaret "June"
Crt 10h55'50"-11d55'
Frederick,Mark Milan
Tau 4h5'14"20d43'
Frederick,Matthew Steven
Ser 16h3'1"10d29'
Frederick,Rob
Dra 14h28'0"64d2'
Frederick,Ron & Diana
Vul 20h39'1"20d17'
Frederick,Shelline Marie
Sge 19h22'50"16d31'
Frederick,Steven Curtis
Her 18h42'17"12d42"
Frederick,Steven K
Cnv 12h33'20"40d43'
Frederick,Teresa Louise
Crb 15h54'25"32d17'
Frederick-Raylin Raymond J
Peg 21h55'23"28d31'
Fredericks,Florence
Sgr 19h33'34"-42d6'
Fredericks,Josh
Umi 16h3'56"76d14'
Fredericks,Jr,James W
Dra 14h56'0"60d18'
Frederickson,Pamela Franklin
Lyr 19h23'59"30d49'
Frederickson,Shirley
And 23h46'13"44d19'
Frederickson,Suzanne E
Mon 8h4'31"-2d10'
Frederiksen,Louise
Aql 19h57'32"14d3'
Frederiksen,Oluf
Aql 18h59'57"16d28'
Frederiksen,The Star of Per
Per 4h5'35"38d19'
Fredi H
Per 3h11'1"55d44'
Fredna
Lmi 9h57'51"34d20'
Fredregill,Sandra
Cyg 21h26'43"41d8'
Fredrich,Gregory J
Aql 19h34'14"-0d50'
Fredricks,Mari
Lyn 8h55'23"46d40'
Fredrickson,Arthur H
Aur 05h50'1"50d36'
Fredrickson,Haley Chinn
Com 13h24'1"25d16'
Fredrickson,Mabel Carolyn
Sornson
And 2h23'21"46d35'
Fredriksen,Bent
Dra 13h25'23"64d56'
Fredriksson,Peter
Her 16h25'35"36d3'
FREDWBJR(KB2CT)
Lac 22h38'1"48d46'
Fredy
Aql 18h57'25"-6d30'
Frederica
Lmi 10h33'42"37d54'
Free Bird
Aql 20h5'33"4d25'
Free,Billy
Cyg 20h22'58"39d26'
Free,Eric
Cet 2h56'37"3d34'
Free,Robert(Bob)
Dra 19h24'24"58d35'
Free,Wayne
Hya 8h43'32"3d44'
Freebairn,Blair
Aur 6h2'34"45d2'
Freeburn,Dee
Cas 0h48'40"71d6'
Freeby,Clarence E
Ser 15h18'13"7d7'
Freed
Uma 9h54'23"53d19'
Freed,Elizabeth & Bennett
Ori 5h55'53"10d52'
Freed,Janci Carlson
And 23h22'1"51d41'
Freed,Morgan Lydia
Lyn 7h51'48"35d38'
Freed,Shawn & Roger
Cyg 21h24'1"41d1'
Freed-Selbo,Nicholas
Per 3h47'21"36d19'
Freedland,Sylvia
Del 20h20'0"10d25'
Freedman's Star,Joshua
Per 3h28'0"50d50'
Freedman,Ilana
Peg 21h39'34"23d48'
Freedman,Jeremy
Oph 18h41'16"8d50'
Freedman,Jonathan
Sex 10h25'28"-0d6'
Freedman,Kenneth S
Per 3h11'52"41d5'
Freedman,Lori
And 1h48'12"37d47'
Freedman,Mark J
Aql 19h3'23"1d36'
Freedman,Rebekka McFarland
Peg 22h45'10"10d50'
Freedman,Richard
Per 2h10'50"57d50'
Freedman,Sheryl P
Eri 2h53'33"-18d41'
Freedom
Pup 8h24'33"-22d42'
Freedom 12-26-80
Aql 18h53'0"-0d2'
Freeland,Barbara
Cas 0h41'60"63d52'
Freeland,Gwen Alicia
Hya 8h46'54"-6d49'
Freeman 6,Matthew
Christopher
Boo 15h45'45"41d47'
Freeman Math Star,The
Michael B
Tri 2h28'51"34d47'
Freeman,Alan
Ori 5h55'20"0d16'
Freeman,Barrie & Tommy
Crb 16h7'0"38d35'
Freeman,Beverly Janeth
And 23h1'36"48d12'
Freeman,Bobbie (Wiffie)
McGlynn
Aql 20h2'9"9d5'A
Freeman,Charles A
Aql 19h55'60"14d8'
Freeman,Chelsey Lyn
Peg 21h55'1"22d47'
Freeman,Cheryl
Cas 1h36'0"68d40'
Freeman,Claire
Cas 0h14'37"46d40'
Freeman,Dennis
Cet 2h50'18"5d5'
Freeman,Doris M
Cmi 7h55'30"0d41'
Freeman,Dr Ronald
Oph 18h38'19"8d60'
Freeman,Eleanor
And 0h22'49"34d45'
Freeman,Eugene A
Dra 10h35'55"78d6'
Freeman,Gloria
And 0h25'14"45d26'
Freeman,Grace M
Cas 22h57'39"55d28'
Freeman,Grandad Harry
Her 18h4'11"40d10'
Freeman,Gregory James
Cep 23h21'15"70d18'
Freeman,J Phil Hawk (Hubbie)
Aql 20h2'9"9d5'B
Freeman,Josh
Ser 18h18'20"-14d28'
Freeman,Jr,Sam
Aur 5h1'50"41d39'
Freeman,Justin Wade
Boo 15h17'58"40d1'
Freeman,Kelly
Cyg 19h47'58"30d18'
Freeman,Leland Francis
Fredrick
Crt 11h15'19"-13d25'
Freeman,Leslie
Tau 5h45'52"27d22"
Freeman,Lisa
Lyr 18h51'22"40d1'
Freeman,Lon
Boo 14h44'0"32d34'
Freeman,Marc Adison
Lac 22h28'32"35d37'
Freeman,Max & Sandi
Aql 19h58'60"10d2'
Freeman,Melvin (Fingerhut)
Cma 6h56'6"-19d2'
Freeman,Michael
Hya 8h33'32"-1d51'
Freeman,Michael
Crt 11h0'51"-12d5'
Freeman,Michael Douglas
Ser 15h39'17"-1d45'
Freeman,Nellie Adams
And 22h59'39"36d7'
Freeman,Sally Adele Bennett
And 23h46'0"47d21'
Freeman,Sarah Catherine
Mon 6h54'1"-0d8'
Freeman,Scott Michael
Boo 13h39'56"15d15'
Freeman,Sr,Donald Timothy
Cap 20h29'19"-10d11'
Freeman,Stacy Ann
And 0h52'1"36d0'
Freeman,Sybil Maureen
Lib 15h43'35"-24d28'
Freeman,Tyrone
Boo 14h53'30"34d43'
Freeman-Compton
Lyr 18h28'20"44d9'
Freemyer,Becky
Dra 9h44'51"73d15'
Freeny,Dolores
Gem 7h2'45"35d12'
Freer Star,The
Cam 7h57'0"60d33'
Freer,Jordan A
Cyg 20h21'44"38d47'
Freese,Adrienne Grace
And 22h55'42"51d24'
Freese,Angela Eve
Cam 4h45'51"68d6'
Freese,Peter
Lyr 19h4'56"28d30'
Freestone,Mack Austin
Cet 2h45'33"-0d5'
Freestone,Maryl
And 0h34'56"45d20'
Freeth,Doris & Bertram
Cyg 20h20'58"41d5'
Freeway
Mon 8h7'55"9d32'
Freeze,Jason & Marjorie
Hya 9h5'58"6d30'
Freeze,Theresa Louise
And 22h58'22"37d57'
Fregeau,Kathleen Louise
Arnold
Cas 2h34'49"57d38'
Fregeau,Marc
Cep 0h55'38"77d28'
Fregeau,Ty
Ori 5h2'56"1d41'
Frehe,Ewald
Vir 13h13'2"-9d9'
Freiberger,Lynne
Vir 13h53'0"1d31'
Freid,David
Gem 7h53'1"31d38'
Freid,Sheila
Vul 19h59'1"23d2'
Freidel,Fritz
Dra 15h6'1"62d6'
Freidel,Michael
Ser 15h57'30"2d37'
Freifeld,Nancy
Lyr 19h23'22"31d18'
Freiherr von Arnim, Hans-
Lothar
Uma 11h56'17"62d34'
Freil,William Flynn
Del 20h35'53"10d45'
Freiman,Denise Joy
Lyn 6h16'39"61d52'
Freimuth,Harald
Hya 9h33'16"-9d15'
Freissinet,Gerard
Sex 9h55'22"-2d51'
Freist,Otto Emil
Per 4h27'34"50d54'
Freistat,Ivan
Aql 20h31'21"0d20'
Freitag,Jr,Paul W
Dra 9h25'1"74d11'
Freitag,Stefan
Cep 22h0'0"68d49'
Freitag,Timothy Wade
Hya 8h12'15"4d9'
Freitag,William
Aql 19h51'1"14d36'
Freitag-Köhler,Gudrun
Gem 7h23'37"35d15'
Freitas "Sunshine", Kim B
Peg 21h57'38"29d45'
Freitas,David Royal
Aql 18h58'25"14d6'
Freitas,John Botty
Hya 10h22'46"-11d24'
Freitas,Kristen Rose
Crb 16h3'39"28d39'
Frelinghuysen,Beatrice P
Cyg 21h12'20"38d8'
Frelly
Aql 18h57'59"-4d12'
Fremaint,Austin Blue
Aur 6h28'37"38d8'
Fremed,Richard & Ellen
Eri 2h49'20"-5d52'
Fremont,Elaine Bruckheimer
Lyn 8h14'29"51d22'
French III, George D
Cap 21h52'0"-21d58'
French Star,Georgina &
Benjamin
Cyg 20h38'43"44d55'
French Star,The Carolyn &
Kathleen
Cas 2h39'29"70d11'
French V Class of 1996
Lyn 7h7'18"51d5'
French,Alecia Irene
Com 12h45'42"20d1'
French,Andrea Annette
Oph 17h43'14"12d57'
French,Angela Marie
And 1h39'22"38d49'
French,Beverly
Sco 16h32'33"-30d0'
French,Dax King
Cet 2h28'12"4d7'
French,Denson Moore
Hya 8h10'56"5d13'
French,Derrick Todd
Aql 19h15'40"19d29'
French,Dorthea Baker
Cas 0h32'19"50d2'
French,Elizabeth M
Lyr 18h59'49"26d25'
French,Erika
Gem 6h59'23"11d32'
French,Janet
Cyg 19h59'45"38d12'
French,Jefferson Taylor
Aql 20h4'55"4d20'
French,Jessica
Gem 6h59'40"12d40'
French,Jessica Laura
Cnc 8h55'44"25d33'
French,John & Diane
Aur 5h5'25"29d17'
French,Jr,"Little Ray" Houston
Mon 6h33'32"8d41'
French,Kathy
Oph 17h24'41"-20d26'
French,Kevin & Shelley
Cyg 21h25'47"30d24'
French,Kirsten Elizabeth
Peg 21h59'22"2d23'
French,La Belle E'Toile Donna
Cas 2h32'59"57d29'
French,Lady Jane
Ari 2h50'42"30d33'
French,Loretta F
Mon 6h23'49"-6d45'
French,Mr & Mrs Paul J
Lyr 18h18'34"41d5'
French,Muriel & Richard
Cam 5h10'15"70d34'
French,Patricia Collins
And 2h19'32"50d29'
French,Peter Alan
Aur 6h27'47"38d15'
French,Peter D
Ser 16h12'44"1d24'
French,R D
Gem 6h59'22"10d4'
French,Tamara Adams
Uma 10h57'30"44d50'
French,Tyrone Clinton
Mon 6h6'15"-8d12'
French,Warren Blake
Cmi 7h19'39"7d54'
French,William Taylor
Lyr 19h14'22"38d46'
French,Zachary Louis
Ori 5h58'19"6d36'
Frenchie
Uma 11h56'44"32d1'
Frendahl,Bruce E
Cap 20h9'13"-10d6'
Frenesia
Cas 1h59'37"58d20'
Frenette,Darlene
Lyn 7h46'15"50d6'
Freni,Valeria
Dra 20h8'44"68d21'
Frenzel,Corina
Aqr 20h38'12"-14d14'
Frenzel,Virginia
And 0h57'16"45d53'
Frerix,Peter
Eri 4h54'34"-8d7'
Frerking,Juliet Rose
Sge 18h55'1"19d16'
Fresch,Joan
Cam 3h52'51"60d57'B
Frese, My Love,Kirsten Marie
Lyr 18h31'45"46d41'
Frese,Dieter
Ari 2h28'12"21d10'
Fresh Tsutom
Lib 15h44'0"-20d2'
Freshman,Nancy
Lyn 9h4'21"45d11'
Freshour,Phyllis
Mon 6h18'37"8d22'
Freshwater,NiceMiaMD: M F
Uma 9h54'16"45d46'
Fresia
Ari 3h6'28"30d38'
Fresoni,Janice & Kenny
Her 18h4'53"38d52'
Fressinet,Veronique
Aql 19h56'46"-10d18'
Freston,Karan Lee
Lyr 18h34'45"31d52'
Fretsky
Uma 9h12'1"60d15'
Fretwell Family Star, The
Aur 5h15'26"47d42'
Fretwell,Michael Peter
Lac 22h2'42"50d37'
Freudenberger,Hans- Dieter
Aur 5h15'29"43d22'
Freudenberger,Wilfried
Lac 22h41'21"56d43'
Freudenstein,Peter John
Ari 2h36'24"20d28'
Freund's Star,The
Dra 16h54'32"71d45'
Freund,Barbara
Sgr 19h0'43"-22d47'
Freund,Christine
Cam 6h35'1"83d39'
Freund,Emily
And 2h15'35"49d55'
Freund,Jerry & Jeannie
Peg 23h15'33"33d9'
Freund,John
Per 3h40'4"38d60'
Freund,Peter
Dra 19h35'28"70d45'
Freundlich,Golda
Cet 1h24'55"-5d52'
Frevert IV,Warren George
Ori 5h48'36"10d12'
Frey,Bryan
Cet 1h19'11"-04d39'
Frey,Bryan
Aql 19h58'34"13d42'
Frey,Donna Faye
Lyn 7h30'14"41d3'
Frey,Dustin Alan
Her 16h30'1"34d58'
Frey,Evelyn Kittleson
Mon 6h29'51"11d31'
Frey,George Fredrick
Hya 8h11'41"2d58'
Frey,Gregory Terrell
Cnv 12h15'0"37d5'
Frey,Hans
Peg 3h33'20"12d50'
Frey,Jon Christopher
Her 16h20'1"26d38'
Frey,Kathleen McGladrey
Cas 0h3'0"50d10'
Frey,Martin
Cam 3h53'15"58d28'
Frey,MD,Harry Bradford
Oph 18h17'42"8d56'
Frey,Michael
Cep 22h9'24"65d5'
Frey,Tresita M
Peg 22h24'47"31d19'
Freya & Frieder
Uma 13h39'11"48d31'
Freya Dylan Joy's Star
Lyr 18h25'23"38d59'
Freyer,Alec Cameron
Sgr 19h23'42"-45d8'
Freyermuth,Barbara
Peg 23h4'42"21d40'
Freytag,Robert P
Her 16h17'46"21d26'
Fricano,Marina
Cyg 21h26'11"40d7'
Fricano,Paul Anthony A Star Is
Born
Ori 5h54'1"14d38'
Frichtl,Joshua Thomas
Boo 15h9'40"32d4'
Frick,H Douglass
Ori 5h55'1"15d17'
Fricke,Sabine
Vir 12h25'46"-6d59'
Fricker,Bill & Cathy Marino
Cam 6h37'1"68d10'
Fricker,Edith L
Cam 4h8'18"71d4'
Fricotin,Chaton Babymou
Tri 1h58'39"34d23'
Friday
Mon 6h21'48"-1d11'
Friday,Jr,Adam A
Aur 4h47'44"50d40'
Friddell,Richard Guy "Rip"
Cnv 13h51'0"40d31'
Fridel Hans
Boo 14h10'25"32d23'
Fridella,Cosmo
Vul 20h44'59"28d30'
Fridley,Marvin & Doris
Boo 14h30'6"35d59'AB
Fridlund,Donald Kenneth
Aur 4h42'24"31d5'
Fridlund,Mark John
Boo 14h10'1"50d32'
Frieband,Morris
Per 3h31'36"40d42'
Friebe,Gisela
Sco 17h54'46"-31d0'
Frieberg,Janet
And 2h11'0"41d48'
Frieberger,Gordon & Miriam
Lmi 9h55'26"37d53'
Friebis,Sr,George
Cep 22h4'0"67d34'
Fried,David
Tri 1h54'54"27d23'
Fried,James Powers
Aur 4h58'13"50d39'
Fried,Larry & Mrs
Crb 16h11'10"34d56'
Fried,Rachel
Cas 1h49'18"75d57'
Fried,Robert B
Del 20h18'0"14d10'
Fried,Sheri & Hal Danzer
Her 16h45'35"30d57'
Frieda
Aql 18h41'28"1d28'
Frieda
Cyg 19h58'33"38d24'
Frieda Carlene
Hya 8h20'23"0d51'
Frieda Mae
Eri 2h53'17"-2d6'
Friedbauer's Fire
Aql 18h55'46"11d42'
Friedberg My Own Star,
Joshua Lee
Ser 15h44'41"24d49'
Friedberg My Own Star, Brian
Per 3h6'53"46d24'
Friedberg Star Muglet,
Matthew Eli
Lac 22h9'46"46d7'
Friedbergs's Personal
Star,Alexander
Her 16h28'0"34d10'
Friede,Michael Ross
Boo 15h26'52"42d3'
Friedekarl
Cap 20h35'26"-18d48'
Friedel,Herbert Jack
Cep 23h12'35"60d20'
Friedel,John W
Cnc 8h58'39"18d17'
Friedel,Konrad
Dra 16h19'30"52d29'
Friedel,Marilyn Neumann
Lyn 7h46'26"42d30'
Frieden,Gina
Mon 6h12'55"-10d11'
Friedenberg,Eva
Vir 12h59'31"-11d33'
Friedkin,Suzie
Hya 8h43'1"2d22'
Friedkin,Thomas Hoyt
Lac 22h20'32"53d54'
Friedl Rosi
Tau 5h53'14"24d16'
Friedl,Hans
Lyn 8h0'44"42d13'
Friedland,Dorothy A
Lyr 19h21m0"30d27'
Friedlander,Angel
Sge 19h0'29"19d26'
Friedlander,Ari Ezra
Cam 6h6'11"56d14'
Friedlander,Bob
Sge 19h19'24"16d60'
Friedlander,Bob
Uma 8h50'10"50d40'
Friedlander,Cherryl Thompson
Cam 13h32'47"78d38'
Friedlander,Harvey Lee
Per 2h26'52"57d15'
Friedlander,Jack
Com 12h9'43"19d29'
Friedlander,Michael H
Dra 17h8'34"62d5'
Friedman's Birhday
Star,Stanley
Sco 16h52'38"-38d28'
Friedman,Alan
Cnv 12h44'0"41d12'
Friedman,Amie Morgan
Dra 15h12'0"63d57'
Friedman,Arthur
Lac 22h52'46"40d40'
Friedman,Benjamin
Oph 17h14'50"-20d51'
Friedman,Bruce
Aur 4h56'14"51d25'
Friedman,Christine Louise
Peg 21h57'1"29d25'
Friedman,Christopher
Benjamin
Her 16h19'35"23d26'
Friedman,David
Hya 8h11'0"6d9'
Friedman,Dr Carl
Cep 21h16'36"68d20'
Friedman,Eileen "Jeckle"
Cam 7h5'37"70d29'
Friedman,Gabrielle Simone
Mon 6h55'22"-5d38'
Friedman,Gerald
Dra 17h1'52"67d38'
Friedman,Harry D
Cet 2h35'0"1d14'
Friedman,Harry R
Cma 6h54'7"-18d25'
Friedman,Henrietta Marks
Psc 22h54'40"5d13'
Friedman,Howard
Her 18h12'14"37d34'
Friedman,Jacqueline Ruth
Eri 3h34'36"-4d19'
Friedman,Jared Benjamin
Cam 6h18'47"80d37'
Friedman,Jason Alexander
Boo 15h7'56"24d60'
Friedman,Jeffrey
Aql 20h7'39"4d58'
Friedman,Jeffrey Arnold
Aur 4h38'42"30d30'
Friedman,Jerald
Ser 15h31'0"1d41'
Friedman,Larry
Sex 10h30'48"-5d3'
Friedman,Leroy Neil
Gem 7h29'44"34d12'
Friedman,Lisa Samet
Psc 1h22'32"12d13'
Friedman,Madeline
Mon 6h19'59"9d41'
Friedman,Mallory Adler
Cyg 19h30'44"34d17'
Friedman,MD,Stephen J
Umi 15h5'21"68d28'
Friedman,Michael B
Cep 23h17'23"64d35'
Friedman,Randolph
Aql 19h36'13"-6d11'
Friedman,Raymond C
Aql 19h53'19"15d36'
Friedman,Richard P
Uma 11h24'1"43d33'
Friedman,Robert H
Per 3h41'58"51d13'
Friedman,Robert Mark
Hya 8h25'17"-10d38'
Friedman,Ruby May
Eri 3h40'31"-13d0'
Friedman,Sadie Wirth
And 1h59'1"47d13'
Friedman,Shelby Paige
Cyg 19h30'42"31d4'
Friedman,Shirlee & Phil
Cyg 20h55'17"53d25'
Friedman,Teal G
Cet 2h59'16"-0d1'
Friednamer,Florence
Lyn 8h28'44"38d31'
Friedrich,Aaron McLean
Per 3h52'53"38d54'
Friedrich,Elijah James
Aur 5h19'23"41d6'
Friedrich,Happy Star Christine
& Karl
Umi 17h27'13"75d1'
Friedrich,Helmut
Dra 16h46'40"73d27'
Friedrich,Herbert
Cnc 8h0'33"10d33'
Friedrich,Monika
Cnc 8h8'45"31d53'
Friedrich,Natascha
Lib 15h11'37"-23d7'
Friedrich,Paul & Thelma
Cam 4h18'19"61d36'
Friedrichs,Clark
Per 3h1'1"46d15'
Friedrichs,Doris Lucina Mauck
And 0h20'53"40d16'
Friedrichsen,Kai
Gem 7h13'0"33d46'
Friedsam,Andrea
Lib 14h48'50"-1d22'
Friedsberg,William
Eri 3h44'27"-0d41'
Frieg,John Henry
Cnv 12h21'47"35d29'
Friel,James O
Ori 5h33'23"-0d23'
Friel,Josephine Mary
Cyg 21h0'23"38d20'
Friel,Pat
Cam 3h50'44"60d8'
Friel,Paul
Lyn 9h38'11"40d33'
Frielingsdorf,Jordan Louis
Per 3h59'38"52d19'
Friemel,Rainer
Cmi 7h17'27"1d52'
Friend
Lyr 18h38'0"37d3'
Friend 11-19-49,James
Joseph
Sex 10h23'60"-5d7'
Friend Boy Springtime
Hya 8h33'22"-0d35'
Friend,Greg
Aql 19h57'36"12d54'
Friend,Ian John
Ori 5h42'1"11d16'
Friend,Michael Paul
Dra 19h4'36"48d2'
Friend,Nicola Kim
And 0h28'15"30d21'
Friend,Paul
Uma 9h9'17"55d39'
Friend,Robert
Per 1h59'60"56d34'
Friendly Ed
Sct 18h52'55"-6d24'
Friends Immortal KSR
Uma 13h43'28"61d9'
Friends of Burlington
Cam 6h45'46"80d31'

Friends Through Eternity
Uma 10h39'1"50d41'

Friends Will Be Friends
Her 16h38'22"11d30'

Friends,Alexander James
Her 17h47'1"40d17'

Friendship
Lyr 19h7'1"38d56'

Friendship 5
Uma 10h23'41"40d37'

Friendship Circle,The
Tri 2h5'0"31d36'

Frierson,Jerri
Scl 23h14'0"-34d59'

Frierson-Johnson,Carol
Peg 22h31'1"27d36'

Friery,Paula Ann
Umi 15h9'0"70'16'

Fries,Darrell William
Equ 21h22'56"3d0'

Fries,Morgan Leila
Crb 15h36'54"26d5'

Fries,Randolph H
Cam 5h52'58"56d14'

Fries,Susan Renee
Vul 20h45'59"28d28'

Friese,Bob
Aql 19h54'29"15d2'

Friese,Chandra
Peg 21h56'1"31d45'

Friese,Dr Klaus
Gem 7h23'36"35d17'

Friese,Hedda
Sco 17h1'49"-30d50'

Friese,Karen A
Mon 6h29'50"-7d22'A

Friese,Katherine L
Mon 6h29'50"-7d22'B

Friesen,Brian
Cep 22h27'21"68d26'

Friesen,Jennifer Marie
Mon 7h16"13"-6d26'

Friesen,Sean Heath
Peg 22h38'13"8d19'

Friesen,Sharyn R
Aql 19h1'37"0d8'

Friess,Gerd
Del 20h13'20"13d13'

Friesz,Bill
Per 3h54'35"31d6'

FRIEVER
Aql 20h32'26"-6d43'

Frige #18
Cyg 20h0'33"38d25'

Frighetto,Joseph Michael
Lac 22h21'33"55d32'

Frignani,Paolo
Tri 1h47'24"31d1'

Frigo,Lopez
Boo 15h10'58"50d51'

Frigon,Lorraine
Cas 0h40'15"76d26'

Friis "Dad" Marty
Ser 15h53'26"17d40'

Frimer,Lone
Cep 23h6'59"67d30'

Frimmer,Helene
Cyg 21h1"46"40d23'

Frimodig,Frank V
Cyg 20h33'1"38d6'

Frimoth,Marshall Rytter
Leo 9h30'47"7d12'

Frimousse
Her 16h55'1"30d13'

Frimout,Dirk
Lyn 6h39'14"59d26'

Frink,Shelby Davis
Uma 11h34'27"32d12'

Frino,Sarah Elizabeth
Cyg 19h30'20"36d27'

Frinsilla,Ulrike
Lib 14h23'25"-22d39'

Frinzell,Melody
Peg 22h45'25"35d9'

Friou,Bernard
Sgr 19h33'7"-31d56'

Frisani,Piero
Lyr 18h38'50"29d47'

Frisbee,Pastor Kenneth L
Uma 10h44'13"52d4'

Frisby My Hero,David
Ori 4h55'0"0d25'

Frisby,Timothy Scott
Her 17h4'0"42d36'

Frisch,Associate Professor Eric V
Tau 4h3'10"21d5'

Frisch,Dr Eric V
Per 2h58'57"32d42'

Frisch,Joel
Lac 22h23'1"37d31'

Frisch,Samuel Thomas
Aql 19h46'28"12d52'

Frisch,Steve
Per 3h52'34"38d56'

Frisch,Wolfgang
Lyr 19h13'30"31d30'

Frische
Cet 3h0'0"0d53'

Frisella,Scott & Tracy Ehlmann
Uma 11h40'1"50d26'

Frishberg,Morton C
Sco 17h27'37"-40d59'

Frishman,Eileen V
Lyr 19h21'23"31d37'

Frisk Pentium Star
Cep 22h45'35"70d49'

Friskney,Gail Alison
Aql 20h2'31"0d27'

Frismavi,Paolozzi
Cyg 20h20'34"40d45'

Frisoska,William R
Her 17h28'30"21d27'

Friss,Dylan Thomas
Her 17h50'22"48d49'

Friss-Speck,Silke
Psc 0h6'44"-5d45'

Frisse,Elizabeth
Cas 2h45'28"70d39'

Frisse,John L
Ori 5h56'20"15d25'

Frith,Benny G
Her 17h0'21"50d49'

Friton,Bernice J
Tri 2h7'21"32d11'

Friton,Carolyn Ashley
Cam 3h22'1"61d12'

Friton,John Francis
Per 3h18'47"41d48'

Friton,Natalie Elizabeth
And 23h20'56"44d29'

Fritsch,Klaus-Peter
Lyr 19h19'60"30d29'

Fritter
Cet 3h5'36"6d55'

Fritton,Virginia Daley
Boo 14h44'1"46d59'

Fritts,Peter & Susan
Cyg 21h56'36"53d3'

Fritz
Psc 23h34'33"-1d23'

Fritz
Lib 15h30'59"-23d57'

Fritz
Ori 6h4'16"20d28'

Fritz
Aur 7h20'45"38d37'

Fritz,Charlene May Zottarelle
Cas 22h56'24"55d17'

Fritz,Dorothy
Uma 11h55'24"52d46'

Fritz,Edward G
Ori 5h47'0"11d1'

Fritz,Hailee Alina
Lyn 9h7'45"38d12'

Fritz,Nicole Kirsten
Lyr 18h45'22"30d7'

Fritz,Patrick Andrew
Psc 1h22'31"33d24'

Fritz-Hermann,Margat & Volker
Uma 8h33'19"47d3'

Fritzi
Aur 6h33'32"35d38'

Fritzy's Star
Tri 2h41'17"31d38'

Friza
Cep 21h19'50"58d34'

Frizell II,Travis Michael
Ser 15h17'46"8d19'

Frizzell IV,Leigh Hartley
Aql 19h4'37"2d51'

Frizzell,Gloria
Eri 4h8'54"-0d1'

Frizzell,Jr,Richard T
Aql 18h42'1"0d3'

Frizzell,Richard T
Ser 15h59'26"10d36'

Frizzell,Richard W
Uma 13h45'32"50d59'

Frizzell,Scott
Aur 7h3'25"37d30'

Frketish,Brenda Withrow
And 1h51'56"47d2'

Froats,AED-HWE Casselman 061234
Eri 4h34'44"-11d56'

Frodo B
Hya 8h54'13"-7d31'

Froehle,Buddy
Cet 2h34'45"0d43'

Froehlich,Charlotte Anne
Cas 0h59'29"62d15'

Froelich,Kelly
Aql 19h56'21"7d40'

Froemming,Bruce N
Cet 1h53'28"-8d50'

Froemming,Elaine
Lyr 19h20'34"38d18'

Froggy
Cep 0h3'1"67d5'

Froggy
Equ 21h2'31"10d27'

Frogman
Eri 2h51'25"-11d30'

Frogtown Storybooks & Stuff
Peg 22h14'51"5d52'

Frogue,Jon Timothy
Oph 17h56'25"-8d45'

Frohlich,Patricia Mier
Sco 17h1'19"-38d46'

Frohn "Mom",Ruby
Tri 2h45'45"32d9'

Frohn,Ulrike
Vir 14h30'39"7d16'

Frohne,Otto
Uma 8h52'0"52d22'

Froio,Glen
Aur 5h18'47"42d33'

Frois 831,J P
Vul 20h39'43"25d27'

Froistad,Papa Richard
Aur 4h47'24"41d13'

Froitzheim 19/12/1995, Lena
Boo 13h38'29"12d47'

Froitzheim,Benjamin M W
Ari 2h0'34"21d5'

Froitzheim,Bettina
Sco 17h30'24"-30d9'

Frojen,"Fluffy"
Cyg 20h57'48"30d53'

Frojo,Roberto
Lep 5h22'11"-24d57'

Frola
Peg 22h34'49"33d56'

Frolic A Star to Steer By
Cet 1h32'48"-10d54'

From Dulcinea With Love
Mon 6h49'60"11d22'

From Mother Reiko to Lovely Sons
Cnc 8h54'27"17d20'

From Tennis To Eternity
Boo 15h39'21"47d45'

From Tomomi To Hiroshi
Cap 20h59'36"-20d38'

From,Theodora
Uma 10h20'21"68d24'

Froman,Steffany Brett
Lyn 7h7'1"59d35'

Fromberg,Mora Eve
Uma 11h44'23"37d51'

Froment,Perrine
Umi 17h21'1"75d12'

Fromentaud,Frédérique
Aur 5h2'19"38d4'

Fromowitz,Jeffrey Stuart
Vul 19h22'24"25d22'

Fronek,Nicole
Lyr 18h33'58"44d54'

Froney,Dean & Jayne
Cyg 19h31'37"33d28'

Fronrath,Dennis
Oph 16h24'22"-6d43'

Frontzak,Irving A
Boo 14h19'44"37d25'

Fronzoni,Alessio
Sgr 19h30'9"-32d53'

Fronzoni,April Elizabeth
Lyn 9h12'43"45d10'

Frorian
Aur 4h47'16"40d43'

Frosch-Furz-Fisch- Polizei
Vir 12h0'38"12d5'

Froschle,Marion
And 0h16'26"38d48'

Frost 30,Jack
Lac 22h30'48"40d51'

Frost II,John W
Cet 3h6'57"6d2'

Frost,Amy Melissa
And 0h4'37"30d52'

Frost,Barbara Ellen
Cas 23h14'49"60d39'

Frost,Betsy A
Eri 3h26'57"-5d41'

Frost,David Richard
Sex 10h1'56"-1d60'

Frost,Douglas Roger
Cnv 12h16'21"40d51'

Frost,H J
Lyr 19h20'55"31d28'

Frost,Ira
Leo 10h1'40"18d37'

Frost,Jack
Cep 21h20'34"85d30'

Frost,Janice
Com 12h49'23"20d2'

Frost,Jeanie Ruth
Mon 7h16'14"-5d27'

Frost,Joan Kelly
Lmi 9h57'1"38d20'

Frost,Jr,Robert S
Aur 6h9'0"37d9'

Frost,Kathleen Ann
And 1h20'54"38d53'

Frost,Laura G
Mon 0h1'30"-1d41'

Frost,Maryanne
And 1h56'39"41d13'

Frost,MD,Allan R
Per 2h53'27"34d15'

Frost,Nadia
Cyg 21h46'51"36d47'

Frost,Sarah
Cnc 8h35'60"18d51'

Frost,Stephen
Lib 15h19'0"-23d5'

Frost,Steve & Erin
Cyg 21h38'39"38d49'

Frost,William
Per 2h50'35"46d35'

Frost,William
Aql 19h44'52"14d10'

Frost-Povlsen,Rikke
Peg 22h27'43"30d14'

Frostick,Lucy
Umi 13h10'28"70d46'

Frosty
Cyg 21h8'52"31d11'

Frosty JPS
Cam 3h22'53"60d46'

Frosty My Love
Lac 22h35'35"56d31'

Frowerk,Ullrich
Cnc 8h9'1"32d23'

Fruchter,David A
Lac 22h55'45"51d12'

Fruchtman,Erin Elissa
Hya 8h17'29"-6d21'

Frueh,ann-Bruckschen, Elsa
Tau 5h53'42"24d3'

Frueh,Reverend Henry Chapin
Per 3h11'28"46d12'

Frueh,Robert Henry
Lac 22h9'52"46d30'

Fruehauf,Eugene
Cam 7h45'47"60d54'

Fruen,Jeanette Lona
Lyr 19h1'40"31d43'

Fruend,Susan
Lyn 8h8'48"58d10'

Fruge
Oph 16h50'1"-28d44'

Frugier,Jean-Marie
Peg 23h34'33"10d51'

Fruity
Cas 0h57'19"60d40'

Frummet,Franz Xavier
Aur 7h5'11"38d36'

Frundel,Summer Madonna
Cnc 7h55'1"11d36'

Fruner,John Stephen
Lac 22h52'39"51d41'

Frusciello,Giusy
Cyg 21h22'1"40d45'

Frush,Kiran Rajender
Vul 19h47'45"28d33'

Frush,Sarina Rajender
Tri 2h25'33"28d25'

Frutchey,Christine
Boo 14h25'11"28d27'

Fruth,Stephan
Uma 13h57'28"50d24'

Fry's Wish-Upon-Star, Susan & Ed
Eri 3h3'44"-12d25'

Fry,Andrew James
Lyn 7h30'1"40d37'

Fry,D C Bunny
Dra 20h17'53"61d52'

Fry,Deborah
And 2h8'39"41d38'

Fry,Denise Ann
Cas 22h56'23"53d54'

Fry,Mary Kelly
And 1h57'51"39d12'

Fry,Nicole Jean
And 2h34'23"39d26'

Fry,Oma Hazel (Lovely) Gillette
Her 17h13'39"48d42'

Fry,Robert M
Her 16h1'1"40d9'

Fry,Tristan James
Cet 1h29'32"-3d14'

Fry,William Grant
Ser 18h5'10"-14d21'

Frycklund,Daniel James
Lac 22h12'22"48d18'

Frycklund,David George
Dra 16h59'13"73d49'

Frydendall,Shelly Marie
Uma 9h8'18"48d6'

Frydrych,Gabriele
Psc 1h43'11"10d27'

Frye,Deryl Eugene
Crb 15h26'16"31d1'

Frye,Lois
Crb 15h47'47"27d3'

Frye,Margaret
Equ 20h59'0"2d56'

Frye,Psalm 19-TM "Pete"
Uma 11h8'14"55d50'

Frye,Tiffany Dianne
And 1h44'1"48d60'

Fryer,Jr,George R
Aur 5h55'29"31d51'

Fryman,Katie
And 23h30'37"46d41'

Fryman,Meggie
Lyn 8h46'34"37d3'

Frymire,Carrie
Peg 21h59'56"22d40'

Frysinger,Todd Michael
Aur 6h10'35"45d29'

Fryz,Linda A
Del 21h4'31"12d46'

Fränznick,Jochen
Tau 4h4'1"0d40'

Fräss-Ehrfeld,Dr Claudia
Cyg 21h31'35"40d6'

Fréderic et Sabrina 7/17/92
Per 3h52'50"37d59'

Frédérik A de St Exupéry's brother
Uma 10h37'16"50d31'

Frédérique
Eri 4h4'27"-18d18'

Frödin,Margareta
Aur 7h6'59"38d38'

Fröhlich,Ariane
And 23h3'30"51d44'

Fröhlich,Dieter
Lac 22h3'25"51d13'

Fröhlich,Jörg
Her 18h5'31"40d4'

Fröhlich,Maria
Cam 4h9'59"67d43'

Fröhlig,Gottfried
Lmi 10h42'0"27d29'

Frömling 1
Sgr 18h47'37"-26d23'

Fröschelein Ute Glatzel
Lib 15h35'12"-20d3'

Frühsammer,Rainer
Ser 15h29'0"18d37'

Frühwirth,Thomas
Vir 12h54'38"-17d52'

Fucci,Pete
Her 16h53'46"27d23'

Fuccy,Robert Frank
Dra 16h18'43"68d18'

Fuchs Star Bright- Mary Ann Rose
And 2h27'0"44d37'

Fuchs,Birgit
Cnc 9h14'30"7d7'

Fuchs,Chris
Cet 2h33'46"9d13'

Fuchs,Don Carol DJ & Timmy
Lac 22h26'31"37d31'

Fuchs,Georg
Lyn 8h0'31"47d16'

Fuchs,Gerard Phillip
Aur 6h0'35"30d17'

Fuchs,Hans-Heinrich
Aur 5h18'37"43d55'

Fuchs,Johnathan Al
Aur 5h29'26"54d36'

Fuchs,Joshua
Peg 22h47'50"31d2'

Fuchs,Michael Wilhelm
Sco 16h59'50"-43d51'

Fuchs-Eternally Your's Hans Dieter
And 23h23'0"43d40'

Fuchsia
Vel 9h43'43"54d8'

Fuchsia Marshmallow
Sge 20h2'11"20d23'

Fuchsmann,Thomas
Aqr 20h59'53"-14d1'

Fudd,Otis
Dra 14h40'37"62d1'

Fudge Brownie
Dra 17h20'33"70d25'

Fudge,Debbroah A
Lyn 9h4'0"44d25'

Fuellgabe,Hans
Oph 17h4'37"1d6'

Fuensanta Hernández Navio
Cep 22h4'0"63d49'

Fuente,David Ira
Hya 9h36'31"0d15'

Fuentes,Elena- Estrella Estupenda
Peg 22h59'12"21d46'

Fuentes,Nancy
And 0h14'29"45d49'

Fuentes,Regan Casey
Peg 22h47'37"34d6'

Fuer immer diis Mautzgl
Cas 3h2'1"57d36'

Fuer Immer Unter Einem Guten Stern
Pho 23h41'29"-40d15'

Fuer Sonny
Cnv 12h22'14"41d47'

Fuerst,Elaine
Vul 20h14'51"22d59'

Fuerst,Tommy
Cnv 12h23'49"50d44'

Fuerstein,Mark
Boo 14h40'20"36d47'

Fuerteventura
Ori 5h18'52"7d54'

Fuessle,Stephen D
Aur 6h1'46"31d34'

FUFA
Ari 3h10'17"18d49'

Fugardi (My Adopted Mom),Star Carrie
And 0h51'25"38d56'

Fugate,Alpha Marie
Del 21h5'33"12d17'

Fugate,Michael Leigh
Dra 17h27'41"73d22'

Fugate,Sasha Luc
Lyn 8h13'19"57d52'

Fugent,Don & Autumn
Sge 19h58'0"18d53'

Fugent,Madonna
Mon 6h56'15"11d30'

Fugent,Sandra
Eri 3h42'55"-16d31'

Fugère,Joey
Her 17h17'56"42d57'

Fuhr,Klaus-Joachim
Ari 2h40'11"28d20'

Fuhrer,Mark & Pat
Uma 11h28'52"40d30'

Fuhrman,Jean
Cyg 20h0'42"31d11'

Fuhrman,Lauren Dina
Cas 1h16'28"66d5'

Fuhrmann,Grandma
Lyn 6h38'0"59d27'

Fuhst,Karin
Psc 0h1'46"-2d1'

Fuip,Jr,John Richard
Oph 17h21'32"10d45'

Fujii,Masao
Aur 4h55'12"41d8'

Fujita,Diana Jeanette
Lmi 10h46'19"30d13'

Fujita,Noriko
Ori 5h6'33"0d54'

Fujito's Star
Sco 16h18'53"-8d54'

Fujiwara,Kazuko
Sco 16h3'20"-15d18'

Fukui,Atsushi
Vir 14h28'6"-2d56'

Fukunaga,Ray & Miyo
Eri 4h56'41"-11d23'

Fukushima,Mel
Eri 3h2'58"-8d40'

Fulcher,C Win
Cam 7h6'43"61d12'

Fulcher,Christopher
Lyn 7h50'35"35d55'

Fulcher,Jr,Gene M
Hya 8h57'0"0d47'

Fulco,Joseph H
Ser 16h15'25"-2d26'

Fulford,Dr Adrian
Per 3h28'1"40d53'

Fulforth,Victoria Rose
And 23h58'34"51d37'

Fulgeri,Iosina
Lyn 7h51'18"50d8'

Fulham-Choen,Olivia Celeste
Cnc 8h27'32"22d13'

Fulio y Eiko
Leo 10h18'42"11d19'

Fulk,Arneda Haney Lopshire
Sge 19h42'32"16d32'

Fulk,Patricia Ann Whitinger
Cet 1h30'42"-11d47'

Fulknier,Lauren Frazier
And 23h30'40"40d10'

Full,David Damian
Her 16h7'22"41d41'

Full-Hearted EH plus CT
Cep 20h26'36"75d58'

Fullam,Anita R
Vul 20h14'26"23d10'

Fullam,Sr,Andrew
Per 3h12'1"48d14'

Fullarton,Amanda-Jane
Cas 0h10'11"64d25'

Fuller's Star of Dynasty,Chuck
Cep 22h9'1"53d9'

Fuller,Angela Dawn
And 0h20'39"33d55'

Fuller,Angelina Marie
Cyg 19h24'58"32d16'

Fuller,Blythe McKenzie
Gem 6h56'39"16d8'

Fuller,Bonnie Hurowitz
Umi 17h12'22"76d22'

Fuller,Charles Jerry
Cet 2h42'14"2d25'

Fuller,Christopher Thomas
Umi 15h59'41"80d4'

Fuller,Daniel
Ser 15h53'50"23d57'

Fuller,David
Aql 20h0'0"10d4'

Fuller,Eric Stephen
Per 2h21'19"58d21'

Fuller,Garrett
Boo 14h26'39"21d55'

Fuller,Gary Someone's Always
Aql 20h10'28"11d36'

Fuller,Hazel L
Lmi 10h18'0"35d20'

Fuller,Jack
Mon 6h23'38"8d4'

Fuller,Jean Grace
Boo 14h58'32"34d51'

Fuller,Jeanette Evonne
Peg 23h19'19"33d22'

Fuller,Jennifer & Lou Sisl
Uma 9h9'0"51d16'

Fuller,John H
Lmi 10h15'41"35d58'

Fuller,John"J J"
Her 17h7'0"41d32'

Fuller,Jon Courtney
Uma 10h24'11"54d1'

Fuller,Jonathon
Cyg 21h32'53"41d8'

Fuller,Jr,Howard Emanuel
Ari 2h56'16"22d4'

Fuller,Karen
Lyn 8h28'1"46d43'

Fuller,Kent & Sandra
Aur 5h54'48"50d21'

Fuller,Lindsey Katherine
Mon 7h24'30"-10d54'

Fuller,Lisa
Mon 6h20'1"7d30'

Fuller,Lloyd "Bud"
Her 17h14'56"27d46'

Fuller,Lynn & Richard
Lac 22h53'16"52d50'

Fuller,Patricia
And 0h8'1"27d40'

Fuller,Pete
Leo 10h50'36"-5d7'

Fuller,Richard Harlan
Oph 17h59'19"8d52'

Fuller,Robert & Judy
Leo 11h4'55"-2d27'

Fuller,Robert W & Carolyn R
Lyn 7h50'21"36d38'

Fuller,Roger
Ori 6h14'10"8d10'

Fuller,Ronnie
Cnv 12h7'0"37d8'

Fuller,Sarah-Beth
Tri 2h11'12"30d16'

Fuller,Stephen J
Per 2h39'1"43d32'

Fuller,Tawny Lee
Lyn 8h23'39"42d5'

Fuller,Will
Aur 5h26'37"40d43'

Fuller,William T
Vul 19h22'28"26d12'

Fuller-My Shining Star Zachary
Gem 6h52'36"12d13'

Fullerton
Uma 9h52'55"51d9'

Fullerton,Catherine H M
Cas 0h37'20"60d15'

Fullerton,Colin Sandlin
Cma 6h51'33"-15d44'

Fullerton,Leon & Zana
Sgr 19h22'7"-44d28'

Fullerton,Timothy G
Cam 4h2'22"68d13'

Fulmer,Christine Daniel
Lmi 10h28'40"28d24'

Fulmer,John R
Tau 5h49'32"28d23'

Fulmer,Jr,Donald W
Uma 10h55'27"71d46'

Fulmer,Tanner
Her 17h3'53"20d53'

Fulmer,Ty
Ori 6h0'11"8d31'

Fulton,Claire
Cyg 19h31'38"33d19'

Fulton,David
Tri 2h22'0"32d26'

Fulton,Donna E
Aur 6h14'23"30d8'

Fulton,Jean A
Her 18h41'3"19d24'A

Fulton,Jeannine Barnes
Com 12h23'45"27d41'

Fulton,Nancy A
Peg 22h7'55"21d42'

Fulton,Pamela J
Crb 16h6'51"30d13'

Fulton,Richard Charles
Her 16h40'13"34d23'

Fulton,Robert A
Her 17h36'42"24d40'

Fulton,Robert B
Her 18h41'3"19d24'B

Fulton,Robert E
Cep 20h52'57"67d49'

Fulton,Robert Kenneth
Her 18h5'36"31d38'

Fulton,Robert Lee
Lyn 6h30'18"60d33'

Fulton,Thomas Edward
Her 17h2'42"50d5'

Fulton,Yolanda Aurora
Cet 0h27'16"-1d47'

Fults,David Michael
Aur 7h23'40"40d31'

Fultz,C B
Aur 6h11'14"32d0'

Fultz,Christopher Joseph
Uma 9h24'30"55d27'
Fultz,David Alan
Mon 7h49'52"-6d39'
Fulvia
Tri 1h54'0"28d54'
Fumagalli,Luisella
Tau 5h25'15"16d23'
Fumagalli,Raffaella
Ant 10h44'5"-33d3'
Fumearis,Katie
Lyr 18h32'41"37d53'
Fumo,Vincent J
Aur 5h58'32"29d59'
Funaro,Kathie
Cam 5h51'44"60d55'
Funaro,Lucilla
Aql 19h34'53"-0d17'
Funcky
Dra 17h58'12"82d51'
Fundaro,Todd
Aql 18h59'44"-2d45'
Funderburg,Blu
Lmi 10h38'50"32d35'
Funderburg,Carl & Aroline
Del 20h20'11"10d26'
Funderburg,Jr,Carl Lee
Sex 10h29'12"-2d4'
Funderburk,Jr,William Albert
Aql 18h58'17"3d1'
Funderburk,Thomas Duncan
Cet 0h27'60"-11d43'
Fundus,KelliAnn
And 0h24'22"37d24'
Fung,Benjamin Chi Cheong
Uma 11h19'1"50d6'
Funk
Boo 14h31'0"30d55'
Funk III I Love You
Erin,William H
Cep 0h19'0"75d41'
Funk,Alan S
Aur 6h32'19"33d24'
Funk,Amanda Dorothy
And 23h17'16"40d13'
Funk,Amy Raub
Lyn 7h57'13"52d15'
Funk,Andrew William
Lac 22h30'56"38d7'
Funk,Byrna
Cet 2h44'48"1d40'
Funk,Ella Frances
Peg 23h45'45"31d17'
Funk,Gerhard
Cmi 7h6'45"5d30'
Funk,Gregory Charles
Ser 15h37'1"21d49'
Funk,Joan
Peg 22h19'33"29d17'
Funk,Jr,Seville S
Aql 20h5'52"7d1'
Funk,Julia "G G"
Mon 6h19'48"6d52'
Funk,Lowell Dixon
Cet 2h35'32"-4d33'
Funk,Matthew Ethan
Cnv 12h50'33"40d22'
Funk,Michael J
Her 17h55'35"31d40'
Funk,Raymond
Aur 7h5'22"41d16'
Funk,Thomas P
Lac 22h53'45"50d27'
Funk,Tristan,Matthew, Nathan & Justin
Peg 22h33'29"21d31'
Funk-50th,Byrna & Joe
Cet 2h25'1"5d46'
Funke,Bernhard
Cmi 7h32'20"1d44'
Funke,H D
Lyr 19h23'56"30d37'
Funke,Jane
Ori 5h57'24"17d55'
Funkhouser,Zachary Andrew
Del 20h52'1"7d54'
Funny,Susie B
Vul 19h19'38"26d41'
Funquest Chabuk
Equ 21h0'46"10d39'
Fuorry,Thomas A
Aql 19h58'53"0d21'
Fuqua,Meg
Com 12h11'10"20d11'
Furbeck,Jr,Howard Rollins
Dra 17h37'42"55d47'B
Furbeck,Rose Victorine
Dra 17h37'42"55d47'A
Furbetta,Gianluca
Cam 7h3'60"61d27'
Furchtsam,Herman "Slim"
Per 2h53'20"40d36'
Fureigh,Kierstin Elaine
And 0h53'14"40d54'
Furger,Heidi
Cam 5h47'56"60d59'
Furgiuele,Teresa & Eugenio
Cyg 19h24'1"31d18'
Furgueson "Poppa", William G
Sex 10h20'43"-5d15'
Furhmann,Eugenia Livaros
Cas 3h8'0"61d51'
Furia
Pyx 8h41'8"-28d30'
Furin,Stephanie Leigh
And 23h1'40"51d5'
Furino,Linda G
Aur 4h58'54"30d52'
Furleiter,Erica
And 2h2'54"40d32'
Furlong,Bryan
Uma 11h9'33"55d50'
Furlong,Lynne
Cas 0h26'21"61d23'
Furmaga,Jacob Stephen
Aur 6h51'54"35d37'
Furman,Eliezer
Aql 18h48'48"11d6'
Furman,George & Martha
Lac 21h55'28"38d47'
Furman,Stanley Albert
Aql 19h50'52"15d46'
Furman,Wendi Hope
And 0h18'49"33d4'
Furmanchik IV,Andrew Steven
Aql 18h55'36"-0d24'
Furnari,Vickic
Uma 11h14'48"58d26'
Furnell,Nicole
Lyn 8h12'33"44d31'
Furnell,Ruth
Vul 19h16'54"25d22'
Furnelle,Pascale
Cas 1h10'25"58d19'
Furness,James
Cep 22h24'48"63d55'
Furney,Mary Craft
And 1h59'46"40d44'
Furnham,Professor Adrian
Uma 14h0'48"48d30'
Furnish,David
Per 3h13'39"50d26'
Furnish,David
Cep 21h20'43"55d42'
Furniss,Francis X
Lyr 18h48'26"30d40'
Furphy,Judith
Ari 2h30'41"24d51'
Furrer,David
Lyn 8h21'41"49d45'
Furrer,Patrick James
Psc 23h0'19"6d50'
Furrer,Paul & Robin & Family
Cep 3h10'1"78d0'
Furrieluv
Cyg 20h22'12"38d53'
Furst,Elaine Ann Zacconi
Lyr 18h30'19"30d18'
Furstenau,Bob & Jeanette
Dra 17h35'28"60d43'
Furtado,Steven Costa
Lac 22h27'17"55d46'
Furtaw,Michael
Aur 6h56'19"37d9'
Furtek Elex
Uma 14h2'50"50d27'
Furth,"MEF" Mari Ellen
Cyg 20h4'42"39d35'
Furuhjelm,Johan
Cep 21h12'23"68d47'
Furukawa,Tyler Hisashi
Cet 0h56'37"-0d37'
Furuyama,Satoshi
Ori 6h16'12"20d16'
Fury Jamar Toti-Stahl
Boo 15h19'33"50d9'
Fury,Brandon J
Her 16h37'21"32d23'
Furzibär
Lmi 10h0'24"30d26'
Fusako
Tau 5h26'29"27d46'
Fusaro,Vincent
Her 18h45'1"12d28'
Fusco,Arnold C
Cep 22h7'35"58d55'
Fusco,Bob
Ori 5h57'1"14d6'
Fusco,Celia Anne
Aqr 21h59'13"0d24'
Fusco,Erna
Uma 11h31'51"32d3'
Fusco,Lorraine
Cas 2h7'57"63d48'
Fusco,Maria
Peg 21h44'48"27d55'
Fusco,Matt
Dra 19h33'30"61d33'
Fusco,Rosemarie
Cyg 20h8'31"40d33'
Fusco,Stephanie A
Aur 5h30'25"38d26'
Fusconi,Elisa
Cas 0h10'14"58d13'
Fuselier,Donald Lee
Eri 2h51'45"-17d12'
Fuselier,Kevin Lee
Eri 2h55'21"-18d13'
Fuselier,Michelle Rene
Eri 2h48'40"-17d6'
Fuson,Gerry & Jeri Ann
Cyg 19h55'21"30d25'
Fuson,Jeremy Neal
Cep 22h7'14"60d5'
Fuson,Jerry & Jeri Ann
Uma 11h1'0"47d8'
Fuss,Harold Grant
Per 2h48'45"46d22'
Fussell,Anna Marie Bourdier
Aql 18h58'1"-5d40'
Fusselman,NU 'Huskers
Fan,Gary E
Aur 5h7'55"44d26'
Futchko,Alyshia Rae
Cyg 20h23'55"38d29'
Futia,Melissa Ann
Cas 0h24'1"72d5'
Futterman,Marlene
Cet 2h30'18"0d19'
Futuraprima
Nor 16h23'35"-55d0'
Future Atsuhiro
Cap 20h58'45"-16d38'
Future Magic
Lib 14h43'29"-13d33'
Fuursted,Camilla
Umi 13h25'50"71d20'
Fuzzy Haired Doo-Doo
Ind 20h29'33"-54d2'
Fuß,Helmut
Lyr 19h21'32"31d51'
Fwsco,Rocky,Jennifer and Anthony
Eri 2h54'54"-7d12'
FWX
Dra 12h42'42"71d13'
FXK
Cam 6h16'52"68d2'
Fydean
Uma 9h27'14"47d51'
Fyfe,Richard Alan
Oph 18h41'14"10d10'
FYI
Peg 23h6'22"21d55'
Fylemans Anniversary Star,Margaret & Geoff
Cyg 20h42'52"45d28'
Fylnns Love 1991, William & Kitreana
Cnv 12h38'53"50d1'
Fährenkämper,Horst
Peg 23h34'32"11d21'
Föge,Günter
Aqr 22h46'15"2d16'
Fölster,Herr
Boo 14h23'23"10d41'
Förster,Michael
Aqr 21h0'25"-14d30'
Für DeLyse
Sgr 19h32'24"-43d37'
Für Immer Deiner
Boo 14h48'27"50d30'
Für Immer Liebe
Cam 3h55'0"60d8'
Fürnsinn,Karin
Eri 2h57'34"-7d46'
Fürst,Dieter
Uma 9h19'35"45d33'

G

G & E's Bus Stop
Eri 3h45'52"-5d18'
G & G Basket Player
Psa 22h6'51"-26d58'
G & J delacy
Cyg 20h19'0"39d31'
G & Lee
Peg 22h42'12"34d57'
G A-J P
Ori 6h6'54"6d6'
G AKA:Mr Confidence
Her 18h4'22"31d10'
G AN 3,1415 ERO
Aur 7h8'12"38d30'
G D S 1
Aur 7h7'1"40d32'
G David K
Aqr 23h28'32"-18d54'
G E M
Peg 22h42'43"34d25'
G E P
Dra 20h24'1"64d43'
G E V W
Cep 21h50'20"61d42'B
G Force Production Video Light Sound
Boo 14h21'30"29d59'
G G
Eri 3h14'1"-2d37'
G G
Cep 23h11'35"80d29'
G G Davy's Star
Cet 1h18'1"-13d23'
G G G Woobie
Uma 12h2'51"36d32'
G Money
Cnv 13h29'27"41d11'
G O D
Aql 19h47'48"13d11'
G P
Uma 9h27'46"48d47'
G R Informatica
Pyx 8h51'51"-29d52'
G R M L
Crt 11h2'33"-18d41'
G R S Loves Anne Marie Weizant
Sge 19h5'5"19d1'
G R Star,The
Per 1h32'51"53d27'
G Star
Del 20h17'38"14d14'
G T C III
Lac 21h58'1"38d51'
G W Y:Paradigm
Per 2h54'38"31d23'
G&M "Soulmates"
Gem 7h5'56"21d34'
G-Man 052168
Lyn 7h58'58"36d18'
G-Man/"Star Glaser"
Sex 9h51'12"-6d59'
G-N-G Bear
Dra 14h18'40"63d60'
G-Star Pipe Dreams & Air Castles
Crb 16h4'28"31d27'
Ga St Xa
Per 3h48'15"39d48'
Gaanderse,Rene
Lyr 18h31'1"30d11'
Gaasch,Megan
Aql 20h31'22"0d45'
Gab's Spot Light Forever
Cam 4h11'53"68d15'
Gabaret,Michel
Her 18h2'41"28d55'
Gabay,Raphael
Cru 12h5'45"-63d30'
Gabay-Te Lintleo, Marlene
Cas 1h46'39"60d19'
Gabbard,Michael
Oph 17h25'23"-24d42'
Gabbert,Franziska Viktoria Regina
Lib 14h18'48"-23d29'
Gabbs,Albert & Anna Belle
Lyn 8h6'12"51d2'
Gabe's Wild Ride
Uma 9h39'13"44d59'
Gabel,Laura Stewart
Peg 23h46'0"15d27'
Gabella,Eugenio
Lyn 8h21'34"49d48'
Gabesajo
Mon 6h21'55"-0d36'
Gabilka-Huehnergarth
Aqr 21h4'52"-1d31'
Gabin
And 0h19'27"40d25'
Gabirondo,Claudia Gili
Ser 15h19'33"3d56'
Gabirondo,David Gili
Ser 15h19'45"4d57'
Gabka,mit Dir will ich leben!,Ilse
Ori 5h56'50"18d12'
Gabl,Dave
Uma 11h15'49"53d23'
Gable,Judy Lynn
And 22h57'28"37d33'
Gable,Julia Ann-Lyle Grant Gable
Lyr 19h21'11"30d42'
Gable,Shelly K
Mon 7h43'6"-1d42'
Gablehouse,Linda Diane
Aqr 23h33'59"-11d60'
Gableman,Michael
Boo 13h35'14"20d55'
Gabler,Gabriela
Cnv 12h42'55"39d45'
Gabor,Georgia
Cas 23h38'37"50d26'
Gabrache,Marie Marthe Avedian
Cas 22h57'40"54d54'
Gabrella
Cas 0h56'40"66d38'
Gabrial Louis
Mon 7h41'49"-8d58'
Gabriel
Ori 5h56'31"15d25'
Gabriel
And 0h54'38"41d10'
Gabriel
Aur 4h55'22"38d35'
Gabriel
And 2h20'57"38d57'
Gabriel
Uma 8h55'0"57d43'
Gabriel 93
Her 18h3'36"38d4'
Gabriel Ross
Cas 0h17'0"61d54'
Gabriel Thomas
Peg 21h52'20"32d0'
Gabriel's Lighthouse, Jorge
Lac 22h54'12"55d35'
Gabriel,Aaron
Ser 15h21'39"2d41'
Gabriel,Eric
Her 16h40'17"29d18'
Gabriel,Glenn
Lac 22h16'44"38d46'
Gabriel,Lisa Helen
Lmi 9h55'36"40d1'
Gabriel,Megan Kate
And 0h30'59"22d24'
Gabriel,Pauline
Ser 15h13'1"18d8'
Gabriel,Pierre Jean
Lyn 7h57'58"40d44'
Gabriel,Samuel
Aqr 20h53'57"-6d24'
Gabriel,Sir Scott Edward
Aql 19h30'1"10d34'
Gabriel,Walter
Ori 5h52'48"14d33'
Gabriela
Sct 18h56'0"-4d38'
Gabriela
And 2h27'51"42d31'
Gabriela & Michael
Cyg 21h3'1"30d48'
Gabriela 51760
Cas 1h7'34"61d41'
Gabriela Martina
Per 2h58'28"34d5'
Gabriele
Leo 10h59'36"21d8'
Gabriele
Cnc 9h13'27"8d5'
Gabriele
Cmi 7h20'56"8d52'
Gabriele e Yacopo
Dra 17h58'48"50d30'
Gabriele,Agnes Sylvie
Lyr 18h30'14"33d2'
Gabriele,Bruno Domenico Giovanni
Boo 15h8'15"48d49'
Gabriele,Tibor & Tagath
Vir 11h38'54"-2d47'
Gabriele,Valerie Angelina
And 23h6'33"44d49'
Gabriella
Ant 10h45'38"-36d32'
Gabriella
Eri 3h44'43"-1d27'
Gabriella
Lyr 18h47'50"42d3'
Gabriella
Lyr 18h27'17"44d51'
Gabriella
And 23h0'30"51d33'
Gabriella
Cae 4h50'13"-31d12'
Gabriella
Her 16h46'55"48d17'
Gabriella
Cas 1h59'36"58d14'
Gabriella & Danilo 1994
Vel 9h21'58"-40d53'
Gabriella & Edmondo
Boo 15h4'53"47d38'
Gabriella & Eric-My Stars Forever
Peg 22h51'26"27d22'
Gabriella Jennifer "BFF"
Mon 6h56'13"10d24'
Gabriella Nicola
Cas 0h29'38"60d9'
Gabriella Racheal
Cyg 20h40'16"45d35'
Gabriella's Light
Lyn 7h56'1"58d26'
Gabriella,Emilie
Peg 21h47'42"33d32'
Gabrielle
Dra 15h49'0"60d29'
Gabrielle
Mon 6h57'1"-1d45'
Gabrielle
Cam 5h4'22"67d48'
Gabrielle Alexis
Peg 23h41'31"30d14'
Gabrielle Ana
And 2h25'0"39d29'
Gabrielle Ann
Lyn 7h44'1"50d14'
Gabrielle B
And 1h23'46"35d8'
Gabrielle LL-1996
Peg 22h56'57"20d30'
Gabrielle's D'Letoile
Cyg 21h7'44"31d17'
Gabrielle's Star of Peace
Tau 5h53'0"27d15'
Gabrielle's Testimony
Crb 16h13'32"37d56'
Gabrielle's Twinkling Star
Uma 10h34'54"59d59'
Gabrielle,Joseph Florian
Her 16h38'34"36d26'
Gabrielle,Joshua James
Aur 7h24'55"41d20'
Gabrielle-Antoinette
Cas 2h43'32"61d38'
Gabrieller
Leo 10h57'44"10d18'
Gabriels Sixteenth Birthday Star
Boo 14h0'39"20d32'
Gabrielson,Mike & Glenda
Mon 7h7'14"-0d10'
Gabrielson,Shirley
Com 12h46'0"20d34'
Gabrio
Cam 4h54'43"58d59'
Gabry
Del 20h13'1"10d5'
Gabry Merci
And 2h7'37"40d6'
Gabry,Mercy
Peg 0h4'41"21d13'
Gabus,Cindy L
Aur 5h6'18"37d38'
Gabus,Louis J
Aur 5h50'4"37d19'A
Gabus,W Grace
Aur 5h50'4"37d19'B
Gabut,Jean Pierre
Ori 6h5'44"20d39'
Gaby & Bernd
Cas 0h18'27"60d12'
Gaby & Frank
Uma 11h50'0"50d38'
Gachira,Lois Wanjiro
Psc 23h27'0"5d5'
Gack Star
Per 2h1'0"50d30'
Gadarowski,JD,James J
Cep 0h51'39"80d5'
Gadbaw,Matt
Cyg 21h20'19"40d31'
Gadd,Ernest
Uma 10h4'12"58d52'
Gaddis,Charles Andrew
Aql 20h3'32"4d59'
Gaddis,Connie Lynn
Peg 22h6'34"4d47'
Gaddy,Cheryl L
And 0h20'19"33d22'
Gaddy,Connor Joshua
Cep 22h18'14"55d46'
Gaddy,Irene Mae
Uma 9h3'12"48d42'
Gade,Stefan
Per 3h50'27"38d52'
Gademsky,Janis
And 2h21'0"46d2'
Gadenne,Dominique
Sgr 18h53'1"-26d56'
Gadermann,Hans-Otto
Lyr 19h19'56"31d18'
Gadolini,Daniela
Cnv 13h4'36"51d0'
Gadrow,James Carle
Sex 9h50'26"-0d6'
Gadsden,Jr,Dr Thomas
Cmi 7h28'31"8d25'
Gae
And 1h19'20"40d7'
Gael Marie
Sgr 18h2'45"-28d19'
Gaer,Kira
Cnv 13h21'48"41d18'
Gaertner,Emily Brianna
Gem 7h22'42"30d23'
Gaertner,King for Christopher
Cep 22h9'38"70d4'
Gaetan,Jr,Gaetan
Per 3h22'45"41d4'
Gaetane pour Christian
Tri 1h58'48"31d14'
Gaetanina,Allocca Felice
Leo 9h35'34"8d12'
Gaetano Sr
Lac 22h9'33"46d27'
Gaetano,June
Lyr 18h58'12"45d30'
Gaetano,Mike
Ser 15h54'12"18d19'
Gaeth,Lola Maxine
Cas 1h21'30"67d36'
Gaeth,Ronda Lea
And 23h27'13"40d58'
Gaetjens,Prof Carol Adele
Cas 0h27'16"61d9'
Gafert,Harro
Cet 2h40'31"2d20'
Gaffey,Suzanne
Sco 15h57'60"-25d51'
Gaffin,Jonathan
Dra 9h26'17"81d10'
Gaffin,Reginald L
Aqr 22h2'42"-8d53'
Gaffney,Bernadette
Lyr 18h19'47"42d11'
Gaffney,Bernard Robert
Ser 15h56'38"4d33'
Gaffney,Carol
Cas 0h0'11"54d42'
Gaffney,Chris
Hya 8h13'26"1d11'
Gaffney,Christopher B
Cet 0h58'48"-10d51'
Gaffney,Denice R
Del 20h52'59"4d33'
Gaffney,Helen Katherine
Eri 4h32'55"-12d17'
Gaffney,James Gerard
Aur 7h5'33"41d8'
Gaffney,John J & Lynn R
Cas 2h8'41"67d39'
Gaffny,Tim
Dra 16h44'45"68d4'
Gaffrey,Eric
Aur 6h33'20"40d59'
Gafner,Mike
Sge 20h14'35"17d49'
Gaga
Cnv 12h54'42"48d57'
Gaga's Star
Vul 21h22'23"26d52'
Gagarin,Michael A
Cam 5h57'27"61d26'
Gage,Albert Thomas
Cep 21h42'53"65d34'
Gage,Kellie Ann
Lyn 7h37'1"50d54'
Gage,Michael
Per 4h0'0"38d56'
Gage,Rachel E
Ser 18h0'58"-13d21'
Gage,Richard G
Dra 16h57'26"68d22'
Gage,Skip & Barbara
Dra 17h37'46"57d35'
Gage,Susan
Cas 23h36'17"61d5'
Gagel,Madeline Sinclair
Lyn 6h12'15"60d50'
Gagern,Jennifer Laurie
Tau 5h22'0"20d24'
Gaggia (Chery),Alfredo
Boo 14h59'27"19d17'
Gaggia,Elena
Ori 5h59'19"6d3'
Gagi
Boo 15h1'26"29d9'
Gagic,Dr Nedush
Sgr 19h33'34"-37d51'
Gagliano,Damien
Her 18h6'24"48d7'
Gagliano,Father Philip J
Umi 13h35'0"70d24'
Gagliano,Salvatore M
Boo 14h41'34"37d4'
Gagliano,Taylor Gray
Uma 10h14'51"72d17'
Gaglianone,Giovanni
Lyn 8h55'27"42d3'
Gagliardi,Eleanor
Cam 4h32'33"67d32'
Gagliardi,Joseph
Cam 4h22'58"69d28'
Gagliardi,Michael Dennis
Lyn 7h52'60"40d30'
Gagliardi,Mike
Aur 7h5'48"41d8'
Gagliardi,Mike
Hya 8h16'52"0d34'
Gagliardi,Prince Mikey Mike-Michael
Lyr 18h44'19"47d14'
Gagliardis Star For Life,Dr John
Uma 12h54'21"57d38'
Gaglio A&J Cheese Company,Jack
Aql 19h59'18"15d30'
Gagne,Janice Claire
Dra 16h53'51"60d36'
Gagne,Joseph Tyrell "Ty"
Cet 1h5'38"-18d34'
Gagne,Kerri Lynn
Eri 4h26'0"-11d12'
Gagnier,Kevin Michael
Dra 16h57'33"74d21'A
Gagnier,Kristin Marie
Dra 16h57'33"74d21'B
Gagnon Pianiste Compositeur,André
Per 3h12'24"41d32'
Gagnon Édouard
Cep 22h20'15"66d27'B
Gagnon,Ann
Umi 14h52'27"66d43'

Gagnon,Claire
Cas 0h45'22"73d28'
Gagnon,Donald & Joyce
Crb 15h53'53"27d57'
Gagnon,Francis
Umi 16h49'16"78d46'
Gagnon,Jaqueline
Leo 10h56'37"20d17'
Gagnon,Jeannot
Her 17h54'25"18d34'B
Gagnon,Jennifer
Peg 22h7'30"29d8'
Gagnon,Kaylie Elizabeth
Cas 0h29'27"75d47'
Gagnon,Pierre
Tri 1h45'54"30d58'
Gagnon,Reynold
And 0h23'54"40d22'
Gagnon,Valérie
Uma 8h57'39"47d34'
Gagné,Guy
Her 16h54'17"33d42'
Gagné,Johanne
Umi 13h49'0"75d19'
Gagné,Réjean
Her 17h12'39"20d49'
Gago,Eileen Londeau
Eri 3h59'32"-6d49'
Gago,Frank José
Hya 8h16'35"1d58'
Gags
Dra 12h19'22"68d9'
Gahagan,Tom
Her 16h5'24"40d20'
Gaharan,Michelle
Peg 22h47'15"21d34'
Gahm,Louise B
Vul 20h15'39"23d5'
Gaia
Cam 8h21'48"80d60'
Gaia
Mon 7h51'58"-4d6'
Gaia,Deborah
Cyg 20h19'49"38d55'
Gaida,Brenda C B
And 23h38'1"44d32'
Gaido,Michael K
Cas 0h13'0"60d30'
Gail
Tri 2h46'22"33d43"B
Gail
Lyn 7h38'1"51d6'
Gail
Cyg 20h45'18"38d39'
Gail
Cas 1h15'22"60d5'
Gail & Gordon
Sge 19h58'15"16d51'
Gail & Marty Forever
Per 2h36'42"40d17'
Gail & Tom
Uma 10h9'29"42d23'
Gail 'N' Mal
Peg 22h17'0"30d41'
Gail Allison
Cyg 21h34'1"42d51'
Gail Ann
Cas 0h58'58"61d8'
Gail Elaine,A Family Prodigy
Mon 6h44'11"6d1'
Gail Lost Star of My Life!
Lyn 7h38'40"51d8'
Gail Lynn
Sco 17h54'30"-30d49'
Gail Margret
Mon 6h47'28"10d13'
Gail Marie
Lyn 8h47'16"42d52'
Gail Marie
Aql 19h58'39"11d13'
Gail Posey
Cet 1h52'24"0d33'
Gail The Magnificent
And 1h37'1"39d39'
Gail's Glory
Aur 6h27'36"34d14'
Gail's Heart
Cas 0h16'0"58d12'
Gail's Marathon Star
Tri 1h56'11"26d54'
Gail's Passion
Dra 19h56'12"60d35'
Gail,Susan
Cas 1h4'28"60d9'
Gail,Walter Park
Lac 22h55'19"54d51'
Gailani,Isa
Cas 0h47'45"72d58'
Gailbreath,Katherine Ann
Bryan
Aql 18h46'15"10d29'
Gaillo,Nadia
Per 3h30'14"38d36'
Gails-Trail
Cas 2h26'32"61d2'
Gailstar
Lyn 7h43'11"50d24'
Gaily,Vicki
Lyn 9h6'19"46d25'
Gailya
Aql 18h53'49"-1d19'
Gaimo,Jr,Gerald J
Boo 14h52'58"34d50'
Gaims,Horace M
Her 16h58'14"48d45'
Gain,Tom & Tina
Lyr 18h25'57"44d16'
Gainer,John I
Vir 13h56'44"-0d53'
Gainer,Trina
And 23h7'55"44d49'
Gaines "Magoo",Mary E
Cam 4h7'58"70d12'
Gaines III,F Pendleton
Aql 19h50'11"15d25'
Gaines XC,Ben B
Ser 15h37'26"-2d53'
Gaines,David L
Aur 6h26'20"31d36'
Gaines,James Henry
Cep 5h30'1"85d54'
Gaines,June
And 0h14'29"38d45'
Gaines,Ken
Boo 14h55'33"35d45'
Gaines,Leviathan Christopher
Silberbach
Aql 19h29'24"-8d38'
Gaines,Tilford Craig
Cep 21h35'1"60d29'
Gainey,Mike
Aur 7h8'37"43d19'
Gainey,Sally & Billy
Cyg 19h50'60"41d4'
Gainey,Scott
Dra 14h49'18"62d26'
Gainsbury,James
Aur 5h33'55"38d42'
Gaiofatto,Giorgio
Per 2h11'40"58d24'
Gaisch,Maria
Sgr 19h55'14"-43d14'
Gaiser,Ingeborg und Oskar
Lyn 8h3'22"49d59'
Gaitan,Adriana
Aql 19h57'22"10d30'
Gaitan,August Alexander
Lac 22h10'0"54d31'
Gaitan,Felipe Antonio
Aql 19h0'18"0d32'
Gaither Jones,Lee Ann
Hya 9h9'0"4d49'
Gaj,Dr Gregory S
Per 1h38'55"52d43'
Gajdica,Michael Shannon
Dra 19h2'43"50d14'
Gajdusek,Corinne M
And 23h18'35"41d52'
Gajewski,Joseph Michael
Cep 21h25'16"55d42'
Gajewski,Sr,Robert J
Dra 16h47'1"70d31'
Gajota
Lyn 8h21'40"46d11'
GAKU HIROSHI & YOKO
Eri 4h5'44"-6d16'
Gala
Boo 14h29'47"53d42'
Gala Star Of Stars, Beall
Lac 22h52'21"55d22'
Gala-Kiwi
Nor 16h19'30"-51d26'
Galadriel
Lac 22h48'0"53d26'
Galafassi,Alice
Boo 15h3'26"41d39'
Galanos,Nick & Jeannie
Aur 5h2'3"29d18'
Galanova,Nadya
Aqr 21h54'59"-6d20'
Galante,Christopher Francis
Lac 22h50'21"54d38'
Galante,Edward M
Boo 14h59'57"51d27'
Galantino,Billy
Aur 6h32'36"38d21'
Galarneau,Marie Paule
Eri 4h53'44"-8d46'
Galas,DC,Dr Robert M
Aql 20h36'25"-5d50'
Galasso '25'
Uma 9h56'59"48d7'
Galasso,Ciro
Boo 15h7'1"50d25'
Galasso,Grace (Nenar)
Cas 0h57'27"63d45'
Galata,Jim "Big Guy"
Lac 22h29'0"38d47'
Galateus
Lac 22h34'25"40d25'
Galati,Joseph Blaise
Dra 16h36'34"66d51'
Galaunder,Barret Lee
Cam 3h30'0"61d43'
Galaunder,Brock Curtis-Otto
Vul 19h21'55"26d18'
Galaviz,Albert M
Hya 8h44'26"5d41'
Galaxie Cuisine
Lac 22h4'38"37d39'
Galaxy Glenn
Ori 4h59'55"5d8'
Galay,Joan
Tri 2h7'36"32d30'
Galazka,Gregory Richard
Per 3h32'57"37d4'
Galban,Craig J
Hya 8h41'52"-6d16'
Galbraith,Bruce I
Hya 8h57'10"-10d57'
Galbraith,Diana Valerie
Cas 0h27'21"61d46'
Galbraith,Gary M
Cyg 21h36'0"42d6'
Galbraith,Linda
Cyg 20h21'54"38d50'
Galbraith,Maxine Jones
Lyn 6h59'51"59d7'
Galbraith,Scott Kenneth
Ori 5h16'57"-5d30'
Galceran,Susana
Mon 7h21'0"-1d50'
Galde,Matthew Nicholas
Cnv 12h55'21"51d38'
Gale Houston TX,Austin
James
Aur 7h17'40"37d6'
Gale's Tales
Cam 3h55'60"55d8'
Gale,Ashley Alyssa
And 0h23'1"30d28'
Gale,Donna
Umi 15h20'54"82d11'
Gale,Douglas E
Ser 15h22'46"24d22'
Gale,Gary
Lac 22h11'58"49d35'
Gale,Jackey
Peg 22h14'19"5d3'
Gale,JoAnn
Mon 6h57'25"1d11'
Gale,Kimberley & Mark
Uma 11h35'14"50d13'
Gale,Robert Alan
Aql 19h46'26"14d58'
Galeano,Traci
Vul 19h21'52"26d43'
Galeazzi,Gabriella Shaye
Aur 4h56'23"40d24'
Galef,Andrew
Eri 4h7'57"-16d33'
Galella,Vincent
Her 15h53'44"46d45'
Galen
Vul 19h44'14"28d20'
Galen
Uma 8h48'55"48d19'
Galen
Aur 6h25'58"38d11'
Galen,Gigi
Tri 2h10'24"30d48'
Galen,Tamara Anne
And 2h33'0"41d55'
Galena "Snow Bunnies 96"
And 0h11'31"37d23'
Galeno's Shining Star, Dr John
A
Per 2h52'24"40d7'
Galeone
Ori 5h52'29"14d31'
Galeucia,Richard David
Lac 22h35'47"54d11'
Galgano,Christine Louise
And 2h11'55'38d3'
Galiani,Pier Paolo
Cyg 21h17'47"37d55'
Galiger,Mary Ann
And 1h26'21"36d41'
Galimberti,Angelo Emilio
Aur 5h6'55"37d14'B
Galina & Oleg
Cyg 19h27'13"31d8'
Galina,Lisa
And 2h29'26"48d58'
Galindo's Dream,John
Leo 9h38'53"7d31'
Galindo,Alexander
Aql 19h5'51"15d4'
Galine's Daughter,Evan
Aur 7h20'27"40d39'
Galinski,John Edward
Per 3h58'16"38d57'
Galinski,Korin Christa
Sge 19h5'49"17d55'
Galinski,Shirley Mary
Ari 2h1'34"26d29'
Galinsky,Angela
Vir 13h18'10"-8d33'
Galinsky,Dina R R Perloff
And 0h49'58"35d2'
Galioto,In Loving Memory Of
Jason
Boo 14h36'12"21d36'
Galipeau,Roger Alfred
Oph 17h18'40"-20d32'
Galkevirse
Gem 6h52'32"31d5'
Gall,Gregory K
Ser 15h39'44"24d17'
Gall,Orsolya
Uma 10h47'13"54d10'
Gall,Stacy
And 2h24'49"47d31'
Gallagher "Babykid", Michael
Edward
Per 2h24'30"55d41'
Gallagher + Taz & Sam, Sandi
Ann
Tri 2h3'11"31d38'
Gallagher,Amanda Michelle
And 2h18'36"49d15'
Gallagher,Andrew Vernon
Ori 5h18'59"1d34'
Gallagher,Aubrey Leigh
Vul 20h1'0"25d51'
Gallagher,Brendan Knoll
Cep 20h56'58"55d23'
Gallagher,Brian Isaac
Oph 17h53'30"12d31'
Gallagher,Charles
Cep 9h52'0"80d23'
Gallagher,Charney Elizabeth
Cas 22h59'51"56d7'
Gallagher,David L
Oph 17h5'17"-24d42'
Gallagher,Dennis Patrick
Cyg 21h17'38"38d49'
Gallagher,Destin Hank
Lac 22h23'53"55d13'
Gallagher,Edward
Per 2h56'20"43d30'
Gallagher,Eric
Aur 6h48'38"37d6'
Gallagher,Erin Kathleen
Mon 7h57'33"-8d7'
Gallagher,Francis
Uma 9h40'57"68d25'
Gallagher,George- Grampy
Uma 10h59'45"68d12'
Gallagher,J K R J
Ori 5h57'33"16d10'
Gallagher,Jack
Per 2h55'47"50d32'
Gallagher,Jada S
Lyr 18h49'25"36d3'
Gallagher,James Lloyd
Cep 2h45'1"77d51'
Gallagher,Jimmy & Kristin
Boo 14h38'27"35d39'
Gallagher,Joanne E
Psc 0h0'46"2d3'
Gallagher,John J
Ori 6h3'18"10d58'
Gallagher,John Joseph
Sgr 19h33'19"-42d22'
Gallagher,Joshua Corey
Her 17h13'1"47d21'
Gallagher,Jr,Harry Joseph
Dra 17h43'1"64d57'
Gallagher,Jr,Thomas Patrick
Aur 6h10'36"37d0'
Gallagher,Kayley Kimberly
And 1h47'39"39d46'
Gallagher,Keith P
Lmi 10h58'29"27d25'
Gallagher,Kevin
Uma 9h49'26"42d53'
Gallagher,Leon
Boo 14h10'37"30d30'
Gallagher,Luisa Anna
Lyr 19h14'1"41d57'
Gallagher,M G
Her 17h26'0"37d30'
Gallagher,Marie
And 0h11'56"31d35'
Gallagher,Mark Peter
Ori 6h9'0"5d58'
Gallagher,Mary
Cas 2h34'41"65d8'
Gallagher,Mary Anne
Peg 22h34'22"35d31'
Gallagher,Michael Lawrence
Oph 17h19'50"-20d26'
Gallagher,Michael Thomas
Aur 7h23'50"36d27'
Gallagher,Pat
Boo 13h41'0"19d29'
Gallagher,Patrick L & Cecilia I
Cyg 19h28'26"34d5'
Gallagher,Paul D
Aur 6h4'35"31d14'
Gallagher,Peter
Uma 11h10'25"50d9'
Gallagher,Rob PB/BH
Dra 14h42'24"62d32'
Gallagher,Sean Kevin
Aur 5h35'16"41d4'
Gallagher,Shannon Leigh
Cyg 19h59'36"30d11'
Gallagher,Shelley A
Cas 0h5'58"63d37'
Gallagher,Shirley Irene
Mon 6h39'24"8d45'
Gallagher,T J & Tricia
Ori 5h59'22"9d33'
Gallagher,Terry
Ori 6h3'11"5d40'
Gallagher,Thomas A
Aur 5h7'56"40d33'
Gallagher,Willoe Maire
Lyn 8h56'15"41d36'
Gallahan,Lance
Aur 6h0'49"38d19'
Gallaher,Mary K
Cmi 7h41'55"5d24'
Gallant Ron
Cap 20h23'59"-26d18'
Gallant,Andrea
Lyn 9h2'36"35d34'
Gallant,Liane
Lac 22h45'56"53d17'
Gallard's Star
Cyg 19h45'29"37d57'
Gallarda,Sir Denny
Lac 22h47'18"38d39'
Gallardi,Marcus Vinicius
Ser 15h42'44"-1d37'
Gallardo,Oscar
Cnv 12h49'24"35d46'
Gallardo,Rico J
Aql 20h1'24"1d47'
Gallas,Ann Marie & James
Larry
Cyg 21h5'38"32d42'
Gallas,Steven
Cap 20h35'38"-10d13'
Gallatin,Adrianne Nell
Mon 6h56'11"1d19'
Gallatin,Pauline
Cas 0h29'57"60d40'
Gallaxer
Peg 22h13'40"30d22'
Gallazzi,Stefano
Pho 0h3'40"-41d6'
Galle Johann
Umi 16h33'28"76d16'
Galle,Kevin
Dra 18h34'45"58d28'
Gallego,Johnny
Cep 21h16'12"61d41'
Gallego,Richard Charles
Her 16h30'15"50d42'
Gallegos,Jordan
Per 2h54'1"50d29'
Gallegos,Linda
Mon 6h55'12"-3d12'
Gallegos,Rose
Cet 2h6'52"1d41'
Gallegos,Rudy
Ser 15h41'32"-1d21'
Gallegos,Sandra J
And 0h22'52"32d6'
Galleni,Julia Louise Ratto
Leo 10h38'46"22d3'
Galler,Lois Joy
Boo 13h36'41"8d15'
Galler,Lorraine M
Cas 23h34'12"61d39'
Galletta,Dr Joseph
Sge 19h43'59"16d39'
Galletti,Jr,Charles
Oph 18h21'25"6d38'
Gallgher,Ian Christian
Per 1h42'55"53d14'
Galli,Anthony Philip
Uma 9h30'11"50d5'
Galli,Brian John
Oph 18h0'60"7d47'
Galli,Jenna Nicol
Cas 1h47'23"61d21'
Galli,Venera Lucretia
Uma 9h38'48"50d51'
Galli,Yolanda
Lyn 7h49'15"38d23'
Galliano,Toni Siri Dharma
Mon 7h18'32"-6d53'
Gallico "What Love Has
Done",Kayla Rose
Cas 0h48'47"61d12'
Galliers,Shannon Lee
And 23h26'11"48d45'
Galligan,Helen Alma
Peg 23h29'27"22d16'
Galligan,Lynn-Patrick
Cam 6h51'44"62d56'
Galligan,Sharon
And 02h24'50'39d28'
Galligani,Grandma Mary
Margaret
Ari 1h47'58"18d57'
Gallina,Elena
Lyn 7h30'28"51d26'
Gallina,Marianne
Cyg 19h58'51"29d33'
Gallina,Virginia
Lyr 19h3'53"28d10'
Gallinghouse-Forever,
Stephen Mark
Lmi 10h36'0"28d56'
Gallistel
Crb 15h49'0"27d42'
Gallo Family,The
Ori 5h54'30"14d44'
Gallo,Barbara
Dra 11h46'28"67d36'B
Gallo,Cheryl Marie Haught
Cyg 19h25'22"30d9'
Gallo,Constance
Vul 19h42'33"26d34'
Gallo,D V
Lac 22h32'59"53d24'
Gallo,David
Lac 22h5'1"47d36'
Gallo,Dean A
Dra 14h56'32"-61d34'
Gallo,Frank
Lac 22h34'47"53d18'
Gallo,Gina
Lyn 8h6'34"58d37'
Gallo,In Loving Memory Of
Loretta
And 1h18'34"33d49'
Gallo,Josephine "Honey"
Mon 7h46'43"-1d46'
Gallo,Josephine & Albert
Ori 5h53'33"7d20'
Gallo,Jr,(Fozzie Bear) Richard
Arthur
Her 16h35'48"32d42'
Gallo,Leonard
Dra 11h46'28"67d36'A
Gallo,Lisa Anne
Aur 4h59'52"31d50'
Gallo,Patricia
Cyg 21h4'11"39d21'
Gallo,Rachael
Cas 0h33'35"66d26'
Gallo,Vincent
Aur 5h59'0"29d25'
Gallo,Vincent
Cnv 12h32'30"32d24'
Gallogly,Aunt Eleanor
Cyg 20d2'44"58d26'
Gallois,Sophie
Lyr 18h45'0"40d3'
Gallot,Laurence
Cep 22h25'47"57d53'
Galloway
Ori 4h41'1"10d20'
Galloway "Andy",Steven
Andrew
Ser 17h36'0"-14d3'
Galloway,James Crooks
Hya 8h25'16"5d45'
Galloway,Kari F
Peg 0h1'42"31d33'
Galloway,Mary
Ori 6h1'1"5d6'
Galloway,MD,George W
Oph 17h30'32"-20d11'
Galloway,Melvin
Lyr 18h37'10"34d16'
Galloway,Rhonda Diane
Cam 3h53'34"69d13'
Galloway,Sarajoyce
Cas 2h15'1"60d7'
Galloway,Torisha Drew
Cnv 12h13'21"45d7'
Galloway,William
Aql 19h1'46"0d1'
Gallucci,Eileen
Peg 23h48'31"28d54'
Galluccio,A Whole New
World-Doug & Lisa
Cyg 21h2'22"31d32'
Gallup,My Maxi Maximillian
Uma 11h9'38"57d41'
Gallus,M F
Tri 1h45'17"27d58'
Galluzzo 1971,Enza
Tel 19h6'0"-48d32'
Galluzzo,Santo
Dra 17h15'26"60d26'
Gally,Victoria
Cam 4h3'33"68d26'
Galmard,Louise
Lyn 9h5'54"33d26'
Galmende,Pshan
Dra 20h23'0"64d49'
Galombeck,Amy Beth
Ori 5h57'0"9d31'
Galotti,Donna
Cas 22h58'57"55d59'
Galotti,Ron
Per 2h59'0"50d20'
Galstonia,The
Dra 20h3'27"67d55'
Galteri,Helene
Cam 6h13'48"68d36'
Galterio
Aur 5h15'29"40d12'
Galton's Galaxy
Com 13h7'21"27d31'
Galuppo,Ron
Cep 0h19'0"80d31'
Galuszka,Linda Louise
And 22h57'0"50d25'
Galutia,Diane
Lac 22h45'57"56d24'
Galvan,Abraham
Cma 6h52'41"-18d10'
Galvan,Jaime
Vir 12h4'33"1d60'
Galvan,Mike
Lac 22h37'0"55d1'
Galvan,P Michael
Cet 3h14'50"0d17'
Galvan,Pierre
Aql 19h55'39"14d40'
Galvani,Patrick
Vir 12h43'29"-9d48'
Galvani,Serena
Tri 1h47'59"34d57'
Galvez,Salvador Lopez
Cet 1h26'40"-2d34'
Galvin (The T Star), Theresa
Lyn 7h51'27"38d41'
Galvin,Alana
Vul 21h23'12"28d22'
Galvin,Mary Jacquelin
Uma 10h55'18"51d37'
Galvin,Princess Elisabeth
And 2h17'1"38d21'
Galvin,Tommy Handsome
Aur 6h33'16"33d39'
Galvin,Vance
Lyr 18h39'0"47d19'
Galway III,John H
Cep 22h54'56"57d6'
Galy,Corinne
Ori 5h57'1"18d10'
Galy,Jean Michel
Sex 10h14'8"-1d31'
Galya,Julius
Lyr 18h59'0"27d36'
Galyon,Rosalind Nancy
Peg 22h33'18"21d46'
Galzy,Franâoise
Per 3h57'33"35d37'
Gama
Mon 6h35'21"11d41'
Gamache,Alfage
Cnv 13h12'23"38d44'
Gamache,Jr,Louis M
Aur 6h31'26"31d46'
Gamache,Martin
And 0h43'24"44d35'A
Gamache,Michael Emile
Dra 9h26'34"73d42'
Gamache,Michelle Rose
Cas 0h40'16"60d6'
Gamache,Raymonde
Cyg 19h27'17"35d19'
Gamalu
Cyg 19h33'39"30d2'
Gamarra,Erick Salvador
Per 2h55'29"55d7'
Gamba,Maria Rosa
Hor 3h29'17"-44d47'
Gambaccini,Annette
Cyg 19h21'43"28d34'
Gambale,Ethan William
Sco 16h6'25"-22d49'
Gambale,Fredric James
Dra 18h59'32"68d34'
Gambale-Cortes,Hernan &
Amy
Cyg 20h38'57"41d15'
Gambardella,Terry
Lyn 7h43'50"52d21'
Gambardella,Vincent Michael
Boo 14h36'54"17d35'
Gambee,Jean Marie
Mon 6h28'35"11d5'
Gamberoni,Graziano
Dra 16h22'44"58d12'
Gambeski,George P
Lac 22h21'1"53d29'
Gambill,Grayson Luke
Cap 20h15'56"-26d51'
Gambino,Anna
Lyr 18h18'17"38d31'
Gambino,Christopher
Her 18h19'0"12d8'
Gambino,Salvatore
Boo 15h20'33"47d42'
Gamble,Anna Marie
And 0h4'1"46d51'
Gamble,Betty Louise
And 0h5'1"31d38'
Gamble,Bridget Erin
And 0h28'1"37d53'
Gamble,Chase Bradford
Ori 5h56'49"13d4'
Gamble,David & Tanya
Per 3h8'17"48d50'
Gamble,Dylan Lowe
And 2h34'20"37d17'
Gamble,Kerry
Del 20h27'22"18d51'
Gamble,Lucas Edward
Boo 15h4'18"26d7'
Gamble,Paul & Genevieve
Eri 4h2'52"-19d5'
Gamble,Rick
Eri 2h57'12"-17d10'
Gamble,Rory
Lmi 9h48'18"33d32'

Gamble,Todd Michael
Per 3h11'53"47d11'

Gamboa,Jose R
Dra 19h30'13"60d22'

Gambrell,Laurie
Mon 7h2'14"4d46'

Gambrill,Kelly Christine & Charles
Cyg 19h30'49"31d43'

Gambsky,Shelley M
Cas 22h57'31"56d15'

Gamelli,Mary
And 0h27'15"27d53'

Gamet,Pierre-Paul
Umi 16h7'1"73d47'

Gamm,Matthew Aaron
Tau 5h1'16"28d15'

Gamm,Why Not?! Matthew Aaron
Vul 19h23'25"26d45'

Gamma "V"
Ori 5h51'33"18d33'

Gamma Sigma Chapter of Alpha Phi
Umi 15h40'53"67d38'

Gammill,A Todd
Aur 5h3'27"38d9'

Gammill,Andrew T
Cnv 13h46'38"31d59'

Gammy-Carole Sue
Uma 9h33'37"48d45'

Gamoonwarden
Peg 23h24'1"15d49'

Gamp Star,The
Sex 10h19'10"-1d13'

Gamper,Hope Ja Yung
And 1h15'56"41d9'

Gampp,Peter
Per 3h16'10"46d13'

Gams
Mon 6h29'43"-10d48'

Gamsby,Kirby Alexandra
Cap 20h35'21"-13d36'

Gamse,LeRoy
Dra 16h57'14"63d27'

Gamze,Michael
Aur 7h25'45"43d44'

Gana,Emily
Cas 0h0'1"63d55'

Ganahl,Jym
Oph 17h37'25"-3d1'

Ganbarg,David
Cnv 12h22'11"45d45'

Gancarlo,H R
Per 3h47'47"35d19'

Gance,Kimberly Ellen
Vul 20h58'58"20d20'

Gance,Sandra & Joe
Peg 22h59'36"29d53'

Ganci,Joe Ann
Peg 21h37'46"27d22'

Ganci,Justine Marie
Lyr 18h38'57"39d33'

Gandara,Maria & Jenny Sanchez
Eri 4h14'0"-19d47'

Gandee
Boo 14h2'24"12d4'

Gandee,Kathrine
And 23h42'40"45d34'

Gandharva
Aur 5h9'20"40d56'

Gandhi,Hemali
Aql 19h57'1"14d52'

Gandolfi
Aql 20h21'51"5d25'

Gandy,Darlene Rose
Cyg 19h32'10"30d43'

Ganeles,Brock
Sct 18h41'53"-4d56'

Ganepola,Sayuri
Lyr 19h24'39"41d24'

Ganet,Frederic
Aur 5h57'41"50d14'

Ganey,"Star Coach" Tanya
Lyn 9h13'55"34d29'

Ganey,Aloysius Joseph
Her 18h19'40"21d28'A

Ganey,Carol B
Cam 7h7'12"68d58'

Ganey,Elaine Veronica
Her 18h19'40"21d28'B

Gang,Jaime Kuriyama
Cas 0h23'26"58d34'

Gangel,Little John
Ser 15h9'28"20d58'

Gangestad,Myron
Her 16h23'19"30d2'

Gangi,David P
Per 1h53'1"50d21'

Gangi,Graham
Dra 18h27'0"50d32'

Gangi,Gregory Peter
Cep 23h7'54"60d10'

Gangi,Jo-Ann
Cyg 21h1'16"40d41'

Gangler's Song,Suzanne
Lyr 18h59'24"37d22'

Gangsei,Virginia
Cet 1h9'0"1d0'

Ganley,Kiki
Cas 3h5'53"57d49'

Ganley,Lynda
Cas 0h34'40"66d20'

Gann,Virginia
Oph 18h40'27"7d30'

Gann,Winfred
Cet 2h8'32"6d19'

Gannarelli,Marco
Uma 13h28'40"61d3'

Gannaway,Byron
Aur 6h3'13"37d42'

Gannaway,Cliff & Lela
Aql 18h45'26"10d13'

Gannon,Daniel
Dra 11h16'54"74d49'

Gannon,Jordan James
Her 17h21'45"27d20'

Gannon,Lilly
Cyg 19h45'10"29d36'

Gannon,Richard & Madlyn
Cet 0h30'10"-12d21'

Gannon,Scott Anthony
Cet 1h49'15"-3d41'

Gannon,Toni
Aur 4h38'28"31d45'

Gano,Kathleen Yvonne
Gem 7h51'57"31d45'

Gano,Laurie Gail
Peg 23h2'50"17d38'

Ganon Star,The
Uma 8h30'33"60d16'

Ganong,Matthew
Aur 6h29'37"38d2'

Ganoudis,Zoe Nicola
And 22h57'34"51d19'

Gans,Howard S
Ser 18h17'25"-1d13'

Gansel,Lelyn Kay
And 1h3'54"47d1'

Gansel,Lelyn Kay
Ori 5h53'41"15d34'

Ganski,Robert "Bobby"
Cnv 12h19'37"34d10'

Gansor Rl
Lmi 10h19'31"36d28'

Gansser
Per 1h45'0"50d17'

Gant,Patricia Owen
Cam 7h28'52"67d35'

Gant,Tanya
Peg 23h26'54"23d45'

Gantoy,Georges
Peg 0h6'49"13d42'

Ganz
Uma 9h37'29"44d1'

Ganz,Andrew
Cas 0h17'57"65d32'

Ganz,Erhard
Dra 18h25'27"67d40'

Ganz,Evan Bradley
Aqr 21h58'0"-1d34'

Ganz,Jordyn Alana
Cyg 20h55'14"39d48'

Ganzer,Chris & Michelle
Ori 6h4'14"-0d33'

Gaoiran,Liza Cortez
Peg 22h46'47"21d42'

Gaquiere,Sabrina
Ser 16h3'1"8d55'

Gar & Dar
Cyg 21h47'18"53d5'

Gar-Ba
Dra 16h2'24"58d10'

Gara Paige
Cnv 13h41'44"34d22'

Garabed,Carlyn Ann
Cnc 8h1'1"11d11'

Garabed,Cheryl Ann
Uma 10h56'18"57d33'

Garabed,Kylie Amanda
Aur 6h32'33"32d25'

Garabedian,John
Sgr 19h36'52"31d43'

Garabrant,Joyce Ann
Cas 0h28'0"68d23'

Garage Mahal Minor,The
Cet 1h8'0"-1d41'

Garaguso,Joan Elizabeth Pugarelli
And 1h32'23"40d15'

Garanan
Ori 4h55'39"4d44'

Garant,Jane Allison
Mon 6h55'36"10d42'

Garant,Stuart Daryl
Ori 5h52'30"-0d13'

Garapetian,Vrej
Per 2h53'37"54d45'

Garapon,Patrick
Umi 16h28'36"70d14'

Garay,Carlos
Her 16h25'17"24d10'

Garb,Margaret
Aql 19h2'15"-0d8'

Garbacki,Ronald
Aur 6h5'1"38d18'

Garbade,Peter
Cmi 7h17'29"5d24'

Garban,Steve & Penny
Cyg 21h2'35"40d30'

Garbarczyk,Edward S
Aur 5h11'26"42d7'

Garbarini,Shirley
And 1h16'13"39d58'

Garber Family,Randy
Ori 5h39'30"11d59'

Garber,Adele
Lyn 8h27'58"40d43'

Garber,Arnie
Uma 9h9'16"58d21'

Garber,Ashley Sayaka
Peg 22h33'1"20d16'

Garber,David Jason
Cam 6h50'23"80d4'

Garber,Samantha
Com 13h1'12"28d35'

Garber,Walter Nevin
Cnv 12h20'17"34d54'

Garber,Wilbur Leroy
Aql 20h33'33"-8d2'

Garberding,Ken
Aur 6h35'11"40d33'

Garberina,Sr,PhD,Dr William L
Her 17h1'50"38d22'

Garberson,Cassy Ann
Eri 3h31'36"-5d27'

Garbett,Yvonne F
Lyr 19h11'19"40d30'

Garbi,Matteo
Cep 21h20'58"61d19'

Garbiack,Shauna
Aql 20h6'59"-6d30'

Garbiel,Lisa
Mon 6h17'46"-5d29'

Garbo,Dr Joseph Phillip
Hya 9h4'60"1d21'

Garbrecht,Jürgen
Sgr 19h17'4"-42d36'

Garbutt,Jr,"Ziggy", Gerald Ray
Aur 6h25'22"38d49'

Garcia
Her 17h13'46"29d13'

Garcia "Last Of The Cowboys",Manuel
Tau 4h40'18"16d21'

Garcia de Quevedo,Rose Marie
Cap 20h34'24"-26d54'

Garcia's Estrella
Peg 23h5'23"10d51'

Garcia,"Star Bright", Vanessa U
Lib 14h59'29"-10d35'

Garcia,Alfred Perez
Umi 17h14'12"75d20'

Garcia,Amanda Michele
Cma 6h30'20"-15d53'

Garcia,Ange
Aql 19h28'15"7d55'

Garcia,Ashley
Peg 21h58'39"2d55'

Garcia,Aurora
Lyr 18h56'0"30d30'

Garcia,Carla
Sge 19h58'56"16d45'

Garcia,Catherine
Peg 23h16'48"31d9'

Garcia,Cindy
Leo 10h6'15"15d18'

Garcia,César
Ori 6h3'15"1d18'

Garcia,Debra Lynn
Cet 3h4'58"-0d16'

Garcia,Donald J
Per 2h10'22"57d27'

Garcia,Elizabeth
Lyn 8h42'44"34d55'

Garcia,Felizabel C
Aur 5h29'51"38d31'

Garcia,Friedrich Austin
Per 4h28'53"51d22'

Garcia,Gina Marie
Peg 22h42'34"20d7'

Garcia,Isabel
Lep 4h59'11"-11d49'

Garcia,Jaime A
Aql 19h18'37"14d58'

Garcia,Janis L
Mon 7h39'3"-3d41'

Garcia,Jen
Ari 3h0'39"23d37'

Garcia,Jeraldine Carmen
Mon 6h28'35"10d3'

Garcia,Jerry
Her 16h41'19"23d39'

Garcia,Jessica
Sex 10h29'14"2d41'

Garcia,Juanita Jordan
Umi 15h37'11"68d34'

Garcia,Manuel
Cep 21h4'50"60d14'

Garcia,Maria Carolina Lopez
Com 12h28'46"26d35'

Garcia,Melissa Isabel
Uma 16h3'1"72d4'

Garcia,Merideth Hardy
Peg 22h17'39"32d56'

Garcia,Michelle Nicole
Cas 3h8'39"71d6'

Garcia,Mike A
Ori 6h3'50"8d21'

Garcia,Nemesio Gabriel Perez
Aql 18h42'17"1d20'

Garcia,Ofelia
Aqr 21h34'12"-1d30'

Garcia,Paul
Ori 5h56'48"6d33'

Garcia,Paul G
Crt 11h51'26"-7d44'

Garcia,Philippe
Boo 15h6'44"7d38'

Garcia,Phillip Blas
Cet 3h3'43"1d50'

Garcia,Rachel Ellen
Cyg 19h27'1"38d44'

Garcia,Raquel Guadalupe
Peg 22h52'26"29d13'

Garcia,Rebecca M Fajardo
Aur 6h56'10"44d24'

Garcia,Ricardo Benjamin
Vul 19h20'30"26d38'

Garcia,Ricardo Benjamin
Boo 14h1'26"20d12'

Garcia,Roberto Torres
Aql 18h59'30"8d27'

Garcia,Roman Miguel
Lac 22h24'33"40d46'

Garcia,Shawn A
Cnv 13h53'28"37d55'

Garcia,Shelbey Reade
Mon 6h29'51"1d38'

Garcia,Suzanne Renee
Vul 19h44'30"28d37'

Garcia,Tanya
Cas 0h51'39"64d2'

Garcia,Ted
Cet 2h28'23"3d34'

Garcia,Tulio
Cam 4h4'53"67d59'

Garcia,Wendy
Vul 21h1'0"20d19'

Garcia,Xavier Salinas
Ser 18h16'36"-14d17'

Garcia,Yesenia Lynn
Uma 11h38'18"43d1'

Garcia-Ball,Juanita I
Peg 22h59'19"25d38'

Garcia-Blizzard, Cecilia Guadalupe
Umi 15h20'11"71d1'

Garcia-Blizzard,Monica Del Carmen
Umi 15h20'48"77d32'

Garcia-Fernandez
Com 12h28'53"20d15'

Garcia-Ribeyro,Gonzalo
Peg 23h22'58"13d40'

Garcin,Gisele Andrea
Sgr 20h17'36"-34d56'

Garcin,Isabelle
Cyg 19h47'25"30d5'

Garczek,Edwin A
Lac 22h15'20"49d8'

García,Marta Alonso
Aur 7h7'22"38d39'

Gard,Melissa Jo
Lyn 7h35'1"44d28'

Gard,Nicole
Cam 4h59'23"60d48'

Garda,Mario
Gru 22h19'26"-51d2'

Gardei,Suzie
Sct 18h52'38"-7d15'

Gardella,Anthony Jacob
Her 17h22'0"46d54'

Gardemal,Dr John Wallace
Cmi 7h15'50"4d58'

Gardener (Bushy), Patricia Ann
Cyg 19h34'19"33d49'

Gardenia,Robert's Star
Boo 13h35'21"16d7'

Gardeux,Sajan Chakraverty
Aql 19h0'46"16d46'

Gardeux-Zanotti,Lorene
Mon 6h25'36"7d42'

Gardey,Kim & Barb
Dra 12h38'52"68d15'

Gardf,Keith "Keico"
Dra 19h46'23"60d35'

Gardia
Lyr 18h22'40"46d14'

Gardiner,Colin R
Ori 6h1'23"3d27'

Gardiner,George R
Per 3h2'56"41d20'

Gardiner,Glenn
Dra 16h23'22"68d36'

Gardiner,Ian Hills
Uma 11h36'30"31d11'

Gardiner,Kevin D
Aur 5h57'45"40d5'

Gardiner,Lisa Jane
Crb 16h21'55"28d8'

Gardiner,Michael
Aur 6h6'15"46d43'

Gardiner,Nancy P
Lyn 6h28'1"58d38'

Gardiner,Susan Mary
And 23h24'30"40d53'

Gardini,Adrianna
Del 20h25'10"20d6'

Gardini,Roseann
Mon 7h30'17"-10d11'

Gardner Star,The Allison P
Ori 5h37'48"-2d12'

Gardner's,The
Uma 10h58'38"56d25'

Gardner,Adrienne K
Mon 7h15'26"-1d26'

Gardner,Alice Rhodes
Cas 0h4'19"56d6'

Gardner,Amanda & Jeffrey
Crb 16h4'40"37d45'

Gardner,Andrew Frank Boyd
Uma 11h5'47"60d28'

Gardner,Ann & Mitchell Karton
Vul 19h23'1"25d18'

Gardner,Audrey McDowell
Mon 6h53'60"-10d11'

Gardner,Ben
Leo 9h22'19"31d8'

Gardner,Betty & Reggie
Lyn 9h9'17"35d51'

Gardner,Charles Allen
Aql 20h11'37"12d53'

Gardner,Charles L
Aur 6h10'23"36d2'

Gardner,Christina Casey
Vul 19h16'40"25d32'

Gardner,Colin
Lac 22h52'51"37d54'

Gardner,Derek Douglas
Aql 18h48'24"11d55'

Gardner,Donna
Mon 8h3'8"-1d37'

Gardner,Donna Maria
Lyn 8h12'13"48d29'

Gardner,Emma
Cep 21h48'47"67d31'

Gardner,Emma Louise Veronica
Cas 23h28'27"61d19'

Gardner,Gail
And 23h27'60"48d35'

Gardner,Hannah Essig
Lyr 19h20'38"42d46'

Gardner,HL "Larry"
Cet 0h0'18"-18d10'

Gardner,Jack Ross
Umi 15h14'60"69d14'

Gardner,Jane Elizabeth
And 0h19'21"38d21'

Gardner,Jennifer Ann
Lyn 6h29'35"58d37'

Gardner,Jerry
Sex 9h43'29"-5d22'

Gardner,Jill Alison
And 0h21'38"30d25'

Gardner,Jill Mary
Lyr 19h3'20"26d25'

Gardner,John D
Umi 14h21'39"68d8'

Gardner,Jonathan Richard
Uma 8h19'14"62d12'

Gardner,Jr,Alton
Boo 15h16'30"40d37'

Gardner,Jr,Paul Andrew
Ori 6h13'13"6d18'

Gardner,Judith C
Del 20h20'37"18d48'

Gardner,Katherine W
And 23h0'26"48d55'

Gardner,Kirk
Dra 14h29'47"62d6'

Gardner,Kristen Lane
And 23h7'0"37d56'

Gardner,Kristen Taylor
Sgr 19h36'50"-34d14'

Gardner,Laura Boyd
Cyg 21h39'36"38d9'

Gardner,Lisa Michelle
Del 20h26'30"11d19'

Gardner,Lukas G
Hya 8h55'33"-18d41'B

Gardner,Marilyn
Vul 20h57'33"28d24'

Gardner,Mary Ann
Crb 15h42'36"26d23'

Gardner,MD,Thank You Lawrence D
Oph 16h48'21"1d38'

Gardner,Michael
Crt 11h28'12"-10d14'

Gardner,Nathan C
Hya 8h55'33"-18d41'A

Gardner,Raelyn
Ari 2h34'32"21d36'

Gardner,Rebecca J
Del 21h5'26"18d58'

Gardner,Richard Thomas Arzdorf
Tau 5h32'54"22d43'

Gardner,Robin
Cep 23h31'26"67d53'

Gardner,Sarah
Eri 2h51'0"-10d15'

Gardner,Stewart Anderson
Ser 15h32'0"10d29'

Gardner,Susan
Ori 5h54'43"16d6'

Gardner,Susan
Cas 0h57'56"63d35'

Gardner,Tara Lyn
Peg 0h2'41"30d37'

Gardner,Thomas
Lac 22h34'30"53d46'

Gardner,Todd Elden
Peg 21h49'12"30d27'

Gardner,Trevor
Peg 23h31'36"32d25'

Gardner,Wayne
Umi 16h6'7"79d36'

Gardner-Natures Angel, Helen Templin
Cyg 20h25'29"40d13'

Gardomikjome
Sgr 19h59'5"41d40'

Gare Bear
Uma 11h21'41"60d15'

Gareau-Feldman,Ginette
Umi 15h12'12"68d18'

Gareis,Christine
Ser 15h30'31"18d19'

Gareth
Mon 7h43'52"-6d47'

Gareth
Uma 9h42'47"68d39'

Garett
Cep 20h32'42"63d51'

Garfein,Carolyn
Aql 18h44'51"-2d47'

Garfield,Catheryne & Eugene
Lmi 10h41'19"32d19'

Garfield,Jennifer Gayle
Mon 6h53'44"0d29'

Garfield,Johnathan George
Peg 23h34'0"18d33'

Garfield,Lindsay Beth
And 23h38'26"47d26'

Garfinkel,Paulyne
Cyg 19h58'20"31d17'

Garfinkle,Steven Lang
Her 16h53'14"40d53'

Gargano,Denise
Sgr 18h59'44"-20d29'

Gargano,Frank
Dra 20h29'0"73d27'

Garguilo,Michael
Lac 22h2'52"47d33'

Gariano,Louis A & Thelma
Cyg 20h20'38"41d2'

Garibaldi,Anne Marie
And 0h17'47"45d5'

Garibaldi,Eugene D
Aur 5h29'24"40d60'

Garibaldi,Garith E
Tri 1h59'20"34d27'

Garibaldi,George J
Aur 5h1'43"50d31'

Garibaldi,Katherine N
Peg 21h53'0"28d50'

Garino,Ryan
Cnv 12h16'0"45d8'

Garis,Greg
Mon 7h1'57"-6d46'

Garko,Paul J
Dra 16h38'37"66d49'

Garlan
Lmi 10h33'1"33d40'

Garland
And 0h4'35"39d22'

Garland
Lyr 19h20'0"37d54'

Garland III,Howard M
Cnv 13h11'0"40d5'

Garland III,Louis Andrew
Cep 22h2'34"54d10'

Garland,Adrian Judson
Ori 5h54'43"12d22'

Garland,Bridgette
And 23h23'46"47d28'

Garland,Bruce E
Lyn 9h9'0"38d3'

Garland,Frank William
Aql 20h0'50"-7d49'

Garland,Glenn G
Aql 18h59'32"16d1'

Garland,Gregory A
Uma 10h17'25"59d41'

Garland,Jason
Ser 15h44'1"20d23'

Garland,Kimberlee
Aql 18h57'0"-3d12'

Garland,Leslie Alice
Cam 7h35'52"83d49'

Garland,Mary J
Com 12h2'36"22d12'

Garland,Nancy Towle
Del 20h20'1"10d55'

Garland,Will
Cep 23h12'52"64d9'

Garland,William
Cep 23h37'38"65d8'

Garlaska,Doris Jayne
Peg 23h47'28"30d22'

Garlin,Arnetta Ann
Uma 9h25'1"48d28'

Garlin,Iven Earl
Uma 9h51'56"48d10'

Garlin,Les
Aql 20h10'0"1d17'

Garlin,Lessie
Uma 9h54'1"48d8'

Garlin,Rod Arnett
Uma 9h56'24"48d25'

Garlin,Zeral RyNell
Uma 9h44'37"48d2'

Garlinger,John & Myrtle
Peg 23h38'43"30d14'

Garlington VII, Christopher Thomas
Oph 17h29'18"-20d50'

Garlinski,Edward
Her 17h22'18"22d50'

Garliss
And 1h54'23"38d35'

Garlock,Dr Victor
Oph 18h17'34"12d26'

Garmain,Matthew Harrison
Lac 22h29'58"55d57'

Garman,James
Her 16h1'50"40d11'

Garman,Phil
Cas 1h31'50"58d10'

Garman,Sean Michael
Her 17h49'29"14d55'

Garmer's Grands
Lac 22h0'24"51d54'

Garmon,Marshall Irving
Del 20h22'17"10d1'

Garmon,Micheal Scott
Eri 2h57'0"-4d30'

Garmond
Uma 8h5'54"68d15'

Garn,MariaElena
Vul 19h45'37"28d26'

Garnaud,Jacques
Per 3h48'17"35d36'

Garncia,Jeffery A
Aur 5h52'44"50'31'

Garneau,Stéphanie
Cas 0h45'1"73d56'

Garner Family
Uma 10h47'1"57d13'

Garner is Loved, Samantha Ryan
Mon 6h28'35"5d29'

Garner,Alvin Joel "Papa"
Aur 5h0'48"41d55'

Garner,Amanda Lee
Peg 0h5'46"17d56'

Garner,Bernice Lea Marie Miller
And 22h59'34"36d48'

Garner,Carolyn
Equ 21h10'28"11d55'

Garner,Carolyn Zajac
Peg 22h17'28"31d54'

Garner,Chris
Aql 19h57'42"0d10'

Garner,Chris & Caree
Lmi 9h53'10"34d53'

Garner,Emily Nichole
Umi 14h53'30"66d14'

Garner,John Lee
Boo 14h32'0"40d36'

Garner,Jon
Aur 6h58'0"43d18'

Garner,Jordan Grant
Cet 0h25'14"-1d50'

Garner,Jr,Donald B
Aur 7h10'46"39d43'

Garner,Marilyn
Cas 3h4'17"73d21'

Garner,Michael David
Aql 20h5'16"4d46'

Garner,Nicholas Lee
Umi 15h8'53"69d20'

Garner,Stephanie Marie
Eri 3h53'25"-5d50'

Garner,Terri R
Leo 9h58'31"19d31'

Garner,Theodore V
Eri 2h52'0"-5d37'

Garner,Virginia S
Eri 2h51'40"-5d32'

Garner,William C
Sex 9h42'0"4d10'

Garner,Zak
Cyg 19h27'26"39d50'

Garnesson,Estelle Honey
Aql 18h56'37"-6d33'

Garnet,Judi L
And 2h33'27"39d44'

Garnet,Steve Ego
Ser 15h10'21"6d5'

Garnett,Delores
Sex 9h51'59"1d28'

Garnich,Diddi
Vir 13h36'22"-10d20'
Garnier,Jean Louis
Tri 2h1'17"35d22'
Garnier,Michel
Cam 3h23'42"60d2'
Garnier,Sylvia
Per 3h53'9"39d25'
Garofalo,Barbara & Ronald
Crb 16h18'13"38d35'
Garofalo,Gary
Her 18h25'47"20d9'
Garofalo,Gary
Lmi 9h48'47"38d10'
Garoian,George
Dra 18h20'14"48d9'
Garoian,Scott Michael
Her 15h56'54"48d43'
Garold
Mon 7h11'20"-6d28'
Garon,Beverly
Leo 9h59'39"8d12'
Garon,Jay
Her 16h27'54"27d18'
Garone,Robert David
Leo 9h56'39"8d13'
Garoni,Gloria
Del 20h19'12"11d11'
Garonzik,Ethan
Oph 16h21'0"2dd51'
Garonzik,Mark
Ser 16h2'11"0d31'
Garoogian,Zakar
Aql 20h9'36"3d50'
Garr Bear
Her 18h8'59"40d4'
Garr,Betsy
Aql 19h0'1"0d49'
Garr,C J
Cet 1h57'0"-1d1'
Garra,Dr Brian
Aur 7h8'38"40d52'
Garra,Jennie DePetro
Cet 0h39'55"-7d21'
Garrabrant
Vul 20h1'49"28d51'
Garrahan,Abigail Marie
Lyr 18h40'12"28d36'
Garrard,Megan Kyla
Cyg 19h24'26"35d51'
Garratt,Mark
Her 16h56'34"50d33'
Garratt,Steve
Aur 5h5'38"41d1'
Garratty,Gemma Rose
And 23h16'17"47d17'
Garraway,J Elizabeth
Cas 0h24'34"56d0'
Garreau,Kelly Lynn
Boo 14h31'49"44d18'
Garreffa,Annamaria
Dra 16h29'38"63d60'
Garren,Maggie Pedersen
Cam 8h24'59"78d34'
Garret Loves Julia Forever
Peg 22h25'58"25d28'
Garrett
Her 18h0'44"30d19'
Garrett Star,The
Sex 9h48'52"0d40'
Garrett"Six"
Vul 19h23'17"25d13'
Garrett's Galaxy
Oph 17h14'50"-20d38'
Garrett's Star-ALGB1920,Albert Leon
Cep 22h35'0"58d51'
Garrett,Amanda Leigh
Eri 3h57'24"-1d1'
Garrett,Amy
Mon 8h3'43"-0d30'
Garrett,Andree A
Aur 5h8'53"40d19'
Garrett,Brian
Ori 5h59'40"11d14'

Garrett,Brittany Nichole
Cyg 19h47'1"37d48'
Garrett,Daphne L
Aur 6h30'13"30d15'
Garrett,David F
Aql 19h58'22"-6d10'
Garrett,Douglas
Aur 5h6'15"40d53'
Garrett,Elke
Vul 20h2'39"28d30'
Garrett,Ernest D
Per 3h7'36"37d56'
Garrett,Eve Felix
Aur 5h53'46"38d51'
Garrett,Inspector
Oph 17h55'24"11d26'
Garrett,Janice
Cet 3h6'38"1d9'
Garrett,Jay
Aql 18h57'22"15d19'
Garrett,Jesse Harrison
Vul 19h46'27"28d33'
Garrett,John M
Cep 22h50'19"71d10'
Garrett,Jonathan Paul
Dra 16h26'47"69d27'
Garrett,Jordan Douglas
Leo 10h38'0"22d27'
Garrett,Joy Ellen
Peg 22h52'26"20d30'
Garrett,Kristen Lynn
Cet 2h30'1"4d59'
Garrett,Mackenzie Elizabeth
And 0h35'20"40d35'
Garrett,Marissa
Peg 22h32'1"24d29'
Garrett,Matthew Lawson
Aur 6h28'1"37d28'
Garrett,Miri
Cyg 20h3'27"31d50'
Garrett,Nancy
Lyr 18h54'13"32d3'
Garrett,Robert & Maria
Uma 11h1'1"35d28'
Garrett,Steven William Longlad
Ser 15h22'18"2d15'
Garrett,Ted
Aql 18h45'52"8d19'
Garrett,Tevin
Uma 9h41'16"62d14'
Garrett,William
Cep 22h15'18"67d60'
Garrett-GF
Aql 20h0'15"9d46'
Garrett-Smith,Sally Ann
Vul 19h47'50"28d23'
Garrido,Brooke Lynn
Mon 6h30'45"3d20'
Garrido,George
Oph 18h17'1"10d57'
Garrigan,Stephen Robert
Dra 16h38'25"61d42'
Garris,G J
Cet 1h36'12"-0d49'
Garris,H Clark
Her 17h52'31"30d58'
Garris,Suzanne M
Cas 2h11'1"71d10'
Garrison
Cam 4h57'37"70d18'
Garrison
Aql 19h19'25"15d10'
Garrison III,P R
Aur 6h30'23"31d37'
Garrison,Betty
And 2h35'0"50d52'
Garrison,Diane & Sue Bundy
Peg 21h57'37"29d50'
Garrison,E Gary
Cet 3h12'44"6d17'
Garrison,Gary Wayne
Her 17h33'37"48d29'
Garrison,Glenn
Cnv 12h58'0"51d20'

Garrison,Haley Lynn
And 23h1'55"51d43'
Garrison,Harvey Bair
Cet 2h30'34"4d26'
Garrison,Jr,Clyde H
Tau 4h37'12"20d32'
Garrison,Jr,Earl Ray
Sct 18h54'57"-4d0'
Garrison,Minor
Lyn 7h5'43"58d27'
Garrison,Missy
Cas 0h32'1"63d5'
Garrison,Philip
Her 17h2'36"21d59'
Garrison,Randy
Boo 14h23'32"17d52'
Garrison,Theodore W
Aur 6h4'15"36d54'
Garrison,Tyler Ryan
Peg 21h57'24"34d9'
Garritano,Ti-Amo Mario-Marc
Lyn 7h2'38"53d22'
Garrity,Jr,Albert Francis Caching
Lac 22h54'0"55d53'
Garrity,Nichole Renee
Lyr 18h36'49"45d3'
Garrity,Patrick James
Boo 15h15'58"32d58'
Garro,Gabriella Mariah
And 23h46'24"43d12'
Garruto,747 Captain Frank Michael
Aql 19h22'58"-1d44'
Garruto,Esq,Andrew F
Per 3h20'42"40d55'
Garry H
Boo 14h41'33"21d32'
Garry III,Joseph A
Dra 18h36'32"68d48'
Garry Lloyd
Ori 5h19'48"15d18'
Garry,James W
Ori 5h31'19"-0d11'
Garry,Molly
And 2h1'46"38d40'
Garry,Sylvia
Equ 21h21'57"10d31'
Garsh,Louis
Mon 7h59'53"-6d2'
Garsh,Trinidad
Cet 0h8'18"-11d24'
Garsick,Aaron James
Cep 21h57'39"58d45'
Garside,Deborah Louise
Lyr 19h0'1"35d0'
Garside,Michelle
And 23h15'1"50d26'
Garside,Scott F
Cet 2h52'1"1d40'
Garson,Bernard
Oph 16h58'57"10d37'
Garson,Stuart Irwin
Gem 7h59'0"20d17'
Garten,R Gray
Ser 15h50'1"19d59'
Garten,Rev Retha
Ori 5h53'18"10d50'
Garth
Ori 5h53'22"-6d7'
Garth
Vel 9h43'56"52d3'
Garthune,Nina Mae Cranston
Cyg 19h24'21"33d22'
Gartland,James Keith
Dra 15h37'50"58d44'
Gartland,Jessica Nicole
And 0h10'54"37d46'
Gartland,Kathleen,"A Memory A Wish A Dream"
Uma 9h33'49"58d23'
Gartman,John A
Cet 1h16'0"-6d33'
Gartner,Christopher Carl
Her 17h21'16"47d19'

Gartner,Jennifer
Cyg 21h34'29"41d13'
Gartner,Joan
Lmi 10h31'58"31d11'
Gartner,Josef
Lyn 9h1'20"40d13'
Gartner,Joseph Cameron
Aur 6h9'1"40d55'
Gartner,Paul
Lmi 10h50'27"31d18'
Garton,Todd
Aql 19h7'13"3d1'
Gartrell,Scott MacFarlane
Oph 16h48'27"11d19'
Garver,Alexandra Rae
Mon 6h20'56"2d56'
Garvey,Josephine Patricia
Cyg 21h0'18"37d36'
Garvey,Judith
Leo 10h56'12"21d24'
Garvey,Marc Devin
Boo 14h26'17"37d50'
Garvey,Nicholas Robert
Per 2h23'20"55d36'
Garvey,Shawn S
Uma 8h19'37"70d16'
Garvey,Vincent Alan
Tau 5h48'20"23d4'
Garvey-McLeish,Judi
Cas 0h40'28"62d21'
Garvie,Gail Ann Purks
And 2h29'12"50d36'
Garvin,Alexander William
Mon 6h20'53"8d55'
Garvin,Dawn
Vul 20h17'17"25d52'
Garvin,Doyle Burgess
Del 20h18'35"10d31'
Garvin,Laura Grace
Aql 20h35'11"0d19'
Garvin,Mallory E
Vul 20h18'17"25d39'
Garvin,Natalie V
And 23h26'56"43d34'
Garvin,Pamela Ann
Mon 8h4'52"-9d21'
Garvy,Neil
Gem 6h48'7"12d13'
Garvy,Vanessa Christine
Vir 13h31'23"-4d39'
Garwood,John Grant
Per 3h3'0"50d17'
Gary
Dra 20h0'54"70d58'
Gary
Per 1h32'34"53d11'
Gary
Lac 22h45'36"52d37'
Gary & Cushy
Uma 8h22'17"61d37'
Gary & Dana
Cyg 21h19'44"37d34'
Gary & Deborah
Eri 4h47'56"-5d53'
Gary & Julia's "Wedding Star"
Ori 6h5'13"8d40'
Gary & Lori
Uma 9h38'23"56d54'
Gary & Michele
Cyg 21h10'11"35d26'
Gary 3
Mon 6h56'56"8d52'
Gary Alan
Cep 22h48'45"70d59'
Gary Andrew
Peg 23h26'47"24d46'
Gary Ann
Lac 22h21'36"50d30'
Gary B
Ori 5h1'24"0d55'
Gary Joseph
Ori 5h56'49"18d25'
Gary Lee
Hya 8h32'1"5d56'

Gary Paul
Ari 1h55'35"17d59'
Gary Starship of the Enterprise
Aur 5h26'25"40d46'
Gary's Dark Star
Cep 20h32'0"63d4'
Gary's Gang
Cep 21h39'14"68d37'
Gary's Glow
Boo 13h38'45"22d58'
Gary's Guardian
Cep 22h12'36"60d52'
Gary's Love
Ori 5h55'40"11d19'
Gary's Love
Her 16h38'25"42d9'
Gary's Nova
Aql 18h56'25"-0d29'
Gary's Star
Aur 5h35'45"38d48'
Gary's Star
Boo 15h11'30"52d42'
Gary's Star
Cap 21h27'10"-23d1'
Gary,Brian Wayne
Aur 6h9'42"38d58'
Gary,Dianne Elaine
Mon 7h44'54"-1d55'
Gary,Dorothy C
And 23h1'18"48d16'
Gary,My Love
Her 16h25'0"48d59'
Gary,Rose Emerson
Mon 7h21'13"-6d55'
Gary,Tanya McTeer
Mon 7h31'23"-10d44'
GaryJules Magic
Uma 12h38'33"56d6'
Garyson,Grace
Ser 18h36'57"5d53'
Garza,Billie
Dra 12h57'33"68d56'
Garza,Catherine
Lac 21h58'13"37d48'
Garza,Enrique
Boo 15h9'12"26d2'
Garza,Jeni
Cas 22h59'30"58d4'
Garza,Joe
Uma 11h42'24"50d3'
Garza,Joel
Ser 15h14'43"9d42'
Garza,John Michael
Dra 17h25'19"72d17'
Garza,Jr,Filemon Sergio
Aql 20h1'13"6d32'
Garza,Kristen Ford
Cas 1h58'41"73d36'A
Garza,Paladio
Lyn 8h5'21"33d38'
Garza,Poli
Cam 7h56'42"60d29'
Garza,Rene C & Enid J
Eri 3h8'40"-7d2'
Garzon,Maria Nelena
And 1h8'41"38d9'
Garzone,Justin Lawrence
Her 16h21'48"20d16'
Garzone,Kristyn Alicia
Cyg 21h24'1"38d33'
Gasbarro,Silvana
And 23h38'14"46d46'
Gasca,Travis Stephen
Oph 17h33'31"-20d43'
Gascoin,Bernard
Her 18h29'51"12d22'
Gaseor,Alyssa Ann
Cyg 20h7'43"41d11'
Gash,Denise Ann
Sge 20h4'39"20d4'
Gasho-Cheek
Aql 20h10'19"4d3'
Gashouse,The
Ori 4h58'37"0d2'

Gasibel
Cam 5h44'52"60d45'
Gasiorek,Scott Brian
Vir 13h27'35"-0d34'
Gaskell,Frances Petti
Mon 6h35'0"-0d21'
Gaskill,Mark J
Her 16h38'53"39d48'
Gaskill,Megan
Lyn 8h13'18"33d44'
Gaskin's Star,Lew
Aql 18h40'13"1d49'
Gaskin,Roger W
Boo 14h58'31"21d56'
Gaskin,Stuart
Aql 20h11'1"10d30'
Gaskins,Terry
Lyn 8h49'59"44d28'
Gaso,Miquel Angel Sanahua
Ori 6h6'35"0d15'
Gaspar Star,The
Lyr 19h21'23"31d2'
Gaspar,Jim
Oph 18h3'54"12d43'
Gaspar,Maria Arlette
Uma 9h18'35"50d3'
Gaspari,Amanda Clare
Tau 4h44'1"20d17'
Gasparik,Jeffrey Florian Sebastian
Cep 23h21'41"64d20'
Gasparino Forever, Martin
Dra 12h11'1"67d55'
Gasparri,Joseph
Aur 6h26'16"33d27'
Gasper,Nicholas Ralph
Aur 5h3'20"51d42'
Gasper,Peter
Tau 3h41'58"7d45'
Gasperi,Renata
Lyn 9h5'1"40d18'
Gasperich,Nathan Matthew
Lyr 18h43'40"34d19'
Gass,Billy Irvin
Sgr 19h43'54"-41d15'
Gassaway,Clyde
Oph 17h8'32"-23d37'
Gasse,Sabine
Equ 21h4'30"3d39'
Gasser,Kathy
And 2h0'16"41d22'
Gasser,Michael Joseph Crawford
Cnv 12h8'12"38d19'
Gassner,Werner
Vul 19h47'0"27d17'
Gasso,Don
Aur 6h14'10"31d2'
Gast,Brandon
Her 17h28'19"40d43'
Gast,Cody
Lac 22h9'37"37d35'
Gastaldin,Dale Leo
Leo 9h53'25"10d52'
Gastaldo,Dona Green
Ser 18h36'18"3d16'
Gastauer,Edward C
Per 2h10'54"56d31'
Gastelu,Marie
And 0h5'58"35d10'
Gastevich,Vladimir
Lmi 10h9'0"40d42'
Gastevich,Vladimir
Her 17h30'43"30d53'
Gastin,Danny
Boo 14h27'40"52d52'
Gastin,Louise Rice
Lyn 7h25'39"58d9'
Gaston,Adeline
Aql 20h53'40"-6d58'
Gaston,Barbara J
And 2h3'40"46d50'
Gaston,Lindsey Erin
Lyn 7h59'13"38d38'

Gaston,Sir Lamont C
Peg 21h25'43"18d56'
Gaston,Theis Hope
Del 20h13'0"14d55'
Gasvoda,Jerome
Lac 22h38'32"38d45'
Gata-Aura,Chane
Cep 22h30'12"63d31'
Gatchel,Chelsee Rose
Lyn 7h36'38"48d51'
Gately,Joshua Tyler
Dra 15h12'35"63d31'
Gater,Jean Margaret
Cyg 19h50'21"58d26'
Gates, James D & Susie D Gates
Crb 15h15'1"32d23'
Gates William
Lyn 7h2'27"52d56'
Gates,Chrysie
Cas 0h23'56"62d31'
Gates,Cody
Crt 10h52'12"-11d42'
Gates,Doreen S
Lac 21h55'19"37d17'
Gates,Dr Kar-La Pierette
Peg 21h58'41"24d13'
Gates,Echo
Cam 3h48'40"68d14'
Gates,Fred & Natalie
Cyg 21h34'44"30d22'
Gates,John Randall
Cep 22h20'56"55d41'
Gates,Jr,Augustus J
Aur 6h4'14"38d37'
Gates,Margaret
And 2h28'0"39d17'
Gates,Meaghan
And 1h26'55"36d23'
Gates,Mike
Lyn 8h13'46"44d10'
Gates,Nicole M
Lyr 18h14'43"38d44'
Gates,Phillip
Cyg 20h38'19"45d51'
Gates,Ralph Jon
Boo 15h3'21"41d22'
Gates,Sarah Jo
Peg 21h57'41"20d1'
Gates,Swanee
Sex 9h58'36"5d25'
Gates,Timothy Lansing
Aql 18h54'21"8d32'
Gates,Vivienne Angela
Lyr 18h16'1"33d57'
Gates,William Frederick
Peg 21h53'54"18d49'
Gates,William H From The CRPG
Uma 8h39'40"70d30'
Gates,Zachary
Gem 8h2'20"31d11'
Gatesy,Jacqueline C
Cas 0h54'52"69d52'
Gath,Craig Stephen
Cep 23h28'17"68d42'
Gathof,Blessed,Karl
Uma 8h23'18"68d26'
Gatine,Joelle
Per 3h28'36"50d41'
Gatliff,Robert Paul
Aur 7h20'27"36d23'
Gatlin II,James Arnold
Lmi 10h15'45"32d5'
Gatlin,Amy
Cyg 20h59'37"28d44'
Gatlin,David Lee
Per 2h51'36"46d38'
Gatlin,Jamie Leslie
Cas 23h41'22"63d16'
Gatlin,Jessie Alberta
Lyr 18h53'15"42d21'
Gatlin,John-Ross Alexander
Her 16h37'59"27d25'

Gatlin,Michael James
Boo 13h55'1"17d7'
Gatling,Paris Savon
Peg 0h2'23"10d53'
Gaton,Michele
Ari 2h31'53"20d11'
Gator
Eri 4h13'20"-17d58'
Gator,Grippy T
Cep 23h5'45"60d38'
Gatsby
Leo 9h25'19"7d10'
Gatschenberger,Marilyn S
Peg 21h25'48"3d33'
Gatson,Emily
Cas 23h6'29"58d11'
Gatta,Douglas Damon
Lac 22h17'0"37d33'
Gatta,John Joseph
Vir 13h24'11"-2d13'
Gatti 26/9/79-4/4/93, Carlos
Dra 17h52'42"60d30'
Gatti,Alexander Vincent
Cma 7h17'1"-15d12'
Gatti,Sergio
Peg 23h0'21"28d31'
Gatti-Damiot,Michele
Sge 19h58'40"20d31'
Gattina
And 23h44'1"38d15'
Gattina
And 0h10'37"40d13'
Gatto,Charles & Kelli Caputo
Lyr 18h38'30"42d3'
Gatto,Joseph P
Eri 3h18'52"-10d56'
Gatto,Joseph P
Dra 19h1'41"54d37'
Gatton,Delbert J & Anna S
Cyg 21h3'17"28d23'
Gatton,M
Umi 15h10'1"67d39'
Gatton,Patsy
Cma 6h57'1"-20d22'
Gattozzi,Claude Eric
Cnv 13h13'0"32d3'
Gattuso,Stephen John
Cep 3h57'19"80d5'
Gatzke,Elizabeth
Cas 1h43'50"70d47'
Gaubatz,Daniel B
Ori 5h15'1"-9d16'
Gaubatz,Walter
Boo 14h20'44"11d21'
Gaubert,Dorothée
Per 3h27'23"38d49'
Gaubert,Marcel
Dra 17h7'26"66d50'
Gaubis,Phil
Dra 20h29'32"70d36'
Gauch,Armin
Ori 5h58'18"18d14'
Gaucher,Jacques
Cam 3h37'21"71d34'
Gaucher,Kenneth A
Cep 0h37'0"77d59'
Gauci,Allan John
Uma 8h47'51"56d57'
Gauci,Vincent Thomas
Dra 17h22'1"60d55'
Gaudenzi,Marcella
Ser 15h52'41"2d12'
Gaudet,Alexandre
Per 3h30'5"37d14'
Gaudet,V A
Cam 14h0'53"81d13'
Gaudette,Marie- Laurence
Gem 6h58'51"18d15'
Gaudia-idem-persempre
Umi 16h23'59"80d26'
Gaudiano (93),Luisa
Sgr 18h50'10"-29d6'
Gaudiano,Samantha Stewart
Ori 5h51'0"20d1'

Gaudin,Donald W
Uma 10h26'52"51d20'
Gaudio,David Allen
Lac 22h35'25"55d36'
Gaudio,Honey
Lyr 19h8'21"38d25'
Gaudio,Mikki C
Cnv 12h41'46"40d14'
Gaudio,Rose M Moskal nee
And 0h10'33"28d58'
Gaudioso,Gloria
Cam 6h17'24"71d31'
Gaudioso-Jackman I
Cyg 19h44'18"29d44'
Gauerke,Juli
Ari 2h43'43"30d57'
Gauerke,Victoria & Ronald
And 0h18'1"31d50'
Gauger,James Erwin
Boo 15h5'23"31d7'
Gauggel,Ingrid
Psc 0h13'1"10d54'
Gaughan,Karen
Ori 5h51'54"14d53'
Gaugler,Gilbert John Luke
Per 4h0'38"40d12'
Gaugler,Jennifer Lynn
Cyg 21h7'51"39d50'
Gaul,Cory Justin
Hya 9h8'54"2d20'
Gaul,Phyllis Elizabeth
Com 12h14'26"20d2'
Gauler,Brian James
Her 16h12'33"21d55'
Gaulke,Scott W
Lac 22h3'57"48d47'
Gaullier,Jean-Pierre
Aur 7h9'20"38d51'
Gaulocher,Gregg
Eri 3h13'33"-6d37'
Gault,Lon
Hya 8h11'39"1d22'
Gault,Morgan Reed
Oph 17h13'20"-10d21'
Gault,Roy V
Her 16h0'11"40d46'
Gault,Stanley C
Vir 11h38'12"2d12'
Gault-Forever Tigers, Jim & Clydie
Boo 15h6'33"19d32'
Gaultney,Camden Elizabeth
Lyr 19h14'11"35d13'
Gaumier,Alain
Aur 4h34'40"31d30'
Gaumond Family
Sct 18h54'55"-7d46'
Gaumond,Joyle
Cma 6h44'32"-13d53'
Gaunce,Maxwell Stewart
Tau 4h3'29"21d48'
Gaurino,Meg & Chuck
Boo 16h11'26"53d51'
Gaury,Lea
Dra 16h46'54"62d38'
Gaus,Michelle B
Del 20h53'29"8d43'
Gause,Jr,Jacob A
Peg 22h42'60"32d27'
Gausepohl,Christel
Dra 18h0'12"67d31'
Gauss,Arno
Leo 10h59'37"11d21'
Gauss,Timothy
Eri 3h30'43"-6d7'
Gaut,Christine
Mon 6h19'26"-1d9'
Gaut,Dr Norman
Per 3h6'57"41d33'
Gautam's Star
Dra 10h15'21"77d52'
Gautam,Dr Anil
Per 2h36'50"37d57'
Gaute,Adele
And 23h16'42"47d36'

Gautheir,Daniel
Per 2h56'29"48d46'
Gauthier 03/30/1978, Marc
Dra 9h30'55"73d25'
Gauthier,Alexandre
Uma 10h53'51"71d30'
Gauthier,Daniel Nelson
Per 1h58'51"50d9'
Gauthier,Jacqueline M
Cyg 21h28'60"41d8'
Gauthier,Jacques-Yves
Cam 5h56'27"65d30'
Gauthier,Jade
Umi 16h54'0"75d52'
Gauthier,John Gates Bourgeois
Oph 18h34'26"10d8'
Gauthier,Laurence
Umi 16h37'45"75d47'
Gauthier,Lucienne Leclerc
Sgr 18h53'36"-32d25'
Gautier,Matthew
Per 3h19'20"50d5'
Gautierrez,Lewis A
Hya 8h56'49"-6d49'
Gautmier,Eric
Mon 5h55'23"-5d40'
Gautschi,Hans & Bridget
Her 17h9'1"49d9'
Gauvin,Hervey P
Aur 7h24'0"38d41'
Gavalick,Anna Rose
Vir 11h45'58"8d59'
Gavazda,Totdeauna Frania
Uma 13h56'37"58d56'
Gavazda,Wlodarczak
Ori 6h5'36"-0d14'
Gavazzi,Cristina
Eri 4h26'0"-6d51'
Gavenas,Mary Lisa
And 23h19'49"50d34'
Gavigan,Brianna Monaghan
And 0h2'28"35d30'
Gavilan,M Claudia Estrada
Cra 18h23'20"-44d30'
Gavilli,Alain et Yoann
Aur 5h2'45"38d11'
Gavin Gloriously Glowing,Stephen D
Cyg 19h31'0"33d23'
Gavin,Brian
Per 3h55'19"38d39'
Gavin,Elaine
Mon 8h6'51"-1d9'
Gavin,Jr Family Star, John Gregory
Lac 22h32'14"49d16'
Gavin,Justin J
Sex 10h48'39"-8d18'
Gavin,Tiffany Victoria
Tau 4h7'3"22d44'
Gavini,Danny
Her 16h56'59"50d7'
Gavino,Rosemary D
Cas 0h29'31"75d26'
Gavito,Dan
Aur 6h17'1"31d45'
Gavlak,Marty
Aur 6h11'11"32d45'
Gavoue,Genevieve
Uma 11h25'1"34d52'
Gavredes,Cheryl Ann
Equ 21h8'0"11d19'
Gavric,Djordje
Aur 4h47'47"50d29'
Gavric,Izabella
Cas 0h35'1"62d13'
Gavrilos,Joanne
Cet 2h5'51"2d8'
Gavrilov Concord Limousine,Alexander
Leo 11h10'19"-5d49'
Gavrilova,Svetlana
Crb 15h33'45"32d23'
Gaw,David
Lac 22h29'41"40d13'
Gaw,Jean E
Cas 23h14'44"60d41'
Gaw,Stephenie
Lyn 8h2'36"35d10'
Gawenda von Rokittnitz Paul
Cnc 9h15'38"32d18'
Gaworzewski,Heidi
And 23h34'22"39d37'
Gawronski"I Love You", Rhonda Lynn
Sge 19h32'8"17d44'
Gawronski,F Matthew
Her 17h4'55"20d54'
Gawronski,Michael Anne
Cas 1h5'37"58d47'
Gawronski,Susan E
Mon 6h35'0"-6d26'
Gay Darlene
Aqr 21h59'35"-18d12'
Gay Little Kicky Feet, Nora Elizabeth
Cas 1h33'51"60d23'
Gay Lynn
Cmi 7h22'58"0d54'
Gay,Allen Thomas
Oph 17h31'24"-23d30'
Gay,Claire Suzanne
Cas 0h25'38"62d25'
Gay,Deborah Ann
Lac 22h1'26"48d47'
Gay,Donna Lyn
And 23h36'49"45d40'
Gay,George
Oph 16h6'11"-6d39'
Gay,Herbert 10-09-1923
Vir 13h35'20"-2d29'
Gay,Jennifer
And 1h19'51"40d59'
Gay,Jr,Leon Roosevelt
Umi 14h44'47"65d34'
Gay,Jr,Walter Lee
Her 17h41'31"18d49'
Gay,Margery Shaer
Lyn 7h35'10"38d48'
Gay,MD,Mary St John
Cyg 20h20'12"30d53'
Gay,R Rondell
Aur 5h58'46"38d2'
Gay,Samantha Rachelle
And 1h23'13"33d42'
Gay,Sandra
Cas 2h52'48"70d33'
Gay,Ted
Lac 22h7'17"51d30'
Gay-Knott was 40!,Nic
Ori 6h4'13"4d6'
Gaybriella Marie
Oph 17h43'57"13d16'
Gayda,Alan
Ori 5h30'21"7d48'
Gayden Stacey
Cap 20h5'13"-12d53'
Gaydos,Andrew
Aur 5h51'25"38d54'
Gaydos,Michael
Cep 22h8'29"53d34'
Gaye & Armando "Don't Go Away"
Peg 22h47'1"11d32'
Gaye,Gregory Andrew
Boo 14h44'20"26d57'
Gaye,John Patrick
Her 17h5'52"48d54'
Gayla
Aql 19h48'44"14d14'
Gayla
Ori 6h11'33"17d55'B
Gayla Jo "Sunshine"
Lyn 6h25'33"58d21'
Gayle
Cyg 21h12'54"38d10'
Gayle & Bob
Per 4h0'37"51d31'
Gayle 171
Tri 2h8'50"31d20'
Gayle 6-4-3
Peg 23h3'41"30d16'
Gayle,Christopher Sean
Cam 3h53'20"57d56'
Gayler,Susan
Sge 20h3'31"20d41'
Gayley,The
Aur 6h6'0"36d42'
Gaynor
Cyg 19h30'21"31d5'
Gaynor,Alex Johnathon
Oph 18h4'57"12d38'
Gaynor,David
Her 17h56'22"40d43'
Gaynor,Jacqui
Umi 14h50'58"69d7'
Gaytan,Elizabeth
Mon 6h58'51"-10d15'
Gaz Star
Uma 8h12'54"61d38'
Gazarek,Cyril "Cy"
Cnv 12h22'0"34d49'
Gazaway,Renee
Cyg 21h45'17"37d39'
Gazda,Sandra Anne
And 23h45'37"38d18'
Gazdik,Kirk Williamson
Dra 15h9'42"56d33'
Gaze,Weston
Her 18h4'53"28d59'
Gazer,Gemini II Taz
Gem 8h3'1"28d6'
Gazlay,Jeannine, Robert,Kyle
Ori 4h54'21"4d2'
Gaztambide,Denise G
Peg 23h44'19"28d58'
Gazza's Glory
Ori 5h40'36"-0d9'
Gazzala
Cas 0h25'38"74d55'
Gazzelli,Laurie Jean
Lyn 8h7'21"36d9'
Gazzillo,Ralph
Her 15h57'31"47d59'
Gaëtan,Gentile
Ori 5h57'0"20d34'
Gaëtane
Lib 14h58'59"2d57'A
Gaßmeier,Gerhard
Peg 23h36'46"11d5'
GB 4247
Uma 9h59'35"51d15'
GC Everlasting
Cet 0h24'1"0d23'
GCD
Sct 18h54'0"-7d25'
Gdula & Kitties,Jerry & Kim
Cyg 19h51'42"41d11'
Ge-h-heim der Com-HERZ-iant
Vir 11h41'24"8d35'
Geach,Carol Marie
Vul 21h12'20"20d16'
Geake Heavenly Star, Christopher M
Aur 6h29'1"37d58'
Gear,Barbara Ann
Vul 21h1'37"28d13'
Gear,Emily
Cyg 19h24'42"33d26'
Gear,Erin "Eggie"
Aql 19h32'17"0d34'
Gear,Timothy W
Aql 20h13'39"5d8'
Gearen,Shane Michael
Cet 1h39'12"-1d35'
Gearhart,Jendy Lynn
Mon 6h50'26"11d22'
Gearhart,Julie Russell
Uma 11h39'0"43d32'
Gearhart,Justin Lee
Cyg 21h15'0"35d5'
Gearity,PhD,Lauree P
Ari 2h1'0"21d37'
Geary Family,Don
Aql 20h11'40"10d35'
Geary Family,The
Aql 20h18'27"7d49'
Geary,Christopher
Aql 20h11'55"10d49'
Geary,Christopher S
Aqr 20h54'1"-0d58'
Geary,Dorothy & Warren
Uma 11h5'47"71d52'
Geary,Louise
Cas 0h23'21"59d27'
Geary,Miriam M
Lyn 8h57'29"46d51'
Geary,Neal
Cet 1h40'48"-5d59'
Geary,Nikki
Cyg 20h1'17"50d9'
Geb
Cae 4h51'21"-32d54'
Geb 12-1-1950,Rudy Lionel Vinck
Dra 10h16'31"81d17'
Gebahrdt,Philippe
Sex 10h14'3"-2d31'
Gebauer,Angelika
Mon 7h52'48"-1d32'
Gebbia,John Thomas
Hya 8h11'0"5d41'
Gebel,Sheila C
Mon 8h8'30"-6d29'
Gebert,Lisa Ann
And 0h9'50"46d31'
Geberth,Melinda
Dra 19h32'32"67d40'
Gebhard 8 23 1958
Lac 22h35'15"50d16'
Gebhard,Kenny
Aql 19h56'35"7d33'
Gebhards,Brandy
Sge 19h53'43"20d28'
Gebhardt,Austin Richard
Boo 14h23'24"37d24'
Gebhardt,Philip A
Tri 2h5'15"32d42'
Gebhardt,Rolph
Cyg 20h42'12"46d44'
Gebhardt,Rosemary
Cyg 20h3'1"50d13'
Gebhart,Gregory Gerard
Per 2h22'49"55d27'
Geboren am 13/04/1960, Markus Setz
Dra 15h51'42"62d27'
Gebrana
Oph 17h55'1"13d60'
Gebura,Karen Ann
Peg 23h7'1"8d11'
Geburtstagsstern,Eva Haberlands
Lyn 8h15'50"49d34'
Gebus,Charles Carter
Ori 5h50'53"18d42'
Geck,Karen Lynn Swistock
And 23h37'46"49d18'
Geck,Russell James
Hya 8h39'59"3d24'
Gedack,Dad Theo
Cep 21h36'0"71d15'
Gedda,Thomas & Amy Desotell
Uma 10h6'28"61d32'
Geddings,Wendy
Vul 19h13'29"22d5'
Gedeon,Theodore P
Cet 2h51'42"5d0'
Gedinsky,Cissy
Cyg 21h26'42"41d9'
Gedney,Annaleigh Crispin
And 23h34'17"43d15'
Gedye,Helen Eloise
Cyg 19h26'35"56d40'
Gee be why
Dra 18h34'1"71d3'
Gee Gee
Lyn 8h26'57"51d18'
Gee Salute Starlight Fifty,Michael
Ant 9h37'58"-33d27'
Gee,Kathy
And 0h21'10"40d24'
Gee,Kenneth
Her 16h54'42"26d31'
Gee,Marilyn & Richard
Lyr 18h30'42"32d33'
Gee,Matthew Daniel
Umi 13h4'45"76d36'
Gee,Mr John
Umi 14h37'19"74d27'
Gee,Nigel
Dra 16h20'42"61d29'
Gee,Paul
Ori 5h48'24"18d31'
Gee,William E & Michele M Gee
Cyg 19h27'1"33d21'
Geeda
Cyg 21h6'56"30d54'
Geekatorius
Uma 10h20'29"53d31'
Geels,Jonathan
Cet 0h59'31"-0d12'
Geels,Marc-Antoine
Peg 22h27'60"30d22'
Geer,Doris M
Com 12h19'12"27d28'
Geer,Mom's Love Shines Eternal-Betty
Cyg 19h24'22"44d23'
Geerlandt,Geneviève
Cyg 21h20'55"41d6'
Geertje
Peg 0h10'36"17d31'
Geesey,Thomas Mitchell
Vul 19h19'1"26d32'
Geest,Thomas von der
Del 20h15'16"13d20'
Geetter,Darya
Lyn 9h4'16"45d0'
Geezer's Rowdy Ramona
Cma 7h22'40"-15d53'
Geffay,Elodie
Ser 15h13'14"8d20'
Geffert,Julia Heather
Aql 18h58'1"-3d6'
GeGe
Crb 15h32'15"31d52'
Gegenheimer,Harold Walter
Del 20h58'35"13d13'
Geheb,Grace Eichenberger
Lyr 18h42'60"33d56'
Gehi,Mickey
Ori 5h27'58"1d45'
Gehin,Jean Claude
Cnv 13h25'14"38d13'
Gehlhaus,Taylor Emma
And 23h7'41"44d52'
Gehlot,Omprakash"Andy" Kalpna Hirani
Aur 5h57'42"31d9'
Gehoski,Esme Marguerite Swan
Sex 10h3'17"2d9'
Gehr,Dooley & Tracy
Cnv 13h50'44"38d20'
Gehrig,Günter-Rainer
Aqr 20h53'49"0d5'
Gehriger,Marietta & Martin
Oph 17h54'55"10d24'
Gehring,Eckhard
Cmi 7h27'45"8d27'
Gehring,Emil
Dra 18h13'49"68d32'
Gehring,Ilfe
Sco 17h31'23"-33d59'
Gehring,Neal
Ori 5h56'20"13d20'
Gehrke,Carsten & Ellen
Leo 10h49'58"2d19'
Gehrke,Charles Robert
Hya 8h13'28"5d44'
Gehrke,Uwe
Cnv 12h28'39'51d0'
Gehrke,Victoria Jade
Cas 0h5'27"63d48'
Gehrsitz,Dawn
Lmi 10h10'16"33d53'
Geib,Sara Catherine
And 0h10'25"47d40'
Geibel,Ronald & Maureen
Uma 9h35'0"51d52'
Geick,Christina
Cyg 19h54'22"47d51'
Geier,Gretchen Leanne
Mon 8h6'22"-1d26'
Geier,James
Lac 22h7'47"49d32'
Geier,Lisa A & Francis M
Boo 14h38'55"33d0'
Geier,Michael
Per 2h41'0"43d13'
Geier,Norbert
Eri 4h49'6"-7d30'
Geiger,Andrew
Ser 15h55'38"17d56'
Geiger,Glenn Warren
Lac 22h7'32"48d55'
Geiger,Jeffrey Scott
Per 3h10'42"46d14'
Geiger,Jr,Mark Eric
Her 16h39'11"35d38'
Geiger,Mary Ann & Vincent
Sge 20h14'0"17d28'
Geiger,Parker
Equ 20h58'19"10d25'
Geiger,T
Eri 3h16'1"-1d51'
Geiler,Jordan G
Sct 18h41'16"-7d21'
Geiman,Steve
Ser 15h14'30"19d44'
Geis
Vul 19h47'11"29d6'
Geis Humanitarian, Billy R
Per 2h53'33"32d57'
Geis,Katie Ann
Cyg 20h4'0"40d53'
Geis,Mary Jane
Cas 2h23'31"70d15'
Geis,Tracie Lynn
Aql 20h17'58"7d56'
Geischläger,Theodor
Umi 15h21'32"70d34'
Geise's Glory
Aur 7h1'52"38d47'
Geisel,Erich
Hya 8h42'36"-8d7'
Geisel,Lynne
Ori 6h6'18"5d57'
Geiselman,Diane Elena
Cas 23h42'13"61d28'
Geiser,MD,B
Peg 22h43'14"30d11'B
Geisler,Claudia
Aql 19h58'36"10d45'
Geisler,Stacy Rebecca
Mon 6h29'43"8d48'
Geissdorfer,Ronald W
Cyg 19h25'20"32d38'
Geissler
Sct 18h43'9"-6d51'
Geissler,Robert Lutz
Leo 10h59'18"18d54'
Geissler,Valarie Ann
Aur 7h23'25"40d30'
Geist,John N
Cmi 7h42'29"4d39'
Geist,Karen Bowlzer
Cas 2h32'49"57d34'
Geist,Keith Grant
Dra 18h26'38"50d15'
Geist,Linda Anne
Aur 5h5'38"38d48'
Geist,Nicole
And 23h10'43"36d58'
Geist,Ulrike
Lib 14h46'25"-23d6'
Geithner,Heidi
Lac 22h2'59"51d40'
Geißenhörner,Rolf
Cmi 7h19'50"8d40'
GEJA
Cen 14h54'47"-33d3'
Gejay,Douglas E
Per 1h30'25"53d26'
Geko
Lyn 7h37'60"39d50'
Gelato,Gregory Vincent
Cep 21h41'25"67d51'
Gelb,Karen
Oph 17h27'53"-6d12'
Gelb,Sean Jason
Aur 5h17'51"46d3'
Gelbart,Rae
Oph 18h39'1"6d20'
Gelber,David Allyn
Cep 20h55'12"60d38'
Gelber,Jeffrey A
Aur 6h16'11"38d34'
Gelber,Marla
Per 3h12'59"43d11'
Gelber,MD,Howard
Uma 11h57'16"30d5'
Gelber,Steven Vincent
Lmi 10h43'1"28d7'
Gelblum,Jeffrey Barton
Her 18h5'0"31d3'
Geldbach,Gregory William
Cet 2h13'14"8d2'
Geldhauser,Gayle Marie
Cyg 19h53'0"38d26'
Gelfand,Daniel Micah
Aur 5h0'35"41d46'
Gelfand,Linda
And 0h49'39"33d42'
Gelfat,Mark & Sandra
Sge 19h31'15"18d39'
Gelfgat-Ziserson, Victoria S
Lyr 18h41'29"37d1'
Geli
Ari 2h4'45"22d15'
Geli
Cam 7h47'41"60d43'
Geliebte Kaiserin - A Straubinger
Gem 6h42'44"13d50'
Gelik,Ronald W
Sex 10h35'28"2d42'
Gelinas,Lisa Marie
And 0h21'1"35d44'
Gelinas,Lynn & Peter
Peg 22h32'58"26d35'
Gelio,Lara
Boo 14h20'35"52d39'
Gell,Jennifer Kaye
Gem 6h58'17"30d4'
Gellatly,Keanu Christian
Ori 5h15'0"-8d6'
Geller,Andrew
Cep 1h33'48"80d3'
Geller,Brian Marshall
Boo 14h20'59"16d48'
Geller,Christina Shalini
Peg 22h42'34"33d21'
Geller,Daniel M
Cyg 19h43'40"30d36'
Geller,Evan
Lyn 9h27'29"40d29'
Geller,Uri
Ori 5h34'37"-0d53'
Gellert,George A
Her 18h2'40"47d53'
Gelles,Stuart R
Her 16h47'50"50d46'
Gellinek,Inge
Uma 9h39'28"50d35'
Gellinek,Philipp Christian
Ori 6h15'17"10d13'
Gelling,Dora Lee
Lyr 18h43'27"45d6'
Gelling,James Burke
Boo 14h9'28"34d13'
Gellings,Jason Walter
Aur 4h46'27"51d20'
Gellings,Linda Marie
And 22h57'39"50d11'
Gellink,Jean Charles
Mon 7h49'35"-4d27'
Gellis,Denise
And 23h36'31"48d5'
Gelner,Leo & Frieda
Cyg 21h28'32"40d57"
Gelnett,Iva P
And 23h42'25"47d23'
Gelnett,Nora
Lyn 8h44'48"36d26'
Gelski,Lucille E
Boo 14h20'29"32d43'B
Gelsomine,Joseph
Lac 22h8'25"47d14'
Geltman,Susannah Sidney
Crb 15h33'59"32d19'
Geltz,Cathleen
Aql 19h56'0"13d8'
Gelven,Andrew Thomas
Aur 6h20'41"32d37'
Gelvin
Cep 22h9'31"54d9'
GEM
Cyg 21h17'25"28d25'
GEM
Cam 3h30'15"63d31'
Gem 3
Com 12h49'37"20d8'
Gembis,Raeann & Doug
Aur 4h51'13"40d48'
Gemini
Umi 13h27'36"73d51'
Gemini Delivery Star, The
Gem 6h57'34"14d34'
Gemini Mom,I Love You More,J J
Vir 13h51'1"0d40'
Gemini Systems Software
Gem 7h18'48"20d60'
Gemino,Grace
And 1h37'29"39d20'
Gemma
Aur 6h32'18"32d44'
Gemma Elizabeth
Aur 5h0'59"47d46'
Gemma T
Lyn 8h21'36"47d6'
Gemma Verde
Nor 16h23'29"-59d4'
Gemma's Little Clown
Cas 0h15'11"47d23'
Gemma,Joshua M
Cnv 12h9'26"43d43'
Gemma,Robert Michael
Cam 3h53'1"71d9'
Gemmel,Hans Rudolf
Boo 13h45'0"10d38'
Gemmell,Kimberly Beth
Mon 6h30'18"0d27'
Gemmingen,Lee Ann
And 0h19'59"30d58'
Gemsiewolf
Cma 6h56'55"-13d23'
Gemski,Mary Hobbs
And 1h33'23"39d60'
Gena
Hya 8h27'50"-6d3'
Gena
Lyr 18h34'32"46d22'
Gena's Smile
Aql 20h0'43"14d30'
Genabus
Sct 18h42'30"-7d28'
Genal,Scot David
Dra 20h7'40"62d7'
Genan
Eri 4h54'25"-4d3'
Genaro & Mara Our First Decade
Crb 15h26'46"31d2'
Genaver,Joy Willow
Cyg 20h4'23"40d12'
Genbet,Hilbert Martine
Sct 18h45'6"-6d38'
Gencarelli,Dawn Noel
Cas 1h5'21"63d42'
Genco,JoAnn
Cnv 13h38'44"45d15'
Gendel,Claudine Jutchen
Lyn 9h14'52"33d21'
Gendel,Lisa
Aql 19h26'48"7d46'
Gendell,Matthew Dare
Boo 15h6'59"21d59'
Gendey,Francis
Ori 5h1'1"14d49'
Gendreau,Jean-Pierre
Her 16h51'21"33d18'
Gendron,Claude
Cep 22h26'32"63d24'
Gendron,Madeleine
Umi 13h27'45"70d15'
Gendron,Patrick
Dra 16h52'24"52d21'
Gendron,Pierre
Ori 6h2'43"6d16'
Gene
Aur 6h5'17"50d19'
Gene
Dra 18h25'0"70d28'
Gene & Bobbie
Eri 4h1'25"-10d27'
Gene & Dina's Cycling Star
Lyn 7h57'13"41d31'
Gene & Maria
Lac 22h52'44"48d53'
Gene & Mary's Star
Cyg 21h5'40"31d4'
Gene Gregory
Her 16h7'28"40d56'
Gene III
Per 2h3'46"56d48'
Gene Our Love Will 4Ever Shine-Sue
Her 18h6'57"28d15'
Gene,Marilyn & BK-LK Class 95-96
Lmi 10h52'0"27d41'
Genender,Jay
Aur 6h21'23"38d48'
Genentech
Ori 5h30'1"-6d51'
General Motors
Uma 11h45'1"33d34'
General Technology Corp
Per 2h2'37"56d52'
Generalli,John
Cet 3h0'10"5d14'
Generalovich,Alyssa
Lyn 7h57'48"50d40'
Generonden
Tri 2h8'1"32d9'
Generoso,Michael
Aqr 21h59'23"0d53'
Genesee
Cyg 19h40d18"40d24'
Genesen,Mac Charles
Ori 5h55'14"11d14'
Genesis (In The Beginning)
Cyg 21h0'54"28d24'
Genesis 1:16
Boo 14h31'26"39d35'
Genest,Antoine
Gem 7h34'43"34d19'
Genest,Claude
Uma 11h33'51"56d24'A
Genest,Francine
Umi 15h13'37"68d43'
Genest,Ghislaine
Uma 11h33'51"56d24'B

Genet,Daniel
Dra 10h6'1"80d56'
Geneteau,Natalia Ernestina
Vul 20h14'57"25d14'
Genetics & IVF Institute
Boo 15h10'0"30d54'
Genetti,Andrew Boucher
Her 16h42'0"33d4'
Geneva,Florence
And 23h1'1"40d11'
Genevieve
And 23h20'49"50d10'
Genevieve
And 23h17'52"40d40'
Genevieve
Aql 19h14'55"15d3'
Genevieve Christine
Cas 2h2'54"59d21'
Genevieve Marie
And 23h33'11"41d40'
Genevieve-Fatema
Dra 10h9'38"74d30'
Geneviève
Cas 0h1'26"61d38'
Genevose,Charles
Aql 19h48'60"14d25'
Genezaret Barron's Aurora,The
Umi 14h36'46"66d46'
Gengo,Jean Marie
Eri 2h59'11"-3d49'
Genice Lynda
Cas 0h45'22"60d23'
Genie
Cnv 13h52'24"37d45'
Genie
Aur 5h20'39"40d3'
Genievich,Christopher John
Aql 19h26'13"15d2'
Genii's Star
Leo 10h30'53"20d9'
Genina Emilo
And 23h27'34"45d51'
Genine Renee
And 0h53'12"36d15'
Genio
Lep 5h24'43"-20d19'
Genise René
Lyn 8h29'47"58d31'
Genitoni,Clara Scarpitta
Uma 11h55'18"32d17'
Genix,Yves
Cam 7h33'39"67d58'
Genna,Tony & Joan
Cyg 20h31'29"40d39'
Gennadiy
Dra 19h4'29"68d58'
Gennarelli,MD,Louis B
Aur 5h58'17"29d1'
Gennatos,Sakis
Aur 4h50'14"50d58'
Genner,Barbara
Cyg 19h22'38"44d55'
Genni's Star
Cas 2h15'55"61d17'
Gennino,LeNette Renee
Lyn 8h42'47"43d44'
Gennums
And 1h5'43"39d45'
Gennuso,Francine
And 0h3'47"46d46'
Genny
Del 20h13'23"14d1'
Genny
Aur 4h58'27"40d20'
Genny Babe
Cyg 20h15'0"38d14'
Geno
Boo 14h49'14"38d58'
Geno
Lyn 9h13'10"41d27'
Geno's Heart
Uma 9h27'10"48d41'
Geno,Michele Nichole
Com 13h7'14"30d47'
Genotropin R Pharmacia
Dra 20h10'39"63d58'
Genova,John & Ann
Cyg 19h40'49"42d47'
Genovese
Aqr 21h58'49"-15d51'B
Genovese,Cara Paige
Cam 3h15'33"61d19'
Genovese,Philomena
Cyg 21h13'39"38d43'
Genovese,Thomas Michael
Vul 20h40'19"20d31'
Genovese,Vincent
Ori 5h53'0"16d29'
Genovesi,Lauren
Her 17h30'40"20d7'
Genshemer,Chase Christopher
Hya 8h54'23"-7d55'
Genske,Angelika
Psc 1h15'20"17d32'
Gensler Family Star
Cyg 19h44'1"30d2'
Gensmer Vons Company, William
Sex 10h42'44"4d46'
Gensorek II,David Anthony
Her 17h11'55"44d50'
Gensorek,Matthew Gerald
Dra 20h6'39"76d43'
Gensorek,Sarah Jayne
And 23h30'23"40d52'
Gent Anne
Leo 9h39'55"10d57'
Gent,Andrew
Umi 14h42'33"68d10'
Gentempo,Jr,Patrick
Aur 6h16'16"38d27'A
Gentempo,Peggy
Aur 6h16'16"38d27'B
Gentet-Jacques,Marine
Cam 5h53'27"68d14'
Gentil,Christian
Lyr 18h58'24"30d16'
Gentilcore,Len Pop-Pop
Lac 22h24'0"54d9'
Gentile(Kid Gentle), Frank A
Dra 12h27'10"71d50'
Gentile,Gina Marie
Cam 3h22'13"61d27'
Gentile,Jessica Madeline Julienne
And 2h13'41"41d52'
Gentile,Kathleen
And 0h7'37"47d13'
Gentile,Marc M
Aur 5h15'16"45d4'
Gentile,Phil
Her 16h18'56"24d59'
Gentile,Sheila & David
Crb 16h2'45"32d26'
Gentile-My Eternal Flame,Chuck
Cep 21h1'59"80d2'
Gentle Thomas & Faithful Ruth
Lac 22h9'48"37d24'AB
Gentleman,Virginia
Vul 19h18'0"26d57'
Gentric,Jean Puc
Uma 11h7'44"50d47'
Gentry
Eri 2h59'35"-1d57'
Gentry,Alexandria
Peg 23h41'42"26d56'
Gentry,Beth Marie
Peg 22h1'59"11d14'
Gentry,Danette
Mon 6h42'36"11d20'
Gentry,Jana Lee
Vul 21h22'54"28d9'
Gentry,Jennifer Amy
Peg 22h1'25"30d34'
Gentry,Jennifer Lee
Vul 19h46'43"28d35'
Gentry,Joan Graham
Mon 6h58'14"10d51'
Gentry,John H
Aur 4h50'33"51d19'
Gentry,Scott
Crt 11h53'24"-10d46'
Gentry,Scott
Sct 18h55'27"-5d52'
Gentry-Eternity,David
Eri 3h24'58"-12d0'
Gentsch,Helga
Sco 16h49'21"-28d14'
Genung,Janell
And 23h32'57"46d41'
Geny
Lac 22h33'46"49d35'
Genzel,Karen Ward
And 2h30'55"39d13'
Geneviève
Per 3h43'20"38d50'
Geo
Cep 4h33'53"80d37'
Geo & Tom
Peg 23h27'0"17d30'
Geo Ken
Eri 4h13'11"-12d42'
Geo Willy
Uma 13h35'28"50d5'
Geo-D
Ori 6h13'1"0d39'
Geobar 1-22
Cam 6h56'52"63d50'
Geoffrey
Cep 20h41'47"60d3'
Geoffrey & Jill
Vir 12h33'59"-8d29'
Geoffrey & Young Forever
Aur 6h30'29"35d46'
Geoffrey C
Ori 5h1'48"14d27'
Geoffrey Scott
Boo 14h7'49"51d57'
Geoffrion,Joyce Marie
And 2h27'19"42d56'
Geofftri
Gru 22h39'36"50d42'
Geoghan,Jim
Cma 7h15'18"-16d16'
Geoleena
Cas 0h51'50"73d22'
Geomela
Uma 9h16'58"53d51'
Geomorphic Matt
Aur 6h0'35"36d9'
Georecca
Cam 4h47'22"68d12'
Georg
Cam 6h14'54"64d26'
Georg
Oph 17h50'15"-7d54'A
Georg Gossen "Schorsch"
Lyn 8h10'34"42d43'
Georgantas,Cassandra Elaine
Peg 22h12'54"34d27'
Georgantas,Panos D & Maxine E
Aql 18h59'40"-1d37'
George
Oph 18h27'1"8d6'
George
Tri 1h48'20"25d38'
George
Peg 22h26'32"31d26'
George
Per 2h56'0"45d13'
George
Leo 10h34'1"13d33'
George
Dra 15h0'0"60d27'
George
Cyg 21h1'50"33d41'
George
Ser 15h23'55"24d50'
George & Bobbie
Cyg 19h47'52"38d46'
George & Cynthia
Peg 22h45'1"29d49'
George & Dawn
Cyg 21h2'22"31d34'
George & Jon
Tau 5h44'41"20d0'
George & Maria
Cyg 20h23'26"38d36'
George & Nancy
Per 3h2'59"57d21'
George & Stacey
Gem 6h44'24"16d27'
George & Toni Ann's "Luckey Star"
Peg 22h27'58"20d6'
George & Veronica, Together Forever
Cyg 19h34'32"38d56'
George C
Cnv 13h24'29"41d12'
George I
Boo 15h13'31"37d38'
George Maciunas aus Litauen
Lac 22h9'17"38d9'
George The Splendiferous
Dra 14h22'39"63d32'
George Thomas
Umi 16h25'55"72d2'
George's Diamond
Ari 3h0'23"21d23'
George's Dream
Ser 18h8'36"-5d38'
George's Gem
Per 2h26'45"56d51'
George's Little Bit of Heaven
Cep 21h0'21"60d38'
George's Peace Of A Dream
Boo 15h29'14"41d54'
George's Secret Garden
Per 1h54'48"48d51'
George's Shining Star
Sex 9h57'27"-1d47'
George's Star
Aql 19h0'1"-8d57'
George's Star
Per 2h36'38"37d43'
George's Sunshine
Her 18h9'22"38d4'
George,Alec Bailey
Vir 13h26'50"-2d12'
George,Angel
Umi 13h32'19"75d51'
George,Anne B
Vul 19h23'27"27d42'
George,Barbara
Eri 2h55'42"-10d44'
George,Bonnie Lee
Lyn 8h10'41"34d42'
George,Carly Jean
Lyr 18h47'36"33d51'
George,Cassie "Light" Ann
And 1h21'56"38d54'
George,Christopher Michael
Per 3h0'40"50d18'
George,Debbie
And 2h25'1"42d18'
George,Deh-L
Boo 14h46'24"38d10'
George,Edward
Dra 18h59'15"48d22'
George,Ella Mae
Cyg 20h2'44"40d29'
George,Emory Jack
Ori 5h54'55"14d4'
George,Evelyn
Ori 6h7'29"4d17'
George,G Chester
Oph 17h6'17"7d48'
George,Hettie Fae
Lyr 18h30'18"32d9'
George,Jane
Lac 22h50'21"38d6'
George,John
Ori 5h47'14"21d28'
George,John W
Psc 23h8'52"6d34'
George,Julie Anna With Love From Vijay
Cas 0h35'42"68d42'
George,Kathryn
Ori 6h8'51"7d42'
George,Keith
Dra 17h29'21"61d4'
George,Lisa
Dra 11h24'19"77d30'
George,Mary Kay
Cet 0h24'14"-1d27'
George,Michael Christopher
Lac 22h12'18"49d50'
George,Michael M
Cep 21h56'58"56d6'
George,Mickey
Tri 1h55'52"25d38'
George,Nicholas T
Her 18h2'37"31d21'
George,Norana
Cmi 7h55'45"4d33'
George,Rajankutty
Aqr 22h36'36"2d5'
George,Rex
Tri 1h48'20"27d11'
George,Richard Charles
Cep 3h27'40"78d11'
George,Rob
Uma 8h16'59"71d28'
George,Robert Lynnwood
Ari 3h3'1"23d53'
George,Stephanie
Mon 6h30'60"8d8'
George,The Majestic
Cma 6h59'38"-18d58'
George-"Beacon of God's Love",Eileen
Cas 3h7'16"57d42'
Georgeanne K
Lyn 8h4'1"38d1'
Georges
Per 3h2'27"37d41'
Georges Acogny
Oph 18h0'28"0d6'
Georges,Helen
Cet 2h41'25"-1d8'
Georges,Jean Marie
Mon 6h22'45"8d34'
Georges,Linda Bernadine
Vul 19h42'24"20d16'
Georges,Mr Blond
Cet 0h57'51"0d48'
Georges,Robyn M
Mon 7h44'0"-0d26'
Georges,Steven John
Cet 1h54'60"-5d36'
Georgesen,Nathan William
Dra 16h31'46"72d27'
Georgette
And 2h31'26"38d47'
Georgette
Uma 11h39'41"48d48'
Georgette Yvonne
Cyg 20h0'35"30d1'
Georgia
Mon 8h3'2"-3d37'
Georgia
Lyr 18h33'33"31d39'
Georgia
Ori 5h56'20"17d38'
Georgia Clare
Peg 22h56'18"21d16'
Georgia May
Cyg 20h1'1"30d41'
Georgia May
Lyn 8h29'47"49d27'
Georgia T
Ori 5h20'14"1d24'
Georgia T
Tri 2h0'43"32d31'
Georgiades,Benjamin
Uma 8h52'40"56d16'
Georgiani,Joseph
Dra 10h45'14"73d22'
Georgie
Sgr 20h6'14"-42d38'
Georgie
Ser 16h7'51"9d58'
Georgie Kitty
Mon 6h33'42"-6d44'
Georgifabi
Mon 7h16'44"0d19'
Georgilas,Peter Anthony
Sct 18h32'52"-6d13'
Georgina
Peg 23h1'59"8d49'
Georgina
And 23h27'46"42d38'
Georgina
Vir 12h58'13"-7d39'
Georgina Light Of My Life
Cyg 20h39'23"30d49'
Georgina Louise
Cyg 19h49'54"50d6'
Georgina Maria
Del 20h16'49"14d21'
Georgina Ruth
Peg 21h22'41"18d45'
Georgina Sarah
Lyr 18h34'32"28d26'
Georgina's Star
Cyg 19h34'0"28d19'
Georgitis,Jason Warren
Dra 15h18'0"57d9'
Georgjane's Eternal Love
And 23h34'48"49d33'
Georgosopoulou (George),Zoe
Vir 13h1'21"11d50'
GeorNat
Lac 22h36'13"52d59'
Georose
Lac 22h18'55"51d31'
Gepner,Dr Ronald
Cep 23h1'19"60d10'
Geppeg
Cae 4h50'42"-31d35'
Geppert,Anthony George
Her 16h59'26"40d1'
Geppert,Birgit
Uma 10h37'33"57d48'
Geppert,Kathryn Elizabeth
Ari 2h34'11"21d45'
Geppert,Paul Joseph
Uma 10h42'0"40d48'
Gerace School 25th Anniv,Stephen
Dra 16h4'53"62d56'
Gerace,Giuseppe
Dra 15h45'32"53d42'
Gerace,Julia & James
Cyg 20h31'40"40d32'
Geraci,Brendan H
Aur 6h0'33"36d8'
Geraghty,Bernie
Her 16h37'27"40d19'
Geraghty,Melissa Anne
Cas 1h6'0"55d13'
Geraghty,Thomas Michael
Her 16h18'32"41d1'
Gerald G
Uma 11h39'51"32d33'
Gerald Martin
Boo 14h47'1"25d45'
Gerald Robert
Cam 13h4'53"81d15'
Gerald's Light
Aql 20h2'1"6d40'
Gerald's Star of Love
Ori 6h6'23"8d4'
Geraldina
Dra 16h22'37"62d18'
Geraldine
Cas 0h13'1"59d51'
Geraldine
Her 18h49'34"18d47'
Geraldine
Cam 7h11'44"67d41'
Geraldine
Lyr 18h36'32"43d53'
Geraldine
And 23h34'21"46d60'
Geraldus
Her 16h49'54"32d0'
Geraldus
Vir 13h5'38"-11d12'
Gerami,Sohrab
Hya 8h46'59"1d47'
Geranebula of Spika
Boo 13h36'0"21d20'
Gerard & Jill Forever
Lyr 18h29'35"38d25'
Gerard Andre Jeambon
Cnv 12h42'47"50d17'
Gerard B, 40
Boo 14h34'46"48d6'
Gerard,Alexander Francis
Cam 6h30'47"83d42'
Gerard,Bjorn Curtis
Hya 8h11'0"2d30'
Gerard,Bradley Lawrence
Dra 17h27'10"60d11'
Gerard,Christopher
Dra 13h46'15"63d34'
Gerard,Dudoret
Aur 5h3'26"50d9'
Gerard,Joseph
Cnv 12h43'0"40d9'
Gerard,Joyce E
Cas 0h29'35"67d17'
Gerard,Quinn Thomas
Peg 0h2'14"18d5'
Gerard,Robert
Per 4h34'18"37d48'
Gerardi Star,The
Lac 22h20'32"53d19'
Gerardi,Alfred Michael
Uma 10h0'45"42d19'
Gerardi,C J
Lyr 18h13'44"38d6'
Gerardi,Clara
Lib 15h32'21"-10d5'
Gerardi,Jr,Frank
Her 17h20'50"42d23'
Gerardi,Loretta
Aur 4h53'0"40d49'
Gerardo,John Robert
Her 16h19'47"28d20'
Gerards,Lothar
Ori 5h49'36"21d49'
Gerardus,Rietveld Johannes
Uma 9h39'54"58d47'
Geraro,Skippy & Manny
Cep 0h1'0"68d28'
Gerasimov,Nelga
Cam 6h13'15"68d6'
Gerasimovich,Dennis
Equ 21h4'51"11d16'
Gerasimovich,Mary
Aql 19h56'41"11d9'
Geray,Rudi
Per 3h56'11"39d9'
Gerb & Lady G,The
Crb 15h35'0"26d38'
Gerber Star,The
Lyn 7h40'55"40d26'
Gerber,Colton James
Cep 21h53'44"63d50'
Gerber,Jenny
Cyg 19h57'0"37d44'
Gerber,Kathy Jean
Boo 14h32'29"31d44'
Gerber,Laurie
Cnv 12h43'45"38d48'
Gerber,Louis E & Sherry L
Uma 9h35'1"56d57'
Gerber,Louise & Aaron
Ori 6h0'36"20d23'
Gerber,Nettie Helen & Morris
Cyg 20h56'42"30d54'
Gerber,Richard
Cep 20h56'43"58d50'
Gerber,Roxy
Cas 1h1'50"53d8'
Gerber,Steven Kenneth
Dra 10h39'22"81d27'
Gerbino,Song Cha
Uma 11h18'52"62d0'
Gerbod,Veronique
Ori 5h20'32"12d7'
Gerbracht,Ernest
Her 18h0'37"28d12'
Gerch,Hans-Reinhard
Uma 10h30'56"40d15'
Gerd
Uma 9h0'23"56d18'
Gerd M
Oph 17h0'10"10d23'
Gerda
Cnc 9h13'50"7d8'
Gerda Maria
Cas 2h26'13"68d5'
Gerda—Günther
Psc 1h0'18"32d52'
Gerdes,Jorene Marie
Crt 11h19'1"-14d24'
Gerdi & Peter "Das himmlische Paar"
Uma 9h15'23"50d44'
Gerding,Heiner
Aur 5h8'27"42d6'
Gereg,Alan "Squeak" P
Aur 6h26'46"32d32'
Gereg,Steve
Dra 16h23'13"68d43'
Gerek,Kevin William
Aur 6h7'57"35d52'
Gerek,Kirsten Florianna
Cyg 20h51'26"39d43'
Geren,Lynn
Ori 5h3'56"-2d37'
Gerette
Cam 7h32'34"61d23'
Gerhard Pieper, Brelingen
Cas 0h42'32"60d35'
Gerhard,Anita L
Aur 6h36'1"41d5'
Gerhard,Claus
Hya 8h53'1"1d37'
Gerhard,Johnny
Pic 4h46'45"-47d43'
Gerhard,Roxanna L
Aur 6h32'1"41d5'
Gerhardstein,Kenneth
Lmi 9h44'40"38d43'
Gerhardt,Gabriele und Werner
Lyn 8h12'19"42d18'
Gerhardt,Sherri Lynn
Vul 21h13'46"20d25'
Gerhardt,Susan McCullough
And 1h14'19"34d38'
Gerhart,Bill
Per 4h32'24"31d28'
Gerhart,Ronald Steven
Boo 15h0'10"30d58'
Gerhart,Susann Gilbert
Com 13h14'50"28d11'
Gerhauser,Stephen
Gem 7h24'15"34d0'
Gerheart,Bob
Tau 5h12'0"20d7'
Geri 1
Peg 23h28'18"21d23'
Geri Elise
Cnv 13h2'42"40d53'
Gerica,The
Uma 11h0'15"56d29'
Gerich,Alfred Adam
Per 2h52'42"36d38'
Gerich,Dr John L
Hya 8h40'15"2d2'
Gerich/Wilburn,Silvia
Lyn 7h41'53"48d15'
Gerick,Alice Wilson
Ari 2h49'50"30d10'
Gerils
Her 18h2'59"30d28'
Gerilynn
Lyn 7h53'1"33d42'
Gerimonte,Dr Dean R
Lmi 10h28'1"30d46'
Gerinemy
Dra 15h37'11"53d59'
Gering,Norbert
Dra 18h4'1"67d44'
Gerisch,Margaret
And 0h5'28"47d7'
Geritano,Frances
Dra 16h25'22"68d28'
Gerjes,Hans
Dra 18h1'1"67d33'
Gerke,Herr
Lyr 19h2'52"31d45'
Gerken,Edward H
Lac 22h10'33"46d4'
Gerken,Grandfather
Boo 15h22'11"37d44'
Gerken,Juliette Lynn
Com 12h26'30"20d6'
Gerkens,Werner
Eri 4h54'38"-7d24'
Gerkis,Catherine Helen
Cam 8h9'1"77d55'
Gerla,Holly
Aur 7h8'13"40d57'
Gerla,Steinen
Cam 13h16'13"77d51'
Gerla,Tawnya
Cas 0h28'12"54d35'
Gerlach,Antje
Cap 20h42'49"-20d1'
Gerlach,Jennifer
Cas 0h47'42"68d53'
Gerlach,Kim
Cnc 9h15'1"30d13'
Gerlach,Ralf
Tau 5h55'38"26d15'
Gerlack,Janet
Com 12h18'10"27d20'A
Gerlack,Jules
Com 12h18'10"27d20'B
Germain,Maureen
Peg 22h26'29"30d35'
Germain,Michel
Oph 17h57'1"11d11'
Germain,Paul & Flo
Cam 5h2'38"65d15'
Germaine
Cyg 20h48'0"38d41'
Germaine & Sarah
Cyg 20h52'24"38d10'
German,Alline
Crb 15h33'23"26d38'
German,Ashleigh Shannon
Vul 19h58'30"22d42'
German,Brittney Noël
Mon 8h5'51"-1d6'
German,Irene Loban
Sco 17h27'23"-40d14'
Germany,Linda
Mon 7h56'44"-1d50'
Germenis,Nicolette Lauren
Peg 0h4'30"20d12'
Germinario,Mariangela
Psc 23h25'54"1d14'
Germinaro,Rosann
And 1h54'41"40d7'
Germroth,Edward F
Uma 9h30'13"68d52'
Gerogia H
Mon 7h3'1"-5d41'
Gerome,Forever Darlene
Uma 11h10'40"68d30'
Geronimo
Aur 5h46'40"31d46'A
Gerou Family Star,The
Cyg 21h18'23"31d29'
Geroux,Robert Maxwell
Vul 19h47'45"28d50'
Gerrard,David
Uma 9h35'0"57d45'

Gerrard,Jon Matthew Mon 6h22'17"-0d8'
Gerrety,Thomas C Cas 23h16'0"62d22'
Gerriets,Gerriet Leo 9h23'54"19d56'
Gerrish,Joy & Fred Uma 9h6'1"54d3'
Gerrit Cyg 21h53'48"41d35'
Gerrity,Katharine Casey Cas 2h51'17"75d20'
Gerritzen,Eugene W Ori 5h51'57"11d32'
Gerry Cyg 21h55'0"52d39'
Gerry Ryan Ori 5h24'32"1d51'
Gerry Star Ori 5h53'23"17d6'
Gerry's Diamond in the Sky-Love Laura Uma 13h20'53"57d39'
GERRY'S LIGHT Uma 9h43'31"55d40'
Gerry,Edward T Hya 8h54'52"2d52'
Gersch,Lisa R Mon 6h19'59"7d38'
Gersenson,Michell & Robert Aqr 20h58'52"-10d10'
Gersh,Emma Louise Eri 4h28'16"-12d9'
Gersh,Gloria Palumbo And 2h24'28"41d37'
Gershberg,Beatrice Peg 22h37'18"33d57'
Gershman,Larry Oph 17h30'0"-0d47'
Gershon,Sam & Esther Lmi 10h2'0"35d11'
Gershona Cam 3h14'43"61d2'
Gersky,The Morning Star Gerald F Aur 4h37'19"30d49'
Gersmann,Sue Tri 1h58'30"27d45'
Gerson Marks Ori 5h55'35"15d7'
Gerson,Elaine Lyn 7h5'47"52d35'
Gerson,Jillian Marissa Vir 11h45'0"8d5'
Gerson,Randolph Robert Lac 22h5'1"40d10'
Gerson,Win Per 2h55'60"31d25'
Gerst,Michael T Aur 4h58'51"40d1'
Gerstein,Alison N And 0h25'0"45d16'
Gerstein,Karla M Lyn 8h49'34"44d9'
Gerstein,Marge Cas 0h3'0"58d52'
Gerstel,Karen Oph 17h53'24"-8d58'
Gersten,Sherri Aql 19h58'34"1d0'
Gersten,Susannah Rebecca Cas 0h29'42"58d17'
Gerster,Frauke Nora Cnc 8h54'59"17d39'
Gerster,Lars Hendrik Lib 15h17'11"-21d12'
Gerster,Sven Thorsten Ari 1h58'1"22d27'
Gerster,Wibke Cora Lib 15h19'15"-22d42'
Gerstley,Paul Hya 8h51'15"-6d22'
Gerstman,Ethan Philip Her 17h44'51"41d15'
Gerstner,Jr,Louis V Uma 8h10'21"70d42'
Gert's Galactic Getaway Tri 2h1'26"31d36'
Gerth,Jo Ellen And 2h14'20"41d12'
Gertje,Lathan Cam 6h55'56"78d45'
Gertrud Cap 22h22'39"-20d10'
Gertrude Peg 22h18'50"34d19'
Gertrude Cam 5h0'20"70d7'
Gertrude's Shining Light Cmi 7h15'47"4d37'
Gertud B Peg 22h43'44"4d25'
Gertz,Annalisa Isabella Peg 22h15'34"7d52'
Gertz,Anne & Edward Crb 16h21'12"31d3'
Gertz,Jeri Lyr 18h45'1"39d6'
Gerughty,Stanley Cet 1h32'40"-12d3'
Gerundino,Consuelo Cyg 19h28'0"40d20'
Gervais,Jarrod & Kelly Eri 4h1'33"-10d26'
Gervase,John & Chiara Stella Cyg 21h32'12"41d18'
Gervasi,Dianne Cas 22h56'46"54d39'
Gervasi,Rosemary Crb 16h17'35"34d45'
Gervino-Ellis, Catherine Rebecca Cas 1h8'18"65d3'
Gervis,Brent Per 2h56'12"32d56'
Gerwing,Klaus Mon 7h1'48"-5d11'
Geryl's Galaxy Bat Mitzvah Star Uma 8h35'51"59d41'
Gerylosharad Her 17h16'13"42d28'
Gerzabek,Olivia Vul 19h17'21"25d28'
Gescheidler's Heavenly Light Del 20h18'53"10d41'
Geschke-Vogler, Maximilian Leonard Umi 15h52'0"71d40'
Geschwentner,Eduard Ori 5h47'60"21d42'
Gesell,Jack Umi 15h15'57"66d50'
Geske,Katja Cnc 9h0'12"32d3'
Gesnouin,Doria And 23h2'12"38d55'
Gesregan,Paul E Her 15h51'21"44d40'
Gessert,The Star Of Earginia Mon 7h18'1"-8d26'
Gessie Poorman's Star, Gerry Aur 4h56'42"40d43'
Gessler,Michelle Elizabeth Aqr 21h0'40"-13d14'
Gessler,R Kenneth Lac 22h38'44"55d43'
Gessler-Carlson Ori 5h56'41"13d16'
Gessner "DAD",William F Aql 18h59'23"15d30'
Gessner,Jennifer Lyr 18h54'53"33d10'
Gessner,Jr,Barry Dean Boo 14h43'28"26d40'
Gessner,Liz Dra 18h11'29"70d23'
Gessner,Thomas W Aur 6h40'16"37d36'
Gessow,Murry "Dads" Cep 0h4'42"69d39'
Gesswein,Kyle Boo 14h57'22"20d11'
Gest,Carla Kay Com 12h3'14"24d50'
Gesualdi Aur 7h8'30"36d29'
Gesualdi,Galor Uma 9h16'37"43d23'
Gesualdo,Eric Michael Dra 14h29'15"62d49'
Gethaiga,Wanjiro Cam 13h19'13"80d11'
Gething,Paul Michael Ori 4h51'0"5d44'
Getlan Family,Michael C Cyg 19h55'12"37d36'
Getman,David N Aur 6h11'46"31d4'
Getrajdman,Joelle Erika And 2h15'24"42d47'
Getson,Philip & Coleen Vul 20h26'59"28d37'
Getsos,John P Her 17h5'1"44d31'
Getterman,Jr,Louis T Sex 10h30'57"-8d42'
Gettinger,Diana And 0h10'48"26d42'C
Gettings,PhD,JD,DABT, Stephen D Boo 15h18'48"47d42'
Gettler,Benjamin Lac 22h22'0"56d47'
Getty,Doris Cas 0h4'42"60d36'
Getty,Karen Frances Hya 10h13'51"-11d48'
Getz,Michael Jon Ser 18h17'14"-13d17'
Getz,Sr,Robert Cep 22h14'51"65d12'
Getz,Walter Uma 10h57'41"70d25'
Getzan,Anna Hope Mon 6h33'26"3d35'
Getzie Aql 20h11'15"5d34'
Gevalt,Jr,Peter Y Psc 13h5'29"2d20'
Gevert,Michelle Cam 4h56'60"68d17'
Gevertz,Justin Matthew Lac 22h24'56"38d45'
Gevirtz,Sidney & Ellen Boo 14h8'1"39d27'
Gewecke,Birgit Psc 23h22'55"1d58'
Gewecke,John Lac 22h19'28"54d49'
Gewertz,Kenneth A Dra 17h2'57"69d13'
Gewirtz,MD,Harold S Psc 1h5'55"27d60'
Gewirtz,Mr Evan & Sharon Crb 16h16'34"30d38'
Gey,Heinrich Dra 18h28'14"67d39'
Geyer,Eberhard Dra 18h26'45"68d54'
Gezon,Judy Lyr 19h7'37"41d3'
Gfeller,Marcus Bradley Aql 20h7'45"4d16'
Gfoeller,Dr Joachim Mon 6h44'37"10d2'
Gfoeller,Shirley Eri 3h52'34"-7d12'
GFR-Da Bully Gem 7h0'32"11d5'
Gfroehrer,Susan C Ari 3h2'43"23d3'
Gfroerer,Paul Cet 3h20'31"1d10'
Gfüllner,Jacqueline Cam 5h6'0"61d21'
Ghahramani-Allen, Johnathan Aql 18h59'24"12d31'
Ghanasyama dasa Psc 1h40'31"28d26'
Ghaswala,DA-NA-SON-KEV Cmi 7h26'33"7d50'
Ghaznavi,Shireen Leo 9h32'0"8d6'
Ghenne,Martha J Vul 19h41'31"22d40'
Gheroni,Anna Maria Boo 15h27'34"38d43'
Ghertner,Frank "5 Star" Aql 19h57'51"10d20'
Ghery Cep 21h22'46"67d55'
Gheysem,Antoine Boo 15h7'60"7d19'
Ghia Cmi 7h40'0"4d41'
Ghilardi,Irene Eri 4h34'0"-0d2'
Ghilardi,Laura Helene Del 20h22'50"20d11'
Ghinnis,Amy & Michael SNC Lyn 7h57'34"48d41'
Ghinos,Cassandra M Aur 6h6'50"31d12'
Ghio,Diane L Psc 23h8'48"0d47'
Ghiorso Gabriella Cep 1h15'23"80d33'
Ghislaine Alajouanine Per 3h13'21"40d54'
Ghlea Eri 3h4'39"-3d36'
Ghlee,Stephanie Aql 20h9'56"7d9'
Ghosio,Caitlin Marie And 2h19'31"48d11'
GHS*AMOG*ESS*97 Cep 23h12'47"78d48'
Ghyslaine Sgr 19h3'15"-28d45'
Ghyslaine,Lartigue And 0h27'37"43d30'
Giabi Lyr 18h38'59"27d33'
Giacalone II,Ronald P Aur 5h6'14"44d35'
Giacalone,Frank Cam 8h23'21"73d21'
Giacalone,Gail M Psc 1h23'19"22d27'
Giacchino,Michael Aql 19h6'45"3d6'
Giaccobi,Linette Cyg 21h4'21"48d45'
Giaccone,Therese D Cam 3h53'33"58d27'
Giacobbi,Garbu Cyg 19h32'54"31d19'
Giacobello,Thomas J Cma 6h51'52"-16d27'
Giacobini,Silvana Cas 23h42'11"60d14'
Giacoletto, Marie G Dra 17h1'16"65d40'
Giacomazzo,Margaret Cas 0h6'51"63d8'
Giacomazzo,Salvatore Lac 22h55'48"55d26'
Giacometti,Bruno Crb 16h10'29"31d12'
Giacomo Vul 19h50'49"20d15'
Giacomo John Boo 14h11'46"30d32'
Giacomo-Santoro 19/9/94 Pic 4h42'10"-48d21'
Giacona,Maureen Eri 4h8'14"-13d3'
Giada And 23h41'49"38d10'
Giada,Albertina Aqr 22h5'14"-0d39'
Giaever,Ivar Og Inger Lac 21h57'49"37d39'
Giaime,Erin Lyr 18h33'41"40d47'
Giaimo Family,The Joseph Ori 5h38'24"15d3'
Gialloreto,Gabrielle And 1h52'45"36d47'
Giambalvo,Leah Cas 1h44'27"58d29'
Giamella,Frank James Ori 4h53'34"-0d0'
Giamella,Victoria Lauren Cas 3h12'29"73d14'
Giamietri-96 Gem 6h38'25"28d33'
Giammarco,James Matthew Cam 3h51'24"70d17'
Giammarco,Joseph Michael Per 3h56'43"51d38'
Giammarese,Carl Aql 19h56'18"1d33'
Giammarino,Johnny Ori 6h17'26"0d35'
Giampapa,Jr,Sebastian Richard Ori 5h7'18"-2d21'
Giampiero Cet 2h58'56"-0d54'
Giampiero Loves Arron Forever Mon 6h24'0"4d51'
Gian,Joey Boo 15h12'0"50d11'
Gianacakos,Nicholas G Oph 18h16'52"12d24'
Gianadda,Ron Chipper Uma 11h13'0"49d53'
Gianantonio,Dominick Her 16h27'1"41d44'
Gianatassio,Liz Uma 13h48'25"60d46'
Gianburrasca,Gianluca Cas 0h10'27"58d17'
Giancarlo 11 febbraio Gem 6h55'57"17d46'
Giancarlo Franco Sgr 19h35'56"41d39'
Giancarlo Simone Uma 10h52'1"48d4'
Giancoli,Jennifer Anne And 0h12'47"31d28'
Gianduso,Dominick Louis Lac 22h41'1"50d4'
Gianduso,Jessica Ann Cas 1h16'1"60d55'
Gianduso,Jr,Daniel Diego Mon 8h1'28"-2d48'
Gianetti,Gary Vul 20h0'0"28d42'
Gianfortoni,Joseph Her 16h53'0"33d47'
Gianfranco,Dolce And 23h17'32"41d27'
Giangiulio,Nicole Olyvia Cas 22h56'1"55d36'
Giangrosso,Roy Aur 6h59'49"35d35'
Gianitsos,Anna Uma 11h54'24"40d3'
Gianluca Cam 9h52'55"82d2'
Gianluca Psa 22h26'0"-26d34'
Gianluca,Samantha Psa 22h33'29"-27d31'
Gianluca-GST Aur 5h53'12"40d39'
Gianlucarana For 2h46'33"-24d52'
GianMarco Pic 5h1'26"-43d15'
Gianna For 2h20'31"-28d51'
Gianna And 2h34'15"37d39'
Gianna Rose Equ 21h5'22"10d60'
Giannakakis,Demetrios Lac 22h27'57"38d33'
Giannakakis,Sam Dra 15h43'17"67d33'
Giannakaris,George Per 2h51'10"40d29'
Giannandrea,Marcus Uma 8h33'56"58d24'
Giannandrea,Rory Uma 8h43'26"57d3'
Giannantonio,Frank Aur 6h27'48"35d1'
Giannantonio,Joseph Anthony Tri 1h47'58"31d13'
Giannatasio,Nicholas Luke Gem 8h5'11"27d3'
Giannecchini,Pauline Vul 19h22'36"26d42'
Gianni F Her 16h10'0"40d47'
Giannie,Joseph P Her 16h27'53"39d56'
Giannina Boo 15h5'2"18d38'B
Giannina Mia Peg 23h30'1"22d2'
Giannina,Muriel Uma 10h42'59"51d35'
Giannini,Judge Peter E Cas 2h33'43"61d26'
Giannini,Michael D Lac 22h30'0"50d15'
Giannini,Nicole Cam 5h51'44"60d55'
Giannioses Galaxy,The Her 16h37'0"36d52'
Giannola,Samantha Marie And 23h3'28"51d5'
Giannone,Susan Equ 21h4'13"2d26'
Giannotto,Rebecca Cam 6h22'10"80d28'
Giannoula Vassilopoulou Dra 17h3'20"68d7'
Giannoutsos,Angelo Uma 11h36'19"51d50'
Gianotti,Margaret Uma 9h3'19"60d9'
Gianotto,Shanen Lee And 0h6'57"47d36'
Gianoulakis,Mindy Lyr 19h3'46"28d43'
Gianpaola Uma 11h44'31"50d52'
Giansoldati,Robert Her 18h13'25"40d38'
Giardina,Doreen Carole And 1h4'34"40d32'
Giarnese,Gordon Uma 10h25'0"59d45'
Giarrusso,Eva Margaret Peg 21h10'56"16d42'B
Giarrusso,Gino Alfred Peg 21h10'56"16d42'A
Giarrusso,Jr,Ralph T Per 2h47'54"40d9'
Gib Dra 17h38'45"70d17'
Gib Gib Her 17h17'26"44d28'
Gibb II,Robert Per 3h4'53"46d51'
Gibb,Dwina Eri 2h58'26"-13d28'
Gibb,Ewan James Uma 14h0'40"52d18'
Gibb,Melissa Mon 8h7'60"-6d24'
Gibb,Noah-Tiffany Vul 19h0'16"25d6'
Gibb,Robin Peg 0h7'59"18d3'
Gibb,Robin-John Mon 8h8'42"-6d21'
Gibb,Spencer Cet 3h8'41"1d1'
Gibbon,Christopher Cep 22h5'16"62d21'
Gibbon,Janice Cas 0h2'20"61d36'
Gibbons Golden Anniversary Star Cyg 20h22'32"39d45'
Gibbons,Billy F Ori 6h2'20"5d21'
Gibbons,Charles W Vul 19h36'0"27d35'
Gibbons,Conor Francis Aur 5h3'0"40d7'
Gibbons,Jo-Ann Lyr 18h36'27"28d17'
Gibbons,Jr,Mark Anthony Hya 9h52'56"-17d30'
Gibbons,Kraig Alan Tri 2h26'12"28d33'
Gibbons,Krysta Cam 13h38'58"80d18'
Gibbons,Laura Lane Martin Peg 23h20'23"32d24'
Gibbons,Leigh-Anne "Tilly" Cnc 8h7'47"20d11'
Gibbons,Martin Leo 11h29'17"-5d8'
Gibbons,Michael David Cep 22h32'1"60d45'
Gibbons,Tyrone C Per 1h30'26"54d22'
Gibbons,Wallace John Cep 22h24'39"61d33'
Gibbons,Wendy Elizabeth And 22h57'60"50d25'
Gibbons,William J Cet 3h15'0"3d27'
Gibbs,Alexander James Ori 6h0'33"20d36'
Gibbs,Andrea Boo 14h31'49"48d16'
Gibbs,Angel Marie Lyr 18h59'50"29d29'
Gibbs,Angela Cas 2h50'52"68d26'
Gibbs,Anne-Marie Mon 7h25'12"-5d44'
Gibbs,Charles S Aur 5h24'17"31d14'
Gibbs,Debbie Del 20h30'15"10d7'
Gibbs,Fran French Cma 6h56'14"-16d60'
Gibbs,Jason Wesley Cmi 8h6'50"6d35'
Gibbs,Jr,Spencer Ori 5h25'51"0d48'
Gibbs,Julian Cra 18h15'59"-39d44'
Gibbs,Julianaa Umi 14h36'35"67d7'
Gibbs,Kathleen S Cas 0h35'60"66d18'
Gibbs,Mark T Aql 19h1'37"3d52'
Gibbs,Michael Aur 6h1'26"31d5'
Gibbs,Sonya Louise Cas 0h44'15"70d41'
Gibbs,Sylvia Peg 22h42'19"22d17'
Gibbs,Tara Emoria Cyg 20h29'48"42d48'
Gibbs,Thelma Lucille Matten Ori 5h25'39"0d21'
Gibbs,Traci "The Dancer" Com 12h52'20"26d60'
Gibby Uma 8h44'29"50d50'
Gibby,Bealin Eugene Vul 19h18'26"22d42'
Gibby-n-Beth Cyg 19h28'0"30d51'
Gibeau,Dr Edward B Uma 8h36'57"61d43'
Giberson,Scott Aaron Oph 17h57'0"8d5'
Gibertoni,Titta Umi 16h17'38"82d47'
Gibgot,Neil A Shapiro Cnv 12h23'0"38d45'
Gibigianna Vel 9h21'41"-41d21'
Gibney,Suellen Muzzy Lyn 8h47'1"36d20'
Gibson III,Richard Wayne Lac 22h6'21"37d34'
Gibson Super Star,The Larry Oph 17h53'19"13d31'
Gibson(Love & Serenity),James Joseph Peg 23h0'36"30d6'
Gibson,Adele And 2h34'12"39d40'C
Gibson,Allen F Aql 20h11'0"10d31'
Gibson,Ansley And 2h34'12"39d40'A
Gibson,Antony Adam Ori 6h7'47"0d24'
Gibson,Brett Thomas Cep 22h28'0"68d57'
Gibson,Brian R Her 16h18'22"24d27'
Gibson,C & C Sge 19h34'17"16d7'
Gibson,Carolyn Cas 23h21'20"60d58'
Gibson,Charles Anton Her 15h50'41"43d13'
Gibson,Charles Edward Gem 7h14'26"21d32'
Gibson,Cleo Patricia Cas 0h37'1"62d11'
Gibson,David & Dawn Uma 9h0'1"54d8'
Gibson,David Keith Aur 5h17'1"48d38'
Gibson,David Revere Her 17h33'1"20d53'
Gibson,David Wayne Aql 19h58'40"15d12'
Gibson,Janet Milburn Cas 1h31'0"68d11'
Gibson,Jaydene Elizabeth Umi 16h50'0"76d40'
Gibson,Jean & Billy Lmi 10h2'42"35d2'
Gibson,Joan Aql 19h47'11"13d56'
Gibson,John Roger Umi 16h10'20"80d32'
Gibson,Jr,James O "Duke" Boo 14h39'59"46d18'
Gibson,Kathryn Lyn 8h20'45"48d30'
Gibson,Kendall And 2h34'12"39d40'B
Gibson,Kirsty Lyr 18h32'18"37d17'
Gibson,Lauren Cas 23h33'10"60d15'
Gibson,Lilian Cas 0h17'56"62d7'
Gibson,Mark G Pyx 9h8'42"-23d58'
Gibson,Mark Robert Per 4h6'27"38d1'
Gibson,Martha Jean Cnc 8h27'18"10d7'
Gibson,Marty (Two Steps Too Slow) Per 1h26'38"53d12'
Gibson,Mary Jill Cas 0h25'13"75d21'
Gibson,Matthias W Leo 10h54'17"10d21'
Gibson,Michael A Per 1h49'1"52d54'
Gibson,Pearl Marie & George Cole Cyg 21h4'35"36d28'
Gibson,R G Cra 18h5'21"-37d25'
Gibson,Richard Davis Aql 19h56'52"11d18'
Gibson,Richard James Cnc 9h12'23"31d47'
Gibson,Richard William Oph 18h15'30"0d9'
Gibson,Ron Peg 21h29'33"2d17'
Gibson,Rondal Scott Aql 19h0'14"1d11'
Gibson,Ruthie Ives And 23h2'47"48d52'
Gibson,Shannon Alexander Uma 11h39'49"52d55'
Gibson,Stacy Lee Peg 22h5'53"3d5'
Gibson,Stephanie Cyg 20h39'47"53d48'
Gibson,Thomas Francis Aql 19h18'40"13d47'
Gibson,Todd Her 17h21'1"40d43'
Gibson,Tom Cet 2h28'18"3d51'
Gicquec,Olivier Lac 22h23'53"56d41'
Gidaley,Max C Cep 2h2'25"78d42'
Gidaly,Keith Oph 17h39'20"-21d60'
Giddens-Pedziwiatr, Melba Amanda Del 20h25'19"7d45'
Giddings,Abbie Cet 3h9'48"2d12'
Giddings,Dawn Cyg 19h47'20"30d58'
Giddings,John Longmore Aur 5h2'46"41d22'
Giddings,Megan Marie Cas 3h10'47"75d46'
Giddings,Pat & Mike Sge 19h2'1"20d26'
Giddings,Ruby Ori 5h53'41"14d24'
Gideon Sge 20h17'43"18d47'
Gidget Peg 22h44'52"34d48'
Gidlow,Rachel Umi 13h19'52"73d46'
Gidman,Charlotte Com 12h27'29"24d22'
Gidwitz,Alexander Cep 20h44'37"68d15'
Gidwitz,Scott Boo 14h7'30"38d27'
Giebel,Emil 06-06-1920 Gem 6h53'57"16d56'
Gieda's Gallimaufry Uma 11h43'25"48d47'
Giegerich,Edmund Per 4h43'20"51d54'
Giegerich,John R Cyg 19h34'19"32d4'
Giegerich,Melody Cyg 20h4'42"39d47'
Giegling,Karla Lib 14h59'0"-23d58'
Giel,Kimberly Lyn 8h9'55"33d33'

Gienko,Allison
Peg 22h33'46"26d13'
Gienko,Jr,Robert C
Leo 10h1'40"8d1'
Gienko,Randall C
Gem 6h29'20"14d36'
Giercke,Jan
Sco 15h50'32"-21d52'
Giere,Emily Louisa
And 0h11'1"35d5'
Gierer,Jr,Vincent A
Per 3h3'24"50d10'
Gierke,Christopher W
Her 16h25'60"28d49'
Gies,Brent James
Cnv 13h40'53"33d20'
Gies,Herbert Francis
Cet 0h45'14"-7d6'
Gies,Patrick John
Lac 22h34'27"52d34'
Giesbrecht,Gerald & Jane
Mon 5h59'20"-4d39'
Giese,Ellen Patricia Worst
Cas 23h37'23"61d40'
Giese,Klaus
Lyr 19h16'25"30d46'
Gieseking,Josef
Cmi 7h31'1"7d43'
Gieslak,Edith
Sgr 18h56'58"-26d45'
Giessauf,Primarius Dr W
Ori 5h34'26"-0d23'
Gietz,Gordon W
Aur 7h12'27"39d56'
Gietzen,Christopher John
Leo 10h54'60"12d47'
Gietzen,Meghan Rae
Cap 20h21'17"-12d28'
Gievers,Franz
Cap 20h8'10"-8d11'
Giffard,Angelique
Ser 16h6'58"7d50'
Giffen,Robert B "The King"
Her 17h24'43"47d54'
Giffin,Duane M
Vir 13h30'2"-7d11'
Giffin,Mary J
Tau 3h43'12"1d31'
Gifford,Beth
Aql 19h55'20"10d21'
Gifford,David
Her 16'56"41d40'38'
Gifford,Dawn L
And 1h20'35"37d48'
Gifford,Elizabeth A
Peg 22h56'11"18d46'
Gifford,Eric Kelly
Dra 18h44'34"68d12'
Gifford,Graham
Cyg 19h47'35"30d4'
Gifford,Gregory
Uma 10h10'53"53d38'
Gifford,Jr,Kenneth Eugene
Cam 6h0'0"60d55'
Gifford,Katharine Sarah
Equ 21h20'41"11d24'
Gifford,Kathie Lee
Uma 13h43'44"51d26'
Gifford,Peter Knox
Boo 15h12'26"52d39'
Gift From Above to the Gilbert I Love
Cma 6h57'52"-30d26'
Gift,George William
Boo 13h45'49"18d25'
Gifts Of God,Lindsey, Ryan,Natalie
Mon 7h4'1"4d12'
Gigante,A Ward
Her 16h36'51"51d7'
Giggle-DSB
Sct 18h40'28"-4d16'
GiGi
Lac 22h41'32"38d28'
GiGi
Cnv 13h58'22"45d43'
Gigi
Her 18h29'28"24d52'
Gigi
Ori 5h57'0"17d47'
Gigi
Peg 22h59'42"24d14'
Gigi-Avé
Tau 4h36'20"15d22'
Gigi-The Beautiful & Special
Lyr 19h23'0"37d42'
Gigia
Pyx 8h51'48"-29d22'
Gigiumbi
Boo 14h46'37"50d10'
Gigli
Cnv 12h59'18"38d21'
Giglia,Jian Anthony
Cnc 8h59'12"30d52'
Giglio
Uma 8h30'37"72d2'
Giglio,Alexandra Page
Cas 1h18'19"61d40'
Giglio,Donna
And 1h57'46"38d4'
Giglio,James Jack
Equ 20h57'48"8d37'
Giglio,Lori Ann
Peg 23h6'0"18d10'
Giglio,Melanie Marie
Peg 23h6'0"18d38'
Gigliotti,Joseph & Arlene Lawson
Cam 4h54'0"70d28'
Gigliotti,Mary Madeline
Cyg 20h17'41"31d6'
Gignac,Christine Fontaine
And 2h25'33"45d6'
Gigous,Candice
Lyr 19h23'0"40d13'
Gigous,Roman
Her 18h12'28"47d11'
Gigs
Lyr 18h42'43"42d33'
Giguere,Chantelle Marie
Cas 1h44'1"58d56'
Giguère,Charles- Edouard Imbeau
Uma 12h10'20"61d58'
GIK-WAK 50
Boo 14h49'37"50d33'
Gikofsky "Mr G",Irv
Tau 5h59'0"28d37'
Gil & Donna's Wedding Star
Cep 23h9'45"64d2'
Gil's Choice
Cet 2h17'52"8d1'
Gil,Nuria Lozano
Her 17h3'15"50d4'
Gil,Ralph Lopez & Carmen E Holmes
Cyg 21h19'15"32d23'AB
Gilberg,Ken & Alvaro Lopez-Levy
Cam 3h30'36"60d5'
Gilbert
Uma 10h7'31"50d9'
Gilbert
Aur 6h31'19"38d41'
Gilbert 11-4-44,Donald Ray
Oph 18h42'46"7d15'
Gilbert 16451 South Pacific,John H
Uma 11h43'60"49d35'
Gilbert et Benoit, l'etoile de
Ori 5h59'0"18d42'
Gilbert The Magnificent
Cra 18h24'54"-40d17'
Gilbert's Valentine Heartlight,Darbi
Ori 5h39'11"-1d50'
Gilbert,Alfred Avrum
Cnv 12h56'49"37d42'
Gilbert,Andrew Christian
Crt 11h12'14"-18d23'
Gilbert,Arlene
Cas 0h31'31"58d45'
Gilbert,Bruce
Ser 17h33'60"-13d34'
Gilbert,Candace Michelle
Peg 21h57'39"24d5'
Gilbert,Captain Anthony J
Aql 20h7'53"1d36'
Gilbert,Colin
Aql 19h54'17"15d33'
Gilbert,Connie
Cyg 21h39'12"41d37'
Gilbert,Craig S
Ori 5h56'1"14d20'
Gilbert,Daniel Adam
Per 4h7'0"40d48'
Gilbert,Elana Beth
Lyr 19h18'56"40d2'
Gilbert,France
Cas 0h47'34"64d1'
Gilbert,Francoise & Andre
Her 17h13'15"26d26'
Gilbert,Hal H
Aql 19h1'53"5d30'
Gilbert,John
Ser 16h4'41"22d6'
Gilbert,Jr
Cyg 20h2'52"30d8'
Gilbert,La Yolande
Cyg 19h28'23"33d7'
Gilbert,Laurent
Tau 5h46'32"23d30'
Gilbert,Lawrence
Aur 5h1'1"50d49'
Gilbert,Lillian Berger
Lyn 7h37'0"58d46'
Gilbert,Lydia Grace
Cyg 20h28'18"42d38'
Gilbert,Mark
Her 17h35'15"25d11'
Gilbert,Michel
Lyn 8h25'32"49d44'
Gilbert,Natasha E
Cas 0h49'27"76d18'
Gilbert,Nicholas Krekor
Boo 15h3'34"21d4'
Gilbert,Paul
Cma 6h52'32"-18d5'
Gilbert,Perry
Sge 20h7'1"20d18'
Gilbert,PhD,Dr Francis Charles
Oph 17h25'5"-20d28'
Gilbert,Renaud
Per 3h16'31"41d16'
Gilbert,Richard Michael
Aql 19h47'1"10d11'
Gilbert,Robert H
Her 17h3'55"47d58'
Gilbert,Sandra Y
Aqr 22h25'28"0d45'
Gilbert,Sharon
Cas 0h9'39"60d40'
Gilbert,Sonny
Peg 22h22'48"2d21'
Gilbert,Stephanie Lynn
Mon 7h21'57"-0d20'
Gilbert,Stephen J
Cet 0h27'43"-6d45'
Gilbert,Sue
Tri 2h15'31"31d10'
Gilbert,Tom
Her 18h41'40"12d49'
Gilbert,Warren K
Aql 18h57'43"-4d27'
Gilbert,Wendy Jean
And 23h46'33"45d44'
Gilbert,Willard R
Her 18h17'36"14d55'
Gilbert,Zachary
Mon 7h48'30"-1d37'
Gilbert-Lurie,Gabriel Loren
Cet 1h28'19"-11d24'
Gilberti #1 Star of Husbands,The Gary
Aur 5h25'21"30d43'
Gilberti #1 Star of Moms & Nannies,Mary
Vul 19h22'47"23d48'
Gilberti #1 Star of Dads & Poppies,Gage
Vul 19h23'13"22d35'
Gilblom,Alex David
Aql 19h2'0"-01d16'
Gilblom,Liza Anne
Oph 17h33'1"-23d39'
Gilbreath,Don
Sct 18h34'40"-6d26'
Gilbreath,Thomas A
Hya 8h47'32"-0d31'
Gilby
Uma 12h57'48"58d20'
Gilby,Jacqueline Ann
Lyr 19h21'18"40d8'
Gilchrist,Denise Elizabeth
Uma 8h57'35"72d10'
Gilchrist,Dr Annette
Cas 0h49'1"64d5'
Gilchrist,Joanne
Cas 0h15'30"63d26'
Gilchrist,Mary Diane
Cas 0h56'54"64d6'
Gilda
And 0h33'47"45d21'
Gilda
Pic 4h59'9"-48d41'
Gilda e Angelo
Sco 16h27'52"-35d7'
Gildart,Lee William
Oph 18h1'0"10d48'
Gildea,Darkstar in honor of Pamela
Lyn 8h29'54"40d5'
Gildea,Raymond Paul
Cep 22h46'36"80d15'
Gilden,Ed & Tee
Cyg 19h11'55"48d39'
Gilden,Jeanne
Ori 5h14'0"-8d26'
Gilder,George A
Boo 15h27'0"47d55'
Gilderman,Dr Larry
Tau 4h16'23"23d45'
Gilderman,Ryan Alexander
Dra 16h31'1"60d55'
Gilders,Sean
Boo 15h20'12"34d60'
Gildner,Jesse Scott
Equ 20h59'29"10d58'
Gildo
Peg 22h27'39"4d56'
Gildow Solaris
Cnc 8h56'53"17d17'
Gile,Stephanie LeAnn
Cet 3h16'20"0d36'
Gileno,Ashley Nicole
Peg 22h38'10"33d46'
Gileno,Kathleen T
Lyr 18h31'51"38d43'
Giles
Crb 16h12'47"33d27'
Giles Star,The Alan
Per 2h26'48"56d37'
Giles,Betti Ann
Lyr 18h45'32"40d7'
Giles,Erica Sheree
And 23h20'0"48d16'
Giles,Harriet Norris
Oph 17h12'4"-20d48'
Giles,James
Hya 9h11'0"3d48'
Giles,Jill Diane
Cyg 20h14'57"38d9'
Giles,Kristen Lynette
Lyr 18h50'28"33d53'
Giles,Nancy
Cas 2h43'1"71d8'
Giles,Oley
Tri 1h46'42"25d52'
Giles,R Thomas
Aql 19h57'39"8d58'
Giles,Roger
Dra 11h51'48"74d19'
Giles,Sarah Abbott Davis
And 2h16'1"37d24'
Giles-Wylde,Louis
Per 4h2'21"38d7'
Gilford,Bert
Boo 14h2'19"19d29'
Gilgallon,David
Boo 14h48'33"46d43'
Gilgenberg,Kimberly
Mon 6h44'54"1d44'
Gilgore,Alison Kim
Cas 23h41'51"62d25'
Gilgore,Valerie Paige
Lyr 18h34'52"42d28'
Gilham,Tracy Joe
Cet 1h24'34"1d44'
Gili,Evie & Richard
Tri 1h46'17"28d20'
Gili,Jr,Joseph
Aur 5h3'39"31d9'
Giliberto,J Christopher
Her 16h37'34"50d21'
Gill,Clarence
Boo 14h14'32"35d10'
Gill,Derek
Per 3h23'11"40d33'
Gill,Elaine & David
Lyr 18h14'55"30d58'
Gill,Glenn "Savage"
Aur 6h11'59"31d52'
Gill,Hazel June
Aql 19h57'43"8d57'
Gill,Jonathan James
Lmi 10h33'50"36d39'
Gill,Judge Bob
Cet 0h54'20"-0d7'
Gill,Kevin John
Ori 5h12'0"15d2'
Gill,Maureen
And 23h28'49"45d47'
Gill,Meagan
Tri 2h4'59"31d28'
Gill,Patrick Conner
Oph 18h2'15"12d0'
Gill,Patti & Tim
Per 4h22'53"50d5'
Gill,Samuel Henry
Vul 19h47'34"23d59'
Gill,Sophie
Lyr 18h36'52"33d41'
Gill-Light Of The Heavens,Nicola
Del 20h53'0"4d27'
Gillam,James
Cep 20h26'20"66d27'
Gillam,Zoé 'Poppet'
Lyn 7h30'38"51d35'
Gillan,Heather Redda
Lyn 7h50'27"58d40'
Gillan,Laura D
Equ 20h55'27"4d29'
Gillan,Ryan P
Boo 14h58'21"21d18'
Gilland,Angel of Rockwell,Elaine
Tri 1h55'1"28d30'
Gillard,Colin River
Boo 15h14'58"47d47'
Gillard,Michael J
Oph 17h29'60"8d50'
Gillard,Michel
Uma 12h3'27"61d4'
Gillard,Roger
Per 3h14'59"54d49'
Gillard-Scott UNION
Lyr 18h58'1"34d45'
Gillary,Jack E
Tri 1h57'53"34d15'
Gillas,Gabriel Alexander
Umi 15h10'39"70d42'
Gille,Marianne
Cas 0h28'29"60d28'
Gilleece,Deanna
Cyg 19h53'53"38d32'
Gilleland,Troy & Tracye
Aql 19h57'16"12d48'
Gillen,Jacob James
Tau 4h2'0"30d36'
Gillen,Jeffrey H
Hya 9h21'32"-10d7'
Gillen,Leesa Ann
Mon 6h25'16"8d56'
Gillen,Rita & Jim
Dra 12h50'41"67d55'
Gillen,Shirley
Crb 16h0'34"28d44'
Gillen,Walter Edward
Hya 9h10'35"2d41'
Gillentine,Jordan Malcolm
Per 2h51'17"46d33'
Gillenwater,Gene
Aur 5h9'58"42d23'
Giller,Dr Robert M
Her 16h58'54"26d33'
Gilles Barral
Cep 20h58'0"66d16'
Gilles et Alma
Uma 12h9'30"60d34'
Gilles Papy
Mon 7h58'48"-4d29'
Gilles,Jennifer
Lyn 7h3'0"50d21'
Gilles,Karl-Heinz
Cmi 7h31'43"1d33'
Gilles,Karl-Heinz
Psc 1h17'12"17d33'
Gilles,Werner A
Uma 9h36'53"44d4'
Gillespie
Aur 5h17'1"40d15'
Gillespie "Goddess I", Donna Marie
And 0h14'17"37d44'
Gillespie "The Maverick",Murdoch W
Aql 18h57'15"16d19'
Gillespie 11-27-54, Allison
Lyn 8h51'32"46d38'
Gillespie 5-22-1941, Jeffrey Harmon
Lyn 8h52'40"45d31'
Gillespie,Ashley Erin
Cam 6h14'15"67d45'
Gillespie,Barr
Cet 3h15'31"2d22'
Gillespie,Brandon
Her 17h3'20"42d50'
Gillespie,Brian J
Per 2h37'52"40d30'
Gillespie,Dana Lyn
Uma 11h31'10"41d52'
Gillespie,Daniel
Sex 10h44'1"-0d51'
Gillespie,Dave
Aql 19h15'54"12d30'
Gillespie,Heather Anne Elizabeth
Com 12h58'13"31d31'
Gillespie,Jennifer May
Lac 22h34'21"56d40'
Gillespie,Kevin
Aur 6h7'46"31d6'
Gillespie,Laurita Boyce
Aql 19h26'16"12d18'
Gillespie,Leara Noisette Gantt
Uma 11h46'30"58d10'
Gillespie,Nana & Pa
Uma 9h21'13"50d14'
Gillespie,Neil
Vul 19h19'17"26d22'
Gillespie,Sr,James I
Lac 22h45'11"52d39'
Gillespie,Steve
Per 1h48'47"54d21'
Gillet,Henri
Mon 7h48'58"-4d27'
Gillet,Jean-Pierre
Cyg 19h42'34"37d40'
Gillett,Lisa Maryon
Sco 17h27'25"-30d54'
Gillette,Bruce MacLean
Gem 6h24'41"14d42'
Gillette,C Peter
Dra 16h2'54"66d42'
Gillette,Haley Elizabeth Alexandra
Umi 15h22'27"80d15'
Gillette,John Edward
Cet 3h18'49"6d16'
Gillette,Marge
Mon 6h32'0"0d56'
Gillette,Marianne
Cas 1h34'22"61d23'
Gillette,Pamela Kay
Mon 8h5'0"-3d24'
Gillette,Patricia A
Crb 15h43m50"27d34'
Gillette,Robert
Cmi 7h26'18"8d54'
Gilley,E Glenn
Lmi 10h39'19"25d39'
Gilley,Greg
Aql 18h59'50"-1d41'
Gilley,Jerry Sue
Vir 12h7'1"1d39'
Gillham,Patricia A
Cam 4h48'33"70d54'
Gillham,Vera
Lyr 18h48'18"43d7'
Gilli Pi 58
Aur 5h52'38"31d40'
Gilliam,Amy
Ori 5h55'30"20d31'
Gilliam,Robert
Crb 15h56'53"38d58'
Gilliam,Terri Lee
Peg 22h56'32"27d4'
Gilliam,Tyler David
Cnc 8h12'1"32d32'
Gillian
Uma 11h51'1"32d13'
Gillian
Cam 4h2'22"68d13'
Gillian
Mon 8h6'16"-6d13'
Gillian
Sge 20h6'0"20d49'
Gillian
Cas 23h24'39"60d1'
Gillian
Lyr 18h28'27"37d51'
Gillian Fay
Ori 4h54'58"0d18'
Gillian Marie
Lyr 18h36'41"27d55'
Gillian,"Less Than Tomorrow"
Uma 9h10'0"60d12'
Gillian,The Singing Star Jay
Sco 17h26'0"-33d46'
Gillian-21445-MHL
Cas 2h56'18"58d43'
Gillians Star
Lyr 18h15'37"46d46'
Gilliard,Charleen
Lyr 18h49'45"39d44'
Gillie Family,The Douglas L
Umi 14h22'43"65d45'
Gillie,Carole
Cas 3h5'10"68d37'
Gillie,Geoffery
Aql 20h31'54"0d48'
Gillie,Jamie
Lyn 7h57'10"44d46'
Gillie,Jim
Cep 22h39'54"68d3'
Gillies,Carol
Cas 0h9'17"63d7'
Gillies,Clark
Cam 4h18'19"67d52'
Gillies,Ian & Ena
Mon 6h53'60"-10d11'
Gillies,Natasha Claire
Cyg 20h21'30"38d41'
Gillies,Nicola Jane
Peg 21h56'26"2d45'
Gilligan's Star, May He Always Shine
Dra 19h28'22"58d48'
Gilligan,Carrie
Cyg 19h57'51"40d51'
Gilligan,Christina
Uma 12h13'1"53d55'
Gilliland,Marianne
And 23h32'10"48d20'
Gilliland,Nancy Kay
Cyg 19h17'0"45d19'
Gilliland,Ronald Glenn
Gem 6h40'13"30d13'
Gillin IV,John Joseph
Cet 2h51'45"5d39'
Gilling,Jenny & Wilf
Cyg 21h43'47"38d33'
Gillingham,Victoria Mary
Peg 23h0'42"22d6'
Gillings,Elaine M
Cas 1h54'0"73d9'
Gillioz,Léa
Uma 10h39'50"39d51'
Gillis,Deborah
Lyr 18h15'19"42d22'
Gillis,Máire Ni
Cas 0h2'59"63d56'
Gillis,Patricia
And 2h24'35"42d31'
Gillispie,Delmus Rush
Dra 20h15'35"68d57'
Gillispie,Joan
Lyr 18h31'16"36d44'
Gillispie,Lucas Brandon
Dra 17h59'56"65d11'
Gillispie,Robert
Aur 6h9'23"33d8'
Gillispie,Tiffany Marie
Cas 0h29'18"61d58'
Gillman,Amanda Grace
And 23h3'40"45d52'
Gillman,Gregory Brian
Her 16h38'30"40d57'
Gillman,JoAnn Lisbeth
Crb 15h53'2"34d30'A
Gillman,Nada Linder & Robert
Ari 2h26'30"10d42'
Gillman,Ramsey Howze
Aql 19h55'51"12d51'
Gillman,Stevie "The Beare"
Aql 19h52'18"11d13'
Gillman,Stuart
Oph 17h35'50"-24d19'
Gillmann,Karola
Cnc 8h34'0"30d54'
Gillmore,Greta K
And 0h10'22"47d3'
Gillogly,Kelly Anne
Tri 2h43'1"33d49'
Gills Star,The
Cam 10h5'51"84d48'
Gilluly,Hope M
Lac 22h54'59"55d00'
Gilly 'n' Jon
Cyg 19h28'31"38d24'
Gilly's Way
Peg 21h59'54"10d19'
Gilman,Dana Lee
Peg 22h24'36"33d46'
Gilman,Emma Catherine
Umi 13h14'45"72d23'
Gilman,George Emery
Lac 22h28'0"55d57'
Gilman,Gertie
Cyg 21h54'14"38d6'
Gilman,Marcia
Peg 23h34'21"10d48'
Gilman,Nate
Cet 1h27'1"-12d53'
Gilman,Patti
Lyr 18h21'0"40d57'
Gilman,Sr,Russell T
Oph 18h40'45"8d24'
Gilmartin,John
Per 3h7'25"40d30'
Gilmartin,Sr,Joseph Michael
Boo 13h55'47"17d41'
Gilmer Down Smithfield Lane,Mark
Per 4h7'28"51d56'
Gilmer,Craig Randall
Dra 20h37'24"70d57'
Gilmer,Linda
Eri 4h7'54"-1d36'
Gilmer,Suzanne
And 0h4'15"47d18'
Gilmore Family Star, Robert & Linda
Uma 13h42'50"60d14'
Gilmore Family,The Anderson
Ori 5h57'16"6d42'
Gilmore,Aaron Christopher
Aur 4h55'52"41d15'
Gilmore,Andrew
Cep 22h35'1"70d31'
Gilmore,C Alan
Leo 10h50'46"18d59'
Gilmore,Danny Michael
Dra 19h57'45"70d58'
Gilmore,Dean
Oph 17h8'25"10d18'
Gilmore,Dede
Lyn 7h10'57"58d31'
Gilmore,Dehn W H
Her 17h25'28"21d37'
Gilmore,Hailey Thomas Hume
Cnv 13h5'38"51d43'
Gilmore,Jan
Com 12h29'34"30d5'
Gilmore,Julie
Cas 2h54'1"70d31'
Gilmore,Megan Patrice
Lyr 18h40'48"29d47'
Gilmore,Mildred "Memo"
Cas 3h7'45"60d24'
Gilmore,Nicholas Joseph
Ori 5h54'24"12d19'
Gilmore,Thomas C
Aur 6h8'36"33d1'
Gilmour,Ginger
Uma 13h55'19"60d53'
Gilmour,Lady Alison
Cas 0h17'42"62d4'
Gilmour,Peter Kurtis
Vul 19h18'33"26d56'
Gilmour,The
Dra 17h40'24"71d14'
Giloy,Jon
Ser 15h54'0"1d47'
Gilpen
Cet 2h29'44"4d51'
Gilpin,Holly Jane
Lyr 19h15'50"28d47'
Gilreath
Mon 6h56'16"11d40'
Gilreath Anniversary Star,The
Crb 16h16'39"32d29'
Gilreath,Michael York
Oph 16h58'34"-25d40'
Gilreath,Tessa York
Mon 7h2'10"-1d17'
Gilroy,John & Pauline
Peg 23h40'46"12d9'
Gilroy,Karen Rose
Lyr 18h57'0"30d31'
Gilsela & Miles Star
Cet 1h28'18"0d23'
Gilsenan's Glory
Aur 6h4'43"36d52'
Gilson,Ari Crane
Oph 17h56'14"7d31'

Gilson,Howard
Per 2h41'18"34d47'
Gilson,Malia Ann
Cyg 20h20'33"30d52'
Gilson,Tim R
Ori 5h47'14"18d57'
Gilster,Julie
Cas 0h38'28"71d44'
Giltner,Michael David
Cep 22h2'12"62d41'
Gilvern
Dra 17h53'16"68d25'
Gilytteota
Cet 3h2'23"2d12'
Gim & Shirley's Stormy Point
Dra 17h44'10"65d35'
Gima
Boo 15h3'56"48d33'
Gimakas,Marcia & George
Per 1h44'26"53d17'
Gimbert,Donald
Aql 19h58'16"10d26'
Gimblet,Aaron
Lac 22h25'1"53d2'
Gimenez,"Stud Muffin"
Ser 18h54'1"2d27'
Gimenez,Christine
Uma 13h24'42"61d23'
Gimlin,Lorraine
Mon 6h55'48"11d4'
Gimpel,Elfie
Lyr 18h27'53"42d25'
Gimpel,Mira
Lyr 18h27'27"42d44'
Giménez,Carmen
Aql 19h53'51"11d19'
Gina
Cyg 19h48'0"38d12'
Gina
And 0h44'19"31d41'
Gina
Aql 20h19'36"5d1'
Gina
Leo 11h0'38"10d29'
Gina
Lyn 8h6'36"38d5'
Gina
Peg 22h39'46"30d22'
Gina
Cyg 21h11'17"38d41'
Gina
Cyg 19h26'26"45d38'
Gina
Cas 1h21'59"55d10'
Gina
Peg 22h28'32"31d45'
Gina & Andy's Star
Cyg 19h26'51"56d49'
Gina & Mike
Del 20h53'19"9d40'
Gina & Wayne
Lyn 7h54'0"58d37'
Gina Ann
Sco 17h34'45"-43d8'
Gina B
Sge 20h4'39"20d4'
Gina D
And 0h47'0"33d40'
Gina Denise
Eri 3h21'29"-12d24'
Gina Elizabeth
And 0h21'26"43d27'
Gina Josephine
And 23h43'15"42d17'
Gina Kay
Vul 20h18'57"28d47'
Gina Lois-Anna
Cas 0h46'56"62d54'
Gina Margo
Cap 21h50'13"-14d10'
Gina Maria
Lyr 19h15'38"40d35'
Gina Marie
Mon 7h46'26"-6d47'
Gina Marie
Lyr 18h38'24"38d55'
Gina Marie
Ori 6h2'49"-0d17'
Gina Marie
Ori 6h6'13"20d7'
Gina Marie
Tau 4h44'53"16d7'
Gina Marie
Cyg 21h11'49"38d24'
Gina Marie
And 0h5'39"34d34'
Gina's 'Lil Bit O' Heaven
Lyn 8h40'13"38d50'
Gina's Eyes
Lyn 7h35'0"45d23'
Gina's Star
Ori 4h55'15"5d45'
Gina,The Universe's Art
Umi 15h43'55"76d49'
Gina-Marie
Cam 3h55'21"58d23'
Gina-Queen Of Hearts
Cas 1h1'58"55d54'
Gina-Regina Elaine
Peg 22h1'58"8d39'
Gina-The Powerful One
Cyg 20h19'44"30d52'
Gina-Weena
Com 12h54'58"31d24'
Gina-You Deserve The Milky
Way, Mom
Lyr 19h16'42"28d24'
Ginatta (Italy),Carlo V
Peg 22h38'0"21d52'
Gindele,Reinhold "Stanis"
Tau 3h46'51"2d13'
Ginese Mae
Aql 19h50'16"15d17'
Ginette
Ori 5h42'37"11d13'
Ginette 40
Per 2h12'0"58d52'
Ginette et Réjean
Umi 16h26'40"72d11'
Ginevra,Mary
Mon 6h24'24"-10d25'
Ginger
Uma 11h52'0"51d12'
Ginger
Lyn 8h28'1"37d59'
Ginger
Per 1h45'14"52d34'
Ginger
Cam 4h14'42"67d50'
Ginger
Per 2h53'1"40d49'
Ginger Ann
Lyn 8h14'44"35d16'
Ginger at VSDR
Lyr 18h54'35"33d43'
Ginger's HB Gold
Cyg 21h25'0"37d48'
Ginger, "I Love You."
Com 12h7'0"21d19'
Gingi
Cyg 21h37'26"38d36'
Gingras,Kathrine
Mon 7h39'44"-3d11'
Gingrich,Karen
Peg 22h14'28"10d50'
Gini
Sco 16h34'0"-25d49'
Gini
Cyg 20h18'53"39d9'
Ginn,Clotilde "Pooh B"
Cet 3h18'1"9d53'
Ginne,Leonard & Charlotte
Cmi 7h16'31"4d22'
Ginnia
Mon 6h44'46"11d35'
Ginnie May
Peg 22h27'40"21d33'
Ginny
Cas 23h31'0"60d18'
Ginny
Com 12h24'48"32d40'
Ginny
And 1h37'21"36d4'
Ginny
Equ 20h56'41"4d12'
Ginny
Mon 7h15'41"-5d42'
Ginny Carey
Peg 21h55'22"30d48'
Ginny Mc D
Cas 2h38'47"57d50'
Ginny's Heart
Uma 9h7'37"58d51'
Ginny's Light
Ari 3h0'57"20d42'
Ginny-John
Cyg 19h23'15"44d6'
Gino
Aql 19h50'36"15d11'
Gino
Boo 14h32'44"43d45'
Gino Daniel
Ori 5h55'58"16d34'
Ginorio,Ralph
Cet 3h13'16"2d19'
Ginoulias,Laura
Cas 0h43'20"75d53'
Ginsberg,Aly & Russell
Cyg 19h26'46"30d33'
Ginsberg,Andrew Chase
Tau 4h4'57"11d10'
Ginsberg,Michael Kim
Cet 2h31'0"0d45'
Ginsberg,Minky
Mon 6h57'15"-10d23'
Ginsburg,Alix Faith
Aql 19h59'18"14d51'
Ginsburg,Daniel Issac
Cma 6h53'19"-17d14'
Ginsburg,David H
Boo 15h4'1"29d29'
Ginsburg,Dr Jeffrey Bruce
Cnc 8h26'1"7d38'
Ginsburg,Pamela Beth
Peg 21h59'1"30d30'
Ginsburgh,Robert Sol
Del 20h13'38"11d44'
Ginsy La leologa
Ori 5h55'33"0d54'
Ginther's Dream Star, Kevin
Per 2h28'13"57d26'
Ginther,Milton & Marion
Uma 10h50'36"57d33'
Ginther,Wendy Joanna
And 23h22'54"40d13'
Ginther,Wendy Joanna
Cas 22h58'12"54d29'
Ginty
Lep 15h17'10"40d52'
Ginty,Kathleen
Tri 2h11'16"32d24'
Ginty,Loren Danielle
Cas 0h32'32"64d29'
Gio
Uma 11h11'0"43d58'
Gio Miss Piggy-One Year
Uma 8h22'27"60d3'
Gioconda "Sabris" Yorkshire
Terrier
Cep 3h24'47"80d34'
Gioe,Daniel Anthony
Umi 16h15'25"71d37'
Gioele
Del 20h18'33"15d12'
Gioffredi,Estelle
Umi 16h13'41"70d10'
Gioia
Cas 2h55'0"60d31'
Gioia Starrlight Design,Mary
Aql 20h8'45"0d51'
Gioiosa,Michael Charles
Per 3h10'15"46d3'
Gioisa
Vel 9h23'28"-47d50'
Gioluz
Lyr 19h16'0"26d29'
Giomar
Eri 2h43'49"-1d27'
Giomat
Ant 10h45'45"-31d46'
Giomer
Per 3h32'1"51d6'
Giona
For 2h1'1"-25d5'
Gionfriddo,David
Boo 14h21'0"18d33'
Gioppi
Cae 4h47'24"-33d10'
Gioppo,Julie Margaret
And 0h50'56"38d37'
Gioranne
Her 18h8'20"40d13'
Giordani,Bradley Paul
Cep 22h1'23"61d56'
Giordano & His Ventura
Stars,Mike
Sex 10h31'35"0d29'
Giordano,Alyssa Novella
And 2h23'17"47d26'
Giordano,Ben
Boo 14h35'50"51d56'
Giordano,Casey
Lyn 7h15'47"50d11'
Giordano,Cinzia
Lyr 18h39'44"29d51'
Giordano,Cory
Aur 5h26'0"31d6'
Giordano,Ferdinando
Cam 14h9'59"80d14'
Giordano,Gianni
Cnv 14h54'31"28d5'
Giordano,Jill Irene
Tau 3h41'15"2d28'
Giordano,John
Aur 6h29'47"32d55'
Giordano,Kerry Ann
Ari 2h56'29"21d25'
Giordano,Matthew
Tri 2h36'33"31d22'
Giordano,Maurizio
Ant 10h45'10"-39d32'
Giordano,MD,Anthony J
Oph 17h0'49"11d58'
Giordano,Palma M
Uma 14h9'40"58d30'
Giordano,Ralph & Gerri
And 23h32'1"47d49'
Giorgetti,Gail Kurker
Tri 2h7'0"32d40'
Giorgi,Angela
Cas 0h59'30"60d14'
Giorgi,Frankie
Aur 6h32'33"33d11'
Giorgi-Campbell,Rose
Uma 11h48'0"49d13'
Giorgia
Lac 22h13'1"55d20'
Giorgiangela
Del 20h18'45"12d24'
Giorgiani Family Star Of Love
Lac 22h51'41"56d36'
Giorgina
Lyr 19h19'50"40d44'
Giorgio
Aqr 23h5'20"-6d40'
Giorgio
Cnv 13h29'38"38d21'
Giorgio
Nor 16h21'20"-52d3'
Giorgio
Ori 6h1'21"7d33'
Giorgio & Raffaella
Aur 4h55'0"40d43'
Giorgio e Stefania
Hor 3h25'38"-49d2'
Giota Angel H
Lyn 7h16'50"58d59'
Giouzelis,Evangelos
Cma 7h15'46"-13d29'
Giova
Hor 3h18'39"-48d31'
Giovagnoli Alexis
Cyg 20h48'1"38d37'
Giovane,Bella Amica
Gru 22h18'43"-55d51'
Giovanelli,Eva
Cam 8h24'37"80d24'
Giovanelli,Louis Brandon
Boo 14h46'47"38d18'
Giovanello,Carmella "Aunt
Millie"
Com 13h25'16"26d4"
Giovangrossi,Fiorella
Nor 16h19'57"-52d1'
Giovanna
Psc 22h55'56"5d45'
Giovanna 66
Ori 5h56'20"20d57'
Giovanna Per Sempre, Tony E
Per 3h8'39"41d52'
Giovanna,Ferretti
Per 2h11'55"58d28'
Giovanna,Gaia
Com 12h45'44"21d39'
Giovanna,Sacchetti e Toschi
Nerina
Cyg 20h59'1"30d50'
Giovanna,Schiazza
Cyg 20h41'57"31d21'
Giovannelli Family, Dominick
& Louise
Aur 5h9'20"42d45'
Giovanni e Isabella
Scl 23h24'19"-25d13'
Giovanni Et Patrizia
Aql 19h0'15"16d1'
Giovanni,Joseph "Poppy"
Her 16h52'0"32d21'
Giovanniello,Elaina Marie
And 0h52'17"40d29'
Giovannone,Louisa
Cet 2h34'41"1d57'
Giovannone,Thomas Joseph
Lyr 19h25'16"39d27'
Giovinazzo,Shannon Leigh
Peg 22h52'44"25d2'
Giovine,Raymond
Lac 22h51'51"56d41'
Giovine,Robert L
Gem 7h5'36"21d43'
Giovinetti,Giacomo
Cam 4h59'46"68d40'
Giovingo's,The
Lmi 10h6'30"35d26'
Giovino,Jessica Grace
And 0h13'30"39d29'
Giovino,Justin Michael
Her 17h1'31"20d56'
Gip,The
Lyn 7h29'18"41d1'
Gipling,Andreas L
Cas 1h34'13"61d11'
Gippetti,Andrew J
Aql 18h56'40"10d39'
Gippy
Cyg 21h29'26"40d30'
Gipsirania,Monica
Gem 6h25'50"14d52'
Gipson Star Of My Life,Coretta
Maeife
Mon 6h55'0"-10d32'
Gipstein,Edward
Lib 14h46'0"-18d53'
Gira,Ellen Marie
Tau 3h39'37"23d12'
Girard 22/2/80,Michel Déry
Marie
Umi 16h56'59"77d9'
Girard 8-27-88,Dana Chari
Ori 6h14'46"0d39'
Girard,Alexandre
Oph 18h1'29"10d2'
Girard,Arlyne Holly
Lyr 18h40'30"41d1'
Girard,Brandon David
Her 17h38'56"14d56'
Girard,Brandon Michael
Vul 19h47'46"28d28'
Girard,Colleen
Vul 19h44'16"28d52'
Girard,Daniel F
Aql 20h2'48"1d7'
Girard,Guillaune Clerc
Cep 22h14'32"55d43'
Girard,Jennifer H
Cyg 21h0'1"30d37'
Girard,Jody
Uma 8h45'46"48d18'
Girard,Lauren Marie
Peg 23h25'20"10d18'
Girard,Louise
Sgr 19h33'13"-30d6'
Girard,Louise
Lyr 18h57'33"45d29'
Girard,Madeline B
And 0h5'10"38d35'
Girard,Marie-Line
Per 2h58'14"45d23'
Girard,Michiline Ann
Mon 8h8'43"-1d1'
Girardeau-Dumur,Lucas
For 3h30'13"-32d28'
Girardet "Binnie", Sabine
Dra 10h52'14"73d25'
Girardi,Eric R
Her 18h1'32"31d40'
Girardin,Valérie
Cas 1h59'40"61d52'
Girardo,Reneé Chacho Janier
Daniel
Psa 22h6'41"-27d8'
Giraud,Alain
Sgr 19h26'30"-36d24'
Giraud,Aline
Cep 21h29'28"68d38'
Giraud,Daniel
Dra 15h0'36"64d11'
Giraudon,Benjamin
Boo 15h1'47"8d31'
Girbaud,Franáois
Lac 22h41'31"56d34'
Girette,Michel Annie
Cyg 21h31'14"50d7'
Girgel
Cet 2h54'0"3d5'
Giri,Aaron Kata
Del 20h36'24"10d38'
Girlfriends,The
Lac 22h8'1"46d37'
Girlie Jo
Aur 5h19'53"48d51'
Girling,Alexander
Peg 23h29'54"21d19'
Girmant,Janet Patricia Clark
Mon 7h29'1"-1d4'
Girod,Chantal
Crb 15h51'42"31d46'
Girod,Frederic
Uma 9h52'19"51d35'
Girod,Gabriel
Per 3h57'33"37d15'
Girolami,Cecilia Marie
Cas 0h39'13"65d9'
Girolamo,Jamie Lynn
Eri 3h49'53"-2d25'
Girouard,Frederick
Dra 19h53'26"70d46'
Girouard,John
Aur 5h7'31"41d17'
Girouard,John M
Aql 18h56'54"16d47'
Girouard,Josephine
Cas 0h24'20"50d5'
Giroux,Hélène
Boo 14h51'57"48d12'
Giroux,Marie-Chantal
Cyg 20h23'39"39d4'
Giroux-Daniel Lanouette,Josée
Umi 16h53'1"75d41'
Girshick,Erica Wein
Aqr 21h30'46"-1d19'
Girshick,Frederick
Vir 13h59'1"6d32'
Girt,Joseph E
Cmi 7h32'30"1d34'
Girton,Baby
Boo 14h9'40"40d1'
Girton,James L & Mary Ellen
Per 3h4'56"50d0'
Giry,Olivier
Aur 7h5'41"38d53'
Gisby MCMLXIX,Sharon
Lyr 18h51'22"38d54'
Gise,Rachel
Cyg 20h25'22"40d14'
Gisel,Sébastien
Uma 9h30'13"51d3'
Gisela
Vul 19h23'42"23d57'
Gisela
Uma 10h52'41"70d40'
Gisela
Vir 13h37'47"2d27'
Gisela
Ori 5h34'53"-2d20'
Gisela & Norbert
Dra 17h32'0"60d56'
Gisela Höfer Baden- Baden
Germany
Aqr 21h5'25"-1d36'
Gisela-Me
Ori 5h52'33"21d17'
Gisella
Cae 4h54'1"-32d0'
Giselle
Lac 22h8'46"50d19'
Giselle 95
Aql 19h50'0"14d19'
Giselleous Hoyteous
Tri 1h56'11"28d27'
Gism
Ori 5h56'23"9d12'
Giso,Marie
Cyg 21h17'17"38d45'
Giss,Markus
Umi 13h15'1"70d2'
Gist,William M
Cet 1h21'19"-4d8'
Githens,Deane
Sex 10h15'1"-3d43'
Giti Forever
Sgr 18h48'34"-31d19'
Gitlin,Dolores
And 2h27'14'41d36'
Gitlin,Irving
Cep 21h47'19"58d59'
Gitlin,Robert- Alexandra
Dra 16h24'14"69d4'
Gitnick Family Celebration
Star,The
Dra 20h33'44"68d23'
Gitta
And 23h23'41"50d1'
Gitta
Lyn 8h2'39"42d8'
Gitta
Vir 13h58'28"7d18'
Gitte
Tau 5h51'45"26d26'
Gitzel,Stephan
Psc 23h39'36"-1d15'
Gitzler,George W
Mon 8h5'10"-2d45'
Giudice,Angelo & Debbe
Marshall
Mon 7h59'6"-5d17'
Giudice,Marco
Cam 6h8'25"65d2'
Giuele
Pho 0h36'44"-42d55'
Giuffrida,Caroline
Cas 1h24'1"70d54'
Giuffré,Bob
Oph 16h22'50"-6d36'
Giulia
And 22h58'33"48d57'
Giulia
Pho 0h33'54"-43d30'
Giulia
Vel 9h41'2"-41d29'
Giulia
Cam 7h1'37"61d45'
Giulia
Ant 10h31'13"-35d48'
Giulia
Lup 15h18'25"-45d33'
Giulia
Uma 11h57'30"37d34'
Giulia C
Cep 22h12'26"55d56'
Giulia et Danilo
And 1h2'55"41d3'
Giulia Francesca R M
Vel 9h41'6"-44d47'
Giulia Illuminato
Ari 2h33'25"22d23'
Giulia T D
Cyg 20h4'57"31d49'
Giulia! La stella della mia vita
Cmi 8h4'0"0d29'
Giuliana
Pic 4h53'19"-46d26'
Giuliana
Cep 21h22'17"68d49'
Giuliani Michele
Per 2h10'36"58d34'
Giuliani,Dale Marie
Cyg 21h8'19"31d51'
Giuliani,Victor
Mon 6h30'18"3d10'
Giuliano
Lyr 18h51'13"40d38'
Giuliano
Del 20h17'53"13d5'
Giuliano 1994,Babbo
Uma 11h24'17"56d33'
Giuliano e Luisa
Cyg 20h4'0"31d25'
Giuliano,Christina
Vir 15h6'13"6d21'
Giuliante,Jaclyn
Vir 12h20'34"1d31'
Giulietti,Sharon Ann
Lyn 8h47'20"46d57'
Giulio
Ant 10h45'30"-33d43'
Giulio
Cam 12h19'11"78d55'
Giunchini,Greg Joseph
Crt 11h11'19"-18d52'
Giunta,Carl
Uma 11h26'15"45d45'
Giunta,Frankie
Boo 14h40'28"31d10'
Giunta,James R
Cep 23h6'48"65d36'
Giuntoli,Mara
Aur 6h1'37"30d22'
Giuntoli,Susie B
Lib 15h3'32"-5d26'
Giuseppe
Hor 3h0'28"-49d36'
Giuseppe
Her 18h9'1"30d16'
Giuseppe e Patricia
Vel 9h41'3"-49d54'
Giuseppe,Rocco & Chantel-
Marie
Dra 14h20'31"63d30'
Giussani,Daniela
Cas 23h42'46"64d46'
Giussani,Silvia
Nor 16h19'33"-51d52'
Giust,Rossella
Cep 23h29'18"64d50'
Giusti,Betty J
Uma 10h50'40"48d27'
Giusti,Elia
Ori 5h59'15"15d36'
Giusti,Lars Axel
Aur 5h12'44"42d8'
Giusti,Mariangela
Her 16h42'41"18d46'
Giusto
For 2h33'24"-28d34'
Giustozzi,Antonello
Lib 15h57'52"-8d40'
Giusy
Tel 20h10'37"-48d21'
Giusy Giusy
Mon 8h7'39"-7d39'
Given,Fionnuala Ann
Cyg 21h18'0"35d25'
Givens Golden Star, Roger &
Eva
Mon 7h22'35"-8d40'
Givens,Tammy Eileen
Cam 4h53'17"78d46'
Giver Of Life,Keeper Of
Knowledge
Hya 8h41'10"3d53'
GIVINCY
Uma 11h7'0"70d45'
Givler,Amy & Randy
Uma 13h1'23"54d58'
Givliani,Patrick
Aur 4h47'18"36d32'
Giz,My Valentine
Cam 12h42'52"76d37'
Gize,Walter Gregory
Aql 19h2'33"16d29'
Gizmo
Cyg 21h8'11"31d15'
Gizmo
Aur 5h53'16"38d56'
Gizmo
Cam 3h57'27"58d3'
Gizmo
Cet 2h47'32"1d37'
Gizmo
Dra 14h35'17"64d42'
Gizmo,Debbie
Lyr 18h38'22"42d7'
Gizzi,Lisa Marie
Lyn 7h58'28"50d3'
Giò
Uma 11h36'44"51d54'
Giò
Boo 15h4'24"26d50'
Giò
Cyg 20h5'0"38d8'
GJ-46
Aql 18h44'22"10d27'
Gjerde,Dr William Mem/B/G
Eugene H Beebe
Cep 21h55'47"55d19'
GJW
Per 3h9'40"41d51'
GKGK 150
Aqr 20h45'40"-6d37'
Glackin,Alexandra
And 0h57'54"37d12'
Glad
Lac 22h8'1"37d42'
Glad & Dizzy The Tom & Alex
Star
Cam 3h54'21"55d10'
Gladden,Ann Damsgaard
Lyr 18h43'37"46d29'
Gladden,Jr,Joseph Rhea
Cet 2h16'24"5d1'
Gladding,David (Papa
Shango)
Aql 19h56'0"8d29'
Glade,Ella Johanna Bremer
Cas 0h29'1"70d51'
Gladfelter
Boo 13h35'0"26d14'
Gladin,Joseph R
Cap 21h43'0"-23d38'
Gladiola
Ori 6h8'1"3d5'
Gladish,George
Crb 16h2'52"38d41'

GladRon 40
Aur 5h19'18"45d44'
Gladstone,Elaine Mary
Cas 0h25'1"60d55'
Gladstone,India Kate
Umi 16h47'51"76d57'
Gladstone,Madeline
Vir 11h41'52"2d4'
Gladwell,Jenny
Cyg 21h44'18"38d34'
Gladwin,William Jackson Earlie
Mon 7h58'20"-8d48'
Gladys C-B
Lyr 18h58'19"30d2'
Gladys Violeta
Cas 0h50'40"65d8'
Glaes,James G
Del 20h27'31"5d45'
Glaeschig,Wolfgang
Per 3h20'26"46d13'
Glaholm,John 'Laughing Boy'
Aql 20h1'46"1d14'
Glaiman,Sylvain et Florence
Cyg 21h56'34"50d12'
Glair,Donna Lee Mary Nuessle
Cas 1h53'49"61d2'
Glamm,Ryan Timothy
Peg 0h7'44"21d52'
Glamorous Gladys
Aql 18h58'54"-6d5'
Glamour Shots
Umi 16h39'29"78d11'
Glamp,Rachael Mariah
Cam 3h11'54"60d42'
Glamp,Savannah Reneé
Cas 3h1'0"57d50'
Glance,Norbert & Marge
Lyn 7h54'1"36d12'
Glandon,Brian Patrick
Ser 17h30'38"-14d45'
Glandon,CW-2 David Russell
Aur 6h8'58"46d48'
Glandus,Sandrine
Boo 13h57'1"16d48'
Glantz,Harris
Cet 3h10'52"6d37'
Glantz-Salesky,Karen Grace
Cnv 12h42'14"37d58'
Glao,Caitlin Marie
Peg 23h26'50"15d26'
Glapa Family
Equ 21h1'35"10d30'
Glas,Hervé
Peg 23h31'27"10d21'
Glasberg,Bunni & Barry
Cyg 19h27'29"33d24'
Glascock,Lisa Dawn
Oph 18h5'35"12d9'
Glaser,Alison
And 23h44'1"41d8'
Glaser,Davina
Cyg 19h28'18"34d14'
Glaser,Herbert O
Com 13h14'16"22d6'
Glaser,Jim
Aur 5h6'14"43d29'
Glaser,Mary Denise
Eri 3h55'13"-17d18'
Glaser,Paul M
Aql 18h59'13"11d47'
Glaser,Tonya J Steinbrecher
Mon 6h20'52"2d45'
Glaser,William & Darlene
Sge 20h16'37"17d56'
Glasgow,Jackie
Eri 4h3'1"-14d56'
Glasgow,Marnie Kathleen
Aur 5h26'12"38d53'
Glashow,MD,Jules Lewis
Peg 21h48'44"28d56'
Glasow aka (Lazer), Steven Edward
Cet 2h11'29"4d32'
Glasow,Muffin Vanderpool
Umi 16h25'40"70d10'
Glass Goose,The
Vul 19h57'45"25d46'
Glass Wedding Star,The
Aur 6h8'1"37d39'
Glass, Peter Mark
Dra 12h36'27"68d57'
Glass,Andrew G
Ori 5h57'34"11d38'
Glass,Barbara & Bob
Crb 15h55'34"38d56'
Glass,Billy C
Mon 6h51'38"-2d54'
Glass,D/G
Vul 20h21'53"23d17'
Glass,Eric Lamont
Lmi 11h1'18"32d49'
Glass,Gary
Hya 8h18'49"0d15'
Glass,Josef
Gem 7h57'11"32d4'
Glass,Stephen
Ori 4h44'44"5d12'
Glass,Vivian A
Cas 0h31'1"63d60'
Glassbrenner,Gary
Boo 15h4'44"23d50'
Glasscock,Anthony Francis
Her 16h47'13"47d59'
Glasser,Cathy Ann
And 2h14'32"39d32'
Glasser,Gary Duane
Her 17h56'42"38d54'
Glasser,Jennie
Cyg 21h15'16"28d50'
Glasser,Joshua Alexander
Psc 0h57'46"32d22'
Glasser,Shelley
And 23h15'0"49d36'
Glasser,Ted Lewis
Per 3h8'19"55d3'
Glassglow-MSG
Mon 6h57'16"-0d55'
Glassgow,Sarah Margaret
Cnc 9h5'21"31d54'
Glassgow,Steven Donald
Sgr 20h6'25"-43d30'
Glassman,Harriet C
Uma 8h32'22"51d54'
Glassman,Harry
Hya 8h56'34"4d18'
Glassman,Max 1 & 2
Aur 5h44'32"50d13'
Glassner,Don
Aur 6h42'13"38d55'
Glassow,Lynn V
Sgr 19h9'23"-26d52'
Glatow,Friederike
Her 17h22'25"46d53'
Glatthorn,Isabel Kae
Cas 0h37'1"60d47'
Glaubensklee,Royal Stanley
Per 2h56'48"55d3'
Glauber,Michael A
Uma 12h3'36"31d0'
Glaubitz,Carola
And 23h22'17"40d4'
Glaubman,Michael David
Her 16h35'38"40d23'
Glaum,D
Dra 17h52'50"68d27'
Glaus
Cas 1h21'1"55d14'
Glauth Star,Mark
Her 17h14'14"42d31'
Glavan,Lisa Renee
Ori 5h58'56"18d38'
Glavaris,Alma
Com 12h24'0"27d6'
Glawar,Iris
And 0h1'10"43d31'
Glaz,Jean Claude
Lyr 18h58'21"31d24'
Glazar,Alan
Per 4h42'15"51d21'
Glaze,Geneva-Elzie
Cet 2h39'33"-1d45'
Glaze,Hailey Noquisi
Eri 4h18'42"-18d51'
Glaze,Kipp-Cailean Jones
Mon 5h58'11"-6d47'
Glaze,Sandra Marie
Lyn 7h58'1"45d28'
Glaze,Sylvia-Dean
Mon 7h55'25"-1d12'
Glazell,Barbara
Per 4h5'19"46d5'A
Glazer,Adam Bennett
Her 15h54'51"40d36'
Glazer,Harvey
Uma 8h33'1"60d38'
Glazik,Karen "Special K"
Lyn 9h10'52"42d9'
Glaß,Steffen
Dra 18h55'29"67d43'
GLC
Psc 23h21'1"1d54'
GLD-France 1995
Sge 19h12'0"21d7'
Gleadall,Fiona
Lyr 18h27'23"46d45'
Gleason's Gazer
Aql 18h46'10"10d23'
Gleason,Andee Moriah
Crb 15h59'10"38d1'
Gleason,Angela Beth
Cas 3h6'32"73d14'
Gleason,Anthony Dale
Hya 9h17'12"1d23'
Gleason,Garry
Dra 17h31'26"70d47'
Gleason,Glenn
Ori 5h38'41"15d15'
Gleason,Heidi
And 23h16'50"41d37'
Gleason,Jack
Ori 5h28'21"-0d35'
Gleason,James
Lac 22h36'58"53d51'
Gleason,Kayla Dawn
And 0h43'25"22d25'
Gleason,Paul
Vul 20h58'1"28d48'
Gleason,Scott Michael David
Tau 3h57'26"24d33'
Gleason,Toni Lynn
Uma 13h32'57"58d31'
Gleason,William John
Hya 9h6'53"3d15'
Gleason-James,Wendy
Cas 2h58'37"76d53'
Gleckl,Mark Anthony
Aql 19h15'34"13d50'
Gledhill,James
Uma 9h10'47"55d27'
Gledhill,Joshua Wright
Cam 7h44'1"60d7'
Gledhill,Jr,John Joseph
Dra 17h34'44"61d5'
Gledhill,Norm
Lac 22h8'1"49d12'
Gleed,Michelle Deann
Psc 0h46'42"31d58'
Gleeson,Benjamin
Aql 20h6'13"6d46'
Gleeson,Sarah Jayne
Sgr 19h28'39"-37d17'
Glein,Eugene
Lac 22h55'47"53d28'
Gleinser,Mary T
Del 20h56'12"16d14'B
Gleize,Betty
Lyn 8h32'21"41d45'
Glemnitz,Dylan C
Cam 4h9'26"67d58'
Glemnitz,Riley S
Cep 22h22'18"70d59'
Glen
Boo 14h36'49"7d56'
Glen & Pat
Uma 12h3'39"33d5'
Glen Ridge GOAL Students of 92
Per 4h1'28"48d60'
Glen's Star
Hya 10h12'21"-16d8'
Glen,Barbara Taurean
Lyn 8h1'29"51d35'
Glen,Esteem's Lorraine
Cyg 20h38'60"45d37'
Glen,Peter Craig
Cep 5h14'14"86d39'
Glen,Tania Marie
Peg 23h0'1"21d41'
Glenannee
Mon 6h23'22"-8d4'
Glencross,Stephanie Kathleen
Umi 15h53'0"72d15'
Glenda
Mon 6h21'0"9d1'
Glenda
Vul 19h0'1"22d7'
Glenda For All Eternity
Cam 9h12'1"84d50'
Glenda Granny
Leo 9h35'1"14d7'
Glenda J,The
Peg 23h30'57"22d47'
Glenda The Good Witch
And 23h23'0"41d7'
Glendenning(010944), Raymond
Ori 6h6'18"2d49'
Glendinning,Ralph O
Cet 3h2'20"5d32'
Gleneck,Samuel Bruce
Sex 10h1'52"1d30'
Glengary,Marija Rasa
Lyn 8h27'1"41d19'
Glenn
Hya 8h18'22"6d12'
Glenn
Lmi 10h7'12"31d21'
Glenn
Sex 9h54'50"1d52'
Glenn & Katie
Dra 16h5'17"61d59'
Glenn & Robert's 1st Anniversary Star
Vul 20h0'1"28d50'
Glenn & Syndee
Uma 8h54'1"50d16'
Glenn & Tara
Ari 2h33'32"30d28'
Glenn 50,Tyler
Ori 5h58'19"16d44'
Glenn II,Gerald Marvin "Mike"
Boo 14h33'1"50d27'
Glenn's Jewel
Aur 5h1'43"37d51'
Glenn's Shoes
Ori 5h59'56"16d30'
Glenn*MP-10
Sgr 19h21'55"-29d24'B
Glenn,Amanda
And 0h57'41"33d32'
Glenn,Ann Koltun
Peg 22h42'44"31d1'
Glenn,D Kyle
Aur 5h2'24"38d26'
Glenn,Donald Earl
Cep 22h21'45"56d14'
Glenn,Hilary Anderson
Crt 11h17'34"-10d59'
Glenn,Jeffrey Gene
Cet 3h16'14"1d33'
Glenn,Jennifer M
Peg 22h8'46"29d37'
Glenn,John & Annie
Her 17h29'16"28d14'
Glenn,Judy Marie
Lyr 18h52'35"31d31'
Glenn,Kathy deMarise
Peg 22h50'43"27d36'
Glenn,Lara Diane
Uma 8h59'18"68d28'
Glenn,Mary
Cas 0h48'14"62d32'
Glenn,Mickey
Cas 1h11'21"65d5'
Glenn,Randel Kurtis
Boo 14h56'17"28d57'
Glenn,Shelton William
Oph 17h56'47"12d40'
Glenn,Susan
Lib 14h59'27"-0d29'
Glenn,Wade Matthew
Hya 8h12'15"6d20'
Glenn,William-Hamilton
Cep 22h8'0"70d6'
Glenn-Toby
Cam 3h50'50"56d20'
Glenna
Cma 6h27'18"-16d22'
Glenna
And 2h22'22"49d34'
Glenner,Corrine Joyce
And 2h24'34"39d26'
GlennGary
Sco 17h23'58"-38d10'
Glennon,Betty
Lyn 9h19'34"36d13'
Glenpatrick,Ire 88, Nicholas Tobin
Per 2h22'35"55d57'
Glenwood Landing School Class Of 1995
Ori 6h15'41"-0d3'
Glenwright,Cole Tristan
Peg 22h13'41"32d40'
Glenys
Cas 0h45'40"63d48'
Gless,Michael McCarthy
Sex 9h52'48"-2d4'
Gless,Sharon
Lyr 18h50'40"36d16'
Glezman,Michael
Oph 16h40'47"0d55'
GLG & MDM
Crb 15h49'52"32d17'
Glibby-Meister,The
Sex 9h57'38"1d38'
Glick,Ana B
And 1h18'1"33d46'
Glick,Jr,David L
Cep 5h6'36"80d28'
Glick,Lorenzo Cat
Boo 13h46'0"16d35'
Glick-Nuckolls,Kristin
Aur 5h25'59"40d15'
Glickman,Alec Douglas
Boo 15h29'51"40d49'
Glickman,Henrietta
Uma 9h7'16"52d33'
Glickman,Jeffrey Alan
Per 2h55'24"40d35'
Glickman,Kara
Cas 1h5'1"60d1'
Glickman,Paul
Ori 4h43'14"0d12'
Glidden,Brooke & John
Her 18h1'50"48d28'A
Glidden,John & Brooke
Her 18h1'50"48d28'B
Glidden,Laurn Elizabeth
Cas 1h40'51"58d32'
Glidden,Peter R
Her 17h18'14"22d27'
Glidden,Rob
Cep 21h45'26"55d5'
Glidden,William Flint
Oph 18h1'47"1d1'
Glide
Uma 13h32'58"60d23'
Glienke,Lothar
Her 17h19'57"42d39'
Gliesche,Marita
Aqr 20h38'38"-14d10'
Gliessner,Jeffrey S
Cet 2h8'22"4d27'
Glimcher,Robert I
Boo 15h1'28"18d56'
Glimco,James "Bud"
Aur 7h0'53"40d18'
Glimmer
Lyr 19h6'37"25d55'
Glinche,Charlie
Aur 6h0'39"40d32'
Gline,Alain
Lmi 10h48'10"33d16'
Glinei Et Nelly
Boo 14h46'51"38d17'
Glissan,Paul Raymond
Leo 9h19'27"21d37'
Glisson,Kemper James
Hya 8h11'0"2d41'
Glista's Cosmic Cafe, Joe
Cep 23h26'1"63d39'
Glitsch,Shawn
Boo 14h0'0"21d55'
Glixman,Miles Parker
Ser 18h17'60"-13d14'
Glo's Own
Uma 12h10'42"57d11'
Glo-Le-Ra
Sex 10h27'43"-1d12'
Globo
Cap 21h2'12"-22d46'
GLOBO
Boo 13h32'12"30d12'
Globosky Forever In A Day,Michael
Lac 22h22'42"53d40'
Glock
Ori 4h51'55"0d33'
Glocksen,Bob & Judie Glick
Leo 10h54'30"20d54'
Glocup
Tri 1h58'57"31d50'
Glodek,Jim
Vul 19h51'11"20d35'
Glodek,Jr,Mitchell
Her 17h17'1"45d26'
Glodek,Stella
Cyg 20h16'0"38d33'
Glodek-Neilan,Linda
Lyr 19h26'51"40d6'
Gloeckler,Gustav
Tau 4h1'37"0d7'
Gloede,Kurt Ancil
Aur 5h25'53"30d8'
Gloeggler's Light
Dra 16h35'30"68d39'
Gloff,Jeffrey David
Boo 14h44'14"22d46'
Glogowski,Richard & Janice
Cyg 21h56'16"53d7'
Glois
Lmi 10h4'60"33d19'
Glomb,Diana Marie
Eri 5h6'37"-6d31'
Glomb,Jessica
And 1h53'20"36d25'
Gloria
Uma 11h54'45"33d12'
Gloria
And 0h20'40"32d15'
Gloria
Peg 22h51'31"27d21'
Gloria
And 23h30'41"43d16'
Gloria
Del 20h12'46"11d37'
Gloria
Mon 8h7'24"-1d52'
Gloria
Sco 16h25'0"-40d32'
Gloria
And 0h3'27"38d20'
Gloria - Love Without End
Vul 19h59'25"26d6'
Gloria 143
And 23h2'0"46d37'
Gloria i Antonia
Boo 13h46'41"13d54'
Gloria Jean/K
Uma 8h54'0"50d16'
Gloria Josephine
Ori 5h39'41"15d9'
Gloria June
Sco 16h56'27"-43d43'
Gloria Loretta
Her 17h32'48"20d33'
Gloria Nona C D
Cas 1h9'32"67d43'
Gloria Savage's Star Forever
Ori 5h53'0"15d54'
Gloria Star,The
Eri 2h49'53"-7d21'
Gloria Therese
And 2h26'44"42d20'
Gloria's 'Lil Bit of Heaven
Cap 21h25'55"-24d6'
Gloria's Delight
Cnc 8h26'45"11d3'
Gloria-Sudfass
Lyn 7h59'0"40d29'
Gloria-The Piano Teacher
Ori 6h8'33"4d47'
Gloriana
Cas 0h27'13"60d37'
Glorianne
And 2h23'59"44d18'
Glorianrocktarius
Lyn 6h21'38"59d12'
Glorich
Vul 19h48'25"28d31'
Gloriela
Peg 22h47'12"21d36'
Glorilei Liana
Mon 6h22'42"1d23'
Glorioso,Carol Ann
Com 12h21'11"21d2'
Glorious Light
Her 18h11'1"40d52'
Glorious Love Junie Moon Pony Star
Vul 20h21'16"23d54'
Glorious Starling,The
Eri 2h48'40"-1d60'
Glorius Stellaris
Lyr 19h16'28"35d11'
Glory
Gem 7h0'27"15d35'
Glory
Sex 9h47'27"3d30'
Glory Summer
And 23h20'0"45d44'
Glory's Star
Uma 11h36'44"31d33'
Glosemeyer,Nana Margaret
Cas 0h28'23"61d9'
Glosik,Dennis S
Cam 3h43'59"68d12'
Glotfelty,William Randolph
Aql 19h15'10"12d49'
Glover,Bryan
Umi 15h14'21"68d17'
Glover,David
Lac 22h9'1"47d53'
Glover,Elizabeth Nicole
Cyg 21h12'11"35d48'
Glover,Geoffrey J
Cam 6h16'44"68d15'
Glover,John David
Her 17h10'28"22d17'
Glover,John Gil
Cnc 8h50'12"32d19'
Glover,Julia Roberta
Cas 0h59'30"58d56'
Glover,Kevin
Aur 5h50'60"40d60'
Glover,Linda A
Aur 5h48'46"38d32'B
Glover,Lori Kim
Aql 19h49'0"15d27'
Glover,Melissa Kate
And 0h37'0"41d6'
Glover,Miriam
Cas 0h35'0"67d15'
Glover,Robert William
Per 2h0'30"57d57'
Glover,Sean Marsh
Cnv 12h58'43"33d15'
Glover,Sherry
And 1h25'44"39d39'
Glover,Sylvia
And 2h10'43"50d33'
Glovera
Mon 7h17'20"-7d4'
Glovka,Leo
Sct 18h30'52"-6d56'
Glow
Cam 6h15'2"73d49'
Glow Worm Keith & Gloria
Cyg 21h37'42"41d57'
Glow,Fieldens Amber
Cmi 7h34'45"1d28'
Glowacki's Piece Of Heaven,Samantha
Cas 2h0'0"61d38'
Glowacki,Samantha & Fraser Horn
Cyg 21h2'1"28d37'
Glowczewska,Klara
Cam 5h4'1"68d5'
Glowski,Michael
Dra 19h26'51"56d49'
Glowski,Michael Andrew
Uma 9h6'53"72d27'
Gloyd R A G Star, Robert
Sex 9h53'48"3d14'
Gloyd T A G Star,Tim
Cet 0h51'1"-1d0'
Gluck,Mildred
Vir 13h10'22"-5d26'
Gluckman,Ted David
Hya 9h8'17"2d36'
Glueckert,Michael Scott
Her 16h37'58"8d18'
Gluhanich,Gerald Thomas
Cnv 12h34'22"40d10'
Glumac,Dick Mile
Cet 2h28'1"0d23'
Glunz,Mara Lee
Cam 4h16'31"60d57'
Glusco,Jr,Michael Paul
Per 3h53'10"40d55'
Glusovich,Kelly
And 22h56'23"51d7'
Gluszko,Joseph Andrew
Cet 1h46'0"-1d24'
Gluszko,Vilena
Lyn 8h59'1"44d55'
Gly-Hb
Oph 17h57'12"12d17'
Glyn Charles
Uma 11h45'12"50d60'A
Glyni
Lyr 18h57'28"38d48'
Glynn,Brendan Patrick
Gem 6h33'21"12d8'
Glynn,Cathy Wolf
Lyn 7h38'0"37d29'
Glynn,David J
Per 2h54'55"40d32'
Glynn,Joe
Cmi 8h1'55"6d5'
Glynn,Joseph E
Ser 15h33'59"18d26'
Glynn,Mary Wolstencroft
Aur 6h13'36"31d40'
Glynn,Meagan Mary
Ari 1h47'19"11d7'
Glynn,Michael F X D
Dra 11h34'57"70d14'
Glöckner,Wenzel Veit Sebastian 14.8.43
Leo 11h2'48"12d19'
Glücksstern
Lib 15h27'4"-22d49'
Glücksstern für Sieglinde Koschare
Boo 15h20'0"41d33'
Glücksstern für "Udo & Brigitte"
Uma 9h39'60"46d55'
Glücksstern im Sternzeichen Dephin
Del 20h29'39"10d1'
Glücksstern,Annette und Uwe Fischer
Boo 14h39'24"14d16'
Glücksstern,Dr Martin Schmucker
Boo 15h22'36"38d26'
Glücksstern,Schlommis
Ori 5h55'24"15d30'
Gma & Gpa
Eri 3h34'0"-5d26'
GMC1
Cam 3h42'38"70d29'
GMD Tigger
Aur 7h24'12"38d27'
Gmeiner,Eva
Cet 2h54'26"7d12'
Gmerek,John J
Dra 15h47'54"57d44'
GMP
Cyg 21h14'1"28d37'
Gmyrek,Bryan D
Boo 14h40'53"36d44'
Gnacik,Benjamin Henry
Dra 17h47'1"64d15'
Gnade,Jill
Cas 2h5'24"59d40'
Gnassi,Charles P
Boo 15h2'34"18d46'
GNC-25
Eri 4h1'27"-12d42'
GNDRFL
Eri 3h49'0"-1d6'
Gnehm,Elizabeth A
Cas 1h13'42"63d43'
Gniewek,Roma Marie
Cas 0h53'43"61d13'
Gniot
Oph 17h17'4"-20d38'
Gniot,Kathy Lea
Tri 1h55'22"27d12'
Gnirk,Mark A & Amanda H Fox
Oph 17h35'33"11d49'
Gnirrep,Jack
Uma 9h48'0"68d45'
Gnom Nicola
Cnc 9h16'1"8d49'
Gnossis,Phineas
Mon 6h53'31"-2d8'
Go back
Cnv 13h3'29"32d40'
Go Boy's Royal Blue
Peg 21h25'21"20d13'
Go On Shining Especially For You
Gem 7h12'30"23d36'
Go,Jaclyn Kelly
And 23h43'20"44d53'
Goachee,Marsha
Tau 5h55'19"28d8'
Goachee,Ronald
Sco 17h53'30"-30d21'
Goad,Allison L
Cma 7h12'26"-13d57'
Goad,Andrea Lee
Cnc 8h50'27"30d22'
Goad,Artur Yahola
Cap 20h24'23"-27d7'
Goad,Diann Williams
Vir 13h54'1"-21d55'
Goad,Keaton Arthur
Ari 2h33'18"30d15'
Goad,Robert Arthur
Ari 2h35'37"30d47'
Goad,Robin Pruitt
Lib 14h58'51"-22d47'

Goad,Timothy Owen
Lib 14h58'37"-22d44'
Goak
Lmi 9h42'39"33d42'
Gobeil,Cynthia Alpha- Omega
Dra 17h27'45"50d54'B
Gobeil,Fowler Alpha- Omega
Dra 17h27'45"50d54'A
Goben,George
Aql 19h35'1"0d3'
Gober,In Memory of Ray
Cet 0h58'31"-6d59'
Gobets,J & C
Per 4h41'43"40d1'
Gobi
Uma 10h29'1"50d52'
Goble,Bob "Why"
Boo 15h26'41"40d34'
Goble,Jennifer J
Sct 18h39'39"-6d4'
Goble,Joyy & Wilbur
Cyg 19h52'48"37d53'
Goble,Vernon James
Lac 22h16'28"46d15'
Goblet,Bernard
Sct 18h45'4"-7d16'
Gochnoor,Michael
Mon 7h3'25"4d43'
Gocklin,Jr,USMC,L CPL Ronald
Cyg 21h20'30"40d1'
Goczan,Karen
Lyn 9h12'27"40d6'
God Bless Our Loved Ones M-N-G-F-J
Per 4h24'26"51d4'
God Child,the
Lup 14h38'42"51d27'
God Love N Bless Mary, Robbie & Stevie
Aql 18h49'50"11d0'
God Sent Me The Best Baby He Had "Mando"
Her 17h2'0"41d3'
God Speed
Ori 6h15'11"-0d9'
God's Blessings To Beth & Al
Sge 20h0'51"19d44'
God's Love
Aql 19h9'32"2d32'C
God's Serious Light The Useted Star
Ori 5h53'31"16d58'
Godaire,William A
Oph 17h11'15"-22d56'
Godbey,Bob
Cet 3h4'55"-0d2'
Godbey,Vancel Raymond
Lac 22h0'22"48d58'
Godbey,Vancel Raymond
Hya 10h47'52"-12d1'
Godbout
Ori 5h54'44"10d26'
Godbout,Susan
Cas 23h40'42"64d6'
Goddard Star,The Handel
Ori 6h3'46"1d25'
Goddard,Bethany Jayne
Umi 14h37'57"68d45'
Goddard,Billy
Dra 19h42'17"68d21'
Goddard,Erin
And 23h39'1"47d40'
Goddard,Faith Whitney
Mon 8h2'27"-9d37'
Goddard,Faith Whitney
Lyn 8h10'50"41d45'
Goddard,Frank W
Ser 15h12'35"0d5'
Goddard,Gary Wayne
Aqr 20h58'1"-8d39'
Goddard,Lester
Aql 20h6'11"7d47'
Goddard,Orville Huns
Aur 7h26'57"40d50'
Goddard,Peter
Cep 22h21'38"61d35'
Goddard,Phillip
Aql 19h4'0"1d14'
Goddard,Rachel Louise
Cyg 21h1'34"36d30'
Goddard,Ray
Eri 2h48'0"-7d1'
Goddard,Susan Mary
Sge 19h37'22"16d19'
Godden,Andrew Peter
Ori 5h25'11"1d52'
Godden,Matthew James
Ori 5h52'48"8d59'
Goddess Kim
Cet 0h53'15"0d50'
Goddess Stephanie
Cyg 20h18'57"41d53'
Goddess Wendy-(Wendy Wilson)
Aql 20h1'51"-6d27'
Goddyn,Eric
Uma 11h25'18"34d33'
Godel,Tara M
And 1h21'37"37d8'
Godell,Audrey M
Cmi 7h42'15"0d4'
Goderwis,H
Aql 20h0'30"7d51'
Godfrey,Alan Robert
Cep 23h31'38"68d18'
Godfrey,Carl
Crb 16h10'56"33d59'
Godfrey,David
Uma 12h8'20"46d43'
Godfrey,Eileen M
Del 20h22'0"10d12'
Godfrey,Jace
Umi 17h23'0"75d31'
Godfrey,Jane Louise
Lyr 18h43'19"40d5'
Godfrey,John
Ori 6h5'49"7d27'
Godfrey,Judith A
Mon 6h29'58"8d20'
Godfrey,Lynn
Aur 4h55'15"41d11'
Godfrey,Michael
Ori 4h55'37"0d8'
Godfrey,Millie
Aql 18h52'57"11d6'
Godfrey,Ron
Her 16h58'1"30d4'
Godfrey,Ryan
Dra 20h38'46"67d31'
Godfrey,Susan Jane
Lyn 8h25'42"33d50'
Godfrey,Thomas J
Cet 2h28'12"8d47'
Godfrey,Wendy Lee
Mon 6h23'41"10d22'
Godfrey-Cass,Robin
Ori 5h55'25"12d22'
Godin,Olaf Erik Oswald
Uma 11h56'43"63d19'
Godin,Paulette
Dra 17h32'0"70d20'
Godines,Celeste Marie
Tri 1h58'19"25d45'
Godinez,Greta
Peg 22h5'0"3d33'
Godinez,Tania Lynn
Aur 4h51'50"48d59'
Godino,Alexandra Andrea
Mon 6h52'49"1d22'
Godino,Cindi Parris
Mon 6h52'32"0d20'
Godino,Jr,Scott
Mon 6h53'55"1d3'
Godle
And 0h27'35"40d35'
Godman,Jo
Mon 7h55'11"-5d57'
Godman,Keith
Cep 23h7'34"60d49'
Godoy,DorJun
Uma 10h24'38"68d14'
Godreaux,Odjiou
Cmi 7h27'58"5d55'
Godsave,Ted
Ori 5h56'50"19d55'
Godsey,Justin Lee
Ser 15h54'14"-2d3'
Godsick,Joan M
Aur 5h2'47"38d18'
Godsick,Suzanne
Cas 1h58'30"60d9'
Godston,Peter P
Aur 7h2'14"41d1'
Godwin's Wedding Star, Paul & Ceri
Cyg 21h53'26"36d20'
Godwin(You Are The Everything),Amy
Mon 6h25'14"-10d18'
Godwin,Elizabeth "Bethany"
Peg 22h33'0"21d53'
Godwin,Kim Kristine
Lib 14h59'39"-23d41'
Godwin,Nancy Lee-Ette
Cas 23h41'34"58d8'
Godwin,Otto Augustine James
Boo 14h55'43"25d12'
Godwin,Patricia B
Cma 6h51'27"-16d23'
Godynick,Justin N
Dra 17h13'19"61d15'
Goebel,Karl
Hya 8h25'20"1d5'
Goebel,Siegfried
Vir 11h54'17"-6d11'
Goebels,Herbert
Peg 23h35'22"31d9'
Goebig,Charles Joseph
Sex 10h6'35"0d46'
Goebig,Edward Francis
Lyr 19h1'12"47d23'
Goecke,Susan & Shawn
Uma 11h41'54"32d15'
Goedde,Claudia
Per 4h3'4"37d41'
Goeddemeyer,Adolf
Dra 14h20'19"64d20'
Goedert,Andrew
Uma 9h54'46"46d34'
Goedert,Bishop Raymond
Oph 18h41'48"8d18'
Goedert,Nicholas
Uma 9h43'13"48d17'
Goeggel,Herbert
Per 3h18'54"46d19'
Goehring & Poncho, Robert A
Aur 5h45'23"50d6'
Goehring,Jr,Robert
Lac 22h10'0"54d42'
Goeman,David Louis
Her 18h0'15"28d57'
Goen,Bob & Joni
Cet 1h0'58"-1d34'
Goepel Star
Cam 9h19'27"78d40'
Goepp,Jr,Peter Edward
Aql 19h40'24"10d6'
Goerden,Helmut
Sgr 19h56'10"-42d32'
Goeres,Anke
Cep 22h12'1"68d13'
Goergen,Marc
Ori 6h7'52"1d35'
Goeringer,Ronald
Mon 8h1'32"-9d50'
Goerner,Michael David
Cnv 12h50'37"37d38'
Goers,Cheryl Ann- Kuslak
Vir 12h31'0"2d17'
Goertz,Amanda
Cam 6h2'26"58d46'
Goertz,Michael
Per 3h33'52"38d22'
Goeser,Greg
Aur 5h1'14"48d17'
Goethel,Jr,Charles
Her 16h41'56"33d34'
Goetsch,Victoria Rebecca
And 2h34'22"37d1'
Goetz,Mary Ann
Cyg 21h2'0"31d44'
Goetze,Linda Makofske
And 0h55'52"45d18'
Goff,Billie Gail McMullen
Vul 20h14'53"25d15'
Goff,David & Karen
Sge 18h59'0"19d43'
Goff,Douglas Allen
Uma 9h0'34"48d0'
Goff,Elaine Marie Laplume
And 0h0'50"46d34'
Goff,John Samuel
Ser 15h15'1"20d48"
Goff,Joseph P
Cet 1h33'1"-3d3'
Goff,Matthew Daniel
Boo 15h5'46"8d49'
Goff,Norma Jean Menzes
Cas 0h20'15"60d8'
Goff,Patrick Dugan
Gem 7h21'19"31d44'
Goff,Starla
Boo 14h28'31"44d8'
Goff-Rose,Michele
And 0h20'36"30d14'
Goffe,Adeelia
Aur 5h24'1"50d17'
Goffe,Andrae
Aur 5h51'38"54d59'
Goffin,Clarence & Johan
Crt 11h17'33"-19d48'
Goffin,Galadrielle
Cru 12h14'27"-57d25'
Goffin,Tinevielle
Ant 10h57'4"-36d1'
Goffredo,John Edward
Cyg 21h38'31"38d15'
Goforth,Charles & Joanne
Uma 11h21'53"40d25'
Goforth,Janice G
Lyr 18h46'37"38d52'
Goforth,Melody Diane
Peg 21h59'0"22d31'
Gogan,Joshua
Aur 6h53'40"37d53'
Goggin,A E
Vir 13h38'29"-4d10'
Goggin,Michael McSweeney
Lac 22h12'52"49d46'
Goggins,Bernard Lee
Cet 0h52'0"-2d6'
Goheen
Peg 22h5'47"27d51'
Goheen,Alysha Skelly
Cas 23h24'18"60d50'
Gohier,Lotus
Cma 6h57'16"-16d29'
Gohl,Scott
Boo 13h59'0"26d26'
Gohr Best Dad & Hubby, Foxwoods Freddy
Cep 21h31'50"65d7'
Goi
Lyn 7h1'60"51d7'
Goia,Albert Quillo
Cyg 20h21'51"39d38'
Goia,Jr,Joseph P
Per 3h24'38"40d7'
Goings,Keith
Her 18h4'53"14d41'
Goins,Sheri & Ron
Hya 8h47'2"-10d49'AB
Gois,Paulo Henrique Bakker
Ser 15h59'48"4d1'
Goit,The
Cep 21h50'21"55d19'
Goity,Veronique
Mon 6h19'38"5d31'
Gojanovich,Mark S
Aql 19h57'22"13d56'
Gokson,Gerry
Dra 16h39'37"60d50'
Gola,Jo-Ann
And 23h46'12"45d39'
Golba,Joseph
Aur 7h12'44"39d59'
Golconda
Lac 22h46'57"37d43'
Gold Coasts Emelia
Cma 7h22'41"-16d43'
Gold Star
Cet 2h37'13"3d49'
Gold,Allison
Peg 22h5'48"20d20'
Gold,Andy
Aql 18h59'16"-1d23'
Gold,Bert
Ori 5h5'34"-1d60'
Gold,Christine Lynn
Cas 1h19'52"61d15'
Gold,Debbie
Ori 6h1'26"1d6'
Gold,Diana
And 23h18'0"45d9'
Gold,Dr Jay Alexander
Oph 17h6'20"12d1'
Gold,Dr Victor
Oph 17h3'0"-8d16'
Gold,Frema
Cyg 21h29'54"40d11'
Gold,Harry Nathan
Aur 5h10'43"42d48'
Gold,Joshua Seth
Boo 14h59'0"33d12'
Gold,Judge Alan S
Cet 2h41'42"-10d31'
Gold,Kaye
Cma 6h55'29"-16d40'
Gold,Mark & Colleen
Uma 11h16'47"30d50'
Gold,Matthew David
Aql 18h44'1"10d24'
Gold,Matti
Lyn 8h7'53"34d16'
Gold,Peter
Dra 17h4'17"68d41'
Gold,Randy
Aur 7h3'39"40d53'
Gold,Rebecca
Leo 11h1'32"22d35'
Gold,Ruth Marsha
Cas 0h20'1"64d8'
Gold,Sandra
Cet 0h11'54"-8d3'A
Gold,Sheldon A "50th"
Her 16h20'1"48d55'
Gold,Simon
Cet 0h11'54"-8d3'B
Gold,Susan Dudley
Cas 0h32'25"62d59'
Gold,Tillie
Cyg 21h51'28"40d55'
Golda In Loving Memory
Cnv 13h50'48"40d25'
Goldbaum,Andrew Scott
Cep 21h18'37"70d40'
Goldberg's Star,Reggie Jerry & Larry
Cyg 19h14'0"49d1'
Goldberg,Baby
Eri 3h31'31"-7d20'
Goldberg,Barbara
Cas 0h14'41"58d27'
Goldberg,Barry
Dra 9h30'33"77d40'
Goldberg,Barry
Dra 20h7'60"62d2'
Goldberg,Bernice "Bubby"
Cas 0h21'52"68d42'
Goldberg,Betty
Cas 0h45'24"70d3'
Goldberg,Brian
Lac 22h47'0"38d2'
Goldberg,Chip
Lmi 10h40'1"24d58'
Goldberg,David
Lac 22h39'11"53d46'
Goldberg,Debbie
Cen 11h42'44"-48d47'
Goldberg,Dena Beth
Cas 2h36'55"71d14'
Goldberg,Edward "Eddiebear"
Uma 11h39'51"31d53'
Goldberg,Ellen Rose
Cas 0h4'29"61d21'
Goldberg,Glenda
Cas 2h31'40"58d51'
Goldberg,Gloria
And 1h10'30"40d36'
Goldberg,Gretchen
Cas 1h44'40"61d50'
Goldberg,Hank
Peg 22h36'12"8d32'
Goldberg,Hunter Charles
Boo 14h51'29"28d25'
Goldberg,Jack
Cep 21h22'48"60d17'
Goldberg,Jay
Tau 4h19'15"22d3'
Goldberg,Leigh Ann
Lyn 8h3'21"51d38'
Goldberg,Linn
Uma 12h14'34"57d58'
Goldberg,Lisa
And 0h49'27"39d38'
Goldberg,Mildred & Harry
Cyg 21h11'18"38d44'
Goldberg,Nancy & Gene Rubenstein
Cyg 21h9'33"48d46'
Goldberg,Paul
Aur 5h0'22"50d26'
Goldberg,Rae & Lou
Cyg 21h19'45"37d49'
Goldberg,Richard
Cep 21h53'37"68d17'
Goldberg,Scott & Julie
Uma 10h11'49"72d55'
Goldberg,Scott Mikel
Her 16h40'12"10d22'
Goldberg,Shelley Thelma
Uma 11h10'59"55d53'
Goldberg,Steven
Lmi 9h49'23"34d11'
Goldberg,Steven A
Per 4h39'40"36d56'
Goldberg,Thomas B
Per 1h45m12"52d37'
Goldberg,Warren
Ori 5h36'55"-4d3'
Goldberg,Wendy Gail
And 23h30'14"48d53'
Goldberg-Pop-Pop's Star,David
Tau 3h41'53"28d47'
Goldberger,Linda
Cas 1h28'1"67d33'
Goldberger,Thomas Pedro
Per 2h23'12"54d31'
Golde,Joanne Casey
Tri 2h1'35"31d7'
Golden Barbara & Fred
Aql 18h43'0"10d28'
Golden Bear
Uma 11h41'14"62d44'
Golden Boy
Mon 8h3'1"-5d52'
Golden Fish
Psc 1h24'40"30d31'
Golden Gryps
Leo 10h0'2"15d2'
Golden Horstmann Star
Vir 13h3'0"-12d10'
Golden Orpheus 94
Ori 15h27'43"3d9'
Golden Vercon
Peg 21h55'21"33d12'
Golden Webb
Lyr 19h18'21"40d16'
Golden,"Hope" Heather
And 0h45'21"31d37'
Golden,Alan
Lac 22h53'27"40d58'
Golden,Bonnie
Cap 21h0'18"-20d4'
Golden,Daniel
Cmi 7h57'22"8d45'
Golden,Donnie
Ori 5h16'26"10d9'
Golden,Elly
Dra 15h36'0"65d5'
Golden,Gail Hadison
Lyr 18h21'16"47d36'
Golden,Gary B
Aql 19h5'0"15d54'
Golden,Hollen
Her 16h19'39"10d54'
Golden,Kaitlin
Vul 19h41'1"23d59'
Golden,Laura Gail
Tau 4h44'41"16d40'
Golden,Nancy Sue
And 2h22'22"43d7'
Golden,Paul & Kirsten
Crt 11h43'38"-21d51'
Golden,Richard L
Boo 14h48'11"38d25'
Golden,Susan
Peg 22h47'11"21d53'
Golden,Tami
And 2h10'23"42d21'
Golden,Toni Lee & Greg
Cyg 20h0'1"30d38'
Golden,Veronica
Cas 0h5'34"56d10'
Golden-The Golden Galaxy,Douglas
Dra 19h21'58"56d56'
Golden-The Golden Galaxy,Robert
Dra 19h21'53"56d3'
Golden-The Golden Galaxy,Glen
Dra 19h22'30"56d30'
Goldenbaum,Rebecca
Crt 11h7'55"-18d29'
Goldenberg,Alaina
Lyr 19h18'16"42d4'
Goldenberg,Dana
And 1h35'16"48d58'
Goldenberg,Douglas
Dra 15h13'53"60d51'
Goldenberg,Ellen Jo
Cas 0h30'1"64d2'
Goldenberg,Erica
Umi 15h26'21"80d35'
Goldenberg,Forever Radiant Judi
Lyr 19h17'16"40d10'
Goldenberg,Geri
Lyn 7h45'27"41d15'
Goldenberg,Jason L
Cyg 20h41'55"42d52'
Goldenberg,Jay
Uma 12h2'53"61d15'
Goldenberg,Marina
Uma 8h23'20"61d9'
Goldenberg,Mike
Per 2h19'12"58d55'
Goldenberg,Suki
Boo 15h6'38"10d43'
Goldenberg,Toni
Cas 3h25'18"75d1'
Golder,Carrie Marie
Lmi 10h27'24"34d18'
Golder,Jesse James
Cep 22h39'20"70d43'
Goldfarb (BIG),Barry Irwin
Aql 19h1'42"5d45'
Goldfarb,Ann Zorn
And 23h41'23"47d54'
Goldfarb,MD,Irvin D
Her 16h52'0"50d38'
Goldfarb,Rachel Marie
Uma 8h51'25"62d22'
Goldfarb,Robert
Psc 1h5'1"28d21'
Goldfarb,Samuel "Buddy"
Cyg 20h19'25"41d44'
Goldfarb,Sanford
Peg 22h13'52"7d40'
Goldfeier,Eric Steven
Cep 22h7'31"53d41'
Goldfinger,Sheila B
Vir 11h59'33"0d10'
Goldfuss,Lauren Paige
And 1h21'49"40d22'
Goldhamer,Douglas
Cnv 12h5'44"38d2'
Goldi,Robert Colin
Per 1h59'48"56d36'
Goldie
Cet 0h32'56"-3d1'
Goldie
Dra 17h6'25"69d1'
Goldie
Cep 22h47'19"56d58'
Goldie R
Lib 15h1'26"-6d3'
Goldie Royale
Peg 22h52'42"29d18'
Goldie,Teresa M
Cnv 13h58'22"31d49'
Goldie,The
Crb 16h6'55"38d46'
Goldin,NASA Administrator,Daniel
Umi 14h1'17"77d28'
Golding Star,The John E
Aql 18h49'42"11d0'
Golding,John Zlatna Zvyesda
Pho 0h49'58"-47d34'
Golding,Justyn Patrick
Uma 8h42'34"55d3'
Golding,Michael R
Aur 7h17'28"39d46'
Golding,Peter
Cyg 20h20'31"38d42'
Goldingay,Jacqueline Anne
Cas 0h34'35"62d23'
Golditch,Sydney
Boo 15h2'1"32d26'
Goldklang,Eternal Love Sheila & Marv
Boo 15h5'51"41d1'
Goldkäfer,Heidi
Boo 14h26'37"14d42'
Goldlist,Rosylin
Cas 1h13'25"60d12'
Goldman Rising Star, Sarah
Lac 22h53'20"53d13'
Goldman's Light,Margo
Tri 2h41'45"32d22'
Goldman,Annetta Jeanne Mamet
Uma 8h38'1"47d51'
Goldman,Benjamin B
Aur 6h25'50"37d15'
Goldman,Blanche & Sol
Eri 4h32'59"-6d28'
Goldman,Carole
Lac 22h4'1"38d37'
Goldman,Debbie
Leo 10h58'0"18d33'
Goldman,Diane
Mon 8h6'56"-8d20'
Goldman,Donna Lynn Pezzello
Mon 6h46'34"10d31'
Goldman,Gabrielle Irina
And 23h2'31"51d1'
Goldman,George
Aur 7h8'17"40d37'
Goldman,George N
Psc 0h47'39"20d46'
Goldman,Herbert David
Boo 14h8'42"46d58'
Goldman,Jean-Jacques
Crt 11h17'59"-19d5'
Goldman,Jonathan David
Per 3h11'19"45d7'
Goldman,Jr,Jefferson Briscoe
Oph 17h5'21"-20d20'
Goldman,Leah,Sara & Lewis
Cyg 19h36'46"28d39'
Goldman,Lindsay Brooke
Cap 21h0'15"-20d22'
Goldman,Louise
And 23h15'48"49d29'
Goldman,Marcia
Lyr 18h35'1"41d52'
Goldman,Michael
Aur 5h22'59"41d2'
Goldman,Nigel
Sge 20h4'39"20d5'
Goldman,Paul
Sgr 19h8'9"-21d59'
Goldman,Shelly Beth
And 1h52'16"39d41'
Goldman-Eller,SS Esther
Cap 20h55'23"-20d38'
Goldmann,Debra
Lyr 18h34'59"47d23'
Goldmeier,Elsie Baer
And 1h1'20"38d59'
Goldner,Abe
Aur 6h25'1"40d16'
Goldner,Jeffrey
Aur 4h46'57"38d2'
Goldner,Steven
Gem 7h52'55"31d49'
Goldsborough,Ian Spencer
Tau 3h57'15"11d34'
Goldsborough,Ralinda & Steve
Del 20h24'26"10d60'
Goldsbourough,Amelia Caroline
Leo 10h52'30"-6d28'
Goldsmith,Aaron Daniel
Ori 5h57'50"15d0'
Goldsmith,Amelia
Uma 9h59'22"50d48'
Goldsmith,Bill & Frances
Sge 19h31'58"16d19'
Goldsmith,Garrett & Alexander
Her 16h36'0"32d59'
Goldsmith,Glenn A
Cnv 12h20'57"37d46'
Goldsmith,Peggy
Cet 3h10'16"1d27'
Goldsmith,Robert
Her 17h13'32"50d9'
Goldstein Star,The Henni
Lib 14h20'30"-10d60'
Goldstein's Everlasting Star,Barb
Uma 9h24'13"55d31'
Goldstein,Alvin
Tri 2h27'0"30d33'
Goldstein,Becca Leigh
Cyg 19h17'20"44d35'
Goldstein,Becky Kim
Cyg 21h59'14"44d22'
Goldstein,Beth
And 2h26'41"47d37'
Goldstein,Brian Sorrel
Cnv 12h6'45"36d37'
Goldstein,Burton
Cep 21h7'46"70d60'
Goldstein,Carol Elizabeth
And 1h53'30"38d56'
Goldstein,Caroline Beth
Del 20h53'20"6d5'
Goldstein,Charlie
Eri 3h0'0"-16d16'
Goldstein,David J
Ori 5h53'44"20d20'
Goldstein,Ed & Anita
Crb 15h28'41"31d1'
Goldstein,Elissé Jo
Uma 11h36'36"46d45'

Goldstein,Elizabeth Boo 14h36'54"8d22'
Goldstein,Estelle & David Cep 21h3'44"60d58'
Goldstein,Everest Star Peg 22h44'11"11d23'
Goldstein,George Ian Crb 15h59'29"34d1'
Goldstein,Howard Her 17h53'0"31d11'
Goldstein,Jacquelynn And 1h37'57"38d12'
Goldstein,Jennifer G Cam 3h43'40"72d11'
Goldstein,Joe & Jill Cyg 21h27'46"38d27'
Goldstein,Julia B Peg 23h43'29"10d11'
Goldstein,Justin David Cet 2h51'18"4d5'
Goldstein,Lana Dra 12h56'1"68d13'
Goldstein,Laura And 0h51'0d21d46'
Goldstein,Leon M Cep 22h16'58"62d30'
Goldstein,Mark Per 3h0'27"40d5'
Goldstein,Marty Hya 9h32'39"0d40'
Goldstein,Mimi Lmi 10h35'28'23d55'
Goldstein,Miriam Cas 23h20'58"53d55'
Goldstein,Mona Lyr 18h55'19"34d13'
Goldstein,Myron William Uma 11h21'50"45d13'
Goldstein,Noah Mathew Aql 20h4'10"8d30'
Goldstein,Punchin' Judy Leo 9h21'1"7d38'
Goldstein,Rachel Elizabeth Lyr 18h52'45"30d37'
Goldstein,Robert Uma 10h4'29"51d34'
Goldstein,Robert Paul Tau 4h1'36"1d59'
Goldstein,Ross Samuel Cep 23h26'31"65d54'
Goldstein,Sammy Uma 12h44'29"60d27'
Goldstein,Scott Cnv 12h8'0"37d8'
Goldstein,Steve & Sandy Cyg 21h19'42"30d38'
Goldstein,Tamara Elise And 23h33'33"37d45'
Goldstone,Cynthia & David Cyg 19h28'37"36d54'
Goldstone,Erin K Mon 6h19'34"5d51'
Goldstone,Florence Ruth Uma 12h6'0"56d42'
Goldstone,Hana Monroe Umi 15h41'43"78d38'
Goldstone,Lee Uma 10h48'57"57d36'
Goldstück Brigitte Psc 1h20'45"18d42'
Goldsworth-Arrance Ser 16h0'0"13d18'
Goldsworthy,Ethyle C Uma 10h20'54"40d9'
Goldsztaub Boo 14h20'47"48d42'
Goldup,Amanda Faith Cam 5h48'49"79d47'
Goldwasser,Gail S Peg 22h7'25"24d25'
Goldwater,Anna Rae Peg 22h48'35"21d46'
Goldwing,Fabiani Vul 19h40'28"20d31'
Golebiowski,Betty Barbara And 23h29'26"49d33'
Golec's Spectacular Star,Joe & Rita Cyg 21h57'37"50d31'
Goleh,Fariba Peg 22h35'0"12d11'
Goleskie,Jeanne & John Cyg 19h53'36"40d31'
Golia,Michael Cap 20h42'0"16d32'
Golias,Bernard Joseph Per 2h56'0"40d16'
Golias,Jeffery P Ser 15h24'38"21d17'
Goliat,Eleanor Lyr 18h58'11"38d24'
Goliath Her 16h41'46"8d0'
Golightly,Jaime Major Lyr 18h59'22"34d54'
Golin,June & Al Aur 7h2'0"43d34'
Golina Uma 12h29'33"62d19'
Golinki,Sandy Tri 2h5'1"30d59'
Goliwas,Carol Lyn 8h24'39"51d47'
Golka,Chase Michael Vul 19h48'1"28d24'
Goll,Beverly Ari 2h55'53"28d48'
Goll,Pamela & George Mon 6h56'22"-10d58'
Gollan,Wesley Samuel Aql 18h42'0"-0d51'
Goller,Jeanine E Aql 18h54'45"-2d34'
Goller,Marilyn Cyg 20h28'39"42d46'
Gollnitz,Thomas Lac 22h22'51"40d58'
Gollock,Robert Aur 7h16'44"40d28'
Gollon,Otto Her 18h7'0"18d52'
Golman Star,The Sidney J Aql 18h41'37"-0d5'
Golms,Peter Tau 5h47'45"27d53'
Golob,Martin Tim Her 17h4'23"21d19'
Golodner,Maria And 2h35'0"40d8'
Goloubkin,The Star of Elizabeth Cru 12h2'58"-58d6'
Golsby,Lesley Karen Uma 11h7'37"43d30'
Golson,Aaron Ross Sex 10h40'46"4d28'
Golstein,Allison Mira And 23h3'14"40d59'
Goltze,Yves Tau 4h59'0"16d20'
Golub,Orville Aur 5h19'1"48d53'
Goluba III,John Uma 9h48'58"42d27'
Golubchikova,Ekaterina Sgr 20h4'30"-20d14'
Golubock,Roberta Cyg 19h26'17"45d42'
Gomart,Madame Franáoise Peg 23h46'17"30d16'
Gomel,Jr,Carlos Miguel Uma 9h21'25"67d58'
Gomersall,Judith Cas 0h45'0"63d14'
Gomes,Armando Peg 21h28'32"23d14'
Gomes,Franklin Gilbert Cep 23h21'10"70d38'
Gomes,Joseph Anthony Her 15h58'33"40d7'
Gomes,Nicole Christine And 23h1'20"50d20'
Gomes,Sarah Elizabeth And 0h59'60"45d23'
Gomez & Berry Boo 14h56'1"20d19'
Gomez,Angela And 2h6'51"42d23'
Gomez,Bob Oph 18h6'58"7d41'
Gomez,Brenda & Glenn Mon 7h13'41"-6d54'
Gomez,Carlos Miguel Cep 20h26'1"75d29'
Gomez,Cierra Catalin Aql 19h4'16"-1d23'
Gomez,Colleen Peg 22h38'54"26d57'
Gomez,Daniel James Hya 8h42'48"3d58'
Gomez,Danny Dra 16h18'13"67d29'
Gomez,David Aur 4h57'51"50d55'
Gomez,Eva Del 20h29'21"11d9'
Gomez,Frances Peg 0h6'1"14d45'
Gomez,Gustavo Hya 9h4'45"-1d8'
Gomez,Henry Joseph Aql 19h7'0"0d36'
Gomez,Jaime H Mon 7h15'43"-5d45'
Gomez,James Carlton Cma 6h27'35"-13d31'
Gomez,James F Dra 16h3'60"67d21'
Gomez,José Luis Ser 15h48'0"20d43'
Gomez,Joy Denise Leo 10h52'13"22d8'
Gomez,Luis Alberto Aql 19h17'15"14d14'
Gomez,Marguerite Cnc 7h58'42"10d57'
Gomez,Michael Ori 5h48'57"21d3'
Gomez,Monica Viola Cnv 13h57'48"30d36'
Gomez,Nicholas Jones Mon 7h13'13"-5d22'
Gomez,Paz B Cet 2h7'21"0d16'
Gomez,Roland Her 16h50'48"40d49'
Gomez,Sorelba Hinestroza Oph 17h10'42"-22d48'
Gomez,Thomas Dra 19h28'47"58d37'
Gomez-Con Vstedes Para Siempre Aql 20h0'28"6d36'
Gomiller,Jeremy Todd Sgr 20h1'11"-20d30'
Goncalves,John Cep 22h23'55"70d16'
Goncharoff,Dan Her 17h31'0"40d46'
Gondal,Manzoor Lac 22h15'55"51d32'
Gong's Star Mon 6h55'0"-8d21'
Gongora,Victoria Michelle Cas 0h4'30"62d44'
Gonier,David Chester Aur 7h14'45"35d54'
Gonigan,Viviane Louise Peg 22h16'34"8d41'
Gonnella,Anthony R Boo 15h5'56"30d46'
Gonnsen,Karl-Martin Hya 9h7'30"1d11'
Gonshor,Nicole Diane And 23h37'27"33d8'
Gonsoulin,Whitney Cma 6h54'19"-17d8'
Gontard,Brita Mon 6h24'21"-6d0'
Gonyea,Kim Mary Cas 1h08'54"62d60'
Gonzalas,Marlene Elaine Del 20h23'20"10d14'
Gonzales Galactic Oph 17h58'14"8d10'
Gonzales(Marie's Star) Marie E Ari 1h54'18"13d11'
Gonzales,Alex & Virginia Com 13h15'25"28d28'
Gonzales,Armando Dra 17h54'57"61d5'
Gonzales,Dick Her 16h59'20"29d38'
Gonzales,Dora Erlinda Sandoval Mon 6h30'1"3d3'
Gonzales,Edward Cet 1h21'1"-3d40'
Gonzales,Felicia Leo 10h50'36"-5d7"
Gonzales,Francisco Cyg 20h54'36"30d2'
Gonzales,Hipolito Sanchez Her 16h56'1"50d49'
Gonzales,Lee Del 20h25'1"10d51'
Gonzales,Manny And 0h19'1"33d11'
Gonzales,Maria Hita Boo 15h21'45"38d46'
Gonzales,Maria Theresa Cmi 8h4'43"1d44'
Gonzales,MD,Joseph M Oph 18h42'30"10d10'
Gonzales,Paul A Lac 22h36'56"38d33'
Gonzales,Petra Cet 1h16'59"-13d54'
Gonzales,Rene Lyr 19h19'15"35d24'
Gonzales,Star of Jason Dra 11h29'0"68d31'
Gonzales,Tom Ser 17h33'1"-14d50'
Gonzalez "Wild Heart", Marie Mon 7h19'14"-7d8'
Gonzalez,Amy & David Lyn 8h21'0"49d36'
Gonzalez,Ana And 0h52'48"45d9'
Gonzalez,Anthony Her 16h7'41"47d34'
Gonzalez,Betty J Cyg 20h20'31"40d28'
Gonzalez,Carlos Humberto Sarg Aql 19h2'33"6d40'
Gonzalez,Claudia Mon 6h32'0"3d29'
Gonzalez,Cristina Maneiro Cyg 19h35'17"38d45'
Gonzalez,Don & Katherine Crb 16h12'20"34d18'
Gonzalez,Dorothy Benefield Eri 2h53'0"-18d53'
Gonzalez,Elicia Pun al Cyg 21h59'1"53d40'
Gonzalez,Flaviano Sex 9h56'50"-6d28'
Gonzalez,Gabriol Nunez Her 18h19'19"18d51'
Gonzalez,Hank Dra 18h29'57"70d41'
Gonzalez,Irma Peg 22h48'59"8d46'
Gonzalez,Jeanette Cnv 13h59'0"28d55'
Gonzalez,Jennifer Aql 20h2'30"0d33'
Gonzalez,Joe P Her 16h36'45"35d13'
Gonzalez,Jordi Nualart Her 18h29'0"20d4'
Gonzalez,Jorge Del Villar Cmi 7h56'22"7d37'
Gonzalez,Jr,Joe Vidal Per 4h2'58"37d33'
Gonzalez,Jr,Joseph Dra 17h52'54"65d15'
Gonzalez,Juan Carlos Aql 19h24'0"15d47'
Gonzalez,Judy K And 22h56'30"51d49'
Gonzalez,Lucas T Cma 7h14'13"-16d36'
Gonzalez,Luis "Pelon" Cma 6h31'12"-18d2'
Gonzalez,Marta Maneiro Aur 6h55'31"37d56'
Gonzalez,Mathew B Aql 20h20'42"5d28'
Gonzalez,Milagros Oph 17h29'38"-23d56'
Gonzalez,Nancy Barriga Cas 0h41'25"61d27'
Gonzalez,Nicole E Peg 0h12'23"18d32'
Gonzalez,Patricia A Del 20h53'40"6d56'
Gonzalez,Roy Boo 14h30'15"17d32'
Gonzalez,Steven "Gigantor" Hya 8h42'1"4d47'
Gonzalez,Tony Solomon Kaleioku Cep 21h51'44"55d8'
Gonzalez,Wally Alberto Col 5h7'44"-29d54'
Gonzalez-Sandoval,Rosa Elia Oph 18h1'0"11d5'
Gonzalez-Taylor, Natalie Mon 7h45'31"-2d41'
Gonzo Lib 14h58'57"-23d46'
Gonzo Gem Aql 20h19'15"5d15'
González,Alexandra Rachel Cas 1h8'23"62d6'
González,Arthur E Ori 6h0'36"-1d50'
González,Carmen Sylvia And 0h52'0"33d53'
González,Jr,Raymond Alexander Lac 22h12'32"49d49'
González,Raimundo José Cep 0h10'26"69d37'
González,Raymond Alexander Lac 22h45'45"55d16'
González-Mrs Bunny, Cathy Ann And 0h21'49"31d23'
Goo Umi 14h39'1"78d15'
Goobanoff,Lana Cet 3h2'26"8d25'
Goober Ori 5h54'21"16d39'
Goobie Lmi 11h0'17"33d17'
GoobieandGoobie Mon 6h47'19"11d37'
Goobielou Cma 7h13'32"-15d0'
Gooch,Barbara Luella Peg 22h44'23"10d2'
Gooch,James L Cma 7h13'35"-15d19'
Gooch,Jerad Tudor Per 3h1'29"40d38'
Gooch,Joshua Dwight Tri 2h18'28"28d32'
Gooch,Michael Ori 5h4'0"14d47'
Gooch,The Boo 14h29'26"43d22'
Goocher,Andrea Marie Del 20h39'1"18d53'
Good Company Crb 15h48'57"28d44'
Good Housekeeping Cep 23h8'38"60d19'
Good II,Thomas Jefferson Cet 0h49'13"-8d37'
Good Luck Star-AP For 2h8'21"-24d33'
Good Morning America Ori 5h27'14"0d42'
Good Morning-I Love You Cam 12h45'55"78d60'
Good Night Irene Peg 22h55'51"21d9'
Good Night Irene & George Cyg 21h19'22"39d19'
Good Night Miss Agnes And 22h57'1"51d52'
Good Night, Margie Rose Peg 23h3'12"32d31'
Good Star,The Lyr 19h26'41"38d30'
Good You Are Our Shining Star,Jim F Cep 23h28'15"64d4'
Good You Are Our Shining Star,Jim F Cyg 19h54'0"38d1'
Good,Ashley Jordan Lyr 18h47'46"31d26'
Good,Christiana Denise Peg 23h38'20"25d36'
Good,Daniel C Aur 7h21'60"38d38'
Good,Dave & Marge Cyg 19h24'16"31d53'
Good,David & Marjorie Cyg 19h26'38"31d4'
Good,David E Aur 6h15'41"31d34'
Good,Kathy J Lyr 18h44'37"30d54'
Good,Kevin Dra 10h23'14"73d19'
Good,Kimberly Mon 7h7'1"0d46'
Good,Kyrstie Jo Gem 7h33'30"20d6'
Good,Lori M And 22h57'56"40d5'
Good,Margaret Elizabeth Cas 0h19'51"63d13'
Good,Samuel E Her 17h21'21"21d35'
Goodacre,Lynn Uma 11h5'32"37d34'
Goodale,Fanny Wilson Cyg 19h26'58"30d5'
Goodale,Julie Marie And 23h5'29"37d12'
Goodall,Anne Cas 0h28'31"61d4'
Goodall,David J Sgr 18h51'32"-22d20'
Goodall,Georgia Connie Rose Mon 6h27'1"2d45'
Goodall,Hope Elizabeth Uma 9h19'18"50d44'
Goodall,Keith Uma 8h24'54"62d28'
Goodall,Marc Per 1h52'11"56d42'
Goodall,Samantha Jayne Lyr 18h29'55"31d14'
Goodarzi,Hamid R Tri 2h5'42"32d21'
Goodchild,Christel Psc 23h20'40"2d23'
Goodchild,Lin Cyg 19h19'56"28d23'
Goode,Greg Ser 15h53'23"21d29'
Goode,Gregory Scott Cet 2h31'31"-1d25'
Goode,Jack Cep 21h59'36"60d55'
Goode,Joseph Andrew Aur 5h11'0"42d14'
Goode,Jr,Noel A Mon 7h1'34"-8d15'
Goode,Lari Gem 7h29'52"20d18'
Goode,Rhiannon Rose Cnv 13h28'16"40d44'
Goode,Rosemary Evely Cyg 19h31'55"37d48'
Goode,Sheila Peg 22h5'17"2d20'
Goode,Sofia Marie Cnc 8h37'23"7d26'
Goode,Thomas & Cari Vul 20h2'45"26d8'
Goode,Tyler Eugene Her 18h6'47"48d29'
Goodell,Christopher Austin Her 17h13'37"44d33'
Goodell,David & Caroline Crb 15h49'23"37d30'
Goodell,Jayna Louise Mon 6h57'35"8d12'
Gooden,Charmaine Uma 11h9'0"37d56'
Gooden,Tammy M Eri 3h14'19"-12d56'
Gooderham,Edward H & Margaret M Del 20h50'1"8d13'
Goodfader,Mary Eri 2h55'50"-3d39'
Goodfellow,Leslie Ann Cas 1h22'25"50d22'
Goodgion,Kevin Walt Ser 17h34'57"-14d5'
Goodhart,Carol Ann And 2h24'58"41d26'
Goodheart,Todd John Cep 20h43'42"65d35'
Goodhew,Martin Dudley Per 1h52'19"50d7'
Goodin,Sweet Potato & Lambie Uma 8h52'33"48d50'
Gooding Inc,David Lmi 10h51'35"38d30'
Goodlett,Jim & Erika Umi 15h9'19"67d5'
Goodliffe,Leah Jane Crb 16h5'12"34d24'
Goodlow,Jeanine Marie Umi 15h25'55"80d16'
Goodlow,Kateri Elizabeth Cyg 21h53'22"41d50'
Goodman (Sam),Adele Monihan Cyg 19h26'22"32d41'
Goodman 20th Ann,Fran & Yossi Sgr 19h31'1"-36d21'
Goodman's 30 Peg 22h52'55"27d22'
Goodman,Allan Her 15h48'30"41d54'
Goodman,Andrew Ser 18h43'43"2d39'
Goodman,Bernie Lac 22h44'40"54d45'
Goodman,Betty & Ray Ori 6h7'21"8d48'
Goodman,Brian Chance Dra 16h1'57"60d29'
Goodman,Carrie Cyg 21h29'54"40d11'
Goodman,David Oph 18h2'38"10d42'
Goodman,David Sex 9h48'56"0d47'
Goodman,Elizabeth J Cas 23h12'31"60d23'
Goodman,Eric Lac 22h16'31"37d58'
Goodman,Faith Lyr 18h58'16"32d50'
Goodman,Gail B Lyn 9h11'30"40d36'
Goodman,Geoffrey Stuart Aqr 23h6'0"-19d53'
Goodman,Grandma Jean Vul 20h14'57"23d12'
Goodman,Harlan Equ 21h6'58"3d28'
Goodman,James Calvin Dra 11h31'18"71d50'
Goodman,James Stanley Boo 15h2'59"23d48'
Goodman,Jeff B Aql 19h30'49"7d60'
Goodman,Jillian Belle Lyr 18h40'19"29d18'
Goodman,John Cep 21h1'29"56d1'
Goodman,Justin Deen Lac 22h10'16"47d53'
Goodman,Katherine Ann And 0h5'36"35d15'
Goodman,Laurie Miho Leo 9h55'20"10d5'
Goodman,Marcie Lyn 7h6'40"60d52'
Goodman,Marie M And 0h1'26"38d2'
Goodman,Mark Christopher Ori 5h54'19"7d32'
Goodman,Mary Equ 20h57'16"10d8'
Goodman,Matthew Shen Oph 18h16'48"7d59'
Goodman,MD,Jay "Dr Dad" Oph 18h4'44"0d51'
Goodman,Michael Richard Aur 6h10'14"31d46'
Goodman,Ms Frances R Equ 20h57'30"11d6'
Goodman,Perry Constantin Cep 3h1'0"77d10'
Goodman,Simon Sge 20h2'38"16d48'
Goodman,Sylvia Zeida Mon 6h34'44"0d42'
Goodman,Yvette And 1h45'20"47d6'
Goodman-Davis 850824 Aur 5h55'0"40d41'
Goodman-Lionstar, Howard Michael Dra 19h26'12"56d49'
Goodmom & Gooddad 1947 Tri 2h9'15"31d50'
Goodness,Bonnie Uma 13h4'26"60d22'
Goodney,Christopher James Vul 19h5'37"24d52'
Goodnight,James H Dra 20h28'46"70d10'
Goodniss,Maria Uma 10h13'47"68d21'
Goodrich,Allen C Aqr 22h46'11"-5d26'
Goodrich,B G Mon 8h6'37"-3d3'
Goodrich,Holly Mon 7h56'51"-1d9'
Goodrich,Jackie And 2h29'41"47d39'
Goodrich,Janet T Gould Aqr 22h45'10"-6d39'
Goodrich,Merle H Gem 6h56'48"13d59'
Goodrich,Paul & Clara Crb 15h53'56"31d16'
Goodrich,Robert Lee Per 2h28'16"58d27'
Goodrich-Glenn's Glistener,Glenn S Cmi 7h15'0"1d38'
Goodridge,Mary Ellen Sco 17h25'53"-33d46'
Goodsaid,Ira Jesse And 0h15'33"37d31'
Goodsell,Andrew Equ 21h4'22"3d33'
Goodsell,La Dore And 0h50'1"33d43'
Goodson,Jacqueline & Alexandros Cyg 19h54'30"48d59'
Goodson,Tim,Catharine & Timothy Aql 19h10'19"13d38'
Goodspeed,Caitlyn M Sge 19h11'49"16d28'
Goodstein,Donald Phillip Ser 18h18'41"-13d47'
Goodstein,Geoffrey Cnv 13h45'40"39d33'
Goodstein,Michel Cnv 13h21'0"40d59'
Goodtimes West Boo 13h39'31"16d44'
Goodwell,Robert H Cet 3h15'37"2d2'
Goodwin Fifieth Anniversary Star Mon 6h35'51"8d41'
Goodwin III,Edward M Cep 0h46'1"77d59'
Goodwin,Adele Louise Aur 5h1'27"49d12'
Goodwin,Barry Alan Dra 19h3'15"70d40'
Goodwin,Ben Ori 5h0'34"0d0'
Goodwin,Betti Kessi Cyg 19h42'13"31d22'
Goodwin,Bev & John Cyg 20h51'11"38d38'
Goodwin,Brenda M Cru 12h7'13"-56d45'
Goodwin,Cameron Lac 22h3'57"48d47'
Goodwin,Clarence Lac 22h42'55"53d7'
Goodwin,Erin A Peg 22h21'36"21d46'
Goodwin,Gail Tri 1h51'44"27d44'
Goodwin,Heather Parker Vul 20h1'0"23d37'
Goodwin,Janice Umi 15h52'40"77d1'
Goodwin,Jennifer Com 12h24'45"22d22'
Goodwin,Jr,Charles R Per 1h55'33"56d49'
Goodwin,Lance Lac 22h10'42"48d4'
Goodwin,Laura Leo 10h57'26"-0d15'
Goodwin,Mark S Cru 12h8'51"-58d2'
Goodwin,Maryann Victoria Jarmuzik Cam 3h39'26"71d38'
Goodwin,Patrice Elizabeth Peg 22h2'15"30d32'
Goodwin,Paul Cnv 13h40'0"40d16'
Goodwin,Scott Aur 7h6'0"40d29'
Goodwin,Sean Lac 22h22'40"55d22'
Goodwin,Tyler E Sco 17h31'36"-30d44'

Goodwin,Wade Edward
Her 17h19'36"40d55'
Goodwin-'Sharon Star',
Sharon
Peg 23h17'1"30d20'
Goody's Vermont
Lyn 9h3'13"33d33'
Goodyear,Jane L
And 22h58'28"50d24'
Goodyear,Janice Ann
And 23h45'25"41d6'
Goodykoontz,Dorothy
Oph 18h5'34"13d34'
Goofy
Uma 10h15'16"56d28'
Goofy
Cnv 13h58'26"28d15'
Goold,Sr,Jack E
Her 17h21'16"14d52'
Gooley,Gerald Warren
Cep 2h48'1"86d47'
Goorian,Suzi
Lyr 18h50'56"30d46'
Gooser White
Cet 1h55'51"-4d53'
Gooska,Jonathan TRG
Sct 18h55'29"-6d51'
Gooslin,Robert H
Aql 19h40'12"13d6'
Gootee,Margaret Emma
Laughery
Cyg 21h28'25"37d45'
Gopalan,Champa
Boo 14h58'12"42d21'
Gopalan,Vijayan
Cam 6h0'38"60d30'
Gopodarek,Winfried "Winni"
Sco 17h32'15"-31d23'
Gorab,Lauren Jean
Umi 14h15'18"70d4'
Goral,Barbara M
Lyn 8h33'54"42d2'
Goralnik,Berton "Scotty"
Cet 2h57'26"0d60'
Goranson,Emmett & Thelma
Cyg 21h0'29"31d15'
Gorbachev,Soviet CCCP Amy
Kizmet
Mon 6h31'44"7d2'
Gordana
Lib 14h28'41"-23d49'
Gordanier's Dream
Her 17h34'39"21d58'
Gorde,Robert Harvey
Lyr 18h54'40"35d1'
Gorden,Celeste Pollock
Cas 2h33'1"63d46'
Gorder,Bryan James
Cep 22h3'1"70d1'
Gorder,Charles David
Cnv 13h11'16"41d12'
Gordikens
Uma 12h9'44"60d16'
Gordillo,Dionette
Peg 23h2'1"18d3'
Gordis,Tamar Michal
Sco 17h25'43"-41d6'
Gordo's Blunt Wheezing Yaz
Star
Cet 2h20'23"-8d19'
Gordo,José Maria
Dra 20h22'20"68d57'
Gordon
Ser 18h15'43"-14d20'
Gordon
Oph 17h31'7"-22d16'
Gordon
Uma 8h34'21"55d25'
Gordon & Andre Forever
Cyg 19h33'31"31d31'
Gordon & Christine
Aur 6h19'21"34d8'
Gordon Star,The
Peg 23h40'0"26d24'
Gordon Star,The Brooks
Gilman
Cep 21h58'21"55d43'
Gordon's & Jean's Star
Ori 5h37'46"1d24'
Gordon's Celestial Beacon
Cet 0h58'16"-12d55'
Gordon's Simplex Mundittis
Ori 6h6'18"5d57'
Gordon's Star,Dougie
Boo 14h19'48"18d38'
Gordon's Star,John H
Uma 8h13'57"72d14'
Gordon,Amy Beth & Seth
Aur 6h1'28"35d23'
Gordon,Barbara Brigitte
Peg 23h32'1"20d41'
Gordon,Bernard
Lac 22h18'12"51d38'
Gordon,Bettye Lou
Cas 0h25'18"50d14'
Gordon,Beverly & Victor
Gordon
Cam 8h18'12"74d58'AB
Gordon,Blaine
Uma 9h14'33"49d42'
Gordon,Bob
Per 3h7'0"47d7'
Gordon,Bradley David
Uma 10h13'0"72d3'
Gordon,Bradley Hale
Hya 8h12'38"0d10'
Gordon,Brent Dean
Oph 17h53'49"12d30'
Gordon,Brian Keith
Cet 2h8'26"0d27'
Gordon,Cheryl
And 2h12'1"39d46'
Gordon,Dana
And 0h11'48"37d37'
Gordon,Darren
Dra 20h1'0"63d23'
Gordon,David Michael
Oph 18h0'34"10d10'
Gordon,Deborah Ann
Cam 3h27'31"53d16'
Gordon,Demetra Mickey
Aql 19h14'53"15d4'
Gordon,Dennis Ray
Sgr 18h56'49"-29d23'
Gordon,Diane Michelle
Lyr 19h23'0"30d45'
Gordon,Ed
Cep 22h15'52"60d56'
Gordon,Gary Wayne Roberts
Aur 7h7'40"35d55'
Gordon,Graeme & Sue
Cyg 20h19'1"38d59'
Gordon,Hannah Marie
Sge 19h23'1"18d50'
Gordon,Heather Lynn
Cas 0h3'0"62d45'
Gordon,Heather Noël
Aql 18h44'16"11d12'
Gordon,Ida B
Cas 2h27'39"60d13'
Gordon,James
Cet 3h7'40"2d22'
Gordon,Jane & Arthur
Ori 5h57'1"8d19'
Gordon,Jeanne
Cam 3h51'18"78d21'
Gordon,Jennifer Lynn
And 2h21'21"42d52'
Gordon,Jessica
And 23h2'0"46d8'
Gordon,Jessica Lynne
Cyg 20h30'31"40d22'
Gordon,Jim
Lac 22h17'1"46d6'
Gordon,Jocelyn
Umi 15h43'44"70d15'
Gordon,John
Hya 8h12'11"5d41'
Gordon,Jr,Frederick W
Per 1h46'43"56d59'
Gordon,Julie Ann
Lib 14h24'58"-11d7'
Gordon,Karen
And 1h52'42"41d7'
Gordon,Kathy
And 0h20'58"34d32'
Gordon,Kelsey Chandler
Peg 23h42'41"30d26'
Gordon,Kenneth McDonald
Eri 3h42'51"0d14'
Gordon,Kenneth C
Lmi 10h31'42"33d9'
Gordon,Lee
Aur 6h56'0"37d44'
Gordon,Lillian
Lyr 18h33'1"40d53'
Gordon,Linda Leigh
And 0h39'54"40d54'
Gordon,Mark A
Aur 5h2'47"40d17'
Gordon,Marty
Cet 1h4'41"-0d48'
Gordon,Mary
Mon 7h42'58"-5d58'
Gordon,Michael A
Oph 17h6'22"8d53'
Gordon,Monica
And 23h1'23"51d45'
Gordon,My Possible Dream
Walter
Lac 22h19'18"51d56'
Gordon,Nadia
And 2h24'22"48d32'
Gordon,Nadine
And 23h23'51"51d45'
Gordon,Nathan David
Hya 8h41'40"2d38'
Gordon,Patrick
Aql 18h54'18"-0d12'
Gordon,Paul D
Cep 20h32'1"61d4'
Gordon,Paul Thomas
Aql 20h1'48"1d13'
Gordon,Peter Luckett
Hya 8h16'11"6d20'
Gordon,Pierce Alexander
Briault
Aql 18h57'32"17d21'
Gordon,Rob
Oph 18h38'22"10d17'
Gordon,Robert E
Mon 6h55'16"10d34'
Gordon,Ronald
Boo 14h45'47"27d28'
Gordon,Rosemary
Cas 23h43'23"58d12'
Gordon,Ruth
Cet 3h4'11"1d14'
Gordon,Samantha S
Umi 15h33'44"68d15'
Gordon,Samuel Jacob
Boo 14h56'12"52d18'
Gordon,Sandy
Eri 4h10'19"-11d46'
Gordon,Shelley
Cas 0h33'46"63d11'
Gordon,Shep E
Per 3h11'55"44d1'
Gordon,Shoshana Meira
Lyr 19h25'0"41d7'
Gordon,Stacy R
Ori 4h55'1"1d20'
Gordon,Sue
Cam 3h18'34"61d10'
Gordon,Sweet Mama
Cas 1h52'11"65d22'
Gordon,Terry
Aur 5h4'58"41d32'
Gordon,The Empress Of
Schmooz,Audrey
Cas 0h3'23"62d20'
Gordon-EDC,Myles
Per 4h2'1"52d1'
Gordon-Saker,Paul Declan
Aql 20h18'41"1d34'
Gordon...The Flash David
Matthew
Crt 10h59'22"-12d29'
Gordongebert-G4
Ori 5h54'20"16d19'
Gordy XL
Aql 18h44'34"8d50'
Gordy,Berry
Mon 7h19'38"-6d54'
Gordy,Debbie
Cyg 19h30'42"37d9'
Gore
Aql 19h5'21"-0d58'
Gore (The Ami Star), Naomi
Karen
Cas 0h3'17"61d42'
Gore,B Kenneth
Lac 22h18'37"46d9'
Gore,Christina
Cyg 21h55'29"52d32'
Gore,Clive Anthony
Psc 23h0'17"0d35'
Gore,Dana Lee
Lyn 8h4'14"34d3'
Gore,Fion Thomas
Ori 6h5'51"0d13'
Gore,Genevieve Walton
Cas 1h59'0"58d12'
Gore,US Vice President Albert
Umi 14h56'60"77d25'
Gore-Hickman,Erin Emily &
Christy
Uma 12h53'36"60d52'
Gorelick,Erin Michelle
Cas 0h1'0"64d37'
Gorelick,Maxwell Benson
Mon 8h7'50"-1d46'
Gorelick,Stuart L
Aur 5h58'0"31d26'
Gorelkin,Maryann
Cas 3h1'14"61d19'
Gorell-Bienzobas,Anna
Crb 16h4'12"33d4'
Gorenkoff,Sheppard
Sco 16h54'43"-44d22'
Goretski,David Michael
Lac 22h21'11"50d19'
Goretzki,Dr Günter Geb 23-
07-41
Leo 10h54'28"18d6'
Gorgas,David
Per 4h7'1"51d9'
Gorgeous Chris
Uma 9h46'51"48d3'
Gorgeous Christopher
Uma 10h21'20"54d21'
Gorgeous Eyes
Del 20h13'58"11d43'
Gorgeous Jen
And 0h50'50"33d32'
Gorgeous Joanna
Sge 20h2'1"20d2'
Gorgeous Lady
And 0h10'1"27d44'
Gorges,Anette
Aqr 21h59'41"-3d41'
Gorges,Nina
Lyn 8h17'34"43d5'
Gorges,Ria
Lyn 8h1'43"49d52'
Gorham,Jeanne
Lyn 8h20'59"46d11'
Gorham,John
Uma 8h51'18"56d15'
Gorham,Jr,Shawna J G &
John P
Per 2h25'13"56d31'
Gorham,Kenneth Durand
Uma 15h48'20"81d45'
Gorham-Poole,Jennifer
Sgr 18h51'3"-35d13'
Gori,The Francesca
Cas 1h5'14"55d48'
Goriup,Emmy C
Lyr 19h21'48"38d50'
Gorlewski,Roman J
Cnv 12h22'35"42d48'
Gorlier,Lucien
Cep 22h0'33"54d15'
Gorman III,John J
Cep 21h15'19"84d29'
Gorman Star,James G
Cet 1h3'1"-4d16'
Gorman,Anne Martha
Peg 21h35'1"20d9'
Gorman,Avery Lewis
Uma 9h54'0"60d36'
Gorman,Grant Adair
Uma 8h36'41"58d47'
Gorman,Jean M
Aql 20h10'34"8d52'
Gorman,Joseph Dominic
Dra 16h47'19"62d14'
Gorman,Karyn
Eri 2h55'14"-13d32'
Gorman,Linda D-I Love You
Mom!
Gem 7h21'53"33d47'
Gorman,Martha C
Lyn 7h25'0"58d28'
Gorman,Matthew Edmond
Per 4h0'20"51d44'
Gorman,Megan Catherine
Lyn 7h27'0"40d44'
Gorman,Pat
Crb 15h23'60"30d29'
Gorman,Raymond Vincent
Aur 6h32'25"38d52'
Gorman,Scott Thomas
Her 16h19'25"50d24'
Gorman,Sr,William Carl
Aql 19h0'49"1d15'
Gorman,Thomas S
Ori 5h3'49"15d43'
Gorman,Titi's Pride, James
Salvatore
Cnc 8h11'48"32d13'
Gormley,Craig
Sex 9h39'45"-6d41'
Gormley,John
Oph 16h56'39"0d59'
Gormlie,Miss Catherine Ann
Cas 0h28'22"61d23'
Gornall,Danielle P
Her 17h3'0"37d46'
Gorney,Beth
Cma 6h51'3"-17d4'
Gorniak,Rev Peter J
Cnv 12h20m31"41d31'
Gornisz,Lara-Sophia
Ari 2h32'24"30d31'
Gorno,Mary Lou
Cyg 19h56'11"38d23'
Gornostayeva,Vera
Cet 1h54'0"0d11'
Gornowski,Laura Beth
And 23h24'54"44d15'
Gorodetzer,Harry
Lmi 10h18'0"38d20'
Gorodetzer,Lois
Lmi 10h15'20"36d23'
Gorogias,Elizabeth Williamson
Peg 21h56'1"30d26'
Gorombly,Susan
Cyg 21h20'45"30d10'
Gorovoy "The Tzar" My
Darling,Charles
Per 2h3'0"50d10'
Gorra,Lucia
Uma 12h2'25"42d38'
Gorrell,Joanne
And 0h51'11"36d2'
Gorreri,Daniela
Pic 5h2'5"-45d53'
Gorrie,Cooper
Per 3h5'53"46d45'
Gorrie,Kendra
Cyg 20h36'55"42d26'
Gorrill,Charles H & Era Lea
Crb 15h53'30"28d33'
Gorrill,Douglas M
Equ 21h4'1"10d38'
Gorringe,Jeffery Alan
Aql 18h59'15"14d28'
Gorringe,Joe Alexander
Cep 21h20'1"70d26'
Gorringe,Kaye Marie
Com 12h45'1"30d54'
Gorshel,Arthur D"Papa"
Her 17h23'0"20d0'
Gorski,Brenda
Cas 2h39'0"68d22'
Gorski,Robert Joseph
Her 16h27'38"32d4'
Gorsuch,Leonard F
Lac 22h1'37"38d21'
Gorsuch,Mary M
Lyr 19h3'57"40d50"
Gorsuch,Melanie
And 2h6'0"40d8'
Gorsuch,Todd Michael
Cet 2h34'0"1d44'
Gorton,Richard Kenneth
Aql 20h11'10"10d16'
Gorum,Frederick
Lac 22h8'18"46d33'
Gos
Eri 2h50'31"-11d32'
Gosch,Samuel Adrian
Lac 22h29'49"50d24'
Goscicki,Allison
Vir 11h55'16"-2d32'
Goscicki,Steven
Ori 5h57'14"17d31'
Goscinski,Olga
Cam 12h43'48"80d2'
Goselin My Star Love,
Raef,Patricia
Lyn 7h47'47"38d35'
Gosh Star,The
Umi 15h7'1"81d38'
Goshorn,Chance Edward
Dra 17h16'45"68d3'
Goshorn,Elizabeth Diane
Tri 2h14'1"33d33'
Goshorn,Rebecca
And 0h22'57"43d39'
Goska
Dra 16h15'46"61d5'
Goslar,Jan
Ori 5h54'50"7d55'
Gosling,Ambrose David
Ori 5h58'39"15d8'
Gosling,Susan & Anthony
Cyg 21h31'27"42d10'
Gosmire,Jennifer
Tri 2h12'42"32d23'
Gosnell,Aaron Dean
Lac 22h21'1"50d34'
Gosnell,Molly
Sco 17h29'58"-30d36'
Gosnell,Rosemary
Boo 15h2'34"50d39'
Gosney III,John W
Lyr 18h31'34"30d58'
Gosney,Heather Leigh
Ori 6h6'39"1d46'
Goss,David Charles
Peg 22h1'48"10d20'
Goss,Hunter Lynn
And 1h37'38"37d9'
Goss,Jana Gravitt
Mon 6h38'34"10d55'
Goss,Jaycee
Lyr 18h13'46"38d47'
Goss,Julie Mae
And 2h6'54"40d32'
Goss,Sarah Emily
Lyr 18h17'52"37d38'
Gossage,Debra Ann
Cra 18h20'17"-45d9'
Gosselin,Andre
Cam 6h45'47"80d6'
Gosselin,Eric Patrick
Ser 17h30'1"-10d45'
Gosselin,Hervé Roland
Dra 12h3'0"70d8'
Gosselin,Jo Anne
Tri 1h44'49"31d52'
Gosselin,Rosalie
Per 2h56'21"45d59'
Gossert,Judith C
And 2h19'1"40d35'
Gosset,Larry Thomas
Dra 17h0'0"73d52'
Gossett,Denean M
Aur 4h49'29"50d32'
Gossett,Jackson Spencer
Oph 17h38'28"-20d25'
Gossett,Jr,Robert A
Aql 18h57'12"-6d11'
Gossett,Linda Ann Glenn
Uma 9h43'35"48d43'
Gossett,Roy D
Her 18h50'1"14d28'A
Gossette,Rex Brandon
Cep 21h55'0"60d54'
Gossin,Pamela
Vul 19h41'13"26d31'
Gossman,Joan A
Com 12h19'52"31d39'
Gossman,Jody M
Cyg 20h59'42"38d7'
Gossner,James F
Cmi 7h7'10"5d2'
Gosson,Timothy John
Boo 14h19'35"17d17'
Gostling,George DeVello
Lac 22h13'53"49d29'
Goswami,Raja A
Dra 18h57'51"48d58'
Goswick,Dana Lynn
Cas 2h25'55"60d7'
Gotcher,Charles & Mildred
Ori 5h37'0"11d41'
Goteiner,Beth Eliya
Vul 19h5'41"25d18'
Gotgart,The
Cyg 20h56'33"40d59'
Gotham,Alexis Hunter
Peg 22h38'1"27d39'
Gothe,Jurgen
Ari 2h43'29"28d43'
Gotimer,George Christopher
John
Boo 13h41'1"14d18'
Goto,Kathie
Crt 11h9'50"-18d38'
Gotsis,Alexander C
Her 18h5'24"40d34'
Gott,Mary Alice
Cas 0h42'31"64d19'
Gottesman,Grandma Esther
Sugar
And 1h17'16"40d29'
Gottesman,Ilise & Juleen
Cas 1h51'46"63d58'
Gottesman,Sol Evelyn
Cas 1h54'31"68d27'
Gotthalf,Rebecca
Cnv 12h50'42"45d46'
Gottlieb Family,The
Uma 10h56'54"40d31'
Gottlieb,Debbie
Peg 21h19'38"21d46'
Gottlieb,Dr Norman
Oph 16h48'55"10d6'
Gottlieb,Letty Canalonga
Sex 10h8'47"0d42'
Gottlieb,Molly Rachel
Cyg 19h59'40"38d6'
Gottlob,Gerson B & Florence
R
Cyg 19h58'22"30d8'
Gottron,Kelly Christine
Peg 23h25'40"33d6'
Gottron,Matthew Robert
Vul 19h43'0"20d34'
Gotts,Monica "Mouse"
Aql 19h14'31"15d8'
Gottschalk,Amanda Rose
Uma 12h18'1"53d18'
Gottschalk,Prince Albert
Boo 15h45'58"41d46'
Gottschall,Carl
Peg 21h19'32"22d43'
Gottscheber,David
Peg 23h41'1"13d59'
Gottscheber,Patrick
Peg 22h16'42"35d28'
Gottshall,Kevin Lee
Uma 9h0'53"61d4'
Gottuso,Calicia J
Cet 0h28'46"0d46'
Gottuso,Carisa S
Mon 7h52'24"-5d7'
Goubert,Philippe
Ori 6h20'49"10d29'
Goudie,Monique Olene
Cas 2h49'54"61d2'
Goudreau,Kenneth Allan
Uma 10h59'53"34d10'
Goudreau,Sylvain
Her 16h51'30"33d15'
Goudreault,Arthur R
Lyr 18h42'34"45d13'
Goudreault,Bertrand
Umi 15h32'0"67d58'
Goudsward,Jeffrey
Uma 11h52'1"41d1'
Goudy,Shannon Marie
Gem 7h53'26"32d36'
Goue,Michel
Per 4h2'57"50d42'
Gouge,Catherine Courtney
Vul 20h26'19"28d56'
Gouge,Charles & Barbara
Peg 22h38'37"25d26'
Gouger
Cam 5h44'52"60d45'
Gough,Annie
Lyn 8h2'53"52d21'
Gough,Dr James Stockman
Aqr 23h37'57"-5d17'
Gough,Eric
Cep 1h12'47"77d57'
Gough,Hunter Curtis
Ori 5h56'41"11d19'
Gough,Mark Joseph
Boo 13h45'23"14d21'
Gough,Michael Jeremy
Umi 17h26'18"86d11'
Gougon,Cristina & Juan Luis
Cep 23h7'27"60d56'
Gouin,Margo
Per 2h56'11"46d16'
Gouins
Cnv 12h47'58"38d14'
Goukler,Rick
Lac 21h57'13"42d29'
Goukler,Ryan Matthew
Dra 16h52'58"63d24'
Goula,Dawn Marie
Cnc 9h0'53"30d3'
Goulakos,Peter
Ori 5h51'53"21d17'
Goulart-King,Dianna T
Aql 20h3'0"-6d17'
Gould
Cyg 21h13'26"38d34'
Gould's Guiding Light
Eri 4h1'56"-17d23'
Gould's Star of
Courage,Caroline
Lyr 18h52'18"30d49'
Gould,Bill
Cep 3h28'22"78d43'
Gould,Colleen K
Uma 9h42'40"43d58'
Gould,Darrick H
Lyn 9h3'21"41d44'
Gould,David L
Cet 0h30'47"-5d15'
Gould,David Martin
Ori 5h27'40"-0d24'
Gould,Deborah Romaine
Cas 0h13'41"61d41'
Gould,Don
Per 1h51'21"52d47'
Gould,Ethan Jacob
Per 2h51'36"45d10'
Gould,Herbert E
Peg 21h47'21"34d59'
Gould,Karen Moffatt
And 23h42'0"41d41'
Gould,Marilyn & Maxwell
Crb 16h19'25"32d7'
Gould,Michael
Aql 19h59'40"0d35'
Gould,Michael F
Cep 23h1'24"70d2'
Gould,Nicky
Peg 21h58'28"31d27'
Gould,Randall Dale
Sge 19h30'22"16d44'
Gould,Ronald A
Per 2h59'60"50d9'
Gould,Staci Olivia
Mon 6h28'49"7d39'
Gould,Theresa
Lyr 18h48'37"40d33'
Gould-McElhone,Leona
And 2h23'12"40d32'
Goulding,Karen Elizabeth
Boo 15h4'42"23d21'
Gouldthorpe,Frances
Lyn 7h29'25"40d56'
Goulet,Alexander Patrick
Hya 8h11'36"-1d8'
Goulet,James
Tau 5h49'17"24d19'
Goulet,Jo-Anne
Per 3h19'28"40d54'
Goulet,Kevin Joseph
Oph 17h12'14"-23d16'
Goulet,Lori Ann
Cas 0h11'0"63d9'
Goulet,Marie Apolline Attala
Cyg 20h55'55"38d58'
Goulet-Doré,Gisèle
Cep 22h31'35"69d39'A
Gouline,Jeffrey
Vir 14h3'59"7d23'
Gouloff-Musgrave
Aur 6h55'36"44d7'
Gouls
Cam 3h56'40'61d17'
Goulston,Mark S
Cep 21h22'49"56d9'
Goulston,Maxwell Delano
Aur 6h24'1"32d24'
Goulston,Paul Richard
Per 3h15'21"41d31'
Gounon,Jacques
Sgr 19h28'18"-33d31'
Goupil,Marie Thérèse Louise
Cas 1h19'43"60d12'
Goupil,Martine Morency
Réjean
Cyg 19h24'58"32d30'
Gour,Laurent
Del 20h49'42"9d51'
Gour-Creek
Lyn 7h6'0"58d59'
Gourbil,Yoeline
Boo 14h8'15"41d16'
Gourd,Carol Jean Ayres
Com 13h0'0"27d13'
Gourd,Jack A
Cet 3h1'1"0d12'
Gourde,T J
Aql 19h5'49"2d5'
Gourdier,J & C
Dra 17h12'24"61d21'
Gourdon,Jean-Marc
Aur 4h34'42"31d22'
Gourdouze,Stephan
Aur 6h34'49"33d60'

Gourgeon,Hughes Benoit Adrien
Cnv 13h6'17"40d55'
Gourier,Philippe
Aur 4h35'39"31d19'
Gourkanti,Sarojani
Cyg 20h24'23"39d58'
Gourlay,Paul
Ori 5h56'52"19d18'
Gournay,Paula "Buddie"
Lyr 19h23'41"38d3'
Gourneau,Zhin-Zhaw Claude William
Cet 2h14'58"7d40'
Gournet,Patrick I
Peg 21h52'35"20d34'
Goursky,Michael Francis
Psc 22h55'36"0d28'
Goushaw,Barbara
Lyr 19h15'0"40d57'
Goushaw,Barbara
Lyr 19h12'56"40d37'
Gout,Jayne
Peg 21h52'26"30d40'
Gout,John Trevor
Lib 14h24'43"-8d46'
Gout,Nicholas William
Cnc 8h50'24"32d12'
Gouverneur,Karl G
Her 16h7'59"41d52'
Gouy,Samantha L
Mon 6h20'54"2d41'
Govan,Clyde D
Aql 19h6'50"2d60'
Gove,Curtis
Lmi 9h59'15"30d15'
Gove,Sacha
Aql 19h40'40"10d32'
Govea,Blanca Idilia
Mon 6h36'27"7d42'
Govelovich,George D
Tri 2h44'33"33d59'
Governali,Ketty
Cas 0h41'32"77d7'
Gow
Uma 11h22'39"43d50'
Gowan's Carousel
Cyg 21h11'0"38d21'
Goward,Graham
Lmi 10h46'37"33d45'
Gowdy & Family,John & Viola
Uma 11h18'51"37d43'
Gowdy,Cindy
Uma 11h23'25"47d55'
Gowen,Erica L
Lac 22h22'17"54d28'
Gowen,Kathy A
Vul 19h57'34"28d28'
Gowen,Mrs Joan
Cyg 19h42'10"30d42'
Gowens,Robert
Ser 15h52'37"24d36'
Gower,George Lawrence
Cam 5h47'51"68d5'
Gowers,Jennifer
Aur 4h56'32"40d26'
Gowin,Heather
Uma 12h10'46"53d54'
Goy,Alain
Cnc 8h39'14"17d12'
Goyer,Matthew Joseph
Lmi 10h12'1"32d15'
Goyert,Dorothy Mather
Lyn 7h8'1"51d30'
Goyette,Ellen Glover
Cas 0h36'50"64d21'
Goyette,Helen Mackintosh
Peg 0h7'46"28d8'
Goynes,Janis
Crb 16h9'53"31d31'
Gozon,Richard
Dra 19h0'0"60d45'
Gozonsky,Eileen
Ari 1h55'12"15d40'
Gozs,John
Aql 20h12'0"10d59'
GR 27/1
Per 3h35'37"51d58'
Graae,Niels
Cas 23h39'44"65d21'
Grab A Root & Growl
Uma 8h54'13"58d38'
Grabber,Andrea Nicole
Cyg 19h59'48"41d11'
Grabeel-Stone, Katherine
Aql 19h43'40"11d24'
Grabek,Mariusz Robert
Lac 22h19'39"54d26'
Grabell-Soul Mate,Moe
Uma 8h36'42"53d2'
Graber,Karin
Aql 18h56'17"17d35'
Grabert Günter
Sge 19h6'28"18d2'
Grabher,Karen Ann
Cep 22h16'59"67d39'
Grabinski,Delores Anna
Cam 4h57'19"68d49'
Grabka,Katja Franziska
Vul 20h42'21"20d31'
Grabka,Nina Fiona
Peg 21h55'41"34d52'
Grable,Grace
Uma 8h32'47"59d48'
Grabman,Jim
Dra 16h55'57"68d12'
Grabow,Star of
Crb 15h49'38"38d57'
Grabowicz,Stanley
Boo 14h51'60"23d41'
Grabowska,Jaime
Umi 14h8'25"66d52'
Grabowska,Jola
Lyr 18h33'1"33d2'
Grabowski "Star Romandis",Roman
Cam 11h48'55"82d20'
Grabowski,Kelly L
Peg 23h3'0"21d51'
Grabowski,Kristof
Vir 11h39'25"1d25'
Grabowski,Paul
Aql 19h50'21"15d58'
Graca & Ruslan
Lmi 10h38'54"27d50'
Grace
And 0h8'21"30d38'
Grace
Nor 16h23'40"-57d58'
Grace
Cas 0h26'19"72d55'
Grace
Mon 6h43'44"9d17'
Grace & Joe's P T C
Mon 6h59'12"-10d48'
Grace Elizabeth
Cas 0h34'54"58d37'
Grace Elizabeth's Star
Lyn 8h7'0"45d35'
Grace Ellen
Cyg 19h59'13"58d21'
Grace Family Star,The
Ori 5h10'45"-4d17'
Grace Lynnette
Peg 22h18'33"35d12'
Grace Mari Nicole
Cyg 19h57'16"31d41'
Grace Mond
Mon 6h28'47"-6d46'
Grace,Ann & Dick
Aql 20h3'34"3d55'
Grace,Caitlin
Mon 6h31'23"-1d34'
Grace,Constance L
Ori 5h56'11"10d60'
Grace,Genevieve Claudia
Cas 0h37'51"71d60'
Grace,Jennifer Lynn
Ori 5h45'34"-6d43'
Grace,Joel
Aql 20h34'33"0d8'
Grace,Joseph Parker
Boo 15h7'0"15d6'
Grace,Jr,Richard Lee
Dra 10h8'38"80d5'
Grace,Lydia
Cam 4h57'49"60d48'
Grace,Margaret Isobel (Mig)
Cas 0h33'15"62d7'
Grace,Mariel
Lyn 7h38'0"38d9'
Grace,Nicole Danielle
Peg 22h40'35"21d21'
Grace,Precious Angel Of The Ritz
Peg 21h56'42"29d49'
Grace,Richard A
Cnv 14h0'22"32d27'
Grace,Vincent
Boo 15h22'28"48d11'
Grace-Warrick,Christa
Per 2h38'51"40d23'
Graceanne
Hya 8h25'56"-5d40'
Gracecelia
Umi 16h21'26"70d16'
Graceful Airwalk on Green Day
Peg 22h23'36"4d44'
Graceful Kyoko
Vir 14h4'1"5d13'
Graceful Little Honey Bee-Rofrano
Lyr 19h18'13"35d17'
Grachanin II, Charles
Dra 12h45'35"72d3'
Gracia,Jorge Estadella
Dra 17h17'0"60d4'
Gracie
Peg 23h3'16"31d2'
Gracie,Larry Wayne
Psc 23h5'1"1d10'
Graciela
Lmi 9h46'60"33d33'
Gracielita Star Of My Life
Lyr 18h34'49"27d36'
Gracies' Bit-Of-Heaven
And 1h15'24"36d44'
Gracietta Maria
Cnv 12h40'28"36d36'
Gracin,Lenko Daniel
Ori 5h40'59"10d8'
Gracious Gail
Boo 15h11'25"49d57'B
Grad,Andreas
Oph 17h7'23"1d22'
Gradante,Jessica Lee
Aql 19h5'40"-0d2'
Gradante,Tony
Tau 4h41'17"16d38'
Graddy Luv
Ori 5h55'24"-2d45'
Grade,Sandy Denson
Lyr 18h33'17"40d43'
Gradin,Bruce A
Dra 16h56'26"62d49'
Gradner,Cindy
Ori 6h7'21"8d37'
Gradster,The
Aur 5h18'21"48d9'
Graduate,The
Cam 4h5'46"67d51'
Grady 41358143,Les
Uma 8h33'45"73d9
Grady 9591,Quinton
Uma 8h26'0"70d55'
Grady I
Aur 4h52'18"40d0'
Grady,Beverly C
Cas 1h7'1"65d35'
Grady,Catherine Louise
Peg 23h29'39"33d18'
Grady,Ferdinand M
Cnv 13h48'1"40d16'
Grady,Grover
Boo 14h57'18"27d9'
Grady,Jayne Marathe
Peg 22h27'55"18d52'
Grady,John & Patricia
Crb 15h33'18"31d49'
Grady,Jr,The "Dougie", C Douglas
Dra 16h43'28"61d14'
Grady,Kevin Lee
Boo 14h1'34"16d41'
Grady,Kevin Michael
Umi 20h2'57"88d46'
Grady,Pamela
Cas 1h1'49"61d10'
Grady,Sarah Michelle
Peg 21h51'1"33d40'
Grady,Sean Stafford
Lac 22h10'46"47d4'
Grady,William R
Aql 19h24'37"10d7'
Grady-Hunter Star,The
Cet 3h15'13"7d52'
Graebling P
Lac 22h46'19"38d11'
Graebner,Jasper Cornelius
Cep 22h13'36"70d42'
Graef,Thea
Sgr 19h3'35"-22d9'
Graefingholt,Renate
Peg 0h0'20"28d39'
Graeler,Samuel Walter
Her 18h13'57"30d13'
Graeme
Per 1h47'46"47d31'
Graeme
Ori 5h26'57"-1d12'
Graeme's "Eternal Treasure"
Uma 11h23'40"40d51'
Graeter,M Dorothy
Uma 11h51'0"46d46'
Graeves,Kyle Edward
Boo 14h2'24"23d34'
Graf,Ben
Tau 4h47'45"16d15'
Graf,Brigitte
Tau 5h30'55"28d20'
Graf,Jr,Kenneth L
Her 18h8'50"30d55'
Graf,Luise (Lusele)
Sco 17h27'24"-38d59'
Graf,Pat & King
Cyg 20h0'13"40d10'
Grafenberg,Ann E
Lyn 8h49'59"46d4'
Graff's Star 8-4-82, Andrew
Cam 7h43'16"60d16'
Graff,Bryan Scott
Tri 2h35'26"35d28'
Graffeo
Cam 8h10'51"73d39'
Graffeo (My Black Star),Karen
Leo 10h45'30"15d43'
Graffeo,Joseph
Cep 20h58'55"58d38'
Graffius,Jan
Lib 15h35'10"-28d36'
Graffstedt,Kurt Johann
And 23h15'13"42d6'
Grafton,Michelle E
Cas 1h7'59"75d49'
Grage,Kurt
Hya 9h13'34"-6d50'
Graglia,Cian
Peg 23h30'19"18d28'
Graglia,Kristina
Lyr 18h54'38"31d24'
Graglia,Quintella
Mon 6h18'14"0d6'
Graham
Umi 14h50'21"70d26'
Graham
Ori 5h20'28"15d9'
Graham Michel
Uma 8h18'37"60d48'
Graham Star,The
Lyr 18h51'57"41d27'
Graham Star,The
Cas 0h0'0"58d35'
Graham Star,The
Cam 11h7'18"81d60'
Graham's Celestial Hideway
Mon 7h55'58"-2d49'
Graham's Star
Lyn 8h24'0"47d29'
Graham,50th Anniversary,Ev & Paul
Eri 4h32'45"-10d28'
Graham,Aileen
Peg 21h21'26"23d13'
Graham,Alan J
Ser 15h53'28"1d15'
Graham,Alison M
Crb 16h6'39"31d58'
Graham,Alison Rita
Lyr 19h15'57"26d29'
Graham,Alycia
Peg 22h46'38"32d38'
Graham,Avril & David
Uma 12h42'1"58d29'
Graham,Bill
Dra 11h41'18"68d41'
Graham,Bonnybel M
Per 3h12'1"37d5'B
Graham,Cheryl A
Mon 6h19'59"-6d32'
Graham,Chester
Cam 6h33'24"66d12'B
Graham,Colin
Aur 4h59'24"41d13'
Graham,Craig Ross
Sct 18h53'53"-10d13'
Graham,Derek
Aql 19h40'15"14d45'
Graham,Dillon Gerald Keller
Cet 2h34'59"4d28'
Graham,Donna Marie
Lyn 8h17'0"37d20'
Graham,Donnie E
Cma 7h20'0"-15d26'
Graham,Esther M
Ori 6h17'30"0d27'
Graham,George A
Per 3h12'1"37d5'A
Graham,Gwen E
Cas 0h18'16"64d2'
Graham,Harrison Lloyd
Aql 20h19'48"1d32'
Graham,Hayley Stuart
And 23h42'48"47d37'
Graham,Iraina
Boo 14h46'53"34d31'
Graham,J Mark
Her 17h22'40"41d6'
Graham,J T
Her 17h0'37"18d47'
Graham,Jack
Hya 8h59'45"-7d12'
Graham,Jack
Uma 9h14'22"57d47'
Graham,Jason Z R
Peg 23h1'13"30d9'
Graham,Jr,Paul R
Com 12h28'28"28d45'
Graham,Judith Ruth
Eri 2h58'13"-17d39'
Graham,Karen
Del 20h25'24"11d7'
Graham,Kate Helen
Uma 10h26'39"68d47'
Graham,Katie
Aql 19h3'10"-6d35'
Graham,Kenneth
Sex 9h57'13"-5d24'
Graham,Kenneth
Cnv 12h49'26"32d30'
Graham,Kim
And 0h1'13"41d4'
Graham,Kimberly Blue
Mon 6h23'50"7d13'
Graham,Leigh Taylor
And 0h51'25"38d56'
Graham,Lori Lynn
Aql 19h2'21"15d19'
Graham,Lory B
Eri 3h57'1"-4d54'
Graham,Madeleine
Cet 0h34'41"0d33'
Graham,Maggie
Lmi 9h35'24"37d31'
Graham,Mallory
Peg 22h28'28"7d38'
Graham,Margaret
Cam 6h33'24"66d12'A
Graham,Marilyn
Lyr 19h1'32"26d1'
Graham,Marvin LeRoy
Lyr 19h13'11"38d9'
Graham,Meagan N
And 0h15'45"33d21'
Graham,Patricia K
Equ 21h20'25"11d14'
Graham,Rebecca Faye
Aqr 21h0'25"0d12'
Graham,Richard
Ser 15h51'28"20d39'
Graham,Robert Gordon
Dra 20h3'48"62d24'
Graham,Robert Kennedy
Lac 22h34'47"40d16'
Graham,Robert S
Her 15h55'32"43d17'
Graham,Ruby
Cet 0h27'48"-11d35'
Graham,Sanford Alton
Cam 06h39'0"82d53'
Graham,Sarah
And 2h23'17"49d59'
Graham,Scott Dwyte
Aql 18h50'16"11d3'
Graham,Seymour C
Dra 17h30'39"70d20'
Graham,Sr,David Bolden
Her 17h58'59"40d40'
Graham,Steve
Ori 6h4'33"0d56'
Graham,Tracy
And 0h1'1"40d34'
Graham,Trevor Douglas
Aqr 22h1'26"-5d26'
Graham,Walter & Hilary
Cyg 19h20'37"44d27'
Graham,William David
Per 2h39'37"40d6'
Graham-Gibbs,Damian
Aur 6h1'38"37d54'
Graham-Princess Of Pisces,Patricia
Psc 23h27'18"6d50'
Grahame & Nicky
Cas 0h13'24"59d13'
Grahi,Gary Caterpiller Satya
Lac 22h4'25"38d8'
Grahl,James-Michael
Aur 5h18'0"54d27'
Grahl,Melissa Brook
Aql 19h56'38"13d24'
Grahn,Julia Elizabeth
Cas 0h50'20"64d30'
Graig,Marty
Umi 14h51'53"80d49'
Grain Of Sand,The
Aur 6h6'32"40d33'
Grainger,Beckl Lin
Boo 14h45'13"20d23'
Grainger,Brian
Ori 5h23'1"15d51'
Grajcar,John & Rita
Cep 21h38'34"55d46'
Gralla,Shirley & Milton
Lyr 18h41'31"38d2'
Gram Star,The
Eri 3h34'37"-12d41'
Gram's Places
Lac 22h40'41"54d42'
Gram,Karina
Umi 9h44'3"47d41'
Gramajo,Noemi Carmen
Uma 11h46'48"38d49'
Gramando,Rhody
Lac 22h44'43"54d16'
Grambeck,Holger
Cep 21h36'44"60d44'
Grambsch,Alvin
Dra 14h29'22"60d3'
Grambsch,Anne Elizabeth
Peg 22h24'1"20d17'
Grambusch,Jason
Aql 19h57'49"0d23'
Gramins,Todd
Dra 19h28'13"67d39'
Gramlich,Robert
Tri 1h57'49"27d19'
Gramlick,Jason M
Aur 6h59'43"38d16'
Gramling (ALS) Dreamweaver,Leon
Oph 17h1'29"8d53'
Gramling,Joel Michael
Cyg 19h56'35"40d23'
Gramling,Joel Mickael
Her 16h48'33"40d31'
Gramma Suzie
Mon 6h26'0"8d34'
Grammens,Kevin Casey
Her 15h50'38"44d10'
Grammer II,James Michael
Hya 8h55'24"-0d53'
Grammer,Allen Kelsey
Psc 23h4'21"5d21'
Grammer,Bryan Thomas
Cet 1h23'24"-4d13'
Grammie
Peg 23h6'10"10d47'
Grammie 85
Umi 15h18'36"71d7'
Grammy Jewel
Vul 20h15'36"25d54'
Grammy Star
Eri 3h13'59"-15d33'
Grammy's Star
Cas 1h19'26"62d22'
Grammy's Svnoy Ajil
And 0h2'59"44d18'
Gramp's Star
Aql 19h42'55"10d22'
Grampa & Sweetheart
Aur 4h55'24"51d19'
Grampa Sahs
Boo 14h52'37"50d36'
Gramps
Lyr 18h44'11"36d13'
Gramps Bundle
Boo 14h41'57"28d28'
Grampy
Cyg 21h46'44"30d49'
Grampy
Aql 19h40'20"8d15'B
Grampy
Oph 17h2'59"10d46'
Grampy
Aur 6h54'26"43d19'
Grampy Twinkle
Per 2h3'0"48d55'
Grams
Cyg 20h36'54"40d19'
Grams,Michael H
Lac 21h57'42"36d59'
Gramstad,Ann
Cas 0h54'58"72d32'
Gramza-Uma & Umpa,Earl & Lois
Cep 22h8'11"61d44'
Gran
Uma 13h29'0"61d55'
Gran Canaria
Ori 5h30'30"1d47'
Gran,Peter Rolf
Aur 6h7'25"33d37'
Granada,Paul
Ser 15h43'35"15d40'
Granadillo,Mireya
Peg 22h18'52"20d27'
Granados,Carmen Tello
Cyg 19h37'58"38d12'
Granados,Matthew David
Per 3h2'59"46d44'
Granados,Nicholas Adaman
Hya 9h3'55"5d8'
Granakas,Ruth
Cam 8h23'46"78d52'
Granat,Robyn Lynn
Cyg 20h37'19"47d30'B
Granatelli,Margaret Lynn
Mon 7h39'10"-1d14'
Granath,Alan
Aql 20h10'1"3d48'
Granath,Paula
Per 4h5'45"36d41'
Granatir,Charles Elliot
Per 4h5'1"50d31'
Grancagnolo,Ginger
Cam 5h4'1"68d5'
Grand Adventure
Tri 2h11'1"31d21'
Grand Ashbrook Star, The
Ori 5h55'14"0d59'
Grand Clanton 66,The
Mon 7h1'40"4d30'
Grand Master Sin II Choi
Her 18h10'55"30d6'
Grand Mere Camille
Cnv 12h10'34"40d58'
Grand Passion
Lyr 19h18'29"40d5'
Grand Shanker
Lac 22h15'24"46d51'
Grand Wazoo
Cet 1h48'1"-1d49'
Grand' Mere Amour
Cep 22h52'0"65d14'
Grand,Everett Alexander Lowrance
Ari 2h31'1"22d18'
Grand,Lucie
Oph 17h57'17"11d45'
Grand,Michael
Cep 22h22'52"56d10'
Grandad Bill
Per 2h9'47"57d9'
Grandad Cliff
Cep 4h19'38"80d36'
Grandad's Star
Vul 20h4'48"23d42'
Grandad's Star
Her 16h41'36"26d1'
Grandad's Superstar
Psc 22h55'1"2d24'
Grandchamp,Marie
Cas 23h40'40"63d30'
Grandchamp,Paula Lucille
Peg 22h8'23"3d27'
Granddaughter Tracy Lynn
Gem 6h56'43"18d10'
Grande,Daniel Gregory
Per 4h0'0"51d27'
Grande,Deana & Joe
Cyg 19h50'59"40d0'
Grande,Dr Stephen A
Her 16h26'55"48d35'
Grande,Jacqueline Antonio
Com 12h53'56"24d3'
Grande,Kenny R
Aur 6h33'0"37d53'
Grande,Lothar
Lyn 8h6'52"48d4'
Grande,Maria Rodriguez
Tri 1h59'16"31d13'
Grandel-Galbrun,Annie
Tri 1h59'39"27d48'
Grandest Mütti of Them All,The
Lyr 18h38'20"41d19'
Grandi,Vera
Boo 15h0'41"41d46'
Grandidier,Bruno
Cam 5h4'56"65d36'
Grandimougin,Jean-Luc
Lmi 10h41'30"24d22'
Grandinetti Anniversary Star
Lyn 8h23'0"40d54'
Grandiosus Immobilius R Saubert
Aql 18h43'0"10d44'
Grandjean,Sandro
Cas 0h32'32"64d29'
Grandma & Grandpa K
Cma 6h43'52"-16d4'
Grandma & Grandpa O
Aql 18h57'21"-6d52'
Grandma & Pop,Happy 50th! Love,Sam
Eri 4h34'47"-18d5'
Grandma Andy's Forever In The Sky
Cap 21h55'15"-10d12'
Grandma Barbara Love Brandon Salinas
And 2h23'33"48d31'
Grandma Betty
Com 12h16'40"28d18'
Grandma Charlotte
Uma 11h10'25"43d54'
Grandma Claire
Cas 1h25'59"61d46'
Grandma Dee-Light (G B O)
Cyg 21h5'0"34d27'
Grandma Dorothy
Mon 6h37'23"11d3'
Grandma Dorothy/Moma
Cyg 19h6'59"48d43'
Grandma Elizabeth
Cas 3h22'38"75d13'
Grandma Ev & Papa Bill
Cyg 19h54'39"38d50'
Grandma Eileen
And 23h21'1"45d43'
Grandma Goose
Mon 6h19'25"8d44'
Grandma Ida
Eri 3h10'53"-6d42'
Grandma Ilse - Grandpa Walter
Cyg 19h52'14"40d56'
Grandma Jackie's Grandkids
Aql 19h48'33"13d12'
Grandma Jean
Cas 1h44'0"61d11'
Grandma Jeannie
Uma 11h19'20"47d18'
Grandma Jo
Cas 3h23'41"75d11'
Grandma Johnnie
And 23h2'56"37d40'
Grandma Margaret - Papa Ralph
Lyn 7h12'1"50d9'
Grandma Mary's Star
And 2h0'41"40d12'
Grandma Midge
Lyr 18h46'25"40d16'
Grandma Nancy
Lyn 8h1'48"45d20'
Grandma Norma
Cas 0h57'37"76d9'
Grandma Rollo
And 23h41'60"44d51'
Grandma Rose Jarvis
Cas 23h39'41"64d29'
Grandma Rosemary
Cep 20h27'35"63d5'
Grandma Ruth
Vul 19h48'38"28d28'
Grandma Star
Aql 19h41'1"14d19'
Grandma Teddy
Cam 12h29'0"77d15'

Grandma Tooties Roll
Eri 4h26'60"-11d58'
Grandma Vi
Cas 0h35'0"63d18'
Grandma Wilhelmina's Star
Cas 2h57'22"70d9'
Grandma's Angel Baby Christopher
Peg 21h54'60"28d42'
Grandma's Glow
Cas 0h29'46"65d4'
Grandma's Star
And 23h21'0"43d44'
Grandma's Star
And 2h27'53"40d48'
Grandma's Twinkle Star
Sgr 19h34'38"-42d31'
Grandmacus The Bestus
Cas 1h1'59"61d6'
Grandmother Rit
Cas 0h25'55"65d10'
Grandmother Star,The
Oph 17h18'22"-23d26'
Grandner,Keith J
Lyn 7h52'24"45d11'
Grandom,Verlelaina
Aql 19h15'30"12d6'
Grandominico,Jodi
Lyn 8h1'1"45d9'
Grandpa Bob's Shining Star
Cnv 13h41'42"30d28'
Grandpa Charlie
Psc 1h18'58"22d15'
Grandpa Compton
Per 2h59'30"31d41'
Grandpa Dick's Star
Cep 23h3'1"60d27'
Grandpa Dougald's Star
Cep 22h16'18"55d52'
Grandpa Fred
And 1h50'0"40d42'
Grandpa Harry
Lyr 18h42'25"33d52'
Grandpa Joe
Vir 11h42'28"9d52'
Grandpa Joe & Grandma Theresa
Lyr 18h46'26"31d5'
Grandpa Lee & Mariah
Peg 22h4'21"10d13'
Grandpa Mac
Eri 3h25'7"-13d15'
Grandpa Max
Cet 2h9'42"3d5'
Grandpa Ned
Cyg 19h55'42"38d27'
Grandpa Norman
Pho 23h49'50"48d2'
Grandpa Pete
Aur 6h24'0"33d14'
Grandpa Ralph
Cet 2h43'18"8d57'
Grandpa Ray
Oph 18h2'35"11d38'
Grandpa Rudy
Per 3h33'35"36d5'
Grandpa Star
Ori 5h14'0"12d39'
Grandpa Ted
Ori 6h5'46"0d22'
Grandpa Tino
Dra 10h29'1"80d3'
Grandpa's Harbor Light
Lac 22h20'0"51d18'
Grandpa's Star
Pho 0h32'40"-49d12'
Grandpa's Star
Tri 2h15'20"31d37'
Grandpuppy Suz
Per 3h14'42"40d37'
Grandsard,Jeffrey A
Her 16h42'43"38d34'
Grandstaff,Coleman
Aql 19h36'30"1d10'
Grandstaff,Kelly
Peg 23h27'50"10d48'
Grandt,Vicki
Lmi 10h17'31"31d38'
Grandville
Boo 13h57'26"21d56'
Grandy's Star
Oph 17h35'9"-23d25'
Grandy,Catherin Sterrett
Aqr 21h3'25"-10d26'
Grandy,Neil A
Lyr 18h16'55"38d50'
Grandy,Serry Berry
Cam 3h30'1"58d45'
Granes,Andrev Fabregas
Ser 15h14'30"6d4'
Granes,Anna Fadregas
Ser 15h14'18"8d2'
Granet,Melissa Jacquelyn
Her 17h51'55"28d45'
Graneto,Elizabeth Ann
Peg 23h21'0"11d59'
Graney,Patti Ann
Lib 15h40'49"-24d5'
Granger,Claire
Dra 16h15'1"62d25'
Granger,Claire
And 23h3'41"50d26'
Granger,Gloria
Cas 0h33'54"74d58'
Granger,Heather Victoria
Com 12h28'28"28d45'
Granger,John Phillip
Her 16h25'30"11d1'
Granger,LaNell
Ori 4h48'12"4d15'
Granger,Marissa Blair
Cyg 19h32'14"33d50'
Granger,MD,Carl V
Aur 5h9'54"41d11'
Granger,Tressa
Lyr 19h22'18"37d43'
Granger,William S
Lac 22h17'0"49d45'
Grangie,Hugues
Boo 15h25'25"37d38'
Grangier,Catherine
And 0h27'0"40d59'
Graniel,Rafael Rodriguez
Hya 9h37'41"-18d50'
Granieri,Joseph R
Boo 15h4'1"20d33'
Granito,In Memory of Jane C
Lyn 9h13'39"36d8'
Granitto,Marissa Cecilia
Peg 23h26'33"10d58'
Granitz,Sheridan
Cyg 20h0'16"40d44'
Granitz,Sheridan
Ori 5h56'31"17d13'
Grankowski,Reverend Zbigniew
Cep 1h20'13"80d2'
Grann,Edward William
Uma 11h14'12"53d5'
Granna
Eri 3h43'56"-1d41'
Grannar,Mark Levengood och Hans
Cnc 8h49'54"30d14'
Grannemann,Janet Louise
Aql 19h57'53"10d9'
Grannie
Ori 5h53'15"0d48'
Grannie & Poppie
Ori 5h57'49"14d38'
Grannie Annie
Cas 0h56'12"60d31'
Granny
Peg 23h27'11"31d31'
Granny
And 2h26'1"39d28'
Granny & Popeye
Eri 4h32'0"-11d49'
Granny Moo
Mon 7h56'0"-3d41'
Granola
Vul 19h48'31"23d11'
Grant
Sct 18h55'7"-7d36'
Grant & Suzan's Anniversary Star
Crb 16h2'52"32d12'
Grant 3-15-43
Boo 14h11'36"30d16'
Grant Man
Ari 2h4'1"18d58'
Grant's Star
Ori 4h56'1"15d20'
Grant,Allison Ryan
Tri 1h33'0"30d11'
Grant,Amanda
Uma 14h21'13"57d51'
Grant,Andrew
Per 2h40'28"37d12'
Grant,Angel Of Our Hearts,Kay
Crt 11h14'19"-11d58'
Grant,Betsy
Cyg 21h24'43"38d30'
Grant,Brandon David
Her 16h39'49"25d50'
Grant,Connie McNamara
Sge 18h55'47"19d35'
Grant,Daniel James
Dra 16h7'24"68d55'
Grant,Darren Mark
Ori 5h55'24"8d6'
Grant,Debbie Webb
Equ 21h6'56"11d45'
Grant,Emily & Gene
Boo 14h42'56"35d18'
Grant,Eugene F
Uma 11h9'0"48d19'
Grant,Frederick W
Cep 20h47'32"60d31'
Grant,Gordon
Uma 9h16'25"44d7'
Grant,James P
Aql 19h1'47"10d15'
Grant,James William
Uma 9h28'29"48d6'
Grant,Jean Frances Miles
Boo 14h29'19"18d8'
Grant,Jessica Leigh
Cas 1h2'53"53d4'
Grant,Josh
Lac 22h28'43"50d23'
Grant,Kathleen Marie
Ori 5h18'60"12d41'
Grant,Lois Marsh
Cas 0h2'1"60d17'
Grant,Lori
Lyr 19h26'29"40d26'
Grant,Margaret
Cam 6h7'26"60d7'
Grant,Marie
Cam 6h11'0"67d59'
Grant,Marquerite Erin
Cam 3h49'0"58d53'
Grant,Martin J
Dra 18h46'34"70d60'
Grant,Mary Alexandria Leigh
And 1h28'57"35d36'
Grant,MD,Kingsley Beaconsfield Jacob
Per 3h25'39"38d34'
Grant,Melissa
Cas 0h27'37"67d49'
Grant,Melissa Carol
Uma 9h49'30"55d19'
Grant,Meredith Ashley
Cyg 21h6'54"30d30'
Grant,Michael Barrett
Sex 10h36'1"5d50'
Grant,Michele M
Cam 8h14'56"83d49'
Grant,Paula Dimeo
Cas 1h3'44"55d17'
Grant,Rebecca
Mon 6h19'21"0d18'
Grant,Robert & Patrick
Dra 16h35'17"51d24'
Grant,Sally Ann
Cas 0h10'0"64d42'
Grant,Sr,James
Cep 20h44'39"63d49'
Grant,Staci & Testa Carmine
Mon 6h54'56"8d30'
Grant,Stephanie
Eri 3h20'17"-13d32'
Grant,Tracey Louise
Lyr 19h9'11"47d39'
Grant,William Peyton
Tri 2h5'27"32d12'
Grant-Johnson,Kim A
And 2h28'12"50d22'
Grant-Smith,Phoebe
Peg 23h42'17"11d14'
Grantchester House Star,The
Eri 4h43'23"-0d3'
Grantham III,Daniel Lyndel
Ori 5h54'12"10d38'
Grantham's Star,Daniel
Umi 14h31'0"68d4'
Grantham's Star:Donum de Eius Henderson,Jake
Ori 5h35'43"7d47'
Grantham's Star:Donum de Eius Deo-Patre,Spik
Ori 5h33'31"7d57'
Grants,Lynette
Per 2h56'30"34d51'
Granty
Uma 13h34'1"51d51'
Granville,Larry
Boo 15h16'0"33d5'
Granzer,Fritz
Tau 3h55'30"7d42'
Granzin,Mark
Aql 19h28'54"14d58'
Grapeintine,Kirsta
Ori 5h12'52"8d23'A
Graphmans,Mark & Wendy The
Uma 9h6'53"56d8'
Gras,Dominique
Aur 7h8'60"38d48'
Gras,Remi
Uma 9h36'33"49d25'
Grasing,Neal Vincent
Dra 15h1'26"62d39'
Grasmane,Dace
Cas 0h58'0"61d34'
Grasmick,Alta Caroline Mecham
And 0h59'49"45d17'
Grass,John J
Aur 6h31'25"30d55'
Grassel,K C
Vul 19h19'57"26d56'
Grasser,Achim
Cmi 7h21'34"4d24'
Grasser,Dr Hans
Boo 15h0'1"10d48'
Grasser-A S
And 2h24'39"45d32'
Grasserbauer,Anton
Psc 23h23'0"5d37'
Grasshopper
Cam 3h41'31"61d44'
Grasshopper's Happinesss
Vul 19h58'0"23d16'
Grassi,Elwood Joseph
Aur 6h2'0"32d52'
Grassi,Kelley A
Lyr 19h17'35"40d19'
Grassia,Kathleen L
Mon 7h3'1"4d17'
Grassie,Lachlan Steuart Henderson
Cam 4h48'14"68d40'
Grassl,Steven
Uma 11h7'27"42d2'
Grassl,Wolfgang
Lyr 19h3'50"28d48'
Grassman,Joshua Meyer
Cet 2h23'1"-8d36'
Grasso,Jacob Michael
Leo 10h33'29"14d17'
Grasso,Jason
Her 18h18'20"12d23'
Grasso,Mason Samuel
Umi 14h55'33"67d48'
Grasso,Ron & Ruth
Aur 5h53'17"30d33'
Grasso,Tony,Jamie, Allison,TJ
Dra 16h9'41"67d29'
Grate,Kelly Lynn
Ori 6h16'52"12d18'B
Grath,Edith
Tau 5h47'39"27d52'
Gratias,Franáois- Xavier Alan
Uma 8h8'1"71d5'
Gratias,Jenny René
Cyg 20h56'21"30d29'
Gratius,Günter
Sco 16h38'31"-44d47'
Gratonilue
And 23h1'1"52d47'
Grattan,Meg & Rod
Sge 19h56'17"16d28'
Grattan,Megan Elizabeth
And 1h59'27"41d0'
Grau,Ingeburg
Ser 15h14'31"2d44'
Grau,Kenny
Her 18h16'34"20d18'
Grau,Word Star,Julie
Uma 9h18'1"42d17'
Graue,Herbert
Tau 3h56'0"8d41'
Grauel,Kristopher Scott
Boo 14h49'55"29d3'
Grauer,Gaby
Gem 6h37'55"34d29'
Grauer,PhD-DD,Rev Joanne
Ari 3h22'32"28d53'
Grauer,Wayne C & Barbara A
Cnv 13h23'0"41d3'
Graunke,Grace Dorothy
And 0h11'10"37d8'
Grauso,George Peter
Aur 5h6'59"40d55'
Gravagna,Silvia-Eliana
Vel 9h43'46"-49d23'
Gravalis III,James
Cep 23h25'40"70d22'
Grave,Tina
Vul 20h57'1"28d21'
Gravel,Jr,William (Billy) Henry
Tri 2h44'51"31d53'
Gravel,Lisa Dendre
Cyg 19h26'51"28d51'
Gravel,Pamela
And 23h6'20"40d3'
Gravelle,Chad Michael
Cam 7h51'31"60d1'
Gravenhorst,Anne-Marie
Vir 12h59'6"-9d41'
Graver,T&S
Her 16h19'32"4d17'
Graves "Yurggie",Yuri & Maggie
Peg 23h24'0"15d32'
Graves Celestial Diamond
Ori 5h46'37"10d24'
Graves II,George Faustine
Cep 23h3'1"64d1'
Graves,Amy Elizabeth Marcella
Mon 6h22'36"0d11'
Graves,Douglas Edward
Umi 16h15'0"72d11'
Graves,Eddie
Sex 9h54'33"-6d51'
Graves,Edna
Cyg 21h37'37"41d35'
Graves,Emileigh Ann
Peg 23h16'27"33d27'
Graves,Erin Elizabeth
Cyg 20h8'55"40d46'
Graves,Frances Louise
Uma 9h17'1"53d25'
Graves,Frank Graham
Dra 17h43'19"63d49'
Graves,Gladys Olive
Lyn 9h3'49"38d52'
Graves,Hunter Breanna
Lyn 7h4'43"59d5'
Graves,Jon Gregory
Cep 23h25'29"64d23'
Graves,Jr,Joseph Lee
Per 2h27'27"58d20'
Graves,Julie
Peg 22h11'33"3d56'
Graves,Kristen
Aql 20h5'12"1d0'
Graves,Margaret
Uma 11h9'18"61d3'
Graves,Mary"Owen"
Mon 7h41'3"-2d45'
Graves,Rex & Judy
Cyg 21h30'37"41d33'
Graves,Robert
Uma 11h4'27"50d46'
Graves,Rodney Lorenzo
Ser 15h34'22"18d45'
Graves,Stephanie Anne
Cas 0h9'13"63d35'
Graves,Vera
Cam 3h26'45"60d46'
Graves-Dyana's Carino, Vaughan Channing
Boo 14h33'56"41d30'
Graves-FHA Guiding Star,Margaret
Aql 18h52'49'-1d18'
Gravestock,Heather
Cas 0h9'26"50d14'
Gravestock,Mary Ann
Lyn 6h29'44"59d58'
Gravestock,Matthew
Cep 21h5'24"55d36'
Gravett,Brendan Allen
Aqr 22h21'1"-5d7'
Gravier,Mike & Ann
Cyg 21h13'29"35d31'
Gravilenko,Francois
Uma 9h26'31"53d54'
Gravitt,Zeatelford
Lyn 8h40'15"33d28'
Gravlee,Casey
Oph 17h54'32"12d16'
Gray Beard
Aur 5h2'28"40d23'
Gray Family,Elmer
Aur 6h9'1"38d50'
Gray Family,Gary
Peg 23h20'20"11d8'
Gray Fossil,The
Uma 9h6'42"47d56'
Gray Libras,The
Lib 15h19'1"-21d14'
Gray Star,The Harry and Helen
Peg 21h28'58"21d50'
Gray's Gram Of Gold 25
Dra 16h47'0"62d56'
Gray's Own Shining Star,C G
Mon 7h0'28"5d18'
Gray's Own Twinkling Star,Vanessa
Ori 5h54'1"16d18'
Gray's Star,Douglas
Dra 16h35'35"73d17'
Gray,Adele "Mother Goose"
Cas 1h30'19"65d23'
Gray,Alma
Cas 23h40'10"61d24'
Gray,Alma Marcum
And 0h20'16"36d49'
Gray,Alvin Jerome
Lac 22h13'21"50d1'
Gray,Angela Lynn
Cyg 20h19'0"38d34'
Gray,Avril Claire
Lyr 19h16'48"41d2'
Gray,Ben
Her 17h18'42"41d42'
Gray,Benjamin Kenilworth
Cma 7h17'11"-15d18'
Gray,Beverly
Cmi 7h23'41"1d3'
Gray,Brooke Suzanne
Del 20h24'0"20d1'
Gray,Carolyn
Peg 23h42'18"31d37'
Gray,Cassandra Henry
Mon 7h48'15"-3d34'
Gray,Christian Robert
Peg 22h57'30"28d46'
Gray,Christina
And 0h19'1"31d28'
Gray,Christopher Michael
Her 16h59'19"33d36'
Gray,Constance
Cas 0h57'22"50d33'
Gray,Danalee & Robert
Cyg 19h27'50"31d13'
Gray,David & Dianna
Aql 20h12'40"4d33'
Gray,David Garry
Aqr 23h43'56"-6d52'
Gray,David Waterhouse
Aql 18h58'54"13d26'
Gray,Debra Ann
Mon 7h49'36"-7d50'
Gray,Dee J
Crb 16h4'39"30d33'
Gray,Dickie
Cet 3h9'59"3d0'
Gray,Donna
Mon 7h0'29"5d28'
Gray,Doris Jean
Tri 2H14'17"33d32'
Gray,Douglas B
Aur 5h57'24"31d19'
Gray,Dr Fred Allen
Lyr 19h4'44"47d43'
Gray,Ella Mae
Peg 22h6'1"5d10'
Gray,Frances M
Gem 7h17'1"30d25'
Gray,Gary E
Cet 1h10'12"-6d48'
Gray,George Howard
Sco 15h56'11"-28d12'
Gray,Ginger
Cet 1h35'0"0d40'
Gray,Gladys
Cas 0h16'34"65d56'
Gray,Gladys
Peg 22h31'10"26d6'
Gray,Great Aunt Alma
Cas 1h2'1"53d60'
Gray,Harold
Cep 22h15'0"56d11'
Gray,Harrison
Lmi 9h32'53"37d48'
Gray,Harry Jack
Per 2h45'23"43d48'
Gray,Hayley Maris
Lmi 10h0'19"34d54'
Gray,Heather, Scott & Al's - Linda
Cyg 20h59'27"31d45'
Gray,Helen
Cas 0h50'19"60d33'
Gray,Hester
Uma 12h27'11"61d17'
Gray,Howard
Cep 22h59'34"65d9'
Gray,I'll Always Be With You,Tom
Cnv 13h32'58'42d26'
Gray,Ian Evans
Ori 5h44'53"10d52'
Gray,James Louis
Ser 17h20'25"-13d48'
Gray,Jane Lee
Uma 9h19'0"57d37'
Gray,Jason Wilder
Lac 22h13'35"46d35'
Gray,Jeffrey C
Aur 7h24'34"37d59'
Gray,Jennifer K
Lyn 8h8'23"37d22'
Gray,Jenny
Cas 0h18'11"62d1'
Gray,Jeremy Williams
Dra 19h28'40"65d31'
Gray,Jill
Lyr 18h38'28"26d11'
Gray,John Sullivan
Lac 22h11'47"46d5'
Gray,Jr,Earl Washington
Cep 22h17'14"60d25'
Gray,Jr,Randle Edward
Oph 18h3'0"10d5'
Gray,Jr,Thomas Alan
Cep 21h45'50"68d15'
Gray,Julie
Eri 2h53'37"-5d3'
Gray,Julie Lorene
Lac 22h3'39"46d36'
Gray,Karen Johnson
Peg 21h29'43"20d20'
Gray,Kasey Helen
And 1h45'40"47d4'
Gray,Kenneth Sr and Yolande
Tau 5h22'29"16d50'
Gray,Libby
Sgr 19h32'23"-40d36'
Gray,Lori
Cas 0h26'26"68d55'
Gray,Marcia
Lyr 19h0'34"41d4'
Gray,Marina Bardosy
Cam 4h54'37"67d37'
Gray,Maxwell Martin
Aql 19h56'40"10d26'
Gray,Michael
Lac 21h59'19"36d34'
Gray,Molly Rowena
Lyn 7h41'36"48d17'
Gray,Nancy
Peg 0h1'1"21d6'
Gray,Peter Bristow
Tau 4h8'32"23d6'
Gray,Philip
Cyg 19h19'44"45d20'
Gray,Richard William Bernard
Lac 22h27'12"50d23'
Gray,Riley Alan
Aur 5h3'1"40d53'
Gray,Rosie
Peg 23h34'26"31d27'
Gray,Roslyn
Cas 0h15'1"61d3'
Gray,Ruth Muldoon
Cas 1h30'60"41d55'
Gray,Sally
Cyg 19h35'18"38d5'
Gray,Scott Charles
Psc 22h57'57"0d42'
Gray,Sheila
Cas 2h12'13"67d38'
Gray,Stephen S
Ari 2h30'23"30d56'
Gray,Sterling
Dra 17h21'1"72d27'
Gray,Suzanne
Cyg 21h25'45"40d4'
Gray,Taylor Eston
Cyg 19h32'25"37d40'
Gray,Thomas & Sharon
Peg 21h19'19"22d34'
Gray,Trent Patrick
Oph 18h2'59"10d22'
Gray,Ward
Cep 0h16'43"87d29'
Gray-McCarthy
Oph 16h58'15"-23d36'
Gray-Reyes,Lou
Sge 19h55'49"16d17'
Graybill,David Lee
Ori 5h33'17"-6d25'
Grayce Anne
And 23h3'1"40d60'
Grayley,Cellar Kisses- Simon
Cyg 19h27'20"34d35'
Grayman & Flattop
Tri 1h57'13"34d26'
Grays,Jr,Dr Eddie Lee
Oph 17h15'45"11d51'
Grayslake High School- Class Of 1996
Lyr 18h54'31"31d39'
Grayson Love You Forever,Randal D
Aur 5h0'29"40d26'
Grayson,Charles "Piano Man"
Boo 15h11'52"27d32'
Grayson,George G
Cam 11h58'18"81d44'
Grayson,Linda Laurie
And 1h25'14"35d4'
Grayson,Lori
Cas 0h37'26"60d40'
Grayson,Maralyn
Umi 14h28'47"68d50'
Grayson,Natasha Justina & Eric
Cyg 20h28'1"41d2'
Grayson,Natasha & Eric
Cyg 20h37'40"40d57'
Grayson,Ronald
Ori 4h53'12"5d22'
Grayson,Walter Patrick
Cet 2h25'40"1d59'
Grayson-My Best Friend ,Michael
Mon 7h2'1"4d10'
Grazaitis,Peter
Aur 6h1'58"38d20'
Grazia Anna 1995
Pic 5h0'30"-49d40'
Grazia Bernice Sweetie
Vul 19h41'36"26d41'
Grazia,Crispino
And 23h26'58"40d32'
Grazia,Ettore e Maria
Ant 10h37'36"-30d5'
Grazia,Maria
Pho 0h2'35"-45d27'
Graziana/Roberta
Del 20h18'49"13d3'
Graziani,Gregory P
Dra 16h14'53"66d43'
Graziano
Ind 21h2'44"-52d5'
Graziano,Angelique
Cyg 21h34'32"41d43'
Graziano,Anthony Raphael
Dra 16h5'36"64d28'
Graziano,Marisa
Mon 6h52'50"-5d48'
Graziano,Ronald
Lac 22h5'51"51d11'
Graziano,Steven
Lmi 10h32'42"32d14'
Graziella
Com 12h9'56"21d49'
Graziella,Barbaglia
Cyg 21h53'12"42d56'
Graziella,Mamma
Cnv 13h28'15"51d4'
Grazier,Barry Joel
Ori 5h54'0"7d31'
Grazier,Berry J
Lac 22h34'38"40d56'
Grazioplene-Hahn,Julie Ann
And 2h9'19"39d33'
Grazzini,Alessandro
Cmi 8h8'58"0d26'

Grbesa,Dipl Ing Boris
Cra 18h13'19"-39d48'
Grcich,Eloise H
Aur 6h25'42"38d30'
Grealish,Denis
Oph 17h39'60"-24d3'
Greaney,Donald N
Per 4h28'26"52d10'
Greaney,William Anthony
Dra 17h42'59"68d14'
Great "48",The
Aur 6h44'0"38d4'
Great Adi,The
Ori 4h57'42"1d10'
Great Baldwin,The
Hya 8h28'1"5d44'
Great Balls of Fire
Peg 22h0'18"8d2'
Great Dane D,The
Her 16h30'0"32d1'
Great Expectations
Per 1h53'35"56d21'
Great Gonzo,The
Her 16h31'17"20d24'
Great Gram Nina Marie
And 2h21'22"37d39'
Great Grandmother
Madeline,The
Dra 11h42'39"70d40'
Great Guest Magician
Extraordinaire,The
Leo 11h1'23"23d24'
Great Lakes Shining Stars
Boo 13h55'44"15d5'
Great Pop Star,The
Boo 14h7'30"22d54'
Great Pop-Pop Star,The
Lyn 8h52'40"41d19'
Great Roland,The
Aur 6h21'39"32d31'
Great Stella of Ryota
Lib 14h47'41"-22d8'
Great Ted
Her 17h12'1"28d20'
Great Tomar,The
Lyr 19h24'50"38d20'
Great Uncle Dave
Aql 18h53'59"10d11'
Greater Lessers,The
Dra 16h1'28"67d18'
Greathead,Jesse Morgan
Uma 10h21'25"48d18'
Greathead,Susan
Peg 22h21'13"4d34'
Greathouse,Jeanne P
Ori 4h57'58"4d14'
Greaver,Russell & Irene
Cyg 21h5'46"32d44'
Greaves,Shannan
Lyn 9h14'54"34d45'
Greb,Tricia Ann
Peg 23h16'49"33d55'
Grebe,Uwe
Ori 5h57'22"19d46'
Grebetz,Kevin Louis
Aur 5h3'23"46d50'
Grebin,MD,Burton
Oph 17h56'59"7d44'
Greby,Kenneth John
Hya 8h13'47"1d6'
Grecco,Alyssa Tia
Aqr 21h24'0"-0d22'
Grecco,James Michael
Dra 17h49'57"68d59'
Grech(The Maltese
Princess)Stacey R
Lyn 9h14'41"33d58'
Greco Star,The Anthony E
Aur 7h24'22"35d50'
Greco's Popstar, Charles V
Aur 6h36'43"40d46'
Greco,Anita
Del 20h29'1"18d52'
Greco,Ann & Bob
Aql 19h30'13"10d40'

Greco,Bobby Krystina
Dra 19h3'15"48d3'
Greco,Brianna Nicole
Cam 12h38'40"77d35'
Greco,Caitlin
Cam 12h15'48"76d41'
Greco,Catherine
Uma 9h21'51"51d1'
Greco,Chiara Rina
Cam 5h42'29"60d59'
Greco,Christopher Michael
Cnv 13h31'19"42d26'
Greco,Dennis
Her 15h51'18"40d3'
Greco,Elizabeth Altman
Mon 6h39'28"10d21'
Greco,Gerald R
Aql 20h4'18"3d50'
Greco,Harry
Aql 20h10'41"1d25'
Greco,Jennifer
Vul 19h2'50"24d32'
Greco,Jimmy
Boo 14h52'20"37d22'
Greco,Jr,Stephanie & Michael
Vul 19h23'39"25d33'
Greco,Kathy Del
Ori 5h0'22"1d50'
Greco,Linda & Phil
Mon 7h44'1"-1d44'
Greco,Nicole
Aql 18h41'17"-0d13'
Greco,Salvatore Santo
Hya 8h26'14"-17d12'
Gredal,Ruth
Cas 0h11'0"63d49'
Gredassoff,Denise A
Peg 21h40'0"21d43'
Gredler,Alys
Cas 1h48'28"74d29'
Gredler,MD,G P
Aur 6h17'1"46d31'
Greear,Joy K
Aql 18h52'25"11d56'
Greeden
Peg 22h11'45"4d27'
Greeff,Bernard & Harriet
Mon 6h24'0"-8d37'
Greeley,Donald
Ser 15h15'56"22d22'
Greeley,Kenneth Marshall
Lac 21h58'21"41d60'
Greeley,Mildred Ann
Aql 20h33'21"-6d10'
Green 21,Micky
Her 18h26'19"20d30'
Green Eyes
Cyg 21h30'50"53d6'
Green Eyes Forever
And 23h26'39"40d54'
Green Family,The
Uma 10h46'51"57d12'
Green Frog 1
Eri 5h6'1"-4d59'
Green Gables
Aqr 22h19'12"-9d52'
Green My Star Mom, Arlene F
Lyr 19h23'22"30d42'
Green nee Smulowitz, Henry
"Heshy"
Aql 20h0'23"0d9'
Green Star,Susie
Lyn 9h15'19"41d56'
Green Star,The Tanzania
Nor 16h28'7"-43d35'
Green,"The GreenLight" Dave
& Cindy
Uma 9h47'51"55d40'
Green,Aaron & Jana
Crb 16h8'22"26d17'
Green,Aaron J
Tri 1h59'39"26d37'
Green,Adolph
Dra 14h55'0"55d4'

Green,Alex "Bally"
Her 17h13'0"48d45'
Green,Alyssa Ashly
Cet 0h46'18"-4d31'
Green,Arnie
Eri 2h59'18"-4d21'
Green,Betty Patricia
Cet 2h59'11"0d38"
Green,Bibby
Peg 0h5'39"18d37'
Green,Big Al
Aur 7h18'28"36d17'
Green,Bill F
Her 17h34'23"26d37'
Green,Bonnie Doris & Richard
Uma 9h19'18"60d15'
Green,Bos
Boo 14h39'57"41d3'
Green,Boyd & Ruthanne
Eri 3h1'44"-2d52'
Green,Brian Victor
Her 17h36'33"40d47'
Green,Bryant & Ollie
Eri 2h55'0"-2d34'
Green,Carol & Troy
Lyn 7h31'25"48d58'
Green,Caroline
And 23h41'0"41d59'
Green,Carolyn Dorothy
Lyn 8h13'57"39d56'
Green,Catherine
Lyr 18h55'37"30d37'
Green,Catherine E
Lac 22h3'28"49d51'
Green,Charles Darren
Boo 14h1'37"16d1'
Green,Charlie F
Her 16h25'57"38d6'
Green,Corey & Michele
Cyg 20h14'56"38d19'
Green,Craig Justin
Dra 11h43'37"70d37'
Green,Cynthia E
Ori 5h57'59"17d53'
Green,Danielle
Gem 6h53'57"16d21'
Green,Darwin
Aur 6h2'21"31d13'
Green,David
Uma 11h41'40"38d28'
Green,David A J
Cnv 12h26'0"46d16'
Green,Debbie
And 23h31'39"36d8'
Green,Debbie Faye
And 23h44'15"40d56'
Green,Donald
Tau 4h39'55"28d33'
Green,Doren Wade
Lmi 10h55'13"33d46'
Green,Eda
Cas 1h30'54"61d20'
Green,Eleanor J
Cam 3h41'11"71d60'
Green,Elina
And 0h0'0"38d0'
Green,Ellaquen
And 0h0'60"34d25'
Green,Elliot Ryan
Cet 0h36'13"1d43'
Green,Emily Julianna
Leo 10h49'58"2d19'
Green,Eric
Dra 14h58'1"65d28'
Green,Ernest Alexander
Her 16h53'38"27d55'
Green,Genevieve
Boo 15h18'1"50d18'
Green,George Mac Ewan
Cmi 7h35'21"1d9'
Green,Ginger's Infinite
Mon 8h4'57"-7d15'
Green,Gordon Joseph
Sct 18h44'22"-6d18'

Green,Gregg
Her 16h20'45"20d27'
Green,Henry I
Uma 8h50'45"61d57'
Green,Hunter James Franklin
Leo 10h22'36"13d43'
Green,James F
Dra 18h9'13"58d56'
Green,James Roby
Hya 9h1'25"5d56'
Green,Jane
And 0h24'18"38d42'
Green,Janet H
Cam 6h38'13"67d41'
Green,Jeff
Ori 5h58'15"16d27'
Green,Jenee
Uma 9h2'21"48d17'
Green,Jo Ann
And 0h50'15"38d43'
Green,Joanne Nola
Cas 1h51'53"61d43'
Green,John
Aql 19h57'47"-0d3'
Green,John
Her 17h1'0"46d28'
Green,Joni Grant
Gem 6h47'10"30d8'
Green,Julia McCarthy
And 23h35'48"44d31'
Green,Kaitlyn
Cas 1h32'13"70d36'
Green,Kathleen
Lyn 6h55'40"54d22'
Green,Kathleen Ann
Sgr 19h24'48"-43d59'
Green,Kathryn Ryan
Cap 20h45'46"-26d15'
Green,Kelly Jade
Cyg 20h21'55"31d36'
Green,Kevin & Kate
Aql 19h6'12"15d35'
Green,Kevin James
Uma 8h51'49"51d32'
Green,Kristen Marie
Cas 1h3'29"65d23'
Green,Kristy Lee
Lyr 18h31'29"33d17'
Green,Laura Marie
And 2h5'44"42d53'
Green,Laurie
Hya 8h37'47"-6d45'
Green,Leland J
Dra 16h14'46"62d37'
Green,Lewis U
Her 16h8'44"16d48'
Green,Lisa Carole
Cyg 21h31'35"40d6'
Green,Lorraine & Bob
Crb 16h4'14"32d22'
Green,Lukas
Boo 14h58'31"41d55'
Green,Margaret Mary
Cas 0h57'25"50d32'
Green,Marin
Eri 2h45'40"-15d22'
Green,Mark Thomas
Sex 10h40'1"0d52'
Green,Mary
Cyg 20h23'12"38d22'
Green,Mary Jane
Lyn 7h8'1"44d55'
Green,Michael Lewis
Lyn 6h44'2"55d45'B
Green,Michael"The Bear"
Cyg 19h28'11"50d10'
Green,Millie
Her 16h59'51"25d16'
Green,Naomi Marcus
Lmi 10h37'57"24d39'
Green,Natasha Rose
Sgr 19h41'54"-42d13'
Green,Nickole S
Cas 23h17'56"61d53'

Green,Nicole Martel
Mon 7h24'23"-10d26'
Green,Pamela
Oph 17h53'52"13d1'
Green,Pamela Andrew Michael
Umi 16h10'22"80d35'
Green,Patricia Jane
And 23h1'0"47d54'
Green,Patrick
Psc 23h27'1"6d45'
Green,Patty Jo
Aql 19h21'17"12d13'
Green,Peyton Richard
Per 2h56'52"40d57'
Green,Philip Scott
Ori 5h51'35"16d2'
Green,Rachel
Lyr 18h35'11"40d60'
Green,Raleigh Burton
Cmi 7h39'40"5d7'
Green,Rhiannon Cecily
Aql 19h27'13"10d49'
Green,Robert M
Cet 2h4'27"4d41'
Green,Roger
Gem 7h27'16"31d27'
Green,Ronald W
Cet 3h19'41"8d10'
Green,Sam
Umi 15h56'32"70d3'
Green,Samantha Leigh
Aql 18h57'36"3d51'
Green,Sarah Eliza
Umi 16h0'20"74d5'
Green,Savanah Danielle
And 2h13'21"41d31'
Green,Shelia
Cas 0h51'59"69d14'
Green,Sherry
And 1h14'33"38d36'
Green,Sheryl
And 23h34'34"48d53'
Green,Sinikka
Lyn 7h26'57"40d50'
Green,Sr,Super Star, The
Thomas J
Dra 10h54'13"73d15'
Green,Stacey Ann
Uma 8h41'18"53d43'
Green,Stephanie Denise
Cas 0h22'16"64d43'
Green,Sterling
Crt 11h22'33"-18d51'
Green,Steven
Boo 15h2'54"17d25'
Green,Steven J
Hya 8h54'29"0d5'
Green,T R
Ori 5h57'39"18d59'
Green,Tammy L
And 23h37'50"48d14'
Green,Tammy Margaret
Llewellyn
And 0h11'30"37d45'
Green,Taylor Mae Rhiannon
Boo 15h22'39"34d25'
Green,Taylor Nicole
Peg 23h24'46"25d15'
Green,Terrell L
Lac 22h37'48"38d35'
Green,Thomas Harrison
Lac 22h15'58"54d47'
Green,William M
Cam 3h55'14"74d44'
Green,William TP
Dra 18h40'38"68d33'
Green-Dunlap Star,The
Cyg 21h15'48"35d39'
Green-eyed Lovely Karen
Crt 11h35'0"-18d28'
Green-Paez,Dr Henry Gregory
Ori 5h59'54"15d57'
Greenall,Melanie Joan
Crb 16h4'58"34d30'

Greenan,Julie Francis
Cas 0h15'22"47d21'
Greenawalt
Dra 18h56'38"58d9'B
Greenawalt,Elizabeth Anne
Peg 21h19'0"23d4'
Greenawalt,Jon P
Vul 19h59'49"25d25'
Greenbalt (Elvis), Thomas M
Cam 3h13'53"61d31'
Greenbaum,David Andrew
Ser 16h2'1"4d4'
Greenbaum,Honey
Cas 1h44'60"65d14'
Greenbaum,Rhona
Cas 1h46'32"65d2'
Greenbaum,Ryan Scott
Hya 8h56'31"0d28'
Greenberg,Al
Tau 4h33'55"30d13'
Greenberg,Alexandra Jane
And 0h55'10"40d21'
Greenberg,Alexandra
Dra 15h35'37"65d1'
Greenberg,Benjamin
Alexander
Boo 13h33'21"21d34'
Greenberg,Bob
Sex 10h0'18"-2d4'
Greenberg,Cory John
Cep 21h25'58"85d12'
Greenberg,Gilad
Aur 4h57'46"50d37'
Greenberg,Joren Kye
Uma 10h21'46"52d41'
Greenberg,Kathryn & Alan
Tri 2h21'20"30d19'
Greenberg,Kelley
Peg 23h6'16"11d37'
Greenberg,Leah Morgan
Lyn 9h14'43"38d9'
Greenberg,Marcia & Joel
Cyg 21h52'21"42d20'
Greenberg,Marshall Sonny
Aqr 21h35'10"-1d29'
Greenberg,Paul Adam
Aur 6h17'43"31d56'
Greenberg,Roz & Hank
Ori 5h56'42"15d5'
Greenberg,Sanford
Lmi 10h1'1"35d11'
Greenberg,Zachary Aaron
Uma 11h51'23"31d30'
Greenberger,Alfred
Lac 22h8'13"46d46'
Greenberger,Eva
Vul 20h21'17"28d49'
Greenbrook,Linda
Ori 5h55'49"12d51'
Greenbug
Cyg 21h20'1"28d55'
Greene II,John Keith
Aql 18h43'53"10d2'
Greene Star "For Bass"
Hilliard
Cas 1h6'1"60d3'
Greene,Albert
Lac 22h14'49"38d26'
Greene,Alexander
Ori 5h50'22"14d37'
Greene,Ann Marie
Mon 7h10'41"-10d48'
Greene,Benjamin
Lac 22h5'1"40d10'
Greene,Brenda Lee
And 23h10'43"39d43'A
Greene,Bryan Robert
Per 2h7'29"57d56'
Greene,Bryan Robert
Per 2h10'27"57d2'
Greene,Cecil P
Peg 23h3'50"33d7'
Greene,Charles Deverne
Equ 20h58'1"9d6'

Greene,Charles Michael
Ser 15h25'40"7d59"
Greene,Christopher Lorne
Cyg 21h18'12"38d16'
Greene,Derek
Her 17h4'20"40d15'
Greene,Dorrit & Danny
Crb 15h55'22"28d17'
Greene,Elise
Vul 20h14'27"22d56'
Greene,Gabrielle
Peg 23h5'0"32d2'
Greene,George
Gem 6h56'37"17d41'
Greene,Gloria Friedman
And 0h52'32"45d19'
Greene,Holly
Cet 2h37'0"0d27'
Greene,Jake Marcus
Ori 6h6'42"0d30'
Greene,James Kimo Joseph
Her 17h46'40"14d38'
Greene,Janice A
Peg 21h53'15"34d23'
Greene,Jeremy Michael
Boo 14h50'32"30d56'
Greene,John Logan
Cet 2h37'35"-8d3'
Greene,Jonathan Dwight
Gru 21h51'16"-40d10'
Greene,Lanny L
Aql 19h29'57"0d36'
Greene,Malorie
And 2h14'30"39d54'
Greene,Ms Vicki Lee
"Gorgeous"
Uma 10h9'24"53d37'
Greene,Nathaniel
Aur 6h8'38"46d32'
Greene,Paula Hope
Lyn 7h53'39"38d35'
Greene,Ricky"Ziggy"
Sct 18h42'33"-4d2'
Greene,Robert E
Cep 0h55'10"80d4'
Greene,Roger & Dorothy
Uma 10h45'47"68d2'
Greene,Sandy
Cas 1h26'21"60d41'
Greene,Sr,Fred McLaughlin
Cep 21h16'43"63d59'
Greene,Trey,Chase,
Jordan,Colin
Ori 5h50'31"14d26'
Greene,Valencia P
Vul 19h46'15"20d14'
Greene,Virginia
Cet 2h18'59"2d60'
Greene,William "Sarge"
Leo 10h33'1"21d5'
Greener's Acabar,Chris
Aql 19h14'52"12d46'
Greener,Janine
Cas 1h10'48"60d55'
Greener,Victoria Elizabeth
Cas 0h51'25"74d10'
Greenfeder,Joshua Louis
Vir 13h18'57"-8d56'
Greenfield
Lyn 7h29'53"58d51'
Greenfield,Brett Andrew
Lac 22h10'0"46d58'
Greenfield,Jacob Lokken
Lac 22h32'0"37d48'
Greenfield,James Todd
Aqr 21h59'26"-0d54'
Greenfield,Marlene
Cet 2h39'20"-9d34'
Greenfield,Nettie
Cnv 13h49'30"42d10'
Greenfield,Paul & Phyllis
Cnv 13h48'53"40d40'
Greenfield,Steve
Cep 20h23'55"60d28'

Greenfield,Theresa M
Lyr 18h38'28"40d6'
Greenhalgh,David
Ori 6h1'1"7d36'
Greenhalgh,Gary Alan
Gem 6h29'52"14d2'
Greenhauff,Adela
And 0h59'46"40d2'
Greenhut,Caryn
And 0h10'48"26d42'A
Greenlake
Aql 19h54'13"15d6'
Greenleag,Sean
Dra 14h58'23"61d10'
Greenlee Star,The
Umi 14h11'53"78d18'
Greenlee,Doleres Miranda
Cyg 20h5'11"38d10'
Greenlee,Leanne Marie
Aur 6h57'23"40d38'
Greenlee-Remm
Cnv 13h46'21"48d9'
Greenlees,Kaye
Aur 5h42'5"30d28'
Greenslade,Taran Kevin
Uma 10h4'14"52d11'
Greensmith,Janet Morgan
Ori 6h7'1"5d42'
Greensmith,Joe
Her 16h7'16"40d39'
Greenspan,Tobey Sarah
Cam 3h19'37"67d16'B
Greenstein,Amy Beth
Peg 21h50'21"30d19'
Greenstein,Henry & Frances
Cyg 21h8'32"40d12'
Greenstone,Arthur B
Lac 22h35'54"37d38'
Greenup,Marion T
Cas 0h38'14"64d17'
Greenwald,(Jimmy) James H
Dra 16h42'47"70d4'
Greenwald,Audrey & Stuart
Eri 3h12'51"-5d59'
Greenwald,Brian Ross
Tau 4h43'1"20d8'
Greenwald,Caryn & Thom
Cep 22h6'18"61d49'
Greenwald,Eva E
And 23h18'0"40d50'
Greenwald,Gary
Ser 15h58'11"21d15'
Greenwald,Graig
Her 17h5'18"31d9'
Greenwald,Jessica Lynn
Cam 13h29'49"84d37'
Greenwalt,Tammy
Lyn 8h34'47"41d5'
Greenway,Loie Ann
Lyr 18h52'51"34d2'
Greenway,Reg
Ori 5h59'52"14d50'
Greenway,Tommy Lynn
Hya 10h23'0"-18d52'
Greenway,Tracie
Tau 4h28'32"30d36'
Greenwell,Julie Kay
Cnv 13h54'53"41d19'
Greenwell,Mark Alan
Her 18h13'37"40d28'
Greenwell,Sandra
Aur 5h16'56"40d36'
Greenwell,Steven Michael
Del 20h13'36"10d38'
Greenwold,Jonathan & Louise
Cyg 19h33'45"33d6'
Greenwood Star,The
Ori 5h51'18"7d40'
Greenwood,Barbara J
And 0h18'32"34d39'
Greenwood,Carol
And 1h7'15"37d36'
Greenwood,Dale B
Del 20h13'47"15d37'

Greenwood,Dalton
Her 16h58'11"22d10'
Greenwood,Kevin Paul
Cep 21h43'1"55d41'
Greenwood,Kimberly Anne
Gem 7h56'51"20d2'
Greenwood,Marvin
Lac 22h50'56"56d42'
Greenwood,Michael Scott
Aql 20h0'1"8d60'
Greenwood,Patrice Ellen
Mon 7h57'0"-1d29'
Greenwood,Peggy Sue
And 23h2'18"46d38'
Greenwood,Robert
Aur 6h30'0"38d53'
Greenwood,Ruthanne "Joy"
Mon 6h54'0"-10d44'
Greenwood,Sharon & Jim
Cyg 19h27'1"31d44'
Greenwood,Stefenee
Cma 6h56'43"-15d8'
Greenwood,Vanessa Kathüleen
Cas 22h58'0"55d38'
Greenwood,Wesley
Aql 18h59'39"-6d35'
Greenwood:Man With A
Vision,Eric
Lmi 10h34'27"38d42'
Greer "Star Of Hope"
Eri 4h3'36"-17d17'
Greer,Adam
Lmi 10h5'42"40d46'
Greer,Alisa Howard
Cam 3h19'27"58d37'
Greer,Barbara Ann
Cas 1h45'34"60d34'
Greer,Betty
Cma 6h53'15"-19d0'
Greer,Caroline Alicia
Lmi 9h23'59"34d41'
Greer,Christie Lee
Cyg 19h51'46"38d16'
Greer,Creighton Willis
Cet 3h17'15"1d25'
Greer,D Ryan
Leo 10h51'12"15d42'
Greer,David S
Aur 4h47'0"41d15'
Greer,Diane
Peg 22h44'1"11d28'
Greer,Doey
Peg 23h41'1"30d24'
Greer,Donna
Lyn 6h59'0"59d49'
Greer,Elizabeth Lane
Mon 8h8'1"-5d12'
Greer,Garry
Cep 21h37'20"55d38'
Greer,Gordon Bruce III
And 23h22'19"49d41'
Greer,Haleigh Marie
And 23h20'15"49d53'
Greer,Jesse R
Psc 1h38'17"20d25'
Greer,John David
Cep 23h27'1"65d32'
Greer,Jr,Richard Donald
Mon 6h23'42"8d34'
Greer,Kyle J
Sco 17h1'0"-38d56'
Greer,Lisa V
Del 20h27'15"11d56'
Greer,Margaret
Cam 5h12'56"78d49'
Greer,Paige Nicole
Lyr 19h20'0"33d46'
Greer,Patrick Gerald
Ser 15h49'57"24d19'
Greer,Richard Keith
Ser 15h39'31"8d51'
Greer,Richard M
Aql 19h55'59"10d36'
Greer,Robert E
Dra 17h47'32"64d3'

Greer,Sandra
Del 20h28'40"20d28'
Greer,Vivacious Vicki Marie
Cas 0h20'59"64d25'
Greeson Star One,James Randall
Hya 8h57'43"1d27'
Greeson,Edgar C
Ori 5h46'55"11d29'
Greever,Lesley Ann Marie
Vul 20h19'16"22d37'
Gref,Dietrich
Peg 23h32'28"13d10'
Greg & Davina
Cyg 20h55'0"30d59'
Greg & Debbie's "Engagement Star"
Ori 5h32'33"-1d29'
Greg & Debi "Our Dreams"
Crb 16h8'16"38d14'
Greg & Erika (Best Friends)
Cyg 21h8'16"30d46'
Greg & Erin
Eri 3h49'33"-0d18'
Greg & Karen
Boo 15h14'1"47d42'
Greg & Renee 9-15-92
Aur 5h30'51"50d20'
Greg & Tammys Heavenly Friendship Star
Cyg 19h44'22"30d35'
Greg's Cosmos
Gru 22h41'30"56d25'
Greg H Sweetest Guy in the Universe
Uma 9h11'14"68d23'
Greg Martin
Aql 20h2'1"10d26'
Greg Scott
Aur 4h57'28"41d2'
Greg You Light Up My Life Puddinghead
Ori 6h3'48"-0d59'
Greg's Diamond
Del 20h16'40"14d59'
Greg's Foley
Aql 19h3'41"-5d50'
Greg's Heavenly Body
Uma 11h30'0"43d2'
Greg's Heavenly Body
Boo 14h3'1"13d17'
Greg's Star
Ori 5h15'47"-5d2'
Greg,Darrell Scott
Crb 15h29'21"30d47'
Greg-N-Windy
Dra 20h17'16"73d25'
Greger,Evi
Cas 1h8'32"60d28'
Gregers,Jens Otto Soberg
Lac 22h1'0"49d57'
Gregersen,Charlotte Falk
Cas 2h2'29"68d12'
Greget,Sandra
Per 3h4'0"38d31'
Gregg
Cep 23h1'55"64d59'
Gregg III,Eugene Stuart
Ari 2h38'48"22d19'
Gregg's Grand
Lyn 8h29'13"48d39'
Gregg's Love
Aql 19h23'1"-6d53'
Gregg's Star
Oph 18h19'29"6d45'
Gregg,Anne
Lyr 18h17'39"47d23'
Gregg,Beverly L
Mon 7h29'55"-10d37'
Gregg,Casey Elizabeth
Mon 6h35'1"6d16'
Gregg,Donald Gordon
Oph 17h23'11"-22d14'
Gregg,Dorothy Miller
Sco 17h54'27"-31d46"
Gregg,John Martin
Ori 6h18'0"8d40'
Gregg,Joyce
And 1h3'33"40d44'
Gregg,Jr,Eugene Stuart
Cap 21h19'17"-26d24'
Gregg,Linda Dalton
Psc 1h1'18"18d8'
Gregg,Richard Nathan
Her 16h38'51"35d50'
Gregg,Sean
Ori 6h7'0"0d20'
Gregg,Sr,Robert William
Ser 15h54'23"21d34'
Gregg,Theone
Aur 4h59'50"37d35'
Gregg-Bob & Jeannie- Belle
Aql 19h16'1"15d38'
Gregg-Eheler,Shirley Mirl
Umi 16h2'51"70d6'
Greggie-Joe
Boo 15h1'59"26d29'
Greggo,Anne
Umi 14h15'38"78d55'
GregLori
Vul 20h22'17"25d4'
Gregoire,James Anthony
Aur 5h24'36"31d5'
Gregoreadis,Gregory
Her 17h45'29"40d37'
Gregori,Valerie
Cas 1h0'11"55d47'
Gregorian
Cmi 7h29'50"1d46'
Gregorio,Nicole Lauren
And 23h33'21"39d24'
Gregorio,Steven
Per 3h26'18"52d34'
Gregorius Harrisus
Per 2h57'1"37d54'
Gregory
Her 16h27'43"28d21'
Gregory
Aur 6h16'49"38d32'
Gregory
Cnc 8h11'47"6d52'
Gregory
Aur 7h9'0"36d21'
Gregory
Aql 19h31'41"8d58'
Gregory
Uma 11h19'11"30d10'
Gregory Alexander
Her 18h10'42"31d5'
Gregory Alias Gregorius
Vul 21h22'14"27d30'
Gregory August 27,1950
Aql 19h29'20"8d8'
Gregory Cosimo Tino
Gem 7h11'14"24d40'
Gregory Edmond
Ind 20h25'48"-54d5'
Gregory Eternally
Uma 11h0'25"38d16'
Gregory Forever
Sex 10h29'24"5d56'
Gregory III,E W
Sex 9h52'16"5d59'
Gregory John
Cep 21h12'53"60d21'
Gregory Philip
Dra 16h13'33"64d15'
Gregory Star,The
Dra 15h6'47"60d41'
Gregory To Be
Hya 9h3'20"5d39'
Gregory's Diamond Direct
Cyg 20h38'0"40d52'
Gregory's Dream
Pup 7h28'45"28d0'
Gregory,Alanna Eve
And 23h24'1"45d40'
Gregory,Alice
And 0h26'1"40d10'
Gregory,Alice Loves You
Dra 19h1'13"61d12'
Gregory,Alyson Lea
Uma 12h23'34"61d5'
Gregory,Bette
Vir 11h50'17"7d32'
Gregory,Betty
Crt 11h8'46"-14d19'
Gregory,Bill & Melody
Hya 8h59'48"2d52'
Gregory,Chad M
Cep 22h9'31"61d4'
Gregory,David Charles
Mon 7h42'48"-8d53'
Gregory,Deanna
And 23h1'48"47d45'
Gregory,Deloris
Lyr 18h37'26"28d44'
Gregory,Dhati Changa
Cam 6h45'20"67d60'
Gregory,Doug
Aql 19h52'27"14d16'
Gregory,George G
Her 18h20'39"13d7'
Gregory,Glenn H
Hya 8h18'16"-1d24'
Gregory,Hollie
Cyg 19h22'11"44d36'
Gregory,John R
Per 2h4'37"57d6'
Gregory,Jon
Cma 6h52'11"-18d25'
Gregory,Joseph
Tau 3h43'24"0d57'
Gregory,Kris
Aql 20h10'26"0d30'
Gregory,Lindsay M
Crb 15h36'22"28d33'
Gregory,Michael Leland
Ori 5h49'25"21d31'
Gregory,Mike
Aur 7h2'0"43d34'
Gregory,Mike
Aql 19h57'56"10d3'
Gregory,Rachael Ann
Uma 11h55'1"31d11'
Gregory,Regine Char11'
Cas 1h2'0"61d32'
Gregory,Rush
Hya 8h54'14"0d35'
Gregory,Sallie & Lyle Granger
Cyg 19h44'21"29d43'
Gregory,Sandra Maria
And 0h8'43"44d53'
Gregory,Shannon Dawn
Mon 7h17'48"-7d12'
Gregory,Tamara Le
Crt 11h8'15"-12d15'
Gregos,Terri
Sge 20h16'41"20d22'
Gregovich
Gem 6h50'34"30d19'
Gregson's Star Happy 60th,Ernie
Aur 5h3'47"38d23'
Gregson,Joanne
Cyg 22h0'59"50d1'
Gregson,Nicole Margaret
Cas 0h36'59"60d46'
Gregson,Sherri Diane
Boo 14h3'0"22d31'
Gregson,Simon
Uma 14h1'0"53d55'
Greguras,Wendy
Cyg 19h24'17"31d42'
Greguska,John R
Aur 6h27'40"35d26'
Greif,Friedrich Karl
Sco 15h52'2"-22d12'
Greig,Cynthia
And 0h27'17"38d42'
Greiner
Her 16h16'25"24d57'
Greiner Valentine 1996 Laura Beth
Peg 22h26'38"25d17'
Greiner,Adam Wayne
Hya 9h9'1"4d54'
Greiner,Angelika
Lyr 19h21'38"30d21'
Greiner,Jeremy Christian
Aql 20h4'45"7d14'
Greiner,Michael
Peg 0h7'0"27d47'
Greiss,Georg und Ruth-Else
Uma 11h30'17"33d11'
Greitemann,Josef
Cmi 7h17'39"8d38'
Greitzer,Caleb
Her 16h9'3"47d56B
Grelock,Justin
Cet 2h13'14"2d28'
Grelock,Ryan
Cma 7h12'0"-15d22'
Gremer,Charles Edward
Hya 8h14'11"0d30'
Gremillion,Gina Louise
Mon 7h56'17"-1d29'
Gremillion,The Gracious
Crt 11h52'45"-7d36'
Greminger,Renee
Cas 0h53'54"61d29'
Gremlion & Hootowl 5'21'94
Cam 5h26'12"65d30'
Gremminger,Denise Lynn
Peg 0h1'13"31d30'
Grendon,William
Cep 20h35'45"61d35'
Grenfell,Jackie
Cas 1h1'38"63d26'
Grenier,Christa M
Cas 0h33'40"64d32'
Grenier,Christopher J
Cep 23h10'1"60d32'
Grenier,Denise
Uma 8h45'36"54d52'
Grenier,Dominique
Ori 6h7'52"20d48'
Grenier,Gérard
And 0h51'17"39d26'
Grenier,Louis
Psc 0h2'13"8d59'
Grenier,Misa Ann
And 23h42'20"37d38'
Grenier,Murielle Amyot
And 0h54'14"39d59'
Grenier,Nenette
Her 17h8'0"40d37'
Grenier,Paul Gabriel
Aql 19h6'55"3d32'
Grenier,Philippe- Olivier
Ori 5h40'55"12d39'
Grenier,Steven Alfred
Cap 21h5'25"-22d47'
Greninger,Bob
Dra 19h23'17"58d15'
Grenko,Colleen Ann Bridget
Aur 7h7'15"40d23'
Grenko,Pat
Lyn 7h33'60"38d23'
Grenon Louise
Cep 22h20'15"66d27'A
Grenquist,Carl
Lyr 19h17'52"41d20'
Grenter,Jane
And 0h54'36"41d4'
Grenville,Lucie
Ara 17h54'18"-56d12'
Grenz,Keith James
Boo 15h14'0"28d44'
Grepper,Pierree
Aqr 23h33'23"-8d24'
Greschler,King Edward
Cep 21h43'54"68d25'
Gresham,John Steven
Aur 5h9'38"42d57'
Gresham,Stephanie
Aql 19h41'1"11d31'
Greshik,Konnie & Ted
Lyn 7h10'27"52d30'
Greshler,Abbey
Cas 1h46'0"61d18'
Greshman,Eric Matthew
Cep 21h5'21"61d5'
Gresko III,Joe V
Aql 19h16'48"15d1'
Gresly,Glen Patrick
Aur 6h2'12"31d26'
Gress
Peg 23h43'46"31d3'
Gresser,Helene Michelle
Uma 10h6'51"58d48'
Gresser,John Paul
Uma 10h8'14"51d26'
Gresser,Rose Frances Konicek
Uma 10h17'27"53d20'
Greta
Sgr 18h55'9"-35d57'
Greta
Lyr 18h50'42"36d42'
Gretchen
Ori 6h2'17"7d50'
Gretchen S
Cas 0h17'58"64d20'
Gretchen Star,Karann
Com 12h9'13"26d45'
Gretchen's Star
Cma 6h51'22"-17d17'
Grethe Marie
Lac 22h39'44"20d29'
Grethe-Beaven Family Star,The
Lyr 18h30'55"37d45'
Greto,Kristina Victoria
Eri 3h46'0"-4d4'
Gretori Forever
Ser 18h7'18"-13d28'
Grettenberger,Isolde
Cep 22h22'1"61d0'
Gretzula,Gene Budash
Aql 19h39'53"10d46'
Greuling,Gerald J
And 0h57'29"36d21'
Greves,Nancy S
Cyg 21h5'54"38d49'
Grevile-Heygate,David
Per 3h15'37"42d10'
Grew,Jo
Dra 9h58'17"81d7'
Grewell,April
Crb 16h16'0"30d2'
Grewell,Bill
Her 16h50'42"33d40'
Grey Account Supervisor
Vir 11h52'17"8d30'
Grey,Alice
Aql 19h53'21"14d8'
Grey,Coleen
Com 12h53'26"26d23'
Grey,Donna Anita
And 0h1'23"47d58'
Grey,Helen Claire
Cap 20h42'46"-26d2'
Grey,Hugh M
Hya 8h33'34"-6d47'
Grey,Jess
Cet 1h13'21"-1d44'
Grey,Lauren
Aql 20h0'22"7d41'
Grey,Lorin
Dra 11h33'44"71d48'
Grey,Michael Albert
Boo 15h35'36"50d38'
Grey,Rachel Frances
Cyg 19h47'46"38d50'
Grey,Sara
Ori 5h24'22"1d26'
Greyhawk
Dra 18h3'17"65d22'
Greyson
Cep 21h10'54"68d24'
Greywolfe
Aur 6h32'45"33d7'
Gribben 18KR,Claire
Uma 11h28'0"42d58'
Gribben,Alice S
Cas 3h0'1"57d47'
Gribben,Michael Francis
Aur 7h15'1"41d53"
Gribben,Phyllis
Mon 6h30'37"4d38'
Gribben,Robert & Dolores
Hya 9h30"41"5d36'
Gribbin,Michael D
Her 18h12'57"40d39'
Gribble,Chuck Arline
Mon 7h44'30"-8d29'
Gribi,Tyler Cameron
Boo 15h7'29"22d11'
Gribouche Suprâme
Cam 3h50'11"63d1'
Grice,Delberta C
Cet 2h8'19"-1d43'
Grice,Kim
Lac 22h4'21"37d55'
Grice,Melodi Love East Pacheco
Cyg 21h3'59"33d49'
Grice,Nicky Lynn
And 2h27'26"38d15'
Gridelli,Gerardo
Uma 11h32'46"53d8'
Grider III,Charles H
Dra 14h53'13"64d25'
Gridley,John & Shelley
Uma 10h5'28"68d26'
Grieb,Herbert
Eri 4h48'39"-7d12'
Griebel,Edgar
And 23h9'56"42d32'
Griebel,Steffani Bottas
Eri 3h41'42"-15d2'
Griebling,Sarah
And 23h1'46"50d2'
Grieco,Christine Michele
Peg 21h59'43"28d38'
Grief,Jude Judith
Cas 0h34'27"62d39'
Grief,Kenneth F
Aur 6h55'60"40d55'
Griego II,Larry "Slick"
Aur 5h1'1"49d48'
Griego,Christine V
Mon 6h53'1"-5d42'
Griego,Jeannette B
Aql 19h1'1"0d51'
Griep,Lucy J
Com 12h23'35"27d55'
Grier,Harold N
Leo 10h55'0"12d41'
Gries,Uwe
Ori 5h50'60"20d40'
Griesbach,Eduard
Lyr 19h21'32"31d2'
Griesenbeck,Markus
Tau 3h55'1"8d22'
Grieshaber,Dr Romano
Her 17h8'16"43d5'
Griesmyer,Katherine Louise Elliott
And 0h21'53"37d1'
Grieve,Eleanor Anne Downie
Cas 23h2'34"58d57'
Grieves,Tracey E
Peg 23h40'11"18d59'
Grie ßer,Helmut
Lyn 8h16'40"47d10'
Grifaffe
Umi 13h12'37"71d37'
Grife,Len "Big Daddy"
Uma 12h18'10"54d12'
Griff's Shining Star
Per 1h52'1"50d28'
Griff,John
Hya 9h2'36"4d57'
Griffey,Emma Lynne
Cas 23h29'1"60d56'
Griffey,Jr,Thomas L
Her 17h6'57"42d39'
Griffey,Megan Marie
Del 20h18'17"10d14'
Griffin
Sge 19h19'50"17d39'A
Griffin
Her 16h52'35"32d39'
Griffin (Kooi),Barbara
Lac 22h45'57"54d18'
Griffin 8/29/45 With Love,Paul John
Boo 15h1'13"20d15'
Griffin III,Earl J
Aur 6h35'30"31d36'
Griffin PA/CTP,William F
Boo 15h8'1"21d38'
Griffin's "Wishing Star",Neal
Lmi 9h54'0"37d50'
Griffin,Alan & Barbara
Lyn 8h11'48"50d22'
Griffin,Alexander
Lyn 6h56'14"59d4'
Griffin,Alexander James
Cyg 19h42'1"38d12'B
Griffin,Anne C
Cyg 20h6'49"40d60'
Griffin,Billy G William
Boo 14h48'58"26d39'
Griffin,Chris
And 0h57'12"45d32'
Griffin,Christie
Cyg 20h59'43"38d24'
Griffin,Christine Dawn
Cas 0h2'58"50d29'
Griffin,Christopher D
Cnv 12h22'32"36d3'
Griffin,Claudia E
Sct 18h48'29"-6d39'
Griffin,Cornelia Caufield
Aql 19h29'22"8d22'
Griffin,Curyn Sky
Cet 0h48'13"1d23'
Griffin,Deborah Jean
Peg 23h23'41"8d54'
Griffin,Dominic
Uma 10h10'29"61d53'
Griffin,Dora Darlynn Bowman
Mon 7h0'10"3d48'
Griffin,Edward
Aur 5h24'11"40d0'
Griffin,Edward
Aur 5h0'1"40d5'
Griffin,Ethan Christopher
Per 3h35'28"38d46'
Griffin,Heather Renee
Mon 6h55'14"11d26'
Griffin,Jerry
Mon 7h58'24"-1d21'
Griffin,John
Cep 1h46'48"78d47'
Griffin,John Maxwell Sherrerd
Cet 1h32'48"0d1'
Griffin,Joy Arline
Cet 0h56'53"-2d22'
Griffin,Jr,Jerry M
Boo 15h9'41"30d49'
Griffin,Katheryn Ann
Com 12h12'12"32d43'
Griffin,Kelsey Elizabeth
Cyg 21h51'56"44d41'
Griffin,Lisa Ann
Peg 22h43'0"10d52'
Griffin,Lynn Marie
Peg 21h40'46"24d32'
Griffin,Mammao Catherine
Tau 3h45'14"1d1'
Griffin,Mark Roland
Sex 10h30'54"0d48'
Griffin,Matthew Scott
Per 2h21'58"54d57'
Griffin,Melissa Anne
Uma 9h51'53"50d29'
Griffin,Michael Ross
Cet 1h54'54"0d49'
Griffin,Michele Lynn
Cyg 19h42'1"38d12'A
Griffin,Patricia Ann Le Kron
Peg 22h10'39"27d47'
Griffin,Paul
Lyn 7h55'26"50d46'
Griffin,Richard & Phyllis
Lmi 10h3'38"32d26'
Griffin,Ryan M
Per 4h44'23"51d25'
Griffin,Samuel Evans
Cep 22h49'49"57d26'
Griffin,Sara Jane
Cet 0h52'54"-0d27'
Griffin,Shelby Malee
Peg 23h44'0"31d22'
Griffin,Steve
Lyr 19h5'0"28d59'
Griffin,Steven Craig
Aur 4h56'58"50d51'
Griffin,Susan
Cet 2h42'0"1d20'
Griffin,Tara Ann
Aql 19h31'30"12d17'
Griffin,Thomas Richard
Aur 5h8'31"43d59'
Griffin,William
Cyg 21h18'35"35d18'
Griffin-S PJerry
Aql 18h51'0"11d22'
Griffing,Laura Michelle
Cnv 13h54'28"40d14'
Griffing,Patricia A & John F
Aql 18h52'0"11d50'
Griffis,Janet
And 0h9'13"37d49'
Griffis,Margaret
Cas 1h45'44"75d28'
Griffith Golden Boy, Joseph Arthur
Cet 2h31'1"-10d52'
Griffith Royal Angel Glow,Sharon
Peg 0h7'0"18d27'
Griffith's Star, Graham
Uma 9h56'56"45d38'
Griffith's Star,Jody
And 23h41'57"43d51'
Griffith,Amelia
Vul 19h47'22"28d48'
Griffith,Belinda
And 0h48'1"37d3'
Griffith,Blanche Taylor
Ori 5h53'59"7d44'
Griffith,David
Eri 4h4'37"-19d38'
Griffith,Delbert Gordon
Aql 20h17'53"5d1'
Griffith,Denman Howard
Lac 22h37'27"55d25'
Griffith,Donald Wayne
Ori 5h47'59"20d59'
Griffith,Doris Edith
Peg 22h46'48"25d51'
Griffith,Douglas Leroy
Dra 19h4'56"50d11'
Griffith,Eric R
Dra 20h8'13"62d53'
Griffith,Evelyn Marie
Peg 0h8'45"13d43'
Griffith,Gary B
Hya 8h30'46"-6d24'
Griffith,Gayle Michelle
Del 20h53'19"7d36'
Griffith,Howard
Del 20h21'18"2d27'
Griffith,Jamie
Cet 2h43'1"3d30'
Griffith,Jason Patrick
Boo 14h49'35"32d33'
Griffith,John
Boo 15h8'1"48d31'
Griffith,John D
Cet 3h17'59"8d59'
Griffith,Larry
Aql 19h57'34"-6d31'
Griffith,Larry Dawson
Aur 5h7'23"50d22'
Griffith,Mary
Cyg 19h28'53"38d35'
Griffith,Michael T
Aur 5h25'54"40d59'
Griffith,Michelle
Crb 15h54'0"31d10'
Griffith,Rocky (James)
Cet 2h55'18"8d40'
Griffith,Tonya Rae
Equ 21h2'48"7d50'
Griffith,William Dexter
Aql 20h11'57"1d52'
Griffith-the Big 50!, Patricia Ann
Cas 1h59'54"60d13'
Griffiths III
Aur 6h57'1"40d1'
Griffiths,Anthony Gabriel
Cep 23h28'27"68d46'
Griffiths,Big Rear Admiral
Uma 9h49'51"67d45'
Griffiths,Carolyn
Lyr 18h31'20"46d17'
Griffiths,D J
Aql 19h30'0"10d46'
Griffiths,David Earnest
Cyg 20h21'51"38d29'
Griffiths,David
Ori 6h4'47"0d13'
Griffiths,Emily
Lyr 18h32'1"46d31'
Griffiths,Eriks M
Her 16h24'45"10d32'
Griffiths,Florence Amy
Cas 2h17'25"60d6'
Griffiths,Jenica Laine
And 23h42'15"45d49'
Griffiths,John Malcolm
Aur 5h0'51"45d34'
Griffiths,Joyce
Mon 6h22'16"-1d52'
Griffiths,Matthew Edmund Batho
Uma 12h3'30"40d54'
Griffiths,Morgan
And 23h30'23"42d39'
Griffiths,Ron
Peg 21h52'0"30d44'
Griffiths,Stephen
Lac 22h18'30"51d5'
Grigg,David
Ori 5h41'19"10d53'
Griggs,Camila "Jeff"
Aql 19h57'56"15d35'
Griggs,Holly Marie
Lyn 8h25'15"58d54'
Griggs,Sr & Jr,Maurice Thornton
Gem 6h59'58"14d46'
Grigoli,Jerry
Per 1h51'56"53d39'
Grigor-Taylor,Oona
Cas 2h28'12"60d14'
Grigorov,Alexander
Aur 6h33'14"34d49'
Grigorov,Christa
Aql 19h49'45"13d16'
Grigsby,Howard
Peg 22h45'56"29d19'
Grijp,Docteur Linda
Uma 13h1'14"62d34'
Grill,Jessie Amato
Cma 6h10'28"-16d46'
Grilleas,Peter
Ori 5h52'54"15d10'
Grilley,Nicole Georgette
And 23h30'40"38d52'
Grilli,Eugene
Aqr 20h56'51"-10d36'

Grilli,Mary E
Tri 2h8'35"31d7'
Grilliot Family,The Ken
Aur 4h54'1"50d9'
Grillo,Anne M
Cyg 21h50'40"38d60'
Grillo,James Charles
Her 15h48'18"41d22'
Grills,Anne Darling
Uma 12h1'28"32d26'
Grilü
Mon 7h57'52"-4d9'
Grim,Anita Lee
Cas 1h28'1"58d36'
Grim,Mark Scott
Cap 21h9'9"-15d9'A
Grimadell,Kenneth Michael & Pamela Eliz
Uma 9h16'22"54d40'
Grimal,Jacques
Peg 23h34'0"10d55'
Grimaldi,Charles
Cep 1h24'31"80d12'
Grimaldi,Danielle
Per 1h33'32"53d3'
Grimaldi,Helen
Cas 1h32'0"61d50'
Grimaldos,David G
Per 1h28'24"53d7'
Grimaloli,Francesa
Aur 4h51'21"40d33'
Grimard,Joseph
Gem 7h55'31"30d0'
Grimard,Nicole Franáoise
Cyg 19h24'12"33d32'
Grimberg,Charlotte
Cyg 21h7'33"31d24'
Grimes Love Stacy Kenworthy,To Tina
Cyg 21h14'1"38d6'
Grimes"'Starship Enterprise",C B
Sco 17h7'48"-38d35'
Grimes,Chris & Susan
Vul 19h47'49"26d37'
Grimes,David
Her 17h16'33"43d2'
Grimes,Elizabeth Gayle
Mon 8h4'1"-8d17'
Grimes,Erika Demé
And 2h26'25"37d42'
Grimes,Graham & Marie
Cyg 20h43'42"45d39'
Grimes,Gretchen Kimberly
Sge 19h59'0"18d47'
Grimes,Ish
Peg 22h2'36"31d15'
Grimes,John William Anthony
Aql 20h18'33"0d3'
Grimes,Jr,Charles
Aur 5h16'23"41d3'
Grimes,Matt
Aql 18h42'14"1d30'
Grimes-P J,Paul J
Cap 21h27'28"-23d18'
Grimler,Nancy Jo
Vul 19h17'27"21d30'
Grimlinger,Adrian
Crt 11h12'48"-19d13'
Grimlinger,Adrian
Aur 5h57'33"38d17'
Grimm
Umi 13h45'40"74d25'
Grimm,Colleen Elaine
Lyn 9h10'29"34d3'
Grimm,Jon
Cep 23h21'32"63d10'
Grimm,Kate
Umi 16h19'17"70d54'
Grimm,Rebecca
And 23h45'1"38d50'
Grimmett,Courtney Elizabeth
Mon 6h53'0"-0d55'
Grimmett,Jr,Dominic "Nick"
Cet 2h53'19"3d34'

Grimminger,Helga
Tau 4h6'36"22d24'
Grimmo From Fatima Mansions
Umi 17h27'41"85d14'
Grimshaw,Michael
Cep 22h11'28"60d51'
Grimsley 1942,Norman
Cnv 13h59'54"38d35'
Grindall,David
Aql 20h1'26"1d3'
Grinder,T C
Aur 5h20'0"41d4'
Grindle
Sco 15h59'19"-25d49'
Grindley's Glimmer
Cas 0h51'0"70d2'
Grindrod,John
Ori 6h6'50"0d11'
Grindrod-Feeny,Sophie Olivia
Cas 0h44'14"60d58'
Grindstaff,Nancy Greenly
Dra 12h57'59"72d28'
Grinel,Ronald & Tanya
Mon 6h57'35"-10d28'
Grinham,Lynda Ruth
And 0h37'60"45d58'
Grinkov,In Memory of Sergei
Cep 23h11'31"64d41'
Grinkov,Sergei
Lyr 18h42'49"39d16'
Grinkov,Sergei
Ori 5h57'10"15d17'
Grinnell,Allan Charles
Uma 8h45'0"57d5'
Grinsell,Charlene
Vul 19h47'44"22d56'
Grinsell,Jamie
Aql 19h55'45"13d36'
Grinwis,Sharon Squish
Cyg 19h31'25"38d52'
Griot,JP
Lac 22h15'37"55d20'
Gripp,Margo
Eri 3h47'41"-6d16'
Grippi,Larry
Aur 5h7'59"42d9'
Grippo,Robin
Eri 2h47'1"-18d57'
Gripton III,Charles Edward
Her 15h50'37"41d45'
Gripton,Barbara
Cas 1h28'49"58d33'
Grisak,Bryan Nicholas
Boo 15h46'17"48d49'
Grisanti's Star, Stephen R
Cet 1h37'29"-2d46'
Grisanti,Eugene
Dra 17h26'25"65d34'
Grisanti,Stellar
Peg 23h26'25"15d5'
Grisantis Star, Christopher C
Aql 19h20'50"-0d0'
Griscom,Verne Dowdell
Cyg 21h23'30"37d37'
Grisham,Mary
And 23h36'53"48d57'
Grisham,Taylor Deal
Tau 3h57'18"20d12'
Grisllian,Clare Helen
Lyr 18h32'37"31d47'
Grisse,Walter
Lyn 7h2'1"58d32'
Grisso,Michael Edwin
Uma 10h3'1"47d33'
Grissom,Walt
Cep 22h8'27"71d14'
Grist Star,The
Aur 5h0'13"45d14'
Grist,William J
Eri 3h34'57"-2d17'
Grisvard,Gilbert
Sgr 19h27'13"-34d1'
Griswaold,Kathleen Marzi
Cas 2h33'40"57d40'

Griswold,John Arthur
Sex 9h56'0"5d49'
Griswold,Kelly Elizabeth
Mon 7h14'26"-10d34'
Griswold,Marcia
Vul 19h58'40"29d3'
Griswold,Patricia Gay
Mon 6h55'20"-6d33'
Grit
Cam 5h2'1"58d35'
Gritsch,Cory Joseph
Lyn 8h49'33"38d24'
Gritsch,Gianna Marie
Peg 22h36'0"30d26'
Gritsch,Madison Jade
Vul 19h48'36"28d32'
Gritsch,Zachary Edward
Lac 22h27'0"37d37'
Grittani,Ronald (Gritts)
Uma 11h45'31"32d12'
Grivas,Yvonne
And 2h19'0"42d24'
Griveau,Antoine
Cet 0h52'28"1d6'
Grivetti,Mark Douglas
Boo 13h53'0"15d12'
Grivotet,Antoine
Cas 1h8'24"66d17'
Grizzard,Donna M
Mon 7h52'30"-6d49'
Grizzle,Lee
Cmi 7h55'35"3d55'
Grizzly
Uma 9h29'17"50d20'
Grizzly's Phatom,The
Uma 8h59'15"58d30'
Grlovich,Vincent A
Cma 6h44'38"-13d18'
Grniet,Garrett
Cma 7h24'59"-15d17'
Gro & Jorgen
Umi 16h14'35"80d22'
Groatman,Suzanne
Cet 2h16'23"6d6'
Grobben,John David
Cet 2h31'23"-9d51'
Grobelny,Jordon Charles
Ari 2h38'38"22d12'
Grober,Walter
Peg 23h29'22"12d37'
GroBheim,Gerhard
Gem 6h26'3"13d47'
Grobman,Olga
Lyn 6h32'29"58d13'
Grochman,Joseph F
Cep 0h9'36"69d35'
Grocholski,Brigitte
Sco 17h32'23"-31d21'
Grocock,Wayne
Aur 5h2'16"49d1'
Groder,Sr,Gary Joseph
Oph 18h37'53"6d54'
Groditzky,Bernd
Boo 14h4'1"10d17'
Grodner,Dr Toby Nekris
Aur 6h7'33"31d35'
Grodner,Robyn
And 0h5'15"47d14'
Groen,Joke
Her 16h13'0"48d25'
Groenendijk,Huibert
Cam 4h49'59"68d27'
Groenhout,J-n-D
Cyg 20h4'29"31d41'
Groenier,Diane Elizabeth
Cyg 19h29'56"30d47'
Groesch,Michael W
Cet 0h59'53"-0d42'
Groesch,Peter
Ori 5h52'30"16d29'
Groethe,Amos
Cas 23h14'17"61d41'A
Groethe,Edith L
Cas 23h14'17"61d41'B

Groething,The Star of
Cam 5h49'23"73d50'
Groff,Sarah
Peg 22h59'36"10d37'
Grog B C
Aql 19h44'38"14d32'
Grogan IV,John Patrick
Aql 20h1'56"6d58'
Grogan Star,The
Uma 8h18'30"68d19'
Grogan,Jordan Tyler
Lac 22h21'12"54d59'
Grogan,Kevin
Gem 7h39'28"20d14'
Grogan,Kevin Mackenzie
Ser 15h57'1"1d0'
Grogan,Laurie A
Peg 23h26'54"10d45'
Grogan,Michael Joseph
Cep 21h26'1"58d19'
Grogan,Sandra Leigh
Vul 20h3'21"23d53
Grogan,Thomas Daniel
Oph 17h54'38"12d58'
Groh,Adam MacAlister
Dra 14h28'47"63d3'
Grohol,Robert M
Aur 4h55'59"40d22'
Grohowski,Camille & Michael
Cet 1h30'1"-6d32'
Grohs,Heinz Adolf
Cas 0h40'1"60d37'
Groiselle,Family
Umi 16h10'19"70d53'
Grojsman,Sophie
Cnv 13h10'58"38d7'
Groll
Her 18h19'36"14d25'
Grom,Gerd-Joachim
Ari 2h42'36"28d20'
Gromala,Bernard Anthony
Sex 9h58'49"0d8'
Gronager,Rose Marie
Eri 2h54'43"-5d45'
Grondahl,John
Aql 18h58'53"-1d54'
Grondin,Jacky
Cra 18h33'14"-43d17'
Grondin,Veronique
Crt 11h17'37"-18d4'
Grondin,Wenly L
Cyg 19h28'0"33d52'
Grone,James A
Aql 19h54'20"-0d3'
Groner III,John James
Cep 22h56'16"70d7'
Groner,P J
And 23h38'55"42d51'
Grones,Amy
Mon 8h1'46"-7d36'
Groniarté
Ori 6h17'38"20d27'
Gronich,Maxwell Armand
Cap 20h28'1"-12d48'
Gronich,Susan Leslie
Cap 20h48'44"-16d49'
Gronich,Theodore L
Cap 20h50'1"-17d40'
Groninger,Jennifer Elizabeth
Eri 2h54'39"-11d31'
Gronqvist,Terttu Iellervo
Umi 15h48'2"76d9'
Groom's Wish,Richard L
Umi 14h8'18"69d47'
Groom,Bill
Lyn 9h18'19"40d52'
Groome,Merilyn Ann
And 23h37'49"49d48'
Grooms,Annette Nicole
Her 17h9'58"47d7'
Grooters,Marja Kristiina Raunnos
Lyr 18h16'0"30d11'
Groothuyse,Georgina W
Dra 10h14'13"81d17'

Gropp,David
Ori 5h55'45"14d42'
Gros,Laurence
Crb 16h8'57"38d60'
Grosby,Alfred G
Dra 9h20'52"73d34'
Grosch,Gabriele
Gem 7h23'1"35d14'
Groschupf,Elmar Dr Ing
Leo 10h52'50"2d4'
Grose,Norene E
Eri 3h22'47"-12d8'
Groseth,Jacqueline Ella
Peg 22h43'35"33d49'
Grosh,Richard
Her 18h40'46"18d57'
Groshong,James Chester
Peg 22h42'54"29d23'
Groshong,Stuart & Virginia
Lyr 18h40'52"32d1'
Grosko,Brandon Michael Mallory
Lac 22h16'22"37d36'
Grosnickle,Brian Paul
Per 4h1'45"50d48'
Grospietsch,Dr Thorsten
Cnv 12h37'42"38d11'
Gross (My Eternal Light),Lisa
Peg 22h24'30"3d3'
Gross',The
Oph 16h41'52"2d46'
Gross,Abigail Nicole
Mon 6h19'1"3d54'
Gross,Alyson Joan
Vul 19h48'1"29d3'
Gross,Amy
Uma 11h8'1"33d39'
Gross,Andrew H
Cep 22h59'31"80d28'
Gross,Angela
Sco 16h27'20"-40d49'
Gross,Arthur B
Boo 13h45'12"25d30'
Gross,Barbara Kemper
Ari 1h50'21"12d24'
Gross,Bettina
Crb 16h5'1"26d6'
Gross,Carl Andrew
Aur 5h11'47"43d31'
Gross,Caroline Lord
Umi 16h13'45"71d10'
Gross,Don E
Aql 19h25'12"0d58'
Gross,Drucilla Jule Graf
Cam 4h38'57"59d25'A
Gross,Evan David
Boo 14h35'27"17d23'
Gross,Gerti
Vir 12h0'1"-2d15'
Gross,Harold Eugene
Cam 4h38'57"59d25'B
Gross,Heather Elizabeth
Cas 0h18'20"66d20'
Gross,J R
Uma 8h37'45"56d39'
Gross,Jacquelyn Margaret
And 2h26'16"42d43'
Gross,Jeanne K
Com 12h23'42"32d13'
Gross,Jeffrey Jerome
Aql 19h8'47"3d11'
Gross,Jimmy,Stevie, Stephanie
Tau 5h48'40"26d21'
Gross,John J
Dra 16h33'52"64d0'
Gross,Jürgen
Vul 19h58'57"29d17'
Gross,Kelly René
Com 12h28'39"30d48'
Gross,Kevin John
Hya 9h3'15"2d16'
Gross,Matthew Douglas
Eri 2h51'0"-5d20'

Gross,Melinda
Cas 0h16'44"58d31'
Gross,Michael Joseph
Her 18h0'23"38d48'
Gross,Michael Stephen
Her 18h39'32"12d13'
Gross,Mitchell D
Hya 9h15'36"6d17'
Gross,Ms M
And 1h1'1"40d11'
Gross,Paul Timothy
Dra 20h21'50"68d51'
Gross,Peter
Sct 18h22'27"-6d56'
Gross,Peter A
Boo 13h37'26"11d26'
Gross,Richard A
Lib 15h41'34"-24d28'
Gross,Shelby
Aur 6h54'56"35d31'
Gross,Shelly
Del 20h15'34"10d2'
Gross,Spencer
Her 18h31'0"20d13'
Gross,Tressa A
Peg 21h38'31"24d46'
Grossa,Elvio Q
Mon 8h1'43"-1d60'
Grosse,Christian
Cap 21h38'0"-19d25'
Grosse,Helga
Cnc 9h13'17"32d10'
Grosse,Kimberly J
Tri 2h6'21"32d19'
Grosser JaquesBär & Kleiner Nutzbär
Uma 8h58'57"59d15'
Grosser Karpfen
Psc 23h8'11"0d50'
Grosser Weiser Hase R W Taeschner
Leo 10h28'35"11d47'
Grosser,Janet
Lmi 10h5'22"31d51'A
Grosser,Morton
Lmi 10h5'22"31d51'B
Grosset,Annie
Her 16h18'26"50d57'
Grossey,Patrick
Aur 4h56'1"38d18'
Grossi,John
Tau 3h57'50"2d13'
Grossin,Jacques
Ari 1h57'44"20d15'
Grossklaus,Robert
Boo 13h35'32"14d11'
Grosslight,Samantha Marie
Vul 20h4'42"28d27'
Grossman,Barbara
Cyg 19h22'23"28d60'
Grossman,Beverly
Com 12h50'32"27d8'
Grossman,David Lee
Lac 22h14'21"47d44'
Grossman,Elizabeth
Lyn 7h40'15"41d5'
Grossman,Gwen Michelle
Gem 6h59'12"16d5'
Grossman,Haley
And 23h10'10"36d12'
Grossman,Jay Harry
Aql 19h58'19"-8d7'
Grossman,Jerry
Aql 19h53'48"11d20'
Grossman,John L
Her 17h22'25"46d53'
Grossman,Karen Ashcraft
Cap 21h26'59"-22d54'
Grossman,Kenneth
Dra 10h41'20"74d53'
Grossman,Kira Lee
And 1h18'0"36d32'
Grossman,Landry
And 23h11'43"37d16'

Grossman,Marsha
Aql 19h40'37"11d60'
Grossman,Matthew
Lac 22h22'31"50d2'
Grossman,Micah
Cyg 21h34'22"41d59'
Grossman,Phyllis Madray
Lmi 10h10'37"30d45'
Grossman,Ronald E
Her 17h59'57"14d33'
Grossman,Sheila
Lyr 18h20'30"46d44'
Grossman,Wendie
Cma 7h48'39"4d25'
Grossmann,Friedhelm
Cap 20h23'28"-26d58'
Grossmann,Jeff "Jake"
Cep 2h24'47"77d37'
Grossmen,Janice
Umi 17h27'47"75d16'
Grossnickle,Heather Carole
Cet 1h33'19"0d29'
Grossnickle,Katlyn Ruth
Peg 22h39'49"25d27'
Grosso,Dolores & Richard
Crb 15h51'20"30d59'
Grosso,Michael Vincent
Aql 19h26'38"15d26'
Grosso,Michelle A
Cas 3h34'15"70d19'
Grostic,Catherine H
Uma 10h20'58"48d47'
Groswird,Bev
Mon 8h3'46"-5d36'
Groszek,Dane
Cep 22h13'50"60d45'
Grote,Hermann
Sco 17h29'0"-30d45'
Grote,Kevin
Boo 14h27'0"41d59"
Grote,Martina
Oph 18h1'50"8d27'
Grote,Sabine
Peg 23h33'44"21d44'
Groteboer,Jeffrey Lee
Lib 15h36'51"-20d21'
Groth,Jim
Ori 6h3'19"6d45'
Groth,Rita Helene Miller
Cas 0h46'34"62d1'
Grothe,Norbert
Equ 20h57'38"10d5'
Grotheer,Martin H
Aur 6h23'43"38d24'
Grotnes,Pearl
Com 13h13'41"22d24'
Grotte,Lee Bryan
Uma 10h54'1"37d43'
Grotton,Scott
Cep 21h49'48"58d21'
Grotz,Meaghan
Lyn 7h41'54"44d25'
Grouchie
Her 17h1'23"22d11'
Groulx I,GPJJ
Aur 6h49'0"39d33'
Groumoutis,Leo
Cam 5h41'29"68d26'
Grounds,Charles James
Equ 21h18'39"2d52'
Grounds,Christopher Malcolm
Uma 8h9'33"61d34'
Grounds,Ed
Hya 8h26'27"-6d7'
Grounsell,Scott
Aqr 22h48'12"-5d32'
Group B Strep Association,The
Sex 10h12'27"-9d24'
Group Zodiac (Constellation EmCare)
Uma 8h47'51"50d40'
Group,The
Umi 14h51'15"78d10'

Grove,Bret
Ori 5h59'52"16d55'
Grove,Dr Andrew S
Oph 17h2'38"-5d41'
Grove,Dru Ann
Ori 4h51'23"4d5'
Grove,Frankee
Lac 22h16'0"49d28'
Grove,Jennifer Ann
Vul 20h2'35"23d55'
Grove,Leah Michelle
Leo 10h36'29"14d44'
Grover & Linda's Destiny
Cyg 21h30'51"37d36'
Grover,Joey
Aur 6h11'36"35d34'
Grover,Paul
Cnc 7h57'41"10d59'
Grover,Robert
Ori 6h4'55"0d20'
Grover,Rosemary
Sct 18h55'53"-7d47'
Grover,Shane Paul
Her 17h26'27"21d34'
Groves' Star,The
Eri 4h9'58"-12d21'
Groves,David
Cma 6h56'21"-17d20'
Groves,Dawn
Vul 19h16'0"21d57'
Groves,Eugene
Cet 2h40'31"-0d55'
Groves,James L
Eri 3h41'12"-1d51'
Groves,Jr,William (Bill) F
Her 17h19'0"49d11'
Groves,Vicki "Thumper"
Uma 10h13'0"42d30'
Grovier,N K
Uma 8h48'26"62d3'
Grow,Brittany Manda
Cas 2h13'26"60d33'
Grow,Joseph Bernard
Aqr 22h3'22"-8d29'
Grow,Louis Eric
Mon 6h53'36"-4d9'
Grow,Rosemary
Vul 19h39'24"20d20'
Groß,Wilfried
Ser 16h5'13"2d4'
Grubb,Amy
And 1h47'56"39d23'
Grubb,Gary S
Sgr 19h34'43"-41d55'
Grubb,Jack Warren
Hya 9h12'16"-14d53'
Grubb,Kevin Harrison
Per 3h50'19"39d58'
Grubb,Michelle Jeanette
Lyn 7h35'46"50d51'
Grubba,John L
Aur 6h1'0"34d14'
Grubbs,Felicity Ann & Kenneth Gene
Sge 20h1'1"17d25'
Grubbs,Kim
Mon 6h29'42"-5d42'
Grubbs,Robert Walker
Del 20h13'38"15d23'
Grube,Bryan Keith
Boo 13h46'51"18d35'
Grube,Tara
Peg 22h15'42"35d10'
Gruben,David
Sct 18h55'1"-4d57'
Gruber,Anne Smolowitz
Cyg 20h0'29"38d27'
Gruber,Bogda
Lyr 18h29'51"31d8'
Gruber,Helen
Lyr 19h21'37"40d57'
Gruber,Irene D
Mon 8h5'29"-8d2'
Gruber,Johann
Gem 6h0'15"26d35'

Gruber,Michael Thomas
Aur 6h14'28"46d11'
Gruber,Robert Joel
Cam 5h36'54"65d9'
Grubert,Horst
Sco 17h5'0"-30d22'
Grubisich,Larry
Per 2h36'12"40d47'
Grubman,Emily
Cet 2h48'15"4d55'
Grubman,Margaret
Eri 3h0'51"-3d18'
Grubman,Matthew
Hya 8h43'35"6d36'
Grubmüller,Wolfgang 21/08/1955
Vir 13h32'18"-6d34'
Gruby,Vera Flora Smith
And 23h42'27"47d2'
Gruca,Jim & Lori
Mon 6h53'29"0d56'
Grucella,Dan
Aql 18h57'0"-6d3'
Gruchala,Christopher Thomas
Cyg 19h24'48"33d8'
Grude,Patricia Marie
Cas 0h59'49"64d40'
Gruebel-Lee,David Mark
Boo 14h32'28"52d24'
Grueber,Roy
Tau 4h16'54"20d29'
Gruen,Robert A
Dra 15h0'12"60d27'
Grueneberg,William
Aur 5h17'0"42d56'
Gruening,Michael
Aqr 22h59'45"-18d55'
Grueninger,Al & Sharon
Cep 0h13'11"70d24'
Grueninger,Bob
Cep 20h38'33"60d8'
Grueninger,Jack & Marci
Cep 0h59'32"78d0'
Grueninger,Kara Michelle
Cep 0h12'1"71d1'
Grueninger,Krista Marie
Cep 0h12'0"66d39'
Grueninger,Sherrie Lynn
Cep 0h13'1"70d13'
Gruenwald,Angela Hogan
Com 13h17'60"30d8'
Gruetzmacher,Jeff
Her 17h0'29"42d52'
Grumbach,Helen Sexxy
Mon 6h34'49"-6d41'
Grumell,Kristin
Cyg 21h31'19"30d1'
Grumet,Edward S
Cep 1h11'11"77d56'
Grumich,Mike
Aql 19h58'34"14d11'
Grunberg,Maurice
Dra 19h21'1"67d31'
Grund,Pamela
Lyn 9h3'30"37d50'
Grundblatt,Fanny
Uma 10h58'40"70d20'
Grundborg,Don
And 2h34'33"37d40'
Grundei,Reinhard
Umi 14h7'37"67d31'
Grundhofer/1996 FAF Shining Star,Jerry
Her 16h37'25"36d56'
Grundig,Maria Alexander
Sgr 19h4'2"-29d43'
Grundt,Jr,Vaughan S
Her 16h50'49"32d44'
Grundy III,Vaughan S
Her 16h56'25"32d38'
Grundy IV,Vaughan S
Her 16h42'0"32d8'
Gruner,Carole
Lyn 8h32'56"42d28'

Gruner,Emily Lynn
Cas 1h39'25"61d11'
Gruner,Jean-Georges
And 23h14'23"45d37'
Gruner,Margrit
Com 12h29'1"30d21'
Gruner,Mary Elizabeth
Cas 1h39'49"61d38'
Grunfeld,Gideon
Cet 2h59'21"1d44'
Grunfeld,Miriam
Mon 7h56'33"-2d44'
Grunigen,Forest
Dra 18h13'17"71d1'
Grunke,Detlef
Cmi 7h29'38"8d54'
Grunow
Lac 22h31'0"55d36'
Grunsby,Rosemary T
Cas 0h32'37"61d12'
Grunt
Uma 11h32'48"46d49'
Grunther,Andrew Joseph
Cnc 8h33'44"31d13'
Grupa,Marcus
Sgr 19h20'54"-43d7'
Grupp,Eldon,Susan,Erin
Peg 22h20'0"33d48'
Grupp,John Stephen
Dra 16h7'0"67d53'
Gruppo,Beth Ann
And 2h33'18"44d23'
Gruppo,Sr,Anthony C- Cheryl A Spreen
Crb 16h4'15"32d22'
Grusky,Christian Henning
Leo 11h23'0"-0d41'
Grusq-Chouraqui, Annabelle
Lac 22h18'32"50d16'
Grusq-Chouraqui,Chloe
Lac 22h18'46"54d35'
Grusq-Chouraqui,Sophie
Equ 21h19'33"11d13'
Gruss,Andrea
Lyr 18h15'44"35d15'
Gruss,Heather Lynn
Uma 8h47'1"47d53'
Gruss,Pamela
Cas 3h35'28"73d9'
Gruszka,Luke Max
Ori 5h51'54"7d29'
Gruver,Dave
Oph 18h21'18"10d45'
Gruver,Sr,Morris
Gem 6h51'8"18d18'
Gruytch,Gail
Del 20h13'52"14d42'
Gruzinova,Larisa
Eri 4h29'56"-11d60'
Grygo,Jim
Her 16h17'0"10d47'
Gryniuk-Lifestar, Jonathan Darby
Dra 20h12'39"61d49'
Gryphon Omega
Lac 22h11'21"48d28'
Grzebieniak,Richard
Aur 7h7'0"35d29'
Grzebyta,Michael
Cas 23h37'56"63d57'
Grzech,Georg
Her 17h20'18"42d7'
Grzesik,Kandra Marie
Cas 0h35'47"60d27'
Grzib,Robert Stewart
Per 1h52'0"56d29'
Gräff,Helmut
Cap 20h30'1"-11d7'
Gräfin
Cet 2h55'11"2d59'
Gräfin Sonja Bernadotte
Boo 14h17'15"19d49'
Grégoire,Nathalie Monferini
Cam 4h59'0"58d27'
Grégory et Frédérique
Uma 11h27'38"34d41'
Gröger,Puken
Lyr 18h41'16"41d6'
Grüben's Celestial Pea
Hya 8h13'57"1d2'
Grünberg,Heinz
Cap 20h4'19"-10d8'
Grünberg-Hansen,Elke
And 0h53'41"21d42'
Gründig,Rainer
Lyr 19h19'28"30d49'
Grünewald,Vera
Lib 15h11'26"-22d12'
Grünling-Kater,Jürgen
Cnv 14h2'56"30d36'
Grünn,Pirkko
Uma 8h32'0"68d4'
Grünthal,Rolf Heinz
Cap 20h45'21"-26d29'
Grünthal,Thea
Gem 8h4'57"30d30'
Gráinne
Aql 18h57'32"17d27'
Gsell,Christian
Cam 7h37'44"68d31'
Gstrein,Markus
Lib 15h5'59"-5d38'
GT
Aur 5h30'14"38d43'
GTS-JAM
Sct 18h49'38"-7d11'
GTT
Boo 14h23'31"45d44'
GTX
Sco 17h50'0"-40d32'
Guadagnoli,Luciano
Lyn 8h16'15"35d7'
Guadalupe,David B
Com 12h14'29"20d16'
Guadamillas Cortés, Gabriela R
Gem 6h33'56"26d49'
Guadarrama F, Alejandro
Ori 4h41'30"13d13'
Guadian Forever,Rick
Ori 5h59'0"15d38'
Guagenti,Mary D Battista
Aqr 21h59'44"-1d4'
Guagliano,Christopher Frank
Boo 13h42'51"18d53'
Guagliano,Keith C
Aql 19h40'1"14d25'
Guaraldo,Annamaria
Lac 22h23'0"50d1'
Guarascio,Roberta (Rags)
Cas 1h5'53"60d29'
Guard,Joan W
Cas 1h24'23"70d44'
Guard,Ronald S
Dra 17h0'23"68d28'
Guardia Angel
Peg 22h44'25"32d51'
Guardian Angel
Mon 6h21'56"3d26'
Guardian Angel
Aql 19h8'0"0d47'
Guardian Angel
Aql 19h32'11"0d24'
Guardian Angel Mom
Ari 1h49'53"23d55'
Guardian Angel of Stephen & Sandy
Lyn 7h10'44"53d28'
Guardian Angel Peter
Mon 7h16'13"-6d30'
Guardian Angel,The
Sgr 19h57'38"-42d18'
Guardiola,Maria Trinidad
Lyn 7h31'18"45d58'
Guare,Jayne
Cmi 7h57'15"0d28'
Guare,Shelly
Aql 19h7'16"-1d49'
Guarente,Thomas Arthur
Aur 5h38'30"50d32'
Guarinello,Jayme
Lyn 9h13'18"38d29'
Guarinello,Leigh
Del 20h13'12"11d22'
Guarinello,William R
Cep 22h2'0"61d38'
Guarini,Joseph Paul
Boo 14h44'33"51d27'
Guarino,Antoinette
Cyg 19h30'36"33d46'
Guarino,Antonio et Mimma
Lmi 10h50'51"30d14'
Guarino,Dr Richard A
Oph 17h37'54"-20d51'
Guarino,Katie & Kelley
Cyg 19h32'39"33d11'
Guarino,Norianna
Aqr 22h8'34"0d31'
Guarino,Richard A
Cyg 20h25'21"40d4'
Guarino,Stevie & Alex
Cyg 19h29'22"33d26'
Guarisco,John Michael
Per 2h53'36"54d33'
Guarnera,Joseph
Hya 8h53'45"-6d39'
Guarnieri,Clara
Lyn 7h47'1"50d50'
Guarriello,Jr, Theodore J
Per 2h37'1"50d29'
Guarriello,Jr,Anthony
Hya 8h49'50"5d40'
Guastaferro,Dennis A
Aur 6h11'37"33d13'
Guay,Frédéric
Umi 14h25'13"68d57'
Guay,Ginette
Cas 0h47'37"73d31'
Guba,Valerie Anne
Com 13h24'27"26d24'
Gubala,Maria
Dra 18h11'60"70d42'
Gubar,Susan
Lyn 7h1'57"51d2'
Gubarevich,Yedviga
Her 17h32'60"20d43'
Gubau,Juli Prat
Ser 15h14'1"1d36'
Guber,Elizabeth
Mon 6h40'30"10d59'
Guber,Jodi
Peg 22h15'1"10d40'
Guber,Samuel Dakota
Aql 18h54'1"8d53'
Guberman,Dorothy
Mon 7h56'1"-0d56'
Gubitz,Irving
Her 17h18'1"49d36'
Gucciardo,Stephen
Her 16h33'35"37d1'
Guccione,Luca
Cnv 13h48'22"37d42'
Guckelberger,David & Valerie
Cyg 21h17'15"38d42'
Guckenberger,August
Crt 11h20'44"-8d32'
Guckert,Stephanie Lynn
And 0h24'0"31d47'
Gudapati,Ravi
Cam 5h41'39"60d33'
Gude,Samuel
Boo 14h24'27"21d36'
Guden,Doris
Eri 4h29'31"-10d12'
Guden,Guy
Aql 18h55'20"-0d8'
Guden,Jack C
Aqr 22h10'51"0d55'
Guden,Robert
Sex 10h43'15"2d11'
Guderian,Walter
Cep 23h37'20"64d26'
Guderjohn,Ashley Nicole
And 1h12'30"47d55'
Gudewicz,Julia,Rachel Emma
Cas 2h20'43"73d23'
Gudewicz,Mary Beth & Jeffrey
Peg 23h19'21"30d31'
Gudi & Peter 4-80-19
Lyr 18h30'0"32d49'
Gudines,Mark
Aql 19h31'46"10d25'
Gudobba,James F
Mon 6h15'0"-6d21'
Gudrun
And 23h48'45"37d39'
Gudrun Guenter
Cas 0h44'22"72d49'
Gudrun N 03081953
Leo 9h19'52"10d49'
Gudrun und Carlo -Guc-
Cnc 8h6'55"31d11'
Gudvangen,Lindsay Ann
Lyr 18h32'24"36d1'
Gueary,Henry
Aur 6h3'27"31d46'
Gueck,Helmut
Aur 5h40'12"49d42'
Guedon,Nicole
Mon 6h36'58"10d58'
Guedry's Piece of Heaven,Donna
Sge 19h41'27"16d49'
Guelfi,Cynthia Ann
And 23h21'25"49d42'
Guelfi,Jean-Dominque Ange-Pierre
Lac 22h28'37"40d7'
Guen Du Leen
Sgr 19h10'6"-15d6'
Guendalina
Cae 4h48'0"-33d17'
Guenes,Branka
And 0h54'14"21d37'
Guengerich,Florence W
Lac 22h35'27"56d39'
Guenin,Jon Eric
Lac 22h45'53"35d37'
Guenola,Chevaux
Ori 6h15'59"0d9'
Guenther,Blakely Galdstone
Per 3h19'51"41d27'
Guenther,Janette Lynne
Lyn 8h13'0"36d35'
Guenther,The Geoffrey L
Cet 3h4'0"5d33'
Guenther,Ursula
Per 4h6'5"49d17'
Guenther,Whitney Carol
And 23h23'29"42d9'
Guenthoer,Norma E
And 23h3'41"44d46'
Guerda,Akli
Com 12h11'0"21d31'
Guerin,Aurélie
Uma 12h58'29"58d23'
Guerin,Francois
Cyg 20h29'1"30d11'
Guerinault,Emmanuel
Crb 15h54'34"35d13'
Guerinault,Michel
Crb 15h38'52"35d14'
Guerini,Giuseppe
Cam 6h9'20"68d28'
Guernalec,Andre
Dra 10h12'17"74d43'
Guerndt,Gary "Sweet Cheeks"
Her 17h28'17"21d16'
Guernick,Brittani Amber
Mon 6h58'47"10d19'
Guernsey 40,Tracey
Lyr 18h14'13"34d24'
Guerra
Her 17h21'28"29d6'
Guerra
Ori 5h6'14"8d18'
Guerra "Star Teacher", Josephine
Cas 0h56'15"75d1'
Guerra,Danny
Aur 6h29'41"32d23'
Guerra,Eva
Oph 18h0'52"11d13'
Guerra,Gracie
Lyr 19h13'24"38d12'
Guerra,Heather Nicole
Mon 8h8'29"-4d54'
Guerra,Hector
Aur 7h9'25"43d24'
Guerra,Jamie Daniel
Cet 1h58'49"-4d48'
Guerra,Lisa Brewer
Gem 8h5'34"28d30'
Guerra,Marco
Per 3h2'27"38d44'
Guerra,Michael Patricio
Per 1h56'11"53d27'
Guerrant,Richard
Aql 19h58'40"-5d48'
Guerraz,Humbert
Peg 23h28'0"22d28'
Guerre,Jean Louis
Lyr 18h56'58"31d48'
Guerrero Family Star, Joesph & Debra
Cet 0h59'44"-5d9'
Guerrero Family Star, Miguel & Susan
Mon 6h22'1"3d42'
Guerrero Family Star, David & Ranae
Oph 16h54'49"11d36'
Guerrero Ramily Star, Mark & Donna
Equ 20h58'41"5d19'
Guerrero,David Andrew
Oph 16h32'37"-6d27'
Guerrero,Enrique"Rick"
Her 17h5'16"30d17'
Guerrero,Erica Victoria
Mon 6h44'0"7d39'
Guerrero,Gary
Mon 6h55'21"-10d52'
Guerrero,Julio
Mon 6h54'19"0d28'
Guerrero,Mark C
Per 4h1'36"51d7'
Guerrettaz,Jr,William A
Cep 22h43'1"59d44'
Guerrier,Catherine
Dra 15h44'1"58d4'
Guerrini,Massimo
Tau 4h10'44"0d8'
Guerry,Eleanor Gooding
Leo 10h22'32"14d31'
Guertin,Scott
Mon 6h18'52"8d59'
Guertin,Timothy
Her 17h19'1"18d47'
Guertin,Timothy Peter
Aur 5h29'31"38d38'
Guertler's Wake
Tau 4h43'31"20d25'
Guesne,Sylvie
Sgr 19h27'14"-38d4'
Guess What!
Cet 3h19'19"1d50'
Guess,MD,George
Cet 2h39'48"-7d33'
Guest,Betty Jane
Cas 23h41'16"58d31'
Guest,Daddy Steven Alan
Aur 7h8'48"40d7'
Guest,Francis Mary
Equ 21h7'31"11d57'
Guest,Gloria
Cas 1h1'42"52d52'
Guest,Josephine Ann
Del 20h30'51"10d36'
Guest,Lee
Cmi 7h36'1"11d25'
Guest,Mary Elizabeth
Oph 17h58'0"10d8'
Guest,Melanie
And 0h21'0"40d58'
Guest,Phyllis
Cet 3h8'53"1d56'
Guest,Thomas Haden
Aql 19h10'37"19d58'
Gueudre,Thomas
Cep 20h8'40"60d51'
Guevara,Ben
Cap 20h36'4"-14d56'B
Guevara,Gustavo & Maria
Lyr 18h24'32"46d39'
Guevara,Holly Bennette
Peg 22h35'11"27d53'
Guevara,Jose
Hya 8h15'55"4d25'
Guevara,Leo Anthony
Sct 18h44'51"-6d48'
Guevara,Maria Janeth Irre
Mon 7h56'4"-6d59'
Guevara,Marisa Beth
Peg 22h47'30"25d22'
Guevara,Robert Joseph
Aql 19h0'0"7d52'
Guevara,Theresa C
Tri 1h37'32"35d25'
Guffee,Brady
Hya 8h33'19"5d45'
Guffee,Landis
Hya 8h34'19"5d54'
Guffey,Lloyd V
Per 2h53'18"38d44'
Guffroy,Dominique
Cyg 19h42'55"30d32'
Gugg,Dr Eberhard & Partner
Leo 11h6'56"-5d35'
Gughi
Peg 22h57'12"22d20'
Guglielmi,Angelo
Cas 23h20'0"53d53'
Guglielmi,Daryl
Lac 22h9'45"46d29'
Guglielminotti, Patrizia
Hor 3h19'47"-47d56'
Guglielmo
Pup 8h8'22"-26d45'
Guglielmo,Amedeo
Per 3h18'30"50d28'
Guglielmo,Joseph "Pooh"
Aur 6h26'11"38d44'
Guglielmo,Jules
Eri 3h55'0"-5d13'
Guglielmo,Rocco
Her 18h10'58"48d10'
Guglielmo,Theresa
Cam 7h53'1"60d18'
Gugliotta,S Victor
Aur 5h19'14"44d41'
Gugliotta,Stefano
Sgr 19h2'5"-21d1'
Gugliucciello,Charles Carmine
Dra 10h17'0"81d43'
Gugu
Psc 1h6'55"27d37'
Guhr,Stephan
Peg 23h28'37"10d25'
Guian,Dominique
Peg 23h45'45"31d17'
Guibaud,Marie-Claude
Cnv 13h45'0"47d48'
Guibert,Julien
Lyn 8h3'49"39d22'
Guibée,Christine
Per 3h59'57"36d37'
Guibée,Philippe
Aql 18h58'35"3d11'
Guichaoua,Eric
Peg 21h6'32"12d16'
Guicher,Laurence
Boo 14h28'54"48d57'
Guida Verde Srl Roma
Psa 22h27'56"-25d24'
Guida,Anna
Del 20h18'56"16d51'
Guida,Lisa Ann Haley
Leo 11h28'33"10d12'
Guida,Stephanie R
Lyr 18h27'11"30d37'
Guidera,James Anthony
Peg 22h0'45"28d40'
Guidi,Silvia
Pic 5h1'28"-44d53'
Guidice,Anthony
Per 3h14'59"50d26'
Guidice,Officer Vincent
Her 17h12'19"27d45'
Guidicianne,Diane C
Lyr 18h36'12"40d46'
Guiding Light
Peg 23h21'59"13d52'
Guido
Cas 0h35'1"61d19'
Guido,Aria "Freddy"
Dra 16h25'31"68d7'
Guido,Patrick N
Vir 13h27'36"-8d20'
Guido,Patrick Richard
Dra 19h6'26"50d29'
Guidone,Francesco Antonio
Cep 23h21'1"65d41'
Guidone,Raffaella
Cyg 20h4'40"30d49'
Guidotti,Tarquia
Umi 15h24'21"69d23'
Guidroz,John C
Uma 9h48m0"70d43'
Guidry,Adrienne
Aql 19h30'38"13d13'
Guiducci,Graziano
Cnv 12h11'57"40d15'
Guiffray,B
Cam 5h1'32"60d17'
Guija,Rony
Per 4h21'0"51d33'
Guild,Peter Michael
Uma 11h12'41"30d17'
Guild,Timothy
Her 16h40'52"36d23'
Guilfoile,Lisa
Cyg 19h44'40"29d42'
Guilfoyle,Joan
Cas 0h55'55"75d12'
Guill,Manuela
Her 16h44'28"34d41'
Guillain & Monica
Lyr 18h27'43"35d2'
Guillard,Julianne
Cas 1h16'17"62d3'
Guillaume
Uma 8h58'16"47d28'
Guillaume
Per 2h26'0"56d25'
Guillaume DeFrance de Tersant
Cap 21h24'0"-14d7'
Guillaume,Eric
Cam 6h55'12"64d37'
Guillaume,Patrick
Sex 10h14'60"-1d10'
Guillauminaud, Dominique
Sge 19h0'25"19d35'
Guillebaud,Ginette
Oph 18h15'27"10d46'
Guillemain,Denis
Cet 0h54'49"1d6'
Guillemette,Étienne
Uma 8h57'56"55d55'
Guillermo
Dra 16h20'42"63d32'
Guillermo
Crt 10h53'41"-10d11'
Guillet,Claude
Lmi 11h4'11"33d29'
Guillet,Kevin
Sgr 20h18'22"-33d29'
Guillette,Paul
Oph 17h35'32"14d3'
Guilliot,Jerry "Dakota"
Her 15h55'59"40d23'
Guillmette,Suzanne
And 2h28'10"39d51'
Guillo,Arturo
Ori 6h12'16"8d44'
Guillory,Pete
Oph 18h16'36"8d12'
Guillot,Chris
Hya 8h41'11"3d54'
Guillot,Darren John
Cep 5h14'28"80d26'
Guillot,Pierre Yves
Lac 22h23'1"40d21'
Guillot-Tantay,Thierry
Her 16h58'60"50d41'
Guillou,Jimi Lynn & Bob
Lyr 18h39'31"31d51'
Guillou,Nathalie
Cnv 13h22'52"30d57'
Guilmette,Bryan Michael
Peg 22h12'20"20d52'
Guiloff,Allan Nicolas
Cra 18h30'33"-41d15'
Guimara,Angelina
Lyn 8h45'54"43d1'
Guimara,Neil Philip
Cep 22h38'0"61d16'
Guimez,Michel
Peg 23h31'56"10d17'
Guimond-Deschênes,Prys
Tau 5h46'17"24d19'
Guin,Hillary
Aql 19h56'10"13d57'
Guinard,"Isabel"
Ari 3h8'0"30d16'
Guinasso,James
Boo 13h56'44"22d22'
Guindi,Natalie
And 0h30'0"45d37'
Guinevere
Dra 20h9'51"62d42'
Guiney,Kristin Melissa
Lyn 8h26'22"47d58'
Guinn,Carla Sue
Peg 21h55'0"20d20'
Guinn,Gloria Ann
Hya 8h25'21"0d31'
Guinns' Star
Aur 6h35'40"34d23'
Guiot,Eric
Crb 15h50'4"35d16'
Guirard,Jean-Jacques
Aur 6h5'17"31d1'
Guirc,Jr,Marvin
Sex 9h59'12"-6d55'
Guirgues,Georgette Aziz
Eri 2h43'24"-4d57'
Guirola,Jorge & Camille
Oph 17h36'35"-24d24'
Guisinger,Nancy Norcross
Cam 5h43'56"60d38'
Guitar,John
Hya 9h1'29"-8d16'
Guitard,Leanor
Sct 18h53'9"-7d19'
Guiterman,Seth
Ser 15h30'19"10d22'
Guitjens,Henk
Hya 9h8'25"2d10'
Guittar-Donner,Rebecca
Del 20h13'21"15d10'
Guittard,Charles Francis
Sex 9h46'49"-1d18"
Guitton,Gilles
Aur 7h25'12"40d10'
Guity,Lorna
Cas 1h11'19"75d41'
Guiver,Brian Clarance
Aur 5h51'24"30d33'
Guiza,Fanny Lillian Gaines
Aql 20h1'47"-1d21'
Gula,Brandee Christine
Cas 0h22'0"64d52'
Gula,Elisa Danielle
Cas 0h22'1"61d51'
Gulas,Vic, Maureen & Alex
Cnv 13h3'50"40d56'
Gulbrandson,Pamela L
Cnc 9h2'60"21d33'
Gulden
Sge 20h3'0"18d51'
Gulden,Jack
Boo 15h5'41"22d19'
Guldi,Adea Frances
Lyn 7h32'56"38d49'
Gulick,Christopher G
Crt 10h49'1"-8d57'
Gulickson,Jeremy Ross
Cet 1h49'27"-0d26'
Gulielmetti,John
Aur 5h2'30"48d48'
Gulinello "Star In The Sky",Michael
Per 2h25'40"58d25'
Gullander,Robert
Cep 20h15'20"75d14'
Gullette,Samuel Robert
Per 2h36'1"45d16'
Gulley's Star,Vicky
Eri 3h56'0"-4d36'
Gulley,Dana S
And 23h43'55"43d23'
Gulley,Neale I
Aur 7h24'31"38d60'
Gulliksen,Stian
Ser 15h24'44"18d25'
Gullixson,Sheila
Lyr 19h2'52"31d50'
Gullo,Joanne D
Cap 21h37'51"-24d18'
Gullotta's Galaxy
Leo 9h23'1"18d58'
Gullotto,John & Mary
Hya 8h12'1"0d58'
Gulotta,Thomas S
Cep 22h33'31"68d11'
Gulu-Gulu
Cnv 12h16'46"37d7'
Gum Ball Eddie
Aur 5h42'1"50d12'
Guma,Katharine
Eri 3h55'12"-3d55'
Guma,Linda
Crt 10h50'30"-6d46'
Guma,Lindsay
Cma 7h0'1"-16d6'
Gumby
Cnv 12h19'33"45d25'
Gumina,Anthony
Aur 5h47'39"54d31'
Gumina,Squeeker
Boo 13h45'13"27d7'
Gumm,Robert William
Lac 22h12'55"37d60'
Gummere,John
Cep 4h1'33"80d33'A
Gummersheimer,Father Gary
Aur 7h20'53"40d38'
Gummerson,Barbara Joan
Aql 19h30'16"-1d22'
Gumowitz,Joe
Aur 6h21'24"31d7'
Gums,Patricia
And 23h18'29"49d13'
Guncheon,Carol & Michael
Sge 19h56'1"20d8'
Gundaker,Beth Madonna
Lyr 18h21'19"38d40'
Gundel
Cnc 8h56'23"7d38'
Gundell,Harold L
Boo 14h33'60"38d8'
Gundell,Shirley V
And 0h53'10"33d41'
Gundersby,Eva
Dra 10h12'0"80d9'
Gundersen,Leah
Peg 0h10'36"14d5'

Gundersen-Herman, Amelia Rose
And 0h58'31"38d28'
Gunderson,James Mark
Lmi 10h1'30"28d11'
Gundlach,Daniel Mark
Aqr 21h22'27"-13d14'
Gundlach,Liliane-Edith & Hans-Georg
Uma 10h53'13"61d27'
Gundlach,Sonja
Equ 20h55'33"3d27'
Gundlach,Steve
Oph 17h3'49"10d42'
Gundrum,Yvonne
Com 12h11'1"19d36'
Gundry,Nancy J
Boo 14h32'50"40d28'
Gundry-White,Catherine
Crb 16h2'12"30d5'
Gungurru
Cyg 21h35'19"42d56'
Gunia,Busia & Dzia Dzia
Ori 5h55'31"10d3'
Gunn,Anna S
Lyr 18h35'0"26d3'
Gunn,Christopher William
Boo 14h36'0"20d39'
Gunn,Jaini
Cet 2h48'1"0d14'
Gunn,Robert David
Ori 5h54'57"8d23'
Gunn,Ross Peter
Cet 2h2'12"-0d8'
Gunn,Steve & Elizabeth Murphy
Lyr 18h35'11"40d1'
Gunn,William Ernest
Boo 14h37'18"20d11'
Gunnar
Aql 19h11'29"13d49'
Gunnels,Gary Clayton
Boo 15h21'45"51d23'
Gunner,Samuel
Ari 3h10'11"19d34'
Gunneson,Grant Wood
Per 2h50'22"35d54'
Gunnill,Andrea
Lib 15h36'46"-21d5'
Gunnin,Jr,Claude O
Hya 8h31'60"-8d6'
Gunning,Forever John
Sgr 20h23'27"-28d59'
Gunnion,A True Star, Mark
Per 3h23'1"40d51'
Gunritz,Tammy
Cas 0h15'52"60d52'
Gunsberg,Merry Lou
Aql 19h26'46"10d41'
Gunsolley,Lane
Lac 22h47'52"53d19'
Gunter,Janiece
Hya 9h6'38"-2d12'
Gunter,Kyle
Cep 23h1'19"60d43'
Gunter,Lorri Hanemig Bardezbanian
Com 12h6'0"20d13'
Gunter,Luther Elverton
Aql 19h55'36"14d53'
Gunter,Paul Edwards
Ori 4h54'10"0d37'
Gunter,Rachael Elizabeth
Lyr 19h16'23"28d21'
Gunter,Sean Henry
Cet 3h1'43"7d26'
Guntermann,Dr Karl L
Eri 3h46'3"-10d27'
Guntermann,Pater Paul
Cam 3h53'47"70d1'
Guntert Family,The
Per 3h47'54"37d22'
Gunther,Andrew & Marianne
Cyg 20h3'0"37d53'
Gunther,Jane & James
Cyg 21h28'20"38d38'
Gunther,Marty
Per 2h11'23"58d15'B
Gunther,Nicole
Per 2h11'23"58d15'A
Gunther,Stephen J
Aql 19h54'44"-0d17'
Gunton,Deborah
Peg 22h3'17"28d25'
Guntrum,Edward Joseph
Cep 4h47'50"80d11'
Gunz,Gabriele
Tau 3h55'18"0d52'
Gunzelman,Austin Lee
Cet 3h17'0"0d23'
Gupta,Meena
Lyr 19h25'33"37d57'
Gupta,Umang From GRPG
Cyg 21h22'35"37d35'
Guptil,Chad & Gretchen
Umi 14h42'43"65d44'
Gupton,Howard L
Her 18h2'46"28d46'
Gurden,Karen
Com 12h21'28"21d31'
Gurewitz,Grandma Marcie
And 23h25'55"42d41'
Gurewitz,Grandpa Harvey
Peg 3h19'39"40d32'
Gurganus,Patricia Hartley
Uma 9h17'13"54d21'
Gurgold,Carl
Oph 17h54'59"13d38'
Gurka-1946,J & T
Cam 3h19'55"60d52'
Gurke,Don
Uma 11h15'12"51d22'
Gurley,Mary-Marshall
Cyg 21h3'46"31d28'
Gurman,Amy Cheryl
Cas 0h59'6"68d38'
Gurman,Karen Alyssa
Cas 0h34'15"75d28'
Gurnee,Julie
And 2h24'53"44d58'
Gurney
Boo 14h25'57"39d6'
Gurney,Geoffrey David
Per 1h47'58"47d28'
Gurney,J Thomas
Aur 5h3'12"40d56'
Gurney,Thomas
Umi 14h39'36"68d16'
Gurney,Win
Cam 11h22'47"80d17'
Gurnsey
Cyg 21h9'24"38d47'
Gurr,Socha Elisabeth
Uma 11h3'28"43d37'
Gurriero,Anthony John
Aur 6h26'46"33d8'
Gurrola,Jaime & Faye
Peg 23h1'26"10d50'
Gurruminas De Costa Rica
Oph 17h39'5"-23d2'
Gursky,Jens Daniel
Ari 2h43'10"21d16'
Gursky,Olga
Equ 21h2'19"10d34'
GuRu
Ser 18h18'51"-13d48'
Gurule,Benjamin P
Aql 20h0'24"10d57'
Gurwitch,Janet
Tri 2h15'54"35d42'
Gus
Boo 14h46'48"23d50'
Gus
Cen 11h52'41"-49d55'
Gus' Place In The Universe
Aur 5h15'21"41d34'
Gus-Gus
Her 18h5'16"37d55'
Guselli,Gene
Per 3h12'51"49d42'
Gusenbauer,Renate Gerlinde
Tau 5h49'44"24d11'
Gusfa,Helen Adele
Lyr 19h3'17"38d29'
Gush,Robert
Her 18h12'53"40d35'
Gushiken,Betty K
Col 6h1'3"-38d52'
Gushiken,Zenko
Pho 0h49'3"-44d38'
Gusick,Johnny
Aql 20h12'15"5d37'
Gusick,Skip
Her 16h18'50"25d35'
Gusinda,Jeffrey J
Vir 14h22'36"-7d42'
Guske,Sally
Del 20h14'20"15d1'
Gusler,Virginia Elizabeth Homsher
Com 12h27'12"26d27'
Guslo,!lia
Aur 5h13'46"43d21'
Gusman,Ana Maria Quiroz
Cmi 7h15'1"5d3'
Gusmerotti Star,Jim
Cep 22h17'52"55d40'
Guss,Geraldine Spitzer
Uma 10h26'39"58d7'
Gussarson,Audrey
And 2h32'23"39d5'
Gussie Monsterus Muffinus Maximus
Cas 1h44'1"60d12'
Gust,Jim
Tri 2h0'42"31d22'
Gust,Wild Bill
Equ 21h0'19"7d50'
Gustafson II,Charles "Chaz" William
Her 18h19'35"12d11'
Gustafson,Aaron Alexandar
Cma 6h47'23"-17d21'
Gustafson,Andrew Steven
Vul 19h47'56"27d27'
Gustafson,Andrew
Aur 5h7'0"40d22'
Gustafson,Anna Elizabeth
Dra 19h50'25"71d1'
Gustafson,Chase Clay
Per 3h11'54"48d10'
Gustafson,Laura Kristy
Cyg 21h23'40"40d28'
Gustafson,Tori
Mon 6h44'53"11d57'
Gustafson,Tori D
Ori 5h5'40"8d32'
Gustafson,Warren Howard
Psc 23h22'48"5d24'
Gustavo
Aur 5h25'14"54d47'
Gustavo
Lac 22h47'25"55d26'
Gustin,Marc Alan
Aur 5h24'15"37d60'
Gustin,Michelle A
Lyr 18h49'48"30d59'
Gustke,Patricia Parker
Cet 3h11'13"0d50'
Gusz,Rosemary
Com 12h2'0"27d2'
Guszkowski,Kenneth Matthew
Her 17h9'16"38d39'
Gut,Paul
Cyg 20h25'21"40d34'
Gutchie Star
Boo 14h33'34"22d24'
Gutenkunst,Froggy I In Honor of R L
Lac 22h7'56"51d40'
Gutermuth,Bill
Per 3h47'44"39d46'
Gutermuth,Julie Griffin (Hall)
Lyr 18h59'57"29d50'
Gutfriend,DDS,Joseph
Her 16h40'58"23d38'
Guth,Brian Thomas
Hya 10h12'39"-19d50'
Guth,Nicole
Ari 2h6'23"20d40'
Guth,Pauline Rosenberg
Sco 17h51'15"-38d10'
Guthart,Janet
Cas 1h31'50"63d50'
Gutherie,Kayla Nicole
Peg 22h36'34"27d24'
Guthery,Melissa Marie
Uma 9h59'13"59d11'
Guthery,Sebastian Daniel
Uma 8h47'47"56d51'
Guthier,Steven
Cmi 7h39'38"8d33'
Guthormsen,Steven Dale
Dra 17h31'37"63d47'
Guthrie,Deidra Michele
Cyg 21h3'51"30d45'
Guthrie,George Allan
Cmi 7h15'1"5d3'
Guthrie,Jim
Her 17h36'18"27d50'
Guthrie,Joan
Cas 0h23'1"64d44'
Guthrie,Kenneth W
Her 17h38'38"21d49'
Guthrie,Mary Ann Pryor
Lyn 8h5'38"58d3'
Guthrie,Rebecca J (Buppy)
Lyr 18h35'21"28d39'
Guthrie,Robert
Cyg 21h1'17"39d15'
Guthrie,Scott C
Uma 9h23'33"67d59'
Gutierrez,Dennis Cookiehead
Ser 15h13'10"11d30'
Gutierrez,Edward
Her 17h54'18"31d35'
Gutierrez,Frances
And 0h11'32"37d33'
Gutierrez,Marta B
Cas 22h57'1"55d35'
Gutierrez,Matthew Evan
Boo 15h19'42"48d39'
Gutierrez,Natasha
Eri 2h46'45"-6d50'
Gutierrez,Timothy James
Cet 0h29'0"-1d37'
Gutknecht,Liesl Margaret
Lyn 7h43'1"38d34'
Gutman,Howard
Per 3h16'57"41d38'
Gutman,John
Her 17h32'56"41d10'
Gutman,Ron
Cep 22h50'31"70d20'
Gutowski,Mitchell Martin
Aur 7h8'19"40d30'
Gutrich-Bubulka
Boo 14h47'0"25d6'
Gutsch,Raymond Paul
Lac 22h52'33"51d35'
Gutschow,Steven W
Cep 21h46'38"55d57'
Gutsell,Michael John
Her 17h54'49"18d54'
Gutstein,Jamie Michele
Lyn 8h22'51"42d45'
Guttenberger,Fanny
Eri 4h5'35"-0d51'
Guttendorf,Raymond
Aur 4h52'19"38d27'
Gutteridge,Janice Lynn
Aur 5h1'31"47d24'
Guttman,E Eleanor
And 23h0'36"48d12'
Guttman,Matthew
Cep 22h28'0"70d8'
Guttridge,Carolina Distaulo
Aur 5h3'11"40d45'
Gutwein,Philip Bruce
Cma 6h46'34"-16d9'B
Gutwein,Sieglinde
Uma 10h25'51"60d3'
Gutwein,Sieglinde
Cyg 21h19'48"34d9'
Gutwein,Victoria Lee
Cma 6h46'34"-16d9'A
Guy
Cyg 20h41'44"42d40'
Guy & Megen
Uma 10h17'48"55d21'
Guy et Monique
Uma 13h41'38"60d15'
Guy,Aaron
Lac 22h46'58"55d55'
Guy,Alison Tepe
Lyr 18h46'45"40d10'
Guy,Ashley Elizabeth
Gem 7h7'36"21d48'
Guy,Autumn
Dra 14h53'32"64d52'
Guy,Carol S
Cas 0h33'31"61d19'
Guy,Kirk A
Boo 15h29'32"42d5'
Guy,Priscilla & Denise Sheldon
Cas 1h35'11"75d22'
Guy,Richard Allan
Oph 17h31'40"-20d1'
Guy,Steve
Her 18h28'56"20d36'
Guy-Mennell Star,The
Lyn 8h16'45"44d10'
Guya 2 Tato
Cep 21h21'19"68d58'
Guyan,Jason Bradford
Del 20h17'22"10d10'
Guydish
And 2h12'21"38d43'
Guyer,Chester
Ser 15h45'46"20d20'
Guyett,Jodi Lyn
And 2h17'0"45d47'
Guyett,Sr,David T
Aur 5h55'39"30d39'
GuyJoy
Sex 10h42'14"2d10'
Guylaine
Cas 0h6'55"63d10'
GuyLory
Sge 20h0'13"19d41'
GuyLynda
Uma 10h46'46"56d48'
Guylynn
Aql 18h46'26"11d16'
Guyon,Jocelyn
Sgr 19h33'4"-32d7'
Guyse,Paul
Aql 19h4'15"3d9'
Guyson
Aql 18h56'32"15d3'
Guzaldo,Joseph
Cet 2h51'42"1d15'
Guzelis,Lone Star
Cnv 13h21'33"31d34'
Guzenda,Marla Anna
And 23h23'11"51d22'
Guzie,Andrew Paul
Per 2h37'41"38d55'
Guzman Forever May You Shine,Melanie
Cyg 21h13'1"37d33'
Guzman Star,The
Peg 21h54'38"30d9'
Guzman's "Vibrant Light",Victor M
Cma 6h12'0"-16d13'
Guzman,Andrew John
Hya 8h18'1"4d26'
Guzman,Orietta Maria Patricia Velez
Lyr 19h2'1"38d22'
Guzman,Pedro
Cep 22h52'27"70d7'
Guzman,Starlet Mary
Peg 0h1'25"13d47'
Guzman,Tyler Austin
Cyg 21h14'50"28d56'
Guzman,Victor M Jr & Kevin M
Peg 23h1'40"13d47'
Guzman-Sheth,Debra
Cas 0h50'40"65d54'
Guzofski,Michele
Cyg 19h27'58"30d6'
Guzy,Sacha
Lac 22h47'13"56d41'
Guzzardo,Stephanie Lynn
Peg 21h38'28"23d23'
Guzzetti,Mara
Vel 10h11'54"-49d59'
Guzzo,Mr & Mrs Vincent
Cep 0h54'30"77d47'
Guérette,Lyne
Per 2h58'23"45d31'
Guérin,Michel
Cyg 19h24'23"30d56'
Gu Cyg 19h24'23"30d56'
Dra 13h48'51"68d5'
Gvirtzman,Doris "Gee"
Aur 6h29'57"52d30'A
GW-A
Cyg 20h34'18"40d40'
GW3PYX-Jim
Uma 10h22'1"54d22'
Gwaltney,Jr,John O'Connor
Sgr 19h48'21"-40d50'
Gwaltney,Lamar Anderson "Andy"
Cet 2h31'60"0d55'
Gwaltney,Luke Royston
Boo 14h44'27"37d58'
Gwee,Catherine Shu-Ming
Peg 22h32'20"18d51'
Gweet
Lac 22h55'51"55d15'
Gwen
Leo 9h19'1"7d53'
Gwen
Cyg 20h19'43"39d11'
Gwen
Leo 10h59'49"22d27'
Gwen & Herb's Paradise
Cyg 20h15'30"41d48'
Gwen & Jill
And 23h32'13"44d35'
Gwen & Ken"GwenKen"
Mon 7h3'1"0d1'
Gwen Marie
And 0h23'0"34d16'
Gwen Noel
Cas 0h54'46"72d31'
Gwen-Tino
Mon 7h15'48"-1d3'
Gwenaëlle
Uma 11h51'0"52d53'
Gwendoline
Ori 5h30'12"-6d17'
Gwendoline
Aql 20h31'40"0d17'
Gwendoline
Cnv 12h10'46"38d46'
Gwendolyn
Cam 4h9'57"60d35'
Gwendolyn
Lyn 7h37'12"48d17'
Gwendolyn
Tau 5h44'11"16d21'
Gwendolyn
Del 21h4'39"13d5'
Gwendolyn
And 23h28'20"39d45'
Gwendolyn Gayle
Cyg 21h39'56"37d57'
Gwendolyn Irene,The
Mon 7h2'26"0d51'
Gwenlan,Gareth
Uma 11h36'53"50d6'
Gwiazda Wandzia
Ori 4h41'33"14d34'
Gwiazdowski,Robert (Philly Bob)
Cam 8h7'30"81d37'
Gwin,Christie Lynn
Cas 1h31'41"73d28'
Gwin,My Belove,Sam
Per 1h58'37"47d48'
Gwinn,Jason Emerson
Hya 8h32'1"-8d30'
Gwinn,Matthew Richard
Boo 15h13'49"40d9'
Gwyla
Uma 8h36'47"58d10'
Gwyneth
Uma 10h55'26"47d30'
Gwynn
Crb 16h14'42"37d44'
Gwynn,Maureen
Boo 14h49'59"31d19'
Gwynn,Todd A
Ori 6h0'32"0d1'
Gwynne,Lin
And 22h56'15"51d46'
Gyasi
Lyn 6h59'17"54d1'
Gygi,Audrey Inez
Lac 22h24'18"53d20'
Gyla
Cam 3h15'53"60d58'
Gyland,Percy
Dra 18h16'33"70d24'
Gyllenhoff,Wendy
Cyg 19h30'55"34d6'
Gyllenhoff-Jacobsen, Gabriella M
And 23h23'25"44d38'
GYM 80
Leo 10h57'44"0d17'
Gyore,Steven L
Her 17h28'18"28d38'
Gyori,Robert
Ser 18h4'53"-14d59'
Gypsy
Pho 0h42'54"-45d17'
Gypsy Boy
Cam 4h12'0"68d38'
Gypsydom (To Travel)
Dra 11h51'16"70d38'
Gysin's Forever Star, Bob & Bette
And 2h0'17"38d32'
Gyspy West
Eri 2h48'56"-2d19'
Gyulay,Landon Hull
Her 16h22'15"23d38'
Gyurits,Irene R
Vul 19h57'18"25d2'
Gyöngyi Zahava Pearl
Cas 1h17'1"61d29'
Gyánti,J R (Joe)
Her 16h29'0"40d22'
Gzerwin,Margaux Julia
Tri 2h33'10"33d16'
Gzors
Cyg 20h1'26"31d3'
Gärtner,Julia
Gem 6h26'24"13d11'
Géczy,Michelle
And 0h49'17"22d50'
Gélinas,Christianne
Uma 11h35'14"31d34'
Gérald
Cep 21h35'37"85d54'
Gérome Daniel
Umi 16h42'47"75d14'
Göbel,Heinz
Peg 23h28'51"13d26'
Göhner,Sieglinde
Sgr 19h38'16"-42d41'
Göhring,Rainer
Gem 6h25'49"12d47'
Gölz 8-3-1946,Hans Jürgen
Ori 5h33'28"-1d31'
Görtz,Anja
Ari 2h25'50"26d34'
Görtz,Heinz-Werner
Aur 5h13'40"44d41'
Götting,Doris
Ari 2h42'23"21d38'
Götting,Rebekka
Lyn 2h28'0"40d53'
Götz,Franz
Aql 19h31'1"8d48"
Götz,Martin
Ori 5h37'22"12d21'
Götz,Monika
Ori 5h55'20"0d16'
Gühring,Marliese
Cap 20h29'20"-20d18'
Gülden,Hermann-Josef
Ori 5h50'21"20d59'
Günter
Cas 0h19'33"60d11'
Günter
Boo 14h18'26"10d18'
Günter,Siegfried
Dra 20h28'16"68d6'
Günter,Steinle Schnuckibär
Tau 5h32'44"28d35'
Günther
Cet 2h40'27"1d21'
Günther,Karin
Tau 3h40'31"0d5'
Günzel,Bodo
Peg 23h31'45"16d48'
Gürtler,Franz
Aql 19h58'18"8d36'

H

H & G 21T
Peg 23h23'0"28d59'
H & H
Sge 20h0'47"16d10'
H & S Galaxy of Stars
Cyg 21h24'1"48d51'
H & Y
Cnc 8h49'41"13d29'
H B Always
Uma 10h18'55"50d17'
H B x 2
Tri 2h9'51"31d44'
H B-994
Uma 11h45'48"64d17'
H Barbara B
Cas 1h58'16"65d30'
H C H II
Per 3h16'35"41d30'
H Dietrich Trapp Luce Terram 1935
Psc 22h53'0"3d13'
H G L
Peg 22h33'59"18d58'
H H
Cep 0h39'1"80d9'
H L Always
Boo 15h0'47"41d28'
H M S Tiger Cub
Ori 5h55'46"16d3'
H N B
Eri 3h21'20"-12d52'
H R D K
Lac 22h53'41"50d28'
H&M Star Paradise
Cam 5h52'0"68d41'
H'Ver
Umi 14h23'58"66d26'
H,Nicole
Cap 21h4'27"-23d9'
H,Sabi
Vir 14h42'35"-0d39'
H-MAN Rocks & Shines After Midnight
Oph 17h52'46"8d1'
Ha BS MS JD Esq, Michael Bruce
Crt 11h31'51"-10d35'
Ha Wai
Cru 12h46'24"-63d55'
Haaarika
Cet 1h7'20"0d57'
Haab,Denise Christine
Cyg 21h50'1"42d9'
Haab,Jessica Elizabeth
Cas 0h9'48"58d9'
Haack,Logan
Uma 9h44'0"42d20'
Haack,Monika
Equ 21h3'36"3d32'
Haag,Christoph und Anja
Lyn 8h18'19"42d33'
Haag,Eveline
Aql 20h8'39"6d13'
Haag,Larry Herbert
Her 18h5'30"28d23'
Haag,Reuben Lael
Dra 14h13'53"64d34'
Haag,Robert B-Patricia J Danaher
Cyg 20h13'16"58d26'
Haage,Madame Anne Marie
Mon 7h51'38"-4d33'
Haak,Brittney Aileen
And 23h22'56"43d19'
Haakansson,Anna
Com 12h9'22"25d57'
Haake 11:02,Hunter Dane
Uma 11h35'35"30d4'
Haake,Caroline
And 23h16'19"46d54'
Haake,Horst-Günter
Lyr 19h17'30"30d25'
Haake,Jack
Sct 18h41'26"-6d38'
Haakert,Manfred
Peg 22h15'31"3d32'
Haalen,Elly van
Vir 13h2'16"-8d22'
Haan,Isabel Bonomo
And 2h21'48"46d21'
Haan,Jerome Clifton Troy
Cyg 20h4'0"31d30'
Haan,Valerie "Dirl"
Cas 0h5'34"63d59'
Haapala Forever,Bruce & Cindy
Cyg 19h24'0"31d34'
Haapanen,Olli
Cep 22h25'47"65d32'
Haar,Benjamin Edison
Lac 22h39'0"53d19'
Haardt,Christel
Cmi 7h20'11"0d3'
Haas IV,Arthur
Hya 8h42'43"2d28'
Haas,Adam Einstein
Aur 0h22'10"38d50'
Haas,Alejandro Julian
Aql 19h9'59"5d31'
Haas,Benjamin Edward
Her 17h27'43"26d26'
Haas,Bev
Ori 6h9'14"9d7'
Haas,Christopher Steven
Aur 7h19'19"38d58'
Haas,Christopher
Cep 20h49'49"76d18'
Haas,Eric
Cep 2h31'12"78d8'
Haas,Evelyn
Uma 9h38'18"71d57'

Haas,Guenter
Lib 15h36'44"-20d22'
Haas,Harry & Lucille
Cyg 19h54'11"48d50'
Haas,Jean
Hya 8h33'21"-6d21'
Haas,Johann
Cnc 9h14'48"12d11'
Haas,John
Cet 2h18'27"4d23'
Haas,Mary K
Cet 2h19'15"5d36'
Haas,Michael
Cap 20h4'29"-10d3'
Haas,Paula Lain
Lyr 18h55'51"30d24'
Haas,Suzanne Marie
Cyg 19h33'37"28d26'
Haas,Ted
Aql 19h33'21"0d26'
Haas,Tracey C
Cyg 20h19'34"38d6'
Haas-Hunter,Muffy, Jamie & Morgan
Cyg 21h6'26"31d22'
Haas-We Love You Dad, Dan Willard
Per 3h47'58"35d28'
Haase,Edward
Eri 3h51'31"-6d23'
Haase,Kera Jo
And 1h42'58"40d53'
Haase,Marco
Cmi 7h28'60"8d42'
Haase,Mary Elizabeth
Peg 22h1'11"24d27'
Haase,Tucker F W
Aql 19h31'10"11d33'
Haayer,Sarah
And 23h37'46"47d12'
Habas,Ashley Sydelle
And 2h0'53"42d16'
Habas,Owen Samuel
Her 16h29'53"42d10'
Habash,Louis
Ori 6h8'54"9d14'
Habedank (För All Tid Pernilla),Richard
Boo 14h27'47"22d9'
Habekovic,Darko
Gem 8h4'19"27d52'
Habenaria Radiata Spreng
Gem 7h14'36"13d35'
Haber,Joan "Our Favorite Captain"
And 0h9'14"38d17'
Haber,Sarah Tamar
Cas 0h53'1"63d7'
Haberek,Paige Mary
Lyn 7h37'40"37d19'
Haberern,Jack
Cep 1h45'32"77d21'
Haberkorn,Falko
Vir 12h26'53"-10d21'
Haberkorn,Manfred "Alonso"
Uma 8h39'55"68d28'
Haberman,Howard A
Her 16h41'0"5d14'
Habert,Marie Christine
Per 3h1'45"31d47'
Haberzettl,Horst
Gem 7h15'50"24d29'
Habetz,Astrid
Lyn 8h22'54"42d18'
Habib,Emad E
Dra 13h26'22"68d7'
Habib,Hayley Ann
Oph 17h20'54"10d27'
Habib,MD,Mohsen A
Dra 16h14'25"68d49'
Habiba Bihi
Sgr 18h54'50"-25d28'
Habibi
Del 20h15'45"14d23'

Habibi(Fred Dawli)
Cam 4h19'52"69d37'
Habiby,Jeff-Julie Pinchuk Love Star
Aql 19h6'43"3d14'
Habich,Son Billy (William)
Cep 23h18'26"65d50'
Habict,Marguerite
Cas 0h35'14"70d45'
Habinek,George A
Her 18h8'0"30d58'
Habingreither,Sheri
Cas 23h24'50"61d15'
Hablitzel,Mary
Leo 9h35'13"8d5'
Hach-Price,Craig & Heather
Cyg 21h7'38"31d3'
Hachem,Letitia
And 23h31'55"40d43'
Hachey,Janette
Lmi 10h0'54"38d15'B
Hachey,Joseph
Lmi 10h0'54"38d15'A
Hachey,Tom
Dra 18h55'42"70d9'
Hachiya,Yvonne
Cam 3h50'35"60d57'
Hack's Hideaway III
Uma 11h26'59"38d10'
Hack,Erik
Dra 14h59'21"63d24'
Hack,John W
Cet 2h4'19"0d42'
Hack,Steven Roger
Lmi 10h29'43"38d51'
Hack,Theodore W
Her 17h24'51"47d10'
Hackbarth,Tony
Cap 21h0'59"-23d8'
Hackedorn,Scott Stanton
Mon 6h55'1"-10d10'
Hackemack,Ingrid
Cmi 7h32'0"0d27'
Hackenberg,Mark C
Dra 16h56'40"65d25'
Hackenberg,Robert & Mildred
Cae 4h28'48"-48d5'
Hacker May 18,1976, Daniel James
Tau 3h52'40"20d1'
Hacker,David Mark
Her 16h59'43"25d34'
Hacker,Eleanor Baird
Cas 22h59'38"54d14'
Hacker,Gale
Aqr 22h21'16"-0d9'
Hacker,O W Mike
Dra 16h36'1"64d7'
Hackett,Amy
Lyr 19h23'18"40d15'
Hackett,Busy Ann
Vul 20h0'46"22d49'
Hackett,Eleanor Beatrice
Cas 1h7'0"50d24'
Hackett,Gillian
Cyg 20h18'30"41d20'
Hackett,John Patrick
Cep 22h54'26"63d33'
Hackett,Kayla Anne
Gem 7h14'1"21d30'
Hackett,Lena Michelle
And 0h0'14"40d20'
Hackett,Michele S
Tau 5h18'24"28d50'
Hackett,Victoria
Aql 19h55'14"15d35'
Hacki,Hans
Ori 5h36'1"-6d20'
Hacking,Robert H
Her 18h9'1"38d32'
Hackler,Jason L
Oph 18h6'21"11d58'
Hackler,Leslie Ann
Cmi 7h54'43"0d53'

Hackman Star,The Paul
Her 17h15'56"48d13'
Hackman,Teresa
Cyg 20h15'46"40d56'
Hackmann,Rachel Anne
And 1h55'0"41d1'
Hackney,Stuart Wilson
Per 2h37'1"46d18'
Hackworth,Christopher
Ser 16h1'44"1d15'
Hacquard,Lorraine
Lmi 10h59'46"30d57'
Hadavi,Dr Nouredin
Aur 5h2'23"40d28'
Hadaway,Scott M
Her 16h41'49"33d33'
Haddad,Gabrielle Nicole
Cnc 9h9'25"32d30'
Haddad,Jacques
Cam 3h36'1"61d21'
Haddad,Nicholas James
Boo 15h41'15"48d23'
Haddad,Pascale Gagnon
Lyn 7h27'35"38d38'
Hadden,Carrie Elizabeth
Lib 15h2'39"-11d14'
Hadden,Charles E
Aql 19h57'30"13d0'
Hadden,John John Paul
Lib 14h59'20"-10d20'
Hadden,Sheila
Lyr 18h34'0"31d34'
Haddock,Eugene C
Tri 1h58'48"25d47'
Haddock,Kay I
Cet 2h1'20"-1d39'
Haddon's Star
Dra 16h49'38"60d18'
Haddon,Timothy J
Her 17h5'1"38d43'
Hadelman Star,Lisa
Mon 7h47'11"-2d43'
Haden's Star
Her 16h21'1"26d55'
Hades Lady For Jill
And 23h27'52"49d49'
Hadesty,Beth
And 23h40'37"45d56'
Hadet,Annie
Per 3h58'25"35d32'
Hadi,Mohammed Faisal
Lmi 10h4'29"35d29'B
Hadjikyriacou,Gavriel
Aur 6h15'10"30d14'
Hadjimichael Star,The
Cep 23h7'34"65d28'
Hadjiyaakov's Shelter, David
Dra 12h3'46"71d35'
Hadland,Nichol Anne
Eri 4h55'52"-9d0'
Hadley
Uma 12h35'18"56d17'
Hadley's Love Forever
Sge 20h17'25"20d10'
Hadley,Cheryl J
Cmi 7h35'57"6d6'
Hadley,Donald S
Per 4h20'6"42d18'B
Hadley,George Albert
Cru 11h57'58"-56d0'
Hadley,Janet
Per 4h20'6"42d18'A
Hadley,Jeannie H
Peg 21h55'1"29d27'
Hadley,Kelly Moore
Aql 18h57'36"11d31'
Hadley,Kenneth T
Uma 10h25'32"40d28'
Hadley,Pam
Mon 6h20'0"5d40'
Hadley,Sarah Marie
Eri 3h34'51"-1d46'
Hadley,The Star of Lauren
Cyg 20h31'13"48d56'

Hadley-Olivero,DeLoris
Cas 0h42'12"68d13'
Hadlington,Jay
Her 17h17'50"42d47'
Hadneberg,Traci Marie
Cas 0h33'42"67d30'
Hadrava,Donald W
Aur 7h22'17"41d17'
Hadsall "Teaching Star",Daniel
Cet 2h15'21"2d37'
Haduch,Nicole Lee
And 2h34'12"50d35'
Haeberle,H B
Aur 7h11'30"36d25'
Haeberle,Heinz
Lac 22h1'47"51d44'
Haefen,Heiko von
Oph 18h29'55"8d16'
Haeflinger
Hya 8h47'27"-0d50'
Haefner,Matthew
Aql 20h7'21"7d32'
Haemer,Joeseph Michael
Her 17h56'48"50d28'
Haendel,Glen
Sgr 19h40'10"-44d15'
Haenel,Judith Sara "Moon Child"
Ori 6h6'21"2d30'A
Haenel,Manfred
Her 17h21'41"42d54'
Haenel,Sylvan & Beatrice
Ori 6h6'20"-0d6'
Haertel,Darton Edward
Aql 20h6'0"4d38'
Haertl-Dorr,Barbara
And 1h20'19"48d54'
Haeuser,Hagen F
Sgr 19h4'58"-28d38'
Haeusler,Thomas
Oph 16h20'18"2d30'
Hafer,James Robert
Aqr 22h39'28"-1d45'
Hafer,Thomas Robert
Aur 6h30'0"37d18'
Haff,Jeanne Marie
Cas 0h26'58"63d40'
Haff,Wilt M
Cap 20h20'30"-20d32'
Haffer,Theresa
Lyn 9h16'1"37d19'
Haffner,Christopher Paul
Cnv 13h54'49"46d2'
Haffner,Kristi Jo
Cas 0h51'29"66d5'
Haffner,Mark Louis
Per 2h59'40"35d43'
Haffner,Paul Matthew
Cep 22h54'50"70d6'
Hafke,Lukas Maria
Leo 9h32'12"8d23'
Hafke,Moritz Maria
Sgr 19h17'29"-28d5'
Hafley,Barry
Cet 2h12'53"5d31'
Hafner,Elizabeth Anne
Cyg 20h35'32"40d29'
Hafner,Ernst D
Her 17h30'59"20d51'
Hafstrom,Ivan T & Dorothy P
Dra 16h36'23"70d24'
Hagadone Star,The
Boo 13h37'14"20d41'
Hagama,George G
Hya 8h58'56"-0d14'
Hagaman,Jean
Peg 23h37'13"20d44'
Hagan,Andrew
Ser 15h37'27"9d3'
Hagan,Eric Michael
Cet 3h0'56"-0d57'
Hagan,Lenny & Doris
Uma 12h5'25"59d49'
Hagan,Lisa Lee
Peg 22h6'0"3d51'

Hagan,Roy & Debbie
Cyg 19h28'0"38d35'
Hagan,Shannon Elizabeth
Ori 5h53'26"14d60'
Hagar,Jack Christopher
Peg 22h26'0"26d10'
Hagar,Terrilyn
Lyn 9h2'31"45d18'
Hagar,The Red Rocker Sammy
Aur 6h56'58"43d58'
Hagata,Brent M
Cma 6h47'53"-17d18"
Hagata,Kathryn B
Hya 8h54'1"0d15'
Hagberg-Gow,Carol
Peg 22h53'40"29d39'
Hagedorn,Klaus
Cmi 7h17'53"4d9'
Hagedorn,Klaus H
Peg 23h28'26"12d23'
Hagel,Margaret Helen
And 0h19'12"33d6'
Hagel,Scott M
Aur 4h50'13"38d47'
Hagelberg,Samuel
Peg 22h25'26"4d5'
Hagelin,Jack F
Boo 15h3'48"28d44'
Hageman,Carol A
Cas 23h29'56"58d52'
Hagemann,Gertrud und Michael
Tri 2h4'0"31d12'
Hagemann,Hans
Peg 23h31'0"17d36'
Hagemann,Joerg
Cmi 7h17'57"7d39'
Hagemann,Shannen Michelle
Mon 6h54'52"-10d4'
Hagen
Lyr 19h13'16"37d56'
Hagen 1-27-93,Harlan Stanley
Aur 6h35'16"37d48'
Hagen,Arts Education Wizard Duane
Peg 22h2'12"31d31'
Hagen,Howard Glen
Mon 7h39'1"-0d48'
Hagen,Jörg
Cnc 8h58'56"8d29'
Hagener,Paul
Eri 2h57'34"-7d46'
Hager
Uma 10h17'34"50d20'
Hager's Star,Jim
Dra 19h26'22"67d48'
Hager,Franz
Cap 20h46'1"-20d8'
Hager,Holly Elizabeth
Com 12h1'20"14d1'
Hager,James Richard
Umi 16h14'47"79d20'
Hager,Michelle Marie
Del 20h22'31"10d40'
Hager,Pamela
Cyg 20h52'54"30d9'
Hager,Ron L
Tau 4h38'17"28d50'
Hager,Tracey
Mon 8h8'38"-7d44'
Hagerling,Tina
And 23h39'0"46d37'
Hagerman,Shara Guttu
Ori 5h29'29"1d43'
Hagerstrom,Ed
Hor 2h59'15"-49d57'
Hagerty,Kelly
Ori 4h53'12"4d2'
Hagerty,Kris
Cet 2h30'1"-18d51'
Hagg,Grant Thomas
Cam 4h17'15"69d60'
Haggar,Hector
Dra 13h16'35"67d32'

Haggard,Tim
Boo 15h2'17"30d17'
Haggart,Laurel
Hya 9h54'49"-16d9'
Haggart,Suzanne
Mon 7h50'44"-1d45'
Haggarty,Patrick J
Ori 5h25'16"-1d14'
Haggarty,Scot
Her 16h46'0"7d41'
Haggerty 100,Ann
Her 16h57'25"27d41'
Haggerty,Donald
Per 2h37'19"37d40'
Haggerty,Linda Lee
Sgr 18h48'4"-29d28'
Haggett,Richard
Aur 5h25'1"38d8'
Hagglof,Judith
Uma 11h33'27"41d11'
Haghighi,Foronzan Pessain
Aql 20h35'15"0d27'
Hagihara,Louise & Yasuhiro
Eri 2h49'15"-3d5'
Hagist,Sophie
Cep 21h42'0"58d50'
Hagiwara Family, Stella of
Cnc 8h46'41"9d2'
Haglund,Barbara Ann
Cas 3h4'1"58d5'
Haglund,Endless Love-Georgia & Ben
Crb 16h23'0"32d8'
Haglund,Torben
Uma 9h16'28"48d56'
Hagmaier,Thomas R
Her 16h12'23"5d33'
Hagman,Jodie
Mon 6h20'29"5d41'
Hagman,Trevor "TJ"
Aur 6h35'15"33d48'
Hagood's Shining Star, Margaret B
Lyr 18h22'43"38d58'
Hagood,Shelley
Equ 20h59'0"3d7'
Hagopian,Nancy Kate
Lyr 18h22'13"44d58'
Hagopian,Sara
Lyn 8h55'50"40d25'
Hagsten,Kristian
Sco 17h53'25"-30d56'
Hagstrom-Dennis(KJHD1) Kimberly J
Cyg 19h40'55"28d28'
Hague,Linda
Eri 2h50'44"-6d45'
Hague,Scott Ryan
Crb 15h54'41"28d27'
Hague,The
Ser 15h56'36"18d17'
Hagy's Star,John Addison
Boo 14h48'22"34d52'
Hahm,Ruth & Zion
Aur 4h51'23"40d17'
Hahn II,Gene L
Oph 17h54'24"10d35'
Hahn Star
Aql 20h35'11"0d19'
Hahn, Paul & Betsy
Sge 19h59'1"16d38'
Hahn,Andrea
Gem 7h5'3"21d7'
Hahn,Angela
And 0h11'1"34d27'
Hahn,Anne Elizabeth
Vul 19h47'36"28d20'
Hahn,Annemarie
Peg 23h34'29"15d21'
Hahn,Bruce
Ser 15h53'32"19d21'
Hahn,Dan F
Dra 17h5'22"58d19'
Hahn,Friedhelm
Cmi 7h29'49"7d44'

Hahn,Garth & Pat
Eri 4h56'1"-8d11'
Hahn,George W
Cet 3h12'1"7d26'
Hahn,Hans-Otto
Aqr 21h42'22"0d12'
Hahn,Hedwig
Cyg 19h24'29"33d16'
Hahn,Jane Marie
Cyg 19h57'33"37d33'
Hahn,Julie
Cyg 19h33'29"31d38'
Hahn,Knut Walter
Lyn 7h48'24"48d41'
Hahn,Michael
Psc 1h37'0"21d37'
Hahn,Oswald
Lmi 9h20'54"38d3'
Hahn,Randy
Cep 21h0'21"60d51'
Hahn,Sanford A
Cep 22h27'49"58d20'
Hahn,Steve
Dra 16h4'45"66d23'
Hahn,Walther
Aur 5h1'12"38d58'
Hahn,Wilhelm
Cas 2h2'24"59d42'
Hahne,Cherylyn M K
Mon 6h21'53"7d40'
Hahne,Linda S
And 2h31'29"50d7'
Hahne,Thomas John
Ori 5h58'1"16d19'
Hai,Kwek Leng
Dra 17h54'0"67d49'
Hai,MHP,Mahmood A
Aql 20h13'11"0d58'
Hai,Sheri
Ori 6h2'38"8d21'
Haible,Dr Winfried
Cnv 13h7'0"33d38'
Haid,Larry L
Dra 18h38'49"67d43'
Haid,William & Kathleen
Crb 15h25'16"30d34'
Haida Sioux V
Cet 1h28'37"0d9'
Haiden,Ashley
Umi 13h49'0"79d22'
Haider,Sherina Ferdouis X
Crb 15h27'20"31d5'
Haidé Costa Vilar31d5'
Lyn 8h1'36"43d40'
Haila
Boo 14h54'26"46d2'
Haig,Russell King Of The Whiners'
Her 17h52'41"28d52'
Haigh,Athan Michael
Hya 8h14'26"-5d45'
Haigh,Christopher
Aqr 21h2'37"-5d56'
Haigh,Diane & David
Cyg 20h22'55"39d39'
Haigh,Nicholas Jonathan
Lyn 8h4'48"41d50'
Haight,Jason Lee
Aur 5h0'1"45d20'
Haight,Jennifer Christie
Mon 7h50'8"-5d21'
Haight,Simon
Cep 21h46'56"55d36"
Haight,Stephen William
Hya 9h30'14"2d17'
Hail,Patricia
Gem 7h5'48"21d25'
Haile,Lynn
And 0h1'49"38d8'
Hailes,Anne
Cas 1h1'36"60d0'
Hailes,Shannon
Vir 11h50'40"3d36'
Hailey
Cet 1h57'38"-1d5'

Hailey's Star
Gem 6h46'1"31d38'
Hailey,Gary
Uma 12h3'51"56d3'
Hails,Clay
Lyn 7h52'54"50d45'
Hailstork,Lyria Virginia
Lyr 18h51'58"37d44'
Haimowitz,Ron
Cet 2h2'13"1d49'
Hain,Erhard
Her 18h20'1"12d53'
Haines,Alan Jay
Her 18h10'49"30d48'
Haines,Chris
Sge 19h29'20"17d60'
Haines,Doug & Lori
Crb 16h19'47"28d36'
Haines,Laura E
And 0h56'11"37d28'
Haines,Lisa M
Del 21h0'32"12d33'
Haines,Marjorie Dralle
Cas 1h9'51"63d11'
Haines,Martin L
Sct 18h43'0"-4d22'
Haines,Russell W
Boo 14h31'0"47d22'
Haines,Shaun
Hya 8h58'13"5d39'
Haines,Sheila Lynn
Crb 15h52'47"27d48'
Haines-Ricco,Jan
Cas 1h4'60"60d1'
Haini
Sco 17h24'1"-30d39'
Hainish,David Mirek
Uma 11h26'58"57d44'
Hainish,Suzanne Margaret
Uma 11h25'30"58d38'
Hains'50',Allan
Uma 10h20'0"54d0'
Hainthaler,Richard
Aqr 22h41'28"-1d26'
Hainzl,Sybille
Dra 15h12'1"62d57'
Hair E Canary
Vul 19h18'41"26d30'
Haire
Lib 14h46'47"-8d9'
Hairr,Randy
Aql 19h35'19"-6d57'
Hairston,Bryan
Dra 9h30'18"74d19'
Hairston,Jennifer
Cyg 21h32'48"40d59'
Hairston,Ronald
Lac 22h26'13"37d47'
Hais,Alan Barry
Aur 5h11'39"41d23'
Haislip,Taylor Jo
Boo 14h11'50"30d16'
Haist,Kurt
Mon 7h51'16"-2d18'
Haitt,Brooks Van
Lyn 7h54'45"43d25'
Haitt,Joyce
Vul 20h2'12"23d18'
Haitt,William
Lac 22h39'30"53d3'
Haizelden,Michael
Cep 22h22'37"56d57'
Hajcak,Pamela Ann
Cas 3h7'56"58d37'
Hajime 1994
Tau 5h27'45"22d17'
Hajkova,Ivana
Boo 13h44'33"12d10'
Hajmiragha,Lynne Marie
Cas 2h3'14"67d49'
Hajmiragha,Sholeh Rhiannon
Dra 14h53'14"64d22'
Hajne,Jr,Alex
Dra 16h54'41"63d59'

Hajnis,Richard Vaclav Vladimir
Her 17h29'37"20d25'
Hak,Amy
Lyr 19h5'1"26d5'
Hak,Louis
Her 17h12'26"20d58'
Hake,Frances"Corky"
Cyg 19h22'23"28d15'
Hake,Jr,Earl Eugene
Aur 6h31'1"38d48'
Hake,Kathryn
Peg 22h38'18"8d36'
Hake,Laura E
And 23h29'43"46d53'
Hakel,Bruce John
Mon 7h49'16"-1d13'
Hakes,Juliana Watkins
And 0h51'46"40d14'
Hakim & Erin
Oph 18h39'39"7d13'
Hakim,Lotty
Peg 22h42'30"18d57'
Hakin,Douglas
Uma 9h46'24"48d10'
Hakola,Tuomas-Henrik- Aleksi
Umi 16h39'41"77d18'
Hakuna Matata
Cam 9h34'49"82d2'
Hakuna Matata Amy We Love You
Mon 7h59'16"-8d1'
Hal
Lib 14h58'27"-1d38'
Hal
Per 2h52'23"48d57'
Hal & Joan
Cyg 19h35'34"28d14'
Hal's Shining Star- Lost Golf Ball
Ori 5h4'49"12d27'
Hal-J
Peg 21h56'25"33d10'
Halaas,Jeffrey Eugene
Aur 6h26'20"37d22'
Halambeck,Chris Donahue & Wendy
Cnc 8h48'12"32d59'
Halberg,MD,Gyula Peter
Aur 6h24'29"37d49'
Halbert,Barbara Louise Cox
Aql 18h54'50"7d41'
Halbert,James Darrell
Sex 10h23'18"0d51'
Halborn,Debra A
And 23h23'27"43d29'
Halboth,Doc
Cep 22h34'21"60d51'
Halbweiss,Dr Dipl Werner Wolfgang
Oph 17h31'48"7d59'
Halcyon-Ed Aagaard
Aql 18h39'23"0d39'
Hale "The Mountaineer",H David
Equ 20h59'24"10d16'
Hale 7-14-46,Patricia J & Robert L
Cyg 21h13'30"38d33'
Hale III,Cal-Boy Richard T
Cep 23h0'23"61d43'
Hale,Anna Linda
Lib 15h36'51"-20d21'
Hale,Beckham
Peg 23h2'30"32d30'
Hale,Blake Thomas
Aql 19h7'60"0d57'
Hale,Carlena
Lyn 7h21'1"58d33'
Hale,Carol Sue
Ori 5h54'0"1d3'
Hale,D
Cam 11h20'1"81d42'
Hale,Darrell Robertson
Her 18h36'44"18d57'

Hale,David Ser 15h25'56"8d28'
Hale,Denny Aql 19h31'0"8d30'
Hale,Dillard O Ori 5h15'41"15d49'
Hale,Dr Sharon L Oph 18h7'39"12d59'
Hale,Ernie Cnv 12h42'56"51d9'
Hale,Eva Flowers Peg 23h20'13"30d37'
Hale,Frank L Uma 9h15'1"48d38'
Hale,Georgina Mary Lyr 18h29'23"46d17'
Hale,Iris Sge 19h56'41"19d25'
Hale,James G Cnv 12h56'46"37d42'
Hale,James Henry Aql 18h42'0"-2d9'
Hale,Jeanine Marie Cyg 19h55'48"37d54'
Hale,Kenneth Aql 19h16'1"13d14'
Hale,Lisa Marie And 1h53'26"40d56'
Hale,Mary Hya 9h11'43"2d22'
Hale,Mickey Per 2h56'32"40d23'
Hale,Nicola And 0h23'30"38d42'
Hale,Pam & David Tri 1h56'36"27d46'
Hale,Poet,Mary Boo 14h43'40"33d29'
Hale,Rhea Donann Mon 7h38'15"-6d34'
Hale,Robert Joseph Cet 0h50'24"0d19'
Hale,Shaina Kaylene Del 20h13'13"15d13'
Hale,Suzanne Lmi 10h37'54"25d12'
Hale,W Fred Per 2h5'1"58d40'
Hale,William John Lyr 18h59'49"40d34'
Hale-Thorpe,Alcinda Peg 22h42'1"24d36'
Haleigh's Happiness Lyn 8h24'57"58d58'
Hales Cyg 20h1'27"30d48'
Hales,Alison Marie Cas 0h29'56"60d21'
Hales,Susan Lynda Lyn 8h22'47"44d60'
Haley Cmi 8h7'31"1d31'
Haley Ann Com 13h8'24"20d14'
Haley Ann And 23h24'32"43d3'
Haley Jo Lyr 18h31'54"30d35'
Haley Rose And 23h1'44"50d10'
Haley's Commet Uma 11h37'17"46d36'
Haley,Alexander Aron Her 16h15'34"50d29'
Haley,Bob Sco 17h4'17"-30d32'
Haley,Charles Milton Cmi 7h7'34"7d3'
Haley,D Sue Cas 0h0'49"50d16'
Haley,Dorothy Lib 14h59'22"-7d32'
Haley,Douglas Cet 3h13'54"1d54'
Haley,Eddie & Mel Uma 9h53'30"68d21'
Haley,Genevieve Adel & Grace Marie Del 20h38'15"18d53'
Haley,Glenn Edward Tau 4h18'51"21d29'
Haley,Irene Gem 6h19'14"18d53'
Haley,Jamie Umi 21h11'39"88d25'
Haley,Jennifer And 2h33'27"45d50'
Haley,Martin William Aur 6h3'0"46d36'
Haley,Michael Her 16h53'17"25d9'
Haley,Nettie Opal And 23h32'32"49d36'
Haley,Paul & Jennifer Cyg 21h6'55"30d51'
Haley,Reed James Lac 22h23'22"52d32'
Haley,Shelley And 1h58'19"37d53'
Haley,Shula Com 12h33'56"25d31'
Haley,Suzanne Cyg 20h50'43"38d33'
Haley,William Dalton Ori 4h45'37"4d20'
Halflora Aql 18h44'55"-0d26'
Halfon,Jean-Michel Cas 23h39'59"60d20'
Haliburtan,Ruth Cas 3h10m18"60d8'
Halibut Cove-Solstice '92 Uma 9h40'32"42d20'
Haliczer,Kevin Aur 6h54'54"37d46'
Halie Rachel Uma 10h8'31"50d52'
Halikas,James Oph 18h36'56"7d27'
Halkedis,Semele Evangeline Peg 23h43'27"25d50'
Halko,Tiffany Tri 2h4'40"31d22'
Hall Cet 2h56'16"0d7'
Hall 01/19/47,David Charles Cep 22h54'55"59d28'
Hall 10-1-1970,Thomas Neil Cnv 13h27'10"51d37'
Hall 10-31-95,Kathryn Dayle Peg 22h30'24"26d16'
Hall Family Star,The Kevin & Wendy Cyg 19h34'40"34d29'
Hall Happy Valentines Day,Cline W Boo 15h16'22"40d22'
Hall II,Lyle Edgar Uma 10h27'52"47d53'
Hall Star,Joedy & Hi Cnv 12h30'0"34d20'
Hall Warsaw's Star, Judith E And 2h16'14"48d52'
Hall Wikki-Wikki, Michael W Cam 6h5'36"58d29'
Hall,Adam S Cyg 20h24'42"37d32'
Hall,Albert Mon 7h44'59"-8d50'
Hall,Albert & Diane Cyg 20h44'44"45d34'
Hall,Alexander David Webster Umi 17h26'13"86d3'
Hall,Amy B Hya 9h18'6"-14d53'
Hall,Anthony David Umi 15h5'36"71d49'
Hall,Arline Cet 3h8'28"2d7'
Hall,Ashley Ann Cet 2h42'38"-17d47'
Hall,B Dra 18h8'59"58d48'
Hall,Barry Wayne Lac 22h30'52"53d12'
Hall,Benjamin Brian Her 17h13'24"27d47'
Hall,Beverly Ann Eri 3h50'21"-2d56'
Hall,Bradford W Aur 7h22'1"40d29'
Hall,Brenda Miriam Cas 0h36'43"73d13'
Hall,Brian Man Mon 7h13'44"-9d48'
Hall,C W Ori 5h53'31"12d9'
Hall,Carol Peg 22h38'36"20d13'
Hall,Celestial Bill Cnc 8h23'1"7d22'
Hall,Charles E Peg 21h54'35"28d18'
Hall,Charlie Lac 22h21'14"55d50'
Hall,Charlotte Ruth Webster Umi 16h51'0"76d7'
Hall,Chelsea Alana And 2h21'57"38d21'
Hall,Christine M Cas 0h19'32"63d51'
Hall,Christopher Uma 8h20'32"71d19'
Hall,Christopher Lunn Boo 13h37'0"18d51'
Hall,Darlene & Bobby Lyr 18h41'28"30d38'
Hall,David G Aql 19h14'48"12d47'
Hall,David Jonathan Cyg 21h2'1"39d28'
Hall,Dennis Dwayne Aql 19h56'1"15d41'
Hall,Diane Lyn 9h9'1"40d6'
Hall,Don R Oph 16h28'28"-5d57'
Hall,Donald Clifford McCrorey Hya 8h30'19"-0d11'
Hall,Dorothy Shair Hya 8h31'21"-6d39'
Hall,Dorthy Evelyn Hamilton Aql 19h9'34"13d9'
Hall,Dr Thomas C Oph 17h33'49"-23d16'
Hall,DVM,Susan Lyn 9h16'32"35d17'
Hall,Edward W Dra 16h53'12"66d29'
Hall,Elaine Marie Cnc 9h0'60"8d17'
Hall,Elizabeth Cyg 20h0'13"40d14'
Hall,Elizabeth Ann Com 12h59'46"20d41'
Hall,Elizabeth Jane Lyn 7h58'1"39d27'
Hall,Erwin Ari 2h58'34"28d55'
Hall,Estelle Mon 6h53'39"-3d10'
Hall,Evan Eri 3h34'39"-5d25'
Hall,Fire Fighter/ Paramedic Alba Cma 6h53'23"-18d22'
Hall,Gareth Cyg 19h32'49"39d44'
Hall,Gillian Lyr 18h29'23"46d56'
Hall,Gregory B Aur 5h17'42"45d13'
Hall,Hadley Makay Peg 22h14'20"33d46'
Hall,Hart Aql 19h10'37"18d50'
Hall,Henry Burke Aql 20h19'20"1d29'
Hall,Howard H Ser 15h28'25"18d48'
Hall,Ida Psc 0h58'58"22d2'
Hall,Irene Cyg 21h35'20"40d38'
Hall,J R Tri 2h34'26"35d42'
Hall,Jacquelyn Marie And 1h35'16"36d57'
Hall,Jade Sonny Reebok Uma 8h33'37"56d54'
Hall,James C & Marion F Boo 15h14'1"52d36'
Hall,Janet Lyr 18h57'33"42d42'
Hall,Jared Cyg 19h55'32"37d58'
Hall,Jennifer C Ori 5h52'46"18d50'
Hall,Jennifer Lee Com 12h5'47"24d44'
Hall,Jennifer Lynn And 0h50'32"36d42'
Hall,Joan LaRochelle Cyg 20h16'17"31d48'
Hall,Joe LeRoy Tau 5h45'30"18d49'
Hall,John Her 16h55'1"29d30'
Hall,John David Aur 5h4'50"40d19'
Hall,John W Per 4h46'27"51d19'
Hall,Joseph & Virginia Sge 20h6'57"20d7'
Hall,Joseph I Sct 18h38'54"-5d29'
Hall,Joshua Caleb Cnv 12h30'1"38d58'
Hall,Jr,James Newton Cep 21h41'42"61d1'
Hall,Judy Cas 0h26'1"61d19'
Hall,Kathy J Lyr 18h46'23"40d26'
Hall,Kay Lmi 9h30'46"38d18'
Hall,Kelly Elizabeth Per 4h7'35"40d38'
Hall,Kerry N Del 20h37'20"3d13'
Hall,Kim M And 23h23'0"45d7'
Hall,Kimberly & Deborah Del 20h14'20"11d10'
Hall,L A Uma 12h48'25"53d53'
Hall,Larry Neal Psc 1h40'13"22d19'
Hall,LeAnne D And 2h16m27"38d50'
Hall,Lisa Ann Lac 22h24'1"56d46'
Hall,Lisa Marie Peg 22h34'0"29d42'
Hall,Lonnie Uma 10h20'38"71d52'
Hall,Lori Saunders Cas 0h25'55"69d32'
Hall,Lowell Reuben Ari 2h34'58"30d43'
Hall,Malcolm And 0h51'53"34d60'
Hall,Malinda Crawford Umi 12h35'41"88d45'
Hall,Mallissa L And 2h23'1"44d29'
Hall,Marci & Thierry Coulet Mon 7h42'30"-8d26'
Hall,Marcus Per 1h53'58"54d17'
Hall,Martha Susan Cas 1h38'1"63d52'
Hall,Mary Cam 3h23'15"61d40'
Hall,Mary Uma 11h47'16"49d47'
Hall,Mary Ellen Cyg 21h11'1"38d57'
Hall,Mary N And 0h20'45"35d45'
Hall,Mason Daniel Aql 18h59'26"-6d41'
Hall,Maureen Peg 23h19'17"30d18'
Hall,Megan Eileen Lyn 7h59'55"44d57'
Hall,Meri Boubala Gem 6h59'52"17d30'
Hall,Michael John Aur 7h17'23"37d24'
Hall,Michael Scott Her 18h19'40"12d23'
Hall,Michael Worthy Oph 16h50'0"10d33'
Hall,Mimi Lib 15h32'36"-28d15'
Hall,Morgan Per 1h47'35"48d20'
Hall,Nancy Sue Mon 6h23'16"4d57'
Hall,Nancy Violet Cas 0h59'59"64d1'
Hall,Nicholas John Her 16h58'18"24d48'
Hall,Nicole Kelsey Lyn 6h29'37"59d11'
Hall,Parry Alan Aql 19h57'26"8d27'
Hall,Pat Lyr 18h52'38"31d14'
Hall,Patricia Anne Peg 23h35'1"30d43'
Hall,Peggy Jean Aur 6h32'14"38d42'
Hall,Philippa Anne Mon 7h28'11"-6d31'
Hall,Rachel Elizabeth- Anne Lyn 7h52'51"58d17'
Hall,Reuben Devon Eri 2h58'0"-1d34'
Hall,Robin Lynn & Michael Blaine Per 3h42'28"45d31'A
Hall,Robin Stephen Cep 22h2'45"60d27'
Hall,Rock Leo 10h22'0"14d19'
Hall,Roger Aql 19h52'31"10d32'
Hall,Ronald Kamanna Psc 0h58'38"20d42'
Hall,Russell E & Helen E Cyg 20h3'0"30d38'
Hall,Rusty Lac 22h14'0"38d40'
Hall,S Ann And 23h49'42"40d41'
Hall,Sarah Elizabeth Lyn 7h22'13"58d59'
Hall,Scott Loren Boo 15h14'58"28d34'
Hall,Sharon And 0h50'33"38d56'
Hall,Sharon Cam 3h54'38"77d48'
Hall,Sheri G D Oph 18h6'0"13d47'
Hall,Sophie & Jessica Cmi 7h34'33"0d13'
Hall,Sr,Richard D Cma 6h56'52"-19d32'
Hall,Stacie R Lyn 7h26'25"44d60'
Hall,Stuart Marius James Per 2h25'17"54d48'
Hall,Tammany Lyr 19h10'56"40d39'
Hall,Tania Chantel Lyn 9h6'54"35d36'
Hall,Teri Gail Cyg 20h6'27"40d45'
Hall,Terra Felicia Mon 6h18'20"9d37'
Hall,The Honorable Robert H Aql 18h47'1"10d8'
Hall,Theodore David Ori 5h56'31"17d15'
Hall,Thomas Adrian Ori 5h6'40"1d13'
Hall,Thomas L Cmi 7h26'52"7d37'
Hall,Thomas M Cam 5h50'0"65d33'
Hall,Tina Ann Cas 1h28'16"73d54'
Hall,Toni Marie Lyr 18h19'49"44d30'
Hall,Victoria Louise Mon 6h21'39"8d20'
Hall,Wanda Elizabeth Sarah Simpson Peg 22h0'29"28d21'
Hall,William & Eugenia Cyg 21h1'33"33d44'
Hall-Ford,Maura Elizabeth Cas 0h25'46"69d46'
Halla"Babe",Dorothy Cyg 21h32'13"42d38'
Halladay,Robert B Del 20h13'41"9d28'
Hallam,Ben Cam 3h58'39"63d37'
Hallam,Jacqueline Claire Per 4h41'42"40d29'
Hallam,Maria V And 0h16'46"34d40'
Hallam,Peter Her 16h52'20"40d52'
Hallas,Dorothy Uma 9h50'1"46d34'
Hallas,Joseph Per 4h45'1"48d59'
Hallas,Tammy Uma 9h45'35"48d46'
Hallaxs,Lisa L Lmi 9h29'12"38d28'
Hallbauer,Katherine Elaine Aur 6h26'40"48d49'
Hallbauer,Steven Wilson Cas 23h2'52"53d21'
Hallberg Family Star, The Cep 23h32'19"65d46'
Hallberg,Timothy "Buddah" Cnv 13h49'32"32d2'
Hallchurch,Andy Cam 4h32'18"65d3'
Hallden,Trisha Cyg 20h7'1"38d31'
Halle Cam 6h46'59"68d54'
Halle,Jeffrey Aur 6h0'0"30d53'
Halle,Peyton Page Lee Oph 17h23'36"-20d32'
Hallee Mon 6h56'14"0d58'
Halleh Lyn 8h3'33"40d58'
Hallenbeck Cam 5h9'21"67d55'
Haller,Dr Peter Sco 17h26'56"-38d14'
Haller,Edwin Hya 9h7'29"5d14'
Haller,Lewis Her 17h15'0"42d23'
Haller,Lindsey Cyg 21h52'0"37d26'
Haller,Lothar Ori 5h35'29"7d53'
Hallereau Guillaume And 23h5'28"40d5'
Hallett III,Nez Conrad Umi 16h29'54"75d30'
Hallett,Kathryn J Com 12h29'17"26d16'
Hallett,Sr,William James Peg 23h37'54"11d2'
Halley,Hannah Rose Tri 2h7'31"31d34'
Halliburton,Carolyn & Michael Umi 16h21'0"71d58'
Halliday,Corinne Uma 11h10'18"38d50'
Hallie Eri 3h51'57"-5d45'
Halligan's Wishing Star,Steve Boo 14h26'30"39d29'
Halligan,David Timothy Paul Dra 15h16'58"57d45'
Hallikainen,K E "Dad" Crb 15h32'52"31d2'
Hallinan,Jr,Robert J Per 4h25'37"50d59'
Hallissy,Marianne Peg 22h40'50"27d49'
Hallitt,Dr John William Her 16h35'0"47d49'
Hallman,Sasha Nicole Eri 4h35'35"-11d5'
Hallman,Webb D Aur 4h51'56"50d52'
Hallmann,Charles A Dra 16h36'19"72d47'
Hallmann,Tracy Lynn And 22h56'13"37d55'
Hallock,Jenna Marie Gem 6h49'11"12d23"
Halloran,Devin Patrick Aql 20h0'53"0d28'
Halloran,Dillon Hamilton Boo 14h55'50"42d59'
Halloran,Kathleen And 23h45'47"41d9'
Halloran,Katie Peg 23h29'46"32d35'
Halloran,Kimberly & Joseph Cyg 19h17'48"45d37'
Halloran,Marijane Cas 0h53'60"66d28'
Halloran,Nicole And 0h29'43"44d41'
Halloween Lyn 6h58'29"54d35'
Hallowes,Paul & Mary Eri 3h19'13"-2d24'
Halls,Aaron Lyn 8h14'12"44d9'
Halls,Les & Laura Ori 5h41'25"0d16'
Hallum,Patricia Ann And 0h10'41"31d37'
Hallum,Ryan Hya 9h20'45"2d48'
Hallwood,Claire Louise Cyg 20h21'34"39d2'
Hallyday,Johnny Lyn 8h12'58"39d53'
Halm,Alexandre Per 3h52'48"36d8'
Halmes,Nicholas Anthony Boo 14h6'39"44d51'
Halonen,Sami Dra 17h28'1"72d32'
Halpain,Shelley Cas 0h5'49"63d58'
Halper,Cynthia Peg 0h3'27"27d42'
Halper,Cynthia Boo 14h9'25"50d31'
Halper,Sharron Lynne Uma 10h8'22"52d2'
Halperin,Dan Cep 22h50'56"57d28'
Halpern 50th Anniversary Star,The Cyg 21h26'43"41d8'
Halpern,Janice Henry Cep 23h15'54"70d2'
Halpin's Anniversary Star,Michael Cep 22h49'21"56d51'
Halpin,Harold J Vir 13h12'14"-1d4'
Halpin,Joan Rita Bill Vul 19h58'43"22d48'
Halpin,Joseph Dean Her 16h43'53"39d34'
Halpin,Marisa Ann And 23h32'12"42d20'
Halpine's Star, Roxanne Lyr 18h56'41"30d16'
Halpren,Betty Cyg 21h7'49"30d7'
Hals,Erik David Ori 5h27'25"0d6'
Hals,Pam Cas 0h22'19"72d7'
Halsey,Stewart John Cyg 21h16'58"28d40'
Halstead,Edward Alva Dra 20h14'46"61d51'
Halstead,Melanee Lynn Mon 7h11'15"-5d50'
Halstead,Norma Jean Mon 6h37'0"-1d0'
Halsted Street School Class of 1994 Per 2h24'50"54d30'
Haltzman,Jay"Double J" Per 1h36'11"52d51'
Haluska,Sharon Lynn Vul 19h3'0"21d29'
Haluzak,Sallyann Vul 19h42'0"25d15'
Halverson,Ed Aur 7h13'37"37d17'
Halverson,Jeffery Michael Boo 14h29'41"27d31'
Halverson,Jennifer Michele Crb 15h37'57"26d13'
Halverson,Kris Marie Lyn 7h55'34"38d4'
Halverson,Louise Joyce Cyg 21h23'42"38d50'
Halverson,Lyman G Lac 22h20'57"48d49'
Halverson,Michael Hya 9h0'22"1d10'
Halvorsen,Joann And 23h40'40"45d45'
Halvorson,Patricia J Peg 22h29'31"29d44'
Halvorson,William A Ori 4h53'60"4d7'
Halwany,Mark Badr Cet 1h57'25"-6d52'
Ham,Edwin Earl Lyn 7h27'34"44d6'
Ham,Tae Jun Uma 8h35'45"52d53'
Ham,Tony Scl 23h7'42"-30d34'
Hamacher,Ralf Gem 6h46'41"12d48'
Hamack,Emily Rebecca Cas 23h39'32"61d30'
Hamad,Sarah Metzger Cyg 20h8'30"41d12'
Hamada,Shogo Cap 20h53'8"-15d30'
Hamady,Maya Cet 1h24'20"-1d44'
Hamalainen,Heikki Uma 12h38'17"59d6'
Haman,Gerald Cap 21h25'40"-23d51'
Haman,Thomas W Her 18h54'11"12d47'
Hamand,Rosemary Gem 5h58'54"26d58'
Hamann,Darrel & Karen Dra 14h2'44"64d59'
Hamann,Jens Peter Ari 2h52'14"21d7'
Hamann,Juergen Hya 9h35'59"5d1'
Hamann,Marvin F Aur 5h6'60"42d38'
Hamaoui,Sammer Aur 5h18'35"43d10'
Hambas,Valerie Boo 14h13'0"40d16'
Hamberg,Denise Cas 0h28'30"60d20'
Hamblen,Jr,Lisa & George Per 3h42'48"50d5'
Hamblin,Jack Anthony Cyg 21h17'21"35d9'
Hamblin,Matthew Aur 6h22'24"32d30'
Hambly,Ginger Equ 21h19'58"8d5'
Hambrick,John Aql 18h58'33"-1d4'
Hambright,Marilynn June Oph 18h5'55"13d4'
Hambright-Newman And 1h6'44"39d54'
Hambro's Dark Star,Lee Aqr 21h0m1"-0d14'
Hambro,Marina Isabella Kimberley Psc 0h56'53"31d8'
Hamburg Süd Ori 5h53'59"7d44'
Hamburg,Brandon Allen Peg 22h48'44"7d49'
Hamburg,Chase Christian Dra 16h35'34"60d1'
Hamburger Dad Hya 9h4'30"-6d31'
Hamburger,David A Ser 15h11'42"-2d52'
Hamburger,Dr Ulrich Ori 5h30'60"1d49'
Hamby,Darrell Lynn Ori 5h34'1"1d20'
Hambücker,Hans Peter Vir 11h51'15"3d41'
Hamel,Hans Sco 17h26'48"-43d49'
Hamel,Misti Peg 22h7'28"26d41'
Hamel,Shea Danielle Cyg 21h12'51"38d18'
Hamel,Sr,Roland Joseph Aur 5h12'31"42d53'
Hamer,Robert C Ser 15h16'29"6d5'
Hamera,Anthony Alexander Aur 7h18'0"36d27'
Hamerich,Adolf Tau 5h56'48"23d7'
Hamerling Star,The Karen & Barry Eri 4h13'32"-12d29'
Hames,Olivia Sge 20h3'33"16d11'
Hamid,Aryan Per 1h44'1"53d28'
Hamid,Ashley Ori 4h42'25"14d31'
Hamilton Uma 8h55'49"49d29'
Hamilton & Wells Cyg 21h25'16"40d21'
Hamilton Hall of Fame Star,Gregory Lmi 9h47'36"37d57'
Hamilton III,Sammy J Cet 2h56'0"2d11'

Hamilton Star,The
Cam 8h21'56"74d16'
Hamilton Star,The
Dra 11h13'40"73d33'
Hamilton Star,The
Cet 2h11'56"6d35'
Hamilton's Horizon
Aur 6h21'48"32d42'
Hamilton,Alexander Scott
Cet 1h35'10"-10d30'
Hamilton,Amy
Vir 13h25'17"-1d51'
Hamilton,Anne Lehr
Aqr 21h52'55"-1d35'
Hamilton,Cacey Elizabeth
Ari 3h13'43"28d47'
Hamilton,Crystal R
Cas 23h39'44"64d54'
Hamilton,Daphne
Umi 13h9'24"71d46'
Hamilton,David
Ori 4h43'13"15d12'
Hamilton,David John
Per 3h12'29"42d15'
Hamilton,David Mark
Cet 2h50'27"1d58'
Hamilton,David Wendell
Dra 11h44'31"72d32'
Hamilton,Dennis
Ser 16h6'49"14d26'
Hamilton,Dottie
Peg 22h38'39"20d30'
Hamilton,Elliott
Boo 15h7'32"12d59'
Hamilton,Eternally Diana
Cas 23h14'1"60d15'
Hamilton,Ginette
Com 13h18'34"26d15'
Hamilton,Heather
Aql 20h18'24"7d51'
Hamilton,Jacob
Per 3h25'19"41d0'
Hamilton,James Dickson
Dra 15h8'31"60d50'
Hamilton,Jane Kaski
Mon 6h58'54"8d16'
Hamilton,Janell & Tom
Pledger,Jr
Del 20h51'52"7d34'
Hamilton,Jeannie
Cas 0h36'54"60d19'
Hamilton,Jennifer L & Todd E
Lmi 10h27'51"34d53'
Hamilton,Katie
Peg 22h32'56"25d44'
Hamilton,Kirk Lewtz
Per 2h0'1"50d36'
Hamilton,Larry Jerome
Cep 23h10'47"60d32'
Hamilton,Lynsey Julia
Lyn 7h48'1"48d28'
Hamilton,Marcus
Boo 14h44'26"28d13'
Hamilton,Martha Lauren
Mon 6h54'56"0d36'
Hamilton,Mary Janine
Aql 19h55'27"0d39'
Hamilton,Maureen E
Eri 2h52'20"-15d13'
Hamilton,Morag
Aql 20h10'41"10d19'
Hamilton,Nicole D
Com 12h24'54"23d43'
Hamilton,Peggy
Cmi 8h2'36"1d44'
Hamilton,Philip Alan
Dra 14h58'17"63d6'
Hamilton,Rev Charles &
Covette
Aur 6h9'48"33d7'
Hamilton,Richard D
Dra 17h13'42"61d12'
Hamilton,Sarah Jane
Umi 13h15'1"71d24'
Hamilton,Shannon Antle
Boo 13h46'38"14d10'
Hamilton,Terry
Lmi 10h46'25"23d11'
Hamilton,Theodore (Dick) R
Aql 18h43'51"6d57'
Hamilton,Thomas
Cep 21h56'44"56d13'
Hamilton,Todd
Aql 19h7'26"3d35'
Hamilton,Wesley Dalton
Peg 22h18'0"11d4'
Hamilton,Wilma
Sge 19h57'0"16d33'
Hamilton-Megami,Dina
Mon 8h3'49"-1d16'
Hamiluis,Arnaud
Ori 6h7'1"0d24'
Hamish's Snuffleuppogus
Cen 11h51'4"46d27'
Hamlett,Lloyd Thomas
Ori 5h32'15"-0d51'
Hamlin,Bradley Sean
Cep 22h44'58"78d37'B
Hamlin,Glenda Sue
And 23h3'29"51d35'
Hamlin,Jeff
Aur 5h54'1"31d20'
Hamlin,Kimberly Ann Marie
Cep 22h44'58"78d37'A
Hamlin,Sensei George
Aqr 23h35'49"-12d3'
Hamlisch,Marvin
Her 18h3'54"38d1'
Hamlyn,Helen
Cas 0h48'32"61d17'
Hamlyn,Paul
Cep 21h15'1"58d20'
Hamm
Dra 18h56'38"58d9'A
Hamm aka The Hammster,
John
Her 15h54'20"46d41'
Hamm,Dawson
Lac 22h27'45"56d42'
Hamm,Jr,Joseph N
Dra 16h47'17"68d1'
Hamm,Julia Ann
Vul 20h43'60"20d34'
Hamm,MD,Jeffrey T
Lac 22h0'0"38d26'
Hamm,Michele Nicole
Crb 16h23'57"32d21'
Hamm,Petra
Lyn 8h10'30"47d34'
Hamm,Robert E
Tri 2h15'40"31d25'
Hamma,Brad
Oph 16h56'38"-28d21'
Hammack,Greg & Hatti
Com 12h58'55"17d32'
Hammack,John Daryl
Uma 10h26'0"50d49'
Hammadh,Fatma
Mon 7h48'44"-4d26'
Hamman-Lifesaver,Ernie J
Hya 9h14'36"6d33'
Hammar,Raymond L
Uma 9h41'16"68d11'
Hammarin,Frank & Ross
Cep 22h6'28"60d57'
Hammel,Debbie Brehl
Cnv 12h54'42"40d30'
Hammel,Dr Herbert
Cep 22h58'22"65d21'
Hammer I
Cap 21h4'17"-24d45'
Hammer,Benjamin Scott
Leo 9h52'30"31d30'
Hammer,Birdie
Lyr 18h46'33"30d4'
Hammer,Bruce Edward
Cet 0h43'24"-2d53'
Hammer,Coach
Aqr 22h37'51"-0d11'
Hammer,Joachim
Dra 18h31'1"68d37'
Hammer,Joe
Sex 10h35'21"2d8'
Hammer,Julie Marie
Mon 6h23'13"2d46'
Hammer,Lazarus
Aql 19h0'14"-8d41'
Hammer,Mariana Elizabeth
Aur 6h25'43"37d57'
Hammer,Sandra
Vir 11h53'37"-2d1'
Hammer,The
Leo 10h52'0"20d48'
Hammer-(Movie Star), Casey
Cam 3h57'30"69d23'
Hammerand,Russell John
Boo 14h45'0"52d12'
Hammerle's "Hideaway", Fred
& Emily
Ori 5h53'35"16d45'
Hammerman My Own Star,
Emily A
Tri 2h10'11"33d12'
Hammerman My Own Star,
Leah M
Lmi 10h33'20"31d47'
Hammerman,Beth
Cas 0h36'20"68d39'
Hammershoj,Lisbeth
Cas 2h32'11"61d22'
Hammersley,Robin Lee
And 0h2'24"46d11'
Hammersmith,Gennie
Mon 6h54'54"-0d58'
Hammes 25th Anniversary
Star
Cyg 21h52'1"40d12'
Hammett,Cierra
Per 2h59'47"46d21
Hammill,Kenneth Ellsworth
Oph 17h35'6"-22d0'
Hammink,Norman
Aur 6h7'12"45d22'
Hammock,Tom
Sex 10h28'56"0d51'
Hammond
Del 20h21'21"10d48'
Hammond,Andrew D
Aur 5h16'55"46d16'
Hammond,Andy
Cma 7h47'52"10d56'
Hammond,Charlotte Eva
And 22h59'39"37d17'
Hammond,Dale & Pearle
Crb 16h1'1"32d46'
Hammond,Don Paul
Tau 5h20'1"20d9'
Hammond,Jocelyn K
Lyn 7h37'0"40d23'
Hammond,Kelvin
Umi 15h9'42"66d34'
Hammond,Kent A Dad
Granpap Kenner
Sex 10h26'57"-8d22'
Hammond,Kimberly & Laura
Cyg 21h7'0"37d48'
Hammond,Leah G
Mon 7h3'19"1d40'
Hammond,Michael
Aql 19h18'25"14d15'
Hammond,Michaela Louise
And 0h3'26"31d26'
Hammond,Ms Frances M
Hya 8h44'24"2d59'
Hammond,Neil
Ori 6h6'20"1d36'
Hammond,Richard H
Her 16h6'55"23d50'
Hammond,Richard Lee
Per 2h56'16"38d28'
Hammond,Robert L
Cep 21h10'35"67d48'
Hammond,Stacey
Scl 0h1'34"-27d53'
Hammond,Susan Lesley
Gru 23h14'6"52d8'
Hammond,Susan P
Lyn 8h53'56"44d13'
Hammond,Valerie
Eri 2h59'10"-4d59'
Hammond,Wayne
Lac 22h53'45"54d49'
Hammond,William Earl
Hya 8h52'54"0d31'
Hammond-The Pretty
One,Kathie
Cyg 20h18'53"38d39'
Hammonds,Jr,Gordon
Sct 18h54'33"-6d22'
Hammonds,Kalan Scot
Hya 8h54'21"-1d49'
Hammonds,Patrick V
Aur 6h4'33"38d2'
Hammons,Chasity
Cnc 8h27'12"8d51'
Hammons,Helen Martha
Mon 6h39'58"-10d48'
Hammontree,Rosalinde D
Eri 4h8'54"-0d28'
Hammou,Rabia
Ser 18h30'27"0d43'
Hammoud,Mr & Mrs Sadek
Lyn 6h19'17"61d43'
Hamner,Marian
Lib 15h40'50"-28d37'
Hamnett,Rebecca Ruth
Marjorie
Lyn 8h9'47"44d20'
Hamor,Mae C
Cet 2h5'38"0d57'
Hampe,Joseph K
Per 2h2'37"48d48'
Hampe,Robert
Uma 10h48'59"50d41'
Hampel,Brian David
Dra 19h24'19"56d59'
Hampel,Joe
Cam 3h35'33"63d42'C
Hamper Star,The Fabulous Fay
Cas 2h22'26"65d7'
Hampshire,Dave
Dra 16h38'1"67d10'
Hampson,John Douglas
Lib 15h3'45"-10d26'
Hampstar
Aql 19h25'11"10d26'
Hampton's Wish
Per 1h29'31"52d32'
Hampton,Crystal A
Uma 9h51'38"53d12'
Hampton,Dawn Regene
Cas 2h7'39"62d6'A
Hampton,Denise René
Cas 2h7'39"62d6'B
Hampton,Diane
Lyr 19h23'56"30d37'
Hampton,Jamie
Eri 4h13'40"-12d15'
Hampton,Jeffrey Kirk
Cnv 12h43'34"51d40'
Hampton,Jon Andrew
Lac 22h49'43"35d33'
Hampton,Leah Kathleen
Cam 5h41'57"67d54'
Hampton,Matthew Byron
Aur 4h56'0"50d56'
Hampton,Richard Hal
Ser 18h41'15"4d32'
Hampton,Sandi
Lmi 10h51'1"27d10'
Hampton,Wendy
Ori 5h13'37"-5d18'
Hamre,Brian
Dra 18h56'22"48d51'
Hamre,Henry George
Aur 5h55'0"40d43'
Hamre,Margaret Eileen
Uma 11h3'16"40d5'
Hamrick,James Ray
Ser 16h6'0"5d35'
Hamrick,Narelia D
Equ 21h20'51"7d57'
Hamrick,Ronda Renee
Eri 3h14'22"-1d51'
Hamrick,Roy E
Ori 5h54'24"14d42'
Hamrick,Walter B
Oph 17h33'38"-24d32'
Hamrock,Constance
Cas 0h31'36"61d13'
Hamsa
Cmi 7h44'0"0d12'
Hamson,John William
Dra 11h50'13"70d51'
Hamson,Yvonne DeSala
Cas 0h45'42"64d28'
Hamster
Cep 22h32'28"70d41'
Hamster `93
Cas 0h11'56"61d24'
Hamwi's,John "Little Annie's
Eating House"
Uma 10h17'23"61d24'
Hamza,Jerold A
Hya 8h55'1"1d33'
Han Lien
Boo 14h50'18"24d52'
Han,Alex Alexis Jessica
Ori 6h8'50"8d54'
Han,Corey D
Cas 1h15'29"61d20'
Han,Gabrielle Song-Yi
Cas 1h4'14"70d21'
Han,Nora
Cmi 7h32'18"0d29'
Hana's Mitzvah
Uma 8h40'52"60d31'
Hanaburgh,Donald
Lyn 7h50'55"35d5'
Hanabusa,Kazumi
Aqr 22h2'41"-23d45'
Hanagan,Nancy E
Cnc 9h1'11"7d0'
Hanan,Eric D
Boo 15h7'47"50d9'
Hanan,Joseph
Her 17h13'1"26d21'
Hanbury,Robert K
Her 16h54'0"50d24'
Hanchar,John A
Her 16h34'27"42d10'
Hancher,Earl L
Aur 6h6'14"45d42'
Hanchey,Janet G L G
Vir 12h28'10"-5d48'
Hanckel,Sunny
Sct 18h53'1"-6d40'
Hancks,Dean
Hya 8h47'23"-01d38'
Hancock & Muse, William S
Ori 5h38'53"7d43'
Hancock Family Star, The
Uma 11h57'43"50d3'
Hancock,"Rich"
Oph 17h27'17"10d3'
Hancock,Anthony
Cnv 13h17'21"38d35'
Hancock,Gaia May
Aql 18h58'39"16d29'
Hancock,Graham J
Her 17h13'49"28d17'
Hancock,Jim
Cnv 12h51'20"51d23'
Hancock,Joyce A Matney
Peg 22h33'17"20d39'
Hancock,Jr,Michael Lee
Cma 6h55'51"-17d12'
Hancock,Kimberly Lane
Vaughn
Uma 12h11'48"57d46'
Hancock,Loran
Sct 18h55'42"-7d14'
Hancock,Melissa
Peg 22h41'17"31d6'
Hancock,Michael A & Gidget
C
Mon 6h54'45"10d15'
Hancock,Sarah
Lyr 18h29'23"31d5'
Hancock,Susan
Aql 20h1'26"8d50'
Hancock,Taylor Rice
Uma 11h55'1"37d46'
Hancock-Glass,Cathy Lee
Mon 6h20'1"7d33'
Hancox,Alan Stanley
Ori 6h6'13"1d24'
Hancox,Candy-Jane
And 0h1'43"38d28'
Hanczar,Diane
Cas 0h58'35"64d38'
Hand My Everlasting
Love,Tammy
Cyg 21h22'24"53d27'
Hand Star,Bee
Lmi 9h22'14"38d13'
Hand,Deborah
Crb 16h13'47"31d31'
Hand,Elizabeth E
Mon 6h27'20"10d7'
Hand,Jennifer
Vul 19h45'13"28d59'
Hand,Marian K
And 0h8'26"47d49'
Handel,Evan
Peg 21h29'43"2d38'
Handel-Bailey,Jean
And 2h15'1"38d29'
Handke,Jennifer Amber
Cas 1h1'19"55d56'
Handler
Boo 14h50'16"45d53'
Handler "Our Love Shines
Eternal",BS
Her 17h13'49"48d40'
Handler,Emily K
Boo 14h26'43"25d32'
Handler,Helen-Nathan
Lac 22h0'52"37d59'
Handler,Samuel J
Uma 11h55'56"52d34'
Handley,Raymond
Cma 6h50'41"-18d22'
Handley,Sr,Jimmy G
Cet 3h4'39"5d5'
Handlin,Deanna Kay
Mon 6h56'33"10d1'
Handlon,Eric Martin
Ori 5h3'34"-02d25'
Handlos,Patricia
Umi 13h56'1"79d12'
Handman,Arthur L
Dra 14h27'56"62d27'
Hands,Newby
Dra 16h3'53"62d24'
Handsome Devil
Lac 22h33'0"38d13'
Handsome Harel
Dra 15h55'60"67d3'
Handsome Jack
Cep 23h12'41"65d49'
Handsome Winter Blizzard
Per 1h28'21"50d15'
Handu
Per 3h0'11"50d32'
Handwerker,Pete
Cam 3h44'47"74d20'
Handy,Christopher James
Boo 14h47'58"38d29'
Handy,F P
Boo 14h40'0"50d26'
Handy,Raymond
Her 17h20'38"20d45'
Hanecak,John & May
Mon 7h15'30"-7d4'
Hanel,Colin Edward Emile
Cet 2h17'51"5d5'
Hanes Haven
Cep 21h12'60"65d24'
Hanes,Bill & Blanche
Cet 0h0'1"-12d2'
Hanes,Christopher & Jill
Hanes
Lmi 10h35'57"33d18'
Hanes,Craig
Aur 5h0'46"46d55'
Hanes,Hall W "Hollywood"
Aql 20h53'1"-1d5'
Hanewinkel,Ryan Joseph
Ser 15h12'25"-1d45'
Haney (Nedimyer), Jeanne
Marie
Dra 16h26'24"64d28'
Haney Birthday Star, Erin
Vir 13h11'3"-1d4'
Haney Jr,Thomas F
Dra 11h37'18"71d25'
Haney Sr,Thomas F
Dra 11h37'1"68d19'
Haney,Bubba
Dra 9h21'13"74d4'
Haney,Christopher
Dra 11h37'10"68d22'
Haney,Christopher
Aur 6h34'39"37d43'
Haney,Dona
Crb 16h5'34"27d34'
Haney,Donna Elaine
Cas 0h8'15"64d4'
Haney,Ed
Dra 9h44'52"73d23'
Haney,Eleanor A
Dra 11h37'0"68d31'
Haney,Marie
Dra 9h21'27"74d35'
Haney,Mark Norman
Aur 6h0'24"32d55'
Haney,Michael S
Dra 11h36'42"71d43'
Haney,William
Hya 9h8'17"3d4'
Hanford,Christopher R &
Jeffrey G
Del 20h34'38"10d47'
Hanford,Debbie Marie
Uma 9h54'42"61d28'
Hangauer,Katharine
Lyr 19h13'36"41d2'
Hangen,Linda Marie Thoman
Cyg 19h25'35"44d32'
Hangen,Linda Marie Thoman
Uma 10h14'22"50d9'
Hanhauser,Ruth W
Eri 3h54'50"-16d28'
Hanidania
Cyg 19h30'23"35d37'
Hanie I
Sgr 19h36'38"-36d41'
Haniel,Klaus
Peg 23h32'31"15d51'
Haniford,Helen Rutledge
And 23h19'52"42d43'
Haniford,Starr
Uma 11h22'14"40d50'
Hanig
Hya 8h17'15"2d40'
Hanisch,Steffen
Cep 20h40'33"60d17'
Hanisko,The Star of Patty
Com 12h45'40"23d17'
Haniszewski,John
Peg 21h57'38"34d35'
Hank
Aql 20h18'0"5d16'
Hank
Cep 22h19'1"55d36'
Hank
Per 3h41'16"51d46'
Hank & Irish
Cep 0h54'25"77d32'
Hank & Melissa
Cyg 21h11'51"39d5'
Hank D
Cep 22h15'43"65d52'B
Hank The Yank
Per 3h11'59"42d10'
Hank's Eyes
Her 15h54'34"50d10'
Hank's Green Stone
Oph 18h42'0"6d22'
Hank's Lucky Star
Ori 4h53'30"1d52'
Hanke,Marianne
Mon 7h52'18"-2d14'
Hankel,Alice Rose
Uma 11h41'59"56d37'
Hanker,Stephen
Uma 8h19'25"68d29'
Hankey,Robert E
Her 18h43'14"12d24'
Hankins,Jacob R
Sex 10h27'59"4d9'
Hankins,Mary Frances
Aql 20h13'48"4d7'
Hankins,Roger Glen
Aql 20h7'56"7d27'
Hanks,Bryant Marshall
Sex 10h44'1"-5d29'
Hanks,Lorelei J & William V
Hanks
Dra 16h31'1"52d4'
Hanks,Oliver (Buppy) Lemay
Dra 17h7'59"68d7'
Hanks,Travis
Ori 5h56'17"16d1'
Hanks,Truman Theodore
Her 17h37'31"27d13'
Hanley Family Wishing Star
Peg 23h5'15"33d23'
Hanley,Anne
Lyr 18h19'35"42d2'
Hanley,Barbara Dolan
Del 21h4'36"12d47'
Hanley,Brennan Thomas
Hya 8h44'21"4d36'
Hanley,Douglas M
Aur 6h28'53"32d12'
Hanley,Edmund James
Aur 6h22'19"31d36'
Hanley,Edward & Kathryn
Boo 15h16'17"47d48'
Hanley,Jennifer Mariah
Ser 15h52'51"17d44'
Hanley,John J
Lac 22h6'31"38d37'
Hanley,Jr,Charles Stewart
Cnv 12h31'50"51d11'
Hanley,Nina Louise
Cyg 20h54'1"48d50'
Hanley,Rebecca Noel
Cam 3h58'21"53d56'
Hanley,Sarah
Cas 1h8'0"60d20'
Hanlon,Donald F
Cet 0h32'42"0d4'
Hanlon,Josefina L
Peg 22h45'0"24d37'
Hanlon,Lawrence P
Ori 5h25'31"-4d12'
Hanlon,Mighty Dave
Per 1h25'47"53d31'
Hanlon,My Special Friend
Rose Marie
Lac 22h40'1"51d43'
Hanlon,Pamela J
Leo 11h9'1"1d36'
Hanlon,Patricia
Psc 1h28'0"20d18'
Hanlon,Patrick Jeremiah
Aur 7h4'59"40d6'
Hann,Fran
And 2h7'0"42d7'
Hann,Ralph L
Dra 16h11'54"68d8'
Hann,Thomas Charles
Her 17h11'0"41d4'
Hanna
Crt 11h39'31"-22d35'
Hanna & Janne
Uma 10h10'1"70d40'
Hanna 9/27/54,Mariana 9-27-
54
Cas 23h31'38"61d35'
Hanna's Glow
Peg 22h0'35"23d8'
Hanna,Ashley Elizabeth
And 1h11'48"48d51'
Hanna,Bonnie Jean
Aqr 20h54'0"-6d24'
Hanna,Buddy
Aql 18h40'49"-1d9'
Hanna,C A
Crt 10h54'40"-11d5'
Hanna,Darlene E
And 0h13'0"33d39'
Hanna,Eric Ward
Uma 11h21'31"38d51'
Hanna,Heather Lynn
Cyg 20h16'17"31d2'
Hanna,Hugh & Nona
Aur 6h4'51"46d39'
Hanna,James
Lac 21h56'56"41d12'
Hanna,Kamel Gabreil Assad
Ori 5h44'39"11d58'
Hanna,Kelly & Mark
Cyg 19h41'22"37d33'
Hanna,Kevin Michael
Dra 16h53'1"68d2'
Hanna,Kyle Stewart
Cam 4h0'55"58d52'
Hanna,Rachel O'Donna
Cas 0h46'10"70d58'
Hanna,Sean Patrick
Cep 20h36'13"67d15'
Hanna,Susie A
Lmi 9h55'1"40d13'
Hanna,Trevor Wayde
Per 2h57'28"32d11'
Hanna,Willard & Mary
Mon 6h23'0"4d24'
Hanna-The Sky's the
Limit,Mark Joseph
Per 2h55'30"35d52'
Hannah
Cyg 20h22'1"38d24'
Hannah
Ori 5h47'15"18d49'
Hannah
Cas 1h15'45"62d7'
Hannah
Sco 16h56'1"-37d37'
Hannah Beatrice
Peg 22h18'45"33d20'
Hannah Caitlin
And 2h19'45"38d30'
Hannah Daisy
Cyg 19h31'40"31d50'
Hannah Danielle's Jewel
Ari 3h13'49"22d45'
Hannah Diane
Peg 23h1'1"20d2'
Hannah Elizabeth
Cep 23h19'44"70d9'
Hannah J
Cas 0h58'38"61d17'
Hannah Lee
Cyg 21h4'13"53d47'
Hannah Marie
Peg 23h32'49"18d1'
Hannah Marie
Cas 1h27'21"60d30'
Hannah Mary
Per 2h48'57"43d9'
Hannah May
Lyr 18h30'0"32d17'
Hannah Nicole
And 23h20'1"49d54'
Hannah Paige
Lyn 7h6'30"52d13'

Hannah Rose
And 23h24'28"40d22'
Hannah Teri Marie
Mon 7h17'32"-5d6'
Hannah Victoria
Peg 22h26'38"3d35'
Hannah's Star
Uma 10h32'49"55d2'
Hannah,Charlotte,Ann
Cyg 20h31'31"42d46'
Hannah,Dale
Aur 6h36'58"37d32'
Hannah,Daryl
And 23h15'18"50d7'
Hannah,Kerin Lee
Mon 7h1'18"5d14'
Hannah,Pooh Bear Ryan F
Ser 15h57'48"1d16'
Hannah,Shannon Reneé
Vir 11h41'16"3d7'
Hannah-Transwalton, C47
And 0h17'16"39d21'
Hannan III,John Joseph
Mon 5h55'35"-4d46'
Hannan,James J
Eri 3h39'13"-6d50'
Hannan,Nancy & Niall
Boo 15h46'45"46d47'
Hannan,Roy
Ori 5h34'32"-5d46'
Hannan,Terrance D
Ser 18h6'58"-0d32'
Hannan,Thomas Danes
Aql 20h35'41"0d39'
Hannawa-Ivan The Great,Ivan
Sgr 18h52'18"-34d40'
Hannel,Marsha Z
Vul 19h18'57"23d12'
Hannele
Cas 1h45'26"75d10'
Hannele,Marja
Uma 11h18'37"43d22'
Hannelore
Cas 1h1'54"55d37'
Hanneman,LeRoy
Oph 18h39'41"8d28'
Hannemann,Megan N
Lyr 18h30'31"30d35'
Hannes Number One
Cnc 9h14'26"11d17'
Hannett,Gayle Lyn
And 0h18'45"31d13'
Hannibal,Katie
Mon 8h3'35"-8d43'
Hannig,Guenter
Her 17h22'35"42d6'
Hannigan,Alyson
Cam 4h0'18"67d53'
Hannigan,Christopher James
Lac 22h36'16"53d1'
Hannigan,Grace Louise
Ori 5h38'15"-0d58'
Hannigan,Joe
Sge 20h3'0"18d59'
Hannigan,Patricia
Boo 14h29'38"52d37'
Hannigan,Tom
Her 17h38'25"38d56'
Hanning,Fred
Ori 6h3'1"5d26'
Hanno
Tau 3h55'58"18d48'
Hannon & Chita Rivera, Chuek
Oph 17h38'23"-20d13'
Hannon,Anne M
Uma 9h19'43"71d24'
Hannon,Benjamin James
Lac 22h52'31"52d34'
Hannon,Jr,Richard C
Aur 6h27'0"37d42'
Hannon,Richard
Dra 14h26'17"64d13'
Hannoun,Jacqueline
Dra 10h11'53"74d37'
Hanns,Charles Gregory
Her 17h1'25"40d16'
Hanns,Jr,Stephen Allan
Aur 4h57'30"36d57'
Hanns,Sheila Renee
Lyn 8h58'56"44d7'
Hannsi
Leo 9h58'0"18d35'
Hannum,Haley Katherine
Cas 1h55'52"63d56'
Hannum,James E
Aur 7h23'0"39d43'
Hannum,Joseph Max
Oph 16h59'24"-28d14'
Hanny,Dr Stefan
Ori 5h23'45"1d49'
Hano's 19th Hole
Lyn 8h24'42"33d36'
Hanoon,Rachel
And 0h15'12"38d59'
Hanrahan,Barbara Evelyn
Liddy Hunt
Leo 10h52'1"-0d10'
Hanrahan,Christopher
Eri 3h18'28"-3d55'
Hanrahan,Cynthia Lynn
Lib 14h53'38"-0d28'
Hanrahan,Helen M
Cas 0h13'31"47d32'
Hanrahan,James Bernie
Aqr 22h48'43"0d56'
Hanrahan,Michael Walsh
Dra 18h58'24"48d52'
Hanrahan,Richard William
Mon 7h3'1"-5d50'
Hanrahan,Sharon Mae
Bjornson
Gem 6h41'56"34d17'
Hanratty,Trooper Thomas J
Per 1h32'43"52d34'
Hanreck,Richard
Lac 22h10'40"38d27'
Hans
Cmi 7h19'36"4d17'
Hans
Dra 10h56'27"74d16'
Hans
Aur 6h0'54"30d3'
Hans
Cas 0h13'0"60d50'
Hans Peter
Tau 5h53'24"26d41'
Hans Slade
Oph 16h48'22"10d7'
Hans,MD,Paul
Cyg 21h32'17"37d42'
Hans-Karl
Cnv 13h45'26"46d33'
Hans-Karl Doll
Lib 15h38'27"-20d30'
Hansard,Michael Clayton
Her 16h19'21"48d38'
Hansbarger,Dr Clark
Oph 17h53'37"12d31'
Hansch,Robert William
Sco 16h58'35"-44d10'
Hansel
Cet 1h26'1"-13d56'
Hansel,Kayla & Chad Stewart
Cyg 20h5'12"40d47'
Hansel,Larry
Ser 15h52'35"18d3'
Hansen "Jamie" 1978-
95,James Steven
Peg 21h19'1"21d37'
Hansen & Hansen
Uma 11h38'48"51d16'
Hansen & Miller
Sex 10h39'45"2d18'
Hansen 8-18-21,Donald
Gordon
Leo 10h33'51"13d59'
Hansen Queen of the
Cosmos,Jeannie
Equ 21h10'31"11d57'
Hansen's "Nitelite", Ethan A
Lac 22h14'29"38d8'
Hansen's Dutchmans Secret
Cam 7h29'38"61d16'
Hansen's Star,Jim
Boo 14h5'19"38d25'
Hansen(The Sister's
Star)Victoria&Caroline
Boo 14h47'53"35d15'
Hansen,Anita & Torben
Cyg 19h34'45"30d24'
Hansen,Betty Lou
Cas 1h30'20"76d17'
Hansen,Bradley Blake
Cet 2h29'15"7d27'
Hansen,Brian Kanne
Oph 17h9'43"10d10'
Hansen,Bruce Ivan
Dra 18h4'36"65d17'
Hansen,Carmen S
Cas 0h5'11"63d22'
Hansen,Christy Marie
And 0h18'12"31d24'
Hansen,Dana & Julie
Vul 19h43'0"20d23'
Hansen,Dave
Her 17h8'12"50d24'
Hansen,Diane"Snow Princess"
And 0h13'31"47d32'
Hansen,Dianne
Cam 13h58'1"80d58'
Hansen,Donald Keith
Cet 0h52'18"1d2'
Hansen,Donna
Del 20h51'10"8d54'
Hansen,Dr Raymond D
Oph 18h39'23"6d12'
Hansen,Geoffrey A
Aql 18h43'21"-0d59'
Hansen,Gerald
Boo 14h40'51"38d28'
Hansen,Hilary Burges
Cas 2h19'49"63d60'
Hansen,Howard
Del 20h24'29"10d1'
Hansen,III,Carl Richard
Boo 15h4'17"12d8'
Hansen,Izumi
Kahokuhulaokekai
Dra 16h46'59"62d47'
Hansen,Jennifer
Cas 0h36'23"62d41'
Hansen,Joshua Gregory
Ori 6h1'32"10d40'
Hansen,Kaitlyn Grace
And 1h3'32"37d36'
Hansen,Katrina
Aql 20h11'45"4d49'
Hansen,Keith & Patricia
Lyr 18h43'35"46d32'
Hansen,Kenneth M
Her 16h31'21"50d1'
Hansen,Kirk Richard
Her 16h42'42"48d34'
Hansen,Lisa Dawn
Cyg 21h4'1"36d31'
Hansen,Margaret Bell
Mon 7h52'37"-5d17'
Hansen,Mark Victor
Hya 8h43'0"4d19'
Hansen,Melinda C K
Cas 0h29'17"63d49'
Hansen,Michael Allen
Lib 15h42'45"-23d36'
Hansen,Michael G
Tau 4h2'31"20d41'
Hansen,Pat
And 0h7'0"38d2'
Hansen,Paul & Dorothy
Aql 18h45'15"11d10'
Hansen,Pauline Christine
Peg 22h18'19"35d8'
Hansen,Poul M T
Lac 22h38'13"40d47'
Hansen,Randy
Her 17h33'27"28d56'
Hansen,Richard D & Cathleen
N
Vul 19h46'32"28d48'
Hansen,Robert
Peg 23h30'20"21d44'
Hansen,Roger Lyle
Her 17h0'21"50d49'
Hansen,Ruth Baker
Umi 14h58'51"67d23'
Hansen,Thomas
Per 2h54'52"35d58'
Hansen,Timothy Sean
Peg 0h12'19"13d3'
Hansen,Trey & Christine
Lynch
Ori 5h59'14"10d48'
Hansen,Tricia "Rangy Lil"
Mon 8h6'4"-4d8'
Hansen,William Peter
Hya 9h9'1"4d13'
Hanser,Franáois Roy
Uma 10h23'33"52d18'
Hansford I Love You 119,Gail
Ann
Cyg 21h13'34"28d30'
Hansford,Shirley Kathleen
Aql 20h5'1"0d57'
Hanske,Edward Albert
Eri 4h47'9"-8d54'
Hanske,Lois Marie
Sex 10h10'55"-7d3'
Hanslik,Cynthia R
Cyg 21h14'0"28d37'
Hanslin,Tyler Semolic
Boo 15h0'42"20d14'
Hansman,Sarah Grace
And 0h22'11"36d20'
Hanson "Loving
Superstar",Dee Dee
Aql 19h51'37"14d49'
Hanson,Alexander Day
Ramirez
Boo 15h16'54"41d24'
Hanson,Benjamin James
Tau 5h53'32"24d4'
Hanson,Bobby Charles
Aql 20h1'31"10d3'
Hanson,Brianna Reed
Cas 23h31'0"61d19'
Hanson,Cecelia
Cas 0h49'58"63d37'
Hanson,Clara
Lyr 19h6'46"37d53'
Hanson,Clayton Garvin "Clay"
Lyr 18h59'52"28d8'
Hanson,Constance S
Peg 22h48'36"29d41'
Hanson,Erik James
Lyr 18h38'1"37d42'
Hanson,Erin Elizabeth
Peg 22h16'27"11d13'
Hanson,Heather
Aur 5h3'35"38d21'
Hanson,Ian
Ori 5h37'22"-6d1'
Hanson,James & Belinda
Sco 17h24'45"-43d17'
Hanson,Jane
And 1h16'44"37d35'
Hanson,John J
Aur 5h37'24"40d10'
Hanson,June Margaret
Oph 17h19'39"11d6'
Hanson,June Reid Richards
Cas 2h9'50"59d54'
Hanson,Keith
Ori 5h37'0"-6d6'
Hanson,Kimberly Beth
Eri 3h32'1"-6d49'
Hanson,Korey,Lee
Cet 1h37'14"-12d58'
Hanson,Lowell
Per 3h54'24"37d44'
Hanson,Mary Katherine
Tau 5h47'32"24d4'
Hanson,Meghan Kali
Tau 5h50'31"24d11'
Hanson,Merlen
Aur 5h12'27"42d49'
Hanson,Michael J
Cep 0h3'1"68d23'
Hanson,Nancy
And 2h17'46"39d56'
Hanson,Paul James
Cyg 19h41'34"41d53'
Hanson,Paul James
Cyg 20h38'31"29d37'A
Hanson,Richard Rockwell
Her 16h34'32"20d8'
Hanson,Rob "Precious"
Aur 6h24'58"38d39'
Hanson,Robert Paul
Peg 23h32'30"18d50'
Hanson,Roger A
Lac 22h24'25"54d4'
Hanson,Scott "Sexy"
Aur 7h12'17"39d54'
Hanson,Terry Jay
Tau 5h57'59"24d1'
Hanson,Thom Wayne
Vir 11h36'27"8d57'
Hanson,Tim
Boo 15h14'20"27d6'
Hanson,Tyler Scott
Per 2h51'48"45d3'
Hansse,Patrick
Del 20h36'17"5d29'
Hanssen,Jon G W
Dra 17h1'32"68d27'
Hanssen,Kirsten
Cap 21h35'34"-26d42'
Hanssler,Jennifer
Uma 10h41'45"68d36'
Hansson,Abigail
Tau 3h41'1"28d26'
Hanton,Doug
Ori 5h57'33"8d45'
Hanula,Heather Lynn
Cas 1h45'1"58d18'
Hanula,Nicholas
Cnv 13h12'38"40d8'
Hanus,Barbara Ann
And 1h31'58"39d25'
Hanuscak,Kathryn
Lyr 18h15'21"30d48'
Hanzlik,Christie Cern
And 22h59'0"50d30'
Haobsh,Dima
Ori 5h46'20"21d1'
Hapner,Bryan Keith
Vir 11h58'47"-4d5'
Happach,Manfred
And 0h54'35"21d59'
Happiness
Ori 6h2'0"1d5'
Happiness
Pup 8h8'35"-26d50'
Happiness For Denise
Mon 7h50'9"-1d53'
Happiness Shinji
Tau 5h14'44"17d19'
Happnie III,John J
Lmi 10h51'43"38d35'
Happy 10th Anniversary Jan &
Scott
Aql 19h22'10"10d47'
Happy 20th Bill, Love Debbie
Aql 20h13'0"1d39'
Happy 21st Birthday Peter
1973-1994
Per 2h59'15"32d50'
Happy 25th Birthday, Rubin
Aur 6h12'29"32d8'
Happy 26th Thomas & Erma,
Jean Tracy
Cet 1h19'23"-12d42'
Happy 34th Baby Steve
Per 1h33'26"52d31'
Happy 50th Ben & Lou
Eri 3h2'29"-7d25'
Happy Birthday Bob
Per 2h3'26"56d49'
Happy Birthday Darla 63
Lyn 7h25'25"44d7'
Happy Birthday Don
Hya 8h44'0"5d39'
Happy Birthday Gerald
Vir 11h57'16"-6d59'
Happy Birthday Giekie
Hya 8h19'1"4d47'
Happy Birthday Min!
Lyr 19h1'0"38d27'
Happy Camper Jane
Tau 5h50'28"26d48'
Happy Cat
Lyn 7h21'26"50d21'
Happy Cat
Oph 16h49'34"10d2'
Happy First Birthday to Mizuki
Lib 14h46'32"-19d42'
Happy Holidays
Ori 6h5'0"8d15'
Happy Jack
Dra 19h21'40"56d11'
Happy Marion
Cyg 20h57'60"31d33'
Happy One
Cap 21h40'53"-19d58'
Happy Paws Obedience
School
Cnv 12h5'25"38d4'
Happy Pete
Uma 8h34'37"51d17'
Happy Place,The
Ori 5h53'11"5d26'
Happy Shooting Fellow, The
Ori 6h17'26"0d35'
Happy Stella of Mizuki
Lib 14h41'12"-12d50'
Happy Thought-GEG
Del 20h35'25"10d2'
Happy Times
Aur 4h55'13"51d19'
Happyness
Del 20h12'23"10d42'
Hapsburg,Raymond Marion
Aql 18h56'0"15d20'
Hara,Megumi
Ari 2h28'45"21d26'
Harabin,Andrew & Anna
Peg 22h47'11"35d8'
Harach,Tod D
Uma 10h17'13"40d24'
Harada Tammie
Cet 1h20'0"-11d37'
Harael
Umi 13h40'51"76d39'
Harald & Ilse
Ari 2h42'24"26d1'
Haraldson,Aaron Rene
Her 18h6'0"38d13'
Haralson,Dr Harold H
Oph 17h0'26"11d0'
Haralson,J Ailene
Cma 7h1'13"-16d51'
Haramis,Sonya
Peg 23h28'46"22d56'
Haramy
Peg 22h19'45"20d58'
Haran I,Daniel J
Cep 23h23'0"64d19'
Harary,Edward "The Great"
Aur 6h26'53"33d58'
Harbadin,Nicholas Paul
Aur 4h59'0"38d16'
Harban,Fraser Mark John
Mon 8h3'23"-1d26'
Harbaugh,Maggie
Del 20h49'44"7d43'
Harbeck,Cole Pirie
Aur 5h5'19"51d20'
Harbeck,Conor Nash
Gem 6h59'46"15d6'
Harbeck,Conor Nash
Tri 2h12'54"31d12'
Harberg,Beryl L
Crt 11h30'24"-12d4'
Harbers,Arnold D
Ori 5h26'24"1d35'
Harbert,Malyne
Peg 22h13'39"10d41'
Harbic's Star
Ori 6h5'56"20d32'
Harbick,Stephen Paul
Sgr 18h49'12"-31d2'
Harbin,Courtney Jennifer
Psc 1h4'1"27d49'
Harbin,Jr,Michael
Leo 10h59'51"15d38'
Harbinson,Dorcas
Lyr 18h16'30"40d19'
Harbison,Ciara
And 0h19'11"38d50'
Harbison,Cindy
Lyr 19h6'14"26d6'
Harbison,Kimberly Ann
Maples
Uma 9h58'12"50d34'
Harbor House
Mon 7h5'6"-6d28'
Harbor Light
Cet 2h4'1"1d2'
Harbrecht,Diane
And 0h59'29"36d36'
Harcleroad,Carolyn
Lyr 18h39'0"36d10'
Hard Cocks of Hartcliffe,The
Uma 11h56'11"57d33'
Hardage,Denise
Cet 1h21'49"-7d10'B
Hardaker,Linda
Aql 20h30'30"0d38'
Hardeley,Margaret
And 23h18'30"49d51'
Hardell's Beacon
Uma 9h20'1"47d48'
Harden I Love You Dad
TH,William J
Uma 10h58'49"35d3'
Harden III,MD,Wesley R
Aur 7h23'40"41d36'
Harden III,William Britt
Aqr 21h24'31"-1d43'
Harden,"Desperado" Eric W.
Per 2h53'37"43d49'
Harden,Jr,Monroe B
Hya 9h15'52"5d37'
Harden,Vincent Joseph
Eri 4h14'55"-17d55'
Hardenbrock,Bonnie
Mon 6h41'15"7d42'
Harder,Jeremy Thomas
Peg 22h40'53"26d9'
Harder,Robert J
Per 2h55'0"37d16'
Harder,Timothy Patrick
Hya 8h43'0"5d23'
Harder,William Joseph
Her 16h39'57"36d16'
Hardie,Joseph Ray
Lac 22h33'51"55d27'
Hardie,Julie Johnson
Uma 8h6'55"71d2'
Hardin "Grandba", Margaret
Stone
Cas 0h33'35"66d26'
Hardin,Alexanna Casandra
Mon 6h20'38"1d30'
Hardin,Carolyn Sue
Cyg 19h18'56"28d35'
Hardin,Christina
Eri 3h42'18"-15d29'
Hardin,Courtney
Lyr 19h23'35"35d8'
Hardin,Elizabeth Ann
Oph 17h26'16"-20d25'
Hardin,Jennifer Lyn
And 2h31'27"41d13'
Hardin,Jr,Robert L
Aur 6h15'23"37d7'
Hardin,Ken
Her 18h5'16"14d54'
Hardin,Melinda "Shnookie"
Cas 1h4'53"62d53'
Hardin,Michael Claude
Mon 6h21'51"7d54'
Hardin,Rebecca
Lyr 19h6'24"26d14'
Hardin,Shane Michael
Mon 6h20'1"8d0'
Hardin,Stacie"Grump"
Eri 4h32'1"-8d25'
Harding 29 Nov 1960, James
Wesley
Her 16h58'38"28d8'
Harding High Class of 1946
Char NC
Aql 19h1'56"5d17'
Harding LAF JC,Melanie
"Highlander"
Uma 9h20'18"51d25'
Harding,Cherry
Lyr 18h29'1"46d35'
Harding,Fiona Jane
Cyg 19h32'33"35d19'
Harding,Gregory Lynn
Hya 8h11'42"-1d9'
Harding,Jack
Aql 19h0'34"7d28'
Harding,Jack
Uma 11h45'30"41d58'
Harding,Jane
Peg 23h33'39"21d40'
Harding,Joy & Robert
Uma 11h7'45"38d29'
Harding,Kitt L
Cyg 20h4'42"30d31'
Harding,Mary Lou
Cet 0h45'35"1d15'
Harding,Nancy Lee P
Cma 7h15'24"-16d47'
Harding,Shirley Belinda A
Aql 20h10'51"10d55'
Harding,Stuart
Oph 18h5'40"12d1'
Harding,Timothy Lawrence
Dra 18h19'26"48d38'
Harding,William F
Sco 16h53'32"-40d39'
Hardingham,Jessica Marie
Lyr 18h30'53"38d22'
Hardison,Ray
Boo 14h47'21"31d60'
Hardman,Brandon James
Equ 21h1'17"8d33'
Hardman,Dylan Thomas
Aqr 22h27'1"0d27'
Hardman,Robert
Per 4h3'13"50d51'
Hardman,Robert
Sco 15h59'50"-21d26'
Hardt,Hans-Theo
Ori 5h51'40"21d6'
Hardt,Konrad
Hya 8h46'46"-8d6'
Hardt,Lindsay Leigh
Gem 7h13'14"21d42'
Hardware,Odell
Ser 18h7'12"-14d29'
Hardway,Sasha Diane
Cas 0h23'1"61d6'
Hardwick,Alan
Cep 22h6'50"68d29'
Hardwick,Glen & Anella
Lyr 19h23'46"37d34'
Hardwick,Jill
And 2h29'15"38d45'
Hardwick,Martin
Cep 4h40'25"80d25'
Hardwick,Noreen
Lyn 7h8'11"50d41'
Hardwicke,BC
Dra 17h13'57"68d55'
Hardy 9/10/54-20/1/89, Linda
Ann
Ori 6h4'22"5d14'
Hardy De Visne
Per 2h21'29"55d50'
Hardy IV,Thomas Francis
Oph 17h32'12"-22d40'
Hardy's Wishing Airplane,Bill
Mon 6h55'45"7d56'
Hardy,Amanda Lee
Cas 1h6'34"55d53'
Hardy,Ann Sebrell
Lyn 8h13'40"51d11'
Hardy,CBE,Maurice G
Gem 7h5'8"29d2'
Hardy,David 'Bilbo'
Boo 14h19'51"27d32'
Hardy,Deborah
And 0h52'51"37d56'
Hardy,Evan Perry
Her 17h13'15"40d38'
Hardy,Genevieve H
Com 12h55'35"27d4'
Hardy,Ieny K
Uma 9h47'59"59d49'
Hardy,Jacqueline M
And 23h34'0"40d7'
Hardy,Jaki
Cnv 12h20'29"35d44'
Hardy,James
Peg 23h7'19"12d17'
Hardy,Jeanne M
Cas 1h2'15"61d54'
Hardy,Jeff M
Aql 19h14'6"-11d4'C
Hardy,Joanna
Uma 14h13'18"58d13'
Hardy,Joanne
Cyg 19h25'28"34d18'
Hardy,Jr,Percy
Hya 8h58'13"0d46'
Hardy,Kellie Lynn Apalategui
Gem 7h8'37"23d18'
Hardy,Leroy & Patti
Crb 15h59'1"28d33'
Hardy,Linda
And 23h25'17"49d30'
Hardy,Luther Hensel
Uma 11h28'23"31d4'
Hardy,Michael
Her 16h40'39"33d34'
Hardy,Raymond G
Lac 22h9'1"40d5'
Hardy,Sarah Ann
Lac 22h40'49"56d15'
Hardy,Vergil (Stardust)
Boo 15h5'16"18d60'
Hardy,William J
Her 17h11'18"42d4'
Hardyman,Chris
Aqr 23h8'52"-10d30'
Hardys Christ Eggum
Cet 2h15'25"5d0'
Hare Bear
Aql 20h30'22"0d5'
Hare,Alice Ridley Chambers
Tau 4h21'23"15d24'
Hare,Christopher Collins
Cnv 12h21'16"40d46'
Hare,Clarence Clifton
Ori 5h50'17"9d57'
Hare,Cliff J
Cet 1h14'28"-1d42'
Hare,Peter C
Uma 9h30'0"60d37'
Hare,William Arthur David
Aql 19h53'12"13d30'
Haren,Ginger Ann
Lyr 18h30'1"45d27'
Haren,Kathleen
Sct 18h47'50"-7d58'
Harenberg,Steven Douglas
Umi 16h58'0"78d44'
Harff,Rolph
Aur 6h51'54"40d40'

Harford,Earl
Peg 21h54'24"30d16'
Harford-Fox,Leslie
Tri 2h29'1"28d44'
Hargadine,Craig L
Aql 20h2'21"10d3'
Hargadon,Bernie
Her 18h13'55"41d7'
Hargadon,Joseph Michael
Aur 7h13'0"36d29'
Hargan,Osa & Gerry
Lyn 7h5'51"51d21'
Hargarten,Paul
Oph 17h28'9"-20d5'
Harger,Burt
Dra 17h48'54"64d43'
Harger,D A
Aur 7h6'22"43d50'
Hargett,John Witney
Oph 17h39'50"-28d24'
Hargis,Leslie A
And 2h29'13"49d56'
Hargis,Lisa
Vul 20h16'56"22d56'
Hargis,Patricia Denise
Lyr 18h38'46"38d53'
Hargis,Sam Henry
Her 17h20'46"48d10'
Hargis,Sylvia
Peg 21h42'36"22d52'
Hargiss,Delta
Ori 6h5'60"-2d29'
Hargrave,Julie
Peg 21h52'49"31d18'
Hargreaves,Alison
Cyg 20h15'10"39d27'
Hargreaves,Mavis Scannett
Boo 14h17'21"53d43'
Hargreaves,Shirley
Aql 18h57'41"17d10'
Hargrove aka Nicolo Sturiano,H
Lac 22h7'51"50d57'
Hargrove,Cara H
Del 20h39'25"10d15'
Hargrove,Whitney Jean
Mon 7h14'41"-10d49'
Harinck,Sally
And 2h6'0"40d44'
Haring,Sr,Wayne William
Aur 6h33'1"33d36'
Hariri,Abdollah
Uma 10h40'23"40d35'
Hariri,Bita
Uma 10h28'17"40d9'
Hariri,Laleh
Cas 23h40'25"60d56'
Hariri,Masond
Uma 10h38'26"40d11'
Hariri,Nassrin Gol
Uma 10h42'0"40d48'
Hariri,Rafi
Uma 10h45'0"40d59'
Hariri,Reza
Lyn 9h12'30"33d39'
Harjes,Danny
Lyn 7h15'0"50d12'
Harjo,Gerald & Jeneia
Sge 19h14'51"16d27'
Harjulong
Sex 10h30'26"2d6'
Harkanson,Susan S
And 0h30'1"38d54'
Harker,8-21-94,8:07AM, Ben Speros
Sct 18h55'15"-6d45'
Harker,Cara Elizabeth
Mon 6h52'37"-0d14'
Harker,Howard G
Uma 10h2'58"48d15'
Harkey,Elisabeth Marie
Cas 2h35'0"60d40'
Harkie
Uma 9h49'27"58d0'
Harkin,Charles Bernard
Aur 6h20'14"31d37'
Harkins,Alice Galloway
Lyn 8h4'1"34d47'
Harkins,CF
Psc 0h19'34"10d3'
Harkins,Christopher Allen
Cep 22h8'22"53d58'
Harkins,Corky
Lmi 10h10'1"38d57'
Harkins,Donald E
Per 1h50'25"53d51'
Harkins,George Francis
Aur 5h29'43"37d32'
Harkins,Gregory John
Cet 1h37'11"-2d43'
Harkins,Jennifer
Com 12h28'56"26d33'
Harkins,Jessica Rose
Vul 19h46'1"28d14'
Harkins,Jimmy E
Per 4h1'15"48d54'
Harkins,Melissa Katherine
Aql 19h55'25"10d60'
Harkins,Patrick
Lac 22h26'0"52d46'
Harkins,Peggy & Larry
Mon 7h0'0"1d38'
Harkins,Scott Alexander
Her 16h25'1"36d28'
Harkness,Brian
Lyn 7h36'27"42d3'
Harkness,Dr James J
Oph 17h24'37"-24d43'
Harkness,Gregory Lynn
Cmi 7h57'29"8d31'
Harkness,Pamela Lynn
Uma 10h56'24"70d58'
Harkness,Patrick
Cep 20h3'17"60d0'
Harkness,Shannon Lynn
Uma 10h55'40"71d20'
Harkness,Stephen
Boo 14h47'0"27d25'
Harkness,Stuart
Her 16h13'49"48d32'
HarlamBakis Family,The
Her 16h10'49"25d17'
Harlambos
Boo 15h22'59"51d18'
Harlan
Boo 14h16'38"47d7'
Harlan,Jamison Eldridge
Gem 7h24'41"22d15'A
Harlan,John & Beth
Uma 9h7'49"52d25'
Harlan,Michelle Marie
Peg 23h38'23"20d17'
Harlan,Timothy Eldridge
Gem 7h24'41"22d15'B
Harlan,Timothy M
Aql 20h31'22"0d48'
Harland
Cam 7h31'48"60d2'
Harland,Bernard
Cet 3h19'41"2d19'
Harlea
Umi 15h23'11"68d21'
Harlee
Lmi 10h44'48"33d19'
Harlee & Seth
Del 20h15'53"10d10'
Harles,Ingeburg Susanna
Ari 2h31'46"21d57'
Harley
Her 18h0'37"30d32'
Harley & Mary's Star
Per 1h39'21"52d39'
Harley,Francis
Cep 23h29'44"68d58'
Harley,Joelle Marie
Vul 19h52'41"20d35'
Harley,Jr,Lloyd D
Eri 3h55'43"-6d25'
Harley,Lesla D'ann
Peg 22h27'0"30d26'
Harley,Norah
Aur 5h1'21"45d58'
Harlfinger,Heidi L
And 0h29'0"41d11'
Harlick,Duke Rudolph
Tau 4h38'28"10d47'
Harlig,Forever Harley
Ori 5h4'1"13d54'
Harlin Darlin's Rosebud
Aur 5h38'25"38d55'
Harlin,LESTAT Joseph Anthony
Agr 21h21'43"-8d8'
Harlin,Wade
Crt 11h44'0"-17d31'
Harline,Hayley Ann
Peg 22h33'58"24d34'
Harm,Honey
Mon 7h43'23"-2d28'
Harman,Bob
Ori 5h59'32"9d8'
Harman,Erich Anthony
Gem 7h16'6"21d35'
Harman,Ikabob
Cma 7h24'23"-11d10'
Harman,Jeni
Aqr 21h56'1"-10d55'
Harman,Jr,Robert E
Tau 4h31'34"30d28'
Harman,Kathy J
And 23h15'17"51d44'
Harman,Kevin C
Cep 1h29'53"77d36'
Harman,Mattie Marie
Mon 6h43'54"11d31'
Harman,Neil
Cep 20h31'35"62d59'
Harman,William Sam
Aql 19h5'49"0d19'
Harmanci,Cem
Boo 13h58'21"27d16'
Harmatz,Betsy
Cas 0h32'38"60d37'
Harmatz,Vanessa Brownhurst
Uma 12h23'46"60d36'
Harmen & Constance United
Cyg 19h26'0"33d15'
Harmening,Herr
Lyr 19h21'22"31d19'
Harmer,Madeleine Lorraine
And 23h2'26"49d36'
Harmo III,John Wesley
Dra 18h19'50"70d33'
Harmon Aug 7,1964,Lisa
Hya 9h9'43"0d4'
Harmon Beloved Husband,Billy Joe
Eri 5h7'23"-8d1'
Harmon SSSAS,Michael & Cheri
Lyr 18h58'44"45d8'
Harmon,Bailey Elizabeth
Gem 6h54'54"14d46'
Harmon,Beverly Pines
Peg 22h1'0"28d46'
Harmon,Christopher Booth
Aur 7h3'55"38d35'
Harmon,Clay Hooper
Gem 6h56'52"12d33'
Harmon,Dan'l
Cep 23h0'21"65d32'
Harmon,Donna Kay
Vul 20h15'59"28d18'
Harmon,Edward J
Per 3h53'34"38d59'
Harmon,Gerald
Sct 18h44'22"-6d4'
Harmon,Janet Ann Leskowsky
And 1h1'52"36d13'
Harmon,Kathy
And 2h1'43"45d32'
Harmon,Kristen Marie
And 0h22'1"32d18'
Harmon,Lillian Helen Gates
Cas 1h48'16"58d1'
Harmon,Linda Sue
Peg 22h35'40"25d16'
Harmon,Marina
Lyr 18h36'52"36d49'
Harmon,Sandra L
Aql 19h57'58"14d1'
Harmon,Star James M
Psc 23h1'37"6d51'
Harmon,Terri
Uma 11h35'18"30d10'
Harmon,Theodora
Cas 23h1'45"53d29'
Harmon,Yvette
Cas 2h30'38"59d7'
Harmonica Lynn
Aql 18h41'28"-1d14'
Harmonize The World
Lyr 18h35'1"28d56'
Harmons' Sacred Ground
Uma 8h33'42"54d6'
Harmony
Mon 7h57'14"-0d55'
Harmony
Lyn 8h48'28"34d53'
Harmony
Lyr 18h41'31"32d22'
Harmony
Cas 0h14'40"62d58'
Harmony Charlotte Theresa
Cyg 20h50'33"40d15'
Harmony's Way
Del 20h18'43"10d25'
Harms,Brandi Lynn
Vul 19h22'55"26d48'
Harms,Cassandra Ann
Mon 8h4'38"-9d41'
Harms,Charlotte
Del 20h52'1"8d2'
Harms,Colleen Nicole
And 23h31'21"47d7'
Harms,Dawn Ann & Kenneth John
Crb 15h53'22"37d30'
Harms,Hillary Ann
Uma 10h24'56"40d57'
Harms,Jessica Marie
Uma 10h49'18"62d29'
Harms,Jr,Kenneth John
Peg 22h1'46"31d50'
Harms,Kimberly
Peg 22h58'45"29d3'
Harms,Michele Lea
Peg 22h54'58"28d37'
Harms,Neita Elizabeth
And 0h51'48"35d43'
Harms,Norman E
Oph 17h30'6"-22d46'
Harms,Stephanie A
Peg 21h37'47"25d59'
Harmsen,Maryanne Joan
Lyn 9h16'50"37d24'
Harmston,Wendy Jo
Cas 3h3'20"61d35'
Harmsworth,Vere
Uma 12h4'10"46d36'
Harmsworth,Vere
Ori 6h1'60"0d52'
Harner,Andrew Stephen
Aur 5h30'0"38d26'
Harner,Richard N
Ori 5h58'10"5d36'
Harnett,Rita
And 0h21'53"33d39'
Harney,Alan
Cma 7h15'1"-13d37'
Harney,Robert T
Boo 15h46'41"51d25'
Harnicar,Thomas Albert
Dra 20h15'53"67d33'
Harnist,Albert
Cep 22h16'18"55d30'
Harnoss,Prof Dr Bernd Michael
Cam 10h57'47"82d23'
Harold
Peg 22h32'43"30d32'A
Harold
Her 16h37'59"40d29'
Harold
Cma 7h13'22"-13d54'
Harold & Virginia
Ori 5h25'17"-3d9'
Harold Lee,The
Aur 7h24'47"40d52'
Harold Love Forever, Jessica
Uma 9h25'0"56d54'
Harold's Golden Star
Cyg 21h2'33"35d40'
Harold,Beverly
Cyg 19h34'45"35d47'
Harold,Stacy
Umi 14h2'14"70d48'
Harold-Be-Thy-Name
Del 20h13'1"11d8'
Harootunian,Lee Craig
Cep 21h45'42"61d15'
Haroun,Fahmy
Umi 15h38'1"67d56'
Haroutinian,Araxie M
Ori 5h52'51"14d50'
Harowitz,Deborah P
Mon 6h54'21"-1d48'
Haroyn
Hya 8h53'0"4d27'
Harp,Jerry
Vul 20h37'54"20d12'
Harp,Robert Jonathan
Cet 1h4'0"0d55'
Harper
Uma 10h22'21"58d55'
Harper #7,D W
Her 17h12'24"21d54'
Harper 1993,Cheryl
Sge 19h56'39"18d51'
Harper IV,John
Her 17h34'56"27d10'
Harper's Ferry Kirsten
Cam 4h12'25"58d18'
Harper's Haven
Cyg 21h16'41"28d52'
Harper(Watai-1),Sharon Joyce
Aql 18h50'59"11d18'
Harper,Angelique
Lyn 9h11'13"37d43'
Harper,Arika
Cam 11h33'1"80d56'
Harper,Becky
And 0h58'23"36d15'
Harper,Charles Edward
Aql 19h59'0"10d58'
Harper,Christopher
Uma 10h30'32"52d17'
Harper,Danny
Lac 22h47'31"54d35'
Harper,Gene
Lyr 18h51'1"40d30'
Harper,Henry B
Aur 5h3'18"40d25'
Harper,James
Cet 0h52'32"-3d8'
Harper,Joyce
Ori 5h55'54"16d13'
Harper,Jr,Edward J
Her 16h9'58"24d15'
Harper,Jr,WR
Aql 19h57'1"1d9'
Harper,Khalfani (Asar) KW
Psc 1h31'37"20d11'
Harper,Loni Ann
Cet 1h33'29"-0d47'
Harper,Madeline Y
Leo 9h39'38"18d23'
Harper,Marilyn
Crb 15h48'34"30d23'
Harper,Randall McCambridge
Lac 22h7'51"49d2'
Harper,Ruth
Sgr 19h0'6"-28d47'
Harper,Scott E
Cep 21h56'59"60d22'
Harper,Stephen
Sco 16h51'12"-37d47'
Harper,Suzanne
Lyr 18h27'33"47d17'
Harper,Terence Michael
Ori 6h7'11"0d27'
Harper,Tory
Cnv 13h8'1"40d24'
Harper,William "Billy Ray"
Aql 19h50'10"15d42'
Harpham,Lynn "Fango"
Mon 7h54'19"-5d45'
Harpin,Poopie Chuck
Uma 8h46'43"51d36'
Harpo 7
Tau 4h36'26"28d34'
Harpt,Tamara Jean
Peg 23h46'13"15d29'
Harpum Le 'Maron'- Lord Of Light
Her 18h3'13"28d50'
Harr,James Daniel
Cnc 8h0'51"8d6'
Harrand,Julie
Crb 15h21'0"31d26'
Harrel's Golden Star
Eri 4h48'39"-7d57'
Harrell
Equ 21h2'12"3d6'
Harrell
Aql 19h0'0"13d23'
Harrell,Always Caston
Tri 1h46'54"26d57'
Harrell,Bernadine M
Gem 6h32'49"14d58'
Harrell,Richard S
Her 17h58'51"14d52'
Harrell,Steven Philip
Cet 1h34'19"-10d43'
Harrell-Geisler
Cmi 7h38'35"1d16'
Harrell-Gorham,Beverly
Del 20h22'23"7d54'
Harrelson,Lindsay Erin
And 0h37'0"41d6'
Harrene
Boo 15h16'21"34d43'
Harriet
Sgr 19h1'3"-26d46'
Harriet
Crb 16h10'26"27d4'
Harriet In Heaven
Cyg 19h52'34"37d44'
Harriet's "Lucky" Star
Lyn 7h55'40"40d34'
Harriett,Albert L
Sgr 19h32'4"-41d13'
Harriette
Tri 1h54'21"28d11'
Harrigan,Beth
Sgr 20h7'1"-44d16'
Harrill,Diana
Cet 1h43'45"0d55'
Harrill,Hannah Alexandria
Eri 3h36'34"-6d44'
Harrill,Harrison Alexander
Tri 1h51'0"27d57'
Harriman,Kathy
Lyr 18h17'43"44d9'
Harrington III,Gordon F
Mon 8h8'52"-1d25'
Harrington,Ann
And 23h16'44"50d18'
Harrington,Barbara
Umi 13h38'17"73d50'
Harrington,Brooks Evan
Mon 7h16'35"-6d6'
Harrington,Chuck & Sue's Star of
Uma 10h34'11"48d33'
Harrington,CLU,ChFC, CFP,Michael A
Cep 23h43'14"83d33'
Harrington,Denise
Cyg 19h18'46"28d24'
Harrington,Edna
Cas 0h22'34"75d10'
Harrington,Erin C
Uma 10h14'49"48d57'
Harrington,Francis John
Hya 8h23'44"-10d40'
Harrington,Grace
Psc 1h40'17"28d1'
Harrington,John Michael
Uma 9h58'35"53d34'
Harrington,John P
Crt 11h53'0"-18d23'
Harrington,Jr,Best Dad,Daniel J
Lyn 9h3'19"34d31'
Harrington,Jr,Gordon Franklin
Per 2h47'11"43d26'
Harrington,Kathleen Caroline
Ori 5h36'37"0d16'
Harrington,Lauren
Uma 8h14'39"68d39'
Harrington,Lauren G
And 0h57'16"35d53'
Harrington,Madeleine Elizabeth
Peg 22h26'14"20d38'
Harrington,Michael Thomas
Dra 14h19'24"64d19'
Harrington,Mrs Megin J
Cas 23h22'30"60d9'
Harrington,Patrick Kiernan
Cet 1h38'32"-1d52'
Harrington,Rhoni Jo
Lac 22h55'1"50d8'
Harrington,Robert L
Cyg 19h55'50"40d30'
Harrington,Ross Daniel
Hya 9h13'41"-7d23'
Harrington,Susan June
Cet 2h7'24"1d43'
Harrington,T E Sam
Her 16h57'10"20d19'
Harrington,Teresa
Cas 1h4'50"60d23'
Harrington,Timothy George
Aur 6h20'24"30d28'
Harrington,Todd
Aur 6h26'13"33d53'
Harrington,Tyler Griffith
Mon 8h8'1"-0d3'
Harrington,William
Ori 4h53'22"4d38'
Harrington-Best Mom, Ann Oliver
Cas 0h12'1"60d43'
Harris Birthday Star, Bradley
Cep 22h51'49"65d38'
Harris Greig Family
Aur 6h0'37"30d51'
Harris Laird Of Camster,John Henry
Cep 20h55'15"58d49'
Harris MBBS FRCA, Doctor Richard William
Ori 5h38'27"10d4'
Harris MD,Dr Robert W
Oph 16h55'1"-25d19'
Harris of Peckham,Lord
Cep 21h34'15"70d52'
Harris Star,Lucky
Lyr 18h30'16"37d60'
Harris Star,The Pearlman
Sge 19h54'49"16d10'
Harris Star,The Andrea
Uma 9h19'53"53d10'
Harris Star,The Lou
Uma 11h52'29"30d53'
Harris's Bridge
And 22h59'60"40d0'
Harris's Giant Step, Bob
Dra 16h35'53"36d55'
Harris,Aaron Michael
Sgr 18h49'36"-34d10'
Harris,Adam Julian
Tri 2h37'59"35d36'
Harris,Adam Julian
Cep 1h0'1"84d53'
Harris,Alexandra Rose
Sgr 18h49'36"-34d10'
Harris,Alfred I
Aur 5h1'0"42d44'
Harris,Alison
Mon 6h27'24"10d5'
Harris,Anne
Peg 22h3'1"25d4'
Harris,Anne Marie
Aur 6h6'28"45d7'
Harris,Antonia Kathrine
And 1h16'22"36d53'
Harris,Art
Aur 6h45'58"37d13'
Harris,Audrey Jane
Cas 2h35'20"57d30'
Harris,Barbara
Cyg 21h50'52"38d29'
Harris,Barbara Darlyn
And 1h53'42"38d14'
Harris,Billy Eugene
Gem 6h55'6"16d59'
Harris,Brian
Tri 1h35'17"31d7'
Harris,Brooks
Com 12h32'50"27d9'
Harris,Bryan Thomas
Tau 4h39'1"7d46'
Harris,Burke Reed
Umi 14h54'59"71d20'
Harris,Caitlin Elizabeth
Leo 9h42'11"7d30'
Harris,Caorl Ann
Crb 15h54'56"27d22'
Harris,Caprice
Vul 19h43'16"22d34'
Harris,Captain JB
Cas 1h20'18"72d35'B
Harris,Caryn Emily
Psc 23h30'20"1d60'
Harris,Catherine Evelyn
Lyr 19h14'23"42d55'
Harris,Chancler Trey
Peg 23h4'26"10d5'
Harris,Chanclor Trey
Aql 18h57'46"-6d40'
Harris,Charles Joseph
Aur 7h14'1"40d48'
Harris,Christopher Thomas
Per 2h56'0"38d19'
Harris,Daniel L
Equ 21h1'14"2d28'
Harris,Daniel Tyler
Mon 8h8'41"-7d56'
Harris,David
Cma 6h59'28"-11d13'A
Harris,David E
Her 16h43'50"33d38'
Harris,David S & Alison T
Sge 18h56'58"19d51'
Harris,Della
Ori 6h5'26"0d33'
Harris,Devon Elizabeth
Mon 7h3'30"-1d25'
Harris,Donna Jean
Cnv 12h21'1"40d9'
Harris,Doreen
And 23h3'14"47d6'
Harris,Douglas Roy
Vul 19h22'54"25d10'
Harris,Duncan William
Psc 0h46'6"18d24'B
Harris,Edith Marion Lavina Howe
Lyn 8h0'33"37d30'
Harris,Eloise Dickerson
Gem 6h42'42"30d51'
Harris,Emma Jane
And 0h50'43"33d24'
Harris,Emma Louise
Cyg 19h58'48"58d48'
Harris,Emmylou
Peg 22h3'47"10d24'
Harris,Evelyn K
Crb 15h54'29"31d1'
Harris,Farah-Marie
Cam 9h7'50"81d1'
Harris,FrancesKa
Sco 17h31'36"-30d44'
Harris,Garnett Judson
Ori 5h31'42"-10d33'
Harris,Gary
Ori 6h3'15"1d18'
Harris,Gary Robert
Her 16h59'11"33d52'
Harris,Georgiana P
And 23h48'16"33d17'
Harris,Grace
Cyg 20h9'42"40d57'
Harris,Haydn R
Aur 4h55'27"40d11'
Harris,Howard Hazen
Aur 6h29'10"37d48'
Harris,In Memory of Stephen E
Aur 4h49'56"50d14'
Harris,James Ridout
Ori 5h55'17"20d53'
Harris,Janet Sue
Lyr 19h17'1"42d13'
Harris,Janie Kathryn
Mon 6h54'52"7d37'
Harris,Jason K
Aql 20h7'26"0d2'
Harris,Jeanne Duran
Dra 18h56'17"67d45'
Harris,Jeffrey Steven
Aql 19h18'1"12d31'
Harris,Jeffrey Stuart
Uma 12h22'58"61d12'
Harris,Jennifer Lynn
And 0h51'43"38d45'
Harris,Jenny
Cas 1h50'11"60d14'
Harris,Jesse M
Lyn 7h0'20"51d42'
Harris,John
Her 17h19'33"44d56'
Harris,John & Jonni
Eri 3h36'41"-6d37'
Harris,Joseph Collin
Aql 19h51'0"11d53"
Harris,Joseph Mark
Cyg 20h18'0"39d5'
Harris,Joya Renee
Cyg 21h5'59"44d45'A
Harris,Joyce Lee
Sex 10h27'54"-5d41'
Harris,Jr,Jesse L
Cet 0h30'29"-0d5'
Harris,Judi
Cas 0h10'47"64d58'
Harris,Julie
Cma 6h59'28"-11d13'B
Harris,Kasey Renee
Lyr 18h23'12"42d37'
Harris,Keith Wynn
Lmi 10h10'17"34d30'
Harris,Kelly Suzanne
Cyg 21h2'0"31d13'
Harris,Ken
Her 16h2'43"21d7'
Harris,Lauren
Cam 13h4'53"81d15'
Harris,Lori & Jim
Aql 19h27'0"7d43'
Harris,Love To Marcia
Cyg 19h33'1"30d19'
Harris,Maggie Elizabeth
Peg 21h45'31"22d41'

Harris,Marianne & John
Vul 19h58'55"26d12'
Harris,Mark
Her 17h3'12"50d1'
Harris,Martin
Per 3h2'11"41d11'
Harris,Marty
Dra 15h15'0"64d3'
Harris,Mary
Cas 0h1'34"61d31'
Harris,Mary Anne
Ori 5h55'1"18d45'
Harris,Mary M
Peg 21h56'34"26d23'
Harris,Matt
Cam 4h59'25"60d13'
Harris,Matthew
Vul 19h48'0"27d39'
Harris,Matthew
Her 18h19'41"28d39'
Harris,Matthew Michael
Cet 2h40'1"-5d41'
Harris,Megan
Lyn 6h43'1"50d8'
Harris,Megan M
And 1h0'0"40d43'
Harris,Michael Cole
Ari 2h24'0"11d20'
Harris,Mickie
Vul 21h12'39"20d23'
Harris,Mike
Per 3h20'30"54d45'
Harris,Nathan Stewart
Psc 0h46'6"18d24'A
Harris,Nicole Marie
Mon 7h29'33"-8d29'
Harris,Paul J
Per 3h51'52"38d38'
Harris,Peggy
Eri 2h54'41"-8d1'
Harris,Peggy
Peg 21h38'38"27d51'
Harris,Peter Frederick
Aql 19h55'11"12d5'
Harris,R & M
Cet 2h36'40"-0d47'
Harris,Rachel Macbeth
Cas 3h32'51"69d9'
Harris,Raymond George
Boo 15h0'51"31d10'
Harris,Richard L
Cep 3h3'30"80d24'
Harris,Robert L
Boo 14h31'39"48d25'
Harris,Robert Roden
Ori 4h55'51"5d12'
Harris,Robin Jason
Lyn 8h34'20"45d9'
Harris,Ruth
Cyg 19h47'22"58d49'
Harris,Ryan Andrew
Sex 10h7'39"-2d27'
Harris,Ryan Everett
Aql 19h41'55"13d0'
Harris,Sarah
Boo 14h30'35"51d34'
Harris,Sharlyn Marie Dixon
Mon 7h45'39" 3d10'
Harris,Shawnda Jane
Peg 23h33'30"21d43'
Harris,Skylar M
Dra 19h58'39"68d7'
Harris,Sr,Jeffrey Davis
Cet 2h25'50"4d55'
Harris,Stephanie Meredith
Peg 22h19'42"30d40'
Harris,Stephen
Her 17h19'50"43d2'
Harris,Stuart
Sex 10h26'24"3d3'
Harris,Susan
Sco 16h50'15"-40d38'
Harris,Susan H
Mon 7h0'18"-8d15'

Harris,Susan Mary Ann
And 23h19'51"49d41'
Harris,Suzy
Aql 20h5'26"1d20'
Harris,Sydne Taylor
And 0h59'0"40d50'
Harris,Tammy & Kerry
Peg 23h16'27"30d47'
Harris,Terry Dean
Gem 6h51'4"13d33'
Harris,The Traveler, Cecelia A
Cyg 20h28'1"41d2'
Harris,Theresa
Mon 8h6'22"-8d45'
Harris,Theresa
Mon 8h5'19"-1d23'
Harris,Theresa
Lyn 7h53'10"42d38'
Harris,Theresa
Cas 1h2'23"63d57'
Harris,Tivon Nelson
Aur 6h1'47"33d4'
Harris,Tovi
Cam 4h22'55"60d46'
Harris,Trevor Jay
Her 17h16'13"41d38'
Harris,Trevor Paul
Ara 17h53'50"-55d53'
Harris,Virginia Mulford
And 2h30'22"48d48'
Harris,Wallace Wally C
Vul 19h22'59"26d39'
Harris,Wayne Shelton
Aql 18h57'26"-6d29'
Harris,Will E
Aur 6h16'34"46d49'
Harris,Zoë Alexa
Mon 7h1'55"5d8'
Harris-Forkner,Katy Irene
Lyn 8h13'15"46d27'
Harris-Hawkins
Peg 23h3'0"22d6'
Harrison
Uma 9h26'38"47d33'
Harrison
Cnc 8h21'50"18d54'
Harrison
Cet 0h36'18"-5d24'
Harrison III,Charles William
Cma 6h13'12"-24d30'
Harrison III,Frank W "Billy"
Cet 2h55'20"0d48'
Harrison Star,The
Uma 11h30m46"63d21'
Harrison ZøB,Lullelia W
Peg 22h36'26"21d60'
Harrison,Aime
Peg 23h15'49"30d5'
Harrison,Albert A
Cet 2h15'43"6d28'
Harrison,Alex Lee Medden
Ori 5h58'23"8d41'
Harrison,Alex Mullan
Ser 18h14'43"-9d39'
Harrison,Ann Wicker
And 0h10'56"46d10'
Harrison,Arthur S
Dra 14h59'58"63d40'
Harrison,Barbara Renee
Cas 1h48'23"61d16'
Harrison,Bob & Cindy
Cyg 19h28'21"41d2'
Harrison,Bradi
Cam 7h6'29"61d33'
Harrison,Brian
Aql 19h36'39"-6d34'
Harrison,Brian Geuer
Cmi 8h5'32"6d12'
Harrison,Caterina Maria
Cyg 20h1'37"40d14'
Harrison,Charles
Sct 18h55'1"-6d56'
Harrison,Christopher M
Umi 13h22'15"71d54'

Harrison,Colin
Ori 6h3'0"1d31'
Harrison,Cristina Winter
Mon 8h8'1"-8d29'
Harrison,Dale Allan
Boo 14h27'24"42d56'
Harrison,Daniel Gene
Oph 18h32'50"11d42'
Harrison,Darroll Keith
Dra 16h10'50"61d23'
Harrison,Debra D
Mon 7h29'48"-10d13'
Harrison,Earl Spencer
Cep 22h55'60"63d8'
Harrison,Edward John
Dra 14h44'49"62d36'
Harrison,Eric Jon
Aur 7h3'21"39d42'
Harrison,Frank G
Boo 15h2'1"28d41'
Harrison,Grady Antero Michael
Cep 21h53'40"55d35'
Harrison,Helen
Cas 0h27'1"63d11'
Harrison,Jade Davida Rose
And 23h8'48"37d58'
Harrison,James Martin
Her 16h41'56"4d16'
Harrison,Jamie
Lac 22h25'29"50d26'
Harrison,Jane
And 0h52'28"37d17'
Harrison,Jerry
Cas 0h30'58"61d26'
Harrison,Joan
And 23h22'24"51d56'
Harrison,John David
Cet 1h47'15"-6d15'
Harrison,Joyce
Cyg 21h3'41"34d18'
Harrison,Jr & Family, William A
Ori 5h52'44"18d48'
Harrison,Kathleen Carol
Cas 0h54'55"73d38'
Harrison,Keith A
Aql 19h55'0"15d11'
Harrison,Kevin Patrick
Aql 18h55'24"11d37'
Harrison,L N
Ser 18h16'43"-14'59'
Harrison,Leandra
Lyr 18h49'21"41d13'
Harrison,Leslie
Cep 23h6'22"60d50'
Harrison,Lionel & Georgia
Crb 16h8'42"30d45'
Harrison,Lisa
Sge 19h30'48"16d35'
Harrison,Lynette Ann
Peg 22h29'20"21d7'
Harrison,Mark & Sarah
Cyg 19h18'13"28d20'
Harrison,Martin Palmer
Aql 19h48'36"14d21'
Harrison,Mary Ann Bricker
Lyr 18h45'15"39d23'
Harrison,Maurice
Hya 8h12'13"1d29'
Harrison,Michael
Peg 23h33'10"31d43'
Harrison,Michael John
Cep 21h26'32"78d55'
Harrison,Michelle
Cyg 19h24'1"35d24'
Harrison,Mr & Mrs E V
Ori 5h55'0"18d10'
Harrison,Myrna Gayle
Cet 1h21'49"-7d10'A
Harrison,Nigel, Catriona & Thomas
Uma 11h31'35"50d17'
Harrison,Patricia & Chad
Mon 6h59'41"-1d25'

Harrison,Peter & Joanne
Sge 19h55'19"20d20'
Harrison,Rita R
Aql 19h31'32"12d48'
Harrison,Robert "Grandpa"
Cep 22h52'1"57d2'
Harrison,Sally Lynn
And 1h35'38"39d4'
Harrison,Sam
Ori 5h56'22"8d51'
Harrison,Trenton
Cep 21h19'53"71d10'
Harrison-Dees,Alison & Geoffrey
Cyg 21h0'30"36d42'
Harrison-Ipse,David Frances
Ori 5h31'42"-10d33'
Harrison-Sherman Shannon
Vul 19h47'52"25d43'
Harrisons Eternity
Aql 19h30'1"0d33'
Harrisson,Huguette
Cas 1h6'60"60d23'
Harro 1
Umi 16h54'39"78d55'
Harrod,Garry S
Lac 22h23'27"48d52'
Harrod,Jeff
Cnv 12h11'1"37d56'
Harrod,Neoma Jean
Peg 22h28'55"11d25'
Harrod,Richard Allen
Sex 10h8'32"-0d38'
Harrold,Barbara Lindsay
Lyn 7h56'1"38d32'
Harrop's Comet
Peg 22h32'41"28d32'
Harrop,Raymond Anthony
Uma 11h55'22"43d10'
Harrow,Lisa J
Leo 10h43'58"15d15'
Harrower,Hanon
Aql 19h11'60"10d52'
Harrsch,Gregory
Boo 15h20'58"48d35'
Harry
Dra 16h32'16"72d11'
Harry
Boo 15h1'17"18d47'
Harry & Louise
Cyg 21h46'43"53d14'
Harry & Phil
Ori 5h28'0"-0d4'
Harry & the Fool
Ori 4h59'0"5d49'
Harry B
Her 17h33'43"24d11'
Harry W B und Marion ter Heide
Cep 22h8'44"61d21'
Harry's Autumn Star
Lac 22h29'25"40d42'
Harry's Harbor
Equ 21h20'31"12d10'
Harry's Hide Away
Cru 12h52'55"-60d38'
Harry's Star
Oph 17h21'50"-20d31'
Harry,Gordon W
Sgr 18h53'49"-23d22'
Harry,Jaynece Lee
And 23h46'30"46d45'
Harry,June Sprague
Uma 12h4'21"62d35'
Harsch,Anne H
Vul 20h43'1"28d36'
Harsh,Amy Elizabeth
Lyn 7h37'17"51d52'
Harshbarger,Leonard & Arlene
Aql 18h52'0"11d7'
Harshbarger,Timothy & Christine
Cyg 19h33'47"38d54'
Harshfield,Donald
Hya 8h53'1"2d16'

Harshita
Her 17h0'19"31d30'
Harshman,Jeffery J
Lyn 9h10'38"39d59'
Harson,Katie Elizabeth
Peg 23h24'30"33d60'
Hart To Hart
Uma 10h4'13"68d42'
Hart's Fireballs 93/94 ,Mrs
Lyn 6h59'47"56d4'
Hart's Golden Memory
Mon 7h2'51"0d42'
Hart,Alec
Ari 2h25'29"10d55'
Hart,Alister
Cep 20h49'56"70d50'
Hart,Angela
Ori 6h7'47"9d16'
Hart,Betty & Irving
Cyg 21h19'56"37d54'
Hart,Carl
Uma 11h34'52"32d12'
Hart,Caroll
Lup 15h20'31"-45d40'
Hart,Christian
Aur 5h30'50"38d10'
Hart,Christopher S
Aur 5h7'36"38d41'
Hart,Clarence Vernon
Cet 3h10'25"2d4'
Hart,Cliff
Cra 18h32'16"-44d28'
Hart,Craig Augustus
Per 1h45'19"48d33'
Hart,Daniele
Ant 10h39'47"-34d40'
Hart,Donald
Cam 5h23'44"68d45'
Hart,Dr Donna
Lyn 7h10'0"52d11'
Hart,Elaine Katherine
Lyn 9h13'0"46d34'
Hart,Eleanor Joan O'Mara
Hya 8h26'17"-9d29'
Hart,Elija Z
Lyn 7h3'28"53d20'
Hart,Grandpa
Her 16h49'11"47d33'
Hart,Jacqueline Altmeyer
And 23h2'36"50d28'
Hart,Jean
Cyg 19h28'30"30d9'
Hart,Jessica
Vul 20h2'57"23d56'
Hart,Jim
Boo 14h20'38"28d49'
Hart,John
Ori 6h8'46"9d56'
Hart,John Curtis
Cet 2h31'33"-3d46'
Hart,JoLeene
Peg 0h5'56"21d1'
Hart,Jr,David R
Aur 5h13'55"44d3'
Hart,Katherine
And 2h34'15"45d36'
Hart,Katie
Mon 7h1'44"5d12'
Hart,Katie & Shane
Cas 0h24'51"69d33'
Hart,Kevin Thomas
Her 16h37'34"50d21'
Hart,Laurence
Peg 0h4'57"20d3'
Hart,Lillie
Dra 18h4'0"70d26'
Hart,Margaret
Peg 22h50'30"29d39'
Hart,Marilyn Klotnia
And 0h7'0"47d58'
Hart,Marvin T
Boo 14h50'33"35d13'
Hart,Maxfield Joseph
Her 17h20'29"45d24'

Hart,Melissa Ann
Lyn 9h8'22"44d19'
Hart,Melissa Joan
Aql 19h31'39"14d18'
Hart,Michael
Her 16h29'55"35d60'
Hart,Ray
Aur 5h3'59"42d1'
Hart,Rick W
Cas 2h3'17"73d42'
Hart,Robby
Cam 8h1'36"60d26'
Hart,Robert
Boo 14h5'56"12d38'
Hart,Robert L
Cet 1h37'11"-2d13'
Hart,Sharon & Pat
Lac 22h54'37"38d11'
Hart,Sheldon
Peg 22h50'1"24d14'
Hart,Ted
Aql 19h57'56"-10d2'
Hart,Terry (Momma)
And 0h56'36"36d34'
Hart,Tyron Robert Walker
Her 17h19'30"42d24'
Hart-Daddy,William Michael
Her 16h25'40"33d7'
Hartberger,Elvira
Cas 0h13'1"60d14'
Hartcher,Michael George J
Ser 15h14'1"9d59'
Hartdegen
Hya 9h54'56"-16d48'
Harte-Seed
Lac 22h31'20"53d40'
Harteis,Colleen
Ori 5h32'35"-1d11'
Harteis,Todd
Aur 6h3'26"30d17'
Hartel,Jill
And 0h48'0"36d59'
Hartenberg,Rodica En Wilko
Dra 10h15'51"81d49'
Hartenstein,Günter und Gisela
Cap 20h4'38"-22d47'
Harter,Jeneane
Mon 6h19'52"6d46'
Hartery,Katherine Marie
Dra 11h56'24"66d42'
Harteveld,J
Cas 3h3'1"57d26'
Hartfield,Jennifer
And 23h40'15"40d50'
Hartfield,Marin Leeann
Uma 11h49'23"49d32'
Hartford,Ruth
Uma 10h3'0"50d37'
Harth,Paul
Hya 8h27'38"-0d23'
Hartig,Dr Richard E
Oph 17h36'30"-1d0'
Hartigan,Megan
Peg 23h23'1"33d49'
Hartin,Patricia Susan
Peg 22h1'41"5d43'
Hartjen,Travis
Her 16h41'35"34d22'
Hartjen-Grimm,Astrid
Mon 7h53'37"-1d38'
Hartke,Holly
Lyn 8h40'51"45d36'
Hartl,Joel Alexander
Her 18h4'35"40d22'
Hartl,Roland
Sge 19h57'33"20d6'
Hartle Family,The
Uma 9h50'41"49d50'
Hartle,Mindy Sue
Lyr 19h1'16"25d44'
Hartleben,Michael
Ari 2h38'39"21d15'
Hartlein,Briana Star
Sct 18h19'20"-14d26'

Hartley Family Star, The
Eri 4h34'46"-17d44'
Hartley's Lifelong Wishing Star,Addy
Vul 21h23'13"28d24'
Hartley,Dustin Hollis
Oph 17h15'32"-20d24'
Hartley,Elise & Doc
Uma 11h25'43"32d25'
Hartley,Fred Jack
Sex 9h53'58"1d19'
Hartley,Jena Kaye
Uma 10h58'0"70d20'
Hartley,Jessica Amber
Cyg 20h21'13"38d28'
Hartley,Landon Russell
Lac 22h23'53"50d24'
Hartley,Mark Anthony
Aql 19h12'23"13d2'
Hartley,Roger
Per 3h59'51"38d1'
Hartley,Shea & Chase
Cma 6h41'48"-11d9'
Hartley,UConn's President Harry J
Psc 0h59'29"27d47'
Hartley,Walter L
Per 3h0'23"40d9'
Hartlieb,Sandy
Uma 12h47'35"57d46'
Hartline,Cheryl Lyn
Del 20h29'21"10d28'
Hartlove III,Joseph George
Per 4h4'16"35d5'
Hartlove,Linda Maria
Cas 3h6'0"70d40'
Hartman
Uma 8h48'0"50d6'
Hartman "The Sheik", Charles E
Lac 22h42'1"54d18'
Hartman #1,Carol
Cas 0h32'11"63d10'
Hartman,Aaron
Hya 8h19'39"0d33'
Hartman,Anja
Cam 6h2'41"58d43'
Hartman,Beatrice P
Lyn 7h55'57"45d17'
Hartman,Brandon D
Ser 17h33'46"-14d28'
Hartman,Brent
Oph 16h53'44"-0d23'
Hartman,Cris & Steve
Lyr 19h4'49"28d49'
Hartman,Eliza M
Dra 16h8'3"58d4'B
Hartman,Erik
Oph 16h57'55"1d1'
Hartman,George C
Her 17h45'55"14d27'
Hartman,Gladys Kear
Lyn 8h3'1"39d36'
Hartman,Heather Marie Elizabeth
Leo 11h6'29"13d22'
Hartman,Jason H
Ser 15h12'32"18d35'
Hartman,Jennifer
And 23h44'32"47d8'
Hartman,John McMaster
Aur 6h28'60"33d45'
Hartman,Kali Marie
And 1h32'43"38d46'
Hartman,Kathy & Jeff
Cyg 21h4'19"38d35'
Hartman,Linda/Russell Keene
Lyr 18h31'14"34d55'
Hartman,Lori Ann
And 23h33'20"40d21'
Hartman,Marc Gregory
Lac 22h26'13"54d57'
Hartman,Molly, Christopher,David
Eri 4h11'37"-10d4'

Hartman,Ray
Hya 9h30'15"0d35'
Hartman,Shannon DaVette
Cas 1h3'56"62d9'
Hartman,T Alan
Lmi 9h47'28"40d40'
Hartman,Theresa
Cas 23h24'17"60d41'
Hartman,Tot
Aql 19h50'1"15d41'
Hartman,Wendy Marie
And 2h11'43"42d15'
Hartman,Wesley B
Dra 16h8'3"58d4'A
Hartman,Yvonne
Gem 7h2'16"21d41'
Hartman-Nitromethane, Karl
Ser 15h9'59"20d59'
Hartmann,Christopher
Per 4h0'50"51d34'
Hartmann,Dr Thomas M
Aql 19h0'18"13d14'
Hartmann,Friedhelm
Lac 22h7'60"50d25'
Hartmann,Gerald R
Ori 5h36'56"0d2'
Hartmann,Gunter
Lac 22h5'50"50d48'
Hartmann,Jörg-Dietrich
Cmi 7h17'47"7d32'
Hartmann,Peter
Peg 23h33'38"10d30'
Hartmann,Peter
Oph 17h54'1"8d57'
Hartmann,Robert & Terri
Cyg 20h17'38"38d25'
Hartmann,Torben
Peg 23h31'12"11d4'
Hartmann-URMY*M L B, Gregg S
Lmi 10h24'0"30d12'
Hartmanstorfer,Joseph S & Caroline L
Cyg 20h22'0"38d45'
Hartmark,Kimberli
Cet 0h54'1"0d31'
Hartmon,Arlene
Lyn 7h2'1"51d2'
Hartner,Brad & Christi
Aql 18h45'30"11d41'
Hartness,Brian A
Ser 15h55'33"0d2'
Hartnett III,Sept 7, 1945,Edward Joseph
Cam 3h27'19"67d24'B
Hartnett September 23, 1945,Marie Keefe
Cam 3h27'19"67d24'A
Hartnett,Marjorie Fern
Aql 19h29'51"0d22'
Hartnett,Richard
Uma 11h9'0"48d19'
Hartnett,Tara Marie
Uma 9h2'47"50d24'
Hartnett,William
Dra 16h11'53"63d10'
Hartnett,Winifred E
And 1h3'13"40d34'
HartNoMore
Lyn 7h15'57"59d8'
Hartog,Elisabeth R J
Cas 0h22'21"66d33'
Hartog,Liesbeth L
Lyn 7h38'34"44d19'
Harton,Shannon Michael
Lac 22h14'28"51d21'
Hartopp,Linda
Eri 2h58'17"-5d39'
Hartree,Frederick Barry
Uma 11h4'50"43d13'
Harts,Cynthia
Vir 11h58'34"0d52'
Hartsel,Holly Beth
Cyg 20h23'45"30d26'

Hartsell,David L
Her 18h0'24"28d47'
Hartsells,The
Dra 17h5'58"52d7'
Hartsfield Family Star Tim
Her 16h37'1"33d51'
Hartsfield's,Star is Rising,Heather
Lyr 18h39'33"35d53'
Hartshorn,Patricia Lee
Lyn 8h23'37"40d53'
Hartshorn,Will
Oph 17h25'32"-20d8'
Hartsock,Cody James
Aql 20h34'19"0d12'
Hartson,Nancy Stafford
Com 12h2'58"26d29'
Hartsook,Austin Robert
Her 17h30'55"14d48'
Hartsook,Dave & Cathy
Lyr 18h21'51"42d17'
Hartsough,Vicki
Gem 6h49'49"30d31'
Hartt,Sharon
Cas 0h39'20"64d14'
Hartung III,James E
Aqr 20h45'28"-6d25'
Hartung,Anna DeFalco
Lyr 19h0'33"37d34'
Hartung,Wendy
Aql 19h59'56"10d13'
Hartwell,Barbara
Ori 5h58'25"21d17'
Hartwell,Sharlene
Cyg 19h42'42"41d48'
Hartwick,John L
Dra 16h47'34"66d34'
Hartwig,Barbara
Aur 5h17'1"44d25'
Hartwig,Corina
Boo 13h37'55"10d33'
Hartwig,Hellmut A
Ser 18h54'29"2d41'
Hartwig,Nicole
Cas 0h13'48"62d54'
Hartwig-Michael
Her 16h34'50"33d25'
Harty,Dennis J
Per 2h9'1"56d27'
Harty,Laurie Lyn
Vul 19h37'1"27d11'
Harty,Patricia A
And 0h49'1"36d45'
Hartz,dans le ciel, Harry & Ruth
Lyr 18h25'16"41d13'
Hartz,Heather Diane
Cet 0h27'0"-11d8'
Hartz,Lee
Cet 1h7'32"-4d12'
Hartzband,Dave "The Incredibull"
Sct 18h56'32"-6d43'
Hartzell,Dane Christian
Cas 23h31'12"61d18'
Hartzell,Gene
Dra 18h8'54"58d30'
Hartzell,Margaret & Francis
Uma 9h8'32"58d17'
Hartzell,Mark Andrew
Cyg 20h24'35"40d22'
Hartzman,Ruthie Genen
Cap 21h28'18"-24d18'
Hartzog's Kingdom
Cma 7h14'27"-15d58'
Hartzog,Katie Rebecca
Peg 22h33'25"33d60'
Haruka
Tau 5h29'9"17d29'
HARUKA
Tau 4h12'14"19d59'
Harv's Star
Per 2h10'46"57d12'
Harvard,Julia Nicole
And 0h48'36"36d12'

Harvender,Nelson
Lyr 18h58'0"35d0'
Harvey & McGonagle
Aql 20h2'0"8d16'
Harvey & Reina
Aql 19h57'42"10d20'
Harvey Forever,Curt- Cookie
Lyr 19h0'13"28d50'
Harvey's "Ranle's Wedding Star",The
Peg 22h25'35"5d52"
Harvey,Amber Ellen
Peg 22h46'21"33d3'
Harvey,Ana Recio
Peg 22h32'15"27d4'
Harvey,Ashley Nicole
Peg 23h1'46"30d23'
Harvey,Barbara
And 0h48'26"39d26'
Harvey,Barbara
Vul 19h48'1"27d21'
Harvey,Barbara & Loren
Aql 19h33'36"0d45'
Harvey,Brigadier General John F
Lac 22h38'13"53d5'
Harvey,Christopher S
Ser 15h10'37"12d16'
Harvey,Daniel Joseph
Uma 8h24'35"71d38'
Harvey,Donna Jo
Lyr 18h41'23"28d17'
Harvey,Eileen V
Cas 1h3'57"60d3'
Harvey,Gene Alton
Lac 22h4'31"46d23'
Harvey,Jacob Weston
Her 16h43'35"22d25'
Harvey,James
Per 4h30'58"37d38'
Harvey,James R
Aql 20h6'51"8d59'
Harvey,Kevin Nicholas
Uma 9h12'1"57d14'
Harvey,Laura Sue Rebecca
Cyg 20h33'21"31d41"
Harvey,Margaret Marion
Lyr 18h48'37"32d32'
Harvey,Melissa
Mon 6h58'23"-8d5'
Harvey,Michelle Joanne
Lyn 7h57'41"50d31'
Harvey,Patrick Dale- Verlin
Per 2h36'11"45d9'
Harvey,Paul John
Aql 20h30'51"-0d28'
Harvey,Peter & Ellie
Umi 13h35'45"70d12'
Harvey,Rachel Mignon
Lyr 10h57'53"37d25'
Harvey,Reba
Peg 21h42'40"20d6'
Harvey,Robert Thomas
Lyr 19h4'28"37d34'
Harvey,Samantha
And 0h22'24"38d2'
Harvey,Sharlene
Lyn 8h1'39"34d43'
Harvey,Shauna
Del 20h54'42"9d4'
Harvey,Simon John
Her 16h35'17"16d59'
Harvey,Simone
Vir 13h56'37"2d24'
Harvey,Spencer Ray
Uma 8h24'20"68d31'
Harvey,Sr,Robert G
Cep 21h43'31"55d15'
Harvey,Steve
Cas 2h36'35"57d37'
Harvey,Vivian
Cas 0h51'25"61d28'
Harvey,Zachary Hays
Mon 7h43'7"-3d31'
Harvey-The Great Star, Lynne "Tweetie Pie"
Cas 0h42'30"70d38'
Harvick "PFR",Valerie Anne
Lyr 19h22'35"40d4'
Harvick,Deanna
Lyn 7h36'39"38d46'
Harvie's Star
Per 2h48'53"43d11'
Harvill,James W
Cep 21h0'53"80d8'
Harville,Rachel
Lmi 11h3'1"32d19'
Harvin,Jeri
Psc 0h25'47"7d43'
Harviv 1947
Aql 19h0'43"5d38'
Harwich,Jr,Susan Diane & Murray D
Per 3h11'20"49d1'
Harwich,Mary Belle Taylor Hay
And 2h28'17"40d59'
Harwin,Dr Steven
Gem 8h1'28"28d30'
Harwood,Blaine
Boo 15h5'45"22d34'
Harwood,Forever Stacy
Cas 1h55'56"70d54'
Harwood,Ian
Cyg 21h42'1"38d55'
Harwood,Johnathan
Per 3h38'30"38d59'
Harwood,Paul
Cep 22h34'44"67d43'
Harwood,Richard Kistler
Her 16h16'0"50d14'
Harzan,Jaclyn Grace
And 0h30'31"45d30'
Harzer
Cam 14h23'11"82d20'
Harzig,Herbert M
Vir 13h24'24"-2d19'
Harzinski,Erin Marie
Aql 20h19'40"0d42'
HAS/JRK
Cnv 13h34'57"41d48'
Hasa
Cam 3h54'52"62d38'
Hasamona
Aur 5h55'54"31d48'
Hasan & Leila
Lyn 7h55'40"50d7'
Hasan,Jahanara
Umi 15h43'55"76d21'
Hasan,Jeffrey Khalid
Lac 22h29'35"37d58'
Hasbou,Hugues
Peg 23h8'0"10d15'
Hasbrouck,Louise Julia
Cap 20h9'17" 10d21'
Hasbrouck,Peter & Catherine
Aql 20h3'23"4d40'
Hasch,Ursula und Jürgen
Gem 8h2'39"31d23'
Hase
Psc 0h23'54"-0d16'
Hase,Gerald A
Cet 2h31'10"-3d45'
Hasegawa,Kiyoko
Hya 8h18'56"-1d20'
Hasehuluh
Vir 14h9'0"-0d3'
Hasek III,Anthony Paul
Per 2h57'27"50d20'
Haseldine,Luke Jean-Louis
Uma 11h39'18"38d1'
Haseleu,Marie Louise
Mon 6h7'0"-10d36'
Haseltine,Helga
And 23h23'41"51d15'
Hasenauer's 21st,Corey
Dra 15h37'12"61d57'
Hasenbein,Gordon
Uma 9h18'22"56d43'
Hasenfratz,Sister Celine
Tri 2h18'39"30d12'
Hasenhuhtl,Eva
Cnv 13h21'23"32d16'
Hasenstab,Robert
Aql 18h56'21"-2d12'
Hash,Lisa
Cas 2h42'41"73d33'
Hash,Margi
And 0h13'18"46d25'
Hash,Patricia Lea
Peg 22h56'15"18d60'
Hashagen,Laura Frances
Aqr 22h1'57"-1d4'
Hashagen,Robert Peter
Aqr 21h0'16"-13d56'
Hashimoto,Maki
Cam 6h33'30"65d12'
Hashin III,Michael
Lac 22h27'1"41d12'
Hashyagin
Cma 6h54'21"-18d29'
Hasimoto,Kiyoshi
Vir 12h28'47"8d51'
Hasino,Nourig
Tau 4h18'0"1d11'
Haskell
Aql 18h56'29"13d52'
Haskell's Stellar Legacy,Richard N
Ari 2h56'41"21d30'
Haskell,Jr,James Spencer
Her 16h47'47"50d34'
Hasker,Ryan
Hya 9h14'19"2d29'
Haski,Robert Ruben
Cra 18h20'31"-41d37'
Haskin,John Cody
Boo 14h51'22"35d27'
Haskins,"Honey Bunny" Donald
Ori 6h8'46"9d59'
Haskins,Amy Michelle
Del 20h24'18"11d7'
Haskins,Harry
Per 4h1'58"37d57'
Haskins,Margaret M
Vul 19h39'0"20d30'
Haslam,Jeb
Uma 11h6'53"37d46'
Haslbeck,Reinhold
Cep 20h40'19"67d41'
Hasler,Nikolaus
Leo 9h55'51"12d7'
Hasley,Jordan
Scl 23h16'24"-25d28'
Hasmonek,David & Catherine Chouinard
Lac 22h22'0"40d19'
Haspel,Dori Beth
Cyg 20h31'0"40d6'
Haspel,Jessica Lynn
Cyg 20h59'24"40d10'
Hass,June
Lyr 18h45'17"30d57'
Hass,Robert
Aur 6h15'16"31d34'
Hass,Timothy Burton
Aql 18h44'51"6d28'
Hassa,Andrea
Psc 22h58'1"2d0'
Hassall,Arthur
Cet 2h9'32"4d45'
Hassan Ali Hussain Al-Ni'mah
Vul 20h21'54"22d45'
Hassan,Alexander Jamiel
Cam 5h45'25"68d12'
Hassan,Ebrahim Ali
Her 18h3'48"14d51'
Hassan,Joseph Anthony
Cyg 19h29'14"35d59'
Hassan,Spencer Robert
Cam 5h54'45"68d26'
Hasselbrink,Glenn
Cep 22h4'44"53d10'
Hasselhof,David et Pamela
Uma 13h46'51"60d52'
Hassell,Douglas
Boo 14h57'52"37d23'
Hassell,Linda Anne
And 0h38'25"41d10'
Hassell,Michael R
Cep 3h2'14"78d7'
Hasselmann,Hayo
Lib 14h46'34"-1d52'
Hassen,Yolande
Cru 11h58'53"-58d37'
Hassenberg,Jennifer Leslie
Aur 6h29'21"40d35'
Hassett,Wendy
And 1h23'17"39d41'
Hassinger,Deborah Ann
Oph 17h36'1"-22d34'
Hassinger,Harvey
Cma 7h17'35"-15d39'
Hassler,Bonnie S
And 23h0'56"48d59'
Hassler,DorothyL.
Lyn 7h32'30"51d4'
Hassler,Ferdinand Rudolph
Cep 20h39'59"55d15'
Hassman's Love Star, Bobbie & Sid
Cyg 21h32'15"30d9'
Hasson Family,The
Cam 3h34'1"71d41'
Hassoon,Baslah Tahab
Lyn 6h35'11"56d0'
Hassoun,Michael George
Tau 4h40'1"15d28'
Hastain,Mick & Freda
Cyg 21h53'13"38d4'
Hastings,Brian Nathaniel
Ori 5h49'36"11d52'
Hastings,Clifford & Lucile
Hya 9h32'44"2d8'
Hastings,Dan & Maya
Eri 3h47'15"-6d30'
Hastings,Frances J
And 23h32'0"47d50'
Hastings,Gregory A
Peg 23h1'49"10d33'
Hastings,Jeffrey
Tri 1h59'19"35d15'
Hastings,Lacy Lee Ann
Del 20h49'0"7d49'
Hastings,Larry
Aql 19h34'20"1d31'
Hastings,Tyler James
Cnc 8h54'0"31d33'
Hastings-Sewell, Jessica
Lyr 18h18'43"43d17'
Hastreiter,Helmar
Uma 9h16'36"48d50'
Hasty,Brooks B
Peg 22h32'45"24d36'
Hasty,Elizabeth Freda
Uma 8h15'55"71d58'
Haswell,Jeanne C
Vul 20h19'59"23d3'
Hat 12-12-80
Peg 23h26'59"23d60'
Hatala,Tom
Vir 13h26'47"-8d56'
Hatch
Her 15h49'13"47d7'
Hatch,Gayle Marie
Cas 1h4'19"60d3'
Hatch,Gerald L
Cmi 7h39'38"8d5'
Hatch,Gwyn
Boo 14h15'23"54d46'A
Hatch,Isabel Noonie
Lyn 8h5'1"50d10'
Hatch,John David
Cnv 12h49'41"50d49'
Hatch,Kay Elizabeth
Peg 23h7'0"8d57'
Hatch,Kyle Alton
Oph 17h1'45"11d56'
Hatch,Michelle Marie
Aql 19h30'59"12d53'
Hatcher,Brady James
Eri 4h12'46"-17d0'
Hatcher,Cathy Koche
Peg 23h5'46"31d26'
Hatcher,Charles Phillip
Hya 9h7'46"2d40'
Hatcher,Greg
Aql 19h30'35"13d31'
Hatcher,Joshua
Uma 8h40'1"70d38'
Hatcher,Kathleen Hines
Cyg 21h35'0"40d30'
Hatcher,Maxine Irminger
Uma 12h1'18"31d10'
Hatcher,Oeida
Lyr 18h24'0"37d55'
Hatfield Forever, Darlene
Eri 3h34'59"-2d35'
Hatfield Jr,John H
Cet 1h36'47"-0d44'
Hatfield's Eonian, Philip
Her 15h48'24"41d52'
Hatfield,Christy
Lyr 19h20'24"35d5'
Hatfield,Danny
Cnv 12h26'0"34d21'
Hatfield,Katelyn Michelle
Peg 21h48'13"34d12'
Hatfield,Papa Archie
Aur 6h34'46"34d7'
Hathaitham,Pricha
Sct 18h41'24"-7d37'
Hathaitham,Tiparat
Cma 6h54'55"-18d47'
Hathaway,"Galactic" Ginny
Mon 7h50'55"-3d27'
Hathaway,Deborah Anne
And 2h15'19"46d24'
Hathaway,Eldon H
Hya 8h33'12"-6d43'
Hathaway,Greg
Aql 18h57'42"10d3'
Hathaway,James & Dorothy
Cra 11h15'54"-11d18'
Hathaway,Susan Baggette
Mon 6h2'36"-10d26'
Hathcock,James
Aql 20h14'36"4d37'
Hathcock,Ronald Dale
Ori 5h45'13"10d32'
Hathcock,Sean David
Peg 23h15'7"18d1'B
Hathcox,Rayna Nicole
Peg 22h39'1"20d42'
Hatheway,M D
Lac 22h44'56"56d40'
Hati
Cmi 7h17'14"8d57'
Hatin,Kresten
Lyn 7h55'32"39d36'
Hatleberg,Laurie
And 1h19'41"38d54'
Hatley,Berry-Ruth's Steve Franklin
Cnv 13h47'21"42d20'
Hatley,Estelle Stephanie
Cyg 21h36'18"42d3'
Hatley,Ginny
Cas 0h17'59"64d4'
Hatmaker,Clarence G
Mon 6h5'37"-4d20'
Hatman,III,Wayman D
Aur 5h26'52"48d59'
Hatosy,Tonya
Cam 4h22'28"65d8'
Hatrick,Sophie Olivia
Cmi 8h8'58"6d29'
Hattemer,Volker
Cam 5h6'57"70d55'
Hatten,"Dr" J
Oph 16h56'59"-25d13'
Hatter,Tami
Aql 19h42'39"11d2'
Hatton,Christopher
Aql 19h1'47"4d57'
Hatton,Nicole Dana
Umi 16h28'42"71d15'
Hatzke,Matthew W
Hya 8h18'42"5d52'
Hatzke,Nicholas M
Cet 2h24'45"8d18'
Hatzke,Trevor R
Her 18h26'45"12d14'
Hau,Hermann
Vul 20h44'13"20d19'
Hau,Thomas & Gloria
Lyn 7h54'40"39d26'
Hau-Kan Diana
Del 21h1'0"18d51'
Hauben,Bobbie
Per 3h56'19"51d51'
Hauber,Bernd
Peg 23h28'29"13d59'
Haubert,Michael Venard
Cnv 13h54'0"30d53'
Haubrich,Joyce Marie
Peg 22h45'11"24d37'
Haubrick,Aggie
Cam 4h8'1"60d16'
Haubrick,Don
Her 17h52'0"30d36'
Hauch,Ross Allen
Cyg 21h2'55"39d34'
Hauck,Carmen
Cet 2h54'51"2d45'
Hauck,Dr James P
Oph 18h15'52"0d26'
Hauck,Father Herbert C
Cep 0h2'0"80d13'
Hauck,Heather Lynn
Eri 2h57'26"-12d9'
Hauck,Heather Lynn
Lyr 18h17'34"30d53'
Hauck,Joseph P
Dra 15h7'20"63d25'
Hauck,Victoria Lynn
Cnc 8h25'50"30d24'
Hauenstein,Kim
And 0h9'15"35d3'
Hauer,Elizabeth Kay
Lmi 10h42'24"30d13'
Hauer,Sabine
Lib 14h30'38"-22d45'
Hauer,William Bibs
Aur 5h17'17"45d22'
Haufe,Rainer
Ori 5h37'44"12d19'
Hauff,Jimmy
Aql 19h10'26"13d16'
Haug's Star,Jim & Dawn
Lac 22h46'58"56d40'
Haug,Eugene
Cet 2h53'47"1d15'
Haug,Johanna
Mon 6h54'49"-0d32'
Haug,Peter
Her 17h12'17"43d28'
Haugan,Holly Jo
Peg 22h36'47"20d41'
Hauge,Corrin D
Uma 8h39'31"58d9'
Hauge,Rodney Lee
Cet 1h55'1"-10d45'
Haugeberg,Darrell
Lyr 19h16'26"42d58'
Haugen,Eileen Edith
Cas 0h24'58"72d29'
Haugen,Norman & Marvella
Cyg 20h17'30"38d52'
Haught,Timothy Shane
Lyr 18h43'31"32d51'
Haughton,Thomas P
Per 3h12'51"40d15'
Haugland,Ruben
Lac 22h15'14"38d21'
Hauk,Debora
Cnc 8h56'40"32d16'
Haukaas,Stein
Leo 10h57'23"12d28'
Hauke,Audrey
Boo 15h5'38"27d24'
Haukeli,Keli
Tri 1h59'49"28d27'
Haulenbeek,Glen B
Dra 12h38'58"70d3'
Haull,William E
Lac 22h20'31"41d10'
Haumschild,Kristina
Cet 2h25'22"4d17'
Haun,Patsy Jean Koon
Cyg 19h41'22"31d23"
Haunsberger,Leopold A
Dra 17h18'24"67d50'
Haunschild,John
Per 2h54'50"56d55'
Haupt,Detlev (RD)
Vir 13h36'35"-6d39'
Haupt,Herbert A
Aur 6h43'18"35d40'
Hauptman,Geoffrey Stuart
Dra 20h15'1"68d7'
Hauptman,In Loving Memory of William D
Aur 5h8'0"41d13'
Hauptman,Oscar
Uma 12h29'0"63d17'
Hauptmann,Steven K
Ser 18h41'41"3d50'
Haus,Dr Erhard Leo
Aur 6h28'53"33d37'
Hauser,Elizabeth
Mon 7h46'51"-7d18'
Hauser,Gordon
Ari 2h44'43"21d2'
Hauser,James
Ari 1h48'44"25d13'
Hauser,Jim
Ori 5h54'44"6d35'
Hauser,Julia Ann
Lyn 8h54'53"41d40'
Hauser,Rachel Marie
Mon 6h54'13"11d35'
Hauser,Robbe M
Crb 15h48'55"38d44'
Hauser,Ruth Mary
Com 12h21'0"28d56'
Hauser,Ryan William
Aql 19h58'1"15d57'
Hauser,Sara Drennen
Lyn 8h2'17"38d47'
Hauserman,Amy
Lyn 8h17'42"36d21'
Haushahn
Aql 20h19'33"1d43'
Hauska,Hildegard
And 23h3'1"45d17'
Hausladen,Kathleen Janet
Peg 22h46'31"20d25'
Hauslein,Leslie Ann
Cam 3h53'46"71d57'
Hausler-Cheadle Forever
Uma 9h17'11"59d50'
Hausman,August Paul
Aql 20h4'46"6d59'
Hausman,Clark
Aur 5h3'38"49d38'
Hausman,Jillian Delaney
Aql 18h57'38"-5d17'
Hausman,Larry
Her 18h4'60"14d7'
Hausmann,Gerd & Ingrid & Nicole
Ari 2h41'42"21d37'
Hausmann,Jeffrey Hudson
Cep 1h10'43"78d22'
Hausmann,Jr,William S
Aur 6h32'27"38d50'
Hausner,Jr,Joseph Stanley
Psc 23h1'52"1d23'
Haussler,Horst
Oph 16h48'15"10d16'
Hauswirth-Balsinger, Martin & Claudia
Boo 14h14'21"17d42'
Haux,Krisztina
Dra 18h18'12"65d17'
Havard,The
Lyn 9h0'25"35d25'
Havaux,Bernard
Ori 6h7'22"8d54'
Have a "Brilliant" Eighteenth! Love,Sarah
Cam 5h42'56"80d32'
Have I Told You Today
Her 16h17'0"7d50'
Haven,Clayton Elmer
Ori 6h4'23"7d3'
Havener,Bobbie
Ori 5h54'11"17d57'
Havenick,Alexander
Peg 23h25'27"32d36'
Havens,Hugh Hunt
Dra 14h27'25"60d4'
Havens,Sarah Dolores
And 23h2'57"45d13'
Haver,Annette
Cap 20h27'24"-26d23'
Haverbush,Thomas
Lmi 10h0'29"31d43'
Haverkamp,Robert V
Cam 3h59'39"78d41'
Haverly,Elise Louise
Peg 0h2'20"18d19'
Haverstick,Debra
Cet 3h11'34"0d51'
Havey,Elizabeth Rita
Vul 20h27'36"28d30'
Haviland,Bear
Ori 4h43'14"4d12'
Havill,Jr,Charles Henry
Per 3h11'46"46d22'
Havlick,Milton
Ori 5h3'37"1d34'
Havlik,Stacy
Peg 21h58'30"26d18'
Havner,Robert N
Ser 15h50'35"17d47'
Havourd,Katrina
Uma 13h12'36"60d3'
Havrilla,Kim
Lyr 18h57'42"38d21'
Havrilla,Lora
Lyr 18h55'13"31d33'
Havrilla,Trynity Anne
Cas 0h16'21"60d34'
Havrisik,George
Her 16h10'22"11d47'
Hawa,Famille Sam et Kik
Per 4h6'17"37d22'
Hawaiian Glo
Lyn 7h54'14"43d43'
Hawdon,John
Ori 5h56'14"15d8'
Hawe,Bob & Cheryl M
Dra 18h55'1"58d23'
Hawe,James Holly
Ori 6h9'38"4d51'
Hawes,Daniel
Per 3h2'30"41d2'
Hawes,Forever Karen
Cam 4h33'16"68d43'
Hawes,Joseph A
Hya 8h26'3"1d16'
Hawes,Peter B
Lac 22h53'14"38d7'
Hawes,Thomas
Dra 15h46'58"60d26'
Hawes,Viana
Lyn 9h7'34"38d39'
Hawie,Irmgard Thomas
Lac 22h19'11"37d31'
Hawk
Cnc 8h10'34"31d18'
Hawk,Adrian Matthew
Cet 2h7'21"-5d56'
Hawk,G Wayne
Cnv 12h22'54"43d52'
Hawk,Julia Anne
Vul 20h0'1"28d41'
Hawk-Milliken, Charlotte
And 0h23'46"36d30'
Hawke,Ethan
Cet 2h30'0"4d20'
Hawke,Ian Malcolm-MGL-Ben R Howard
Sct 18h42'36"-6d53'
Hawken,Brett
Cnv 12h49'0"50d48'
Hawkes,Andrew Malcolm
Umi 15h17'31"70d48'
Hawkes,Chesney
Ori 5h31'1"1d16'
Hawkes,Malcolm Tudor
Cep 22h22'40"61d46'
Hawkesley,Ann Mary
And 23h37'22"40d49'
Hawkings,Parish
Lyn 7h53'13"43d43'
Hawkins II,Eldridge Thomas Enoch
Per 1h43'18"52d59'
Hawkins,Alison
Ori 5h53'49"7d59'
Hawkins,Ashley Beth
Mon 6h45'38"11d47'
Hawkins,Caren
Peg 22h37'54"7d60'
Hawkins,Cissy
Peg 21h47'0"31d53'
Hawkins,Corinne Rita
Aql 19h41'1"11d23'
Hawkins,Dawn Danielle
Peg 22h9'14"20d18'
Hawkins,Dianna
Cam 4h23'17"68d9'
Hawkins,Ellie Tyne Rebecca
Crb 16h6'53"34d53'
Hawkins,Frederich
Lac 22h2'54"38d11'
Hawkins,Gladys B
Uma 9h59'17"42d5'
Hawkins,Graham
Ori 5h42'0"10d23'
Hawkins,Hillary Alicia
Lyn 6h30'30"58d20'
Hawkins,Isaac Walter
Ori 6h2'31"4d33'
Hawkins,Jonathan Alan
Aql 19h31'14"-8d52'
Hawkins,Lillian
Cas 1h48'47"68d23'
Hawkins,Lilly Eva
Peg 22h47'44"25d39'
Hawkins,Mary Welch
Tri 1h48'49"27d33'
Hawkins,Melanie
Cas 1h19'11"63d38'
Hawkins,Patrick Michael
Per 2h39'58"38d49'
Hawkins,Paula Marie
Gru 22h10'46"-55d29'
Hawkins,Renee Frances
Cas 1h40'57"58d8'
Hawkins,Richard William
Boo 14h21'44"46d42'
Hawkins,Richard Ernest
Aql 19h31'31"8d40'
Hawkins,Richard L
Aql 18h59'42"-1d39'
Hawkins,Ringmaster Jerry
Lyr 19h8'30"38d16'
Hawkins,Robert Francis
Her 16h37'45"38d16'
Hawkins,Sadie
Uma 11h38'0"38d53'
Hawkins,Stacey Renee
Peg 22h53'48"27d1'
Hawkinson,Ben Allen
Sco 17h52'39"-31d46'

Hawkinson,Jennifer Kjelle
Sgr 19h40'33"-38d21'
Hawkinson,Randy R
Leo 9h49'25"33d5'
Hawkridge,Taylor McKenley
Cap 21h20'37"-18d60'
Hawks,Dana
Peg 21h50'16"34d28'
Hawksley,Jeff
Aur 4h54'33"37d43'
Hawley,Contance
Lyr 18h55'31"33d23'
Hawley,Daniel James
Lac 22h14'20"46d33'
Hawley,Kaitlyn Gray
Peg 23h2'51"11d55'
Hawley,Nancy A
Del 20h13'23"14d38'
Hawley,Timothy Wayne
Lac 22h1'51"40d57'
Hawn,Joachim
Ser 15h41'10"24d28'
Hawn/Precious Baby
Nikki,Nicole Lynn
Peg 0h1'42"21d54'
Haworth PJDA
Cru 12h45'8"-60d47'
Haworth,Carrie Louise
Cas 2h55'41"57d29'
Haworth,Paul
And 23h1'15"50d31'
Hawthorn-Wood,John Norman
Arthur
Leo 10h28'0"11d25'
Hawthorne,Caitlin Marie
Vul 19h18'27"23d4'
Hawthorne,Jessica Hope
Boo 14h37'57"45d19'
Hawthorne,John
Aql 19h5'1"-1d23'
Hawthorne,Jr,Jerry Alan
Ser 16h12'59"2d11'
Hawthorne,Kelsey Anne
Tau 4h22'25"16d27'
Hawthorne,Nan
Cas 1h14'1"63d29'
Hay Es Cuarenta &
Twinkling,Pamela Davis
Cas 1h8'25"63d31'
Hay In A Needlestack
Uma 9h10'43"50d12'
Hay Yang (Feb 6th)
Cas 23h33'39"61d31'
Hay,Charlie & Joanne
Cyg 21h55'26"53d38'
Hay,Donald & Sybil
Cyg 20h38'37"52d57'
Hay,Elizabeth
Hya 8h53'1"0d25'
Hay,Jean Gladys Theresa
Cas 1h5'1"58d38'
Hay,Joan Felice
Lyr 18h33'42"37d11'
Hay,Marion E & Lucille G
"Suzy"
Cyg 21h22'1"39d56'
Hay,Marva
Cet 3h19'28"1d25'
Hay,Preston J
Dra 19h33'24"70d54'
Hay,Steve
Lyn 8h23'1"52d19'
Hay,Trevor
Hya 8h44'18"3d46'
Hayabusawake
Oph 18h3'0"0d57'
Hayashi,Clifford Thomas
Holeski
Boo 14h50'48"39d21'
Hayashi,Joey
Hya 8h53'1"0d46'
Hayashi,MD,T Teruo
Uma 11h59'31"55d31'
Hayashi,Yuka
And 23h12'27"36d41'
Hayat
Aur 4h55'34"50d22'
Haybach,Jonas
Uma 10h36'29"41d36'
Haycock,Ian
Ori 4h44'0"0d22'
Hayda,Roma Maria
Lyn 8h28'40"57d45'
Hayden
Ori 5h49'0"18d5'
Hayden 4:26AM,Anna
Elizabeth
Oph 17h8'36"-16d25'B
Hayden 4:29AM,Troy Daniel
Oph 17h8'36"-16d25'A
Hayden 5-5-69,Jennifer Dawn
Aql 19h10'0"10d9'
Hayden Honey Fear, James
Mitchell
Aur 6h10'36"38d59'
Hayden Star,R
Cep 20h28'50"66d7'
Hayden,Amber Jean
Peg 22h9'16"24d16'
Hayden,Ann
Vul 19h43'1"27d17'
Hayden,Irene F
Cyg 21h17'39"38d23'
Hayden,J David
Lyn 7h53'25"45d20'
Hayden,Jennifer Crummer
And 23h16'53"47d58'
Hayden,Julius
Uma 13h51'47"60d19'
Hayden,Katherine Michelle
Lib 15h31'45"-11d10'
Hayden,Kimberly A
Lyn 8h24'58"47d43'
Hayden,Margaret
Cet 1h27'25"-11d16'
Hayden,Melanie
Cam 5h26'46"65d3'
Hayden,Michael Joseph
Dra 17h42'32"64d35'
Hayden,Michael Vincent
Aql 19h51'33"10d34'
Hayden,Robert Kenneth
Aur 6h10'28"36d16'
Hayden,Ryan
Cet 2h19'40"5d58'
Hayden,Sharla Uman
Mon 6h56'11"10d17'
Hayden,Sister Lawrence Marie
Cyg 20h23'51"40d32'
Hayden,Terri Ann McQuade
Cet 3h6'57"0d57'
Hayden,W Lee
Aql 19h57'59"15d28'
Haydon,Dr Paul R
Oph 18h16'56"12d15'
Haydon,Kurt
Lyr 18h19'53"45d32'
Hayek,Gloria
Ori 6h0'53"0d56'
Hayek,Kelly
Cyg 19h50'28"38d9'
Hayes
Cep 6h42'57"85d1'
Hayes "Baby Doll", Donna
Lynn
Lyr 19h16'21"28d30'
Hayes III,Edward
Aur 6h12'56"33d12'
Hayes Just Love Anne, Warren
S
Ser 18h0'20"-13d25'
Hayes Star,The Chuck &
Peggy
Ori 6h8'55"8d52'
Hayes,Alan
Cnc 8h2'48"7d20'
Hayes,Amanda Taylor
And 0h50'10"36d41'
Hayes,Amy Jo
Lyr 18h47'17"30d2'
Hayes,Angela
Com 13h30'30"26d35'
Hayes,April Foxworth
Mon 6h52'59"-0d25'
Hayes,Ashley & Lynn
Cyg 20h20'11"38d38'
Hayes,Barbara Anne
Lac 22h40'43"56d10'
Hayes,Beryl Cameron
Aql 19h18'50"15d23'
Hayes,Billy
Dra 19h39'23"84d55'
Hayes,Brandon S
Cet 1h18'37"-4d40'
Hayes,Bruce
Aur 7h23'17"40d26'
Hayes,Carol
Oph 17h1'21"1d42'
Hayes,Christopher- Hayesy
Gem 6h5'36"26d32'
Hayes,Clifton W
Sex 10h24'39"-5d11'
Hayes,Darin
Per 2h29'42"58d32'
Hayes,Darrick Michael
Per 3h11'57"50d19'
Hayes,Dihanne
Sge 20h16'37"16d45'
Hayes,Don
Aql 19h45'26"10d20'
Hayes,Donald James
Aql 18h54'45"-0d53'
Hayes,Douglas Scott
Boo 14h44'32"25d60'
Hayes,Edward C "Papa"
Aur 6h9'46"48d51'
Hayes,Emily Lorraine
Peg 21h48'42"30d23'
Hayes,Fred
Dra 15h52'16"51d56'
Hayes,Freelan S
Ser 15h22'40"1d16'
Hayes,III,John Joseph
Cnv 12h54'19"41d8'
Hayes,Iral Edward
Uma 11h4'20"40d11'
Hayes,James
Sex 10h13'54"-1d20'
Hayes,James M
Dra 16h59'44"58d54'
Hayes,Jamie & Andrea M
Cas 1h39'14"58d30'
Hayes,Jared's Nana- Carol
Ann
Cas 0h18'54"60d52'
Hayes,Jareds Pop-Pop John
David
Per 2h10'22"57d43'
Hayes,Jennifer
Peg 22h12'37"33d15'
Hayes,Joan F
Lyn 8h12'41"33d25'
Hayes,Joanne & Howard
Lyr 18h13'48"38d13'
Hayes,Jr,Edward A
Her 18h15'11"48d47'
Hayes,Jr,Thomas Baker
Cnv 12h20'36"38d19'
Hayes,Jr,William S
Boo 14h26'34"38d12'
Hayes,Kenneth
Umi 15h38'52"68d25'
Hayes,Keri
Aql 19h30'53"8d41'
Hayes,Kimberly Lynn
Ser 15h50'35"16d5'A
Hayes,Kimberly Sue
Cas 2h45'56"73d27'
Hayes,Kristen B
Peg 22h16'51"34d11'
Hayes,Leah
Hya 9h18'56"-6d20'
Hayes,Linda Chaganos
Lyn 8h40'47"37d27'
Hayes,Liz
Col 5h41'29"-42d52'
Hayes,Marie "Tootsie"
Crb 16h15'39"28d13'
Hayes,Marie Jean
Cas 22h59'51"55d41'
Hayes,Marjorie R
Cyg 20h2'1"39d46'
Hayes,McKinley Craig
Cnv 13h58'52"28d60'
Hayes,Meryl Jane
Cas 1h53'48"60d36'
Hayes,Michael
Crt 11h42'29"-10d27'
Hayes,Michael Ney
Cep 4h37'26"80d38'
Hayes,Mike
Ser 15h35'31"17d51'
Hayes,Pauline
Cyg 20h52'25"37d55'
Hayes,Peter Madison
Aql 20h35'14"0d26'
Hayes,Rachel
Cas 2h32'43"57d41'
Hayes,Richard Todd
Her 16h8'44"23d60'
Hayes,Robbie
Dra 17h21'34"60d14'
Hayes,Ronald Eugene "Papa"
Aql 19h43'57"12d33'
Hayes,Sean Charles
Aur 6h34'49"31d0'
Hayes,Teresa"Sparkles"
Tri 1h30'17"30d2'
Hayes,Thomas J
Peg 23h28'19"23d58'
Hayes,Thomas Robert
Boo 14h2'25"26d26'
Hayes,Timothy Douglas
Aur 6h20'49"33d11'
Hayes,Tom & Bernice
Peg 21h51'17"31d41'
Hayes,Wade
Aql 20h0'53"0d28'
Hayes,William J
Per 1h27'20"50d26'
Hayes-A Stream In The Desert
Cam 3h26'50"53d3'
Hayes-Donan'Patty
Uma 12h11'58"62d10'
Hayes-Hardin,Robin Lynn
Mon 6h20'33"0d7'
Hayhoe's Pearl
Peg 22h10'43"3d20'
Hayhow,James E
Dra 18h39'20"68d57'
Hayler,Ron & Jean
Cyg 20h22'37"39d22'
Haylett,Robyn L
Cyg 20h53'46"50d19'
Hayley
Cas 2h47'32"65d17'
Hayley Amy
Crb 15h25'24"30d18'
Hayley Ann
Cas 2h1'43"61d24'
Hayley Ku'unani
Aur 5h0'43"48d50'
Hayley's Comet
Aqr 21h4'14"-1d41'
Hayley's Star
Vul 20h2'17"28d40'
Hayley,Ralph
Dra 16h42'29"64d43'
Haylie Beth
Lyn 7h28'41"36d26'
Hayman Star,The Peter
Dra 14h41'0"62d43'
Hayman's"Delicious"
Perfume,Gale
Cyg 19h50'11"40d33'
Hayman,Barbara
Aql 19h33'36"0d28'
Hayman,Stacey
Peg 21h41'59"21d41'
Hayman,Ted
Her 16h54'24"29d30'
Haymes,Gail Ann Lowe
Mon 6h20'37"2d58'
Haymes,Kenneth Fredrick
Cma 7h49'0"11d4'
Hayne,David Alexander
Cep 23h10'41"65d22'
Haynes & Trouble, Everette E
(Gene)
Uma 8h39'26"51d11'
Haynes (Bob & Lucy), Robert
& Lucille
Mon 7h51'45"-5d8'
Haynes 81,Michael
Aql 19h48'28"14d57'
Haynes,A New
Generation,Caitlin
And 2h7'28"42d48'
Haynes,Adam Christopher
Vir 13h6'54"-0d11'
Haynes,Anita
Lyn 8h14'19"34d21'
Haynes,April
Lyn 7h34'11"45d56'
Haynes,Ashley Daryl
Cap 21h27'33"-23d4'
Haynes,Charles & Ana Yepes
Her 17h59'55"40d20'
Haynes,Colleen
And 2h28'1"45d15'
Haynes,Danelle Katherine
Cas 0h3'58"60d12'
Haynes,Diane Lynn
Lyr 18h46'39"30d28'
Haynes,Duane
Cma 6h27'25"-13d30'
Haynes,Edward B
Aur 4h58'30"51d48'
Haynes,Gary
Her 17h2'36"48d42'
Haynes,Gordon H
Aql 19h55'23"7d46'
Haynes,Ken & Sharon
Cam 3h44'58"70d50'
Haynes,Kevin Lee
Uma 9h18'0"62d20'
Haynes,Mark
Her 17h21'28"41d54'
Haynes,Oraleaze
Cet 2h8'22"6d15'
Haynes,Preston Stanford
Ser 18h53'35"2d44'
Haynes,Rachel H
Cas 2h20'42"68d16'
Haynes,Vivien
Crt 11h16'36"-8d40'
Haynes,Yvonne E
Mon 6h22'1"2d49'
Haynie,Teresa Bussells
Vul 19h59'1"25d11'
Haynos,Karen
Lyr 18h44'37"45d48'
Hays M H C 1989, Cynthia Y
Del 20h35'40"20d29'
Hays,Charles Garner
Dra 16h26'36"69d19'
Hays,Elinor Rice
Oph 17h20'1"12d2'
Hays,Evelyn Skippy
Lyn 8h58'28"39d33'
Hays,Marjorie J
Cet 0h24'35"1d43"
Hays,Mary & Charles
Cam 3h32'46"58d18'
Hays,Rachel Sommers
And 22h55'14"50d22'
Hays,Ryan Michael
Cam 3h22'16"58d14'
Hays,Sarah Hulsey
Lyr 19h1'10"30d4'
Hays,Toni McBee
Ori 4h42'1"11d11'
Hays,Wes
Her 17h32'39"26d51'
Hayward III,Robert Gilliam
Mon 7h5'33"-5d28'
Hayward,Aidan Rile
Lyr 18h38'45"43d38'
Hayward,Aidan Rile
Per 2h43'0"40d13'
Hayward,Bailey
Cet 3h15'42"4d46'
Hayward,Elizabeth Morgan
And 23h30'36"41d23'
Hayward,Jackson Douglas
Eri 4h14'38"-12d46'
Hayward,James
Sge 20h1'22"16d48'
Hayward,Joanna L
Cas 0h1'11"62d40'
Hayward,Lawrence J
Her 18h18'34"12d38'
Hayward,Nikki Jayne
Cyg 20h26'54"46d11'
Hayward,Roberta Anne
Gem 7h29'45"28d30'
Hayward-Jones Ruby
Cyg 19h27'1"34d12'
Haywood's Halo
Her 18h4'51"30d38'
Haywood,Alan
Lyn 7h41'0"39d32'
Haywood,Carol
Eri 4h59'38"-10d59'
Haywood,Floyd & Pris
Sge 19h11'16"16d34'
Haywood,Frances Delores
Cet 1h51'43"-8d6'
Haywood,Jeanne
Cyg 20h13'57"38d9'
Haywood,John Norman
Her 16h15'43"20d27'
Haywood,Keeley
Per 2h0'45"57d17'
Haywood,Rick
Aur 6h7'0"46d2'
Haywood,Sharon
Cyg 20h13'36"38d26'
Hayworth,Kimberly Yvonne
Mon 6h31'33"-6d19'
Hazard,Colette Kring
Ori 5h1'1"13d30'
Hazard,Kay
And 2h9'51"38d43'
Hazard,Lori Lynn
Cyg 19h50'31"38d38'
Haze
Vir 13h53'39"-1d28'
Haze's Star
Mon 6h46'58"11d41'
Hazel
Gem 7h6'56"21d7'
Hazel
Cas 2h37'31"63d49'
Hazel
Cyg 19h34'0"39d24'
Hazel
Hya 9h14'13"1d30'
Hazel & Andrew
Cyg 20h38'10"45d48'
Hazel & Haig's Precious Little
Star
Lyr 19h11'21"37d48'
Hazel & Jim
Gem 6h33'33"14d55'
Hazel Elaine
Cyg 20h37'31"53d14'
Hazel Grace
And 23h21'22"51d10'
Hazel Victoria
Cyg 20h39'32"40d2'
Hazel's Casa Del Sol
Sct 18h41'39"-7d27'
Hazel,Debbie
Lyr 19h18'55"40d20'
Hazel,Jennifer Marie
Cam 4h59'17"68d34'
Hazel,Melissa
Lyn 8h1'56"34d2'
Hazel,Michael Dustin
Aql 19h29'51"13d45'
Hazel,Stanley Frances
Dra 17h30'41"64d15'
Hazelett,Betty Lewrena
Uma 13h1'28"61d6'
Hazelett,Catherine Wanedia
Uma 12h58'45"62d11'
Hazelett,Liddie May
Uma 13h12'0"60d25'
Hazell TogetherForever
EternalLove,Jim&Marion
Cyg 20h16'44"38d47'
Hazell,Amy
Lyr 18h29'51"32d23'
Hazell,Luke Jeffrey
Cep 21h59'16"65d22'
Hazell,Skip
Peg 23h35'18"20d12'
Hazelnis,Michael
Per 2h50'59"40d60'
Hazelton-My Brightest
Star,Shelly
Vul 19h57'47"28d57'
Hazelton-Star Gazer, James
Ryan
Boo 14h49'14"26d25'
Hazelwood,Dianne C
Lyn 7h3'60"53d31'
Hazelwood,Kenneth A
Her 17h59'47"50d3'
Hazen,Anita
Cet 3h2'22"3d32'
Hazen,Jennifer Lynn
Cas 22h58'14"54d44'
Hazen,Karlene
Cas 1h19'48"64d44'
Hazle,Madeline Elise
Aql 20h2'13"7d38'
Hazlehurst,Peter F
Gem 6h48'43"19d43'
Hazlett,Catherine Louise
Lyn 7h50'46"42d36'
Hazlett,Dorothy
Cyg 20h34'1"42d55'
Hazlett,Lana Kay
Mon 6h32'35"-0d47'
Hazlett,Marie Louise
Cyg 21h35'19"30d43'
Hazlewood,Leo
Oph 17h36'52"-3d21'
Hazlewood,Richard
Lac 22h32'12"55d18'
Hazouri's Nova
Cma 6h53'54"-18d41'
Hazy
Peg 23h6'24"8d59'
Hazzard,Grace Harriet
Peg 22h42'37"5d27'
Hazzard,John C
Oph 16h55'54"-5d57'
Hazzard,Michelle
And 1h2'28"39d36'
Haïgo-Garo
Lyn 8h43'24"42d48'
HB
Lac 22h50'12"55d41'
HBB #1 & HBB #2
Lyr 18h18'23"41d14'
HDC+JMW=TBWFAFT
Cas 0h59'27"61d30'
HBS Punches
Vir 12h2'25"7d30'
HDD 50
Aur 5h58'1"38d36'
HDH-92
Cnv 12h46'25"48d52'
He Stands Tall
Aql 18h56'31"8d53'
He's Tall & Next Door Eternally
Cliff
Peg 22h1'20"10d56'
Heaber Dd
Ori 5h39'15"-0d57'
Heacox,Paul Robert
Per 2h55'1"40d22'
Head Angel,The
Com 12h19'20"19d24'
Head III,Richard Henry
Leo 11h31'42"2d9'
Head IV,Richard Henry
Ari 2h38'23"22d2'
Head Sot
Crb 16h4'49"26d50'
Head Star
Cyg 21h51'36"53d17'
Head,Andrea Kathryn
Mon 6h37'19"10d58'
Head,David Paul
Aur 5h12'58"40d54'
Head,Elizabeth"Betsy"
Com 12h29'42"28d37'
Head,Jacqueline Renee
Sco 16h32'1"-44d29'
Head,Janice Leigh
Lib 14h45'1"-0d35'
Head,Jeff
Cep 21h29'45"78d58'
Head,Jennifer Rachael
Ari 2h38'31"22d10'
Head,John & Ruth
Uma 9h20'57"58d19'
Head,John Edward
Her 16h58'41"36d40'
Head,Jr,Floyd N
Aur 5h53'34"38d47'
Head,Keirsten
Leo 10h43'1"15d46'
Head,Linda
Per 3h18'22"41d28'
Head,Roy
Oph 17h10'54"-22d47'
Head,Shirley Ann Jaeger
And 1h43'48"39d20'
Headford,Steve
Her 17h50'36"14d47'
Heading,Ian,June, & Darren
Ori 5h5'0"1d8'
Heading,Karen Sander
Com 12h30'1"30d44'
Headley,Alexandria Rosa
And 1h55'10"40d16'
Headley,Robin Lyne
Ori 5h56'44"5d33'
Headley,William F
Uma 13h1'60"62d31'
Headrick,Gary Leon
Hya 8h44'43"5d21'
Headstrom's 25th,Bob & Kathy
Oph 18h7'29"0d29'
Heady,Brian Kenneth
Hya 9h8'22"0d37'
Heady,Gregory S
Hya 8h30'14"0d57'
Heafey,Audreye Eve
Cam 3h52'26"60d30'
Heafey,Nathalie
Ori 5h23'5"-2d50'
Heafey,Pierre
Cep 22h58'53"70d52'
Heafner,Braden Lee
Dra 20h13'52"68d23'
Heafner,Elise
Lac 22h54'51"50d3'
Heafner,Joshua David
Dra 20h39'22"68d40'
Heakin,Erin Frances Mary
Lyr 19h1'1"30d12'
Heakin,Scott Morgan
Lac 22h23'16"52d59'
Heal The World-Doc B's Star
Uma 8h51'1"62d20'
Heald,Bradley Alan
Cep 20h27'57"62d5'
Heald,Timothy "Tim"
Her 17h24'21"20d32'
Healey 1-21-75,Kieran
Kathryn
And 23h32'18"48d54'
Healey,David L
Cra 18h24'23"-41d2'
Healey,Joseph Eugene
Boo 15h4'10"10d17'
Healey,Kathy Damico
Sge 18h57'44"20d16'
Healey,Kristine Couto
Peg 21h52'56"31d22'
Healey,Mary Beth
And 23h13'0"36d45'
Healey,Nathan John
Dra 16h53'1"68d2'
Healey,Roseanne S
And 0h47'57"38d33'
Healey,Sophie
Cas 0h54'20"64d2'
Healey,Steven J
Aur 7h6'15"43d40'
Healing-Star *Chaman*
El'Ca'Sa
Ari 2h51'27"21d49'
Health Tour- Established-1989
Cma 7h24'20"-15d5'
Healy III,Daniel Alphonsus
Uma 9h19'11"57d59'
Healy,Brennan Michael
Cnc 8h56'1"31d24'
Healy,Daniel R
Crb 16h0'12"33d39'
Healy,David Adam
Boo 14h20'1"37d39'
Healy,Evelyn M
Cyg 20h22'23"39d32'
Healy,Francis T
Cet 0h10'39"-17d43'
Healy,Johnie & Dawn
Cyg 21h19'20"39d34'
Healy,Kathleen Ann
Cyg 19h50'17"38d39'
Healy,Kathy
Lyn 9h32'14"40d47'
Healy,Marilyn "Zap"
Cam 7h42'45"67d46'
Healy,Megan
Cyg 19h28'0"34d9'
Healy,Michelle
Vul 21h24'59"27d32'
Healy,Rachel Louise
Umi 15h17'14"66d4'
Healy,Sean
Cmi 7h33'1"1d18'
Healy,Tim & Mary
Crb 15h49'22"38d11'
Healy,Tom & Kathi
Uma 8h50'24"62d19'
Healy,Viola
Cma 7h12'55"-13d19'
Healy-Lucciola,Alex
Ori 5h26'0"-5d25'
Hean,Robert Scott Arnold
Her 16h49'42"37d37'
Heaney,B E M,Mary
Cmi 8h8'37"6d26'
Heaney,Elyse
Cet 1h53'22"0d48'
Heaney,Jennifer,Anne
Lyn 8h6'59"51d50'
Heaney,Joseph
Per 3h9'37"40d33'
Heaps,Sr,Harry D
Dra 10h31'28"74d57'
Hear,George William
Gem 7h8'34"21d28'
Heard The Great, Anthony
Baron
Boo 14h6'40"23d59'
Heard,Bettye Olivia
Equ 21h5'59"10d55'
Heard,Reginald K
Sct 18h52'36"-5d24'
Hearn,Loretta Ann
Cyg 19h27'54"30d31'
Hearn,R Peter
Aqr 21h23'15"-5d53'

Hearnden,Derek
Cep 22h51'31"56d53'
Hearne,Bonnie Lou
Vul 19h48'32"25d45'
Hearst,Kevin
Cet 2h41'11"-18d3'
Heart Of Gold
Cep 22h25'48"60d46'
Heart of Haruhi
Ari 2h15'17"13d12'
Heart of Travis,The
Cep 22h21'0"59d10'
Heart Star
Cyg 20h33'12"46d46'
Heart to Heart
Boo 14h17'54"48d42'
Heartso,Grandpa & Grandma
Aql 19h28'34'0d14'
Heary,Brendan John
Boo 14h26'11"22d17'
Heasley,Rachel C
Cas 23h6'1"54d21'
Heasley,Tina
Uma 9h24'0"49d12'
Heaslip,Daniel Bug
Aql 19h57'45"15d56'
Heasman,Sally Jane
Cep 21h20'0"55d0'
Heater
Cas 2h4'57"73d11'
Heath
Aur 6h35'12"37d47'
Heath
Vul 20h0'33"22d35'
Heath Family,The Darren & Laurie
Aql 20h7'51"8d52'
Heath III,Charles Daniel
Dra 12h44'54"68d12'
Heath's Harbour
Dra 14h49'55"55d32'
Heath,Bobby & Deborah
Lyr 18h58'16"45d17'
Heath,Cullen Dean
Psc 0h5'0"8d7'
Heath,Dorothy & Charlie
Cep 23h9'47"65d33'
Heath,Edward P
Leo 9h36'0"10d50'
Heath,Emily Margaret
Com 12h54'0"25d36'
Heath,Hellen
Aql 20h33'41"0d22'
Heath,Ivor Bernard "John"
Ori 5h58'16"0d13'
Heath,John R
Cam 3h27'49"61d29'
Heath,Jr,Calvin
Lyr 19h15'29"40d46'
Heath,Morgan Victoria
Lyn 7h2'45"58d26'
Heath,Rebecca Grace
And 1h0'52"39d40'
Heath,Richard B
Aql 19h3'42"-1d6'
Heath,Tory Seebeck
Cyg 21h8'18"31d36'
Heather
And 2h30'28"50d25'
Heather
Boo 14h10'43"40d0'
Heather
Del 20h55'55"8d27"
Heather
Cyg 21h39'44"41d57'
Heather
Uma 10h38'32"48d4'
Heather
Uma 10h2'0"51d13'
Heather
Cas 1h12'24"65d34'
Heather
Cas 22h57'21"54d18'
Heather
Uma 9h36'51"70d57'
Heather
Peg 22h2'23"10d57'
Heather
Lyn 7h33'1"41d10'
Heather
Del 20h39'43"18d53'
Heather
Cas 3h0'18"60d21'
Heather
Boo 13h45'58"20d35'
Heather
Cet 3h8'0"5d58'
Heather
Lyn 7h49'1"38d46'
Heather
Eri 2h53'41"-3d33'
Heather
Cas 2h29'57"61d43'
Heather
Crb 16h7'13"30d21'
Heather
And 0h11'49"34d38'
Heather
Aql 19h31'41"10d10'
Heather
And 23h35'36"44d31'
Heather
Mon 6h42'18"10d46'
Heather & Ben,Eternal Love
Cnv 12h35'0"50d27'
Heather & Bill
Cyg 19h48'26"29d46'
Heather & Bill
Crt 11h37'34"-11d19'
Heather & Jordan
Sgr 18h52'17"-34d42'
Heather & Kristian
Cyg 21h39'39"40d18'
Heather & Marc's Love Star
Uma 9h26'23"50d24'
Heather & Robert
Cen 11h32'26"46d10'
Heather & Tim
Sge 20h17'18"20d8'
Heather 101994
Cas 0h43'14"68d56'
Heather 318
Hya 8h32'23"-0d56'
Heather Amanda
And 2h32'58"41d11'
Heather Andrina
Peg 23h48'26"27d52'
Heather Ann
Cas 1h56'1"61d25'
Heather Ann
Cas 0h22'12"61d43'
Heather Ann
Aql 19h31'31"14d6'
Heather Breann
Sco 16h20'1"-40d58'
Heather Diana
Lyn 9h12'36"45d20'
Heather Diane
Uma 10h5'44"48d7'
Heather Elaine
Cas 2h40'47"61d50'
Heather Elizabeth
Cam 4h48'44"67d24'B
Heather Elizabeth
And 23h34'0"45d4'
Heather Joy
Cma 6h50'60"-16d9'
Heather Joyce
And 2h23'32"42d16'
Heather Lanai
Hya 9h42'30"-18d48'
Heather Lee
Mon 6h56'43"-10d36'
Heather Linda
Lyn 9h16'18"40d44'
Heather Love
Cas 2h11'18"68d42'
Heather Lyn
Aql 19h18'1"13d2'
Heather Lynn
Dra 20h0'31"61d12'
Heather Marie
Cas 1h3'17"56d1'
Heather of the Heavenlies
Mon 7h55'24"-2d29'
Heather Patrice
Peg 23h29'0"21d57'
Heather Renee
Ori 6h17'34"-1d30'
Heather Ruth
Cas 0h46'19"61d13'
Heather S
Psc 22h53'46"2d11'
Heather Salon
Ori 5h56'0"15d55'
Heather Star
Cyg 20h2'1"40d6'
Heather's Cradle
Vul 20h2'55"23d22'
Heather's Dazzling 18th
Gem 8h1'27"31d5'
Heather's Delight
Lyr 18h44'15"32d13'
Heather's Light
Sct 18h52'30"-6d52'
Heather's Pooh Star
Eri 3h17'0"-13d43'
Heather's Star Holdrige
Vul 20h45'53"28d22'
Heather's Star
Sgr 18h3'20"-28d9'
Heather's Star
Vir 15h3'45"6d15'
Heather's Star
Lyr 19h15'20"42d29'
Heather's Wish
Uma 9h7'37"67d58'
Heather,Kendall Lynn
Hya 8h19'48"3d23'
Heatherdell,Duchess of
And 23h48'54"33d33'
Heatherdell,Duke of
Boo 14h52'16"22d37'
Heaths,Leland Valerie
Tau 3h54'1"0"60'
Heatley,James Spencer
Boo 15h25'31"38d52'
Heatley,Pauline Andrey
Cas 23h14'13"60d37'
Heatwole,Phyllis L & Marion G
Cyg 20h1'35"58d19'
Heaven
Cyg 19h8'47"47d55'
Heaven
Del 20h15'50"10d22'
Heaven
Cam 7h16'46"83d37'
Heaven for the Lawsons I John 4:19
Cet 0h4'0"-10d54'
Heaven In A Life Time
Cyg 20h58'56"30d45'
Heaven In Your Eyes
Cam 4h11'42"67d59'
Heaven On Earth
Cyg 19h44'50"30d29'
Heaven Sent
Per 2h38'1"38d32'
Heaven Sent
Peg 22h59'50"8d44'
Heaven Sent Dora
Cas 23h46'17"50d12'
HEAVEN'S KISS To Toni
Mon 6h53'55"-1d23'
Heavenleigh Hi
Vul 19h18'0"26d55'
Heavenly Angel-Nanny
Cas 0h6'22"61d4'
Heavenly Beth
Com 13h4'38"30d18'
Heavenly Carol
Ori 5h52'34"9d8'
Heavenly Cheryl
Cyg 20h10'27"39d12'
Heavenly Christine
Cas 1h54'1"73d45'
Heavenly Deana
Aur 5h53'52"31d52'
Heavenly Hans
Cmi 7h15'23"4d14'
Heavenly Lady
Cyg 19h43'17"30d29'
Heavenly Lisa
Sgr 18h55'23"-25d45'
Heavenly Liz
Cas 1h40'46"75d13'
Heavenly Love
Lyn 8h18'21"36d1'
Heavenly Love
Oph 18h17'31"12d14'
Heavenly Love,Heather & Todd
Aql 19h56'37'14d32'
Heavenly Luna Strauss, The
Lyr 19h23'34"31d27'
Heavenly Mary
Mon 8h3'47"-10d14'
Heavenly Niece Josette
Gem 6h58'19"30d11'
Heavenly Peace
Lyr 19h12'0"33d52'
Heavenly Philip
Peg 21h53'33"33d30'
Heavenly Sheila
Crb 16h16'47"33d46'
Heavenly Union
Ser 15h35'42"8d16'
Heavenly Wendy Ann
And 1h40'32"40d48'
Heavenly-Becky & Mark
Cyg 20h9'42"41d9'
Heavens A Step Away Together,Diane
Cyg 21h0'1"38d10'
Heavens To Grace Lingers At 50!!!
Cyg 20h50'43"40d24'
Heavey,Robert J
Her 17h34'50"26d50'
Hebbel,Daniela
Gem 7h19'46"24d10'
Hebbend
Cen 11h32'30"43d0'
Hebbert,Matthew Anthony
Dra 16h38'18"68d12'
Hebenbrock,Dieter
Cnc 8h27'25"31d10'
Hebenstreit,Joseph William
Her 16h58'0"23d13'
Hebenstreit,Sabine
Gem 7h14'52"24d32'
Heberle,Julie Et Clemence
Ser 15h59'23"11d1'
Heberlein,Jr,G Erich
Dra 16h36'42"64d55'
Hebert,Brennan Douglas
Vir 11h41'24"5d35'
Hebert,Darryl "Chief"
Cma 6h59'11"-16d35'
Hebert,Denis Bertrand
Cnv 13h32'50"51d21'
Hebert,Dylan Francis
Dra 19h32'31"61d18'
Hebert,Joe & Anne
Aur 5h2'32"38d15'
Hebert,Lisa Ann
Her 18h6'0"30d52'
Hebert,Louis Curtis
Aur 6h19'15"45d26'
Hebert,Michael Spencer
Hya 8h19'48"2d43'
Hebert,Robert J
Lac 22h21'43"55d21'
Hebert,Sarah Mae
Lyr 18h34'19"46d51'
Hebert,Sidney Joseph
Mon 7h0'51"4d11'
Hebert,Tasha
Ori 6h21'59"16d5'A
Hebert,The Heavenly
Hya 9h0'14"0d40'
Hebert,Trace Michael
Aqr 22h1'14"-5d17'
Heboian,Diane M
Gem 7h36'1"34d11'
Hebranko,Michael (Choo)
Lac 22h54'46"54d30'
HeBrRaRê
Leo 9h19'0"18d20'
Hecht,Johnny
Vir 12h8'38"-5d51'
Hecht,Julie-Mon
Peg 22h39'25"21d4'
Hecht,Martin der Eisläufer
Uma 8h44'29"51d54'
Hechtman,Elizabeth C
Ori 5h49'0"21d5'
Hechtman,Isabelle Lynn
Vir 11h40'52"9d31'
Heck Family Star,The Charles E
Lyn 8h47'0"35d54'
Heck,Erminia B
Mon 8h3'13"-5d33'
Heck,Johann
Per 4h5'37"45d53'
Heck,Keri Koepke
Sge 19h57'28"16d27'
Heck,Kristen
Aql 19h30'19"8d41'
Heck,Mona
Lyr 18h50'55"38d29'
Heck,Scott Vincent
Oph 18h5'13"12d60'
Heck,Sr,Russell T
Hya 8h11'42"1d37'
Heck,Wendy C
Hya 8h16'56"0d29'
Heckel,Tatjana
Cmi 7h17'0"8d53'
Heckenbach,Diane
Tri 2h4'39"32d29'
Heckendorn,Christoph
Sgr 19h3'23"-29d35'
Heckenlively,Josephine
Mon 6h53'29"-0d9'
Hecker,Pam
Com 12h0'24"28d48'
Heckman Star
Vul 19h52'22"20d7'
Heckman,Brandon
Peg 0h12'51"14d46'
Heckman,Isabelle Eshleman
Sco 16h52'38"-38d22'
Heckman,John H
Boo 13h44'34"23d28'
Heckman,Karen Sue
Peg 22h9'48"24d27'
Heckman,Mark
Ser 15h24'30"20d17'
Hector
Her 18h2'0"18d51'
Hector & Cheryl Forever
Aql 18h59'35"14d52'
Hector,Will
Sct 18h20'45"-5d10'
Hedden,Susan Walker
Cas 0h2'50"50d29'
Hedderly,Anngela Lynne
And 23h41'48"33d8'
Hedderman-McNichols, Mary Kay
Cam 13h29'49"78d34'
Hedgepeth,Preston"PJ"
Uma 10h57'48"72d39'
Hedger,Lori
And 2h25'32"39d57'
Hedger,Matt
Aur 6h29'36"34d30'
Hedges,Dennis E
Her 16h18'51"10d22'
Hedges,Grahame Alan
Ori 5h59'1"14d11'
Hedges,Jane Susan
Uma 13h5'34"53d47'
Hedges,Stephen Henry
Ori 5h55'56"18d31'
Hedgins,Jewgie Odessa
Sex 9h55'36"0d39'
Hedgpeth,Elle Grace
Cas 2h55'22"68d42'
Hediger,Peter
Lyn 7h0'26"60d2'
Hedley,Mark Harry
Mon 7h2'52"5d30'
Hedlund Love Your Husband,To Dorothy
And 1h38'25"40d5'
Hedon-Star
Uma 11h46'30"32d5'
Hedouin,Jean-Pierre
Cas 1h35'55"64d36'
Hedrich,Christof
Lac 22h53'13"56d25'
Hedrick,Christine M
And 0h50'58"34d17'
Hedrick,Daniel T
Her 17h17'32"44d18'
Hedrick,John
Per 1h59'32"48d26'
Hedrick,Tammy
Uma 10h59'53"58d43'
Hedström,Jan
Peg 22h43'21"34d33'
Hedtke,Joan Marie
And 23h27'50"48d23'
Hedum,Timothy J
Aur 5h7'42"42d15'
Heduschka,Jörg
Sge 19h58'35"20d36'
Hedwall,Robert"Bob" E
Aqr 22h5'52"-8d29'
Hedwig 23-08-1957
Vir 13h36'22"-12d10'
Heeg,Georg
Aql 19h57'37"10d8'
Heeg,Kilian Peter
Lib 14h41'1"-5d43'
Heekin,Sixten
Lyn 8h5'28"35d39'
Heekin,William Moore
Lyr 18h44'55"38d37'
Heeley,Christopher Charles
Gem 7h20'21"30d57'
Heen,Victor Frederick
Aur 6h26'56"37d36'
Heer,Ray Chris
Cet 2h43'43"1d42'
Heer,Roxane Alexandra
Tau 5h23'25"16d3'
Heeres,James "Red"
Her 15h49'50"48d22'
Heery,Rev Cornelius J
Uma 10h58'16"36d30'
Heese,Nancy
Peg 23h52'29"28d21'A
Heese,Ned
Peg 23h52'29"28d21'B
Heeter QTZ,Ron
Per 2h24'0"55d57'
Heffele,Petra
Ari 2h58'23"21d34'
Heffelfinger,Beth Suzanne
Aql 19h52'30"-6d29'
Heffer,Sophie Jane
Uma 9h6'0"68d20'
Heffernan,Edna
Tri 2h18'1"32d43'
Heffernan,Sarah Grace
Mon 6h49'1"11d21'
Heffington,Catherine Dichtel
Aql 18h48'35"10d36'
Heffler,Roberta
Vul 20h16'14"25d13'
Heffley,Dennis A
Ser 18h35'19"3d40'
Heffley,Elaine
Cyg 20h29'1"38d31'
Heffley,Richard D
Aur 6h6'1"54d24'
Heffner,Betty Scheibner
Aql 19h47'28"13d39'
Heffner,Brent Nathaniel
Hya 8h53'27"0d43'
Heffner,Joel Evan
Aql 19h52'27"14d35'
Heffner,Sarah
Cmi 7h52'53"1d30'
Heffron,Carl J
Boo 15h3'59"51d49'
Hefley,James Michael
Cet 2h45'49"3d14'
Heflin,Bob & Linda
Boo 14h14'29"45d57'
Heflin,Patrica
Mon 7h38'59"-0d33'
Hefner,Bette S
Cyg 19h17'32"44d11'
Hefner,Christie Ann
Cas 0h18'14"64d40'
Hefner,James Richard
Oph 17h55'15"12d1'
Hefner,Robert
Peg 23h35'14"13d7'
Hefner,Stephen Paul
Her 16h23'56"50d34'
Hege & Staale
Peg 22h27'33"31d51'
Hege,Edwin Collins
Her 16h19'42"11d18'
Hegedus,Geraldine & Steven
Eri 3h56'1"-0d55'
Heger,Darlene Lynch
Lyn 8h18'48"51d58'
Heger,Jennifer
Cyg 21h50'31"41d35'
Hegforce,Ann & Ray McAllister
Cyg 20h23'38"30d29'
Hegg,Anna
And 23h1'57"48d37'
Hegg,Anne Marie
And 0h48'15"36d57'
Heggs,Alvin
Peg 21h56'34"30d40'
Heggs,Owen L R
Aur 5h25'13"30d0'
Heggstar,The
Lmi 10h54'17"28d8'
Hegmann,David Anthony
Boo 14h24'41"50d49'
Hegmann,Margaret Moore
Boo 14h20'12"50d59'
Hegmann,Richard Anthony
Boo 14h23'21"50d18'
Hegmann,Robert Alan
Boo 14h31'57"50d21'
Hegwood,LeiLani Picoll
Aur 6h12'16"36d9'A
Hegwood,Michael Keane
Aur 6h12'16"36d9'B
Hegy,Gloria
Aqr 22h12'1"0d58'
Hegy,Jr,Eugene A
Aqr 22h11'0"0d29'
Heib,Andreea
Cas 1h34'51"64d55'
Heibbel,Sue Les
Aur 5h39'42"40d9'
Heid,Charlotte Kay
Tri 2h28'22"28d49'
Heid,George A
Tri 1h29'1"34d24'
Heid,Guenther
Hya 8h11'41"-6d36'
Heidbreder,Yorck André
Cep 22h24'48"63d55'
Heide, Günter
Lyr 19h19'48"31d8'
Heidel,Gregory & Jean
Aqr 23h2'45"-6d56'
Heidel,Michelle Amanda
Lyr 18h19'0"38d48'
Heideman,Martha Lynn Estelle Blakely
Peg 21h58'49"23d34'
Heideman,Richard D
Cnv 12h15'14"42d2'
Heidemarie
Cap 20h5'36"-10d42'
Heidemarie-Heidchen
Tau 5h54'23"27d54'
Heiden,Christiane auf der
Cap 20h30'54"-21d6'
Heidenberger,Stellar Cory
Her 18h20'12"13d5'
Heidenreich "Silver Jubilee",Fr Bob
Uma 11h12'1"33d15'
Heidenreich,Peter
Cep 22h0'35"61d50'
Heider,Thomas
Aql 19h55'41"10d11'
Heiderman,Pauline
Cyg 21h8'47"48d53'
Heidersdorf,Rosanne Marie
Vul 19h48'32"29d8'
Heidhues,Joy
And 0h2'24"44d36'
Heidi
Leo 9h56'0"11d14'
Heidi
Aqr 21h57'1"0d23'
Heidi
Eri 3h14'0"-10d18'
Heidi & John's Star
Aql 19h29'25"11d6'
Heidi & Norbert
Cap 21h39'0"-19d15'
Heidi & Peter We're Always With You
Umi 14h31'37"88d52'
Heidi & René
Lib 15h15'53"-28d47'
Heidi Dawn
Cas 0h30'17"70d5'
Heidi Elizabeth-70
Sgr 18h49'37"-20d35'
Heidi Justina
Aqr 22h48'27"0d38'
Heidi Lynn
And 23h19'12"45d8'
Heidi Lynn
Mon 6h39'12"1d31'
Heidi Marie
Oph 17h57'24"11d32'
Heidi MCMXCII
Cas 0h15'18"58d26'
Heidi Renée
Cas 1h13'1"64d59'
Heidi Ruth
Lyr 18h56'30"33d9'
Heidi Sue
Cas 2h2'1"70d26'
Heidi's Chance
Del 20h13'47"10d10'
Heidi's Comet
Lyn 7h53'29"44d39'
Heidi's Evening Star
And 1h11'28"40d13'
Heidi's Hope
Aql 19h27'41"10d52'
Heidi's Light
And 1h44'0"39d33'
Heidi's Star
Lyn 7h50'31"39d53'
Heidi's Star
Lyr 19h20'53"40d14'
Heidinger,Peter
Mon 7h51'33"-1d28'
Heidy
Ari 3h13'16"25d51'
Heidy,Jason
Aur 6h9'28"37d44'
Heier,Gary
Cam 5h33'21"61d14'
Heiermann,Andre
Lyn 8h9'30"49d35'
Heifer,Bertie Mae
Umi 13h27'35"71d57'
Height,Melody
Aql 19h24'32"15d52'
Heighway,Stephen
Aur 5h5'23"41d21'
Heihupper
Ori 5h44'12"8d35'
Heik,Erica
Vir 13h36'50"-8d15'
Heik-Wauzz
Her 17h19'14"42d29'
Heike
Dra 16h8'27"63d2'
Heike
Sco 17h29'58"-30d36'
Heike & Mathias
Ori 5h13'27"-8d45'
Heike & Thomas
Cam 6h14'20"64d50'
Heike Stefanie
Eri 2h56'12"-3d51'
Heike's Liebesstern
Sge 19h57'13"20d13'
Heikenen-Weiss,Eilee
Cam 4h4'0"60d17'
Heikes III,Dana
Dra 17h2'22"62d11'
Heikes,Melanie A
Aql 20h10'44"1d34'
Heiko
Vir 13h28'24"-5d14'
Heiko-Rubin
Cnv 12h38'1"35d40'
Heil,Adelbert
Psc 23h2'29"6d10'
Heil,Del
Aql 19h9'1"5d31'
Heil,Donald & June
Aql 19h49'11"13d35'
Heil,Tanner McKenzie
Ser 15h39'0"24d46'
Heil,Terry
Umi 14h22'16"66d16'
Heilbrunn,Dr Karl E
Psc 0h56'0"21d13'
Heile,Best Grandpa Bernie
Dra 16h14'20"67d35'
Heilgeist,Warren & Pauline
Crb 15h27'46"32d11'
Heiliger,Douglas G
Her 16h58'16"24d51'
Heiliger,Nancy L
Lyn 8h45'35"40d47'
Heilman,Carl E
Cnv 12h56'57"33d13'
Heilman,Charles Haines
Ori 5h3'26"12d47'
Heilmann,Tobias
Dra 19h2'23"58d4'
Heim's Guiding Star, Jan M
Cam 3h48'34"72d3'
Heim,James
Cep 20h49'25"65d18'
Heim,John Joseph
Her 18h5'34"30d57'
Heim,Lester "Ltd"
Aur 6h9'28"48d56'
Heim,Lillian Joan Bertha Heavenly Light
Sco 17h29'11"-30d12'
Heim,Margie L
Gem 6h56'47"12d25'
Heim,Max-Dieter
Aur 5h11'26"43d48'
Heim,Michael
Cap 21h45'24"-22d38'
Heim,Nicholas Fain
Lyr 19h20'0"38d3'
Heim,Paige
Peg 22h2'37"4d34'
Heim,Shelly
Ori 5h56'37"12d28'

Heiman,Brett Nicholas
Aql 18h45'41"10d55'
Heimann's Haze
Col 6h0'15"-34d22'
Heimann,Bernadette & Walter
Leo 10h55'42"18d2'
Heimann,Heinz-Peter
Mon 7h52'6"-2d47'
Heimann,John Gaines
Dra 16h47'39"68d11'
Heimbach,Paul
Cnc 8h26'0"10d12'
Heimberg,Ron
Her 16h41'51"20d30'
Heimer,Keith L
Her 16h37'1"50d27'
Heimerle,Christopher M
Her 18h39'45"12d33'
Heimos,George Adam
Boo 15h14'1"28d1'
Heimroth,Silke
Eri 4h48'22"-5d50'
Hein (Salatiel),Paul
Ori 5h56'1"17d9'
Hein,Anne
Cas 0h9'42"50d30'
Hein,Belinda
Lyr 18h46'24"31d51'
Hein,Dominique Elyse
Lmi 10h37'0"30d37'
Hein,Kimberly Alecia
Uma 8h58'27"49d35'
Hein,Luca Alexander
Umi 17h4'1"80d20'
Hein,Maretes Dawn
Lyn 8h46'26"36d37'
Hein,Meggan
And 1h13'32"38d46'
Hein,Melissa Ashley
Cep 21h29'23"77d42'B
Hein,Michael Garrett
Cep 21h29'23"77d42'A
Hein,Nancy
Mon 7h41'11"-1d45'
Hein,Richard A
Cam 3h56'24"61d0'
Heindl,Anna Theresia Maria
Leo 11h15'50"-1d55'
Heindl,Franz & Gertrude
Dra 19h9'41"67d55'
Heindl,Pat & Dorothy
Cyg 21h22'46"40d5'
Heine,Bernd
Cmi 7h29'14"8d10'
Heine,Bruce
Hya 8h14'48"-6d23'
Heine,Christian
Lyr 19h13'23"35d24'
Heine,Marten Norman
Boo 14h39'1"37d41'
Heine,Renaud
Aql 19h2'0"16d40'
Heineman,Frances
Uma 12h10'26"60d8'
Heinemann,Linda
Vul 19h15'32"25d1'
Heinemann,Lois
Vul 20h5'20"28d23'
Heinemann,Walter
Lyr 19h20'39'31d16'
Heines,Dr Karl-Dieter
Lyr 19h8'37"38d20'
Heinicke,Herr
Boo 14h24'27"11d3'
Heinle,Klaus
Aql 19h58'1"10d55'
Heino,Arto
Hor 3h1'11"-49d53'
Heinonen,Keijo
Cam 3h46'52"60d37'
Heinrich's Brilliance W W
Gem 6h50'35"13d54'
Heinrich,Evelyn Krueger
Tau 5h46'40"27d19'
Heinrich,Joyce
Sge 20h16'40"20d29'
Heinrich,Jr,Gary L
Aur 6h6'15"33d21'
Heinrich,Jr,Ward D
Aql 20h31'54"0d48'
Heinrich,Karylyn
Mon 7h36'0"-0d4'
Heinrich,Lynn Ann
Cyg 20h22'20"38d37'
Heinrich,Sara Aimee
And 2h24'56"44d53'
Heins,Berle Eugene
Cep 22h44'54"80d25'
Heins,Carlton D
Psc 23h1'1"2d23'
Heins,Phoebe Elise
And 0h57'11"37d29'
Heins,Richard Marriott
Per 2h44'21"40d27'
Heins,Stacy Marie
Peg 22h41'28"21d55'
Heins,Zoë Leigh
Hya 9h11'23"-14d56'
Heinsen,William
Cep 20h25'34"76d28'
Heinsius,Ernst W Dr
Mon 7h36'46"-4d46'
Heintze,Catherine Ann
Cas 1h34'30"60d29'
Heinz
Lac 22h33'33"40d10'
Heinz & Richarda
Lyr 18h41'58"30d13'
Heinz,Christopher
Peg 21h6'51"18d49'
Heinz,Karl
Dra 18h28'49"65d16'
Heinz,Michael W
Lib 14h21'16"-23d3'
Heinz,Woody
Aur 7h19'44"38d41'
Heinz-Günter
Cas 0h13'11"60d39'
Heinze,Tonya Bordier
Lac 22h46'54"56d2'
Heir,Robin & Jeff
Ori 5h53'57"21d33'
Heise,Grant Randall
Her 17h56'1"40d32'
Heise,MD,Carl Warmington
Sco 17h53'43"-30d26'
Heise,Stephanie Lynn
Lyr 18h41'0"36d19'
Heisel-Star of Loveland,Betty
Cam 3h56'11"74d57'
Heiser,Mary & Stephen
Aql 19h41'33"10d29'
Heiser,My Brother Andy
Lac 22h21'45"55d37'
Heiser,Robert Francis
Boo 14h45'13"23d13'
Heisinger-Dede, Elizabeth Ellen
Peg 21h55'15"21d33'
Heisinger-Uncle Jerry, Gerald H
Aql 19h51'48"14d46'
Heisler,Audrey Olive
And 23h42'29"46d34'
Heisler,Joshua Charles
Boo 14h37'20"20d4'
Heisley,Dawn
Aql 18h44'36"8d22'
Heisner,Jerry
Per 3h1'20"50d18'
Heisner,Patricia A
Cet 1h45'31"-2d14'
Heiss,Justin
Aql 20h14'46"0d3'
Heistand,Richard Michael
Ser 18h16'28"-14d59'
Heiston,Dillon Cole
Cnv 12h36'36"51d29'
Heiston,John Thomas
Her 18h3'29"40d11'
Heiting,Bud
Tau 5h44'56"28d19'
Heitkamm,Hans
Cnc 8h57'35"11d49'
Heitland,Bonnie J
Cas 0h59'25"75d46'
Heitman,James Robert
Dra 11h49'56"67d12'
Heitner,Arthur B
Lac 22h35'1"54d7'
Heitshusen,Lena
And 23h9'37"40d33'
Heitshusen,William
Dra 20h23'1"68d11'
Heitz,Esq,William R
Aur 4h47'1"48d50'
Heitz,Lenore G
Peg 22h22'58"29d37'
Heitz,Megan
Peg 23h19'41"30d44'
Heitz,Olivia True Burgess
Mon 6h37'60"1d30'
Heitzeberg,Isaak
Ari 2h32'23"21d37'
Heitzenrater,Krista
And 2h24'26"45d49'
Heius
Umi 15h3'52"66d18'
Heizmann,Arnold
Boo 14h6'18"11d1'
Hejde,Daniel Ryan
Dra 14h20'0"63d52'
Hejde,Richard
Her 16h34'50"32d43'
Hel Bell
Peg 23h28'21"17d48'
Helander,Milja
Dra 14h20'58"64d14'
Helbing,Heire
Ori 5h24'43"-6d13'
Helbling,Beat
Vir 13h58'44"5d29'
Helbling,Chris
Aql 18h58'10"-1d18'
Helbraun,Rebecca Gardner
Uma 11h29'10"48d23'
Helcia
Tau 5h46'1"23d59'
Held,Franz
Lib 14h57'0"-22d54'
Held,Hendrik
Uma 10h25'30"50d14'
Held,Jeff
Cep 1h0'43"77d42'
Held,Johanna Geborene Struck
Boo 14h17'48"19d39'
Held,Karen
Cas 1h42'46"70d14'
Held,Steven M
Aqr 21h0'39"-1d8'
Helder Legacy Star, Earl Claude
Dra 16h4'52"64d20'
Heldmann,Erwin J
Lib 15h4'12"-6d3'
Heldner,Jürg
Cas 0h19'33"60d11'
Heldt,Peter
Dra 16h51'41"66d33'
Hele,Jill
Cyg 20h20'42"40d8'
Helen
Cas 1h18'23"70d2'
Helen
Cas 23h22'48"60d2'
Helen
Umi 15h46'34"78d21'
Helen
Ori 5h12'54"12d49'B
Helen
Tau 3h43'42"28d46'
Helen
Cas 0h18'13"61d38'
Helen
And 23h41'22"43d34'
Helen
Cyg 19h32'49"33d36'
Helen
Com 12h25'1"21d10'
Helen
Cam 3h14'32"63d41'
Helen
Lyr 19h4'58"40d9'
Helen
Com 12h21'30"19d54'
Helen
Sge 19h55'53"19d8'
Helen
Mon 7h44'35"-2d26'
Helen
Cyg 19h45'37"29d30'
Helen & Alfie
Oph 18h0'56"10d45'
Helen & Dave Sparkler,The
Cyg 21h57'54"50d3'
Helen & Dave 25
Cyg 21h31'58"31d26'
Helen & Gene
Boo 15h3'51"40d43'
Helen & Harry's Christmas Star
Cyg 21h47'59"37d41'
Helen & Jordan
Uma 8h44'1"50d45'
Helen & Len
Her 16h22'56"22d51'
Helen & Reberta
Umi 14h11'53"66d14'
Helen & Seymour
Mon 7h14'25"-5d29'
Helen Alexandra
And 0h48'45"36d9'
Helen Ann
Lac 22h26'59"38d32'
Helen B & Howie G
Aur 5h59'33"38d46'
Helen Elaine
And 02h25'55"39d29'
Helen Elizabeth The Bright One
Lyr 18h14'11"38d39'
Helen Elizabeth,The
Ori 5h52'41"14d34'
Helen Josephine
Crb 16h10'17"37d34'
Helen L
Cas 0h1'50"60d14'
Helen Lee-Best Mom in the Universe
Cas 0h27'46"50d15'
Helen Louise
And 23h18'43"41d41'
Helen Louise
Lyr 18h46'30"33d5'
Helen Marie
Lac 22h33'25"55d2'
Helen Marie
Mon 6h38'43"1d47'
Helen Marie
Cam 3h27'20"61d39'
Helen Marie
Vir 11h37'20"1d56'
Helen Mary 4/29/64
Cas 0h12'36"64d39'
Helen Mildred
Cyg 19h15'22"44d15'
Helen Millennium Star
Com 12h39'53"21d40'
Helen of Fowey
Cas 0h33'12"74d38'
Helen Vivian
And 0h51'0"35d36'
Helen's First
Vul 19h44'1"28d35'
Helen's Heaven
Eri 3h26'47"-2d0'
Helen's Light
Lyr 18h29'15"31d41'
Helen's Rainbow
Lyr 18h15'29"30d31'
Helen's Star
Cas 2h6'21"58d16'
Helen's Star
Hya 8h49'45"5d31'B
Helen's World
Sco 17h53'16"-38d12'
Helen-Ruth
Aql 18h53'34"-2d11'
Helena
Cas 0h15'52"60d26'
Helena
Cyg 19h47'22"31d5'
Helena
Sge 20h0'49"16d14'
Helena
Aql 19h1'1"16d34'
Helena
And 0h36'1"45d31'
Helena Elia
And 0h53'10"36d22'
Helena's First Christmas Star
Cyg 19h34'55"33d3'
Helena-45th Birthday
Cyg 19h29'59"58d2'
Helena-Dolce/Dear Old Dad
Eri 4h56'21"-11d48'
Helena-L
Cas 1h42'32"75d39'
Helenbrook,Howard & Shannon
Crb 15h55'22"28d17'
Helendrick
Cyg 21h18'59"39d46'
Helene
Dra 17h55'50"63d59'
Helene
Lac 22h7'28"38d6'
Helene
Peg 23h31'41"33d15'
Helene
Cyg 19h50'35"40d40'
Helene
Per 2h52'23"32d58'
Helene C
Lyr 18h35'11"40d60'
Helene,Yvette Maria
Crb 16h9'15"32d8'
Helenius,Agnes
Vul 19h59'17"23d31'
Helenius,Dorthe
Lyr 18h17'20"30d26'
Helenka
Cam 7h41'55"61d15'
Helens Asterism
Cyg 20h17'51"39d6'
Helenė
Cyg 19h54'33"48d31'
Helfand,Joe
Boo 14h10'0"32d24'
Helfano,Joshua Bryan Inouye
Umi 14h50'27"71d34'
Helfenstein,Mary
And 0h38'24"45d35'
Helfferich,Barbara M L
Boo 15h4'12"48d45'
Helfgott,Matthew
Cet 0h59'51"2d1'
Helfman,Timmy & Stuart
Uma 11h14'44"40d16'
Helfmann,Loryn Emily
Cas 2h39'1"60d42'
Helfrich,Anja Cook
Cyg 20h50'55"40d15'
Helga
Vir 13h52'15"-1d44'
Helga
Cnc 8h9'45"30d38'
Helga
Gem 6h25'59"13d28'
Helga
Lib 15h5'51"-2d41'
Helga
Sgr 18h52'56"-24d50'
Helga
Boo 15h13'10"38d16'
Helga & Hans
Uma 11h30'44"33d7'
Helga I
Psc 22h49'31"5d56'
Helga Maria
Cap 20h34'59"-18d47'
Helgans,Elliott Conley
Uma 11h52'59"42d31'
Helgemo,Alice
Cas 1h5'1"63d60'
Helgen,Bill
Her 16h6'36"25d38'
Helgeson,Daniel Quinn
Lac 22h29'13"50d24'
Helgesson,Ada
Per 2h36'24"56d26'
Helgren,Elsie
Lyr 18h32'29"41d11'
Helhowski,A Star For Heather
And 1h39'38"48d54'
Helhowski,Rachel Marie
Cam 3h17'52"60d26'
Helicher,Ellen
And 0h8'10"37d39'
Helin,Hannu
Uma 8h10'56"71d36'
Helin,Matthew
Cnv 12h58'25"41d17'
Helinski,Rachel Davey
Lyr 18h18'1"38d50'
Helinski,Rebecca Helen
Lyr 18h22'57"38d35'
Helix de l'Orbital
Aqr 23h31'24"-18d58'
Helka,Mirka
And 23h19'57"41d11'
Helker,Velma
Eri 2h48'53"-16d20'
Helkowski,Margaret
Cam 3h48'21"58d47'
Hell,Mario
Lib 15h8'6"-21d37'
Hella
Cap 20h35'38"-11d13'
Helldoerfer,Rosemarie
Cyg 20h27'43"30d51'
Hellen,Argiros Bill
Dra 19h41'30"70d46'
Heller Star
Umi 15h27'49"70d57'
Heller,Becky
Aql 18h39'59"-2d3'
Heller,Daniel
Aql 19h8'52"5d8'
Heller,David T
Aqr 23h5'21"-6d36'
Heller,Hal Scott
Boo 14h24'51"30d14'
Heller,Heidi
Cas 0h32'18"71d35'
Heller,Joan Wood
And 22h59'57"50d5'
Heller,Kitt & Ryan Perry
Mon 7h15'7"-7d8'
Heller,Marjory A
Cas 0h32'14"64d12'
Heller,Neil
Ser 15h20'20"9d24'
Heller,Richard
Cep 3h16'18"78d39'
Heller,Robert Dixon
Dra 19h4'57"47d56'
Heller,Ross
Lac 22h49'13"53d22'
Heller,Sherri Ann
Cas 3h34'60"70d22'
Heller,William C
Boo 14h26'60"25d19'
Helles,Betina
Per 2h59'31"35d16'
Hellgren's Stellar Gem
Peg 22h58'19"20d9'
Hellie Jade
Cyg 20h20'1"39d29'
Hellier,Jr,Gilbert L
Aur 6h0'51"37d1'
Hellin
Uma 10h21'30"62d23'
Hellin,Wiltelmina Moser,Dr
Per 3h10'22"54d47'
Helling,Julia
And 23h27'43"41d46'
Helling,Mark D
Cep 22h3'36"62d3'
Hellings,Dulcinea Marie
Mon 6h44'22"7d49'
Hellman,David Jay
Aur 6h2'13"32d16'
Hellman,Judy
Cyg 21h17'49"28d55'
Hellmann,Bernie
Oph 17h15'34"-20d13'
Hellmann,Brigitte
Leo 9h19'1"11d26'
Hellmann,Ralf
Cnc 8h9'37"32d36'
Hellmayr,Renate
Cnc 8h48'30"32d28'
Hellmers,Christian David
Cnc 7h54'56"11d16'
Hellmers,Ryan Patrick
Ari 1h55'1"15d38'
Hellmich,Rene
Lyr 19h12'47"31d35'
Hello,Love
Cyg 20h33'31"31d51'
Hellwig Mechanical Co Inc,Ray L
Dra 16h54'27"66d8'
Hellwig,Karina
Mon 7h52'41"-2d9'
Hellwig,Kathryn Marie
Cas 0h49'41"61d11'
Hellwig,Mareile
Sco 16h13'49"-22d16'
Hellwing,Peter
Ori 5h44'19"0d17'
Helm,C
Cyg 21h27'45"28d47'
Helm,George & Margery
Mon 7h48'8"-1d51'
Helm,Paul Herbert
Her 18h2'32"30d41'
Helm,Randy Caroline Katy Lacey
Uma 10h21'24"56d22'
Helm,Rosann
And 1h49'56"36d7'
Helm,Zachary Emile
Ori 5h54'10"14d55'
Helman,Dorothy
Crb 15h56'12"31d18'
Helme,Andrew W
Dra 16h41'16"68d2'
Helme,Garrick
Del 20h24'34"10d54'
Helme,Kevin & Judith
Cyg 20h40'39"44d25'
Helmer
Lyn 8h12'17"39d42'
Helmer,Claus
Cam 5h0'33"68d26'
Helmer,Claus
Psc 23h8'27"1d14'
Helmer,Helen
Vul 20h4'55"20d53'
Helmer,LPN,Theresa
Lyr 18h57'19"32d58'
Helmers,Herman
Her 16h30'53"32d20'
Helmers,John F
Dra 16h52'23"62d56'
Helmig,Ida Marie
Leo 10h58'34"11d46'
Helmig,Laura Lee
Cyg 21h33'0"38d40'
Helmlinger,Jessica Leigh
Cas 3h2'0"60d11'
Helmold,Kenneth Paul
Her 17h32'41"50d11'
Helmore,Francesca Tersina
And 0h1'44"47d35'
Helms Genuine Glatz Lite
Boo 13h43'21"15d34'
Helms,Alana Rose
Mon 6h26'37"8d24'
Helms,Chloe Rose
Sgr 18h51'35"-35d47'
Helms,David R
Cet 2h8'39"3d25'
Helms,Jr,James Reding
Sex 9h58'51"-8d47'
Helms,Nancy Ann
Lmi 10h8'24"34d1'
Helms,Rogan William
Cet 2h31'14"-9d41'
Helms,Taylor Nicole
Gem 6h32'28"12d30'
Helmsley,Leona & Harry
Aql 19h2'1"-6d11'
Helmsley,The
Cyg 19h22'0"44d13'
Helmut "Wuschel" Stern FRANZ
Ser 15h9'31"-1d24'
Helmut,Hans
Per 2h2'48"57d48'
Helmuth
Cep 22h10'39"61d51'
Helona
Cyg 19h57'38"44d18'
Helping Hands
Oph 17h59'58"8d24'
Helplessly Hoping 72995
Lyn 9h18'59"40d40'
Hels Belles
Her 16h30'47"36d41'
Hels Bels
Lyr 19h9'36"47d28'
Helsing,Ola
Cam 7h56'31"60d49'
Helsley,Gary
Sex 10h39'34"-6d40'
Helstrom,Dane & Sean
Hya 8h20'1"-10d29'
Helt,Rachel Lillian
Cyg 19h28'47"30d60'
Heltau,Michael
Cnc 8h30'11"30d60'
Helton III,Kathleen & Teddy
Cyg 20h6'20"40d9'
Helton,Daniel William
Cet 1h33'51"-12'30'
Heltzer,Debby
Lyn 8h14'10"39d41'
Heltzer,Mildred M
And 0h30'49"45d47'
Heluain,Franáois
Aur 6h5'31"31d48'
Helvellyn
Lmi 10h57'47"25d58'
Helwagen,George Robert
Aur 6h24'1"31d51'
Helwig,Glenn A
Ser 16h2'12"14d20'
Helwig,Susanna
Aqr 22h16'31"-1d13'
Hely,Mary Alice
Mon 6h34'21"-6d40'
Helyer,Sara
Cyg 19h34'38"33d16'
Helynn
Cas 23h25'16"61d33'
Helzerman,Leah Marie
Lac 22h42'17"56d30'
Helène-Andrée
Lac 22h23'34"40d22'
Heman,Jim
Ori 6h6'16"0d23'
Hemani,Mohammed
Cnv 13h20'50"38d11'
Hembd,Biggi
Vir 13h26'54"-4d46'
Hembd,Russell & Carol
Cyg 21h24'34"38d24'
Hemberger,Eva Dr
Leo 11h2'27"-1d48'
Hembree,Faye Robinson
Dra 17h18'34"63d46'B
Hembree,Sr,Jerry David
Dra 17h18'34"63d46'A
Hembrough,Julie
Lyr 18h20'13"38d47'
Hemeryck,Joseph L
Her 16h21'28"28d54'
Hemig,Jane L
Lyr 19h16'38"41d38'
Heminger,Hashim
Uma 10h6'1"57d23'
Heminger,Hazimah
Ori 6h3'1"6d2'
Hemko,Leeann Lindsay
Cyg 20h2'24"40d25'
Hemko,Leeann Lindsay
Peg 23h43'31"27d11'
Hemley,Philip
Per 1h55'1"56d54'
Hemm,Dieter
Eri 4h48'32"-5d49'
Hemm,Kristen
Del 20h37'52"6d15'
Hemmel,Dieter
Cyg 20h26'17"30d42'
Hemmen,Kimberly Anne
And 23h47'20"47d17'
Hemmer,Birgit
Cas 3h35'52"74d19'
Hemmer,Rosemary C
Uma 8h44'56"48d5'
Hemmerich,Bruce
Aur 6h0'51"30d14'
Hemmerich,Ilse
Cam 4h7'58"70d12'
Hemmick,Marguerite A
Lyn 7h58'55"35d29'
Hemmingsen,Hilary
Oph 18h0'23"13d17'
Hemmingsen,Michael C
Cep 23h4'56"64d30'
Hemmingsley, Hieronymus Troy
Umi 17h15'25"85d13'
Hemmo,Valérie
And 0h21'1"40d38'
Hemmrich,Erich
Cyg 20h27'33"37d35'
Hempel,Chester Walter
Dra 20h12'0"67d53'
Hempel,Hattie
And 23h34'24"38d4'
Hempel,Nathan
Dra 19h1'30"56d42'
Hemphill's Destiny, Claudia
Del 20h13'40"12d16'
Hemphill,Jenny
And 0h56'38"37d59'
Hemphill,Judy
Sgr 18h52'50"-34d23'
Hemphill,Mary Lynn
Psc 1h2'48"21d9'
Hemphill,Roby
Cyg 19h57'2"39d57'
Hemphill,Samuel Patrick
Leo 11h9'23"26d50'
Hempnall
Gem 7h14'28"30d55'
Hen The Handsome
Dra 19h16'28"67d41'
Hen-Ei
Leo 9h59'47"21d44'
Henas,Janet
Lyn 8h33'57"40d10'
Henault,Emile J
Lyn 8h9'0"46d51'

Henceroth,Justin
Her 16h41'44"33d13'
Henckens,Johan
Mon 6h12'2"-9d1'B
Hencsie,James W
Aur 6h1'47"34d1'
Hendel,Kenneth Bruce
Hya 9h41'60"-18d59'
Henderling,Terry
Aql 19h53'39"15d20'
Henderson (Michael & Marjory)
Peg 22h53'39"28d47'
Henderson 07/25/93,S L
Crb 16h2'20"31d49'
Henderson 4/9/1954, Mary L Sanders
And 0h27'0"40d22'
Henderson Family Star, W Peter
Oph 17h39'47"11d50'
Henderson's Dog Star
Cmi 7h52'41"0d17'
Henderson's Heavenly Body
Cyg 19h29'51"32d58'
Henderson,Adam McCall Thompson
Per 4h2'1"34d8'
Henderson,Alexandra Eugenia
And 0h56'1"39d36'
Henderson,Archie
Cet 2h38'43"0d17'
Henderson,Barbara Jane
Lyr 18h56'24"31d6'
Henderson,Bob & Erma
Eri 4h14'40"-17d0'
Henderson,Bruce
Lac 22h51'1"54d16'
Henderson,Carol & Jim Thompson
Cnc 8h52'34"30d6'
Henderson,Carole
Vul 19h13'13"21d41'
Henderson,Catherine Claire
Peg 23h4'23"10d53'
Henderson,Christian Le Travis
Her 17h38'46"27d32'
Henderson,Christine B
Ori 5h55'51"12d0'
Henderson,Danny "Boone"
Tau 3h41'1"28d28'
Henderson,Danny Louis
Dra 17h57'32"63d58'
Henderson,David
Hya 8h44'43"4d26'
Henderson,Deanne Lynn
Lyr 19h1'38"26d4'
Henderson,Diana Melamid
Lyr 18h20'47"40d35'
Henderson,Earline L N
Cas 1h58'27"70d59'
Henderson,Elizabeth
Sex 9h59'35"-6d41'
Henderson,Emily Meredith
And 1h56'0"40d3'
Henderson,Erin
Sct 18h53'23"-5d29'
Henderson,Gail
Eri 4h0'53"-7d37'
Henderson,Grace Marie
Peg 23h50'25"11d38'B
Henderson,Graham Leslie
Per 2h2'42"57d7'
Henderson,Gregory B
Dra 17h9'23"52d10'
Henderson,Griffin A
Aur 6h34'55"32d47'
Henderson,Hal K
Cyg 21h36'1"44d13'
Henderson,Jacqueline Jeannette
Peg 23h50'25"11d38'A
Henderson,James Harris
Dra 12h2'54"71d16'
Henderson,Jamie Lynn
Cnc 9h8'25"32d25'
Henderson,Jeff
Per 3h1'28"56d45'
Henderson,Jesse C
Hya 8h34'19"0d50'
Henderson,John
Uma 8h53'1"56d39'
Henderson,Johnee
Dra 16h52'11"69d31'
Henderson,Jr,Harry Elmont
Lib 15h37'25"-28d46'
Henderson,Lawrence E
Her 16h6'1"41d25'
Henderson,Linda Faye Gottschalk
Mon 6h34'31"8d7'
Henderson,Louise
Aur 5h8'36"40d27'
Henderson,Lynn Elizabeth
Peg 21h26'26"10d51'B
Henderson,Mary Lynne
Peg 22h39'0"27d41'
Henderson,Matthew Thomas
Ser 16h2'52"13d20'
Henderson,Michael
Hya 9h2'22"3d13'
Henderson,Michael Ryan
Sct 18h47'11"-8d15'
Henderson,Nancy
And 23h0'13"45d17'
Henderson,Natalie Victoria
Mon 6h43'0"1d30'
Henderson,Nicole Reed
Peg 21h49'31"34d2'
Henderson,Norma Ivah Felcyn
Cas 1h23'23"60d35'
Henderson,Paul & Louise
Umi 15h21'12"70d26'
Henderson,Provost Donald M
Cam 7h55'28"68d5'
Henderson,Raechel "Roach"
And 23h30'31"39d38'
Henderson,Ray A & Donna C
Oph 18h6'30"7d50'
Henderson,Robert & Catherine
Cyg 21h32'1"50d1'
Henderson,Robin M
Uma 11h45'29"41d7'
Henderson,Sarah Eleanor
Cyg 21h33'0"41d27'
Henderson,Stephen Bruce
Ser 15h53'39"20d44'
Henderson,Steve
Boo 15h3'32"18d27'
Henderson,Trisha
Ser 18h41'25"2d54'B
Henderson,Vivian & Cecil
And 0h55'1"38d56'
Henderson-Evening Light,Eve D
Cyg 21h30'15"42d12'
Hendersonium Sheilaquarian
Peg 23h24'0"8d55'
Hendijani,Jahan Riz
Cet 2h38'55"-3d55'
Hendin,Clark Leland
Her 18h16'50"38d26'
Hendon,James C
Vir 11h56'1"-6d5'
Hendren,Wayne
Cet 2h45'37"3d10'
Hendrick,Grant Kerr
Aur 6h18'30"37d45'
Hendrick,Kurt J
Lyn 6h52'21"60d44'
Hendricks,Ann
And 2h26'13"40d26'
Hendricks,Barbara
Lac 22h6'58"51d43"
Hendricks,Barbara
Sge 19h52'1"16d23'
Hendricks,Barbara
Cas 0h38'23"73d9'
Hendricks,Frank F
Cma 7h14'60"-16d46'
Hendricks,Gina Rose
Lyn 6h58'1"52d54'
Hendricks,Gregg L
Cam 3h56'35"57d41'
Hendricks,Hilary Hope
Peg 21h57'0"28d27'
Hendricks,James B
Cep 0h17'10"73d29'
Hendricks,Jay
Cet 0h30'23"0d40'
Hendricks,Jr,David C
Ser 15h10'10"6d60'
Hendricks,Lorraine
Cam 5h30'52"61d27'
Hendricks,Michael O'Farrell
Aur 5h11'0"40d26'
Hendricks,Priscilla B
Mon 7h13'0"-10d38'
Hendricks,Rodney Joseph
Cet 0h28'20"-12d21'
Hendricks,Sam-Jessica- Ron
Cyg 20h38'48"42d41'
Hendricksen
Peg 23h30'37"10d39'
Hendrickson,Annetta
Lac 21h59'56"37d5'
Hendrickson,Cynthia
Peg 21h57'47"30d17'
Hendrickson,Jerry Lou
Her 18h4'27"31d19'
Hendrickson,Jr, Benjamin
Lyr 18h44'32"30d36'
Hendrickson,Judy
Lyn 8h15'1"52d5'
Hendrickson,Leif & Karen
Cyg 21h26'1"40d11'
Hendrickson,Mary E
Sco 16h32'1"-28d30'
Hendrickson,Michael
Hya 8h14'20"-5d40'
Hendrickson,Mike K
Ori 6h10'56"8d17'
Hendrickson,Penny
Peg 0h0'49"30d20'
Hendrickson,Willy Psychedelic Relic
Per 1h45'52"52d55'
Hendrie,Donald Franz
Hya 8h56'29"-8d35'
Hendrie,Eric William
Uma 8h24'27"61d56'
Hendriks,William
Aur 4h56'34"38d9'
Hendriksen,Susanne
Equ 20h57'36"10d40'
Hendrikszs,Dad's Star For Paul J
Ser 15h9'31"10d44'
Hendrix,James David
Equ 21h10'0"10d48'
Hendrix,Jonathan Howard
Lac 22h4'1"51d40'
Hendrix,Lisa Marie & Carl I Gable III
Lac 22h27'53"52d9'AB
Hendrix,Michael Don
Ori 5h50'49"13d7'
Hendrix,Richard
Hya 8h53'53"5d43'
Hendrix,Tim
Vul 19h53'18"20d34'
Hendrix-Miller,Heather
And 0h13'34"36d33'
Hendry,Alexander Nathanial
Per 2h43'51"43d19'
Hendry,Dale
Her 17h24'0"27d4'
Hendry,Duncan & Ruth
Lyr 19h1'13"28d60'
Hendry,Jr,Robert M
Vul 19h42'40"23d25'
Hendry,Lindsay Dawn
Cet 2h7'0"0d21'
Hendry,Michael S
Mon 8h5'23"-2d7'
Hendry,Tiffany
Cas 23h27'39"63d22'
Hendy,Cheryl A
Cas 0h44'56"63d48'
Hendzel,Carol
Cas 0h3'47"58d30'
Henebry,Mamie
Uma 11h57'47"46d8'
Henegar,Albert L
Her 17h13'1"44d58'
Henenberg,Jinny M & Delilah D
Cas 0h37'55"65d45'
Henerson's Star,Casey
Her 16h14'55"47d45'
Hengehold,Eric Daniel
Cyg 20h25'52"42d46'
Hengehold,Sabrina Ann
Lyr 18h46'38"37d36'
Hengst,Friedrich
Cnv 12h15'24"35d18'
Hengst,Maximilian
Lmi 10h44'51"27d56'
Hengst,Wolfgang Horst
Lup 15h28'2"-42d13'
Hengsteler,Aris
Boo 15h3'16"10d57'
Henika,Dick & Bea
Aql 19h43'21"13d6'
Heningburg,Chiara Angelique
Lyn 7h48'36"48d10'
Heningburg,Dylan Bodie
Dra 17h19'1"71d58'
Heningburg,Jules Nehemiah
Cam 6h18'23"83d39'
Heninger,Brian T
Dra 18h30'12"70d33'
Heninger,Max Von
Lyn 9h11'37"36d55'
Henk en Esther
Cnv 13h32'1"47d55'
Henke's Eidolon
Aql 19h7'33"3d28'
Henke,Ludger
Lyr 19h19'44"30d33'
Henkel,Albert
Eri 2h54'0"-5d7'
Henkel,Ardell
And 2h0'45"45d33'
Henkel,Colleen K
Cas 0h32'0"64d4'
Henkel,Douglas
Her 17h30'10"40d4'
Henkel,Kathryn L
Com 12h25'47"30d50'
Henkel,Lester
Gem 6h56'24"13d27'
Henkel,Manuel
Cas 0h19'15"60d15'
Henkel,Megan Renne
Com 13h16'32"20d21'
Henkel,Melanie
Mon 6h37'44"10d46'
Henkel,Pia
Eri 4h49'55"-9d33'
Henkel,Ronald
Her 16h40'45"32d40'
Henkel,Sr,Thomas B
Per 2h57'37"46d18'
Henkels' Heavenly Body
Aql 18h57'21"12d4'
Henkle,Marylyn
Del 20d14'45"10d42'
Henley,Annabel Olivia
Peg 22h21'27"28d25'
Henley,Jill
Sge 19h32'32"19d23'
Henley,Jim
Cet 3h15'0"3d41'
Henley,Laurel
And 2h25'22"45d4'
Henley,Margaret
Cas 23h41'0"50d25'
Henley,Sara
Cet 3h15'0"4d47'
Henmann,Clara Annika
And 0h29'21"38d53'
Henn,Colleen Marie
Cas 0h30'27"63d0'
Henn,Dr Fritz
Cnv 13h33'33"46d29'
Henn,Kimberly Ann
Lyn 7h37'49"50d46'
Henn,Michael Scott
Per 3h56'46"38d28'
Henn,Michael Scott
Per 3h56'46"38d28'
Henn,Michelle
Uma 11h57'43"43d0'
Henn,Ulrich
Umi 13h3'27"75d8'
Henne,Emil
Peg 23h30'0"13d27'
Henne,William Gregory
Her 18h2'29"37d57'
Henneberger,Lawrence F
Per 2h24'25"55d2'
Henneberry,John L
Lac 22h53'16"50d9'
Henneka,Carol Ann
Uma 9h0'38"71d40'
Henneka,Joseph
Cep 21h7'48"56d0'
Hennelly,Edmund Paul
Ari 2h30'22"25d52'
Hennes,Willi 090852
Leo 9h25'1"18d28'
Hennessey
Her 16h20'1"42d29'
Hennessey,Brian Michael
Per 2h1'45"56d49'
Hennessey,Michael W
Umi 15h14'18"70d53'
Hennessey,Paul Nicholas
Dra 9h28'55"74d8'
Hennessey-Kalousdian
Cyg 20h30'32"42d43'
Hennessy A Special Friend,Marcus A
Aqr 23h4'49"-19d37'
Hennessy,Catherine
Lyn 7h59'40"35d1'
Hennessy,Ludivine
Cnv 12h8'15"38d57'
Hennig,Curt
Aql 19h4'0"0d17'
Hennig,Erna
Lib 14h20'40"-23d24'
Hennig,Thomas
Sgr 19h37'37"-42d36'
Henniger,Nana
And 0h59'23"36d12'
Henning,Alicyn
Cas 3h7'16"75d56'
Henning,Erin Layne
Cam 3h25'24"55d29'
Henning,James P
Aur 6h30'0"30d18'
Henning,Robert
Lac 22h54'28"54d1'
Henning,Sicyne
Cam 13h48'1"81d57'
Henning,Thomas Fred
Ori 5h1'17"7d32'
Henning,William H
Boo 14h10'52"35d52'
Henninger,Hannelore
Cap 20h46'54"-20d8'
Henninger,Robert
Vir 11h59'22"-0d10'
Henningson,Sr,Ray
Dra 19h1'1"47d52'
Hennyandy
Sge 19h12'39"20d21'
Henon,Michel
Aql 19h56'11"14d59'
Henri et Christine
Aur 6h4'10"31d58'
Henri et Muriel
Aur 4h47'16"40d25'
Henri et Muriel
Tri 1h58'37"33d0'
Henri-Joyce
Oph 18h16'11"12d54'
Henrich,Herr
Cmi 7h6'15"4d0'
Henrich,Larissa Valerie
Mon 6h41'13"7d37'
Henrich,Steven John
Ser 15h21'36"-0d6'
Henrich,William
Uma 10h57'30"61d0'
Henrichs,Jessica Dean
Aql 19h25'16"-7d56'
Henricksen-Porter, Delaine May
And 2h30'21"50d38'
Henrie,Jade Michelle
Lyr 18h53'37"34d53'
Henrietta Claire
And 23h45'41"42d23'
Henriette
Gem 8h1'13"30d56'
Henriksen,Dorothea L Schmude
Cas 0h31'19"60d48'
Henriksen,James Ralph
Aur 5h5'34"40d16'
Henrion,Martine
Dra 9h22'57"81d45'
Henry
Hya 9h1'1"0d25'
Henry
Cep 20h29'47"65d36'
Henry & Elsie- Embracable You
Cyg 21h6'45"31d35'
Henry & Mimi
Cyg 21h50'35"42d15'
Henry B & Anna Marie "Dazzler",The
Cnv 12h5'35"44d17'
Henry Bradley
Cam 12h43'39"82d14'
Henry Family,The Rev Amos
Ori 5h49'10"10d14'
Henry III,Jim
Peg 23h35'15"30d11'
Henry III,Thomas Joseph
Boo 14h18'28"39d9'
Henry The Educator
Her 17h37'1"26d18'
Henry's 326
Per 3h6'39"46d35'
Henry's Brilliance
Per 1h57'30"48d52'
Henry's Hope
Mon 7h47'49"-2d10'
Henry's World,K W
Cnv 13h23'48"41d5'
Henry, der Stern meines Lebens
Lyn 8h4'1"49d41'
Henry,Alice
Uma 12h45'55"57d48'
Henry,Alice & Dick
Eri 4h34'48"-1d40'
Henry,Amber Christine
Mon 7h17'51"-5d53'
Henry,Bob & Mary
Her 16h44'53"11d7'
Henry,Bruno Pierre
Uma 8h42'50"52d23'
Henry,Carmel
And 23h18'53"49d42'
Henry,Cathy
Vir 11h51'41"2d4'
Henry,Christopher
Aql 20h7'14"8d49'
Henry,Christopher Robin
Cmi 7h39'16"5d33'
Henry,Curtis
Dra 18h59'0"58d57'
Henry,Dale & Michele
Boo 15h0'49"50d30'
Henry,Darren Michael
Oph 17h54'39"10d12'
Henry,David & Karen
Sge 19h55'46"20d12'
Henry,Denis & Mary
Lyr 19h1'20"28d32'
Henry,Diane Kraft
And 2h18'58"49d27'
Henry,Elizabeth
Cas 22h57'11"56d17'
Henry,Elliott William
Boo 13h38'22"15d56'
Henry,Erin
Del 20h35'49"10d56'
Henry,Ernest
Cet 0h33'1"-6d59'
Henry,Francis
Cyg 21h0'51"36d16'
Henry,Geoffrey Walker
Aql 19h30'17"13d19'
Henry,Harold Robert
Her 16h51'1"50d27'
Henry,Helen E
And 0h5'17"47d2'
Henry,Jacquelyn
Equ 21h3'25"8d56'
Henry,Jeff W
Oph 16h27'1"-1d5'
Henry,Jessie Grace
Mon 8h0'23"-0d54'
Henry,Jim & Betty
Sge 19h53'55"18d48'
Henry,John Doty
Aqr 22h3'1"-11d27'
Henry,Keira Elizabeth
Aql 18h58'49"-5d19'
Henry,Kevin M
Lac 22h27'0"50d27'
Henry,Marshal
Sex 9h55'56"-2d17'
Henry,Martha
Cas 0h27'25"54d59'
Henry,Mary Ann
Ori 5h47'0"21d22'
Henry,MD,Robert Raymond
Oph 17h54'1"-5d51'
Henry,Michael
Aur 4h55'20"50d42'
Henry,Neil Patrick
Sct 18h53'1"-6d44'
Henry,Patrick
Aur 4h38'24"34d34'
Henry,Pierre Michel
Umi 14h39'11"67d34'
Henry,PJ
Hya 9h14'19"1d38'
Henry,Ray Glenn
Dra 16h30'19"52d7'
Henry,Ricahrd Charles
Per 4h24'44"51d30'
Henry,Robert A
Her 17h55'23"38d47'
Henry,Samuel C
Ori 5h55'60"15d44'
Henry,Seddon Wilson
Mon 6h50'0"10d29'
Henry,Steven L
Leo 11h1'21"1d18'
Henry,Tonya
Lyr 18h14'37"31d24'
Henry,Trish
Mon 7h3'24"1d3'
Henry,Vanessa A M
Cyg 20h21'29"41d40'
Henry,Vickiann
Cas 0h46'46"61d15'
Henry/Nancy
Lyn 8h44'58"44d2'
Hens
Uma 10h45'1"40d39'
Hensch,Kathrin
Ori 6h3'12"9d11'
Henscheid,Robert Bayard
Aql 19h5'18"15d0'
Henschel,Diane
Aql 19h31'0"8d15'
Henschel,Nicole
Cmi 7h16'31"4d27'
Henschell,Matt
Cam 4h11'42"67d59'
Henschen,Randy
Crt 11h47'33"-10d12'
Hense,Josef
Oph 18h1'19"8d56'
Hense,Sandra
Leo 10h20'49"13d58'
Hensel,Peter
Ari 2h5'15"20d42'
Hensel,Steven
Cnv 13h53'1"41d26'
Henselman,Wife of Tom, Carol Jeane
And 0h3'25"47d44'
Henshaw,Cyril Ford Middleton
Ori 5h57'1"16d27'
Henshaw,Derek Gordon
Dra 12h37'44"67d43'
Henshaw,Lucy & Phil
Oph 18h40'52"8d49'
Henshaw,Steven A
Ser 15h28'26"10d26'
Hensinger
Lac 22h13'54"54d29'
Henske,Albert J
Per 1h27'1"50d22'
Henslee,Chella
Tau 5h21'0"28d10'
Hensley's Wishing Star
Lyn 8h31'46"52d23'
Hensley,Adam L
Tau 3h58'48"20d29'
Hensley,Amanda Marie
Mon 7h47'43"-2d56'
Hensley,Amber Danette
Peg 0h3'48"31d1'
Hensley,Emma Lynn
Lyr 19h19'21"40d9'
Hensley,Evan Michael
Her 16h38'53"32d26'
Hensley,Kathryn
Cap 21h38'48"-24d8'
Hensley,Lauren Michelle
Aql 20h35'51"-1d25'
Hensley,Leslie Diane
Cas 0h21'18"63d1'
Hensley,Mark William
Ser 15h28'12"24d47'
Hensley,Martha
Mon 8h6'35"-3d17'
Hensley,Mildred
Ori 5h55'33"0d54'
Hensley,Sr,George
Her 18h8'0"38d52'
Hensley,Violette I
Vul 19h38'29"20d14'
Henslin,John Norris
Hya 8h30'37"0d7'
Henson,Anna
Com 12h37'44"20d55'
Henson,Chelsea Megan
And 0h26'1"21d59'
Henson,Cory Lynn
Sex 9h56'0"-6d59'
Henson,Dereck Marsh
Sct 18h42'36"-6d34'
Henson,Dirk "D T"
Cyg 19h29'41"30d2'
Henson,Douglas Oates
Cet 1h10'19"-3d34'
Henson,Jean Francis
Eri 4h14'53"-14d51'
Henson,Nicole
Peg 0h1'0"18d36'
Henson,Reiley Marie
Aqr 22h45'1"-6d21'
Henson,Rhea
Ori 5h26'23"0d16'
Henson,Robert Taylor
Aur 5h2'53"46d20'
Henson,Rosa
Ori 5h58'27"10d39'
Henson,Rose
Cam 3h30'17"60d6'
Henson,Susan Jane
Tri 1h45'12"34d12'
Henson,Tara Marie
Aql 19h4'55"-6d28'
Henszey,Daniel Roberts
Aql 18h50'28"9d0'
Henszey,Sara Denise
And 2h5'22"42d17'
Henszey,Sara Denise
Cas 0h49'46"64d23'
Hentic,Yves
Cam 4h0'41"61d4'
Hentrich,Sonya J
Tri 1h40'0"33d48'
Hentschel,Dieter
Equ 20h56'16"7d38'
Hentschel,Ida E
Leo 9h54'1"7d14'
Hentz,Tracy & David
Cyg 21h57'0"48d50'
Henwood,Bryan
Her 16h27'57"40d2'
Henwood,Lalania
Peg 21h21'0"23d52'
Henze,Anissa 5-4-77
Lyr 19h14'43"31d46'
Henze,Dietlind 9-10-94
Lyr 19h14'48"31d5'
Henze,Horst 15-6-46
Lyr 19h14'13"31d44'
Henze,Oliver 12-12-67
Lyr 19h16'1"31d10'
Henze,Tobias 10-1-72
Lyr 19h14'38"31d30'
Henzel,Greg
Boo 13h40'14"14d6'
Henpen,Nicole
Lyr 18h26'26"47d51'
Heon,Gregory David
Peg 22h58'40"20d11'
Hepburn,Matthew
Umi 16h22'0"76d26'
Hepenstrick,Dr Heinrich
Sco 15h52'7"-21d56'
Hepfer,Anmarie & Steven
Uma 8h44'1"52d2'
Hepfner,Van Frederick
Aur 5h15'22"45d38'
Hepler,Hazel
Per 3h14'14"42d29'B
Hepler,Jr,Michael
Dra 14h42'49"61d18'C
Hepler,Michael
Dra 14h42'49"61d18'A
Hepler,Pamela
Dra 14h42'49"61d18'B
Hepler,William
Per 3h14'14"42d29'A
Hepner,David Michael Eric
Aql 20h12'0"11d21'
Hepp,Michael Arthur
Cep 21h25'28"60d46'
Heppenstall IV,Walter Leonard
Per 2h38'16"34d39'
Hepple,Thomas
Her 16h53'22"40d2'
Heppner SIU USMC 1992, Vincent Bernhard
Per 2h58'44"40d17'
Hepps-Pun Intended, "Sonny"
Cam 8h3'42"70d15'
Hepworth Family Star, Cogatotcka
Uma 14h19'18"61d55'
Her Father's Diapason Vessel
Cep 0h10'41"75d57'
Her Royal Highness Princess Chulabhorn
Cra 18h20'10"-45d9'

Her Royal Majesty Queen Kelley Deer
Cas 0h58'0"64d46'
Herakhan
Cyg 20h22'50"39d12'
Heraklith
Per 4h28'43"51d22'
Herald,Caroline Marie
Sco 17h2'31"-31d33'
Herald,Keith
Her 16h37'22"26d13'
Herald,William Rara Rudolph
Lmi 10h4'0"31d30'
Herald,Winifred "Nana" Kirkham
Com 12h12'53"26d15'
Heras,Ibis A
Lyn 7h7'17"50d51'
Herb
Cam 5h5'57"63d32'B
Herb
Sct 18h55'45"-7d36'
Herb & Gale's "Lodestar"
Crb 16h9'12"30d9'
Herb & Sarah
Cyg 21h26'36"41d14'
Herb,Jake Thomas
Her 17h32'29"21d38'
Herber & Carder
Lac 22h3'0"51d15'
Herber,Elizabeth F
Tau 4h42'44"16d31'
Herberg,Darby Marie
Vir 13h43'36"5d21'A
Herberger,Bob & Kax
Eri 3h16'47"-12d22'
Herberholz,Karl Heinz
And 23h34'30"37d14'
Herbert
Cam 12h54'11"78d8'
Herbert
Dra 17h24'30"73d51'
Herbert
Cnc 8h56'38"8d1'
Herbert One
Ari 2h51'1"30d25'
Herbert Star,The Jack
Cet 1h50'52"-1d1'
Herbert's Star Over New Mexico,Tracy Boyce
Cyg 21h2'24"30d47'
Herbert,A P
Lyn 7h46'40"39d29'
Herbert,Avril "Christmas Morning"
Lyn 8h21'42"40d44'
Herbert,Derek Jerome
Uma 10h24'11"48d3'
Herbert,Jennifer
Boo 14h8'49"53d31'
Herbert,Lillian
Cas 0h44'33"62d53'
Herbert,Mary Catherine
Eri 2h59'45"-13d4'
Herbert,Richard Michael
Lac 22h34'44"40d31'
Herbert,Roger
Dra 18h12'22"67d39'
Herbert,Saxon Taylor
Aur 6h23'38"37d7'
Herbert,William Michael
Aql 20h7'30"1d50'
Herbi's Lucky Star
Sco 17h29'1"-30d54'
Herbie
Her 17h31'1"38d26'
Herbie Baby
Boo 14h5'60"22d37'
Herbon,Vicky
Cyg 21h19'20"37d48'
Herbowy,Anne
Tri 1h38'27"28d47'
Herbst Star,The
And 2h25'43"41d7'
Herbst,Jr,Arthur L
Gem 6h51'22"21d14'A
Herbst,Nicole L
Gem 6h51'22"21d14'B
Herbst,Star of Leonard & Thelma
Uma 9h34'13"49d40'
Herbstritt,Norma Jean
Cas 0h59'20"60d15'
Herbus
Hya 8h44'46"6d1"
Herchel,Eckard
Ari 3h0'41"21d35'
Herchenbach,Lisa Karen
Lyr 18h47'16"32d54'
Herchenbach,Nellie Kathryn
Cyg 20h26'37"38d27'
Hercher,Jean
Cam 3h40'1"61d51'
Hercot,Agnes (Bunis)
Cas 1h37'28"73d34'
Hercule le chef- D'oeuvreux
Her 16h52'58"40d56'
Hercules
Her 17h38'34"40d40'
Hercules,Brenda
Psc 1h36'0"28d12'
Hercules,Frank E M
Her 17h36'18"21d12'
Herd,Kathy Jo
Lyn 7h53'21"34d47'
Herda
Aql 19h17'23"14d57'
Herdegen,Kristina Louise
Cas 0h23'27"56d30'A
Herdegen,Stacey Marie
Cas 0h23'27"56d30'B
Herdman III
Cnv 12h11'52"39d57'
Herdman,Jay
Dra 14h54'1"62d15'
Herdman,Paul N
Per 2h58'1"50d10'
Herdt,Jon R
Ser 15h55'27"-1d17'
Here Comes King Joseph!
Cep 23h33'31"60d37'
Here's Mud In Your Eye,Frank
Ori 5h54'57"15d30'
Heredia,Elisabeth Baeza
Cep 22h23'46"67d37'
Hereth,Cristine R
Sct 18h56'35"-6d5'
Herfurt,Joachim
Oph 17h57'29"8d37'
Herger,Patrick
Vir 13h15'53"-8d25'
Herilla,Laura
Cyg 20h37'25"30d36'
Hering Star 8/6/94, The Kim & Karl
Aql 18h41'52"0d48'
Hering,Brenda D.
Cam 3h56'13"70d50'
Hering,Elfie
Cet 2h54'28"8d10'
Hering,Jörg
Uma 10h46'0"40d45'
Heringklee,Ingeborg
Aql 18h59'17"16d59'
Herion,Mattie
And 0h11'54"39d58'
Heritage,Barbara Amy
Cyg 21h0'49"38d60'
Herka Doodle
Vul 19h1'27"21d43'
Herkey,Jason Daryl
Aur 6h34'24"38d10'
Herklotz,Brett
Boo 14h37'0"40d53'
Herkner,David Matthew
Lac 22h18'11"54d39'
Herlic,Claire
Dra 15h43'22"54d42'
Herlice
Cam 3h49'1"60d19'
Herlihy,Caitlin
Cyg 21h3'53"28d49'
Herlihy,Kaz
Lyr 19h25'22"41d5'
Herm Boy Diggle III, Eddy
Her 16h22'15"25d34'
Herman C
Uma 11h53'38"41d36'
Herman,Alan Paul
Cet 2h10'41"7d44'
Herman,Ann
Sct 18h53'56"-13d13'
Herman,Charlotte
Cas 2h26'13"61d50'
Herman,Christopher Sean
Lib 15h39'32"-28d30'
Herman,Dr
Uma 10h43'1"50d30'
Herman,Dr William
Oph 18h18'15"10d38'
Herman,Dylan Sky
Lac 22h54'49"52d56'
Herman,Hilde
Lyn 7h56'1"34d41'
Herman,James A
Aur 6h0'45"37d38'
Herman,Kelly Eileen
Cas 0h32'16"63d52'
Herman,Larry
Her 17h31'12"14d45'
Herman,Lesley Gene
Lyn 9h1'21"40d45'
Herman,Noah Ross
Aur 6h34'36"32d50'
Herman,Norman
Her 17h32'53"42d7'
Herman,Robin
Peg 21h25'17"10d59'
Herman,Scott H
Cep 1h1'43"77d32'
Herman,Shari
Cam 3h51'12"58d43'
Herman,Sharon L
Cas 1h6'14"65d34'
Herman,Steve
Aur 5h1'58"48d7'
Herman,Willard Henry
Dra 15h15'17"61d23'
Herman,William Morris Belknap
Cet 2h4'34"-2d5'
Herman-Mach
Lac 22h51'17"55d9'
Hermanas
Del 20h13'1"10d44'
Hermance,Edward Raymond
Boo 15h14'0"29d12'
Hermann
Aur 5h56'11"31d23'
Hermann & Hanna
Cmi 7h7'17"3d48'
Hermann,Adam Peter
Mon 7h52'52"-1d1'
Hermann,Al & Evelyn
Uma 11h12'32"51d53'
Hermann,Christa A
Sco 17h0'46"-30d10'
Hermann,Darja
Cap 20h30'7"-26d12'
Hermann,Ida Mae Heakin
Lyr 19h22'32"35d34'
Hermann,Michael
Peg 23h30'17"10d54'
Hermann,Paula & Ron
Eri 4h55'50"-8d13'
Hermann,Peter
Aql 18h59'0"17d15'
Hermann,Ralph Joseph
Lac 21h55'14"38d46'
Hermann,Robert Edward William
Car 7h29'39"-59d27'
Hermann,Willhelm
Sgr 18h57'15"-24d27'
Hermans,Liduine
And 0h29'46"40d5'
Hermanson,Aaron Christian
Ori 5h52'36"15d25'
Hermanson,Erik Christopher
Peg 23h25'25"17d22'
Hermes,John
Aur 7h9'1"36d0'
Hermes,Martina
Aqr 21h55'31"-3d24'
Hermida (Chan),Jose A
Oph 17h34'44"-23d27'
Hermine
Cet 2h40'52"3d38'
Herminghaus,Trisha & Joe Kurtak
Cyg 21h27'0"40d21'
Hermione
Cas 0h16'45"47d48'
Hermosa,Alejandra
Cyg 19h29'19"37d35'
Hermosa,Maria Luisa
Peg 22h17'50"4d44'
Hermosa,Yolanda Estrella
Peg 22h7'58"29d37'
Herms
Cnv 13h20'56"40d5'
Hermsdorfer,Joel
Ori 4h57'37"4d17'
Hermstein,Ramona
Aqr 21h59'37"-3d40'
Hern,Robert
Cnv 12h28'12"32d42'
Hernandez Birthday Star,The Lou
Aur 5h28'17"37d45'
Hernandez"Magnificent" Andrew Joseph
Ori 5h32'19"-2d28'
Hernandez,Aida
Cnv 12h50'21"39d35'
Hernandez,Aime
Peg 23h3'1"8d55'
Hernandez,Albert
Per 3h55'1"52d36'
Hernandez,Andrea Carmen
Lyr 19h17'25"41d46'
Hernandez,Barbara Jean
Aql 19h24'21"-10d1'
Hernandez,C & C
Oph 18h19'0"6d28'
Hernandez,Charles
Boo 14h57'34"27d33'
Hernandez,Dan
Cet 3h14'30"1d7'
Hernandez,Dave Allen
Cnv 12h26'43"50d44'
Hernandez,Davie
Cet 1h38'0"-5d40'
Hernandez,Donna
Oph 18h37'35"6d14'
Hernandez,Edward Victor
Mon 7h54'2"-8d58'
Hernandez,Francisco
Lyn 7h52'36"33d24'
Hernandez,Guadalupe
Ori 5h39'0"10d15'
Hernandez,Gustavo A
Aur 5h33'26"38d26'
Hernandez,Gustavo A & Fern Gelb
Dra 17h37'42"55d47'AB
Hernandez,Hector & Christy
Sex 9h58'45"-4d32'
Hernandez,Isa Peirato
Umi 16h53'50"75d44'
Hernandez,Jesica
Cas 22h59'0"54d49'
Hernandez,John Patrick
Sex 10h21'45"-5d18'
Hernandez,José
Aql 19h58'48"10d17'
Hernandez,Juano
Cet 2h32'38"-3d18'
Hernandez,Julie A
Mon 6h24'16"-8d50'
Hernandez,Katherine
Cyg 20h19'21"38d11'
Hernandez,Leah Marie
Hya 8h43'39"2d23'
Hernandez,Mariano E
Gem 6h48'47"30d32'
Hernandez,Mary
Lyn 7h57'24"39d50'
Hernandez,Michael
Cet 1h59'51"-6d50'
Hernandez,Nelson
Crt 11h20'46"-8d40'
Hernandez,Paula
Peg 23h16'0"33d51'
Hernandez,Ricky
Oph 18h41'52"10d46'
Hernandez,Robinson Rodriguez
Ori 6h0'45"8d47'
Hernandez,Scott
Mon 8h6'17"-3d11'
Hernandez,Steve L
Peg 21h47'41"31d45'
Hernandez-Ponti,Jihan
Aur 4h55'53"40d46'
Hernanutz,Birgit
Tri 2h29'43"31d24'
Herndon,Kristina Sedej
And 0h10'0"37d31'
Herne,Unni
Sgr 18h48'3"-30d4'
Hernesniemi,Sari Marjaana
Cas 23h38'45"63d57'
Hero
Per 4h20'52"50d55'
Hero Jeff
Aql 18h43'52"11d57'
Herold
Dra 18h26'34"67d56'
Herold C537
Com 13h32'0"21d31'
Herold Raises 6 & Gets A Star,Ellen
Crb 15h48'34"28d46'
Herold,Ethel Mae
And 2h15'1"48d9'
Herold,Heike
Equ 21h3'27"3d43'
Herold-Schumacher, Margaret
Lyr 18h32'14"32d50'
Heron,George
Cyg 20h18'18"38d46'
Heron,Lee "Jester"
Mon 6h47'1"11d28'
Heronemus,Seth Martin
Her 17h3'25"21d32'
Heroshi Chesu No Kamisama
Her 15h50'21"42d1'
Heroy,Sasha Gabriel
Peg 22h35'45"21d39'
Herpers,Martin
Aql 19h7'43"3d34'
Herpers,Oliver
Eri 4h49'14"-6d30'
Herpolsheimer,Arthur B Sally A
Crb 15h46'57"27d55'
Herr Binder
Cmi 7h19'48"8d33'
Herr Mazalka
Mon 7h54'58"-1d27'
Herr Thiel
Lyr 19h21'54"31d15'
Herr,Allison
Lyn 8h5'19"37d25'
Herr,Brandon H
Mon 6h34'27"-0d37'
Herr,Mary Jean O'Brien
Umi 15h22'30"71d7'
Herrald V,Fletcher Harold
Aql 19h30'15"12d22'
Herrald,Ashley Brooke
Peg 22h47'57"27d10'
Herrberg,Jane A
Com 13h32'29"25d55'
Herrbold,Cindy L
Lyn 6h54'30"60d58'
Herrel,Daniel K
Her 16h32'37"13d5'
Herrell,Alma Lee
Mon 8h7'9"-6d35'
Herren's Star
Aur 4h56'13"40d39'
Herren,Erica
Lyr 18h56'0"40d58'
Herren,Lilian & Lester
Leo 10h2'30"15d6'
Herrera,Alesha
Lyr 18h54'29"42d60'
Herrera,Alicia Maria
Peg 22h8'0"4d53'
Herrera,Alma
Sgr 18h55'55"-33d58'
Herrera,Andrea
Lyn 9h8'28"40d14'
Herrera,Andrew Evan
Cep 21h50'1"68d11'
Herrera,Daniel Albert
Ori 5h49'1"18d59'
Herrera,Kelcey Marie
Lac 22h28'54"48d48'
Herrera,Manuel R
Her 16h38'0"27d59'
Herrera,Melissa
Peg 23h46'33"31d32'
Herrera,Miyelina
Equ 21h23'38"3d34'
Herrera,Movia Paz
Per 1h34'56"53d52'
Herrera,Robert Lee & Dayna Leanne
Mon 6h45'55"10d1'
Herrera,Shining Star Of Christopher Patrick
Peg 22h1'40"21d32'
Herrera,Suzanne D
Cyg 20h15'0"41d29'
Herrera-Markovic,Amara Felice Linda
Cyg 19h19'23"45d47'
Herrero,Tere Casals
Cma 7h21'0"0d21'
Herrewig,Evelyn LaVerne (Toni)
Vul 20h19'31"22d54'
Herrhausen,Traudl
Cyg 21h22'33"48d57'
Herrick
Cnv 13h47'50"40d30'
Herrick,Andrew Lee
Her 17h56'40"14d27'
Herrick,Benjamin Richard
Her 17h33'1"20d56'
Herrick,Brian
Dra 11h5'46"74d40'
Herrick,Craig Whitehead
Psc 0h0'0"1d20'
Herrick,Elayne S
Cnv 13h47'11"46d15'
Herrick,Evan
Cnv 12h14'43"37d48'
Herrick,Jacob Anthony
Hya 8h39'18"0d32'
Herridge,Dave
Cep 22h30'49"70d35'
Herries,Linda
Lyn 7h46'45"45d51'
Herriges,Diane
Mon 7h8'40"-6d60'
Herriges,Greg
Her 16h59'30"33d21'
Herriman
Crb 15h27'20"31d5'
Herriman,Amanda P
Eri 2h47'1"-17d38'
Herrin,Dawn J
Cas 0h52'47"64d48'
Herrin,Joseph
Her 17h49'36"14d52'
Herring,Ann
Uma 11h57'18"32d22'
Herring,Diane Elaine
Leo 11h52'47"21d17'
Herring,James C
Cyg 21h17'24"38d30'
Herring,Jim
Dra 15h50'12"62d31'
Herring,John Brandon
Ori 5h26'20"-0d8'
Herring,Otha
Mon 6h3'49"-8d15'
Herring,Samuel Robert
Umi 13h17'34"76d35'
Herring-Smith Wagon Hitch
Aur 5h3'56"30d11'
Herrington,Carol Christine
Cas 1h57'51"58d32'
Herrington,Chad Alan
Lac 22h8'54"50d50'
Herrington,Curtis
Cep 22h58'25"57d49'
Herrington,Garrel
Oph 18h2'1"7d40'
Herrington,Karen
Vul 19h18'51"20d25'
Herrington,William Glen
Tau 4h37'27"18d51'
Herriot,Mhairi
Umi 15h15'1"66d4'
Herrman,Esther L "Granny"
Cas 0h27'27"75d42'
Herrman,Nell E
Cam 6h6'40"80d19'
Herrmann I
Cap 21h19'34"-14d3'
Herrmann,Barbara
Cap 21h14'56"-26d41'
Herrmann,Constanze
Ari 2h30'30"26d12'
Herrmann,Cornelia
Ari 2h38'32"21d44'
Herrmann,Dirk
Tau 4h59'59"16d49'
Herrmann,Eric
Ser 15h14'60"0d6'
Herrmann,Eric Michael
Ori 5h26'1"-0d3'
Herrmann,Gabriel J C
Lyr 18h54'42"30d36'
Herrmann,Gustav E
Lyr 19h3'29"28d55'
Herrmann,Hannah Elisabeth
Tau 5h56'1"27d10'
Herrmann,John Patrick
Sct 18h48'42"-8d54'
Herrmann,Jonathan, Gail
Cnv 13h18'0"51d48'
Herrmann,Juergen
Mon 7h51'18"-1d42'
Herrmann,Pamela Rae
Umi 15h17'35"71d12'
Herrmann,Paul J
Per 4h42'42"51d38'
Herrmann,Renato
Ari 3h0'31"28d57'
Herrmann,Robert Friend
Boo 13h54'31"15d3'
Herrmann,Tracey & Karl
Eri 3h29'11"-8d5'
Herrmann,Wolfgang
Her 17h20'50"42d23'
Herrmann-Lisi
Lyn 8h16'18"47d57'
Herro,Robert Michael
Cas 0h40'55"67d43'
Herro,Stephen Robert
Per 1h44'0"53d39'
Herron,Arthur
Per 2h1'0"50d30'
Herron,Daniel David
Leo 11h54'26"26d32'
Herron,Marjorie A
Lyn 7h35'22"39d35'
Herron,Mindy
Eri 4h9'12"-15d37'
Hersam,Beth
Lyn 9h9'16"41d23'
Hersch,Dr Lee
Mon 8h2'5"-9d14'
Hersch,Rachel Lynn
Com 13h7'17"28d45'
Herschdörfer,Gabriel Stefan Mimis
Cap 20h4'49"-10d22'
Herscher,Adam Joseph
Oph 17h59'58"11d5'
Hersey,Geoffrey Alan
Aur 5h0'35"45d57'
Hersh,Allan
Lac 22h45'13"52d40'
Hershaft-Ross,Talia Suzanna
Sco 17h55'50"-38d31'
Hershenhart,Elise J
Cas 0h11'0"64d6'
Hershey Class Of 2000
Uma 8h28'27"72d31'
Hershey,John E
Aql 19h23'28"14d16'
Hershey,Rebecca
Cyg 20h23'1"39d30'
Hershkowitz,Susan Debra
Vir 11h51'33"6d24'
Hershman,Amanda Haley
Cyg 19h28'24"32d16'
Hershman,Bryan G
Dra 17h3'1"63d4'
Hershman,Jerome & Dee
Crb 16h11'44"32d3'
Hershon,Lawrence M
Cep 21h39'59"68d23'
Herson,Richard Matthew
Her 18h21'34"12d32'
Hersovie
And 2h7'48"40d6'
Herssens,Walter R
Psc 1h0'14"20d8'
Herstein,Max & Mizzi
Lyn 8h37'0"43d2'
Herstein,Morris
Cam 7h30'30"80d9'
Herstemarsven
Cet 2h40'30"1d4'
Herta & Hans-Joachim "für immer"
Boo 15h40'45"48d9'
Hertan,Bill "Puff"
Hya 8h39'14"1d36'
Hertel,Adam James
Cet 1h54'23"-3d41'
Hertel,David
Aql 20h30'29"0d47'
Hertel,Eberhard
Her 17h20'40"42d50'
Hertel,Ina 17 40 h
Cap 21h22'46"-20d13'
Herter II,Everit A
Hya 9h11'46"1d3'
Herter,Carolyn Andrea
Mon 7h51'29-5d14'
Hertgers,Harry & Cheryl
Cyg 21h5'53"36d2'
Herth,Jonathan Robert
Uma 11h2'23"45d27'
Hertweck,Jochen
Tau 4h3'50"1d21'
Hertz,John Edward
Her 16h29'23"47d39'
Hertz,Kelly Leigh
Sct 18h33'35"-06d20'
Hertzendorf,Brenda & Jay
Cnc 8h34'57"30d13'
Hertzler,Russell John
Aur 4h52'1"48d47'
Hertzler,William Winfield
Ser 16h2'26"10d15'
Hervatin,Giselle
And 0h21'59"36d55'
Herve B
Lac 22h19'19"51d40'
Herve Maryvonne
Cam 10h5'52"82d26'
Herve,David
Cep 22h15'0"58d45'
Hervetaollier
Cas 22h57'47"57d35'
Hervey,David Steven
Mon 6h22'1"-1d23'B
Hervey,Douglas John
Mon 6h22'1"-1d23'A
Hervey,Mark Loves Dr Debbie DO
Sge 19h13'51"16d23'
Hervey,Sr,Joseph
Aur 6h53'43"37d4'
Hervey,Treasure Fay
Cas 1h21'13"53d55'
Hervy,Jean-Claude
Tri 1h58'34"33d24'
Hervé,Vicogne
Per 3h57'40"39d19'
Herwig,Fred M
Her 16h59'18"38d22'
Herzeelle,Juanita
Lyr 18h56'20"32d23'
Herzens-Didel
Ori 6h17'45"10d13'
Herzfeld,Steven
Cnc 8h58'11"20d26'
Herzig,Abby Jana Kleiner
Cnc 8h57'23"28d6'
Herzig,Leigh K
Sgr 18h59'12"-23d46'
Herzig,Robert Alexander
Cnc 8h57'0"22d0'
Herzler,Gary,Ann & Cory
Lyn 7h43'0"40d30'
Herzlich
Lyr 18h15'0"38d33'
Herzog Hans Eberhard
Psc 22h54'17"2d20'
Herzog,Ann
Uma 13h20'52"57d39'
Herzog,Jane
Aur 6h22'40"37d25'
Herzog,Jason Carter
Lyn 8h16'60"58d31'
Herzog,Mary Lou "Wimpy"
Mon 6h55'1"0d18'
Herzog,Peter
Sgr 19h3'56"-29d14'
Herzog,Rosita & Werner
Uma 12h0'32"51d22'
Herzog,Shelly Ann
Peg 21h19'15"23d20'
Herzog,Susan Lynn
Aql 19h53'12"15d24'
Hesbrook,Denise Alynn
Lac 22h21'38"54d22'
Hesch,T M
Per 4h2'19"50d7'
Heselton's Golden
Sge 20h4'0"20d36'
Heseman,Marta
Lyn 0h26'50"52d26'
Heshmaty,Ramean Aaron
Boo 13h44'16"18d39'
Heske,Frank
Boo 14h10'46"31d30'
Hesketh,Howard & Joyce
Uma 15h0'51"70d32'
Heskiel,Israel
Cep 21h44'59"60d10'
Hesler,Daniel James
Her 16h57'16"23d2'
Heslin,Patricia
Cet 0h28'47"0d43'
Heslip,Malcolm Farnsworth
Aur 5h15'12"41d15'

Heslop,Brian
Aur 6h11'34"32d37'
Heslop,Christopher R
Aur 4h59'19"40d19'
Hespel,Maria Soledad Domenech
Oph 17h55'46"7d42'
Hespell,Laura"Bright Eyes"
Cas 2h33'32"57d34'
Hesperides
Lyr 18h31'25"38d44'
Hess For Xmas 1994, Daddy Heiri
Boo 13h45'36"26d13'
Hess I,Jorja Louise
And 2h30'47"39d31'
Hess,Alexander Joseph
Cnc 8h41'0"17d37'
Hess,Cynthia
Mon 6h19'40"7d40'
Hess,Deidre Marie
Vul 21h3'50"20d36'
Hess,E Edmund
Gem 7h2'45"35d12'
Hess,Gilla
Tau 4h14'38"22d25'
Hess,Gordon
Aql 19h2'35"-0d24'
Hess,Heather Rochelle
Cas 0h25'13"60d36'
Hess,Helen L
Lyn 7h47'25"50d28'
Hess,Jonathan Louis
Her 18h5'0"28d32'
Hess,Lauren Andrea
Cas 1h38'55"65d22'
Hess,Lisa Wolfson
Lyn 7h37'41"43d17'
Hess,Macy Lee
Peg 0h0'21"30d28'
Hess,Martha
Peg 21h55'34"6d6'
Hess,Meredith
Cas 2h8'0"59d4'
Hess,Michael Anton
Aql 20h0'35"10d1'
Hess,Paul
Her 16h39'36"35d16'
Hess,Peter Clark
Her 16h58'48"31d33'
Hess,Preston Edward
Lac 22h19'1"50d24'
Hess,Ray
Ori 5h52'56"20d54'
Hess,Robert D
Boo 15h12'43"50d14'
Hess,Robert David
Cep 22h59'21"68d22'
Hess,Sarah Elizabeth
Cas 1h50'0"60d11'
Hess,Steve
Cnv 13h55'39"32d27'
Hess,Sue E
And 23h15'14"41d8'
Hess,Therese
Crb 15h17'49"30d32'
Hess,Timothy W
Leo 11h32'21"-2d6'
Hess-Fennell's Crossroads,Peg
Lyn 8h11'31"42d16'
Hess-Kerner,Michelle
Dra 19h7'24"61d19'
Hesse,Claes
Cam 8h46'41"81d39'
Hesse,Jean
Lac 22h29'30"54d7'
Hesse,Martha
Cet 1h19'15"-04d44'
Hessel,Volker George Hotstepper
Cnc 8h29'17"7d30'
Hesselbacher,Artur
Hya 8h51'45"-1d46'
Hesseltine,Deanna Jean
Lyn 8h23'14"50d23'
Hessen,Arnold
Aql 19h4'17"-1d35'
Hessen,Susan Marie
Cet 0h28'55"0d23'
Hesser,Corey
Aur 5h15'48"40d16'
Hesser,Jr,Peter L
Sco 17h1'0"-31d32'
Hesser,Lesley Woodhouse
Ari 1h51'16"10d42'
Hesser,WW
Aql 19h56'25"-0d0'
Hession,Lanell
Mon 6h36'39"11d34'
Hessions Hope
Uma 11h46'1"40d5'
Hessler,Barbara & Leroy
Cma 6h31'34"-20d25'
Hessler,Gabi Winklbauer
Cmi 7h35'23"6d20'
Hested-Gallerani
Uma 11h6'55"44d18'
Hester
Uma 9h9'14"67d49'
Hester God's Little Angel,Josie M
And 2h32'34"37d35'
Hester,John
Sct 18h41'49"-5d53'
Hester,Lisa
Tri 2h14'55"30d27'
Hester,Matthew John
Aur 7h4'1"43d20'
Hester,Stefan
Lyn 8h0'52"43d40'
Hester,Teri Lynn
Del 20h15'11"10d18'
Hesterberg,Gregory Xavier
Dra 17h42'54"64d10'
Heston,Miss Amber
Tau 3h56'42"28d15'
Hestvik,Erik
Sge 18h59'56"20d13'
Heth,Leonard R
Lac 22h32'24"55d6'
Hethea
Uma 8h35'21"58d26'
Hetherington,Barney
Dra 17h20'14"61d17'
Hetherington,Irene Laura
Lyn 7h25'0"58d27'
Hetherington,Jacquelin & Sebastian Gooch
Cyg 21h31'1"42d16'
Hetherington,Paul W
Boo 15h21'16"33d13'
Hetman,Cynthia
Ori 5h56'30"15d43'
Hetman,Milton R
Ori 5h51'0"15d38'
Hetmanski,Mary Catherine
Del 20h26'43"11d49'
Hetrick,Ray
Ori 5h55'43"16d48'
Hetrick,Ruth
Ori 5h55'15"18d15'
Hett,Wendy S
Peg 22h35'40"29d19'
Hettarachchi,Parry
Sge 19h56'0"18d46'
Hettinger,Ashley Suzanne
Vul 20h4'20"25d24'
Hettmannsperger,Günter & Marie Luise
Cyg 20h3'16"31d49'
Hettmansberger,Harlan
Oph 18h5'12"10d40'
Hetz,Dorothy
Cas 0h52'32"61d6'
Hetzel,Judy O
Lyn 6h38'0"59d26'
Hetzel,Ray
Aql 20h4'0"3d57'
Heuberger,Sam
Ser 18h2'31"-14d17'
Heuchert,Mike
Ser 15h30'42"24d37'
Heuer,Ausbern
Sct 18h43'13"-6d51'
Heuer,John L
Sge 19h54'59"16d4'
Heuer,June
Sge 19h54'42"16d14'
Heuer,Robert
Aur 6h29'28"32d43'
Heuer,Rolf-Rainer
Sgr 19h7'18"-25d17'
Heuer,Susanne
Sgr 19h3'47"-22d54'
Heuk,Kelly McKee
Lyr 18h58'0"30d52'
Heukrath,Diane
And 0h7'56"31d25'
Heule,Claire
Cas 0h34'42"68d44'
Heumann,William Anton Wicki
Cam 6h39'18"68d36
Heun,Heinz
Ser 16h5'32"2d44'
Heureka
Ori 5h53'52"15d9'
Heuring,Lori Anne
Com 12h59'39"20d59'
Heuschkel,Steven M & Paula A
Lyn 8h44'27"36d1'
Heuser,Werner
Aur 5h12'38"43d16'
Heusser,Jr,John F
Boo 14h48'0"51d5'
Heussner,George
Her 18h6'23"38d49'
Hevey,Jillian
Peg 0h1'24"22d17'
Hevia,Debbie
Del 20h13'1"10d54'
Hevly,Robbie
Gem 6h43'27"34d44'
Hewatt,Ira Mae
Cet 2h4'27"6d9'
Hewer,Marian Jane
Cas 23h26'40"60d58'
Hewer,Paul William
Cep 4h29'20"80d20'
Hewett,Helen & Jim
Cnc 8h57'17"32d54'
Hewett,Kathleen
Cyg 19h40'27"30d37'
Hewgley,Kevin
Ori 4h56'24"15d11'
Hewitson,Jane
Equ 21h20'55"11d3'
Hewitt,Allison Rebecca
Vul 19h45'0"28d26"
Hewitt,Balin M
Cet 2h7'27"4d51'
Hewitt,Bradley James
Cnv 12h21'0"47d31'
Hewitt,Dana A
Cmi 7h26'58"8d49'
Hewitt,Heather Eleanor
Tau 3h54'12"20d36'
Hewitt,Heidi Elizabeth
Tau 3h54'1"20d28'
Hewitt,Henry Cromwell
Lac 22h2'53"47d22'
Hewitt,Jackie Joy
And 2h25'33"46d51'
Hewitt,Jamey & Beth
Sge 20h2'48"-26d50'
Hewitt,Lois
Mon 7h48'45"-3d7'
Hewitt,Michael & Rebecca
Peg 22h46'0"34d24'
Hewitt,Nancy
Oph 17h33'0"8d7'
Hewitt,Nancy M
Cas 1h0'0"62d32'
Hewitt,Rebecca "Becca"
Peg 22h23'23"4d19'
Hewitt,Richard Alexander
Cep 20h16'39"60d56'
Hewitt,Richard Daddles
Her 16h48'58"8d33'
Hewitt,Roger Glenn
Dra 20h35'24"75d5'
Hewitt,Sandra A
Com 12h54'54"30d29'
Hewitt,Simon J
Per 3h2'25"40d58'
Hewitt,Stirling Alexander
Aur 7h17'1"36d29'
Hewitt,Sunny N Martin
Tau 3h54'41"22d52'
Hewitt-Widmyer,Connie Marrie
Cas 0h43'49"69d45'
Hewlett Star,The Steve
Aql 20h11'20"10d6'
Hewlett,Clifford Thomas
Aql 19h26'20"13d26'
Hewlitt,Sarah Elizabeth Cowtland
Lyr 18h35'18"43d18'
Hewson,Chappie
Aur 4h55'53"40d49'
Hewson,Marsha
Cas 1h45'31"56d27'
Hewston,Jordan James Mark
Umi 14h36'23"68d20'
HEX FF
Del 20h34'43"10d26'
Hey Baby
Sex 10h8'32"-0d40'
Hey Gorgeous
Boo 15h1'1"53d18'
Hey Princess,Don't Forget About Me
And 0h5'38"35d53'
Hey,Nanci Marie
Ori 5h18'26"-8d41'
Hey,Roxylanie Hummel
Cas 23h40'0"64d33'
Heydeck,Gloria & Georg
Cyg 19h29'1"33d22'
Heydel,Heidrun
Lyr 19h17'12"30'41'
Heyden,Gunter
Per 3h27'33"42d9'
Heyden,Randall G
Cet 3h6'26"1d58'
Heydt,Francis(Go Blue)
Peg 23h44'11"31d22'
Heyduk,Meghan Lindsey
Aqr 23h10'35"-5d10'
Heyer 6th Grade 1995-Waukesha,WI
Dra 20h6'32"64d11'
Heyl,Thomas
Aqr 23h42'25"-5d22'
Hoyma Ayma-`95
Lmi 10h36'0"24d15'
Heyman,Daniele
Lmi 9h58'34"30d17'
Heymann,Alliana
Cnc 8h57'24"28d55'
Heymann,Charles
Aqr 20h59'51"-0d13'
Heymann,Christine
Cap 20h6'49"-20d34'
Heymann,Daniel
Cnc 8h55'45"22d25'
Heymann,Leo
Sgr 19h14'23"-26d34'
Heymann,Leon
Sco 16h49'19"-28d40'
Heymann,Peter
Equ 20h57'0"8d14'
Heymann-Deutsch,Herta
Tau 3h45'28"0d46'
Heymer-Chervenyak, Janet
And 1h23'17"39d42'
Heyse,Donald G
Ari 1h45'13"22d3'
Heyse,Robert W
Ari 1h46'1"21d53'
Heyward,Peter & Kathryn Maher
Her 16h27'19"20d6'
Heywood,Dennis
Per 1h52'41"47d38'
Heywood,James David
Cep 22h50'10"63d38'
Heywood,Mr & Mrs John Adams
Crb 16h6'20"26d36'
Heß,Gerda
Vir 11h51'32"-4d53'
HG Liebe meines Lebens um 89
Lyn 4h28'33"40d12'
hgll
Cam 3h19'37"67d16'A
HH-1
Her 17h8'14"50d12'
Hi
Ser 15h12'29"7d59'
Hi Beautiful
Mon 6h54'16"-10d6'
Hi-Co
Ori 4h57'29"-0d33'
Hi-Steppers & Krew
Dra 16h41'46"62d44'
Hiatt,Amy D
And 23h39'19"48d24'
Hiatt,David
Sex 9h55'14"-5d57'
Hiatt,Ronald William
Per 3h7'22"40d5'
Hiatt,Susan
Del 20h16'41"11d50'
Hibbard,Courtney
Peg 21h42'48"23d2'
Hibben,Hajo Karl Ludwig
Dra 16h14'27"66d38'
Hibbert Star,The
Eri 4h2'56"-8d17'
Hibbert,Pop Pop
Lac 22h52'44"54d23'
Hibbett,Laverne
Mon 8h2'19"-4d9'
Hibbett,Mary Rachel Fasci
Del 20h32'46"6d6'B
Hibbits,Carol A
And 0h20'57"33d31'
Hibdon,Gregory Edward
Oph 17h36'33"-24d1'
Hibdon,Jr,Charles Edward
Hya 8h16'1"-6d50'
Hiben,Jim & Chris
Cyg 20h39'22"53d7'
Hibler,Ralph Edward
Her 18h10'34"37d39'
hibus der spiegelfeste Querbinder
Cep 22h2'36"68d39'
hibus der spiegelfeste Querbinder
Her 17h19'22"48d51'
Hice,Patricia
Cyg 21h3'50"38d36'
Hichens,Jenni
Cas 2h43'15"60d46'
Hickam,Wayne
Ser 15h57'42"24d56'
Hickel,Wilhelm
Cep 21h32'27"55d29'
Hickerson's Star,Doc
Per 1h47'53"47d38'
Hickey
And 0h57'46"45d34'
Hickey's Poo,Karen
And 23h18'58"44d6'
Hickey,"Gus" & Michael R
Cam 5h10'48"68d24'
Hickey,Alan Christopher
Cep 1h23'26"78d20'
Hickey,Arlene
Vir 14h1'0"-2d30'
Hickey,C J
Cam 3h52'32"56d34'
Hickey,Christopher John
Lyn 8h13'52"51d0'
Hickey,Christopher Lawrence
Aur 6h19'11"38d15'
Hickey,Cindy Robin
Lyn 7h31'0"58d23'
Hickey,Denis
Her 17h19'44"48d22'
Hickey,Gary Martin
Pho 23h43'55"-40d57'
Hickey,John
Her 17h28'33"38d6'
Hickey,Jr,Ronald J
Aur 6h0'0"38d50'
Hickey,Laura Smilanich
Peg 22h14'17"32d32'
Hickey,Marie E
Lyr 18h13'36"35d8'
Hickey,MD,Ann
Lyn 7h50'52"37d18'
Hickey,Nell
Cas 1h8'41"75d16'
Hickey,Steve & Nancy
Aur 5h1'44"29d6'
Hickey,Stewart Glenn
Peg 22h25'45"17d0'A
Hickey,Thomas J.
Cet 1h33'22"-12d4'
Hickie,David John
Lyn 9h8'37"37d37'
Hickli,John A
Ori 6h7'45"20d30'
Hickman Star
Boo 14h20'54"29d10'
Hickman,Bertha P
And 23h11'23"39d57'
Hickman,Inshallah Delores
Uma 10h49'10"51d53'
Hickman,Jennifer S & Richard J
Uma 11h58'1"43d8'
Hickman,Johanna
Mon 6h19'44"7d60'
Hickman,Jr,Barnes B
Hya 8h52'1"-0d8'
Hickman,Kayleigh
Vul 19h45'38"23d6'
Hickman,Liberty Ella
And 23h24'17"40d47'
Hickman,Patty
And 0h11'14"47d12'
Hickman,Samuel
Lac 22h38'52"54d21'
Hickman,Stephen
Hya 9h8'44"4d35'
Hickman,Susan
Lyr 18h23'26"41d10'
Hickman-Sunshine,Jim
Hya 9h4'12"3d52'
Hickmann,Rick Edward
Ori 05h37'29'-00d56'
Hickok,Adam
Mon 6h6'18"-8d46'
Hickok,Kate
Aql 20h9'32"6d1'
Hicks Family,The
Cma 6h54'54"-19d49'
Hicks' Halo
Uma 9h2'41"51d6'
Hicks,Alayna
And 0h11'18"33d7'
Hicks,Brian Lewis
Lac 22h31'58"55d22'
Hicks,Caitlin Marie
Peg 21h20'29"23d8'
Hicks,Carey Neal
Sco 16h37'20"-30d22'
Hicks,Carol Ann Manafort
Lyr 18h33'0"34d29'
Hicks,Carroll
Lyr 18h58'1"47d38'
Hicks,Catherine Elizabeth
Cnc 8h31'49"8d10'
Hicks,Christie N
Lyr 18h37'53"40d9'
Hicks,Edna
Cyg 19h17'38"45d16'
Hicks,Edwin Michael
Dra 14h20'31"63d30'
Hicks,Emily
Lyn 9h6'32"38d20'
Hicks,Frank William (Billy)
Vul 19h39'1"27d5'
Hicks,James Marshall
Oph 18h17'48"11d20'
Hicks,Jason
Per 2h30'53"56d54'
Hicks,Jeannie B
And 0h22'13"36d19'
Hicks,Jessica Ann
Lyn 7h3'14"60d55'
Hicks,Jim
Cep 22h50'56"57d28'
Hicks,John Phillip
Her 16h23'60"38d28'
Hicks,John William
Cyg 19h33'21"39d29'
Hicks,Kate
Lyr 18h50'56"34d6'
Hicks,Kathryn
Mon 7h0'15"4d48'
Hicks,Larnce D
Aql 19h39'53"14d44'
Hicks,Larry Wayne
Oph 17h56'28"12d6'
Hicks,Margaret
Peg 22h39'30"25d37'
Hicks,Margaret L
And 0h21'39"30d43'
Hicks,MD,Jesse Robinson
Oph 17h38'52"-16d47'
Hicks,Pat
Mon 8h5'20"-6d60'
Hicks,Robert Merritt
Her 17h2'57"48d48'
Hicks,Robin Lee Moran
Peg 23h30'56"31d52'
Hicks,Ron
Cet 0h27'48"-11d35'
Hicks,Sarah
Cas 1h1'52"61d26'
Hicks,Snuffy
And 2h25'1"41d32'
Hicks,Stanley
Aql 18h57'54"-0d29'
Hicks,Susanne
And 0h1'0"46d12'
Hicks,T C
Aur 4h50'32"40d23'
Hicks,Tamara Lyn
And 0h58'11"37d7'
Hicks,Thelma
Cyg 21h18'23"31d18'
Hicks,Thomas Ollis
Equ 21h4'45"2d33'
Hicks,Victor
Her 14h48'11"48d46'
Hickson,Brandon Wayne
Peg 22h25'25"20d5'
Hickson,Raymond
Cep 22h19'47"62d17'
Hidaji,Rhonda & Fred
Cyg 20h21'25"38d23'
Hidaka,Yoshimasa
Cet 3h14'25"2d57'
Hidalgo,David Augusto
Per 2h44'24"40d49'
Hidalgo,Glena Valdivieso
Com 12h45'0"30d2'
Hidalgo,Miguel Angel Gonzalez
Her 16h28'36"40d45'
Hiddleston,Carla & Craig
Peg 23h0'0"8d29'
Hide,Stephanie
Aql 18h56'17"8d5'
Hideaki & Yoko
Vir 14h25'23"1d32'
Hidejo,Kanzaki
Cas 22h56'58"55d54'
Hideki
Cap 20h53'42"-24d36'
Hideki,Mori
Sco 16h17'51"-22d28'
Hideyuki
Cap 20h57'27"-23d54'
Hieatt-Dalton,Barbara Ann
Cmi 7h19'31"1d0'
Hiebert,Peter
Oph 17h18'3"-20d8'
Hiebert,Sylvia
Sge 19h43'56"16d43'
Hier,William
Aql 20h0'35"4d15'
Hierlmeier,Judie & Larry
Crb 15h27'40"31d60'
Hieromniman,Paul
Hya 8h10'0"-0d5'
Hieronimus,Heink
Oph 17h29'34"8d15'
Hieronymus,Lynn
And 2h8'52"50d56'
Hierosolyma
And 0h16'58"39d25'
Hiers,Nicholas Ryan
Sex 10h22'32"-5d33'
Hiers,Susan Lea Bowman
Peg 23h1'35"14d39'
Hiestermann,Sylvia
Lib 15h29'51"-28d17'
Hietala,Justin Michael
Umi 16h0'35"71d46'
Hietamies,Heikki
Cam 4h18'19"67d52'
Hiett,NaNaMay
Peg 23h46'27"17d5'
Higdon 296,Charlotte
Cet 0h52'23"1d50'
Higdon,Keith
Cmi 7h29'16"0d32'
Higdon,USFA,RET,Major Thomas N
Ori 5h0'34"0d0'
Higgenbotham,Mickey
Hya 8h9'39"0d37'
Higginbbottom Mizpah, Colin
Aur 6h28'34"38d51'
Higginbotham,Hollie
Mon 8h0'11"-10d33'
Higgins,Andrew Colin
Ari 2h56'26"22d10'
Higgins,Antony B
Boo 13h58'45"12d10'
Higgins,Barbara Elizabeth
Uma 9h8'12"49d28'
Higgins,Bob
Her 18h1'15"14d6'
Higgins,Brian
Cep 22h13'30"62d30'
Higgins,Butch
Aur 6h27'0"37d51'
Higgins,Carolyn Thanasea
Mon 6h37'1"10d22'
Higgins,Catherine & Michael
Cyg 19h24'0"31d49'
Higgins,Claire Anne
Mon 6h23'58"-10d42'
Higgins,Eileen
And 2h5'32"37d15'
Higgins,Jr,William Clyde
Uma 9h9'11"49d24'
Higgins,Kearsley
Cam 3h28'50"58d37'
Higgins,Kyle
Cet 2h11'51"1d44'
Higgins,Malachi Edward
Her 16h47'26"18d48'
Higgins,Patrick W
Cep 21h19'1"55d12'
Higgins,Robert John
Uma 9h10'0"49d48'
Higgins,Ronald F
Aql 19h0'21"12d30'
Higgins,Scott Allen
Boo 15h20'45"38d7'
Higgins,Scott E
Dra 20h6'1"64d53'
Higgins,Sharon Louise
Uma 9h8'1"49d51'
Higgins,Shawna Ann
And 23h33'44"35d31'
Higgins,Tad
Aur 5h14'19"40d25'
Higgins,Tara
Peg 23h33'1"30d56'
Higgins,Tara Marie
Cyg 19h28'15"38d51'
Higgins,Trudy
Cyg 20h29'49"40d18'
Higgins,Vera
And 23h38'33"40d41'
Higgins,Vincent J
Cnv 13h45'58"37d5'
Higgins,William Clyde
Uma 9h9'57"49d39'
Higgins,Wilson P & Betty Lou
Cyg 20h54'39"39d45'
Higgins-50 not out!, Christopher
Cep 22h21'59"56d57'
Higgins-Witch II, Donna
Tau 5h45'20"28d20'
Higginson - Oso Blanco Gary
And 23h35'60"33d50'
Higginson,Maria Kimberley
Lyr 18h29'50"47d31'
Higginson,Muir
Cyg 21h19'50"35d31'
Higginson-Kansas Girl, Marcia
And 23h48'0"33d1'
Higgs,Jim David
Ori 5h22'31"0d27'
Higgs-"Drew",Andrew Allan
Leo 10h0'14"10d1'
High Hopes
Peg 23h0'0"18d7'
High Passion
Lyr 19h23'1"38d32'
High,Garry Philip
Aql 19h4'1"0d0'
High,Lesley
Com 13h12'24"28d33'
High,Roberta
Peg 22h22'45"21d33'
Higham,David
Aur 6h21'41"32d22'
Higham,Matthew
Cep 23h11'26"62d53'
Highest Laurel
Ori 5h59'26"10d49'
Highfill,Angie"Panda"
Tri 1h50'33"25d52'
Highlander Racer
Peg 21h10'20"10d57'
Hight,Courtney Elise
Mon 7h48'45"-2d30'
Hight,Joan & Norton
Crb 16h11'38"32d9'
Hight,Shirley Anne
Sgr 18h3'59"-28d57'
Hightower Family Star, The
Uma 13h58'13"60d17'
Hightower,Doris Irene
And 23h17'48"47d23'
Hightower,Teratha
Uma 10h10'11"52d20'
Hightower,Thomas Stephen
Cet 2h59'0"1d22'
Higley,Moran Michael
Aql 19h20'53"15d1'
Higley,Steven J
Cet 2h55'11"6d46'
Higney,Paul A M
Per 3h12'37"50d19'

Hignite,Debra S
Cas 0h34'59"69d57'
Higson,Gregg Duncan
Ori 5h32'38"-1d10'
Higuera,J Gilberto
Per 1h48'52"53d19'
Higy,Chantal
Uma 13h41'14"61d35'
Higy,Stella
Mon 7h43'11"-4d9'
Hijjawi,Ayman Hisham
Hya 9h20'24"-9d37'A
Hikaru
Tau 5h26'38"19d59'
Hikaru Genji
Cap 20h50'26"-23d0'
Hilarion & Jennifer
Leo 9h22'37"19d6'
Hilary
Peg 22h37'0"11d12'
Hilary Ann's Nova
Vir 13h38'48"-4d22'
Hilary Kristen
Sge 19h5'42"19d46'
Hilary Star
Cet 1h37'32"-10d41'
Hilary-S 'Nur'
Cas 0h39'23"72d34'
Hilaryflo
Lyr 19h0'17"26d50'
Hilber,Hans Reiner
Cas 1h56'45"60d46'
Hilber,J Alison
Cnv 13h1'46"37d46'
Hilbert,Jeremy Tyler
Sex 10h7'1"-6d11'
Hilbert,Steven Austin
Boo 15h2'31"10d13'
Hilbun,Dr William
Boo 13h41'18"17d50'
Hilburn,Socrates Theodore
Ori 5h46'15"20d44'
Hilda
Lyn 9h6'36"44d25'
Hilda Maude
Crb 16h0'11"27d35'
Hilda Mildred
And 23h6'0"37d49'
Hildebrand,Diana Kaye
Oph 16h50'26"-26d2'
Hildebrand,Don
Per 3h30'1"51d55'
Hildebrand,Dr Wilbur
Dra 11h26'46"70d20'
Hildebrand,Drew & Jenny
Cyg 21h19'17"36d49'
Hildebrand,Jana
Aql 18h40'1"-2d2'
Hildebrand,Joyce B
Peg 21h58'50"33d24'
Hildebrandt 94
Aur 4h36'1"34d32'
Hildebrandt,Alena
Lyr 18h58'0"42d1'
Hildebrandt,Heide
Uma 11h4'36"50d55'
Hildebrandt,Marilyn
Cam 3h31'20"61d20'
Hildebrandt,Rüdiger
And 23h33'27"38d55'
Hildegard
Cnc 8h32'57"8d41'
Hildegard 01 06 1964, Ganser
Per 4h3'32"50d46'
Hildegard-Stern Der Liebe
Cap 21h16'1"-26d56'
Hildegarde
Cam 3h17'30"67d52'
Hildenbrand,Dennis J
Lac 22h54'44"50d20'
Hildenbrand,Eugen
Mon 7h54'10"-1d14'
Hildenbrand,Helmut
Cyg 19h52'38"48d39'

Hildenbrandt,Russel Victor
Dra 16h51'0"73d55'
Hilderbrandt,Alexander
Dra 17h37'12"60d15'
Hilderbrandt,Gregory
Lac 22h22'19"56d49'
Hilderic
Cnv 12h10'18"40d18'
Hilderman Sterne
Cep 20h20'0"60d13'
Hilditch,James Stephen
Umi 13h57'24"70d30'
Hildner,Lynna
And 0h3'59"38d33'
Hildreth,Adam Charles
Her 17h9'1"49d38'
Hildreth,Chris
Per 4h27'38"51d31'
Hildreth,John F James P &
Grace M
Lac 22h1'36"49d39'
Hildreth,Mayor Patrick A
Her 16h41'57"29d20'
Hildreth,Richard
Aql 19h9'1"15d02'
Hildy
Mon 6h24'31"4d56'
Hildy
Cet 1h32'24"-14d54'
Hilferty,Christine E
Crb 15h43'47"27d45'
Hilfiker,Wesley Holden
Aql 20h6'15"4d47'
Hilgarth,Eduard
And 23h11'33"40d16'
Hilgers,Angelika
Sco 17h27'35"-31d14'
Hiljus,Wallace (Curly)
Hya 8h13'46"3d44'
Hilkert,Hunter
Ori 4h42'35"14d48'
Hill
Vir 11h55'14"-3d6'
Hill "President Of A C C,"Patti
Com 13h32'21"22d22'
Hill "Trinkle Little Star",Trink
Sgr 20h3'30"-26d50'
Hill 9-3-95,Maribeth & Mike
Ori 5h53'12"9d4'
Hill Family,J David & Virginia
J
Aql 19h51'52"15d58'
Hill's Wishing Star, Skip &
Bari
Crb 16h15'29"28d56'
Hill(Ycnan),Nancy Jane
Lyr 18h49'48"41d41'
Hill,Addison & Irene
Cas 1h16'14"67d58'
Hill,Amanda Jane
And 0h43'1"31d49'
Hill,Amanda Louise
Aqr 23h2'27"-20d32'
Hill,Ariel
Lyn 8h52'23"41d52'B
Hill,Arthur Christopher
Her 17h7'22"20d37'
Hill,Arwood
Oph 17h4'29"11d33'
Hill,Baird
Hya 8h56'37"4d43'
Hill,Betsy Hatch
Umi 15h34'20"70d59'
Hill,Bobby Joe
Boo 15h0'44"47d31'
Hill,Brenda Kaye
Peg 21h44'54"24d17'
Hill,Brooke Arden
Boo 14h55'54"45d59'
Hill,C P & Betty
Peg 23h7'39"14d2'
Hill,Candace "Candy"
Cam 6h53'0"70d46'
Hill,Carol C
Peg 22h58'13"29d0'

Hill,Catherine Louise
Cyg 21h33'26"41d7'
Hill,Charles James
Lac 22h47'18"37d46'
Hill,Charlotte Michelle
Lyr 19h16'49"25d50'
Hill,Charlotte (SAM)
Eri 3h49'0"-3d4'
Hill,Chris
Uma 9h56'17"54d24'
Hill,Chuck & Pat
Eri 4h17'31"-11d54'
Hill,Cindi
Mon 8h5'1"-1d47'
Hill,Claire B
Cas 2h58'38"61d8'
Hill,Clinton
Gem 7h17'3"21d13'
Hill,Craig
Ser 18h18'1"-0d7'
Hill,Darcy L
Dra 19h33'17"60d30'
Hill,David Devereux
Her 18h42'12"12d31'
Hill,David Eugene
Her 18h15'28"48d34'
Hill,Davy Lee
Aql 20h12'26"14d9'
Hill,Dawn & Seth's Joy
Eri 3h57'42"-16d51'
Hill,Diane R
Aql 19h53'53"11d49'
Hill,Doreen Millicent Marina
Cyg 19h29'58"35d15'
Hill,Eileen
Lyr 19h26'23"38d13'
Hill,Eleanor Estelle
Com 12h30'39"30d23'
Hill,Elizabeth Lee
Oph 18h39'0"6d26'
Hill,Elizabeth U
Lyn 8h9'46"57d35'
Hill,Emma
And 1h21'10"39d11'
Hill,Eric
Boo 14h28'28"52d42'
Hill,Ernie
Ori 5h30'14"-8d46'
Hill,Etta
Mon 6h19'32"7d42'
Hill,F Charles
Dra 16h12'26"63d6'
Hill,Faith
Cas 3h10'55"65d26'
Hill,Fawn
Cyg 19h7'0"50d25'
Hill,Gareth Griffin
Aql 18h57'26"-4d31'
Hill,Gary
Aur 4h50'44"41d4'
Hill,Gary Dean
Her 18h17'15"14d30'
Hill,Gcoff D
Cyg 21h34'3"29d49'B
Hill,Gisela
Cnc 8h37'18"18d12'
Hill,Grandma
Eri 3h55'45"-8d0'
Hill,Heather
Mon 6h2'0"-10d17'
Hill,Helen H
Sex 10h48'32"1d20'
Hill,Herbert Ray
Aur 6h14'26"31d54'
Hill,J Michael
Her 17h29'35"21d35'
Hill,James & Marion
Cnv 13h52'55"40d21'
Hill,James Allen
Aql 19h4'18"0d24'
Hill,Janie Alice Arnett
Cas 1h31'58"61d33'
Hill,Jean Shevis
Eri 4h25'29"-1d51'

Hill,Jennifer Lynn
Peg 0h3'13"31d15'
Hill,Jessica Harley
Lyr 18h29'56"46d1'
Hill,Joseph E
Leo 9h20'41"27d58'
Hill,Jr,Dr Arthur Ted
Her 17h2'10"51d10'
Hill,Justin James
Lib 15h17'29"-22d21'
Hill,Katherine Marie
Eri 4h9'56"-10d16'
Hill,Kathleen Margaret
And 0h45'50"28d32'
Hill,Kathryn C
And 23h31'50"44d37'
Hill,Keith F
Sex 9h59'50"-1d44'
Hill,Keith Richard & Ann
Elizabeth Hill
Cyg 21h5'41"30d31'
Hill,Kenneth Richard
Cam 3h52'21"61d28'
Hill,Kevin Patrick
Cet 0h53'25"-6d1'
Hill,L Jay
Hya 9h31'14"-2d24'
Hill,Landon Chloe
Oph 18h6'35"13d10'
Hill,Leigh B
Cyg 21h34'3"29d49'A
Hill,Lill
And 23h36'48"48d36'
Hill,Lisa A
Vul 19h43'48"23d35'
Hill,Lori Elizabeth
Ori 5h54'46"16d26'
Hill,Lucille
Aur 4h36'52"34d60'
Hill,Mallory
Lyn 7h27'21"37d3'
Hill,Mara Lee
Mon 6h54'15"7d54'
Hill,Martynn & Robert
Eri 3h29'30"-7d10'
Hill,Michael
Cnv 12h24'54"42d37'
Hill,Michael Christopher
Her 17h37'16"14d45'
Hill,Michael G
Aur 5h28'14"40d7'
Hill,Mike
Her 16h20'45"40d26'
Hill,Mildred Dawe
Aqr 21h28'1"-1d39'
Hill,Naomi Dorothy
Aql 20h0'1"-5d40'
Hill,Nicholas
Aur 7h22'23"40d40'
Hill,Nicole
Lyn 8h52'23"41d52'A
Hill,Nina Jane
Cas 0h50'25"63d3'
Hill,Patricia & Sanford
Eisenberg
Sct 18h31'54"-4d45'
Hill,Perry Mack
Sex 10h29'33"5d22'
Hill,Perry Wayne
Her 18h12'32"30d2'
Hill,Peter
Boo 13h40'60"20d4'
Hill,Phyllis & Brecke
Cyg 20h8'55"38d53'
Hill,Pippa
Cyg 20h0'38"30d10'
Hill,Rene
Cas 1h26'38"75d52'
Hill,Richard Anthony
Ori 6h5'59"4d57'
Hill,Richard B
Cet 1h1'58"-1d56'
Hill,Robert C
Mon 7h5'55"-5d7'

Hill,Robert E
Aur 6h17'2"31d4'
Hill,Ronni
Cep 20h16'22"76d24'
Hill,Ryan
Lac 22h20'31"52d38'
Hill,Ryan Daniel
Mon 6h22'57"8d56'
Hill,Sam & Pegg
Cyg 19h27'13"31d6'
Hill,Samuel Craven
Aql 19h28'36"1d23'
Hill,Sarah Kate Monica
Cyg 19h33'59"34d8'
Hill,Sena
Del 20h56'29"10d22'
Hill,Shannon
And 23h22'0"40d48'
Hill,Sharon Michelle
Cas 0h29'29"50d31'
Hill,Sheila N
Leo 9h27'19"28d12'
Hill,Starr
Lmi 10h41'46"25d24'
Hill,Stephen Ashley
Boo 13h38'37"20d38'
Hill,Steven G
Lac 22h18'17"47d35'
Hill,Tanner
Hya 8h57'41"2d35'
Hill,Terry
Per 2h16'20"58d21'
Hill,Tim
Ser 15h28'46"10d27'
Hill,Timothy
Per 3h36'13"38d38'
Hill,Valerie C
Uma 11h22'19"41d56'
Hill,Vanessa Dyanne
Cet 2h41'15"-1d52'
Hill,Victoria Sue
Cam 6h31'24"80d30'
Hill,Wanda Diane
Sct 18h54'45"-10d44'
Hill,William Christopher
Cep 0h20'53"78d22'
Hill,William E
Boo 14h57'0"43d4'
Hill,William Jesse Perry
Oph 18h2'40"-8d45'
Hill,Wind Songs Bonnie
Krystal
Mon 6h40'52"10d50'
Hill-Bruder
Vul 20h26'38"28d58'
Hill-Crim,Margaret Angela
Peg 22h42'26"27d52'
Hill-Quimby,Karen Elizabeth
Uma 10h25'27"51d35'
Hillabrandt,Jr,Larry
Cep 21h56'22"56d1'
Hillard,Bill
Hya 8h14'34"4d41'
Hillard,Charlie Love
Ser 15h15'40"19d24'
Hillard,Jeffrey Lewis
Cas 0h25'1"63d57'
Hillary
And 0h6'0"40d23'
Hillary Cheré
Lyr 18h25'37"40d24'
Hillary Paige
Lyr 19h19'55"41d14'
Hillary The Brittany
Sge 19h30'42"16d22'
Hillbilly A,The
Ser 15h14'18"8d54'
Hille,Wolfgang
Uma 9h28'43"46d36'
Hillebrand,Rolf A
Cet 0h53'54"0d20'
Hillegas,Sarah Brian Delbert &
Cathy
Eri 4h37'25"-0d16'

Hillemeyer,Jean & Pattie
Cyg 20h38'14"42d4'
Hillenbrand,Jordan Michael
Aur 6h23'24"30d1'
Hiller,Autocratos Georg
Gem 8h3'12"30d32'
Hiller,Roz
Crt 10h55'54"-8d23'
Hiller,Susan Margaret Anne
Uma 8h36'1"47d8'
Hillery & Chris Forever
Crb 16h5'15"38d44'
Hillery,Maria Francesca
Cas 23h19'52"60d16'
Hilley II,Ronald Lamar
Gem 6h49'23"14d32'
Hillger,Jan & Lin
Lyn 7h49'1"50d14'
Hillhouse I,Pamela
Mon 6h55'23"8d6'
Hillhouse,Professor Edward
Uma 9h48'22"46d48'
Hilliar,Grace
Cas 2h0'21"61d30'
Hilliard,Harold I
Del 20h36'21"20d23'
Hilliard,John Mauk
Her 17h15'20"46d27'
Hillier,MAH Marjorie Arlene
Uma 8h39'46"56d50'
Hillier,Susanne
Peg 23h33'12"10d0'
Hillier,Whitman
Cam 6h11'32"60d31'
Hillila,Lee Oliver
Ori 6h15'11"-0d9'
Hillis,Michael Guy
Lac 22h17'31"55d13'
Hillis,Susan Diane
Mon 6h28'1"11d12'
Hillje,Jurgen
Eri 4h56'9"-8d15'
Hillman,Amber Marie
Cyg 21h31'14"34d14'
Hillman,Donovan Bruce
Oph 17h4'32"8d10'
Hillman,Hayley Paige
Del 20h18'19"18d51'
Hillman,Jennifer
Peg 21h19'13"22d45'
Hillman,Kendra Terese
Peg 22h31'54"8d32'
Hillman,Richard & Betty Jane
Cyg 21h27'13"38d4'
Hillman-DeWitt,Emily
Oph 18h0'54"13d21'
Hillmer,Thomas Patterson
Her 17h5'1"38d43'
Hillner,Jeffrey Charles
Aql 19h46'48"10d29'
Hills (30),Kay
Cas 0h25'57"50d26'
Hills,David Farrell
Lac 22h14'41"48d29'
Hills,Sandra Kent
Lyn 8h59'47"41d47'
Hills,Tracie Michelle
Mon 7h51'10"-2d54'
Hills,William Gregory
Lyn 8h11'23"43d11'A
Hillson,Edward Nicholas
Oph 16h59'38"10d44'
Hillstrom,Cur
Cma 6h27'0"-18d53'
Hilltop Peppercorn
Cmi 8h8'1"4d12'
Hilly
Peg 23h0'32"21d10'
Hillyard,Don
Sex 9h52'0"-0d38'
Hillyard,Robert Dean
Aql 18h54'55"8d11'
Hillyer,Alexzandra
Lyn 7h44'47"44d29'

Hilsee's Star,Scott R
Uma 10h36'1"40d10'
Hilsenbeck,John
Dra 14h49'15"61d6'
Hilson,Gary
Aur 5h17'16"41d58'
Hilson,Stephen S P
Dra 14h23'1"64d18'
Hilston,James & Sharon
Lyn 7h27'59"50d6'
Hilton
Her 17h7'29"41d16'
Hilton
Cyg 20h38'1"45d57'
Hilton & Paula
Pho 0h46'55"-47d41'
Hilton's Halo
Uma 11h40'44"43d54'
Hilton,Andrew"Fluffy"
Per 4h2'27"50d41'
Hilton,Angela Kay
Mon 6h42'49"11d33'
Hilton,Arthur & Lorraine
Cyg 21h36'22"41d14'
Hilton,Bonnie Sue
Ori 5h57'10"14d5'
Hilton,Edward R
Cep 21h55'28"55d7'
Hilton,Jr,Donald James
Ori 6h17'4"8d15'
Hilton,Kathleen
And 23h1'18"47d43'
Hilton,Marilyn
Peg 23h0'24"20d25'
Hilton,Pat
Ori 6h3'51"20d18'
Hilton,Paula
Mon 6h22'55"4d36'
Hilton,Randy
Ori 4h54'27"-0d5'
Hilton,Sarah
Lyr 18h16'50"38d26'
Hiltrud
Sco 17h26'52"-31d26'
Hilzenkopp,Claude
Mon 7h58'43"-4d28'
Himanen,Harri Ja Eija
Cam 3h22'44"60d50'
Himawari
Sco 16h6'0"-29d10'
Himebrook,Leslie S
Mon 7h55'8"-1d29'
Himelfarb,Steven
Dra 16h51'14"70d43'
Himer's Shining Star
Umi 15h56'31"80d13'
Himer's Spell
Lyr 19h22'40"41d1'
Himer,Etsi
Hya 8h24'0"-0d51'
Himes,Cecelia Constance
Canning
Mon 8h5'15"-1d37'
Himes,Cody
Mon 7h43'46"-2d1'
Himie's Hideaway
Sct 1h55'53"-6d47'
Himmel
Peg 22h45'35"29d6'
Himmelmayer,Renate Turtle
Uma 10h29'43"51d47'
Himmelsstern für Wolfgang
Cas 0h3'56"60d45'
Himmelwright,MD, James P
Cep 23h13'46"60d22'
Himmlische Wölfin,Die
Cet 0h35'37"-4d44'
Himsworth,Rebecca Anne
And 23h19'53"41d14'
Himsworth,Victoria Alison
Cyg 20h1'0"30d37'
Hince,Lieutenant Colonel
Geraldine M
Cyg 20h21'1"39d30'

Hince,Natalie Jane Bough
And 1h42'54"40d36'
Hince,Ronald A
Per 4h0'53"50d41'
Hincelin,Marilyne
Lyn 7h52'53"58d25'
Hinch,Jr,Star of
Andrew/Andrew Lewis
Sct 18h55'15"-6d47'
Hinch,Louise
Cyg 20h24'55"41d15'
Hinch,Star of Joan- Joan
Christine
Leo 11h31'26"-2d13'
Hinchcliffe,Jean M
Cam 6h36'37"68d39'
Hinchey,Brian E
Cep 21h1'1"60d38'
Hinchey,Michael
Her 17h18'29"41d32'
Hinchliffe,Bob
Ori 6h15'36"8d32'
Hinchliffe,Gary
Aur 6h9'32"45d13'
Hinchliffe,Marie & Joe
Cyg 21h30'0"38d24'
Hinchmiffe My Guiding
Light,Forever Robert
Cep 21h3'1"55d60'
Hinckley,Philip David
Dra 15h55'27"52d28'
Hind,David Vincent
Ori 5h55'41"15d46'
Hind,Peter
Lyr 18h32'15"33d3'
Hinde,Lucy
Lup 15h4'17"-42d23'
Hindle,Debra Louise
Aql 19h13'20"15d16'
Hindman,Sherry Lynn
Cet 3h14'18"0d26'
Hindmarch,Clementine C
Lyr 18h31'19"34d26'
Hindmarsh,John Werner
Per 2h24'17"54d57'
Hindra
Ser 15h16'50"8d16'
Hinds Love Forever, Wrey &
Connie
Sge 19h35'0"16d24'
Hinds,Claudia-Donald-
Marion & Claude
Mon 6h23'50"-1d40'
Hinds,Janice (E B)
Del 0h16'1"10d24'
Hine-Phillips,Sarah Jane (5-
15-34)
Mon 7h16'29"-6d0'
Hinegardner,Michael
Aql 19h0'12"5d5'
Hines A Loving Husband-
Dad-Gramps,Bob
Lyn 7h55'1"43d29'
Hines MD,David
Sgr 19h31'47"-42d17'
Hines,"Teddy-Mumble Bear"
Kevin
Ori 4h53'50"1d38'
Hines,Alice Snow
Mon 8h3'13"-1d38'
Hines,Caleigh Jo
Peg 23h0'30"25d20'
Hines,Christopher
Oph 16h53'55"10d53'
Hines,Dana Michelle
Lyn 8h40'55"37d27'
Hines,Elibear Logan
Uma 9h38'37"52d45'
Hines,Gail
Lyr 19h6'30"28d25'
Hines,Gregory
Peg 21h22'27"18d50'
Hines,Heidi R
Crb 15h24'14"32d14'

Hines,Joni Lynn
Cet 1h54'41"-0d9'
Hines,Kelly M
Mon 6h27'59"11d59'
Hines,Leslie Anne
Sge 20h1'55"20d29'
Hines,Peter
Umi 15h36'19"71d44'
Hines,Rachel Morgan
Tau 3h54'15"30d38'
Hines,Susan Jean
Lyn 6h15'54"58d54'
Hines,Tammi Lynn
Lyr 18h59'45"31d7'
Hines,Thomas Fritz
Gem 7h2'43"30d1'
Hines,William Edgar
Her 16h33'34"20d30'
Hiney,Barbara
Cyg 21h1'0"28d46'
Hininger,Ann
Peg 22h12'1"34d40'
Hink,III,Charles "Chaz" Levi
Aur 5h1'12"41d12'
Hink,Janis Wheeler
Lyr 19h4'53"38d60'
Hinkebein,Michael
Ori 5h57'56"20d56'
Hinkel,Christopher James
Dra 17h4'32"62d59'
Hinkeldeyn,Lothar
Aur 5h11'41"43d7'
Hinkes,Rick
Her 18h7'35"14d49'
Hinkle,C Austin & Heather
Hartley
Lyr 19h0'26"30d20'
Hinkle,Curt
Sex 9h44'12"2d21'
Hinkle,David O'Neal
Cmi 7h23'37"0d23'
Hinkle,Jack M
Nor 15h47'12"-44d54
Hinkle,Monty Darden
Ser 18h0'13"-14d1'
Hinkle,Steve
Aur 6h30'58"38d31'
Hinkley,Annie
Mon 6h57'12"10d20'
Hinkley,Janet Ann
Peg 22h49'53"8d21'
Hinks,Larry A
Aql 19h57'36"0d51'
Hinksman,James Lowe
Cep 22h45'14"63d37'
Hinlay,Michael P
Cnc 8h35'22"15d13'
Hinman,Alexander Nicholas
Lac 22h3'32"46d38'
Hinman,Helen Boyle
Mon 6h56'48"-5d56'
Hinman,Jr,John Harold
Her 17h4'0"42d36'
Hinman,Ken
Uma 11h25'37"40d2'
Hinnant,Kai William
Vir 13h31'59"12d24'
Hinojosa,Martin
Sge 19h53'13"18d53'
Hinoto
Sgr 19h10'9"-27d59'
Hinrichs "Tommi 40", Thomas
Dra 16h45'12"65d58'
Hinrichs,Corie Eliese
Peg 0h8'39"22d8'
Hinshaw,Donna L Kaneshiro
Lyn 8h22'54"51d23'
Hinshaw,Richard
Aur 5h52'56"38d42'
Hinshaw,Stephen Leon
Dra 18h32'22"58d15'
Hinskton,Jeremy
Per 3h24'13"38d44'
Hinson,Dana R
Cas 3h4'0"60d38'

Hinson,Jr,Sherwood F
Aur 5h8'0"44d57'
Hinson,Sr,Sherwood F
Aur 5h8'40"44d58'
Hinterman,T Andrew
Boo 15h20'0"38d31'
Hinton,Alice J
Sge 19h58'20"20d25'
Hinton,Barbara
Mon 6h58'21"10d56'
Hinton,Claude
Per 2h52'11"37d4'
Hinton,Jackie
Cas 3h7'25"57d22'
Hinton,Karen L
Eri 3h45'56"-7d28'
Hinton,Michael J
Aur 5h16'1"45d19'
Hinton,My Very Best Friend,Brad
Sct 18h32'11"-6d25'
Hinton,Roderick
Lyr 19h13'56"40d26'
Hinz,Manuel
Cnv 12h40'37"34d30'
Hinz,Marcus J
Aur 4h54'58"40d58'
Hinz,Matthias
Aqr 21h21'21-8d32'
Hinz,Ulrich
Cep 23h17'35"68d26'
Hinze,Christina Lynn
Lib 15h18'1"-23d4'
Hinze,Marc
Dra 19h3'35"48d53'
Hinzmann,Renee Michele
Mon 7h6'58"-6d26'
Hip '93
Oph 18h4'13"8d10'
Hipelius,Michael Colin
Her 16h7'15"48d34'
Hipelius,Patrick James
Ori 5h57'21"16d34'
Hipkiss,Andrew Phillip
Umi 17h9'48"80d20'
Hippe,Nick Alan
Her 17h37'49"25d59'
Hippe,Silvia
Ari 2h59'56"21d25'
Hipperson,Tom
Her 15h54'31"40d59'
Hippler,Rochelle Lenore
And 2h29'13"50d29'
Hippopotamus,Dr
Tau 5h26'53"19d58'
Hipps,Byron Nathaniel
Her 16h56'33"32d48'
Hippy
Del 20h13'59"11d39'
Hipworth,Reg
Uma 10h20'36"54d0'
Hirakochi,Stella of Rie
Leo 10h1'5"19d10'
Hiram Loves Judy
Uma 9h55'48"58d58'
Hirano,Miduki
Uma 10h50'16"70d22'
Hirao,Kasha
Cas 1h32'12"72d43'
Hirata,E & P
Boo 13h58'17"24d33'
Hirata,Kathy & Ed
Cam 3h40'54"61d7'
Hirayama,Beni
Oph 17h54'42"12d43'
Hirchert,Lutz
Hya 9h0'30"5d31'
Hird,Christine
Cyg 21h6'36"31d25'
Hirko,Dianne & Jim
Lyn 7h1'37"58d18'
Hiro & Fumi
Cap 20h54'8"-17d42'
Hirofumi & Haruka
Ari 2h16'3"12d46'

Hirohata,Tyler Makoa
Dra 19h29'11"58d24'
Hiroi Kokoro
Leo 10h16'27"14d53'
Hiroki
Sgr 19h11'53"-17d6'
Hiroki Birthday Star
Leo 9h19'21"10d44'
Hiromu
Ari 2h27'30"23d2'
Hiron Star,Carolynn
Uma 9h12'36"70d18'
Hiroshi
Cmi 7h14'59"8d15'
Hiroskey,Doug
Her 17h38'15"14d11'
Hiroyuki & Noriko
Boo 15h4'15"25d37'
Hirsch III,Leon
Aql 19h28'44"-8d47'
Hirsch,Anne
Psc 0h52'49"8d17'
Hirsch,Gerald
Per 2h57'13"34d35'
Hirsch,Harry
Uma 11h21'41"43d37'
Hirsch,Jade
Lyn 6h27'56"59d22'
Hirsch,Jeffrey K
Ori 5h57'20"15d20'
Hirsch,Judith A
Cam 4h5'58"58d24'
Hirsch,Larissa
Mon 6h35'11"8d56'
Hirsch,Laurence E
Cet 3h9'19"2d45'
Hirsch,Ludwig
Psc 22h50'46"1d52'
Hirsch,Rhoberta Reah
Com 12h0'13"27d59'
Hirsch,Roberta
Cep 22h45'17"57d45'
Hirsch,Roxanne C
And 23h46'0"32d34'
Hirsch,Steven Alan
Aql 20h0'1"7d20'
Hirsch,Uve
Gem 6h24'53"12d32'
Hirschberg's Star, Adam Conrad
Aur 6h27'46"30d57'
Hirschberg's Star, Claire Evalyn
And 1h49'57"40d0'
Hirschberg's Star, Valerie Rose
Cra 18h30'9"-43d30'
Hirschberg,Eric & Lorraine Stanhope
Cyg 20h51'53"30d2'
Hirschberg,Wendy
Dra 18h12'20"70d5'
Hirschberger,Harry
Cep 1h42'13"80d1'
Hirschboeck,Rob
Peg 23h19'44"11d53'
Hirschcorn,Nick
Ori 6h2'11"8d27'
Hirschfeld,Betty
Cas 0h34'0"75d52'
Hirschfeld,JoAnn O
Mon 8h8'1"-0d3'
Hirschfield,Steven
Lib 15h32'58"-28d57'
Hirschhaut,Jaclyn Barrett
Cas 23h22'17"60d38'
Hirschkoff,Cie
Ori 5h51'58"7d4'
Hirschler,Eberhard
Uma 13h37'31"48d49'
Hirschovits,Asne & Sam
Crb 16h7'16"30d47'
Hirschovits,Fred
Oph 16h53'13"10d19'
Hirschy,Marthe
Eri 3h54'17"-5d34'

Hirsey,Kezia
Cyg 20h9'1"40d43'
Hirsh,Haley Kimbell
Aql 18h55'27"8d15'
Hirshberg,Sidney S
Cas 2h47'21"73d47'
Hirshfield,Nancy
Crt 11h11'16"-10d31'
Hirson,Richard
Cep 22h40'32"56d59'
Hirst,Clive
Uma 9h35'15"53d33'
Hirst,James
Lyn 8h17'48"44d46'
Hirte,Catherine Mary Grealish
Peg 22h38'58"20d49'
Hirter,Rhoberta
Eri 3h32'1"-2d60'
Hirth,Klaus
Mon 7h51'28"-1d30'
Hirtz
Ari 2h3'1"20d57'
Hirtzel,Stephanie Marie
Lyr 18h50'0"31d9'
His Hughness #40
Oph 18h8'36"13d27'
Hisashi,Star of Yumiko
Per 2h9'59"56d28'
Hisatomi,Stella Of Katsuhiko
Vir 14h6'12"2d19'
Hisch,Savannah Jean
Lyn 8h3'11"39d55'
Hiscock,Elizabeth N M
Uma 9h4'24"58d6'
Hiscock,Emily
Peg 22h6'22"4d23'
Hiscock,Ruth Clayton
Boo 14h16'1"46d59'
Hiser,William R
Aur 5h29'0"30d56'
Hiserodt,Jamie
Mon 7h23'43"-1d8'
Hisey's Heavenly Body, Bo
Aur 6h20'29"31d34'
Hisey,Nancy
Peg 23h27'33"12d3'
Hislod,Marian Hansen
Lyn 7h46'42"51d3'
Hislop,James
Uma 9h58'31"42d48'
Hislop,Jesse Thorne
Aur 6h57'1"37d39'
Hislop,Valerie
Tau 4h19'33"20d32'
Hitch,Robin Elizabeth Ann
Mon 6h59'41"-0d15'
Hitchcock,Alexandra Lisette
Cas 0h22'41"58d36'
Hitchcock,Alfie
Ori 4h49'1"0d18'
Hitchcock,Naomi Ruth
Cyg 19h33'13"33d8'
Hitchcock,Nicholas Erasmus
Aur 4h56'42"50d44'
Hitchcock,Stephany L
Cas 0h35'1"65d29'
Hitchcock,Timothy Michael
Peg 23h1'58"10d34'
Hitchens,Derek Kirk
Aql 20h10'31"10d22'
Hitchins,Denise E
And 23h43'37"47d41'
Hitchman,Myrna Millar
Lyr 19h25'56"38d3'
Hite 6-27-65,Karen Kay
Cas 23h32'22"61d9'
Hite,James R
Sex 10h35'39"0d39'
Hiter,Dick
Aql 19h6'52"3d47'
Hites,William George
Boo 14h39'1"17d18'
Hito
Tau 5h33'17"19d35'

Hitte,James Eldred
Psc 0h54'49"21d12'B
Hitte,Robert Willard
Psc 0h54'49"21d12'A
Hittenmark,David
Lac 22h40'47"51d52'
Hittle,Leslie
Eri 2h53'43"-12d28'
Hittman,John Desimone
Per 2h29'41"56d19'
Hitzig,Zoe Ketlar
Uma 10h54'36"40d50'
Hitztaler,Michael Alan
Hya 8h55'58"-1d22'
Hiu Yuen Wu
Aql 19h5'40"1d18'
Hiube's Starlight
Cep 2h17'41"77d22'
Hively,John
Hya 9h16'31"2d21'
Hivon,Grégoire
Ori 5h58'38"14d60'
Hix,Jr,William
Sco 17h52'52"-38d59'
Hixon's Star
Lmi 10h17'47"33d20'
Hixon,Larry R
Boo 14h23'13"27d8'
Hizel,Jordan Lee
Uma 9h9'29"58d11'
HJ
Lac 22h44'0"54d29'
Hjeltnes,Arne
Cam 7h37'22"67d32'
Hjorth,Jens
Dra 12h59'27"71d26'
Hjorth,Josephine Toft
Lyr 19h3'25"28d24'
HKT
Sct 18h46'18"-7d15'
Hlabse,Beth
Cas 0h53'1"68d26'
Hlad,Eric Jerome
Cma 6h56'14"-18d45'
Hladky,Karen
Lyr 19h22'38"37d59'
Hlavac,Ashley Sarah
And 2h19'11"45d41'
Hlavaty,Vince
Cet 0h34'53"-5d10'
Hlavenka,Celeste E
Mon 6h55'0"10d15'
Hlavsa,Krista
Lyn 6h55'56"58d8'
HLEAHTOR
Aqr 22h0'19"0d37'
HLH
Cyg 21h33'43"36d40'
Hlinka,Edward & Jessica
Cyg 19h46'46"38d3'
Hlubik
Peg 23h3'58"30d50'
Hlubik,Nancy
Tri 1h47'18"34d16'
HMB 022842
Aql 19h47'54"14d25'
Hmiel,Mrs
Crb 16h21'47"27d46'
Hmmm Really
Ser 15h55'17"21d11'
HMS Loving Heart
Uma 11h6'21"31d13'
Hnat Star,The Gerald & Elizabeth
Sge 19h11'28"16d49'
Hnatow,Peter
Aur 6h21'20"30d22'
Hnatuk,Norbert
Cnc 7h55'1"10d37'
Ho Chi
Aql 19h5'39"15d30'
Ho,Cyndi A
Eri 3h23'45"-14d30'
Ho,Don
Lyn 7h12'40"57d44'

Ho,Ivan
Cep 23h38'16"65d52'
Ho,Janis Diane
Uma 10h54'44"38d25'
Ho,Jovita
And 23h46'0"40d37'
Ho,Thomas
Cma 6h52'52"-18d55'
Ho,Trevor
Cma 6h52'39"-19d3'
Ho,Valesia
Cet 2h3'47"1d29'
Ho,Veronica Klein
Cet 2h3'27"1d36'
Ho,Wildes Green Hastings
Cyg 19h16'60"49d37'
Ho-John
Her 16h38'0"8d26'
Hoadley,Janet Marie
And 0h59'14"45d47'
Hoadley,Wendy Diane
Vir 14h58'32"6d2'
Hoag,Diane
Cas 0h35'51"54d35'
Hoag,Violet Anderson
Mon 6h15'1"-10d16'
Hoagland,Charles Raymond
Per 2h55'21"35d7'
Hoagland,Justin Micah
Ori 5h28'32"-3d37'
Hoagland,Lisa
Cas 23h36'42"61d29'
Hoague,Abigail
Cam 3h28'23"53d48'
Hoak,Darell
Dra 19h6'14"48d36'
Hoak,Jonathan Hale
Aur 5h10'42"41d1'
Hoana
Aql 19h2'38"-6d32'
Hoang Ha
Lyn 8h11'37"36d43'
Hoangvy
Crb 16h10'33"38d35'
Hoarau,Remi
Cru 12h14'19"-56d7'
Hoard,Claudia
Crt 10h59'30"-10d31'
Hoard,Mike
Cet 3h1'49"1d45'
Hoard,Phaedra
Uma 12h10'37"53d22'
Hoard,Suzanne
Com 12h20'14"20d16'
Hoare,George Arthur
Peg 22h42'19"31d33'
Hoback,Sunday Elizabeth
Cyg 20h36'1"38d5'
Hoban
Ori 5h41'47"10d56'
Hoban,Justin
Ori 5h0'1"15d11'
Hoban,Ryan
Her 16h22'1"26d42'
Hobart L,The
Cyg 21h37'1"44d9'
Hobart,Billie
Ari 3h19'7"24d13'
Hobart,James
Her 18h13'25"45d11'
Hobart,Teddy
Ori 5h35'47"-0d5'
Hobaugh,Claude & Tessie
Lmi 10h11'48"31d4'
Hobbes,The Slick
Boo 14h57'51"47d45'
Hobbit,Susie
Lyr 18h27'14"46d40'
Hobbs Mein Liebling, Peter W
Vir 14h59'55"7d1'
Hobbs,Angela
And 2h21'30"49d4'
Hobbs,Ashley
Peg 22h34'48"24d2'

Hobbs,Barbara B
Uma 11h33'0"32d50'
Hobbs,Casey
Her 17h59'21"38d53'
Hobbs,Connie J
Psc 0h23'49"-5d38'
Hobbs,Craig Alan & Jodi Lynn
Crb 15h18'0"30d20'
Hobbs,Donald
Aql 20h36'25"0d19'
Hobbs,Doug
Per 3h25'47"40d24'
Hobbs,Elmdale
Oph 19h0'46"8d13'
Hobbs,Joanne Elizabeth
Umi 13h58'13"77d26'
Hobbs,John H.
Ser 18h35'16"2d24'
Hobbs,Johnny F
Ari 1h46'43"23d45'
Hobbs,Joy C
Cet 2h54'13"1d17'
Hobbs,Rachel P
Ser 15h58'16"13d9'B
Hobbs,Ray
Boo 13h34'58"10d48'
Hobbs,Rebecca
Cas 3h1'25"60d1'
Hobbs,Robert Bruce
Cet 3h20'45"0d21'
Hobbs,Susan K
Leo 10h51'23"22d9'
Hobbs,William (Billy)
Dra 20h0'15"62d11'
Hobbs,William I
Ser 15h58'16"13d9'A
Hobbs,William Joseph
Hya 8h53'19"2d26'
Hobbs-Rice,Sheralon Mae
Lyn 7h54'60"58d58'
Hobby H21BD,Jennifer Mae
Sge 19h12'0"20d9'
Hobby,Elizabeth
Peg 22h47'1"4d55'
Hobby-Lehr,Allison
Aql 19h57'54"11d18'
Hobday,Paul Anthony
Equ 21h11'27"11d48'
Hobden,Simon
Her 17h34'14"25d25'
Hobdy,Robert Shaun
Ori 6h13'18"1d11'A
Hobdy,Roland Shaine
Aur 6h26'22"37d11'
Hoberman,Bradley Thomas
Boo 14h12'45"38d22'
Hoberman,Roni Victory
And 0h57'0"34d52'
Hobgood,Mary Beth
Peg 22h5'35"4d31'
Hobie
Ser 15h10'31"7d25'
Hobson From Physics 1994-95,To Mr
Dra 17h14'0"60d41'
Hobson Star,The Kathleen
Cas 22h56'47"54d33'
Hobson,Alison S E
Del 20h22'29"4d44'
Hobson,Charles "Air"
Cam 3h30'29"53d49'
Hobson,Michael A
Cep 22h27'50"80d17'
Hocevar,Anthony & Laura Morgan
Lac 22h54'31'54d47'
Hoch,Donald & Marion
Cep 22h7'19"54d5'
Hoch,Hans-Elmar
Oph 16h54'21"1d35'
Hoch,Jordon
Uma 10h13'48"61d21'
Hoch,Mary
And 23h48'1"42d31'

Hoch,The Leroy & Emily
Eri 2h49'49"-10d45'
Hochberg,Leslie
Cam 3h53'59"60d11'
Hochhaus,Stephen
Cma 7h0'13"-13d8'
Hochhauser,Sophie Claire Bedel
Cas 3h4'26"75d55'
Hochheiser,Elaine
And 23h17'12"46d19'
Hochman,David Samuel
Aql 9h15'33"12d1'
Hochman,Lawrence-Heidi & Leah
Mon 6h23'49"8d5'
Hochron,Cole Daniel
Aql 18h46'54"11d39'
Hochstein,Robert Gary
Cnv 12h47'38"50d9'
Hochwarter,Franz & Brigitta
Cyg 20h2'29"58d10'
Hochzeitsstern,KariOns
Boo 14h27'1"14d37'
Hock,Jr,Frederick Charles
Sco 16h25'56"-40d42'
Hock,Jr,John
Per 3h22'1"40d52'
Hock,Tim
Boo 15h8'13"22d34'
Hock-Hussey,Deborah
Aql 19h26'0"8d49'
Hockenbury,Gary
Cep 0h3'45"86d22'
Hockert,Cathlene Ann
Lyn 7h32'35"51d40'
Hockstader,Avery Diane
Lac 22h0'34"40d31'
Hockun,Edith
Her 17h10'21"48d20'
Hoczmann,Anna-Sophie
Sco 17h54'29"-31d46'
Hoda,Kay
Peg 22h7'27"4d57'
Hodam,Shannon
Lyn 8h24'20"41d21'
Hodder,Harold Francis
Her 16h8'24"20d30'
Hodder,Kenneth R
Aql 19h4'37"3d53'
Hodder,Lucille Coty
Her 16h12'25"20d26'
Hodder,Paul P
Cet 3h5'31"2d7'
Hodel,Diane Marcia
Lyn 7h37'32"51d22'
Hodel,Duncan Hill
Aql 19h54'58"8d45'
Hodel,George Hill
Aql 19h56'53"8d16'
Hodel,George Kelvin
Dra 11h4'38"73d35'
Hodel,Mark Denis
Dra 11h3'39"73d32'
Hodel,Ramon Allen
Her 17h49'27"14d55'
Hodel,Steven Kent
Lmi 9h59'1"33d41'
Hodel-Malinofsky, Teresa Vasanta
Com 11h57'31"19d57'
Hodge,Arthur A
Lac 22h2'14"49d15'
Hodge,Chad E
Lac 22h18'26"46d59'
Hodge,Chris
Sct 18h55'51"-6d44
Hodge,Cindy
Peg 22h1'43"31d16'
Hodge,David Hellyer
Cep 3h27'25"80d37'
Hodge,David William "Bill"
Aur 6h23'0"30d22'
Hodge,Judy L
Com 12h1'31"25d24'

Hodge,Kelli
Equ 21h23'15"7d44'
Hodge,Lauren Carol
Mon 6h43'27"7d49'
Hodge,Linda
Mon 7h59'22"-8d5'
Hodge,Mary Ellen
Eri 2h52'1"-18d36'
Hodge,Megan Eve
Vul 19h46'35"28d9'
Hodge,Nicole C
Peg 22h0'47"21d41'
Hodge,Ronald A
Ori 5h4'32"8d4'
Hodge,Shirley Ann
Lmi 10h3'19"33d19'
Hodge,Sr,Don G
Aql 19h6'1"4d19'
Hodge,Tim G
Cnv 12h29'59"48d29'
Hodgerney,Michael David
Dra 16h26'26"68d38'
Hodges I,Neil C
Dra 17h40'20"60d40'
Hodges I,Timothy B
Cep 0h17'41"75d7'
Hodges II,William G
Lac 22h34'27"40d43'
Hodges,Bartley Sevier
Cet 1h3'53"-1d3'
Hodges,Beverly A
And 23h24'31"44d57'
Hodges,Bob
Boo 15h0'57"31d22'
Hodges,Charles "Chuckie Cheese"
Mon 7h2'48"-5d0'
Hodges,Christopher
Per 2h49'55"48d21'B
Hodges,David H
Cmi 7h42'1"4d5'
Hodges,Diane
Per 2h49'55"48d21'A
Hodges,Hugh Dorsey
Aur 6h3'27"36d31'
Hodges,Jewel
Mon 6h26'20"-8d40'
Hodges,John
Lyn 8h17'18"44d8'
Hodges,Kara-Marie Germaine
Cet 3h6'14"0d49'
Hodges,Karen Leanne
Peg 23h21'55"30d14'
Hodges,Kayla Leigh Ann
Cam 8h44'16"81d34'
Hodges,Kimberly K
Aql 20h12'45"11d41'
Hodges,Kimberly Y
And 1h36'1"38d43'
Hodges,Kynda Leigh-Anne
Peg 23h28'11"24d24'
Hodges,Linda
Equ 21h19'51"3d4'
Hodges,Marjorie Sousa
Cas 23h29'0"61d58'
Hodges,Sidney Lee
Cam 6h20'20"78d50'
Hodges,Star of Adam
Aql 20h10'52"10d40'
Hodges,Star of Dan
Eri 2h55'44"-11d8'
Hodges,Star of Rick
Cet 3h0'1"6d8'
Hodges,Thomas E
Aql 19h55'0"13d16'
Hodges,Thomas Wayne
Ser 15h10'25"10d28'
Hodges,Wallace G
Aql 19h27'38"10d55'
Hodges,Walter Anthony
Her 17h16'51"45d51'
Hodgeson,Hannah
Lyn 7h27'55"50d36'
Hodgetts,Lisa Catherine
Mon 6h41'31"3d39'

Hodgkin,Pete & Debbie
Cyg 21h30'35"50d32'
Hodgson's South Wold, David
Ori 4h55'0"0d48'
Hodgson,Anne Edna
Lyr 18h28'0"38d51'
Hodgson,Helen
Cyg 19h51'34"58d25'
Hodgson,John F
Ori 5h17'34"15d17'
Hodgson,Lindsay
Umi 15h19'29"80d7'
Hodgson,Timothy Scott
Boo 13h37'55"16d32'
Hodian
Gem 7h58'59"28d60'
Hodkin,Joseph
Her 17h18'45"42d33'
Hodková,Petra-Barbora
Cam 3h20'31"60d4'
Hodnett,Bobbie
Ser 16h0'12"9d48'
Hodnett,Dana Martin
Lac 22h12'21"50d2'
Hodnett,Debra
Cas 0h25'28"61d29'
Hodnett,Martin Frank
Mon 7h17'32"-5d28'
Hodo,Flora
Cyg 21h25'0"40d53'
Hodo,James Michael
Leo 9h54'45"8d22'
Hodorff,Janelle Josephine
Uma 9h59'10"47d37'
Hodson,Molly Louise
Cam 3h24'59"63d36'
Hodson,Paula Maxine
Lyr 18h58'0"42d50'
Hodson,Robin Elizabeth
Mon 6h19'54"7d30'
Hodum,Günther
Ori 5h49'46"21d9'
Hoebel,Sarah Monroe
Eri 3h10'1"-3d22'
Hoechst,Christian Mark
Boo 14h37'58"50d4'
Hoeckner,Fritz
Aur 6h45'41"35d59'
Hoefflin Star,The
Aur 5h52'57"30d12'
Hoefle,Brandon Michael
Aur 7h19'49"35d47'
Hoeft,Daniel
Aur 6h59'1"36d42'
Hoeft,Dr Thea M
Oph 17h14'42"11d42'
Hoehing,Sr,Charles Alfred
Cep 22h8'27"58d13'
Hoehle,Sibylle
Cnv 12h39'21"36d47'
Hoehn,Douglas William
Oph 18h19'42"7d37'
Hoehnemann
Her 17h56'23"14d22'
Hoehner,Christine Marie
Lyr 18h58'33"46d21'
Hoekel,Helen
Eri 2h55'15"-14d33'
Hoekenga,Blythe Inge
Peg 22h40'16"21d50'
Hoekstra,Stephen Charles
Cma 6h12'1"-13d40'
Hoelen,Robin Stacy
Mon 6h41'40"8d35'
Hoelen,Suzeriek
Ori 5h8'46"-1d28'
Hoelle II,William John
Cep 0h43'56"80d4'
Hoelle,Christopher John
Aur 6h7'0"40d42'
Hoelscher,Christopher Scott
Aql 20h30'33"0d58'
Hoenig,Michael David
Aur 6h29'53"38d2'

Hoenscheid,Joe "Alexandra's Daddy"
Her 17h3'22"20d33'
Hoepfl,Michelle Lisa
Peg 0h6'14"27d44'
Hoerauf,Beth Navarrette
Cas 23h36'41"61d45'
Hoerig,Jaime Diane
Peg 22h38'20"20d31'
Hoerl,Evan Arthur
Cep 22h8'21"61d50'
Hoerl,Shannon Colleen
Mon 6h41'10"8d49'
Hoes,Hannah Hutton
Her 17h9'46"40d60'
Hoesch,Hayley McKenzie
Cyg 21h28'60"41d8'
Hof,Marguerite M
Uma 12h57'25"57d56'
Hofbauer,Georg
Lyn 8h19'31"47d19'
Hofem
Ori 6h12'28"10d16'
Hofer's Silver Star
Cyg 21h42'47"29d40'
Hofer,Al
Aur 6h5'49"46d44'
Hofer,David L
Cet 3h1'47"9d27'
Hoff,Gladys Nellie
Cam 6h14'42"68d32'
Hoff,Madison Ivy
Ser 15h10'0"0d32'
Hoff,My Shining Star Marisa Ann
Cyg 20h38'55"42d26'
Hoff,My Shining Star Serena Lin
Peg 22h11'1"29d5'
Hoff,Nadine Anne
Cas 3h34'1"74d56'
Hoff,Siegbot von
Leo 10h52'1"0d57'
Hoff,Ted,F Faggin, S Mazor, M Shima
Umi 14h22'29"75d41'
Hoffacker,Chloe'Luree
Aur 5h26'21"40d12'
Hoffellner,Gertrud
Cyg 19h32'34"37d49'
Hoffenberg,Robert
Ser 15h49'49"23d16'
Hoffer II,William M
Aur 6h7'1"30d45'
Hoffer,Abby
Lac 22h54'39"38d24'
Hoffer,Gregory Joseph
Her 17h25'52"40d31'
Hoffer,Jacob Alan
Cam 3h31'51"60d28'
Hoffer,Terry
Peg 22h43'33"35d29'
Hoffman Star,The Sharon
Cyg 21h7'44"30d23'
Hoffman's Cottage Point
Aql 18h43'1"1d46'
Hoffman,Adele
Cnv 13h51'60"38d44'
Hoffman,Alan & Judy
Crb 15h57'15"30d21'
Hoffman,Alexandra
Ori 4h58'47"4d29'
Hoffman,Alexandra Jane
Lyr 19h5'1"37d49'
Hoffman,Alice Marie Margaret
And 22h56'54"50d28'
Hoffman,Amy Lynn
Cas 0h53'57"65d39'
Hoffman,Arlan James
Per 4h24'48"50d8'
Hoffman,Barbara Ellen
Cas 3h4'50"60d56'
Hoffman,Beate Reichel
Psc 22h55'49"5d37'

Hoffman,Bob & Judie
Uma 10h43'49"56d3'
Hoffman,Carolyn
Cas 1h2'54"62d15'
Hoffman,Charlene
Lmi 10h31'57"28d32'
Hoffman,Christopher Michael
Hya 9h7'47"3d59'
Hoffman,Cinnamin K
Ori 5h44'37"-0d22'
Hoffman,Conrad & Rhoda
Uma 13h59'47"53d9'
Hoffman,David Wesley
Cet 3h2'1"0d15'
Hoffman,Dennis & Patti
Uma 9h15'42"51d41'
Hoffman,Deon Warren
Aur 6h20'33"38d19'
Hoffman,Dorothy Matilda Kutil
Lyn 8h6'56"50d29'
Hoffman,E Michael
Boo 15h3'18"12d26'
Hoffman,Edwin Earl- Doris
Hoffman Prichard
Umi 15h52'32"78d55'
Hoffman,Elliott
Lib 15h46'0"-8d24'
Hoffman,Ethan Walter
Hya 9h6'19"2d39'
Hoffman,F P Elaine Whitaker
Oph 17h1'12"-21d36'
Hoffman,Gary
And 2h28'50"41d31'
Hoffman,Gary Walter
Sgr 19h37'27"-38d55'
Hoffman,Georgianna N
Lyn 7h46'18"42d50'
Hoffman,Gertrude "Peggy"
Tri 2h5'15"31d40'
Hoffman,Harold J
Aur 6h2'29"38d47'
Hoffman,Hunter
Lac 22h13'13"46d20'
Hoffman,In Memory of George
Her 18h5'33"38d28'
Hoffman,Jack & Ginny
Cyg 19h6'14"48d36'
Hoffman,Jason
Boo 14h36'21"30d40'
Hoffman,Jeanette & Paul
Uma 8h43'0"50d37'
Hoffman,Jeffrey
Cet 0h48'42"-6d32'
Hoffman,Jeffrey Geno
Aur 6h6'35"38d16'
Hoffman,Jerry S
Cep 22h22'46"68d10'
Hoffman,Jessica
And 23h35'41"42d5'
Hoffman,Jillian
Lyr 18h43'10"35d49'
Hoffman,Joanne Carol Rose
And 2h28'40"38d48'
Hoffman,Joel
Tri 1h48'58"27d7'
Hoffman,John
Cet 0h31'12"-1d29'
Hoffman,Jr,James A
Aql 20h12'11"1d45'
Hoffman,Jr,William Edgar
Cam 4h26'0"68d24'
Hoffman,Julie
Lyr 18h37'0"46d18'
Hoffman,Larry
Cep 23h4'35"64d26'
Hoffman,Lauren Elizabeth
Cas 0h29'10"63d52'
Hoffman,Lynn
Lyr 18h30'59"40d11'
Hoffman,Madelene Haspel
Cas 1h5'55"75d29'
Hoffman,Mark & Laura 3-7-1986
Her 17h19'0"22d43'

Hoffman,Mark Charles
Leo 9h37'0"8d23'
Hoffman,Marlene Jean
Cas 0h46'31"67d4'
Hoffman,Megan Alicia
Cas 1h46'22"61d47'
Hoffman,Melissa
Cas 1h15'15"60d13'
Hoffman,Micahel B
Aur 6h32'21"32d56'
Hoffman,Nicholas Michael
Boo 15h25'31"42d19'
Hoffman,Noelle Paige
Peg 21h47'13"34d11'
Hoffman,O F P Kris
Mon 7h20'14"-6d42'
Hoffman,Patricia Ann
Cet 1h42'1"-1d48'
Hoffman,Paul
Cnv 13h47'34"30d45'
Hoffman,Rachel Ann
Cyg 19h47'56"29d28'
Hoffman,Richard J
Per 3h4'1"46d59'
Hoffman,Robert
Ori 4h58'18'4d25'
Hoffman,Ronald F & Louise
Umi 14h10'1"69d45'
Hoffman,Sarah (Suranovitz)
Aql 18h43'1"10d11'
Hoffman,Sheila E
And 0h10'1"31d31'
Hoffman,Sr,John R
Per 2h53'42"55d35'
Hoffman,Stacie
Mon 8h0'17"-8d11'
Hoffman,Steven John
Aql 20h19'17"5d25'
Hoffman,Susan Elizabeth
Cas 0h34'26"65d55'
Hoffman,Tara Lynn
Lyr 19h2'43"37d34'
Hoffman,Tim
Cet 0h32'15"-0d57'
Hoffman,Tristan
Uma 8h44'43"50d4'
Hoffman,Virgil
Dra 14h45'48"62d14'
Hoffman,William Douglas
Her 17h42'43"40d18'
Hoffman,Wolfgang Peter
Ser 15h53'21"-0d14'
Hoffmann
Per 2h36'20"57d15'
Hoffmann,Billy John
Sex 10h17'41"-5d38'
Hoffmann,Christoph
Dra 20h28'38"68d47'
Hoffmann,Elfie
Sco 17h29'1"-38d39'
Hoffmann,Gary W
Dra 19h26'28"68d35'
Hoffmann,Jane W
Crt 11h12'4"-18d12'
Hoffmann,John L
Aur 6h0'0"30d38'
Hoffmann,Jürgen
Lyn 8h0'10"42d22'
Hoffmann,Jürgen
Uma 9h28'44"47d34'
Hoffmann,Luc
Uma 9h47'26"50d28'
Hoffmann,Marlies
Vir 11h41'17"8d55'
Hoffmann,Niclas
Lib 14h43'35"-23d10'
Hoffmann,Nolan Christopher
Her 17h11'40"48d38'
Hoffmann,Olivia Carolynn
Cas 0h50'20"64d55'
Hoffmann,Oskar
Ori 5h39'42"8d38'
Hoffmaster,Robert
Cet 3h8'53"1d56'

Hoffs,Michael S
Ori 6h2'14"1d1'
Hofmann "Lurchi",Peter
Her 17h22'0"46d54'
Hofmann,Bettina
Cnv 13h6'59"38d39'
Hofmann,Brita
Sgr 19h27'-28d11'
Hofmann,Carol
Oph 17h6'28"-22d56'
Hofmann,Deborah Ann
Vir 13h38'15"1d49'
Hofmann,Gabriele
Cap 21h52'53"-14d41'
Hofmann,Gernard
Peg 23h30'18"18d45'
Hofmann,Gert-Maria
Peg 23h29'39"21d41'
Hofmann,John William
Aql 19h52'57"12d35'
Hofmann,Keith Alan
Ori 6h5'24"8d0'
Hofmann,Kerstin
Leo 10h51'57"8d2'
Hofmann,Max Herbert
Cnv 13h25'38"48d58'
Hofmann,Norbert
Her 16'20"49"50d26'
Hofmann,Richard
Hya 9h7'33"2d17'
Hofmann,Werner
Del 20h19'12"10d31'
Hofmann,Winfried
Aql 19h55'58"10d36'
Hofme/Tus Wallau-Massenheim
Ori 4h59'38"10d12'
Hofmeister,David Lynn
Dra 16h1'12"60d32'
Hofmeister,Elke & Uwe
Dra 15h8'45"62d34'
Hofmeister,Mary B
Cas 22h56'53"55d18'
Hofner,Peter
Peg 23h34'0"28d44'
Hofschuster,Henry
Cep 22h11'23"55d50'
Hofstaedter,John
Her 15h49'45"41d57'
Hofstee-Ran,Margot
Cam 4h58'59"58d42'
Hofstetter Family Star
Per 1h36'1"53d34'
Hofstetter,Harvey & Evelyn
Peg 23h47'46"10d56'
Hofstetter,Reinhard
Eri 4h49'9"-8d32'
Hogan (Dad),John Patrick
Cep 20h10'41"60d18'
Hogan (Mame),Mary Patience
Mic 21h14'33"-34d17'
Hogan 1935-94 Luvd life S,G B
Per 3h10'59"47d16'
Hogan's Halo
Dra 13h50'42"68d55'
Hogan's Halo,The
Uma 9h20'53"68d11'
Hogan,Anna Rose
Cas 1h33'51"60d5'
Hogan,Ben
Cmi 7h34'49"0d47'
Hogan,Bobby
Umi 16h59'15"79d35'
Hogan,Cecelia
Cas 23h32'47"60d27'
Hogan,Charles Gerald
Vir 11h38'46"6d28'
Hogan,Christina Marie
And 1h53'1"36d50'
Hogan,Dorothy K
Uma 11h54'31"51d33'
Hogan,Dylan Thomas
Her 17h49'58"40d10'

Hogan,Garnette D
Crb 16h3'31"27d11'
Hogan,George "Puppetman"
Ser 15h21'12"2d44'
Hogan,Gerald William
Cep 21h39'31"80d8'
Hogan,Hannah Lyn
And 23h34'41"39d51'
Hogan,Harry
Uma 9h38'1"53d37'
Hogan,Jack & Kathleen
Lyr 18h39'44"34d37'
Hogan,James Wayne
Dra 16h54'44"61d48'
Hogan,Jenny Marie
And 2h17'53"39d29'
Hogan,John
Ori 6h9'23"1d9'
Hogan,Joseph C & Betty C Hogan
Cyg 19h33'19"37d32'
Hogan,Kathy
And 1h27'1"34d54'
Hogan,Kristin Carol
Cas 0h1'44"64d9'
Hogan,Loretta T
Uma 12h48'28"53d13'
Hogan,Margaret
Hya 9h10'1"5d40'
Hogan,Mary Ellen Murphy
Cas 0h19'45"63d43'
Hogan,Michael Joseph
Dra 18h38'29"58d35'
Hogan,Michael Joseph
Per 3h1'44"50d7'
Hogan,Penny S
Cam 3h57'57"57d15'
Hogan,Ray
Boo 15h2'1"28d41'
Hogan,Rich
Lac 22h34'30"40d7'
Hogan,Susan
Crt 11h19'15"-19d14'
Hogan,Ted
Aur 4h55'29"48d60'
Hogarth,Jenny
Cyg 19h31'30"33d21'
Hogarth,Suzanne Linnea
Lyr 19h1'1"37d31'
Hoge,Thomas A
Sge 18h59'20"19d42'
Hogenson,Helen
Cyg 21h6'34"48d53'
Hogestyn Family,The
Aql 18h59'39"-0d2'
Hogeweg,Desiree Ingrid Mirjam
Cas 0h10'49"63d50'
Hogey Bear
Aql 20h3'48"1d42'
Hogg,Ian Buchanan
Mon 7h43'14"-5d47'
Hogg,John M
Peg 0h2'21"26d22'A
Hogg,Robbye Lee
Peg 0h2'21"26d22'B
Hogg,Simon
Uma 11h15'1"55d47'
Hoggarth,Jr,Joseph
Cep 0h10'39"66d45'
Hoggarth,Velma Frazier
Cas 3h10'14"63d43'
Hogge,D
Oph 16h55'49"11d12'
Hogrewe,Laurie
Lyn 9h6'13"44d33'
Hogstrup,Josephine
Lyr 18h41'0"27d60'
Hogue,Donald M
Her 17h54'55"40d12'
Hogue,John H
Dra 20h2'1"80d19'
Hogwood,Leigh Holly
Lyr 18h52'21"38d33'

Hoh,Matthew Paul
Dra 19h53'54"60d48'
Hoh,Melissa Brietta
Mon 7h6'28"-6d30'
Hohe,Dieter
Cmi 7h29'23"7d33'
Hohensee,Clif
Ori 6h7'0"11d11'
Hohert,Robynn N
Com 13h31'34"19d32'
Hohing,Douglas Kurt
Per 2h3'0"58d55'
Hohl,Jordan John
Aur 6h31'55"35d50'
Hohlbein,Daniel O
Per 3h33'11"36d11'
Hohlfeld,Covernor Al
Hya 8h41'36"5d48'
Hohm,Harold
Her 16h50'0"38d34'
Hohman,Colleen & Erin
Cyg 21h3'16"33d45'
Hohman,Jeanie
Crb 15h50'32"37d55'
Hohmann,Dr Peter Franz-Josef
Tau 4h3'39"0d6'
Hohmann,Tom
Dra 19h30'1"58d40'
Hohmeister,Elizabeth Kollasch
Cas 1h0'0"68d14'
Hohn,Michael William
Oph 17h32'46"-0d0'
Hohner,Frederick S
Her 16h42'45"25d16'
Hohorst,Jennifer Joan
And 23h19'49"44d22'
Hohrmann,Trine
Cet 1h26'1"1d8'
Hoidale,Kevin Watson
Dra 19h34'37"61d22'
Hoisington,Heather Hennessy
Lyn 8h16'11"51d53'
Hojnacki,Frank & Wilhelmina
Vir 13h38'24"2d15'
Hokamp,Rich
Dra 19h0'49"50d37'
Hokans,Afton Janet
Peg 23h3'40"11d18'
Hoke,John Riley
Per 2h13'44"58d19'
Hoke,Sr,Patrick
Her 16h59'38"32d56'
Hokey
Aql 19h14'1"10d11'
Hokie,The
Vul 19h0'46"25d10'
Hoku Lani O Pihana
Oph 17h1'21"-23d2'
Hokuf,Myron Bernard
Aql 19h58'12"-0d60'
Hokuf,Rosebel Jayn
Crb 15h53'31"27d22'
Hokumakalani
Mon 6h30'11"8d58'
Holaday,Mary Adeline Maguire
Aql 19h46'14"13d48'
Holaday,The Honorable T W (Bill)
Aql 19h51'18"13d40'
Holahan,Maureen
Cyg 20h3'53"40d10'
Holahan,Patricia F
And 23h18'53"40d33'
Holbay,Ashley Rose
Aql 20h7'49"1d41'
Holberg,Bud
Aur 4h46'58"38d3'
Holbert,Clark
Per 2h54'29"55d23'
Holbert,Kairo Graham
Eri 3h20'40"-5d30'
Holbert,Timothy Wayne
Sct 18h56'30"-6d4'

Holborn,Robert
Ori 5h10'23"-6d59'
Holbrook's Christmas Star
Aur 4h54'18"41d1'
Holbrook,Dennis
Aql 19h37'12"-0d3'
Holbrook,Jessie B
Eri 4h4'58"-11d24'
Holbrook,John E
Dra 17h58'39"64d5'
Holbrook,Marion & Jim Gudenrath
Cyg 19h29'1"48d58'
Holbrook,Steve & Bonnie
Uma 11h43'60"51d47'
Holcado,Elena Perez
Lyn 8h0'15"39d60'
Holck,Mildred Wray
Lyr 18h32'29"30d45'
Holcomb "My Shining Star",Mark
Boo 14h38'56"37d2'
Holcomb Star,The
Cyg 21h55'0"52d46'
Holcomb,Crystal Ann
And 2h18'43"48d29'
Holcomb,Garen Ashley
Vul 20h42'42"28d20'
Holcomb,Heather
Peg 22h24'42"21d56'
Holcomb,Jennifer Lynn
Cas 1h26'1"58d60'
Holcomb,Jodi Alain
Eri 3h27'1"-5d20'
Holcomb,Lynn Adele
And 22h57'52"51d14'
Holcomb,Michaela Ann
Cas 0h58'1"66d26'
Holcomb,Ryan
Aql 19h9'21"4d18'
Holcomb,Steffan
Cet 2h25'12"4d21'
Holcomb,Susie Starr
Crb 15h20'35"30d6'
Holcombe's Heaven, Peter
Cep 20h54'50"78d57'
Holcombe,Robert Craig
Aur 6h25'15"37d2'
Holcombe,Roderick
Ori 5h25'21"1d22'
Holcombe,Rowland L
Dra 19h5'21"48d45'
Hold Fast To Dreams
Lyr 19h26'56"38d44'
Hold,John Martin
Oph 17h54'14"13d26'
Holden
Per 3h15'15"41d8'
Holden Noel
Boo 14h11'12"43d13'
Holden Star,Ann
And 23h42'52"41d18'
Holden,Alec
Her 16h41'46"53d44'
Holden,Brian Andrew
Her 17h54'37"37d55'
Holden,Charles Kellett
Lmi 10h59'31"31d7'
Holden,Christopher
Mon 7h29'48"-5d47'
Holden,Colin J
Her 16h55'60"26d34'
Holden,Donald George
Peg 23h45'17"30d55'
Holden,Forrest Robert
Aur 6h33'0"31d25'
Holden,Frances Mary
Crb 15h48'46"26d24'
Holden,Hale Hobart
Peg 22h41'29"10d53'
Holden,Helen Valentine
Cas 0h13'25"47d35'
Holden,Jeffrey Bryce
Boo 14h0'29"27d5'

Holden,John F
Dra 19h54'23"70d12'
Holden,Louise
Lyr 18h57'42"30d2'
Holden,Nichele Yvonne
Cyg 20h52'29"38d8'
Holden,Sheri
Cas 0h31'50'67d35'
Holden,Stanley
Aur 6h28'28"35d6'
Holden,Stanley
Boo 14h37'54"32d43'
Holden,Thomas Clinton
Her 16h56'0"47d49'
Holden,Thomas Raymond
Cep 0h57'30"87d22'
Holden,Tracy Michelle
And 0h29'26"38d27'
Holden-60,Joseph L
Cet 3h18'1"8d3'
Holdener,Don R
Uma 11h55'58"52d57'
Holder Star,The
Her 16h39'28"34d59'
Holder,Bertram
Cnv 12h28'0"32d12'
Holder,Edward
Dra 14h44'27"60d55'
Holder,Hazel Zella Lee
Mon 7h23'36"-8d40'
Holder,Mary Dee
Vul 19h48'35"27d23'
Holder,Richard
Aql 19h43'13"13d45'
Holder,Roya Soroush
Lyr 18h58'51"31d24'
Holder,Shelli Loren
Cet 3h19'38"1d44'
Holder,Sr,Judson A "Mostly Wonderful"
Mon 7h46'49"-8d58'
Holder,Tony Mark Rne
Per 2h8'58"57d58'
Holderman,Brooke Ashlie
Sge 19h7'17"19d53'
Holderman,Terra Jayne
Sco 17h51'1"-38d51'
Holderness,Austen
Boo 14h54'28"33d26'
Holderness,Richard Thurston
Aql 19h31'33"12d33'
Holdforth,Ian Robin
Lyn 7h56'41"51d59'
Holdman,Beth
Del 20h14'25"15d22'
Holdmann,Bob
Dra 19h15'33"70d44'
Holdren,Marlo
Mon 7h4'34"1d8'
Holdsworth Happy Star, The Jane
And 2h2'28"43d0'
Holeman,Irene Ethel
Ori 5h55'0"14d45'
Holeva,Stephen Marcus
Cma 7h19'14"-15d1'
Holewinski,Ann Lauren
And 23h28'0"45d3'
Holey,Rüdiger
Sgr 19h56'47"-43d12'
Holger,Andreas
Peg 23h31'0"11d8'
Holiday,Colleen M
Peg 22h18'21"22d48'A
Holiday,Steve's Dream- Steven Earl
Her 16h1'51"47d48'
Holiday,Steven Lane
Lmi 10h9'16"40d22'
Holimon,Thomas Monroe
Aql 19h3'1"15d35'
Holin,Carole
Cyg 19h44'36"38d33'
Holladay,Amy Louise
Cas 0h29'56"65d34'

Holladay,Charles R
Aql 20h3'0"4d28'
Holladay,Evelyn Canne
Cap 20h16'23"-20d20'
Holland III,Reuben
Aql 19h44'56"10d10'
Holland IV,Anne & Peter
And 1h59'25"36d58'
Holland Sincerity, Kimberly
Cam 3h34'13"61d13'
Crt 10h54'14"-11d24'
Holland Star,Claudia Noel
And 1h59'25"36d58'
Holland's Star,Brady
Cet 0h55'15"-5d49'
Holland,Aaron
Hya 9h4'43"2d43'
Holland,Alyssa's Angel
And 0h21'21"33d48'
Holland,Anthony Mark
Oph 17h34'19"-6d59'
Holland,Bevin
Hya 8h44'56"1d5'
Holland,Carl R
Her 18h8'0"38d12'
Holland,Charla
Aql 19h31'0"13d19'
Holland,Christopher
Her 17h30'28"50d4'
Holland,Christopher Charles
Cmi 7h57'21"8d33'
Holland,Donna Kay
And 2h29'1"38d14'
Holland,Doug
Boo 14h48'11"35d45'
Holland,Edith
Sge 19h54'39"16d12'
Holland,Gladys K
Cas 0h6'23"62d48'
Holland,Hilda
Cyg 20h0'35"30d25'
Holland,J M
And 22h56'10"37d13'
Holland,Jack
Ser 15h33'32"-1d26'
Holland,Jacquie Ruth
Sco 17h43'1"-31d6'
Holland,Jens Arne
Cep 22h2'15"60d11'
Holland,John Norman
Dra 12h29'41"64d6'
Holland,Jonathan Peter
Per 1h53'55"50d16'
Holland,Joseph Howard
Umi 14h24'27"61d26'
Holland,Jr,William P
Boo 15h2'46"24d34'
Holland,Kathryn"Kathy"
Vul 19h47'21"28d20'
Holland,Kitty
Aql 20h12'29"11d51'
Holland,Lareta
Oph 18h7'13"12d1'
Holland,Leonard & Gloria
Cnv 13h33'41"41d42'
Holland,Lucy In The Sky
Lyn 7h31'52"38d40'
Holland,Michael
Uma 10h2'43"55d51'
Holland,Nancy Lee
Cas 0h16'51"63d11'
Holland,Patrick MacIntyre
Aur 5h13'28"44d25'
Holland,Paul Dean
Per 1h56'26"56d31'
Holland,Penelope
Cas 0h35'0"60d43'
Holland,Randy
Cep 22h59'32"80d28'
Holland,Saul
Ori 5h57'45"16d30'
Holland,Steve
Dra 15h52'41"60d53'
Holland,Wildon M
Lyn 8h42'0"45d16'

Holland/Syrstad,John W
Her 18h13'48"40d3'
Hollander,Aaron D
Boo 15h22'0"41d25'
Hollander,Adam
Aur 6h57'11"37d22'
Hollander,Becky L
And 2h30'43"42d37'
Hollander,Brad
Aql 19h54'59"8d22'
Hollander,Brad
Ser 17h15'52"-13d39'
Hollander,David Avi
Aur 5h54'45"29d2'
Hollander,Helaine
Lyr 19h19'38"42d56'
Hollander,Rachel S
Lyn 7h31'13"50d39'
Hollander,Ralph "Poppy"
Cet 1h57'0"-6d44'
Hollander,Robert "Poppy"
Oph 17h34'31"-0d46'
Hollander,Thomas
Aur 6h34'1"35d7'
Hollando,Roger & Debby
Cyg 19h27'11"33d15'
Hollands,Sylvia E
Cas 1h5'16"55d27'
Hollanender,Sr,Robert P
Cyg 19h27'33"33d58'
Hollar,Jr,John S.
Cep 22h14'58"58d33'
Hollaway,Jeremy Richard
Cmi 7h27'26"8d2'
Holle,Donald F
Cnv 12h23'21"35d34'
Holle,Janine Marie
Dra 20h18'25"68d14'
Holle,Kevin Lee
Dra 20h19'25"73d11'
Holle,Lena Slattery
Dra 20h2'55"76d29'
Holle,Michael Raymond
Dra 20h36'0"74d58'
Holle,Reese Robert
Dra 20h28'14"74d41'
Holle,Robert Alan
Dra 20h27'11"84d34'
Holle,Stephen Douglas
Dra 20h32'55"75d6'
Holle,Vicki Marquardt
Dra 20h1'17"75d23'
Hollebecq,Christiane
Ori 6h0'24"8d53'
Hollenbeck,Dorothy
Cas 0h33'1"63d32'
Hollenbeck,Judith A
Aql 19h40'29"11d1'
Hollenbeck,Pixie
Mon 8h7'1"-5d48'
Hollender,Michael
Her 17h32'56"41d10'
Holler,Blue Max-Blake
Mon 8h4'33"-9d38'
Holler,Jochen
Eri 4h48'56"-7d36'
Holler,Kathrin
Lyn 8h1'11"57d42'
Holleran,Amy Alexandria
Psc 1h23'11"31d46'
Holleran,Matthew
Leo 10h30'33"21d45'
Hollerbach,Michael Robson
Her 18h7'41"31d24'
Hollerbach,Michael
Her 18h17'54"14d31'
Holley
Mon 6h26'51"-6d46'
Holley SMS,Michelle J
Lyr 18h29'0"46d24'
Holley's Star
Lib 15h15'59"-20d20'
Holley,Frieda Mary
Cas 1h38'43"75d16'
Holley,Gary C
Aur 5h9'23"43d4'
Holley,Gregory Scott
Vir 12h39'11"0d43'
Holley,Lempi E
Aur 4h55'1"38d37'
Holley,Scott R
Umi 16h42'59"78d18'
Holley,Sr,The William S
Boo 14h15'56"18d60'
Holliday "China Doll",
Elizabeth
Com 12h54'53"25d57'
Holliday Star,The Silver
Lyn 7h55'47"50d34'
Holliday's James &
Rosemary,40th
Cyg 21h39'0"37d41'
Holliday,Adam
Lyr 18h34'47"33d25'A
Holliday,Alexander Scot
Lmi 10h2'16"32d10'
Holliday,Ashley
Lyr 18h34'47"33d25'B
Holliday,Clint
Lyr 19h1'51"28d44'
Holliday,Jeanna
Del 21h2'52"12d27'
Holliday,Jim
Equ 20h56'33"2d34'
Holliday,Lucille
Cas 23h26'25"61d7'
Hollie Ann
Sge 19h14'54"16d30'
Hollie,Darcy & Gary
Cyg 19h46'56"38d4'
Hollieanna
Ori 4h57'28"15d4'
Hollier "Heavenly
Teacher",Anne S
Lyn 7h47'17"48d26'
Hollier,Cindy
Cas 0h3'1"61d21'
Holliman,Earl
Ser 16h4'0"1d12'
Holliman,Linda Ann Lanier
Cyg 19h45'42"29d38'
Holling,Richard
Ori 6h9'55"5d47'
Hollinger,"Mean Gene" Edgar
Eugene
Cet 1h53'51"-2d6"
Hollings,Janine
Cyg 21h36'17"38d4'
Hollingshead,Ron
Per 1h50'42"50d4'
Hollingsworth,Bill
Cmi 7h41'1"8d11'
Hollingsworth,Carol Ann
Lyr 18h44'43"40d24'
Hollingsworth,Clay Todd
Sex 9h54'35"5d8'
Hollingsworth,Heath
Umi 14h36'20"57d47'B
Hollingsworth,Heather Lee
Del 20h23'40"10d30'
Hollingsworth,James P
Her 17h52'21"14d23'
Hollingsworth,SW
Cet 2h44'38"1d19'
Hollingworth,Joan
And 2h34'49"40d25'
Hollingworth,Ryanne R
Mon 7h3'17"0d24'
Hollinsworth,Raymond Donald
Boo 15h22'34"51d41'
Hollis
Ori 4h42'50"4d23'
Hollis,Adrian Colin (ACE)
Ori 5h5'15"0d47'
Hollis,Brooke Ashley
Peg 22h22'1"21d18'
Hollis,Cameron King
Dra 14h17'0"63d28'
Hollis,Kaitlyn Lee
Umi 14h29'1"68d8'
Hollis,Sharon
And 1h53'30"36d13'
Hollis,Tami Jo
Mon 6h21'59"3d5'
Hollis,Trish & Tim
Uma 9h15'38"48d41'
Hollister Leasing Star,Heather
And 2h25'0"49d57'
Hollister,Andy
Cnv 13h10'1"40d40'
Hollister,Jane Q
And 23h18'50"45d58'
Hollman,Florence Scott
Cnv 13h53'46"41d6'
Hollman,Jennifer Lyn
Cas 1h3'47"60d50'
Hollo,Barrett Christine
And 0h21'47"44d18'
Holloman,Brian
Aur 6h33'49"37d36'
Holloran,Edward J
Cnv 13h29'56"40d53'
Holloway (Hollie), Abigail
Cas 3h5'1"65d0'
Holloway,Billy John
Ori 4h40'27"7d59'
Holloway,Bryan John
Hya 8h54'0"4d50'
Holloway,D'Lorah Ann
Ori 5h38'55"12d43'
Holloway,Kelly Jo Marie
Boo 15h7'0"38d20'
Holloway,Kevin Michael
Aur 5h58'23"37d52'
Holloway,Lorinda Jill
Ori 5h38'60"12d3'
Holloway,Mary
And 23h20'41"41d6'
Holloway,Mr & Mrs Shirley F
Cyg 21h18'55"31d42'
Holloway,Nancy
And 23h16'25"40d43'
Holloway,Philip
Ser 15h13'54"10d37'
Holloway-Heggbloom,
Johanna Katherine
Ori 5h37'59"12d20'
Holloway-Heggbloom, Erin
D'Lorah
Ori 5h38'21"12d23'
Hollowell,Mack
Sgr 18h48'18"-22d4'
Hollowell,Preston Mavor
Peg 21h58'46"10d43'
Hollrah,Mark R
Aql 20h12'60"4d30'
Holly
Lyn 8h54'54"44d29'
Holly Holly
Lyn 8h58'34"44d49'
Holly
And 22h57'18"51d38'
Holly
Eri 3h1'19"-15d19'
Holly
Cet 0h50'1"1d13'
Holly
Aql 19h54'30"11d41'
Holly
Tri 2h16'24"30d23'
Holly
Sge 20h1'51"16d51'
Holly
And 2h35'13"48d54'
Holly
Cet 2h57'39"0d34'
Holly & Joe
Mon 8h6'17"-8d13'
Holly Ann
Cas 0h19'44"61d12'
Holly B
Peg 23h6'0"20d5'
Holly Beth
Uma 12h18'57"60d22'
Holly Bianca
And 0h22'16"30d26'
Holly Eileen
Vul 21h14'31"20d31'
Holly Elizabeth
Eri 2h48'54"-10d31'
Holly Josephine
Cyg 19h30'15"33d47'
Holly Lee
Peg 22h44'27"29d1'
Holly Louise
Umi 15h24'0"70d25'
Holly Raven
Lyr 18h15'1"38d51'
Holly Regina
Peg 22h36'23"29d53'
Holly Robin
Lyr 18h28'30"34d14'
Holly Ruth
Cyg 19h30'27"32d25'
Holly Star,The
And 23h32'12"47d44'
Holly's Rock
Cmi 7h57'21"8d33'
Holly's Sanity
Peg 23h5'59"21d57'
Holly's Star
Cyg 21h30'38"48d52'
Holly's Star
Cas 22h58'44"56d18'
Holly's Star
Com 12h2'1"27d12'
Holly,Michael
Aql 19h31'52"13d7'
Holly,Shawn
Ari 1h56'28"11d47'
Holly,Tiffany
Cyg 20h35'17"30d39'
Holly-Annali
Cas 0h59'12"71d11'
Hollyn,Jessica
Lyn 7h51'14"43d25'
Hollywood
Aql 19h3'41"2d49'
Hollywood Hayes
Uma 10h20'58"68d7'
Hollywood John
Lac 22h7'12"51d38'
Hollywood,Gary
Per 2h57'11"48d49'
Holländer,Iris
Dra 17h52'47"68d21'
Holm,Dolores
Cyg 20h34'18"31d25'
Holm,Eric
Ori 4h59'51"0d31'
Holm,Ian Martin Sutter
Aql 19h2'1"4d46'
Holm,John K
Aql 20h14'0"0d38'
Holm,Kevin James
Ori 5h59'59"11d21'
Holm,Kristina Marie
Lyr 19h16'36"26d54'
Holm,Sebastian Eske
Ari 2h25'25"10d42'
Holm,Tana Marie
Cyg 19h58'0"58d48'
Holm-Jensen,Ditte
Umi 17h9'12"85d46'
Holm-Jensen,Johanne
Cas 2h17'33"67d57'
Holman,Alice Marie Stinson
Sgr 18h51'11"-22d57'
Holman,Austin Ray
Ori 5h49'1"18d19'
Holman,Darlene
And 23h24'43"45d41'
Holman,Joy
Cyg 19h51'15"41d5'
Holman,Michael
Tri 2h17'48"33d35'
Holman,Pam
Peg 23h22'51"13d48'
Holmberg,Lorayne & Bob
Per 3h2'26"43d30'AB
Holmberg,Sandy
Oph 17h0'17"8d41'
Holme,Andrea Elizabeth
Lyr 18h36'44"27d38'
Holme,Julian's
Uma 9h23'43"51d19'
Holmes III,Dr Charles W
Cet 2h29'46"0d25'
Holmes III,The Tom
Oph 18h3'53"1d32'
Holmes,Alec Edward
Aur 6h53'43"38d34'
Holmes,Carolyn Ethel
Dra 14h54'44"64d11'
Holmes,Chandra Elizabeth
And 2h25'30"44d48'
Holmes,David Matthew
Ser 15h51'28"24d2'
Holmes,Dr Robert
Her 15h48'38"40d52'
Holmes,Gene
Cet 3h0'39"4d57'
Holmes,Grace
Cam 4h7'45"60d13'
Holmes,James Jonathan
Ind 20h49'17"-56d56'
Holmes,Jason William
Oph 17h45'59"14d6'
Holmes,Joyce Carrie
Lmi 9h22'41"33d49'
Holmes,Kenneth"Bunny"
Her 17h2'43"37d37'
Holmes,Lawrence Kennedy
Cmi 7h6'20"9d47'
Holmes,Mary Bridget
Cam 3h58'19"55d27'
Holmes,Nancy
Eri 4h58'1"-4d54'
Holmes,Natalie Rebecca
Lyr 18h29'28"32d46'
Holmes,Nellee A
Peg 22h26'46"31d29'
Holmes,Patti
Cru 12h43'1"-57d46'
Holmes,Russell Stephen
Cnv 12h50'11"50d14'
Holmes,Skyler
Eri 4h8'15"-19d54'
Holmes,Sr,Charles
Boo 14h20'48"16d21'
Holmes,Vanessa
Cas 0h44'40"62d32'
Holmes,Willie
Cyg 21h4'15"37d36'
Holmes-Laserpoint,Andy
Cep 21h0'43"55d19'
Holmgren,Joshua
Crb 16h0'18"32d13'
Holmlund,Pam
Lib 15h31'55"-8d34'
Holmquist,Elizabeth Marina
Cas 0h4'0"61d22'
Holmquist,Lance & Sherry
Rogers
Sge 19h53'55"16d10'
Holmquist,Mollie
Cas 1h13'32"68d10'
Holmsen,Anita
Eri 3h56'0"-16d47'
Holmstrom,Johan
Per 2h0'32"50d22'
Holofcener,Nicole & Ben
Allanoff
Crb 16h22'44"38d47'
Holomb 1968 Love,Josh
Charles
Cet 1h38'11"1d12'
Holomon,Jack C
Eri 4h12'26"-19d36'
Holomshek,Joan & Charles
Cnv 13h54'20"40d2'
Holopikian,Charles M
Dra 16h3'38"63d22'
Holopikian,Pauline
Dra 16h19'1"62d32'
Holotik,Jim
Crt 11h52'36"-7d56'
Holowka,Donna Marie
Peg 22h37'23"14d17'A
Holowka,Michele Lee
Peg 22h37'23"14d17'B
Holsbeck,Mark Van
Boo 14h30'19"22d3'
Holsenbeck,Wade Davis
Ori 5h52'22"16d19'
Holsomback,Stewart E
Lac 22h22'0"52d53'
Holst,Howard Norman
Dra 10h50'21"74d9'
Holst,Wolfgang
Lyr 19h21'50"30d18'
Holstead,William Drury
Gem 7h15'2"28d34'
Holstein,Elizabeth Rebecca
Vir 12h32'18"-10d11'
Holstein,John J
Lac 22h3'10"40d5'
Holstein,Peter
Aql 19h13'1"13d6'
Holsten,Kyle Alexander
Per 3h42'36"50d27'
Holsten,Stephen E
Per 3h23'46"40d58'
Holster-Bob Borealis
Cep 22h43'0"70d8'
Holstman,Melissa Duncan
Del 20h29'48"20d13'
Holsworth,Sara Ellen
And 0h23'24"37d0'
Holt Love Story
"Piccadilly",Max
Cma 6h12'16"-15d10'
Holt Special Education Staff
93-94
Cep 21h9'32"55d45'
Holt,Aaron
Cet 1h37'0"-13d2'
Holt,Amanda Mary
And 2h25'37"38d19'
Holt,Austin Jesse
Aur 7h7'43"41d6'
Holt,Barbara Van
Cet 2h17'51"4d25'
Holt,Bruce Anton
Ser 15h53'20"0d37'
Holt,Carolin Ruth Jokemca
Sublette
And 23h3'1"46d16'
Holt,Catherine L
Tri 1h36'47"30d20'
Holt,Chad
Dra 15h10'1"61d29'
Holt,Daniel
Aql 20h7'47"1d15'
Holt,Darren Thomas
Crt 11h19'12"-19d21'
Holt,Ed & Pat
Per 1h46'1"53d44'
Holt,Ellen C
Cet 0h27'29"-17d54'
Holt,Gary W
Cet 0h50'28"-5d52'
Holt,Georgina
Lyr 18h29'49"46d34'
Holt,Glenn & Yvonne
Cyg 20h8'35"41d2'
Holt,Harold L
Her 18h23'12"28d47'
Holt,Janet S
Crt 11h14'31"-19d20'
Holt,Johnnie
Cet 1h59'20"-6d7'
Holt,Jr,James Arthur
Sgr 19h24'52"-43d34'
Holt,Jr,John D
Cmi 7h45'37"8d42'
Holt,Kathi
Lyr 18h22'37"37d48'
Holt,Laura Elizabeth
Cyg 19h44'1"30d2'
Holt,Laura Margaret
Peg 22h3'43"27d22'
Holt,Lillian
Peg 22h14'48"5d18'
Holt,Linda Jayne
Boo 14h31'0"50d40'
Holt,Margaret Eleanor
Ori 6h4'35"5d9'
Holt,Marjorie M
Mon 7h39'25"-3d38'
Holt,Mary Elizabeth
Equ 20h58'0"5d20'
Holt,Paul
Tri 2h19'1"31d12'
Holt,Rick
Dra 17h13'0"60d20'
Holt,Robert Douglas
Aur 4h48'55"40d30'
Holt,Robert R
Del 20h37'46"4d51'
Holt,Stanley Ray
Cnc 9h2'31"30d49'
Holt,Sue & Steve
Ori 6h7'29"8d54'
Holt,Susan N
And 23h3'19"51d45'
Holt,Tommy
Aur 7h22'51"41d33'
Holt,Wilma Lee
Vul 20h39'24"28d58'
Holt,Zachary Aaron
Lyn 9h15'30"39d47'
Holt-Hills,Alison Gail
Lyn 8h11'23"43d11'B
Holtan,Erik Joseph
Her 16h59'59"31d35'
Holtby,Bill & Lorraine
Eri 4h53'42"-6d13'
Holten,Bud
Per 1h47'46"50d31'
Holtham,Frank C
Cep 22h17'36"60d30'
Holthaus,Fusako
Cam 3h46'45"57d57'
Holthaus,Rebecca
Lyr 18h14'28"31d45'
Holthe,Hanne
Dra 17h39'12"64d20'
Holtman,Carrie Elizabeth
Leo 9h51'22"17d35'
Holtman,Harry
Her 16h19'1"4d29'
Holtman,Joseph A
Dra 17h48'33"76d2'
Holtman,Martha Lucille
Peg 22h37'34"26d41'
Holtman,Violet
Cet 2h56'43"8d10'
Holton (EP),Patricia Mavis
Lyr 18h16'42"31d21'
Holton,Burch
Lac 22h3'38"51d37'
Holton,Katie & Eilish
Lyr 19h26'13"38d12'
Holton,Maria Ellen
Cyg 20h17'25"39d28'
Holton,Robert D & Bonnie M
Eri 3h38'21"-7d3'
Holton-AILAT,Harvard S
Her 16h42'46"32d41'
Holts,Scott
Dra 16h43'19"72d16'
Holtsclaw,James William
Ori 5h51'10"14d53'
Holtus Bearialis
Uma 10h36'11"70d42'
Holtz Family Star,The
Boo 14h59'26"29d59'
Holtz Rubenstein & Co, LLP
Tri 2h15'20"31d55'
Holtz,Chris & Janet
Cyg 21h2'57"31d12'
Holtz-Hanefeld,Dorothe
Psc 1h1'23"21d30'
Holtzman,Dr Herbert
Lac 22h35'18"55d55'
Holub,Diane Clayre
Cas 23h39'44"61d7'
Holvick,Brenda Jean
Vul 19h16'12"25d37'
Holvie
Lac 22h33'30"53d59'
Holway,Frances
Cnc 8h9'1"30d6'
Holweg,Konrad
Cam 6h6'15"70d24'
Holwell,Agnes
Uma 11h6'0"60d52'
Holwell,Mervin
Her 16h21'17"24d22'
Holycross,Kathleen
Cas 1h47'30"60d18'
Holycross,M Alexander
Mon 6h27'51"-0d21'
Holyoak,Heather Kathleen
Peg 22h57'29"30d2'
Holzbaur,David R
Sgr 19h7'54"-25d45'
Holzberger,James A
Ari 3h4'47"30d54'
Holzer,Craig
Dra 12h26'1"71d39'
Holzer,Tiffiny Linda
Peg 22h18'17"5d57'
Holzinger,Deborah Ann
Crb 15h33'22"26d18'
Holzkaemper,Maxine
Lib 15h8'0"-20d1'
Holzknecht,Günther
Leo 11h14'0"-5d17'
Holzleitner,Günter
Uma 11h54'27"64d53'
Holzman,Philip Andrew
Cep 22h4'54"68d60'
Holzman,Shirley & James
Cyg 21h29'49"40d4'
Holzmann Star,The
Lac 22h31'27"54d8'
Holzmann,Britta
Sgr 20h3'57"-42d12'
Holzmann,Julia- Katharina
Cap 21h55'11"-22d30'
Holzmann,Jörg Anton
Cap 21h52'21"-22d23'
Homa Star,My Legendary C
William
Ori 4h42'57"7d53'
Homan,Adam Dean
Gem 7h38'37"20d21'
Homan,Santana
Ori 5h27'20"1d30'
Homburg,Allysa
Cas 0h37'18"63d37'
Homburg,Eric & Jodie
Crb 15h29'49"31d4'
Homburger Juwel
Uma 9h52'35"57d45'
Home
Cet 3h2'49"8d30'
Home
Uma 11h50'1"32d21'
Home News
Boo 14h57'11"48d59'
Home Of The Jean Marie
Umi 14h5'0"66d40'
Home On The Range
Dra 16h4'49"68d15'
Home Show,The
Cyg 21h59'54"50d37'
Homeboy
Per 3h5'50"46d33'
Homelsky,Isidore Norman
Sex 9h52'47"5d40'
Homelsky,Tinnie Venetek
Sex 9h53'26"2d24'
Homenko,PhD,Donna
Cep 23h13'23"64d12'
Homer #3
Tau 5h51'13"28d41'
Homer,Cheryl Ann
Lyn 8h20'47"58d19'
Homer,Rachel Louise
Com 12h30'30"21d49'
Homer,Steven J
Sge 19h59'40"16d47'
Homewood,Amanda
Cas 0h57'56"70d27'
Homey
Lyn 7h22'24"44d36'
Homi
Cnc 8h32'51"7d0'
Homily
Aur 5h0'57"46d38'
Homm,Debbie
Cas 1h17'44"64d45'
Homme D'Asie
Lyn 9h5'50"40d14'
Hommell,Lauren Leann
Peg 23h46'54"29d53'
Hommer,Mark Philip
Cep 20h54'0"60d40'
Hommet,Marie-Laure
Lyn 7h57'53"40d23'
Homminga,Alphonsus
Peg 23h21'1"33d30'
Hompe,Phyllis L (Stone)
Cas 3h13'17"70d55'
Homthal,Becca
Peg 22h5'54"29d0'
Hon
Lyr 18h31'20"41d26'
Hon,Amy
Cas 22h59'58"55d60'
Honadel,Alexandra Lauren
Ari 2h30'46"20d11'
Honaker,Shannon
Lyr 18h30'48"30d13'
Honaki
Per 3h33'11"36d14'
Honan,The Rita
Cyg 19h45'35"29d30'
Honda Rhonda
Uma 10h11'30"57d20'
Hondros,John Paul
Her 16h57'46"28d11'
Hone,David & Anne
Aql 19h24'49"15d40'
Honecker,Ernest
Per 1h50'42"52d35'
Honegger,1992- Centenaire
d'Arthur
Uma 9h10'58"52d5'
Honegger,Barbara
Peg 21h57'58"34d7'
Honegger,Elaine S
Lyr 18h55'30"40d59'
Honegger,Kathryn Elizabeth
And 2h22'21"47d41'
Honer,Joseph Anthony
Aqr 22h31'0"-1d36'
Honesty
Peg 21h57'9"28d51'
Honesty
Oph 17h38'16"-20d36'
Honey
Pup 8h8'42"-27d42'
Honey
Her 16h58'1"32d45'
Honey
And 1h40'43"39d28'
Honey
Cep 21h5'44"58d24'
Honey
Del 20h23'23"10d34'
Honey
Del 20h49'12"7d52'
Honey
Oph 17h15'54"-20d35'
Honey
Cyg 19h58'52"30d41'

Honey
Del 20h17'0"11d14'
Honey
Lyn 7h51'23"48d17'
Honey & Brandy's Mom
Mon 6h39'33"6d60'
Honey & Melvin
Per 2h53'57"32d21'
Honey Bear
Uma 13h31'15"57d34'
Honey Bear Brian
Her 17h55'31"28d50'
Honey Bee
Dra 17h35'59"65d27'
Honey Bunkins Star
Uma 11h45'11"43d58'
Honey Bunney
Uma 8h18'30"68d19'
Honey Bunny
Peg 22h9'38"21d3'
Honey Divide
Ori 5h45'45"11d6'
Honey Girl
And 1h18'22"38d31'
Honey Girl-94
And 1h24'16"33d32'
Honey Hogan
Equ 21h15'0"3d30'
Honey in my Heart
Dra 19h53'25"67d44'
Honey Phiphi
Per 3h52'46"38d25'
Honey Star,The
Peg 22h22'19"35d5'
Honey's Haven
Cyg 20h9'24"39d48'
Honey's Spirit Of Pig
Aur 7h22'36"40d29'
Honey's Star
Aqr 21h31'39"-1d38'
Honey,Frederick J
Boo 14h15'57"36d12'
Honey,Patricia M
Cas 3h5'0"65d18'
Honey-Du
Uma 8h52'41"61d39'
Honeybear
Aur 6h1'17"30d17'
Honeycass
Cnv 12h17'39"50d36'
Honeycutt,Barry & Wendy
Del 20h15'37"13d5'
Honeycutt,Julie
Peg 22h43'1"29d24'
Honeyman,Jeb Cory
Mon 7h42'41"-1d3'
Honeysuckle
Lyr 18h20'17"42d1'
HoneySuckles "Love Bunny"
Boo 14h19'41"37d4'
Hong,A Ryung
Aql 19h56'28"12d7'
Hong,Grace
Lmi 10h47'43"32d45'
Hong,Master Y K & Mrs
Onsook Hong
Uma 16h46'17"82d27'AB
Hongisto,Mike
Dra 13h6'36"76d27'
Honhon,Yves
Cyg 20h23'1"39d31'
Honi
Lyn 8h16'14"36d12'
Honickman,Mauri Elizabeth
Cyg 20h17'19"49d16'
Honickmann,Harold
Lyr 18h58'48"38d12'
Honig's Wishing Star, Marc
Per 3h6'36"47d23'
Honingh,Karin
Uma 13h23'53"62d41'
Honkanen,Raimo
Cam 12h9'23"77d7'
Honneffer's Love, Daniel
Peg 22h36'23"8d59'

Honor Bright Bob
Eri 4h55'23"-4d59'
Honor,James
Per 2h52'49"45d25'
Honorine
And 2h24'0"45d22'
HonorJeanne
Lib 15h4'12"-0d16'
Honschopp,Kyle Jacob
Cep 21h29'58"63d57'
Honschopp,Tyler Ross
Dra 16h53'0"60d43'
Honse,Samuel Greer
Dra 16h58'45"68d48'
Honsey,Jon & Sue
Lyr 19h0'25"31d44'
Honston,Alexandra Jane
And 0h7'51"37d57'
Hontz,Lillie Jane
Mon 6h42'33"11d49'
Honymoon
Peg 23h30'26"15d50'
Hoo "Going Home", Jonathan
Per 1h47'1"56d40'
hoob & poo
Vul 19h36'45"20d16'
Hood
Umi 15h4'40"71d25'
Hood II,John A
Peg 22h1'18"10d53'
Hood III,John A
Cep 23h9'13"60d39'
Hood,Alberta
Cas 0h5'1"56d6'
Hood,Andrew Christopher
Per 3h2'26"41d1'
Hood,Corey Tyler
Cyg 19h36'19"28d31'
Hood,Corrie Lorraine
Vul 20h4'14"28d33'
Hood,David Ian
Oph 18h34'44"10d42'
Hood,Dennis & Shannon
Eri 2h50'40"-1d49'
Hood,Emily Ellen Thacher
Cas 0h53'48"64d59'
Hood,Grant
Cyg 20h26'23"40d2'
Hood,John A
Vul 19h36'59"20d9'
Hood,Lynn Jennifer
And 2h3'45"40d26'
Hood,Marsha C
Ari 2h2'17"18d7'
Hood,Michele
Cas 2h40'25"61d51'
Hood,Paul
Cet 2h42'43"4d30'A
Hood,Robin E
Oph 17h1'16"8d46'
Hoodem Magic
Boo 14h52'30"36d47'
Hoodless,Elain Margrette
And 23h38'22"41d57'
Hoof,Ryan Michael
Mon 6h15'21"-6d23'
Hoofnagle,Alice Margaret
Murdach
Lyr 18h37'46"45d40'
Hooge Troll & Magic Fairy
Princess
Lyr 18h37'20"40d50'
Hook Muench
Aql 18h43'23"6d34'
Hook's Rising Sun
Cam 5h59'55"67d39'
Hook,Alice Marie
Lyr 18h17'26"44d50'
Hook,Betty & Brian
Cyg 19h48'33"37d37'
Hook,Jabez & Carol
Sge 19h19'56"16d47'
Hook,Jerry
Boo 15h6'29"10d39'

Hook,Jr,John Stanley
Lac 22h12'1"54d37'
Hook,Kari A
Cnc 8h22'14"15d12'
Hook,Miriam
Uma 10h57'52"40d15'
Hook,Neva A
Cam 5h58'53"58d50'
Hook,Randolph C
Ori 5h56'18"18d20'
Hook-Meade,Kyle Joseph
Uma 17h8'21"76d13'
Hooker's Delight
Mon 6h54'54"-0d58'
Hooker,David A
Her 18h9'15"30d21'
Hooker,Dr Andrew Neil
Uma 9h14'14"68d12'
Hooker,E G
Aql 20h1'59"14d13'
Hooker,Julie
Cas 2h56'38"61d33'
Hooker,Katelyn Ann
And 2h4'18"37d15'
Hooker,Ryan Patrick
Cep 22h43'59"70d4'
Hooker,Sarah
Gem 8h5'24"28d47'
Hooks,Audrey
And 1h25'28"36d3'
Hooks,Morgan Rachel
Ari 1h45'35"13d54'
Hoonan,Rosemary April
Tau 5h28'10"28d29'
Hooper,Bernard L & Cheryl
Ireland
Vir 13h55'25"-21d34'
Hooper,Beth Laura
Cas 2h54'33"68d39'
Hooper,Desmond
Per 3h21'44"40d48'
Hooper,Elizabeth
Crb 15h16'32"30d49'
Hooper,Erin L
Aql 19h2'14"1d54'
Hooper,Evelyn Day
Eri 3h48'50"-2d2'
Hooper,Helen
Tri 1h41'59"34d9'
Hooper,Jeffrey Alan
Ari 2h37'28"21d23'
Hooper,Mary Judith
Cas 2h22'58"70d40'
Hooper,Rebecca Leigh
Mon 8h2'28"-6d9'
Hooper,Steven C
Aur 6h4'27"31d8'
Hooper,Suzanne
And 23h1'28"50d19'
Hooper,Timothy Scott
Cma 7h14'24"-16d49'
Hooper,Vearl
Sct 18h45'5"-7d50'
Hoops,Jerry Clark
Cap 21h52'58"-20d21'
Hoos,Gary
Cam 4h56'42"60d40'
Hoos,Suzanne
Vul 19h44'1"28d24'
Hooser,Sr,In Memory Of Carl
Ori 6h16'25"-1d39'
Hootie
Cam 8h7'46"80d14'
Hooton,Brittany
Peg 22h46'19"4d26'
Hooton,Ray
Uma 11h30'27"38d41'
Hoover (Pop),Ralph
Aql 20h30'60"0d23'
Hoover Forever Star, Molly &
Cliff
Peg 21h55'44"24d33'
Hoover,Bryan Dale
Peg 23h29'19"23d47'

Hoover,Carolyn Pemberton
Ori 5h50'1"16d47'
Hoover,Charles
Boo 14h25'0"29d0'
Hoover,Charlie
Ser 18h4'58"-14d49'
Hoover,Chris & Allison
Dra 18h43'0"70d57'
Hoover,Cindy Lea
Dra 12h3'2"69d4'AB
Hoover,Colin H
Her 16h19'36"23d28'
Hoover,Emily Savannah
Mon 6h42'1"10d3'
Hoover,Ernie
Her 18h3'20"48d49'
Hoover,Henry J
Ser 18h9'41"7d18'B
Hoover,James B
Cet 2h1'36"0d55'
Hoover,Linda
Lyr 19h24'1"37d58'
Hoover,Linda Strait
Crb 15h56'58"31d18'
Hoover,Linda Strait
Mon 8h5'17"-0d23'
Hoover,Matthew R
Cnv 12h47'0"51d29'
Hoover,Mildred B
Ser 18h9'41"7d18'A
Hoover,Missy
Dra 16h7'42"62d23'
Hoover,Ronald Grant
Peg 22h25'45"17d0'B
Hoover,Stephen Paul
Cnv 13h56'33"40d28'
Hoover,Taylor Rhiannon
Mon 6h54'1"0d34'
Hoover,Terri
Ori 5h54'57"13d4'
Hoover,William L
Cma 7h17'13"-15d40'
Hooverville Elementary School
(Waynesboro,PA)
Uma 8h28'55"71d0'
Hop Step Australia
Cru 12h53'16"-60d50'
Hop,Steven J
Cyg 20h3'47"38d46'
Hopcroft,Dorothy Margaret
Mon 6h54'36"-10d57'
Hope
Peg 22h58'1"22d1'
Hope
Cyg 21h21'47"38d43'
Hope
Ori 5h57'58"8d11'
Hope
Uma 10h12'0"68d40'
Hope
Aql 19h45'18"13d53'
Hope
Peg 22h44'0"26d15'
Hope
Ser 15h35'16"7d49'
Hope
Uma 10h25'12"50d50'
Hope
Mon 6h20'33"6d46'
Hope
And 0h14'58"38d41'
Hope
Cnv 12h7'26"42d14'
Hope
Cas 0h21'39"56d7'
Hope
And 1h47'22"40d57'
Hope
And 23h39'1"38d59'
Hope & Nate-Friends Eternally
Peg 22h8'1"3d35'
Hope Kelly
Cas 1h5'60"61d28'
Hope Megan
Lyr 18h59'39"28d51'

Hope Star
Cyg 21h35'26"38d26'
Hope Star,The
Peg 21h23'44"23d4'
Hope's Heart Eternally 143
Ori 5h3'28"14d58'
Hope's Star
Uma 10h34'14"56d2'
Hope's Twilight
Mon 7h59'35"-1d33'
Hope,Anita
Peg 22h20'42"5d6'
Hope,Ann
Lmi 9h26'19"37d35'
Hope,Bob
Per 2h30'25"57d46'
Hope,Bob
Boo 14h3'36"13d25'
Hope,Karen
Cas 0h37'1"63d55'
Hope,Katherine Jeanne
Mon 6h54'47"0d1'
Hope,Susanne Devik
Dra 11h0'13"73d55'
Hopely,Stephanie Jean
Schmauk
Mon 7h51'11"-3d28'
Hopeman,Robert Mark
Sex 10h29'33"-8d35'
Hopf,Christl & Horst
Lib 14h31'28"-20d12'
Hopf,Herr
Boo 14h0'15"10d4'
Hopf,Martin
Cam 5h50'31"70d24'
Hopke,Fred
Cep 21h35'19"55d3'
Hopkins Master
Metaphysician,Arthur M
Aur 6h4'46"48d54'
Hopkins Star
Her 18h2'47"28d58'
Hopkins Star of The
Universe,Marie
Peg 23h32'27"33d50'
Hopkins,Adam Noah
Cnc 8h39'42"17d48'
Hopkins,Adam Noah
Cnc 8h39'53"17d11'
Hopkins,Amanda Carol
Mon 6h39'39"1d50'
Hopkins,Amanda Jane
And 23h39'0"41d55'
Hopkins,Andrew Brian
Her 16h24'27"39d40'
Hopkins,Baby
Ori 5h34'0"-0d4'
Hopkins,Catherine
Cas 0h58'57"63d36'
Hopkins,Charles Cooper
Aur 5h14'19"40d22'
Hopkins,Charles Eason
Alexander
Her 16h43'0"27d19'
Hopkins,Colin Noel
Sgr 20h21'38"-28d21'
Hopkins,Cyril
Aur 5h25'29"30d8'
Hopkins,Daniel D
Uma 12h8'58"53d42'A
Hopkins,Danielle C
And 23h34'43"39d34'
Hopkins,Della May Solmon
Cas 0h8'26"60d35'
Hopkins,Earl Mortimer
Aql 19h53'47"14d32'
Hopkins,Edris Jo
Peg 22h31'40"21d11'
Hopkins,Everett A
Peg 23h43'57"20d8'A
Hopkins,Geraldine
Lyr 18h54'12"42d29'
Hopkins,Geraldine Ruth
And 2h31'50"37d40'

Hopkins,Gordon & Irene
Uma 10h11'1"47d31'
Hopkins,Harland
Her 16h54'38"38d44'
Hopkins,Jane
And 1h43'30"40d38'
Hopkins,Jedi Master, Keith
Alexander
Cnv 12h21'60"51d20'
Hopkins,Jennie
Umi 13h11'10"71d16'
Hopkins,John
Cep 22h28'44"70d58'
Hopkins,John Robert
Per 2h52'46"48d58'
Hopkins,Joyce "Our Staar"
Mon 6h19'43"6d58'
Hopkins,Leo M
Psc 0h55'54"28d17'
Hopkins,Marilyn "Perk"
Uma 9h7'12"71d18'
Hopkins,Rachel Nicole
Mon 6h33'20"4d34'
Hopkins,Raymond
Her 18h1'39"14d23'
Hopkins,Ryan Barnard
Lyr 19h0'46"40d1'
Hopkins,Ryan Blair
Hya 8h33'39"0d49'
Hopkins,Sharalynn Melissa
Lyr 18h47'48"39d55'
Hopkins,Sheila
Lyr 18h20'26"38d44'
Hopkins,Sir Anthony
Cep 21h0'1"65d21'
Hopkins,Sr,Mark Anthony
Boo 14h35'14"8d1'
Hopkins,Susan B
Peg 23h43'57"20d8'B
Hopkins,Thomas
Tri 2h41'16"35d6'
Hopkins,Victoria Claire
Cas 0h50'55"71d32'
Hopkins,William Gregory
Tau 4h33'48"30d39'
Hopkins-Connors- Medallis
Cam 7h26'35"61d28'
Hopman-Star
Cet 2h57'11"2d47'
Hopp,Jane Anna
Lyr 18h18'28"43d9'
Hopp,Kristen Johanna
Lac 22h24'48"54d48'
Hopp,Mary Lou
Lyn 7h1'55"50d6'
Hopp,Nancy S
Lmi 9h50'0"38d17'
Hoppe,Donald L
Cep 0h10'48"67d11'
Hoppe,June L
Gem 7h32'45"31d21'
Hoppe,Karen
Cas 23h22'25"60d25'
Hoppe,Manfred
Uma 10h31'14"48d46'
Hoppe,Mark & Darlene
Cet 2h43'23"6d56'
HOPPER
Uma 10h51'33"37d40'
Hopper Nr vedby
Denmark,Sten
Ori 5h26'0"0d14'
Hopper's Twilight
Ori 5h57'28"16d29'
Hopper,Barbara
Vul 19h44'59"23d54'
Hopper,Drew
Eri 3h19'31"-17d14'
Hopper,Dustin William
Her 16h59'26"48d16'
Hopper,George R
Hya 8h19'42"1d2'
Hopper,Graham
Cep 20h54'50"71d7'

Hopper,Jeff
Cet 2h59'11"0d38'
Hopper,Jennifer
Com 13h7'11"27d20'
Hopper,John P
Cma 6h47'53"-19d23'
Hopper,Lewis Dewayne
Cet 0h25'36"-18d52'
Hopper,Linda
Eri 4h1'31"-19d14'
Hopper,Samuel Richmond
Sco 17h5'54"-38d44'
Hopper,Steven A C
Ori 5h36'44"-0d18'
Hopper,Thomas & Pauline
Cyg 19h14'51"44d22'
Hopper,Toni Rae
Lyr 18h27'19"38d48'
Hopper-United Forever, Jeff &
Kelli
Cyg 19h26'29"33d10'
Hopps,Steven "My Brother"
Boo 14h37'43"47d55'
Hops,Sam
Cam 5h38'48"61d6'
Hopson,Leonard
Lac 22h7'27"50d16'
Hopson,Louise Androff
Eri 3h17'30"-12d23'
Hopson,Robert Earl
Hya 8h46'21"-8d14'
Hopson/Meriwether
Cyg 21h3'28"40d20'
Hopta,Christopher David
Per 2h45'1"43d9'
Hopta,Matthew David
Peg 22h24'26"20d32'
Hopta,Samantha Lynn
Lmi 9h50'49"34d37'
Hoptoma-Reviens-Moi
Boo 14h52'1"48d14'
Hopuful Stella of Yukari
Leo 10h4'5"18d2'
Hopwood,Martin M
Dra 18h58'0"58d37'
Hopwood,Robert W
Uma 9h20'18"54d54'
Hoque,M D Mozammel
Her 18h27'40"13d3'
Horacek III "Jody", Joseph
Cep 20h44'46"67d36'
Horak's Heart
Ori 5h2'56"1d41'
Horan,Ellen
And 23h31'31"48d2'
Horazdovsky
Her 17h37'41"14d55'
Horbath,Andrea Kay
And 2h33'42"44d7'
Horbina
Cet 2h57'24"0d58'
Horcher,Joyce Lynn
Peg 21h38'49"26d42'
Horchuck,Michael & Helene
Cyg 21h11'13"37d43'
Hord,Mary Colette
Tri 2h1'43"31d48'
Hord,Nichole Marie
Cas 0h47'0"61d41'
Hord,Ronald Dale
Per 2h55'13"32d43'
Hordynski,Stephen N
Aqr 21h27'53"-0d29'
Horel,Eugène
Umi 16h53'13"79d38'
Horeman,Audrey Beryl
Cas 0h37'59"60d39'
Horina
Dra 16h12'1"66d57'
Horkan,Michael
Dra 17h19'36"68d40'
Horlacher,Klaus
Tau 4h18'14"1d5'
Horlick,Gavin
Cnv 12h53'0"47d29'

Horling,Sonja
Lib 14h19'0"-20d18'
Horloger Des Etoiles, Phillipe
Cyg 21h23'46"50d2'
Hormann,Hans-Peter
Leo 10h57'0"11d34'
Horn,Carolyn Ann
Mon 7h47'46"-9d3'A
Horn,Casey
Cyg 21h3'30"31d27'
Horn,Christina
Lmi 10h24'19"31d19'
Horn,Dianne
Lyr 19h18'22"42d35'
Horn,Dr Alfred
Cap 21h24'20"-23d18'
Horn,Dr George A
Oph 17h19'27"10d31'
Horn,Gary
Dra 17h0'28"67d52'
Horn,Jason S
Aur 6h11'41"33d54'
Horn,Jeanne
Cam 3h27'12"53d20'
Horn,Joachim
Lyr 19h21'49"31d17'
Horn,John Michael
Psc 1h36'1"18d33'
Horn,Jr,Claude R
Lmi 10h10'11"32d54'
Horn,Judith
Ari 3h21'43"30d4'
Horn,Michael
Her 17h1'26"45d36'
Horn,Sara A & Taylor B
Gem 8h0'57"33d10'B
Horn,Stacy
Crb 15h48'35"30d59'
Horn,Trevor
Dra 16h4'16"64d0'
Horn,Vince Penny Jake &
Lizzy
Umi 15h33'18"68d26'
Horn-L'il Angel, Zachary
Christopher
Aur 5h11'46"41d13'
Hornaday,Charles C
Her 17h20'15"14d27'
Hornaday,Elizabeth H
Peg 23h5'41"22d20'
Hornbeck,Carol Elizabeth
Raynor
Gem 6h41'47"18d44'
Hornbeck,Doug & Betty
Ori 5h6'32"1d39'
Hornbecker,Lindsay Claire
Cam 6h13'49"56d4'
Hornbek,Michael Andrew
Thomas
Cet 3h16'24"8d0'
Hornberger,Reed & Carolyn
Cmi 7h6'37"9d8'
Hornberger,Rita Mae
Peg 23h19'21"28d31'
Hornbrook Star,The Jill
And 22h58'1"51d26'
Hornbuckle
Dra 17h40'47"67d35'
Hornburg "Jolly John", John
Webber
Boo 14h11'28"37d29'
Horne Star,The Bill
Her 17h7'35"49d18'
Horne's Heavenly Hound Dog
Cma 7h17'17"-15d29'
Horne,Clyde B
Hya 8h43'0"6d3'
Horne,Connie
Eri 3h38'17"-4d13'
Horne,Deborah Norris
Cet 2h47'21"0d28'
Horne,Derrick
Her 17h31'28"20d15'
Horne,Jeff
Dra 16h29'57"63d19'

Horne,Jordyn Elizabeth
Cyg 21h33'54"30d5'
Horne,Kary
Aql 20h11'34"10d23'
Horne,Lillian
Cyg 20h37'19"47d30'A
Horne,Mark
Ori 6h0'24"3d25'
Horne,Paige Veronica
Umi 16h24'28"75d16'
Horne,Shelley Dionne
Del 20h15'28"9d44'
Horne,Stacy
Cas 2h3'49"60d1'
Horne,Terry
Lac 22h5'10"49d54'
Horne,Tierney
Cnv 13h8'15"33d36'
Horne,Walter William
Cyg 20h26'45"38d17'
Horned Beast,The
Dra 20h13'59"62d43'
Hornek,Dr Herbert
Aqr 22h40'31"-5d50'
Horneman,Jillian Rae
Gem 5h58'20"26d46'
Horner BA,PGCE,Carl Peter
Jonathan
Cep 23h6'27"68d0'
Horner,"Rocky"
Cet 0h31'47"-3d3'
Horner,Bill & Margot
Mon 7h57'29"-9d2'
Horner,Bruce
Lac 22h11'0"54d51'
Horner,Evelyn Ann
Eri 3h56'11"-17d58'
Horner,Forever Jack
Aur 4h54'55"50d59'
Horner,J P
Crb 15h32'55"30d4'
Horner,Joseph F
Aur 5h31'42"30d10'
Horner,Paul L
Per 2h52'37"45d58'
Horner,Phillip F
Cap 21h22'11"-25d16'
Horner,Sally Melton
Ori 4h57'55"15d6'
Horner-Hawkins
Per 2h57'37"38d51'
Hornfeldt,Javier
Cas 0h4'24"60d46'
Horni,Clint & Peg
Cyg 21h2'13"38d25'
Hornig,Brigitte
Crb 16h5'41"26d5'
Horning,Joseph Francis
Lac 22h9'30"38d15'
Hornreich,Sean
Mon 7h16'11"-6d2'
Hornsby,Leo's William &
Margaret
Peg 21h58'45"20d22'
Hornsby,Lesley E
Cet 0h50'24"-1d6'
Hornsby,Marian Pauline
And 0h3'0"40d34'
Horny
Boo 14h58'31"40d32'
Hornyak,Jr,John Michael
Aur 7h1'54"36d56'
Horobin,Lynne
Lyr 18h35'36"39d32'
Horowitz 12,Liza
Mon 7h20'13"-8d3'
Horowitz's Star,Robert &
Eleanor
Cet 1h33'18"0d53'
Horowitz's Star,Steve &
Melinda
Ori 6h9'19"7d2'
Horowitz,Austin N
Aql 18h42'52"-2d4'

Horowitz,Dr Howard (Howie)
Oph 17h57'29"8d37'
Horowitz,Dr Stuart Elliot
Cnc 8h37'23"18d33'
Horowitz,Evelyn Stella
And 0h29'0"45d2'
Horowitz,Hun 1/10 Ronald J
Oph 17h9'13"-23d27'
Horowitz,Jill
Crb 15h21'35"31d2'
Horowitz,Marlene Elyse
Eri 2h58'40"-4d11'
Horowitz,Peaceful Nicholas
Boo 14h52'57"53d32'
Horowitz,Rudy J
Dra 11h5'24"73d33'
Horowytz,Bruce Aaron
Uma 10h19'48"60d45'
Horridge,Richard John
Umi 15h9'36"69d23'
Horridge,Richard M & Claire H
Cyg 21h57'1"50d12'
Horrigan,Tony
Lyn 7h48'32"47d49'
Horrobbin,Seana P
Boo 14h28'31"21d12'
Horrobin,Karla M
Lmi 9h39'43"38d7'
Horrock,Victoria
Mon 7h1'60"1d51'
Horsburgh,Barrie & Sandra
Cyg 21h2'1"38d47'
Horschel,Gwendolyn Rae
Mon 6h19'30"3d24'
Horscroft,Claire Elizabeth
Peg 22h30'60"11d26'
Horsefield,Gregory Steven
Her 17h21'50"48d13'
Horseshoe & Pearl, Somewhere In Time
Lyr 19h20'1"38d36'
Horsey,Henry Ridgley
Aur 5h0'18"52d17'
Horsley,Hilary Elizabeth
Dra 15h36'16"58d53'
Horsley,Luke Ellison
Oph 18h40'1"8d13'
Horsley,Rhian Marie
Lac 22h5'39"51d42'
Horsley,Scott
Ori 6h3'30"8d29'
Horsman,Teresa Kay
Uma 9h41'25"50d40'
Horst
Mon 7h35'41"-3d25'
Horst
Sgr 18h52'25"-28d48'
Horst,Eric & Mandy
Cyg 19h17'37"44d45'
Horst,Lesley Anne
Com 12h7'21"27d32'
Horst,My Shining Star, Jay
Ori 5h41'1"-0d30'
Horst,Sandra K
Lyr 18h15'14"31d46'
Horstick,Joan Piersen
Crb 15h16'11"31d25'
Horstmann,Herr
Cmi 7h17'35"0d48'
Horswell "Blade", Robert Daniel
Gem 7h35'51"20d31'
Hortense (For Dorothy Tittley)
Lmi 9h24'52"38d41'
Horting,Reed
Crt 10h51'16"-12d25'
Hortmann,Sonja
Leo 10h58'45"10d57'
Horton II,David Walter
Peg 21h57'59"21d55'
Horton's Dream Star, James Monroe
Lmi 10h32'14"30d14'
Horton,Alvin Everett
Her 18h44'23"12d19'
Horton,Doris May
Mon 7h0'21"0d40'
Horton,Eric Scot
Cnv 12h57'22"32d59'
Horton,Herbert & Jo Ann
Eri 4h58'38"-10d18'
Horton,Janie Rouse
Sgr 18h49'51"26d27'
Horton,Jude & Larry
Dra 16h47'0"71d41'
Horton,Karley Taylor
And 1h49'56"37d47'
Horton,Ken
Mon 6h58'0"11d7'
Horton,Louis Jackson
Cam 3h54'0"55d32'
Horton,Norman Michael
Oph 17h5'34"10d31'
Horton,Robert William
Her 17h38'56"27d1'
Horton,Roberta Lorraine
Aql 19h47'16"14d48'
Horton,Sandra
Uma 12h7'33"47d15'
Horton,Suzanne
Peg 23h2'33"13d14'
Horton-91443,Georgia Louise
Cam 4h0'41"58d47'
Hortus
Cnc 8h58'17"10d8'
Horvath,Andy
Hya 9h17'28"-6d46'
Horvath,Charlyn Rita
And 23h38'0"48d16'
Horvath,Christina & Julius
Lyn 8h37'0"40d23'
Horvath,Christine
Del 20h13'48"14d51'
Horvath,Jenny Elizabeth
And 23h36'55"43d24'
Horvath,Jr,Paul John
Dra 16h57'26"68d22'
Horvath,Karen
Lyn 9h10'1"34d53'
Horvath,Kerry Lynn
Eri 3h21'11"-16d14'
Horvath,Krystin Eileen
Lyr 18h46'0"32d51'
Horvath,Michael (Mikey)
Cep 21h31'13"58d36'
Horvath,Michael
Boo 14h7'0"34d48'
Horvath,Richard Edward
Aur 5h22'0"38d25'
Horvath,Stephanie Ann
Cma 6h53'6"-19d8'
Horvath,Thomas
Her 18h8'52"31d34'
Horvatich,Rudolph L
Aur 5h25'30"54d33'
Horwitz of Ridgewood, Theresa
And 2h23'0"43d0'
Horwitz,Aaron Martin
Tau 5h47'52"28d46'
Horwitz,Deborah
Cet 1h32'33"-0d20'
Hosansky,Aaron Jason
Oph 17h56'34"8d6'
Hosbach,Jennifer
Cyg 20h31'14"38d55'
Hoschek,Uli
Aqr 22h0'32"-0d35'
Hosek,Walter
Boo 13h40'18"26d7'
Hoselton,Charles
Tau 3h42'41"22d10'
Hosemann,Josef
Dra 15h48'26"67d46'
Hoser' Pickle Suit
Per 2h41'37"35d12'
Hoser,Albert
Cam 13h20'27"77d39'
Hosey,John P
Equ 21h23'0"2d38'
Hosford,Lindsey
Psc 1h1'58"32d23'
Hosford,Marcus
Ari 2h39'43"30d50'
Hoshauer,Kimberly Ann
Cas 0h48'13"62d59'
Hosking,Nick
Cam 3h10'50"60d50'
Hoskins Reach For The Stars,"Wendala"
Uma 11h12'25"50d33'
Hoskins,Arnold Wilson
Cet 2h13'36"3d59'
Hoskins,Carl Bryan
Cet 0h5'13"-11d30'
Hoskins,Dean Marie Dorsey
Crt 11h12'48"-12d1'
Hoskins,Donna
Cyg 19h50'46"38d33'
Hoskins,Elizabeth Melanie
Mon 6h36'0"10d41'
Hoskins,Sandra
Cas 1h32'48"63d52'
Hoskins,Stacey Lynn
Com 12h3'30"27d1'
Hosley,Forever Lynnie P
And 0h23'20"37d48'
Hosmanek,Andrew J
Ori 4h54'58"-0d12'
Hosmer,Julie Ann
Aql 18h48'40"11d36'
Hosna,Andrew Joseph
Aql 19h48'40"11d18'
Hosna,Evan Joseph
Oph 17h56'13"13d31'
Hosny,Said Al
Her 18h4'26"14d27'
Hosomi,Keiichi
Aql 20h9'1"1d24'
Hoss,Daniel Dryden
Vir 13h36'22"-11d44'
Hoss,Ronald Talbott
Vir 13h57'1"-20d7'
Hosseini,Naaz
Ari 2h1'13"18d23'
Hosselet,Stephen
Lac 22h48'51"56d11'
Hossman,MD,James Patrick
Per 2h29'45"57d35'
Host,Jennifer M
Cnv 13h14'1"38d18'
Hoster,Grace Marie Zumstein Lerdy
Umi 15h4'39"69d45'
Hoster,Hot Rod
Cnv 13h0'43"50d11'
Hostetler,Christopher & Beth
Cyg 21h28'0"41d12'
Hostetler,Miriam Olive
Dra 14h52'32"56d39'
Hostetter,Henry & Barbara
Umi 13h15'49"70d18'
Hostetter,Micah Evan
Cam 6h23'16"83d22'
Hostetter,T & H Wells
Umi 13h59'39"72d2'
Hosto,Ronald E
Boo 14h19'48"12d57'
Hot Glowing Surprise-Derrick's Star
Cma 6h57'12"-17d36'
Hot Rod
Her 16h6'57"48d25'
Hot Shot
Cma 6h50'47"-18d12'
Hot Stuff Too
Boo 13h55'55"21d36'
Hotchkiss IV,Adair John
Lac 22h49'20"54d19'
Hotchkiss,Alice Sharp
Cam 7h43'34"70d37'
Hotchkiss,Eric Alan
Aur 5h35'51"50d8'
Hoti
Vul 20h22'1"22d47'
Hotin,Richard
Lyn 6h57'54"59d19'
Hotman,Les
Dra 18h18'45"80d34'
Hotson,Barbara
Per 2h48'16"43d37'
Hotspur Plantagenet
Dra 17h29'26"60d25'
Hott Family Star Henry Lee
Dra 18h55'59"56d40'A
Hott Family Star Aaron Michael
Dra 18h55'59"56d40'B
Hott Family Star Linda Sue
Dra 18h55'59"56d40'C
Hott,Cristina Kay
Lac 22h8'50"51d4'
Hotta,Mrs Ikuka
Uma 12h38'27"59d58'
Hottel,Ellen & Rob
Mon 7h53'50"-6d19'
Hotton,Jean-Michel
Aql 20h4'15"0d45'
Hotwagner,Mag Gerhard
Her 18h0'27"47d31'
Houbary,Rose
Com 12h4'58"24d13'
Houbre,Catherine
Sex 10h13'50"-9d41'
Houchen,Evelyn L
Vul 19h43'20"25d29'
Houchens,Brandy Elizabeth
Aql 18h53'59"11d46'
Houchin,Jennifer Gwen
Cam 4h0'22"68d34'
Houck,Matthew Eric
Vir 15h0'32"0d54'
Houck,Michelle & Richard
Mon 7h56'52"-3d23'
Houck,Nana's Star AKA Mildred
Cas 3h7'17"61d44'
Houck,S J
Lac 22h41'52"53d29'
Houck-Fecteau 25
Lyn 8h44'18"37d45'
Houda,Bud & Lor
Uma 10h34'32"57d26'
Houdayer,Yves
Ser 15h17'35"11d43'
Houde,Gabrielle Gagnon
Lyn 9h2'0"34d46'
Houde,Jon Christian
Mon 7h50'20"-3d50'
Houde,Thomas J
Aur 6h10'30"31d51'
Houdmon,Therese
Peg 21h51'45"34d18'
Hough (Harmony Girl), Angela
Boo 14h31'41"48d26'
Hough (sar),Sara
Lyr 19h16'56"26d36'
Hough,Archie D
Tri 2h5'11"31d18'
Hough,Charles Robert
Lac 22h7'39"50d11'
Hough,David Lee
Aur 4h49'0"40d36'
Hough,Dr Travers Eugene
Tau 5h23'52"18d50'
Hough,Helen Lukes
Peg 22h14'22"4d50'
Hough,Jim
Cep 0h47'45"78d3'
Hough,Julia M
Vul 19h58'0"25d37'
Hought,Steven Andrew
Her 16h39'35"21d10'
Houghtaling,J Myles
Ori 5h55'43"11d2'
Houghton III,Lee E
Aql 18h59'35"-8d50'
Houghton's Haven
Dra 16h46'0"67d40'
Houghton,Connor Starr
Peg 23h27'18"28d12'
Houghton,Dean G
Cet 2h39'30"-0d28'
Houghton,Gary Sean
Uma 9h57'60"71d53'
Houghton,Hilda
Ori 5h56'47"7d46'
Houghton,James DeK
Ori 6h13'48"8d46'
Houghton,Joanne Louise
Cas 0h10'1"60d32'
Houghton,Kitten
Aur 5h0'20"46d27'
Houghton,Msgr Francis J
Vul 19h15'32"25d9'
Houghton,Roger E
Cep 0h6'43"73d59'
Houghton,Ruth & Jerry
Cyg 21h38'60"42d44'
Houghton,The Julie & Brian Orchard Star
Cyg 21h23'42"28d20'
Houille,Didier
Ori 5h6'0"10d47'
Houk,Erica
Lmi 10h8'47"34d45'
Houk-Seeger
Cyg 21h15'43"35d2'
Houlditch,Allison Shaned
Eri 4h43'21"-1d4'
Houldsworth,Georgann E
Cas 0h48'42"69d25'
Houle,Denis
Lmi 10h55'40"28d4'
Houle,Earl Walker
Cnv 12h46'15"40d9'
Houlette,Beth
And 2h23'0"39d44'
Houlette,Brandon
Lyr 18h50'11"42d15'
Houlette,Christopher
Boo 13h58'17"20d40'
Houlette,David L
Aql 20h2'0"1d9'
Houlette,Dylan
Tri 2h22'48"34d46'
Houlette,Gregory L
Lyn 8h56'14"44d33'
Houlette,Sarah
Cam 7h44'14"61d26'
Houlette-Leahy,Ramona
Lyr 18h59'0"40d18'
Houlihan (1-4,7-9), Greg
Cep 20h51'40"61d50'
Houlihan,Ryan Patrick
Ser 15h40'50"13d49'B
Houlihan,Timothy J
Ori 5h55'26"18d50'
Houliston,Paula
Boo 14h9'44"53d46'
Hoult,Ellen Parks
Com 12h19'25"28d23'
Hound Dog (from hell)
Her 18h5'25"38d40'
Hound of the Baskervilles,The
Tau 5h12'57"19d53'
Houpt,James Andrew
Cma 6h26'25"-15d35'
Houriet,Dominique
Cet 2h0'13"1d56'
Hourin,Suzy
Cas 23h15'0"61d51'
House of Astrology,The
Uma 9h8'41"49d3'
House of Susi,The
Cas 23h23'52"53d56'
House,Arthur B
Aql 20h4'57"1d48'
House-Lantto,James
Aur 6h13'31"33d3'
Houseman,Hunter
Sgr 19h37'6"-30d22'
Houser,Carol
Lyr 18h31'19"30d31'
Houser,David William
Sco 16h9'1"-38d5'
Houser,Mark Gerard
Oph 17h53'34"13d14'
Houser,Parker
Cnv 14h3'35"38d11'
Houshmand,John & Charlotte
Uma 12h12'1"58d58'
Houssais,Patrick
Sgr 19h27'7"-35d9'
Houssu,Jacques
Cas 0h2'48"62d29'
Houston's Star Power, Paul
Uma 9h35'57"70d40'
Houston,Amanda Jane
Aql 19h56'60"12d44'
Houston,Barcley
Aql 20h12'59"1d42'
Houston,Bill & Anna
Cyg 19h32'42"33d9'
Houston,Bryan
Cet 3h11'53"3d60'
Houston,Catherine
Cas 1h23'33"58d41'
Houston,Darrell C
Lyr 18h24'38"38d37'
Houston,R Clinton
Oph 17h9'20"7d57'A
Houston,Regina
Tri 2h17'33"32d60'
Houston,Robert
Sco 17h53'42"-40d39'
Houston,Stephen
Aur 6h25'40"34d14'
HoustonLand
Aql 20h13'27"1d3'
Houstoun,Katherine Nicholson
Leo 11h31'21"-2d18'
Houstoun,Martha Anne
Cap 21h23'21"-15d42'
Houstoun,Patricia Banker
Leo 11h8'39"2d5'
Houstoun,Rock Nicholson
Gem 6h28'46"14d13'
Houstoun,Sallie Banker
Sco 16h32'36"-43d43'
Houstoun,William Patrick
Leo 11h31'24"2d26'
Houtchens,Bruce Allen
Cet 0h59'1"0d27'
Houtman,Cristi M
Peg 22h41'45"20d47'
Houtz,Michael Lamar
Aur 5h8'28"40d7'
Houx et Tom sans cesse
Per 3h20'25"54d26'
Houy,Jean
Peg 21h23'51"22d60'
Hovanec,George
Her 18h8'1"28d25'
Hovanec,Robert M
Cet 1h21'14"-1d7'
HOVDA
Uma 9h43'14"48d29'
Hovda,Stephanie
And 0h12'51"39d55'
Hoven,Louise
Cyg 20h25'24"42d43'
Hoversten,Christopher David
Hya 8h51'1"-1d41'
Hovey
Her 15h54'20"41d40'
Hovey,Thomas Blumer
Umi 14h4'31"71d23'
Hovis,Kristen
Mon 7h0'32"0d16'
Hovland Esq,Carl M
Aur 5h3'32"44d48'
Hovland,Carl Michael
Aur 6h26'34"35d4'
Hovland,Michael
Cma 7h14'40"-16d44'
How I "Wanner" What You Are...
Lac 22h50'44"37d44'
How,Alan
Her 16h27'35"40d47'
Howar,Wind Beneath My Wings,Ray
Her 17h5'40"21d8'
Howard & Lowenstein
Her 17h18'49"14d28'
Howard III,Bucky
Sgr 18h49'32"-22d37'
Howard III,Charles Henry
Sex 9h50'0"-5d29'
Howard Jeffrey
Boo 13h52'21"18d27'
Howard Sailor Jim, James Oral
Del 20h24'38"8d29'
Howard's Bar Mitz-Star
Her 18h4'18"38d27'
Howard's Fastline,Bill
Aur 5h4'30"30d56'
Howard's Star
Boo 15h4'20"13d57'
Howard's Star,Robert W
Cmi 7h13'6"7d42'
Howard,Adam Charles
Hya 9h10'17"0d21'
Howard,Albert Benjamin
Cep 21h45'55"55d6'
Howard,Barbara
Lmi 9h46'18"38d31'
Howard,Barrie L
Vul 19h48'32"28d43'
Howard,Benjamin Daniel
Lyr 19h5'0"40d1'
Howard,Bill
Her 16h43'37"21d30'
Howard,Brigid
Ori 6h3'60"20d12'
Howard,Bruce Reid
Aql 20h5'26"4d23'
Howard,Bryan
Cma 6h27'26"-13d27'
Howard,Bryon L
Aur 5h20'44"38d30'
Howard,Carol
Peg 23h46'22"13d10'
Howard,Carole Anne
Lib 14h22'18"-18d45'
Howard,Christopher
Boo 15h0'20"12d23'
Howard,Constance B
Mon 7h41'24"-2d54'
Howard,David
Ser 15h17'1"21d10'
Howard,David & Elsie
Lyr 19h42'54"40d25'
Howard,David W
Aql 19h30'45"8d44'
Howard,David W
Hya 8h9'43"6d25'
Howard,Debbie & Marty
Cyg 20h0'34"50d26'
Howard,Drew Louis
Cep 22h9'10"61d31'
Howard,Edward James
Aur 6h59'11"37d28'
Howard,Eileen T
Cyg 20h16'57"38d55'
Howard,Elizabeth Boggan
Peg 21h37'51"25d42'
Howard,Ellen Patrice
Del 20h56'44"12d58'
Howard,Emily Renee
And 23h22'53"50d7'
Howard,Emma Kate
Col 6h36'1"-33d51'
Howard,Essie
Del 20h14'15"10d38'
Howard,Eva Mae
Uma 11h47'26"41d40'
Howard,Fiona
Sge 20h3'27"20d24'
Howard,Floyd Wayne
Mon 7h58'17"-1d53'
Howard,Gerald Lee
Uma 9h55'12"51d55'
Howard,Jack E
Ser 15h9'23"11d46'
Howard,James
Aur 5h1'32"48d20'
Howard,James Michael
Cet 2h53'11"3d5'
Howard,Janine
Hya 9h10'41"3d8'
Howard,Jason Michael
Cet 2h28'12"1d39'
Howard,Jeffrey Charles
Boo 14h13'26"48d16'
Howard,Jerry A
Cet 2h31'1"8d23'
Howard,Joanna Starone
Lib 14h44'26"-23d15'
Howard,John Lee
Cep 0h6'14"70d4'
Howard,Joseph D
Dra 17h35'46"64d16'
Howard,Julie
Com 12h22'1"31d2'
Howard,Juwan Antonio
Her 17h38'21"27d26'
Howard,Kathleen & Hugh
Crb 15h25'24"30d18'
Howard,Kathleen & Hugh
Cyg 19h27'57"30d7'
Howard,Keighla Ashli
Eri 3h16'29"-2d6'
Howard,Kenneth E
Cet 1h52'32"-10d28'
Howard,Kimberly Christine
Peg 0h2'30"18d47'
Howard,Kit
Aql 18h58'38"4d7'
Howard,Lisa
Aur 5h0'10"40d41'
Howard,Lois Catherine Sullivan
Lyr 18h38'1"31d34'
Howard,Mark A
Uma 13h30'44"61d42'
Howard,Mary
Her 17h19'54"22d3'
Howard,Mary
Com 12h16'19"26d56'
Howard,Mary Catherine
Mon 7h3'36"0d41'
Howard,Michelle
Lyr 18h47'57"30d6'
Howard,Mr Mike
Pyx 8h33'40"-28d18'
Howard,Muriel & Ed
Lyr 18h54'57"30d11'
Howard,Nancy & Dan
Aur 6h12'41"46d42'
Howard,Patrick James
Cyg 21h3'11"33d41'
Howard,Patti F
Cam 6h4'49"60d50'
Howard,Paul & Delores
Cyg 20h40'53"38d9'
Howard,Raymond A
Her 16h48'0"30d28'
Howard,Robert A
Dra 18h23'18"80d22'
Howard,Roger Dean
Aql 19h12'55"12d45'
Howard,Ruth
Crb 15h29'21"30d47'
Howard,Sally P
Eri 2h53'43"-12d28'
Howard,Sarah Galen
Mon 8h1'57"-9d16'
Howard,Tammy
Cyg 20h2'22"31d25'
Howard,Terry
Boo 14h16'52"51d36'
Howard,Tina
Lyn 7h8'49"58d34'
Howard,Torin Anthony
Equ 21h8'1"11d31'
Howard,Wendy Joy
Cas 0h18'43"60d4'
Howarth,Bruce Nobel
Per 3h6'11"47d31'
Howarth,Joan
Lyr 18h19'13"42d5'
Howarth,Peter
Ori 4h55'0"0d49'
Howarth,Philip James
Cmi 7h26'45"0d46'
Howarth,Rob
Her 16h2'0"16d18'
Howden,Richard William
Her 16h12'49"40d21'
Howe AKABH,Barbara
Cyg 19h47'58"29d30'
Howe,Allison Clark
Com 13h7'31"20d46'
Howe,Amber Lee
Com 13h12'35"20d7'
Howe,Austin Kristopher
Lac 22h17'35"49d34'
Howe,Christina
Vir 11h43'0"0d6'
Howe,Donna L
Lyn 7h13'0"58d56'
Howe,Doris S X
And 23h17'28"45d53'
Howe,Eleanor Cameron
And 0h2'20"40d36'
Howe,Jim & Pat
Sge 19h31'0"19d12'
Howe,Lisa Evelyne
And 23h50'1"33d38'
Howe,Lucinda Faye
Eri 3h46'59"0d47'
Howe,Mary Martha
Aqr 21h58'57"-21d38'
Howe,Michael
Ori 5h18'21"11d6'
Howe,Michael David
Cnc 9h1'51"21d3'
Howe,Natalie Elizabeth
Eri 3h56'0"-19d5'
Howe,Norma Kay
Cas 1h59'22"60d40'
Howe,Penny
Cas 0h25'57"50d26'
Howe,Scott
Sgr 18h50'19"-21d47'
Howe,Sr,Jack
Her 17h9'22"48d43'
Howe,Valmai
Uma 9h0'1"61d14'
Howell
Psc 1h36'50"28d4'
Howell & Vernita
Cet 0h44'43"-6d8'
Howell,Andrew Dickinson
Cet 2h51'47"3d23'
Howell,Brian
Oph 17h36'23"-23d59'
Howell,Carolyn
Cmi 7h56'33"0d56'
Howell,Carolyn
Her 16h11'31"10d16'
Howell,Chris William
Uma 9h25'47"70d27'
Howell,Dr
Ori 5h37'14"-0d44'
Howell,Earle & Carole
Hya 9h35'18"2d24'
Howell,Florence
Hya 8h23'57"-7d18'
Howell,Floyd K
Sex 10h24'0"-5d18'
Howell,Gerald A
Aql 19h24'39"-1d49'
Howell,Jack
Per 1h37'20"53d6'
Howell,Jamie
Lyr 18h55'30"30d17'

Howell,Jonathan D
Cep 22h13'37"80d22'
Howell,Jr,William L
Del 21h0'1"13d3'
Howell,Karol Marie
Cas 0h20'54"70d53'
Howell,Linda C
Cas 23h21'18"63d21'
Howell,Lowell Thomas "Tom"
Cmi 7h15'30"3d55'
Howell,Mark
Hya 9h53'22"-19d51'
Howell,Megan Sioux
Cas 0h43'21"60d7'
Howell,Pamela Arnell
Mon 6h33'38"-1d7'
Howell,Paul Michael
Cet 3h9'53"3d2'
Howell,Richard E
Her 16h6'42"42d29'
Howell,Robert Wayne
Hya 9h3'0"0d33'
Howell,Roxie A
Cam 5h37'20"61d13'
Howell,Tammy
Del 20h17'39"10d8'
Howell,Todd R
Uma 11h41'30"56d8'
Howell,Uncle Bill (Billy)
Lac 22h9'52"46d29'
Howells GN,Bethselamin
Ori 6h4'15"2d24'
Howells,Richard William
Lyn 6h27'40"54d32'
Howen,Bob
Sgr 18h55'3"-24d40'
Howes III,EC
Mon 8h3'59"-9d23'
Howes,Jill Raye
Hya 8h54'57"-1d5'
Howie
Gem 7h27'16"30d54'
Howie
Peg 22h18'41"34d23'
Howie's Protector
Boo 15h2'24"50d55'
Howie,Bob
Ser 17h31'28"-10d20'
Howie,Brian Hugh
Per 1h56'23"48d5'
Howie,Donald Scott
Sgr 20h4'23"-26d40'
Howie,Jayne Helen
Cas 0h37'46"60d4'
Howie,Kimberly Heather
Lyr 18h40'16"41d1'
Howie,O G
Boo 14h18'51"32d21'
Howie-"pulcher.. amans..uni-cus"
Dra 15h7'23"57d56'
Howison,Barbara
Aqr 20h58'1"-1d7'
Howit,Kimberly Diane
Eri 4h5'21"-16d39'
Howit,Matthew Allan
Cmi 7h19'50"8d11'
Howk,Todd
Her 16h56'56"25d16'
Howland,Aaron Matthew
Ser 15h11'59"2d16'
Howland,Elinor
And 1h16'54"40d58'
Howland,Ward Wilson
Per 2h26'13"55d37'
Howle,Julie Anne
Cas 2h7'34"68d31'
Howlett,Ashley Grace
Cas 23h41'0"61d15'
Howlett,Susan
Cas 0h29'37"61d20'
Howlett,Susan Smoleski
Psc 1h30'49"20d29'
Howlett-Star!,Micheal C
Peg 22h38'16"21d2'
Howley,Dr Joan Ertel
Aql 19h43'27"4d7'A
Howley,Kaye M
Cas 2h2'43"58d8'
Howley,Philomena
Cas 1h32'25"72d48'
Howley,Scott C
Hya 9h1'34"0d35'
Howley,Sean Michael
Lac 22h38'27"55d27'
Howley,Sr Helen Patrick
Her 18h6'1"30d1'
Howrani,Adeeb
Uma 9h24'10"42d10'
Howsam,Sandy
Peg 21h53'47"2d59'
Howse 95
Aur 7h15'56"36d17'
Howser,Joy Ann
Lyr 19h8'42"37d56'
Howser,RC
Tri 2h3'0"31d10'
Howton
Cmi 7h46'10"8d39'
Hoy,Blair Kenneth
Her 16h6'27"21d50'
Hoy,Joyce Jean
Ori 5h55'30"18d47'
Hoyack,Vern
Cet 1h34'1"-11d22'
Hoyer,Adrienne
Cnv 13h13'39"32d14'
Hoyer,Elizabeth Ann
And 1h13'1"38d47'
Hoyer,Thomas
Cas 0h9'0"60d41'
Hoyle,Stephen Lewis
Aur 6h2'12"46d17'
Hoyt,Doreen Elizabeth
Lyn 7h51'36"34d54'
Hoyt,Grandpa Pete
Her 17h16'0"46d38'
Hoyt,Gregory F
Aql 19h53'44"14d32'
Hoyt,Milton E
Sgr 19h9'15"-23d42'
Hoyt,Sue
And 23h37'54"46d28'
Hoyt,The Star Of Betina Coffey
Uma 10h1'40"50d15'
Hoyt,The Star Of Dolph Graham
Uma 9h44'46"49d50'
Hoyt,Tony
Aur 4h36'14"31d3'
Hrabal,Stacie
Lyn 9h18'30"38d33'
Hranicky,Kim
And 0h45'23"34d12'
Hrast,1-4-3 Forever Julie L
Ori 6h7'27"7d26'
Hrazdina,Helga M
Boo 14h11'10"31d49'
Hrehorova,Silvia
Vir 12h52'10"-0d34'
Hreys & Fryk
Umi 15h9'24"71d12'
Hribernigg,Bernd
Aur 6h21'30"33d4'
Hribernik,Curt S
Dra 11h31'33"72d47'
Hribko,Rick
Lac 22h25'57"53d60'
Hripcsak,Michael J
Dra 19h28'29"58d44'
Hritz,Dolores
Peg 23h4'12"21d53'
Hritz,R Thomas
Dra 15h19'16"63d3'
Hrobak,The Star
Mon 6h58'47"11d51'
Hromadka,Steven A
Cet 3h9'30"1d33'
Hromalik,Mark & Elizabeth
Her 17h29'13"41d14'
Hron,Allen Edwin
Lac 22h35'0"54d10'
Hrovat,Claire Ann Marie
And 2h27'32"49d35'
Hrovat,John
Aqr 23h8'30"-4d5'
Hrstich,Marko
Crt 11h11'0"-19d39'
Hrubi,Felix Montgomery
Leo 11h14'22"-5d26'
Hruska
Vul 18h57'23"24d32'
Hruska,Heather
Peg 22h8'12"26d1'
Hruska,Louise Napolitano
And 2h32'0"49d49'
Hrynkiewicz,Sarah
And 0h35'34"40d49'
Hsi,Yutz & Wai-Ying
Ori 6h4'0"6d10'
Hsia #1 Lady In The World,Hanna
Eri 3h26'10"-6d36'
HSN & GBN's Star
Cyg 21h7'22"30d39'
Hua Hua
Uma 10h19'0"56d4'
Huan,Fion Tan Khoon
Lup 15h20'31"45d31'
Huang,Mr & Mrs Wei Lun
Cam 3h57'17"69d58'
Huang,Paul
Aur 6h39'43"35d22'
Huang,Pei-Ying
Cam 4h18'18"68d56'
Huang-Ku,Herching & Shun
Lyn 7h55'34"42d49'
Huang-Wang,HouYang & Diwha
Peg 21h30'12"20d36'
Huard,"Pete"
Boo 15h4'12"20d50'
Huard,Michael
Cyg 19h27'20"33d44'
Huard,Michael & Terri
Eri 3h42'0"-12d7'
Huard,Patrick
Per 2h52'21"38d25'
Huart,Hedwig G
Cnc 9h5'32"32d24'
Huart,Ina Marion
Lyr 19h25'33"38d23'
Huart,William
Lib 15h3'15"-1d18'
Huascar,Robin
Dra 17h5'1"60d11'
Huba,Jr,Walter A.
Cam 4h3'27"68d40'
Huband,Franklin
Dra 14h17'18"63d60'
Hubba Bubba J C W
Uma 8h54'58"53d14'
Hubbard 80
Aql 20h0'28"3d59'
Hubbard Dream Star, Louis E
Mon 6h14'39"-10d22'
Hubbard's Harlequin
Peg 23h22'17"28d20'
Hubbard's Nova,Holli
Peg 23h8'28"12d22'
Hubbard,Abbott Green
Cnv 12h24'50"37d21'
Hubbard,Addie
Tri 1h34'13"35d3'
Hubbard,Alan William
Per 2h4'34"56d33'
Hubbard,Bridgett Rose
Peg 21h28'26"21d2'
Hubbard,Claire
Lyr 18h19'25"40d18'
Hubbard,David
Ori 5h55'15"6d47'
Hubbard,David Alan
Aur 6h0'0"32d12'
Hubbard,Dixie Christine
Equ 20h55'0"3d51'
Hubbard,Laura
Lmi 10h17'33"28d38'
Hubbard,Lena Howell
And 0h57'26"36d34'
Hubbard,Lillian
Lyr 19h19'19"41d12'
Hubbard,Robert Page
Her 17h35'15"40d57'
Hubbard,Sandra
Ori 5h53'11"9d51'
Hubbard,Shaun
Lyn 7h53'27"48d60'
Hubbard-Rejika 1992 & 1993,Bobbie
Lyn 9h13'20"33d55'
Hubbell Family,The
Ori 6h14'20"8d49'
Hubbell,Brian J
Per 4h20'25"52d13'
Hubbell,Kenneth Tyler
Umi 15h34'1"68d4'
Hubbell,Ric
Equ 20h57'18"2d48'
Hubbell,Stan
Vir 11h51'14"9d42'
Hubbell,Windy
Oph 17h4'47"11d15'
Hubbs Double H Ranch, Homer J
Uma 11h46'22"31d31'
Hubby Love Bug
Lac 22h42'59"35d35'
Hubcap
Cmi 7h58'25"0d16'
Hube,Klaus-Jürgen
Ori 5h49'18"20d38'
Huber
Lib 15h16'58"-23d3'
Huber MA,MFCC,Lauren
Del 20h17'28"9d34'
Huber's Heaven Anchor, Daniel James
Dra 17h1'45"65d34'
Huber,Abram Paul
Uma 11h23'37"58d18'
Huber,Alfons
Aql 18h57'46"17d39'
Huber,Ashley Nicole
Lyn 6h38'58"59d22'
Huber,Cathy
Psc 22h54'13"6d3'
Huber,Gus
Uma 10h42'0"47d41'
Huber,Jane C & Norman V
Cyg 19h25'38"31d6'
Huber,Kenneth Dale
Sct 18h50'29"-7d3'
Huber,Megan Nichole
Sct 18h55'17"-7d44'
Huber,Mr Ulli
Aur 4h55'12"50d20'
Huber,Nicole Renee Elizabeth
Lyr 19h4'51"38d19'
Huber,Peter
Lyn 8h14'21"48d43'
Huber,Robert
Cmi 7h17'59"8d45'
Huber,Robert
Aur 5h18'12"48d56'
Huber,Rudolf
Ser 15h30'36"-2d59'
Huber,Walter & Marie
Aql 19h29'0"-6d1'
Huber,William
Aur 4h52'42"40d43'
Huber,Wolfgang
Uma 9h18'1"45d26'
Hubert
Boo 15h21'27"38d38'
Hubert
Cma 6h50'5"-19d23'
Hubert & Una
Eri 2h49'58"-17d54'
Hubert et Emmanuelle
Cyg 20h17'0"39d8'
Hubert MD,Gary D
Oph 18h38'39"10d35'
Hubert,Catherine D
Peg 23h39'0"30d37'
Hubert,David E
Boo 14h55'16"45d16'
Hubert,Frances
Oph 17h38'33"-20d36'
Hubert,Frank
Psc 22h50'20"5d11'
Hubert,Shawn
Lac 22h6'17"47d57'
Hubert,Vernie & Kathleen
Sge 18h59'39"19d47'
Huberth,Connolly Rose
Cet 1h55'37"-18d53'
Hubertus,Bobby
Lyn 6h53'51"59d0'
Hubick,Arnold N
Aur 7h23'35"39d35'
Hubick,Gloria B
And 2h20'1"47d56'
Hubio,Francisco Javier Alonsio
Cam 5h6'17"65d12'
Hublet,Valerie
Boo 13h43'29"16d17'
Hubley,Bert
Lyn 7h7'47"51d19'
Hubner,Wilhelm
Eri 4h10'15"-15d51'
Hubnik,Gary "Nicholi"
Her 16h36'29"11d40'
Hubsch,Sharon
Cas 0h24'27"67d54'
Hubuck
Ori 5h54'24"21d54'
Huck,Alain
Mon 7h45'57"-4d9'
Huck,Fred
Lyn 8h0'52"38d54'
Huck,Mimi
Del 20h14'0"15d30'
Huck,Paul E
Aur 6h0'41"30d54'
Huck,Samuel Paul
Com 12h37'36"30d13'
Huck,Tom & Judy
Cep 0h10'21"69d45'
Huck,Wolfgang
Lyn 8h15'0"47d3'
Huckanourus,Marcus
Boo 14h36'51"20d31'
Huckle,Peter
Hya 8h36'32"-0d53'
Huckleberry
Cmi 7h46'24"8d42'
Huckleberry Markus
Boo 15h24'28"41d27'
Huckleberry,Hobo
Lac 22h10'16"37d55'
Hucks,Arnold
Sco 17h31'29"-38d29'
Huckstadt,Wilma
Ori 5h48'1"12d5'
Huckstep,Gene E
Boo 14h15'52"37d17'
Hudak,Cortney Janelle
Ori 5h40'0"-1d12'
Hudak,Lorraine
Cnv 13h30'1"42d12'
Hudd,Lea Justine
Umi 15h5'11"67d3'
Hudd,Mike
Ori 6h3'21"2d19'
Huddleston,Ann Wallace
Mon 8h6'57"-8d50'
Huddleston,Don & Michelle
Cma 6h51'19"-16d50'
Huddleston,Mary Tomashek
Cas 0h20'16"60d20'
Huddleston,Teresa M
Lyn 6h45'43"60d58'
Huddleston,Zakkary Keith
Dra 17h47'37"68d52'
Hudelson,Laura
Vul 20h15'35"23d14'
Hudes' 1st Birthday, Claire
Aqr 23h12'29"-5d0'
Hudgens,Gretchen
Aql 20h1'55"11d36'
Hudgens,Major Edward M
Aql 19h46'1"12d8'
Hudgins,Drew Ben
Sct 18h43'40"-6d32'
Hudiak,D
Cam 4h18'59"65d9'
Hudiburg,Robert J
Cet 3h5'57"1d57'
Hudoklin,Daax
Uma 11h21'25"62d13'
Hudson & Saleeby With Chris & Doug
Aur 5h52'54"50d36'
Hudson's Dancing Star, Gill
Ari 2h26'34"21d38'
Hudson,Ariana
Mon 6h23'14"3d39'
Hudson,Audrey Ann
Lyn 8h59'1"36d27
Hudson,Betty Boop
Cas 0h33'24"63d20'
Hudson,Betty J
Lyr 18h56'33"30d11'
Hudson,Brian E
Cet 2h26'52"6d7'
Hudson,David & Brooke
Cet 3h11'59"0d30'
Hudson,Deborah C
Aql 20h18'42"8d15'
Hudson,Deborah C
Aql 19h31'38"-1d3'
Hudson,Gayle
Cyg 19h17'13"45d25'
Hudson,Gena Ann
Cyg 19h23'39"44d53'
Hudson,Harry
Uma 9h58'22"45d31'
Hudson,Henry L
Aql 19h44'22"10d24'
Hudson,Jean Lesley
Cas 0h37'19"60d25'
Hudson,John Vernon
Dra 14h9'13"63d31'
Hudson,Jr,James Edward "Hudstar"
Aqu 20h54'38"05d56'
Hudson,Micah
Ser 16h3'16"-0d1'
Hudson,Molly
Cru 12h39'26"-62d41'
Hudson,Nathan John
Ori 5h55'0"7d57'
Hudson,Robert Charles
Aur 5h52'30"51d22'
Hudson,Sharon Lee
Tri 1h51'47"27d59'
Hudson,Willis Jason
Ori 4h53'0"-2d53'
Hudson-Jones,Kathy
Cyg 19h21'33"44d15'
Hudson-Tottle,Renda Carol
Lyn 8h8'28"40d29'
Hudspeth,John
Cep 0h10'1"80d10'
Hudspeth,Larry Lloyd
Boo 13h48'42"20d29'
Hudspeth,Vernelle
Aql 19h59'49"11d42'
Hudy
Lac 22h12'16"47d56'
Hudziak,Daniel
Cnv 12h41'44"36d2'A
Hudziak,Ginny
Cnv 12h41'44"36d2'B
Huebner's Star,Pete
Lac 22h39'26"50d4'
Huebner,Candace
Cyg 19h45'57"31d48'
Huegel,Len J
Cep 22h52'15"58d55'
Huelin,Lawrence
Her 17h19'1"49d37'
Huellen,Paul T
Aur 6h1'29"32d54'
Huelsbyrne
Cam 6h5'53"71d49'
Huemann III,Alvin Schroeder
Vir 11h41'55"7d29'
Huemmer,Mr & Mrs James
Cyg 21h16'59"38d14'
Huensch,George E
Her 17h57'55"40d37'
Huertas,Joe "The Admiral"
Aql 18h59'48"-8d9'
Huertas,Martine
Del 20h36'16"8d34'
Huerter,Marie
Cet 2h37'15"9d29'
Huestis II,Jesse Logan
Hya 9h25'10"-0d26'
Huet,Aleese Marie
Del 20h24'0"10d45'
Huett,Diana Charlene
Cma 6h41'47"-13d38'
Huetter,Mr & Mrs Robert F
Cyg 20h24'0"38d41'
Huettig,Morgan
Leo 11h32'49"-5d48'
Huettig,Taylor
Lib 15h17'1"-22d51'
Huettner,Glenn Arthur
Sex 10h37'11"-0d27'
Huey James
Sct 18h42'16"-6d8'
Huey,Martha Thurston Gram
Mon 7h7'21"06d39'
Huey,Price & Elaine
Sge 19h4'7"18d59'
Huey,Shannon Marie
Eri 2h50'0"-17d45'
Huey,Ted & Flo
Uma 11h41'55"56d22'
Huff's Rio Bravo
Aql 19h30'0"10d27'
Huff,Barbara
And 0h14'33"35d51'
Huff,Chyrel Sue
Aql 19h5'16"6d58'
Huff,David
Aql 19h7'21"3d44'
Huff,Earl
Sge 19h57'58"16d14'
Huff,Eliza Lena
Del 20h53'0"9d41'
Huff,Gladys
Cas 0h27'11"58d30'
Huff,J Winston
Ser 15h40'53"-2d21'
Huff,Kariel & Jean
Cyg 20h41'22"30d49'
Huff,Margo
Peg 21h43'41"24d49'
Huff,Muriel Emily
Cyg 19h20'59"34d54'
Huff,Olen & Wynona
Sge 20h13'29"17d6'
Huff,Patricia G
And 0h4'18"41d1'
Huff,Susanna Elise
And 22h56'24"40d54'
Huff,Terry
Lyr 19h6'41"28d17'
Huff,William D
Aur 5h6'30"42d28'
Huffa
Oph 17h22'8"-22d2'
Huffine,Deadra Glyn
Cam 5h54'34"71d11'
Huffines,Nadine
Cas 0h57'0"63d16'
Huffman Family,The Jeffrey A
Her 17h1'19"40d11'
Huffman,Aimee
Ori 6h9'38"7d47'
Huffman,Amanda Lauren
Lib 15h15'42"-18d14'
Huffman,Jr,Domer Jones
Boo 13h40'49"22d59'
Huffman,Judy & Doug
Vir 11h38'0"8d47'
Huffman,Kacia Marie
Tau 4h27'1"28d56'
Huffman,Larry Lee
Lac 22h23'40"52d55'
Huffman,Tianna Lee
Cnc 8h29'12"10d27'
Huffnagle,Cleda M
Vul 19h2'42"25d32'
Hufham,Linda L
Sco 17h1'14"-31d23'
Hufnagel,Elizabeth Sherwood Newman
Cas 0h21'26"61d27'
Hufnagel,Michael L
Eri 4h8'17"-8d31'
Hufnagel,Petra
Lyn 7h46'19"40d9'
Hufner,Eileen
Vul 20h0'52"28d11'
Hufsmith,Justin Robert
Dra 20h25'55"63d34'
Huge Morgan
Tri 1h47'42"26d7'
Huggles Forever
Mon 8h3'16"-7d26'
Huggs,Gerald E
Psc 22h59'59"-2d51'
Huggy Bear
Uma 11h53'36"64d32'
Huggy Bear
Uma 8h27'53"62d30'
Hugh
Cep 20h36'48"75d17'
Hugh
Ori 6h7'28"0d43'
Hugh III
Oph 17h28'1"-0d59'
Hughes
Uma 11h8'21"58d23'
Hughes 24 Jan 1937, Glynne
Ori 5h57'16"15d32'
Hughes born July 12, Marilyn Lea
Eri 2h51'13"-8d20'
Hughes Forever,Tim & Diane
Crb 15h38'21"29d53'
Hughes III,Michael Arthur
Aql 19h56'37"10d14'
Hughes III,Philip Carl
Aur 5h2'0"48d53'
Hughes' Hopes,Dreams & Inspiration,Connor D
Ori 5h55'36"19d20'
Hughes,A Blair
Aur 6h25'53"38d15'
Hughes,Aija
Aql 19h2'1"-5d47'
Hughes,Alexandra Helene
Eri 4h12'25"-18d5'
Hughes,Amy Lynn
Cas 0h3'35"58d54'
Hughes,Benjamin Clarke
Psc 23h21'0"6d3'
Hughes,Buddy
Aur 5h59'17"37d50'
Hughes,Calvin H
Dra 16h45'19"60d60'
Hughes,Cameron McKay
Oph 17h58'22"8d4'
Hughes,Carole
Crb 16h20'52"27d9'
Hughes,Casey
Mon 7h1'22"4d20'
Hughes,Casey Jane
Lyn 9h5'1"35d46'
Hughes,Chad C
Her 17h4'40"40d43'
Hughes,D R
Aql 19h0'1"16d49'
Hughes,Dennis
Her 18h4'25"31d9'
Hughes,Diane
Lyr 18h19'0"40d50'
Hughes,Donna T
Uma 11h29'50"30d9'
Hughes,Dorothy J
Cas 23h30'33"63d11'
Hughes,Eleanor Elizabeth
And 22h55'34"50d10'
Hughes,Eric
Boo 14h41'57"34d11'
Hughes,Eric Dwayne
Uma 8h26'21"71d23'
Hughes,Eric Timothy
Lac 22h17'38"54d45'
Hughes,Erin
Lyr 18h44'42"47d44'
Hughes,Eva & Virgil
Mon 7h12'1"-10d51'
Hughes,Francis
Lyn 8h9'37"58d20'
Hughes,Gail Marie & Joseph William
Cma 6h53'28"-19d48'
Hughes,Gary D
Dra 16h2'45"65d5'
Hughes,Gary Everett
Hya 8h35'0"-10d0'
Hughes,Gary Wayne
Tri 2h45'27"31d10'
Hughes,Glenn Michael
Aql 20h8'1"4d44'
Hughes,Hannah Lucy
Umi 17h4'38"80d30'
Hughes,Heather
And 0h57'50"35d25'
Hughes,Heather
Boo 13h54'0"17d38'
Hughes,Hunter
Ser 18h21'1"0d43'
Hughes,Jack & Myrta
Cyg 20h5'39"40d1'
Hughes,Janet Sue
Com 12h25'17"23d41'
Hughes,Jason Christopher
Eri 4h4'42"-18d42'
Hughes,Jennifer
And 2h19'26"40d10'
Hughes,Jennifer Tracey
Peg 21h38'0"25d27'
Hughes,Jim
Her 15h54'26"48d38'
Hughes,Joelle J
Lyn 7h47'0"44d57'
Hughes,John
Uma 11h6'57"30d9'
Hughes,John
Cep 20h43'44"75d28'
Hughes,John Francis
Aql 19h1'16"12d36'
Hughes,John-Philip Mishler
Uma 10h6'0"48d52'
Hughes,Joseph
Cmi 7h45'22"8d51'
Hughes,Joseph Pierre
Aql 19h52'57"12d57'
Hughes,Joyce
And 23h5'44"42d37'
Hughes,Jr,Ray
Uma 11h29'50"32d13'
Hughes,Jr,Thomas Robert
Aql 19h0'18"16d32'
Hughes,Justin Mark
Ori 5h23'1"1d18'
Hughes,Kathleen
And 2h1'33"47d20'
Hughes,Katie
Peg 22h2'26"32d35'
Hughes,Lauren Noelle
Uma 11h55'53"52d50'

Hughes,Letty J Uma 11h4'41"62d30'
Hughes,Lindsey Peg 23h32'20"31d25'
Hughes,Lisa K Cas 0h49'24"63d41'
Hughes,Lori Ann Peg 22h21'49"26d13'
Hughes,Lyndon Her 18h0'48"40d37'
Hughes,M C Curtis Cep 21h4'1"55d37'
Hughes,Marian Louise Cas 1h44'1"77d27'
Hughes,Mary Lyr 18h14'14"46d22'
Hughes,Mary Hya 8h40'0"-6d29'
Hughes,Michael Peg 22h13'1"10d9'
Hughes,Michael David Gem 7h27'50"28d19'
Hughes,Morgan Brian Aur 6h28'15"30d31'
Hughes,Mrs Lori Cet 1h23'58"-18d53'
Hughes,Nicholas Peg 23h0'36"18d55'
Hughes,Nora Cyg 20h53'55"39d45'
Hughes,Patricia S And 0h56'56"37d4'
Hughes,Paula Clare Cyg 19h29'53"35d21'
Hughes,Philip Charles James Per 1h50'26"53d8'
Hughes,Richard E Aql 20h5'0"6d13'
Hughes,Ricky Hya 8h39'35"-11d6'
Hughes,Rita Louise Lanaux Peg 23h24'59"33d23'
Hughes,Robert & Carol Cnv 12h27'19"32d0'
Hughes,Robert John Cmi 7h33'37"1d21'
Hughes,Ronnie Cyg 19h33'57"30d26'
Hughes,Sarah Catherine And 23h13'21"40d23'
Hughes,Shane R/ Kimberly A Hamilton Cyg 21h16'52"28d47'
Hughes,Shelby Nicole Mon 6h52'57"-6d39'
Hughes,Smokey & Laura Cam 3h19'0"61d11'
Hughes,Stephen Paul Ori 5h37'31"10d18'
Hughes,Susan And 23h19'35"51d53'
Hughes,Tanya M Lyn 8h49'48"45d58'
Hughes,Thomas Fisher Aur 5h30'20"38d14'
Hughes,Tony Cyg 19h33'42"31d50'
Hughes,Violet Mon 6h44'55"10d57'
Hughes,Wendy Jane Davies Lyr 19h2'29"26d24'
Hughes,William Uma 9h10'15"54d36'
Hughes-Rill,Emily And 1h58'30"37d15'
Hughes-Wistow Cyg 21h0'15"38d16'
Hughey Family,The Kevin Lac 22h15'19"49d20'
Hughey,Elizabeth Hightower Mon 7h40'12"-8d56'
Hughey,Jennifer Gayle Cas 0h35'24"62d54'
Hughitt,Jeremiah K Boo 15h3'37"17d39'
Hughs,Daniel Dra 11h53'53"72d25'
Hughs,John Dra 11h54'54"66d4'
Hugin Cnv 12h15'57"47d27'
Hugkulstone,Simon William Cep 21h15'0"55d33'
Hugo Vul 20h15'19"22d57'
Hugo Vul 20h16'19"28d48'
Hugo Ori 5h55'14"12d13'
Hugo Per 3h2'21"41d42'
Hugo "Wild Star" Cam 12h24'45"81d46'
Hugo,Benjamin V Aur 5h16'0"48d44'
Hugonin,Jenny Vul 20h0'35"23d38'
Hugs "N E W" Cap 20h23'58"-13d31'
Hugs & Huggy Peg 23h20'23"16d47'
Hugueley,Cordelia N Aql 19h53'45"11d51'
Huguenin "Luciolle", Florence Uma 10h34'58"39d54'
Huguenin,Jay Aur 6h23'38"33d33'
Huguette Crb 16h14'56"38d39'
Huh,Heidi Ori 5h54'13"14d24'
Huhman,Amanda Nicole Lyn 7h15'21"59d53'
Huhn-Heidrich,Martina Cep 0h4'11"81d55'
Hui,Pirus Pho 23h54'47"-42d43'
Huiras,Cletus & Phyllis Crb 16h21'1"32d6'
Huizenga,Dr Gary N Aur 5h58'20"31d36'
Huizinga,Steve Dra 17h32'55"60d9'
Hukezalie 4-16-93, Robert & Pamela Aql 19h48'22"10d59'
Hukka,Kyllikki Boo 14h31'35"50d16'
Hulbert Boo 14h8'49"30d4'
Hulbert,Cheryl Cas 0h38'53"74d2'
Hulbert,Dustin Marion Aur 6h17'34"33d10'
Hulbert,Erin Marie Cyg 19h27'32"32d39'
Hulbert,Sheila Marie Lyn 8h5'12"33d37'
Hulburd "Big Jazz", Chip Crb 16h11'41"37d19'
Hulen,Tarrace Boo 15h9'54"31d30'
Hulet,Joshua Per 4h28'56"50d48'
Hulet-The Jeweler Star,B David Aql 19h55'23"7d46'
Hulett,Gerald Cam 11h20'1"81d42'
Hulett,Thomas William Boo 14h29'21"25d58'
Hulick,Jr,Charles Edwin Dra 14h18'40"63d60'
Huline,Lauren Anne Lyr 18h14'14"30d26'
Huling,Jr,John Reynolds Per 2h55'0"37d13'
Hulitar,Renée Cosmy Cas 22h56'1"54d2'
Hull,Angelina B Cyg 20h58'51"38d7'
Hull,Barton H Cet 2h5'1"-1d59'
Hull,Carl J E Aql 19h7'52"4d2'
Hull,Christina L Aql 20h10'15"0d21'
Hull,James S Aql 20h18'31"8d22'
Hull,Janet Sue Gem 6h45'14"31d34'
Hull,John Tillman Cap 20h8'57"-10d22'
Hull,Lorraine Julie Cyg 21h58'23"52d34'
Hull,Marvin W Her 18h6'53"40d1'
Hull,Mary J Cyg 19h42'21"28d14'
Hull,Randy Peg 23h5'0"33d46'
Hull,Rev Eileen M Peg 23h35m1"18d15'
Hull,Richard & Mary Cyg 19h58'39"30d23'
Hull,Stephanie Cas 23h41'35"60d7'
Hull,William Sherman Cet 3h12'56"6d41'
Hull-Kimball,Leslie Vul 20h0'0"23d49'
Hullana,Lisa Lyr 18h37'52"27d40'
Hullar,Stephanie Cas 23h47'0"50d7'
Hullhorst,David Hya 8h43'25"3d11'
Hullihan,Bill Her 18h19'56"20d7'
Hullinger,Judith E Cyg 19h28'11"38d16'
Hullinger,Kurt Aur 6h2'41"31d41'
Hullingshead,Kelly, Barry,Emily,Melina Per 1h41'34"52d51'
Hulme,Alexandra Nicole Com 13h2'28"20d28'
Hulme,Cerissa Lauren Dra 17h32'1"64d2'
Hulme,Dawn L Umi 17h21'1"76d31
Hulme,Diana Margaret Ori 4h59'30"5d55'
Huls,Stacy Cas 23h20'13"60d22'
Hulse,Fred & Carol Lac 22h2'0"50d36'
Hulse,George A Per 2h6'52"57d49'
Hulse,Sarah Margaret And 0h31'51"41d1'
Hulsey,Imogene Vul 20h3'33"28d45'
Hulsey,Terry Cet 1h27'44"-12d50'
Hulsey,Tracy Leanne And 0h52'27"40d16'
Hulson,Geoffrey Paul Vir 11h56'0"2d18'
Hulstrom,Carol & Martin Elliot's Star Cyg 19h30'39"35d21'
Hultgren,John B Aur 6h30'1"38d8'
Hulton,David Nakamura Del 20h25'36"18d49'
Hults,Kelly Lyr 18h41'13"32d49'
Hum,Carol L Aql 19h42'1"11d31'
Hum,Jr,John Per 4h19'19"51d14'
Humaira Ori 6h11'37"1d46'
Humann,Erin And 2h22'28"41d35'
Humann,Susan & Richard Cyg 21h31'19"44d10'
Humar,Eleanor And 23h22'46"41d27'
Humberside-Stella Nova Cuan Luke Cnv 13h39'37"44d52'
Humbert,Marie Dra 17h5'18"60d4'
Humbert,Natalie Kay Nichole Mon 7h1'15"8d57'
Humbert,William Cristopher Vul 20h15'59"28d11'
Humble Efterskole Umi 15h32'24"77d40'
Humbles,Margaret Kimberly Cas 0h38'42"60d37'
Humbs,Adrian Cmi 7h30'25"1d24'
Hume,Ashley Nicole Peg 22h19'0"30d2'
Hume,Janis Lyn 7h46'46"40d21'
Humenny,Marco John Per 3h27'46"40d45'
Humera Boo 14h20'21"50d20'
Humes,Joyce Mon 7h35'17"-2d28'B
Humes,Marci Uma 9h42'34"52d18'
Humes,Norman Mon 7h35'17"-2d28'A
Humfleet,Mysterious Sarah Cmi 7h32'41"1d2'
Humiston,Vivian & Jack Eri 3h44'37"-11d48'
Hummel's Lucky Star, Brad Hya 8h44'29"1d3'
Hummel,DO,Andrew E Cep 22h47'1"58d6'
Hummel,Frazier Cep 20h59'56"61d48'
Hummel,Giff Cep 22h53'42"59d22'
Hummel,Irving "Music" Lib 15h17'31"-23d57'
Hummel,Keith Patrick Her 17h26'19"21d13'
Hummel,Mitchell Robert Aur 6h11'25"54d45'
Hummell,John Her 16h17'34"40d59'
Hummer,Allan Cep 20h51'26"63d33'
Hummer,Caroline Elizabeth Peg 0h11'32"14d31'
Hummer,Erika Cas 23h32'17"53d51'
Hummer,Sharon Lyr 19h16'0"35d2'
Hummingbird Uma 9h24'44"59d32'
Humnisk,Alara Breanne Aur 5h32'0"38d33'
Humperdinck,Engelbert Her 17h59'1"38d58'
Humphrey,Beth Ann Cyg 21h26'45"38d20'
Humphrey,Betty Templeton Eri 3h25'21"-1d59'
Humphrey,Catherine Marie And 23h8'1"41d7'
Humphrey,Elisabeth A Peg 22h19'58"25d22'
Humphrey,Jill Elizabeth Her 16h24'36"28d40'
Humphrey,Joby Aql 18h53'35"-0d34'
Humphrey,Jocelyn Suzanne Lyr 18h54'26"31d56'
Humphrey,Jr,William Merritt Oph 17h27'25"10d34'
Humphrey,Kathleen And 0h11'35"31d41'
Humphrey,Kelly Christine Oph 18h6'40"10d58'
Humphrey,Maureen & Rodney Lac 22h11'15"49d23'
Humphrey-Guerra,JoAnn Cas 0h56'27"75d17'
Humphreys Infinite spirit,Schultz Cmi 7h58'37"5d30'
Humphreys,Anna-Marie Umi 14h16'41"67d21'
Humphreys,Jr,James Fraser Del 20h55'54"9d20'
Humphreys,Lucy Connell Moore Mon 7h44'31"-2d14'
Humphreys,May Rose Cyg 21h21'58"28d17'
Humphreys,Roberto Robert Dra 16h59'1"69d49'
Humphries,David C Uma 9h43'47"47d5'
Humphries,Forever-Owen & Ginar Lyr 18h50'33"41d36'
Humphries,Fred Ward Peg 0h5'29"13d8'
Humphries,Harry B Aql 20h22'60"0d9'
Humphries,John Cep 22h9'46"62d59'
Humphries,Lu Ann Hya 8h18'47"-1d26'B
Humphries,Patrick Andrew Dra 16h45'0"68d3'
Humphries,Randielle Marie Umi 14h13'27"66d24'
Humphries,Robert Michael Cep 2h3'54"80d10'
Humphries,Robert R Hya 8h18'47"-1d26'A
Humphries,Sara Walker Cyg 20h5'54"31d8'
Humphries,Tina Ruan Peg 22h40'47"24d22'
Humphries,William C Cep 22h11'58"60d42'
Humphry,CPA,Charlotte Del 20h17'23"9d22'
Humphry,Wallace Boycott McNabb Uma 8h12'13"60d16'
Humrich,Paul & Ruth Cet 2h49'0"7d7'
Hun 1 Cmi 7h30'55"0d39'
Hun,Angelo T Her 17h24'32"40d39'
Hunault,Benjamin Oph 17h3'54"10d3'
Hunault,Olivier Her 17h53'0"48d55'
Hung Hung & Yan Yan And 1h46'39"38d56'
Hung,Nai Uma 9h26'1"55d4'
Hung,William Kwok Peg 21h56'37"31d28'
Hunger,James "Pumpkin" Her 17h9'54"41d33'
Hungerford,Ellen Com 13h14'49"27d24'
Hungerland,Udo Boo 13h40'59"13d11'
Hungridge,Bryan Robert Dra 16h34'34"58d51'
Huning,Allan Cet 3h13'54"0d27'
Hunke,Jürgen Eri 4h50'43"-7d47'
Hunkins,Mauri Peg 23h19'0"32d45'
Hunneybell,Jean Barclay Uma 8h51'15"56d21'
Hunnicutt,Hap Uma 10h15'23"50d50'
Hunnicutt,Jennette Uma 10h10'10"57d36'
Hunnicutt,Tom Ori 5h58'17"6d2'
Hunny Bunny Cet 1h36'18"-13d42'
Hunny Bunny (Adam) Ori 4h55'26"5d47'
Hunsaker,Brooke Elizabeth And 0h50'51"35d39'
Hunsaker,Donald Ray Aur 6h47'19"37d44'
Hunsberger,Christopher B Per 4h4'35"51d52'
Hunsberger,Colin Troy Cam 3h20'46"61d37'
Hunsberger,Galen Tyler Her 16h55'37"38d23'
Hunsberger,Logan Anders Her 17h29'30"26d55'
Hunsberger/Arimura Lac 22h53'17"53d5'
Hunsinger,Franz Dra 16h11'16"51d47'
Hunsinger,Jan Scott Cam 5h42'1"61d5'
Hunsinger,Sean Aur 6h31'44"33d26'
Hunsucker,Bobby Joe Her 17h34'32"20d14'
Hunt (Spud),Charlotte Susan Lyr 19h17'47"41d42'
Hunt Et Al,Howard & Ida Lac 22h8'42"46d30'
Hunt For Joy Lyn 8h19'27"40d58'
Hunt Star,The Per 3h5'35"40d50'
Hunt Star,The Jim Cnv 12h30'15"32d16'
Hunt's Millennium Star Fiona Com 12h40'19"21d33'
Hunt's Star,Debbie Jane Cas 0h5'55"61d2'
Hunt,Agnes Erika Cas 0h45'17"73d37'
Hunt,Airborne Johnny Aql 19h25'1"-7d54'
Hunt,Al & Charlie Cas 23h20'1"53d18'
Hunt,Aleisha And 1h17'21"38d56'
Hunt,Alexander David Peg 21h20'17"12d55'
Hunt,Allyson P Com 12h2'27"21d28'
Hunt,Amy Lee Leo 11h53'41"21d19'
Hunt,Angela Lyn 9h22'52"41d20'
Hunt,Anthony James Aql 20h2'0"6d39'
Hunt,Asia Cam 7h34'26"61d40'
Hunt,Austin Cep 20h33'0"67d2'
Hunt,Azarm M Uma 12h41'59"62d12'
Hunt,Barbara Ann Lyr 18h15'59"35d16'
Hunt,Bill & Claire Lyn 8h41'1"40d53'
Hunt,Chris Aql 20h4'1"8d21'
Hunt,Christine Ellen Aql 19h52'1"13d49'
Hunt,Claire Dra 17h37'42"71d2'
Hunt,Cotton & Billie Cma 6h52'46"-19d6'
Hunt,Craig "Fatboy" Per 3h4'33"43d26'
Hunt,Cynthia L Ori 4h42'27"4d33'
Hunt,Dave Her 16h7'44"40d6'
Hunt,David James Frederick Cyg 20h22'37"41d8'
Hunt,David P Dra 14h40'48"61d28'A
Hunt,Donna M Peg h21h58'35"28d50'
Hunt,Doris & Max Cyg 20h55'41"30d0'
Hunt,Dr G & Stephanie Neurohr Gem 5h58'25"27d22'
Hunt,Edward Buddy Cmi 7h56'12"0d3'
Hunt,Eric T Aql 19h18'34"15d10'
Hunt,Glenn Oph 18h0'37"11d53'
Hunt,Glorine Cas 0h58'37"58d28'
Hunt,Graham Reginald Her 18h17'31"14d55'
Hunt,Harley J Aur 5h9'0"38d55'
Hunt,Hazel McCubbins Kearney Tau 3h27'44"30d30'
Hunt,Heather Ann Aql 18h54'40"-0d1'
Hunt,Isabell Cyg 19h28'19"38d50'
Hunt,Jr,John Michael Her 17h53'52"31d32'
Hunt,Jr,Robert Carl Aur 5h59'37"38d45'
Hunt,Julia Cas 2h33'32"57d34'
Hunt,Julida "Judy" F Cas 0h39'58"67d7'
Hunt,Keith W Dra 19h31'24"61d16'
Hunt,Kelsey Kristine Mon 8h8'10"-9d53'
Hunt,Margaret Cyg 19h30'16"37d44'
Hunt,Martha L Com 12h54'54"24d3'
Hunt,Martin Ori 5h36'60"1d35'
Hunt,Mary Ann Lyr 18h49'29"42d20'
Hunt,Michael Terrell Cma 7h12'39"-16d31'
Hunt,Mitchell S Per 1h48'59"48d18'
Hunt,Myron M Her 17h34'54"21d17'
Hunt,Phillip Ori 5h54'28"18d41'
Hunt,Randall Glenn Lep 5h2'43"-11d38'
Hunt,Rexford R Vul 19h40'57"23d26'
Hunt,Richard Her 18h1'47"40d18'
Hunt,Robert William Aur 6h15'1"45d4'
Hunt,Rowena Francis Vul 19h46'20"28d52'
Hunt,Ruth Ann Cas 23h35'54"61d27'
Hunt,Sean Edward Joseph Cam 3h55'56"1d6'
Hunt,Sharon Lyr 18h16'46"44d59'
Hunt,Shaun G Ori 5h54'1"11d13'
Hunt,Sherry L Cyg 21h9'28"39d51'
Hunt,Shirley Nancy Sct 18h50'4"-7d48'
Hunt,Simon T Ori 4h51'47"5d35'
Hunt,Stevie Peg 23h18'58"18d13'
Hunt,Teke Cyg 20h18'60"40d17'
Hunt,The Ashley Umi 14h30'32"68d46'
Hunt,The Little Man- James Daniel Umi 14h16'11"65d52'
Hunt,Thomas Ori 4h41'46"12d46'
Hunt,William Michael Lac 22h35'1"53d18'
Huntemann,John H Her 16h29'47"39d41'
Hunter Cet 3h18'44"3d40'
Hunter Lac 22h18'40"49d60'
Hunter B Dra 15h54'0"61d53'
Hunter Den Goat Cma 7h12'47"-15d12'
Hunter Esquire,H Joe Crt 10h53'24"-7d44'
Hunter ME,Leslie Martin Sgr 19h37'10"-31d23'
Hunter Wray Leo 10h33'42"14d35'
Hunter's "Guiding Light",Diana Sge 19h57'0"20d37'
Hunter's Star Dust Cma 6h55'48"-15d7'
Hunter's Treasure Cma 7h12'13"-16d15'
Hunter,Adam Duval Ori 5h55'20"16d24'
Hunter,Adam Robert Cep 2h33'24"78d45'
Hunter,Aileenna Danielle Mon 6h24'31"1d39'
Hunter,Barbara Joyce And 2h3'43"40d58'
Hunter,Benjamin Robert Ori 5h31'50"-1d54'
Hunter,Billie Anne Cet 0h55'39"0d36'
Hunter,Bob & Pea Cyg 20h20'48"39d21'
Hunter,Bonnie Mon 6h35'24"6d28'
Hunter,Bradley Cet 1h27'1"-2d15'
Hunter,Carol Gem 6h49'19"14d46'
Hunter,Catherine Ellis Lyn 7h56'17"40d35'
Hunter,Charles Dale Aql 20h8'0"4d2'
Hunter,Christopher Randolph Her 17h32'16"31d20'
Hunter,D Lloyd Aur 5h20'44"54d36'
Hunter,David E K Cyg 21h18'32"38d26'
Hunter,David Matheson Uma 9h46'47"46d37'
Hunter,Devon Bradfield Per 3h59'52"51d45'
Hunter,Diane Susan Uma 10h59'41"71d4'
Hunter,Dorothy Cyg 19h58'1"30d23'
Hunter,Elaine Cas 1h40'23"73d9'
Hunter,Gordon & Theresa Ori 5h29'49"1d16'
Hunter,Gordon Murray Lyn 8h20'45"47d39'
Hunter,Heather Mon 7h7'19"-1d24'
Hunter,Heidi Ori 6h0m36"-1d50'
Hunter,Helen Brand Lyn 6h12'53"59d23'
Hunter,Ian Michael Curtis Her 16h57'2"15d33'
Hunter,Jackie R Uma 11h53'58"40d59'
Hunter,Jacqueline Rose Cep 22h7'20"61d55'
Hunter,James Russell Boo 14h20'31"18d9'
Hunter,Jerry E Sex 10h31'24"-6d25'
Hunter,Jonathan Marlin Peg 22h34'17"11d25'
Hunter,Joy A Uma 9h48'27"50d10'
Hunter,Jr,Donald Glenn Sex 9h49'33"2d60'
Hunter,Julia Anne Tatum Cet 3h5'35"1d48'
Hunter,Lynsey Michelle Lyr 18h34'21"35d3'
Hunter,Margaret S And 23h34'28"36d6'
Hunter,Nancy Umi 13h44'0"72d28'
Hunter,Nathan Allon Samuel Her 17h16'34"48d53'
Hunter,Patricia Ori 5h48'48"10d11'
Hunter,Rachel R And 1h8'48"40d22'
Hunter,Richard Charles Boo 14h58'22"17d45'
Hunter,Robert Heybyrne Ori 5h24'1"15d23'
Hunter,Robert J Umi 14h52'32"71d40'
Hunter,Robert James Cnv 12h21'59"43d47'
Hunter,Ryan Matthew Aur 5h54'1"50d14'
Hunter,Sandra Jean Mon 8h1'14"-4d50'
Hunter,Stephen Cep 23h6'44"61d47'
Hunter,Stephen Charles Boo 14h30'25"41d9'
Hunter,Tracia Lyr 19h18'15"42d21'
Hunter,Tristan John Cet 2h36'44"0d14'
Hunter,William & Allison Her 18h19'38"12d27'
Hunter-Kysor,Riley Schmidt Eri 3h51'34"-10d29'
Hunters Diamond Cyg 21h54'52"53d30'
Huntingford,Jonathon Peg 23h41'12"18d15'
Huntington,PhD,Gordon Oph 17h0'42"10d36'
Huntleigh's Superstar Uma 12h11'33"57d44'
Huntley,Howard Per 2h41'58"37d20'
Huntley,Verna Grace Cas 0h15'16"65d17'
Huntoon,Dr Carolyn Leach Oph 17h32'0"-1d27'
Huntoon,Jim Cep 20h58'22"68d5'
Huntowski,Frank Cap 21h20'22"-16d21'
Huntrakoon,Chavee Lac 22h4'34"49d22'
Hunts Heaven Umi 13h56'47"76d25'
Huntsman,Karen & Bret Mon 7h39'28"-3d43'

Hunwick,Bernard Barton
Aql 19h58'57"7d38'
Hunyard,Mark
Her 17h10'48"46d16'
Huo,Rico S K
Ori 5h39'21"10d27'
Huos,Michel Coustille
Cnv 13h12'38"50d52'
Huot,Christian
Aur 6h55'1"44d21'
Huot,Patty
Lyn 7h39'0"58d23'
Hupper,Merry Christmas Julie L
Lyr 18h46'53"41d52'
Huppertz,Anne Marilyn Hess
And 1h59'16"38d58'
Huquet,Guy
Cep 22h25'40"57d47'
Hurant,Jeffrey David
Per 2h33'48"57d2'
Hurant,Leslie M
Peg 23h19'52"25d10'
Hurant,Robert Michael
Dra 16h3'32"61d22'
Hurb
Cam 6h9'36"58d11'
Hurd 40th Birthday, Laura
Ari 2h57'39"20d51'
Hurd,Caroline Stirling
And 0h11'57"36d36'
Hurd,Donna
Boo 15h23'36"47d34'
Hurd,Francis Jackson
Leo 10h14'52"10d24'
Hurd,Frederick Ernest
Cep 21h24'10"68d23'
Hurd,Gary & Deborah
Uma 10h59'58"50d14'
Hurd,Myya Naomi
Lmi 10h28'0"32d8'
Hurd,William Michael
Cep 4h27'40"80d2'
Hurd,Zachary
Per 3h4'0"47d55'
Hurder,Greg
Aur 5h15'48"45d6'
Hurdisandron
Cep 21h53'10"60d22'
Hurer,Rashide
Lac 22h36'58"55d41'
Hurich
Cyg 19h24'22"33d16'
Hurkamp,Mark Shelton
Boo 13h44'43"17d25'
Hurlburt,Betty Latty
Sco 16h4'14"-26d5'
Hurles,John J
Cep 20h41'35"65d31'
Hurley,Chip
Peg 22h34'20"7d51'
Hurley,Dan
Aur 6h58'44"38d43'
Hurley,Deborah Lynn
Cam 12h48'57"78d20'
Hurley,Dr,William
Lyr 18h25'25"40d52'
Hurley,Fred
Dra 18h42'35"78d57'
Hurley,George
Aql 20h1'47"9d23'
Hurley,Graham Thomas
Cep 21h30'52"70d20'
Hurley,Jonathan Michael
Boo 14h59'29"20d14'
Hurley,Kevin
Cyg 21h1'18"38d25'
Hurley,Lynn C
And 1h31'23"40d9'
Hurley,Mary
Cas 2h21'50"73d24'
Hurley,Niall Thomas
Per 1h28'21"52d38'
Hurley,Ruth E
Lac 22h7'36"46d14'
Hurley,Susan J
Equ 20h58'12"8d5'
Hurley,William Thomas
Cet 2h45'1"-0d21'
Hurley-Felt,Patricia
And 23h47'60"42d30'
Hurn,Hollie Victoria
Mon 6h28'50"2d52'
Hurney,Dorothy
Boo 14h29'13"10d45'
Huror,Viking Ken
Ori 5h56'22"19d46'
Hurowitz,Richard
Boo 14h18'49"15d52'
Hurrell,Deborah
Peg 22h12'27"30d1'
Hurricks,Maurice Edward
Cep 20h31'24"61d14'
Hurst
Peg 23h38'23"26d32'
Hurst's Heavenly Body
Boo 14h55'1"36d55'
Hurst,Angela Diana
Cnc 8h55'1"22d10'
Hurst,Ashley Rae
And 2h22'27"42d8'
Hurst,Candy
Equ 20h58'21"5d20'
Hurst,Cathy
Eri 4h7'28"-14d56'
Hurst,Cheryl
Her 18h6'31"14d52'
Hurst,Edna
Cmi 7h43'38"4d26'
Hurst,Greg
Per 1h50'51"52d42'
Hurst,James Harrison
Ori 5h57'34"14d52'
Hurst,Jason Jeffery
Hya 8h42'14"4d1'
Hurst,Jeffery M
Tri 2h18'24"31d53'
Hurst,John
Peg 22h13'56"10d11'
Hurst,Jr,Charles G
Dra 13h56'21"68d11'
Hurst,Mary Beth
And 23h37'0"47d3'
Hurst-Bryant
Tau 4h44'44"16d28'
Hurt of Pontyclun
Her 18h5'42"45d27'
Hurt,Chance Jon
Hya 8h14'26"3d55'
Hurt,Jr Nathan H
Her 17h54'11"14d58'
Hurt,Jr,David Hadley
Aur 5h5'60"50d18'
Hurt,Kristi
Cet 1h30'12"-10d39'
Hurtado,Tiffany Lynn
Cam 6h0'0"60d54'
Hurteau,Vincent
Uma 10h6'10"47d43'
Hurtig,Etta
Hya 9h33'41"-8d44'
Hurtig,Henry
Umi 17h1'45"76d9'
Hurtig,Juila
Umi 17h1'36"76d10'
Hurtig,Shirley
Tau 4h32'22"18d54'
Hurtubise,Paul
Cep 21h7'33"58d35'
Hurwitz,Ben
Per 2h55'25"43d39'
Hurwitz,James Daniel
Lyr 19h6'1"41d29'
Hurwitz,Michael
Per 2h54'1"38d40'
Hurwitz,Ronald & Andrea
Cyg 21h38'1"40d44'
Hurwitz,William Charles
Hya 8h28'1"-6d14'
Huryk,Jessica
And 1h56'44"38d1'
Husa,Jr,William J
Oph 17h29'16"-20d35'
Husayni,Deanna Lena Jad
Uma 11h2'36"46d19'
Husband,Gaylon
Aql 18h57'31"17d35'
Husband,May Elizabeth Lloyd
Ori 5h14'48"-6d12'
Husch-Husch Helga
Cep 22h7'27"61d8'
Huscusson,Eric Taylor
Boo 14h59'30"53d10'
Huse,Hu H Timothy
Tri 1h49'24"28d18'
Huseby,Bob & Jane
Equ 21h17'39"3d41'
Huselton,Bob
Lib 15h4'37"-0d40'
Huseman,Alexander Arons
Hya 10h32'0"-11d44'
Husgburg-Drach,Rose
Lmi 9h53'38"38d28'
Husher,Sherie Shutterly
Aqr 21h43'58"0d48'
Hushion,Casey
Aur 7h2'14"39d40'
Huskalange Star,The
Aql 19h53'44"15d18'
Huskey,Erik C
Mon 7h5'29"-6d24'
Huskey,Jr,Michael David
Aql 20h11'58"10d30'
Huskey,Romantic Roy
Cyg 20h56'17"40d56'
Husmer,Buzz & Judy
Sct 18h41'34"-8d5'
Huson I,Bradley Upton
Aur 6h13'47"35d19'
Huson's Wishing Star, Kimmy
Tau 5h28'16"28d12'
Huson,Linda S
Lyn 9h6'14"39d35'
Huss,Chris
Cnv 12h22'1"45d38'
Huss,Kerri Lynn
Mon 6h53'45"-0d53'
Huss,Leland Henry
Uma 11h1'32"47d60'
Hussain,Michele
Crb 15h35'38"28d53'
Hussar,Frank Lawrence
Cmi 7h55'26"8d4'
Hussein Salah Hussein
Her 18h28'46"18d57'
Hussein,Metin
Ori 5h59'22"16d57'
Hussey,Betty Lee Trafton
And 1h54'25"38d54'
Hussey,Bobby
Dra 15h59'31"67d60'
Hussey,Eileen
Cam 3h58'51"68d27'
Hussey,Kathleen
And 23h45'1"44d55'
Hussey,Roy
Ser 15h15'35"11d45'
Hussman,Merrily Joan
Peg 23h31'28"21d27'
Hussong,Mark A
Lac 22h16'16"38d56'
Hustad,Janet
Cas 1h30'0"58d49'
Hustead II,James Miller
Aql 20h5'52"1d47'
Husted,"Betsy", Elisabeth
Cyg 21h5'38"40d36'
Husted,Betty
Lyn 9h32'0"41d28'
Husted,Bruce H
Dra 15h51'49"60d50'
Hustlin' Henry
Mon 6h20'24"8d18'
Huston's Star
"Jitterbug",Jenny
Crb 15h54'29"38d48'
Huston,Jacob Blair
Mon 8h1'27"-3d17'
Huston,Janet Sue
Cas 0h11'22"54d41'
Huston,Margo
Mon 7h21'0"-6d22'
Huston,Michael G
Her 18h5'25"40d12'
Huston,Sr,John A
Her 17h16'44"18d55'
Huston,Stephany Ann
Tri 1h59'23"30d12'
Huston,Taran
Lmi 9h46'1"40d25'
Husty,Super-Dave Rick
Sct 18h43'22"-6d40'
Hutchence,Michael
Aur 4h54'57"41d6'
Hutchenreuther,Robert William
Her 16h57'30"34d6'
Hutchens,Jr,James
Her 18h9'58"38d41'
Hutchens,Robbie
Cet 2h52'37"3d20'
Hutchens,William V
Cep 21h48'12"58d59'
Hutcherson,Joshua Ryan
Lib 15h4'51"-8d49'
Hutchings
Umi 15h9'14"66d24'
Hutchings,Dion Jude
Boo 14h47'37"22d36'
Hutchings,Mary Louise Frisch
Peg 23h23'43"11d47'
Hutchins III,William Tate
Her 16h37'60"38d32'
Hutchins,Alma Albert
Lyn 7h5'25"58d23'
Hutchins,Blanche Francis
Ori 6h6'42"0d30'
Hutchins,Donna
Ori 5h57'1"15d35'
Hutchins,Gary
Lmi 9h48'24"33d52'
Hutchins,Gregory
Eri 3h1'16"-10d20'
Hutchins,Magdalene M
Cyg 21h11'37"35d8'
Hutchins,Susan
Lyr 18h17'32"42d2'
Hutchins,Thomas Benton Onas
Uma 11h14'46"57d38'
Hutchins,Tom
Aql 20h10'37"1d2'
Hutchinson Polack's
Uma 9h9'57"49d39'
Hutchinson,Audrey
Com 12h52'36"26d35'
Hutchinson,Candace Lee
Mon 7h42'38"-3d32'
Hutchinson,Cayla
Mon 7h13'22"-6d34'
Hutchinson,Connie
Cas 0h44'24"63d58'
Hutchinson,Diane E
Oph 18h19'15"7d44'
Hutchinson,Elizabeth Stafford
Cas 0h33'59"72d18'
Hutchinson,John
Cep 22h28'0"70d57'
Hutchinson,Joseph Fulton
Lac 22h17'41"54d56'
Hutchinson,Nelson C
Aql 18h59'52"11d11'
Hutchinson,Robert
Umi 14h12'53"78d43'
Hutchinson,Robert
Sct 18h56'12"-6d57'
Hutchison,Carrol
Ori 5h23'21"0d15'
Hutchison,David Arthur
Ori 6h4'44"8d40'
Hutchison,Jennifer
And 2h6'20"38d34'
Hutchison,Jerry E
Aql 19h4'0"-0d37'
Hutchison,Julie K
Crt 11h22'1"-21d37'
Hutchison,Kelli Jo
Cet 1h31'33"-11d12'
Hutchison,Robert
Dra 14h58'40"64d18'
Hutchison,William "Bill"
Dra 20h18'20"68d7'
Huter,Irmgard
Peg 23h32'25"10d37'
Huth,Aaron Michael
Ori 5h39'41"10d21'
Huth,Harald
Peg 23h28'31"12d45'
Huth,Jerome Richard
Cma 7h29'23"0d27'
Hutkins,Lil & Sol
Boo 14h7'56"40d36'
Hutsell-33 Years,Wayne
Leo 9h42'51"7d17'
Hutsko,Bradley Kent
Her 17h21'15"48d28'
Hutson,Lynne Ellen
Peg 22h16'49"34d38'
Hutson,MD,Dr J J
Oph 18h1'48"12d43'
Hutson,Oliver
Umi 14h37'58"68d36'
Hutson,Pete(Frank)
Oph 17h57'32"8d5'
Hutt Star,The Steven
Lyn 7h48'22"44d24'
Hutt,Nesy
Del 20h13'37"15d37'
Hutt,Walter Philip
Uma 13h40'29"48d45'
Huttner,Fred J
Cma 6h31'1"-15d17'
Hutto,Sherry
Peg 22h38'44"29d1'
Hutto,Tim
Aql 19h57'24"14d40'
Hutton III,Dr Howard David
Per 3h51'27"37d32'
Hutton,Elizabeth
Cas 0h44'14"71d12'
Hutton,George Draper
Del 20h55'47"6d40'
Hutton,Greg
Aql 19h27'55"15d15'
Hutton,Kathleen
Com 12h22'43"20d7'
Hutton,Lauren Elizabeth
Cyg 20h2'41"30d52'
Hutton,Matthew
Aur 7h9'43"40d1'
Hutton,Michael J
Lac 22h16'1"51d34'
Hutton,Nic
Cas 0h33'11"62d30'
Huttram,30/50 Mark
Cep 21h29'0"60d28'
Hutwagner,Brett E A
Aur 5h14'54"41d39'
Hutz,Joshua William
Lac 22h9'21"47d1'
Hutz,Sarah Allane "Ladybug"
Cas 0h9'46"64d56'
Hutzell,Glenn
Aur 6h0'1"30d37'
Hutzell,LaVon
Cyg 19h39'56"28d41'
Huw James Star,The
Cep 22h43'28"56d57'
Hux,Tyler Carlton
Aur 4h55'16"40d2'
Huxhage,Jutta
Lmi 10h0'55"35d29'
Huxtable,Karen
Lyr 19h6'1"28d56'
Huyard,Wayne
Aql 20h3'15"7d43'
Huygh,Baptiste Etienne
Cyg 21h8'22"30d21'
Huyghe,Cecile
Cas 2h30'39"77d12'
Huyghe,Fernande
Ser 15h9'36"7d35'
Huyin,Nancy Jean
Uma 11h50'1"40d19'
Huynh,Loc Patrick
Uma 11h50'11"46d34'
Huynh,Myduyen
Ori 5h22'57"0d59'
Huynh,Ran
Aql 19h31'44"12d13'
Huzzey,Kent
Aql 19h34'29"1d20'
Hvezda-Jandak Star, Jandakova
Per 2h52'41"40d59'
Hvidsten,Ross
Her 16h42'30"33d26'
Hvidt,Flemming
Ori 6h1'21"0d44'
Hvitfeldt,Mogens
Ser 15h22'18"18d55'
Hvizda
Cas 1h43'54"65d8'
Hwang,Andy Jin
Cep 0h18'14"78d59'
Hwang,Shirley
Lyn 7h44'42"48d56'
Hy-Lu
Dra 11h45'19"72d12'
Hyams,David Samuel
Crt 11h12'44"-19d27'
Hyams,Debbie
And 0h1'44"47d36'
Hyatt's Dream,Ron
Aql 19h57'44"11d32'
Hyatt,Edward T
Aql 20h11'33"0d35'
Hyatt,Jennifer Nicole
Peg 23h4'16"10d54'
Hyatt,Sumner
Cyg 21h17'53"37d36'
Hyatt,Victoria
Cyg 21h17'41"38d7'
Hyatt,Wanda M
Aql 20h7'20"1d21'
Hyatte,Daniel
Her 17h3'37"48d29'
Hyde"Mi Amor-Ilywamh", Sir JC
Sco 17h29'29"031d30'
Hyde,Anne
Cas 2h32'35"57d32'
Hyde,Carol Ann
Peg 22h16'0"20d10'
Hyde,Charles R "Chip"
Cet 3h10'44"1d20'
Hyde,Daniel Nicholas
Cam 3h40'11"77d54'
Hyde,Frankie
Lyn 7h26'23"41d3'
Hyde,Frazer
Ori 6h6'58"3d43'
Hyde,Jack
Cyg 21h53'13"41d42'
Hyde,James
Per 2h59'55"45d49'
Hyde,Joseph
Ori 5h3'26"-2d0'
Hyde,Judith Ann
Sex 9h48'0"4d42'
Hyde,Karen
Cas 0h58'52"59d33'
Hyde,Mildred Rosalie
Aql 19h1'53"17d48'
Hyde,Richard
Cyg 20h0'59"30d20'
Hyde,Rosalie
Mon 7h51'30"-3d20'
Hyde,Sandra
Lyr 18h29'10"40d33'
Hyden,Olive S
Cas 2h1'38"61d26'
Hyder,Linda Wilson
Eri 4h2'26"-10d1'
Hyder,Walter Benton
Mon 7h29'56"-5d55'
Hydorn,Gerri Lynn
Cas 0h57'25"50d32'
Hydrick,Paul Anthony
Cet 2h6'32"0d23'
Hyer,Dr Reginald Clinton
Oph 17h33'36"-23d21'
Hyer,Rachael Loyola
Cas 1h39'56"61d3'
Hyla
Uma 12h42'41"62d22'
Hyland,Barbara
And 23h31'0"37d38'
Hyland,Daniel Francisco
Gem 6h50'50"12d23'
Hyland,Daniel Breen
Gem 6h50'25"12d6'
Hyland,Dylan T
Her 17h3'37"20d59'
Hyland,Ed & Kathy
Cyg 21h18'16"36d44'
Hyland,Jason M
Per 3h17'30"40d45'
Hyland,Jr,Thomas T
Aur 7h3'1"36d52'
Hyland,Kevin
Cep 20h43'36"70d17'
Hyland,Lisa Marie
Mon 6h22'41"-6d34'
Hyland,Marion
Uma 9h18'42"58d12'
Hyland,Nicola Danielle
Cyg 21h31'36"52d38'
Hyland,Pamela Jeanne
Cas 1h56'27"75d38'
Hyles Family Star,Jack & Beverly
Lac 22h26'38"56d46'
Hyles-Daniel 12:3,Dr Jack
Her 16h57'36"50d53'
Hylton,Kelly Jeanne
Mon 6h54'32"0d35'
Hylton,Warren James
Peg 22h40'58"26d13'
Hylton-A Star To Follow,Bud
Aur 5h6'17"41d24'
Hyman,Curtis Van Hook
Her 18h3'1"31d21'
Hyman,Fredric S
Tau 4h21'30"16d19'
Hyman,Irving
Cnv 12h57'36"38d10'
Hyman,Matthew E
Aqr 22h4'1"0d43'
Hyman,Merv
Aur 7h12'31"40d20'
Hyman,Mimi
And 0h10'55"39d54'
Hyman,Nicola Jayne
And 0h28'0"38d26'
Hyman,Pessi Gitta
Lyn 7h0'47"53d53'
Hyman,Rick & Jill
Cyg 19h28'59"31d3'
Hyman,Ruth
Cam 10h50'28"82d25'
Hyman,Stellar Eurus Joe
Cam 3h15'52"60d5'
Hymans,Jeff
Her 17h28'33"26d43'
Hynan,In Memory of Edward Vincent
Her 18h5'51"28d27'
Hyndman,Peter
Per 3h0'41"47d49'
Hynds,Patricia Marie
Uma 10h10'45"70d23'
Hynes,Brendon Michael
Sct 18h34'43"-4d54'
Hynes,Jessica Rene'
Cam 3h46'43"55d15'
Hynes,John G
Cnv 12h38'1"51d49'
Hynes,Karen C
Peg 21h40'23"21d34'
Hynes,Mary
Cas 0h32'27"66d52'
Hynes,William M
Aur 6h40'47"38d54'
Hyon,Chong
Boo 14h47'44"32d14'
Hyperion
Cam 4h54'11"61d34'
Hypher,Benedict John
Cap 21h52'56"-17d52'
Hysham,Ellen
Sct 18h33'1"-6d55'
Hyter,Joanna
Cas 1h10'21"64d25'
Hyun,Lee Ji
Uma 9h56'50"56d2'
Hyun-Sook Kim
Del 20h52'58"2d15'
Hyzy,Brian Spencer
Lyr 18h58'10"31d45'
HèLéne
Cam 6h31'25"68d18'
Häcker,Angelika "Angie"
Leo 10h57'52"8d0'
Hägen Christian
Cep 0h33'39"78d40'
Hähnlein,Margarete
Aql 19h59'16"10d19'
Hämel,Horst
Ori 5h50'39"20d28'
Hänel,Gisela
Tau 5h55'43"27d49'
Hänsch,Bärbel
Vir 13h23'17"-8d5'
Härtel,Rudolf K
Vir 14h8'25"-5d50'
Hässner,Manfred
Sgr 19h44'27"-41d33'
Häusler,Nicole
Mon 6h41'22"6d58'
Hélène
Col 6h34'57"-33d14'
Hélène
Per 3h7'36"41d13'
Hélène et Jean-Marie
Peg 22h58'19"24d1'
Hélène et Raymond, Amour Infini
Cyg 19h59'60"30d31'
Hélène et Stéphane
Cyg 19h28'15"33d30'
Héon,Gisèle Lamothe
Lyn 9h38'60"40d27'
Hétu,Nicole
Ori 5h41'38"12d5'
Hitol Mövenpick Genève
Per 4h2'2"39d15'
Höka Ernö
Gem 6h21'26"19d6'
Höka Mihály
Gem 6h32'41"18d48'
Höeg,John Anthony
Cmi 7h30'41"7d41'
Höfer,Susanne
Uma 9h38'20"46d39'
Höfflin,Eva
Boo 14h25'46"14d47'
Högfors,Anders Birger
Cam 5h21'41"80d36'
Högl,Agnes
Lyn 8h11'31"48d14'
Höhl,Karin
Leo 9h55'47"11d19'
Höhn,Alexander
Tau 5h45'51"21d28'
Höhner,Sabine
Aqr 22h44'25"-2d15'
Höllmich,Christoph
Umi 16h28'20"70d4'
Hölzl,Horst
Sgr 18h52'40"-20d56'
Höner zu Siederdissen, Detlev
Uma 9h20'48"47d37'
Höpfner,Gurine-Karin
Lib 14h42'0"-23d14'
Höpoltseder,Jürgen
Vir 13h52'16"-1d30'
Höppner,Birgit
Lyn 7h51'10"43d9'
Hösl,Dr Ernst
Her 17h39'47"42d4'
Hübel,Sonja
Tau 5h53'41"23d27'
Hübler,Jens
Lac 22h5'0"51d47'
Hübner,Nicolai
Uma 12h45'16"53d7'
Hühnke,Andreas
Lib 15h7'23"-20d30'
Hüppi,Cornelia
Lib 14h59'23"-10d35'
Hüppi,René Jean
Cas 1h5'54"60d9'
Hüppi,Tobias Roberto René
Lib 14h57'51"-22d58'
Hütgens,Thomas
Cnc 8h25'10"32d54'
Hüther,Johannes
Cam 6h1'15"61d30'
Hüther,Julius Alfred-Meinrad-P
And 23h9'40"42d2'

I

I
Lyn 7h27'17"37d8'
I C One Star
Dra 9h53'47"77d49'
I Carry Your Heart With Me
Uma 9h16'25"51d43'
I Castorina
Del 20h18'0"11d36'
I Claudius Ricktus
Umi 16h18'38"71d38'
I Could Love You Like That 12/95
Tri 2h3'46"31d57'
I D E A Company I994
Aur 4h54'13"51d20'
I D F Flower Poppy
Aur 6h24'36"32d30'
I Don't Know Why, But I Do
Aqr 22h34'37"-11d23'
I Dreamed I Painted Living Music
Lyr 19h23'31"40d54'
I For An Angel
Cas 0h52'1"61d43'
I Knew You Were There, Patricia Ann
And 0h52'1"41d11'
I L U
Ser 17h34'43"-13d36'
I L U F E
Cyg 19h27'35"35d13'
I L U-Menchie
Lac 22h38'53"55d7'
I Love Libby C-Randy P
And 0h8'0"38d17'
I Love Lucy
Cmi 7h58'24"1d13'
I Love Sarah
Lyr 18h41'50"26d53'
I Love You
Sge 20h6'55"16d49'

I Love You Bigger Than The World
Uma 11h58'39"46d32'
I Love You Dad...Gabby
Vir 13h4'2"-5d16'
I Love You Daddy,Orren
Ori 5h7'16"1d15'
I Love You Davids
Per 3h23'60"40d21'
I Love You Dawn
Mon 7h0'19"4d32'
I Love You Dearly
Cnc 8h46'53"12d21'
I Love You Forever, My Brother Joe
Aur 6h33'0"34d54'
I Love You Jo
Lyr 19h6'33"38d0'
I Love You More
Ori 5h27'0"0d25'
I Love You More!
Uma 9h13'49"49d21'
I Love You Star
Cap 21h42'30"-22d35'
I Love You Sue
And 0h30'0"45d9'
I Love You This Much, Scott!
Per 2h5'1"48d50'
I Love You Tracy
Lmi 10h29'11"31d42'
I Love You! Love, Jamie
Cas 2h38'1"58d53'
I Love You, Abel
Boo 15h6'60"22d25'
I Love You,Amy
And 1h25'22"37d22'
I Love You,Garen
Vir 13h39'39"-8d31'
I Love You,Hoona
And 2h24'58"47d3'
I Love You,Joanne
Cas 1h45'0"61d4'
I Love You,Matthias
Sge 19h14'44"16d41'
I L U
Gru 23h12'31"56d17'
I Luv Mom & Dad
Cnv 13h26'16"48d51'
I Luv Wu Star,The
Lyr 18h25'1"37d40'
I Muvrini
Ori 5h55'60"10d39'
I of Many
Lyn 8h0'0"45d28'
I Once Had A Bear Named Fred
Umi 16h51'31"77d54'
I Swear To You
Ori 5h58'37"17d50'
I Swear...Completely
Eri 3h37'59"-2d8'
I TAI
Her 16h18'35"23d25'
I Wish I Could Give You More
Uma 9h0'42"59d10'
I'Dabole,Janboulat
Aur 5h9'53"40d55'
I'Dabole,Julianna
Cam 5h39'51"68d33'
I'll Always Love You Spence & Marianne
Cyg 19h54'1"48d25'
I'll Loveth You Always Eyeore
Cas 0h38'39"66d7'
I-n-I Living Irie
Peg 22h48'11"3d29'
Iaccio-Nardoza
Cam 4h44'46"68d5'
Iacino,Edward M
Her 16h58'0"40d1'
Iacobus
Aur 6h29'1"34d44'
Iacona,Nora Iacona,B Joseph
Cyg 19h53'50"50d4'
Iacopelli,Rosina M
Lyn 7h53'0"44d39'
Iacopetti,Luigi
Cam 14h11'44"81d23'
Iacopo,Capriotti Daniele Marino
Gru 22h31'33"-54d26'
Iacovazzo,Iolanda
Cap 21h2'41"-23d39'
Iacovelli,Andrea
Lyn 8h49'13"34d32'
Iacoviello,John G
Per 4h5'38"37d42'
Iacoviello,Myriam
Pyx 8h51'44"-24d6'
Iacuessa,Alice J
Dra 15h56'1"63d13'
Iacuessa,William M
Dra 15h55'1"67d27'
Iadevaio,Gloriann
Vir 13h6'11"-0d38'
Iafaie
Peg 21h58'39"31d33'
Iafrate,Monte
Her 16h58'41"34d25'
Iafrati,Matteo
Cam 3h39'21"74d25'
Iago,Ann
Peg 22h14'40"4d10'
Iah,Y H W H Michele
Nor 16h18'48"-52d14'
Iain's Happy Place
Umi 15h17'23"68d27'
Iain's Star
Ori 5h59'42"10d54'
Ialongo,Cristiana
Dra 14h52'24"64d36'
Iamah
Peg 22h51'47"24d20'
Iamsean,P J
Oph 18h7'0"0d3'
Ian
Equ 21h21'56"10d37'
Ian
Sct 18h53'52"-9d16'
Ian "The Real Deal"
Aur 7h17'13"36d36'
Ian & Bagpuss In Eternity
Cyg 19h24'10"31d20'
Ian & Dawn
Crb 16h2'38"27d23'
IAN (Immortal Amado Novio)
Cep 0h0'1"70d53'
Ian 50
Lyn 7h36'55"36'5'
Ian Flemming's Daisy
Cas 0h7'31"63d60'
Ian One
Cam 3h12'29"58d29'
Ian Patrick
Dra 15h5'33"60d54'
Ian the Twelfth
Cep 20h40'1"63d60'
Ian's Illumination
Sct 18h55'1"-4d43'
Ian's Justice
Ori 4h43'42"15d12'
Ian's Starbryte
Aur 6h12'39"31d5'
Ian,Janis
Eri 4h14'5"-19d31'
Iannace Family Star, The
Cnc 8h27'14"7d27'
Iannacito,Vince
Per 3h14'60"54d36'
Iannacone,Gianluca
Vir 12h2'23"8d1'
Iannapollo,Robert
Dra 11h31'48"70d26'
Iannarone,Anthony J
Aur 6h0'50"32d3'
Iannarone,Ruth A
Cas 1h19'29"63d21'
Iannece,Alfredo
Lac 22h14'12"48d59'
Iannelli,Deanne
Dra 17h10'42"68d27'
Ianniello,Shelby Lee
Peg 22h13'53"5d17'
Iannitto,Ann Elizabeth
Mon 6h34'41"-1d9'
Iannitto,Matthew Albert
Aur 6h26'35"38d41'
Iannone,Louis
Aur 4h37'16"33d52'
Iannotti,Joan Bouthillier
Cyg 20h23'1"40d14'
Iannucci,Anita
Dra 20h39'27"73d49'
Iannucci,David Michael Jozo
Her 17h1'39"30d2'
Iannuzzi,Dominic Frank
Boo 13h45'23"24d20'
Iannuzzi,Ernest Christopher
Her 16h18'0"23d15'
Iansito,Donna Michelle
And 0h51'43"45d16'
Iantha
Del 20h35'46"10d24'
Ianuzzi,Leonora Lombardo
Lyr 18h31'35"32d21'
Iapoce,My Light,Kyle Patrick
Aql 18h45'0"11d28'
Iattarelli,Carol A
Cnc 7h56'32"11d20'
Iattarelli,Kristen L
Cnc 7h55'47"10d45'
Iavarone,Steven Joseph
Lac 22h18'40"49d51'
Iazzetta,Nicholas Alexander
Ori 5h31'1"-0d54'
Ibarra,John
Sco 17h26'58"-30d6'
Ibarra,Silvia
Mon 6h57'60"-8d2'
Ibberson,Heather Mary
Lyr 18h31'1"34d6'
Ibberson,Nicola Michelle
Uma 8h27'16"61d22'
Iberia
Ori 4h53'30"1d52'
IBG
Tau 5h50'56"26d38'
Ibing,Louisa
Cam 14h9'59"80d14'
IBM 1993 Golden Circle The Majestic West
Uma 10h15'34"71d51'
IBM 1993 Golden Circle Main Street,USA
Uma 10h8'14"70d55'
IBM 1993 Golden Circle America's Crossroads
Uma 10h6'55"70d2'
IBM 1993 Golden Circle Gateway To Freedom
Uma 10h15'20"70d47'
IBM 1993 Golden Circle Atlantic Seaboard
Uma 10h20'56"70d5'
IBM 1993 Golden Circle Sunshine South
Uma 10h17'57"71d6'
IBM 1993 Golden Circle Big Sky Country
Uma 10h20'59"71d7'
IBM/JMP Star,The
Cyg 21h51'21"42d56'
IBN Luqmaan
Psc 22h51'39"5d32'
Ibold,Charlotte
Lyr 18h34'45"40d36'
Ibone
Peg 22h16'39"4d23'
Ibrahim,Khalil
Lyn 8h24'47"51d14'
Ibrahim,Mary
Her 16h54'49"23d18'
IbOHer 16h54'49"23d18'
Cep 22h50'15"58d1'
IbáCep 22h50'15"58d1'
Hya 8h54'13"-5d42'
Icasson,Theresa
And 0h25'16"21d51'
Icayan,Pat
Peg 23h29'25"21d33'
Ich Liebe Dich
Lac 22h22'24"35d36'
Ichaso,Mari R
Tri 2h5'56"30d43'
Ichel,Steven Adam
Per 2h5'1"58d38'
Ichiban Star for Tony & Sharon
Cyg 21h20'43"38d48'
Ichibanboshi Star Blessing 1995
Cru 12h34'27"-56d30'
Ichihara-Robinson
Oph 18h2'53"7d47'
Ichikawa,Martha
Ori 4h52'46"0d41'
Ichikawa,Yusuke
Lyr 18h58'40"26d32'
Ichinose,Alan I
Aql 20h4'35"4d18'
Ick
Com 13h30'47"26d31'
Ick Star,The
Mon 7h17'26"-1d42'
Ickenroth,Gregor
Gem 6h45'60"18d44'
Ickes,Anne & Merle
Crb 15h55'20"31d26'
Icky
Vul 20h40'46"28d50'
Ida
Ari 1h58'55"25d17'
Ida For Ever
Pup 8h24'40"-21d27'
Ida Wray
Cyg 19h25'21"31d19'
Idalina
Hya 9h13'15"2d17'
Iddings,Frank
Cep 22h1'42"68d11'
Iddings,Paul
Aql 19h59'0"10d44'
Iddoya
Crb 16h4'55"30d56'
Ide,Richard
Eri 3h13'51"-6d54'
Idelicato,Randy & Kim Ratteree
Lyn 7h26'46"44d32'
Iden
Scl 23h17'1"-32d29'
Iden,Don & Madelyn
Ser 18h19'25"0d44'
Idgie
Peg 23h46'1"11d50'
Idgie
Cas 3h10'16"67d48'
Idha & Andy's Love Star
Ori 6h5'58"5d9'
Idir,Pascal
Mon 6h23'1"-1d52'
Idit
Peg 22h35'16"11d39'
Idle,Joanne
Sge 20h16'38"20d18'
Idoni,Mayor Timothy C
Her 16h59'23"35d15'
Idris Michael
Gem 6h55'11"18d43'
Idriss
Cep 20h28'11"75d17'
Idzal,Alexandra Rose
Peg 23h28'13"18d19'
Ienan's Star,Jac
Lyn 7h30'23"40d22'
IEP
Mon 6h27'25"7d54'
Ierley,Kemp
Cmi 7h57'1"1d17'
If Wishes Came True
Lyn 8h48'55"41d51'
If you see a chance take it
Ori 5h57'25"15d45'
ifert
Peg 21h55'23"21d46'
Ifigenia
Pic 4h55'1"-43d30'
Iggy
Uma 12h9'0"48d44'
Igi,George
Ser 18h22'21"-1d36'A
Igi,Sharon
Ser 18h22'21"-1d36'B
Iglehart,Sasha
Dra 18h5'37"70d3'
Iglesias,Jessica
Tau 4h31'45"15d38'
Iglesias,José
Cnv 12h34'0"38d1'
Iglesias,Rachel
And 23h45'47"45d52'
Iglhaut,Daniel Mark
Aur 5h30'24"31d39'
Iglikowski,Mary Florence
Cep 22h20'41"68d32'
Ignacio,Ana Tamun- Island Star
Aur 4h55'53"40d49'
Ignacio,Ayoroa
Uma 10h52'56"40d32'
Ignacio,Benham
Cep 21h57'27"61d25'B
Ignatius,Bomar
Aur 6h22'14"34d17'
Ignatowicz III,Vincent Francis
Boo 14h17'23"38d26'
Ignazio
Lyn 7h55'58"58d37'
Ignis
Lyn 7h51'23"36d16'
Igo,John
Sex 9h44'52"-5d44'
Igor
Crt 11h7'35"-17d34'
Igor & Christine
Cyg 21h28'41"41d12'
Igou,Colleen Bridget
Peg 21h54'57"28d19'
Igou,Jr,William
Her 16h52'50"50d54'
Iguaran,Roberto
Dra 11h20'59"77d37'
Ihanblapi Waste
Cnv 13h30'43"48d15'
Ihde,Mike
Aur 5h25'0"54d38'
Ihde,Timothy Jay
Dra 12h32'11"64d34'
Ihlenfeldt,Chad M
Her 16h21'49"10d10'
Ihnat,Kathrine Susan H
Vul 20h14'38"22d37'
Ihrie in the Sky
Cnv 13h53'16"38d21'
Ihrig,Christa
And 0h20'55"30d59'
Ihry,Robin
Cyg 19h16'51"47d5'
Iizuka,Dorothée
Boo 14h22'11"13d55'
Ike The Road Runner
Uma 11h8'0"58d19'
Ikebana
Pic 4h59'29"-46d28'
Ikeda,Marion
Cam 8h19'39"78d29'
Ikeda,Stella of Hisatoshi
Lib 14h44'20"-2d46'
Ikezawa,Wendy H
Aql 20h16'10"5d2'
Iki's Dream
And 22h57'46"51d36'
Ikins,Rachael Zacov
Com 12h23'50"26d41'
IKKITOOMI
Aur 5h11'1"40d39'
Ikonen,Eero Augustus
Pho 23h49'25"45d49'
Ikuko
Sgr 18h50'26"-30d15'
Ikuno,Ginger L
Ori 5h53'49"11d36'
Ikuno,Gloria T
Ori 5h54'28"14d26'
Il Etait Une Fois
Boo 14h47'21"50d47'
Il mio amore durerá per Sempre
Cyg 20h33'29"45d24'
Il orso e la colomba
Uma 8h37'57"68d32'
Il Pescatore AJD
Per 2h29'51"56d19'
IL Richard FATS
Her 17h2'45"20d15'
il Viaggio di Nozze
Cam 10h1'48"82d16'
Il VO Gero Viareggio
Gru 22h21'51"-54d3'
Ila
Aql 18h57'24"0d44'
Ila Zee
Cnv 12h31'51"33d44'
Ilan
Aql 20h31'36"0d46'
Ilan & Dolce
Cyg 19h43'27"30d9'
Ilana
Cyg 21h59'35"52d55'
Ilanamysugy
Cet 2h56'47"3d46'
Ilareguy,Suzon
Lac 22h52'53"52d55'
Ilaria
Col 6h36'36"-34d6'
Ilaria Micai
Lep 5h23'14"-25d23'
Ilaria,Valerio Nikita
Peg 23h30'49"11d14'
Ilaria,Valerio Nikita
Equ 21h11'26"11d56'
Ilaria/S
Aql 18h59'16"-6d23'
Ilas,Elgina
Mon 6h30'46"7d32'
Ilasat,Tamora
Boo 14h37'12"46d56'
Ilbak,Mehmet
Uma 10h48'1"55d29'
Ildhuso,Jr,Forever Gunnar
Uma 13h56'1"51d32'
Ileana
Mon 7h13'0"-10d30'
Ileana
Lyr 18h26'30"46d57'
Ilene/Tito
Sge 19h34'22"18d49'
Iles,Anthony
Her 18h13'37"38d41'
Iles,Jeff
Cyg 21h21'56"40d49'
Iles,Robert C
Sex 10h7'55"5d23'
Ileta,Marcella
Cas 23h14'15"60d12'
Ilex,Gabriele M
Cnc 9h17'45"30d56'
Ilg,William John
Dra 19h3'29"67d33'
Ilic,Dean(Monchki)
Hya 8h13'23"-0d2'
ILic,Lynette M
Cyg 21h47'27"37d20'
Iline,Stephanie
Lyn 9h9'35"37d4'
Iliopoulos,Gina
Lyn 8h40'50"44d29'
Illavsky,Jerry
Dra 14h46'30"62d34'
Illig,Rebecka Jane
Peg 23h21'60"10d5'
Illing,Peter A
Her 17h13'38"26d35'
Illinois Lottery Second Chance
Her 18h4'56"38d46'
Illouz,Michel
Oph 17h11'58"11d9'
Illuminate Darkness: Dan A A & Joan E C
Lyr 19h3'37"28d13'
Illuminating Companion
Uma 11h13'0"49d53'
Illuminating Love
Ori 5h2'57"-1d47'
Illusions
Her 16h25'38"28d47'
Ilo
Psc 1h22'44"21d56'
Ilona S
Cnc 8h26'1"30d7'
Ilse's Schaf
Gem 7h16'36"24d30'
Ilsfe
Lyr 18h17'0"32d40'
Ilsley,Rowan
Cyg 21h0'18"38d2'
ILUTOO
Her 16h9'0"10d24'
Ilves,Jan Peter
Lac 22h16'42"54d34'
ILY
Dra 18h56'38"58d9'A
Ily
Cnv 12h13'45"44d15'
Ily
Uma 8h45'59"47d40'
Ily Ashley K
Cyg 21h48'50"37d19'
Ily,Toni Marie
Per 4h24'43"50d3'
ILY2
Dra 18h56'38"58d9'B
Ilyana,Allyson
Mon 6h31'0"10d47'
Ilys
Cnv 13h30'37"40d9'
Ilyse & Glen
Mon 8h2'30"-7d31'
Ilyssa's Star
Lyr 18h29'25"30d31'
Ilze Ruta
Cam 11h48'21"80d34'
Imad
Uma 13h18'27"61d28'
Image Media Productions
Lyr 18h17'29"45d21'
Image Uniquity
Equ 21h19'48"7d58'
Imaginari
Cap 21h53'36"-21d56'
Imago(T D Ellis)
Uma 9h18'41"51d21'
Imagoa
Cep 22h6'0"68d47'
Imamura,Ian Drinkwine
Boo 15h3'0"27d32'
Imanari,Naomi
Peg 22h37'50"11d5'
Imani,Malak
Sex 10h44'54"5d40'
Imbeault,Stevens
Cyg 20h24'37"39d25'
Imber's Photo Star, Jerry
Aql 19h9'1"3d41'
Imbert,Emile
Ori 6h21'0"10d40'
Imbert,Jean Marie
Ori 6h21'23"10d5'
Imbert,Les
Oph 17h6'18"7d50'
Imberti,Paola
Aur 5h57'48"50d3'
Imbimbo,Jim
Sex 10h10'40"-7d43'
Imbimbo,Victor
Dra 20h23'57"62d40'
Imboden,Tris
Dra 17h43'0"65d29'
Imbriani,Ralph Edward
Aql 19h4'13"5d19'
Imdahl,Helmut
Lac 22h5'40"51d52'
Imfeld,Michael Dennis
Leo 9h22'0"17d36'
Imgor
Cnv 12h15'0"38d5'
Imhoff,Jake
Aur 5h36'11"50d23'
Imhoff,Michael
Cnc 8h6'20"32d48'
Imhoff,Michael
Cnc 7h53'16"10d32'
Imhoff,Special Principal,James
Boo 15h22'11"40d24'
Imi
Ori 5h21'58"15d16'
Imke,Lauren Elizabeth
And 23h5'1"44d50'
Imler,June L
Crt 11h9'0"-13d27'
Imma
Eri 3h54'15"-1d7'
Imma
Nor 16h23'26"-50d17'
Imma
Lep 5h55'56"-24d51'
Imma Soler Pons I Conrad Enllac
Cap 20h34'30"-8d57'
Immaculate Conception School
Uma 11h4'26"35d47'
Immanence
Ori 5h59'59"9d27'
Immer
Lyn 8h26'39"58d36'
Immeraufblickenzufinde nunserliebe
Ser 18h37'31"4d22'
Immerman,Isidor S
Cas 1h15'1"61d1'
Immonen,Heidi Tuulia
Cas 2h38'59"57d48'
Immordino,Sam
Her 16h17'31"42d24'
Immortal Beloved
Cam 3h53'13"60d50'
Immortal Dawn LAF JC
Cas 23h5'25"54d15'
Immortal Love-James & Lisa
Ori 6h3'13"20d24'
Immortal Martin
Cam 3h19'28"67d60'
Immortal Michael
Cep 22h57'35"58d2'
Immortale,Vivlamore
Uma 12h0'31"36d9'
Immu
Cam 5h49'17"61d44'
Imogen
Cas 23h2'29"58d50'
Imogen
Ant 10h34'0"36d19'
Imogen Iona
And 23h16'49"50d31'
Imondi,Beth
And 23h45'25"45d4'
Imp's Heavenly Garden
Her 16h56'52"26d20'
Impack Forge,Inc
Uma 11h45'24"53d23'
Imperatore,Frank
Per 4h26'0"51d17'
Imperial 3,Leopold John
Dra 11h42'0"68d46'
Impett,Royce
Boo 15h15'27"34d41'
Improta,Filomena
Lep 5h24'1"-25d47'
Imrie,Donald Hamilton
Cep 20h47'55"78d49'
Imsiecke,Hermann
Aur 5h16'52"43d58'
Imson,Katherine D
Peg 23h24'47"31d22'
Imzadi
Lyr 19h2'40"26d44'
Imzadi
Uma 12h57'38"60d52'
Imzadi-John & Yolande 102983
Crb 15h15'0"27d19'
In Celebration of Shelby
Del 20h18'52"10d41'
In Celebration of Sarah & Roger
Cyg 19h24'19"44d47'
In Hoc Signo Aeterna Spes Mea
Uma 8h5'38"67d58'
In Honor Of: Victim Witness
Aur 4h58'1"31d9'
In Lancelot's Fashion
Lmi 10h24'52"28d22'
In Liebe Deine Mama "Claudia"
Boo 13h58'56"10d42'
In Liebe für meinen Schatz Ralf
Com 13h5'11"28d56'
In Love of Theresa
Ori 5h49'45"10d1'
In Memory
Sgr 19h49'15"-43d54'
In Memory Of Bob & Susan's Love
Peg 22h3'47"10d30'
In Memory Of Dad
Her 17h21'56"22d39'
In Memory Of Grandma Marian
Cyg 20h12'38"37d38'
In Memory Of Grandpa Herman
Uma 11h52'0"40d9'
In Memory Of Mort Stern
Cep 23h25'31"64d53'
In Memory Of Our Special Grandpa
Ori 4h43'52"4d53'
In Memory of Our Dear Gaby
Lac 22h3'34"46d13'
In Memory Of Princess & Silver Cloud
Lyn 7h31'55"45d4'
In Memory of Ruth Lillian
Cyg 21h29'22"40d18'
In Sook
Lyr 18h38'30"41d46'
In your eyes
Cyg 20h55'35"38d28'
Ina
Cam 8h3'54"72d23'
Inaba,Star of Harumi
Leo 10h2'36"7d22'
Inacio,Luis
Per 3h57'48"38d1'
Inagaki,Goro
Sgr 19h19'42"-29d10'
Inamorata
Peg 21h6'24"12d8'
Inamorata
Peg 22h9'10"25d31'
Inamorato
Col 5h50'9"-40d42'
Inamorato Jerome
Per 2h53'11"40d22'
Inara
And 23h19'10"43d49'A
Incaini,Sarah Marie Crocco
Lyn 7h37'1"51d54'

Incandescent Rozzie 5544
Lyn 9h22'34"40d57'
Inch,Gail Nystrom
Lyn 6h35'52"60d28'
Inch,Kelsey Malia
Lyn 6h35'21"60d3'
Inchaffray
Uma 10h52'45"38d48'
Inchul Lee
Uma 9h7'0"57d24'
Incitti,Jay
Cmi 7h35'22"11d32'
Incollingo,Dana Nicole
And 23h42'22"46d57'
Incomparable Duo David &
Irene,The
Lmi 10h6'27"34d4'
Incredible Games
Ori 5h26'36"-4d12'
Indano,Marilyn
Cas 1h23'34"67d48'
Independent Pipe and Supply
Cnv 13h23'37"37d36'
Inderbitzin,Ron & Kim
Crb 16h10'50"26d11'
Inderhees,Sr Carol Louise
Cep 21h33'1"60d55'
India Rose
And 0h56'0"37d30'
Indian & The Princess, The
And 0h57'43"34d0'
Indian Paul
Hya 8h12'16"-5d45'
Indian Summer
Aur 5h53'43"38d13'
Indiana
Vul 20h21'42"28d57'
Indidarius
Lyn 9h12'1"42d40'
Indigo-blue Ocean -AIMI-
Aqr 22h1'39"-14d42'
Indio
Ari 2h58'0"30d7'
Indochina
Cnv 12h35'30"38d16'
Indu
Cet 3h5'27"6d17'
Ines' wunderbarer Zauberstern
Cap 20h22'50"-26d49'
Inez
Boo 15h4'29"31d1'
Inez's Star
Hya 8h59'31"5d12'
Inez,Carmela
And 0h53'43"41d4'
Inez,Kimberly
And 23h0'1"51d32'
Infante Star,The
Lac 22h23'56"54d5'
Infante"Queenie", Nicole A
Aqr 21h7'1"0d34'
Infinate + 2-Julian & Joanna's
Sge 20h6'19"16d33'
Infinite Jaap
Ori 5h54'36"21d14'
Infinitely Eric
Sct 18h54'7"-7d38'
Infinitesimal
Lyr 19h16'46"40d12'
Infinito
Dra 19h53'11"60d58'
Infinity
Crb 15h47'22"38d49'
Infinity
Lmi 9h54'27"34d30'
Infinity
Lyn 9h8'25"33d52'
Infinity
Aur 5h6'1"40d6'
Infinity
Ori 6h6'53"0d59'
Infinity
Tri 2h2'35"31d8'
Infinity
Ori 5h55'41"20d54'
Infinity
Lac 22h14'12"38d7'
Infinity
Uma 8h44'21"48d9'
Infinity
Lyn 7h38'60"43d1'
Infinity
Sgr 20h3'27"-26d52'
Infinity
Gem 6h44'40"31d28'
Infinity
Uma 9h15'35"50d40'
Infinity
Vul 20h20'13"25d50'
Infinity Peter & Marcia
Cyg 21h3'23"38d39'
Infinity Infinity
Uma 8h46'1"47d56'
Infinity Plus Infinity
Ori 5h28'40"-6d26'
Infinity Plus Infinity
Tri 2h35'15"31d42'
Infinity Plus One
Vul 20h46'42"28d57'
Infinity Ron
Per 3h11'26"45d30'
Infinity Squared
Peg 21h58'32"26d47'
Infinity I
Eri 2h44'18"-1d55'
Infinity-RM
Boo 15h47'1"41d16'
Infinity/Niles & Jennifer
And 23h38'35"46d24'
Infozino,Joseph
Dra 14h27'18"60d22'
Infusium 23
Peg 23h29'19"23d47'
Ing,KateLynn Marie
And 0h25'10"43d15'
Inga
Col 6h29'29"-34d36'
Inga To Matthias To The End
Of Time
Psc 1h20'52"21d44'
Inga's Glow
Cmi 8h8'1"4d60'
Ingalls,Eve
Mon 7h58'33"-0d37'
Inge
Aqr 21h21'50"-13d33'
Inge E
Cet 2h57'11"6d17'
Ingebor Ida
Lyn 7h28'0"40d11'
Ingeborg
Lib 15h14'56"-28d26'
Ingeborg-Christiane- Sophie
Boo 14h4'20"13d15'
Ingela Birgitta
Cas 1h58'40"61d46'
Ingels
And 0h20'16"32d7'
Ingenheim-Molitor, Marianne
Gräfen v
Sco 17h3'36"-31d32'
Ingerman,Joanne & Jeffrey
Cyg 20h36'28"42d31'
Ingersoll,Alan-Cindy
Lyn 9h18'43"39d50'
Ingersoll,Margot Ann
Tau 4h5'20"20d9'
Ingersoll,Paul M
Cep 20h22'0"78d53'
Ingham,Andrew & Nilgün
Cyg 21h0'44"36d31'
Ingham,Gail Robert
Mon 6h29'25"11d20'
Ingie J
And 1h47'27"37d15'
Ingino,Sean Michael
Boo 14h25'1"39d53'
Ingino,Vincent John
Boo 14h25'41"38d44'
Ingison,Peggy Sullivan
Cas 0h21'50"73d59'
Ingle,Grady & Ditter
Sge 19h7'19"16d16'
Ingle,Jerry
Lac 22h36'55"50d36'
Ingle,Larry
Dra 15h56'1"62d58'
Ingle,Ruth
Sge 19h9'47"19d48'
Ingle-Ostendorf,Paige
Sct 18h54'23"-9d26'
Ingles,Eric
Oph 18h38'58"6d13'
Ingles,Joseph
Peg 23h45'21"26d27'
Ingles,Tasia
Sgr 19h39'23"-32d21'
Inglesby Family,The
Vul 19h44'1"20d23'
Inglese Xmas 1995, Toni-Nikki
Lyr 18h41'55"30d37'
Inglis,Elizabeth Jean Malcolm
Cnv 13h37'41"28d15'
Ingmire,Charles
Lyr 19h20'34'42d55'
Ingo
Lyn 8h14'35"48d16'
Ingolia,Jim
Aql 18h58'1"-1f14'
Ingraham,Claude Raymond
Del 20h22'21"3d27'
Ingraham,Daniel B & Gayle L
Cyg 20h58'28"30d58'
Ingraham,Herbert John
Cep 21h0'16"68d6'
Ingram Micro b v
Cnv 13h38'32"47d35'
Ingram,Andrea
Lyn 7h50'40"52d1'
Ingram,Brian
Cet 2h27'0"2d29'
Ingram,Caitlyn Mariah
And 0h57'25"45d55'
Ingram,Catriona M
Cep 23h8'22"68d51'
Ingram,Connie
Peg 21h41'32"28d7'
Ingram,David
Boo 14h59'39"20d10'
Ingram,Debra Kaye
Aql 20h6'38"0d20'
Ingram,Dori
Mon 6h25'22"-5d45'
Ingram,Evan Alexandra
Leo 09h21'23"20d22'
Ingram,Fredrick(Rick)
Hya 9h3'1"2d22'
Ingram,Heather
Mon 6h56'30"7d31'
Ingram,Jacqueline Ann
Eri 3h11'1"-2d9'
Ingram,Jane E
And 23h1'1"42d57'
Ingram,Lew
Cep 2h41'56"77d57'
Ingram,Lowell & Ruth
Uma 9h11'17"50d30'
Ingram,Mary Alexandria
Com 12h9'30"20d1'
Ingram,Matthew Dennis
Ori 5h51'45"5d34'
Ingram,Nicolas
Ori 4h50'49"0d37'
Ingram,Phillip James
Cam 12h16'0"78d4'
Ingram,Sam
Eri 3h10'19"-5d46'
Ingram,Stabler Maris
Mon 6h51'49"11d2'
Ingram,Thomas Garry
Lac 22h5'26"49d21'
Ingram,Thomas P
Eri 3h26'18"-2d19'
Ingram,Timothy Ray
Sgr 19h38'1"-32d59'
Ingram,Tyler Kieffer
Aql 19h42'30"14d5'
Ingrassia,Sal
Cet 1h56'22"-2d24'
Ingrid
Peg 22h0'19"2d4'
Ingrid
Cas 0h19'0"60d2'
Ingrid
Cnc 8h28'24"10d55'
Ingrid
Cam 5h2'1"58d13'
Ingrid
Aql 20h10'29"12d45'
Ingrid
Cyg 20h23'29"38d20'
Ingrid
Com 12h43'39"20d36'
Ingrid
Lyr 19h22'0"30d39'
Ingrid
Aur 6h4'53"46d41'
Ingrid "Ingi"
Sgr 19h18'53"-24d28'
Ingrid "Star of Poellau"
Leo 10h53'31"0d58'
Ingrid B
Lyn 8h18'30"51d50'
Ingrid Ilona
Boo 13h59'31"13d15'
Ingrid K
And 1h38'30"40d11'
Ingrid Mäuschen Mania
Cnc 8h1'27"8d27'
Ingrid,My Queen
And 23h28'46"45d38'
Inhelder,Gina Marie
Cet 1h9'39"-1d52'
Inhelder,Heidi Suzanne
Aql 19h30'13"7d46'
Inhelder,Kenneth
Hya 8h19'34"-1d18'
Inhoffen,Karin Frigger
Lyr 18h47'58"35d23'
Ini
Uma 11h54'21"60d41'
Inka
Boo 13h59'43"10d2'
Inke + Norbert-Maus + Maus
Sgr 18h53'27"-29d27'
Inkiest Pooh-Bear Of Lamb
Chop Dogs
Cnv 13h57'1"38d47'
Inma H M
Lyn 8h5'23"43d18'
Inma S M
Lyr 18h16'49"38d28'
Inman,Barbara Canterbury
Mon 6h18'54"7d54'
Inman,Elizabeth Zehntbauer
Cam 3h58'31"60d18'
Inman,III,S C
Uma 14h2'25"59d11'
Inman,Robert C & Robert S
Peg 23h18'0"25d15'
Inna
Cas 2h52'60"60d60'
Inna
Mon 6h56'58"11d20'
Innace,Arthur G
Her 18h17'44"14d56'
Innaconi,Geri
Sex 10h38'0"0d12'
Innamorato,Anthony
Cep 22h22'52"70d57'
Innamorato,Jennifer
Uma 13h57'12"48d1'
Innamorato,Louie
Per 3h58'24"50d16'
Innes,Lauren
Cas 0h54'32"66d29'
Inness,Wendy
Oph 17h53'40"12d29'
Inniss "Mamy",Yvonne
Richards
Mon 8h0'33"-6d33'
Innocent
Aql 20h1'39"-1d33'
Innocente,Marina
Lyr 18h58'23"46d46'
Innocenti,John Charles
Dra 20h32'45"70d34'
Inns,Clare
And 0h59'52"39d42'
Inocenia
Cyg 21h2'36"38d22'
INOFINA Consulting &
Management GmbH
Ser 16h7'11"9d12'
Inoue Family,Stella of
Cap 20h54'35"-16d38'
Inoue,Noriyuki & Yumi
Psc 0h7'30"0d53'
Inprint Systems U S A
Psc 1h38'32"13d42'
Insalaco,Sherrie & Nicholas
Mon 6h30'14"-0d2'
Insecon
Sct 18h52'22"-7d7'
Inserauto,Massimo Gabriele
Sgr 19h30'13"-30d11'
Inserra (Pino), Giuseppe
Lac 22h12'55"51d29'
Insinga,Lee Ryan
Leo 10h42'57"15d11'
Inskeep,Alyssa Nicole
And 23h20'37"40d25'
Insley,Amber
Aql 19h56'33"10d1'
Insley,Patricia A & Charles W
Lockwood
Eri 3h3'0"-14d18'
Insley,Walter A
Aur 6h19'24"33d17'
Insogna,Daniel
Cep 21h49'39"85d50'
Insoll,Rowena Mair
Cyg 19h39'26"30d5'
Inspiration
Eri 2h55'0"-3d5'
Inspiration
Lyr 18h30'40"35d15'
Inspiration
Lyn 9h2'21"34d9'
Inspiration
Uma 9h40'40"52d11'
Inspire
Lac 22h26'48"53d11'
Institut f Pflegewis- senschaft
SS96
Boo 14h39'50"14d9'
Int'l Inst of Practical
Psychology
Tri 2h5'44"32d16'
Integral-10
Cyg 21h14'41"36d28'
Intel
Boo 14h17'26"48d56'
Intention Of Kato
Sgr 18h54'21"-33d7'
Inter American Data, Inc
Aur 5h30'0"31d32'
Interface
Umi 13h24'1"75d19'
Intergalactica
Aur 5h51'46"38d18'
Interlink EFS
Cra 18h7'0"-39d56'
Interlochen
Uma 8h46'44"70d50'
Intermediate Multiage Class
1996
Lyr 18h35'47"38d7'
Interrante,Sr,Charles Cosmo
Per 3h56'0"31d25'
Interrante,Vincent
Per 1h48'22"50d22'
Interstellaroonski
Peg 22h0'54"20d54'
Intiso,Andrea
Cet 1h53'11"-10d23'
Intracorp's EA/El Dallas
Cmi 7h6'49"4d14'
Intravaia's Lucky Star,Tom
Her 16h14'20"48d22'
Intrepid Star of Beth U S A
Crb 15h44'24"28d49'
Intrinsic Elegance
Cas 0h4'52"50d19'
Introcaso,Kate
Lmi 10h22'42"28d18'
Intromasso,Joseph A
Aur 6h41'15"38d15'
Introzzi,Fabio
Oph 17h53'36"11d47'
Intveld,Julie J
Lyn 6h40'25"50d22'
Invicta,Stuart Silver
Ori 5h57'59"7d26'
Invictus
Cam 7h7'20"61d20'
Invincible
Cam 3h15'14"60d15'
Inwood,Luke
Ori 5h57'1"9d11'
Inzani,Victor
Aql 20h30'45"-1d11'
Inzelbuch,Marsha
Boo 14h56'58"31d40'
Inzerillo,Robert A
Per 1h59'47"50d29'
Io Amore Voi
Cet 2h56'1"-0d38'
Ioanna
Dra 19h51'11"61d34'
Ioannou,Alexandra A Summer
Cnv 12h44'56"38d7'
Iodice(TDC),Scott Crispin
Boo 15h15'0"38d13'
Iodice,Victoria Marie
And 23h1'54"50d3'
Iole
Aql 19h1'1"13d10'
Ion
Boo 14h28'22"50d42'
Iona
Cyg 19h48'42"38d8'
Ione Rose
Cnv 13h56'0"38d22'
Ionica
Umi 15h9'19"67d5'
Ions,Catherine
Cas 0h5'58"64d6'
Ionta
Umi 16h22'24"77d33'
Iorio,Gina
Ori 5h32'34"-1d22'
Iosto
Psc 22h54'21"6d31'
Ioviero,Andrea Robert
Peg 22h2'28"2d28'
Iovino,Christina "Bina"
Lyr 18h34'15"33d11'
Iovino,Christina Lee
Lyn 8h58'43"40d50'
Iovino,Joe
Leo 10h52'32"-0d58'
Iovino,Leslie A
Her 17h19'22"48d51'
Iowa's Night
Aql 20h19'35"7d42'
Iozia,Jerry A
Ori 5h19'0"1d4'
Iozze,Gina
Lyr 18h59'29"26d3'
Ip,Angelina
Per 2h58'1"48d57'
Iparraguirre,Alvaro & Ximena
Cyg 19h56'1"38d51'
Ipema,Ed & Vicky
Cyg 20h55'41"38d27'
Iphigenia
Cep 1h44'14"80d38'B
Ippen,Stephanie Trude
Her 16h27'37"30d46'
Ippolita
For 2h38'36"-25d15'
Ippoliti,Roseanne
Ori 5h27'20"1d27'
Ippolito,Jessica Lauren
And 0h18'47"31d59'
Ippolito,Marion M
Mon 7h59'40"-0d13'
Ipser Star,Sue B B Hampden
Eri 3h14'40"-18d46'
Ira's Night Light
Her 18h2'46"14d49'
Ira's Star
Cmi 7h43'52"3d60'
Irace,Luke Forester
Cap 21h0'13"-23d51'
Iraci Your Love Shines
On,Dave:David K
Gem 7h17'40"28d42'
Iraine,Emma
Boo 14h9'29"52d59'
Iram 33
Cep 22h13'48"61d21'
Irani,Joan D
Mon 7h38'48"-4d0'
Irbicella,Giulia
Gru 22h31'24"-55d30'
Irby & Family,Mr Jeff
Peg 22h52'36"27d34'
Irby,Dave
Aql 18h55'12"-0d16'
Irby,III Alton Fernando
Lac 22h22'35"54d19'
Irby,Mary Katherine
Lyr 19h21'52"30d30'
Ireland,Carol
Cas 0h56'35"62d33'
Ireland,Gladys & Angus
Aur 5h1'1"47d7'
Ireland,Gwendoline
Cyg 19h28'20"38d54'
Ireland,Harold
Cyg 19h29'0"38d4'
Ireland,Jeanne Brunetti
Psc 1h0'1"20d30'
Ireland,Keeley Shannon
Eri 4h44'48"-6d1'
Ireland,Kim
Eri 3h41'14"-10d53'
Ireland,Kim-Kev
Crb 16h10'51"37d49'
Ireland,M Sue
And 0h11'57"37d36'
Ireland,Minerva & Jay
Mon 6h47'24"-0d15'
Ireland,Philip John
Uma 9h0'28"54d6'
Ireland,Stargazer Waz
Sco 16h23'12"-35d26'
Irene
Tri 2h4'41"31d11'
Irene
And 23h2'49"51d46'
Irene
Sgr 18h6'36"-28d55'
Irene
Cam 8h58'15"74d8'
Irene
Cas 1h16'37"70d50'
Irene
Pho 0h43'11"-43d24'
Irene
Tau 5h55'38"23d8'
Irene
Cap 20h21'23"-26d10'
Irene
Cas 2h24'16"67d60'
Irene
Cas 0h15'59"62d39'
Irene & Baby
Umi 13h11'59"71d15'
Irene & Joe
Peg 22h6'21"2d21'
Irene & Josef
Uma 10h3'32"58d44'
Irene F
Pic 4h56'6"-43d6'
Irene Isabel
Eri 2h55'26"-6d50'
Irene Katharina Sophia
Vir 13h33'29"-6d46'
Irene Lou
Aqr 22h33'25"-0d19'
Irene The Perfect Mom
Mon 6h55'15"-2d40'
Irene's Baby Star- Mary Ann
Lib 15h55'44"-6d5'
Irene's Beam
Mon 7h25'21"-8d46'
Irene's Inspiration
Aql 19h12'30"15d40'
Irene's Star
Aur 4h49'40"40d37'
Irene/Joanne/Rosanna
Cma 6h28'44"-20d10'
Irenes's Star
And 23h38'29"37d40'
Irenius
Psc 23h39'31"-2d51'
Irenpetraganz
Cyg 21h5'40"33d36'
Irfan & Saiqa
Cyg 20h36'0"45d32'
IRI
Cam 9h21'57"81d53'
Irick,Ashlyn Juanita
Lyn 9h2'12"37d36'
Irick,Rebecca
And 1h58'47"38d9'
Irimajiri,Kana
Cet 1h45'31"-0d22'
Irina
Peg 23h32'18"20d39'
Irina
Dra 10h43'13"74d52'
Irina
And 2h35'15"40d29'
Irina
And 2h17'19"47d13'
Irina & Thilo Just Married
Lib 15h5'39"-8d22'
Irini
Cyg 20h54'39"31d36'
Irinia
Mon 7h11'50"-6d40'
Iris
Pup 7h28'49"22d9'
Iris
Ser 16h9'44"2d34'
Iris
Peg 23h23'18"13d46'
Iris
Aql 20h0'27"9d49'
Iris & Wheaties, Soulful
Spirits
Uma 11h12'27"55d30'
Iris Edythe
Cas 23h0'0"58d16'
Iris Joy
Leo 10h42'14"15d50'
Iris Lovebundle
Lyr 19h17'0"26d40'
Iris Marie
Ori 5h57'49"20d18'
Iris Star
Lyn 7h59'45"39d39'
Iris' Love Star
Sge 19h53'36"16d22'
Irish Charlene
Mon 7h41'27"-2d35'
Irish Eyes
Cyg 21h30'40"42d46'
Irish Tail Twister,The
Cam 5h5'19"60d0'
Irish,Carody Elise
Vul 19h41'1"23d59'
Irish,Joan H
Lyr 19h24'38"37d31'
Irish,Rosie
And 23h1'17"50d48'
Irizarry,Dr Luis
Aur 5h28'11"41d13'
Irizary-Principus Stupendous
Aql 19h27'59"-6d57'
Irka
Boo 14h12'41"51d39'
Irlbeck,Keith & Joleen
Umi 15h7'58"68d50'
Irle,Beth Ann
Cam 4h1'0"58d32'
Irle,Chip
Dra 20h6'27"67d55'
Irle,Chuck
Cnv 12h26'43"33d36'
Irle,Linda
Uma 10h26'26"62d13'
Irle,Mark
Aur 5h5'24"40d27'
Irle,Sara
Cas 1h41'50"58d14'
Irma
Lep 15h18'44"45d54'
Irma Julia Jean
And 23h37'11"46d10'
Irmalex
Aqr 21h7'1"0d27'
Irmer,Andrea
Cap 20h8'13"-11d0'
Irmgard "Laba-Laba"
Lib 15h40'33"-20d29'
Irmiere,Trystan Gabriel
Vul 20h16'29"26d2'
Irminger,Verda Maxine
Eri 3h34'18"-6d16'
Irmler,Günter
Dra 15h57'13"62d5'
Iron Horse,The
Vir 11h44'29"0d30'
Irons,Leon
Her 17h32'11"27d8'
Irons,Paul Simon
Dra 16h13'30"63d3'
Irons,Zachary Cole
Her 17h35'27"41d11'
Irpino,Michael
Ori 5h53'30"14d12'
Irre,Lars
Aql 19h13'43"14d38'
Irrgang,Erik
Oph 17h13'57"-22d40'
Irvando
Hor 3h7'45"-49d12'
Irvin,Emily Sheppard
Aur 5h50'29"31d48'
Irvin,Gladys & Ernie
Crb 16h16'50"37d57'
Irvin,Merrill P
Hya 8h54'0"5d1'
Irvin,Robert
Her 16h58'28"29d29'
Irvine
Uma 8h58'34"56d23'
Irvine,Andrew A
Cam 6h15'42"62d24'
Irvine,Anthony Roberts
Dra 17h40'53"68d23'
Irvine,Betsy
Cas 23h4'40"53d3'
Irvine,Carly-Ann
Dra 19h20'22"80d20'
Irvine,George Young
Cet 2h12'51"4d28'
Irvine,Norbert White
Boo 14h42'59"31d35'
Irvine,Whitney Cantrell
Mon 6h6'1"-4d44'
Irvine-Stienessen
Dra 20h4'44"62d40'
Irving's Star
Cep 22h40'14"70d18'

Irving,Andrew & Machaela
Cyg 19h29'47"31d8'
Irving,Angela Kaye
Lyr 18h59'34"46d20'
Irving,Fraser Deckland
Dra 16h24'42"60d8'
Irving,Jane Ann
Cyg 19h25'43"33d13'
Irving,Jim
Uma 11h16'37"52d60'
Irving,MD,Albert Stoddard
Oph 17h16'26"11d18'
Irving,Michael Johnathan
Per 3h2'11"41d11'
Irwin Wedding Star, Carl & Sue
Cyg 21h51'49"40d22'
Irwin's "Fine"
Cnv 12h10'13"34d6'
Irwin,A G
Lac 22h4'46"47d4'
Irwin,Birdie Marie Cassidy
Lac 22h24'32"37d11'B
Irwin,Brent Eric
Cep 1h9'19"78d19'
Irwin,Brian Michael Jude
Her 17h3'43"38d47'
Irwin,Briana Jean
Aur 6h27'16"38d47'
Irwin,Charles Finney
Her 17h23'1"38d35'
Irwin,Clifford Lukens "Luke"
Lac 22h24'32"37d11'A
Irwin,Craig T
Her 16h6'36"50d29'
Irwin,Dylan Haynes
Ser 15h13'31"8d49'
Irwin,George S
Oph 16h57'15"-25d45'
Irwin,Jeanne
Peg 23h26'19"10d52'
Irwin,John M
Her 18h13'15"41d7'
Irwin,Jr,James R
Ser 16h1'36"-0d32'
Irwin,Linda J
Umi 15h6'23"80d23'
Irwin,Lora Darlene Amelia
Ori 6h2'56"5d46'
Irwin,Marjorie M
Com 12h1'34"26d36'
Irwin,Mary Place
Cyg 20h8'26"40d60'
Irwin,Patrick Vincent
Cam 13h1'59"78d47'
Irwin,William E
Her 18h1'20"38d21'
Is This Heaven? No, it's Gooch
Aur 6h35'42"31d51'
Isaac Forever UDI,Aby
Cas 23h27'12"58d31'
Isaac's Inspiration
Her 16h51'44"48d54'
Isaac,Fred C
Sgr 18h2'20"-28d9'
Isaac,Mose Azar
Boo 15h3'28"22d5'
Isaac,Sydney Anne
Cas 0h59'0"64d17'
Isaacs,"Circle I Star" Steve
Per 2h25'34"56d34'
Isaacs,Alessandro Marcello
Dra 17h46'57"64d19'
Isaacson,Laurence
Lac 22h2'0"50d5'
Isaacson,Richard
Boo 14h49'30"36d56'
Isaak,Franz
Hya 9h28'4"3d3'
Isabel
Del 20h15'41"14d6'
Isabel
Peg 23h47'32"11d16'
Isabel
Lyr 19h3'1"28d32'
Isabel
Cyg 19h25'24"44d7'
Isabel
Lyr 19h16'16"26d53'
Isabel
Lyn 7h54'39"34d17'
Isabel
Cep 22h8'48"68d37'
Isabel
Cyg 19h28'16"36d55'
Isabel 831 TL
Lyr 18h37'37"30d28'
Isabel Rose
Peg 22h59'1'11d3'
Isabel und Ralph
Ari 2h32'14"20d49'
Isabell,George
Her 17h53'29"14d44'
Isabella
Cyg 20h59'42"30d2'
Isabella
Peg 23h42'50"12d8'
Isabella
And 22h56'0"40d58'
Isabella
For 2h10'31"-28d15'
Isabella & Family Vincent & Jeannie
Ori 4h59'23"4d53'
Isabella One LSG
Cyg 19h32'31"33d50'
Isabella-A Star Is Born
Peg 22h55'1"21d31'
Isabelle
Cnv 13h22'28"50d50'
Isabelle
Crb 16h15'36"34d60'
Isabelle
Peg 21h51'54"19d21'B
Isabelle
Cnv 12h27'24"37d36'
Isabelle
Cam 7h50'17"68d59'
Isabelle Ann
Tau 5h43'39"28d15'
Isabelle Blanche Jeannin
Aur 5h1'46"38d46'
Isabelle C-Benoit B
Per 3h18'33"42d14'
Isabelle et Hossein
Oph 18h36'48"7d12'
Isabelle Of Paris
Ari 2h43'60"21d46'
Isabelle Salle
Umi 16h30'1"76d1'
Isabelle Éternelle
Cam 3h29'21"61d43'
Isadomad 7
Dra 15h43'37"58d1'
Isadora
And 0h19'24"31d4'
Isadore Magnifico
Cam 9h12'14"84d28'
Isakson,Ronda
Lyn 7h58'1"35d36'
Isaline et Yann
Peg 22h31'37"31d32'
Isbell,Dawn Lee
And 23h0'0"45d18'
Isbell,Dr William M
Ori 6h12'33"7d33'
Isbell,Eve Jolie
Cas 0h57'28"60d4'
Isbell,Jordan Michel
Her 16h20'13"42d12'
Isbell,Stephanie Ann
Tri 1h57'57"27d49'
ISC/EDS Sportacular Cybernetics
Aur 5h34'42"38d38'
Isel,Paul M
Oph 17h27'24"-5d44'
Isele,Olaf Eric Alexander
Vul 19h6'32"24d5'B
Iseman,Kenny
Gem 6h51'43"18d44'
Isenberg,Hans-Georg
Psc 23h1'54"5d38'
Isenberg,Stanton J.
Ser 15h13'19"11d23'
Isene,Geir
Cam 3h25'24"63d43'
Isenhour,J Bradley
Per 3h37'12"38d26'
Isern,Jordan Nadine
Mon 8h6'48"-1d52'
Isern,Jordan Nadine
Lib 14h44'24"-20d3'
Isetta
Nor 16h23'34"-57d38'
Ishak,Monir
Uma 12h4'52"57d47'
Isham,In Memory of Bob
Lyr 18h42'21"40d54'
Isherwood Domus Stella
Her 17h13'47"28d28'
Ishida,Tatsuhiro
Cnv 12h58'35"32d48'
Ishikawa,Neil
Cet 2h38'1"5d12'
Ishka
Eri 3h38'45"-17d0'
Ishoy,Kimberly Clay
Hya 8h53'56"5d53'
Isiah
Uma 12h11'46"63d24'
Iside
Lep 4h59'1"-19d33'
Isidenanetta
Vel 9h18'44"-43d42'
Isik,Umit
Ori 6h17'38"7d43'
Isis
Uma 9h49'36"48d0'
Iskalis,Thomas J
Dra 16h50'36"52d6'
Iske,Ronald Charles
Lac 22h10'59"49d51'
Islam,Aheela
Leo 11h0'24"17d52'
Islano,Drummond
Eri 4h8'1"-18d16'
Islar,Richard"Ice"
Her 18h0'32"28d28'
Isle of McCaskill
Cet 3h7'39"2d21'
Isleoftamburino
Del 20h17'20"14d20'
Isley,Kyle W
Oph 17h20'28"10d18'
Islu,Ines Herrnleben
Leo 9h36'35"7d54'
Ismail,Yasmin
Aql 20h11'0"14d29'
Ismay,David Robert
Ori 6h3'1'10d30'
Ismerie
Gem 7h17'40"28d51'
Ismerio,Alejandro
Ser 15h32'47"-1d31'
Ismet
Lyr 18h41'45"30d30'
Isoard,Agnès
Aur 5h8'45"37d41'
Isobel
Per 3h6'51"56d49'
Isobel,Ena
Uma 12h41'34"62d53'
Isolde
Sex 10h17'4"-8d51'
Isom,John Patrick
Cnv 12h43'57"42d23'
Isom,Karen Lee
Mon 6h42'34"10d5'
Ison,Linda K
Mon 8h1'43"-2d46'
Isops
Eri 3h42'30"-5d34'
Israel
Aql 20h4'0"6d6'
Israel,Charles
Her 17h37'48"40d30'
Israel,Craig Myles
Her 18h31'1"20d3'
Israel,David Nissan
Boo 15h2'0"50d22'
Israel,James Donovan
Ari 2h31'20"21d59'
Israel,Jesse
Hya 9h1'39"0d16'
Israel,Mark & Lori
Sge 20h1'24"19d19'
Israela,Eva
Cas 0h4'1"64d43'
Israelson For Eternity,Ken & Bev
Uma 10h53'40"39d21'
Isreal,Phyllis A
Umi 15h32'46"71d20'
Iss-Hogai
Her 16h47'23"11d2'
Issa,Lisa
Cas 0h32'38"50d36'
Issacs,Martha
Cas 0h27'0"72d3'
Issei "Starman"
Sgr 19h5'54"-14d32'
Issenjou,Corinne
Boo 13h54'12"16d47'
Issho Ni Rei
Aql 19h28'22"0d34'
Issorat,Rudy
Ser 15h53'17"2d58'
ISTV?
Cam 10h51'15"82d26'
István Sänger 50
Ari 2h27'54"26d36'
Iszczalowicz,Daniele
Aur 5h1'32"37d54'
It Had To Be You
Aql 18h57'0"7d25'
It had to be you
Boo 15h0'38"53d50'
It's About Time
Eri 4h8'1"-14d56'
It's Magic
Lmi 10h35'57"32d5'
IT's"SEARCH" Dream Team Love,Suz
Eri 5h4'13"-8d7'
Itema
Hor 3h8'56"-49d20'
Ithaque
Cam 5h56'44"67d52'
Itkin,Anne Grace
Lyn 7h36'25"44d37'
Itkin,Lynne S
And 0h10'14"32d12'
Itkin,Olga
And 2h2'26"41d38'
Itkonen,Unto Johannes
Uma 11h10'29"32d38'
ITT Hartford Regional Office
Aur 6h2'1"30d6'
Ittap 50
Eri 4h48'8"-5d5'
Itten,Markus
Cep 21h1'36"58d52'
itti
Gem 7h6'48"21d39'
Ittrich,Gerd E A
Dra 14h29'18"62d54'
Itzenthaler,Edeltraud "Wangerl"
Cap 20h34'35"-18d56'
Itzkowitz,Carol Brooks
And 23h38'45"47d8'
Iu-Taylor,Christine
Peg 22h9'1"5d12'
IU2
Aql 19h6'20"15d38'
Iurato,Joseph
Umi 16h46'27"78d55'
Iusi,Roberto Richard Anthony
Her 16h28'23"50d39'
Iuzzolino Milken Fingers,Michael
Dra 16h56'0"65d44'
IV Ann
And 0h59'25"38d34'
IV Mi Donna
Lyn 9h9'25"40d27'
Ivalle
Crt 11h16'7"-19d44'
Ivan
Cas 0h47'15"77d13'
Ivan
Scl 0h2'13"-27d4'
Ivan & Judi
Uma 10h27'48"41d41'
Ivan & Tiffany
Ori 5h52'45"1d13'
Ivan e Paola
Hor 3h7'38"-49d10'
Ivan,Sarahlyn Twinkie
Lyn 7h42'14"58d13'
Ivana
Mon 7h21'31"-3d52'A
Ivana
Cma 6h11'56"-15d21'
Ivana
Cap 21h21'15"-24d10'
Ivana
Eri 3h57'38"-6d1'
Ivana
Ori 5h1'48"12d55'
Ivana Comoli di Novara
Sco 16h36'1"-35d6'
Ivana e Goffredo
Col 6h35'20"-34d45'
Ivana Rita
Cas 0h4'17"65d19'
Ivana-9
Sct 18h31'55"-6d26'
Ivanho & Dame Ed-Nah
Dra 16h27'24"64d44'
Ivanjack,Thomas Robert
Lmi 10h35'24"27d15'
Ivanka
Lyr 18h48'54"38d30'
Ivannova,Marika
Sex 9h51'43"1d27'
Ivanoff,Keith Andrew
Aql 18h58'35"3d11'
Ivans,Danielle
Vul 20h0'1"22d46'
Ivans,Kelly
Peg 22h23'44"27d12'
Ivansimone
Lac 22h16'51"38d23'
Ivatury,Mani
Mon 6h22'45"8d6'
Iven,Theodore Robert
Cap 21h24'45" 24d52'
Ivens,Derek Jason
Dra 15h27'38"65d28'
Iverna"Abiding In His Grace"
Cyg 19h40'47"40d40'
Ivers,Betty
Cas 23h39'11"61d16'
Ivers,Carol
Ori 5h28'16"-6d38'
Ivers,Joseph G
Per 3h8'26"46d48'
Iversen,Rosemary
Cas 23h34'1"60d28'
Iverson,Aimee Therese
Lyr 19h21'44"30d39'
Iverson,Anne-Marie
Cam 14h24'25"81d42'
Iverson,Camio V
Lyn 8h54'48"36d1'
Iverson,Cimone A
Lyn 8h10'42"37d4'
Iverson,Clayton Robert
Ori 5h15'1'-9d40'
Iverson,David Allen
Boo 15h6'53"16d14'
Iverson,David G
Cep 23h33'49"66d41'
Iverson,Eric Patrick
Ser 15h54'27"0d22'
Iverson,Shirley Nelson
Mon 8h8'29"-0d55'
Iverson,Therese Rose
Cyg 20h38'42"53d18'
Iverson,Trent Michael
Aql 19h47'59"10d19'
Iverson,Winston David
Aur 5h35'18"50d2'
Ives,Grant Stephen
Per 3h2'26"40d28'
Ives,Hannah
Aql 19h52'45"15d13'
Ives,Melissa
And 0h52'57"40d43'
Ives,Rebecca (Becky)
Cas 2h35'15"61d24'
Ives,René
Mon 7h15'56"-6d20'
Ives,Richard
Per 1h38'0"53d15'
Iveson,Kyle
Lyr 18h14'54"30d26'
Ivette
Com 12h22'59"24d59'
Ivey,Ann
Cet 1h23'50"-2d58'
Ivey,Dennis
Boo 15h0'23"31d19'
Ivey,Dianne
Aql 19h4'48"0d26'
Ivey,J Winston
Lac 22h11'57"54d35'
Ivey,John R
Oph 16h58'45"-28d48'
Ivey,Kiva Orion
Lyn 8h36'0"41d40'
Ivey,Pat
Sex 9h56'41"2d23'
IVIA
Cyg 19h46'38"30d4'
Ivie,Jerry
Uma 8h26'29"62d25'
Ivimas,Jose Babaloo & Miriah
Hya 10h46'30"-18d21'
Ivonne
Cas 0h39'42"62d51'
Ivonne
Lyn 8h5'13"33d60'
Ivor
Ori 5h47'59"21d42'
Ivor & Lucy
Cyg 21h14'22"28d18'
Ivory Paula
And 0h44'33"38d57'
Ivory,Lucille
Cas 2h2'30"58d29'
Ivory,Rae Beth
Cas 1h11'16"63d51'
Ivory,William H
Cet 0h22'14"-17d49'
Ivry & Young
Her 18h29'34"12d41'
Ivy
Del 20h55'24"10d23'
Ivy 1904
Cas 1h29'0"58d33'
Ivy May
Lyn 8h30'0"45d9'
Ivy's Outer Space Island
Uma 11h59'26"40d8'
Ivy's World
Mon 7h57'2"-9d1'
Ivy,Caitlin
Mon 7h2'0"4d59'
Ivy,Ernest
Hya 8h11'42"1d37'
Ivy,Roger Dale
Aql 19h30'18"8d49'
Ivy-Taylor,David Peach
Oph 18h29'14"8d19'
Iwanick,Bill
Aql 19h58'38"10d42'
Iwanicki,Edward P
Aql 20h13'0"1d32'
Iwanowski,Thomas
Cnc 8h2'33"8d43'
Iwen,Kristine Charity
Aql 19h26'16"10d15'
Iwicki,Gary
Dra 19h54'39"60d25'
IYELAK
Cnv 12h31'50"40d35'
Iyer,Azul
Lac 22h49'0"54d15'
Iyer,Krishna S
And 1h26'57"33d55'
Izawa,Star Of Tomomi
Vir 13h17'16"-1d52'
Izer,Ruth
Lyn 8h22'37"43d37'
Izetta
And 1h36'25"38d45'
Izolda's Inspiration
Peg 23h33'23"23d53'
Izquierdo,Chary
Cep 22h41'38"68d16'
Izzi,Steve
Aur 5h30'0"30d35'
Izzo,Luann Schingo
Peg 23h22'24"11d40'
Izzo,Sabina F
Cas 0h8'34"54d36'
Izzo,Sr,Robert "Bob"
Cma 4h52'41"10d40'
Izzy
Ori 5h56'1"16d18'
Izzy,Ernie
Boo 14h55'12"45d32'

J

J
Uma 9h50'34"68d12'
J
Oph 17h18'12"10d36'
J Siempre esta conmigo J
Aql 19h3'0"4d11'
J & B
Ser 18h54'11"2d25'
J & B Star
Ori 5h51'55"14d30'
J & B's Celstial Hideaway
Cyg 19h47'16"29d37'
J & J
Cam 6h24'52"80d38'
J & J Superior Landscaping
Umi 15h18'0"69d58'
J & K Always
Lmi 10h11'1"38d32'
J & M Promise,The
Lyr 18h49'42"42d1'
J & M Sparkle,The
Cep 1h26'24"78d12'
J A D-Crystal- Christopher
Vul 20h58'54"28d45'
J A M My Shining Star
Uma 10h17'32"70d23'
J B & Don
Cyg 20h15'1"31d48'
J B 7-11
Aur 6h20'43"31d16'
J B G
Oph 17h29'33"8d56'
J B M
Uma 11h11'16"43d35'
J B's Triangle
Ser 15h36'0"9d8'
J C
Cyg 19h57'38"40d40'
J C
Cyg 20h0'20"40d60'
J C
Aur 6h37'21"38d56'
J C 22
Cet 3h0'27"2d11'
J C B
Peg 0h8'47"17d44'
J C C
And 2h31'12"44d58'
J C M Pepe
Aur 5h3m53d37"53'
J C Nina
Cnv 12h9'12"38d48'
J C The Little Wolf
Lac 22h36'23"52d51'
J C-Always-M J
Cet 2h46'54"6d48'
J D A
Aur 7h13'17"38d37'
J D M Big Bear
And 23h39'28"47d20'
J D's Colts
Dra 20h23'0"68d9'
J D's Star
Aur 6h27'14"31d21'
J E B's Angel Eyes
Umi 14h10'35"65d60'
J E H D
Ori 5h58'0"17d50'
J E S '94
Cam 4h32'48"68d0'
J F K
Pic 4h41'31"-49d4'
J F R
Del 20h52'0"9d53'
J Forever N
Ori 5h56'36"13d52'
J G & D W
Lyn 8h49'21"36d27'
J G IV
Tau 5h49'58"28d53'
J I S Star
Ser 16h1'24"14d20'
J J
Cnv 13h59'49"48d4'
J J
Boo 14h49'47"52d13'
J J
Cam 11h34'23"81d3'
J J
Lac 22h18'17"51d37'
J J F-B 9:12:25
Ser 15h56'26"22d35'
J J G R 1970 10743
Hya 9h34'14"-9d47'
J J Island
Hya 10h43'58"-18d52'
J J M's Elite Dreams
Vul 19h6'58"25d10'
J J's Eternal Flame
Uma 10h4'0"56d10'
J K Sex God
Her 17h16'14"46d48'
J L C
Del 20h58'22"16d39'
J L J M
Eri 4h7'16"-8d33'
J L O - R M G
Del 20h22'29"18d48'
J M
Her 17h10'55"47d37'
J M - L C N
Cam 6h40'1"68d9'
J M R Y "Muffstar"
Lyn 7h1'13"60d38'
J M T -Jeri
Tri 2h36'35"35d10'
J M Y Babe
Cyg 19h24'1"35d24'
J Michael
Her 17h58'17"40d55'
J My Beautiful Boy from Ohio
Aql 19h50'0"13d42'
J O C
Sct 18h43'27"-6d42'
J P
Uma 8h55'20"48d29'
J P
Hya 8h11'26"4d28'
J P B
Uma 11h41'41"53d17'
J P M
Uma 11h58'47"63d15'
J P M
Cas 0h7'58"60d6'
J P S
Boo 15h5'1"42d5'
J R
Lac 22h33'50"40d45'
J R
Aur 6h6'29"37d43'
J R
Oph 17h0'0"11d26'
J R 50
Psc 0h49'24"31d31'
J R B,Jr
Ser 15h33'1"-1d38'
J R K "Against All Odds"
Per 2h58'11"40d47'
J R Promise
Ori 5h57'49"1d30'
J R T
Cyg 21h25'12"40d10'
J S B L
Cam 3h17'57"60d10'
J S C H
Per 3h0'46"47d0'
J Scott & Nori Sue 1st Anniv Star
Lyn 7h57'0"40d58'
J T
Dra 18h34'15"70d3'
J T 93
Aql 19h23'41"-5d54'
J T Feb 6,1985 to Nov 15,1993
Aql 18h59'11"-5d3'
J T Master of the Universe
Umi 15h17'1"67d17'
J T T Made For Comfort Not Speed
Her 17h7'1"20d55'
J T's Jewel
Uma 8h34'47"60d6'
J V B
Peg 0h6'1"13d13'
J V H
Cyg 19h34'33"38d30'
J W B
Cnc 9h2'57"7d24'
J W C-Corbin
Cet 3h1'11"9d34'
J W S
Leo 9h55'26"10d48'
J&B Till The End of Time
Cmi 6h58'45"-26d14'
J&S Love Forever Star
Sge 19h3'13"16d39'
J's 1st 40
Sge 19h59'24"16d17'
J's Promise:Forever Friends
Lac 22h3'29"51d47'
J's Special Place in the Sky
Peg 23h0'31"30d21'
J-Bird
Cyg 21h0'12"30d43'
J-Dain
Cep 22h30'23"61d52'
J-R's Star
Vir 13h39'48"1d40'
J-Weeks 12784
Cam 5h57'40"70d18'
J.P.
Cep 21h5'0"55d18'
J.T. (Tommy) III
Aur 5h10'23"40d54'
J1395
Per 4h7'11"50d5'
J8-2-Lourdes-7
Peg 23h32'42"10d41'

J9
Mon 7h13'1"-10d32'

Jaab 2/5/93,Ben Robert
Peg 23h32'1"18d37'

Jab 4 Eva
Uma 10h43'35"40d29'

Jabaily,Nicole Ariel
Vul 19h45'21"28d36'

Jabal,Abdo Zuhair
Ori 5h57'28"17d57'

Jabal,Noelle
Ori 6h2'43"5d1'

Jabara
Lac 22h21'56"55d47'

Jabba
Cyg 19h29'49"40d50'

Jabbour,Carole
And 2h21'16"46d9'

Jabbour,Nicolette Marie
Cet 1h24'16"-10d29'

Jabes
Uma 10h28'36"57d8'

Jablon-Fowler
Cam 12h9'53"80d31'

Jablonka
Del 20h14'50"12d20'

Jac-Clé
Cam 3h35'12"63d16'

Jac-Cor-Laine
Her 18h1'11"30d10'

Jacalyn Carol
Cas 2h37'47"63d56'

Jacalynn
Cyg 21h38'40"38d23'

Jacang,David R
Cma 7h13'1"-16d11'

Jacarol & DinLitz
Cma 6h50'26"-17d22'

Jacelyn Noelle
Sge 19h11'33"19d12'

Jacey My Prince
Her 17h57'1"40d28'

Jach
Cma 6h31'28"-15d12'

Jach,Marek Maciej
Mon 7h47'46"-9d3'8

Jachim,F Dick
Sex 10h18'14"-1d27'

Jachimczyk,Patricia N
Cyg 21h33'51"41d1'

Jachlewski,Daniel B
Peg 21h41'49"27d58'

Jachowicz,Hanna Alice
Com 12h54'21"28d47'

Jachowicz,MD,Robert
Oph 17h17'55"-21d10'

Jaci Rae
Peg 22h0'35"35d2'

Jacinta
Aql 19h20'34"12d17'

Jack
Peg 23h15'21"33d31'

Jack
Per 2h59'36"45d31'

Jack
Ori 6h5'40"0d4'

Jack
Cep 21h50'22"65d31'4

Jack
Her 16h45'46"34d55'

Jack
Cep 22h51'31"57d33'

Jack
Lyr 19h17'18"40d16'

Jack
Cet 3h11'28"4d26'

Jack "N" Skip
Per 3h31'0"31d11'

Jack & Debbie
Sge 20h0'12"18d6'

Jack & Donna
Cyg 19h41'0"38d45'

Jack & Gayle 711
Cam 10h58'56"80d29'

Jack & Imo
Ori 6h8'48"9d17'

Jack & Jim-The Eagle's Nest
Aql 19h36'26"1d27'

Jack & Jo
Umi 16h16'46"73d13'

Jack & Laurel
Crb 15h56'35"30d48'

Jack & Mitzi, "Will Shine Forever"
Aql 19h24'17"10d50'

Jack & Rori
Her 18h19'52"20d36'

Jack A Lucky Star 4U I Luv U Maggie
Eri 2h48'45"-4d22'

Jack Alexander
Aur 7h17'12"35d57'

Jack Joseph 411
Uma 10h55'0"35d57'

Jack Norman
Her 18h54'30"12d49'

Jack of Hearts
Dra 19h27'26"70d28'

Jack The Rapper
Mon 6h20'17"7d52'

Jack The Rebel (Farenheit IV)
Per 4h41'32"50d28'

Jack's & Karen's Star
Cet 2h21'52"-5d42'

Jack's Beamer
Boo 14h35'58"8d21'

Jack's Diamond
Lyn 7h36'52"38d30'

Jack's Hole in One
Cep 21h47'56"70d48'

Jack's Journey to Eternity
Dra 13h11'41"68d53'

Jack's Lookout
Aur 5h10'36"40d51'

Jack's Place
Cep 23h34'36"64d49'

Jack's Point
Per 4h9'52"38d15'

Jack's Rolex In The Sky
Boo 13h54'29"20d45'

Jack's Serenity
Ser 15h22'46"8d12'

Jack's Star
Aur 7h9'57"38d30'

Jack's Star
Cam 8h19'47"77d45'

Jack's Wish
Lac 22h4'59"47d26'

Jack,Cathy & Jacey
Aql 18h56'33"10d13'

Jack,Danny Lee
Hya 9h47'26"-19d32'

Jack,Janet Kyle
Eri 3h53'1"-3d29'

Jack,MD,Robert A
Oph 17h15'0"12d38'

Jack,Ralph Dottie
Hya 8h54'47"-7d11'

Jacka,Klaus
Vul 19h44'24"29d20'

Jackel,Denyse Geneviève
Her 17h10'41"43d1'

Jackes,Courtney Elizabeth
And 0h54'37"40d36'

Jackfool
Ori 6h6'48"-0d3'

Jacki
Lyn 7h50'1"33d46'

Jacki
Cap 21h52'0"-17d54'

Jacki's
Peg 22h44'21"32d52'

Jackie
Mon 7h29'31"-10d15'

Jackie
Cyg 19h25'32"50d26'

Jackie
Cas 1h12'18"61d42'

Jackie
Lyn 8h49'56"34d21'

Jackie
Cyg 20h34'58"40d13'

Jackie
Lyn 9h9'55"45d50'

Jackie
Tri 1h38'31"31d41'

Jackie
Sge 19h59'57"20d32'

Jackie
Cyg 19h26'31"30d42'

Jackie
Aql 20h11'25"5d13'

Jackie & Alan
Aql 19h43'1"10d44'

Jackie & Chris
Uma 12h52'15"60d46'

Jackie & Otis
Cyg 21h2'43"50d37'

Jackie Ann
And 23h48'16"38d4'

Jackie Blue
And 0h58'50"33d42'

Jackie Boy
Psc 0h44'0"20d31'

Jackie Le
Lmi 9h55'18"33d29'

Jackie O
Cyg 19h42'16"31d30'

Jackie OH !
And 2h18'20"41d9'

Jackie's Birthday Jewel
Peg 21h55'36"20d14'

Jackie's Delight
Del 20h14'43"14d36'

Jackie's Hope
Mon 6h19'36"9d2'

Jackie's Jewel
Boo 14h53'16"33d56'

Jackie's Precious Adventure
Ori 6h6'35"9d45'

Jackie's Shooting Star
Lyr 18h34'34"42d37'

Jackie's Star
And 1h3'51"39d60'

Jackie's Star
Lyn 9h3'33"39d53'

Jackie's World
Cnv 13h47'45"46d13'

Jackie,"Light of My Life",Love Jack
Umi 16h58'54"76d21'

JackieMike
Cyg 19h21'31"28d33'

Jackman,Janice
Lyr 18h31'56"30d17'

Jackman,Richard Wayne
Sex 10h1'0"2d25'

Jackos
Vul 19h34'9"24d3'

Jackrel,Justin
Aql 19h21'37"13d56'

Jacks,Darren
Ori 5h10'0"-8d3'

Jacks,Mabelle
Boo 13h40'55"19d17'

Jackson Birthday Star, The Peter
Ori 5h22'1"15d25'

Jackson Celestial Body D A
Ori 5h55'21"16d46'

Jackson Dean
Cep 22h8'44"53d4'

Jackson Did This
Aql 19h30'60"13d52'

Jackson Heal The World Star,Michael
Vul 19h43'11"26d45'

Jackson Highland Luminary,Margaret
And 0h15'16"32d40'

Jackson III,M Lewis
Mon 8h2'56"-0d56'

Jackson The Angel
Uma 9h56'28"49d57'

Jackson"PJ",Patricia Kay
Sgr 18h7'23"-28d59'

Jackson's Childrens Angel,Michael
Uma 9h32'26"56d35'

Jackson's Dream,Scott
Cet 2h17'1"1d15'

Jackson's Eternal Flame
Del 20h53'36"7d40'

Jackson's Nova
Per 3h39'18"32d46'4

Jackson's Wedding Star,Brian & Tamara
Crb 15h53'15"27d44'

Jackson,Abner & Dainty
Ori 5h9'34"-5d33'

Jackson,Alexander Edward
Mon 6h25'57"2d52'

Jackson,Alexandra Gloria
Lyn 8h26'37"52d30'

Jackson,Alexandra Jane
Lyr 19h19'30"42d33'

Jackson,Amanda Grace
Aur 5h38'17"37d52'

Jackson,Amanda Renae
Lyr 19h25'36"38d58'

Jackson,Amy Beth
Lyr 18h45'19"42d29'

Jackson,Andrew M
Boo 14h14'33"37d0'

Jackson,Andy
Cep 23h5'1"65d30'

Jackson,Arlene Rita
Cas 0h8'1"62d52'

Jackson,Bailey
Eri 3h42'36"-18d42'

Jackson,Benjamin Edward
Per 1h47'1"50d22'

Jackson,Bradley Jay
Her 17h57'41"14d26'

Jackson,Carol Janis
Cas 1h2'1"58d59'

Jackson,Carter
Oph 16h40'22"2d20'

Jackson,Charlotte Elizabeth
And 0h1'20"38d25'

Jackson,D K
Cas 0h55'55"63d59'

Jackson,Derek Robert
Peg 22h40'54"11d21'

Jackson,Don & Betty
Mon 6h56'18"-1d15'

Jackson,Donna Faye
Sge 19h21'47"16d11'

Jackson,Dorothea May
Cyg 21h1'0"30d26'

Jackson,Dr James W
Oph 18h1'40"10d1'

Jackson,Dr Stanley E
Cep 0h16'30"76d34'

Jackson,Duane Leo
Mon 6h38'36"10d22'

Jackson,Elizabeth
Ori 5h59'16"10d4'

Jackson,Elizabeth Powers
Cas 0h0'24"56d0'

Jackson,Ellis Carter
Cep 22h38'51"63d59'

Jackson,Empyrean Dancer Michael
Vir 11h57'32"-1d15'

Jackson,Forever Michael
Uma 11h8'59"56d7'

Jackson,Francis
Mon 6h24'0"-0d33'

Jackson,Frank Eastwood
Ori 6h3'58"1d37'

Jackson,Franklin A
Gem 6h35'14"26d30'

Jackson,Glenn & Robin
Lac 22h27'31"53d20'

Jackson,Glyn
Ori 6h1'22"1d1'

Jackson,Harry Benjamin
Her 18h1'53"28d55'

Jackson,James & Laura
Cyg 21h12'0"35d17'

Jackson,Jane Purvis
Peg 22h16'0"34d13'

Jackson,Jenny K
Aql 19h42'23"14d44'

Jackson,Jeremy Cain
Her 18h1'1"14d30'

Jackson,Jeremy Joseph
Ori 4h44'15"15d13'

Jackson,Joanne
Peg 22h8'48"24d25'

Jackson,Joe
Boo 13h49'50"19d57'

Jackson,John Cat
Cep 22h50'53"57d37'

Jackson,John Matthew
Umi 15h12'0"68d27'

Jackson,Jon
Crt 11h42'12"-10d46'

Jackson,Jr,Glenn Bluford
Cnv 12h24'35"37d56'

Jackson,Jr,William J
Aql 18h56'47"-1d12'

Jackson,Julia Thomas
Cyg 20h55'0"31d12'

Jackson,June S
Cyg 20h38'1"40d42'

Jackson,Kara Celeste
Peg 23h44'21"25d43'

Jackson,Kathryn
Cas 0h57'54"64d33'

Jackson,Keesha Naté
Vir 12h58'1"-17d41'

Jackson,Kellye Hudson
Mon 6h30'54"7d40'

Jackson,Kevin
Aur 7h15'48"41d38'

Jackson,Kevin
Boo 14h6'22"50d44'

Jackson,La Toya
Boo 15h26'23"33d39'

Jackson,Lady Brenda
Lyr 18h49'0"30d56'

Jackson,Laura
And 1h29'25"39d43'

Jackson,Laura
Sge 19h54'45"19d26'

Jackson,Lee William
Vir 13h34'29"-8d19'

Jackson,Lela
Oph 17h4'33"11d51'

Jackson,Lt Col Mo
Aql 19h26'56"-7d49'

Jackson,Lynda
Del 20h14'0"12d50'

Jackson,Mark Lee
Ser 15h31'53"-2d40'

Jackson,Mary & Willis
Cyg 19h31'59"36d32'

Jackson,Michael
Cep 22h8'21"61d4'

Jackson,Michael
Peg 21h56'57"26d18'

Jackson,Michael
Cma 6h54'51"-17d18'

Jackson,Michael
Vir 13h34'46"-2d22'

Jackson,Michael "Michael"
Vir 12h33'0"-8d28'

Jackson,Michael Joseph
Aur 6h56'0"38d23'

Jackson,Moira
Cas 0h14'59"63d50'

Jackson,Nancy June Lane
Lyn 7h3'20"44d40'

Jackson,Patricia Ann
Equ 20h55'0"5d54'

Jackson,Paul William
Her 17h1'1"42d51'

Jackson,Peter & Dorothy
Cmi 8h8'21"3d52'

Jackson,PhD,Cliffordum Ray
Oph 17h1'32"8d18'

Jackson,Philip H
Ori 4h55'0"0d48'

Jackson,Rex Steven
Dra 18h59'34"78d55'

Jackson,Rhonda Sue Conover
Cas 22h59'15"58d14'

Jackson,Robert E
Cet 3h16'40"3d13'

Jackson,Robert E
Ser 15h10'14"5d40'

Jackson,Ron
Cap 21h19'34"-14d23'

Jackson,Ron & Sue
Cma 6h44'36"-15d35'

Jackson,Roy
Her 17h16'46"45d30'

Jackson,Samuel Lear
Aql 19h0'58"13d18'

Jackson,Sandra E
Lyn 8h59'29"35d39'

Jackson,Sarah
Her 16h16'20"41d42'

Jackson,Scott & Emma
Sge 19h32'20"16d26'

Jackson,Scott Garrison
Umi 15h33'52"70d7'

Jackson,Sean Michael
Peg 23h42'19"30d29'

Jackson,Shari Renee
Lyr 18h30'16"38d21'

Jackson,Sharon
Del 20h39'14"11d1'

Jackson,Simon
Aur 6h2'57"30d34'

Jackson,Sophie
Aql 18h43'32"6d35'

Jackson,Stacy
Lyn 8h23'15"43d22'

Jackson,Stephen
Per 2h56'41"38d18'

Jackson,Steve
Cet 2h33'36"0d33'

Jackson,Stuart B
Cet 2h0'56"-1d52'

Jackson,Tara
Vul 19h48'0"26d32'

Jackson,Terry
Aur 6h24'36"40d39'

Jackson,Terry & Janice
Cyg 21h59'35"52d59'

Jackson,The J Bryan
Cmi 7h46'43"8d5'

Jackson,Theresa V
And 0h26'34"40d50'

Jackson,Thomas B
Aql 18h53'30"-1d18'

Jackson,Tim & Beth
Eri 3h5'21"-2d40'

Jackson,Timothy Ian
Cep 0h57'39"80d29'

Jackson,Vera Lockwood
Peg 23h35'47"31d52'

Jackson,Vincent Ray
Aur 6h10'45"37d37'

Jackson,Wendy
Cyg 21h32'21"53d5'

Jackson,Whitney Erin
Tau 5h48'30"26d39'

Jackson,William B
Hya 9h39'45"-6d41'

Jackson,William C
Oph 17h52'34"-0d1'

Jackson,William Dean
Cet 2h34'11"-12d23'

Jacksons,The
Lyn 8h7'28"42d22'

Jackstar
Sct 18h48'21"-7d35'

Jackster,The
Boo 15h46'29"44d9'

Jacky
Sgr 18h50'4"-34d40'

Jacky & Alan
Cyg 19h26'19"31d28'

Jacky's Jewel
Cyg 19h33'0"39d39'

Jaclaire '48
Aql 19h31'27"12d14'

Jaclyn
Cas 0h24'34"56d0'

Jaclyn Paige
Cas 0h33'16"54d27'

Jaclynn Rae
Cas 0h4'21"64d41'

Jaco,Charles Dennis
Eri 3h2'60"-8d23'

Jaco,Kimberly
Psc 23h20'38"5d3'

Jacob
Dra 18h15'54"53d10'

Jacob
Uma 10h18'31"68d1'

Jacob
Per 1h44'0"53d58'

Jacob
Oph 16h30'23"-6d50'

Jacob Alexander
Cet 3h0'40"0d12'

Jacob's Hope
Cep 21h12'57"68d45'

Jacob's Point
Uma 11h51'49"50d34'

Jacob,Christian
Her 16h27'1"33d29'

Jacob,Dr Stanley W
Oph 17h33'45"-23d27'

Jacob,Duane "Jake"
Lac 22h39'0"53d39'

Jacob,Elizabeth Rachel
Vul 20h16'34"25d38'

Jacob,Kenneth A & Michelle M P
Cam 4h12'45"67d31'

Jacob,Kensey
Cmi 7h33'58"6d40'

Jacob,Lisa Marscene
Lac 22h19'17"51d18'

Jacob,Marella Rita
Peg 23h22'21"12d15'

Jacob,Travis
Sex 10h30'52"4d54'

Jacobellis,James Anthony
Aur 5h12'34"40d54'

Jacobi,Derek Anthony
Cmi 8h6'38"6d4'

Jacobi,Lacey Marie
Mon 7h5'28"-5d48'

Jacobi,Robert Anthony
Cet 0h2'42"-10d19'

Jacobina,Irene
And 23h30'21"47d59'

Jacobowitz,Caryn
Psc 23h28'17"1d23'

Jacobs"Serenity Star", Max R
Cyg 21h2'1"40d16'

Jacobs,Albert
Umi 15h42'38"77d12'

Jacobs,Andrea Isabell
Cap 20h46'52"-26d59'

Jacobs,Anita
Dra 18h50'24"68d59'

Jacobs,Dale
Per 3h26'29"40d35'

Jacobs,Dana Kay
Peg 23h3'1"12d57'

Jacobs,David W
Her 16h33'0"48d32'

Jacobs,Eddie
Hya 9h53'33"-15d5'

Jacobs,Edward George
Cep 21h47'1"55d25'

Jacobs,Felice
Aur 4h48'35"48d52'

Jacobs,Gayle
Uma 11h41'30"56d8'

Jacobs,Jacqueline Amanda
Cyg 21h55'11"53d26'

Jacobs,James S
Her 16h40'13"8d7'

Jacobs,Jared T
Her 17h12'37"27d4'

Jacobs,Jo
Cam 5h2'49"61d29'

Jacobs,Joy
Mon 7h39'46"-6d46'

Jacobs,Justin Richard
Cep 22h20'20"60d37'

Jacobs,Karin
Aqr 22h59'39"-22d6'

Jacobs,Keith Brian
Aql 18h57'15"8d49'

Jacobs,Kenneth D
And 2h14'16"50d42'

Jacobs,Lenny
Leo 11h6'46"-5d55'

Jacobs,Maria
Lyn 7h55'47"47d37'

Jacobs,Mark
Her 17h30'1"14d54'

Jacobs,MD,Brad J
Lac 22h3'49"47d26'

Jacobs,Melanie A
And 0h12'53"30d8'

Jacobs,Michael Buckley
Per 1h28'27"53d47'

Jacobs,Nicole
And 0h59'34"40d36'

Jacobs,Randa E
Uma 10h41'43"47d6'

Jacobs,Randy
Aur 6h12'25"31d18'

Jacobs,Ruby Janet
Aql 20h15'54"1d38'

Jacobs,Sean Randall
Aql 20h0'0"14d42'

Jacobs,Stacey Irene
Vir 13h58'25"-1d40'

Jacobs,Stephanie
Psc 1h0'59"21d38'

Jacobs,Stephen
Her 18h1'0"30d58'

Jacobs,Steven Michael
Uma 9h33'49"51d27'

Jacobs,Suzy
Sge 19h28'57"16d24'

Jacobs,Tammy Carol
Leo 11h29'33"11d38'

Jacobs,Tim
Sge 20h3'16"20d20'

Jacobs,William & Doreen
Cyg 21h0'46"38d13'

Jacobs,Yvonne
Mon 6h23'53"8d29'

Jacobs-Williams,Judy
Aur 6h29'30"39d49'8

Jacobsen,Amber Fallon
Com 13h8'54"21d43'

Jacobsen,Barbara J
And 2h6'27"38d43'

Jacobsen,Captain Harry Arthur
Cep 23h9'55"61d23'

Jacobsen,Danny
Uma 10h40'56"40d33'

Jacobsen,Dave Otto
Vul 19h58'53"23d14'

Jacobsen,Edward R
Aur 4h38'19"31d51'

Jacobsen,Helen P
And 23h24'0"49d1'

Jacobsen,In Memory of Shirlee
Eri 5h8'1" 6d48'

Jacobsen,Jan
Cyg 19h29'10"38d17'

Jacobsen,Loretta
Cam 10h53'1"81d38'

Jacobsen,MD,Arthur Jacob
Uma 10h33'53"47d15'

Jacobson 80,Nathan & Ethel
Dra 14h19'0"64d10'

Jacobson,"Annie"
And 0h55'1"34d54'

Jacobson,Alexander Leaff
Peg 22h36'25"24d5'

Jacobson,Bernel Adil Morton
Cma 7h50'0"11d37'

Jacobson,Carmen
Mon 6h23'31"8d52'

Jacobson,Carol
Cyg 20h21'54"30d17'

Jacobson,D J
Boo 13h43'26"24d17'

Jacobson,David
Her 17h19'29"27d13'

Jacobson,Deb
Cmi 7h30'26"8d6'

Jacobson,Devora L
Gem 7h24'53"31d40'

Jacobson,Diane
Cyg 20h48'15"37d56'

Jacobson,Jaake
Lmi 10h15'31"38d9'

Jacobson,John
Lac 22h6'45"51d6'

Jacobson,Michael Aaron
Cnv 12h54'34"51d45'

Jacobson,Nathan & Florence
Dra 15h59'0"63d34'

Jacobson,Peyton
Aql 19h2'0"0d23'

Jacobson,Ralph
Aql 20h3'60"4d3'

Jacobson,Richard
Aur 6h28'56"37d52'

Jacobson,Stephanie Ann
Vul 20h13'58"23d35'

Jacobson,Stephen Eric
Dra 14h3'28"63d29'

Jacobson,Tali
Cnv 12h30'55"38d45'

Jacobson,Vanessa
And 0h21'0"40d28'

Jacobson,Vivian M
Peg 22h43'16"3d34'

Jacobus
Per 3h30'46"37d55'

Jacoby III,MD,John Z
Vul 19h22'19"23d45'

Jacoby,Alyssa Jane Thomson
Peg 22h0'41"21d57'

Jacoby,Meagan Anne
And 0h17'53"34d13'

Jacoby,Phil
Her 16h52'48"47d45'

Jacome,Francine
And 23h15'55"42d5'

Jacop "The Brightest Star Of All",Bill
Uma 11h52'26"46d30'

Jacopino,Marcella
Dra 17h50'14"64d9'

Jacopo
Uma 11h47'30"50d14'

Jacopo
And 23h4'28"52d34'

Jacot,Cheryl & Stan
Uma 8h50'55"50d28'

Jacox,Ada
Cam 5h44'54"67d34'

Jacquart,Jared Joseph
Uma 12h11'33"62d50'

Jacquart,Rachel
Aur 6h0'21"38d10'

Jacquault Philippe
Cnv 13h58'24"37d54'

Jacque
Lyn 6h28'20"56d11'

Jacque's Star
Peg 23h6'39"17d38'

Jacque,Michael"I.O." Ronald
Cet 2h3'60"0d8'

Jacque-B
Cet 3h15'1"2d50'

Jacquelin Gail
Eri 3h42'29"-1d51'

Jacqueline
Cnc 8h23'46"29d50'

Jacqueline
And 0h51'41"38d22'
Jacqueline
Cas 0h24'10"63d49'
Jacqueline
Aqr 21h56'1"-6d8'
Jacqueline
Cam 3h15'13"60d5'
Jacqueline
And 2h29'15"50d46'
Jacqueline
Cet 1h31'27"-10d20'
Jacqueline
Mon 6h25'41"-5d50'
Jacqueline
Oph 18h16'23"11d38'
Jacqueline
Ari 2h50'45"21d10'
Jacqueline 26-3-67
Lyr 18h54'46"30d27'
Jacqueline A
Ind 20h49'10"-54d45'
Jacqueline Amber
Cas 0h23'45"66d49'
Jacqueline Anne
Cas 0h6'19"59d1'
Jacqueline Elizabeth
And 0h8'24"31d43'
Jacqueline Marie
And 2h30'27"50d7'
Jacqueline Mary
Cas 0h22'0"58d55'
Jacqueline Rose
Sex 10h15'35"-2d46'
Jacqueline Star
Mon 7h1'13"0d17'
Jacqueline's Star Rising
Gem 7h55'32"20d15'
Jacqueline's Star
Lyr 18h20'38"38d45'
Jacqueline's Star
Lyr 19h13'35"42d33'
Jacqueline,Therese
Ori 5h25'40"1d38'
Jacquelyn "My Nova of Love"
Lyr 19h2'52"25d52'
Jacquelyn Ann
Del 20h21'36"20d27'
Jacquelyn Dawn
Oph 17h52'42"7d49'
Jacquemet,Stéphanie
And 23h9'33"40d33'
Jacquemin-Clark,08-30-1950,Jane L
Lyn 8h49'29"43d26'
Jacques Arthur
Cyg 21h8'52"30d14'
Jacques et Marie Andrèe
Uma 11h44'1"46d55'
Jacques Philippe
Cyg 19h47'1"30d2'
Jacques' Nova
Per 4h9'16"38d56'
Jacques' One
Dra 15h59'27"68d21'
Jacques,Dominique
Aur 6h26'43"33d37'
Jacques,James A T
Cyg 21h19'48"35d9'
Jacques,Magdelaine
Umi 16h42'44"76d44'
Jacquestine Amour
And 23h8'53"40d33'
Jacquet,Beatrice
Peg 22h12'34"30d1'
Jacquet,Heather
Peg 22h19'15"30d6'
Jacqui
Her 17h0'0"46d60'
Jacqui Ann
Cas 2h54'0"58d13'
Jacquie
Peg 22h42'46"25d15'
Jacquie & Lisa
Cyg 20h8'18"38d58'

Jacquotte Et Momo
Per 2h56'34"56d54'
Jacton
Ori 6h7'30"3d20'
Jade
Vel 10h5'41"53d0'
Jada
Cas 0h1'37"54d32'
Jade
Cam 7h38'17"68d42'
Jade
Peg 22h58'57"32d32'
Jade
Cyg 19h29'42"35d24'
Jade's Promise
Leo 10h52'1"15d15'
Jaden
Aur 5h48'59"54d46'
Jadoux,Martine
Per 3h1'12"37d59'
Jadzak,Lori A
Peg 21h56'0"30d8'
Jae's Star
Cam 3h37'42"60d1'
Jaeb,Joel Thomas
Cmi 7h43'1"8d60'
Jaecker,Saskia
Cep 22h6'53"61d35'
Jaeger (July 6th 1914) ,James
Per 2h54'36"40d45'
Jaeger,Eric
Her 16h12'36"10d9'
Jaeger,John (Jackie) J
Her 18h0'40"38d25'
Jaeger,Thomas Karl
Ori 5h52'31"1d45'
Jaekel III,"Lover", Carl J
Cep 5h2'40"80d11'
Jaffa,Margaux Lauren
Cas 1h31'48"60d34'
Jaffari,Cyrus Afshar
Cnv 12h6'1"40d56'
Jaffe 1994,Steven
Dra 17h35'42"60d11'
Jaffe,Abigail
Dra 20h11'22"73d48'
Jaffe,Betty
Ori 6h13'32"8d32'
Jaffe,Helias Cody
Cam 3h57'51"52d41'
Jaffe,Julianna
Cas 0h35'31"70d23'
Jaffe,Lew
Boo 14h35'56"45d50'
Jaffe,Martin Jay
Cep 21h14'51"84d35'
Jaffe,Rachel Leigh
Lyn 8h5'20"40d14'
Jaffe,Sandra Kay
And 0h39'19"40d59'
Jaffe,Sci-Fi Star, Matthew
Ori 5h39'34"7d33'
Jaffer,Adrian
Mon 8h4'22"-2d35'
Jaffer,Nadia
Sge 20h16'11"20d35'
Jag
Her 16h20'0"24d55'
Jager,Frank
Aur 5h6'18"43d7'
Jager,Jon A
Hya 8h28'23"-0d22'
Jager,Madeleine
Cyg 19h47'18"30d9'
Jaggers,Caroline
Ori 5h55'43"16d1'
Jaggers,Karen E
Cas 0h29'37"58d20'
Jaggers,Terri
Del 20h14'40"15d1'
Jaggs,Angela
And 0h50'33"33d23'
Jagiel-Douthit's Love Star,D Renee
Cyg 19h25'43"33d40'

Jago,David William
Dra 18h32'1"50d31'
Jago,Ted & Jayne
Cep 23h7'36"64d8'
Jagoda,Dorothy L
And 1h3'54"38d22'
Jagu
Cnc 7h55'0"10d45'
Jaguar's Den
Cnv 13h52'36"40d10'
Jahn "Big Bill", William E
Boo 14h55'60"38d48'
Jahn,Jr,Joseph
Per 3h24'54"54d37'
Jahn,Jürgen
Cap 20h34'11"-21d9'
Jahn,Michael
Vir 12h30'42"-8d2'
Jahn,Michael C
Vul 20h16'32"26d11'
Jahn,Miriam
Sgr 20h3'51"-41d57'
Jahn,Norman Barry
Boo 15h8'1"29d9'
Jahn,Roland
Hya 8h50'31"-8d4'
Jahn,Susanne A F
Cra 18h0'1"-37d28'
Jahnel,Thomas
Sge 19h8'38"18d41'
Jahnka,Wilma
Boo 15h37'54"41d17'
Jahnke,Dan & Lisa
Dra 18h2'49"58d28'
Jahnke,Jutta Martina
Sgr 18h54'31"-36d22'
Jahnke,L P
Cyg 21h32'1"42d11'
Jahnke,Suzanne Joy
Sct 18h48'36"-6d46'
Jahnke,Ursula
Sgr 19h59'49"43d3'
Jahns,Rudolf
Sco 16h39'18"-40d38'
Jahr,Alexandra
Sco 17h29'50"-30d37'
Jahut Ole M
Aql 19h16'36"12d20'
Jai,Joni
Eri 3h45'23"-3d17'
Jai,Ken
Ser 15h38'19"3d35'
JAI-HYL
Uma 10h12'0"55d25'
Jaillet,Gordon Craig
Ori 4h47'60"4d27'
Jaime Elana
Mon 6h19'49"5d4'
Jaime Lark
And 23h49'21"37d36'A
Jaime Marie
Cas 0h42'33"60d21'
Jaimse
Ori 6h10'18"8d36'
Jain,Saraswati Hiralal
Peg 23h27'48"18d29'
Jain,Urvashi Praveen
Ori 4h43'21"4d24'
Jaino
Lyr 18h53'56"42d48'
Jaisa
Sco 16h36'1"-30d4'
Jak & Karen 1996
Boo 14h49'36"31d48'
Jakacki,Benhard
Her 17h25'0"18d60'
Jakacki,Violet Wagner
And 2h22'52"38d34'
Jakala,John & Lynda Woodham
Sge 19h55'29"16d39'AB
Jake
Per 2h31'38"57d14'
Jake
Her 16h59'51"21d25'

Jake
Ori 5h52'1"18d47'
Jake
Cet 0h50'59"-4d2'
Jake & JoJo's Dream
Lac 22h18'0"50d6'
Jake Thomas
Oph 17h34'58"-1d38'
Jake's Star
Her 16h22'31"42d23'
Jake's Star
Lac 22h6'10"48d52'
Jakeko
Peg 23h39'39"30d24'
Jakel,MacKenzie Chase
Aql 19h45'1"10d23'
Jakes II,James Warren
Lac 22h28'30"54d20'
Jakes,Pauline Anne
Peg 22h2'26"10d36'
Jaki
And 23h17'30"49d15'
Jakk's Pet,19 March 1946
Psc 1h1'25"21d38'
Jakob
Sco 16h12'41"-24d49'
Jakob,Ann Wilbraham
Cas 23h39'33"50d7'
Jakob,Annette
Cas 22h57'31"55d1'
Jakob,Wolgang
Aql 20h9'32"1d45'
Jakobeit,Margitta
Vir 12h56'54"-18d35'
Jakobowski,Michael
Aql 19h24'41"8d59'
Jakobsons,Astrida
Lyn 8h55'14"41d13'
Jakovac,Donald
Cep 21h58'51"55d58'
Jaksola,Jaakko
Uma 10h12'40"47d45'
Jaksztat,Sandra Barbara
Cnc 9h0'29"12d13'
Jakubczak,Hans-Josef
Del 20h58'33"12d32'
Jakubczyk,Kimberly
Crb 15h55'29"28d35'
Jakubecz,Brigitte
Ari 2h52'29"21d41'
Jakubzick,Elsa
Mon 6h57'46"7d57'
Jal Rustom Dadabhoy
Peg 22h32'42"14d21'B
Jale
Aur 4h56'27"38d6'
Jalenak,Jr,L R
Uma 11h45'52"31d59'
Jalensky,Nicolas Joseph
Ari 3h14'16"28d59'
Jallot,Liliane
Aql 20h35'46"0d21'
Jalon,Francisco Montunoi
Her 18h7'16"20d19'
Jalonski,Jim
Aql 18h59'36"-6d12'
Jalotar
Cam 13h33'24"80d59'
Jalowiec,Maryann R
Cas 0h57'35"59d48'
Jalpipom
Ori 5h34'10"-0d16'
Jam
Cam 7h54'19"70d57'
Jam 64
Aur 5h21'39"38d16'
Jam Blazing Ad Infinitum
Per 1h45'29"5012'
Jam,Jimmy
Her 16h46'22"47d43'
Jam,Jimmy & Lisa Harris
Uma 8h44'21"48d5'
Jama
Peg 22h9'26"29d49'

Jama
Uma 9h28'1"48d15'
Jama-Bootsy
Hya 8h17'36"5d15'
Jamail I
Aql 18h58'52"-5d11'
Jamain,Michel
Sgr 19h30'34"-36d58'
Jamain,Sabine
Tri 2h0'35"30d24'
Jamais
Cyg 21h0'32"28d29'
Jamal et Nathalie
Lyr 18h54'29"30d29'
Jamarin
Dra 19h40'15"70d45'
Jamason,Jr,Gregory Carl
Ser 15h21'0"6d58'
Jambor,Katie Louise
Cas 0h22'59"62d25'
Jame,Claudine
Boo 14h31'0"40d34'
Jamerson,Tony Gene
Cyg 19h29'54"30d56'
James
Her 17h0'55"42d27
James
Aur 6h7'48"35d50'
James
Boo 15h2'30"27d14'
James
Tau 3h36'46"24d48'
James
Aur 6h7'15"33d10'
James
Aur 6h20'53"30d44'
James
Per 3h25'56"46d45'A
James
Boo 14h5'49"8d16'
James
Cam 11h39'25"84d46'
James
Ori 6h11'33"17d55'A
James "Against the Flow",Brooke
Per 3h10'23"41d57'
James & Jennifer
Sge 19h58'30"16d16'
James & Kelly
Her 17h33'51"21d1'AB
James & Lillian
Cyg 20h2'30"31d28'
James & Marcie
Mon 7h19'32"-8d17'
James & Marjorie
Del 20h14'14"15d10'
James & Michaela's Place
Sge 19h13'0"16d12'
James Andrew
Lac 22h42'0"53d1'
James B (Precious)
Aql 20h0'54"7d56'
James C
Ser 15h12'50"19d25'
James Christopher
Boo 15h40'28"47d35'
James Craig
Per 1h57'59"52d33'
James Dale
Her 17h24'17"27d7'
James David
Aur 4h47'12"41d10'
James H
Cma 6h52'56"-16d47'
James Howard
Aur 6h25'23"38d60'
James Joseph
Tri 1h58'0"27d17'
James Matthew
Cnc 9h6'11"30d3'
James Oliver
Oph 18h35'37"10d36'
James Richard
Psc 22h54'47"6d60'

James Richard
Per 3h2'1"40d52'
James Scott
Uma 8h58'23"56d24'
James Star
Hya 8h41'47"2d32'
James Star To Steer Her By,Jesse
Uma 13h33'20"60d38'
James Star,The
Cyg 19h26'55"50d1'
James Stuart
Uma 10h39'28"47d46'
James The Jersey City Squirrel
Gru 22h9'5"-55d46'
James The Protector
Aql 19h27'55"13d28'
James Tracy
Boo 15h1'53"50d38'
James Tracy
Boo 15h1'53"50d38'
James V & Erica V
Hya 9h33'14"-19d21'AB
James Way
Her 17h39'44"37d32'
James We Love You!, Okie Bill
Per 2h53'12"40d24'
James We Love You!, Mildred Mom
Lyn 7h50'43"37d10'
James William
Her 16h36'33"34d3'
James' Alderan
Cet 2h7'18"0d53'
James' Diamond
Aur 6h9'45"37d32'
James's Melody
Her 18h22'26"13d1'
James's Sparkle
Her 18h2'31"28d19'
James,Adam Gustav
Dra 18h30'36"58d37'
James,Andrew
Oph 17h2'34"-23d23'
James,Anne & Bob
Ori 5h51'46"15d18'
James,Anthony B
Her 17h56'52"30d35'
James,Arthur
Cep 20h38'56"60d45'
James,AV80R Jeff
Cam 3h49'59"52d40'
James,Bennett Warren
Oph 17h53'33"12d49'
James,Bernadette
Cas 0h7'16"63d59'
James,Betty
Cas 0h29'41"61d56'
James,Betty Claire (Tyler)
Aql 19h1'13"16d3'
James,Bill
Crb 15h55'27"26d53'
James,Brandon
Psc 1h27'13"30d4'
James,Brenda Carole
Hya 9h31'19"5d31'
James,Cambell
Sex 10h26'12"-1d33'
James,Carolyn Cheryl
Cyg 20h27'23"58d17'
James,Clara
And 0h11'42"47d42'
James,Daniel Paul
Uma 11h48'7"59d15'
James,David
Cep 22h12'1"68d1'
James,Debbie
Hya 8h58'0"-5d49'
James,Deborah
Cam 8h48'31"78d1'
James,Dorothy
And 1h15'51"40d16'
James,Dr Crafton Davis
Oph 17h58'24"7d30'

James,Emerson
Cmi 7h58'33"4d35'
James,Erin Edward
Hya 9h3'39"4d55'
James,Goddess Sherri
And 0h51'43"40d20'
James,Graham Andrew
Gem 6h55'54"17d53'
James,Harold Irving
Cep 23h2'16"64d1'
James,Henry Clay
Boo 14h36'56"50d20'
James,Irma L
And 0h6'48"31d15'
James,Jai Edward
Ori 6h6'59"5d23'
James,Janette
And 0h1'20"47d24'
James,Jeanette Lesley
Ari 2h30'59"20d20'
James,Jeffrey
Her 17h59'26"31d40'
James,Jennifer
Dra 19h40'57"80d35'
James,Joann Janine
Mon 6h54'48"0d14'
James,Jobeanna
Cyg 21h33'35"42d9'
James,John
Aur 4h59'18"51d29'
James,Judith Kay
Cas 1h48'33"60d15'
James,Mai & Phil
Eri 4h33'32"-19d44'
James,Mary
Vul 20h16'40"25d5'
James,Megan P
Tri 2h19'0"32d37'
James,Meghan Elizabeth
Lyn 8h38'40"41d15'
James,Meghan Elizabeth
Aql 19h31'28"12d57'
James,Melissa
Cas 1h54'57"74d26'
James,Michael
Cyg 20h57'46"48d52'B
James,Nannie Beatrice
Peg 22h26'33"3d3'
James,R Spencer
Aql 19h3'29"-6d40'
James,Ramey
Oph 16h59'14"0d58'
James,Rita
Cyg 22h0'16"51d13'
James,Robert G
Her 17h49'40"14d45'
James,Robert Lynn
Lac 22h32'38"53d13'
James,Rose
Lyn 8h42'18"33d33'
James,Sandra Lee
Lyr 18h53'52"31d43'
James,Sr,Morgan David
Lyr 18h44'45"40d46'
James,Star Of R
Dra 16h25'29"65d46'
James,Steven Elden
Her 16h42'12"29d59'
James,Sydney & Eileen Ann Turner
Her 16h18'27"23d12'
James,Tina
Crt 11h7'35"-15d9'
James,William R
Cap 20h35'59"-11d11'
James,William Walter
Cma 6h49'40"-18d39'
James,Zenisha Lynnell
Equ 21h6'49"10d32'
Jameson,Adam
Cnv 12h24'55"36d2'
Jameson,Bradley Morgan
Lib 15h3' 29"-10d20'
Jameson,Bretton
Uma 11h0'46"58d36'

Jameson,Jan & Michele
Cam 5h49'41"58d29'
Jameson,Kim Erickson
Umi 15h17'23"71d19'
Jameson,Mary
And 0h58'23"39d55'
Jameson,Patrick Gary
Aur 4h55'15"40d2'
Jameson-Burns,Michael
Dra 13h1'32"70d39'
Jamesson,Steven
Hya 8h44'38"-1d16'
Jametta's Light
Lyn 9h18'55"35d8'
Jamey-The Twinkle Of My Eye
Tau 5h51'1"23d28'
Jami
Com 12h27'53"21d21'
Jami
Umi 16h29'30"70d18'
Jami
Oph 17h36'1"11d50'
Jami Alexander
Uma 11h55'36"40d40'
Jami Catherine
Tri 2h15'36"32d42'
Jamie
Lyr 18h37'23"43d1'
Jamie
Tri 2h23'1"34d29'
Jamie
Uma 11h31'27"32d37'
Jamie
Ori 5h34'1'0d25'
Jamie
Per 2h8'23"57d12'
Jamie
Lyr 18h43'35"36d36'
Jamie
Vul 20h2'42"25d14'
Jamie
Cyg 19h32'42"39d39'
Jamie
Cnv 12h7'36"40d59'
Jamie
Cap 20h34'57"-10d16'
Jamie
Cas 0h47'45"66d8'
Jamie
Gem 6h54'27"17d33'
Jamie
And 1h33'17"41d10'
Jamie "The Oz" 1021
Tri 2h15'59"33d28'
Jamie & Kim
Cyg 20h5'54"40d45'
Jamie & Marianne's Star
Del 20h24'53"8d13'
Jamie Ann
Per 4h3'21"36d27'
Jamie Annette
Cnc 8h48'30"30d33'
Jamie Christopher
Her 16h18'25"48d28'
Jamie Dawn
And 0h11'32"37d33'
Jamie Dawn
Aur 6h17'28"32d33'
Jamie Dean
Umi 13h30'18"70d49'
Jamie Grayce
Aur 5h0'0"41d36'
Jamie Lee
Mon 6h55'53"0d59'
Jamie Lee
Ser 15h56'35"20d27'
Jamie Loves Jose
Peg 23h43'56"11d29'
Jamie LSB
Lyr 19h22'16"31d12'
Jamie Lynn
Cma 6h53'36"-19d31'
Jamie Lynn
Aql 19h54'0"12d32'

Jamie Mae
Cas 23h18'30"62d12'
Jamie Robin S
Cyg 20h18'57"39d13'
Jamie Robinette
Aur 5h14'1"40d34'
Jamie Star,The
And 2h29'49"42d27'
Jamie's & Jouni's October Star
Lyr 18h43'28"38d15'AB
Jamie's Love Forever
Del 20h54'20"8d50'
Jamie's Lucky 7
Vul 19h48'22"26d11'
Jamieson III,William Brian
Dra 14h55'0"59d38'
Jamieson,Camilla Anne
Peg 21h43'17"27d30'
Jamieson,Dorothy A
Cyg 19h30'57"31d18'
Jamieson,Lawrence Gerard
Cet 1h31'0"-0d39'
Jamieson,Rebecca
Lyr 18h29'54"46d4'
Jamieson,Terry L
Sgr 19h9'35"-24d30'
Jamieson,Wendy
Lyr 18h38'18"33d7'
Jamieson,William & Mary (Warner)
Boo 14h36'48"18d48'
Jamis
Aur 5h1'17"41d56'
Jamison,James Brian
Oph 17h1'0"10d7'
Jamison,Judith
Cyg 21h4'0"35d38'
Jamison,Little Grandma Susan C
Uma 8h56'58"68d0'
Jamison,Ronald Michael
Aur 4h48'14"40d44'
Jamison,Tim
Lac 22h40'0"54d52'
Jammerman,Jill
Cas 1h25'1"50d31'
Jammet,Marjorie
Sgr 20h17'56"-36d4'
Jamnik,Roni
Cam 6h1'59"82d41'
Jamoca Swirl
Ori 5h51'55"15d15'
Jampac
Nor 16h18'27"52d18'
Jamplis,Robert Warren
Ori 5h59'12"-1d45'
Jamri,Yasmeen Abdul Hameed Ali Al
And 23h40'12"38d44'
Jamryl
Ori 5h53'17"21d4'
Jamstar
Ori 6h14'59"0d22'
Jan
Cyg 19h58'49"31d38'
Jan
Cet 0h28'54"-8d52'
Jan
Lyn 9h1'15"40d35'
Jan
Lyn 8h10'57"38d39'
Jan & Attmore-Oct 2, 1981
Lyn 7h52'32"40d46'
Jan & Kerstin
Lyn 8h2'1"43d55'
Jan & Lou's 10th Anniversary Star
Cyg 19h32'58"38d56'
Jan & Silke's Glücksstern
Cas 0h14'12"60d41'
Jan & Steve's July Sparkler
Uma 11h56'17"57d59'
Jan & Walt
Vul 19h48'42"28d41'

Jan Ellen
Cet 2h36'48"0d29'
Jan Jan
And 23h17'46"49d28'
Jan Jans
Cas 1h41'1"75d11'
Jan of Caerphilly `68
Uma 11h25'42"42d32'
Jan R-I Will Always Love You
Gru 22h8'58"-54d54'
Jan's ED-VENTURE
Uma 10h48'12"47d45'
Jan's Friendship Star
And 0h4'24"35d4'
Jan's Star
Cet 3h8'1"5d12'
Jan's Star
Aur 5h51'29"31d10'
Jan-Doris
Crb 16h12'56"26d14'
Jan-Erik
Cam 3h27'19"63d27'
Jan-Lynn's Eternal Star
Cet 1h3'38"0d50'
Jan-n-Jim
Crb 15h57'32"29d5'
Jana
Lyr 18h53'32"42d56'
Jana
Crt 11h16'34"-16d52'
Jana
Vul 19h18'0"23d8'
Jana
Del 20h54'55"7d16'
Jana & Mark
Lyr 19h19'1"41d46'
Jana Belinda
Vul 19h21'40"26d42'
Jana Loves Kevin
Mon 7h14'37"-5d22'
Jana Marie's Star
And 23h1'26"51d23'
Jana Rose
Mon 7h47'18"-5d13'
Jana's Star
Uma 8h37'18"49d15'
Janaboo 13
Uma 10h25'41"70d53'
Janal
And 1h47'0"38d15'
Janas,Adeline & Adam
Dra 15h53'17"60d59'
Janavice,Justin Dean
Tau 4h6'19"22d41'
Janaway,Timothy
Per 4h7'1"51d12'
Jandorf,Karen KJ
Lyn 8h5'56"51d49'
Jandt (John & Tracie)
Cyg 20h20'45"38d43'
Jandziol,Tereza
Cyg 20h58'37"53d29'
Jane
And 0h34'0"31d15'
Jane
Ari 2h53'50"30d6'
Jane
Cyg 20h22'58"39d26'
Jane
Lyn 8h6'20"41d38'
Jane
Peg 22h21'45"18d53'
Jane
And 23h0'0"47d50'
Jane & Dallas
Lyn 8h44'22"35d19'
Jane & Danny Forever as One
Cyg 20h59'50"40d45'
Jane & Hansi
Cyg 21h2'27"31d22'
Jane & Karl
Cyg 19h21'44"28d12'
Jane & Manos (Eternity)
Cyg 20h58'25"30d31'
Jane & Michael
Gru 22h48'26"54d37'
Jane & Mike
Lyr 18h20'33"42d9'
Jane & Roman's Dreams
Aql 19h11'50"19d11'
Jane 56
Eri 3h49'46"-1d46'
Jane Alice
And 2h6'40"41d57'
Jane Deborah Liz Caroline
Judy Jane
Uma 8h35'1"57d49'
Jane Elizabeth
Aql 19h31'26"10d22'
Jane Elizabeth
Cas 0h39'1"64d56'
Jane H
Peg 0h9'33"13d28'
Jane Louise
Cyg 19h27'37"35d31'
Jane Mara
Ori 6h4'29"8d27'
Jane Mara
Lyr 18h47'32"34d42'
Jane Marie
And 23h34'1"48d12'
Jane Marie
And 1h37'15"38d32'
Jane Marie Francis
And 0h17'35"38d47'
Jane of Arnprior
Cam 3h55'14"57d52'
Jane Rachel
Lyr 18h38'24"38d58'
Jane XL
Cas 0h10'51"63d47'
Jane's Ba-Wana Star
Her 16h9'21"40d17'
Jane's Bright Smile
Cnc 7h56'0"11d26'
Jane's Family Star Of
Happiness
Lyr 19h18'19"33d58'
Jane's First Star
Cas 1h59'22"60d40'
Jane's Hope
Umi 15h8'13"68d17'
Jane's Star
Aql 20h0'27"1d24'
Jane's Star (Plain Jane)
Cnc 9h4'44"31d6'
Jane,Jane "The Butterbean"
Cyg 20h52'31"30d59'
Janecek,Lois E
Del 20h14'18"9d25'
Janecka,MD,Ivo P
Gem 6h57'0"20d17'
JANEIL
And 2h13'42"50d25'
Janel
Uma 8h41'40"47d9'
Janel Kay
Lac 22h25'0"50d34'
Janel Lynn
Vul 20h1'55"22d43'
Janell
Vul 20h25'0"28d16'
Janelle
Mon 7h6'0"-1d33'
Janelle
Lyr 18h40'0"29d51'
Janelle
And 2h10'24"39d32'
Janelle
Lyr 18h48'0"33d46'
Janelle,Tiffany
Cas 23h2'17"58d59'
Janeltis Karaine
And 0h56'46"34d18'
Janer
Mon 7h4'21"-6d33'
Janer Ba-ba-94
Mon 7h2'12"4d51'
Janert,Betty Ann
Mon 7h58'20"-6d38'
Janes,Colin
Peg 22h58'44"10d13'
Janes,Janina Z Piorkowska
And 23h15'54"48d60'
Janes,Melinda
Aql 19h30'33"1d19'
Janes,Michael D
Aur 5h26'31"38d39'
Janes,Samantha
Eri 3h22'33"-4d57'
Janesville Gazette
Cyg 20h17'31"38d35'
Janet
Tri 2h34'0"31d8'
Janet
Cas 23h13'54"63d2'
Janet
Crb 15h32m24"28d12"
Janet
Cas 0h39'42"65d15'
Janet
Cyg 19h20'10"30d10'
Janet
Cet 0h55'25"1d42'
Janet "CACA"
Com 12h51'19"30d59'
Janet & David
Crb 16h0'29"31d40'
Janet & Marie
Cyg 21h32'33"38d41'
Janet & Mike's Magic Star
Mon 6h54'10"11d15'
Janet & Ron's White
Illumination
Cyg 19h59'36"30d11'
Janet & Steve's Star
Cyg 19h39'50"30d57'
Janet 812
Cas 0h57'27"67d52'
Janet Anne
And 23h44'41"45d38'
Janet Gayle
Uma 11h46'29"55d57'
Janet Heather"Unto Eternity"
Cyg 19h31'0"38d28'
Janet Irene
Lyn 9h0'30"42d1'
Janet Kathrein
Cnc 8h27'51"8d1'
Janet Lea
And 2h24'22"50d18'
Janet Lynn
Eri 3h56'40"-0d24'
Janet Lynn
Cas 23h0'25"53d29'
Janet Marie
Cas 1h47'40"58d32'
Janet Marie
Lyn 7h23'51"44d21'
Janet Marie
Cas 1h29'43"65d25'
Janet Marie
Lyr 18h57'1"41d38'
Janet Ruth
Cet 2h18'23"-10d5'
Janet's "Star-Cloud"
And 0h7'44"40d29'
Janet's Star
Lyn 8h0'59"50d3'
Janet's Star
Del 21h5'51"18d60'
Janet's Star
Lyn 7h25'59"58d48'
Janetta
Ari 2h30'23"30d35'
Janette
Cas 2h7'41"65d30'
Janette's "Jem"
Cas 0h56'11"59d38'
Janevicius,Lori
Cas 23h14'1"60d11'
Janevicius,Raymond
Her 16h45'19"35d35'
Janey K's World
Lac 22h24'52"49d28'
Janey,Gerald Laurence
Uma 11h52'39"44d32'
Jang,Michael
Cas 1h4'40"60d29'
Jangula,Victor Thomas
Lib 15h30'27"-10d19'
Janhonen,David
Ori 5h48'1"11d10'
Jani
Lyr 19h26'41"38d30'
Jani "Mother of Love"
Cas 0h29'54"68d23'
Janiak,Jeffrey
Cyg 19h57'37"58d16'
Janica's Dream
And 1h10'1"39d31'
Janice
Cam 5h5'57"63d32'A
Janice
Cyg 21h5'46"40d15'
Janice
Leo 10h43'26"15d53'
Janice
Mon 6h45'20"10d53'
Janice
Com 12h28'10"21d8'
Janice
Ori 5h55'47"15d23'
Janice
Lyr 18h31'27"30d46'
Janice
And 23h35'35"40d48'
Janice
Cas 22h58'12"54d29'
Janice "Ubiquitous"
Lyr 18h41'30"40d8'
Janice & Berny,Our Golden
Light
Cas 1h6'1"63d13'
Janice & Spooky's Star
Peg 22h22'31"5d15'
Janice 66
Vul 20h15'21"25d40'
Janice Ann
And 23h39'0"46d52'
Janice Anne
Vul 20h39'46"23d50'
Janice ASD
Del 21h5'35"12d31'
Janice Brittany
CMA 6h53'38"-17^9'
Janice Elaine
Lyr 18h15'1"31d18'
Janice G
And 23h48'16"45d55'
Janice G
And 23h30'25"43d3'
Janice G
Cyg 19h55'43"38d24'
Janice Kay
Cas 23h41'39"60d57'
Janice Lew
Peg 22h4'23"3d45'
Janico M J
Lup 15h42'53"46d19'
Janice Noel
Vul 20h2'42"23d42'
Janice"The Challenger"
Lyn 6h55'23"60d17'
Janice's Luminous Love
Cma 6h53'37"-17d57'
Janice's Star
Aur 4h48'37"40d21'
Janice(July 26th 1934)
Cas 23h32'45"61d18'
Janice,Edda
Ori 5h2'27"12d40'
Janice,Lovely Spirit By The
Sea
Cas 1h35'40"60d28'
Janice-Queen of the Lobster
Nebula
Vul 19h58'9"26d2'B
Janicke,Mark
Per 2h54'45"40d49'
Janicola,Lewis Steven
Aur 6h23'14"31d40'
Janie
Uma 11h5'24"31d31'
Janie
And 23h29'54"48d12'
Janie
Tau 4h31'22"30d21'
Janie
Vul 19h47'23"22d48'
Janie
Peg 0h4'38"11d16'
Janie 143
Aqr 22h48'36"-1d15'
Janie's Fireball
Vul 20h15'1"23d1'
Janie-Jim
Crt 11h36'1"-10d23'
Janik,Jerzy
Aur 6h16'1"31d44'
Janik,Kathleen
Lyr 18h15'0"37d39'
Janik,Sonja
Boo 14h19'29"14d59'
Janin,Matthew
Her 16h54'49"23d41'
Janine
Cas 0h48'20"72d13'
Janine
Cas 22h58'44"56d20'
Janine
Lyn 8h33'1"37d58'
Janine & Christopher
Cyg 20h41'55"44d7'
Janine & Jerry "Star of Love"
Sge 19h59'56"20d2'
Janine (Star Of Australia)
Cru 12h44'39"-59d53'
Janine 10
Uma 9h48'54"46d29'
Janine,Callie
Uma 11h24'0"61d47'
Janis
Cma 6h27'41"-11d11'
Janis
Umi 13h40'48"73d22'
Janis
Umi 16h29'48"71d58'
Janis
Aql 19h55'0"10d42'
Janis
Ori 5h43'10"-0d41'
Janis & Mike
Cyg 20h51'12"30d33'
Janis' Star,Byron & Maria C
Boo 14h48'49"25d15'
Janis,John Paul
Per 1h41'26"50d33'
Janis,Pam R
And 1h8'23"38d29'
Janis-Wight
Ori 5h2'13"8d44'
Janisio
Cnv 12h14'12"37d59'
Janisse,Jill Haamen
And 1h56'13"38d2'
Janissen,Liane
Sgr 18h47'39"-22d19'
Janiszewski,Lois Jean
Tri 2h5'0"31d21'
Janiszewski,Michelle Corrinne
Cas 0h34'49"64d37'
Janiszewski,Walter
Aql 19h39'54"13d42'
Janito,Jodi Beth
And 1h49'31"38d17'
Janitschke,Ethel & Robert
Cyg 21h35'39"37d45'
Janiuk,"Autumn" Sandy
Mon 7h13'50"-10d29'
Janjigian,Edward
Ser 15h18'19"7d60'
Jank,Christopher Edward
Aur 7h23'0"40d3'
Jankamar
Crb 16h13'1"34d56'
Janke,Jay Wayne
Her 18h9'1"38d35'
Janke,Kay
Her 16h11'52"41d5'
Janke,Steven Michael
Her 18h1'58"30d47'
Janklow,Juliet Faye
Peg 0h9'27"20d13'
Janko,Michael Stephen
Aql 19h49'39"14d43'
Jankowski Spot,The
Cep 23h29'51"71d3'
Jankowski,Jessica Elise
Cas 23h40'0"61d19'
Jankowski,Manuela
Mon 7h53'43"-1d49'
Jankowski,Nicholas Grant
Hya 9h36'53"-10d27'
Jankura's 80th,Ethel K
Psc 0h9'46"-3d3'
Janlyn & David Soulmates
Forever
Lyr 19h21'12"38d41'
Janman,Denis W
Aur 5h19'29"47d46'
Janmey,Alia Marie
Cas 3h11'38"58d43'
Janmoore,The
Cam 4h59'53"69d39'
Jann's Star
And 0h58'1"39d41'
Janna L
Lyr 19h22'1"30d45'
Janna"Paragon of Love"
Cyg 21h3'22"39d26'
Jannaccio,Bruce &
Caltabiano,Mike
Oph 18h0'22"7d46'
Jannace,Louis J
Mon 7h45'55"-2d29'
Jannarellé,Marcello
Uma 13h0'30"61d35'
Jannazo,Jerry
Her 16h41'52"48d49'
Jannell Joy
Peg 22h5'25"24d7'
Jannet
Cas 0h12'42"59d37'
Janney,Alison
Lyr 18h23'12"38d23'
Janney,Bryan
Her 18h11'28"37d43'
Janni
Lib 15h11'13"-20d23'
Janni,Mary Anne
Cas 1h41'30"54d16'
Jannicelli,Dena
Lyn 8h4'57"38d24'
Jannick,Caroline
Uma 12h11'50"61d21'
,Jannik-,I,I-No 1
Cep 23h18'39"64d59'
Janniks Stjerne
Cep 23h0'18"68d36'
Jannke,Bevan Ann
Ori 5h57'50"17d47'
Jannke,Cyrus Paul
Mon 6h56'1"11d46'
Jannke,Olin Ernst
Aql 19h11'48"15d29'
Janny Lynn
Lmi 10h32'0"34d50'
Janoff,Allan
Uma 11h58'47"64d7'
Janoff,Debbie & Allan
Crb 15h53'0"37d33'
Janofsky,Michael
Cep 21h5'54"58d56'
Janosek,Joan M
Boo 14h28'58"22d46'
Janota
Aql 18h56'33"16d32'
Janousek,MD,James M
Aql 20h5'22"6d10'
Janov,Daniel
Cma 7h20'46"-15d17'
Janovec,Conner Matthew
Cyg 21h17'29"28d18'
Janovic,Ted
Her 16h27'22"47d30'
Janovsky,Julie
Cas 23h35'18"58d8'
Janowitch,Lawrence A
Lac 22h4'11"46d33'
Janowski,Andrew
Lyn 8h26'33"35d8'B
Janowski,Leah
Lyn 8h26'33"35d8'A
Janowski,Paul Andrew Vincent
Cnv 12h45'19"51d57'
Janquin Star,The
Lmi 10h32'33"28d26'
Jans,Sue
Crb 15h49'13"31d47'
Jans,Tom & Deborah
Del 20h37'60"6d14'
Jans,Virginia
And 23h37'18"49d46'
Jansen "Baby Snooz", Judy
Aur 5h26'41"31d9'
Jansen 4-0,Etty
Ori 5h42'54"11d8'
Jansen,"Odessa" Bert
Cam 5h3'16"58d45'
Jansen,Darron Lee
Boo 14h51'17"22d35'
Jansen,David
Lac 22h23'42"38d3'
Jansen,Ernst Günter
Uma 9h27'35"47d47'
Jansen,Jeroen
Per 3h3'0"41d16'
Jansen,Jörg
Cep 20h40'0"65d16'
Jansen,Michael Christopher
Aur 7h20'0"38d18'
Jansen,Michele Marie Theresa
Ori 5h43'0"11d6'
Jansen,Nicole C
Lyn 7h37'28"51d5'
Jansen,Patricia Ann
Lyr 19h2'20"40d9'
Jansen,Sonja und Oliver
Ori 4h41'57"10d1'
Jansen-Berkley Superlight
Cmi 8h8'49"3d47'
Jansma,Dean Pierce
Cam 13h1'14"81d47'
Janson,Eric J
Boo 14h21'50"31d5'
Janson,John
Her 18h3'13"28d50'
Jansse,Adriaan Tieleman
Cyg 20h55'56"31d5'
Janssen,Barbara Anne
Ori 5h52'27"8d57'
Janssen,Carl Arthur
Aqr 23h1'45"-5d20'
Janssen,David
Ari 3h22'48"30d12'
Janssen,Delia
And 1h41'25"38d29'
Janssen,Erich
Lac 22h0'37"51d28'
Janssen,Heyo
Lyr 19h18'31"31d35'
Janssen,Jacobus (Sjaak)
Franciscus
Ori 6h22'41"18d45'
Janssen,James G
Ori 5h56'54"16d1'
Janssen,Peter
Uma 13h58'57"48d54'
Janssens,Alicia Lynn
Vu1 9h58'1"23d7'
Janssens,Christina Louise
Cep 22h10'14"55d47'
Jansson From Sweden,
Kerstin
Uma 11h23'27"43d42'
Jansson,Frank
Cep 22h38'58"63d22'
Jantho,Deborah
Cas 0h40'1"65d52'
Jantzen,Ariel Elizabeth
Mon 7h41'36"-8d25'
January
Cas 1h54'1"58d41'
January 7,1995
Lyn 8h29'42"44d48'
Janus Clinic,The
Oph 18h22'1"6d54'
Janus,Banie
Cam 3h16'44"58d37'
Janus,Julie Anne
Cyg 20h33'44"42d14'
Januszewski,Tadeusz J
Lmi 9h52'22"34d20'
Jany
Per 3h2'33"38d42'
Jany
Boo 15h15'30"37d54'
Janzer,Dennis John
Dra 20h17'22"63d50'
Jan ßen,Sabine
Lyn 7h27'0"40d44'
Jaouen,Marc Isabelle
Com 13h13'0"28d38'
Japhet,Valérie
Tri 1h59'24"34d32'
Jaq
Cru 12h49'26"-57d56'
'Jaqueline-Jane
Cnc 8h28'1"7d3'
Jaques,Caroline Allison Marie
Cas 0h50'20"62d46'
Jaques,Glenda Jo
Peg 22h3'35"26d38'
Jaquier,Chantal
Aur 4h47'26"40d15'
Jaquith,Michael L
Her 16h6'48"50d47'
Jaquith,Patricia E
Aql 18h57'1"17d58'
Jaquith,Philip A
Aql 18h57'21"17d60'
JAR
Lyn 6h12'41"58d24'
JAR Forever
Uma 11h26'20"48d58'
Jaracz,Daniel L
Peg 21h57'46"28d34'
Jaramasbabolas
Ori 4h43'11"0d51'
Jaramillo,Elsa
Eri 3h57'60"-12d50'
Jaramillo,Robert Allen
Mon 7h0'55"-6d37'
Jaramillo,Steve
Aur 6h0'41"54d50'
Jarboe's Star,Grandma &
Grandpa
Aql 19h4'10"3d50'
Jardene
Aql 19h40'17"14d46'
Jardin,Yves
Per 3h3'37"40d29'
Jardine
Lyr 18h36'26"28d23'
Jardine,Tom
Boo 15h4'59"52d27'
Jarecki,Kevin & Diane
Sge 19h56'44"20d35'
Jared Lee
Cru 12h34'31"-60d31'
Jared Thomas
Aur 5h16'27"48d0'
Jared,Matthew
Ser 16h1'11"1d36'
Jareds One
Leo 10h25'36"10d47'
Jarena
Cyg 19h15'18"45d0'
Jarest-Prince,Carolyn Mae
Cam 4h53'13"60d14'
Jarett,Carly Blair
Aql 20h5'59"-1d45'
Jarka,Edward F
Dra 16h18'56"66d21'
Jarlet,"Boupy" Philippe
Cmi 7h28'49"5d1'
Jarmolinski,(Angel
Star),Wendy
Aql 19h58'0"7d60'
Jarmuschek,Karin
Lac 22h2'25"51d6'
Jarnagin,Sierra Colby
Mon 7h42'48"-1d42'
Jarnstedt,Bo
Her 16h5'35"47d39'
Jarosch,Martina
Lyn 8h11'14"43d24'
Jaroslawski,Keith
Dra 17h31'27"60d9'
Jaroslawski,Steven
Lyn 8h41'13"38d23'
Jarosz,Ashley
Lyn 7h33'15"35d21'
Jarosz,Kathleen Cookie
Mon 8h7'10"-6d20'
Jarosz,Michele
Cas 2h17'58"68d03'
Jarosz,Sandra Kay
And 2h20'32"40d51'
Jarot,Christopher Paul
Her 16h26'20"30d58'
Jarowski,Charles Ignatius
Dra 14h47'18"64d9'
Jarrand,Daniel
Cyg 19h47'0"30d37'
Jarratt
Peg 22h58'43"31d46'
Jarre Star,The
Lyr 19h17'37"40d57'
Jarreau,James Burke
Hya 8h11'49"5d59'
Jarreau,Jordan Elizabeth
Mon 7h12'33"-5d22'
Jarred-Westwick, Katherine
Cas 2h56'58"60d34'
Jarrell,Clint
Ser 15h10'21"21d18'
Jarrell,Lucille Lipp
Lyr 19h24'55"38d32'
Jarrett
Cnv 12h16'0"47d34'
Jarrett BF,Kim
Peg 23h45'49"28d11'
Jarrett,Franklin
Aur 7h0'17"37d27'
Jarrett,Ginny
Vul 19h45'20"20d18'
Jarrett,John Crow
Aur 6h23'54"30d18'
Jarrett,Kathryn S Vierra
Tau 3h39'33"28d17'
Jarrett,Nicholas
Per 3h3'59"41d51'
Jarrod's Star
Aur 4h50'0"40d52'
Jarry,Francine
Cyg 21h1'58"39d6'
Jarry,Franáois Premier
Lac 22h49'46"53d33'
Jarschke,Jr,Carl Martin
Hya 8h16'28"0d54'
Jarski,Ashley R
Uma 10h21'49"52d26'
Jarski,W Andrew
Uma 10h30'49"52d26'
Jarvi,Jennifer
Lyn 8h20'37"40d45'
Jarvid Excellent
Uma 10h52'21"54d32'

Jarvis,Andrew Robert Dra 16h38'27"69d26'
Jarvis,Chad Ryan Cmi 7h43'0"5d37'
Jarvis,Mary Lmi 10h43'49"26d50'
Jarvis,Matthew Ori 5h12'42"-6d29'
Jarvis,Michael A Ori 5h55'20"16d40'
Jarvis,Natalie Elizabeth Mon 6h36'16"-6d5'
Jarvis,Timothy L M Aql 19h57'57"14d31'
Jarvis,Victoria Lmi 10h38'44"26d20'
Jary,Patrick Aql 19h0'26"-8d2'
Jas Ser 15h15'37"21d17'
Jasbec,John Aur 5h53'53"29d54'
Jasch,Erin Julia-Riley And 1h31'19"39d56'
Jaschowez,Andreas Her 17h9'58"48d5'
Jasen Scott & Christopher Todd Eri 3h48'24"-6d27'
Jasen Star of Bright Dra 16h24'42"60d15'
Jasen,Elizabeth Anne Cas 23h32'54"62d36'
Jashar Aur 6h21'0"31d57'
Jasiak,Jean-Marc Cep 22h0'10"54d3'
Jasiewicz,Theodore M Vul 20h56'31"20d11'
Jasikowski,Eugeniusz Dra 16h45'43"73d59'
Jasionowicz,Jodi John Mon 7h39'51"-6d59'
Jasko,Ashley Victoria Crb 15h17'1"31d27'
Jaskson,Gary C Aur 6h45'1"37d19'
Jaskulske,Donald Dra 19h7'47"48d25'
Jasma Cyg 21h54'43"52d36'
Jasman,Jason Her 17h54'33"40d2'
Jasmer "Pumpkin",Gary Aur 5h6'47"42d50'
Jasmin Mon 6h38'36"-7d31'B
Jasmin Cap 20h6'21"-11d2'
Jasmina And 0h33'21"40d32'
Jasmine And 23h46'37"38d58'
Jasmine And 0h23'17"32d10'
Jasmine Lyn 7h14'26"56d42'
Jasmine Aur 7h17'32"39d23'
Jasmine Lyr 18h37'54"39d53'
Jasmine Umi 15h32'39"75d42'
Jasmine Boo 14h33'48"11d28'
Jasmine 25 Lyr 18h36'41"33d7'
Jasmine Linn Cas 2h11'15"59d34'
Jasmine Renee Mon 5h57'54"-5d50'
Jasnon Eri 3h49'52"-1d56'
Jasnosz,Diane Marci Cyg 19h28'32"30d28'
Jason Her 17h5'57"42d11'
Jason Dra 16h19'11"67d29'
Jason Cma 6h44'52"-13d57'
Jason Eri 3h2'57"-3d15'
Jason Crt 11h0'16"-18d59'
Jason Ori 5h36'15"-0d25'
Jason Aur 6h19'53"38d48'
Jason Uma 9h37'0"57d45'
Jason Boo 14h31'54"39d10'
Jason Aql 19h55'44"12d46'
Jason Her 16h12'11"21d14'
Jason Boo 15h2'55"52d33'
Jason Ser 15h33'34"19d4'
Jason Dra 20h27'54"74d57'
Jason & Cindy Now & Forever Sge 19h36'23"16d51'
Jason & Krista Umi 16h28'58"70d59'
Jason & Margarita Cyg 20h21'26"38d39'
Jason & Susan Lyn 7h47'14"50d50'
Jason B Her 17h24'33"46d35'
Jason Elliott Her 16h8'57"47d46'
Jason F S Lmi 11h4'12"31d14'
Jason Gilbert Her 16h58'17"33d23'
Jason K Aql 20h2'55"7d17'
Jason Lee Aql 19h24'52"8d32'
Jason Loves Stacey Del 20h18'42"10d28'
Jason Robert Her 17h31'28"31d12'
Jason Robert Aql 18h59'60"-4d51'
Jason Robert Her 16h56'42"25d41'
Jason Scott Psc 1h6'54"27d41'
Jason The Great Tri 1h52'15"28d58'
Jason Tsvi Her 18h3'56"30d7'
Jason's Christmas Star Boo 13h36'12"17d56'
Jason's Destiny Aur 6h26'0"31d40'
Jason's Enterprise Ser 18h16'36"-5d30'
Jason's Fire Cet 3h17'58"1d3'
Jason's Journey Lac 22h40'20"51d33'
Jason's King Soul Aqr 20h37'1"0d30'
Jason's Light Cnc 7h55'47"10d45'
Jason's Miracle Per 1h59'0"47d38'
Jason's Smile Lmi 10h58'12"26d19'
Jason's Star Per 2h30'33"56d53'
Jason's Wish Tri 2h35'1"33d47'
Jason's Zootin' Star Aql 19h0'0"13d42'
Jason-n-Kristi 4Ever 7-6-94 Sge 20h0'40"18d11'
Jasonium & Kerbearium, Forever Ori 6h1'11"1d47'
Jaspan,"Little Lady" Myrna Jane Leo 11h47'24"25d25'
Jasper Aql 20h10'26"13d8'
Jasper Boo 13h59'1"8d9'
Jasper Lyn 7h57'58"33d37'
Jasper Samuel Dra 19h37'0"60d46'
Jasper,Baby Aur 7h21'18"35d46'
Jasper,Barbara Jean Uma 10h54'20"41d9'
Jasper,Charles Lee Lac 22h18'47"46d12'
Jasper,Dorrena V And 23h19'1"40d46'
Jasper,JoAn Marie Aqr 21h43'44"0d36'
Jasper,William Marcus Dra 16h59'30"67d17'
Jaspers,Roberta Crb 15h21'21"32d25'
JASPN Peg 23h26'43"28d57'
Jasra Cas 0h24'0"61d42'
Jastrow,Kary E Mon 7h0'58"4d17'
JASYPOES Tri 2h4'32"30d59'
Jaszmine Peg 22h7'44"24d38'
Jaubert,Nadine Per 3h59'44"38d9'
Jaume Ministral Ori 5h58'46"19d43'
Jaume MS Uma 10h54'42"40d40'
Jaume Rigau i Vilata Cmi 7h20'23"4d3'
Jauregui Oph 17h34'51"-24d35'
Jausten Bad Bear Umi 16h29'33"70d52'
JAVA Morelle Lyr 18h17'20"31d51'
Javate-My "Shining Knight",Alan Lac 22h55'35"51d7'
Javelle,Joyce And 0h40'56"38d36'
Javerliat,Marie Vul 19h18'43"27d16'
Javier Her 17h42'57"31d9'B
Javits,Jr,Eric M Per 2h50'18"32d30'
Javoric,Anthony V Cet 0h56'57"2d4'
Javorski,Jiri Peg 23h35'0"11d38'
JAW Cam 4h32'39"60d50'
Jawbeer und Puschel Uma 9h58'19"50d60'
Jawereksmjckbbrtyptla, Emmanuel Her 17h13'56"48d56'
Jawitz,Beverly And 2h1'43"38d17'
Jawork,Lisa Lib 15h29'43"-8d54'
Jaworowski,Edward Dra 17h39'32"60d31'
Jaworski,Malgosia & Robert Cyg 21h17'11"38d32'
Jaworski,Patricia Marie Piccione Mon 6h20'45"-6d36'
Jax Cyg 20h20'14"38d24'
Jax 'n' Rich Dra 16h25'24"60d40'
Jaxon Del 5 Delroro Salvador Lumbrias Cep 21h39'37"58d57'
Jaxon's Pass Leo 9h52'1"31d18'
Jaxx,Jill And 2h22'1"42d29'
Jay Gem 6h1'37"26d44'
Jay Tri 1h52'37"28d2'
Jay Mon 6h55'17"-6d3'
Jay Ori 5h47'18"20d56'
Jay Aql 19h53'39"13d10'
Jay Ori 6h7'45"9d12'
Jay & Angela's Star Aql 20h2'12"4d22'
Jay & Barb to"Win, Place & Show" Cet 2h39'24"1d5'
Jay & Bea Aur 5h53'21"31d37'
Jay & Carolann Sge 20h0'35"18d13'
Jay & Erin Uma 9h56'50"42d13'
Jay & Jackie's Star Hya 9h2'21"-6d29'
Jay & Julie Cyg 19h28'1"30d58'
Jay & Papa's Star Ser 15h9'16"22d8'
Jay Allen Aur 6h30'0"38d53'
Jay Always the Shining Light within our Lives Cyg 21h12'32"34d7'
Jay André Ori 6h2'0"20d17'
Jay Bird Tau 4h45'56"16d45'
Jay David Cmi 7h26'32"0d7'
Jay Dee & Jolie Uma 11h49'40"44d48'
Jay Kee & Cheri's Star of Love Del 20h17'19"14d29'
Jay Man Who Walks Among The Stars Ori 5h56'46"13d2'
Jay Robyn Hya 8h54'50"4d35'
Jay Star Of My Life I Love You Kris Uma 11h13'30"30d9'
Jay Warren Forever Her 18h32'58"24d56'
Jay's "Daddy" Nova Cep 22h9'0"67d35'
Jay's Future Ori 6h7'20"1d37'
Jay's Melody of Light Tau 5h29'37"20d3'
Jay's Reflection Cep 22h15'44"61d52'
Jay's Skinny Dipper Ser 18h16'27"-7d51'
Jay's Spirit In The Sky Hya 8h52'14"-7d7'
Jay's Star Cas 0h36'39"60d55'
Jay's Star Munyang Tau 4h28'53"4d15'
Jay's Twinkler Uma 10h25'43"40d13'
Jay,Arthur Jacobowitz Aur 6h25'20"33d51'
Jay,Dr Erwin Her 17h29'45"42d59'
Jay,Dr Sian E Uma 9h20'54"56d12'
Jay-Hunter Aql 18h41'45"1d42'
Jayaraman,Balachandar Uma 9h13'23"56d50'
Jayavarthanavelu, Rajyalakshmi Tau 5h56'19"24d4'
JayBear Her 17h16'13"42d28'
Jaybird Per 3h1'46"56d18'
Jaycie & Jeff Com 12h54'18"24d27'
Jaycox,Donald Patrick Cnc 8h56'53"18d1'
Jaycox,Jack McCarthy Ser 17h30'41"-13d41'
Jaye,Rochelle Mon 6h21'51"7d55'
Jaylan Peg 22h24'58"30d2'
Jayleigh Lib 15h13'0"-20d26'
Jaylou's Maximum Maggie May Cma 7h17'34"-16d43'
Jayme Pooh Cep 23h49'57"75d15'A
Jaymi Lyr 18h50'60"42d8'
Jayna * True Love Endures All Forever Cyg 21h10'54"35d44'
Jayne Cam 3h27'23"61d15'
Jayne Cas 1h37'1"73d12'
Jayne Lyr 18h36'49"31d42'
Jayne Cas 0h38'27"58d32'
Jayne & Larry Cma 7h19'40"-16d38'
Jayne & Steven Sge 20h2'34"20d11'
Jayne Elizabeth Cas 1h1'17"60d31'
Jayne Victoria Lyr 18h58'46"41d10'
Jayne's Star Lyr 18h23'0"38d17'
Jayne, Emma, Laura Lyr 18h16'27"32d56'
Jayne-n-Steve Lyn 7h49'28"37d1'
Jaynee Mon 8h3'3"-8d22'
Jaynes,Max High Oph 17h1'12"10d58'
Jaynes,Pamela L Cet 1h25'26"-1d33'
Jayo Lyn 9h5'58"35d11'
Jayroe,Brandon Oph 16h59'59"10d57'
JAZ-Z-AG Cam 3h50'40"60d34'
Jazzdare Boo 14h56'58"18d5'
Jazzmin Cyg 19h22'12"28d6'
Jazzy Lac 22h15'0"50d19'
Jazzy's Star Uma 13h31'59"53d43'
JB Ori 5h56'12"15d2'
JB Aur 6h29'0"31d55'
JB Baby Lac 22h15'28"48d52'B
JB's Hope Cam 5h34'1"61d12'
JB1 Cmi 7h34'1"0d37'
JBJ Star,The Uma 9h52'14"57d36'
JBRJA Tansey Star,The Cam 3h22'39"61d50'
JC Dra 17h7'56"61d44'
JC-The Inca Princess Leo 9h57'48"18d26'
JDB Jr Dra 19h6'26"68d29'
JDG & PB Forever Dra 19h5'39"54d51'
JDL - Hand Surgeon Cep 0h11'55"80d18'
Je T'aime Cyg 21h1'0"30d30'
Je t'aime aussi Ori 5h25'52"-0d13'
Je t'aime Marion Cyg 19h55'49"58d15'
Je Taime Mon Choisi Oph 16h54'23"11d52'
Je' T'aime Cathie Lynne Lib 15h43'40"-22d23'
Je'Rod Cherry Dra 19h28'0"68d6'
Jeal,Nicola Cnv 13h13'0"32d3'
Jeambon,Gerard Andre Cnv 12h42'47"50d17'
Jeambon,Laury Ford Cam 4h46'30"68d45'
Jean Cas 0h58'28"58d32'
Jean Dra 19h43'32"80d20'
Jean Aur 4h50'15"40d35'
Jean Vir 11h44'39"2d9'
Jean Cet 2h58'0"4d4'
Jean "Sissy" Cyg 21h54'1"40d59'
Jean & David Cyg 21h28'50"38d49'
Jean & Edgar Cep 21h57'0"58d43'
Jean & Roy Lyr 19h26'19"38d39'
Jean 317 Psc 23h37'1"1d42'
Jean Alexandria Per 2h56'11"37d57'
Jean Alice Lyn 8h21'12"49d33'
Jean Andre Aur 4h35'47"30d58'
Jean Claude et Sabine Per 3h32'13"39d38'
Jean Delia Aqr 21h2'1"0d42'
Jean Elizabeth Cas 0h44'17"60d4'
Jean Ellen Lyr 19h17'37"41d35'
Jean et Angela Peg 22h28'42"31d25'
Jean et Renée Cas 1h55'24"61d15'
Jean Franáois et Sandrine Peg 21h54'54"21d21'
Jean Louise Cas 0h47'1"65d9'
Jean Luc Per 1h51'54"50d11'
Jean Marie Ori 5h57'58"8d11'
Jean Marie Lyn 7h52'12"36d10'
Jean Marie Cmi 7h40'38"8d55'
Jean Marie C Equ 21h19'31"7d53'
Jean Marie Elizabeth Lyn 9h11'21"45d9'
Jean Michelle And 23h34'17"47d46'
Jean Noël And 2h6'53"40d29'
Jean The Apache Tear Boo 13h35'24"19d11'
Jean's Delight Eri 4h4'41"-8d17'
Jean's Jubilant Stardancer Hya 9h16'25"5d34'
Jean's Paradise Per 4h6'22"37d25'
Jean's Star Uma 10h23'14"60d51'
Jean,Floyd H Her 17h9'18"45d25'
Jean,Gloria And 2h31'0"38d48'
Jean,Leslie Ori 5h56'31"5d23'
Jean-Baptiste Her 18h23'19"12d19'
Jean-Baptiste,Germano Cnv 12h52'33"40d31'
Jean-Benoât Her 17h11'50"20d49'
Jean-Charles,Betsie And 1h21'21"39d20'
Jean-Christophe Uma 13h22'20"60d35'
Jean-Luc Et Fabienne Vul 19h21'20"23d45'
Jean-Marc Her 17h31'25"20d56'
Jean-Marc Diana Cyg 20h28'31"30d42'
Jean-Marie And 23h38'35"47d38'
Jean-Maxime Cam 6h38'40"67d57'
Jean-Paola Umi 14h59'32"69d38'
Jean-Philippe Une Voix Cnv 12h26'39"40d20'
Jean-Pierre Cnc 8h34'42"31d59'
Jean-Pierre Cyg 20h30'43"30d23'
Jean-Pierre et Francine Margot Per 4h0'26"36d18'
Jean-Pierre Leuis Her 17h20'1"41d0'
Jean-Piérre Aur 6h5'1"35d60'
Jeanalis,Frankus Dra 16h48'1"62d37'
Jeanbug Cyg 21h34'36"40d12'
Jeandot,Martine Dra 11h56'27"71d0'
Jeanelle Umi 16h27'37"80d20'
Jeanette Cas 23h32'28"58d18'
Jeanette Dra 9h38'51"81d6'
Jeanette And 2h31'1"42d38'
Jeanette Peg 22h43'17"21d23'
Jeanette And 23h3'52"35d42'
Jeanette "Tabatha" And 2h22'22"42d3'
Jeanette DeLuca Shines Com 12h28'0"21d19'
Jeanette Lee Lyn 7h7'48"44d12'
Jeanette of Eternal Goodness Crb 15h42'41"29d36'
Jeanette Rachelle Vir 13h8'29"-17d54'
Jeanette's Radiant Light Lyn 8h10'58"43d20'
Jeanette's Stella De Amor Del 20h54'35"8d53'
Jeanette,James W Lyn 8h22'23"57d35'
Jeanguenat,Cheryl Ari 3h1'42"23d6'
Jeanie Lyr 18h53'49"30d31'
Jeanie Marie And 1h33'51"38d57'
Jeanie Weenie And 1h31'0"37d48'
Jeanie's Dreams Lyr 18h39'29"32d10'
Jeanine Lyn 8h51'51"34d47'
Jeanine Vul 21h2'0"27d55'
Jeanine Cam 3h57'0"53d0'
Jeanine "My Eternal Love" Dra 15h6'31"63d51'
Jeanine & Gene Crb 15h24'33"31d51'
Jeanine Collette Lyn 7h50'34"40d53'
Jeanine Marie And 1h45'30"38d47'
Jeanine Marie & Vasili Cyg 19h56'20"41d11'
Jeaninne Lac 22h45'60"56d30'
Jeann Amour Peg 23h6'18"20d10'
Jeanna & Jessica Sisters For Enternity Cam 3h38'34"59d48'B
Jeanna P And 0h48'16"36d46'
Jeannathan Aql 18h44'18"7d40'
Jeanne Sgr 18h51'52"-22d1'
Jeanne Eri 4h44'24"-5d14'
Jeanne Lyr 18h40'13"45d18'
Jeanne Lyr 18h55'35"40d55'
Jeanne Lyr 18h38'1"31d44'
Jeanne & Michael Cyg 20h30'35"50d11'
Jeanne & Tom Aur 5h3'10"40d22'
Jeanne & Walter's 50th Anniversary Crb 16h18'50"30d49'
Jeanne & Whit's Angel Star And 23h25'14"46d55'
Jeanne - Sophie Cas 0h26'31"61d16'
Jeanne Alice Cnc 8h55'59"32d15'
Jeanne Ann And 23h24'38"43d48'
Jeanne au Sourire Joyeux Crb 15h27'14"31d57'
Jeanne B And 23h33'1"45d23'
Jeanne Faye Peg 21h39'36"23d21'
Jeanne I'll Look Up To You And 23h40'0"45d38'
Jeanne Kathryn Peg 21h32'21"20d29'
Jeanne Louise Cyg 20h20'54"37d58'
Jeanne M And 1h21'55"40d19'
Jeanne Marie Lyr 19h21'16"30d15'
Jeanne Marie Ori 6h7'27"-2d56'
Jeanne Z Umi 15h11'39"68d31'
Jeanne's Christmas Star '94 Uma 8h57'0"49d9'
Jeanne's Guiding Light Lyn 8h9'11"38d19'
Jeanne's Jewel De Noel Hya 9h16'52"5d6'
Jeanne's Wish Del 20h17'51"10d3'
Jeanne-Claire-Celina Lyn 8h32'45"41d30'
Jeanne-Todd Tri 2h6'0"32d49'
Jeanneret,Paul Richard Cet 3h4'30"1d4'
Jeannes' Journey And 23h33'14"46d30'
Jeannette And 23h49'11"44d33'
Jeannette Et Jean- Philippe Cam 4h39'42"58d48'
Jeannette P Vul 19h57'41"25d30'
Jeannette,Judith K Lyn 8h23'11"58d18'
Jeannick Peg 22h30'15"30d29'
Jeannie Eri 4h9'49"-10d10'
Jeannie Cet 1h37'1"0d49'
Jeannie Marie Cas 0h48'1"65d26'
Jeannie's Twinkle Cas 2h33'59"60d3'
Jeannine Cas 1h40'1"60d41'
Jeannine Lib 14h20'1"-23d53'
Jeannine"T"Star Com 12h11'60"21d47'
Jeannine's Future Crt 11h26'39"-17d47'
Jeanron Eri 4h7'1"-8d39'
Jeantheau,Gabriel G Psc 23h6'49"1d47'
Jeather Eri 3h11'28"-5d1'
Jeb & Rom Love Star Lyn 8h17'38"50d47'
Jecevicus,Ashley Anne And 1h54'26"39d58'
Jed"The Rock Star" Lac 22h20'29"50d18'
Jedar,Maurice Lyr 19h22'0"30d11'
Jedon Aur 4h47'18"40d48'
Jedrzejak,Gail Cet 2h38'15"1d42'
Jeep's Jarman Boo 13h37'1"22d30'
JEF Jennifer's Eternal Favorite And 23h29'29"44d30'
Jeff Lac 22h21'18"53d26'
Jeff Cmi 7h42'27"5d15'
Jeff Aur 6h24'56"31d56'

Jeff
Boo 14h47'24"30d28'
Jeff
Her 18h55'1"18d46'
Jeff & Brenda-25
Crb 16h17'26"28d11'
Jeff & Debbie
Lac 22h53'24"53d44'
Jeff & Gina
Cyg 21h6'49"31d46'
Jeff & Jill
Ori 5h26'1"1d28'
Jeff & Leslie Star
Cas 0h51'0"64d1'
Jeff & Lilia
Lyn 8h6'0"58d14'
Jeff & Misty
Eri 4h7'23"19d35'
Jeff & Shelly
Cmi 7h59'37"0d9'
Jeff Jamais en amour
Ori 6h5'56"7d57'
Jeff N' Leese
Uma 10h9'12"58d6'
Jeff's "GoobStar"
Per 3h30'45"38d14'
Jeff's 19th Hole
Her 17h55'14"18d49'
Jeff's Hole-In-One
Aql 19h53'14"15d34'
Jeff's Lucky Stars
Uma 10h42'42"40d27'
Jeff's Pagoda
Per 1h50'0"56d20'
Jeff's Passion
Aur 5h11'48"41d27'
Jeff's Place
Her 16h46'1"20d22'
Jeff's Place
Uma 8h38'35"50d27'
Jeff's Spirit
Ser 15h9'21"1d23'
Jeff's Star
Her 17h8'45"44d25'
Jeff's Star
Cmi 8h8'0"2d52'
Jeff's Star
Cet 3h1'1"1d34'
Jeff,My Shining Star, My Love Forever
Uma 9h13'30"61d41'
Jeffares,Brett Gentry
Peg 23h49'1"13d32'
Jeffcoat,Carl T
Sex 9h55'0"0d24'
Jeffelle
Aur 6h35'54"37d44'
Jefferds III,Joseph Crosby
Sex 10h49'0"3d34'
Jefferies,A K
Ori 6h8'27"1d9'
Jefferis,Marion Falconer
Lyn 7h11'0"59d21'
Jeffers,Harold Bridgwater
Cen 12h36'52"-51d59'
Jeffers,Jessica
Sge 20h3'26"18d49'
Jeffers,Victor
Aql 20h1'48"4d41'
Jefferson
Sco 15h50'6"-22d8'
Jefferson Star,The Martha
Lyr 18h31'45"32d59'
Jefferson's Place
Aur 5h32'41"38d32'
Jefferson,Bakari Kwame
Per 1h46'29"53d54'
Jefferson,Frances Christine
Eri 3h47'27"-4d29'
Jefferson,Gloria
Ori 5h56'44"17d30'
Jefferson,Leon W
Her 17h48'15"40d25'
Jefferson,Mark
Ori 5h56'18"18d18'
Jefferson,Pauline
Lyr 18h33'43"38d60'
Jefferson,Sr,Edward "Reds"
Aur 5h18'12"48d50'
Jefferson,Stella
Lyr 18h52'52"32d19'
Jeffery Allen's Nova
Aur 5h25'1"54d58'
Jeffery's Star
Tau 3h59'57"28d10'
Jeffery,Christine
Equ 21h9'18"11d4'
Jeffery,Colin & Noreen
Cyg 19h23'12"44d16'
Jeffery,Eileen
Lyr 18h34'44"28d37'
Jeffery,Jessica N
Cas 1h13'59"61d28'
Jeffie
Cep 23h3'1"70d31'
Jeffords,Joyce Virginia
Cyg 20h5'29"41d7'
Jeffrey
Per 3h23'41"40d37'
Jeffrey
Her 16h58'0"40d41'
Jeffrey
Ori 5h47'25"21d11'
Jeffrey
Peg 22h27'16"30d44'
Jeffrey
Vul 21h16'1"28d16'
Jeffrey
Aur 5h9'42"40d42'
Jeffrey
Aql 19h40'18"14d16'
Jeffrey
Aur 6h34'1"33d44'
Jeffrey
Aur 6h41'26"37d19'
Jeffrey & Dorianne
Lac 22h1'0"49d7'
Jeffrey & Sharon
Cep 3h24'50"78d8'
Jeffrey & Stacy
Dra 14h36'58"63d47'
Jeffrey 25
Aur 5h56'51"31d6'
Jeffrey Alan
Per 2h53'39"37d56'
Jeffrey Curtis
Dra 19h1'0"48d43'
Jeffrey Dean
Dra 20h27'16"70d20'
Jeffrey Kevin Psyche of Theseus Kimberly's
Ori 5h34'10"-0d16'
Jeffrey R
Vul 19h22'32"23d54'
Jeffrey S S
Dra 12h18'21"68d26'
Jeffrey Thomas
Aur 6h27'16"38d47'
Jeffrey with Love Always Chrissie xxx
Umi 15h27'41"68d17'
Jeffrey's Alan's Light
Cmi 7h7'40"5d53'
Jeffrey's Susan
Lyr 18h40'16"28d18'
Jeffrey,Jay Raleigh
Her 15h59'24"48d40'
Jeffrey,Linda Stacy
And 23h34'26"43d17'
Jeffrey,My Precious Son
Peg 23h36'21"18d42'
Jeffrey,Rose Marie
Lyr 18h48'43"31d41'
Jeffrey-Kathryn
Vul 19h59'1"23d31'
Jeffries,Andrew McCullough
Aql 19h31'43"13d44'
Jeffries,Andrew McCullough
Cyg 19h31'50"34d5'
Jeffries,Christopher E
Aur 6h14'15"37d49'
Jeffries,J J
Lac 22h1'44"37d45'
Jeffries,Janene (Pope)
Cas 0h37'17"58d46'
Jeffries,Jeremy
Per 3h47'38"52d39'
Jeffries,Mr Michael Sydney
Cep 21h35'18"70d51'
Jeffries,Rosemary
Cyg 19h49'30"37d40'
Jeffries,Timothy Burns
Hya 9h14'0"0d15'
JEH & SKG Light-O'- Love
Uma 10h24'42"67d38'
Jehn,Shelly
Ori 5h13'34"15d16'
Jehu
Uma 10h52'0"38d17'
Jeitler,Michael
Peg 23h30'19"18d39'
Jeitner,David Michael
Cyg 21h50'21d40"28'
JeJean I
And 0h8'58"46d55'
Jekel,Kevin Blake
Her 16h59'16"15d1'B
Jeker,Krista Marie
Cyg 21h10'1"38d44'
Jelenic,Andrea
Cas 0h13'1"58d1'
Jelf,Christopher
Uma 10h14'0"47d39'
Jelinek,Barbara Marie
Aur 5h43'0"54d50'
Jelinek,Carolyn Marie
Aur 5h43'50"54d58'
Jelinek,Kathryn Marie
Aur 5h33'42"54d37'
Jelinek,Kristen Alexandra
Lyr 18h42'1"31d46'
Jelinek,Michael Lester
Aur 5h33'52"54d39'
Jelinek,Otto Harrison
Aql 19h48'43"13d39'
Jelinek,Phillip Lester
Aur 5h33'32"54d55'
Jellema,Alexander Stuart
Lib 14h53'28"-0d23
Jellicoe (a Star), Diane
Cas 0h27'60"50d36'
Jelliffe,Raymond Charles
Boo 14h9'56"40d4'
Jelly Bean
Boo 13h41'49"14d10'
JellyBean
Cam 5h21'13"67d49'
Jellybean
Dra 19h20'52"58d39'
Jellybean
Sge 19h21'58"16d13'
Jellyfish-Michelle
Peg 22h45'15"32d36'
Jelmi,Sandra
Cas 0h21'13"72d41'
Jelnicky,Stephanie Grace
Cmi 7h26'18"7d47'
Jelus,DDS,Paul
Dra 17h4'35"63d11'
Jemalcan
Cnv 12h26'48"33d11'
Jemison,Sabrina Cole
Vir 13h0'0"-20d1'
Jemma 21
Cyg 21h54'32"52d30'
Jen
Cas 0h37'26"65d14'
Jen
Sge 18h56'18"19d14'
Jen
Mon 7h3'0"-0d28'
Jen
Mon 7h34'14"-0d27'
Jen & Dick Forever
Crb 15h58'16"30d42'
Jen & Don Forever
Uma 9h33'31"50d9'
Jen & Drex
Cyg 21h15'12"35d18'
Jen & Joe
Lyr 18h40'58"41d4'
Jen & Joe
Lyn 7h29'36"38d51'
Jen & Will
Cyg 19h45'60"38d24'
Jen 318
Ori 5h44'1"1d26'
jen oublierai jamais
Cma 6h12h44"-15d20'
Jen Star
Com 12h6'23"21d7'
Jen Star
Vir 12h55'39"-9d1'
Jen Sue Ram-z
Lyn 8h0'36"39d40'
Jen X
Mon 6h42'38"11d49'
Jen's Unicorn
Mon 6h42'49"2d15'
Jen-A-Bone 1775
Uma 9h35'50"48d56'
Jen-R-Stern
Cam 3h25'33"58d33'
Jena
And 0h15'25"39d48'
Jena's "Secret" Star
Del 20h20'39"10d39'
Jenae Noelle
Mon 8h0'29"-1d32'
Jenart
Sex 10h34'0"0d10'
Jenavik
Peg 22h51'40"24d10'
Jencius,Will Patrick
Eri 3h40'13"-5d9'
Jencks,Donald E
Oph 18h18'25"6d42'
Jencks,Robert A
Ori 6h15'14"1d16'
Jencks,William S
Per 3h2'25"38d4'
Jenco,Michael
Lac 22h53'23"54d17'
Jenda
Cam 8h19'38"81d44'
Jendi's
Mon 7h22'28"-0d44'
Jenean's Star
Lyn 9h19'0"33d49'
Jenelle
Peg 23h43'39"31d31'
Jenerator
Lyn 6h35'40"54d57'
Jeneva-Aurora
Lyn 6h19'47"59d46'
Jeneé
Crb 16h2'51"28d28'
Jeneé's Star
Peg 23h40'30"25d37'
Jenga Major
Peg 22h19'32"3d34'
Jeni
Psc 1h2'0"17d56'
Jeni
Cyg 20h4'1"30d9'
Jeni
Cas 23h28'41"60d24'
Jenian 13
Pho 23h43'6"-44d55'
Jenica
Boo 14h29'46"41d12'
Jenicek,Joan Marie
Lyn 8h35'59"45d4'
Jenifer Rose
Cyg 21h27'21"31d6'
Jenifer S021864 HB
And 22h56'13"50d23'
Jenik,Alison
Ori 4h58'51"4d30'
Jenindary Visgus
Dra 16h56'20"61d59'
JENJIM
Lyr 18h41'31"31d33'
Jenkel,Anne Edlo
Tri 2h5'31"33d7'
Jenkins II,Robert L
Ori 6h4'0"-0d29'
Jenkins, Roku-Dan, Shihan
Cas 1h12'36"75d58'
Jenkins,Alice Irene
And 2h24'31"49d31'
Jenkins,Always & Forever;Emily
Mon 6h22'49"1d26'
Jenkins,Andrew J
Cep 1h28'46"78d4'
Jenkins,B Ann
Lmi 10h8'19"41d2'
Jenkins,Brennan S
Crb 15h27'37"32d11'
Jenkins,Carol
Lyn 7h33'25"58d44'
Jenkins,Casey
Hya 8h32'1"-8d30'
Jenkins,Charles B
Cep 20h56'49"62d10'
Jenkins,Charles L
Ori 5h56'35"11d35'
Jenkins,Claire Elizabeth
Cyg 19h28'28"36d9'
Jenkins,Daniel Kenneth
Per 1h57'12"47d36'
Jenkins,Dawn S
Dra 17h59'25"58d15'
Jenkins,Dennis
CMI 7h15'1"5d32'
Jenkins,Earl J
Boo 14h7'28"31d34'
Jenkins,Elaine Kathleen
Peg 22h38'1"33d54'
Jenkins,Elizabeth Catherine
Tri 2h37'46"35d57'
Jenkins,Elizabeth K
Uma 11h56'24"53d50'
Jenkins,Eric Timothy
Cet 0h27'1"-8d56'
Jenkins,Frances Maureen
Tau 3h55'17"28d48'
Jenkins,Garth
Ori 5h56'29"19d17'
Jenkins,Gerald P
Mon 6h56'30"7d31'
Jenkins,James Michael
Aql 18h59'28"12d9'
Jenkins,Jeanine Marie
Mon 6h7'0"-4d36'
Jenkins,Jeanne Elizabeth Cross
Lyn 8h7'44"37d3'
Jenkins,Jeff
Per 2h58'57"38d45'
Jenkins,John L
Oph 17h31'20"-8d39'
Jenkins,Jr,Harold Francis
Lac 22h35'37"50d5'
Jenkins,Julie
Mon 6h57'42"-8d23'
Jenkins,Julie-Anna
Cas 0h46'16"68d16'
Jenkins,Karen Fawn
Eri 2h50'31"-18d30'
Jenkins,Karn
Mon 6h18'1"0d37'
Jenkins,Kent
Per 1h51'16"50d5'
Jenkins,Khanssaa
And 23h41'48"38d31'
Jenkins,Laura
Hya 8h13'57"-1d15'
Jenkins,Leslie H
Eri 4h56'6"-8d58'
Jenkins,Linda
Lac 22h51'0"38d24'
Jenkins,Mark W
Cap 21h34'56"-21d19'
Jenkins,Matthew
Peg 22h35'0"26d49'
Jenkins,Michael F
Ori 6h0'41"8d21'
Jenkins,Mrs Ceri Diane
Cyg 19h25'29"35d11'
Jenkins,Nathaiel L
Cep 23h33'20"71d13'
Jenkins,Parry
Ori 4h48'26"0d16'
Jenkins,Patti F
Peg 22h39'10"20d37'
Jenkins,Paul Kevin
Aur 7h3'26"38d37'
Jenkins,Phillip Kendall
Boo 14h38'52"43d45'
Jenkins,Shirley Cathleen
Cas 1h26'28"60d50'
Jenkins,T C A
Ori 5h56'1"16d39'
Jenkins,Thomas Quinn
Vir 14h2'56"-1d43'
Jenkinson,Abigail Rose
Cas 1h51'42"75d40'
Jenkinson,Kathryn Louise Russell
Sge 20h4'0"16d5'
Jenkis,Robert Elmar
Cet 2h43'25"2d59'
Jenks,David Alden
Her 16h18'38"8d32'
Jenks,Mark Story
Dra 12h57'60"68d23'
Jenks,Wickliffe
Ori 5h55'52"15d10'
Jenksin,Nancy
Cap 20h6'55"-18d47'
Jenn
Mon 6h26'1"-10d53'
Jenn
Lyr 18h47'43"40d17'
Jenn T
Lyn 8h36'19"41d37'
Jenn's Celestial Sky
Aql 18h43'15"10d7'
Jenn's Shining Star
Peg 23h23'39"31d39'
Jenna
Cas 0h16'16"65d6'
Jenna
Boo 14h28'31"8d38'
Jenna Lynn
Cas 1h31'40"58d38'
Jenna Lynn
And 2h14'22"41d46'
Jenna Michelle
Cas 23h44'8"64d1'C
Jenna Suzanne
Tau 4h38'38"15d8'
Jenna's Shining "Sweet 16" Star
And 23h45'57"38d49'
JennDav's Everlasting Love
Peg 22h0'0"5d28'
Jennen Schrum,The
Mon 6h43'53"6d1'
Jenner,Alexander
Aql 20h0'45"1d35'
Jenner,Stephen
Aql 20h0'15"1d3'
Jenner,Theodore
Cep 21h43'17"85d23'
Jennett,Jimmy
Dra 16h32'21"61d55'
Jennette
Cam 7h28'30"67d35'
Jenney Bird
Cas 1h45'0"75d55'
Jenni
Eri 2h53'57"-3d46'
Jenni
Cas 2h53'24"65d17'
Jenni & Matt
Sge 20h15'14"17d3'
Jenni & Tim's Eternal Love
Uma 10h21'13"72d32'
Jenni 99
Equ 21h21'11"12d44'
Jenni Star
Cep 21h51'18"85d6'
Jenni's Star
Sgr 19h26'40"-40d17'
Jenni,Mary Anne
Uma 11h2'0"48d44'
JenNicks
Aql 19h0'12"13d4'
Jennie
Cas 0h24'23"50d30'
Jennie
Cas 0h16'46"46d56'
Jennie
Lib 15h34'32"-28d20'
Jennie
Cyg 19h31'44"33d53'
Jennie
Cyg 20h11'0"40d55'
Jennie Lee
Peg 22h49'27"21d24'
Jennie Lynn
Cas 2h0'42"59d59'
Jennie Lynne
Del 20h25'42"20d20'
Jennie Marie
Peg 22h37'60"24d40'
Jennie Rose & Steven Vincent
And 0h5'34"31d6'AB
Jennie Twenty-one
And 0h38'33"31d31'
Jennie Victoria
Lyr 18h57'0"46d12'
Jennifer
Cam 4h15'17"61d2'
Jennifer
And 23h19'10"43d49'B
Jennifer
Lyn 7h32'37"50d40'
Jennifer
Cas 23h5'49"58d18'
Jennifer
Per 1h34'34"53d8'
Jennifer
Mon 6h29'0"1d45'
Jennifer
Boo 15h30'53"41d42'
Jennifer
Dra 11h25'17"68d3'
Jennifer
Lyn 8h23'15"43d4'
Jennifer
Cam 3h19'26"58d48'
Jennifer
Cas 0h26'55"67d41'
Jennifer
Cas 0h21'53"66d10'
Jennifer
Peg 23h25'46"33d52'
Jennifer
Lyn 7h35'10"45d19'
Jennifer
And 1h38'36"39d3'
Jennifer
Mon 6h25'26"8d40'
Jennifer
Ser 15h22'36"8d36'
Jennifer
Peg 21h55'0"22d44'
Jennifer
And 23h29'17"48d25'
Jennifer
And 23h35'0"43d33'
Jennifer
Peg 22h25'36"24d10'
Jennifer
Lyn 8h9'43"33d33'
Jennifer
And 1h8'32"38d48'
Jennifer
And 1h17'17"39d24'
Jennifer
Lyn 8h20'48"40d26'
Jennifer
Mon 6h47'0"11d11'
Jennifer "My Little Princess"
And 23h0'50"45d44'
Jennifer & Andrew
Cyg 20h20'34"38d35'
Jennifer & David
Uma 14h15'42"59d34'
Jennifer & Fernando
Uma 10h4'54"51d16'
Jennifer & Garry
Cyg 21h17'1"38d21'
Jennifer & Jason
Cyg 19h58'21"29d33'
Jennifer & John
Lac 22h46'38"56d49'
Jennifer & Lovio
Sge 19h53'11"16d42'
Jennifer & Mason Forever
Crb 16h9'54"34d14'
Jennifer & Mike
Crb 16h15'29"30d33'
Jennifer & Stephen
Ari 3h0'48"25d14'
Jennifer & Tim
Cam 3h55'38"78d56'
Jennifer & William
Cyg 20h23'1"40d39'
Jennifer (Precious Princess)
And 23h46'20"43d7'
Jennifer 021689XO
And 2h28'19"50d3'
Jennifer 032271
Cam 8h3'59"71d58'
Jennifer 5794
Lyn 9h23'58"41d35'
Jennifer Always Have, Always Will
Peg 23h47'41"27d19'
Jennifer B
Vul 20h14'46"23d48'
Jennifer Carole
Umi 13h44'20"86d3'
Jennifer d'Abo
Cas 0h24'14"60d5'
Jennifer Diane
Peg 22h5'41"5d60'
Jennifer Ellen
Crt 11h15'28"-18d58'
Jennifer Gail
And 22h56'32"51d6'
Jennifer H
Peg 22h41'56"10d14'
Jennifer H-(The Camper Star)
Lyn 7h32'0"39d23'
Jennifer Hope-My Eternal One
Ori 4h50'31"4d8'
Jennifer I
And 23h0'18"45d49'
Jennifer J P
Mon 6h41'20"-10d55'
Jennifer Kay
Lyn 6h32'24"60d13'
Jennifer Kelly
Peg 22h25'15"29d52'
Jennifer Lara
Peg 22h3'57"2d20'
Jennifer Lauren
Lyn 8h10'55"37d28'
Jennifer Lee
Aql 20h21'13"0d4'
Jennifer Louise
Del 20h18'26"16d30'
Jennifer Louise
Uma 12h2'24"31d17'
Jennifer Louise(Hugga)
And 1h1'12"39d1'
Jennifer Loves Scott
Tri 2h9'45"32d46'
Jennifer Lynn
Sge 19h39'1"16d29'
Jennifer Lynn
Aql 18h57'22"12d30'
Jennifer Lynn's Lucky Star
Cyg 21h17'23"28d34'
Jennifer Lynn's Star
And 0h55'1"38d35'
Jennifer Marie
And 0h58'40"38d35'
Jennifer Marie
Cam 10h57'47"82d23'
Jennifer Mechelle- My Strength
And 23h3'36"51d15'
Jennifer Michelle
And 2h7'16"38d42'
Jennifer Michelle
Mon 7h4'0"4d12'
Jennifer O
Vul 21h21'1"27d55'
Jennifer Paige
Peg 21h27'1"23d40'
Jennifer Paige (Snooky)
Cyg 19h19'51"45d7'
Jennifer Rebeca
Vul 20h15'0"28d16'
Jennifer Renae
Vul 20h0'59"22d53'
Jennifer Renee
Del 20h14'28"15d37'
Jennifer Rose
Eri 3h14'25"-1d55'
Jennifer Rose
Uma 8h7'13"68d20'
Jennifer Shinning Brightly
Oph 17h54'14"10d38'
Jennifer Starburst
Tri 2h25'48"31d45'
Jennifer Sue
Uma 8h56'0"71d11'
Jennifer Sue
Lmi 10h29'26"31d43'
Jennifer Susan
Cas 0h57'0"61d35'
Jennifer Suzanna
Aqr 21h3'30"-8d35'
Jennifer The Long Awainted Friend
Del 20h52'14"9d39'
Jennifer "Gentle Spirit"
Ori 5h59'0"16d25'
Jennifer ~ Douglas
Peg 22h16'1"34d26'
Jennifer's (Tuck)
Cyg 20h24'1"30d15'
Jennifer's Angel
Cas 0h28'0"63d36'
Jennifer's Dream
And 0h39'59"45d6'
Jennifer's Fire
Cas 2h51'16"61d36'
Jennifer's Joy
And 1h5'11"41d14'
Jennifer's Leap of Faith
Cas 0h6'1"54d57'
Jennifer's Star
Sct 18h49'32"-6d24'
Jennifer's Star
And 2h35'28"40d34'
Jennifer's Star
And 2h28'0"38d58'
Jennifer's Star
Peg 21h59'1"30d27'
Jennifer's Star
Lyn 8h6'20"35d47'
Jennifer's Star Of Hope
Cas 3h5'24"68d17'
Jennifer's Towishupon
Peg 21h33'27"20d12'
Jennifer(Kitty Kat)
Tri 1h47'44"26d9'
Jennifer,"Our Star"
Cas 3h25'32"75d35'

Jennifer,143 Always Mom
Cas 23h30'23"60d50'
Jennifer-John
Cam 4h59'28"65d34'
Jennifer-Our Bright Star
Com 12h49'11"21d18'
Jennifer-Ryan
Cnv 12h52'34"40d6'
Jennifer...Jim's Shining Star
Cyg 21h32'60"30d35'
Jennifer/Travis
Lyn 6h30'27"59d9'
Jennimar
Mon 7h0'33"-0d12'
Jennine
And 0h8'25"47d46'
Jennine
Cet 2h23'20"1d53'
Jennings Eternal
Eri 3h24'52"-5d10'
Jennings II,Charles Wallace
Oph 17h57'53"12d38'
Jennings II,Evan Daniel
Ori 5h12'16"-4d37'
Jennings III,Buddy James Gordon
Aql 20h2'40"-0d39'
Jennings' Star in Heaven,Lillian D
Crt 11h16'39"-21d15'
Jennings(Princess), Jessica Sue
And 1h39'49"40d10'
Jennings,Alpha
Dra 12h31'11"68d52'
Jennings,Amylea Zipporah
And 23h2'0"51d0'
Jennings,Barbara L
Peg 22h1'23"30d10'
Jennings,Brian Patrick
Cnv 12h26'0"48d54'
Jennings,C Eugene
Aql 20h2'20"11d54'
Jennings,Cara
Lyr 18h14'0"40d60'
Jennings,Casey
And 2h16'16"47d43'
Jennings,Colleen Catherine
Crb 15h51'30"38d56'
Jennings,David H
Her 17h32'53"20d25'
Jennings,Dee 1
Cnc 8h26'54"8d20'
Jennings,Edwin Francis
Dra 16h32'1"71d8'
Jennings,Gertrude "Dolly"
Aur 6h4'1"31d37'
Jennings,Gina
And 0h24'51"37d18'
Jennings,Grace Marlene Hughes
Cnc 8h28'15"7d48'
Jennings,Jackie
Uma 11h57'40"63d55'
Jennings,Jacob William
Aur 5h28'34"31d43'
Jennings,James Theodore
Ori 5h56'1"13d24'
Jennings,James Richard & Laura K
Ser 15h52'47"-0d26'
Jennings,Jill
Aur 5h15'12"40d26'
Jennings,JoAnn
Umi 16h51'0"75d33'
Jennings,Karen A
And 23h40'57"44d49'
Jennings,Katie
Aqr 22h33'23"-1d37'
Jennings,Kayleigh Ann & Shiela Robin
Aql 19h31'37"10d14'
Jennings,Louis Bryan
Umi 15h46'13"70d9'

Jennings,Matt
Boo 14h53'1"31d48'
Jennings,Mike
Aql 18h55'31"-0d39'
Jennings,Paulette Mae
Umi 16h49'25"78d57'
Jennings,Ralph W
Oph 17h10'27"-23d35'
Jennings,Rhonda
And 2h28'1"44d4'
Jennings,Robert & Lynnae
Cyg 21h33'15"41d34'
Jennings,Shannon
Eri 3h41'24"-14d44'
Jennings,Shona
Cas 22h56'34"57d13'
Jennings,Sr,Star,The Raymond D
Crt 11h19'22"-20d29'
Jennings,Susan Marie
Aql 19h25'17"8d30'
Jennings,Terrance Michael
Dra 11h7'18"73d13'
Jennings,Tina Ann
Mon 6h24'21"5d53'
Jennings,William Randall
Boo 14h48'56"35d30'
Jennings,Zackary Allan
Uma 11h29'0"40d58'
Jennison,Charlotte
Cas 0h25'38"62d25'
Jennison,Kimberly
Lyr 18h36'27"28d20'
Jenniess
Aql 20h2'26"7d36'
JennTina
Peg 22h10'38"3d7'
Jenny
Eri 3h49'26"-2d9'
Jenny
And 0h16'20"38d1'
Jenny
Peg 22h4'60"20d28'
Jenny
And 23h40'16"43d48'
Jenny
Crb 16h1'42"28d3'
Jenny
Cas 0h6'0"54d33'
Jenny
Lac 22h50'38"38d0'
Jenny
Peg 0h12'45"13d19'
Jenny & Evan Friends Forever
Vul 19h47'21"27d26'
Jenny & Jon
Ori 5h55'36"7d8'
Jenny & Mike's Love
Uma 12h13'59"56d43'
Jenny & Motty
Aql 19h28'41"-8d35'
Jenny & Rich (Waters)
Equ 20h58'37"9d42'
Jenny & Steve
Aur 5h29'53"31d3'
Jenny 12 21
Oph 18h0'28"13d34'
Jenny 18
Cyg 19h28'59"32d8'
Jenny 458
And 2h34'46"44d31'
Jenny C-29
Peg 21h19'26"21d40'
Jenny Eternal
And 0h13'25"47d41'
Jenny G
Uma 9h59'22"60d56'
Jenny K 95
Pho 0h31'57"-47d57'
Jenny Lee
Eri 2h52'21"-8d59'
Jenny Lynn-16
Peg 23h24'30"32d51'
Jenny Lynne
Cas 23h34'54"63d11'

Jenny Och Björns Stjärna
Uma 12h6'17"46d29'
Jenny Rebecca
Eri 3h22'49"-14d19'
Jenny's Angelica
Mon 6h57'13"10d3'
Jenny's Eyes
Peg 21h21'20"23d34'
Jenny's L'Etoile
Mon 7h17'58"-5d28'
Jenny's Light
Peg 0h5'58"14d2'
Jenny's Light
Sco 16h40'46"-28d54'
Jenny's Rehab
And 0h40'14"40d36'
Jenny's Star
Cas 0h23'24"70d40'
Jenny's Star
Vir 14h59'55"7d1'
Jenny's Star
Cas 2h13'49"68d16'
Jenny,Ann,Nickel
Vul 19h13'45"21d28'
Jenny-Light of My Life
Peg 23h16'0"32d33'
Jenny-Love Eternal
Cma 6h55'11"-19d49'
Jennyfer,My "Funny Face",My Love
And 2h34'29"49d3'
Jennyjump
Lac 22h1'0"47d24'
Jens,Rachel Katherine
Uma 10h14'12"55d57'
Jensean
Ori 5h53'24"12d30'
Jensean I
Ori 6h2'45"7d37'
Jensen 309048 Mother of Six,Yolanda
And 0h42'21"31d30'
Jensen Star,The
Lyn 6h46'15"60d26'
Jensen,Alexander Gulnar
Hya 8h41'28"2d27'
Jensen,Austin & Keely
Aql 18h44'0"8d21'
Jensen,Betty
And 1h23'1"39d37'
Jensen,Bjarna Askov
And 23h23'12"47d15'
Jensen,Brian Maslak Frisendahl
Aur 6h4'53"45d2'
Jensen,Carrie Ann
Uma 8h41'35"47d21'
Jensen,Carroll E
Cet 2h37'18"-1d57'
Jensen,Cassandra Lynne
Lyr 18h48'0"38d41'
Jensen,Charlotte
Crb 15h46'18"34d46'
Jensen,Charlotte
Cas 1h18'12"68d31'
Jensen,Chris
Aql 18h58'18"12d57'
Jensen,Christopher E
Crt 11h22'29"-17d38'
Jensen,Colleen
Lyn 7h3'58"44d54'
Jensen,Darrell Anthony
Cam 3h17'13"61d19'
Jensen,Deborah
Lyr 18h21'55"41d13'
Jensen,Dennis
Tri 1h54'46"27d9'
Jensen,Derek Thomas
Lac 22h12'33"48d47'
Jensen,Derek William
Her 16h12'46"50d25'
Jensen,Don
Umi 16h15'45"79d45'
Jensen,Donald
Uma 11h41'34"57d3'

Jensen,Elsa H
Aql 20h0'44"-6d36'
Jensen,Finn & Mary 50th Anniversary
Crb 15h29'21"30d47'
Jensen,James Trevor
Dra 19h21'1"56d20'
Jensen,Jane Lynn
Vir 13h32'24"-21d27'
Jensen,Janet Lynn
Cam 12h56'0"77d33'
Jensen,Jeri
Lac 22h4'57"46d22'
Jensen,Jorgen
Uma 10h13'20"42d11'
Jensen,Judy A
Lyr 18h35'39"28d25'
Jensen,Kate-Marie
And 1h57'26"37d39'
Jensen,Katherine Margaret
Aql 18h43'11"-1d17'
Jensen,Katie Marie
And 23h33'31"40d7'
Jensen,Kelsey Marie
Cap 21h13'30"-22d11'
Jensen,Kenneth H
Per 2h56'15"40d39'
Jensen,Kenneth Lynn
Eri 3h25'48"-13d9'
Jensen,Kim
Umi 15h9'24"71d56'
Jensen,Kristen
Cnc 8h50'11"31d6'
Jensen,Lars
Ori 5h55'50"15d1'
Jensen,LaVonne
Cam 6h10'51"68d16'
Jensen,Leonard M
Boo 14h36'32"40d55'
Jensen,Louisa
Lyn 6h38'40"58d12'
Jensen,Mette
Boo 15h24'1"51d10'
Jensen,Nancy Rhe
Cas 2h57'59"57d50'
Jensen,Pamela
Vir 15h7'60"7d19'
Jensen,Randall David
Per 2h56'12"34d52'
Jensen,Ray Swinyard
Cnv 13h5'55"39d0'B
Jensen,Rebecca Ann
Cas 0h55'0"60d46'
Jensen,Rebecca Ann
Cas 0h55'0"60d46'
Jensen,Robert O
Cet 3h7'0"3d53'
Jensen,Rogene
Boo 14h38'52"40d37'
Jensen,Sarah B
Vul 21h2'30"20d0'
Jensen,Shirley Joan
Cnv 13h5'55"39d0'A
Jensen,Steven James
Cma 6h54'55"-11d11'
Jensen,Susan & David
Cyg 19h54'33"38d32'
Jensen,Tanya Sonne
Cam 3h31'23"53d40'
Jensen,Tavs
Umi 14h49'41"68d31'
Jensen,Traci A
Aql 19h53'22"10d51'
Jensen-Our Special Star,Harlan Peter
Cet 0h56'16"-1d15'
Jenson III,Gust
Cep 22h26'24"60d15'
Jenson,Bruce Gerhardt
Cep 22h13'0"61d20'
Jenson,Ella Gerhardt
Cep 22h28'17"59d14'
Jenson,Natalie Christine
Vul 19h59'0"26d6'

Jenson,Ruby Delores
Cep 20h38'55"76d29'
Jenstar
Gem 6h36'59"13d28'
JenT LVB KALA
Lyr 19h9'18"41d6'
Jentel,Florence
And 2h32'1"45d52'
Jentine
Gem 6h41'30"14d14'
Jentoft,Ebba Helen
Aql 19h29'0"12d3'
Jentsch,Conrad
Dra 17h1'11"66d59'
Jentsch,Elke
Lib 14h55'31"-6d52'
Jenuss,Frank
Cep 21h32'28"55d26'
Jené
Peg 21h54'1"30d23'
Jeong,Yoo Mi
Lmi 10h51'43"38d35'
Jephanie
Tri 1h59'54"27d22'
Jephcott,Jess
Mon 6h52'34"10d54'
Jeppesen,Lisa
Umi 14h6'17"71d1'
Jepsen,Gene Richard
Ori 5h57'29"16d43'
Jepsen,Kristin
Cap 20h22'22"-13d4'
Jepsen,Linda
Lyr 18h57'36"32d50'
Jepson,Keith Eugene
Cep 22h20'51"58d25'
JER
Uma 11h44'1"48d58'
JER,3rd & MSR,50th Wedding Anniversary
Crb 15h53'22"34d44'
Jer-Woman
Uma 10h5'44"48d7'
Jerah
Peg 22h51'44"24d5'
Jeran,Jason Matthew
Aur 6h11'60"30d50'
Jeran,Jeffrey Joseph
Uma 9h36'1"51d33'
Jerby
Mon 6h56'26"-10d16'
Jerdo
Uma 13h31'40"54d24'
Jerelyn,Annette
Uma 10h12'27"57d39'
Jeremiah
Mon 6h55'11"-8d59'
Jeremiah Love Mali
Cyg 20h24'12"40d38'
Jeremiah Terrance, Child Of Wonder
Hya 8h52'28"0d54'
Jeremiah's Star
Ari 2h0'0"20d20'
Jeremy
Aur 5h52'34"30d28'
Jeremy
Cet 2h31'1"1d34'
Jeremy
Aur 6h42'51"38d56'
Jeremy
Boo 14h18'48"33d16'
Jeremy
Dra 11h31'1"66d3'
Jeremy
Lac 22h46'33"38d53'
Jeremy
Ori 6h14'57"11d3'B
Jeremy
Equ 20h58'15"5d42'
Jeremy
Sct 18h41'10"-6d47'
Jeremy & Bridget
Cyg 19h34'0"37d53'

Jeremy Tyler Andrew
Cep 22h39'53"70d32'
Jeremy's Courage
Lib 15h33'41"-8d35'
Jeremy's Wishing Star
Aql 19h44'21"12d22'
Jeremy, Meredith's Shining Star
Cyg 20h36'21"40d41'
Jeremy,Cyndi & Venture
Cyg 19h28'56"40d9'
Jerez,Rose
And 23h35'39"49d11'
Jerez,Sonia
Boo 15h25'59"38d12'
Jerger,Christopher Elwin
Ori 6h2'22"8d52'
Jeri
Ori 5h53'59"10d55'
Jeri
Ori 5h50'40"21d30'
Jeri Lynn
Sgr 18h58'11"-28d53'
Jeri Lynn
Cmi 7h58'17"0d56'
Jeri Lynn
Crb 16h1'34"27d47'
Jeri's Star To Dream On
Leo 11h5'0"-1d16'
Jeri-J
Vul 20h20'28"25d1'
Jerilee
Mon 8h4'18"-9d27'
Jerilynn & Jeffrey Donald
Aur 6h45'42"35d45'
Jeris,Aurora Mierendorf
And 0h14'31"39d36'
Jerjan
Lyn 7h52'21"44d2'
Jerkenstad,James F
Aqr 21h13'20"0d50'
Jerkins,Jesselyn Lauretta
Gem 6h47'44"14d37'
Jerkins,Jillianne Rosemary
Gem 6h47'52"13d9'
Jermaine & Shannon
Sge 19h54'16"16d48'
Jerman,Ann Marie
Cas 1h5'1"50d29'
Jermdogandpea
Cnv 13h28'37"40d57'
Jernberg,Frances G
Her 18h25'1"12d44'
Jernegan,Gary William
Aur 5h53'10"30d43'
Jernigan,Beverly
Peg 22h2'0"21d19'
Jernigan,Joshua Carter
Dra 17h18'40"69d28'
Jernigan,Marsha & John
Aql 18h45'52"11d32'
Jernigan,Tyler
Ser 18h23'52"1d46'
Jero
Cep 22h2'1"61d21'
Jerome
Her 18h31'43"18d45'
Jerome,Ellen
Peg 22h19'1"21d35'
Jerome,Gabrielle
Cyg 21h16'20"38d22'
Jeron
Ori 6h0'1"-0d1'
Jerrad's Star
Uma 10h43'20"53d27'
Jerram IV,James Edward
Uma 10h49'58"60d6'
Jerram,Katherine Cate
Uma 10h50'55"61d58'
Jerrell,Leslie Kent
Oph 18h1'21"10d48'
Jerri
Lyn 7h32'41"44d34'
Jerri Annette
Sct 18h51'57"-7d35'

Jerri Lynn
Vul 19h44'54"22d59'
Jerrold,Rhya Hope
Aur 5h52'54"38d53'
Jerrolyn
Per 4h6'44"50d46'
Jerrom,(Mum & Dad) Harry & Maureen
Cyg 20h58'1"53d55'
Jerry
Ori 4h41'11"5d40'
Jerry
Lyr 18h44'57"45d38'
Jerry "Forever Special"
Leo 9h54'59"7d58'
Jerry & David
Boo 14h5'59"14d2'
Jerry & Lucille Star, The
Uma 9h1'42"50d16'
Jerry & Robin
Mon 6h56'59"0d46'
Jerry 80 & Adelle 75
Ori 6h6'38"8d24'
Jerry Forever
Dra 11h10'17"74d5'
Jerry Harold
Ori 5h52'31"10d59'
Jerry Lynn
Vir 11h40'0"6d26'
Jerry's & Kathee's Wedding Star
Crb 15h48'1"32d20'
Jerry's AstroGizmo
Sct 18h42'19"-7d17'
Jerry's Blackjack
Aur 6h21'30"33d1'
Jerry's Blessing
Oph 17h52'37"12d36'
Jerry's Destiny
Oph 18h19'22"7d29'
Jerry's Dream 12/09/41
Cet 1h41'59"-2d26'
Jerry's Love Light
Cmi 7h25'43"1d7'
Jerry's True Love
Cep 23h33'51"65d5'
Jerry's Wish
Uma 11h4'40"38d32'
Jerry, Ted & McGee
Uma 10h34'12"54d31'
JerryRose
Cyg 21h1'57"30d31'
Jersey Girl Kir
Lyn 8h50'54"35d35'
Jersey Joanne
Del 21h2'28"13d4'
Jersey,Caryl D
Dra 16h45'53"73d53'
Jersey,Katelin Jean
Com 12h29'15"26d54'
Jersey,Victoria
Lyr 18h55'35"41d27'
Jersha,Jamie
Cyg 19h41'47"37d41'
JeRu-Lady of Avalon
Hya 8h31'43"5d59'
JerVera
Ori 6h17'41"-1d53'
Jervis,Kenneth
Aur 6h34'59"34d18'
Jervis,Logan Troy
Lac 22h40'38"54d23'
Jervis-Guiding Light, Ruby P
Lyr 18h44'1"40d49'
Jervis-Heaven's Jewel, Nancy A
Cas 0h48'24"69d21'
Jes Eryn 16
Aqr 23h3'52"-5d27'
Jeschke,Maik
Cep 22h3'32"68d8'
Jeshiva,Carole Andrea
Cas 0h19'47"61d27'
Jesiolowski,Carole
Uma 12h4'34"34d4'

Jeska,Todd
Cet 0h55'12"0d56'
Jeske,Karl-Wilhelm
Lac 22h1'0"51d57'
Jeske,Wolfgang
Lyn 8h10'15"42d39'
Jespergard,Norma Damiano
Cyg 21h51'24"40d21'
Jess
Cas 0h6'39"50d33'
Jess
Del 20h18'54"20d8'
Jess & Kristine's Star
Aql 19h6'32"3d56'
Jess & Sara's Star
Mon 6h45'46"-10d14'
Jess & Syndle's Star
Cam 5h24'46"68d52'
Jess & Tova's Star
Mon 7h0'20"-6d25'
Jess Lara
Aql 20h2'52"0d46'
Jess' Joy
Uma 10h8'1"60d51'
Jess,Melanie
Gem 6h26'44"13d1'
Jessami Irish
Aql 20h1'20"-8d51'
Jesse
Cma 7h30'1"8d44'
Jesse
Cet 1h59'57"-5d25'
Jesse
Aql 20h18'55"0d17'
Jesse
Cnv 13h19'47"32d18'
Jesse
Boo 13h37'44"15d45'
Jesse
Her 18h7'0"18d59'
Jesse & Denise
Cam 3h56'58"70d56'
Jesse & Tommy
Aur 6h31'21"38d50'
Jesse B
Vul 19h45'0"28d19'
Jesse's Heavenly Body
Uma 9h18'26"48d0'
Jesse's Lodestar 50
Cap 20h25'42"-13d28'
Jesse's Wishing Well
Aur 7h3'15"43d54'
Jesse,Neil Wilson
Cap 21h25'38"-19d2'
Jessen,Donald
Aur 5h51'27"31d35'
Jessen,Gary
Boo 14h14'41"37d41'
Jessen,Irene Mary
Peg 23h21'57"32d13'
Jessen,Joseph
Ori 5h46'55"10d59'
Jessen,Karen
Cas 0h18'19"58d15'
Jessep,Jane Nordli
Com 12h12'49"20d48'
Jessica
Cas 0h33'11"68d44'
Jessica
And 2h30'38"42d14'
Jessica
Eri 4h8'15"-14d57'
Jessica
Vul 20h4'0"25d15'
Jessica
Lyr 18h35'50"26d1'
Jessica
Umi 13h46'33"71d57'
Jessica
And 2h1'22"46d5'
Jessica
And 2h19'48"46d26'
Jessica
Cet 0h33'12"-3d35'

Jessica
Cam 3h34'52"60d37'
Jessica
Ori 5h33'15"-0d55'
Jessica
Cas 1h44'43"76d48'
Jessica
And 0h29'27"22d9'
Jessica
Ari 2h6'14"22d8'
Jessica & Jeanna Sisters For Eternity
Cam 3h38'34"59d48'A
Jessica & Luke
Cmi 7h53'12"4d8'
Jessica & Rhoda
And 2h33'19"40d36'
Jessica & Roger Love Shining Forever
Umi 17h52'49"89d2'
Jessica 19-10
And 23h46'31"41d56'
Jessica Ann
Mon 7h30'44"-1d47'
Jessica Ann
Cas 3h3'1"75d21'
Jessica Anne
Crt 10h57'56"-22d28'
Jessica Caitlin
And 23h22'42"41d42'
Jessica Dawn
Lyr 19h3'33"37d43'
Jessica Dawn
Lyr 19h16'55"28d44'
Jessica Dionne
Ori 5h53'40"21d31'
Jessica Erin
Lmi 10h17'58"30d58'
Jessica Fay
Peg 22h0'40"30d11'
Jessica Francis
And 2h25'54"42d49'
Jessica Jean
Lmi 10h10'27"33d47'
Jessica Jeanne
Com 12h0'0"26d39'
Jessica Lee
Tri 1h41'1"30d36'
Jessica Lee
Cas 0h39'10"67d50'
Jessica Leigh
And 1h16'58"39d22'
Jessica Lin
Cas 0h30'12"61d20'
Jessica Lynn
And 2h32'13"40d31'
Jessica Lynn
Lyn 8h52'23"44d47'
Jessica Marie
Peg 23h28'26"21d49'
Jessica Marion
Lyn 8h14'26"38d36'
Jessica May
Crb 15h39'34"26d7'
Jessica R (Reynolds)
Mon 7h15'12"-10d59'
Jessica Rae
Tri 1h45'52"34d51'
Jessica Rae
And 1h52'49"37d41'
Jessica Starbright
And 23h40'39"38d10'
Jessica Suzanne I
Lyr 19h5'0"38d50'
Jessica's Brilliance
And 1h26'55"34d4'
Jessica's Halo
Aql 20h32'24"0d59'
Jessica's Heartbeam
Cas 0h42'18"70d48'
Jessica's Jewel
Equ 20h59'24"10d46'
Jessica's Miracle
Boo 15h5'31"38d12'

Jessica's Special Star
Lyr 18h40'16"30d56'
Jessica's Star
Ori 5h56'16"8d25'
Jessica-Our Bright & Shining Star
Mon 6h44'1"7d57'
Jessie
And 23h11'1"41d53'
Jessie
Umi 13h56'28"78d49'
Jessie
Lyn 7h52'32"37d8'
Jessie
Aur 5h18'51"46d4'
Jessie
Cyg 20h18'25"39d8'
Jessie
Lac 22h48'14"38d12'
Jessie
Lyr 19h5'34"38d25'
Jessie (F A I L S)
Ori 5h34'49"-8d17'
Jessie Allyn-Our Miracle
Lib 14h27'48"-22d35'
Jessie Leigh
Uma 12h5'0"60d25'
Jessie Rose,The
Peg 21h49'45"7d19'
Jessie's Dream
Eri 3h50'31"-1d46'
Jessie's Star
Cam 3h34'18"61d3'
Jessie's Star
Aql 19h55'34"15d11'
Jessie-Danny-Mandy- Mikey
Vul 19h47'28"28d53'
Jessie-Our Little Angel
Lyr 18h42'36"30d14'
Jessika's Dream
And 1h19'34"38d37'
Jessiman,Sarah
Cnc 8h34'24"30d31'
JessiMorgan
And 23h18'14"50d28'
Jessop Star,The Ian
Ori 4h48'20"0d43'
Jessop,Mary
Umi 16h12'38"77d16'
Jessop,Rosemarie
And 0h25'38"40d29'
Jessup Star,The
Cyg 19h34'58"33d59'
Jessup Star,The Ryan
Per 3h19'30"40d34'
Jessup,Alyssa Lynne
Aql 19h7'0"-0d54'
Jessup,Dawn
Del 20h12'49"11d1'
Jessup,Our Noble Star, Mary Blanche
Aur 5h1'36"38d13'
Jessup,William Brian
Aur 7h4'33"38d9'
Jester Mother of Universe,Charlotte
And 0h58'16"34d35'
Jester,Charles Aaron
Cmi 7h57'26"8d33'
Jester,Edward EO
Vir 13h56'0"-0d24'
Jester,Melinda
Del 20h54'54"8d22'
Jesus-Schultz
Lmi 10h4'39"32d28'
Jet
Lyn 7h9'35"52d24'
Jet"If I Was Your Girlfriend"
Gem 7h1'14"10d42'
Jet'amine
Lac 22h11'49"49d12'
Jeter,Andrea Marie Cordell
Cas 2h59'20"57d50'
Jeter,Michael
Per 4h26'33"50d16'

Jeter,Mrs Janet
Aql 19h7'19"4d11'
Jeters,Harold A
Her 18h0'26"18d45'
Jethro
And 0h15'54"25d51'B
Jethro Tull
Ori 5h56'21"19d56'
Jetorian
Dra 11h30'29"68d33'
Jett,Rhea (Honey)
Eri 3h35'24"-16d1'
Jett,Starke (Honey)
Lmi 10h37'46"30d33'
Jette
Uma 8h46'41"50d49'
Jette Bean
Peg 21h51'27"28d59'
Jetti-Mercedes
Sco 15h57'9"-22d3'
Jettie & Chris 1994 I Love You
Peg 23h39'36"30d60'
Jetté,Dr Pierre
Per 3h6'27"40d17'
Jetzinger,Eva Maria
Cas 2h6'52"68d37'
Jeu D'Eau
Oph 18h27'27"7d58'
Jeunai,Jacquelyn
Lyn 9h5'44"37d24'
Jevec,Tom
Cep 0h19'31"70d47'
Jevons,Brendan Alec
Dra 16h50'0"68d5'
Jevons,Christopher William
Aur 5h4'1"41d54'
Jevons,Miss Abigail
Cas 23h5'26"58d23'
Jewdricks,The
Oph 18h0'48"8d46
Jewel
Lyr 18h17'18"30d11'
Jewel
Ori 5h59'12"12d6'
Jewel "The Brightest Star"
Uma 10h59'0"57d9'
Jewel Of Geael
And 0h7'24"46d45'
Jewel's Diamond
Cet 0h42'1"-4d35'
Jewel,Bobbie
And 23h42'0"43d6'
Jewel,Lela
Mon 6h26'57"-6d49'
Jewell
Lyr 18h40'55"41d29'
Jewell's
Lyn 8h46'22"45d20'
Jewell,Arthur
Her 16h52'45"32d37'
Jewell,Cheryl M
Peg 21h57'23"24d46'
Jewell,Debi
Lyr 19h19'24"35d17'
Jewell,John Clayton
Hya 8h46'22"-7d30'
Jewell,Lisa Schuyler
Aur 5h29'53"38d20
Jewell,Moire
Cas 0h38'0"60d42'
Jewell,Stuart Valuable
Aur 4h36'57"34d10'
Jewell,Tim
Cep 21h22'17"68d56'
Jewell-Thomas,Stephen
Her 16h8'14"42d4'
Jewels
Per 1h54'26"53d5'
Jewels
Cas 0h52'16"68d3'
Jewels
Sge 20h2'1"20d46'
Jewett,Andre T
Ori 4h42'34"11d19'

Jewett,Kathryn M
Peg 22h2'1"24d15'
Jewett,Robert O'Neal
Crb 15h57'22"28d23'
Jewett,Robert Roy
Sex 10h46'1"-6d17'
Jewett,Vera P
Peg 22h4'24"24d3'
Jezae
Uma 9h29'1"42d5'
Jezebel's Fire
Lyn 8h55'26"40d46'
Jezequel,Martin
Mon 7h49'48"-4d11'
Jezidija,Katherine A
And 2h20'14"38d50'
Jezischek,Heinz M
Peg 23h31'18"17d15'
Jezyk,Susan
Peg 22h37'53"26d18'
JFJ II-The Encourager
Aql 19h58'29"7d47'
JGH + LJM
Aql 20h20'26"0d20'
JGS - The Always & Forever Star
Aur 5h3'35"40d5'
Jhangiani,Shuja
Boo 14h15'20"17d53'
Ji,Ge Xui
Her 17h38'32"40d16'
Jia
And 23h0'47"51d19'
Jian,Hovik Kulke
Dra 17h0'34"66d45'
Jianoran,Eleno Gonzales
Equ 20h59'26"3d2'
Jianoran,Khristie Cunanan
Peg 21h59'29"21d1'
Jibilian,Aram
Lib 15h1'37"-28d48'
Jibilian,Gary
Crb 16h11'1"32d25'
Jicky
Ori 5h58'41"7d23'
Jiddo
Aql 19h57'58"10d34'
Jiggs Is Up,The
Boo 15h4'35"17d42'
JIGLAN
Ori 5h18'55"1d12'
Jigs "The Love God"
Per 3h33'34"37d11'
Jikasei's Chewbacca Inu
Uma 11h6'32"45d7'
Jill
Umi 14h10'13"70d38'
Jill
Mon 7h39'31"-1d43'
Jill
Cam 5h30'53"60d1"
Jill
Peg 23h27'30"33d44'
Jill
Cas 0h38'18"62d50'
Jill & Chris-Our Infinite Love
Umi 14h21'39"68d8'
Jill & Eric's Eternal Love Light
Cyg 19h44'22"29d33'
Jill & Jeff
Cam 6h1'41"67d48'
Jill & Jennifer,Mother & Child
Cas 1h38'57"58d22'AB
Jill & Joe
Lyr 18h53'11"31d41'
Jill & Rob
Uma 10h21'11"61d52'
Jill & Steve
Ari 1h57'35"25d15'
Jill & Steve Eternally
Ori 5h50'0"16d52'
Jill Anne
Peg 23h33'33"30d28'
Jill At The Elkavox
Psc 0h16'42"20d39'

Jill Bonnie
Del 20h15'25"10d31'
Jill Christine
Eri 4h3'38"-13d13'
Jill Diane
And 23h43'12"43d51'
Jill Elaine
Cyg 21h51'13"42d25'
Jill Louise '95
Cas 0h59'59"72d3'
Jill Marie
Umi 13h7'16"70d0'
Jill Marie
Cyg 21h4'1"36d13'
Jill Marion
Cas 0h20'20"61d35'
Jill Suzanne 03/08/75
Peg 22h13'13"5d52'
Jill's Delight
Aur 4h59'0"34d35'
Jill's Star
Aql 18h59'1"-5d2'
Jill's Star
Crt 10h54'30"-23d18'
Jill's Star
Eri 3h8'44"-3d54'
Jill's Star
Lmi 10h36'38"32d9'
Jilletella
Peg 22h43'29"25d22'
Jilli Bear
Uma 11h24'52"33d25'
Jillian
Cet 2h41'49"5d1'
Jillian
Cap 21h25'42"-23d11'
Jillian
Crb 15h56'49"27d58'
Jillian Elizabeth
Cas 1h8'59"60d1'
Jillian Marie
And 23h30'50"45d25'
Jillie,Jr,Don W
Vul 19h44'21"23d1'
Jillstar
Lyn 8h57'43"40d20'
Jilly
Boo 14h8'23"41d17'
Jilly & Jimi
Cyg 21h12'0"38d54'
Jilly & Ted
Boo 15h16'13"53d3'
Jilly Beans Golden Birthday Star
Dra 17h31'33"60d30'
Jillyanno
Lyn 8h26'34"43d58'
Jillyn Rae
Cyg 20h41'55"30d49'
Jily110292HB
Peg 21h55'37"28d24'
Jilya
Ori 6h0'1"20d21'
Jim
Aql 18h55'0"-0d2'
Jim
Ori 5h39'21"10d40'
Jim & Anne-Marie
Uma 9h45'38"44d12'
Jim & Barb
Aql 18h56'40"-7d58'
Jim & Bridget-A Perfect Love
Lac 22h46'46"37d45'
Jim & Cheryl 1995
Sge 20h5'56"20d12'
Jim & Christi
Lyn 7h50'31"42d31'
Jim & Christina's Wedding Star
Ori 6h6'0"0d53'
Jim & Cindy
Dra 19h30'20"68d18'
Jim & Diane
Crb 16h2'45"32d16'

Jim & Dina
Cam 3h54'48"61d39'
Jim & Eleanor 50 years
Cyg 21h14'29"28d48'
Jim & Geo
Lyr 18h59'58"37d30'
Jim & Joan,1022 Enchantment
Cyg 21h36'18"40d20'
Jim & Kelly Forever
Cyg 19h36'0"28d48'
Jim & Leslie
Uma 9h3'22"51d59'
Jim & Nikki's Normandy Nuptials
Lyr 18h32'0"40d30'
Jim & Nora
Cyg 19h41'34"37d54'
Jim & Shel's Guiding Lite
Cyg 19h58'11"41d6'
Jim & Tina
Aql 20h2'1"4d38'
Jim All
Per 2h55'31"40d44'
Jim Cosmo
Sex 10h15'14"-1d54'
Jim E B
Cnv 13h52'40"37d50'
Jim Irish Star
Aur 5h2'1"48d52'
Jim Karen Jerome
Dra 17h0'12"63d12'
Jim Loves Mary Lou
Lyr 18h29'23"31d5'
Jim My Shining Star
Cep 0h58'45"78d19'
Jim's "Malibu Light"
Psc 0h57'23"20d28'
Jim's Brown-Eyed Girl
Cas 0h47'30"71d42'
Jim's Fantasy Island
Del 20h26'49"10d23'
Jim's Fire Within My Heart
Aql 20h0'49"10d29'
Jim's Gaseous ORB
Per 4h44'17"41d8'
Jim's Jungle
Cet 0h37'0"-1d13'
Jim's Light
Her 15h56'59"48d27'
Jim's Lone Star
Ori 5h57'1"14d14'
Jim's Quiet Flight
Ori 5h56'0"13d51'
Jim's Rainbow
Dra 14h34'21"61d48'
Jim's Train In The Sky
Uma 9h50'1"48d25'
Jim's World Connection
Del 21h2'60"18d51'
Jim, Yes Yes Yes!
Lib 14h58'32"-20d29'
Jim,My Love For All Eternity,Woo
Uma 8h38'26"49d14'
Jim,Shirley, & Stephen
Peg 23h29'50"18d20'
Jim bob
Per 2h25'15"54d35'
Jim-Shar
Cam 3h59'33"55d17'
Jim-The King of Layfield
Her 17h14'35"29d51'
Jimarcia
Her 16h45'15"38d33'
Jimarjie
Gem 6h4'9"18d1'
Jimbabwe
Boo 14h0'30"18d41'
Jimbo
Peg 23h5'52"33d56'
Jimbo
Cnc 9h19'0"32d34'
Jimbo/Neister
Uma 11h20'34"45d16'

Jimbobaybabadoo Cluster,The
Uma 10h21'40"47d52'
Jimdeshermuff Star
Oph 16h57'47"-28d17'
JIMEDITH
Peg 22h44'13"32d20'
Jimenez "Charlie", Charles Albert
Aql 20h12'29"1d0'
Jimenez,Antonio Adriano
And 2h0'49"45d12'
Jimenez,Astrid
Lyn 7h3'11"52d55'
Jimenez,Gaspar
Cet 2h15'23"5d15'
Jimenez,Joaquin Lopez
Cas 23h2'47"58d27'
Jimenez,Ramiro & Kathleen
Cap 20h5'43"-12d9'
Jimenez,Victor "Vic" M
Cyg 21h7'52"30d21'
Jimenny
Vul 19h22'23"26d42'
Jimerson,George Dewey
Boo 14h19'0"16d11'
Jimi
Aur 7h23'29"37d15'
Jimi & Shelia
Aql 19h29'57"-6d5'
Jimithins-"feels Like Rain"
Aql 19h48'30"14d51'
JimJesAlextar
Lac 22h5'27"46d29'
Jimmares "For Your Wishes"
Boo 14h21'0"38d49'
Jimmer Glimmer
Ori 6h7'21"20d12'
Jimmiemma
Dra 11h37'52"70d43'
Jimmifer
Uma 9h27'23"49d28'
Jimmimax
Cet 0h51'46"-4d9'
Jimmy
Dra 17h40'27"61d20'
Jimmy
Lac 22h16'56"38d19'
Jimmy
Her 17h5'13"20d53'
Jimmy
Boo 15h30'25"41d33'
Jimmy
Her 15h58'23"43d17'
Jimmy
Boo 15h2'32"21d5'
Jimmy
Lac 22h32'34"40d26'
Jimmy
Boo 15h6'56"48d38'
Jimmy & Anna
Boo 14h22'30"26d27'
Jimmy & Lucille's Star
Peg 0h0'48"31d6'
Jimmy & Michelle
Uma 9h54'27"44d42'
Jimmy & Micki's Tranquility Star
Sex 10h27'30"1d15'
Jimmy (15699)
Cma 7h19'27"-15d8'
Jimmy Bryan
Ori 5h22'27"15d46'
Jimmy G
Cet 1h52'1"0d43'
Jimmy J,"Scooter Pie in the Sky"
Cap 21h52'53"-14d41'
Jimmy K
Lac 22h1'1"47d60'
Jimmy Lee
Aur 6h22'23"38d46'
Jimmy Lee's Baby Blue Ice
Aur 5h1'59"40d31'
Jimmy Mack
Peg 23h2'37"12d15'

Jimmy the One
Cep 5h31'16"86d5'
Jimmy's 50th
Her 17h4'0"42d42'
Jimmy's Comet
Psc 1h15'26"20d40'
Jimmy's Star
Ser 16h1'54"14d11'
Jimmy's Star
Cep 21h49'0"70d26'
Jimmy's Star
Oph 17h27'29"-20d11'
Jimmy-"My Piece Of Heaven"
Per 2h56'42"45d7'
JimmyDon
Ori 5h32'13"1d17'
Jimmyv Always Reach For Your Dreams
Her 16h45'23"22d27'
Jimnjune
Uma 11h53'53"63d49'
JimPin's Equinox
Ori 5h52'39"14d40'
Jimster
Oph 18h32'49"11d26'
Jimtom 12/95
Aql 19h28'59"13d47'
Jimtown Jr High Faculty
Lac 22h52'57"51d45'
Jiménez,Ana Parro
Lyn 8h16'40"43d48'
Jindle,Vijay
Cep 22h1'46"70d28'
Jinene,A
Ori 5h39'18"15d14'
Jingle Bells
Lac 22h18'42"49d44'
Jingle,Marty Dziegel
Her 17h51'13"50d6'
Jini Jo
Cas 0h9'57"50d31'
Jinkins,Susan Lee
Lyr 18h45'31"34d51'
Jinks,Bill & Juliet
Cyg 21h3'58"53d54'
Jinks,Darlane S
Eri 2h47'1"-2d58'
Jinks,Jerry
Lac 22h30'21"37d31'
Jinks,Marcus Odell
Aql 20h8'0"1d32'
Jinnie
Ori 6h7'31"20d31'
Jinsk
Aur 6h6'15"50d5'
Jinx
Lyr 19h1'59"33d46'
Jinx Spille
Cam 8h33'18"73d37'
Jipps,Katie Emma
Cyg 21h56'21"52d36'
Jirdd
Dra 18h51'37"68d46'
Jirous,Marvin (Bud) D
Dra 18h8'30"58d27'
Jirousek,Georgianne
Ari 1h48'20"17d7'
Jirousek,Kimberley Ann Theresa
Cyg 19h26'0"30d28'
Jirsa,Laura Lee
Cas 0h7'12"58d46'
Jitosho Tomoko
Sgr 19h12'45"-14d0'
Jizz Glop FEZ
Uma 11h14'24"42d8'
JJ * Maggie
Peg 22h47'50"34d55'
JJ's "Lucky Star"
Uma 8h56'0"58d34'
JJM Sparks R/O
Aur 7h6'48"38d58'
JK
Cep 22h14'1"60d34'

JK My Special Sweetheart
Peg 0h0'53"18d2'
Jkekimo
Her 18h18'17"12d34'
JL's Star
Cma 6h56'44"-20d19'
JLC TO KFC 5-21-94
Per 2h51'57"35d35'
JLM Reserve I
Ser 15h13'14"7d30'
JMB #1
Hya 8h46'19"-7d6'
JMB 111063
Hya 9h4'23"5d54'
JME My Magician LPS
Dra 11h43'24"68d49'
JMK's Own Special Star
Lmi 10h10'25"33d14'
JMK-50
Crb 16h11'37"26d8'
JMR" 25
Eri 3h30'33"-11d10'
JMS
Boo 15h14'56"26d18'
JMS '93
Mon 8h3'0"-9d39'
JO
Cam 4h6'1"67d58'
Jo
Tri 2h36'34"34d11'
Jo
Hya 10h18'42"-16d50'
Jo
Cam 6h59'1"80d0'
Jo
Sge 20h2'19"18d60'
Jo & Armando
Uma 10h36'26"56d28'
Jo & Mike
Aur 6h29'12"35d4'
Jo & Ruth 25
Cyg 21h45'31"31d37'
Jo & Trevor- Celebration of Love
Gru 22h9'28"-56d39'
Jo (Blinkus Flippius)
Peg 22h38'11"11d60'
Jo and Bubba
Hya 9h2'24"5d60'
Jo Ann
Cas 0h3'0"63d44'
Jo Ann
Psc 1h2'31"20d6'
Jo Ann
Lyn 8h13'18"37d37'
Jo Ann
Aqr 22h24'29"-0d4'
Jo Ann 45986
Cas 23h2'24"58d16'
Jo Anna
Mon 7h0'1"8d27'
Jo Anne
And 0h58'23"39d55'
Jo Barnhill Brown Brown
Ori 6h6'13"1d24'
Jo Jo 1
Ari 2h36'43"20d31'
Jo Jo's Sunshine
Aql 19h30'53"13d27'
Jo Wadsworth's Piece Of Heaven
Lib 15h3'30"-11d15'
Jo's (Soulfullness)
Ser 16h16'11"-0d57'
Jo's und Lieses Kürbisstern
Peg 23h34'0"13d13'
Jo-Anne
Cru 12h44'35"-57d10'
Jo-el & Dylan Forever & Always
Cep 2h44'33"78d49'
Jo-Jo
Lmi 10h2'56"34d5'
Jo-Jo
Peg 21h34'18"20d33'

Jo-Jo
Aur 6h0'59"37d8'
Jo-Jo & Hoochie (4-29-92)
Crb 16h19'1"38d13'
Jo-jo's Star
Cas 1h36'31"73d30'
Jo-Simply The Best
Peg 21h48'0"31d22'
Jo-Su Vice Versa
Cam 5h46'40"61d37'
Joachim
Cep 22h5'36"65d16'
Joachim Labudde
Ori 5h58'1"10d28'
Joachim,Heather Michelle
And 2h26'50"50d27'
Joachim,Linda
Lyn 8h4'19"38d30'
Joan
And 0h11'0"46d41'
Joan
Cep 23h14'39"70d50'
Joan
Cas 0h12'46"54d26'
Joan
Cam 9h1'47"80d19'
Joan
Cas 1h43'30"75d4'
Joan
Cas 23h19'0"60d24'
Joan
Peg 22h38'27"22d9'
Joan
Cyg 21h42'21"29d25'
Joan "Mother Of Elaine"
Peg 23h2'19"30d31'
Joan "Orchid Mattone"
Mon 7h3'0"5d6'
Joan & Joe
Cyg 20h37'20"45d33'
Joan & Paul
Mon 7h58'27"-3d45'
Joan & Richard
Cyg 21h58'0"48d54'
Joan & Sammy's Endless Love
Cyg 19h32'59"39d5'
Joan Elaine
Lyr 18h31'45"33d45'
Joan Ellen
Ari 1h46'51"17d42'
Joan Margaret
Cas 0h45'0"64d28'
Joan Marie
Gem 6h54'8"18d49'
Joan Mary For P F Stal Kelley
Boo 14h35'52"31d53'
Joan Shirah
Cma 7h0'0"-11d2'
Joan"Our Caring Star"
And 2h30'0"48d35'
Joan's Eternal Star
Lmi 9h22'48"33d23'
Joan's Guy
Cep 23h22'0"71d4'
Joan's Jewel
Peg 22h31'1"26d60'
Joan's Star
And 0h22'25"28d31'
Joan-N-Bernie
Cnv 13h24'31"42d6'
Joanie
Aql 20h0'26"1d4'
Joanie
Cas 1h43'0"60d32'
Joanis,Claudia
Lac 22h48'54"53d17'
JoAnn
Mon 6h24'21"2d41'
JoAnn
Cam 4h13'29"70d18'
Joann
Com 12h18'54"20d3'
Joann
Uma 10h52'32"53d15'

Joann
Peg 22h41'23"29d36'
Joann
Cas 1h30'0"60d45'
JoAnn "Fumerelle I"
Mon 7h1'0"-1d43'
Joann & Allan 40th Star
Cyg 21h0'38"40d56'
JoAnn & Frank-Love Forever
Lyn 8h55'13"40d7'
Joann & Tom
Cyg 21h55'35"52d60'
Joann 82395
Equ 21h21'42"10d13'
Joann Felicia
Vul 19h4'23"25d25'
JoAnn's Love Light
And 2h8'30"40d29'
JoAnn's Sparkling Love Light
And 2h9'1"38d18'
JoAnn,Julie Ann,Karen, Kristine
And 1h57'51"37d53'
JoAnn-ica J J
Mon 7h40'48"-8d13'
JoAnna
Uma 11h27'27"47d1'
Joanna
Cyg 20h57'38"38d23'
Joanna
Cas 23h35'53"61d18'
Joanna
Cet 2h3'53"-0d17'
Joanna
Umi 13h12'26"70d49'
Joanna
Lyn 8h22'59"47d33'
Joanna
Uma 10h13'45"50d13'
Joanna Alice
Uma 11h45'44"53d15'
Joanna Beautiful Bird
Cyg 19h40'59"40d28'
Joanna Eleese & Jacqueline Elizabeth
Mon 6h55'47"-4d38'
Joanna Leigh
Oph 17h25'32"-20d42'
Joanna Marie
Lyr 19h23'10"40d2'
Joanna Marie
Cas 23h33'13"61d13'
Joanna of Hutton Magna
Cas 1h24'37"52d33'
Joanne
Cyg 21h3'36"28d53'
Joanne
And 23h38'31"43d15'
Joanne
Cas 1h32'48"75d53'
Joanne & Bill All Our Love Always
Cyg 21h35'20"40d38'
Joanne & Edwin Forever
Eri 4h22'16"-6d47'
Joanne & Scottie
Mon 6h22'53"-0d51'
Joanne 1
Uma 11h9'10"61d21'
Joanne Dawn
Lyr 18h18'47"43d39'
Joanne Elizabeth
And 0h59'38"39d22'
Joanne Hippy Chick
Lyn 8h5'37"41d30'
Joanne Kathryn
Crb 16h13'0"32d56'
Joanne Marie
Cas 2h33'11"57d33'
Joanne Marie
Lyr 18h59'53"37d40'
Joanne Marien
And 2h22'32"45d23'
Joanne May 18th
Cam 5h50'1"70d36'
Joanne Nicole
And 23h0'12"41d55'
Joanne Star,The
Cas 0h44'45"71d18'
Joanne's Dream
Peg 21h40'22"26d28'
Joanne's Star
Peg 23h49'12"30d59'
Joanne,Gracious Gift of God
And 2h22'38"40d4'
Joannides,Stan
Boo 14h50'59"34d58'
Joannot,Yves Et Madeleine
Cam 3h26'54"60d48'
Joaquin
Cam 4h29'25"65d22'
Joaquin & Sharon
Boo 14h27'0"46d29'
Joaquin,Dr Nicanor
Aqr 23h12'1"-6d4'
Joaquin,Gary
Aur 4h54'27"51d21'
Job's Daughter-Orange Bethel #337
Crt 11h38'29"-21d38'
Job's Daughters of Mrs K Pitts
And 22h55'39"50d56'
Job,Kerry
Cet 2h31'58"1d15'
Jobe,Anne
Uma 9h44'11"45d54'
Jobe,Carrie B
Peg 21h39'1"25d33'
Jobe,Joël
Lyn 8h1'54"33d29'
Jobe,Lance Klile
Per 4h2'23"34d3'
Jobes,Bremembery
Lac 22h52'27"56d33'
Jobin
Lib 14h40'1"-18d54'
Joblon,Jr,James Thomas
Sex 9h59'47"-6d30'
Jobman,Sheila
Cyg 20h0'39"31d38'
Jobmann,Dierk
Oph 18h3'29"7d42'
Jobo
Cep 23h13'14"71d12'
Jobobly Joe
Cnv 12h19'34"50d0'
Jobson,Andy
Ori 5h51'54"20d29'
Joby
Eri 4h14'13"-13d12'
Joby
Eri 3h0'11"-18d35'
JOCALOSAJEBEARTH
Oph 18h19'55"8d51'
Jocelyn
Lyr 18h28'1"42d5'
Jocelyn Elise
And 22h59'0"38d12'
Jocelyn Rachel
Ori 5h24'0"0d41'
Jocelyn's (J C)
And 1h4'1"47d2'
Jocelyn,Sari
Uma 11h59'16"48d26'
Jocelyn/JB
Peg 22h13'56"34d49'
Jocelyne
Mon 6h24'37"10d46'
Jocelyne
Uma 10h57'49"51d57'
Jocelyne
Boo 14h2'0"12d22'
Jochen
Cet 2h59'29"5d19'
Jochum Family Star,The
Uma 10h55'41"36d56'
Jochum,Dieter
Lyr 19h18'1"30d2'
Jock
Her 16h19'56"26d28'
Jockey's Light
Cyg 21h4'0"39d48'
Jockie
Ori 6h4'1"2d46'
Jocobsen
Cet 0h47'49"-8d27'
Jocque,Magdalen
Cas 1h17'1"64d55'
Jodamist
Per 2h57'11"40d49'
Jodan
Per 2h59'0"31d7'
JoDave Eternity
Ori 4h48'0"5d41'
JoDee Elaine 7-27-70
Cas 0h54'15"61d51'
Jodei
Del 20h13'49"14d58'
Jodi & Jay
Crb 16h18'59"33d3'
Jodi Catherine
And 0h16'49"36d49'
Jodi-Praised & Beloved Of God
Ori 5h51'41"10d2'
Jodie Ann
And 2h26'21"41d59'
Jodie's Dad
Ori 5h35'19"0d15'
Jodie's Joy
Vir 15h5'29"0d36'
Jodie's Light
And 0h21'58"31d17'
Jodie's Wishes
Aql 20h11'12"5d9'
Jodieann My Lady
And 23h33'35"41d4'
Jodush,Stephen Thomas
Lac 22h8'1"51d27'
Jody
Leo 10h52'34"21d50'
Jody
Cet 1h1'1"-4d2'
Jody
Hya 8h30'30"1d18'B
Jody
Cet 2h56'39"5d34'
Jody & Al's Place
Tau 4h10'29"1d5'
Jody Ann
And 1h27'1"37d43'
Jody Lynn
Lyr 18h30'17"40d53'
Jody Lynn
Lyn 9h5'48"40d15"
Jody's Smile
Vul 20h15'34"23d5'
Jody's Star
Com 13h28'30"25d28'
Joe
Cnv 13h29'53"38d2'
Joe
Aur 6h30'1"38d53'
Joe
Dra 16h51'32"61d34'
Joe & Catharine
Cyg 21h24'1"53d8'
Joe & Colleen's Wish
Peg 22h21'45"27d42'
Joe & Diane
Uma 9h5'30"59d55'
Joe & Elyse Star,The
Her 16h51'33"34d14'
Joe & Gloria
Cam 4h20'24"70d14'
Joe & Loretta's "Nite Lite"
Crb 15h57'45"26d28'
Joe & Lynn
Aql 19h25'54"-0d54'
Joe & Michelle
Mon 6h30'55"-0d56'
Joe & Renee,Eternal Passion
Sge 20h2'1"16d3'
Joe & Sandi's "Annus Mirabilis"
Sge 20h6'1"18d45'
Joe & Teresa's Love
Vul 20h39'59"20d31'
Joe B
Lmi 10h29'57"30d25'
Joe Baby
Dra 19h53'35"61d12'
Joe Bets
Cep 21h56'34"55d54'
Joe Loves Karen Forever
Sge 19h54'48"20d32'
Joe S
Tau 4h5'58"20d49'
Joe T
Her 16h37'55"32d44'
Joe Thomas
Cnv 12h9'0"42d55'
Joe Z
Dra 17h4'18"58d49'
Joe Z,The
Uma 10h47'31"68d9'
Joe's
Cam 7h53'22"60d48'
Joe's "Great Spirit" Star
Uma 10h6'0"47d53'
Joe's "New Vision"
Vir 13h33'46"-6d57'
Joe's Eternal Flame
Cnv 12h18'0"34d13'
Joe's Guardian Angel
Boo 13h39'19"20d30'
Joe's Legacy
Oph 17h53'60"11d19'
Joe's Papa
Uma 11h27'36"40d10'
Joe's Star
Per 1h54'27"50d21'
Joe's Star
Mon 7h19'17"-0d57'
Joe's Star Of No Coincidences
Lac 22h25'33"52d36'
Joe's True Gravity
Per 2h58'10"50d33'
Joe's Way
Ser 18h43'20"2d27'
Joe,Amicus Carus
Cam 5h43'37"70d15'
Joe,Marcia"Forever"
Cma 6h59'32"-18d58'
Joe-Dee
Lyn 6h39'40"59d55'
Joe-Man 5
Lac 22h15'38"51d45'
Joe-Wes
Per 4h45'29"50d44'
Joean Aeternus
Ori 5h32'25"-0d43'
Joebamb Natdor
Aur 5h18'45"47d13'
Joebges,Heribert
Peg 23h36'18"15d39'
Joeffrey
Lac 22h43'25"38d54'
Joekel,Charles
Hya 8h12'34"0d20'
Joekel,Linda Kelley
Eri 3h53'25"-5d50'
Joel
Her 17h59'34"40d49'
Joel
Aur 6h30'37"33d14'
Joel
Cyg 19h32'38"35d9'
Joel Anthony
Uma 10h11'33"60d30'
Joel Christopher
Cep 0h10'1"70d23'
Joel's Jupiter
Boo 15h9'0"29d59'
Joel's Light in the Night
Cep 22h18'14"55d31'
Joel's Place
Hya 8h58'21"3d37'
Joel's Polaris
Uma 12h5'34"57d24'
Joel,Tom & Michelle
Cam 5h46'40"61d37'
Joelbarbwendy
Uma 10h22'44"40d2'
Joeli
Ori 5h49'1"21d50'
Joell's Anniversary Super Star
Eri 2h47'54"-2d46'
Joella Marie
Ari 1h55'46"11d2'
Joelle
Ori 5h48'57"10d2'
Joelle
And 0h10'29"31d5'
Joelle B XVI-190187
Uma 11h31'58"31d32'
Joelle's Familiar
Sge 19h55'24"16d16'
Joelle's Music In The Heavens
Lyr 18h35'1"27d15'
Joelle's Nicholas
Per 1h56'0"54d18'
Joelle,Sweet Stephanie
Cas 23h13'25"60d42'
JoEllen My Love
Psc 1h2'44"18d31'
Joels,Herman Eduard
Lyr 18h56'37"34d24'
Joelson,Yasmina
Mon 6h39'31"-0d6'
Joelster
Sex 9h59'27"-6d24'
Joely Rebecca
And 22h56'51"50d8'
Joerg,Markus
Cas 23h2'12"53d23'
Joesy & Les
Cyg 19h21'1"30d4'
Joey
Boo 13h44'26"14d22'
Joey
Ori 5h50'56"12d43'
Joey
Cep 22h34'1"60d59'
Joey
Boo 14h59'25"21d51'
Joey
Peg 23h25'52"17d13'
Joey
Sct 18h31'11'-6d25'
Joey
Aql 19h6'39"3d11'
Joey
Aql 19h50'12"14d49'
Joey
Lac 22h16'0"54d58'
Joey
Her 17h30'49"14d44'
Joey
Aur 4h52'32"52d17'
Joey "Dx"
Aur 4h47'47"40d58'
Joey "Mr Wonderful"
Her 17h3'28"21d36'
Joey "of the BIG VII"
Aur 6h6'28"46d27'
Joey & Diane
Crb 16h9'13"31d59'
Joey & Lonon's Wedding Star
Cyg 21h55'38"53d34'
Joey & Maggie
Del 20h13'20"9d51'
Joey & Roseanne
Aur 5h24'17"40d52'
Joey & Tara's Star For Eternity
Cyg 20h18'35"31d34'
Joey & Tara's Star For Eternity
Cyg 20h18'35"31d34'
Joey D
Del 20h25'18"8d35'
Joey D
Dra 16h38'21"63d53'
Joey Girl
Cyg 19h41'10"41d43'
Joey Scott
Per 2h55'21"50d26'
Joey V
Boo 15h11'29"53d32'
Joey's "Celestial Song"
Aqr 21h53'29"-0d15'
Joey's Joy
Her 16h56'31"32d56'
Joey's New York Rangers
Her 17h43'23"40d59'
Joey's Night Light
Uma 9h59'0"51d44'
Joey's Star
Cep 21h5'0"65d30'
Joey's Star Forever Shines
Per 2h29'53"51d39'
Joey"
Aur 6h52'7"37d58'AB
Joff's & Lauren's Piece of Heaven
Sco 17h55'52"-38d0'
Joffre
Leo 11h55'17"23d36'
Joggl und Dorli
Tau 4h5'40"20d4'
Johal,Tanisha
And 0h31'11"40d3'
Johal,Tindy
Per 2h57'16"34d56'
Johann,Konrad
Cap 21h38'47"-20d31'
Johanna
Ori 6h17'12"12d38'
Johanna
Cam 7h56'1"60d39'
Johanna
Gem 8h4'19"30d56'
Johanna
Psc 23h34'39"-1d19'
Johanna
Sco 16h13'45"-20d43'
Johanna
Peg 22h26'15"25d11'
Johanna & Tobias
Umi 17h13'16"85d23'
Johanna Clare
Equ 21h0'15"2d13'
Johanna Michelle
And 0h14'57"39d51'
Johanna Ruth
And 23h48'1"40d56'
Johanna's Dream
Peg 23h42'0"26d28'
Johanna's Star
And 23h35'14"42d11'
Johanna-Maria
Vir 13h59'1"-6d37'
Johanne
Cas 0h48'1"73d22'
Johanne-Lucette
Cas 22h59'50"55d11'
Johannes
Lyn 8h1'55"49d21'
Johannes & Adrian Forever Love
Eri 3h11'28"-5d1'
Johannes Maximilian
Lyn 7h55'1"50d34'
Johannesen,Nicholas
Her 18h12'1"47d38'
Johannesen,Pat
Leo 9h56'36"11d57'
Johannessen,Dorothy Claire
Lyn 9h14'59"34d29'
Johannesson,Taylor Ellenne
Aur 6h54'29"40d18'
Johanning,Jr "Jungle Bob",Robert
Aql 19h27'30"15d32'
Johansen,Billy & Michelle
Uma 9h56'22"42d38'
Johansen,Bob & Briana
Aql 19h57'44"1d46'
Johansen,David
Aur 4h50'26"40d26'
Johansen,Elizabeth Whitney
Com 12h32'1"14d20'
Johansen,Jonathan C
Per 4h1'14"50d33'
Johansen,Kimberly
Lib 15h42'57"-23d53'
Johansen,Stephanie A
Cas 0h16'44"64d9'
Johansen-His Very Own, Kenneth
Vul 19h48'1"28d8'
Johanson,Wesley Eugene
Aql 19h5'1"3d5'
Johansson,Dagmar
Cnv 13h20'44"40d41'
Johantgen,Jessica E
Uma 9h7'23"54d6'
Johmann,John
Her 16h35'27"50d1'
John
Cep 22h26'27"59d53'
John
Aql 19h52'38"1d44'
John
Cas 2h12'16"68d1'
John
Cep 22h32'60"60d50'
John
Lac 22h6'41"51d45'
John
And 0h5'7"36d54'A
John
Cyg 21h0'24"37d27'A
John
Ori 4h46'1"4d47'
John & Barbara
Peg 21h59'53"30d7'
John & Betsan
Cyg 19h29'48"36d21'
John & Betty Star
Aql 19h20'38"15d2'
John & Camille's Eternal Love
Mon 6h19'13"8d47'
John & Catherine
Uma 9h0'0"58d54'
John & Charlotte
Per 3h33'39"35d21'
John & Chloë Will Always Love Gabriela
Vir 11h41'53"9d33'
John & Christie's Star
Dra 18h15'47"65d2'
John & Dana,Always & Forever
Cnv 12h48'57"39d22'
John & Denise
Crb 15h20'32"30d21'
John & Dorothy's Star Of Golden Love
Cyg 20h16'25"41d45'
John & Edel 1996
Aql 18h44'44"10d7'
John & Erin
Cyg 21h19'0"38d9'
John & Florence
Her 16h42'36"34d39'
John & Illia
Pho 0h49'7"-46d39'
John & Irene
Ori 5h59'12"11d25'
John & Janes "Wedding Light"
Cam 6h8'1"68d5'
John & Janet's Star of Mt Whisper
Com 12h29'37"23d22'
John & Joan
Lyn 7h12'37"56d13'
John & Joann's Lucky Star
Sct 18h44'19"-8d16'
John & Joette
Uma 8h57'47"54d35'
John & Julia
Cyg 20h53'30"31d36'
John & Julia's Wedding Star
Ori 5h38'55"-0d5'
John & Karen
Cet 2h38'15"1d42'
John & Laura
Cyg 19h26'19"31d13'
John & Linda
Cyg 20h5'1"40d0'
John & Lori's Star
Cyg 21h3'25"38d36'
John & Michele Forever
Lac 22h15'36"37d57'
John & Pat
Cyg 21h31'0"37d42'
John & Rachel
Uma 11h6'19"32d1'
John & Rebecca's Soul- Star
Mon 6h53'32"-1d43'
John & Regina's Special Star
Crb 15h53'1"27d18'
John 1990
Tau 5h46'14"24d14'
John A "Mac"
Boo 14h41'31"21d5'
John A True Believer
Uma 11h37'37"50d16'
John Adam
Cyg 21h14'57"28d46'
John Alexander
Uma 9h57'37"42d11'
John Andrew
Boo 15h13'1"52d39'
John Andrew
Aur 5h40'22"50d32'
John Anthony
Cep 21h26'45"56d2'
John Anthony
Sge 20h17'57"20d37'
John B
Vir 11h44'31"7d34'
John B,The
Cep 20h20'11"80d3'
John Benjamin
Per 3h11'41"44d4'
John Boy
Cma 6h54'18"-19d57'
John Boy
Sct 18h49'37"-7d38'
John Bradley Jacques
Per 3h15'0"41d15'
John Britta
Equ 21h4'33"12d1'
John Callie
Cep 0h29'19"78d4'
John Charles
Cas 2h10'50"59d49'
John D
Per 1h56'28"47d51'
John Daniel
Ori 4h50'16"5d13'
John David
Oph 18h4'30"10d40'
John David
Aqr 22h48'0"0d53'
John E
Her 17h38'23"26d46'
John Eddie
Ser 15h56'10"21d20'
John Eric
Peg 23h28'57"31d25"
John Forever
Her 17h59'1"40d42'
John Forever Dianne
Aur 7h24'12"38d27'
John James & Sheryn Eve
Cam 8h49'39"74d32'
John Jane Laura Victoria
Umi 15h23'0"68d41'
John Jim Randy Laurie VCHS Class of 1973
Uma 9h28'36"48d49'
JOHN JOLLY PA-PA'S PINNACLE
Aql 18h45'15"8d11'
John Lee
Aql 20h7'34"1d25'
John Loves Janet Forever 6-27-87
Eri 2h49'0"-7d23'
John M
Uma 11h29'42"31d12'
John Michael
Oph 17h2'37"-20d47'
John My Father My Hero
Del 20h22'27"19d7'B
John Nichol Mark
Oph 17h17'0"10d20'
John Of Tysnes
Lac 22h38'37"50d25'
John Our Love Burning Eternally Lynette
Ara 17h56'26"-56d16'
John P
Per 4h19'23"51d42'
John Paul
Cep 23h9'59"64d1'
John Peter
Eri 3h4'17"-2d34'
John Philip
Dra 15h12'52"63d59'
John Randolph
Dra 16h13'56"60d10'
John Robert
Tau 4h4'18"22d34'
John Robert
Aur 4h56'59"50d34'
John Star,The
Aql 18h49'0"11d5'
John The Elder
Her 16h11'1"40d11'
John Thomas
Her 17h35'36"40d15'
John Vincent
Her 16h20'12"8d42'
John Walter
Lac 22h33'50"52d32'
John William & Sarah Marie
Uma 9h55'24"47d35'
John's Arizona Star
Lac 22h11'29"38d55'
John's Charisma
Sgr 18h55'59"-33d1'
John's Dream
Aql 19h55'56"12d32'
John's First Sun
Cep 22h58'45"58d7'
John's Joy
Cam 4h0'48"61d4'
John's Light
Mon 7h50'49"-2d54'A
John's Light
Her 15h49'1"40d57'
John's Little Bit of Heaven
Aql 18h56'36"11d27'
John's Lucky Star
Per 2h59'58"32d53'
John's Noel 92
Aur 06h26'58"41d7'
John's Pass Christian Dream
Aql 19h1'0"8d30'
John's Promise
Boo 13h36'37"20d36'
John's Realm
Uma 11h37'0"30d15'
John's Secret Star
Boo 14h57'57"48d36'
John's Shooting Star, William
Her 18h0'54"40d19'
John's Star
Cep 22h9'0"67d43'
John's Star-Forever And Always
Cnv 12h39'26"51d45'
John's Ugheli
Aql 19h53'34"13d46'

John,Benjamin
Her 18h17'47"20d21'
John,Daniel
Cep 21h33'22"58d38'
John,Elton
Lac 22h1'57"40d39'
John,Fischer Jordan Fledder
Vul 19h20'31"27d16'
John,Kurt Anthony
Lac 22h24'38"38d47'
John,My Love
Per 2h3'0"48d55'
John,Par Amour Eternity
Cyg 21h33'22"30d8'
John,Peter
Aur 6h50'43"38d22'
John,Rachael Elizabeth
Lyr 18h29'43"46d31'
John-Forever Loved
Cet 3h7'58"5d23'
John-Paul
Aql 19h18'45"19d33'
John-Roger
Sex 10h46'51"-6d3'
John-Wes-Howie
Per 2h59'51"46d21'
Johna Catherine
Oph 17h13'27"-18d56'
Johncox II,William L
Cep 22h55'20"57d5'
Johnecheck,Cheryl L
Cet 0h24'36"-11d50'
Johnecheck,Marjorie H
Eri 4h56'3"-7d47'
Johnecheck,Steve & Colette
Ori 5h56'51"18d59'
Johnes,Eric James
Her 16h37'16"38d34'
Johnetta,Celeste
Cyg 20h33'45"42d18'
Johnheather
And 23h40'38"44d8'
Johnke,Jenny Ray
Cam 3h28'17"61d33'
Johnnie R
Eri 3h13'37"-4d10'
Johnnie,Love's First Kiss
Lac 22h21'21"50d37'
Johnnifer
Lyn 8h40'20"36d24'
Johnny
Her 15h53'46"41d13'
Johnny
Aur 5h21'48"54d51'
Johnny "Angel"
Cep 0h59'32"78d0'
Johnny & Gioconda Forever Love
Eri 2h54'31"-17d50'
Johnny & Gwendarlin'
Cyg 20h36'41"45d26'
Johnny & Shanna
Cet 2h41'39"-12d22'
Johnny Ang,The
Per 4h46'1"51d1'
Johnny Angel
Uma 8h39'0"50d3'
Johnny Angel
Peg 21h53'47"18d55'
Johnny Angel
Her 18h4'34"31d30'
Johnny Angel
Her 18h12'1"31d16'
Johnny Angel
Her 17h39'11"21d30'
Johnny Angel
Aur 6h21'1"31d40'
Johnny Boy
Mon 7h0'42"1d9'
Johnny Boy
Peg 23h16'27"30d47'
Johnny D
Ser 18h5'49"-1d6'
Johnny D
Hya 8h37'59"-10d3'

Johnny McC
Aur 5h17'43"42d33'
Johnny O
Per 4h5'44"50d52'
Johnny S
Per 3h23'56"54d26'
Johnny S
Aqr 21h4'45"-8d26'C
Johnny Star,The
Boo 14h13'44"16d34'
Johnny V
Lac 22h3'33"49d38'
Johnny V
Aur 5h7'34"42d11'
Johnny Z
Dra 18h11'1"65d24'
Johnny's Star
Cep 20h18'29"75d22'
Johnny's Star "Dream Come True"
Ori 5h55'57"15d11'
Johnny,No Stranger to Danger
Ori 6h6'0"5d6'
Johnny-Mary-Jinny & Benji
Ori 5h56'11"9d3'
Johnny-O
Aur 7h2'14"36d37'
Johns,Agnes M
Cyg 19h18'58"45d21'
Johns,Aristotle Christopher
Cep 21h19'0"65d31'
Johns,Bethany Leigh
Peg 23h15'55"33d13'
Johns,Bobbie
Per 3h4'12"50d23'
Johns,Brandon Robert
Sco 16h26'11"-30d24'
Johns,Charles Clarence
Ori 4h42'40"0d37'
Johns,Charles Robert
Oph 17h31'34"-22d53'
Johns,Christopher Constantine
Cam 10h53'28"80d32'
Johns,Constantine Christopher
Cam 10h56'0"80d4'
Johns,Emily Kristin
Mon 8h8'46"-0d33'
Johns,Ian Michael
Aql 19h5'32"-0d16'
Johns,Joe Lel Duce
Per 1h36'36"53d49'
Johns,John L
Lac 22h37'1"50d14'
Johns,Kathleen Lynn
Cas 3h22'22"75d35'
Johns,Lynn P
Mon 6h17'0"-5d39'
Johns,Rachael Katrina
Mon 6h6'41"-4d10'
Johns,Rosemarie
Crt 10h59'16"-17d49'
Johnsen,Betty Jane Bernier
Cas 2h34'23"70d29'
Johnsen,James Warren
Cmi 7h16'22"4d17'
Johnsen,Petter
Psc 0h22'15"7d59'
Johnsen,Thomas Alan
Ari 2h56'11"30d49'
Johnsey Star,The
Cyg 19h29'12"41d4'
Johnsey,Mona
Peg 0h2'58"30d10'
Johnson
Vul 19h19'19"26d22'
Johnson
Her 17h22'40"29d6'
Johnson "Dr Hooks", David L
Tri 1h34'58"35d14'
Johnson 1/27/53-12/13/94,Scott Alden
Ori 5h15'32"7d30'

Johnson 1993,Frieda Lou
Oph 17h10'25"-20d41'
Johnson 1994,Charley & DJ
Cep 22h19'41"65d10'
Johnson Architect, Samual J
Her 17h7'1"50d13'
Johnson Class Of 94, Melanie Arlene
Cas 23h40'23"60d47'
Johnson Dy N 2 Live, Phyllis Ann
Tau 3h57'0"10d10'
Johnson Family,Everett
Aur 6h14'41"46d50'
Johnson H5OTHA,Art & Dolly
Her 16h23'49"50d40'
Johnson HB,Thom
Aur 5h55'10"29d43'
Johnson I, '94
Umi 17h23'55"76d24'
Johnson II,Fredrick Palmer
Aql 18h59'46"-4d42'
Johnson II,Robert Edward
Sgr 18h3'52"-28d25'
Johnson II,Thomas
Peg 23h16'12"31d22'
Johnson III,Frank
Per 3h27'41"31d16'
Johnson IV,Ernest K
Hya 8h56'41"0d23'
Johnson Light Of My Life,Cassandra
And 23h36'38"46d43'
Johnson My Guiding Light,Robert Gordon
Her 16h45'28"48d56'
Johnson Of Omaha,NE, Linda & Perry
Peg 21h52'42"31d40'
Johnson Star,The J W
Lac 22h37'39"53d46'
Johnson Super Star, The Michael D
Boo 14h27'10"26d1'
Johnson Sweet 16,Tasha Michelle
And 0h57'47"34d55'
Johnson"Jay Llska Dae",B Bruce
Del 20h21'38"8d8'
Johnson's Eastern Star Yvonne
Crt 11h16'30"-20d28'
Johnson's Jewel
Uma 9h34'53"57d55'
Johnson's Pennsylvania Dog Star,D P
Uma 10h47'29"48d33'
Johnson(Pappy),Samuel Eugene
Aur 5h53'48"38d50'
Johnson,"My Dad" Jeffrey M
Oph 18h0'40"13d57'
Johnson,Adam Brandley
Cet 3h5'1"0d43'
Johnson,Adam Casey
Aur 7h23'49"38d47'
Johnson,Adam Ross
Vir 11h51'41"7d42'
Johnson,Albin "Al"
Aur 6h52'57"35d30'
Johnson,Alek Linwood York
Dra 17h47'53"68d42'
Johnson,Alexander Miles
Per 2h56'23"37d31'
Johnson,Alice H & Frederick
Lyn 7h47'37"41d36'
Johnson,Allan Robert
Aur 6h28'1"30d52'
Johnson,Amazing Jason
Cmi 7h52'47"10d30'
Johnson,Amy
Cas 0h56'26"61d7'
Johnson,Andrew Chuck
Dra 17h17'34"63d60'

Johnson,Andrew D
Leo 11h28'32"-2d3'
Johnson,Andrew Ross
Aur 4h57'30"37d51'
Johnson,Angela Janine
Peg 22h57'45"22d54'
Johnson,Anita
Del 20h24'44"7d39'
Johnson,Ann Barrett
Del 20h21'11"8d46'
Johnson,Anna
Cas 2h32'17"61d33'
Johnson,Anthony William
Cnv 12h43'0"51d16'
Johnson,Anthony Godby & Jack Godby
Cep 21h10'49"68d18'
Johnson,Anthony Godby
Uma 10h46'46"47d4'
Johnson,April Angel
Peg 22h41'1"10d14'
Johnson,April E
Cas 23h1'44"53d8'
Johnson,Ashley Elizabeth
Aql 18h58'58"-6d58'
Johnson,Barry James
Com 13h21'32"26d27'
Johnson,Bennie Grace Haley
Lyr 18h53'22"32d15'
Johnson,Bernadine Ann
Peg 23h5'28"33d7'
Johnson,Bernard Allan
Lac 22h44'45"54d9'
Johnson,Bernard W
Aur 5h0'31"31d4'
Johnson,Bernett L
Lyn 8h55'35"38d43'
Johnson,Betty
Cyg 19h45'0"38d38'
Johnson,Beverly
Lyn 7h45'20"58d8'
Johnson,Bill
Hya 8h57'11"5d38'
Johnson,Blake & Marissa
Mon 6h32'26"1d9'
Johnson,Bob
Her 16h2'24"20d3'
Johnson,Brent Allan
Cet 2h31'11"8d7'
Johnson,Brett Alan
Hya 9h36'17"1d28'
Johnson,Brett Curtis
Aql 18h58'1"16d55'
Johnson,Brett D
Cnc 8h40'24"18d27'
Johnson,Brian
Uma 10h26'35"54d32'
Johnson,Brian
Per 1h56'26"56d44'
Johnson,Brian L
Aur 6h11'32"31d29"
Johnson,C Farrell
Lmi 9h24'1"37d41'
Johnson,Candace Susan
And 23h36'19"45d40'
Johnson,Candice Marie
Cas 22h59'38"54d33'
Johnson,Carina
Lyr 18h14'1"32d4'
Johnson,Carl
Uma 8h15'12"72d33'A
Johnson,Carol
Lib 14h58'16"-10d26'
Johnson,Caroline Mary
Uma 8h33'41"67d60'
Johnson,Carolyn
Del 20h18'11"18d52'
Johnson,Carter Vincent
Cma 7h17'22"-15d38'
Johnson,Cary Lee
Aql 19h0'13"-6d15'
Johnson,Catherine
Eri 3h36'13"-6d57'
Johnson,Cecelia & Guy
Sge 19h58'21"16d44'

Johnson,Chad
Her 15h50'40"42d4'
Johnson,Charles
Per 4h33'51"33d56'
Johnson,Charles Preston
Aql 19h52'31"10d39'
Johnson,Charles "CJ"
Cyg 19h23'51"44d51'
Johnson,Charles & Susan
Ori 5h16'56"-8d49'
Johnson,Chelen
Cep 22h11'24"55d27'
Johnson,Chelen Hey
And 0h13'21"46d11'
Johnson,Christina Ann
Cyg 21h1'33"30d53'
Johnson,Christine Elizabeth
Ori 6h13'0"8d5'
Johnson,Christine
Cas 0h46'31"66d12'
Johnson,Christopher
Cep 0h18'14"69d57'
Johnson,Christopher Randolph
Crb 16h18'43"30d15'
Johnson,Christopher Armando
Her 16h58'1"38d13'
Johnson,Christopher G
Boo 15h0'52"11d7'
Johnson,Chuck "Lucky"
Eri 4h13'20"-15d8'
Johnson,Cindy
Ori 4h43'22"12d52'
Johnson,Cindy Michelle
Ori 5h51'25"14d26'
Johnson,Cindy Rose
Lyn 9h14'0"35d13'
Johnson,Claudia Jane
And 23h27'15"45d4'
Johnson,Colin C
Aur 5h2'31"40d39'
Johnson,Connor Ross
Cep 20h20'50"75d40'
Johnson,Coyote Man- Erik David
Lyn 7h51'17"40d8'
Johnson,Craig R
Aur 6h42'23"38d54'
Johnson,Curtis Alan
Boo 14h3'18"14d16'
Johnson,Cynthia A
Eri 3h36'11"-2d49'
Johnson,Cyrus E
Cep 22h33'52"58d10'
Johnson,D Neil
Dra 14h14'49"64d13'
Johnson,Dale Dean
Boo 14h56'0"28d2'
Johnson,Dalton Robert
Sex 10h35'43"2d23'
Johnson,Daniel Galen Arrowsmith
Umi 15h11'31"67d20'
Johnson,Daniel Gerald
Cnc 7h57'27"11d2'
Johnson,David
Dra 17h21'14"64d40'
Johnson,David
Aql 19h55'28"15d16'
Johnson,David Evon
Ser 16h18'29"0d24'
Johnson,Dawn "Goofy"
Mon 7h38'31"-5d40'
Johnson,Deborah A
Cyg 21h19'38"37d50'
Johnson,Dennis Anthony
Cet 2h33'18"7d29'
Johnson,Diane Elizabeth
And 1h21'1"41d8'
Johnson,DiAnne
And 22h57'30"36d36'
Johnson,Don
Uma 10h1'42"57d17'
Johnson,Don & Letha
Cyg 20h20'12"40d22'

Johnson,Donald
Aur 6h4'25"31d6'
Johnson,Donald G
Dra 19h49'58"84d50'
Johnson,Donald T
Aql 20h14'32"0d3'
Johnson,Donna & Phillip
Cyg 21h8'11"30d49'
Johnson,Donna Jo
Lyn 7h45'43"41d21'
Johnson,Donna Marie
Lmi 9h20'19"34d7'
Johnson,Donna Marie
Com 13h30'33"26d19'
Johnson,Donovan
Cep 21h14'1"55d42'
Johnson,Doris
Dra 11h34'50"78d56'
Johnson,Doris J
Mon 5h56'13"-6d47'
Johnson,Dorothy
Peg 21h55'40"26d23'
Johnson,Douglas
Cep 21h49'1"85d52'
Johnson,Dr Krista
Oph 17h59'51"7d45'
Johnson,E K
Crt 11h39'32"-8d41'
Johnson,Earl
Uma 8h15'12"72d33'B
Johnson,Earl R
Ori 5h57'17"14d52'
Johnson,Earl Riley
Per 4h6'1"51d3'
Johnson,Edgar & Eleanor
Sge 19h10'23"16d60'
Johnson,Edmund Lloyd
Oph 17h39'21"-28d20'
Johnson,Edward Perry
Per 2h45'13"43d52'
Johnson,Edward William
Aur 5h44'53"29d34'
Johnson,Eileen W
Sge 19h53'37"19d19'
Johnson,Eleanor H
Lyr 19h16'19"40d44'
Johnson,Elliott Thomas
Mon 6h39'0"10d40'
Johnson,Ellsworth H
Her 17h13'1"48d59'
Johnson,Elois Westbrook
Cet 2h19'41"7d17'
Johnson,Emily Dawn
Lyr 18h47'53"40d36'
Johnson,Eric M
Boo 14h52'43"23d37'
Johnson,Eric P
Her 16h18'52"10d8'
Johnson,Eric Robert
Lac 22h26'20"55d5'
Johnson,Eric Russell
Leo 11h12'40"27d50'A
Johnson,Eric Samuel Tecumseh
Aql 19h24'38"10d39'
Johnson,Erica Simone
Lyn 8h59'18"46d13'
Johnson,Ervin C
Cet 3h2'45"4d16'
Johnson,Ethan Mark
Ori 6h4'12"10d36'
Johnson,Evelyn
Lyr 18h40'42"26d10'
Johnson,Fisherman's Dream,Irv
Hya 8h51'24"-5d49'
Johnson,Flying-Ace "Daddy" Don
Aql 19h47'16"12d49'
Johnson,Fran
Oph 17h21'45"-20d17'
Johnson,Frank James
Boo 14h36'31"52d10'
Johnson,Frederick T L
Ori 6h15'14"1d16'

Johnson,Freedom For Emma
Aql 20h5'22"1d23'
Johnson,Gail
Leo 9h21'27"18d59'
Johnson,Gary
Per 1h38'19"54d18'
Johnson,Gary A
Ser 15h59'25"-1d30'
Johnson,Gary J
Per 1h50'51"56d37'
Johnson,Gayla Michelle
Sct 18h53'18"-7d29'
Johnson,Gemma Leigh
Cas 0h17'33"62d38'
Johnson,Gene Paul
Aql 20h2'46"7d34'
Johnson,Genny
Cmi 7h35'24"0d32'
Johnson,George
Aur 5h41'19"50d37'
Johnson,George F
Cnv 12h52'20"40d9'
Johnson,Gerald Herbert & Ruby May
Cyg 20h51'26"40d55'
Johnson,Gladys
Cyg 20h7'12"41d13'
Johnson,Glenn R
Cap 21h2'41"-15d49'
Johnson,Gregory Colin
Del 20h21'21"7d57'
Johnson,Harold D
Cnv 13h55'14"41d33'
Johnson,Harriet Truland
Leo 9h55'21"33d8'
Johnson,Haven
Aql 20h1'45"9d26'
Johnson,Hazel
Aql 19h39'48"12d2'
Johnson,Heather Lee
Lyn 7h58'14"44d15'
Johnson,Henry
Uma 12h56'26"53d42'
Johnson,Herbert C
Cet 0h36'42"-5d47'
Johnson,Howard K
Cep 3h29'25"77d35'
Johnson,Ian Andrew
Her 17h34'46"21d55'
Johnson,IEJ Janet
Cas 1h2'1"62d45'
Johnson,Ingrid
Cep 22h56'42"80d28'
Johnson,Jack
Her 18h11'34"40d13'
Johnson,Jacqueline
Umi 15h36'10"74d49'
Johnson,Jake Thomas
Boo 15h16'32"52d41'
Johnson,James
Her 18h13'25"37d37'
Johnson,James Alfred
Boo 13h44'32"27d49'
Johnson,James Alfred
Ser 15h59'19"0d17'
Johnson,James E
Vul 19h36'38"22d59'B
Johnson,James Eric
Her 17h12'30"48d40'
Johnson,James H
Oph 18h5'30"13d14'
Johnson,Jamie
Cep 22h57'18"68d9'
Johnson,Jamie William
Uma 10h41'53"47d36'
Johnson,Janet Ann
Oph 17h2'30"7d53'
Johnson,Janet Eileen
Eri 3h3'43"-3d12'
Johnson,Janet I
Ori 6h12'0"1d35'
Johnson,Jasmine Yvonne
Sco 16h58'42"-37d40'
Johnson,Jason
Aur 6h27'48"31d35'

Johnson,Jason George
Her 17h50'22"48d49'
Johnson,Jayne Ann
Cyg 19h36'56"28d25'
Johnson,Jean
Lyr 18h17'43"42d16'
Johnson,Jeffrey
Cam 6h12'47"61d49'
Johnson,Jeffrey Alan
Cet 3h16'54"9d31'
Johnson,Jenni & Wayne
Cyg 21h16'55"38d55'
Johnson,Jennifer
Aur 5h26'58"40d44'
Johnson,Jennifer A
Cam 6h15'13"65d5'
Johnson,Jennifer Ann
Cas 0h10'1"50d10'
Johnson,Jennifer Lee
Equ 21h16'28"2d41'
Johnson,Jenny E
Vir 11h38'37"5d40'
Johnson,Jessica Grace
Lyn 8h31'0"43d14'
Johnson,Jimmie Lee
Hya 9h37'59"0d13'
Johnson,Joan & Curtis
Cam 13h19'0"77d54'
Johnson,Joan Delores
Lyn 9h2'0"37d53'
Johnson,Joanne
Mon 6h57'24"10d0'
Johnson,Joanne
Sco 16h56'23"-40d17'
Johnson,Joe
Hya 9h7'57"3d51'
Johnson,Joel Gregory
Boo 15h9'53"40d53'
Johnson,Joel Scott
Tau 4h2'55"22d49'
Johnson,John David
Aur 5h57'17"31d50'
Johnson,John R & Socorro M Johnson
Lac 22h28'36"55d10'
Johnson,Joseph
Aur 7h16'43"40d54'
Johnson,Josh
Lyn 8h53'43"45d40'
Johnson,Jr,Guy Russell
Aur 6h13'43"38d58'
Johnson,Jr,James D
Her 17h11'51"46d29'
Johnson,Jr,Johnny Emanuel
Cep 22h24'58"58d50'
Johnson,Jr,Joseph J
Cet 0h39'12"-2d49'
Johnson,Jr,Loyd C
Sex 9h55'54"-0d8'
Johnson,Jr,Marvin John
Aur 7h15'1"37d3'
Johnson,Jr,Morris I
Cnv 13h42'39"40d40'
Johnson,Juanita May
Eri 4h4'15"-6d52'
Johnson,Judy Kay
Cas 1h17'32"61d24'
Johnson,Julie
And 2h10'0"40d52'
Johnson,Julie Dawn
Cmi 7h33'34"7d39'
Johnson,Kandi Joy
Uma 9h22'58"52d25'
Johnson,Karen
Lyr 18h40'12"36d19'
Johnson,Karen
Cas 0h23'50"74d2'
Johnson,Karen & David Goodrich
Cam 7h48'1"83d30'
Johnson,Karin Ashe- Taylor
Lyr 18h47'11"35d4'
Johnson,Kathleen
Uma 9h52'32"57d36'

Johnson,Kathleen Ann
Lyr 18h47'0"31d27'
Johnson,Kay
Cas 2h33'14"60d1'
Johnson,Keith & Linda
And 0h36'0"40d39'
Johnson,Kelli Brooke
Cmi 7h22'1"8d53'
Johnson,Kelsey
Lyr 18h51'19"40d17'
Johnson,Ken & Eileen
Crb 15h48'38"30d54'
Johnson,Kenneth Bradley
Oph 16h59'50"-25d47'
Johnson,Kenneth
Her 17h29'40"20d3'
Johnson,Kenneth G
Aur 7h14'44"36d19'
Johnson,Kenneth M
Lib 14h59'34"-10d15'
Johnson,Kenneth R
Aur 5h2'56"40d2'
Johnson,Kent
Aql 20h5'42"0d39'
Johnson,Kevin & Angie
Dra 19h4'37"80d17"
Johnson,Kim
Vul 19h59'20"25d13'
Johnson,Kimberly
Cap 20h32'1"-13d0'
Johnson,Kimberly Lynn
And 1h13'1"38d30'
Johnson,Kona
Sco 17h54'11"-38d38'
Johnson,Kristen Danielle
Mon 7h48'29"-1d51'
Johnson,Kristy
And 2h23'20"44d26'
Johnson,Kyle Wilde
Boo 14h6'19"25d35'
Johnson,L Robert & Reneé N DuPont
Cyg 19h28'1"56d53'
Johnson,Lance & Roberta
Mon 6h37'0"1d46'
Johnson,Larry & Beth
Lyn 7h50'56"44d40'
Johnson,Laura M
Cas 23h39'27"62d22'
Johnson,Lauren Marie
Uma 10h42'52"51d44'
Johnson,Lawrence Earl
Cmi 7h38'35"4d0'
Johnson,Lawrence R
Ori 6h16'33"7d49'
Johnson,Lawton C
Per 2h0'42"47d27'
Johnson,Leah Danielle
Equ 21h11'25"10d59'
Johnson,Lee
Peg 22h37'50"29d14'
Johnson,Lee J
Cep 21h33'22"68d49'
Johnson,Lei Selu
Lyn 8h13'27"50d21'
Johnson,Leigh
Uma 9h12'33"48d21'
Johnson,Len
Cmi 7h45'56"8d15'
Johnson,Lesley
Uma 12h8'38"47d35'
Johnson,Leslie Marie
Boo 14h45'33"51d11'
Johnson,Leslie Roberta Beatty
And 0h5'31"38d12'
Johnson,Lillian
Cmi 7h27'0"1d50'
Johnson,Linda
Lyr 19h15'57"28d33'
Johnson,Linda F
Cmi 7h55'13"7d46'
Johnson,Linda P
Lyn 9h14'46"39d2'
Johnson,Lindsey Sue
Uma 10h44'25"52d57'

Johnson,Lisa
Mon 6h33'1"7d17'
Johnson,Liza M
Peg 21h57'24"35d52'
Johnson,Logan Kenneth
Cep 0h43'30"78d41'
Johnson,Logan Wayne
Uma 13h24'51"61d1'
Johnson,Lois Joy
Uma 11h8'1"37d49'
Johnson,Lonny G
Cyg 19h29'16"30d0'
Johnson,Lorayne
Cet 1h51'19"-3d7'
Johnson,Lyndsay Michelle
Cas 1h4'23"61d4'
Johnson,M David
Her 16h30'17"42d10'
Johnson,Mabel S
Cas 1h19'13"61d29'
Johnson,Marc Douglas
Ser 15h40'0"6d30'
Johnson,Marian M
Aur 4h56'31"40d41'
Johnson,Mark
Her 17h0'1"18d50'
Johnson,Mark A
Ser 15h14'16"8d41'
Johnson,Mark Gaylene Kara & Brian
Dra 20h19'43"68d14'
Johnson,Marlee Rae
Mon 7h43'12"-8d36'
Johnson,Marsha S
Com 12h25'0"26d21'
Johnson,Mary Alma
Cam 5h10'28"68d28'
Johnson,Mary E
Cas 1h20'14"53d1'
Johnson,Mary Elizabeth
Cas 23h39'55"61d13'
Johnson,Mary Patricia
Peg 22h38'18"27d41'
Johnson,Matthew Gregory
Per 4h43'17'40d45'
Johnson,Matthew Adam
Ser 16h19'51"0d50'
Johnson,Maura Elizabeth
And 23h1'55"50d2'
Johnson,Maxwell Lewis
Ori 5h55'21"7d7'
Johnson,McKenzie Lee
Lyn 9h3'1"44d33'
Johnson,Megan
Sge 19h40'36"18d57'
Johnson,Melanie Roberts
Mon 6h44'50"10d49'
Johnson,Melanie Ann
Ori 5h52'0"12d26"
Johnson,Melinda Q
Peg 22h2'20"31d39'
Johnson,Michael
Cet 1h29'28"-14d23'
Johnson,Michael
Lyr 19h23'37"31d16'
Johnson,Michael J
Cyg 19h48'19"38d30'
Johnson,Michael James
Her 17h14'54"49d33'
Johnson,Michele Dyan
Boo 14h31'0"10d44'
Johnson,Michelle
Cas 23h16'0"61d1'
Johnson,Michelle Anne
Aql 19h46'48"10d19'
Johnson,Midnightstar- Shar Ryan
Dra 16h52'30"64d57'
Johnson,Mona
Peg 0h6'29"14d51'
Johnson,Monica Sue
Lyr 18h44'48"33d1'
Johnson,Mousie
Mon 6h52'36"10d11'

Johnson,Ms Patricia
Cas 0h28'42"58d51'
Johnson,Muriel
Mon 7h44'25"-1d39'
Johnson,Nadine Victoria & Aldridge
Ori 5h36'27"-0d1'
Johnson,Naida W
Mon 6h30'31"-0d49'
Johnson,Nancy G
Aur 5h53'12"29d21'
Johnson,Nancy H
Cam 7h55'55"61d43'
Johnson,Nathan Rustin
Dra 14h43'37"62d35'
Johnson,Nathaniel Paul
Aur 6h16'35"45d33'
Johnson,Nicole Sylvia Rose
Aqr 21h1'39"-5d52'
Johnson,Our Grandad
Uma 10h6'0"48d30'
Johnson,Pamela Trosper
Uma 10h9'48"52d22'
Johnson,Patrice L
Cas 0h23'57"66d7'
Johnson,Patricia G
Uma 12h57'54"58d22'
Johnson,Paul
Ori 4h56'32"15d25'
Johnson,Paul Edward
Ori 4h58'49"13d57'
Johnson,Paul William
Sex 9h44'28"-0d30'
Johnson,Perry Edward
Per 3h9'1"38d56'
Johnson,Perry Lee
Aql 19h56'57"8d27'
Johnson,Perry Lee
Sct 18h44'43"-5d25'
Johnson,Perry M
Boo 14h52'43'23d36'
Johnson,Peter Craig
Ori 5h37'22"-4d46'
Johnson,Peyton
Aql 20h1'23"6d53'
Johnson,Peyton Selena
Uma 9h29'53"56d9'
Johnson,PhD,Daniel Palmer
Aql 19h47'33"14d60'
Johnson,Philip
Ori 5h35'13"-1d41'
Johnson,Phillip Renford Anthony
Cep 0h19'19"68d44'
Johnson,R J
Sgr 18h55'9"-32d36'
Johnson,R K
Ser 16h14'22"1d14'
Johnson,Rachel
Mon 7h43'20"-3d24'
Johnson,Randall Lee & Betsy
Boo 15h35'35"40d9'
Johnson,Randy
Per 1h50'37"48d7'
Johnson,Raymond Peter
Lac 22h27'1"50d16'
Johnson,Rebecca
And 2h31'56"44d41'
Johnson,Rebecca Ashley
And 1h43'18"40d32'
Johnson,Reg
Her 17h14'1"29d26'
Johnson,Richard Rienhold
Per 2h59'1"46d22'
Johnson,Richard
Oph 17h4'48"-23d50'
Johnson,Richard
Aur 6h37'48"37d51'
Johnson,Richard Charles
Sex 10h3'28"5d21'
Johnson,Richard D
Aql 20h11'15"12d18'
Johnson,Rick
Cmi 7h23'47"7d44'

Johnson,Rita "Coo"
And 2h13'57"39d38'
Johnson,Robert
Uma 9h28'45"55d10'
Johnson,Robert
Cep 22h40'21"65d58'
Johnson,Robert Alan
Her 16h19'15"20d10'
Johnson,Robert Allen
Vul 19h18'53"23d6'
Johnson,Robert Andrew
Cmi 7h23'38"9d50'
Johnson,Robert C & Ruth M
Boo 14h17'58'36d20'
Johnson,Jr,Robert Dudley
Ori 6h3'58"10d51'
Johnson,Robert Edward
Sgr 18h3'0'-28d22'
Johnson,Robert L
Dra 19h0'59"50d5'
Johnson,Robert LeRoy
Ori 0h11'25"0d58'
Johnson,Robert Russell
Cet 3h10'25"2d4'
Johnson,Robin Lee
Cam 3h14'59"61d39'
Johnson,Rodney M
Sex 10h20'16"-5d54'
Johnson,Ron "Jules"
Her 16h29'58"40d38'
Johnson,Ronald L
Cnv 12h5'1"34d48'
Johnson,Ronald Wayne
Per 3h57'19"38d37'
Johnson,Rosalina Angelina Pepe
Aur 5h11'18"40d35'
Johnson,Rosemarie Brady
Mon 6h59'32"-5d45'
Johnson,Ruby Jean
Hya 8h43'55"3d5'
Johnson,Rubye Spain
Cyg 19h12'20"48d12'
Johnson,Rudy
Cam 3h59'10"58d2'
Johnson,Russell Edward
Ari 2h59'47"28d33'
Johnson,Ruth
Aql 19h20'43"15d50'
Johnson,Ryman Gordon
Sct 18h49'30"-7d17'
Johnson,Samuel David
Aur 5h18'38"48d56'
Johnson,Sandra Marie
Uma 14h14'15"60d46'
Johnson,Sandra Marie
Vir 13h15'1"-15d11'
Johnson,Sativa
Oph 17h51'0"13d22'
Johnson,Scott Christian
Lyr 18h58'0"30d41'
Johnson,Shane Myboolah
Per 4h7'12"51d11'
Johnson,Shannon M
Cep 22h14'25"70d28'
Johnson,Shendelle Marie
Ori 6h1'0"-0d33'
Johnson,Sherri Lynn
Cyg 20h15'13"38d15'
Johnson,Siobhán
Lyr 18h36'40"33d24'
Johnson,Sky
Uma 10h19'34"50d42'
Johnson,Sonja Lee
Peg 22h43'1"4d26'
Johnson,Spinner
Per 3h26'52"51d53'
Johnson,Sr,Wayne Perley
Cam 7h7'30"60d51'
Johnson,Staci Hilene
Peg 21h48'1"35d59'
Johnson,Stamichael
Aur 4h56'41"40d4'
Johnson,Starlett
Aql 19h55'45"13d59'

Johnson,Stephanie Louise
Uma 8h45'31"62d29'
Johnson,Stephen William
Cet 2h14'0"6d25'
Johnson,Stephen Alan
Her 17h7'29"41d52'
Johnson,Steve A
Gem 7h0'24"35d17'
Johnson,Steven
Per 2h58'57"43d18'
Johnson,Steven M
Uma 12h51'50"53d10'
Johnson,Steven Wayne
Boo 14h33'54"48d5'
Johnson,Sydney Reed
Leo 10h49'1"-1d49'
Johnson,Sylvia Jewel
Lyn 7h55'16"44d33'
Johnson,Terence
Cep 23h7'40"64d52'
Johnson,Terri
Uma 12h11'24"61d34'
Johnson,The Star Lainey
Lyn 9h14'0"36d5'
Johnson,The Star of
Boo 14h50'35"28d47'
Johnson,Thelma
Cam 9h31'22"82d4'
Johnson,Theresa Carol
Mon 6h43'1"11d17'
Johnson,Thomas M
Crt 11h22'29"-10d15'
Johnson,Tish
Eri 3h44'1"-2d36'
Johnson,Todd G
Aur 7h0'42"41d9'
Johnson,Tom
Cma 6h13'28"-13d16'
Johnson,Trett McCabe
Aur 4h54'1"40d8'
Johnson,Tyler Lauren
Lac 22h55'45"51d16'
Johnson,Tyler Scott
Crt 11h13'33"-17d5'
Johnson,Vanessa
And 23h1'58"51d16'
Johnson,Vera
Com 12h28'20"23d58'
Johnson,Vera Emily
Ori 4h54'1"5d19'
Johnson,Vernon
Mon 6h53'18"-1d17'
Johnson,Victor L
Her 17h22'21"46d45'
Johnson,Victoria Ann
Cas 1h5'53"53d31'
Johnson,Virginia Frieda
Lyr 19h21'1"37d37'
Johnson,Warren M
Cet 0h35'19"1d9'
Johnson,Weston Lee
Aql 20h3'51"6d15'
Johnson,Will
Boo 14h40'38"20d0'
Johnson,William Clifford
Aur 7h24'38"35d39'
Johnson,William
Aur 6h25'47"36d6'
Johnson,William H
Aur 5h23'32"40d5'
Johnson,Wunderman Cato
Her 16h42'13"22d30'
Johnson,Zachary Bernard
Boo 15h43'11"42d7'
Johnson,Zachary McFarland
Dra 17h16'48"64d48'
Johnson,Zelma A
Lac 22h41'30"56d25'
Johnson-A Real Star!, Thirl
Aur 4h48'19"40d6'
Johnson-Beard,Glynda Louise
Crt 11h8'53"-18d48'
Johnson-Kennon,Wren
Sco 17h1'28"-31d30'

Johnson-Loving Mother, Lou
Ori 6h8'19"5d52'
Johnson-Michaels, Sabrina Mary
Peg 22h14'1"31d0'
Johnsons' Evans T & Ellen H,The
Ori 5h38'0"7d34'
Johnspon,Kevin
Her 17h57'10"28d45'
Johnsson Underbar éngel,Louise
And 0h7'47"38d29'
Johnstar
Ser 15h57'45"23d10'
Johnston Constellation
Uma 10h33'46"52d5'
Johnston(me),Lee
Ori 5h50'40"20d52'
Johnston,"Elliot"
Boo 15h3'23"22d14'
Johnston,"Little" Luke Taylor
Vul 19h43'53"27d45'
Johnston,Alan Albert
Cep 21h23'11"70d8'
Johnston,Alfonso "Pelon"
Ari 1h57'50"22d14'
Johnston,Alison
Lyn 7h43'26"47d49'
Johnston,Andrea J
Cas 22h56'57"55d21'
Johnston,Andrew
Gem 6h29'43"13d14'
Johnston,Andrew Rhett
Cet 0h30'18"-0d13'
Johnston,Ann C
Peg 22h45'22"32d14'
Johnston,Anne
Sgr 18h23'58"-20d6'
Johnston,Anne & Pat
Mon 7h1'34"-6d39'
Johnston,Ansley
Peg 22h6'32"25d27'
Johnston,Benjamin
Her 17h27'33"18d52'
Johnston,Betty
And 22h56'42"51d2'
Johnston,Betty Marie Foyston
Peg 23h26'26"10d33'
Johnston,Blakely James
Eri 3h4'49"-6d3'
Johnston,Carson
Lac 22h38'46"40d9'
Johnston,Christopher
Sco 17h52'24"-30d4'
Johnston,Cole Walker
Ori 5h48'36"11d39'
Johnston,Curious George
Oph 18h16'20"1d22'
Johnston,Daryl Peter
Aql 19h54'58"10d28'
Johnston,Eileen
And 2h22'60"45d35'
Johnston,Elise Langdon
Cas 0h56'33"58d39'
Johnston,Gordon
Her 17h12'15"42d0'
Johnston,Grace Allen
Equ 20h57'43"7d25'
Johnston,Isobel
Lyr 18h14'57"46d15'
Johnston,James Edward
Aql 18h45'42"7d45'
Johnston,Janice
Lyn 8h6'0"39d12'
Johnston,Jeffrey Howard
Peg 21h58'53"2d21'
Johnston,Jessica
Lyr 18h37'53"37d33'
Johnston,Joan Minett (Star of Stars)
Lyn 7h57'52"38d32'
Johnston,Jodi P
Sex 10h15'3"-0d37'

Johnston,Joe
Per 2h56'29"54d51'
Johnston,JR
Boo 15h2'28"41d34'
Johnston,Jr,Joseph P
Ori 5h58'46"-0d5'
Johnston,Kane Michael
Cap 21h25'44"-23d35'
Johnston,Karen Ann
Ind 20h29'14"-58d16'
Johnston,Kathy N
Vul 19h17'11"22d20'
Johnston,Katie
Peg 21h24'17"13d28"A
Johnston,Kelley
Cet 2h1'45"-5d14'
Johnston,Kelly Rae
Equ 20h54'20"3d47'
Johnston,Kenneth Wade
Sex 9h58'43"-6d56'
Johnston,Kristi Renee
Vul 20h3'52"25d50'
Johnston,Larry & Susie
Cyg 21h38'46"42d49'
Johnston,Margaret A.
Johnston,Margaret A.
Vul 21h21'48"24d4'
Johnston,Mark
Ser 18h21'36"-1d32'
Johnston,Meagan Leigh
Peg 23h5'50"20d49'
Johnston,Megan
Peg 21h24'17"13d28'B
Johnston,Michael Earl
Aur 5h4'23"38d37'
Johnston,Michael Hal
Boo 13h49'38"15d1'
Johnston,Millicent & Craig
Tri 2h46'19"32d37'
Johnston,Muy Man
Lac 22h36'47"41d8'
Johnston,Mysta
Her 16h30'11"38d29'
Johnston,Neal W
Uma 9h54'46"43d55'
Johnston,Paul
Sct 18h55'15"-6d41'
Johnston,Paul
Her 18h9'27"40d13'
Johnston,Pickens Brian
Cnv 13h20'26"41d20'
Johnston,Rory
Aur 6h57'27"38d51'
Johnston,Ross
Per 2h57'48"34d5'
Johnston,Shane
Aql 19h31'51"-8d52'
Johnston,Shannon Clair
Peg 21h21'37"23d6'
Johnston,Sr,"Our Dad", Eugene J
Boo 14h30'0"27d13'
Johnston,Sr,Howard Eugene
And 1h22'49"40d38'
Johnston,Sr,Vernon V
Equ 20h57'15"3d17'
Johnston,Steven Allen
Sex 9h59'20"-1d57'
Johnston,T Jane & G Doug Daughtry
Peg 0h2'14"31d44'
Johnston,Thomas Quintin
Cma 7h1'46"-30d24'
Johnston,Vickie E
Uma 10h9'51"57d12'
Johnston,William John
Per 2h26'41"57d7'
Johnstone Family Star, The
Boo 15h1'13"16d10'
Johnstone,Anthony Lennox
Cep 0h12'42"75d12'
Johnstone,Paul George
Ori 6h7'14"0d40'
Johnstone,Robert Roy
Cep 20h55'22"59d39'

Johnstone,Shona Ann
Crb 15h52'14"6d42'
Johnze,Jean
Mon 6h25'27"8d44'
Joie
Ori 5h58'33"6d43'
Joignon,Jose
Mon 6h28'24"-6d16'
JOILOVUSO
Her 15h57'58"42d3'
Joiner,Ashley
Cas 1h3'24"62d44'
Joiner,Donna Marie
Eri 3h38'14"-13d28'
Joiner,Irene
Cyg 19h59'60"30d8'
Joiner,Jennifer
Lyn 9h15'21"35d34'B
Joiner,Wendy Chaney
Peg 22h11'12"5d13'
Joiner,William
Cnv 12h8'57"36d57'
Joines,Jerry Thomas
Aql 19h25'46"0d34'
Joinnides,Art
Lac 22h23'13"38d43'
Jointél
Cmi 7h15'51"8d57'
Jojake
Cas 0h7'47"59d33'
Jojo
Boo 14h38'56"14d0'
JoJo
Cnv 12h51'37"40d56'
JoJo
Gem 6h38'30"28d34'
JoJo I
Uma 10h56'18"40d14'
Jojo Luvs Suzy
Cet 0h27'54"1d38'
JoJoe
Dra 12h24'0"64d30'
Jokhu,Urmilla Neeta
Ori 5h58'57"12d38'
Jolain,Jean-Claude
Aur 7h1'41"38d22'
Jole
Pup 8h24'23"-22d55'
Joleil
Uma 11h35'22"33d28'
Jolene
Cyg 20h35'11"37d38'
Jolene Marie
Com 12h59'1"21d44'
Jolene's Light
Tau 4h4'35"23d2'
Jolibois,Zoé
Lac 22h3'23"37d43'
Jolicoeur,Marcel George
Ari 2h54'55"22d29'
Jolie Gabrielle
Umi 16h4'44"79d21'
Jolie Marie
Vul 19h57'54"28d20'
JoLin
Uma 9h25'60"50d40'
JoLisa Dawn
Cnc 8h49'0"32d6'
Jolivette,Mary Alisa
Cas 23h24'15"62d28'
Jolley,Inara Illona
Com 12h44'59"21d26'
Jolley,Irene
Lyr 19h15'39"40d25'
Jolley,Lori Kay
Lyn 8h18'60"50d56'
Jolley,Marianne Kephart
Aql 19h31'1"12d5'
Jolley,Scott
Uma 9h36'25"52d51'
Jolley,Stefani Senae
Sex 10h26'54"-1d2'
Jolley-McKune,Lisa Ann
Cet 2h39'11"-11d32'

Jolliffe,Steven Michael
Oph 17h57'27"13d1'
Jolly Buana
Aql 18h57'46"-4d13'
Jolly Julie
Lyn 8h22'51"49d50'
Jolly Roger
Aql 18h57'10"-8d50'
Jolly Too
Mon 6h48'24"11d19'
Jolly,Alan Frank
Aql 18h59'25"14d54'
Jolly,Anita
Lyr 18h36'49"43d15'
Jolly,Bruce
Aqr 22h20'17"0d15'
Jolly,George
Her 16h59'13"38d7'
Joly,Celine Rebecca
Com 13h2'11"26d40'
Joly,Charlotte
Mon 7h57'53"-5d43'
Joly,Christophe
Lac 22h41'31"38d43'
Joly,Eric
Cas 23h38'38"61d37'
Joly,Marie-Line
Her 18h24'55"12d58'
Jomard,Mireille
Ori 5h55'58'16d58'
JoMarie
Cyg 20h29'60"38d0'
Jome
Her 16h45'0"10d29'
JoMi
Lib 15h30'56"-18d47'
Jominy,Jack J
Aql 20h4'58"1d51'
Jompole,Irving
Cam 12h58'0"80d7'
Jon
Uma 10h14'17"61d8'
Jon
Sex 9h53'13"2d27'
Jon
Aql 18h58'20"-6d51'
Jon & Mandy
Umi 15h7'55"69d55'
Jon & Rachael Forever
Cyg 19h29'37"32d7'
Jon & Stacy
Cyg 21h5'49"35d18'
Jon Joe Mac
Oph 16h55'43"11d42'
Jon Jon "1994"
Lac 22h34'54"53d40'
Jon Karl
Dra 17h0'1"73d37'
Jon Sparkling Above All
Boo 13h35'47"17d59'
Jon, Loving Son & Brother
Cep 22h14'53"61d55'
Jon-Ber-Ton 50
Cet 2h17'53"1d4'
Jonaitis,Robert W
Her 16h36'11"48d34'
Jonak,Dorothy
Cam 3h59'0"69d8'
Jonakin,Rachel Marie
Cyg 19h35'59"38d18'
Jonas
Boo 15h4'1"21d10'
Jonas,Klaus
Cep 0h5'6"83d5'
Jonas,Margaret
Cas 0h27'36"62d2'
Jonas,Stacie Leigh
Cam 12h21'52"78d53'
Jonash Bok/Bok/Bok
Lac 22h31'39"50d18'
Jonathan
Dra 11h28'0"68d30'
Jonathan
Boo 14h45'57"38d16'

Jonathan
Leo 10h34'1"21d59'
Jonathan
Aur 6h14'37"31d16'
Jonathan
Cep 21h5'26"68d50'
Jonathan
Uma 12h18'1"60d18'
Jonathan & Alexander
Aur 6h3'27"38d1'
Jonathan & Dana
Her 16h17'0"42d21'
Jonathan & Leila
Sgr 18h58'49"-27d10'
Jonathan & Sushma
And 0h8'42"31d31'
Jonathan Andrew
Per 4h24'45"51d2'
Jonathan Brett
Aur 4h49'58"41d6'
Jonathan Christopher
Psc 1h0'48"32d41'
Jonathan E
Cep 22h18'28"62d5'
Jonathan Joseph Coom World
Uma 9h36'0"47d42'
Jonathan Our 2nd Shining Light
Her 16h37'10"48d47'
Jonathan Scott
Cma 7h24'12"-15d3'
Jonathan Wayne
Hya 10h1'47"-15d41'
Jonathan You Will Always Shine
Aql 19h9'29"0d11'
Jonathan's Brownie
Lac 22h17'59"50d10'
Jonathan's Fire
Her 16h38'13"21d50'
Jonathan's Lucky Star
Cnv 12h28'0"38d56'
Jonathan,Judith & Paula
Ori 6h7'15"0d19'
Jonathans Star
Lib 15h53'57"-5d44'
Jonathon
Aur 6h7'32"31d45'
Jonathon's Celestial Body
Aql 20h30'38"0d56'
Jonathon's Star
Her 16h26'37"28d35'
Jonathroban
Boo 14h28'34"47d16'
Jondreau,Caryn Patrice
Aql 19h42'35"12d60'
Jondreau,Denise Danielle
Peg 22h58'27"30d9'
Jondreau,Michelle Marie
Mon 6h26'1"3d31'
Jones (Carrie & Dei), Caroline & David
Cyg 20h54'0"37d51'
Jones 18091967-12061994,Jeremy Paul
Ori 5h21'1"15d13'
Jones Aka Tootsie, Evelyn Florence
Mon 6h55'52"8d60'
Jones Eternal Light, Julie
And 2h23'42"38"13'
Jones Forever,David & Laura
Aur 6h26'0"37d10'
Jones Greatest Dad in the Galaxy,Ole
Ori 4h41'29"5d50'
Jones II,Bob
Ori 5h3'17"-2d56'
Jones II,Russ Harris
Tau 3h57'19"10d18'
Jones III,W Hampton
Cet 2h5'38"-0d57'
Jones IV,Leonard S
Lac 22h38'42"55d15'

Jones Loving Brother, Ernest Wayne
Lac 22h24'53"52d43'
Jones Star,The Bobbi
Dra 16h24'20"68d44'
Jones Star,The Douglas A
Uma 10h19'28"40d53'
Jones Star,The LC
Cyg 19h28'23"39d24'
Jones the Shelf
Her 16h10'1"5d22'
Jones Twenty First, Elaine
Cas 0h37'52"60d30'
Jones WGP Ontario 1994-95,Donald
Her 16h59'56"24d40'
Jones"Susie",Marilyn Sue
Cmi 7h24'57"8d36'
Jones' White Light, Barry
Per 2h51'13"48d47'
Jones,"Bambar" Janet Merris
Lib 15h31'29'-8d58'
Jones,Adrienne
Uma 10h6'47"56d25'
Jones,Alan William
Ori 4h57'59"5d14'
Jones,Alice R
And 2h15'47"42d39'
Jones,Alun
Leo 9h22'39"18d39'
Jones,Alysia Ann
Mon 8h5'25"-3d8'
Jones,Amanda Jo
Peg 23h6'36"8d37'
Jones,Analu Beth
Peg 23h27'21"33d46'
Jones,Andrew Edward
Oph 17h23'10"-20d41'
Jones,Andy Dalziel
Aur 5h0'15"48d59'
Jones,Anna M
Peg 21h55'52"28d37'
Jones,Arlene "Boopski"
And 23h36'55"38d43'
Jones,Barbara
Lmi 10h51'0"30d53'
Jones,Barbara J
Lyr 18h59'34"42d56'
Jones,Barbara Licata
Lyr 18h31'31"45d37'
Jones,Becky
Eri 2h51'12"-1d31'
Jones,Bernice
Cyg 19h29'32"30d47'
Jones,Bert
Aql 19h53'17"13d39'
Jones,Bessy
Cas 0h52'48"74d15'
Jones,Beverley Linda
Cyg 19h32'47"37d13'
Jones,Blaine
Hya 9h7'40"5d16'
Jones,Blaine Tillman
Aur 6h53'13"44d23'
Jones,Brandon W
Boo 15h6'27"20d14'
Jones,Brenda
And 0h4'23"46d20'
Jones,Brennan Alan
Ori 6h17'23"0d37'
Jones,Britt Robert
Tri 1h47'28"27d16'
Jones,Brynley Owen
Her 16h42'1"50d56'
Jones,Brynn Althea
Aql 20h9'47"0d6'
Jones,C M
Lac 22h8'0"51d56'
Jones,Callum Robert
Ori 4h47'1"0d41'
Jones,Carl Warren
Per 2h53'25"50d33'
Jones,Carla
Sge 20h14'10"17d3'
Jones,Carol Ann
Mon 6h28'33"-6d58'
Jones,Carolyn D
Cma 6h54'48"-18d37'
Jones,Carolyn Dae
And 2h22'20"37d42'
Jones,Carolynne Deanne
Dra 15h3'1"58d15'
Jones,Casey Oceans
Cnv 12h27'26"44d51'
Jones,Celeste
Lyr 18h40'21"45d47'
Jones,Charles
Aur 5h32'0"30d37'
Jones,Charles Edward
Aur 4h47'49"40d24'
Jones,Charles Scott
Boo 15h1'40"53d47'
Jones,Chas P
Tau 3h45'34"28d38'
Jones,Christina Helen Frances
Cas 1h57'1"60d41'
Jones,Christine Anne
Uma 10h30'33"48d30'
Jones,Christopher
Cam 5h41'53"73d23'
Jones,Christy
Peg 23h1'1"33d28'
Jones,Clinton Christopher
Aql 18h46'1"11d34'
Jones,Codi Marie
Lmi 10h51'1"26d20'
Jones,Colin Hunter
Cma 7h12'37"-16d36'
Jones,Constance Virginia Perry
Equ 21h5'38"3d41'
Jones,Cora Jean
Lyr 18h42'51"41d60'
Jones,Cynthia
Peg 23h25'40"15d45'
Jones,Daniel Ryan
Dra 17h13'26"60d27'
Jones,David
Ori 5h50'16"16d42'
Jones,David Debellis
Her 16h53'22"27d40'
Jones,David Mac
Aur 6h21'0"33d24'
Jones,David R
Sct 18h55'25"-14d14'
Jones,David Wayne
Her 17h56'49"41d10'
Jones,Debora Lynn
Mon 7h17'34"-5d18'
Jones,Deborah
Umi 13h32'45"72d34'
Jones,Dennis
Her 16h57'42"48d8'
Jones,Dennis
Cep 21h22'36"60d32'
Jones,Deputy Kenobi
Uma 11h35'33"33d34'
Jones,Des & Rita
Uma 11h14'58"57d7'
Jones,Diane Jane
Lyr 18h29'30"30d7'
Jones,Donald P
Aql 19h2'22"3d9'
Jones,Donna Lee
Cas 0h49'21"71d4'
Jones,Dr George
Dra 20h15'0"63d15'
Jones,Dr Howard
Oph 17h19'1"-23d45'
Jones,Duane Davy
Boo 15h38'21"41d33'
Jones,Dub
Vul 19h52'46"20d26'
Jones,Duncan McArthur
Uma 8h24'1"60d37'
Jones,Dustin Harold
Her 16h32'0"40d18'
Jones,Edward D
Sex 10h40'57"5d50'
Jones,Edward Raymond
Aur 4h54'1"41d8'
Jones,Eli F
Lac 22h9'35"40d44'
Jones,Elizabeth Jean
Cam 4h46'46"68d50'
Jones,Elsie
Peg 22h2'32"32d11'
Jones,Eric & Megan
Cyg 21h55'1"53d31'
Jones,Eric N
Crb 15h58'30"27d59'
Jones,Eryl Lee
Per 1h46'44"52d56'
Jones,Evalyn,Sadlier
Cet 1h31'14"-3d38'
Jones,Eve
And 0h11'28"38d24'
Jones,Evelyn Marlene
Eri 3h41'33"-18d26'
Jones,Everett Arnold
Per 3h30'56"38d58'
Jones,Everette Ruark
Cep 22h2'21"53d2'
Jones,Frances Agnes
Hya 9h12'37"3d50'
Jones,Frances Fontaine
Peg 23h22'42"33d18'
Jones,Frederick C
Cma 6h52'56"-17d1'
Jones,Gailene
Ant 9h37'49"-37d29'
Jones,Geoffrey Alan
Hya 8h40'12"-6d28'
Jones,George T
Cnv 12h44'1"36d40'
Jones,George Thomas
Sct 18h41'15"-6d29'
Jones,Geri-Lynn
Tri 2h9'0"32d13'
Jones,Gil
Cmi 7h38'38"4d22'
Jones,Glyn
Lyn 8h16'1"46d21'
Jones,Graham & Michel
Aql 18h58'32"10d43'
Jones,Gregory
Her 17h8'46"45d2'
Jones,Gregory L
Boo 14h39'34"21d40'
Jones,Gregory M
Per 1h43'16"53d38'
Jones,Griffin Tyler
Aql 20h8'35"0d19'
Jones,Gwenan Mary Jenny
Lyr 19h20'18"42d7'
Jones,Helen
Lyr 19h21'51"38d21'
Jones,Hollis W
Cma 6h53'18"-19d36'
Jones,Honor & Max
Cet 3h16'42"-0d29'
Jones,J Mark
Boo 14h19'52"29d22'
Jones,James Edward
Peg 23h49'38"11d6'
Jones,James Ethan
Tri 1h50'13"34d15'
Jones,James Lamon
Sgr 18h48'3"-27d2'
Jones,James Michael
Her 17h20'38"44d49'
Jones,James Richard
Peg 23h3'48"33d47'
Jones,Jamie Leanne
Aql 20h12'35"12d23'
Jones,Jamie Thomas
Umi 13h9'23"70d51'
Jones,Jana Marie
And 0h19'22"30d32'
Jones,Janie
Cas 23h40'13"50d30'
Jones,Jason Riddick
Her 17h35'23"24d30'
Jones,Jay
Sct 18h45'18"-9d58'
Jones,Jean
And 0h4'10"31d17'
Jones,Jean S
Aur 6h55'11"40d4'
Jones,Jeanette
Vul 20h23'32"28d14'
Jones,Jeffery
Boo 14h48'54"27d32'
Jones,Jeffrey Ronald
Her 16h19'27"10d43'
Jones,Jeffrey William
Oph 17h30'22"-20d56'
Jones,Jennifer
Sex 10h10'34"-8d13'
Jones,Jennifer
Cas 2h2'35"58d56'
Jones,Jennifer Marie
Equ 21h9'58"11d54'
Jones,Jenny Kathleen
Lyr 18h58'12"33d39'
Jones,Jeri Riha
Mon 7h29'29"-0d57'
Jones,Jim
Ori 5h58'1"10d40'
Jones,Jimmy Sue
Uma 10h12'38"47d44'
Jones,Joe Steven
Sex 9h56'53"5d59'
Jones,Joel T
Cet 2h33'41"5d60'
Jones,John Roberts
Aur 4h59'39"52d30'
Jones,Jonathan
Dra 18h49'27"68d9'
Jones,Jonathan V
Aur 7h14'0"40d39'
Jones,Joshua Daniel
Her 15h50'12"46d6'
Jones,Jr,Max Edward
Hya 9h54'25"-17d45'
Jones,Jr,Robert J
Cnc 9h11'0"31d40'
Jones,Julia Claire
Per 4h1'19"52d26'
Jones,Julie
Cas 0h32'51"62d18'
Jones,Karin R & Joel L
Peg 23h47'51"15d54'
Jones,Katharine Spencer
Gem 6h47'0"30d37'
Jones,Katherine Dawn Coe
Peg 23h45'13"11d49'
Jones,Kathleen
Lyr 18h35'1"37d53'
Jones,Kathleen Lander
Mon 6h21'1"0d0'
Jones,Kathy A
Sge 19h53'37"19d6'
Jones,Kathy A
And 23h22'53"49d42'
Jones,Kathy Sue
And 2h21'39"41d23'
Jones,Keith & Whitney
Mon 8h8'4"-6d47'
Jones,Kelly
Sex 9h50'48"-2d22'
Jones,Kelly Diane
Cas 23h28'41"63d28'
Jones,Kelly Louise
And 23h2'29"47d51'
Jones,Ken
Hya 9h34'30"-6d11'
Jones,Kenneth Frederick
Ser 18h43'10"3d49'
Jones,Kevin Wayne
Lac 22h23'0"37d38'
Jones,Kiersten Nicole
Oph 17h54'46"10d56'
Jones,Kristie Lynn
And 1h27'38"39d41'
Jones,Larry Dexter
Sex 9h49'44"1d50'
Jones,Larry Martin
Tri 2h20'30"34d15'
Jones,Laura
Lyr 18h16'40"32d26'
Jones,Lauren Christina Kaye
Lyr 18h19'12"40d12'
Jones,Lauren Taylor
Mon 7h14'38"-5d3'
Jones,Lawrence Alymer
Ser 15h50'31"20d34'
Jones,Lee & Lisa
Lyn 6h37'11"54d32'
Jones,Leslie
Cyg 20h52'0"30d41'
Jones,Lindsay
Ser 16h17'32"2d13'
Jones,Lindsey Marie
Ori 6h17'19"8d51'
Jones,Linus P "Princess"
Psc 23h20'33"1d21'
Jones,Lynn
Mon 6h41'36"-10d57'
Jones,Lynne Katrina
Lyr 18h23'0"45d19'
Jones,Mabel T
Cas 0h45'26"70d54'
Jones,Mace Di Grandi
Lyr 19h17'44"28d28'
Jones,Madeline & Philip
Lyr 18h30'30"44d21'
Jones,Malcolm Eugene
Sge 19h56'57"16d22'
Jones,Mandy
Ori 5h58'52"15d7'
Jones,Mandy Jo
Uma 11h29'32"47d19'
Jones,Marie
Cas 0h45'52"64d24'
Jones,Marie Hucks
Peg 22h42'44"27d12'
Jones,Mark
Sct 18h42'1"-6d32'
Jones,Mark A
Cnv 12h54'26"32d8'
Jones,Mark Byron
Ori 6h4'0"20d25'
Jones,Mark Thomas
Per 4h40'1"38d55'
Jones,Martha Kate- Darlin
Com 13h6'51"20d28'
Jones,Mason & Alice
Cyg 21h6'50"38d46'
Jones,Master Zachary D
Dra 12h25'0"64d54'
Jones,Maximilian Jordan
Per 3h1'48"40d23'
Jones,McKenna
Hya 10h1'11"-18d25'
Jones,Meredith
Eri 2h46'55"-5d54'
Jones,Michael Churchill
Oph 17h2'38"8d43'
Jones,Michael Dan
Lmi 10h2'34"31d51'
Jones,Michael L
Her 18h12'33"30d42'
Jones,Michael Thomas
Aql 20h35'14"0d26'
Jones,Michael William
Gem 7h27'7"24d55'
Jones,Michelle Andrea
Gem 6h37'58"31d49'
Jones,Michelle Denise
Aql 18h56'27"8d0'
Jones,Michelle Rae
Peg 22h29'1"31d10'
Jones,Miss Emma
And 23h3'25"50d28'
Jones,Mollie
Crt 11h17'44"-8d26'
Jones,Molly T
Lyr 19h25'43"41d24'
Jones,Mr Keith
Aqr 22h7'42"-6d52'
Jones,Nancy L
Vir 13h0'32"12d44'
Jones,Natasha
Uma 8h53'45"55d6'
Jones,Nicholas Andrew
Cep 21h41'29"61d13'
Jones,Nicky
Lyn 7h36'23"51d35'
Jones,Oswald & Mary Louise
Uma 17h24'0"80d28'
Jones,Owen Stanley
Ori 4h47'0"15d24'
Jones,Patricia Lynn
Eri 2h58'19"-6d44'
Jones,Patricia Violet
Mon 6h52'48"10d13'
Jones,Patrick Casey
Dra 20h11'16"73d39'
Jones,Patti
Cyg 21h9'46"40d35'
Jones,Paul & Mary
Lyr 19h11'44"41d14'
Jones,Paul Edward
Aur 5h32'1"48d48'
Jones,Peggy
Aqr 21h2'56"-0d28'
Jones,Pete
Vul 19h2'18"24d24'
Jones,Peter A
Uma 8h59'21"61d11'
Jones,Phillip Alan
Cma 6h54'6"-19d35'
Jones,Phyliss F
Cas 0h52'33"63d27'
Jones,Pierson Russell
Her 16h30'14"48d48'
Jones,Pres Save Our Rivers,Margaret (Peg)
Cet 2h53'1"-1d16'
Jones,Quincy
Sct 18h41'47"-5d29'
Jones,Rafe
Ser 15h36'32"-2d32'
Jones,Richard
Cep 22h21'1"63d49'
Jones,Richard
Her 18h15'14"48d32'
Jones,Richard Edwin
Cet 2h10'36"6d24'
Jones,Richard J
Her 16h7'14"21d18'
Jones,Richard Louis
Lac 22h16'17"46d34'
Jones,Rita
Peg 22h59'59"34d38'
Jones,Rob
Uma 10h55'0"40d22'
Jones,Robert
Hya 8h58'21"2d20'
Jones,Robert Edwin
Boo 14h34'56"31d33'
Jones,Robert/Marie
Crb 15h54'9"34d34'
Jones,Rodney Dugdale
Mon 6h22'17"-0d32'
Jones,Ron
Per 2h58'0"38d18'
Jones,Ron
Ser 15h55'58"24d10'
Jones,Rowland
Cep 23h34'56"64d45'
Jones,Roy Whitman
Cmi 7h18'43"0d5'
Jones,Roz
And 23h1'56"47d31'
Jones,Ryan Jacob
Cnv 12h11'18"41d21'
Jones,Sabrina Noel
Vul 19h45'19"25d23'
Jones,Sahi M
Boo 14h17'22"50d0'
Jones,Scott
Tri 2h38'37"31d23'
Jones,Scott Michael
Her 17h20'1"48d32'
Jones,Sean Byron
Aur 5h10'28"42d2'
Jones,Sean M
Hya 8h53'30"2d26'
Jones,Sharon
And 21h30'13"3948'
Jones,Shawn Scott
Aql 20h19'39"0d52'
Jones,Shelda
Mon 6h19'21"1d48'
Jones,Sir Ryan Kyle
Her 16h36'29"28d31'
Jones,Stanley George
Dra 14h57'55"56d16'
Jones,Star Romer
Hya 8h43'53"6d33'
Jones,Starship Jimmy & Judy
Vul 19h18'0"22d45'
Jones,Starship Thomas Lee
Aql 19h55'0"7d44'
Jones,Stephen James
Per 3h1'24"47d28'
Jones,Sterling Michael
Cet 2h9'57"6d14'
Jones,Steven R
Aql 19h54'59"13d16'
Jones,Stéphanie
Umi 15h3'33"77d21'
Jones,Sue
Lyn 7h55'30"58d48'
Jones,Susan
Cas 0h15'48"62d15'
Jones,Susan
And 0h18'25"32d59'
Jones,Susie
And 22h56'41"48d55'
Jones,Suzi
Ori 5h31'43"-0d0'
Jones,Tanya Tereena
Cas 1h8'28"61d42'
Jones,Teri
Vul 19h53'56"20d21'
Jones,The Danangela
Crb 15h50'41"26d6'
Jones,The Gerald W
Umi 16h23'0"71d20'
Jones,The Light of Alyce W
Aql 19h30'37"10d34'
Jones,Thomas Jefferey
Cet 2h30'44"6d8'
Jones,Thomas John
Hya 10h45'41"-11d45'
Jones,Thomas Michael
Uma 10h7'22"48d46'
Jones,Thomas W
Del 20h21'26"3d11'
Jones,Thomas W
Ori 5h54'46"14d1'
Jones,Timothy
Cet 2h57'12"5d55'
Jones,Timothy James
Her 18h7'0"47d33'
Jones,Tina
Vul 21h3'0"20d22'
Jones,Todd B
Per 2h50'53"43d20'
Jones,Tom
Per 3h3'30"47d27'
Jones,Tony
Peg 22h15'53"33d40'
Jones,Tracy E
And 2h32'50"48d10'
Jones,Troy
Crt 11h18'46"-8d57'
Jones,Venita
Sco 16h23'48"-40d38'
Jones,Vera J.
Cas 2h12'40"70d16'
Jones,Verna Mae
Eri 4h46'17"-5d22'
Jones,Victoria Ann
And 0h14'52"38d9'
Jones,William Arthur
Ari 1h56'30"11d45'
Jones,William Harrison
Oph 6h59'24"11d56'
Jones,William Hugh
Her 16h28'13"36d16'
Jones,William Lynn
Per 3h1'47"41d42'
Jones,Willie Aurelia
Umi 16h2'42"70d9'
Jones-"My 5'11" Hero", - 100% Man,E Rodney
Her 17h39'15"14d15'
Jones-Bratest Star In Da Sky!,Terri
Sge 20h17'56"17d36'
Jones-CJ,Aaron
Sge 18h57'37"19d56'
Jones-Galaviz
Her 17h29'46"28d0'
Jones-Grandma,Helen C
Del 20h18'23"20d29'
Jones-Love From Mary, Spike Y
Uma 11h12'12"70d36'
Jones-Love From Spike, Mary C
Cyg 20h20'40"37d51'
Jones-Love Is Forever, Amber
Peg 0h6'13"28d27'
Jones-Roguska,Peggy J
And 23h23'46"48d50'
Jones-Thompson
Dra 19h58'50"70d37'
Jones-With Love, Miranda Harmony
Cyg 19h25'51"45d48'
Joney Pony
Boo 13h39'48"23d2'
Jonfy & Lali
Cyg 19h32'0"56d38'
Jong,Sylvie et Georges
Boo 13h50'35"16d48'
Jongebreur,Mireille D
Cas 0h5'1"63d50'
Jongerius,Geertje Marie
Cyg 19h50'1"58d29'
Jongewaard,Todd Christian
Cep 22h59'41"65d21'
Jonglez,Guillaune
Per 3h27'50"50d52'
Joni
Cas 0h42'16"68d48'
Joni
Peg 22h46'49"31d43'
Joni & Charlie's Everlasting Light
Cam 3h29'16"60d11'
Joni Lynn
Mon 6h56'54"1d4'
Joni's Jottingstar
Lyn 6h57'21"60d32'
Joniec,Angelina Dawn Costa
Ari 2h33'26"20d51'
Jonjer
Aur 6h22'28"33d23'
Jonker,T J Tracey
Del 20h14'31"15d29'
Jonmonica
Aql 20h0'58"0d29'
Jonna H
Lac 22h54'27"53d44'
Jonnie
Aql 19h49'36"11d31'
Jonnie
Lmi 10h53'41"31d46'
Jonnie Angel
Her 16h14'12"20d26'
Jonnie Dee
And 2h0'24"40d43'
Jonny
Del 20h16'20"11d49'
Jonny Adventure
Hya 8h29'21"1d33'
Jonny Healing Soulmate Bear Star
Aur 7h2'56"40d8'
Jonny's Star
Psc 1h5'29"28d6'
Jono
Cep 21h32'49"71d10'
Jonser
Her 16h24'13"32d57'
Jonsha Aka J J Headhunter
Dra 13h53'38"69d47'B
Jonson,Love & Sunshine Mimi
Vir 14h9'59"-8d8'
Jonsson
Vul 19h46'41"28d12'
Jonsson,Thröstur
Boo 15h4'19"27d41'
Jonworld
Cyg 21h37'35"38d9'
Jonzina
Uma 11h7'51"44d3'
Joolie Rea
Sge 20h1'0"17d14'
Joolz
Peg 22h29'42"28d25'
Joop,Wolfgang
Lib 15h11'36"-23d21'
Joost Eeftink
Ori 5h10'0"-1d17'
Jope,Alan
Uma 11h11'37"55d20'
Jope,Peter S
Umi 13h9'24"71d46'
Jopiti,Karin & Florindo
Ori 6h8'1"1d22'
Joplin,Chaz William
Ori 5h38'0"8d12'
Joqua
Aql 18h46'45"10d12'
Jordahl,Felicia & Dick
Mon 7h45'27"-5d51'
Jordahl,Judy
Crb 15h21'14"30d9'
Jordahl,Lois M
Cas 0h40'1"67d33'
Jordan
Per 4h23'54"50d6'
Jordan
Cep 0h50'48"77d30'
Jordan
Her 16h40'38"32d38'
Jordan
Per 3h20'57"41d6'
Jordan Alexander
Ser 15h34'41"8d30'
Jordan Alexandra
Aur 4h38'0"31d20'
Jordan Andrew
Her 18h0'39"38d15'
Jordan Kelley's Good Luck Star
Cyg 19h43'18"37d55'
Jordan Noel
Uma 11h33'1"64d49'
Jordan Paul
Dra 19h14'51"68d27'
Jordan Rose
Tau 5h34'23"28d36'
Jordan The Star of My Life Luv KKN
Uma 10h3'1"72d37'
Jordan"Free Spirit", Judy
Sgr 19h25'3"-41d41'
Jordan's Eye
Lac 22h33'1"53d46'
Jordan's Star
Her 16h53'33"27d57'
Jordan's Star
Lyn 7h49'22"51d54'
Jordan,Alexander Scott
Aql 20h12'45"11d33'
Jordan,Angela
Cas 0h16'46"46d56'
Jordan,Brittany Nicole
Lyn 8h50'43"37d6'
Jordan,C Wallace
Cam 4h17'45"69d33'

Jordan,Chelsea
Umi 17h45'49"80d3'
Jordan,Christopher Shane
Per 3h4'1"40d48'
Jordan,Dana Allen
Cas 0h12'55"58d13'
Jordan,David Charles
Aur 6h0'54"34d4'
Jordan,Douglas
Oph 17h10'37"-23d35'
Jordan,Ed
Cnv 12h29'33"40d60'
Jordan,Elise M
And 23h49'43"44d32'
Jordan,Gayla
Mon 7h6'29"-1d39'
Jordan,Hilde
Sgr 19h19'14"-29d3'
Jordan,Ian Stewart Robert
Uma 9h44'45"44d42'
Jordan,James
Uma 12h12'39"57d59'
Jordan,Jamie Thomas
Per 1h47'29"56d18'
Jordan,Jayce Lyn
Cyg 19h21'12"54d43'
Jordan,Jenny
Peg 23h47'57"30d36'
Jordan,Jo Lenore
Aur 6h57'38"40d32'
Jordan,John & Susan
Aur 4h52'37"37d49'
Jordan,JP
Aql 18h57'49"-7d48'
Jordan,Kaitlyn A
And 1h9'1"38d11'
Jordan,Kristen
Cet 1h32'42"-1d39'
Jordan,LaVerne
Com 12h55'1"28d60'
Jordan,Linda
Gem 6h58'21"35d12'
Jordan,Lucy Christian
Uma 10h57'14"48d29'
Jordan,Marc
Cam 7h30'30"80d9'
Jordan,Marc James
Ser 17h53'41"-14d45'
Jordan,Mary Cynthia
Mon 7h40'56"-5d38'
Jordan,My Bashert
Gem 7h56'48"31d35'
Jordan,Naomi Renee
Peg 22h32'29"21d24'
Jordan,Patricia
Peg 22h25'44"30d38'
Jordan,Phyllis "Grandma"
Cas 1h3'40"64d32'
Jordan,Polly
Eri 2h46'1"-4d8'
Jordan,Rachel Yvonne
Cmi 7h23'59"0d9'
Jordan,Ray & Christine
Bickerdyke
Cyg 20h24'19"40d34'
Jordan,Sally
Lyr 18h32'1"37d20'
Jordan,Sandra Diane
Lyr 19h13'59"42d42'
Jordan,Scott & Terry
Eri 3h55'43"-10d17'
Jordan,Shannon
Lyn 7h27'1"40d42'
Jordan,Sondra
Cas 1h51'38"61d9'
Jordan,Sr,James N
Aur 6h0'16"30d6'
Jordan,Stuart R
Cep 22h28'48"59d8'
Jordan,Terri Anne
And 2h24'43"45d33'
Jordan,Thad
Aql 20h2'20"1d22'
Jordan,Thomas
Per 3h7'14"40d2'
Jordan,Victor Bradley
Boo 15h26'39"37d30'
Jordana
Aql 18h56'53"17d49'
Jordana Lauren
Uma 11h57'58"47d4'
Jordanek,Melissa Sunshine
Cas 1h6'35"60d13'
Jordanhazy,Robert
Cep 23h35'45"63d26'
Jordann Leandra
And 2h2'0"35d53'
Jordanna
Peg 23h31'27"18d58'
Jordanne
Aur 4h48'1"40d45'
Jorden
Equ 21h6'23"3d25'
Jorden's Valentine Star
Uma 11h43'1"43d6'
Jorden,MacKenzie
Cam 8h32'43"74d57'
Jordheim,Nancy,Neil, Brent & Erik
Uma 12h27'28"61d19'
Jordiphinelangielle
Hya 8h13'0"1d20'
Jordon & Family,Lynn
Peg 0h9'23"22d14'
Jordon,Arthur Anderson
Cet 2h2'54"-1d45'
Jordon,Joann J
And 0h7'39"35d60'
Jordon,Nicole
And 23h38'16"45d54'
Jordy
Aqr 22h4'42"0d42'
Joret,Aghate
Oph 18h7'27"10d15'
Jorg,The Pamela Marie
Cyg 20h20'35"31d51'
Jorgensen,Annie Howarth Smith
Cas 1h2'28"58d60'
Jorgensen,Colleen Kay (Harris)
Cas 0h57'22"67d52'
Jorgensen,Forever Star
Aql 19h25'38"-1d38'
Jorgensen,James Robert "Bob" & Mary Lou
Cyg 21h30'40"34d46'
Jorgensen,Kayla Marie
Gem 7h0'14"35d15'
Jorgensen,Kyndi Ann
Gem 7h0'24"35d17'
Jorgensen,Lisa
Umi 15h11'44"66d56'
Jorgensen,Marilyn
Mon 6h57'53"-10d60'
Jorgensen,Monty
Aur 5h0'26"49d25'
Jorgensen,Nicholas Lord
Lac 22h35'1"37d50'
Jorgensen,Richard & Betty
Cyg 21h0'55"53d7'
Jorgensen,Ruth "Rufie" Ann
And 2h6'28"38d12'
Jorgensen,Tina H
Lyn 9h5'38"33d39'
Jorgensen,Torben
Ori 5h31'1"0d31'
Jorgenson,Thomas P
Lac 22h5'0"46d48'
Jorgenson,Timothy
Oph 17h26'33"-0d7'
Joriot
Per 3h2'58"37d47'
Jorja
Cet 0h52'24"-1d6'
Jorkasky,Dr Diane
Vul 20h58'25"28d47'
Jorndt Birthday Star, Patricia M
Cas 0h35'50"64d25'
Jos & José
Boo 14h30'38"46d36'
Josda-Curhe
Cet 1h4'44"0d29'
Jose & Jacky
Sge 19h12'11"21d3'
Josef,Helen
And 0h55'1"45d20'
Josef,Helene
Cas 0h54'32"56d9'
Josef,Philip
Per 1h43'19"52d41'
Joseff,Jeffrey René
Her 18h2'38"30d47'
Joseff,Jr,Jeffrey Rene
Her 18h8'57"31d50'
Josefine Louise
Her 17h56'32"42d46'
Josep Maria-Magda
Cas 0h18'50"60d55'
Josep-Antoni Fabregas Julia
Ser 15h14'48"8d32'
Joseph
Aql 18h57'26"-2d19'
Joseph
Lyn 7h53'27"41d4'
Joseph
Aur 4h46'15"40d59'
Joseph
Dra 19h42'0"60d39'
Joseph
Her 15h49'45"41d57'
Joseph
Dra 14h19'38"63d13'
Joseph
Cep 22h21'31"55d52'
Joseph & Kellie & Rico
Aur 6h7'29"38d6'
Joseph & Laura's Eternal Love Light
Cyg 20h52'52"40d7'
Joseph & Polly
Lyn 8h27'59"50d49'
Joseph A J-Computer Wiz
Aql 20h21'51"8d11'
Joseph Angel
Per 2h32'26"56d47'
Joseph Harvey
Her 18h1'0"30d7'
Joseph Henry
Dra 16h27'46"61d38'
Joseph Loves Christina 4-21-91
Uma 10h44'39"68d47'
Joseph The Just
Aur 6h33'0"37d53'
Joseph's Dream
Cep 21h22'37"61d26'
Joseph's Little Corner of Heaven
Equ 20h56'40"2d53'
Joseph's Vision Quest
Cnv 13h29'1"50d13'
Joseph,Curtis Johnson
Boo 14h29'0"20d12'
Joseph,David
Aur 7h15'34"37d1'
Joseph,David Neil
Boo 15h0'42"10d33'
Joseph,Ellen
And 23h33'0"38d12'
Joseph,Forever My Love,Elizabeth
Vul 19h18'24"20d26'
Joseph,Jamal E S
Lyn 7h24'59"50d37'
Joseph,Jr,John F
Her 18h17'25"14d13'
Joseph,Lawrence D
Aur 6h37'48"37d51'
Joseph,Meghan Kelsey
Cas 2h59'35"61d37'
Joseph,Sammie Jacqueline
Aql 20h2'59"-6d26'
Joseph,Stanley
Aql 19h17'15"15d52'
Joseph,Violet
Lyr 18h44'38"31d50'
Josephine
Mon 8h3'10"-2d39'
Josephine
Lyn 8h14'59"39d36'
Josephine
Cnc 8h56'35"30d12'
Josephine
Oph 18h4'49"12d36'
Josephine
Eri 4h30'13"-6d31'
Josephine Clare & Andrew Charles
Per 3h15'36"40d45'
Josephine Infinite
Del 20h55'1"10d2'
Josephs,Roberta
Vir 11h42'0"3d30'
Josephson,Ashley Nicole
Peg 23h47'46"16d14'
Josephson,Dr Donald
Uma 10h14'12"51d16'
Josephson,Judith Coates
Eri 4h11'3"-14d38'
Josephson,Mark Eric
Dra 20h38'1"70d4'
Josephson,Rosebjorg
Umi 15h17'17"68d33'
Josephson/Myra Rogers, Susan
Mon 7h0'24"5d15'
Joseppibergin
Ori 5h3'30"8d42'
Josetteet Jean
Com 12h22'43"21d23'
Josey
And 23h17'49"50d43'
Josh
Cep 20h31'40"61d47'
Josh
Cep 21h18'58"70d17'
Josh & Stacey
Cyg 21h4'46"38d45'
Josh Loves Mindi
Aur 6h19'45"38d31'
Josh's Star
Uma 9h51'14"48d49'
Joshi,Christine
Eri 3h50'31"-1d47'
Joshi,Eileen
Cas 0h34'18"62d26'
Joshi,Hetal R
Aql 19h55'51"0d44'
Joshi,Mukund R
Aql 19h58'29"8d51'
Joshroglin Trio Star For Joy & Music
Dra 16h21'30"60d13'
Joshua
Boo 14h13'0"42d0'
Joshua
Dra 16h3'44"64d49'
Joshua
Lac 22h5'38"48d59'
Joshua
Ori 5h53'41"15d11'
Joshua
Her 17h22'24"47d56'
Joshua
Aur 6h0'43"30d1'
Joshua 1:9
Peg 0h2'12"17d59'
Joshua 2
Lya 18h59'55"30d22'
Joshua Allen
Aql 19h57'35"14d30'
Joshua Charles
Ori 6h9'51"3d57'
Joshua Hahn-Pauls
Tau 5h31'50"20d26'
Joshua Jeffery Michael
Dra 16h3'18"63d23'
Joshua The Bold
Cep 22h4'1"61d50'
Joshua's Jewel
Lyn 7h56'33"58d45'
Joshua's Love
Cet 3h3'39"-0d3'
Joshua's Star
Her 17h13'46"22d25'
Joshua's Star
Equ 21h20'29"11d24'
Joshua's Sun
Per 2h41'0"43d52'
Joshua,By George
Uma 11h29'22"31d18'
Joshua,Steven
Psc 22h59'53"5d19'
Joshua,The
Hya 9h37'47"-10d58'
Joshua,Travis,Allison
Cet 0h55'35"-4d42'
Joshuamanda
Per 3h33'44"38d17'
Joshy & Kath
Dra 12h35'37"70d11'
Josi
Psc 22h56'18"1d17'
Josi Renee
Lyn 7h33'51"45d17'
Josiassen,Joseph
Dra 13h40'53"63d47'
Josie
Del 20h24'49"7d57'
Josie
Lyn 9h15'15"39d50'
Josie "Please Say Yes"
Lib 14h33'23"-18d48'
Josie Ho
Cap 21h21'48"-22d40'
Josie's Star
Cam 3h49'1"53d11'
Josiger,Kurt
Dra 14h30'42"63d6'
Josipa,Gordana
Boo 14h29'43"50d39'
Joslin,David Andrew
Gem 6h25'3"13d23'
Joslin,Douglas M
Her 18h11'35"46d59'
Joslin,Shelby M
Cas 23h36'56"63d1'
Joslin,Thomas Dillon
Cyg 20h9'57"40d22'
Josling,June
Lyr 18h44'56"42d37'
Jospeh,Michael
Her 16h42'54"21d4'
Joss,Jerome
Ori 5h53'0"17d29'
Jossa,Angela Marie
Peg 22h49'34"8d14'
Jostameling,Bianca
Cnc 8h34'26"30d36'
Josuah
Cyg 20h49'20"38d15'
Josund,Duncan John
Aql 19h45'50"10d1'
Josusie
Mon 6h18'51"8d29'
José
Com 12h26'44"21d35'
Joséke,Lief
Per 4h29'31"50d7'
Jota Jota
Aql 20h8'50"8d19'
Joubaud,Henri
Dra 11h56'18"72d45'
Joules
Peg 22h37'30"11d55'
Joumeaut,Youam
Ser 15h9'0"12d20'
Jouni & Leea
Cam 6h31'25"68d18'
Joura Anniversary Star,Frank & Mary
Sge 20h17'54"16d9'
Jourdain,Jacques
Lmi 10h48'12"32d13'
Jourdain,Laurent
Sge 20h2'24"20d37'
Jourdan,Pierre
Gru 22h8'58"-55d13'
Journey,The
Her 16h49'39"35d32'
Joussard,Michelle
Com 12h27'58"21d58'
Jouleux,Marie-Alice
Com 13h9'54"30d36'
Jouve,Magali
Ori 6h20'34"10d13'
Jouven,Francois
Sgr 19h26'20"-34d53'
Jovain
Gru 22h29'21"-56d13'
Joval
Cyg 21h30'11"37d19'
Jovaria Ali's Star
Cyg 19h32'60"39d23'
Jover Flix Benito
Dra 16h49'50"65d44'
Joverner
Per 1h54'0"56d21'
Jovi,Federico
Cmi 8h7'1"1d47'
Jowads
Ind 20h54'40"48d23'
Jowyk
Ori 5h53'27"12d22'
Joy
Ori 5h52'49"14d52'
Joy
Lyr 19h24'26"37d45'
Joy
Pup 8h8'34"-22d15'
Joy
Vel 10h5'19"54d19'
Joy
Lyr 18h41'35"34d9'
Joy
Aur 5h0'10"47d33'
Joy
Mon 7h12'12"-6d50'
Joy
Cyg 19h22'20"54d31'
Joy
Uma 9h28'11"50d13'
Joy
Del 20h16'0"9d23'
Joy & Andy's Star Of Booth
Mon 7h22'28"-8d56'
Joy Alene
And 2h30'55"41d48'
Joy Diane
Cap 20h35'13"-20d24'
Joy Kelly
Mon 6h42'35"7d14'
Joy Noël
Uma 8h36'15"56d50'
Joy Peace
Lyr 18h50'16"34d51'
Joy To The World
And 0h24'49"45d59'
Joy"To The World"
Cmi 7h58'57"8d22'
Joy,Austin William
Mon 7h56'38"-8d13'
Joy,David Baxter
Boo 15h2'37"29d58'
Joy,David Sidney
Her 18h5'1"28d32'
Joy,Evan Tyler
Uma 10h51'51"47d36'
Joy,Joshua Evan
Hya 9h11'28"0d33'
Joy,Nicki
Lyn 7h9'34"58d11'
Joy,Samantha Agnew
And 1h0'45"34d5'
Joy,Sandra
Peg 23h46'27"8d16'
Joy-18
Eri 3h23'43"-11d33'
Joyal 454
Dra 20h6'1"70d7'
Joyal,Nicolette
Lyn 6h45'53"60d10'
Joyce
Cet 2h44'17"1d4'
Joyce
Lyn 7h33'14"39d56'
Joyce
Cet 2h47'32"3d6'
Joyce
Cas 0h42'50"74d18'
Joyce "The Dancer"
And 0h54'27"40d33'
Joyce & Jerry
Dra 20h6'1"62d45'
Joyce 1
Peg 22h3'50"5d35'
Joyce Ann
And 0h11'54"39d26'
Joyce Ellen
And 1h26'19"36d22'
Joyce Gwendolyn
Peg 21h42'0"23d7'
Joyce L
Mon 6h15'40"-4d54'
Joyce Margaret (Blanche)
Crb 16h22'13"28d60'
Joyce Marie
Lyn 8h53'42"44d34'
Joyce Stacey 70th Birthday Star,The
Cas 0h7'50"61d16'
Joyce Star,The
Cyg 21h29'27"30d55'
Joyce Suzanne
Cnv 12h42'12"33d33'
Joyce's Dream
Lyr 19h2'27"33d53'
Joyce's Star
Crb 15h52'20"31d18'
Joyce's Stellar Tutu
And 0h11'39"34d39'
Joyce*Beautiful Mother Partner, Friend & Star
Lyr 18h53'26"40d56'
Joyce,Andrew Sprouse
Aql 19h7'0"-1d52'
Joyce,Arlene Mary
Com 12h53'35"27d37'
Joyce,Catherine M
Lmi 10h11'13"38d6'
Joyce,Jeanine Santerre
And 23h47'13"45d26'
Joyce,Jeffrey A
Uma 11h14'50"30d33'
Joyce,Joshua Mark
Cep 21h2'46"58d9'
Joyce,Jr,Charles Bryant
Sex 9h58'23"2d13'
Joyce,Karen
Lyn 8h1'59"41d39'
Joyce,Kerry Michael
Boo 13h35'51"22d3'
Joyce,Mortimer John
Aur 5h15'35"45d3'
Joyce,Our Family Star
Mon 6h22'30"4d44'
Joyce,Patrick & Crysaly Aviles
Lmi 10h18'31"34d28'
Joyce,Phyllis Secrest
Lyr 19h22'56"30d58'
Joyce,Quinlan Campos
Cam 3h21'37"60d26'
Joyce,Rachel Morrison
Lyr 19h0'35"26d13'
Joyce,Randy
Per 3h15'1"54d36'
Joyce,Susan M
And 23h45'0"42d59'
Joyce,Thomas Michael
Cnv 13h5'36"50d32'
Joyce,Tina Marie
And 0h20'31"34d5'
Joyce,Vincent,Jessica, Ariel S O T M
Dra 16h40'0"64d49'
Joyce,William Kenneth
Her 18h4'18"14d51'
Joyce-Jake & Stella's Little Angel
And 23h42'30"47d5'
Joycelyn Star,The
Aur 5h39'34"37d47'
Joycenken
Lyn 8h29'46"43d1'
Joycie:My Sweet Princess
Peg 21h19'45"21d51'
Joye Noelle
Aql 19h58'17"10d4'
Joyer,Sophia Ann
Cap 21h23'54"-16d26'
Joyful
Mon 7h6'19"-6d49'
Joyle,Wendy
Peg 23h7'25"11d57'
Joyner
Uma 10h0'14"58d43'
Joyner,Alberta
Lyr 18h40'0"32d1'
Joyner,Fred
Cet 3h5'29"2d34'
Joyner,Vickie
Gem 7h56'37"32d25'
Joyner-Siebrecht, Jewell Josephine
Uma 10h59'59"58d18'
Joyous Song of Happiness
Lyr 19h20'18"38d18'
Joyous Victory
Lyn 7h27'0"45d8'
Joz
Cet 0h56'1"-2d47'
Jozefin
Cep 23h1'42"61d14'
Jozefowicz,Richard Michael
Cep 22h36'22"60d33'
Jozkowski,Michael
Aqr 22h0'14"-11d35'
Jozsef & Debra
Uma 10h5'38"60d52'
Joëlle
Peg 22h44'29"5d59'
JP
Cam 3h23'30"61d15'
JP
Cyg 21h35'1"41d31'
JP
Uma 11h59'39"56d8'
JPB4ever (Joanne P Beck)
Com 12h20'0"19d42'
JPPP Oostvogels omdat ik van je hou
Lac 22h51'44"40d18'
JPR & HLR-Always & Forever
Lyn 8h18'54"45d2'
JPR 19/95
Lyn 7h30'30"38d35'
JPS
Uma 11h33'43"66d52'
JPV "Always" JRV
Cam 4h58'1"67d47'
JR
Aur 6h51'38"37d27'
JR
Cep 23h7'0"68d39'
JR Image
Aql 20h4'0"-5d54'
JR-ER Eternally
Dra 19h44'48"71d1'
JRA
Lac 22h19'56"49d29'
JRDII/1 + RVF-F =JRD"
Crb 16h20'55"52d29'
Jrecki,Maxson Drew
Boo 15h11'33"29d7'
JSP Esq
Her 15h48'54"42d13'
JSPUT
Lac 22h5'20"49d23'
Jss Ply-Gem
Ori 6h6'50"8d60'
JS"
Ori 5h57'31"15d31'
JT2
Boo 15h8'56"18d26'
JTH-60
Cam 6h9'36"71d8'
JTKT
Umi 15h17'54"70d1'
Ju Ju Jasmine
Vul 19h18'47"25d57'
Juah
Dra 19h0'55"70d11'
Juan
Peg 23h16'0"31d17'
Juan & Annette's Destiny
Crb 16h14'23"37d48'
Juan Jose
Pho 0h42'2"-46d15'
Juan Y Maria
Ori 4h56'27"0d43'
Juan,Justin J
Cep 22h21'39"56d10'
Juancar
Cep 22h40'32"68d29'
Juanita
And 0h25'27"21d42'
Juanita
Lyn 7h44'13"50d4'
Juanita
Cyg 20h20'11"37d55'
Juanita Clare
Lyr 18h18'0"40d20'
Juanita Little Brewer
And 23h29'0"45d48'
Juanita Louise
Peg 22h37'33"18d49'
Juanita's Love
Boo 15h36'38"50d28'
Juanma Y Rosa
Ori 5h38'33"7d53'
Juarbe,Anthony M
Equ 21h20'33"3d1'
Juarez,Santiago
Gem 6h54'16"18d3'
Juarez,Willibaldo
Cep 22h18'29"58d37'
Juba (Jutta Bachmann)
Sco 17h29'43"-30d52'
Jublee Jo
Lyr 19h6'39"28d40'
Jubu
Del 20h17'48"13d50'
Jud
Crb 16h13'57"33d46'
Jud's Song
Lyr 18h19'55"38d6'
Judah's Star
Cep 21h25'24"55d3'
Judah,Charles
Aql 19h13'48"19d3'
Judah,Julian
Aql 19h5'28"15d30'
Judastar
Psc 1h19'18"17d43'
Judd,A-Yo-Na-Kit-Ne- Shi William W
Her 16h50'30"34d42'
Judd,Janet
Aql 19h7'21"0d18'
Judd,Rex Langdon
Ori 5h33'25"-1d44'
Jude
Lmi 10h27'52"30d57'
Jude
Boo 14h37'31"18d8'
Jude In The Sky with Diamonds
And 1h22'54"48d59'

Jude's Star
Lib 15h57'49"-11d8'
Jude-N-Malisa Forever
Lyn 8h5'49"38d12'
Jude-Terese
Sex 9h53'21"0d50'
Judf
Nor 16h18'39"53d45'
Judge,Jane E
Dra 17h52'40"76d18'
Judge,Joan S
Peg 23h2'1"10d36'
Judge,Martha Turner
Crb 15h51'46"27d35'
Judges Star,The Christopher
Cmi 7h15'1"1d42'
Judges,Alexandra
Sge 20h5'52"16d49'
Judging Doctor Star, The
Dra 17h35'37"64d30'
Judi
Cas 2h10'0"67d33'
Judi Christine "Shawnee"
Peg 23h43'37"31d42'
Judi Princess of Quite-A-Lot
And 1h1'0"37d2'
Judi Star
Vul 19h44'59"20d27'
Judi's Wishing Star
Cap 21h1'18"-27d9'
Judice,Fatima Cotta
Lyr 19h23'27"41d14'
Judiczew,Alexander
Cnc 8h59'16"7d36'
Judiczew,Efrosina
Cnc 8h58'56"8d29'
Judie
Lyn 7h47'49"40d42'
Judie & Rolf
Aql 19h5'37"5d4'
Judilyn
Mon 6h29'45"8d33'
Judit Rigau i Cortal
Her 18h19'25"14d46'
Judith
Eri 3h54'41"-5d39'
Judith
Cyg 20h33'25"37d59'
Judith
Cas 0h14'21"63d13'
Judith
Cyg 19h28'33"39d52'
Judith
Lib 14h57'22"-10d17'
Judith
And 0h16'45"45d56'
Judith
Vul 19h22'52"25d16'
Judith
And 23h40'51"38d1'
Judith
Hya 8h13'49"1d2'
Judith
Mon 7h24'51"-10d10'
Judith
Cyg 21h7'17"31d33'
Judith
Cas 1h49'37"73d19'
Judith & Edwin
Eri 4h1'0"-18d15'
Judith & Joff
Cyg 19h43'12"29d52'
Judith & Leigh
Eri 5h1'14"-4d14'
Judith Ann
And 23h38'41"46d51'
Judith Ann
Peg 23h24'57"32d22'
Judith Anne
Cam 8h28'0"80d2'
Judith B
Ori 5h59'48"20d57'
Judith Ellen
Peg 23h1'26"11d53'
Judith Jean-Marie
Cyg 20h24'18"39d17'
Judith Mae
Cyg 20h36'54"38d11'
Judith P
Cas 23h41'11"61d39'
Judith's Angel
Peg 21h53'49"21d20'
Judith's Love Shines Upon Us
Mon 6h57'42"-10d52'
Judith's Star
Mon 7h43'11"-1d17'
Judkins,Madeline Blanche
Cam 3h52'54"71d58'
Judson,Julie Anne
Lyn 7h16'24"50d29'
Judson,Stephen Francis
Her 17h2'31"22d28'
Judy
Lyn 7h58'55"37d21'
Judy
And 2h30'32"42d16'
Judy
Aql 20h10'15"4d34'
Judy
And 0h11'58"39d34'
Judy
Peg 21h43'58"21d23'
Judy
Sct 18h35'17"-6d32'
Judy
Cyg 21h5'19"39d43'
Judy
Mon 7h27'11"-1d7'
Judy
Mon 6h53'25"-1d28'
Judy
Vul 19h43'36"22d58'
Judy & David
Cyg 19h26'48"30d35'
Judy & Louis
Cyg 21h4'17"38d51'
Judy 91225
And 2h11'50"50d46'
Judy Ann
Peg 23h3'31"21d59'
Judy Ann
Cas 0h52'24"66d25'
Judy Ann
Cyg 19h39'19"37d54'
Judy Ann
Lmi 10h7'52"32d35'
Judy B's Inspiration
And 23h0'53"46d36'
Judy Diana "40"
And 0h22'0"45d29'
Judy I Love You,Stu- Happy
Birthday
Leo 9h26'0"7d17'
Judy Kay
And 23h42'23"43d24'
Judy Kelly
Uma 9h45'23"58d52'
Judy Lee
Uma 11h23'37"40d8'
Judy M
Cas 0h8'18"61d23'
Judy Marie
Vul 21h22'35"27d41'
Judy Maureen
Cet 0h29'26"1d28'
Judy Nora
Aql 19h55'11"10d7'
Judy Rose
And 23h33'0"41d22'
Judy"I Love You" David
Cyg 19h54'37"38d2'
Judy's
Umi 13h51'50"76d12'
Judy's (Dash)
Cet 3h5'0"1d55'
Judy's (Nip)
Mon 6h55'19"-2d1'
Judy's 50th Birthday
Lmi 9h41'36"37d44'
Judy's Gem
And 0h8'16"38d53'
Judy's Inspiration
Lyr 18h30'47"44d4'
Judy's Light
And 23h41'1"33d38'
Judy's Star
Ori 4h59'29"14d22'
Judy,Alisa Elizabeth
Mon 6h53'57"1d31'
Judy,Bruce, & Eric
Cyg 21h25'44"31d9'
Judy,Drew William
Aql 18h57'14"-6d19'
Judy,Thomas Robert
Cep 21h44'0"55d19'
Judy-Doo
And 0h44'53"22d22'
Judy/David
Sge 20h13'0"17d33'
Judybeans
Cyg 19h53'27"38d8'
Juelly,Alexander
Ori 5h55'6"9d46'
Juen,W Elisabeth
Cap 20h27'42"-26d42'
Juer-Vonna
Sge 19h53'17"16d36'
Juergens,Mary
Peg 22h6'20"20d20'
Juhase-Brown September Star
Ori 5h46'42"10d54'
Juhasz,Carla & Richard
Sex 10h12'16"2d43'
Juhasz,Debra Sue
And 23h3'17"37d50'
Juhasz,Zachary Paul
Peg 21h59'48"28d39'
Juhe
Vir 13h51'45"1d54'
Juhl,Marilyn
And 2h30'1"42d36'
Juhl,Mary L
Cyg 19h33'0"32d29'
Juill,Yannick
Uma 12h57'25"57d56'
Juillet 1263
Sgr 18h50'54"-28d40'
Juin,Michel
Lac 22h42'0"38d35'
JuJu Bean
Aur 6h32'50"35d19'
JuJuBe I
Mon 6h30'29"-0d13'
Jujubee
Per 2h57'58"34d26'
Jukebox Dave
Dra 19h47'30"61d26'
Jukes,David
Lyn 8h10'37"44d37'
Jukkala,Andrew Kenneth
Equ 21h21'1"10d60'
Jul
Eri 3h41'1"3d41"
Jula,Jim F
Hya 8h50'0"-0d26'
Jule
Cam 4h11'18"70d19'
Jule
Lyr 18h26'11"47d6'
Jule
Aur 5h16'34"43d46'
Jule's Planet
Aql 19h28'36"13d51'
Julee
Mon 8h1'15"-10d0'
Julee
Cas 0h59'22"68d26'
Jules
Uma 11h54'50"57d24'
Jules
Aur 5h1'58"40d56'
Jules
Cas 1h8'26"61d43'
Jules' & Treva's Eternal Flame
Cep 20h52'12"68d52'
Juletta
Crb 16h5'0"26d49'
Juli-Please will you marry
me?
Cyg 20h20'40"40d59'
Julia
Cyg 19h57'1"30d44'
Julia
And 1h43'23"40d44'
Julia
And 23h16'14"50d19'
Julia
Lyr 18h 45'39"41d19'
Julia
And 2h34'50"38d26'
Julia
Cyg 21h0'37"31d10'
Julia
Vul 21h25'14"24d12'
Julia
Cyg 19h27'52"35d10'
Julia
Lyr 19h2'14"30d14'
Julia
Cyg 20h7'22"40d13'
Julia
Lyr 19h16'1"41d31'
Julia
Cam 7h51'17"68d47'
Julia & Gene
Cam 4h4'41"61d46'
Julia 1004;The Guardian
Angel
Lyn 7h36'32"39'33'
Julia 2
Aql 19h0'35"7d30'
Julia A
Cyg 21h37'42"41d57'
Julia Ann
And 0h58'1"45d9'
Julia Cecille
Cas 2h18'29"70d54'
Julia Diane
And 1h23'26"34d57'
Julia G
And 23h35'44"47d60'
Julia Joy
Ori 4h59'15"0d19'
Julia K
Mon 6h29'15"11d31'
Julia K
Vul 19h51'51"20d12'
Julia Katharina Dee
Lyn 8h28'0"33d28'
Julia Lee
Cyg 19h48'0"38d18'
Julia Louise
And 1h16'0"37d6'
Julia Maria's Star
Cet 2h47'38"3d49'
Julia Rae
Cyg 20h7'16"38d28'
Julia Rosemary
Cas 1h50'47"60d21"
Julia star
Ori 5h55'31"7d5'
Julia T
Oph 18h40'31"7d2'
Julia's (Taco Attack)
Cas 2h8'40"61d21'
Julia's Jewel
Cas 0h34'39"69d37'
Julia's Star
Lyn 8h14'14"36d52'
Julia's Star
Cas 0h3'30"54d46'
Julia's Star From Shayne
Aql 19h27'23"10d45'
Julia-21
Cnc 8h32'0"10d41'
Julian
Her 16h59'35"13d2'
Julian
Peg 22h52'38"25d21'
Julian
Boo 13h48'1"16d56'
Julian
Cep 20h39'1"75d9'
Julian
Cnv 12h18'20"48d55'
Julian (Julie & Ian)
Lyr 18h31'20"34d17'
Julian James
Hya 9h1'0"5d19'
Julian Poet in Arms
Per 4h0'54"31d29'
Julian's Wisher
Lyn 6h28'50"54d59'
Julian,Ann
And 0h44'1"38d59'
Julian,Cornie F of
Accokeek,MD
Oph 18h42'44"6d38'
Julian,Donna Jean
Psc 0h0'12"1d46'
Julian,James Philip
Oph 17h4'29"10d42'
Juliana
Cyg 21h5'48"31d32'
Juliana
Uma 13h5'21"52d40'
Juliana
Cyg 19h30'42"36d53'
Julianna
Ori 6h2'42"8d20'
Julianna
And 0h46'37"22d37'
Julianna & Jay's Love Star
Eri 2h52'23"-17d32'
Julianne
Cas 1h52'12"58d27'
Julianne
Uma 10h54'14"56d9'
Julianne 1970
Cam 9h7'29"80d19'
Julianne, Anthony, Michael
Hya 8h42'55"-5d49'
Juliano,Donna & Rick Bryan
Uma 11h35'12"33d41'
Juliano,Joseph
Mon 6h18'28"-0d42'
Juliano,Nick
Her 18h24'52"20d25'
Julich,James Joseph
Aur 5h13'0"40d41'
Julie
Cam 8h22'47"78d51'
Julie
Cas 0h1'30"60d29'
Julie
Peg 22h30'44"8d38'
Julie
And 2h5'33"37d46'
Julie
Ori 5h51'46"15d45'
Julie
Aql 19h25'31"11d12'
Julie
Uma 10h32'31"47d45'
Julie
Cyg 20h5'57"40d45'
Julie
Cas 1h28'54"60d3'
Julie
Lyn 6h55'23"54d58'
Julie
Boo 14h14'32"48d16'
Julie
Boo 15h11'10"48d59'
Julie
Cas 3h9'52"60d45'
Julie
And 2h29'0"44d24'
Julie
Lmi 10h49'25"26d46'
Julie
And 23h23'15"42d8'
Julie
Cyg 19h33'28"39d55'
Julie
Cas 0h23'21"74d18'
Julie
Crb 16h13'37"33d13'
Julie
Ser 16h5'16"13d11'
Julie & Christine's Road Trip
Star
And 23h26'57"49d55'
Julie & David
Aql 20h0'56"9d50'
Julie & Derrick
Eri 4h36'12"-1d26'
Julie & Don:The Chips Have
Fallen
Uma 11h0'43"34d5'
Julie & Fred
Cyg 21h3'41"38d21'
Julie & Julie
Ori 5h57'23"10d39'
Julie & Kimberly's Union Of
Love
Eri 4h9'38"-11d35'
Julie & Michael
Lac 22h15'22"46d52'
Julie & Terry's Star
Peg 22h31'48"21d35'
Julie & Tom are Married
Sge 20h1'0"18d38'
Julie Aline
Vul 19h46'27"22d40'
Julie Ann
Com 12h7'1"27d1'
Julie Ann
And 0h50'48"40d47'
Julie Ann
Aql 19h55'34"1d7'
Julie Ann
And 2h3'20"40d32'
Julie Ann
And 2h7'38"38d54'
Julie Ann
Cyg 20h15'0"38d56'
Julie Ann D
Cas 0h35'53"64d3'
Julie Anna
And 23h39'35"37d56'
Julie Anne
Peg 22h27'46"35d15'
Julie Arlene in Spring:393.26
Ori 5h27'13"-6d33'
Julie Christine
Peg 21h58'50"33d20'
Julie Christine & Arnaldo
Sco 17h32'0"-30d21'
Julie Cristine
Vul 20h40'1"28d10'
Julie Dawn
Cyg 20h27'16"30d46'
Julie Elise
Lib 15h15'49"-19d18'
Julie Elizabeth
Lyr 18h42'10"41d15'
Julie Elizabeth
Gru 22h39'58"51d36'
Julie Ellen
Cas 0h29'20"58d63'
Julie Helaine
Peg 22h42'21"27d32'
Julie Lynn
Peg 21h41'0"20d7'
Julie Lynn
Sco 16h57'1"-31d20'
Julie Lynn
Umi 21h34'51"89d13'
Julie Marie
And 2h32'16"48d16'
Julie Marie
Lyn 9h5'33"41d28'
Julie Mary-Marjorie
And 1h16'37"39d34'
Julie Susan
And 1h41'41"38d58'
Julie Suzanne
Mon 6h27'38"11d28'
Julie T
Cas 1h57'40"71d15'
Julie's Belle Star
Crb 15h42'56"28d43'
Julie's Blue Sky Happiness
Cas 1h22'31"58d17'
Julie's Gem
And 1h46'36"40d0'
Julie's Mini Dipper
Peg 22h28'40"26d2'
Julie's Shining Little Star
And 2h28'50"47d45'
Julie's Star
And 0h45'14"38d45'
Julie's Star
And 2h2'38"38d59'
Julie's Star
Peg 22h24'59"31d23'
Julie's Whatever
And 0h10'1"27d44'
Julie's Wish
Peg 22h51'44"20d8'
Julie's Wishing Star
Lyr 19h16'53"40d33'
Julie,Daniel
Dra 15h46'17"61d46'
Julie,David
Uma 9h16'14"51d34'
Julie,Love Andrew
Sge 19h28'17"17d16'
Julie-Joe
Eri 3h4'18"-1d53'
Julie-Sidney
Crb 15h20'30"30d22'
Julien
Cas 2h21'24"77d12'
Julien
Aql 20h0'21"10d24'
Julien
Her 16h24'59"23d22'
Julien et Annie
Umi 16h27'57"74d3'
Julien M V M Sc
Vir 13h54'49"2d23'
Julienne A
And 23h1'51"51d47'
Juliep
Cyg 21h8'18"30d45'
Juliet
And 23h3'1"49d44'
Juliet in the Night
Del 20h26'12"11d32'
Juliet Letita
And 1h52'27"37d35'
Juliet Marie
And 1h44'1"40d55'
Juliet Star
Psc 1h0'51"31d13'
Julieta
Eri 3h53'40"-1d1'
Juliets' Phoenix
Com 12h30'46"20d52'
Juliette
Uma 11h43'15"38d12'
Juliette
Com 12h14'42"27d53'
Juliette
And 23h25'45"38d36'
Juliette
Ori 5h48'38"19d14'
Juliette
Aur 5h0'46"45d3'
Juliette 1996
Peg 22h44'0"31d19'
Juliette et Marc
Vir 14h20'47"-7d37'
Juliette Rose
Cyg 20h24'31"37d44'
Juliette's "Natural Born" Star
Aur 4h59'1"36d31'
Juliette-Dawn
Eri 3h43'55"-10d23'
Julio & Paul
Lyr 19h7'45"40d33'
Julio Is Three-Oh!
Mon 8h0'44"-6d60'
Julissa
And 23h46'18"44d10'
Juliusz & Lulu
Crb 16h7'1"31d52'
Juljea
Lyr 19h21'24"41d18'
Jullian Ingrid
Vul 20h16'41"28d49'
JulNat
Cyg 20h37'55"40d7'
Julric
Lyn 7h51'42"36d51'
July Candia
Uma 10h19'32"58d25'
Julye Ann
Crt 11h27'55"-10d11'
Jumes,Dorothy
And 0h22'25"34d57'
Jump,Teri
Uma 10h11'0"42d3'
Jun
Cep 22h9'57"53d27'
Juncar,Stephen Andrew
Ori 5h54'14"10d33'
Juncke-Williams, Jeannie
Sgr 20h4'32"-44d20'
Juncker,Grant
Cet 0h22'35"-12d18'
Junction,Kristine Zaback
Aql 18h50'52"11d34'
June
Cas 0h16'1"62d1'
June
And 23h38'34"43d44'
June
Ser 16h0'48"13d38'
June
And 0h9'53"28d42'
June
Uma 11h2'0"38d34'
June
Eri 3h52'12"-1d10'
June & Dan Forever
Sge 19h31'45"16d13'
June & Poo
Cas 2h29'51"59d59'
June 'n Hilda's Charley
Lyr 18h56'32"42d18'
June 17,1994 IPO
Cyg 20h0'60"38d14'
June 70
Cas 0h6'45"64d6'
June Ann
Mon 6h42'59"6d46'
June Bug
Hya 8h38'26"-5d40'
June Fox,The
Pho 0h17'34"-42d24'
June Lyn
Cep 20h56'40"78d55'
June Patricia
Cas 0h1'45"61d36'
June Pearl
Lyr 19h0'15"40d55'
June Star Light '94
Boo 15h7'25"47d46'
June Tiffany
Cnc 8h9'47"6d49'
June's Daughters
Lyr 18h14'51"30d53'
June,Karl
Eri 4h54'58"-6d53'
June-June
Aql 18h58'1"15d3'
Juneau,Joe
Aur 4h57'58"52d27'
Junebug
Cyg 21h20'0"40d2'
Junewick,Charlotte Ann
Cas 0h24'53"60d14'
Junewick,Ellen Jensine
Psc 0h23'22"-0d16'
Jung,Anton
Uma 11h17'54"41d26'
Jung,Cannie
Boo 14h57'33"32d26'
Jung,Heidi
Sgr 19h11'1"-21d17'
Jung,Josef
Ori 6h17'57"10d45'
Jung,Joseph
Lyr 18h18'1"47d1'
Jung,Mara
Cas 0h10'30"61d55'
Jung,Patricia
Lyr 18h49'55"37d11'
Junga,Terry
Aur 7h26'1"40d26'
Junge,Christian F
And 1h56'48"40d60'
Junge,Ulrich
Cep 22h5'19"61d35'
Junger Pfennig
Dra 15h12'1"63d9'
Junger,Kristina
Peg 22h8'1"5d27'
Junghanns,Hans-Werner
Lyn 8h15'0"42d5'
Junghans,Stefan
And 23h1'1"43d11'
Jungle Green
Per 3h3'49"40d54'
Jungle Joe
Cmi 7h41'54"4d28'
Jungle Storm
Leo 10h38'45"13d2'
Jungseniorinnen TC RW
Sobernheim
Uma 11h36'31"38d27'
Jungslager,Anouk
Ori 5h50'16"21d55'
Jungwirth,Manfred
Lac 22h6'1"50d42'
Junia Gail
Cas 0h23'24"56d10'
Junie
Sct 18h49'60"-7d2'
Junie Ann
Umi 15h10'41"70d60'
Junigeoff
Cyg 21h28'0"53d31'
Junior
Psc 22h51'26"5d30'
Junipat
Aql 18h45'1"10d46'
Juniper Thomas
Boo 14h18'11"37d55'
Juniran
Uma 9h11'1"67d43'
Junji Katsuhara
Sco 16h2'2"-19d1'
Junk 11,Jack
Dra 16h58'37"68d21'
Junker,Gerhard
Vir 12h34'33"-8d56'
Junker,Patrick
Leo 10h23'43"11d17'
Junkermann,Dietmar
Lyr 19h23'24"30d33'
Junko's Star
Sgr 19h8'6"-15d10'
Junnier,George
Aql 19h55'39"15d31'
Junod,Kelly
Lyn 6h26'23"58d41'
Jupa,Diana Lynn
Cas 23h2'22"53d5'
Jupiler
Aur 5h27'1"40d21'
Jupiter,Horton
Uma 8h40'19"62d10'
Juranek,Missy
Uma 11h0'50"60d7'

Jurata,John A & Linda E Wagner
Per 4h5'55"48d38'

Jurblum,Marc
Cra 18h29'10"-45d12'

Jure,Richard C
Dra 15h52'48"67d39'

Jurek,Charles & Joy
Cma 6h56'12"-18d24'

Jurek,James L
Crt 11h47'0"-12d27'

Jurek,Mark R
Cma 7h6'2"15d35'

Jurek,Roger
Cep 23h4'31"65d19'

Jurgeit,Harald
Boo 14h6'11"14d24'

Jurgen's Giggle Star
Umi 14h54'42"66d15'

Jurgens,Talia Dewl
Aql 18h59'39"-5d41'

Jurgensen,Robert Edward
Ori 5h11'26"-4d30'

Jurich,Michael Joseph
Aur 6h16'34"31d8'

Juriga,James Andrew
Cep 22h22'46"59d25'

Jurilli,Jose
Cyg 19h47'27"31d2'

Jurim,Barbara Bernadette
And 0h33'48"38d14'

Jurist Family,N J & Flo
Lyn 9h5'35"45d36'

Jurist,Ceil
And 1h42'46"40d59'

Jurist,Theodore
And 1h50'52"40d37'

Juristo,Natalia
Leo 9h20'14"8d48'

Jurka,Karl
Umi 13h44'27"74d12'

Jurkacek,Carol
Del 20h18'39"10d20'

Jurkoic "Sal",John Douglas
Cmi 7h21'49"8d38'

Jurkovac,Jay Thomas
Dra 15h50'0"67d4'

Jurkovic,Rachel
Aur 5h32'0"41d3'

Jurkovich,Cindy
And 0h21'38"30d25'

Jurnak,Joseph J
Her 17h14'17"46d47'

Jurney 6-26-95,Troy Weston
Dra 15h56'29"64d30'

Jury,Jacob C
Oph 17h6'16"10d33'

Jury,Lori
And 23h44'31"47d58'

Juryoanne
Per 1h29'13"53d33'

Jurysta,Joshua A
Her 15h56'37"42d46'

Jurysta,Sarah
Cas 23h4'25"58d31'

Juska,Joe & Jurate
Cyg 21h8'16"31d26'

Juskevich-Boyer, Barbara
Cyg 21h58'30"53d43'

Jusko,Emily Carol
Peg 0h6'52"18d55'

Jusko,Matthew
Oph 17h23'59"-20d23'

Jusko,Michael P
Cet 1h12'32"-2d48'

Jusko,Nathan William
Aur 4h57'29"34d15'

Jusko,Oma Lajuan
Cam 6h58'15"75d35'AB

Jusko,Tammy
Peg 21h39'1"28d15'

Jusko,William R
Hya 8h53'31"2d17'

Jussiau,Philippe
Ori 6h21'57"10d55'

Just A Tine Star
Vir 13h21'54"-0d52'

Just Another Day In Paradise
Sge 18h58'44"20d27'

Just Because-BCAA
Cyg 21h38'0"40d50'

Just David
Lyr 18h42'53"45d60'

Just Don
Sct 18h54'19"-6d37'

Just For Kicks
Lac 22h6'55"48d41'

Just Jer
Tri 1h29'0"30d22'

Just Judy
Cas 0h28'33"72d17'

Just Look Up Paul
Aur 6h41'21"37d50'

Just Me
Peg 23h0'58"11d54'

Just Peachy
Aql 19h25'21"-6d45'

Just Plain Jim
Cep 21h33'44"80d36'

Just Plain Phil
Uma 11h18'46"48d35'

Just,Eleanor "Peaches"
Lmi 10h37'53"32d37'

Justess Rose
Mon 6h20'35"8d53'

Justice
Cru 12h9'6"-63d5'

Justice
Per 2h26'29"56d54'

Justice
Cru 11h58'10"-57d41'

Justice
Ari 2h36'13"20d18'

Justice
Lmi 11h2'21"32d25'

Justice,David O
Her 16h25'11"40d44'

Justice,Debra Lynn
Aql 20h14'17"1d34'

Justice,Dixie Wright
Mon 6h53'33"-4d48'

Justice,Georgina
Cyg 19h33'0"39d44'

Justice,Joe & Amber
Peg 21h43'49"20d37'

Justice,Sandra Jean
Mon 8h4'57"-1d48'

Justico
Oph 18h8'15"12d0'

Justin
Her 16h46'22"34d9'

Justin
Aur 6h20'31"38d13'

Justin
Peg 22h3'15"29d54'

Justin
Del 20h16'52"10d27'

Justin "Little Man"
Mon 7h44'51"-2d32'

Justin 143 Traci
Aql 19h3'55"5d44'

Justin Christopher
Vir 11h51'0"9d25'

Justin David
Dra 11h10'17"74d5'

Justin E
Cet 0h59'44"-5d9'

Justin James
Cam 11h7'45"81d48'

Justin Michael
Boo 15h1'27"11d14'

Justin's Star
Ori 5h50'39"14d19'

Justin,Ryan & Victoria
Cyg 21h36'15"30d48'

Justin,Sharon,& Rex
Aql 20h16'43"5d4'

Justina Nicole
Del 20h14'0"15d33'

Justine
Vel 9h43'56"55d30'

Justine
Lyr 18h35'16"27d5'

Justine
Cas 0h27'18"50d19'

Justine
And 1h54'31"37d30'

Justine
Ori 5h48'21"19d25'

Justine
Ori 5h56'14"14d0'

Justine
Aur 6h10'55"54d52'

Justine
Tau 5h43'44"29d39'

Justine & Mark
Her 16h25'53"50d37'

Justine & Mark- Together Forever
Cyg 19h25'53"34d35'

Justine Marie
Peg 22h29'54"8d15'

Justine Randall
Lyn 8h7'22"51d29'

Justine Twinkle Twinkle Little Star
Cas 0h48'58"62d44'

Justine's Jewel
Lyn 8h13'43"40d27'

Justine's Redemption
Eri 4h0'37"-7d36'

Justus,Arthur
Lyn 8h8'41"47d0'

Justus,Donna J
Mon 7h53'41"-1d35'

Justus,Lisa Case
Cma 6h55'24"-17d1'

Justyn & Rachael Star, The
Cyg 21h1'53"30d13'

Justyn,Lynette Marie
Vul 19h48'17"23d3'

Justé,Irène
Cas 23h42'0"62d3'

Jutka-Pinci
Mon 7h15'44"-6d53'

Jutt,Genevieve & Edgar
Sge 20h17'58"17d27'

Jutta
Lyn 7h6'10"53d44'

Jutta
Her 17h48'12"46d59'

Jutta & Andi
Uma 8h59'52"60d27'

Jutta & Hans-Werner
Ori 5h58'58"20d33'

Jutta I
Ari 2h55'20"20d49'

Jutta Spalzi
Lib 14h41'31"-20d15'

Juusela,Berit
Aur 7h5'42"40d48'

Juvema
Cet 2h57'12"1d49'

Juvimi
Aql 19h56'25"13d40'

Jwar,Karmar
Lyn 7h21'57"58d15'

JWB 21 For Uncle Al's 50th
Aur 5h2'13"49d8'

JWF II,A Son That Shines Lake A Star
Cep 2h6'0"77d12'

JWV 1359
Cyg 19h26'52"30d39'

JXHSO4B/VRGV87A Eternal Star,The
Ori 4h42'0"11d33'

Jylvean
And 23h30'54"39d53'

Jyoti-Kumar-Ra
Lib 14h42'1"-22d47'

JYOTSNA
Uma 10h59'38"39d32'

Jyrki Torni
Ori 6h1'11"0d16'

Jyrkkä,Reijo Aukusti
Cam 4h55'11"65d34'

JZ1
Lyn 9h30'35"41d39'

Jzendolyn
Crt 11h16'14"-8d57'

Jäger,Detlef
Cas 0h4'51"60d40'

Jäger,Herr
Peg 23h31'13"16d7'

Jämtenius,Erik
Ori 5h48'39"18d5'

Jänisch,Julius
Lmi 9h31'38"38d30'

Jérard
Her 16h43'0"40d51'

Jérome
Peg 23h34'32"20d4'

Jérémie
Cam 3h36'15"63d18'

Jöri
And 2h24'58"45d14'

Jûèsçmé
Cam 5h0'16"54d26'

Jöchle,Beate
Ari 2h32'28"30d31'

Jöel et Aurelie
Vul 19h21'13"23d52'

Jörg
Her 17h50'52"43d1'

Jörg Jung * Dresdner Bauspar AG
Ari 2h29'19"20d19'

Jörg,Drexler
Sgr 18h57'7"-20d52'

Jöris,Stephan
Lyn 8h36'27 42d57'

Jügü
Vir 11h43'15"7d59'

Jünkermann,Dietmar
Peg 23h29'51"13d0'

Jünti
Cet 2h55'22"2d3'

Jüptner,Matthias
Psc 23h22'1"2d29'

Jürgen
Ari 2h38'48"21d55'

Jürgen
Sge 19h36'37"16d41'

Jürgen
Cep 22h7'21"67d36'

Jürgen
Uma 9h10'59"50d17'

Jürgen
Cas 0h6'40"60d43'

Jürgen's Schatzi
Cet 0h29'26"-2d21'

Jürgens,Mechthild
Sco 16h39'58"-43d15'

Jürgensen,Ursula
Leo 9h19'0"18d19'

Júlia Galera i Cuffi
Aur 5h6'53"43d18'

Jørgensen,Lena
Aur 5h12'46"29d12'

K

K
Peg 22h59'15"32d10'

K & D "Until The End Of Time"
Her 16h57'58"20d17'

K & G Together For Eternity
Lac 22h10'45"49d33'

K & J-10
Cam 3h41'0"74d23'

K & K's LoveLight 143
Lac 22h34'26"41d13'

K & M Forever
Crb 16h10'34"38d26'

K & O Woisetschläger
Aql 19h5'31"10d7'

K & P
Uma 12h38'25"60d55'

K 8
Cyg 20h6'1"40d10'

K A D T A
Lyn 7h31'23"35d28'

K A G E L L
Peg 21h55'50"22d45'

K B
Tau 5h51'19"23d54'

K C Masterpiece
Lac 22h1'1"48d45'

K C's Fifth
Ori 5h30'50"0d47'

K C's Voice Of The Stars
Lac 22h6'35"47d49'

K D
Lac 22h8'49"49d5'

K D C A
Sco 17h29'29"-40d56'

K D J "Red"
Cyg 21h15'50"38d10'

K D Shaka Goodfoot
Peg 23h6'0"11d33'

K D's Forever
Cam 4h45'0"68d56'

K E M's Gem
Uma 11h55'18"32d17'

K E W
Uma 9h44'55"58d0'

K G G
Dra 12h4'0"71d34'

K I I
Cyg 21h45'41"30d35'

K J
Umi 15h18'25"69d20'

K J MAC
Boo 13h37'23"19d23'

K J's Twinkle
Cyg 21h31'43"37d31'

K K
Cam 3h58'38"58d28'

K K A P S
Tri 2h6'37"31d3'

K K Express
Cnc 9h8'1"31d30'

K K H Star,The
And 0h23'39"30d38'

K K J & Daphnis
Leo 9h56'19"12d9'

K L Katherine's Light
And 1h24'59"39d48'

K M A X
Lac 22h36'40"52d45'

K P & Bella
And 1h18'20"39d45'

K Rose
And 23h3'0"45d3'

K T & Me With Love
Uma 9h26'53"51d4'

K T K I
Uma 12h9'58"60d11'

K T Topping Moon Star
Vir 13h21'13"-8d13'

K T's Love
Oph 18h41'57"8d41'

K W
Aur 7h25'48"40d25'

K&S in Eternity
Gem 7h24'6"20d51'

K's Star
Eri 3h37'32"-16d20'

K,Antje
Ari 2h53'14"22d10'

K,Lollo
Ari 2h53'55"20d14'

K-Boy Wishing Star
Aql 19h57'1"15d3'

K-Swiss
Eri 3h12'32"-5d30'

K27 Narozeninám Petrovi Zverinskému
Boo 13h37'59"10d18'

K2PJ
Aur 7h23'0"40d25'

K6
Cnv 12h53'0"40d9'

Ka-Brice,The
Vul 19h44'55"28d21'

Kaakimaka,Hanwell & Carol
Eri 3h33'42"-11d38'

Kaarresalo,Anneli
Cas 1h1'0"60d5'

Kaaya,Benedict Wood
Aql 19h55'18"13d32'

Kabakoff,Ellen
Vul 20h19'0"23d38'

Kabel,Christopher
Dra 15h2'34"61d52'

Kaber
Del 20h15'43"14d46'

Kabira Cora
Gem 7h26'19"28d26'

Kabitzky,Alexandra
Her 17h59'20"46d58'

Kabler,Daniel J
Her 15h53'17"48d3'

Kabnick's Star
Uma 8h43'0"70d7'

Kabob
Lyr 18h50'46"34d51'

Kaborvi
Lmi 10h14'54"40d28'

Kacaba UR Light 143
MBK,Michael R
Aql 18h58'38"-6d40'

Kacalek,Deena Michelle
Mon 6h35'15"-0d15'

Kacey & Mel
Crb 15h41'1"27d23'

Kacey's Komet
Cam 4h31'44"68d19'

Kachellek,James
Ori 6h1'41"1d36'

Kacher,Patti
Del 20h33'57"11d1'

Kachorsky,Nanny
Oph 18h22'22"6d52'

Kachur,Helen Elizabeth
Ari 2h4'45"22d15'

Kachur,Kyla Alayna Brietta
Uma 13h48'28"48d42'

Kachura,Sandra
And 2h30'43"44d31'

Kacie Marianne
Cas 1h35'47"61d17'

Kacinski,Kristen Anne
Mon 6h23'26"1d40'

Kacludis,Jessica Marie
Uma 11h34'48"31d57'

Kacludis,Jr,Paul Steven
Uma 11h33'23"33d40'

Kacmarek,Regina Helen
And 23h20'41"51d32'

Kacner,Mario
Lib 15h4'31"-5d47'

Kacprowicz,Kimberly Paige
Lib 15h19'35"-21d58'

Kacprzak,Henry & Irene
Cyg 20h32'56"42d31'

Kacvinsky,Andrew & Mildred
Crb 15h55'43"26d28'

Kaczmar,Gregory Kenneth
Lyr 18h32'23"42d39'

Kaczmarski,Donna
And 23h1'58"51d36'

Kaczor,Kimberly A
Peg 21h49'1"28d59'

Kaczor,Leslie A
Aql 18h58'0"16d15'

Kaczynski,Victor
Dra 17h20'51"61d37'

KADA 1
Tau 5h45'42"23d22'

Kada,Jon Kazumi
Mon 6h55'1"0d18'

Kadash,Michele
Vul 19h47'13"28d39'

Kader,Mehabob Abdul
Her 18h7'1"14d46'

Kaderabek,Marilyn Parks
And 0h10'45"46d59'

Kadi Jade
Cru 12h41'23"-57d5'

Kadi,David C
Lac 22h2'1"48d52'

Kadi,Karen N
Cas 0h14'43"58d28'

Kadien,Jo Ann Marie
Tri 2h39'11"32d56'

Kadilak,Daniel Andrew
Dra 17h5'12"64d33'

Kadish,Susan
And 0h9'1"46d33'

Kadlac,Robert Louis
Ser 18h41'44"4d35'

Kadlubowski,Sophia
Crb 15h28'19"31d14'

Kae
Lac 22h44'1"56d50'

Kaech,Chloé
Peg 22h14'1"32d57'

Kaefer,Donna
Cas 0h29'53"58d19'

Kaelberer,Dwayne P
Sco 17h1'59"-37d46'

Kaeleigh's Wish Come True
Tri 1h34'42"28d59'

Kaelin,James
Sex 10h45'0"0d45'

Kaemmerer,Paige Lynette
Lyn 7h46'53"47d56'

Kaemmerlen,Aurelien
Sgr 19h31'37"-32d54'

Kaeppel,Peter
Lib 15h15'0"-28d51'

Kaer,Walter
Aur 4h54'44"51d3'

Kaese,Kory
Dra 18h53'21"48d47'A

Kaestner,J Adair
Cet 1h57'36"-2d49'

Kafel,Melanie Lynn
Boo 15h18'49"41d9'

Kaffenberger
Uma 9h56'54"44d19'

Kafka,Gertrude Thomas
Cas 1h7'21"61d13'

Kafka,Ronald W
Cnv 13h50'27"38d55'

Kafran
Gem 6h53'45"30d20'

KAG & BWM
Psc 1h2'0"32d30'

Kagan's Place In The Universe,Russell A
Her 17h28'21"31d32'

Kagan,Barbara
Umi 17h16'32"75d26'

Kagan,Dori L
And 23h17'1"41d6'

Kagan,Florrie
Mon 6h29'18"10d43'

Kagan,Jonathan
Her 18h54'52"13d0'

Kagan,Marie Louise
Lyn 8h50'1"46d11'

Kagan,Paulette
Com 13h12'0"30d3'

Kagan,Stephen "Steal Your Face"
Aql 19h3'16"-1d45'

Kagan,Vickie
Cyg 19h11'45"47d15'

Kagawa,Kathy
Cas 23h45'24"50d35'

Kagayaki 1995
Vir 11h39'0"6d31'

Kagayaku-Mirai
Cnc 9h19'16"31d27'

Kage
Umi 15h45'0"71d40'

Kage
Equ 21h2'23"10d54'

Kageyama,Shigeko & Yusaku
Psc 0h7'56"0d47'

Kagney
Cmi 7h55'1"5d7'

Kahan & Brown DDS
Uma 11h34'44"31d59'

Kahan,Barry & Holly
Uma 11h36'14"30d31'

Kahan,Michael
Dra 15h10'50"62d2'A

Kahane mon amour, Carrie
Umi 15h20'24"67d54'

Kahea
Cnv 13h52'21"40d58'

Kahekili
Cnv 13h49'44"41d2'

Kahikina,Elizabeth
Peg 22h51'17"27d35'

Kahl,Barbara Hahn
Lyr 19h13'16"42d1'

Kahl,Christmas Glory-Dorothy Mae
And 0h1'1"47d24'

Kahl,Linda K
Cas 0h43'19"64d17'

Kahl,Reinhold
Sgr 19h19'16"-22d10'

Kahle,Cheryl
Com 12h28'22"20d46'

Kahle,Lisa
And 2h35'31"41d54'

Kahler,Christian Robert
Cmi 8h6'0"6d55'

Kahler,Ross Charles
Mon 6h29'11"10d33'

Kahlmeyer,Julia Ann
Cas 0h51'49"68d33'

Kahlmeyer,Justin Carl
Cet 3h11'39"1d23'

Kahn
Vul 19h42'1"23d2'

Kahn 4-24-59,Leslie Adams
Lyr 18h57'19"46d45'

Kahn 7-8-59,Dr Bruce R
Cep 20h56'54"60d1'

Kahn's Star,Larry
Dra 16h30'1"67d14'

Kahn,Dr Arthur
Her 17h11'36"46d14'

Kahn,Hedy
And 0h17'0"33d57'

Kahn,Jamie
Hya 8h19'38"1d57'

Kahn,Jasper Jake
Her 17h12'20"22d49'

Kahn,Jules Sumner
Aur 7h6'27"40d41'

Kahn,Louis
Oph 17h37'48"11d24'

Kahn,Martin Jay
Oph 16h52'24"-28d29'

Kahn,Megan
Lyn 8h6'42"58d56'

Kahn,Rachel Esther
Cas 2h45'12"75d34'

Kahn,Samuel Alexander
Aur 6h11'36"45d29'

Kahn,Sol Barak Averom Shepsell ben Jacov
Gem 7h57'49"30d26'

Kahn-Keating Wedding Star,The
Cyg 20h20'12"30d19'

Kahnk,Ray Kent
Cet 2h44'1"6d44'

Kahns,Mitchell-Adrian
Lyr 19h25'18"42d25'

Kahoolawe-Sept 25, 1992,Lanai
Uma 11h3'11"38d50'

Kahouadji,Anne Sophie
Per 3h52'48"36d17'

Kai
Eri 4h14'12"-10d12'B

Kai & Josh
Crb 15h52'11"38d18'

Kai & Julie:Bridging Friends Forever
Cam 3h31'58"61d34'

Kai Liu
Boo 14h32'42"43d34'

Kai Lyn
Lyn 8h41'35"43d45'

Kai Sun
And 23h20'1"44d42'

Kai's Star
And 23h12'35"35d39'

Kaia's Love
Aqr 23h8'52"-6d36'

Kaibas,Cherilyn Mary
Cyg 20h18'45"41d10'

Kaigreor
Cnv 12h14'30"45d18'

Kaila
Aqr 22h19'19"0d23'

Kaila
Cet 2h16'54"3d33'

Kailey
Cas 23h37'29"60d22'

Kailing,Alexandra
Lyr 18h52'1"43d2'

Kailing,Penny
Cyg 20h4'25"30d5'

Kaim "Eddie",Edward Henry
Her 18h2'39"28d52'

Kaim,Brian Andrew
Her 16h49'26"34d2'

Kain,Carolyn
Cyg 19h58'24"40d60'

Kaine Baptist
Uma 8h39'40"70d30'

Kaipo,Mary
Lyn 8h3'11"34d16'

Kairey,Micki
Cnc 8h7'23"31d39'

Kairis,Rachel Ann
Lyr 18h36'39"35d26'

Kairos,Giselle
And 23h39'40"48d25'

Kaiser
Her 17h23'18"29d24'

Kaiser "Dolly",Dorothy T
Crb 16h2'18"28d14'

Kaiser Family's Star
Uma 9h57'41"56d42'

Kaiser Planet,Sam
Sgr 19h58'15"-42d56'

Kaiser,Caleb
Uma 11h21'11"61d13'

Kaiser,David Harrison
Dra 16h1'41"67d16'

Kaiser,Emily
And 2h16'19"48d59'

Kaiser,Eugen Fredrick
Ori 4h59'0"5d23'

Kaiser,George Robert
Hya 8h56'58"5d32'

Kaiser,George Robert
Hya 8h49'4"-0d54'B

Kaiser,Gottfried
Per 3h56'29"37d49'

Kaiser,Gregory K
Del 20h25'12"8d1'

Kaiser,Heinz
Cnc 9h16'0"18d17'

Kaiser,Herbert J
Her 17h28'19"20d55'

Kaiser,Kristofer Fredrick
Cyg 21h59'1"53d5'

Kaiser,Linda Marie
Hya 8h49'4"-0d54'A
Kaiser,Louise
Ori 5h30'34"1d0'
Kaiser,Marissa Alaina
Peg 22h35'14"26d25'
Kaiser,Mark Alan
Peg 22h0'30"25d53'
Kaiser,Marlo Ann
Peg 22h22'1"25d58'
Kaiser,Mary Ann
Lyr 18h59'51"33d5'
Kaiser,Michael Anthony
Peg 22h7'20"26d26'
Kaiser,Milas A
Hya 8h21'16"-1d40'
Kaiser,Molly Ann
Lyn 6h39'12"54d20'
Kaiser,Rudi
Aql 18h47'22"10d37'A
Kaiser,Ruth
Aql 18h47'22"10d37'B
Kaiserman,Joe
Aql 19h35'39"-5d45'
Kaitlin
Cas 3h28'23"70d27'
Kaitlin
Peg 22h31'12"29d1'
Kaitlin Ann
Lyn 7h57'42"42d18'
Kaitlin Rose
Lyn 8h25'11"46d3'
Kaitlineel
Uma 8h56'46"53d13'
Kaitlyn
Vir 11h42'0"0d42'
Kaitlyn
Peg 23h32'47"18d52'
Kaitlyn
Cam 7h55'1"60d8'
Kaitlyn Elizabeth
Cas 2h35'1"57d36'
Kaitlyn Marie
Cyg 21h14'21"38d8'
Kaitlyn Sarah
Mon 7h17'16"-6d31'
Kaitlyn's Folly
Lyr 18h39'0"27d53'
Kaitz,Betty & Larry
Aur 5h19'27"40d37'
Kaj & Allison-25 Years
Uma 10h47'33"47d32'
Kajiya,Stella of Yoshihito
Sco 16h16'59"-24d15'
Kajkowski,Edward
Sco 17h5'52"-38d6'
Kajsa 68-29-01 Forever
Uma 9h24'21"47d39'
Kajun Always Wanted To Be Star
Cyg 19h21'1"28d49'
Kakad,Reshma
And 0h22'23"45d56'
Kaki Ke
Cam 3h56'16"60d43'
Kakic
Per 3h29'28"39d8'
Kakie
Per 3h29'50"51d32'
Kal
Boo 14h30'30"41d48'
Kal's Kismet
Sgr 19h31'7"-36d9'
Kal Res
Cra 18h19'48"-42d26'
Kalai I
Equ 21h10'14"10d5'
Kalail,Fred
Aur 4h47'44"38d52'
Kalail,Pat
Lac 22h8'53"38d24'
Kalamaha,Kipton
Aur 5h2'13"41d20'
Kalangis,Peter G
Aur 6h48'42"37d14'
Kalas,Bob
Aur 6h30'0"38d53'
Kalash,G D
Eri 2h57'13"-2d33'
Kalata,Mark Thomas
Per 3h11'30"45d53'
Kalb,John V
Sct 18h53'23"-6d48'
Kalbacher,Adrienne
Aur 6h1'17"30d17'
Kalberloh,Barbara & Ralph
Crb 15h57'21"28d8'
Kalbfleisch,Coralice
Cet 0h41'41"-4d28'
Kalbfleisch,Myrna
Oph 16h50'14"11d22'
Kalbfleisch,Robert
Aur 7h8'60"38d48'
Kalbfleisch,Timothy
Ori 5h11'27"-5d32'
Kalebich,Maureen
And 1h17'27"38d57'
Kaleia
Vul 21h22'42"26d22'
Kaleigh
Cyg 20h36'60"30d46'
Kaleigh Ann
Lib 14h24'29"-10d0'
Kalen
Cas 0h12'50"65d52'
Kalen
Cam 3h36'1"60d6'
Kalenda
Lyn 8h17'60"37d23'
Kalene Rae
Lmi 10h24'0"30d52'
Kalentyr
Cam 3h40'12"70d0'
Kaleonahe-"Soft Expression"
Eri 2h59'57"-3d6'
Kalesia
Cae 4h47'23"-27d51'
Kaleva Marja
Ori 5h51'27"19d59'
Kalex
Aur 5h3'41"44d34'
Kaley
Cas 0h23'58"70d15'
Kalfen,Lester
Per 1h56'42"48d55'
Kalff-Wentink,Henny
Uma 10h45'14"40d7'
Kalgon
Her 16h37'60"7d47'
Kalha,Mr Sanjeev
Ori 5h10'59"-6d22'
Kaliful
Lep 5h57'17"-20d27'
Kalil,Elyse Jessie
Cyg 21h41'37"38d26'
Kalin Jesse
Cam 5h57'45"70d13'
Kalin's Comet
Aql 19h25'26"-10d22'
Kalin,Alexia
Cep 21h20'47"68d7'
Kalin,Randy
Lac 22h28'39"55d39'
Kalina
Eri 3h25'11"-3d32'
Kalinowski,Frank S
Aql 20h11'25"5d3'
Kalish,Adam Michael & William Frank
Her 18h11'28"40d24'
Kaliski,Pow Wow L
Lyr 18h25'0"46d2'
Kalitkin,Stephen Joseph
Uma 9h19'0"54d55'
Kalkhurst
Tau 4h0'18"1d23'
Kalkwarf,Beverly Jane
And 0h14'37"30d15'
Kallas,Scott
And 1h6'11"40d16'
Kallatch,Grandmom
Cas 0h53'26"64d9'
Kalle,Marie E
Uma 11h52'26"50d21'
Kallen,AKA P C,Edward P
Her 16h14'12"20d26'
Kallergis,Kristyna
Lib 14h57'1"-7d39'
Kalli
Eri 3h17'33"-15d5'
Kallie Christine
Peg 21h58'0"2d5'
Kallio,Markku Mac
Uma 9h11'0"58d6'
Kalliste
Mon 7h11'54"-6d43'
Kallisti Elena
And 0h33'1"45d42'
Kallivatu
Uma 10h22'0"57d15'
Kallmeier,Barbara I
Cas 0h31'50"58d50'
Kallqvist,Veronica
And 1h23'0"38d53'
Kalm,Richard W
Boo 14h27'20"22d39'
Kalman,Joshua Jonathon
Cyg 20h51'1"39d37'
Kalman,Kim
Uma 10h2'17"56d14'
Kalmans,Melvin
Cmi 7h53'45"1d49'
Kalmbach,Ute
Gem 6h39'41"31d7'
Kalmer's Golden Star Light,Glen
Boo 13h45'58"20d2'
Kalnasy,Melissa & Steven
Peg 22h23'37"25d9'
Kalodemas,Alexis Ann
And 2h23'30"39d13'
Kalogeros,Andreas
Lyr 18h38'34"36d54'
Kalogridis,George A
Mon 8h4'9"-2d9'
Kalomira,Alifanti
Aur 6h7'0"54d28'
Kaloos,MD,PA,Nilza P L
Eri 3h56'29"-13d20'
Kalpakis 33rd Ann Star,Bill & Faye
Del 20h37'1"10d56'
Kalpana
And 0h4'30"38d17'
Kaltenbach,Neva
Aql 19h23'11"12d36'
Kaltenmeier
Dra 18h14'26"68d38'
Kalter,Alan R
Cep 22h21'54"61d37'
Kalthoff,Alois
Lib 14h58'30"-6d49'
Kalthoff-Hamann,Petra
Gem 6h34'56"26d44'
Kaltman,Jessica Michelle
Cam 6h11'22"61d5'
Kaltmann,Brigitte
Lib 14h18'26"-20d29'
Kaltner,George
Cep 21h18'38"67d41'
Kalupa,Glenn B
Equ 20h57'35"3d10'
Kaluric
Umi 16h18'40"70d14'
Kalus,Gwendolyn
Cyg 21h54'1"40d59'
Kalusa,Ray
Uma 12h0'51"32d4'
Kalustian,Dr Lynne
Uma 11h2'39"67d49'
Kaluzny,Alfons
Psc 1h15'18"17d48'
Kaluzny,Evelyne
Sgr 18h54'14"-23d53'
Kalvin,Hannah Leah
Cas 23h34'11"61d21'
Kalvoe,Peer Gunnar
Per 2h3'22"50d24'
Kalwag
Cyg 19h19'16"28d38'
Kalyanpur,Anil
Cas 1h12'41"61d52'
Kam
Cep 22h53'57"58d11'
Kam,Jon & Diane
Cyg 20h22'0"38d59'
Kamada,Stella Of Akira
Leo 10h2'54"13d11'
Kamala Cara
Per 1h35'0"53d41'
Kamat,Star Jameson
Aql 18h53'31"-2d25'
Kambal,Nasir Mohammed
Her 18h26'1"12d53'
Kambeitz,Louis Anselm
Hya 9h5'1"-6d12'
Kambers,Emily
And 22h58'42"51d50'
Kamcheff,M
Uma 9h21'48"68d22'
Kamees,Kelly Marie
Mon 6h42'32"1d20'
Kamel,Ihab Roushdy
Lyr 19h1'1"41d11'
Kamen,Libby
Aur 4h57'36"44d16'
Kamen,Nick
Uma 10h30'15"41d40'
Kamenir,Robert H
Her 17h50'46"40d22'
Kameno,Chelsea
Cas 1h34'51"67d45'
Kamenova,Masha
Uma 9h2'28"51d15'
Kamens,Cathy
Lyr 18h30'58"35d11'
Kamermayer,Patricia
Cyg 19h53'57"37d35'
Kamielaar,Helane
Lyn 9h12'27"37d17'
Kamikawa,Lindsay Nicole
Aql 20h4'28"1d11'
Kamil,Stephanie & Michael
Uma 13h40'36"50d35'
Kamin,Judy
And 23h21'49"40d33'
Kamin,Tommy
Her 18h4'1"37d40'
Kaminer
Lyn 8h24'52"44d13'
Kamins,Alexander S
Oph 16h57'17"-25d50'
Kaminski Billy The Kid,Williams A
Lac 22h34'56"55d40'
Kaminski,Angela Rose
And 1h46'11"39d37'
Kaminski,Big John
Cam 6h10'12"61d47'
Kaminski,Gabi
Ori 5h57'52"16d51'
Kaminski,Kalina Maclan
Dra 17h23'27"60d41'
Kaminski,Kristen Mari
Cyg 19h51'33"38d19'
Kaminski,My Husband My Love Gregory
Dra 19h6'16"48d7'
Kaminski,My Son Pumpkin Puss Jacob
Dra 19h5'1"50d23'
Kaminski,My Son Honey Bunny Kevin
Dra 19h3'55"48d1'
Kaminski,Nicole
Uma 10h19'11"53d35'
Kaminsky,Alan
Uma 10h20'27"48d17'
Kaminsky,David B
Sgr 18h51'28"-22d2'
Kaminsky,Dawn
Aqr 22h5'21"0d40'
Kaminsky,Killer
Her 16h48'0"50d23'
Kaminsky,Mathew
Uma 10h0'44"54d35'
Kaminsky,Pam
Aql 18h56'0"-1d19'
Kaminsky,Sally
Uma 10h50'0"48d0'
Kaminsky,Simone
Uma 10h19'34"50d49'
Kaminsky,Thomas Joseph
Ori 5h42'1"11d7'
Kaminsky,Yvonne Waldram Marchel
Sge 19h56'13"16d31'
Kaminstein,Mr & Mrs Philip
Uma 8h37'48"51d0'
Kaminstein,Wendy & Dan
Ori 4h54'43"0d6'
Kamis Misericordia Dulcis Somnium
Aur 5h6'56"42d47'
Kamisato,Ryan Michael
Crt 11h14'50"-16d34'
Kamiya,Yusuke & Takafumi
Cru 12h11'0"-60d55'
Kamlage,Jens
Vir 12h7'38"-5d53'
Kammeier 1923 Alpha, Hesse
Mon 7h0'50"-10d5'
Kammenis Magnus Donum, Robert
Aql 20h0'53"4d45'
Kammer,Katherine "Kit" N
Cam 5h1'22"54d39'
Kammerer's Star,Sally Kathleen
Lyn 8h0'1"37d45'
Kammerer,Christopher John
Per 2h55'38"43d43'
Kamont,Thomas L
Cnv 12h10'28"43d12'
Kamoreau
Cep 2h47'45"80d16'
Kamots
Sct 18h52'32"-4d47'
Kamp
Her 17h24'48"29d41'
Kamp,Franz-Josef
Peg 21h21'37"13d3'
Kamp,Peter
Lib 14h40'51"-6d20'
Kamp,Tanja
Aql 19h25'12"15d24'
Kampa,Kim
Cam 7h20'43"73d10'B
Kampa,Kornelia-Barbara
Cnc 8h24'11"20d4'
Kampas,Sean Robert
Cep 22h41'30"57d8'
Kampe's Star,Deb
Lyr 18h35'31"29d37'
Kampe,John Arthur
Dra 13h48'32"68d24'
Kampen,Diana
Oph 18h29'36"8d33'
Kampen,Gerd-Christian/ Ute
Sgr 18h54'12"-23d51'
Kampen,Johanna van
Gem 6h38'30"34d25'
Kamper,Werner
Sgr 18h50'57"-29d8'
Kampf,Jill Marie
Cyg 19h21'44"28d12'
Kampf,Michelle Lynne
Cas 0h12'29"54d46'
Kampf,Serge
Leo 9h58'39"10d58'
Kampfe,Sofia Annalise
Peg 22h3'16"20d33'
Kamphuis,Gina
Boo 14h44'26"47d1'
Kampmann,Sandra
Lac 22h52'28"38d6'
Kampmeier,Ulrich
Cnv 13h6'39"38d36'
Kamppinen,Allie Jane Marie
And 0h17'30"36d49'
Kampwirth
Uma 9h29'20"47d33'
Kamradt,F R
Cmi 7h33'60"7d35'
Kamuda
Aur 6h29'27"32d43'
Kamyck,ILU4,Kirsten"My Bunches"
Lyn 7h39'59"51d57'
Kan,Joey
Hya 8h18'1"0d2'
Kanady,Kirk Edward
Ser 15h50'16"18d32'
Kanakis,Margaret L
Peg 22h43'33"3d37'
Kanal,Shobhana Laveen
Tau 3h57'30"20d13'
Kanaley,Jenny
Tau 4h41'1"16d49'
Kanarr,Paul & Chris
Dra 17h42'26"60d15'
Kanas,William Peter
Dra 19h48'21"71d2'
Kanavos,Jr,Peter J
Her 17h31'0"40d46'
Kanczuzewski,Kristel Marie
Sco 16h52'49"-40d58'
Kanda,Davinder
Per 2h41'7"34d54'B
Kanda,Ruby
Per 2h41'7"34d54'A
Kandahar
Dra 17h2'58"63d52'
Kander,Sandra
Cas 2h14'44"75d37'
Kandes,Jr,John A
Aur 7h5'24"40d41'
Kandi R S
Lmi 10h32'52"31d31'
Kandler,Jennifer Schindler
Lyr 19h22'55"31d44'
Kandra,Kelly Marie
And 0h12'0"37d47"
Kane
Uma 10h48'21"47d36'
Kane
Cnv 12h24'18"35d54'
Kane "Super Star",John
Aql 18h56'57"-6d37'
Kane 7-17-72,Lynn Suzanne
Tri 2h10'23"32d21'
Kane Star,Grace Marie
Cam 7h8'1"60d60'
Kane Star,Sheridan
Cyg 21h37'0"41d53'
Kane's Star
Cyg 19h56'41"37d57'
Kane,Alan Scot
Cyg 21h25'33"38d28'
Kane,Ann & Andy
Eri 2h50'23"-5d0'
Kane,Arnold T
Aql 19h52'0"12d21'
Kane,Arvilla M
Peg 0h11'22"13d16'
Kane,Bruce A
Per 1h48'51"56d39'
Kane,Debbie Penner
Mon 7h2'22"4d59'
Kane,Donna
Crb 16h9'0"27d38'
Kane,Douglas Eternal Wave
Sex 10h12'9"5d23'
Kane,Eliza
Cas 0h25'18"50d35'
Kane,Jonathan Dane
Per 2h46'43"40d20'
Kane,Jr,Robert Whitson
Her 17h57'1"14d52'
Kane,Kathy
Lib 15h41'0"-22d41'
Kane,Kimberly C
Cas 0h31'1"62d14'
Kane,Lilianna Tobey
Peg 22h7'1"3d15'
Kane,Martin
Dra 10h15'18"73d47'
Kane,Matthew Eric
Aur 5h15'17"48d45'
Kane,Matthew Thomas
Per 4h8'23"37d31'
Kane,Maureen V
Cam 7h7'19"68d60'
Kane,Michael Patrick
Aur 5h3'19"46d8'
Kane,Patrick
Cep 20h39'43"55d1'
Kane,Paul Edward
Per 2h2'41"50d34'
Kane,Raymond C
Aql 19h0'45"16d9'
Kane,Richard & Rose
Peg 23h22'31"28d12'
Kane,Riley Edward
Dra 20h1'15"63d33'
Kane,Rob
Hya 8h15'56"4d31'
Kane,Robert Kohler
Aql 20h11'1"4d48'
Kane,Roxanna Wright
And 23h39'41"42d18'
Kane,Sean Lyle Stratton
Sco 17h51'12"-31d3'
Kane,Sean P
Aur 5h15'29"45d23'
Kane,Sheryll Ann
Boo 14h39'25"50d25'
Kane,Stella Rae
Umi 17h34'28"84d36'
Kane,Steve A Star Strategic Partner
Per 2h41'14"35d54'
Kane,Suzanne
Mon 7h43'49"-6d52'
Kane,The Star of
Ori 6h3'50"8d59'
Kane,William E
Dra 18h11'27"58d38'
Kane-Grimes,Katie
Uma 11h20'0"48d55'
Kanehl,Sr,George A
Cep 21h54'12"56d7'
Kaner,June
And 0h59'44"39d55'
Kanesaka,Nanami
Cyg 19h32'15"28d11'
Kang,Kyong-Ja & Chang-Wuk
Cam 8h6'49"80d3'
Kangas "The Boss", Marilyn
Tri 2h46'57"32d7'
Kanhaeuser,Peter
Her 18h12'16"30d27'
Kania,Dorothy M
Cyg 19h59'16"30d24'
Kania,Edward Joseph
Aur 6h0'0"30d30'
Kanis Kelly The Hearing Dog
Cma 6h40'51"-16d9'
Kanka,Megan Nicole
Peg 23h25'50"18d4'
Kankowski,Dr Stanley J
Per 4h22'10"51d7'
Kann Splendrous,Deanna Francis
And 0h21'1"36d41'
Kannaan
Cam 4h32'30"68d23'
Kannapell,J
Vul 19h38'19"20d29'
Kanner,Brian David
Mon 8h1'13"-6d47'
Kanner,Karlee & Keith
Cma 6h54'26"-18d37'
Kanno,Calvin Taro
Ori 5h42'0"12d21'
Kanno,Lindsey Maurine Aiko
Ori 5h50'23"18d48'
Kanno,Sophie-Claire Michiko
Peg 22h33'51"30d54'
Kannon,Cheryl
Com 12h34'0"20d20'
Kannon,Matthew
Aur 6h40'49"35d52'
Kanode,Keith
Aql 20h21'53"5d28'
Kanode,Mark Kalep
Ser 17h33'13"-14d0'
Kanofsky,Adena B
And 23h19'54"43d45'
Kanoo,Adel Salman
Leo 9h28'49"27d36'
Kanoo,Alia
Eri 4h2'1"-16d15'
Kanov,Gary Wayne
Aur 5h15'15"49d32'
Kansa,Pamela E
Uma 10h55'0"72d16'
Kansanaho,Veikko
Uma 12h12'45"56d24'
Kansas Surgery & Recovery Center
Uma 10h50'31"72d50'
Kansier,Roger C
Aql 19h6'53"3d58'
Kansteiner,Mark
Aur 5h16'30"41d18'
Kant,Mark Jay
Cam 12h51'34"77d15'
Kanter,Ayran & Teddy
Umi 15h15'36"69d42'
Kanter,Lauren
Lyn 9h25'45"41d2'
Kantner,Zachary
Peg 23h23'59"15d40'
Kantola,Margaret Jody
And 23h31'18"48d49'
Kantola,Robyn Marta
Mon 8h7'47"-5d60'
Kantor,Charlotte Jacobson
Uma 10h55'39"38d29'
Kantor,Eileen Dolores Connoly
And 0h55'33"39d44'
Kantor,Joshua Matthew
Cet 2h1'43"1d55'
Kantorski,Joseph & Patricia Baiardi
Uma 11h48'43"53d19'
Kantounatakis, Sieglinde
Cyg 20h34'42"48d56'
Kantounis,Dr Stratos Liberty
Aur 6h3'12"31d4'
Kantovowitz Star, The David
Her 16h16'28"23d32'
Kanza,Raphael
Peg 22h54'15"24d29'
Kao's 1993 Birthday Star,Hsiao C
Cam 8h5'20"74d22'
Kaori Ten Age Now
Cnc 7h55'27"10d25'
Kaough,Karen Lee
Mon 7h52'5"-8d58'
Kapa,Loretta
Sgr 18h50'18"-24d38'
Kapala,Janet C
Vir 13h39'48"-8d4'
Kapcsandi,Carlene Lane
And 23h1'13"50d27'
Kapelczak,Grzegorz
Cnv 12h5'12"41d47'
Kapferer,Christel
Vir 14h38'60"-0d20'
Kaphan "SSSK",Stephen Scott Stacey
Psc 0h31'35"21d37'
Kaphing,Sara Elizabeth
Cas 0h51'34"61d21'
Kapka,Autumn Marie
Cas 0h25'29"61d31'
Kapke,Eric James
Cet 3h7'12"2d10'
Kapko,Jr,Nicholas
Cep 23h11'14"64d24'
Kapko,Peter & Valerie
Eri 4h55'29"-6d3'
Kapla,Bruce
Lac 22h54'35"54d15'
Kaplah-To Steven
Psc 1h2'55"30d42'
Kaplan,Alexa
Lyn 7h30'46"45d33'
Kaplan,Annalee Rose
Peg 22h10'50"3d59'
Kaplan,Arlene Lynn
And 0h51'39"39d50'
Kaplan,Arnold M
Ser 17h32'43"-14d18'
Kaplan,Auren
Aql 19h3'23"11d11'
Kaplan,Carl & Susan & Jana
Lac 22h30'21"54d22'
Kaplan,Darren J
Lac 22h1'51"46d59'
Kaplan,Dr Louis L
Oph 18h32'55"6d15'
Kaplan,Dr Robert J
Aur 4h59'42"36d56'
Kaplan,Elyssa Stephanie
Lyn 8h41'16"38d35'
Kaplan,Etta Jehkins
Cyg 21h6'1"39d22'
Kaplan,Gadi
Cnv 12h47'16"46d42'
Kaplan,Gene
Lmi 10h8'26"31d17'
Kaplan,Gregory A
Ari 1h54'28"11d5'
Kaplan,Harry Arthur
Dra 20h16'12"62d39'
Kaplan,Helen
And 0h49'11"36d31'
Kaplan,Iggy & Irma
Crb 16h14'31"32d7'
Kaplan,James Scott
Dra 15h20'1"53d26'
Kaplan,Janet A
Uma 10h8'13"51d22'
Kaplan,Jerry Lee
Per 2h55'1"40d22'
Kaplan,Jonathan
Lac 22h17'58"54d28'
Kaplan,Julie
Uma 11h40'22"40d51'
Kaplan,Ken
Per 1h54'37"56d40'
Kaplan,Kyle Shauna
Uma 10h37'14"47d48'
Kaplan,Lewis
Cep 22h11'54"55d57'
Kaplan,Lisa Joy
Cyg 19h59'36"29d38'
Kaplan,Lisa S
Mon 6h21'1"8d37'
Kaplan,Louis Irving
Lyn 7h48'1"35d58'
Kaplan,Marjorie & Leo
Col 6h0'46"-3d9'
Kaplan,Marlene
And 23h32'53"41d42'
Kaplan,Mary Anne
Per 2h55'22"48d51'
Kaplan,Maurice & Charmaine
Mon 6h29'32"-10d42'
Kaplan,Max & Ava Jodi Holodak
Aur 5h13'45"44d6'
Kaplan,MD,Joel
Tau 3h27'1"20d2'
Kaplan,Michael
Cnv 12h56'49"43d2'

Kaplan,Paul
Per 1h30'32"53d24'
Kaplan,Robert & Virginia
Lyr 19h4'20"26d29'
Kaplan,Sara & Richard
Vul 19h36'49"20d32'
Kaplan,Sidney H
Cep 3h45'14"86d35'
Kaplan,Stan
Cet 2h6'52"1d41'
Kaplan,Steven Frederick
Dra 16h52'52"69d9'
Kaplan,Sylvia Brody
Mon 6h27'37"-0d16'
Kaplans Golden Luminary
Aql 18h59'0"13d31'
Kapler,Johann
Aqr 21h54'59"-6d20'
Kaploe,Burton
Per 1h48'20"47d58'
Kapner,James
Psc 23h22'12"6d45'
Kapono 143
Cyg 20h8'30"41d12'
Kapoor,Melanie
Lyn 7h50'51"36d53'
Kapoor,Tarun
Hya 8h19'26"4d38'
Kapoun,Margie J
Lyn 8h47'57"34d11'
Kapp,Brian Michael
Hya 8h41'1"2d51'
Kapp,Justin Wesley
Gem 6h52'9"18d18'
Kapp,Marie J
Aqr 20h36'1"-0d6'
Kapp,Shelley Ann
Cas 1h22'43"58d14'
Kappa Alpha Jackie
Aqr 20h46'45"-1d22'
Kappaeffe
Nor 16h23'41"-52d21'
Kappel,Pat
And 0h19'47"32d44'
Kappel,Robert Frank
Cet 2h25'18"4d8'
Kappele,Cora
Hya 8h19'35"4d47'
Kappelmann,Dixie Mae Martin
And 0h19'45"45d54'
Kapper,Victoria Beth
Eri 3h30'40"-7d20'
Kappler,Sebastian
Aqr 22h9'26"2d15'
Kapplow II,Arthur Leon
Cam 4h52'22"70d39'
Kapplow III,Arthur Leon
Sco 17h54'13"-30d13'
Kappos,Elizabeth Ann
Lyr 19h6'42"28d23'
Kappy's Comet
Boo 14h39'13"30d24'
Kaprielian 5-2-54, Donna
Aivazian
Cet 2h37'55"2d5'
Kaps,Winfried
Cep 20h39'11"60d5'
Kapushion,Betty Joyce
Cam 5h43'28"71d13'
Kapushion,Leonard Joseph
Her 15h52'32"48d11'
Kapushion,Randal Anthony
Cas 0h13'21"58d20'
Kapusta,Donald Joseph
Her 16h2'0"18d48'
Kapusta,Mario
Eri 4h36'34"-0d29'
Kapustka's Dream
Eri 3h30'0"-6d54'
Kar Rob
Cnv 12h53'28"50d28'
Kar-Sun
Oph 17h53'40"10d2'
Kara
Lyr 18h40'31"40d22'

Kara
Per 4h22'46"51d12'
Kara
Lyn 7h44'50"44d35'
Kara
Peg 22h34'50"10d4'
Kara
Cnv 12h50'52"48d56'
Kara
Cam 5h10'50"70d18'
Kara
Mon 6h18'39"8d17'
Kara & Brian
Eri 4h18'48"-18d11'
Kara Beth
Cyg 19h45'32"30d3'
Kara Leigh
Mon 6h51'58"10d18'
Kara Lynn
Del 20h13'59"10d6'
Kara Megan
Vul 19h37'49"20d5'
Kara Rea
Mon 6h30'40"0d11'
Kara's Hope
Peg 22h45'20"29d22'
Kara's Loving Star
Del 20h54'0"7d6'
Kara's Star
Hya 9h37'16"-9d39'
Kara-Kar
Cas 2h9'39"67d46'
Karabin,Jeffrey
Gem 7h24'57"31d33'
Karada,Susannah Katherina
Cas 0h13'55"60d23'
Karagas,Tony
Vul 20h56'27"28d30'
Karahalios,Alexandria
Demetrius
Com 13h14'1"31d16'
Karahalios,Alexis
Sco 16h57'44"-40d30'
Karahalis,Great
Uma 9h59'0"59d24'
Karahalis,Sam Andrew
Lac 22h2'21"38d7'
Karahalis,Sandra
Peg 22h36'1"27d8'
Karakache,Mirella
And 1h57'50"36d21'
Karalla,Malak
Oph 17h5'34"-9d39'
Karam,Kristin Hartel
Dra 19h23'48"68d4'
Karam,Ted
Dra 19h32'53"68d30'
Karamesines,Chris
Lmi 10h58'1"26d47'
Karamu
Cam 5h7'33"70d42'
Karan Elizabeth
Cas 0h53'52"65d52'
Karanian,Edna
Aql 19h37'16"-6d26'
Karant,Niko
Aur 7h10'0"38d17'
Karanzias,Kellan
Mon 6h30'0"8d4'
Karas,Charles David
Lyr 19h13'0"46d57'
Karas,Jason Aaron
Aur 6h13'0"38d48'
Karas,Jean & Robert
Uma 11h29'37"40d30'
Karas,Jeffrey Scott
Her 16h6'1"42d27'
Karas,Theodore S
Cam 6h6'23"61d44'
Karasek,Dara
Tri 1h41'40"34d52'
Karatony
Uma 10h1'40"59d21'
Karatwys
Cyg 20h21'1"38d58'

Karatzas,Niki
And 23h2'23"51d8'
Karavas,Fred A
Lac 22h27'11"50d30'
Karavish,Aaron
Lac 22h47'0"37d47'
Karayalcin,Gungor
Boo 15h7'46"18d28'
Karber,Edward & Helen
Uma 9h54'48"53d30'
Karberg,Kelley
And 23h31'46"48d5'
karbil-94
Tau 4h20'1"16d25'
Karch,Craig J
Boo 14h34'23"21d5'
Karcher 7
Mon 7h19'44"-6d59'
Karcher's Star
Cam 13h48'32"80d21'
Karcher,Tom & Ann
Uma 10h4'1"59d32'
Karchers' Beam
Tri 2h42'58"31d51'
Kardelan
Cnv 13h30'51"50d22'
Kardell,Heather Lynn
And 23h37'1"39d58'
Kardis,Andrew Pulver
Psc 1h26'15"11d42'
Kardis,Michael Steven
Ari 1h46'18"24d22'
Kardis,Suzanne Pulver
Psc 1h28'18"11d38'
Kardlec,Hunter Aaron
Cet 0h2'34"-10d43'
Kardong,Gloria Marie
Vir 13h34'21"2d18'
Kardos,Gary
Cep 0h57'35"77d25'
Kardys One
Ori 5h59'11"18d7'
Kardén,Jorma
Cep 21h50'0"65d24'
Kare Star
Sge 20h17'41"16d53'
Karel
Pup 8h24'35"-22d50'
Karelesian Memoriam
Mon 6h32'51"4d1"
Karen
Lib 15h13'33"-22d22'
Karen
Lib 15h15'17"-28d44'
Karen
Cas 0h32'24"62d38'
Karen
Cmi 7h43'30"4d37'
Karen
Cam 5h38'48"71d8'
Karen
Lyr 18h28'35"31d27'
Karen
Lyr 19h20'25"42d54'
Karen
Peg 21h57'49"34d37'
Karen
Cam 3h35'58"60d48'
Karen
Cas 0h38'57"58d22'A
Karen
And 1h56'56"37d48'
Karen
And 1h19'14"39d8'
Karen
Cyg 20h25'37"40d11'
Karen
Leo 10h27'43"15d56'
Karen
Lyn 6h17'14"60d12'
Karen
Cyg 19h51'0"37d36'
Karen & Andrew
Sge 18h56'35"19d56'

Karen & Andy
Cyg 21h33'37"42d46'
Karen & Colin's "Wedding
Star"
Cyg 21h50'30"40d13'
Karen & Curwen
Cyg 19h24'1"31d18'
Karen & Garys Star
Cyg 21h20'42"39d34'
Karen & Jack
Cyg 20h40'54"42d6'
Karen & Nichole Forever
Friends
Lyn 8h35'13"41d26'
Karen & Paul
Cyg 19h47'25"30d5'
Karen & Tad's "Nitelite"
Peg 22h36'12"10d30'
Karen & Terry
Her 17h36'52"18d54'
Karen & Warren
Cyg 20h20'0"38d20'
Karen 40
Cas 2h42'55"73d50'
Karen A
Lyn 8h31'56"50d22'
Karen Alice
Cas 0h36'45"54d38'
Karen Alisa
Mon 7h0'17"-10d13'
Karen Ann
Lyr 18h31'1"41d13'
Karen Ann
And 23h38'33"48d4'
Karen Anne
Cyg 20h24'55"38d33'
Karen Anne's Star
Lyn 8h23'24"51d34'
Karen B
Uma 9h41'31"47d6'
Karen CC
Lyn 8h42'36"46d52'
Karen Diane
Aql 19h2'46"1d30'
Karen Elena
Peg 22h5'25"20d21'
Karen Elise
Cyg 21h50'39"41d12'
Karen Elizabeth
Lyr 19h15'0"41d31'
Karen Elizabeth
Lyn 6h58'53"54d27'
Karen Elizabeth
And 0h0'32"47d50'
Karen Ellen
And 0h8'56"43d38'
Karen Jane & J Star
Lyr 18h50'1"31d11'
Karen Jean
Cas 1h10'55"58d6'
Karen Jean
And 1h54'31"40d58'
Karen Joy
Peg 22h3'10"8d57'
Karen Krystal
And 2h8'45"38d24'
Karen L & Mike G
Peg 22h31'31"27d37'
Karen la Merveille
Cas 0h15'32"46d41'
Karen Lee
Cas 0h42'21"67d15'
Karen Lee
Peg 22h14'54"5d8'
Karen Lis 1938
Vul 19h22'50"25d41'
Karen Louise
And 1h58'19"38d32'
Karen Louise 25
Cas 2h31'0"61d50'
Karen Lynn
Eri 2h46'0"-6d58'
Karen Lynn
Oph 18h5'36"2d12'B

Karen Marie
Lyr 18h47'14"36d14'
Karen Marie
And 0h54'29"39d47'
Karen Marie & Hailey My Two
Girls
Aql 19h31'20"11d49'
Karen Michelle
Lyr 19h21'55"35d3'
Karen O'
Cam 3h55'40"60d2'
Karen Precious
And 23h24'31"40d19'
Karen Pretorius 1
Cas 1h50'0"58d15'
Karen Rae
Leo 10h51'39"-0d39'
Karen Rose
And 23h40'29"46d3'
Karen Ruth
Cyg 21h32'17"38d31'
Karen Sue
Ori 5h20'17"13d4'
Karen Super-Star
Cru 12h39'22"-62d43'
Karen The Great (KTG)
Cas 23h26'0"61d7'
Karen Tisci's "555"
Mon 8h0'46"-0d42'
Karen Virginia
Equ 21h22'47"10d24'
Karen W
And 2h18'26"46d30'
Karen's Comet
And 1h19'33"38d3'
Karen's Dream
Uma 11h46'30"58d10'
Karen's Hope
Cyg 19h52'44"40d55'
Karen's Jewel
Cyg 19h21'12"28d46'
Karen's Light (Woobie-1)
And 0h22'47"33d51'
Karen's Light
Mon 6h43'30"-4d47'
Karen's New Beginnings
Sco 17h3'47"-31d18'
Karen's Radiance
And 23h3'23"41d59'
Karen's Radiance
And 1h1'54"39d38'
Karen's Sapphire
Lyr 19h9'27"40d53'
Karen's Star
Cam 3h21'20"61d42'
Karen's Star
Vul 19h42'27"26d44'
Karen's Star To Wish Upon
Lyr 19h22'40"31d44'
KarenAl 96
And 0h20'51"32d17'
Karenannhahn
Eri 3h17'20"-13d29'
Karenelisa
Cas 23h5'0"58d46'
Karenella
And 0h16'53"39d36'
Karenlee
Cyg 20h3'42"40d41'
Kareotes,Dena
Cyg 19h28'25"32d21'
Karess,Maia
Cas 0h49'28"75d38'
Karfgin,Kristin
Cyg 19h26'10"32d10'
Karg,Sonja
Uma 9h41'11"57d53'
Kargatis,Marsha L
And 23h36'34"49d7'
Karge,Cmdr R
Sex 10h3'0"-2d24'
Kargl,Carmen Marianne
Leo 9h56'59"28d17'
Kargul,Angela Lynn
Boo 14h31'27"30d23'

Kari
Tri 1h49'31"27d26'
Kari
Umi 14h41'1"81d54'
Kari Ann "Sparkie"
Mon 6h27'13"11d18'
Kari Beth
And 1h45'57"39d36'
Kari Forever
And 1h17'1"36d46'
Kari Lynn
Cas 0h59'21"62d29'
Kari's Beacon
Peg 22h43'18"32d30'
Kari's Hope
Eri 3h52'54"-6d33'
Kari's Light
Cas 1h14'14"61d16'
Karibenas,Theresa
And 23h24'44"37d49'
Karician,Elaine
Peg 23h29'1"20d2'
Karicori
Vul 19h22'14"27d40'
Karie
And 1h35'15"38d12'
Karim,Atel
Ori 5h57'32"20d33'
Karim,Daniel
Aur 6h4'44"36d7'
Karimito
Lmi 9h56'52"37d33'
Karin
Cyg 21h20'45"39d28'
Karin
Vir 12h28'38"-8d17'
Karin
Lib 14h43'52"-22d44'
Karin
Sgr 18h53'23"-27d22'
Karin
Lib 15h32'27"-10d24'
Karin
Cyg 19h59'21"41d37'
Karin
Psc 1h0'0"12d6'
Karin
Scl 23h18'1"-32d24'
Karin
Her 17h27'25"42d16'
Karin "She Is The Best "
Psc 1h15'51"21d16'
Karin A Walter Prien/ Munich
Omega
Sco 16h31'19"-44d56'
Karin Adrian
Cyg 21h6'18"30d54'
Karin Forever
Boo 14h26'28"23d56'
Karin I
Cet 2h55'36"5d51'
Karin M
Aur 5h26'36"31d4'
Karin und Werner:Viel Glück!
Cas 0h49'48"60d46'
Karin's Asta
Hya 8h43'42"2d28'
Karin's Star
And 0h7'24"47d2'
Karin-Matt Wedding Star,The
Cyg 21h54'18"42d39'
Karin-Mausi
Cnv 12h39'51"38d29'
Karina
Cas 0h54'32"63d50'
Karina
Peg 21h58'39"21d4'
Karina Anastassia
Uma 10h1'39"72d11'
Karina Ann
Cam 5h7'32"68d14'
Karina et Marc
Crb 16h18'0"38d51'
Karina Margarita
Eri 4h59'50"-4d49'

Karine
Cas 23h39'20"61d50'
Karine et Yan
Cas 22h58'1"55d4'
Karine L O M L
Cyg 19h29'55"35d29'
Karis,Anna & Eva
Cas 0h30'28"68d9'
Karis,Peter & Birgit
Cam 3h41'0"60d13'
Karissa Elaine
And 0h24'34"44d2'
Karisumma Vivastrum
Hya 8h27'43"-17d7'
Karkhanis,Kiran & Priya
Aur 5h8'0"40d29'
Karl
Her 18h5'36"14d18'
Karl & Dan 94'
Lac 22h49'48"38d41'
Karl Jeffery
Lyn 8h9'10"49d10'
Karl O
Aur 4h56'31"40d41'
Karl's Viola
Ori 5h56'54"15d21'
Karl,Irv
Gem 7h0'14"35d15'
Karl,Larry W
Crt 11h50'12"-11d31'
Karl,Max H
Equ 21h6'29"2d55'
Karl,Shayna Michelle
Umi 15h12'18"67d5'
Karl,Thomas Joseph
Tri 1h37'45"33d56'
Karl,W R A
Her 17h12'12"48d21'
Karl-Heinzi
Sgr 19h39'35"-40d13'
Karl-Schuele
Tau 4h1'15"0d29'
Karla
Equ 21h5'31"2d33'
Karla
Uma 9h43'37"59d19'
Karla Jean
Mon 6h55'49"-2d2'
Karla Rose
Lmi 10h14'57"30d30'
Karla S
Lyr 18h46'29"41d13'
Karla Vera
Mon 6h22'42"4d37'
Karla with A "K"
And 0h12'49"39d46'
Karla's Guardian Angel
Eri 4h3'26"-16d60'
Karla's Rising Star
Sge 19h33'1"16d33'
Karla's Wishing Star
Peg 22h35'16"20d33'
Karlas,Karen
Lyn 7h38'14"50d48'
Karle,Dyland Charles
Gem 6h47'41"12d27'
Karle,Kathleen A
Aur 6h27'41"31d2'
Karlee Rae
Cyg 20h52'52"40d6'
Karlen's Dreams
Cam 11h28'58"80d40'
Karlene E
Cas 0h2'35"65d5'
Karli Marie
Cas 0h49'29"70d27'
Karli Michelle
Vul 19h43'59"20d22'
Karli-Hasi
Boo 14h5'29"13d28'
Karlik,Donna
Del 20h51'0"8d17'
Karlin
Uma 8h33'54"60d39'

Karlin
Aql 19h11'46"12d9'
Karlin,Kim
Dra 19h51'51"61d48'
Karlingers' Star
Cnc 8h32'43"8d8'
Karlmari Kristi I
Cam 3h55'37"52d51'
Karlos
Pyx 8h31'29"-25d34'
Karlsen,Donald
Cep 0h14'13"80d1'
Karlsson,Tom
Dra 16h46'38"67d38'
Karlstadt,Robyn
Aql 20h11'29"14d48'
Karlyn & Bruce
Cyg 21h2'29"39d31'
Karman,Kristy Ann
And 2h23'25"49d0'
Karman,Larry Steven
Cet 3h5'45"0d31'
Karmazyn's Mushin
Her 18h3'0"48d5'
Karmel,Marianne
Cnv 13h9'41"32d21'A
Karmela
Vul 21h22'0"27d28'
Karner,Stefan
And 23h9'36"43d4'
Karney "J F K",Jules F
Ori 6h7'48"9d19'
Karnezis,Peter Steven
Boo 14h52'32"22d44'
Karno,Art
Eri 4h30'47"-10d56'
Karno,Gary
Cet 2h38'44"-0d32'
Karno,Zelda
Mon 7h50'37"-7d49'
Karns,Daniel & Heather
Cyg 21h16'43"35d45'
Karns,John William
Cep 23h13'1"61d26'
KaRo
Umi 14h31'29"68d55'
KaRoJ
Per 2h10'20"57d54'
Karol Anet
And 0h27'51"45d47'
Karol Lee
Lmi 10h35'43"30d30'
Karol Suzanne
Cmi 7h27'36"8d53'
Karol,Paul Eugene
Aql 18h43'12"11d40'
Karola
Cap 20h27'5" 26d30'
Karoly,Simonyi
And 23h5'45"44d17'
Karon
Vul 20h0'37"28d26'
Karon Elizabeth
Cyg 20h20'44"40d27'
Karon K
Cas 0h42'14"67d10'
Karon,David & Libby
Cra 18h13'33"-39d24'
Karon,Thomas Willard
Boo 14h37'16"39d40'
Karoon,Kayvan
Cnv 12h24'52"51d27'
Karp"Bat Mitzvah"
Star,Rebecca Shira
And 0h47'46"45d31'
Karp's Star,Elodie & Jerry
Eri 2h53'29"-6d17'
Karp,Cathy
Cas 0h5'1"63d50'
Karp,Dr Louis
Aur 7h3'17"40d26'
Karp,Essie
Sco 16h26'46"-30d18'
Karp,Evan Craig
Cet 2h9'57"6d3'

Karp,Stephen
Uma 11h28'45"31d1'
Karpick,David
Vir 11h35'50"-2d9'A
Karpick,Greg
Vir 11h35'50"-2d9'B
Karpie,Jennifer Ann
And 2h25'41"39d39'
Karpinski's Quest
Cam 9h7'50"81d1'
Karpinski,Jr,John D
Aur 6h8'47"31d55'
Karpy (T K),Timothy Roger
Lac 22h15'44"46d54'
Karr,Charlotte Miller
Lyr 18h52'29"40d17'
Karr,Mary M
And 0h50'1"45d46'
Karr,Rachel
And 0h48'47"45d20'
Karra,Kristel
Lyn 8h25'0"43d40'
Karram,Lynne
Cyg 20h27'33"37d32'
Karram,Zachary George
Cep 22h17'27"56d4'
Karras,Alexis Ann
Lyn 7h31'53"50d9'
Karras,Tyler John
Lac 22h23'16"54d10'
Karrie
And 0h1'18"35d55'
Karrissa
Cyg 20h16'25"38d55'
Karron-Davis Our
Friend,Susan
Cyg 20h5'1"40d0'
KarRos
Eri 4h3'1"-18d25'
Karschnia,Kristin
Cyg 19h44'39"38d43'
Karsdorf's Stern
Uma 13h38'0"48d26'
Karsh,Myra & Brad
Uma 11h17'1"42d46'
Karst,Alison Marie
Cas 22h57'21"56d25'
Karsten,Dorothy
Cam 8h8'16"77d54'
Karstens,Matthew Ian
Lyr 18h57'45"31d47'
Karstensen,Ashleigh Marie
Ori 5h58'13"20d60'
Karstensen,Darren Scot
Ori 5h58'10"21d57'
Karstensen,Tyler Scot
Ori 5h57'52"20d24'
Karthika,Bindu
Cyg 20h0'34"30d5'
Kartono,Teddy
Lyr 19h12'13"31d10'
Kartorie II,Willis Cordell
Cnv 12h23'31"34d54'
Kartorie,Alicia Marie
And 0h34'59"40d7'
Kartrude,True Love 30 Kathy
Tau 5h53'27"28d53'
Kartye,David C
Peg 0h2'49"13d58'
Karvelas,Angelo
Vul 19h43'0"23d17'
Karvin
Cas 0h26'45"62d8'
Karvinen,Ilsa-Hanna & Petri
Cyg 19h32'16"39d41'
Karvinen,Joni Kultapää
Uma 8h25'59"62d29'
Karyn
Lyn 7h25'0"58d13'
Karyn
Sex 10h37'14"5d45'
Karyn's Planet
Lyr 18h57'30"30d56'
Karz,Richard John
Cep 0h4'11"70d17'

Kasai,Jane Cas 22h57'36"54d43'
Kasal,Christine Cma 6h53'17"-19d7'
Kasala,Kisto Cas 0h5'29"61d15'
Kasanoff,Lawrence Ser 16h16'32"1d25'
Kasarda,Nicole Com 13h5'43"21d12'
Kasari Tau 5h17'0"24d20'
Kasarsky,DDS,Jason S Oph 17h15'38"11d1'
Kasarsky,Emma Louise Uma 11h43'37"53d26'
Kasarsky,Sarah Joy Uma 11h49'29"52d58'
Kasbee,Marion Ruth Hohenstein Lyr 18h38'48"40d8'
Kaschak,Edward & Kathleen Lyr 18h32'17"34d58'
Kasey Umi 13h42'23"76d39'
Kasey Lynn Lyr 18h44'59"45d1'
Kasey,Bruce Cep 0h59'48"87d53'
Kash,Samuel Shea Lac 22h31'44"55d6'
Kashare,Jonathan Alexander Tau 5h49'26"23d6'
Kashauer,Ethel M Vir 11h45'37"9d13'
Kashee & Schaumee Lyn 7h26'1"58d38'
Kashetta-Konzman, Gloria Cyg 20h53'43"30d28'
Kashick Boo 15h4'0"24d1'
Kasia Sge 20h1'15"20d16'
Kasia's Joy Aql 19h18'28"15d32'
Kasiewicz,Belle And 23h44'42"47d35'
Kasik's Cosmo,Kurt Aur 7h6'15"43d40'
Kasinger,Mary Ann Peg 22h49'34"27d52'
Kasirobus Dra 17h1'23"62d33'
Kaskoun,Kelsey Cas 3h7'0"57d42'
Kaskoun,Neil Finn Boo 14h42'37"37d50'
Kasky,Cle Her 17h31'34"30d24'
Kasler,Yvette Eri 3h35'57"-2d8'
Kasmenealitagine Cep 23h18'0"70d48'
Kasnic,Anna B Cyg 19h55'16"37d34'
Kasnic,Tyler Cep 23h0'0"64d8'
Kasparek,James J Her 17h3'53"47d53'
Kasper's Mark Aql 19h6'26"-0d8'
Kasper,Anique Jacqueline Cam 7h47'23"68d42'
Kasper,Barry David Cep 23h8'0"65d7'
Kasper,Colonel John C Cep 22h14'29"62d19'
Kasper,Hans-Rüdiger Lib 14h20'17"-22d43'
Kasper,Paul David Aur 4h49'47"40d51'
Kasperina Leo 9h59'20"7d23'
Kasperowicz,Brianna Debra Cmi 7h22'35"9d0'
Kasprowicz,Friedrich Ori 5h58'55"19d28'
Kasprzak,Irene Maskaly Cet 0h26'14"-12d23'
Kasprzak,Kelly Cet 1h30'38"-1d48'
Kasprzycki,Edyta Anna Lyn 7h24'0"50d19'
Kasrel,Bruce Aql 18h44'22"6d40'
Kass & Doug Beyond the Edge... Cyg 19h57'30"29d33'
Kass,Beverly S Lyr 19h15'51"28d56'
Kass,Dr Ethan Ben Oph 17h53'52"12d49'
Kass,Edwin A Cam 12h52'13"77d42'
Kass,Jeremy Morris Aur 5h29'51"31d5'
Kass,Our Star Alison Elizabeth And 23h3'1"49d50'
Kassab II,Thaddeus K Per 2h56'28"46d16'
Kassab,Julius Her 16h43'24"33d39'
Kassan,Lee E Her 18h16'55"14d45'
Kassay,Marie Gabriel And 2h9'1"40d48'
Kassel,Guenter Cep 0h5'34"84d30'
Kassel,Sylvia & Art Umi 14h22'12"66d14'
Kassel,Victor Aur 5h11'51"42d34'
Kassell,Jack M Dra 14h18'11"63d52'
Kassidy,Joseph & Rebekka = Vul 20h2'57"23d21'
Kassie Ann Umi 14h23'18"66d57'
Kassigkeit,H Curt Cnv 12h47'40"37d50'
Kassin,Alec Robert Vul 20h4'49"28d34'
Kassirer,Lynne Cyg 21h33'20"40d31'
Kassis,Shirley "Darlin" Cas 1h23'31"53d38'
Kassner Fantasy,Linda Cet 3h11'58"1d13'
Kassoff,Thomas Dra 18h45'11"70d40'
Kasson Uma 8h59'1"51d32'
Kasson,Paul "Doc" Boo 14h22'1"33d23'
Kassorla,Jackie Mon 6h40'39"10d35'
Kast,Lothar und Susanne Cnc 8h1'34"8d37'
Kastelic,Craig A Ser 15h43'14"-2d48'
Kasten,Klaus Ari 3h22'25"30d43'
Kastenbaum,Jr,James Eugene Dra 12h59'50"68d6'
Kasting Celestial Legacy Aur 4h52'30"51d22'
Kastley,Jillian Margaret Vir 12h59'1"7d49'
Kastner,Viola Hage Kelley Mon 7h9'16"-5d19'
Kastroll,Betty Jane Peg 22h6'40"5d55'
Kastrup,Karl W Sex 10h46'22"-8d7'
Kaszowski,Monica Lynn Cyg 21h3'0"40d41'
Kat Vul 21h0'1"22d34'
Kat Lyr 19h15'13"41d12'
Kat Lyr 18h49'3"32d4'
KAT Lyn 8h52'15"37d59'
Kat & Bad Old Dog Cyg 20h21'22"41d50'
Kat's Kolestar Ari 2h37'48"22d11'
Katajamäki,Tuula Peg 21h56'53"33d16'
Katalan Eri 4h4'44"-16d25'
Katalin,Várszegi Ma'kos Ori 5h55'1"20d57'
Katalina Peg 22h16'10"34d46'
Katanja,Latrice Peg 22h13'51"5d7'
Katara,Tatiana And 23h0'33"45d57'
Katariina Cas 2h34'45"60d55'
Katarincic's Celestial Body,Amy Dee Cam 3h14'0"58d25'
Katatonia Aur 4h49'0"40d51'
Katavolos,Marianne & Peter Cyg 19h14'44"47d45'
Katayoun Cnc 9h9'15"32d27'
Katchmer,Lillian Peles Mon 6h18'32"5d36'
Kate Sge 19h28'5"17d48'
Kate Lyr 18h40'13"42d56'
Kate Cyg 20h22'58"30d54'
Kate Lyn 9h12'40"38d16'
Kate Cyg 19h32'1"35d15'
Kate Ori 5h14'23"-9d47'
Kate And 0h10'45"47d17'
Kate And 0h29'1"43d36'
KATE And 23h23'50"50d52'
Kate & Bruce 1993 Christmas Aql 20h8'56"1d26'
Kate & Gary Cyg 20h21'0"38d28'
Kate & Jon's Engagement Star Cyg 21h14'20"28d50'
Kate & Miles Always Sge 20h1'35"16d36'
Kate 50 Sgr 19h39'21"-35d1'
Kate Helene And 23h26'20"42d8'
Kate Michael Cap 20h34'31"-20d8'
Kate Rachel Mon 6h23'34"4d45'
Kate's Dream Cyg 19h25'42"35d56'
Kate's Dream "Star Of James" Dra 11h36'42"70d24'
Kate's One Uma 9h21'0"68d22'
Kate's Special Star Lyr 18h33'25"31d3'
Kate's Star Uma 9h20'19"51d45'
Kate's Star Tau 5h57'52"23d48'
Kate-My One Shining Moment Ori 5h57'0"16d57'
Kateins Sge 20h0'1"16d3'
Katelyn Peg 0h2'42"31d16'
Katelyn Abigayle Cyg 20h23'0"38d57'
Katelynd Marie Mon 6h50'33"11d48'
Kater Jane Mon 6h54'57"-10d15'
Kater Kasimir Ori 5h56'54"7d56'
Katerbau,Gabriella Gem 7h28'56"20d45'
Kateric Lyr 18h28'0"42d24'
Katerina Cyg 19h57'36"29d38'
Katerina Sgr 20h1'47"-33d53'
Katerinos,Bernardine Victoria Bogucki Lyn 8h6'28"40d52'
Katerinoula Cnv 12h21'19"40d52'
Kates,Marcel Felix Vanderlaan Per 3h3'37"38d5'
Kath Eri 4h9'32"-5d0'
Kath Cyg 21h8'18"30d4'
Kath Cas 0h36'0"60d25'
Kath Cyg 21h8'18"30d4'
Kath & Tony Cyg 20h39'12"45d58'
Kath Marie Umi 14h1'56"71d29'
Kath's Star Cas 0h52'16"73d16'
KATH-ALLISON,6190 Mon 7h3'17"0d31'
Kath-o-leen & Allen Lac 22h39'35"53d43'
Kathaleen Hilda And 23h19'32"41d38'
Kathaleen Mary Cyg 21h35'1"44d6'
Kathamegos,Sandra Lee And 23h36'17"48d44'
Katharina Anna Maria Sgr 19h44'43"40d34'
Katharine's Star And 23h38'0"50d38'
Katharine's Star Cas 1h59'10"60d33'
Kather,George Richard Vul 19h48'1"28d11'
Katherine And 1h56'14"37d47'
Katherine Lmi 9h49'52"40d44'
Katherine Cas 23h23'53"60d59'
Katherine Aql 19h16'50"13d48'
Katherine And 23h22'38"49d52'
Katherine & Anthony's Star Peg 2h34'32"35d23'AB
Katherine & Bob Lyr 18h57'57"33d8'
Katherine & Rubin, Together Forever Mon 7h1'56"-6d60'
Katherine Alexandra Lyn 7h54'14"40d42'
Katherine Cecella Gem 7h30'53"34d9'
Katherine Loves Gill Mon 7h5'44"-6d33'
Katherine Lydia Aqr 22h23'18"0d12'
Katherine Mary Lyr 18h15'32"32d41'
Katherine Maureen Tri 1h53'19"27d44'
Katherine Rose Cas 0h32'45"62d11'
Katherine S Lyr 19h5'23"26d28'
Katherine Sophia Cam 6h40'47"67d36'
Katherine Sue And 1h33'18"36d44'
Katherine's Eternal Light And 1h48'26"40d34'
Katherine's Shooter Lyn 8h12'13"34d59'
Katherine's Spirit Equ 21h12'19"7d15'
Katherine's Star Cas 1h1'19"63d56'
Katherine's Star Lyr 18h58'43"27d45'
Katherine,Stuart Dra 16h11'1"68d2'
Katherine-KMSB Ori 6h3'57"6d41'
Katheryne Mary And 1h52'10"37d56'
Kathi Cas 1h23'51"60d55'
Kathie Aql 19h24'0"-1d34'
Kathie Cyg 19h28'45"30d1'
Kathie Ann And 0h30'57"45d33'
Kathie's Star Lyn 8h2'47"51d7'
Kathleen Del 20h23'47"16d29'
Kathleen Cam 7h15'46"68d48'
Kathleen Cas 1h21'44"55d45'
Kathleen Cyg 19h18'0"44d17'
Kathleen Eri 4h34'45"-5d38'
Kathleen Uma 9h51'1"42d13'
Kathleen Cas 1h44'52"61d2'
Kathleen Cnc 9h3'41"12d19'
Kathleen Cas 1h50'24"60d20'
Kathleen Peg 22h54'50"29d52'
Kathleen Lyn 9h14'1"38d36'
Kathleen Ori 5h51'20"8d13'
Kathleen & Jack Vul 20h47'1"28d23'
Kathleen Ann Peg 21h57'16"20d28'
Kathleen Anne Coogan Aql 20h12'30"10d35'
Kathleen Beatrice Cas 0h24'14"61d19'
Kathleen C Lyn 7h27'44"45d10'
Kathleen Carol Aqr 22h0'37"-5d40'
Kathleen Ellen Lyr 19h2'24"30d11"
Kathleen Florence Sco 16h20'13"-38d54'
Kathleen Grace Del 21h5'29"12d39'
Kathleen Indra Magdalena Cas 1h19'12"70d1'
Kathleen Joy Aqr 22h2'56"-11d34'
Kathleen Kelly And 2h20'50"42d57'
Kathleen Luella Lmi 9h36'1"38d49'
Kathleen Marie And 23h43'0"44d29'
Kathleen Mary And 23h40'21"41d44'
Kathleen Mary,The Crb 15h48'22"26d46'
Kathleen Michele Vir 11h51'53"8d8'
Kathleen Star Cnv 13h6'30"40d41'
Kathleen Star,The Uma 11h11'42"44d13'
Kathleen Sue - Kase 71153 Cnc 8h12'18'31d38'
Kathleen's Shining Star Lyr 19h5'45"25d39'
Kathleens Eternal Light Aql 20h1'54"9d25'
Kathlene Dawn And 1h40'25"40d29'
Kathrin & Kirk's Pustefix Tau 5h54'58"26d54'
Kathrin-Schubo Her 17h27'42"42d37'
Kathrina Uma 10h4'56"61d46'
Kathrine Marie And 1h49'28"39d42'
Kathryn Cet 3h16'1"1d46'
Kathryn And 2h32'1"42d55'
Kathryn Cyg 20h22'1"41d29'
Kathryn Uma 10h54'1"54d32'
Kathryn Cas 3h2'21"71d10'
Kathryn Alice Cas 1h37'23"73d10'
Kathryn Amelie Aql 19h56'59"14d34'
Kathryn B And 2h27'16"42d18'
Kathryn Elise And 22h55'27"38d11'
Kathryn Elizabeth Lyr 18h22'33"38d36'
Kathryn Jane Cas 0h28'1"61d43'
Kathryn Jane Com 12h51'0"20d6'
Kathryn Lenice Eri 4h0'50"-18d28'
Kathryn Louise Uma 9h4'49"50d52'
Kathryn M Cyg 21h52'60"52d35'
Kathryn Madeline And 2h17'18"48d8'
Kathryn S Lac 22h35'18"56d21'
Kathryn Teresa 4/29/69 Cyg 21h17'41"37d34'
Kathryn's Enchanted Place Peg 22h34'41"24d30'
Kathryn,Steven & Temina Her 18h8'40"41d3'
Kathy Uma 11h7'24"71d5'
Kathy Peg 22h25'11"31d3'
Kathy Crb 15h36'11"28d9'
Kathy And 1h27'40"37d23'
Kathy Aql 19h44'0"14d33'
Kathy & Amy Uma 9h43'43"58d21'
Kathy & Anne Uma 8h47'27"70d22'
Kathy & Jeff's Star Eri 3h40'56"-0d56'
Kathy & Leah Aur 5h59'54"31d12'
Kathy & Pete 12/31/94 Lyn 7h57'24"38d25'
Kathy & Tom Cyg 21h7'20"30d31'
Kathy 10/7/71 Umi 16h23'12"70d17'
Kathy A Lyr 18h43'59"32d13'
Kathy Ann Vul 19h5'0"25d3'
Kathy Ann II Sco 16h54'1"-41d14'
Kathy B And 23h37'1"47d33'
Kathy J And 0h49'56"37d47'
Kathy K And 0h56'0"45d39'
Kathy K Cam 11h2'48"80d55'
Kathy Lee Eri 3h11'46"-3d4'
Kathy Lynn Cet 3h19'60"4d39'
Kathy M Vir 13h14'17"-19d47'
Kathy M And 22h56'14"40d44'
Kathy Regina Cam 5h55'19"58d44'
Kathy Road Runner Lyn 8h36'12"44d40'
Kathy Ruth And 23h40'36"38d47'
Kathy's 40 And 2h8'0"41d13'
Kathy's Angel In The Sky And 1h49'0"39d38'
Kathy's Celestial Body And 2h36'19"38d54'
Kathy's Dream Boo 14h36'50"11d31'
Kathy's Forever;B M B! Ori 6h4'15"8d48'
Kathy's Gift Boo 14h28'52"20d7'
Kathy's Love Star And 23h37'0"49d22'
Kathy's Own Peg 23h32'41"10d15'
Kathy's Purple Vision Aur 5h1'17"50d10'
Kathy's Rose Vul 19h52'32"20d4'
Kathy's Shamrock Tau 3h58'25"30d38'
Kathy's Star Peg 22h24'53"29d54'
Kathy,I'll Love You Always,Tommy Lyn 6h36'19"60d45'
Kathy-J Cyg 20h31'34"38d8'
Kathy-Meow-Alice Lyr 19h20'41"38d55'
Kathykato Cmi 7h46'35"7d36'
Kati,John Boo 13h34'22"30d29'
Katia And 0h45'60"45d21'
Katica Nikic Gem 7h0'59"31d18'
Katie Lyr 18h57'35"45d8'
Katie Cyg 21h3'36"38d3'
Katie Cas 23h29'14"63d26'
Katie Lyr 18h44'31"40d44'
Katie Ori 6h0'45"20d53'
Katie Cas 1h58'35"60d11'
Katie Lyn 8h2'22"33d26'
Katie And 23h41'23"33d42'
Katie Peg 22h59'48"28d40'
Katie Cyg 21h54'50"36d26'
Katie Cet 0h9'28"-8d29'
Katie Vul 21h1'13"28d9'
Katie Cyg 21h30'37"41d33'
Katie Cas 1h45'33"61d15'
Katie Cet 1h39'16"-11d34'B
Katie "The Beautiful" Lyn 7h26'42"41d11'
Katie & Alan's Star Sge 19h33'24"16d40'
Katie & Dennis And 23h35'18"45d40'
Katie 16 And 2h0'23"38d23'
Katie Anne And 1h43'55"38d55'
Katie Beth And 2h18'38"40d1'
Katie Beth Eri 3h37'43"-14d29'
Katie Bug Lyn 6h25'54"60d58'
Katie Dean Ant 10h53'40"53d45'
Katie Ellen Cas 0h18'52"62d47'
Katie J Peg 23h21'28"33d49'
Katie Jane And 0h29'37"30d35'
Katie Jane's Star Mon 6h12'54"-10d33'
Katie Jo And 1h55'15"37d14'
Katie K Lyn 9h7'56"44d18'
Katie Kat Cyg 21h55'1"53d33'
Katie Kate And 2h31'56"40d37'
Katie Lauren,The Eri 4h5'34"-13d55'
Katie Laurie And 23h44'58"47d22'
Katie LoRe' Cmi 8h2'45"6d45'
Katie Lou Peg 22h47'50"33d31'
Katie Lynn Peg 21h45'32"23d0'
Katie Lynn Eri 2h55'23"-8d18'
Katie M Uma 8h57'19"57d9'
Katie Mae Lyr 18h31'15"32d25'
Katie Marie Lyr 19h21'59"42d39'
Katie O Aql 19h36'18"3d15'AB
Katie R Star Aqr 23h17'59"-4d4'
Katie Rene Equ 21h15'26"2d35'
Katie Rose And 1h48'0"39d50'
Katie Rose Leo 9h57'22"10d9'
Katie's Diamond Peg 22h32'48"27d13'
Katie's Dreams Aur 7h3'18"40d54'
Katie's Endeavour And 23h45'20"41d16'
Katie's Eye Lyn 9h12'1"35d30'
Katie's Guiding Light And 23h32'21"47d9'
Katie's Night Light Peg 22h55'50"29d47'
Katie's Smile And 23h17'27"35d15'
Katie's Star Cyg 21h8'18"31d32'
Katie's Twinkle Lyn 8h9'49"38d5'
Katie's Wishing Star Mon 6h43'43"3d23'
Katie-Boots Cas 0h32'20"62d17'
Katiestar Uma 10h15'29"50d30'
Katinka-KB230295 Tau 4h44'15"16d48'
Katish And 1h57'14"36d46'
Katja Uma 11h2'30"70d26'
Katja Uma 11h52'56"62d19'
Katja & Frank Uma 10h9'1"70d22'
Katja-Bill Her 17h32'45"41d10'
Katjas Private Universe KBTH Cnv 12h42'1"39d15'
Katka And 2h29'37"49d45'
Katka Cam 9h10'36"77d45'
Katkish,Andrea Lynne Cyg 20h9'40"40d21'
Katkish,Christopher John Ser 16h6'10"13d17'
Katkish,Lauren Angela Mon 7h22'40"-1d34'
Katlin Elizabeth Cyg 20h39'52"52d55'
Katlin Leigh And 0h10'53"47d41'
Katlyn Mon 6h35'47"6d13'
Katmar Tel 20h22'55"-46d13'
Kato,Hiroko Cet 2h26'0"3d19'
Kato,K C Peg 21h59'49"28d20'
Kato,Kenichl Aur 5h4'16"43d31'
Kator,Family Star Of Uma 8h59'1"57d50'
Katrin Marie Eri 2h48'50"-5d54'
Katrina Eri 3h55'0"-7d7'
Katrina Com 12h44'27"21d42'
Katrina Cas 3h13'15"77d6'
Katrina Cas 1h17'53"60d42'
Katrina Cyg 19h32'46"37d51'
Katrina 01-26-63 Mon 6h23'57"0d30'
Katrina Loves Robert Sex 9h55'13"-6d40'
Katrina's Star Lyn 7h45'43"38d34'
Katschthaler,Hans Dr Psc 1h16'1"21d38"
Katsoulis,Alexandra Cyg 20h57'52"30d2'

Katsoulis,Esq,Timothy E
Her 17h32'27"40d2'

Katsoulis,Theodore
Cnv 12h59'50"46d39'

Katsuhara,Hidehiro
Sco 16h1'5"-14d58'

Katsumi Miwa
Cet 2h39'45"-3d14'

Katsutoshi & Meiko Love Forever
Ser 15h37'42"20d27'

Katt,Malcolm E
Lmi 10h56'53"32d1'

Katt,William
Cet 1h59'0"-3d9'

Kattan,M
Mon 7h10'41"-6d22'

Katten,Daniel Rudy
Cam 4h12'36"58d53'

Katten,Deanna
Cas 3h9'16"57d17'

Katter,Hugh
Eri 2h57'0"-15d3'

Katterson,Carol & Lyman
Gem 6h49'5"14d20'

Katty-O
Ori 5h55'1"13d44'

Katuscak,Christopher A
Uma 11h39'32"44d20'

Katy
Cam 5h59'24"73d43'

Katy
Cyg 20h39'21"31d2'

Katy
Tau 5h53'15"28d23'

Katy Joanna
Sgr 18h56'36"-21d53'

Katy Laurel
Cyg 19h28'24"31d57'

Katy Liz (For Katarina E Walker)
Mon 6h57'14"-10d55'

Katy Marie
Hya 9h11'44"2d22'

Katy's Jewel
And 1h8'53"39d9'

Katy,Loveliest forever
Cas 1h53'13"70d52'

Katy-Lady
Mon 8h4'45"-8d54'

Katya
Lyn 8h27'22"41d10'

Katya
Cen 13h24'0"-58d1'

Katya Amelia
Cas 0h0'31"61d18'

Katz Forever,Joshua Dustin
Sex 9h54'31"-1d46'

Katz Star,The
Uma 13h41'31"51d8'

Katz,Adam Jeffery
Cnc 8h36'17"11d42'

Katz,Arthur Valentine
Dra 18h19'26"48d22'

Katz,Benjamin
Dra 13h31'37"64d56'

Katz,Bob "Santana"
Oph 17h38'11"-23d9'

Katz,C Susan
Lyr 19h0'10"26d0'

Katz,Caroline
And 0h51'25"38d28'

Katz,Catherine & Reuven
Cnv 13h52'1"40d39'

Katz,Craig J
Leo 9h20'43"18d43'

Katz,David
Cam 5h39'57"68d31'

Katz,David Michael
Cyg 21h50'44"52d55'

Katz,Declan Charles
Tau 5h55'57"24d13'

Katz,Dorothea A
Cnv 12h13'36"44d21'

Katz,Everett E
Aql 19h58'38"10d32'

Katz,Guy
Cet 2h33'10"-4d56'

Katz,Harry
Her 18h7'38"30d12'

Katz,Jack P
Per 1h46'42"56d20'

Katz,Jaclyn Michelle
Uma 10h4'1"48d30'

Katz,Jason Barett
Dra 18h3'15"58d16'

Katz,Jeff
Boo 14h5'45"33d39'

Katz,Johann
Aur 5h15'39"43d11'

Katz,Lauren
Lac 22h26'19"56d33'

Katz,Lauren Ann
Cet 2h49'20"2d5'

Katz,Marjorie Lewis
Lyn 6h56'25"52d43'

Katz,Mark
Aur 5h5'27"42d18'

Katz,Marni Aileen Smith
Crb 15h49'52"38d8'

Katz,Martin H
Her 18h0'34"50d21'

Katz,Matt
Aql 19h4'25"0d7'

Katz,Mia
And 23h48'13"46d56'

Katz,Norman B
Cam 3h46'31"61d5'

Katz,Philip
Cma 6h57'24"-17d21'

Katz,Philip Sean
Boo 15h8'48"50d19'

Katz,Ralph
Boo 14h55'34"29d58'

Katz,Ray
Oph 17h18'52"12d27'

Katz,Renate
Cnc 8h57'30"18d24'

Katz,Rita Starr
Leo 11h4'34"-1d18'

Katz,Sandra
Per 3h59'13"39d15'

Katz,Seymour,& Lea
Sge 19h24'27"16d37'

Katz,Sharon Helene
And 2h24'1"39d47'

Katz,Sid
Aql 29h16'42"8d7'

Katz,Steve & Roberta
Crb 16h10'34"32d7'

Katzander's Smile
Lyn 7h8'30"59d58'

Katze & Viech
Sgr 19h17'32"-44d46'

Katzel,John & Janet
Mon 6h18'29"2d55'

Katzen,Matalie
Del 20h13'0"10d41'

Katzenbeisser,Helmut
Vir 11h52'57"-1d16'

Katzenberg,Samuel Robert
Aur 6h37'30"38d41'

Katzenberger,Gladys
Cas 0h3'36"61d11'

Katzer,Benjamin Lee
Ser 15h29'50"10d49'

Katzman,Karen Block
Lyn 7h31'16"50d3'

Katzmarczyk,Andrea
Eri 3h40'59"-16d32'

Katzmarczyk,Bernhard
Ori 6h1'53"4d46'

Katzmarczyk,Jan
Aql 19h47'47"13d12'

Katzmarczyk,Marie- Therese
Mon 6h56'48"10d17'

Katzner,Iris Tanenbaum
Lmi 9h56'50"33d46'

Kauai Kookie
Cet 0h46'51"-6d6'

Kauan,John
Psc 23h9'38"1d33'

Kauffer,Jean-Franáois
Aur 7h1'36"38d42'

Kauffman
Dra 19h51'56"70d42'

Kauffman,Alan & Melanie
Aur 6h50'46"35d33'

Kauffman,Barbara
Cyg 19h29'18"31d43'

Kauffman,Lauren Diehl
Lyr 18h32'55"30d55'

Kauffmann,Chris
Her 18h9'25"46d58'

Kaufhold,Erika Eva
Peg 23h27'26"27d44'

Kaufhold,Eugen
Lib 15h6'32"-6d11'

Kaufka,Lawrence M
Boo 14h19'27"38d31'

Kaufman Star,The
Oph 17h19'36"-18d57'

Kaufman,Aryeh
Per 1h32'0"52d33'

Kaufman,Christine
Her 16h1'42"48d59'

Kaufman,David
Her 18h6'54"48d17'

Kaufman,Donna M
Crt 11h14'39"-18d33'

Kaufman,Harrison Roig
Dra 16h57'59"58d59'

Kaufman,Helen
Uma 8h10'60"72d28'

Kaufman,James Francis Matthew
Boo 15h17'0"50d35'

Kaufman,Jennifer
Cam 4h42'22"67d58'

Kaufman,Julian
Ari 2h0'17"25d33'

Kaufman,Kristina Anne
Lyn 6h13'35"60d10'

Kaufman,Laura & Kimba & Elsa
Lyr 18h38'45"42d46'

Kaufman,Louis
Dra 16h51'21"51d36'

Kaufman,MD,Stuart
Vir 13h30'18"-8d58'

Kaufman,Midnite L
Lyr 18h44'56"33d46'

Kaufman,Pat & Bob
Mon 8h8'30"-6d29'

Kaufman,Pim
Aur 6h53'21"48d47'

Kaufman,Scott Kenneth
Cet 1h25'14"-0d29'

Kaufman,Sherry
Aql 18h59'20"-5d24'

Kaufman,Wanda(Tanner)
Cas 23h29'0"60d30'

Kaufman,Wendy
Uma 11h57'51"40d56'

Kaufmann,Bryan David
Cep 22h16'0"70d10'

Kaufmann,Karen Lossing
Cyg 21h55'28"53d3'

Kaufmann,Kathleen
And 1h7'43"40d19'

Kaufmann,Raymond
Per 2h42'21"40d43'

Kaugher,Chester Edward
Equ 21h5'12"3d26'

Kaulana & Baby Kalau
Oph 17h54'30"0d4'AB

Kaulius-Barry,Aldona
Per 2h48'44"40d52'

Kaun,Dennis
Cyg 20h7'58"40d5'

Kaun,Jenny
Gem 7h18'54"24d45'

Kaupa,Michael
Sco 17h28'0"-38d55'

Kaur,Amarjit
Cap 21h22'55"-15d19'

Kaur,Har Darschan
Psc 1h14'53"21d10'

Kaur,Kuljit
Aql 19h2'21"-1d46'

Kausch,Siegfried
Peg 23h36'56"15d18'

Kauski,Valerie A
Cap 20h57'22"-16d43'

Kausler,Nicole
Cet 2h55'20"0d48'

Kaut
Mon 7h3'30"-6d56'

Kaut,T
Ori 5h54'42"7d51'

Kaut,William Alexander
Sex 10h26'45"-6d21'

Kauth,James Frederick
Del 20h16'44"14d46'

Kauth,Stephan William
Oph 17h32'8"-24d5'

Kautsch,Irmgard A
Boo 15h11'30"50d43'

Kava,Kendal Kay
And 23h5'24"40d56'

Kavalauskas,Mark
Lac 22h48'1"55d56'

Kavalich,Allan
Per 2h53'0"34d13'

Kavanach,Charles Leon
Hya 9h10'39"0d9'

Kavanagh Star,The Niall J
Peg 23h28'31"27d47'

Kavanagh,Brian Patrick
Aur 6h29'46"38d24'

Kavanagh,Ginette
Cma 6h46'46"-16d16'

Kavanagh,Jackie
Cru 12h52'38"-60d12'

Kavanagh,Jim
Her 17h49'0"48d50'

Kavanagh,Melanie
And 1h56'34"38d36'

Kavanagh,Michael John
Ori 5h0'22"8d26'

Kavanaugh,Duane & Jessica Darling
Cyg 20h53'35"40d39'

Kavanaugh,Leslie & Jack
Mon 6h29'10"2d47'

Kavanaugh,Sharon Marjorie
Aql 20h0'40"11d47'

Kavlock,Steven E
Aur 4h55'25"40d55'

Kawa,Dru
Dra 16h53'34"63d53'

Kawadler,Holli Sharyn
Cyg 20h13'0"37d37'

Kawadler,Jaeson
Aur 6h4'48"37d38'

Kawaguchi,Lance
Her 17h24'1"43d37'

Kawaguchi,Shigenori
Leo 10h9'48"19d6'

Kawaharada,John
Cet 2h38'0"5d14'

Kawaja,Nellie
Peg 21h26'25"22d49'

Kawakami,Vivien
Del 20h14'38"15d12'

Kawakita,Stella of Naomi
Vir 14h33'37"5d35'

Kawal,Chuck
Boo 15h22'1"41d22'

Kawczynski,Yvonne Mendoza
Cas 0h28'57"63d50'

Kawesch,Alex Nicole
Cas 3h10'16"67d48'

Kawiecki,Richard
Per 3h2'30"47d6'

Kay
Lib 15h13'24"-21d40'

Kay
And 0h3'16"30d27'

Kay
Cas 1h15'0"61d39'

Kay
Uma 10h16'41"52d56'

Kay
Peg 22h0'39"29d3'

Kay
Sco 17h25'31"-38d42'

Kay
Sge 19h55'44"19d6'

Kay & Arthur's Golden Star
Cru 12h45'8"-60d4'

Kay & Mitch's Love-Light#5,Shine On
Crb 15h55'1"28d47'

Kay & Stan
Mon 7h41'14"-8d54'

Kay Ellen's "Fire"
Cyg 20h7'53"40d39'

Kay Happiness Twinke, Bret
Gem 7h55'1"30d17'

Kay Kay
Sco 16h8'48"35d33'

Kay Marie
Cma 6h41'12"-16d17'

Kay My Angel,Brett
Aql 19h58'28"15d14'

Kay The Eternal Light
Cyg 20h23'27"39d33'

Kay's EBTG
Cam 5h49'55"67d55'

Kay's Eternal Star of Hope
And 23h15'17"48d12'

Kay's Kind Heart
Lyn 7h53'27"41d3'

Kay's Special Star
Peg 21h55'59"30d43'

Kay's Star
Lyn 7h52'45"50d6'

Kay's World
Lmi 10h33'21"30d52'

Kay,Av N Virginia
Cyg 21h0'33"39d7'

Kay,Barney
Sco 16h26'19"-31d12'

Kay,Barry
Cam 11h19'26"80d15'

Kay,Courtney A
Mon 6h43'34"10d4'

Kay,David
Cep 22h6'15"60d28'

Kay,H B
Cyg 21h46'42"53d30'

Kay,Kevin
Ori 5h25'41"-01d54'

Kay,Leah
Vul 20h14'0"22d55'

Kay,Peter
Ori 4h58'56"15d14'

Kay,Richard & Christine
Cyg 20h28'0"50d10'

Kay,Robert Henry Christopher
Oph 18h28'27"8d25'

Kay,Robert Jeffrey
Per 4h0'27"51d55'

Kay,Sr,Robert Lyle
Aur 5h9'17"42d10'

Kay,Teresa
Cyg 19h55'35"58d22'

Kay,Teresa L
Per 4h2'56"43d17'A

Kay-Bob
Aur 4h59'30"51d29'

Kay-of-the-Light
Ori 4h56'39"0d1'

Kay-two Forever In Time!
Ori 6h20'15"16d32'B

Kaya
Mon 7h2'26"1d10'

Kaya
Aql 20h31'28"0d56'

Kaya & Marut
Sco 17h25'1"-31d42'

Kayachanian,Ernesto Gutierrez
Aql 19h57'60"15d42'

Kayalee
Cet 2h47'40"1d36'

Kaydee
Cas 1h46'29"74d2'

Kaye
Lyr 18h34'59"29d9'

Kaye
And 1h54'57"37d58'

Kaye
Lyn 6h27'23"58d33'

Kaye & MacDonald
Boo 14h52'58"26d33'

Kaye's (Ultimate)
Cas 2h48'10"75d53'

Kaye,Alan
Aqr 23h5'35"-8d6'

Kaye,Daniel
Oph 17h54'24"-8d21'

Kaye,Melvin S
Her 18h10'29"50d20'

Kaye,Pamela
Cyg 21h50'16"37d37'

Kaye,Richard E "Snuggles"
Ori 5h4'59"13d4'

Kayfman,Lauren Marie
Cyg 21h8'12"31d6'

Kayl,Eric
Boo 13h50'29"18d56'

Kayla
And 2h25'18"38d41'

Kayla
Leo 11h54'56"20d22'

Kayla
Hya 9h22'41"-10d14'

Kayla
Cas 0h21'27"61d44'

Kayla 22
Cas 1h46'24"60d54'

Kayla Ann
Mon 6h44'25"10d40'

Kayla I
Oph 18h39'31"10d2'

Kayla Lyn
Tau 5h45'22"27d46'

Kayla Lynn
Mon 6h24'50"7d41'

Kayla Lynn
Lyr 19h23'26"35d29'

Kayla Marie
Mon 7h0'39"5d15'

Kayla Marie
Cep 20h10'30"76d35'

Kayla Nycole
Ori 5h52'15"16d29'

Kayla's Sweetness
Uma 11h48'1"45d42'

Kayla,Robert F
Aur 4h57'17"40d8'

Kaylani,Safouh
Lac 22h16'57"54d34'

Kaylee Frances
Peg 22h45'32"27d58'

Kayleen Lisa
Umi 15h47'8"77d19'

Kayleigh's Comet
Peg 23h36'0"18d31'

Kayler,Gene J
Aqr 23h8'1"-5d58'

Kaylin
Cyg 20h40'42"30d7'

Kaylin Michelle
Uma 10h54'13"51d42'

KayLynn's Arcadia
Leo 11h21'38"-2d11'

Kayne,Alex
Aur 6h13'0"30d19'

Kayne,Jenny
And 23h48'49"40d21'

Kayne,Maya
Cyg 21h38'40"41d48'

Kayne,Sara
Vul 20h57'1"20d28'

Kayrene
Cyg 20h1'1"41d9'

Kaysen,Janice
Sgr 19h47'58"-43d16'

Kayser,Darrin
Sex 9h50'42"-0d48'

Kayser,Ing Helmut
Umi 14h7'1"67d55'

Kayser,Nicolas (Fils)
Uma 12h56'54"59d42'

Kaysinger,Gary
Cep 21h57'34"58d18'

Kaystar
Uma 12h17'27"59d39'

Kaystar
And 0h24'13"45d20'

Kaytes,Adrienne
Uma 11h47'1"43d34'

Kazakoff,Christian
Vul 19h45'10"28d22'

Kazdin,Miriam
Cma 6h44'55"-15d30'

Kaze,Princess
And 2h1'22"38d39'

Kazelis,John "Mr Ditka"
Gem 6h38'21"20d9'

Kazeminy Star,The Yasmine Yvonne
Cam 3h15'26"60d13'

Kazenmayer,Ulrich
Peg 23h30'46"18d53'

Kaziulina,Natalia Vladimirovna
Umi 13h2'44"71d41'

Kazlauskas,Katherine
Tau 5h50'34"27d36'

Kazmierczak,David Michael
Her 18h1'36"14d42'

Kazmierski,James
Cet 2h29'18"4d55'

KaZoo (Chantale D'Aoust)
Psc 1h34'30"21d32'

Kazu & Ake
Psc 1h24'48"6d8'

Kazuhiro & Ruriko's Wish
Lib 14h40'2"-20d33'

Kazuki
Peg 22h57'1"32d13'

Kazuki Asao
Ari 2h11'53"18d50'

Kazulin Boshi
Crt 11h16'36"-19d60'

Kazutaka 1994-9-1
Vir 14h28'44"5d45'

Kazuyo
Ari 1h54'49"23d36'

Kazuyoshieg,M
Tau 5h29'45"20d51'

Kazzano,Anthony J
Dra 18h29'12"70d27'

Kazzer
And 0h51'16"39d10'

KB's "Big Wave"
Hya 9h22'17"-10d59'

KB4BLU
Aql 18h51'23"11d4'

KBF One
Peg 22h4'31"7d45'

KC & Kim Forever
Lyr 19h2'49"31d6'

KC 143
Per 2h39'25"40d59'

KC,Inc
Cet 0h45'40"0d44'

KCD 923 95
Lyr 18h45'24"34d52'

KCH8368
Lyr 18h33'1"35d8'

KC
Peg 21h54'37"28d38'

Kea,William Paul
Aql 19h25'29"-6d37'

Keables,Mark
Her 17h57'48"37d44'

Keady,Kathleen
Gem 7h15'55"21d28'

Keahey,Jessica Nicole
And 1h46'54"47d8'

Keais,Kenneth Charles
Oph 17h34'22"-23d4'

Keal,Don & Watson, Jeanni & Jordan
Cep 22h40'34"68d56'

Kealey,Corinna
Tri 2h5'16"32d46'

Kealey,Lori Cathrine
And 0h54'59"34d57'

Kealohapau'ole
Cyg 19h52'48"38d45'

Kean,Katie Lynn
Boo 15h13'0"30d0'

Kean,Martin Morph
Cas 0h11'11"61d48'

Kean,Maurice William
Uma 9h53'49"53d3'

Kean,Nicholas James
Aql 20h2'27"10d7'

Kean,Ray
Uma 11h42'42"38d6'

Keane,An Eternal Star For Eddie
Uma 11h24'45"60d18'

Keane,Colleen
Peg 22h28'39"11d25'

Keane,Donald E
Tri 2h40'22"34d27'

Keane,Kristen Emily Ashley
And 0h23'50"31d24'

Keane,Megan Paige
Cyg 19h31'15"32d5'

Keane,Patsy
Cas 1h57'11"61d18'

Keaney's Nodligh Realt John
Uma 8h13'54"61d34'

Keany,"Creeka"Erica Symmons
Lyr 19h18'11"42d23'

Keany,Erica
Cas 0h35'0"58d53'

Kear,Adam Conner
Ori 5h12'15"-7d7'A

Kear,Debbie L
Sge 19h53'35"16d41'

Kear,Mary Lou
Uma 13h28'9"60d36'B

Kear,Michael T
Aur 6h21'1"33d36'

Kear,Pamela
Pho 23h44'45"-49d53'

Kear,Suzanne
Cyg 19h32'21"32d30'

Kear,Thomas B
Uma 13h28'9"60d36'A

Keara "Cosmic Child"
And 0h20'24"38d25'

Keara's Gem
And 23h29'35"48d14'

Kearly "50",Donald Murriel
Lyr 19h24'0"38d16'

Kearnes,JoAnne
Cma 6h55'16"-15d17'

Kearney Star,The
Aql 19h2'0"11d13'

Kearney,Brian Richard
Cep 22h34'44"61d32'

Kearney,Ceil
And 0h55'0"36d10'

Kearney,Iris Gayle
Crb 15h31'19"31d39'

Kearney,Jack Sinclair Carmichael
Oph 17h36'53"-22d42'

Kearney,James M
Cep 0h58'1"87d16'

Kearney,Joseph
Per 1h45'11"52d39'

Kearney,Kate A
Leo 11h28'60"-1d51'

Kearney,Pauline Florence
Cas 1h44'13"58d15'

Kearney,Robert R
Her 17h53'11"20d21'

Kearney,Samantha Jacqueline
Cet 1h57'0"0d46'

Kearney,William R
Vul 21h16'43"20d4'

Kearns,Beth Ann
Cas 0h3'32"62d24'

Kearns,David
Hya 8h37'28"5d45'

Kearns,Helen Ilene
And 1h52'34"41d9'

Kearns,Matthew Thomas
Dra 11h26'28"78d31'

Kearns,Orless B
Her 18h17'48"18d56'

Kearns,Robert P
Her 18h3'17"28d54'

Kearns,Vickie
Lyr 19h22'1"31d6'

Kearsley,Edward
Cep 0h4'20"69d58'

Kearsley-Brown Star, The James R
Her 16h48'14"30d19'

Keas,Michael Scott
Cet 1h34'50"-2d45'

Keasey,Conner Reid
Aql 18h39'25"0d40'

Keasey,Sarah Elizabeth
And 0h17'46"34d39'

Keast,Adriana Patricia
Cas 1h13'1"71d13'

Keast,James D
Aur 6h0'44"37d2'

Keast,Matthew Andrew John
Ori 5h58'15"0d17'

Keast,Samantha Christine
Cas 2h2'18"61d2'

Keates,Patricia A
Dra 13h1'47"68d28'

Keates,Robert L
Dra 13h19'22"67d45'

Keates,Teresa A & Douglas J Sortino
Dra 17h58'1"60d51'

Keathley,Sr,MD, Franklin Burr
Uma 9h26'58"68d14'

Keating's Legacy,Bob
Aql 20h22'24"2d1'

Keating,Brent Cole
Oph 17h56'24"8d53'

Keating,Charles
Aql 18h41'16"-1d55'

Keating,Christine Marie
And 0h45'59"22d13'

Keating,Deanna
Equ 21h22'10"8d46'

Keating,Dolores
Com 12h2'44"25d36'

Keating,Finian
Umi 16h26'39"80d18'

Keating,Karen Kestel
And 0h25'19"37d50'

Keating,Kevin Christopher
Her 17h57'1"50d13'

Keating,Marge
Cyg 20h19'11"31d45'

Keating,Maureen & Jimmy
Cyg 19h34'27"39d54'

Keating,Peggy Jo
Oph 17h58'58"7d43'

Keating,Sean Francis Scannell
Cnc 8h52'48"31d48'

Keating,Tammy-Larry
Cnv 14h3'51"46d10'

Keating-Rice
Cam 3h30'53"55d1'

Keatings,Penelope Jane
Crb 15h28'19"31d14'

Keatley,Joanne
Psc 0h4'30"2d16'

Keaton,Dexter Dean
Ori 6h6'48"8d53'

Keator,Mary & Ray
Dra 12h1'50"70d2'
Keats' Quasar
Aql 20h2'28"4d50'
Keavney,Robert Darrell
Lyn 7h59'34"50d40'
Keavy
Aql 20h2'24"12d51'
Keay,Super Star Stephen
Cep 22h29'39"70d2'
KEBA-RIN
Lmi 10h14'52"39d26'
Keck II,William M
Cet 2h39'0"-9d32'
Keck II,William Myron
Oph 18h7'50"12d47'
Keck III,William Matthew
Ser 15h21'16"3d16'
Keck,Kathryn
Mon 8h5'50"-10d3'
Keck,Leslie E
Peg 22h41'11"10d15'
Keck,Libby Ann
Cas 1h10'14"58d30'
Keck,Mandy Lynn
And 0h21'22"43d26'
Keck,Michael Andrew
Per 1h55'36"53d18'
Keck,Stephen Myron
Cet 2h33'53"4d12'
Keck,Theodore James
Aql 19h31'35"12d7'
Kecklik,Ashley Corinne
And 2h20'26"46d6'
Kedmi,Solomon
Sco 16h54'26"-40d49'
Kedy
Cet 0h38'41"-5d48'
Kedzierski,James Stanley
Her 17h20'1"41d0'
Kedzior,Mary Kathleen Ardis
Cyg 21h59'15"53d21'
Kee,Hilary Davis
Cnc 9h0'1"17d48'
Kee,Mae Elise & Jeffrey Wing-Kit
Aql 20h5'0"1d50'
Kee,Pete
Gem 6h45'54"31d45'
Keebaugh,Foster Woolett
Ori 6h9'57"7d19'
Keeble,Jenni
And 0h29'30"40d12'
Keeble,Robert
Ori 5h57'40"16d2'
Keedy,Michael H
Aur 7h24'23"37d3'
Keefe 01-15-94,Rachel Lillian
Com 12h53'57"28d9'
Keefe,Jack "Mister Ed"
Aur 6h1'37"37d53'
Keefe,John
Cru 12h24'31"-61d56'
Keefe,John Lawrence
Sct 18h54'13"-7d46'
Keefe,Maxwell Thomas
Aql 19h46'20"14d12'
Keefe,Megan Ann
Vul 19h53'54"20d31'
Keefe,Peter
Cma 6h54'5"-19d58'
Keefe,Sheila
Lyr 18h41'23"40d7'
Keefer,Barbara J
Cyg 21h23'53"39d29'
Keefer,Donald D
Aur 4h34'30"34d7'
Keefer,Donna S
Cas 0h17'22"58d32'
Keefer,James Bruce
Ser 15h20'1"2d19'
Keefer,Linda S
Cyg 21h24'1"39d54'
Keefer,Loren Ralph
Ser 18h9'44"-5d38'
Keefer,Robert S
Mon 8h5'60"-4d40'
Keefer,Shannon Marie Nadine
And 23h5'45"40d19'
Keegan
Cep 22h17'44"62d28'
Keegan Shea
Cyg 20h42'18"42d1'
Keegan,Derrell Mark
Uma 10h0'45"71d52'
Keegan,Elizabeth K
Lyr 18h15'51"38d59'
Keegan,Gina
Peg 22h23'0"25d4'
Keegan,Jennifer Catherine
Col 6h0'58"-30d24'
Keegan,Lexus
Eri 5h8'1"-8d13'
Keegan,Lily
Per 2h36'54"56d35'
Keegan,Lorraine & Joseph
Crb 16h10'21"32d3'
Keegan,Michael & Gay
Cyg 21h16'27"28d30'
Keegan,Philip M
Per 3h5'27"40d12'
Keegan,Raymond
Ori 5h55'53"17d37'
Keegan,Sean & Tim
Per 4h20'1"50d29'
Keegan,Sr Muriel
Lyr 19h0'25"25d37'
Keegans Star
Cam 7h28'37"61d6'
Keegans,Lee & Wayne
Boo 14h7'11"26d26'
Keehn,Paul
Ori 5h53'43"15d15'
Keel,Samantha Elizabeth
Cas 0h11'31"62d16'
Keelan,Pauline
Cyg 19h55'51"58d55'
Keeler,Bobby Allen
Hya 8h40'38"-10d50'
Keeler,Christopher Sean
Eri 4h2'1"-11d57'
Keeler,Daniel Joseph
Cep 1h51'38"80d22'
Keeler,David Thomas
Per 4h36'16"36d39'
Keeler,Elaine
Lyr 19h6'51"25d38'
Keeler,Irvin & Darlene
Uma 10h8'35"58d47'
Keeler,Joan
Cyg 21h21'58"28d22'
Keeler,Joseph B
Her 16h41'54"38d6'
Keeler,Michael Scott
Per 1h26'56"53d41'
Keeley
Cas 0h14'21"47d42'
Keeley,Colleen Aisling
Peg 23h1'1"30d38'
Keeley,Daniel Patrick
Her 16h50'13"41d14'
Keeley,Timothy James
Cet 3h0'50"1d0'
Keelin 423673968,Peter W
Leo 9h52'40"27d38'
Keeling,Anita Loyce
Equ 21h1'56"11d17'
Keeling,Robert Allen
Hya 8h49'16"-0d33'
Keely's Comet
Cnv 13h54'35"31d11'
Keely,Ed (Mr Britt)
Lyn 9h2'33"39d49'
Keely,Kristen Lynn
Vul 19h2'58"25d30'
Keely,Michael
Lyn 6h35'18"59d8'
Keely,Robert Charles
Vul 19h6'13"25d0'
Keen Resplendence, Michael E
Aur 7h10'55"37d26'
Keen's Perennial Star, Dick & Betty
Uma 11h6'30"47d0'
Keen,Fred & Mary
Cyg 20h16'34"39d11'
Keen,Henry
Cnv 12h5'23"37d41'
Keen,Lila
Eri 2h55'1"-3d16'
Keen,Mark Douglas
Cnv 12h16'37"34d13'
Keen,Melissa Hope
Lyr 18h29'11"46d1'
Keen,Monte Steven
Lyr 18h43'13"47d40'
Keen,Rex Conrad
Crt 10h55'1"-12d12'
Keenan,Abbigail
Cas 1h33'52"73d33'
Keenan,Alexis
Cnv 13h41'1"37d19'
Keenan,David J
Ori 6h6'17"8d53'
Keenan,Dr Thomas A
Lac 22h32'43"55d22'
Keenan,Francis William (Bill)
Boo 14h15'1"31d19'
Keenan,Joyce
Cas 2h32'30"61d18'
Keenan,Megan
Cas 1h47'28"73d31'
Keenan,Patti
Cas 0h8'0"50d14'
Keenan,Uncle Poodlahah Rev Hugh D
Aur 6h47'56"38d59'
Keene,Evelyne D & Joseph R Keene
Mon 7h31'31"-6d28'
Keene,Franny
Aql 20h5'17"1d34'
Keene,Joel Anson
Her 16h53'41"32d52'
Keene,K M
Lac 22h4'28"47d1'
Keene,Michelle
Cas 1h17'50"62d10'
Keene,Patricia Marjorie
Peg 23h29'36"15d3'
Keene,Travis Maxwell
Lac 22h13'47"46d28'
Keene,Tricia "Light of My Life"
And 1h22'49"39d31'
Keener,Carl R
Aql 18h43'55"11d31'
Keener,Emily Virginia
Vul 19h45'27"28d59'
Keener,Robert L
Her 16h43'1"22d5'
Keener,Taylor Elizabeth
Mon 6h47'19"10d6'
Keeney
Uma 8h53'24"53d22'
Keeney,Erin M
And 0h28'0"41d10'
Keeney,Gail Billerman
Vul 20h22'0"23d45'
Keeney,Grace
Cet 3h4'13"-0d42'
Keeney,Judy Clymer
Lyr 18h32'52"46d20'
Keeney,LaVonne
Peg 23h32'58"33d43'
Keeper Of The Stars - 114
Lyr 18h57'22"31d53'
Keeper,The
Lyn 8h54'20"40d22'
Keers,Sharon
Aur 4h52'0"51d21'
Kees,Ludwig
Aql 18h59'0"17d49'
Kees,Wayne G
Ori 6h4'41"-2d15'
Keese,Robert Eckhard
Peg 22h56'18"17d33'
Keese,Siegfried
Hya 9h35'55"1d40'
Keesha
Cyg 19h15'30"44d7'
Keesling,Nancy C
Aql 19h14'1"15d51'
Keeta
Eri 2h56'21"-16d55'
Keetch,Megs
Cyg 19h33'51"39d42'
Keeter,Donald C
Her 18h9'15"38d58'
Keeter,Norma Jean
Cnv 12h28'12"40d29'
Keeth,Jeannette & Monty
Cyg 19h57'12"30d19'
Keeton,Kelsea
Cyg 20h35'26"45d18'
Keeton,Ruby
Ori 4h42'47"14d37'
Keeton,Simon David
Ori 6h1'22"1d1'
Keevert,Phillip Dean
Uma 10h45'37"71d22'
Keevill,Ann Marie
Uma 11h42'19"64d8'
Keezer,Jenny Grace
Com 12h57'19"25d60'
Kefalas,Dylan Zachary
Her 18h8'51"48d51'
Kefalas,Sasha
Cas 1h45'21"75d52'
Kefalonitis,Barbara Anna
Cyg 20h30'16"38d39'
Kegans Star,The Chad, Kristin & Cole
Her 16h52'24"37d54'
Kegel,Reiner
Aur 5h12'28"43d51'
Kegelman,Jessica Ann
Lyn 9h14'52"46d52'
Kegelman,Julia May
And 0h48'1"39d33'
Kehaly,Paniz
Cyg 20h38'27"39d40'
Kehl,Amy Jean
Cas 3h9'47"61d50'
Kehl,Joseph John
Aql 19h59'33"10d2'
Kehn,Jason
Aur 7h20'53"40d38'
Kehoe,Colleen Mary
Cyg 21h6'59"30d54'
Kehoe,Jr,John M
Per 2h58'25"32d51'
Kehoe,Jr,Miles J
Aql 18h56'41"7d4'
Kehoe,Patrick Kevin
Boo 14h58'39"22d24'
Kehrer,Sandy
And 1h20'53"38d2'
Kehrl,Howard H
Peg 0h4'26"13d56'
Kehrman,Hugh R.
Ser 18h18'40"-14d12'
Kehry,Walter
Lib 14h42'1"-1d23'
Kei,Jr
Cnc 8h45'20"8d24'
Keibler,Matthew
Oph 17h25'30"-22d17'
Keifer,Devon Matthew
Ori 5h57'40"16d50'
Keigley,Lacey & Kevin
Uma 9h16'35"43d30'
Keiichi & Kumiko With Tatsuya
Cap 20h52'27"-19d8'
Keil,Elda
Vul 20h0'17"28d54'
Keil,Katrina
Tri 1h54'15"27d32'
Keil,Peter Alfred
Cnv 12h27'18"33d10'
Keil,Ronesha Noel
Umi 15h16'1"70d4'
Keil,Shanon L
Peg 22h58'58"20d5'
Keilan's Wish
Aql 19h52'10"10d16'
Keilholz,Klaudia
Mon 6h34'24"5d17'
Keiling,Gregory Paul
Cma 6h58'59"-18d20'
Keilman,Kristin Elaine
Del 20h13'23"14d10'
Keilty,Sean Michael
Lmi 9h51'1"33d33'
Keilwitz,Connie Ann
Mon 7h28'0"-10d56'
Keim,Virginia Moyer
And 1h15'56"37d7'A
Keim,Wesley Everett
And 1h15'56"37d7'B
Keiper,Kevin-Corrine & Eion
Equ 20h58'49"8d47'
Keiper-Quinn,Linda
Lyn 6h58'11"58d41'
Keippel I,John R
Her 16h29'19"32d60'
Keippel,Erica L
Cyg 21h31'1"42d16'
Keir
Her 16h58'25"50d48'
Keiran's Light
Cep 20h39'1"75d57'
Keirnan Space Star 3-2-3-2=10,Nancy
Lyr 18h40'50"40d35'
Keirsten Marie Romance &Enchantment
Aur 4h47'0"51d55'
Keirstin Haire
And 0h14'48"36d55'
Keiser,Willis G
Her 16h34'53"35d34'
Keisling,Frederick Carter
Cet 1h7'21"-0d33'
Keissi
Lac 22h17'36"49d13'
Keister,William S
Cep 22h29'0"58d56'
Keith
Leo 10h59'51"11d54'
Keith
Aur 5h18'31"45d2'B
Keith
Ari 2h58'41"30d8'
Keith
Her 16h58'55"32d44'
Keith
Ori 6h5'59"1d38'
Keith
Cep 22h5'40"60d46'
Keith & Ann
Aql 19h1'1"1d22'
Keith & Chris New Orleans
Cap 21h3'33"-14d59'
Keith & Jen Love Texas!
Com 13h17'33"26d50'
Keith & Laura Go Bragh
Uma 11h17'21"40d58'
Keith & Mariann
Lyr 18h46'13"39d38'
Keith & Mary Ann
Sge 19h19'0"16d10'
Keith & Michele Forever 1/29/94
Eri 2h55'59"-12d0'
Keith & Nan's "Wedding Star"
Cyg 20h39'0"41d9'
Keith & Priscilla Forever
Lyr 18h23'1"38d37'
Keith & Raquel
Lyr 18h54'53"30d25'
Keith & Sandra
Sge 20h2'15"20d6'
Keith & Sandra
Boo 13h35'41"22d4'
Keith & Wendy
Cyg 21h5'34"30d43'
Keith III,Stewart T
Her 17h14'20"27d4'
Keith Immortal Optimist
Ori 6h8'45"3d53'
Keith Jr Star Valentine 1993
Her 16h55'14"38d52'
Keith K2 Kime
Aql 18h56'21"16d49'
Keith Reach for The Stars
Cep 22h31'57"59d55'
Keith Ryan
Aur 6h17'25"32d47'
Keith the Spirit of love
Per 2h1'42"57d50'
Keith Thomas
Sex 10h13'54"-3d27'
Keith's Cosmic Kaleidoscope
Ori 5h59'31"15d2'
Keith's Great Nebula
Ori 5h29'51"-5d49'
Keith's Guiding Light
Her 16h51'22"50d56'
Keith,Audrey Ellen
And 23h3'41"44d44'
Keith,Beverly McGaffin
And 1h9'23"37d59'
Keith,Bonnie Joy
And 2h20'0"40d60'
Keith,Colette
Del 20h37'18"9d24'
Keith,Elle
Peg 22h42'49"22d2'
Keith,Francine
Cas 1h4'0"60d59'
Keith,Judy Gail
Boo 13h39'11"19d12'
Keith,Katie Lynne
Mon 6h56'27"8d33'
Keith,Lou Fulton
Her 17h22'49"44d46'
Keith,Messenger Of Light
Ori 5h33'27"-6d30'
Keith,Nicholas Scott
Ori 5h27'11"-0d12'
Keith,Todd Lewis David
Uma 13h53'15"51d16'
Keith,William Richard
Her 17h12'17"44d14'
Keith,Woodrow-Victoria
Lyr 19h19'0"38d35'
Keith-Clio
Dra 10h37'58"74d54'
Keith-Elaine
Cet 2h37'52"2d12'
Keithley,Zoie
Mon 8h0'23"-0d54'
Keizer/Angel Plane, Jeff
Uma 11h45'26"30d55'
Kejzlar's Wayward Wind Vratislav V
Aql 20h18'1"1d5'
Kekaula,Mele Kalani
Cyg 21h4'38"37d23'
Kekina,Wayne
Mon 6h55'1"-1d48'
Keko
Aql 19h8'22"15d30'
Kel Bel
Vul 20h15'11"25d41'
Kel's Irish Way
And 22h56'19"50d28'
Kel-Artinian,Sarkis
Sge 20h17'27"16d56'
Kel-Ron
Cyg 21h17'47"28d46'
Kel-Une
Aur 5h44'37"29d12'
Kelbel,Dennis Ray
Aql 20h35'38"0d45'
Kelcey"The World Is Yours"Tamara Lynn
And 1h15'40"33d37'
Kelch,Anthony W
Hya 8h14'39"-5d52'
Kelch,Lynn
Dra 19h5'3"59d12'A
Keldevash
Peg 23h17'44"33d32'
Kelegian,Haig Todd
Aql 18h58'26"17d7'
Keleher,Cecelia Rose
Cyg 19h50'56"38d1'
Keleher,Jan
Oph 17h55'1"12d3'
Keleman,Bruce A
Her 17h54'0"14d14'
Keljaz
Cas 1h39'13"60d53'
Keljo
Lyr 18h22'55"37d49'
Kell
For 3h28'50"-32d40'
Kell,Jennifer Arlene
Cnc 8h57'24"31d20'
Kellam,Edgar
Boo 15h8'55"50d12'
Kellar,Jeffrey E
Del 20h23'58"10d2'
Kellar,Lois Ann Ellercamp
Cas 1h3'42"68d21'
Kelleghan,Jason
Cet 2h25'30"5d11'
Kelleher
Cam 5h7'41"68d17'
Kelleher, Richard & Rivituso, Maria
Cyg 19h47'47"29d30'
Kelleher,David Christopher
Per 1h25'41"50d28'
Kelleher,Georgia Mae Bakken
Aql 19h29'0"7d52'
Kelleher,James
Dra 17h59'55"68d57'
Kelleher,Kyle Stacy
And 0h0'32"46d20'
Kelleher,Leah
Cyg 19h29'16"33d16'
Kelleher,Matthew Sidney
Per 2h23'12"54d26'
Kelleher,Roark Shannon
Uma 11h55'13"52d51'
Kelleher,Ronald Donald
Boo 15h2'30"12d5'
Kelleman,Ginger
Cas 0h0'26"62d40'
Kellen Jay
Lmi 11h2'13"33d9'
Kellenberg-Eggmann
Cam 4h59'0"61d23'
Keller,Alexis Reneé
Del 20h20'59"10d33'
Keller,Alice Florella
Cma 6h23'23"-19d57'B
Keller,Anthony E
Cet 0h29'20"-2d51'
Keller,Bobbi
Peg 21h52'20"31d38'
Keller,Bobbie
Cam 9h32'12"84d58'
Keller,Brian Michael
Boo 14h27'19"38d32'
Keller,Christa
Mon 8h2'0"-1d48'
Keller,Daniel
Uma 10h53'37"54d26'
Keller,David Chester
Aur 5h3'45"40d22'
Keller,Eck
Lac 22h43'47"38d17'
Keller,Elaine
Tau 4h15'19"22d54'
Keller,Elisabeth Ch Friedenslicht
Lib 15h38'15"-20d19'
Keller,Elizabeth Zerbe
Peg 0h5'31"18d21'
Keller,George Albert
Lac 22h7'13"51d40'
Keller,Hank-Dottie Speed
Umi 16h52'1"77d53'
Keller,Helmut
Cyg 20h27'58"30d29'
Keller,Ilona
Cas 0h45'45"67d27'
Keller,Jerry
Boo 14h37'55"35d36'
Keller,Joanne L
Cyg 19h40'45"31d52'
Keller,Joseph & Rose
Cnc 7h54'60"11d40'
Keller,Karisa & Gage
Crb 15h52'54"30d53'
Keller,Kimberle
Mon 6h56'14"8d42'
Keller,Kimberly Anne
And 2h6'49"38d19'
Keller,Klaus Peter
Gem 6h40'0"34d19'
Keller,Lynn
Eri 4h41'51"-6d16'
Keller,Lynn Hayhurst
Peg 23h23'34"8d36'
Keller,Lynne
Lyn 7h53'34"39d56'
Keller,Madison Christine
And 0h22'46"40d31'
Keller,Marc Alexander
Aur 5h15'29"49d28'
Keller,Mathew & Adam
Lac 22h9'55"47d16'
Keller,Max Gordon
Her 16h38'21"32d40'
Keller,Mel
Mon 7h54'14"-6d51'
Keller,Mildred
Cyg 21h33'35"42d9'
Keller,Moirnja
And 23h41'21"40d41'
Keller,Nicole Marie
And 23h36'55"39d50'
Keller,Penny
And 2h33'45"37d3'
Keller,Ronald L
Lac 22h13'49"49d46'
Keller,Sandrine
Aql 19h55'46"14d56'
Keller,Sara
Sex 10h26'55"-5d53'
Keller,Scott Robert
Ori 6h6'49"8d32'
Keller,Thomas Jefferson
Cma 6h23'23"-19d57'A
Keller,Todd
Aql 20h14'17"5d18'
Keller,Todd Jay
Lac 22h20'31"50d35'
Keller,Valerie Lauren
Uma 11h52'42"45d36'
Keller,Victoria Lee
Aqr 21h28'42"-6d2'
Keller,Zoe Amelia
Peg 21h49'25"33d25'
Kellerman Star,Terry & James
Cep 21h11'13"58d19'
Kellermann,Bianca
Lyn 8h13'39"47d45'
Kellermann,Thomas
Lac 22h32'22"37d45'
Kellett,Kathy
Lyn 8h29'19"40d39'
Kelley & David
Lyn 7h52'59"58d15'
Kelley 2-10-73,Sharon J
Peg 22h36'33"22d46'
Kelley and Jason
Del 20h28'31"10d45'
Kelley Elizabeth
Cmi 7h58'47"8d51'
Kelley Marie "My Bright Star"
Uma 10h42'25"48d23'
Kelley's Star System, Catherine
Vul 19h0'49"21d9'
Kelley,Alyssa Francis
Peg 23h2'50"10d48'
Kelley,Bernest
Mon 7h59'55"-8d19'
Kelley,Blake
Sge 19h58'0"16d9'
Kelley,Brian Francis
Hya 8h12'1"0d12'
Kelley,Christopher Michael
Vul 19h19'26"27d4'
Kelley,Craig
Del 20h14'31"14d29'
Kelley,D.T.
Uma 11h5'18"48d29'
Kelley,Don & Geri
Peg 22h21'25"24d14'
Kelley,Elijah Judd
Peg 0h4'48"28d9'
Kelley,Harold E
Aql 18h43'58"10d36'
Kelley,Jarrett Benjamin
Vul 19h22'40"25d48'
Kelley,Jennifer Neil
Vul 20h14'36"23d35'
Kelley,Jim
Cyg 20h4'1"40d6'
Kelley,Joanne
And 23h48'42"44d21'
Kelley,Joseph Thomas
Lac 22h38'46"37d32'
Kelley,Jr,E Rick
Gem 7h56'53"32d19'
Kelley,Jr,Edward J
Lac 22h1'1"46d17'
Kelley,Jr,Esq,Irish I James E
Her 16h41'11"21d59'
Kelley,Karen Rose
Lyr 18h57'53"30d20'
Kelley,Kevin N
Per 2h55'43"31d33'
Kelley,Laura D
Cam 5h20'26"68d46'
Kelley,Pamela J
Cyg 21h34'29"41d49'
Kelley,Peter T
Lac 22h32'50"50d36'
Kelley,Rebecca Anne
Cyg 20h4'45"40d23'
Kelley,Richard Edward
Dra 15h41'52"67d56'
Kelley,Robert & Shirley
Dra 17h46'31"63d49'
Kelley,Robert J
Her 17h34'12"38d35'
Kelley,Robert T
Oph 16h59'20"-25d5'
Kelley,Rose Adda Johnson
Cas 23h39'29"61d19'
Kelley,Shayla Rae
Cyg 20h24'22"40d23'
Kelley,Sophie Ann
Lyr 19h22'38"38d48'
Kelley,Taylor Paige
Del 20h20'20"10d13'
Kelley,Thomas J
Aql 18h54'12"-2d13'
Kelley,Timothy Alan
Cmi 7h30'23"8d46'
Kelley,Virginia
Uma 11h19'34"41d41'
Kelley,William J
Aql 19h58'30"0d19'
Kelley-Angel
Cas 0h33'1"61d22'
Kelli
Lac 22h44'40"56d32'

Kelli Rachel XJ 220
Crb 16h15'31"32d45'
Kelli's Life
Ori 6h7'34"20d2'
Kellie
Lyr 18h29'57"38d52'
Kellie Noel
Cmi 7h42'57"1d42'
Kellie's (Gleam)
Aql 19h55'42"8d52'
Kelliher,Sharron
And 23h0'13"48d52'
Kellimar
Hya 8h48'22"-1d18'
Kellin,Olivia D
Mon 5h57'20"-5d59'
Kelling,Debbie
And 1h26'52"38d57'
Kellish,Alexander Ross
Ori 5h42'23"3d59'B
Kellish,Lisa
Ori 5h42'23"3d59'C
Kellman,Dorothy Marie
Scl 23h10'35"-31d47'
Kellman,Joel Robert
Uma 10h30'35"52d53'
Kellman,Russell Albert
Scl 23h9'40"-30d55'
Kellner,Tara
Umi 15h6'25"66d12'
Kellogg,Carroll Atkinson
Cyg 19h18'26"28d59'
Kellogg,Christine
Cet 1h52'1"-7d51'
Kellogg,Denise Rae
Peg 21h53'19"31d35'
Kellogg,Ritt
Peg 23h2'31"17d55'
Kellogg,Tina Marie
Lyr 18h41'18"31d44'
Kellom,Kimberly
Peg 21h55'1"30d15'
Kellom,Susan
Vul 19h59'23"23d39'
Kellringer,Alois
Lib 14h23'22"-23d20'
Kelly
Cas 0h35'47"58d54'
Kelly
Peg 23h28'15"22d11'
Kelly
Lac 22h26'25"48d17'B
Kelly
Leo 11h50'54"22d23'
Kelly
Del 20h38'58"10d50'
Kelly
Psc 23h27'21"5d42'
Kelly
Peg 23h23'14"10d4'
Kelly
Sex 9h56'48"1d21'
Kelly
Cyg 19h30'22"31d19'
Kelly
Eri 2h52'1"-5d51'
Kelly
Cyg 19h32'1"34d45'
Kelly
Lmi 10h45'40"24d47'
Kelly & Donmichael
Ori 5h28'28"1d18'
Kelly & Mark Forever
Aql 20h7'34"0d9'
Kelly 01-17-74
Cam 3h57'0"68d56'
Kelly An
Lyr 18h47'44"32d18'
Kelly Ann
Peg 23h46'1"28d5'
Kelly Ann
Cas 1h55'54"76d7'
Kelly Ann
And 22h59'50"37d12'
Kelly Ann
Mon 6h36'47"-0d11'
Kelly Consellation-JBH
Lac 22h52'31"53d54'
Kelly Constellation- Yvonne
And 2h17'30"40d8'
Kelly Constellation- Pat
Lyr 18h45'43"40d57'
Kelly Constellation- Debbi B
Dra 17h17'0"61d42'
Kelly Constellation- Kathi H
Uma 11h38'23"50d1'
Kelly Constellation- Kevin H
Lac 22h51'54"53d21'
Kelly Constellation- Sean H
Boo 13h46'18"20d22'
Kelly Constellation- Mike G
Per 2h57'50"43d58'
Kelly Constellation- Jack
Lac 22h51'50"53d9'
Kelly Constellation- Melody G
And 2h16'40"40d56'
Kelly Dawn
Lyn 6h29'18"56d12'
Kelly Edwards Elementary
Peg 22h26'32"20d6'
Kelly Friendship Star, Jackie
Ori 5h55'45"6d42'
Kelly II,David M
Her 16h9'25"8d15'
Kelly III,John B
Per 4h42'27"37d51'
Kelly III,Robert D
Uma 11h7'54"71d35'
Kelly Jean
Vul 21h25'31"24d14'
Kelly Jo
Lyn 7h49'50"36d55'
Kelly Jo
Lyr 18h17'46"44d43'
Kelly Jo Lynn
And 1h50'29"40d14'
Kelly Joël
Aql 19h0'52"13d21'
Kelly Kandy
Lib 15h34'24"-8d2'
Kelly Kaye Star
Aql 20h17'24"5d12'
Kelly Lee
Mon 6h54'43"-0d3'
Kelly Louise
And 23h41'31"41d22'
Kelly Louise
Ori 4h54'35"0d4'
Kelly Lynn
Peg 21h55'1"23d41'
Kelly Lynn
Lyn 8h3'39"34d30'
Kelly Lynn
Lyn 6h59'46"54d47'
Kelly Lynn
Cyg 19h46'49"38d56'
Kelly N' Kevin
Cyg 19h25'49"44d42'
Kelly R & Cheri N Match
Made In Heaven
Cas 1h18'54"62d53'
Kelly Rachel
And 1h3'36"37d32'
Kelly Ruth-Sweet Sixteen
Cyg 20h36'29"31d30'
Kelly Shines Bright, Erin K
Lyn 8h0'35"40d4'
Kelly Star
Lmi 11h3'41"32d53'
Kelly Star,The
Peg 22h47'42"10d5'
Kelly Star,The Tony
Cap 20h57'57"-15d58'
Kelly Wed Aug 8,1946, Leo &
LaVern
Cyg 21h3'59"36d56'
Kelly's Bear
Uma 9h51'18"43d33'
Kelly's Comet
Uma 11h34'49"33d14'
Kelly's Comet
Lyn 9h15'1"34d38'
Kelly's Dream
And 23h49'11"40d15'
Kelly's Golden Knight
Lyr 18h59'18"36d29'
Kelly's Heart
Eri 2h50'44"-6d45'
Kelly's Heavenly Light
Peg 22h56'0"18d23'
Kelly's Kismet
Lib 15h18'47"-21d13'
Kelly's Light
Pho 0h49'12"-48d5'
Kelly's Light
Gem 6h58'21"15d53'
Kelly's Peace
Aur 5h10'44"54d41'
Kelly's Smile
Lyn 9h3'41"46d6'
Kelly's Star
And 23h34'1"47d30'
Kelly's Star
Cas 0h8'44"62d47'
Kelly's Star Of Dreams
Eri 3h51'1"-7d6'
Kelly's Very Own Star
Cas 3h8'28"61d24'
Kelly,A Leah
And 2h6'19"40d34'
Kelly,Adrienne J
And 1h47'16"40d48'
Kelly,Aileen Frances
Lyn 8h4'15"38d13'
Kelly,Anna
Mon 6h44'51"1d42'
Kelly,Arthur Francis
Lac 22h26'27"55d24'
Kelly,Bernadine A
Lac 22h4'0"50d22'
Kelly,Bernard Joseph
Oph 18h7'11"12d38'
Kelly,Bryanna C
Lib 14h6'1"39d55'
Kelly,Bud
Per 3h14'12"56d17'
Kelly,Cameron Finn
Cnv 13h59'52"37d49'
Kelly,Carol Diane Gardner
Leo 10h42'43"14d30'
Kelly,Cathy
And 0h17'35"38d35'
Kelly,Chris E
Cet 0h33'52"-2d15'
Kelly,Claudia Lacy
Cas 1h38'56"58d9'
Kelly,Clayton
Her 18h21'0"12d39'
Kelly,Cynthia Johnson
Equ 21h23'37"8d43'
Kelly,Daniel
Lac 22h14'49"49d15'
Kelly,Daniel Edward
Leo 9h35'0"8d56'
Kelly,Daniel J
Aur 6h29'1"31d4'
Kelly,Daniel J
Lyr 18h28'39"45d3'
Kelly,Darcy
Lyr 18h17'21"34d4'
Kelly,Dave & Ann Marie
Cam 12h47'0"77d34'
Kelly,Debby
Peg 23h20'56"25d7'
Kelly,Deborah J
Lyn 7h53'24"44d39'
Kelly,Deborah R
Mon 6h38'47"6d59'
Kelly,Denise M
Cas 2h7'54"67d50'
Kelly,Devon Marie
Psc 23h20'22"5d28'
Kelly,Don
Cet 2h32'20"0d38'
Kelly,Doreen Ann Callow
And 23h20'20"46d3'
Kelly,Douglas
Lyn 6h30'21"60d39'
Kelly,Dr Maryann
Oph 17h17'4"-23d18'
Kelly,Elizabeth A
Gem 7h31'11"20d32'
Kelly,Erin Katherine
Cas 1h1'22"64d57'
Kelly,Erin Kathleen
Com 12h12'30"27d12'
Kelly,Fiona
Lyr 19h0'25"26d51'
Kelly,Frank B
Lmi 10h6'46"30d44'
Kelly,Fredrick
Her 18h55'52"12d54'
Kelly,Gail Marcina
And 23h41'23"40d34'
Kelly,Grandpop
Dra 17h28'56"72d8'
Kelly,Heather Kathleen
Crb 16h11'1"30d57'
Kelly,Jack
Boo 14h8'45"32d6'
Kelly,Jack
Lac 22h13'24"38d33'
Kelly,Jackie
Lyn 7h55'34"40d36'
Kelly,Jacob Ian
Cep 22h39'46"61d24'
Kelly,James Jacob
Aur 6h28'1"37d53'
Kelly,James Patrick
Dra 18h19'29"71d15'
Kelly,Jennifer
Com 12h21'59"19d44'
Kelly,Jericho
Nor16h18'25"56d6'
Kelly,John
Her 16h29'19"48d59'
Kelly,John C
Boo 15h2'44"24d37'
Kelly,John Ennist
Aur 6h43'38"38d49'
Kelly,John Fitzgerald
Dra 15h59'26"68d36'
Kelly,John J
Aur 5h6'41"40d19'
Kelly,Jon Stuart
Eri 2h47'30"-6d1'
Kelly,Karen S
Vir 11h50'24"7d5'
Kelly,Kathryn
Cas 0h7'33"63d10'
Kelly,Kathy
Lyn 8h48'39"41d18'
Kelly,Kay Elizabeth
Peg 23h23'33"15d46'
Kelly,Kay Marie
Cyg 19h34'0"33d13'
Kelly,Keith
Eri 3h25'31"-6d4'
Kelly,Kevin Brian
Lac 22h50'18"37d58'
Kelly,Kevin Leo
Cyg 21h33'42"44d49'
Kelly,Kimberly Ann
Cas 1h27'26"58d37'
Kelly,Kimberly Loves Marc
Maison
Aur 4h51'56"50d57'
Kelly,Kristen
Peg 21h53'19"30d55'
Kelly,Kristina Marie
Cnc 8h59'16"20d33'
Kelly,Kurt
Cyg 19h34'53"28d0'
Kelly,Lauren Shipley
Vul 20h21'41"22d43'
Kelly,Lenore Christine
Lyr 19h1'28"31d50'
Kelly,Linda Darlene
Mon 7h0'1"8d43'
Kelly,Marcella A
And 0h11'39"37d59'
Kelly,Megan Elizabeth
Crb 15h58'1"26d57'
Kelly,Michael T
Hya 9h32'50"-2d15'
Kelly,Mrs Irene
Cas 0h17'31"62d35'
Kelly,Neil James
Boo 15h2'31"31d30'
Kelly,Nellie Ann
Peg 23h40'14"8d36'
Kelly,Nick
Cam 10h51'32"80d34'
Kelly,Patrick John
Uma 10h53'52"38d4'
Kelly,Paul & Paulette
Mon 7h2'1"1d3'
Kelly,Peggy Foster
Cam 3h59'25"62d8'
Kelly,Peter James
Cep 20h35'47"76d15'
Kelly,Peter Thomas
Lac 22h3'20"46d25'
Kelly,Rachael Lynn
Cap 21h1'29"-15d14'
Kelly,Ralph William
Cnc 8h35'22"18d25'
Kelly,Richard W
Sex 10h18'23"-2d40'
Kelly,Robert J
Her 17h32'0"30d44'
Kelly,Robin Michelle
Eri 4h7'58"-16d42'
Kelly,Rodger
Dra 13h51'13"68d2'
Kelly,Sally Payne
Cnv 12h40'39"40d14'
Kelly,Sandy
Her 17h14'0"18d49'
Kelly,Sarah Marie
Lyn 8h59'37"41d16'
Kelly,Sarah Scrogham
Eri 2h47'30"-6d59'
Kelly,Sr,Larry
Per 3h15'51"50d32'
Kelly,Stephen A
Aql 19h32'45"0d18'
Kelly,Stephen D
Aur 6h46'43"38d60'
Kelly,Steve
Her 16h21'26"28d43'
Kelly,Timothy Patrick
Tri 2h3'53"32d37'
Kelly,Timothy Peter
Uma 9h56'28"51d28'
Kelly,Todd
Her 18h31'56"24d51'
Kelly,Tommy
Boo 15h3'21"14d26'
Kelly,Virginia M
Cas 0h38'32"63d1'
Kelly,Walter
Cep 22h47'18"56d58'
Kelly,William
Hya 8h38'30"5d45'
Kelly,Zoie Margaret
Uma 11h48'15"43d45'
Kelly-Belle
Lyr 18h59'35"47d17'
Kelly-Lieberson,Sandy & Rob
Vul 19h48'45"26d8'
Kelly/Jeff
Peg 22h6'37"3d53'
Kellye
Peg 22h18'59"31d6'
Kellymatt 6593
Peg 22h19'32"4d1'
Kellystella
Cam 3h26'1"61d1'
Kelm "Bobbie",Roberta Lynn
Lyn 8h56'47"46d44"
Kelm,Leatrice & Roger
Cyg 21h32'29"41d53'
Kelman,Glenn Arthur
Lac 22h17'34"47d12'
Kelman,Jan
Eri 2h58'11"-10d27'
Kelman,Judi
Cyg 19h26'26"35d13'
Kelmiski,Timothy
Her 17h22'37"46d53'
Kelsa
Uma 9h42'53"42d43'
Kelsay,Bruce
Ari 2h56'22"28d51'
Kelsch's Kooky Kilonnie
Lyn 8h4'16"43d8'
Kelsea
Peg 23h36'19"18d18'
KelseaP93
Lyr 18h46'13"34d8'
Kelsey Alexandra
Peg 22h45'45"21d11'
Kelsey Carin
Uma 9h46'27"51d38'
Kelsey Forever,Russell Walter
Equ 21h6'37"11d32'
Kelsey Lynn
Cas 0h25'20"60d7'
Kelsey Marika
Equ 21h10'51"10d23'
Kelsey May
Cyg 19h28'17"39d23'
Kelsey Nicole
Cam 6h8'1"60d45'
Kelsey Rae
Uma 12h32'14"61d59'
Kelsey's Krystal
And 0h37'23"40d54'
Kelsey's World
Boo 14h41'23"39d45'
Kelsey,Forever Susan, Miles
Del 20h28'1"10d47'
Kelsey,Kathryn
Crb 15h42'19"27d26'
Kelsey,Morgan
And 2h29'54"44d34'
Kelsey,Stephen F
Dra 19h55'17"68d30'
Kelsey,Vivian
Lyn 6h56'27"54d4'
Kelsh,Laurel
Lmi 9h43'49"34d24'
Kelsi
Lyn 8h0'41"54d53'A
Kelso,Connor Daniel
Her 18h7'0"30d23'
Kelso,Dana Marie
Mon 7h48'35"-1d18'
Kelso,Sharon
Vul 19h22'34"26d47'
Kelsoe,Allison Leigh
Ori 5h46'55"11d11'
Kelson,Jane-Gorgeous
Girlfriend
Sgr 19h29'18"-33d0'
Keltner I,Shirley & George
Lyn 7h57'26"41d1'
Keltner,Brian Charles
Aur 6h32'53"31d6'
Kelton,John Allen
Cma 7h7'27"5d13'
Kelvin
Sex 10h47'1"-0d1'
Kemak
Her 17h25'30"29d58'
Kemenosh,Mak
Cnv 13h56'47"42d26'
Kemens,Trisha
Cas 22h57'18"55d52'
Kemist,Adam G
Her 15h48'1"42d32'
Kemm-Er-Bees
Ori 5h48'36"20d42'
Kemmer,Philip C
Per 1h41'0"52d31'
Kemmerer,Michelle
Lyn 8h10'0"40d54'
Kemmet,Klaus
Sgr 20h4'13"-41d9'
Kemmling,Kerstin
Vir 12h4'23"5d18'
Kemnitz,Dee
Cam 7h47'6"71d41'
Kemo
Cas 22h57'41"56d37'
Kemosabe Tom
Ser 15h19'1"10d20'
Kemp,Bo
Aql 19h5'29"3d6'
Kemp,Brandon James
Cep 20h47'0"61d11'
Kemp,Catherine
Del 20h14'52"11d52'
Kemp,Christine
And 23h3'22"46d23'
Kemp,Danielle Christine
Lmi 10h52'46"26d17'
Kemp,Dorothy Laurel
Mon 7h29'30"-10d18'
Kemp,Jan
And 0h23'51"37d33'
Kemp,Kara
Equ 21h2'44"10d54'
Kemp,Mark Steven
Boo 15h3'33"30d20'
Kemp,Michelle Clare
Crb 15h29'36"30d4'
Kemp,Paul
Lac 22h55'29"52d33'
Kemp,Paul Wesley
Lmi 10h52'1"28d26'
Kemp,Philip Walter
Dra 15h7'34"61d32'
Kemp,Richard A
Eri 4h13'14"-17d46'
Kemp,Robert E
Dra 19h6'30"54d50'
Kemp,Ross
Ori 5h35'40"-8d5'
Kemp,Rosser Sterling
And 0h49'27"38d58'
Kemp,Scott Stephen
Lmi 10h52'47"27d58'
Kemp,Sherran Joy
Lmi 10h56'11"25d47'
Kemp,Siena Lee
And 23h37'27"44d28'
Kemp,Stefanie No 1
Lib 14h40'26"-0d13'
Kemp,Sue Ellen
Lmi 10h56'32"31d14'
Kempa,Walter
Cep 0h18'34"73d36'
Kempczynski,Patricia
Lyn 7h50'28"33d51'
Kempen,Christopher
Cyg 21h52'13"37d55'
Kempen,Dennis
Cas 23h42'18"61d15'
Kempen,Jeffrey
Cyg 21h52'20"37d39'
Kempen,Larry
Cyg 21h52'43"37d29'
Kempen,Lisa
Cas 23h42'30"61d30'
Kemper,Linda
Lyn 7h26'54"45d13'
Kemper,Lindsay
Ori 5h19'1"-6d43'
Kemper,Ross Hilton
Lmi 10h50'31"28d22'
Kemper,Sydney & Isobel
Umi 15h11'56"68d1'
Kemper,Tamerah Kay
Boo 14h7'35"10d11'
Kemper,Wolfgang
Com 13h28'0"26d40'
Kempf,Allen Jacob
Aql 18h58'24"0d53'
Kempf,Arno
Cam 4h9'1"58d33'
Kempf,Kathy
Peg 22h52'11"29d40'
Kempf,MacDonald John
Mon 6h40'39"11d3'
Kemph,Deborah R
Cas 0h28'17"75d47'
Kemph,James E
Cet 0h36'59"1d35'
Kemph,Lisa M
Lyn 7h6'30"50d35'
Kempka,Glenn
Cam 3h56'50"74d51'
Kempler,Joanie
Lyr 19h19'0"38d59'
Kempner,Sheila & Edward
Crb 15h18'0"30d58'
Kempnerward,Sheila & Ed
Aur 5h22'1"41d2'
Kempr,Wienand
Lac 22h5'39"51d42'
Kempter,Devon George
Francis
Uma 8h17'29"67d47'
Kempton,Randy P
Her 17h32'13"26d50'
Kempy
Ori 4h49'31"0d1'
Kems,Mary & Harry
Uma 10h35'39"70d52'
Ken
Cyg 21h3'12"30d50'
Ken
Cnv 12h26'16"51d58'
Ken
Uma 9h50'30"47d31'
Ken
Uma 9h32'60"49d32'
Ken "Our Daddy"
Ori 5h59'1"-2d51'
Ken & Annabelle's Star
Cyg 19h30'1"35d53'
Ken & Barbie
Cyg 19h21'11"28d15'
Ken & Deven Always
Per 2h50'38"40d42'
Ken & Donna
Aql 19h42'41"10d27'
Ken & Jenny's Star
Aql 19h40'41"11d2'
Ken & Marsha's "New
Beginnings"
Cyg 20h36'15"58d27'
Ken & Megan
Lyr 19h17'46"40d34'
Ken & Sharon's Star Forever
Eri 3h20'59"-16d46'
Ken George
Cep 3h49'28"80d4'
Ken Mar 50
Cep 21h11'22"68d47'
Ken Star,The
Per 2h29'43"57d46'
Ken's Blue Jay
Dra 9h56'33"77d53'
Ken's Dream
Ser 16h14'23"0d11'
Ken's Drean
Aql 19h50'1"14d48'
Ken's Figment
Aur 6h29'24"34d44'
Ken's Princess
Del 20h35'31"20d15'
Ken's Promise
Per 2h38'1"37d26'
Ken's Star
Aur 5h57'26"29d53'
Ken's Star
Per 0h9'50"18d48'
Ken's Star
Ser 15h27'44"1d25'
Ken's Way
Aur 7h19'20"41d12'
Ken-Ann
Cnv 12h24'50"46d44'
Ken/Teresa
Boo 13h58'46"26d7'
Kenan
Aql 19h17'58"12d19'
Kenars Wooden Gallery
Mon 8h3'13"-1d8'
Kench,Jayne & Wes
Cma 6h59'12"-19d23'
Kendal
Cyg 19h25'43"30d20'
Kendall
Aql 18h44'17"10d40'
Kendall
Cas 2h35'14"61d48'
Kendall Alison
Lyr 18h30'50"30d15'
Kendall's Rae
Peg 22h21'44"35d17'
Kendall's Star,Brett
Boo 15h3'28"25d7'
Kendall,Amara Sok
Aqr 23h4'42"-4d6'
Kendall,Ben R
Vir 13h29'32"-4d26'
Kendall,Brandt
Lyr 18h40'1"33d2'
Kendall,Charles Arthur
Cep 21h43'1"68d35'
Kendall,Cody
Aur 7h0'49"40d4'
Kendall,Erick R
Per 2h56'50"40d48'
Kendall,Glenn
Cep 0h12'41"80d8'
Kendall,Ion
Umi 15h55'34"77d3'
Kendall,Marcene Marilyn
Cas 0h32'53"54d47'
Kendall,Susan
Nor 16h18'20"56d18'
Kendall,Tela
Peg 22h54'40"29d36'
Kendall,Theresa Joan
Cas 0h35'0"72d24'
Kendall,Virginia Kay Rozich
Uma 11h52'22"54d60'
Kendall-Smith,Lou-Lou
Cas 0h37'27"60d26'
Kenderes,Lisa
And 23h0'47"51d7'
Kendra
Mon 8h7'10"-8d26'
Kendra
Vir 13h54'1"1d18'
Kendra Lyn
Cyg 19h25'0"32d40'
Kendra's Paradise
Cet 1h29'58"-4d42'
Kendra's Twinkle
Leo 9h46'33"19d43'
Kendra's Wish
And 0h5'35"46d14'
Kendra,Anne D
Cyg 21h8'47"47d29'A
Kendra,John S
Cyg 21h8'47"47d29'B
Kendrew
Lyn 8h39'53"44d59'
Kendrick's Kindle
Vul 19h17'27"25d26'
Kendrick's Star,Mum & Dad
Peg 23h47'35"30d57'
Kendrick,Barbara
Peg 22h50'19"20d23'
Kendrick,Don
Cet 2h51'0"0d28'
Kendrick,Elwood
Boo 15h0'1"22d42'
Kendrick,Evelyn Douglas
Leo 9h33'60"10d49'
Kendrick,Kim David
Mon 6h18'28"5d18'

Kendrick,Rachael Marie Cas 2h31'13"67d37'
Kendrick,Robert Dennis Aql 18h59'57"-6d55'
Kendrick,Samuel Stuart Cam 3h54'21"68d25'
Kendrick,William Rodney Del 20h17'41"14d25'
Kendrioski,Susan "Suzy Q" Lyr 19h24'25"38d41'
Kendryna,John P Sex 9h55'25"2d36'
Kendzior,Steven J Her 17h24'46"45d46'
Kenefick III,Peter Oph 17h39'22"-23d42'
Kenefick,Alexander Her 16h11'38"11d30'
Kenefick,Captain Dave Aur 6h0'11"50d20'
Kenefick,Elizabeth Ann Mon 7h0'53"0d8'
Kenehan-Stanchak,Alice Regina And 0h46'45"37d40'
Kenfield,Kathleen Aql 20h12'47"11d51'
Kenigsberg,Dara Cas 0h15'55"65d6'
Keniry,Lena Theresa Peg 22h54'21"27d38'
Kenison,Lisa Dianne Cam 9h2'53"82d23'
Kenitzer,Stephanie N And 23h30'15"49d38'
KenJean Cyg 19h14'34"45d47'
Kenji,Curran Lac 22h12'39"47d52'
Kenkel,Andrew James Dra 15h45'45"62d33'
Kenlee & Kevin in the Heavens Umi 16h19'27"71d41'
Kenlon,Virginia Minogue Lyn 7h3'57"50d13'
Kenmar Ori 4h53'0"0d33'
Kenmar '95 Her 16h10'51"21d28'
Kenn,Andrew Aur 5h29'14"31d37'
Kennamore,William Earl Aur 5h19'43"46d6'
Kennan,Amanda & Tyler Sge 19h38'1"16d37'
Kennan,Elizabeth Topham Cas 0h58'52"61d1'
Kennan,Melanie Anne Peg 23h46'30"12d21'
Kennard,John Cet 0h52'55"-6d59'
Kennaro,Janice Cnv 13h19'41"40d45'
Kennaugh,Alicia Aql 18h55'42"-1d23'
Kenneally,Mary Josephine Joyce Uma 13h49'20"48d15'
Kennedy Hya 9h39'20"-18d54'
Kennedy III,John Patrick Cep 20h59'0"59d48'
Kennedy IV,"Billy" William Patrick Hya 8h13'13"3d20'
Kennedy Love Gerald, Shirley Ann Sge 19h23'58"16d41'
Kennedy Star,The Erin Uma 9h27'16"50d45"
Kennedy's Celestial Mustang,Craig Leo 11h0'12"8d55'
Kennedy(Hek),Hillary Elizabeth And 23h40'26"33d27'
Kennedy(KLK),Katherine Louise And 23h36'26"33d2'
Kennedy,Adam C Her 18h3'20"14d56'
Kennedy,Alyssa Rose Tau 5h54'35"28d49'
Kennedy,Amy Jo Oph 17h50'14"12d1'
Kennedy,Barbara Ann And 0h20'15"31d46'
Kennedy,Blake Ryan Uma 9h47'11"62d12'
Kennedy,Bryce Boo 14h2'1"18d53'
Kennedy,Cam Mon 8h7'24"-8d33'
Kennedy,Catherine Eileen Eri 3h57'37"-4d16'
Kennedy,Charles Lmi 10h37'57"24d47'
Kennedy,Christine And 0h16'40"31d47'
Kennedy,Christine & Chris Sge 20h1'1"16d23'
Kennedy,Daniel Cnv 12h22'55"37d26'
Kennedy,Daniel Joseph Aql 19h9'0"4d30'
Kennedy,Darlene Paige Darleeno Vul 20h17'34"25d36'
Kennedy,Darren Robert Uma 10h51'59"70d30'
Kennedy,Delany Aql 19h31'36"12d18'
Kennedy,Evelyn Cyg 21h3'38"37d47'
Kennedy,Florence Judith Mon 6h58'12"7d36'
Kennedy,Glenna Sge 19h39'15"16d14'
Kennedy,Helen & James Cam 6h12'38"58d10'
Kennedy,Helen Leona Lyn 8h11'32"34d14'
Kennedy,Jacob Michael Lyr 18h32'0"32d46'
Kennedy,James Corcoran Cep 7h59'1"86d47'
Kennedy,James Patrick Her 16h14'52"48d22'
Kennedy,Jeanne Peg 21h28'37"20d25'
Kennedy,Joe Aql 20h3'15"6d43'
Kennedy,John Uma 8h46'53"73d11'
Kennedy,John & Ronee Cyg 21h34'1"42d51'
Kennedy,John David Cet 0h58'31"-1d4'
Kennedy,John F Her 18h7'35"37d48'
Kennedy,John J Aur 6h25'15"38d2'
Kennedy,John M "Dodger" Cet 1h26'1"-13d39'
Kennedy,John R Boo 13h56'12"22d1'
Kennedy,John Stewart Cmi 7h22'25"4d38'
Kennedy,John Walter Oph 18h35'29"10d52'
Kennedy,John-Wesley Stephens Her 16h10'0"24d44'
Kennedy,Jr,Alfred Parker Cep 23h12'56"62d3'
Kennedy,Jr,James G Her 16h58'60"20d26'
Kennedy,Jr,Jerry Riley Aql 19h48'52"14d25'
Kennedy,Karen A Del 20h14'52"15d14'
Kennedy,Katelyn Cas 3h7'0"58d33'
Kennedy,Kathleen Peg 21h40'31"28d18'
Kennedy,Kathleen E Aur 4h56'17"40d60'
Kennedy,Keith Ori 5h14'15"-5d14'
Kennedy,Kelly Ann Cas 0h29'40"50d4'
Kennedy,Kelly Patricia Vul 20h14'34"23d55'
Kennedy,Kenneth Aql 19h32'28"1d16'
Kennedy,Lachlan Gardner Lyn 6h12'31"60d6'
Kennedy,Lauri Ann Peg 0h2'27"14d27'
Kennedy,Linda J Aql 18h58'56"-6d60'
Kennedy,Maj Thomas J Aql 19h26'46"-1d25'
Kennedy,Mammaw's Star Shirley Eri 3h54'39"-5d34'
Kennedy,Marie Leighann Eri 4h55'9"-5d14'
Kennedy,Marsha Umi 16h25'51"70d26'
Kennedy,Mary Jane Sontheimer Cyg 20h4'26"31d48'
Kennedy,Nikki Simone Sge 20h16'1"20d13'
Kennedy,Olive Dyson Vul 19h48'37"29d13'
Kennedy,Pam And 23h39'30"41d2'
Kennedy,Robert J Tau 4h10'8"24d11'
Kennedy,Robert J "Pip" Aql 19h28'0"8d9'
Kennedy,Roy Her 16h47'12"10d18'
Kennedy,Ruth W And 1h6'59"38d56'
Kennedy,Scott M Aur 6h3'32"46d42'
Kennedy,Sid Umi 15h41'14"68d48'
Kennedy,Sophia Elizabeth Sge 19h21'22"18d49'
Kennedy,Tania Ant 10h39'57"-38d6'
Kennedy,Tania Marie Lyr 18h42'11"41d3'
Kennedy,Tessa Rae Lyn 7h42'39"42d49'
Kennedy,Tim "Tammi" Lac 22h41'12"50d3'
Kennedy,Vera E Cas 1h4'17"53d6'
Kennedy,Vince,Mary, Eric,Rita Lyr 18h27'35"31d40'
Kennedy,Wendy Cas 0h21'31"65d0'
Kennedy,William Sex 10h32'50"2d3'
Kennedy-"Bobby",Robert S Cnc 8h11'12"32d6'
Kennedy-Kardas,Psy D, Dr Dorothy Cyg 19h37'40"28d18'
Kenneke,Michael Gerard Cep 0h8'17"68d15'
Kennell,Ken Dra 18h55'59"68d49'
Kennemer,Janice Crt 11h36'35"-11d51'
Kennemer-Mr Moon Over Texas,Wayne Cet 3h14'17"9d8'
Kennerknecht,Bob Oph 17h39'53"10d29'
Kennerly,William Gordon Lmi 10h12'44"40d57'
Kenneth Cmi 7h42'29"4d44'
Kenneth B Aur 6h55'1"41d13'
Kenneth Gerard Del 20h21'1"7d59'
Kenneth James Her 17h11'59"44d9'
Kenneth Laurance Aql 19h49'51"11d59'
Kenneth Patrick Boo 15h4'34"11d20'
Kenneth's Event Horizon Aql 19h4'53"-0d9'
Kenneth:Shakespeare Sonnet XXIII Lmi 9h20'25"34d10'
Kennett,I Love Sharon Cyg 19h59'11"45d8'
Kenney,Bernie Cep 21h32'35"61d47'
Kenney,Charles William Boo 14h37'15"36d11'
Kenney,Daniel M & Sheryl A Lac 22h14'15"55d16'
Kenney,Docia Jo Bowers Peg 21h26'31"23d6'
Kenney,Dr Howard W Oph 18h17'22"11d54'
Kenney,Francis & Grace Ori 5h51'54"10d24'
Kenney,Jo & Peter Umi 14h55'34"66d10'
Kenney,Joseph W Cam 3h37'26"60d58'
Kenney,Jr,Lawrence Michael Lac 22h31'43"55d22'
Kenney,Julia & Francis Sge 20h3'48"20d7'
Kenney,Kathleen And 23h50'39"33d36'
Kenney,Marion H Cas 3h11'35"58d34'
Kenney,Martha D Uma 10h16'58"48d50'
Kenney,Michael A D Lac 21h56'18"42d36'
Kenney,Pamela A And 23h11'56"41d54'
Kenney,Phoebe K D Cnv 12h39'54"40d58'
Kenney,Shannon Lyr 19h0'53"37d58'
Kenney,Sue & Ed Lewis Lyn 7h57'48"33d50'
Kenning,Quinn Dermot Aqr 20h52'45"-1d48'
Kennington,Jack Oph 17h32'59"7d30'
Kennon,Marcia Ellen Vul 19h49'31"20d36'
Kennosuke Psc 1h21'23"2d27'
Kenny Per 2h29'41"57d41'
Kenny Per 1h54'32"54d22'
Kenny Dra 18h29'31"58d35'
Kenny Aur 5h22'36"41d0'
Kenny "The Lonesome Cowboy" Peg 23h3'1"11d37'
Kenny & Kristen's Valentine Love Star Cyg 20h32'14"50d22'
Kenny & Laura"Light To The Future" Boo 15h18'1"52d41'
Kenny & Lori Since 7-10-90 Cam 13h9'0"78d40'
Kenny & Marc's Star Aur 6h35'16"34d53'
Kenny Dog,Rockstar,The Cnv 12h28'0"46d2'
Kenny III,Thomas F Ori 5h59'24"6d46'
Kenny's 1st Birthday Cam 3h58'59"53d49'
Kenny's Forté Dra 16h15'0"61d48'
Kenny's Good Luck Star Ori 5h5'34"-1d3'
Kenny's Place Aur 6h6'53"31d29'
Kenny's Place Ori 4h53'42"-1d10'
Kenny's Star Cnv 12h6'53"38d32'
Kenny's Star Aur 5h17'1"49d31'
Kenny's World Ori 5h43'33"11d3'
Kenny,Brigid Ann Uma 9h41'36"57d21'
Kenny,Dr Kevin J Her 16h54'21"39d27'
Kenny,Frances Gates Cas 1h3'41"55d27'
Kenny,Henri Bernard Ori 5h56'47"20d22'
Kenny,Jr,Liam Joseph Sgr 19h16'31"-20d31'
Kenny,Kaitlin D Mon 7h16'0"-1d38'
Kenny,Kerry Cam 5h41'0"60d51'
Kenny,Kristen Adele Tri 2h7'20"33d58'
Kenny,Margaret Hanson Cyg 19h26'23"30d47'
Kenny,Michael Eric Vul 20h3'34"22d45'
Kenny,Paula Louise Cap 21h37'1"-19d44'
Kenny,Rachel Ann Vul 19h23'0"25d22'
Kenny,Sydney Margaret And 0h53'0"36d30'
Kenny,William John Boo 14h19'28"51d52'
Kenny-Stith,Sharlande Eve Lyn 8h10'54"57d40
Kenrick,James M & Amy Oksner Tri 2h11'24"30d29'AB
Kensandra Peg 22h36'55"25d7'
Kensinger,Amanda Marie And 0h48'31"38d47'
Kensington Academy Super Star Uma 8h30'11"70d56'
Kenstar Uma 12h4'29"33d3'
Kent Dra 17h41'1"68d10'
Kent & Judy Lyn 7h33'30"38d56'
Kent & Laura Cet 1h39'60"-14d52'
Kent & Susanne's Wedding Star Crb 16h12'22"31d58'
Kent Michael Per 3h2'21"41d42'
Kent,D C,Ray Aur 5h19'24"45d44'
Kent,Desmond Per 2h39'52"40d7'
Kent,Fred Ori 6h21'59"16d5'B
Kent,I will love you forever Martyn,Tracey Ori 5h53'42"7d17'
Kent,Isabel Finley Cas 2h43'0"70d37'
Kent,John Kramer Sex 10h17'25"-9d35'
Kent,Karen Lyr 18h39'15"30d55'
Kent,Kennith Dale Aql 19h55'12"7d55'
Kent,Lesley Lyr 19h17'15"42d33'
Kent,Melissa Lee Sex 9h53'23"2d37'
Kent,Mike & Mary Cyg 19h26'24"31d49'
Kent,Naomi Cma 6h59'16"-20d4'
Kent,Peter E Boo 14h37'33"48d31'
Kent,Polly Hya 8h51'59"0d32'
Kent,Rodney Douglas Aql 19h0'53"12d19'
Kent,Scott Her 17h12'28"21d0'
Kent,Theresa Marie Psc 0h53'25"11d9'
Kent,William Joseph Lac 22h23'32"37d40'
Kentammi Uma 10h8'33"50d27'
Kenton And 23h18'39"47d39'
Kenton,Roger Ser 15h38'34"4d2'
Kentucky Aur 5h0'43"44d28'
Kentucky SIDS Chapter Uma 11h57'25"51d1'
Kentwood Uma 10h6'0"48d52'
Kenty,Mary Sge 19h13'55"20d36'
Kenworthey-Male Cra 18h8'48"-39d48'
Kenworthy,Betty Uma 8h41'52"61d13'
Keny,Jane Cas 0h18'1"61d28'
Kenyatta,Floyd Her 18h3'11"40d53'
Kenyon II,Arnold O Aur 6h1'44"30d38'
Kenyon's Choral Constellation,Deb Lyr 19h18'22"38d48'
Kenyon,Courtney L And 1h28'12"38d21'
Kenyon,Peter & Jill Boo 15h8'0"52d12'
Kenzelmann,Elizabeth Teresa Peg 22h37'26"27d49'
Kenzie Aql 18h52'0"11d18'
Kenzil 5-24-64 Bernard And 0h31'54"40d8'
Keo Lib 15h2'31"-20d18'
Keo Per 4h24'15"50d57'
Keogler-Happy 50th Birthday,Bill Ori 5h21'14"0d6'
Keohane,Elizabeth And 1h20'20"38d12'
Keohane,Susan Riggs Rice Del 20h15'47"14d27'
Keoki Cadiz Cnc 8h35'34"16d2'
Keoki Laka Honi Mai Ke Moana Ahe Aiu Cyg 20h22'1"38d7'
Keong,Cheang Wai Per 2h29'26"57d23'
Keough Lib 14h57'29"-0d11'
Keough,Barbara "Doodle" And 23h37'33"38d39'
Keough,Kathleen M Cyg 21h20'50"28d17'
Keough,Kristen Cnv 13h21'37"32d18'
Keough,Scott Boo 14h29'24"28d31'
Kep Cet 1h31'54"-12d32'
Kepashosta 100 Umi 15h16'33"68d2'
Keplar,Mary P Uma 10h19'0"59d13'
Kepler-Whiteford, William Her 18h38'1"18d52'
Keppel,Berthold J Kreuz von Stift Sgr 18h50'23"-22d56'
Keppel,Paul Her 16h31'53"38d49'
Keppeler,Brita Dra 17h40'0"61d17'
Kepple,Courtney Elizabeth Tri 1h52'36"26d38'
Kepple,Lawrence Richard Aur 4h46'35"31d18'
Keppler,Louise Jane Umi 16h15'48"74d31'
Ker Marianne Aur 5h54'54"50d31'
Kerasavich,Joseph Ori 5h53'18"10d58'
Kerasotes,George Ser 15h42'35"4d2'
Kerber,Jason "Jason's Smile" Cnv 12h49'59"38d13'
Kerby,Kristin Lyn 7h28'1"38d48'
Kerby,Mae Mon 8h4'1"-5d48'
Kercher,David Matthew Per 2h56'41"40d10'
Kerchner,Jr Star,The Charles F Aur 5h8'20"40d48'
Kerege,Anna Ashley Psc 1h24'34"18d30'
Kerek,Constance Cas 0h40'45"61d0'
Keresey,Kelley Eri 4h32'21"-11d46'
Kereta,Joseph (King) Ser 17h55'20"-14d28'
Kerfin,Carol Lmi 9h51'31"38d54'
Kerho,Michael Fort Her 17h21'35"45d50'
Keri Umi 15h13'17"81d37'
Keri & Shaun's Star Lyr 19h4'46"37d44'
Keri's Guiding Light And 23h27'41"47d13'
Keri's Love Lyr 19h22'55"38d5'
Kerins,Jane M Del 20h19'32"16d21'
Kerlagon,Gary Stephen Aur 5h12'24"41d12'
Kerley Star,The Jackie Peg 23h29'44"16d57'
Kerlova,Olga Aur 5h1'33"29d46'
Kerman,Elliott Stuart Hya 8h9'28"-9d2'
Kermit Eri 4h4'16"-8d10'
Kermit's Light Vul 20h39'33"25d59'
Kermode,Tom Dra 17h7'28"68d38'
Kern"Family Star", Diane-Lorri-Rich-Al Sex 9h48'50"2d35'
Kern,Alexis Victoria Cas 1h38'42"75d25'
Kern,Ann Marie Cyg 21h7'0"30d45'
Kern,Eddy And 1h22'1"33d31'
Kern,Emily Ann Vir 13h28'39"-7d31'
Kern,Jenny Elizabeth Psc 0h55'50"32d20'
Kern,Marcella And 0h11'21"37d52'
Kern,Otto "Wauii" Aqr 21h24'1"-0d1'
Kern,Reinhard Lyr 19h23'48"30d25'
Kern,Richard Cep 0h0'56"69d59'
Kern,Sara Rose And 1h44'39"37d1'
Kernagis,Amy McFall And 2h32'12"42d20'
Kernan,Barb Tri 1h52'18"28d4'
Kernan,Ryan Finley "Balobster" Cnc 8h48'25"32d40'
Kerner,Ulrike And 23h39'13"37d57'
Kernetzke,Kathy Cyg 20h30'41"39d50'
Kernit Ori 5h40'28"-6d49'A
Kernodle,Cindy I Sct 18h56'19"-4d28'
Kernot,Kevin Ori 5h59'32"14d29'
Kerns,Kameron Klay Cnv 13h10'13"32d36'
Kerns,Matthew Richard Her 18h16'40"20d20'
Kerns,Sandra J And 23h36'47"46d22'
Kerns,Stephen Aql 20h19'1"5d34'
Kerns-Nathan,Nickolas Her 17h37'41"27d51'
Kerola & Karola's Enchanted Sojourn Uma 9h46'0"71d6'
Kerr 50 Year Star, Baine & Mildred Cra 18h28'30"-41d9'
Kerr Family Star Lyn 8h56'56"34d4'
Kerr's Passion Mon 6h53'33"-0d57'
Kerr*Mother Star,Anne Lyr 18h49'1"42d12'
Kerr,Bess & Bob Oph 18h17'33"11d14'
Kerr,Bob Peg 23h48'0"8d8'
Kerr,Cameron Thomas Peg 23h43'50"27d38'
Kerr,Carey Eri 3h44'40"-11d15'
Kerr,David T Per 3h33'27"38d16'
Kerr,EP (Trip) Cet 1h19'18"-0d31'
Kerr,Howling George Cnv 13h49'44"32d1'
Kerr,Jean Aur 6h27'45"34d46'
Kerr,Jr,James R & Karen L Cloud Cyg 21h28'49"41d7'
Kerr,Julia Vul 20h39'30"23d35'
Kerr,Kelly Cet 2h55'46"2d17'
Kerr,Laura Ashley Lyr 18h56'58"40d58'
Kerr,Lynn And 0h9'24"38d27'
Kerr,Margie Lyn 7h6'16"52d35'
Kerr,Michael Ray Per 1h42'57"50d35'
Kerr,Morag Dra 17h58'41"58d10'
Kerr,Otho Vul 20h39'16"23d16'
Kerr,Philip Per 3h15'51"40d41'
Kerr,Princess Martha Terrell And 23h44'12"46d49'
Kerr,Rob Del 20h52'45"9d12'
Kerr,Ross Patrick Her 16h27'36"38d7'
Kerr,Sally Cam 4h59'0"70d5'
Kerr,Sarah Aqr 22h26'32"-17d43'
Kerr,Shannon D Lyn 7h54'0"43d23'
Kerr,William R Cam 12h53'0"77d21'
Kerr,Wilma Jean Mon 8h5'8"-8d50'
Kerri Mon 7h24'49"-8d10'
Kerri Peg 22h15'18"8d22'
Kerri Cap 20h30'50"-21d20'
Kerri And 23h45'56"47d13'
Kerri & Tom Forever "96" Aql 18h59'19"12d58'
Kerri Lee Umi 15h48'58"77d20'
Kerri Lyn Cam 12h42'40"76d38'
Kerri,Raymonde And 0h17'55"40d50'
Kerrick's Star Uma 10h19'33"50d58'
Kerrie Uma 14h3'54"54d38'
Kerrie Gru 22h10'44"-55d32'
Kerrigan "Hangin with the Stars",Tammy Mon 7h6'58"-5d22'
Kerrigan,Judith Ann Oph 17h35'35"-23d23'
Kerrigan,Sandra Ann Lota Cam 5h54'17"71d5'
Kerrin - Number One Cep 22h8'43"65d17'
Kerry Sge 19h23'35"18d10'
Kerry Cas 0h49'46"66d29'
Kerry Lyr 18h14'41"45d19'
Kerry And 23h29'57"48d10'
Kerry Lyn 9h9'25"33d21'
Kerry Uma 11h2'20"48d29'
Kerry & Nina,A Heavenly Match Boo 15h47'25"46d5'
Kerry & Ray Vul 20h0'0"25d19'
Kerry - Todd Cma 6h42'33"-13d43'
Kerry Ann & Jason Forever Always Lyr 18h41'30"45d15'
Kerry Clair Cas 0h25'60"61d24'

Kerry E
Vul 19h2'50"22d13'
Kerry Lynn
And 1h34'28"48d51'
Kerry Tod
Cyg 19h24'52"34d2'
Kerry's Astral Hub
Lac 22h1'34"51d31'
Kerry's Beauty
Cas 1h20'47"55d57'
Kerry's Place of Peace
Dra 17h41'47"73d59'
Kerry's Star
Cyg 20h16'14"39d8'
Kerry's Wishing Well
Boo 14h27'53"17d52'
Kerry,Frank
Cyg 20h39'35"30d38'
Kerry,Mandy Jayne
Cyg 19h32'47"37d57'
Kerry-Sacha
Cas 2h34'0"65d9'
Kerryland
Mon 7h4'0"1d4'
Kerryn
Lyr 18h34'26"28d44'
Kersavage,Jeffrey Jon
Dra 16h6'1"65d27'
Kersch,Tom
Dra 12h22'24"71d32'
Kerschbaum
Peg 23h24'31"16d21'
Kerschner,Helen Kale
Oph 18h2'26"11d15'
Kersey,Aubrey Timothy
Boo 15h44'26"50d9'
Kersey,Daniel Hendrik
Hya 8h16'0"3d29'
Kershaw Elysium
Cyg 19h34'60"39d9'
Kershaw,John Andrew Charles
Umi 15h17'14"66d4'
Kershaw,John Hugh D'Allenger
Cep 1h4'14"78d49'
Kershaw,Sammy
Ori 6h4'59"8d11'
Kershaw,Sammy
Hya 9h37'30"-9d47'
Kershaw,Vernon B
Cep 20h26'58"76d39'
Kershner,Bill
Aql 20h0'10"12d15'
Kershner,Jim
Cet 3h15'56"6d13'
Kerstan,Angelika
Cam 12h15'0"80d20'
Kerstin
Tau 4h3'11"0d18'
Kerstin & Kay
Leo 10h0'27"10d55'
Kerstin's Star
Ori 6h11'59"10d12'
Kersting,Heinz
Umi 15h34'0"80d54'
Kersting,Sophie
Leo 11h17'55"-1d57'
Kert,Lillian
Lyr 18h33'27"40d59'
Kertanis,Helena
Sge 19h4'51"19d40'
Kerting,Thomas
Her 17h21'22"40d5'
Kerttula,Jim & Charmayne
Per 3h12'0"54d52'
Kertz,Rose
Aql 19h0'50"-6d42'
Kervian,Barbara
Lyr 18h41'43"42d11'
Kervick,William John
Aur 5h1'11"38d1'
Kervin,Kate & Kellie
Peg 22h14'38"32d55'
Kervio,Chloé
Crb 16h15'0"38d39'

Kerwin"Mr Picker",Kay
Cep 21h22'59"80d13'
Kerys' Star
And 2h31'49"50d33'
Kerzius Major
Per 2h59'38"56d19'
Kerznar,Tricia Ellyn
Cyg 20h55'15"31d47'
Kerzner,Judi
Aur 5h9'59"40d23'
Kes
And 0h45'0"38d39'
Kes Ming
Lyn 7h46'32"45d11'
Kes-Shine Like My Love Forever
Cyg 19h26'1"31d24'
Keseley,Marie
Vul 19h15'30"21d41'
Keshner,Derrie Ann
Vir 13h7'16"-8d30'
Keshock,Charles Robert
Boo 14h17'19"15d58'
Kesler & Family,Steve & Kim
Cet 0h58'58"-5d25'
Kesler,Duane
Her 16h39'29"48d3'
Kesner,Daniel & Denise
Cyg 19h56'14"38d6'
Kesner,Leslie Rubin Andrew & Nicole
Tri 1h54'34"25d54'
Kesolits,Thomas J
Aur 5h27'59"38d24'
Kess,Hans
Cep 20h39'16"60d34'
Kessel,Bernd
Del 20h53'49"3d34'
Kessel,Rachel
Cyg 19h30'55"38d15'
Kesselring,Beatrice & Joseph
Peg 22h4'16"2d28'
Kesselring,Rick
Per 1h44'58"50d10'
Kessey,Kathy
Aql 19h28'1"10d23'
Kessinger,Kristin Marie
Peg 22h49'50"20d13'
Kessler (Pidge),Jaclyn Nicole
And 1h15'34"33d44'
Kessler's,"Wedding Star",Drew & Aimee
Cyg 20h58'11"40d11'
Kessler,Christina
Lyn 8h5'56"58d9'
Kessler,Christopher A
Per 3h8'42"46d9'
Kessler,Daniel Timothy
Her 16h54'0"41d7'
Kessler,Denis
Cam 13h10h36'81d31'
Kessler,Elizabeth Nicole
Aur 7h20'45"37d37'
Kessler,Erika Gabrielle
Cep 23h6'1"61d43'
Kessler,Fred Al Cliss
Dra 17h54'1"63d50'
Kessler,George "Snuckems"
Cnv 14h0'13"42d16'
Kessler,Kimberly Ann
Cyg 21h31'52"38d47'
Kessler,Lawrence Jonas
Cnc 8h59'31"32d28'
Kessler,Margrit
Mon 7h1'51"-6d26'
Kessler,Mary L
Tau 4h12'49"6d4'A
Kessler,Nicholas
Cnv 13h48'30"40d26'
Kessler,Norbert
Sge 19h8'9"17d32'
Kessler,Paul
Cep 21h33'15"55d11'

Kessler,Peter - One In a million
Cen 11h53'37"-49d35'
Kessler,Richard Callie
Peg 22h0'54"2d21'
Kessler,Richard Gary
Dra 16h52'1"61d57'
Kessler,Robert Charles
Peg 23h46'0"31d27'
Kessler,Sr,Lawrence
Tau 4h12'49"6d4'B
Kessler,Vance E
Cam 3h34'47"61d6'
Kessler,William J
Boo 14h36'56"51d52'
Kesslering,MF
Uma 9h24'32"48d27'
Kest,Alan R
Ser 16h17'59"1d11'
Kesteloot,Roger
Her 18h3'39"31d17'
Kestenbaum,Alice
And 2h3'36"45d23'
Kestenbaum,Joshua
Hya 10h23'54"-18d58'
Kester
Cyg 20h24'10"39d51'
Kester,Anthony David
Sgr 18h51'40"-35d11'
Kesterman,Michelle
Tri 2h4'34"35d9'
Kestersone,Linda & Steve Saferin
Aql 19h34'29"1d20'
Kestner,James
Dra 16h47'33"68d23"
Ketan
Aur 7h2'13"36d0'
Ketaŷ,Kenneth
Ori 5h58'57"10d38'
Ketch
Cep 20h42'43"76d46'
Ketcham,Shauna Leigh
Aql 19h30'25"7d46'
Ketchie,Robert
Sct 18h37'50"-4d27'
Ketchum,Hal Michael
Per 2h36'41"50d4'
Keth,Willy
Cnc 8h31'32"30d6'
Kethelen
Ori 5h19'57"11d26'
Keti
Dra 18h10'23"67d43'
Keton
Crb 16h20'0"32d41'
Kett,Margaret
Lyr 18h19'41"43d48'
Kett,Paw
Aur 5h4'38"37d52'
Kett,Peter Wilson
Ori 6h1'41"1d36'
Kettaneh,Marina Grumpetta Smilodona
Lyr 18h30'10"43d7'
Kettelhut,Hans K & Alice H
Boo 14h35'1"51d15'
Kettelwell,Matthew Dustin
Ser 15h37'42"8d19'
Ketteman,John Robert
Peg 21h59'0"7d49'
Ketterer,Tom, Christine,Ryan
Del 20h18'39"10d20'
Kettering,Donald L
Cet 1h17'43"-5d53'
Kettie
Lyn 7h10'1"58d45'
Kettle,Vic
Cyg 19h27'31"36d25'
Kettlekamp,Kate Emily
Lyn 7h44'13"40d26'
Kettler,Carol Lynn
Cas 1h9'47"61d17'

Kettler,Ulrike
Sco 17h51'48"-31d7'
Kettler-371704, Brandon Douglas
Dra 19h4'20"65d26'
Kettley,Amy
Umi 13h25'51"72d24'
Kettlitz,Steffen
Uma 9h28'44"47d30'
Kettunen,Shaun Thomas
Aur 6h15'10"35d12'
Ketz,John Douglas
Cnv 12h43'15"51d56'
Ketzer,Robert A
Boo 14h32'36"46d33'
Keuler,Ron
Ori 5h56'58"10d37'
Keune,John K
Dra 12h19'1"76d25'
Keuntje,Barbara
Cas 0h25'1"60d56'
Keupen,Manfred
Aql 18h56'38"16d36'
Kev D
Boo 14h39'32"40d33'
Kevelly Pro35
Aql 20h6'36"1d12'
Keven/Khalil
Cep 20h22'44"60d35'
Kevie,Richard Monroe
Cap 21h1'26"-23d3'
Kevifer
Sge 19h14'17"16d21'
Kevin
Dra 17h58'35"61d19'
Kevin
Her 17h31'52"27d5'
Kevin
Cmi 7h56'30"8d31'
Kevin
Her 16h31'32"30d36'A
Kevin & Angie Always
Cyg 19h47'11"29d51'
Kevin & Gina
Cam 13h0'49"78d28'
Kevin & Heather
Cyg 21h33'24"36d43'
Kevin & Jenny
Cyg 19h18'18"48d46'
Kevin & Kristi Forever
Cyg 21h22'33"41d8'
Kevin & Melissa
Sge 20h0'17"19d19'
Kevin & Sarah
Cyg 19h28'0"35d53'
Kevin & Theresa's Star of Eternity
Per 3h25'51"51d4'
Kevin 85
Col 6h26'29"-33d50'
Kevin Christopher
Lac 22h13'20"49d15'
Kevin Joseph
Ori 5h55'24"15d5'
Kevin Little Angel
Cam 7h54'45"61d18'
Kevin Michael George
Lac 22h9'28"40d28'
Kevin Star I Love You More
Her 17h25'55"37d30'
Kevin Thomas
Vir 13h31'22"-7d41'
Kevin's Dream
Lac 22h2'41"49d59'
Kevin's Millennium Star
Uma 10h5'0"58d40'
Kevin's Sparkle
Aur 6h7'29"38d6'
Kevin's Star
Uma 8h39'25"68d40'
Kevin's Tomorrow
Cep 20h42'49"68d13'
Kevin's Wave
Cmi 7h25'0"1d13'

Kevin's Wish
Aur 5h1'34"49d13'
Kevin's Wish
Lac 22h37'0"50d31'
Kevin's World
Dra 17h30'1"61d26'
Kevin,Michael J
Her 16h44'56"32d37'
Kevin,The Holder Of My Heart
Dra 15h7'24"57d21'
Kevina,Klaus Christiane Johanna
Sco 17h5'49"-31d49'
Kevlar
Per 3h7'1"50d23'
Kevlin III,James Courtney
Per 2h24'1"58d37'
Kevlin,Joseph James
Cam 5h58'25"68d32'
KevMar
Umi 16h24'0"70d60'
Kevorkian,David
Aur 7h18'29"40d13'
Kevorkian,Ron Cary Kristopher & Dean
Aur 6h39'22"38d18'
KevSu- May We Always Shine As One,The
Dra 18h32'18"50d9'
Kewin,Gary
Umi 14h36'31"68d56'
Kewin,Simon
Per 4h3'7"35d27'
Kewley,Eve Suzanne
Eri 5h5'39"-10d32'
Kewley,Nicholas Dahlman
Aur 6h12'47"35d44'
Key West
Cyg 19h57'29"40d28'
Key,Annelle Hadley
And 0h32'52"38d35'
Key,Barbara Clair Thaxton
Lac 21h59'38"37d46'
Key,Christopher
Tri 1h32'0"31d15'
Key,Dr
Cep 1h19'39"77d44'
Key,Elizabeth Russell
And 0h29'42"38d43'
Key,Jeremy Shane
Dra 19h41'21"68d9'
Key,Josephine
Cru 12h35'15"-59d38'
Key,Lisa
Cyg 20h18'25"38d40'
Key,Mr & Mrs V E
Aql 19h46'10"14d19'
Key,Robert
Ser 15h25'0"1d45'
Key-Keizer,José
Cam 4h29'25"58d22'
Keyes,Amanda Nicole
Mon 7h7'2"-5d5'
Keyes,Ashley Stanton
Mon 7h37'52"-6d2'
Keyes,Carolyn M "Carrie"
Cas 0h29"59"60d18'
Keyhani,Dr Tina
Aur 5h26'0"40d50'
Keylanshel
Aql 20h5'20"6d55'
Keys,Margaret
Cas 1h2'52"61d46'
Keys,Roger,Mary, Matthew,Megan
Crb 15h39'27"29d1'
Keyser,Pooh's Star For Gloria Joan
Del 20h14'60"15d32'
Keyser,Sr,Robert Louis
Per 4h45'47"41d2'
Keystal
Cam 3h55'25"58d14'
Keystone TLC Duo
Hya 8h12'17"4d50'

Keyt,Diane
Lyn 7h39'42"58d45'
Keyt,Donna
Cas 0h58'56"50d33'
Keyt,Natasha
Cas 0h9'55"58d28'
Keyt,Ruth
Lyn 7h39'0"52d5'
Keywan,Pamela
Lyn 8h43'0"38d57'
Kezar,Jr,Thomas F
Dra 20h16'42"63d15'
Kezlarian,Barbara Jo
And 23h43'37"47d33'
Khalaf,Faride
Ori 5h53'27"15d9'
Khalaf,Tracy
Cas 23h35'57"61d57'
Khaleel,Richard
Ori 6h6'18"6d27'
Khaleghl's Star
Eri 5h1'0"-4d31'
Khalid Mohammad Al-Ali
Ori 6h6'28"8d13'
Khalifeh,Alexander
Dra 14h47'15"60d56'
Khalil,Mariam J
And 23h24'16"48d0'
Khalili,Daniel
Uma 11h13'12"53d2'B
Khalili,Hadassah
Uma 11h13'12"53d2'A
Khalsa,Ann-Julie Kirkpatrick
Peg 22h40'48"34d53'
Khamis-"Lilly Marlene" Lillain Eiwaz
Peg 22h47'19"35d12'
Khan,Ayub
Cyg 21h46'1"36d28'
Khan,Azura
Ori 5h26'37"-2d26'
Khan,Derek
Umi 13h42'23"76d39'
Khan,Josephine Anna
Lyr 19hh22'32"40d8'
Khan,Lailah N
Peg 22h0'21"33d22'
Khan,Samina Phoenix
Cyg 21h7'30"39d27'
Khan,Shazia
Cep 20h35'19"76d45'
Khan,Waqqas H
Uma 9h59'30"55d59'
Khan,Wena
Uma 12h0'53"43d32'
Khan,Xenos
Umi 14h12'41"69d50'
Khan,Zacharia Jamal
And 2h0'51"37d12'
Khan,Zafar Ali
Cep 21h10'1"61d35'
Khan,Zainool
Peg 23h5'1"33d9'
Khandekar,Sanjeev
Cet 3h9'34"1d36'
Khanna R2RS2S,Anish K
Cam 3h27'24"61d5'
Khantalu
Cam 5h5'39"65d31'
Khanzadian,Lisa Ann
Cas 1h39'0"70d36'
Khashayar
Uma 10h34'15"57d30'
Khashoggi,Adnan
Ori 5h18'37"8d38'
Khatchadourian,Lois
Cas 0h55'48"67d5'
Khatibi,Niloofar
Mon 8h6'3"-3d21'
Khayman
Cam 14h14'1"81d10'
Kheel,Richard
Her 18h9'30"31d43'
Kheit,Robert
Her 18h7'18"31d40'

Khenaffou,Daniel et Paulette
Boo 15h0'13"8d32'
Khinda,Philip & Pamela
Crt 11h3'47"-11d52'
Khlystov,Anatoli
Uma 10h35'1"57d20'
Khong,Michelle Yoon Chee
Cra 18h17'21"-39d24'
Khoonyam,Catherine
Lup 15h19'56"-41d36'
Khorshid Al- Balushi
Lac 22h0'29"37d37'
Khoury,Angelina Agnese
Leo 9h23'1"30d8'
Khoury,Aurore
Lyr 18h58'0"34d44'
Khoury,Carole Louise Sacco
Gem 6h51'25"31d31'
Khoury,George M
Cet 2h32'35"6d6'
Khoury,Imad Nemr
Aur 4h49'36"40d5'
Khoury,Julianna Allegra
Peg 23h26'15"11d53'
Khoury,Khalil Paul
Ori 5h57'42"10d14'
Khoury,Laura Beth
Aur 5h28'0"40d45'
Khoury,Paul
Her 16h51'1"40d23'
Khoury,Yves-Laurent
Peg 22h39'1"21d8'
Khun,Ariane
Ori 6h6'34"10d20'
Khuner,Janine
Aql 19h0'42"-8d58'
Khushdil,Aman
Cnv 13h59'13"46d36'
Ki Ki
Ori 4h59'24"12d30'
Kia
Eri 5h4'22"-10d9'
Kian
Cnv 12h16'11"37d16'
Kianjah 72892
Boo 15h13'44"50d28'
Kianoucatioun
Oph 18h17'11"10d14'
Kibat,Marce
Peg 23h1'31"20d45'
Kibat,Nicole
Aql 19h4'33"-5d51'
Kibbey in Orbit
Aql 19h55'1"0d27'
Kibbey,Richard D
Hya 8h31'47"-8d22'
Kibler,MD,Gordon Eugene
Ori 5h56'12"15d44'
Kibria,Masud
Per 4h23'0"50d40'
Kic's Favorite
Cet 3h7'0"2d8'
Kicak,M
Uma 12h51'43"58d30'
Kicca & Giovanni
Scl 23h24'42"-27d43'
Kicinski,Georgana M Kreidler
And 1h40'21"40d15'
Kicker
Boo 15h20'1"52d51'
Kicki,A & E
Cyg 20h21'30"38d41'
Kicklighter,Willis
Cnv 13h44'41"30d58'
Kiczula,Leah Nicole
Cma 6h59'27"-19d4'
Kid
Cyg 20h7'33"40d21'
Kid Coleman
Uma 11h27'49"40d31'
Kid Cowboy
Per 3h5'40"47d26'
Kid,The
Del 20h15'20"10d21'

Kidanan
Cap 21h44'0"-22d52'
Kidby,Samuel Robert
Mon 8h5'34"-6d26'
Kidd,Caskie Dalton
Ori 6h7'30"20d34'
Kidd,David Justin
Ori 6h7'20"3d38'
Kidd,Gene
Cet 2h34'35"4d59'
Kidd,Gordon
Per 3h2'51"40d35'
Kidd,Jane (JBS)
Umi 14h19'18"67d29'
Kidd,Judson
Her 15h51'35"46d44'
Kidd,Lauren Nicole
Boo 13h59'20"19d46'
Kidd,Lori Dare
Cet 1h16'55"0d43'
Kiddell,Sydney Arthur
Her 18h9'15"45d42'
Kidder"Star Of Love", Irene
Cyg 19h24'0"33d5'
Kidder,Ray Skylstead
Oph 17h15'53"-20d39'
Kidder,Steven
Mon 7h11'39"-10d42'
Kidder,The Love Of Mary & Hartwell
Vul 19h45'0"28d30'
Kidder,Trudy
Boo 14h53'54"32d30'B
Kiddo
Uma 10h15'57"53d8'
Kido,Melanie Emiko
Uma 10h1'26"58d14'
Kiecker,Pam
Peg 22h0'58"33d44'
Kiecolt,Georgia
Sex 10h33'1"2d54'
Kiefer,Andrew Stephen
Gem 6h44'36"14d9'
Kiefer,Helmut
Lac 22h5'39"51d20'
Kiefer,Hermann
Lyr 19h18'25"30d48'
Kiefer,John F
Her 16h54'39"32d25'
Kiefer,Maxine E
Cyg 21h32'42"41d10'
Kieffer,Thomas
Lac 22h18'57"50d9'
Kiefner,Anthony & Kellie
Cyg 21h24'41"40d56'
Kiehle,Brittany
Tau 4h38'32"1d28'
Kiehle,Danique
Sgr 19h38'27"33d35'
Kiehle,David & Natalie
Cnc 8h51'12"11d45'
Kiehle,Fred & Rosalie
Leo 11h5'17"-0d52'
Kiehn,Daniel James
Ori 5h55'45"15d10'
Kiel Star-Robert's Pal,Matthew Thomas
Lac 22h13'23"54d42'
Kiel,Fern Oquist
Hya 9h4'1"4d1'
Kiel,JoAnn
Cas 0h29'55"66d45'
Kiel,Peter
Peg 23h28'1"11d20'
Kiel,Renate
Lib 14h59'19"-22d52'
Kiel,Richard Philip
Hya 8h54'21"0d47'
Kiel,Walter Richard
Hya 9h3'14"3d41'
Kielceski,Katherine Mary
Cam 4h12'0"68d38'
Kieley,Carol Marie
Lyn 8h30'22"45d49'

Kielian,Scott A
Dra 16h53'14"70d16'
Kielsmeier,Arnold F
Aql 19h47'1"12d53'
Kielty III,John L
Boo 14h11'0"32d24'
Kieltyka's Star,Pat
Aur 4h42'28"30d31'
Kieltyka,Bruno
Aur 6h29'0"31d26'
Kiely,Jennifer
Cas 1h17'29"62d25'
Kiely,Stephen J
Her 16h31'14"41d34'
Kiener & Wittlin
And 1h23'58"42d33'
Kienholz,Michael B
Dra 11h45'29"72d15'
Kienzl,Friedrich
Ari 2h29'25"22d29'
Kiepfer,Dr Richard
Oph 17h21'10"11d33'
Kier,Nesandra C
Tri 2h4'58"32d26'
Kiera Anne
Lyn 7h55'58"40d7'
Kiera Michele
Cas 0h32'11"61d40'
Kiera The Dancer
Cyg 21h32'17"31d37'
Kieran
Crt 11h3'57"-13d57'
Kieran
Aur 6h0'11"36d46'
Kieran's Krystal
Aur 6h26'18"37d57'
Kieran,Karen E
And 0h13'24"47d57'
Kierantis,Melaniesha Devonicus
Cyg 20h15'1"39d30'
Kiermeier,Erwin
Sco 16h39'19"-44d39'
Kiernan III,Daniel Edward
Dra 16h32'25"72d50'
Kiernan,Alison Louise
Lyr 18h14'36"43d19'
Kiernan,Charles F
Aur 6h20'0"30d16'
Kiernan,Florence L
Cam 5h34'52"68d20'
Kiernan,Peter
Lac 22h17'0"51d36'
Kiernan,Phyllis
Com 13h7'15"20d36'
Kiernan,Taylor Skye
Peg 23h31'50"21d38'
Kiersky,Miss Loretta J
Cas 0h59'23"61d26'
Kierstead,Carl Walter
Peg 22h10'39"28d4'
Kiersten Marie Romance & Enchantment
Cyg 20h28'1"50d32'
Kiersten's Inferno
Lyr 18h59'12"27d52'
Kierstyn,William Allen
Lmi 10h23'55"28d37'
Kies,Bill
Ser 15h14'15"9d7'
Kies,Joshua Michael
Oph 17h36'34"-0d45'
Kiesche,Marlene & Stephen
Psc 1h36'33"27d60'
Kiesel,Carson L
Cet 2h26'17"3d3'
Kiesel,Grace Therese
Lyn 8h50'12"35d5'
Kiesel,Hertha
Psc 23h22'34"1d45'
Kieser,Mary Ann
Cas 23h24'20"61d37'
Kiesewalter,Sabine
Gem 7h33'40"20d28'

Kiesewetter,Thomas Aur 6h32'32"38d42'
Kiesgen,Michael Peg 0h12'12"18d45'
Kieslich,Otto J Cep 22h26'1"58d45'
Kiesling,Jenny Yohanna Peg 22h52'15"35d18'
Kiessling,Anthony Joseph Lac 22h36'42"53d4'
Kiessling,Dr Louise Cas 0h48'45"69d9'
Kiester,Kevin Her 16h19'46"24d45'
Kieswether,Jr,Leo M Umi 16h23'1"71d20'
Kieth & Dagny's Celestial Home Sge 19h53'40"16d13'
Kietzke,Christopher Boo 15h3'28"32d11'
Kiewel,Kurt J Aur 6h31'57"38d7'
Kiffen Lyn 6h55'32"58d11'
Kiffer Hya 8h52'30"-0d28'
Kiger,Bruce Gem 6h53'3"18d21'
Kiggins,Lawrence Robert & Deborah Anne Mon 7h44'1"-3d24'
Kigmalis,Christos Cam 4h0'57"67d35'
Kihara,Yoko Mon 7h6'0"0d51'
Kihiczak,Nadia Cas 0h55'15"60d20'
Kijek,Al Lac 22h28'57"53d56'
Kijorski,Kathy Cas 0h30'28"67d48'
Kijowski,Dennis Aql 19h14'6"-11d4'B
Kik & Leo Cyg 20h57'1"37d57'
Kiker Aql 20h12'38"11d6'
Kiki Umi 16h14'45"74d27'
Kiki Cra 11h14'25"-17d44'
Kiki's Star Mon 6h30'31"-8d7'
Kiki-Chris Umi 17h4'36"76d48'
Kiki-le Stelle Italiano Lmi 10h55'13"33d30'
Kilagallon,Kameron Cyg 19h28'1"50d2'
Kilberg,Scott Robert Her 16h58'25"30d32'
Kilbert,Florence Aql 18h54'53"8d32'
Kilburn,Janet Cam 8h4'58"71d27'
Kilburn,Mark D Lmi 10h35'40"28d36'
Kilburn,Paul A Ori 4h55'24"0d52'
Kilburn,Roland "Rocky" Sex 9h55'57"-0d23'
Kilbury,Phillip E Aur 7h10'53"36d26'
Kilby's Quest Oph 17h38'3"-21d33'
Kilcarr,Logan Cnv 13h55'10"40d17'
Kilcourse,Yvonne Lac 22h4'60"51d40'
Kilcourse-Vrondos, Kathleen Cas 0h29'54"60d16'
Kilcrease,Janet Del 20h35'54"18d58'

Kilcullen,Jr,John Cep 23h2'0"70d9'
Kildale,Bryan Austin Uma 10h41'53"47d36'
Kilduff,R David Per 1h44'24"53d34'
Kilduski,Dad Lawrence Hya 8h34'56"5d48'
Kile,Jr,William H Boo 14h24'17"39d20'
Kile,Matthew Her 16h47'18"35d59'
Kiley II,Kathleen Eri 3h19'0"-6d54'
Kiley,Kelsey Joanmarie Vir 13h30'55"-3d8'
Kiley,Nadine Vul 19h6'47"24d42'
Kilfoil,Bradley Boo 13h38'24"20d8'
Kilfoyle,Crawford Hurst (Smiles) Oph 17h31'1"-22d13'
Kilfoyle,Jason Michael Hya 8h13'58"5d24'
Kilgarriff,Lynn Sge 20h3'1"20d12'
Kilgore,K L Peg 23h1'29"10d10'
Kilgore,Maggie Irene Aur 6h29'18"38d34'
Kilgore,Peggy R Lyn 7h44'50"41d60'
Kilgore,Rhonda Ori 6h9'22"6d3'
Kilgore,Thomas Paul Cep 22h38'13"63d37'
Kilian,Andrea Duncan Dra 19h55'51"61d1'
Kilian,James C Her 16h46'11"32d43'
Kilian,Roy A Boo 15h6'0"40d2'
Kilishek,Gary Thomas Cep 22h34'44"61d32'
Kilkenny,Jack Mon 6h43'54"10d55'
Kilkenny,Patrick Oph 16h56'1"11d13'
Killebrew,Mitchum Cet 2h15'30"5d60'
Killeen,Ashley Nicole Cas 1h3'57"50d14'
Killeen,Evan Patrick Psc 1h21'39"20d43'
Killeen,Robert Leo Aur 6h32'21"31d42'
Killeen,Sean B Ser 15h53'56"-2d18'
Killen Star,The Ken & Cathie Boo 14h28'39"40d15'
Killen,Buddy Peg 22h42'0"34d4'
Killer Uma 9h15'0"70d46'
Killer aka Ann's Shnookerdoodle Ori 5h56'11"8d59'
Killer's Star-Griffin Cet 2h31'0"-0d36'
Killgore,Georgey Cnv 12h22'15"47d3'
Killian,Charles Sct 18h41'12"-7d26'
Killian,Dr Samuel Theodore Lac 22h10'22"50d25'
Killian,Jack Hubert Anne Hya 8h11'19"4d45'
Killin,Dolores Ann And 23h38'15"46d34'
Killinger,Shawn Marie Cyg 20h28'41"37d48'
Killingsworth,Cam Uma 10h55'47"60d44'

Killingsworth,Greg Her 18h5'39"28d37'
Killmeier,Ike Mary Carly Eddie Aql 19h59'43"15d41'
Killper-Krause, Gertraud Vir 12h5'48"-2d7'
Kilmer,Jr,Harry Her 18h7'57"38d22'
Kilnes,Dag Einar Ori 5h46'1"10d8'
Kilolo Eva-Marie Cam 6h47'36"68d37'
Kilpatrick,Sheila Ann And 0h50'25"34d49'
Kilpatrick,Wayne R Cet 0h33'1"0d18'
Kilroy,Paul Seamus Lac 22h19'14"49d35'
Kilt,Thomas Francis Crt 11h21'33"-16d36'
Kilter,Rita Ann Lyn 7h40'21"39d42'
Kiltie Lmi 9h57'29"34d37'
Kilyanek,Frank Joseph Cmi 7h23'25"8d24'
Kilzhen,Dennis Cep 23h19'51"80d7'
Kim Lyn 7h50'0"39d43'
Kim Cet 3h18'0"3d19'
Kim Uma 11h59'49"51d37'
Kim Peg 23h30'26"18d34'
Kim Mon 6h36'32"6d3'
Kim And 0h45'43"28d22'
Kim Peg 22h35'0"8d35'
Kim Cas 0h41'15"64d48'
Kim Cam 12h43'48"80d2'
Kim Lac 22h0'34"40d31'
Kim & Andy Cyg 19h29'56"36d9'
Kim & Anthony-True Love Always Sge 20h17'45"16d9'
Kim & Greg's Star Cra 18h20'3"-43d51'
Kim & Jimmie Crb 15h57'49"29d31'
Kim & Jon's First Valentine's Day 1994 Mon 8h3'36"-0d4'
Kim & Ken's Pathways 1 Cyg 21h7'18"30d48'
Kim & Nichole,Forever Gem 7h0'44"11d18'
Kim & Scott Always Cyg 19h47'52"30d16'
Kim Alessandra Cam 5h55'26"67d59'
Kim Alexsandra Ori 5h56'27"15d36'
Kim Andrea And 1h51'11"36d11'
Kim Anh Cas 1h16'30"60d34'
Kim Ellen Lyr 19h7'44"37d36'
Kim et Elsa Cam 5h46'55"67d45'
Kim I Only Have Eyes For You Tau 5h53'1"23d57'
Kim June 25 Lyn 8h58'51"43d5'
Kim M 6-14-93 Peg 22h16'34"7d47'

Kim Marie Cyg 20h6'0"39d56'
Kim Tav Peg 23h3'11"33d55'
Kim'N Dale Uma 12h1'0"38d26'
Kim's "Wishing Star" Cam 5h43'29"61d29'
Kim's Birthday Star And 22h59'0"51d48'
Kim's Burning Gem And 23h34'48"44d19'
Kim's Diamond Peg 22h0'0"28d14'
Kim's Diamond In the Sky Leo 10h0'14"14d5'
Kim's Heavenly Dreams Crb 15h30'58"30d37'
Kim's Karat Cam 8h15'51"81d20'
Kim's Piece of Heaven Cet 0h54'60"1d23'
Kim's Place Cas 0h55'58"58d33'
Kim's Special Star Cas 0h54'39"70d46'
Kim's Star And 0h54'17"22d23'
Kim's Star Cas 0h25'54"61d28'
Kim's Star Aql 19h16'48"15d1'
Kim's Star Cyg 19h33'15"34d18'
Kim's Wishing Star Peg 22h5'44"27d24'
Kim,Cynthia Tri 2h1'37"35d11'
Kim,Dr Jaemin Aur 4h49'59"51d24'
Kim,Hanna Marie Vul 19h46'0"23d14'
Kim,Helen Lyr 18h36'12"42d51'
Kim,Jean H And 2h31'23"50d19'
Kim,John Augustus Ori 5h3'26"-2d0'
Kim,Kristine Cas 1h10'50"60d37'
Kim,Kwan Cam 9h6'0"80d27'
Kim,Masami Gem 6h58'16"13d51'
Kim,Mr & Mrs Andrew B Lyr 19h23'42"37d35'
Kim,Sharon Sylvia Cas 1h46'42"71d2'
Kim,Soo Yeun Cas 2h20'47"73d16'
Kim-Eck Peg 23h3'43"31d59'
Kim-Star Crt 11h23'5"-13d43'
Kim-The Little Goober Cas 1h24'24"71d11'
Kim-Zak 93 Cma 6h50'15"-19d48"
Kimba Lyn 0h27'15"47d10'
Kimba And 1h1'49"37d56'
Kimba Lmi 9h52'13"34d16'
Kimball 2-10-96,Clare Coughlin Tri 2h8'36"33d29'
Kimball,Albert V Uma 11h43'19"33d36'
Kimball,Arthur Aur 5h11'55"42d47'
Kimball,Edward Her 17h7'38"20d49'
Kimball,Jr,Paul Clark Aur 5h31'40"38d43'

Kimball,Jr,Ralph E Lyn 6h55'38"59d21'
Kimball,Julius Her 17h14'53"44d56'
Kimball,Kathleen Patricia Cet 2h45'47"1d9'
Kimball,Matthew James Per 3h1'54"40d45'
Kimball,Patricia And 1h16'1"38d24'
Kimball,Sara Elizabeth And 0h56'54"37d33'
Kimbell,Wayne Leo 10h35'21"15d19'
Kimber Mon 7h1'56"-1d24'
Kimber Leigh Lib 15h27'20"-10d11'
Kimber,Hakkedesh Tri 2h19'47"30d60'
Kimber,Ivy & Bill Cyg 20h20'26"40d36'
Kimberlea Mon 7h4'17"-5d12'
Kimberlee Nicole Mon 6h57'12"-10d36'
Kimberley And 23h3'20"51d26'
Kimberley Abigail Lyr 18h18'51"44d49'
Kimberley's Star Cam 8h5'10"80d25'
Kimberli Del 20h12'20"10d35'
Kimberly And 2h25'12"44d58'
Kimberly Ori 5h55'45"17d52'
Kimberly And 0h22'34"44d32'
Kimberly Equ 21h0'17"10d38'
Kimberly Lyn 7h11'36"50d18'
Kimberly Psc 1h0'28"18d54'
Kimberly Mon 6h24'34"8d56'
Kimberly Lyr 18h17'1"44d22'
Kimberly Peg 23h37'41"10d55'
Kimberly & Anthony Loving Forever And 0h6'55"37d34'
Kimberly & Chris Cyg 21h12'55"28d46'
Kimberly & Dan Aur 5h18'0"42d18'
Kimberly (Goober), Joshua Per 2h0'1"56d38'
Kimberly Ann Lyn 7h47'10"40d48'
Kimberly Ann Uma 9h20'46"56d4'
Kimberly Ann Crb 16h2'44"33d7'
Kimberly Ann 6/25/93 Cyg 20h55'0"39d58'
Kimberly Anne Peg 22h42'29"26d33'
Kimberly Danielle Cas 2h3'1"67d33'
Kimberly Dawn Del 20h20'47"10d22'
Kimberly Dawn Mon 6h57'26"0d28'
Kimberly Dorrien Cyg 21h14'0"38d20'
Kimberly Hope And 1h34'25"36d14'
Kimberly Ingrid Cas 0h18'50"59d59'
Kimberly J Vul 19h43'17"24d12'

Kimberly Jill Gem 7h6'26"21d33'
Kimberly Joy(Trouble Maker) Lyn 7h58'22"40d25'
Kimberly K Ori 5h33'56"-0d15'
Kimberly K Cas 0h28'23"62d59'
Kimberly K 1-13-67 Lyr 19h17'1"42d48'
Kimberly Mae Cyg 19h59'19"58d35'
Kimberly My Love Cyg 20h24'58"41d39'
Kimberly Nicole Cas 0h20'20"70d6'
Kimberly Rene Leo 10h26'60"10d25'
Kimberly's Image Com 12h23'1"32d4'
Kimberly's Star Ori 5h10'39"-5d24'
Kimberly,Bryan Aur 6h26'41"37d48'
Kimberly,Jill Cas 1h13'1"61d52'
Kimberly-Campaspe Tau 4h7'48"22d38'
Kimble Star,The Brigitte Nicole Lyn 7h57'60"35d16'
Kimble,Diane Marie Wilson Ori 6h1'55"8d44'
Kimbo Cam 6h11'27"68d25'
Kimbo "40" Boo 14h36'10"7d57'
Kimbri Cyg 21h31'33"37d45'
Kimbrough,Dr Richard Walter Cnv 12h18'44"36d58'
Kimbycoot Leo 10h51'16"-5d38'
Kime,Aaron Maxwell Uma 9h15'15"51d60'
Kime,Devin Whitney Vul 19h21'22"25d41'
Kime,Kevin Aur 6h54'17"38d38'
Kime,Megan Leslie Vul 21h23'1"26d31'
Kimi Peg 23h29'1"28d42'
Kimi Sue Cas 2h59'28"58d49'
Kimi's Fire And 23h18'30"41d40'
Kimiaki,Kuroki Lib 15h14'36"-20d32'
Kimich,Michael Ray Cet 2h10'21"3d1'
Kimihiro & Tomoko Ari 2h28'27"21d51'
KimJoe Peg 21h57'0"22d31'
Kimley,Elizabeth Frances Aur 7h3'17"40d26'
Kimlin "Lukie",Gloria Mae Uma 11h3'55"46d8'
Kimlyn Lyn 7h10'53"50d23'
Kimm,Alice Y Cap 20h39'27"-16d21'
Kimmel,Charles Cnv 12h23'21"34d53'
Kimmel,Kris A Aql 20h17'35"0d57'
Kimmel,Larry C Her 16h44'42"32d60'
Kimmel-Our Beloved Dog Kimberly (Butch) Cnv 14h0'24"31d24'
Kimmell,Kristin S Cnv 13h36'35"46d17'
Kimmelle Com 12h7'1"26d53'

Kimmer Tri 2h8'59"31d21'
Kimmer,The Lmi 9h59'0"40d11'
Kimmidge Peg 23h2'0"31d32'
Kimmie Cas 0h42'20"69d45'
Kimmie Peg 23h49'43"10d37'
Kimmie 26 Oph 18h2'24"11d43'
Kimmie's Star Dra 19h59'0"60d14'
Kimmy And 2h18'34"46d46'
Kimmy Bug Cam 10h7'41"82d22'
Kimmy-Kim-Kim Lmi 10h30'18"38d32'
KimmyDorothyBuckyJeanA Cyg 20h44'1"38d26'
Kimo Her 16h39'0"48d58'
Kimo Cru 12h54'48"-60d32'
Kimo Kaua Noi Mon 6h25'0"3d36'
Kimpel,Herr Boo 14h24'26"10d44'
Kimpel,Sue Lyn 7h53'14"42d36'
Kimpel,Wayne Allan Lac 22h19'17"37d45'
Kimrey,Sterling Drew Aql 19h53'37"15d31'
Kimrob Aql 18h57'33"-6d28'
Kims Angel Peg 22h57'53"27d49'
Kimsey 1/4,Asterion Umi 13h18'47"71d40'
Kimsey's,The Dra 17h24'1"72d24'
Kimsey,Evelyn Lyr 19h23'0"35d31'
Kimsey,Tyler Vul 19h3'32"21d38'
Kimura Twins,Tasuku & Kanaini Oph 18h5'48"11d4'
Kimura,Masayuki Cam 10h51'34"81d19'
Kimya Her 17h9'33"42d5'
Kina And 22h56'47"50d6'
Kinard,Fred Scott McFarling Aql 20h11'33"10d32'
Kinard,Kenneth J Dra 17h4'42"68d51'
Kincaid,Barry Mon 8h7'28"-1d19'
Kincaid,Jim Sct 18h55'37"-6d8'
Kincaid,June Sge 18h55'47"19d35'
Kincaid,Macy Lynn Peg 21h57'24"31d44'
Kinch,Carissa M Lyr 18h49'16"41d54'
Kinch,Rochelle L Lyn 7h45'54"44d46'
Kind,Matthew James Aur 6h25'22"34d37'
Kinder,Carl Ori 5h29'29"-0d35'
Kinder,Ford Tau 4h14'26"20d9'
Kinder,Nicole Lyn 9h9'28"43d9'
Kinder,Paula Kaye And 0h20'34"36d20'
Kinder,Richard Duval Lmi 10h32'18"30d14'

Kindermann,Oliver Tau 5h45'12"20d44'
Kindinger,PhD,Paul E Boo 15h1'0"27d3'
Kindle,Connie Marie Hall Oph 16h39'55"2d13'
Kindle,Julie Equ 21h0'11"10d4'
Kindle,Keith Earl Peg 21h58'23"30d18'
Kindred Spirit Cam 7h16'14"68d40'
Kindred Spirits Mon 7h23'47"-1d29'
Kindscher,Hendel Uma 10h22'11"50d57'
Kindt,Charles Per 3h44'36"50d35A
Kindt,Jean Per 3h44'36"50d35B
Kineen,Cindy & Jerry Cyg 21h39'29"41d33'
King (Dib),Elisabeth Cas 0h16'41"46d46'
King Amfortas Cyg 20h22'53"38d56'
King Anne Cyg 19h44'50"29d38'
King Arthur Star,The Her 18h53'40"12d3'
King Beak Cep 2h35'0"87d21'
King Boy Aql 20h10'1"-8d47'
King C Stice,The Per 4h26'30"50d50'
King Chickie Cep 21h5'29"58d56'
King David Cet 2h52'1"0d50'
King DME,The Uma 10h33'56"52d16'
King Edward Lyn 7h38'21"37d32'A
King Emmy Cep 22h7'54"61d32'
King Francesco Cep 22h36'44"60d10'
King Gary Aur 5h17'17"41d45'
King HPRK,Virginia Elizabeth And 1h37'46"38d12'
King In My Heart,The Cep 21h51'18"85d6'
King Jerry Aql 20h13h35"1d29'
King Ketonakai Noble Bare Cyg 19h19'0"28d25'
King Marc Jeffrey Cep 2h49'1"77d52'
King of Hearts Cep 22h53'0"70d52'
King Paul Cep 0h3'33"73d46'
King Rabbit's Far Out Wedding Star Gem 6h25'40"12d10'
King Richard Cep 6h37'21"85d21'
King Sprout Cep 23h10'55"80d17'
King Tai Shan Cameron Cep 23h3'46"64d44'
King Trinary ABC,The Larry Cas 2h24'54"67d10'ABC
King V,Thomas Caldicott Cep 21h23'32"65d33'
King William Hya 9h2'30"-0d41'
King Wish,The Her 17h19'22"29d8'
King's Birthday Star, Mel Her 18h10'50"38d11'
King's Little Dream, Kelly Lyr 18h43'58"33d27'

King's World,Michael Her 18h6'31"40d55'
King's"Star",Harold S Sco 16h50'0"-40d2'
King,1935: Ruth & Marshall Cyg 21h33'1"31d46'
King,Alexa Elizabeth And 0h11'57"47d2'
King,Alice Aql 19h29'0"-6d47'
King,Alison Marie And 22h58'11"50d12'
King,Alles Traei Mon 7h54'53"-5d21'
King,Ambrose Morrison Peg 22h25'14"21d40'
King,Ann Com 13h17'19"26d18'
King,Ann Cyg 19h43'51"38d5'
King,Anna Com 13h3'12"16d43'
King,Anne Elizabeth Cas 0h53'52"66d1'
King,Barbara Lyn 6h58'55"52d40'
King,Barbara Hya 8h39'50"-10d13'
King,Barbara Lynn And 23h16'0"46d23'
King,Bill Cep 23h0'28"64d49'
King,Bill & Dinah Oph 18h15'50"11d37'
King,Bob Per 1h42'18"53d0'
King,Bob Ori 5h56'16"19d37'
King,Bonnie Tau 4h41'20"16d23'
King,Brian Boo 14h16'31"17d35'
King,Buford & Fran Mon 7h44'21"-5d26'
King,C L Dusty Her 17h2'42"40d29'
King,Catherine Robin Vir 11h35'51"8d4'
King,Charles Oph 17h27'51"10d44'
King,Christine Robin Cas 2h37'16"58d24'
King,Christopher John Dra 9h32'27"77d40'
King,Constance & Warren S Peg 21h53'1"20d34'
King,Cynthia(Cyndi) Aql 18h55'20"-0d8'
King,Dave Cep 22h54'60"56d56'
King,David E Aur 6h4'51"38d17'
King,David F Ser 15h38'17"8d50'
King,David J Aur 6h14'24"31d30'
King,David John Lac 22h17'32"47d39'
King,Debbie Umi 14h57'46"67d16'
King,Delroy Cep 23h11'59"64d19'
King,Dennis George Uma 9h46'15"45d42'
King,Diane And 0h29'57"30d47'
King,Dorothy Laird Peg 23h4'26"32d3'
King,Dwandalyn Reece And 0h29'1"44d26'
King,Edward & Elizabeth Uma 12h11'0"60d12'
King,Elizabeth Jane Eri 2h57'36"-5d18'

King,Ellis F
Gem 7h29'38"34d38'
King,Erika-Lee
Equ 21h5'17"11d51'
King,Erin & Bruce
Eri 4h0'41"-13d28'
King,Everett
Aql 19h50'53"13d9'
King,Ferrall Bennett
Mon 6h21'15"3d54'
King,Frances & Francis
Aur 5h4'15"40d25'
King,George Henry
Cet 1h24'52"-3d35'
King,Gloria
Cet 2h58'57"1d46'
King,Gov Bruce
Uma 9h10'48"48d18'
King,Grady
Cmi 7h11'15"9d7'
King,Graham Peter
Ori 6h3'15"8d33'
King,Issac E
Aur 7h17'0"39d42'
King,James C
Oph 17h2'30"-20d39'
King,James G
Oph 17h8'21"-23d3'
King,James Henry
Aur 6h36'25"37d51'
King,James Royce
Aql 18h59'17"-5d37'
King,James William
Her 18h0'12"38d23'
King,Jamie Leigh
Aqr 23h13'33"-6d0'
King,Janine L
Del 20h29'43"18d49'
King,Jeffrey Paul
Cet 2h34'35"4d59'
King,Jennifer Leigh
Peg 22h41'32"22d50'
King,Jeri
Cet 2h30'30"5d14'
King,John M
Eri 2h59'58"-6d8'
King,John Robert
Aql 19h45'0"12d52'
King,Jonathan Philip
Ser 18h4'16"-0d32'
King,Joseph Lucas
Cet 0h47'16"-3d36'
King,Joshua Oakes
Cep 22h54'15"57d34'
King,Joyce L
Cas 0h6'34"64d40'
King,Judy Ann
Ser 15h35'46"-2d18'
King,Kassandra
And 23h39'11"45d34'
King,Keith & Virginia Lea
Cas 1h38'35"70d7'
King,Kevin
Hya 8h44'22"5d38'
King,Kevin Scott
Cet 2h0'35"-0d54'
King,Krista A
Peg 21h43'1"23d17'
King,Krystal M
Mon 6h29'1"3d35'
King,Kyle William
Uma 9h50'33"68d12'
King,Lani Spencer
Uma 9h41'14"52d2'
King,Lawrence
Del 20h52'32"3d23'
King,Leila
Mon 6h56'17"10d20'
King,Leo
Cep 20h25'30"63d28'
King,Linda
Aql 19h55'53"12d58'
King,Little Mo Maureen
Cep 21h50'53"61d13'

King,Lloyd Keith
Tri 2h38'26"31d8'
King,Ma & Pa
Vul 21h26'53"24d18'
King,Marissa A
Hya 9h5'37"-11d26'
King,Mark D
Sct 18h32'31"-6d44'
King,Mary Christine
Cmi 7h57'29"8d31'
King,Mary L
Peg 23h20'24"15d36'
King,Mary Wheeler
Cas 1h28'17"60d22'
King,Mathew
Cnv 13h37'54"45d13'
King,Matthew Scott
Lac 22h28'49"55d26'
King,McKenzie Gemma
Cyg 20h20'1"40d46'
King,Megan Katherine
Cam 8h39'49"73d48'
King,Michael
Per 3h27'51"40d37'
King,Michael H
Cep 22h53'27"70d16'
King,Michael Pa Ying
Cnv 13h55'59"42d11'
King,Michael Ross
Lac 22h9'46"47d14'
King,Michelle Erin
Del 20h18'1"9d27'
King,Michelle Rose
Del 20h26'20"11d42'
King,Molly
Ori 6h6'51"20d32'
King,Mrs Marie
Aql 19h46'39"11d58'
King,N Reid
Sgr 19h40'5"-41d13'
King,Patti
Cyg 20h39'50"38d13'
King,Peter & Dorothy
Lyr 19h21'24"41d23'
King,Phyllis Alma
Lyn 7h57'14"44d57'
King,Preston William
Her 18h6'11"41d9'
King,Princess Jennifer Anne
Cyg 19h27'31"37d58'
King,Rachel
Lyn 8h4'44"44d32'
King,Ray
Cep 22h11'48"61d31'
King,Rev Dr Barbara Lewis
Mon 6h18'0"0d37'
King,Richard "Casey"
Ori 5h13'29"-8d47'
King,Roger
Cet 2h8'12"0d40'
King,Ronald & Gail
Aql 19h29'27"8d28'
King,Sally B
Lyr 19h0'25"40d53'
King,Sarah Kathleen
Cas 0h17'16"64d58'
King,Selina
Uma 11h28'13"66d33'B
King,Sheldon Andrew
Aqr 23h3'32"-11d55'
King,Shelly Bundy
Vul 19h19'0"22d37'
King,Sherrie Lynn
Mon 6h56'44"-10d34'
King,Shirley
Cep 0h4'26"67d6'
King,Shona Mary
Cyg 20h20'11"40d51'
King,Stephanie
Peg 0h10'17"14d5'
King,Stephen Andrew
Aqr 21h38'11"0d57'
King,Steve & Shauna
Boo 14h59'26"51d52'

King,Steven Lee
Cep 21h0'20"70d23'
King,Steven Michael
Dra 13h43'12"64d18'
King,Tanya
Boo 15h3'60"32d4'
King,Thomas James
Oph 18h24'30"12d15'
King,Thomas Richard
Aql 18h58'34"16d54'
King,Tom & Cherie
Sge 18h57'38"19d30'
King,Virgil W
Cam 3h18'1"60d3'
King,Virginia Ann Garber
Uma 9h27'36"55d4'
King,Wes & Doris
Uma 11h30'45"41d4'
King,William David
Ari 1h51'25"12d6'
King,William David
Cep 23h10'11"61d52'
King,William E
Her 18h43'12"12d35'
King,Zachary John
Dra 17h0'38"68d42'
King-Shaw 130595, Juliet
Umi 16h51'14"76d10'
King-Star 16952, Richard F
Per 2h20'36"55d16'
Kinga
Dra 18h52'19"65d8'
Kinga & Kristin Kindred Spirits
And 2h29'26"41d39'
Kingdom Of Coombe
Psc 23h7'0"2d1'
Kingdom,Christopher
Cas 0h50'12"62d31'
Kingera,Richard Nicholas
Dra 20h15'47"68d58'
Kingery,Todd
Vir 13h32'41"-4d29'
Kingfish
Ori 5h57'15"15d39'
Kingham,Clara
Cas 0h26'0"62d10'
Kingham,Mary Wilson
Cyg 19h48'21"38d52'
Kinghorn,Emma J
Cas 0h7'0"63d58'
Kingloff,Amanda
Ori 5h5'21"10d34'
Kingloff,Amanda Leigh
Cmi 7h56'43"8d29'
Kingma,Jackie J
Mon 5h55'32"-6d49'
Kings 50
Cyg 19h58'52"58d39'
Kingsbury PW,Robert B
Her 17h16'56"46d38'
Kingsbury,Audrey
Tri 2h4'1"32d47'
Kingsbury,Fred
Tau 3h51'48"22d14'
Kingsbury,Jacob
Mon 6h57'16"10d16'
Kingsbury,Lindsey Renee
Peg 21h44'51"24d60'
Kingsbury,Maxime
Tri 2h4'11"32d4'
Kingsbury,Michaël
Tri 2h4'43"32d3'
Kingsbury,Virginia Suzanne
Crb 15h47'1"26d37'
Kingsley Shooting Star
Cnv 12h23'37"36d51'
Kingsley,Bob
Aql 20h6'43"4d53'
Kingsley,Lucy Clinton
Hya 9h36'32"-5d52'
Kingsley,Nanette
Ari 1h54'22"12d37'
Kingsley,Zachary Jordan
Aur 5h24'16"31d1'

Kingsmill,Timothy
Sct 18h44'0"-5d7'
Kingston,Steve
Dra 17h56'49"64d47'
Kingston,Susan Jeanne McCahill
Cas 0h3'53"64d14'
Kini
Uma 8h54'35"48d13'
Kini
Aql 18h44'48"7d42'
Kinkead-Weekes,William
Cma 6h12'55"-28d23'
Kinkel,Father Bob
Uma 10h52'56"40d46'
Kinkel,George Andrew
Aql 20h4'57"7d33'
Kinkel,Jack
Her 17h59'35"50d20'
Kinley,Sarah Louise
Cas 23h0'51"58d23'
Kinlin,Patrick Thomas
Per 3h42'47"37d58'
Kinloch,Anne Doreen
And 23h21'21"50d53'
Kinloch,Robert
Mon 7h0'30"8d57'
Kinman,Laurin Dean
Sge 19h38'14"16d44'
Kinman,Peter & Renee
Lyr 18h25'52"37d54'
Kinman,Rachel Fairlight
Equ 21h33'33"8d16'
Kinnair I-J * 1954
Aur 6h2'0"30d1'
Kinnaird,Robert
Umi 15h18'11"68d3'
Kinne,Richard
Cep 20h53'55"58d54'
Kinnear,Hilary Andrea
Cyg 19h29'30"35d34'
Kinnear,Mary
Lyr 18h59'1"33d43'
Kinnear,Richard Michael
Her 16h39'45"29d22'
Kinner,Josef
Uma 9h25'11"47d34'
Kinney,Bill & Teresa
Cyg 20h33'14"42d1'
Kinney,Elizabeth "Beth"
Cyg 21h23'56"40d1'
Kinney,Gregory Allan
Mon 7h23'0"-8d38'
Kinney,Kevin
Aql 19h58'55"12d48'
Kinney,Margaret
Cas 1h39'39"58d48'
Kinney,Michael Christopher
Tri 2h11'28"33d47'
Kinney,Roland F
Aur 5h41'41"50d6'
Kinney,Ryan Zachary
Her 17h19'31"48d7'
Kinney,Seth
Dra 19h36'36"68d2'
Kinney-OCD,Fr Donald
Lyr 18h43'11"45d55'
Kinnie,Desirae Regina Ann
Uma 9h3'58"56d57'
Kinnie,Raymond J
Boo 15h34'17"42d10'
Kinny,Paul R
Hya 8h37'0"-0d54'
Kinou,Karine
Boo 13h41'49"16d51'
Kinsel,Jennifer
Lib 15h43'49"-23d55'
Kinsella,Austin
Aql 18h58'57"6d27'
Kinsella,Elizabeth
Peg 22h25'55"25d42'
Kinsella,Gene Loughlin
Cep 22h15'12"63d52'
Kinsella,Martha Ann
Vul 20h56'20"28d16'

Kinsella,Nicholas
Cma 6h59'41"-24d53'
Kinser Star,The Michael
Aql 19h0'19"5d17'
Kinsey,Jennifer & Mark
Cyg 19h34'16"30d0'
Kinsey,Rev John C
Cep 21h49'55"58d50'
Kinsey,Robert
Oph 16h54'13"1d32'
Kinsey,Starshine David Leon
Per 1h53'38"53d26'
Kinseys Day Dream
Uma 10h38'49"58d43'
Kinsky,Monika-Stand By Me
Psc 1h20'0"18d39'
Kinsley,Marjorie Claire Thompson
Ori 5h54'18"19d27'
Kinsley,William Kyle
Her 17h30'1"14d43'
Kinslow,Amy A
Lyr 19h8'14"40d50'
Kinsman,Deborah
Aql 19h2'51"16d39'
Kinsman,Dr Ben A & Christina L
Cyg 21h23'20"40d10'
Kinsman,Ops Beloved Ollie Harry M
Uma 14h12'52"61d1'
Kinsman,Robert E
Crt 11h38'18"-8d32'
Kintoki
Sco 16h21'14"-27d1'
Kinton,Randi Singer
Equ 21h19'30"12d14'
Kintzler,Constance
Cmi 7h43'19"0d46'
Kinzie,Raymond Wyant
Dra 13h39'21"64d45'
Kinzler,Helga
Ser 15h30'17"18d2'
Kinzler,Jimmilee
Her 17h14'54"18d56'
Kiolbassa,Onufry I
Oph 17h6'48"0d39'
Kip
Boo 15h13'18"47d41'
Kip
Sex 10h30'33"1d2'
KIP
Cet 3h17'35"0d50'
Kip & Elaine
Cnv 13h57'1"40d31'
Kip2
Cyg 19h56'58"37d55'
Kipfmüller,Heidi und Erwin
Uma 11h50'44"33d12'
Kipke,Hermann
Cmi 7h20'39"4d38'
Kipke,Tom
Cmi 7h20'0"5d15'
Kipp's Red Chiffon Memoir
Aur 5h5'47"38d47'
Kipper
Cnv 13h23'14"32d9'
Kipper
Aql 18h58'40"-6d4'
Kipper,Abigail Marie
And 0h38'34"40d31"
Kipper,Tamar Judith
Lyn 7h57'15"50d4'
Kipreos,Theophilos Haralambos
Oph 18h17'35"11d41'
KIPRIJANOW,JAQUELINE von Frank (FF)
Gem 8h5'1"31d9'
Kir Kum
Aur 6h22'11"38d30'
Kira
And 2h3'0"47d1'
Kira Alexandra
Cas 2h2'1"68d29'

Kira Angelique
Lyn 7h50'12"51d48'
Kira Brianne
Cet 3h8'41"1d1'
Kiracofe,Victoria
Cas 2h51'49"68d53'
Kirakosjan,Edgar
Cep 22h17'0"67d32'
Kiraly,Mr Ernest L
Cyg 21h3'45"48d49'
Kiramastzelle
Ind 21h10'11'-50d48'
Kirby
Her 17h13'22"21d41'
Kirby
Dra 19h34'15"61d5'
Kirby
Cma 6h16'13"-24d29'
Kirby & Carolyn
Peg 22h0'22"28d25'
Kirby Star,Taylor- Elsie
Peg 23h17'60"31d32'
Kirby's Love
Sgr 18h47'58"-22d41'
Kirby,Charles Ross
Aur 6h16'54"38d26'
Kirby,Cheryl
And 23h3'1"47d2'
Kirby,John
Her 16h28'40"41d1'
Kirby,John Dixon
Ser 15h15'0"10d20'
Kirby,John Leo
Her 17h38'26"14d52'
Kirby,Kara Louise
Cam 4h9'20"58d20'
Kirby,Karolyn
Mon 7h25'30"-8d16'
Kirby,Kevin Scott
Cet 3h19'37"2d40'
Kirby,Michelle
And 0h6'16"31d51'
Kirby,Preston R
Tri 1h46'34"28d21'
Kirby,Scott Dale
Ser 15h53'59"18d24'
Kirby,Terry
Aur 6h30'12"37d37'
Kirchenbauer,John William Kirkpatrick
Ari 1h45'42"15d51'
Kircher,Madison Malone
Ori 5h25'44"-4d33'
Kirchhofer,Kory Tyler
Vul 19h2'11"25d15'
Kirchhofer,Tamara L
Cyg 19h28'49"30d29'
Kirchhoffer,Tracy Anne
Ant 10h39'15"-32d39'
Kirchmann,Larry
Equ 21h15'57"2d18'
Kirchmeyer,Bill
Cnv 13h47'0"31d49'
Kirchmeyer,Klaus
Cmi 7h18'31"0d5'
Kirchner,Amanda Susan
Cnv 12h42'25"33d6'
Kirchner,Beth Ann
Cas 23h36'1"60d1'
Kirchner,Frederick Roy
Cyg 20h4'12"31d46'
Kirchner,Jason & Christine Campbell
Ori 6h4'57"8d51'
Kirchner,Jürgen
Peg 23h30'23"10d37'
Kirchner,Karen Ann
Equ 21h2'12"8d58'
Kirchner,Sylvia
And 2h5'47'40d20'
Kiriazis,Judith Ann Hodges
Cep 20h57'52"78d54'
Kirilloff,Sally Morrow
Eri 3h37'39"-5d41'

Kirimura,Yoshie
Ori 5h52'7"9d45'
Kirk
Del 20h13'36"14d36'
Kirk
Sct 18h56'21"-4d25'
Kirk & Mindy
Hya 9h2'1"-6d46'
Kirk Most Brilliant of Them All Xoash
Aql 19h30'1"10d34'
Kirk's Cosmic Chrome Dome
Her 17h1'56"40d21'
Kirk's Gift
Her 17h3'42"48d5'
Kirk,A L
Oph 16h54'55"-25d27'
Kirk,Abigail
Lyr 18h38'27"34d54'
Kirk,Adam
Ori 6h16'1"-2d51'
Kirk,Capitaine
Umi 15h17'24"77d31'
Kirk,Captain
Cnc 9h4'51"32d29'
Kirk,Captain
Uma 10h31'0"41d37'
Kirk,Christine Marie
Cyg 21h4'0"39d24'
Kirk,Helen & Bud Much
Her 16h39'25"13d8'
Kirk,James C & Wendy E
Eri 4h58'48"-8d44'
Kirk,John Russell
Per 4h27'52"50d42'
Kirk,Mary R
Uma 12h44'0"61d21'
Kirk,Noah Jay
Peg 22h20'35"21d34'
Kirk,Philip W
Lac 22h16'56"49d12'
Kirk,Robert Chapman
Cet 3h14'48"0d40'
Kirk,Ruth Evelyn
Eri 3h4'20"-5d44'
Kirk,Wesley David
Aur 6h11'18"37d52'
Kirkbride,Carol Lee Macolly
Lyn 6h31'54"59d3'
Kirkbride,Jr,Jon Mahlon
Per 3h0'32"41d25'
Kirkbride,Linda
Boo 14h49'18"32d31'
Kirkby,Martin & Ann
Cam 7h56'1"67d50'
Kirkeide,Jane L
And 22h58'55"50d1'
Kirkell,Michael
Aql 19h58'59"12d45'
Kirkham,Howard Douglas "Cactus"
Her 16h49'54"50d55'
Kirkham,Quinn M.
Her 18h5'0"45d46'
Kirkham,Richard
Uma 9h25'26"47d57'
Kirkhart's Star
Mon 7h2'44"4d43'
Kirkland,Austin Michael
Aql 19h1'44"5d3'
Kirkland,Claude Murray
Aql 18h41'19"-2d9'
Kirkland,Dave
Lmi 10h57'1"32d43'
Kirkland,Jason S
Her 16h51'53"40d56'
Kirkland,Laura M
Cas 23h14'53"62d50'
Kirkland,Margo J
Cyg 21h0'27"28d46'
Kirkland,Mary
Cas 3h2'17"67d57'
Kirkland,Mary Faye Perdue
Eri 2h58'21"-12d7'

Kirkland,Meghan Beth
Uma 10h4'47"70d20'
Kirkland,Olga Lynn
Mon 6h4'34"-5d11'
Kirkland,Pat
Ori 4h42'55"4d0'
Kirkland,Purnell A
Ori 5h57'39"10d43'
Kirkman,Albert
Hya 8h46'50"-0d47'
Kirkman,Amy Caroline
Peg 23h47'0"12d17'
Kirkman,David Olinger
Uma 11h24'36"37d53'
Kirkman,Holly
Peg 23h21'1"18d43'
Kirkman,Leona Genieve
Lyn 8h28'0"41d27'
Kirkman,Rick
Crb 16h9'53"37d51'
Kirkner,Billy
Aur 4h56'43"38d54'
Kirkness,Laurie
Cas 2h29'35"67d46'
Kirkpatrick - Star Voyager,Zephaniah
Sco 17h28'16"-30d15'
Kirkpatrick F L C, Richard David
Aql 19h47'55"11d15'
Kirkpatrick,Bessie Mae
Peg 21h54'16"20d31'
Kirkpatrick,Betty J
Cyg 20h58'46"31d33'
Kirkpatrick,Donald Tiberius Kirkback
Per 2h42'11"35d52'
Kirkpatrick,Dr Robert Thomas
Oph 18h21'13"10d38'
Kirkpatrick,James William David
Ser 15h29'60"9d35'
Kirkpatrick,John & Kristine
Cyg 19h46'45"29d24'
Kirkpatrick,Teri
Vul 20h56'0"28d16'
Kirkpatrick,Vilma Rita
Crt 11h33'0"-18d43'
Kirks,Louise P
Uma 10h8'36"51d32'
Kirksey,Pam
Cyg 21h20'36"37d55'
Kirkwood,Andy
Ori 4h49'1"0d32'
Kirkwood,Dolores & John
Tri 1h42'26"34d19'
Kirkwood,Gina Rachel
Com 12h5'27"27d55'
Kirkwood,Jack
Per 1h57'28"56d33'
Kirkwood,John
Dra 16h6'15"64d26'
Kirkwood,John Robert
Her 16h17'57"22d48'
Kirkwood,PJ
Boo 14h53'38"40d4'
Kirkwood,Stuart Kennedy
Ari 2h1'34"21d32'
Kirley,Judith Lea
Umi 15h47'40"80d58'
Kirouac,Diane
Umi 16h8'43"77d49'
Kirpach,Catherine
Cas 2h25'0"61d52'
Kirra's Star of Hope
Peg 0h12'57"14d6'
Kirrane-50yrs,John & Louise
Cyg 21h23'37"31d10'
Kirree
Ori 6h0'45"8d47'
Kirsch,Cynthia Ann
Peg 22h1'51"31d35'
Kirsch,Dr Ann
Cam 8h34'44"77d49'

Kirsch,John Edward & Paula
Per 2h59'47"38d1'
Kirsch,Julie
Cam 5h37'41"78d49'
Kirsch,Louis Allen
Aur 5h19'43"52d52'
Kirsch,Ray
Eri 3h1'19"-13d0'
Kirsch,Stan
Cep 22h45'44"57d14'
Kirschbaum,Bianca Geb Plum
Tau 5h33'36"26d6'
Kirschbaum,Hans
Aur 6h8'43"45d3'
Kirschbaum,Martha
Aur 6h56'38"44d23'
Kirschner,Alexis
Peg 22h31'44"8d5'
Kirschner,Christopher André CAK1971-72
Her 17h13'38"41d37'
Kirschner,Dr Marc
Oph 17h22'34"-6d57'
Kirschner,Jessica
Aql 20h11'1"0d22'
Kirschner,Melanie Anne
Tri 1h58'0"25d28'
Kirschneseit,Siegrun
Crb 16h5'18"28d19'
Kirshheim,Ingeborg
Uma 11h46'18"45d53'
Kirshpaularon
Mon 6h22'18"8d15'
Kirshy,Virginia Marie McGowan
Mon 8h8'60"-8d26'
Kirsi & Jari
Cyg 21h56'43"53d10'
Kirsimaria
Cam 6h32'37"80d14'
Kirstan Lily
Ori 5h51'59"16d5'
Kirsteen
Cas 0h25'24"50d33'
Kirsten
And 2h24'54"45d49'
Kirsten
Cas 23h24'39"53d2'
Kirsten
Cas 1h4'1"55d31'
Kirsten & Tom "Forever One"
Gem 7h29'18"20d16'
Kirsten B
Cas 1h23'12"58d25'
Kirsten Helena
Lyr 18h49'15"36d48'
Kirsten Loves Robin Forever
Uma 11h42'1"60d17'
Kirsten's
And 23h37'43"45d34'
Kirsten,Virgil J
Cnv 12h18'29"51d20'
Kirstie
Mon 6h26'29"-0d57'
Kirstie
Aql 19h29'24"8d55'
Kirstin
Mon 8h3'39"-8d44'
Kirstine,Nancy L
Mon 7h43'45"-5d7'
Kirsty
And 0h23'30"40d30'
Kirsty S
Cyg 20h15'14"39d9'
Kirti
Lib 15h38'10"-28d8'
Kirton,Gavin James
Peg 23h1'32"30d44'
Kirts,Rosemeiry Chaves
Cma 6h43'0"-13d9'
Kirvan-Zweig
Uma 11h53'54"31d2'
Kirwan,Dominic
Aur 5h0'41"48d48'

Kirwan,Katherine
Cyg 20h27'25"37d40'
Kirwin,Jr,Patrick Owen
Del 20h34'38"20d14'
Kirykowicz,Robert Alexander
Her 16h27'0"39d36'
Kisa
Cyg 21h32'35"30d4'
Kisch,Martha C
And 2h21'0"44d11'
Kiseliova,Natasha
Cep 22h23'23"70d7'
Kiselow,Mark C
Per 3h5'36"40d15'
Kiser's Star,Thelma & Curt
Mon 7h39'37"-4d7'
Kiser,Archus
Cet 3h19'31"8d41'
Kiser,Bruce David
Hya 9h34'47"1d36'
Kiser,Cara Lynn
And 2h26'18"48d12'
Kiser,Chad Erik
Boo 14h42'33"52d23'
Kiser,Emily Rebecca
Cas 23h38'28"64d35'
Kiser,Joseph Paul
Dra 16h4'53"52d9'
Kiser,Kim
Boo 14h56'31"36d4'
Kiser,Nagiko Sato
Mon 7h21'25"-1d22'
Kiser,Nathan
Aur 5h18'36"45d54'
Kiser-Hyde,Karen
Crt 11h21'40"-16d24'
Kish,Stephen John
Boo 15h46'18"47d56'
Kishel,Kendra
Cas 0h56'1"66d52'
Kishline,Gregg Alan
Lac 22h14'28"50d2'
Kishore & Rama
Cyg 20h59'1"28d60'
Kisin,Marcia
And 23h29'23"49d9'
Kisler,Anne
Lyr 18h49'55"31d13'
Kismet
Lac 22h9'57"49d38'
Kismet
Cep 0h1'0"66d41'
Kismet
Cam 6h34'51"83d14'
Kismet
Cam 10h42'14"82d18'
Kismet,Douglas
Oph 17h0'49"-28d38'
Kismit-Jorge & Terry
Umi 16h10'40"72d0'
Kisowetz,Klaus Dieter
Cmi 7h18'1"7d48'
Kiss of Peach Blossom
Leo 10h2'35"16d42'
Kiss,Alexandre
Aql 19h59'47"10d35'A
Kiss,Daniel C
Her 17h36'18"20d42'
Kiss,Michael Andrew
Ari 3h0'23"25d1'
Kiss,The
Cep 21h1'27"65d5'
Kissane,Joseph K & Dorothea M
Aur 5h4'24"40d1'
Kissen,Jennifer Joy
Mon 6h32'27"-0d14'
Kissiah,Leslie Carol
Per 3h0'1"55d10'
Kissinger,Jean Dale
Her 17h21'14"42d49'
Kissinger,Samuel Fulton
Cnc 9h0'49"27d60'
Kissinger,Sir Henry
Uma 9h40'12"58d8'
Kissinger,Sophia Frances
Tau 4h3'22"10d28'
Kissling,Christopher Brian
Lac 22h46'27"54d47'
Kistemann,Leo
Sge 19h6'11"17d42'
Kit's Dream
Lmi 10h40'26"26d12'
Kit-Cal
Lyn 7h56'21"43d22'
Kit-My Shining Star In Frosthaven
Ari 3h14'58"28d16'
Kita,Mahal
Mon 6h57'0"-0d59'
Kitakaze Familys
Ari 1h50'32"23d2'
Kitakaze-Stella Of Macky
Tau 5h29'24"17d42'
Kital,Mark Nihal
Cmi 7h16'29"5d31'
Kitamura,Akemi
Cmi 7h54'57"8d59'
Kitamura,Junko
Cyg 20h22'14"30d20'
Kitch,Kara Lee
Lyr 19h23'28"31d4'
Kitch,Michael F
Dra 10h15'57"78d46'
Kitcharoen,Kenzie
And 1h11'26"38d19'
Kitchen "Daddy's Star" Robert David
Tri 2h16'42"30d33'
Kitchen,Elda
Ori 5h56'11"15d26'
Kitchen,Emilee
Aql 20h10'27"14d27'
Kitchen,Matthew Brian
Lmi 10h45'37"23d42'
Kitchen,Robert N
Ser 15h40'54"-1d48'
Kitchener,Stephen A
Aql 20h6'1"6d35'
Kitchens,Gordon
Aql 19h56'32"13d26'
Kitching,Bronwen
Cra 18h20'13"-40d29'
Kite,Brianna
Peg 21h59'60"20d25'
Kite,James E
Per 1h47'51"54d12'
Kitnick,Alexander Lawrence
Cet 1h21'31"-5d13'
Kito
Boo 15h18'14"48d54'
Kito Star
Sco 17h29'43"-40d33'
Kitou,Evi
And 0h18'47"38d26'
Kitowski,Dina
Lac 22h54'34"55d21'
Kitron,Marshall
Aur 6h23'53"35d39'
Kitson,Chloe Louise
Cnv 13h45'45"32d21'
Kitsos,Dr George
Boo 14h17'0"37d15'
Kitsos,George & Jean
Uma 11h23'1"33d32'
KITT
Uma 12h17'41"58d21'
Kitt,Brian Lee
Peg 23h37'23"30d6'
Kitt,Gilbert
Sct 18h30'26"-6d51'
Kitt,Karl A
Aql 20h12'25"11d37'
Kittel,Scot
Aur 5h1'39"51d8'
Kittel,Ulla und Daub, Hans
Uma 9h51'41"52d12'
Kittelberger,Carl
Uma 11h44'21"64d40'A
Kittelberger,Ruth
Uma 11h44'21"64d40'B
Kitten
Peg 22h31'21"26d6'
Kitten
Lyr 18h58'22"30d24'
Kittinger,Jeffrey Lee
Aur 7h23'0"38d49'
Kittle,Clare Adams
Crb 16h5'27"31d55'
Kittle,Jr,Elmer
Her 16h54'1"34d12'
Kittle,Jr,Frank Louis
Per 3h45'35"33d54'A
Kittle,Ruth
Per 3h45'35"33d54'B
Kittleson
Sge 19h16'0"20d1'
Kittredge,Ronald Earle
Boo 14h9'50"41d23'
Kittredge,William E
Dra 16h23'19"60d2'
Kittrell,Deborah Mary
Cas 0h51'21"64d35'
Kittrell,Mary Ruth
Peg 23h27'51"10d8'
Kitts,Rita
Cmi 7h5'12"3d58'
Kitty
And 0h41'53"30d1'
Kitty
Lyr 18h15'52"40d31'
Kitty
Equ 20h55'58"2d43'
Kitty
And 1h19'52"38d3'
Kitty
Cas 2h29'26"59d52'
Kitty & Jason
Cyg 21h23'48"28d52'
Kitty And The Marlboro Bunny
Lyn 8h6'0"36d29'
Kitty Angel
Cyg 20h36'49"38d38'
Kitty Kelly
Dra 17h32'39"60d34'
Kitty Lindsay
Mon 6h56'18"11d0'
Kitty Samantha
Lyn 7h29'30"50d15'
Kitty's Star
Cas 0h15'11"63d39'
Kitty,The
Boo 13h55'48"20d39'
Kitty-Bunny
Hya 9h20'37"-0d41'
Kittybear
Equ 21h9'51"11d28'
Kituna
Ori 5h34'44"-5d59'
Kitz,Angela Marie
And 23h2'48"51d42'
Kitzi's Honor
Ori 6h4'33"0d56'
Kitzis,Mrs Pauline
Tri 1h49'20"28d16'
Kitzmiller,Catherine Caroline
Uma 14h16'21"60d6'
Kitzmiller,Tad Stuart
Lac 22h22'43"56d21'
Kivell,Wanda Lucy Korqul
Cas 2h16'32"65d36'
Kiviniemi,Aimo & Vi
Vul 21h1'25"20d32'
Kivlon,Don
Peg 23h5'4"32d33'B
Kivlon,Jean
Peg 23h5'4"32d33'A
Kiwiet,Eva
Cam 8h31'2"74d53'B
Kiwus,Donald P
Oph 16h33'16"-6d35'
Kiwwelschisser
Cet 2h58'0"3d48'
KIX Country 104
Her 18h30'1"24d29'
Kiyner,Zisi
Cmi 7h21'12"4d1'
Kiyo
Ori 4h45'58"4d51'
Kiyo,Camille
Cas 1h1'48"60d4'
Kiyomi
Cnc 8h53'32"16d59'
Kiyomi & Toshio
Ori 6h8'31"9d20'
Kiyoshi Doi
Psc 1h24'3"10d38'
Kiza,Shawn
Aur 5h59'32"50d29'
Kizer,Jr,Jerry D
Per 3h59'52"38d52'
Kizer,Richard L
Per 1h49'48"48d34'
Kizer,Todd
Cep 22h46'14"57d32'
Kißmer,Bodo
And 23h45'15"38d34'
Kjaer,Erik
Ori 5h46'51"21d20'
Kjaerbye,Henrik
Dra 11h58'36"68d56'
Kjar,Michael E
Aql 19h55'50"13d52'
Kjellerup,Knud
Ori 5h24'43"1d44'
Kjenes,Anita
Cas 0h40'53"60d45'
Kjentvet VII,Henry Henry
Aur 5h2'58"42d48'
Kjerulf,Stig
Cep 21h12'46"68d42'
Kjolner,Dana
Cru 12h25'35"-60d26'
KKNTS Where Dreams Come True
Sct 18h45'23"-7d12'
Klaas,Forever Polly
Cyg 19h28'19"40d51'
Klaas,Polly
Peg 23h26'39"23d32'
Klaas,Polly Hannah
Lyr 18h57'46"40d43'
Klaas,Polly Hannah
Tau 4h4'40"21d13'
Klaban,Maria & Walter
Cyg 19h46'50"38d35'
Klabin,Dan
Lac 22h37'27"38d22'
Kladek,Franz
Ser 15h12'49"8d35'
Klaes,Dianne
And 0h21'43"35d34'
Klaes,Robin Thomas
Lyn 8h31'36"52d25'
Klages,Jacob A
Cma 7h37'24"6d32'
Klages,Jacob B
Mon 7h41'11"-4d23'
Klainbaum,Yohanna
Mon 6h56'52"0d30'
Klajda,Dolores Ann
Ori 6h8'44"17d23'B
Klancer,Richard J
Dra 16h55'38"58d53'
Klanderud,Timothy Lewis
Her 16h42'56"28d59'
Klang,Eileen Marie
Cas 0h23'11"62d35'
Klanjac,John A
Per 2h9'34"57d26'
Klanke
Lyr 19h19'1"30d26'
Klanke,MD,Charles W
Lac 22h22'0"40d52'
Klapper,"Cute" Regina
Vir 13h34'32"-10d14'
Klar,Dawn Anne
Lyn 8h13'27"50d21'
Klar,Emily Marie
Cas 1h6'23"52d56'
Klar,Michael Hager
Cep 21h56'60"60d27'
Klar,Sara Ann
Peg 23h3'16"32d3'
Klar,Waltraud
Psc 1h22'30"18d8'
Klara da Kuáova
Ind 21h2'30"-53d6'
Klarinka i Robertka
And 0h19'15"34d41'
Klarska (Kov), Christian
Ori 5h24'43"1d44'
Klas Goran Ek
Eri 4h27'0"-17d50'
Klase,Mary N
Cam 5h51'48"71d14'
Klasen,Heinz
And 23h10'18"40d13'
Klaska,Joanna
Mon 8h0'1"-6d26'
Klate,Ozzy
Her 16h47'47"10d7'
Klatt,Pauline R
Cas 0h45'12"63d58'
Klau,Norma
Cas 1h49'1"73d8'
Klaudia-Geggo
Cnc 8h28'32"7d28'
Klaus Fritz
Cma 7h21'44"1d43'
Klaus Marko
Gem 6h43'24"14d1'
Klaus,Bernhard
Per 2h6'29"58d53'
Klaus,Dr Tremmel
Sge 19h57'0"20d5'
Klaus,Louis B
Lyn 7h2'39"59d43'
Klaus,Mark
Her 16h24'16"40d47'
Klaus,Reinhold Walter
Lyn 8h18'0"43d12'
Klausner,Jospeh Marshall
Her 17h55'35"35d40'B
Klausner,Shirley Brook
Her 17h55'35"35d40'A
Klaustermeier,Amy
And 2h21'46"42d36'
Klavon,David "Maqtar"
Lac 22h0'44"37d58'
Klawunn,Russell Edward
Cet 2h10'48"7d53'
Klay D's Star
Aur 5h10'0"41d0'
Klay,Johanna Rachel
Cyg 19h28'39"30d37'
Klayder Together Forever,Velma Urquhart
Uma 12h54'6"54d22'B
Klayder,Jr Together Forever,Paul A
Uma 12h54'6"54d22'A
Klazstar I
Cyg 19h43'49"30d37'
Kleber,Carol Ann
Lyn 7h59'1"34d57'
Klecka,Marc Robert
Dra 16h25'11"69d39'
Kleckley,Jr,Edgar
Crt 10h58'27"-17d35'
Klee,Gerhard
And 23h18'31"50d14'
Klee,John
Ori 5h54'0"9d0'
Klee,Mary Ellen
Cam 3h37'40"71d49'
Kleeblatt,Josef
Vir 12h35'20"-10d52'
Kleeman,Tammron Jay
Uma 10h23'56"40d18'
Kleemeier-Bambach, Christine
Leo 9h21'39"33d12'
Kleemeyer,Wolfgang
Dra 18h28'0"65d20'
Kleene 06
Ari 2h53'23"22d11'
Kleespies,Kathy
Uma 8h50'34"50d38'
Kleffner,Lisa
Lyr 18h40'30"28d23'
Kleich,Thomas Eric
Her 16h33'24"37d44'
Kleid,Richard Maxwell
Per 1h47'1"56d39'
Kleiman,Amy & David
Cyg 19h18'10"28d28'
Kleiman,Linda & Ted
Umi 14h29'0"65d51'
Kleiman,Matt
Cep 22h41'18"57d34'
Kleiman,Zachary Harrison
Peg 23h35'29"10d46'
Kleimenhagen,Chris A
Cyg 21h36'23"42d28'
Klein
Aql 20h8'15"8d54'
Klein "Super Star", Sam
Sex 10h42'15"-6d49'
Klein's Destiny
Aur 6h26'48"35d44'
Klein,Alex Charles
Her 17h14'58"47d52'
Klein,Betty
Cet 2h40'17"-12d10'
Klein,Bob
Cep 20h35'0"61d17'
Klein,Bruce & Hillary
Cyg 20h50'1"37d36'
Klein,Carly Ruth
And 1h4'12"39d12'
Klein,Charles Marcel
Cet 2h35'31"-4d53'
Klein,Cheryl Goodrich
Gem 6h32'49"13d42'
Klein,Christopher Sean
Aur 6h11'0"46d28'
Klein,Dana Anne
Cas 3h10'44"68d26'
Klein,David R
Per 2h47'0"41d5'
Klein,Dieter
Uma 9h33'10"45d18'
Klein,Dipl-Ing Horst
Ori 5h37'53"12d53'
Klein,Don
Oph 17h59'29"8d17'
Klein,Dr Betty Crothers
Cma 6h44'17"-15d48'
Klein,Emilee Suzanne
And 22h58'30"38d23'
Klein,Fay
Mon 6h54'38"-10d32'
Klein,Gene
Her 16h46'51"30d2'
Klein,Gerd W
Vir 11h38'14"2d52'
Klein,Gisela
Cas 0h4'27"60d14'
Klein,Grace Marie
Cmi 7h41'1"8d52'
Klein,Hans-Martin
Per 3h56'0"37d27'
Klein,Harlan Matthew
Aql 19h24'28"10d53'
Klein,Hayden Charles
Cep 21h15'0"61d17'
Klein,Hope M
Eri 2h46'32"-10d9'
Klein,Irene
And 23h38'31"33d25'
Klein,Jade Hellen Rose
Cas 1h25'44"73d25'
Klein,Jaques Paul
Her 17h43'0"18d46'
Klein,Jarvis
Eri 4h13'40"-18d26'
Klein,Jeffrey
Boo 15h6'15"40d13'
Klein,Jon Anthony
Sex 10h45'26"-1d25'
Klein,Jr,Walter C
Cnv 13h40'1"40d10'
Klein,Karen Elizabeth
Mon 6h30'56"-6d57'
Klein,Katherine Mary "Sisty"
Cam 4h55'25"60d54'
Klein,Lauren Jean
And 23h7'60"44d54'
Klein,Lori Beth
Peg 22h45'27"5d48'
Klein,Maria
Sct 18h54'55"-6d33'
Klein,Mark A
Boo 14h56'1"47d32'
Klein,Martin
Lac 22h1'53"50d14'
Klein,Max
Aur 5h0'41"48d48'
Klein,Michael
Sgr 19h0'59"-2d1'
Klein,Michael
Cnc 9h15'11"30d21'
Klein,Michael
Lyr 19h17'35"30d18'
Klein,Michael & LeeAnn
Cyg 21h8'12"31d0'
Klein,Mildred
And 23h27'0"43d21'
Klein,Monika
Cap 20h37'17"-11d4'
Klein,Morton
Peg 21h58'13"3d17'
Klein,Paul E
Cet 2h51'13"1d54'
Klein,Philip E
Per 1h50'55"56d25'
Klein,Philip John
Gem 6h32'48"12d37'
Klein,Rachael Joy
Mon 6h54'58"8d50'
Klein,Randy D
Cep 23h9'12"60d21'
Klein,Renate
Lac 22h7'56"51d40'
Klein,Robert J
Hya 8h31'14"1d12'
Klein,Sally Rountree
Eri 2h52'1"-4d57'
Klein,Samantha Brooke
Lyr 18h21'31"46d55'
Klein,Shannan Fay
Lyn 8h47'49"37d51'
Klein,Shlomo Ilan
Lib 15h3'21"-2d3'
Klein,Sonia
Cyg 21h3'44"41d15'
Klein,Star of Steven S & Ruth Kamin
Aur 5h52'53"38d55'
Klein,Susan & David
Cyg 20h22'55"30d10'
Klein,Thomas Alan
Aur 5h1'27"52d8'
Klein,To The Very Lovely Ave Nicole
Lyn 6h33'1"60d43'
Klein,Travis Austin
Dra 11h55'12"72d14'
Klein,Walter J
Aql 19h31'50"0d32'
Klein,Yves A M
Lyn 8h38'0"41d22'
Kleinath,Kathrin
Gem 6h43'15"14d24'
Kleinbaum,Rabbi Sharon Anne
Crb 16h18'17"37d32'
Kleinberg,Banjo Dave Doc
Ori 5h54'38"6d16'
Kleine Maus Gabi
Ser 16h16'1"2d18'
Kleine Moira
Mon 8h3'15"-8d43'
Kleine,Aaron David
Her 17h35'1"28d53'
Kleine,Alice & Lou
Cam 6h11'15"60d29'
Kleine,Sharlene Renee
Ori 6h8'28"8d19'
Kleine-Weischede,Horst
Her 17h39'10"42d9'
Kleiner Karpfen
Psc 23h9'33"0d43'
Kleiner Our Star In The Sky,Timmy
Her 18h0'34"38d1'
Kleiner Prinz
Ari 2h52'47"20d54'
Kleiner,Garrett William
Cet 2h28'12"4d7'
Kleiner,Helen
Peg 23h16'48"33d31'
Kleiner,Milka J
Aql 20h12'37"4d48'
Kleiner-Krümel-Birte-Abraham-300595
Cep 22h8'59"60d53'
Kleinginna,Mark D
Aql 19h37'22"-6d18'
Kleinheinz,Captain James
Aql 20h11'42"13d19'
Kleinhempel,Hans
Aql 19h53'52"1d22'
Kleinhenn,Charles
Lmi 10h6'28"30d5'
Kleinke,Michael
Her 16h19'41"50d49'
Kleinman,Cathy G
Cnc 7h57'19"10d2'
Kleinman,King
Sct 18h45'59"-7d59'
Kleinman-All Star Loren
Her 17h39'29"40d59'
Kleinmann,Daniel Isaac
Del 20h50'45"9d10'
Kleinpell,Peter J
Del 20h34'35"20d19'
Kleinschmidt,Alberta
And 23h15'0"49d19'
Kleinschmidt,Peter
Her 17h12'45"48d27'
Kleinsmith,Audrey Lynn
Peg 22h0'22"20d59'
Kleinstes
Gem 6h53'56"17d18'
Kleinweber,Bob & Kay
Umi 14h23'1"71d28'
Kleist,Bonnie
Vul 20h13'57"23d34'
Klekowski,Helen
Peg 21h45'51"23d9'
Klemens,Jamie
Uma 10h57'23"72d44'
Klement,Heinz
Sgr 19h57'29"-42d9'
Klemm,Sami
Boo 14h6'48"40d56'
Klemm,James R
Lyn 9h12'15"42d2'
Klemm,Randy Michael
Cam 3h59'28"67d38'
Klemmer,Kandace Lauren
Peg 22h3'28"18d52'
Klendrou,Ioanna
Uma 9h18'21"51d8'
Klenert,Amy
Tau 3h59'39"20d27'
Kleoppel,Bill
Aur 6h4'43"31d50'
Klepetko Grandpa, Anthony R
Boo 14h19'14"12d15'
Klepodlo,LaVerne
Uma 8h13'15"68d60'
Klepper-Windham,Ernae
Aur 6h7'54"43d48'A
Klepsic,Peter
Per 4h0'0"50d6'
Klesert,Amy
And 0h55'35"45d5'
Klesert,Vanessa
And 0h55'46"40d47'
Kleszczewski,John
Ser 15h35'22"9d44'
Klett,Angelika
Leo 9h24'47"17d49'
Klett,Francis Joseph
Psc 23h7'17"2d1'
Kletz,Albert E
Cam 9h0'57"82d27'
Kletz,Allison Hope
Cam 3h54'52"68d5'
Kletz,Cathi Lynn
Cam 8h33'20"77d52'
Kletz,MD,Michael Robert
Cam 3h47'57"61d50'
Kletz,Michelle Brooke
Cam 8h3'37"60d56'
Klevans,Joshua Goheen
Lac 22h5'0"49d25'
Klevemann-Frystak, Paula
Cyg 21h28'0"41d12'
Klevickis,Justin Charles
Dra 17h6'45"62d42'
Klevinsky,Thomas J
Dra 15h51'49"67d18'
Kleydorff,Ludwig Frhr V
Lac 22h16'1"46d31'
Kleyn,Lisette Henk
Cas 1h14'0"60d8'
Klics,Dr Richard
Oph 16h40'24"2d35'
Kliemann,Erica
Aqr 21h19'21"-10d58'
Klienfeldt,Jill
Lyn 8h48'11"46d24'
Klimala,Walter Edmund
Cnv 13h3'10"38d55'
Klimchak,Justin Robert
Vul 21h2'55"24d8'
Klimczak,Henry J
Her 18h3'41"50d7'
Klimek,Petra
Lib 14h22'57"-20d4'
Klimek,Winfried M
Ari 2h58'37"22d18'
Klimek-Kayser,Gerda
Cap 20h35'22"-10d14'
Klimes,Candy
Peg 0h2'1"18d20'
Klimes,Peter M
Cma 6h59'50"-16d29'
Klimisch,Helga
Sgr 19h1'56"-24d44'
Klimo,Peter
Tau 5h32'38"28d22'
Klindt,Jason Michael
Lyr 18h48'45"37d2'
Kline,Andrew T
Cmi 7h24'5"0d18'
Kline,Caren Joanne Whitman
Vul 19h19'10"20d27'
Kline,Daniel Howard
And 2h21'27"44d7'
Kline,David John
Aur 6h22'50"30d18'
Kline,Donald Gerald
Aur 5h11'60"40d4'
Kline,Dr Michael J & Patricia A
Cyg 21h3'39"33d27'
Kline,Eleanor J
And 23h40'50"40d13'
Kline,Garrett
Dra 20h35'1"68d8'
Kline,Gerard
Sge 19h5'20"16d40'
Kline,Gregory
Dra 17h27'18"64d9'
Kline,James David
Ori 6h4'48"7d16'

Kline,Jimmie
Cnv 13h32'9"39d2'A
Kline,Johanna Kathleen
Lac 22h30'50"38d49'
Kline,Jonathan Richard
Dra 16h0'30"61d55'
Kline,Leah
Cyg 19h14'45"45d26'
Kline,Norma
Cnv 13h32'9"39d2'B
Kline,Paul & Deirdre
Lyr 18h33'17"42d41'
Kline,Rachel Stacy
Boo 14h28'0"26d24'
Kline,Robert Jerome
Cet 1h38'45"-0d53'
Kline,Roberta
Peg 22h52'26"27d41'
Kline,Robin
Cas 0h48'0"62d32'
Kline,Ruth
Cyg 19h37'39"38d44'
Kline,Sarah
Lyr 18h58'15"31d31'
Kline,Stanley
Cet 2h54'52"3d14'
Kline,Ted
Her 17h36'34"26d33'
Kline,Tom
Aur 5h7'1"41d18'
Kline,Tyler
Hya 9h32'51"2d13'
Klinect,Bryan Bradford
Cnc 8h48'28"32d29'
Klinefelter,Brian David
Uma 12h3'10"39d3'
Kling,Daniel
Lac 22h51'25"53d6'
Kling,Michael John
Dra 12h11'13"75d52'
Klingaman
Umi 15h25'12"77d49'
Klingberg,Karl LUDWIG
Uma 10h29'53"70d48'
Klingemann's Star, William H
Aql 19h53'42"15d0'
Klingenberg,Clifford
Per 3h17'56"40d35'
Klinger,Beigitte
Lac 22h2'57"51d58'
Klingert,Janet Marie
Cyg 19h37'1"28d45'
Klinges,Dr
Oph 16h50'44"11d4'
Klingler,Logan Stuart
Her 16h40'0"50d36'
Klingler,Megan Mackenzie
And 0h23'45"36d21'
Klingler,Staci Beth
Cyg 19h32'0"36d2'
Klingman,Glenn Evan
Boo 13h34'29"21d23'
Klingsberg,Marc Allen
Dra 17h33'51"60d49'
Klink,Bruce
Dra 12h55'58"76d24'
Klink,Lisa "Me"
Crb 15h50'33"27d42'
Klink,Melissa
Uma 11h9'14"60d12'
Klinka,Shirley
Uma 9h49'55"68d54'
Klinker,Jacob Michael
Ori 5h57'28"12d37'A
Klinker,Lucas Andrew
Ori 5h57'28"12d37'B
Klinkhammer,Peter
Oph 17h20'54"10d30'
Klinnert,Karl
Her 16h24'46"36d25'
Klintworth,Ralph
Eri 4h54'47"-8d59'
Klisura,Dominic
Aql 18h59'43"-6d58"

Kliton,Andy
Uma 9h36'0"48d32'
Klitta,Uwe
Gem 6h43'59"14d54'
Klitus,Thomas & Denise
Cyg 20h24'26"38d36'
Klock,Felix S
Per 3h0'1"40d5'
Klockner,Paul D
Aur 5h5'1"40d26'
Kloeting,Nora
Ari 2h32'57"22d6'
Klokner,Ryan Lee
Sct 18h50'14"-7d40'
Kloo,Matthew David
Dra 18h55'59"68d49'
Kloo,Matthias
Lyn 8h7'33"48d60'
Klooster,Amber Noel
Lyn 6h55'43"59d58'
Klopfer,Benjamin David
Aur 6h11'33"30d7'
Klopfer,Evan Michael
Aur 6h5'43"30d5'
Klopfer,Ralf
Sgr 19h19'57"-26d49'
Klopp,Kenneth Henry
Aqr 21h51'59"-3d32'B
Klopp,Linda
Cam 5h7'12"70d47'
Kloppenburg,Manfred
Cmi 7h29'22"1d6'
Klor,William M
Boo 14h55'26"46d16'
Klora,Dipl-Ing Eckhardt
Her 17h4'19"46d54'
Klos,Angelika
Gem 6h37'1"28d15'
Klos,Sylvain
And 2h34'0"39d21'
Klose,Annette
Lib 14h22'46"-20d25'
Klose,Juergen
Lac 22h5'44"51d30'
Klosinski,Lynda Ann
Cas 3h23'60"75d34'
Kloski,Gerald Lloyd
Her 17h38'1"22d20'
Kloss (CAT),Nathalie
Lyr 18h35'44"40d9'
Kloster's Magical Wishing Star,Mrs
Mon 6h32'50"-6d32'
Klosterknecht,Uwe Frank
Cap 20h6'11"-20d7'
Klosterman,Natalie Amber
Oph 18h3'16"8d2'
Klosterman,Sammy K-
Hya 9h6'45"1d21'
Klotvich "Bird-Man", Michael
Aql 19h7'11"2d42'
Klotzek,Ralf
Her 17h43'35"42d57'
Klovsky,Sidney B
Cyg 19h32'45"36d5'
Kluczynski,Mary Ann
Peg 22h3'14"5d8'
Kluetz,Amy
Cam 8h2'35"82d7'
Kluft,David
Vir 13h14'14"-1d54'
Klug,Beth
Cet 2h16'47"6d20'
Klug,Christine M
Mon 6h19'21"5d6'
Klug,Gram & Pop
Aql 19h47'38"12d9'
Kluge,John W
Dra 17h38'23"64d23'
Kluger,Harald H J
Psc 23h23'56"2d18'
Kluka,Miranda Nicole
Mon 6h30'51"-6d46'
Klumph,Bruce Stewart
Cet 0h7'31"-10d42'

Klumpp,Kurt
Per 4h6'53"47d7'
Klumpp,Patricia Anne Hunter
Vul 19h48'46"23d12'
Kluss,Tara Ellen
Cas 0h23'44"61d34'
Klussmann,Sonja
And 1h42'39"41d25'
Kluttz,Austin Elizabeth
Ori 5h59'50"8d45'
Kluz,Krystyna
Cyg 20h0'57"40d28'
Klyber,Raymond
Per 1h47'43"53d52'
Klym,John Michael
Lac 22h4'43"46d38'
Klyne,Richard Patrick
Boo 14h50'13"48d59'
Klären,Josef
Lyr 19h2'16"31d43'
Klöser,Herr
Cmi 7h17'19"1d27'
Klöver,Gerhard
Cmi 7h6'23"4d34'
KMC 50/50
Cyg 19h26'54"30d0'
Kmetz,Cynthia J
Oph 17h27'49"-1d1'B
Kmiecik,Marikay
Cas 2h30'14"59d43'
KMS 1
Uma 8h41'33"52d51'
Knadler,Patricia
Cyg 20h32'17"42d44'
Knapköien,Amina Lovscell
Ari 3h23'35"30d33'
Knapp's, Rosedale Star
Cnv 12h7'49"36d13'
Knapp,Arthur Earl
Cam 3h50'1"58d58'
Knapp,Darryl Daniel
Cet 1h19'22"-10d32'
Knapp,David "Shameless"
Gem 6h44'56"14d34'
Knapp,Dr Robert C
Cep 23h4'39"64d46'
Knapp,Eva Madeline
Mon 6h19'1"2d42'
Knapp,Felix
Sgr 19h47'2"-43d54'
Knapp,Helen E
Mon 7h27'49"-5d48'
Knapp,Helga
And 2h17'24"46d23'
Knapp,Jennifer
Cas 23h29'0"60d48'
Knapp,Joan Frances Green
Cyg 20h6'58"41d6'
Knapp,Julie Ann
Cmi 7h24'22"1d0'
Knapp,Nancy
Lyn 9h0'43"44d46'
Knapp,Patricia Anne
Mon 7h18'34"-6d45'
Knapp,Rachel
Cyg 20h24'48"40d31'
Knapp,Sabrina Sampson
Psc 23h32'57"0d11'
Knapp,Sharon
Lyr 19h19'23"42d39'
Knapp,Thomas Hunt
Ser 15h9'31"-1d24'
Knapp,William H
Aql 20h0'59"13d54'
Knapp-Brightest Star, Douglas C
Aur 4h55'26"48d49'
Knappmann,Wilbur J
Eri 2h54'53"-5d4'
Knapstein,Harald
Boo 14h52'0"25d9'
Knauer Star,The Lumpy
Aur 7h7'26"36d33'
Knauer,Ian
Boo 14h33'13"47d17'

Knauer,Scott Carroll
Aur 4h47'24"51d2'
Kneale,Jean Anne
Aql 19h27'46"7d58'
Knecht,Elmer H
Her 18h17'0"14d20'
Knecht,Kathi A
Cas 1h40'47"73d32'
Knecht,Leo Valentine
Lib 15h43'11"-20d21'
Knechtle,Andrew
Dra 11h53'52"70d3'
Kneebone,Paul Dean
Ser 15h20'12"20d35'
Kneerim,Arthur
Dra 16h36'37"61d16'
Kneeter's Golden Star
Cyg 21h51'24"40d4'
Kneisel,Christina
Lib 14h51'32"-1d9'
Kneisley,Grayson Baird
Cep 21h3'37"61d24'
Kneisley,Jonathan Reid
Boo 14h56'51"38d45'
Knepp,Miriam & Perry
Vir 11h56'12"-3d51'
Knepper,Barbara
Lyr 18h33'41"36d3'
Knepper,James
Her 16h30'43"39d35'
Knepper,Jo Ann
Aur 7h0'21"40d23'
Knetz,Matthew T
Lac 22h15'1"46d6'
Knetzger,Jr,Edwin L
Aur 5h2'58"42d20'
Kneubühl,Britta
Dra 20h29'53"68d9'
Kneubühl,Lina
Dra 19h51'39"86d5'
Kneupper,Video Star Karl-Dieter
Ser 15h55'13"-0d6'
Knez,Brandon Cole
Sge 20h16'24"17d54'
Knick,Miss Jennifer
Cnc 8h37'11"18d28'
Knieriem,Kyle McKinley
Mon 7h45'20"-7d13'
Kniesner,William Stephen Freund
Aur 6h26'0"33d56'
Kniess,Robert P
Aql 20h11'39"8d33'
Knieter,Lyn
And 1h56'56"38d55'
Knifton,Eleanor
Umi 17h7'60"85d59'
Knight
Cma 6h36'5"-16d53'
Knight Anniversary Star,Ibb & Jim
Peg 22h58'45"10d12'
Knight Family, Blessings,Burton Lee
Lyn 8h52'26"36d31'
Knight of Day
And 23h1'36"51d37'
Knight Star Albert
Umi 15h45'49"80d12'
Knight Star,The
Uma 10h5'14"42d14'
Knight Star,The
Uma 12h9'47"62d1"
Knight `Life',Peter
Per 4h2'1"40d58'
Knight's Lair
Ori 5h59'13"12d54'
Knight,Anita Samantha
Lyr 18h15'56"31d45'
Knight,Brittany Lynn
And 1h18'35"33d53'
Knight,Christel Ursula
Cas 22h56'37"54d46'

Knight,Christopher Charles
Uma 9h7'17"52d51'
Knight,David Francis
Cep 23h5'52"60d45'
Knight,Forest Boyd
Hya 8h35'55"-0d4'
Knight,Francis B
Aql 19h26'19"8d18'
Knight,Garrett Roulston
Hya 9h8'54"1d41'
Knight,James Patrick O'Hara
Per 1h47'36"56d28'
Knight,Jan
Cyg 20h22'1"38d55'
Knight,Jimmy Lee
Aql 20h5'21"6d30'
Knight,Joan & Robert
Peg 23h0'11"13d47'
Knight,Jonathan Rashleigh
Lyn 9h1'11"40d53'
Knight,Jordan
Peg 22h33'14"10d19'
Knight,Judith N
Lyr 19h21'0"42d14'
Knight,K R M-White
Sct 18h29'18"-4d59'
Knight,Kevin W
Hya 8h34'1"-10d32'
Knight,Kimberley
Del 20h19'25"18d49'
Knight,Lynn R
Cas 0h32'50"64d4'
Knight,Mark D
Boo 14h18'56"15d36'
Knight,Matthew Calvin
Ori 5h59'45"17d9'
Knight,Michelle
Mon 6h38'39"7d6'
Knight,Michelle Catherine
Uma 9h0'32"68d27'
Knight,My Love,Happy 15 Anniv,Carol
Uma 11h52'46"33d43'
Knight,Pamela
Peg 22h19'25"5d11'
Knight,Paul
Ori 6h2'42"4d52'
Knight,Phoenix Rising
Cet 3h0'23"0d29'
Knight,Professor Walter
Ori 6h0'41"4d16'
Knight,Rachel Giessinger
Crb 15h19'33"31d21'
Knight,Richard E
Crb 15h56'47"38d55'
Knight,Robert Lester
Hya 9h2'0"-8d16'
Knight,Sandi
Mon 6h43'0"10d20'
Knight,Sara Rochelle
Del 20h27'11"11d31'
Knight,Shalta
Peg 0h11'50"18d1'
Knight,Shannon Michelle(Knoefel)
And 23h3'45"46d59'
Knight,Simon Cosmo
Cyg 21h0'28"35d28'
Knight,Starlene
Mon 7h47'22"-2d4'
Knight,Stephanie
Cyg 21h58'52"52d57'
Knight,Steve & Michael Goodman
Mon 7h56'49"-2d31'
Knight,Stuart & Robin
Ori 6h5'44"7d24'
Knight,Tami Lee
Mon 6h54'19"-8d42'
Knight,Tina
Lyr 18h47'0"31d23'
Knight,Valorie
Lyn 7h7'46"44d44'
Knight,Vi & Jim
Cyg 21h5'36"40d7'

Knight,Yvonne Jean
And 23h16'34"51d52'
Knight,Zanna
Aqr 21h32'52"-5d48'
Knight-Burke,Rhonda
And 2h26'41"47d59'
Knight-Ruffin 50
Equ 21h1'1"3d36'
Knights In Shining Armor
Sex 10h0'19"0d9'
Knights,David Ronald
Ori 5h56'0"15d0'
Knights,Geraldine
Aqr 23h8'48"-12d21'
Knights,Roberta
Mon 7h59'41"-1d49'
Knights,Ross Dexter
Cep 20h34'0"75d43'
Knip
Aur 6h3'0"31d45'
Knipp,Big John
Uma 9h31'22"48d23'
Knippa,Gary Louis
Aql 19h30'52"11d2'
Knippa,Jerry Paul
Cet 1h2'42"-3d28'
Knis,Laurie
Lyr 18h42'29"32d0'
Knise,Jr,Joseph M
Cnv 12h44'17"51d56'
Knittel,David L
Lac 22h39'15"53d50'
Knittel,Joan & Jeffrey C
Peg 22h46'54"34d6'
Knitter,Constanze
Cas 23h42'26"61d34'
Knitter,Linda & Rolf
Mon 6h43'34"-1d55'
Knitzer,Steven
Her 17h59'0"50d18'
Knizner,Diane
And 23h37'13"45d35'
Knobel,Hans-Jürgen
Ari 2h32'48"30d57'
Knobel,PhD,Roland
Aur 5h3'21"40d51'
Knobel,Simon Bennet
Peg 21h54'32"34d3'
Knoblauch,Anne Duncan
And 2h23'57"45d14'
Knoblich,Scott
Her 18h2'13"31d13'
Knobloch 143,My Special Mom,Shirley
Lac 22h12'19"47d11'
Knobloch,Fred W
Aur 5h11'39"42d48'
Knoblock,Heather Lee Bechtold
Ori 4h43'12"12d47'
Knock Out,Bernd Freier
Vir 12h55'42"-12d3'
Knode,Herr
Lyr 19h21'10"31d21'
Knodt,Karl-Heinz
Sgr 18h57'20"-21d49'
Knodt,Leo
Sgr 19h49'47"40d24'
Knoebl,F
Lac 22h50'46"40d14'
Knoedler,Bonnie
Peg 22h37'39"26d30'
Knoeffel,Alfred & Lotte
Cet 0h57'20"-1d49'
Knoepfler,Gayle S
Cam 5h37'16"61d45'
Knoerr,Adam
Her 16h12'0"50d42'
Knoerr,Randy
Boo 14h1'30"26d42'
Knofel,Dan
Dra 14h52'45"61d17'
Knoll,Matthew John
Dra 17h10'0"51d14'

Knoll,Rainer
Lyn 8h8'13"47d46'
Knoller,Eric
Aql 19h39'49"10d47'
Knoop,Amanda
Eri 4h17'32"-12d50'
Knoop,Kristin Anne
Vir 11h56'1"0d59'
Knop,Günter
Dra 16h55'60"72d41'
Knopes Best Bear,Keith
Cep 22h14'17"59d58'
Knopf,Sascha
Uma 9h22'34"58d50'
Knopf,Wally MacKenzie
Cnv 13h22'44"51d58'
Knopp III,John W
Her 16h26'20"22d34'
Knopps 25th,Maureen & Peter
Lyr 18h32'0"32d48'
Knops,Kenneth J
Per 2h48'41"40d25'
Knorr,Richard A
Boo 14h36'54"8d22'
Knorzer,Joachim Kurt
Del 21h5'31"12d8'
Knott Star,Shelley Bernice
Boo 14h22'1"53d53'
Knotts,Tristan Emery
And 0h59'0"36d49'
Knouse,"The Duchman"
Oph 18h0'26"8d59'
Knowle & Athena Joy
Lyr 19h11'40"38d38'
Knowles & Grandma-PaPa James Raymond
Ori 5h17'59"1d37'
Knowles Eternal Light, The Jack O
Cet 2h26'38"5d6'
Knowles,Charles Patrick
Her 17h12'35"45d57'
Knowles,Christine & Warren
Cyg 19h43'30"29d37'
Knowles,Gary W
Sex 9h59'0"-10d12'
Knowles,Greg
Oph 16h2'31"-5d51'
Knowles,Jacqueline Rogers
Aql 19h54'31"11d42'
Knowles,Jennie Mae
And 1h24'33"37d31'
Knowles,Justin
Sex 10h10'7"-9d10'
Knowles,Katrina
Aql 19h7'0"-6d50'
Knowles,Kelsy Anne
Peg 21h58'34"32d31'
Knowles,Robert Michael
Cep 21h21'1"67d48'
Knowles,Valerie
Cyg 21h57'18"48d46'
Knowlton Star,The Peter
Cep 20h21'22"78d59'
Knowlton,Beth Ann
Cas 1h13'38"61d40'
Knowlton,Joanne
Cam 6h9'47"71d49'
Knowlton,Mary Andreé
Sct 18h54'23"-8d10'
Knowlton,Samantha
Ori 5h51'51"16d60'
Knowlton,Spencer Brooks
Gem 6h30'26"12d13'
Knowlton,Stephanie Michelle
Umi 13h10'50"75d59'
Knowlton,Susan
Leo 10h0'22"7d56'
Knox,Andrew
Cep 21h18'1"65d27'
Knox,Carol Ann Ewart
Com 12h29'58d14d35'
Knox,Colin Morton
Her 16h51'26"40d46'

Knox,Corrin F
Oph 17h28'36"-23d58'
Knox,Joanna
Cas 1h55'54"61d8'
Knox,John
Dra 18h51'37"65d26'
Knox,Jr,George C
Dra 11h39'51"68d37'
Knox,Laura
Mon 7h48'49"-1d50'
Knox,Pearl
Cas 1h49'41"73d48'
Knox,Rodney
Umi 15h10'0"78d38'
Knox,Sorie & Marsha
Cyg 19h15'18"48d57'
Knox,Victoria Lynn
Sct 18h51'41"-6d51'
Knuddelfloh
Lib 15h1'24"-28d24'
Knudi
Cap 21h39'49"-19d25'
Knudsen,Alexandria Nicole
Cyg 21h7'0"31d6'
Knudsen,Bob & Muriel
Del 20h20'21"10d23'
Knudsen,Chris
Ser 15h19'44"-1d2'
Knudsen,Douglas
Aql 20h34'52"0d40'
Knudsen,Edith & Kenneth
Hya 8h10'11"1d24'
Knudsen,Harnold
Mon 6h38'55"7d26'
Knudsen,Linda
Del 20h13'18"14d58'
Knudsen,Lisa Pauline
Boo 14h29'37"50d17'
Knudsen,Travis K
Dra 19h47'57"61d1'
Knudson 7/31,Sue
Boo 14h27'54"22d57'
Knudson,Andy & Harry
Aur 6h16'39"46d34'
Knudson,Robert Patrick
Cep 22h13'42"68d39'
Knudson,Terra Lynn
Mon 6h59'47"8d39'
Knudten,Richard David
Cep 22h39'46"61d24'
Knudulu
Leo 9h19'13"12d2'
Knue 061074,Jeanie
Com 12h58'0"14d38'
Knuffel
Sco 16h27'25"-40d44'
Knupfer,Nikolaus
Sco 16h48'52"-28d37'
Knupp,Suzanne Dale
Ari 3h23'23"28d51'
Knust,Reinhard
Dra 18h21'34"65d5'
Knute
Cam 5h46'29"60d46'
Knuth,Melissa G
Psc 0h46'23"27d58'
Knutsen JM,Diane 1693
And 0h10'1"38d49'
Knutson,Agne Cattaneo
And 23h46'1"45d38'
Knutson,Constance Faye
Cap 20h57'1"-15d43'
Knutson,Kevin
Her 17h45'23"14d26'
Knutson,Leonard & Jeanene
Lyr 18h37'35"29d15'
Knutson,M O
Mon 6h39'51"10d59'
Knutson,Machele Kathleen
Cyg 21h44'42"29d37'
Knutson,Tryc
Tau 3h59'13"1d31'
Knödel 22/04/1926, Gustav
Cep 21h36'43"61d7'

Knüppel,Melanie
Cap 20h21'49"-26d31'
Ko,Chong Min
Mon 6h23'19"8d46'
Ko,Derek F Y
Umi 13h4'29"74d25'
Ko,Eric F K
Umi 13h3'44"74d34'
Ko,Eric S C
Umi 13h3'11"74d5'
Ko,Jason S Y
Umi 13h12'45"74d21'
Ko,Leon S T
Umi 13h6'52"74d2'
Ko,Lucilla
Umi 13h10'51"74d39'
Koality's Dream
Per 2h38'25"43d52'
Kobacker II,Alfred J
Ori 5h55'16"-1d37'
Kobbernagel,Christina
Cyg 20h43'35"45d24'
Kobel,Sheri
And 2h28'20"47d15'
Kober,Karin und Robert 20-09-93=30
Sgr 18h57'2"-23d7'
Kober,Sabine
Cap 20h26'38"-26d22'
Koberg,Gwyneth Nadine
And 0h4'48"46d16'
Koberg,Kathea Elsbeth
Vir 13h56'22"5d54'
Kobernusz,Julie Ann
Peg 0h2'59"28d8'
Kobi,Heidi J
And 23h27'0"40d28'
Kobilke,Monia
Gem 7h32'57"34d0'
Kobishop,Ed & Mavis
Dra 20h20'30"62d8'
Koblitz,Norbert Franz
Aql 19h3'56"3d6'
Kobra,Mr-Davide Sassi
Cep 21h35'24"65d36'
Kobrzycki,AnnMarie
Uma 11h48'42"63d48'
Kobs,Angela
Cas 2h5'1"59d5'
Kobs,Jared Ashton
Cet 2h19'59"5d36'
Kobylarz,Jr,Michael Adam
Aur 5h28'56"30d38'
Kobylarz,Michael Albert
Aur 6h3'20"46d26'
Kobylski
Lyr 18h49'45"32d50'
Kobzeff,Matthew
Her 15h48'39"46d24'
Kobzon,Iosif
Umi 16h46'59"75d42'
Kocanda,Zachary John
Cyg 20h2'37"31d25'
Koceski,Sandy
Eri 3h38'18"-3d39'
Koch's Star,The
Vul 20h4'17"28d27'
Koch,Alexandre
Lyn 7h32'29"51d17'
Koch,Andrea Lee
Mon 7h4'21"3d45'
Koch,Arthur F
Dra 17h3'36"52d3'
Koch,Barry Stephen
Hya 8h13'36"2d4'
Koch,Best Mom In The Galaxy Dorothy
And 1h51'54"36d11'
Koch,Chris J
Sct 18h41'52"-5d23'
Koch,Christine
Psc 23h21'11"1d50'
Koch,Conrad B H
Lac 22h28'26"41d13'

Koch,Curtis
Uma 10h40'44"70d33'
Koch,Dominique
And 0h15'47"38d10'
Koch,Donald B & Sherry L
Dra 16h8'13"58d27'
Koch,Gabriele Maria
Com 13h10'31"20d11'
Koch,Günter
Sge 19h38'34"17d33'
Koch,Harald
Aqr 22h24'59"-1d58'
Koch,Hayden Matthew
Cnc 7h55'0"11d43'
Koch,Hermann
Her 16h38'30"38d18'
Koch,James Gregory
Cnv 12h32'48"37d52'
Koch,Jeffrey
Dra 16h7'35"67d18'
Koch,Jeffrey Paul
Per 4h28'30"52d16'
Koch,Jessica Ashley
Aql 19h50'57"15d60'
Koch,Johann
Lyr 19h12'16"31d7'
Koch,Josephina Anna Baeni
Cma 6h50'5"-17d5'
Koch,Jr,Robert Earl
Leo 10h40'54"12d28'
Koch,Kim
Uma 10h12'20"71d34'
Koch,Manfred
Psc 23h2'53"1d2'
Koch,Mareen
Aqr 22h42'10"-5d24'
Koch,Pari
Oph 16h55'23"0d47'
Koch,Patricia Lee Moore
Mon 6h43'1"3d13'
Koch,Robert
Uma 9h18'19"47d24'
Koch,Robert W
Leo 10h59'25"11d22'
Koch,Shaun
Boo 15h26'26"50d55'
Koch,William F
Cep 22h42'1"70d8'
Kochan,Jennifer Rae
Cnv 13h54'36"32d15'
Kochan,Sarah
Cas 0h57'17"60d45'
Kochania,Ania
And 2h33'11"50d19'
Kocharian,H E
Ari 2h40'44"25d4'
Kochav Haliallen
Uma 9h39'26"43d57'
Kochav Robin Sue
Lyn 8h9'41"46d56'
Koche,Missie
Vul 19h48'25"23d33'
Kochems,Karl
Ori 5h51'24"21d58'
Kochheiser,Joseph Michael
Her 17h3'25"46d20'
Kochiesen,Joseph
Her 16h46'11"33d32'
Kochkin,Jeri
Uma 10h50'1"41d4'
Kochsiek,Prof Dr Med Kurt
Uma 10h53'1"40d42'
Kochubka,Anne
And 2h4'50"39d51'
Kocik,Grandma
Ari 1h48'14"13d17'
Kocina,Jefferey Scott
Her 17h28'40"28d9'
Kock am Brink,Ulla
Cnc 9h0'1"8d59'
Kocks,Klaus Dr
Psc 23h37'27"-2d52'
Kocozom,Andrew et Danielle
Cep 20h58'38"62d50'
Kocsis,John M
Her 17h49'35"41d1'
Kocsis,Star
Lac 22h4'11"46d27'
Kocul,Jack
Lac 22h52'47"37d38'
Kocur,Kristian Anne & Kendal Reed
Umi 16h28'39"72d44'
Koczan,Wayne Joseph
And 0h9'35"38d27'
Koczwara,Joseph
Dra 16h18'54"61d29'
Kodak Portrait-Partner
And 1h41'38"41d33'
Koder,Michele
Peg 21h40'50"27d49'
Kodis,Theresa
Uma 9h52'59"71d10'
Kodish,Alan E
Her 17h35'27"27d30'
Kodman,Robert A
Her 16h19'1"28d25'
Koduri,MD,Sailaja
Uma 10h52'43"52d17'
Koebel,Gerlinde
Uma 10h40'1"55d11'
Koebnick,Chad Robert
Uma 8h24'30"71d1'
Koeferl,Michael Toyota
Boo 14h50'33"38d51'
Koegel,Jane
Mon 6h36'53"-0d20'
Koehl,Joel
Boo 13h59'59"25d42'
Koehler,Beth & Walter
Per 4h6'40"50d52'
Koehler,Fred W
Ser 18h25'35"1d35'
Koehler,Jerry
Gem 7h2'33"33d54'
Koehler,John G
Her 17h10'13"45d8'
Koehler,Kurt Allen
Her 17h32'29"21d14'
Koehler,Linda S
Peg 22h44'1"29d53'
Koehler,Nancy Wells
Aql 20h12'16"12d24'
Koehler,Tracy Dee
Mon 6h42'32"7d40'
Koehler,Valerie Christine
Peg 22h45'20"27d56'
Koelbel,Brian & Mary Ellen
Cnv 13h48'59"42d3'
Koeller,Georgene
And 1h32'36"36d7'
Koelling,H Fred
Ser 18h17'26"-14d54'
Koelling-Louvar
Uma 10h56'51"35d55'
Koellisch,Joseph & Valerie
Uma 11h12'55"53d55'
Koellner,Rick
Her 16h57'36"30d34'
Koen,Amber Terrell
Ori 5h31'10"-1d46'
Koeneman,Jenna
Lyn 9h4'14"46d3'
Koeneman,Nancy
Lyn 7h53'46"40d6'
Koenig II,Robert H
Cma 6h58'50"-30d37'
Koenig IV,Henry A
Her 17h21'42"22d55'
Koenig,Amy Michelle
Peg 23h42'1"31d36'
Koenig,Esq,Robert Evans
Lac 21h58'43"38d57'
Koenig,Heike
Aqr 22h23'11"1d10'
Koenig,Joshua August
Dra 17h22'18"68d36'
Koenig,Joshua John
Aur 4h55'31"51d1'
Koenig,Julia W
Lyn 9h4'30"37d29'
Koenig,Lana Vacha
Cyg 21h54'1"41d57'
Koenig,Lindsay Jane
Lyn 9h14'39"35d15'
Koenig,Lucas Daniel
Aur 7h11'48"40d3'
Koenig,Marianne
Peg 21h24'14"18d55'
Koenig,Marilyn
And 2h17'53"46d22'
Koenig,Michelle Kathleen
Vul 19h42'51"26d2'
Koenig,Robert
Aur 6h29'43"38d1'
Koenig,Rueben
Ori 4h52'57"4d38'
Koenig,Sara Beth
Vul 19h40'48"26d22'
Koenig,Scott Anthony
Cet 2h49'55"7d26'
Koenig,Star C S C E
Vul 19h47'44"28d49'
Koens,Ann
Hya 8h58'54"-1d5'
Koepke,Beverly
Eri 3h17'58"-16d24'
Koepp,Brett
Aql 20h2'1"10'9'
Koepp,Martha
Tri 1h56'49"27d54'
Koepp,Mitch
Cmi 7h56'1"8d18'
Koeppel,Renee
Uma 9h0'57"50d52'
Koeppel,Robert Paul
Cyg 21h37'27"41d7'
Koeppen,Kirk,Jackson & Devon
Aql 19h55'53"10d7'
Koeppen,Mary Jo
Peg 22h51'60"27d10'
Koerner,Clement Benjamin Charles
Per 3h1'47"41d42'
Koerner,Clive Nicolaus
Eri 3h38'20"-5d18'
Koerner,James Gordon
Cep 20h44'37"68d36'
Koerner,Natalie Marie
Cyg 19h45'58"31d19'
Koerner,Volker
Boo 13h37'26"12d0'
Koerper,Chelsea Diane
And 23h26'1"50d24'
Koerper,Ryan Scott
Her 15h48'30"42d56'
Koers,Vincent & Kitty
Her 18h2'45"18d57'
Koesoemawardani,Renie
Uma 11h13'36"50d10'
Koessler,Einahpets L
Tau 3h37'31"24d30'
Koester
Del 20h54'48"6d8'
Koester,Brittany
Lyn 7h55'10"38d8'
Koester,Caroline Rose
Lyn 7h27'54"43d26'
Koester,Don N
Her 18h18'1"20d35'
Koester,John
Her 17h11'49"43d17'
Koester,Karl E
Per 3h58'24"50d33'
Koester,Kelsey A
Cam 9h14'0"81d16'
Koester,Lance J
Her 17h20'0"14d22'
Koester,Pete
Ori 6h2'29"-0d51'
Koester,Sean Michael
Lac 22h12'44"49d6'A
Koester,Smith Matthew
Lac 22h39'39"50d4'
Koester,Thomas Jame
Lac 22h12'44"49d6'B
Koestler,Courtney Ann
Cet 2h11'56"6d8'
Koetting
Ari 2h1'36"25d26'
Koferl,Fredrick
Dra 10h54'50"73d9'
Koff,Mitchell Todd
Uma 10h48'42"40d36'
Kofler,Anna
Aur 5h17'53"43d0'
Kofoed Supreme Worthy Advisor,Margaret
And 23h1'31"50d26'
Kofsky,Charles
Cam 3h52'1"52d42'
Kofsuske,Marianne
Dra 11h57'27"72d6'
Kogan,Elana
Cyg 21h7'31"30d33'
Kogel,Fred
Uma 11h17'59"41d53'
Kogelschatz,Robert C
Boo 14h29'11"47d21'
Kogen,William H "Papa"
Aql 19h31'55"10d33'
Kogl,Celeste Louise
And 2h35'29"41d1'
Kogo Amaranthine
Uma 12h9'29"58d56'
Kohagura,Mark Shigeru
Dra 18h57'1"58d35'
Kohaku
Ari 2h12'29"22d21'
Kohan,Cherly Launa
Lyn 7h54'16"40d40'
Kohan,Jaime
Cma 7h16'38"-16d18'
Kohan,Jr,John Harry
Her 15h53'0"50d36'
Kohankie,Amy & Hubert Rust
Uma 10h6'57"62d1'
Kohfink,J-Michael
Aqr 21h5'29"-1d10'
Kohl,Dagmar
Cnc 8h7'38"30d0'
Kohl,David
Cnv 13h10'13"41d13'
Kohl,David Lawrence
Her 18h16'24"14d54"
Kohl,Deanna Marie
Crb 16h4'34"27d13'
Kohl,Hannelore
Lmi 11h4'45"30d10'
Kohl,Hans-Dieter
Vul 20h44'12"20d12'
Kohl,Katja
Cam 7h53'58"73d3'
Kohl,Michael
Psc 0h59'28"21d44'
Kohl,Monica Leigh
Lyr 19h19'25"40d55'
Kohl,Stefan
Lmi 9h30'32"38d14'
Kohl,Stefan
Ari 2h41'30"21d49'
Kohlbeck,Nancy
Uma 12h20'1"62d38'
Kohlberg,Margaret E
Tri 1h58'0"30d47'
Kohlenberger,Charles William
Her 16h38'26"50d55'
Kohler,Erica Lyn
Cas 0h25'13"61d29'
Kohler,Heidi
Cas 1h46'0"71d29'
Kohler,Kelly Patricia
Cet 1h24'49"0d18'
Kohlhof,Eckert
Hya 9h14'38"-6d35'
Kohlhoff,Erick
Cam 7h51'0"61d20'
Kohlman,Robin Jean
And 23h25'1"49d58'
Kohlmann,Kurt
Boo 14h28'0"26d8'
Kohlmeier,Justin V
Dra 19h1'35"48d6'
Kohlrieser,Douglas George
Boo 14h56'19"30d34'
Kohls,Dr Jürgen
Leo 9h57'51"11d59'
Kohlschmitt
Cet 2h5'13"-1d50'
Kohn,Amy
And 0h56'0"35d34'
Kohn,Dr Jerry
Oph 17h55'14"12d39'
Kohn,George M
Vir 13h32'0"-20d57'
Kohn,Lori
Boo 14h10'37"44d7'
Kohn,Richard
Lac 22h14'44"51d33'
Kohn,Robbie
Cyg 19h56'1"38d31'
Kohnhorst,Tarrah-Shea
Peg 23h39'19"28d54'
Kohr,Pat & Tish
Crb 16h9'0"28d47'
Kohrherr,Barbara J
Vul 21h14'23"20d21'
Koiv,Erika Anne
Cas 1h1'29"50d13'
Koivu,Timo
Cam 3h42'20"60d26'
Koivuniemi,Jacquè
Vul 19h59'0"23d37'
Koivuniemi,Kalle
Cam 3h24'10"60d9'
Koji 20 Age Now
Ari 1h45'27"21d33'
Koji Nakamura
Sgr 19h19'36"-29d39'
Kokabh-Hannah
Umi 14h47'51"68d46'
Kokak,Cetin
Uma 9h14'41"48d49'
Koki
Ser 16h8'11"6d49'
Koko
And 2h21'1"46d59'
Kokoschka,Richard
Dra 18h21'14"65d15'
Kokum,Robert Henry Peter
Her 16h55'38"32d27'
Kol-Hallel,Lorna
And 23h20'1"42d22'
Kola
Peg 22h41'56"32d38'
Kolaci,Nicholas Jan
Aur 5h3'53"42d6'
Kolakowski,Jolene Catherine
Crb 15h49'1"32d13'
Kolambage,Sumith
Crb 15h15'15"31d3'
Kolanich,John James
Aur 6h22'41"34d12'
Kolanko,Stanley J
Lac 22h38'50"52d44'
Kolanowski,Franz-Josef
Cas 23h30'27"63d9'
Kolar,Helen Georgina Tesmer
Peg 23h4'1"8d17'
Kolasa,Thomas S
Cep 22h10'19"55d9'
Kolator,Alice
Equ 21h0'36"10d22'
Kolb II,John N
Cep 22h9'12"61d20'
Kolb,Arthur Benjamin
Cnv 13h58'45"40d1'
Kolb,Hans Martin
Lac 22h3'37"50d4'
Kolb,Jennifer
And 23h1'51"51d47'
Kolb,Letitia
Uma 13h27'7"56d29'A
Kolb,Linda Manges
Sex 9h45'33"4d50'
Kolb,Michael
Uma 13h27'7"56d29'B
Kolb,Roger
Aur 6h55'44"38d16'
Kolba,Bobby
Ser 15h59'50"1d13'
Kolbe,Brigitte
Cnc 8h27'1"32d43'
Kolbe,Gerd
Ari 2h24'55"21d36'
Kolber,Elizabeth P
And 0h0'38"40d49'
Kolberg,John
Boo 14h49'45"25d48'
Kolberg,Keith
Aur 6h11'30"38d58'
Kolberg,Lynn Christine
Peg 22h4'57"4d26'
Kolbert,Steven Eric
Lib 15h39'51"-23d33'
Kolby,Robert C
Boo 13h36'57"25d53'
Kolde,Sasha Schuyler
Cas 2h2'32"58d40'
Koldobsky,Ms Ana Maria
Cas 0h28'19"70d29'
Kole B's Star
Uma 11h45'53"38d50'
Kole's
Lyn 7h53'22"48d37'
Kole's Star,Jennifer
And 2h24'1"42d9'
Kole,Cecilia Guzman
Cyg 20h54'52"40d19'
Kole,Princess Teresa L
Cas 1h4'19"50d31'
Kolega,William J
Cet 2h59'31"1d53'
Kolek,Linda L
Lyn 8h26'16"44d4'
Kolessy,Jean Marie
Cas 2h0'25"58d53'
Kolestar,Joseph Stephen
Vir 13h56'38"-5d28'
Kolestar,Joseph John
Cnc 8h24'19"18d56'
Kolestar,Lisa Rena
Cnc 8h5'58"18d60'
Kolestar,Paul Michael
Cnc 8h56'41"31d23'
Kolestock,Faith Michael
Lyn 7h4'43"50d3'
Kolhhammer,Inge
Lmi 10h4'0"31d11'
Kolhoff,Madison C
Lac 22h6'13"38d47'
Kolinski,Danette "Danie"
Aqr 20h59'19"0d41'
Kolitsch,Andreas
Peg 23h36'27"13d4'
Kolk,Susan
Cas 2h35'38"68d52'
Koll,Bill
Her 16h47'15"13d20'A
Koll,Monique
Mon 6h21'19"8d25'
Koll-Nesher,Issac
Vir 11h39'1"8d49'
Kollar,Jeffrey George
Aql 20h32'54"-6d48'
Kollar,Pam
Cas 23h31'26"60d5'
Kollarik,Arthur & Frances
Her 18h4'27"48d6'
Kolleen
Aqr 22h22'12"0d25'
Koller,Christiane
Lyn 8h56'34"45d32'
Koller,Frank & Julie
Cyg 20h40'1"38d29'
Kollhof & Family,Jan Klaas
Hya 9h35'57"1d16'
Kolligs,Werner
Aqr 22h42'1"0d29'
Kolling,Mary A
Cyg 19h28'32"31d13'
Kollman,John RE
Per 3h59'59"31d29'
Kollmar,Stephanie Lea
And 0h10'37"35d47'
Kollo,B Joan
Cam 12h56'37"76d58'
Kolman,Patsy
And 0h54'42"45d2'
Kolmparis
Aur 5h0'38"50d25'
Kolodny,Nancy Harrison
Peg 23h3'12"31d21'
Kolodny,Sherrill David
Peg 21h8'39"13d1'
Kolodziej,Edwin A
Boo 13h58'58"22d30'
Kolodziej,Kent
Boo 14h6'30"40d52'
Kolodziejczak,Olek
Aur 5h54'47"30d2'
Kolodzinski,Victoria Dolores
Cas 1h1'43"52d58'
Koloff,Joshua Beau
Her 18h1'14"28d31'
Koloff,Kevin
Hya 9h16'59"5d9'
Koloko,Eric
Umi 6h49'40"88d57'
Koloms,Bradley A
Cap 20h35'15"-13d9'
Kolonko,Bettina
Lyr 18h35'60"40d52'
Kolonko,Nadine
Boo 13h42'37"10d23'
Kolosvary,Stephen Daniel
Per 3h19'11"40d10'
Kolowski,Mary
Peg 23h46'22"30d32'
Kolozsváry,Victor
Aur 5h13'15"44d44'
Kolsky,Charles L
Psc 1h4'0"23d45'
Kolsrud,Rachael Alexis Lagemann
Peg 22h24'48"29d46'
Kolter,John Francis
Hya 9h16'51"-6d33'
Kolter,Michael Joseph
Oph 17h54'55"10d24'
Kolton,Jeffrey E
Her 16h43'15"32d37'
Kolyno Kadesh-Kasey's Kimah
Leo 11h54'31"26d16'
Kolysko,Matthew Stephen
Aql 19h29'57"-7d46'
Komanicky,Alexander
Aql 20h9'56"0d15'
Komar's (Krog)
Cmi 7h59'15"8d53'
Komar,Chris
Her 16h27'33"25d9'
Komar,Harold
Sgr 19h38'49"-37d59'
Komarek,Jon & Cynthia
Aql 18h57'0"-5d20'
Komarek,Kari Joyce
Uma 9h58'15"51d25'
Komars Light
Mon 7h58'13"-3d56'
Komel
Cma 6h54'38"-18d54'
Komer,Myron
Uma 9h57'57"51d20'
Komerex,San drA
Cyg 19h27'1"32d18'
Kominko,John Michael
Lmi 9h59'55"38d26'
Komisar,William L
Cep 23h38'0"64d46'
Kommert,Hans Ludwig
Cnv 13h57'1"46d29'
Kommerzialrat Franz Denner
Ori 6h3'50"8d21'
Komnenich,Michael J
Sgr 19h40'14"-40d55'
Komninos,Margery Ann
And 2h20'0"41d26'
Komolvajana,=Jht= Chusri
Oph 17h55'18"10d12'
Komorinezumi
And 0h13'14"34d25'
Komorn,Diane Gail Bachman
Cas 23h32'24"58d49'
Komornik,Patricia
Cas 0h13'24"47d57'
Komp,Laura Mae
Vul 21h25'0"24d5'
Koncius,Esq,Jeffrey A
Cam 3h17'46"55d30'
Koncius,Jura
Aur 5h26'25"38d17'
Kondanen
Ori 5h1'0"0d28'
Kondek,Amanda Nicole
Cas 0h40'41"65d58'
Kondschak,Hannelore
Lib 14h46'32"-23d15'
Kondysar,Nicholas Peter
Aur 6h27'40"30d22'
Konecky,Sean & Sibylle
Cep 23h9'48"64d22'
Konecky,William S
Cep 22h10'30"60d37'
Konestabo,Barbara G
Tri 2h9'33"33d40'
Konfederak,Bernice
Lyn 9h1'47"35d42'
Kong,Betty M
Cyg 19h59'15"31d40'
Kong,Jeremy Alexander
Lmi 10h52'51"31d13'
Kongsuwan,Sansern & Michelle
Lmi 10h43'15"33d41'
Koni & Jim(K&J Ranch)
Her 17h28'43"20d3"
Konicek,Donald
Her 16h56'33"23d58'
Konicki,Mike
Per 4h23'24"50d25'
Konieczny,Karl-Heinz
Peg 23h29'41"13d45'
Konieczny,Olaf
Boo 13h36'32"12d27'
Konigsberg,Sidney S
Lyr 18h51'17"41d50'
Konior,Eugene R
Peg 23h28'11"24d24'
Konior,Hulda M
Cas 0h44'34"66d17'
Konior,Joseph E
Aql 19h31'15"8d55'
Konishi,Shoji
Lyn 7h51'33"41d11'
Koniuto,Mary E
Com 12h29'0"28d59'
Konkol,Aleta
Aql 19h30'0"1d21'
Konkol,Robin & Brian
Cyg 20h23'12"38d22'
Konmo,Süß
Lyn 7h48'0"40d48'
Konni
Cyg 19h26'27"32d13'
Kono,Mic
Dra 16h26'41"64d46'
Kono,Nicholas Francis
Per 3h42'44"50d32'
Konopa,James E
Dra 19h25'12"56d32'
Konopka,Justin Alexander
Lac 22h51'18"54d41'
Konopka,Stephen P
Boo 14h52'35"30d46'
Konopka,Uschi Kleiner Bär
Sgr 19h52'33"43d48'
Konopnicki,William
Her 18h3'16"48d6'
Konrad,Manfred "Tiger" aus Wien
Cnc 8h57'56"10d5'
Konrad,Mary Jane
Cas 1h22'32"68d45'
Konrath,Kate
Cep 23h19'26"65d36'
Konrath,Kathryn J
Com 12h25'11"19d42'
Konroy,Joseph P
Oph 17h13'59"-24d55'
Kons Forever,Jim & Katie
Lyn 8h10'36"50d12'
Konstant,Susan
And 2h15'24"40d19'
Konstantina
Cam 3h42'0"70d56'
Kontogianni Star,Vasos Alex
Cyg 21h34'1"44d58'
Kontonickas,Neysa
Per 4h4'17"38d45'
Kontul's (Lucky Seven) Erma & Ed
Tau 4h44'51"16d19'
Kontusch,Wolfgang
Cap 20h58'18"-26d21'
Konz,Julie
Cas 1h3'48"62d27'
Koob,Paul Stephen
Aur 5h2'11"46d21'
Koob,Peter James
Dra 16h15'34"67d25'
Koob,Rosemary
Cyg 19h45'0"30d5'
Koogle,Paris Ann
Cam 7h58'48"70d9'
Kooistra,Tara
Aur 7h47'46"40d37'
Kook,Maximilian
Per 2h1'13"50d6'
Kookie
Cnc 8h29'1"10d10'
Kookie B
Lyr 18h55'53"41d23'
Koomouma,Lea
Ori 5h27'12"0d8'
Koon,Jennifer Rae Patterson
Cyg 19h54'32"40d36'
Koon,Melissa R
Lyn 8h8'22"46d10'
Koonce,Paul D
Aql 19h54'16"1d43'
Kooney,George P
Her 16h39'11"26d48'
Koons,Rick
Aql 19h30'45"12d23'
Koonse,Monty W
Per 2h37'25"41d4'
Koontz,Nickolas Tanner
Ser 15h58'55"24d36'
Koontz,R Stephen Bennett
Per 3h18'46"41d9'
Koontz,Robert Parker
Aur 7h24'51"39d33'
Koontz,Roger & Shari
Crb 15h49'37"37d38'
Koontz,Shirley
Aql 20h30'27"0d8'
Koop,In Loving Memory Of Juanita D
Cyg 21h4'0"39d27'
Koop,Thomas Abraham
Dra 17h50'32"64d58'
Kooperman,Todd
Per 3h14'57"50d27'
Koopman,Pam & Simon Buchman
Lyn 7h34'22"40d15'
Koopmann's Star
Her 17h10'1"20d25'

Kooros
Cam 3h31'41"60d34'
Koos
Cap 21h4'51"-22d23'
Koos,Jr,Richard H
Uma 15h58'57"77d0'
Koosmann,Jill & Chuck
Aur 5h2'50"29d24'
Kopack,In Memory Of George
Her 16h34'28"37d11'
Kopec 1
Cep 23h26'37"64d38'
Kopec,Lindsey Nicole
Cas 2h20'0"60d30'
Kopec,William Anthony
Oph 18h6'27"11d19'
Kopecky,Tracy Marie
And 0h28'40"41d13'
Kopeky
Lyr 19h17'1"30d24'
Kopel,Larry
Aur 5h6'0"41d20'
Kopelousos,John
Aql 19h53'25"13d39'
Koper,Chris
Lac 22h44'20"53d37'
Kopervos III,Joseph Michael
Tau 4h7'16"22d48'
Kopetski,Vondene
Equ 21h6'52"11d59'
Kopff,Gary
Dra 19h48'59"70d43'
Kopicki,Allison Nicole
Cyg 20h21'55"38d22'
Kopidakis,Victoria Theodora
And 0h58'42"40d59'
Kopija,Mary Kathleen
Com 13h0'1"18d56'
Kopinski,Wendy
Peg 22h41'34"27d45'
Kopko,James B
Boo 13h40'25"14d11'
Kopleton,Jack
Her 16h39'36"34d18'
Koplowitz,Arthur & Janet
Lyn 7h6'15"58d13'
Koplowitz,Joseph Rae
Cnv 12h12'21"46d34'
Kopnicky-25-FDATV
Peg 23h38'58"11d32'
Kopooshian,Lisa
Cas 2h10'11"59d33'
Kopp,Andrew Jeffrey
Per 2h50'40"46d5'
Kopp,Auntie Carol
Com 12h14'20"27d49'
Kopp,Jerry F
Her 16h50'34"38d39'
Kopp,Judith L
And 23h42'1"45d25'
Kopp,Megan Eileen
Eri 3h31'1"-5d37'
Kopp,Robert Lowell
Her 17h20'26"46d43'
Koppa,Kathleen Ann
And 23h24'27"40d52'
Koppel,Felicia Christina
Tau 4h33'32"18d58'
Koppel,Uri Mordechai
Ori 5h31'1"0d31'
Koppelman,Brian & Amy
Cyg 20h34'54"40d50'
Kopperman,Leigh Ellen
Cas 0h53'22"76d15'
Koppie,Douglas A
Cep 22h22'0"55d33'
Koprin,Sr,William
Her 16h55'51"26d5'
Kopriva,Thomas
Cet 1h18'57"-13d2'
Koptcho,Lisa
And 23h26'34"37d52'
Kor Al
Uma 11h56'44"64d58'

Korak,Milan & Sophie
Cyg 19h21'54"44d59'
Koral,Esqs,Margi & Mark
Boo 15h8'39"51d6'
Koral,Phyllis M
Lyn 8h57'58"45d52'
Koraleski-My Love, Darren R
Per 3h11'51"46d12'
Korali,Dann
Ori 5h55'22"15d28'
Korbonski,Duncan
Aql 19h48'0"11d43'
Korbus,Virginia King
Oph 18h39'32"8d59'
Korchak,Jerome R & Cynthia A
Leo 10h13'32"17d59'AB
Kordas,Alexandra
Uma 11h58'19"58d59'
Kordaszewski,Tom
Boo 14h48'32"51d29'
Kordic,Yvette Leticia
Eri 4h53'52"-8d0'
Kordowicz Star,The
Her 16h22'3"11d2'
Korec,Ashley Christine
And 23h33'14"42d60'
Korell's Superstar, Christ is Kay
Lyr 19h21'26"30d35'
Koren
Crb 16h21'54"32d29'
Koren,Jena Marie
Lib 14h26'56"-18d52'
Koren,Roger James
Per 2h37'17"38d22'
Korendor
Car 7h34'40"-56d56'
Korenyi-Both,Tyler Ethan
Lac 22h2'14"47d51'
Korff,Toni Lee
Oph 16h56'11"-28d32'
Korgen Star,The James Matthew
Hya 10h41'23"-18d27'
Korhonen,Otto Kalervo
Umi 15h35'30"81d6'
Korhonen,Seppo Jaakko
Dra 16h9'17"61d27'
Korhonen,The Star of Seppo
Cam 5h47'42"58d21'
Korhummel,William
Aur 6h28'39"48d48'
Koribaki
Cnv 13h58'22"28d19'
Korich,Tracy Ann
Cap 21h1'32"-26d12'
Korinchock,JoAnn
Lyn 8h6'39"40d49'
Korinko,Deborah
Del 20h18'54"10d49'
Korinne
And 1h8'21"39d33'
Korka,Craig Stephen
Ori 5h55'31"15d26'
Korman's Star
Lac 22h35'44"53d27'
Korman,Clyde
Cma 6h54'52"-19d21'
Korman.Lee
Her 16h24'40"37d54'
Korn,Alyssa Lynn
Cas 23h27'25"61d60'
Korn,Barbara Lee
Sct 18h52'53"-8d51'
Korn,Carl
Boo 15h38'57"40d58'
Korn,Estelle
And 2h26'1"42d36'
Korn,Karl
Del 20h13'54"10d8'
Korn,Michael T
Dra 16h28'1"69d35'
Korn,Sally Wood
Cet 3h11'47"5d46'

Korn-Rainbows & Stars, Jennifer
Lyn 8h43'14"41d18'
Kornacki,Timothy F
Her 18h5'21"38d57'
Kornatowski,Brian
Lmi 10h19'30"36d31'
Kornbluth,4-2-75, Jennifer Carlie
Cam 3h51'17"52d40'
Kornbluth,Herman
Cet 2h19'13"5d41'
Korneisel Littlest Angel,Keith
Hya 8h13'45"-6d49'
Kornelia
Cnc 8h6'49"30d55'
Korner,H Calvin
Uma 8h33'27"58d36'
Korner,Helen & Jack
Cyg 20h38'0"40d40'
Kornet,Gloria A
Uma 9h1'34"50d3'
Kornet,Michele L
Uma 9h2'1"52d7'
Kornet,Robin Michelle
And 0h52'16"33d35'
Kornexl,Kenneth "Lobo"
Ser 15h38'51"6d30'
Kornhauser,Lee Richard
Tau 4h15'52"7d46'
Kornhiser
Aql 20h12'37"4d21'
Kornman,Matthew James
Umi 14h56'0"70d41'
Kornmann,Christopher
Boo 14h2'18"22d14'
Kornmann,Felix
Cep 22h29'18"67d44'
Kornmeyer,Matthew P
Aur 4h52'23"40d39'
Kornswiet,Debi
Cyg 19h25'22"30d27'
Korol,T
Aql 20h7'19"-5d55'
Korolija Star,The Goren Antonio
Aql 19h31'30"10d47'
Korosec,Michael J
Sco 17h25'42"-33d59'
Korot,Marvin
Ori 6h7'52"8d28'
Korpela,Joe
Ori 4h48'26"10d59'A
Korpela,Julie
Ori 4h48'26"10d59'B
Korpi-Cloninger,Carly Shea
Peg 23h2'15"33d51'
Korpinen,Neil
Cma 6h30'19"-18d3'
Korse,Barbara
Uma 12h11'0"60d12'
Korsgaard Family Star, The
Ori 5h52'39"18d55'
Korsgaard,Rosalie
And 0h10'48"46d42'
Kort,Susan Michelle
And 2h34'12"50d35'
Kortanek,Apryll Leigh
And 1h55'29"47d23'
Korte,Ursula
Cyg 19h25'18"30d19'
Korth,Jason
Mon 6h47'38"10d45'
Korthas,Kelly L
Uma 10h3'1"56d5'
Korus,Agnes Lucy
Com 12h28'21"28d2'
Kory's Star
And 0h30'43"40d55'
Kosa's Bank & Trust Me Dan
Cnv 12h24'20"34d55'
Kosako,Glen Takeo
Her 16h54'19"22d31'
Kosanovich,Nicholas Eli
Aur 4h50'34"38d32'

Kosar,Ruksana
And 23h20'41"42d29'
Kosarek,Amadeus Joshua Wolfgang
Crt 11h12'29"-19d49'
Kosbau,Tamera Kay
Lmi 9h56'40"34d18'
Kosch,Kathleen & Theodore
Cyg 19h43'52"31d7'
Koschoreck,Kenneth R
Peg 23h36'20"30d34'
Koschwitz,Thomas
Umi 13h32'25"74d9'
Koscinski,Frank S
Uma 11h23'48"43d43'
Kosciol,Jessica Lee
Lyn 8h40'26"38d10'
Kosco,Bernard
Dra 17h44'40"67d57'
Kosczowsky,Manuela Marie
Boo 13h46'11"12d0'
Kosek,Dennis & Susan
Cyg 19h32'50"33d11'
Koser,Margaret E
Peg 22h0'32"35d13'
Koshivas,Matthew K
Sco 16h31'1"-31d22'
Kosia
Cyg 21h30'48"34d46'
Kosiba,Camille
Vir 13h52'35"-5d55'
Kosik,Glen
Dra 20h3'22"62d46'
Kosim,Johnny
Her 17h49'54"50d23'
Kosinski,Larry
Her 16h56'34"39d27'
Kosinski,Patricia Ann
Peg 23h23'1"32d26'
Kosinski-vonBotefuhr's Wishing Star
Cam 5h45'44"61d13'
Kosiorek,Ryan
Uma 12h24'31"56d1'
Koska,Joseph
Aur 5h19'31"52d41'
Koski,Kristina Michelle
And 2h15'52"39d37'
Koskinen,Heikki
Cam 6h16'22"62d13'
Koskinen,Osmo
Dra 15h44'0"62d15'
Koslick,Jürgen
Cep 20h39'59"65d15'
Kosloff,Rebecca Amy
Lyn 8h12'19"40d53'
Kosmic Kim
Cas 0h16'17"63d36'
Kosmische Sonnenrose Helga Strecker
Sco 17h54'33"-33d48'
Kosner,Carol Verdelle
Umi 16h27'35"72d3'
Kosner,Michael A
Cet 0h49'41"-5d37'
Kosnitzky "The Magician",Gary
Ori 5h11'29"-5d54'
Koss Family Star
Lac 22h6'30"47d16'
Koss,Janier Rose
And 0h36'37"41d3'
Kossak,Jennifer Rachel
Lyr 19h11'0"38d36'
Kossak,Michael Timothy
Lyr 18h40'53"37d39'
Kossdorp's Welt
Peg 21h38'36"27d25'
Kossen,Juanita
And 0h11'1"41d14'
Kossoff,Esther & Arnold
Crb 16h1'47"32d14'
Kossoff,Irwin & Edith
Crb 15h59'49"32d18'

Kossoff,Mitchell Hal
Per 1h51'16"52d60'
Kossoff,Phyllis & Burton
Cnv 12h54'29"41d8'
Kostakis,The Star Of Nick & Maria
Cep 22h1'18"67d39'
Kostecki,Daniel Henry
Hya 8h47'4"-7d20'
Kostek,Pierre
Mon 7h59'54"-4d15'
Kostel Ray Of Light, Mary Elaine
Aql 18h56'1"4d17'
Koster,Helen W
And 0h19'34"31d21'
Koster,Rosemary Ann
Eri 2h44'36"-6d21'
Koster,Tracey Louise
And 23h20'0"42d4'
Kosters,Kody James
Aur 5h11'59"40d12'
Kosters,Kyle Joseph
Lac 22h45'57"54d50'
Kostival,Jessica Ann
And 23h39'11"39d55'
Kostoff,James R & Patricia A
Umi 16h29'19"70d27'
Kostycz,Gayle C
Lyn 9h16'52"37d42'
Kostyk,Michael
Aur 5h0'1"42d12'
Kosut,Linda
Peg 22h14'43"34d16'
Kotagal,Kalpana
Mic 20h25'20"-32d22'
Kotarski,PhD,Michael A
Per 2h46'59"43d9'
Kotas,Cassandra
Cam 7h56'44"70d20'
Kotas,Jakob
Dra 19h11'20"71d8'
Kotchick,Claudia
Mon 6h55'0"0d47'
Kotecki,Florian Roy
Cam 3h47'56"57d43'
Koten,John A
Boo 14h38'34"40d48'
Kothe,Jackson Patrick
Peg 22h43'47"11d24'
Kothmeier,Siegfried
Sgr 19h57'10"-43d15'
Kotinsky,Tom Cat
Gem 7h16'55"28d50'
Kotkins,Katherine Marion
Lyr 18h29'24"30d45'
Kotler,Barry Michael
Sco 16h58'29"-30d14'
Kotonias,Nancy M
Cas 1h11'43"68d31'
Kotora,Marjorie Lucile Smith
And 1h28'1"35d18'
Kotowski,Donny
Oph 17h1'11"11d52'
Kotowski,Toni
Ori 6h16'10"-1d21'
Kotrofi,John Michael
Ori 5h4'37"10d57'
Kotsen,Duane & Wendy
Uma 8h14'0"68d12'
Kotsifas,Heidi Marie
Peg 21h54'28"23d55'
Kotsubka,William, Rudy & John
Cma 7h4'19"-11d12'ABC
Kott 93',F J
Uma 11h53'10"44d42'
Kott,Charlene Emiko
Lyr 18h54'18"30d4'
Kott,Debra Kellerman
Peg 22h22'57"30d57'
Kott,Jacob Daniel
Cnv 13h17'56"40d55'
Kott,Julia Marie
Crb 15h49'57"38d4'

Kott,Kelly Burton
Com 12h4'0"20d14'
Kott,Stephen Jacob Burton
Her 16h12'25"42d4'
Kott,Stephen James
Vul 19h47'55"26d2'
Kott,Stephen Jay
Lyr 19h17'1"28d31'
Kott,Tara Dawn
Crb 15h31'31"30d46'
Kottisch,Hans-Juergen
Ari 2h37'0"21d50'
Kottmeier,Kathleen June Ainsworth
Uma 10h13'13"50d54'
Kottmeier,Klaus
Vir 13h59'49"-1d9'
Kotula,Jr,Richard Steven
Boo 14h29'18"41d37'
Kotula-Russ, Christopher
Lac 22h28'42"56d35'
Kotula-Russ,Robert
Boo 13h37'14"27d41'
Kotwal,Zarina H
Lmi 9h52'0"34d23'
Kotwasinski,Kari Raquel
Mon 7h58'14"-0d45'
Kou
Vir 12h35'0"8d48'
Kou-Kou
Psc 1h18'57"24d58'
Koubek,Melissa
hya 9h16'1"6d16'
Koubek,Vlastimil
Dra 12h42'46"75d2'
Koucheck,Joe & Jean
Crb 16h20'0"30d1'
Koudriachova,Lilia
Cas 0h26'26"68d55'
Kough,R Michael & Barbara
Boo 14h9'57"30d28'
Kougher,Kamron Casey
Eri 4h56'5"-6d37'
Kouicem,Malek
Ori 4h50'26"5d19'
Koukatsikas,Stavros A
Aql 19h16'56"12d21'
Koukoulas,James Michael
Cep 23h32'40"65d38'
Koulouris,Julia Jacqueline
Cyg 19h25'1"35d16'
Kountz,Angela C
Cet 0h37'39"2d6'
Kourbagh,Allison Marie
And 23h44'1"37d52'
Kouremenos,Zoe Trifona
Lac 22h14'36"49d2'
Koures,Kristi & Bill
Lyn 6h24'19"60d43'
Kouri,Robbie F
Ser 15h38'41"4d6'
Kourofsky,Allen
Aur 5h11'42"41d56'
Kourofsky,Margaret Rae Wood
And 0h9'24"38d27'
Kourosh
Cep 0h12'34"68d9'
Koury,Jeanette Loretta Albert
Lib 14h19'34"-11d14'
Koutcher,Nancy
Vul 19h59'48"26d1'
Koutsakos,Nicholas
Lac 22h30'13"40d54'
Kovac,8/21/93 Forever, Mr & Mrs Ron
Crb 15h59'30"26d36'
Kovacevic,Janeen
And 23h32'39"41d31'
Kovach,Ann
Cas 0h42'48"64d34'
Kovach,Joan
And 0h18'19"33d28'
Kovach,Mrs Gwen
And 1h35'41"40d1'

Kovach,Sarah
Peg 23h32'31"23d8'
Kovacic,George & Frances
Cyg 21h33'34"41d48'
Kovacs,David
Her 16h2'32"40d24'
Kovacs,Robin
Aur 6h50'0"40d32'
Koval,Christopher Ryan
Cet 2h28'23"3d34'
Koval,John & Helen
Lac 22h41'1"51d32'
Koval,Sherrie Ann
Lyr 18h25'11"45d42'
Kovalik,Nicholas John
Dra 14h55'20"61d40'
Kovall 1-25-74,Kristen Elizabeth
Cyg 19h51'49"41d0'
Kovalski's Wishing Star 1993
Cam 14h12'18"80d28'
Kovalsky,Christopher Joseph
Cet 1h52'41"-7d40'
Kovanda,Kimberly
Vul 20h2'58"28d28'
Kovarik,Jerry
Per 1h50'0"50d31'
Kovarik,Sondra Pfutzenreuter
Cyg 21h18'32"38d12'
Kovarik,Susan & Dale
Sex 10h11'41"-2d26'
Kovarsky,Norm
Her 16h9'30"50d8'
Kovesdi,Ramona Anne
Cyg 20h18'15"41d57'
Kovian,Theresa Maria
Cas 23h13'59"61d54'
Kovic,Joe
Ori 5h2'23"0d39'
Kovner,Sidney
Aql 18h56'31"13d22'
Kowal,Jessica
And 2h22'1"44d24'
Kowal,John
Per 4h20'0"51d33'
Kowal,Zachary
Cep 22h27'48"60d59'
Kowalchik,Amy Joan
Cas 1h39'50"58d48'
Kowalcyk,Kimberlee Meg
Cam 3h43'57"70d8'
Kowaleski,Anthony Joseph
Aur 5h6'1"42d31'
Kowaleski,Lindsay Catherine
And 2h15'60"42d26'
Kowaleski,Todd Michael
Her 17h15'20"46d43'
Kowalewski,Anthony Vincent
Her 17h59'44"28d52'
Kowalewski,Nicole Danielle
Lyn 8h51'41"43d31'
Kowalski"69",Sue
And 23h38'0"46d14'
Kowalski,Allen R
Psc 23h4'46"5d32'
Kowalski,Debra Ann
Psc 23h9'1"5d46'
Kowalski,Dolores F
Aur 5h1'39"41d21'
Kowalski,Emil
Cas 23h26'15"54d13'
Kowalski,Eugene S
Uma 10h15'30"42d23'
Kowalski,Mathew Aaron
Vul 19h48'14"28d12'
Kowalski,Raymond L
Aur 6h28'41"30d33'
Kowalski,Roman P
Oph 18h19'47"8d0'
Kowalski,Ronald
Cep 20h46'59"73d59'
Kowalski,Sandy
Lyn 6h55'46"59d4'
Kowalski,Susan Marie
Cas 1h33'27"73d58'

Kowalski,William A
Aur 5h7'1"42d27'
Kowalsky-06/19/1926AD, John T
Aur 6h4'0"37d28'
Kowamiet
Boo 14h37'51"17d37'
Kowarsh,George
Cam 3h31'11"53d50'
Kowatsch,Michelle
Aql 19h7'20"15d1'
Kowlessar,Dira
Uma 9h44'47"71d55'
Kowzun,Andrew et Danielle
Per 3h59'9"39d60'
Koz
Cnv 12h33'29"51d56'
Koza,Bethany
And 0h28'48"38d60'
Koza,Josephine
Cas 1h32'54"60d20'
Koza,Mitchell P
Uma 8h53'36"60d56'
Kozak,Daniel Kevin
Peg 22h1'1"28d48'
Kozak,Ellette Johns
Ser 15h39'1"4d7'
Kozak,Mary Imogene
Peg 22h30'24"24d3'
Kozaka,Alexa
Cas 3h9'43"58d52'
Kozaka,Jordan Mikaela
Cet 2h31'57"4d34'
Kozakiewicz,Carol
Cyg 19h29'14"32d4'
Kozal,Karen
Com 12h47'48"20d17'
Kozal,Robert
Aur 5h38'9"29d28'B
Kozar,John M
Aur 5h48'59"54d56'
Kozar,Ruby Willich
And 2h21'1"40d17'
Kozarits,Josef
Cmi 7h20'32"8d56'
Kozicki,V Michael
Per 3h37'1"40d35'
Kozie,Ma-Pa
Cyg 21h5'42"30d23'
Kozielski,Debra Elaine
Vul 20h57'34"28d34'
Kozik,Heather Lynn
Cas 2h12'30"59d54'
Kozik,Mami
Sco 16h13'9"-23d5'
Kozin,Dr Arthur
Dra 17h3'50"69d4'
Koziol,John A
Aur 7h4'35"36d30'
Koziol,John Edward David
Aur 4h57'1"31d43'
Kozlik,Susan R
Cas 0h0h22'61d17'
Kozlowski,Matt
Lyn 7h3'54"51d59'
Kozman,Edward J
Dra 20h21'21"68d34'
Kozmetsky,IC² Institute,Dr George
Umi 14h4'54"73d53'
Kozsukan,Kathy
Per 3h2'46"40d31'
Kozub,Amy Elizabeth
Cas 1h53'0"68d23'
Kozuka,Ted
Uma 12h11'26"57d58'
Kozyra,Suzanne R
Uma 11h10'37"33d43'B
KP Nina
Aur 5h16'42"43d20'
KPM
Lmi 10h3'48"34d31'
Kraal,Professor Georg
Lyn 7h6'39"60d18'

Kraay,Dana
Ori 5h44'58"10d27'
Krabach,Joseph Albert
Cma 6h14'32"-20d20'
Krabbe,Werner
Peg 23h31'15"18d37'
Krabbenhöft,Kristian
Gem 6h26'39"12d6'
Kracht,Anja
Ind 20h25'54"-55d15'
Kracht,Elizabeth Anne
Cas 23h21'1"61d8'
Kracht,Martina
Mon 8h7'11"-1d41'
Kracht,Stefan
Per 1h47'0"56d23'
Krackenberger,Mike
Aql 18h55'16"-1d9'
Kraemer,Peter
Cyg 20h25'23"30d50'
Kraemer,Todd
Ori 5h45'26"10d48'
Kraemer,William H
Sct 18h40'39"-4d22'
Kraff,Love & Dreams, Kelly,David G
Lmi 10h55'33"27d55'
Krafft,Hans
Uma 10h45'46"40d6'
Krafft,Maryann
Com 12h5'23"21d40'
Kraft "1",Douglas Edward
Dra 16h48'44"69d3'
Kraft,Armand M
Per 2h48'18"40d17'
Kraft,Dean
Boo 13h34'20"16d22'
Kraft,Gilman
Boo 14h7'0"38d10'
Kraft,Gloria R
Del 20h20'32"10d24'
Kraft,Hermann
Aur 5h14'55"43d20'
Kraft,Horst
Peg 23h29'33"10d17'
Kraft,Jackie
Lyr 19h16'25"38d23'
Kraft,John A
Uma 11h40'24"37d55'
Kraft,Kim
Uma 10h53'20"71d52'
Kraft,Ruth
Mon 6h51'31"10d29'
Kraft,Ruth N
Ori 6h0'37"-0d16'
Kraft,Sy
Dra 14h19'42"64d29'
Kraft,Vickie
Cam 7h7'42"60d15'
Krafthefer,Troy
Her 16h33'18"35d21'
Krafton,Michael
Cep 21h23'45"55d31'
Kraftsow,Carole & Stan
Cet 2h31'12"-10d36'
Kragh,Daniel
Boo 14h42'15"21d15'
Krahenbuhl,Nora Rose
Vul 19h20'21"26d45'
Kraisinger,Regis F
Dra 17h42'12"61d32'
Kraisky,Matthew James
Lac 22h7'16"49d58'
Krajenke,Mackenzie Clare
And 23h39'50"37d58'
Krajick,Kent
Aql 19h44'51"13d53'
Krajnc,Ralf & Christina
Uma 10h54'27"71d24'
Krakowski,Jennifer
Com 12h57'20"25d27'
Kral I Lv U W/All My Heart,Billy
Hya 8h55'12"0d44'

Kral,John & Nancy Sct 18h55'55"-6d2'
Kral,Marc Peter Aur 5h17'42"43d12'
Kral,Margaux Cas 23h1'14"58d15'
Kralapp,Jacob Per 1h42'14"53d17'
Kralik-Godwin,Mary A Uma 11h1'49"36d41'
Krall,Veronica L Cas 0h56'32"60d16'
Kraly,Ed Dra 16h46'49"52d24'
Kram,Eve And 1h45'41"47d7'
Kram,Florence Cyg 20h22'13"41d5'
Kramar,Erin Marie Tau 4h0'13"10d47'
Kramarz Uma 8h5'54"68d58'
Kramer Peg 23h27'55"13d31'
Kramer Uma 10h23'22"48d13'
Kramer Sex 9h42'30"2d8'
Kramer Remarkable,Fred Aql 19h28'58"13d50'
Kramer Star,The Rubin & Miriam Eri 3h40'55"-18d24'
Kramer,April Lyr 19h3'26"28d11'
Kramer,Barbara Lyr 19h15'42"28d30'
Kramer,Barton Jay Vul 19h23'1"25d15'
Kramer,Blanche Virginia Vir 11h38'21"2d45'
Kramer,Brian James Cam 6h13'0"83d47'
Kramer,Carissa Jayne Mon 6h35'1"0d33'
Kramer,Christopher Hugh Hya 8h42'52"5d50'
Kramer,David & Carolyn Cet 0h57'30"-2d17'AB
Kramer,Deborah Sgr 19h40'17"-43d33'
Kramer,DeeDee Cyg 20h35'36"39d45'
Kramer,Denise Noel Rosenfeld Cas 0h8'16"58d45'
Kramer,Dennis Michael Vul 19h48'31"26d5'
Kramer,Dr Elliott Cma 6h43'17"-15d43'
Kramer,Earl Uma 12h1'16"42d55'
Kramer,Erica Jean Peg 22h29'55"21d21'
Kramer,Glenn Thomas Aql 19h24'58"8d44'
Kramer,Jeffrey Scott Hya 8h27'15"1d39'
Kramer,Joseph Ori 5h56'11"8d59'
Kramer,Judith A And 1h21'49"48d00'
Kramer,Justin Her 18h12'47"31d13'
Kramer,Kathleen Cas 1h34'47"62d5'A
Kramer,Kristen Marie And 23h48'58"40d55'
Kramer,Mackenzie Leigh Dra 13h12'58"67d44'
Kramer,Marcel Tri 2h4'0"35d45'
Kramer,Margie Lyn 7h40'20"38d49'
Kramer,Mark Allan Her 18h31'13"24d30'
Kramer,Melanie Deason Aql 19h31'29"10d16'
Kramer,Norbert Boo 14h6'21"10d13'
Kramer,Rachel Molly Com 13h5'35"21d37'
Kramer,Spencer Sterling Uma 8h39'24"68d29'
Kramer,Tammy Mon 6h41'30"-10d25'
Kramer,Wanda Aql 20h2'23"6d58'
Kramers,The Boo 14h28'24"48d5'
Kramm,Douglas J Her 16h49'44"33d34'
Krammer,Ingrid Cmi 7h20'28"1d32'
Krampe,Mein Liebling Matthias Her 15h52'11"46d53'
Krandev,Todor Aur 6h6'22"45d20'
Krane,Sasha And 0h41'10"33d2'B
Kranendonk,Rene Uma 9h9'49"50d19'
Kranichfeld,Harry Edward Cet 2h3'21"-1d5'
Kranichfeld,Jane Fritche Leo 10h51'52"-5d47'
Krank Aur 5h57'48"31d37'
Krantz,Michael Cmi 7h31'18"1d49'
Kranz,John David Aur 5h1'1"50d4'
Kranz,Othmar Sco 17h24'1"-30d39'
Kranzusch,Terry L Cam 3h34'16"61d54'
Krapff,Jennifer Peg 23h1'46"30d23'
Kras,Lindsay Renee Lyr 19h2'1"31d8'
Krasaway,John Cma 6h56'51"-18d40'
Kraska,Kimberly Kay Cyg 19h12'53"49d40'A
Krasne,Thatcher Boo 14h26'41"44d7'
Krasner,Gerald & Heather Uma 12h12'17"57d6'
Krasniqi,Ruzhdi Cmi 7h44'14"4d58'
Krasnomowitz,Adam Michael Gem 7h51'33"31d52'
Krasnow,Marc Aur 5h8'14"43d28'
Krasovskis,Osvalds Uma 12h9'45"48d33'
Krasowski,William John Cam 3h57'11"57d45'
Kratky,Joan & John Ori 5h45'19"10d6'
Kratky,Tyler Takeo Dra 17h1'1"67d33'
Kratochvil,Amie Lyn 8h12'0"40d31'
Kratochvil-From Dana & Nicole,Family Uma 8h32'33"52d24'
Kratovil Daughter, Nicole Uma 8h49'16"57d41'
Kratovil Mother,Dana Uma 8h35'58"56d36'
Kratovil,Mark Mon 6h45'13"0d24'B
Krattli,Robert William Per 3h10'1"47d45'
Kratz,Donna Lee Lyr 18h54'48"33d7'
Kratz,Julia Katherine Vul 19h58'11"28d18'
Kratzat,Louise Victoria Mon 6h55'52"-3d8'
Kratzer,Emily Elizabeth Cas 0h47'38"61d22'
Kratzert,Jan Aur 4h53'35"41d1'
Krau,Linda Sue Davis Cas 2h6'51"63d45'
Krau,Michael Paul Dra 16h35'24'51"58d
Krauer,Michael Per 2h10'18"57d60'
Kraus,Ashley Lyr 18h40'23"41d53'
Kraus,Cameron Allan Uma 9h58'29"58d36'
Kraus,Diana Cushing Mon 8h6'31"-1d47'
Kraus,Gerry Per 1h54'33"48d19'
Kraus,Josef Ser 16h18'27"0d51'
Kraus,Kyle Per 4h27'34"50d23'
Kraus,Mitch Her 16h49'29"50d35'
Kraus,Patricia Peg 22h44'40"32d10'
Kraus,Stephen Richard Sex 10h10'59"-6d39'
Kraus,Steven Cep 22h20'16"61d11'
Kraus,Thomas W Vul 19h22'36"22d38'
Krausch,Melodie Hya 8h31'17"-0d36'
Krause 5-20-38 to 3-1-89,Erwin D Lyr 18h35'35"40d56'
Krause Fertility Star Sex 10h4'10"-1d35'
Krause,Angels' Breath- Angela M Cas 0h56'1"72d8'
Krause,Anne Vir 13h15'54"-5d27'
Krause,Cheryl Lee Aur 6h37'31"40d20'
Krause,Christa Lac 22h2'55"50d5'
Krause,Dave & Andrea Crb 16h21'45"32d53'
Krause,Donald Her 17h30'55"20d56'
Krause,Dr John Oph 17h53'42"10d57'
Krause,Eternal Umi 15h43'0"70d25'
Krause,Eva-Maria und Bernd Her 18h20'46"12d4'
Krause,Frauke Cap 20h57'36"-20d2'
Krause,Horst Dra 18h55'28"65d36'
Krause,John V Lac 22h52'1"56d16'
Krause,Mary Cas 2h54'55"75d35'
Krause,Pamela Cmi 7h41'20"5d7'
Krause,Rosena Augusta Cep 0h34'23"78d13'
Krause,Sabrina Marie And 0h9'34"35d11'
Krause,Stephan Ori 5h50'56"21d2'
Krause,The Light Of My Life,Judy A And 23h29'21"42d35'
Krause,Yvonne Uma 10h6'11"58d10'
Krausen,Franz Boo 14h19'51"10d31'
Krauske,Kristin Nicole Uma 10h51'15"57d35'
Krauss,Alan J Aql 19h4'39"17d35'
Krauss,Dorothy Aqr 20h57'19"-13d48'
Krauss,Gladys Victoria Leo 11h5'21"25d16'
Krauss,Howard Ori 5h58'15"16d9'
Krauss,Jessie Uma 9h43'15"56d1'
Krauss,John Aur 5h53'1"31d43'
Krauss,Lobo Hank Cap 20h27'5"-26d1'
Krauss,Maxi Leigh Mon 6h35'43"0d21'
Krauss,Samara Lynn Peg 23h23'46"8d55'
Krauss,Sarah J Vir 13h37'48"-2d16'
Krauss,Seth L Sex 9h46'28"4d17'
Krausse,Dolores Peters And 0h56'17"38d10'
Krausse,Tommy Aql 20h1'30"-8d28'
Kraut,Alan J Vul 21h23'14"24d12'
Krauthamer,Charles Cam 3h56'40"57d52'
Kravcov,Casey John Boo 14h34'0"41d43'
Kravik,Linn Cas 1h19'37"68d31'
Kravis,George Tri 1h55'12"26d60'
Kravis,Henry Aur 6h47'23"38d11'
Kravitsky V,Michael Per 1h55'52"50d1'
Kravitsky,Savannah Jules Lyn 7h56'1"38d55'
Kravitz,Joseph Daniel Lac 21h58'28"40d17'
Kravitz,Michael Cet 3h18'28"09d54'
Krawczak,Bob Aur 6h5'42"35d40'
Krawczyk,Sheila Marie Sex 10h25'1"-10d39'
Krawczynski,Aaron A Tri 2h11'1"33d40'
Krawiec,Carl Cnv 12h57'19"40d6'
Krawitz,Ruth Cam 3h26'20"58d15'
Kray,Rosemary F Aur 5h56'55"31d15'
Kraybill,Gary Cep 0h11'56"78d55'
Krayenhagen,Rachael Leigh Aqr 22h31'12"2d13'
Kraymer Angel Star Ori 6h0'45"8d47'
Kraz Cep 22h20'41"68d36'
Kreamer Cam 4h36'44"68d24'
Krebs III,Floyd Aloysious Cmi 7h32'57"2d37'
Krebs,Dr Heinrich Cap 21h38'55"-21d20'
Krebs,Ernst Aql 19h6'15"10d49'
Krebs,Jackie Sge 20h16'13"20d27'
Krebs,Jordan Joseph Ori 5h59'17"13d4'
Krebs,William & Teresa Cyg 21h25'0"40d13'
Krebsbach Mi Amor El Magnifico,Kurt Per 3h7'19"41d5'
Krebsbach,MD,Richard James Her 18h2'23"30d44'
Krech,Friedhelm Peg 0h12'49"18d46'
Krecko,Robert Ellsworth Per 1h41'54"53d57'
Kreder,Joseph F Cnv 12h12'0"34d21'
Kreeger,Theodore W Lib 15h30'50"-10d19'
Kreek,MD,Mary Jeanne Peg 23h43'24"31d53'
Kregar,Simon Aql 19h5'53"15d2'
Kregn Uma 11h33'47"32d40'
Kreha,Janet Cas 0h2'56"54d32'
Kreibich,Ralf Lyr 19h14'58"42d49'
Kreidel,Kelly Uma 9h34'18"67d59'
Kreidel,Kelly Cas 1h0'35"61d1'
Kreidler,Linda M Mon 6h53'26"-4d39'
Kreilein,Thomas Peg 23h5'28"14d53'
Kreiling,David Duane Mon 7h53'4"-3d40'
Kreimer,Heron Lyn 7h52'21"37d59'
Kreimer,Lori Cam 7h46'44"82d40'
Krein,Adele Mellender Oph 17h37'23"-0d36'A
Krein,Bernhardt Leo Oph 17h37'23"-0d36'B
Kreiner,Sylvia Eri 4h27'25"-19d7'
Kreinhop,Hilda Joyce Hollingsworth Mon 6h18'38"3d7'
Kreinhop,Paul Aur 5h43'38"50d30'
Kreis,Barbara Ann Lib 14h39'32"-10d20'
Kreisberg,Billy Cep 21h10'1"60d6'
Kreisch,Siegfried Gem 7h15'60"24d41'
Kreiser,Joe & Dawn Crb 15h36'21"29d34'
Kreisher,Hailey Breana Cyg 20h17'17"30d53'
Kreisman,Lottie Lyn 8h53'53"41d8'
Kreitler,Isabel & William Aur 6h1'0"45d53'
Kreitz,Karin Lib 15h8'42"-22d25'
Kreitzberg,Skip Aur 5h56'20"41d6'
Krejci,Tatiana Mon 6h30'0"10d43'
Krekus,Charles A Her 16h7'21"40d30'
Krelitz,Bridget Cam 7h27'18"60d21'
Krell,Dr Ted Oph 17h4'46"7d59'
Krell,Marlene Peg 22h20'29"27d27'
Krell,Robert D Lyr 19h10'36"47d44'
Krell,Sylvia Cnv 13h23'1"41d59'
Kremer,Selena Maria Eugenia And 1h53'0"36d49'
Kremer,Shirley Cyg 19h29'14"31d34'
Kremers,Coleen Kenner Ori 5h52'1"18d40'
Kremlicka,Elisabeth Lyn 8h16'24"48d36'
Kremmer,Chad Elliott Hya 9h13'39"6d30'
Krempels,Ettamae Lyr 19h0'41"40d3'
Krems,Thomas Ori 5h10'43"-6d32'
Krenitsky,Lucas Sica Ori 5h56'30"14d31'
Krenitsky,Nicholas Sica Sex 10h33'39"4d4'
Krenz,Angeline Mary Cyg 21h54'47"52d47'
Krenz,Madalyn Michael And 0h5'25"34d57'
Krenz,Thomas Cyg 21h1'31"30d55'
Kreppel,Bob Sct 18h56'43"-6d10'
Kresge,Lela Lyn 7h25'46"58d45'
Kresge,Serena Lea And 2h23'54"43d5'
Kresken,John J & Margie Com 13h3'1"21d55'
Krespan,Jon C Umi 16h25'22"70d5'
Krespan,Lawrence P Umi 16h24'53"71d58'
Kress,Herr Boo 14h24'32"11d44'
Kress,Kristina Cas 0h49'26"61d47'
Kress,Mary Christina Cas 0h27'0"61d16'
Kress,Philip I Aur 6h4'56"37d34'
Kresser,"Big Turtle" Michaela Cnc 8h27'60"31d15'
Kresser,Rachel Catherine Cam 3h26'25"53d33'
Kressler,Kenneth Per 2h51'39"32d39'
Kressler,Virginia Per 2h51'22"32d13'
Kresson,Brandon Robert Per 3h4'46"50d25'
Kretan Cam 5h37'24"70d47'
Kretly,Elisabeth Lac 22h42'18"38d49'
Kretman,Michael Leo 11h55'42"21d19'
Kretovic,Michael Aaron Vul 19h28'36"19d35'
Kretsch,Mary Carol Uma 11h38'1"32d34'
Kretschmer,Doktor Helmut And 23h9'36"40d14'
Kretschmer,Stephen Linwood Cep 21h56'51"55d31'
Kretz,Cara & Tom Peg 23h3'0"11d22'
Kretz,Robert Karl Dra 1651'60"63d59'
Kretzschmar,Mark Aur 6h0'24"32d14'
Kreuch,Peter Pollard Boo 14h48'16"35d51'
Kreusch,Michael Psc 0h59'26"21d43'
Kreuscher,John Walter Cep 22h3'14"61d30'
Kreutzberg,Reese Oph 17h4'4"-20d10'
Kreutzer,Jason Matthew Her 16h46'47"32d35'
Kreutzer,Wolfgang Sco 17h28'60"-38d58'
Kreuz,Gary Aur 6h29'32"33d47'
Kreuz,Joachim Cyg 20h25'23"30d32'
Kreuzberg,Sandy Sgr 19h3'24"-28d43'
Kreuzberg,Stefanie Cap 21h24'47"-26d59'
Kreuzwieser,Mark Henry Ser 15h54'48"4d2'
Krevens,Skip Dra 17h56'29"64d20'
Krey 11-25-74,Matthew Paul Peg 23h40'15"31d21'
Krich,Karen Equ 21h1'1"8d31'
Krichiver,Donna & Joel Cyg 19h25'0"30d33'
Kridelbaugh,MD, LeAnn Eri 2h47'41"-7d23'
Kridle"Nana",Dora M Boo 14h6'0"22d51'
Kriebel,Paula Aql 18h54'38"11d51'
Krieck,Dona Lyn 8h15'25"40d14'
Kriedt Family Star, The Karl Cyg 19h46'43"30d2'
Krieg,Carolyon And 23h38'54"43d29'
Krieg,Juergen Cyg 20h25'52"39d59'
Krieg,Tish Tri 1h51'43"26d40'
Kriegel,Al L Dra 15h4'20"60d23'
Krieger The Golden BEAR,Claus Peter Ori 5h39'40"8d8'
Krieger,Brittany Allison Uma 12h33'1"60d37'
Krieger,Carly Rachel Aqr 21h5'29"-0d11'
Krieger,Henni Cas 23h40'58"61d19'
Krieger,Jeffrey James Her 17h31'42"28d15'
Krieger,Lois Hill Mon 7h0'59"5d7'
Krieger,Mary Ann Lac 22h38'58"44d44'A
Krieger,Rhonda Cas 1h26'58"61d50'
Krieger,Sonja Gem 6h44'5"13d49'
Krieger,Suzanne Marie Mon 7h39'41"-8d5'
Krieger,Terry Aql 19h56'46"15d37'
Krieger,Wayne Lac 22h38'58"44d44'B
Krieger,William (Bill) Ari 2h43'42"30d52'
Kriegeskotte,Hans Sge 19h8'52"18d47'
Kriegsman,Beatrice Crb 15h47'19"38d35'
Kriegsman,Minnie Lyn 7h48'40"48d7'
Kriegsmann,Danielle Cas 0h40'44"69d27'
Krien,Chase Aql 19h43'1"10d44'
Krier,Patricia "Starwish" Cyg 20h40'41"30d19'
Kriescher,Raymond Cas 0h3'48"66d11'
Kriesel,James H Aql 19h53'0"-5d59'
Kriesel,Steven C Boo 14h48'11"50d12'
Krieter,Jennifer Dean And 0h10'0"37d31"
Krietsch,Christopher Allen Aur 6h3'27"31d41'
Krietsch,Geoffrey Michael Equ 21h0'36"10d6'
Krigbaum,Rick E Tri 2h17'19"28d35'
Krijnen,Jacques Sge 19h57'39"20d8'
Krik Per 2h54'1"31d22'
Krikalo Uma 11h37'29"50d23'
Kriket Eri 2h55'41"10d29'
Krikles 612 Her 18h19'1"14d36'
Krimbacher,Bernhard Ser 15h18'53"22d29'
Krimminger,Anthony Eugene Her 16h45'42"16d10'
Kring,Kristin Obrien Cet 1h44'59"-1d4'
Kringlee,Betty Cline Cas 1h10'19"68d15'
Krinick,Selma Tri 2h41'37"31d22'
Krinsky,Charles H Aur 6h9'33"46d51'
Krinsky,Rachael Lyr 18h38'28"41d6'
Krinstine J Lyn 7h48'47"51d16'
Krinzman,My Hero- Edward A Per 1h43'19"53d22'
Kripke,Dan Cet 1h35'59"-6d51'
Krippendorf,Nicholas Lac 22h10'52"49d29'
Krippinger,John & Dot Cyg 20h57'1"31d31'
Kris Ari 1h45'18"22d49'
Kris Ori 6h6'0"6d54'
Kris Cam 3h55'41"72d13'
Kris "K H & John "Gas" Cyg 21h12'21"39d21'
Kris & Marci's Anniversary Star Umi 13h55'48"71d41'
Kris' Cosmo Cma 6h46'0"-11d13'B
Kris's Goodluck Star Cas 0h52'0"68d30'
Kris's Jewels Cam 8h8'23"80d40'
Kris,Donna Schulze And 2h8'57"42d23'
Kris,My Shining Star And 23h23'18"42d10'
Krisby,Paul William Boo 14h36'50"32d11'
Kriscintilla XXIV Lyn 7h31'42"35d30'
Krisdathanon 1 Cep 21h18'0"60d27'
Kriser,Robert Andrew Dra 16h24'30"60d35'
Krish,Jeff Cnv 12h45'18"50d25'
Krishna Cam 4h34'1"61d26'
Krishna Mary Cet 2h42'27"2d46'
Krisi And 0h25'13"45d3'
Krisman,Adele And 0h45'27"38d42'
Krissa & David Oph 18h17'36"12d33'
Krissy Cas 0h30'52"61d47'
Krissy Lyn 7h43'48"38d49'
Krissy L K Aql 20h5'53"-0d53'
Krista Mon 6h48'21"-6d39'
Krista And 0h7'17"40d33'
Krista Lyr 18h30'25"45d26"
Krista Vul 20h17'11"28d37'
Krista Cmi 7h56'51"0d18'
Krista & David Uma 10h22'0"48d24'
Krista & George Uma 10h35'44"48d16'
Krista Ann Sgr 18h52'13"-20d19'
Krista Kerr Peg 22h24'0"29d54'
Krista Lynn Peg 22h0'20"23d0'
Krista Lynn Aql 18h58'38"-1d59'
Krista Lynn Cas 1h5'25"48d52'
Krista Marie Vul 20h39'47"20d28'
Krista Zipper Peg 0h4'59"13d10'
Krista,Star Of Goodness & Beauty And 23h21'16"46d8'
Kristal And 1h17'52"40d4'
Kristapiazzi Mon 7h0'35"-1d3'
Kriste:Keeper Of The Flame Cas 0h26'45"75d55'
Kristel Mon 8h7'0"-1d29'
Kristel Kaye Lyn 9h8'14"44d15'
Kristen Lyr 18h39'51"44d44'
Kristen Cma 7h15'26"-15d24'
Kristen Com 12h3'38"27d52'
Kristen Lyr 18h41'34"27d56'
Kristen Anne Lyr 18h49'1"42d21'
Kristen Anne And 2h16'11"38d52'
Kristen Casey Jeffrey Aur 7h2'1"38d55'
Kristen Eileen Cas 23h32'46"61d1'
Kristen Elizabeth Cas 0h54'28"59d31'
Kristen Lynn And 22h57'48"37d29'
Kristen Michele Del 20h55'11"7d56'
Kristen Nicole Cet 0h35'57"1d21'
Kristen Patricia 3-6-94 And 23h3'12"44d48'
Kristen R Aql 19h28'43"10d23'
Kristen Star,The And 2h29'43"42d31'
Kristen's (School- daze) Cas 2h12'40"61d42'
Kristen's Cella System Ori 6h12'45"18d18'B
Kristen's Light Lyr 19h7'23"37d42"
Kriston's Twenty first Com 12h28'39"30d48'
Kristen-J Del 20h55'39"9d8'
Kristens Star 12-25-94 Lmi 10h0'0"34d50'
Kristensen,Rikke Lyr 5h27'41"29d56'
Kristensen,Stephanic Nystrup Cyg 20h1'41"31d28'
Krister,Charles Oph 18h19'59"6d39'
Kristi Mon 7h59'25"-0d3'
Kristi Aql 20h11'46"5d35'

Kristi & Charles Forever
Cyg 19h40'18"38d5'
Kristi's Star
Cyg 20h53'1"31d36'
Kristian 27
Cam 3h57'28"71d26'
Kristie Lynn
Cas 23h36'52"62d39'
Kristie,Glen
Eri 4h4'19"-10d30'
Kristien,Dale
Ori 5h22'46"0d13'
Kristin
Cas 1h44'14"61d10'
Kristin
Cas 2h26'25"77d25'
Kristin
And 2h19'30"46d32'
Kristin Ann
Cas 0h5'1"54d49'
Kristin Anne
And 0h8'52"31d39'
Kristin Elizabeth
Cyg 20h3'54"40d57'
Kristin Kathleen
Cas 2h59'12"57d38'
Kristin Leigh
And 23h26'46"48d2'
Kristin Lindsey
Mon 6h25'0"10d33'
Kristin Lori
Cas 0h22'0"75d49'
Kristin Loves Glenn
Uma 11h18'58"40d17'
Kristin Marie
Tri 2h20'25"35d24'
Kristin My Shining Star
Del 20h14'29"14d4'
Kristin R
Cyg 21h9'12"40d19'
Kristin's Little Star
Ori 5h51'0"17d58'
Kristin's Star
And 23h22'18"46d44'
Kristin's Star For Wishes & Dreams
And 0h57'54"41d2'
Kristina
Cas 23h24'1"58d59'
Kristina
Uma 10h53'36"40d56'
Kristina
Umi 14h54'0"68d41'
Kristina
Cyg 19h48'34"37d49'
Kristina
Lyr 18h36'11"40d13'
Kristina
Lyn 8h17'0"33d48'
Kristina Elizabeth
Cas 1h12'32"63d35'
Kristina L
Cas 23h21'26"60d22'
Kristina Michelle
Uma 9h39'15"44d55'
Kristina Ulrike St
Uma 13h50'40"50d24'
Kristina's Light
Eri 2h48'27"-18d60'
Kristina,Bona Libra Puella
And 0h7'32"46d3'
Kristine
And 0h4'37"46d27'
Kristine
Uma 11h17'30"57d26'
Kristine & Frank
Lyr 18h37'48"35d4'
Kristine & Michael Our Light Of Love
Her 18h17'18"20d6'
Kristine Brooke
Cet 1h29'58"-11d45'
Kristine Lee
Del 20h57'40"10d23'
Kristine Robed In White-Song 4:7
Cyg 20h22'34"38d4'
Kristine Rose
Cra 18h15'36"-39d47'
Kristine's (Sweetness)
Vul 19h48'11"20d17'
Kristine's Star
Mon 7h45'52"-1d11'
Kristine's Star,Love Always Dad
Eri 4h6'0"-18d11'
Kristine,Princess of Light
And 1h56'20"40d13'
Kristofer Jon
Her 16h44'31"29d41'
Kristoph,Katie
And 23h42'40"42d44'
Kristy
Peg 21h43'13"24d9'
Kritsch,Rich
Boo 14h20'45"27d27'
Kritter
Hya 9h5'33"-11d51'
Kritzler (Milky Way), Robert August
Dra 16h4'45"66d23'
Krivich,Tyler
Lmi 10h35'30"30d30'
Krize"Your Special Star",Tara Lani
Cas 23h25'0"61d10'
Krizmanic
Uma 10h40'1"51d59'
Krnak,Adelaide
Crb 15h42'40"26d53'
Krnc Star,The
Cnv 13h21'47"41d17'
Krneta,Mariateresa Bonezzi
Cyg 21h22'51"28d47'
Krob,Kathleen Kay O'Hara
Cas 1h42'34"61d48'
Krobot-Stone,Susan Marie
Mon 6h19'29"-0d50'
Kroeger,"Kando"Kandis S
Cyg 20h25'29"40d13'
Kroeger,Jan & Gary
Aql 19h27'46"1d7'
Kroeger,Paul
Peg 0h0'40"28d6'
Kroeger,Rosemarie
Psc 0h55'1"2d25'
Kroeger,Shannon
Aql 20h1'54"11d54'
Kroely,Jean-Louis
Aur 7h0'38"38d20'
Kroeplin,Jochen
Del 21h4'32"12d5'
Kroese,Emma Miriam
And 23h19'1"45d48'
Kroese,Roeland Martôn
Aur 5h12'30"42d8'
Krofta,Dr Milos
Cmi 7h15'26"9d30'A
Krofta,Marla
Cmi 7h15'26"9d30'B
Krog,Evald
Eri 5h6'59"-4d11'
Kroger,Matt
Mon 8h2'42"-2d10'
Kroger,Rickle
Boo 14h44'30"50d26'
Krogh,John Russell
Cep 21h29'59"65d19'
Krogh,Patricia Lee
Peg 23h5'54"33d57'
Krogmann,Norbert
Uma 10h24'37"48d16'
Krogull,Steven Robert
Lmi 10h2'1"36d51'
Krohm,Sabrina
Uma 14h22'0"60d4'
Krohmann,Rolf
Peg 0h10'27"18d5'
Krohmer,Addie
Aql 18h55'1"-1d15'
Krohn Star,The
Uma 9h45'50"51d14'
Krohn,Jules
Aur 6h17'33"35d20'
Krohngold,Morgan James
Cma 7h15'14"-16d18'
Kroin,Lawrence & Kelly
Dra 16h41'55"70d4'
Krol
Ori 6h16'44"7d57'
Krol,Jan
Uma 10h4'29"50d52'
Krol,Joanne
Aql 20h1'28"11d15'
Krol,Rosemary
And 0h9'50"27d39'
Krolak,Willa
Lyn 7h30'33"41d8'
Krolik,Kaitlyn Kelly
Cyg 20h30'23"37'53'
Krolikowski,Debbra A
Cas 23h27'20"61d13'
Krolikowski,Nicholas Thaddeus
Vul 19h41'21"26d46'
Kroll,Klaus
Cap 21h32'3"-20d40'
Kroll,Phyllis
Del 20h18'9"10d47'
Kroll,Werner
Cmi 7h29'23"7d42'
Krollfeifer,Jr,Howard
Aur 5h15'1"45d26'
Krom,Marianne
Cas 2h43'23"61d47'
Krombach,Matthew Stedman
Peg 23h16'1"32d40'
Kromer,James F
Cnv 12h12'14"50d5'
Kromer,Megan Lea
Cyg 20h22'40"40d6'
Krompier,Ashley Diane
Lyn 7h46'1"42d31'
Kromwall,Adrian
Cet 1h58'45"0d49'
Kron,Carolyn
Sge 19h35'29"16d52'
Kron,Colleen
Cas 0h37'34"68d9'
Kron,Erin
Mon 8h6'58"-0d26'
Kron,Kevin
Her 18h16'40"14d56'
Kron,Michelle
Lyr 19h6'20"28d16'
Kron,Robert Devin
Lac 22h25'47"55d47'
Kron,Tara
Lyn 8h11'54"57d31'
Kron,Uwe
Ori 5h51'24"21d8'
Krona,Judith Ann
Mon 7h16'0"-6d49'
Kronberg,Steve
Her 18h2'12"30d21'
Kronborg,Karina
Umi 14h56'54"70d44'
Kronengold,Matilda
Mon 6h30'25"8d4'
Kronenthal,Eric Ilan
Ser 15h27'47"22d2'
Kronk,Michael & Janet
Cet 2h28'51"-10d59'
Kronlund The Legend, Leif
Ori 5h52'14"9d11'
Kroon,Captain James R
Cep 22h26'13"59d5'
Kroon,Ronnie & Bob
Umi 14h52'42"65d45'
Krop,Emily
Mon 6h0'50"-5d48'
Kropf,Claus
Per 3h50'29"35d10'
Kropf,Robert A
Per 3h53'11"38d18'
Kropff The Beast ta-ta,Jake
Her 17h59'44"14d51'
Kropik,Renee M
Oph 17h54'42"13d27'
Kropornicki,Jackie
Psc 1h31'34"20d3'
Kropp,Albert
Cyg 19h52'43"45d28'
Kropp,Alice Marie
Dra 11h34'35"78d3'
Kropp,Bill Durante
Dra 16h37'42"70d39'
Kropp,Charles A
Uma 10h0'52"50d8'
Kropp,Jennifer Mauschild
Ori 5h33'30"-6d46'
Krosnick,Morton & Diane
Del 20h13'17"10d1'
Krotz,Judith
Cas 0h15'1"61d44'
Krouse,Mary & George
Cap 21h20'46"-18d56'
Krouse,Mr Jim
Cep 22h27'17"60d4'
Krout,Pamela
Equ 21h18'37"2d54'
Krouzil,John J
Psc 1h2'38"30d39'
Krowitz,Israel Murray
Tri 2h20'32"28d40'
Krsna's Cornelia
And 0h32'56"45d30'
Krstevski
Cam 4h50'59"69d32'
Kruas,Dagmar
Leo 9h55'0"12d25'
Kruchoski
Ser 15h33'60"20d24'
Kruchten,Karl
Her 17h21'39"42d20'
Kruckenberg,Erik Allen
Uma 11h13'1"32d51'
Kruckman,Floyd F & Harriet E
Dra 13h24'36"68d12'
Kruczyna,Roger
Cep 22h4'34"60d27'
Krueger 10/26/71, Tracy S
Mon 7h47'24"-3d26'
Krueger's Komet
Dra 17h33'0"61d28'
Krueger,Amy
Cyg 21h35'57"41d26'
Krueger,Anna
Cas 1h7'21"50d16'
Krueger,Annette H
Cas 1h5'11"65d18'
Krueger,Dave
Dra 11h3'14"73d45'
Krueger,David & Pam
Mon 5h59'58"-8d16'
Krueger,Diane Stevens
Equ 20h58'53"6d13'
Krueger,Douglas Michael
Her 17h7'1"21d33'
Krueger,Harold & Patricia
Aql 18h51'35"11d1'
Krueger,Keith John
Leo 11h6'18"-2d23'
Krueger,Michelle Linn
Cam 5h2'14"60d43'
Krueger,Reinhold Richard
Cet 2h55'0"2d47'
Krueger,Ron
Dra 16h20'25"64d30'
Krueger,William
Uma 11h53'23"31d25'
Krueger-Orten
Her 17h55'14"14d20'
Kruegers,Ten
Ori 5h56'44"14d7'
Krug,Bobby
Aur 4h49'33"50d38'
Krug,Erwin
Cep 23h10'1"63d44'
Krugar II
Dra 20h6'55"67d39'
Kruger 12-09-1995, Richard
Dra 16h6'40"64d26'
Kruger,Brandon Scott
Ori 5h4'1"10d24'
Kruger,Charles J
Cap 20h39'33"-16d48'
Kruger,Diane Marie Eide
And 0h9'1"35d29'
Kruger,Jack
Sct 18h35'39"-6d21'
Kruger,Jim I
Aql 19h12'11"10d30'
Kruger,Joey Bobo
Aur 6h50'57"38d26'
Kruger,Kevin Gary
Peg 21h59'54"31d22'
Kruger,Layne & Dania
Crb 16h17'1"30d34'
Kruger,Leslie Diane
Vul 19h19'19"26d45'
Kruger,Mitchell Kai
Boo 13h40'36"24d39'
Krugler,Matthew Samuel Russell
Oph 18h8'28"13d43'
Krugman,Doug Shmoo
Uma 10h50'1"50d16'
Kruis,Lynn Margarethe
Ori 5h55'28"17d17'
Kruise,Frank,Debbie, Christine & Anthony
Cam 07h41'1"68d6'
Kruk Arrigoni, Alessandra
Gem 6h40'50"30d43'
Krukowski,Manfred
Peg 23h39'59"18d22'
Krul,Scott
Her 16h41'0"4d5'
Krulan,Clare Christina
Lyn 8h8'53"39d60'
Krulig,Dr & Mrs Eduardo
Cam 9h23'24"81d34'
Krull,Angela
Vir 13h1'47"-5d16'
Krull,Darlene V
Peg 22h14'15"33d44'
Krull,Janelle Jodi
Ser 15h35'12"18d50'
Krum,Howard
Per 4h39'49"36d26'
Krum,Lester William
Per 2h54'55"38d36'A
Krum,Marian Trexler
Per 2h54'55"38d36'B
Krum,Penny L
Aql 20h1'1"14d19'
Kruman,Howard & Chris
Boo 14h35'58"8d21'
Krumes,Willi
Cap 21h21'25"-23d51'
Krumm,Thomas
Ori 6h17'0"8d48'
Krumme,Irene
Cam 7h59'52"73d46'
Krumme,Michael
Sge 19h8'11"17d24'
Krumnow,David Lee
Per 1h39'0"53d18'
Krump,Emma Rose
Cyg 20h22'34"38d4'
Krumpe,Edward Georg
Ori 5h54'21"16d29'
Krumrie,Corliss Anne
Cas 0h45'28"70d57'
Krupa,Bozena Joanna
Cas 1h16'0"61d37'
Krupa,Samantha Lee
Lyr 19h18'46"38d2'
Krupen,MD,Jeffrey
Uma 8h44'47"72d16'
Krupinsky,Steven Scott
Ori 5h30'34"-1d33'
Krupnick,Ida & Jack
Dra 15h54'24"61d53'
Krupnick,Tammy
Vul 19h20'48"25d15'
Krupnikoff,Jerry
Aur 6h6'12"30d22'
Krupp,Gina Marie
Lyr 19h15'31"28d13'
Krupp,Meline
And 23h46'50"44d24'
Kruppenbacher,Matthew Charles
Aql 20h4'52"8d46'
Kruppenbacher,Michael Francis
Hya 9h10'18"2d23'
Kruppy's Star
Cet 2h14'23"9d16'
Kruse,"Annie" Anne Elizabeth
Sge 18h58'20"19d13'
Kruse,"Kasey" Katharine Marie
Umi 14h18'56"66d15'
Kruse,Brigitte
Tau 3h44'0"11d3'
Kruse,Caleb Knox
Aql 18h57'38"13d51'
Kruse,Darold
Hya 8h43'28"2d28'
Kruse,David Lee
Aur 6h22'23"38d46'
Kruse,Harley
Cru 12h39'18"-60d22'
Kruse,Hubert & Karen
Cyg 1h20'47"37d42'
Kruse,James Richard
Aur 4h53'43"38d47'
Kruse,Marie Lindal
Cep 23h9'49"60d20'
Krusemark,Dawn Marie
Aur 4h47'1"40d58'
Kruspig,Volkmar
Ori 5h49'49"21d25'
Krust
And 0h59'1"39d34'
Kruszynska Star,Danka
And 23h47'17"41d57'
Krut,Brook Deanne
Ser 15h17'53"18d44'
Kruta,Jason
Tau 3h39'1"28d50'
Kruty,Kimberly Ann
Gem 6h35'56"12d4'
Kruz
Cma 6h44'37"-11d7'
Kruzel,Charlotte
Lyr 18h38'11"45d30'
Kruzel,Joseph J
Her 16h49'36"41d7'
Kruzic,Jacquelyn S
Vul 20h14'33"28d19'
KRV Loves MMP
Aql 19h31'31"14d44'
Kryger,Albert
Boo 15h0'56"23d55'
KRYJOZ
Sct 18h53'54"-4d39'
Krypel Standard Star, Michael
Aur 6h13'19"45d41'
Krypel,Busia
Cam 3h23'58"60d25'
Krys,Ms
Mon 6h20'56"2d56'
Krys,Stan & Carol
Aur 5h58'33"30d60'
Krys-BJ's Eternal Light
Lyn 8h31'28"43d46'
Kryshak,John A
Her 18h19'28"12d29'
Krysick,William
Sgr 20h0'52"-45d10'
Kryss,Caitlin
Del 20h14'42"15d40'
Kryss,Gretchen
Equ 21h11'0"10d41'
Krystal
Umi 15h11'27"70d16'
Krystal & Jeff's Wedding Star
Cyg 20h21'42"41d40'
Krystal Lynn
Peg 23h5'0"33d46'
Krystal Tears
And 0h5'37"44d52'
Krystar
And 23h43'1"32d27'
Krystyna das geliebte Maiglöckchen
Ser 15h25'17"2d11'
Krywulak,Alexandria Tina
Uma 8h37'55"68d1'
Krzak
Uma 9h7'1"50d6'
Krzciok's Star,Joni Bayle
Lyr 18h49'38"37d9'
Krzeminski,Wally
Cep 20h54'0"65d3'
Krznaric,Jr,Larry Charles
Her 16h31'20"47d54'
Krzysztof,Stefan
Ser 15h52'46"-2d23'
Krzywiecki,Sharon Ann
Lac 22h8'36"51d45'
Krzyzaniak,Together Always Ken & Carol
Lac 22h26'25"54d2'
Kräupel,Louis
Cet 2h40'12"4d18'
Kröber,Jürgen
Ori 5h49'1"20d59'
Kröger,Uwe und Christa
Lyn 8h8'1"49d14'
Kröner,Wolfgang
Gem 6h42'30"12d14'
Krösser,Johny
Ori 5h50'53"21d22'
Krüger,Ingrid
Psc 0h28'29"8d7'
Krüger,Monika
Eri 3h34'47"-7d10'
Krüger,Philipp Dagobert
Leo 11h51'17"28d36'
Krüger,Ulrice
Lib 15h1'24"-2d24'
Krüger-Hasenmüller, Olaf
Psc 0h28'1"8d39'
KSA-1
Sct 18h55'10"-4d32'
Ksander,George (Pop)
Lmi 9h34'29"37d44'
KSB & VLL (Fool's Rushin)
Uma 9h59'41"71d26'
Kship,Donna
And 0h45'47"40d57'
KSS#1DC
Del 20h26'48"11d28'
KT CARUANA-RANDOLPH
Boo 14h22'0"25d59'
KT3Y
Lmi 9h24'56"37d57'
Ktema Eis Aei
Oph 18h2'42"10d11'
KTY
Aqr 20h10'34"10d13'
Kuan,Mary
Cnc 9h8'17"30d19'
Kubach,Karl
Dra 16h46'22"73d5'
Kubacz,Joseph L
Her 18h2'30"31d49'
Kubacz,Joseph L
Her 18h2'51"31d15'
Kubala,Jane Banaszak
Cyg 20h4'33"58d36'
Kubala,Joel Frank
Ori 5h25'58"-4d8'
Kubale,Fritz
Sgr 19h51'43"43d25'
Kubeck,Ali
Cyg 21h7'13"31d36'
Kubes,Ellen A
Lyn 8h58'34"40d29'
Kubes,Mark J
Cnv 12h21'38"36d27'
Kubiak,Daniel Roman
Aur 5h36'47"54d27'
Kubicki,Ed
Uma 15h26'42"84d51'
Kubicki,Edward Lawrence
Uma 15h22'18"81d28'
Kubicki,Elizabeth Sharon
Uma 14h35'45"80d47'
Kubicki,Henry Thomas
Uma 15h42'15"82d26'
Kubicki,Ian Edward
Uma 14h42'24"81d2'
Kubicki,Irene Zelenyuk
Uma 15h24'17"80d14'
Kubicki,Jacquie Dean
Uma 15h32'28"81d8'
Kubicki,Jeanne Svelik
Uma 15h28'52"82d0'
Kubicki,Johanna Moore
Uma 15h2'15"81d11'
Kubicki,Marc Alan
Uma 15h2'42"82d16'
Kubicki,Marge
Uma 15h39'30"84d53'
Kubicki,Matthew Joseph
Uma 15h38'14"81d0'
Kubicki,Nicolas Edward
Uma 15h44'17"80d31'
Kubicki,Patricia Ann Stevens
Uma 15h42'0"81d33'
Kubicki,Paul Christopher
Uma 15h36'51"80d16'
Kubicko,Sharon Ann
Cas 0h53'47"58d26'
Kubik I,Klaus
Ari 2h41'18"21d42'
Kubik,Claire
Cas 23h39'35"50d11'
Kubisiak,Tiffany
And 22h55'0"50d43'
Kubisova,Ingeborg
Tau 5h56'46"26d51'
Kubitz,Abby S
Boo 14h59'1"50d23'
Kubitz,Jack & Tracey
Cnv 12h50'42"41d34'
Kubitz,Jack A
Dra 17h57'38"68d49'
Kubitz,Jeremy Allen
Boo 15h6'59"12d35'
Kubitz,Kyle L
Per 3h25'16"40d27'
Kubitz,Nancy L
Uma 12h32'13"61d28'
Kubitz,Sandra L
Aur 4h53'0"50d9'
Kubler,Devin Michael
Uma 10h52'39"56d48'
Kubler,Rick
Dra 13h2'1"67d41'
Kubokura-Strawn,Asako
Dra 16h49'0"62d33'
Kuborek,Grandpa Walter
Cam 5h49'0"60d11'
Kuby,Stuart
Her 15h50'28"41d38'
Kucera,Edna Lee
Lyn 7h40'24"40d31'
Kucera,Gregory
Her 17h1'36"41d9'
Kuchar,Karin
Aqr 22h44'36"-2d17'
Kucharik,Pastor Joseph
Her 16h8'12"20d3'
Kuchenbecker, Kristopher
Cet 3h20'55"1d6'
Kuchenbecker,Karalyn
Mon 6h11'1"-10d47'
Kuchenbecker,Katherine
Peg 23h4'1"13d10'
Kuchenbecker,Ralf
Cep 22h2'0"61d46'
Kucher,David J
Aur 6h0'51"32d54'
Kuchie's Cosmic Connection
Psc 0h44'26"20d33'
Kuchta,George B
Dra 14h58'0"62d37'
Kuchta,John F
Ari 2h44'27"30d16'
Kuchy,Lisa M
Mon 6h57'42"-8d23'
Kuchy,Seth M
Sex 9h51'43"1d27'
Kucij,Miriam
Uma 8h59'11"52d43'
Kucine,Milton
Lac 22h8'0"40d44'
Kucinski,Vickie Marie
Crt 10h55'1"-10d14'
Kuckuck,Robert Wade
Sco 15h54'39"-26d0'
Kucler,Joyce Ann
Lyr 18h23'45"38d2'
Kuczj,Benjamin John
Ser 15h38'43"21d45'
Kuczkowska,Gerald Fabien
Dra 11h1'19"73d50'
Kuczma,Philip Andrew
Her 17h11'20"20d19'
Kuczwara,Glen A
Boo 14h30'0"26d37'
Kuczynski,Alice Borzym
Sco 17h5'25"-38d53'
Kuczynski,Andrew Lucas
Sco 17h34'15"-38d56'
Kuczynski,Les Stanislaw
Sco 17h6'20"-38d11'
Kudebeh,Krevis Jude
Uma 11h30'14"44d22'
Kudela,Tanner Ashley
Peg 22h31'59"8d10'
Kudla,Todd Andrew
Her 17h0'1"48d2'
Kudler,Carrie & Harley
Uma 11h9'32"30d17'
Kudo,Irma S
Cas 23h0'32"53d49'
Kudzia,Sr,Edward A
Per 3h43'51"37d18'
Kuecher,Howard & Mary Alice
Peg 22h13'36"20d57'
Kuefnoris Intravaneous
Boo 15h1'0"15d42'
Kuehl,Anna Ashleigh
Mon 7h29'28"-6d44'
Kuehl,Elkelev
Gem 7h9'58"28d39'
Kuehling,Bryan James
Boo 14h15'42"32d8'
Kuehn,Frank W
Cnc 8h31'57"30d45'
Kuehn,Patricia G
Cyg 21h1'23"37d39'
Kuehn,Stephen Joseph
Aql 19h48'25"14d42'
Kuehne,Johannes
Gem 6h46'9"16d33'
Kuehne,Nicole Beth
And 2h9'11"41d31'
Kuehner,Thomas
Aur 5h39'46"48d15'
Kuehnle Family Star, Wes & Grace
Del 20h13'1"15d11'
Kuenstle,Billy
Aur 6h19'43"32d32'
Kuenzi
Uma 9h44'45"50d43'
Kuenzinger,Robert Joseph
Aql 19h3'13"4d57'
Kueper,Charles
Aur 5h21'46"38d5'

Kueppers,Elke
Cnc 8h27'22"31d29'
Kues,Henry August
Boo 14h56'10"46d12'
Kuester,Brigitte
Oph 18h39'22"8d22'
Kuester,Gerd
Oph 16h20'19"1d39'
Kuffell,Daniel James & Laura Lynn
Mon 7h53'41"-6d49'
Kufro,Carolyn & Elmer
Cyg 21h38'1"40d44'
Kugel,George
Cep 22h16'27"62d1'
Kugel,Helga
Aqr 22h40'1"-2d3'
Kugler,Christian
Uma 13h34'0"48d52'
Kugler,Frank
Cep 0h5'37"80d20'
Kuhlin,Vicki Sue
Lyn 8h47'0"46d57'
Kuhlman,Cael A
Ser 15h59'54"0d12'
Kuhlman,Sally
Cas 2h4'50"59d21'
Kuhlman,Scottie Coblentz
Cam 7h36'29"70d25'
Kuhlmann,Friedrich- Karl
Sgr 19h11'2"-23d16'
Kuhlmann,Nicholas Alexander Gray
Her 16h19'48"10d36'
Kuhlmann,Taylor Anne
Tri 1h49'31"28d42'
Kuhn II,Gerard Raphael Joseph
Cet 0h52'26"-4d34'
Kuhn,Becky
Lyn 7h56'43"45d54'
Kuhn,Berthold
Psc 1h21'54"10d54'
Kuhn,Brian James
Per 3h3'1"41d33'
Kuhn,Carol
Uma 10h25'28"67d60'
Kuhn,Carol
Cyg 21h16'41"36d21'
Kuhn,Carol
Crb 15h56'54"31d44'
Kuhn,Carol
Cam 11h50'38"78d12'
Kuhn,Charles
Ori 6h6'36"20d35'
Kuhn,Charles Richard
Lyr 18h52'26"34d56'
Kuhn,Eugenia
Sco 17h21'44"-38d21'
Kuhn,Genevieve Marie
Cam 4h3'1"67d59'
Kuhn,George David
Dra 14h54'30"60d30'
Kuhn,J C
Ori 5h52'54"19d0'
Kuhn,Jerilynn Hope
And 23h21'32"48d40'
Kuhn,Marie B
And 0h4'17"38d52'
Kuhn,Reagan
Cyg 19h33'33"30d37'
Kuhne,Kay C
Uma 11h59'50"45d49'
Kuhnen,Molly Mae
Cas 1h16'28"68d6'
Kuhner,Dr Susan M
Leo 10h30'34"21d48'
Kuhnke,William & Wilda
Crb 15h53'0"38d40'
Kuhns Star,Prince Charming The Carl
Lac 22h19'20"50d19'
Kuhns,Anne Duvall
Lib 15h18'13"-20d2'
Kuhns,James Edward
Cet 2h56'50"0d40'
Kuhns,Joan A
Uma 10h51'31"48d48'
Kuhrman & Family
Boo 15h7'1"18d56'
Kuhrtz,Meff & Jane
Crb 15h48'1"38d5'
Kuhta,Jr,Richard Thomas
Cyg 21h27'14"40d3'
Kui,Ellen Bo Yee
Lyn 8h52'39"40d39'
Kuipers (TAL),Derek R
Uma 10h1'19"42d28'
Kuipo
Del 20h18'20"14d58'
Kuitunen,Timo
Cam 3h41'35"60d50'
Kujala,Amy Jo
And 0h46'31"36d10'
Kujawa,Andre Kasmir
Cet 0h52'0"0d27'
Kujawa,Haley Marie
Lyr 19h17'18"38d28'
Kukay,Norman John
Lac 22h52'12"53d16'
Kukiela,Richard
Dra 16h55'44"70d42'
Kukla,Linda Marie
Aql 20h16'33"5d12'
Kuklinski In Memorium 1933-1992,George
Cyg 20h36'1"40d14'
Kuklish,Nick & Adella
Cet 1h28'0"-10d47'
Kukor-Merry Christmas 1994,Randy
Dra 17h7'0"62d47'
Kukuk,Norbert
Del 21h5'17"12d57'
Kulak's Class RP,Joe
Aur 6h9'27"31d9'
Kulak,Brianna Maria
And 0h17'1"38d32'
Kulak,Taylor Ann
Gem 6h39'33"35d6'
Kulas,Andreas Anton
Cma 6h55'43"-18d2'
Kulbida,Carol Frances
Cas 2h2'1"59d10'
Kulcsar,Tünde Téczely, & Lászl
Dra 20h28'44"62d49'A&B
Kulczycki,Thomas A
Crb 15h19'0"30d46'
Kulczyk,Virginia
Lyn 8h4'36"40d27'
Kulewicz,Leslie Anne Crisafulli
Cas 0h28'20"63d43'
Kulich,Vlada
Lyn 6h24'50"54d22'
Kulick III,John William
Ori 5h26'25"15d13'
Kulik,Gene J
And 1h13'58"40d39'
Kulikov,Hanya
Uma 8h46'35"56d30'
Kulikowski,Matthew A
Aur 6h26'58"38d12'
Kulk,Ilene Yarmark
Mon 6h33'12"0d33'
Kulk,James
Cet 3h13'48"5d31'
Kulk,Kenneth
Equ 21h21'55"3d27'
Kulk,Marguerite
Cra 18h19'36"-37d11'
Kulka,Patricia Iovieno
Lyn 7h48'31"38d16'
Kulowski,Holly Caroline
Vul 19h22'12"23d36'
Kulp,Mary Anne
Uma 10h54'11"50d17'
Kulpinski,Edmund
Dra 18h51'37"68d46'
Kulwicki,Alan
Dra 16h12'47"66d33'
Kulyk,Elizabeth Rogers
Cet 2h14'58"3d48'
Kulyk,John William
Cet 2h54'17"5d24'
Kuma,Marcia
Aur 7h7'38"40d12'
Kumano,Ralph
Aql 19h8'13"0d19'
Kumar
Cam 3h50'1"69d22'
Kumar,Gita
Aql 19h10'14"12d38'
Kumar,Nimi
Peg 23h26'51"24d11'
Kumar,Ravimdar
Her 18h3'40"14d31'
Kumari
Sge 20h7'51"18d52'
Kumari,Asha
Lyr 18h28'0"31d15'
Kumataro
Her 17h53'49"14d21'
Kumbernuss,Astrid
Sge 20h13'46"17d53'
Kumble,The Celestial Body of Roger
Sct 18h55'0"-5d40'
Kume,William John
Cyg 19h33'39"35d50'
Kumichan Stella
Tau 5h23'45"24d57'
Kumiko,A Precious Glory Of Rosemary
Cap 20h54'26"-22d34'
Kumm,Matthew Scott
Cep 3h18'0"78d14'
Kummerfeld,Theodosia "Tia"
Cas 1h50'56"60d4'
Kummerman,Alex et Mathilde
Cyg 22h0'39"51d56'
Kummerow,Claudia
Sgr 18h47'44"-29d59'
Kump,Howard James
Lac 22h31'35"53d18'
Kumpf,Niklas
Vir 13h37'52"2d21'
Kumstel,Otto
Uma 10h2'30"57d38'
Kumud
Lyn 8h59'47"45d52'
Kun-Ja
Uma 10h52'39"71d33'
Kunash,Andrew Raymond
Dra 17h41'35"60d18'
Kunce,Marc Lee
Tri 1h46'15"28d14'
Kuncl's George Alby Gloria,Glorious
Boo 15h6'49"31d50'
Kundrath,Stephen Joseph
Lac 22h55'54"52d40'
Kuneck,Stacie Jane
Cas 2h3'43"71d3'
Kunert,Eva
Cep 22h9'52"60d58'
Kunert-Woofie,Edward Robert
Cnv 12h53'25"43d9'
Kunesh,Mai Tai
Cam 3h38'0"61d58'
Kunetz,Caroline Elizabeth
Lyr 18h28'36"35d1'
Kunhikrishnan P P
Eri 4h8'12"-5d3'
Kuniko
Cet 3h1'27"02d8'
Kunin,Jake Lewis
Tau 4h17'47"7d48'
Kunisch,Dr Rolf Compliments La Prairie
Lac 22h37'43"37d50'
Kunkel,Elisabeth
Mon 6h23'14"4d35'
Kunkle,Eileen
Cas 2h6'55"68d56'
Kunkle,James
Cam 5h2'29"65d1'
Kunkle,Marian
Mon 8h1'11"-9d47'
Kunkle,Mikell R
Lyn 9h0'20"46d31'
Kunkle,Steven J
Her 17h18'1"20d48'
Kunn,Jean-Pierre
Lac 22h6'19"37d50'
Kunter,Eva
Cnc 9h18'14"30d58'
Kuntner,Sonja
And 2h4'0"40d30'
Kuntsch,Ewald
Hya 9h9'8"-9d42'
Kuntz Family Star
Crt 11h15'50"-18d42'
Kuntz,Leon C
Aur 6h32'25"41d7'
Kuntze,Elisabeth
Ari 2h20'58"21d25'
Kuntzelman,Carol
Com 12h9'33"27d59'
Kuntzman,Eugene & Blanche
Ori 6h9'1"8d51'
Kuntzsch,Thomas Fredrick
Her 17h0'55"42d27'
Kunz,LaRae
Equ 21h19'18"10d38'
Kunz,Monique
Sgr 19h34'30"-34d18'
Kunz,Reneé
Peg 23h4'35"22d22'
Kunz,Richard William
Leo 10h25'32"15d53'
Kunz,Rudolph
Peg 23h4'42"20d12'
Kunz,Sr & Jr,William
Dra 18h11'1"65d24'
Kunz,Tom
Tau 4h18'24"20d46'
Kunze,Christian
Lyr 19h12'44"31d30'
Kunzelmann,Heike & Frank
Vir 13h52'0"-2d12'
Kunzman's Half Century William A
Lib 15h39'17"-28d13'
Kunzmann,Karl-Heinz
Mon 7h51'29"-2d27'
Kupczewski,Kimberly
Cas 0h43'28"69d24'
Kuper's Star "Woo's World",CJ
Aql 20h7'19"0d23'
Kuper,Paul Joseph Michael
Gem 6h49'57"19d48'
Kupferberg,Ian Seth
Dra 16h7'1"61d39'
Kupiec,Marcyanne
Aql 19h53'1"11d51'
Kupilik,Margaret
Cyg 19h39'15"30d54'
Kupkowski,Diane Dermott
Lyr 18h38'54"30d59'
Kuppers,Daniel Alexander
Her 18h12'19"41d4'
Kuprevich,Benjamin J J
Her 17h38'22"26d50'
Kurapati,Sudha Rani
Vul 20h3'49"25d4'
Kurcz,Tamma
Cap 20h41'15"-26d57'
Kurek's Star,Dolores
Cyg 21h21'16"40d3'
Kuren Dell'Aquila, Phyllis J
Sgr 18h59'53"-27d25'
Kurgan,Glen R
Her 17h5'48"45d33'
Kuri"Forever as One", Carl & Olga
Crb 15h58'33"26d37'
Kurian,Barbara
Peg 22h21'27"33d46'
Kurie Star of Principals,The Mannie
Ori 6h17'50"-1d6'
Kuris,Cary & Sonya
Uma 11h31'0"49d21'
Kurkjian,Diane
Cyg 19h43'38"29d17'
Kurkjian,Jeffrey Patrick
Peg 23h37'40"30d20'
Kurkjian,Rudolph
Sgr 18h0'56"-28d43'
Kurkowski,Julie P
Lac 22h29'29"53d36'
Kurkowski,Kevin
Dra 14h36'23"62d14'
Kurkowski,Michael Frank
Aur 6h16'59"38d13'
Kurland,Robert Duff
Aur 5h4'0"40d55'
Kurlemann,Monica
Com 12h24'47"31d22'
Kurlok & Sharlo
Boo 15h4'54"15d40'
Kurnava,Stacey Ellen
Boo 14h31'18"10d49'
Kuroda,Daien
Hya 9h8'17"3d58'
Kuroda,Junyu
Eri 4h1'43"-12d44'
Kuroda,Shunyu
Ori 5h40'1"11d49'
Kuronen,Eeva Ja Sulo
Umi 15h17'51"78d17'
Kuropatkin,Paul Jacob
Lyr 19h2'13"28d50'
Kuroski,Alyson
And 0h58'16"45d27'
Kuross
Cyg 19h28'22"37d38'
Kurpinsky,Lawrence & Joan
Cnv 13h45'56"39d57'
Kurpjuweit,John R
Sge 19h1'0"18d59'
Kurran,Kirsten
And 23h30'43"47d44'
Kursch,Mark Allan
Lib 14h57'51"-1d29'
Kurt 40
Boo 14h38'15"39d16'
Kurt in the Sky
Ari 2h4'53"20d2'
Kurt Paul
Per 3h28'52"40d48'
Kurt's Future
Her 16h3'53"41d25'
Kurt's Little One
Vul 21h25'21"27d37'
Kurt,Kamil
Vul 20h46'12"20d27'
Kurth,Beth & Casey
Cyg 21h23'21"39d54'
Kurth,Cydney Taylor
Lyr 18h15'21"33d50'
Kurth,Edward
Ser 15h31'1"9d28'
Kurti
Boo 15h39'1"48d55'
Kurti
Boo 14h39'19"14d59'
Kurtz,Casey
Dra 19h43'30"61d27'
Kurtz,Daniel Victor
Lac 22h25'45"52d46'
Kurtz,Jackson Hunter
Lac 22h9'52"37d44'
Kurtz,Julien
Cam 3h25'13"61d47'
Kurtz,Mary
Lyn 7h49'20"50d5'
Kurtz,Stacy
Cas 23h22'33"63d17'
Kurtzal
Lyr 18h37'20"26d9'
Kurtzrock,Lisa Dawn
And 23h44'22"41d2'
Kurutz,Gary Allen
Dra 19h18'53"67d49'
Kurwicki,Robert Alan
Aur 6h29'56"37d17'
Kuryluk III,Edward Charles
Lac 22h14'21"50d6'
Kurz "Special K", Robert J
Aur 6h8'24"35d34'
Kurz,Inge
Eri 4h48'56"-6d20'
Kurz,Margaret
Mic 20h52'24"-44d46'
Kurz,Rudolf
Cep 22h0'12"61d25'
Kurzejewski,In Memory of Josephine
And 1h0'57"41d12'
Kurzeka,Susan R
Mon 7h56'4"-3d38'
Kurzner,Peter Meyer
Hya 8h41'0"1d53'
Kurzok,Paul
Her 16h13'24"8d31'
Kurzweil,Elaine & Brett
Cyg 19h28'1"34d1'
Kurzynski,Master Mike
Aur 6h32'14"38d42'
Kus,The
Ari 2h14'32"18d6'
Kusas,Areti Evangelia
Aql 19h30'13"0d12'
Kusch,Meghan
And 1h58'38"37d32'
Kusch,Sr,Robert David
Aur 7h4'25"40d44'
Kuschel
Com 13h8'39"31d13'
Kuschelbrchen Uwe
Ser 16h5'53"2d2'
Kuschelmäuschen
Boo 15h5'45"62d17'
Kuschill,Terri
Peg 22h0'12"20d12'
Kuscho
Uma 9h4'56"54d12'
Kuse,Michael James
Aql 19h30'0"10d46'
Kusel,Robert & Lois
Sge 19h56'22"16d43'
Kusevich,Valerie
Cyg 21h22'18"38d1'
Kushner,John
Uma 11h5'20"48d20'
Kushner,Kathleen "Kay"
And 1h12'29"35d55'
Kushner,Marian McCarthy
Peg 23h1'1"18d28'
Kuske,Anne Greta
Ori 5h49'54"21d1'
Kuskowski,Helen W
Tau 4h39'38"10d59'
Kusmierz,Allyn Matthew
Sgr 20h2'50"-43d21'
Kusnyer,Michael C
Eri 5h5'53"-8d11'
Kusrow,Dorothy
Aql 20h5'43"3d47'
Kuster,Alfons
Ori 5h32'30"-8d21'
Kuster,Lineli
Ari 2h43'16"30d10'
Kuster,René
Cam 7h47'15"61d31'
Kustka,Glenn
Cep 22h14'26"68d10'
Kustra,Adam
Her 16h36'16"10d52'
Kusumagraj
Gem 7h18'1"31d52'
Kutash,Sarah Joan
Psc 0h55'29"27d53'
Kuter,Jason Loves, Liz Witmer
Cep 20h36'22"65d5'
Kutik,Zachary James
Lac 22h42'54"37d55'
Kutlik,Jr,Lawrence Edward
Her 16h48'41"48d23'
Kuttner,Ilse Ruth
Com 12h1'56"27d47'
Kutz,Uwe
Lib 15h54'18"-18d48'
Kutzner,Christine und Georg
Lyr 18h34'12"27d8'
Kuuipo
Umi 16h47'30"76d3'
Kuuipo
Cam 5h50'59"73d10'
Kuusik,Andres & Maureen
Cru 12h49'27"-58d28'
Kuykendall,Captain
Lyn 8h9'20"36d24'
Kuykendall,Greg
Hya 9h37'51"-9d6'
Kuypers,Roger
Aur 7h0'1"38d32'
Kuzara,Eric James
Aur 5h22'22"37d57'
Kuzdenyi,Larry
Uma 11h21'35"33d3'
Kuzel,Nicholas Emil
Boo 14h1'58"11d56'
Kuzel,Peter
Sgr 17h58'56"-30d36'
Kuzemko,Annabel
Cep 21h58'1"68d53'
Kuzemko,Matthew
Cep 21h43'49"71d1'
Kuzera,Monica Genevieve
Cyg 19h40'12"30d13'
Kuzia,Sister's Star: Michelle & Cathy
Boo 14h37'46"45d19'
Kuziak,Mr & Mrs Michael J
Lyn 8h18'0"41d14'
Kuzma,Katharine Celeste
Mon 7h42'4"-1d46'
Kuzma,Stephen Joseph
Cnv 12h58'57"40d36'
Kuzminski,Stephen Rawlins
Sgr 18h56'12"-27d14'
Kvasnack,Kristopher P
Hya 8h43'46"5d8'
Kveton,Nicholas James
Lac 22h25'45"53d30'
Kvirkvelia,Tamara
Uma 11h58'55"31d57'
Kwakie,My Forever Seoul Mate
Lac 22h14'35"38d8'
Kwaller,Michael Kelly
Cnc 8h22'60"15d47'
Kwan,Ms Maxine
Mon 7h30'45"-8d22'
Kwan,Timothy A
Ori 6h0'46"-1d3'
Kwapis,Henry David
Cap 21h56'23"-24d7'
Kwas,Linda
Lyn 9h5'58"46d49'
Kwasniak,Ron
Aur 6h32'52"38d46'
Kwasny,Joshua Stanley
Uma 11h30'0"32d44'
Kwasny,Nana Helen
Uma 11h30'0"30d10'
Kwasny,Popi Stanley
Uma 11h30'0"33d33'
Kwaszkiewicz,Kimberly
Tri 2h36'60"35d27'
Kwayne
Per 1h52'51"53d40'
Kwentus,MD,Joseph
Hya 8h56'33"5d40'
Kwet,Claudia
Aqr 22h40'55"-5d42'
Kwiatkowski,Mary
And 1h53'13"40d38'
Kwiker,Louis A
Hya 8h15'45"-5d51'
Kwitny,Cassandra Ilene
Tau 4h45'48"18d53'
Kwitny,Jackson Charles
Tau 4h43'1"18d59'
Kwitowski,Juliet
And 23h27'44"49d29'
Kwon,Byung K
Cam 8h7'1"80d27'
Kwong,Arthur
Sco 17h28'40"-30d11'
Kwong,Elizabeth
And 1h9'24"39d29'
Kydd,Frances
Aur 4h56'50"41d7'
KyF
Lac 22h42'53"54d11'
Kyla
And 2h32'51"41d4'
Kyla Madonna
Lmi 10h17'40"31d16'
Kylane
And 2h3'0"39d60'
Kyle
Ori 6h6'48"-0d3'
Kyle
Cas 0h33'48"63d24'
Kyle
Cep 23h17'0"64d43'
Kyle
Cet 1h39'16"-11d34'A
Kyle & Josh
Umi 13h40'51"79d26'
Kyle Christopher Star, The
Cap 20h20'31"-26d58'
Kyle Family,The Roger
Aql 20h11'44"10d20'
Kyle Francis
Lyr 19h21'29"41d19'
Kyle Matthew
Uma 12h3'36"31d24'
Kyle Steven
Vul 19h48'30"20d11'
Kyle Thomas
Aur 5h55'51"31d13'
Kyle's Lantern
Uma 9h10'41"56d19'
Kyle's Light
Cyg 20h50'30"40d21'
Kyle's Miracle
Aql 19h10'56"13d5'
Kyle,April Michele
Aql 19h54'1"15d12'
Kyle,Brooke Denee
Peg 22h27'0"20d9'
Kyle,Dana Marie
Mon 7h40'0"-10d11'
Kyle,Emily Virginia
Peg 23h21'1"13d21'
Kyle,Michael A
Boo 14h38'39"37d53'
Kyle-The Great
Hya 8h41'19"2d28'
Kylee
And 23h23'36"43d49'
Kyler
Ori 4h58'46"10d7'
Kyler,Brenda June
Lyn 7h7'23"50d4'
Kyler,Katherine June
And 0h51'11"36d12'
Kylie Ann
Cas 2h2'18"59d38'
Kylie Loves Andrew
Boo 15h5'11"26d34'
Kylisa
Aur 5h9'12"40d32'
Kymberley Anne
Lyr 18h40'39"30d27'
Kymberly
And 0h14'45"37d35'
Kynara
Sco 16h1'42"-40d32'
Kyong Suk
Uma 9h28'1"52d5'
Kyra
Mon 7h40'29"-8d4'
Kyra Marie
Lyn 7h52'56"50d21'
Kyrc,Cassie
Peg 22h57'21"31d58'
Kyriakopoulou-Vafia, Melina
Cas 1h16'0"60d9'
Kyrian
Cam 5h36'27"61d46'
Kyrianna Marie
Del 21h2'35"12d39'
Kyrie
And 2h21'57"47d18'
Kythira
Ori 6h2'45"6d55'
Kytrina
Eri 2h56'50"-10d25'
Kählert,Uwe
Lib 14h40'57"-23d4'
Kämmerer,Hugo
Aqr 22h5'54"2d3'
Kämpf,Eleonore
Aqr 22h57'1"-6d49'
Kärn,Diane
Sco 17h51'48"-31d6'
Kästner,Sylvia
Cas 0h2'0"63d56'
KÜrstad,Thorstein
Cep 21h11'0"67d35'
Köck,Christian
Peg 22h16'36"4d44'
Köckinger,Dr Othmar
Ori 5h55'51"7d45'
Köder,Eva
Ser 15h29'53"8d21'
Köhlbach,Birgit
Cmi 7h21'45"0d2'
Köhler,Hedda-Ursula
Tau 4h3'29"20d6'
Köhler,Wolfgang
Ori 5h56'1"0d53'
Köhler,Wolfgang
Mon 7h51'56"-2d1'
Köhlmoos,Peter
Aqr 21h44'52"-3d39'
Köksal,Berrin
Lyr 18h18'20"38d45'
Köllmann,Jürg E Köllmann-Gruppe
Cap 21h26'19"-23d43'
Kölln,Dr Hans- Christian
Peg 23h41'25"17d16'
Köllner,Herbert
Ari 2h28'13"26d52'
König,Emmerich
Psc 22h56'34"5d34'
König,Gero
Uma 9h38'24"44d9'
König,Nadine
Tau 4h15'16"0d22'
Könner,Claudia
Cnc 8h12'38"31d20'
Köpcke,Petra
Gem 6h46'44"13d43'
Köpke,Klaus W
Cmi 7h17'1"8d21'
Körbelin,Jan
And 0h22'0"36d58'
Körner,Carsten
Lyn 8h1'11"43d6'
Kübler
Dra 20h18'25"62d55'
Kück,Peter
Vul 19h44'19"29d6'
Küdde,Christian
Lib 14h22'28"-20d28'
Kügelgen,Bernhard
And 23h13'52"42d7'
Kühl,Dorle
Psc 23h9'44"5d44'
Kühlen
Lyr 19h19'0"31d3'

Kühne,Barbara
Cep 22h8'35"63d52'
Kühne,Mareike
Sgr 18h51'1"-21d53'
Kühnel,Werner
Aqr 22h42'24"-0d48'
Kühner,Miriam
Cmi 7h21'26"4d33'
Kühner,Willi
Cas 0h9'34"60d10'
Künnecke,Otto
Psc 22h55'13"0d34'
Küpper,Heinz
Vul 19h47'0"29d5'
Kürschner,Jürgen
Cmi 7h17'31"8d7'
Kürschner,Jürgen
Aur 5h19'40"43d45'
Küster,Inge
Vir 14h0'32"7d13'
Küttler
Her 17h41'56"14d24'
K´
Lyn 6h34'0"59d20'

L

L A
Lyn 7h47'10"40d48'
L A's Star
Aql 19h31'53"11d33'
L B V K "Leenda"
Cas 0h21'33"51d44'B
L E P
Tri 2h0'25"31d9'
L H S
Peg 21h45'1"22d34'
L J
Sgr 19h0'39"-27d47'
L J
Dra 10h25'46"73d53'
L J
Aur 4h59'56"51d37'
L J Lite
Lyn 8h8'16"37d12'
L J R
Ori 5h52'58"20d22'
L L S
Uma 9h11'54"59d4'
L M C
Uma 11h31'0"53d34'
L M G K 143
Peg 23h24'26'25d59'
L O
Cyg 21h24'1"48d51'
L O M L
Cep 23h10'0"62d50'
L P G Mama I,9-27-41
Umi 17h23'0"76d24'
L T
Cma 6h22'2"-12d56'A
L T
And 2h1'0"42d33'
L T B Jean
Psc 1h32'52"20d13'
L U L A
Aur 7h23'47"40d4'
L Z K
Cam 3h34'21"71d2'
L'Aigle De Feu
Her 17h25'39"40d22'
L'Amore Di Jan e Bill Maietta "25 years"
Vul 19h23'1"26d58'
L'amour sans frontière
Tri 2h38'56"34d7'
L'Ange
Per 3h7'16"40d52'
L'Annee Marie Jean
Cep 21h39'54"67d56'
L'Arc En Ciel De Renata
Cyg 21h55'58"53d53'
L'Arc-en-Ciel
Aqr 21h58'41"-1d19'
L'Chaim,Lillian & Walter
Del 20h16'43"13d58'
L'Ecqosa
And 0h10'29"40d56'
L'Espérance,André
Her 16h28'28"18d44'B
L'Espérance,Francine
Her 16h28'28"18d44'A
L'essence
Del 20h18'51"12d1'
L'Estel
Aql 20h1'38"3d50'
L'etoile d'Yvonne
Cas 0h56'0"73d45'
L'Etoile de bonheur
Cyg 21h34'43"38d42'
L'Etoile De Cremeans- Joie De Vivre
Aql 19h2'60"-1d12'
L'etoile de Danielle
And 0h21'18"35d53'
L'Etoile de Guillaume
Dra 16h19'25'61d8'
L'etoile De La Paix
Lac 22h45'18"38d39'
L'etoile de la vérite et la grÈce
Vir 14h32'31"5d47'
L'Etoile de Mark et Josette
Ori 5h59'25"-2d38'
L'Etoile de mere
Lyn 9h18'0"38d10'
L'Etoile De Mon Amour
Leo 10h37'26"18d3'
L'Etoile De Notre Vie
Uma 11h6'38"57d32'
L'etoile de Rija
Uma 9h19'0'50d29'
L'etoile de Suzette
Eri 3h50'0"-5d27'
L'etoile de Troy
Cam 8h6'12"80d37'
L'Etoile De Voeux
Lac 22h15'18"54d43'
L'Etoile Dennis Y Ann Toujours
Uma 9h51'1"70d30'
L'Etoile Diamant
Mon 7h39'11"-4d32'
L'etoile Louise
Tri 2h39'58"34d20'
L'Etoile Nicholson
Cam 4h7'0"61d8'
L'Etoile Shana Marie
Gem 8h4'1"28d42'
L'Etoile Vincent
Hya 8h58'11"-5d59'
l'Heautontimoroumenos
Aur 5h3'19"50d37'
L'heureux RSS Grad '94,Daniel
Lac 22h54'57"38d38'
L'Heureux,Cathy (Smoochy)
And 2h10'40"40d34'
L'Heureux,Ken & Carole
Cyg 20h42'35"46d49'
L'Huillier,Liisa Beth
Cas 0h0'12"62d8'
L'Huillier,Lisa
Lyn 8h33'39"44d2'
L'il Sue
Cas 0h4'34"62d49'
L'incontro con Paolo al Parisimare
Dra 16h2'37"62d39'
L'Indefinissable
Cet 2h34'0"3d33'
L'Insondable
Cas 0h30'47"70d26'
L'Opal Du Paix
Dra 17h40'58"61d8'
L'étoile Anne Elizabeth
Cas 0h18'17"62d6'
L'Étoile Bé
Lyn 9h36'15"41d32'
L'Étoile D'Hema
Umi 14h30'51"66d53'
L'Étoile de Billie
Ari 1h55'1"19d41'
L(U)=CA 12/12/'66: What A Star!
Vul 19h48'44"28d44'
L-19
Cam 10h50'28"82d25'
L-Is For Lisa
Cas 1h22'32"55d39'
LA
Uma 11h25'1"41d2'
La Bella Gay
Dra 17h8'39"71d31'
La Bella Rosa
Cet 2h54'52"1d42'
La Bella Stella Rebecca
Peg 21h50'29"31d34'
La Belle Etoile Sally
Cyg 20h33'31"40d22'
La Belle Kimberley
Lyn 6h19'1"60d2'
La Bets
Lib 15h38'49"-28d29'
La Boheme
Lyn 8h28'43"39d9'
La bonne fée Malie
Ori 5h56'15"16d32'
La Brillante Estrella De Norm
Mon 7h57'39"-1d19'
La Brosse
Peg 0h5'0"14d47'
La Carmeline
Cyg 19h48'0"38d21'
La Carrubba,Diane L
Lyr 18h30'48"31d41'
La Clotterie Des Clottes
Ori 5h4'26"10d17'
La Curandera
Cam 4h53'37"67d33'
La Dolce Vita
Cnv 13h24'25"37d44'
La Duchesse de Langrai's
Cas 1h0'42"60d7'
La Estrella Bonita Tina Marie
Lyr 18h48'27"35d36'
La Estrella De Catherine
Tri 2h5'27"31d11'
La Estrella de Armando
Lac 22h14'44"49d36'
La Estrella de Jani
Cet 1h15'60"0d53'
La Estrella de Janiero
Lyr 19h11'28"37d41'
La Estrella de LA-Kev
Tau 4h43'11"28d20'
La estrella de mi coraz
Uma 12h3'54"30d30'
La Estrella de Rafaela
Aql 19h9'43"3d17'
La Estrella de Santos
Aur 6h3'14"37d36'
La Estrella Esmeralda
Ori 5h19'25"0d22'
La ESTRELLA eterna
Lyr 19h13'48"31d37'
La Felichironi
Her 18h22'50"12d18'
La Flair You're A Star Love Jeff,Karen
Psc 1h0'51"31d56'
La Fonta,Gonzague
Ori 5h57'32"11d15'
La Forza Del Destino
Cas 1h58'31"60d9'
La Fountaine,Daniel
Per 4h34'31"34d16'
La Gazelle
Her 17h34'13"22d36'
La Giacone
Cnc 8h22'37"7d2'
La Gigi
Cyg 21h35'20"38d59'
La Gioia Di June
Boo 15h15'36"34d23'
La Giornata del Destino
Sgr 19h18'9"-29d53'
La Gomera
Ori 5h54'15"1d5'
La Gould
Boo 14h15'59"48d46'
La Grace
Mon 8h6'43"-9d45'
La Ina
Sgr 18h54'16"-20d19'
La Juene,David
Cyg 21h53'18"37d10'
La La Jane
Cnc 9h0'53"31d27'
La Licata,Fabrizio
Lyn 7h9'30"44d46'
La luce del padre e della madre
Cae 4h52'33"-27d47'
La Lumaca
Col 6h29'7"-39d3'
La Lumiére De Thérése
Sex 9h49'39"0d38'
La Lustre De Elisa
Uma 1h21'27"48d15'
La Luz de Fahy
Lyn 8h5'29"40d19'
La Madeleine
Lib 15h16'38"-22d27'
La Maravilla Blanca
Lyr 19h16'26"31d25'
La Menestrelli, Keldrille
And 0h11'30"40d4'
La Merle
Cra 18h0'39"-37d4'
La Montagne,Rich
Her 17h7'0"42d46'
La Nae
Tri 2h20'30"31d49'
La numero 9
Aur 5h25'54"31d26'
La Palma
Ori 6h16'47"5d44'
La Pergola,Nick
Aql 19h36'47"1d5'
La Petite
Sco 17h51'14"-30d22'
La Petite
Psc 23h23'13"0d9'
La Petite Claude
Cas 0h49'33"73d43'
La Petite Je T'Aime Etoile De David
Cam 3h54'46"57d32'
La Petite Souris (Y D & D D)
Her 18h47'11"25d59'AB
La Pine
Ori 5h2'0"0d6'
La Plage De La Soleil
Cam 3h31'40"61d40'
La Plant,Timothy
Sct 18h56'57"-6d37'
La Plante,Mary Ellen
Cas 2h4'18"61d34'
La Plume,Larry
Aql 20h11'0"0d18'
La Polaire
Lac 22h55'1"53d16'
La Principessa
Pho 0h6'23"-47d33'
La Principessa
Cam 7h56'23"73d25'
La Rocca,Anthony
Per 2h39'18"38d51'
la Rosa-Enzo
Peg 23h32'23"18d46'
La Rose,Garry James
Aql 19h20'0"15d9'
La Sagrada Familia
Cep 22h17'57"55d56'
La Samaritaine
Per 3h16'1"41d16'
La Star
Cyg 19h34'50"30d1'
La Stella Di Angie
And 2h15'15"46d50'
La Stella Di Gregorio
Peg 23h27'26"8d11'
La Vague Eternelle
Crt 11h23'10"-15d52'
La Vette
Cnc 8h26'23"31d51'
La'Nette,Tanisha
Uma 8h38'55"47d48'
La-Di's Centauri Starhawk
Mon 6h3'35"-8d55'
Laabs,John
Oph 18h22'21"10d35'
Laarhoven,Andrew,Adam & Alexander
Ori 6h4'53"6d58'
LaBa
Lac 22h36'0"55d19'
Laban
Umi 17h12'53"86d10'
Labanca,Angela
Dra 15h43'5"53d31'
LaBarbera,Jr,Justin
Gem 6h38'1"20d32'
LaBarrere,Florence & Joseph
Cyg 20h22'0'40d52'
Labash,Zoe Anya
Aqr 23h29'1"-18d47'
Labate,Shelley
Mon 6h24'40"1d35'
LaBaw,Claudine
Hya 9h6'46"0d9'
LaBayne"My Shining Star",Gerald Nelson
Boo 14h32'28"20d8'
LaBeau,Jr,Joseph R
Lib 15h0'22"-10d31'
Labedz-Tcheresky,Luba
Vul 19h59'37"27d9'
LaBella,Julia "Coolio"
And 0h20'37"30d6'
Labella,Miguel Perez
Her 18h20'51"18d47'
LaBelle Cortney
Ori 5h51'1"16d16'
LaBelle,Linda J
Cam 3h45'54"77d9'
Labelson,Hannah Sylvia
Psc 1h0'13"21d36'
Labhart,Billie
Lac 22h54'22"56d0'
Labia
Com 12h12'55"21d20'
Labisch,Anna Elizabeth
Cam 8h47'20"74d3'
Labit,Tina
Lyr 18h42'27"39d12'
LaBombard,Nancy & Tom
Cam 3h28'0"61d24'
Labonte,Leonard H
Dra 11h2'47"74d6'
Labonté,Fabienne "La Jolie"
Aqr 23h45'58"-6d35'
Labonté,Raymonde
Cep 21h17'47"60d5'
Laboratoire Pfizer
Her 18h21'44"12d33'
Laborde,Pierre
Aur 7h9'31"38d37'
Laboureur,Olivier
Dra 10h5'20"80d28'
Labouré,Dominique
Umi 15h54'53"73d21'
Labow,Judy
Cam 4h56'60"68d17'
Labowitch,Petranella Adele
Mon 7h5'59"-6d12'
Labozzetta,Kristi Jean
Psc 1h33'0"21d22'
Labozzetta,Mark Anthony
Cep 23h5'0"70d33'
Labozzetta,Matthew Patrick
Aur 5h12'15"41d3'
LaBranche,Don
Dra 11h51'50"70d30'
LaBranche,Sylvie
Umi 15h19'18"65d48'
Labre,Edmond
Cep 22h15'17"55d57'
Labrecque,Roger L
Aur 6h13'17"31d49'
LaBree,Michael Robert
Per 2h7'45"58d46'
Labrenz,James Robert
Ori 5h50'35"18d19'
Labrenz,Julia Marin Fanjoy
Peg 23h43'23"30d33'
LaBrie,Daniel V
Lac 22h21'10"55d60'
Labrosse,Maxime
Aur 5h57'26"50d19'
Labrot,Martine
Dra 10h11'52"73d51'
LaBrucherie,Anthony Stirling
Aql 20h11'36"11d12'
LaBrucherie,Dawn Marie
Peg 23h47'35"15d4'
Labrugnas,Ghislaine
Cam 3h24'38"60d42'
Labuda,Kristen
Peg 23h46'23"32d28'
LaBusier,Sara Danielle
Cnv 13h27'45"41d26'
Labuzan,Bruno
Cyg 19h39'27"38d59'
Labyak,My Handsome Silver Fox Bob
Lyn 9h11'31"37d54'
Lacanaud,Jacques
Per 3h59'52"39d34'
Lacap,Eden
Ori 5h56'49"15d15'
LaCapra,Rocco V
Her 16h23'27"47d53'
LaCarubba,Paul
Her 18h5'54"38d22'
Lacasa,Carlos A
Aql 19h26'46"11d6'
Lacasse,Antoinette
Dra 20h6'55"65d2'B
Lacattiva,Anna-Lisa
Cyg 20h58'21"41d10'
Lacau,Claude
Cyg 19h39'33"38d13'
LaCava,G R
Uma 11h36'31"38d36'
LaCava,Nancy & David
Cyg 20h45'10"38d16'
LaCava,Sally
Cas 0h9m33"60d29'
Lacave,Cathy Stoll
Tri 2h0'39"30d57'
Lacaze,Alain
Cyg 19h47'59"30d38'
Lacchini,Michele
Lyr 19h3'1"37d31'
Lace,Charles Leavers
Boo 14h11'39"41d0'
Lace,Ronald E
Uma 11h7'26"30d46'
Lacerra,Eugene & Nicolene
Cyg 19h40'23"38d13'
Lacerte,Aunt Patty
Eri 3h45'18"-4d3'
Lacey
Peg 23h48'22"27d56'
Lacey
Lyr 18h38'12"44d6'
Lacey Meus Donum Ab Caelestis
Ori 5h54'50"7d55'
Lacey Nalani
Peg 22h40'1"24d52'
Lacey,Jessica
Del 20h21'0"10d20'
Lacey,Miranda
Aql 18h59'53"14d24'
Lacey,William Morgan
Her 16h42'1"26d59'
Lachal,Claudie
Mon 6h22'55"-6d3'
Lachance,Emelene
Oph 17h15'52"-22d11'
LaChance,Gayle
Cyg 19h51'0"40d29'
Lachance,Michael Roy
Ori 5h58'31"14d44'
Lachance,Myriam
Lyn 8h45'14"42d57'
Lachance,Pascal
Lac 23h50'45d50'
LaChapelle,Helen Thorsen
Sct 18h55'37"-4d50'
Lachelle,Denise
Lyr 18h48'45"36d51'
Lacher Meiertöns,Uta
Cep 22h10'43"61d38'
Lacher,André Marcel
Her 16h26'0"35d6'
Lacher,Barbara
Cam 5h52'0"58d38'
Lacher,John & JoAnne
Cyg 19h47'0"31d7'
Lachermair,Wilfried & Regina
And 23h20'25"45d29'
Lachin,Linda Beth
Cas 0h2'24"62d48'
Lachman,Brooke Nina
Psc 23h5'54"2d10'
Lachman,Pam
And 23h26'1"43d37'
Lachmann,Christine
Psc 23h22'12"6d45'
Lachmann,Frank
Sco 17h31'20"-31d19'
Lachowiec,Joseph
Hya 8h9'26"1d32'
Laciak,Anthony M (Yontek)
Her 16h5'40"50d15'
Lacik,J I
Lyr 18h55'26"31d50'
Lacinda Dawn
Oph 17h16'17"-20d26'
Lack,Jr,Andrew Richard
Boo 14h42'37"31d3'
Lackey,Linda
Eri 4h1'24"-12d7'
Lackey,Mattie L
Lyn 9h10'36"38d42'
Lackie,M Linda
Uma 10h2'44"60d36'
Lacklear,Pam
Sct 18h54'0"-4d30'
Lackmann,Ute
Psc 22h50'20"5d11'
Lackner,Bud & Toni
Cyg 20h36'30"42d4'
LaClair,Lucille
Mon 6h29'45"-8d31'
LaComb,Christy "Jardell"
Eri 3h8'32"-1d33'
LaCombe,Andrew J
Per 1h27'57"53d8'
LaCombe,Barry
Boo 15h6'38"12d19'
Lacombe,Hugues
Ori 5h55'23"17d1'
Lacomblez,Eric
Peg 21h51'19"20d31'
Lacomme,Thibault Et Loic
Aql 20h4'0"1d44'
Lacorte,Maria
Peg 21h48'0"35d57'
LaCoste,Christa Fleischer
Eri 4h13'0"-10d22'
LaCour,Alice Marie
Peg 23h32'12"32d35'
LaCovara,Michael Kevin
Per 2h25'27"52d3'
Lacrambe,Fanny-Pruvost
Cnc 8h52'1"30d2'
Lacrambe,Henry
Sgr 19h58'24"-41d8'
Lacrambe,Paul-Henry
Tau 5h50'44"23d31'
Lacroix,Clément
Her 17h31'43"20d44'
LaCroix,Elizabeth Ashley
And 2h0'25"42d20'
LaCroix,Gary Terrill
Cam 3h51'44"70d30'
LaCroix,James Matthew
Aur 6h18'32"45d5'
LaCroix,Julienne
And 0h10'0"37d17'
LaCross,Nancy Ann
Crt 11h12'18"-21d23'
Lacy's Todd
Uma 8h45'41"50d41'
Lacy,Jason Thomas
Cet 2h49'30"0d12'
Lacy,Matthew Wilson
Lac 22h15'36"49d54'
Lacy,Michael Jon
Hya 9h17'0"4d56'
Lacy,Sandy
Cas 0h53'33"63d43'
Lacy,Sharon Rose
And 1h24'32"39d38'
Lad
Cas 0h31'40"73d19'
Lad's
Cma 6h50'49"-16d16'
Ladagona,Stephen T
Aur 5h19'50"43d13'
Ladas,Tina
Ori 5h55'34"9d3'
LaDawn
And 0h2'49"44d49'
Ladd,Bimini Lee
Lyn 7h47'0"44d57'
Ladd,Eldred Bridges
Tri 1h57'0"30d8'
Ladd,Kevin E & John L Finlon
Her 16h45'42"48d25'
Ladd,Lara
Cet 2h55'52"9d28'
Ladd,Nolan John
Per 4h2'44"52d25'
LadedaLee
Eri 3h18'1"-2d60'
Lademan,Lawrence J
Aur 6h12'35"38d17'
Ladendorf,Doug & Ingrid
Aur 7h4'43"41d51'
LadiDwagon
Ori 5h53'17"15d8'
Ladies Who Sang With The Band,The
Mon 6h24'20"7d41'
Ladin,Cindy
Sex 10h11'27"-8d9'
Ladonna
Ori 6h12'15"4d25'B
Ladonna Kay
Mon 7h17'1"-1d52'
Ladson,Shonna
Del 20h17'0"10d35'
LaDuca,Jacob Daniel
Boo 15h14'58"47d43'
LaDue,James
Her 16h48'55"18d57'
LaDuke,Jean Elizabeth
Sct 18h43'22"-6d57'
Ladwig,Heinz
Aql 19h59'0"10d56'
Ladwig,Todd & Elease
Eri 3h55'34"-10d12'
Lady
Boo 13h38'31"12d19'
Lady 4-29-91
Lyn 7h53'40"34d10'
Lady Adrienne
Lyn 8h22'15"43d4'
Lady Anne of Wellesly Centre Glass Lake
Cas 23h33'1"63d16'
Lady Ashley
Cas 1h53'19"61d17'
Lady Ashley
Uma 8h35'49"54d59'
Lady Barbara Salina
Dra 17h50'0"64d47'
Lady Beth
Cap 21h10'3"-22d37'
Lady C
Sge 19h56'41"20d34'
Lady Catherine,The
Peg 23h1'23"32d31'
Lady CEO of Seacroft Centre Studio
Cas 0h30'28"61d38'
Lady Charlene
Cas 0h57'11"60d47'
Lady D'Arbanville
Lyn 8h1'4"49d13'
Lady Della
Peg 22h44'19"18d56'
Lady Delpapa
Cas 2h29'16"59d27'
Lady Di
And 1h14'35"48d51'
Lady Di
Cas 0h35'1"66d40'
Lady Diedre
Vul 20h43'27"20d14'
Lady Donna
And 1h54'18"39d35'
Lady Elaine Suzanne
And 23h34'30"45d31'
Lady Elizabeth
Cas 1h28'40"60d27'
Lady Gayle
Cas 22h57'32"54d14'
Lady Godiua
Dra 11h42'24"71d18'
Lady Grace
Cet 2h15'25"1d31'A
Lady Hawke Snoopy
Cet 0h55'30"-11d7'
Lady In Red
Eri 3h36'24"-17d32'
Lady Jake
Cnv 16h4'0"35d8'A
Lady Jane
Cyg 20h37'56"31d37'
Lady Jane
Tri 1h40'27"33d49'
Lady Jean
Lyr 19h23'1"35d32'
Lady Jean-Love shines in your eyes
And 23h9'34"41d20'
Lady Johanna
Cas 1h24'1"53d35'
Lady Julia
And 23h43'24"42d1'
Lady Kathryn A
Cyg 21h0'1"40d27'
Lady Kim
Cas 1h2'28"61d50'
Lady Kristine,The
And 1h59'1"39d22'
Lady Lavinia's Anchor
And 0h15'38"38d7'
Lady Lee
Uma 9h15'53"46d41'
Lady Linda
Crb 15h35'47"30d50'
Lady Linda
And 1h54'41"40d50'
Lady Lisa Anne Of The Crescent Cove
And 1h58'34"41d5'
Lady Love
Lyr 19h12'1"35d2'

Lady Luv
Lyn 7h53'57"38d30'
Lady Margret
Cet 0h56'37"-18d52'
Lady Nairobi
Cas 0h17'59"64d30'
Lady of Avon
Lyr 19h6'42"26d21'
Lady P
Ori 5h25'0"-0d5'
Lady Petra
Com 13h31'55"20d8'
Lady Rebecca,One Who Shines
Peg 23h1'0"18d21'
Lady Ruth
And 23h49'28"41d9'
Lady Susan
Mon 6h20'22"8d40'
Lady Tanya
Cmi 7h35'1"0d56'
Lady Victoria
Cas 23h38'0"62d57'
Lady Wize
Cas 0h18'15"61d0'
Lady,The
Lyr 18h25'44"38d44'
Ladye Britannia
Cas 1h34'1"73d38'
Ladye-British Ocean, The
Uma 11h44'0"57d20'
Ladyga,Kelley
Peg 22h39'53"22d36'
Ladyman,Kate
And 23h2'46"51d38'
Ladzinski,Ryan Y
Per 3h56'42"50d22'
Lael
Cam 8h3'1"67d33'
Laetitia
Cas 23h30'44"60d15'
Laetitia
Peg 22h44'41"4d51'
Laetitia
Ori 6h0'49"20d50'
Laetitia out of Stereolab
And 0h7'35"30d0'
Laeuger,Michael Joseph
Cnv 13h48'49"35d41'
LAF
Uma 9h35'40"68d45'
LaFalcia,Austin Joe
Dra 15h4'12"61d60'
Lafargue,Bertrand
Lac 22h43'59"38d50'
Lafata,Pat
Com 13h15'29"27d41'
Lafave,Diana Jane
Cyg 19h47'38"29d22'
LaFavor,Dana Kay
And 0h11'12"46d49'
LaFavre,Ann B & Robert E
Aql 19h26'23"1d46'
LaFavre,Charles E & Ruth A
Sct 18h54'22"-6d47'
LaFavre,Deborah A, Karen R,Mary J
Vul 21h2'45"28d12'
LaFawn-Kevin 1986
Cam 3h59'28"77d39'
Lafayette,David Edward
Aur 4h59'1"38d28'
Lafayette,John Harold
Dra 16h42'0"63d18'
Lafebre Brightest Star Ingrid H
Cet 1h49'0"0d43'
LaFerla,Michael & Alexandria
Cyg 19h57'1"30d3'
Laferriere "The Slick Star",Richard
Dra 14h59'38"64d33'
LaFever,John William
Her 17h25'41"20d16'
Lafever,Mack
Per 2h50'39"32d36'
Lafferty 10-20-77,Max Christopher
Lyr 18h42'35"36d4'
Lafferty,Daniel James
Lac 22h2'1"40d36'
Lafferty,Daniel James
Cmi 8h5'53"0d53'
Lafferty,Meredith Paul
Cap 21h3'42"-16d28'
Lafferty,W H K
Ori 5h2'20"0d22'
Lafflie,Elaine
Peg 21h54'19"32d41'
Laffman,Kevin
Aur 6h9'23"37d20'
Laffon,Océane Jibault
Cyg 20h22'24"40d7'
Lafler,Claude
Cnc 8h13'41"30d25'
Lafleur,Herve
Her 18h2'39"28d54'
Lafleur,Katherine Lavoie
Cam 3h30'0"63d29'
Laflin,Molly
Leo 11h1'48"23d57'
LaFoe,Margaret
Lyn 8h27'46"44d52'
LaFoe,Maryann
Cas 0h32'44"62d1'
Lafollette,Amanda K
Cyg 19h51'52"40d42'
LaFollette,Katherine
And 23h21'1"45d24'
LaFontaine,Ben
Boo 14h29'28"23d14'
Lafoon,Karen M
Mon 8h3'33"-5d60'
LaForest,Genevieve Elizabeth
And 0h31'33"28d27'
Laforest,Joël
Lyn 8h28'0"49d46'
LaForest,Michelle Alexandra
And 0h31'51"28d42'
LaForge,Loren
Lyn 8h13'35"39d1'
LaForge,Rick
Ori 6h8'57"9d40'
LaForgia's Horizon Star,Donna
Crb 16h22'54"27d28'
LaFortune,Tiffany Angelina
Uma 12h56'29"62d2'
Laframboise,Elise C
Lyr 19h3'29"28d29'
Laframenta,Casey
Cmi 7h27'19"0d36'
Laframenta,Meryl
Ser 15h10'28"9d5'
Lafrance,Evelyne
Cas 0h53'31"75d15'
Lafrance,Jean-Pascal
Umi 16h59'32"78d27'
LaFrance,Paul & Pauline
Mon 6h54'47"-1d11'
Lafranceschina,Giusy
Dra 16h19'0"62d24'
Lafuente,Carlos
Her 15h49'50"40d56'
Lafuente,Francisco
Ori 5h57'44"15d50'
Lagace,Suzanne
Cas 0h7'36"65d36'
Lagacé,Jean Guy
Cep 23h31'44"64d11'
Lagacé,Noella
Uma 11h34'0"42d40'
Laganiere,Tim
Aur 5h34'20"50d31'
Lagarde,Christine
Cam 7h26'22"80d14'
Lagarde,Christine
Cam 7h26'22"80d14'
LaGasse,Anthony James
Aur 5h25'37"41d7'
LaGena,Cathy,Kim, Heather
Mon 7h1'30"-5d49'
Lagenberg The Big Ma, Mary B
Tri 2h12'58"30d25'
Lagerfeld,Karl
Aur 4h58'34"30d36'
Lagergren,C R
Dra 13h34'26"64d57'
Lages,Anja
Sgr 18h48'31"-23d11'
Laget
Vul 20h16'1"28d40'
Lagies,Sabine
Sco 17h21'0"-38d38'
Lagler,Ingenieur Franz
Uma 9h23'12"47d41'
Lago,Carol Ann
Cam 3h33'21"60d30'
Lagorce
Crb 16h19'34"37d58'
Lagos,Bonnie
Lyn 8h45'13"37d1'
Lagrange,Elliot
And 23h18'0"50d42'
LaGrange,Guy
Per 1h43'33"53d39'
Lagrasta,Fred D
Aur 7h6'33"41d4'
Lagraulet,André
Lyn 8h27'10"34d47'
LaGrossa,Lisa
Lyr 18h39'60"41d54'
LaGruth,Francine Ann
Vul 21h2'1"20d11'
LaGuardia Star,The Leslie
Cyg 19h43'27"30d9'
Laguathn,Karen A
Cas 23h5'48"53d2'
Laguerre,Jacqueline Guiselin
Cam 6h57'42"64d13'
Laguna,Orlando
Uma 11h8'47"48d29'
Lagunowich,Corey
Peg 21h53'40"30d17'
Lahah,Jacob Justin
Boo 14h56'40"33d8'
Lahaie,Marielle
Cyg 21h1'47"39d14'
LaHavich,Bruce
Her 18h1'51"37d39'
LaHerran,William J
Cet 3h15'16"2d26'
Laherty,Patrick James
Aur 5h31'22"30d23'
Lahiff,Mark Andrew
Cep 0h6'40"67d11'
Lahl Sharr
Cyg 21h29'0"40d39'
Lahmers,Christopher
Sco 17h30'0"-31d28'
Lahmers,Daniel S
Cep 7h10'58"86d49'
Lahmers,Danielle
Lib 14h25'20"-11d6'
Lahmers,Diane Fay
Cam 13h15'42"78d10'
Lahmers,Jennifer
Psc 1h26'47"11d43'
Lahmers,Kay
Sgr 18h53'51"-29d27'
Lahmers,Matthew
Psc 1h27'1"10d1'
Lahmers,Nancy Kay
Cas 2h56'28"70d22'
Lahmers,Richard B
Her 15h50'12"40d31'
Lahmers,Robert E
Sgr 18h57'41"-20d41'
Lahmers,Robert F
Lac 22h3'19"49d59'
Lahmers,Robin Guthrie
Sgr 18h54'47"-21d48'
Lahmers,Rose
Uma 10h4'23"48d19'
LaHorgue,Frank
Oph 18h42'18"7d3'
Lahourcade,Christian
Aur 6h52'20"48d54'
Lahoussaye,Marie Laure
Per 3h43'5"37d45'
Lahtinen,Maija
Aql 20h8'58"4d12'
Lahue,Bruce
Her 17h11'0"22d36'
Lai Lin Lee
Mon 8h2'42"-1d52'
Lai Mei
Lyr 18h14'27"30d12'
Lai,Olivia K Y
Cyg 20h26'40"46d19'
Lai,Yin
Dra 17h55'55"67d54'
Laia y Nacho
Peg 23h32'43"17d41'
Laible,Sylvia
Boo 14h32'1"52d6'
Laico's on the Square
And 23h37'10"44d23'
Laidig,Gary Wayne
Lyn 8h33'46"42d7'
Laidley
Cam 3h32'56"62d5'
Laik,Thierry
Cmi 7h29'54"5d59'
Laikola,MD,Dr Leslie Alexander
Uma 9h35'57"44d11'
Laila,Marzia Monia
Cam 13h33'24"80d59'
Laile
Boo 15h10'52"31d40'
Laimishader,Solus
Boo 14h50'55"34d59'
Laina,Sr,Alan
Aur 5h11'21"43d32'
Lainati 131
Tel 20h14'12"-46d6'
Lainesse,Sherley
Lyr 18h47'35"36d16'
Lainey
Mon 6h54'43"10d34'
Laing,Christopher K
Aql 19h19'41"15d7'
Laing,Cooper
Lyn 8h23'31"47d50'
Laing,Gary
Cap 21h26'11"-24d29'
Laing,Mary Grace
And 2h17'29"45d26'
Laino,Robert
Lac 22h5'56"47d9'
Lainsbury 7-25-81, David & Debi
Crb 16h2'59"26d53'
Laipple Reunion Eternity Star Base
Crb 15h57'53"38d29'
Lair,Greg
Boo 14h42'48"35d7'
Lair,Jr,Paul
Cam 3h57'33"61d30'
Lair,Tim & Stephanie
Aql 18h49'56"11d23'
Lair,Troy Lee
Per 2h4'59"58d41'
Laird,Bart William
Cet 2h45'32"1d37'
Laird,Elizabeth
Sgr 18h51'59"-31d23'
Laird,Jackie
Mon 7h22'33"-5d40'
Laird,Ria Clara
Umi 15h20'37"66d22'
Laird,Vanessa Anne
Cyg 21h22'26"37d40'
Lais,Peter
Boo 13h44'18"12d0'
Laisne,Pierre
And 23h0'1"38d48'
Laithwaite,Simon James
Cep 20h30'35"61d38'
Laiv
Tri 1h41'20"28d50'
LaJeana Dawn
Mon 7h25'0"-0d45'
Lajiness,Daryl
Her 18h6'17"38d49'
LaJudice,Glenn & Kim
Cyg 21h20'39"38d21'
Lak,Caitlen
And 2h20'34"44d46'
Lak,Courtney
Cnv 12h45'1"38d8'
Lakan
Boo 14h27'1"48d51'
Lakarosky,Jeffrey Philip
Cet 3h17'33"0d18'
Lakatamitis,Alexandros
Umi 15h15'19"65d58'
Lakatamitis,Polikarpos
Umi 15h17'25"65d60'
Lakatamitou,Tatiana
Lyr 19h2'37"28d30'
Lake,Antonia Camilla
Cyg 20h19'58"39d58'
Lake,Christopher Edward
Dra 17h49'54"60d7'
Lake,Corey Thomas
Uma 10h34'51"40d12'
Lake,Geoff
Cep 21h50'17"61d32'
Lake,Perry
Her 16h49'0"32d26'
Lake,Ricki
Cep 21h15'0"61d5'
Lake,Timothy C
Her 18h0'45"31d34'
Lakeram,Yvonne Denise Ann
Cyg 21h59'34"53d54'
Lakeside High School Star 2000
Ori 5h57'17"17d47'
Lakewood Lutheran School 4th Grade
Peg 22h25'44"2d18'
Lakey (Lakestar), Christopher
Dra 16h37'41"62d27'
Lakin,Shiela & Len
Cyg 21h15'17"36d3'
Lako,Marie
Cas 0h32'11"63d49'
Lakota Blue
Aql 19h7'45"1d26'
Lakshmi,MD,Dr
Per 2h51'31"31d18'
Lal,Rohini
Aql 18h58'1"12d34'
Lala
Psc 1h18'24"18d3'
Lala
Per 4h1'0"34d34'
Lalage
Aql 19h54'47"-6d36'
LaLande's Lantern
Lyn 9h2'26"41d60'
Lalande,Louis-Michel
Cam 14h23'15"82d41'
LaLanne,Jack & Elaine
Ori 5h27'15"15d3'
Lalela
Lyn 7h40'15"41d5'
Lalelu
Cnv 14h2'19"35d19'
Lali & Rosie
Dra 17h46'56"61d12'
Laliberte-Coppa
Cyg 20h47'45"37d53'
Laliberté,Daniel
Uma 9h5'48"60d24'
Lalich,Robert Joseph
Aql 18h46'27"11d38'
Laline
Cas 0h33'31"62d42'
Lalitaji
Crt 11h20'30"-10d50'
Lalko,Amy
Cas 2h6'30"59d9'
Lalko,Jon Francis
Dra 14h24'1"63d22'
Lall,Sumita
Dra 15h45'39"67d33'
Lalla 36 15
Cyg 21h56'24"53d26'
Lallave,Gilbert
Aur 4h50'21"38d5'
Lalley,Robert Wayne
Crb 16h14'48"28d54'
Lalli,Eileen Marie
Mon 7h24'42"-5d47'
Lalli,Jessica M
Lyr 19h12'1"40d39'
Lalli,Michele Marie
Peg 22h16'48"7d49'
Lallo's Dream,William & Barbara
Umi 15h43'43"70d26'
Lally
Her 18h27'26"24d23'
Lally Wedding Star,The
Cyg 19h32'34"38d54'
Lally,James R
Cep 22h27'12"60d7'
Lally,Jr,Jack
Aur 6h0'27"34d53'
Lally,Kenneth T
Cep 1h56'56"77d53'
Lally,Paco
Cet 0h49'59"-3d17'
Lalonde,Aleric Martin
Ori 5h2'51"1d14'
Lalonde,André
Ori 5h58'11"17d32'
Lalonde,Francis
Uma 11h34'50"30d42'
LaLonde,Ka-Rin Elizabeth
Vul 19h48'0"28d10'
LaLonde,Kaidin Edward
Vir 11h42'39"0d6'
Lalonde,Maurice
Cep 3h10'15"80d2'
Lalonde,Pierre
Umi 16h40'38"79d49'
LaLonde,Rosemary J
Peg 21h59'1"32d1'
Lalonnier,Karl
Cmi 7h39'20"4d12'
Lalou
Dra 17h3'0"70d40'
Laloyau,Claude
Cyg 19h39'48"38d40'
Lalu,Jacqueline & Pierre
Her 17h53'38"14d27'
Lalumia,Midge-Jessica
Eri 3h25'58"-5d1'
Lam
Ori 5h31'27"-1d29'
Lam,Christopher "Hedgehog"
Lac 22h6'11"49d58'
Lam,Jenna Elizabeth
Cas 1h0'45"68d21'
Lam,Lori
Uma 13h50'28"51d57'
Lam,Mandi Dawn
Del 20h28'31"10d19'
Lam,Mimi Siu Mui
Mon 8h7'36"-1d36'
Lama,Serge
Ori 4h43'14"0d12'
Lama,Thaddeus Joseph
Aql 19h53'0"15d47'
LaMacchia,Anthony M
Cet 1h37'1"-5d43'
Lamaison De Sagesse
Cnv 12h26'31"40d57'
Lamantia,Christopher M
Cep 21h47'42"61d1'
LaMantia,Shelia Frances
Cet 3h16'45"0d53'
LaMar,Casey Ambrose Runnman
Cyg 20h5'12"40d47'
Lamar,Silver
Umi 15h31'59"69d15'
LaMarca I II III, Fredrick Benjamin
Uma 9h34'54"71d57'
Lamarche,Amy
Cyg 21h4'25"32d25'
Lamarche,Audrey
Per 2h58'39"37d60'
Lamarche,Denis
Uma 9h58'19"50d60'
LaMarche,Elaine Ann
And 2h22'25"45d39'
Lamarche,Sr,Normand H
Lyr 18h57'37"31d25'
LaMarine,Cherry
Lyn 8h59'47"41d48'
LaMarsh Star,The William J
Aur 5h27'60"47d35'
Lamas,Alexander Charles
Her 17h43'12"40d40'
Lamas,Lorenzo
Aql 18h43'1"6d46'
Lamason,Cherie
Aur 6h26'46"33d4'
LaMassa,Ralph
Her 17h58'1"41d9'
LaMay,Cindy
And 23h42'0"43d13'
Lamb Family,The
Lyr 18h30'1"30d34'
Lamb,Brian
Cyg 21h1'10"37d57'
Lamb,Brian Thomas
Boo 15h5'56"14d36'
Lamb,Dennis
Dra 11h52'36"72d27'
Lamb,Edmund
Cas 0h58'11"69d5'A
Lamb,Gayla & Richard
Sco 17h29'29"-30d8'
Lamb,Gregory Robert
Aur 5h20'1"40d11'
Lamb,Heather Marie
Ori 5h32'35"-0d59'
Lamb,James H
Ori 5h10'10"-0d43'
Lamb,Jennifer Yvonne
Crb 16h5'1"34d53'
Lamb,Joy J
Peg 23h20'48"10d3'
Lamb,Marie
Lyn 8h23'22"43d58'
Lamb,MaryAnn
Peg 22h3'39"24d2'
Lamb,May You Soar With The Angels,Jim
Her 18h12'44"41d2'
Lamb,Michael
Dra 16h11'20"60d11'
Lamb,Nicholas Christopher
Com 13h29'18"26d40'
Lamb,Peter Franklin
Uma 14h0'1"58d33'
Lamb,Scott & Andrew
Cyg 19h43'23"30d31'
Lamb,Suzanne Michelle
And 23h24'13"44d43'
Lamb,The
Ori 5h39'1"-6d29'
Lamb,Theano & Chuck
Cam 9h42'27"82d6'
Lamb,Thelma & Leslie Foust Forever
Lyn 7h44'20"41d27'
Lamb,Timothy M
Cnv 13h41'20"36d47'
Lambchop
Aur 5h53'56"30d27'
Lambda 1,Julie
Cyg 21h39'44"41d57'
Lambe,Frank & Nicholas
Dra 20h13'20"63d23'
Lambe,Nicholas Anthony
Mon 6h51'30"10d54'
Lamberg,Eric & Heidi
Crb 16h13'18"28d59'
Lambert
Umi 16h29'19"74d58'
Lambert Raynald
Cet 2h21'1"1d2'
Lambert Star,The
Ori 5h57'11"18d29'
Lambert's Lucky Star
Lac 22h39'18"53d30'
Lambert,"Our Soul" Jerry & Susie
Ser 17h56'10"-13d12'
Lambert,Anna K
Aql 19h57'39"0d0'
Lambert,Barbara L
Del 20h14'1"15d17'
Lambert,Bradley M
Dra 15h56'52"61d0'
Lambert,Brandon "Justin"
Ser 17h30'43"-10d52'
Lambert,Cedric
Lyn 8h41'33"45d2'
Lambert,Christopher Joseph
Peg 22h41'1"10d20'
Lambert,Christopher
Vul 20h38'23"20d3'
Lambert,Cynthia Marie
Cma 6h55'3"-18d8'
Lambert,Edward E
Cep 21h3'19"58d18'
Lambert,Elisabeth Ann
Del 20h26'26"11d8'
Lambert,Elizabeth Carroll
Aql 20h32'26"-5d58'
Lambert,Fred & Helen
Peg 23h3'33"18d27'
Lambert,Halley Alexandra
And 2h0'25"38d21'
Lambert,Ian Christopher
Cmi 7h44'16"4d8'
Lambert,Jackie
Lyr 18h16'0"31d44'
Lambert,Jessica Ann
Vul 19h23'19"26d49'
Lambert,Jr,William M
Per 2h1'1"50d7'
Lambert,Julia
Cep 20h58'32"65d58'
Lambert,Karen & Mark Menezes
Cyg 20h20'0"38d49'
Lambert,Kimberly Ann
Lyn 8h9'39"34d29'
Lambert,Kristina Lynn
Cas 0h30'44"58d58'
Lambert,Laura
Cep 20h58'31"66d11'
Lambert,Marie
And 23h20'1"38d35'
Lambert,Megan Elizabeth
And 23h13'58"37d6'
Lambert,Melinda Sue
Ori 6h7'12"8d44'
Lambert,Michael William
Cep 4h16'16"80d31'
Lambert,Michel
Per 3h58'33"35d20'
Lambert,Nicola Anne
Lyr 18h37'44"43d36'
Lambert,Patricia
And 1h14'38"40d42'
Lambert,Patrick Michael
Per 2h41'26"43d59'
Lambert,Peter John Bowes
Del 20h28'39"10d10'
Lambert,Peter Thomas
Cyg 19h33'47"35d48'
Lambert,Precious Baby Joseph
Her 17h8'1"45d43'
Lambert,Randy
Uma 12h14'23"62d35'
Lambert,Rich("Big Rich")
Per 2h6'1"56d52'
Lambert,Ronald Jack
Sgr 19h6'44"-21d30'
Lambert,Thomas Richard
Eri 3h44'50"-10d1'
Lambert,Timothy Ryan
Cep 21h49'29"68d3'
Lamberta,Peter Charles
Dra 15h32'47"65d24'
Lamberth,Skyler Jacob
Aur 4h47'1"38d39'
Lamberti,Forever My Love Elias
Boo 14h59'21"50d12'
Lamberto
And 23h22'0"52d52'
Lamberts,Elizabeth
Lyn 8h47'33"37d11'
Lambertus,Patrick
Cet 2h18'41"7d16'
Lambeth,Julius "Jay"
Aql 19h55'41"13d53'
Lambie Pie
Psc 23h1'0"5d23'
Lambinicio,Kathie
Lyn 7h59'59"58d34'
Lambke,Sheila Suzanne
Lyr 19h15'41"41d22'
Lamblin-Boulanger,Anne - Marie
Cas 0h57'40"73d19'
Lambotte,Claudine
And 23h14'19"41d17'
Lambourne,Emma-Fleur
Cyg 20h0'43"31d41'
Lambourne,Josié
Cyg 20h1'17"30d58'
Lambrecht,Evan Richard
Aql 18h59'50"-6d25'
Lambright,Connie
Lyr 19h15'53"40d54'
Lambright,Connie
Aql 19h31'23"10d60'
Lambright,Ray
Ori 5h35'35"-0d51'
Lambrinides II, Nicholas James
Per 2h49'0"45d52'
LambStar
Per 2h53'0"40d47'
Lambuth,Alan L
Per 3h7'15"41d30'
Lamczyk-Mas,Jennifer
And 23h33'47"42d23'
Lameira,Cristina F
Uma 11h15'1"57d45'
Lamela,Josephine
Com 12h49'59"23d18'
LaMendola Family Star, The
Cyg 19h29'27"33d53'
Lamens,Andre
Lyr 18h57'18"47d8'
Lamet,Lynn
And 0h21'54"30d41'
Lametta,Lou Lou
Tau 4h4'37"23d58'
LaMew,Adam Elzo
Sco 16h53'34"-40d0'
Lami,Kathryn Rose
Cmi 7h58'17"8d0'
Lamiri,Akim
Cas 23h23'54"54d16'
Lamitie Star,Lee- Lynda-Lelyn-Lynsie
Cet 1h48'26"-8d15'
Lamm,Ellery Rose
Mon 6h53'54"10d43'
Lamm,Michael E
Ori 5h10'0"-6d32'
Lammerding,Nancy Kay
Eri 2h58'10"-18d12'

Lammers
Aur 5h6'0"40d0'
Lammers's Folly, in honor of D Keith
Her 17h47'55"40d41'
Lammie
Uma 10h31'44"51d31'
Lammie,Melissa G
Sge 19h59'0"16d1'
Lamminpohja,Jarmo Juhani
Cam 5h47'58"58d47'
Lammlin,Jon V
Peg 21h59'33"7d56'
Lammons,Barbara
Peg 23h19'13d53'
Lammons,Lesly
Cyg 21h51'15"44d6'
Lamon,Michelle Tina (MTL)
Lyr 18h47'40"42d49'
Lamond,Emeline
Dra 13h4'54"67d54'
Lamond,George
Her 18h12'52"45d3'
Lamond,Rich "T M"
Her 17h5'57"45d33'
Lamont,Anthony D
Uma 11h3'11"45d18'
Lamont,Charles
Leo 9h53'12"30d47'
Lamont,Dylan Mathew
Uma 11h21'0"41d50'
Lamont,Ferrin
Aql 18h57'17"13d55'
Lamont,Ruth Marie
Gem 6h35'50"13d14'
Lamont,Taylor
Ser 15h21'1"4d33'
Lamontagne,René
Umi 15h18'14"68d30'
Lamontagne,Yves
Gem 6h36'35"13d14'
Lamonte,Greg S
Tau 5h44'59"16d38'
Lamonte,John & Chyna
Crb 15h21'21"32d25'
Lamore,Lisa
And 2h0'59"40d17'
Lamothe,Kate
Lac 22h11'31"50d13'
Lamothe,Marc
Cep 3h6'1"80d6'
Lamothe,Sylvie
Cep 22h10'6"59d28'B
Lamoureaux,Edward T
Aur 6h27'26"32d59'
Lamourenita,Lady Byron
Peg 21h24'19"8d30'
Lamouret,Benjamin
Lyr 18h51'43"34d60'
Lamoureux,Franáois
Per 2h58'0"34d16'
Lamoureux,Gilbert
Per 3h2'50"41d33'
Lamoureux,Melaine Jeanne
Cet 2h9'53"6d17'
Lamoureux,William
Cep 22h25'52"63d24'
Lamp'l,Bill
Cet 0h42'34"-5d37'
Lamp,Sue
Lyr 18h40'20"26d13'
Lamp,Tony
Tri 2h20'26"32d46'
Lampa,Stan
Dra 16h48'21"67d10'
Lamparter,Cara
Lyr 18h48'19"39d49'
Lampe,Cathy
Aql 20h1'21"1d37'
Lampe,Debbie
Aql 20h16'12"0d32'
Lamper-The Spirit Of The Industry,Lew
Aql 18h59'44"11d45'
Lampert,Thomas K
Lmi 9h20'37"34d15'
Lampi,Dovie (Sponaugle)
Oph 16h56'57"-28d24'
Lampi,Ruby Kathryn
Cas 0h0'39"56d0'
Lampi,William C
Aur 4h52'44"40d49'
Lampkin,Terrence Lloyd
Ori 6h1'26"1d6'
Lampl,Brian K
Per 3h10'28"46d28'
Lampman,Donna (Lampy)
Cam 4h9'50"60d39'
Lampman-My Super Star, Aaron
Lac 22h7'48"51d50'
Lamporella,Calogera Cavasecnno
Aur 5h32'10"41d13'
Lamport,Forever Allan
Ser 18h16'7"-7d19'
Lamrock,Colleen"Bean"
Crt 11h3'33"-13d36'
Lamson "Willow",Mike
Vul 19h58'29"22d32'
Lamson,James E
Ori 5h52'23"12d56'
Lamson,Judith Lee
Mon 7h25'1"-0d53'
Lamson,Tara
And 2h30'0"50d14'
Lamune
Cap 20h56'24"-20d38'
Lamy,Aurore
Ori 6h7'11"20d14'
Lamya
Del 20h19'50"14d37'
Lan Neo,Alice Chee
And 23h15'44"50d18'
Lana
And 2h6'39"39d26'
Lana
Mon 6h57'32"10d59'
Lana & Jim's Endless Love Star
Aql 18h43'26"10d53'
Lana-Marie
Cas 0h50'49"62d25'
Lanahan,James F
Tau 3h54'12"20d16'
Lanak,Elizabeth Christina
Lyn 7h43'11"44d38'
Lancaster III,James Joseph
Mon 6h58'26"7d46'
Lancaster,Daisy Hochanadel
Equ 21h18'49"3d43'
Lancaster,David R "Smooches"
Uma 11h27'14"49d5'
Lancaster,Grace
Peg 21h42'15"23d54'
Lancaster,Jr,Ralph Ivan
Ori 5h49'1"18d48'
Lancaster,Kimberly Jonelle
Lyn 8h41'17"46d49'
Lancaster,Leonard Alfred
Eri 4h3'35"-11d2'
Lancaster,Lynch Pryor
Aur 6h56'11"43d57'
Lancaster,Perry
Cet 2h41'1"-1d34'
Lancaster,Tonya Elizabeth
Sge 20h5'0"20d5'
Lancaster/Webb,Ann Marie
Lyr 19h1'12"25d51'
Lance
Boo 13h51'1"20d10'
Lance
Aur 6h9'58"45d33'
Lance
Sct 18h42'49"-6d55'
Lance
Psc 1h20'33"22d43'
Lance
Dra 14h46'0"65d9'
Lance
Uma 10h39'44"60d13'
Lance & Taryn
Lyn 7h21'26"58d45'
Lance & Yandell
Cra 11h8'3"-19d2'
Lance Romance
Psc 1h2'26"21d42'
Lance's Star,Erin Rebecca
And 2h20'25"39d9'
Lance,Donna M
Eri 3h55'0"-15d29'
Lance,Jessica Ann
Lyr 19h1'60"37d46'
Lance,Marian Marie
Cas 0h43'23"69d51'
Lance,Marsha
Lyn 7h7'28"52d23'
Lancellotti,Teresa
Lac 22h25'55"54d34'
Lancelot
Per 2h6'54"47d23'
Lancelot
Cep 22h14'17"59d23'
Lancelot Franáois
Umi 16h30'29"78d53'
Lancer
Cet 2h28'33"5d37'
Lancer,Brett
Sct 18h54'12"-6d37'
Lancer,David
Eri 2h57'28"-10d17'
Lanchak,Lillian C
Lyn 8h10'36"40d57'
Lanchester
Uma 13h10'45"60d3'
Lanci,Antoinette
Lyn 7h28'24"36d12'
Lancieri,Lara
Cyg 21h52'21"42d20'
Lanciloti,Kim
Vul 19h48'18"28d18'
Lanciloti,Louis
Aur 5h17'0"43d17'
Lancry,Jean Jacques
Cam 3h47'18"67d51'
Lanczki,Amber Joyce
And 0h53'40"37d15'
Land 9-3-91,Stacy Lynett
Cnc 8h53'60"30d53'
Land of Good
Aql 19h50'17"12d16'
Land,Bette & Don
Lyn 7h56'23"42d52'
Land,James Cameron
Cep 20h14'57"60d19'
Land,James Christian
Cet 2h26'17"3d3'
Land,Jr Harry L
Aql 20h11'37"1d45'
Land,Kelly Kristine Weisenberger
Cep 20h14'0"60d31'
Land,Kevin Mark
Cep 20h14'26"60d29'
Land,Michael John
Cep 20h14'18"60d21'
Land,Mike & Trudy
Crb 15h30'41"30d27'
Land,Nicholas Daniel
Cep 20h14'28"60d36'
Land,Roy Daniel
Cep 20h15'0"60d1'
Land,Roy Samuel
Cep 23h11'0"60d45'
Land,Ute Gudrun Schöne
Cep 23h10'13"60d21'
Land,Vincent Scott
Cep 20h14'0"60d49'
Land,William O
Ser 17h30'29"-14d54'
Landa,A Toby
Lac 22h25'47"55d32'
Landa,Kelsey Summer
Ori 5h49'34"20d56'
Landa,Lisa "Lil Puddin"
And 23h38'1"38d48'
Landa,Philip
Oph 17h6'20"7d46'
Landa-Dancing With The Angels (18)
Uma 9h8'18"49d5'
Landahl,Elise Victoria
Lib 14h50'60"-1d23'
Landaluce,Adam Alton
Cep 23h1'49"60d57'
Landau"Noah Nick", Noah Nicholas
Uma 10h0'46"70d2'
Landau,Barbara & W Loeber
Cyg 20h53'13"39d51'
Landau,Karin
Uma 9h36'1"50d47'
Landau,Nicholas J
Her 16h28'15"33d19'
Landau,Ralph
Tau 4h0'32"2d8'
Landau,Sol
Eri 3h1'10"-5d20'
Landauer,Kurt "Opi"
Oph 17h33'46"-24d20'
Landauer,Star of Rebecca,Amy & Jerry
Aqr 20h58'12"0d21'
Landaw,Jackie & Bob
Crb 16h19'32"32d19'
Landay Victoria
Umi 15h37'0"70d5'
Landay,David
Lac 22h33'27"40d40'
Lande,Jeri Roth
Dra 12h10'22"64d49'
Lande,Neil Bronfman
Ser 15h43'53"4d39'
Landenberger,John H
Her 16h33'1"38d13'
Lander,John William
Her 16h32'15"36d5'
Lander,Millie
Ori 5h58'55"9d43'
Lander,Paul Davidson
Cep 23h9'38"65d6'
Landergren,Jr,Edward Curtis
Peg 22h33'1"26d6'
Landerman,Jon Taylor
Umi 6h48'55"89d5'
Landeros,Ernesto (Neto)
Sct 18h55'56"-4d24'
Landers,Dale
Cet 0h34'53"-7d12'
Landers,Gloria J
And 23h21'31"44d21'
Landers,Paul
Lac 22h16'46"51d28'
Landes,L B
Aur 5h38'33"38d51'
Landesberg,Ella Mery
Cas 1h44'32"73d44'
Landesberg,Noah Samuel
Lac 22h13'19"51d34'
Landess,Nancy Ryerson
Uma 11h41'41"44d39'
Landgarten,Jo Ann & Harris
Ori 5h28'15"-1d59'
Landgarten,Michelle
And 0h24'35"43d12'
Landgraf,Evi
Lib 14h43'42"-22d30'
Landgraf,Stefan
Vir 13h28'24"-11d38'
Landgren,Beverly
Ori 5h57'32"16d41'
Landheim,Elliot H
Her 16h4'22"50d47'
Landie,Cordon B
Cep 22h56'41"59d31'
Landini,Louis R
Aur 5h58'41"30d39'
Landino,Albert
Lac 22h32'10"55d33'
Landino,Lori
Del 20h14'1"15d13'
Landis
Uma 10h59'37"38d51'
Landis Clan,The
Uma 13h59'19"58d23'
Landis,Barbara
Uma 9h32'23"61d10'
Landis,Daniel Scott
Cam 3h55'38"78d56'
Landis,Doris Jean Archer
Per 4h1'50"50d19'
Landis,Dorothy & Fred
Cyg 20h53'0"38d6'
Landis,Erica Staci
Uma 8h33'49"51d57'
Landis,Fred
Dra 10h17'0"74d40'
Landis,Jeffrey Clayton
Her 18h6'32"28d29'
Landis,Judith
Peg 23h44'37"31d40'
Landis,Justin Wayne
Aql 19h45'32"11d46'
Landis,Leah Rose
Cyg 19h16'1"45d50'
Landis,Samuel I
Cma 7h2'1"-28d27'
Landis,Scott Louis
Aql 20h2'39"4d58'
Landis,William Edward
Dra 14h41'51"65d8'
Landler,Monika
Her 17h4'10"49d60'
Landm,Christophe
Lyn 7h52'13"50d43'
Landman,Kasper Henry
Her 16h19'52"8d37'
Lando,Andrea
Lyr 18h58'55"47d15'
Lando,Galina & David
Sct 18h42'16"-7d15'
Landon,Jack
Tri 1h57'0"27d38'
Landor,Edna
Lyr 18h37'0"39d55'
Landovitz,Steven G
Per 1h44'22"50d28'
Landre Sonia
Uma 11h4'27"50d4'
Landreau,Dominique
Lac 22h43'15"38d14'
Landress,Joshua
Gem 7h1'47"24d42'
Landriault,Alain Guy
Cam 3h50'19"61d58'
Landriault,Sylvie
Cas 0h51'50"61d3'
Landrichi,Rachid
Oph 18h3'25"10d6'
Landrum,"My Black Magic Man" Keith M
Dra 19h0'52"48d6'
Landrum,Graham Gordon
Boo 15h6'24"12d43'
Landrum,J Michael
Per 1h58'39"48d58'
Landrum,Jr,Benjamin
Cam 8h26'29"82d10'
Landrum,Lisa
Lyr 19h18'37"38d11'
Landry's Ocean,Robin Marie
Umi 14h58'23"80d5'
Landry,Adam Roy
Cet 2h43'27"0d22'
Landry,Angela Diane
Oph 18h2'16"10d53'
Landry,Ashley Marie
Cas 3h6'0"73d22'
Landry,Blake Douglas
Hya 8h31'53"-8d14'
Landry,Chad
Dra 18h38'19"80d18'
Landry,Daryl T
Sct 18h55'15"-6d60'
Landry,Jim
Cyg 20h6'20"32d13'
Landry,Marcus Joseph
Mon 6h43'26"10d19'
Landry,Pam
Cas 1h22'13"71d6'
Landry,Paul
Eri 4h35'31"-1d26'
Landry,Serge
Her 16h46'1"20d27'
Landry,Thérèse
Lyn 8h27'0"46d1'
Landsberger,Peter Leslie Herbert
Aur 6h5'48"35d9'
Landschoot,Gloria Dawn
Equ 20h59'29"6d50'
Landsman,Allison T
Lyn 7h16'59"58d31'
Landsman,Stefanie T
Peg 23h1'55"21d26'
Landstorfer,Andreas
Cap 21h27'49"-23d56'
Landström,Airi Marjatta
Cam 10h57'0"80d1'
Landström,Matilda
Umi 14h38'32"69d1'
Landun,Coach Howard
Oph 17h59'52"-5d49'
Landvik-Larsen,Deborah Joy
Lyn 7h49'23"48d10'
Landwehr,Father & Grandfather,Bob
Dra 19h15'43"70d7'
Landwehr,Kyle Dennis
Uma 10h16'51"59d57'
Landwehr,Sharlaine Smart
Uma 10h52'42"40d38'
Landwehrmann,Rainer
Sgr 18h50'33"-28d47'
Lane & Children,Ashara Cindi
Aql 19h58'28"13d34'
Lane (LX),John
Cep 21h1'53"56d6'
Lane Spell
Sge 19h58'1"16d7'
Lane Star,The Susan
Cnv 12h43'23"34d7'
Lane"The Little Goat", Hana Umlauf
Aur 5h27'19"29d36'
Lane's Birthday Star, Hank
Aqr 21h53'0"-1d9'
Lane's Jewel
Boo 14h24'0"45d47'
Lane,Agnese
Cas 23h43'36"58d24'
Lane,Amber Dawn
And 1h40'36"36d58'
Lane,Anthony
Ori 5h26'17"15d13'
Lane,Beau Suzanne
Per 2h38'0"37d31'
Lane,Brett A
Vir 12h4'34"1d48'
Lane,Carol Christine
Aur 7h9'30"42d44'
Lane,Charles
Dra 17h11'36"72d2'
Lane,David Charles
Aur 7h12'13"40d5'
Lane,Deborah
Peg 21h57'36"36d19'
Lane,Diana Gail
And 2h24'11"45d55'
Lane,Doil Keith
Lyr 18h58'51"34d35'
Lane,Ellen Elizabeth
Tri 1h48'45"27d35'
Lane,Ethel L
Eri 3h12'42"-17d7'
Lane,Frederic Mirick
Aur 7h23'25"40d30'
Lane,Genevieve
And 0h22'1"32d13'
Lane,Georgina
Mon 7h42'57"-8d47'
Lane,Gillian
Lyr 18h33'31"29d51'
Lane,Jane Earl
Boo 14h11'25"30d5'
Lane,Jeana
And 23h2'12"41d23'
Lane,Jennifer
And 2h31'48"40d53'
Lane,Jill M
Cas 0h54'42"68d39'
Lane,Jimmese
Lyr 18h41'0"34d17'
Lane,John Kade
Dra 14h53'46"62d53'
Lane,Jonathan Edmund
Cep 21h37'18"71d2'
Lane,Josie Gale
Lyr 19h20'49"42d52'
Lane,Kenny
Ori 5h46'0"21d59'
Lane,Kerry Scott
Ori 5h59'31"17d26'
Lane,Kevin
Cnv 12h59'12"32d35'
Lane,Linda Faye
Oph 17h35'19"8d3'
Lane,M Pauline
And 1h47'50"39d54'
Lane,Madeline B
Uma 11h53'17"46d14'
Lane,Mary Freda
Cam 13h14'0"78d34'
Lane,Megan Ann
Cnc 8h5'54"18d56'
Lane,Michael Alan
Cet 3h13'48"5d38'
Lane,Molly Ann
And 0h13'16"47d42'
Lane,Rachael Anne
Lyr 18h58'22"38d21'
Lane,Rachel
Boo 14h45'0"51d24'
Lane,Robert B
Hya 8h14'19"1d4'
Lane,Roberta
Aql 20h32'58"-6d37'
Lane,Robin
Hya 8h54'59"-1d51'
Lane,Rocky B
Ori 6h7'43"8d31'
Lane,Ronnie
Peg 21h39'1"25d36'
Lane,Rosa
And 23h38'10"47d8'
Lane,Rosa Bell
And 0h53'38"35d2'
Lane,Sarah A
And 0h56'25"39d53'
Lane,Sean Edward
Ori 5h4'48"8d30'
Lane,Stan
Dra 18h7'12"58d34'
Lane,Susan
Cyg 20h36'25"60d41'
Lane,Sutton
Umi 16h48'30"75d48'
Lane,Tami K
Eri 2h47'18"-18d50'
Lane,Terri L
Aql 19h30'26"12d5'
Lane,Thomas W
Dra 19h56'1"67d33'
Lane,Travis
Sex 10h15'17"-4d9'
Lane,Valerie
Cam 8h27'22"74d6'
Lane,Vienna Rose
Mon 8h2'54"-6d8'
Lane,William
Her 18h10'14"47d49'
Lane,William Denis
Cmi 7h54'48"0d51'
Lane,William Gregory Wilson
Tau 4h2'42"21d30'
Lane,Willis
Cyg 20h35'47"41d12'
Lane-Prater,Cassie Renee
Eri 3h18'0"06d57'
Lanese,Michelle
Leo 10h2'12"12d50'
Lanette's Dream
Uma 10h38'28"47d25'
Laney
Oph 17h33'28"-24d46'
Laney
Aur 5h0'12"47d40'
Laney,Jessica Erin
Mon 6h4'1"-8d19'
Laney-2/14/95,Patrick Owen
Eri 4h4'56"-15d27'
Laney-5/20/90,Nicholas Robert
Hya 8h15'58"2d31'
Lanfranco S171.256
Cae 4h58'24"-33d27'
Lanfranco,Marco
Psc 22h53'11"1d15'
Lanfranco,Massimo
Cnv 14h55'23"28d53'
Lang's Star,Susan
Mon 6h55'39"-0d42'
Lang,Adam James
Her 17h18'45"48d44'
Lang,Brennan
Aql 18h58'48"13d17'
Lang,Brian Christopher
Uma 11h53'30"55d12'
Lang,Carol
And 0h4'25"46d22'
Lang,Carole
Peg 23h46'52"31d6'
Lang,Carrie
Peg 22h8'41"3d46'
Lang,Carrie Elizabeth
Ori 6h8'50"9d14'
Lang,Chunky
Per 2h57'52"46d51'
Lang,Eileen
Dra 18h11'0"70d57'
Lang,Eugene
Aur 7h2'20"43d19'
Lang,Gabriele Helene
Aqr 23h32'50"-12d25'
Lang,George
Boo 14h53'50"48d37'
Lang,Hans-Jürgen
Cap 20h39'7"-26d42'
Lang,Heather Kristen Elizabeth
Cas 1h45'0"68d57'
Lang,Illuminating Lana
Com 13h18'29"22d4'
Lang,Jeannie
Lyn 7h27'28"45d4'
Lang,KD
Lyn 7h44'32"40d56'
Lang,Mary Alice
Psc 1h25'18"17d35'
Lang,Norman Edward
Per 2h57'15"46d30'
Lang,Pauline
Cas 3h34'49"73d39'
Lang,Rainer
Aur 5h54'24"48d26'
Lang,Renee Lynn
Mon 6h44'17"11d26'
Lang,Roger
Dra 19h5'55"48d23'
Lang,S
Vir 13h36'22"-8d3'
Lang,Stacy
Mon 5h57'53"-4d58'
Lang,Stan
Cnv 13h58'35"37d41'
Lang,Tayln Gabriel
Uma 11h56'16"32d14'
Lang,Volker
Peg 23h31'22"15d23'
Lang,Zachary Randall
Ori 5h1'38"1d34'
Langan's Quest
Lac 22h18'17"54d25'
Langan,John Scott
Tri 1h58'45"27d3'
Langan,Patti & John
Cas 23h13'33"63d10'
Langdon Star,The
Cet 1h26'27"0d29'
Langdon,Adam Michael
Cep 2h54'0"80d29'
Langdon,Allison
Cas 0h3'55"65d6'
Langdon,Lorraine
Cas 0h7'0"64d33'
Langdon,Marie
Aql 18h57'39"12d3'
Langdon,Ruth
Lmi 9h32'41"38d20'
Langdon-Sivak,Carla
Sge 19h28'34"16d22'
Lange,Alex Christopher
Her 17h59'55"14d21'
Lange,Alex Peters
Cmi 7h21'45"3d45'
Lange,April Lynn
Uma 9h1'35"48d32'
Lange,Beau
Cet 2h20'1"-0d10'
Lange,Bianca
Ori 5h56'0"7d44'
Lange,Christy
Eri 3h18'55"-2d58'
Lange,Dietrich Günther
Cap 20h29'1"-26d50'
Lange,Edward A
Her 18h19'35"12d10'
Lange,Elizabeth "Oma", Schlobohm
Peg 22h32'58"35d14'
Lange,Gerson & Katia
Lmi 10h26'23"31d21'
Lange,J Douglas
Cet 2h55'41"5d46'
Lange,Jeff
Lib 14h55'48"-22d12'A
Lange,Josie
Mon 6h29'40"10d21'
Lange,Jürgen
Aur 5h4'19"43d41'
Lange,Kelly
Ari 3h0'13"11d40'
Lange,Linda
Peg 21h54'50"23d37'
Lange,Patrick Anthony
Cma 6h52'31"-16d30'
Lange,Robinn
Lac 22h35'48"56d31'
Lange,Roxanne Marie
And 1h20'36"38d40'
Lange,Sr,Charles G
Cep 23h29'56"70d53'
Lange,Stuart Michael
Mon 7h11'12"-6d46'
Lange,Tom D
Hya 8h16'33"3d54'
Lange,William Paul
Ser 15h57'0"0d38'
Lange,Wolfgang
Umi 13h57'22"73d22'
Lange-Kelly,Gordon Henry
Cap 21h54'17"-20d20'
Langel,Nancy
Vul 19h13'53"21d38'
Langelier,Guy
Uma 10h5'48"54d19'
Langenbach,Sr,Michael P
Her 16h39'22"38d12'
Langenberg The Big O, Oliver M
Her 17h1'51"20d34'

Langendoen,Kerrie Anne
And 0h33'27"40d53'
Langenmayr-Superstar, Ken
Ori 5h23'53"-6d35'
Langensiepen,Angelika
Lib 15h16'36"-21d23'
Langer,Al & Betty
Ori 6h6'13"8d48'
Langer,Ann M
Cyg 20h36'1"38d39'
Langer,Doris
Dra 16h46'32"65d43'
Langer,Marion
Ari 2h57'48"21d33'
Langer,Peter und Dorthy
Vul 19h40'32"20d32'
Langer,Sonny
Cep 22h0'33"61d26'
Langevin,Maryann
Cyg 20h8'47"39d49'
Langfelder,Diane
And 2h21'18"41d13'
Langfelder,Jeffrey W
Ori 5h38'0"0d58'
Langford
Cas 0h29'33"58d3' C
Langford
Boo 14h38'5"22d11'C
Langford
Peg 22h49'2"26d7'C
Langford's Star
And 22h55'19"40d28'
Langford,Lane
Ser 15h53'37"0d16'
Langhans,Tanja
Vir 13h26'46"-2d2'
Langille,Allan Harding
Boo 15h28'51"47d48'
Langjahr,Karen Gail
Uma 11h13'13"71d51'
Langkil,Sheera(P O P)
Lyn 7h49'15"38d15'
Langlais,Brandon Izaiah
Cep 7h45'25"86d41'
Langley "Shorty", Shelia
Peg 22h40'50"31d23'
Langley,Andreas und Amy
Cap 21h4'55"-24d25'
Langley,Bess
Lyr 19h16'20"41d26'
Langley,Brandi
Mon 6h39'46"10d28'
Langley,Glen Martin
Aur 6h0'42"38d38'
Langley,Jancy Eskew
Mon 6h53'59"-6d38'
Langley,Jennifer Drais
And 1h48'58"39d19'
Langley,Jonathan "Elmo"
Eri 3h12'57"-6d14'
Langley,Margaretta Polly
Lyr 18h45'0"31d44'
Langley,Patricia
Lyn 7h48'1"50d13'
Langley,Russell Lawrence
Per 3h20'0"40d26'
Langley,Thomas Aaron
Tau 4h44'51"16d19'
Langlois III,Herbert J
Uma 9h38'47"52d9'
Langlois,Franáois
Uma 9h23'31"60d1'
Langlois,Jean-Michel
Uma 9h33'58"60d46'
Langlois,John
Ori 4h59'18"8d41'
Langlois,Jullen
Uma 9h29'1"60d52'
Langlois,Lucette Dubois
Uma 9h21'44"60d33'
Langlois,Mary Allie
And 0h55'11"22d7'
Langlois,Patrice
Uma 12h8'16"60d21'
Langlois,Patrick
Cep 20h28'55"75d16'
Langlois,Roland
Her 17h39'25"22d46'
Langlois,Ronald
Uma 9h1'35"60d43'
Langlois,Shayne Jonathan
<<Keats>>
Dra 17h8'58"52d20'
Langlois,Stéphane
Uma 10h25'45"52d21'
Langlois,Sue
And 1h12'12"39d5'
Langlume,Patrice- Michel
Cyg 19h39'13"37d42'
Langman,Andy
Cep 22h38'29"56d56'
Langmead,Donna
Leo 9h36'1"7d51'
Langmead,Eileen Hamilton
And 23h26'26"40d14'
Langmead,Michael
Ari 2h23'28"10d40'
Langnead,Sheryll
Cyg 19h51'0"40d19'
Langner,Brooke Marie
And 22h59'32"51d17'
Langner,Thomas
Gem 7h6'45"24d37'
Langner,Wayne
Aur 6h8'52"38d13'
Langpap,Rainer
Cep 0h4'57"81d41'
Langson,Nicole Marie
Mon 7h40'0"-6d30'
Langstar
Aur 4h55'15"41d11'
Langsted,Clyde Regnar
Aur 5h10'0"44d57'
Langston,Leighton
Ori 5h49'59"11d43'
Langton,Erin Margaret
Cas 0h34'27"66d21'
Langton,Lawrence Iain
Hya 8h58'54"-10d53'
Langton,Ryan Joseph
Per 3h5'20"50d5'
Langton,Timothy Michael
Cep 22h30'0"63d24'
Languille,Jean-Pierre
Cyg 19h47'13"37d49'
Langweiler,Joseph Paul
Boo 15h40'28"41d49'
Langworthy,Stanton Barber
Aur 5h32'0"30d33'
Langzemis,"Patches"
Dra 15h35'1"53d22'
Lani
Aur 4h48'22"40d35'
Lani
Cyg 19h46'51"38d13'
Lani
Uma 11h2'42"42d43'
Lanier Construction Co,Robert
E
Cet 0h51'41"-1d60'
Lanier III,Bruce N
Aql 18h59'0"15d36'
Lanier's Eastern Star, Alice
Mon 7h44'43"-5d52'
Lanier,Jillian Nicole
Eri 2h56'21"-5d5'
Lanier,Maryalice
Cmi 7h40'46"4d18'
Lanier,Shannon I
Aql 18h56'20"13d48'
Lanier,Susie & Bruce
Aql 18h57'0"11d48'
Lanigan,Paul Augustine
Dra 19h0'15"71d13'
Laningham,Larry
Sco 17h25'1"-31d42'
Lanjohnette
Peg 23h33'1"30d56'
Lankford,Norce Gene
Aql 19h27'11"-10d55'
Lankford,Pamela
Ari 1h46'0"20d16'
Lankford,Shellie
Peg 21h50'28"34d21'
Lanky
Her 17h36'13"40d53'
Lannan,Kirk
Cep 22h17'1"62d24'
Lanni,Jr,Orlando L
Cma 6h31'8"-16d52'
Lannigan,Stephenie
Lyn 7h58'20"35d27'
Lanning,Amelia
Com 13h13'0"27d52'
Lanning,Moria & E J
Peg 0h5'0"13d26'
Lanning,Robert J & Lillian R
Cyg 21h15'40"38d59'
Lanny
Sct 18h44'57"-6d44'
Lanny
Cam 8h49'19"74d7'
Lano,Charles
Lac 22h47'37"52d44'
Lanoff,Lawrence Gregory
Com 13h6'51"28d36'
Lanoue,Michael
Dra 16h2'19"61d28'
Lansburgh,Cheryl Strickler
Sgr 19h9'28"-23d7'
Lansche,Jr,John E
Dra 15h7'0"58d10'
Lansdell WGM Ontario 1994-
95,Jan
Cas 0h44'51"75d5'
Lansdorf,Andrew William Paul
Aur 7h23'42"38d7'
Lansford Star,The Jack
Aql 20h12'23"4d3'
Lansing,Joanne Engel
Lyr 18h58'32"31d26'
Lansing,Melvin Everett
Cnv 12h32'42"40d45'
Lansing,Sherry
Mon 6h41'0"7d25'
Lant,Skott
Lac 22h22'50"55d14'
Lant,Vicky
Lac 22h4'21"50d0'
Lanthorn,Craig W
Boo 15h32'1"40d28'
Lanthorn,D Adam
Cep 22h11'1"58d32'
Lanty,Michael
Lac 22h12'48"46d22'
Lanty,Rosalie
Aql 19h52'38"12d2'
Lantz,Douglas
Her 17h24'33"38d56'
Lantz,Laura
Peg 22h33'40"8d41'
Lantz,Nada
Cam 7h29'16"80d21'
Lanum,Kent
Ori 6h16'53"-2d24'
Lany
Vul 20h4'35"23d4'
Lanza III,Charles
Her 18h39'12"12d6'
Lanza,Edward
Her 16h53'19"38d16'
Lanza,Joseph Dominic
Tri 2h12'39"31d50'
Lanza,William Conrad
Dra 13h8'1"68d22'
Lanzarini,Claire
Cep 22h26'31"57d46'
Lanzarone,Dana W. & Sherri A
Vischio
And 2h9'26"38d59'
Lanzarote
Ori 4h54'47"1d35'
Lanzetta,Stanley & Fortuna
Boo 14h10'50"39d44'
Lanzi,Mr & Mrs Marco
Gem 7h53'41"30d18'
Lanzillotta,Jamie Leigh
Ari 1h55'36"11d21'
Lanzillotta,Kimberly A
And 23h32'12"46d57'
Lanzilotta III, Frank James
Sex 10h26'0"-5d32'
Lanzotti,Josephine
Aql 19h31'1"10d4'
Lanzrath,Mary K
Cyg 19h32'49"34d59'
Laopandria
Gru 22h5'18"-56d21'
Lapa,John & Odette
Cyg 21h35'32"40d33'
Lapacek,Don M
And 2h0'15"42d34'
Lapacek,Les
And 2h23'25"42d42'
Lapacek,Michael
And 2h7'16"41d3'
Lapacek,Sue
And 2h8'35"42d4'
Lapadula,Nicholas Francis
Lac 22h11'10"46d14'
LaPage,Joyce,Roger &
Rosemary
Cas 2h33'42"60d38'
Lapalorcia,Angela
Cyg 20h58'52"38d24'
LaPardo,Dawn
Cas 2h1'15"61d36'
LaPenta,Bob & Anne
Crb 16h21'45"38d23'
LaPenta,R J
Cnv 12h16'35"41d51'
Lapera Star,The Nicole
Uma 9h26'1"57d12'
Lapetina,Michael Lee
Oph 18h25'24"7d36'
Lapeyre,Jr,Philip Frances
Ori 5h21'59"1d21'
Lapham,Peter
Boo 14h56'0"22d46'
Lapham,Warren Vail
Cep 22h21'54"56d14'
Lapides,Alexander Moore
Dra 19h26'1"56d43'
Lapides,Benjamin Moore
Lac 22h15'11"48d53'
Lapides,Olivia Bette
Peg 23h39'21"26d13'
Lapides,Star of Albert
Hya 8h32'28"1d31'
Lapidos aka Honest Abe,Jack
M
Lmi 9h47'42"38d12'
Lapidus,Letitia Rose Suzanne
And 0h1'25"46d27'
Lapidus,Stanley
Aur 6h6'11"31d33'
LaPierre,Eric T
Hya 8h10'29"1d41'
Lapierre,Jacqui M
Uma 9h16'0"42d57'
Lapierre,Nathalie et Robert
Racicot
Cep 23h39'36"64d39'
Lapierre,Olivier
Cyg 21h57'0"50d16'
Lapinski,Estelle
Dra 17h50'21"58d27'
Lapinski,Thomas
Dra 17h37'18"60d44'
Lapiska,Hayden Alexander
Aql 20h12'26"4d40'
LaPlaca,Joan
Cas 0h7'21"50d19'
LaPlante,Billy
Aur 5h0'47"51d14'
LaPlante,Leo
Per 3h18'12"40d42'
LaPlante,Marianne
Lac 22h12'1"49d2'
Laplante,Nancy
Cas 0h53'28"74d31'
Lapo
Her 16h11'34"10d18'
Lapo
Cnv 14h1'0"45d52'
Lapo
Mon 7h54'1"-4d5'
Lapo
Cyg 20h41'19"31d1'
LaPoe,Don
Per 3h43'53"51d22'
LaPoe,Donald W
Boo 15h18'25"33d37'
Lapointe,Anne-Julie
Cyg 21h1'35"39d10'
LaPointe,Daine
Aur 7h22'0"40d43'
LaPointe,Karl Joseph
Tri 1h58'36"26d41'
Lapointe,Roland
Lyn 8h38'54"42d7'
Lapointe,Valou-Valérie
Leo 11h31'58"-1d47'
Lapole,Jacob Ryan
Cyg 21h12'18"39d23'
Lapolla,Jennifer C
Del 20h21'0"20d35'
Lapolla,Matt
Hya 9h5'10"5d20'
Lapolla,Molly
Mon 7h26'46"-1d28'
LaPonte,Maribeth
Psc 23h1'54"5d38'
LaPort,Jeanne
Aql 18h57'0"3d14'
LaPorta,Lauren Joy
Aql 19h56'28"13d59'
LaPorta,Margaret Kadrich
Lmi 9h46'50"38d9'
Laporta,Philip Charles
Cep 7h12'0"86d44'
Laporte,Beth Erin Schrock
Lmi 10h55'34"31d45'
Laporte,Chantal
Cyg 19h25'47"30d38'
Laporte,Erin Michelle
Umi 14h26'43"68d15'
Laporte,Jean-Franáois
Lyr 18h40'40"29d30'
Laporte,Jeanne S
And 0h44'0"40d56'
LaPorte,Linda Louise
And 0h51'11"35d37'
Laporte,Rejeanne Dulude
Cam 3h54'1"52d34'
LaPorte,Victor
Aql 18h54'37"-1d9'
Lapostolle-Rogers, Marie
Christine
Cas 0h58'21"54d26'
Lapp,Jr,Charles E
Cnv 12h55'1"32d52'
Lappan's Of Gaylord, Inc
Aur 6h52'38"35d40'
Lappen,Catherine Lee
Cas 1h6'1"55d28'
Lappin,Marie & Gareth
Sge 20h3'16"20d20'
Lappin-Weiser,Eileen
Uma 9h2'1"51d16'
Lappy
Aur 5h8'35"40d59'
Laprade #722
Uma 10h2'47"61d45'
LaPrade,Crista & Stephen
Lac 22h26'18"54d22'
LaPreciosa Christy
Ori 5h51'60"9d9'
Lapriore,Justin John
Cam 12h6'51"77d19'
Laprune,Jean Ollé
Cep 22h24'47"57d50'
Lapsley,Alan
Her 16h28'29"36d20'
Lapsley,Ryan Patrick
Boo 14h40'54"22d22'
Lapstar
And 1h25'26"33d53'
Lapychak,Marilynn M
Com 12h20'0"24d39'
Laquatra,Idamarie
Vul 19h18'24"20d26'
Lar
Per 3h19'1"50d21'
Lar & Nora
Uma 11h22'24"50d52'
Lar Star
Cam 4h9'28"60d28'
Lar-Bear
Aur 5h1'48"40d20'
Lar-Bear
Cyg 21h31'11"44d51'
Lara
Cmi 7h27'32"0d31'
Lara
Sco 17h35'20"-33d1'
Lara
Pic 5h7'4"-45d27'
Lara
And 0h54'57"35d39'
Lara
Uma 9h37'47"46d29'
Lara & Alex
Lyr 18h21'18"40d47'
Lara & Dave's Eternal Love
Peg 23h3'16"33d58'
Lara Evonne
Cas 1h23'53"55d60'
Lara K
Cas 0h30'0"68d36'
Lara Karina
Eri 3h21'24"-11d56'
Lara my Angel
Tau 4h31'1"30d40'
Lara's Star
Mon 7h19'16"-5d18'
Lara,Gracie Zarate
Boo 14h48'50"35d7'
Lara,John A
Ori 5h36'38"-6d25'
Lara,la Piccola
Mon 7h15'52"0d41'
Laraki,Princesse Laïla
Cyg 20h59'49"31d26'
Laramee,Adelord
Lyn 8h17'1"45d35'
Laramis Joansie
Lyn 7h52'58"50d5'
Laramore,Marie H
Aql 20h2'43"0d11'
LaRatro,Raeann
Com 13h3'56"30d38'
Lardas,Antigone
Tri 1h33'45"28d30'
Lardini,Roberta
Lyn 7h54'29'41d4'
Lardner,Janet Thomson
Cas 0h48'53"66d51'
Lardy,Christian David
Cep 2h40'12"77d44'
Lare's Pluto
Cnv 14h2'27"30d21'
Lare,Joanna M
Peg 22h45'20"5d56'
LaReau,Gretchen Anne
Cyg 21h33'17"30d47'
LaRee
Aur 5h0'12"40d12'
LaRee
Aql 19h53'32"14d8'
Larew III,John William
Ori 4h57'56"4d33'
Largailloli
Lyn 7h45'38"48d43'
Large,Bonnie Marie
Cyg 19h29'0"31d45'
Large,Eleanor
Peg 22h4'24"5d37'
Large,Erin Stacy
Cet 3h16'53"-1d28'
Large,James Daniel
Aur 6h2'49"32d56'
Large,Jonny
Ori 5h35'37"-0d30'
Large,Wilbur R
Aur 6h54'0"38d40'
Largen,Jim & June
Cyg 20h32'51"42d36'
Largeron,Isabelle
And 23h3'39"38d11'
Lari Leane
Aql 19h4'52"4d2'
Larich,Paula A
Ori 6h15'1"7d48'
Larimer Star,The
Ori 4h55'46"-1d41'
Larin,Vitali
Aur 5h14'11"43d34'
Laris,Linda
Del 20h25'30"7d45'
Larisa
Peg 22h23'20"35d4'
Larisa Ananda-Kojro
And 23h26'1"44d21'
Larisa Desireé
Cas 2h26'43"59d22'
Larison,Emilie Elizabeth
Cas 1h17'51"62d41'
Larissa
Aql 20h12'0"11d37'
Larissa
And 0h10'23"46d53'
Larissa Dawn
Cnc 8h52'41"31d15'
Larissa's Star
Cas 0h39'52"58d47'
Larivee,Eric S
Gem 6h52'0"31d9'
Lariviere,Chuck
Cam 6h4'1"61d6'
Larivière,Isabelle
Cyg 19h28'15"33d56'
Larizadeh,Darya
Ori 5h51'21"16d53'
LARJAPAR
Ori 5h58'51"16d55'
Lark VI,Callen
Cep 22h58'54"58d15'
Larke,Lily Sze-Man
Sco 15h55'13"-22d54'
Larkin
Aur 5h3'12"42d24'
Larkin "Big Ed",Edward H
Cep 23h32'56"70d46'
Larkin I Love You, Richard &
Judy
Cyg 21h54'19"52d38'
Larkin Star,The Robert Nelson
Aql 19h9'19"0d22'
Larkin,"Old No 7" James Keith
Per 3h26'0"41d2'
Larkin,Araxie Hairabedian
Aur 5h17'33"40d24'
Larkin,Brittany Katherine
Cep 20h27'21"61d48'
Larkin,David Scott
Boo 14h17'10"39d21'
Larkin,Gabriel Andre
Her 17h36'45"25d22'
Larkin,Grant
Cep 22h39'20"59d14'
Larkin,John Stephen
Peg 23h23'1"17d28'
Larkin,Lucy
Vul 20h18'54"28d53'
Larkin,Meredith
Cyg 21h50'27"41d55'
Larkin,Michael P
Dra 16h30'18"71d15'
Larkin,Morgan Frances
Aur 5h9'14"40d17'
Larkin,Peter
Aql 18h58'42"17d16'
Larkin,Samantha
Peg 22h27'25"31d46'
Larkin,Sean Francis
Aql 19h58'54"13d56'
Larkin,Staci "Crystal of
Beauty"
Tri 2h23'0"34d3'
Larkin,Theresa Anne
Lyr 19h25'26"37d58'
Larkin-30 Years, Patricia &
James
Lac 22h17'49"54d30'
Larkin-Lorbeer,Ashley
Com 12h13'18"20d53'
Larkin-Tuma,Cyndi
Eri 2h43'57"-3d5'
Larko,Eddie J
Sex 10h25'42"-0d48'
Larmand,Andrew James
Lac 22h4'32"40d2'
LaRocca,Arthur
Cyg 19h27'41"36d10'B
LaRocca,Peter Paul
Cep 22h46'45"68d49'
LaRocco,Anna
Cyg 19h27'41"36d10'A
Larocco,Kathryn L
And 0h12'21"37d48'
LaRocco,Robin
Boo 13h35'44"19d59'
Larochas,Florence
Lyr 18h28'20"38d3'
LaRoche,Adrienne Vivian
Cas 2h49'45"61d14'
Laroche,Anne-Alix
Cas 23h39'60"61d24'
Laroche,Julie
Lyr 18h35'58"39d25'
Laroche,Roland
Lac 22h49'33"53d39'
LaRoche,Stéphanie Raymonde
Lillian
And 23h3'1"48d40'
LaRochelle,Christine Ann
Eri 3h35'25"-12d43'
Larocque,Carmelle
Lac 22h54'49"38d59'
LaRocque,Cassandra Doris
Vul 19h46'49"28d41'
Larocque,France
Cyg 21h4'53"31d24'
Laron's Kick
Aur 7h6'22"43d50'
Larosa,Boby
Lac 22h11'36"54d25'
LaRosa,Freddie
Per 2h58'35"38d23'
Larosa,James
Per 3h8'0"38d58'
LaRosa,Joe
Aql 20h1'34"12d18'
LaRosa,Robert A
Dra 12h28'1"71d48'
LaRosa,Scott John Michael
Her 17h19'1"20d53'
Larose 50,Ken
Uma 11h32'26"38d34'
Larose,Kathryn Ann
Peg 22h34'20"21d14'
LaRose,Ronald W
Aqr 21h2'20"-1d23'
LaRose,Sharon
And 23h45'48"47d57'
LaRossa,Dana
And 0h59'1"34d44'
LaRouere,Jessica
Cas 0h1'18"63d30'
LaRovere,Sheba
Uma 12h2'47"57d42'
Larr,Shannon
Uma 9h53'0"56d10'
Larrabee,Elizabeth
And 23h46'40"45d5'
Larrabure,Anne-Marie
Eri 3h13'25"-16d53'
Larrianne
Cas 0h29'1"75d36'
Larrick,Jon Pentland
Dra 19h26'32"61d46'
Larricq,E Pete
Lyr 19h24'34"37d42'
Larripa,Maria-Cinta
Lyr 18h34'22"29d29'
Larrissa Maxine
Lyr 18h29'39"31d31'
Larrivée,Denise Normandin
And 23h56'47"34d44'B
Larrivée,Normand
And 23h56'47"34d44'A
Larrivée,Simon
And 23h33'58"41d31'
Larrivée,Sophie
And 23h28'43"41d12'
Larrivée,Yves
And 23h29'41"41d38'
Larroque,Laurence
And 2h30'53"40d59'
Larrouturou,Alain
Aql 19h30'11"-10d29'
Larry
Her 16h33'41"21d16'
Larry
Uma 8h55'29"68d8'
Larry
Per 1h58'43"50d21'
Larry
Per 2h59'47"37d33'
Larry
Aur 5h42'53"50d4'
Larry
Ori 5h52'15"19d6'
Larry & Audrey
Aur 6h12'0"45d41'
Larry & Christine
Lyr 18h28'22"47d14'
Larry & Liz
Cyg 19h41'21"31d18'
Larry & Lynn
Dra 19h1'58"65d28'
Larry & Nica
Cyg 21h37'11"42d45'
Larry & Sheri
Cma 7h15'58"-11d5'
Larry B,The
Boo 14h28'53"25d38'
Larry Goodenough
Sex 9h54'42"5d48'
Larry Victor
Per 2h39'24"57d14'
Larry's Dream
Oph 17h57'49"12d11'
Larry's Dreams
Per 2h58'15"32d38'
Larry's Shining Star
Aql 19h3'12"0d12'
Larry's Star
Boo 15h6'21"48d22'
Lars
Aur 5h15'54"40d51'
Lars
Ser 17h35'56"-10d35'
Larsen
Cep 22h6'24"61d29'
Larsen
Peg 0h6'1"27d6'
Larsen Star,The
Dra 12h0'60"67d52'
Larsen Star,The
Cyg 19h34'22"28d4'
Larsen,Ana
And 2h7'14"42d19'
Larsen,Anne
Cmi 7h20'20"7d45'
Larsen,Audrey R
Lyn 8h10'25"34d17'
Larsen,Britt
Umi 13h26'0"72d27'

Larsen,Gloria
Psc 23h0'20"2d26'
Larsen,Jack & Suzanne
Cyg 20h37'11"50d23'
Larsen,Jason K
Aur 5h11'30"41d52'
Larsen,Jennifer L
Lyn 7h34'27"45d14'
Larsen,Jenny
Ori 5h31'40"8d1'
Larsen,John "Basketball"
Uma 10h3'58"50d4'
Larsen,Laura
Aql 20h6'33"1d6'
Larsen,Lori
And 23h23'10"51d8'
Larsen,Lorraine
And 2h17'15"48d25'
Larsen,Niels
Cep 23h4'60"61d5'
Larsen,Olaf Eugene
Cma 6h52'21"-18d42'
Larsen,Rafe Anders
Lac 22h14'0"37d57'
Larsen,Verna LaRue
Peg 21h57'29"30d49'
Larson II-Star Of Bubba,Glenn
Lac 22h21'26"52d36'
Larson My Special Girl Jessica
Com 12h9'48"20d55'
Larson Star,The
Cma 6h43'40"-15d44'
Larson, Kathy D
Ori 5h52'21"10d05'
Larson,"Aunt Ann" Ann Hicken
Aql 19h50'12"15d39'
Larson,Abigail E
And 1h12'22"35d58'
Larson,Beth
Umi 14h56'0"71d9'
Larson,Cassie
Lyn 7h31'13"50d39'
Larson,Craig
Aur 4h56'58"50d51'
Larson,Danielle
Cyg 19h45'0"31d23'
Larson,David Campbell
Per 2h23'36"58d40'
Larson,David Eric
Her 17h32'50"42d37'
Larson,David Michael
Cnv 12h14'31"41d1'
Larson,David Thomas
Ori 5h52'50"15d5'
Larson,Dorothy
Cyg 21h38'19"38d11'
Larson,Erica
Cas 23h22'1"60d48'
Larson,George
Lac 22h8'56"47d8'
Larson,George P
Sgr 18h52'0"-20d21'
Larson,James M
Her 16h44'0"32d46'
Larson,Jennifer Annette
Lyn 8h11'59"48d22'
Larson,Joe
Boo 14h15'23"54d46'B
Larson,Jr,Robert Lee
Cmi 7h55'11"8d0'
Larson,Kari
Cyg 20h38'41"31d35'
Larson,Kelly
Com 12h34'58"14d30'
Larson,Kevin Ronald
Dra 18h56'0"70d30'
Larson,Laurie Wood
Mon 6h43'53"10d13'
Larson,Lawrence E
Vul 20h14'23"26d2'
Larson,Leann
Cas 0h10'22"62d28'
Larson,Lori Smidt
And 23h34'1"38d38'
Larson,Mark & Linda
Dra 14h2'19"64d36'
Larson,Mary Adele
Lyr 18h52'55"31d25'
Larson,Mike
Lmi 10h5'14"30d50'
Larson,Sandy
Aql 20h5'58"4d51'
Larson,Scott A
Lyn 7h55'22"50d55'
Larson,Shanda
Uma 10h59'41"50d10'
Larson,Sr,In Memory Of Arthur H
Aur 5h57'53"30d17'
Larson,Stella Goddard
Mon 6h25'43"3d37'
Larson,Susan Booth
Cas 1h58'16"60d19'
Larson,Suzanne
Peg 21h25'39"2d29'
Larson,Thea
Peg 21h27'51"2d19'
Larson,Tropic of Irmama
Del 20h17'18"12d13'
Larson-Love & Tenderness,Margaret
Lac 22h12'16"48d44'
Larsson,Jessica Elisabet
And 2h32'46"37d30'
Larsson,Ulrika E
Cnc 9h3'31"32d21'
Larsuskaler
Tau 4h46'48"16d5'
Larter,Kathleen
Lyr 19h15'60"25d59'
LaRue,Amy Elizabeth
Tau 3h29'56"30d59'
LaRue,James David
Tau 3h56'36"30d20'
LaRue,Jason J
Cep 22h18'46"80d4'
LaRue,PhD,Rodney M
Her 17h8'26"42d22'
LaRussa,Michael
Dra 17h27'25"60d34'
LaRussa,Nicholas
Aur 7h18'29"39d58'
LaRusso,Jeannine
And 1h57'37"40d1'
LaRusso,Melissa
Uma 10h24'39"40d16'
Larvin,Brian
Cyg 21h44'10"31d22'
Larvin,Kristin
Cam 5h47'38"56d15'
Las
Uma 10h22'21"54d21'
Las Raqueles
Uma 9h23'14"55d5'
Las,Lauren Lucy
Lyn 7h52'33"35d25'
Lasagni 28/11/1987, Giulia
Sct 23h41'16"-27d24'
Lasala,Carole
Ari 1h58'0"11d57'
Lasala,Stephen F
Cam 3h53'29"52d38'
LaSalandra,Frankie
Dra 17h50'48"60d44'
Lasalle,Frederic
Peg 23h5'16"31d30'
LaSalle,Lisa Marie
And 0h50'12"39d42'
Lasapio,Frank E
Eri 2h44'0"-2d59'
Lasarcyk,Wilfried
Cap 21h17'57"-21d46'
Lasater,Colton Dane
Cam 7h57'0"61d45'
Lasater,Tim
Leo 10h56'57"-0d51'
LaSaxon,Jessica Reneè
Sct 18h54'1"-6d33'
LaSaxon,Rebecca Anne
Sge 19h53'11"16d41'
LaScamme,R Anna of Charles Francis
Cas 0h36'25"64d57'
Lascelles,Victoria
Cas 0h30'29"60d14'
Lasch,Alan P
Her 18h42'38"12d21'
Lasecki,Judy
Lyn 8h53'28"41d20'
Lasey's Christmas Star
Per 2h53'13"38d31'
Lash,Jessica Alyson
And 23h38'51"40d37'
Lash,Megan Marie
Cnc 9h5'42"27d35'
LaShawn,My Forever Love
Peg 22h27'40"21d30'
Lashelle-Bunkfeldt Family,The
Cyg 20h34'59"58d42'
Lasher of Ft Johnson NY,Donald F
Uma 9h54'40"68d37'
Lashley-Galvan,Barbara
Mon 7h47'38"-5d58'
Lashmit,Sandra
Mon 8h4'43"-4d27'
Lashon,Christy
Lyr 18h32'27"42d49'
Lashua,Jerry
Uma 11h29'0"40d25'
Lasinski,Charlotte
Com 12h26'20"20d34'
Laska,Laurel Jean
Cyg 20h3'54"40d57'
Lasker,Alexandra Baxter
And 23h47'0"45d53'
Lasker,James Edwin Baxter
Aur 7h15'42"36d13'
Lasker,Margo Baxter
Lyn 8h48'0"38d18'
Lasker,MD,Harold
Her 16h28'48"33d1'
Laskey,Anne
Peg 23h8'17"13d13'
Laskey,Carol Maureen
Peg 23h19'41"12d13'
Laskey,Ian William
Peg 23h8'1"11d29'
Laskey,Maureen
Peg 23h0'47"10d39'
Laskey,Pam & Ken
Cyg 20h17'45"39d8'
Laskey,Thomas Gair
Peg 23h0'52"13d19'
Laskin,Fern
Cam 4h59'23"60d48'
Laskow,G Blaise
Ori 5h17'21"-8d45'
Laskowitz,Daniel Todd
Cep 2h26'1"77d11'
Lasky,Carla
Cyg 19h28'59"30d23'
Lasky,Connor Robert
Eri 3h7'21"-2d59'
Lasky,Dara S
Uma 9h43'35"67d39'
Lasky,Dora Henrietta Liberman
Com 12h14'50"21d47'
Lasky,Kenneth Paul
Sex 9h41'27"0d32'
Lasky,Michael J
Ori 5h31'30"0d5'
Laslo's Light
Lyn 7h52'35"41d12'
Laslo,David Nathan
Dra 16h36'56"61d58'
Lasnier,Dame Cécile
Uma 8h57'54"55d51'
LaSorsa,Evangelia
Lac 22h5'1"49d53'
Lasoski,Pat & Ken
Lyr 19h20'35"37d58'
LaSpina,Lawrence
Aql 19h58'45"15d31'
Lass,Joan
Crb 15h30'10"38d45'
Lass,Peter
Lac 22h15'45"37d32'
Lass,Tracy
Aql 20h34'28"-7d32'
Lassabe,Mathieu
Cam 4h5'47"58d45'
Lassalle,Jean Pierre
Ori 6h1'35"11d2'
Lassaso,Maureen & Scott
Tri 1h58'0"30d33'
Lassen,Christian Riese
Hya 9h16'14"1d28'
Lassen,Susanne
Cas 0h56'45"50d23'
Lasseter,Gina-Lena
Mon 7h15'3"-6d52'
Lassignardie,Amandine
Mon 6h24'45"-5d44'
Lasswell,Pamela
Del 20h57'0"10d19'
Last Class Of 240
Her 17h57'49"14d18'
Laster,Jon & Jenny
Uma 8h22'37"68d39'
Laster,Nancy President Hospice of El Paso,Inc
Ori 6h10'40"0d7'
Lasting Love- Linda & Hap Cole
Tri 2h25'20"28d45'
Laszko,Larry
Per 2h42'1"43d40'
Laszlo
Cet 2h28'0"7d46'
Laszlo,Ronald Jay
Her 16h38'23"22d31'
Laszlo,Titli
Peg 23h37'55"22d1'
Latai,Mojdeh
Cnv 13h22'14"41d37'
Latanzio,Alan & Denise
Cam 3h38'22"61d4'
Latapie,Dominique
Sgr 19h27'15"-31d55'
Latashia
And 1h19'1"36d22'
Latchford,Erich Andrew
Aur 4h47'20"36d25'
Latchère,J P
Tri 1h55'52"30d35'
Lateana,Angelena
Cyg 19h17'34"44d6'
Laterzo,Gary
Uma 11h11'1"55d22'
Latham,David
Aur 6h52'0"41d11'
Latham,John D & Patricia D
Aql 18h58'52"6d56'
Latham,Lee R
Sex 9h57'14"2d27'
Latham,S Duane
Gem 7h54'0"20d33'
Lathan,Arielle
Aur 6h1'11"38d36'
Lathan,Liliane
Uma 11h35'43"31d59'
Lathbury,Sharon
Lyr 18h30'39"30d38'
Lathlean,Melissa Ann
And 0h38'51"38d36'
Lathrop,Sherry Lynn
Del 20h37'36"18d55'
Lathwesen,Jörg
Tau 5h50'11"26d16'
Lathwood,Edward
Ori 4h59'55"0d32'
Latibeaudiere,Timothy Phillip
Oph 18h17'41"12d57'
Latif,Fatima
Lyn 8h43'41"36d12'
Latif,Salha
Uma 9h3'11"51d27'
Latif,Shahab
Ori 5h36'21"15d15'
Latika-Heavens Flower, Madeline
Cyg 20h37'43"40d24'A
Latimer,Kathryn J
Cas 0h4'58"65d7'
Latimer,Teresita Paniagua
Aur 4h58'26"30d43'
Latina An Honoured Star,Robert
Car 7h29'20"-54d23'
Latini,David
Aur 5h39'38"47d58'
Latiolais,Janet
Mon 6h54'20"-6d11'
Latishas Light
Lac 22h45'46"37d49'
Latka,Viktor
Lyr 18h56'0"34d35'
Latori,Claudio
Boo 14h33'54"18d16'
LaTour,"Frenchman"
Cam 4h1'20"67d30'
Latour,Benjamin Ivy
Aql 19h31'18"10d23'
LaTour,Daniel Trigg
Boo 14h19'41"34d22'
Latour,Jean Franáois
Sgr 19h33'41"-36d3'
Latour,Justin
Cyg 19h23'43"44d52'
Latour,Mademoiselle Kristina
Sgr 18h2'26"-28d47'
Latour,Yves
Cyg 20h42'13"47d2'
Latray,Jack
Hya 9h17'50"1d1'
Latremouille,Neal E
Her 16h29'20"35d40'
Latrisha
Lyn 7h16'45"50d33'
Latta,Dale
Aur 5h13'1"41d59'
Latta,Jason M
Dra 16h34'1"68d33'
Latta,Marion E
Lyn 7h20'55"50d1'
Latta,Michele Della
Mon 6h29'34"11d59'
Lattanzi,Jason
Boo 14h52'53"46d41'
Lattarulo,Delores
Cas 1h52'16"65d1'
Lattavo,Michael & Denise Vander Vorst
Cyg 21h33'20"40d31'
Latterman,Eryn Rose
Aqr 23h14'43"-6d43'
Lattig (Teddy Bear), Theodore Anthony
Cam 7h6'29"61d9'
Lattimer,Billy
Hya 9h10'49"1d49'
Lattimer,Camille Odessa
Cam 12h40'44"76d59'
Lattimer,William Ervin
Boo 13h53'25"17d28'
Lattimore II,Anderson Joseph
Her 17h14'37'20d50'
Lattman Star,The
Mon 6h55'0"0d21'
Latulippe,Lucien N
Aur 5h29'1"37d51'
Latuszek,Joseph P
Cmi 7h23'25"0d12'
Latuszek,Mark J
Ser 15h30'19"1d13'
Latuszek,Paul D
Sex 9h51'57"2d48'
Latuszek,Veronica F
Dra 9h39'41"77d47'
Latvis,Donald P
Lyn 7h52'47"41d6'
Latzka,Kurt P
Per 3h12'12"48d33'
Latzke,Elsie
Umi 13h51'33"70d43'
Lau Star
Mon 7h49'23"-5d54'
Lau,Doris
Vir 13h26'9"-2d2'
Lau,Estelle Tsui
Mon 8h5'46"-7d40'
Lau,H T & Pauline Au
Uma 10h47'60"53d44'
Lau,Ian
Cma 6h28'34"-24d51'
Lau,Jennifer Lynn
Lyn 7h48'1"48d28'
Lau,John
Cet 3h19'41"2d19'
Lau,Kate Rose
And 23h34'22"38d43'
Lau,Randall H
Her 17h28'19"41d4'
Lau,Rodger
Lac 22h19'25"48d53'
Lau,Rose
Cas 0h17'1"61d20'
Lau,Smiling Sue
Lyr 18h32'18"41d12'
Laub,Arnold
Sct 18h30'19"-6d30'
Laub,Donald E
Boo 13h58'34"21d54'
Laub,Lelia Montague
Lyn 8h31'1"45d52'
Laub,Louis Solomon
Cet 1h22'14"-4d50'
Laub,Marika Daël
Cyg 20h10'47"58d41'
Laubach,James Blake
Cma 6h59'1"-15d16'
Laubach,Matthew William
Aur 6h6'58"45d32'
Laube,Gifford
Uma 10h22'45"51d37'
Laube,Robert B
Aur 4h52'44"51d57'
Laubis,Silke
Sgr 19h24'23"-42d39'
Lauble,Harry E
Her 17h19'12"40d28'
Laubner,F Paul
Her 17h8'24"22d26'
Lauby,Nancy J
Ori 6h0'28"1d26'
Lauck,Michael Joseph
Her 17h28'0"31d24'
Laudadio,Robert Anthony
Dra 16h33'1"72d54'
Laudani,Alexander Steven
Per 1h58'51"56d46'
Laudenbach,Bonnie L
Lyn 8h35'1"37d42'
Lauder,Gordon
Cep 22h24'0"68d9'
Laudet,Chippy
Per 2h31'40"52d18'
Laudig,Misty
Cas 1h1'17"60d14'
Laue,Nikolai
Peg 22h49'27"20d21'
Lauer,Astro Andy
Aql 18h56'24"15d14'
Lauer,Brandon Kent
Ori 4h49'57"4d25'
Lauer,Courtney
Lyr 18h33'23"38d33'
Lauer,Dorothy Lee
Cam 5h51'17"58d24'
Lauer,Kirsten
Cep 1h23'28"78d1'
Lauer,Matthew
Aur 5h18'19"40d35'
Lauer,Nathan
Her 18h38'42"18d58'
Lauer,Rosemarie
Lyn 8h9'47"45d14'
Lauersen (God's Helper),Dr Niels
Oph 18h5'0"11d8'
Lauf,Dipl Ing Gerhard
Cep 22h38'59"61d50'
Laufenberg
Uma 9h36'1"45d36'
Laufenberg,George H
Cep 0h14'30"66d39'
Lauffer,Donna Jean
Uma 9h24'45"56d55'
Laughhunn,Alexis Sue
Lyn 7h53'1"43d27'
Laughhunn,Paige
Cam 3h18'41"61d24'
Laughing Loving Generous Kind Helen
Cas 1h17'37"61d24'
Laughing Star
Boo 14h48'27"50d30'
Laughlin,Renee Denise
Lyn 6h20'55"54d13'
Laughman,Courtney Rae
Cyg 20h23'47"30d1'
Laughter Always
Lyn 7h8'21"52d9'
Laughter,Judson C
Mon 6h18'1"8d34'
Laughton,Jake John
Ori 5h55'31"8d60'
Laughton,John David
Her 16h30'52"40d41'
Laugier,Aurore
Per 2h51'13"50d16'
Laugier,Lola
Vul 21h18'18"24d4'
Laugier,Maryline
Per 2h51'1"50d9'
Lauhon,Alexander Paul
Oph 18h19'18"8d55'
Lauk,Michaela
Ari 2h2'18"22d1'
Lauletta,James Daniel
Aur 4h48'46"40d0'
Lauletta,Margaret Anne
And 2h1'33"38d42'
Laulhere,Cherese Mari
Cyg 21h10'45"34d21'
Laumelnatant Star
Cmi 7h34'12"10d52'
Laumen,Anette
Sgr 19h35'58"44d51'
Laumeyer,Richard
Lac 22h48'48"54d28'
Launderville,Juanita Ann
And 2h17'10"40d55'
Launer,Lisa
Cep 23h28'50"64d33'
Launi Marie
Lyr 18h50'20"30d22'
Launt,Catherine Dewees
Peg 23h2'42"33d49'B
Launt,Walter Taylor
Peg 23h2'42"33d49'A
Laupretre,Jean-Pierre
Aql 19h3'52"16d29'
Laura
Pho 0h4'0"-44d3'
Laura
Pho 0h6'53"-40d59'
Laura
Crb 15h23'34"30d1'
Laura
And 2h32'23"38d28'
Laura
Cam 3h40'11"68d51'
Laura
And 2h12'25"38d31'
Laura
Lyr 19h9'12"47d35'
Laura
Sge 20h17'0"18d57'
Laura
Her 16h47'10"10d12'
Laura
Cas 0h38'57"58d22'C
Laura
Sge 19h14'36"18d52'A
Laura
Lyn 8h16'27"35d12'
Laura
Psc 22h51'39"6d51'
Laura
Lyn 7h48'49"40d56'
Laura
Del 20h17'30"9d30'
Laura
And 0h9'1"46d56'
Laura
Cam 12h40'56"77d31'
Laura
And 0h51'14"33d25'
Laura
Cas 2h52'18"70d33'
Laura
Sge 19h35'38"19d20'
Laura
Ind 20h50'7"-50d24'
Laura
Cas 0h10'22"56d1'
Laura
Lmi 10h10'53"32d11'
Laura
Psc 23h2'35"5d28'
Laura
Cap 20h7'54"-10d12'
Laura
Cas 3h9'33"70d42'
Laura
Uma 13h41'13"50d10'
Laura
Per 2h12'42"58d29'
Laura "The Garden Star"
Ori 5h5'15"0d14'
Laura & "Bitsy"
Crt 11h19'11"-17d5'
Laura & Corey
Cyg 19h16'31"49d49'
Laura & Joe
Aur 6h10'32"45d21'
Laura & Matthew
Peg 23h43'50"28d45'
Laura & Michael
Crb 15h34'20"30d44'
Laura & Mike Forever
Ori 5h59'36"10d18'
Laura & Susan
Cyg 20h53'24"31d13'
Laura & William
Col 6h30'51"-33d7'
Laura & Yong-Ho
Uma 9h56'45"43d24'
Laura 30
Lac 22h0'24"51d52'
Laura Alice
Lyn 8h19'14"57d52'
Laura Angel
Aur 5h19'55"47d51'
Laura Anne
And 1h2'14"35d27'
Laura Anne
Sge 18h56'14"18d50'
Laura B
Crb 15h58'59"27d46'
Laura Beth
Peg 22h54'44"29d31'
Laura D
Cyg 20h23'14"41d5'
Laura Dawn
Cas 1h44'60"58d36'
Laura e Marcello
Eri 3h3'46"-2d25'
Laura Elisa,The
Cma 6h53'59"-18d37'
Laura Elizabeth
Cas 1h53'32"58d2'
Laura Elizabeth
Cas 1h45'51"60d19'
Laura Elizabeth
And 23h21'30"48d54'
Laura Elizabeth
Del 20h14'31"11d13'
Laura H & Benito G Q
Sge 19h36'27"16d25'
Laura Helen
Peg 22h0'0"26d12'
Laura Holly
Mon 8h4'51"0d20'
Laura Iacullo,Gambino, Anthony
Peg 21h7'48"12d3'
Laura J
Cet 3h11'0"1d16'
Laura Jacqueline
And 0h38'18"40d12'
Laura Jane
Lyr 18h59'52"26d3'
Laura Jane
Lyr 18h28'0"46d11'
Laura Janine
Cyg 20h48'11"37d54'
Laura Jean
Lyr 18h27'1"41d14'
Laura Jean
Cas 1h47'50"61d10'
Laura Jean
Peg 22h33'56"21d44'
Laura Jean Precious Starbeam
Cyg 19h26'14"33d42'
Laura Jennifer
Cyg 20h25'49"40d17'
Laura Kay
Peg 23h4'33"8d30'
Laura Kay
And 0h18'50"30d24'
Laura Lee
Aql 19h29'26"0d32'
Laura Lee
Uma 12h16'1"58d18'
Laura Lee
Ori 4h56'24"0d6'
Laura Lee
And 0h10'23"31d51'
Laura Li
And 1h39'28"40d12'
Laura Lynn
And 0h21'1"33d49'
Laura Lynne M
Cas 0h2'29"58d28'
Laura M
Psc 0h55'1"32d18'
Laura Mae
Lyr 18h32'11"32d28'
Laura Mae-jor
Mon 7h27'1'-10d34'
Laura Marie
Lyr 18h59'1"42d32'
Laura May
Cam 3h40'30"71d25'
Laura MHB
Cma 7h21'36"-18d48'
Laura Michelle
Lyr 18h27'42"37d41'
Laura Michelle Number One
Ori 5h59'49"6d51'
Laura Mon Amour
Per 2h59'47"32d58'
Laura Noel
Lyr 19h17'20"37d38'
Laura Rose
Mon 7h48'55"-5d21'
Laura Rose
Cet 1h53'16"1d36'A
Laura S
Lac 22h41'16"56d10'
Laura '76
Cnv 13h5'38"51d43'

Laura's Beacon
Psc 0h48'43"32d25'
Laura's Ice Cream Star
Uma 11h2'46"48d5'
Laura's Light
Lmi 9h25'19"38d17'
Laura's Light
Cet 3h16'43"0d55'
Laura's Panacea (Sweetleaf)
Lyr 19h0'21"38d55'
Laura's Shining Star
And 23h31'58"45d4'
Laura's Star
Peg 23h30'0"10d17'
Laura's Star
Peg 23h31'34"18d29'
Laura's Star
And 0h10'50"37d8'
Laura's Star
Ori 6h5'1"3d52'
Laura's Wish
Cnc 8h31'40"8d55'
Laura,Edioige
Cyg 20h23'46"39d37'
Laura,My Star On Earth
Lyn 9h14'46"33d28'
Laura,The Light of
And 1h7'24"38d5'
Laura,Yoko
Cas 0h51'1"61d8'
Laura-Forever Yours I
Tri 2h18'46"33d59'
Laura-Jon
Lyr 18h53'11"30d28'
Laura/Joan
Per 2h52'26"37d31'
Laura/PMF
And 0h0'38"43d9'
LauraCanadoreMatt
And 23h23'45"48d55'
Laurajean
And 23h29'40"49d57'
Laurano,Anthony F
Peg 23h23'52"18d21'
Laure
Lac 22h4'52"37d56'
Laure,Annie
Per 2h59'43"38d15'
Laure,Louis
Cam 13h52'58"82d16'
Laureano,Alexa Dorothy
Lyn 8h3'48"40d55'
Laureca
Aql 19h3'1"16d20'
Lauree
And 23h38'36"40d37'
Laureen
Cyg 21h20'1"31d25'
Laureen
Lyr 18h20'53"45d12'
Laurel
Dra 16h19'15"62d1'
Laurel
Lmi 10h0'19"32d32'
Laurel
Uma 12h0'19"61d56'
Laurel
Cas 0h58'36"50d34'
Laurel
Del 20h35'59"20d4'
Laurel Christine
And 23h19'41"35d16'
Laurel Elizabeth
Cmi 7h35'30"0d14'
Laurel Jean
Lyn 7h25'37"44d42'
Laurel Jean
And 0h4'22"35d18'
Laurel Olivia
Mon 7h30'41"-5d39'
Laurel Rose
Ari 2h44'1"24d29'
Laurel,Michele H
Cyg 20h23'39"39d4'
Lauren
Dra 18h13'48"56d34'B
Lauren
Sco 17h32'0"-30d21'
Lauren
Cyg 20h18'0"41d50'
Lauren
Gem 7h3'18"24d56'B
Lauren
Cyg 19h53'28"40d21'
Lauren
Cnv 12h14'0"44d20'
Lauren
Aql 20h8'17"0d52'
Lauren
Aql 19h32'48"-6d59'
Lauren
Vul 19h18'42"26d49'
Lauren
Cas 0h53'0"62d41'
Lauren "The Luminous Bat Mitzvah"
Ori 5h55'1"11d57'
Lauren Alexandra
Cet 2h42'46"-8d51'
Lauren Alison
Cas 1h6'21"60d51'
Lauren Ann
Cas 23h14'46"60d50'
Lauren Ashley
And 0h54'14"21d37'
Lauren Christine
Uma 9h29'31"50d58'
Lauren Danielle
Cyg 21h1'0"38d40'
Lauren Emily
Peg 23h17'1"30d38'
Lauren Haley
Oph 18h3'23"11d59'A
Lauren Kate
Mon 6h58'0"11d14'
Lauren Louise
Uma 11h50'45"51d49'
Lauren Marie Illuminati
Cas 23h14'33"62d51'
Lauren Marie
Cyg 21h35'1"38d7'
Lauren Michele
Cas 23h14'11"62d56'
Lauren Michele
Ori 5h14'58"-10d0'
Lauren Nicole
Tau 5h48'35"23d42'
Lauren Nicole's Star
And 23h7'26"43d7'
Lauren of the East
And 23h5'19"42d17'
Lauren Rebecca
Aql 20h6'10"1d21'
Lauren Renee
Oph 18h1'1"12d51'
Lauren Rose
Uma 13h15'34"61d57'
Lauren Victoria
Lyn 8h2'14"35d47'
Lauren's Daddy's Wishing Star
Cep 22h14'47"61d5'
Lauren's Dream
Dra 10h11'52"80d5'
Lauren's Eternal Love For Emanuel
Lyn 6h56'47"58d14'
Lauren's Gaze
Cas 0h9'1"58d57'
Lauren's Gift
Cet 1h50'0" 10d14'
Lauren's Legacy
Peg 22h23'1"31d48'
Lauren's Star
Lyn 8h10'42"51d4'
Lauren,Kelly
Del 20h18'59"10d52'
Lauren,Kelly & Nathan
Aur 6h26'46"36d37'
Laurence
Cep 4h15'50"80d32'
Laurence
Tau 4h10'43"0d6'
Laurence C
Peg 21h54'0"21d44'
Laurence,Florrie Leigh
Mon 6h50'15"10d12'
Laurence,Je t'aime
And 1h3'11"41d4'
Laurence,Lynne Griffin
Lyn 7h35'0"45d22'
Laureni,Joe
Boo 14h49'25"27d45'
Laurens,Didier
Cam 13h51'1"80d3'
Laurenson,Robert Charles
Psc 0h52'48"33d17'
Laurent
Aur 5h56'16"38d23'
Laurent & Lubna
Lmi 10h3'13"41d10'AB
Laurent Et Oliver
Lyr 19h22'40"31d7'
Laurent,Arcay
Oph 18h16'40"10d46'
Laurent,Chloé Marie
Cas 23h31'29"61d23'
Laurent,Fiona
Cyg 19h39'27"30d30'
Laurent,Jean Franáois
Aql 19h34'51"-6d27'
Laurent,Patricia
Lac 22h43'59"38d35'
Laurent,Pierre
Ori 5h56'0"20d43'
Laurent-Brisart,Jean- Philippe
Aql 19h1'0"-10d5'
Laurenti's Haven For Joe & James
Ori 5h38'10"-3d57'
Laurentine
Eri 3h42'45"-18d40'
Lauret,Marie & Philippe
Col 6h1'23"-36d37'
Lauretano,Patrizia
Pic 4h32'18"-48d11'
Laureth
Aur 5h4'37"38d45'
Lauretta
Sge 20h1'56"20d26'
Laurette,Meris
Boo 13h33'24"20d27'
Lauri,Otto
Cra 18h19'12"-42d37'
Lauria,David P
Dra 14h20'27"64d41'
Lauria,Nicky & Micky
Cyg 21h10'22"39d4'
Lauria,Vin & Marie
Cyg 19h56'27"40d45'
Laurian 96
Peg 22h32'19"21d52'
Lauriandy
Cyg 20h43'46"45d38'
Laurice,Mike
Sct 18h55'36"-5d31'
Laurich,Stephanie A
And 2h31'35"41d10'
Laurie
And 22h59'52"37d41'
Laurie
Cas 23h16'31"60d13'
Laurie
Eri 3h32'43"-2d53'
Laurie
Cyg 20h39'1"30d22'
Laurie
Lyn 8h54'14"34d58'
Laurie
Psc 1h34'42"22d14'
Laurie
Cas 23h21'21"53d53'
Laurie
Del 20h14'46"11d1'
Laurie
Peg 23h43'30"16d25'
Laurie & Carmen's Taurean Moon
Cyg 20h42'1"37d31'
Laurie & Christopher
Cyg 21h51'35"53d33'
Laurie & Helmut Forever & Ever
Crb 15h50'40"27d11'
Laurie & Ian Joined In Love
Lyn 8h17'47"40d28'
Laurie & Mickey
Aql 18h56'1"13d1'
Laurie 63
Mon 7h3'23"-06d52'
Laurie Anne
Vir 14h1'51"-6d24'
Laurie Beth
Hya 9h42'26"-18d58'
Laurie Day
Crb 16h3'52"34d58'
Laurie H 30th
And 0h3'0"40d34'
Laurie Jo
Cas 0h56'22"58d27'
Laurie Joan
Tau 5h28'29"28d51'
Laurie Kay
Cyg 20h53'30"30d3'
Laurie's Heart
Cas 2h9'19"70d56'
Laurie's Heavenly Body
And 2h6'46"38d58'
Laurie's Star
Lyn 6h28'0"61d15'
Laurie's Star 50
Cas 0h57'24"54d35'
Laurie,Al
Cam 5h18'17"67d31'
Laurie,Barbara
Cyg 20h33'58"38d43'
Laurie,Diane Frances
Uma 10h14'46"52d32'
Laurie,Julie & Rodney
Eri 4h53'21"-7d18'
Laurie,Neva
Lyn 8h54'34"44d48'
Laurila,H & P
Cyg 21h13'46"28d58'
Laurilania
Aql 19h30'45"12d12'
Laurisch,Tyler James
Her 18h1'20"38d21'
Laurita Marie
Lyn 7h28'55"42d38'
Lauritano,Robin Marie Constance
And 1h1'36"35d59'
Lauritson,Kasey Marie
Mon 6h24'60"0d0'
Laurna & Paul
Cyg 20h20'1"40d34'
Laurrell 93345
Umi 15h41'54"67d42'
Laursen,Carly Marie
Eri 2h45'35"-6d8'
Laursen,Julie
Del 20h18'28"13d10'
Laursen,Søren
Aur 5h12'25"29d54'
Laury
Lyr 19h0'60"30d48'
Laury
Ori 5h55'43"19d13'
Lauryn's Friendship Star
Lyr 18h56'59"30d53'
Lausch,Adrian Leander Thassilo
Dra 16h23'57"58d22'
Lausch,Melissa
Lyr 18h39'12"42d1'
Lausi,Justin Pickering
Aql 20h1'38"13d20'
Lausike,Darmon
Lmi 10h34'32"31d34'
Lausmann,Vance
Cma 6h59'58"-19d37'
Lauson,Nicole
Cas 0h33'42"62d14'
Lausten,Andrew
Hya 9h11'1"5d4'
Lauten-Hayes,Monica
Tau 4h41'50"28d18'
Lautenbach,Daniel James
Aur 4h48'59"50d42'
Lautenschlager,Harley -N- Elizabeth
Cas 1h25'19"64d58'
Lautenshlager,Rene Alden Yumi
Peg 22h7'28"27d44'
Lauter,Alfred
Aql 19h55'1"10d56'
Lauterbach,Gerd
Cmi 7h19'35"5d14'
Lautner Irrigation, Inc
Per 3h15'0"41d2'
Laux,Ernest & Evelyn
Cam 4h59'16"69d50'
Laux,Jeffrey J
Her 17h48'37"14d46'
Laux,Mathew James Christopher
Lac 22h19'1"54d24'
Lauxmann,Axel
Cep 22h43'13"57d23'
Lauxmann,Kristof
Lib 15h12'10"-20d19'
Lauzon (Canada),Gaëtan N
Peg 22h38'44"21d59'
Lauzon,Cheryl
Mon 6h38'60"3d34'
Lavado,Janylove Espinosa
Aur 7h22'1"43d25'
Lavallee,Caitlin
Cas 0h49'31"75d38'
LaVallee-Davidson
Cyg 20h16'20"41d44'
Lavallée,Hans-Hugo
Boo 14h3'0"10d28'
Lavane
Cas 0h53'43"67d0'
LaVangie,Mark S
Dra 13h1'22"64d15'
LaVelle
Umi 16h29'39"70d21'
Lavelle,Jr,Glenn Thomas&Glenn Thomas,Sr
Uma 10h46'41"51d49'
Lavelle,Robin
Sgr 18h55'13"-24d49'
Lavely,Mark & Nadine
Umi 7h13'55"88d21'
Lavender Ladie's & Gent's
Lyr 19h19'1"41d48'
Lavender's Love
Ori 5h56'40"16d8'
Lavender,Karen Plischke
Cas 0h35'48"66d55'
Lavender,Rick
Boo 14h14'30"38d13'
Lavendo
Vir 14h6'0"-8d13'
LaVenture,"Mr Ed"
Per 1h48'0"53d3'
LaVerde,Aaron Joseph
Lac 22h14'56"48d55'
Laverdiere,John
Per 4h37'0"37d38'
Laverdière,Liliane
Uma 10h14'22"58d46'
LaVergne,Kyle Stephen
Peg 0h12'37"18d6'
Lavergne,Léo
Cyg 19h29'60"35d8'
Laverne,Mae
Uma 8h37'52"57d54'
Laverne
Aql 19h7'1"3d36'
Laverne
Mon 7h44'27"-3d24'
LaVerne (JJ) Mae
Sco 17h27'20"-40d58'
Lavernock Point
Cyg 21h32'1"42d11'
Laverty,Wayne J
Dra 15h39'49"58d54'
Lavery
Cas 1h14'16"60d3'
Lavery,Charles
Uma 8h49'1"56d26'
Lavery,Joan Carol
Mon 6h36'1"-0d20'
Lavery,Jonathan Thomas
Leo 10h33'33"15d51'
Lavey,Elliott B
Aql 20h4'13"1d34'
Lavezzo,Richard
Her 16h19'31"5d27'
Laviani,Adeline
Tau 5h20'48"16d32'
Laviani,Dolores
Vir 11h39'40"3d26'
Laviani,Mary
Sco 16h37'51"-31d31'
Lavieille,Estelle
Cam 4h31'45"58d16'
Lavigna,Michael P
Cas 0h33'30"62d36'
LaVigne,Amanda Katherine
And 0h8'35"46d40'
Lavigne,Ameuf
Sgr 20h17'51"-31d28'
Lavigne,Andrée
Ori 5h57'12"16d39'
LaVigne,Ian Wayne
Cep 21h9'42"61d22'
Lavigne,Pierre
Cyg 21h37'18"38d6'
Lavin,Reid Alexander
Peg 22h30'36"27d8'
Lavin,Theresa Ellen
Peg 23h35'11"31d40'
Lavina Jaan
And 0h32'28"45d23'
Lavine,Sean Paul Francis Mohandas
Uma 11h46'38"33d21'
Lavinia
Lyn 8h15'45"58d19'
Lavinia
Pup 7h57'0"-29d26'
Lavino,Henry Dixon
Boo 14h1'1"26d13'
Laviola,John Carl
Aur 6h12'1"46d47'
Lavis
Vul 20h21'60"25d12'
Lavis,Mike
Uma 11h32'12"50d58'
Lavish,My Star,Sharon
Cas 0h53'10"68d42'
LaVista,New Canaan Summer Stars Tony
Boo 14h37'16"9d30'
Lavlle,Lois
Aql 19h29'40"-1d1'
Lavoi,Greg
Hya 10h40'0"-18d2'
Lavoie,André
Ori 4h54'22"0d42'
Lavoie,Denys
Lyr 18h34'59"34d11'
Lavoie,Étienne
Lyn 9h2'1"34d19'
Lavonne
Eri 4h7'53"-16d58'
Lavorerio,Matthew Charles
Lac 22h18'36"46d57'
Law & Darcey
Sge 20h5'21"20d6'
Law & St Evelyn
Aur 4h58'20"34d17'
Law,Amy
Eri 4h9'28"-10d32'
Law,Beth Ann
Ori 6h0'59"1d11'
Law,Debby Joy
Vul 20h15'20"28d23'
Law,Douglas Watson
Her 16h15'47"4d47'
Law,Fredy
Uma 11h50'14"31d58'
Law,Graham
Per 3h15'12"40d9'
Law,James "Jay" Arnold
Oph 17h9'36"-23d59'
Law,James Roland
Oph 17h50'20"13d15'
Law,Jude
Mon 6h27'56"8d59'
Law,Julice Bowen
Peg 22h42'54"10d11'
Law,Kathy
Cet 0h6'45"-12d10'
Law,Kristen Munson
Cmi 7h57'43"5d23'
Law,Scott
Ori 5h44'54"10d19'
Lawanda Rose
Cet 1h34'45"-5d15'
Lawatsch,Wilfried
Aqr 21h58'33"-10d53'
Lawburgh,Bryce Eric
Cyg 21h13'0"39d52'
Lawes,Desmond Staubin
Per 2h55'18"35d34'
Lawler,James
Per 2h6'1"56d38'
Lawler,Kaitlyn Francis
Peg 0h12'0"20d27'
Lawler,Philip
Her 17h4'32"30d22'
Lawler,Thomas Lewis
Her 17h16'0"40d12'
Lawles,Lynn
Mon 6h55'41"-6d15'
Lawless,Shawn
Aur 6h1'50"45d16'
Lawley Star,Lucky
Lyr 18h31'0"37d5'
Lawlor,Caitlin Elizabeth
Com 12h15'59"32d46'
Lawlor,Jr,Michael
Lac 22h13'0"55d15'
Lawlor,Ken & Leeann
Vul 19h15'37"22d24'
Lawman,Mary Lyn
Boo 14h43'40"45d24'
Lawn,Ashley Ann
Aql 19h0'14"13d10'
Lawn,Bob
Sco 16h51'0"-38d3'
Lawpel
Hya 8h33'0"1d14'
Lawrang
Uma 9h6'28"54d44'
Lawren's Place
Mon 8h6'20" 1d10'
Lawrence
Her 15h57'27"45d41'
Lawrence
Del 20h25'1"10d0'
Lawrence
Dra 11h46'57"67d57'
Lawrence
Aur 5h14'22"41d40'
Lawrence Alexander
Psc 0h56'13"32d18'
Lawrence Dean
Cep 21h55'46"61d3'B
Lawrence ET Angela Fur Immer
Cyg 19h34'1"35d53'
Lawrence Family
Lac 22h3'13"50d31'
Lawrence Robert
Sco 17h54'51"-30d38'
Lawrence's Star
Per 1h51'24"56d52'
Lawrence, Irene, Christina Auriana
Umi 14h30'41"68d20'
Lawrence,Adam David
Boo 15h18'12"38d12'
Lawrence,Alexandria Elizabeth
Mon 7h42'36"-2d56'
Lawrence,Amanda S
Lyr 18h13'18"38d28'
Lawrence,Andrew
Her 18h31'32"17d41C
Lawrence,Ann G
Lyr 18h40'54"32d24'
Lawrence,Ann-Marie
Cas 1h46'0"75d52'
Lawrence,Anthea Mary
And 0h16'11"39d37'
Lawrence,Barbara
Cas 0h25'33"50d12'
Lawrence,Barry
Per 3h39'34"38d1'
Lawrence,Brett A
Cep 22h25'15"59d26'
Lawrence,Brian
Dra 11h51'48"68d58'
Lawrence,Caitriona
Cas 0h43'36"74d3'
Lawrence,Curtis Dale
Aql 19h57'58"11d6'
Lawrence,Dave
Lyn 7h46'58"39d35'
Lawrence,David Matthew
Boo 14h3'0"22d31'
Lawrence,David Matthew Lennon
Hya 8h25'58"-2d21'C
Lawrence,Douglas & Tammy
Cyg 21h17'11"38d32'
Lawrence,Gary
Aur 5h14'35"40d16'
Lawrence,Geoff & Cathi
Crb 16h6'27"34d18'
Lawrence,Gertrude
Ori 4h42'22"0d15'
Lawrence,James
Cep 21h1'19"63d55'
Lawrence,Jean Karen Lennon
Hya 8h25'58"-2d21'A
Lawrence,Jeremy John
Uma 8h38'55"57d42'
Lawrence,Joey
Her 18h31'32"17d41A
Lawrence,John David
Lac 22h53'52"55d9'
Lawrence,Kayla Marie
Peg 22h47'55"3d36'
Lawrence,Kelly Alisa
Hya 8h25'58"-2d21'B
Lawrence,Lee John
Aur 5h3'22"40d19'
Lawrence,Linda & Robert
Uma 10h49'1"51d23'
Lawrence,Linda Kay Shanklin
Sge 19h1'21"21d4'
Lawrence,Martha C
Lyr 18h52'29"40d33'
Lawrence,Mary
Cas 1h8'35"60d44'
Lawrence,Matthew
Boo 14h51'38"38d10'
Lawrence,Matthew
Her 18h31'32"17d41B
Lawrence,Nichola Sharon
And 23h16'1"40d57'
Lawrence,Ninon Dupras
Peg 22h18'0"3d25'
Lawrence,Patrick
Cep 22h35'15"58d17'
Lawrence,Philip
Ori 5h34'53"-0d11'
Lawrence,Richard Gale
Cam 5h49'13"58d44'
Lawrence,Robert William
Cep 23h5'45"60d44'
Lawrence,Robert Joseph
Her 18h9'32"31d13'
Lawrence,Ronald Louis
Tri 2h0'46"33d38'
Lawrence,Ryan
Lac 22h17'16"50d14'
Lawrence,Sandra & Carl
Cyg 19h23'51"30d7'
Lawrence,Sarah
Mon 6h42'57"7d44'
Lawrence,Shaun & Kimberly
Eri 2h48'20"-6d9'
Lawrence,Sheldon
Ser 16h5'16"13d11'
Lawrence,Sonia
Per 3h15'30"41d17'
Lawrence,Sr,Lamon
Aql 20h1'39"12d56'
Lawrence,Stacye Kaye
And 0h45'37"33d36'
Lawrence,Steven Todd
Per 2h59'0"34d8'
Lawrence,Susan
Lyr 18h59'56"29d44'
Lawrence,Susan
Vul 20h15'0"25d12'
Lawrence,Susan-Deborah Leigh
Mon 6h54'15"-10d23'
Lawrence,Susie
And 2h1'35"45d19'
Lawrence,Thomas Joseph
Ari 2h30'1"20d11'
Lawrence,Timothy W
Per 2h36'57"45d11'
Lawrence,Tracy
Crb 16h6'11"30d35'
Lawrence,William Jackson
Lac 22h14'40"49d33'
Lawrence,William "Bill"
Oph 17h15'10"-21d38'
Lawrence-Jones,Sophie
Cyg 19h29'46"39d27'
Lawrence-Molloy,Colin Kenneth Neville
Equ 21h20'41"11d1'
Lawrencejoy Star,The
Uma 8h46'23"52d51'
Lawrencio I
Lmi 10h46'22"27d5'
Lawrenz,Dean
Cet 1h4'36"-3d14'
Lawruk,Nickolaus
Per 4h20'20"50d58'
Laws,Hannah Alice Cook
And 0h3'35"31d13'
Laws,Helen
Cas 0h24'1"68d24'
Laws,Irene Evelyn
Uma 9h44'43"55d32'
Laws,Mary & Harry
Cyg 19h28'59"38d14'
Laws,Michael Liam
Cep 21h23'1"58d49'
Laws,Patricia A
Cam 5h5'50"68d35'
Laws,Patricia Ann
Peg 22h5'44"25d58'
Laws,Sr,Carl Robert
Uma 8h55'54"55d27'
Lawson
Boo 14h11'33"50d37'
Lawson Most Radiant Star,Hal A
Her 18h20'35"28d25'
Lawson,Amy L
Cyg 20h55'44"30d4'
Lawson,Andrea & James
Lyr 19h18'41"38d52'
Lawson,Anna B
Lyr 18h31'27"30d46'

Lawson,Barbara
Del 20h13'59"15d32'
Lawson,Brigitte
Cam 3h25'0"59d46'A
Lawson,Celia
Mon 7h48'45"-5d13'
Lawson,Dawn Marie
Tri 1h48'22"28d46'
Lawson,Dorothy
Mon 7h21'45"-1d7'
Lawson,Forest & Rose
Vul 19h46'34"23d5'
Lawson,Heather Lynn
And 0h12'0"46d26'
Lawson,James D
Per 2h35'29"56d28'
Lawson,Jennifer
Com 12h17'28"19d50'
Lawson,Jerry Wayne
Cam 3h15'36"61d11'
Lawson,Jess
Mon 7h0'1"8d39'
Lawson,John C
Her 16h54'33"38d23'
Lawson,Jonah Steven
Lac 22h28'54"53d11'
Lawson,Joseph James
Crb 15h18'25"32d50'
Lawson,Juanita I
Cyg 20h9'58"40d33'
Lawson,Katherine
Lyr 18h44'13"39d8'
Lawson,Kendell & Elizabeth
Cyg 20h52'48"38d51'
Lawson,Lillian
Vul 21h27'38"24d16'
Lawson,Mark A
Cyg 19h55'23"37d41'
Lawson,Mary R
And 0h6'36"43'38'
Lawson,Matthew T
Boo 14h50'30"35d31'
Lawson,Melissa Kaye
Lyn 7h49'20"36d56'
Lawson,Merritt Allen
Lac 22h15'57"49d22'
Lawson,Michael Sean
Cyg 20h7'22"39d38'
Lawson,Noah Thomas
Boo 14h16'57"16d40'
Lawson,Steven Robert
Eri 3h7'1"-5d59'
Lawson,Suzanne "Sweetheart"
Lyn 7h10'7"48d35'B
Lawson,Thomas M
Dra 18h25'1"80d28'
Lawson,William L
Cep 22h18'57"60d20'
Lawson,Yvonne & F H (Rudy)
Eri 4h30'57"-8d41'
Lawton,Courtney
Mon 6h20'15"0d36'
Lawton,Daniel James
Lac 22h27'36"37d43'
Lawton,Elizabeth Jean
Cas 1h22'28"53d40'
Lawton,Jacob Andrew
Cas 0h57'26"67d37'
Lawton,Julie
Mon 7h0'40"8d55'
Lawton,Sherry
And 0h7'16"35d26'
Lawton,Stephen Thomas
Peg 21h58'48"28d1'
Lawton,Wonderful Parents
Gerald & Joan
Crb 16h14'10"37d40'
Lay Worthy
Vul 19h41'1"26d16'
Lay,Julie Ann
Umi 15h3'29"68d33'
Laybourn,Nicole Marie
And 2h35'34"40d36'
Layburn,Margaret Rhome
Cas 1h8'25"61d18'
Layden,W H
Boo 14h38'51"30d4'
Laydon,Clara-May Mercedes
Aql 19h4'49"5d14'
Layer
Lac 22h39'56"56d42'
Layer,Joy D
Uma 10h1'1"71d51'
Laygue,S
Cam 5h54'13"61d1'
Layla
Cam 5h59'1"61d27'
Layla Ann
And 0h39'28"40d38'
Layla Francesca
Lyn 8h6'18"50d42'
Layman,Barbara
Com 13h1'43"22d7'
Layman,Frederick- Petronella
& Larry
Oph 18h2'56"11d21'
Layman,James
Cam 3h31'29"60d34'
Layn,Susan M
Cru 12h37'12"-63d10'
Layn,Toni Michelle Elizabeth
Ori 5h54'52"21d11'
Layne
Umi 15h30'27"65d60'
Layne
Tri 2h30'30"31d29'
Layne Marie
Oph 18h2'40"12d48'
Layne,Alberta Marion
Cas 1h58'30"77d12'
Layne,Sydney Taylor
Cep 22h54'56"57d45'
Layne,Tory
Mon 7h41'13"-3d24'
Layton
Lyr 19h26'13"38d12'
Layton,Brooke Elizabeth
Mon 8h0'20"-8d4'
Layton,Katy Marie
Peg 22h40'24"20d24'
Layton,Marlene
Lac 22h6'51"51d31'
Layton,Max Alexander
Dra 20h0'39"68d39'
Layton,Peter
Sge 20h17'30"16d13'
Layton,Star David Ross
Per 4h1'51"50d29'
Laz,René
Eri 3h32'24"-2d17'
Lazaar,Louis B
Cas 0h31'38"64d42'
Lazar,Aline-Nicole
Lib 14h40'16"-6d33'
Lazar,Bev
Cyg 19h43'20"30d16'
Lazar,Jason A
Per 3h46'40"35d35'
Lazar,John
Aur 5h39'10"50d30'
Lazar,Robert & Kellie
Cam 3h13'23"61d35'
Lazar,Vera
Cam 9h13'35"73d39'
Lazar-Mr West-"Mr Apple
Pie",Aaron
Boo 15h12'31"28d5'
Lazarcheff,Mason James
Vul 19h57'43"28d22'
Lazare's Shining Star, Paul
Per 3h14'56"40d55'
Lazare,Paul
Aur 7h15'51"41d25'
Lazaretti,Margaret Louise
Dra 11h51'24"72d12'B
Lazaretti,Raymond Frank
Dra 11h51'24"72d12'A
Lazarin
Cet 1h52'0"0d39'
Lazarow,Linda Lee
Sco 16h55'34"-38d3'
Lazarson,Rosalie
Cas 23h31'55"61d19'
Lazarte,Adriane
Cnc 8h38'0"18d17'
Lazarus Legacy,The
Sco 17h54'14"-30d20'
Lazarus,Aaron Jacob
Gem 7h4'52"28d31'
Lazarus,Corinne & Lou
Crb 15h50'0"38d9'
Lazarus,Ethan Benjamin
Cap 20h29'33"-26d3'
Lazarus,Jordan
Lyn 8h5'27"43d50'
Lazarus,Lois Jane
Tau 4h26'52"28d57'
Lazarus,Marysia
And 23h28'39"42d6'
Lazarus,Monte
Tau 4h24'0"28d55'
Lazarus,Taylor
Aql 19h33'29"0d6'
Lazenby,Gail
Uma 10h52'14"53d8'
Lazenby,Jeff S
Her 17h14'44"47d43'
Lazenby,Lorrie
Cas 0h30'31"75d14'
Lazers Family Star,The
Hya 8h26'15"-11d3'
Lazio,Frank Paul
Boo 13h54'0"20d2'
Lazo,Debe
Mon 6h54'16"1d38'
Lazy J Acres
Cnv 12h30'35"38d2'
Lazy Liza
Lyr 18h36'41"37d57'
Lazy S O B Ranch "Satellite
Branch",The
Lac 22h1'23"47d31'
Lazzara,Rita Maria
Uma 8h47'1"71d28'
Lazzari,Richard
Per 2h52'1"32d27'
Lazzarini,James A
Aur 7h2'52"36d34'
Lazzaris e Manu Molini Max
Col 6h33'6"-34d47'
Lazzeri,Patty & Jon
Cyg 19h33'11"31d41'
LB Sweet Memorys Future
Hope Love N
Her 18h42'29"12d43'
LB Valentine 93
Lyn 8h4'56"47d52'
LDMV-1
Uma 11h9'34"61d28'
LDS
Lyn 7h50'0"48d20'
Le Baigue,Linda
Cyg 20h19'47"38d58'
Le Baron
Peg 21h26'26"10d19'
Le Bastard,Ludovic
Ser 15h42'46"2d24'
Le Batard De Dieli
Cyg 20h56'1"31d19'
le Bec Hellouin, Edouard
Umi 16h36'27"78d60'
Le Bihan
Peg 22h28'51"20d7'
Le Bihan-Capelle, Christophe
Umi 13h56'47"76d25'
Le Blevec,Philippe
Cyg 19h52'30"47d53'
Le Boulanger 6/19/64, Sabine
Lyn 8h28'37"39d2'
Le Brasseur,Guy
Peg 23h30'16"15d26'
Le Cam,Anne Gaëlle
Sex 10h12'57"-3d50'
Le Champagne,Philippa Aime
Lyn 7h51'34"36d6'
Le Clainche,Luc
Peg 21h58'30"11d11'
Le Compte,Donna
And 1h3'14"40d2'
Le Conte,Sandie Coppola's La
Contessa
Ari 2h37'0"21d50'
Le Couer D'isa
Uma 10h44'31"60d28'
Le Deux Cour
Ori 5h7'18"0d51'
Le Dieu,Bernard
Umi 16h24'42"71d7'
Le Diouron,Régine & Alain
Guillaume
Dra 15h42'0"53d44'
Le Duc,Iain & Alexis
Cep 4h8'12"80d15'
Le Dep 4h8'12"80d15'
Ori 5h57'23"8d22'
Le Fave II,William Barry
Crt 11h30'60"-10d12'
Le Fave,Patricia Kristine
Eri 2h56'49"-12d20'
Le Faverollet,Robert
Cas 1h4'51"65d7'
Le Fevre,Marius
Cnv 13h18'0"50d47'
Le Gousse,Veronique
Del 20h20'30"16d31'
Le Grain Josette
Lac 22h3'13"38d59'
Le Grys,Val & Phil
Cyg 20h34'14"40d13'
Le Grégoire
Cma 7h12'45"-13d19'
le Guenic,Didier
Cnv 13h21'11"38d8'
Le Guillou,Bertrand
Lac 22h43'21"38d42'
Le Gurun,Monique
Peg 21h9'19"12d16'
Le Lorain,Claude
Cyg 19h38'1"38d15'
Le Marrec,Laurence
And 0h16'58"40d11'
Le Moal,Alexia
Ser 15h17'33"11d36'
Le Montelli
Del 20h17'1"10d29'
le Page,Laurence
Per 3h29'43"39d27'
Le Petit Prince
Per 2h53'48"46d15'
Le Petit Prince Nicolas
Umi 16h3'0"72d42'
Le Petite Fleur
Cas 1h43'27"68d44'
Le Pèlerin-18/08/1946,
Michel
Her 17h13'29"20d25'
Le Quesne,Sukica
Aql 19h5'46"1d30'
Le Rhun,Ghislaine
Dra 15h37'36"60d29'
Le Scanff
Her 17h24'0"41d5'
Le Shuttle
Lyn 7h26'13"51d33'
Le Star De Sam
Ori 5h24'24"0d53'
Le Subelight
Cam 5h52'20"80d31'
Le Tallec,Michele
Ser 18h29'14"0d53'
Le Tellier,Marta W
Crb 15h38'0"28d40'
Le Tellier,Michelle M
Cyg 20h33'47"37d33'
Le Tutour,Martine
Cyg 19h44'58"37d45'
Le Uierstra
Mon 7h43'48"-1d19'
Le Vaillant,Sabrina
Del 20h36'16"7d46'
Le Vines' Eastern Star Mary
Sex 9h52'50"4d2'
Le Vision du Cendre
Cas 1h6'0"60d44'
Le Vu Thi Kim-Chi
Ser 15h20'59"2d30'
Le Winter,Nancy
Vul 19h42'40"26d37'
Le,Duke
Dra 18h59'28"50d17'
Le,Madeleine Thu-Hang
Equ 21h21'41"2d18'
Le,Tiffany
Cet 1h59'45"0d32'
Lea
Vir 13h34'51"-14d2'
Lea
Cam 13h10'24"77d13'
Lea
Vir 14h38'33"-5d42'
Lea & Simon
Cyg 20h25'0"38d43'
Lea A
Her 16h24'0"48d21'
Lea Ann
Crb 15h30'14"38d43'
Lea Loves Doug
Boo 15h18'19"50d24'
Lea,Kathryn Jane
Cas 0h50'16"73d19'
Leach (Gray's Jack), John
Cep 22h4'10"60d7'
Leach,Adrienne marie
Oph 18h4'1"12d57'
Leach,Agnes Irene Blake
Borough
Mon 7h53'20"-6d6'
Leach,Allan
Per 4h32'40"31d47'
Leach,Amanda Leigh
Cas 3h9'22"73d59'
Leach,Anita
Cyg 21h3'14"37d30'
Leach,Carlee Patricia
Cas 23h33'25"60d30'
Leach,Dennis O'Keefe
Cet 2h51'16"5d5'
Leach,Eirik
Her 18h10'16"38d56'
Leach,Eleanore Krajewski
Lyn 7h57'38"44d11'
Leach,Everett Leland
Cam 6h18'12"65d32'
Leach,Gary
Cnc 8h39'36"17d28'
Leach,Ian Miller
Cma 7h1'43"-16d16'
Leach,Jeanne Shaw
Cas 0h37'14"68d22'
Leach,Robin
Dra 17h27'1"68d48'
Leach,Sarah
Uma 9h20'18"57d22'
Leach,William M
Ser 15h28'45"21d36'
Leachman,Jamie Hooper
Aql 20h34'18"0d24'
Leacy,Heather J M
Cas 23h12'29"60d8'
Leadbeater
Oph 18h42'52"10d13'
Leader,Connie J
Lyn 8h17'14"39d37'
Leader,Jordan Colby
Boo 14h8'22"22d55'
Leader,MaryJane Strickler
Cyg 21h7'47"30d57'
Leaf,Dale "Pete"
Dra 17h21'39"71d37'
Leaf,N C
Del 20h15'38"10d4'
Leaf,Robert
Lac 22h12'36"49d51'
Leago,Betty
Cyg 21h48'31"36d34'
League,Richard Edmund
Her 17h27'16"28d56'
Leah
Peg 22h12'56"10d45'
Leah
Eri 3h27'27"-2d41'
Leah
Peg 22h40'15"22d55'
Leah & Chip's Wedding Star
Crb 15h30'38"32d24'
Leah Anita
Sge 19h38'41"16d18'
Leah CXLIII
Hya 8h18'18"5d14'
Leah Faith
Cas 1h33'22"60d40'
Leah Jane's Eyes
Vul 19h43'0"22d53'
Leah Marie
Ori 5h27'27"-3d60'
Leah Marie
Sex 10h0'34"1d26'
Leah Michelle
And 23h40'27"38d59'
Leah Of Yoda
Cnc 8h34'32"7d27'
Leah René
Lyn 8h5'26"38d5'
Leah Scarlett
Cas 1h2'59"58d49'
Leah Star Shamana
Tau 3h36'24"24d44'
Leah's Light
Cyg 21h32'53"33d45'
Leah's Star
Mon 7h39'14"-3d3'
Leah's Star
Peg 23h20'52"25d28'
Leahey,M S
Lyn 9h7'36"34d35'
Leahy,Elisabeth V
Cet 1h29'39"-11d16'
Leahy,Joseph M
Lib 15h32'39"-11d11'
Leahy,Jr,Sidney M
Dra 13h41'11"63d26'
Leahy,Michael Richard
Hya 8h14'41"0d32'
Leahy,Pat & LeAnn
Uma 10h11'42"53d2'
Leahy,Patrick
Cet 0h44'14"-6d36'
Leahy,Paula
Peg 23h2'1"21d3'
Leak,Kyle J
Hya 9h34'37"5d39"
Leal,Annette M
Cas 1h39'57"59d30'
Leal,Mr Alan
Uma 8h19'25"68d29'
Leallyn Fay
Peg 22h42'10"30d59'
Leaman,Christopher M
Aur 6h42'11"35d56'
Leaman,Jeanne
Her 17h23'30"41d37'
Leaman,Tom
Her 15h50'1"44d33'
Leamon,James David
Dra 14h57'1"65d8'
Leamy,Ginny
Psc 1h26'56"22d34'B
Lean,Frederick D
Ori 5h18'21"1d50'
Leana
And 1h45'1"38d41'
Leanda,Jane
Car 7h31'54"-51d36'
Leane
Uma 9h31'19"61d37'
LeAnJaZé Dance Family
Mon 6h28'16"7d26'
LeAnn
Del 20h20'12"10d26'
Leann
Lyr 18h24'43"38d42'
Leanna-Brooks
Psc 0h20'58"20d36'
Leanne
And 0h31'30"28d8'
Leanne
And 0h1'1"46d55'
Leanne
Sex 10h29'55"-0d57'
Leanne 22
Vir 11h36'10"2d3'
Leanne's Wish Upon A Star
Cyg 20h19'23"31d1'
LeAnne,Hollie
Lyr 18h54'36"31d42'
Leanora
Mon 6h23'41"3d6'
Leanos
Cyg 21h51'14"40d50'
Leanza's Light,Laura
Lyn 9h0'0"45d9'
Leanza,Lillian Karin
Lib 15h16'49"-23d8'
Leap Of Faith
Vir 11h55'25"-6d13'
Leap Star 1 Sax
Tri 2h36'24"35d40'
Leap,Jennifer Margaret
Lyr 19h5'36"38d23'
Leaper,Percy F
Lac 22h27'34"54d11'
Lear's Star
Umi 15h27'0"81d27'
Lear,Brianna Elizabeth
Peg 22h56'2"9d5'B
Lear,Carol
Dra 19h5'3"59d12'B
Lear,Donna
Cas 2h56'59"73d12'
Lear,George William
Per 2h0'0"57d53'
Lear,Harris Edward
Oph 18h31'1"10d19'
Lear,Madeline Rose
Peg 22h56'2"9d5'A
Lear,Nicholas
Cep 23h10'26"61d22'
Lear,Peter David
Aql 19h0'58"13d18'
Lear,Susan
Gem 6h15'10"25d56'A
Lear,William Alpheus
Per 3h12'47"45d21'
Leardini,Andrew
Vul 19h46'50"28d51'
Leardini,Christopher
Cep 22h41'44"58d53'
Leardini,Matthew
Lyn 8h28'40"51d50'
Learman,Trudy & Dave
Dra 18h56'16"58d25'
Learnard,John N & Mary
Perkins
Ori 4h59'19"13d39'
Learoyd,Cathryne Ann
Cyg 19h59'1"47d42'
Leary's SYBRON
Vul 20h20'49"28d46'
Leary,Derek
Uma 10h37'0"68d46'
Leary,Eddie
Aql 19h30'1"12d3'
Leary,Edward
Ari 2h59'40"28d10'
Leary,Janet Elizabeth
Cas 0h36'18"62d21'
Leary,Jean
Lyr 18h32'15"35d47'
Leary,Jessica Lynn
Cas 23h31'0"61d25'
Leary,Monica
Cas 1h13'0"61d30'
Leas...Soul Mates, Robin &
Annette
Cyg 21h17'52"37d44'
Lease,Baby Boy
Per 3h56'46"52d20'
Lease,Pamela Sue
Uma 10h11'43"57d12'
Leasing Dynamics Inc
Ori 6h5'34"8d4'
Leath,Eric Cordell
Per 2h53'17"40d34'
Leath,Jordan
Aur 6h31'55"31d55'
Leatham,Donald Lee
Her 17h30'48"27d37'
Leather,Donald
Dra 16h53'1"61d04'
Leather,Hilary & Peter
Cyg 21h58'23"53d45'
Leatherland,David
Lyn 8h2'0"44d10'
Leatrice
Peg 21h55'0"23d44'
Leauitt 1st Anniversary
Star,The
Cyg 19h21'35"44d41'
Leaumont,Cyndy & Noel
Sge 20h0'34"18d17'
Leavell,Connie Rae
Aql 20h17'42"8d6'
Leaver,Grandad's Star- Richard
Cyg 21h2'1"37d10'
Leaver,Todd Jeffery
Per 2h22'11"58d54'
Leaverton,Thelma "Zelm"
And 23h26'58"38d38'
Leavey,Susan Ann
Sgr 19h24'6"-40d52'
Leavey,Susanne Teresa
Sco 17h27'49"-30d50'
Leavid
And 22h57'53"40d22'
Leaving Norcross
Vul 19h42'0"20d1'
Leavitt,Briant Ralph
Aql 18h58'51"11d18'
Leavitt,David Charles
Lac 22h8'46"40d28'
Leavitt,Derek
Cep 22h52'1"70d10'
Leavitt,Hugh Lewis
Per 3h3'45"46d58'
Leavitt,Jessica Nova
And 23h35'52"39d46'
Leavitt,Lawrence of Arabia
DBA Michael
Ser 15h20'23"2d38'
Leavitt,Shawna
Vul 19h47'53"25d24'
Leavy Love-Saved The Best
For Last
Uma 9h11'20"60d14'
Leaza,Valentina
Peg 21h52'54"33d53'
Lebard,Marie
And 0h18'18"40d38'
LeBarre,Charley
Dra 20h23'1"80d14'
Lebarrois,Marie-Pia
Oph 17h58'59"10d15'
LeBarron
Aur 6h0'0"48d52'
Lebbad,Brenda
Lyn 8h13'53"58d12'
Lebeau,Chantel Claudette
Lib 15h40'51"-22d45'
Lebel "Wuuff",Jon D
Aur 6h6'21"32d30'
Lebel,Marie-Josée
Cas 0h20'0"73d35'
Lebel,Marie-Pierre
Cas 22h57'23"54d32'
LeBel,René
Umi 15h18'11"68d3'
Lebell,Judo Gene
Her 16h53'1"28d16'
Leben,Joshua Scott
Cep 21h44'1"63d52'
Lebental,Carole & Jack
Oph 17h6'44"11d15'
LeBer,Nicole Erin
Cam 7h35'35"61d6'
Leber,Robert Lloyd
Aur 6h21'0"37d49'
Lebert,Philippe
Aur 4h47'33"40d19'
Lebeuf,Dominique
Gem 6h51'33"13d34'
Lebkuechner,Wolfgang
Aur 5h5'0"43d20'
Leblanc
Umi 13h19'53"70d60'
Leblanc,Albert Yvan
Her 16h53'22"23d1'
Leblanc,Andrew Tyler
Cep 21h54'43"61d18'
LeBlanc,Curtis E
Aql 19h58'0"8d56'
LeBlanc,Denise & Michael
Uma 9h49'24"42d6'
LeBlanc,Dr P Dudley
Oph 17h18'51"10d39'
LeBlanc,Erin Elizabeth
Aql 19h13'36"19d41'
LeBlanc,James David
Cma 6h50'20"-18d28'
LeBlanc,Katherine Marie
Cas 0h49'32"68d19'
LeBlanc,Lorrie Ann
Cyg 21h20'46"39d27'
LeBlanc,Michael D
Dra 16h28'41"64d14'
LeBlanc,Mitchell T
Peg 22h44'30"33d3'
LeBlanc,Monique
Peg 21h43'43"24d38'
LeBlanc,Omer A
Peg 22h1'23"21d3'
Leblanc,Pierre
Per 3h7'36"40d15'
LeBlanc,Sheryl W
Lyn 7h19'24"58d33'
LeBlanc,Susan
Cyg 19h34'22"34d56'
Leblanc,Suzanne
And 23h18'21"47d16'
LeBlanc,Val Joseph
Cas 2h39'1"61d12'
Leblond,Romuald
Dra 16h11'25"60d21'
Leboeuf,Audrey Dominique
Sylvain
Cas 0h48'31"73d55'
Leboeuf,Kathleen
Uma 10h9'1"54d22'
Leboeuf,Paul Louis
Uma 11h22'0"38d55'
LeBoff,Jared Benjamin
Uma 10h13'31"52d2'
Lebot,Annie
Lup 15h12'36"-44d10'
LeBov,Leonard
Per 3h1'29"40d51'
Lebovitz,In Loving Memory of
Robert
Aur 7h23'33"38d57'
Lebow,Stephanie
And 2h29'51"41d38'
Lebowitz,Elise
Lyn 8h13'40"38d12'
Lebowitz,Marvin
Per 1h32'44"52d52'
Lebowitz,Stephen J
Cnv 14h0'37"38d34'
LeBrecht,Marcia
Eri 3h41'30"-13d36'
Lebreton,Lucien
Ser 16h6'39"7d49'

LeBreux,Gaëtan
Her 16h58'14"20d11'
Lebrun,Jean-Guy
Uma 8h56'53"47d17'
LeBrun,Vivien I
Tri 2h1'56"32d13'
Lebski,Helga Maria
Cnc 9h13'29"32d22'
Leca,Yann
Peg 23h3'14"31d44'
Lecallard,Eric
Per 3h56'47"38d18'
Lecavalier,Frédérick
Sco 16h33'11"-28d43'
Lecchini,Jean Pierre
Dra 12h35'31"70d4'
Lecerf,Lola
Uma 11h22'26"38d37'
Lech,Joyce R
Cas 1h25'47"61d6'
Lechardeur,Claude
Aql 18h57'48"3d39'
Lechler,Inge
Cas 0h40'30"60d8'
Lechner,Franziska
Aqr 22h40'34"-6d41'
Lechnir,Jake Laurence
Psc 22h50'56"2d8'
Leckie,Ann Marie & Michael
Lyn 6h57'46"48d55'
Leckrone,W Thomas
Oph 18h7'23"1d16'
LeClair
Aql 20h9'1"1d30'
LeClair,Joseph Henry
Boo 14h25'1"28d20'
LeClair,Nancy
Mon 6h15'50"-6d10'
Leclerc,Danielle
Uma 12h57'0"60d38'
Leclerc,Patricia
Dra 12h16'13"64d49'
Leclerc,Rusty
Aql 19h53'24"15d24'
Leclercq,Bruno
Ser 18h30'0"1d41'
Leclere,Albert & Evelyn
Dra 17h2'0"62d56'
LeCompte,Kitsy
Peg 21h41'39"21d45'
LeCompte,Patrica
Cas 23h28'1"53d15'
Lecompte,Serge
Per 2h57'18"52d9'A
Lecosse,Cyril
Cnv 12h50'35"40d34'
LeCouffe,W J
Hya 8h26'1"-8d44'
Lecrique,Pascal
Aql 19h29'28"7d51'
Lecry,Christian Brigitte
Uma 13h42'42"61d58'
Leda 27-6-72
Lep 5h23'27"-25d29'
Leda Maria
Cyg 19h22'14"44d29'
Ledah
Lyn 8h44'56"45d22'
Ledbetter's Star Wade & Susan
Cyg 19h47'11"29d25'
Ledbetter's Star,Paula Gay
Mon 7h52'45"-4d24'
Ledbetter,Alice M
Lyr 19h2'46"25d51'
Ledbetter,Emory-David
Lmi 10h32'12"32d14'
Ledbetter,Mark Wayne
Hya 9h4'34"0d40'
Ledbetter,Michael D
Lyr 18h37'19"41d5'
Ledbetter,Michael Paul
Equ 21h0'48"10d48'
Ledbetter,Sierra Brooke
Aql 19h55'42"15d2'

Ledbetter,Taylor Nicole
Eri 3h45'40"-5d19'
Ledden,Cathy Ann
Tri 2h23'42"34d47'
Ledder,Ted & Mildred
Uma 11h44'13"45d5'
Ledeboer
Uma 12h16'59"60d30'
Ledebrink,Hermann
Peg 23h34'40"12d14'
Ledell,Susan
Cam 5h19'1"68d60'
Leder,Laura Schur
Peg 23h25'50"11d15'
Lederer,Georg
Boo 14h30'31"20d9'
Lederer,John Henry
Her 17h8'53"48d53'
Lederer,Mary Regina
Lyn 8h46'37"36d29'
Lederman,Alex Abraham
Aur 5h3'0"31d52'
Lederman,Charles
Lmi 9h57'49"34d28'
Lederman,David Jay
And 0h4'28"38d1'
Lederman,Evan & Jill
Lac 22h12'56"37d59'
Ledes,Bayly
Lyn 7h54'1"41d6'
Ledes,John
Cep 22h50'21"58d14'
Ledesma 123137-011994, Richard
Cam 4h17'45"70d20'
Ledesma,Enrique
Aur 5h29'43"37d47'
Ledford,The Guiding Star Of Mitch
Ori 6h4'1"9d24'
Ledgerwood,Margaret
Lyr 18h55'33"40d17'
Ledgerwood,Stuart
Peg 22h0'23"32d43'
Ledgerwood,Tom C
Hya 8h10'51"0d1'
Ledig,Marc
Aql 19h55'30"13d20'
Ledingham,(Belle) Mrs Isabel
Uma 9h2'60"48d15'
Ledinsky,Susanne
Psc 0h17'1"11d3'
Lednicky,Allison & John
Peg 22h58'31"26d42'
LeDorothy Charlene
Cnv 12h16'13"45d57'
LeDoux,Chris
Cep 22h8'15"61d32'
Ledowsky's Light
Pup 7h26'17"-27d27'
Leduc,André
Uma 9h2'31"60d11'
LeDuc,Edward John
Aql 20h0'55"-1d36'
Leduc,Esther et Pierre Soucis
Cep 23h37'29"64d2'
Leduc,Ghislaine
Uma 8h59'41"59d23'
Leduc,Jacques
Cep 22h43'55"63d1'
Leduc,Pierre
Per 3h0'49"40d36'
Ledwell,Sharon E
Lib 15h27'34"-17d15'
Lee
Lyr 19h24'30"42d59'
Lee
Aql 20h15'52"5d17'
Lee & Dianne
Cyg 19h24'42"30d25'
Lee & Nancy's Corner Star
Uma 9h30'57"52d29'
Lee (God's Star Hope) Karen Darlene Parker
Vir 13h12'29"-8d29'

Lee Amerley Fio Amarteitio
Lyn 8h25'55"42d49'
Lee Ann
Mon 8h2'39"-8d54'
Lee Ann
Lyn 7h42'51"38d16'
Lee Ann
Ori 5h27'46"1d2'
Lee Ann
Cyg 21h39'14"37d34'
Lee Ann's Distant Guiding Light
Del 20h19'1"20d2'
Lee Ashley
Cmi 7h22'37"1d9'
Lee County AIDS Task Force
Ori 5h2'1"1d28'
Lee Dennis
Cyg 20h44'58"45d54'
Lee Forever,I Will Love Pamela
Peg 22h3'23"21d8'
Lee Kong Memorial Star
Cyg 21h11'22"34d42'
Lee Lee
Lac 22h35'0"56d27'
Lee Out Of This World Dad,Kenny
Vul 19h22'28"26d46'
Lee Star,The Simon (Sad Bastard Society)
Uma 8h42'34"57d45'
Lee William
Boo 13h52'1"21d42'
Lee's Dream
Crb 16h9'58"26d15'
Lee's Focus
Aur 7h22'0"37d7'
Lee's Lovelight
Aql 20h2'12"11d6'
Lee's Shining Star, George & Nona
And 23h19'48"43d28'
Lee,Annie
Tau 4h32'28"30d46'
Lee,Annie & Jack
Cyg 20h54'49"38d27'
Lee,Arthur D
Del 20h24'37"8d3'
Lee,Benjamin
Aur 5h3'22"38d8'
Lee,Bob & Lucy
Lac 22h4'47"40d15'
Lee,Bonnie K
Vul 19h47'27"28d12'
Lee,Booker T
Cet 1h51'36"-10d18'
Lee,Brian Yoon
Hya 9h5'47"3d55'
Lee,Catherine E
Peg 22h23'29"33d53'
Lee,Charlotte M
Aql 20h0'0"13d19'
Lee,Cherry
Vul 19h18'29"20d0'
Lee,Christopher Bruce
Ori 5h32'41"-1d49'
Lee,Cindy Marie
Lyr 10h33'10"30d50'
Lee,Claude J
Cnv 12h56'49"40d29'
Lee,Clinton Boyd
Aql 19h30'34"10d32'
Lee,Damon Andrew
Ser 15h18'31"11d42'
Lee,Dana
Lmi 10h17'52"30d38'
Lee,Daniel Steven
Ori 5h48'25"20d7'
Lee,Dave Eric & Allan
Cyg 19h58'38"30d50'
Lee,David Kien Ping
Sge 20h17'19"17d8'
Lee,Devon Ray
Uma 11h11'1"48d52'

Lee,Dolores H
Cas 1h34'33"60d58'
Lee,Douglas Charles
Ori 5h48'29"20d60'
Lee,Edwin H
Lyr 19h20'60"41d25'
Lee,Elizabeth Ann
Hya 9h0'28"2d23'
Lee,Eric Robert
Lac 22h41'0"56d1'
Lee,Erin Alexandra
Cyg 20h23'18"38d50'
Lee,Ester
Lyn 8h30'0"43d2'
Lee,Ethan Carter
Her 17h1'25"46d4'
Lee,Gene
Lyr 19h1'35"42d16'
Lee,Gertrude Electa D'Elosua
Cam 3h39'43"61d22'
Lee,Gil L
Dra 12h6'0"70d22'
Lee,Gina
And 0h47'1"39d17'
Lee,Helen May
Eri 3h55'1"-0d20'
Lee,Ian Lyp
Aur 5h1'43"50d14'
Lee,In Memory Of Linda
Mon 6h30'31"-6d2'
Lee,Jack Alan Colin
Cep 2h48'0"80d22'
Lee,Jacob Michael James
Her 17h35'12"27d33'
Lee,Jaime K
Uma 9h56'21"50d7'
Lee,Janene & Buddy
Cyg 19h56'44"50d1'
Lee,Janet
Ori 5h42'23"3d59'B
Lee,Jeanette Elizabeth
Aql 20h6'51"0d55'
Lee,Jeffrey K
Per 1h48'34"53d15'
Lee,Jennifer & Phil
Lyn 6h35'39"54d38'
Lee,Jennifer Brianna
Lyn 7h53'33"45d4'
Lee,Jennifer Lynn
Sge 19h53'59"19d21'
Lee,Jesse
Aql 19h55'1"14d45'
Lee,Jessica E
Uma 10h32'56"50d19'
Lee,Jet
Uma 10h2'24"71d23'
Lee,Jillian Noelle
Per 3h51'1"38d54'
Lee,Johnathan
Cet 1h34'16"-0d10'
Lee,Jonathan Richard
Lac 22h34'33"55d22'
Lee,Josephine
Mon 7h51'13"-2d3'
Lee,Joyce
Umi 15h59'1"77d45'
Lee,Jr,Scott Edgar
Ori 5h57'33"8d45'
Lee,Juan Miguel
Ori 6h17'0"-2d51'
Lee,Judith Kennedy
Cyg 21h32'48"40d15'
Lee,Juel
Crt 11h11'1"-14d47'
Lee,June Victoria
Crb 16h9'56"28d17'
Lee,Kara Adele
And 23h44'1"40d6'
Lee,Katelyn Patricia
Cas 0h30'35"60d19'
Lee,Kathleen Carrol
And 2h24'1"50d3'
Lee,Kelly
Cma 7h3'25"-16d49'

Lee,Kelsey Cody
Peg 21h54'35"34d39'
Lee,Ken
Uma 11h23'21"50d43'
Lee,Kevin
Her 18h7'51"30d46'
Lee,Kimi
Aql 20h3'1"0d46'
Lee,Kit Tzing
Psc 23h5'55"0d38'
Lee,Kori Reneé Kinslow
Uma 11h2'29"55d23'
Lee,Larry
Aur 5h22'1"37d59'
Lee,Laura
And 0h17'25"46d7'
Lee,Lenore
Aql 19h0'56"8d41'
Lee,Leo Tung Hai
Lyn 7h57'17"43d28'
Lee,Mabel & Y C
Lyr 19h0'23"26d43'
Lee,Margaret Ann
Cnc 9h8'43"33d4'
Lee,Margot Renee
Mon 6h56'29"-10d54'
Lee,Marissa Celeste
Sge 20h6'21"20d1'
Lee,Mark
Per 2h36'14"40d5'
Lee,Mark Avery
Aql 20h2'40"0d28'
Lee,Marty & Kristina
Hya 9h17'41"5d9'
Lee,Marvin J
Aur 6h35'41"32d44'
Lee,Mary Ann
Lyn 6h53'30"60d46'
Lee,Mary Barbati
Cas 1h36'58"67d46'
Lee,Matthew F
Ori 5h52'11"14d46'
Lee,Michael & Kasie
Cyg 21h50'1"42d9'
Lee,Michael R
Her 16h32'53"34d29'
Lee,Monika Lucia
And 1h32'16"36d57'
Lee,Ms Mona
And 1h40'27"40d41'
Lee,Nancy Jane McCleary
Peg 21h53'0"36d10'
Lee,Nettie
Cas 23h39'39"65d51'
Lee,Nick
Cyg 21h6'1"30d36'
Lee,Noni
Mon 7h6'23"-1d33'
Lee,Paul Royce
Sco 16h39'40"38d43'
Lee,Pearl
Mon 7h15'43"-0d33'
Lee,PhD,Dr Diane W
Lmi 10h28'36"28d60'
Lee,Princess Audrey
Vul 19h21'47"26d35'
Lee,R K
Peg 22h32'57"29d36'
Lee,Rachel Elizabeth
Mon 6h57'56"-10d53'
Lee,Robert Milne
Per 2h55'21"43d12'
Lee,Rose A
Uma 11h53'1"41d24'
Lee,S S,William J
Crb 12h50'32"20d6'
Lee,Sandra
Mon 6h54'29"-6d48'
Lee,Sandra Gayle
Ori 5h28'44"1d12'
Lee,Sarah Ann
And 0h2'10"47d48'
Lee,Scott Walter
Cam 5h50'4"70d21'

Lee,Shauna
Boo 14h22'23"28d42'
Lee,Soon Sung
Mon 7h0'39"0d42'
Lee,Special Wishes Forever-Tara
Ori 5h30'50"-0d57'
Lee,Stephanie Nicole
Eri 3h42'47"-15d9'
Lee,Sun Me
Boo 15h6'0"13d20'
Lee,Susan G
Sge 20h16'24"16d33'
Lee,Tawnie
Lyr 18h17'43"44d2'
Lee,Taylor Harrison
Her 17h39'1"21d12'
Lee,Toy
Cyg 19h31'21"36d28'
Lee,Tyler
Ori 6h0'1"1d3'
Lee,Van T
Vir 14h7'17"-7d37'
Lee,Wayne
Vul 20h15'26"23d53'
Lee,William M & Patricia E
Aql 19h30'35"-0d44'
Lee,World's Greatest Dad-Daniel
Boo 15h1'39"24d46'
Lee-Lee
Lyn 7h43'38"51d56'
Leeanddina
Cam 8h27'42"77d53'
LeeAnn
And 23h37'58"43d48'
Leeann
Eri 4h9'21"-19d33'
Leeann
Aql 19h53'17"11d19'
LeeAnna
Cas 0h57'49"50d6'
Leeanne
Uma 10h18'0"54d33'
Leeb,Dr Alvin J
Cep 23h13'0"60d24'
Leeb,Tanja
Cep 22h0'53"60d49'
Leeburg,Linn T
Aql 19h58'1"0d9'
Leece,Robert B
Dra 18h2'17"58d23'
Leech's Light Lovingly Illuminating
Psc 1h3'58"20d12'
Leech,Judith
Eri 2h58'13"-4d9'
Leech,Marlin
Mon 6h49'21"10d4'
Leeder,Hermann
Lib 14h30'0"-23d58'
Leedom,Craig Paul
Aur 6h14'27"37d52'
Leedom,Lester & Jean
Cyg 19h17'40"44d58'
Leeds,Ingrid
Tri 1h59'0"25d55'
Leeds,Joe & Shirley
Her 17h23'48"20d17'
Leeds,Karen Elizabeth
And 23h3'26"45d56'
Leeds,Naomi Rachel
Cas 0h36'44"58d10'
Leedy,Kathleen
Aql 19h32'57"-6d52'
Leef,Anna Josephine Dundon
And 1h16'0"41d5'
Leef,Thomas Marvin
And 0h32'7"34d41'
Leefe,William Robert
Dra 9h43'1"78d14'
Leeker,Tom
Ser 15h24'48"7d42'
Leela-Janakinath
Her 16h45'0"10d3'

LeelaBruce
Ori 6h0'34"-0d4'
Leeming,David Paul
Lyn 8h14'55"46d45'
Leeming,Kate Victoria
Cas 1h37'57"74d7'
Leeming,Taren
Lyr 18h33'23"38d1'
Leemon,Mark
Her 18h18'34"18d53'
Leenhouts,Kallin Harrison
And 23h37'27"47d30'
Leeny
Lyn 7h0'46"50d12'
Leep,William Reed
Cep 21h25'17"55d25'
Leeper III,John Edward
Sex 10h2'48"1d6'
Leeper,Audrey L
Vul 20h1'28"22d49'
Leerex
Mon 6h56'12"-10d11'
LeeRoy Loves Deanna
Ori 5h32'4"-2d6'
Lees,Cathryn Jane
Lyr 18h37'55"33d25'
Lees,John
Cep 22h37'44"60d22'
Lees,John Kinler Tanya Stogsdill Lees
Aql 18h47'44"10d28'
Lees,Karen Lynn
And 2h29'41"44d15'
Lees,Madison Kate
Aur 4h54'0"41d3'
Lees,Richard Lawrence
Cep 21h6'45"55d24'
Lees,Roger
Per 2h55'60"31d25'
Lees,Samuel Charles
Uma 10h12'52"71d43'
Leesa
Peg 23h27'0"18d52'
Leesemann,Dr Peter
Sgr 19h4'18"-25d19'
Leeser,Audrey Kuethe
Aql 18h44'35"10d30'
Leeson,Norma Lee
Boo 15h46'1"51d14'
Leetch,Deanna K
Cma 6h51'19"-15d11'
Leetham's Place in the Stars,Sandra
And 1h23'21"40d16'
Leeves,Barbara
Cyg 20h18'28"41d31'
Leeves,Craig
Aur 5h5'17"41d14'
Leeves,Danny Michael
Aur 5h1'1"45d4'
Leeves,Kirk David
Aur 6h2'38"45d41'
LeeVin
Aql 19h8'49"4d1'
Leewood star,the
Aql 20h8'56"1d19'
Lefcoe,Amy
Peg 21h30'36"20d31'
Lefcoe,Kevin
Crb 16h1'10"31d40'
Lefcovich,My Love,Mark
Per 2h41'16"40d23'
Lefebure,Chantal
Her 16h58'55"50d42'
Lefebvre,Daniel
Ori 6h0'18"7d23'
Lefebvre,Gilles
Lmi 10h55'12"28d17'
Lefebvre,Patrick Anthony
Per 1h39'43"52d21'
Lefehocz,Margaret Volanski Heikkila
Vul 20h15'1"23d21'
Lefekne,Mauria et Laura
Peg 0h6'49"13d42'

Lefeuvre,Jean Noël
Lyr 18h58'41"31d42'
Lefeuvre,Sebastien Laurent
Aur 6h8'10"48d59'
Lefever,Margaret Briggs
Cas 3h29'31"73d54'
Lefevere,Jean M
Crb 15h59'14"35d20'
Lefevre,Bastien
Boo 15h26'10"48d7'
Lefevère,Monique
Crb 15h38'1"35d57'
Leff,Deborah
Uma 9h25'27"61d10'
Leff,Forever Sharon
Lac 22h14'43"50d33'
Leff,Morgan Paige
Cam 7h15'1"67d37'
Leffert,Jenalyn
Lyr 19h3'44"40d48'
Leffert,John Robert
Gem 7h6'47"28d30'
Leffler,Holly Kristen
Hya 9h34'36"1d40'
Lefke,Barbara L
And 0h50'19"45d10'
Lefkovitz,Jay Robert
Her 15h55'51"41d18'
Leftin,Lawrence Joseph
Dra 16h39'43"61d7'
Lefton,Lester
Dra 16h32'0"68d59'
Lefty
Oph 17h5'58"11d39'
Lega,M Grazia
Oph 17h59'25"10d50'
Legacy Of Love Marie's Glowing Star
Del 20h24'39"8d40'
Legacy,Eileen M
And 23h56'56"33d26'A
LeGACY,Mark Turner
Aur 5h26'36"40d50'
Legacy,Roy W
And 23h56'56"33d26'B
Legal,Marine
Aur 5h51'1"30d14'
Legault,Guy
Uma 11h38'14"38d6'
Legault,Pierre
Per 3h12'46"42d50'
Legend
Uma 9h52'36"67d53'
Legendary Laubach Luminous
Hya 9h20'39"3d42'AB
Legendre,C
Cam 5h58'32"60d52'
Legends
Lyn 8h57'54"45d39'
Leger,Lauretta Loretta Marie LeBlanc
Vul 20h40'16"28d57'
Leger,Maureen Janice Veronica
Peg 23h34'1"31d60'
Legere,Marcia
Peg 21h56'22"24d34'
Legg,John
Cet 1h29'49"-3d20'
Legge Deirdre
Lac 22h24'0"56d51'
Leggett,Louise
Del 20h18'54"14d15'
Leggett,MD,Philip L
Oph 17h15'21"-22d31'
Leggett,Sarah Corren
Cet 2h14'21"6d2'
Leggett,Saskia Noel
Ari 2h42'1"25d19'
Leggie
Peg 23h41'32"15d49'
Leggio 1948,Frank & Anne Thérèse
Cyg 21h50'38"40d17'

Leggio,Jr,Tommy John
Aur 5h2'42"41d36'
Legnini,Famiglia
Tau 5h53'1"23d52'
LeGrand,Carl
Aur 4h57'21"37d2'
LeGrand,Dwayne Joseph
Sct 18h52'40"-5d60'
Legrand,Sylviane
Cnv 13h18'15"38d10'
Legris,Luc
Lmi 10h7'52"38d25'
LeGros"Headstar", Leighton
Lmi 10h17'15"38d14'
LeGros,Dawn Lee
Lyn 6h59'51"54d42'
Legros,Yves
Per 3h30'19"37d8'
Legué,Juliette
Lyr 18h13'23"39d24'
Leh,Jens
Peg 0h9'59"18d46'
Leh,Robert Donald
Umi 15h41'26"68d35'
Leha
Mon 7h1'1"-5d11'
Lehamn,Donald L
Ori 5h50'13"11d10'
Lehan,Diane Mary
And 0h52'44"37d51'
LeHerve,Laurence
Per 3h27'52"51d22'
Lehew,Steve W
Aur 7h5'13"37d60'
Lehman,Alan M
Sct 18h51'42"-9d45'
Lehman,Charles & Paige
Crb 16h12'0"30d35'
Lehman,Dallas W
Dra 16h57'29"63d49'
Lehman,Dana Mei
Dra 15h47'0"62d42'
Lehman,Dancer
Cet 1h12'35"-4d8'
Lehman,Ethel I
Aur 5h52'26"36d55B
Lehman,Imogene
Cam 4h21'57"60d55'
Lehman,Jean
Dra 16h47'12"68d4'
Lehman,Jeffrey
Her 17h38'60"23d3'
Lehman,Jerry
Ser 15h57'43"24d54'
Lehman,John W
Aur 5h52'26"36d55A
Lehman,Kayla Ann
And 2h25'36"44d52'
Lehman,Kyle
Dra 16h25'54"67d20'
Lehman,Leslie
And 0h29'29"45d15'
Lehman,Martin A
Peg 23h38'36"10d29'
Lehman,Mary R
Lyn 6h19'0"60d58'
Lehman,Melinda A
Cam 3h18'29"60d15'
Lehman,Sean Kristopher
Aur 5h39'26"38d3'
Lehman,Sheldon "King"
Ori 5h3'15"-0d9'
Lehman,Stephanie Fran
And 0h56'40"45d25'
Lehmann,Alan David
Per 2h55'53"46d51'
Lehmann,Donald
Lac 22h47'34"56d37'
Lehmann,Edgar
Mon 7h54'1"-2d33'
Lehmann,Edward Byrnes
Lac 22h11'26"48d2'
Lehmann,Gabriella Brigitte
Cas 0h20'0"62d50'

Lehmann,Heinrich
Cmi 7h16'48"4d19'
Lehmann,Monika
Cmi 7h16'54"5d5'
Lehmann,Nina
Cmi 7h16'55"4d58'
Lehmann,Sally Yapp
Cas 0h31'38"58d37'
Lehmkuhl,Herr
Peg 23h31'39"17d5'
Lehn,Jacqueline
Aql 19h53'24"12d14'
Lehnda
Cyg 21h21'1"38d45'
Lehnen,Birgit
Vir 14h41'24"-6d28'
Lehner Danny & Remy, The
Hya 8h42'48"6d29'
Lehner,Hans
Ari 2h3'0"20d13'
Lehner,Jean Donovan
Lyr 18h27'51"30d16'
Lehnert,James Edward
Lac 22h46'46"53d24'
Lehnes,Ellie
Lyr 18h51'50"33d22'
Lehnhoff,Erwin
Boo 13h37'28"10d36'
Lehoisky,Michael William
Vul 19h58'58"23d58'
Lehovitis,Christina Kassiani Siarres
Leo 11h8'57"1d35'
Lehr (Heaven's Light), Leslie Alison
And 1h52'22"39d13'
Lehr,Dr Max I
Per 3h37'31"37d50'
Lehr,Pauline Grabill
Lyn 6h15'55"60d8'
Lehr,Samuel Gordon
Her 16h43'25"27d16'
Lehr-My Love,Douglas Charles
Aql 19h50'35"10d46'
Lehrer,Rick
Cap 20h25'33"-12d34'
Lehrhoff,Sunya "Sunny"
Cas 0h3'24"63d24'
Lehrman,Stephen J
Her 16h57'39"26d37'
Lehtinen Matti Veikko Tapani
Uma 11h3'55"68d11'
Lehtinen,Riitta
Peg 22h49'34"20d13'
Lehto,Danielle Suzanne
Aur 5h16'11"40d56'
Lehto,Pentti
Cam 5h42'1"60d30'
Lehtola
Eri 3h30'0"-6d54'
Lehtonen,Mika
Cam 3h44'16"60d0'
Lehtonen,Sweet Pea Scott
Peg 22h39'18"29d15'
LEI-I+MAV=MAV"
Aql 20h2'37"0d34'
Leia's Paradise
Mon 6h4'20"-6d7'
Leibenguth III,Albert A
Lac 22h9'22"50d42'
Leiber,Dirk
Uma 9h36'41"50d38'
Leiber,Miss Margaret
Vul 20h21'1"26d10'
Leiber,Miss Margaret
Cnc 8h58'42"18d34'
Leibman,Julie Kantrowitz
And 0h0'21"41d14'
Leibowitz,Elise Hannah
Cyg 21h22'43"31d37'
Leibowitz,Maxine
Lyn 7h34'40"37d12'
Leibrock,Don
Aur 4h53'24"38d46'
Leiby,Betty
Lyr 18h55'37"32d35'
Leiby,Marilyn
And 0h58'59"33d40'
Leicester City
Dra 19h40'24"60d48'
Leichner,Lorry
Lyn 8h32'59"43d51'
Leichsenring,Dana
Dra 16h59'39"66d6'
Leichtag,Toni & Lee
Lmi 10h41'13"32d35'
Leidel,Alice Ann
And 23h41'28"38d50'
Leiden's Radiant Gold Heart,Steve
Cep 22h14'53"65d29'
Leidheuser,Star of Horst & Regine
Uma 11h23'1"41d36'
Leidig,Peter Germany Mannheim
Gem 6h40'12"31d19'
Leidner,Alice
Cyg 19h25'31"33d52'
Leiendecker,Lindsay Allison
And 0h7'1"37d45'
Leif's Life
Ori 4h41'30"14d49'
Leif's Wish
Ori 5h58'55"16d50'
Leifer,Jennifer Bess
Lyn 6h14'45"60d28'
Leifer,Kelly
Gem 6h56'39"17d25'
Leigh
Cma 6h50'58"-18d33'
Leigh
Mon 8h5'15"-1d42'
Leigh
Lyn 7h42'44"40d14'
Leigh
And 0h32'28"36d33'A
Leigh
Crb 16h8'50"33d55'
Leigh & Claire
Sge 20h16'34"16d6'
Leigh & Forrest
Cyg 19h47'1"29d59'
Leigh & Tim
Lac 22h54'22"52d57'
Leigh Ann's "My Teddy Bear" Star
And 0h46'29"33d31'
Leigh Loves Jay
Uma 9h7'24"51d28'
Leigh,Hana
And 0h37'0"40d14'
Leigh,James J.
Cep 22h48'51"57d20'
Leigh,Jessica Morgan
Equ 21h0'26"3d38'
Leigh,John & Sandra
Uma 9h27'57"54d27'
Leigh,Kiera
Peg 21h52'42"3d17'
Leigh,Tara
Cas 1h56'20"60d28'
Leigh,Terri
Vul 19h23'0"25d42'
Leigh,Tim
Mon 7h0'17"4d28'
Leigh-Anne
Mon 06h19'13"-1d51'
Leigha Rose
Ori 5h32'27"-2d22'
Leighfield,Lisboa '94- James Benjamin
Aql 20h18'39"1d49'
Leighric 94
Aql 19h31'32"11d5'
Leight,Paul (Bright)
Crt 10h52'12"-11d42'
Leighton 100 Years, Lucas Arvo
Uma 10h46'18"61d56'
Leighton,Christopher J
Aur 6h35'30"37d30'
Leighton,LuAnn
Com 13h4'26"19d58'
Leighton,Margaret Elizabeth
Cas 2h54'16"57d31'
Leighton,Tracy Lee
Lyn 8h7'11"35d6'
Leighton,Virginia Woodruff Hubbell
Leo 11h32'12"0d43'
Leikem,Carmen Gomez
Peg 22h34'17"22d57'
Leiken,Stanley
Aur 4h35'20"31d41'
Leiker,David C
Lmi 10h3'0"32d9'
Leila
Cnc 8h35'6"18d23'
Leila
Aql 19h57'54"15d12'
Leila's Blooming Star
Peg 22h42'17"10d32'
Leila's Star
Peg 23h43'1"18d49'
Leila-Star Of Rapidan
And 23h34'35"49d50'
Leilah D
And 2h24'40"42d11'
Leilani Mia
Lyr 18h27'49"41d3'
Leimer,Bill
Ser 15h35'1"19d7'
Leimkühler,Dieter
Peg 23h30'47"12d14'
Leimone,Dave
Cmi 8h4'11"0d7'
Leimuvirta,Olavi
Uma 8h39'24"68d29'
Lein 082695,Saleki
Cam 3h33'31"61d51'
Lein,Mila Nina
Cyg 19h52'35"38d3'
Lein,Timothy Allen
Aur 7h12'22"35d40'
Leinadracir
Cep 22h10'1"60d15'
Leiner Light
Aql 18h57'21"-4d7'
Leiner,Michael R
Oph 18h36'23"11d42'
Leinhos,David Andrew
Per 2h48'24"32d17'
Leininger,Christopher Robert Eugene
Cap 20h21'19"-26d15'
Leininger,Robert
Lyn 7h31'41"39d41'
Leinster,Thomas Stanley
Uma 9h29'17"48d14'
Leiper-Peterhead,Baby Steven John
Umi 16h27'17"75d18'
Leis,Kevin LeRoy
Vul 19h22'33"26d35'
Leisenring,R Craig
Aur 5h2'23"40d26'
Leishman,Aaron Crawford
Dra 16h57'11"71d10'
Leishman,Jeremy Wyatt
Ori 5h51'15"8d11'
Leising My Guiding Light,James L
Sgr 19h32'11"-45d15'
Leisky,Gailen Robert
Lac 22h33'44"54d1'
Leisman,Cheri
Hya 8h18'27"4d15'
Leiss,Robert H
Aql 19h54'10"15d14'
Leist,EG
Lac 21h56'19"37d44'
Leisy,Aimee Christine
Eri 3h15'54"-18d57'
Leitao,Judith
Cas 2h30'45"61d4'
Leitao,Paulo de Lima
Tau 4h56'30"16d25'
Leitch,Donald Richard
Lib 15h16'42"-23d11'
Leitch,Eric B
Cnv 12h42'17"32d20'
Leitch,J R
Ori 5h50'1"14d17'
Leitch,Phil
Cam 3h29'46"63d33'
Leitch-Campbell, Jennifer
Lyr 18h44'13"41d17'
Leite,Diane
Peg 23h37'29"30d17'
Leitenmaier,Peter
Cep 22h38'49"65d36'
Leiter,Jack
Uma 11h57'15"57d7'
Leitermann,Horst
Cyg 20h26'43"40d13'
Leith,Patricia
Lmi 9h57'27"34d25'
Leith,Steven
Cep 1h1'39"77d55'
Leitheiser,Agnes
And 1h59'30"38d20'
Leithner,Uli & Christian
Uma 9h37'29"46d38'
Leitner
Cet 2h13'37"0d46'
Leitner's Legacy
Peg 22h41'12"32d39'
Leitner,Herbert
Cas 0h15'48"60d40'
Leitz,Edward
Dra 17h42'32"64d35'
Leitz,Gabriele
Sco 17h20'58"-38d33'
Leitzel,Doreen
Cyg 21h51'53"42d13'
Leiva,Larry
Lyr 18h34'58"40d54'
Leivermann North Star, The
Per 2h5'22"48d41'
Leizear,Sharon Louise
Boo 14h27'20"25d51'
LeJarde,Victor J
Ori 5h58'22"16d53'
Lejman's Love
Cyg 20h23'28"38d47'
Leka,Christopher Allen
Cma 6h52'40"-16d37'
Lekas,"Star Of Yanni" For John C
Her 16h18'59"48d13'
Lekawski's Golden Wedding Star
Sge 19h54'54"18d49'
Leker,Jim & Mary
Sco 17h52'18"-38d11'
LeKites,Cole Aaron
Boo 15h0'33"14d32'
LeKites,Morgan Patrick
Aur 6h20'22"33d4'
Lekus,Melissa
Vul 20h56'26"20d3'
Lekutis,Cody Taylor
Dra 10h21'54"74d34'
LeLa
Lyr 19h9'46"37d31'
Lela Mae
Com 12h16'0"31d56'
Lela-N-Letha The Twins
Lyn 7h59'0"50d45'
Lelakis Star,The
Vel 10h8'24"44d12'
Leland,John
Mon 7h40'28"-1d37'
Leland,Maria M
Uma 8h56'52"49d59'
Leland,Wayne A
Aql 20h7'57"8d21'
Lele
Sgr 18h50'8"-23d25'
Lele e Peppino
Her 16h53'55"41d5'
Leleiko,MD,PhD,Neal
Per 1h50'45"52d36'
Leliveld,Ron
Lac 22h49'45"38d42'
Lella
Cyg 20h30'0"58d12'
Lella
Cyg 20h37'15"48d49'
Lella per sempre
Col 6h25'38"-39d36'
Lellarosella
Ori 6h5'1"8d26'
Lelli,Paola
Uma 11h58'57"30d25'
Lello,James M
Per 3h58'16"38d42'
Lelly
And 23h27'1"40d5'
Lelouch,Claude
Sco 17h53'58"-33d54'
Lema,Eileen
Cas 0h23'60"60d15'
LeMahieu,Bethene
Lyn 9h15'43"34d13'
LeMaire,Brooke B
Ori 5h59'28"9d18'
Lemaire,Gilles et Sylvie Fortier
Tau 4h40'1"15d45'
Lemaire,Kathleen
Cas 1h42'53"60d46'
Lemaire,Nadine
Aur 4h48'41"40d59'
Lemaitre Alain
Mon 7h48'35"-4d21'
Lemal,Pierre
Per 3h58'36"39d29'
Leman,Elizabeth Anne
Lyn 6h26'52"61d41'
Lemansky,Marko y Rosi
Her 17h19'50"43d2'
Lemar
Peg 21h18'46"22d56'
Lemarquand,Jeannette
Per 3h4'48"41d0'
Lemaréchal Family Star,The
Uma 10h15'19"54d20'
Lemaster,Cinnamon
Mon 7h50'23"-1d54'
LeMaster,Eternally Chyree'-Chyree'
Vir 13h34'51"-21d44'
Lemaster,Matthew
Lac 22h41'0"54d40'
Lemay,Bernice G
Cnc 8h57'21"31d34'
Lemay,Hughie C
Leo 9h58'0"15d48'
Lemay,Karine
Ori 4h50'31"5d28'
Lemay,Laurent et Véronique
Com 13h12'50"28d57'
Lemay,Lynda
Per 2h56'48"34d11'
LeMay,Patricia
And 1h59'1"37d36'
LeMay,Patricia Filkin
Lyr 18h58'41"47d35'
LeMay,To My Star-Gary
Aql 19h21'29"15d48'
Lemberg,Aira Annikki
Uma 9h47'19"51d42'
Lemberger,Regina C
Mon 6h58'0"11d52'
Lembke,Jana M
Tri 1h51'51"28d3'
Lembke,Romel Hess
Dra 18h58'14"48d34'
Lemen,Ted
Cep 21h46'25"60d58'
Lemercier,Jacky
Mon 7h54'55"-4d23'
Lemere,Kenneth & Teresa
Peg 21h51'0"35d45'
Lemettre,Barbara
Aql 18h57'59"-6d15'
Lemieux,Hilaire
Dra 20h6'55"65d2'A
Lemieux,Marie Danielle
Cam 3h51'1"62d21'
Lemieux,Nicolas
Umi 13h23'19"70d19'
Lemieux,Pierre
Ori 5h55'57"14d55'
Lemieux,Renée
Her 17h18'57"48d29'
Lemieux,Sylvain
Aqr 23h47'1"-6d53'
Leming,Cathy
Sct 18h54'56"-8d41'
Leming,Gene
Peg 21h59'22"4d31'B
Leming,Jean
Peg 21h59'22"4d31'A
Lemire,Curtis & Victoria
Cyg 20h57'20"30d42'
LeMire,Judith Lynn
Peg 21h22'11"23d55'
Lemiska,Erin Kanani
Eri 3h56'45"-15d24'
Lemière,Jean Michel
Per 3h43'40"38d28'
Lemke,Bridget
And 2h23'21"44d17'
Lemke,Jacqueline
Peg 23h0'24"20d25'
Lemley,Daniel Callihan
Oph 18h21'0"6d18'
Lemm,Bob & Mary
Sag 19h58'41"16d21'
Lemmer,Elizabeth (Betsy) Mary
Cyg 20h25'0"42d56'
Lemmon,Jr,Chris
Aur 5h43'16"54d58'
Lemmon,Sydney
Lyn 7h29'23"39d39'
Lemmons,John Mark
Lac 22h12'57"50d37'
Lemoal,Solweig
Per 3h32'27"39d30'
Lemoine,Frèderique
Cep 23h0'10"60d56'
Lemoine,Frédérique
Per 3h59'0"35d20'
Lemoine,Kari K
Peg 22h29'36"29d13'
Lemoine,Kristi F
Mon 6h40'1"11d40'
Lemoine,Richard
Per 3h58'60"36d13'
Lemoine,Richard
Cep 21h26'27"68d31'
Lemoine,Richard et Nicole
Per 3h58'32"36d24'
Lemoine,Richard et Nicole
Cep 20h57'32"63d41'
Lemon Forever,James & Clarita D
Oph 18h41'18"10d2'
Lemon Sensei Jerry, Gerald L
Boo 14h6'1"34d51'
Lemon,Bryan
Cet 1h34'37"-6d32'
Lemon,Corbyn
Mon 6h53'11"-6d23'
Lemon,Wolfman Mark R
Cma 7h16'35"-13d18'
Lemoncelli,Craig V
Lac 22h28'55"40d5'
Lemond,Judith
Mon 6h44'47"11d18'
Lemons,Debbie
Eri 2h59'21"-18d27'
Lemoor
Peg 22h42'17"33d33'
Lemorande "Lemo", Thomas S
Dra 16h53'49"61d28'
LeMoullec,Carole Aimeé
Mon 6h55'29"11d29'
Lemée,Viviane
Mon 7h47'19"-4d14'
Len,Su Ling
Mon 6h42'58"3d56'
Lena
Cma 6h50'2"-15d51'
Lena
Lyn 8h27'23"40d20'
Lena
Ori 5h56'36"8d38'
Lena
Lyr 19h23'51"40d4'
Lena
And 0h40'44"40d20'
Lena 13-11-1992
Lyr 19h16'40"31d30'
Lena Joannes
Del 20h36'11"9d21'
Lena Kay
Aql 18h58'56"17d54'
Lena S
Cet 2h57'36"6d26'
Lena W
Cyg 20h15'47"39d2'
Lena's Dream
And 0h21'27"40d14'
Lena's Light
Uma 13h37'28"58d26'
Lenaburg,Steve
Aql 19h25'57"7d33'
Lenaki
Uma 10h53'11"53d35'
Lenartowicz,Cesira Petrucci
Cyg 21h31'1"40d5'
Lenartowicz,Michal
Cam 4h47'21"68d0'
Lenartowicz,WGM Of MI, Lyene
Vul 21h16'17"28d13'
Lence,Alejandro Pimpo
Pup 7h59'43"-29d41'
Lenczewski,Michael
Boo 15h2'0"21d4'
Lender,John Samuel
Hya 8h19'0"1d42'
Lender,Mark David
Cmi 7h7'17"1d47'
Lender,Richard Christopher
Oph 18h1'57"11d2'
Lendy,Talon
Uma 11h40'34"30d30'
Lene's
Cas 3h0'33"60d34'
Lenehan,David
Cnv 12h29'18"33d18'
LenElf
Lyn 8h30'0"41d35'
Lenelle's Star
Cyg 19h22'30"44d6'
Lenert,Jr,Dennis M
Boo 14h38'35"17d22'
Lenert,Kimberlee Rose
Com 12h0'1"21d28'
Lenerz,Mark Edward
Cam 6h35'53"68d54'
Lenfant,Christi Michelle
Vul 21h3'11"20d16'
Lenfestey,Kyle Ford
Mon 6h29'18"8d12'
Leng,Bernard
Ori 5h57'35"7d35'
Lenga-Balk,Edith
Boo 14h4'0"11d54'
Lengacher,Bill
Her 18h13'37"30d46'
Lengauer,Paula
Cap 20h26'1"-26d40'
Lenger,Barbara Marie Grubbs
Mon 7h50'44"-5d21'
Lenhard III,Peter Paul
Peg 21h53'44"2d31'
Lenhard,Dieter
Cmi 7h19'15"8d36'
Lenharr,Urtha
Lmi 9h53'57"34d52'
Lenhart Love Star, Henry & Marilee
Mon 6h19'50"8d52'
Lenhart,Emma Rose
And 0h12'1"33d27'
Lenhart,Harold F
Aql 19h26'1"-0d40'
Lenhelen
Peg 23h0'53"14d28'
Leni,Omait Franco
Her 18h9'41"31d52'
Lenia
Cet 2h57'37"1d34'
Lenick-Watts,Shelley
And 1h58'52"40d44'
Lenihan,"Apropos" Carol
Her 17h4'1"20d49'
Lenihan,Denise
Cas 2h8'48"70d20'
Lenik,David Lee
Her 17h0'47"31d22'
Lenik,Debra Lynn
Lyn 7h3'23"44d30'
Lenk,Christopher
Aur 6h10'35"38d26'
Lenk,Heidi und Rainer
Her 17h56'59"43d5'
Lenka
Uma 10h41'0"60d23'
Lenke,Frau
Boo 14h23'23"10d41'
Lenkov,Peter Mitchell
Cyg 19h59'11"29d30'
Lenkowski Lumina
Dra 18h18'0"68d53'
Lenner,Jessica Suzanne
Cyg 20h17'43"38d37'
Lennert,Bill
Cet 1h35'7"-9d39'B
Lennert,Vicki
Cet 1h35'7"-9d39'A
Lenni,The
Aql 19h30'26"12d44'
Lennig,Eckhard
Peg 22h15'24"5d14'
Lennon's Light,Bobby
Boo 14h59'42"35d6'
Lennon,Donna Adrienne
Mon 8h2'18"-1d36'
Lennon,Frank & Barbara
Uma 11h30'1"56d53'
Lennon,Helen Schneider
Dra 14h36'54"60d31'
Lennon,Henry Maxwell
Cet 2h34'21"-6d23'
Lennon,Isolde Concetta
Gem 6h6'0"26d45'
Lennon,Jake & Barabas
Her 16h27'1"35d32'
Lennon,James"Jim" William
Cnv 13h2'12"50d31'
Lennon,Jr,John Francis
Gem 7h14'4"21d37'
Lennon,Justin R
Gem 7h40'0"28d47'
Lennon,Karen Michelle
Cyg 20h22'55"39d39'
Lennon,Kelly Michelle
Aql 20h14'25"1d36'
Lennon,Lindsay Hope
Vul 19h23'19"23d44'
Lennon,William James
Boo 14h46'53"27d47'
Lennox,Dave
Cma 6h49'9"-17d60'
Lennox,Jesse Levi
Cep 20h16'32"60d52'
Lennox,Systems Anarchist-Craig
Boo 14h56'0"51d21'
Lennstrom,Carly Marie
Cam 8h5'33"82d40'
Lenny
Aur 5h54'14"31d20'
Lenny
Her 16h38'1"11d53'
Lenny
Her 17h2'26"31d24'
Lenny
Aur 5h6'54"43d13'
Lenny & Deborah Our Love Forever
Aql 19h5'19"3d6'
Lenny Illuminata
Aql 20h11'47"1d31'
Lenny M
Lac 22h9'51"40d47'
Lenny Zee's 60th
Cnv 12h11'43"40d32'
Lenny's Light
Boo 15h2'34"20d36'
Lenny's Own
Vul 19h22'38"25d36'
Lenoci,Jr,Alfred G
Aur 6h16'39"33d19'
Lenoir,Christian
Cyg 19h38'0"38d7'
Lenora
Gem 5h58'55"27d55'
Lenore
Del 20h18'0"14d23'
Lenore
Uma 11h46'11"45d52'
Lenore & Donna
Uma 8h57'38"54d55'
Lens,Maria
Aur 7h25'14"38d31'
Lensing,Thomas James
Her 17h10'58"44d33'
Lent,Bob
Hya 8h59'0"3d4'
Lent,Sheldon
Her 18h4'0"31d35'
Lenthall,Allan
Her 17h19'19"46d36'
Lentin,Lorraine
And 2h17'25"38d41'
Lentner,Jacob Joseph
Peg 21h59'32"28d6'
Lenton,Dean George
Lyn 8h5'32"41d50'
Lentz,Christiane
Boo 14h38'25"14d30'
Lentz,Colleen
And 23h23'1"43d8'
Lentz,Connor Bayer
Cam 3h42'40"70d21'
Lentz,Gundi
Vir 12h1'27"-2d13'
Lentz,Paul
Per 3h58'52"52d37'
Lentz,Rebecca Anne
Sco 16h57'27"-40d41'
Lentz,Stephen
Ori 4h53'31"1d43'
Lentzer,Albert H
Oph 17h55'14"11d43'
Lenz 1962,Mark Steven
Cet 3h0'24"4d44'
Lenz Family Star,The
Uma 11h12'35"44d25'
Lenz,Clara Erma
Cas 23h24'41"53d30'
Lenz,Cynthia R
Lyn 7h9'19"58d32'
Lenz,Deborah Jane
Cas 0h31'25"75d23'
Lenz,Dorothy Smith
Cam 10h8'30"84d29'
Lenz,Gregory Alan
Ser 15h58'44"23d52'

Lenz,Irving & Mabel
Cam 8h11'24"78d16'
Lenz,Jessica Wolken
Eri 3h59'1"-14d57'
Lenz,Travis Austin
Per 3h11'51"41d16'
Lenz,Volker
Psc 23h9'22"0d52'
Lenz-"The Fisherman", Arthur Paine
Cet 0h8'57"-10d7'
Lenz-Pietschmann, Siegrid
Lac 22h4'30"51d28'
Lenzi
Mon 6h23'27"8d36'
Lenzner,Allan Joseph
Her 16h33'56"41d15'
Lenzner,Benjamin Hardy
Per 3h22'45"40d31'
Lenzner,Guy
Sgr 19h28'28"-36d3'
Lenzner,Laura Adasko
Aur 4h54'29"50d7'
Lenzo
Tau 5h23'0"20d29'
Leo
Cet 2h57'28"4d37'
Leo
Cet 1h53'16"1d36'B
Leo "Mr Continental"
Cep 22h15'54"80d33'
Leo III
Leo 11h31'41"10d30'
Leon
Her 18h18'23"18d57'
Leon
Hya 9h7'32"0d55'
Leon III
Cet 2h8'15"5d23'
Leon's Way
Dra 14h57'27"56d57'
Leon,Francisco Lucena
Cam 5h51'36"68d37'
Leon,Geoffrey Patten
Eri 4h56'19"-11d35'
Leon,George R
Ori 5h55'40"16d53'
Leon,Keegan David
Leo 10h32'21"11d30'
Leon,Stephanie Andrews
Uma 11h26'1"56d4'
Leon-Gutherie,Abigail Jolene
Leo 10h24'44"10d48'
Leona
Cas 2h55'30"61d12'
Leona "Shorty"
Sgr 19h36'46"-30d2'
Leona & Dennis' Light Of Love
Mon 7h56'0"-6d57'
Leonard
Cam 4h11'38"50d16'
Leonard
Aur 7h3'23"36d24'
Leonard 1
Aql 20h0'27"-6d45'
Leonard Star,The Roger
Aur 6h29'31"34d39'
Leonard Unit
Aur 4h35'0"33d50'
Leonard's Enterprise
Her 17h12'58"18d53'
Leonard's Eternal Star Jennifer Ann
Aql 19h43'27"12d23'
Leonard's Forever
Cep 22h27'50"80d17'
Leonard,Ali-Oop
Ori 5h25'45"0d43'
Leonard,Allison Elizabeth
And 23h42'55"43d17'
Leonard,Bandit Joe
Boo 15h1'1"47d54'
Leonard,Brandy Akasha
Lyn 7h35'0"41d17'
Leonard,Brent Michael
Boo 14h55'23"28d53'
Leonard,Brian
Cet 3h17'39"3d46'
Leonard,Christine M
Cma 6h52'46"-18d18'
Leonard,Cindy A
Cnc 9h1'28"30d5'
Leonard,Cody Mark
Lac 22h3'19"38d15'
Leonard,Dominique
Mon 6h27'17"0d34'
Leonard,Frank Gerald
Dra 15h6'29"63d27'
Leonard,James E
Cam 6h2'28"58d44'
Leonard,James Wharton (Jamie)
Aql 20h17'49"5d14'
Leonard,James-Galen
Peg 23h43'1"31d30'
Leonard,Jean Marlene
Cas 0h11'22"58d37'
Leonard,Jesse Edward
Lac 22h5'1"49d46'
Leonard,John C
Hya 9h38'12"-1d32'
Leonard,Johnna Amato
Uma 9h53'53"45d0'
Leonard,Jr,Joe
Aur 4h53'28"51d47'
Leonard,Jr,Robert Jonathan
Her 16h42'51"5d1'
Leonard,Kim M
Lyn 8h58'0"46d48'
Leonard,Kyle Joseph
Aur 6h27'54"31d59'
Leonard,Lisa Rene
Cyg 20h22'1"41d56'
Leonard,Mark
Aur 6h55'19"44d8'
Leonard,Mary
Cyg 19h43'25"30d13'
Leonard,Matthew Raymond
Her 18h8'18"28d35'
Leonard,Pamela J
And 1h34'56"38d60'
Leonard,Paul
Sct 18h21'59"-13d18'
Leonard,Stella Estes
Ori 5h56'44"17d7'
Leonard,Steve
Boo 14h2'17"20d39'
Leonard,Susan M
Vul 20h15'0"23d3'
Leonard,Thomas Joseph
Dra 15h1'50"62d10'
Leonard,Trey Komel
Uma 10h32'11"53d22'
Leonard,Troy L
Aql 19h58"1"13d31'
Leonard,Valerie Ann
23h 23h36'40"46d53'
Leonardi Forever,John & Barbara
Lyr 18h45'53"42d22'
Leonardi,Paul
Aql 18h59'1"-0d46'
Leonardis,Deborah
And 23h5'0"44d33'
Leonardo
For 2h31'0"-27d60'
Leonardo
Eri 2h43'45"-7d22'
Leonardo (B-Promised 1-4-3),Charles E
Aur 7h25'20"37d44'
Leonardoandreamarco car-locristina
Cep 23h30'13"70d23'
Leone,Erica & Meghan
Com 12h28'15"22d49'
Leone,Goddotter Susie (Nee Smith) & Vince
Uma 10h54'24"48d16'
Leone,Jr,Richard Paul
Cep 23h2'45"68d22'
Leone,Marcy & Tony
Lac 22h12'45"46d2'
Leone,Reynold
Cnv 13h23'22"50d50'
Leone,Tony & Mary
Cam 6h12'26"65d2'
Leonetti's,Forever Yours!,Lynda
Sco 16h31'45"-44d48'
Leonetti,Nick
Peg 22h40'16"27d59'
Leong,B James
Dra 18h1'13"58d11'
Leong,Miss Michelle
Mon 7h15'2"-6d31'
Leong,Pauline
Sgr 19h58'47"-43d37'
Leonhard der Dritte
Psc 0h43'38"20d5'
Leonhard,David & Evelyn
Uma 11h35'40"41d32'
Leonhard,Edward Anthony
Per 2h59'57"37d54'
Leonhard,Linda
Cet 1h30'41"-1d60'
Leonhard,Timothy Wayne
Per 2h37'23"37d21'
Leoni,Hugh
Cam 9h12'1"73d57'
Leoni,Oliver
Dra 18h50'0"70d31'
Leonides Miaz Tomas, Maria
Uma 12h42'1"58d36'
Leonie & Robert Eternal Lover
Ori 5h56'34"16d35'
Leonie Rae
Peg 22h34'41"28d27'
Leonie,Deborah
And 0h29'0"28d50'
Leonis,Trinty Oksana
Peg 23h17'11"32d44'
Leonor
Ori 5h29'23"8d39'
Leonora
Peg 21h48'26"35d52'
Leonora & Jorge
Cam 5h44'30"60d45'
Leonore
Uma 12h4'29"58d34'
Leontiades,Louisa
Ori 5h56'47"7d15'
Leopoldi,Norbert
Lac 22h28'45"38d59'
Leopoldine
Lyn 8h29'39"47d10'
Leopoulos,George E
Her 16h39'51"7d44'
Leora,A Child of the Commandments
Cam 11h57'39"77d58'
Leoria
Vel 10h5'18"55d41'
Leos,David Corey
Peg 22h16'55"10d45'
Leos,Kyera and Kyleigh
Vir 11h50'40"3d09'
Leota Rae Bernadette
Cam 8h7'1"80d27'
Leove
Lyr 19h21'26"30d35'
Lepadatu,Cornelia & Florin
Cyg 20h25'21"42d18'
LePage,Joel
Per 1h54'58"54d19'
Lepage,Michel
Uma 10h1'12"61d37'
Lepage,Nicole
Del 21h2'47"12d16'
Lepage,Stéphane
Cep 0h33'51"78d8'
LePage,Theodore
Per 2h54'20"43d55'
LePage,Tyler
Aur 5h45'28"54d33'
LePain,Jerry
Aur 6h7'41"45d58'
Lepante,Paul L
Cep 21h6'20"60d2'
Lepard,Diana Catherine
Cas 0h35'49"66d55'
Lepard,Willaim Paul
Per 2h26'22"54d41'
Lepaumier,Marie-Claire
Mon 6h27'34"8d23'
Lepesqueux
Cyg 21h59'50"50d35'
Lepeutrec,Catherine
Ori 6h1'42"8d47'
Lepey's Christmas Star Fred
Dra 16h48'19"70d11'
Lepiorz,Manfred B
Cam 7h36'23"74d46'
Lepley IV,Gordon Taylor
Lyr 18h40'39"39d3'
Lepley,Matthew
Aql 19h55'17"14d60'
Lepola,William Matthew
Per 1h28'28"54d9'
Lepore,Desiree C
And 23h16'33"40d6'
Lepore,Kass
Lyn 9h10'22"38d24'
Lepore,The Beautiful Linda
Dra 13h6'49"68d36'
Leporini,Viviane
Tri 1h29'11"28d60'
LePoutre,Monique
Cam 3h42'14"68d13'
Lepp,Papa Joe
Lyn 8h13'32"36d10'
Leppard 5/7/70,Geoff
Aur 5h53'36"40d19'
Leppard,Adam George
Peg 22h20'19"31d22'
Leppek,Janet Louise
And 0h10'27"37d29'
Leppere,Caitlyn Michelle
Mon 6h48'27"10d7'
Leppert,Fats
Cam 7h38'43"60d2'
Leppla,Lucille
Cam 8h55'23"81d21'
Leprechaun
Umi 15h19'39"69d51'
Leprince,Suzie
Per 3h3'31"38d34'
Leprotta Innamorata
Lyn 7h0'48"44d48'
Leprottino
Per 2h10'28"58d54'
Leps,Richard J
Her 16h17'49"41d28'
Lepschi,Gabriela
Ari 2h0'1"22d5'
Lepucki,Ted & Karen
Cyg 19h52'47"40d24'
LeQuatte,Rosalie Hope
Aql 19h52'59"-6d33'
LeQuatte,Willis E
Aql 19h31'36"-6d48'
Lercho Margot
Gem 6h34'24"26d45'
Lererway To Heaven
Per 3h21'34"38d58'
Lericheux,Jacques
Oph 18h7'26"10d53'
Lerman,Alexandra
Lyn 6h56'26"44d50'
Lermen,Christel
Sco 16h36'14"-43d34'
Lerner
Cap 21h45'1"-24d2'
Lerner,Claude George
Ori 5h56'17"16d48'
Lerner,E Brooke
Lyn 7h30'0"41d44'
Lerner,Heidi Ingrid
Uma 8h57'28"71d35'
Lerner,Jeffrey Steven
Her 17h14'14"27d47'
Lerner,Michael "Micky" Raymond
Ori 5h11'1"-4d52'
Lerner,Mitchell
Her 18h9'55"33d26'AB
Lerner,Patsy Lee
Cyg 21h52'18"40d6'
Lerner,Peter Alan
Her 16h6'1"48d45'
Lerner,Ruth Geyer
Cas 2h11'46"59d10'
Lerner,Sol
Aur 5h0'52"30d8'
Lerner,Thomas D & Connie R
Her 17h32'59"21d17'
Lernihan,Brian
Uma 11h17'11"51d29'
Lerouge 17-7-71, Famille
Boo 13h51'58"16d9'
Leroux,"Mamoush" Thérèse
Lyn 9h35'1"41d1'
LeRoux,Brenda Lee
And 23h39'19"44d59'
Leroux,Catherine
Ari 3h13'23"19d9'
LeRoux,Danielle Elice
Vul 20h38'0"20d24'
Leroux,Lynnda
Lmi 10h53'14"32d18'
Leroy
Cmi 7h42'40"5d27'
Leroy & Anita Lewis
Cam 13h19'59"78d7'
Leroy Eternally Yours
Per 2h57'45"50d22'
LeRoy Leslie Center
Oph 17h26'49"-22d21'
Leroy R
Cep 21h37'17"58d33'
LeRoy,Carri Jeane
And 22h57'39"38d51'
Leroy-Beaulieu,Terence
Ori 6h9'1"4d4'
Lerro,Alessandro Maria
Pyx 8h45'33"-24d9'
Lerro,Niccolò
Her 16h43'49"18d52'
Les
Aur 6h28'20"31d23'
Les & Laura
Aur 6h4'53"38d50'
Les & Vicki's Silver Star
Boo 14h22'0"54d52'
Les Amis Toujours
Uma 8h6'29"60d29'
Les Artistes Laura & Ken
Lac 22h17'21"47d28'
Les Jyrans
Lyr 19h22'35"30d12'
Les M
Uma 8h16'0"71d0'
Les Poissons
Lac 22h30'27"38d6'
Les Trois Anges
Peg 21h25'26"3d30'
Les' Miracle
Sgr 19h37'30"-35d59'
LeSaar Star,The
Uma 10h38'26"48d39'
Lesage,Michel
Crb 15h55'54"35d20'
LeSage,Steve
Uma 10h48'52"68d27'
Lesagor,Lauent
Per 1h53'53"48d35'
Lescaze,Marie Claire
Per 4h5'27"37d19'
Lescelestialley
Cas 2h13'28"68d38'
Lescop,Jeanine
Her 16h16'0"50d39'
Lesczinski,Albina Soczynska
Lyr 19h23'29"38d1'
Lesczinski,Robert L
Her 18h4'20"38d27'
Lese,Jr,William D
Aqr 20h36'23"-6d46'
Lesenechal,Yves
Cyg 19h44'22"37d35'
Leser,Linda
Peg 23h37'43"16d51'
Lesh Family Star,The
Cyg 19h30'20"37d58'
Lesho,Daniel Lee
Ser 15h53'33"20d25'
Lesieur-Zelie
Uma 12h20'34"60d12'
Lesinsky,PFC Jonathon Vincent
Aur 6h11'42"38d10'
Leske,Robert
Lmi 10h26'14"30d29'
Lesko,Angela Reneé
Sco 16h24'21"-29d10'A
Lesko,Helen
Cyg 20h31'60"42d39'
Lesko,John
Her 17h31'0"30d19'
Leslea Ann
Cet 2h41'29"-0d53'
Lesley
Aur 5h17'58"49d20'
Lesley
Peg 22h17'0"34d46'
Lesley
And 0h55'54"38d22'
Lesley Elizabeth's Nova
Sco 16h21'23"-30d15'
Lesley Fay
Eri 3h5'42"-3d44'
Lesley Kate
Lyr 18h33'54"33d11'
Lesley May
Cyg 19h27'42"36d5'
Lesley Mc's Star
Ori 5h51'33"9d14'
Lesley-Anne
And 23h21'28"40d4'
Lesley-Ray-Carter
Boo 14h58'27"31d6'
Leslie
Psc 0h58'26"32d17'
Leslie
And 0h20'35"34d18'
Leslie
Lyr 18h55'13"30d53'
Leslie
And 2h23'15"47d18'
Leslie
Tau 3h46'1"1d24'
Leslie
Cyg 20h35'15"50d7'
Leslie
Vir 13h52'56"-5d37'
Leslie
Lyr 18h51'19"40d29'
Leslie
Com 12h18'26"20d8'
Leslie
Mon 6h57'0"7d48'
Leslie "International Superstar",David
Per 3h26'27"40d17'
Leslie & Christian
Lac 22h54'58"51d23'
Leslie & Kevin,The Marriage of
Uma 11h42'52"56d10'
Leslie & Leo Means Love Forever
Cyg 19h29'25"40d33'
Leslie Ann
Peg 23h6'0"20d5'
Leslie Anne
And 23h43'30"40d12'
Leslie Colleen
Lyn 8h13'45"37d48'
Leslie David
Cas 0h31'0"75d45'
Leslie Earlene
Del 20h13'54"10d35'
Leslie Erin
Mon 6h4'0"-4d55'
Leslie F
Aql 19h27'22"13d38'
Leslie H
Aur 4h57'15"36d57'
Leslie Lynn
Peg 22h40'21"20d38'
Leslie Miranda
Hya 10h31'39"-11d60'
Leslie René
Mon 7h39'23"-3d30'
Leslie the Shining Star
Lyn 7h0'42"58d34'
Leslie's Eternal Star
Ori 4h58'19"0d45'
Leslie's Heart
Mon 6h53'41'-0d56'
Leslie's Love
Lyn 7h0'28"50d2'
Leslie's Loving Heart
Mon 7h16'42"-8d7'
Leslie's Music
Lyr 18h51'28"30d4'
Leslie's Play-N- Peekaboo
And 0h18'41"36d13'
Leslie,Brian Russell
Cep 22h6'59"61d29'
Leslie,Burton Warren
Per 3h26'21"38d60'
Leslie,Denise
Hya 8h24'38"0d50'
Leslie,Eugene E
Dra 19h56'42"84d42'
Leslie,II,MD
Ori 4h58'46"-1d39'
Leslie,James C
Uma 12h1'0"57d2'
Leslie,Margaret Mary
Cyg 20h25'25"46d42'
Leslie,Michael Patrick
Dra 19h29'29"68d37'
Leslie,Neal
Hya 8h24'26"0d57'
Leslie,Tina Marie
Ari 2h57'20"28d20'
Leslie,W L,Lottie & Jessica
Lyn 8h24'53"46d52'
Leslye's Light Of Love
Hya 8h44'0"3d41'
Lesmeister,William A
Per 3h1'45"50d15'
Lesne,Cyrus
Per 3h59'28"36d17'
Lesniak,Matthew Brian
Dra 15h6'25"63d55'
Lesnick
Per 4h45'18"51d11'
Lesnick,Stephen Matthew
Hya 8h11'1"2d53'
Lesnicky,Jr,Jennifer & Frank
Cyg 19h31'33"34d13'
Lespagne,Claude
Aur 6h52'20"49d17'
Lespagnol,Marie Piesse
Cam 3h36'1"60d57'
Lessard,Daniel
Gem 7h34'43"30d49'
Lessard,Lawrence
Her 16h9'22"11d5'
Lesser Magic Star,The Eugene
Cet 1h58'56"-6d50'
Lesser,Amy Rachel
Umi 14h2'15"70d1'
Lesser,Leonard
Cep 21h1'42"85d31'
Lesser,Martin V
Cap 20h59'1"-18d4'
Lesseski,Forever My Love David C
Her 16h53'25"40d27'
Lesslie,Gordon
Tau 5h56'26"23d18'
Lesslie,Mona
Ari 1h59'24"17d41'
Lessly,Star Of
Leo 11h33'31"1d21'
Lessner,Maria & Cory
Peg 23h22'16"25d24'
Lessor,Robert
Aur 5h28'14"30d9'
Lestat
Boo 15h0'42"19d7'
Lester
Cep 1h13'38"80d32'
Lester "The Honest Confidant",S J
Ori 6h17'48"7d40'
Lester,Bradley Marc
Gem 6h51'43"13d18'
Lester,Dr Michael George
Cep 23h10'28"62d50'
Lester,Janis
And 0h2'11"37d57'
Lester,Jeremy
Peg 22h40'0"34d54'
Lester,Juanita Lynn
Peg 23h38'0"18d4'
Lester,Layde Ashley
Com 12h17'59"21d27'
Lester,Leanne
Per 3h11'13"46d55'
Lester,Lydia
Cas 1h46'50"74d59'
Lester,Michele
Lib 15h02'0"-8d55'
Lester,Sara Jane
Peg 23h18'42"18d48'
Lester,Simone
Mon 7h0'54"0d39'
Lester,Vivien
Peg 21h59'32"33d4'
LeStrange Family Star, The
Lac 22h22'37"53d56'
LeSueur,Wilma W
Lyr 19h4'57"28d51'
Leswing,Carolyn Nye
And 2h28'20"45d31'
Lesyna,Jonelle C
Lyn 8h6'27"51d56'
Leszinsky,Alexander
Cnv 12h35'16"36d34'
Let The Children Shine-Riv Presb PS
Vul 20h20'40"28d30'
Leta
Lyr 18h17'17"30d54'
Leta Joan
Peg 22h29'13"27d18'
Letasi,Marianne
Lyn 9h14'0"35d16'
Letayf,Jose y Raquel
Crb 15h50'27"30d50'
Letendre,Nicolle C
Ori 5h2'41"-0d10'
Leticia
Cru 12h53'30"-16d3'
Letitia-Goddess Of Love
Com 13h5'40"14d33'
Letizia
Gru 22h5'5"-50d21'
Letizia
Ori 6h4'32"1d16'
Letizia,Maria
Pup 8h8'23"-21d48'
Letley,David John
Cep 20h31'1"62d17'
Leto,Susan Buffum
Cas 23h0'15"53d48'
Letortu,Stephane
Sgr 19h29'49"-30d41'
Lett,K C
Dra 14h55'29"62d45'
Lett,Lacy
Mon 6h21'13"7d34'
Lettau,Dennis S
Dra 18h27'34"58d53'
Letterel,Nicholas Ralph
Aur 5h13'40"40d37'
Letterman,David
Uma 10h52'40"48d36'
Letterman,David
Sex 10h24'28"-1d48'
Letterman,Kathryn
Com 12h29'31"20d48'
Letterman,Ron Lynn
Ori 5h50'0"21d52'
Letters,Frank Gilbert
Aur 6h31'58"32d12'
Lettieri,Andrew
And 0h21'15"36d9'
Lettieri,Nicholas M
Lyr 18h30'23"30d11'
Lettington,Letha Maxine Hook
Cyg 21h2'32"31d50'
Letton,Robert W
Ori 5h7'25"1d15'
Lettow,Duane E
Gem 6h32'43"13d16'
Letty
Vir 14h16'56"-7d43'
Letz,Peggy Ann
Lyn 9h12'28"36d37'
Leu,David Jonathan
Aur 5h2'0"48d53'
Leubner,Katryn Christine
Del 21h1'47"12d45'
Leudway,Iris und Michael
Peg 23h16'14"31d29'
Leue,Judy L
Lyn 7h55'50"44d29'
Leuenberger,Travis Chase Nicholas
Lac 22h47'47"54d6'
Leuis,Jean-Pierre
Per 4h2'26"50d57'
Leukart,Petra Maria
Leo 9h25'1"18d17'
Leukothea-Meredith
Lyn 8h57'20"41d6'
Leung,Bonnie
And 23h35'0"40d3'
Leung,Jessica
Her 16h42'38"38d5'
Leung,William Wai Kit
Ori 4h41'1"10d7'
Leung,Wong Chin
Aql 19h24'23"-6d27'
Leupold,Brigitte
Lib 15h39'26"-20d12'
Leupold,Lisa
And 2h25'26"45d29'
Leurance,Leslie
Cas 0h38'12"61d20'
Leus Arthur
Cam 10h50'59"81d51'
Leuschen,Sam
Sct 18h53'17"-6d26'
Leuschner,Hans
Lyr 19h21'1"31d1'
Leuschner,Manfred
Her 17h19'44"43d14'
Leutfeld,Annekäthe
Boo 13h34'48"10d15'
Leuthold,Johannes Walter
Her 17h48'0"40d1'
Leutner,Peter
Cyg 20h24'28"31d36'
Leutrol
Dra 19h1'15"48d22'
Leutza,Baby
Cnv 13h2'49"50d59'
Lev,Alice Raful
Cyg 19h56'15"38d23'
Leva,Peggy
Com 12h3'27"20d1'
Levac,Barbara
Cas 2h48'18"70d29'

Levac,Chantal
And 23h35'37"49d27'
Levack,Alistair David
Ori 5h20'0"15d37'
Levaggi,Sandberg & Taylor
Oph 18h15'32"1d10'
Levandosky,Chester
Cyg 20h19'19"30d51'
Levangie,Cynthia Diane
Lyn 8h35'25"34d19'
Levant,Natalie K
Cam 7h8'36"71d5'
Levantino,Florence
And 2h17'55"45d1'
Levantino,Lisa Gaye
Lyn 7h35'60"39d55'
Levantino,Peter
Boo 14h3'20"25d41'
Levasseur,Gerald
Crb 16h1'52"36d13'
Levasseur,Lynn Lee
And 23h33'44"43d40'
Levasseur,Marci Kay
Lyn 9h8'1"39d50'
Leveille II,Dennis William
Aur 5h2'18"47d48'
Leveille,Benedicte
Cyg 19h42'48"30d3'
Leveille,Cynthia H
Lyn 7h56'19"43d39'
Levell,James
Lac 22h22'38"56d48'
Levene,Karen A
Lyr 18h18'28"38d33'
Levengood,Jennifer
Aql 19h5'60"14d60'
Levengood,Mark
Cnc 8h49'47"31d12'
Levenknight,Stefany Erika
Psc 1h2'28"23d7'
Levenston,Stella Jonathon L
Cap 21h19'37"-22d42'
Leventhal Jacqueline C
Eri 3h18'46"-12d50'
Leventhal,Drew Robert
Psc 0h55'58"32d10'
Leventhal,Ronald M
Cet 0h45'39"-3d60'
Leveque,Jean
Umi 16h24'13"70d16'
Lever,Allison Welborn
Eri 4h4'24"-17d9'
Leverence-Soul Mate, Thomas P
Per 2h9'20"58d35'
Leverett,Kathy
Mon 7h19'42"-6d32'
Levering,Kimberly
Uma 9h51'0"50d6'
Levers,Alicia Cathryn
Cas 3h1'0"61d7'
Levesque,Jeanne Lillian
And 1h0'45"40d27'
Levesque,Miranda
Lyr 18h44'1"46d33'
Levett,Susan C
Hya 9h35'18"2d9'
Levey,Forever Robert David
Lyn 8h29'41"33d25'
Levi,Connie & Jim
Cam 6h34'0"67d46'
Levi,Jason B
Lac 22h27'39"55d11'
Levin 30,Adam
Lac 22h23'18"50d32'
Levin Family,Jess S
Boo 15h5'0"11d1'
Levin Hahn-Pauls
Tau 5h32'11"28d40'
Levin's Star,Keithe
Dra 19h4'1"61d17'
Levin,Amanda Nicole
Peg 23h44'1"30d56'
Levin,Collin Mckanzee
Aql 18h54'35"10d42'

Levin,Grandma Sarah
Cyg 20h36'22"42d15'
Levin,Gregory Cole
Her 16h54'1"32d26'
Levin,Karl-Johan
Lyr 18h15'22"31d1'
Levin,Khea
Ari 3h6'44"30d44'
Levin,Leanna Rachel
Eri 5h8'42"-6d5'
Levin,Mart
Cyg 20h25'11"56d28'A
Levin,Matilda Kertzer
Aqr 20h57'1"-8d49'
Levin,Maxine"Aunt Mac"
Mon 6h42'22"2d59'
Levin,Sharyn
Cyg 20h25'11"56d28'B
Levin,William
Sct 18h52'51"-5d39'
Levine "Prine",Ronald
Dra 16h15'50"60d46'
LeVine & Kipp,Robert
Boo 14h57'54"31d29'
Levine,Abby
Ari 2h38'51"30d32'
Levine,Adam David- Friend Extraordinaire
Crb 16h5'40"38d17'
Levine,Alicia Celeste
Lmi 9h39'27"38d29'
Levine,Andrea
Cas 1h22'15"72d7'
Levine,Annabel
Ser 15h37'41"8d51'
Levine,Anne Rose
And 2h0'1"38d44'
Levine,Asher
Mon 6h23'46"8d4'
Levine,Barbara
Cnv 12h43'28"37d58'
Levine,Bruce E
Cep 2h48'0"80d22'
Levine,Carl Max
Per 2h3'38"50d14'
Levine,Dean E
Per 2h38'1"40d16'
Levine,Dorothy Brun
Lmi 10h41'23"26d49'
Levine,E Jean
Cas 0h34'16"60d29'
Levine,Edward Jay
Cam 3h24'31"60d41'
Levine,Elaine
Peg 21h30'57"20d15'
Levine,Elise & Harvey
Lac 22h40'27"48d45'
Levine,Harold
Ori 5h56'27"14d43'
Levine,Harriet "Bigfoot"
And 1h21'33"37d59'
Levine,Hilary Brooke
Peg 22h46'1"20d12'
Levine,In Memory Of Filmore
Cyg 21h53'29"42d43'
Levine,Jack
Psc 1h3'38"22d7'
Levine,Jason Charles
Oph 18h38'25"8d51'
Levine,Jeff
Aur 6h9'13"46d32'
Levine,Jeffrey M
Her 17h36'59"25d43'
Levine,Jesse Aaron
Cep 23h2'48"64d28'
Levine,Joel R
Her 16h45'53"50d30'
Levine,Julius "Poppi"
Sgr 19h10'51"-22d1'
Levine,Katherine
And 1h29'56"38d23'
Levine,Larry
Aur 6h22'26"37d36'
Levine,Laura Alexandra
Uma 10h13'24"67d49'

Levine,Linda
Aql 18h46'1"10d12'
Levine,Manfred
Boo 14h44'50"30d38'
Levine,Mark H
Hya 9h8'50"1d6'
Levine,Martha
Boo 14h43'49"51d16'
Levine,Nathan & Irene
Umi 15h13'19"70d55'
Levine,Neil
Del 20h21'17"11d7'
Levine,Philip
Cep 23h1'32"71d1'
Levine,René Ruth Bond
Aql 20h4'42"1d48'
Levine,Richard Andrew
Hya 8h59'32"1d25'
Levine,Ruth
And 2h5'15"38d37'
Levine,Seymour
Per 3h33'13"52d14'
Levine,Sgt Norman
Her 16h32'0"32d24'
Levine,Staci
Cma 7h38'0"10d56'
Levine,Sylvia "The Z"
Aur 4h49'0"40d12'
Levine,Zachary Nathan
Her 16h17'40"10d43'
Levine,Zöe
Psc 1h22'37"31d4'
Levinger
Cyg 21h35'54"41d32'
Levingston,Alfred A
Oph 17h27'41"-4d19'A
Levingston,Vivian K
Oph 17h27'41"-4d19'B
Levins,Chester
Cep 20h45'12"68d53'
Levins,Ellen
Aur 4h36'11"31d49'
Levinsohn,John
Hya 8h55'0"-6d53'
Levinson Star,Jerome H
Aql 19h27'45"15d9'
Levinson's,Paul
Ori 6h6'13"20d24'
Levinson,Dr Larry
Her 17h57'41"40d39'
Levinson,Emelie
And 2h11'54"43d1'
Levinson,Jeff
Hya 8h56'50"4d13'
Levinson,Joshua Kyle
Vul 19h16'30"22d21'
Levinson,Karen L.
And 0h58'39"40d38'
Levinson,PhD,Ronald N
Aql 20h11'30"11d13'
Levinstein,Robert F
Cma 7h51'38"11d46'
Levison,Dick
Aur 4h55'31"48d46'
Levison,Donna
And 2h26'50"38d27'
Levison,Henry Michael
Aur 6h35'44"32d30'
Levison,Michelle
And 0h10'57"47d12'
Levitan,Charan
Aql 19h55'32"13d30'
Levitan,Michael
Per 2h55'26"40d35'
Levitan,Safrah Angelica
And 0h24'48"44d41'
Levitas,Sasha Hope
Mon 6h59'41"8d4'
Levite's Dream 030962, A
Psc 0h53'1"11d46'
Levitsky,PhD,Dr Debra
Cas 0h41'13"64d25'
Levitt,Andrew
Aur 7h5'13"37d59'

Levitt,Ann Lee
Cyg 19h33'23"30d11'
Levitt,Art & Jamie
Cyg 21h30'11"44d30'
Levitt,Francine Bara
Cas 0h4'17"62d50'
Levitt,Mat
Boo 14h2'1"7d57'
Levitt,Philip
Cet 2h1'25"1d30'
Levitt,Rose
Lyr 19h20'1"34d43'
Levitt,Uncle Harold
Crb d15h25'0"31d60'
Leviy
Eri 3h56'34"-17d35'
Levon
Her 16h23'52"10d43'
Levon,Florence
Lyr 19h1'32"37d58'
Levora,Clarence
Her 16h41'47"32d5'
Levoyer,Patrick
Ori 5h1'49"10d25'
Levrero,Giulia Martina
Cam 14h4'50"82d27'
Levy Star Of The Protector,Jason
Aur 6h19'0"37d5'
Levy,Abby J
Peg 21h56'0"2d25'
Levy,Alice Victoria
And 23h17'38"48d55'
Levy,Anne
And 2h30'56"40d4'
Levy,Bernice & Sanford
Cam 3h53'0"57d19'
Levy,Carlos
Cep 20h39'13"65d31'
Levy,Danny
Lac 22h16'59"51d10'
Levy,Elliott Louis
Her 15h53'22"50d36'
Levy,Fred
Aqr 21h35'29"-5d54'
Levy,Glen I
Her 16h26'48"48d12'
Levy,Helen & Marty
Aur 6h39'22"38d18'
Levy,Hellmut Edward
Aur 5h52'18"31d2'
Levy,Hilma & Walter
Eri 4h5'32"-18d38'
Levy,Jessica Randi
Gem 6h42'30"30d10'
Levy,L Robert
Crb 16h1'29"37d35'
Levy,Lucille & Robert Geist
Vul 20h19'41"28d19'
Levy,Peggy
Lyn 8h25'40"34d37'
Levy,Rebecca
Cas 0h27'48"61d22'
Levy,Robert
Lac 22h26'54"48d53'
Levy,Robert-Judith- Seth-Rachel
Eri 3h22'23"-6d23'
Levy,Rose
Cas 0h24'19"56d2'
Levy,Rosemarie
Cas 2h4'19"61d47'
Levy,Teddi Q-Tip Woo Tip
Mon 7h13'56"-5d4'
Levy-Bailey,Star
Cam 4h38'0"78d51'
Levy-Richard,Dominique
Aql 19h58'54"-10d43'
Lew
Dra 18h26'56"50d16'
Lew,Linda
Tri 1h41'50"30d40'
Lewandowski,Charlie
Aur 5h24'34"40d54'

Lewandowski,Jr,Mark Anthony
Vul 19h18'32"26d58'
Lewandowski,Michael
Cru 12h7'49"-60d50'
Lewandowski,Ute
Cas 0h11'30"60d31'
Lewandowski,Violet
Gem 7h32'27"20d24'
Lewandowsky,Rüdiger
Lmi 9h20'1"38d29'
Lewbart
Sgc 18h57'25"18d36'
Lewen,Hermann
Aql 20h9'15"4d55'
Lewenhaupt,Lotta
Uma 11h7'44"52d17'
Lewert,Pretendiente Favorito!Ronald
Aqr 23h12'3"-10d57'AB
Lewicki,Reinhard
Lib 15h35'47"-28d55'
Lewicky,Yuri Michael
Del 20h25'13"8d43'
Lewie
Del 20h20'23"13d11'B
Lewin,Don
Cep 21h8'1"70d1'
Lewin,Dr Bruce
Lyn 7h57'41"43d54'
Lewin,Karen
Cyg 19h28'59"38d23'
Lewin,Rachel Marie
And 0h4'33"30d26'
Lewing,Debbie J
Com 13h7'38"16d18'
Lewinski,Dennis J
Cep 22h52'11"70d47'
Lewis
Cet 3h0'28"-0d18'
Lewis
Ori 5h34'53"0d10'
Lewis
Her 16h50'58"39d10'
Lewis #1 Teacher,Carla Kotas
And 22h58'53"50d13'
Lewis & Heidi's "Wedding Star"
Cyg 9h40'49"40d15'
Lewis Derick
Umi 16h59'37"76d54'
Lewis II,John McIntyre
Cep 2h28'46"78d49'
Lewis In The Sky
Boo 14h20'20"27d59'
Lewis MD,David Arthur
Llewellyn
Oph 18h5'49"0d56'
Lewis Of Independence Oregon,Kathy
Uma 10h5'57"60d47'
Lewis of Kegil
Cyg 21h19'21"39d23'
Lewis Star,The Edward
Cyg 21h26'22"50d26'
Lewis Together Always, James & Karen
Cyg 21h38'50"37d32'
Lewis"Blackie"Richard
Dra 16h43'32"71d5'
Lewis' Wedding Star, Jeff & Meredith
Sge 20h7'34"18d50'
Lewis,Abigail
Vul 19h59'0"28d59'
Lewis,Addison Hetzer
Cnc 8h57'30"22d6'
Lewis,Al & Solly
Eri 2h46'53"-6d29'
Lewis,Angela
Lyr 19h16'11"26d32'
Lewis,Ann Carol
Cas 1h9'41"61d18'
Lewis,Ann M
Ori 5h55'0"14d44'

Lewis,Ashley
Aql 20h17'56"7d38'
Lewis,Asia Inez
Uma 10h15'50"68d17'
Lewis,Betsy Sands
Mon 7h38'45"-0d36'
Lewis,Betty
Cet 0h1'41"-17d42'
Lewis,Bill
Cet 0h1'31"-15d23'
Lewis,Billy
Hya 8h15'38"2d9'
Lewis,Brad
Her 18h1'40"40d46'
Lewis,Bradly
Equ 21h4'0"11d57'
Lewis,Bree Star
Peg 23h25'1"18d13'
Lewis,Brenna Katherine
Cas 1h19'20"65d35'
Lewis,Brian David
Ori 6h1'22"0d35'
Lewis,Brian Wesley
Cmi 7h42'55"5d19'
Lewis,Brittney Nicole
And 23h22'19"43d29'
Lewis,Carolynn K
Vul 19h41'40"26d2'
Lewis,Cathy
Cas 0h13'52"63d45'
Lewis,Cecil Leslie
Boo 14h29'11"53d54'
Lewis,Christie Lynn
Mon 6h26'0"-10d9'
Lewis,Christina Lynn
Uma 12h9'46"57d20'
Lewis,Christopher Jordan
Dra 11h47'25"74d33'
Lewis,Claudia
And 0h36'60"45d48'
Lewis,Constance
Cas 0h58'41"61d4'
Lewis,Courtney Michele
Cas 0h57'1"63d45'
Lewis,Craig
Her 17h35'0"40d25'
Lewis,Daniel
Hya 8h16'46"2d32'
Lewis,Daniel Ryan
Her 17h12'1"43d45'
Lewis,David
Aur 5h23'43"50d20'
Lewis,David H
Lac 22h25'29"50d22'
Lewis,David Trent
Her 17h35'34"40d24'
Lewis,Dean
Her 17h20'32"38d39'
Lewis,Dean
Ori 4h50'40"4d46'
Lewis,Debbie
Cas 1h14'0"64d39'
Lewis,Denise R
Lyn 7h2'33"50d35'
Lewis,Diane
Mon 7h38'55"-4d30'
Lewis,Diane
Cyg 19h24'41"35d4'
Lewis,Diane P
Aql 18h59'25"-6d54'
Lewis,Donald D
Lyn 7h51'22"44d35'
Lewis,Donald H "Dinky"
Aur 6h13'14"37d36'
Lewis,Dorene
Cep 23h0'24"80d30'A
Lewis,Doris J
Cas 23h21'34"54d15'
Lewis,Dr Edward Leon
Her 16h52'22"48d59'
Lewis,Dr Steven
Oph 17h53'58"10d4'
Lewis,Dusty
Tri 2h41'42"31d31'

Lewis,Edward C
Aur 6h40'1"35d50'
Lewis,Eleanore M
Cet 1h33'1"-5d0'
Lewis,Erika
Vul 20h37'1"20d34'
Lewis,Francine Mary
Cas 0h20'58"62d57'
Lewis,Gail Boreali
Cas 1h7'57"61d12'
Lewis,Gary Ray
Lyn 6h35'0"58d40'
Lewis,Gaynor
Aql 20h5'56"4d15'
Lewis,Grady Arrasmith
Aql 19h57'14"10d53'
Lewis,H Frederick
Lmi 10h35'45"23d58'
Lewis,Hannah R
Vul 19h46'50"28d32'
Lewis,Heather Irene
Peg 22h0'27"30d10'
Lewis,J D
Lac 22h16'24"51d33'
Lewis,Jac
Her 16h0'51"47d51'
Lewis,James
Cep 20h42'58"60d38'
Lewis,James W
Aur 4h55'45"51d12'
Lewis,James Warren
Her 16h58'1"39d3'
Lewis,Jan
Uma 11h11'59"31d9'
Lewis,Janet
Boo 13h42'12"15d45'
Lewis,Jared Brett
Umi 15h35'27"77d56'
Lewis,Jean
Com 12h4'31"21d12'
Lewis,Jeanne
Eri 3h58'59"-13d6'
Lewis,Jeanne McCartney
Lyn 9h1'28"40d27'
Lewis,Jeffrey Peter
Ori 5h7'19"0d8'
Lewis,Jennifer A
Cyg 19h27'34"30d23'
Lewis,Jessica Louise
Cas 0h52'45"71d29'
Lewis,Joan Joel
Psc 1h24'23"32d26'
Lewis,John C
Peg 23h39'10"31d39'
Lewis,John Lawrence
Lyr 18h49'57"37d13'
Lewis,John Vaughn
Per 1h38'56"53d38'
Lewis,Jonathan George
Oph 17h36'1"11d17'
Lewis,Jr,Carroll A
Cet 1h9'0"1d24'
Lewis,Jr,Reginald
Cyg 20h4'36"31d33'
Lewis,Jr,Robert L
Boo 15h14'44"26d24'
Lewis,Jr,Tilford
Cmi 7h43'58"5d30'
Lewis,Justin Coulter
Uma 8h53'44"51d60'
Lewis,Karlee Lynn
Umi 16h0'36"70d6'
Lewis,Katharine L
Tri 2h9'0"32d58'
Lewis,Katherine
And 23h40'33"45d52'
Lewis,Kathryn
Uma 12h52'23"58d35'
Lewis,Kathy Lynn
And 23h41'49"46d59'
Lewis,Kim
Uma 9h28'42"68d37'
Lewis,Lance
Her 15h58'30"50d27'

Lewis,Leo "The Legend"
Leo 10h38'40"21d8'
Lewis,Lillian
Lyr 18h48'1"36d33'
Lewis,Linda Carol Landon
And 23h33'19"45d41'
Lewis,Linda Christine
And 0h12'38"39d22'
Lewis,Linda Elaine
And 23h0'27"49d20'
Lewis,Linda K
Ori 5h51'33"18d33'
Lewis,Lisa
Mon 7h1'0"0d1'
Lewis,Loren Kay
Lyr 18h55'23"40d18'
Lewis,Lorna
Peg 21h55'57"24d56'
Lewis,Lorre Lea
Peg 23h40'54"8d9'
Lewis,Madison Kaye
Cas 22h57'0"56d48'
Lewis,Margaret Elizabeth
Peg 22h27'38"20d32'
Lewis,Mark
Ori 5h31'47"-1d36'
Lewis,Marla
Eri 3h38'0"-14d51'
Lewis,Martha
Vul 20h55'52"28d36'
Lewis,Martin
Her 17h19'17"40d9'
Lewis,Mary Elizabeth
Dra 15h15'35"65d16'
Lewis,Mary Lynn
Cam 6h32'0"68d25'
Lewis,Mary M
Lyn 7h1'18"58d29'
Lewis,Melony Janina
Aql 19h32'33"0d48'
Lewis,Memrie
Uma 10h37'10"47d40'
Lewis,Michael
Her 17h6'20"20d7'
Lewis,Michael A
Lmi 10h4'47"39d50'
Lewis,Michael David
Dra 15h54'45"57d47'
Lewis,Miss Natalie
Crb 16h15'0"32d42'
Lewis,Mitzi & Harvey
Cyg 19h27'14"32d36'
Lewis,Monica A
Lyn 8h4'58"38d42'
Lewis,Patricia Ann
Cas 23h28'52"53d38'
Lewis,Peter B
Aur 6h56'13"43d54'
Lewis,Pricilla
Aur 4h58'29"40d17'
Lewis,R Clark
Boo 14h25'21"38d40'
Lewis,Reggie
Per 4h2'52"51d46'
Lewis,Rex
Aur 5h4'54"41d40'
Lewis,Richard P
Cep 21h6'58"55d5'
Lewis,Richard Peter
Per 1h52'1"56d36'
Lewis,Robert
Cep 21h4'20"65d21'
Lewis,Robert David
Her 17h50'22"48d51'
Lewis,Robert Edison
Cmi 7h57'20"3d53'
Lewis,Robert Harry
Aql 19h56'37"10d44'
Lewis,Robert John
Cep 22h26'41"63d43'
Lewis,Robert P
Dra 16h6'13"52d19'
Lewis,Robert R
Uma 10h50'27"40d23'

Lewis,Rodney F
Cnv 12h17'26"34d42'
Lewis,Ron G
Eri 2h45'32"-4d31'
Lewis,Russell Scott
Aql 19h9'54"3d14'
Lewis,Ruth Ellen
Cas 0h46'40"60d32'
Lewis,Samantha Allisa
Gem 6h46'57"30d37'
Lewis,Sandra Fitzpatrick
Peg 22h0'59"23d49'
Lewis,Sarah Wilcox
Mon 6h29'51"8d26'
Lewis,Scott
Aur 5h9'0"40d49'
Lewis,Sesha Dea
Cyg 19h37'1"40d3'
Lewis,Sharon (Ashara)
And 2h21'32"42d37'
Lewis,Solina
Lyr 18h46'28"35d26'
Lewis,Sondra
And 0h8'18"38d40'
Lewis,Sonya
Lyr 18h15'1"40d8'
Lewis,Sr,Richard
Aql 19h51'56"10d50'
Lewis,Stanley J
Aur 6h24'42"33d25'
Lewis,Stephanie
Cyg 20h23'34"38d54'
Lewis,Steven
Lac 22h49'15"53d25'
Lewis,Steven
Cyg 21h1'52"37d58'
Lewis,Sylvia
Per 2h38'56"35d56'
Lewis,TC
Cam 13h34'39"77d48'
Lewis,Terry
Aql 19h3'0"-0d36'
Lewis,Terry & Karyn
Cyg 21h19'0"31d10'
Lewis,The Star Of David John
Pho 0h43'13"-42d49'
Lewis,Thomas J
Oph 18h7'29"0d1'
Lewis,Thomas Jason
Her 17h12'52"44d45'
Lewis,Thomas William
Per 2h2'45"57d25'
Lewis,Tilson & Charlotte
Leo 9h38'26"10d19'
Lewis,Trevor William
Dra 16h4'57"61d28'
Lewis,Umpire Thomas
Ori 6h3'0"1d11'
Lewis,Vaughan K
Aur 6h52'30"37d3'
Lewis,Walter L
Per 4h22'26"50d55'
Lewis,Wilburn Elzo
Mon 6h43'0"7d36'
Lewis,William & Dawn
Lyr 19h0'37"38d3'
Lewis-Rakuljic,Shirley
Peg 22h34'17"29d57'
Lewis-The Pumpkin Star
Sarah Taylor
Cnc 9h2'1"31d4'
Lewistar
Dra 17h49'37"60d53'
Lewman,Todd A
Cet 2h6'0"2d24'
Lewoczko,Susan Stumpf
Uma 8h52'20"50d31'
Lewter,Jennifer Lynette
Tri 1h48'45"28d58'
Lewter,Melvin Dan
Cet 1h30'11"0d9'
Lewter-Flood,Sharlene
Cyg 21h4'38"36d44'
Lewton,Ida Alice
And 0h5'1"35d34'

Lex 21
Eri 3h15'11"-14d28'
Lex I
Lyn 8h0'20"39d28'
Lex,Ekkehard
Cep 22h5'20"60d58'
Lex-Tugaw,Heidimarie
Mon 7h45'7"-3d14'
Lexi July
Eri 3h28'0"-5d25'
Lexi Nicole
Equ 21h6'41"2d56'
Lexo,Dianne E
Boo 14h18'27"48d44'A
Lexo,Joan F
Boo 14h18'27"48d44'B
Lexvold,Terry
Cam 7h20'43"73d10'A
Ley Family Star,The
Uma 11h25'40"71d14'
Ley Star,The
Cyg 19h32'51"39d14'
Ley,Ralph Frederick
Cep 21h47'21"56d9'
Ley,Thomas Cooper
Cnv 12h50'31"51d5'
Leya Elizabeth
And 0h53'15"40d7'
Leyden,Erin
Lmi 10h42'50"27d21'
Leygriff,Becema
Lyn 8h38'0"45d17'
Leygue,Fredrique
Dra 20h27'58"75d5'
Leyla
Com 12h47'51"20d45'
Leyli
Cep 21h28'57"61d40'
Leyrer,Freda Beatrice
Boo 14h33'22"41d21'
Leyshon,Danielle Marie
Cam 4h56'29"67d42'
Leyva,Aurora L9"67d42'
Ori 5h28'1"0d12'
Leyva,Lisa
Ser 16h15'18"-0d46'
Leyva,Mario Roberto
Aur 6h1'56"32d45'
Leyva,Martha
Cyg 19h59'59"40d7'
Leyva,Martha
Crb 16h0'28"27d2'
Lezeray,Jean-Claude
Cyg 20h33'15"46d57'
Lezlie
And 23h1'18"36d13'
Lezotte's Legacy
Ari 1h45'52"17d46'
Lezotte,Ellen Webb
Cam 5h36'48"61d49'
Lezotte,Patrick Jeffery
Peg 22h35'56"20d14'
Le,Peg 22h35'56"20d14'
Sge 19h57'28"16d8'
Le Sge 19h57'28"16d8'
Dra 17h18'14"70d9'
Lhota,Tamra Roberts
Crb 16h3'49"32d8'
Li Mei
Lac 22h3'35"49d10'
Li,Agnes Akane
Cyg 20h22'53"38d44'
Li,Arthur K C
Umi 13h3'22"70d4'
Li,Chun
Dra 15h50'30"65d16'
Li,David
Hya 9h39'51"-10d16'
Li,Sharron
Mon 7h9'34"-5d49'
Lia
Pic 4h34'32"-46d27'
Lia
Gru 22h12'5"-54d13'
Lia
Vir 14h11'25"-7d43'
Lia Victoria
And 0h22'0"37d57'
Lia,Kari
Lib 15h41'0"22d26'
LiaBraaten,James O
Cma 7h23'40"-18d58'
Liakas,Markos
Ori 5h53'51"18d7'
Liaki
Lyr 18h37'35"40d32'
Liam & Lauren
Cyg 20h25'25"42d1'
Liam's Love For Skylar
Mon 7h4'28"-6d48'
Liam's Star
Hya 8h43'17"3d25'
LiAn
Cas 0h34'27"66d21'
Liana
Dra 15h5'16"60d42'
Liang Star
Aql 18h58'13"-7d55'
Liano
Ori 5h58'25"21d3'
Liao,Maguy
Cam 5h5'0"65d9'
Liapes-My Love, Catherine Ricci
Cyg 21h33'34"40d59'
Liardi,Roseanne
Vul 21h0'56"24d19'
Lias,Jim
Her 17h53'21"14d56'
Libbey
And 23h24'37"45d12'
Libbey,Michelle
Lyn 9h1'47"33d21'
Libby
Umi 15h6'0"65d59'
Libby
And 22h59'39"50d28'
Libby "The Scorpio"
Sco 17h51'17"-40d17'
Libby & Sarah
Per 3h46'13"39d32'
Libby K
Oph 18h0'1"8d52'
Libby's Wish
And 2h35'43"41d15'
Libby,John Willis
Cep 20h15'21"75d46'
Libby,Rick
Her 16h39'16"35d33'
LiBentley,Patricia Ann
And 23h24'46"48d31'
Liberatore,Anne Lauren
Cam 5h8'0"60d11'
Liberatore,James F
Aur 6h26'42"31d20'
Liberatore,Jr,R J
Lac 22h10'29"51d47'
Liberman,Caryl & Larry
Vul 20h0'33"28d39'
Liberman,Saul
Cnv 13h49'18"31d40'
Liberman,Saul
Her 17h14'31"48d19'
Libert,Marie-Astrid
Sgr 19h33'34"-38d55'
Libert,Thomas M
Aur 5h1'46"46d12'
Liberti,Gabe O
Aur 6h26'32"35d22'
Liberti-Morandina, Sylvia C
And 2h7'22"40d4'
Libertini,Chris
Umi 16h48'59"77d46'
Liberty Check Printers
Aql 19h54'42"15d13'
Liberty,Thomas A
Aur 7h2'28"36d39'
LibertÖ,Amore Michele
Boo 15h5'29"38d32'
Libiez,J
Sct 18h47'60"-7d44'
Libman,Alan David
Ser 15h49'0"23d59'
Libra Music Ltd
Lib 14h20'17"-8d48'
Libra Rita
Lib 14h46'34"-8d58'
Librandi,Angela
Lyr 18h28'46"38d14'
Liburti,Hallie
And 2h30'51"48d3'
Licameli,Joseph
Cep 23h3'37"64d17'
Licare,Thomas Michael
Uma 12h59'43"54d1'
Licari,Tess
Cas 0h2'56"62d46'
Licata,Carole June
Cas 1h10'23"63d10'
LiCata,Michele
Vul 19h48'22"23d13'
Licata—Happy 50th!, Pauline & John
Uma 11h15'40"42d11'
Liccardi,Michael Enoq
Aur 6h9'0"38d54'
Licciardi,Roland
Lyr 18h56'35"31d42'
Lich,The Legendary Little Guy
Her 18h17'25"14d10'
Lichnousky,Albert Lee
Ser 18h36'47"3d2'
Lichodziejewski,Ian Reece
Cet 1h57'50"-8d41'
Licht,Judy Elaine
Mon 6h20'60"-5d44'
Licht,Jürgen
Tau 5h56'49"27d36'
Licht,Ronald Paul
Aql 20h0'25"11d43'
Licht,Star
Aur 5h18'25"42d57'
Lichtenberg,Brett R
Uma 9h2'60"48d15'
Lichtenberger III,John
Aur 6h31'39"33d60'
Lichtensteiger,Manfred
Her 16h54'22"38d20'
Lichtenstein,Gayle West
And 2h1'36'41d27'
Lichtenstein,Gerald
Cnv 12h24'1"42d27'
Lichtenstein,Jessie Paige
Lyr 19h21'42"30d43'
Lichtenstein,Landon Amory
Her 16h58'25"24d13'
Lichtenstein,Landon Amory
Lac 22h26'40"52d56'
Lichtenstein,Steven
Cep 22h39'44"70d39'
Lichtenwalter,Jeff C
Ori 5h9'46"-5d14'
Lichter,Kelly Lynn
Vul 20h21'0"23d28'
Lichtman,Scott
Boo 15h3'14"26d24'
Lichtner,Michael
Uma 10h4'18"58d18'
Licia
Cnv 12h15'24"51d58'
Lick,Nancy L
Lyr 18h36'0"38d7'
Lickel,Magdalena Long
Mon 6h27'57"8d44'
Licker,Nancy
Vul 20h1'12"22d48'
Lickey,Jackie
Peg 22h15'0"33d11'
Licklider,Matthew Alan
Dra 17h2'39"52d9'
Licona,Laura Cristina
Cas 1h11'43"75d41'
Licopoli
Cam 4h3'35"70d56'
Lida's Star
Del 20h23'15"7d57'
Lidagoster,Mark Ian
Sgr 18h55'35"-28d1'
Liddell,Jared
Dra 12h52'50"70d58'
Liddle,Wayne
Peg 23h42'43"30d41'
Liddy,Kevin Gerard
Sge 19h58'57"20d18'
Lide,Cody Tyler
Cep 22h54'39"57d27'
Lidiard,Emma
Sge 20h4'48"18d60'
Lidman,Fred
Cep 22h45'57"70d43'
Lido Elementary School
Uma 9h7'58"53d56'
Lidsky,Anita & Simon
Mon 6h54'33"-6d48'
Lidston,Evelyn K D
Cas 1h51'1"61d42'
Lidstone,William John
Boo 14h37'40"40d14'
Lidwina
Cyg 21h29'19"40d36'
Lieb 22 02 1938,Denise
Cas 1h42'30"60d6'
Lieb,Briana Michelle
Tau 5h45'0"26d30'
Lieb,J Michael
Cap 21h11'52"-22d50'
Liebau,Karl-Friedrich
Tau 5h31'1"20d29'
Liebe,Amy Ewige
Peg 22h57'15"33d46'
Liebe,Lust & Leidenschaft-M & J
Cmi 7h16'14"4d16'
Liebe,Pookie
Eri 4h0'31"-18d8'
Liebenow,Franklin & Katherine
Eri 3h52'47"-2d27'
Liebens Feuer Stern Sieben
Aur 6h28'23"37d34'
Liebenthal,Mayaba
Sge 20h1'17"20d5'
Lieber Star,The Ben & Rachel
Uma 9h51'18"52d54'
Lieber,Danielle Kara
Sco 17h4'31"-31d5'
Lieber,Josephine E
Mon 7h45'1"-2d11'
Lieber,Sam
Sgr 19h6'43"-25d15'
Liebergot,Harry
Cep 22h56'13"65d7'
Lieberman,Cassandra (Civi)
Tau 3h55'34"20d19'
Lieberman,Janine "The Queen"
Cas 1h48'24"60d3'
Lieberman,Keith Erica
Cyg 20h1'29"50d35'
Lieberman,Melysa Ann
And 23h2'24"37d31'
Lieberman,Rachel Janet
Peg 21h27'1"2d19'
Liebers,Alexander William
Vul 19h20'12"26d41'
Liebert,MD,Peter Selig
Lac 22h31'21"54d21'
Liebert,Midge & George
Peg 22h30'51"27d41'
Liebgen,Brigitte & Manfred
Sge 20h7'41"20d44'
Liebgen,Charlotte
Cas 2h15'49"61d37'
Liebhaber,Viki
Cas 0h15'1"62d59'
Liebich,Mason
Her 16h59'48"18d48'
Liebig,Astrid
Tau 5h55'50"24d4'
Liebisch,Peter
Cnc 8h24'17"30d31'
Liebl,JoAnna
Umi 16h16'43"72d8'
Liebl,Max
Aur 5h1'12"38d58'
Liebling Rainmar
Cap 20h43'15"-26d39'
Liebman,Ralph
Aql 19h24'33"7d49'
Liebovich,Lou Bonnie Alyse Sheri
Ori 5h26'47"15d18'
Liebowitz,Patsy Jean
Cas 1h49'25"73d46'
Liebowitz,Sholem
Cyg 19h29'22"33d13'
Liebscher,Karin
Lac 22h2'21"50d1'
Liebscher,Lawrence Arthur
Lib 15h57'37"-8d4'
Liebzeit,Jason
And 0h23'11"43d43'
Liechti,Annah C
Mon 7h37'1"-5d54'
Liedel,George & Judy
Sge 19h58'0"16d21'
Liedgens
Sge 19h52'30"16d29'
Liedholm,Hakan
Leo 11h32'39"-5d6'
Liedig,Thomas F
Ser 15h12'18"20d35'
Liedl,D C C
Aur 6h26'49"37d47'
Liedtke,Katja
Cap 21h1'51"-26d21'
Liefde,Hedy & Ron
Cyg 20h43'23"42d29'
Lieftenchaus
Cet 3h17'26"5d49'
Liegl Star,The Peter
Cep 21h56'56"55d14'
Lien,Jean & Dave
Cam 6h10'23"70d35'
Lien,Sydney Marie
Cas 23h34'24"62d39'
Lienhart,Alexandre
Aur 7h9'57"38d30'
Lientz,Delores
Cyg 20h32'0"50d1'
Liepold,Klaus
Cyg 20h27'40"30d57'
Lier,Patrick
Ser 15h18'58"8d34'
Lierman,Chad Eric
Aur 6h16'28"30d39'
Lies,Anita
Boo 14h42'23"30d28'
Lieselotte
Lyr 19h23'44"31d1'
Liesl
Cyg 19h46'38"30d4'
Liesl & Loli
Cnv 12h14'13"40d50'
Liesl Bean
Cam 4h32'40"61d37'
Liesl Marie
Peg 22h16'0"21d23'
Lietard,Marie
Cas 1h0'32"56d9'
Lietmann,Walter
Lyr 19h20'43"31d29'
Liette,"Dougie"
Dra 16h18'53"60d0'
Lietz,Penny & Artie
Eri 2h55'57"-12d5'
Lietz,Volker
Lib 15h20'0"-23d34'
Lieval,Robert M
Cet 2h31'0"1d35'
Liew,Mary
Cyg 19h24'47"35d34'
Life After Peanut World
Per 4h41'58"51d54'
Lifshen,Susen
Mon 8h4'23"-6d47'
Lifter,Les
Her 18h4'44"41d3'
Lifter,Marcy Robin
Cas 1h21'59"61d51'
Lifton,Zelma Sophia Koehne
Aql 20h15'42"1d40'
Ligatti,Tony
Boo 15h0'18"27d42'
Ligeros,Mike
Cet 0h37'0"1d1'
Liggett,Margaret Ann
Aql 19h30'51"13d25'
Light 9-27-95,Skyler Madison
Aql 19h55'38"10d46'
Light of Bram
Dra 20h37'1"70d52'
Light Of Bryan,The
Per 2h39'28"40d59'
Light Of Candyce,The
Mon 7h3'31"5d4'
Light of Christine
Lib 15h0'16"-2d30"
Light of Chrystal,The
And 23h0'57"37d14'
Light Of Consciousness -SMG
Aur 5h5'24"40d22'
Light Of David's Love, The
Her 16h59'40"32d15'
Light of Dawn
Ari 1h51'56"10d37'
Light Of Dawn,The
Del 20h16'17"14d48'
Light of Earl-Wechsler
Aur 5h28'11"54d45'
Light of Eleanor
Sgr 19h34'11"-41d34'
Light Of Erin's Smile, The
Aur 6h20'60"35d3'
Light of Friendship
Cyg 21h57'28"52d34'
Light of Healing & Peace for All
Uma 8h54'1"56d34'
Light Of Love-Dee Greg Brad Clint
Tau 3h56'13"20d10'
Light Of Lucy Clare
And 23h15'52"50d16'
Light Of My Life Carson & Grandpa
Eri 3h12'37"-16d17'
Light of my Life,The
Lac 22h54'41"54d2'
Light of Paris
Cam 5h52'18"60d59'
Light of Robyn & Havier
Cyg 21h3'22"40d23'
Light of Sean,The
Boo 15h4'0"20d58'
Light Of The George's
Dra 14h52'24"62d8'
Light of the Medicine Gypsy
Psc 1h3'17"22d15'
Light Of Truth Ben & Amy '95,The
Sge 19h59'56"16d30'
Light,Ann Ruth
Cas 0h52'48"68d58'
Light,Antonia Lauren
Peg 22h16'60"34d37'
Light,Kellie Lynn
Cmi 7h26'17"0d44'
Light,Leigh
Cam 10h54'57"80d44'
Light,Mike & Linda
Uma 9h34'18"47d24'
Light,Rosemary
Cas 0h28'16"60d16'
Light,The
Cet 2h2'1"-0d17'
Light,Wanda Adams
Peg 22h15'57"5d36'
Light-Of-My-Life
Eri 5h3'25"-8d10'
Lightbody,Thomas William
Dra 14h52'0"64d10'
Lightcap,Roberta
Cyg 21h1'43"30d22'
Lighten Fawn
Vul 20h21'1"25d3'
Lightfoot,Teddi
Aql 19h13'47"14d41'
Lightheart (The Gog Star),The Fiona
Cas 1h20'48"53d22'
Lighthorse,Randall H
Her 17h24'1"47d35'
Lighthouse Bill
Peg 23h1'59"10d11'
Lightkep,Joan Marie
Cyg 20h31'18"37d31'
Lightner-Walters,Tina Leigh
Com 12h32'14"28d36'
Lightning
Uma 11h20'47"33d28'
Lightowler,Nicky Detlef
Ori 4h59'49"0d39'
Lightowler,Rachel
Cas 22h59'48"54d22'
Lights Of Lorna
Del 20h18'23"13d22'
Ligman,Jordan Rae
Dra 17h6'34"61d42'
Lignell,Patricia Ann
Hya 8h54'27"1d47'
Lignon,Antonio
Col 6h1'9"-35d57'
Lignum
Boo 14h20'26"12d55'
Ligon,Catherine
Cyg 21h31'37"52d37'
Ligon,Leslie
Sco 16h2'17"-40d43'
Ligon,Mark G
Aur 6h0'1"38d30'
Ligon,Marty Parish
Per 3h18'36"41d3'
Liguori,CFP,Frank C
Cep 23h51'10"82d53'
Liguori,Daniel Cristopher
Uma 9h16'55"44d26'
Liguori,John
Aql 19h4'29"2d45'
Liguori,Lisa Avigdor
Com 12h10'31"20d7'
Liguori,Tom & Sue
Aql 18h59'29"13d49'
Lih,June
Cas 1h50'0"60d10'
Liisa & György
Uma 9h1'37"50d56'
Lija-The Star That Shines On Us
Sct 18h38'44"-4d20'
Lijay
Cnc 9h4'46"30d59'
Like Lisa
And 0h47'53"38d40'
Likens ABR
Ori 5h50'51"10d36'
Likens,Dickie
Aur 6h32'15"37d48'
Likos,Billy Boy
Leo 9h58'27"11d27'
Lil Egg
Boo 14h58'21"40d47'
Lil G
Oph 18h40'50"7d40'
Lil Hub
Lyr 19h4'29"38d49'
Lil Jimmy
Uma 10h18'48"67d51'
Lil Re
Cyg 21h44'23"28d33'
Lil Roo
Lyr 19h25'36"41d55'
Lil St Nicholas Merry Xmas Star 94
Hya 8h59'49"3d48'
Lil' Angelface
Peg 22h20'0"33d47'
Lil' Doots
Sco 17h52'46"-38d46'
Lil' E
Cam 4h19'1"70d17'
Lil' Kristopher
Sgr 18h50'49"-35d48'
Lil' Paul
Dra 16h9'25"68d10'
Lil' Red
Lyr 18h43'0"34d45'
LIL' RIF
Lyr 19h21'52"31d34'
Lil' Sister,Carol Jean
Cam 5h36'17"61d6'
Lil'Hug
Peg 22h43'47"33d37'
Lil(Rose)
Cas 0h14'25"63d59'
Lil-Lee-Pat
Cam 5h44'1"60d28'
Lil-Pooter
Lyr 18h20'28"47d17'
Lila B
Lyn 7h46'24"38d38'
Lila,Marja-Leena
Sge 18h57'15"19d9'
Lila-Lloyd's Li'l Love Forever
Lyn 8h5'17"46d18'
Lilalalila
Lyr 18h55'1"31d7'
Lileikis,Molly
Peg 22h41'18"27d56'
Lileikis,Robin
Cyg 20h35'1"42d55'
Lili
Dra 17h44'34"61d40'
Lili
And 23h37'0"48d55'
Lili-Ruffie
Oph 17h54'39"13d22'
Lilia e Stefano '87
Cam 7h5'29"61d23'
Lilian
Uma 8h34'56"58d47'
Lilian Anita
Lyr 18h29'39"31d31'
Liliana
Peg 22h59'14"34d1'
Liliana
Aql 19h31'11"12d17'
Liliana
Ori 5h56'47"20d38'
Liliane
Lac 22h3'34"49d53'
Lilin's Stjerne
Mon 7h16'0"-1d16'
Liling Hu
Mon 6h45'24"11d29'
Lilita Anne
Cet 2h15'54"5d10'
Lilja,Dorothy Lillian
Vul 19h2'19"25d23'
Liljedahl,Joan & Louie
Lyr 18h59'56"29d21'
Liljedahl,Wendall Brian
Cyg 20h40'36"42d42'
Liljegren,Dana
Gem 6h52'19"30d2'
Lill,Mary-Margaret Lorraine
Cyg 19h35'53"38d47'
Lill-Sue-Ann-Rob & Imogen's Star 95
Aur 6h26'1"34d57'
Lilley,Jennifer Ann
And 1h58'0"40d47'
Lilley,John & Virginia
Cyg 21h55'40"53d15'
Lilley,Leslie Arthur
Boo 14h28'50"23d42'
Lilley,Mary
Tri 1h30'45"28d36'
Lilli,Norbert
Aql 18h55'29"17d53'
Lillian
Sgr 18h55'51d-25d12'
Lillian
Aql 20h31'40"0d17'
Lillian
Vul 19h45'48"28d13'
Lillian
Lyr 18h31'47"32d51'
Lillian
Cam 3h42'54"74d30'
Lillian
Cas 23h21'50"57d15'B
Lillian Leigh
Mon 6h3'58"-10d1'
Lillian's Hope
Lmi 10h50'51"31d41'
Lillianne Grace
And 23h12'1"38d32'
Lillianus,Eugene
Her 17h14'36"42d9'
Lillie
Her 17h2'34"20d23'
Lillie,Heather Beth
Cas 0h54'41"61d59'
Lillie,Tori Anne
Lyr 18h59'1"33d43'
Lillingston,Hugh & Cat
Uma 9h5'32"68d8'
Lilliput
Peg 23h32'0"17d37'
Lillis,Charles J
Hya 9h6'0"-5d58'
Lillis,Kacey
Lyn 9h5'30"39d43'
Lillis,Lauren Elizabeth
Cyg 19h14'39"45d33'
Lillith
Cas 1h17'45"60d6'
Lilliumblù
Gru 22h12'55"-53d48'
Lillo-Massimiliano Di Mambro
Dra 18h10'41"51d4'
Lilly
And 2h28'54"42d54'
Lilly
Cet 3h1'1"4d47'
Lilly,Anna Michelle
Cam 6h51'59"64d53'
Lilly,Brenda Lee
Mon 6h28'46"-6d41'
Lilly,Cameron Hastings
Oph 16h51'27"-25d3'
Lilly,Earl H
Lmi 9h45'25"38d40'
Lilly,Lance Warren
Cet 2h23'16"-5d29'
Lilly,Lindsey
And 1h35'51"37d37'
Lilly,Loren Wayne
Cas 0h42'57"60d28'
Lilly,Sheila
And 23h46'14"45d19'
Lilly,Virginia Smith
Uma 8h37'19"47d49'
Lilly-Mattia-Clarissa
Uma 11h20'0"50d48'
Lilo
Ser 16h1'46"1d5'
Lilo
Cas 1h58'30"60d9'
Lilou
Dra 11h0'51"73d17'
Lilou
Uma 11h39'32"61d7'
Lily
Aql 19h26'22"8d18'
Lily Leigh
Sge 20h16'1"18d51'
Lily Rose
Peg 23h5'0"32d20'

Lily's Star Lyn 8h54'39"42d39'
Lim,Clifford We Sun Hya 9h7'29"3d39'
Lim,Marissa & Mikayla Aur 6h7'53"43d10'AB
Lim,Stephanie Nicole Cas 23h22'19"53d44'
Lima Aur 6h59'25"37d7'
Lima,Chris Mon 7h1'1"0d27'
Lima,Father Orlando H Oph 17h3'51"-22d48'
Lima,Jean Garcia Ser 15h17'25"7d40'
Lima,Ruthy Cas 1h15'1"60d1'
Lima,Victoria Rose Gem 7h57'44"32d53'
Lima-Duret,David Freddy Cep 21h50'18"68d39'
Lima-Duret,Sarah Cas 2h17'24"61d12'
Limabean Com 13h6'50"28d36'
Limare,Jacques Cam 3h27'48"60d60'
Limbaugh III,Rush H Aql 20h2'40"0d28'
Limbaugh,Mega Dittos By & For Rush Cnv 13h57'26"45d38'
Limbaugh,Patriot on Loan,Rush Ser 15h49'22"21d44'
Limbaugh,Rush Uma 10h56'59"70d2'
Limbaugh,Rush Her 17h14'50"22d33'
Limberg,Claudia Oph 18h7'11"0d49'
Limberg,Helen Cyg 19h32'1"32d21'
Limberger,Adore Uma 9h20'54"56d12'
Limitless Dream Sgr 19h5'6"-15d18'
Limmer,Elisabeth Umi 15h5'38"67d14'
Limon,Mathieu Ser 15h17'18"11d8'
Limozon,Jean Claude Dra 15h5'49"62d30'
Limpach,Jr,Ronald M Her 17h19'54"45d26'
Limprecht,Nellie Cma 7h0'15"-11d5'
LimCma 7h0'15"-11d5' Ser 15h34'53"7d60
Lin Star Ori 5h56'55"16d20'
Lin's Star Per 1h37'51"54d13'
Lin,Andrea And 0h54'59"40d44'
Lin,Fred Shar-Chaun Lac 22h4'60"46d14'
Lin,Paul & Eileen Ori 5h57'39"20d2'
Lina Cas 0h40'51"63d55'
Lina Cam 8h24'37"80d24'
Linaka "Woman Who Keeps The Stars" Mon 6h30'1"-10d30'
LinaMark Crb 16h3'47"32d18'
Linares,Victoria C Psc 1h43'29"21d37'
Linares,Véronique Cas 1h29'59"58d52'
Linbea Cyg 19h26'59"35d52'
Linblad,Jennifer Hya 8h58'26"2d30'
Lincalis,Rosemarie Cyg 20h8'51"40d51'
Linck III,Charles Fredrick Vir 13h6'23"-1d54'
Lincoln,Anita Lyn 7h52'13"44d8'
Lincoln,Jeffrey Scott Mon 7h6'52"-0d37'
Lincoln,Nancy E Tri 1h50'17"26d34'
Lincoln,Sarah Peg 21h40'30"21d3'
Lind's Star,Andrew Keenan Utkilen Dra 17h7'60"51d35'
Lind's Star,Peter Christian Utkilen Per 1h39'56"53d26'
Lind,Barry J Her 17h59'55"40d20'
Lind,Christian Myles Ori 6h3'42"-1d35'
Lind,Emma And 1h57'14"38d7'
Lind,Eric Thomas Lac 22h21'28"37d52'
Lind,Harold Edward And 23h34'49"45d43'
Lind,Joanna Rose Cas 0h58'0"75d38'
Lind,Kathryn Daniels Becker Sco 17h9'35"-38d38'
Lind,Keith J Aur 5h39'33"40d57'
Lind,Light Of Louise Cas 0h48'34"72d50'
Lind,Roy E Dra 20h23'29"64d14'
Linda Lyr 18h29'1"32d15'
Linda Lyr 18h23'34"40d7'
Linda Cas 0h59'29"61d0'
Linda Lyn 8h47'51"44d4'
Linda And 02h31'51"39d51'
Linda And 23h37'1"48d36'
Linda Peg 22h16'1"30d27'
Linda Mon 6h42'55"6d6'
Linda Mon 6h41'32"11d58'
Linda Lyr 18h46'37"35d22'
Linda Cas 0h39'18"74d13'
Linda Cmi 8h6'51"0d13'
Linda And 23h4'13"37d55'
Linda Mon 8h1'33"-3d13'
Linda Cet 2h26'36"-2d1'
Linda Uma 9h45'33"50d26'
Linda Mon 7h4'30"-5d43'
Linda Lyn 6h28'1"58d16'
Linda Cas 0h44'57"62d45'
Linda "Kizzy" Jean Cnv 12h24'54"35d41'
Linda & Brandi Friends Forever Ori 4h59'21"10d3'
Linda & Daniel Cyg 21h3'19"39d10'
Linda & James Friends Forever Uma 10h26'16"56d37'
Linda & Jeffrey Per 2h41'22"34d50'
Linda & Jim's Star Cyg 21h4'56"38d10'
Linda & Louis Cyg 19h17'37"44d45'
Linda & Paul Uma 11h52'24"32d24'
Linda & Peter Forever Ori 6h5'37"7d15'
Linda & Scott Aur 5h20'1"37d49'
Linda & Scott's First Anniversary Crb 15h47'1"28d23'
Linda 50 Com 12h42'45"23d11'
Linda 69 Aql 19h54'13"12d15'
Linda Agness Lyr 18h52'0"42d50'
Linda Amey And 2h33'34"40d6'
Linda Ann Tau 5h23'25"16d3'
Linda Ann Del 20h18'1"10d12'
Linda Ann Tau 4h16'0"7d56'
Linda Aurora,The Aql 20h0'51"14d18'
Linda B Com 12h38'21"23d3'
Linda B Tri 2h11'57"32d2'
Linda Bug Cas 0h8'55"58d53'
Linda Carol Cas 0h48'40"66d28'
Linda Christine Uma 9h45'19"50d48'
Linda Claire And 0h12'1"47d9'
Linda D Sge 19h31'54"18d59'
Linda Diane Peg 21h54'37"20d19'
Linda Diane Com 13h27'29"25d1'
Linda Elizabeth Peg 22h54'18"27d33'
Linda Eve Cas 23h20'55"58d16'
Linda Faye T I Leo 10h49'35"-2d26'
Linda Felice Lyr 18h45'34"36d10'
Linda Forever Lyr 18h44'38"32d39'
Linda G Cyg 19h42'1"31d26'
Linda Gail Lyr 18h29'28"37d36'
Linda Grace Vul 20h39'30"26d11'
Linda Honey Bah Eri 2h49'13"-5d40'
Linda Jean Com 13h4'16"17d38'
Linda Jeanne Aql 18h58'43"11d30'
Linda Jo Peg 22h51'29"27d26'
Linda K Peg 22h17'11"8d51'
Linda K G M Aql 19h42'41"10d34'
Linda Kaye Peg 23h42'43"30d41'
Linda Lee Del 20h13'23"15d44'
Linda Lee Mon 8h2'1"-6d30'
Linda Lindsay Vul 19h47'43"28d49'
Linda Lindsay Cyg 20h5'20"58d55'
Linda Lou Lyn 7h39'13"58d26'
Linda Lou Peg 23h28'47"31d44'
Linda Lou Aql 19h48'1"13d27'
Linda Lou II Cet 3h8'49"4d33'
Linda Louise Cet 2h51'43"0d33'
Linda Marie Psc 0h46'1"32d3'
Linda Marie And 0h59'48"37d60'
Linda Marie Hya 9h1'1"5d41'
Linda Marie Lyr 18h41'13"26d2'
Linda Marie Vul 20h4'60"25d46'
Linda Marie Ara 17h54'14"-55d9'
Linda Marie-Twenty Five Cyg 21h28'0"37d34'
Linda My Love Lyr 18h43'58"32d50'
Linda Sue Lyr 18h48'30"42d32'
Linda Sue Mon 6h30'54"0d43'
Linda Sue Mon 7h56'25"-3d22'
Linda Sue And 23h7'36"38d52'
Linda U Cam 5h51'20"68d28'
Linda V Lyr 18h51'0"42d26'
Linda's Destiny Umi 15h52'31"82d50'
Linda's Destiny Cyg 20h5'24"38d4'
Linda's Dream Gem 6h55'43"18d34'
Linda's Hand Cyg 19h53'40"37d47'
Linda's Heart Del 20h17'58"13d23'
Linda's Hideaway Oph 18h5'30"7d46'
Linda's Light Cet 1h18'52"-10d18'
Linda's Ram-Apsides One Ari 2h37'45"21d15'
Linda's Star And 0h41'52"38d14'
Linda's Star And 1h41'1"36d17'
Linda's Star And 2h28'16"47d10'
Linda's Star Cas 1h18'13"60d52'
Linda's Star Sapphire Cra 18h3'30"-37d27'
Linda's Wish Place Cyg 19h29'37"31d38'
Linda-Rose Cas 0h27'1"60d38'
Lindabery,Rik J Aql 18h44'17"8d27'
Lindabury,Florence Evelyn Cet 2h15'21"7d30'
LINDALE Vir 13h22'2"12d41'AB
Lindas' Star Cas 1h7'59"61d31'
Lindasmine Lac 22h23'30"40d19'
Lindau,Pär Cam 6h18'0"67d32'
Lindavid Her 16h49'1"50d46'
Lindberg,Judy E Lyn 7h50'37"36d2'
Lindberg,Lars Arthur Dra 17h12'2"54d11'A
Lindberg,Lauren Allison Vul 19h3'35"25d39'
Lindberg,Neil & Debbie Uma 9h11'12"56d26'
Lindberg,September Polk Dra 17h12'2"54d11'B
Lindblom,Annika Birgitta Cas 1h18'12"75d43'
Lindblom,Mark Per 2h1'25"50d18'
Lindbloom,Roy Edward Dra 15h19'30"61d53'
Lindborg,Carl Emmanuel Per 2h54'13"35d31'
Lindburg,Diane C Mon 6h20'52"9d28'
Linde's Special Star, Teresa Lynn Lyr 19h23'46"35d8'
Linde,Lauren G Cet 2h19'13"0d32'
Lindeberg,Sven Oph 16h20'51"2d49'
Lindell,In Memory of Jeff Mon 8h0'48"-6d2'
Lindell,Roy Sct 18h52'14"-6d37'
Lindeman"We Love You", David J Aur 5h32'1"30d28'
Lindeman,Amanda Mon 6h21'16"2d53'
Lindeman,Paul G Cyg 19h24'42"32d25'
Lindemann,Cecelia Lyn 8h13'1"44d41'
Lindemann,Paul Hya 8h55'12"0d44'
Linden,Cara Melissa Psc 0h57'12"27d60'
Linden,Joan Lyn 7h2'52"50d44'
Linden,Maria Mon 6h24'35"9d55'
Linden,Marjorie A Cas 0h36'15"75d46'
Linden,Pauline Cyg 19h53'15"44d28'
Linden-Shinji Masayo Tau 5h25'41"21d27'
Lindenfelser,Betty Cam 3h42'26"68d15'
Lindenmayer,Chip Lyn 8h27'25"50d46'
Linder (Lion),Graeme Uma 11h21'52"50d38'
Linder,Ben Sex 10h3'16"2d20'
Linder,Carolyn Rochelle And 23h5'10"44d38'
Linder,Derrick Micheal Per 3h5'28"38d57'
Linder,Jennifer Rene And 23h43'29"33d44'
Linder,Leslie Aql 20h1'0"0d8'
Linder,Steve Cam 11h3'35"82d14'
Linderman's Light Uma 10h11'45"48d14'
Linderman,Brian Cep 21h1'54"58d33'
Linderman,Elaine Leonora Uma 14h21'33"58d59'
Linderman,Susie Cra 18h16'45"-43d54'
Linders,Leentje Aur 7h3'10"38d21'
Lindgren "God's Little Star",Edward A Tri 1h47'20"26d11'
Lindgren,Paige Cas 0h5'35"63d31'
Lindgren,Roy Shannon Cmi 7h50'4"3d30'A
Lindgren,Ruby Ethel Cmi 7h50'4"3d30'B
Lindholm,Denise Catherine Vir 11h40'16"8d35'
Lindholm,Eric Boo 14h58'24"28d8'
Lindholm,Kristen Lyn 7h55'49"44d20'
Lindholm,Sandra A Ori 5h55'21"18d60'
Lindini's Light Sex 9h40'47"0d33'
Lindisa Eilenee Vul 19h58'1"26d8'
Lindjord,Harald & Mary Cyg 21h10'57"48d50'
Lindlar,Colleen Cyg 21h54'19"38d15'
Lindlbauer,Ben Gay Aur 4h51'17"50d21'
Lindley 4,T J J J Peg 22h59'38"26d4'
Lindley,DW / Geraldine Ori 5h57'16"16d12'
Lindley,George William Jacob Psc 1h36'35"22d5'
Lindner 0377,Gene, Betsy,Scott & Amber Uma 9h56'0"51d39'
Lindner,Barbara M Cas 0h34'0"56d10'
Lindner,Doris Cas 2h4'37"68d25'
Lindner,Paul Cyg 20h27'54"31d6'
Lindner,Wolfgang Peg 0h6'58"28d14'
Lindo Aql 19h5'1"1d11'
Lindo,Cymelin "Mae" Mefania Eri 4h11'23"-12d43'
Lindop,Suzanne And 0h4'33"30d56'
Lindquist II,John Aur 6h29'57"33d53'
Lindquist(Auld Lang Syne),Kari C Tri 2h21'28"35d49'
Lindquist,Aubrey Nicole-Rivers Uma 11h12'30"47d45'
Lindquist,Jr,Robert B Aql 18h59'0"17d26'
Lindquist,Link-N-Dandy Cyg 19h58'36"38d12'
Lindquist,Wayne-N- Sandy Cyg 19h58'1"37d42'
Lindra Lyr 18h38'24"35d45'
Lindros,Stanley Morrison Aur 4h53'20"40d37'
Lindsay Com 13h7'55"27d24'
Lindsay And 0h52'53"36d18'
Lindsay Uma 8h34'39"47d45'
Lindsay Ann Peg 23h2'1"22d21'
Lindsay Belle Lib 14h24'15"-10d48'
Lindsay My Love Lyn 8h48'11"38d27'
Lindsay's & Ashley's Star Of Angels Lyr 18h31'36"32d20'
Lindsay's Twinkle Cyg 20h42'50"45d29'
Lindsay's Valentine's Day Star Lyr 18h21'44"40d39'
Lindsay,Allison Taylor And 1h1'56"40d2'
Lindsay,Avis Vul 19h18'29"20d0'
Lindsay,Carl Cyg 20h38'12"45d37'
Lindsay,Cassie Renee And 23h29'27"48d35'
Lindsay,David Jeffery Her 15h49'29"41d56'
Lindsay,David McDougal Uma 10h46'60"40d2'
Lindsay,David Parker Per 2h44'25"40d20'
Lindsay,Eunice Mary Cyg 21h21'23"28d50'
Lindsay,Francis Alexander Ori 6h5'13"0d17'
Lindsay,George & Doris Mon 6h33'42"1d18'
Lindsay,Janet E Lmi 10h56'1"31d33'
Lindsay,Jr,Steven Kelley Peg 23h48'1"12d46'
Lindsay,Liam Tau 5h51'25"24d16'
Lindsay,Melinda Ann Peg 21h53'40"30d32'
Lindsay,Michele Gem 7h11'45"21d26'
Lindsay,Muriel Opopie And 23h0'45"44d32'
Lindsay,Robert & Virginia Sex 10h8'0"1d38'
Lindsell,Robert Per 2h11'53"57d44'
Lindsey And 2h34'42"41d42'
Lindsey Cyg 20h40'52"38d49'
Lindsey Peg 23h3'16"33d58'
Lindsey Uma 9h32'0"52d22'
Lindsey Mon 6h47'47"10d36'
Lindsey & Adam's Star Uma 9h17'39"52d40'
Lindsey & Katie Star And 22h57'36"50d27'
Lindsey Alan Robinette Mon 6h19'38"8d17'
Lindsey Anne And 0h9'20"39d8'
Lindsey Daniell Oph 18h2'50"13d19'
Lindsey Erin Ari 1h53'48"25d3'
Lindsey Jane Cnv 12h25'50"34d14'
Lindsey Paige Psc 0h52'39"32d27'
Lindsey's Shooting Star Lyn 8h59'50"34d2'
Lindsey,Alima Peg 22h25'42"30d4'
Lindsey,Carol Ori 5h50'48"14d32'
Lindsey,Christina Marie Psc 1h3'36"23d40'
Lindsey,David Lawrence (Sweatbelly) Cep 20h50'10"60d52'
Lindsey,Diane Marie Leo 9h32'15"7d31'
Lindsey,Elissa Ann Lyn 7h1'22"58d30'
Lindsey,Erin Lyr 19h14'48"42d51'
Lindsey,Henry Carter Cnv 12h18'1"34d40'
Lindsey,Jace Thomas Gem 6h53'24"31d28'
Lindsey,Jacob Wesley Peg 23h7'45"11d31'
Lindsey,Jeannie Robert Dra 10h25'30"80d10'
Lindsey,Jeffrey R Dra 15h2'37"63d17'
Lindsey,Jim & Pat Eri 3h39'23"-7d29'
Lindsey,Kit Del 20h19'32"10d21'
Lindsey,Mari Anne And 0h3'44"39d7'
Lindsey,Martha L & Edward A Bia,Jr Hya 9h28'0"1d28'AB
Lindsey,Matthew David Tau 4h16'5"21d54'
Lindsey,MD,George T Her 17h19'0"40d5'
Lindsey,Stephanie B Mon 7h1'54"4d56'
Lindsey-Michael Peg 22h58'29"31d30'
Lindsie Marie Mon 7h6'40"0d47'
Lindstrom,Buzz Dra 14h35'25"65d15'
Lindstrom,Chandler Hya 8h17'16"4d15'
Lindstrom,Gerald R Aur 6h8'0"45d12'
Lindstrom,Jason Tau 5h53'43"27d12'
Lindstrom,Maire Uma 8h34'38"67d44'
Lindstrom,Rachel Mon 6h47'36"10d41'
Lindstrom,Steven R Uma 11h52'24"32d24'
Lindström,Jacqueline Becker Lyr 18h56'24"33d5'
Lindström,Matti Cnv 13h47'44"42d10'
Lindståml,K Cnv 12h31'47"38d11'
Linduska,Coach Cnv 12h22'0"35d30'
Lindy Gem 7h52'0"32d54'
Lindy Lyr 18h54'1"31d2'
Lindy Babey Cas 0h9'1"65d3'
Lindy's Light Cyg 19h57'16"45d33'
Lindy,Annabel & Philip Lmi 10h39'0"32d53'
Lindzy,Louise & Andy Crb 16h0'47"28d49'
Line Bager Cas 1h28'37"64d58'
Lineberger,Jerome Baldwin Cnv 12h25'51"33d19'
Lineberry,Billy Bluegrass Vul 19h48'0"25d3'
Linee,Kathie Cyg 19h27'16"31d18'
Lineham,Nigel Rory Alan Cyg 19h24'39"35d60'
Linehan,Jr,Raymond E Boo 14h56'0"51d47'
Lines,Shane Uma 8h35'0"50d17'
Lines,Thomas "Warren" Aur 4h59'17"51d14'
LinFante,Charles Vincent Lac 22h3'24"49d1'
Ling,Huang Hya 9h7'0"6d31'
Ling,John Howard Cep 23h41'51"66d45'
Ling,Jr,Robert T Hya 8h53'45"-6d39'
Ling,Kao Yu Aur 5h17'22"46d51'
Ling,Peter Hya 9h1'11"0d36'
Lingaard,Knute Cet 3h16'23"5d35'
Lingeman,Robert Jacobus Cet 2h4'22"-6d27'
Lingenberg,Walter Com 13h9'1"28d41'
Lingenfelter,Michael P Per 3h50'14"35d5'
Lingg Family Star,The Uma 10h6'43"59d54'
Linh,My Guiding Light, Pepe Cas 23h15'24"61d13'
Linhart,Diane Lac 22h20'27"54d40'
Linhoss Dra 16h59'48"63d29'
Linic,Gerda J Cet 1h35'19"-2d42'
Linington,George Edward Dra 15h5'1"60d18'
Link,Jeffrey Stevens Cep 21h52'1"84d32'
Link,Jennifer Erin And 0h8'32"43d55'
Link,Jessica Katherine "Baby J" Cap 20h21'5"-26d40'
Link,Johannes Willi Aql 18h43'28"8d45'
Link,Jr,Gottlieb Dra 16h32'0"62d37'
Link,Kendall Cyg 21h20'29"40d20'
Link,Margrit Psc 23h22'42"0d6'
Link,Nina Jane Cam 8h3'25"80d40'
Linkage Guru Ser 15h41'10"15d22'
Linke,Alex Richard Cep 20h46'26"68d53'
Linke,Brian Avery Aql 20h6'1"-0d54'
Linke,Mascha Cnc 8h27'15"31d38'
Linker "Wunderkind", Goerge S Cet 2h36'54"-6d51'
Linker,Lavender Gerrie Cyg 20h1'0"40d9'
Linker,Lorie Lalaine Dra 13h37'29"67d51'
Linker,Robert C Aur 6h57'11"40d47'
Linkiewicz,Judith Leo 10h14'42"13d44'
Linkous,Rodney L Lyr 19h19'19"41d12'
Linley,James Cep 3h29'10"80d12'
Linmerwin Cyg 19h32'11"35d8'
Linn,Bob Dra 16h53'52"63d24'
Linn,David Leander Cet 2h27'12"0d9'
Linn,Elaine Jewel Cas 1h43'36"61d16'
Linn,Gary & Deborah O'Keeffe Cyg 19h33'11"30d2'
Linn,Ginny Uma 10h32'22"68d0'
Linn,Jr,Lester Uma 10h31'48"68d14'
Linn,Mary Kay Mon 6h38'18"-10d6'
Linn,Regen B Uma 11h41'12"63d19'
Linn,Robert K Lac 22h10'32"54d26'

Linn,Robert Steven
Aql 19h4'35"-1d14'
Linn,Theresa N
Cas 1h46'39"60d8'
Linnaloo
Lyn 7h52'29"43d41'
Linnartz-Eternity, Duane & Gwen
Mon 8h4'47"-3d43'
Linnarz,Chase Matthew
Cep 23h6'30"70d22'
Linne,Merideth Ann
Cyg 21h34'0"38d26'
Linnelli,Frank Capozzi
Lac 22h7'0"48d54'
Linnelli,Gregory Frank Joseph
Lac 22h52'11"52d59'
Linnemann,Gina Cecelia
Cas 0h41'1"73d34'
Linnemann,Juergen
Cep 0h5'23"83d14'
Linnemann,Klaus
Lac 22h4'1"51d40'
Linnenbrink,H
Her 18h20'24"12d17'
Linnert,Sylvia & Karl- Heinz
Uma 8h44'0"50d45'
Linnerud,Peter
Cas 1h55'36"61d18'
Linnhe
Umi 14h44'0"65d31'
Linnie's Star
Com 12h57'35"31d23'
Linny
Aql 19h1'18"-6d24'
Lino
Cyg 20h4'12"38d51'
Lino
Scl 0h58'40"-26d49'
Lino E Cristina
Lac 22h7'43"37d35'
Linore
Boo 14h6'1"11d12'
Linsdall Light,The
Cas 0h46'42"64d27'
Linse,Ruth Alice
Sct 18h51'52"-6d24'
Linsell,Love You Sally
Cyg 21h57'33"52d41'
Linson,Jeremy Michael
Her 18h1'42"28d52'
Lint,Candy Ranee
Mon 7h30'16"-0d19'
Lintgen,Mary Martha Rose Daley
Peg 21h56'58"34d45'
Linthicum,Anthony
Her 16h13'30"48d55'
Linthicum,David
Oph 17h53'34"13d7'
Linthicum,MD,William R
Dra 16h29'1"61d18'
Linthicum,Ryann
Dra 16h45'10"64d6'
Linton,Jennifer Stolte
Sco 16h38'14"-31d4'
Lints,Ashleigh Paige Turner
Del 20h21'57"11d3'
Linwood
Cet 0h24'44"-5d9'
Linwood,Frank
Her 18h12'39"30d19'
Linz,Brigitte
Sco 16h52'0"-40d15'
Linz,Leona E
Cas 3h5'46"61d40'
Lioba
Uma 10h24'21"57d4'
Lioi,Justin J
Cnv 12h43'37"38d44'
Liolios,Nicolas
Dra 17h8'11"63d38'
Lion
Cma 7h20'11"-15d10'
Lion Heart
Lmi 10h21'34"30d44'
Lion's Den
Aql 19h5'1"5d1'
Lionard,Christiane
Umi 16h19'21"78d55'
Lionel
Crb 16h7'1"38d2'
Lionel,Angelillus
Cyg 20h48'32"38d11'
Lionheart
Leo 11h34'40"16d46'
Lioy,Jennifer
Cam 5h43'21"68d60'
Lipa,Jeremy Ryan
Per 3h8'35"47d17'
Lipa,Mary Helen
And 23h38'22"42d35'
Lipari,Anthony
Aur 6h26'1"38d54'
Lipera,Jacqueline
And 23h0'27"51d51'
Lipetz,Erik
Her 16h49'19"37d17'
Lipiec-"Our Angel", Nichole Colleen
And 2h16'14"42d28'
Lipinfki,Little Christine
Lac 22h45'58"56d47'
Lipinska,Anna Alexandra Elizabeth
Lyn 9h1'17"38d29'
Lipinski,Cecelia Marie Borkowski
Cas 3h6'59"68d35'
Lipinski,Sheree B
Aql 19h31'19"10d46'
Lipinsky,Stuart
Boo 14h29'60"28d16'
Lipira,Kimberly Ann
Aqr 20h57'45"-10d32'
Lipka,Delane & Harry
Cyg 20h56'55"30d33'
Lipke,Michael Gerard
Per 3h7'24"38d19'
Lipman "Lovey-Dovey" Florence
Lyr 18h55'0"31d58'
Lipman,Barry Ross
Cep 22h24'46"59d52'
Lipman,Beverly Lynn Feinstein
Cas 0h25'43"73d24'
Lipman,Coleman
Per 1h46'58"56d41'
Lipman,Dr Ivan
Oph 18h42'0"6d39'
Lipman,Macey
Equ 21h2'50"11d57'
Lipman,Marie
Vul 19h45'0"20d15'
Lipman,Meimei
Aql 20h30'17"0d3'
Lipous,Jean-Franáois
Lyn 7h58'55"40d52'
Lippa,Marcia & Edward
Cyg 19h31'38"36d44'
Lippe-Hagedorn,Doris
Leo 11h28'30"12d25'
Lippert,Eva
And 0h24'42"34d55'
Lippert,Ursula
Lyr 19h18'34"30d22'
Lipphardt,Sandra L
Tri 2h7'21"31d3'
Lippi,The Louis
Cet 3h15'31"9d18'
Lippil
Equ 21h21'16"11d19'
Lippitt,Charles Chandler
Per 3h1'11"46d32'
Lippitz,Beate
Sgr 19h55'41"-43d32'
Lippman,Cyras C
Aql 20h10'0"27'
Lippman,Ellen M
Hya 8h57'60"4d32'
Lippold,Thomas Henry
Umi 16h40'41"84d45'
Lipps,Stephen H
Dra 9h58'52"74d14'
Lipschutz,Allen
Boo 14h26'47"47d33'
Lipscomb III,John Walling
Her 16h37'32"38d23'
Lipscomb,J Hardy
Aql 19h26'0"15d45'
Lipscomb,Lorna
Cas 1h59'44"61d42'
Lipscomb,Margot Walker/Matthew
Aql 20h11'42"11d0"
Lipscomb-Ross,Bruce & Feeny
Hya 8h10'51"0d1'
Lipset,Ruth Fishman
Tau 3h35'28"30d30'
Lipshutz,Ellen & Raymond
Cyg 21h7'45"31d3'
Lipski,Alyson E
Cas 23h13'1"61d26'
Lipsky,Alex
Cma 7h2'59"-15d16'
Lipsky,Dr Peter E
Aql 20h14'56"04d48'
Lipsky,Evan James
Her 18h18'54"14d10'
Lipsky,Ross Evan
Lib 15h39'32"-20d12'
Lipsky,Stephen
Oph 17h8'19"-23d30'
Lipson,Carol Ann
Lyn 7h32'11"38d46'
Lipson,Stephen Casey
Tri 2h12'30"31d16'
Liptack,William E
Lac 22h2'0"50d22'
Liptak,Karen
Uma 14h24'27"61d26'
Lipton "Harmony Star", Margie & Jamie
Cam 12h17'39"80d47'
Lipton "Miracle Star", Harrison
Aur 6h13'1"38d51'
Lipton "The Star",Adie
Peg 22h45'34"32d8'
Lipton,Janice
Mon 7h1'1"-0d17'
Lipton,Lauren
Cas 1h2'42"56d11'
Lipton,Richard
Cyg 21h3'1"52d11'B
Lipton,Richard
Cep 0h47'1"77d49'
LiPuma,Matthew Anthony
Dra 16h31'31"61d38'
Lirio,Jason Philip
Her 17h39'60"38d10'
Lis
And 23h35'41"40d58'
Lis Anth
Cyg 20h26'44"58d32'
Lis Family Star,The
Mon 6h43'52"4d13'
Lis Georgo
Cyg 19h34'31"34d8'
Lis,Frank C
Aur 7h8'31"40d58'
Lisa
Her 18h30'46"18d58'
Lisa
Gem 6h51'32"13d54'
Lisa
And 2h17'30"41d52'
Lisa
Lyn 7h58'1"39d26'
Lisa
And 2h2'38"38d59'
Lisa
Lyr 19h22'42"33d49'
Lisa
And 2h3'29"37d13'
Lisa
Lyr 19h22'33"31d29'
Lisa
And 1h46'24"37d25'
Lisa
And 23h27'21"48d47'
Lisa
Tri 2h4'40"31d26'
Lisa
Lyn 9h16'22"36d43'
Lisa
Cyg 19h28'12"38d16'
Lisa
Del 20h42'31"12d7'B
Lisa
And 0h11'0"47d4'
Lisa
Lmi 9h46'42"38d33'
Lisa
Vul 19h53'40"20d32'
Lisa
Hya 9h1'43"2d2'
Lisa
Lyn 6h33'11"59d14'
Lisa
Cet 0h25'57"-17d31'
Lisa
Uma 10h35'24"50d31'
Lisa
Mon 7h5'0"-1d37'
Lisa
Cnv 13h12'26"51d41'
Lisa
Peg 23h48'20"17d41'
Lisa & Doug
Crb 16h5'41"28d7'
Lisa & Eric
Peg 22h41'30"35d11'
Lisa & John's Wedding Star
Crb 16h4'0"32d44'
Lisa & Mike Forever
Lyn 7h42'53"40d59'
Lisa & Precious
Mon 6h24'43"3d17'
Lisa & Raymon's Love Light
Lyn 7h56'29"58d21'
Lisa & Tom
Cyg 19h40'31"42d47'
Lisa & Travis
Equ 20h58'26"4d38'
Lisa & Wally
Cyg 19h32'38"34d36'
Lisa & Yana
Peg 23h44'15"15d38'
Lisa 83
Vul 19h38'34"20d1'
Lisa Amanda
Lyr 19h16'22"25d46'
Lisa Angelica
And 22h58'14"51d19'
Lisa Ann
Oph 17h11'35"-23d21'
Lisa Ann
And 1h54'36"39d20'
Lisa Ann
And 0h8'1"35d42'
Lisa Ann
Vir 13h2'57"-1d30'
Lisa Ann
Cas 1h45'50"60d32'
Lisa Ann
Cas 0h48'1"62d9'
Lisa Ann & Peter Joseph
Cyg 21h0'33"28d42'
Lisa Chris
Peg 22h46'49"25d18'
Lisa Christine
Peg 21h43'0"24d16'
Lisa D
Cnv 12h50'46"37d60'
Lisa D
Lyn 6h31'46"58d8'
Lisa Danielle
Eri 2h53'27"-15d54'
Lisa Dawn
Cas 2h46'43"70d59'
Lisa F S'Own
And 0h50'53"37d11'
Lisa Forever
Aql 20h18'58"5d3'
Lisa Glowing Softly A Guiding Light
Del 20h50'40"8d59'
Lisa H
Umi 16h13'34"74d51'
Lisa J M
Uma 11h58'1"40d35'
Lisa Jane
Cnc 8h32'28"30d45'
Lisa Jean,My "B"
Lyr 18h24'0"45d41'
Lisa Jo
Cyg 21h5'60"38d42'
Lisa Jo
Her 17h34'30"27d14'
Lisa K C
Peg 21h29'0"25d48'
Lisa Kay
Com 12h23'0"19d19'
Lisa la trentaine éternelle
Leo 10h36'13"21d7'
Lisa Lee
Sco 16h26'26"-37d54'
Lisa Lee
Del 20h24'31"10d4'
Lisa Lisa
And 23h0'23"50d27'
Lisa Lou
Mon 7h1'32"1d46'
Lisa Lou
Cyg 19h28'59"37d33'
Lisa Lynn
Cas 0h51'39"66d54'
Lisa Lynn
Mon 7h1'12"7d48'
Lisa Lynne
Lyr 19h12'41"37d58'
Lisa Marguerite
Lib 15h2'52"-20d7'
Lisa Marie
Aqr 21h56'24"-6d41'
Lisa Marie
Cas 0h1'0"58d26'
Lisa Marie
Cas 0h53'35"64d48'
Lisa Marie
Lyr 18h34'18"46d34'
Lisa Marie
Uma 13h54'36"61d22'
Lisa Marie
Lyr 18h44'55"34d6'
Lisa Marie
Lyr 18h42'52"34d9'
Lisa Marie
Lyr 18h43'10"31d53'
Lisa Marie's Eternal Guiding Light
Mon 6h15'12"-5d44'
Lisa Marie's Own Star
And 0h29'18"40d27'
Lisa Michelle
Mon 7h6'23"-0d56'
Lisa Mon Amour
And 23h13'34"35d53'
Lisa My Everlasting Star
And 0h27'30"37d57'
Lisa My Shinning Star
And 23h6'41"38d43'
Lisa Nicole
Lyn 8h46'50"36d19'
Lisa of the Angels
Uma 10h0'1"56d22'
Lisa Renee
And 1h38'12"38d19'
Lisa RW 1994
Cas 23h44'35"58d59'
Lisa Star
Cyg 20h17'59"38d31'
Lisa Star
Eri 3h41'53"-5d56'
Lisa Star,The
Cas 23h31'55"54d0'
Lisa T's Own
Com 12h18'58"27d51'
Lisa Venuti's Little Feather
And 2h5'52"38d7'
Lisa Victoria
Uma 11h17'26"38d57'
Lisa's Dream Light
Lyn 6h58'1"58d29'
Lisa's Guiding Light
Lyn 9h3'23"34d16'
Lisa's Hanukkah Kids
Umi 16h11'57"70d51'
Lisa's Heart For Brion
Cyg 19h31'12"34d23'
Lisa's Lathe of Heaven
Crb 16h15'59"31d39'
Lisa's Light
Cas 0h58'0"58d11'
Lisa's Light
Peg 22h42'55"33d47'
Lisa's Little Star
Peg 23h38'11"10d21'
Lisa's Lonestar
Gem 6h41'40"12d24'
Lisa's Lucky Star
Sge 18h56'17"18d48'
Lisa's Lucky Star
Aqr 23h37'31"-19d37'
Lisa's Quirk
Lyr 18h29'1"44d45'
Lisa's Secret
Peg 22h51'52"25d23'
Lisa's Star
Psc 23h20'14"1d35'
Lisa's Star
Cyg 19h30'19"33d19'
Lisa's Star
Lyr 18h56'54"41d7'
Lisa's Star (LLR)
Mon 6h45'19"10d6'
Lisa's Star Of Hope
Eri 4h11'42"-12d5'
Lisa's Totally Awesome Shining Star
Lyn 7h47'45"40d25'
Lisa,Edward,Michael, Matthew
Per 3h58'25"50d22'
Lisa,LaEstrella DeOro
Lyr 19h14'1"42d33'
Lisa,Robert J
Cnv 13h30'1"50d18'
Lisa—The Little Woman
Sge 19h56'37"16d8'
Lisa-Katharina
Tau 5h49'23"23d35'
Lisa-Lisa
Tri 2h0'44"32d26'
Lisa-Paul-Voula-Michos
Cam 5h35'1"60d32'
Lisa/Brad
Lyr 19h24'16"41d44'
LisaAnn
Uma 11h46'27"49d59'
Lisac,James N L
Lac 22h22'47"38d12'
Lisante,April Danielle
Cas 0h33'21"63d30'
Lisante,James Patrick
Ari 1h46'0"23d24'
Lisanti,Angeline Obie
Aur 5h1'16"49d11'
Lisaro 7
Cas 0h3'1"62d37'
Lisashley
Cyg 21h1'41"37d1'
Lisbeth
Cyg 19h44'29"38d18'
Lischer III,Alan
Peg 23h18'26"26d7'
Liscia,Adam Nicholas
Per 2h51'0"46d13'
Liscia,Mark Benjamin
Aur 6h12'38"45d6'
Lisciandrello,Andrew Carl
Aql 20h3'51"8d17'
Liscio,John F
Boo 14h14'46"17d50'
Liscomb,Clark M
Ori 6h16'44"7d57'
Lise
Umi 13h51'59"76d13'
Lise Ann & Andrew
Ori 5h36'28"-1d45'
Lise et Jules
Ori 5h58'23"17d42'
Lise, Beautiful Mountian Flower
And 23h18'44"47d27'
Lise-Hélène
Lyn 8h30'23"42d50'
Lisel Star
Boo 13h45'18"13d51'
Lishamer,Justine
Cnv 12h17'34"34d28'
Lisiewski,Nancy
Vul 19h59'30"23d6'
Liska,Anne
Psc 1h1'0"17d34'
Liska,Cassandra
And 2h28'37"49d1'
Liska,Courtney
And 2h30'15"49d40'
Liska,Jamie
And 2h27'39"49d56'
Liska,Kaitlin
And 2h23'13"47d37'
Liska,Nevada Star Richard & Marcia
Lmi 9h56'27"33d26'
Liska,Robin L
Cas 1h43'31"68d59'
Liske,Debbie L
Ari 2h57'27"30d36'
Lisker,Thorsten
Del 20h54'31"2d19'
Lisle,Monica
Aur 5h52'36"29d36'
Lislie's Brillance
Umi 9h37'21"38d15'
Lisman,Michelle
And 0h28'33"38d30'
Lisnic,Thomas
Cep 21h39'49"61d23'
Lisnness,Niles
Aql 19h12'13"13d55'
Lison
Lyr 18h48'43"35d17'
Liss 80th Birthday Star,Joseph R
Lac 22h49'35"54d29'
Liss,Dr Ron
Her 16h59'1"34d16'
Liss,Ilana Dara
And 23h5'34"40d41'
Liss,Justin Phillip
Her 17h27'15"38d50'
Lissa
Gem 7h18'30"28d45'
Lissadrun
Peg 21h20'1"23d36'
Lissette,Jill
Cam 14h15'1"80d56'
Lissner,Louise Auf der Heide
Lyn 8h40'19"39d6'
Lissy Bear
Mon 7h7'11"-5d5'
List,Keith A & Christina A Rule
Lyn 8h10'18"47d10'
Listenberger,Corrie Ann
Vir 13h18'56"-9d2'
Listenberger,Laura Lynne
And 1h58'13"37d35'
Lister,Alison
Eri 4h35'39"-11d50'
Lister,Christopher K E
Aur 5h3'0"45d23'
Lister,David Howard
Cep 20h36'18"76d47'
Lister,James M
Dra 15h51'20"67d27'
Lister,Jason R
Crb 15h50'45"29d19'
Lister,Joedy
Cmi 7h15'0"1d35'
Lister,Michelle Karen
And 0h12'51"39d36'
Listermann 10,Kimberly Ann
Lyr 18h42'53"38d48'
Listo-Star
Per 4h3'10"37d16'
Liston,TLP,Douglas C
Her 16h49'47"33d46'
Listorti,Joseph"Pip"
Aur 5h54'1"40d37'
Lisy
Aql 19h59'56"12d21'
Lisz,Gary
Ori 5h56'15"17d49'
Liséric
Per 3h11'0"43d6'
Lita
Cyg 21h31'26"53d47'
Litchfield,Lauren
Lyr 18h41'28"32d14'
Litchfield,Sarah
Sge 20h5'24"16d17'
Lite,Cecile
Mon 6h23'33"-1d52'
Litfin,Lisanne
Aql 20h7'15"8d44'
Litherland,Paul John
Sge 20h5'52"16d42'
Litherland-Todd,Janet
Cyg 21h18'21"36d34'
Lithgow,Diana I G
Aql 20h6'45"1d18'
Lithgow,Douglas Michael
Hya 9h8'1"5d4'
Litke,Doug
Lyr 18h36'58'34d18'
Litke,Jackie
Lyr 18h36'0"44d0'
Litke,Rosemarie
Vul 19h46'59"28d50'
Litke,Vernon
Uma 11h43'25"47d57'
Litsa
Uma 10h37'38"57d42'
Litsky,Susan Albina
Cyg 21h55'1"53d33'
Litt,Joan
Lac 22h4'0"49d51'
Litt,Sandi
Cas 0h23'18"70d21'
Littau,Gunther Keenan
Dra 17h37'45"68d21'
Littau,Sally R
Peg 22h19'46"21d38'
Littell-Herrick,Ame
Uma 12h1'36"61d44'
Littell-Herrick,Ray Thomas
Cnv 13h32'17"48d13'
Litten Lasthaus-der Süsse,Katarina
Uma 9h2'11"50d44'
Litteral,James Arnold
Hya 8h44'43"1d57'
Litterer III,William E
Cep 22h54'38"59d36'
Litterscheid,Claudia
Her 17h17'50"42d47'
Little
Lac 22h14'52"49d21'
Little Henry W E W
Cnv 13h52'0"37d40'
Little "G"
Cnv 12h31'39"51d28'
Little "T"
Lmi 10h6'36"39d55'
Little Abby Dabby
Lyn 6h24'41"58d54'
Little Amy's Star
Cyg 19h45'19"29d22'
Little Angel
Peg 23h33'43"18d26'
Little Angel
Aqr 21h26'58"-1d7'
Little Angel
Mon 7h26'50"-1d33'
Little Angel Amy
And 23h4'22"40d19'
Little Anne
Cnv 12h23'22"40d33'
Little Annie
Peg 22h0'18"35d8'
Little Bart
Sco 16h39'1"-37d38'
Little Bear Claire
Umi 16h41'11"86d3'
Little Beauty
Ari 2h25'51"26d33'
Little Bee
Dra 16h31'53"61d50'
Little Bess
Cmi 7h34'17"0d25'
Little Beth
Sco 16h58'14"-40d16'
Little Betty
Uma 14h43'0"82d12'
Little Bevie
Lyn 7h29'48"37d25'
Little Big Mann
Uma 9h25'39"48d25'
Little Bit of My Heart
Cam 12h5'1"76d37'
Little Blue Eyes
Aur 6h1'41"45d20'
Little Blue Spirit
Cam 3h38'26"60d59'
Little Boo
Lac 22h9'45"48d55'
Little Bop
Ori 6h2'17"0d56'
Little Boss
Aur 6h24'22"40d41'
Little Brad
Cyg 21h0'42"30d26'
Little Brian
Mon 6h54'14"1d8'
Little Brother
Per 2h51'17"40d55'
Little Chad
Cyg 20h16'58"39d27'
Little Chicken
Aql 18h58'40"-10d10'
Little Chief- Hookociihit Neecee
Sco 17h51'36"-31d18'
Little Chipper 94,The
Aur 6h15'12"38d52'
Little Clary
Umi 16h9'35"70d20'
Little Clyde
Umi 15h38'13"76d48'
little debbie
Aqr 22h24'22"0d10'
Little Dicky
Her 17h5'1"49d26'
Little Doreta
Peg 23h16'48"33d12'
Little Freddy
Peg 21h52'54"40d58'
Little Fuentes
Lyn 7h53'34"43d46'
Little Goe
Eri 4h12'53"-14d54'
Little Grace
Umi 14h27'0"82d28'
Little Guy
Dra 15h30'29"57d47'

Little H
Per 4h3'0"52d11'
Little Heaven (Brian & Mary)
Cyg 21h53'14"38d3'
Little Helen
And 2h24'18"44d44'
Little Honey
Lyn 8h22'37"48d50'
Little Janie - Big Janie
Del 20h15'59"10d44'
Little Jean
Gru 22h42'34"53d45'
Little John
Umi 16h45'26"76d14'
Little Juder
Cyg 19h25'34"30d29'
Little Kahuna,The
Uma 11h26'35"44d25'
Little Katie
Umi 13h6'37"70d57'
Little Kell-7/3/63
Lac 22h5'1"49d24'
Little Lauren
Lyr 18h19'25"37d41'
Little Legend with Tatsuya
Cap 20h57'47"-16d1'
Little Leo
Leo 11h0'37"2d1'
Little Lewis
Umi 16h53'19"76d25'
Little Ligenza
Aqr 23h43'59"-5d12'
Little Lilli-o
Tri 1h54'59"26d42'
Little Lisa
Uma 9h17'31"68d5'
Little Little Little Shadow
Lyn 7h49'24"44d16'
Little Lois
Dra 18h59'37"70d35'
Little Lueder-Butt
Umi 15h15'45"72d34'
Little M
Cyg 19h28'20"38d39'
Little Mac
Cap 21h20'42"-14d38'
Little Maggie
Lyn 9h14'58"41d46'
Little Man
Aur 7h22'15"40d17'
Little Marge
Ori 5h53'46"11d55'
Little Marie
Cyg 21h0'33"28d22'
Little Marty
Lac 22h13'1"50d26"
Little Mate (CDKT)
Cyg 20h33'37"48d47'
Little Mermaid,The
Ori 6h8'0"7d39'
Little Michelle
Uma 11h26'56"49d51'
Little Michie
Uma 10h53'1"67d38'
Little Miss
Aql 19h7'31"15d17'
Little Miss Amanda
Mon 7h59'17"-2d55'
Little Miss Can't Be Wrong
Del 20h48'60"7d40'
Little Miss Ella-Marie
Peg 22h57'1"22d14'
Little Miss Hannah
And 1h7'32"38d34'
Little Miss Muffet
And 2h30'48"40d57'
Little Mom
Cas 0h1'0"60d28'
Little Monica Girl
Cet 3h19'52"4d49'
Little Moo
Lac 22h24'1"53d40'
Little Mother
Cas 0h27'13"61d20'
Little Niffer
Tau 3h57'30"20d13'
Little One
Uma 8h40'0"60d35'
Little One
Vul 19h44'39"23d54'
Little One
Peg 22h2'28"2d28'
Little One
Ser 15h54'24"12d37'B
Little One Luminance Of Love
Uma 8h33'51"48d44'
Little Oscar
Cep 21h39'52"55d28'
Little Package,The
Uma 12h23'46"53d52'
Little Panney Our Shining Star
Uma 10h57'36"38d21'
Little Pasquale
Lyr 19h14'45"41d9'
Little Paul
Umi 14h22'1"68d22'
Little Pea
Cas 0h14'39"65d58'
Little Pete
Per 1h27'34"52d50'
Little Prince
Del 20h18'15"10d53'
Little Protector Spirit
Cam 3h38'1"67d41'
Little Red
Per 3h19'24"50d17'
Little Richard
Per 2h50'23"40d47'
Little Ron
Umi 13h3'1"72d24'
Little Schmitka
Uma 10h47'41"48d36'
Little Seal
Ori 5h51'59"10d57'
Little Shadow
Cam 3h35'43"71d59'
Little Sheba G
Sge 19h38'18"16d45'
Little Shining Fay
Del 20h55'55"8d27'
Little Sister
And 2h28'44"45d53'
Little Skipper,The
Lyn 9h4'15"40d40'
Little Squirt
Dra 15h19'46"63d47'
Little Steffer
Peg 21h58'1"30d13'
Little Stevie's Star
Cet 1h32'0"-2d59'
Little Sunshine
Ori 5h29'40"1d24'
Little Susie
Cas 0h48'0"69d27'
Little Susie,The
Lyr 18h59'56"36d57'
Little Sweetie
Aql 19h27'20"-0d44'
Little T
Aur 6h21'32"38d15'
Little T,The
Sct 18h50'50"-7d17'
Little Terry
Peg 0h0'43"30d49'
Little Tonia
Peg 23h5'1"30d53'
Little Treasure of Love
Vul 19h45'51"25d32'
Little Walter, Our Poodleoopie
Cmi 7h59'52"8d39'
Little White Buffalo
Uma 10h26'0"56d40'
Little Witch,The
Vir 13h59'35"-12d3'
Little Wonder
Mon 6h22'51"7d37'
Little's Star
Ori 5h56'16"47d48'
Little,Alister
Aql 19h2'0"-5d49'
Little,Allen William
Cma 6h52'1"-16d30'
Little,Anne
Uma 9h41'40"61d8'
Little,Billie
Sct 18h55'54"-6d48'
Little,Bob
Hya 9h6'1"6d9'
Little,Cameron Royal
Mon 7h47'35"-1d22'
Little,Darren Lee
Uma 12h11'26"57d58'
Little,David Scott
And 23h49'51"40d23'
Little,Debra L & Edward J
Lyn 8h11'54"40d37'
Little,Devin Tanner
Peg 22h25'26"21d41'
Little,Donna
Cas 0h41'58"65d17'
Little,Emily Suzanne
Cet 3h1'54"8d38'
Little,Isla Dean
And 0h8'32"28d45'
Little,Jeanne Graves
Oph 17h59'29"8d17'
Little,Jr,Robert Stanley
Lac 22h6'53"51d48'
Little,Judith Ann
Mon 8h2'1"-6d30'
Little,Judy
Hya 9h6'46"3d26'A
Little,Julie
Vir 12h35'51"0d36'
Little,Linda K
Oph 18h42'31"7d2'
Little,Mark
Per 2h30'49"57d59'
Little,Mark
Eri 3h42'14"-3d13'
Little,Mary (Big M)
Cas 23h19'0"61d59'
Little,Mindy & Tony
Peg 23h46'36"11d7'
Little,Polly
Mon 8h2'37"-8d20'
Little,Sammy Joshua
Gem 6h49'11"14d55'
Little,Sean Patrick
Aur 6h3'43"35d31'
Little,Serah
Aql 19h29'40"8d19'
Little,Simon Leslie
Col 6h31'23"-36d15'
Little,Willie D
Cas 2h18'39"65d1'
Little-Bit
Uma 8h53'32"51d27'
Littlefield Star,The
Aql 18h58'57"-6d50'
Littlefield,Douglas
Mon 8h7'0"-1d29'
Littlejohn,Leslie & Angus
Per 3h37'59"39d59'
Littlejohn,Nancy
Mon 6h44'27"0d25'
Littlejohns,Peter Thomas
Dra 18h1'27"70d14'
Littlest Angel,The
Vul 19h48'42"28d22'
Littlest Bear
Umi 16h0'40"78d31'
Littleton,Mark-Luke & Jackie
Uma 9h24'52"57d43'
Littleton,Martha Morrow
Lyr 18h37'0"43d42'
Littman,Barbara
Sco 16h56'41"-41d11'
Littman,Carrie Wenzer
Cas 0h45'49"61d14'
Littmann-45 Years, Walter & Jane
Cep 22h48'16"67d41'
Litton,Susan Catherine
Crb 15h59'19"37d47'
Litton,Teresa
Cet 1h0'33"-1d49'
Littrell,Barbara
Lac 22h35'33"38d13'
Litwack,Florence
Lyr 18h33'40"38d25'
Litwak,Katherine Fiske
Mon 7h42'39"-5d20'
Litwhiler,Austin Gunner
Her 17h4'26"38d39'
Litwiller,Theodore Ronald
Ser 17h59'42"-14d21'
Litwin,Irene
And 23h39'58"42d27'
Litzenblatt,Seth Nathanial
Eri 4h36'31"-1d18'
Litzinger,Katherine Anne
Crb 16h10'1"34d7'
Liu,Chao-Li & Susan
Ori 5h6'0"1d45'
Liu,Crystal
Ori 5h13'13"-8d33'
Liu,Helen
Cas 1h36'20"60d6'
Liu,Kelly & Ben
Lyn 9h17'26"38d56'
Liu,MD,Jason
Oph 17h29'45"7d34'
Liubava
Cep 21h47'36"60d18'
Liuccia
Col 6h31'29"-37d46'
Liuzi Lou
Com 12h29'15"25d1'
Liuzzi,Rosaria
Cnv 13h35'40"28d18'
Liuzzo,Francesca
Psc 23h24'34"5d57'
Liv
Ori 5h52'36"1d50'
Livanou,Irene
Cas 2h58'54"61d36'
Livdahl,Dale
Cam 13h22'24"81d20'
Livengood,Jeffery Dallas
Boo 14h24'56"20d23'
Livengood,Ryan Michael
Uma 12h3'58"30d27'
Liver,Jay
Aql 19h57'21"1d2'
Liverakos,Ruth M
Lac 22h13'46"51d1'
Liverman,Lois & Bill
Cyg 19h6'55"48d30'
Liverman,Sandra Friedman
Aql 19h31'44"13d17'
Liverman,William Todd
Her 18h16'45"14d26'
Livermore,Jr,Gordon Dexter
Sco 16h7'23"-33d49'
Livermore,Robert & Tammie
Her 16h37'38"40d30'
Livernoche,Marie- Audrey
Cas 22h59'25"55d46'
Livesay,Betty Annalee
Vir 13h34'11"0d6'
Livesey,Brian Harold
Ori 6h2'39"7d56'
Livesey,John
Cap 21h36'50"-21d10'
Livesey,Patricia Ann
Lyr 18h32'28"47d14'
Livesey-Hulme
Lyr 18h31'24"37d56'
Livezey,Jessica Suzanne
And 0h53'21"35d37'
Livezey-Lawson Mary Alyce
Del 20h50'37"9d6'
Livia
Lyr 18h35'1"37d40'
Livia
Scl 23h14'10"-33d3'
Liviana
Hor 3h21'13"-49d57'
Livigne,Matthew Altello
Lac 22h39'0"56d46'
Livingston 1994, William Tyler
Aql 11h59'46"5d42'
Livingston 1994,Lisa & Randy
Eri 2h50'10"-2d41'
Livingston,Anthony L
Her 17h7'49"20d33'
Livingston,Dalton Kelly II
Cet 0h24'25"-10d39'
Livingston,David Russell
Aql 19h48'40"10d41'
Livingston,David William
Lmi 10h41'52"31d33'
Livingston,Debby
Peg 22h3'52"25d33'
Livingston,Douglas
Her 18h1'17"28d39'
Livingston,Douglas
Sgr 20h12'20"-29d21'
Livingston,Haley
And 0h14'49"38d50'
Livingston,J & J
Cet 2h5'22"4d45'
Livingston,Jaclyn
Uma 10h6'1"61d48'
Livingston,Jodi Schneider
Mon 6h22'23"7d54'
Livingston,John & Cynthia
Cyg 21h21'44"30d60'
Livingston,Jonathan
Cyg 21h8'19"30d33'
Livingston,Justin
Cep 22h18'0"58d43'
Livingston,Lori
And 2h18'22"45d13'
Livingston,Myra Cohn
Peg 22h20'43"5d23'B
Livingston,Richard Roland
Peg 22h20'43"5d23'A
Livingston,Richard G
Hya 9h24'42"-10d30'
Livingston,RN,Carolyn
Gem 5h57'19"26d60'
Livingston,Ryan
Uma 10h6'17"62d13'
Livingston,Steve & Brenda
Mon 7h44'20"-5d34'
Livingston,Susan Briana
Lyr 18h41'52"32d37'
Livingston,Terry Lynn
And 23h32'48"49d32'
Livingston-Goldman, Sally
Lyn 8h6'45"40d6'
Livingstone
Ori 4h59'24"1d32'B
Livingstone,Bonnie
Vul 20h57'38"28d44'
Livio,Cristiano
Cnc 9h1'1"8d23'
Liviu
Umi 16h34'40"78d1'
Livornese,Joseph Domenick
Aur 6h5'31"38d14'
Livowsky,Jean Michel
Aur 5h2'42"30d16'
Liz
Cyg 21h7'13"30d8'
Liz
Aur 6h14'18"54d4'
Liz
Aqr 20h54'0"0d6'
Liz
Cas 0h25'27"62d17'
Liz
Peg 23h5'0"33d46'
Liz
Mon 6h39'14"3d31'
Liz & Bernie
Cyg 19h17'34"44d6'
Liz & Daryl's Place
Tau 3h36'13"20d17'
Liz & David
Aur 5h13'0"41d52'
Liz & Dwayne "Children of Light"
Tau 4h4'59"20d19'
Liz B
Cet 1h36'0"-6d50'
Liz W
Crb 16h12'30"33d56'
Liz' Lasting Lustre
Aql 19h28'46"14d38'
Liz's Star
Crb 16h4'0"32d56'
Liz's Wish
And 23h1'0"50d3'
Liza
Aur 4h59'24"52d1'
Liza
Umi 16h43'54"76d44'
Liza's Eyes
Cyg 20h29'32"50d6'
Lizabeth
Vul 19h58'1"22d56'
Lizabeth
Lyn 9h25'50"40d59'
Lizabetta
And 0h14'14"45d41'
Lizakowski,Pat & Scott
Uma 10h47'1"48d31'
Lizarazo,Alex G
Cam 3h41'56"70d17'
Lizard's Worm
Lac 22h41'0"56d32'
Lizbeth
Cyg 19h26'48"35d26'
Lizlea
Peg 21h57'37"22d37'
Lizpigui
Umi 15h18'56"68d7'
Lizundia,Laura
And 1h16'43"39d22'
Lizy 6-12-93
Com 13h18'24"26d48'
Lizzard
Lac 22h50'23"56d51'
Lizzard II
Lac 22h27'53"55d57'
Lizzie
Cas 0h32'0"70d31'
Lizzie Lovedrops
Cyg 20h17'10"38d25'
Lizzy
Cyg 19h21'17"28d45'
Lizzy
Cas 0h52'43"68d4'
Lizzy May Lou-Lou
Cas 1h38'23"73d27'
Lizzy's Light
Aql 19h56'15"0d3'
LJ
Vir 11h52'15"0d15'
LJD 100
Lmi 10h38'57"25d6'
LJL 91095
Vul 20h15'22"23d48'
Ljuba
Del 20h12'53"14d36'
Ljubaliza
For 2h8'36"-26d18'
Ljutich,Paul Calvin
Her 17h14'1"49d45'
LKM 12
Cas 0h3'31"62d18'
Llafet,Gail
Uma 10h55'17"48d29'
Llapitan,Tyler William
Hya 8h41'57"0d27'
Llarena,Lori Lynn
And 0h49'29"41d7'
Llenroc
Lyn 7h56'30"58d52'
Llerena,Giannina Gabriela
Mon 7h45'51"-8d45'
Llewellyn,Bill
Ori 5h32'46"0d28'
Llewellyn,Brian
Cep 23h8'1"60d16'
Llewellyn,Corey Anthony
Per 2h48'16"32d29'
Llewellyn,Don & Barbara
Cam 7h56'37"61d9'
Llewellyn,Judy Brejot
And 2h29'39"44d25'
Llewellyn,Justine
Lyr 18h52'0"42d43'
Llewellyn,Laura
Cas 23h14'15"61d51'
Llewellyn,Lee
Eri 4h2'17"-18d19'
Llewellyn,Marie
Cas 0h5'51"65d32'
Llewellyn,Nan
Dra 17h42'20"60d53'
Llewellyn,T David
Cep 1h21'29"77d53'
Llewelyn,Rachel
And 23h15'46"51d41'
Llinas,Jaume
Per 4h5'56"37d1'
Llinos
Cas 2h0'36"61d23'
Ill and Rising
Dra 18h58'15"48d15'
Llohis,Juan Solaz
Her 17h50'1"41d2'
Llorens,Ariadna
Boo 14h44'1"46d16'
Llorens,Eduard Boixet
Cas 0h5'55"60d18'
LLorin,Brittany Nicole
Mon 8h7'9"-9d29'
Lloveras,Rosa Oliva
Leo 9h32'0"8d26'
Lloyd
Uma 11h0'52"60d56'
Lloyd
Umi 13h15'0"76d4'
Lloyd
Aur 7h22'15"41d10'
Lloyd & Kate
Cyg 21h6'43"30d48'
Lloyd David
Lac 22h19'0"48d30'
Lloyd II,Richard A
Lac 22h7'23"37d46'
Lloyd Jr,Charley
Per 2h39'12"35d48'
Lloyd's Light
Aur 5h50'52"50d6'
Lloyd,Barbara
And 23h23'25"40d45'
Lloyd,CJ
Aur 5h0'59"45d18'
Lloyd,Donald
Dra 19h49'1"61d41'
Lloyd,Elwyn Peter
Cep 4h0'28"80d13'
Lloyd,Erin Lee
Del 20h31'2"5d16'A
Lloyd,Flora Berry Shreve
Cas 0h13'1"61d32'
Lloyd,Harry J
Aur 6h10'47"31d32'
Lloyd,James B
Cma 6h57'3"-16d31'
Lloyd,Jane
Peg 22h28'46"18d56'
Lloyd,Jeffery Kenneth
Oph 17h57'0"12d12'
Lloyd,Joan
And 2h26'55"44d10'
Lloyd,Jr,Theodore W
Aql 19h17'45"12d14'
Lloyd,Judith
Cas 0h54'57"56d4'
Lloyd,Kay Dawn
Peg 22h34'35"10d15'
Lloyd,Lane
Cep 22h15'23"56d0'
Lloyd,Lillian
Cas 0h38'27"69d46'
Lloyd,Lillian Mae
Vul 19h46'48"26d11'
Lloyd,Marion
Lyn 7h52'31"58d18'
Lloyd,McKenzie
Equ 21h3'1"10d53'
Lloyd,Nicci
Cas 0h48'34"73d37'
Lloyd,Ralph W
Cep 21h57'49"56d12'
Lloyd,Rebecca
Lyr 18h36'0"30d50'
Lloyd,Rebecca Caroline
Lyr 18h15'37"38d56'
Lloyd,Ronda
Ori 5h57'1"17d58'
Lloyd,Sarah Jeanne
Aur 6h2'32"48d57'
Lloyd,Shannon Marie
Cmi 7h55'27"8d10'
Lloyd,Sheila & Corinne
And 23h20'22"42d36'
Lloyd,Tim
Oph 17h37'2"-21d26'
Lloyd-Carr,Patricia
Umi 16h19'39"71d9'
Lloyd-Owen
Ori 5h23'13"1d9'
Lloydoris
Cam 5h49'41"60d48'
Lloydy
Per 2h11'41"57d47'
Lloydy
Uma 10h10'44"42d12'
LMB 4
Mon 8h3'12"-9d35'
LMB Microcomputers
Cyg 21h29'26"40d30'
LMF/RKB
Cet 2h55'39"1d11'
LMV:Till Time & Times Are Done
Uma 9h59'51"48d58'
Lo Conte,Antonio
Lyn 7h53'1"41d25'
Lo Curto,Nancy A
Lyr 18h39'51"40d27'
Lo,Eric Kenneth
Dra 16h34'31"64d20'
Lo,Peter Pak Shiu
Mon 6h28'31"3d44'
Lo-Go-31-07-1991
Leo 11h53'48"20d49'
Loague My Brightest Star,Nancy Louise
Tri 2h0'19"31d19'
Loaiza,Peter
Cap 21h56'34"-20d43'
Loalbo,John M
Aur 5h15'0"41d15'
Loar,Sr,Richard & Ethel L
Aur 6h26'53"34d48'
LoBasso,Dominick S
Her 16h39'60"38d8'
Lobaugh,Lindsey
Cas 1h17'23"63d6'
Lobben,Denny
Hya 8h17'16"3d47'
Lobberecht,Janet
Cyg 20h39'57"52d43'
Lobbestael,Colonel Wayne J
Her 18h12'17"45d35'
Lobdell,Nana & Pampau
Vir 13h34'45"-8d16'
Lobdell,Scott & Amie
Cyg 19h30'22"31d11'
Lobe,David Roger
Leo 9h41'25"8d15'
Lobelson,Harvey
Sex 10h30'0"3d52'
Lobengerg,Kelly
Lyr 19h6'28"28d51'
Lober,David Meyer
Per 3h25'57"40d36'
Loberg,Rigmor
Cet 2h27'22"0d57'
Loberg,Thomas
Cet 2h27'57"0d24'
Loberger,Gordon
Cnv 12h56'29"32d53'
Lobl-Guiding Stars, Leona & Albert
Crb 15h59'19"27d52'
Lobo
Uma 9h55'12"45d9'
Lobo Love
Boo 14h15'1"54d13'
LoBoda's,Frenchie, Miraculous 80
Cyg 21h20'26"39d28'
Lobosco,Dana Lyn
Lac 22h6'1"46d29'
Lobsinger,Nancy
Cas 0h30'24"66d1'
Locascio,Amber Marie
And 0h4'22"34d45'
Locascio,John & Mary
Ori 5h56'30"5d22'
Locascio,Lisa
Cyn 20h24'1"37d32'
LoCascio,Valerie
Tri 2h43'0"31d40'
Locascio,Kyle
Lyr 18h37'59"36d56'B
Locascio,Travis
Lyr 18h37'59"36d56'A
Locatelli,Mario
Ari 3h4'0"28d58'
Locatelli,Sergio
Lyn 7h3'41"60d31'
Lochart,Olivia Kim & James
Cam 4h45'5"66d26 AB
Lochbaum,J
Uma 10h13'1"51d23'
Lochbigler,Doris Angel
Com 13h4'30"26d16'
Locher,Gunther
Ori 5h37'28"12d19'
Lochhead,Gladys Ellen
Lyr 18h18'1"47d5'
Lochiatto,Dale Ann Kennedy
Cas 1h7'24"61d43'
Lochinvar's Lucky Star,Alexandar
Lib 14h28'19"-22d53'
Lochinvar,D.H.F.
Uma 8h45'39"71d2'
Lochmann,Dieter
Peg 23h35'24"10d56'
Lochner,Sally Soth
Cam 4h55'54"58d49'
Lochridge,Scott
Dra 16h50'38"68d54'
Locicero,Dorothy
Lyr 18h33'60"37d39'
LoCicero,Lisa
And 23h15'0"45d32'
Lock,Brittany
Com 12h6'13"21d57'
Lock,Derek
Uma 9h4'21"58d6'
Lock,Harry R
Boo 14h54'54"28d31'
Lock,Terry
Boo 13h39'31"21d8'
Lockamy,Denise Foster
Oph 17h19'31"-23d45'
Lockamy,Elizabeth Joy
Peg 23h18'18"33d58'
Lockamy,Patrick Ryan
Aql 20h2'20"10d9'
Lockard,Doris
Lyn 8h11'41"50d27'
Lockard,In Memory Of Merle
Lyn 8h11'44"57d57'
Lockard,Laramie Isabelle
Hya 8h44'16"2d54'

Lockard,MD,Vern M
Aur 5h9'26"41d57'
Lockart,John Robert
Dra 19h2'42"58d44'
Locke The Doc- Renaissance Man,J
Boo 13h36'30"21d59'
Locke,Al-Sharon- Adriana-Christina
Per 4h26'55"51d26'
Locke,Alexander
Cru 12h9'60"-64d3'
Locke,Attica
Cet 2h41'25"4d2'
Locke,Brittany Elizabeth
Cas 0h51'1"70d48'
Locke,Deborah G & Lenox P
Her 16h20'20"40d5'
Locke,Eva Marie
Lyr 18h47'36"31d15'
Locke,Garry A
Dra 18h39'43"80d5'
Locke,James D
Boo 15h3'52"10d31'
Locke,Katherine
Col 6h1'41"-38d52'
Locke,Lisa Jane
Cas 1h48'25"75d32'
Locke,Martha
Cas 1h1'45"61d40'
Locke,Nicki
And 0h45'13"38d59'
Locke,Robert
Her 18h6'51"14d37'
Locke,Stephen & Christina
Peg 22h26'27"21d59'
Locke,Toby Peter
Ori 5h57'13"15d24'
Locker,Janice
Sge 19h11'18"19d57'
Lockett,G Jeffrey
Dra 18h59'46"48d1'
Lockhart
Cyg 19h13'20"48d17'
Lockhart,Celeste
Cam 7h57'45"61d38'
Lockhart,Elaine
Uma 9h48'36"47d39'
Lockhart,Jr,George F
Aql 20h7'1"6d24'
Lockhart,LCDR Mark Stephen
Dra 18h41'28"68d15'
Lockhart,Stacey Lee
Mon 8h4'23"-8d38'
Lockhart-God's Oncologist,Dr Sharon
Oph 17h58'1"-8d52'
Lockin,Rose Anna
Cas 1h11'36"60d42'
Locklear
And 23h51'31"39d0'B
Locklin,Dr Walter Kaye
Dra 19h32'37"68d15'
Lockman,Susan
Cas 1h9'0"62d50'
Lockner,Francis John
Uma 9h24'58"61d17'
Lockrem,Kristi
Lyn 7h55'1"41d25'
Lockridge,Mark Alan
Cnv 12h10'37"37d45'
Lockwood,Cheryl
Cas 23h32'15"60d25'
Lockwood,Chester
Sct 18h55'27"-6d54'
Lockwood,Jr,Love Forever Mom,Dale H
Aur 7h0'44"36d31'
Lockwood,Linda
Lyn 8h56'28"42d12'
Lockwood,Lisa
Peg 21h55'32"34d39'
Lockwood,Mark Andrew
Boo 13h36'1"19d37'
Lockwood,Mark Andrew
Her 16h14'33"4d59'
Lockwood,Rachel Sloane
Cam 12h26'31"80d41'
Lockwood,Samuel Bridger
Ari 3h2'18"25d2'
Lockwood,Terri & Jeff
Cyg 21h6'32"30d30'
LoCosa Five,The
Tri 2h4'41"31d38'
Locquet,Daniel
Dra 9h35'37"80d34'
LoCurto,Stephanie Christine
Lyr 18h28'39"31d49'
LOCUS ISTE
Sco 16h36'51"-41d13'
Lodato,Nicholas Matthew
Boo 13h33'20"22d11'
Loday,Yves
Peg 23h30'58"21d33'
Lodders,Hans
Peg 23h17'1"31d57'
Loddie Star,The
Dra 18h13'38"80d18'
Lodes,Amanda
Cet 1h8'0"-3d39'
Lodes,Cathy
Aql 19h30'1"0d0'
Lodes,Evan
Ser 18h54'23"3d17'
Lodes,Tom
Sex 9h53'28"2d3'
Lodestal 4
Psa 22h33'29"-27d31'
Lodewyck,Sandy
Lyn 8h18'17"40d8'
Lodi,Bruno
Pho 0h27'12"-43d59'
Lodi,Feroz
Lac 22h51'54"41d9'
Lodise,Genny
Cas 0h45'33"64d1'
Lodl,James Anthony
Lyr 19h23'13"37d48'
Lodon,Dominique
Uma 9h35'24"62d21'
Lodovisi,Gianluca
Her 18h9'18"31d43'
Lodoza,Charles
Ori 5h25'0"1d38'
Loeb,Mort R
Her 18h45'17"12d39'
Loeb,Walter M
Ser 17h54'41"-14d32'
Loebenberg,Kenneth
Boo 13h45'17"14d40'
Loebig,Bill
Dra 16h54'31"67d14'
Loeffelman,Betty Jane
Lyn 9h13'0"45d9'
Loeffler,C Helen Pease
Aql 19h45'52"13d56'
Loeffler,Kate
And 0h45'0"39d29'
Loeffler,Sharie
Cas 0h39'0"64d54'
Loegan,Mattie
Lyr 18h58'32"45d19'
Loel
Mon 6h54'51"0d32'
Loenhart,Anne-Sophie
Gem 8h2'53"30d21'
Loer,Harald Arthur Karl
Lyr 19h22'56"30d31'
Loera,Baley Katherine
Mon 6h54'22"-0d28'
Loera,José Leyva
Her 16h59'1"28d24'
Loera,Jr,Luis Miguel
Per 3h6'22"38d28'
Loery,Jay
Hya 9h7'33"2d17'
Loerzel,Randy Joe
Lac 22h14'33"51d38'
Loesch,Michael
Cet 1h58'0"-8d33'
Loesl,Kristina Marie
Lyn 7h23'15"50d7'
Loetscher,Mirjam
Aql 19h59'40"12d7'
Loevsky,Louis
Lac 22h53'37"55d27'
Loew,Daniel
Per 3h56'5"37d26'
Loew,Dieter
Peg 23h34'21"18d58'
Loewchen=Mechthild Bermpohl-Nolden
Leo 10h30'26"18d32'
Loews,Carl
Lac 22h14'38"51d4'
Lofaro's Lucky Star
Aur 7h13'15"40d11'
LoFaro,Guillaume
Ori 5h32'26"-6d56'
Loff,My Everlasting Love,Marta
Cas 1h43'12"60d47'
Loffredo,Carlo A
Crt 10h55'56"-21d49'
Lofgren,Getrude D
Cyg 21h5'23"34d50'
Lofquist,Katherine Ann
Boo 14h34'20"7d38'
Loft,Marta Bell
And 2h18'59"38d15'
Loftie's Light Bulb
Cam 6h1'19"60d42'
Loftin,Jenifer Leigh
Vul 19h22'25"25d22'
Loftin,Robin
Peg 22h39'56"25d29'
Loftis,Kelly Harasti
And 0h2'0"34d25'
Lofton,Dot Caldwell
Aqr 22h20'47"0d25'
Lofts,Margot Patricia
Peg 23h31'39"11d30'
Loftus,"Cece" Sienna Lee
Lyr 19h18'16"40d57'
Loftus,Allison Jayne
Cas 2h46'28"68d46'
Loftus,Allison M
Del 20h12'1"15d20'
Loftus,Catherine Grace
Cas 2h47'48"61d9'
Loftus,Edward Roland
Eri 4h57'40"-8d59'
Loftus,Michael Lee
Cet 2h1'44"1d55'
Logan "The Amphibian Queen",Teri J
Cas 0h32'58"61d16'
Logan Alpha,Sean
Cet 0h41'47"-6d24'
Logan's Star
Ser 15h30'16"0d34'
Logan's Star Sallee
Her 17h4'0"42d42'
Logan's Starasaurus
Per 1h47'54"50d26'
Logan,Brian David
Lac 22h23'34"40d59'
Logan,Brian P M
Cnv 12h23'33"46d41'
Logan,Donna
Lyr 18h47'51"41d17'
Logan,Elaine
Cas 23h17'52"62d8'
Logan,Evelyn Stark Pedrazzi
Lyn 7h37'42"36d12'
Logan,Jacqueline Anne
Cas 2h56'45"65d13'
Logan,Jon Paul
Boo 15h7'34"11d16'
Logan,Joseph Lee
Aur 6h2'41"45d26'
Logan,Kim Lorain
Leo 10h38'0"15d19'
Logan,Loren L
Peg 23h47'55"15d47'
Logan,Matthew
Aur 4h54'52"40d13'
Logan,Megan
Mon 6h56'26"-10d16'
Logan,Megan Kristen
And 0h28'22"27d42'
Logan,Morgan
Cam 3h53'13"61d29'
Logan,Robert Faid Bell
Cma 6h59'31"-18d27'
Logan,Sharon
And 23h11'23"37d32'
Logan,Sue
Com 12h43'13"20d5'
Logan,Vanessa
And 0h2'1"35d59'
LogaNancy
Ori 5h47'35"11d12'
Logerot Merel
Per 3h42'37"39d59'
Logette,Lilia M
Mon 6h54'26"-1d25'
Loggins,Pamela L
Aqr 22h0'12"0d45'
Loggins,Patsy
And 23h0'1"51d32'
LoGiudice,Anthony
Cnc 9h13'54"31d32'
Logiudice,Susan
Cas 0h51'1"64d20'
Logiudice-Davis 54112ø
Cnv 13h55'40"40d28'
Logory,Dustin Joseph
Aur 7h5'34"36d52'
Logrono,Mary
And 23h20'35"51d24'
Logsdon,Bob E
Oph 17h1'17"11d12'
Logsdon,Frank Lee
Her 17h9'33"43d15'
Logsdon,J Daane
Sgr 19h38'26"-38d14'
Logsdon,Margie
Mon 6h57'51"11d16'
Logsdon,Mary Ann
Cam 3h16'56"61d47'
Logsdon,Teresa
Peg 22h45'48"25d7'
Logsy La La!!
Psc 23h8'41"5d58'
Logue,Carol Gigante
Peg 23h34'29"23d49'
Logue,Clare T
Peg 22h22'31"27d39'
LoGuirato,John
Aur 7h22'43"40d15'
Loh,Yen Lin
Boo 14h53'32"42d4'
Lohbauer,Donna M
Eri 5h2'49"-5d59'
Lohbauer,Tom
Her 17h53'27"40d29'
Lohbrandt,Amber
And 23h37'0"46d52'
Lohengrin
Ant 10h44'46"-32d51'
Lohman,Ray
Cnv 13h32'44"47d45'
Lohman,Robert E
Oph 18h32'26"11d13'
Lohmann,Austin Gregory
Hya 8h39'43"-10d56'
Lohmann,Ulrich
Cam 7h38'34"74d36'
Lohmeier,Peter
Aql 18h54'26"-1d3'
Lohnes,C C
Aur 6h39'28"38d37'
Lohnes,William Edward
Per 3h10'1"47d14'
Lohr,Heather
And 1h50'11"39d55'
Lohr,Jack
Oph 17h19'33"-21d22'
Lohr,Linda
Peg 23h43'16"31d3'
Lohr,Richard E
Boo 15h3'25"50d37'
Lohrke,Helmuth
Del 20h31'57"3d49'
Lohse,Eric
Uma 11h16'38"51d12'
Loiacano,Sr,James M
Aql 18h58'38"-0d5'
Loiacono 1947-1994, Nicholas Peter
Aqr 20h45'47"-6d33'
Loiacono,A J
Her 18h3'24"38d48'
Loiacono,Blaine
Lyn 7h37'23"50d52'
Loiacono,Nicholas Dominic
Boo 15h29'51"48d4'
Loiacono,Philamena
Equ 21h2'23"11d11'
Loika,Liz
Com 12h16'24"31d11'
Loiko,Kara
Cyg 19h42'46"31d23'
Loiola,Amerigo
Vir 13h2'1"-20d6'
Lois
Mon 8h6'24"-3d46'
Lois
Mon 6h20'25"8d33'
Lois
Ori 5h44'32"7d47'
Lois
Aur 4h21'9"34d12'B
Lois
And 1h22'15"37d23'
Lois & Carl's Happy Day Star
Equ 20h58'1"6d12'
Lois & Jerry
Cyg 19h56'18"41d3'
Lois & Lisa Daycare Star,The
Uma 11h6'49"30d38'
Lois & Lorne
Aur 6h38'0"41d7'
Lois & Rick's Star
Uma 11h4'47"57d10'
Lois Ann
And 2h22'42"47d14'
Lois Anne
Cas 0h51'24"64d29'
Lois' Light 2-19-41
Per 1h47'1"53d56'
Lois's Star
Lyr 18h25'36"38d15'
Lois's Star
And 23h15'41"50d49'
Loiseau,Alex Jean
Boo 14h43'30"35d44'
Loiseau,Todd Allan
Dra 16h13'59"60d32'
Loish,Tanya Lee
Lyr 19h14'26"37d50'
Loisi
Mon 6h53'23"0d2'
Loiska
Cep 2h13'60"78d18'
Loizou,Nasia
Eri 3h41'32"-19d54'
Loizou,Stavriny
Ori 5h56'20"20d36'
LOJAK
Oph 17h17'51"-23d15'
Lojo,Anthony James
Her 16h55'51"34d9'
Lojo,Nicholas
Boo 13h43'0"21d15'
Lokay,Craig
Cnv 13h43'14"35d26'
Loken
Ser 18h7'39"-14d9'
Loken,Orlin G
Aql 20h17'1"0d45'
Loki Devildog Terra Nova
Cma 7h2'50"-31d35'
Lokke,Devonshire Grace
Dra 17h26'0"73d30'
Lokke,Hirum & Lolita
Lmi 10h52'22"26d19'
Lola
Dra 17h3'28"66d39'
Lola
And 23h21'31"50d18'
Lola
Lyr 18h29'30"31d51'
Lola
Ori 6h5'48"7d16'
Lola Aznarez
Ori 5h54'22"1d20'
Lola Maya
Cas 0h41'38"61d22'
Lola-Eugenie
Lac 22h4'33"37d60'
Lolangelauren
Del 20h14'16"12d17'
Loletta Lee
And 23h3'29"38d57'
Loli
Aur 5h38'48"50d6'
Lollipop
And 23h45'11"45d53'
Lollis,Leila B (Bennie)
Aql 19h30'43"10d52'
Lollis,Mary Catherine
Peg 22h46'45"33d55'
Lolly
Sgr 19h33'36"-42d55'
Lolly
Peg 22h4'23"20d5'
LoLo
Mon 7h0'1"5d10'
Lolo
Aur 5h2'30"37d34'
Lom-Ajan Star,Silvia
Vul 19h18'5"26d33'B
Loma
Ori 5h51'51"14d48'
Loma
Equ 20h55'39"2d50'
Lomacis
Hya 8h16'45"-1d50'
Lomas,Shanine Leigh
Mon 6h29'45"2d53'
Lomascolo,Charlie
Cep 20h26'53"75d9'
LoMastro,Carol & Victor
Lyr 18h44'17"41d33'
Lomax
Ori 5h0'49"15d27'
Lombard
Peg 0h2'0"20d9'
Lombard Junior Woman's Club
Lyr 18h20'19"44d5'
Lombard,Christina Cameron
Mon 7h49'27"-5d6'
Lombard,D Sterling
Cet 2h19'58"0d26'
Lombard,Elizabeth Ann
Hya 9h6'10"2d28'
Lombard,Frank James
Oph 18h31'0"10d50'
Lombard,Michael Francic
Aur 5h1'1"44d32'
Lombard,Paige
Sex 10h28'39"4d32'
Lombard,Zoe
Ori 5h17'11"-9d30'
Lombardi,Antoine
Peg 23h35'18"10d10'
Lombardi,Donna Marie
Cas 2h23'29"70d1'
Lombardi,Francesco
Lyn 7h52'53"41d48'
Lombardi,Georgia
Lyn 8h2'14"35d56'
Lombardi,Gerardo & Katerina
Cyg 19h25'27"33d11'
Lombardi,Joan Elizabeth
And 2h29'43"38d28'
Lombardi,Kim Marie
And 2h25'43"40d52'
Lombardi,Larry
Lac 21h59'20"42d49'
Lombardi,Mark Anthony
Uma 11h40'43"41d10'
Lombardi,Michael G
Dra 19h25'52"67d46'
Lombardi,Mike
Dra 17h26'1"70d8'
Lombardi,Mr E F
Cet 2h32'40"0d16'
Lombardi,Quentin
Ori 5h56'1"20d17'
Lombardi,Silvestro
Dra 16h24'60"62d42'
Lombardo Misura/Re Amore,Anita
Cas 0h28'30"63d42'
Lombardo,Alice Ann
Lyn 8h11'41"57d34'
Lombardo,Christian
Uma 13h35'36"57d42'
Lombardo,Chuck
Ori 5h56'49"14d31'
Lombardo,Dr Thomas A
Oph 17h34'41"8d28'
Lombardo,Eric James
Cep 0h46'18"86d29'
Lombardo,Michael
Lac 22h15'58"54d24'
Lombardo,Scott
Aur 6h8'1"30d1'
Lombart,Marcel
Her 17h53'41"40d38'
Lomberg,Maryanne
Cyg 19h34'42"32d0'
Lombini,Emanuele
Aql 19h53'55"-5d44'
Lombino,Glenn & Leslie
Lyn 7h44'58"40d44'
Lombre,Anthony V
Lib 14h58'59"-1d50'
Lomeli,Anthony M
Vul 20h38'48"20d12'
Lomeli,Marcela A
Mon 6h54'43"-5d44'
Lomicky,Carolyn & Craig
Lyn 7h58'28"38d15'
LOML
Lyn 9h8'52"33d55'
Lomon,Dominique
Cyg 19h47'1"38d20'
Lomot,Michael
Lmi 10h47'51"30d1'
Lomotan,Mike (Stubbs)
Vul 20h14'54"25d41'
Lomovsky,Alice
Cas 23h4'55"58d24'
Lon Lee
Her 16h38'0"41d53'
Lon's Alternative Beacon
Aql 18h42'52"0d60'
Lon's Light
Del 21h1'28"12d4'
Lon's Reach
Per 4h41'35"51d12'
Lonardo,Elizabeth
Sco 17h50'42"-38d50'
Lonardo,Enrico
Tau 5h53'49"28d16'
Lonbom,Chelsie Ann
And 2h33'1"40d21'
Loncto,Thomas Jeffrey
Boo 15h7'15"53d53'
Londa
Del 20h30'14"20d27'
London Star,The
Oph 17h21'16"-23d39'
London,Lauren Elizabeth
Cyg 19h29'42"31d0'
London,Ryan Lawrence
Per 1h32'46"53d15'
London/Buschbacher
Ori 5h56'0"19d29'
Londoner,Daniel
Cet 2h41'29"5d37'
Londry,Cal
Lyn 8h26'20"46d24'
Londry,Joel
Cyg 20h57'53"50d16'B
Londry,Karen
Cyg 20h57'53"50d16'A
Londry-Loose,Laura Jean
Uma 11h15'1"40d22'
Lone Larsen
Uma 11h31'11"50d12'
Lone Vanessa
Cam 3h18'52"60d26'
Lone Wolf
Ori 6h8'38"8d27'
Lonergan,Matthew & Jennifer
Vul 20h4'58"25d24'
Lonergan,Tim
Aur 4h53'40"40d18'
Lones,Jane Louise
Cet 0h54'49"1d1'
Lonestar Mike
Her 17h29'22"31d37'
Lonez,Renee
Cmi 7h18'31"3d60'
Long "Rusty",Russell Ernest
Oph 18h19'0"6d17'
Long 10,Christopher Gene
Cet 0h10'17"-18d52'
Long Island Star For The Gifted
Uma 12h12'0"54d55'
Long,Adam C
Cyg 19h30'35"39d25'
Long,Albert E
Per 2h56'1"45d14'
Long,Allen F
Cam 13h19'13"80d11'
Long,Amanda Marie
Peg 23h39'1"31d18'
Long,Amy Jo
Cmi 8h4'47"6d29'
Long,Anthony
Mon 6h51'39"11d45'
Long,Arthur
Lyn 7h53'18"44d2'
Long,Bonnie L
Cas 0h0'1"62d46'
Long,Brandon Allen
Dra 19h23'49"67d58'
Long,Bruce D
Aur 6h57'1"35d38'
Long,Charles Abell
Boo 15h9'57"28d15'
Long,Charles Andrew
Lac 22h18'20"51d8'
Long,Charles Anthony
Per 2h56'42"40d8'
Long,Charles Kimberly
Uma 11h11'40"44d32'
Long,Chiquita Maria
Ori 5h39'9"-2d55'A
Long,Christopher
Uma 9h19'48"53d44'
Long,Christy
Cas 0h6'0"65d37'
Long,Cynthia Galey
Peg 22h42'18"27d56'
Long,David
Peg 0h10'18"14d7'
Long,Dennis Brian
Aql 18h41'16"-1d55'
Long,Diana Janice
And 23h16'33"46d32'
Long,Doris Lee
Ori 5h39'9"-2d55'B
Long,Dorothy Grace
Peg 22h19'1"31d41'
Long,Dr James E
Her 17h50'42"40d11'
Long,Emily Ann Foley
Aql 19h30'0"13d12'
Long,Florence Jean Castleberry
Cyg 21h1'28"50d3'
Long,Freda Lee
Peg 0h1'26"11d52'B
Long,Holly
And 0h1'0"44d9'
Long,Holly Beth
Peg 23h44'60"30d20'
Long,Jan
Aql 19h53'24"-0d28'
Long,Jeffrey R
Hya 8h16'49"4d26'
Long,Jessica
Cnv 13h21'36"41d3'
Long,Jr"Alimal",Alan Richard
Uma 8h47'53"48d28'
Long,Jr,James Edward
Ori 5h33'12"0d52'
Long,Kay E
Del 20h13'25"10d56'
Long,Keri L
Sco 16h25'51"-40d37'
Long,L'étoile de l'amour
Cyg 21h4'50"37d16'
Long,Lacey Marie
Mon 8h3'41"-8d19'
Long,Laura Lea
Peg 23h44'0"26d4'
Long,Leigh Robin
Uma 13h37'36"61d32'
Long,Lesley
Lyn 7h52'28"43d16'
Long,Lynda L
Aql 19h57'24"10d29'
Long,MA-KD2Q,Gary Robert
Crt 11h14'59"-15d44'
Long,Marianne
Lyn 9h0'19"34d14'
Long,Mark Allen
Tau 3h37'26"21d14'B
Long,Mark D
Dra 17h31'22"64d56'
Long,Matthew Aaron
Boo 15h10'55"47d37'
Long,Melissa Jill
Lyn 8h47'33"45d38'
Long,Michele
Cas 1h34'0"77d3'
Long,Nancy Crawley
Tau 3h37'26"21d14'A
Long,Peggy Joyce
Cas 2h3'0"59d12'
Long,Robert James
Boo 15h8'38"28d6'
Long,Samuel Jonathan
Dra 19h6'20"58d31'
Long,Sr,Kenneth Richard
Peg 0h1'26"11d52'A
Long,Steven
Hya 8h43'1"3d35'
Long,Thomas Lloyd
Cet 2h35'55"-0d58'
Long,Timothy Shawn
Cmi 7h25'17"0d4'
Long,Willis F
Dra 12h3'40"68d36'
Long,Woodrow A
Aur 5h27'58"30d16'
Long,Zoe Ann
Lyr 18h16'56"43d38'
Long-Angel Forever, Nicole Danielle
Lyr 18h36'32"40d55'
Long-Havrilla "Honey", Agnes Cecilia
Uma 12h3'18"43d47'
Longacre,"Yak"
Uma 8h10'57"68d17'
Longaker-Plaisance '94 Cara L
Lyn 6h28'41"56d2'
Longano,Jr,Alfred A
Tri 2h18'50"33d19'
Longbody Daddy
Uma 9h28'1"58d2'

Longenberger,Edward
Aqr 23h37'56"-12d30'
Longenhagen,Jeanne Wolfe
Cas 1h52'50"71d9'
Longfellow,Jack M
Cet 3h7'55"4d18'
Longfellow,Joshua D
Cyg 20h27'35"40d32'
Longgerth
And 1h5'11"38d57'
Longhelt,Christine M
Vul 20h15'13"23d13'
Longhi,Andreina
Lyr 19h16'12"42d39'
Longhi,Letizia
Hor 3h23'34"-48d6'
Longhran,Sarah Louise
Sge 20h4'18"16d40'
Longino,Harry Hugh
Aql 19h56'49"10d55'
Longley,Joan
Her 15h51'48"43d46'
Longley,Michael Sandman
Per 3h59'28"51d9'
Longley,Michael Owen
Her 18h0'34"38d45'
Longmore,Scott
Hya 8h15'48"0d43'
Longmuir,Jacquelyn Marie
Lyr 19h22'51"33d56'
Longo's Star,Tony
Aur 6h24'19"32d43"
Longo,Angelina Ragonese
Com 13h2'13"17d48'
Longo,Carmen
Her 17h35'51"38d46'
Longo,Christopher
Dra 14h21'12"63d40'
Longo,James S
Aur 6h0'21"38d10'
Longo,Jessica Fern
Lyn 7h57'0"50d39'
Longo,Raquel Nicole
Lyr 18h14'36"46d33'
Longo,Steven Nicholas
Boo 15h3'19"22d20'
Longo,Victor W
Tri 2h16'21"30d20'
Longo-Dente,Camilla
Boo 15h5'15"23d8'
Longoni,Ambrogio
Lyr 18h37'0"28d11'
Longoria,Karen
Mon 6h58'22"-5d48'
Longoria,Ricardo Elliott
Aql 18h57'52"12d39'
Longran,Kevin David
Cyg 21h0'45"38d8'
Longsden,Charise Lynn
Lyr 18h58'44"27d18'
Longshaw,Alpha Ted
Cep 1h11'31"77d60'
Longshore,Amy Marie
And 1h52'58"39d41'
Longshore,Andrea J
Eri 3h27'13"-4d29'
Longson,Marky
Her 17h38'43"40d1'
Longstaff,SamKate
Cyg 21h36'30"42d28'
Longstaff,Sarah Louise
Aql 18h57'28"-10d5'
Longstreth:My One Wish
Come True,Rob
Boo 15h16'50"51d51'
Longthorn,Ann Radband
Crb 16h14'24"32d46'
Longthorne,Joe
Aql 20h5'45"4d29'
Longtin,Michael
Cep 21h44'24"58d34'
Longuehaye,Kayleigh
Aql 19h0'13"-8d47'
Longuépée Géraldine
Mon 7h46'23"-4d32'

Longview,Lucile Kitson
Schuck
And 0h24'0"34d21'
Longwell,Connie Dee
Mon 7h7'19"-6d34'
Longwell,Erika N
And 2h32'33"49d18'
Lonica's Star
Lyn 7h 32'19"52d1'
Lonier,Aaron B
Aur 5h14'15"41d36'
Lonigro,John Richard
Psc 23h4'40"1d2'
Lonigro,Patricia Jane
Cas 23h36'43"62d5'
Lonija's Zvaigzne
Lyr 18h56'17"47d18'
Lonner's Star,Matthew & Adam
Crb 16h20'1"38d2'
Lonnie
Vir 13h56'49"-20d51'
Lonnie
Aql 18h58'42"4d25'
Lonnie Lee
Uma 11h0'0"62d24'
Lonnie's Light 1993
Peg 22h2'43"10d49'
Lonnie,Lynnne & Judy
Uma 8h35'31"68d34'
Lonny
Ori 4h57'29"-0d25'
Lonschein,Eric P
Cep 22h5'1"61d59'
Lonsdale,Antony
Aql 19h55'26"10d43'
Lonsdale,Matthew Paul
Her 17h18'37"49d44'
Lonser,MD,Roland E
Aql 19h44'51"14d59'
Lonsinger,Jazz "Her
Dogliness"
Lmi 10h48'22"30d33'
Lont,Nicholas
Ori 5h58'0"1d9'
Lonyay,Barbara
Ori 5h57'40"16d32'
Loo
Aql 19h50'30"11d3'
Loo,Leo
Lyn 7h55'34"50d50'
Loock,Brett Thomas
Her 15h51'14"41d39'
Looges,Adrian
Peg 21h10'0"12d27'
Look at Us-Mom & Dad -Curt
Uma 11h22'50"49d12'
Look JTB
Cru 12h43'7"-57d0'
Look,Bennett Markwell
Her 18h13'43"47d1'
Look,Edith Amanda
Uma 9h25'45"58d24'
Loomes,Roy Laurence
Ori 5h50'41"10d11'
Loomis II,Clifford "Barry" N
Aql 20h0'44"7d42'
Loomis,Timothy
Sct 18h49'18"-9d57'
Loomis,Toni
Uma 8h52'1"59d14'
Loon Lake 092395
Aql 19h57'28"12d48'
Looney,Barbara
Ari 2h7'30"18d8'
Looney,Brenda Kay
Aql 18h54'35"10d7'
Looney,Francis William
Ori 5h36'29"0d3'
Looney,Joseph R
Aur 6h26'44"35d28'
Looney,Robert
Dra 17h31'46"70d2'
Loonis,Jacques
Cyg 19h42'39"30d2'

Loonopollis
Dra 17h1'0"68d41'
Looofie
Ori 6h6'50"6d20'
Loos,Herbert
Cmi 7h18'50"8d4'
Loos,Linda
Dra 17h28'41"75d8'
Loose,Amanda Lynne
And 0h23'1"37d34'
Loose,Jessica Anne
Lyr 19h4'1"47d33'
Loose,Nancy
And 0h50'19"36d40'
Loosemore,Tracy
Lib 15h2'32"-0d59'
Looyenga,Donna
Peg 23h35'35"30d9'
Lopaka
Ori 5h55'54"13d19'
Lopardo,Sharon Lee
Aql 19h44'47"11d53'
Lopas,Larry Arthur
Lyn 7h29'14"50d4'
Lopeman,Joshua C
Boo 13h41'43"14d10'
Loper,Stanley Lewis
Vul 19h18'17"26d45'
Loperena,Jordan
Dra 17h32'45"61d21'
Lopes,Anivet
Cnv 13h55'43"31d2'
Lopes,Joaquina Maria
Cas 1h14'11"64d60'
Lopes,Matthew Edward
Cep 20h39'15"55d23'
Lopez "A Star at Fifty",Judith
Ann
Mon 6h27'15"8d33'
Lopez III,Eduardo
Aql 19h58'36"11d20'
Lopez"Love You",Cindy Moya
Peg 23h23'0"17d23'
Lopez,"My Babe" Jorge
Armando
Dra 11h39'41"72d0'
Lopez,"Yo Joe" Joseph
Sct 18h46'33"-7d2'
Lopez,Adrian
Cas 0h25'12"68d11'
Lopez,Albert B
Ori 6h16'36"8d37'
Lopez,Alejandra
Uma 10h9'27"51d35'
Lopez,Andrea Rae
Uma 11h48'48"62d23'
Lopez,Annette
Lmi 10h49'1"26d17'
Lopez,Beatriz
And 23h38'0"46d11'
Lopez,Berto
Uma 11h21'33"38d35'
Lopez,Carlos
Uma 11h6'21"41d8'
Lopez,Carolyn
Cyg 19h23'16"30d8'
Lopez,Catherine
Hya 9h34'19"-1d23'
Lopez,Chale
Dra 18h15'42"70d54'
Lopez,Darrin Lee
Aur 5h0'13"49d11'
Lopez,Davey Luis
Cet 2h59'54"0d28'
Lopez,David Zachary
Oph 17h0'12"11d1'
Lopez,Eduardo & Eusebia
Sge 19h31'1"16d37'
Lopez,Eli-Elleen
Vul 20h0'32"23d55'
Lopez,Fatima
Aqr 22h22'46"0d8'
Lopez,Gary Daniel
Aql 19h1'19"5d27'

Lopez,George & Ann Marie
Sge 19h9'25"19d51'
Lopez,Gloria Christine
Lyr 19h23'0"30d48'
Lopez,Hilda
Cas 1h59'55"68d19'
Lopez,Ivan A
Cet 2h42'47"1d28'
Lopez,Jessica
And 23h27'1"49d59'
Lopez,Joseph Gregory
Aur 7h15'46"36d45'
Lopez,Karina
Lyn 7h14'28"58d9'
Lopez,Kelly Deanne
Tri 2h1'37"33d51'
Lopez,Kenneth Paul
Ori 5h54'52"8d1'
Lopez,Kyle Brodie
Cas 0h34'38"72d37'A
Lopez,Laura Bichillo
Cyg 21h5'29"40d3'
Lopez,Maria Louisa
Cas 2h4'31"68d57'
Lopez,Marisol
Lyn 9h11'32"34d59'
Lopez,Matthew Mitchell
Cas 0h34'38"72d37'B
Lopez,Melissa
Equ 21h6'39"11d23'
Lopez,Michele
Peg 22h39'24"10d16'
Lopez,Pamela Sue
Cet 1h52'18"0d52'
Lopez,Peter (L M)
Dra 17h11'20"65d25'
Lopez,Phillip & Jan
Dra 18h55'20"70d6'
Lopez,Ramon
Hya 9h38'38"0d19'
Lopez,Ramon Elszy
Her 16h54'14"40d2'
Lopez,Ramon S
Eri 4h46'46"-9d49'
Lopez,Raquel
Mon 6h37'21"3d3'
Lopez,Remembering Yvonne
Mon 6h44'49"0d34'
Lopez,Rick A
Lac 22h27'23"41d6'
Lopez,Roy
Sex 9h53'18"0d59'
Lopez,Sergio
Aur 6h25'53"52d57'
Lopez,Sheila Grace
And 0h23'10"36d7'
Lopez,Tammy Lynn
Sct 18h55'58"-6d5'
Lopez,Tom & Janet
Magnusson
Lyn 9h10'49"37d32'
Lopez,Trent Nathan- McClure
Sex 9h50'31"-5d38'
Lopez,Victoria S
Mon 6h30'48"-6d55'
Lopez,Y
Cam 5h0'0"61d19'
Lopez,Yaneth Yvonne Pacheco
Lyr 19h8'34"40d30'
Lopez,Yvonne
Tri 2h18'31"31d8'
Lopez-Flor,Adriano
Per 2h31'46"56d27'
Lopez-Meyers,Mitch
And 0h54'35"37d22'
Lopezalles,Alex
Her 16h24'40"33d26'
LoPiano,Matthew
Her 17h15'47"18d49'
LoPiccolo,Phyllis
And 23h22'31"48d57'
Lopinto,Arlene
Vul 19h46'48"23d34'
LoPinto,Liam Quinn
Mon 6h54'59"-6d0'

Lopizzo,Antonio
Lep 4h58'41"-19d35'
Lopopolo,Nicoletta
Eri 3h40'22"-19d21'
LoPresti,Conor James
Per 2h25'11"55d4'
Loprete,Darlene
Lyr 18h55'40"40d40'
LoPrete,Ruth P
Dra 17h56'28"73d47'
Lopuch,Manfred
Aur 7h15'41"37d3'
Lopus,Bill
Oph 18h17'1"11d22'
Loquacious Harold
Dra 19h20'44"70d30'
Lora
And 0h8'0"47d40'
Lora "My Honey"
And 23h42'20"43d40'
Lora Jayne
Lyn 7h46'1"38d38'
Lora Star,Christina Anthony &
Andrea
Del 20h23'21"11d54'
Lorac
Del 20h53'14"9d54'
Loracher,Erwin
Lyr 19h20'50"31d52'
Loraine
Lyn 8h26'14"43d40'
Loralee
Cas 0h38'49"68d58'
Loralee-"But Honey"- Steve
Cyg 19h32'0"33d44'
Loranger,Larry
Del 20h16'1"12d30'
Loranger,Lucas
Del 20h16'17"15d10'
Loranger,Mary
Del 20h15'44"14d3'
Loranneagh Jewel
Eri 4h12'31"-19d2'
Loratta,Nicholas James
Per 2h22'0"55d18'
Lorayne
Mon 7h48'52"-3d4'
Lorber,Valerie
Del 20h17'15"13d45'
Lorch,John T
Lac 22h12'28"51d12'
Lorchel,Philippe
Tri 1h59'11"31d1'
Lord Alge
Tau 3h58'33"1d39'
Lord Arian Shining Star
Blessed Be
Aur 6h31'29"31d13'
Lord Bertram
Cam 5h17'1"68d33'
Lord Daniel Defender US
Silvertere
Aql 19h27'37"-8d46'
Lord Gerlad
Aql 19h53'34"12d24'
Lord Knifton
Per 2h51'11"40d44'
Lord Mhoram's Victory
Her 17h10'18"40d19'
Lord Sethx Grimme Of
ShatterMind Hold
Per 4h44'23'38d41'
Lord Seymour & Lady Lillibet
Cas 0h53'29"62d48'
Lord Spooney
Dra 16h9'53"64d13'
Lord's"Master-Star", Ken
Ori 5h38'36"-0d24'
Lord,Amy
Umi 14h58'42"65d47'
Lord,Beverley
Lyn 9h0'17"37d57'
Lord,Carrie Melissa
Cyg 21h55'48"44d51'

Lord,Christina
Lyn 9h6'39"37d31'
Lord,Conor Adams
Uma 9h24'31"57d45'
Lord,David G
Crt 11h53'20"-18d46'
Lord,Helen Renee
Mon 6h57'37"7d53'
Lord,Jacob Sylvan
Aql 18h43'31"8d24'
Lord,Jacqueline
And 2h34'38"40d41'
Lord,Joan & Ernest
Aur 5h48'0"50d6'
Lord,John Thomas
Aur 6h2'35"34d58'
Lord,Louis-Philippe
Per 2h58'45"38d0'
Lord,Lynne
Hya 9h25'32"-1d30'
Lord,Marion
Ori 6h2'42"10d46'
Lord,Penny Lei
Lyr 19h4'16"28d22'
Lord,Philip John
Ori 5h55'13"15d15'
Lord,Rebecca Jean
Cas 22h57'29"56d44'
Lord,Shirley Rosenthal
Lyn 7h54'49"40d53'
Lord,Thomas Edward
Lac 22h23'1"52d39'
Lord,Tom
Aur 6h34'11"30d37'
Lord,Yves-André
Umi 13h32'12"70d52'
Lord-Anthony
Her 17h35'1"27d27'
Lordan,Jack & Mary
Aur 6h54'50"44d23'
Lordo,Jacob Anthony
Cep 22h40'33"57d26'
Lore,Imme
Gem 6h26'22"14d53'
Lore,LInda
Uma 10h54'43"40d40'
Lore,Linda
Mon 7h19'50"-8d53'
Loredana
Lyn 8h23'21"50d12'
Loredana
Ind 21h1'35"-54d0'
Loredana
Lyr 18h39'1"29d55'
Loredana
Scl 23h44'49"-26d31'
Loredana
Lyn 7h27'16"40d53'
Loredana
Cas 2h34'28"50d52'
Loredana
Dra 16h1'13"62d50'
Loredana Simone
Cyg 21h5'0"32d24'
Loredana,Tatiana Isabella
Lyn 7h0'30"44d42'
Loree
Dra 17h54'42"68d42'
Lorek,Michael
Lac 22h12'40"48d38'
Lorella-Matteo
Lyn 7h29'26"48d54'
Loremm
Lyn 7h28'26"40d54'
Loren & Adele's
Sct 18h55'20"-6d2'
Lorena
Cas 0h5'56"50d30'
Lorena
Umi 16h22'41"72d19'
Lorena
Cyg 21h59'33"53d38'
Lorena
Lep 6h2'16"-18d57'

Lorena
Mon 6h35'28"11d18'
Lorena
And 1h1'0"37d37'
Lorena
Cyg 20h16'35"38d39'
Lorenc Colonia
Gem 6h24'54"12d1'
Lorencene-Marquis- O'Neal-
Blair
Boo 14h46'59"46d51'
Lorenda
Cam 7h30'43"78d52'
Lorenson,David Leslie
Cam 3h48'19"67d53'
Lorentz,Joan C
Uma 11h37'22"42d56'
Lorenz III
Aur 5h9'45"40d47'
Lorenz,Amelile
Lyn 8h24'21"42d6'
Lorenz,Clayton
Ser 17h57'15"-13d8'
Lorenz,Debra Sue
Lyn 7h26'46"58d56'
Lorenz,Dr Rudolf
Lib 14h21'0"-23d15'
Lorenz,Eva
Leo 9h20'25"18d2'
Lorenz,Gerd
Ori 5h57'42"7d48'
Lorenz,Joshua Andrew
Sex 10h26'13"1d9'
Lorenz,Kathryn Elizabeth
Cmi 7h58'0"4d6'
Lorenza
Eri 3h47'0"-1d27'
Lorenzen's Star,D
Dra 9h32'0"78d20'
Lorenzen,Joern
Dra 20h12'22"68d18'
Lorenzen,Peter J
Per 1h55'50"50d31'
Lorenzina
Tel 18h9'1"-45d42'
Lorenzini,Helen
Cas 23h18'28"62d5'
Lorenzini,Silvia
Lup 14h29'31"-44d53'
Lorenzo
Her 16h15'46"47d40'
Lorenzo
Cae 4h58'38"-27d51'
Lorenzo
Her 16h45'22"47d42'
Lorenzo
Aql 20h2'1"3d50'
Lorenzo 1988
Cnv 13h25'46"37d55
Lorenzo,Dino,Mario
Per 3h31'36"36d55'
Lorenzo,Kari
Peg 22h2'39"20d27'
Lorenzo,Paola E
Aql 18h58'56"-6d13'
Lorenzo-Gardini
Per 3h7'17"37d35'
Loretan,Ann
Tau 5h21'45"16d15'
Loreth,Gordon Albert
Uma 8h34'26"67d54'
Loreti,Joseph Raymond
Dra 18h57'1"48d16'
Loretta
Vel 10h12'7"-48d28'
Loretta J
Mon 8h0'29"-1d32'
Loretta Rose
Lyr 19h24'41"38d43'
Loretta's Star My Sweety
And 23h21'57"44d47'
Loretta,Raphael
Lyr 18h43'39"32d34'
Lorette
Ori 6h3'50"-0d20'

Lorette
Sco 17h3'1"-31d42'
Lorey,Dr Robert P
Aur 4h48'34"50d40'
Lori
Cyg 20h51'20"30d1'
Lori
Lyr 19h19'17"41d18'
Lori
Lyr 19h22'15"42d13'
Lori
Cet 3h19'33"6d57'
Lori
Lyn 8h56'33"45d32'
Lori
Vul 20h16'20"23d3'
Lori
And 1h36'0"37d17'
Lori
Aql 19h42'14"12d14'
Lori
Peg 22h2'12"34d28'
Lori
Cet 2h18'47"2d41'
Lori "Boots"
Sco 17h52'33"-38d10'
Lori & Adam
Cyg 20h19'34"31d51'
Lori & Fabir
Eri 4h3'27"-17d45'
Lori & John
Cnv 12h21'11"46d17'
Lori & Sean's Love Star
Del 20h36'12"10d17'
Lori Ann
Cas 0h1'30"56d3'
Lori Ann
Lyn 8h12'30"41d13'
Lori Beth
Gem 6h55'58"16d10'
Lori Beth "The Boss"
And 2h6'20"40d36'
Lori Denise & Phillip Arthur
Lyn 8h17'28"35d9'
Lori Iris
Del 20h39'0"10d39'
Lori Kristen
Lyn 8h11'31"37d31'
Lori Lacey
Peg 21h42'32"20d12'
Lori Lee
Cas 2h2'32"58d52'
Lori Lynn
And 23h35'50"49d1'
Lori Lynn
Lyn 9h1'0"38d38'
Lori My Love
Uma 8h9'26"61d42'
Lori Patti
Cyg 20h36'18"58d48'
Lori Sue-129
Lyr 18h47'27"34d53'
Lori Susan
Cas 23h16'33"62d32'
Lori's "Sparkle Star"
Cyg 19h40'0"30d18'
Lori's Hollywood Star
Eri 2h51'0"-16d10'
Lori's Hope & Faith
Cyg 20h3'35"31d32'
Lori's Love
And 22h56'47"50d48'
Lori's Peace of Heaven
Dra 12h2'0"72d1'
Lori's Personal Star
Cam 12h40'40"82d22'
Lori's Radiance
Lib 15h41'0"-25d10'
Lori's Shining Knight
Ori 5h0'0"10d23'
Lori's Star
Peg 22h57'13"18d16'
Lori's Star
Umi 14h59'1"66d6'

Lori,Jennifer M
Peg 21h18'60"22d41'
Loria,Grace Marie
Lyr 18h21'17"42d44'
Loria,Mike & Olive
Sge 19h10'32"19d25'
Lorial 60
Cam 4h26'52"68d50'
Loriaux,Bridgette Yvonne
Aur 5h9'60d37d46'
Loriaux,Charisse Lela
Cas 1h46'22"75d39'
Loricchio,Susan
Sex 10h14'22"-7d45'
LoRicco,Millie
And 23h48'0"40d56'
Lorie Harriet
Eri 3h51'14"-6d36'
Lorie Kay
Peg 21h39'54"27d37'
Lorik,Leslie Kay
Cas 1h47'1"60d15'
LoriLee
Vul 19h44'19"28d47'
Lorilla
Cyg 19h19'47"49d59'
Lorimar
Mon 7h1'22"0d54'
Lorimer,Dominica & Vincent
Cra 18h1'1"-37d29'
Lorimer,Marnie
Peg 22h37'23"25d3'
Lorimier,Frédéric Thibault de
Umi 16h37'23"75d43'
Lorin
Del 20h13'1"11d30'
Lorina Lynn II
Aql 18h57'55"2d2'
Loring,Asha
Tri 2h28'50"30d28'
Loring,Noelle M
Ser 15h49'53"15d5'
Loring,Robert Binford
Sct 18h44'50"-6d38'
Lorion,Michael Jerome
Lac 22h26'25"55d4'
LorJan "Friends Forever"
Sex 10h8'24"-6d56'
Lorkay
Peg 23h39'14"15d29'
Lorna
Mon 6h14'46"-10d14'
Lorna
Cas 0h4'58"64d33'
Lorna
Lyr 18h36'32"27d15'
Lorna
Sco 16h34'56"-25d14'
Lorna
Cyg 19h33'28"35d7'
Lorna Beverly
And 0h12'1"47d5'
Lorna June
Cyg 19h28'18"36d4'
Lorna's Light
Cas 1h33'30"60d39'
Lorna's Smile - 93
Uma 14h1'1"48d56'
Lorne's Passion
Cnv 13h22'19"41d36'
Lornik II
Boo 14h23'11"31d26'
Lorracher,Erwin
Ori 5h57'38"19d38'
Lorraine
Lyr 18h16'1"46d58'
Lorraine
Aql 18h59'55"-5d46'
Lorraine
Vul 21h17'32"20d10'
Lorraine
Sge 20h0'18"16d48'
Lorraine
Uma 9h37'1"54d57'

Lorraine
Vul 20h21'55"25d47'
Lorraine
And 2h0'46"38d3'
Lorraine
Aur 6h1'15"50d1'
Lorraine
Gem 7h38'1"31d47'
Lorraine J
Peg 22h42'54"29d21'
Lorraine Lee
Cyg 20h55'55"31d18'
Lorraine Marie
Cas 0h9'38"63d22'
Lorraine Mary
Mon 6h28'35"-10d25'
Lorraine of Minnesota
Ori 6h0'49"0d13'
Lorraine Teressa
Cas 0h33'45"62d14'
Lorraine's Crown
Peg 22h45'44"27d46'
Lorraine's Fish Face
Ori 6h0'18"10d48'
Lorraine's Harte
Lmi 10h54'0"30d7'
Lorraine's Star
And 1h17'25'35d16'
Lorraine's World
Cas 0h17'35"60d57'
Lorraine,Aaron
Cmi 7h44'1"7d33'
Lorraine,Andrew
Lac 22h26'1"52d51'
Lorraine,Susan
Mon 7h47'27"-3d59'
Lorraine-Jean
And 1h13'33"39d57'
Lorrannette
Cas 1h19'28"64d56'
LorRay
Peg 22h19'34"5d5'
Lorrayne
Cmi 7h21'58"7d52'
Lorre Lee
And 1h33'47"48d56'
Lorreta Lynn
And 2h0'29"38d23'
Lorrie
Aur 7h24'59"41d2'
Lorrie
Tau 4h53'34"28d27'
Lorrie's Star
Cas 0h25'25"69d56'
Lorrin B
And 0h23'19"27d36'
Lorry
Mon 6h36'52"1d19'
Lorry
Cet 1h58'26"0d46'
Lorry
Cyg 19h31'54"39d29'
Lorry & Will Forever
Cyg 20h35'36"40d57'
Lortie,Joseph Robert Bernard
Uma 12h11'36"60d4'
Lortie,Marie Jeanne Caroline
Uma 12h59'26"60d11'
Lorusso,Dawn And Valerio
And 0h54'19"45d29'
Lory,Alain
Sex 10h13'54"-8d6'
Loryn
Psc 0h57'26"32d35'
Loryn Beth
Lyr 18h59'34"27d16'
Loréda
Cam 5h1'58"61d47'
Los Fontanez Estrella Del Amor
Boo 14h49'37"50d33'
Los Star
And 2h29'28"50d48'
Losa,Claudia
And 0h20'59"38d42'
LoSardl,Rosemarie
Lmi 9h57'0"38d11'
Losavio,William
Per 2h59'46"31d32'
LoScalzo,Vincent T
Boo 14h50'1"23d56'
Loscher,Jennifer Christine
Ori 5h56'57"13d23'
Loscher,Jennifer
Oph 18h39'21"7d59'
Loscocco,Teresa Marie
Cas 0h32'47"64d2'
Loserl,Robert
Her 17h43'53"43d2'
Losh,Linelle
Vul 19h46'20"20d10'
Loshin,PE,Steven M
Aur 6h4'18"31d42'
Loshonkohl,Donald Dean
Cet 1h50'26"-5d51'
Losier,Brent A
Leo 10h49'59"15d12'
Losier,Martine
Lyr 18h59'14"34d2'
Losito,MD,Francis C
Uma 10h32'25"57d60'
Loskot,Nancy
Cam 13h10'24"77d13'
LoSquadron,Eugene Robert
Lyn 7h42'22"51d38'
Lost Horizon
Per 2h59'19"48d53'
Lostfogel,Frances
And 0h25'46"40d3'
Lostritto,Dana Andrea
And 23h21'34"37d59'
Lota,Salvatore
Cam 14h27'55"81d13'
Lote,Diana Vagueiro
Lyr 18h58'1"41d30'
Lothamer Star,The
Ori 5h21'28"12d43'
Lothar
Per 1h55'50"47d41'
Lotherington,Bruce
Cam 6h36'37"68d39'
Lothridge III,Billy Gene
Cet 1h37'28"-1d2'
Lotoro,Matteo
Aur 5h15'0"41d15'
Lotsolove
Sge 19h53'28"16d1'
Lott,Alexandra
Cet 2h40'12"2d26'
Lott,Celestial
And 0h55'13"36d19'
Lott,Duncan René
Cyg 21h7'57"30d29'
Lott,Gladys
Aql 19h2'44"17d52'
Lott,Hubert E
Gem 6h53'23"30d58'
Lott,K C L Kelley Charles
Psc 1h22'40"17d37'
Lott,Steven W
Aql 19h31'0"8d29'
Lotter,Gerard
Uml 15h16'31"78d48'
Lotterik's Star
Cas 1h55'36"61d18'
Lottero,Dennis
Cep 23h11'29"78d48'
Lotti,Joseph
Per 2h54'49"43d49'
Lottie
Her 15h45'44"44d45'B
Lottie,Iziah David
Vul 21h26'39"28d11'
Lotz,Ruth Pauline
Lyr 19h22'42"33d55'
Lou
Col 6h33'55"-33d26'
Lou
Uma 10h8'26"59d56'
Lou
Cmi 7h59'20"18d30'A
Lou & Stella In Heaven
Lyr 19h1'29"38d10'
Lou & Susan
Aql 20h7'39"0d2'
Lou 2-20-93
Mon 7h0'57"-6d44'
Lou 21
Uma 9h48'23"48d18'
Lou C
Aql 19h56'56"13d54'
Lou Et Ninou
Boo 15h1'35"20d52'
Lou Simon
Aql 19h25'50"-6d49'
Lou's Vision
Oph 17h8'55"-21d20'
Lou,Kin
Lac 22h27'15"40d20'
Lou,Leenee
And 2h29'59"39d18'
Lou-Lou
Lyr 19h2'28"26d54'
Lou-Lou
Umi 14h56'21"67d22'
Louan,Sandra
Per 4h42'31"51d57'
Loubar
Lac 22h0'43"51d48'
LouCa
Cyg 21h5'1"48d49'
Louck,Jr,Harold Eugene
Ser 15h10'48"6d51'
Loucka,Lenora L
Lyr 18h58'55"37d8'
Loud,Adam Curtis
Her 17h44'29"48d55'
Loud,Christopher S
Cnv 13h35'41"46d12'
Louden,Kane S
Cmi 7h25'40"1d38'
Louden,Olwen
Her 16h18'50"26d36'
Louder,Lisa Marie
Crt 11h14'52"-17d5'
Loudermilk's 20th, Anthony Ray
Cnc 8h13'25"30d38'
Loudmer,Pierre
Lac 22h23'56"53d41'B
Loudon,John Hugo
Dra 16h13'24"62d3'
Loudon,Karen
Lyn 6h24'1"54d20'
Loudon,Kevin
Lyr 19h24'30"41d8'
Louella
Cas 23h25'58"58d51'
Louganis,Darius Ryan-Gregory E
Aql 20h6'57"0d30'
Louganis,Greg E
Del 20h20'0"20d16'
Lough,Suzanne
And 23h42'51"37d56'
Loughlin Forever,Jim & Helen
Cyg 19h49'49"38d56'
Loughlin Star,The
Cyg 20h42'1"45d21'
Loughlin,Heather Lynn
Cyg 20h21'28"41d27'
Loughlin,Judy
Del 20h52'43"9d39'
Loughran,Margarita
Cma 6h25'1"-15d7'
Loughrey,Frank & Kathy
Peg 22h44'32"3d17'
Lougina
Umi 15h51'1"78d45'
Lougue,James Keith
Boo 14h31'11"8d20'
Louie
Cma 6h25'54"-15d10'
Louie & Barbie
Uma 9h49'30"47d42'
Louie & Erin
Ori 5h43'57"-1d25'
Louie B
Per 2h7'46"47d18'
Louie's Italian Grill & Bar
Cyg 19h56'16"38d53'
Louie,Corinne Elizabeth
Mon 6h31'42"11d9'
Louie,Lisa
Sct 18h42'16"-7d44'
Louie,Thomas Bing
Aql 19h57'51"15d12'
Louis
Uma 10h17'44"48d43'
Louis
Her 15h59'51"41d5'
Louis & Esther"Put it in your pocket"
Cyg 21h5'24"37d13'
Louis & Laura,50th Anniversary
Cyg 21h5'31"31d51'
Louis My Love
Cas 0h54'57"69d38'
Louis T
Lac 22h4'52"49d12'
Louis Torrence Star #1
Peg 21h24'28"22d37'
Louis's Light
Cas 0h31'44"64d2'
Louis,Andrew D
Aql 19h3'57"-0d7'
Louis,Donna Blackwell
Mon 7h16'38"-6d49'
Louis,Gertrud
Her 47h10'13"50d12'
Louis,John Paul
Oph 17h56'34"8d6'
Louis,Jr,John Frank
Sex 10h4'1"0d10'
Louis,Sherry Renee
Crt 10h54'11"-8d30'
Louis,Trudy
Vul 20h18'12"22d45'
Louis,Viola Mihalich
Aql 19h55'57"-8d39'
Louis-Charles,Joseph
Aur 5h57'1"31d21'
Louis-Dreyfus, Margarita
Cas 2h33'11"57d33'
Louis-Dreyfus,Eric
Her 16h14'55"42d20'
Louis-Dreyfus,Robert
Per 1h55'13"52d47'
Louis-Prescott,Leah
Vul 19h18'40"25d6'
Louisa
Crb 15h55'30"26d56'
Louisa Boo
Cet 2h41'36"3d28'
Louisano
Her 17h15'18"20d42'
Louise
Cyg 19h57'25"38d39'
Louise
Lyr 10h30'14"40d1'
Louise
Lyr 18h54'29"32d43'
Louise
Ori 5h56'37"17d56'
Louise Claire
Cyg 20h1'14"38d49'
Louise G
Cas 2h52'16"60d5'
Louise K P
Boo 15h4'0"51d34'
Louise Kelly
Dra 15h44'21"62d16'
Louise my Beautiful Princess
Lyr 18h16'56"44d54'
Louise Of Small Bones & Curly Hair
And 0h9'0"47d46'
Louise Rebecca
Lyr 18h17'37'30d56'
Louise Superstar
Mon 7h28'55"-6d31'
Louise,Cairo
And 1h18'20"37d5'
Louise-Eva
Cas 0h31'52"61d46'
Louiselle,Allan Richard
Her 18h12'32"38d14'
Louisolo,Gina
Crt 11h8'42"-19d9'
Louisolo,Gina
Eri 3h53'13"-6d57'
Louize
And 0h32'1"31d40'
Loukakis,Elefteria
Cas 0h20'52"65d28'
Loukas,Rosalie Camileri
Lyn 7h24'55"50d21'
Louks,Mary
And 2h13'46"40d4'
Loula Belle
Cyg 19h33'56"35d10'
Loulabelle
Cas 1h59'19"71d38'
Loulette
Umi 15h45'2"73d8'
Loulou
Tau 3h58'39"28d13'
Louman
Boo 14h6'11"37d56'
Loup,France Anne Marie
Aur 5h58'25"50d11'
Loupus Major
Lup 14h20'35"-45d26'
Loupy,Emmanuel
For 3h29'18"-30d41'
Lourdelle,Michele
Dra 10h6'30"73d24'
Lourdes
Cep 22h9'57"67d52'
Lourdes
Lmi 10h4'39"32d28'
Lourdes Avellana
Peg 22h3'56"10d43'
Lourdes Bravo "Luli"
Lib 15h3'36"-28d59'
Lourdes Bravo "Luli"
Lib 14h19'1"-22d55'
Lourie,Hadlee Richardson
Oph 17h55'30"8d47'
Loux,Diane
And 1h45'17"40d40'
Lovable Lou
Aur 5h53'19"30d19'
Lovable Lou & The Big A
Peg 23h19'14"32d49'
Lovable Old Geez
Del 20h15'46"14d31'
Lovaglia,Diane Palla
And 2h17'24"38d16'
Lovaglio,Alfredo
Dra 14h58'18"61d47'
Lovallo,Joe
Cnv 12h27'1"33d7'
Lovati,Renato
Aur 5h3'1"30d47'
Lovato,Lil & Joe
Cyg 21h31'39"37d40'
LovaU Mike
Her 17h33'1"40d36'
Love
And 0h15'54"25d51'A
Love
Lyn 7h30'17"37d14'
Love
Cyg 20h22'23"39d32'
Love
Cyg 20h21'54"39d15'
Love "The Navigator", Jertha O
Ori 5h52'44"14d6'
Love & Faith Eternally ,BCS & JMW
Vul 19h23'38"25d11'
Love & Happiness
Uma 11h43'34"44d43'
Love & Happiness
Sge 20h17'13"20d28'
Love & Peace
Tau 3h45'18'24d43'
Love Always
Sct 18h44'10"-5d22'
Love Always
Her 18h4'44"30d12'
Love Always & Forever
Uma 9h18'1"61d48'
Love At First Sight, Forever Eric
Ari 1h50'1"12d2'
Love Bug
Uma 9h23'31"55d43'
Love By The Lake
Cam 5h10'17"68d26'
Love Conquers All
Crb 16h10'48"31d33'
Love Dove Star,The
Boo 13h38'22"18d25'
Love Fiendish
Cas 1h2'53"48d51'
Love For David
Uma 11h35'43"31d59'
Love Fore Mike
Aql 20h34'56"0d32'
Love Forever Heather
Boo 15h6'13"38d7'
Love Forever You
Cyg 20h19'19"30d51'
Love From Above
Uma 9h0'40"48d5'
Love Glub
Boo 14h6'0"47d60'
Love Is Always & Forever-Matt & Lisa
Cyg 19h33'57"30d12'
Love is for Always
Cyg 19h27'39"36d2'
Love Kimiko
Ari 1h45'60"23d49'
Love Knotts
Peg 22h2'48"3d33'
Love Never Fails-Love I Give to You
Uma 10h58'47"36d43'
Love of a Lifetime
Uma 11h48'34"37d60'
Love of Clara
Vul 19h45'24"28d17'
Love of Hope
Boo 13h34'14"21d20'
Love of John & Camille The
Peg 22h3'1"10d48'
Love Of Joy,Six Men And A Lady,The
Sge 19h55'34"18d55'
Love Of My Life
Cet 2h6'0"-5d41'
Love of My Life Jeffrey
Her 16h53'44"47d59'
Love Of My Life Bobbie
Uma 8h35'0"51d16'
Love of Natalie & Jalal,The
Ari 2h25'41"12d6'
Love Of Sarah Kate
And 23h2'36"50d3'
Love Shoko
Cap 21h40'1"-19d39'
LOVE STAR PPP 5937
Vir 13h4'23"-11d42'
Love Star,The
Cyg 20h9'35"40d44'
Love Star/Brooke- Brandon-Aunt Sandy
Cyg 20h0'57"40d10'
LOVE TWINKL Stella Yoshihiro & Mari
Vir 12h5'19"2d25'
Love Ya Linda
Cas 3h23'20"75d3'
Love You Forever
Crb 15h22'43"32d3'
Love You More
Cyg 19h30'35"38d11'
Love You Too Much
Sco 17h55'1"-38d2'
Love's Darcy
Aql 19h30'21"13d18'
Love's Illusion
Ori 5h44'27"11d26'
Love's Promise Scott & Susan
Umi 16h3'22"72d28'
Love,Alexandra Elizabeth
Mon 7h0'12"-1d31'
Love,Alice K
Mon 8h1'20"-0d40'
Love,Amanda Carol
Lyn 8h56'1"36d17'
Love,Brian Patrick
Aur 4h55'0"38d34'
Love,Conner Jameson
Ori 6h4'54"-1d49'
Love,David
Ser 15h20'48"-2d54'
Love,David
Cyg 19h41'52"38d13'
Love,David James
Ori 4h45'17"0d6'
Love,Elan Nathaniel Jacob
Oph 18h2'47"12d8'
Love,Fred & Bertha
Lyr 18h39'33"41d44'
Love,Holly
Lyr 19h4'0"37d38'
Love,Jamie Erin
Cyg 21h30'20"37d39'
Love,Janice Mae
And 2h7'19"38d11'
Love,Jr,Floyd W
Her 18h18'55"12d16'
Love,Judith Ann
Lyr 19h5'51"26d48'
Love,Kaitlin Elaine
And 2h24'12"42d59'
Love,Kim
Cyg 19h27'1"31d15'
Love,Lilli An
Aql 18h59'52"10d28'
Love,Lorna
Tri 2h1'0"31d44'
Love,Marguerite Cameron
And 0h17'51"38d10'
Love,Nathan
Lyn 8h10'58"35d58'
Love,Nigel
Uma 12h3'38"41d8'
Love,Oliver
Cmi 7h22'0"8d5'
Love,Patrick Donald
Per 1h45'0"52d45'
Love,Penny & Sol
Aql 20h2'34"1d2'
Love,Sr,James
Mon 6h29'56"-6d8'
Love,That's Lillian
Sco 17h29'23"-31d49'
Love,Timothy O C
Per 3h11'58"46d50'
Love,Tonya Melissa
Gem 6h59'16"11d30'
Love,Vistina Marie Barger
Cas 2h47'31"61d3'
Love-Davies,Christine Joy
Cyg 20h0'48"30d39'
Love-Life
Lyr 18h55'11"34d53'
Loveable Harry
Per 4h1'16"50d59'
Loveartus Musicis
Cmi 7h34'19"1d46'
Lovebud
Ori 6h3'38"4d19'
LoveBug
Cet 2h11'28"4d7'
Lovebug Star,The
Ori 5h48'48"11d13'
Lovecat
Peg 23h37'11"32d9'
Lovecchio,Lucia
Cas 0h10'28"50d10'
LoveCheer-800
Cyg 21h18'50"37d40'
Loved you always Joe
Cyg 21h2'12"37d12'
Loveday
Uma 8h36'49"51d55'
Lovegren,Mildred Effie
Cam 9h53'14"82d5'
Lovegren,Norman Victor
Cap 21h22'18"-20d6'
Lovejoy,Deven Bickford
Aur 5h14'39"50d13'
Lovejoy,Louis Leroy
Aur 6h14'0"33d48'
Lovelace III,Dallas W
Vul 20h14'22"22d35'
Lovelace,Kay
Lyn 7h27'31"42d59'
Lovelace,Samantha Lou (Sammie)
Peg 21h38'47"26d53'
Loveland
Peg 21h28'37"22d52'
Loveland,Daniel & Cheryl
Cyg 21h53'27"41d45'
Loveless,Melinda Doris
Mon 6h20'49"8d54'
Lovelet,Star
Uma 11h9'45"42d39'
Lovelett,Jon
Tri 2h15'39"30d23'
Lovelight 81889
Cyg 19h40'39"41d29'
Lovell,Amanda Nicole
And 0h39'11"40d46'
Lovell,Golda
And 0h58'36"39d35'
Lovell,John R
Per 2h50'46"45d16'
Lovell,Jr,Olen Ira
Aur 5h1'33"38d48'
Lovell,Robert Edmund
Cas 23h15'12"61d5'
Lovell,Victoria
Cam 7h49'29"68d57'
Lovella,Gay
Uma 8h50'20"48d2'
Lovelle
Cap 21h3'1"-20d19'
Lovelock,Ian J
Mon 6h19'59"7d38'
Lovelock,Robert Ian
Uma 9h41'12"46d59'
Lovely Angela
Cyg 20h37'59"45d6'
Lovely Edith
And 0h15'39'38d53'
Lovely Hutty
Cas 0h31'41"74d18'
Lovely Jimmie Dean
Uma 11h49'51"51d4'
Lovely Laura
Lyn 7h42'0"38d35'
Lovely Laura
Cas 1h24'31"58d17'
Lovely Laurie
Lyn 8h11'25"37d15'
Lovely Lesley's Light
Boo 15h21'33"33d8'
Lovely Leslie's Light
Cas 1h13'41"61d2'
Lovely Linda
Lyr 19h21'16"41d6'
Lovely Lisa
Cas 0h31'0"64d29'
Lovely Lisa
Nor 16h23'30"48d50'
Lovely Lisa
Aqr 23h4'13"-3d47'
Lovely Lorianne,The
Cas 0h37'45"75d40'
Lovely Lorna
Cas 22h56'23"53d36'
Lovely Lucille
Lyr 19h7'41"37d31'
Lovely Luscious Lady Lillian
Lyr 18h59'0"28d49'
Lovely Lynda
Mon 6h38'38"-10d27'
Lovely Maura Star Over County Clare
Cas 2h56'0"57d32'
Lovely Monika
Sgr 19h18'52"-29d43'
Lovely Ms B,The
Lyn 7h52'23"40d6'
Lovely Roseann,The
Del 20h16'12"9d39'
Lovely Rox Star
For 2h44'30"-25d33'
Lovely Spry & Vociferous Rachel
Eri 2h54'1"-6d27'
Lovely Trouble
And 0h17'53"39d24'
Lovely-Bwana Jim
Lyr 18h58'43"38d18'
Lovenduski,Alysha Marie
Cam 8h27'22"82d16'
LOVER BOY
Ser 16h18'25"0d53'
Lover Dover in the Wover
Aql 18h59'11"-6d55'
Loverbean
Lac 22h10'33"47d55'
Lovercheck
Mon 6h33'1"1d26'
Loverd,Christel Ibsen
Cyg 21h1'58"33d48'
Loveridge,Erin Lee
Sge 19h12'18"20d5'
Loveridge,Lauren Lee
Sex 10h45'51"2d23'
Loverly Woman
Lac 22h26'0"55d35'
Loverman Despain
Cam 6h10'51"60d23'
Lovers of Longer Than Forever
Aql 19h50'36"12d27'
Lovers,The
Cas 1h34'4"58d22'AB
Loverso,Frank
Her 17h2'22"13d2'
Lovestar
Cyg 19h26'44"56d45'
Lovett,John & Elizabeth
Sge 19h26'30"18d54'
Lovett,Sophia Myers
Peg 22h23'25"21d24'
Lovett,Tina
Cas 1h28'46"60d19'
Lovette Light of Love
Peg 23h30m1"32d2'
Lovette,Camille
Peg 21h57'52"29d59'
Lovey
Sex 10h26'50"0d53'
Lovey Princess
And 23h0'24"49d31'
Loveys,Kristopher
Cam 6h3'42"60d2'
Loveys,Ruth H
Peg 21h37'57"26d44'
Lovgren,Jordyn Nicole
Cas 23h41'20"64d29'
Lovig,William N
Aur 7h22'53"40d14'

Lovin'
Aql 19h22'24"10d40'
Lovin,Cathy
Cam 7h0'1"82d31'
Loving Daddy of Mine
Her 16h2'47"16d3'
Loving Michael
Her 17h39'15"14d23'
Loving Paul Eternally
Gem 7h5'54"28d44'
Loving Wife Michele's Special Star
Ori 5h56'1"19d50'
Loving Wife Nancy
Peg 21h54'43"30d48'
Loving,Cheryl Ann
Cas 0h47'22"61d17'
Loving,Diane
And 23h48'16"47d21'
Loving,Ellen Keith
Cyg 20h18'45"30d49'
Lovinger,Mitchell Scott
Sex 9h57'40"-0d25'
Lovinger,Sheldon
Dra 9h51'57"77d42'
Lovingfoss,Ferne Louise
Sge 19h3'47"19d7'
Lovingly Monica
Per 3h38'26"38d11'A
Lovins,Lois Marie Hanson
And 1h50'60"38d36'
Lovins,Richard
Ori 6h0'32"0d1'
Lovis,Paul
Sgr 19h18'36"-24d10'
Lovitch,Derek Joshua
Dra 18h31'32"50d7'
Lovre,Nancy"The Bear"
Eri 3h22'56"-4d7'
LoVullo
Umi 7h28'40"88d38'
Lovvorn-Rodriguez, Danielle René
Mon 6h54'57"-6d13'
Low,Christine Grace
Com 12h14'60"22d50'
Low,E Holland
Aql 20h10'56"10d18'
Lowary Lumina,The
Cyg 19h30'11"34d57'
Lowas
Cyg 21h8'47"30d14'
Lowden,Arthur Douglas
Uma 9h49'41"45d51'
Lowder,Jann K
Lyn 6h33'32"59d53'
Lowe #1 Teacher,Ken
Psc 1h41'25"27d45'
Lowe 2nd Jan 1971, Christine
And 23h15'23"50d19'
Lowe,Alan Wallace
Sex 9h41'10"3d28'
Lowe,Angela L
Cam 5h59'49"60d17'
Lowe,B J
Ori 5h52'38"12d60'
Lowe,Bob L
Aur 6h18'18"32d1'
Lowe,Brian Thomas
Boo 14h57'0"29d23'
Lowe,Carl M
Her 17h38'51"21d25'
Lowe,Cheryl
Cyg 20h20'23"39d23'
Lowe,Damian LaGary
Gem 6h52'49"31d41'
Lowe,Danny Harrison
Lac 22h16'0"51d31'
Lowe,Denise
Uma 11h29'0"30d59'
Lowe,Dorothy
Aql 19h25'38"13d49'
Lowe,Eric Michael
Her 15h52'15"44d5'
Lowe,Erica Lea
Psc 23h5'19"0d5'
Lowe,Gary R
Vul 20h4'1"23d40'
Lowe,Georgia Olivia
Ari 2h33'17"20d8'
Lowe,Heather
Peg 23h47'49"8d6'
Lowe,Janice
Cas 1h18'31"60d17'
Lowe,Joseph G
Cep 21h0'0"61d44'
Lowe,Joseph H
Sct 18h30'57"-5d8'
Lowe,Joshua
Her 18h6'18"47d57'
Lowe,Justeen Helen
Lyr 18h52'21"38d27'
Lowe,LaGary Otis
Vir 12h39'49"2d5'
Lowe,Lisa Q
Peg 0h9'1"14d19'
Lowe,Malcolm
Uma 9h1'35"48d32'
Lowe,Matthew Joseph
Cyg 21h4'38"37d55'
Lowe,Michael
Hya 8h54'1"0d15'
Lowe,Michelle & Michael Porte
Crb 16h14'10"28d59'
Lowe,Ralph Craig
Her 17h0'1"42d38'
Lowe,Rebecca Anne
Del 20h13'31"14d50'
Lowe,Richard J
Cnc 8h54'57"12d19'
Lowe,Robert
Her 16h43'11"23d27'
Lowe,Sarah
Ori 5h57'49"9d32'
Lowe,Stephen
Uma 10h35'0"42d15'
Lowe,Sue
And 23h6'48"44d56'
Lowe,Tim Karen James Ryan Chelsea
Ori 5h57'51"8d47'
Lowe,Tommy
Aql 20h14'39"1d21'
Lowe,William Michael
Ori 5h30'0"-1d28'
Lowell
Cet 2h6'17"8d31'
Lowell & Jean
Cyg 21h6'37"31d42'
Lowell High School Class Of 1995
Uma 10h13'1"67d45'
Lowell Kaye,Humanus Maximus
Aql 20h19'57"8d34'
Lowell's Light
Cet 2h42'17"1d52'
Lowell,Clark E
Hya 8h30'45"0d27'
Lowell,Jennifer Pearce
Cyg 19h59'35"30d37'
Lowell,Jenny
And 2h1'21"37d21'
Lowell,Mary Elizabeth
Peg 22h24'44"26d15'
Lowell,Nesha Diane
Peg 23h37'21"15d47'
Lowell,Russell Ivan
Cmi 7h38'20"3d57'
Lowenberg Star,The
Uma 8h41'19"62d22'
Lowenbraun,Vienna
Sge 19h37'56"16d13'
Lowenstein,Lori
And 1h21'39"38d57'
Lowenstein,Lynda
Psc 1h0'47"23d57'
Lowenstein,Nicole
And 23h1'51"51d53'
Lowenthal,Richard Elliot
Sex 9h55'23"-2d6'
Lower,Charles Bates
Aur 6h27'1"32d40'
Lowery Super Star,Nick
Her 16h34'56"32d39'
Lowery `96 Superstar
Oph 17h37'58"11d30'
Lowery,Bill
Dra 16h54'42"67d10'
Lowery,Bryan Keith
Ori 5h47'30"11d23'
Lowery,Corey French
Aql 18h54'15"8d9'
Lowery,David L
Uma 12h2'47"31d25'
Lowery,Dennis Love U Beyond 4ever
Per 2h56'0"32d11'
Lowery,Kim Marie
Cyg 20h28'13"38d21'
Lowery,Nancy Ellen
Cnv 13h0'30"33d21'
Lowery,Richard Arthur
Sex 10h23'32"-6d21'
Lowin,Paul
Ori 4h42'10"0d14'
LOWK/Andrew's Star
Lyn 8h31'17"43d45'
Lowles,Chris
Cam 4h37'57"67d43'
Lowman 50th Anniv, Raymond & Olga
Aql 19h58'13"11d13'
Lowman,Jacqueline Renee
Mon 6h39'45"8d56'
Lowman,Jr,(Joey)George R
Per 4h42'16"37d43'
Lowmaster,Wendy Rae
Cas 1h13'58"60d51'
Lown,Georgina & Peter
Cru 12h9'36"-60d26'
Lown,Tosha L
Uma 9h35'46"51d20'
Lownds,Norm
Boo 14h59'57"42d5'
Lowrance,Harold & Cybil
Mon 6h22'1"8d35'
Lowrey
Boo 15h19'1"41d30'
Lowrey,Anna Cathleen
And 0h55'49"33d38'
Lowrey,David Thomas
Leo 11h44'43"27d21'
Lowrey,Jim & Sheila
Per 3h13'1"55d36'
Lowrey,Shannan Pegine
Lyr 18h49'19"39d40'
Lowrie,Roz
Sge 20h5'1"20d48'
Lowrie,Terry
Lyr 18h46'43"30d40'
Lowrie,Theodore Mon Etoile Brillante
Aur 4h54'45"37d47'
Lowrimore,Glenn
Ser 15h15'37"18d25'
Lowry
Per 3h10'16"41d12'
Lowry,Ashley R
Aql 19h7'11"2d57'
Lowry,Governor Mike
Uma 8h13'46"71d1'
Lowry,Jeneane
Aql 19h34'1"1d48'
Lowry,Kelly Ann
Ori 5h59'53"12d47'
Lowry,Kristina
Cmi 8h6'45"0d5'
Lowry,M. Eleanor C.
Cam 10h50'1"81d31'
Lowry,My Angel My Light Daddy Robert
Aur 7h22'27"40d4'
Lowry,RSM,Maureen
Sgr 19h10'44"-22d0'
Lowther #1,Beverly S
Lyn 8h7'31"48d59'
Lowther,Joshua James Marple
Umi 13h9'18"76d46'
Lowther,Ray T
Cma 6h55'54"-19d18'
Lowy,Michelle Hope Abels
Lyr 18h25'21"45d44'
Lox & Mox Forever
Ser 15h58'53"21d12'
Loxton,Neville Coleman
Ind 20h50'17"-53d36'
Loxton,Princes Helen
Crb 16h17'32"37d33'
Loy
Hya 9h2'3"-12d53'
Loy,Avery & Pauline
Cep 22h55'23"78d3'AB
Loy,John A
Aql 20h30'42"-5d48'
Loya
Aur 6h53'22"40d41'
Loya,Arthur A
Oph 17h5'29"-20d57'
Loyal,Jr,Benjamin Charles
Hya 8h15'1"3d58'
Loyd,Inge
Lmi 10h39'16"25d43'
Loyd,James F
Cam 6h8'0"56d15'
Loyd,Jerry H
Her 17h21'42"21d14'
Loyd,Robert
Per 2h40'0"40d5'
Loyez,Monique
Lac 22h29'0"38d5'
Loyle,Nicholas
Her 18h40'51"12d18'
Lozada,Michelle Lee
Lyr 18h47'48"35d22'
Lozada,Tricia
Oph 17h53'37"12d29'
Lozak,Lauren
Cas 0h30'1"58d33'
Lozano,Dianne E
Mon 6h20'27"5d7'
Lozano,Rene
Cet 2h14'0"9d48'
Lozano,Sophie
Dra 16h16'19"62d22'
Lozier Pierre Marie
Del 20h36'2"8d39'
Lozier,Logan Jeffrey
Hya 8h14'59"1d42'
Lozinak,John
Aur 5h52'23"31d50'
Lozito-Monnecka,Fran
And 0h56'32"39d57'
Loòs-Alex
Umi 15h54'52"73d11'
LP-62
Cet 2h1'22"-5d22'
LPK-40
Boo 14h11'25"30d5'
LQCHO
Peg 22h46'14"12d29'
LT H B
Lep 5h31'27"-11d12'
Lu
Ori 5h53'54"0d59'
Lu Lu
Aql 20h0'0"7d44'
Lu-ka
Uma 11h55'0"30d13'
Luana
Her 16h41'16"18d56'
LuAnn
Lyr 19h18'58"35d17'
LuAnn
Eri 3h14'17"-15d48'
Luann
Lyn 8h20'14"43d55'
Luann
Aur 5h16'33"43d34'
Luanne Marie
Cam 3h54'0"60d0'
Luanne's Raven Hill
Lyn 8h30'0"41d58'
Lubach,Donna
Lyr 18h58'58"26d48'
Lubapa
And 23h9'44"40d16'
Lubas,Rev Michael H
Cyg 21h39'12"41d37'
Lubben,Heather Beath
Cas 0h7'56"64d8'
Lubell,Kenneth David
Vir 12h57'46"-11d16'
Lubenova,Adriana Hristoslavova
Leo 11h51'50"26d22'
Luber,PhD,La Chiam Marilyn
Lyn 8h59'0"41d16'
Lubert,Hollace & William
Vul 20h56'41"28d40'
Lubetkin,David
Aql 20h0'0"11d42'
Lubi
And 23h11'49"41d25'
Lubian,Andres Sanchez
Her 16h19'40"50d10'
Lubich,Joseph W
Oph 17h4'0"8d38'
Lubin (60th Birthday), Barry
Ori 4h59'34"-0d16'
Lubin,Joseph "Joe"
Aql 19h53'30"14d24'
Lubin,Roger
Oph 16h34'20"-6d3'
Lubinsky,Jim
Dra 14h20'48"64d4'
Lubke,MD,Bernard William Joseph
Cep 23h1'0"60d17'
Lubos
Boo 14h48'47"48d58'
Lubov
Lac 22h43'0"54d43'
Lubow,Hal William
Her 17h39'1"40d3'
Lubow,Mary Ellen
Lyn 7h5'36"50d14'
Lubow,Michael
Ari 1h48'10"15d3'
Lubowicke,Donald
Aur 4h49'53"50d32'
Lubrano,Jonathon
Cep 21h49'29"58d34'
Lubrano,Juanilla Franchesca
Mon 6h21'0"3d28'
Lubs
Cet 0h29'10"-0d40'
Lubus,Officer Shellee
Lmi 9h48'49"40d34'
Luby Sunshine
Ori 4h52'35"0d23'
Luc
Cep 21h47'59"58d29'
Luc,Jeannette
Oph 17h41'1"10d3'
Luca
Boo 15h11'44"38d40'
Luca
Her 16h28'45"18d53'
Luca
Aur 4h50'18"40d47'
Luca
Cas 2h49'0"60d34'
Luca
Cas 23h3'46"53d55'
Luca & Nicole
Uma 11h29'31"61d21'AB
Luca Manuel
Pup 8h8'32"-23d14'
Lucadoro
Aur 7h20'45"38d37'
Lucarelli,Florence & Bruno
Lyn 6h21'27"61d45'
Lucari,Rachel Briana
Del 20h48'54"9d33'
Lucari,Sherry Dea
Mon 7h44'39"-8d54'
Lucariello,Flavio
Lep 5h53'33"-20d16'
Lucas
Her 16h19'0"26d46'
Lucas
Ari 3h2'12"28d31'
Lucas B
Boo 14h37'50"16d15'
Lucas Construction
Per 3h8'11"48d55'
Lucas V,Luke-Jean Baptiste Charles
Hya 8h19'34"0d50'
Lucas'"Light of the World"
Mon 7h29'14"-1d19'
Lucas,Amanda Em
Aql 20h10'17"14d41'
Lucas,Billy
Lyr 18h36'53"37d0'
Lucas,Darrell Blaine
Cep 4h1'33"80d33'A
Lucas,Denice R
Del 20h13'23"12d57'
Lucas,Dorothy Carl
Cep 4h1'33"80d33'B
Lucas,Edward
Aur 4h49'39"51d8'
Lucas,Frank N
Cam 4h14'43"68d29'
Lucas,James Travis
Oph 17h27'10"10d21'
Lucas,Jr,Harry
Hya 8h58'33"-7d12'
Lucas,Judith
Peg 22h9'17"5d2'
Lucas,Kelly "Sweet Cheeks"
Lyn 7h7'32"44d18'
Lucas,Kyle Grant
Aql 19h42'32"14d48'
Lucas,Laraine
Mon 7h2'42"4d19'
Lucas,Lore'ta
Equ 21h0'18"2d60'
Lucas,Mary Carol
Lib 14h53'11"-0d45'
Lucas,Mary Grab
Ori 5h55'36"6d35'
Lucas,Maureen Beryl Robertson
Sgr 20h0'37"-44d2'
Lucas,Michael W
Cet 2h58'49"0d24'
Lucas,Michelle Marie
And 23h14'1"40d58'
Lucas,Pam
Cam 3h28'16"61d52'
Lucas,Patrick
Cyg 19h38'46"38d52'
Lucas,Pauline
Cas 1h41'28"73d1'
Lucas,Philip E
Lyn 8h57'1"40d42'
Lucas,Sara Lynn
Mon 7h30'10"-8d15'
Lucas,Stephanie
Cyg 19h20'0"28d24'
Lucas,Susan Anne
ANd 0h41'17"40d17'
Lucas,Virginia
Lyn 8h14'28"57d59'
Lucassiopeia
Uma 9h45'23"62d8'
Lucca,Joyce & Pasquale
Cas 1h13'35"68d28'
Lucca,Violet Veva
Mon 7h50'23"-3d49'
Lucchi,Darcy Wilde
Per 1h27'43"50d31'
Lucci,Jerry
Boo 14h37'1"12d17'
Lucci,Louis C
Aql 19h16'36"15d37'
Lucciola
Psa 22h29'36"-27d35'
Luce II,Chester Walter
Cam 3h31'37"60d26'
Luce,Laura
Cas 23h25'1"63d9'
Luce,Ryan "Bunny"
Hya 8h13'46"0d6'
Luce,Suzanne
Eri 4h18'13"-14d35'
Lucenaquarantadue
Pic 4h36'20"-48d23'
Lucenta "Per Sempre", Dominic
Lac 22h6'16"51d55'
Lucente,Nick & Mary
Com 12h8'29"33d17'
Lucerito de la Noche
Ant 10h31'24"-30d46'
Lucero
Her 16h39'55"36d23'
Lucero,John Michael
Aql 19h9'58"1d45'
Luces,Johnny
Her 18h15'45"31d48'
Lucette
Per 3h7'16"40d52'
Lucey V,John Anthony
Per 2h53'17"43d45'
Lucey,Bitsy Lee
Mon 6h11'31"-10d37'
Lucey,Tim & Anne Marie
Eri 3h27'41"-8d26'
Lucey,Tina
Her 17h32'11"14d48'
Luchesa,Giorgio
And 1h4'35"46d59'
Lucht,Hans-Jürgen
Psc 0h18'17"10d12'
Luci,Edward
Cet 1h38'57"-12d34'
Lucia
Lyn 8h28'47"34d25'
Lucia
Ari 2h4'1"18d51'
Lucia
Cas 0h31'55"67d49'
Lucia
Cam 5h30'18"65d28'
Lucia
Per 2h8'41"58d53'
Lucia
Lyr 18h54'0"38d31'
Lucia
Lup 14h45'0"-45d27'
Lucia
Peg 21h59'47"29d18'
Lucia & James
Cyg 20h26'49"48d52'
Lucia & Scibona
Lyn 8h41'20"35d17'
Lucia Sofia
Aql 18h43'32"11d56'
Lucia's Prosperity
Aql 20h14'0"0d12'
Lucia's Star
Com 13h1'11"27d28'
Lucia,Antonia "Jean"
Cas 23h5'22"58d24'
Lucia,Dolly
Del 20h13'43"14d38'
Lucia,la stella di Hallowen
Lup 15h0'32"-37d56'
Lucia,Maria
Cyg 21h57'24"50d8'
Lucian
Psc 1h39'41"28d17'
Luciana
Dra 15h46'59"53d6'
Luciana Basilico Merigo
Leo 10h59'24"17d53'
Luciano
Her 18h10'23"38d56'
Luciano
Boo 15h13'51"38d13'
Luciano,Bob & Lori
Uma 10h3'50"52d10'
Luciano,Cella
Lyr 18h53'38"38d11'
Luciano,Daniel A
Ser 15h56'20"0d30'
Luciano,Franzoni Filiberto Gilberto
Cas 23h20'40"53d12'
Luciano,Louise
And 2h2'44"40d40'
Luciano,Maccari
Cep 23h33'11"65d37'
Luciano,Nicoletta
Lyn 8h11'28"58d59'
Luciano,Rosalie
Boo 14h30'35"43d21'
Lucidine
Cep 21h20'54"68d59'
Lucidity
Tri 2h6'14"31d56'
Lucido,Nina May
Cas 23h42'1"58d16'
Lucie
Uma 9h55'53"51d30'
Lucie
Cam 4h31'43"58d21'
Lucie
And 0h1'52"46d58'
Lucie Catherine Marie
Cep 21h39'19"63d53'
Lucie Florence
Mon 6h19'0"8d38'
Lucie-Fabien
Uma 12h21'21"61d52'
Luciea
Her 16h10'12"48d20'
Lucier,James Henry
Aur 5h3'53"37d53'
Lucier,Kenneth Richard
Cyg 19h30'22"31d50'
Lucile Elizabeth
Cam 3h22'16"60d02'
Lucilla
Cae 4h41'0"-33d29'
Lucille & Buddy Forever Loved
Cyg 20h41'27"38d56'
Lucille Martha of St C
And 22h59'53"50d18'
Lucille's Star
Oph 17h54'14"13d31'
Lucin,Sebastien
Ser 15h55'1"2d7'
Lucinda
Lyn 8h36'18"40d36'
Lucinde
Pyx 8h50'25"-28d48'
Lucio,Victor (NMI)
Her 17h53'25"28d13'
Lucius,Harald
Lac 22h7'51"38d3'
Luck
Col 6h35'54"-37d54'
Luck (A Man Who Groks),Dale
Ser 15h18'1"20d29'
LUCK and LOVE "good luck charms
Aur 5h53'55"30d40'
Luck,Jennifer Karen
Aql 19h54'22"14d42'
Luck,Melodie Elizabeth
And 0h59'36"37d27'
Luck,Peter
Cep 21h2'0"70d60'
Luck,Philippa Jane
Lyr 18h57'28"38d46'
Lucka,Robert William
Boo 15h7'60"21d21'
Luckasen,Sara Noel
Cyg 20h46'25"38d2'
Lucke,Adrianne
Cam 5h44'21"62d47'A
Lucke-Schaefer
Peg 21h55'42"34d54'
Lucken,Liberty Sol- Marda
Cyg 19h32'1"33d46'
Luckenbill,Terry
Aur 6h25'23"38d31'
Lucker II
Cam 5h44'50"61d18'
Luckern,Miss
And 23h1'56"48d18'
Luckey,Alexis Nicole
Eri 3h50'47"-2d55'
Luckey,Kyle Michael
Boo 14h14'43"33d4'
Luckey,Sherman C
Cyg 21h38'37"28d17'
Luckhurst,Deborah
Aql 20h10'27"10d46'
Lucki,Stacia
Cmi 8h3'11"6d6'
Luckman,Jr,Chuck
Boo 15h1'35"50d19'
Luckman,Stewart & Diane
Cyg 19h20'0"44d13'
Luckmann,Kath Heide
Uma 10h7'1"51d57'
Luckow,Guenter
Lac 22h4'37"50d43'
Lucksted,Tracie
Cyg 20h17'27"30d58'
Lucky
Her 16h12'55"41d56'
Lucky
Dra 19h1'1"48d37'
Lucky
Cet 2h20'47"0d40'
Lucky
Cnv 12h43'22"37d56'
Lucky
Vul 20h46'44"28d14'
Lucky Boy
Aur 6h26'32"37d31'
Lucky Eva Star,The
Cyg 21h17'46"35d21'
Lucky Hutchy
Boo 15h18'0"34d56'
Lucky Lindy
Lac 22h15'28"48d52'A
Lucky Mike
Hya 8h25'1"-0d36'
Lucky Rosemarie
Gem 6h57'20"18d18'
Lucky Seven
Lmi 10h42'34"24d47'
Lucky Star
Her 17h50'0"42d29'
Lucky Star Heide
Lyn 8h17'37"47d43'
Lucky star of Andrea & Hans,The
Leo 9h18'34"11d33'
Lucky Stella of Yuichiro & Hiroko
Sgr 19h10'53"-16d5'
Lucky-Future
Sco 17h33'48"-38d54'
Lucky-Star
Sco 17h51'35"-30d57'
Lucretia
Umi 13h5'47"76d17'
Lucrezia
Del 20h18'1"10d12'
Lucy
And 2h20'49"40d59'
Lucy
Peg 21h55'11"33d42'
Lucy
Cmi 7h56'32"4d38'

Lucy
Cas 0h41'30"70d56'

Lucy
Del 20h16'26"11d36'

Lucy & Irvings Star
Uma 11h21'21"40d37'

Lucy A J
Ori 6h4'31"7d9'

Lucy Ann
Vir 13h56'0"-1d50'

Lucy Diamond in the Sky
Ori 5h56'16"14d5'

Lucy in the Sky
Psc 1h24'19"11d27'

Lucy in the Sky
Uma 11h28'11"32d13'

Lucy In The Sky With Diamonds
Mon 7h30'31"-0d38'

Lucy in the Sky with Diamonds
Vel 9h18'47"-47d21'

Lucy Jane
Lyr 18h16'21"38d47'

Lucy Liz
Cyg 19h33'19"34d16'

Lucy Mae Claire
Aql 18h58'0"-10d14'

Lucy Mira
Ori 5h53'30"7d8'

Lucy's Daddy
Cam 3h33'11"61d21'

Lucy's Diamond
And 0h1'54"43d45'

Lucy's Light
Cyg 20h25'21"40d4'

Lucy's Light In Pam
Leo 11h47'55"22d8'

Lucy's Little Liberty Belle
Cma 7h14'59"-13d27'

Lucy's Sgnauss
Ser 16h4'59"10d41'

Lucy's Star
Mon 7h22'34"-1d50'

Lucy's Star
Lyr 18h37'33"35d25'

Lucy's Star
Lyr 18h50'1"36d1'

Lucy's Tiny Bubble of Light
Cas 2h1'36"63d46'

Lucy-Heaven's New Light
Lyr 18h41'21"45d5'

Lucynka
Cyg 21h10'48"39d6'

Luczak,Halina
Ori 5h51'17"15d9'

Luczkowski,Douglas
Lyr 19h20'43"40d33'

Luczny,Stephanie
Cet 1h34'43"-2d25'

Ludden,Leo V
Her 16h25'1"22d54'

Ludeman,Paul Gene
Psc 1h43'23"21d58'

Ludeman-Hopkins, Elizabeth Ann
Ori 5h14'21"11d16'

Ludewig,Andrew
Cep 20h58'26"65d35'

Ludgate Rubyann
Cyg 19h49'24"58d43'

Ludgate,Gemma Anne
Aql 20h10'29"13d25'

Ludington,Trevor Allen
Aur 5h2'12"51d12'

Ludivine
Lyn 8h59'54"42d22'

Ludjylbi
Aur 5h56'59"30d60'

Ludkin,Kenneth David
Aur 5h58'33"30d5'

Ludlam,Susan
Peg 22h22'43"26d9'

Ludlow,Andrea
Lyr 19h1'14"28d55'

Ludmann,Linda M
Cas 1h26'35"61d39'

Ludovic,Danel
Cam 7h37'45"68d50'

Ludovic,Garcia
Aql 19h27'1"7d59'

Ludovic,Laurent
Per 3h43'11"36d54'

Ludovica
Uma 11h52'57"50d29'

Ludovica
Del 20h12'51"10d27'

Ludshott,Wynlyn
Peg 21h56'1"33d6'

Ludtke,Erik
Sct 18h56'1"-5d12'

Ludwick,Andrew
Per 2h51'1"46d9'

Ludwick,John & Marjre
Crt 11h3'5"-14d53'

Ludwig
Boo 14h22'25"33d57'

Ludwig
Ori 6h4'18"3d1'

Ludwig,Anthony- Antonio's Veritas
Tau 4h12'60"21d58'

Ludwig,Betty Ruth Lundberg
And 1h33'56"40d20'

Ludwig,Bob
Dra 16h50'23"52d18'

Ludwig,Hannah Nicole
Peg 21h59'56"34d44'

Ludwig,Happy Birthday Eddie
Boo 14h41'53"32d8'

Ludwig,Jennifer Ann
Peg 23h30'0"23d53'

Ludwig,Josef
Ser 15h14'23"4d51'

Ludwig,Kurt E
Cep 22h4'40"53d14'

Ludwig,Mark A
Cep 22h12'54"50d1'

Ludwig,Renee Lynn
And 23h0'34"48d7'

Ludwig,Rose
Cas 0h13'49"61d49'

Ludwig,Sara
Mon 6h22'32"8d42'

Luebbers,Venita
And 2h30'27"50d7'

Luebke,Dolly & Bud
Cyg 20h27'1"40d28'

Luebs,Ahna Marie
Sgr 19h2'57"-26d32'

Luebs,Erin
Mon 6h25'35"10d34'

Lueck,May & Martin
Uma 11h12'53"31d30'

Luedders,Judy & Corey
Crb 15h51'0"27d53'

Luedeking,Heinz & Marianne
Eri 2h54'1"-14d53'

Luehr,Reinhold-Walter
Her 17h8'26"42d22'

Luehrs,Harry & Dona Lee
Equ 21h2'56"10d8'

Luehrs,Lesley Marie
Cyg 19h28'52"37d40'

Luening,Corinna
Peg 23h32'46"22d5'

Luense's Hopes & Dreams,Alice
And 1h26'57"40d44'

Luesing,Jr,Richard Turner
Her 16h36'40"33d15'

Luesing,Lauren Ashley
Lyn 8h44'11"38d56'

Luetticken,Stefan
Peg 22h56'28"8d56'

Luf,Fiona Smith
Uma 10h20'56"70d9'

Luff the 1st,Mark
Her 16h50'0"41d15'

Luff,Charles A
Lib 15h16'29"-22d20'

Luffey,Angela Angel
Del 20h19'1"11d4'

Luffman,Bryan Carter
Eri 4h8'38"-19d29'

Luffman-Cook,Joanne
Cyg 21h45'20"31d12'

Lufkin,Carole & Lee
Peg 22h19'14"5d37'

Luft,A G Astra
Boo 14h21'55"18d23'

Luft,Sharron
Mon 6h20'18"1d9'

Lugara,Lynda Edy
Mon 7h0'1"4d48'

Lugaric,Meg
Mon 6h10'53"-10d40'

Lugaric,Meg
Mon 6h10'53"-10d40'

Lugay,Marina McRae
Cas 0h4'21"58d48'

Lugero,Monique
Lyn 8h7'36"51d60'

Lugero,Sophia
Cas 1h5'48"50d19'

Luggi,Karen
Uma 11h13'34"52d39'

Luginbühl,Simon
Lac 22h9'58"50d9'

Lugli,Barbara
Cma 6h23'23"-13d27'

Lugo,Anne Marie
Dra 17h59'52"71d5'

Lugo,Aramis
Her 16h1'30"41d24'

Lugo,Carlos A
Ori 4h42'44"10d55'

Lugo,Isabel
Boo 15h23'49"38d40'

Lugo,Jacqueline Rose
Aql 18h45'0"11d41'

Lugo,Javier
Aur 6h34'12"38d48'

Lugo,Rikki
Hya 8h18'48"0d39'

Lugui
Cyg 20h49'48"38d37'

Luhman,Randall S
Sex 9h51'18"2d33'

Luhrman,T
Cnv 13h39'28"47d46'

Luhrs,Patricia A
Cap 20h21'31"-20d26'

Luigi
Lac 22h50'23"38d53'

Luigi
Hya 8h21'10"0d37'

Luigi et Luke
Uma 9h36'1"54d34'

Luigi Friends Forever Claudia
Cet 2h55'35"1d15'

Luigi,Toto & Patruno Maria
Scl 23h44'17"-28d59'

Luigina
Lac 22h23'37"50d35'

Luikart,Elizabeth Kathleen
Gem 6h42'29"30d9'

Luis & Michelle "Love Star",The
Dra 17h15'19"60d52'

Luis Felipe
Nor 15h48'0"-44d27'

Luis y Begona
Umi 15h22'1"66d16'

Luis,Jake Ryan
Aql 19h20'25"15d1'

Luis,Jerry A
Cet 0h29'28"-11d1'

Luisa
Cyg 20h52'16"31d9'

Luisa
Cma 6h10'48"-15d36'

Luisa
Pic 4h37'41"-48d3'

Luisa Z
Cep 22h17'13"55d30'

Luisella
Lyr 18h18'43"40d37'

Luisella,Mamma Sublime
Aur 7h24'0"43d22'

Luisi,Stephanie
And 0h1'1"46d20'

Luiz,Jonathan Michael
Oph 16h51'47"-25d59'

Luiz,Michael Robert
Tri 1h57'15"34d6'

Luizard Meryl
Cet 0h53'17"0d49'

Lujack City
Aur 5h5'1"38d53'

Luka
Aqr 21h1'52"-13d51'

Lukacs's Starr,Lydia
Lyn 9h11'59"34d28'

Lukacs,Lisa
Mon 6h33'40"0d42'

Lukas,Cary & Judy
Ori 5h46'28"20d35'

Lukas,Edith
Cmi 7h20'49"3d59'

Lukas,Peter
Her 16h6'32"50d1'

Luke
Pic 4h40'15"-48d15'

Luke
Peg 21h49'0"34d27'

Luke
Cmi 7h9'11"5d44'A

Luke
Cyg 19h30'24"38d24'

Luke
Cnc 8h53'41"21d12'

Luke 16
Cet 2h16'38"5d29'

Luke Star,The
Cam 4h4'52"68d17'

Luke The Babe
Boo 15h4'11"30d26'

Luke's Star
Her 16h15'0"22d33'

Luke,David Matthew
Ser 18h15'37"-1d50'

Luke,Forever
Per 3h36'24"37d33'

Luke,G W
Eri 4h31'58"-18d52'

Luke,John Anderson
Her 16h20'14"41d58'

Luke,Melissa
Uma 10h55'30"70d11'

Luke,Naoise
Umi 13h40'42"78d50'

Luke,Sheena
And 22h58'17"37d40'

Lukemire,Eddie
Per 1h45'33"53d19'

Lukemire,Meredith
Vul 20h29'50"28d25'

Lukens,Zachery J
Per 4h6'40"48d7'

Luker,Dr John A
Uma 8h33'32"58d45'

Luker,Steven Daniel
Tau 4h40'52"7d57'

Lukes,Dean L
Dra 17h29'51"64d38'

Lukes,La Van A
And 2h20'51"37d54'

Lukhurst,Corinne Marie
And 23h45'0"44d1'

Lukitsch,Joey
Aql 18h59'15"-2d46'

Lukos-Bradt,Andrew Christopher
Ori 5h48'12"21d1'

Lukosavich,Annie
Mon 6h26'58"11d33'

Luks,Kraemer & Christi
Eri 4h43'17"-0d24'

Luksic,Richard Grant
Boo 13h51'38"18d33'

Lulich,SM
Cam 7h57'46"60d15'

Lull,Rena A
And 1h57'27"37d35'

Lull,Ruth
And 23h35'16"44d33'

Lulu
Uma 11h51'1"40d17'

Lulu & Olli
Cnc 8h48'30"32d28'

Lulu's Estrella Verde
Lyn 9h2'1"37d20'

LuLu-Henry's True Love
Psc 1h1'1"23d19'

Lulù
Lac 22h15'40"38d53'

Lum,Aileen
Vul 20h39'36"28d34'

Lum,Carla
Peg 22h0'1"25d26'

Lum,Jonie
Hya 9h10'19"4d58'

Lum,Richard C H
Dra 19h0'41"56d39'

Lumadue,James W
Per 2h8'0"56d18'

Luman's RC Warbird World,Kreis
Lib 15h12'1"-3d38'

Lumarh,Thalita
Ant 10h32'29"-31d5'

Lumb,Basil
Dra 9h46'26"73d23'

Lumberjack
Dra 17h56'52"50d32'

Lumen de Aetermus Sodalitas
Lyr 18h33'36"33d5'

Lumi
Uma 12h8'17"56d3'

Lumia,Glen Patrick
Ori 5h32'10"-0d39'

Lumiere de Nuit
Peg 22h26'21"5d58'

Lumina,Erica Regan
Cas 0h25'11"50d12'

Luminar,Elizabeth
Umi 15h5'25"69d41'

Luminari,Holly Theresa
Cas 23h4'46"58d52'

Luminescent of Valued Expression
Ser 15h32'1"18d8'

Luminous
Ori 5h28'49"1d48'

Luminous Alma
Tri 2h6'37"33d34'

Luminous Lee
Umi 15h29'21"68d14'

Lumley,John Stuart Penton
Cnv 14h1'25"30d28'

Lump
Uma 10h40'48"52d26'

Lump,JuliJohn
Uma 11h1'0"67d56'

Lumpkin,Caleb D
Peg 23h6'22"8d53'

Lumpkin,Sr,Dr Lee R
Oph 17h55'48"0d59'

Lumpkins Family Star, The David
Aql 19h22'30"15d12'

Lumsden,John
Leo 9h59'1"20d25'

Luna
Eri 4h3'47"-16d15'

Luna Fina
Psc 1h37'49"22d13'

Luna,Alexander
Her 16h55'44"38d33'

Luna,Carlos
Per 2h59'20"35d10'

Luna,Ramon Melindez
Boo 15h17'0"50d37d

Lunae D N,Isabella Noctis
Gem 6h47'1"17d46'

Lunan,Mandy
Cyg 20h16'46"38d38'

Lunardi,Maria Cristina
Lyn 7h2'33"60d31'

Lunceford,Charles
Cam 4h7'18"70d14'

Lunckenbein,Hans
Peg 23h28'17"11d49'

Lund,Alan "Paul"
Her 18h5'14"31d39'

Lund,Barbara
Lyr 18h35'36"29d46'

Lund,Briana
Aql 8h56'57"4d20'

Lund,Clarence E
Her 17h25'30"21d2'

Lund,Deanna
Aql 19h3'0"-0d53'

Lund,Lisa Lynne
Mon 6h24'11"1d10'

Lund,Miles
Lac 22h32'29"55d30'

Lund,Tone Haug
Cnv 13h32'12"38d21'

Lund,Tracy
Vul 20h57'0"28d48'

Lunday,Claudia
Cas 0h22'38"69d45'

Lundberg,Alanna Lee Noelle
Tau 5h45'35"16d7'

Lundberg,Charlie
Dra 20h21'18"63d41'

Lundberg,Christina Marie
Vir 12h48'42"-6d14'

Lundberg,Erik Paul Sven
Tau 5h25'36"16d48'

Lundberg,Joan Ellen
Tri 1h34'1"35d16'

Lundblad,Lea Elizabeth
Peg 21h26'50"2d16'

Lundbom,Carl & Pat
Eri 2h52'0"-16d52'

Lunde,George
Leo 11h22'25"-1d56'A

Lundeberg,Thomas Dean
Leo 10h53'55"17d48'

Lundeen,David Alexander
Vir 13h24'9"-2d29'

Lundeen,Kent
Cnv 13h53'14"38d3'

Lundell,Pamela Jo
Com 13h26'43"26d45'

Lundemo
Eri 2h53'0"-6d22'

Lundemo,Jamison Leigh
Aql 18h46'0"11d49'

Lundgard,Phyllis
Cas 0h45'44"72d5'

Lundgren,John Donald
Dra 20h21'12"80d22'

Lundgren,Robert Erik
Ser 15h11'0"1d41'

Lundgren,Uncle Harry
Oph 18h7'14"12d22'

Lundgrun,Bernard
Oph 17h58'55"10d45'

Lundi,Jacky
Per 3h57'49"35d21'

Lundi,Valérie
Per 3h57'20"37d25'

Lundon 40,Thomas J
Aur 5h5'57"42d37'

Lundquist Love Nickolas,Mary V
Cyg 19h28'36"41d1'

Lundquist,Charles Michael
Ori 4h56'46"4d10'

Lundquist,Darrell G
Boo 14h59'35"40d32'

Lundquist,Hortense
Lyr 18h52'45"43d1'

Lundquist,Karen B
And 1h26'11"40d21'

Lundquist,Sylvia Jean
Aql 19h18'52"15d8'

Lundstrom,Carl
Aur 4h58'34"48d55'

Lundstrom,Karin Irene Margareta
Mon 7h56'38"-8d13'

Lundy,Abigail Lamet
Uma 9h43'46"47d21'

Lundy,Albert & Ruth
Aur 4h49'0"51d19'

Lundy,Jack
Ori 5h51'38"21d32'

Lundy,Ken & Keetja Yee Lundy
Uma 11h41'13"60d17'

Luneau,Gabriel D
Mon 6h36'1"-6d5'

Luneburg
Per 2h13'59"58d46'

Lunetta,Aaron Paul
Cet 1h30'57"-11d58'

Lunetta,Henry Cole
Aur 5h7'1"40d37'

Lunin,Martin
Vir 13h22'13"-6d43'

Lunquist,David
Ori 5h57'53"-0d4'

Lunsford,Debbie
Peg 22h17'25"34d28'

Lunsford,Walter Bishop
Sco 16h21'1"-30d43'

Lunsway,Forrest Edward
Ori 5h52'1"17d37'

Lunt,Jennifer Michelle
Mon 7h38'38"-6d52'

Lunzer,David Anthony
Oph 17h53'45"10d5'

Lunøe,Henning
Cam 3h29'50"61d52'

Luoieneuse
Per 3h38'1"50d29'

Luong,Eugene
Cam 5h10'51"70d21'

Luongo III,Bernard Carducci
Vul 19h22'34"26d47'

Luongo,Devon
Mon 7h5'1"-5d47'

Luongo,Kelsey
Del 20h24'16"10d17'

Luongo,Rhonda
And 1h52'1"46d56'

Luosetam
Com 12h19'1"20d47'

Luostarinen,Tuula Margid
Cam 6h30'23"68d6'

Luoto,Marja Katriina
Uma 9h54'58"45d31'

Lup,Jeff R
Aql 19h0'0"-6d46'

Luparello,Steven
Aur 6h23'52"37d41'

Lupel,Warren S
Cnv 12h42'19"36d56'

Lupescu,Hedwig
Leo 9h18'33"18d45'

Lupien,Andrée
Lac 22h22'33"55d14'

Lupien,Claire
Lac 22h23'47"55d53'

Lupien,Diane
Lac 22h23'44"55d50'

Lupien,Fernande
Lac 22h23'25"55d5'

Lupien,Francine
Lac 22h44'16"55d18'

Lupien,Gilbert
Lac 22h22'0"55d39'

Lupien,Gilles
Lac 22h24'23"55d31'

Lupien,Gisèle
Lac 22h24'44"55d21'

Lupien,Jean
Lac 22h44'27"55d55'

Lupien,Josée
Lac 22h44'54"55d56'

Lupien,Marielle
Lac 22h44'40"55d15'

Lupien,Micheline
Lac 22h23'40"55d42'

Lupien,Réjean
Lac 22h23'42"55d18'

Lupien,Thérèse Bleau
Lac 22h22'56"55d20'

Lupien,Eloïse
Cam 3h36'0"63d32'

Lupili
Sco 16h59'28"-41d10'

Lupin,Elyse Hilary
Com 12h58'58"30d55'

Lupinacci Assunta Zanzot,Anna Grace
Hya 9h14'19"1d38'

Lupinacci,Anna
And 23h3'17"48d55'

Lupino
Boo 15h9'31"27d39'

Lupis,Alyssa Nicole
Mon 6h56'19"-8d45'

Lupo,Andrew Michael
Cyg 19h33'32"28d29'

Lupo,Maria
And 0h45'18"40d35'

Lupoli,S
Ori 6h9'28"8d30'

Luprado,Scott Joseph
Aur 6h8'58"30d14'

Lupton,Margaret
And 23h1'0"51d7'

Lupton,Mary Irene
Lyr 18h58'47"40d1'

Lupus Patrie
Cep 23h5'21"64d50'

Luquet,Lionel
Tri 2h21'2"31d17'

Lura Danae
Uma 11h33'57"62d35'

Lurch,Hermann
Del 21h3'49"12d58'

Lurie,Cathy Jane
Lyn 7h29'13"37d8'

Lurie,Dan
Aur 5h26'38"48d57'

Lurie,Julian M J
Uma 10h13'56"60d12'

Luris,Samantha
Aql 19h40'54"10d3'

Lurot,Guy
Cam 3h42'19"68d40'

LuRoyce
Cet 2h16'1"8d4'

Lurtz,Jane
And 0h33'37"45d31'

Lusakan Blaine Six Actual
Cnv 12h19'28"41d59'

Lusardi,Anthony Ernest
Boo 14h30'11"48d31'

Luschilaus
Gem 5h58'29"27d13'

Luschniki
Umi 14h52'0"68d15'

Luscinia M
Lyn 6h37'43"54d22'

Luscious Lou
Sco 17h20'52"-37d19'

Luscri,Giovanni
Umi 17h11'21"75d24'

Luse,Albert William
Aql 19h43'29"10d8'

Luse,Blake
Her 16h42'25"27d41'

Lush,Jacob Douglas
Cep 0h23'58"78d38'

Lushnycky,E Clyde
Dra 19h36'50"68d20'

Lusia
Cas 2h20'33"60d7'

Lusignan,Jacques
Ori 6h0'42"20d25'

Lusk,Pat & Larry
Ori 5h56'19"11d27'

Lusk,Tommy
Sct 18h44'51"-6d11'

Lusk,Victoria Suzanne
Cyg 21h6'0"40d1'

Lussiaud,Jean Henri
Sgr 20h17'33"-37d51'

Lussier Star,P S
Cam 7h35'1"61d29'

Lussier,Catherine
Peg 22h34'22"26d50'

Lussier,Richard Roy
Aql 19h51'26"11d57'

Lussier,Ruth Ann
Lyn 8h9'35"40d44'

Lust Ful Lady Di
Lyn 7h59'45"40d12'

Lust,C James
Her 16h56'23"22d34'

Luster,Billie Gayle
Sco 16h35'54"-40d2'

Luster,Marc Loves Marilyn DeLores
Vul 19h47'1"20d25'

Lustig,Marlene Jo
And 23h34'29"49d20'

Lustrous Lauren
Crb 16h11'16"36d8'

Lusty,Gabriel David
Cam 5h49'16"70d32'

Lusty,Patricia Ruth
And 0h31'1"28d42'

Luszcz,Heather Connell
And 2h13'25"39d32'

Lute,Carey R
Tau 4h38'42"0d45'

Luter,Thomas Theron
Equ 21h20'32"3d17'

Lutge,Albert Thomas & Shirley Faye
Cyg 21h8'10"40d1'

Lutgendorf,Mira
Cas 0h21'40"61d19'

Luther & Steven Rosemary & Danny
Eri 4h8'15"-5d0'

Luther's Star
Oph 18h42'44"8d18'

Luther's Star
Crb 15h22'25"31d44'

Luther,Constance
Cas 23h3'56"54d12'

Luther,Dr Craig W
Her 18h19'48"13d7'

Luther,Kristin
Eri 3h44'53"-4d57'

Luther,Michael
Oph 17h29'34"8d59'

Luther,Ronald
Dra 9h49'0"73d17'

Luther,Shirley A
Cas 1h15'12"66d40'

Luthern,Bill
Aql 19h37'19"1d26'

Luthi,James Reed
Boo 15h7'51"23d40'

Luthier,Jean
Uma 10h34'35"68d33'

Luthman,William H
Uma 10h36'26"50d6'

Lutke,Melissa
And 2h14'35"42d3'

Lutkowitz,Kitty
Del 20h17'26"10d46'

Lutman,Liza Yvonne
Lyr 18h40'47"33d14'

Lutman,Sr,Darrell Elwood
Her 16h50'31"32d28'

Luton,Holli
Cet 1h16'21"0d10'

Luttenegger,Greg
Aur 6h25'53"38d53'

Lutterbel,Harald
Cmi 7h17'37"1d44'

Lutton,TC
Hya 9h39'47"-18d22'

Luttrell,Richard
Lyn 8h11'21"47d35'
Luttrell,Roger Clark
Hya 8h56'39"1d29'
Lutts,Ray
Boo 15h15'1"33d8'
Luty,Derick
Cyg 20h21'48"38d43'
Lutz
Lyr 19h4'47"28d26'
Lutz III,Joseph W
Aur 5h5'55"40d57'
Lutz,Ann
Cas 0h27'1"54d51'
Lutz,Caroline
Cas 1h12'41"61d13'
Lutz,Catherine
Peg 22h0'27"30d7'
Lutz,Christine Margaret
Cas 23h0'54"53d31'
Lutz,Cindy
Peg 23h23'33"26d2'
Lutz,Edward
Aqr 23h17'22"-5d1'
Lutz,George
Boo 15h11'46"52d46'
Lutz,Gretl Otto
Peg 23h34'29"12d9'
Lutz,J Dean
Dra 11h49'1"67d17'
Lutz,John Hartwell
Aur 6h28'19"31d38'
Lutz,John Peter
Aur 4h59'44"30d9'
Lutz,Joseph William
Aur 5h24'37"31d32'
Lutz,Manfred
Lyr 19h18'1"31d43'
Lutz,Mathieu
Cep 22h26'29"58d16'
Lutz,Melissa Ann
And 0h28'59"41d13'
Lutz,Robert E
Lac 22h33'50"56d43'
Lutz,Steven & Sharon
Cas 0h8'20"62d43'
Lutz,Valerie Melissa
Gem 7h0'15"11d13'
Lutz,Yves
Cyg 19h47'13"31d51'
Lutze,Peggy Jane
Peg 23h2'17"18d49'
Lutzky,W
Cam 4h10'54"58d46'
Luukka,Dr Minna-Riitta
Umi 16h49'44"77d53'
Luv U Nic
And 23h18'44"51d22'
Luvaul,Russell Lynn
Vul 19h44'22"25d25'
Luvus,C L
Eri 3h41'39"-13d52'
Lux Originis
Cap 20h33'37"-10d41'
Lux,"Doris May" E Tenebris
Lyr 18h18'52"42d15'
Luxae,Stella
Cyg 21h8'1"30d29'
Luxardi,Lorenzo
Ori 4h55'56"5d34'
Luxenberg,Jacqueline Paige
And 23h6'43"44d55'
Luxor
Pup 8h24'2"-22d54'
Luxton,Clyde C
Dra 16h56'28"68d29'
Luxton,Michael
Hya 9h52'53"-16d47'
Luxusburg,Rosa
Lac 22h25'48"40d25'
Luyke,William Kurt
Aur 5h31'35"50d17'
Luyten,Amaryllis
Lyn 7h45'53"47d34'
Luzetta S W
Cas 3h11'0"68d38'
Luzorgues,Jean-Claude
Mon 6h28'28"-1d46'
Luzzi,James M & Victoria S
Luzzi
Cnc 9h10'13"31d1'
Luzzi,Larry & Roberta Pon
Cap 21h16'14"-26d48'
LWB Prometheus
Eri 3h51'43"-5d2'
Lyall,Gavin
Cet 0h27'20"-4d55'
Lyanne
Uma 8h56'58"47d17'
Lyanne et Enzo
Cyg 20h23'50"39d28'
Lybar
Peg 23h4'46"30d35'
Lybarger,Brenda & Randy
Lyn 8h47'0"36d27'
Lybarger,Craig
Lyn 7h51'17"35d22'
Lyckberg,Doris
Mon 6h19'31"0d11'
Lycke,Keil Pience
Aql 20h32'33"0d14'
Lyde,Johnny
Lyn 7h59'53"33d26'
Lyden,Francis
Sco 17h28'42"-41d10'
Lyders,Helen M
Cnc 8h1'27"8d27'
Lydford,Cynthia
Cas 1h46'41"61d13'
Lydia
Lyn 9h37'20"40d1'
Lydia
Mon 6h52'43"-1d48'
Lydia
Lac 22h20'19"54d27'
Lydia
Tel 20h21'34"-45d28'
Lydia
Aqr 23h4'47"-12d13'
Lydia
And 0h55'50"33d57'
Lydia Anne
Cyg 20h0'56"31d26'
Lydia C
Lac 22h38'16"53d41'
Lydia C
Aql 18h51'1"11d5'
Lydia Claire
Cas 0h54'29"71d7'
Lydia Mary
And 0h36'59"41d10'
Lydia Ruth
Peg 23h46'32"8d19'
Lydia's Miracle:The Wishing Star
Uma 8h33'35"51d58'
Lydia's Place In the Stars
Leo 10h57'16"-5d38'
Lydia's Star
Lyn 7h30'42"45d4'
Lydia's World
Uma 14h16'1"60d10'
Lydia/David
Sge 20h17'50"20d18'
Lydianus
Per 4h30'43"38d54'
Lydiard,George "Chip"
Ser 15h19'1"1d4'
Lydiat,Janet
Peg 23h33'14"33d57'
Lydie
Cep 22h17'11"55d53'
Lydon,Christa
And 23h20'12"43d47'
Lydon,Steve
Aqr 23h8'31"-10d7'
Lyell,Amanda Valerie
And 2h16'32"38d39'
Lyell,Bob
Equ 21h1'52"8d59'
Lyell,Charleen Marie
Cas 0h48'41"66d52'
Lyell,Elaine Marie Boyd
Cyg 21h5'48"30d11'
Lyell,Robert
Her 18h25'46"12d18'
Lyford,Matthew Adam
Gem 6h52'7"13d52'
Lyke
Umi 15h7'38"66d46'
Lykins,Clint & Cindy
Cyg 19h17'29"49d50'
Lykins,Jimmy
Aql 18h44'10"10d50'
Lykowski,David Lee
Dra 16h39'41"72d54'
Lylah
And 2h0'43"38d17'
Lyle & Angie
Uma 13h19'29"63d4'
Lyle Family,The
Boo 14h9'54"31d24'
Lyle III,Herman Oscar (Chad)
Tri 1h59'34"27d26'
Lyle,Amanda
Eri 3h43'24"-0d49'
Lyle,Elizabeth Lane
Peg 21h29'1"26d12'
Lyle,Loren Williams
Dra 20h23'57"62d40'
Lyle,Patsy R
Sge 19h53'15"19d0'
Lyle-Chavez 8-2
Aql 18h49'26"11d45'
Lyles,Pamela
Cas 3h3'13"75d30'
Lyman,Marcia
Del 20h53'37"2d46'
Lyman,William V
Aql 18h54'1"11d32'
Lymburn,Byron Hugh
Dra 16h30'33"70d49'
Lyme
Cam 5h44'37"73d38'
Lyn
Lyn 7h47'22"51d23'
Lyn
Lyn 7h42'0"44d48'
Lyn
Com 12h0'34"27d58'
LYN & BOB (vingt-cinq)
Lyn 8h49'1"40d7'
Lyn & Pete
Aqr 23h3'11"-11d1'
Lyn & Rebecca
And 1h11'1"37d38'
Lyn,Dennis John
Sco 17h27'25"-30d54'
Lyn,Lloyd & Gabriel
Uma 11h47'32"50d11'
Lyn-Jim 7-31-93
Crb 16h21'20"38d39'
Lynah,Gary J
Per 2h12'45"57d49'
Lynaire
And 23h20'38"40d26'
Lynan,Mary Elizabeth
Oph 18h20'13"8d35'
Lynas Marriage Star, Stephen & Karen
Ori 5h14'27"-0d5'
Lynaugh,Daniel F
Her 17h8'56"47d20'
Lynbarbra
Uma 10h0'19"51d55'
LynCamet-Awakening Of Cosmic Ecstasy
Lyn 8h26'1"51d25'
Lynch "Star of Recnat" Merle
Boo 15h2'17"30d17'
Lynch II,Stan
Lac 22h24'33"52d35'
Lynch III-Born 07/31/46,John W
Lmi 10h56'1"26d46'
Lynch,Amanda Bree
Eri 3h40'0"-4d51'
Lynch,Brian N
Her 18h12'59"47d19'
Lynch,Cecilia A
Peg 21h32'51"20d34'
Lynch,Charles & Kathleen
Ari 1h51'45"25d4'
Lynch,Christopher
Oph 17h54'43"12d51'
Lynch,Connor Peter
Umi 14h11'49"68d30'
Lynch,Cynthia
Vul 20h17'1"23d7'
Lynch,Darryl
Oph 17h56'17"8d55'
Lynch,David L
Uma 11h51'15"50d57'
Lynch,Donn
Ori 5h29'41"-8d41'
Lynch,Dorothy
Cam 3h27'33"60d21'
Lynch,Drew Clay
Vul 19h42'0"26d24'
Lynch,Emily Catherine
And 1h36'59"37d32'
Lynch,Ethyl Maye
Lyr 18h54'0"41d59'B
Lynch,Irene
Lyr 18h56'21"34d21'
Lynch,James & Trina
Cyg 19h36'10"38d29'
Lynch,Jennifer J
And 23h5'44"38d52'
Lynch,John F
Per 3h8'0"48d45'
Lynch,John Paul
Peg 0h7'23"27d33'
Lynch,John R "Jack"
Aql 20h8'26"6d37'
Lynch,Jr,John Joseph
Dra 19h3'11"58d11'
Lynch,Jr,The Chief- Lawrence M
Eri 2h58'1"-7d25'
Lynch,Kaitlin
Lyr 19h7'25"40d4'
Lynch,Kara Ann
Lyr 18h53'50"35d26'
Lynch,Linda
And 1h58'11"37d40'
Lynch,Luanne
And 2h32'1"40d36'
Lynch,Margaret Rose
Cyg 21h44'44"28d45'
Lynch,Marnie Christi
Mon 6h54'31"-10d30'
Lynch,Mary Ann
Lyr 18h50'19"41d52'
Lynch,Mary K
Lyn 7h43'51"52d26'
Lynch,MD,E Gene
Del 20h13'37"10d46'
Lynch,Mervyn
Cas 0h12'39"61d20'
Lynch,Mr & Mrs Paul Lamar
Aql 20h31'39"0d24'
Lynch,Nora Barbara
Cas 0h55'36"58d18'
Lynch,Opal Lee
Mon 8h2'36"-1d34'
Lynch,Ozzie
Lac 22h27'59"54d9'
Lynch,Patrick
Boo 14h21'16"50d35'
Lynch,Robert
Aql 19h56'58"15d49'
Lynch,Robert Allen
Oph 17h6'52"-0d30'
Lynch,Sachi
Cmi 7h53'47"1d20'
Lynch,Shane
Cep 23h12'29"60d8'
Lynch,Sharon Rae
Mon 8h7'47"-5d60'
Lynch,Sr,Edmund E
Cam 3h51'53"70d41'
Lynch,Sr,Paul E
Cep 22h38'38"58d37'
Lynch,Thomas C M
Cep 6h9'1"85d44'
Lynch,Thomas David
Lyr 18h54'0"41d59'A
Lynch,Thomas Joseph
Cnv 12h4'53"38d42'
Lynch,Toni
Mon 6h21'59"3d5'
Lynch,USN,Ret,LCDR FM
Sex 10h2'40"5d47'
Lynch,Wendy June
Cra 18h18'35"-42d37'
Lynch,William O
Boo 15h3'0"27d32'
Lynch,Yvette
Cas 0h29'33"61d44'
Lynch-Fuer,Bonnie
Cas 0h21'27"64d56'
Lynch-Holman,Carole
And 23h29'12"48d42'
Lynchmills,S F J
Lyn 6h36'45"59d46'
Lynda
Aur 5h4'47"40d56'
Lynda 200647
Cyg 20h19'1"39d6'
Lynda A
Cas 23h1'1"53d34'
Lynda Carol
Pup 7h39'53"22d14'
Lynda Faye
Cam 3h32'1"60d36'
Lynda Lou
ARI 1h54'0"25d15'
Lynda Marie
Cas 0h1'11"58d18'
Lynda Star
And 1h29'33"40d23'
Lynda's Heart
Lyn 9h1'32"38d8'
Lynda's Limelight
Cas 0h35'1"73d47'
Lynda's Sky Diamond
Lyr 19h14'35"33d59'
Lynden,My Grandson Jamie
Umi 15h15'52"66d15'
Lyndsay
Lyr 19h17'17"40d43'
Lyndsey
Mon 7h42'33"-1d6'
Lyndsey,Charise & James Lesher
Crt 11h22'40"-19d28'
Lynes,Jeffery Alan
Per 4h7'31"38d13'
Lyness,Scott Alistair
Her 16h20'58"41d17'
Lynette
Lyn 9h14'47"44d8'
Lyngdal,Quentin E
Lac 22h25'22"53d29'
Lynhen
And 1h30'41"39d51'
Lynja-I
Cet 2h59'27"1d5'
LynLar
Cyg 21h37'56"40d29'
Lynmarie
Mon 6h27'19"11d31'
Lynn
Cyg 19h59'57"45d38'
Lynn
Peg 22h52'42"21d44'
Lynn
Uma 8h32'0"68d4'
Lynn
Vul 21h1'16"27d33'
Lynn
Vul 19h58'11"22d56'
Lynn
Lyr 18h19'59"40d26'
Lynn & Allen Forever
Her 18h12'15"31d24'
Lynn & Michael
Mon 7h30'0"-5d46'
Lynn & Paul
Cyg 21h38'1"40d54'
Lynn & Warren's Cup of Tea
Sgr 18h56'34"-32d27'
Lynn 69333
Aql 19h50'23"10d52'
Lynn Ann The Great
Uma 11h17'48"47d6'
Lynn Carole
Uma 10h39'26"67d58'
Lynn Ellen-The Schnooting Star
Cet 2h55'40"6d14'
Lynn Lynn
Peg 23h32'51"23d47'
Lynn Marie
Sct 18h52'22"-9d33'
Lynn Marie
Lyr 18h50'13"41d40'
Lynn Marie
Cas 0h34'48"64d37'
Lynn Marie
Mon 6h43'31"11d2'
Lynn Mi Cari'31"11d2'
Ori 5h55'52"16d39'
Lynn Michele
Ori 5h59'34"20d43'
Lynn Stacey
Cyg 22h0'0"51d28'
Lynn The Wunnerful
Mon 6h20'52"2d45'
Lynn Your Time Is Now
Umi 15h23'12"88d9'
Lynn's Aura
Com 13h2'28"28d58'
Lynn's Light
Tri 1h56'42"28d3'
Lynn's Own
And 23h41'51"46d55'
Lynn's Star
Cam 6h6'27"72d53'
Lynn's Star
Umi 16h23'24"75d0'
Lynn's Star, From Shayne
Cyg 21h13'23"28d22'
Lynn's Valentine Star
Cam 12h42'53"78d12'
Lynn's Wishing Star
Vul 19h15'35"25d3'
Lynn,Carl
Her 18h3'42"31d20'
Lynn,Elizabeth Kloo
Cyg 20h2'0"30d21'
Lynn,Jospeh Anthony
Boo 15h21'42"40d10'
Lynn,Sandy
Cas 0h23'0"60d49'
Lynn,Sommer
Aql 19h54'42"13d12'
Lynn,Ted
Lmi 10h49'51"33d46'
Lynn-50th Anniversary, Raoul
Ser 16h3'42"13d27'A
Lynn-50th Anniversary, Nancy
Ser 16h3'42"13d27'B
Lynn-August I995
Lyn 9h13'0"40d55'
Lynn-Rogers,FLOGAN
Ori 5h58'41"17d56'
Lynn/Michelle
Tri 2h12'0"32d42'
Lynnann
Tri 1h52'44"26d25'
Lynndee
And 2h26'41"49d55'
Lynne
Umi 15h5'20"66d46'
Lynne & Darren
Cyg 21h24'17"28d36'
Lynne & David
Uma 12h42'22"53d6'
Lynne & Lorrie
Uma 10h2'0"73d7'
Lynne & Mark
Eri 3h54'24"-2d51'
Lynne A
Lmi 9h57'32"34d29'
Lynne Jeanne
And 2h32'28"38d21'
Lynne Marie
Cep 21h53'57"55d42'
Lynne's Light
Cyg 21h3'11"36d57'
Lynne's Star
Crb 15h55'28"31d36'
Lynne,Barbara
Cyg 20h3'58"40d56'
Lynnea's Wishing Star
Lmi 9h46'30"37d57'
Lynneray
Vul 19h22'1"22d33'
Lynnette & Jim
Peg 21h50'55"34d14'
Lynnie
Lyn 6h57'56"60d20'
Lynnie Little Butt
And 23h15'38"41d9'
LynnMarie
Mon 6h28'22"-10d8'
Lynnse's Star
And 1h22'11"40d25'
Lynnsong
Cyg 20h35'39"37d31'
Lynsey & Martin
Umi 13h3'22"70d4'
Lynsey Kara
And 0h38'57"30d37'
Lynsey,Siobhan
Cmi 8h8'22"3d55'
Lyn's Light
Cen 11h50'29"49d43'
Lynster,Diana Hope
Lyn 7h51'20"40d48'
Lynwood Star,The
Peg 23h0'31"18d16'
Lyon's Star,Patricia
Uma 10h55'16"40d32'
Lyon,Anna Marie Adams
Mon 6h52'44"-1d40'
Lyon,B Carl
Cet 0h49'59"-3d17'
Lyon,Charles S
Boo 14h12'1"39d21'
Lyon,Christopher McTee
Ser 18h16'19"-11d35'
Lyon,Dee Lilley
Ori 6h13'38"1d28'
Lyon,Donna T
Mon 7h54'3"-3d49'
Lyon,Gail
And 23h36'14"39d55'
Lyon,Henry Jesse
Dra 17h13'34"63d50'
Lyon,Hilary
And 23h17'58"37d53'
Lyon,Jr,Michael David "Mick"
Uma 11h49'53"55d5'
Lyon,Laurie
Lyr 19h3'40"28d26'
Lyon,Leslie P
Peg 23h1'34"31d21'
Lyon,Marjorie
Cyg 20h21'0"38d28'
Lyon,Rabbi David A
Boo 14h2'0"18d50'
Lyon,Rachel A
Umi 16h37'52"77d17'
Lyon,Scarlet Ruby
Cas 0h10'40"58d36'
Lyon,Todd
Dra 16h8'33"66d60'
Lyon-Behre,Janet
And 0h53'51"40d40'
Lyon-N-Hale,Love
Crb 15h32'15"31d15'
Lyons,A P N D Alphonse D'Abernon
Uma 11h41'30"37d51'
Lyons,Ann Elizabeth
Psc 0h53'24"18d25'
Lyons,Carol & Jayel
Cyg 19h42'26"42d3'
Lyons,David
Hya 9h12'19"5d50'
Lyons,Emily Marie
Cam 7h52'19"83d26'
Lyons,Gerald A
Cet 2h18'55"1d1'
Lyons,Gerald Thomas
Hya 9h7'0"5d36'
Lyons,Gingus,Attilla Vaugn
Uma 12h13'34"58d31'
Lyons,Jennie Lee Helen
And 2h28'26"47d29'
Lyons,Kevin John
Cmi 7h35'21"0d15'
Lyons,Kody Grant
Her 16h38'20"32d13'
Lyons,Kyle Albert
Boo 14h30'20"45d59'
Lyons,Leo
Dra 16h33'49"60d58'
Lyons,Lorraine
Lyr 18h14'19"41d5'
Lyons,Mildred Bethania
Cet 3h1'52"0d39'
Lyons,Molly
Cyg 20h3'12"31d37'
Lyons,Mrs
Umi 13h39'26"70d41'
Lyons,Msgr Joseph
Crb 15h52'19"30d4'
Lyons,Pamela A
Ori 5h54'39"0d7'
Lyons,Paul Lightner
Uma 13h22'18"62d34'
Lyons,Paula
Lyn 8h27'39"40d36'
Lyons,Sarah Jane
Lyr 19h6'1"26d59'
Lyons,Tammy
Mon 7h44'35"-5d29'
Lyons,W Terry
Lac 22h29'25"52d38'
Lyons,Wayne
Dra 17h48'54"64d43'
Lyovina
Peg 23h1'28"21d44'
Lyran
Cam 3h57'36"70d14'
Lyric
Lyr 18h57'52"40d6'
Lysaght,Kathleen M
Uma 9h7'25"48d6'
Lysandre
Ori 5h40'14"12d56'
Lysandrou,Lillian Thalia
Cas 0h54'17"60d4'
Lysenko,Victor
Cnc 8h59'23"7d53'
Lysle,Jennifer
Mon 7h8'14"-6d53'
Lysohir,Stephen A
Boo 14h51'31"34d46'
Lystad,Kevin Henrik
Psc 1h27'19"12d17'
Lytell,Roxanne Patricia
Del 20h48'37'9d20'
Lythgoe,(my love),Pam
Cyg 21h44'28"31d7'
Lythgoe,Clare Louise
Cas 2h29'30"68d23'
Lytle II,Marshall B
Mon 6h1'10"-8d9'
Lytle,Clint
Aqr 20h58'50"-14d2'
Lytle,Linda "Pumpkinmuffin"
Sex 10h40'36"-0d28'
Lytle,Lynda
Cas 1h11'41"62d17'
Lytle,Rick Aye
Aur 6h7'39"31d45'
Lyttleton,Danielle
And 23h33'17"40d55'
Lytwynuk,Richard
Per 2h56'30"45d33'
LYZ
Cas 1h8'54"68d18'
Lässig,Dipl-Ing Rainer
And 23h13'14"40d30'
Léa
Cnv 12h8'35"38d5'
LÉcuyer,Monique
Ori 5h40'1"11d6'
Légaré,Franáois
Umi 16h12'20"73d13'
Légaré,Michaël
Cep 22h9'42"60d7'
Léger
And 2h17'33"45d22'
Léger,Gentle Jill
Lyr 18h30'48"45d46'
Léger,Réjean
Tri 2h41'54"34d12'
Léocécile
Leo 10h20'34"14d22'
Léona
Umi 14h29'44"67d51'
Léopardina
Sgr 19h28'33"-30d1'
Lépine,Diane
Cas 0h53'36"74d46'
Létourneau,Denis
Lmi 10h1'38"39d60'
Létourneau,Gilles
Per 3h5'0"41d53'
Létreux
Vul 21h20'54"24d15'
Lévesque,Jean-Pierre
Cyg 21h7'1"30d14'
Lévesque,Michel
Umi 15h34'20"67d41'
Lévesque,Olivier Lamarre
Uma 8h42'43"55d32'
Lévesque,Sylvie
Cyg 21h5'59"31d51'
Löber,Bernd Eric
Cap 20h26'15"-26d38'
Löblein,Melanie
Cap 20h41'27"-23d49'
Löblein,Rene
Sco 17h30'1"-30d42'
Löffelmann,Günther
Psc 1h16'1"18d15'
Löffler,4-12-47 Christine
Sgr 19h17'8"-40d4'
Löhn,Peter
Dra 17h52'0"68d10'
Löneke,Horst
Tau 4h4'3"24d8'
Löser,Holger
Vir 12h33'59"-8d29'
Löw,Herrmann
Ori 5h57'18"1d39'
Löwe Anja Walter 12/08/77
Lyn 8h11'16"48d33'
Lüchtefeld,Till
Ori 5h56'0"0d5'
Lück,Katharina Lisa
Cas 0h59'57"63d49'
Lückerath,Gerhard
Del 20h17'58"14d39'
Lüdeke,Friedhelm
Her 17h40'36"14d18'
Lührs,Rainer
Psc 1h19'58"21d43'
Lünse,Anke
Cep 22h23'31"65d8'
Lüthy,Gottlieb
Lyn 8h11'14"42d46'

Lüttgens,Waltraud
Tau 5h33'57"26d8'
L Tau 5h33'57"26d8'
Hya 8h53'43"2d40'
L Hya 8h53'43"2d40'
Boo 14h49'29"32d27'
Løvbjerg,Maibritt
Del 20h35'1"3d54'

M

M
Cyg 20h24'18"39d17'
M & M
Tau 3h55"47'20d17"
M & M
Cam 4h6'50"70d13'
M & M
Gem 6h32'29"14d30'
M & M
Uma 9h32'35"50d8'
M & M
Lmi 10h46'28"31d26'
M & M
Lmi 11h2'48"32d44'
M & M
Aur 5h53'0"30d47'
M & M Twiggy
Ser 17h33'17"-14d10'
M & R W
Cyg 19h58'49"31d38'
M * A * C
Cam 3h45'0"68d58'
M A C
Ori 5h10'42"-6d57'
M A C-50
Dra 20h7'48"73d17'
M A Decot-"40"
Aur 6h57'0"44d7'
M A T
Lyr 19h22'41"31d18'
M B Lock Lips
Cam 13h22'24"8d20'
M C C Class of 1996
Per 3h14'54"41d12'
M C L J M B
Mon 8h2'27"-9d20'
M C Robin
Sco 17h5'0"-31d15'
M D D Stephanie
Mon 6h19'46"8d45'
M E M
Vir 13h20'27"-2d7'
M Eden's Star
Cyg 19h28'51"35d18'
M I C J S T
Her 17h16'21"40d29'
M I T
Aur 6h17'60"50d24'
M I T '72
Cnv 13h52'16"46d19'
M J 10
Her 17h9'10"45d36'
M J C
Tau 4h22'25"15d41'
M J C R
Peg 23h50'47"31d29'
M J Lea (Leander)
Cyg 21h6'0"30d45'
M J S
Her 16h22'57"47d35'
M J T
Uma 13h33'18"53d59'
M J's Always,Always M J's
Ori 6h5'56"-2d42'
M Jeanne
Ori 5h33'55"-0d6'
M K Rebecca
Cnc 8h0'0"10d45'
M L V & E L K
Cas 23h30'50"53d11'
M Lou
Crt 11h21'44"-18d13'
M Lydia
Aqr 22h40'20"-1d56'
M M Mark
Oph 17h32'1"11d54'
M N M 2
Cam 9h33'31"82d9'
M P
Cam 6h59'26"83d29'
M P T
Ori 4h55'56"5d50'
M R G Studmuffin
Dra 20h18'42"64d7'
M R K
Her 16h55'12"18d54'
M R P Loves Woody
Boo 14h54'21"43d56'
M R Pa 2T
Her 17h21'11"14d39'
M T
Lyn 7h29'45"58d60'
M T H
Mon 7h41'40"-3d42'
M W H Bear
Cam 5h3'16"61d5'
M&M
Sge 20h2'24"20d17'
M&M Crouch
Oph 17h36'33"8d44'
M&M Star,The
Vul 21h4'1"27d38'
M&M Twinkle Star
Psc 1h20'27"16d1'
M'Lady's Hope
Cyg 20h21'28"40d29'
M'Marina
Cap 20h57'57"-23d51'
M-123 Torrance,CA
Dra 16h27'1"69d9'
M-235 Temecula,CA
Aql 19h57'35"13d51'
M-Kon Al Halilaina
Cas 24h5'22"53d58'
M-V
Psc 1h3'1"21d47'
Ma & Pa Forever
Lmi 10h0'46"32d19'
MA & Rocky
Cyg 19h28'37"30d34'
Ma Bonne Etoile
Per 4h3'34"35d17'
Ma Bonne Etoile
Cas 1h58'38"60d6'
Ma Bonne Etoile
Equ 21h16'5"11d21'A
Ma Cherié Diane
Per 2h22'43"57d0'B
Ma Conroy
Cyg 21h51'24"40d4'
Ma Douce
Uma 10h5'29"54d23'
Ma Doudou
Ori 5h56'0"7d53'
Ma Mere Joni
Lyn 7h31'14"44d42'
Ma Puce
Ori 6h16'1"18d48'
Ma Teresa-Ivet
Lyn 8h12'29"48d11'
Ma Toche
Ori 5h55'60"18d19'
Ma,Cathy
Umi 16h43'50"77d52'
Ma-Pi
Uma 10h52'45"38d48'
Maack,Arlene Kozyra
Cmi 7h40'53"3d36'A
Maack,David Ralph
Cmi 7h40'53"3d36'B
Maack,Dominique Marie
Gem 8h2'1"28d30'
Maack,Emily Tyra
Uma 8h34'45"68d36'
Maack,Jessica Kristen
And 0h47'35"35d17'
Maae,Lars
Uma 12h5'29"56d42'
Maahs,Theresa
Uma 8h43'1"68d49'
Maak,Eric David
Cep 20h58'34"60d24'
Maas
Peg 23h46'27"8d1'
Maas,Helen
Mon 7h39'39"-3d30'
Maas,Mahalo
Eri 3h41'30"-10d38'
Maas,Natascha
Lyn 8h3'46"47d39'
Maas,Scott C
Ori 6h6'50"8d9'
Maas,Tammy Christine
Crt 11h38'46"-18d28'
Maaskant,Nicole
Crb 15h48'18"38d50'
Maass,Evelyn
Sco 16h59'0"-40d24'
Maass,Werner
Cmi 7h17'19"0d20'
Maastricht,Namens Restaurant L'Entree
Cas 1h48'0"68d50'
Maat,Tineke
Cnv 12h44'0"40d15'
Maatta,Bert Ian
Boo 14h22'28"16d44'
Maazouz,Moulay
Lyn 7h39'22"39d52'
Maaβ,Herr
Boo 14h39'0"11d21'
MAB 28
Sgr 18h56'35"-26d4'
Mab Aster Fortune Ztm
Cam 6h39'18"67d38'
Mabe,Kathleen
Del 20h13'51"14d4'
Mabel & Louise
Cyg 19h45'25"30d51'
Mabel K
Lyr 18h56'31"31d34'
Mabel-Rose,We Love You
And 1h0'44"37d55'
Maben,Buster
Cep 22h54'34"68d32'
Maben,Debra
And 2h3'34"45d35'
Maber,Elsie Annetta
Cas 22h57'20"55d32'
Mabes
Cam 5h57'29"68d29'
Mabire,Pierre
Lmi 10h47'24"23d58'
Mabry,Cynthia L
Equ 21h3'10"10d41'
Mabry,Karen Lisa
Aql 20h6'47"0d55'
Mac
Hya 8h14'35"6d36'
Mac
Dra 14h50'12"60d40'
Mac
Per 3h4'50"40d8'
Mac
Per 3h19'27"41d32'
Mac
Cma 6h59'21"-19d4'
Mac & Sheryl Love Everlasting
Ori 5h55'27"17d17'
Mac & Suzy's World
Her 16h20'54"40d38'
Mac & Tommie
Aql 20h20'1"5d12'
Mac Daddy
Boo 15h25'16"30d34'
Mac Kenzie,John W
Boo 14h29'13"30d2'
Mac Lean Star
Aql 19h34'37"-0d18'
Mac the Knife
Her 16h31'12"35d52'
Mac the Light
Cep 20h36'34"75d18'
Mac's Millennium
Cet 2h19'1"9d3'
Mac-Myra
Mon 8h6'40"-9d38'
MAC-us-DANA
Hya 8h28'30"-17d13'
Maca
Sge 20h17'54"18d52'
Macade,Alain
Cam 3h26'53"61d49'
MaCafee,PhD,Patrick
Cma 6h55'11"-19d34'
Macak-Sima
Cap 20h4'24"-10d33'
MacAlister,Ian Scott
Hya 8h56'45"0d24'
Macall,Thomas F
Psc 1h43'14"27d54'
Macaluso & Burton
Cyg 19h40'14"40d35'
Macaluso,Jean Ann
Del 20h14'44"14d4'
Macaoay,Stacy Leigh
Psc 0h44'26"20d33'
Macario,Anthony
Her 16h42'15"48d52'
Macario,Katheryn L
Cyg 19h57'10"38d17'
MacArthur,Jet
Cet 3h0'25"0d49'
Macary,Ann Marie (Nellie)
Miele
Tri 2h5'29"31d45'
Macary,James
Aur 6h2'12"31d26'
MacAskill,Ryan Peirce
Per 1h44'58"54d14'
Macatee,Midnight Flight Patricia
Peg 21h42'20"21d39'
Macaulay,Helen
Cas 0h33'51"74d15'
MacBurne,Jamie Fe
Vir 11h38'25"9d4'
MacCallum,Archibald
Uma 11h48'0"30d22'
MacCallum,Douglas C
Her 16h45'51"33d11'
MacCallum,Robert
Per 2h48'57"40d57'
Maccani,Thomas R
Lac 22h07'55"50d30'
Maccarini,Paola Maria
Gem 6h49'36"17d12'
Maccario,Elizabeth J
Lyr 19h2'21"40d36'
Maccario,John M
Her 15h58'25"40d18'
MacCarn,Neale & Kenneth
Lmi 10h10'1"33d10'
Maccarrone,Christine
Cnv 13h5'41"51d11'
Maccarrone,Susan Linda
Cap 20h41'52"-20d32'
Macchi,Ercole
Cas 0h11'54"66d22'
Macchia,Giuliano
Lep 4h59'43"-19d21'
Macchia,Greg
Her 15h48'1"45d31'
Macchia,John & Gale
Cyg 21h25'19"28d46'
Macchia,Michael Brian
Boo 13h59'19"18d52'
Macchia,Sr,David A
Dra 16h56'30"64d46'
Macchiavelli,Jr,Joseph Anthony
Dra 15h58'46"51d40'
Macchiavelli,Mary R
Tri 2h17'40"31d7'
Macchio,Gino Steven
Boo 14h46'1"48d58'
Macchio,Julia Rose
Cam 3h50'0"60d22'
Macchion,Barbara
And 0h2'48"40d55'
Maccia,Sir Gregory Charles
Her 17h28'43"28d32'
MacCollum,Laurie
Lyn 8h45'31"37d6'
MacConchie,Cory Griffen
Lac 22h24'57"52d40'
MacCormack,Karen
Vul 20h16'0"26d2'
MacCready,Thelma Louise
Sco 15h59'1"-25d43'
MacDaddy
Her 15h53'32"41d20'
Macdak
Cam 3h42'19"61d29'
MacDermaid,David John
Umi 10h13'1"88d16'
MacDermid,Robert G
Peg 21h55'43"2d27'
MacDonald February 19th,1987,Leslie W
Cas 23h39'51"65d19'
MacDonald III,William Shaw
Eri 4h24'50"-24d11'B
MacDonald,Alexander
Oph 16h54'49"0d24'
MacDonald,Allie Marie
Uma 8h14'33"60d56'
MacDonald,Amber Anne
Uma 8h28'17"62d10'
MacDonald,Ann Elisabeth
And 23h39'11"48d5'
Macdonald,Ann E
And 2h28'25"47d28'
MacDonald,Betsy
Peg 21h57'38"33d51'
MacDonald,Brandi
And 0h17'0"46d6'
Macdonald,Christian
Cas 2h15'25"60d21'
MacDonald,Damien Lee
Cet 2h52'11"1d47'
Macdonald,Ellen Kristine
Lib 15h2'47"-10d42'
MacDonald,Fiona Onio
Cas 3h1'59"65d9'
MacDonald,Isaiah Todd
Aur 7h10'14"39d32'
MacDonald,Janyn Darnell
Del 20h12'1"9d7'
MacDonald,Jean Alison
Crb 15h52'20"26d57'
MacDonald,John
Sct 18h55'1"-6d50'
MacDonald,John
Ori 5h43'23"10d48'
MacDonald,John & Lynn
Lyn 7h36'11"43d14'
Macdonald,Kay
Uma 8h43'15"53d34'
Macdonald,Laura
Vir 11h42'43"5d54'
MacDonald,Laurel Ann
Lyr 18h27'16"42d3'
MacDonald,Maria Rosario
Eri 4h24'50"-24d11'A
MacDonald,Mary Alison
And 1h19'1"40d1'
MacDonald,Maureen
Peg 22h54'31"29d7'
MacDonald,Neil
Aur 4h48'47"41d6'
Macdonald,Nicola Claire
Cas 1h5'24"53d28'
MacDonald,Raymond Harrell
Aql 19h8'20"5d58'
MacDonald,Rev Kenneth
Sct 18h42'27"-5d20'
MacDonald,Ritchie
Cam 9h31'22"82d4'
MacDonald,Thomas Avis & Jason
Del 20h34'31"20d7'
MacDonald,Victor
Cru 12h25'2"-61d45'
Macdonald,W Malcolm
Sct 18h52'2"-6d53'
MacDonald,William & Maria
Cru 12h9'42"-62d40'
MacDonnell,David
Lac 22h40'56"55d60'
MacDormand,Nellie
Lyr 18h32'44"38d32'
MacDougald,Doug
Cep 21h59'11"55d57'
MacDougall I
Cep 21h52'0"60d59'
MacDougall,Colin Kennedy
Boo 13h45'35"21d29'
MacDougall,Elizabeth Clare
Aql 18h57'19"0d10'
MacDougall,Mary Katherine
Aql 20h18'24"7d51'
MacDuff,Leslie
And 2h33'41"40d19'
MacDuffee
Cyg 21h15'0"37d41'
MacDulla,Frank & Marianne
Her 16h38'1"24d29'
Mace"The Alexis", Ralph E
Sct 18h54'41"-4d42'
Mace, Phebe
Cam 5h56'56"61d40'
Mace,Allan Keith
Dra 16h53'59"68d29'
Mace,Barbara J
Com 12h23'55"24d15'
Mace,Daniele
Cam 3h27'33"60d31'
Mace,Doris Mae
Vul 19h39'1"20d4'
Mace,Kent
Her 16h24'0"34d52'
Mace,Lady Pamela
Aql 19h53'19"15d55'
MacEachern,Charlie
Aql 20h12'49"0d13'
Macedo
Her 17h26'18"29d39'
Macedo,Monica Marie
Equ 21h7'0"3d29'
Macedonia,Gloria "Butterball"
Cas 1h13'51"67d50'
Macejko II,Joseph Andrew
Cnv 12h44'0"40d50'
Macek 50th Anniversary,Ruth & Les
Ori 5h58'23"8d41'
Macemore,Charlie "Moe" Monroe
Hya 8h54'31"-7d58'
Maceoin,Beth
Cas 22h58'41"55d47'
Macera's,"Fascinating Star",Monica T
Cet 2h47'0"3d2'
Maceren,Myra M
And 0h12'1"37d36'
Macesich,Matthew Robert
Boo 14h46'19"45d14'
MacEwen,Amanda
Mon 6h55'1"-6d6'
MacFadyen,Dr Edward C
Cas 2h45'50"64d25'A
MacFadyen,Elizabeth C M
Ori 4h52'52"0d26'
MacFadyen,Evelyn Mary
Cas 2h45'50"64d25'B
MacFarlane IV, Alexander
Boo 14h56'56"44d25'
MacFarlane,Carleton R
Cet 1h36'60"-6d12'
Macfarlane,Duncan Clayton
Dra 16h30'55"62d21'
MacFarlane,Laura Catherine
Com 12h41'16"21d44'
MacFerren,Sr,Oscar Dean
Lmi 11h0'29"26d4'
MacGeorge,Jr,Douglas David
Lac 22h16'19"38d42'
MacGillivary,Jordan Alexia
Sex 10h28'1"-1d17'
Macgowan,Stefan
Ori 6h8'1"8d56'
MacGowan,William Stewart
Mon 6h29'56"7d54'
MacGregor,Ian Cline
Cnv 13h37'25"40d14'
MacGregor,James Alexander
Aur 6h35'40"34d43'
MacGregor,Joyce
Cam 5h50'28"58d35'
MacGregor,Julia Nichole
Lyr 18h41'0"36d44'
MacGuinness,William Michael
Her 17h17'50"42d3'
MacGyver
Her 15h48'59"48d28'
Mach
Pup 8h24'38"-29d36'
Mach,Stellar Marie- William
Lyr 18h59'14"30d9'
Machacek,Joseph William
Uma 11h3'16"44d13'
Machado,Ashley
Cas 0h24'33"60d54'
Machado,Norbert
Boo 13h45'22"26d19'
Machado,The
Boo 14h48'49"51d36'
Machamer,Mary-Jean
And 0h59'0"33d46'
Machan,Richard Edwin
Hya 8h22'56"5d44'
Machel,Richard
Dra 15h56'41"52d7'
Machenbach,Dr Brigitte
Vir 11h54'46"-3d5'
Machi
Mon 6h20'12"5d17'
Machi,Fran
Cas 0h24'36"59d46'
Machiaverna,Antoinette
And 1h15'45"40d55'
Machiko
Lib 14h31'26"-22d12'
Machill,Rosemarie
Lyn 8h7'54"48d23'
Machin (Tiger), Kenneth G
Lyn 8h2'39"46d38'
Machin,Eddie
Her 16h45'25"10d45'
Machnik,Elizabeth Spik
Uma 11h33'39"49d6'
Machnik,Mäusel
Psc 23h22'43"5d51'
Machnik,Thaddeus Stephan
Lac 22h18'52"50d25'
Machowski
Cyg 19h43'38"31d5'
Machradt
Lib 14h46'11"-20d20'
MACHRI
Ari 1h46'32"20d55'
Macht,Anna Irene
Vir 11h39'48"2d46'
Macht,Julia Irene
Mon 6h20'23"0d46'
Machtinger,Dory Diamant
Umi 14h58'36"66d3'
Machtinger,Lawrence Arnold
Sex 10h36'19"3d41'
Machuca,Pilar
Cnv 12h46'34"51d1'
Machulis-Thaell, Jessica
Del 20h33'52"10d38'
Machurek,Jim (Santa) & Kathy
Cyg 20h28'11"42d53'
Macias,Charles L
Lmi 10h32'0"28d7'
Macias,Enrico
Crt 11h17'56"-19d18'
Macias,Mark
Per 2h39'13"34d33'
Macias,Meganne Maureen
Uma 8h37'18"48d25'
Maciel,José
Cyg 19h51'29"44d35'
Macintire,Cordelia
Cas 23h35'42"61d11'
MacIntosh,Arlene
Cas 0h31m34"50d8'
MacIntyre,Doug "Roy-El"
Aur 6h10'58"37d59'
Macintyre,Julie Anne
Vul 19h46'16"26d8'
Macintyre,Lyndsey Victoria
And 0h52'29"34d59'
MacIntyre,Ruthelaine
Cas 1h19'1"65d35'
Maciolek,Donald
Per 2h59'25"55d45'
MacIsaac,Joseph Bruce Patrick
Oph 17h31'42"11d39'
MacIver,Norman
Uma 8h55'30"50d34'
Mack
Leo 9h37'32"7d57'
Mack
Her 18h18'0"13d7'
Mack
Cma 6h25'38"-15d23'
Mack 1994,Kara Michelle
Del 20h24'48"10d38'
Mack III,Lawrence John
Dra 17h40'58"59d38'
Mack III,Wallace R
Hya 9h38'58"2d11'
Mack's Monoceros
Mon 7h3'51"-6d47'
Mack,"Pixie" Dorothy
Cap 21h21'1"-26d10'
Mack,Adam
Dra 10h12'48"80d2'
Mack,Ben
Aur 6h52'59"38d8'
Mack,Charlie
Lmi 9h42'0"34d49'
Mack,Dale
Cam 4h56'52"60d4'
Mack,Danny
Uma 10h28'54"51d43'
Mack,E A
Aqr 23h43'51"-5d58'
Mack,Emily Marie Eve
Cas 0h1'28"54d37'
Mack,Jeffrey C
Lac 22h17'11"38d49
Mack,Jimmy
Tri 1h53'35"27d52'
Mack,Joe
Aur 7h2'39"40d15'
Mack,Karl
Mon 7h51'21"-2d26'
Mack,Kathy
Lyr 19h18'52"35d10'
Mack,LaDonna Faye
Leo 11h16'40"-0d49'
Mack,Larry
Ser 15h52'41"0d21'
Mack,Michael
Oph 17h58'51"14d15'
Mack,Norma Edith
Boo 14h26'23"46d0'
Mack,Norman
Per 2h59'54'45d22'
Mack,Patti
Eri 4h11'41"-17d33'
Mack,Ralph & Petra
Lyr 19h3'40"38d37'
Mack,Robert
Lac 22h15'55"48d51'
Mack,Robert & Debby
Aql 19h10'52"12d11'
Mack,Tiffany D
And 1h17'53"39d17'
Mack,Tracy Lorraine
Cnc 8h38'33"18d46'
Mack-You are One in a Million,Kevin
Ori 6h14'37"9d4'
Mackail,Emily Clair
And 0h53'35"41d7'
Mackall Star,The Glenn
Mon 8h0'33"-6d33'
Mackall,Rosemary
Mon 8h7'1"-8d0'
MacKavanagh,Jill Anne
Cas 3h25'59"73d24'
Mackay
Dra 17h49'39"73d26'
Mackay Star
Uma 14h1'55"48d58'
Mackay's Dream
Uma 8h42'1"70d10'
MacKay,Anne Glaze
Aql 19h26'45"0d29'
Mackay,Arlene
Uma 12h34'58"60d31'
MacKay,Carolyn F
Mon 8h7'36"-1d47'
MacKay,Carrie
Mon 7h42'31"-1d51'
Mackay,Duncan MSP
Umi 17h4'41"76d32'
Mackay,John William
Ori 5h43'29"10d58'
Mackay,Trudy L
Umi 14h59'38"71d19'
Mackbarth,Lynn
Cas 0h34'0"66d26'
MacKeil,John & Mary
Cyg 21h11'22"39d32'
MacKellar,Stuart"Doc"
Oph 17h0'52"-25d26'
MacKenna
Mon 6h33'1"8d4'
MacKenna,Juliann
Lyn 9h1'38"33d59'
Mackenzie
Peg 23h17'13"33d55'
Mackenzie Leigh
And 23h19'27"35d31'
Mackenzie's Light
Oph 18h18'34"8d41'
MacKenzie,Brooke
Tau 4h3'3"22d39'
MacKenzie,Colin C
Vul 20h1'14"28d56'
MacKenzie,Corrina
And 23h0'47"51d7'
MacKenzie,Daniel Finlay
Per 3h4'51"41d50'
MacKenzie,David
Cep 20h44'18"76d40'
Mackenzie,Donald Hector
Ori 5h56'13"15d43'
MacKenzie,Ellen Lee
And 2h20'53"47d58'
MacKenzie,Forever John
Cep 20h41'0"76d3'
MacKenzie,James Robert
Dra 14h21'0"64d38'
MacKenzie,My Cosmic Ray-Raymond
Ori 6h4'1"0d8'
Mackenzie,Sophie Nicole
Tri 1h45'40"31d44'

MacKenzie,Sr,Edward P Boo 13h40'52"26d18'
Mackenzie,Whitney S Cam 7h16'45"84d3'
Mackeprang,Jr, Stephen J Per 2h38'12"37d8'
Mackereth,Erin Boo 15h14'0"38d19'
Mackes,Faith Kristan Cas 0h53'40"61d5'
Mackesy,Sarah And 23h0'25"35d25'
Mackewich,Susan Tilghman Ball Tau 5h45'0"23d29'
Mackey & Family,Don Uma 10h36'38"50d30'
Mackey,Barbara Pia Haile Eri 3h20'50"-6d11'
Mackey,Carla Cyg 21h15'1"37d41'
Mackey,Carson Uma 13h46'12"51d46'
Mackey,Jerry Her 16h10'29"24d7'
Mackey,Ka-Ma-Ly Lyn 7h56'15"40d44'
Mackey,Kathryn Cas 0h48'34"65d1'
Mackey,Kellie Peg 21h56'13"24d7'
Mackey,Shawna And 0h15'0"38d35'
Mackey,Shelagh Lyr 18h23'52"37d47'
Mackie,Doug Dra 14h54'48"60d26'
Mackie,Laura Marie Vul 20h3'49"23d43'
Mackie,Mo Ghaoil Catriona Cas 2h32'35"61d32'
Mackie,Sue Lyr 18h32'33"43d12'
Mackie,William Lac 22h9'39"49d17'
Mackin,Clifford & Georgina Dra 12h40'0"71d22'
Mackin,Kenneth John Sco 16h21'41"-30d53'
Mackin,Mary Crb 16h18'24"27d16'
Mackin,Sr,Peter J Per 1h54'25"56d43'
MacKinen,Lily Sylvia Peg 22h44'0"5d27'
MacKinley Aur 5h4'60"40d24'
MacKintosh,Christian David Boo 14h59'55"25d55'
MacKintosh,Paige,Renae Lac 22h23'36"54d20'
Mackius,Ken Aql 18h56'32"-5d10'
MACKIW Boo 14h8'60"39d47'
Mackley,Janet Louise Cam 6h7'16"68d35'
Mackley,Meme Sge 19h18'0"16d19'
Macklin,Christopher Bear Per 2h36'10"45d30'
Macklin,John Joseph Cep 22h19'25"61d42'
Macklin,Nina Aql 19h3'27"1d19'
Macklin-Broad,Allan Thomas Dra 14h45'47"64d30'
MacKnicki,Jessie Umi 15h13'46"67d14'
Macko Dra 14h41'59"64d50'
Mackrell,James Cnv 13h21'20"28d49'
Macksanna And 2h22'26"47d29'
Macku Eri 3h29'59"-8d25'
Maclachlan Uma 10h9'0"48d27'
MacLachlan,John Douglas Cam 14h3'54"80d45'
MacLaren,Anthony Merrill Tri 2h27'42"30d37'
MacLean's Star,Cameron Cam 5h54'13"61d1'
MacLean,John Patrick Lac 22h6'10"48d52'
MacLean,Katherine Michelle Hya 8h11'15"2d49'B
MacLean,Margaret Harris Hya 8h11'15"2d49'A
Maclean,Mary Weeks Cam 3h24'56"55d27'
MacLean,Suzanne Lmi 10h50'46"32d55'
MacLellan 45th Anniversary Lmi 10h35'54"25d29'
MacLellan,George & Myrtle Crb 15h34'21"31d11'
MacLellan,Jean Cas 1h38'57"72d4'
MacLellan-Schwartz, Louise De Marillac And 0h15'1"45d46'
MacLeod Dra 19h26'47"80d13'
MacLeod,Allan Edward Uma 11h49'46"46d37'
MacLeod,David Lawrence Tri 1h44'43"30d39'
MacLeod,Heather Gem 7h12'37"31d3'
Macleod,John Alexander Boo 14h29'53"21d35'
MacLeod,Joyce Grob Eri 3h43'46"-2d20'
MacLeod,Kevin Bruce Eri 4h11'0"-15d24'
MacLeod,Lorraine Vir 11h51'33"6d24'
Macleod,Marge Tau 4h18'1"18d59'
MacLeod,Scott Braden Uma 11h31'45"30d7'
MacLeod,Stephen Sgr 18h4'25"-28d41'
MacLeod,William Lib 15h16'44"-23d26'
Maclin,Ernest Dra 13h55'1"63d21'
Maclot Denis Aql 20h22'31"1d13'
Maclyn 21470 Her 17h32'46"42d8'
MacMahon,Elaine Eri 4h29'56"-11d60'
MacMartin,Douglas Cam 5h55'0"68d23'
MacMillan's Fire Aur 6h53'1"37d35'
MacMillan,Andrew Charles Aur 5h3'0"42d38'
MacMillan,Bill Alan Aur 6h26'31"31d51'
MacMillan,Grace Ethelwynne Cas 0h48'20"70d50'
MacMillan,Laura Cas 1h9'15"70d30'
MacMillan-Scott,Flora Cyg 20h20'28"38d42'
MacMonagle,Pat & Gerri Cyg 19h25'28"30d20'
MacNamara,Ian Boo 14h54'15"23d49'
MacNaught,James S Lac 22h21'56"55d30'
Macnee,Patrick & Baba Cyg 21h2'50"38d57'
MacNeil,Donalh Uma 8h43'55"60d41'
Macneil,Neil & Frances Peg 22h4'39"26d10'
MacNeil,Sharon Rosemary Lyr 19h25'19"42d58'
MacNeill,Rebekah Kathleen Lyr 19h6'0"37d53'
Macnow,J M H Crb 16h1'47"32d3'
MacNutt,Robert Per 2h53'23"40d26'
Macomber,Laura Vul 19h59'53"28d53'
Macomber,Trevor Cep 23h0'1"65d22'
Macon,Karen M M Cet 1h25'44"-1d46'
Macondray,Alex Ari 1h46'1"10d28'
Maconi,Alyssa "Alabama" Uma 10h25'33"59d45'
Macosajual Mon 7h57'35"-6d18'
Macphail,Colette Dra 16h51'30"68d32'
MacPhail,Donald J Dra 16h10'28"64d31'
MacPherson's,The Lyn 9h13'35"36d43'
MacPherson,Allan E Per 3h5'25"50d32'
Macpherson,Don Cma 6h55'53"-18d50'
MacPherson,Donald Cnc 8h12'56"32d30'
MacPherson,Edna Cas 23h32'27"54d5'
MacPherson,Joan Cnc 8h13'51"30d29'
Macpherson,Jody Hya 9h3'25"5d51'
MacPherson,Karen And 23h1'43"50d18'
Macpherson,Mark Aur 4h49'37"40d56'
MacQuarrie,Rosanne Cas 0h34'10"60d9'
MacRae,Gary J Pho 23h42'59"-48d57'
MacRae,Samantha Jean And 0h20'0"30d35'
Macri (Little Mac), Richard Her 15h50'23"42d5'
MacRobbie,Martha & Jim Aur 6h55'30"37d24'
Macron,Jeanne Patricia Umi 17h1'28"84d54'
MacSoft Cma 6h55'19"-15d40'
MacThréinfhir,Marial And 0h15'28"39d49'
MacVay,Kerry Lyn 8h27'14"41d29'
MacVicar,Richard Ian Umi 14h34'58"67d45'
MacWilliams,William Ori 5h31'54"-1d34'
Macy 45,Mary Aqr 20h56'17"-0d54'
Macy,Deborah Jon Cam 4h23'41"61d30'
Macy,Mike Ori 5h4'39"8d56'
Macy,Timothy John Dra 11h46'35"70d14'
Mad Max Her 17h27'55"42d19'
Mad-Mar Uma 11h5'47"30d13'
Madagan,Dr Nigel George Aql 18h59'1"17d39'
Madaio,Jerry Cmi 7h41'42"0d47'
Madalyn & York 25 Lyn 8h8'24"34d21'
Madamau Ant 10h44'20"-31d41'
Madasa Ori 6h8'16"1d13'
Madaschi,Danilo Cap 21h22'23"-16d3'
Maday,Matthew Ari 2h48'38"30d34'
Madckie Uma 8h33'1"52d42'
Madcurnier,Gerard Her 17h26'46"40d28'
Maddalena Cam 6h10'1"68d1'
Maddalone,Deborah Athenas Cas 2h0'29"61d28'
Madden "Little Bear", Brendan Barrett Umi 15h35'56"66d41'
Madden & Olivia,Jack Cyg 20h2'29"39d57'
Madden Family Star,The And 1h51'34"40d60'
Madden,Ashley Patricia Cas 0h33'30"63d9'
Madden,Caitlin M Eri 3h38'0"-14d3'
Madden,Carol V Umi 15h50'0"74d24'
Madden,Cathy And 23h41'55"37d53'
Madden,Claire A Eri 3h15'39"-12d2'
Madden,Cynthia Uma 9h18'58"44d30'
Madden,Darrian Edward Cep 22h48'20"70d34'
Madden,Diane Kathleen Vul 19h46'0"28d35'
Madden,Dolores A Dra 17h39'12"68d6'
Madden,James D Ser 18h53'44"2d35'
Madden,Jane Her 17h35'25"18d49'
Madden,Jane J Her 17h34'42"14d59'
Madden,Jane LeLaurin Del 20h18'26"9d28'
Madden,John Sge 19h58'51"20d1'
Madden,Phil Cet 1h28'34"-3d53'
Madden,Phyllis Com 13h7'1"22d29'
Madden,Ryan Aur 6h57'27"38d51'
Madden,Sammantha Lyr 18h39'16"36d9'
Madden,Sara Elizabeth Eri 2h51'30" 2d5'
Madden,Sean Her 16h52'26"34d41'
Maddi Lynn Ser 17h33'17"1d1'B
Maddie & Katie Peg 23h33'18"31d46'
Maddison,Tom Her 16h12'0"14d60'
Maddix,Louise Uma 10h40'38"47d38'
Maddock,Greg Hya 8h27'0"-1d38'
Maddock,Greg Equ 21h21'33"12d42'
Maddock,Tara Lee Mon 7h1'15"8d38'
Maddocks,Robert D Aql 18h44'26"10d2'
Maddox Richard Pagga Oph 18h20'12"8d57'
Maddox,John Equ 21h20'25"12d11'
Maddox,LauraBeth Lyr 18h58'49"30d16'
Maddox,Sophie Louise Lyr 18h30'0"37d33'
Maddsion Rosie Peg 23h30'30"18d7'
Maddux 06011961-070519 94,Darwin Hughes Ori 5h52'56"16d23'
Maddux,John William Hya 8h12'16"5d46'
Maddy Lep 4h58'50"-18d51'
Maddy I Eri 2h54'51"-14d0'
Maddy's Cyg 19h33'31"34d10'
Made Balbat Uma 8h39'1"68d25'
Madeja,Glen Joseph Sct 18h51'21"-6d43'
Madelaine Cam 6h22'32"70d33'A
Madelaine Claire Cas 0h23'18"60d20'
Madeleine Mon 6h40'48"11d21'
Madeleine And 0h52'21"41d0'
Madeleine And 0h54'53"45d49'
Madeleine Chez Les Mayas Aql 18h47'42"10d54'
Madeleine Elizabeth Cas 0h54'37"65d40'
Madeline Cyg 21h21'1"40d33'
Madeline Sge 19h8'14"16d29'
Madeline Cnc 8h50'57"31d11'
Madeline Alice Cru 12h3'56"-57d58'
Madeline Florence Lyr 18h30'13"36d31'
Madeline Zara And 1h13'44"39d37'
Madeline's Ray Lyn 7h6'23"50d33'
Madelon Of Sonoma And 0h59'27"40d2'
Madelyn Vul 19h2'1"21d51'
Madelyn Com 13h31'21"15d44'
Madelyn & Homer Eri 2h58'59"-17d19'
Madelyn Virginia Lyr 18h56'53"41d1'
Madelyn's Eternal Love Sge 20h2'15"20d6'
Mademoiselle And 0h30'50"40d37'
Madenkönig Leo 10h2'53"18d44'
Madenski,Jack A Dra 20h2'37"64d18'
Mader,Charles Wiggins Hya 8h44'57"0d49'
Mader,Dr Josef Cnv 14h0'42"39d9'
Mader,Harald Umi 13h51'51"74d9'
Mader,Heinz Cam 6h59'34"70d51'
Madera,Bartholomew Louis Aql 19h2'51"15d6'
Madewell,William D Sct 18h55'51" 5d13'
Madey,Myles Lac 22h12'25"46d8'
Madey,Tyler Dra 19h31'53"65d15'
Madgar,Donna Jean Marie And 0h34'51"40d15'
Madi Ori 5h34'0"-2d23'
Madigan,Mary Aur 6h10'46"37d54'
Madigou,Xavier & Roland Cra 18h32'40"-43d25'
Madikatzi,Brigitte Mon 7h49'39"-9d15'
Madinina Uma 12h25'39"62d7'
Madio,Paul Lib 15h31'22"-28d15'
Madison Sex 9h54'0"-1d27'
Madison Alexandra Eri 3h0'45"-13d45'
Madison Austin's Magic Ori 5h27'33"1d13'
Madison Mae Aql 18h44'45"-2d32'
Madison Nemesis Her 16h12'46"41d22'
Madison Scott Mon 6h50'21"11d17'
Madison Zöe Tri 1h59'46"31d33'
Madison,Bonita W Vul 19h46'25"28d19'
Madison,Chelsea Vir 15h4'1"6d21'
Madison,Jennifer Crb 15h15'34"30d48'
Madison,Lisa J Cas 0h56'36"64d10'
Madison,Rachael Tau 5h46'26"23d7'
Madison,Rebecca Anne And 2h25'48"48d16'
Madison,Tracy Lee Cma 6h50'44"-19d3'
Madkap Dazzler Lyn 6h37'23"54d50'
Madland Star,The Larry Cep 22h50'25"70d21'
Madlinger,Adam David Lac 22h22'35"52d31'
Madoka Nor 16h18'29"54d31'
Madole,Merideth Gay Aql 19h15'58"15d4'
Madonia,Jacqueline Lib 15h42'28"-22d19'
Madonna Leo 9h21'16"28d4'
Madonna C Cas 23h35'50"61d49'
Madonna,Brian Aur 6h18'10"31d52'
Madoogali Dra 11h28'1"68d7'
Madorma,Camille And 0h58'0"39d31'
Madrid 091570,Michael Allen Her 16h1'50"20d20'
Madrid,Rachel Oph 18h1'34"8d8'
Madrigal,Bianca Marcel Cas 0h32'58"61d16'
Madsen,Beth Uma 11h55'24"63d28'
Madsen,Dana Earl Cam 3h50'26"63d17'
Madsen,Deborah Ann Aur 5h58'0"54d35'
Madsen,Gracie Ivaline Hardee Lyr 18h36'39"30d9'
Madsen,Michael Hya 10h42'19"-17d39'
Madsen,Wayne Cep 22h43'0"68d30'
Madson,Daniel James Her 16h18'39"24d15'
Madson,Jonathan Gardner Lac 22h21'32"53d59'
Madson,Manta Damian Uma 11h47'25"38d17'
Madston Family(C+C+T) Peg 21h54'57"28d19'
Madura,Robert Boo 15h46'17"48d49'
Madzik,Paul Sct 18h56'22"-4d29'
Mae Ori 5h23'46"-0d57'
Mae 121558 Com 13h24'44"25d44'
Mae 21 Crb 16h21'43"37d60'
Mae Yee Umi 16h43'40"76d59'
Mae,John Cet 0h29'38"-0d7'
Maebao Gem 7h3'3"21d34'
Maechtlen Her 16h26'24"18d54'
Maeck,"Wilmac II" for William J Oph 18h15'44"0d5'
Maeck,Elizabeth Cam 3h40'45"71d53'
Maed Lac 22h16'40"38d13'
Maedel,Reinhard Cmi 7h19'42"4d6'
Maeder,André E Uma 11h57'1"32d10'
Maeder,Julian And 0h18'19"31d18'
Maeder,Sophie Louesa Vir 14h33'52"7d11'
Maeder,Ursula Tau 4h18'53"21d55'
Maenhout,John Joseph Dra 18h12'12"70d38'
Maenza,Frank Ori 5h30'57"-3d20'
Maerten,Flora Ori 6h5'14"10d36'
Maerzendorfer,Andreas Cap 21h24'36"-14d14'
Maes,Justyn Umi 13h58'0"70d36'
Maes,Phyllis Gast Peg 22h28'26"21d43'
Maese,Molly Chavira Mon 6h59'13"-10d56'
Maestas May 1st 1972, Christopher A Aql 19h2'35"-0d24'
Maestranzi,Lauren Peg 22h25'43"29d12'
Maestro Vul 19h15'37"24d24'
Maestro,Jose Martin Oph 17h58'23"10d2'
Maetoogi Cet 2h31'12"-10d36'
Maeurer,Melinda Vul 19h23'27"22d59'
Maeva Cyg 20h48'59"38d18'
Maeva,Laura Per 1h44'18"53d8'
Maeve Eugenia Aql 20h10'1"0d30'
Maezumi,Hakuyu Taizan Sex 9h51'42"0d17'
Maf Boo 14h10'54"50d9'
Maffei,Christopher Nicholas Boo 14h7'15"47d59'
Maffei,Nicholas James Lac 22h2'16"40d2'
Maffeo,Nicholas Aur 6h15'34"45d51'
Maffessanti,Nancy Per 2h12'19"58d37'
Mafra Lyn 7h54'32"40d0'
Mag Amicitiae Immortales Cas 1h2'41"58d51'
Mag,Connie Lyn 8h0'22"50d23'
Maga,Marsha And 23h48'52"40d51'
Magaddino,Tony Cet 1h59'0"-2d51'
Magaletta,Rose Eri 4h5'31"-2d16'
Magali Peg 22h59'50"32d39'
Magali P T Bright Eyes Peg 22h7'48"3d39'
Magalie Aur 5h38'59"50d34'
Magalnick,Steven Allan Dra 9h46'34"73d13'
Magaly Vianna Sco 16h54'39"-40d41'
Magan Nor 16h22'49"-59d23'
Magan,Mary E And 1h33'0"37d59'
Magana,Antonio D Aur 6h6'42"32d36'
Magar,Linda E Mon 6h22'0"-1d0'
Magari,Claude Gignac alias Anthony Lac 22h54'41"38d54'
Magarity,Jackie Vul 19h22'36"25d24'
Magarity,William Cet 2h37'41"-4d52'
Magathan,Jr,Gen & Mrs Wallace C Ori 5h49'50"18d59'
Magau Tri 1h46'30"30d32'
Magau Aql 20h5'1"1d52'
Magaw,Michael Cam 4h58'55"70d25'
Magaziner,Badiene Sge 19h0'46"20d9'
Magda Ferres i Ainsua Cyg 21h51'39"52d46'
Magda M Uma 13h46'50"48d21'
Magdahl,Dirk Her 16h40'39"23d42'
Magdalene And 0h23'52"36d25'
Magdalene Umi 16h57'24"75d5'
Magdanz,Carla Jane And 1h58'16"38d18'
Magdanz,Jane And 23h49'49"44d10'
Magdelena,Mercedes Ori 5h54'48"16d6'
Magdolene Aql 19h45'0"14d42'
Magdziarz,Raymond Oph 18h24'50"7d50'
Magee,Allison And 23h21'12"50d0'
Magee,Margaret Del 20h20'33"10d60'
Magee,Rocky & Heidi Umi 13h19'56"71d30'
Magee-Milbourne, Brandon Her 17h27'38"20d30'
Magellan Cyg 20h34'20"37d57'
Magellan Boo 14h1'44"16d53'
Magellan,William Lac 22h13'26"51d22'
Magelssen,Gladys Consuelo Aql 19h23'35"14d25'
Magennis,Katherine & James Crb 16h2'22"33d55'
Mager,Charlotte Aql 19h58'0"0d15'
Mager,David Cnv 12h48'51"38d15'
Mager,Marcus 21/11/81 Boo 13h58'39"10d35'
Mager,The General- Judge Gerald Aql 20h30'10"0d10'
Magersuppe Com 12h7'58"22d32'
Mages,Theresa & Nicole Del 20h57'36"10d2'
Maggard,Tamara Blair And 23h32'55"45d48'
Maggi,John & Cecilia Cyg 19h51'19"41d7'
Maggie Umi 15h52'54"74d33'
Maggie And 0h5'41"41d4'
Maggie Cyg 20h3'58"40d56'
Maggie Peg 22h32'43"30d32'B
Maggie Lyr 18h44'59"41d31'
Maggie Cas 23h42'43"62d39'
Maggie Gem 6h59'15"14d10'
Maggie Lac 22h34'1"56d40'
Maggie Cam 3h54'1"72d36'
Maggie Tri 2h13'38"33d53'
Maggie Ori 6h5'26"0d29'
Maggie Cyg 21h2'37"38d39'
Maggie G Cas 23h23'1"53d50'
Maggie May Lyr 18h53'12"33d21'
Maggie May (Titch) Umi 14h34'49"70d12'
Maggie May B Lyr 18h32'38"33d12'
Maggie Sarah Uma 11h34'12"30d43'
Maggie's Magic Com 13h13'17"20d14'
Maggie's Magnificent Loving Magic Cas 0h19'43"61d10'
Maggie's World Eri 4h53'32"-6d34'
Maggiotto,Donna Vul 18h55'58"24d47'
Maggipinto,Maria Cyg 21h27'57"37d44'
Maggs Her 17h56'47"14d38'
Maggs,Steven Michael Cep 1h2'17"78d22'
Maghakian,Erika Ann Cas 2h9'28"59d18'
Maghakian,Kevin John Lac 22h11'1"47d58'
Maghavi Peg 22h22'44"2d24'
Magher,Jackie Aql 19h54'42"14d19'
Magi Cyg 19h47'18"30d9'
Magia Del 20h15'35"15d26'
Magian Prince Cam 3h32'52"60d52'
Magic Uma 9h55'11"52d26'
Magic Dra 19h45'43"67d36'

Magic
Peg 21h24'32"18d56'
Magic
Cas 1h38'43"75d25'
Magic & Teddy's Star of Love
Gem 6h49'25"12d2'
Magic Andy
Psc 1h0'31"30d9'
Magic Man
Lyn 8h28'35"33d52'
Magic Marc
Aql 19h29'35"-7d54'
Magic Marc Anthony
Hya 9h53'29"-15d16'
Magic Milton
Cep 23h20'59"70d50'
Magic Star,The
Ori 6h2'21"8d17'
Magic Stone
Pyx 8h50'21"-24d33'
Magica
Psa 22h3'23"-28d35'
Magica Luce
Col 6h29'51"-33d57'
Magical Light of Tim, The
Ori 4h55'40"5d40'
Magical Tia
Aql 19h25'58"10d31'
Magicfairy
Sco 17h51'36"-30d36'
Magick Ariel
Cam 9h10'16"80d43'
MagicMan
Peg 23h22'1"11d30'
Magid,Kenneth S
Oph 17h53'24"-8d58'
Magid,Marc
Oph 18h17'19"1d45'
Magid,Selina Rae
Equ 21h5'0"3d21'
Magie
Gem 7h23'44"35d14'
Magiera,Hannah Elizabeth
Uma 11h33'41"30d20'
Magill,Jr,Walter James
Boo 14h10'57"46d28'
Magillism
Crt 11h7'23"-17d43'
Maginity,Keith E
Cet 0h3'16"-10d60'
Maginnis,Dr Maria C
Oph 18h16'36"0d2'
Maginski,Walter
Cep 0h50'48"78d32'
Magister Bernhard Weihs
Cnc 8h29'44"8d45'
Magister Huhglings Himwich
Per 3h14'1"41d47'
Magistra,Clare
Lyr 18h57'54"37d50'
Maglia,Louis Henry
Her 18h53'16"18d49'
Maglich,Michael G
Lib 15h3'1"-6d3'
Maglio Family Star,The
Uma 11h46'39"48d35'
Magliocco,Joseph Michael
Sct 18h55'43"-4d18'
Maglione"Faster Lion", Teresa
Leo 11h3'29"23d23'
Maglouyrk,Heather
Peg 22h0'1"10d19'
Magluilo,Tom"Dog"
Lyr 18h57'20"31d27'
Magnan
Ori 6h0'28"7d6'
Magnan,Mireille
Cet 2h59'41"2d17'
Magnani,Randy
Aur 6h0'1"30d37'
Magnaud,Maryse
Ori 5h57'20"20d10'
Magne,Nick
Her 16h42'48"21d18'
Magner,Tony
Her 17h22'32"38d21'
Magness,Addison Lee
Sct 18h53'9"-6d48'
Magness,Carole Suzanne
Cma 6h53'44"-18d51'
Magness,Evan Turner
Oph 17h18'1"10d54'
Magness,Konrad
Tri 1h56'56"26d12'
Magness,Lisa
Mon 7h15'25"-1d26'
Magness,Marcia L
Lyr 19h3'39"40d24'
Magness,Sallyann Camille
Del 20h39'1"10d25'
Magni
Cet 0h24'0"-10d43'
Magni,Giancarlo
Boo 15h5'38"41d52'
Magni,Robert David
Lac 22h43'1"53d24'
Magnier,Jean Louis
Lyn 9h5'56"37d30'
Magnificatus,Elainus
Equ 21h23'50"8d5'
Magnificence of Ray, The
Mon 7h14'22"-10d14'
Magnificent J R C
Uma 9h19'0"44d41'
Magnificent Max
Hya 9h8'0"1d14'
Magnificent Mikel
Aur 6h11'46"37d56'
Magnificent Millie
Cnc 8h59'55"17d17'
Magnificent Molly
Cma 6h14'39"-18d53'
Magnificent Moriello, The
Sco 17h51'48"-31d35'
Magnigicent Mark
Aqr 21h42'51"0d28'
Magnin,Vonn R
Ser 18h20'26"-0d44'
Magnum
Per 2h25'37"51d34'
Magnus Amor
Lyn 8h37'36"41d7'
Magnus,Fredrick
Dra 14h0'55"68d58'
Magnus,Teresa
Sex 10h26'54"-0d2'
Magnusom,Richard W
Aur 6h5'36"45d58'
Magnuson's Star
Ser 15h39'1"21d7'
Magnuson,Brad
Boo 14h18'12"35d46'
Magnuson,Lorretta
Cas 0h28'12"58d43'
Magnusson,Carl Eric & Jenni
Aur 7h16'40"37d13'
Magnusson,Gordon David
Cet 2h5'15"4d9'
Magoo,Bonnie Chris
Cyg 20h42'36"47d40'
Magowan,Norman
Cep 21h44'18"55d20'
Magpoc,Mary Ann D
Cyg 20h1'0"31d24'
Magrah
Peg 23h0'29"11d37'
Magrane,Penney
Aql 19h43'27"4d7'B
Magrini,Gus
Hya 8h12'49"-5d54'
Magrit Klein-Wadegotia I
Vir 13h3'12"-6d7'
Magro,Ms Janet
Lmi 10h34'12"32d9'
Magro,Nancy
Del 20h13'0"13d29'
Magruder,Laura
And 23h15'56"50d14'
Magruder,Walt & Barb
Cyg 19h43'23"30d31'
Maguire Star,The
Uma 11h52'55"48d48'
Maguire,Brian Jeffrey
Lac 22h9'17"40d18'
Maguire,Caroline
Lyr 18h37'36"43d15'
Maguire,Charles
Eri 4h0'0"-8d35'
Maguire,Ciara
Cep 0h11'12"73d50'
Maguire,Craig James 17/4/1975
Ari 2h36'43"22d24'
Maguire,Dennis James
Dra 16h6'20"66d13'
Maguire,Eilish Mary born 9/6/1995
Col 6h36'4"-34d15'
Maguire,Jennifer
Lyn 8h11'14"50d6'
Maguire,John
Cnc 8h34'16"7d28'
Maguire,Luke Kane
Aqr 23h28'40"-6d20'
Maguire,Mary Loyola McCarthy
Cam 5h49'46"71d2'
Maguire,Michael
Cet 2h56'47"4d29'
Maguire,Ritzi
Cma 7h12'1"-15d37'
Maguy & Barry
Cyg 21h27'16"38d51'
Magyar
Cam 3h54'60"55d21'
Mah,Corine
Aql 19h44'35"10d58'
Mah-Tat
Uma 9h52'58"71d32'
Mahaffey,Erin Elizabeth
Cyg 19h33'56"32d29'
Mahaffey,Steffon
Sct 18h44'0"-5d23'
Mahal Kita Jocelyn
Peg 22h48'0"29d51'
Mahalo Mrs G With Aloha
Eri 3h59'31"-19d44'
Maham,Lincoln
Cep 22h19'26"78d60'
Mahan,Deborah
Dra 16h56'60"68d35'
Mahan,Joyce & Kerrigan
Cyg 19h16'23"44d8'
Mahan,The Star Of Eric
Boo 15h2'52"28d50'
Mahan-Best Friend & Lover,Sean E
Aur 6h46'49"37d2'
Maharaj
Cyg 19h59'56"31d28'
Maharisha Trisha
Aur 4h47'35"40d49'
Mahatha,Janine
Vul 20h1'25"28d43'
Maher,Arthur
Cam 4h9'0"68d23'
Maher,Brian David
Ori 5h52'26"11d54'
Maher,Brianne Irene
And 0h52'50"34d19'
Maher,Cassandra
Vul 19h59'14"28d54'
Maher,Cecily
And 23h41'15"33d23'
Maher,Emily
Ori 5h21'41"13d3'
Maher,Heather Marie
Cas 2h37'30"73d21'
Maher,Heather Marie
And 2h21'1"42d1'
Maher,Jerritt Philip
Boo 15h5'1"28d14'
Maher,Jr,Kevin Paul
Lyr 18h59'32"45d48'
Maher,Jr,Lawrence A
Lib 15h16'59"-17d43'
Maher,Kevin F
Boo 14h18'0"36d3'
Maher,Larry
Cep 21h6'1"68d22'
Maher,Leonard
Aur 4h35'12"31d28'
Maher,Margaret
Cap 21h1'49"-18d50'
Maher,Mark Terence
Umi 17h22'57"76d48'
Maher,Michael E
Aur 5h0'55"47d20'
Maher,Peter Douglas
Her 16h38'50"34d2'
Maher,Samantha
Com 12h53'54"23d13'
Maher,Summer Scarbrough
Sge 19h54'15"20d34'
Maheux,Hélène
And 0h51'17"39d51'
Mahfouz,Michael Firmin
Cyg 19h43'40"38d27'
Mahlah Naomi
Aql 20h12'30"1d4'
Mahlau,Peter Claus
And 2h3'57"40d22'
Mahle,Joachim
Sge 19h16'37"20d2'
Mahle,Ryan
Her 16h34'25"41d47'
Mahler,Ethel
Peg 22h15'58"33d15'B
Mahler,Louis
Dra 17h1'1"72d48'
Mahler,Paul
Peg 22h15'58"33d15'A
Mahler,Stefan
Sco 17h27'45"-33d55'
Mahler,Steven A
Aql 19h17'56"12d60'
Mahlstede,Robert & Nancy
Uma 9h6'46"51d30'
Mahlstedt,Robert Charles
Ari 2h23'54"12d28'
Mahlstedt,Spencer Patton
Aql 19h57'53"14d20'
Mahmood,Susan A
Lyn 8h3'12"50d13'
Mahmoud Abu-Qudais
Equ 21h6'46"3d2'
Mahmoud,Maryam A
And 23h17'57"51d10'
Mahn,Gertrude I
Cyg 21h32'0"41d45'
Mahnich,Christine Marie
Ori 6h4'15"8d43'
Mahnke,Doug
Aur 6h29'30"33d10'
Mahnken,David Michael
Sco 16h54'28"-43d14'
Mahnken,David William
Per 2h53'13"35d47'
Mahnken,John"Faithful"
Per 2h3'0"56d30'
Mahnkopf,Monika
Uma 11h4'10"53d13'
Mahon,Deborah Ann
And 22h57'14"38d41'
Mahon,Margaret
Cas 1h26'17"74d33'
Mahon,Shawna Lynn
Peg 22h7'0"24d19'
Mahon,Sr,James F
Boo 15h5'45"12d48'
Mahon,Susan
And 1h51'23"39d39'
Mahon,The
Per 3h27'46"52d31'
Mahoney OSA,Father Thomas M
Hya 8h18'24"3d21'
Mahoney's On Cotton Hill
Dra 11h15'45"74d1'
Mahoney's World
Cyg 21h54'34"37d6'
Mahoney,566-33-2401, Menos Leslie
Gem 7h27'55"30d37'
Mahoney,Anna
Oph 17h6'53"-22d11'
Mahoney,Anne Marie
Cyg 20h24'52"38d40'
Mahoney,Brenna B
Lac 22h25'41"41d8'
Mahoney,Connor Patrick
Lac 22h35'1"50d18'
Mahoney,Darcy Shannon
Cet 2h41'31"4d32'
Mahoney,J R
Boo 15h6'26"19d55'
Mahoney,Janice
And 1h27'32"41d13'
Mahoney,Jean Thorp
Cam 4h14'45"60d16'
Mahoney,John Gary "Bone"
Aql 18h57'0"-6d1'
Mahoney,June M
Crt 11h12'24"-13d28'
Mahoney,Linda Marie
Aql 19h1'46"0d1'
Mahoney,Matt
Uma 13h12'53"62d11'
Mahoney,Matthew C
Cep 20h33'29"62d60'
Mahoney,Matthew Walter
Boo 14h49'54"32d27'
Mahoney,Richard
Cet 3h2'58"6d4'
Mahoney,Sean
Cmi 7h54'23"1d2'
Mahoney,Shane
Hya 9h5'55"3d54'
Mahoney,Thomas Gary
Boo 14h41'30"31d3'
Mahony,Daniel & Diana
Uma 9h6'46"51d30'
Mahony,Nathan Richard
Sge 19h52'33"19d6'
Mahood,Alexis Glenfield H
Cyg 21h1'37"33d35'
Mahr
Aql 19h53'57"10d28'
Mahr,Erik
Cet 0h45'14"-3d50'
Mahr,Ludwig
Ori 5h57'20"19d37'
Mahsus,Akdamar Bazisey
Uma 8h29'27"61d12'
Mai
Cyg 20h28'56"31d43'
Mai,Joyce Cynthia
Uma 12h8'43"63d1'
Mai,Louise Elizabeth
Cyg 19h27'19"35d49'
Mai,Margaret
Mon 6h29'38"-8d6'
Mai,My Dearest Friend
Sgr 18h54'57"-34d36'
Maia
Cep 1h36'52"78d25'
Maia Marie
Peg 22h4'11"10d54'
Maia Samantha
Cas 1h26'25"77d22'
Maia,Lorna
Lyr 18h37'56"39d57'
Maibaum,Kayla
Mon 6h56'59"10d1'
Maidhof,Angelique Farrah
Cas 0h57'54"66d6'
Maids,The
And 2h23'0"42d34'
Maiello,Dominik
Lib 15h3'1"-0d32'
Maiengruen
Her 17h27'50"42d42'
Maier,Bryan
Cmi 7h22'10"0d36'
Maier,Christian
Ori 5h5'35"13d21'
Maier,Dr Ferdinand
Vir 13h30'7"-3d9'
Maier,Howard Simon
Per 4h2'24"37d52'
Maier,Jason Robert
Hya 8h43'55"0d42'
Maier,Kathleen
Cnv 13h57'34"41d45'
Maier,Kurt William
Aur 5h0'0"29d57'
Maier,Margaret Immaculata Cuomo
Cas 0h43'42"60d11'
Maier,Millicent
Cam 9h16'12"77d32'
Maier,Reimund
Lmi 9h27'26"38d34'
Maier,Roman
Sge 19h5'55"18d20'
Maier-Stoll,Gerda
Vir 13h29'55"-11d30'
Maiers Twinkle Star, Olivia Monet
Mon 6h48'58"2d38'
Maierson A Shining Star, Anne Fell
Peg 22h40'49"22d41'
Maiese,Anne
Cyg 19h30'45"36d34'
Maietta,Darrell
Boo 14h52'43"23d54'
Maietta,Samuel
Uma 10h58'28"40d9'
Maiga
Uma 10h15'12"42d8'
Maigné,Joëlle
Uma 12h3'36"61d36'
Maika
Lyr 19h20'22"42d40'
Maike
Lyr 18h47'1"30d49'
Maiken
Crb 15h27'23"30d26'
Maile,Helmut
Her 17h37'24"50d15'
Mailender,Kevin
Tri 2h15'38"33d55'
Mailey,John W.
Lyn 7h27'16"50d22'
Mailfert,Anne Laure
Lyr 18h50'28"38d57'
Mailfert,Martha
Lyr 18h59'16"38d31'
Maillard 95,Valerié
Cas 0h34'43"74d5'
Maillho,Alexander C
Dra 9h57'12"80d30'
Maillho,Jr,Alexander C
Hya 9h38'58"-10d26'
Maiman,Brooke Kristine
Cyg 20h40'56"42d31'
Maimes Happy Anniv, Dave & Debbie
Cyg 21h14'45"35d58'
Maimes,Ashley Faye
Aql 19h6'14"1d57'
Maimes,Danna
Aql 19h2'58"2d54'
Maimes,Jack
Cet 0h57'49"0d28'
Maimes,Michael
Aql 19h1'49"3d11'
Maimone,Danny
Ori 5h17'46"-9d30'
Main,Ian C
Cnv 12h24'15"36d13'
Main,Jeri Lynn
Boo 14h21'54"29d40'
Main,Nancy & Tony
Peg 23h8'21"14d25'
Mainardi,Manuela
Dra 9h27'0"80d40'
Maine,Brant Addison
Aur 4h57'24"38d6'
Mainey,Mary Ann
Cas 0h8'0"50d14'
Maini,Mina Kumari
Lyn 9h0'0"37d45'
Mainio,Mary
Cam 4h5'58"58d19'
Mainio,Michelle
Cas 0h26'48"60d53'
Mainone,Wilhelm Dietger
Aur 5h16'40"43d3'
Mains,Louise & Alan
Lyn 8h22'24"46d43'
Mainz-Carusone Light of Love 1995
Eri 2h47'40"-3d4'
Mainzelmännchen
Cep 22h9'27"60d15'
Mainzer,Firefighter John (Jack)
Per 2h7'0"57d27'
Maiolo Family Star,The
Sge 19h29'59"16d34'
Maioriello 4th,Joseph
Her 18h16'24"14d54'
Maioriello,Ann
And 2h33'10"49d50'
Maiorka,Mary
Mon 7h32'30"-0d45'
Maiparamuake
Sct 18h41'21"-7d47'
Mair,Edith
Sco 17h20'28"-38d5'
Mair,In Honor of Margaret R & Ross C
Peg 21h58'39"3d11'
Mair,Jane
Cyg 21h52'45"38d48'
Mair,Klaus
Gem 7h12'44"21d9'
Mair,Lynne Allison
Boo 15h26'39"48d52'
Maire,Bruno
Cep 21h27'14"60d24'
Mairead,Alainn
Aql 19h2'1"1d27'
Mairéad
Sco 17h20'55"-31d6'
Maisel,Bernd
Sgr 18h47'34"-22d52'
Maisie
Cma 6h14'47"-29d37'
Maisie Florence
Her 17h4'1"20d46'
Maislin,Sydney
Ori 5h51'0"20d6'
Maister,David Hilton
Her 18h0'32"40d4'
Maistrellis,Maria
Aql 20h35'15"0d37'
Maistrellis,Phillip
Cep 21h48'21"68d6'
Maith,Michael
Hya 8h46'60"-8d13'
Maitre
Cyg 19h50'28"44d47'
Maiuri,Lucia
Vel 9h43'3"-49d46'
Maiwand,Soraya
Cas 0h58'31"70d57'
Maja
Leo 10h55'41"20d16'
Majak
Vel 9h43'12"54d51'
Majansky,Linda
Peg 23h48'15"27d38'
Majcher,Susan Maria
Com 12h53'50"27d43'
Majd,Megan & Kam
Eri 2h48'0"-3d6'
Majeed,Abdul Majeed Mamdoo Abdul
Eri 3h44'14"-5d12'
Majella's Star
Umi 15h32'19"86d14'
Majer,Joann Marie
Ori 5h57'42"16d27'
Majerle's Star
Cep 23h6'24"64d49'
Majerle,Karl & Rebecca
Uma 15h4'12"69d7'
Majeski,Jr,Joseph S
Her 17h41'30"14d26'
Majeski,Michelle
Cas 1h8'0"61d42'
Majestic Kimberjus,The
Lyn 7h36'15"51d16'
Majestic Mhyles
Cyg 19h35'34"28d53'
Majestic Michael
Aql 19h26'23"-0d2'
Majestic Michelle! Love,Gary
Mon 7h1'18"4d3'
Majestic Star,The William
Dra 9h22'34"74d19'
Majestika,Azura
Aql 19h8'51"1d3'
Majich,Leo Anthony
Aql 18h59'57"10d24'
Majkowski,Majik & Don
Cep 20h30'29"76d27'
Majlak,Theresa Marie Rossi
Cam 5h58'20"70d37'
Majoarp
Peg 23h31'22"11d14'
Major (La Majorienne), Marcel
Uma 10h3'20"53d55'
Major Chill
Leo 11h31'28"-0d19'
Major Rut
Ori 4h57'30"-0d13'
Major Tom Baba
Aql 18h54'12"11d40'
Major,"Bonze" Ralph
Uma 9h33'24"48d45'
Major,Albert
Boo 14h51'13"26d28
Major,Berlian
Aql 18h58'44"17d23'
Major,Brittany Lynn
Cas 1h44'16"61d16'
Major,Cameron Michael
Cnc 9h5'15"28d3'
Major,Carol
Cyg 19h56'21"37d52'
Major,Jacqueline Doris
Cas 0h25'24"69d38'
Major,Jordan Alexander
Peg 22h33'45"30d10'
Major,Jr,John Edward
Tri 2h39'30"31d17'
Major,Juanita Mary
Lyn 7h48'26"39d39'
Major,Mariette
Uma 9h45'26"49d30'
Major,Scott & Kenia
Uma 9h51'50"67d39'
Major,Tabitha
Tau 5h19'28"16d49'
Major,Toni R Barron
Mon 7h31'50"-10d37'
Major,Trudith
Cet 0h51'47"-5d35'
Majora,Punkinicus Pieabus
Aql 19h53'45"-1d22'
Majors,Aaron James
Ori 5h28'34"-1d50'
Majors,Joyce
And 2h26'47"50d35'
Majors,Kirk Steven
Her 17h4'17"37d44'
Majors,Sherri L
Cas 23h26'40"58d25'
Majorè,Elizabeth A Parascandolo
Lyr 19h25'27"38d19'
Majsterski Minor
Lmi 11h0'1"33d30'
Majure,Margaret
Cma 6h54'21"-19d37'
MAK & BABS
Cyg 21h50'13"53d16'
Mak,Norma Jean
Com 13h11'20"30d4'A
Mak,Warren K
Com 13h11'20"30d4'B
Makaela Marie
Peg 21h42'34"28d24'
Makara,Mary Sinak
Mon 7h4'58"-0d6'
Makarewicz & Servedios Engagement Star
Crb 15h54'28"38d14'
Makaru 33
Dra 20h22'25"63d19'
Make A Wish Chris Love Sue
Vir 11h59'55"8d56'
Makewell,Derrick Herbert
Per 3h13'58"50d13'
Makewell,Thelma Elizabeth
Crb 15h19'33"31d21'
Makhortov,Denis Alexeevich
Aur 5h13'19"44d37'
Makhortova,Olga Jurjevna
Cas 0h59'40"60d7'
Makhrani,Natoushka- Halyet
Cam 3h34'1"61d57'
Maki,Danika Dionne
Her 18h7'1"30d37'
Maki,Janell Marie
Cyg 21h39'39"40d18'
Maki,Marjatta
Cam 3h58'48"58d24'
Maki,Richard H
Her 15h56'32"46d10'
Maki,Taimi
Lyr 19h16'45"38d16'
Makihara,Noriyuki
Tau 5h27'57"19d44'
Makila
Lac 22h2'27"49d48'
Makin,Holly
Sgr 19h13'18"-25d25'
Makita
Aql 19h53'25"13d52'
Makkai,Sr,John
Cmi 8h9'20"6d6'
Makkapati,Shreya
Leo 10h31'0"15d30'
Makko
Pho 0h4'31"-41d45'
Makkonen,Kelsi Jordan
Sco 17h29'28"-41d13'
Makos,Angela
And 2h27'0"45d20'
Makoski,Betty
Cas 1h2'22"61d29'
Makov,Oren
Cam 13h12'19"82d9'
Makowski,Andrew L
Ori 5h36'22" 1d12'
Makowski,Holly Marie
Cas 0h34'35"60d36'
Makowski,Meagan Elizabeth
Cas 1h27'12"60d45'
Makowsky,Ellen Joy
Boo 13h34'56"8d57'
Makowsky,Reinhard
Sgr 19h1'40"-21d5'
Makowsky,Sharon Mildred
Vul 19h42'0"20d12'
Makowsky,Wyndam Isaac
Aur 7h22'1"43d25'
Makris,Julie
And 1h46'0"39d38'
Makrounis,Christa
Cam 3h19'30"58d16'

Maks,Brian J
Ori 6h6'46"7d54'
Maksimovic,Wilma Spicker
Vul 21h16'28"20d1'
Maksimuk,Michael
Cep 22h18'42"58d10'
Maksoudian,Areg
Cep 20h36'57"61d2'
Maksuta,Raymond T
Lac 22h25'1"53d53'
Maksymkow,Andrea Beth
Mon 8h7'11"-6d19'
Maksymkow,Erin M
Cyg 20h19'26"41d26'
Makuch,Leah Beth
Cas 0h51'31"62d3'
Makuch,Suzanne Vivian
Mon 6h43'19"11d27'
Mal Vsa 36-41A
Cra 18h22'46"-45d10'
Mal-de-Rez
Cmi 8h8'47"2d23'
Malacaria,André & Nancy
Cyg 19h40'41"38d20'
Malacaria,Hector
Dra 19h2'27"47d50'
Malacaria,Hector Marcel
Cru 12h3'17"-61d2'
Malacaria,Hector Vicente
Cru 12h3'51"-60d37'
Malacaria,Liliana Andrea
Cru 12h0'10"-59d46'
Malacaria,Maria Rachel
And 23h33'16"40d56'
Malacaria,Sergio Victor
Cru 12h0'21"-61d37'
Malacaria,Vito & Antonia
Cyg 19h40'53"37d33'
Malacaria-Burns, Alejandra Claudia
Cru 12h1'57"-62d28'
Malacaria-Tolmie,Aiden Joshua Sergio
Cru 12h0'44"-60d59'
Malachowski,William E
Her 16h49'52"37d58'
Malachowsky,Michael William
Cnc 8h54'34"20d34'
Malachuk,Kristin April
Cas 23h4'1"58d13'
Malacina,Matthew
Lyr 19h8'16"38d57'
Malacina,Olga
Lyr 19h8'25"38d6'
Malaika
Umi 15h1'40"67d4'
Malakeh
Aql 19h9'18"15d27'
Malamas,Anne Marie Jolley
Aql 19h55'27"13d33'
Malamas,Katherine Alexis
Del 20h53'45"9d29'
Malamas,Ted
Psc 1h27'36"21d52'
Malanchuck,Maggie
Uma 10h37'27"70d6'
Maland,Marissa Renee
Sct 18h50'36"-8d47'
Malandra
Sco 17h52'47"-30d25'
Malandra
Uma 10h24'0"58d18'
Malandra,Mr & Mrs Frank
Cep 22h38'55"63d37'
Malaney,Cynthia Grunder
Gem 7h2'37"35d24'
Malani's Wishing Star
Lmi 10h0'24"31d0'
Malat,David A & Doris
Eri 2h58'58"-6d45'
Malatich,Nick & Helen
Uma 10h1'58"52d16'
Malatino,Maria Jo
Cyg 19h50'0"37d42'
Malaussena,Jean
Cep 23h21'12"70d19'
Malbin,William
Uma 12h10'20"59d31'
Malbon,Erica Faith
Mon 7h52'25"-8d56'
Malbon,John & Lois
Vul 19h23'20"26d27'
Malbone,Frederick L
Per 2h28'40"57d48'
Malbos,Andre
Tri 1h40'54"28d56'
Malcolm
Tri 2h40'37"34d10'
Malcolm Allen
Dra 17h17'18"65d29'
Malcolm's Smiling Eyes
Ori 6h6'40"2d20'
Malcolm,Alice
And 0h59'20"33d47'
Malcolm,Alison Marie
Lyr 18h27'27"45d27'
Malcolm,Claire & Leigh
Crb 16h14'47"32d46'
Malcolm,Katherine Greer
And 0h10'41"36d10'
Malcolm,Lyndsay
Cas 0h49'19"71d3'
Malcolm,Maggy
Hya 8h10'26"-6d28'
Malcolm,Ronnie R
Her 16h9'12"40d12'
Malcomb,Lorena Lynn
Dra 16h28'59"60d42'
Malcon,Anita & Bruce
Cnv 13h5'36"50d32'
Malcus,Kerstin
Ari 2h39'54"21d13'
Maldonado,Candyce Alexes
Cet 1h44'44"-0d30'
Maldonado,Caroline Brandy
Mon 6h35'36"10d19'
Maldonado,Dana Lynn
Sco 16h49'45"-28d60'
Maldonado,Falyn Star
Sco 17h51'1"-37d40'
Maldonado,J Richard
Ori 5h50'25"15d1'
Maldonado,Tammy
Uma 8h49'42"70d5'
Male Mouse
Sct 18h53'24"-7d25'
Male,Donald Warren
Cep 21h3'51"60d37'
Malec,Cincuenta Martin
Aur 6h19'43"34d57'
Malec,Michael
Her 16h57'39"33d29'
Maleck,Annette Marie
Aqr 23h14'54"-5d22'
Maleck-Whitcley,Karon Louise
Sco 17h29'57"-41d3'
Malecki,Barbara"Wooba"
Aql 20h11'56"13d46'
Malecki,Benjamin Paul
Cep 22h54'0"57d34'
Maleeny,Star of Peace, The Robert
Boo 15h8'26"38d44'
Maleezia Star,D
Boo 15h17'59"50d49'
Malek,Janet
Cas 1h50'0"58d33'
Malek,Paula & Tim
Lac 22h3'1"49d1'
Maler,Paige
And 1h51'22"39d7'
Males,Kim Elizabeth
Cyg 21h15'39"35d32'
Malesieux,Bertrand
Ori 6h5'27"10d16'
Malet,Grégoire
Ori 6h20'35"10d46'
Malet,Suzanne
Lyn 9h4'21"45d11'
Malette,Joey Robert
Cep 23h33'22"64d20'
Malette,Korey Alexandre
Cam 3h54'36"57d25'
Maletzki,Christian
Cep 20h39'25"65d3'
Malfait,Christophe
Uma 12h18'55"60d57'
Malfas,Evangelos
Uma 10h10'47"51d12'
Malfitano,Joseph
Her 15h50'31"48d41'
Malgee,Paul F
Dra 16h8'20"58d25'
Malgren,Jeff & Linda
Aql 20h10'30"12d48'
Malherbe,Cassandre
Lyr 18h56'17"31d39'
Malhotra,Devan James Edward
Dra 18h42'20"68d16'
Malia's Day
And 0h13'19"35d53'
Malia,Steven & Lue Ann
Dra 19h53'42"60d56'
Malian,Debra Kay
Vul 20h0'57"22d40'
Malibu Katie
Mon 6h20'37"3d37'
Malibu Segal
Aql 18h58'50"-6d30'
Malice
Pic 4h32'25"-48d56'
Malick,Loretta Theresa
Lyn 8h6'28"39d31'
Malies,Pamela J
Del 20h26'13"18d59'
Malik "You're the Greatest",George E
Cep 22h29'23"80d13'
Malik,Paula
Dra 17h0'23"61d51'
Malik,Rubina
Lac 22h12'40"38d45'
Maline,Michael
Cep 23h31'42"66d18'
Malinoff,Tony
Boo 14h21'36"32d34'
Malinowski Family,The
Lyr 18h55'46"34d1'
Malinowski,Alexander
Leo 10h52'57"7d58'
Malinowski,Joe
Boo 15h47'51"50d22'
Malinverno,Armand
Lmi 10h47'23"28d24'
Malis,Peter
Gem 6h47'52"19d53'
Malishkevich,Anna
Cyg 21h6'23"31d20'
Malisiak,Stella
Lyn 8h7'21"38d53'
Maliszewski,Beth A
Lyr 18h30'0"31d19'
Maliszewski,John Adam
Lib 15h19'0"-28d44'
Malita
Aur 5h37'42"38d42'
Malizia,Louie
Oph 17h16'3"-20d19'
Malkanthi Shinanthi Lecamwasam
Lyn 8h43'1"40d21'
Malkenson,Carol
Aqr 21h4'42"-5d45'
Malki,Omran
Cam 3h52'27"61d19'
Malkin,Bill
Cep 20h55'40"70d56'
Malkin,Clu ChFC, David B
Her 17h54'49"20d2'
Malkin,Jonathan Scott
Cmi 7h17'44"0d2'
Malkin,Judd
Aur 4h57'40"40d35'
Malkin,Nicole Anne
Lyr 18h25'0"38d27'
Malko,Buddy
Dra 17h7'3"50d54'A
Malko,Theresa
Dra 17h7'3"50d54'B
Malkov,Tigerman Jacky
Lyn 8h33'11"41d18'
Malkuch,Virginia J
Mon 6h30'41"1d52'
Mall,Cyndi
Lyr 18h15'17"43d50'
Mall,Michael und Margarete
Uma 11h41'1"33d26'
Mall,Patty & Paul
Cyg 20h13'17"37d59'
Mallam,Brendan
Equ 21h17'12"9d18'A
Mallam,Clark
Equ 21h17'12"9d18'B
Mallan,Stella John Francis
Cep 5h31'39"85d21'
Mallard,Airey Aultman
Eri 4h54'24"-9d41'
Mallard,Emmett Earl
Cet 0h4'41"-12d16'
Mallard,Jean-Claude
Peg 23h45'46"26d53'
Mallard,Odile & Jerome
Ant 10h39'41"-34d48'
Mallari,Lorilynn
Com 12h18'25"32d1'
Mallas,Ray
Per 4h23'49"51d20'
Mallen,Joe
Lac 22h15'24"37d52'
Malleo,Ann
Com 12h11'54"20d24'
Mallet,Etienne
Sgr 19h26'30"-37d8'
Mallett,Deborah
Lyn 8h26'11"41d4'
Mallett,James K
Cet 1h34'1"-1d41'
Malleus,Elizabeth
Cnv 12h26'36"43d57'
Malley,Charles Patrick
Ori 5h36'38"-0d51'
Malley,Suzanne
Del 20h53'0"7d59'
Mallia,Carol E
Lyn 6h49'44"60d31'
Mallia,Christi Anna
Eri 2h55'13"-2d32'
Mallick,Andrew
Oph 17h36'19"-24d12'
Mallin Star,The Joel & Sherry
Lyn 8h52'17"36d15'
Mallin,Dr Robert
Per 2h26'15"54d38'
Mallinson,Polly Jane
Cgy 20h43'34"45d30'
Mallion,John Richard
Per 3h8'41"37d32'
Mallock & His K S, Robert D
Her 16h51'59"37d55'
Mallon,Patricia
Peg 0h1'1"20d43'
Mallonee,Susan
Lac 22h10'28"54d32'
Mallory,Bobby
Dra 16h39'33"60d54'
Mallory,Cheryle Sue
Peg 22h17'1"8d50'
Mallory,Don & Betty
Peg 22h9'1"2d19'
Mallory,Jennifer
Cas 0h52'24"62d5'
Mallory,M S
Peg 23h46'14"30d20'
Mallory,Ryan
Lac 22h41'32"53d43'
Mallory,Stephen
Lyn 8h10'23"40d43'
Mallory,Trish
And 2h27'41"42d51'
Mallory,Whitney Jo & Weston Dale
Peg 23h21'0"25d50'
Mallow,Elizabeth Rose
And 2h33'23"40d32'
Malloy,Brendan J
Dra 16h38'1"52d7'
Malloy,Derry
Lyn 6h32,49"58d43'
Malloy,Eugene
Per 2h50'53"37d49'
Malloy,Michael T
Aql 20h20'36"7d53'
Malloy,Ronald Egan
Aur 5h29'49"40d6'
Malloy,Tim
Dra 11h46'16"72d16'
Mally,Erin Lee
Lyn 7h2'0"50d59'
Malmberg,Chad
Lac 21h56'36"37d43'
Malmberg,Jessie
Vul 19h58'18"28d28'
Malmed,Dan
Per 2h44'1"40d47'
Malmer's 40th Anniversary Star
Uma 8h55'22"58d3'
Malmston,Sarah Lynn
Cyg 21h4'45"38d45'
Malmstrom,Ivar
Cnv 13h8'54"38d0'
Malnoy,Jean-Marc
Peg 23h45'36"25d13'
Malo,Jean Marc
Ori 6h5'0"8d57'
Malone 1985,Anna-Marie
Lyr 18h39'21"38d54'
Malone Feb 2,1981, Julia Ann
Umi 17h19'53"75d10'
Malone Heavenly Light, Josephine B
Del 20h23'0"20d15'
Malone Jan 6,1982, Daniel James
Umi 14h48'40"77d47'
Malone to Shine Forever,Sara Lore
Cet 2h37'18"4d21'
Malone's Millennium
Her 16h53'23"38d49'
Malone,Barry
Hya 8h14'0"1d8'
Malone,Cory Anderson
Tri 1h35'1"30d19'
Malone,Diana Mancel
Vul 20h15'50"23d2'
Malone,Edith J
Cet 0h56'15"1d13'
Malone,James L P
Cam 3h17'1"61d17'
Malone,Jim
Hya 8h12'59"0d18'
Malone,John
Aql 20h17'0"0d18'
Malone,John (Moe)
Cyg 19h33'1"33d10'
Malone,Josephine Elizabeth
Peg 22h41'59"12d13'
Malone,Joshua Isreal
Lac 22h6'18"46d55'
Malone,Jr,John Franklin
Cep 22h4'28"61d12'
Malone,Laberta
Cet 1h34'51"-2d33'
Malone,Lisa A
Peg 21h41'39"23d41'
Malone,Mary Katherine
Psc 0h53'25"32d20'
Malone,Mary Lou
Umi 14h14'40"69d10'
Malone,Mary Phyllis
Oph 17h34'45"8d50'
Malone,Melanie Bryant
And 1h42'31"48d59'
Malone,My Darlin' Bill W A
Boo 14h20'23"38d13'
Malone,Noelle
Lyn 8h51'16"46d22'
Malone,Patrick C
Cmi 7h18'15"3d27'
Malone,Thomas Patrick
Dra 19h17'39"70d44'
Malone,Wayne Ware
Aql 19h55'55"14d52'
Maloney
Boo 14h12'0"32d16'
Maloney,Anne Frances
Uma 11h58'34"61d55'
Maloney,Chick
Uma 15h23'17"71d49'
Maloney,Dorothy Marie Ford
Vul 20h3'27"28d21'
Maloney,Faith Bowker
Cas 1h11'38"61d2'
Maloney,Gerald Scott
Boo 14h40'35"33d54'
Maloney,Jeremiah
Hya 8h15'27"2d60'
Maloney,Jona Lee L
And 1h43'24"37d53'
Maloney,Julie Danielle L
Com 12h13'51"24d58'
Maloney,Karen Beth
Ori 5h56'20"11d39'
Maloney,Mary Kathleen
Crt 11h16'12"-17d42'
Maloney,Mary Therese Cecelia
Mon 6h19'52"8d45'
Maloney,Miss Margaret
Lyr 18h42'43"32d12'
Maloney,Molly
Mon 6h58'27"10d31'
Maloney,Patty
Eri 3h45'56"-3d53'
Maloney,Samantha Suzanne L
Crb 15h59'25"28d34'
Maloney,Shannon Michael L
Cas 1h56'53"70d40'
Maloney,Susan Marjorie
Lyn 6h32'27"59d51'
Maloney,Sylvia B
Uma 9h50'0"48d58'
Maloney,Tess Danielle
Cas 2h58'47"70d59'
Maloney,Tim
Cmi 7h28'26"0d7'
Malongo
Leo 9h59'13"19d56'
Maloof,Katherine
Cet 0h30'38"-18d39'
Malosse,Franáoise
Equ 21h3'1"8d0'
Malotke,Kris
Uma 11h42'27"52d25'
Malou
Ori 5h53'33"14d43'
Maloustra,Matthew Sean
Lac 22h52'38"55d23'
Maloy Family Star,The Ken
Umi 14h47'0"65d53'
Maloy,Jack
Her 16h35'22"42d16'
Maloy,Lucy
Lyr 18h16'29"37d50'
Malroy,Mark A
Cas 1h32'51"77d7'
Malson,Bryan
Ori 4h58'54"10d33'
Maltais,Félix Cøté
Sgr 18h54'10"-33d33'
Maltais,Gary
Dra 9h42'36"78d21'
Maltais,Silvie
Cyg 19h26'31"30d12'
Maltby,Richard Alan
Aur 7h14'20"41d5'
Maltempi,Camille Justine
And 2h21'1"41d22'
Malter,Werner
Aql 18h56'21"17d50'
Maltese,Christopher John
Lac 22h1'1"51d59'
Maltese,Jerry
Ser 15h15'56"20d27'
Maltese,Peter & Debbie
Cet 3h2'34"2d12'
Maltson,Thomas Ward
Ori 5h43'0"10d17'
Maluone-Carire
Uma 11h51'18"45d7'
Maluto,Francine
Per 4h5'49"36d32'
Malvasia
Tri 1h57'19"31d49'
Malveaux-Freeman, Brenda
Cyg 19h33'1"32d45'
Malvick,Danyelle Renee Welch
Peg 22h40'11"20d41'
Maly,Betty Joan & John Meyers
Mon 7h11'46"-8d15'
Maly,Jessica Dawn
Cyg 20h40'36"38d33'
Maly,Marilyn McGarry
Tri 2h8'55"33d28'
Maly,Suzanne Marie
Cma 6h50'7"-19d19'
Malynda Kathleen
Peg 23h25'18"10d41'
Malzone,Joseph
Aur 6h4'53"37d39'
Mam Sixty
Lib 14h57'11"-6d55'
Mam-Maw
And 23h38'20"43d30'
Mama
Pho 0h28'36"-43d37'
Mama
And 23h19'55"43d57'
Mama "G" & Papa "G"
Ant 10h32'51"-30d4'
MaMa Bou-Bou
Com 12h49'12"30d53'
Mama Els 70 Kinderen Kleinkinderen
Uma 10h56'29"40d15'
Mama Jeanne
Uma 9h40'56"50d29'
Mama Millie
Lib 15h43'52"-8d48'
Mama Nina
Uma 10h59'54"61d33'
Mama's Love
Cyg 19h17'1"45d21'
MaMa-n-PaPa Bear
Cyg 20h38'30"31d4'
Mamala
Cas 0h1'11"61d9'
Mamalins,Barnet
Uma 8h51'41"50d47'
Maman Henriette d'amour
Ori 6h2'14"7d27'
Maman Jule & Dada
Uma 11h38'1"50d23'
Maman Loulou d'amour Poitras
Ori 5h41'47"11d3'
Maman,Daniele N
Ori 5h55'23"7d39'
Mamay
Peg 23h4'1"21d15'
Mamer,Megan Marie
Mon 7h46'49"-2d28'
Mamer,Trevor James
Her 16h45'24"48d24'
Mamet,Clara
Cas 2h45'28"61d28'
Mami
Lib 14h30'2"-22d47'
Mamie Blue 8-32
Cyg 21h36'29"40d52'
Mamie Boo
Del 20h16'25"14d49'
Mamie Justine
Umi 13h16'0"71d42'
Maminou
Uma 11h46'11"57d10'
Maminski,Mimi & Fafa
Cyg 19h44'43"30d0'
Mamma Valeria
Nor 16h21'29"-51d25'
Mamma-Adele
And 2h21'56"42d15'
Mammam
And 23h12'24"35d14'
Mammano,Sara Alisha
And 1h25'50"38d56'
Mammele,Jeanne
Cam 3h40'57"60d59'
Mammina Star
Peg 21h53'34"21d1'
Mammolito,Dominick M
Cnc 8h41'0"17d17'
Mammolito,Emilio Jerome
And 1h22'25"38d16'
Mammuth,Lindsey Nauren
Lyn 8h23'32"49d16'
Mamo Big Daddy
Cep 22h18'57"61d9'
Mamo,Andrew
Aur 5h22'45"54d34'
Mamobea
Oph 17h55'40"11d39'
Mamoth
Ori 6h6'35"-2d27'
Mamouchka
Dra 17h17'1"60d5'
Mamoulides,George
Ser 18h6'46"-14d26'
Mamoune
Boo 14h26'48"50d28'
Mamousse et Sidula B
Cam 4h59'51"58d50'
Mamta 31078
Cyg 19h20'20"28d59'
Mamula,Joe
Aur 7h11'35"36d6'
Mamus
Umi 15h51'16"83d36'
Mamychat
Peg 21h59'38"21d27'
Man from Avalon
Boo 14h46'26"50d19'
Man Hill
Gem 7h32'54"20d20'
Man Tora Bishtar Doost Daram Joy
Sex 9h53'55"2d9'
Man Toura Doost Daram Sunshine
Sex 9h42'44"3d51'
Man,Ron
Sct 18h53'24"-6d30'
Mana
Leo 10h1'21"11d51'
Manafort's Summer Sunday
Uma 12h8'44"63d5'
Manai,Roberto
Aur 5h29'57"31d34'
Manaloto,Juanito Q
Cnv 12h54'15"43d34'
Manami No Ohoshisama
Lib 14h43'38"-10d54'
Manansala,Sara Haley
Vul 19h43'51"26d58'
Manar
Cyg 19h24'46"30d53'
Manasco,Rob
Boo 15h6'58"16d59'
Manasseh,Gary
Aql 19h6'39"15d12'
Manassero,Bob
Mon 6h54'40"0d5'
Manastar
Lup 15h4'54"-40d38'
Manaugh,Dean & Merle Tigert
Boo 14h51'2"15d54'AB
Manay May
And 23h18'16"50d13'
Manbeck,Master Chief Robert R
Lac 22h5'29"40d18'
Manchester,Carol Hebold
Cap 20h51'47"-27d52'
Manchester,Laurie
And 2h31'24"37d52'
Manchester,Lawrence
Psc 1h26'0"10d58'
Manchester,Lydia Charlotte
Tau 5h53'55"26d51'
Manchild
Aql 18h47'35"11d54'
Manci,Nicolø
Peg 23h18'51"18d2'
Mancill,Robert Bruce
Aur 5h2'1"48d52'
Mancilla,Oscar Enrique
Eri 3h31'39"-6d46'
Mancillas,Jane
Cas 1h33'1"58d8'
Mancinelli,Roz & Rick Smith
Vul 19h19'13"26d18'
Mancini 95,Marisa
Cet 3h4'30"1d4'
Mancini,Andre
Cet 1h52'18"0d52'
Mancini,Elena
And 1h24'29"40d18'
Mancini,Luciano
Per 2h51'26"50d11'
Mancini,Michelle Elizabeth
Lyr 19h17'22"38d47'
Mancino,Thomas Harry
Leo 11h4'15"-5d10'
Mancourt,Sarah Kathleen
Cas 3h0'21"60d34'
Mancow
Ori 5h34'35"-0d8'
Mancuso 10-July-1991, Antoinette Rose
Ori 5h31'38"1d40'
Mancuso 20-Sept-1985, Vincent Francis
Ori 5h29'31"1d30'
Mancuso,Anthony N
Oph 16h22'20"0d13'
Mancuso,Dianne Nicole
Boo 14h58'48"44d24'
Mancuso,Joan
And 0h10'23"39d6'
Mancuso,Mary T
Cas 1h42'1"67d39'
Mancuso,Nancy
Cyg 21h32'48"41d21'
Mancuso,Peter V
Leo 9h45'12"7d4'
Mancuso,Stacy L
Eri 3h6'1"-7d10'
Mancuso,Vincent
Uma 9h8'19"62d12'
Mancuso-Heitman,Denise
Lyn 7h5'25"58d29'
Mancy Forever,Yonnie
And 2h20'56"40d47'
Mand,Melissa Christine
Lyr 19h18'48"37d47'
Manda Panda
Aur 6h27'27"34d38'
Manda-You Lift Me Up-Chuck
And 0h14'44"45d29'
Mandaglio Family Star
Leo 9h21'19"33d6'
Mandala
Lmi 10h19'20"30d14'
Mandallaz,William Alexandre
Per 4h5'25"36d28'
Mandana & Pat's "Miracles & Memories"
Cyg 20h0'39"30d30'

Mandara,Gina Marie
Umi 14h17'19"77d51'
Mandara,Michael Anthony
Cmi 7h44'21"8d23'
Mandarino,Dorothy M
Cyg 21h52'40"40d6'
Mandarino,Gene
Aql 18h52'39"-1d40'
Mandarino,Michele
Peg 22h5'56"25d35'
Mandato,Claudia
Ori 6h16'11"7d45'
Mandaud,Volodia
Cyg 20h4'54"31d5'
Mandee
Lyr 18h58'54"40d27'
Mandel,David Andrew
Cam 5h57'1"58d46'
Mandel,Manfred
Mon 7h36'48"-4d56'
Mandel,Marvin & Marilyn
Com 13h13'32"21d45'
Mandel,Oren
Per 3h2'35"38d16'
Mandel,Ray of Light
Cep 23h23'1"70d37'
Mandelbaum,Bari Ruth
Mon 6h19'16"3d20'
Mandelbaum,Bryan J
Her 17h16'0"18d55'
Mandelbaum,MD,Bernard
Dra 19h15'58"70d18'
Mandelhro
Aql 19h23'1"-6d53'
Mandelli,Natale
Peg 21h58'15"20d37'
Mandelstam,Paul
Oph 17h57'45"7d59'
Mander,Sr,Richard H
Dra 20h1'31"68d53'
Manderino,Salvatore
Her 18h7'17"31d8'
Manderson,Byron
Uma 11h57'25"31d43'
Mandery,Michael
Aqr 22h56'0"-5d9'
Mandery,Willy
Crb 15h33'33"38d3'
Mandes
Lyn 7h28'36"40d18'
Mandi Ann
Peg 22h46'15"31d22'
Mandi J D
And 0h19'36"38d30'
Mandi Jo's Rock
Cet 0h35'27"1d55'
Mandie
Cas 23h0'53"53d23'
Mandl,Beth
Lyn 8h10'19"38d4'
Mandracchia,Bonnie
Cyg 21h50'31"2d3'
Mandri,Nicholas Daniel
Ser 17h55'18"-13d14'
Mandrun
Cam 6h18'13"68d30'
Mandurrago,Nathan Anthony
Cet 1h24'50"-2d39'
Mandusky,Edwin
Mon 5h55'53"-4d39'A
Mandusky,Frances
Mon 5h55'53"-4d39'B
Mandy
Vul 21h2'0"20d6'
Mandy
Cas 0h52'26"77d19'
Mandy
Cas 23h39'60"10'
Mandy
Cas 23h4'0"58d17'
Mandy
Cas 1h17'59"63d29'
Mandy & Gary's 2nd Anniversary Star
Cyg 20h22'15"39d26'
Mandy & Randy's Star of 1995
Cet 2h11'34"3d56'
Mandy Beth
Tri 1h52'44"26d30'
Mandy C
Hya 9h32'28"-0d28'
Mandy Jane
Boo 15h3'25"31d31'
Mandy Larae
Aql 19h25'0"15d20'
Mandy loves Grey
Cyg 21h43'1"30d2'
Mandy's Star
Lyr 18h55'40"30d37'
Mandy's Star
Cyg 19h26'12"35d32'
Mandés,Taina del Mar Feliciano
Ser 18h2'35"-14d31'
Manee's Stingie
Peg 22h43'1"10d54'
Manek,Nikolai
Her 17h10'54"42d21'
Manela,Helene
Lyr 18h44'19"32d32'
Manelis,Julia & Edward
Cyg 21h1'13"30d51'
Manento,Joseph Anthony
Aur 6h33'17"31d31'
Manes,Kirk Ryan
Aql 19h54'17"12d27'
Manesberg,Jean & Maury
Dra 20h13'45"67d36'
Maness,Jimmy-Melissa-Timmy & Matthew
Aur 6h8'10"48d59'
Maness,Kevin
Dra 16h10'18"68d9'
Maness,Mary Ed & Eddie
Peg 23h39'50"17d13'
Maness,Michelle
Lyn 6h38'54"58d40'
Maness,Rita Little
Vul 19h0'21"24d41'
Maness-Ash,Edith
Cyg 21h1'1"28d6'
Maney My Friend Love Maw,Paw Ronnie
Crt 10h58'0"-7d51'
Maney's "Shining Star" Mary & Bob
Lyr 19h4'45"28d12'
Manforti
Lyr 18h42'18"40d50'
Manfred
Sgr 19h36'44"-40d59'
Manfredi,Andrew Phillip
Her 16h59'1"32d8'
Manfredi,Zachary-John
Dra 17h7'13"63d21'
Manfredo,Jr,Joseph Anthony
Boo 14h27'46"38d16'
Manfredotti,Manuela
Cyg 21h29'45"48d58'
Manfredoux
Sgr 19h18'58"-27d10'
Manfrino,André
Dra 14h57'43"62d24'
Mang,Barbara
Cap 20h10'12"-8d13'
Mangan,Ross Edward
Aur 6h19'52"33d41'
Manganaro,Victor
Her 16h56'55"31d51'
Mangane
Dra 16h16'35"62d38'
Manganiello,Richard
Lac 22h22'23"55d39'
Manganiello,Victoria & Lily
Cep 3h0'10"80d9'
Mangas,Robert Eugene
Aql 19h23'19"-0d3'
Mangelsen,Sonja
Del 20h22'52"11d12'
Mangeno,Angela Rose
Psc 1h22'0"10d41'
Manger,Martha Ann
Lyn 8h0'16"38d33'
Manger,Siegfried
Hya 9h1'47"2d27'
Mangham,Cain Michael
Mon 6h46'36"10d53'
Mangham,Mandi Nicole
Vir 12h29'22"1d59'
Mangieri,Dawn Marie
Cas 0h46'14"61d4'
Mangieri,Scott Michael
Boo 13h46'1"17d3'
Mangili,Ria
Cep 23h33'46"65d49'
Mangin,Melissa
Equ 20h55'36"2d34'
Manginelli,Ralph & Angie
Peg 21h52'27"33d23'
Mangini,Joseph Louis
Dra 20h3'0"76d48'
Mangini,Michael
Boo 14h51'39"33d38'
Mangini,Suzanne P
Cas 3h8'37"58d34'
Mangino,Erica Lynn
Lyn 9h7'47"38d21'
Mangione,Jack
Oph 17h55'0"10d11'
Mangione,Patricia
Cyg 19h33'38"28d22'
Mangle Star
Mon 6h19'37"8d55'
Mango,Frank Donald
Her 17h3'46"31d11'
Mango,Joyce Enright
Cyg 20h31'18"39d41'
Mangola,Mary C
Mon 6h29'54"7d21'
Mangoni,Jennifer S
Crb 15h38'58"28d39'
Mangot,Alexander Samuel
Aur 5h1'27"30d32'
Mangretta,Renee
Cnc 9h7'45"31d22'
Mangrum,Marguerite
Equ 20h57'21"8d1'
Mangrum,Ray
Aql 19h9'21"13d36'
Mangrum,Richard
Ori 5h58'1"12d44'
Mangrum,Valerie
Cas 0h33'48"71d58'
Mangual,Hector
Per 4h5'42"50d45'
Mangum,99 Minutes by William T
Cet 2h16'16"7d34'
Mangum,J Kevin
Aur 6h29'28"37d58'
Mangum,Margaret
Eri 3h40'34"-16d1'
Mangurten,Julie
Peg 23h36'10"21d45'
Manhire,Jack
Aur 6h16'53"45d53'
Mani,Nicolas Christian
Equ 20h57'0"8d38'
Mani,Peter
Dra 10h26'1"80d41'
Maniam,Sheila
And 0h10'19"30d35'
Manieri,Carlo
Hor 3h23'19"-46d38'
Manieri,Clifton Stephen
Dra 12h6'18"70d17'
Manieri,Paul
Her 16h39'38"11d19'
Maniet,Heather Patricia
Lyn 7h34'17"52d26'
Manig,Lena
Sgr 19h49'47"45d13'
Manilath's Wish
Uma 11h36'42"47d2'
Manildi,J Stephen
Cmi 7h36'11"8d7'
Manilow Magic
Aql 20h30'43"0d44'
Manilow,Barry
Per 2h5'25"58d56'
Manilù
Psa 22h8'1"-27d30'
Manin,Grams & Gramps
Cyg 19h53'15"48d34'
Maningas,John
Her 18h12'37"37d51'
Manion,Billie Margaret
Ser 18h16'11"-14d54'
Manion,Lisa Kay
Lmi 10h41'36"26d18'
Manion,Thomas
Dra 18h59'16"67d39'
Maniord
Del 20h54'47"6d10'
Manipulator
Cyg 19h21'50"28d40'
Manis,Sanford Joe
Per 2h36'13"45d1'
Maniscalco,Giuditta
Umi 16h36'38"82d38'
Maniscalco,Margaret & Michael
Per 3h23'1"54d31'
Manisha
Cnc 8h34'25"30d21'
Maniske,Nellie Isabelle Wood
Umi 16h21'49"78d4'
Manius,Cheryl
Lac 22h29'37"52d48'
Manj-V"
Crb 15h16'19"30d4'
Manjoras,Janice
Lyn 8h29'57"38d23'
Mankato,MN,T-663
Cep 23h39'44"65d21'
Mankins,Megan
Cma 6h10'25"-30d39'
Mankowsky,Toni & Alan
Cyg 20h19'38"30d48'
Manktelow,Georgina Rosie
Lyr 18h14'45"38d35'
Manlapas,Mike
Cnv 12h23'15"37d54
Manley's Sun
Lac 22h25'58"53d20'
Manley,Bruce W
Cet 1h12'16"-3d7'
Manley,Jessica
Cyg 19h26'60"32d35'
Manley,Karen
Peg 22h5'1"25d48'
Manley,Lane Nicole
Leo 10h52'0"-0d22'
Manley,Lawrence B
Per 1h53'0"54d13'
Manley,Matt & Liz
Cyg 20h42'15"45d8'
Manley,Sarah
Vul 19h34'27"26d23'
Manley,Yvonne Marie
Lyn 7h51'38"38d47'
Manly I
Her 15h51'31"48d12'
Mann's Star,Jim
Cnv 12h36'48"34d43'
Mann,Adam David
Aur 4h50'41"41d2'
Mann,Angel
And 2h19'18"48d1'
Mann,Ashley Elizabeth
Lib 15h31'1"-28d46'
Mann,Bill & Earline
Aql 20h21'1"2d13'
Mann,Brendan
Aur 4h52'34"40d55'
Mann,Brian Charles
Per 2h25'17"52d22'
Mann,Dessa Jo
Cas 0h0'52"62d30'
Mann,Duane & Leighann
Aql 20h9'23"8d43'
Mann,Erinn Jennifer
Cas 0h10'31"61d42'
Mann,James
Cep 22h20'26"59d24'
Mann,Jason Patrick
Per 2h45'48"43d30'
Mann,Jay
Mon 6h23'0"-10d41'
Mann,Jennifer
Lyn 7h49'15"44d13'
Mann,Jessica Rose
Peg 23h40'19"28d51'
Mann,John G
Tri 2h35'39"34d48'
Mann,Joseph Gabriel
Aur 6h0'23"34d19'
Mann,Joyce
Boo 14h2'31"21d35'
Mann,Karen
Umi 15h16'42"69d51'
Mann,Kevin
Cyg 19h21'0"28d59'
Mann,Kylee Jeanne
Lyr 19h7'40"37d57'
Mann,Laurence A
Boo 14h54'28"26d60'
Mann,Lisa Lynn
Umi 16h20'51"78d50'
Mann,Mary Jo "MJ"
And 23h22'60"41d41'
Mann,Rachel Elizabeth
Cep 22h4'54"70d56'
Mann,Rodd
Per 1h56'34"54d13'
Mann,Rosangela
Gem 6h44'41"35d15'
Mann,Rose
And 1h41'1"40d14'
Mann,Russell B
Cap 20h17'30"-20d5'
Mann,Sally Jo
Del 20h18'18"9d32'
Mann,Shirley Jo
Ari 2h52'47"30d59'
Mann,Sonia
Mon 7h23'16"-8d23'
Mann,The Star Of
Uma 8h56'29"53d6'
Mann,Tierra
Lac 22h1'27"49d37'
Manna,Alexander
Per 3h4'35"57d10'
Manna,Caroline
Cyg 20h44'29"45d55'
Manna,Maria
Mon 6h57'56"-1d50'
Mannal,Anne
Mon 6h5'41"-8d1'
Mannan,Willow
Aur 5h24'0"40d35'
Mannanici,Sofia
Umi 14h42'0"78d41'
Mannaravalappil,Babu
Lyn 9h14'57"33d20'
Mannarelli-Manzane
Ori 5h2'29"10d28'
Mannas,Regis "Dad"
Leo 11h4'34"-0d43'
Mannell,Adesue
Ori 4h56'11"5d23'
Mannering,Adam Hunter
Ser 15h53'39"0d13'
Mannering,Mac Lester
Mon 7h46'13"-1d9'
Manners,Kim
Aql 20h5'51"6d17'
Manners,Marlene
Vul 20h40'33"28d26'
Manners,Raymond F
Oph 18h27'49"8d8'
Mannes,David Alexander
Cnc 8h11'39"32d29'
Mannes,John
Aur 4h49'49"51d59'
Mannes,Julie
Cas 2h2'56"59d26'
Mannetta,Lianne
And 1h25'35"39d0'
Manngard,Kevin Joseph
Per 3h27'53"51d20'
Mannheims,Karl
Aqr 22h42'27"-6d32'
Mannie's Guiding Star
Psc 23h4'50"0d5'
Mannina,Anthony E
Lac 22h34'30"38d11'
Manning Star,The
Per 2h51'47"45d60'
Manning(Starbright), Joseph C
Cam 13h52'19"81d59'
Manning(Starlight), David J
Dra 10h3'12"81d31'
Manning,Alpha Mary
Ori 6h1'16"8d5'
Manning,Barbara
Aql 18h44'34"-1d41'
Manning,Barbara Jean
Lib 15h58'0"-8d26'
Manning,Conley
Oph 17h18'42"11d42'
Manning,Dawnielle Saness
Mon 7h40'33"-8d13'
Manning,Don
Cep 23h2'58"64d36'
Manning,Father's Fancy Roy Eugene
Sex 10h27'0"2d57'
Manning,Frank & Ruth
Vir 14h0'32"-6d23'
Manning,Gabrielle
And 22h56'0"40d8'
Manning,Ian Paul
Ori 5h56'21"12d42'
Manning,Jeff
Her 16h26'34"41d29'
Manning,Joshua John
Cep 2h0'33"78d41'
Manning,Monica
Cet 1h3'35"-2d30'
Manning,Pat
Aql 20h30'10"0d10'
Manning,Patricia Anne
Lyn 8h19'31"52d8'
Manning,Paula
Mon 7h56'9"-2d45'
Manning,Sarah Marie
Mon 7h0'54"8d17'
Manning,Sr,Richard L
Cep 23h3'12"65d20'
Manning,Stacy Ann
Cas 1h51'1"75d32'
Manning,Stuart
Ori 5h41'1"10d3'
Manning,Tempie
Mon 7h22'38"-6d47'
Manning,Vivian LaVerne Wagner
Tau 3h55'35"22d47'
Manning,William Thomas
Cep 23h14'35"64d20'
Manning=13-1-14-14-9- 14-7
Lyn 7h7'0"51d60'
Manningstar
Oph 17h18'23"-23d46'
Mannino,Lynda Christine
Lyn 7h9'39"59d14'
Mannis,Bethany Jane
Cyg 21h21'16"39d49'
Mannix,Frank Lee
Her 15h50'0"41d23'
Manno,Anthony
Aqr 20h59'1"-10d17'
Manno,Joan C
Mon 7h39'53"-1d1'
Manno,Rosina M
Per 3h23'13"40d27'
Mannuel,Heather
Cam 5h48'24"68d44'
Mannum,Justina
Aql 18h59'45"11d5'
Manny
Psc 1h24'58"12d10'
Manny
Lyr 18h43'38"39d7'
Manny & Debby
Cyg 19h32'47"33d53'
Manny Q
Aql 19h31'33"13d32'
Manoa Niama
Lep 4h59'11"-11d49'
Manocchio,Kelly & Mike
Lyr 18h37'19"44d26'
Manocherian,Lindsay Margaret
Cas 1h19'1"63d52'
Manocherian,Sarah Ann
Leo 10h4'31"7d13'
Manolito
Uma 13h38'57"50d41'
Manolov,Valtchan & Valka
Cet 1h27'25"-1d55'
Manolva,Maria
Vir 13h58'15"-10d26'
Manon
Cnv 13h48'24"42d23'
Manon et Laurie-Ann
Uma 9h39'20"60d13'
Manon,La Belle
Cam 6h10'0"80d0'
Manon,Ryan Alan
Her 16h30'43"48d54'
Manooch
Cyg 19h31'54"34d60'
Manooch
Del 20h18'43"11d13'
Manos,Kelly Peter
Mon 6h56'1"8d58'
Manos,Pamela Amelia
Boo 13h57'48"14d26'
Manos,Pete
Dra 10h24'30"78d43'
Manos,Steven Mitchell
Boo 13h34'18"14d26'
Manoukian,Brigitte
Mon 6h26'22"-5d47'
Manoutschehr,Sanei
Ori 5h49'28"21d4'
Manr 50
Peg 22h16'10"32d43'
Manresa,Howard E
Cep 2h1'50"77d57'
Manriel,Laura Elizabeth Brown
Peg 23h23'42"15d59'
Manring,Charles D
Aur 5h58'51"38d41'
Manring,Yvonne A
And 23h43'40"42d0'
Manross,David
Cet 3h17'1"-0d15'
Mansberg,Roy
Lmi 11h2'10"27d55'
Mansell,Laurie Ruth
Cas 23h48'40"50d12'
Mansell,Vanessa
And 23h31'39"41d11'
Mansfeld,Heimke
Leo 11h23'35"7d42'
Mansfield's Star
Aql 20h1'53"8d43'
Mansfield,Allison Marie
And 23h0'34"50d5'
Mansfield,B Mans- Robert
Lmi 10h38'1"23d20'
Mansfield,E Blaine
Aql 19h29'30"0d50'
Mansfield,Ellen
Lyn 7h58'36"44d59'
Mansfield,Francis DeeDee
Lyr 18h43'18"32d49'
Mansfield,George
Aql 18h59'32"12d55'
Mansfield,Jerry Dean
Cap 20h19'4"-19d2'A
Mansfield,Kathy Jean
Mon 6h42'27"10d25'
Mansfield,Lisa Maree
Cap 20h19'4"-19d2'B
Mansfield,Marsia Sue
Ari 2h58'24"21d44'
Mansfield,Mary Gay
Sct 18h51'18"-7d55'
Mansfield,Megan
Uma 9h38'20"57d56'
Mansfield,Mike
Uma 8h41'58"62d1'
Mansfield,Paul James
Her 17h19'52"41d43'
Mansfield,Tim
Ser 18h7'14"-13d30'
Mansi,Kate Elizabeth
Sex 10h6'0"2d23'
Manson,Alasdair N
Per 2h24'0"54d42'
Manson,Elizabeth
Cas 2h39'25"73d21'
Manson,Hannah Ruth
Cyg 19h23'31"44d30'
Manson,Hortense Peterson
Aur 4h57'18"48d56'
Mansour,Paul
Per 2h54'31"55d53'
Mansoura,Monique K
Lyn 8h4'21"41d15'
Manstein,Laurens
Uma 10h15'0"48d48'
Mansur,Peggy
Cnv 13h13'59"37d33'
Mansvelt,Gèrald
Cam 6h4'17"56d7'
Mansvelt,Viviane
Cas 3h2'18"57d22'
Mantei,Braden H A
Her 16h38'28"23d55'
Mantelli,Maria & Anthony
Cyg 19h25'34"31d55'
Manternach,Stephen Lowe
Lac 22h16'51"37d42'
Manternach,Wendy
And 2h13'56"41d3'
Manteuffel,Adam Dustin
Eri 4h47'31"-8d28'
Manteuffel,Travis Jay
Cet 1h35'17"-2d22'
Manteuffel,Wolfgang Richard & Sandra Lee
Per 2h43'26"36d31'
Manthe,Barry & David
Aql 19h52'23"15d18'
Manthe,Rebecca Erin
Aqr 22h20'59"0d38'
Manthei,Marie
Dra 16h55'48"68d23'
Manthey,Mark W
Eri 4h14'47"-19d7'
Mantilla Family
Eri 4h1'36"-3d20'
Mantini,Carla e Pier Luigi Gianquitto
And 23h3'0"52d53'
Mantis/4M
Dra 17h5'57"65d36'
Mantos,Vasilies
Lyn 7h42'16"58d29'
Mantouani,Sandro
Peg 23h45'52"28d38'
Mantoux,Jean-Claude
Cyg 19h45'37"30d2'
Mantovani,Andrea
Cnv 12h52'58"40d12'
Mantovi,Linda Marie
Cas 0h29'33"63d48'
Mantykow,Nestor R
Aur 7h23'33"38d57'
Manu
Eri 3h46'58"-1d27'
Manu
Cet 2h57'35"2d56'
Manu
Sgr 19h1'55"-27d35'
Manu The Best Cousin
Boo 14h24'24"50d15'
Manue
Cam 5h53'16"68d24'
Manuel
Boo 15h5'2"18d38'A
Manuel,Angelia Maria
And 1h46'1"40d23'
Manuel,Carolyn
Lyr 18h48'58"38d55'
Manuel,Christopher Mark
Cep 21h26'0"55d50'
Manuel,Elinor A
Cmi 7h15'16"4d24'
Manuel,Katriel Joy
Cyg 20h55'17"31d40'
Manuel,Luthgard Luthy
Ori 6h1'12"4d26'
Manuel,Marie S
Peg 22h0'40"31d50'
Manuel,Tammy
Cas 0h38'10"73d44'
Manuel,Tammy
Del 20h53'29"2d16'
Manuel,Tommy J
Cmi 7h35'26"0d16'
Manuel,Tony
Cmi 7h15'32"4d29'
Manuela
Com 12h18'0"21d15'
Manuela
Gru 22h12'45"-56d19'
Manuela
Her 16h11'50"10d44'
Manuela
Cnv 13h14'1"38d18'
Manuela
Cet 2h33'52"1d19'
Manuela
Sge 19h28'32"17d47'
Manuela
Lac 22h3'55'51d6'
Manuela
Cnv 13h2'52'32d38'
Manuela
For 2h36'1"-25d6'
Manuela
Cep 21h13'52"68d41'
Manuela & Herbert Forever
Aql 18h40'49"-1d9'
Manuela Cristina
Gem 6h43'44"18d43'
Manuele & Federica
And 23h17'44"51d18'
Manuelle(Berthaud)
Cam 7h35'29"68d7'
Manulek
Uma 10h54'51"60d45'
Manuri,Barbara Ann
And 0h47'17"35d45'
Manusevitz,Hanna
Per 2h30'6"55d10'A
Manuwald,Ray
Cep 0h34'33"80d35'
Manuzzi,Stefano
Dra 16h39'47"66d21'
Manville,Lesley Ann
Cas 1h37'20"74d41'
Manwaring,Ian
Cep 0h6'40"73d23'
Manwell,Edmund R
Boo 14h24'59"21d1'
Many Kisses
Ori 6h6'46"8d40'
Many,M Hepburn
Mon 7h16'3"-6d28'
Manz,Joachim Hugo Hermann
Ari 3h21'1"20d28'
Manz,Louis & Wilhelmina
Boo 13h53'44"20d10'

Manza,Roseann
Cam 6h12'16"56d5'
Manzanares-Amor De Mi Vida,Viviana
Sgr 20h3'15"-26d38'
Manzanedo,Tomas
Lac 22h43'43"38d12'
Manzano,Nola
Com 12h13'12"18d59'
Manzi,Enrico Harry
Dra 16h2'54"66d42'
Manzo,Fred V
Peg 23h0'1"31d31'
Manzo,Jessie
Mon 8h5'25"-9d16'
Manzo,Mark
Ori 5h57'20"15d1'
Manzo,Nino
Uma 11h50'0"41d33'
Manzoli,Nick
Uma 8h46'42"68d30'
Manzones Hidden Treasures
Lyr 18h54'50"37d51'
Manzoor,Tahrah Wasim
Crb 15h48'18"26d36'
Maola,Stephanie
And 0h22'58"27d35'
Maparila
Cam 6h38'25"67d50'
Mapel,Brian
Lmi 9h58'1"34d38'
Mapes,Stephen Alexander
Peg 22h49'1"29d8'
Mapko
Uma 10h33'1"42d21'
Maple,James
Her 17h56'55"41d3'
Maple,Sid
Lac 22h22'21"56d44'
Maples,Carl Steven
Cmi 7h35'48"11d20'
Maples,Douglas Charles
Her 16h25'10"11d12'
Maples,Janet Gail
Com 12h55'33"30d11'
Maples,Julia Pack
Aql 19h29'38"-0d8'
Maples,Tony Lawrence
Her 18h53'41"12d49'
Mapot
Cam 10h23'37"81d47'
Mappa
Ori 4h50'19"0d42'
Maquettes Modèle Aclùolités
Dra 9h37'49"74d50'
Mar Pedroviejo 1993
Lyn 8h35'24"42d10'
Mar,Natalie Tian
Pho 0h42'0"-47d16'
Mar-Lou
Aur 6h52'19"48d58'
Mar-Lyn
Cyg 21h57'1"53d2'
Mara
Cyg 19h35'43"29d40'B
Mara
Mon 8h8'11"0d43'
Mara
Cam 5h0'0"60d45"
Mara
Uma 11h22'0"52d13'
Mara
Tri 1h46'1"34d43'
Mara
Mon 7h15'54"0d56'
Mara
Uma 11h49'37"53d49'
Mara Bara
Lyr 18h36'0"40d26'
Mara Del Pilar,Pilar, José
Lup 15h11'30"-44d7'
Mara's Star
Lyn 8h9'23"50d33'
Mara,Kathleen K
Aql 19h2'1"0d38'

Mara,Victoria L
Com 12h1'41"25d38'
Mara-Lisa
Tri 2h18'1"32d5'
Maraache,Ahmad
Cam 3h38'20"74d4'
Marabeth
Cyg 20h53'26"38d36'
Marabeth
And 1h53'18"39d46'
Maracich,Rosemarie
And 1h10'10"47d30'
Maraconi,C Mad
Lac 22h43'42"52d35'
Maragos,Tasos
Cnc 7h54'29"11d13'
Maragozidis,Haralabos
Uma 9h45'53"70d60'
Maraillat,Louis
Lyr 18h50'37"39d54'
Marais,André & Marietjie
Gru 22h30'20"-51d14'
Marak,Tommy
Boo 15h44'39"40d9'
Maran,Harriet
Aql 19h57'18"-8d36'
Maranda
Ori 5h36'7"-1d1'
Marandet,Christian
Equ 21h21'14"8d51'
Marandi,Katel
Aql 19h55'25"13d26'
Marando,Ivana
Lup 15h17'31"-42d54'
Marando,Kelly
And 0h57'22"45d11'
Marando,Maria
Boo 15h6'49"8d22'
Maranesi,Marco
For 2h33'39"-27d13'
Maranesi,Silvia
And 23h1'59"51d4'
Maranga,Thomas Rocco
Aur 6h22'32"32d56'
Marangolo,Paola
Cyg 21h5'1"48d49'
Marangon,Marcus
Aql 18h44'24"8d60'
Marannetyn-Entwined Forever
Cyg 20h20'49"39d1'
Marano,Barbara A
Lyn 7h34'17"50d41'
Marano,Marie
And 0h15'48"33d56'
Marant,Charlie
Ser 18h3'48"-13d15'
Maranto,Anthony R
Aur 6h31'1"33d52'
Maranto,Gina
Cyg 19h29'32"34d8'
Mararian,Jeffrey Starr
Per 3h11'0"44d38'
Maras
Lyr 18h51'1"41d8'
Marasa,Sabrina
Cas 0h22'45"66d55'
Marasa,Thomas
Aur 5h15'38"40d29'
Marasca,Madison Elizabeth
Cas 0h29'45"63d36'
Maraschiello,Kelly
Ori 5h34'1"-2d24'
Maraschino #072576-92-16,C R
Sex 10h18'32"-7d42'
Marasco,George
Boo 15h3'13"11d26'
Marasco-Mudd
Lyr 18h57'57"33d8'
Marassi,Monica
Tau 4h10'46"0d39'
Maraubaldo
Pyx 8h50'18"-28d1'
Marautious
Cnv 12h52'1"40d12'

Marazio N2YXX,Dom
Aur 4h50'35"41d4'
Marazzi,Carolina
Peg 23h30'1"33d37'
Marazzi,Silvio e Bianca
Cyg 21h35'0"40d37'
Marback,Patricia Dee Winston
And 23h40'43"46d20'
Marbaugh,Corinne
And 2h9'58"42d56'
Marbaugh,Corinne
Cyg 20h53'43"48d59'
Marber,Ian
Her 16h55'36"50d60'
Marble's Star,Jeff
Sco 16h8'10"-40d57'
Marble,Amanda Ruth
Tri 1h43'31"33d56'
Marbuger,Patsy Marie
Mon 7h57'46"-3d34'
Marburger,Gary
Aur 5h56'52"40d16'
Marbury,Lillie Mae
Lyr 18h37'45"41d15'
Marc
Cas 23h1'22"53d17'
Marc & Kathleen's Beau Ideals
Per 1h47'23"53d31'
Marc & Linda
Cyg 21h23'26"30d53'
Marc ed with Laura's Love
Aql 19h58'30"8d56'
Marc II
Her 18h31'1"24d23'
Marc Ivan
Peg 23h22'33"33d58'
Marc Jordan
Cam 7h30'30"80d9'
Marc Philipp P W
Aur 4h55'33"50d53'
Marc Rigau i Cartal
Cmi 7h20'43"8d44'
Marc The Magnificent
Uma 9h59'31"70d51'
Marc's Destiny
Del 20h28'42"10d42'
Marc's Piece Of Heaven
Ori 5h16'15"1d22'
Marc's Star
Aur 6h34'46"34d36'
Marc,My Dream Came True
Cap 21h52'54"-18d60'
Marcail
Aql 20h0'50"14d9'
Marcalus,Heather Rebekah-Sue
Lyr 18h40'59"46d32'
Marcan's Girl
Aur 6h5'43"33d2'
Marcarelli,Gary Peter
Boo 15h3'39"14d17'
Marcarelli,Gary Peter
Lac 22h46'29"55d47'
Marcase,Mariah Barrett
Lyn 6h25'0"58d36'
Marcault,Jean Claude
And 2h7'22"40d54'
Marcault,Michel
And 2h9'16"40d37'
Marceau,Roger
Lyr 18h58'1"39d24'
Marceau,Sister Therese
Cyg 19h26'0"56d59'
Marceca,Anna
Ind 21h2'4"-52d2'
Marceile & John
Vul 19h48'0"25d56'
Marcel
Per 4h5'50"36d40'
Marcel
Cep 23h15'26"68d47'
Marcel & Silvia
Her 17h24'26"42d42'
Marcel Janos Deszö I
Gem 6h46'40"16d36'

Marcel,Raven & Taylor
Psc 0h1'58"0d8'
Marcel,Teresa (Pooh Bear)
Peg 22h8'24"26d13'
Marcela
Lyn 8h56'1"42d7'
Marcela
Del 20h20'52"9d49'
Marcella
Aur 4h58'0"36d7'
Marcella
Cas 2h13'54"70d13'
Marcella
Lyr 18h40'32"41d27'
Marcella
Uma 10h0'20"51d17'
Marcella Maria
Peg 23h39'15"10d0'
Marcella Meini di Rosignano
Lyr 18h39'11"28d9'
Marcella,Fred
Boo 14h56'55"29d49'
Marcella,Melissa
Lyr 18h59'54"29d40'
Marcelle Louise
Peg 23h44'1"30d44'
Marcelli,Jean Luc
Lyr 18h57'18"31d35'
Marcellin,Ron
Her 17h53'23"18d46'
Marcellino's
Uma 11h12'59"31d36'
Marcellis,Jill
Mon 6h53'28"-3d36'
Marcello
Cet 2h10'19"6d32'
Marcellusi,Federico D
Cep 21h41'31"85d55'
Marcelo
Aur 6h12'45"30d17'
Marcengill,Albert
Cmi 7h38'49"5d11'
Marcey's Star
Lyn 9h14'41"34d36'
March,Leo Walter
Cet 3h16'44"1d34'
March,Mateu Mascarco
Boo 13h43'27"15d46'
March,Mryna
Cyg 21h7'15"40d10'
March,Pat Light
Boo 14h28'38"44d30'
March,Wendy J
Gem 6h43'53"34d38'
Marcham,Gabrielle Adrianne
Umi 15h56'20"78d59'
Marchand,Alyssa Nicole
And 0h1'46"38d36'
Marchand,Claude
Ori 6h5'57"8d51'
Marchand,Deanna
And 1h13'56"39d51'
Marchand,Olivier
Peg 23h23'53"11d42'
Marchand,Pierre
Per 2h56'57"35d42'
Marchant's Magic
Lib 15h18'18"-21d2'
Marchant,Elyse Nicole
Mon 6h27'57"-5d59'
Marchant,Kate
Cyg 21h27'12"39d57'
Marche Camilla
Ori 5h56'42"15d5'
Marche Toujours Avec Moi
Cet 0h45'56"0d30'
Marchela & Billy
Cyg 21h33'48"40d27'
Marchesani,Stephen Vincent
Aur 5h0'58"42d6'
Marchese,Bella
Cas 2h42'21"70d1'
Marchese,Carmelo T
Aql 19h29'14"-0d27'

Marchese,James & Dianne
Cam 9h19'41"81d6'
Marchese,James A
Dra 17h4'39"62d5'
Marchese,Raphael E
Aur 5h6'0"40d34'
Marchesi,Daniela
Uma 12h2'36"41d48'
Marchesi,Lynn
Mon 6h54'40"-6d50'
Marcheski,Daniel Harold
Aur 5h0'1"38d41'
Marchi's Wishing Star, Rhonda & Jesse
Cyg 21h8'34"30d6'
Marchione,Joseph
Lyr 18h54'27"34d23'
Marchionna,Kristin
Lyr 18h42'58"42d28'
Marchisotto,Cynthia
Lyr 19h16'12"42d21'
Marchitelli, Christopher
Dra 17h36'23"64d45'
Marchitello,Miguel Angel
Cep 23h38'46"65d24'
Marchitto,Nicholas Victor
Per 2h29'45"57d35'
Marchl,Helmuth
Cnc 9h11'34"8d48'
Marchman,Calder Jedediah
Aql 18h59'34"-2d5'
Marchman,Jamee Charlene
Mon 6h30'30"10d44'
Marchou,Laura Aurore
Peg 0h5'26"20d16'
Marci G
Vul 19h20'40"26d33'
Marcia
Cam 5h25'19"79d48'A
Marcia
Cas 0h31'16"64d40'
Marcia
Com 12h38'44"23d18'
Marcia
Cma 7h23'35"-18d50'
Marcia
Lyr 18h45'27"33d38'
Marcia
Cmi 7h28'18"8d0'A
Marcia 10-23
Del 20h17'12"14d9'
Marcia Ann
Cyg 21h30'38"34d36'
Marcia Ann
And 2h24'0"46d17'
Marcia Ellen
And 1h0'11"40d11'
Marcia Erotic Goddess of the Night
And 23h25'46"45d15'
Marcia Forever
Uma 10h53'1"40d42'
Marcia G
Del 20h26'41"20d24'
Marcia L G L R
And 2h16'59"48d21'
Marcia Lee
Mon 7h44'10"-2d1'
Marcia's Dandelion
Crt 11h13'11"-19d4'
Marcia's Star
Cas 0h21'48"70d33'
Marcial & Smiles
Cet 1h34'0"-1d5'
MarciAnn
And 2h21'10"44d15'
Marciano,Gabriella Camille
Peg 22h22'0"3d34'
Marciante,Alfonso Albert
Lac 22h21'24"50d10'
Marciante,Paula & Peter W Weiler
Del 20h53'19"2d43'
Marcie
Eri 2h48'46"-8d46'

Marcie
And 2h29'26"44d53'
Marcie
Mon 6h33'43"0d8'
Marcie The Forget-Me- Not Of Angels
Uma 14h17'52"61d55'
Marcie's Own Star With Love,Larry
Cyg 19h27'22"32d53'
Marcie's Star
Eri 3h37'42"-12d33'
Marcie's Star 1992
Lib 14h57'0"-18d55'
Marcil,Dany
Ori 5h52'48"8d59'
Marcil,Paul H
Her 17h6'1"22d17'
Marcil,Timothy R
Her 16h55'0"38d5'
Marcille,Tom
Aur 6h1'30"36d6'
Marcillina
Ori 5h57'16"8d28'
Marciniak,Pat & Carl
Lyn 7h6'60"51d2'
Marcinkevicius-Edney
Oph 17h34'6"-22d46'
Marcinko,Joline A
Cnv 13h47'28"46d23'
Marciscano,Edgar
Cet 2h27'26"3d3'
Marckess,Denise
Boo 13h57'48"30d36'
Marco
Her 15h49'39"42d30'
Marco
Sgr 19h28'24"-38d36'
Marco
Umi 17h19'18"78d49'
Marco
Col 6h32'4"-36d11'
Marco
Pho 0h1'32"-45d5'
Marco
And 0h5'2"39d52'B
Marco
Ori 5h56'19"0d16'
Marco
Uma 11h58'27"31d51'
Marco
For 2h26'15"-27d26'
Marco
Cnv 12h52'10"50d9'
Marco
Lac 22h10'40"46d45'
Marco
Nor 16h19'42"-52d8'
Marco Anthony (Tony)
Lyr 18h26'20"42d56'
Marco C
Hor 3h17'46"-49d0'
Marco et Monica
Cas 1h44'40"58d10'
Marco M
Cas 2h10'30"60d24'
Marco Pra Monego
Hor 3h25'24"-49d53'
Marco T
Ori 5h31'1"0d37'
Marco The Greatest
Ser 15h41'59"20d41'
Marco Zanker Germany
Leo 9h58'32"19d1'
Marco's Little Darling Forever
Boo 14h8'1"41d26'
Marco's Star,Great- Grandma
Cyg 19h28'33"34d31'
Marco,Chris
Sge 19h59'34"16d6'
Marco,Marri San
Aql 19h58'36"12d33'
Marco,Max
Gem 7h6'60"21d46'

Marco,Nicole
Aqr 23h24'29"-6d42'
Marcocci,Todd A
Cam 4h50'40"68d21'
Marcolino
Nor 16h22'37"-59d39'
Marcon,Lorella
And 0h11'28"30d60'
Marcon,Simonetta
Dra 17h4'0"71d48'
Marconcini,Leno
Her 16h26'20"33d17'
Marconcini,Mary Jo
Lyr 18h14'53"38d26'
Marcone,John
Lac 22h12'27"46d21'
Marcone,Quin-Joann Dana-Jamie
Cep 22h14'19"67d45'
Marconi Family,The
Dra 18h59'16"67d39'
Marconi's Family Star Melanie
Cet 0h31'56"-4d49'B
Marconi's Family Star Marisa
Cet 0h31'56"-4d49'C
Marconi's Family Star, Karen Galloway
Cet 0h31'56"-4d49'A
Marconi,Luxia by Luca e Lucia
Cae 4h59'17"-32d48'
Marconi,Vanessa
Aur 6h6'43"45d6'
Marcor
Umi 16h12'0"70d43'
Marcosa,Michael Thomas
Hya 8h16'10"5d37'
Marcotte,Christy
Cap 21h13'57"-26d50'
Marcotte,Linda Jane
Mon 6h21'40"2d53'
Marcotte,Melissa Ann
Equ 20h58'0"9d40'
Marcotte,Robert L
Aur 6h1'40"36d21'
Marcoux,Brandon Scott
Ser 17h51'20"-11d19'B
Marcoux,Jeffrey Paul
Cet 2h29'55"3d1'
Marcoux,Robert
Ori 5h12'1"-6d30'
Marcoux,Ron
Per 1h51'20"56d58'
Marcove,Laura
Vul 21h25'37"24d9'
Marcovich,Isaac & Raquel
Aql 20h34'11"-0d32'
Marcus & Victoria "Forever Shining"
Cyg 21h35'1"38d7'
Marcus Allen
Aur 7h14'54"40d25'
Marcus Aurelius
Lac 22h29'25"40d42'
Marcus Aurora
Oph 17h54'32"12d16'
Marcus Julian
Her 17h17'0"42d13'
Marcus Kai Chi Lian
Cyg 20h1'35"37d57'
Marcus Lee's Star
Lac 22h30'29"55d23'
Marcus Magnificus
Aur 5h54'44"31d39'
Marcus Superbus
Her 16h56'36"27d38'
Marcus's Wish
Her 17h20'24"49d46'
Marcus,Burton S
Her 17h2'1"18d48'
Marcus,Carrie Anne
Mon 6h32'31"4d44'
Marcus,Catherine Reid
Lib 14h39'26"-23d45'
Marcus,Charlotte Alexis
Peg 22h42'40"20d41'

Marcus,Edwin
Cap 21h54'14"-18d49'
Marcus,Ellen Levin
Aqr 22h18'51"0d33'
Marcus,Elyssa June
And 0h0'55"40d37'
Marcus,James L
Her 18h5'39"14d59'
Marcus,Kyle Elizabeth
Mon 7h46'55"-1d1'
Marcus,Marty
Her 16h37'47"4d46'
Marcus,MD,Harold S
Dra 17h29'13"64d26'
Marcus,Michael E
Cet 0h58'1"2d0'
Marcus,Mitchell Charles
Uma 11h9'31"43d36'
Marcus,Nancy C
Mon 6h34'39"1d36'
Marcus,Raven
Lyr 18h35'47"40d36'
Marcus,Raymond
Aql 19h8'1"15d15'
Marcus,William
Ori 5h41'37"10d16'
Marcy
Ori 5h54'35"14d24'
Marcy
Cam 7h49'39"60d13'
Marcy,Christine L
And 23h40'55"40d15'
MarcyAnn
Gem 6h52'43"14d56'
Marcyn
Aql 20h7'58"1d22'
Marcèle-Nanou
Cyg 19h28'11"31d32'
Mardarescu,Ilinca
Cas 1h20'1"53d19'
Mardelle
Ari 1h59'1"12d22'
Marden,Scott
Cmi 6h54'55"-15d11'
Marder,Marvin A
Ori 4h58'40"0d58'
Mardirosian,Arax
Lyn 8h10'26"41d46'
Mardit,Rose Julianne
Cyg 20h54'0"30d49'
Mardon 95
Lac 22h7'16"47d57'
Mardy
Cam 5h27'0"65d36'
Mare Nostrum
Peg 22h23'45"35d20'
Mare's Star
Cet 0h57'45"-0d25'
Mareck,Birgit
Sge 19h27'22"16d38'
Marecki,Emma Cordelia
Tau 4h15'1"20d46'
Mared's Paradise
Uma 11h49'1"63d57'
Mareena
Lyn 9h14'20"45d52'
Marek 6
Hya 8h12'36"-5d40'
Marek Star,Tom's
Cet 3h18'23"8d56'
Marek,Anne
Cas 0h45'32"62d9'
Marek,Cynthia
And 1h21'11"40d34'
Marek,Elizabeth Lillian
Cas 0h47'36"61d35'
Marek,J Dennis
Cep 22h17'0"61d27'
Marek,William Joseph
Cep 20h59'30"55d15'
Marel
Tri 2h8'37"32d23'
Marella
Lyr 19h17'42"26d55'

Marelle's Magic
Ori 5h29'30"0d49'
Marelli,Maddalena e Enrico
Pyx 8h31'22"-25d15'
Maren (Perkins)
And 2h20'45"49d0'
Maren,Lisa,Betty, Sergio
Cyg 19h53'26"44d18'
Marengo "Teddy Rose", Deanna Lee
Vul 19h46'45"28d27'
Marengo,J J
Uma 10h48'16"57d12'
Marengo,Kim
Cet 2h37'1"0d36'
Marengo,Marina J
Oph 17h53'54"11d14'
Mares Love
Dra 16h42'40"61d17'
Mares,Don
Her 16h8'42"16d3'
Mares,John W
Per 2h10'22"58d15'
Mares,Johnny Prep
Hya 9h15'25"2d56'
Maresa's Star
Ori 5h27'0"-0d10'
Maresca,Maureen
Cas 0h23'41"75d3'
Marescaux,Nathalie
Cyg 19h42'0"38d43'
Maresco,Eileen Mary
Cam 5h51'36"68d37'
Marese
Ori 6h15'58"1d2'
Maresh,Edw F
Lyr 18h40'51"32d22'
Maretta
And 2h28'0"48d31'
Marfac,Inc
Mon 7h44'44"-8d42'
Marfell,Tony
Cep 21h46'10"70d27'
Marfeo,Deana
Dra 10h36'59"80d5'
Marfeo,Wendy
Dra 9h32'36"80d46'
Marga's Shining Star
Cyg 19h39'1"31d24'
Margalski's Star
Peg 22h58'21"27d60'
Margamobriano
Oph 17h15'37"-20d51'
Margaret
And 0h29'55"30d48'
Margaret
And 2h10'25"40d23'
Margaret
Psc 1h28'55"30d19'
Margaret
Cas 1h2'55"60d49'
Margaret
And 0h18'1"31d37'
Margaret
Sge 19h54'30"19d48'
Margaret
Aql 19h1'28"-6d43'
Margaret
Cyg 19h25'1"44d33'
Margaret
Equ 21h2'16"3d40'
Margaret
Cam 3h32'40"60d34'
Margaret "Magoo"
Uma 9h47'53"71d49'
Margaret & Bob "Forever"
And 1h19'32"34d44'
Margaret & Nicole
Lyr 18h24'23"40d24'
Margaret & Patrick Forever
Aqr 23h11'1"-5d2'
Margaret & Tina Best Friends 4ever
Hya 8h59'54"-0d30'

Marina
Lib 15h1'59"-28d11'
Marina
Umi 16h30'35"76d45'
Marina
Sgr 18h53'19"-27d34'
Marina
Umi 14h52'0"68d15'
Marina
And 23h44'59"38d18'
Marina
Tri 1h59'30"34d14'
Marina
Cae 4h40'54"-33d4'
Marina
And 0h5'7"36d54'B
Marina
Ori 6h4'42"20d19'
Marina
Cyg 20h17'52"38d39'
Marina
Pyx 8h45'26"-20d36'
Marina & Charlie
Crb 15h34'21"30d44'
Marina B
Peg 0h4'17"20d19'
Marina M F
Aur 5h2'33"30d40'
Marina's Star
Aql 19h25'56"-8d15'
Marinan,Suzanne Esther
Cas 2h6'34"68d41'
Marinaro,Janis
Lyr 18h44'30"47d10'
Marinaro,Nicki
And 23h37'1"49d22'
Marine
Cas 0h29'36"68d54'
Marine
Equ 21h19'15"11d56'
Marine B Champetier de Ribes Christolle
Uma 13h12'0"61d25'
Marine MGA Inc
Uma 8h11'0"60d1'
Marine Office Of America Corporation
Cam 5h47'55"61d25'
Marine's Dream
Uma 9h1'46"68d51'
Marine,B B
Equ 21h10'52"11d54'
Marine,Darrin
Cet 1h19'50"-2d31'
Marine,Katie
Lyr 19h1'57"31d0'
Marinella
Cep 2h14'57"77d53'
Marinelli Longaretti
Lyr 18h56'17"31d39'
Marinelli,Anna
Ant 10h46'10"-31d35'
Marinelli,Francesco
Cnv 13h1'22"51d36'
Marinelli,Lou
Vir 13h15'5"-8d26'
Marinelli,Mandy
Lyn 8h53'52"42d48'
Marinelli,Mario
Umi 16h24'26"82d50'
Marinelli,Maurizio
Umi 16h12'10"82d40'
Marinelli,My Great Love Michael
Del 20h52'1"8d16'
Marinelli,Vincent
Dra 12h9'0"75d35'
Marinello,Paulina Clotilde Severino
Cam 3h58'51"56d47'
Mariner,Jennifer Lynn
And 0h5'29"46d29'
Marinho,Rosita F
Cyg 19h19'24"28d51'

Marini,Len
Aur 5h8'25"40d29'
Marini,Leocadia
Aur 5h19'27"42d37'
Marinier,Anna Marie
Cam 4h53'41"61d25'
Marinier,Lisa Marie
Cas 0h24'31"50d0'
Marinne
Aur 4h52'16"40d16'
Marino's Star,Gary Evan
Her 17h11'11"46d39'
Marino,Amalia
Lyn 7h37'47"52d21'
Marino,Amy
Cas 0h31'50"67d35'
Marino,Anna Marie
Peg 21h59'46"20d2'
Marino,Elio
Cam 14h15'1"80d56'
Marino,Florence
Cyg 21h25'0"40d13'
Marino,Frances
Cet 2h25'27"1d3'
Marino,Joseph F
Aur 6h31'34"31d9'
Marino,Mary Michael Graviss
Mon 7h54'56"-9d2'
Marino,Michael
Ser 16h19'38"2d46'
Marino,Michael
Cyg 20h24'16"38d13'
Marino,Nancy
Lyr 19h20'40"35d27'
Marino,Ralph
Cet 2h26'0"1d55'
Marino,Roselyn
Cet 2h28'55"0d43'
Marino,Tony
Cet 2h27'40"1d46'
Marino,Trudy
And 0h5'55"30d36'
Marino-Aaron,Gina
Cas 1h48'21"75d28'
Mario
Ori 5h55'17"13d46'
Mario
Vir 14h54'28"0d15'
Mario
Her 16h38'20"38d4'
Mario
Her 16h37'12"24d26'
Mario
Aql 18h43'47"6d39'
Mario - Der Schmusebär
Cep 22h7'53"61d0'
Mario de los Cobos
Hya 8h51'25"-6d31'
Mario Star,The
Her 17h22'27"44d11'
Mario's Happiness
Cep 22h12'48"61d10'
Mario's Inspiration
Ori 5h55'36"16d30'
Mario's Star
Per 2h57'37"38d15'
Mariolina
Aqr 21h23'55"-13d54'
Mariolina
Aql 20h34'43"-1d8'
Marion
Aqr 22h20'30"0d30'
Marion
Uma 9h18'28"51d47'
Marion "Schieti"
Vir 13h11'10"-8d21'
Marion & Bill
Cyg 20h24'55"38d33'
Marion & Ron
Cyg 19h41'0"38d37'
Marion 2-10-65
Lib 14h20'51"-20d12'
Marion Elizabeth
Ori 6h5'1"0d23'

Marion From Skien
Lmi 9h57'16"33d53'
Marion L
Lyr 18h37'17"31d16'
Marion's "Wishing Star"
Cas 1h10'19"63d55'
Marion's Brilliant Gem
Peg 22h59'0"10d44'
Marion's Galaxy Ripple
Lyr 19h2'25"26d56'
Marion's Integrity
Cyg 19h26'20"34d20'
Marion's Moonbeam
Cyg 21h8'32"40d12'
Marion,John
Aur 5h8'49"40d41'
Marion,Joseph Edward
Psc 0h47'13"20d25'
Marion,Tracy Short
Ori 6h2'49"8d3'
Mariotti,Brian
Dra 19h57'58"61d5'
Maripat
Cet 0h59'44"-0d34'
Maris Multimedia
Ori 5h57'15"17d58'
Maris,Melanie Lynne
Lyr 18h47'0"38d20'
Marisa
Cmi 8h5'29"0d58'
Marisa
And 23h48'0"33d59'
Marisa
Cyg 21h1'32"37d44'
Marisa
Leo 10h8'48"21d55'
Marisa
Crb 16h16'55"37d51'
Marisa
Ori 5h56'1"8d28'
Marisa
And 1h33'45"38d21'
Marisa
Cyg 19h43'1"38d41'
Marisa
Lyr 19h15'35"41d5'
Marisa J G J
Lyr 19h2'11"31d38'
Marisa Johana
Cas 23h23'51"58d37'
Marisa Lelia
Lyr 19h23'17"38d48'
Marisa Louise
Uma 11h29'24"45d3'
Marisa P G G
Cap 21h0'1"-26d33'
Marise Amelia
And 23h48'31"47d18'
Marise Nuelle
Mon 6h25'14"-8d56'
Marisel
Uma 10h50'36"57d6'
Marisela
Peg 22h10'52"24d4'
Marish,Jacqueline The Winged One
Cas 1h13'27"60d37'
Mariskovic,Karen I
Uma 9h55'0"50d10'
Marisol GG1294
Her 18h19'42"14d4'
Marissa
Aql 20h9'56"8d36'
Marissa
Cyg 20h36'17"39d55'
Marissa & Randy
Mon 6h55'29"8d43'
Marissa Dawn
And 23h42'34"44d8'
Marissa Nichole
Peg 23h24'36"8d12'
Marissa's Light
Peg 22h59'27"20d10'
Marissa-Our Shining Star
Tri 2h45'26"31d15'

Maristella
Her 16h10'30"10d5'
Marisuccia
Sco 17h52'1"-30d58'
Marisue
Hya 8h59'33"-7d59'
Marit
Cet 0h5'28"-18d35'
Marit
Uma 9h56'15"58d14'
Marita
Per 2h56'25"48d57'
Marital Bliss
Cas 1h37'15"60d17'
Maritan,Alessandro
Vel 10h11'58"-46d29'
Maritin und Iris
Tau 3h43'45"7d36'
Maritta
Dra 18h8'34"70d41'
Maritza
Mon 6h27'17"11d19'
Maritza Y Linda "Romeo Y Juliet"
Ori 6h0'34"-0d4'
Maritza's Earcake
Ser 15h54'21"4d25'
Mariucci,Anne
Uma 11h32'53"40d42'
Marius & Anjelika
Vul 20h26'34"28d17'
Marius,Charles Philippe
Per 1h31'39"52d30'
Marivic
Cnv 13h43'0"39d46'
Marja
Umi 15h25'0"67d46'
Marjac
Cas 2h30'52"77d17'
MARJACK
Dra 17h33'0"68d9'
Marjan & Navid
Ori 6h4'32"1d16'
Marjanian,Mickey- Michael G
Per 3h13'42"56d1'
Marji V
Lyr 18h46'24"35d31'
Marjo
Cyg 21h0'29"41d12'
Marjo the Dream Catcher
And 23h16'50"51d44'
Marjolaine
Per 2h58'16"46d28'
Marjoram
Aur 6h30'60"34d33'
Marjorie
Peg 22h12'31"32d58'
Marjorie
Sco 17h51'12"-40d45'
Marjorie
Boo 14h27'41"32d0'A
Marjorie Mae
Cyg 21h21'47"38d34'
Marjory Hart
Aur 5h39'48"33d50'B
Mark
Her 17h14'42"20d35'
Mark
Aql 19h26'20"13d16'
Mark
Peg 22h45'38"33d54'
Mark
Sgr 19h39'24"-39d36'
Mark
Her 17h44'57"14d40'
Mark
Dra 18h58'38"48d33'
Mark
Psc 23h1'28"0d33'
Mark
Lac 22h26'25"48d17'A
Mark
Cet 1h20'26"-3d32'
Mark
Uma 8h40'18"51d46'

Mark
Aur 5h0'1"42d7'
Mark & Anita
Sge 20h5'31"20d14'
Mark & Chris Forever March 10,1984
Her 16h10'52"48d30'
Mark & Clare
Cyg 21h28'0"38d22'
Mark & Claude
Aql 19h2'37"16d16'
Mark & Debbie
Lyn 8h1'58"34d6'
Mark & Diane 1994
Eri 2h58'25"-7d4'
Mark & Jennifer Forever
Aql 19h50'23"12d28'
Mark & Jennifer
Cnv 12h20'18"47d1'
Mark & Lainie
Uma 9h52'39"47d39'
Mark & Laura
Cyg 19h56'25"38d17'
Mark & Mary
Eri 4h56'37"-10d13'
Mark & Patty
Cyg 21h13'34"35d47'
Mark & Sarah Star,The
Cyg 19h27'48"33d8'
Mark & Stuart-Lee
Uma 9h44'46"67d50'
Mark & Susan
Cyg 21h0'24"39d21'
Mark & Terese
Cma 6h54'9"-18d36'
Mark & Tracy's Li'l Bit of Heaven
Crb 16h17'1"37d40'
Mark 1
Cep 21h47'25"65d0'
Mark Allen,Knight In Shining Armor
Her 17h34'14"20d26'
Mark Andrew
Ari 1h59'43"19d25'
Mark Andrew
Dra 11h35'24"72d12'
Mark Andrew/Cindi
Lac 22h44'37"53d37'
Mark Anthony
Her 16h21'25"21d51'
Mark Anthony
Cep 0h49'45"86d18'
Mark Bann Dark Star, The
Per 3h15'26"50d22'
Mark Christopher
Lac 22h1'43"50d32'
Mark E
Boo 14h23'33"31d29'
Mark et Stefany
Gem 6h50'26"18d43'
Mark Forever
Lyr 19h21'21"30d57'
Mark Forever Beloved Deborah
Cyg 21h23'21"30d57'
Mark Harrison Billy
Cnv 12h20'32"48d46'
Mark I
Hya 8h40'43"2d52'
Mark of Love - Nick & Aim
Sge 19h31'17"17d13'
Mark Scott
Mon 7h25'14"-0d26'
Mark Soul & Spirit
Cep 0h18'45"68d47'
Mark Star-zak
Boo 14h57'33"31d24'
Mark the Fitter
Mon 6h47'20"10d3'
Mark Thomas
Ori 5h57'45"17d34'
Mark To The Top-Love Karin
Lyn 7h28'57"45d18'

Mark Vincent
Vul 19h46'35"28d23'
Mark"Huggie"30
Cet 1h33'0"1d24'
Mark"The Dragonslayer"
Dra 17h56'22"58d13'
Mark's "Heavenly Body"
Sco 16h31'30"-44d19'
Mark's "Starlight, Starbright"
Dra 17h40'15"60d25'
Mark's Blue Butterfly
Peg 21h23'44"8d57'
Mark's Destiny
Lyr 18h56'36"32d25'
Mark's Dream
Cet 1h21'42"-14d45'
Mark's Fish-Umbra
Psc 0h45'26"20d11'
Mark's Infinity Squared
Per 2h35'20"50d54'
Mark's Magic
Her 15h52'36"41d46'
Mark's Magic
Lac 22h6'37"38d26'
Mark's Molly
Peg 22h1'28"31d51'
Mark's Myriad
Her 17h19'0"29d38'
Mark's Shining Star
Ori 5h33'28"-2d30'
Mark's Star
Dra 16h3'25"66d38'
Mark's Star
Hya 8h49'45"5d31'A
Mark's Star
Hya 8h54'27"1d47'
Mark's Star
Ori 5h55'52"15d3'
Mark's Star Guitar
Aur 4h59'1"33d57'
Mark's Starlight
Dra 14h0'1"63d41"
Mark,Allen David
Her 16h41'25"21d10'
Mark,Alyson Leigh
Peg 22h17'58"5d46'
Mark,Anne Isabel
Cam 10h24'0"81d41'
Mark,Camilla
Cam 7h48'12"71d59'
Mark,Chelsea Rayna Edelle
Cas 1h27'31"60d43'
Mark,David R
Her 18h4'17"38d27'
Mark,Riona Delaney
Mon 7h46'10"-2d49'
Mark,Tina
Crb 15h27'48"32d32'
Mark,You're My Piece of Heaven 143
Aql 19h59'20"10d25'
Mark,Zachary Willis
Cep 21h56'18"56d10'
Mark-And-Margarita 082090
Dra 17h1'29"72d34'
Mark-King of the Lobster Nebula
Vul 19h58'9"26d2'A
Mark-Mackey
Uma 12h54'0"53d50'
MarkAnne
Crb 16h6'23"26d12'
Markar
Tri 1h39'29"34d17'
Markar,John William
Her 16h44'1"30d0'
Markarian,Sawyer David
Per 2h9'45"57d10'
Markate
Ori 6h2'56"8d50'
Markay
Aur 6h54'0"38d40'
Markay
Lyn 8h36'0"43d32'

Markcon B
Vul 19h44'35"28d57'
Markealli,John A
Boo 14h51'11"50d38'
Markegard,Paul
Boo 14h5'11"32d27'
Markel,Bonnie
And 23h5'46"40d44'
Markel,George Sydney Junior
Hya 8h58'59"3d59'
Markelon,Michele
Tri 2h27'49"31d22'
Marker,Eugene F
Cma 6h54'23"-17d52'
Marker,Linda
Aql 19h14'14"13d23'
Markert,Ezekiel Allen
Boo 15h11'18"30d38'
Markertek
Per 2h36'1"46d51'
Markey's Folly-Fonz Bon
Ori 5h35'17"-2d32'
Markey,Jamie Ian
Her 17h21'18"27d24'
Markfield,Gloria & Tony
Tri 2h8'38"31d56'
Markfield,Josh & Barbara
And 23h40'48"44d14'
Markham,Celest
Lyr 18h36'54"41d18'
Markham,Charlotte Lucy
Cyg 19h34'59"33d4'
Markham,E H
Gem 7h8'8"21d37'
Markham,Eugene
Leo 10h31'2"23d36'B
Markham,Johanna
Leo 10h31'2"23d36'A
Markham,O P,Donna 25th Jubilee 1968-1993
Cnv 13h52'1"41d18'
Markham,Robin
Cyg 19h30'14"39d41'
Markham,Stella
Crb 15h59'41"28d22'
Markhoff,Peter
Cmi 7h18'11"8d30'
Markie C
Mon 7h54'19"-3d16'
Markiewicz,Steve
Ori 5h59'15"14d47'
Markitell,James N
Cyg 20h9'14"41d8'
Markkanen,Kalle John
Dra 16h46'60"62d32'
Markl,Terri
And 1h24'52"38d34'
Markland,Mercedes & John
Lyr 19h1'15"37d47'
Markle III,Kenneth L
Ori 5h55'17"6d34'
Markle's Star,Perri Murphy
Ari 2h48'31"30d15'
Markle,Brian
Cet 2h18'18"7d57'
Markle,Constance Abigail
And 2h34'26"40d3'
Markle,David Gates
Per 2h26'48"56d37'
Markle,Gilbert Scott
Cnc 9h1'19"32d23'
Markle,Halle
And 22h56'10"36d10'
Markle,Kendra Lys
Cyg 19h26'1"56d43'
Markle,Linda Marie
Peg 0h5'17"13d35'
Markle,Rae Lynn
Vir 11h58'36"8d26'
Markle,William Gates
Dra 15h15'41"57d49'
Markle,William Scott
Ori 6h16'20"-1d22'
Markley,Jen
Cas 2h33'32"57d34'

Marko
Aur 7h23'0"38d40'
Marko
Cam 12h37'1"80d22'
Markofer Elementary School,Florence
Uma 9h18'58"62d20'
Markoulis
Mon 7h11'44"-6d60'
Markovich My True Love,Jim
Aur 6h3'14"37d31'
Markovich,Paul Joseph
Lac 22h3'58"46d26'
Markovich,Steven Rosich
Peg 22h5'1"10d31'
Markovits,Helene
Uma 11h48'36"40d32'
Markovitz,Esq,Alan F
Her 18h4'22"38d22'
Markow,Rosemary
Lyn 7h53'45"43d17'
Markowitz,Christine
Cnc 8h5'43"18d48'
Markowitz,Deborah
Vir 13h17'4"-8d32'
Markowski,Gregory Joseph
Ori 5h33'50"-8d5'
Marks (Shish),Sharon
Cma 6h50'22"-19d49'
Marks,Alex James McCall
Her 17h52'24"20d13'
Marks,Alexandra Marella
Mon 6h48'35"10d36'
Marks,Allison
Sge 20h0'1"20d30'
Marks,Allison
Sge 20h0'53"19d7'
Marks,Ann Morwenna
Lyr 18h35'21"31d0'
Marks,Bert & Lou
Lyn 7h4'12"44d50'
Marks,David S
Aur 6h0'21"45d44'
Marks,Diana
Cas 0h50'0"62d4'
Marks,Dr Richard K
Oph 18h42'53"6d52'
Marks,Edward & Sylvia
Eri 4h10'55"-23d15'AB
Marks,Elisa Robbin
And 23h0'49"47d39'
Marks,Emily Louise
Mon 7h58'27"-4d15'
Marks,Esther & Irv
Lac 22h24'35"50d32'
Marks,Gertrude
Ori 5h52'37"20d18'
Marks,Irwin
Her 16h55'38"33d46'
Marks,Jacob Keith
Oph 17h6'59"10d51'
Marks,Jeffrey
Aql 19h0'50"14d49'
Marks,Kelby R
Cra 11h9'54"-18d56'
Marks,Laurie Martin
And 23h46'31"42d51'
Marks,Leslie & Lucille
Hya 8h25'43"0d8'
Marks,Lewis R
Cam 3h50'1"56d29'
Marks,Marla L
Sct 18h44'40"-6d9'
Marks,Mary Jo
Lyr 18h47'53"35d34'
Marks,MD,Richard A
Aql 20h7'54"3d46'
Marks,Miss
Del 21h5'47"12d11'
Marks,Morgan Marie
Mon 6h58'57"-10d29'
Marks,Phil
Boo 14h53'54"32d30'A
Marks,Richard E
Aur 5h0'28"41d53'

Marks,Spencer
Vul 19h1'19"24d28'
Marks,Staci D
Lmi 10h46'22"27d5'
Marks,Stephanie Elizabeth
Cyg 19h56'49"44d8'
Marks,Stephen Douglas Haig
Aur 7h14'19"41d24'
Marks,Virgil R
Aur 6h58'44"38d43'
Marks,Walter B
Per 2h53'18"40d19'
Marks-My Sunshine, Denise Cathleen
Lyr 18h59'27"29d13'
Marksity,Burke
Oph 17h19'34"11d47'
Marksity,Drew
Oph 17h19'44"11d34'
Markst
Uma 9h23'47"54d56'
Markstar
Cep 0h7'30"68d21'
Markstar
Lac 22h8'43"51d21'
Markt Control
Cnc 8h33'17"30d46'
Markunas,Mary
Sco 17h55'19"-40d18'
Markus
Cas 1h46'28"61d12'
Markus Linke
Aur 4h53'15"50d4'
Markus' "Thirtystar"
Dra 17h28'60"50d34'
Markus,"Sparky", Kathleen Alice
Cma 7h19'14"-16d18'
Markus,Frances & Fred
Aql 19h42'54"12d24'
Markus,Karl
Her 17h20'39"46d54'
Markus,Linda J
Cas 0h29'1"61d50'
Markus,Ronald Allyn
Dra 17h47'32"61d15'
Markus 23
Ind 21h19'23"48d40'
Marky & Jessy
Dra 19h47'11"60d15'
Marky Boy
Cma 6h56'3"-16d4'
Marky Star
Vul 19h47'22"28d49'
Marla
And 1h34'57"38d4'
Marla Lee
Tau 5h53'35"23d24'
Marla Loves Frank
Uma 9h55'53"45d12'
Marla Mi Amor
Mon 6h14'0"-10d34'
Marla's Heart
Cas 0h23'28"71d10'
Marla's Shining Star
Aqr 21h33'38"-0d46'
Marla's Star
Peg 0h1'57"13d11'
Marlaine
Cas 23h4'32"54d16'
Marland,Philippe
Lmi 10h49'18"30d39'
Marlatte,Valerie
Cet 1h45'0"0d39'
Marlaw
Cam 7h7'18"61d25'
Marlee
Dra 16h48'18"61d44'
Marlena Ann
Cap 21h53'48"-14d38'
Marlena's Birthday Wish
Peg 0h3'51"18d27'
Marlene
Cas 1h21'39"64d59'

Marlene
Cyg 21h56'17"44d47'
Marlene
Lyr 19h17'52"40d37'
Marlene
And 1h13'34"40d44'
Marlene
Leo 11h14'54"-5d11'
Marlene & Richard Always
Cyg 19h43'47"30d5'
Marlene Ann
Cam 9h15'0"80d1'
Marlene Louis
Vul 20h59'17"20d13'
Marlene's Cosmic Connection
Peg 21h27'11"22d47'
Marlenes Stellina
Aur 5h16'56"43d54'
Marler,Boyd
Her 17h21'0"22d23'
Marler,Jean
And 23h20'44"49d48'
Marler,Reece
Ori 5h51'30"14d55'
Marler,Winston
Cep 21h57'27"61d25'A
Marlett,Amanda K
Lyn 7h3'57"50d10'
Marlette,Jr,Walter Floyd
Per 2h59'25"48d56'
Marley
Ori 5h28'30"0d57'
Marley III,Raymond C & Terri
Weaver
Sge 20h1'43"19d17'
Marley,Allyson & David Turner
Aql 19h42'24"12d1'
Marley,Boyd Paul & Anne
Lovejoy
Cyg 19h17'53"44d16'
Marley,Hendrix
Uma 10h24'0"54d50'
Marley,Loretta Lu
Lyn 7h19'17"48d50'
Marlier,Christian
Del 20h36'44"4d38'
Marlier,Edmond
Cyg 20h42'48"47d43'
Marlin,Albert Lee
Lyr 18h58'21"34d9'
Marlin,David Michael
Aur 6h7'20"50d18'
Marlin,Elaine
Lyn 7h4'18"44d45'
Marlin,Marilyn
Cep 0h4'0"68d53'
Marlin,Stephanie Diane
Peg 22h38'42"22d21'
Marlin,William Michael
Her 17h23'36"21d14'
Marline
Lyn 9h34'17"40d52'
Marlo
Aql 18h57'53"3d16'
Marlo
Cyg 19h45'40"38d49'
Marlo Forever
Uma 11h24'53"58d20'
Marlo Marie
Aql 20h14'11"0d24'
Marlo,Ron
Per 2h55'60"50d5'
Marlou
Lyr 18h33'1"36d54'
Marlou & Karen
Aql 19h58'0"11d25'
Marlou (Nelen)
Cam 5h48'0"56d12'
Marlous
Uma 9h43'11"46d45'
Marlow,Brad
Cet 1h27'42"0d59'
Marlow,Brenne K
Uma 8h33'42"56d40'
Marlow,Dennis
Sex 10h35'53"2d40'
Marlow,Nancy
Cyg 21h0'28"31d17'
Marlowe PhD,Cynthia Marie
Oph 17h53'43"12d11'
Marlowe,Kimberly Dawn
Vul 19h19'17"22d38'
Marlowe,Sr,Thomas & Elaine
Cyg 20h14'12"37d52'
MarLyn Crew
Uma 10h10'43"67d40'
Marlyse
Peg 23h25'40"33d24'
Marmar Helen
Cyg 21h3'51"34d57'
Marmey45
Lyr 18h43'50"38d25'
Marmie
Cyg 19h30'19"35d41'
Marmotte Chef et Girafe
Cam 5h46'55"67d45'
Marmy
Lyr 18h47'20"42d13'
Marnace Star,The
Her 16h11'56"8d40'
Marne-Johnson
Lyr 19h17'1"41d25'
Marnell,Sue
Psc 1h26'56"22d34'A
Marnell,Todd & Tori
Eri 4h7'0"-18d35'
Marni
Cet 2h41'20"5d3'
Marni
Uma 12h1'25"31d36'
Marni
Cam 3h57'35"61d10'
Marnie
Aur 4h47'17"51d25'
Marnie Charlotte
Ori 4h45'52"5d1'
Marnie's Light
Aql 20h3'57"0d41'
Marnik
Del 20h21'1"11d11'
Marnina
Peg 21h58'27"30d19'
Marni
Uma 9h43'0"57d44'
Marny Rae
Cam 4h5'41"61d5'
Marny Star
Aur 4h53'21"51d2'
Marnzari,Sonja
Tau 3h40'25"10d4'
Maro,Gérard
Sgr 19h30'42"-30d1'
Marocchi,John Louis
Peg 23h29'50"33d10'
Marocchi,Sory
Uma 12h47'22"53d21'
Marocco,Isabel
Vul 20h14'46"25d1'
Marocco,Susan
Lyn 7h24'33"44d31'
Marold,Dennis Lee
Per 2h6'33"58d26'
Marolla,Daniel
Cep 23h12'58"65d26'
Marolla,Maura
Cyg 20h34'39"48d56'
Marolleau,Mireille
Ser 15h9'25"10d32'
Marolt,Ronald
Vul 19h48'1"20d17'
Maron,"Duffy"
Boo 14h22'43"37d57'
Maron,Anne-Claire
Cas 23h20'36"58d11'
Maron,Elizabeth Suzanne
Lac 22h5'54"50d17'
Maron,Ingrid Barbara
Cnc 8h34'30"31d32'
Marona,Danny
Lac 22h6'1"38d34'
Marone,Jr,Nicholas
Gem 6h54'26"30d18'
Maroni,Graziano
Lac 22h24'30"54d50'
Maroonie,Junie
Cas 1h52'46"58d11'
Marootian,Jeff
Boo 14h22'51"15d48'
Marosi,William Adalbert
Cnc 9h14'47"31d32'
Marot,Kelly Ann
Peg 23h22'43"15d52'
Marotta Shines Over Us,Ralph
Her 17h26'10"31d44'
Marotta You Belong To
Me,Michael
Her 17h20'47"37d36'
Marotta,Betty
Cyg 20h31'14'41d18'B
Marotta,Diane Elizabeth
Cas 0h16'1"58d18'
Marotta,Grace
And 2h19'35"41d3'
Marotta,S Daniel
Cyg 20h31'14"41d18'A
Marotta,Silvia
Gem 6h44'46"16d12'
Marotta,Vincent
Per 2h58'0"31d13'
Marotte,Joseph & Lorna
Crb 15h31'57"30d18'
Maroulis,Madonna Sue Rice
Uma 11h14'23"47d32'
Marovitz,My Passionate Star
Billy
Her 17h56'29"30d22'
Marovitz,William A
Cas 0h7'53"64d50'
Marowski,Maria P
Aql 20h33'50"0d50'
Marozsan,John & Anne
Cyg 21h31'44"36d23'
Marozzi,James Lee
Hya 8h55'11"1d16'
Marozzi,Maurizio
Umi 16h32'59"83d23'
Marpa
For 2h2'32"-24d30'
Marple,Susan Lorena
Lyn 8h49'1"34d58'
Marple,William Edward
Her 18h2'51"28d31'
Marples,William Dennis
Her 17h19'1"14d32'
Marquard,Hermann
Sgr 19h16'55"-40d13'
Marquardt,Jr,Allen Eric
Per 3h56'30"50d32'
Marquardt,Lisa M
Cas 0h54'42"59d50'
Marquart,Andrew Stuart
Dra 20h4'1"68d39'
Marquart,Christopher Louis
Aur 5h8'33"43d26'
Marquart,Marisa Dee
And 23h19'1"42d59'
Marquart,Matthew Christopher
Lyn 7h31'15"35d52'
Marquart,Matthew
Aur 7h22'15"40d17'
Marquefave Bajile Clémentine
Cam 3h36'12"60d16'
Marques,Vasco
Peg 23h26'59"23d60'
Marquessa
Equ 21h22'11"3d35'
Marquet,Donna
Scl 0h55'15"-25d36'
Marquette Turner's Wildlife
Preserve
Uma 11h53'0"40d12'
Marquez, Rosanna
Del 20h30'0"18d51'
Marquez,Bella Maria
Aql 19h13'14"13d25'
Marquez,Bianca Sofia
Aql 18h43'20"10d59'
Marquez,Eli
Cnv 13h33'20"42d17'
Marquez,Ernie
Dra 19h2'33"58d12'
Marquez,Jr,Jose Franco
Boo 14h2'30"20d16'
Marquez,Lori
Lyr 18h30'41"46d41'
Marquis,Gerald & Eileen
Sge 19h36'25"16d48'
Marquis,Gregory D
Cyg 19h21'10"28d59'
Marquis,Robert W
Boo 14h47'37"38d56'
Marquis,Rollin Hilary
Boo 15h6'12"18d58'
Marquise,Geoffrey
Aql 20h0'39"8d28'
Marquisee,Madalyn "Maddie"
Jaffee
Sco 16h55'41"-40d9'
Marquitta Rose
Uma 9h28'29"68d6'
Marr 1929,Ronald Joseph
Aql 20h7'42"4d33'
Marr,Barbara
Cet 2h24'29"-1d9'
Marr,Connie McCandless
Cyg 20h5'15"31d9'
Marr,Linda
Lyr 18h45'56"35d52'
Marr,Mrs Marguerite E
Lyn 9h6'31"39d53'
Marra,Andrea
Eri 3h49'46"-1d46'
Marra,Ashley Nicole
And 0h52'44"34d41'
Marra,Diomira
Oph 18h5'38"11d24'
Marra,Gary
Her 17h0'1"48d12'
Marra,Jennifer Lynn
And 1h11'24"38d43'
Marra,Joe
Per 2h1'0"57d24'
Marra,Patti
Cas 23h2'18"53d51'
Marrazzo,Dean Anthony
Cas 0h22'48"75d18'
Marregio
Ori 5h30'31"8d29'
Marrella,Lillian
Cas 1h3'0"60d30'
Marren,Mike
Oph 17h16'36"-24d28'
Marri
Sge 20h17'35"20d10'
Marriner,Lily
Cas 0h6'48"61d50'
Marrinson,Ivan Dawes Beach
Peg 22h26'21"30d0'
Marrinson,Leopold Cutler
Beach
Per 4h27'34"52d16'
Marriott Clan Star System,The
Arch
Ori 6h5'42"9d46'
Marriott,Carol & Graham
Sge 20h17'37"20d56'
Marriott,Lindsay Jane
Cas 2h5'0"65d19'
Marrocco,Alexie Nicole &
Erica Rae
Uma 11h36'0"63d16'
Marron,Tony
Cra 18h19'16"-45d12'
Marrone,Jr,Charles
Cep 22h59'30"65d28'
Marrone,Maureen Leahy
Lyr 18h53'13"33d54'
Marrs,Chris C
Lyr 19h23'0"31d24'
Marrs,Matthew Roy
Somerville
Umi 14h38'0"71d51'
Marrs,Twila J
Aql 19h58'54"8d16'
Marrs,Yvette
Ori 5h56'0"21d10'
Marrufo,I Love You, Kathy
Aql 23h37'30"-5d18'
Marry Me
Lyr 18h59'32"34d41'
Mars,David Michael
Boo 14h4'34"27d4'
Mars,Linda
Gem 6h35'22"26d45'
Marsalis,Bryan A
Oph 17h34'44"-22d3'
Marsalisi,JoAnne
Lyn 8h56'57"33d58'
Marsalisi,Lisa
Cas 0h6'13"58d52'
Marsan Systems Inc
Boo 14h25'21"45d7'
Marsan,Sabrina Marie
Lyn 8h50'57"45d21'
Marsande,Anne Laure
Vul 19h46'58"20d24'
Marsco,Rosemary Donna
And 1h39'12"36d55'
Marsden,Barry J
Cep 21h12'32"67d38'
Marsden,Leslie Kenneth
Mon 6h19'51"-6d24'
Marsden,Neil
Ori 6h4'13"4d27'
Marselli,Giorgia
Cae 4h52'26"-32d60'
Marsey,Edward Robert
Boo 14h59'15"40d9'
Marsh "Michelle", Patricia
Ann
Lyn 8h9'0"49d7'
Marsh III,Sherman
Cet 2h7'1"3d56'
Marsh,Arnold "Dad"
Per 3h54'27"41d4'
Marsh,Benjamin
Ori 4h44'20"15d25'
Marsh,Claire Coleman
Eri 3h35'43"-4d1'
Marsh,Claire Victoria
Per 4h7'35"40d53'
Marsh,Daniel C
Dra 17h0'38"64d9'
Marsh,Dave
Cep 21h45'57"70d49'
Marsh,Donald & Nancy
Sct 18h55'14"-8d14'
Marsh,Dorothy
Lyr 18h14'35"42d52'
Marsh,Ernestine J
Aur 6h19'48"33d52'
Marsh,Gerry
Vul 20h58'19"28d15'
Marsh,Glenda
And 0h12'17"39d40'
Marsh,Gregory Joseph
Per 1h47'14"56d30'
Marsh,Jon
Boo 15h5'20"21d48'
Marsh,Julianna
Lyr 18h29'50"38d46'
Marsh,Loy Ann
Peg 21h58'30"23d41'
Marsh,Nancy Lynn
Lyr 19h20'46"33d46'
Marsh,Olive Rose
Dra 14h50'42"56d5'
Marsh,Patsy
Mon 7h1'0"4d20'
Marsh,Paul David
Mon 8h2'57"-2d7'
Marsh,Richard E
Uma 9h6'39"47d7'
Marsh,Robert & Jean
Per 3h9'0"46d36'
Marsh,Robert L
Boo 13h37'53"22d29'
Marsh,Rosie Maria
Cas 0h19'33"63d52'
Marsh,Stephen David
Ori 5h53'24"18d18'
Marsh,Stuart Robert
Per 3h9'19"37d39'
Marsh,Timothy B
Aur 4h49'43"51d34'
Marsh,Wendy & Peter
Uma 11h42'23"40d29'
Marsha
And 2h20'0"45d5'
Marsha
Cyg 20h22'15"39d26'
Marsha
Boo 14h58'60"11d26'
Marsha & Dave
Lyn 7h35'26"40d17'
Marsha's Happy Place
Cyg 19h19'1"44d33'
Marshack,George & Pam
Cam 5h56'23"58d52'
Marshal,Julie Dennehy
And 0h53'53"36d39'
Marshalabra
Cam 5h55'19"60d17'
Marshall
Her 16h25'12"27d44'
Marshall
Aql 19h12'31"19d45'
Marshall (Bali Hai), Maralyn
Cas 0h18'1"61d35'
Marshall Star,The Glenda
And 1h6'58"39d28'
Marshall's Milky Way
Del 20h24'13"10d11'
Marshall's Star,Baby Megan
Ori 5h57'1"20d42'
Marshall,Allan Leslie
Boo 14h20'57"51d38'
Marshall,Anna-Marie
Cas 1h58'37"63d48'
Marshall,Bill
Hya 9h5'22"4d2'
Marshall,Brenda M
Del 20h13'39"12d54'
Marshall,Bud & Martha
Cyg 21h27'32"40d32'
Marshall,Candace Louise
Pup 7h40'49"-50d16'
Marshall,Catharine A
And 2h23'29"40d52'
Marshall,Charles Henry
Her 17h20'49"27d14'
Marshall,Daniel Alexander
Dra 16h57'16"62d44'
Marshall,Diana
And 1h42'47"41d4'
Marshall,Diana Lynne
Eri 2h58'44"-18d16'
Marshall,Donald S
Her 16h14'38"24d48'
Marshall,Dr Tom
Uma 8h45'15"52d26'
Marshall,Earline
Lyr 18h58'50"38d53'
Marshall,Ed & Rita
Eri 2h49'39"-1d58'
Marshall,Elaine D
Cyg 19h34'55"32d55'
Marshall,Eric
Boo 15h5'18"20d4'
Marshall,Erin Kathleen
Lyn 6h25'29"60d35'
Marshall,Florence
Equ 20h58'55"9d12'
Marshall,Frank
Boo 14h3'27"27d29'
Marshall,Fred
Aur 7h12'45"41d54'
Marshall,Geoffrey & Evelyne
Aur 5h1'50"29d9'
Marshall,George Anderson
Sex 10h1'13"-0d43'
Marshall,George Henry
Aur 5h52'0"38d53'
Marshall,Greg
Lyr 19h16'37"25d53'
Marshall,Hellen Victoria
Lyn 7h57'1"35d28'
Marshall,Henry E
Ori 5h12'27"-4d38'
Marshall,Irma
Cas 1h54'0"67d57'
Marshall,James Richard
Her 18h39'41"18d54'
Marshall,Janet Knox
Lyn 7h48'12"50d18'
Marshall,Jean & Joe
Cyg 21h38'27"30d43'
Marshall,John
Lac 22h25'22"52d35'
Marshall,Jill
Lyn 6h37'54"54d18'
Marshall,Jim
Hya 9h10'58"3d56'
Marshall,John
Cet 0h0'30"-12d24'
Marshall,John
Dra 16h41'29"61d41'
Marshall,John
Aur 6h11'0"37d1'
Marshall,John (Chick)
Lyn 9h14'50"40d3'
Marshall,Julie
Lyr 18h46'12"42d30'
Marshall,Katherine
Vir 12h1'0"0d49'
Marshall,Katherine Louise
Cas 0h30'0"60d12'
Marshall,Marc Steven
Aql 19h5'26"2d23'
Marshall,Mary S
Lyn 7h55'48"35d2'
Marshall,Myia Renee Ellyson
Eri 5h0'43"-5d53'
Marshall,Oliver
Ori 5h57'38"15d2'
Marshall,Phyllis Jean
Lyr 18h30'41"31d13'
Marshall,Richard
Aql 19h44'18"12d4'
Marshall,Rifka"Rikki"
Cas 0h21'39"58d36'
Marshall,Rob
Sge 20h17'41"16d45'
Marshall,Robert
Her 15h50'0"46d7'
Marshall,Robert
Her 10h14'1"30d40'
Marshall,Saint Edward
Douglas
Aqr 22h48'1"-6d5'
Marshall,Sharon Kathleen
Aqr 22h2'0"-5d44'
Marshall,Smantha Kay
Lyn 7h 26'47"44d32'
Marshall,Sr,Thomas A
Ori 5h3'23"14d57'
Marshall,Stephanie Anne
Mon 6h54'0"1d30'
Marshall,Stevie
Umi 15h30'17"68d35'
Marshall,Susan Jean
Lac 22h52'0"56d49'
Marshall,Susan Kelly
Peg 22h11'30"4d43'
Marshall,Tamara Louise
Eri 3h6'31"-2d29'
Marshall,Tony
Uma 10h47'1"40d9'
Marshall,William Ross
Boo 15h24'1"50d47'
Marshall,William W
Oph 17h27'38"-20d30'
Marshall-Day,Cindy
Mon 6h22'17"3d40'
Marshall-Moore,Jeanne Anne
Ari 2h32'44"21d51'
Marsham-Townshend,June
And 23h0'34"51d51'
Marsharita 95
Peg 21h41'43"22d41'
Marshello,Alan L
Cyg 21h1'24"39d48'
Marshmallow Land (Andrew)
Her 16h26'22"36d40'
Marshman,Philip Pipit
Per 4h5'32"38d38'
Marshman,Terry
Peg 21h56'23"21d11'
Marshoff,Susan
Cyg 19h36'23"28d3'
Marshér
Lyn 7h10'21"53d50'
Marsic,Kimberly A
And 23h42'45"40d13'
Marsicano,Nancy
Eql 21h0'52"10d44'
Marsiglia,Paola
Aur 4h59'57"40d22'
Marsilio,Louis James
Uma 8h57'41"68d48'
Marsjo
Sgr 19h57'18"-40d20'
Marsland,Betty
Cas 3h11'13"58d51'
Marsocci,Steven
Cnv 13h41'1"45d9'
Marsollier,Daniel
Mon 7h46'39"-4d10'
Marson,Kyle Alexander
Boo 13h34'0"15d28'
Marsteen,Clarice M & John B
Per 3h35'56"40d51'
Marston,Barbara K
Lyn 7h41'30"45d55'
Marston,Claire Angel
Aql 19h42'59"11d20'
Marston,Michael
Hya 8h11'1"1d51'
Marston,Paul Francis
Boo 14h40'34"31d29'
Marston,Roxann
Uma 10h4'53"48d19'
Marston,Sande
Com 13h22'0"26d59'
Marston,William
Cep 20h48'35"61d12'
Marsura,Mae Janet
Cam 11h36'0"80d5'
Marszalek,Gwiazda
Cmi 7h41'36"5d13'
Marszalek,Joseph Anthony
Her 18h5'19"19d39'B
Marszalek,Marilyn M & Greg
C Jones
Crb 15h57'42"37d38'
Marszalek,V Jan
Tri 1h30'16"33d39'
Mart Anthony
Her 18h43'43"12d30'
Marta
Cas 1h37'47"58d54'
Marta
Eri 3h55'46"-6d55'
Marta
Lyr 18h48'21"41d29'
Marta
Lyn 7h54'1"40d13'
Marta
Del 20h17'15"12d26'
Marta Lluch Seyda
Ser 15h15'28"1d46'
Marta P K 83
Lyn 7h8'36"44d60'
Martarella's Star,Joe
Cnv 13h56'0"38d13'
Martas 27121991
Peg 21h52'19"30d44'
Martasin,Ed
Aql 19h47'28"14d52'
Martay,Dolores
And 0h31'20"41d9'
Marteau,Marie Carole
Ori 6h2'57"1d15'
Martedi
Pyx 8h46'4"-21d9'
Marteen, Anna
Lyr 18h44'51"45d38'
Marteka Wedding Star, The
Cyg 21h22'22"31d39'
Martel T Q T A 9621, Jolie
Annie
Ori 5h56'42"17d43'
Martel,Carol "BSITU" & Larry
Cyg 21h14'25"37d27'
Martel,Normand
Per 3h19'30"43d40'
Martel,Priscille
Uma 8h38'26"58d33'
Martel,Victoria Renee
Aur 4h59'52"40d9'
Martell,Edward
Cet 1h14'51"-0d13'
Martella,Bill
Uma 11h20'1"33d44'
Martella,James A
Cep 22h34'33"48d48'
Martelli,Debbie Paschetti
Cas 1h47'41"60d21'
Martelli,Gino
Peg 22h26'12"30d39'
Martello,Pamela Ann
Cas 0h58'36"70d18'
Martello,Peter James
Cnv 12h28'32"40d2'
Marten,Brocklyn,Dustin &
Saige
Her 17h39'43"15d58'A
Marten,Wesley Levi
Her 17h14'10"41d35'
Martens aka "The Alien",Ernie
B
Per 3h13'0"40d53'
Martens,Elin Christina
Vir 13h52'50"2d16'
Martens,Frances
Boo 14h7'48"40d36'
Martens,Gregory
Cmi 7h44'26"1d6'
Martens,Gregory Alan
Aur 5h12'20"41d30'
Martens,Hazel M
Aql 20h0'55"12d12'
Martens,Maria Elana
Sgr 20h18'4"-31d12'
Martens,Michael Johannas
Uma 10h35'0"58d3'
Martens,Mike & Lynette
Cyg 19h50'24"41d13'
Martens,Robert William
Her 16h58'19"31d18'
Martes,Bailey Scott
Cnv 13h46'51"32d38'
Martes,Melanie
Lyn 8h12'36"52d2'
Martesio,"Pop-Pop" John
Her 16h24'39"31d60'
Martey-Gems Who Sparkle,A
& M
Uma 11h57'44"32d18'
Martfeld,Levi
Her 18h6'47"30d56'
Marth,Chase
Her 17h24'45"50d14'
Marth,Christopher
Aql 19h7'30"15d55'
Martha
Tri 1h59'0"27d57'
Martha 12-17
Cas 1h18'36"61d23'

Martha Anne
Hya 8h55'1"6d10'
Martha Dear
Cas 0h8'29"64d9'
Martha Elena
And 0h59'27"45d32'
Martha Louise
Uma 9h39'0"50d1'
Martha Margarita
Com 12h24'21"24d53'
Martha P
Cyg 21h27'1"28d31'
Martha's Porch
Crb 16h8'17"30d26'
Martha's Star
Her 16h42'22"48d28'
Marthey,Ed
Ari 1h46'0"17d59'
Marthony
Cam 5h28'0"68d22'
Marti
Cep 0h43'34"86d52'
Marti
Dra 19h42'22"61d40'
Marti "The Dancing Star"
Uma 10h41'41"40d13'
Marti & Ron's "Wedding Star"
Cyg 21h25'54"30d26'
Marti,Beatriz Mur
Cyg 21h36'1"40d48'
Marti,Timothy & Lauren
Cnc 8h37'15"17d37'
Martianich
Tri 1h58'30"33d12'
Marticus
Hya 9h25'4"-2d27'
Martignetti,Ferdinand
Lac 22h11'1"51d48'
Martil,Neil Alexander
Her 15h57'19"40d11'
Martin
Hor 3h7'44"-49d9'
Martin
Pup 7h27'30"25d35'
Martin
Aur 5h1'1"50d40'
Martin
Her 18h19'35"28d31'
Martin
Aqr 21h31'40"-1d34'
Martin
Ser 15h30'8"18d8'
Martin "Mim",Marion G
Aql 19h51'44"15d9'A
Martin & Sheila
Cyg 20h15'57"39d34'
Martin 3
Aur 5h25'28"50d36'
Martin 5/8/65-12/9/90, Tana Wynette
Lyn 7h45'27"41d37'
Martin Albert
Eri 3h31'24"-6d34'
Martin II,Michael L
Oph 18h39'22"8d44'
Martin III,George Elliot
Aur 5h0'38"40d9'
Martin III,Harry F
Boo 14h40'33"21d27'
Martin IV,Dee Elmo
Uma 11h7'47"31d16'
Martin J
Cep 21h38'49"80d3'
Martin Lee
Ori 6h1'10"8d1'
Martin Lynn
Uma 14h3'0"51d13'
Martin M
Cep 22h29'20"59d3'
Martin One
Per 2h40'59"34d60'
Martin Star,Charles Albert
Aur 6h28'22"33d2'
Martin Star,Jesi Danielle-The
Ser 18h37'0"3d5'

Martin Two
Per 2h40'34"34d37'
Martin"Preacher", Jennings E
Lyr 18h58'11"31d14'
Martin's Gallow Pole
Cam 3h57'22"62d30'
Martin,5/13/67, Michelle Renae
Peg 23h37'0"20d53'
Martin,Ace Joseph
Boo 14h38'32"37d49'
Martin,Adam Parker
Ari 2h23'44"12d25'
Martin,Adison B
Uma 8h36'18"48d55'
Martin,Alice
And 2h4'36"43d2'
Martin,Alton Lamar
Hya 8h10'29"1d41'
Martin,Ann Marie
And 2h25'43"42d44'
Martin,Annabelle & Andy
Cyg 21h51'21"42d56'
Martin,Anne Marie Hughes
Cas 3h5'48"58d7'
Martin,Arleen Russo
Vul 20h39'37"20d32'B
Martin,Ashley
Cnv 12h12'22"40d25'
Martin,Bethany
Ori 5h14'1"-4d34'
Martin,Betty Frances
Tau 4h6'27"20d59'
Martin,Brad T
Boo 13h34'52"20d38'
Martin,Brandon
Her 17h34'0"26d19'
Martin,Bryan Lee
Vul 20h39'37"20d32'A
Martin,Bud & Dort
Her 16h54'29"48d10'
Martin,Byron Keith
Eri 3h30'30"-6d35'
Martin,Caroline E
Lyn 9h15'23"40d47'
Martin,Carolyn P
Peg 21h40'45"24d55'
Martin,Cathy Ann
Mon 6h52'48"-5d58'
Martin,Catlin Marie
Del 20h18'40"10d40'
Martin,Cedric
Sgr 19h26'55"-34d51'
Martin,Christopher
Ser 16h5'58"10d36'
Martin,Christopher
Sex 10h41'0"1d58'
Martin,Cindy
Cyg 19h44'1"37d57'
Martin,Claire
And 0h31'51"45d27'
Martin,Colette
Aur 5h33'21"29d39'
Martin,Coley
Aur 6h39'43"37d25'
Martin,Corinna Gayle
Lyn 8h9'0"39d13'
Martin,Crickett Nicole
And 23h45'1"43d53'
Martin,Dan
Hya 9h14'19"-13d55'
Martin,Dana
And 2h28'1"38d14'
Martin,David M
Vul 20h43'57"20d16'
Martin,Debbie
Aql 19h57'49"-5d49'
Martin,Debra
Del 20h15'17"13d51'
Martin,Dominique
Aur 6h2'27"38d30'
Martin,Donald
Lac 22h24'15"40d47'
Martin,Doris
Cyg 20h35'25"38d24'

Martin,Dorothy
Oph 17h55'55"13d59'
Martin,Dorothy Everett Matson
Leo 10h3'38"15d12'
Martin,Dorsey(Beau)
Uma 13h55'12"61d21'
Martin,Dr Ruth S L
Aur 7h22'33"37d59'
Martin,Dr Sam
Oph 17h50'20"13d13'
Martin,Dylan John
Her 18h5'21"46d55'
Martin,Edna "Mimi"
Cas 1h7'33"58d5'
Martin,Elena
Cyg 19h33'39"39d32'
Martin,Elena
Cep 22h6'37"60d49'
Martin,Elliot William
Per 1h56'14"50d22'
Martin,Eric
Cam 13h3'31"81d31'
Martin,Erikka
Lyn 7h57'58"40d52'
Martin,Erin F
Lyn 7h34'31"35d47'
Martin,Fiona G
Lyr 18h45'1"31d13'
Martin,Fred
Peg 22h10'25"28d26'
Martin,Freda May
Crb 16h11'26"28d50'
Martin,Freddie
Cep 0h8'19"80d10'
Martin,Garrett Quinn
Aur 6h17'47"45d3'
Martin,Gary L
Dra 19h22'29"70d24'
Martin,George & Vicki
Cyg 20h21'48"38d43'
Martin,Ginger
Cmi 7h6'44"3d49'
Martin,Gloria
Crt 11h17'5"-11d50'
Martin,Greg
Hya 8h14'45"-6d34'
Martin,Guy
Ori 5h57'0"14d50'
Martin,Gérard
Sex 10h14'29"-1d5'
Martin,Hap
Ori 4h43'32"4d44'
Martin,Hectorin Carlos Gomez
Boo 15h4'51"40d20'
Martin,Herbert Albert
Uma 11h51'41"64d14'
Martin,Ivan Gomez
Sct 18h50'55"-6d59'
Martin,J R
Lac 22h51'41"53d3'
Martin,Jack Cochran
Her 17h50'1"40d43'
Martin,Jacob Tanner
Uma 9h12'38"49d14'
Martin,Jacquelyn E
Peg 21h59'45"31d51'
Martin,James
Her 16h57'51"24d58'
Martin,Jamie-Lee Serena
Umi 15h55'52"71d14'
Martin,Janina Elyse
Mon 6h44'53"8d56'
Martin,Jay
Her 17h22'44"48d7'
Martin,Jean-Philippe
Per 4h1'55"51d50'
Martin,Jeannine & George
Cnv 13h22'0"41d9'
Martin,Jeffrey D
Peg 23h29'20"18d35'
Martin,Jeffrey W
Dra 20h3'48"62d24'
Martin,Jennifer
Cas 1h49'14"58d16'

Martin,Jessica Whitney
Lac 22h21'59"50d30'
Martin,Jeute
Oph 18h2'22"10d53'
Martin,Jill S
Cas 0h32'1"54d49'
Martin,Jimbo
Ser 15h28'35"0d26'
Martin,Joan Huguet
Cas 23h1'29"53d43'
Martin,Joeffrey
Per 3h51'43"37d41'
Martin,John Edward
Hya 09h36'58"-18d47'
Martin,John Kenneth
Mon 7h15'49"-5d5'
Martin,John R
Her 17h6'27"21d25'
Martin,John Robert
Tau 5h56'20"26d54'
Martin,John Thomas
Cnv 13h50'60"30d30'
Martin,John William
Uma 11h51'17"53d8'
Martin,Jr,Francis P
Lyn 7h59'36"44d53'
Martin,Jr,Paul Lawrence
Aql 19h29'17"10d37'
Martin,Jr,Thomas L
Boo 13h54'0"20d2'
Martin,Jr,William Raymond
Cyg 20h57'34"38d32'
Martin,Jr,William H
Cep 22h18'59"61d7'
Martin,Karla W
Cas 0h29'41"61d27'
Martin,Katherine Elizabeth
Cyg 19h59'59"38d19'
Martin,Katherine
Lyr 18h48'43"35d13'
Martin,Kathleen P
Vul 20h43'1"20d37'
Martin,Katie
And 23h34'26"45d50'
Martin,Katie Lee
Lyr 18h32'1"44d13'
Martin,Kay & Robert
Crb 16h16'45"32d8'
Martin,Kelly
Aql 19h28'43"-06d14'
Martin,Kevin
Boo 14h21'43"25d28'
Martin,Kim
Oph 18h25'45"7d35'
Martin,Kyle Arthur
Vir 12h4'43"2d7'
Martin,Leo
Aql 19h0'19"1d15'
Martin,Libby
And 23h2'43"41d8'
Martin,Lillian Marie
Uma 11h33'52"63d10'
Martin,Linda
And 0h18'16"35d16'
Martin,Lisa Kay
And 0h19'58"31d58'
Martin,Lori M
Cyg 21h0'43"30d28'
Martin,Mama & Charlie
Cyg 21h18'41"34d31'
Martin,Mandy Jane
Cas 0h32'50"73d56'
Martin,Maria De Las Nieves Navarroy
Col 6h0'3"-37d36'
Martin,Marie Alder
Tau 5h53'1"28d33'
Martin,Marie Paule
Lyn 8h28'0"34d0'
Martin,Marilyn K
Cas 0h24'31"61d41'
Martin,Marlise Marie
Mon 6h55'13"-3d6'
Martin,Mary
Cas 0h8'29"65d22'

Martin,Mary Chanel
Peg 22h58'43"34d53'
Martin,Mary Louise
Aql 18h56'47"13d33'
Martin,MD,"Tom",John T
Aql 19h51'44"15d9'B
Martin,Meghan Elizabeth Ann
Cas 1h6'26"52d43'
Martin,Melody
Cas 1h13'59"63d32'
Martin,Michael A
Ori 5h59'13"15d58'
Martin,Michael Douglas
Umi 15h17'1"81d9'
Martin,Michael J
Equ 20h59'55"3d49'
Martin,Michel
Ori 4h43'0"9d26'
Martin,Michel
Dra 15h13'44"62d9'
Martin,Mikael
Cet 0h32'17"-5d7'
Martin,Milton Foy
Tri 1h50'13"27d6'
Martin,Miranda Marie
Uma 10h12'60"68d11'
Martin,Nancy
Cyg 21h7'15"40d10'
Martin,Nancy & Mike
Peg 23h42'29"27d42'
Martin,Nancy Casteel
Umi 15h10'1"66d22'
Martin,Nancy S
Cma 7h1'0"-16d17'
Martin,Nicholas Richard
Tau 3h52'15"18d56'
Martin,Nicola
Cyg 19h39'43"30d12'
Martin,Nicole
Cyg 20h28'12"38d40'
Martin,Nora A
Lyn 7h49'50"50d58'
Martin,Norma Helen
Cyg 21h48'12"52d34'
Martin,Oliver/Karen Myers
Lyr 19h12'39"37d35'
Martin,Olivia
Lyr 18h26'0"42d14'
Martin,Orlando & Robert Noreen
Cet 2h46'43"2d34'
Martin,Palmira Mary
Ori 5h53'42"16d21'
Martin,Pamela "Bobo"
Cet 1h9'22"-1d42'
Martin,Patches Anthony
Lac 22h17'57"49d25'
Martin,Patricia Claflin
Cas 23h2'1"53d27'
Martin,Peter
Boo 14h53'48"23d51'
Martin,Rebecca Ruth
Vir 12h27'42"-7d35'
Martin,Rhonda
Lyn 8h29'53"49d42'
Martin,Richard Frederick Kohl
Lib 15h32'37"-28d11'
Martin,Richard John
Aur 7h0'0"36d14'
Martin,Richard W "R-Dub"
Dra 18h16'18"80d28'
Martin,Rick
Boo 14h56'33"44d27'
Martin,Robert Ashby
Boo 14h43'54"26d5'
Martin,Robin
Mon 8h3'43"-0d30'
Martin,Robyn Marie
Mon 7h44'11"-2d10'
Martin,Ron
Uma 12h3'50"31d50'
Martin,Russell David
Ori 5h52'0"19d35'
Martin,Ryan F
Lyn 7h33'1"44d48'

Martin,Ryan Thomas
Per 1h50'59"50d16'
Martin,Régis
Sex 10h13'56"-7d22'
Martin,Sam L
Cep 21h23'37"56d14'
Martin,Samantha
Mon 6h55'0"-10d3'
Martin,Samantha Marie
Cas 1h4'58"60d47'
Martin,Samuel
Psc 1h0'32"20d24'
Martin,Sandra Lee
Cas 1h44'0"60d13'
Martin,Sean,Keith & Ryan
Boo 14h44'18"26d15'
Martin,Shareen
And 1h16'32"40d9'
Martin,Sharon Ann
And 2h23'0"45d21'
Martin,Shaun & Stefan
Cyg 21h0'16"31d34'
Martin,Sherrie Anne
Hya 9h17'22"0d17'
Martin,Sonia,Trent & Molli
Crb 16h5'55"37d35'
Martin,Sr,Mearis Massie
Pho 0h30'5"-44d20'
Martin,Stanley J & Vernell Peterson
Uma 9h58'0"59d47'
Martin,Steve
Aur 5h6'40"40d25'
Martin,Stuart
Cyg 19h39'30"30d15'
Martin,Sue Ann Stricker
Cyg 21h13'58"28d45'
Martin,Susan J
Aql 19h31'44"10d10'
Martin,Susan R
Uma 9h28'0"51d48'
Martin,Tara Ann
Eri 4h8'28"-14d7'
Martin,Thomas Edward
Cnv 12h35'31"37d36'
Martin,Tim
Cep 23h20'46"65d17'
Martin,Timothy John
Ori 4h43'17"10d56'
Martin,Tobias Philipp
Aur 5h10'19"40d56'
Martin,Tommy
Lmi 10h44'31"23d46'
Martin,Tony
Cep 20h25'45"63d25'
Martin,Tracy Genelle
Aql 19h53'33"15d13'
Martin,Trudy D
Crb 16h2'22"27d35'
Martin,Vlasta
Cmi 7h6'38"4d25'
Martin,Walter E
Cma 7h20'20"-16d20'
Martin,When You Wish Upon A Star Paula Lee
Crb 15h33'1"31d13'
Martin,William R
Oph 18h2'25"14d0'
Martin-Aldrich
Oph 17h36'13"8d31'
Martin-Gustin,Lisa
Dra 20h29'40"73d55'
Martin-My One & Only Star,Keith E
Ser 15h35'0"8d34'
Martin-Neuville, Pauline
Cam 5h52'43"60d23'
Martin-Soltau,Wesley C
Her 18h11'35"41d9'
Martina
Hya 8h31'28"-11d11'
Martina
Tau 3h54'13"2d5'
Martina
Lyr 19h2'16"38d28'

Martina
Gem 6h57'59"18d30'
Martina (Kleinpuh)
Cnc 8h29'35"11d10'
Martina Holiday Star
Cas 3h1'58"60d46'
Martina Stephani
Ori 5h39'53"8d31'
Martina The Most Beautiful Star
Ser 15h14'16"1d58'
Martina Theresia
Lyr 18h31'27"34d11'
Martina,Dr John
Hya 8h11'1"0d42'
Martinalbert,William H & Pamela N
Cyg 20h51'19"37d33'
Martindale,Allen & Barbara
Eri 3h16'20"-12d43'
Martindale,Wendy
Lyr 19h4'11"26d6'
Martindelcampo,Joe
Aur 5h58'32"48d57'
Martine
Cam 3h32'38"53d60'
Martine et Stephanie
Lyn 7h52'33"58d18'
Martine Moura
Lyn 8h10'52"40d26'
Martine,Alessi
Mon 7h47'57"-4d6'
Martine,Inaccessible Reve
Ori 6h15'53"11d2'
Martine,Louis et Marc Arnold
Sgr 19h31'31"-35d51'
Martine,Martin
Crb 15h53'24"35d19'
Martine,Michaela
Cyg 20h8'49"58d14'
Martineau "Star Albumn"
Cam 6h49'21"68d8'
Martineau,Etoile d' Amour,Eric
Her 17h30'54"21d31'
Martineau,Helen Evelyn Resendiz
Aql 20h6'20"4d22'
Martinek,Andrea
Vir 12h55'35"-11d28'
Martinell,Scott Thomas
Per 2h59'1"56d25'
Martinelli,Anthony
Cep 0h8'39"70d30'
Martinelli,Inger Olsen
Aur 5h59'24"30d3'
Martinelli,Jean Elliott
Per 3h13'35"40d5'
Martinelli,Louise
Eri 5h5'54"-8d15'
Martinelli,Patricia
Cam 4h19'26"68d26'
Martinet,Charles Andre
Aur 7h5'49"38d28'
Martinez Cinco de Mayo Star,Roberto
Tau 5h51'42"28d47'
Martinez Forever Remembered,Daniel
Crt 11h0'51"-8d49'
Martinez Forever, Carrie & Steve
Lyn 9h17'38"34d10'
Martinez,"TCM" Tech Carlos M
Aql 19h55'43"15d4'
Martinez,(RLMF)Richard Glen
Equ 21h6'59"10d41'
Martinez,A
Cep 21h24'26"55d9'
Martinez,Algela De Las Mercedes
Cam 5h55'19"60d17'
Martinez,Anna Mae
Cyg 21h2'30"30d58'

Martinez,Annie
Ori 6h1'13"18d45'
Martinez,Aurora
Hya 9h3'44"2d8'
Martinez,Blas P
Hya 8h50'58"1d32'
Martinez,Carla P
Lyn 7h27'0"44d6'
Martinez,Christelle
Cep 21h20'54"68d59'
Martinez,Dakota Lee
Mon 6h12'40"-10d23'
Martinez,Dalinda
Peg 23h43'31"31d12'
Martinez,Damian Munoz
Per 3h29'51"31d18'
Martinez,Daniel Andrew
Her 18h13'26"48d59'
Martinez,David Gabriel
Lac 22h7'28"48d14'
Martinez,Derrick David
Aql 19h55'38"10d40'
Martinez,Deserée Nicole
Cyg 20h27'15"48d54'
Martinez,Devon Makena
Mon 6h12'49"-10d34'
Martinez,Emmanuel Alvelo
Eri 4h8'48"-11d34'
Martinez,Fernando
Cma 7h49'0"10d18'
Martinez,Fernando
Aur 6h10'47"30d23'
Martinez,Frances
Mon 6h54'16"-10d6'
Martinez,Frank Andrew
Per 1h55'16"53d3'
Martinez,Gene
Aql 20h1'16"6d11'
Martinez,Grandpa Thomas
Lyr 18h42'15"31d4'
Martinez,Jesse
Hya 9h7'0"3d47'
Martinez,Jesus "Jesse"
Cep 3h21'37"78d13'
Martinez,Joe Frank
Aql 18h59'18"14d4'
Martinez,Jolan
Her 17h18'1"43d24'
Martinez,Jorge Alberto Cabastida
Cep 21h11'14"58d50'
Martinez,Jorge F
Ori 5h48'12"21d1'
Martinez,Jose Cela
Cas 1h18'48"50d31'
Martinez,Josie
Cam 3h17'57"58d21'
Martinez,Jr,Juan
Lac 22h26'1"50d11'
Martinez,Kathleen Stevens
And 2h27'17"50d34'
Martinez,Kelly Lynn Hartley
Cnc 8h11'24"30d1'
Martinez,Kevin Richard
Her 17h11'53"22d38'
Martinez,Lacey
Ori 5h57'39"17d38'
Martinez,Leo Michael
Lac 22h52'51"55d54'
Martinez,Leslie
Mon 7h24'51"-10d10'
Martinez,Lizzette
Lyr 18h20'54"40d43'
Martinez,Louis Manuel
Oph 17h50'54"12d25'
Martinez,Maria
Cas 2h34'11"60d1'
Martinez,Maria Cires
Lyr 19h14'16"31d4'
Martinez,Maria Jose
Cyg 19h39'0"30d29'
Martinez,Mark Anthony
Sct 18h35'12"-4d33'
Martinez,Michael Einstein
Her 16h17'57"41d1'

Martinez,Michael Luis
Aql 18h56'1"4d50'
Martinez,Michelle
Aql 18h56'51"2d41'
Martinez,Mr & Mrs Donald
Cep 22h38'34"61d19'
Martinez,Peter A
Dra 16h54'41"63d59'
Martinez,Rachel
Uma 9h48'11"56d30'
Martinez,Ralph & Dora
Cyg 19h31'19"30d14'
Martinez,Randolph"Ray"
Lac 22h11'17"49d10'
Martinez,Ray
Per 3h15'0"56d31'
Martinez,Ren Farren
Cam 6h10'51"68d16'
Martinez,Reneida
Uma 10h59'18"50d20'
Martinez,Robi
Mon 7h16'46"-7d12'
Martinez,Ruth
Eri 3h14'24"-2d48'
Martinez,Shelley D
Mon 6h27'1"-10d44'
Martinez,Trudy
Lyr 18h32'1"40d22'
Martinez,Trudy E
Ari 2h0'1"26d1'
Martinez,Vicky
Sct 18h37'46"-6d22'
Martini,Sergio
Cet 1h30'23"-18d46'
Martinick,Melanie
Mon 6h56'26"-10d16'
Martiniere,Madelynn
Cyg 20h54'53"38d33'
Martinko,Dan
Aql 19h59'40"0d48'
Martinkus,Rita
Peg 23h29'59"20d51'
Martino
Per 3h18'13"40d13'
Martino's "Magical Glow",Michael
Her 15h50'28"46d20'
Martino,Anjelica Helene
Lyn 7h28'24"44d38'
Martino,Christopher
Per 3h33'13"52d14'
Martino,Gina Marie
Cas 0h31'55"50d12'
Martino,Nicholas
Crb 16h5'11"38d43'
Martino,Rick
Per 2h24'54"58d9'
Martinovic,Janja
Ori 5h51'32"21d18'
Martins,Leigh Santos
Peg 22h59'56"20d40'
Martinsburg Christian Acad (Martinsburg,WV)
Uma 8h27'31"67d37'
Martinson,Kevin & Kris
Aql 20h4'41"0d37'
Martinson,Richard A
Aql 20hl'40"12d5'
Martinson,Violet
Sag 19h7'33"-20d45'
Martinus,Mathias
Uma 9h29'13"48d3'
Martiomairino,Vivian
Sge 18h57'41"20d36'
Martirosian,Rosa
Vul 19h8'53"22d32'
Martita Mia
Sgr 19h34'26"-44d1'
Martius,Sabine & Jens
Cas 0h17'46"60d59'
Martner,John & Hildy
Boo 14h20'28"18d49'
Martolina
Peg 22h6'0"20d34'

Martong,Gertrud "Trudi"
Dra 15h0'0"61d30'
Martorana,Anna
Aur 6h5'31"31d48'
Martorana,Carol
Com 12h53'15"24d3'
Martorell,L Jeffrey
Crt 11h11'55"-18d26'
Martorella,Kyle Bruce
Peg 22h1'1"33d49'
Martorello-Benificial 1955-1994,C
Aql 20h18'19"5d35'
Martti
Cep 20h49'52"61d24'
Marttinen,Anu
Cam 3h28'36"53d57'
Martuccio,Gene
Sex 10h31'37"2d18'
Marturano,Amy Melissa
Eri 2h48'55"-18d30'
Marturano,Anthony Joseph "TJ"
Ori 5h52'20"14d7'
Martwick,Andrew W
Her 16h48'16"40d22'
Marty
Cam 5h7'31"76d24'A
Marty
Dra 19h29'0"68d9'
Marty
Boo 14h6'16"50d30'
Marty
Cam 8h55'51"77d43'
Marty
Oph 17h25'27"-20d4'
Marty
Cep 21h20'4"80d8'A
Marty
Tri 2h41'19"31d50'
Marty & Cindy
Cam 8h27'28"78d5'
Marty Stuart
Cmi 7h44'42"8d0'
Marty Van D
Her 16h40'53"4d51'
Marty's Dreams
Uma 12h11'13"60d29'
Marty's Special Effect
Aql 19h27'22"15d33'
Marty,Claire
Mon 7h50'30"-4d10'
Marty,Nicole
Lmi 10h48'55"32d40'
Marty-My Shining Star (M"S")
Crt 11h49'32"-7d55'
Marty-Scott,Jacob Allen
Mon 7h47'23"-5d14'
Marty-Scott,Pamela Laeigh
Peg 23h6'11"11d42'
Martyn,Brett David
Cep 22h5'52"62d19'
Martyn,Captain Ron
Uma 11h51'19"40d12'
Martyn,Jennifer G
Ori 5h34'46"-0d58'
Martyniak,Ellie
Dra 17h3'55"66d33'
Martyniuk,Taniushka
Eri 4h14'34"-17d9'
Martz,David Allen
Lmi 11h3'26"27d47'
Marté,Denise
Mon 7h55'48"-3d38'
Martín,José G
Cam 8h8'37"81d14'
Martínez,Adán Gonzalez
Her 16h22'21"4d39'
Maru
Eri 2h59'53"-11d49'
Maru,Frisco
Cyg 20h3'25"39d47'
Maruca,Attilio
Cep 3h27'31"77d30'
Maruca,D J
Ori 6h6'56"6d27'
Maruffo,Vanessa L
Gem 6h51'55"12d51'
Maruja
And 23h15'54"51d30'
Maruk
Uma 10h2'54"53d38'
Maruschak,Fayth Dawn
Pup 7h46'30"-23d19'
Maruska
Cae 4h59'29"-33d35'
Marut,Mich Michele
Cas 23h38'31"50d34'
Marv & Marilyn
Lyn 8h46'1"41d12'
Marva Dawn
Cas 1h28'38"60d34'
Marva,Innocent Aloysius
Vul 20h28'53"28d53'
Marvel,William Craig & Judith Stafford
Lac 22h53'24"36d5'AB
Marvel-Jean
Mon 6h53'10"-2d27'
Marvell,Haynes Cruce
Mon 6h38'10"7d35'
Marvella
And 23h2'41"50d16'
Marvellously Magnificent Mallon
Aur 6h50'56"41d5'
Marvelous Marvin
Cep 21h8'42"85d14'
Marvelous Marvin Star
Peg 0h1'30"18d50'
Marvelous Mr Terry
Boo 14h31'27"30d23'
Marvez,Forrest J
Lac 22h17'1"38d53'
Marvi,Ellen Wasilewski
Ori 5h33'0"-0d0'
Marvie
Boo 14h32'0"40d36'
Marvil
Cas 2h34'45"57d55'
Marvin 61556
Lac 22h24'22"55d16'
Marvin Mark
Aqr 21h26'19"-1d44'
Marvin Star,The
Aql 18h43'0"10d54'
Marvin's Star Forever
Cep 22h0'34"61d51'
Marvin,Frederick Wilson
Ori 6h1'42"5d16'
Marvin,Margrete Elizabeth
Del 20h23'56"11d4'
Marvin,Michael Edward
Eri 3h49'47"-4d30'
Marvita!
Tau 3h54'0"1d39'
Marx,Captain Betty Lee
Ori 5h26'16"0d9'
Marx,Captain Duke
Ori 5h26'20"0d48'
Marx,Dietmut-Heike
Cnc 8h29'1"7d50'
Marx,Frank
Uma 11h54'37"46d14'
Marx,Gilda
Cmi 8h3'11"6d38'
Marx,Helen
Cas 23h30'24"53d43'
Marx,Joann
Mon 6h36'25"10d54'
Marx,Raymond J
Aql 19h31'15"12d6'
Marx,Sherran Kay
And 0h51'1"38d17"
Marx,Victoria Paico
Crt 11h15'34"-10d56'
Marx,William J
Cnc 9h13'53"33d16'
Marxer,Michel
Per 3h57'46"35d17'
Mary
Cyg 20h2'1"40d15'
Mary
Peg 21h47'56"34d35'B
Mary
Lyr 19h20'51"40d45'
Mary
Pho 23h37'30"-45d19'
Mary
Col 6h35'10"-35d6'
Mary
And 2h19'0"45d18'
Mary
Boo 14h19'49"30d13'
Mary
Cyg 21h0'24"37d27'B
Mary
Ant 10h44'25"-39d33'
Mary
Lyr 18h13'47"38d48'
Mary
Eri 3h14'33"-15d3'
Mary
Del 20h26'15"11d18'
Mary
Dra 15h11'1"62d45'
Mary
Del 20h14'0"15d8'
Mary "Mountain Puppy" Eileen
Aur 6h7'8"31d24'
Mary "Stretch" Katherine
Ori 6h3'1"8d28'
Mary & Cindy's Guiding Light
Peg 21h19'18"21d31'
Mary & Dale's Wishing Star
Cyg 19h17'49"49d22'
Mary & Floyd
Cyg 19h18'32"28d20'
Mary & Joe
Sge 19h56'50"16d1'
Mary & Lynn
Cnv 12h51'47"37d49'
Mary & Richard
Cyg 21h4'58"48d55'
Mary & The Big House
Lmi 10h27'19"30d54'
Mary Alice
Ori 5h56'47"15d6'
Mary Alice
Cyg 19h25'22"30d17'
Mary Alice
Lyr 18h32'56"38d49'
Mary Allison
Mon 7h0'48"8d18'
Mary Alyce's Guardian
Tau 4h32'43"30d45'
Mary Andy Star
Pup 8h24'30"-23d0'
Mary Angela
Lyr 18h49'57"40d51'
Mary Ann
Mon 6h32'51"0d4'
Mary Ann
Peg 22h43'39"21d13'
Mary Ann
Lyr 18h49'25"33d54'
Mary Ann
Cnv 13h56'48"30d7'
Mary Ann
Vul 19h58'1"28d32'
Mary Ann
Cas 1h6'21"60d12'
Mary Ann Star of Scotts Love
Peg 21h39'52"26d22'
Mary Ann,Illuminating The Darkness
Del 20h24'37"10d5'
Mary Anne
And 2h31'23"38d31'
Mary Anne
And 23h45'26"40d58'
Mary Beth
Cas 0h21'43"67d56'
Mary Beth
Peg 23h48'45"12d39'
Mary Beth's Eternal Star
And 2h32'38"49d46'
Mary Beth-Rachel
And 23h28'38"43d28'
Mary Carol
Eri 3h58'14"-12d48'
Mary Catherine
Lmi 10h9'19"40d23'
Mary Catherine,The
Cyg 19h46'19"30d18'
Mary Christine
Lyr 18h35'16"41d19'
Mary Christine
And 2h18'48"49d7'
Mary Claire
Cas 1h0'13"61d50'
Mary Claudia
Cma 7h15'20"-13d52'
Mary E
Dra 18h31'25"70d37'
Mary Elizabeth
Peg 22h6'22"21d20'
Mary Elizabeth
Cas 1h3'57"58d3'
Mary Elizabeth
Cas 0h54'16"59d3'
Mary Elizabeth
Oph 18h39'17"7d43'
Mary Elizabeth
Cas 1h30'1"61d34'
Mary Elizabeth
Lyn 7h35'0"45d59'
Mary Elizabeth Flo
Peg 22h40'17"20d20'
Mary Ellen
Peg 22h25'22"26d6'
Mary Ellen
Lyr 18h42'41"35d21'
Mary Ellen
Aql 19h54'1"14d8'
Mary Ellen
And 23h33'16"40d40'
Mary Ellin K
Cas 1h5'19"58d25'
Mary Evelyn
Cyg 21h25'39"38d22'
Mary Forever in My Heart
Cas 23h14'38"60d55'
Mary Frances
Cyg 19h26'11"30d31'
Mary Francis
Hya 9h17'43"0d23'
Mary Francis Star,The
Lyn 7h27'53"44d17'
Mary Isobel
Cyg 20h21'1"39d15'
Mary Jacquelyn
And 1h14'56"40d11'
Mary Jane
Eri 3h13'23"-15d58'
Mary Jane
And 23h42'14"43d47'
Mary Janna
Mon 6h22'20"5d17'
Mary Jean
Uma 10h21'41"50d18'
Mary Jo
Vul 19h5'45"25d25'
Mary Jo
Peg 23h10'49"21d48'B
Mary Josephine
Lyr 18h30'17"40d53'
Mary Joy
And 2h10'10"40d56'
Mary K
Eri 2h47'34"-17d43'
Mary Katherine
Gem 6h58'38"13d29'
Mary Kathleen
Cnv 13h21'19"40d31'
Mary Kathryn
Equ 21h7'35"10d19'
Mary Kathryn
Eri 3h26'38"-7d5'
Mary Kathryn
Peg 21h42'54"23d30'
Mary Kay
And 22h59'39"41d3'
Mary Kay Courageous
And 0h7'0"46d3'
Mary Kay,My Love
Mon 6h23'60"7d57'
Mary Kelly
Cas 23h31'44"61d48'
Mary Lee
Peg 22h50'50"29d7'
Mary Lee
Cam 3h57'39"60d47'
Mary Lee
Cas 2h20'50"63d54'
Mary Lee-Bobbi-Becka- Max
Umi 14h49'54"65d51'
Mary Lenore
Eri 3h22'51"-13d36'
Mary Lindsay
Leo 11h30'16"-0d4'
Mary Lou
Vul 19h45'60"20d32'
Mary Lou
Cas 2h51'23"61d36'
Mary Lou
Mon 7h1'1"4d45'
Mary Lou
Tri 2h11'45"31d46'
Mary Lou
And 1h14'0"35d34'
Mary Lou's"Magic Lights"
Ori 6h4'48"-0d39'
Mary Louise
Lyn 7h41'0"58d20'
Mary Lucille
Mon 6h49'14"10d13'
Mary Lucille
Cyg 19h17'1"45d13'
Mary Lynn
And 2h22'27"41d57'
Mary Mac,Our Shining Star Of A Mom
Tri 2h10'18"33d38'
Mary Margaret
Eri 3h52'47"-0d56'
Mary Margaret
Cas 0h40'48"67d33'
Mary Marjorie
Cas 23h29'0"62d14'
Mary Meg
Gem 6h52'25"18d5'
Mary Michelle
Lyr 18h37'1"29d26'
Mary O
Cep 22h39'51"67d53'
Mary O'
Boo 14h32'1"48d1'
Mary P R I
Peg 22h40'13"26d45'
Mary Rebecca
Dra 10h29'28"78d54'
Mary Rose
Cyg 19h42'44"28d49'
Mary Rose,Mary Rose, Mary Rose
And 2h23'11"42d58'
Mary Stan
Cyg 20h18'42"38d41'
Mary T
Cas 2h12'1"60d11'
Mary The Honey Bunch Star
Sge 19h55'11"19d5'
Mary Theresse
Aur 6h37'30"38d41'
Mary Valerie
Tau 4h17'38"20d57'
Mary's "Shining Light"
Mon 6h21'0"-6d41'
Mary's Christmas Star
Cet 2h29'54"-11d7'
Mary's Destiny
Aur 7h8'37"43d19'
Mary's Happy Birthday Star
Com 13h1'11"21d20'
Mary's Light
Ori 5h31'1"-6d41'
Mary's Night Light
Mon 6h19'53"8d53'
Mary's Star
Lyn 8h0'55"38d54'
Mary's Star
Peg 22h5'40"24d3'
Mary's Star
And 1h20'48"48d46'
Mary's Star
Aql 18h55'1"-0d19'
Mary's Star Brings Us Light Forever
Uma 11h13'34"60d21'
Mary's Star Of Companionship&Firework
Leo 9h37'40"8d1'
Mary,Roberta
Gem 7h3'54"21d43'
Mary-Faith
Lyn 6h52'1"60d21'
Mary-Luz
Cam 7h0'58"60d56'
Mary-Orise
Aur 5h2'33"40d34'
Maryam & Bizhan
Eri 4h42'30"-0d59'
Maryam Jami
Eri 3h55'1"-12d27'
Maryan & Tony A Child Is Born
Aql 19h30'10"12d18'
Maryann
Cas 1h2'0"53d15'
Maryann
Cyg 20h9'40"40d45'
Maryann & Judy
Cas 1h51'39"58d12'
Maryann 1
Cas 2h38'20"70d52'
Maryann's Wish
Tau 5h54'51"27d2'
Maryanna
Peg 22h11'1"25d51'
MaryBelle
Aur 6h38'18"38d20'
Marygold,Iris
And 23h15'23"46d31'
Maryjane
Cnc 9h1'58"28d32'
Maryjane
Lyn 7h24'21"44d33'
MaryJo
And 23h29'31"47d48'
Maryland Small Business
Uma 10h7'27"72d41'
Maryland's Drug-Free Workplace Stars
Dra 19h21'40"56d11'
Maryland,Floyd John Steven
Aql 19h57'16"-5d59'
Marylene
Aql 19h29'1"0d41'
Marylene DB
Vul 19h19'37"25d42'
Marylyn Victoria
Peg 22h45'52"29d8'
Marynapuakailu'ulu`uma `ema`ekamea`ohi
Eri 3h22'1"-15d53'
Marynette
Sex 9h50'55"4d6'
Maryrose
Lyr 18h14'35"37d37'
Maryse
Cam 14h21'28"85d29'
Maryse et Marc
Tau 4h40'40"15d10'
Maryse Nicole
Mon 6h26'32"-8d52'
Marystella
Pic 4h42'5"-47d5'
Marz
Uma 8h47'56"52d42'
Marzahn,Diana
Sco 17h3'27"-30d45'
Marzaroli,Louise
Lyr 18h16'32"31d2'
Marzec,Richard David
Lmi 10h7'47"31d55'
Marzella
Cyg 19h58'51"48d49'
Marzendrew
Dra 12h23'44"64d28'
Marzett
Boo 14h5'32"22d55'
Marzetta,Allan L
Lyn 8h45'38"36d56'
Marzia
Tel 20h10'43"-47d36'
Marzia
Lep 5h20'14"-11d41'
Marzluff,Herr
Boo 14h1'44"10d26'
Marzocca,Vincent
Dra 10h42'1"73d37'
Marzola,Joan
Lyn 6h56'21"54d45'
Marzorati,Giovanni
Her 18h9'53"40d50'
Marzulli,Nenell "Ne"
Lmi 10h43'46"23d15'
Marzullo,Gregory Phillip
Per 4h3'0"41d3'
Maréchal,Grégoire
Tri 1h55'49"30d35'
Masae
Lib 14h29'29"-10d44'
Masaharu,Emiko
Tau 5h22'55"20d16'
Masahiko & Shinko
Tau 5h19'50"25d2'
Masak,Barbara Ann
Cyg 20h54'27"31d20'
Masako
Mon 6h20'1"3d23'
Masamitsu & Nami
Psc 1h19'39"26d20'
Masanori & Rika
Ari 2h13'8"23d18'
Masaracchio,Ron
Her 17h25'48"38d33'
Masaraki,Gian Lorenzo
Tri 1h56'36"27d46'
Masaru & Miwa
Sgr 19h12'8"-16d6'
Masaru-Hoshi No Kagayaku,Kawashima
Peg 23h31'35"31d30'
Masat,Agnes Simpson
Lmi 9h27'24"38d13'
Masayo - 93
Lmi 9h53'39"38d15'
Mascal,M B
Cam 6h45'49"78d57'
Mascali,Allyson
Cyg 20h38'31"40d6'
Mascarenas,Jason
Her 18h0'1"50d23'
Mascari,Jimmy
Aur 6h28'37"38d8'
Mascaro,Patricia Anne
Leo 9h33'22"8d13'
Mascaro,Valerie,Jean
Uma 11h0'50"58d60'
Mascavage,Jeanne Patroni
And 0h32'17"45d2'
Maschek,Karyn Ann
Cyg 19h29'46"35d33'
Mascheri,Luca
Lyn 8h8'1"42d7'
Mascherina
For 2h42'25"-25d53'
Mascherino,Dominick James
Aur 6h2'3"48d15'B
Mascherino,Emily Fiore
Aur 6h2'3"48d15'A
Maschino,William David
Cet 3h0'0"2d56'
Maschke,Christy
And 1h46'58"41d12'
Maschke,Steven Douglas
Aur 6h32'15"38d56'
Mascio,Sr,William Michael
Her 17h5'42"21d54'
Mascunana,Christian Alan
Her 17h46'0"14d55'
Mascunana,Rebecca Rose
Mon 6h57'56"10d22'
Maseas
Lyn 9h19'15"34d47'
Masek III,Oldrich
Per 3h33'25"38d39'
Maselko,Jeremy K
Tri 2h5'36"31d50'
Masella,Phyllis
Del 20h57'31"11d1'
Masenheimer,David
Ori 5h29'13"1d49'
Masetta,A Lyn
Cam 5h10'48"68d24'
Masevice,Anthony
Aql 19h30'31"8d45'
Masha
Cet 1h17'30"-16d4'B
Masham,Jenny
Cnv 13h33'21"50d31'
Masi,Monika
Aqr 20h56'23"-14d18'
Masi,Pat
Aql 19h59'56"-1d23'
Masi,Vito
Her 17h16'0"40d34'
Masie,Barbara Ann
Mon 6h53'1"-1d47'
Masie,Estelle Jane
Mon 6h33'39"-1d32'
Masiello,Vincent
Dra 19h25'46"56d60'
Masin,Zachary Mansfield
Gem 6h43'5"18d31'
Masingill,Miss Phyllis
Ari 2h43'32"21d11'
Masiques,Nelia & Alexander Hall
Cnv 13h21'12"31d37'
Maskell,Leticia D
And 2h3'13"47d28'
Maskew,Gregory D & Karen D
Cru 12h37'47"-60d38'
Maskey,Christopher Alan
Uma 11h8'16"41d28'
Maskey,Kelly Alexandra
Cam 10h50'1"81d31'
Maskowitz,Dr Trudy
Com 12h54'19"27d5'
Maslanka,Kazmier
Ori 5h46'36"11d25'
Maslanka,Ronald Valentine
Per 4h35'12"52d10'
Maslansky,Michael
Aur 6h7'52"32d44'
Maslansky,Scott
Aur 6h20'21"32d55'
Maslauskas,Darlene L
Mon 6h36'14"6d18'
Maslen,Wendy Grace
Cyg 20h1'1"40d1'
Maslin,Jamie
Cet 2h52'22"7d9'
Maslow,Emilie Taylor
Cas 0h58'58"68d16'
Maslow,Liza Hayes
Cam 5h23'45"68d41'
Maslow,Lois
Com 13h7'33"26d29'
Maslowski,Mary Marie
Cas 22h56'31"53d46'
Maso,Francesco
Aql 18h43'19"0d30'
Mason
Her 17h0'1"40d43'
Mason's Double M
Aur 5h38'15"37d32'
Mason's Star,Pat
Lyn 7h48'15"44d37'
Mason,"Mostest"
Mon 6h53'28"-0d35'
Mason,Alexsandra
Cas 0h20'0"58d10'
Mason,Bob
Eri 3h21'46"-13d28'
Mason,Bonnie
Cet 1h51'17"0d42'
Mason,Carla Ann
And 0h10'43"31d32'
Mason,Charlie Loran
Ori 5h47'42"11d60'
Mason,Charlotte von Jena
Lyr 18h38'0"45d5'
Mason,Christopher Barrie
Ori 6h2'28"1d32'
Mason,Cynthia Neumann
Cep 20h49'1"60d19'
Mason,Dana
Mon 8h1'20"-0d40'
Mason,Daniel Evan
Ori 5h49'48"19d25'
Mason,Don (Poppop)
Aur 6h29'0"37d46'
Mason,Dr Thom
Her 16h40'1"48d30'
Mason,Elizabeth
And 2h23'53"44d39'
Mason,Elizabeth Anne
Ori 5h25'34"0d20'
Mason,Elsie
Tri 2h2'20"28d51'
Mason,Ernest L
Lac 22h13'29"47d49'
Mason,Eva
Lyn 6h29'14"54d44'
Mason,Fran
Cas 3h11'25"70d26'
Mason,Franklin J
Mon 7h51'8"-2d5'
Mason,Garry
Cep 4h40'35"80d17'
Mason,George
Aql 19h59'0"0d22'
Mason,Greg
Hya 8h48'49"0d20'
Mason,Heather
Uma 11h31'27"63d51"
Mason,Jane
Cas 2h11'0"59d58'
Mason,Jeannene Rice
Lmi 10h1'31"36d46'
Mason,Jennifer
And 2h32'19"38d36'
Mason,Jesse
Cnv 12h12'0"38d47'
Mason,Jr,Raymond K
Hya 8h56'16"0d10'
Mason,Kathleen
Peg 22h30'1"25d6'
Mason,Kathy
Cam 3h27'20"61d39'
Mason,Kendra
Vul 19h58'58"22d51'
Mason,Kevin L
Peg 23h5'1"18d46'
Mason,Kim Lee
Com 12h1'0"27d21'
Mason,Kimberly Dawn
Cyg 19h36'15"28d7'
Mason,L Barto
Oph 17h12'55"9d45'A
Mason,Laura
Cnv 13h43'1"33d57'
Mason,Leanne
Crb 16h10'0"36d53'

Mason,Lois Vars
And 23h48'0"42d45'
Mason,Margaret
Boo 14h5'55"51d48'
Mason,Mary Jane
Lyn 6h14'24"60d54'
Mason,Mary Reynolds
Lyn 7h48'43"36d50'
Mason,Raymond & Jody
Aql 18h56'57"13d18'
Mason,Roseann & Dane
Lyn 7h55'50"42d22'
Mason,Ruth
Oph 17h12'55"9d45'B
Mason,Sacha
Lyr 18h57'0"30d7'
Mason,Shelonda
Equ 21h7'50"10d8'
Mason,Susie
Tau 3h52'47"1d8'
Mason,Tanner Keith
Cep 22h43'30"78d54'
Mason,Tara
Eri 4h12'0"-17d31'
Mason,Theresa Savarese
Cam 5h54'31"60d7'
Mason,Willard
Ori 5h55'41"20d5'
Mason,Zachary Alexander
Aql 19h30'0"14d8'
Mason-Oliver
Ser 15h54'52"20d40'
Masoner,Nancy Lee
Com 12h27'39"28d54'
Maspens,Carrie Beth
Aur 4h55'58"41d14'
Masquerade
Crb 15h29'39"30d56'
Masri,Lina Horchani
Lyr 18h29'23"46d56'
Masri,Rana Al
Hya 9h20'24"-9d37'B
Mass "Trystar",Kathryn &
Robert
Cyg 21h6'51"30d28'
Mass 4ARookie Of The
Year,Candice L
Cas 2h58'1"61d15'
Mass,(75 Shining
Years),Miriam Yasgur
Peg 23h45'24"27d5'
Massa,Barbara
Cas 0h45'12"64d12'
Massa,John M
Ser 15h10'19"24d48'
Massa,Louis S
Uma 9h18'19"55d24'
Massa,Patrizia
Dra 19h3'52"49d50'B
Massabuau,Jean Claude
Sex 10h16'26"-2d50'
Massai,Elisa
Lyn 6h54'1"49d52'
Massali,Rachel
Boo 15h6'44"7d38'
Massand,Pradeep K
Cnc 9h12'1"32d40'
Massara,Darren C
Her 16h52'39"30d45'
Massara,Manica
Cam 13h55'15"82d5'
Massarelli,Shelby Barr
Cas 1h57'12"73d34'
Massaro,Daniel Joseph
Sco 17h24'51"-31d34'
Massaro,Jillian Kerry
Leo 10h14'30"14d39'
Massaro,Nicolette Victoria
And 23h43'0"47d20'
Massaro,Philip P
Dra 19h6'1"54d29'
Massart,Olivier
Her 18h11'53"31d45'
Massaux,Barb
And 0h4'16"40d3'
Massaux,Barb
Cam 11h50'42"77d49'
Massaux,Barb
Tau 4h23'11"16d29'
Massaux,Barb
Cyg 21h19'39"35d4'
Masse,Patricia
For 3h29'38"-31d25'
Masseguin,Leslie
Her 18h54'48"18d53'
Massen,The
Mon 6h47'17"11d24'
Massengill,Bob
Ser 15h28'56"10d4'
Massett,Edward R
Her 17h11'1"48d57'
Massetti,Mauro
Nor 16h19'18"-54d31'
Massey
Cas 0h27'26"64d2'
Massey "Redbreast", Robin
Lea
Mon 8h8'47"-4d26'
Massey,A D
Cmi 7h54'31"1d3'
Massey,Adam Joseph
Lmi 10h1'29"38d49'
Massey,Annabel Christina
Jewel
Aqr 21h36'1"-6d55'
Massey,Brian & Janet
Sge 19h3'11"18d50'
Massey,Crane Douglas
Per 3h4'20"42d4'
Massey,Cynthia Ann
Mon 7h46'4"-5d16'
Massey,David Vard
Sex 10h22'29"2d24'
Massey,Donita
Mon 6h25'13"8d3'
Massey,Eamonn
Her 16h56'34"38d56'
Massey,Jerry
Aql 20h5'27"4d2'
Massey,Joanna
Eri 4h5'1"-7d36'
Massey,Michael & Billee
Sge 20h16'23"17d35'
Massey,Michael Wayne
Cam 3h54'1"57d58'
Massey,Robert Tyler
Her 16h42'26"27d24'
Massey,Sandra Kay
Del 20h14'32"15d17'
Massey,Stephen M
Ser 15h39'50"6d53'
Massey,Sue
Cas 0h37'25"60d32'
Massicotte,Ashley
Uma 12h39'25"61d54'
Massicotte,Franáois
Her 18h55'1"12d46'
Massicotte,Joseph
Ori 5h18'28"11d20'
Massicotte,Marcel
Her 16h53'45"37d45'
Massie
Cas 0h1'1"58d59'
Massie's Esoterica, Jack
Her 16h34'21"34d38'
Massie,Brian
Lac 22h29'29"55d50'
Massie,Jacquelyn Nicole
Peg 23h44'33"31d35'
Massie,Jennifer Kristin
Peg 23h44'24"31d42'
Massie,Julienne
Cas 0h40'58"65d40'
Massimi,Danielle Marie
Lyn 8h5'55"40d55'
Massimiliano
Cet 2h51'32"-0d51'
Massimiliano,Giuseppe
Michele
Per 3h30'51"51d23'
Massimo
Ori 6h7'27"7d56'
Massimo
Lyn 7h53'37"41d57'
Massimo
Del 20h13'36"14d58'
Massimo
Cap 20h31'50"-13d39'
Massimo
Cyg 21h36'1"40d53'
Massimo 63 Segue Lettera
Uma 12h7'14"60d23'
Massimo e Monica
Cyg 19h48'15"38d21'
Massimo,Margaret Caroline
Cyg 20h50'33"50d32'
Massimo,Stella
Aur 6h12'23"31d29'
Massimo,Strada
Ind 21h2'19"-52d7'
Massin
Mon 7h47'9"-5d44'
Massing,Elizabeth
Cas 2h12'31"68d21'
Massing,Louis
Aur 4h56'30"40d13'
Massini 1914,Harry
Her 17h9'28"42d49'
Massive Attack
Ori 5h40'56"1d3'
Masson,Jean
Ori 5h5'56"14d5'
Masson,Marie & Lloyd
Aur 5h39'19"37d35'
Masson,Pierre
Sex 10h16'60"-2d21'
Masson,Robert K
Aql 19h9'36"3d37'
Masson,Victor
Cnc 8h27'28"10d55'
Massoutier,Claudine et
Bernard
Cyg 20h22'44"39d50'
Massé,Mariette Julienne
Cyg 19h27'48"35d18'
Mast,Andrew
Lib 15h14'0"-28d50'
Mast,Kassandra Beth
Del 20h35'49"20d16'
Mast,MD,Jeffrey Welling
Her 18h19'49"14d34'
Mast,Rodney C
Ori 5h56'58"14d43'
Mastandrea,Lorenzo Maria
Edoardo
Ari 3h13'21"23d20'
Mastantuno,Carli
Tau 5h45'30"28d29'
Mastantuno,Samantha
Tau 5h52'54"24d6'
Mastel,Willie
Her 15h59'53"46d20'
Masteller,Tracie L
And 1h46'43"36d51'
Mastenbaum,Rita Book &
Anna Book
And 0h4'41"31d34'
Master
Cam 4h50'40"68d21'
Master Craig & Mistress
Cyndie
Uma 9h29'33"47d48'
Master Ismael
Aql 20h0'14"6d14'
Master Kieren's Star
Peg 21h57'46"30d17'
Master Young Sun Lim
Her 17h28'34"28d10'
Master 'Holden'
Equ 21h1'16"8d32'
Master's Shining Tygre
Her 18h42'55"18d56'
Master,Amy Lynn
Lyn 8h47'26"34d20'
Master,Dawn E & Stacey M
Stone
Cam 3h20'58"60d9'
Master,Maxwell David
Her 16h35'58"37d45'
Master,Todd Rockwell
Lyr 19h12'31"38d27'
Masterov,Boris
Lac 22h3'57"41d3'
Masters "Il Souls, I Love",Ed
& Carol
Sge 18h58'59"18d45'
Masters,Andrew Paul
Cep 20h34'53"76d18'
Masters,Anthony Shane
Sgr 19h41'25"-44d34'
Masters,Bernard Earl
Aql 20h6'14"4d27'
Masters,David
Uma 9h27'45"54d49'
Masters,Don
Per 2h51'0"37d20'
Masters,Dr Paul L
Lmi 10h5'0"31d26'
Masters,Hugh Clarendon
Ensor
Lyn 8h25'30"47d33'
Masters,Irene
Aql 19h54'47"14d46'
Masters,Kathryn & Martin
Tau 3h53'22"20d7'
Masters,Mark Christopher
Her 17h27'26"20d30'
Masters,Paul
Cmi 7h43'42"8d51'
Masters,Phillip Robert Terry
Dra 19h13'20"67d46'
Masters,Rebbeca S
Vul 19h48'33"23d6'
Masters,Ruth/Tom, Lynne,
Chris Jones
Cyg 21h40'55"38d6'
Masters,Terry McDaniel
Cma 6h55'18"-18d8'
Masterson III,Harris
Aql 19h3'1"0d2'
Masterson,Julie
Vul 20h4'18"28d24'
Masterson,Kathleen "Katie"
Cyg 19h58'31"37d42'
Masterson,My Daddy
Cornelius
Cep 22h26'14"63d24'
Masterson,Patrick
Per 2h37'0"40d13'
Masterson,Robert Patrick
Cep 21h58'22"55d49'
Masterson,Tara Jo
And 2h31'58"39d53'
Masterson,Thomas Charles
Peg 22h42'58"10d29'
Mastifino,Lesley M
Sge 20h2'1"20d24'
Mastin,John Alfred
Cet 2h4'28"3d4'
Maston,Melanie Jayne
Cyg 19h33'48"33d45'
Mastra
Boo 15h0'46"37d56'
Mastracchio,Dawn
Aur 5h0'0"29d7'
Mastrangelo,Michael A
Per 1h56'54"47d55'
Mastriforte,Karla J
Psc 0h22'14"-0d57'
Mastriforte,Keith A
Tau 5h44'0"28d8'
Mastrocola,Peter "Little Mack"
Lyr 18h56'0"34d29'
Mastrogany,Margie
Vir 11h45'35"3d53'
Mastrogany,Theodora
Leo 10h54'39"20d22'
Mastrogiovanni,Mary &
Pasquale
Mon 6h26'37"10d48'
Mastrolia,Lucia
Aur 5h25'58"31d27'
Mastronardo,John Vito
Aur 4h59'47"51d43'
Mastronardo,Maria Angela
Com 12h4'20"27d1'
Mastroviti,A Gianni
Uma 13h13'1"60d19'
Mastrup,Elke & Frithjof
Ori 6h4'20"20d54'
Masturzo,Bill
Aur 4h49'56"37d36'
Masuda,Naomi
Del 20h17'1"9d6'
Masuda,Scott
Ser 17h32'23"-10d20'
Masuda,Shigeru & Youki
Cyg 21h19'27"38d40'
Masuda,Stella of Takanobu
Psc 1h17'3"20d59'
Masui & Nozawa Family's
Prosperity
Tau 5h28'23"23d56'
Masumi's Tear Drops
Gem 7h12'57"29d21'
Maszk's Marvel
Psc 1h14'40"20d40'
Mata,David
Ori 6h17'23"-2d24'
Matar,Ashraf
Mon 7h4'32"4d58'
Matarazzo,Donna Marie
Aqr 20h47'46"-0d14'
Matarazzo,Dorothy Ann
Peg 23h20'37"33d34'
Matasovsky,Ryan K
Uma 9h31'1"54d18'
Matata,Hakuna
Mon 7h2'16"-0d43'
Matcham,John J
Oph 18h18'19"10d56'
Matchett,Charles L
Leo 10h39'0"20d46'
Matchinsky,Matthew
Ori 5h50'13"11d17'
Matcorjon
Umi 15h59'8"74d31'
Mateiro,Antonio
Peg 21h6'51"18d49'
Matejuk,Nadine F Honey
Bunch
Peg 22h58'10"31d27'
Matel,Mart
Cyg 20h38'31"40d6'
Matelle,Lisa & Joe
Lyr 18h32'40"36d6'
Mater,Joshua Whitworth Wing
Aur 5h12'0"54d55'
Mater,Regina
And 1h33'53"37d6'
Matera,Anthony D
Dra 18h43'1"70d37'
Matera,Ashley Ann
Umi 14h53'18"67d25'
Matera,Forever My Heart
Paula L
Cyg 20h27'40"40d41'
Matera,Guido
Cep 22h42'58"70d54'
Matera,To My Hurricane Chloe
Glenne
Vul 19h13'38"22d15'
Matera,To My Precious One
Kyrié Elyse
And 0h53'50"45d52'
Materese Kelly
Lyr 19h9'52"37d58'
Mates,Alisha
Aql 19h29'30"-10d0'
Mates,Constantin George
Cep 23h4'36"64d26'
Matesic,Maria
And 0h57'34"41d1'
Mateu,Jose
Aql 19h30'36"-10d58'
Mateyaschuk,Nicholas Gabriel
Her 17h48'14"40d50'
Mathees
Ori 5h58'0"1d9'
Matheis,Dottie
Aql 19h3'48"-6d13'
Matheis,Jerry
Cmi 7h44'31"5d35'
Matheison,Sandra Trovillion
Sgr 18h59'39"-24d26'
Matheny III,Edward Taylor
Tau 4h1'44"23d0'
Mather
Boo 15h46'1"42d56'
Mather's Miracle
Umi 15h18'15"69d55'
Mather,Jack
Lyn 8h38'57"44d28'
Mather,Janet
Peg 22h6'0"5d53'
Mather,Lesley
Cas 0h28'58"60d22'
Mather,Marlene
Dra 19h56'1"80d4'
Mather,Matthew R
Cet 2h7'55"8d38'
Mather,Neil
Cep 21h5'58"70d5'
Mather,Sophie Elizabeth
Umi 14h43'14"67d39'
Mathers,Clay B
Lyr 18h43'15"33d28'
Mathers,Gary Ames
Cet 3h9'57"42d15'
Mathes,Jr,Andrew Joseph
Aql 18h53'22"-0d35'
Mathew
Her 18h0'13"30d25'
Mathew Adrian
Per 2h31'12"57d10'
Mathew's Everlasting Light
Lyr 18h53'22"34d39'
Mathews "Our Mom", Myrtle
Cyg 19h58'1"31d11'
Mathews 10-16-81, Andrew
Scott
Aql 19h55'41"10d11'
Mathews Star,The
Aur 5h37'50"40d31'
Mathews,Ann & Robyn Lingen
Peg 23h36'0"27d48'
Mathews,Deborah Ann
Com 12h18'41"18d51'
Mathews,James R
Ori 6h6'1"8d53'
Mathews,Jerry
Del 20h26'0"20d20'
Mathews,Kevin Douglas
Per 2h49'1"50d28'
Mathews,Nicole Marie
Umi 15h8'1"71d12'
Mathews,Nikolas
Sex 10h8'33"2d25'
Mathews,Nikole
Mon 6h33'40"-1d25'
Mathews,Paige Erin
Sco 17h4'32"-31d42'
Mathews,Rob & Diane
Lac 22h26'24"53d49'
Mathews,Sixty Sparkles,Mary
I
Eri 2h45'44"-18d2'
Mathews,Tristan D
Cet 1h17'0"-5d58'
Mathewson,Donald Joseph
Cam 5h40'1"60d56'
Mathey,Vicky
And 2h28'17"38d26'
Mathias,Bruno
Crb 15h46'36"35d43'
Mathias,Dieter
Cnc 9h16'38"10d54'
Mathias,Juanita
Peg 22h34'1"28d11'
Mathias,Kenneth Alexander
Gem 5h59'0"27d14'
Mathias,Paul
And 0h28'1"44d9'
Mathiasen,Martin Andrew
Cep 2h47'34"77d40'
Mathieson,Simon John
Aur 6h27'47"34d10'
Mathieu
Dra 11h33'0"67d5'
Mathieu
Ser 15h20'24"8d4'
Mathieu,Francois-Louis
Per 2h57'30"37d3'
Mathieu,Jon Paul
Her 17h20'49"44d54'
Mathieu,Mary O'Brien
Peg 23h6'16"10d41'A
Mathieu,Michel
Sgr 19h33'58"-30d29'
Mathieu,Patrick
Sex 10h24'49"5d52'
Mathieu,Raymond F
Dra 19h46'49"58d55'
Mathieu,Ross Charles
Uma 12h25'55"61d40'
Mathieu,Wilfred
Peg 23h6'16"10d41'B
Mathilde
Boo 14h32'28"47d60'
Mathilde
Her 17h40'12"42d28'
Mathilde,"Ingrid" Ingeborg
Aur 5h9'57"42d15'
Mathis,David T
Dra 20h21'51"63d46'
Mathis,Desse
Umi 14h36'20"57d47'A
Mathis,Dreamboy Richard
Aaron
Cet 1h4'17"-2d4'
Mathis,Jean-Philippe
Vir 13h6'35"-5d25'
Mathis,Johnny
Ori 5h56'27"19d4'
Mathis,Katherine Ann
Mon 6h35'0"-1d45'
Mathis,Margaux Dumas
And 23h23'20"45d6'
Mathis,Victor Dumas
Her 16h36'42"38d31'
Mathis-Bresnan,Thomas
David
Ori 5h52'27"14d51'
Mathison,Mr & Mrs William A
Cnv 13h42'12"40d7'
Mathison,Our Special Star,Jim
Aql 18h55'23"-2d26'
Mathosian,Amanda Katherine
Cas 2h4'31"61d52'
Mathosian,Jillian Elizabeth
And 0h18'16"36d49'
Mathu
Tri 2h41'35"34d23'
Mathy,Sandy
Uma 8h48'30"73d4'
Mati
Oph 16h59'24"-25d35'
Matias,Linda
Lyn 8h18'59"36d48'
Matier,John Robert
Her 15h56'15"47d49'
Matijaca,Ivica
Uma 10h56'55"50d39'
Matilda
Cet 1h17'14"-11d9'
Matilda
Eri 2h57'29"-4d27'
Matilda Lee
Tri 1h51'33"28d13'
Matilde
Lyr 19h16'22"40d42'
Matilde
And 0h36'20"45d46'
Matis'Shining Star, Maria
Cas 0h0'1"54d26'
Matis,Bill
Per 4h44'1"37d34'
Matis,Diane Louise
Cas 0h27'53"73d30'
Matishak,Laurence W
Leo 11h41'37"25d30'AB
Matisoff,Ann-Lynn
Psc 1h2'31"18d32'
Matisz,Jr,John A
Boo 14h29'19"21d44'
Matjac
Lac 22h28'38"53d23'
Matla,Hans & Franny
Cam 5h44'50"61d18'
Matlack,Paul Anderson
Dra 19h42'17"61d8'
Matlavage,Nicholas E
Gem 8h1'22"27d56'
Matlick,Dayton
Cet 3h7'1"5d36'
Matlick,Fran
Tri 2h5'21"32d48'
Matlick,Phil
Tri 2h6'20"31d44'
Matlin,Gerald L
Dra 15h54'13"62d15'
Matlock,C D
Cyg 20h34'48"44d5'
Matluk,Sr,Nicholas
Lac 22h1'49"48d52'
Matney,Hannah Elaine
Vul 19h16'38"25d4'
Mato Forever
Lyr 19h1'35"37d40'
Matook,Alen
Cep 3h17'0"78d19'
Matook,Florence Alma
Uma 10h15'37"57d10'
Matos,Robert
Lyn 7h5'33"51d2'
Matousek,Stephanie Anne
Gem 6h59'47"10d44'
Matousek,Thomas Andrew
Lac 22h13'1"49d20'
Matrejean,Eleanore
Lyn 7h38'31"45d12'
Matrimonium Monaghanorum
Cyg 19h34'0"38d12'
Matrisciano,Todd G
Her 16h41'43"11d23'
Matrokebian
Aur 5h19'13"42d51'
Matros,Richard Texas
Lyr 19h18'57"41d37'
Matson's Mercy
Aql 20h8'54"1d49'
Matson,Barb
Cet 1h47'38"-6d42'
Matson,Donna Jean
And 1h18'37"34d30'
Matson,Emily Marie
And 0h20'25"32d27'
Matson,Muriel
Aqr 22h0'0"-5d51'
Matson,Robert Dee
Tau 4h59'33"16d3'
Matson,Sheila & John
Uma 8h45'1"49d41'
Matsuda,Satoru
Vir 12h4'16"8d25'
Matsuda,Yuko
Cap 20h51'14"-26d1'
Matsukawa,Ryuichi
Tau 5h10'17"16d50'
Matsumoto,Alexandria René
And 2h7'35"42d48'
Matsumoto,Ian
Eri 2h51'50"-16d30'
Matsumoto,Jae Toshifumi
Cma 6h59'49"-18d50'
Matsunae,Stella of Motoko &
Tohru
Sco 16h22'32"-29d47'
Matsunaga,Guy
Aql 19h41'27"14d38'
Matsunaga,Stella of Tsubasa
Cap 20h53'54"-16d2'
Matsuo,Fumihiko
Hya 8h11'24"6d18'
Matsuyama,W Brian
Dra 17h52'36"75d37'
Matsylbrumar
Cep 21h39'16"61d37'
Matt
Per 4h27'24"51d30'
Matt
Aur 6h56'31"37d19'
Matt & Beth Forever
Mon 7h14'31"-10d7'
Matt & Char SF
Cyg 20h53'0"38d43'
Matt & Jenny Star,The
Uma 8h55'23"49d39'
Matt & Jeny-"The Silver
Spoon"
Umi 15h43'31"81d12'
Matt & Judy's Star
Lac 22h13'37"51d17'
Matt & Lucy
Aur 5h0'14"50d0'
Matt & Patti's Magic
Per 2h57'40"48d46'
Matt & Stacy
Lyr 18h45'36"42d33'
Matt Loves AnneMarie
Cyg 19h39'42"28d34'
Matt N' Jennie
Dra 16h8'54"58d19'
Matt& Deborah
Cyg 21h0'43"39d1'
Matt's China Cat
Lyn 7h42'55"44d39'
Matt's Destiny
Aql 19h6'33"-6d56'
Matt's Rock'n'Roll Star
Per 2h54'23"35d25'
Matt,Elizabeth
Cas 3h3'53"57d38'
Matt-Man
Per 2h12'1"56d56'
Matta,Ann
Aql 19h26'45"7d48'
Mattaboni,Veronica
And 2h28'43"39d57'
Mattaliano,James
Her 16h18'48"40d51'
Mattaliano,Leanne Rose
Her 16h19'56"42d2'
Mattana,Francesco
Peg 23h29'54"21d19'
Mattanja
Ari 2h43'14"30d47'
Mattarelli,Isabella
And 0h20'35"40d25'
Mattas,Mike & Violet
Eri 4h56'44"-5d49'
Matte,Bernard
Ori 4h50'23"5d39'
Matte,Denny C
Umi 16h7'32"77d50'
Mattei,Nathieu
Lyn 8h27'54"40d58'
Matteini,Isabelle
Cyg 21h8'28"30d42'
Matteis,Aris
Aql 19h31'31"12d51'
Mattenkerl,Volker
Her 17h21'50"42d21'
Matteo
Cep 1h16'37"80d19'
Matteo
Del 20h12'55"12d52'

Matteo 11,Melissa
Cyg 21h23'38"30d56'
Matteo,Joseph
Her 18h12'11"40d32'
Matteo,Joseph
Her 16h57'37"30d17'
Matteo,Jr,Albert Paul
Boo 15h6'13"22d39'
Matteoni,Jim
Lac 22h2'12"50d29'
Matter,Jennifer L
Mon 8h6'30"-5d44'
Mattera,Dana
Lyn 7h35'1"41d49'
Mattern,Christoph
Psc 22h50'1"1d13'
Mattes,Kerry Ann
Vir 13h29'32"-6d38'
Mattes,Sabine
Lac 22h2'13"51d40'
Matteson,Joshua Paul
Cam 4h3'17"58d30'
Matteson,Patricia Ann
Ori 05h50'01"18d52'
Matteson,The Most Beautiful,Susan
And 0h13'19"30d43'
Matteson,Violet
Hya 9h10'50"2d26'
Matthaes,Theresa
Cas 0h7'33"64d14'
Matthaeus,Renate
Aur 5h2'36"40d50'
Matthaious
Cmi 8h6'17"6d24'
Matthau,Walter & Carol
Aql 18h45'1"10d24'
Matthes,Marilyn Judith
And 1h44'0"39d24'
Matthew
Hya 8h58'25"6d1'
Matthew
Boo 14h15'33"39d24'
Matthew
Her 17h18'36"49d27'
Matthew
Aur 6h39'25"37d32'
Matthew
Lac 22h8'21"40d12'
Matthew
Peg 22h31'45"11d1'
Matthew
Peg 23h37'20"31d48'
Matthew
Boo 14h38'5"22d11'A
Matthew
Her 17h6'6"31d16'B
Matthew
Per 1h40'51"52d58'
Matthew
Oph 18h41'28"8d31'
Matthew & Erika
Lyr 18h37'0"29d19'
Matthew & Joshua Star
Peg 23h0'0"22d26'
Matthew & Robin Always & Forever
Mon 6h22'24"8d32'
Matthew & Robin's Burning Love
Sge 19h34'32"16d27'
Matthew & Robindale Star
Dra 17h22'15"64d8'
Matthew & Sara
Cyg 20h43'44"45d32'
Matthew (Froggie)
Boo 14h12'53"52d30'
Matthew 44
Her 17h31'10"27d34'
Matthew A P"
Gem 6h44'0"31d38'
Matthew Angel
Ser 18h42'45"4d28'
Matthew B
Boo 14h55'35"52d49'
Matthew Box My Fate Mate For Life
Aql 19h5'56"-5d45'
Matthew Christopher
Peg 21h56'55"34d16'
Matthew Christopher
Lac 22h55'42"52d34'
Matthew Craig
Boo 15h28'1"50d45'
Matthew Jon
Cep 21h56'47"58d23'
Matthew L
Per 4h3'21"51d19'
Matthew S
Vul 21h24'32"26d26'
Matthew Stephen
Boo 15h6'52"11d7'
Matthew Thomas
Boo 15h6'35"31d17'
Matthew Tod
Aur 5h0'55"49d38'
Matthew Wayne
Mon 7h43'9"-5d13'
Matthew's 13th
Dra 20h33'20"68d52'
Matthew's Christmas Star
Oph 18h19'0"10d19'
Matthew's Diamond
Her 18h28'32"24d47'
Matthew's Guardian Angel
Dra 17h57'23"83d3'
Matthew's Little Bear
Umi 15h54'41"76d46'
Matthew's Lucky Star
Ori 6h1'51"1d37'
Matthew's Magic
Cep 23h3'54"64d36'
Matthew's Moon Beam
Aql 20h2'33"3d59'
Matthew's Providence
Lyr 18h35'44"37d51'
Matthew's Star
Oph 17h23'54"7d38'A
Matthew's Star
Dra 19h1'0"56d59'
Matthew's Star
Aur 5h4'19"42d43'
Matthew's Star, Harrison C
Per 4h43'58"51d36'
Matthew's Star,Samuel A
Cep 23h37'53"64d11'
Matthew's Wishing Star
Uma 12h14'0"58d23'
Matthew,Blair
Mon 6h53'55"-0d43'
Matthew,Christopher
Cet 0h29'26"-2d21'
Matthew,Fern Etta Williams
Vul 19h2'57"25d32'
Matthew,Karie Anne
Aql 19h16'44"10d24'
Matthew,Kenneth & Bonnie
Sge 20h3'43"20d11'
Matthew,Son of David
Per 3h35'31"31d48'
Matthew,Thomas Howard
Her 17h12'21"45d50'
Matthew,Twinkling Forever
Mon 7h41'0"-8d56'
Matthews Family
Peg 22h45'54"32d15'
Matthews,"Papi" Dana W
Oph 17h43'50"13d46'
Matthews,Alan
Per 2h10'1"57d2'
Matthews,Alice E
Cnc 8h57'24"28d55'
Matthews,Brandon"Bean"
Hya 8h12'1"-6d27'
Matthews,Bryan
Per 4h5'19"46d5'B
Matthews,Cary James
Aql 19h42'1"12d48'
Matthews,Charles W
Dra 16h35'1"61d52'
Matthews,Daddy
Aur 5h3'56"40d41'
Matthews,Darren Wayne
Uma 11h6'33"30d35'
Matthews,Dawn
Lyr 18h19'33"41d4'
Matthews,Dorothy Thompson
Uma 8h33'54"60d39'
Matthews,Dorothy "Dot"
Cas 1h57'46"60d40'
Matthews,Dr Martin H
Per 3h42'39"39d50'
Matthews,Eleanor F
Uma 15h35'49"77d13'
Matthews,Elisabeth K
Peg 21h28'25"21d34'
Matthews,Frances P
Vul 20h3'56"25d54'
Matthews,Francis T
Her 17h19'49"44d11'
Matthews,Gary
Uma 8h58'56"56d6'
Matthews,James K & Farrell J
Cyg 21h6'44"31d5'
Matthews,John Burton
Mon 6h56'19"3d40'A
Matthews,Kaimana Marie
Cam 4h58'0"68d31'
Matthews,Kenneth W
Per 2h54'1"38d33'
Matthews,Kyle
Dra 18h7'38"65d33'
Matthews,Lorna-Jane Theresa
Lyr 18h16'34"31d16'
Matthews,M J
Her 18h19'40"20d16'
Matthews,Mackey
Oph 18h31'17"11d39'
Matthews,Marc
Her 17h17'14"40d32'
Matthews,Michael J
Sco 16h58'1"-44d0'
Matthews,Mommy
Aur 5h0'46"46d42'
Matthews,Paul
Ori 5h53'0"10d19'
Matthews,Richard
Gem 7h12'14"21d42'
Matthews,Robert J
Cet 0h24'34"-6d3'
Matthews,Robert W & Jeffrey Gorczynski
Aur 6h15'59"45d14'
Matthews,Robin
Oph 17h31'28"-20d49'
Matthews,Steven F
Cet 1h25'1"-1d4'
Matthews,Susie Jo
Mon 6h33'46"-6d32'
Matthews,Thomas Ira
Sgr 19h27'20"-40d33'
Matthews,Victoria
Cam 6h23'17"65d14'
Matthews,William Wallace
Ori 5h57'38"14d48'
Matthi
Peg 22h27'41"31d25'
Matthias
Cnv 12h41'0"35d29'
Matthias's Star
Mon 6h53'19"-1d29'
Matthies Star,The One & Only Toni
Tau 5h49'0"28d23'
Matthies,Carl Prince
Her 18h19'59"18d47'
Matthieu
Lac 22h39'56"38d25'
Matthieu,Romaen
Cas 3h2'29"58d45'
Matthieu-Benoit, Aguessy
Oph 17h5'34"10d12'
Matthow,Henry
Hya 8h14'33"0d1'
Matti & Gloria's Paradise
Cyg 20h53'31"50d28'
Mattia
Uma 11h48'50"50d0'
Mattick III,Augustus Herman
Cyg 19h52'18"37d53'
Mattick,Sondra Therese
Cas 0h49'13"68d30'
Mattie
Lyn 9h17'16"36d9'
Mattighello,Anna
Umi 17h15'30"76d29'
Mattila's Shining Star,Al
Uma 10h1'14"54d26'
Mattila,Arja
Dra 16h7'46"60d58'
Mattila,Pekka
Cam 3h47'57"58d26'
Mattina,Maria
Sgr 18h56'14"-21d18'
Mattingly III(Little Bit),Vince
Cep 23h2'19"65d11'
Mattingly(Prissy Missy),Melissa
Lyn 6h21'13"60d35'
Mattingly,Charles Q
Mon 6h31'30"-6d14'
Mattingly,Christine
Tri 2h13'50"32d24'
Mattingly,Louise & Sandy
And 0h25'0"43d18'
Mattioni,MD,Thomas A
Oph 18h42'0"10d13'
Mattison,Lauren
Lyr 19h10'13"40d39'
Mattison,Lisa
Peg 22h24'45"30d60'
Mattistar
Cyg 20h59'46"28d26'
Mattivi,Mariella
Aur 6h53'1"37d48'
Mattivi,Robert E
Her 16h39'54"34d29'
Mattli,Paul Martin
Cep 22h12'54"61d21"
Matto,Don W
Her 17h1'53"21d9'
Mattock,Julie
And 0h48'13"40d20'
Matton Star,The Robbie
Aur 6h32'27"34d8'
Mattoon,Bryan Tyler
Boo 15h38'53"41d12'
Mattoon,Carol
Boo 14h26'53"22d59'
Mattoon,Richard ("Matt")
Lmi 10h36'45"26d27'
Mattos,Reed Aualino
Lyn 6h26'0"60d41'
Mattos,Richard A
Oph 17h35'32"-22d29'
Mattox,Chris
Her 16h36'44"25d14'
Mattox,Daryle
Cep 22h32'24"60d14'
Mattox,Molly
Umi 15h17'12"69d42'
Mattox,Wade Robertson
Sgr 15h20'36"20d38'
Mattrose
Cnv 12h38'28"50d21'
Mattsei Porkka
Lmi 9h58'19"37d40'
Mattson,Daryl & Bonnie
Uma 10h18'27"61d16'
Mattson,John G
Uma 12h25'38"61d37'
Mattson,Jonathan
Cma 6h48'30"-19d25'
Mattson,Raymond Erik
Sge 20h7'42"18d58'
Mattster
Lac 22h37'18"41d13'
Mattthew
Peg 21h58'11"31d32'
Mattulat,Gustav & Emmi
Cyg 20h8'13"37d34'
Mattull,Thomas & Therse
Crb 15h52'39"30d31'
Mattus,Joseph
Ser 15h28'14"24d52'
Matty
Cet 0h45'39"-3d60'
Matty,Bea
Sex 10h25'30"3d34'A
Matty,Connie
Mon 6h30'12"-5d18'A
Matty,Newt
Sex 10h25'30"3d34'B
Mattyn
Peg 21h56'41"35d54'
Matuchovi,Ludvig & Ludmila
Vir 11h52'1"9d22'
Matula,John F
Aur 5h5'1"38d53'
Matura,Elnora Ann
Aql 19h3'50"-5d52'
Maturi,Michael David
Cnv 12h28'40"32d1'
Matuszewski,Wanda
Lyn 9h12'0"35d37'
Matuzas,Ben
Cet 2h42'31"6d2'
Matuzas,Dan
Cet 2h42'57"7d49'
Matuzas,Mark
Cet 2h42'58"8d15'
Matviyak
Boo 14h14'42"38d8'
Matwaly,Nagy
Aql 20h0'42"12d18'
Matyas,Daniel Kalman
Aql 20h7'40"8d50'
Matyczynski Love Uncle David,Kaven
Lyn 7h32'16"38d40'
Matyja,Verlyria
Crt 11h0'28"-22d12'
Matys-Hartmann, Claudia
Leo 9h32'40"7d52'
Matysko,Harriet I
And 23h17'42"44d48'
Matz,Anastasia Tasha Marie
And 1h32'46"40d43'
Matz,Gary D
Dra 9h26'38"78d37'
Matz,Genna
Cyg 21h21'1"39d55'
Matz,Joshua
Per 4h40'21"41d5'
Matz,Merri
Cas 0h36'1"58d37'
Matzke's 95 Wildcat Star
Eri 4h1'13"-10d12'
Matzke,B Elaine
Sco 17h24'0"-38d39'
Matzkofz-Eternity
Crb 15h26'48"30d7'
Mau-Estelle
Cam 3h27'32"61d31'
Maubert Aurélien
Mon 7h41'44"-4d12'
Maubon,Jean-Claude
Aql 19h30'24"-10d60'
Mauceri,Angelo Jack
Leo 10h31'1"26d48'
Mauceri,Nancy & Frank
Crb 16h0'20"33d43'
Mauchamp,Isabelle
Lyn 8h55'56"45d4'
Mauck "Forever With Kim",Gregory
Uma 10h51'0"40d31'
Mauck,Lois Anne
Del 20h23'25"11d41'
Mauck,Rhett Adam
Hya 8h11'21"0d49'
Mauclere,Alix
Boo 14h32'54"31d28'
Maudcorkytaflanrodkevz arich-phil
Boo 14h56'1"33d44'
Maudie
Lyn 8h4'60"37d16'
Mauerman,Tod
Boo 14h6'38"25d60'
Mauge,Amaury
Cas 1h36'28"64d44'
Maughan,Michael M
Cep 2h1'44"80d16'
Maughan,Susan
Sgr 18h57'44"-25d46'
Maughon,Landy Phillip
Oph 17h58'1"-8d49'
Mauglan,Breanne Helen
And 2h29'11"48d40'
Mauimar Siempre Dia Monte Brillante
Ari 2h36'24"29d49
Maul,Robert
Per 3h58'12"37d45'
Mauldin's Ace Man in The Moon
Tri 1h41'11"28d47'
Mauldin,G Scott
Ori 5h59'16"10d24'
Mauldin,Joshua Craig
Per 3h14'38"41d48'
Maule,Agnes Mary
Umi 14h33'57"78d6'
Maule,James Harvey
Ori 6h1'11"1d47'
Maulorico,Carmen Anthony
Her 17h0'33"38d51'
Maultsby,Jordan Hilton
Ori 5h16'20"7d36'
Maumoynier,Jr, Alexander & Doris
Oph 17h59'26"12d7'
Maunder,John F
Dra 16h22'58"63d43'
Mauni
Peg 22h20'1"10d29'
Maunsell,Dr Charles Dudley
Oph 18h18'26"6d34'
Maupetit,Indy
Dra 16h59'28"61d28'
Maupin,Emily
Uma 8h33'11"62d6'
Maupin,Joel Keith
Dra 17h0'24"51d44'
Maupin,MD,B Kent
Oph 17h32'36"-6d16'
Maur One
Uma 12h3'17"46d26'
Maura
And 23h41'27"47d20'
Maura's Realta Go Bragh
Uma 9h1'43"56d58'
Maura's Starry Night
Cet 2h31'26"4d29'
Mauran,Michel
Cet 0h58'52"1d36'
Maurea,William Conrad
Ori 5h59'27"11d44'
Maureen
Cas 0h20'53"61d31'
Maureen
Uma 11h54'11"56d58'
Maureen
Umi 15h18'40"69d45'
Maureen
Vul 19h48'19"22d40'
Maureen & Anthony
Eri 2h50'15"-1d58'
Maureen & Marty
Cyg 21h7'30"30d58'
Maureen Alice
Lyr 19h22'25"40d60'
Maureen Ascending
And 1h52'28"40d50'
Maureen B
Cam 8h1'18"61d41'
Maureen Diana
Cas 0h46'32"63d57'
Maureen Elizabeth
Cas 1h58'42"61d8'
Maureen Margaret
And 0h8'21"35d57'
Maureen Patricia
And 0h38'13"40d26'
Maureen's Follies
Sge 19h52'47"16d38'
Maureen's Lucky Star
Aql 20h11'0"13d47'
Maureen's Magic
Peg 0h0'39"18d29'
Maureen's Paul
Peg 22h35'11"24d55'
Maureen's Star
Vir 12h51'6"-9d45'
Maureen's Star
Cyg 19h30'48"35d4'
Maureen's Star
Cyg 20h44'51"42d5'
Maureen(Mo)
Cyg 21h27'18"38d1'
Maurer,Amanda Marie
Lmi 10h46'1"25d37'
Maurer,Cynthia Gwen
Boo 15h47'14"46d4'
Maurer,Earl W
Cep 23h11'19"60d22'
Maurer,J W
Cet 2h11'18"2d17'
Maurer,Katelyn
Cas 23h16'42"62d59'
Maurer,Laura Smith
Lyn 7h42'42"42d31'
Maurer,Michael Raymond
Vul 20h56'1"28d36'
Maurer,Paul M
Cet 1h28'60"-0d34'
Maurer,Richard W
Oph 18h19'1"7d1'
Maurer,Sandy
And 22h58'28"37d46'
Maurer,Susan Faith
And 2h16'11"45d47'
Maurer,Thomas
Ind 20h50'6"-55d46'
Maurer,Werner
Mon 7h51'17"-2d56'
Maurey,Michelle
And 2h27'13"42d46'
Mauri
Equ 20h55'38"2d27'
Mauri,Toby
Her 16h7'1"24d21'
Maurice
Dra 17h20'28"61d43'
Maurice
Cet 2h16'47"2d29'
Maurice
And 0h30'1"22d55'B
Maurice & Jean Love Forever
Cyg 20h25'13"38d6'
Maurice Heart Spirit
Cru 12h8'58"-58d27'
Maurice Richard
Dra 19h59'40"80d19'
Maurice,Julia
Uma 8h33'15"50d58'
Mauriello,"Robbie"-R L
Her 17h10'0"43d52'
Mauriello,Bradley Johnathan
Her 16h49'31"30d10'
Mauriello,Eric
Cet 1h27'24"-6d9'
Maurin
Ser 15h38'24"4d50'
Mauritz,Chiquita & Maurizeo Quintor
Peg 22h0'31"30d28'
Mauriz
Per 4h1'36"50d3'
Maurizi,Kristen Marie
Cam 4h10'35"58d39'
Maurizi,Lauren Ann
Uma 10h44'52"61d50'
Maurizi,Rudy
Boo 14h3'31"23d36'
Maurizio E Antonella
And 0h46'18"40d1'
Maurizio, Persona Meravigliosa?
Cnv 13h0'0"32d0'
Mauro
Col 6h34'39"-39d12'
Mauro Iacomelli
Peg 22h28'38"30d44'
Mauro,Don
Aql 19h46'58"14d22'
Mauro,John Doric
Lac 22h35'54"55d24'
Mauro,Lisa G
Lyn 8h25'24"41d34'
Mauro,Our Star In Heaven Joe
Boo 15h1'11"17d15'
Maurose,Ruth
Cas 1h32'46"58d41'
Maurovic,Natasha Wellesly
Cas 22h58'35"55d44'
Maurutto,Ann
Peg 23h49'16"31d38'
Maus,Carie Leigh
Lyr 18h52'38"33d55'
Maus,Emily Jean
Ori 4h56'22"15d7'
Maus,Frederick Eli
Aql 19h52'30"14d33'
Mausar,Dream Star Owned by Barbara
Tri 1h53'23"27d45'
Mausenflop
Uma 10h4'29"62d20'
Mauser,Ingeborg
Sgr 18h52'48"-27d4'
Mausi,Sylvia
Vir 12h2'32"-1d34'
Mausinsk & Mausowsk
Lyn 8h6'0"42d9'
Maust,Keith
Aur 7h2'55"43d48'
Maust,Lori Lynn
Lyn 8h26'25"40d36'
Mautner,Bryce Alison
Gem 6h54'51"15d26'
Mautz,"Jano" Alexandro R
Hya 9h21'20"-0d54'
Mauz,Karl-Heinz
Aur 5h4'47"43d10'
Mauzy,Ron/Kim Enrenfried
Cyg 19h56'38"41d9'
Mavelikera Godavarma Raja Varma
Ori 5h56'11"7d53'
Maver,Devon Francesca
Lyr 18h55'17"40d16'
Maverick
Dra 16h28'0"66d21'
Maverick
Cae 4h48'37"-32d6'
Maverovi
Cep 21h26'15"61d35'
Maviglia,Mammer-Doris Angelina
Vul 20h16'32"23d12'
Maviglia,Papper-Frank Augustus
Dra 16h41'46"61d37'
Mavis
Lyr 18h34'27"36d9'
Mavis
Per 1h34'51"54d20'
Mavis Elaine
Tau 4h45'38"16d22'
Mavis Iris
And 1h45'13"38d59'
Mavis,Evol Agnes
Com 12h40'23"21d31'
Mavity,Thomas,Love Denise Shisler
Aur 6h52'31"37d60'
Mavourneen
Lyn 7h54'56"58d9'
Mavournin Maria
Cas 1h17'50"63d36'
Mavromatis,Achilles Louis
Per 1h39'38"53d23'
Mavrommatis,Demetrios
Lyr 19h5'10"38d16'
Maw
Lac 22h34'35"55d50'
Mawahib
Her 16h12'58"4d31'
Mawdsley,Kayleigh Sylvia Margaret
And 23h19'38"49d10'
Mawford,Philip
Aur 5h11'34"42d25'
Mawson,Harry E,Jr
Mon 7h39'34"-1d26'
Max
Tau 5h22'29"16d50'
Max
Oph 17h14'60"-20d25'
Max
Uma 10h51'0"40d31'
Max
Sct 18h43'0"-6d21'
Max
Her 18h43'36"18d46'
Max
Cma 6h55'28"-13d25'
Max
Umi 14h44'52"80d16'
Max
Aur 7h13'41"38d41'
Max
Ori 6h5'51"0d13'
Max
Ser 15h38'10"3d9'
Max & Erin
Lyn 9h7'42"35d54'
Max & Helen's Star
Peg 21h45'36"20d7'
Max David
Ser 15h28'55"21d12'
Max Elf Q My
Per 3h14'11"55d7'
Max Kristy
Aql 18h43'42"6d10'
Max My Love
Lmi 9h57'27"28d23'
Max Nathan
Per 2h36'11"38d56'
Max Paul
Lac 22h2'20"37d32'
Max Regis Et Chris Rius
Lyr 19h22'22"30d15'
Max's Moon TA
And 0h11'47"38d56'
Max's Wish
Lyr 19h1'29"28d22'
Max,Cricket Malherbie
Uma 9h2'21"48d17'
Max,Lenny
Cet 2h39'11"-4d40'
Max,Mary
Boo 14h58'18"7d19'
Max-Carl "M-C M"
Leo 10h58'42"1d42'
Max-Million
Umi 15h16'28"68d19'
Maxakan
Uma 9h24'17"57d9'
Maxamillion Perry Blue
Her 18h24'29"20d18'
Maxen,Anna
Umi 15h31'37"77d48'
Maxey,David James
Uma 13h5'18"53d29'
Maxey,Michael Coy
Dra 14h57'1"58d3'
Maxey,Robert A
Mon 8h8'55"-4d53'

Maxfield,Scott & Esther Cas 1h28'53"58d21'
Maxfield,Tom Lmi 10h1'31"39d32'
Maxham,James Edward Cnv 12h49'60"39d11'
Maxi & Betty's 30th Wedd Anniversary Star Peg 23h31'57"11d34'
Maxi und Clemens Lyn 7h47'28"48d8'
Maxie 51 Cyg 20h7'0"58d17'
Maxim Uma 8h57'53"47d46'
Maxim Alexander Per 3h12'51"40d15'
Maxim,Jaclyn Crb 15h18'0"31d12'
Maxima Pic 4h35'33"-47d15'
Maxime Lyr 19h5'32"40d55'
Maxime-Deborah Cas 23h29'42"61d35'
Maximilian Her 17h53'58"42d52'
Maximilian Leo 10h36'14"21d52'
Maximilian Alexander Ser 15h9'51"-1d31'
Maximum Bart Security Cma 6h59'25"-19d13'
Maximus,Fudgy Cnv 12h7'46"43d8'
Maxine Lyn 8h23'23"50d41'
Maxine Gem 6h52'51"31d10'
Maxine And 2h26'44"41d51'
Maxine Lyr 19h24'29"38d34'
Maxine Ari 2h51'1"30d45'
Maxine Cyg 19h34'34"33d25'
Maxine And 0h50'55"39d24'
Maxon,George S Aur 5h0'39"45d13'
MaxPeters Cap 20h34'45"-10d54'
Maxphil Mon 8h8'29"0d21'
Maxson,Barbara Mon 7h22'24"-8d52'
Maxson,Forever John Per 2h7'14"47d24'
Maxson,Robert T Dra 15h22'32"58d34'
Maxted,Freija Robin Umi 16h8'39"71d8'
Maxwell Charles Ser 15h52'52"-1d48'
Maxwell's Very Own Boo 14h22'13"33d32'
Maxwell,A J Lac 22h14'28"51d14'
Maxwell,Andrew & Una Hya 9h11'53"3d36'
Maxwell,Billy Dra 19h6'42"68d4'
Maxwell,Carolyn Cyg 20h35'17"38d3'
Maxwell,Connie Hall Peg 21h53'16"21d56'
Maxwell,Crystal Lynn Cas 1h17'59"63d55'
Maxwell,Debbie Cet 2h17'52"3d42'
Maxwell,Donald Cep 4h21'1"80d26'
Maxwell,Esther & Robert Uma 9h21'48"56d3'
Maxwell,H Lovell Gibby Lyn 7h55'28"58d52'
Maxwell,Jacqueline L And 22h57'31"51d47'
Maxwell,James Reed Sct 18h48'30"-7d2'
Maxwell,Jennifer Cyg 19h32'0"36d39'
Maxwell,Kasha Cnv 13h7'55"33d6'
Maxwell,Kelly And 23h0'0"47d45'
Maxwell,Lindsay Umi 14h56'48"66d4'
Maxwell,Martin Boo 14h39'13"15d18'
Maxwell,Max Sex 10h43'52"0d17'
Maxwell,Nicki Dee Lyn 8h1'17"37d43'
Maxwell,Nikolas Richard Umi 13h59'41"72d28'
Maxwell,Peter James Cep 22h18'59"56d11'
Maxwell,Phillip J Ser 15h10'52"-2d21'
Maxwell,Rick Aur 5h55'0"31d23'
Maxwell,Robert Her 17h53'0"38d4'
Maxwell,Ruth Hille Mon 6h45'27"11d35'
Maxwell,Ryan Vul 19h43'0"23d35'
Maxwell,Sheri Umi 13h32'40"72d4'
Maxwell,Speedy Cmi 7h59'1"0d49'
Maxwell,Steven J Aur 5h51'53"41d7'
Maxwell,Teri Vul 19h44'53"23d35'
Maxx & Carley Cyg 20h17'56"38d34'
May Cas 1h29'54"60d3'
May (Aged 32),Anthony James Ori 6h7'1"3d44'
May AKA Pooh Bear, Nancy Jane Uma 9h5'13"56d12'
May Everybody You Love Be Happy Tau 5h23'41"26d55'
May Madeleine Lib 15h40'10"-23d41'
May Meus Amare Inaetemtas Mon 7h49'43"-5d28'
May Star,Wayne F Dra 17h1'58"64d26'
May Your Dreams Come True Alexandra Sct 18h52'26"-8d49'
May's Star,Andrea Cyg 21h39'58"28d50'
May,A K & Julia May Eri 2h55'0"-13d32'
May,Adabel Mon 7h14'34"-5d17'
May,Al Dra 12h1'48"70d25'
May,Alfred Aur 6h8'31"46d34'
May,Andrew Michael Per 2h41'33"35d56'
May,Babe & Ida Cyg 21h41'18"37d32'
May,Becky & Ike Cyg 21h41'31"37d43'
May,Bill Per 2h51'50"41d10'
May,Brendan Stephen Aur 5h25'15"31d16'
May,Bruce Sco 16h54'1"-38d47'
May,Caleb Elijah Cet 0h10'1"-10d43'
May,Carol Ann Uma 11h13'0"40d11'
May,Caylah McKenzie Peg 22h8'0"5d10'
May,Christopher David Ara 17h57'0"-51d25'
May,Craig J Her 15h50'39"47d15'
May,Delmar & Bertha Crb 15h29'26"30d9'
May,Forever Eddie Cep 20h28'21"67d15'
May,Franklin Eugene Her 18h22'1"13d6'
May,Grace Ellen Peg 21h19'58"22d35'
May,Haley Rhianna Aql 20h1'55"10d42'
May,Harry R Crb 15h56'12"31d18'
May,Howard Michael Psc 0h50'34"31d22'
May,I Love You,Richard G Eri 3h32'24"-2d17'
May,James Leo Ser 16h1'14"-1d34'
May,Jessica Ellen Lyn 8h8'41"50d54'
May,John E Her 17h32'36"23d36'
May,Joshua Anthony Aur 5h0'58"41d17'
May,Jr,James L Cap 21h6'27"-20d9'
May,Juli S And 23h2'50"50d48'
May,Julia Cyg 19h40'52"40d43'
May,Katina Williams Com 12h22'30"28d6'
May,Kelsey Jordan Leo 10h36'27"13d59'
May,Kevin Cep 4h29'59"80d36'
May,Kimberly Allison Cam 12h17'47"80d44'
May,Krista M Lyn 7h54'22"35d57'
May,Margaret "Molly" Gem 6h40'1"30d24'
May,Mary Lou Uma 10h30'0"52d9'
May,Michael Dra 16h58'49"71d28'
May,Michael Raymond Oph 18h39'56"7d23'
May,Paul Peg 22h0'47"32d44'
May,Pauli Peg 23h42'56"26d55'
May,Peter Boo 15h29'41"42d4'
May,Peter Michael Cam 8h29'30"83d12'
May,Rolf-Dieter Uma 8h30'59"68d8'
May,Shane W Aql 19h6'49"3d16'
May,Shelby Lynn Cas 23h30'11"61d18'
May,Stephen Michael Dra 18h53'23"71d5'
May,Stephen P Her 17h38'49"42d6'
May,Steven F Her 17h32'22"23d6'
May,Trent E & Lisa F Uma 12h0'0"30d39'
May,Vivian Marie Pierce And 0h9'12"44d43'
May-Babe,Jonathan Boo 15h0'45"30d22'
Maya Cas 0h46'18"61d44'
Maya Oph 18h7'18"11d54'
Maya Uma 11h11'31"41d29'
Maya Margaret Oph 17h1'25"-18d54'
Maya's Heart And 1h42'46"38d29'
Maya's Star And 2h22'15"48d57'
Mayalice Com 12h9'32"21d44'
Maybe,Happy 20th! Marg & Rich Cyg 21h54'48"38d25'
Maybee Family,The Peg 23h42'22"28d35'
Maybelle Cyg 21h32'33"53d48'
Maybelle & Stuart Cyg 20h3'58"41d6'
Maybin,Howard J Cep 0h41'17"78d55'
Maybruck,Eryn A Ori 5h54'54"8d26'
Maybruck,Lynda Johnson Oph 17h57'54"12d25'
Maybruck,Lysa K Peg 22h56'18"20d53'
Maybury,Bruce Robert Ori 5h10'25"-5d12'
Maybury,Gordon Her 16h6'0"22d54'
Maybury,Loraine Cas 2h49'39"68d40'
Mayce,Joseph Lac 22h19'59"48d43'
MayCo Umi 13h57'50"75d32'
Maycumber Star, "Volcano Sam"-The Aql 19h31'13"14d57'
Mayda Cam 4h15'10"68d1'
Mayela Eri 2h46'0"-5d14'
Mayer Umi 14h45'17"66d43'
Mayer Jean Claude Umi 15h30'0"71d51'
Mayer,Abriana Elizabeth Sgr 18h51'57"-32d31'
Mayer,Alain et Brigitte Per 3h5'56"40d24'
Mayer,Anneliese 06/04/1938 Ari 2h0'1"22d21'
Mayer,Artur Lac 22h1'30"51d29'
Mayer,Bonnie Ann Lyn 7h15'47"55d0'A
Mayer,Christian F Cam 3h51'13"55d14'
Mayer,Erin Michelle Mon 6h20'1"-1d14'
Mayer,Gary Paul Cyg 20h20'47"38d18'
Mayer,Heidi Lyr 18h42'48"42d30'
Mayer,Irene M And 0h10'28"33d5'
Mayer,Jeff Aur 5h29'48"54d54'
Mayer,Jeffrey Karl Lyn 7h15'47"55d0'B
Mayer,Jr,Charles Bennett Lac 22h20'24"55d13'
Mayer,Karl Wayne Boo 15h17'1"53d12'
Mayer,Konsul Peter Ari 1h46'36"23d18'
Mayer,Lee F Hya 8h22'31"0d13'
Mayer,Margaret R Lac 22h31'12"56d45'
Mayer,Marty Aql 19h7'17"15d4'
Mayer,Michael & Lenora Lyr 18h30'1"30d59'
Mayer,Michelle Carolyn Cyg 20h55'45"30d3'
Mayer,Rachael Marie Aqr 21h4'30"0d57'
Mayer,Randy Dra 17h2'46"68d58'
Mayer,Richard Louis Cet 2h10'36"5d50'
Mayer,Ro Cam 7h57'32"61d18'
Mayer,Roger R Aur 4h50'23"40d18'
Mayer,Simon J Lac 22h28'33"41d6'
Mayer,Stefany P-Robin D Roberts Uma 12h10'24"55d56'
Mayer,Stephanie G Tri 2h29'32"30d31'
Mayer,Thomas D Per 2h59'39"46d3'
Mayer,Ursula und Hans Mon 7h54'34"-1d15'
Mayer,Vanessa Anne Lyr 18h16'0"30d11'
Mayer,William L Aur 5h25'43"30d45'
Mayerle-Looman Sge 19h54'0"18d50'
Mayers,Jim Dra 17h1'29"65d26'
Mayes,Darrell Wayne Tri 2h26'53"28d11'
Mayes,Mark Bryan Her 17h13'1"49d22'
Mayfield,Jennifer Cas 1h43'54"70d47'
Mayfield,Lisa J Boo 15h5'32"31d9'
Mayfield,R T Cet 2h19'29"2d27'
Mayflower Enterprises Boo 14h43'40"34d9'
Mayheath Peg 21h58'36"33d9'
Mayhew,Millie "Eykis" Aqr 20h59'34"-0d14'
Mayhew,Trevor Paul Edward Cep 22h51'0"56d56'
Mayjor I,Judy Lyr 19h17'40"26d46'
Maykemper,Dr Bernd Lyr 18h34'37"26d0'
Maykrantz,Teresa Marie Aql 19h31'10"12d22'
Maykuth,Toby Boo 14h10'14"30d10'
Maylam,Tony Her 16h34'1"40d15'
Mayle,Aundra Lopez Aql 20h10'1"0d42'A
Mayling,Carol Ann Eri 4h57'55"-5d58'
Maylor,Clifton Arthur Her 16h41'44"32d31'
Maynard Cam 3h57'32"56d4'
Maynard Vul 19h23'25"23d38'
Maynard 1960-1991, Stephen Oph 17h3'22"8d40'
Maynard II,Daniel Putnam Cam 4h0'34"61d49'
Maynard,Albert E Boo 15h5'25"22d12'
Maynard,Arthur Boo 14h37'45"33d3'
Maynard,Dorothy E Cyg 20h29'1"37d54'
Maynard,Dorothy L Cas 0h7'16"62d43'
Maynard,Elaine Rae Aqr 21h35'19"-1d42'
Maynard,Eva Irene Lyr 19h17'30"42d55'
Maynard,Greg Ser 15h58'12"21d44'
Maynard,Ryan David Cet 0h24'55"1d55'
Maynard,Timothy Shawn Boo 14h48'0"30d23'
Mayne,William H Cyg 20h6'17"39d39'
Mayo,Brenna Peg 22h15'10"31d5'
Mayo,Cathy And 2h31'54"49d53'
Mayo,Esq,James C Boo 15h13'41"28d23'
Mayo,Mark Aur 6h14'51"33d1'
Mayo,Matthew Edmund Uma 11h53'1"38d22'
Mayo,Michael Cleve Her 18h54'23"12d49'
Mayo,Michael David Per 2h40'49"40d33'
Mayo,Michelle Dawn Eri 3h44'13"-5d51"
Mayo,Pat Lyn 7h52'55"51d25'
Mayo,Sheri Vul 20h0'53"23d19'
Mayo,Stephen David Aur 5h15'48"45d22'
Mayo,Thomas Carl Umi 14h39'56"67d20'
Mayo,Vicki Leo 9h21'37"19d41'
Mayo-Smith,John Farrington Boo 14h46'55"35d44'
Mayor,"J" Ori 6h0'1"8d19'
Mayor,Bubba Ori 6h0'24"8d53'
Mayor,Gayle Ori 6h0'19"8d9'
Mayor,Hélène Dra 12h0'30"71d38'
Mayor,Jason Ori 6h0'1"1d3'
Mayor,Katie Ori 6h0'11"8d31'
Mayr,Karl & Olga Boo 14h20'13"45d59'
Mayr,La Eterna Estrella de Aql 19h32'28"1d16'
Mayr,You Beauty Doan Ind 20h55'48"-53d55'
Mays,Avery Earlene Lyn 7h28'43"58d8'
Mays,Barbara J Peg 22h0'46"23d57'
Mays,Carole Lyn 7h43'60"50d51'
Mays,Ellen Patricia And 0h7'33"37d37'
Mays,Harriet Cam 5h38'34"67d50'
Mays,Paquita Robin Lyr 18h31'1"38d21'
Mays,Terry Sge 20h2'21"20d13'
Mays,Thomas A Cru 12h25'36"-61d47'
Mays-Angel,Lori Lee And 2h31'28"41d38'
Mayte Cae 4h46'1"-33d39'
Maytum,Harry Rodell Aql 19h30'35"0d27'
Mayu Peg 23h26'50"16d60'
Mayumi Lyr 18h43'52"41d30'
Mayuri And 2h32'42"40d13'
Maywald,Carlo Cet 2h32'44"1d46'
Mayworm,Bob & Mary Ann Her 17h25'19"21d31'
Mazandi,Yousof A Vir 13h4'22"-8d32'
Mazansky,Janet And 0h18'39"45d38'
Mazansky,Ruth Crb 16h11'41"37d8'
Mazel Lyn 8h9'39"45d17'
Mazel Tov Danny & Amy 2.20.94 Lmi 9h45'45"34d51'
Mazen Cet 2h41'54"-11d28'
Mazen & Adeena (Mazeena) Vul 20h26'52"28d49'
Mazen Alam Cam 4h51'54"65d19'
Mazet,Pauline Battoni And 0h16'46"45d55'
Mazich,Laura-Ashley Ori 6h3'23"8d45'
Mazie 1 Uma 10h36'16"71d33'
Mazier,Carol & Simon Cyg 19h24'16"44d40'
Maziere,Jean-Noel & Catherine Ind 20h26'25"-51d56'
Mazorra,Mary Beth And 2h16'19"45d48'
Mazumder's Star,Sonia Cas 0h51'0"70d52'
Mazur,Julie Sco 17h54'20"-30d46'
Mazur,Ladislaus Lmi 10h21'57"31d48'
Mazurek Aur 6h38'57"38d49'
Mazurek,Faith Cyg 21h24'25"30d46'
Mazurek,Kit Aur 6h28'28"38d26'
Mazurka Leo 9h53'59"12d19'
Mazurka,Alan David Umi 14h44'55"80d52'
Mazurkiewicz,Jennifer Peg 22h18,20"29d43'B
Mazurkiewicz,Kris Uma 11h24'43"42d54'
Mazuroski,Barbara & Edward Cep 23h7'23"60d54'
Mazz Uma 8h56'16"47d59'
Mazza Lyr 18h16'22"41d13'
Mazza,Christopher Dra 14h23'50"63d41'
Mazza,Deb Cyg 20h24'1"38d14'
Mazza,Frank J Her 16h56'60"51d20'
Mazza,Jr,John A Aql 19h30'45"10d34'
Mazza,Nicole Com 13h0'48"27d48'
Mazzacarallo,Anthony Patrick Psc 0h46'28"21d34'
Mazzalla,Theresa Cam 8h28'52"74d44'
Mazzara,Nicolette Lyn 8h0'10"51d15'
Mazzarano,Marie Elaine Com 12h18'57"32d46'
Mazzarol,Xavier Lmi 11h2'45"31d8'
Mazzaron,Maurizio Ori 6h17'36"8d16'
Mazzei-Trongale, Nicholas Giovanni Uma 9h11'44"56d18'
Mazzenga,Chara Vir 12h53'47"1d47'
Mazzeo,Marlett Cyg 20h21'51"41d55'
Mazzeo,Victor Cnv 12h49'45"48d56'
Mazzer,Caroline Joy Cas 2h44'43"61d28'
Mazzesi,Marco Lyr 18h38'58"27d31'
Mazzetta,Anne Lyr 19h26'56"40d59'
Mazzilli,Franco Francesco Lyn 7h41'50"41d16'
Mazzilli,Lee Louis Boo 14h5'21"38d49'
Mazzilli,Sally Cam 3h31'23"53d39'
Mazzocco Vul 21h14'30"28d15'
Mazzocco,Denise Mon 7h10'39"-5d21'
Mazzocco,Lisa Marie And 2h27'25"48d53'
Mazzocco,Tina Uma 10h26'40"51d1'
Mazzola,Dino Tri 2h20'0"32d33'
Mazzola,Gene & Dawn Lyn 8h37'26"40d45'
Mazzola,Jane Cyg 21h11'34"39d6'
Mazzola,Joann & Paul Cyg 20h23'48"39d6'
Mazzola,John Lyn 7h5'33"51d2'
Mazzola,Jr,Joseph Anthony Per 1h52'0"56d53'
Mazzola,Julian Joseph Gem 6h54'5"15d16'
Mazzoli,Thea Hor 2h43'50"-49d30'
Mazzon,Anouchka Boo 14h31'33"50d16'
Mazzone,Michael Francis Per 1h49'0"48d34'
Mazzone,Savanah Kathryn Lyn 8h10'1"57d52'
Mazzoni,Princess Deidre And 1h29'57"37d50'
Mazzotti,Stevie Caryl Boo 15h17'10"51d43'
Mazzuca,Dr Douglas E Ori 6h5'55"7d31'
Mazzucchelli,Gail Susan Cyg 19h42'39"41d2'
Mazzucchelli,Gail Susan Cyg 20h38'31"29d37'B
Mazzurco,Mary E Cep 22h27'45"58d52'
Mazzuto,John D Aql 19h26'51"10d53'
Mazzvillo,Lynn Ori 5h57'0"18d29'
Mazzy Cma 6h27'58"-13d46'
Maïwenn Et Fabrice Lyr 19h21'50"30d48'
Maü Lyn 8h23'38"42d20'
MA" Uma 10h1'35"48d54'
MBA Group Her 17h55'36"50d4'
MB" Ori 5h48'39"10d8'
Mc 93 Dra 20h54'18"80d32'
Mc Cain,Peter Cas 3h5'48"58d59'
Mc Comb,Rita Gladstone Cyg 19h41'49"41d43'
Mc Coy,James Raphael Her 18h5'53"45d54'
Mc Dalton Uma 9h16'20"70d19'
Mc Gregor,Alison Sge 20h6'42"20d56'
Mc Innie Mackiny Eri 4h42'30"-0d59'
MC Kemmer's Sunrise Uma 8h18'59"68d50'
MC's Twinkling Reminder Aur 6h40'45"35d42'
McAdam,Debbie Cyg 20h25'28"40d21'
McAdam,Eileen And 23h31'56"40d51'
McAdams,Buttons Equ 20h58'46"10d46'
McAdams,Jaime Vul 19h46'36"22d34'
McAdams,James Michael Cet 2h43'43"0d33'
McAdams,Patty Cas 0h59'24"60d16'
McAdams,Richard Her 16h14'41"24d30'
McAdams,Tyler Kurtis Cyg 19h57'59"31d37'
McAfee,Bryant Prentice Boo 15h21'35"33d17'
McAfee,Cathy(Gramma) Lyr 18h31'30"35d5'
McAfee,Gena Jayne Uma 9h58'15"51d25'
McAfee,Gerald Brent Lib 14h45'40"-1d5'
McAfee,Kate Cyg 19h40'11"30d48'
McAfee,Millie Mon 7h24'19"-10d25'
McAffe,Dan(Poppa) Lyr 18h38'39"35d58'
McAlear OMI,Rev Richard Ori 6h16'22"7d51'
McAleer,Eugene Russell Tau 5h50'40"28d36'
McAleese,Clark Boo 14h38'18"15d10'
McAlevy,Margaret Cas 23h3'50"54d3'
McAlister,Adriane Lynn & Bryan Charles Boo 13h41'35"14d27'
McAlister,Jasmine K Mon 6h54'1"-1d52'
McAlister,Maurice L Uma 11h25'30"63d27'
McAllister My Love Forever Marti,Ollie Mon 6h19'38"8d16'
McAllister,Annalisa Jane Ari 1h59'46"12d31'
McAllister,Carol Michelle Lmi 9h57'34"33d34'
McAllister,Jamie Aql 20h2'30"0d18'
McAllister,Michelle Leigh Cyg 19h57'34"50d3'
McAllister,Patti Lynn Ori 5h58'53"16d40'
McAllister,Peter "Sluggo" Mon 7h59'17"-6d6'
McAllister,William Per 4h21'13"50d59'
McAloon,Louis H Cam 6h2'48"61d43'
McAlpine (Rigby),Megan Louise Umi 14h34'22"67d32'
McAlpine,Don Cyg 21h0'10"36d20'

McAlpine,Jackie
Peg 22h34'27"29d12'
McAlpine,Ken
Peg 22h21'18"29d45'
McAlpine,Marina
And 23h26'51"40d59'
McAlpine,Mark Paul
Dra 17h7'54"63d32'
McAmis,Leah
Ori 5h27'1"15d7'
McAnany,Francis X
Per 2h56'22"37d12'
McAnany,Magaret A
Aur 6h43'54"38d15'
McAnany,Robert Emile
Cep 21h49'54"60d12'
McAndrew,Susan V
Cyg 20h23'43"38d29'
McAndrews,Kathleen & James
Del 20h16'17"13d30'
McAndrews,Richard J
Gem 6h33'14"13d44'B
McAndrews,Theresa Christine
Gem 6h33'14"13d44'A
McAnerney,Eula D
Ori 6h6'53"-0d7'
McAnish,Sandie Lynn
Eri 2h48'40"-1d60'
McAnulty,Jr,John Adams
Ori 6h9'54"7d29'
McAra,Ed T
Her 16h59'30"22d45'
McArdle,Carol-Loraine
Lyr 18h38'41"33d24'
McArdle,Eva
Cyg 20h37'26"48d47'
McArdle,Joseph Emmett
Aql 19h9'13"1d19'
McArdle,Margaret A
Ori 5h49'11"11d32'
McArdle,Poppy
Aur 6h23'0"40d60'
McArdle,Terrance J
Oph 17h7'34"-10d51'
McArdle-Pettee Family, The
Boo 15h3'27"17d59'
McArthur Star
Her 18h2'0"20d27'
McArthur,Andy
Ser 15h19'13"21d34'
McArthur,Double D
Cyg 21h35'11"34d3'
McArthur,Floral
Com 12h59'27"20d21'
McArthur,Karen & Linda Keith
Cyg 19h24'55"33d16'
McArthur,Katrina Marie
Aqr 22h1'33"-2d28'
McArthur,Tracy D
Crb 16h9'42"28d6'
McAtee,Harold L & Vivien S
Cyg 20h55'41"30d59'
McAtee,Joan
Cam 3h27'56"58d59'
McAtee,Sharon Louise
Cyg 21h48'58"37d35'
McAteer,Autherine
Lmi 9h21'41"38d37'
McAuley,Angela
And 0h34'0"38d7'
McAuley,Robert
Cep 22h11'23"70d24'
McAuliffe,Cal Michael William Robinson
Ori 4h58'19"1d36'
McAuliffe,Charles R
Lac 22h8'0"40d21'
McAuliffe,Frank & Gwendolyn Summers
Peg 23h50'44"30d16'
McAuliffe,James
Aql 19h15'45"10d34'
McAuliffe,Kera Nicole
Lyn 9h13'1"35d56'
McAuliffe,Maria G
Cam 3h47'38"56d55'
McAulisse,Sherrie Lynn
Lyn 6h29'53"60d4'
McAusland Minor
Cyg 20h21'0"39d28'
McAvey,Dorthy
Peg 23h29'39"21d41'
McAvinchey,Donald F
Dra 15h9'37"64d10'
McAvoy,Gary
Cam 6h54'54"68d51'
McAvoy,Jack
Ori 5h28'0"-4d32'
McAvoy,Kathleen
Cyg 21h35'12"31d20'
McBain IV
Dra 16h28'27"69d53'
McBain,Scott
Aql 18h43'49"-1d51'
McBane,Wendy L
Aql 19h57'58"14d36'
McBean,Nealie Jean
Cyg 20h55'31"30d7'
McBee,Bailey Hardy
Mon 8h1'51"-1d52'
McBee,Denise
Cas 0h13'21"47d44'
McBee,Donald E
Her 16h9'0"10d37'
McBratney,Marc Louis
Her 18h5'23"30d54'
McBrayer,Scott
Tri 2h1'42"28d21'
McBreaty's Star,Mike
Dra 14h57'49"64d37'
McBride
Boo 14h31'23"47d42'
McBride "Himself",Ken
Cnc 8h48'50"32d1'
McBride (Ziba),Karen L
Cyg 19h33'27"32d21'
McBride,Brian
Boo 15h45'39"50d32'
McBride,Brian Scott
Her 17h26'46"28d23'
McBride,Doris May
Peg 23h8'33"14d54'
McBride,Eugene King Brolo
Aur 5h0'53"42d3'
McBride,Jr,Arthur Francis
Uma 11h33'1"64d35'
McBride,Lisa Kay
Cam 3h33'0"61d45'
McBride,Matthew
Boo 13h51'11"18d39'
McBride,Maureen Carolyn & Thomas J
Aur 6h13'31"46d25'
McBride,Myrrl
Ori 5h6'28"-0d10'
McBride,Paul Collier
Lyn 7h27'16"41d49'
McBride,Robert W
Her 18h16'39"20d29'
McBride,Sean Walker
Lyn 8h17'31"52d10'
McBroom,Elizabeth Rose
Cet 3h6'46"5d14'
McBroom,Robert W
Aql 19h24'24"-7d48'
McBryde,Sarah
And 0h12'43"39d20'
McBurney,Richard Eric
Aur 7h9'39"40d55'
MCC Class of 1993
Uma 9h43'45"50d36'
McCabe's Majesty
Her 17h24'33"31d28'
McCabe,Cameron M
Mon 7h2'0"-6d14'
McCabe,Carolyn
Com 12h53'57"26d51'
McCabe,Charles Edward
Her 17h30'41"27d59'
McCabe,Colonel Edward B
Per 3h13'41"40d8'
McCabe,Dennis P
Vir 11h42'27"8d27'
McCabe,Janis
Lyr 19h21'20"40d52'
McCabe,Joe & Peggy
Cam 7h37'55"80d36'
McCabe,Judith E
Cas 23h39'31"65d22'
McCabe,Lauren D
Com 13h18'38"20d3'
McCabe,Marshall
Uma 8h35'37"56d46'
McCabe,Michelle
Sco 17h54'13"-38d13'
McCabe,Tee
Dra 19h17'36"78d56'
McCabe,Teresa Bernadette
Cyg 20h33'40"46d3'
McCabe,Thomas Brell
Lac 22h54'37"51d18'
McCafferty,Bobby
Aur 6h2'24"33d55'
McCafferty,Joanne Layte
Crb 16h3'1"33d11'
McCafferty,William & Winifred
Cma 7h15'0"-11d10'
McCaffrey 50
Cam 5h50'19"58d55'
McCaffrey Star,The Michael
Cyg 20h35'44"40d53'
McCaffrey,C Elizabeth
Lyr 18h25'15"45d15'
McCaig,Jasmine
Cet 2h47'23"0d35'
McCain,Cheryl
Eri 3h41'11"-2d50'
McCain,Julie
Cyg 19h52'24"38d50'
McCall,Alyce
Hya 8h16'38"0d11'
McCall,Bob
Cam 3h17'13"61d24'
McCall,Brian Edward
Cyg 21h30'1"52d50'
McCall,Charles Arthur Patrick
Her 16h57'19"33d55'
McCall,Cheryl
Mon 8h54'45"-5d56'
McCall,Elyn Honts
Aql 20h2'1"6d45'
McCall,Gary
Aur 5h31'1"40d22'
McCall,Harold & Marlene
Sge 19h38'28"18d53'
McCall,Linda Marie
Lyr 18h58'20"40d1'
McCall,Mark W
Cep 22h34'16"61d38'
McCall,Oren Wayne
Per 02h36'53"43d12'
McCall,Robert Terry Gayda
Uma 13h52'52"51d35'
McCall,ScotSue
Uma 11h48'0"53d30'
McCall,T J
Sco 15h58'37"-21d11'
McCalla,Charles Franklin
Vir 11h44'42"5d16'
McCallion,Michael T
Cma 6h52'39"-18d33'
McCallum,Jenna
Lyr 18h21'0"46d21'
McCallum,Kadi
Lyr 19h16'31"28d23'
McCallum,Karen
Cyg 21h7'15"8d52'
McCallum,Kirsty
Cmi 7h33'59"1d35'
McCallum,Linda
Cam 3h56'32"61d55'
McCallum,Patrick
Dra 12h56'1"70d42'
McCallum,Pauline A
Aql 20h34'52"0d40'
McCalmont,Andrew
Lmi 11h0'17"33d7'
McCalmont,Joshua
Peg 21h47'0"33d52'
McCamants Comet
Ori 5h38'55"15d18'
McCambridge,Madeline
Lyn 9h15'0"37d58'
McCambridge,Roger M
Dra 17h50'0"76d37'
McCambridge,Shane
Her 16h41'10"5d10'
McCamic,Earl
Oph 18h38'46"8d18'
McCamic,Lisa
Del 20h13'33"10d9'
McCamish,Joseph L
Lep 5h0'32"-18d59'
McCance,Christopher John
Aql 20h5'14"1d28'
McCandless,Frederick Steele
Aql 19h17'58"13d59'
McCandless,Iris
Mon 6h21'0"7d20'
McCandless,Marion
Cas 0h26'56"63d12'
McCandless,Robert
Cmi 7h54'27"4d45'
McCandless,Taylor DeLane
Cet 0h40'13"1d1'
McCandlish,Stuart Alexander
Lyr 19h21'27"35d5'
McCandless,Nancy
And 23h41'49"38d48'
McCann
Eri 2h50'39"-1d44'
McCann's Bright Light
Cyg 21h51'0"38d46'
McCann,Betty Jo
Mon 7h15'48"-1d11'
McCann,Christopher
Cnv 13h46'51"32d38'
McCann,Darrin Edward
Oph 18h17'25"12d0'
McCann,David F
Aur 6h13'19"31d47'
McCann,Elizabeth
And 23h46'57"33d15'
McCann,Jane
Cas 1h30'57"75d51'
McCann,John & Phyllis
Cyg 20h20'38"41d7'
McCann,Kelly A
Ori 4h58'59"1d29'
McCann,Kevin & Cathy
Ori 5h53'12"15d46'
McCann,Kevin Alan
Gem 7h30'21"20d47'
McCann,Maggie May
And 0h5'36"44d19'
McCann,Megan
Lyr 19h19'44"33d47'
McCann,Olivia Anne
And 23h30'58"37d39'
McCann,Rachelle
Lyn 6h35'32"60d29'
McCann,Robert Ellsworth
Aql 19h56'25"15d21'
McCann,Sr & Jr,Edward
Cnv 12h17'52"50d48'
McCanna,Sarah Wade
And 1h22'32"34d34'
McCardell,Donald
Cam 7h54'29"60d19'
McCardle,Janine
Lyn 8h4'0"50d30'
McCarley,George H
Ari 2h36'58"20d32'
McCarley,Jimmy
Her 16h17'56"10d21'
McCarley,Robert Leo Walter
Cet 3h6'0"2d55'
McCarney,Benjamin F
Her 16h42'28"32d58'
McCarney,Claire B
Lyn 9h12'26"42d28'
McCarney,Maggie "Maggie's Farm"
Cyg 19h20'15"28d21'
McCarney,Rachel "Shiara's Star"
Cyg 19h21'13"28d30'
McCarra,Beverly
Sgr 20h2'21"-20d14'
McCarrel,Nancy Kay
Cas 1h10'34"66d14'
McCarrick,Andrew James
Vir 13h38'43"5d41'
McCarrick,Annie Bridget
And 23h9'47"41d0'
McCarrick,Rosemary Lyons
Cam 12h39'16"80d54'
McCarrick,Shawn Patrick
Lac 22h30'1"38d40'
McCarroll,Patrick Winston
Her 16h24'19"18d51'
McCarron,Keri
Lyr 18h59'42"41d9'
McCarry,Ian
Uma 9h49'54"73d5'
McCartan,Brian
Uma 10h2'1"55d26'
McCarter,Beauty Star- Claire
Peg 23h32'1"31d28'
McCarter,Erica Lee
Peg 0h3'48"30d55'
McCarter,Jennifer Lorene
Crb 16h0'57"38d54'
McCarter,Jessica Elise
And 0h20'0"33d14'
McCarter,Jr,My Son, Rick Neal
Her 18h12'58"41d6'
McCarter,Lisa Kaye
Uma 8h31'17"58d28'
McCarter,Teresa Faye
Peg 21h30'51"20d13'
McCarthy "Dot",Dorothy Frances
And 2h3'19"38d43'
McCarthy 50,Geoff
Uma 8h12'45"62d26'
McCarthy I,John Patrick Michael
Aql 20h3'25"1d47'
McCarthy III,Joseph F
Her 16h39'12"33d57'
McCarthy Wedding Star,The
Cyg 19h43'29"29d9'
McCarthy's Star
Ori 5h30'59"1d17'
McCarthy's Star,Robert
Aql 19h55'30"7d44'
McCarthy,Alison
Tau 5h28'42"28d35'
McCarthy,Andrea L
Mon 6h40'57"10d44'
McCarthy,Andrew
Lac 22h9'0"38d26'
McCarthy,Ann
Cam 4h5'58"58d47'
McCarthy,Bob
Her 15h53'25"40d9'
McCarthy,Briana C
Com 12h12'58"24d29'
McCarthy,Cassidy Rose
And 23h36'37"47d24'
McCarthy,Christine
Oph 18h1'47"12d54'
McCarthy,Christine
Tri 2h41'58"31d28'
McCarthy,Christopher J
Dra 17h18'42"60d3'
McCarthy,Cynthia Ruth
Peg 23h48'24"28d47'
McCarthy,Dennis (Denny) Joseph
Cep 23h32'38"66d21'
McCarthy,Donal Joseph
Mon 6h42'17"7d57'
McCarthy,Donna
Cyg 21h0'34"30d49'
McCarthy,Eileen
Cam 11h33'1"80d56'
McCarthy,Elenor
Mon 6h56'39"-6d14'
McCarthy,Florence
Cas 1h49'28"58d25'
McCarthy,Heidi
Mon 6h54'55"-10d34'
McCarthy,Jacqueline E
Lac 22h29'53"56d30'
McCarthy,James
Cep 21h52'33"55d7'
McCarthy,Janet C
Uma 12h25'56"62d24'
McCarthy,Jennifer Taylor
Lyr 19h1'32"38d37'
McCarthy,John F
Aur 6h27'14"31d36'
McCarthy,John M
Her 17h2'49"49d14'
McCarthy,Joseph & Marian
Uma 10h7'16"50d24'
McCarthy,Joyce
Equ 21h2'49"7d52'
McCarthy,Jr,Jim B
Boo 14h18'29"17d29'
McCarthy,Jr,Robert Sean
Per 2h24'50"55d13'
McCarthy,Kevin Michael
Lmi 10h32'53"30d33'
McCarthy,Lauren Olivia
Lyr 18h30'0"30d8'
McCarthy,Margaret
Lac 22h32'27"53d47'
McCarthy,Meghann Lynne
Mon 6h58'19"0d22'
McCarthy,Mike
Her 18h2'19"30d27'
McCarthy,Paul
Uma 11h31'58"45d44'
McCarthy,Paul T
Aur 5h1'24"51d19'
McCarthy,Sandrea Lynn
Peg 22h31'24"27d11'
McCarthy,Scott
Cnv 13h58'31"40d8'
McCarthy,Sue
Tri 1h51'54"26d42'
McCarthy,Theresa M
Cyg 19h29'48"30d16'
McCarthy,William
Her 17h27'1"26d18'
McCarthy,William & Rita
Lyn 8h44'50"36d41'
McCarthy,William P
Oph 16h50'45"-26d11'
McCartney,Brigid
Cep 22h35'21"60d17'
McCartney,Diane Barrett Marcase
Del 20h14'40"14d18'
McCartney,H Thomas
Cep 20h27'57"76d50'
McCartney,Janet & Dan
Cyg 20h40'46"37d43'
McCartney,Kara Jo
Del 20h21'12"20d11'
McCartney,Kate Eun Ha
Mon 6h25'17"8d44'
McCartney,Reed Franklin
Her 08h11'30"45d35'
McCartney,Shirley Leone
Peg 21h30'40"20d11'
McCartney,Zeke Robert
Lac 22h24'51"50d9'
McCartt,Pat D
Aur 5h37'1"54d55'
McCarty Mustang
Boo 15h7'15"30d54'
McCarty Star,The Thomas
Sex 9h53'22"0d40'
McCarty,Charles H
Umi 16h23'53"71d52'
McCarty,Jackson Connor
Mon 6h44'44"8d56'
McCarty,James Michael
Eri 3h28'10"-2d9'
McCarty,Lynn Marie
Per 3h10'31"50d17'
McCarty,Molly Kampas
Uma 11h55'41"41d26'
McCarty,Pat & Jean
Vul 19h48'0"28d18'
McCarty,Ruth G
Cas 1h53'1"58d7'
McCarus,Cheryl
Peg 23h36'28"31d30'
McCary,Brandy Christina
Lyn 7h48'55"58d46'
McCaskill,Mary Kathryn
Peg 22h54'30"29d45'
McCaslin,Jack W
Dra 16h1'22"63d34'
McCaslin,Jay
Aql 20h20'26"5d13'
McCaslin,Jimmy
Aur 6h10'12"30d50'
McCaul,Hiroko Takemura
Peg 23h43'1"31d30'
McCauley,Barbara
Aql 19h57'0"11d49'
McCauley,Carole
Cyg 19h27'26"33d42'
McCauley,Christopher Mowbray
Lac 22h1'44"47d41'
McCauley,Colonel Mark Gregory
Her 16h32'58"38d44'
McCauley,Elizabeth Proctor
And 23h44'13"47d18'
McCauley,Joann
Cam 3h47'17"72d15'
McCauley,Joseph G
Aur 6h8'54"46d27'
McCauley,Julie Marie
Com 12h19'44"32d12'
McCauley,Marian
And 23h24'29"47d44'
McCauley,Matthew Joseph
Aur 7h22'31"35d54'
McCauley,Patricia McSweeny
Lib 14h17'31"-8d11'
McCauley,Peter
Her 16h59'59"48d18'
McCauley,Robert John
Peg 21h38'19"25d54'
McCauley,Rosemary Barlow
Lyn 8h1'50"37d47'
McCausland,Irene
And 2h21'43"45d1'
McCausland,Kirsten Aine
Uma 13h26'37"60d12'A
McCausland,Linda
Oph 18h40'51"6d41'
McCausland,Mary Louise
Uma 13h26'37"60d12'B
McCay-Twin # 1,Renee Hadges
Sgr 19h59'4"44d21'
McChesney,Jimmy
Lyn 9h12'13"37d8'
McClain's Magic
Ori 5h57'11"14d5'
McClain's Valentine Star,Linda
Cas 23h33'29"61d18'
McClain,Barbara
Cas 1h32'24"60d39'
McClain,Danielle
Ori 5h56'20"17d34'
McClain,Dr Barbara
Cas 3h0'1"61d35'
McClain,Florence Dotson
Com 12h11'16"20d10'
McClain,Larry French
Equ 21h0'11"8d31'
McClain,Lisa Annette
Aql 19h59'21"14d50'
McClain,My Husband, Steven
Her 16h23'0"37d55'
McClain,Nicole
Mon 6h12'30"-10d20'
McClain-Kramer,Jessica Kimberly
Peg 22h32'35"8d36'
McClain-Paradise Found,Joan
Mon 6h58'57"11d45'
McClane,Brian Phillip
Lac 22h4'35"47d22'
McClane,Donald & Elva Dean
Mon 6h22'24"8d4'
McClaran,Lance
Hya 8h57'30"3d50'
McClard,Dennis
Aur 5h53'0"30d39'
McClary,Lucille
Sge 20h5'33"20d1'
McClatchy,Elizabeth J
Cyg 21h3'45"30d29'
McClay 50,Pegandrod
Lmi 10h10'43"39d47'
McClean,Celia & Michael
Uma 10h39'1"47d15'
McClean,Jeff
Her 17h53'0"30d35'
McCleary,Grandpa
Cep 2h54'30"77d53'
McCleary,Sharon
Lac 22h6'58"51d36'
McCleery,Stephen & Colleen
Eri 4h35'12"-12d5'
McClellan 7-24-93, Nicole Elizabeth
Cas 0h43'52"61d47'
McClellan,Chris
Cas 23h13'21"63d47'
McClellan,Dylan Burke
Aur 5h14'35"44d31'
McClellan,Jacob Michael
Her 18h42'19"12d13'
McClellan,James H
Sct 18h41'25"-4d21'
McClellan,Jason
Aur 6h51'57"37d59'
McClellan,Kathryn Keeton
Eri 3h7'0"-4d7'
McClellan,Lewis Velton
Cet 3h15'33"7d45'
McClellan,Michael"Mac"
Sct 18h51'10"-9d2'
McClellan,Mike
Her 17h55'1"40d51'
McClellan,Robert Lyle
Cnc 8h54'12"31d30'
McClellan,Shawn Royle
Dra 19h23'34"56d16'
McClellan-Burrows,Spot
Oph 17h37'53"-16d44'
McClelland
Her 17h27'1"29d33'
McClelland,Andrew Charles
Oph 17h10'59"-23d0'
McClelland,Bill
Equ 21h6'17"10d20'
McClelland,Cheryl Anne
Cap 20h23'25"-13d34'
McClelland,Dorrie
Cas 0h27'28"62d8'
McClelland,J R
Lyn 7h47'0"39d54'
McClelland,Jazmin J Lange
And 2h7'1"42d59'
McClelland,Kristen E
Peg 23h42'39"27d19'
McClelland,Mac & Peg
Crb 16h8'48"30d10'
McClelland,Teresa Ann
Peg 21h29'1"24d5'
McClements,George
Lyn 7h30'1"40d37'
McClements,Tara-Jane
Crb 16h17'42"34d15'
McClendon,Clarence Hughe
And 23h7'30"39d11'A
McClendon,Curtis James
Mon 7h45'8"-6d48'
McClendon,Kellen
And 0h12'16"33d51'
McClendon,Phyllis Mae
And 23h7'30"39d11'B
McCleskey Commitment
Lib 15h42'50"-25d37'
McCleskey,Carrie
Lyn 8h9'22"50d34'
McClintock Star,The Gregory R
Ori 4h54'57"0d20'
McClintock,Campbell McCabe
Peg 22h24'1"20d17'
McClintock,Helen Hancock
Lyn 6h38'10"56d10'
McClintock,Robert Stone
Per 4h44'0"50d38'
McClintock,Sallie Lou Colvin
Cmi 7h41'40"1d33'
McClintock,Sharon M
Cas 2h33'32"73d57'
McClinton,Alexander Knox
Aur 6h38'0"40d56'
McClinton,Tanner Day
Lac 22h1'54"37d49'
McClorey,James Howard
Per 2h52'0"32d42'
McClory,Michael
Uma 12h0'58"56d4'
McCloskey,Daniel Robert
Lac 22h12'27"47d19'
McCloskey,Kevin Michael
Her 16h35'47"47d51'
McCloskey,Kiernan Michelle
Mon 6h54'40"-6d50'
McCloskey,Robert
Aur 6h29'13"31d50'
McCloskey,Shane Troy
Oph 18h21'33"6d53'
McCloskey,Tony
Per 3h3'55"40d30'
McCloud 7-24-58,Rhonda Linn
Cep 22h43'33"82d0'A
McCloughlin,Megan Jane
Umi 14h39'36"68d16'
McCloy,James & Mary Ellen
Cnv 13h58'59"41d51'
McClun,Christine Ann
Peg 21h37'41"20d14'
McClung,Wallace Riley
Hya 8h12'10"1d31'
McClure III,James
Gem 6h54'23"15d55'
McClure Shawnee, Oklahoma,Betty Jo
Crb 16h3'32"32d25'
McClure's Miramar Royale-25
Lyn 8h53'41"41d28'
McClure,Amy Jo
Cep 21h49'0"68d58'
McClure,Austin Taylor
Lac 22h43'58"54d33'
McClure,Cathryn M
Peg 21h25'42"23d24'
McClure,Dan H
Hya 9h5'11"1d26'
McClure,Glenn
Psc 1h26'1"33d8'
McClure,Harry Adrian
Ori 5h55'34"17d48'
McClure,James B & James B. McClure,Jr
Cet 2h58'0"2d14'
McClure,Jeanne Marie
And 23h33'44"43d56'
McClure,Kelley Ann
And 23h1'17"50d48'

McClure,Lori Board Cyg 21h52'14"38d17'
McClure,Maggie Alayne Ari 1h55'0"16d44'
McClure,Margaux Odette Cam 13h19'12"80d49'
McClure,Melody Star Vul 19h48'30"22d43'
McClure,Pamela Seymour Mon 6h4'0"-8d12'
McClure,Trina Com 12h29'0"27d54'
McClure,Zachary Shannon
McClaughry Lyr 18h47'43"40d36'
McClurg,Robert & Lisa Cnv 12h5'44"40d56'
McClurkan,Connor Rowe Cep 1h31'21"87d30'
McCluskey,Brendan Sean Per 3h53'13"35d45'
McCluskey,Mel Cru 12h44'42"-58d34'
McClusky,Catherine A Cyg 19h42'11"31d15'
McCoig,Robert Keith Hya 8h53'12"6d33'
McColgan,Charles Ori 4h44'10"15d8'
McColgan,My Mother Florence Uma 11h9'38"48d10'
McColgan,William "BBB" Lac 22h14'43"51d40'
McColl,Cameron Schuyler Boo 14h24'23"45d58'
McColl,Nathan Alexander Boo 14h52'41"52d17'
McCollin,Mr M Uma 8h31'1"55d56'
McCollough,Anthony Lain Mon 7h19'31"-6d57'
McCollough,Kenji Todd Aur 5h52'41"30d8'
McCollow,Timothy J Aur 7h4'14"43d49'
McCollum Star,The Peggy Cas 0h57'28"64d23'
McCollum,Conor Per 1h45'17"52d43'
McCollum,Jennifer Peg 22h28'52"21d35'
McCollum,Jim Lac 22h20'25"38d4'
McCollum,Maxine & Ed Eri 4h5'21"-5d7'
McCollum,Patrick Lac 22h5'24"47d19'
McCollum,Shane Cep 20h56'0"55d58'
McCollum,Stuart Lee Dra 16h0'0"68d19'
McCollun,James Cet 0h25'40"-5d56'
McColm,Mike S Sct 18h49'53"-7d44'
McComas,Captain Ed Sex 10h4'13"-5d1'
McComas,James Robert Cet 2h27'26"0d45'
McComas,Mesina Cam 13h28'23"80d21'
McComb,Jim Aql 18h59'22"-3d1'
McComb,John Cnv 12h8'52"34d5'
McComb,John K Tau 3h55'47"20d17'
McComb,Martin Dra 16h12'39"63d4'
McCombs,Lilian Lance Mon 6h25'0"11d21'
McComis,William Ser 16h2'1"9d44'
McComish,Rachel Cas 0h25'0"62d36'
McConahy,MD,The "Great" John Glass Crb 16h17'60"38d35'
McConarty,Linn Lyn 7h53'29"44d39'
McCone,Cody Her 16h57'27"21d16'
McCone,Samantha Cas 0h34'51"62d46'
McCongray,James Anthony Ori 4h55'1"1d20'
McConn,Rick Sct 18h37'56"-4d45'
McConnell Née Moore, Gwendoline May Cas 0h54'1"73d21'
McConnell,Allisdair Hunter Tomken Gem 7h32'60"33d47'
McConnell,Andy Ori 4h53'1"5d16'
McConnell,Barbara Ann Wright Cas 1h27'25"61d23'
McConnell,Carley Vul 19h53'26"20d25'
McConnell,George Phillip Lac 22h51'49"53d18'
McConnell,Harry Neil Cma 6h58'46"-24d31'
McConnell,Joey Boo 14h30'0"20d57'
McConnell,Karen L Equ 20h59'0"10d33'
McConnell,Keith Dra 19h2'44"58d40'
McConnell,Larry Cep 21h57'37"55d43'
McConnell,Leslie Caroline Lyn 7h48'1"58d13'
McConnell,Mark & Barbra Cyg 19h51'1"41d12'
McConnell,Matthew Tri 2h33'29"35d51'
McConnell,Michie Cnc 8h52'55"31d5'
McConnell,PhD,H Keith Cet 2h11'29"3d58'
McConner Star,The Vul 21h21'1"24d6'
McConnon,Michael Joseph Boo 15h2'34"20d36'
McCoo,Jr & Sr,Harold White Her 17h5'57"42d11'
McCooey,Marie Venus Umi 14h56'10"69d33'
McCooey,Timothy Joyce Vul 20h41'18"20d15'
McCook,Andrea Bahmann Mon 7h21'32"-8d2'
McCool Moon Cet 0h44'35"-2d14'
McCool,Ryan William Lib 15h17'53"-23d23'
McCoomb,Alexandra Katherine Uma 9h49'55"73d5'
McCord,Harold Scott Dra 20h17'59"67d55'
McCord,Jacob Woodberry Her 17h28'25"26d52'
McCord,Nora Lyr 19h22'21"35d35'
McCord,Sloan Cnv 12h49'35"35d4'
McCord,Terri(T-Bird) Lyn 7h26'1"58d38'
McCorduck,Judy Rae Peg 21h59'56"24d2'
McCorgray,Anthony James Ori 4h47'33"1d8'
McCorgray,Christine Catherine Ori 4h56'1"1d51'
McCorkle,Jr,Charles Philip Ori 5h30'33"-8d54'
McCorkle,Kenneth Dra 18h1'54"68d8'
McCorkle,Rae Rae Lyn 9h25'31"40d49'
McCorley,Rosaleen And 2h7'41"47d17'
McCormac Star,The Stephen Her 18h25'53"28d51'
McCormach,MD,William M Her 18h12'55"40d19'
McCormack,Don Her 18h5'19"38d4'
McCormack,G Scott Hya 9h25'0"-1d6'
McCormack,Gavin Blair Dra 17h6'16"60d50'
McCormack,Jessica Peg 23h44'13"25d10'
McCormack,John Edward Hya 9h32'1"2d7'
McCormack,Kellie Ashlyn Peg 21h57'45"23d52'
McCormack,Madeline Cyg 19h58'0"31d41'
McCormack,Robert A Lac 22h34'43"53d13'
McCormack,Rosemary Cyg 20h17'16"38d34'
McCormack,The Ori 6h0'52"0d32'
McCormack,William Elbert Cmi 7h39'43"5d20'
McCormack,Win Boo 14h56'53"42d0'
McCormick Cep 22h35'56"61d0'
McCormick 9 Uma 10h4'0"48d55'
McCormick Star,The Lyn 7h56'59"44d14'
McCormick,Alton L Per 3h50'19"36d30'
McCormick,Bertha Margaret Dra 11h33'48"68d19'
McCormick,Betsy Cas 0h56'0"60d20'
McCormick,Billy(Mc) Frank John Cet 3h1'58"0d44'
McCormick,Bonnie Ott Cas 0h25'1"69d34'
McCormick,Cameron Lac 22h5'28"47d35'
McCormick,Carolyn Mon 8h5'11"-9d24'
McCormick,Cheryl & George Crb 15h50'1"30d58'
McCormick,Corky Her 17h11'13"42d53'
McCormick,Dick Per 2h2'36"50d1'
McCormick,George & Lois Boo 15h15'1"33d18'
McCormick,George G Lac 22h29'13"53d58'
McCormick,James Patrick Aql 19h58'0"1d27'
McCormick,James Boo 14h43'34"52d4'
McCormick,James Barry Aql 19h16'1"15d51'
McCormick,Janet Roxana Cas 23h40'25"61d11'
McCormick,Jeremy Patrick Lac 22h8'52"49d11'
McCormick,Jr,Ens John W Tau 4h23'40"15d57'
McCormick,Jr,Paul Christian Her 16h54'0"47d37'
McCormick,Katherine Jean Cas 0h6'15"64d22'
McCormick,Kathleen Mary Tau 4h58'50"16d47'
McCormick,Kathryn G Com 12h14'23"31d13'
McCormick,Kenneth Michael Lac 22h8'12"48d55'
McCormick,Kevin Aur 7h20'20"35d30'
McCormick,Kevin F Her 18h19'18"14d49'
McCormick,Kevin James Boo 15h21'12"52d44'
McCormick,Kyle Lindsey Lyn 7h56'28"58d24'
McCormick,Marcy Cas 0h13'40"58d38'
McCormick,Mary Lou "Ricque" Cas 1h42'40"61d47'
McCormick,Michael D Aur 4h52'10"48d48'
McCormick,Michelle Cas 0h24'44"59d44'
McCormick,Morgan Amelia And 23h13'18"41d56'
McCormick,Patrick Arthur Lac 22h23'38"50d21'
McCormick,Ryan Matthew Ori 5h33'37"8d54'
McCormick,Sabrina Peg 23h40'24"15d5'
McCormick,Sabrina Cara Lyr 19h2'57"28d23'
McCormick,Shane Her 18h14'55"48d7'
McCormick,Sr,Bill Boo 14h49'1"22d41'
McCormick,Sr,John Wayne Aur 5h7'40"38d38'
McCormick,Todd Christopher Eri 2h53'0"-11d38'
McCormick,Tracy Lyn Cas 1h37'1"60d56'
McCormick,Valerie And 1h26'42"40d27'
McCormick,Vicki Lynn And 1h7'10"41d1'
McCorquodale,Felix Uma 9h24'17"48d44'
McCorry-Crazy Diamond, John Cep 23h27'52"68d46'
McCortney,Ruth L Vul 19h14'0"22d7'
McCorvey,Jr,William H Boo 14h19'17"16d10'
McCosker,Leila Margaret Cra 18h24'36"-43d40'
McCoul My Guiding Light,MaryAnn Mon 6h18'27"4d29'
McCowan,Jim Aql 18h43'49"-1d46'
McCowan,Tamara And 1h44'1"40d15'
McCown,Conner Scott Peg 23h4'60"17d32'
McCoy & Sons,Brian & Christopher,Ronald Ori 5h32'17"-0d18'
McCoy Boys,The Cet 2h15'34"3d35'
McCoy Happy 25th, Patirck & Shirley Her 17h52'38"37d47'
McCoy,Arlene Karen Eri 4h13'60"-12d22'
McCoy,Brent Boo 14h58'47"30d6'
McCoy,Cece Jones Uma 14h4'29"61d9'
McCoy,Christopher Lee Eddings Cet 0h59'57"0d40'
McCoy,Craig Ori 5h12'52"8d23'B
McCoy,Dennis Merle Lac 22h50'55"37d47'
McCoy,Dorothy,Louise Lyr 18h55'51"42d9'
McCoy,Elaine Boo 14h48'12"34d47'
McCoy,Gina Mon 6h30'0"8d25'
McCoy,Kathryn Marita Elena Sco 16h57'41"-40d32'
McCoy,Katlin Nicole Ori 5h48'32"21d47'
McCoy,Khristopher J Her 16h57'1"26d34'
McCoy,Kimberly J Oph 17h2'54"8d43'
McCoy,Laura M And 0h26'47"40d59'
McCoy,Loren And 0h7'18"30d23'
McCoy,Marjorie Mae Mon 6h18'41"4d60'
McCoy,MDM Peg 21h31'59"20d17'
McCoy,Megan Enger Vul 19h46'47"26d2'
McCoy,Michael David Evan Aur 5h0'48"50d25'
McCoy,Morgan Seykora And 0h16'29"38d7'
McCoy,Paul Aql 20h12'21"1d26'
McCoy,Richard Austin Boo 13h48'45"15d24'
McCoy,T J Ari 1h48'50"14d2'
McCoy,Tuesday Lynn Peg 21h23'12"23d11'
McCoy,William Jack Her 15h54'44"46d46'
McCoy-Chen,Joenne Uma 12h54'6"54d22'A
McCracken,Craig John Oph 17h6'20"11d51'
McCracken,D Scott Aur 6h11'53"46d24'
McCracken,Melissa René Lyn 6h13'18"54d2'
McCracken,Wayne Battle Lac 21h58'55"36d31'
McCracken,William Cep 21h0'34"65d32'
McCrady,Star of Stephanie Skalska Lyr 18h48'29"30d54'
McCrain,Chewy Cnv 13h40'56"35d3'
McCraney,Letha H Ori 6h9'12"8d45'
McCranie,Marissa Nicole Lib 15h16'28"-18d56'
McCrary,Kelli F Eri 4h2'11"-17d16'
McCrary,Leon Uma 10h24'27"47d56'
McCrath,Dorothy Jean Vul 19h47'31"22d56'
McCraw's Star,Cindy Cma 6h59'25"-13d47'
McCray,Arie E Sge 19h58'38"16d14'
McCrea,Bob & Carol Umi 15h30'41"69d21'
McCrea,Carol Her 17h21'35"14d12'
McCrea,Stephen Gregory Sct 18h45'54"-6d44'
McCreadie Cep 21h41'33"58d52'
McCready,Paulette Cas 2h34'31"58d50'
McCreery's Star Aql 19h56'38"14d37'
McCreight,Michael & John Krzykoski Cep 21h53'1"55d36'
McCreight,Stephen & Marci Cyg 19h44'43"30d42'
McCrimmon,Ali Peg 23h43'27"25d34'
McCrink,Mary Frances Aqr 21h52'47"0d34'
McCrocklin,Laura Beth Eri 2h50'0"-17d45'
McCrone,Lorie Diane Lyr 19h21'0"31d21'
McCrory,Charles Manget Aur 5h15'36"41d10'
McCrory,Marion And 23h27'53"44d42'
McCrossan,Thomas Walter Her 18h7'30"47d11'
McCrumb,Leila Cas 1h7'1"60d3'
McCrumb,Marchael Boo 15h17'27"48d26'
McCrumb,Spencer Cyg 21h19'0"28d43'
McCrystal,Paul James Her 17h19'54"27d40'
McCuaig,Stephanie And 2h29'54"50d35'
McCue IV,Leonard A Per 2h58'58"40d49'
McCue,Arlene Cas 1h29'48"58d36'
McCue,Cynthia Roberts Mon 7h3'27"-4d26'
McCue,David Boo 14h1'15"26d11'
McCue,Hannah Massie Wolford Aql 18h57'25"-6d30'
McCue,James Ori 5h54'42"9d52'
McCue,Jeraldine Mon 6h20'1"8d20'
McCue,Jessica Lyn 7h58'27"35d56'
McCue,Kitty Lyn 7h45'0"50d22'
McCue,Lucille Cyg 21h30'22"38d59'
McCue,MD,Kathleen Patricia Oph 17h59'55"12d31'
McCue,Meghan Lyn 7h57'31"35d30'
McCue,Russell Aur 6h29'1"35d54'
McCue,Timothy Alan Dra 18h10'23"67d43'
McCuiston,Forrest & Joan Aql 18h57'1"16d37'
McCullaugh,Jim Cep 20h46'45"71d2'
McCullen,Michelle Peg 22h35'23"30d3'
McCulley,Dustin Hya 9h21'60"-18d58'
McCulley,Robert Cmi 7h37'0"10d40'
McCulley,Walter Kenneth Hya 8h40'17"3d7'
McCulloch,Cory Del 20h18'29"20d26'
McCulloch,Donald B Cep 22h27'0"58d57'
McCulloch,Douglas H Cam 4h5'29"70d52'
McCulloch,Gwendolyn Eunice Cyg 21h29'56"41d6'
McCulloch,John Halliday Peg 22h12'35"32d44'
McCullough,Ashley Uma 11h10'54"41d21'A
McCullough,Carol Ann And 2h3'46"41d32'
McCullough,Caroline Lyn 7h9'0"58d56'
McCullough,Donna Mazza And 0h54'58"38d41'
McCullough,Galway (Spider) Cam 7h27'39"83d37'
McCullough,Harry Per 2h28'51"56d21'
McCullough,Kerry Ori 5h1'58"10d13'
McCullough,Lee Aur 6h34'43"38d40'
McCullough,Lindsey Uma 11h10'54"41d21'B
McCullough,Meagan A Psc 1h3'23"22d34'
McCullough,Michael Per 1h57'57"53d33'
McCullough,Quinn Boo 14h53'21"35d42'
McCullough,Richard Cmi 7h34'0"1d19'
McCullough-Paragon Mom,Purdie Nelson Uma 10h58'1"59d26'
McCullum,Delbert Aur 6h35'14"37d39'
McCully,Martha Lyn 7h56'26"40d9'
McCune Paper Co Ori 4h55'1"-1d13'
McCune,Bud Per 1h51'46"47d31'
McCune,David Alan Ser 15h37'33"18d40'
McCune,Joseph Albert Cnv 13h14'27"38d41'
McCurdy,Elizabeth Rice Cas 3h6'34"63d59'
McCurdy,Philip R Mon 6h0'18"-5d38'
McCurley,David R Cyg 20h0'0"40d4'
McCurley,James John Dra 16h49'1"69d57'
McCurrach's Eternity Uma 11h18'32"30d40'
McCusker,Erin Cam 5h36'13"70d53'
McCusker,Robert Matthew Uma 11h12'28"46d25'
McCutchen,Mark Aur 6h28'1"31d43'
McDade,Constance Irene Cma 6h59'48"-18d40'
McDade,Madison Lind Uma 13h36'55"61d5'
McDade,Tara Marie Lyn 7h27'55"39d42'
McDaid,Alan & Shirley Cyg 21h31'1"41d14'
McDaneld,Brent Cet 2h55'1"1d53'
McDaniel,Beth T Com 13h20'12"26d22'A
McDaniel,Christian Valentine Peg 23h37'25"18d54'
McDaniel,Dustin L Dra 11h45'1"70d21'
McDaniel,Jesse L Aql 20h8'53"4d1'
McDaniel,Kasey Nicole Aql 20h2'34"0d57'
McDaniel,Robert & Frances Peg 0h0'23"27d49'
McDaniel,Russell A Aur 5h0'18"42d38'
McDaniel,Scott W Com 13h20'12"26d22'B
McDaniel,W Dwayne Eri 2h54'34"-16d22'
McDaniels,Nora Jane Cam 5h53'59"60d20'
McDaniels,Steven Dwight Cep 23h2'1"60d35'
McDarby's Celestial Trek Lyr 18h28'23"32d60'
McDavid,Radford Kersey Cet 2h2'32"-11d5'
McDavid,William Terry Sex 10h40'22"4d3'
MCDCKXIIIVIIMCMXCVI Lac 22h3'39"48d44'
McDee Cyg 20h52'32"38d41'
McDermed,Duane Hya 8h26'59"1d36'
McDermid,Chande Aql 19h55'22"8d5'
McDermitt,Naida Uma 9h47'18"57d54'
McDermot,Maxwell Blake Leo 9h33'0"8d28'
McDermott Star,The Ori 4h59'45"14d55'
McDermott,"Miracle Baby",Audrey Rose Vul 19h20'48"23d33'
McDermott,Alexander Graham Selkirk Per 3h4'60"43d14'
McDermott,Brian Michael Aqr 21h24'27"-0d27'
McDermott,Clara Cyg 20h53'1"31d4'
McDermott,Dylan Otis Cam 12h40'52"77d50'
McDermott,Gina And 1h47'44"41d12'
McDermott,Harold Vincent Dra 19h0'14"48d37'
McDermott,Jennifer Marcella Lyr 18h45'25"39d45'
McDermott,Jerry Vir 13h51'12"-1d43'
McDermott,Jimmy Her 17h14'38"47d47'
McDermott,John Howard Boo 13h55'25"20d1'
McDermott,John P Cep 21h34'16"65d21'
McDermott,Judy J Com 12h20'12"30d14'
McDermott,Marianne Mon 6h18'19"5d59'
McDermott,Paul Michael Patrick Dra 14h24'1"63d22'
McDermott,Sharry Peg 23h26'17"32d10'
McDermott,Thomas Cas 0h27'31"61d49'
McDermott,Traci Charlene Cyg 19h28'21"34d34'
McDermott,Tracy Mon 7h3'53"0d46'
McDevitt,Bette Joyce Mon 6h25'30"11d6'
McDevitt,Chris Cyg 20h20'24"38d54'
McDevitt,John Neal Oph 16h58'51"10d4'
McDevitt,Keegan Peg 21h52'0"34d19'
McDevitt,Robert Ferguson Uma 10h23'43"40d48'
McDevitt,Sr,George N Lac 22h32'52"55d29'
McDevitt,Wyatt Blake Aql 19h50'46"15d6'
McDolly Tri 2h22'58"35d45'
McDonagh,Francis Joseph Cep 4h14'25"88d21'
McDonald 7-24-44, Patricia Ann Cook Lyr 19h22'13"41d23'
McDonald Family Star, The Cep 22h6'1"60d48'
McDonald Nova,The Alex Cas 22h55'36"55d38'
McDonald,24,Michael "Mike" Uma 10h32'54"42d29'
McDonald,Arielle Rosemarie 30/3/94 Ari 2h32'13"20d19'
McDonald,Astrig Abadjian Mon 6h19'36"-6d37'
McDonald,Barbara E Cyg 21h16'31"28d13'
McDonald,Bill & Candace Cyg 19h23'16"47d41'
McDonald,Bryan Ser 17h57'56"-13d24'
McDonald,Catherine J Boo 14h43'58"37d21'
McDonald,Charles H Lac 22h19'21"51d17'
McDonald,Cheryl Lyr 18h58'34"33d6'
McDonald,Christian Robert Ori 6h5'18"1d5'
McDonald,Cynthia Dianne And 23h49'56"40d52'
McDonald,Daniel J Dra 16h30'59"62d41'
McDonald,Dawn Jackson Cas 1h41'37"58d15'
McDonald,Dell Joseph Boo 13h50'58"19d59'
McDonald,Diana Equ 21h18'19"3d20'
McDonald,Diane Cyg 21h52'22"48d53'
McDonald,Diane & Chris Lmi 10h15'46"30d46'
McDonald,Erin Clark Del 20h13'0"14d54'
McDonald,Gary Lloyd Hya 8h27'19"-6d38'
McDonald,Harry Aur 4h35'34"31d48'
McDonald,Harry & Karen Crb 16h21'34"30d15'
McDonald,Hartland Rose Eri 4h8'45"-5d37'
McDonald,Howard Aql 19h50'15"13d46'
McDonald,James Cet 0h27'16"-1d47'
McDonald,Jane G Peg 22h0'18"28d19'
McDonald,Jeffrey D Aur 6h17'26"37d39'
McDonald,Jim Her 18h3'0"30d27'
McDonald,John Cep 22h6'58"60d33'
McDonald,John D Her 17h35'0"20d45'
McDonald,John McLean Ori 6h0'16"20d35'
McDonald,John T Lac 22h22'34"55d8'
McDonald,Josephine Russo Cas 22h56'0"57d28'
McDonald,Julia Lee Lyn 8h50'42"43d37'
McDonald,Kelly Lynn Cas 1h8'26"61d41'
McDonald,Kenneth Paul Dra 10h26'58"78d24'
McDonald,Kyle Somerled Lyn 6h54'37"59d19'
McDonald,Lark Scott Lac 22h45'27"37d52'
McDonald,Lenny Cet 2h8'38"1d58'
McDonald,Lilburn "Libby" Del 20h25'51"20d19'
McDonald,Lisa Aur 5h25'12"38d58'
McDonald,Mary Driscoll Del 20h26'13"11d12'

McDonald,Matthew James
Lac 22h19'44"49d45'
McDonald,Neil Anthony
Uma 9h45'37"44d24'
McDonald,PJNA
Cam 3h47'11"58d13'
McDonald,Richard M
Aql 20h3'0"4d15'
McDonald,Robert F
Lac 22h33'20"55d16'
McDonald,Sandy
Cet 2h28'36"8d17'
McDonald,Sarah Ashley
Aur 6h43'54"38d24'
McDonald,Sherry Lynn
Com 12h18'19"30d55'
McDonald,Steven
Boo 14h58'30"48d26'
McDonald,The Great
Aur 5h18'39"40d2'
McDonald,Tracey Lee
Aur 6h13'1"33d15'
McDonnell,Alfred D &
Elizabeth L
Aql 19h31'52"10d30'
McDonnell,Brian Christopher
Aur 7h23'51"35d29'
McDonnell,David Victor
Her 18h30'16"24d34'
McDonnell,James
Boo 14h31'33"22d3'
McDonnell,Lynda
Cyg 19h35'25"28d38'
McDonnell,Max
Ori 4h45'45"15d38'
McDonnell,Norann
Per 2h51'7"47d56'B
McDonough,Allison Anne
Cas 22h58'60"55d60'
McDonough,Bert
Per 1h28'39"50d35'
McDonough,Charles E
Hya 9h10'1"1d44'
McDonough,Cherie
Mon 7h49'11"-5d37'
McDonough,Doris Richards
Cas 0h34'43"61d1'
McDonough,James Allen
Cet 0h27'0"-6d3'
McDonough,Kaelin Sean
Cleary
Ori 5h45'24"11d29'
McDonough,Linda S.
Tau 4h11'54"21d49'
McDonough,Margaret
Uma 11h46'54"33d26'
McDonough,Maryanne
Per 2h41'32"40d18'
McDonough,Megan Erminia
Cas 0h23'34"62d40'
McDonough,Michael James
Aql 20h19'1"5d31'
McDonough,Norman
Aur 5h58'44"38d1'
McDonough,Ruthella O &
Daniel J
Lyn 8h2'58"41d1'
McDonough,Thomas
Per 2h51'56"40d22'
McDonough-Ballandras
Eri 2h56'1"-1d33'
McDoody,Rudy
Sco 17h28'26"-30d44'
McDougal,Bryan T
Boo 14h0'22"7d54'
McDougal,Sharon N
And 23h39'60"33d0'
McDougall,George
Cep 21h48'27"61d50'
McDowall,James Pontin
Her 18h8'27"45d4'
McDowall,William Robert
Cap 20h53'43"-22d43'
McDowell Happy 50th, Jack
Uma 11h28'29"64d51'
McDowell,"Mac"
Peg 23h29'10"31d26"
McDowell,Annie
Cas 23h40'24"61d11'
McDowell,Benjamin H
Aur 6h29'53"35d7'
McDowell,Bruce William
Her 16h12'25"18d47'
McDowell,Catherine Patricia
And 23h19'55"40d58'
McDowell,Mary Elizabeth
Cet 1h51'27"-8d30'
McDowell,Peter Adams
Hya 9h39'1"0d30'
McDowell,Ron
Ser 17h18'1"-11d8'
McDowell,Ronald Dean
Her 18h7'46"48d19'
McDowell,Ronnie
Cyg 21h16'60"38d10'
McDowell,Scott
Aur 5h17'45"45d40'
McDowell,Sharon
Sge 18h58'45"19d54'
McDowell,Susan Lynne
Eri 4h14'21"-18d25'
McDowell,Virgil
Cyg 19h32'20"30d16'
McDuade,Dolores & Larry
Uma 13h48'11"48d4'
McDuff,Louis Ralph (Red)
Lyr 18h57'28"37d9'
McDuffie,Amy Elizabeth
Peg 22h35'19"12d16'
McDuffie,Dr Ernest L
Dra 16h57'36"62d33'
McDuffie,Gary D
Gem 6h41'10"18d58'
McDuffie,Michael Eugene
Cet 2h35'19"-0d30'
McEachern,Kelly Cook
Cyg 21h26'42"41d9'
McEachern,Ronald R
Aql 20h13'45"0d1'
McElgunn IV,John Edward
Her 18h12'34"40d33'
McElgunn,Patrick
Cet 0h24'39"1d50'
McElhaney,Diane
And 0h57'46"39d41'
McElhatton,Christopher
Aql 19h52'31"15d15'
McElhill,Charles J
Dra 16h54'12"67d31'
McElligott,Baby
Cyg 20h51'38"39d27'
McElligott,Gloria Reems
Cet 2h28'22"-10d47'
McElligott,Mitzi & Sean
Cyg 21h23'47"30d51'
McElligott,Owen Roe
Ser 18h53'48"2d38'
McElrath,Matthew Sebastian
Uma 10h4'0"50d24'
McElrath,Michael D
Aql 19h36'17"-6d17'
McElrath,Mr Gayle W
Dra 16h27'1"66d12'
McElreath,Lisa L
Del 20h23'51"18d48'
McElroy,Emma
And 0h29'31"43d40'
McElroy,Jr,Donald Robert
Aql 18h39'34"-2d6'
McElroy,Julia Clare LaMuniere
Cas 2h44'30"61d31'
McElroy,Lindsey Rae
Hya 8h13'53"-6d58'
McElroy,Richard
Cet 1h13'44"-3d18'
McElroy,Richard
Aql 19h57'37"8d25'
McElvery,Keith Michael &
Heather Laurie
Peg 23h20'17"28d16'
McElwain,John Patrick
Lac 22h6'36"51d19'
McElwee,Cody Lee
Her 18h2'43"28d47'
McElwee,Dan
Cnv 12h47'12"51d20'
McElyea,"Cyndi" Cynthia
Gayle
Uma 8h50'21"71d12'
McEneamey,Sophie
Lyr 18h16'52"44d36'
McEnerney,Janine
Lyn 7h7'55"58d37'
McEnerney,Mary Katherine
Mon 6h32'40"-0d22'
McEnery,John Silky
Aur 5h57'38"30d6'
McEnroe,Katie
Cas 0h22'54"61d4'
McEntee's Bear
Cam 6h19'42"67d54'
McEntee,Diedre
Cet 1h13'42"-3d20'
McEntee,Mrs Jennifer
Peg 21h54'31"34d6'
McEntire,Reba
Lyn 8h11'44"40d40'
McEntire,Reba
Aql 18h46'28"10d54'
McEntire,Reba
Cam 14h19'41"82d13'
McEntyre,Virginia Elizabeth
Scott
And 0h21'12"36d39'
McEvilley,Christopher
Nicholas
Dra 9h42'0"81d1'
McEvoy III,John Edward
Vir 14h3'21"-0d33'
McEvoy,Aaron Allen
Leo 10h58'0"8d48'
McEvoy,John Mac an Bui
Uma 10h47'1"47d35'
McEvoy,Sean
Gem 7h29'1"30d20'
McEwan,Bernard Christopher
Her 17h6'43"49d34'
McEwan,James
Lac 22h41'42"37d31'
McEwan,Lenny
Cep 22h25'58"61d48'
McEwan,Ryan
Tri 2h5'39"31d4'
McEwan,Stephen Joseph
Aur 5h50'48"40d5'
McEwen,Jeff
Dra 13h29'55"64d6'
McEwen,Mark
Oph 18h17'12"11d16'
McEwen,Michael Clinton
Boo 15h4'17"20d38'
McEwen,Shay Huntington
Uma 8h32'36"56d54'
McFaddem III,John
Dra 17h57'27"64d13'
McFadden,Amber
And 23h25'40"48d48'
McFadden,Anne
And 1h51'47"36d33'
McFadden,Chelsea Lauren
Cyg 19h33'28"36d27'
McFadden,Corrie
Cet 2h4'12"-6d53'
McFadden,Dana Erin
Vul 20h20'39"22d44'
McFadden,Denise
Umi 15h8'41"82d30'
McFadden,Flow
Eri 2h51'56"-15d49'
McFadden,Frances
Cas 0h37'50"64d52'
McFadden,George
Leo 11h49'21"27d5'
McFadden,Jacalyn & Steven
Cyg 21h51'14"40d50'
McFadden,Jeep
Hya 8h54'54"3d37'
McFadden,John/Johnny Mac
Per 1h55'11"54d1'
McFadden,Kathleen
Cas 2h48'21"61d9'
McFadden,Kelly
Lyr 18h59'14"34d21'
McFadden,Madison Ann
And 23h32'14"41d47'
McFadden,Mickey
Ori 4h54'27"-0d5'
McFadden,Moira
Cyg 19h30'41"33d37'
McFadden,Neil
Dra 16h3'19"67d38'
McFadin,Joshua Cole
Ori 5h56'45"16d59'
McFadyen,Alexander &
Margaret
Cyg 21h16'17"28d40'
McFall,Cary Michael
Gem 8h4'20"28d58'
McFall,Connor Jesse
Boo 14h18'52"31d31'
McFarland,Brad Keith
Sct 18h20'23"-10d44'
McFarland,Brian
Per 4h41'43"50d39'
McFarland,Eckhoff
Eri 3h43'27"-16d24'
McFarland,John
Per 4h31'51"34d51'
McFarland,Lynne & Mark
Boo 14h24'35"23d47'
McFarland,Mae
Eri 4h59'54"-10d46'
McFarland,Megan
Scl 0h8'57"-25d48'
McFarland,Monica Marie
Cam 4h44'24"68d23'
McFarland,Siobhan C
Lyn 7h26'23"40d6'
McFarland,Steve Edward
Boo 13h41'32"14d49'
McFarlane,Eric John
Ori 4h51'42"5d48'
McFarlane,Keri
Vul 18h57'52"24d34'
McFarlane,Mary Ellen
Ori 4h52'59"5d1'
McFarlane,Salvatrice Louise
Calarco
Vir 12h59'49"-21d23'
McFarlane,Steven Bradley
Aur 6h4'46"34d40'
McFarlin,Asa Harris
Cnv 13h49'44"32d1'
McFarling,Deanna & Brad
Aql 20h0'29"7d24'
McFaul,Sarah Catherine
Ori 5h54'59"15d10'
McFee,Beaman
And 2h23'31"40d47'
McFellin,Sharon Lynn
Cam 3h27'12"52d52'
McGaffigan,Kim
Cam 7h55'59"60d35'
McGahan,Charles H
Cep 22h41'18"57d24'
McGahan,Paul
Per 3h0'50"46d1'
McGalliard,Krysten Serena
Cet 2h49'33"3d31'
McGann,Karen Ahearn
Cas 0h57'40"77d16'
McGannon,Don
Lac 22h28'11"48d51'
McGannon,Dr,Robert F
Aur 5h59'57"37d34'
McGannon-Kearney, Patricia
Peg 23h39'51"30d29'
McGarey,Robert Morgan
Cet 1h56'32"-2d8'
McGarigle-Gemanis
Eri 2h46'1"-3d13'
McGarity III, Augustus
Courtney
Peg 22h25'13"21d20'
McGarity,Cornelius J
Cep 22h18'31"58d32'
McGarrity,Jane Andrews
McCulloch
Lyr 19h20'43"42d26'
McGarry,David T
Dra 20h20'48"67d17'
McGarry,Donald P David
Boo 14h45'25"31d20'
McGarry,Hedwig Mueller
Boo 15h7'19"53d30'
McGarry,Jaye
Cyg 19h32'0"33d35'
McGarry,Sean
Lmi 10h27'39"28d25'
McGarry,Suzanne Marie
Christine
Psc 0h15'46"20d0'
McGarvey,Mike The Wook
Boo 13h40'58"20d37'
McGaughey,Molly
Com 13h0'22"26d46'
McGaw,George D
Cas 2h38'15"58d24'
McGeachy,Maddison Elaine
Peg 22h9'43"25d27'
McgEE
Uma 10h0'53"48d49'
McGee,Ann Clark
Crb 16h6'10"37d30'
McGee,Billie
Aql 20h16'59"5d29'
McGee,David
Eri 4h32'50"-9d50'A
McGee,Debbie
Boo 14h9'1"52d30'
McGee,Debbie
Aur 4h57'59"50d20'
McGee,Frank Starr
Eri 4h11'42"-16d52'
McGee,Hannah Elaine
Peg 22h28'1"24d2'
McGee,Jacob Charles
Aql 19h18'48"13d56'
McGee,Kim Elizabeth
Cyg 21h17'35"38d57'
McGee,Lisa Noelle
Mon 7h10'48"-10d31'
McGee,Marguerite
Peg 22h13'29"30d52'
McGee,Mary Tara
Cas 0h1'50"61d37'
McGee,Marybeth Hofmann
Crb 16h21'56"33d25'
McGee,Marykate
And 0h0'21"34d49'
McGee,Nora Ann
And 23h12'49"36d53'
McGee,Wayne
Ori 5h13'12"-5d17'
McGehee,Robert Burdick
Uma 10h13'31"56d44'
McGeorge,Kirk
Aur 4h59'12"48d52'
McGeough,Rose
Cas 2h35'20"57d30'
McGhan,Annabel & Louis
Lyr 18h55'37"30d2'
McGhee
Her 17h28'41"29d37'
McGhee,Bobby
Lac 22h24'44"53d48'
McGhee,Jeramiah Lewis
Cma 6h50'32"-16d29'
McGhee,Katie
Lyr 19h5'1"37d51'
McGiff,Kelly
And 0h9'31"34d5'
McGill,Carli Ann
Lmi 9h50'13"38d6'
McGill,Heather Lynn
And 2h19'18"38d17'
McGill,Janet H
Cnv 12h47'36"38d5'
McGill,Jr,David Bruce
Lyn 7h29'36"35d32'
McGill,June C
Crb 15h35'57"28d38'
McGill,Kahan
And 23h44'21"47d40'
McGill,Stephanie
Vul 19h47'12"22d41'
McGillicuddy,Andrew
Aur 6h14'0"31d33'
McGilligan 40,Patrick
Uma 8h41'24"57d8'
McGillivray,Phyllis
Cyg 20h3'13"39d53'
McGillvary,Tammy M
Aql 19h57'36"1d37'
McGinlay,Gordon
Per 1h45'26"47d52'
McGinlay,Joanne
Cam 12h28'42"81d22'
McGinley,Hannah Christine
Cas 1h19'45"66d15'
McGinley,Michael Edward
Crb 16h21'1"38d13'
McGinley,Thomas C
Aql 20h4'51"0d56'
McGinly,Sr,Robert J
Aur 6h15'16"31d19'
McGinn,Anna
Peg 23h45'16"30d56'
McGinn,Collin Robert
Cyg 20h42'41"42d8'
McGinn,Mary
Ori 6h7'37"4d31'
McGinness,Logan W
Cep 20h38'19"58d56'
McGinnie,Joseph Benjamin
Her 17h35'19"38d31'
McGinnis III,William J
Per 2h36'10"43d53'
McGinnis,Allan & Kathie
Sct 18h48'39"7d60'
McGinnis,Frank
Cet 0h25'51"-10d27'
McGinnis,Frank
Eri 4h55'45"-8d56'
McGinnis,Jarred Patrick
Cnc 8h10'54"30d48'
McGinnis,John & Susan
Cam 6h11'19"60d42'
McGinnis,Joseph Benjamin
Her 17h35'19"38d31'
McGinnis,Jr,Roy C
Uma 12h43'46"53d3'
McGinnis,Keith W
Lac 22h55'0"51d30'
McGinnis,Megan Irina
Lyr 19h4'49"38d55'
McGinnis,Sr,Roy C
Uma 12h50'1"58d15'
McGinniss,Jennifer Ann
Lac 22h7'1"46d52'
McGinnity,Sacha & Janet
Cyg 20h21'30"39d34'
McGinnity,Tim
Hya 8h9'39"0d37'
McGinty,James Arthur
Lac 22h6'42"50d20'
McGinty,James Arthur
Boo 14h10'53"46d11'
McGinty,Richard Gerard
Francis
Cam 7h36'55"70d56'
McGinty,Sandra
Cas 0h56'27"59d26'
McGladdery,Moria
Cru 12h9'24"-58d44'
McGlathlin,Linda
Aql 19h36'0"0d4'
McGloin,Mary Ellen
And 23h23'48"42d36'
McGlone,J
Hya 9h30'36"-0d18'
McGlumphy,Mara Kristine
Cnc 8h22'46"15d22'
McGlynn,Matthew James
Her 16h8'27"50d40'
McGlynn,Peter Andrew
Gem 7h17'12"31d49'
McGoey,Chris
Peg 22h46'6"30d50'A
McGoey,John Thomas
Her 16h34'53"21d22'
McGoldrick,Frances Marie
Cavallo
Peg 21h57'56"29d19'
McGoldrick,Kathleen T
Lyr 18h38'45"41d2'
McGonigle,Dr William R
Lyr 19h4'1"41d11'
McGonigle,Maureen
Cas 0h58'49"75d35'
McGonigle,Michael Dennis
Sge 20h0'45"20d32'
McGoo
Lac 22h30'43"50d32'
McGoo
Uma 10h3'48"42d31'
McGoron,David L
Ori 5h28'6"-4d47'
McGough,Christy
Aql 18h58'13"16d20'
McGough,Roger
Aql 20h13'13"0d20'
McGoveran's Destiny, David
Ornan
Ori 6h9'36"6d2'
McGovern,Anne M
Cas 0h52'17"64d33'
McGovern,Becki Kay
Cas 0h30'39"65d28'
McGovern,Caitlin
Lyr 18h42'34"38d31'
McGovern,Christopher Daniel
Dra 20h25'15"68d38'
McGovern,Corene A
Cam 8h8'25"81d49'
McGovern,Fergus
Cyg 19h25'18"34d48'
McGovern,Frances & Michael
McGee
Lyr 18h30'60"31d5'
McGovern,Kristen & Edward
Uma 9h17'34"58d6'
McGovern,Laura Lee
Cas 23h42'0"63d51'
McGovern,Linda
And 1h0'19"34d26'
McGovern,Maria Elena
Peg 21h40'45"27d42'
McGovern,Matthew T
Her 17h48'1"14d21'
McGovern,Maureen T
Leo 9h24'48"17d53'
McGovern,Tony
Cep 21h5'38"70d22'
McGowan,Andrew
Ori 5h8'34"-0d22'
McGowan,Audrey Nell
And 2h24'40"46d40'
McGowan,Brian Michael
Her 16h59'58"35d46'
McGowan,Cecil
Aur 6h2'3"48d15'B
McGowan,David Lee
Cep 23h2'26"62d44'
McGowan,Eunice Nettie
Cas 0h56'1"54d29'
McGowan,Harold
Lac 22h26'13"53d41'
McGowan,Jean Boone
Aur 6h2'3"48d15'A
McGowan,John P
Uma 8h37'35"68d60'
McGowan,Joyce & Mickey
Equ 20h55'12"2d31'
McGowan,Jr,James P
Boo 15h2'1"32d27'
McGowan,Kimberly Kaye
Cyg 21h36'23"42d28'
McGowan,Linda Parker
Peg 22h28'45"27d51'
McGowan,Paul Dorden
Dra 16h55'44"66d19'
McGowan,Paula
Tri 2h29'0"35d52'
McGowan,Peggy
And 0h48'16"45d21'
McGowan,Sean
Lmi 10h41'13"24d54'
McGowan,Shannon Lynn
Peg 22h46'21"34d25'
McGowan,Thomas & Patricia
And 23h13'1"37d43'
McGowan-Lees,The
Aur 5h16'20"47d28'
McGowin's Star,Charlie &
Berniece
Peg 21h28'0"21d8'
McGrady,Christopher Dwain
Boo 14h45'57"26d8'
McGrady,Patrick Blaine
Per 4h25'36"50d50'
McGrail,John Michael
Boo 14h7'36"52d59'
McGrail,St Elmo's Fire Kelly M
And 0h53'47"36d58'
McGranahan,Brian Richard
Aur 6h25'23"37d10'
McGranahan,Erin
Lyr 18h51'1"41d29'
McGrane,David
Aur 5h0'42"44d5'
McGrane,Judy
Tau 3h53'28"20d6'
McGrath III,Francis Robert
Lac 22h43'0"37d45'
McGrath(Annie),Annette
Lmi 9h53'52"34d12'
McGrath,Ann Elizabeth Camp
Peg 21h55'1"21d48'
McGrath,Conor Paul
Dra 16h7'35"67d18'
McGrath,Don Marie Lisa &
Dorothy
Boo 14h51'59"35d24'
McGrath,Donna Ann
Lyn 7h19'1"59d53'
McGrath,Elizabeth Anne
Cyg 19h52'22"37d45'
McGrath,Gram
And 1h24'36"41d4'
McGrath,Grandma Jo
Cas 0h54'38"70d26'
McGrath,Itchie Richie
Boo 14h49'52"48d13'
McGrath,James W
Cep 22h45'36"57d56'
McGrath,Mary Elizabeth
Eri 4h33'29"-8d41'
McGrath,Patrick Francis
Aur 5h19'48"46d15'
McGrath-Regan Joined
Forever in Space
Uma 8h34'31"67d59'
McGraw
Uma 9h23'0"68d9'
McGraw & The Dancehall
Doctors,Tim
Oph 17h17'1"11d35'
McGraw,Andy
Dra 16h31'48"72d1'
McGraw,Betsy Catherine
And 2h20'28"50d25'
McGraw,Denise E
Cyg 21h4'33"31d16'
McGraw,Leslie Marguerite
Cyg 19h31'0"33d43'
McGraw,Sean Patrick
Ori 5h2'22"-0d36'
McGreevy,Brennan
Equ 21h6'27"11d6'
McGreevy,Casey
Cet 0h58'1"0d5'
McGregor,Brooke Ashten
Eri 3h0'38"-3d24'
McGregor,Bulah Marie
Eri 3h19'11"-10d56'
McGregor,Catherine & Alistair
Cyg 20h1'52"31d26'
McGregor,Christopher
Peg 21h47'44"30d39'
McGregor,Michael David
Per 1h48'48"56d46'
McGregor,Rena Kirsty
Cyg 19h48'19"38d26'
McGregor,Russell
Ori 5h41'28"10d4'
McGregor,Sean Michael
Cep 23h14'47"65d32'
McGregor,William Higgins
Aur 7h1'23"43d50'
McGrew,Richard
Dra 19h55'58"68d34'
McGriff,Steven Bruce
Pup 6h59'17"-36d33'
McGriffen Vincennes,
USA,Margaret Ann
Cas 0h17'1"58d10'
McGroarty,Fortune Jim &
Heather
Cyg 19h27'1"38d40'
McGrogan,Darette Stoker
Crb 16h21'45"30d21'
McGrogan,Patrick M
Crb 16h23'19"30d2'
McGruder,Suzi's Tudor
Lyr 19h9'55"38d14'
McGuckian,Desmond
Lyn 8h4'1"44d25'
McGuffey,Kevin M
Boo 15h0'37"20d33'
McGuigan,Thomas E
Cyg 21h18'60"38d9'
McGuiness,William J
Dra 17h57'16"58d41'
McGuinness IV,William
Vincent
Uma 10h6'45"42d11'
McGuinness,Clifford Merrill
Hya 8h19'56"3d30'
McGuinness,Edward W &
Mary P
Eri 3h4'54"-5d8'
McGuinness,Kelly
Lyr 19h19'0"35d22'
McGuinness,Monika Lynne
And 0h48'22"45d51'
McGuinness,Ryan
Aur 6h5'46"46d11'
McGuinness,Sean Edward
Sct 18h53'13"-7d21'
McGuinness,Tim
Per 3h13'25"56d4'
McGuinness-Thurston, Geri
Ann
And 2h18'39"41d43'
McGuire "Dreamkeepers"
Chris & Carol
Eri 3h41'22"-1d53'
McGuire's "Wedding
Star",Kevin & Helen
Ori 5h34'12"0d52'
McGuire,"Max" Henry B
Cam 12h24'45"81d46'
McGuire,Ann G
And 0h23'59"44d8'
McGuire,Casey Jane
Crb 16h11'0"32d23'
McGuire,Dean
Lac 22h11'26"49d24'
McGuire,Denise
Gem 6h58'0"35d23'
McGuire,Emily E
Mon 7h42'16"-1d25'

McGuire,Fraser Leslie John
Cep 20h40'37"60d41'
McGuire,James J
Cep 22h20'34"63d32'
McGuire,Jason
Crt 11h52'28"-17d43'
McGuire,John J
Cep 22h48'40"58d17'
McGuire,Kathy
And 2h1'17"45d34'
McGuire,Kent & Mitch-Heavenly Love Forever
Del 20h26'37"11d4'
McGuire,Kevin
Oph 18h0'0"11d38'
McGuire,Kevin Thomas
Ori 5h56'46"15d18'
McGuire,Liz & Dan "Wedding Star"
Cyg 21h30'44"30d32'
McGuire,Mary & Bill
Dra 17h38'0"64d30'
McGuire,Michael Francis
Lac 22h27'57"40d27'
McGuire,Molly Christine
Peg 22h13'29"3d4'
McGuire,Patrick Gwvan
Per 2h43'19"41d7'
McGuire,PhD,Noreen
Vir 13h35'34"6d3'
McGuire,Robert & Mary
Cyg 21h7'40"37d37'
McGuire,Sarah K
Eri 2h50'1"-10d41'
McGuire,Shirley K
Mon 6h21'1"5d58'
McGuire,Thomas J
Aur 6h8'58"32d59'
McGuire,Tyler Joseph
Aql 18h58'26"15d9'
McGuires's,The
Cnv 13h32'24"48d53'
McGuirk,Stella Marie
Cas 2h52'1"70d8'
McGwan,John Robert
Aur 6h13'27"37d14'
McGwire,Morgan Danielle
Lyr 18h56'14"34d32'
McHale,James A
Dra 16h50'21"70d32'
McHale,Krista
And 2h26'54"44d39'
McHale,Tara
Cyg 20h52'19"38d54'
McHalski,Maureen
And 23h15'55"48d22'
McHarg,James William
Her 16h56'40"32d16'
McHough,John R
Her 16h37'52"47d40'
McHugh,Evan Daniel James
Aur 7h24'15"38d14'
McHugh,James Allen
Dra 20h20'24"63d5'
McHugh,James John
Cep 0h12'41"80d8'
McHugh,John & Mary
Lyr 19h0'44"33d60'
McHugh,Jr,Little Angel Afar Paul
Cam 13h16'17"77d51'
McHugh,Kevin T
Aur 5h53'34"37d39'
McHugh,Mary Suzanne Kennard
Cam 3h45'22"60d39'
McHugh,Paddy
Aqr 20h53'16"-5d42'
McHugh,Paul
Cep 21h25'1"68d31'
McHughes,Donnie & Chris
Uma 12h9'30"60d34'
McIlhenny,The Star of Scott David
Uma 9h55"30d46d57'
McIlmoyle,June
Aql 19h3'41"0d0'
McIlrath,Barbara J
Cyg 20h47'27"38d50'
McIlrath,Scott
Her 15h55'42"40d5'
McIlroy,Andrew Wilson
Cyg 19h30'0"37d43'
McIlvaine III,Joseph F
Uma 11h10'37"33d43'A
McIlwain,Jack & Marjorie
Crb 16h10'54"28d58'
McInally,Karen
Cep 20h30'25"65d39'
McInerney,Matthew
Per 2h2'57"50d26'
McInerney,Megan Shay
Ori 5h58'47"16d29'
McInerney,Sr,Kenneth Joseph
Per 2h11'33"56d47'
McInerny,Daniel Francis
Lac 22h52'52"55d52'
McInnes,Edward
Per 4h9'46"38d1'
McInnestar
Aur 6h7'37"36d31'
McIntee,Katherine Amy
Lyr 19h8'12"38d29'
McIntire,Barbara E
Lyn 8h57'11"41d8'
McIntire,David
Oph 17h31'22"-23d45'
McIntire,Jay
Hya 8h36'57"5d39'
McIntire,Madison Leigh
Peg 22h28'50"21d48'
McIntire,Susan Elizabeth
Peg 22h41'0"20d43'
McIntosh
Uma 12h12'40"60d7'
McIntosh
Per 3h7'56"37d50'
McIntosh,Brian
Cyg 21h0'1"35d12'
McIntosh,Colin Sanders
Mon 6h30'1"8d19'
McIntosh,Dorothy Marie Brinton
Com 12h16'47d26d16'
McIntosh,Gerald M
Del 21h5'1"12d51'
McIntosh,Lisabeth T
Cet 2h27'22"-8d52'
McIntosh-Smith,Margaret
And 0h43'25"30d54'
McIntosh,Mark M
Cep 20h50'49"60d54'
McIntosh,Mike D
Boo 14h22'24"39d26'
McIntosh,My Love,Craig G
Sge 20h17'43"16d29'
McIntosh,Quinlan Earl
Mon 6h34'48"7d29'
McIntosh,Sherri Dawn
Eri 3h19'32"-2d1'
McIntosh,The
Sct 18h53'30"-7d26'
McIntosh,Tiffany Breann
Mon 8h3'12"-9d47'
McIntosh,Trevor
Per 2h24'1"55d40'
McIntryre,Claudette
Lyn 7h46'14"52d3'
McInturff,John Andrew
Per 1h48'57"50d7'
McInturff,Michael Shane
Hya 9h8'0"3d50'
McIntyre
Aur 6h33'13"34d20'
McIntyre Family Star, The
Lyn 8h17'17"37d9'
McIntyre,Alyssa Maureen
Cnc 9h3'55"20d6'
McIntyre,Gavin
Uma 11h21'32"55d2'
McIntyre,Goompabean
Boo 14h28'13"38d37'
McIntyre,James Winfield
Cap 21h16'21"-23d29'
McIntyre,Jeannie
Peg 21h55'11"33d29'
McIntyre,Joe
Aur 6h10'15"33d26'
McIntyre,Jr,John C
Lac 22h53'24"54d32'
McIntyre,Kate Marie
Mon 7h54'2"-3d3'
McIntyre,Kathleen
Cet 0h53'19"-1d53'
McIntyre,Ken
Cnv 13h59'28"37d44'
McIntyre,Lauren S
Cas 1h33'10"70d2'
McIntyre,Mary Susan
Mon 7h0'17"7d40'
McIntyre,Michelle
And 1h46'0"36d4'
McIntyre,Sheila
Cyg 20h39'0"37d59'
McIntyre,The
Aql 18h48'59"1d1'
McIntyre,Vera Ellen
Mon 6h52'14"11d59'
McIntyre,Wyatt
Sex 10h23'1"5d35'
McIssac,Michael Shea
Dra 18h49'27"68d9'
McIver,Barbara J
Cas 1h1'11"63d55'
McIver,Eileen Louise
Dra 10h32'20"81d16'
McIvor,Angela
Aql 20h1'37"9d32'
McIvor,Betty
Uma 10h5'20"58d27'
McIvor,Donald Kenneth
Lac 22h46'1"55d12'
McJohn,Mary Ann"Molly"
Lyr 19h22'32"38d25'
McKaig 10/15/55, Richard & Maureen
Her 16h6'26"25d6'
McKain,Todd
Uma 11h54'27"40d1'
McKalip,Diana
Uma 9h53'30"42d5'
McKalip,Keith
Uma 11h28'52"62d56'
McKane,Anna
And 23h4'1"42d59'
McKane,Mark
Cmi 7h41'36"3d52'
McKaughan,Michelle Lyn
Peg 22h39'53"26d26'
McKay
Nor 16h18'38"51d44'
McKay,Ann & Jack
Uma 12h4'28"55d28'
McKay,Erin Lea
Peg 22h0'0"21d7'
McKay,Gay Lynne
Cas 2h59'51"68d33'
McKay,Gloria
Aql 19h3'29"0d16'
McKay,Jr,James Ross
Mon 6h37'52"8d29'
McKay,Kazimiera Irene Constance
And 23h34'13"37d24'
McKay,Lindsay Morag
Uma 10h43'54"70d50'
McKay,Lyndsey Gay
Cas 2h34'30"75d7'
McKay,Marilyn E
Peg 22h37'0"30d23'
McKay,Mary Matilda
Eri 4h13'15"-11d52'
McKay,Myong
Peg 23h36'21"28d49'
McKay,Pamela
Crb 16h5'59"30d51'
McKay,Peter & Vicki
Cmi 7h6'31"4d28'
McKay,Trish
Mon 6h24'33"5d40'
McKayla,Alexis Ann
Cas 1h8'34"60d2'
McKeague,Barbara Jeanne
Lib 15h27'58"-8d4'
McKean & Bridgette Ludden,Stephen
Ori 5h54'20"10d13'
McKean Wish Star, Sherry Lynn
Lyr 18h27'20"35d12'
McKean"Con Amore", Josepina
Mon 7h17'17"-8d36'
McKeand,Michele
Ser 15h12'49"19d46'
McKechnie,Alison
Leo 10h1'56"21d29'
McKechnie,Heidi
Cyg 21h55'1"53d15'
McKechnie,Suz
Lyr 18h31'1"31d28'
McKee Star
Aql 19h56'11"0d20'
McKee,Arthur J
Ori 6h17'40"18d56'
McKee,Brice "Meisterman"
Cet 2h51'43"0d33'
McKee,Catherine Elizabeth
Cas 0h14'49"63d53'
McKee,Don't Forget
Umi 15h21'10"69d26'
McKee,Gordon J
Uma 10h52'33"51d51'
McKee,Ila Jean
Lyr 18h58'24"33d34'
McKee,J & C
Ser 15h28'0"0d59'
McKee,Jared "Booger"
Aql 20h5'0"-5d40'
McKee,Jr,C Roger
Aql 19h44'1"10d57'
McKee,Katherine Anne Celeste
And 23h21'20"45d31'
McKee,Kyle Wayne
Sco 17h53'20"-30d20'
McKee,Mark W
Ori 5h52'37"14d45'
McKee,Mark Wayne
Peg 23h20'40"10d16'
McKee,Mary Taylor
Gem 7h2'5"21d41'
McKee,Melissa
Crb 15h48'16"31d40'
McKee,Michael Scott
Sct 18h42'27"-6d55'
McKee,Sharon
Ori 6h1'60"-0d0'
McKee,Stephen & Patricia
Eri 4h44'26"-8d56'
McKee,The Miracle of Miles
Oph 16h53'59"0d47'
McKee,Victoria L & David D Moon
Umi 14h33'12"84d34'
McKeehan,Jr,Howard "Cricket" L
Mon 6h54'26"1d14'
McKeel,Dena Marie- Smith
Del 20h48'59"8d54'
McKeel,Millard Filmore
Sex 10h17'36"-4d59'
McKeeney,Alice
Peg 22h2'32"30d55'
McKeever,Captain Bob
Tri 1h48'1"25d32'
McKeever,Jr,Lacy Scott
Uma 11h9'24"70d27'
McKeever,Jr,Michael Joseph Peter
Her 17h21'37"21d22'
McKeever,Kathleen Patricia
Cas 1h25'20"61d30'
McKeever,Laurel
Sct 18h31'25"-5d44'
McKeever,Michael Ryan
Dra 19h24'54"56d13'
McKeever,W Bernard
Cep 21h28'57"67d44'
McKeith I,Adam Thayer
Aur 4h51'25"38d16'
McKeith I,Trevor Nicholas
Dra 17h50'33"65d31'
McKellar,Carolyn
Lib 15h32'27"-20d13'
McKellar,Katherine Gail Tabler
Eri 4h36'31"-19d18'
McKelvey,Rachel
And 0h49'31"38d19'
McKendrick,Cheryl Kay
Cgy 20h43'22"45d33'
McKendrick,Ian Paul
Ori 6h3'0"1d11'
McKendry,Brian
Cep 23h31'28"66d19'
McKendry,John
Boo 15h7'22"11d4'
McKendry,Kevin Barry
Boo 14h33'32"50d45'
McKenna "My Darling", Thomas
Ori 6h8'46"8d5'
McKenna Family Star Barney,Eileen,Ann,Tim
Dra 19h47'27"70d40'
McKenna,Annie & Billy
Uma 10h59'51"59d52'
McKenna,Brian
Lac 22h2'16"40d2'
McKenna,Diane E
And 0h18'1"34d58'
McKenna,Edna & Andrew
Cas 1h39'57"54d16'
McKenna,Elise
Vul 19h51'31"20d30'
McKenna,Elsie
Lyn 6h57'17"58d13'
McKenna,Gerard
Aur 6h7'1"38d13'
McKenna,James Michael Joseph
Ari 2h5'22"18d30'
McKenna,Jim & Chris Perryman
Cam 8h23'47"74d6'
McKenna,Karalyn
Cyg 19h59'1"31d28'
McKenna,Katherine
And 0h42'12"45d48'
McKenna,Kevin
Dra 18h54'51"58d23'
McKenna,Lady Monique
Cas 0h30'29"61d2'
McKenna,Mark
Lac 22h1'57"40d36'
McKenna,Martin
Aql 19h48'20"11d50'
McKenna,Mr & Mrs John
Cam 4h59'0"70d5'
McKenna,Patricia
Cas 0h27'29"70d32'
McKenna,Rebecca Susan
Lyn 6h14'28"60d40'
McKenna,Robert Patrick
Aur 6h31'36"38d59'
McKenna,Ron Alexander
Uma 11h0'56"41d21'
McKenna,Sandra
Cas 1h38'14"58d29'
McKenna,Terence Patrick
Vir 11h38'48"2d7'
McKenney,Doreen
And 23h23'1"41d6'
McKenney,Jacqueline Lee
Cas 0h32'36"62d55'
McKenrick,Jessica Elayne Dylan
Cyg 21h23'1"40d42'
McKenzie,Ann Louise
Her 16h43'59"35d43'A
McKenzie,Anna
Cyg 20h26'51"39d57'
McKenzie,Catherine Morrison
Mon 7h53'35"-3d50'
McKenzie,Henry Dai Keong Moran
And 1h48'14"38d11'
McKenzie,James Kent
Aur 5h40'27"50d30'
McKenzie,John H
Ori 5h47'28"21d31'
McKenzie,Jr,James Joseph
Her 16h43'59"35d43'B
McKenzie,Kerri
Lyn 7h52'37"44d6'
McKenzie,Kwaiki
And 23h42'28"47d41'
McKenzie,Lyn
Mon 8h2'0"-0d49'
McKenzie,Mad Mary
Dra 16h8'16"60d37'
McKenzie,Marilyn Marie McKenney
Mon 7h48'2"-1d9'
McKenzie,Marjorie
Cas 23h4'49"54d10'
McKenzie,Mark Daniel
Ori 4h59'37"-0d59'
McKenzie,Megan
Cyg 19h28'35"39d38'
McKenzie,Rod
Per 2h25'11"54d24'
McKenzie,Thomas Warren
Ori 5h25'42"15d18'
McKenzie,Toujours Barbara
Peg 23h24'52"13d53'
McKeon,Cynthia Ann
And 23h27'37"47d26'
McKeon,Elia
Aql 19h9'18"12d42'
McKeon,James J
Ori 6h16'29"-1d7'
McKeon,Karen & Siobhan Begley
Uma 11h44'34"32d31'
McKeon,Kelly J
Lyr 18h58'19"33d31'
McKeon,Kimberlee Diane
Eri 4h28'37"-12d9'
McKeon,Siobhan Marie
Cam 6h28'20"68d5'
McKeon,Thomas Francis
Per 2h50'1"32d17'
McKeon-I love you always,Alison
Cyg 20h21'52"39d32'
McKeown,Arthur
Aur 7h24'32"41d2'
McKeown,Denise
And 0h18'34"35d24'
McKeown,Katharine Blair
Aql 19h12'25"14d16'
McKeown,Rosemary C
And 23h37'42"48d49'
McKewen,Susan Leitzman
Lib 14h39'26"-11d2'
McKie,Alexander James
Aql 19h54'48"-6d0'
McKie,Betty Lou Stinson
And 2h0'0"45d39'
McKie,Gillian
Cas 0h57'31"50d20'
McKie,Nathaniel
Cmi 7h36'13"0d6'
McKim
Her 18h3'42"38d11'
McKim,Dick
Eri 3h56'32"-10d3'
McKim,Joshua
Hya 9h15'17"4d26'
McKinlay,Morag
Cep 21h1'0"58d36'
McKinley,Jacob Todd
Aql 19h48'39"14d6'
McKinley,Michael
Ser 15h32'58"10d56'
McKinley,Ronald E
Cas 3h10'12"63d20'
McKinley,Thomas Michael
Umi 14h21'39"68d8'
McKinney 27,Sherry B
And 1h21'19"34d10'
McKinney I Love You Robert,Tamara S
Lyr 18h54'56"40d42'
McKinney,Alexis
Eri 4h10'50"-09d38'
McKinney,Archie Bruce
Ori 6h8'33"8d44'
McKinney,Brian
Oph 17h26'28"0d40'
McKinney,Christopher M
Peg 22h23'1"30d34'
McKinney,Dallas & Dillon
Lyr 18h17'28"38d1'
McKinney,Deborah
And 23h1'21"44d39'
McKinney,Don Hunt
Cet 2h50'29"2d25'
McKinney,Dr Donald
Boo 15h29'12"40d8'
McKinney,Geri
Sco 16h9'20"-37d45'
McKinney,Helen & Arthur
Lyr 18h29'53"30d37'
McKinney,J Robert
Lyr 19h1'22"25d39'
McKinney,James I & Sherri A Schmidt
Cyg 21h16'59"38d0'
McKinney,Joan
Crb 15h47'15"38d43'
McKinney,Kevin P
Vul 19h47'11"28d51'
McKinney,Our Star, Caitlyn Alaina
Eri 4h41'31"-1d9'
McKinney,Robert
Sct 18h55'23"-5d34'
McKinney,Thomas Murdock
Aql 20h11'20"5d1'
McKinney-Smith,Weaks
Uma 13h43'25"60d29'
McKinnon Star,The Aaron
Cet 0h52'1"-5d12'
McKinnon,Darlene
Mon 7h47'2"-9d0'
McKinnon,John Robert
Cep 20h38'45"76d15'
McKinstry,Beatrice
Cas 0h30'55"70d52'
McKinstry,Carlene
Uma 13h46'11"50d9'
McKinstry,Darcy
Boo 13h39'52"27d18'
McKinstry,Frederick H
Lmi 10h19'10"31d46'
McKinstry,Raymond
Lmi 10h57'21"30d60'
McKirachan,C David
Uma 12h49'17"54d41'
McKirachan,Joan M
Peg 21h57'13"28d44'
McKissic,Jr,Mary Esther Leon M
And 23h35'20"42d46'
McKitterick,David C
Per 4h0'0"51d34'
McKittrick,Catherine
Cyg 19h30'27"36d34'
McKittrick,DeeDee & Al
Cyg 20h9'30"37d57'
McKneely,Jr,Roland V
Per 2h26'49"57d50'
McKnight '96,Alex
Aur 6h31'38"31d17'
McKnight,"JAM"Joanne A
Cyg 20h41'56"30d54'
McKnight,Carolyn
Peg 23h6'23"12d36'
McKnight,Charlie
Aur 6h1'56"31d30'
McKnight,Charlotte
Cas 22h56'46"54d39'
McKnight,John Daniel
Psc 22h57'56"0d59'
McKnight,Lorna
And 1h18'59"37d47'
McKnight,Robert Eugene
Per 3h15'60"50d34'
McKnight,Stephanie Marie
Crb 16h8'21"30d45'
McKnight,Stephen J
Dra 11h25'19"77d41'
McKnight,Thomas Alexander
Cep 22h8'39"61d19'
McKune,James L
Cep 1h14'18"77d56'
McLafferty,Joanna
Mon 6h22'0"3d2'
McLafferty,Mason
Aql 19h53'19"15d20'
McLain "Matthew 2:10", J D
Her 17h20'16"22d22'
McLain,Calvin West
Cyg 21h21'48"28d38'
McLain,Marci
Cas 2h50'32"75d2'
McLain,Robert Snyder
Her 16h21'59"41d41'
McLaine,Colleen Ann
Peg 22h42'53"27d51'
McLamb,Ronald
Vul 19h43'42"22d54'
McLane,Adam Keith
Aql 18h58'38"-1d59'
McLane,Coie
Psc 23h6'45"6d40'
McLane,Frances
Gem 7h9'42"27d18'A
McLane,Gerald
Gem 7h9'42"27d18'B
McLane,Jr,Richard Francis
Per 3h52'7"36d27'
McLane,Judy
Cas 2h2'47"59d21'
McLaren
Per 3h25'56"46d45'C
McLaren,Althea B
Hya 9h5'33"0d25'
McLaren,Ian
Per 2h53'36"41d8'
McLaren,Joan Patricia
Umi 13h54'33"72d39'
McLaren,Kathleen
Lyr 18h52'17"34d2'
McLaren,Pauline
Cyg 21h1'33"38d15'
McLaren,Sean & Deb
Cnc 8h50'35"30d28'
McLarty,Donald William (Bunny)
Cyg 21h5'1"32d1'
McLauchlan,Kathryn
And 23h15'17"46d38'
McLaughlin Family,The
Lyr 18h30'0"30d48'
McLaughlin Star,The Nana & Grampie
Dra 14h5'32"68d28'
McLaughlin"Superstars" Anthony & Marla
Vir 11h59'30"1d57'
McLaughlin's Star, Michelle & Brandon
Uma 9h51'60"52d31'
McLaughlin,Carol
Cyg 19h49'15"58d16'
McLaughlin,Cayla Loren
Peg 0h0'32"20d54'
McLaughlin,Erik & Becca
Umi 15h7'50"72d17'
McLaughlin,Erin Elizabeth
Gem 6h47'13"16d31'
McLaughlin,Helen Madonna
Cas 0h5'40"60d23'
McLaughlin,Jack Edward
Sct 18h34'44"-4d59'
McLaughlin,Jane Clarkin Gallogly
Cas 0h36'0"67d39'
McLaughlin,Joan S
And 1h30'52"38d27'
McLaughlin,John Donald
Oph 17h18'20"-20d10'
McLaughlin,Kate
Lyn 7h37'30"38d40'
McLaughlin,Kelly Anne
Lyr 18h41'14"28d49'
McLaughlin,Kelsey Paige
Lyr 18h41'48"33d31'
McLaughlin,Madelyn
Cyg 20h12'50"40d36'
McLaughlin,Mary Moran
Uma 9h31'23"47d11'
McLaughlin,Patrick
Hya 9h1'59"0d55'
McLaughlin,Patsy N
Mon 7h2'39"-6d55'
McLaughlin,Rick & Loved Ones
Umi 15h53'34"74d35'
McLaughlin,Sami
Ori 5h59'1"11d42'
McLaughlin,Sandy Irene
And 0h13'0"47d37'
McLaughlin,Star,The Marlene
And 0h51'38"39d26'
McLaughlin,Stephanie Sissy
Del 20h55'23"9d49'
McLaughlin,Tina & Marlon
Aql 19h9'28"1d35'
McLay,Douglas Carlyle
Oph 17h10'32"-20d37'
McLean's Star 9-9-95, Connal
Cra 18h18'40"-37d7'
McLean,Beverly
Aur 7h3'23"40d44'
McLean,Carol
Crb 15h17'48"30d6'
McLean,Carolyn Lippard
Sco 16h36'57"-30d28'
McLean,Christopher Zachery
Cnv 12h35'1"32d49'
McLean,Cophie
Equ 21h7'12"2d27'
McLean,Daly James
Cyg 19h42'23"28d1'
McLean,Dorothy Morefield
Peg 21h54'42"20d18'
McLean,Fori
Cet 2h56'38"1d45'
McLean,Hugh
Cyg 19h27'44"35d20'
McLean,Jack & Pat
Lyn 6h44'15"50d24'
McLean,James A
Lac 22h31'14"55d19'
McLean,Katherine Finch
Cyg 19h50'53"40d18'
McLean,Maxine
Umi 14h45'51"67d11'
McLean,Megan K
Ori 6h0'28"1d26'
McLean,Morgan Scott
Aur 6h4'27"35d54'
McLean,Nigel
Ori 5h21'30"11d33'
McLean,Patricia
Cas 0h42'23"70d39'

McLean,Paul G
Cet 1h29'44"-1d14'
McLean,Phyllis J
Lyn 7h51'58"47d51'
McLean,Robert Allingham
Lmi 10h10'46"38d3'
McLean,Stuart Nicklaus
Cyg 19h26'48"35d59'
McLean,Warren
Ari 2h36'24"21d37'
McLees,John J
Per 1h40'59"53d0'
McLeish,Catherine M
Umi 13h15'58"71d30'
McLeish,Emma Louise
And 2h35'20"40d3'
McLeish,James Douglas
Dra 16h41'50"62d35'
McLeish,Sally & Richard
Mon 6h23'16"5d24'
McLeish,Shawn Douglas
Cam 4h13'0"71d8'
McLeish,Todd Andrew
Cep 23h13'50"68d5'
McLellan 1
Ori 4h59'21"5d23'
McLellan,Jeffrey Charles
Aur 7h12'36"41d43'
McLellan,Rosemary Valerio
And 23h30'19"43d37'
McLelland,Madeleine
Mon 8h1'48"-8d41'
McLendon III,John Harvey
Aur 7h13'17"41d8'
McLendon,Dr Irwin C
Sct 18h42'51"-5d15'
McLendon,Timothy
Dra 19h29'37"68d11'
McLennan,Jon G
Ser 17h55'42"-14d51'
McLennan,Judy
Boo 13h36'58"12d15'
McLennan,Kimberly
Boo 13h57'35"11d21'
McLennan,William James
Cyg 19h27'27"33d15'
McLennan-April 25,
1963,John Avery
Boo 14h34'1"45d12'
McLenon,Andy
Ser 17h32'49"-14d60'
McLeod III,Victor C
Cet 0h36'42"-1d55'
Mcleod,Baby
Umi 10h34'43"86d4'
McLeod,Cheri
Peg 22h2'15"32d17'
McLeod,Donald
Her 16h48'0"51d10'
McLeod,Jay
Cas 0h15'10"47d40'
McLeod,John
Boo 15h28'58"42d13'
McLeod,Laina
Cyg 21h5'25"31d14'
McLeod,Lisa Ann Adams
And 23h24'1"48d21'
McLeod,Maggie & Brady
Sge 19h23'42"16d30'
McLeod,Vicki Lynn
Hya 9h0'23"5d27'
McLeod-Williams, Alexander
N
Ari 1h59'11"17d53'
McLeods,T J
Boo 15h4'11"28d23'
McLernon,Mary Margaret
Crb 16h4'12"27d29'
McLeskey,Linda M
Aqr 22h4'0"0d9'
McLinden,James K
Uma 11h28'13"57d33'
McLoughlin,Caitland Ann
Cyg 21h6'21"30d4'
McLoughlin,Laurie Anne
Lyn 8h2'55"43d30'
McLoughlin,Lori
Eri 4h32'55"-7d56'
McLoughlin,Noel "The
Professor"
Cep 0h10'0"75d8'
McLouth,Don & Effie
Tri 1h49'43"26d53'
McLucas,"Heavenly" Nevin
Her 16h25'37"23d32'
McLucas,Charles & Myrle
Cet 3h1'42"0d16'
McLuckie,Peter Butler
Ori 6h18'52"18d52'
McMahan II,Charles Edward
Lac 22h6'51"51d34'
McMahan Phenomenon,
Richard Lamar
Aql 19h56'33"0d38'
McMahan,D Bruce
Her 18h2'36"31d42'
McMahan,Michel & Jason
Camara
Dra 16h41'15"51d38'
McMahan,Roland Jack
Hya 8h11'53"-1d41'
McMahan,Sharon & Stephen
Cyg 21h0'1"38d37'
McMahan,Tamera L
Mon 6h56'36"10d21'
McMahon III,Edward Joseph
Aur 6h8'59"30d10'
McMahon's"Wishing Star"
Crb 16h0'11"31d43'
McMahon,Barry John
Cmi 7h44'47"5d21'
McMahon,Brent J
Per 1h53'31"50d11'
McMahon,Brian Robert
Christian
Aur 4h58'1"40d24'
McMahon,In Memory of
Robert
Tri 2h14'55"28d40'
McMahon,James Joseph
Per 1h54'26"50d33'
McMahon,Jason Ken
Equ 21h6'34"10d23'
McMahon,Jennifer
Dra 16h7'0"67d53'
McMahon,John
Gem 7h25'11"33d47'
McMahon,John P
Lac 22h47'15"52d36'
McMahon,John William
Aur 6h33'0"34d47'
McMahon,Kathleen Dierdre
Cyg 20h55'0"39d35'
McMahon,Kimberly Anne
Cet 2h41'0"5d10'
McMahon,Mary K
And 23h26'15"44d30'
McMahon,Nicholas
Aur 7h21'12"38d3'
McMahon,Ruth
Vul 20h16'19"22d41'
McMahon,Stillman Dillon
Ori 5h38'1"0d40'
McMahon,Valerie
Cas 0h27'1"68d46'
McMains,Colbert
Aql 19h35'23"-5d52'
McManis,(4 Star), Frances
Fern
And 23h39'35"38d15'
McManis,Catherine Christine
Mon 6h39'15"7d1'
McMann,Sr,Daniel L
Her 15h56'23"41d46'
McMannus,Desmond William
Lyn 7h48'11"47d59'
McManus Jobies "Shine
Forever"
Cyg 21h54'48"38d25'
McManus,Ann O'Brien
Eri 4h31'13"-11d17'
McManus,Ardelle I
Aql 18h57'29"-05d57'
McManus,Brian
Boo 14h19'14"37d56'
McManus,Christopher
Aur 6h25'50"38d3'
McManus,Jerome L
Cep 23h33'0"68d15'
McManus,Kerri Lynn
Vul 20h20'58"25d0'
McManus,Michael John
Cmi 7h23'28"0d24'
McManus,Miriam Catherine
McCarthy
And 0h28'0"27d54'
McManus,Neal Colin
Cet 0h0'54"-18d51'
McManus,Padraig B
Cyg 21h30'15"42d12'
McManus,Rosemary
And 0h47'1"21d30'
McManus,Virginia
Mon 8h1'18"-9d29'
McManus,Wendell
Hya 8h12'20"2d58'
McMarlin,Lois
Cyg 19h23'27"28d37'
McMaster,Douglas Matthew
Cet 0h27'1"-10d15'
McMaster,Paul Joseph
Lac 22h27'1"38d38'
McMaster,Scott K
Her 18h22'47"28d25'
McMaster,Valerie-Anne
Heather
Psc 1h0'28"11d18'
McMeekin,Emma
And 0h23'49"40d15'
McMenamiln,Stephen J P
Ori 6h17'1"18d50'
McMenamin,Daniel James
Ori 6h6'1"9d11'
McMenamin,Elsie
Cas 0h50'22"60d15'
McMenamin,Michelle Louise
Aqr 22h2'31"-18d58'
McMenamin,Ryan Joseph
Per 4h1'49"51d31'
McMenemy,Irene
Cas 1h20'55"55d44'
McMichen,Rocky H
Her 16h17'22"5d1'
McMiddle
Sct 18h53'3"-7d54'
McMilion,Heidi Jo
Peg 22h35'41"21d51'
McMillan,Carol "Irish"
Lyr 18h42'57"35d19'
McMillan,David Evan "Jesus"
Her 16h59'30"13d6'
McMillan,Kristina
Cyg 19h56'32"38d56'
McMillan,Michael Hayden
Sex 10h13'2"-2d2'
McMillan,Rick
Aur 5h45'48"50d23'
McMillan,Sandra Tyers
Crb 15h16'18"31d37'
McMillen,Annette Katherine
Mon 6h26'1"8d17'
McMillen,Charles
Mon 6h56'31"7d32'
McMillen,James
Her 16h30'14"38d29'
McMillen,Lisa Joanne
Hya 9h3'32"0d26'
McMillen,Lisa Joanne
And 23h19'0"40d38'
McMillen,Paula JeanMarie
Lyn 9h0'24"36d50'
McMillen,Raymond
Boo 14h25'24"20d18'
McMillen,Sr,In Memory Of
John William
Eri 2h49'15"-18d38'
McMillian,Roy
Her 15h48'52"44d48'
McMillin,Donald"Papa"
Aur 6h6'26"31d51'
McMillion,Scott
Her 16h8'1"40d35'
McMindes,Beverly Jean
Uma 12h57'50"61d39'
McMindes,David Michael
Uma 12h59'52"61d53'
McMinn,Shyanne
And 0h13'1"33d34'
McMoran,Judy
Lyr 28h14'54"35d6'
McMorrow,Jessica Marie
Vul 19h45'17"20d7'
McMorrow,Katherine Grace
Uma 11h43'20"48d21'
McMorrow,Patricia Mae
Uma 11h0'11"41d7'
McMorrow,Patrick Gerald
Cep 21h39'19"55d11'
McMorrow,Shirley Marie
Cas 1h39'19"60d40'
McMullen,Beatrice Kathleen
Lyr 18h58'30"42d42'
McMullen,Bob
Cep 21h54'32"60d42'
McMullen,James
Cep 22h33'0"59d42'
McMullen,Lori Ann
Mon 6h36'24"6d31'
McMullen,Louis Jerome
Ori 6h2'15"7d40'
McMullen,Nancy
Leo 10h14'51"10d37'
McMullen,Rick
Aql 19h58'15"14d45'
McMullin MSE,Linda
Del 20h38'25"11d9'
McMullin,Anna Elizabeth
Mon 6h24'36"1d41'
McMullin,Laura Nichole
Lyn 9h14'1"41d12'
McMullin,Sarah Jordan
Del 20h14'26"11d24'
McMurray,Brandon B
Cet 2h51'42"1d15'
McMurray,Carissa
Vul 19h23'42"26d57'
McMurray,Franáois
Ori 6h1'21"6d19'
McMurray,John
Her 16h33'39"34d40'
McMurray,Kim R
Mon 6h59'39"-8d21'
McMurray,Lindsey
Cet 1h18'45"-1d3'
McMurtry,Thomas Lee
Ori 4h55'45"-1d41'
McNab,Don
Boo 14h37'19"18d30'
McNab,Mattie Elizabeth Ayres
Cam 3h47'37"58d8'
McNabb,Jane Elizabeth
Cas 2h56'47"68d13'
McNabb,Jennifer Lynn
Lyr 19h5'45"40d54'
McNabb,Nootsie
Cam 7h37'59"80d27'
McNaghten,Thomas Cantor
Her 16h57'36"50d53'
McNail,Deborah S
Ori 5h3'59"14d32'
McNail,Erma
And 23h18'0"44d57'
McNair Star,The
Mon 6h47'22"11d16'
McNair,Mel
Vir 13h9'25"-5d3'
McNairy,Jack H
Lib 15h1'13"-28d25'
McNally's
Vul 19h57'18"26d48'
McNally,Dennis
Lac 22h26'56"50d22'
McNally,Ian Charles
Boo 14h18'28"31d43'
McNally,John F
Per 1h53'31"56d39'
McNally,Jonathan Ryan
Dra 16h48'1"68d6'
McNally,Judith
Mon 6h26'1"-6d26'
McNally,Kevin
Per 2h25'37"55d3'
McNally,Margaret "Meg"
Psc 0h47'34"28d40'
McNally,Matthew Robert
Ori 5h53'16"-10d42'
McNally,Timothy
Sgr 19h1'11"-20d52'
McNally,William J
Vir 13h2'15"-5d18'
McNamara & Boo Star, The
Margaret
Lyn 7h30'27"50d49'
McNamara III,Francis "Dub" L
Aql 18h46'29"10d23'
McNamara,Barry
Her 17h35'20"21d47'
McNamara,Bernadine
Psc 23h39'37"1d30'
McNamara,Blair Bennett
Her 16h34'0"48d52'
McNamara,Dawn
Lyn 7h5'19"44d50'
McNamara,Elizabeth C
Peg 22h17'40"20d28'
McNamara,Gregory M
Boo 15h18'53"50d12'
McNamara,Janet Ray
And 1h24'32"37d17'
McNamara,Jimmy
Uma 9h9'1"47d34'
McNamara,Kathleen Marie
Aqr 22h46'49"-2d3'
McNamara,Kathy
Peg 21h18'42"22d55'
McNamara,Kevin Michael
Cep 21h40'45"56d14'
McNamara,Laura King
And 23h30'55"45d53'
McNamara,Maggie Mae
And 0h54'32"36d34'
McNamara,Margaret
Cas 1h1'25"66d26'
McNamara,Meghan Elizabeth
Tri 1h59'47"34d53'
McNamara,Michelle Kathleen
Uma 12h2'19"32d47'
McNamara,Neil S
Cam 4h4'0"68d32'
McNamara,Robert J
Sco 17h9'0"-38d26'
McNamara,Sarah Anne
Cam 7h34'0"61d46'
McNamara,Suellen & Warren
Cam 3h54'55"57d19'
McNamara,Suzanne Elizabeth
Lyr 18h34'1"26d3'
McNamara,Teresa
And 2h32'17"38d7'
McNamee,Jeanne
Peg 22h27'41"21d35'
McNamee,Jennifer
Cas 23h47'20"50d30'
McNamee,Thomas Gary
Cnv 12h34'15"50d7'
McNames,Penny Lou
Cas 0h4'15"63d40'
McNannay,Robert E
Lac 22h28'60"38d18'
McNary,Nathaniel Whey
Hya 8h45'9"-7d57'
McNatt,BW
Hya 9h11'1"2d39'
McNaught,Jacoby
Vul 19h45'49"20d5'
McNaughton
Ori 5h31'1"8d45'
McNaughton Family
Cam 4h57'1"58d58'
McNaughton,Alisha
Peg 23h3'0"21d51'
McNaughton,Amanda
Boo 14h27'45"51d1'
McNaughton,Ken
Lyn 8h5'44"58d36'
McNaughton,Thomas
Ori 5h48'22"10d36'
McNeal Star,The
Uma 8h42'45"54d49'
McNeal,Dorothy Fay
Peg 22h45'14"26d0'
McNeal,Jr,Laird Charles
Cep 23h39'39"77d53'
McNeal,Marty
Peg 23h39'36"31d50'
McNeal,Michael Ray
Dra 16h50'10"68d48'
McNeel
Aql 19h6'1"15d3'
McNeel,Alexandra
Eri 3h36'22"-18d30'
McNeel,Barry Wayne
Aql 20h12'56"0d16'
McNeel,Morgan
Equ 21h2'59"10d2'
McNeeley,Alison
Cyg 20h16'23"38d59'
McNeely,Anthony Dale
Cam 4h52'42"61d40'B
McNeely,Deborah A
Sct 18h22'1"-5d17'
McNeely,John Edward
Dra 9h33'56"81d27'
McNeely,Mistelle Therese
Cyg 19h24'47"31d5'
McNeely,Tina Marie
Cam 4h52'42"61d40'A
McNees,Maxine B
Lyr 18h33'0"38d32'
McNees,Stephanie
Cas 1h41'17"76d22'
McNeice,Barry Y
Dra 14h11'23"64d59'
McNeil ØB‰,Alvin J
Oph 18h31'0"10d17'
McNeil,Amy Dominica
Cmi 7h26'55"8d2'
McNeil,Dr R J
Lyn 9h12'17"38d32'
McNeil,Hilda Frances
Eri 3h20'39"-14d25'
McNeil,Jennie
Aql 18h56'0"13d33'
McNeil,Mary Joyce
Peg 22h52'28"22d24'
McNeil,Meghan
Aql 18h53'55"-1d30'
McNeil,Rosanne
Lmi 10h32'26"31d47'
McNeil,Troy Scott
Sex 9h52'43"-1d27'
McNeil-I Love You,Rose
Lyr 18h49'12"40d9'
McNeill Our Shooting
Star,Bernie
Aql 18h57'52"17d0'
McNeill,Malcolm
Aur 6h36'36"30d29'
McNeill,Ronnie
Cet 2h33'32"2d21'
McNeill,Susan Ann "Susie-Q"
And 23h13'29"37d36'
McNeilly
Uma 12h58'57d49'
McNeilly Star,The Alexander
Lac 22h6'18"47d47'
McNeish,John Louis
Cet 1h22'47"-1d24'
McNellis,Janet Mills
Vul 21h3'25"28d18'
McNemar,Don & Britta
Cyg 20h58'47"30d56'
McNemar,James W
Aur 5h1'58"41d16'
McNemar,Kelly Colleen
And 1h57'1"37d53'
McNemee,Debbie
Del 20h13'57"10d38'
McNerney's "Star Lite Star
Bright"
Cet 4h47'23"4d28'
McNerney,Mary Kate
Sge 19h58'18"16d0'
McNerney,Patrick & Angela
Sge 19h53'0"18d52'
McNerney,Timothy Andrew
Sge 19h52'25"16d22'
McNevin,Jr,Gerald
Vul 19h36'17"27d7'
McNichol,Fiona
Cyg 20h58'31"37d40'
McNickle,Miriam
Aql 18h43'30"11d45'
McNicol,Claudia Reigh
Lyn 8h58'46"46d8'
McNicoll,Jean-Michel
Ori 6h2'45"6d12'
McNicoll,Stéphane
Ori 4h45'51"5d48'
McNinch,Judy
Aql 19h30'52"7d48'
McNish,SW SL,Yona
Cet 2h49'51"3d47'
McNulty,Ashlyn Bree
Cnc 9h17'55"31d22'
McNulty,Charles A
Aur 6h8'56"50d17'
McNulty,Esther M
Cas 2h12'27"67d34'
McNulty,Karen
Dra 16h11'18"67d56'
McNulty,Lorraine
Lyr 18h29'24"30d55'
McNulty,Michael Augustine
Cep 22h8'27"54d7'
McNutt,Beth Ann
Aql 19h2'1"0d10'
McNutt,Kelly Michele
Lyr 19h15'37"42d4'
McNutt,Pat Starks
Cam 11h58'18"81d44'
McNutt,Sally Ann
Peg 22h13'55"3d40'
McOmber,Corinne Mador
Sgr 18h49'28"-31d2
McOmber,Herbert T
Tri 1h52'33"25d26'
McOmber,Russell H
Boo 14h2'37"20d54'
McOsker,Christopher D
Ori 5h47'17"18d18'
McOwen,Douglas Charles
Umi 15h1'58"68d12'
McParland Three,E F C The
Dra 17h32'51"63d51'
McParland,Michelle
And 2h1'58"41d13'
McPartland,Jane Helen
Crb 15h50'14"38d29'
McPartlin,Ruth E Brasie
Cas 0h20'51"58d38'
McPeak,Christopher Lloyd
Leo 10h36'37"15d36'
McPetrie,Diana
Cam 6h10'33"72d22'
McPhail III,John Finley
Oph 16h17'30"-6d41'
McPhail,Christy Lynn
And 1h53'36"37d58'
McPhail,Courtney Brooke
Sge 20h5'54"20d10'
McPhail,Patrick & Misako
Cyg 21h35'16"37d40'
McPhail,Sharon M
Lyr 19h22'44"38d42'
McPhearson (JSM), Jeffrey
Scott
Cnc 8h36'58"18d26'
McPhee,Norman
Peg 22h40'41"23d44'
McPhee,Robert Lloyd
Ori 5h53'1"10d33'
McPherson
Cyg 19h19'59"49d1'
McPherson Family,The
Vul 19h47'13"22d45'
McPherson I972,Aileen
And 2h19'23"47d3'
McPherson William N & Betty
J
Lac 22h17'52"46d13'
McPherson,Debra
Peg 22h52'1"27d36'
McPherson,Doris Mae
Eri 2h48'48"-6d59'
McPherson,Edgar "Big Daddy"
Her 16h2'36"45d4'
McPherson,Edward Sidney
Aur 6h34'60"31d27'
McPherson,Elizabeth
Alexandra
Sct 18h52'5"-7d32'
McPherson,John Duncan
Per 2h58'11"38d43'
McPherson,Julie
Peg 23h27'56"18d26'
McPherson,Lynn Carol
Lyr 18h36'31"43d24'
McPherson,Mark S
Her 17h10'23"49d48'A
McPherson,Mary Ann
Vul 19h22'16"25d43'
McPherson,MD,Alice R
Aql 20h2'33"10d47'
McPherson,Michael Joseph
Ser 15h45'59"24d14'
McPherson,Raven Storm
Uma 9h54'27"45d6'
McPherson,Robert Michael
(Mike)
Aql 20h13'42"4d12'
McPherson,Scott A
Psc 1h2'40"21d28'
McPherson,Sybil
Peg 22h59'53"10d10'
McPherson,Terri E
Her 17h10'23"49d48'B
McPherson,Thomas F
Tau 4h5'51"20d15'
McPhetridge,William Bryon
Tau 5h15'29"16d19'
McPipkin,Ashley Kay
Aql 20h31'43"0d23'
McPoyle,Kirk
Dra 16h32'57"67d34'
McQuade,Bobbie
Uma 9h0'24"67d33'
McQuade,Travis Jordan
Boo 14h54'12"27d59'
McQuaid,Annette Marie
Cam 6h58'32"63d19'
McQuality,Larry
Sgr 19h4'41"-29d7'
McQuarrie,The Silver Fox-
Ronald M
Per 1h34'51"54d14'
McQueen,JoAnne
Tri 2h23'20"34d43'
McQueen,Mako
Cet 0h59'26"0d37'
McQueen,Shari Lynn
And 0h56'30"40d3'
McQueeney,Marion Wall
Cnv 13h31'17"40d47'
McQuerrey,Shirleen M
And 2h10'19"39d59'
McQuiggin,Brock Matthew
Aur 7h21'41"41d8'
McQuilkin,Jr,Thomas J
Her 16h29'15"50d0'
McQuillan,Emma
Cap 21h22'0"-14d55'
McQuillan,Francis T & Anna F
Her 16h13'16"21d21'
McQuinley,Sheila Rae
And 23h21'17"49d29'
McQuinn,My Endless Love-
Mike
Dra 17h13'52"65d19'
McQuire,Michelle Lynn
Uma 11h21'17"43d16'
McQuiston,Sr,Charles W
Per 1h54'1"50d21'
McRae
Cam 4h11'16"70d55'
McRae,Alan
Ori 5h55'45"16d12'
McRae,Lisa Christine
Aql 19h45'35"14d48'
McRae,Scott
Ori 5h54'44"1d19'
McRee,Noah Lance
Ser 15h16'1"21d20'
McRee,Taliann Nicole
Eri 4h43'50"-8d15'
McReynolds,Ryan John
Cam 13h55'42"80d24'
McReynolds,Susan
Cyg 19h42'19"40d16'
McRit,Susan
Cet 3h12'0"2d38'
McRobbie,Gordon
Her 16h7'1"48d53'
McRobbie,Ian G
Ori 5h23'1"0d37'
McRobbie,Stuart G
Per 4h42'32"40d42'
McRoberts,Christine
Uma 9h47'1"57d58'
McRoberts,Imogene
And 0h53'0"22d5'
McRoberts,William J
Aur 5h5'40"40d18'
McRoy's Forever Star, Jim &
Paula
Cyg 21h4'34"39d18'
McSarland,Gary D
Oph 17h31'59"-22d58'
McSarland,William
Lac 22h35'0"37d41'
McSarlane,James Joseph
Aur 4h50'41"40d2'
McShane,Biff
Lac 22h4'47"46d18'
McShane,Jennifer
Cyg 19h24'37"33d38'
McShane,Margaret
Cas 0h28'59"58d48'
McShea,Gregory
Aur 6h13'26"35d44'
McSherry,Anna
Del 20h14'45"10d8'
McSherry,Brian
Lac 22h20'55"53d52'
McSherry,Kevin
Dra 15h4'53"60d52'
McSorley,Bryan
Her 17h9'29"22d4'
McSorley,Chris & Carrie
Crb 16h20'0"32d29'
McSorley,Jack R
Aur 5h49'28"50d22'
McSorley,Jerry
Lyn 7h58'24"40d35'
McSorley,Kristen Ann
Com 12h42'11"20d31'
McSorley,Peter
Dra 14h14'37"64d20'
McSpadden,Anne
Aur 5h15'55"34d50'A
McSpadden,Warren W "Mac"
Aur 5h15'55"34d50'B

McSparron,John William Edgar
Per 2h6'11"57d10'
McStoddart
Boo 14h55'16"28d18'
McStroul,Geoffrey
Uma 11h43'19"47d32'
McSurdy,Jane
Cet 2h56'0"0d35'
McSwain,Michael J
Hya 8h51'31"-1d8'
McSwaney,Allison Lindsay
Psc 23h33'54"0d33'
McSween,Jr,Harry Younger
Cet 0h45'29"-3d31'
McSweeney,Mykalah
Crb 15h53'35"27d0'
McSweeney,Mykalah Ann
Ori 5h27'30"0d30'
McSweeney,Shannon
Cyg 21h0'36"38d15'
McSwiggan,Jr,James J
Peg 23h8'35"10d59'
McTaggart,Philip Patrick
Lyn 8h17'0"45d0'
McTague,Lenore H
Mon 7h2'11"-0d36'
McTague,Nora
Uma 12h52'55"60d15'
McTague,Robert
Uma 12h50'20"60d16'
McTamney,Colleen S
Boo 13h46'45"20d5'
McTavish,Shamus
Aur 4h48'1"48d54'
McTeague
Lyn 6h58'0"44d52'
McTernan,Lynsey
Lyr 18h31'27"35d23'
McThomaich,Ian Tearlach
Aql 18h58'53"11d38'
McTiernan,Michael
Cet 0h32'59"0d40'
McTigue,HRA Thor Poseidon David
Cyg 21h8'12"48d50'
McTigue,HRA Thor Pluto Tom
Aur 6h5'36"31d51'
McTigue,Michael John
Ori 5h2'40"12d18'
McTurk,Ian Rutherford
Psc 0h46'56"27d59'
McV
Oph 18h42'19"7d16'
McVay 32,E
Sge 20h4'49"20d24'
McVay,Michael
Cyg 19h44'52"38d25'
McVay,William Horace
Cma 6h55'59"-17d1'
McVean,Donald L
Her 16h9'38"10d23'
McVean,Forever Mine Steven Clark
Sct 18h53'42"-7d26'
McVeigh,Coach
Ori 5h56'0"21d50'
McVeigh,Jr,Thomas John
Dra 20h15'55"67d35'
McVeigh,Mary Angela
And 23h0'23"51d2'
McVeigh,Thomas James
Her 16h15'21"47d59'
McVey's Flying Star, David
Cam 10h7'41"82d22'
McVey,Drew D
Aql 20h10'1"0d21'
McVey,Jr,Robert M
Aql 19h55'0"13d16'
McVey,Lisa E
And 23h16'0"41d39'
McVicker,Diane
Cet 1h22'54"-4d39'
McVicker,Jack Marshal
Lac 22h9'57"38d46'
McVickers-Thomas,Dania
Cas 1h49'22"60d12'
McVinney,Erin
Uma 10h38'31"47d37'
McWeeney,Meghan Kathryn
Mon 7h17'7"-7d7'
McWhirter,Maurice J
Ser 15h26'0"9d37'
McWhirter,Norma Lee
Mon 6h23'33"-6d49'
McWhorter,John V
Per 2h25'34"57d57'
McWilliam "Pride of Scotland",Colin
Cep 20h7'12"60d43'
McWilliam,Gordon Crawford
Her 17h30'54"27d39'
McWilliam,Ronald
Her 18h3'1"28d59'
McWilliams,Barry John
Mon 6h32'1"3d42'
McWilliams,Bryan Michael
Cap 21h28'58"-23d9'
McWilliams,Colin M
Peg 21h28'1"23d23'
McWilliams,Darren Eugene
Tau 5h13'43"20d12'
McWilliams,Erica Lee
Mon 6h18'59"7d17'
McWilliams,James Daniel
Her 18h28'1"20d24'
McWilliams,Jennifer
Lyn 7h15'57"59d58'
McWilliams,Linwood
Ori 5h57'58"14d10'
McWilliams,Nancy Rae
Cyg 19h46'59"56d25"
McWilliams,Patricia
Mon 8h7'40"-10d13'
McWilliams,Robert Vincent
Aql 18h59'13"-5d38'
McWilliams,Robert Andrew
Equ 21h22'0"2d59'
MD-D-PB-S-M-O-J-J.
Crb 16h8'11"30d26'
MD2
Ant 10h32'48"-31d33'
MDD-143! Kt
Aur 4h36'36"30d31'
MDG
Tau 5h52'40"27d19'
MDL RJ
Boo 15h7'30"20d21'
ME
Lac 22h6'16"51d60'
Me
Her 17h13'56"48d56'
Me & My R C
Hya 8h39'44"-0d11'
ME Fishman-Planet Schmanet Figaro
Dra 16h42'0"70d11'
Me Sue
Lyn 9h11'46"40d58'
Me, Lord?
Boo 13h36'47"14d16'
Me-Linh
Sge 19h14'37"16d36'
Me-Maw
Hya 8h54'20"3d9'
Mea
And 2h23'35"39d1'
Mea Aloha Aimee
Uma 8h10'10"72d5'
Mea,Al
Uma 9h2'11"48d25'
Meacham
Ori 6h12'15"4d25'A
Meacham (Organon), Jeffrey
Aql 19h24'48"11d12'
Meacham,Jo Ann
Aql 19h31'22"12d37'
Meacham-Moore Eternity
Sge 20h17'25"16d47'
Meachum,L W
Cam 3h14'19"57d52'
Mead III,Robert Winslow
Ori 6h6'41"-2d60'
Mead,Alan James
Aur 6h3'21"30d44'
Mead,Anne Marie
Peg 23h46'40"8d27'
Mead,Catherine S
Per 3h41'39"48d2'A
Mead,Christopher J
Per 3h59'23"50d31'
Mead,John G
Per 3h41'39"48d2'B
Mead,Jr,Earle E
Dra 17h35'23"63d49'
Mead,Judith T
Tri 2h6'51"32d20'
Mead,Kim
Lyn 8h23'54"48d3'
Mead,Lynn
Cas 0h10'34"58d17'
Mead,Michael
Lyr 19h25'49"38d51'
Mead,Sheldon
Her 16h53'58"41d5'
Meade Clan
Lyr 18h42'50"32d29'
Meade's Corvette Heaven,Donnie
Dra 12h23'38"64d28'
Meade,Barbara
Cam 6h11'41"68d6'
Meade,Bill
Dra 19h34'34"60d45'
Meade,Catherine Barry
Lyn 7h28'18"38d4'
Meade,Gerard
Uma 9h32'46"47d31'
Meade,Jr,Mr & Mrs Anthony Carl
Lyn 8h55'53"41d42'
Meade,Leslie
And 23h30'15"42d59'
Meade,Lorraine
Lyr 18h34'50"36d47'
Meade,Monica
Uma 11h13'1"60d50'
Meade,Roger
Her 16h35'0"21d6'
Meade,Susan
Lyn 8h50'0"35d22'
Meadelga
Ori 5h39'54"-2d57'
Meaders,Jewel Hogan
Mon 7h4'28"-1d40'
Meaders,Robert Hogan
Oph 17h29'1"-23d20'
Meador Men,The
Vul 19h15'38"21d40'
Meador,Diana Jean
Uma 11h18'35"55d45'
Meador,Donald
Crt 11h9'49"-13d53'
Meador,Gabriel Paul
Cas 1h22'15"50d28'
Meador,Norma
Equ 21h20'1"11d51'
Meador,Tony & Julie
Dra 15h1'17"55d30'
Meador,Walter
Boo 14h30'0"10d28'
Meadors,Crystal Rea
Cyg 19h59'1"40d8'
Meadors,Jessica Aubrey
Peg 23h25'21"15d43'
Meadow,Lynne
And 23h2'32"44d54'
Meadows (Snow White), Deborah Marie
Mon 6h35'33"-0d19'
Meadows,Alyssa Jeannine
And 0h22'12"33d48'
Meadows,Byran Ray
Aur 6h12'1"32d58'
Meadows,Cody Richard
Mon 7h17'24"-6d57'
Meadows,Joyce
Mon 6h38'47"11d46'
Meadows,Jr,Charles Samuel
Her 15h51'60"41d2'
Meadows,Kenneth Mark
Aql 19h28'26"-8d58'
Meadows,Misty Summer
Mon 6h21'15"2d59'
Meadows,Sara Marie
Cas 0h38'59"58d45'
Meadows,Sarah Alaina
Eri 3h36'55"-17d59'
Meadows,Tamara Lynn
Peg 23h30'34"11d22'
Meadows,Thomas J
Aql 19h36'27"0d42'
Meagan
Lyn 7h35'18"36d28'
Meagan Emily Ann
Mon 7h24'24"-5d39'
Meagan Kathleen
Aql 19h7'10"1d40'
Meager,Christopher Dennis
Her 17h7'27"20d29'
Meager,Martin John
Oph 17h14'20"10d4'
Meager,Norman A
Ori 6h10'19"8d54'
Meaghan
Cas 2h13'40"70d24'
Meaghan Elizabeth
Psc 1h18'21"20d14'
Meagher's Star,Joseph
Lyn 8h44'57"36d59'
Meagher,Andy & Julie
Aql 19h23'1"10d7'
Meagher,Anne Parker
Uma 14h1'1"58d24'
Meagher,Jamie Lee
Lac 22h54'28"38d36'
Meagher,Mary Elizabeth
Cnc 8h52'60"31d45'
Meaker,Craig
Boo 14h55'29"23d5'
Mealey,Judith M
And 0h15'35"33d16'
Mealey,Maureen Coleman
And 1h50'14"36d26'
Mealey,Michael P
Her 17h38'21"21d50'
Mealey,Sr,Charles
Dra 18h58'42"70d5'
Mean Streak Of Comiskey,The
Per 3h8'0"38d32'
Meana,Elias
And 0h9'50"28d1'
Meaney,Robert P
Ser 15h11'50"1d13'
Meanor,Captain Dave
Aql 19h5'40"15d39'
Means,Arnold E
Boo 15h8'0"21d6'
Means,Pryce Hilton
Boo 15h7'13"11d10'
Meara,Catherine Desleyn
Lep 15h28'47"44d39'
Meares,David G
Uma 8h49'45"52d38'
Meares,Jonathan Micheal
Uma 10h57'52"56d56'
Meares,Jr,Walt
Cma 7h19'58"-15d56'
Meares,Michael Bredon
Uma 10h57'1"56d40'
Mearl's Navigae
Uma 10h52'39"60d47'
Mears,Deborah
Cyg 21h4'12"40d5'
Mears,Dennis
Ori 5h57'0"16d5'
Mears,Dick
Uma 12h58'13"53d3'
Mears,Fawn Marie
Uma 12h54'25"53d35'
Mears,Greg Kahuna
Cam 3h14'52"61d57'
Mears,Irene Gyurik
Umi 16h11'15"80d26'
Mears,John Gordon
Umi 16h27'48"80d15'
Mears,Marie
Uma 12h50'1"53d40'
Mears,Sarah Helen
And 0h43'30"30d7'
Mease,Nan & Roland
Crb 16h19'33"32d22'
Measures,Verl T
Del 20h15'18"10d11'
Mecca of the Llano
Cmi 7h16'1"1d38'
Mecca,Melissa
Lyn 6h39'22"59d48'
Mecca,Michael
Dra 17h40'41"68d49'
Mecca,Patrick Joseph
Boo 15h11'14"30d15'
Meccher,Kenneth J
Cmi 7h56'1"0d53'
Mecey,John
Oph 17h57'27"13d1'
Mecham,Larry Jay
Aur 4h54'49"51d15'
Mechanic,Bill
Eri 3h27'58"-8d32'
Mechanical Marvel
Peg 23h30'13"18d54'
Mechel & Spygie Star, The
Cyg 21h23'15"37d39'
Mechtold,Michael
Cnv 12h16'48"38d26'
Meck,Bill
Equ 20h58'18"10d13'
Meck-Our Star,Chris-Al Angel-Jason
Lmi 10h10'28"32d12'
Mecke,Chandra
Cet 2h56'43"6d27'B
Mecke,Robin
Cet 2h56'43"6d27'A
Meckfessel,Kimberly Christine
Mon 6h23'27"6d47'
Meckstroth,PHD,Wilma K
Cas 1h21'31"53d46'
Meco The Great
Lyn 8h18'45"33d43'
Mecrissey,James R
Per 2h36'50"50d21'
Mectex
Col 6h32'57"-35d38'
Mecum,Aileen
Oph 16h56'54"-23d21'
Mecum,Keith
Hya 9h0m1"4d18'
Mecum,Linda Colleen
Aql 20h9'0"3d46'
MED-FED
Aur 7h4'39"38d21'
Medak,Vi
Cam 7h49'57"61d38'
Medalie,Randolph M
Her 18h7'49"40d20'
Medberry,Thirza
Lyr 18h57'16"41d35'
Medby,Michael Christopher
Cet 3h12'36"3d19'
Medders,Brendan Avery
Peg 23h46'1"18d3'
Medders,Morgan Lee
Peg 23h46'40"17d53'
Meddock,Shannen Doreen
Sco 17h4'1"-31d17'
Medea Cornflake Celeste
And 1h20'49"33d42'
Medegaard
Cas 2h18'22"60d9'
Medeiros' Star,Jeffrey
Oph 18h24'14"8d39'
Medeiros,Heath
Dra 16h53'29"62d1'
Medeiros,Julie
Mon 7h0'11"8d39'
Medeiros,Teressa Deneen Welolani
And 22h56'27"51d34'
Medeiros,Trinity Dawn Kalia'akapu'uwi
Lac 22h47'20"53d55'
Medeiros,Veronyca Muniz Veras
Col 5h43'29"-40d13'
Meder,Robert M
Aur 6h28'12"31d24'
Medetbekov Here Comes The Sun,Janys
Gem 8h3'58"30d19'
Medford
Umi 13h7'34"75d21'
Medgie's Magic
Ori 5h28'1"-1d6'
Mediaction
Oph 18h4'35"11d46'
Mediate's Star,Beverly Gaye
Aql 19h7'0"15d47'
Medical Tribune
Ori 5h40'13"7d39'
Medici,Anthony Michael
Aur 5h14'58"41d56'
Medici,Catherine Elaine
And 2h23'58"42d59'
Medici,P & B
Uma 10h7'41"59d39'
Medigovich,Catherine
And 2h19'60"38d48'
Medina 2 Jonathan
Aur 6h56'19"43d21'
Medina,Edda
Lyr 18h21'59"38d6'
Medina,Ezra
Peg 22h51'28"8d52'
Medina,G G
Ori 5h15'15"13d2'
Medina,Jr,Francisco
Cyg 19h21'23"28d45'
Medina,Juli Elvira
Lac 22h7'32"38d18'
Medina,Karen Tami
Cas 1h8'22"61d38'
Medina,Tony
Cet 2h52'47"1d44'
Medlen,Mrs Ida
Aql 19h1'52"4d49'
Medler,Victoria L
Eri 2h58'33"-11d27'
Medley of Musical Madness
Lyr 18h57'1"31d40'
Medlin,Günter
Tau 4h20'56"20d13'
Medlin,Günter
Tau 4h21'59"20d34'
Medlin,Leslie
Eri 4h9'58"-0d29'
Medlin,Lib
Peg 21h58'30"28d50'
Medlock,Marlene
Peg 23h32'23"32d23'
Mednick,Mitchell Stefan
Aur 4h55'19"40d35'
Medoff,David
Tri 1h47'0"30d55'
Medors,Peggy Lee
Cmi 7h52'51"1d52'
Medrano,Adriana Perez
Del 20h30'48"10d10'
Medrano,Anita
Eri 4h12'37"-11d9'
Medrano,Erica
Eri 3h58'57"-6d41'
Medrano,Laura Lynn
Aql 19h53'31"0d55'
Medrano-Stegall, Virginia Lupe
Ser 16h13'44"-1d31'B
Medsker,Orum
Dra 9h34'51"77d54'
Medved,Maxx
Gem 6h39'11"31d3'
Medvedeva,Ekaterina
Peg 22h34'39"12d4'
Medwed,Wendy
Lyr 18h29'39"37d48'
Medwin,Harvey
Aur 7h18'12"39d24'
Mee,Tammie "Momma"
Eri 4h57'40"-11d28'
Meece,Jonthan Karl
Aql 19h58'24"15d25'
Meece,Patricia
Lyr 18h48'54"35d37'
Meech,Alexis Wenning
Ori 5h55'41"17d14'
Meechan,Thomas & Eloise
Cyg 19h40'24"42d59'
Meeder,Bob
Aur 5h4'32"42d19'
Meedom,Viggo
Gem 5h58'21"27d12'
Meeek
Mon 7h3'52"0d60'
Meegan II,Joseph Thomas
Sco 17h27'19"-30d26'
Meegan,Patrick G
Her 16h26'48"39d40'
Meehan
Cnv 13h26'38"40d1'
Meehan,Gertrude
Eri 2h50'30"-2d29'
Meehan,Lee Allen
Ser 15h11'18"24d0'
Meehan,Mary T
And 0h11'17"46d29'
Meehan,Terence Michael
Cep 21h43'17"85d23'
Meehan,Thomas J
Tri 1h50'13"27d25'
Meek,Nadine & Robert
Cyg 21h5'29"31d32'
Meeke,David Roy
Aur 5h15'53"41d10'
Meeker
Ari 1h46'48"15d44'
Meeker,Camelot Love 14 Mary Anne
Cas 1h26'38"61d28'
Meekma,Jim J
Hya 9h53'39"-15d52'
Meekma,Shelly Joy
Cet 3h15'47"2d28'
Meeks,James A
Her 17h32'1"20d7'
Meeks,James Robert David
Cet 2h49'44"6d16'
Meeks,Jayne-Annette
Leo 10h59'49"7d60'
Meeks,Karee A
Del 20h50'59"9d6'
Meeks,Linda L/Alan C DeWolf
Cyg 20h9'17"40d34'
Meeks,Rev James T
Per 3h19'56"40d26'
Meeks,Thomas Jan
Cam 7h55'30"60d50'
Meeley,Walter D
Per 2h56'0"43d19'
Meem
Vul 19h48'11"25d11'
Meemaw & Peepaw
Cyg 19h59'13"40d29'
Meemaw's Star
Cas 0h41'45"66d11'
Meenan,Eithne
Lyr 18h47'12"36d25'
Meenan,William F
Uma 11h14'11"32d11'
Meep Meep3
Uma 9h52'19"56d36'
Meer,Steven Anthony
Cnv 12h55'34"31d49'
Meera
Uma 11h51'54"40d51'
Meernout,Jean Luc
Per 3h32'44"35d3'
Meertje,Kliene
Boo 15h5'54"50d23'
Mees,Penelope
Lyr 19h4'0"28d27'
Meese-A Star Forever, Darlene B
Peg 23h32'59"33d38'
Meesher & Tor
Cyg 21h48'47"38d54'
Meeson,Vicki
Lyr 18h36'19"27d24'
Meeting Place,The
Ari 1h55'14"17d11'
Meetre,Steven
Boo 15h8'49"28d39'
Meewes,BRIS 1992 aka Brian & Chris
Per 3h35'24"40d5'
Meffe,Bill
Boo 15h8'29"29d30'
Mefford,Micah Sinclair
Cnc 8h52'35"7d28'
Mefford,Tanner Anthony
Aql 18h55'45"10d56'
Meg
Cas 0h13'43"58d35'
Meg
Cas 0h31'48"58d24'
Meg
And 0h12'40"37d43'
Meg & Jason
Crb 16h9'52"26d56'
Meg Ann
Eri 3h19'25"-12d30'
Meg Ellen
Cas 1h51'59"70d38'
Meg L
Aur 7h4'48"38d37'
Meg's Dream
Lyr 18h49'41"40d6'
Meg's Star
Cas 0h18'11"58d29'
Meg's Very Own Wishing Star
And 23h35'34"49d27'
Meg-A-Roonie
Peg 22h54'38"8d0'
Mega "Watts" Star,The
Uma 12h2'0"37d41'
Mega-Bährchen Manfred Bähr
Cap 20h40'41"-20d10'
Megadis
And 2h5'57"38d41'
MegaGratis TTP
Ori 5h32'11"8d59'
Megalera
Cas 0h34'32"60d55'
Megalli,MD,Maguid R
Dra 17h52'46"61d46'
Megalu
Psa 22h36'1"-25d12'
Megan
Cas 23h24'0"60d46'
Megan
Cas 0h24'1"65d18'
Megan
Peg 0h5'27"21d55'
Megan
Peg 22h5'56"20d51'
Megan
Lyr 18h49'23"33d55'
Megan
Uma 10h55'59"55d6'
Megan 1013
And 2h14'52"40d34'
Megan 17
Crt 11h11'47"-17d5'
Megan A
And 0h2'41"40d33'
Megan Ann
And 0h50'52"35d19'
Megan Ashley-76
Psc 23h2'50"0d45'
Megan Elizabeth
Cas 3h1'45"60d24'
Megan Emily's Star
Sge 19h57'28"19d58'
Megan Jade
Cyg 21h9'29"37d37'
Megan Joy
Cyg 19h55'11"48d23'
Megan Leah Marie
Cam 4h51'33"68d6'
Megan Leigh
Aqr 23h7'10"-8d31'
Megan Leigh
Aur 5h16'34"40d60'
Megan Leigh
And 23h28'46"39d36'
Megan Lynn
And 0h16'55"37d21'
Megan Marie
Uma 9h22'14"51d48'
Megan May
Lyn 6h32'49"60d46'
Megan Sarah
Uma 14h12'57"72d2'
Megan's - Rags & Me
Cas 0h35'1"75d45'
Megan's Celestial Light
Uma 10h43'52"40d48'
Megan's Light
And 1h43'56"40d11'
Megan's Quasar MB13
Ari 1h55'39"11d51'
Megan's Shining Star
Vul 19h21'1"26d38'
Megan's Star
Peg 23h24'51"10d5'
Megan's Star
Hya 8h41'19"6d26'
Megan,Brittine,Maggie, Rafael
Mon 7h41'6"-5d39'
Megan-Cat-Cliff (Mc")
Lyn 6h39'17"60d15'
MeganP84
And 23h30'36"42d45'
Megans Star
Crb 16h10'39"34d2'
Megara,Barbara
Eri 3h22'16"-3d34'
Megdal,Myles Glenn
Per 2h58'17"32d46'
Megelsh,Samantha Jean
And 2h33'50"49d49'
Meggie
Hya 8h45'51"1d48'
Meghan Loiselle
And 0h58'1"39d41'
Meghan,John Meghan & Joseph
Com 13h5'44"20d3'
Meghana & Aditya
Ser 15h53'19"-2d1'AB
Megias,Gabrielle Alexis
Sex 10h10'33"-6d49'
Megna,Stephen
Aur 6h31'26"31d16'
Megnin,Benedicte
Equ 21h1'25"7d32'
Megoutoftheblue
And 0h57'35"37d34'
Megraw,Eric Charles
Cep 23h1'1"78d45'
Megreg
Ser 15h39'36"18d57'
Megstar,The
Psc 1h43'55"21d40'
Megumi
Cnv 13h30'35"35d9'A
Megumi
Lib 15h8'13"-1d16'
Mehaffey,Bev & Andy
Sge 19h41'57"16d45'
Mehdi
Aur 4h58'52"40d3'

Mehdizadeh,Delores Jane
Crb 15h35'53"26d20'
Mehegan,Andrea H
Lyr 18h54'41"31d20'
Mehegan,Patrick C
Tau 4h31'40"20d31'
Meher
Tri 1h47'40"25d35'
Mehkal
Per 2h52'56"38d30'
Mehl,Carter
Ser 15h14'45"22d18'
Mehl,David Ryan
Cet 2h53'25"0d1'
Mehl,Ghislain
Ser 18h28'29"1d45'
Mehlert,Anna
Cnc 9h13'54"8d44'
Mehlhaff,Edgar
Umi 13h39'42"74d27'
Mehling,M Randel
Umi 14h43'42"66d55'
Mehlman,Bridgette
Uma 12h0'31"31d32'
Mehlman,Rena & Harry
Peg 23h19'35"30d32'
Mehm,Claudia Jean & Joseph Francis
Lyn 8h58'20"34d15'
Mehner,Lieselotte
Gem 8h1'43"30d42'
Mehnert,Alma
Sgr 19h6'41"-21d32'
Mehra,Rahim Dominic
Cep 22h20'27"58d19'
Mehrak
Sct 18h19'34"-6d21'
Mehring,Lothar Dr
Sgr 19h0'45"-24d0'
Mehringer,Karl
Her 18h19'37"14d22'
Mehringer,Raymond E
Cnv 12h28'22"32d53'
Mehrwald,Andreas
Leo 11h33'27"-5d46'
Mehta,Aspy
Mon 7h2'57"-6d40'
Mehta,Madhu
Lyr 19h17'46"30d17'
Mehta,Madhu Kaur
Cmi 7h54'10"1d30'
Mei's Place
Aur 6h32'1"32d55'
Mei-Li,Lily
Cyg 21h49'0"53d53'
Meibaum,Brian Patrick
Aql 19h7'55"4d14'
Meibaum,Julie Anna
Lyn 7h35'26"40d17'
Meichenbaum,Michelle
Sge 19h5'14"18d16'
Meichler,Klaus Uwe
Boo 14h7'33"11d47'
Meidel,Bob
Hya 8h27'45"-8d48'
Meidinger,Ben Ros Susanne Scott Cindy
Mon 6h5'13"-8d11'
Meidl's Satellite Of Love
Lac 22h38'46"56d38'
Meidl,Christopher Donald
Sex 10h10'58"-7d47'
Meidl,Hollis Ann Dorsey
Cmi 7h9'40"9d19'
Meier (H B),Deborah
Lyr 18h57'40"33d34'
Meier,Elke
Aql 18h43'20"7d46'
Meier,Ewin
Cyg 20h28'1"31d42'
Meier,Heiko
Cam 7h7'33"60d34'
Meier,Lynn
Uma 8h21'48"72d8'
Meier,Malene
Aur 5h12'25"29d54'
Meier,Marion
Sge 20h15'55"17d42'
Meier,Susan
And 23h16'0"40d14'
Meiers,Theobald Josef
Aqr 23h33'33"-8d21'
Meijer,Lyn Rust
Uma 12h58'51"60d36'
Meik
Aqr 22h59'22"-20d37'
Meiki S I
Leo 10h6'22"7d16'
Meikle,Robert Burns
Cep 22h43'25"58d2'
Meilenstein 30,Martina
Crb 16h3'50"30d45'
Meimaroglou,Sophia
And 2h29'31"48d55'
Meimbresse,Elke
Equ 20h56'19"7d47'
Mein Stern vom 9 Mai `93
Tau 4h14'31"1d51'
mein treuer und ehrenhafter Freund
Aur 5h29'51"38d27'
Meinbresse,Kazu Amy
Aql 19h46'21"11d2'
Meine liebe Maus Nicole
Ser 17h35'12"-14d14'
Meinecke,Jason Daniel
Sex 9h44'39"-6d5'
Meiner grossten Liebe Frank "Haubi"
Lyn 8h3'35"49d36'
Meiners,Elmo
Aql 20h2'0"7d53'
Meiners,Flora Anna Wilhelmine
Cas 23h15'1"60d25'
Meinhard-1956,Richard & Mary
Cyg 21h10'42"39d44'
Meinhardt,Annalis
Lyn 7h4'11"53d58'
Meinhardt,D-Man; J-Girl; Coo-Coo & Bug
Uma 10h45'31"40d54'
Meinhardt,Jacqueline Diane
Cas 1h9'41"61d10'
Meinhold,Mary
And 23h28'24"45d55'
Meinhold,Sandra Lee Goodin
Del 20h50'0"7d37'
Meinhold,Todd William
Cep 21h20'13"70d56'
Meinke III,Joseph Frederick
Ser 17h54'51"-14d22'
Meinke,Gerald & Allene
Sct 18h42'32"-6d58'
Meinl,Lothar
Peg 23h34'43"16d21'
Meins,Ilse A H
Vir 11h51'37"2d27'
Meints,Les
Cmi 7h6'57"1d31'
Meinunger,Hermann
Gem 7h32'0"33d53'
Meinz,Brecht
Uma 14h24'51"60d19'
Meinzinger,Mathias
Ser 15h52'33"1d37'
Meir,Rachel
Cas 1h45'45"74d45'
Meiseinger,Vernon L
Cep 6h59'42"85d9'
Meisel,Dr Frank RJ
Per 4h7'60"38d18'
Meiser,Darren
Lac 22h54'52"51d11'
Meiser,Derrek
Ser 16h2'18"2d53'
Meishka 12
Cyg 20h17'21"38d47'
Meisinger,Eric Carl
Tri 2h35'36"34d46'
Meisl-1995,Mizes Getaway Kevin W
Cmi 7h56'13"8d29'
Meisler,Evelyn Mae
Hya 9h0'48"1d57'
Meisner,Christina Pike
Oer 3h32'52"38d57'
Meissner,Josef
Mon 6h33'0"4d45'
Meissner,Mark H
Umi 16h10'28"71d27'
Meissner,Thomas
Ser 15h30'58"18d15'
Meissner,Uwe
Peg 23h32'12"17d5'
Meister
Ser 15h13'47"4d23'
Meister - My Kind Of Star,Barbara
And 1h17'57"33d49'
Meister Familys Star, The
Cyg 20h40'12"38d17'
Meister,Carl Nicholas
Her 16h11'19"20d21'
Meister,Eunice
Tau 4h31'33"30d39'
Meister,Laurel G & Tripp
Ser 15h54'19"4d31'
Meister,Robert A
Aur 7h13'20"35d37'
Meiswinkel,Frank G
Gem 7h16'44"30d9'
Meitg,Stéphanie
Per 3h59'10"39d49'
Meixner,Erwin Johann
Lac 22h1'51"40d57'
Meixner,John Reed
Cet 0h45'38"0d49'
Meixner,Stephanie
Cap 20h36'0"-10d46'
Mejak,Larry
Cep 22h36'41"61d10'
Mejanes,Ishtar
Aur 6h57'23"37d58'
Mejean,Gilles
Aur 4h47'36"40d2'
Mejia Family,The
Sge 19h24'27"16d39'
Mejia,Ashley Christine
Aql 19h7'1"-6d57'
Mejia,Brittany Herminia
Tri 1h47'45"27d6'
Mejia,Megan Renee
Mon 7h56'37"-8d18'
Mejjati,Abdelhak
Lmi 10h45'56"25d26'
Mekeel,Linda
And 23h43'0"44d1'
Mekler,Joan
Crb 15h16'57"30d53'
Mekush,Margaret Anna
Mon 6h28'58"-1d34'
Mel
Ori 4h42'41"12d41'
Mel
Vul 20h17'17"23d5'
Mel
Her 17h38'0"14d60'
Mel & Judi
Cyg 21h50'42"53d48'
Mel & Kris
Cyg 19h18'55"28d33'
Mel & Paul's Star
Sge 20h0'54"20d11'
Mel's Mesmerizer, Melanie Frank
Lyn 8h40'58"39d7'
Mel's Star
Her 17h32'38"20d11'
Mel-N-Richie
Cyg 21h19'40"35d10'
Melaffo,Vivienne
Lyr 19h20'12"41d40'
Melancon Whoopytwirl Star,RiseRalph
Her 18h5'30"30d21'
Melancon,Andrée
Uma 12h7'1"46d54'
Melander,Gregory David
Uma 9h32'11"61d57'
Melandri,Vittorio
Cep 21h24'25"68d13'
Melanie
Cas 2h58'59"57d48'
Melanie
Mon 6h56'36"-6d55'
Melanie
Cas 0h21'0"58d18'
Melanie
Cas 23h22'0"54d6'
Melanie
Cam 3h25'30"52d50'
Melanie
Peg 0h0'29"28d31'
Melanie
Cyg 20h20'1"40d46'
Melanie
Oph 18h0'24"12d58'
Melanie
And 23h46'18"47d31'
Melanie
Mon 6h54'19"10d36'
Melanie & Joann's Laughing Star
Cam 5h55'1"58d57'
Melanie & Mark
Peg 23h30'0"17d38'
Melanie & Robert
Cyg 21h42'18"30d9'
Melanie Anne
And 23h35'15"49d57'
Melanie Jane
And 0h32'0"45d5'
Melanie Jane
Eri 3h49'27"-0d2'
Melanie Jane
Uma 9h18'0"42d15'
Melanie Katharine
Com 12h24'58"22d18'
Melanie Lee's Sweetness & Light
Mon 6h29'1"10d39'
Melanie's "Magic"
Cet 1h53'20"-4d50'
Melanie,With Love, Nouno Paul
Cyg 20h2'21"41d5'
Melanie-Ever My Love
Uma 10h1'31"51d20'
Melanie-Susanne
Her 17h27'12"42d34'
Melanis Crowning Glory,Kristi
Cas 1h47'22"60d48'
Melanson,David
Aql 19h53'60"12d46'
Melant,Christine
Cas 0h58'59"54d46'
Melanye,Bennardo
Lyr 18h50'41"41d42'
Melanáon,Julie Bissonnette
Lmi 10h51'20"32d0'
Melaragni,Lena M
Lyn 8h14'60"48d22'
Melaro/Kuzak Together Forever
Uma 8h56'38"54d42'
Melarry
Her 15h52'24"41d45'
Melay,Sandra-John
Cyg 19h30'31"35d44'
Melbob
Eri 2h54'15"-8d29'
Melbourne,Kathleen
Cam 5h7'0"71d1'
Melbourne,Lori H
Eri 4h33'31"-3d42'B
Melby,Marilyn Meldrum
And 23h36'15"48d42'
Melching,Jeffry Shane
Ori 5h56'17"14d18'
Melchiona,Joe
Sct 18h43'59"-6d31'
Melchior Mahot de la Quarantounais
Uma 13h0'25"60d54'
Melchior,Joseph W & M Patricia
Uma 11h55'40"48d57'
Melchiorre,Michael Raymond
Aql 20h1'25"10d57'
Melchisedec "I Love You"
Cyg 19h46'46"31d20'
Melcor
Lmi 10h43'21"23d18'
Meldrem,Samantha Lynn
And 23h18'60"45d42'
Meldrum,Alison
Lyr 19h22'27"40d7'
Mele,Steven Peter
Aur 5h18'57"42d54'
Mele:My Wife & Lover, Dr Celina
Cas 0h12'22"64d25'
Melea,Tanya
And 0h20'15"30d27'
Melemenis,Nikki Ann
And 23h31'50"41d50'
Melena Gayle
Aql 19h59'17"14d22'
Melendez,Alex Nicholas
Tri 2h28'1"28d31'
Melendez,Helen & Bill
Ori 5h54'11"14d58'
Melendez,Joseph Anthony
Dra 9h29'1"78d59'
Melendez,Wanda I
Com 12h11'0"27d55'
Meleni,Gianni
Pic 5h0'4"-47d10'
Melette
Cnc 8h32'42"8d47'
Melgar,Carmen Roxana
Mon 6h40'48"11d30'
Melgarejo Family Star, The
Cyg 20h37'20"30d27'
Melgies
Cyg 19h50'51"44d45'
Melhorn,Jonathan
Aur 4h45'1"30d34'
Melia,Donna
Uma 13h32'34"58d5'
Melia,James Anthony
Mon 8h4'44"-8d17'
Melidis,Anastasios (Sternchen)
Ari 2h1'42"20d14'
Melilli,Joe
Aql 18h58'47"3d19'
Melillo Housing Oversight,Andrew
Per 1h38'33"53d45'
Melillo,Dom Martin
Lyn 7h6'32"50d14'
Melillo,William
Uma 10h17'0"51d56'
Melina
Cyg 19h28'0"40d31'
Melinda
Aql 19h3'32"11d40'
Melinda
Cnv 12h47'36"39d30'
Melinda
Cas 0h30'44"67d60'
Melinda
Cyg 20h42'1"42d53'
Melinda Jane
Tau 4h11'14"22d52'
Melinda Kay
Del 21h4'53"12d15'
Melinda Lee
And 2h30'57"41d35'
Melinda Marlene
Eri 4h11'1"-12d6'
Melinda Sue
Peg 21h49'1"34d4'
Melinda's
Lib 15h19'50"-23d14'
Melinda's Wishing Star
Eri 2h50'30"-2d29'
Melinda-Rob `95
Eri 4h2'32"-17d35'
Melis,Barbara
Uma 11h56'13"37d44'
Melis,Jean-Pierre
Cyg 19h38'57"37d33'
Melisa
Aql 18h59'1"17d39'
Melisa
Sex 10h11'14"-2d55'
Melisande
Peg 23h28'17"22d16'
Melisande
Cep 22h49'39"65d9'
Melisi,Janice Joy
Peg 23h19'24"18d1'
Melisi,Zachary
Peg 23h35'19"30d4'
Meliski,Devon Michael
Cap 21h45'14"-23d4'
Melissa
Boo 15h3'0"22d5'
Melissa
Cam 6h6'41"60d39'
Melissa
Cet 3h12'19"2d55'
Melissa
Cas 0h36'25"69d24'
Melissa
Cam 9h14'0"81d16'
Melissa
And 2h17'18"46d16'
Melissa
Peg 22h0'1"28d49'
Melissa
Com 12h48'38"20d7'
Melissa
And 23h40'51"47d56'
Melissa
Lmi 10h2'36'30d17'
Melissa
And 23h47'18"41d18'
Melissa
And 0h58'52"39d51'
Melissa
Peg 22h25'21"20d21'
Melissa
Cet 2h58'0"7d23'
Melissa
And 1h57'12"40d57'
Melissa
Lyr 18h58'10"30d9'
Melissa
Mon 7h5'34"-6d47'
Melissa
Vul 19h42'27"22d57'
Melissa
Lyn 8h10'44"58d47'
Melissa
Cet 1h24'45"-5d59'
Melissa & Donald's "Lovelight"
Cyg 19h50'43"58d55'
Melissa & Gary
Peg 23h5'0"8d26'
Melissa & Kyle
Dra 13h42'1"64d31'
Melissa & Naum
Cam 3h17'42"61d17'
Melissa 17
Cas 1h23'56"75d2'
Melissa A Santa Maria
Vir 14h8'37"7d24'
Melissa Ann
Uma 10h58'36"38d5'
Melissa Anne
And 1h30'39"36d49'
Melissa Dawn
Eri 2h47'22"-15d36'
Melissa Gene
And 23h29'0"40d36'
Melissa I
And 22h58'40"50d48'
Melissa Jan
Crb 15h56'21"30d6'
Melissa Jayne
Lep 15h9'7"40d44'
Melissa Jeanne
Aql 18h43'43"10d31'
Melissa Joy
And 23h46'23"47d43'
Melissa Kay
And 0h57'57"39d32'
Melissa Laine
Cmi 7h42'0"0d39'
Melissa Lee
Lyn 9h14'0"42d55'
Melissa Lynn
Sge 20h14'18"17d40'
Melissa Marie 21
Leo 10h42'23"15d56'
Melissa R S
Lmi 10h35'1"23d39'
Melissa Renae
Mon 6h36'23"7d10'
Melissa Sue
Mon 6h24'38"11d37'
Melissa's Birthday Star
Cam 5h42'47"80d18'
Melissa's Dream
Cnv 12h51'10"43d6'
Melissa's Memory (Melissa Ann Coffman)
Uma 8h57'1"49d16'
Melissa's Star
Mon 6h25'46"10d5'
Melissa's Window
Mon 6h25'56"10d2'
Melissa's Wish
Equ 21h1'19"10d31'
Melissa's-High Note
Cyg 19h28'54"28d9'
Melissa, My Heaven On Earth
Lyn 9h21'1"44d59'
Melissa-James
Gem 6h41'22"13d57'
Melissa-My Shining Bright Star
And 1h24'15"35d26'
Melissah
Cas 2h43'38"61d31'
Melisse Bruso
Tri 2h14'29"32d59'
Melissen,Peter
Aur 5h4'1"43d27'
Melisummar-4729
And 23h38'46"48d14'
Melita
Vir 11h45'40"2d5'
Melita
Ori 5h53'52"15d9'
Melita,David Benjamin
Tri 1h58'0"30d47'
Melita,Jessica Marie
And 22h56'0"36d0'
Melizan,Daniel Luke
Dra 14h43'58"64d46'
Melkonian,Roe & Bill
Peg 23h35'0"18d44'
Mellar,Jayne Margaret
And 23h6'48"38d46'
Mellas 10-30-60,Maria & William
Lyr 19h17'36"42d51'
Mellas,Big Nick
Her 16h44'15"33d52'
Melldean
Leo 9h38'17"10d55'
Mellen,Carol
Cas 0h58'15"69d31'
Mellen,Jimmy J
Her 17h53'0"37d56'
Mellendick Star, Robert & Carolyn
Cyg 20h4'45"40d23'
Mellerstain
Cyg 21h7'35"31d21'
Melli Sonne
Ser 15h40'31"1d27'
Melling,Brian Stuart
Per 4h22'33"51d53'
Mellini,Patricia
Psc 0h56'49"20d44'
Mellino,Irene
Cas 1h38'59"58d43'
Mellix's Saxiphone Dream,Jerry
Oph 16h54'49"0d24'
Mello Gabriels Hunter, Debbie Granata
And 0h30'21"45d3'
Mello's Milestone
Oph 18h5'42"12d8'
Mello,Douglas J
Lac 22h26'14"37d52'
Mello,James
Oph 18h41'48"8d7'
Mello,Jr,Wayne F
Hya 8h24'51"6d20'
Mellon,Mary Kathleen
Equ 21h2'0"10d34'
mellonhollyandtheinfin itehappiness
Lyr 18h33'19"38d51'
Mellor III,Jesse Lynn
Mon 6h56'12"10d13'
Mellor,Baby
Lyr 18h19'20"40d60'
Mellor,My Beautiful Brandi Nicole
Oph 18h0'46"10d26'
Mellor,Nigel Steven
Per 1h55'26"47d29'
Mellor,Pomoro & Steven
Eri 3h46'26"-1d43'
Mellor,Steven Kevin
Cyg 21h18'50"39d51'
Mellors Star,The
Cep 21h19'14"58d39'
Mellors,Oscar
Cap 21h20'57"-24d4'
Mellors,Tim
Ori 5h37'16"10d57'
Mellott,Oriana
Sge 20h3'1"20d22'
Mellow-Flanagan
Gem 6h55'17"34d31'
Melmac
And 23h25'36"41d8'
Melman,Richard
Aur 6h25'53"38d38'
Melmed,Mrs Barbara
Lyn 7h34'40"36d58'
Melnick,Robert S
Uma 10h13'0"53d36'
Melnikoff,Julia & Art
Lyn 7h59'0"34d18'
Melnikoff,Sue
Cyg 19h50'56"38d50'
Melnyk,Sean Michael
Aur 6h8'0"35d38'
Melnyk,Stephen
Lib 15h42'0"-28d53'
Meloche,Brian Andrew
Aur 4h45'18"30d35'
Melodee's Dream
Mon 7h50'49"-2d54'B
Melodey,R Chellerawn
Lyn 7h55'56"40d27'
Melodia
Uma 11h10'11"56d16'
Melody
Lyr 18h28'44"38d24'
Melody
Uma 9h45'35"48d37'
Melody
Cyg 19h42'28"30d21'
Melody 1
Vul 21h23'53"24d13'
Melody Ann
Cas 0h11'0"64d19'
Melody Anne
Uma 9h56'52"51d0'
Melody Grace
Lmi 11h0'28"31d38'
Melody Lee
And 0h57'30"36d15'
Melody Mike
Lyr 18h13'13"40d8'
Melody-Leigh
Lyn 9h26'52"41d11'
Melon,Ruth
Cyg 21h1'20"30d3'
Melony Jane 469295422
Cyg 20h7'35"40d20'
Meloy,Hatsie
And 0h11'22"32d32'
Melrose #920 Oes, Alberta Mann
Mon 6h54'60"0d3'
MELS 3 AR
Her 16h24'0"41d35'
Melson,Ingo
Tau 5h32'41"20d14'
Melson,Miohael
Gem 6h42'40"13d23'
Melstar 2000
Aur 6h9'16"35d3'
Melstone
Boo 15h5'24"31d60'
Meltoby
Lyr 19h18'28"40d43'
Melton,Floyd W
Her 17h23'22"20d55'
Melton,Grady
Her 16h22'0"41d20'
Melton,Marion
Cas 1h34'17"73d11'
Melton,Stephen Roy
Oph 17h6'11"8d35'
Meltzer,Dr Milton
Boo 13h45'19"14d37'
Meltzer,Eleanor Kikumi
And 0h14'18"37d57'
Meltzer,Sandra Sue "Wa-Wa"
Vul 19h46'33"28d9'
Meltzer,Steven Joel
Cep 21h7'58"68d26'
Meltzer,Susan Massie Wood
Sct 18h36'50"-6d58'
Meltzer,Victoria Rose
Mon 6h58'24"1d51'
Meluvig,Raymond
Lyr 18h59'16"39d49'
Melveney's Romp
Boo 14h9'54"45d34'
Melvie
Lyr 19h3'0"28d27'
Melville,Westley
Peg 23h42'39"30d10'
Melvin
Boo 14h38'34"48d47'
Melvin,Joseph E
Dra 13h49'53"64d25'
Melvin,Sue
Gem 6h47'32"19d16'
Melvins,Diane & Maxwell
Com 13h32'0"20d34'
Melvoin,Sandra "Baby Girl"
Mon 7h0'0"-0d13'
Melzner 55555555555, 55555555 Andrea
Cep 23h18'0"68d4'
Membrino's Spark of Life,Gina
And 0h51'57"45d36'
Membrino,Rosemary Bernard
Cyg 20h8'25"39d31'
Membré,Jean-Julie-Bernadette
Her 16h24'57"48d39'
Meme
Vul 19h48'1"28d43'
Memel,Iris
Vul 19h18'46"27d19'

Memmer-Topmiller
Mon 7h3'18"-5d44'
Memmi,Dominique
And 0h18'18"40d38'
Memmie
Cas 1h18'48"61d3'
Memmo's Star,Peter & Cathy
Cyg 20h2'32"30d39'
Memole
Boo 15h10'56"32d18'
Memon,Dr Nazir A
Oph 17h8'24"10d49'
Memories of Crescent City
Hya 9h6'49"0d28'
Memories of Delia that Sparkle
Uma 11h44'43"37d38'
Memories of Denise
Cam 3h40'44"60d11'
Memories of Lou that Sparkle
Tri 2h46'44"31d47'
Memories Of Mark That Twinkle
Her 17h5'50"38d57'
Memories Stella of Takeshi
Psc 1h15'48"3d56'
Memory,Joanne
Cyg 20h34'58"40d4'
Memuz
Tri 2h23'27"35d54'
Menacher,Rick
Her 16h9'43"20d26'
Menagh,Douglas Paul
Aur 5h56'1"40d46'
Menagh,Nancy Louise
Lyn 8h53'29"41d45'
Menahka,Kito Esto
Boo 14h45'41"34d7'
Menalis,Anthony
Dra 16h4'24"51d54'
Menand,Ian
Cep 21h11'22"58d21'
Menard,Cheryl
Lyr 19h7'25"37d47'
Menard,David A
Her 16h44'34"27d44'
Menard,Elizabeth Ann
Ori 5h55'30"15d27'
Menard,Joanne M
Her 16h56'23"29d16'
Menard,Martine
Per 3h58'25"37d9'
Menard,Pierre
Mon 7h45'30"-3d56'
Menary,Cheryle
And 0h21'30"43d28'
Menasion,Kim
Lyn 6h19'52"54d15'
Mencaer
Ori 6h7'21"8d48'
Menchaca,Grace
Mon 6h55'52"8d50'
Menchaca,Rosie Canales
Mon 6h34'48"-0d58'
Mende,Astrid
Vir 12h58'13"-7d39'
Mende,Dr Reinhold
Lib 14h44'23"-22d52'
Mendel
Cyg 21h33'37"42d46'
Mendel,Herbert Donald
Lac 22h31'16"53d45'
Mendelis,Peter S
Hya 8h12'40"-1d11'
Mendell-Happy"65th" Birthday
Uma 8h49'33"52d21'
Mendelsohn,Brian
Aql 19h51'54"14d36'
Mendelsohn,Sheila & Joel
Cyg 21h1'28"31d43'
Mendelson,Beverly Joanne
Equ 21h10'54"10d24'
Mendelson,Harvey & Florence
Cyg 20h24'52"38d40'
Mendelson,Jordan Edward
Her 17h46'59"40d56'
Mendelson,Paul "Pablo"
Ser 17h21'41"-14d47'
Mendelson,Steven Paul
Ori 5h52'38"14d57'
Menden,Kayla Ann
Cas 23h15'39"61d34'
Mendenhall,Chris
Hya 9h4'40"-7d45'
Mendenhall,Elliott
Her 16h32'56"48d45'
Mendenhall,Jay & Ruth
Uma 8h18'48"68d7'
Mendenhall,Matt
Aql 19h2'0"3d59'
Mendenhall,Monica Ann
And 0h6'56"47d2'
Mendenhall,Robert H
Cmi 7h25'3"0d26'
Mendenhall,Sarah Kaitlin
Peg 22h6'49"4d48'
Mendenhall,Sharon Lorraine
Aqr 21h42'59"0d53'
Mendenhall-The Artist, Ragen Kay
Ari 1h59'24"13d5'
Mendeola,Joe
Cnv 12h20'28"42d13'
Mendeola,Martha
Uma 11h51'16"44d18'
Mendes,Ana
Umi 15h10'0"67d39'
Mendes,Antonio Montiero
Mon 7h34'42"-5d57'
Mendez,Cordelia Fuller
Gem 7h5'46"21d40'
Mendez,Joaquin
Lmi 9h50'40"37d32'
Mendez,Lisa M
And 1h57'49"39d55'
Mendez,MD,Paul Edward
Oph 17h33'4"-23d32'
Mendez,Philip G
Dra 18h59'0"58d34'
Mendez-Star Dancer, Ricardo M
Cet 2h33'51"8d32'
Mendez-Walden,Joelle Diane
Lyn 8h32'58"40d31'
Mendheim,Harris
Aur 5h56'40"41d9'
Mendi B
Peg 23h36'44"22d16'
Mendibles,Lucy B
Vul 19h18'24"23d25'
Mendicino,Felix Gerard
Mon 6h54'16"01d38'
Mendieta,Deana
Mon 6h54'0"0d48'
Mendieta,Hector Homer
Aql 18h57'36"-5d29'
Mendieta,Maritza Liliana Canon
And 0h59'12"34d6'
Mendieta,May
Mon 6h53'0"1d20'
Mendillo,Priscilla
And 0h14'0"45d2'
Mendiola
Oph 18h17'34"13d28'
Mendiola,Hose
Cma 6h44'17"-18d57'
Mendl,Daniel E
Uma 13h19'0"60d56'
Mendlowitz,Claire
Aur 7h6'21"37d5'
Mendolia,Patricia Elizabeth
Cyg 20h25'1"37d55'
Mendolia,Richard
Boo 14h12'1"38d12'
Mendoza
Mon 7h52'55"-6d18'
Mendoza,Emilio & Julie
Sge 18h57'0"18d31'
Mendoza,Jordan B Hamilton
Cma 7h18'15"-18d53'
Mendoza,Jose L
Mon 6h42'22"-10d0'
Mendoza,Joshua
Aur 6h8'32"50d25'
Mendoza,Jr,C Cliff
Sct 18h42'49"-5d44'
Mendoza,Kelly
Equ 21h16'1"2d26'
Menduni,Charlie
Aql 19h57'38"0d15'
Mene Star,The Tom
Ser 15h58'14"-0d29'
Meneely,Veronica T
Lyn 7h30'38"52d8'
Menefee,Robert Dakota
Cyg 20h42'12"42d55'
Menell,Aidan
Sct 18h53'38"-6d45'
Menemteau,Roland
Aur 4h53'18"40d2'
Menendez,Carolina Sofia
Eri 4h30'59"-11d13'
Menendez,Jill
Crt 11h42'32"-10d59'
Menendez,Luis
Cep 22h11'11"63d53'
Menendez,Melanie Noel
Cap 20h59'48"-16d38'
Menendez,Nancy
Hya 8h59'0"5d22'
Menetrey,Jean-Paul
Dra 16h25'28"62d45'
Menett,John Charles
Cyg 21h20'10"28d27'
Menezes,Joseph Charles
Dra 17h4'1"63d41'
Meng,Christine Margaret
Cam 8h20'48"78d9'
Meng,Donald Linzy
Per 1h55'50"56d47'
Menghetti,Peter A
Dra 17h52'32"58d36'
Menghini,Sandro
Her 16h46'34"48d18'
Menig,Brenda Sue
Del 20h27'57"11d8'
Menissier,Dominique
Sex 10h24'60"5d1'
Menk,Eternally,Michael & Niki
Dra 16h6'54"52d2'
Menke,Jessika Gabriele
Lyn 8h9'0"47d17'
Menke,Johannes
Cam 4h8'23"58d11'
Menke,Katherine Irina
Mon 6h36'54"8d48'
Menke,Patricia Herney
Oph 18h2'46"8d6'
Menke,Tina
Sco 16h54'50"-40d3'
Menken
Cam 12h42'24"77d18'
Menken,Andrea Leigh
Umi 15h29'1"68d29'
Menkes,Russell
Cet 0h21'47"-18d58'
Menne,Dr Thomas J
Her 18h11'25"47d42'
Mennette
Peg 21h57'43"34d2'
Mennie Family X-Mas Star,The Scott
Umi 14h55'47"65d39'
Mennillo,Anna
Cnv 13h39'31"32d14'
Menninger,Holly
Her 16h40'1"38d16'
Menniti,Sarah Ashley
Lyn 9h11'32"39d46'
Menold,Sandra
Mon 6h54'58"-1d32'
Menon,Akshay
Cmi 7h55'50"1d9'
Menouar,Najad
Equ 21h3'34"10d53'
Menozzi,William J
Her 18h5'1"30d6'
Menphis
Cas 2h10'36"60d1'
Mensch,Homer
Sgr 18h0'41"-28d25'
Mensch,Sunny
Lyn 7h54'18"51d29'
Mensching-Kayser,Donna Elizabeth
Vir 12h39'0"0d22'
Menser,Daniel
Per 2h20'19"54d43'
Mensing,William D "Bill"
Dra 16h56'50"64d31'
Mensler,Meghan Louise
Tau 4h56'49"16d8'
Menster,Tammy M
Her 18h18'1"13d3'
Mental Ben
Ori 5h13'1"15d46'
Menter,Marsha
Lmi 10h48'19"28d49'
Menton,Gregary Ivar
Aur 6h7'10"38d4'
Menture,Tricia
And 22h55'13"40d35'
Mentzer,Abigail Mabel
Uma 9h55'0"57d18'
Mentzer,Ronald Lee
Aur 6h34'0"38d21'
Mentzer-Zero Defect, R Jordan
Lmi 10h5'0"30d55'
Meny,Danielle Nicole
Peg 22h19'32"30d20'
Meny,Katherine Anne
And 23h21'47"40d19'
Menze,Henning
Aqr 22h59'41"-22d24'
Menzel,Gerald & Patricia
Cyg 21h46'12"38d44'
Menzel,Helga
Vir 12h55'13"-18d33'
Menzel,Horst
Aql 20h9'29"4d34'
Menzel,Judith Anne
Tri 2h0'0"31d8'
Menzel,Jutta Und Wolfgang
Lib 14h22'31"-23d21'
Menzel,Ronald C
Lyr 18h17'56"31d26'
Menzel,Taylor Diane
And 0h6'38"44d37'
Menzell,Deborah Lynne
Cyg 21h3'1"52d11'A
Menzies
Boo 15h12'24"53d53'
Menzies,Catherine
Boo 15h2'35"31d11'
Menzies,Lauren
And 1h11'52"39d18'
Menzler,Megan Elizabeth
Aqr 22h1'49"-10d27'
Meo,Sandra
Cas 1h8'15"60d30'
Meo,Vincent
Aur 6h12'55"31d19'
Meoak,Julia
Crb 15h49'13"38d38'
Meogrossi,Franco
Boo 13h45'0"25d2'
Meola Star
Lyr 18h46'1"33d31'
Meola,Cindy Ellen Trager
Peg 23h2'16"20d31"
Meon,Marlyse
Cyg 19h45'36"31d11'
MER
Aur 5h16'37"49d12'
Meralpes Vacances
For 2h14'49"-24d51'
Merati,Shohreh
Peg 22h0'28"26d24'
Meraviglia,Julia- Fiore Di
Lac 22h32'1"37d33'
Meravigliosa Cristina dei Miei Sogni
Cyg 21h2'27"36d34'
Merbaum,Ira F
Cas 0h18'11"59d56'
Merbt,Margot
Leo 9h23'12"18d12'
Mercadante Star, Elvira & Carmen
Uma 11h53'32"47d46'
Mercadante,Kenneth J
Aur 6h17'41"30d52'
Mercadante,Maria Lee
Cyg 21h50'1"53d41'
Mercadante,Mark L
Aur 5h25'19"40d26'
Mercadante,Sophia Anastasia
Cyg 20h29'36"41d5'
Mercado,Glenn
Dra 16h50'31"52d21'
Mercado,Jose
Cam 3h59'18"78d19'
Mercado,Lee Iveliss
Cnv 12h45'14"40d9'
Mercala,Mark R
Ori 6h1'57"0d33'
Mercaldo,John
Dra 12h39'27"71d48'
Merce,Sue Braithwaite
And 23h42'0"37d48'
Merced,Melissa
Cas 0h8'0"58d4'
Merced,Sonny
Lac 22h32'42"52d47'
Mercer's Waterbug Star 1996
Cas 1h7'11"60d38'
Mercer,Dorothy J
Uma 15h40'55"76d13'
Mercer,Eleanor
Cas 0h45'35"62d5'
Mercer,Gina
Vul 20h4'39"28d38'
Mercer,Harris McNeil
Cnv 13h27'47"50d15'
Mercer,Jasmine Elaine
Mon 7h50'8"-1d20'
Mercer,Jerry
Boo 14h16'21"36d42'
Mercer,Jr,Thomas Lee
Cma 7h19'14"-15d39'
Mercer,Keelan Ashlee
Eri 3h52'56"-5d57'
Mercer,Kevin
Dra 19h25'12"56d41'
Mercer,Linda
Cas 0h56'45"50d26'
Mercer,Rosemary
Mon 7h6'54" 5d55'
Mercer,William Franklin
Cet 2h10'53"5d33'
Merchant"Wheel",Jim
Dra 16h49'1"73d41'
Merchant,Betsy "BeBe"
Vul 20h39'40"25d16'
Merchant,Jaime Danette
Peg 23h0'53"26d5'
Merchant,Kerry Leigh
Cas 3h10'16"61d39'
Merchant,Kristian
Lac 22h55'0"55d57'
Merchant,Moni
Sct 18h55'11"-7d4'
Merchant,Patricia
And 2h28'16"45d9'
Merchant,Richard K
Boo 14h43'49"21d26'
Merchant,Shilpa
Ori 5h59'39"6d48'
Mercier
Uma 9h52'0"47d41'
Mercier,Catherine
Umi 15h46'48"82d40'
Mercier,Ella
Mon 6h19'41"7d58'
Mercier,Mariane
Cnc 8h42'1"17d26'
Mercier,Mélissa
Umi 13h40'46"76d35'
Mercier,Pierre
Per 2h18'17"58d53'
Merciful
Peg 22h1'50"7d47'
Mercin
Her 17h31'29"47d55'A
Merck,Kenneth Michael
Aur 7h2'19"38d58'
Mercorelli,Francesca
Uma 10h28'22"47d36'
Mercuri,Gina
And 0h58'40"38d8'
Mercuri,Luciano
Cam 3h18'23"60d36'
Mercuri,Vincenzo
Aur 5h0'23"48d54'
Mercuria
Hor 2h53'36"-49d51'
Mercurio,Annie
Cnv 13h18'12"50d34'
Mercurio,Ersilia Anna
Umi 13h15'33"72d25'
Mercurio,Michelle
And 1h29'55"50d3'
Mercurio,Norman Charles
Cep 23h13'50"64d32'
Mercurio,Randy
Dra 16h8'25"68d25'
Mercurio,Rebo Micciche
Aur 4h51'49"40d25'
Mercurio,Rosemarie
Cyg 20h35'15"45d23'
Mercurio,Sam
Cap 21h1'1"-26d3'
Mercurio,Tiago
Eri 4h29'13"-1d30'
Mercury,Freddie
Per 3h4'37"50d10'
Mercy's Star
And 0h10'25"37d38'
Merdanian IV,Roy Eugene
Cep 21h11'30"63d50'
Meredith
And 0h39'57"40d34'
Meredith
Cyg 20h35'39"30d39'
Meredith
Lyn 9h10'14"44d35'
Meredith "Daddy's Girl"
Peg 0h2'0"30d60'
Meredith,Alexis K
Del 20h21'53"20d17'
Meredith,Bryon
Per 2h36'40"38d29'
Meredith,Charles & Kathryn Kleiner
Dra 18h13'31"70d14'
Meredith,Jonathan
Oph 17h15'23"10d15'
Meredith,June Mascho
Hya 9h15'41"4d15'
Meredith,Kim Nyegaard
Aql 19h41'20"10d5'
Meredith,Michael
Ori 4h44'48"0d13'
Meredith,Patricia A
And 2h22'22"45d6'
Meredith-Mer
Cas 1h32'57"72d60'
Meredyth
Aql 18h45'12"11d9'
Merel,Leonard
Per 2h52'41"40d51'
Merell,Theresa Lowe
Cnc 8h13'50"31d55'
Mereness,Kelly Lynn
And 22h58'59"50d33'
Meresman,Stephen
Aql 19h58'43"1d25'
Meretedanny
Lyn 7h42'54"44d56'
Merew,Michelle (Sweets)
And 2h11'1"38d39'
Mergenthal,Kerstin & Raphael
And 23h49'32"37d51'
Merhar,Stephanie
Her 16h37'11"38d33'
Meri
Uma 11h56'11"45d4'
Merian,John
Aur 5h26'1"30d58'
Mericle,Tim
Aur 6h3'52"34d52'
Merideth,Elinor
And 2h28'10"44d10'
Meridier,Wilfred F
Lyr 19h7'58"38d2'
Meridith
Aql 19h31'49"10d44'
Merigold,Frank & Margaret
Cyg 19h41'48"31d0'
Merika
And 23h20'56"51d20'
Merino,Ainhoa Izquierdo
Lac 22h3'28"48d10'
Merino,James Anthony
Leo 9h53'48"27d38'
Merino,Stephen Miguel
Boo 14h49'1"22d48'
Merisi,Michelangelo- Davide
Aql 19h5'11"0d30'
Merita
Lyr 18h48'14"35d39'
Merita,Claudia
Lyr 18h54'44"31d46'
Meritai,Mark Edward
Peg 23h37'20"18d57'
Meriwether,Doris
Eri 3h58'41"-10d3'
Merja
Lyr 18h59'40"30d1'
Merk,Edna & Howard
Cas 0h25'48"67d26'
Merk,Nancy Geraldine
Vul 20h2'25"22d36'
Merkee
Eri 4h7'0"-10d10'
Merkel,Dolores R & Christopher R
Cyg 20h59'40"31d18'
Merkel,Herr
Cmi 7h18'26"3d46'
Merkel,James Robert
Aur 7h3'52"39d37'
Merker,Marianne
Psc 23h0'44"1d45'
Merkert,Jessica Valetta
Tau 4h17'0"0d18'
Merkhofer,Jerry
Boo 14h35'56"12d9'
Merkie
Aqr 22h30'21"-2d19'
Merkl,Luke Joseph
Her 16h56'60"51d20'
Merkle,Elisabeth
Cap 21h27'24"-23d6'
Merklin,Maria & David
Uma 9h58'21"59d27'
Merkstern
Cet 3h11'40"2d9'
Merkwan,Sandra L
And 1h43'54"39d11'
Merle
Cas 1h57'19"75d28'
Merle
Tri 1h59'59"30d25'
Merle
Com 13h17'47"28d50'
Merle's Everlasting Point Of Light
Uma 9h19'59"42d29'
Merle,Alice
And 0h26'46"44d45'
Merle,Christopher
Hya 8h43'1"-1d28'
Merle,Claudia
Lyr 19h20'51"30d26'
Merle,Sylvie
Cyg 19h47'43"30d13'
Merli's Star
Dra 16h0'10"63d36'
Merlin
Aql 20h18'11"1d21'
Merlin 1974
Boo 15h6'56"14d47'
Merlin I
Oph 17h5'34"-24d20'
Merlin,Austin Kelly
Mon 6h53'27"1d52'
Merlin,Elisabetta
Her 16h51'43"47d44'
Merlin,Rose of Sharon
Lyr 18h57'51"31d19'
Merlin-The Cat Who Would Be King
Lyn 6h15'31"59d6'
Merlinge,Isabelle
Uma 11h28'29"36d1'
Merlino,John
Eri 2h44'27"-6d7'
Merlino,Richard Edmond
Aql 19h54'0"14d47'
Merlino,Sandy
Lyn 8h24'39"43d10'
Merlisa Kaye 03-31-60
Aql 19h53'38"15d32'
Merlo,Sandra
Ant 10h44'41"-35d49'
Merlot
Cam 3h29'13"60d9'
Merlucci,Susan
Lmi 10h27'36"34d28'
Merly Marina
Mon 6h22'33"8d34'
Merlyn's Star
And 23h22'60"42d9'
Merman,Richard J
Per 2h54'36"43d44'
Mermelstein,Karen & Howard
Cyg 20h49'58"37d38'
Mern,Alexander Jordan
Psc 0h5'54"8d50'
Merna
Mon 6h43'22"10d26'
Mernie
Cas 23h31'57"61d11'
Merok,Robert
Aur 4h52'0"50d56'
Merola,Christopher Erroll
Per 3h25'1"53d13'
Merola,Claire
And 1h7'20"38d5'
Merola,Matthew
Cep 21h54'56"55d53'
Merola,Robert
Boo 14h11'1"39d32'
Meron,Michael
Her 18h43'18"12d41'
Meron,Neil
Aur 6h17'1"37d33'
Meron,Yael
And 23h3'36"50d43'
Meros Steam
Vul 19h48'49"23d9'
Merouc
Cep 21h25'29"67d37'
Merrell,Christopher II
Her 16h14'41"26d45'
Merrell,Jonathon D
Cet 3h19'15"-0d4'
Merrell,Richard S
Aql 20h20'28"5d32'
Merri Lee's Sparkle
Peg 21h58'0"33d23'
Merriam Star,The
Dra 17h48'23"61d40'
Merribeth
Mon 5h57'16"-4d42'
Merrick,Diane
Cma 6h50'18"-15d16'
Merrick,Graham Alan
Cnv 13h24'13"48d56'
Merrick,Joyce E
Cyg 21h7'58"40d2'
Merridee
Uma 11h46'35"47d3'
Merrideth,Holly
Cas 2h5'14"61d52'
Merrie's Heavenly Delight
Cas 22h56'51"54d26'
Merrifield,Gregory John
Her 16h55'42"25d26'
Merrifield,Taylor Connie
Cet 0h40'25"-3d30'
Merrigan,Anthony David
Cep 22h0'40"62d20'
Merrigan,Maribeth
Cyg 19h52'51"38d23'
Merrigan,William & Lillian
Lyn 7h27'23"40d38'
Merrilee
Cas 0h1'0"64d27'
Merrill
Cyg 21h3'25"38d7'
Merrill Lynch Pittsburg
Tau 5h2'0"16d35'
Merrill's Meteor,Mark
Uma 11h27'27"64d58'
Merrill,Caroline June
Mon 7h3'52"0d11'
Merrill,Dan M
Her 16h48'12"39d26'
Merrill,Dorothy May
Cam 3h53'57"58d58'
Merrill,F James "Sonnyboy"
Aur 5h36'45"37d45'
Merrill,Helen
Mon 7h7'17"0d22'
Merrill,James
Aqr 21h30'0"-6d37'
Merrill,Leslie Olmstead
Mon 6h5'25"-8d18'
Merrill,Maxine
Cas 0h43'49"72d44'
Merrill,Sean L
Dra 12h21'1"70d5'
Merrill,Stuart
Mon 7h1'19"3d56'
Merrilyn Sweet
Aql 19h5'40"-6d33'
Merriman
Cyg 19h30'44"36d0'
Merriman,Brett Panda Tai Pan
Ant 9h37'41"-35d51'
Merriman,Casey
Ori 5h16'18"0d57'
Merriman,Lynda C
Cas 0h36'18"62d23'
Merriman,Sarah Alexandra
Uma 9h15'27"62d29'
Merriman,Shannon Elise
Crb 16h10'37"31d28'
Merring,Rachel
And 23h20'53"44d53'
Merrit,Frederick John & Doreen Rose
Peg 22h12'0"4d60'
Merritt,Austin Komor
Ori 5h56'18"11d7'
Merritt,Constance Williams
Cyg 20h23'47"40d37'
Merritt,June Lee
Peg 0h9'16"13d37'
Merritt,Mark
Lyn 8h19'23"49d13'
Merritt,Paul Dean
Cet 2h27'0"6d31'
Merritt,Ryan Michael
Cep 21h53'57"55d39'
Merritt,Sally
Hya 8h47'54"-4d0'A
Merritt,Stephanie
Cas 0h55'30"62d6'

Merritt,Steve Sct 18h44'28"-6d30'
Merritt,T J Dra 18h14'35"70d52'
Merritt,Tammy Lyr 19h13'54"38d18'
Merritt,Tammy Del 20h53'37"2d46'
Merritt,Tiecha D Tri 1h45'12"26d24'
Merritt,Tyrone Khalife Per 3h9'40"41d51'
Merritt,Valerie & Bruce Crt 11h36'46"-21d56'
Merrov Lup 15h15'0"-44d39'
Merry Carla Peg 22h30'1"24d20'
Merry Christmas for Hiroaki from Yukiko Vir 14h2'16"2d8'
Merry Dawn Cas 0h31'17"66d3'
Merry Legs And 0h0'59"47d37'
Merry"Happy Star", Aubrianna Rose Lib 15h32'39"-8d9'
Merry,Mary Ellen Cam 9h11'14"73d27'
Merryl And 23h35'54"49d1'
Merrylees,Ian Umi 14h26'0"68d47'
Merryman In Memory Of, Scott Aur 7h2'0"40d28'
Merryman,Dustin Per 3h11'0"40d56'
Merryman,Mica Dra 16h11'11"68d2'
Mersch,Jens "Watchtawe" Vir 12h5'43"-5d41'
Mersch,Joseph-Michelle & Cody Lyn 9h5'53"42d22'
Merschat,Arlene Lmi 9h21'43"38d52'
Merschat,Arthur H Boo 14h14'0"52d44'
Merschat,Carl E Vul 19h40'31"23d18'
Merschat,Kurt Arthur Oph 17h17'20"-21d14'
Mershon,William Her 17h6'41"38d13'
Mersil Her 16h27'17"36d34'
Mersinger,Ross & Tish Aur 5h1'50"29d44'
Mersiowsky-Otte Peg 22h14'59"8d56'
Mersky,Sue M Cyg 21h33'42"41d52'
Mersman,Douglas Patrick Dra 12h59'59"68d22'
Mersman,Judith Lynn Cma 7h1'13"-18d53'
Mersmann,James F Oph 18h3'24"13d1'
Mersmann,Ulrike Vir 11h37'33"-3d38'
Mersova,Inna Lyn 7h39'54"51d50'
Mert Gem 7h5'27"28d44'
Mert's Nova Cet 1h12'60"-0d15'
Merten,Friedlinde Cmi 7h18'49"5d56'
Merten,H Peter Lac 22h41'48"38d59'
Merten,Herbert P Aur 6h26'48"31d5'
Merten,Robin Dopke Lyr 19h13'26"42d34'
Mertens,Christopher Wayne Her 16h57'0"18d55'
Mertens,Hal Oph 16h49'60"10d14'
Mertens,Susanne Her 17h9'28"42d49'
Mertes,Ewald,Ingrid, Birgit & Monika Mon 7h43'34"-2d29'
Mertes,Paul Mathew Aql 19h2'29"5d34'
Mertins,Devon James Boo 14h25'39"28d4'
Merton,Dorothy Uma 11h49'52"56d55'
Mertz,Carolyn R Lyn 7h39'41"50d56'
Mertz,Claudia Lmi 10h52'59"30d42'
Mertz,Marijane Ori 5h23'32"-6d31'
Mertz,Nicole Lynn Cet 2h52'21"1d16'
Merus Amare Pax Ori 6h7'34"8d1'
Merveilleuse Colette Peg 22h56'58"24d31'
Mervis,Alice Cas 2h25'40"71d9'
Mervish I Sex 9h39'1"0d30'
Mervyns Per 2h7'48"57d29'
Merwil Aql 19h59'38"14d55'
Merwyn Uma 11h53'0"33d22'
Meryl Lee Vul 19h5'58"25d2'
Meryle-Eric 9-13-92 Boo 15h16'58"53d39'
Meryn,Dr Siegfried Cep 21h20'59"56d12'
Merz,Alessia Col 6h31'39"-38d3'
Merz,Michael Joseph Aql 19h55'52"15d36'
Merzedes Dra 15h9'21"62d16'
Merzenlich,Michaela Cnc 7h58'10"20d0'
Merzoug,Hadesh Uma 10h54'28"37d59'
Meschy,Franáois Ori 6h5'53"20d59'
Meserve,Rocky Edward Dra 15h21'0"58d55'
Mesh ki no Ma Cyg 21h57'34"52d34'
Mesh,Jessica And 23h3'14"44d48'
Meshka 3 Lyn 8h16'40"39d50'
Mesick,Holly Susan And 0h44'30"40d8'
Mesick,Jacob Peter Cam 6h54'46"80d10'
Mesin,Dimitry Tri 2h15'47"32d36'
Meskimen,Nicholas Lee Her 16h11'56"50d50'
Meskin,Kira Ann And 0h20'17"36d40'
Mesnil,Christelle And 2h4'45"40d15'
Mesoris,Matthew Joseph Cam 4h40'12"67d35'
Mess,Paul Uma 9h33'20"46d9'
Mess,Walter Lansdale Cet 2h18'55"8d51'
Messengale,Alan Lee Lyn 7h42'1"50d54'
Messenger,George Eric Uma 11h45'13"38d51'
Messens,Ingrid et Gary Oph 18h0'51"11d39'
Messer,Carol J Eri 3h23'13"-5d44'
Messer,Edward M Ori 5h56'17"11d57'
Messer,Marcy And 0h14'53"45d37'
Messer,Robert James Hya 8h14'0"1d34'
Messer,Ronald James Sco 16h50'25"-37d50'
Messerli Anthony Per 4h5'1"37d11'
Messerschmidt, Alexandra Ley Sco 16h56'1"-40d47'
Messerschmidt, Katherine Elisabeth Cas 0h2'40"63d35'
Messersmith,Amy J And 0h21'0"31d1'
Messex,Danny Douglas Boo 15h2'54"30d42'
Messick,Don J Oph 16h30'29"-6d26'
Messick,Gina Lynn Aql 19h47'29"13d17'
Messick,Judith Cyg 21h21'33"40d1'
Messier Her 17h29'22"29d58'
Messier,Flora Tri 2h33'0"31d44'
Messier,Virginia R Lyr 19h23'55"40d24'
Messig,David Walter Cam 4h6'51"70d16'
Messina III,William Allen Cnc 8h34'39"31d18'
Messina's Lucky Star, Steven Alexander Ori 5h57'1"21d35'
Messina,25 years, Charles & Shirley Lyr 19h18'31"42d40'
Messina,Christopher Paul Ser 16h6'17"14d4'
Messina,Gerry Oph 16h40'42"2d35'
Messina,Joseph Per 1h36'48"52d48'
Messina,Kathrine Aql 19h47'30"14d55'
Messina,Mary & Frank Carelli Cyg 21h18'27"28d18'
Messina,Miranda Maria Cyg 20h17'0"38d34'
Messina,Paul A Aql 19h3'26"2d21'
Messina,Vito Aur 6h19'47"32d33'
Messina,Vittorio Alexander Aur 5h1'1"41d10'
Messing-Penella Rebecca Tri 1h46'36"28d50'
Messinger,Justin Lyn 9h9'43"44d24'
Messinger,Nathalie Ori 6h5'48"2d4'
Messinger,Sandy Lyn 9h8'44"44d18'
Messinger,Scott Lyn 9h22'58"41d36'
Messinger,Scott Ryan Lyn 9h8'44"45d22'
Messmer,Heather Marie Cas 22h56'1"55d37'
Messner,Dr Heinz Gem 7h32'48"33d48'
Messner,Dr Heinz Gem 7h33'49"20d24'
Messnick,Friend Dorothy And 1h33'10"36d29'
Messy Dessie Peg 23h3'35"18d47'
Mester,Uschi Cnc 8h1'34"8d53'
Mesterton Aql 19h56'0"12d49'
Met,Patrick Cas 2h36'1"50d10'
Metacarpa,Jeffrey Trent Oph 17h33'60"-24d20'
Metal,Z Cyg 21h3'45"48d49'
Metaxas,Harry Oph 17h16'46"-16d13'
Metaxia Cnv 12h48'29"39d16'
Metcalf Elementary Cowboy,The Cet 2h5'38"0d31'
Metcalf,Bridget Mon 6h23'16"1d52'
Metcalf,Charles Lyr 18h59'22"47d3'
Metcalf,David Uma 13h4'50"53d49'
Metcalf,Donald I Oph 17h3'21"-18d47'
Metcalf,Elizabeth Cody Peg 23h40'33"8d24'
Metcalf,Elvis Anne Cam 8h48'25"73d58'
Metcalf,Kristy Cas 0h30'36"67d4'
Metcalf,Lena Ongel' And 23h1'32"44d52'
Metcalf,Linda Eri 4h52'57"-5d43'
Metcalf,Robert E & Elizabeth M Cnv 12h44'11"33d8'
Metcalf,Stephanie L And 1h54'21"39d1'
Metcalf-Harrison,Linda Lyr 18h24'18"42d20'
Metcalfe,"EV-RO-AR" for Evan Robert Arthur Cam 6h15'29"65d14'
Metcalfe,Angelika Lyn 7h45'20"50d41'
Metcalfe,Richard Keith Gem 6h51'46"12d12'
Metch Skater Scooter Tri 2h22'23"32d11'
Mete,Anne L Cas 0h36'20"68d39'
Meteoric Mervyn Ori 6h5'23"3d16'
Meter,Peter Cmi 7h6'29"4d29'
Meth Family Sct 18h42'53"-6d59'
Meth,Richard Her 18h5'20"18d46'
Metha,Meena Cyg 20h0'37"38d31'
Mether,Ted Dra 18h36'15"58d11'
Methner,Mark M Dra 19h4'16"48d50'
Methven Love Kaila,MD '94,Lisa Aql 18h54'26"8d59'
Metier Inc,M & K Cep 20h44'44"61d10'
Metlicka,Scott D Boo 14h59'48"27d57'
Metois,Mickael Crb 16h13'37"38d3'
Metropolis Lyn 8h46'21"44d45'
Metry,Dean Charles Her 16h13'13"41d25'
Metsker,Gary Wayne Dra 20h21'11"62d28'
Mette Ori 5h58'39"20d11'
Mette-Finn Tri 1h48'17"25d56'
Mettembrink Family Star Aur 4h57'15"36d57'
Metten,Dirk Aur 5h14'33"43d14'
Mettendorf,Brigitte Dra 15h13'0"63d4'
Mettler,Greg R Vul 19h2'1"21d25'
Metts IV,Dr Vergil L Leo 10h18'41"12d28'
Metz's Odyssey Cnv 12h13'10"43d49'
Metz,Anja Cam 7h57'56"73d18'
Metz,Edith Ori 6h0'48"8d50'
Metz,Kathy Kimmel Cam 6h14'20"58d32'
Metzdorf,Mario Cyg 20h35'55"48d59'
Metzen,Henni Die Liebe Psc 23h8'30"0d39'
Metzger (aka "Canyon") Robert Lake Boo 14h26'27"12d59'
Metzger,Brynn Kelly And 0h58'33"36d27'
Metzger,Caryn Sex 10h26'31"-2d1'
Metzger,Edith & Vincent Irving Cyg 21h2'39"50d35'
Metzger,Elaine Uma 11h57'22"32d57'
Metzger,Fredrick C "Fritz" Tau 4h10'40"20d25'
Metzger,Germaine Leo 10h36'29"18d17'
Metzger,Haleigh A And 23h1'29"48d3'
Metzger,John Aql 20h0'49"-6d54'
Metzger,Karlee Nikole Peg 23h21'1"25d7'
Metzger,Matthew Ryan Cep 21h8'34"67d35'
Metzger,Michelle Lee Com 12h19'13"20d26'
Metzger,Sondra & Allan Lyr 18h37'56"39d43'
Metzger,Suzanne And 0h21'56"37d58'
Metzger,Thomas Dean Ser 18h15'32"-13d23'
Metzler,Jutta "Wuschel" Her 17h4'26"50d24'
Metzler,V Louise Morgan Tri 2h22'58"35d49'
Metzler,William & Karen Leo 9h59'12"19d26'
Metzner,Kenneth Lee Dra 18h40'51"68d19'
Metzner,Sabine & Marc Uma 11h46'50"52d21'
Metzner,Warren H Tau 5h57'28"23d50'
Meuer,Horst Hya 8h11'41"1d1'
Meulemans,Jamie Cep 23h8'58"70d34'
Meuley,Eric Cam 7h37'59"80d27'
Meunier Sx 34 Dra 11h36'0"68d4'
Meunier,Gerald E "Jerry" Cet 2h18'1"1d49'
Meunier,Jean Franáois Boo 14h59'40"11d34'
Meunier,Lucie Uma 8h43'34"68d46'
Meunier,M J Lucie And 23h20'20"51d27'
Meunier,Tami Lee Cep 21h58'14"55d43'
Meurer,Doris Tau 5h50'1"23d40'
Meurer,Tilda Lyr 19h14'60"42d35'
Meuric Laëtitia Cet 0h56'46"1d57'
Meusert,Georg Aql 19h59'12"10d51'
Meuti,Enrico Peg 23h0'16"21d28'
MeviBen,Andreas und Helene Gem 7h21'28"20d57'
Mevorach,Samuel S Ari 2h1'14"25d14'
Mew,Stacey Denise Sge 19h59'41"16d5'
Mewborne,George Peg 22h33'55"7d40'
Mewller,Susan M Mon 6h21'43"-6d0'
Mey,Michel Mon 7h51'41"-5d45'
Meya,Klaus-Adolf Gem 6h24'51"12d43'
Meydman,Justin Ian Her 17h14'58"47d52'
Meyer - Greenwood, Henry John Cap 20h31'58"-13d30'
Meyer Star,The Nikki Cnc 8h55'51"31d27'
Meyer The Star Of Our Family,Jackie Cyg 21h34'29"41d13'
Meyer"The Boss",Bruce Ori 5h58'33"5d26'
Meyer's Star,Patty Peg 22h51'45"27d34'
Meyer's Star,Sue Eri 3h20'47"-3d10'
Meyer(Little Bo),Bob Lac 22h28'31"50d37'
Meyer,Adam Per 2h37'25"37d28'
Meyer,Amanda Elizabeth Lyn 7h50'33"48d35'
Meyer,Barbara Ann Cyg 21h4'57"40d9'
Meyer,Benjamin Joseph Aql 19h13'40"14d42'
Meyer,Brigitte Lmi 10h44'31"25d50'
Meyer,Carson Oph 17h4'32"8d59'
Meyer,Catherine W Hya 8h58'18"5d42'
Meyer,Cecelia P Cyg 21h32'29"41d53'
Meyer,Celia Cyg 21h24'23"37d51'
Meyer,Christi And 23h34'36"48d47'
Meyer,Christian Rudolf Dra 10h50'19"74d3'
Meyer,Christine K Vul 19h39'38"20d12'
Meyer,Craig A Her 17h6'0"40d14'
Meyer,Dirk Cam 5h39'47"70d44'
Meyer,Dona Jean Cas 0h9'58"61d41'
Meyer,Donald P Ser 15h42'50"0d38'
Meyer,Dr Eric Thomas Fndr/Meyer Telescope Aur 5h20'18"48d57'
Meyer,Dr Herbert Cam 4h18'18"68d56'
Meyer,Ed Cnv 13h30'1"47d44'
Meyer,Elaine Uma 11h4'15"50d54'
Meyer,Erik Del 20h21'55"16d48'
Meyer,Erin H Cap 21h50'17"-8d16'
Meyer,Evelyn Her 18h19'13"13d2'
Meyer,Francie Mon 7h10'14"-10d56'
Meyer,Frank C Her 16h24'1"48d4'
Meyer,Grant Anthony Boo 14h57'19"51d0'
Meyer,Günter Per 2h3'58"58d56'
Meyer,Hans Dra 15h13'0"63d37'
Meyer,Holger Lac 22h7'36"38d51'
Meyer,Horst Uma 9h28'40"46d50'
Meyer,Jack Preston Lyr 19h4'50"28d36'
Meyer,Jacquelyn A Uma 10h43'12"51d18'
Meyer,James D Cet 3h3'45"1d34'
Meyer,Jamie Marie Lyn 8h18'21"37d60'
Meyer,Janet A & Arthur L Sgr 19h15'43"-28d44'
Meyer,Jeanne Lyn 8h20'47"48d24'
Meyer,Jeffrey Her 16h33'37"51d14'
Meyer,Jim Her 18h7'23"40d31'
Meyer,Johanna Jacoba Cyg 20h1'37"40d38'
Meyer,John E Her 16h0'57"50d24'
Meyer,John Stephen Lyr 18h20'0"38d56'
Meyer,Joseph D Aur 5h57'44"54d33'
Meyer,Jr,William Michael Aql 19h31'36"11d50'
Meyer,Karen Legrand Oph 18h4'24"11d32'
Meyer,LAM Lindsey Allison Cyg 20h21'59"38d10'
Meyer,Marilyn,Chad, & Brendan Hya 9h35'55"-5d27'
Meyer,Melody Ann Vul 19h47'40"28d20'
Meyer,Michael J Sgr 18h58'9"-27d37'
Meyer,Millie Dra 16h24'45"68d22'
Meyer,Nancy Barbara Com 12h11'1"19d15'
Meyer,Paul Douglas Cnv 12h22'17"48d26'
Meyer,Petra Lco 11h40'59"20d36'
Meyer,Petra Dra 18h52'19"65d8'
Meyer,Richard Lyn 9h4'0"39d25'
Meyer,Rick Her 17h54'11"14d58'
Meyer,Rolf Pcr 3h56'49"31d51'
Meyer,Rolf Gustav Cap 20h29'17"-24d14'
Meyer,Ron Oph 16h24'14"-6d41'
Meyer,Sandra Doreen Lyr 18h59'1"27d21'
Meyer,Stefanie Anne Peg 21h58'55"10d50'
Meyer,Steven Lyn 9h3'48"34d27'
Meyer,Terry & Larry Lyr 18h20'59"47d4'
Meyer,Thomas Joshua Ser 15h9'22"0d36'
Meyer,Tresa Mon 6h34'46"-6d10'
Meyer,Urban F Dra 13h30'21"64d45'
Meyer,Willy Cyg 20h26'29"31d17'
Meyer-"Fritz", Frederick Lac 22h55'59"54d58'
Meyer-Herbst,St fl Res Commander,I Cam 5h6'35"58d53'
Meyer-Moon Shadow,Kohl Cam 3h50'12"56d9'
Meyer-Reißenweber,Lutz Uma 10h39'24"41d58'
Meyer-Scharenberg,Kurt Ori 5h51'41"20d14'
Meyerhoff,Albert Henry Lac 22h44'17"54d0'
Meyerhoff,Cole Benjamin Uma 11h51'0"47d37'
Meyerhoff,Susan Mon 7h2'42"4d34'
Meyerholt-Top Gun, Colonel William H Her 16h23'13"37d47'
Meyerhöfer,Günter Lib 15h6'49"-1d18'
Meyers The Star, Marlene Silverman Peg 22h2'36"33d20'
Meyers,Alexis Diane Lyr 19h3'17"25d55'
Meyers,Allison Robin Cas 0h33'49"67d56'
Meyers,Amy Elizabeth Uma 8h32'1"51d45'
Meyers,Bill Oph 17h11'36"-20d54'
Meyers,Bill Ser 17h57'1"-10d58'
Meyers,Carole Ann Lyr 18h47'23"33d25'
Meyers,Clarissa Ann Cyg 19h28'0"34d44'
Meyers,Collin Daniel Equ 21h7'16"3d19'
Meyers,David Shawn Per 3h11'1"45d49'
Meyers,Diana Cas 0h46'53"68d3'
Meyers,Faye Ann Uma 10h39'30"58d7'
Meyers,Fred,Sue,Alex & Zachary Uma 9h8'10"57d50'
Meyers,Gary Brian Lmi 9h49'22"40d54'
Meyers,Karen Na Mee Mon 6h35'22"0d52'
Meyers,Laurel Elizabeth Cyg 20h4'39"40d51'
Meyers,Malcolm Peg 22h26'47"31d38'
Meyers,Marvin E Dra 17h2'26"67d32'
Meyers,Michael & Jennifer Lyn 9h8'11"41d17'
Meyers,Norma Del 20h13'35"10d1'
Meyers,Rick Ser 15h58'58"10d18'
Meyers,Samuel Louis Leo 10h26'30"12d5'
Meyers,Sara & Lee Cyg 21h7'58"30d53'
Meyers,Zachary Scott Aur 7h11'13"37d16'
Meyerson,Andrea Joy Cam 5h48'47"61d4'
Meyniel,Philippe Lyr 18h57'14"30d14'
Meyring,Charles Anthony Aur 6h25'25"33d22'
Meyrowitz,Colette Cas 0h24'59"61d28'
Meza,Augusto "Coffee Bean" Peg 23h37'41"12d33'
Mezera,George & Irene Cnv 12h51'17"38d6'
Mezey Sge 20h1'52"16d39'
Mezyk,Sarah Vul 19h59'10"22d45'
Mezzanzanica,Roberto Cyg 20h4'10"30d48'
Mezzari,Robert F Her 16h47'24"33d56'
Mezzetti,Maurizio Lyr 18h38'28"40d6'
Meß,Siegbert Her 17h22'37"46d53'
MF & LD Her 17h53'38"18d49'
MG 90 Tel 19h4'35"-49d23'
MGM 102994 Cam 6h16'0"67d56'
Mhairi Cas 0h53'54"61d3'
Mi Amor Peg 23h49'1"15d44'
Mi Amor de mi Vida Ser 15h16'0"9d5'
Mi Amor Eterno Jose Luis Uma 9h58'12"48d26'
Mi Amore Cyg 20h8'20"40d57'
Mi Amore Dra 14h18'38"64d8'
Mi Cora Zon Mon 7h48'47"-5d14'
Mi Coraz7h48'47"-5d14' Her 16h53'33"40d38'
Mi Estrella A John Uma 11h2'14"38d1'
Mi Hwa & Joseph Cyg 20h35'49"41d12'
Mi Mi Lyr 18h21'11"46d31'
Mi Nina Bonita Ori 5h48'32"-1d13'
Mi Queca Lyr 18h21'0"37d50'
Mi Rey David Eduardo Ori 5h55'21"13d11'
MI STAR Cap 21h55'52"-18d46'
Mi Teresita Equ 20h58'36"7d23'
Mi Tesoro Aql 20h6'28"1d32'
Mia Uma 10h9'11"50d0'
Mia Eri 4h9'53"-11d20'
Mia Cet 2h59'39"3d35'
Mia & Yasu Cru 12h37'6"-57d28'
Mia Amore Per Luigi Cyg 19h41'1"31d10'
Mia Christiane Col 6h27'0"-38d32'
Mia Dolci MetÖ Lyr 19h1'54"30d27'
Mia Pia Cas 23h16'45"60d51'
Mia Zada Uma 11h5'58"30d15'
Mia's Star Ori 6h7'24"8d54'
MiaDon Cyg 19h19'37"50d17"
Miadé Lyr 18h30'20"30d10'

Miailovich,Patricia
And 23h43'1"45d11'
Miakisz,Jr,Thomas E
Lmi 10h35'25"28d23'
Miale,Drew & Kristen
Lyr 18h57'22"37d0'
MiaLovesHugh
Lyr 18h47'38"38d18'
Miani,Marco
Pup 7h56'51"-28d54'
Miano,Ashley
Lyr 18h58'48"26d4'
Miaoulis,Rachel Elizabeth
Lib 14h20'0"-18d54'
Miata,Helen
Cas 2h10'55"70d41'
Miau bau
Cyg 21h55'21"52d46'
Mibs
Boo 13h59'22"25d14'
Mic & Mac
Aql 19h59'39"0d6'
Mic Tak
Vir 14h41'39"-8d4'
Mic,Kyle C
Tri 2h30'27"31d45'
Mica
Peg 23h30'24"16d48'
Micaela & Marcus
And 23h35'5"39d7'
Micah's Christmas Star
Uma 12h10'51"60d9'
Micah's Wish
Aur 5h54'59"38d0'
Micah,Cheryl
Peg 22h58'19"30d48'AB
Micahel Paul
Her 17h28'58"21d12'
Micale
Cnv 13h33'37"37d36'
Micali,Giovanni
Vel 9h40'56"-48d41'
Micalizio,Alessandra
And 23h1'26"44d59'
Micalizio,Veronica
Cyg 20h21'20"41d60'
Micallef,Michelle Marie Naylor
Cas 0h23'48"75d7'
Micallef,Stephanie & Ramon
Tau 3h53'11"1d56'
Micanmarflopier,Abry
Sgr 19h26'40"-36d51'
Miccariello,Dana
Vul 19h22'0"25d16'
Miccichc,Pauline F
Vul 20h17'20"23d42'
Miccio,Dr Joseph V & Lillian
Oph 16h45'47"2d53'
Miccio,Vera
Umi 13h25'16"70d29'
Micco,Peggy Turl
Cas 0h18'37"66d16'
Miceli's Rising Star, Stephen
Dra 19h0'46"58d25'
Miceli,Danielle "Hon"
Cas 23h38'24"62d26'
Miceli,Jr,Joseph
Psc 1h3'57"20d12'
Micey
Cru 12h9'35"-60d3'
Mich
Dra 15h50'35"51d50'
Micha Rae
Sco 16h58'35"-44d7'
Micha Son Of Dana
Lyn 9h9'0"44d38'
Michael
Aql 20h5'0"8d11'
Michael
Lmi 10h9'32"38d6'
Michael
Ori 5h26'1"-5d10'
Michael
Uma 8h35'0"60d50'
Michael
Aur 7h22'37"35d55'
Michael
Cam 3h50'52"75d1'
Michael
Aur 6h6'30"35d7'
Michael
Lac 22h19'38"47d1'
Michael
Hya 8h13'32"4d11'
Michael
Aql 19h53'11"15d13'
Michael
Uma 9h15'42"53d37'
Michael
Lyr 18h57'57"34d12'
Michael
Cnv 12h38'37"35d9'
Michael
Her 17h50'26"42d9'
Michael
Per 3h13'54"40d18'A
Michael
Uma 11h52'51"37d52'
Michael
Per 2h0'45"56d30'
Michael
Peg 21h47'56"34d35'A
Michael
Cep 21h44'49"55d14'
Michael
Lac 22h11'57"54d31'
Michael
Aur 5h31'52"37d46'
Michael
Dra 13h7'44"64d9'
Michael
Peg 22h58'57"11d27'
Michael
Her 17h6'52"48d3'
Michael
Sco 16h38'16"-44d55'
Michael
Dra 17h8'12"61d18'
Michael "50"
Ori 4h53'50"1d38'
Michael "A Heavenly Body"
Her 16h57'58"28d5'
Michael "Burning Brightly Always"
Aql 19h56'1"0d60'
Michael "Light of My Life"
Hya 9h29'58"-0d34'
Michael "My Eternal Love"
Cep 6h30'1"85d17'
Michael "Tenerifa '92"
Vir 13h6'47"-5d9'
Michael "Zeek" Star
Cnc 8h9'14"6d57'
Michael & Alex
Peg 23h19'29"33d30'
Michael & Ann
Cam 5h42'51"73d38'
Michael & Ann's Wedding Star
Cyg 19h25'33"33d7'
Michael & Annie August 12,1988
Crb 16h19'46"38d34'
Michael & Barbara Ad Infinitum
Ori 5h36'30"-0d36'
Michael & Belinda
Uma 13h1'26"53d48'
Michael & Brenda
Lyr 19h12'16"38d53'AB
Michael & Candida
Eri 4h36'29"-18d11'
Michael & Carole
Crb 16h2'51"38d41'
Michael & Christine
Lmi 10h7'18"32d31'
Michael & Cindy's Magic Star
Aur 6h6'1"45d13'
Michael & Co2
And 1h29'20"39d30'
Michael & Danielle
Cyg 21h23'29"50d11'
Michael & Danielle- Destined To Be
Mon 6h52'43"0d28'
Michael & Deborah
Lyn 8h1'33"51d17'
Michael & Dina
Cyg 20h54'20"5242'
Michael & Gary-GET A BIG GOLD STAR
Ori 5h47'0"10d28'
Michael & Gia's Love Star
Sge 20h6'57"20d7'
Michael & Ian
Ori 6h16'24"-2d55'
Michael & Irene 527
Cyg 19h24'30"30d21'
Michael & Janet
Lyr 18h15'51"34d34'
Michael & Jenny
Lyr 18h39'57"31d50'
Michael & Joyce Forever
Lyr 19h3'50"25d38'
Michael & Kimberly
Peg 22h34'12"8d23'
Michael & Kimberly
Gem 7h21'47"33d51'
Michael & Kirsten
Boo 14h6'42"32d49'
Michael & Laurel
Lyn 6h54'21"52d49'
Michael & Lisa
Cyg 21h0'14"33d33'
Michael & Lisa auf ewig Okt 96
Cas 0h6'50"60d55'
Michael & Lorri
Uma 11h56'58"61d37'
Michael & Marnie
Cyg 19h36'15"28d44'
Michael & Melissa
Cmi 8h3'19"6d8'
Michael & Michelle's Star
Cma 7h15'35"-13d33'
Michael & Mina
Cyg 21h7'33"31d52'
Michael & Nicole Forever
Vul 21h2'18"27d13'
Michael & Parish
Aql 19h52'26"15d42'
Michael & Robin 12/17/1994
Sge 19h43'35"16d31'
Michael & Sherri
Lac 22h24'20"50d16'
Michael & Steph's Precious Moments
Aur 5h13'20"41d9'
Michael & Tamela TLF
Lyn 8h21'0"41d40'
Michael & Tarra Forever & Forever
Cyg 20h56'28"31d1'
Michael & Tish
Eri 4h5'20"-18d45'
Michael & Trina
Mon 6h52'32"-6d13'
Michael & Trisha, Forever
Cyg 21h35'28"40d0'
Michael & Wendy Star
Boo 14h14'33"15d26'
Michael (Hard Luck)
Lac 22h45'47"54d0'
Michael 143
Her 17h0'25"28d18'
Michael 40
Cet 1h46'35"-2d35'
Michael Aaron
Cnv 13h21'11"31d53'
Michael Alan
Ori 5h58'17"10d1'
Michael Angelo
Dra 16h42'45"67d59'
Michael Anthony
Per 2h45'46"40d31'
Michael Anthony
Her 17h24'0"41d5'
Michael Anthony
Pho 23h49'13"44d47'
Michael B
Aur 5h2'0"41d32'
Michael B's
Sgr 19h40'33"-40d33'
Michael Cherri
Cma 6h54'55"-18d25'
Michael Clark
Her 17h59'19"48d52'
Michael Clark & Kandis Sue Eternal Love
Per 1h45'22"53d20'
Michael David
Aur 7h23'16"43d23'
Michael David
Aql 20h2'44"6d7'
Michael Dean
Ori 5h31'30"1d21'
Michael Don & Marilyn Grace
Cet 2h25'24"1d44'AB
Michael Edward
Per 2h42'42"43d15'
Michael Forever
Aqr 21h23'21"-1d42'
Michael Francis
Per 1h59'35"52d41'
Michael Francis
Hya 8h58'12"2d13'
Michael Frank
Cam 12h19'0"81d59'
Michael Fred
Cep 22h37'49"59d3'
Michael George
Per 2h3'0"48d12'
Michael George
Her 16h23'57"32d21'
Michael J
Lmi 9h52'55"33d57'
Michael J
Oph 17h3'54"-23d40'
Michael J Geist My Light In The Dark
Boo 14h57'26"28d22'
Michael Jacques
Dra 17h24'31"64d43'
Michael James
Her 17h11'13"41d17'
Michael John
Dra 19h30'50"67d36'
Michael John
Her 17h26'19"38d49'
Michael Keith
Boo 14h8'12"31d54'
Michael Kent
Boo 15h18'16"40d11'
Michael Lee
Per 2h37'13"37d42'
Michael Loves Richard
Eri 3h55'35"-19d54'
Michael Loves Tiffany
Sct 18h48'25"-7d42'
Michael Luc
Her 16h47'55"35d48'
Michael My Knight In Shining Armor 1st Ann.
Sex 10h18'1"-8d7'
Michael My Love
Her 18h14'35"37d37'
Michael My Lucky Star
Per 1h51'23"53d2'
Michael O
Cep 0h50'28"77d11'
Michael P
Per 3h1'42"41d19'
Michael Patrick & Bridget Lee
Com 12h18'16"18d51'
Michael Peter
Cep 1h14'53"78d38'
Michael Ray
Aur 4h49'0"48d60'
Michael Raymond "Mickey"
Mon 7h2'33"-7d2'
Michael Ruddy
Lac 22h6'44"51d36'
Michael T
Her 15h50'34"44d51'
Michael The Great
Ori 5h31'31"1d11'
Michael Travis
Vul 20h20'54"26d12'
Michael V
Lac 22h54'52"50d31'
Michael Victoria
Cas 0h57'53"61d53'
Michael Vincent-Her Bumble Bee
Cyg 20h37'43"40d24'B
Michael Wayne Charles
Peg 21h49'36"30d20'
Michael's Childhooddream
Cas 0h9'26"60d33'
Michael's "Celestial Nightlight"
Aur 4h51'17"50d20'
Michael's & Paula's Dream-A Reality
Cyg 21h7'55"38d40'
Michael's 50th Birthday Star
Dra 14h23'0"64d35'
Michael's Battle Star
Cap 21h52'36"-21d43'
Michael's Best
Per 4h0'30"51d33'
Michael's Changes
Aql 20h1'23"12d6'
Michael's Destiny
Cep 0h2'27"67d42'
Michael's Destiny
Aql 19h59'40"15d35'
Michael's Dream
Per 2h52'30"40d22'
Michael's Dream
Boo 15h0'53"25d26'
Michael's Eyes
Per 3h7'12"37d51'
Michael's First Little Man
Cep 21h56'23"60d11'
Michael's Forever Spirit
Her 16h40'22"34d56'
Michael's Guiding Light
Her 16h38'1"26d13'
Michael's Guiding Light
Her 15h54'41"50d59'
Michael's Hope
Uma 11h39'36"41d24'
Michael's Knight Light
Ori 5h56'42"15d5'
Michael's Light
Per 2h31'38"56d34'
Michael's Light
Per 2h2'41"50d34'
Michael's Light
Lac 22h50'27"53d17'
Michael's Little Star
Her 16h37'46"33d39'
Michael's Magic
Her 16h44'44"33d29'
Michael's Magnificent Sun
Boo 13h48'47"15d58'
Michael's Malibu Kisses
Mon 8h6'40"-3d31'
Michael's Mark
Per 3h11'59"46d35'
Michael's Muse
Oph 17h36'28"-16d52'
Michael's Myth
Cam 4h3'37"67d36'
Michael's Old Too
Her 16h10'20"8d18'
Michael's Other Piece Of Heaven
Per 3h18'49"40d53'
Michael's Pa On
Her 16h30'39"41d49'
Michael's Passion
Dra 19h39'48"65d4'
Michael's Piece of Heaven
Per 1h52'0"53d2'
Michael's Power Of Love
Lac 22h37'17"55d34'
Michael's Princess
Lyr 18h52'36"40d37'
Michael's Smile
Mon 6h46'41"11d50'
Michael's Splendrous Aurora
Ser 15h57'36"23d56'
Michael's Star
Crt 11h13'35"-12d51'
Michael's Star
Aur 6h7'19"31d21'
Michael's Star
Mon 8h6'43"-9d31'
Michael's Star
Ari 2h58'50"28d19'
Michael's Star
Peg 23h4'53"12d16'
Michael's Star
Per 2h59'21"37d18'
Michael's Star
Aql 18h58'1"4d10'
Michael's Star
Lac 22h55'39"38d34'
Michael's Star
Lac 22h29'27"53d56'
Michael's Star
Oph 17h32'1"11d25'
Michael's Star
Her 17h36'1"20d51'
Michael's Star 7
Ori 5h39'46"-0d16'
Michael's Star Forever Shining
Aur 6h18'0"38d52'
Michael's View
Per 3h17'58"50d16'
Michael's Yankee Choice
Boo 14h49'12"22d58'
Michael,Carol
And 1h32'42"40d35'
Michael,Carroll Florizal
Uma 9h41'52"44d35'
Michael,Christian
Ori 5h54'31"15d32'
Michael,Elizabeth Viola
Cas 2h47'34"75d23'
Michael,GJ
Ser 18h21'0"-2d46'
Michael,Günther
Boo 14h16'21"17d3'
Michael,Sabine Hildegard
Eri 4h8'22"-16d60'
Michael,Sarah
Del 20h32'34"20d13'
Michael,Sherry Lynn
Lyn 7h36'18"50d16'
Michael,Siegfried
Uma 9h39'21"43d48'
Michael,Sven
Cep 20h39'0"65d50'
Michael,The Star of
Oph 17h15'59"12d39'
Michael,Thomas P & Martha Awais
Cyg 19h33'44"36d4'
Michael,Tyler
Her 17h15'10"26d35'
Michael,Wesley Elisabeth
Dra 16h9'43"66d54'
Michael-Angel In The Northern Sky
Her 17h7'1"48d33'
Michael-Master of God's Heart
Lyr 18h56'26"47d9'
Michael-My Inspiration
Boo 14h9'48"40d1'
Michael-Tina
Cyg 21h34'22"41d59'
Michael-To Be One With God
Ori 5h56'12"13d10'
Michaela
Crt 11h9'34"-18d24'
Michaela
Lyn 7h55'52"38d27'
Michaela
Vul 20h14'39"25d59'
Michaela 2000
Crb 16h18'54"33d12'
Michaela 24-6-94
Cnc 9h16'1"11d34'
Michaela Dawn
And 0h51'15"41d2'
Michaela Funny Bunny
Cyg 19h34'23"39d18'
Michaela Janina's Star
Aql 18h59'44"14d12'
Michaela Shea
Peg 23h30'26"33d37'
Michaela und Paul
Umi 16h34'48"75d18'
MichaelAmy 191993
Her 18h3'58"40d12'
Michaelangelo
Per 1h52'34"47d16'
Michaelangelo
Per 3h0'25"41d23'
Michaelangelo diPax
Cnv 13h53'32"41d56'
Michaelene The III
Lyr 18h33'35"45d56'
Michaelinda
Cap 21h18'0"-26d36'
Michaelis,Karsten
Oph 18h1'24"7d56'
Michaelis,Marvin Albert & Nancy Mae
Eri 2h49'14"-6d58'
Michaelisa
Cyg 19h41'1"37d37'
Michaelke
Aqr 21h4'11"-0d52'
Michaelouise
Cyg 20h38'0"40d22'
Michaels,Adam Leigh
Uma 11h20'32"45d5'
Michaels,Anthony G
Her 18h18'1"14d27'
Michaels,Darren
Cyg 21h40'15"37d56'
Michaels,Everett
Dra 20h21'21"68d34'
Michaels,James Walker
Gem 7h36'58"34d23'
Michaels,Katherine Ann
And 2h28'14"50d9'
Michaels,Lindy Rollyson
Hya 9h8'0"2d26'
Michaels,Lori
Uma 12h1'36"32d28'
Michaels,Margaret Mary
Mon 8h3'23"-3d13'
Michaels,Norma
And 23h22'51"41d44'
Michaels,Robert
Her 16h18'59"10d36'
Michaels,Tara L
Del 20h14'0"15d39'
Michaels,Tiana Lauren
Sco 16h37'35"-31d25'
Michaels,Tyler David
Ori 5h54'52"20d43'
Michaels-After & Forever,Lori Ann
Eri 4h17'56"-1d26'B
Michaelson "Golden", Janet & Bertram
Lac 22h39'57"52d49'
Michaelson,Grandpa Max
Aur 6h11'47"38d3'
Michaelson,John
Aur 5h16'41"45d39'
Michaelson,Larry Nolan
Dra 18h44'51"68d40'
Michaelviv
Lyn 8h14'46"37d49'
Michalak,Eric Michael
Her 16h46'58"21d31'
Michalak,Jennifer L Dini
Uma 11h15'53"70d25'
Michalchuk,Diane Rose
Lyn 7h17'1"59d49'
Michalek,Jane P
Aql 19h57'27"0d32'
Michalek,Peter
Tau 5h34'17"28d32'
Michalewsky,Craig
Dra 17h11'60"61d11'
Michalka,Robert E
Per 2h2'57"50d26'
Michalovic,Michael J
Cet 2h4'59"4d19'
Michalowski,Bob
Aur 6h20'49"30d36'
Michalski,Daniel W
Cep 22h50'40"57d53'
Michalski,Dieter
Aqr 21h57'30"-11d18'
Michardière,Sylviane
Aur 5h52'1"30d13'
Michaud,Clare Gallagher
Com 12h54'58"26d39'
Michaud,Cynthia
Lyn 6h58'19"52d58'
Michaud,Elisabeth Joe
Cas 1h2'56"61d35'
Michaud,Guylaine
Cam 7h55'51"61d42'
Michaud,Jared Daniel
Lac 22h2'52"50d24'
Michaud,Jeannette
Lyn 8h30'56"43d12'
Michaud,Robert
Aur 6h8'1"37d39'
Michaudet,Gerard
Cam 3h23'35"66d23'
Miche & Noah
Sge 20h17'11"17d43'
Micheal & Melissa's Ebullient Nebula
Del 20h35'54"10d57'
Micheal Rachel
Oph 16h48'56"11d41'
Micheal"Micheal David Shine"
Ser 15h18'36"20d53'
Michealson,Charlene B
Mon 6h18'11"5d56'
Michel
Umi 13h44'52"74d28'
Michel
Uma 11h33'47"31d24'
Michel
Cam 5h5'30"67d37'
Michel Alex
Del 20h23'18"16d39'
Michel LX
Oph 17h31'53"7d33'
Michel Paul T J
Cep 20h56'47"62d55'
Michel Your Forever, Renate Brigette
Uma 11h35'57"63d46'
Michel,Alexander Stephen
Lac 22h39'17"56d35'
Michel,Babin
Cnv 12h54'0"40d28'
Michel,Cynthia Glenn
Aur 4h54'59"40d55'
Michel,Jean Cédric
Uma 9h12'40"50d31'
Michel,Jim & Joan
Cyg 19h35'0"28d59'
Michel,Jr,Gilbert
Aql 19h5'54"3d12'
Michel,Mallory Glenn
Lyn 6h34'0"59d1'
Michel,Meïr
Ori 5h13'42"15d15'
Michel,R Brian
Lac 22h27'43"41d12'
Michel,R Bryce
Lmi 9h37'56"34d5'
Michel,Stephanie
Ser 18h27'36"1d27'
Michela
Eri 3h56'43"-5d36'
Michela
Peg 22h29'27"28d34'
Michela
Cam 8h26'13"80d27'
Michela P 13/9/94
Boo 15h5'38"32d26'
Michele
Mon 6h54'52"-10d4'
Michele
Cyg 19h44'21"30d13'
Michele
And 1h47'19"36d8'
Michele
Peg 22h48'53"27d15'
Michele
Mon 6h24'25"3d23'
Michele
Cam 4h3'17"61d13'
Michele
Cmi 8h2'39"6d22'
Michele
Cas 0h45'38"64d20'
Michele
Peg 23h16'0"31d17'
Michele
Cnc 8h58'10"12d18'
Michele & Don
Aur 4h58'45"50d57'
Michele & John
Her 18h55'23"12d3'
Michele & Mauricio
Aql 20h18'22"5d11'
Michele (Sparkler)
Uma 11h17'53"68d21'
Michele K
Eri 2h48'23"-4d29'
Michele Louise
Vul 20h57'0"28d47'
Michele Mary
Cyg 20h54'33"50d3'
Michele Renae-W/1L - L A R
And 23h20'46"43d55'
Michele Tara
Peg 0h4'28"27d36'
Michele's Star
Cas 0h22'18"61d44'
Michelen,Dr Nasry
Oph 18h2'35"10d36'
Michelfeit,Walter
Per 4h6'52"47d5'
Michelfelder,Raymond Clifford
Peg 22h57'26"33d53'
Michelin,Wanda Cooper
And 23h44'57"47d14'
Michelina
Pyx 8h40'45"-27d35'
Micheline
Gem 6h51'32"13d15'
Micheline
Cam 3h33'30"63d50'
Micheline
Tri 2h37'31"35d50'
Micheline Colette
Ori 5h57'10"15d20'
Michelini,Aimee Lynn
Lib 15h32'11"-10d21'
Michell's Star,Karen Rosalind
Cas 2h22'16"61d33'
Michelle
Mon 6h53'46"7d46'
Michelle
And 23h2'39"45d55'
Michelle
Hya 9h9'13"3d30'
Michelle
Cas 2h56'38"70d36'
Michelle
Leo 10h55'52"10d31'
Michelle
Cyg 20h24'1"38d57'

Michelle
Sco 16h20'11"-23d48'
Michelle
Cyg 20h58'32"37d35'
Michelle
Cyg 19h33'39"32d60'
Michelle
And 1h25'55"34d2'
Michelle
Psc 0h59'36"22d8'
Michelle
And 1h47'40"41d10'
Michelle "Beavis"
Mon 6h20'0"6d2'
Michelle & Andrew's Wedding Star
Cyg 20h20'56"38d4'
Michelle & Dale
Crb 16h2'53"32d27'
Michelle & David
Uma 11h55'32"32d17'
Michelle & David
Cyg 20h3'40"37d51'
Michelle & Jeff's Star of Hope
Lyr 18h56'50"31d37'
Michelle & Kevin
Eri 3h21'56"-6d50'
Michelle & Mark Night Light
Peg 22h30'46"26d41'
Michelle & Michael
Com 13h2'12"20d2'
Michelle 1994
Cas 1h26'0"54d2'
Michelle 4-16-72
Cam 4h13'0"68d20'
Michelle Ann
Cyg 19h49'56"38d5'
Michelle Anne
And 23h20'41"51d32'
Michelle Anne F D
Uma 14h20'19"60d21'
Michelle B
Lyn 8h37'39"37d56'
Michelle Dawn
Peg 22h2'42"4d46'
Michelle Dianna
And 23h20'32"42d4'
Michelle Elizabeth
Mon 6h53'44"7d58'
Michelle Holly
Ari 2h58'18"22d24'
Michelle I Love You Stars
Uma 9h10'13"62d25'
Michelle Jane
Lyr 18h15'38"38d20'
Michelle June
Cam 3h37'15"61d41'
Michelle L's White Knight "Jay"
Ori 5h0'0"14d23'
Michelle Leigh & James Thomas Dec 95
And 23h34'42"47d10'
Michelle Lynn
And 23h27'15"49d33'
Michelle Lynn
Sco 16h54'39"-40d43'
Michelle Lynn
Vul 19h47'57"28d58'
Michelle Margaret
Cam 12h53'20"77d15'
Michelle Marie
Peg 22h29'56"29d39'
Michelle-Mark
Vel 9h57'37"50d9'
Michelle Nov 1 1992
Lyn 6h51'48"60d13'
Michelle Rae
Cas 2h5'51"59d32'
Michelle Rae
Tri 2h6'27"30d23'
Michelle Rene
Cas 2h6'43"61d5'
Michelle Rose
Lyr 18h18'57"44d35'
Michelle S M
Lyn 7h56'12"58d59'
Michelle Stella Mary
And 23h35'50"40d31'
Michelle 11
Vel 9h59'17"56d14'
Michelle's 600 Smiles
And 23h7'1"36d12'
Michelle's Dream Catcher
Cyg 21h52'45"52d42'
Michelle's Flame
Uma 8h16'32"62d22'
Michelle's Heaven
Lyn 8h15'12"39d13'
Michelle's Light
Lmi 10h39'43"28d21'
Michelle's Star
Cam 5h10'40"70d10'
Michelle's Star
Mon 6h58'28"-5d57'
Michelle's Star
And 0h31'42"45d49'
Michelle's Twinkle
Per 4h6'22"36d58'
Michelle's Wish Come True
Ori 4h56'28"-0d49'
Michelle,My Heavenly Daughter
Ori 5h55'57"17d53'
Michelle,présent de le ciel
Cyg 20h59'19"37d45'
Michelle,Twinkle Littl Star 21 is what U R
Ant 9h38'4"-32d21'
Michelle-31
Lyr 18h15'55"30d29'
Michelle-Lucie Cecile
Eri 3h50'36"-7d19'
Michellemylove
Mon 6h35'40"-6d15'
Michelon,Jon Bradford
Oph 16h49'34"-4d21'
Michelotti
Cet 1h6'1"1d42'
Michels,Barbara
Dra 18h12'48"68d29'
Michels,Katja
Lyn 8h8'24"48d20'
Michels,Larry
Tri 1h51'24"28d12'
Michels,Sarah Eileen
Lyr 18h48'50"34d18'
Michelsen,Edgar "Mike"
Uma 11h10'32"30d31'
Michelsen,Joan
Uma 11h14'0"30d58'
Michelsen,Shirley
And 2h26'60"39d50'
Michelson Family Star, The Selma
Cam 7h37'13"61d50'
Michener,A Horace
Dra 11h28'56"78d22'
Michener,Mary Christine
Uma 9h2'36"57d43'
Michette
Cyg 19h40'47"31d3'
Michetti,Giada
Umi 16h39'0"83d16'
Michi e Nico
Sco 17h30'1"-30d42'
Michi's Pooky Bear
Vul 20h41'46"20d24'
Michiaki
Ari 2h21'12"11d41'
Michiaki & Ikuko
Sgr 18h50'25"-31d11'
Michico,Yuasa
Com 12h7'46"30d1'
Michielini,Denise
Cam 12h47'41"80d51'
Michigan Hope
Uma 11h46'59"45d49'
Michigan,Annie
Per 3h22'41"54d39'
Michiko
Lib 14h56'23"-5d53'
MICHIKO
Ser 15h20'32"4d53'
Michnewich,Alexander
Her 16h29'39"39d34'
Michnowicz,James Casimir
Boo 14h29'49"29d50'
Michnowski,Alan
Cet 0h42'49"1d45'
Michota,Bruce
Aur 7h13'42"40d47'
Michotte
Cra 18h12'44"-42d9'
Michropageo-Owen Family Star
Her 18h18'53"12d42'
Michuda,Jay
Uma 9h43'59"50d46'
Michuda,Mimi
And 1h15'14"39d32'
Michuda,Nic
Vul 20h17'47"26d5'
Michuda,Tony
Dra 16h18'30"63d18'
Michy
Her 16h22'17"10d12'
Michél JaTamé
Aur 4h49'20"40d0'
Mick
Cnv 12h28'38"33d45'
Mick A
Uma 11h9'25"46d59'
Mick Cluster-Suzie,The
Cet 1h4'37"-1d59'B
Mick Cluster-Wally,The
Cet 1h4'37"-1d59'A
Mick's Star
Boo 15h12'0"51d53'
Mick-Where You Go My Love Follows,M
Uma 11h2'46"35d25'
Mickayla's Magic
Crb 15h59'19"38d59'
Mickella
Cyg 19h46'13"30d24'
Mickelsen Sept 30,1976 Warren Lee
Cma 7h0'56"-15d35'
Mickelson,Brittney A
Cyg 21h4'1"28d44'
Mickelson,Dorothy
Lac 22h47'13"56d39'
Mickelson,Edwin
Dra 16h20'0"68d25'
Mickelson,Virginia
Lyr 18h49'55"42d45'
Mickelwaite,Elaine
Gem 6h51'43"31d23'
Mickens,Forever Stanley
Per 3h21'49"40d23'
Mickey
Boo 15h25'40"50d54'
Mickey
Sex 9h50'26"1d35'
Mickey
Dra 14h49'35"59d42'
Mickey
Oph 17h56'21"8d34'
Mickey
Umi 16h38'48"75d53'
Mickey B
Eri 4h1'60"-6d44'
Mickey's First Christmas
Peg 22h21'51"31d9'
Micki
Ori 5h56'13"6d43'
Micki Lynne
Lyr 18h22'14"38d3'
Micki's Lone Star
Tau 5h45'35"16d7'
Mickie
Uma 10h41'48"51d7'
Mickie & Donald
Cam 11h3'33"81d10'
Mickievicz,John
Per 1h47'1"53d22'
Mickle,Jr,Gerald St Claire
Peg 22h30'30"8d17'
Micklem,Eleanor Sarah
Cyg 20h19'14"41d49'
Mickles,Ellen Marie
Cep 21h12'31"76d6'A
Mickles,Nevada Anne
Cep 21h12'31"76d6'B
Mickley,Frank
Her 16h20'16"20d26'
Mickly Mockly Mikely
Umi 15h49'43"80d42'
Mickstorm Adamantium 21
Cra 18h11'1"-39d25'
Mickus,Don
Ori 5h33'32"-1d53'
Micky
Ori 5h59'18"15d26'
Micon
Mon 6h51'21"11d28'
Micoucou,Fanny
Boo 15h15'1"37d37'
Midani,Huda
Uma 9h39'1"42d58'
Middlebrook,Meredith Ann
Aql 18h56'1"-1d40'
Middleton November 1 1994,Fred
Sgr 18h53'22"-36d26'
Middleton Star,The
Lyr 18h53'21"37d33'
Middleton,Alfred
Lac 22h27'36"56d37'
Middleton,Dawnie Lee
And 0h24'0"45d19'
Middleton,Edwina Davidson
And 23h20'47"51d29'
Middleton,Eugene Ronald
Dra 16h55'1"67d21'
Middleton,Fiona Alexandra
Cyg 19h45'15"50d37'
Middleton,Jenny
Cyg 19h45'48"50d26'
Middleton,John
Hya 8h33'20"-6d3'
Middleton,John
Ser 17h31'28"-10d43'
Middleton,John S
Per 3h37'46"38d5'
Middleton,Kent
Sco 16h56'59"-38d45'
Middleton,Melissa Marie
Peg 21h48'50"36d23'
Middleton,Norma & Everett
Cyg 21h10'11"35d6'
Middleton,Pamela Lynn
And 22h57'37"51d56'
Middleton,Robert Wilmot-40
Gru 22h30'32"-51d38'
Middleton,Sara
Peg 23h27'38"23d8'
Middleton,Sarah Shaw
Cyg 19h26'0"40d35'
Middleton,Suzanne
Cyg 19h26'22"34d31'
Mideiros,John James
Oph 18h34'50"7d29'B
Mideiros,Ramona Rae
Oph 18h34'50"7d29'A
Midene
Oph 16h59'13"10d16'
Midge
And 1h26'39"41d2'
Midge
Cep 20h55'51"60d5'
Midge"The Magnificent"
Com 12h13'30"30d6'
Midheaven Vision Quest
Aql 20h1'0"14d51'
Midkiff,Benjamin Joseph
Boo 14h23'0"28d17'
Midkiff,Candy Reiter
Cnv 12h10'11"42d54'
Midkiff,Mary Ann & Jack
Mon 7h50'1"-2d19'
Midkiff,Melissa Aimeé
Peg 22h29'21"21d3'
Midknight Destiny
Equ 21h6'57"10d33'
Midland,CU,Astra Merck
Uma 10h35'56"56d15'
Midnica,Bonnie
Aur 5h6'29"50d8'
Midnica,Carl
Aur 5h3'21"51d29'
Midnica,Gloria
Aur 5h1'18"50d42'
Midnight Diamond
Uma 9h55'42"55d46'
Midnight Marauder,The
Sge 19h16'20"16d19'
Midori
Sgr 19h11'15"-26d42'
Midori,Valerie
Lyr 18h59'16"45d17'
Midura,Todd Andrew
Dra 16h31'1"69d13'
Miechi
Cnc 8h32'31"8d54'
Mieko's Rose
And 0h17'29"38d4'
Miel de Botton
Lmi 11h2'0"32d15'
Mielcarek,RJ
Her 18h15'57"14d17'
Miele,Antoinette Rosso
Cas 23h25'38"62d12'
Miele,Philip John
Dra 17h4'11"63d5'
Miele,Sr,Joel A
Cam 13h19'0"77d54'
Mieles,My Friend Isis
Cas 0h4'0"60d6'
Mielke,Howard O
Gem 7h16'21"30d41'
Miell,Lisa "Bug"
And 0h29'15"28d13'
Mielnicki,Alpha-Linda
Lyn 8h2'1"34d28'
Mienville,Nicolas
Her 18h23'31"18d49'
Mierau,Sylvia
Vir 13h36'47"-6d33'
Miers,David Wayne
Dra 15h42'15"62d39'
Miesseler,Bellinda
Cep 22h6'34"60d31'
Miet,Michel
Umi 15h47'40"80d58'
Miezz
Lyr 19h20'50"38d6'
Mif-143
Tri 2h6'21"33d56'
Mifflin,William B
Lmi 10h35'14"27d57'
Mifsud,Laura Lee
Cyg 19h26'24"34d20'
Mig
And 2h15'47"38d22'
Migas,Bruno
Cma 7h27'48"3d24'A
Migas,Catherine
Lyn 7h9'0"58d22'
Migas,Violet
Cma 7h27'48"3d24'B
Migasi,Dino Ronald
Aur 6h39'37"38d9'
Migawa,Caitlyn Marie
Lyn 9h9'22"35d53'
Migdal,Jeff
Mon 8h4'19"-10d4'
Miggy-42
Cet 2h55'1"-0d11'
Mighty Al
Mon 7h58'5"-1d52'
Mighty Gra
Uma 9h59'1"73d8'
Mighty H Man,The
Sex 10h44'20"3d1'
Mighty Isis
Aql 19h4'37"-0d27'
Mighty Mite
Cyg 21h0'30"28d46'
Mighty Pepper Epiphany Jewell
Cas 0h23'58"70d15'
Miginiac,Jean Pierre
Mon 7h51'31"-4d22'
Migliaccio,Concetta A
Gem 6h57'26"13d55'
Migliaccio,Joseph
Uma 10h3'44"59d3'
Migliaccio,Lillian
Per 2h42'58"35d14'B
Migliaccio,Pasquale Nicholas
Per 2h42'58"35d14'A
Migliaccio,Samantha Nichole
Cyg 19h19'1"45d11'
Miglio,Joan
Lyn 6h44'0"60d14'
Migliorati,Alessandro Beccaro
Psa 22h21'41"-26d39'
Migliore,Brian James
Cam 4h54'54"61d10'
Migliore,Elizabeth
Cyg 20h53'43"39d36'
Migliore,Joseph
Cep 21h50'59"61d13'
Migliore,Kristopher
Cma 6h54'50"-19d22'A
Migliore,Matthew
Cma 6h54'50"-19d22'B
Migliore,Richard T
Aqr 21h7'21"-6d34'
Migliori,Francesca
Leo 9h58'0"20d58'
Mignano,Frank
Lyr 19h20'48"41d10'
Mignano,Gerard
Per 4h3'15"51d49'
Migneault,Richard
Lyn 7h37'1"39d56'
Mignerey,Peter
Lac 22h27'0"50d32'
Mignon,Gerry
Aur 4h50'50"40d25'
Mignone,Mary Alyce
Uma 11h19'51"42d39'
Migoletta
Del 20h38'60"10d44'
Migoya,Raisa L
Cas 0h43'29"61d48'
Miguel
Eri 2h54'48"10d41'
Miguel & Ana
Uma 12h2'44"34d17'
Miguel,Joy
Vul 19h47'50"27d8'
Miguel,Lawrence Andrew
Aur 5h27'0"30d48'
Miguez,"Jo's Joy"-Joy Rita
Peg 22h19'45"31d13'
Migut,Jr,James J
Cnv 13h26'23"41d16'
Migyanka,Tina
And 2h6'1"38d50'
Mihaich,Amanda Lee
Aur 7h24'37"38d19'
Mihaich,Brian John
Cam 3h49'20"53d6'
Mihaich,John
Aur 6h53'43"38d34'
Mihalakos,Anastasia Socrates
Sge 19h16'33"16d29'
Mihalko "Toots",Cheryl Lynn
Cas 23h32'29"62d52'
Mihami,Jean-Paul
Umi 15h47'35"82d6'
Mihaylo,Steven G
Vul 19h40'22"20d29'
Mihia,Oneina
And 0h44'15"30d28'
Mihkels,Arvo
Ori 5h55'0"20d29'
Mihkels,Peter
Sex 10h7'31"-1d59'
Mihkels,Robert
Ori 4h54'29"0d29'
Mihok,Suzanne
Cas 23h41'12"61d27'
Mihorean,Philip F
Uma 9h41'32"51d42'
MiJa
Ori 5h4'29"7d60'
Mijardus
Boo 15h32'34"48d4'
Mijn Lieverd Sohail Asad
Eri 4h59'33"-6d22'
Mijonbo
Ori 5h47'30"11d9'
Miju
And 1h59'1"37d33'
Mik
Ori 5h40'28"-6d49'B
Mika's Dream
Ari 2h14'53"18d12'
Mika,Frank
Lib 14h56'55"-5d46'
Mikaelian,Alex & Suzi Pilavdjian
Cas 1h43'18"58d43'
Mikaelian,Hope
Cas 0h10'55"54d51'
Mikajag
Del 20h26'12"20d12'
Mikalic,Jackson Riley
Lac 22h7'28"49d40'
Mikayla
Cam 5h59'1"61d27'
Mike
Dra 19h49'51"70d22'
Mike
Aur 6h4'53"38d50'
Mike
Dra 14h38'27"62d28'
Mike & Alicia
Peg 21h29'37"20d0'
Mike & Alyson
Per 4h0'44"31d7'
Mike & Amy Wishing Star,The
Lyr 19h4'46"28d48'
Mike & Ashley
Peg 22h2'41"28d34'
Mike & Denise
Dra 17h54'45"65d28'
Mike & Graham
Her 17h0'59"28d16'
Mike & Kristina
Sge 20h7'1"20d18'
Mike & Linda
Peg 21h27'58"23d7'
Mike & Lisa "Eternal Love"
Aql 20h14'41"0d5'
Mike & Lucy
Sge 19h28'21"16d21'
Mike & Mandi:Best Friends Forever
Lyr 18h42'27"30d35'
Mike & Mary "Always"
Lyn 8h29'58"49d36'
Mike & Mary's Big Adventure
Cet 2h35'31"5d16'
Mike & Mary's Star
Oph 17h17'44"-23d33'
Mike & Reesi
Cyg 21h12'35"38d4'
Mike & Wendy's Blitz Night
Uma 13h31'49"54d26'
Mike B
Cet 1h58'17"0d57'
Mike D
Per 2h31'18"51d36'
Mike Loves Lori
Lyr 18h58'30"33d0'
Mike the Love of My Life
Cnv 13h46'55"31d33'
Mike The Mechanic
Dra 16h0'40"52d26'
Mike's Barber Light
Cet 1h22'1"0d37'
Mike's Big Hoohna In The Sky
Cmi 7h16'40"5d12'
Mike's Eternity
Aur 6h24'56"38d40'
Mike's Harp
Com 13h19'54"21d51'
Mike's High Hopes
Her 17h31'34"26d44'
Mike's Love For Lori 4-25-87
Lyn 7h55'1"44d23'
Mike's Nova
Cmi 7h19'54"1d39'
Mike's Penguin
Per 3h28'17"38d52'
Mike's Place
Dra 19h6'1"56d21'
Mike's Silver AIS
Aur 4h53'34"41d13'
Mike's Star
Gru 22h6'12"-52d8'
Mike's Star
Boo 13h47'20"18d38'
Mike's Twinkle
Uma 9h52'25"42d20'
Mike,I Will Love You Forever-Pam
Crb 16h16'35"31d11'
Mike-My Love
Cam 4h5'1"70d60'
Mikel,Joseph
Per 3h59'52"36d30'
Mikel,Knight Of Wands
Dra 20h0'33"68d3'
Mikel-Lari Danielle
Uma 10h9'23"56d9'
Mikela
Psc 0h59'11"28d29'
MIKELL 10/92 A Diamond In The Sky
Sgr 19h14'38"-23d0'
Mikell,Julia LeCraw
Peg 22h41'43"24d31'
Mikelly
Cam 5h43'56"60d38'
Mikeska Star,The Susan
Uma 11h36'0"42d20'
Mikey
Eri 2h59'0"-6d48'
Mikey
Lyn 6h29'51"59d8'
Mikey
Ser 15h20'29"3d0'
Mikey & Al's Light
Uma 10h0'0"57d35'
Mikey & Jackie
Mon 6h57'1"0d40'
Mikey & Minx
Lac 22h49'12"54d19'
Mikhael Stephen
Cep 21h23'19"58d59'
Mikhail's Destiny
Ori 5h52'1"11d15'
Mikhail,Ramzy N
Aur 6h28'51"38d40'
Mikhailsworth
Boo 15h19'1"40d11'
Miki
Uma 8h59'56"70d23'
Miki
Oph 17h38'28"-16d40'
Miki
Cam 5h57'15"58d16'
Miki Finn
Uma 10h4'59"47d36'
Miki Queen of Much-too-Much
Cas 0h38'51"70d38'
Mikie
Lac 22h49'1"56d47'
Mikie III
Boo 14h30'48"30d55'
Mikiko & Kenji Forever
Vir 14h26'33"2d2'
Mikilani
Ori 5h29'18"0d59'
Mikki
Aql 18h59'26"12d56'
Mikkie Leigh
Lyr 18h50'58"30d34'
Mikkola,Katja
Umi 13h55'26"70d15'
Mikla,Frank J
Crb 15h45'1"28d37'
Miklja,Terri & Marinel
Umi 14h4'1"66d46'
Miklos,Peggy
Cyg 21h26'58"50d26'
Miko
Mon 8h6'0"-9d9'
Miko
Uma 10h45'25"59d7'
Mikols,Mark
Cam 3h58'58"60d29'
Mikota,Lubomir
Cep 0h26'1"77d52'
Mikota,Polly
Ori 4h42'54"14d1'
Mikovits,Monica
Lyn 8h46'26"34d37'
Mikrut,Joey Dog
Her 16h27'48"32d50'
Miksa,Lauren
Cyg 21h20'57"38d31'
Miksa,Ronald Lee
Her 17h14'36"42d9'
Miksic,Evonne Marie
Peg 23h5'32"33d7'
Miksic,Paul Richard
Cep 22h39'19"70d4'
Miksuvius
Uma 8h38'53"56d43'
Mikula,Amela
Cas 1h47'16"60d14'
Mikulski Nov 95,Sheila & Dennis
Cyg 19h26'60"32d19'
Mikulsky,James Robert
Cma 7h15'34"-16d11'
Mil & Ray Always
Lmi 10h56'41"27d37'
Mil,Danile
Aql 18h43'1"11d51'
Mil-Ron
Mon 7h4'13"-0d27'
Mila
Ori 5h40'42"-1d10'
Mila
Aur 6h13'55"33d27'
Miladinovich, Christopher J
Cet 2h36'46"2d3'
Milady
Lyn 7h21'60"44d37'
Milagros(Sonia Milagros Sanchez)
Peg 21h28'35"21d47'
Milagros,Daisy
Psc 1h0'29"21d29'
Milam,Daisy Jean
Peg 22h25'40"27d8'
Milam,Justin
Sgr 18h50'47"-31d15'
Milan
Cyg 21h1'59"37d43'
Milan Held Star
Her 18h12'58"40d53'
Milan's Beginning
Uma 11h30'58"64d33'
Milan,Eddie
Lac 22h10'1"46d51'
Milan,James
Hya 8h9'22"-9d15'
Milan,Julius G
Aql 20h36'13"-0d44'
Milan,Sheri
Sex 9h56'39"-5d52'

Milana,Theresa
Leo 11h0'27"23d15'
Milanak,John George
Aur 5h56'12"31d30'
Milander
Ori 5h57'1"18d38'
Milanese,Charles Marie Rose
Her 16h15'28"41d57'
Milanesio,Oscar
Per 3h6'36"31d18'
Milanesio,Simone
Pyx 8h51'51"-27d32'
Milani,Jr,John "Precious"
Dra 17h48'48"58d39'
Milani,MD,Robert J
Oph 17h53'1"1d7'
Milankov-21 03 1988, Nina
Cas 1h49'38"68d55'
Milankov-3 02 1960, Stevan
Uma 9h38'22"58d5'
Milano,Regina Marie
Ori 5h50'17"17d0'
Milapake
Boo 15h20'39"42d12'
Milauskas,MD,Albert T
Cma 6h26'1"-24d39'
Milazzo,David A
Ori 6h3'0"6d50'
Milazzo,Dorothy Mary Mary
Lac 22h42'49"53d30'
Milazzo,Marcia
Milazzo,Marcia
Cas 0h10'51"58d34'
Milazzo,Stacy Marie
Lyr 18h36'49"30d34'
Milbank,Jeremiah
Dra 16h24'39"68d30'
Milbank,Mark L
Cep 20h46'1"70d20'
Milbauer,Michael Jude
Per 2h55'33"40d55'
Milbeck,Palmerina & Jerome
Cyg 20h4'52"31d5'
Milbredt,Kathryn Eleanor Wood
Umi 16h30'17"79d7'
Milburn,Fiona Elizabeth
Cas 0h48'1"63d57'
Mild,Markus Birkenbach
Boo 14h20'1"10d13'
Milde,Johannes Franz
Ari 2h2'13"20d1'
Mildenberger,Frank
Her 17h20'43"38d32'
Mildner,Maryetta & Karl
Uma 9h59'56"51d6'
Mildred & Rose-Women of Valor
Cet 1h53'30"-0d32'
Mildred 90
Tau 4h20'44"15d21'
Mildred Marie
Mon 7h13'37"-6d53'
Mildred-Jules
Cas 0h28'59"62d11'
Mildren,Mitchell Timothy
Cet 1h42'44"-2d39'
Milecofsky,Margo
Uma 10h50'29"40d48'
Mileham,Matthew
Ori 4h44'18"0d35'
Milena
Cas 2h59'35"60d25'
Milena's Star
Uma 10h48'27"52d3'
Miles
Vul 19h58'16"23d12'
Miles 1930,Donald Keith
Lac 22h1'14"38d34'
Miles Away From Alisal
Lac 22h35'51"54d20'
Miles Into 1derland
Uma 8h50'46"62d19'
Miles Rom 8:13
Oph 18h0'36"10d31'
Miles,Abigail J
Cas 0h37'37"56d12'
Miles,Alison
And 23h21'23"50d26'
Miles,Alison
And 2h34'0"40d18'
Miles,Allan Richard
Ori 5h58'1"15d44'
Miles,Alma
Cet 1h24'31"0d9'
Miles,Bettie Rochelle
Cet 3h10'48"1d10'
Miles,David
Cep 22h5'42"68d40'
Miles,Denise
Cet 0h6'39"-17d38'
Miles,Donna Matthews
Cyg 21h56'40"53d13'
Miles,Dr David Neale
Oph 16h56'36"-25d7'
Miles,Edna
Cyg 21h31'42"41d48'
Miles,Elizabeth Joy
Lyr 18h29'49"30d20'
Miles,Frederick B
Her 18h29'49"20d35'
Miles,James P
Lac 22h30'42"56d32'
Miles,Linda Jean
Uma 8h17'17"62d28'
Miles,Maria `The Brightest Star'
Cyg 19h30'14"33d11'
Miles,Matthew Daniel
Aur 7h9'43"40d57'
Miles,Megan Louise Mooney
Umi 15h16'21"66d59'
Miles,Nicholas John
Ser 16h1'47"14d23'
Miles,Rebecca R
Lyn 6h33'14"58d24'
Miles,Robert
Per 2h12'27"57d22'
Miles,Robert
Cep 20h55'1"61d33'
Miles,Sasha
Sge 20h6'43"16d45'
Miles,Seraphine
Cas 1h24'25"52d50'
Miles,Shysti
Cet 3h19'1"9d29'
Miles,Steve
Sex 9h54'54"-5d22'
Miles,Suprise Matthew R
Per 1h53'39"48d48'
Miles,Tiffany Lynn
Com 12h32'12"24d48'
Miles-Taylor,Sandra
Peg 21h56'48"33d37'
Mileston,Jan Muroff
Sex 10h27'13"-1d32'
Milewicz,Felicia J
Lac 22h44'47"38d20'
Miley,Patrick & Juliet
Aql 19h29'20"11d12'
Miley,Paul
Cep 23h8'45"64d32'
Milford,"Muffie"
Lyn 8h40'1"37d51'
Milford,Alexis Andrea
Peg 23h3'56"8d15'
Milford,Barbara
Cas 2h47'19"70d20'
Milford,Joanna
Cyg 19h54'56"45d6'
Milford,Louise H
Cas 0h30'1"61d16'
Milfred
Cyg 19h27'22"35d44'
Milhorat,MD,Thomas H
Oph 17h52'42"7d49'
Milhouse,Ian Patrick
Boo 14h40'56"22d8'
Milhouse,Rebecca McBain
Lac 22h36'0"56d20'
Milidrag,George D
Dra 11h57'49"71d3'
Milis,Taisce
Aur 5h1'25"45d23'
Militello,Maria Tomasino
Boo 15h2'18"31d26'
Milizio,Salvatore Sweetie Precious
Her 17h20'55"38d50'
Milja
Cas 2h43'59"61d35'
Milkey,Vickie Lou Patrick Reid
Lmi 9h55'0"40d23'
Milkins,Graham
Lyn 9h16'40"40d48'
Milkis,Janice Marie
Cas 1h32'46"67d31'
Milkovich,Anna
Lyn 7h29'1"50d24'
Milkovich,Rachel Mait
And 1h39'58"38d24'
Milkowski,Carol McKee
Crb 15h30'60"38d10'
Mill's Lighthouse,Troy
Aur 5h34'15"29d13'
Mill,Barbara
Cam 3h14'42"63d6'
Mill,Nicholas Joseph
Vul 19h23'27"26d60'
Millan,Alex
Her 18h18'1"14d10'
Millan,Robyn
Mon 8h6'57"-9d58'
Millar,Alan Stevenson
Uma 12h3'51"33d38'
Millar,Andrew Hatty
Cyg 20h22'31"39d19'
Millar,Billy
Uma 12h52'37"58d46'
Millar,Bruce
Ori 5h40'55"11d19'
Millar,Iain
Ori 6h3'56"2d13'
Millar,Thomas E
Dra 19h33'0"65d23'
Millard,Billy
Her 16h40'58"50d37'
Millard,Edmund Lloyd
Aur 6h18'0"37d6'
Millard,Jr,Alfred
Cep 22h30'48"63d48'
Millard,Popperdeesma Robert J
Ori 6h1'31"-2d16'
Millard,William Owen
Sco 16h54'33"-40d32'
Millauer,Dr Gerhard
Sgr 18h50'39"-21d6'
Millband,Charles
Cyg 21h19'47"39d48'
Mille Fois Merci
Per 3h9'14"40d20'
Millen Star-DWM & IWM At 40
Cyg 21h18'0"34d14'
Millen,Jack David
Lmi 10h42'1"25d59'
Milleo,Theresa
Lyn 8h1'15"57d35'
Miller
Cmi 7h23'54"7d37'
Miller "21",Kathy
Cet 3h16'1"6d18'
Miller "Dad",Tim
Cet 3h15'30"1d30'
Miller (Mr M),James J
Uma 9h15'40"71d43'
Miller (Turtle II), Paul J
Lac 22h12'41"46d4'
Miller 01/06/44-01/06/94,Ivor
Uma 10h18'32"71d3'
Miller 62693,Stephen & Debbie
Eri 4h58'21"-10d1'
Miller Borealis (Constellation EmCare)
Uma 8h54'57"49d30'
Miller Boys-Milt, Larry & Samuel,The
Per 3h36'21"38d15'
Miller Forever,David & Elizabeth
Mon 7h0'0"8d39'
Miller GOD's Little Angel,Courtney
Lyn 7h4'13"44d43'
Miller II,George Willie
Hya 8h42'19"4d7'
Miller III,Allen F & Janet Saville
Eri 3h20'55"-2d48'
Miller III,Ernest St Clair
Peg 22h0'14"10d54'
Miller III,John Francis
Ori 6h5'39"20d52'
Miller Is My Valentine,Robin
Cyg 20h17'33"31d21'
Miller IV,Allen Reynolds
Aql 18h58'49"15d8'
Miller Star,Freddy "Bubba"
Sct 18h38'1"-6d27'
Miller Star,Jerry
Aql 20h18'22"7d59'
Miller Star,R D
Cnv 12h20'29"42d36'
Miller Star,Susan & Ed
Cyg 20h41'12"38d4'
Miller VII,Andrew Galbraith
Hya 8h30'56"0d43'
Miller"Super Grandma", Norma J
Mon 6h23'41"5d14'
Miller's Destiny, David
Dra 20h6'1"62d19'
Miller's Lite
Mon 7h33'50"-1d26'
Miller's Mark
Cep 21h59'24"70d12'
Miller's Personal Star,Natasha
Uma 10h33'55"50d31'
Miller's Wish,Gabé Jillian
Crb 16h3'12"27d47'
Miller's Wish,John Jay
Aur 4h59'55"51d54'
Miller,"Dub"
Cyg 19h29'24"32d26'
Miller,A J
Dra 11h55'1"71d52'
Miller,Abby Marie
Peg 22h8'60"29d25'
Miller,Ada
Vul 19h40'30"23d18'
Miller,Alan Michael
Aur 5h17'42"49d46'
Miller,Alan R
Aql 19h11'53"12d59'
Miller,Alejandra Careaga
Pup 7h56'41"-22d52'
Miller,Alexander Herald
Dra 16h49'60"61d33'
Miller,Alisha Marie
Cas 0h46'1"75d31'
Miller,Alvin A
Peg 23h2'53"18d52'
Miller,Amanda
Cyg 19h27'55"35d43'
Miller,Amber Dawn
Cet 2h53'38"0d3'
Miller,Amy
And 23h48'44"40d42'
Miller,Andrew M
Lmi 11h3'0"30d47'
Miller,Ann M
And 23h0'27"43d47'B
Miller,Anna
Aql 19h28'14"0d22'
Miller,April Meherlene
Peg 22h23'33"21d51'
Miller,Arielle Sharon
And 23h2'24"41d15'
Miller,Arlene
And 0h2'42"46d19'
Miller,Arlene
Cyg 20h52'37"50d31'
Miller,Avery Eric
Boo 15h2'60"30d6'
Miller,Barbara
Lyr 18h19'18"38d16'
Miller,Barbara Faith
Eri 4h3'47"-16d53'
Miller,Barry
Her 17h20'40"42d50'
Miller,Belinda
Peg 23h43'37"31d42'
Miller,Ben W
Lac 22h54'43"53d42'
Miller,Benjamin
Cyg 20h36'39"40d27'
Miller,Benjamin Carmichael
Sct 18h55'43"-6d51'
Miller,Benjamin Joseph
Per 3h58'48"51d13'
Miller,Benjamin Worth Bingham
Boo 13h37'54"26d0'
Miller,Bill & Bettylou
Uma 9h31'39"57d6'
Miller,Bobby
Cam 3h48'48"73d59'
Miller,Bradley J
Aqu 21h35'0"-1d2'
Miller,Brandon H
Aur 6h53'47"40d34'
Miller,Brandon Scott
Sct 18h32'35"-4d55'
Miller,Brian Chadwick
Cet 3h8'57"7d29'
Miller,Brittany Lauren
Peg 22h17'18"20d12'
Miller,Carol
And 23h35'36"43d45'
Miller,Caroline
Lyr 19h18'0"40d26'
Miller,Chad
Cep 21h52'54"63d53'
Miller,Charles Martin
Boo 13h48'20"15d53'
Miller,Charles W
Cyg 20h31'31"43d11'B
Miller,Chelsea Lauren
Cas 2h45'28"61d10'
Miller,Cheri
And 2h30'0"45d29'
Miller,Cheryl Jayne
And 2h34'1"37d60'
Miller,Chris & Bill
Boo 15h24'55"42d5'
Miller,Christa Buendgen
And 2h6'0"41d42'
Miller,Christiane
Lyr 18h22'18"38d19'
Miller,Christina Patricia Elizabeth
Tau 5h52'56"28d33'
Miller,Christina Marie
Dra 9h44'25"74d10'
Miller,Christopher Michael
Sge 19h57'53"20d32'
Miller,Chuck
Del 20h50'0"8d10'
Miller,Cindie Lee
Lyn 7h53'5"38d19'
Miller,Clayton
Ser 15h40'23"6d57'
Miller,Clem F
Uma 9h35'21"67d29'A
Miller,Colby Anne
Aur 6h20'43"33d55'
Miller,Coleman Chandler
Boo 14h28'39"40d15'
Miller,Connie Bean
Mon 7h36'32"-0d6'
Miller,Cortney
Peg 22h46'19"3d13'
Miller,Courtney Ann
Mon 7h47'14"-2d26'
Miller,Dana Sorrenne
Peg 21h30'0"20d33'
Miller,Daniel Fredrick
Oph 16h58'52"11d53'
Miller,Danielle Nicole
And 0h12'33"33d11'
Miller,Danny Ray
Umi 15h4'23"70d10'
Miller,Daryl Paul
Her 18h12'17"48d10'
Miller,David Daryl
Cam 6h39'20"82d31'
Miller,David E
Oph 17h27'49"-1d1'A
Miller,David Harry
Uma 10h14'56"68d1'
Miller,David Lloyd
Cnc 8h1'24"8d44'
Miller,David Sewell
Her 16h33'45"42d21'
Miller,David Ulderico
Psc 22h54'25"0d25'
Miller,David W
Her 16h11'1"40d54'B
Miller,Deborah Ann
Gem 7h53'55"31d3'
Miller,Debra Lee
Lyr 18h41'20"29d21'
Miller,Deevid Richard
Ser 15h40'0"-1d46'
Miller,Denise
Lyn 8h11'23"45d34'
Miller,Dennis John
Per 2h45'21"36d29'
Miller,Dennis V
Boo 13h45'25"22d56'
Miller,Diana Jeanne
Cas 1h2'30"56d5'
Miller,Diane
Cas 0h38'48"58d15'
Miller,Diane Elizabeth
Mon 6h37'32"10d37'
Miller,Diane K
And 2h4'48"40d13'
Miller,Diane M
Lyr 18h42'53"32d56'
Miller,Don & Pam Leonard
Lyn 8h4'32"45d55'
Miller,Donald William
Per 1h53'55"47d49'
Miller,Donna M
And 1h17'36"48d52'
Miller,Donovan St Aubyn
Aur 6h25'1"30d49'
Miller,Dorothy J
And 1h2'52"38d31'
Miller,Dorothy Laila
Cas 0h18'55"60d23'
Miller,Dot
Lyn 7h34'0"41d12'
Miller,Douglas S
Per 2h40'56"35d24'
Miller,Dr Aaron
Lyr 19h22'57"40d57'
Miller,Dusty & Neat
Aql 19h0'41"-8d29'
Miller,Edward Sinn
Cep 22h15'56"63d56'
Miller,Edwin C
Per 4h24'1"51d47'
Miller,Elaine E
Ori 5h51'56"17d22'
Miller,Eliane Vivace
Mon 7h21'13"-8d33'
Miller,Ellen
Tau 5h44'33"28d31'
Miller,Emily Patricia
Lyr 18h39'28"38d30'
Miller,Emma Louise
Cyg 20h40'52"44d32'
Miller,Eric Wayne
Mon 6h55'0"-10d35'
Miller,Ernest C
Ser 15h43'39"4d12'
Miller,Esther Alice Rohrer
Cam 6h13'25"58d51'
Miller,Evelyn Rogers
Del 20h15'30"9d20'
Miller,Extra Special Teacher, Barb
Lyr 19h11'16"38d38'
Miller,Farris James
Hya 8h52'45"2d55'
Miller,Florence
Ori 5h1'13"12d11'
Miller,Florence Y
Sex 10h10'47"-0d54'
Miller,Floyd
Gem 6h22'20"25d12'A
Miller,Frank G
Gem 6h32'51"22d38'
Miller,Garcon Leoni aka Gary M
Aql 19h26'43"-0d29'
Miller,Gary L
And 23h0'27"43d47'A
Miller,Gary Wayne
Per 3h33'53"38d35'
Miller,Geo W
Ari 1h51'44"25d2'
Miller,Geoff
Uma 11h17'1"53d52'
Miller,George
Ari 1h44'41"21d42'
Miller,George H
Aur 5h7'38"40d18'
Miller,Gerald
Aql 19h1'43"15d58'
Miller,Gina
Cyg 21h53'20"53d20'
Miller,Glenn
Per 2h52'42"41d6'
Miller,Greg
Ser 15h57'0"2d46'
Miller,Greg L
Dra 14h36'27"63d54'
Miller,Gregory T
Her 16h28'0"38d37'
Miller,Guy & Grace
Crb 15h59'11"37d33'
Miller,H Beonita
Cam 7h56'36"61d37'
Miller,Hal J
Her 18h17'38"14d18'
Miller,Harold J
Aur 5h16'55"47d54'A
Miller,Harriet
Del 20h40'30"10d8'
Miller,Harry
Her 17h35'27"38d58'
Miller,Hartwell William Johnnie
Uma 11h37'1"48d6'
Miller,Heather
Mic 21h20'57"-43d49'
Miller,Helen Ann Kitchen Eide
Mon 6h18'58"3d3'
Miller,Helen Celeste Järossi
Cam 3h17'23"60d1'
Miller,Helen Louise
And 1h19'43"35d51'
Miller,J D
Sct 18h31'18"-6d44'
Miller,J D & Patty
Dra 17h21'0"60d22'
Miller,Jackie
Cas 1h46'13"61d8'
Miller,Jacqueline
Uma 12h9'44"60d16'
Miller,Jacqueline R
And 23h39'15"38d1'
Miller,Jaime Lee
Lyn 7h56'1"41d40'
Miller,James L
Her 17h33'1"18d57'
Miller,Janette
And 23h2'58"37d8'
Miller,Jason
Cet 2h19'20"8d35'
Miller,Jean
Aql 20h31'48"-0d18'
Miller,Jeanne
Lac 22h28'41"54d10'
Miller,Jeannine
Lyr 18h17'45"40d11'
Miller,Jeff T
Aur 5h8'37"40d27'
Miller,Jefferson M
Dra 19h31'15"68d32'
Miller,Jeffrey
Sco 17h29'43"-30d52'
Miller,Jeffrey Calvin
Cep 22h19'10"60d5'
Miller,Jeffrey Neil
Her 17h20'56"14d55'
Miller,Jennifer & Richard Yarid
Com 12h15'28"19d45'
Miller,Jennifer Louise
Aqr 20h58'47"-13d29'
Miller,Jesse Raymond
Ori 5h55'39"8d55'
Miller,Jessica K
Mon 6h38'29"11d53'
Miller,Joan
Aql 20h1'28"10d58'
Miller,Joan & Stan
Cap 20h54'9"-27d43'
Miller,Joan Magdaline
Cyg 20h31'38"42d52'
Miller,Joanie
Cyg 21h29'14"40d17'
Miller,Joann Clingler
Lmi 9h50'54"38d17'
Miller,John
Ser 18h7'20"-14d22'
Miller,John Bradley
Cet 2h32'0"-1d22'
Miller,John F
Crt 11h14'39"-14d26'
Miller,John L
Cet 0h25'40"-17d34'
Miller,John Patrick
Vul 20h18'1"26d6'
Miller,Jonathan Hunter
Oph 17h18'20"-22d51'
Miller,Joseph Michael
Her 16h57'34"25d18'
Miller,Joyce
Cas 3h0'39"61d0'
Miller,Joyce Ann
Cas 2h3'59"73d20'
Miller,Jr"D Jaye", Donald M
Aql 18h56'28"10d20'
Miller,Jr,Happy 4th, Von Loves Charles D
Cyg 21h21'56"40d49'
Miller,Jr,Justin McCarthy
Sex 10h30'24"-2d25'
Miller,Jr,Russell Keith
Cmi 7h7'23"5d16'
Miller,Jr,Terry Hollis
Peg 21h26'0"2d31'
Miller,Jr,Ward M
Per 2h5'11"47d10'
Miller,Judith Ann
Cma 7h0'32"-30d14'
Miller,Judy
Peg 23h8'21"10d14'
Miller,Julian Jay
Lac 22h7'0"37d30'
Miller,Julianne Marie
And 0h24'37"34d42'
Miller,Justin C
Aql 20h34'13"0d54'
Miller,Justine
Cep 20h53'58"58d20'
Miller,Kaley
Hya 8h50'23"1d25'
Miller,Karen Judith
And 0h37'53"30d24'
Miller,Karen Renée
Peg 21h54'12"34d32'
Miller,Kathryn Mae
Psc 1h6'57"23d32'B
Miller,Kendra Christine
Aqr 21h41'29"0d5'
Miller,Kenneth Laurence
Per 2h4'42"57d25'
Miller,Kenneth M
Aql 20h0'49"-6d26'
Miller,Kent
Oph 17h13'30"-20d14'
Miller,Kerry L
And 2h19'50"42d54'
Miller,Kim Marie
Peg 23h5'50"20d49'
Miller,Kimberly Anne
Uma 9h21'28"51d47'A
Miller,Kyler Bryce
Ori 5h53'1"8d57'
Miller,Laura Crawford
And 0h15'17"38d22'
Miller,Laura Jean
Cet 0h53'27"0d54'
Miller,Laura Marie
And 2h26'24"45d41'
Miller,Lawrence Elzie
Psc 0h0'34"1d4'
Miller,Lee Amos
Dra 17h18'1"60d5'
Miller,Leif Aaron
Dra 16h27'22"60d3'
Miller,Lenore Anita Sacco
Cas 0h57'41"60d40'
Miller,Leonard M
Cma 6h55'4"-19d47'
Miller,Les C
Vul 19h14'47"22d27'
Miller,Leslie "Red"
Cma 6h52'17"-18d28'
Miller,Linda
Crb 15h57'0"37d42'
Miller,Linda Braithwaite
Lyn 7h49'23"48d10'
Miller,Linda Irene
Dra 10h53'57"74d13'
Miller,Lisa
Lmi 10h59'25"31d5'
Miller,Lisa Jo Ann
And 2h34'18"38d17'
Miller,Little Eddie
Ori 5h16'11"11d15'
Miller,Logan Anthony
Dra 19h4'23"54d60'
Miller,Lori
Lyn 8h2'30"47d29'
Miller,Louis Dadant
Ori 5h6'18"1d6'
Miller,Luanne
Mon 6h32'42"-0d52'
Miller,Luke Bernard
Dra 18h45'38"70d32'
Miller,Lynette
Com 13h31'43"21d9'
Miller,Lynsey Peyton
Peg 22h21'38"29d39'
Miller,Madelyn Keenyn
Cas 1h12'28"61d21'
Miller,Maggie
Lyr 18h35'41"46d13'
Miller,Margaret
And 0h50'32"37d48'
Miller,Margaret & Wm Ray
Eri 2h56'0"-4d18'
Miller,Marilyn Ann
Aur 5h52'12"29d2'
Miller,Marilyn Jean
Cam 6h38'39"83d6'
Miller,Marjorie Cleveland
Sct 18h38'19"-4d44'
Miller,Mark
Cet 0h55'23"-4d17'

Miller,Mark A
Tri 2h0'33"32d4'
Miller,Mark George
Sct 18h51'59"-6d44'
Miller,Marsha & Bob Stilp
Eri 3h55'15"-6d35'
Miller,Martin
Per 4h3'13"37d37'
Miller,Martin(Marty)
Psc 1h25'55"31d57'
Miller,Marvin Elton
Aur 5h58'23"30d20'
Miller,Mary Elizabeth
Cyg 20h3'42"40d41'
Miller,Mary Lou
Peg 23h26'34"32d5'
Miller,Maryanne
Cas 0h33'53"60d26'
Miller,Maxwell Jonathan
Cnv 12h30'18"34d19'
Miller,McKenna Lauren
Mon 6h38'33"1d12'
Miller,Megan
Lyn 7h45'52"52d15'
Miller,Melinda Rose
And 0h2'22"35d21'
Miller,Meredith K
And 1h58'12"37d33'
Miller,Merle & Ophelia
Umi 17h0'47"80d11'
Miller,Mervyn
Boo 13h34'55"22d37'
Miller,Mia Sue
Cas 0h49'14"70d16'
Miller,Micah Christopher
Peg 22h36'11"30d23'
Miller,Michael
Per 4h7'12"50d16'
Miller,Michael
Leo 9h21'31"10d35'
Miller,Michael Pierrino
Cet 2h18'1"8d58'
Miller,Michael Anthony
Leo 11h0'50"18d25'
Miller,Michael J
Peg 23h20'44"10d17'
Miller,Michael Reed
Dra 16h21'28"58d3'
Miller,Michael Robert
Boo 14h5'57"34d36'
Miller,Michael Wayne
Aql 19h57'52"8d44'
Miller,Michael William
Lyn 7h42'0"50d4'
Miller,Micheal Anthonie
Ori 6h6'50"8d9'
Miller,Miseigh
Lyn 8h13'46"44d45'
Miller,Misty Noelle
Cyg 21h22'58"40d41'
Miller,Molly Jo
Peg 22h4'57"3d23'
Miller,Monroe Grant Scott
Aql 18h56'54"2d28'
Miller,Moonlady,Sandra M
Cas 0h27'19"67d8'
Miller,Morgan Haden
Lmi 9h57'0"38d11'
Miller,Morris
Eri 4h5'17"-11d56'
Miller,Nancy D
Umi 14h31'36"80d16'
Miller,Nicholas Andrew
Sco 17h3'40"-30d56'
Miller,Nicholas Clarke
Ori 5h57'18"11d58'
Miller,Nina
Ori 5h51'44"17d52'
Miller,Norman Lawrence
Psc 1h6'57"23d32'A
Miller,P George
Cyg 20h31'31"43d11'A
Miller,Page Layne
Ori 5h55'55"14d55'

Miller,Paige Elizabeth
Lyr 19h2'16"33d56'
Miller,Patricia
Ser 18h16'1"-14d60'
Miller,Paul
Cep 22h51'25"58d33'
Miller,Paul
Boo 15h40'50"40d49'
Miller,Paul
Lib 14h24'56"-8d10'
Miller,Paul Julian
Ori 6h5'1"-0d19'
Miller,Paul Richard
Aur 6h13'25"46d19'
Miller,Peter Joseph
Vul 18h55'23"24d34'
Miller,Philip
Sge 19h3'53"20d31'
Miller,Phyllis
Cyg 21h20'19"38d45'
Miller,Pierre
Tri 1h58'22"33d17'
Miller,Quinton Bryce
Cnc 8h52'14"7d54'
Miller,R Dudley
Peg 22h6'16"4d2'
Miller,Raegan
Dra 13h30'47"67d50'
Miller,Ray
Sex 9h54'34"1d12'
Miller,Raymond
Boo 15h5'60'14d22'
Miller,Raymond
Dra 10h40'56"73d32'
Miller,Raymond V
Per 1h28'24"53d45'
Miller,Rebecca Ann
Woodberry
Del 20h13'0"14d30'
Miller,Rebecca Suzanne
Cap 20h35'11"-12d3'
Miller,Reg
Ori 5h51'13"13d7'
Miller,Rema Sutton
Uma 9h35'21"67d29'B
Miller,Ricchard "My Hero"
Her 16h58'18"31d24'
Miller,Richard
Aur 6h2'58"36d45'
Miller,Richard & Lynn
Aql 18h56'49"8d58'
Miller,Richard Arthur & Nona
Michele
Peg 22h19'31"8d12'
Miller,Richard F
Hya 9h8'16"2d23'
Miller,Richard K
Boo 14h47'13"28d32'
Miller,Richard M
Boo 13h42'13'26d5'
Miller,Richard P
Boo 15h1'47"28d59'
Miller,Riley Alexandra
And 0h23'40"37d47'
Miller,Rita
Ari 1h58'37"25d4'
Miller,Robert A
Boo 13h45'33"23d41'
Miller,Robert Donald
Cet 2h3'38"-1d4'
Miller,Robin L
Cas 23h40'58"61d19'
Miller,Ronald Binks
Her 18h18'1"12d13'
Miller,Ruby
Lyr 18h38'45"40d21'
Miller,Russell
Cet 2h32'22"1d1'
Miller,Ruth
Lyn 8h43'47"33d40'
Miller,Ruth & Philip Horowitz
Cyg 20h39'43"45d38'
Miller,Ryan J
Aur 5h10'51"41d59'

Miller,Ryan Joseph
Boo 14h34'53"8d39'
Miller,Ryan Patrick
Per 3h27'41"31d39'
Miller,S Joseph
Aur 6h4'1"31d37'
Miller,S William "Man In The
Moon"
Ori 6h6'21"2d30'B
Miller,Sally C
Mon 6h54'52"-0d10'
Miller,Sam Allen
Oph 17h22'55"0d52'
Miller,Sarah Elizabeth
Mon 7h16'1"-1d31'
Miller,Sarah Louise
Peg 21h53'0"30d6'
Miller,Seth Morgan
Cet 1h24'18"-1d27'
Miller,Sharon-Elliott
Del 20h56'25"18d53'
Miller,Shawn Christopher
Cnv 12h49'20"39d2'
Miller,Shelby Alexandra
Lyn 8h39'58"44d21'
Miller,Shelley
Cyg 19h43'51"38d7'
Miller,Sherie Alise
And 0h50'11"35d59'
Miller,Sherri
Uma 9h38'27"50d9'
Miller,Shirley
Equ 21h2'15"11d45'
Miller,Simon Christopher
Cmi 7h53'35"3d57'
Miller,Skip
Uma 13h14'0"62d5'
Miller,Spencer Scott & Janet
Finley
Cyg 21h26'57"28d23'
Miller,Sr,Albert August
Uma 11h20'35"43d49'
Miller,Stacey Leigh
Mon 6h57'43"-1d6'
Miller,Stephanie Megan
Cas 1h16'50"61d37'
Miller,Stephen M
Cnv 12h19'38"35d54'
Miller,Steve Michael
Her 18h1'37"30d5'
Miller,Steven
Uma 11h53'51"46d21'
Miller,Steven W
Her 17h21'16"40d44'
Miller,Susan
Cas 1h57'29"70d20'
Miller,Susan G
And 0h20'52"32d46'
Miller,Susan Hope
And 23h46'0"46d9'
Miller,Susanna Marie
And 23h37'36"48d51'
Miller,Suzanne
Mon 7h0'20"4d9'
Miller,Ted
Aql 18h41'59"-2d31'
Miller,Teri
Cam 7h15'27"83d43'
Miller,Terrence Reul & David
Ori 5h52'51"20d29'
Miller,Terri Thompson
Gem 6h44'24"18d25'
Miller,Terril "The Big O"
Peg 23h5'20"10d56'
Miller,Terry Glenn
Aur 6h26'11"37d34'
Miller,Theodore
Peg 23h34'1"30d52'
Miller,Thomas Gerard
Cet 3h16'46"9d9'
Miller,Todd Michael
Aql 19h12'16"12d49'
Miller,Trevor David
Ser 15h35'22"24d32'

Miller,Trisha
Gem 6h22'20"25d12'B
Miller,Tron Lawrence
Ori 5h59'30"21d2'
Miller,Truman & Marie
Ori 5h53'45"5d20'
Miller,Tyson Sean
Oph 18h0'38"10d15'
Miller,Valerie
Eri 2h55'21"-4d16'
Miller,Valerie
Lac 22h34'0"38d53'
Miller,Victoria
And 1h2'1"36d21'
Miller,Victoria Donato
Peg 23h26'0"33d15'
Miller,Vince
Aql 19h50'28"11d51'
Miller,Virginia Ginny
Lyr 19h6'45"37d54'
Miller,W T C
Lyn 8h9'26"51d30'
Miller,Wallace W
Per 2h59'50"45d3'
Miller,Walter Michael
Aql 20h2'26"10d36'
Miller,Warren V
Aur 6h32'35"37d60'
Miller,Wes
Per 3h11'49"50d20'
Miller,Wesley Warren
Ori 5h54'17"14d0'
Miller,William
Boo 15h4'39"27d49'
Miller,William & Wendy
Peg 22h19'52"3d7'
Miller,William Gerard
Cam 3h55'50"78d15'
Miller,William"Willie"
Per 2h39'27"40d58'
Miller,Willie I
Hya 9h5'55"2d30'
Miller,Zachary Charles
Kenneth
Cep 2h36'1"78d7'
Miller,Zoe Adele
Aqr 21h56'40"-17d38'
Miller-Capece,Sheila Maureen
Peg 23h33'1"21d37'
Miller-OSF,Sr Rosemarie
Cnv 12h16'43"45d37'
Miller-Zbieroski,Avery Joel
Lac 22h24'47"38d23'
Millerd,James Elstin
Hya 9h37'39"-9d53'
Millerick,Ronald Paul
Aur 6h24'17"38d18'
Millers's Star,Heidi & Scott
Umi 16h54'50"79d2'
Millers,The
Uma 10h59'44"68d9'
Millership,Stephen John
Aql 20h32'34"0d56'
Millert,Olaf & Juta
Cyg 21h20'17"37d57'
Millesogni
Pyx 8h41'31"-28d28'
Millett,Rufus Rand
Aur 4h53'1"40d54'
Millette,Jimi
Ser 17h56'38"-13d12'
Millholland V,Lewis Curtis
Boo 15h7'10"11d47'
Millhone,Gentleman Mac
Tri 1h59'35"30d41'
Millhouse,Dr Felix G
Aur 4h49'33"41d3'
Millhouse,Quinci Maia
Elizabeth
Peg 23h29'42"20d3'
Millian,Diane Nicole
Cas 0h34'0"71d24'
Millicent
Del 20h25'1"11d28'

Millie
Cam 8h36'29"78d30'
Millie
Lyn 7h40'1"52d17'
Millie
Peg 23h23'0"30d40'
Millie & Willie
Lyn 6h49'0"60d57'
Millie J
And 0h47'39"45d58'
Millie's Mazel Tov
And 0h21'49"30d15'
Millie's Star
Lyr 18h29'1"40d55'
Millier,Samantha
And 2h15'48"45d10'
Milliet,Jacques
Uma 12h25'30"60d47'
Millieymiguel
Peg 22h14'53"31d10'
Milligan,Danielle M
And 23h33'24"48d46'
Milligan,Dillon
Gem 6h42'20"24d44'B
Milligan,Evan Knight
Hyd 8h39'1"-0d36'
Milligan,John Steven
Aql 18h44'18"-2d16'
Milligan,Karen Jean
Leo 9h44'1"7d24'
Milligan,Kyle
Gem 6h42'20"24d44'A
Milligan,Lawrence
Aur 5h3'46"30d31'
Milligan,Lucy
Eri 2h57'59"-12d12'
Milligan,Robert & Catherine
Uma 8h20'53"68d15'
Milliken,David
Ser 15h12'19"20d35'
Milliken,Zachary
Ori 5h43'43"12d5'
Millin,Edward Lewis
Tri 2h17'19"31d8'
Milling,Ariana
Mon 6h45'1"10d45'
Millington,Carol Ann
And 0h5'59"47d28'
Millington,Cerys Rowena
Lyr 18h27'50"46d17'
Millington,Steve
Cnc 9h13'0"7d29'
Millington,Susan Vanessa
Cap 21h26'15"-19d11'
Million Dollar Kiss, The
Peg 22h46'42"5d53'
Million Dollar Maggie
Mon 8h4'59"-8d8'
Million Kisses for You
Sgr 19h7'27"-15d17'
Million,Earl
Mon 6h32'14"-5d44'
Million,Lisa Michelle
Mon 6h26'50"-0d57'
Million,Peter Michael
Sex 10h10'56"-9d50'
Million DeBruce
Sge 19h41'22"18d51'
Milliron,Molly Kay
Cas 1h3'19"58d48'
Millis,Heather Kayliegh
And 23h3'33"49d49'
Millis,Kimberly Austin
Lyr 18h31'1"41d13'
Millis,Marvin Yates
Aur 5h7'14"42d1'
Millis,Norman McDole
Eri 3h20'1"-4d31'
Millman,David Benjamin
Aur 5h24'52"38d52'
Millman,Jack & Elinore
Her 15h48'17"51d5'
Millman,Norman A
Per 2h57'43"38d40'

Millman,Stacey
Vul 20h14'55"25d18'
Millns,Peter
Crb 16h13'33"33d16'
Millon,Theodore
Per 2h51'37"37d58'
Mills II,Robert Mason
Cep 23h35'38"67d50'
Mills My Own True Star
Dennis
Cyg 19h22'26"44d54'
Mills Star,Jane
Cyg 19h34'34"39d39'
Mills,"Lady Bug" Christina
Renee
Lac 22h37'59"37d43'
Mills,Aaron Isadore
Her 16h35'49"36d12'
Mills,Abigail Jinnette
Aql 20h12'41"10d18'
Mills,Alexander
Aur 6h3'16"38d1'
Mills,Amber Lynn
Cmi 7h25'57"8d45'
Mills,Amy
Cyg 20h0'44"30d27'
Mills,Arthur Warren
Cep 22h19'37"58d54'
Mills,Ashly Eva
Lyn 6h44'42"60d22'
Mills,Barbara & Edward
Ori 5h16'16"0d47'
Mills,Blondie
Lyr 18h25'52"38d28'
Mills,Catherine Margaret
Cep 20h53'42"55d41'
Mills,Cathy
Umi 14h8'0"72d26'
Mills,Cecilia
Mon 6h54'20"10d52'
Mills,Chloe Nicole
Cas 2h22'11"67d56'
Mills,David
Cyg 19h40'0"42d4'
Mills,David
Cep 21h1'34"60d9'
Mills,David F G
Uma 8h15'19"61d4'
Mills,David Martin
Cyg 20h27'40"40d41'
Mills,Delbert S
Crt 11h22'37"-10d59'
Mills,Denise
And 23h45'55"44d7'
Mills,Frank & Hilda
Mon 6h20'54"-5d39'
Mills,Georgia Irene
Aql 19h42'21"10d24'
Mills,Haley Rose
Gem 6h56'21"14d4'
Mills,Hayley
And 0h59'28"45d36'
Mills,Hayley
Mon 6h58'39"-10d1'
Mills,Holly
Cma 6h55'34"-18d56'
Mills,James Lowell
Aql 19h8'15"5d59'
Mills,Jennifer Ann
Aql 19h18'31"15d57'
Mills,Jessica Jean
And 23h8'13"40d47'
Mills,Jim
Ser 15h38'29"7d11'
Mills,John Thomas
Lyn 6h53'0"60d5'
Mills,Jolene & Paul
Mon 6h36'29"-0d19'
Mills,Judy
And 22h56'1"36d22'
Mills,Justin Aaron
Aur 5h28'20"40d9'
Mills,Kandice & Russell
Eri 3h25'46"-5d47'

Mills,Katie
Cas 2h44'0"67d53'
Mills,Kaylyn Rachael
Cas 0h11'32"61d43'
Mills,Lillian Laura
Cam 6h9'45"60d48'
Mills,Lindsey Nicole
Crt 11h13'25"-19d28'
Mills,Lois
Cas 2h37'46"61d45'
Mills,Lord Theo
Dra 14h17'12"64d18'
Mills,Marianne
Vul 19h44'51"23d49'
Mills,Matthew Lee
Cmi 7h54'52"1d2'
Mills,Maureen J
Cas 1h8'34"64d53'
Mills,Melissa
Cyg 19h18'0"28d25'
Mills,Patricia
And 0h22'0"37d45'
Mills,Randy
Sex 9h55'0"-5d15'
Mills,Rebecca Klem
Vul 19h57'33"23d47'
Mills,Robert
Per 2h25'56"56d59'
Mills,Ronda Carol
Mon 7h14'33"-10d15'
Mills,Scott & Cindy
Oph 17h16'15"-18d51'
Mills,Scott & Kim
Cyg 20h29'34"42d28'
Mills,Simon
Cep 0h6'23"75d37'
Mills,Sr,Richard Harlan
Dra 17h46'48"68d22'
Mills,Stephen
Per 2h11'27"57d56'
Mills,Stephen
Hya 8h49'57"-7d2'
Mills,Steven Alexander
Dra 13h8'54"64d25'
Mills,Tracy
Cas 1h46'0"73d6'
Mills,Vivienne
And 0h37'31"45d44'
Mills,Wendy
Peg 22h14'58"32d50'
Mills,William John
Cep 21h56'21"71d5'
Mills,Zoe
Peg 22h2'28"10d8'
Millsap,Rex B
Boo 14h36'12"31d33'
Millson Forever,John & Lori
Peg 22h36'28"21d37'
Millspaugh,Sandy & David
Lyn 8h55'59"37d1'
Millstead,Jr,Wendel M
Dra 20h18'42"68d53'
Millstone 49th Int'l
N'siah,Meka Ann
Lyr 19h1'22"33d51'
Millstone,Jennifer & Sam
Ori 5h47'31"20d43'
Millward,Daniel Joseph
Her 16h54'0"48d11'
Millward,Nicollette Noelle
And 22h58'0"36d5'
Milly
Cas 1h34'18"72d56'
Milly's Star
Tau 4h6'33"20d23'
Milly,Nancy
Mon 6h42'26"0d38'
Milman
Aur 6h58'26"41d14'
Milman,Morris
Cnv 12h22'30"40d50'
Milmar
Her 17h31'35"21d59'
Milmeister,Franci
Ari 2h36'47"21d47'

Milne,Aud Helene
And 1h21'43"39d36'
Milne,Bruce Robert
Ori 5h28'34"0d46'
Milne,Florence Lilian
Aql 19h31'0"12d13'
Milne,Frank
Ser 15h25'55"8d24'
Milne,Jack & Florence
Sex 10h47'40"-6d49'
Milne,Lesley Avril
Cyg 21h32'16"53d36'
Milne,Norma Morrison
Oph 18h6'0"11d22'
Milne,Robin
Cas 0h31'21"61d36'
Milne,Tyler & Carina
Oph 18h42'16"8d46'
Milner,Adrian
Equ 21h10'57"11d34'
Milner,John & Martha
Peg 22h32'55"10d40'
Milner,Morgan Daisy
Cas 0h0'0"54d60'
Milnes,Alexander
Ori 6h2'55"7d25'
Milnes,Carolyn K
Sge 20h7'40"20d3'
Milnes,Sadie
Crb 16h16'14"33d57'
Milo Star,The
Aql 20h12'35"5d8'
Milo's Luck
Per 1h56'10"56d38'
Milo,Gerry
Ser 18h4'33"-13d36'
Milone,Colin B
Aur 7h11'16"39d29'
Milos,Janet
Lyr 18h24'30"47d12'
Milos,Pamela Susanne
Oph 18h42'42"8d46'
Milosevic,Nicolas
Mon 6h24'55"7d44'
Milota,Peter Everett
Leo 10h16'34"13d48'
Milota,Sarah Louise
Psc 0h20'28"0d31'
Milovanovitch, Alexandre
Peg 21h55'56"21d26'
Milroy III,Robert B
Cnv 13h57'0"40d36'
Milroy,Karl A
Sex 9h49'40"-5d46'
Milshteyn,Sasha
Cyg 20h19'27"31d36'
Milstead,Charles
Hya 8h29'1"-8d36'24'
Milstead,Otto & Margie
Aql 18h56'20"13d29'
Milstein,Aaron David
Per 2h24'24"55d21'
Milstein,Beatrice
Mon 6h19'24"4d41'
Milstein,Dylan
Aur 6h14'42"35d22'
Milt & Letty
Crb 16h19'44"31d17'
Milton
Her 16h28'24"41d36'
Milton,Bob & Mary
Cyg 20h23'14"39d33'
Milton,Daisy
Lyn 8h10'59"38d59'
Milton,Henry Oscar
Lyr 18h32'30"37d10'
Milton,Joseph Payne
Peg 23h29'18"18d11'
Milton,Lorraine Anne
Cas 0h19'34"62d52'
Milton,My Buddy
Aur 5h0'48"45d13'
Milty 1
Uma 10h48'37"55d52'

Milu,Mary Grace
Sex 10h47'33"0d53'
Miluk,Brianna Rae
And 23h5'1"38d11'
Milum,Zindy
Lyn 8h47'43"34d31'
Miluski,Edward
Dra 16h47'42"68d12'
Milva
Boo 15h14'38"47d44'
Milward,Stanley Frost
Cep 22h46'48"59d31'
Milù
Hor 2h52'18"-49d33'
Milù
Vel 9h43'11"-49d15'
Mim & Gregg
Tri 1h33'55"28d50'
Mim,Heidi Lisa
Aql 19h58'1"8d11'
Mima Mi Amor
Sex 9h59'59"-6d13'
Mimi
Mon 7h55'31"-2d50'
Mimi
Mon 7h56'59"-2d6'
Mimi
Lyn 8h27'54"40d58'
Mimi
Cas 0h40'51"60d55'
Mimi
And 23h19'20"51d18'
Mimi
Lyr 18h24'35"47d39'
Mimi
Cas 2h3'59"70d19'
Mimi
Peg 23h19'34"32d54'
Mimi
Cma 6h59'45"-16d3'
Mimi
And 1h16'58"33d33'
Mimi
Cas 0h27'21"63d22'
Mimi
Mon 6h22'27"4d5'
Mimi
Cnc 8h22'56"15d5'
Mimi & Martin Forever
Aql 19h56'18"15d11'
Mimi & Pop-Pop's Golden
Szavis
Crt 11h48'1"-8d8'
Mimi Memo
Vul 21h16'29"20d22'
Mimi Tony Forever
And 2h18'37"47d36'
Mimi's Star
Cam 5h59'56"58d11'
Mimi-Mae
Cam 7h31'33"68d42'
Mimitran
Cam 5h45'30"67d48'
Mimma
Peg 21h52'33"20d26'
Mimma
Com 13h12'0"21d58'
Mimmo Zito
Mon 6h57'30"6d16'
Mimms,Georgeana
Cmi 7h22'36"8d35'
Mims,Del M
Ari 2h6'1"17d54'
Min Jorgen
Uma 8h51'21"50d14'
Min,Merry Jacque
Cnv 13h20'40"30d45'
Min,Shin Hyun
Ori 5h44'55"11d48'
Min,Xiao
Cma 6h10'11"-18d55'
Mina
Cam 8h3'25"80d40'
Mina
And 0h21'0"35d47'

Mina di Sospiro, Gaetano
Eri 3h19'52"-16d45'
Minaldi 12-15-28, Billie
Eri 3h48'10"-0d53'
Minaqua
Lyn 7h50'53"39d56'
Minard,Eternal Lee
Psc 1h22'46"30d11'
Minard,Susan Hedley
Lib 14h58'17"-8d42'
Minardi,Maria Teresa
Montevecchi
Cyg 20h59'12"30d4'
Minari,Fernand
Dra 17h1'55"60d26'
Minarich
Boo 14h55'33"26d4'
Minarik,Andrew
Ori 6h1'44"3d50'
Minartz,Franz
Uma 9h31'48"45d51'
Minatoya,Cheryle Marie
Aql 19h0'11"12d15'
Minawa
Sgr 19h31'40"-38d15'
Mincer,Shane Anthony
Oph 17h6'35"-20d20'
Minchella,Chiara Raffaella
Del 20h17'52"10d21'
Minchilli,John
Ser 15h37'22"18d55'
Minchong,Kelli
Eri 2h53'33"-18d41'
Mincy,Billie Jeral Hall
Mon 8h5'21"-6d16'
Mincy,Billie Jeral Hall
Uma 9h55'48"56d23'
Mind of Hisayo
Tau 5h26'48"27d2'
Minda
Aql 20h2'25"13d8'
Mindel,Lorenzo
Ori 6h0'18"-0d20'
Mindel,Pamela Watson
Mon 6h47'28"10d3'
Mindel,Phillip Bruce
Cmi 7h23'54"8d37'
Minder,Sean
Per 3h20'39"40d35'
Mindling,Teresa
Peg 23h45'11"11d58'
Mindszenty Jo'zsef
Hercegprima's
Ori 5h55'14"21d5'
Mindy
And 0h39'14"45d53'
Mindy
And 22h59'0"38d53'
Mindy
Lyr 18h43'52"37d59'
Mindy
Lyr 18h59'59"34d29'
Mindy & Doug
Cyg 21h19'46"38d43'
Mindy & Gary's Star
Uma 10h18'47"48d39'
Mindy Bear
Aql 18h58'28"17d10'
Mindy F My Shining Star
Cas 0h4'13"61d51'
Mindy Marie,My Love Forever
Cyg 19h41'47"30d23'
Mine
Aql 19h7'14"2d39'
Mine
Per 3h10'0"40d33'
Mine Alone RDC,Jr
Boo 15h12'56"50d43'
Mineau,Harriet Jean
Vul 19h40'38"26d47'
Mineau,Luc
Per 3h7'52"46d17'
Minece
Uma 13h39'38"60d1'
Mineiro,Kiva
Cam 3h56'46"57d17'
Minello,Jillian Ruth
Lyn 7h59'19"44d37'
Mineo,J C
Hya 10h41'52"-18d17'
Mineo,Sebastiano
Vir 13h14'3"-1d3'
Miner Mpls Mn,Jmom
Lmi 9h50'16"38d33'
Miner,Theresa Z
And 0h9'43"46d29'
Minetta,Karen
And 1h25'30"40d52'
Minette Ginette
Peg 22h26'0"21d1'
Minette,Michael Matthew
Her 17h16'16"45d22'
Ming-Hsialee,MD,PhD,
Catherine
Her 16h42'58"10d59'
Mings' Alpha-Centroy, William
J
Aql 19h4'43"2d27'
Mingshing
Lyn 7h46'39"42d50'
Mingus,Richard David
Dra 18h37'18"58d55'
Minicone Family,S
Dra 16h6'17"62d46'
Minicozzi,Mark J
Her 18h1'18"37d52'
Minier,Pierre
Uma 10h26'0"70d20'
Miniere-Wilson, Jacqueline
Com 12h34'28"21d54'
Minihold,Dipl-Ing Horst
Cam 6h52'30"70d36'
Mininger,Mark Edward
Cep 21h31'10"55d12'
Minis & Abundisalge
Cyg 19h36'25"28d58'
Minissale,Loris
And 23h23'43"51d27'
Miniufabissi
Com 13h10'29"30d49'
Mink,Frances
Cyg 20h44'0"38d24'
Mink,Melissa
Lyr 18h48'0"39d27'
Minkey's Minkey 40
Dra 15h48'20"57d33'
Minkler,Jason L
Her 17h12'52"27d17'
Minkner,David P
Her 18h45'21"12d19'
Minkoff,Mary E
Lyn 8h29'42"41d2'
Minkos,Michael
Cet 1h14'54"-0d53'
Minkow Star,Norman H
Aql 20h2'58"1d42'
Minks,Edward John Christian
Her 18h5'33"40d37'
Minkster "PVA",The
Ori 5h0'0"14d3'
Minkus,Ann
Mon 7h38'52"-2d0'
Minky
Cnc 8h24'11"32d58'
Minna
Com 12h26'21"20d37'
Minne,James (Jimbo)
Her 16h53'37"33d30'
Minnelli,Liza
Lyn 7h1'0"44d19'
Minnelli,Liza
Aur 5h55'47"54d49'
Minnesota Crystal Star
Cep 23h7'34"65d28'
Minnich,Jane Orth
Cet 2h55'21"1d19'
Minnick,Joseph Marcley
Aql 19h26'24"15d37'
Minnick,Krista Therese
Com 12h2'53"20d13'
Minnick,Love You,Lisa
Peg 21h41'0"23d25'
Minnie
Boo 15h21'16"38d18'
Minnie & Chapman
Uma 10h24'20"61d13'
Minnie 88
Mon 6h52'54"-6d9'
Minnie N Me
Sex 9h40'1"3d48'
Minnier 12/25/94, Meryl O
Lyr 18h59'50"28d11'
Minnigerode,Joann Burns
And 2h17'1"39d45'
Minnis,Keith William
Her 18h5'0"38d46'
Minniti,Cristiano
Lac 22h18'43"38d45'
Minniti,Dina Cultrera
Cmi 7h25'45"0d22'
Minniti,Elenamaria
Psa 22h28'0"-28d15'
Minns,Ernie
Cep 22h39'53"78d45'
Minns,Paula R
Cyg 20h29'43"50d28'
Minnucci,Joe
Her 16h23'39"24d57'
Mino,John David
Hya 8h41'43"3d22'
Minock,Jason Robert
Dra 16h2'1"61d36'
Minoff,Robert J
Sct 18h53'35"-6d42'
Minogue,John & Malinda
Aql 20h1'15"8d45'
Minogue,Johnny "Grandpa
Joe"
Dra 15h16'1"55d42'
Minogue,Patrick
Per 2h32'55"56d40'
Minogue,Thomas John
Peg 23h29'11"17d40'
Minor's Lovelight, Yvonne &
Robert
Cyg 21h16'25"28d57'
Minor,Bob
Cam 3h17'29"56d11'
Minor,Jeanne
Lmi 10h2'16"35d53'
Minor,John David
Ser 15h10'1"6d32'
Minor,Jr,Claudie D
Hya 8h27'0"-17d5'
Minor,Mandel
Gem 6h40'19"31d33'
Minor,Mary Anne & Michael
Cyg 20h50'56"40d15'
Minor,Rachel Marie
Ori 6h3'27"0d30'
Minor,Rana
Cmi 7h17'0"8d44'
Minor,Sierra Elisa
Mon 6h33'34"9d20'
Minor,Susie
Cas 0h54'0"54d37'
Minora,Roberta
Cas 23h21'45"53d41'
Minors,John Wesley
Aur 6h54'16"41d9'
Minott's Miracle
Ori 6h16'22"7d51'
Minott,Samantha
Peg 22h11'1"25d18'
Minou
Cyg 20h24'37"39d25'
Minouche-Dominique
Ari 2h1'14"25d14'
Minoughan,Patrick Timothy
Per 2h30'17"51d35'
Minshall
Vir 13h15'41"-1d51'
Minshew,Ann
Ori 4h55'1"-2d43'
Minsker,John Henry
Aur 7h2'41"35d29'
Minson Superlative
LXXV,Sidney
Mon 6h57'36"0d16'
Minta's Dream
Uma 9h12'21"68d30'
Minten,Alice
Eri 4h32'34"-18d8'
Minten,Dorothy
Mon 7h44'60"-6d45'
Minten,Ester
Cet 2h39'44"-6d20'
Minten,Janie
Sct 18h45'20"-6d27'
Minter,Andrew
Her 17h5'38"38d26'
Minter,Donna Kaye
Mon 6h18'59"6d3'
Minter,Michael Christopher
Aql 19h8'57"13d39'
Minter,Vishal Sean
Aql 19h47'43"11d31'
Minterr,James Kerr
Per 2h58'18"35d5'
Minto,Edward
Lyn 7h13'38"50d38'
Minton,Sadilla Ann Petty
Lyr 19h20'49"38d58'
Mintrim,Michelle
And 0h29'43"30d10'
Minty,Pierrette
Sag 20h2'41"-44d29'
Mintz's Star,Rick & Cindy
Oph 17h39'55"11d8'
Mintz,Elliot
Mon 7h4'2"-7d4'
Mintz,Elliot
Mon 7h19'26"-1d0'
Mintz,Natalie & Ezra
Lyr 19h3'43"26d52'
Mintz,Raymond H
Lyn 7h30'43"41d33'
Mintz,Rosemarie
Lyr 18h58'10"33d7'
Mintz,Sandra Jane Earl
Lyn 9h15'60"39d34'
Mintz,Tina
Cyg 21h11'28"37d25'
Minuet
Com 12h33'59"14d41'
Minutella,Jr,Joseph C
Cyg 20h46'23"52d13'A
Minutella,Marcella "Macy" I
Cyg 20h46'23"52d13'B
Minutolo,Michel
Per 3h57'32"37d10'
Minwell,Alison
Cyg 19h31'55"36d32'
Minx,The
Lyn 7h46'0"39d50'
Minyard,Liz
Oph 18h5'24"13d6'
Minzner,Daryce K
And 23h29'28"40d55'
Mio,Colleen Amoro
Gem 6h39'38"12d56'
Mio,Massy
Cas 2h21'27"60d49'
Mioara-Seres,Centro estetico
Del 20h18'19"10d17'
Miodonski,Marilyn Francis
Claire Desch
Com 13h13'0"30d54'
Mione,Alexa Marie
And 0h55'40"45d33'
Mione,Camille
Cas 23h4'0"58d36'
Miozzi,Rita
Aur 5h53'21"31d38"
Miquiz
Pup 7h56'56"-20d12'
Mir-Gouiram,Anne-Marie
Boo 14h46'21"39d23'
Mira
Lmi 9h45'0"33d42'
Mira & Goran's Star Of Peace
Aqr 20h45'57"-1d10'
Mira 0123
Ori 6h7'36"20d36'
Mira Estrellita Tot6'
Pyx 8h35'28"-30d19'
Mira II
Cas 1h57'35"77d21'
Mirabal,Lupe
Aql 18h59'28"11d10'
Mirabella
And 1h20'0"38d29'
Mirabella,Grace
Vul 19h42'37"20d27'
Mirabella,Star,Olivia Michelle
Cas 1h29'58"58d21'
Mirabello,Robert Primo
Per 2h20'11"55d36'
Mirabilevisu,Linda
Cas 0h30'48"65d29'
Miracle,Carolann Linda
Sgr 18h58'9"-27d39'
Miracle,Clarence & Lassie
Boo 14h34'58"19d48'
Miracle,Haley Marie
Eri 3h21'20"-11d50'
Miracle,Scott Andrew
Lac 22h51'40"54d55'
Miracolo,Andrew Scott
Aur 6h25'45"37d31'
Miraculous Megan
Uma 8h47'35"52d16'
Mirai
Aqr 22h4'12"-10d38'
Miral,Fabrice
Cnv 13h25'48"38d48'
Miramontes,Andy & Barbara
Ori 5h15'23"1d21'
Miranda
Uma 10h48'29"48d27'
Miranda
Lyn 8h10'15"33d57'
Miranda
Vir 11h48'17"7d46'
Miranda
And 1h19'50"39d43'
Miranda J
Sge 20h14'20"17d39'
Miranda Jule
Lac 22h53'1"52d58'
Miranda,Ben
Cet 3h7'58"3d29'
Miranda,Brendy Lynette
Sco 17h28'23"-33d54'
Miranda,D'Ann
Peg 22h53'0"20d44'
Miranda,Eddie
Per 3h55'53"40d46'
Miranda,Ella
Com 13h33'20"22d11'
Miranda,Ernest
Lyn 8h54'44"41d55'
Miranda,Jarid Alexander
Cmi 7h18'1"0d13'
Miranda,Michael
Peg 23h19'51"11d34'
Miranda 30/07/73
Cen 11h51'4"48d42'
Mirandi,Eleanor
And 0h45'35"22d15'
Mirando,Barbara
And 23h3'15"44d53'
Mirando,Catherine Joyce
Mon 6h42'18"10d49'
Mirando,Janine
Sco 17h36'1"-38d26'
Mirando,Michelle
Psc 0h56'12"28d43'
Mirandola,Clementina
Lep 5h25'39"-24d48'
Mirante,Alyssa Clare
Mon 6h43'1"7d35'
Mirante,Chantal
Uma 10h6'45"61d51'
Mirante,Lara Marie
Eri 4h37'32"-0d16'
Miravitlles,Luis
Her 16h39'42"41d37'
Mirch,Christopher Thomas
Aql 19h43'34"10d56'
Mirecki,Eddie Pat
Per 1h32'46"52d52'
Mirecki,P M Philip
Aur 5h0'44"40d60'
Mireille
And 23h22'27"50d3'
Mireles,Miranda Paige
Aql 19h44'15"10d41'
Mirella
Lup 15h20'42"-45d46'
Mirella
Cnv 14h53'47"28d4'
Mirella K
Sgr 20h9'20"-33d29'
Mirer,Rick & Stephanie
Uma 11h5'23"32d21'
Mirethomas
Uma 10h53'51"71d30'
Miretsky,Alexander "Sasha"
Ari 2h36'53"20d4'
Mireya
Dra 19h29'56"61d5'
Miri
Peg 23h4'35"30d18'
Miriam
Lup 15h11'28"-37d30'
Miriam
Cas 0h16'26"61d2'
Miriam
Vul 20h4'0"23d47'
Miriam
Lyr 19h3'25"28d43'
Miriam
Del 20h12'1"10d59'
Miriam
Ori 5h56'49"16d29'
Miriam
Mon 8h8'47"-5d57'
Miriam
Cam 5h57'51"68d53'
Miriam
And 2h21'1"45d4'
Miriam Ivette
And 23h12'54"40d22'
Miriam W
Peg 21h23'36"22d55'
Miriam y Gregorio
Lyn 8h1'15"42d43'
Miriam-Apple,The
Sge 19h57'16"20d34'
Miriana,Mary Lou
Cet 1h30'30"-11d21'
Mirilovich,Joseph & Lillian
Equ 21h2'30"3d10'
Miritello,Christopher John
Boo 14h54'1"27d42'
Miritello,Paula & Julian
Cintron
Cas 2h29'57"71d3'
Mirizio,Jr,Jack Frank
Ori 5h56'0"10d31'
Mirjam
Lup 15h10'42"-41d16'
Mirjam
Tau 5h34'18"20d16'
Mirjana (Slatkica)
Uma 11h30'11"43d47'
Mirkes,Jürgen
Dra 20h7'56"62d13'
Mirmelstein,Rona Becker
Ari 1h57'52"18d7'
Mirner,Beatrice Stein
Cas 0h43'0"64d10'
Mirocco,Richard John
Her 18h28'31"20d10'
Miroph
Lyn 7h30'31"46d15'A
Mirosa
Cep 21h25'0"68d31'
Mirosa
Lup 14h44'31"-42d27'
Miroslav
Dra 17h22'0"64d37'
Mirra,John
Sco 17h41'58"-35d10'
Mirro,Donna Ann
And 2h23'28"47d6'
Mirro,Mike
Cas 0h38'47"61d21'
Mirror To You
Boo 14h38'49"33d41'
Mirsky,Barbara & Bob
Cyg 19h24'1"33d10'
Mirt,Clarence G "Dude"
Cyg 19h26'36"33d25'
Mirta
Pup 7h56'37"-29d31'
Mirtha
Equ 21h3'35"10d49'
Mirti,Josephine
Equ 21h2'32"10d40'
Mirtilli
Dra 16h13'23"60d47'
Mirus,Marypat
Lyn 7h31'12"42d21'
Mirza,Gaby
Peg 22h40'1"31d37'
Misan
Ori 5h32'18"8d6'
Misarti,Gabriel J
Tri 2h6'10"31d53'
Mischka
Cyg 21h19'47"37d17'
Mischler
Cet 1h4'48"0d22'
Mischnick,Wolfgang
Lib 15h7'53"-23d57'
Misciagna,Brian Keith
Aur 5h2'45"50d48'
Miserendino,Peter
Vul 19h22'1"23d16'
Misetic,A Robby
Aql 20h5'58"6d0'
Misfund-Mum,Cynthia
Cas 0h1'1"58d59'
Mish,Ira P
Mon 7h43'52"-2d48'
Mish,Michael
Lac 22h36'1"50d17'
Misha
Peg 23h47'43"30d24'
Misha
Tau 4h30'15"30d43'
Misha
Vul 21h26'25"27d11'
Misha
Cyg 20h35'26"50d11'
Misha 16
Hya 9h9'28"2d25'
Mishaan PhD is A "Star
!",Marilyn Haddad
Com 12h24'52"27d25'
Mishal
Mon 6h25'16"11d15'
Mishala
Uma 11h18'55"47d18'
Misher's,Jennifer & Chris
McClean's
And 23h33'12"47d54'
Mishiu
Tau 3h30'50"30d11'
Mishka & Bluebird Woman
Hya 8h59'38"-7d50'
Mishkin,Mike
Per 3h37'52"38d49'
Mishler II,Dale J
Aur 6h8'23"31d34'
Mishler,Alanna Joy
Mon 6h12'40"-10d23'
Mishler,Donica
Lyr 18h58'0"31d37'
Misiak,Elizabeth
Cam 4h14'1"71d13'
Misiurewicz, Christopher J
Lac 22h26'20"52d51'
Miska's Star,Jason
Cep 22h15'26"55d42'
Miske,Lenord
Uma 10h31'1"59d27'
Misner,Dan
Ari 2h32'37"20d27'
Miso,Mrs
Cas 1h18'0"75d42'
Mison,Hilda & Arthur
Cyg 21h32'48"40d15'
Misra,Karan Jay
Cap 21h53'51"-19d36'
Miss Alice
And 0h30'1"22d55'A
Miss Alison
Cas 2h54'25"68d52'
Miss Barbara
And 23h49'35"32d57'
Miss Barbara's Star
And 23h39'1"47d28'
Miss Beth
Uma 11h23'25"38d1'
Miss Blue
Eri 3h14'31"-7d18'
Miss Bobbie
Cas 2h2'30"59d52'
Miss Brenda
Cam 7h7'24"68d39'
Miss Catherine
Uma 9h3'53"60d51'
Miss Chatelaine
Lyr 18h22'35"38d43'
Miss Daryl
Lib 15h42'1"-22d27'
Miss Diane
Mon 6h57'12"8d26'
Miss Dorothy
Crt 11h16'47"-19d55'
Miss Emily
Lyn 6h58'21"60d24'
Miss Emily
Mon 8h7'12"-5d42'
Miss Georganne
And 0h13'0"34d9'
Miss Hazel Two Crows
Sco 17h53'56"-31d11'
Miss Ida
Lyn 6h53'17"44d44'
Miss J
Eri 2h53'37"-15d30'
Miss Jackie
And 2h30'0"39d26'
Miss Jenny
Cnc 8h56'38"30d3'
Miss Joyce
And 23h41'15"37d50'
Miss Kitty
Mon 7h41'3"-2d39'
Miss Laetitia
Uma 9h21'50"54d37'
Miss Laurie
And 1h54'41"40d50'
Miss Marzi Pan
Eri 4h47'30"-5d20'
Miss Pat
And 23h5'54"38d11'
Miss Patricia
Mon 6h55'46"10d41'
Miss Priss
And 2h12'0"41d56'
Miss Protocol
Vul 20h46'54"28d55'
Miss Reba
Mon 6h21'44"8d47'
Miss Rebecca Kathlynn
Vir 13h27'58"11d42'
Miss Sarah
Mon 7h29'35"-10d30'
Miss Sheila
Mon 6h48'13"10d0'
Miss Shirley
Peg 22h2'57"2d30'
Miss Sue's Star
Lyn 8h57'13"38d11'
Miss Teedie
Lyn 7h32'33"50d12'
Miss Vivian
Hya 9h16'0"6d21'
Miss Winnie
Cap 21h54'1"-21d41'
Miss Young's "Super Snail"
Cas 23h28'50"60d3'
Missano,Theresa
Cyg 21h47'22"38d52'
Missant,Agnes Antionette
And 2h15'17"38d46'
Misseck,Jaroslawa A
Oph 18h42'17"10d36'
Missett,Eric John
Her 17h10'28"45d55'
Missi
Lyr 18h37'26"28d55'
Missi Lea
Cyg 19h28'57"32d15'
Missing Line,The
Sex 9h59'1"-5d4'
Missing Link
Cam 3h54'24"57d17'
Missing Lynk
Sex 10h41'0"-0d9'
Missing Piece
Uma 10h46'60"40d2'
Missori,Mauro
And 23h23'27"52d36'
Missy
Lib 15h42'52"-20d15'
Missy
Uma 11h10'29"58d8'
Missy
Mon 8h8'41"-3d30'
Missy
Cmi 8h5'11"6d7'
Missy
Peg 22h2'13"2d17'
Missy
Lyr 18h37'17"29d7'
Missy
Cas 0h51'18"62d51'
Missy
Cas 22h59'29"56d51'
Missy
Lyn 9h12'26"44d29'
Missy 33
Cnv 12h28'57"40d3'
Missy 5491
Cas 0h55'34"70d57'
Missy Baker
Dra 10h39'22"81d27'
Missy Star
Com 12h1'53"14d55'
Missy's Honey Bee
And 1h39'32"37d3'
MissyMister,The
And 0h7'47"38d5'
Mist,Candy
Cas 2h55'43"68d29'
Mister
Dra 12h22'13"64d50'
Mister B
Her 15h45'44"44d45'A
Mister Colin
Umi 14h12'46"68d2'
Mister Happy
Her 17h53'46"20d21'
Mister Mister
Mon 6h42'40"-10d21'
Mister Sof-T
Uma 10h49'38"68d14'
Misti
And 1h8'27"38d48'
Mistina,Barbara
Aur 5h27'33"29d57'

Mistique
Hya 8h14'0"-5d57'
Mistler,Haley A
Mon 6h43'17"3d3'
Misto My Wish Come True,Edward Gene
Aur 5h0'29"41d13'
Mistress Kage
Equ 21h3'12"2d41'
Mistretta,Frank
Lac 22h2'35"51d53'
Mistretta,Victor,S
Tri 2h5'18"35d41'
Mistrytta,Thomas J
Her 17h25'12"30d45'
Misty
Cam 5h4'15"70d15'
Misty
Uma 10h39'0"40d13'
Misty
Cas 1h11'30"62d59'
Misty
Aql 19h49'53"15d7'
Misty
And 0h8'38"34d13'
Misty Jean
And 2h24'30"42d6'
Misty Loves Dale
Cam 13h21'14"77d50'
Misty Marie
And 1h54'58"38d11'
Misty Michaelmus,The
Cam 3h5'18"70d44'
Misty Paul
Ori 5h26'37"0d3'
Misty R S
Her 18h5'46"40d51'
Misty Sunshine
Dra 17h43'53"63d49'
Misty's Heart
And 0h43'19"38d29'
Misty's Light
Peg 21h7'10"12d55'
Misu
Tri 2h43'0"31d27'
Misukanis,David Allen
Cep 6h28'15"86d19'
Miswaki,Hokuto
Sco 16h24'21"-27d9'
Misztal,Agata
Com 12h34'51"22d40'
Mitakides,Nick
Her 16h53'38"34d43'
Mital
Uma 11h8'50"44d33'
Mitarai,Stella of Kiyoshi
Sgr 19h19'3"-27d15'
Mitch
Cnv 13h43'44"37d52'
Mitch
Ori 4h51'0"0d9'
Mitch
Dra 17h38'44"68d21'
Mitch & Yvonne
Cyg 20h56'42"52d53'
Mitch and Kim 1434
Crb 16h3'49"37d41'
Mitcham Magic
Aur 5h29'38"40d19'
Mitchamore,R D
Hya 8h36'24"-1d47'
Mitchel,Jodi
Lyn 9h2'0"38d27'
Mitchell
Cyg 21h27'27"53d17'
Mitchell
Per 1h39'55"54d15'
Mitchell "Queen Of Everything",Laura
And 0h57'28"45d59'
Mitchell 1993,Eugene Chase
Lyn 7h22'1"44d48'
Mitchell 50,Peter Ian
Aur 6h16'55"30d12'
Mitchell At 21,Susan
And 0h22'37"34d30'
Mitchell Crawford My Love & Passion
Mon 7h4'31"-6d48'
Mitchell Curtis
Cep 3h4'59"78d9'
Mitchell II,William
Lmi 10h16'27"39d30'
Mitchell III,Argentry
Uma 10h34'0"48d48'
Mitchell Lee
Peg 0h6'0"20d45'
Mitchell Marc
Boo 14h36'51"7d34'
Mitchell U R My Sunshine,Love Kat
Mon 6h54'57"8d45'
Mitchell's "Twinkle"
Lyn 7h54'13"39d45'
Mitchell's Meritdome
Cnv 12h15'40"42d10'
Mitchell,"E J"
Dra 9h40'54"80d22'
Mitchell,Alexandra
Peg 0h8'11"17d51'
Mitchell,Amy Renee
Aur 5h27'43"31d13'
Mitchell,Andrew LaMounte
Her 15h58'20"17d20'
Mitchell,Ann Christine
Cet 0h38'27"1d59'
Mitchell,Barbara S
Peg 21h25'1"23d5'
Mitchell,Benjamin Laurence
Her 17h35'19"21d52'
Mitchell,Bill,Karina, Kiley,Karsey & Brook
Her 16h50'46"40d19'
Mitchell,Carolyn (Dwyer)
Cas 2h53'12"70d47'
Mitchell,Chad
Aqr 23h8'43"-8d21'
Mitchell,Chad A
Her 16h41'40"11d53'
Mitchell,Charlee Ann
Uma 10h42'37"68d8'
Mitchell,Christine
Cas 0h39'17"60d50'
Mitchell,Clara Belle
Ori 5h54'0"21d12'
Mitchell,Clare Lindsay
Umi 14h14'16"66d3'
Mitchell,Connie Ann
Tri 2h22'0"32d36'
Mitchell,David
Cnc 8h27'55"7d28'
Mitchell,David Milan
Her 17h13'30"18d48'
Mitchell,Debbie
Cas 0h36'0"75d48'
Mitchell,Dennis James
Sco 17h8'0"-38d2'
Mitchell,Dick,Susie & Brooke
Aur 5h24'26"40d5'
Mitchell,Donna
Lac 22h6'22"46d52'
Mitchell,Dorothy K
Aql 19h53'1"11d42'
Mitchell,Doug
Aur 7h4'1"43d50'
Mitchell,Dr Dan
Oph 17h53'0"-3d44'
Mitchell,Drew
Dra 12h42'55"75d60'
Mitchell,Dustin San
Peg 23h24'54"16d11'
Mitchell,Earl F
Per 4h21'0"50d30'
Mitchell,Elizabeth Louise
And 1h44'26"37d1'
Mitchell,Emma Estelle
And 0h7'24"47d2'
Mitchell,Eugene Harold
Aur 4h54'40"40d57'
Mitchell,Fred
Sgr 18h58'7"-29d23'
Mitchell,Glenda L
And 1h34'59"40d37'
Mitchell,Grace Liddy
Lyr 18h47'49"38d44'
Mitchell,Ilene
Dra 19h52'49"60d49'
Mitchell,Irene Elizabeth
Peg 22h1'15"30d6'
Mitchell,Jackie
Aur 6h3'23"31d41'
Mitchell,Jayne
Cyg 19h32'58"37d12'
Mitchell,Jennifer
Aql 18h42'60"-1d26'
Mitchell,Jennifer Dawn
Lyr 18h58'26"31d51'
Mitchell,Jerry W
Her 17h28'47"20d47'
Mitchell,Joan
Ori 6h14'26"1d32'
Mitchell,JoHanna
Gem 6h27'41"13d7'
Mitchell,John G
Cep 23h13'42"62d9'
Mitchell,John H
Cep 22h13'48"55d44'
Mitchell,John Iridium
Her 17h20'1"44d47'
Mitchell,John L
Per 2h5'23"56d57'
Mitchell,Joshua Richard
Crb 16h7'14"37d35'
Mitchell,Joy
Peg 0h10'1"17d35'
Mitchell,Jr,Herb
Mon 7h53'11"-5d44'
Mitchell,Julia Christine
And 23h15'36"50d50'
Mitchell,Kate & Mac
Gem 6h44'29"18d53'
Mitchell,Katherine
Vul 21h4'0"27d21'
Mitchell,Katherine
Eri 3h41'0"-14d9'
Mitchell,Katherine Lee
Lyr 19h20'18"35d17'
Mitchell,Kaylene L
Mon 8h8'33"-7d36'
Mitchell,Kelly
Cas 23h5'19"53d15'
Mitchell,Lauralee
And 2h33'49"45d35'
Mitchell,Lee
Cnc 8h26'1"8d4'
Mitchell,Leslie Gramps EmyLu Nana
Mon 6h39'10"11d14'
Mitchell,Linda
Peg 21h19'41"20d1'
Mitchell,Liz & Bernie
Cyg 21h7'23"31d7'
Mitchell,Lynne Mattern
Eri 3h49'26"-2d9'
Mitchell,Madeline Getz
Ori 5h1'15"11d10'
Mitchell,Maria
Mon 7h48'15"-1d52'
Mitchell,Mary Jane Gordon
Lyr 18h32'29"43d31'
Mitchell,Mary Lou
Cas 0h55'0"58d24'
Mitchell,MD,Steven P
Hya 8h13'36"1d47'
Mitchell,Michael
Cet 2h5'10"-0d0'
Mitchell,Michael
Aur 5h8'46"41d49'
Mitchell,Michael Alton
Cep 22h13'36"65d33'
Mitchell,Michael John
Her 16h59'51"41d11'
Mitchell,Michaela
Cyg 20h37'52"45d29'
Mitchell,Morgen Ruth
Peg 22h53'23"29d33'
Mitchell,Ms Jordan Elizabeth
Sge 20h2'1"20d19'
Mitchell,My Cosmic Dad Michael V
Per 2h7'45"58d20'
Mitchell,Neil
Per 3h31'48"51d43'
Mitchell,Nicholas Michael
Cam 3h17'0"60d10'
Mitchell,Nicole Kimberley
Ori 5h55'39"15d22'
Mitchell,Nigel
Equ 21h6'34"10d23'
Mitchell,Oscar
Ser 15h20'37"6d3'
Mitchell,Pamela Ann
Cyg 19h26'30"33d46'
Mitchell,Pastor Tim
Aql 18h56'53"-2d24'
Mitchell,Pat E
Cet 2h4'0"-18d55'
Mitchell,Patty
Per 3h0'48"40d47'
Mitchell,Paul & Lori Read
Lyn 9h1'36"45d52'
Mitchell,Paul Mark
Aur 5h1'39"45d2'
Mitchell,Randall "Darrion" Blake
Aur 4h55'1"40d56'
Mitchell,Rava
Eri 3h34'57"-2d17'
Mitchell,Rebekah T
Lyn 9h0'38"37d43'
Mitchell,Renee Lea
Lyr 18h19'24"38d22'
Mitchell,Richard
Dra 17h8'23"64d51'
Mitchell,Richard & Melinda
Peg 22h35'1"8d53'
Mitchell,Rickey
Aql 19h58'56"1d6'
Mitchell,Robert
Lac 21h58'1"41d11'
Mitchell,Roman A
Her 15h49'36"41d37'
Mitchell,Rosemary
Ori 6h7'54"2d54'
Mitchell,Ryne Abigal
Cam 3h50'0"61d2'
Mitchell,Sasha
Cet 0h56'27"0d42'
Mitchell,Savannah Kate
Vul 19h21'23"26d21'
Mitchell,Scott Davidson
Ori 6h1'11"20d26'
Mitchell,Shelley
Per 2h38'42"43d12'
Mitchell,Shelley
Oph 18h5'59"13d33'
Mitchell,Sloan
Aur 5h55'1"38d7'
Mitchell,Stacey Anne
Sge 20h17'18"20d2'
Mitchell,Steven Sheldon
Ser 15h19'34"20d9'
Mitchell,Steven
Hya 9h11'21"2d10'
Mitchell,Susan
Lyn 8h41'36"41d50'
Mitchell,Taylor K
Ori 5h30'24"0d36'
Mitchell,Thomas
Dra 11h18'52"73d43'
Mitchell,Tom
Cet 0h51'17"-0d32'
Mitchell-1993,Marc Stephen
Per 2h37'36"41d13'
Mitchell-40,Sarah Diane
Lyr 18h28'54"30d32'
Mitchell-Forever & Always Yours,Tony
Ser 15h18'45"2d52'
Mitchell-Shining High Above,Susan
Mon 6h24'44"-8d57'
Mitchella
Lyn 8h23'32"49d18'
Mitcheltree,Wayne
Eri 3h30'41"-3d8'
Mitchem,Carmen
Aur 5h21'23"40d16'
Mitchem,John
Boo 15h8'52"22d30'
Mitek,Ashley Elaine
Peg 23h28'34"28d57'
Mitek,Jordan Charles
Boo 14h5'13"23d4'
Mitere
Per 3h0'27"38d25'
Mithat
Ori 4h51'52"0d14'
Mithera
Aqr 23h30'20"-18d8'
Mitjans,Dolors Redondo
Mon 8h6'0"-6d23'
Mitnick,Gustave
Her 17h20'38"26d44'
Mitolo III,Rocco
Dra 19h13'54"70d57'
Mitorotondo,Michael James
Cep 20h57'0"56d9'
Mitorotondo,Sr,Michael J
Cam 4h55'45"67d43'
Mitra,Anthony Nath
Ori 5h34'0"-1d43'
Mitrakas,Lightning Gregory George
Vir 12h50'58"1d46'
Mitrione,Bernadette
Cyg 20h17'12"38d11'
Mitrovich,Michael & Michelle Lehman
Crb 15h15'0"31d30'
Mitry
Uma 10h56'23"71d20'
Mitry,Barry
Her 17h25'29"38d53'
Mitsakos,Anthony (Tony)
Per 1h39'57"52d59'
Mitsch,Darelyn Darr
Cet 2h5'30"6d12'
Mitschke,Claudia
Boo 14h18'22"14d2'
Mitsios,Mary
Cas 0h57'43"66d57'
Mitsuko
Psc 0h7'19"0d42'
Mitta,Vinay
Mon 6h47'21"10d37'
Mittelbach,Debbie
Cyg 19h30'56"38d15'
Mittelbach,Tannis D
Equ 21h6'29"10d49'
Mittelman,Harriet
Cas 0h5'0"61d24'
Mittelstaedt,Ron
Boo 14h1'18"27d54'
Mittelsteiner,Lucy
Cap 20h35'1"-10d15'
Mitten,Leigh
Uma 11h10'30"30d51'
Mitten,Valerie & Randall Ray
And 0h44'51"41d6'
Mittens
Cyg 20h50'17"39d36'
Mittermüller-Hutter Bernstein
Vir 13h5'30"-12d22'
Mitternight Star,The
Aql 19h51'44"15d25'
Mittkap Star,The
Oph 16h59'14"1d50'
Mittl,Brian Andrew
Aur 5h1'58"41d19'
Mittl,Gregory Stephen
Her 18h55'22"12d4'
Mittler,Art
Dra 18h43'34"67d56'
Mittmannsgruber,Heinz Michael
Cnc 8h25'56"31d9'
Mitton,Lisa Ann
Cyg 20h35'21"39d38'
Mitty
Uma 8h48'1"62d2'
Mitty
Lyr 18h38'10"32d39'
Mitura 5-18-85,Michael Jason
Cep 20h57'12"56d3'
Mitura 6-28-87,Daniel Joseph
Aur 4h46'23"31d1'
Mitura-Stuart
Ori 5h59'52"15d5'
Mitus,Elliot Joseph
Cnc 9h0'0"31d4'
Mitz,Pierre
Oph 18h1'38"11d31'
Mitze,Kaspar
Ser 15h53'48"1d51'
Mitzi
Aql 18h43'0"6d51'
Mitzolli
Lac 22h53'49"50d37'
Mitzy
Cam 8h27'28"78d5'
Miuccio A
Uma 12h2'36"47d45'
Miville,Elaine
And 0h52'48"37d55'
Miwako My Love Forever
Leo 10h17'5"14d24'
Mix,Jonah Patrick
Per 1h58'21"54d1'
Mix,Keith Allen
Dra 16h11'18"67d56'
MIYABI
Aqr 22h2'54"-18d55'
Miyagawa,Kenichi
Aql 18h43'30"7d12'
Miyahira,Ayako
Cyg 21h18'1"31d12'
Miyahira,Joy Yukie
And 22h58'38"37d52'
Miyake,Teruko
Aql 19h8'0"3d59'
Miyako
Peg 22h5'56"27d46'
Miyako-Stephen
Uma 9h52'16"55d53'
Miyamaru
Gem 7h2'56"17d29'
Miyamoto,June
Psc 23h8'17"0d56'
Miyuki
Cap 20h54'21"-16d21'
Miz Marissa
Cam 4h0'57"67d35'
Miz Nick
Del 21h1'40"12d56'
Mizan,Darren Charles
Cyg 19h29'1"38d31'
Mize,Lori Ann
Vul 21h12'57"28d13'
Mizen Star,The
Cyg 19h26'35"35d21'
Mizerski,Elizabeth
Lyn 6h54'43"49d22'
Mizuguchi,Colin
Hya 0h58'1"-10d14'
Mizuhara,Hiroshi
Aur 5h1'25"44d9'
Mizzell,Margaret H
Mon 6h21'0"7d20'
Mizzer Neal
And 2h28'42"40d54'
MJ 17
Aur 6h1'45"38d5'
MJ CEB MCMXCIII
Umi 14h48'55"67d51'
MJH & PMH "To Have and To Hold"
Cyg 20h19'21"30d20'
MJOYN 33
Aqr 23h12'12"-6d45'
MJS Fergus Forever
Per 1h54'58"54d20'
MJW-A Heavenly Body
XXOO,2E
Lac 22h31'14"38d39'
MK Forever
Cyg 19h59'59"29d53'
MKSA
Ori 5h24'53"-4d23'
MLA Bandit
Cyg 21h27'1"30d57'
MLAA
Uma 9h4'13"56d59'
Mlax
Cas 1h38'12"58d31'
MLE
Uma 8h34'46"71d24'
Mlechick,Mary Komlos
Cas 0h21'0"62d17'
Mlechick,Sr,John Paul
Uma 8h33'50"54d22'
Mlensky,Peter
Her 17h18'12"42d55'
MLV
Lac 22h9'13"50d34'B
Mlyniec,Emma Fannie
And 0h54'13"40d52'
MML Love Star
Her 16h26'32"23d26'
Mnaircika Forever
Peg 23h32'28"18d39'
MND Res Ipsa Locquitur
Uma 9h38'1"58d56'
Mo
Umi 14h38'31"78d12'
Mo
Mon 8h6'58"-6d38'
Mo
Hya 8h37'59"-6d54'
Mo Mo I
Lyr 18h31'26"37d58'
Mo's Glow
Peg 0h11'22"13d46'
Mo-Hee
Del 20h18'14"11d14'
Moat,Jr,James Alton
Cet 2h25'41"3d44'
Moats,Glenda Ellen
Cet 3h0'49"1d19'
Moats,Karen Lynn
Lyr 18h46'32"39d30'
Moauro,James Angelo
Aur 6h0'0"37d57'
Moauro,Mary Beth Page
Lyr 19h14'40"41d55'
Moberg,Kimberly Ann
Cam 4h18'27"61d4'
Moberly,Daniel
Lac 22h20'53"54d43'
Moberly,George Everett
Cmi 8h2'35"6d25'
Mobilia III,Star Tony
Aql 19h11'44"19d5'
Mobistar
Uma 10h3'46"47d44'
Mobley,Christine
Cas 1h25'1"50d26'
Mobley,Jean & Wade
Aur 04h52'12"40d55'
Mobley,John Christopher
Her 16h3'27"21d6'
Mobley,Lindsy Elisabeth
Del 20h21'21"10d2'
Mobley,Marta
Eri 3h15'53"-5d6'
Mobley,Shawn B
Her 16h48'31"51d19'
Mobyed,Jeannette Carol
Vul 20h5'31"25d38'
Moche,Ingrid
Vir 13h21'6"-1d38'
Mochica
Peg 22h6'0"20d26'
Mock,Doug & Barb
Mon 6h31'17"0d49'
Mock,Jamie Elizabeth
Ori 5h26'13"-1d43'
Mock,JoAnn D
And 0h49'59"45d39'
Mock,John
Peg 23h8'1"8d58'
Mockaitis,Debra
Lyn 8h23'1"50d2'
Mockaitis,Edward & Ann
Cma 6h55'17"-19d25'
Mockett Star,The Robert John
Per 1h48'57"56d26'
Mockridge,Thomas Michael
Ori 6h22'10"18d52'
Mocny,Joseph
Cmi 8h3'45"1d4'
Moctezuma,Julio Cesar
Crt 11h22'26"-19d13'
Moczko,Franz
Dra 18h11'34"68d27'
Moczygemba,Michael Anthony
Aur 6h29'27"37d38'
Moda
Lyr 19h20'48"30d54'
Modahl,William B
Per 2h54'58"34d28'
Modano,Mike
Gem 6h48'2"19d12'
Mode,Olivia
Lyn 8h35'0"44d56'
Modeen,"Sweetpea", Annie Louise
Mon 7h43'59"-2d16'
Moden,Irene
Lyr 18h17'13"30d32'
Modene,Florence
Uma 10h47'31"68d9'
Modesitt,T E B
Uma 11h17'16"61d27'
Modica,Samuel Stephan
Uma 8h33'30"48d21'
Modjeski,M Jane Carroll
And 23h23'25"51d25'
Modlin,John R
Aur 6h7'29"45d18'
Modro,Harold W
Lac 22h48'55"55d21'
Modro,Jr,Harold W
Uma 10h2'46"50d6'
Modroo,Anthony T & Ashley B
McBride
Uma 11h16'24"37d44'
Modugno,Marnie
And 23h3'12"49d35'
Modula,Shawn
Sgr 18h56'39"-23d46'
Mody's Navjyot Star, Bahman
Her 16h18'1"10d8'
Moe
Cma 6h54'26"-18d60'
Moe
Cyg 19h21'51"28d48'
Moe
Uma 8h22'1"60d44'
Moe,Darrell Ivan
Dra 19h21'50"68d43'
Moe,Tommy
Her 16h15'46"50d29'
Moebus Star,Captain
Ser 17h56'29"-14d39'
Moed,The Roy
Her 16h51'12"40d27'
Moeders in Nood
Aur 7h23'1"38d40'
Moege's Dipper,Debbie
Peg 22h1'57"8d51'
Moegelin,Gernot
Lib 15h36'18"-20d23'
Moehring,Ethel Ida Louise
Sedo
Cyg 21h20'12"31d11'
Moehring,Jodi
Scl 23h24'42"-27d43'
Moehrle,Dieter
Lyn 9h1'12"45d58'
Moeller,Christa
Sgr 19h8'54"-20d27'
Moeller,Jim
Sct 18h45'23"-7d8'
Moeller,Marc Victor
Aur 7h12'23"39d57'
Moeller,Stephanie Lee
Cas 2h12'48"59d35'
Moellering,Aaron Paul
Dra 16h27'58"69d22'
Moellering,Mark Christopher
Boo 13h42'53"14d13'
Moellman,Janice
Aql 20h7'25"4d57'
Moellman,Marlene
Aql 20h4'1"0d45'
Moellman,Richard
Aql 20h3'25"7d13'
Moellman,Robert
Aql 20h7'44"1d7'
Moellman,Sarah
Aql 20h4'1"3d58'
Moellman,Staci
Aql 20h4'32"1d11'
Moellon,Karin
Peg 22h30'54"33d48'
Moen,Alexandria Lily
And 1h39'0"40d30'
Moench,Krystal
Lac 22h49'0"56d44'
Moenik-O
Psc 0h50'51"32d15'
Moening,Ted Louis
Per 2h26'1"55d34'
Moennich,Ann
Cas 1h7'44"58d52'
Moens,Virginia E
Mon 6h24'21"7d30'
Moeran,Fenner Orlando
Her 18h7'35"28d8'
Moerdyk,Julia Patricia
And 2h30'28"48d1'
Moersch,Brandon
Uma 10h27'56"51d56'
Moersch,Tyler
Dra 12h5'0"71d52'
Moeser,Brigitte
Psc 23h0'58"0d46'
Moeser,Deborah E
Leo 10h35'22"20d48'
Moeser-Jackson, Peggy Sue
Peg 22h6'20"24d23'
Moessner,Gerhard
Per 3h50'36"39d1'
Moet
Tau 5h44'26"29d2'
Moffat,Jack
Cep 5h10'46"87d11'
Moffat,Kerry Anne
Lyr 19h0'1"38d12'
Moffatt,Helen Draudt
Mon 6h26'20"-0d39'
Moffatt,Steven Oliver
Oph 17h21'30"-22d34'
Moffet,Cora Elizabeth Jeffreys
Equ 21h0'16"2d44'
Moffett,Jerry D
Aur 4h54'58"38d12'
Moffitt Family Star, The
Ori 4h43'45"5d11'
Mogan,Natalie Ann
Cam 5h51'33"60d45'
Mogan,Sarah Marie
Cas 1h44'51"61d6'
Mogavero,Louis & Julia
Dra 16h1'33"52d14'
Mogensen,Budd
Uma 8h51'54"49d43'
Mogensen,Elaine Anne
Cyg 20h26'40"58d47'
Mogford,Kevin
And 22h57'56"51d6'

Moghimi,Mohamad Cma 6h57'36"-30d21'
Moglia,Carole Ann Cas 0h54'35"69d39'
Mogol,Gran Boo 15h9'23"47d30'
Mogren,Angela Nicole Lyr 18h38'21"42d19'
Mogs Pixius Pho 0h50'16"-45d37'
Mohamed,Ian Peter A Hya 8h33'10"-5d54'
Mohamed,Ms Sabira Com 12h31'11"24d5'
Mohammady,Renate Aqr 23h29'40"-5d23'
Mohammed's Night Uma 11h23'14"40d58'
Mohammed,Hafizullah Her 18h27'17"12d6'
Mohan,Jim Cmi 7h5'42"4d8'
Mohan,Jr,Bernie Cet 0h52'54"-0d38'
Mohange,Carole Paulette Antoinette Cas 0h15'31"61d29'
Mohar,Joseph J Lac 22h25'36'50d8'
Mohat,Pete Mon 6h44'42"11d18'
Mohawk,Essra Cet 3h12'51"4d49'
Mohen,Conor Joseph Boo 14h18'25"17d28'
Mohler,Bradford Douglas Lyr 18h58'33"47d39'
Mohn,Janet And 2h6'29"38d35'
Mohomer,Chakir Aur 5h1'28"37d33'
Mohr,Anna Becker Lyr 19h24'47"38d8'
Mohr,Bo Aur 6h54'42"40d29'
Mohr,Carol & Chuck Sge 19h33'13"17d5'
Mohr,David Arlo Her 18h38'44"12d39'
Mohr,David L Ori 5h52'38"15d5'
Mohr,Lorenz Aql 20h8'49"6d1'
Mohr,Meribeth Margret Peg 21h56'37"22d45'
Mohr,Michael Donald Ori 5h10'55"-8d18'
Mohr,Urban J Ori 6h3'23"8d40'
Mohrhauser,Whitney Allison Mon 7h19'21"-8d22'
Mohring,Doktor Jan Boo 14h22'0"10d41
Mohrman,Jon Eric Aur 6h16'13"38d41'
Mohsinaly,Hamza & Massemane Ind 20h32'51"-56d55'
Mohundro,Geneva Bill (Hinchey) Lyn 7h29'50"38d42'
Mohwinkel,Jennifer Anne Lyn 7h38'26"42d16'
Moien Et Tilou Uma 8h37'0"54d43'
Moinder Aql 20h1'51"0d34'
Moine,Gérard Cnv 13h22'40"38d46'
Moinet,Glynis Carol Aqr 20h58'0"-13d22'
Moir II,J Daniel Aur 5h16'0"45d14'
Moir,Benjamin Andrew Her 17h47'40"48d49'
Moira Cam 5h54'55"80d28'
Moira Cas 0h45'35"74d7'
Moira's Magical Star Cas 0h55'1"70d21'
Moira's Star Cas 0h46'57"67d12'
Moise,Gene Uma 10h8'31"70d11'
Moiseenko,Ilya Valerievich Cnc 8h11'33"31d10'
Moises Santa Lmi 11h4'51"33d43'
Moist-Sorenson,Shelly Lorraine Cyg 19h43'1"50d10'
Moitié,Douce Umi 13h54'1"76d47'
Moj Damir Od Nebo Uma 9h53'30"42d5'
Moj Zalti Bob Cnv 13h55'22"38d46'
Moja,Suzy Zvezdica Lyr 18h20'41"41d11'
Mojarro,Martin Aql 19h31'29"14d3'
Mojden,W W Hya 9h3'0"5d24'
Mojica,Jesse Her 17h57'44"30d14'
Mojica,Rosa Tau 4h7'15"22d9'
Mojo Ori 5h3'0"0d55'
Mojo's Eternal Love Aql 19h58'15"8d55'
Mok,Wendy Umi 14h50'11"81d42'
Mokay Mother,Julianne Vir 11h38'1"4d19'
Mokey Love Vul 21h19'60"24d3'
Mokhtarzada Cyg 19h55'51"38d51'
Mokros Cam 4h9'59"68d43'
Mol,Bradley John Lyn 7h50'0"48d17'
Molaei,Roya Lmi 10h4'50"30d23'
Moland,Asbjorn Cam 13h14'12"63d15'
Molcard,Sylvie Aur 6h2'43"38d55'
Mold III,William F Cep 20h52'27"68d52'
Moldehnke,Gerd Lyr 19h19'46"30d53'
Moldovan,Karen Lynn And 23h29'1"43d21'
Moldovan,Laci & Edith Aql 19h27'15"8d46'
Moldt,John Morrison Sge 19h54'17"16d28'
Mole,Joseph & Eleanore Lac 22h55'13"53d43'
Molenbeek,Michele Tri 2h37'26"34d4'
Molendinarius,Petro Lup 15h28'38"-42d54'
Molengraft III,Edward Cornelius Her 17h0'0"50d18'
Molfetas,Rita Com 12h28'47"22d54'
Molina,Marco Sct 18h30'51"-5d12'
Molina,Matthew Ser 18h1'59"-13d41'
Molina,M¶Teresa Ori 5h52'37"21d57'
Molina,Raul E Boo 15h0'25"53d11'
Molina,Roger Sgr 19h33'35"-32d11'
Molina,Sydney Rose Boo 13h43'16"20d16'
Molina,Veronica Aql 20h12'0"11d52'
Molinari's Destiny Dra 18h31'11"58d20'
Molinari,David Ori 6h1'33"4d19'
Molinari,Marco Ari 1h15'0"30d20'
Molinario,Amy Mae Cyg 20h48'13"46d28'B
Molinario,Patricia Cas 1h32'22"61d46'
Molinario,Robert D Cyg 20h48'13"46d28'A
Molinaro's Magic Dra 11h25'55"78d4'
Moline,Abby & David Cyg 20h2'10"40d6'
Moline,Lawrence Anthony Cam 13h20'11"80d16'
Molineaux,The Uma 8h14'0"71d25'
Molineux-Summer Night Star,Pat Doyle Crb 16h14'13"30d5'
Molinia,Rosemarie Susan Lyr 19h23'49"38d43'
Molinier,Dominique Cam 3h34'57"60d45'
Molinsky,Bert & Donna Cet 3h16'0"9d36'
Molitch,Matthew Cas 2h3'36"60d8'
Moliterno,Gilda And 2h31'34"37d51'
Molitor,Anneliese & Mario Uma 9h50'49"52d7'
Molitor,Dr Roland Peg 22h34'36"18d55'
Molitor,Inge Peg 22h39'0"18d53'
Molkosan,Angel Peg 23h16'0"33d51'
Moll And 2h35'29"40d1'
Moll (Soulmates Forever),John & Donna Uma 8h40'19"48d33'
Moll,Arthur A Per 1h53'37"53d37'
Moll,Denise & Howie Lyr 18h40'31"30d28'
Moll,Friedrich Hya 9h9'43"2d11'
Moll,Lisa Marie Cyg 20h15'44"30d42'
Moll,Peggy Mon 6h46'0"-4d49'B
Moll,Philipp Ori 6h0'45"8d2'
Molla Uma 11h24'26"67d44'
Mollach,Laura Ann Cyg 20h18'1"38d45'
Molle,Callan Elizabeth Tau 4h36'47"20d15'
Molle,Marcia Sue Cyg 20h54'47"30d7'
Molle,Tony Boo 14h10'50"52d11'
Molle,Zachary Scott Cap 20h33'25"-10d53'
Moller-Petra,Bigs Lyr 19h23'22"42d11'
Mollie Mon 6h58'0"-6d33'
Mollie Joan Lyr 18h14'52"45d49'
Mollie Rae Mon 7h43'26"-2d38'
Mollohan,Alice Sydney (Sami) Legvold Aql 18h59'33"12d4'
Molloy,Anne And 0h38'36"40d18'
Molloy,Becky Aql 20h8'0"8d23'
Molloy,Geoffrey Cep 20h58'26"70d43'
Molloy,Karen Aqr 21h4'17"0d7'
Molloy,Sherril Ant 10h42'32"35d4'
Mollway,Tyler James Dra 17h4'26"50d30'
Molly Leo 9h26'59"28d10'
Molly And 23h29'0"41d14'
Molly Uma 8h17'0"67d48'
Molly And 23h39'22"38d52'
Molly Del 20h19'50"18d53'
Molly And 1h21'58"40d47'
Molly Mon 6h30'35"8d46'
Molly And 2h11'46"39d24'
Molly Lyr 18h41'32"31d45'
Molly Uma 9h11'41"50d28'
Molly Cyg 19h28'16"37d14'
Molly Del 20h14'33"15d7'
Molly & Jake Lyn 8h9'34"35d8'
Molly Ann Peg 22h42'39"34d23'
Molly Elizabeth Cas 2h3'42"59d9'
Molly Gwen Com 13h4'11"27d44'
Molly I Love You Forever-Jason And 1h36'1"38d1'
Molly Laraine Uma 9h33'52"44d16'
Molly Louise And 0h43'31"40d29'
Molly Lynn Cet 3h14'29"2d13'
Molly Lynn Cap 21h25'35"-24d31'
Molly M P Cas 1h35'12"60d23'
Molly Marie And 23h42'20"46d52'
Molly Rebecca Cyg 19h34'51"39d23'
Molly's Heavenly Body Lmi 9h53'0"38d35'
Molly's Lucky Star Eri 2h50'59"-2d44'
Molly's Star Aqr 22h7'1"0d39'
Molly's Star And 23h34'43"45d56'
Molly's Star Del 20h53'20"2d50'
Molly's Star Lyn 7h42'33"51d20'
Molly-Amie-Hazel-Allen Peg 21h49'27"30d1'
Mollycoddle Lyn 8h41'16"36d52'
Molnar,Elizabeth Vir 13h2'0"-20d39'
Molnar,Jessica Olivia Cas 3h0'56"57d31'
Molnar,Joan Cam 5h40'35"73d16'
Molnar,Jr,Charles Cet 3h10'48"4d57'
Molnar,Libi Cas 1h13'31"62d10'
Molnar,Paul C Per 3h4'43"41d23'
Molnar,Princess Judy And 2h23'0"42d55'
Molnar,Sister Beth Cyg 19h30'45"36d7'
Molohan,John Joseph Her 17h4'52"49d29'
Moloney,Ann Cole Cas 23h41'46"50d29'
Moloney,Phyllis M Eri 4h54'17"-4d31'
Moloney,Tom Her 18h11'21"38d17'
Molony,Evelyn Elizabeth Cyg 20h54'54"37d34'
Molowa,Andrea Marie Cyg 20h25'54"40d35'
Molowa,Mark Tucker Cep 21h40'52"60d59'
Mols,Frieda C MacArthur Del 20h53'10"4d20'B
Mols,Pierre M Del 20h53'10"4d20'A
Molski,Thomas M Sex 9h53'48"2d10'
Molsky,Beth Ann And 1h49'20"38d44'
Molstar,The Dra 17h22'59"68d26'
Molta,Sr,Daniel J Sct 18h54'18"-6d19'
Molter,Valerie Lynn Lyn 7h15'29"58d34'
Molter-Serrano,Hope Peg 22h26'0"30d54'
Molthrop,Morgan And 0h31'29"31d0'
Moltnd 0h31'29"31d0' Lyn 8h27'46"34d26'
Molyneaux,Brittany Lyr 18h48'56"39d48'
Mom Lyn 7h52'52"38d30'
Mom And 0h25'58"21d49'
Mom & Bob Cyg 19h25'40"30d27'
Mom & Dad Never Lose Faith Umi 16h19'55"88d46'
Mom & Dad-"My Guiding Star" Cyg 21h37'59'42d55'
Mom & Pop Lmi 10h14'30"30d52'
Mom Channah Leah Peg 22h26'44"3d14'
Mom Dee Dee The Bird Cas 2h1'37"60d55'
Mom Star Eri 3h0'41"-11d42'
Mom The Petite Woman Cap 20h50'28"-26d34'B
Mom's Ori 5h34'53"0d10'
Mom's Octave Tri 1h55'48"26d42'
Mom's Star "Jeri" Cas 0h34'41"68d44'
Mom-Fantastic Eri 3h4'28"-1d32'
Mom-Mom Cyg 21h8'47"48d53'
Mom-Mom Cas 2h21'12"70d10'
Mom-The Wind Beneth My Wings,Denise Cma 6h51'16"-17d28'
Mom/Gran Cas 0h7'56"63d29'
Moma Tonga Her 17h34'56"40d48'
Moman,Timothy Robert Oph 17h57'1"13d36'
Mombou Cyg 21h29'53"38d10'
Momie Lyr 18h51'48"42d8'
Mominee,Katrina Marie Peg 22h25'35"20d2'
Momma Sue Aql 19h56'39"14d28'
Momma,You Are My Shining Star Cyg 21h50'0"53d41'
Mommen,Paula Lyr 19h21'1"30d14'
Mommens,Charlotte & Coralies Lacroix Cru 12h14'43"-59d47'
Mommer,Walter Cnv 12h16'0"37d1'
Mommie Doris Peg 22h32'18"35d20'
Mommie's Lyn 8h11'33"50d26'
Mommie,The Cas 1h43'13"60d34'
Mommy Leo 11h44'35"23d33'
Mommy Nancy Mon 7h3'11"4d9'
Mommy O & Daddy O Uma 10h7'13"57d35'
Mommy Star-I'll Always Be With You And 2h11'43"42d57'
Mommy's Baby Girl Little Ms Mary Jo Peg 22h45'27"33d53'
Momndad Aql 20h0'57"7d32'
Momokabo Forever Star Leo 10h18'38"9d4'
Mon Ami Del 21h5'14"18d58'
Mon Amor Uma 9h7'26"58d6'
Mon Amour Pyx 8h50'28"-29d41'
Mon Amour Dinesh Per 2h22'43"57d0'A
Mon Ange Ori 5h27'1"0d48'
Mon Beau Ideal Lyn 8h40'0"44d58'
Mon Cher Paul Del 20h15'39"14d27'
Mon Chère Martin Cas 1h20'53"71d11'
Mon Etoile,Mon Amour, Cécile Boo 14h31'17"12d26'
Mon Frere Sylvain Dra 17h6'25"73d30'
Mon Petit Ange Lyr 19h5'48"28d24'
Mon Petit Prince Michel Del 20h53'11"9d24'
Mon Petit Rayon De Paradis Diane Mon 7h0'24"-0d6'
Mon Ptit Bébé Equ 21h16'5"11d21'B
Mon Soeur Que J'Aime Isa Uma 10h42'59"62d11'
Mon Ti-pet Nini Cep 23h37'25"64d38'
Mon Étoile D'amour, Steven M Hya 8h24'24"-1d18'
Mon,Maly Cas 0h34'14"66d42'
Mona Cyg 19h39'14"31d37'
Mona Lyn 8h15'54"47d52'
Mona Cas 1h28'21"63d58'
Mona Cas 23h41'1"50d3'
Mona "Monumental Brat" S Uma 8h44'0"52d52'
Mona & Gyp Aur 5h44'58"30d31'AB
Mona B Cet 2h57'42"6d7'
Mona B Cyg 19h51'35"40d20'
Mona G Cas 23h34'56"61d59'
Mona K Uma 9h25'28"47d53'
Mona Lee And 23h15'53"49d12'
Mona Mausi Tau 5h54'40"26d22'
Mona My Love And 2h32'54"50d53'
Mona Star Com 13h14'42"20d48'
Mona's Bear And 1h17'25"39d45'
Mona's Mom Tau 5h45'39"23d53'
Monack,Jack Dra 18h12'49"68d29'
Monaco Of The Rock, Princess Steffi And 23h32'0"43d8'
Monaco,Anthony & Josephine Vul 20h5'0"28d41'
Monaco,Eileen McCann Peg 21h38'1"24d45'
Monaco,Frank Peter Lac 22h21'29"54d41'
Monaco,John "Elvis" Her 16h15'17"41d2'
Monaco,John Del Sco 17h32'26"-38d20'
Monaco,John Joseph Cep 20h12'11"76d1'
Monaco,Kathleen Mary Lyn 8h28'55"50d39'
Monaco,Marcus Aur 6h30'16"35d16'
Monaco,Michael Cnc 8h22'1"15d15'
Monaco,Nicholas Robert Aur 5h18'28"43d58'
Monaco,Tina And 1h40'16"39d9'
Monaghan,Auedrey Peg 22h13'51"31d15'
Monaghan,Dave Ori 5h52'28"8d53'
Monaghan,David Walker Aur 6h8'0"31d25'
Monaghan,Edmund Per 2h23'26"54d54'
Monaghan,Jr,John M Hya 8h24'10"1d3'
Monaghan,Monty A Sct 18h53'52"-6d33'
Monaghan,Patrick Aur 5h37'16"40d54'
Monago,Frank A Ser 15h52'37"-0d36'
Monahan,"Dancing Eyes" ,James F Peg 23h23'32"16d55'
Monahan,David M Lac 22h4'17"49d2'
Monahan,Edward Cnv 12h28'39"33d45'
Monahan,Gregory Cep 22h18'0"55d54'
Monahan,Joan And 2h28'41"41d18'
Monahan,Mark Dra 17h7'38"61d50'
Monahan,Mildred Mary Aqr 20h55'14"-6d51'
Monahan,Pat Hya 8h15'56"0d2'
Monaillon,Pierre Cyg 21h34'14"40d41'
Monarch,The Cep 22h31'58"68d50'
Monarco,Tammy Mon 8h4'23"-6d25'
Monas,Matthew & Jeanette Boo 14h36'57"39d22'
Monast "Papa-Bubba", Albert L Uma 13h57'1"51d23'
Monast,Jeanette A Aur 6h8'34"30d50'
Monast,John D Cas 23h20'54"54d6'
Monast,Pamela M Peg 23h22'29"12d34'
Monastero,Rose Carmella Lyr 18h28'21"38d10'
Monchablon,Maurice Uma 11h7'16"51d7'
Monck,Alec Dahlton Umi 16h17'21"71d57'
Monclova,Blodwen Cyg 19h33'25"30d50'
Moncrief,Adelaide Deborah Aqr 21h18'41"0d10'
Moncrief,Celia Elizabeth Aqr 22h43'23"0d25'
Moncrief,Gloria Marie Mon 6h35'1"7d29'
Moncrief,Kit Tennison Tau 4h0'12"10d34'
Monday,Molly Lyn 6h34'29"54d12'
Mondejar,Noel Aql 18h58'57"10d24'
Mondello,Michael Per 3h11'38"50d16'
Mondello,Tiffanie Lyn 7h32'24"36d30'
Mondenkind Nina Stripling Uma 13h36'55"48d31'
Mondik-50th Wedding, Frank & Stella Cyg 21h2'57"33d11'
Mondloch,Tyler & Ashley Palumbo Uma 8h51'53"57d25'
Mondock,Emma Cas 0h26'0"62d16'
Mondomania Ind 21h9'7"-59d31'
Mondrag-on,Jr,Richard Aur 4h48'52"40d54'
Mondschein Lib 15h11'9"-23d18'
Mondschein,Mrs Cas 23h35'36"60d5'
Mondshine,Michael Lmi 10h10'15"32d22'
Mondshine,Raymond & Shirley Aql 19h23'26"15d19'
Monel,Sylvie Cyg 21h36'11"40d58'
Monet,Elizabeth Peg 22h25'29"27d24'
Monetta,Francis W Boo 15h4'19"40d56'
Monette Dra 16h15'18"62d41'
Monette,Jean-Yves Umi 15h10'48"77d59'
Monette,Kathy Cyg 19h29'0"37d32'
Money's Star Vul 19h47'44"26d10'
Money,Jason Blaine Uma 11h11'30"30d17'
Moneypenny Family Lyr 18h47'47"30d54'
Mongeau,Dr Gilbert J Ori 5h46'1"10d45'
Mongelli,Gabriela De Medeiross Sge 19h35'31"16d47'
Mongelli,Thomas John & Jaclyn Amber Cyg 21h1'34"30d8'
Monges,Lizette Peg 21h29'43"20d21'
Mongey,Eamonn Dermot Cep 22h51'39"65d59'
Mongiello,Donna Gem 6h58'3"18d30'
Mongillo,Michael William Hya 9h26'26"-10d4'
Mongioi,Charles Aur 6h32'54"32d55'
Moni And 1h20'31"35d14'
Moni Ori 6h5'49"5d24'
Moni Psc 1h0'29"21d29'
Moni G Sex 9h47'13"2d40'
Monia Vir 13h1'1"10d8'
Monia Her 16h46'1"10d44'
Monica Per 2h9'10"58d11'
Monica Cam 6h23'13"68d55'
Monica Nor 16h19'31"-51d23'
Monica Mon 7h5'10"-1d41'
Monica And 2h0'12"38d1'
Monica And 1h31'49"37d54'
Monica Cas 22h58'12"54d6'
Monica Del 20h13'56"10d49'
Monica Del 20h17'15"10d27'
Monica Lep 5h56'0"-11d57'
Monica Uma 8h35'1"58d49'
Monica Cam 7h47'49"60d47'
Monica Aql 19h32'55"-1d22'
Monica Oph 18h34'55"7d22'
Monica Peg 22h18'32"3d33'
Monica And 2h20'57"42d28'
Monica & Alicia Cet 1h52'0"0d32'
Monica & Chris Uma 11h47'42"51d10'
Monica & John Cyg 21h1'20"37d41'
Monica & Kathy Del 21h2'49"13d1'
Monica & Michele Cyg 21h3'42"36d37'
Monica 3491 Peg 23h1'50"21d35'
Monica Ann Vul 20h56'20"28d25'
Monica e Giulio Uma 11h26'1"51d2'

Monica J
Cas 23h23'0"60d23'
Monica Lyn
Cma 6h55'8"-19d10'
Monica Mae
Lyr 18h59'0"26d23'
Monica Mazzini,B
Massimiliano per sempr
Cas 1h58'33"58d54'
Monica R's Own Sparkler
Lyr 19h22'21"31d11'
Monica's Light
Cas 1h20'16"61d7'
Monica's Osita
Cas 2h33'24"57d34'
Monica's Star
Mon 6h53'48"-8d45'
Monica's Star
Tri 2h36'49"31d34'
MONICA,First Again!
Ori 5h37'18"-0d9'
Monicamanenti
Cep 22h53'41"57d31'
Monie,Dusty Autumn
Boo 14h37'22"18d33'
Monie,Konrad Apollo
Aql 19h45'24"14d40'
Monie,Sunny Spring
Aur 6h28'31"34d33'
Monigan,Edward T
Aur 6h4'53"38d51'
Monihart,Adolf
Gem 8h4'28"31d13'
Monika
Boo 14h39'46"47d6'
Monika
Aqr 20h57'24"-0d44'
Monika
Sco 16h59'53"-40d0'
Monika
Peg 23h30'13"15d24'
Monika
Her 17h19'12"50d2'
Monika & Alex S
And 2h25'28"37d32'
Monika & Thomas
Leo 11h8'18"-5d8'
Monika & Thomas
Cam 7h57'41"73d30'
Monika Eva
Sge 19h15'33"16d26'
Monika und Kemper Helmut
Retzer
Sgr 20h5'58"-44d3'
Monika's Way Out There
Vul 19h44'56"20d21'
Monika-Ellen
Cap 20h45'33"-23d28'
Monin,Patricia et Xavier
Ser 15h20'41"8d34'
Moniowczak,Sandy
And 2h24'22"39d48'
Monique
Peg 23h1'32"11d43'
Monique
Boo 14h31'14"12d28'
Monique et Guy
Cep 20h15'0"60d22'
Monis Stern
Psc 0h19'47"12d18'
Monita Star
Pho 23h44'48"-41d46'
Moniter
Uma 10h2'10"58d10'
Monitto-30,Molly Todd
Aql 19h13'31"10d36'
Monizc,Astrid Rochelle
Peg 21h35'53"20d5'
Monk,Allan William
Uma 8h45'0"59d0'
Monk,Daddy
Cep 3h15'23"80d22'
Monk,Doug
Aur 5h37'34"54d44'
Monk,Jonathan Ellis
Her 16h58'27"51d13'
Monk,Lesley Anne
Cam 5h3'59"70d53'
Monk,Lesley Anne
Cas 0h23'16"75d58'
Monk,Randall D
Cam 7h1'19"68d58'
Monk,Roy S
Lac 22h38'49"50d17'
Monkey
Aur 4h39'24"31d2'
Monkey Boy
Sge 19h5'8"16d51'
Monkhouse,Valerie Sanger
Cam 3h26'36"61d11'
Monlezun,John Kammer
Dra 16h51'36"51d38'
Monnet,Isabelle
Cnv 13h16'38"38d18'
Monnie Kississippi
Per 2h52'41"31d3'
Monnie,Benjamin
Ori 6h4'40"7d50'
Monnier,Erika Lynn
Lyr 19h17'32"35d12'
Monnier,Jean
Per 3h0'27"48d49'
Monnier,Jean Bertrand
Per 3h29'47"39d53'
Monnier,Megan
Sco 15h56'16"-25d45'
Monoki,Balazs
Ser 16h7'40"13d53'
Monoki,Veronika
Ser 16h0'48"13d38'
Mononen,Alpo
Dra 16h37'28"62d14'
Monosexualis
Ori 5h48'23"19d35'
Monotheistic Morses
Lyr 18h32'24"36d4'
Monrad,Julie Lynn
Lyn 7h56'58"43d5'
Monroe 063-38-9592,
Rosemary Janet
Ori 5h27'13"-0d3'
Monroe,Charles
Cet 2h10'24"2d11'
Monroe,Darrin
Her 16h55'33"25d34'
Monroe,Harris One-in-a -
Zillion
Her 17h57'16"40d29'
Monroe,Heather
Uma 11h46'24"32d57'
Monroe,James Daniel
Aql 19h24'48"8d4'
Monroe,James Hinson
Aur 6h35'56"37d46'
Monroe,James Stewart
Per 2h39'23"40d18'
Monroe,Jr,Garfield "Fat Cat"
Aur 6h35'0"34d9'
Monroe,Judy Kay
And 0h11'26"46d33'
Monroe,Kitty
Lyn 7h55'32"43d27'
Monroe,Kris
Cyg 20h2'58"31d31'
Monroe,Leila
Peg 23h43'54"26d5'
Monroe,Lindsay Lee
Mon 6h36'58"7d2'
Monroe,Mark
Aur 4h52'35"41d8'
Monroe,Melanie Kate
And 0h10'49"30d41'
Monschau,Gabi
Cnc 8h57'53"18d44'
Monsees,John & Ellen
Ori 5h53'30"20d32'
Monsees,Kristen Lee
Umi 17h2'23"75d31'
Monsen,Beverly Jean
Mon 7h46'0"-2d40'
Monsen,Judith Sara
Cet 3h13'11"2d0'
Monsen,Margie
Eri 2h50'60"-6d27'
Monsi
Del 20h30'49"20d28'
Monsieur Jean Baptiste Le Big
C
Cnv 12h17'40"47d11'
Monsieur Pellae Jean-Maurice
Cep 2h42'1"80d7'
Monsita
Lyr 19h17'49"40d39'
Monson,Finns for Dig Robert
H
Oph 17h20'26"12d55'
Monson,Rudy
Dra 17h51'54"75d38'
Monson,Sarah Elizabeth
Eri 3h14'1"-16d36'
Monson,Thomas L
Cep 21h39'49"68d18'
Monsour,Kimberly
Oph 17h16'21"-20d37'
Mont,Richard
Aur 6h35'24"33d27'
Montacute,Sylvia
Crb 15h52'19"26d55'
Montagne,Jean-Paul
Cyg 19h42'40"38d41'
Montagne,Michel
Cyg 19h42'13"37d32'
Montagu,Julia
Lyr 18h36'20"37d36'
Montague,Ben
Aql 19h47'56"14d51'
Montague,Bryan Leslie
Aql 18h59'16"-4d44'
Montague,Donna-Marie
Cyg 19h34'10"35d29'
Montague,Douglas McKenzie
Vul 20h3'0"22d56'
Montague,Stephanie
Cas 0h15'28"47d28'
Montalbano,Christopher
Philip
Boo 14h35'21"18d25'
Montalbano,Frank B
Leo 11h9'23"2d20'
Montalbano,Gregory Anthony
Cep 0h2'14"67d25'
Montalbano,Phil
Urm 11h11'42"61d46'
Montalbetti,Claudio
Her 16h40'45"18d55'
Montalto,Vito & Penelope
Cam 7h3'20"70d24'
Montalvo "Beloved
Mother",Isabel
Cas 0h4'23"62d57'
Montalvo,Cindy
Cas 0h47'10"66d39'
Montamaro,Kayla Augusta
Del 20h17'1"9d52'
Montana
Tau 3h56'57"11d52'
Montana
Equ 21h19'43"2d38'
Montana "Superstar", Joe
Her 16h27'23"48d54'
Montana Big Sky
Cam 6h16'12"62d11'
Montana Tom
Ori 6h3'40"-1d12'
Montana Wedding Star, The
Mon 8h8'35"-9d25'
Montana,Claude
Cnc 8h54'11"32d49'
Montana,Heather McAlpine
And 2h26'55"41d17'
Montana,Zeke
Her 16h33'45"38d1'
Montana-Caceda,Eden Ruben
Gru 22h9'24"-50d31'
Montanagro,Cyndi
Aql 19h43'51"11d6'
Montanari,James William
Lac 22h7'39"49d37'
Montanaro,Michael T
Aur 6h50'56"35d31'
Montanelli,Mario
Ori 6h0'41"8d1'
Montanelli,Massimo
Uma 12h0'20"41d24'
Montano,Carl Vincent
Ser 18h27'53"1d24'
Montano,Eftihia
Aur 7h4'43"43d30'
Montant,AP
Aur 6h2'35"41d15'
Montaroup,Nadia
Ori 5h54'36"19d21'
Montauti,Alyssa
And 2h18'41"49d26'
Montauti,Elena
And 2h29'23"49d30'
Montad 2h29'23"49d30'
Aql 19h24'43"-0d12'
Montbleau,Jr,Joseph A
Dra 18h57'48"48d48'
Monte Ann
Aur 6h28'30"38d56'
Monte Ray Veal Know What
Boo 14h57'12"43d8'
Monte's Hot Flash
Cra 18h14'29"-39d48'
Monte's Night Light
Cet 2h27'26"3d3'
Monte,Anthony John
Lac 22h24'46"54d52'
Monte,Mary Bella
And 0h7'0"34d52'
Monteagudo,MD,Ana
Umi 13h49'52"70d11'
Monteagudo,Victor Leon
Her 16h20'52"41d23'
Montecalvo,Allan
Lac 22h31'48"50d13'
Montecalvo,Robert
Eri 4h1'60"-10d1'
Monteclaro,Stephen Raymond
Cet 1h29'22"-11d30'
Montee,Richard Edward
Ori 6h15'58"1d2'
Monteferrante,Angelo
Cam 4h8'1"69d44'
Monteferrante,Helen
Com 12h0'57"19d54"
Montefusco IV,A
Aql 19h24'58"8d31'
Montefusco,Tony
Ori 5h31'55"-3d48'
Monteiro,Dulce C
Uma 9h17'55"51d13'
Monteiro,Henry F T
Aql 19h51'30"11d47'
Monteiro,Virginia Genevieve
Cas 0h33'1"58d56'
Monteith,Harry Lee
Aur 6h48'24"37d35'
Monteith,Tania
Peg 22h20'17"27d43'
Montel-Anderson
Aql 20h31'60"0d11'
Monteleone,Albert G
Per 3h9'55"40d44'
Monteleone,Joe
Uma 11h17'59"50d38'
Monteleone,Ray
Sgr 18h52'7"-24d4'
Montells,Lisbel
Lac 22h9'38"51d59'
Montemarano Family Star
Cma 6h53'24"-19d52'
Montemayor,Esiquiel G
Aql 18h59'43"14d27'
Montemurro,Michael
Uma 9h4'27"62d29'
Montera,Debra Marie
Sgr 19h9'13"-28d13'
Monteressi,Christopher
Aql 19h9'1"1d46'
Monterie,Coenraad Raymond
Nicolas
Tau 4h16'47"23d11'
Montero,Janie Jo
Tri 2h5'1"30d59'
Montes,Enriqueta Peir
Cas 23h24'0"61d36'
Montes,Estrellita Regina
Cam 6h20'24"80d15'
Montes,Greta
Uma 8h31'20"70d44'
Montes,Manny
Cep 21h11'59"63d56'
Montes,Marcos J
Aql 20h17'56"0d57'
Montesano,Anthony J
Her 17h35'25"28d23'
Montesanti,Richard Clement
Cet 3h7'43"4d55'
Montesion,Thomas Michael
Aur 5h1'1"46d23'
Montessori Academy of
Mobile
Lmi 10h35'18"30d12'
Monteverdi,George John
Tau 5h45'30"16d32'
Montez,The Majestic
Oph 16h56'36"-23d13'
Montezin,Albert
Ori 6h4'0"0d26'
Montgomery
Umi 14h57'59"65d33'
Montgomery's Star
Cep 2h55'45"87d20'
Montgomery,"Anita B" Anita B
Hare
Cas 1h6'30"56d5'
Montgomery,A Christopher
Uma 13h33'47"60d59'
Montgomery,Anne
Lyr 19h16'0"40d27'
Montgomery,Annette
Cet 2h53'49"8d3'
Montgomery,Beverly
Cyg 20h12'18"58d51'
Montgomery,Bob
Vul 19h55'17"20d32'
Montgomery,Brooke Elizabeth
Aql 20h35'0"0d36'
Montgomery,Diane
Aur 6h1'57"30d4'
Montgomery,Ed & Sue
Cnv 13h18'47"28d38'
Montgomery,Ellen
Mon 7h3'51"4d24'
Montgomery,Florence
Lac 22h33'47"56d42'
Montgomery,Guy Timothy
Ori 4h44'54"1d4'
Montgomery,James Terry
Cet 2h8'58"0d14'
Montgomery,Jeffrey
Cet 1h23'27"1d45'
Montgomery,Jennifer
Cas 0h24'20"61d47'
Montgomery,John Leroy
Oph 18h42'11"7d59'
Montgomery,Jr,Robert L
Her 17h6'0"41d32'
Montgomery,June A
And 0h4'11"35d58'
Montgomery,Kate Hutson
Com 12h55'28"27d41'
Montgomery,Lawrence Calvin
Crt 10h50'0"-17d47'
Montgomery,Lawrence
Dra 14h42'0"62d19'
Montgomery,Linda Kay
Eri 3h23'0"-4d37'
Montgomery,Martine
Lyr 18h54'0"38d39'
Montgomery,Mary Meeler
Uma 12h12'55"61d47'
Montgomery,Naomi
Lyr 18h48'1"41d3'
Montgomery,Nathan
Her 16h22'29"41d2'
Montgomery,Patricia Henry
Peg 22h55'59"22d53'
Montgomery,Penn
Ori 6h2'28"4d22'
Montgomery,Randy
Ori 5h53'44"20d19'
Montgomery,Rep Gillespie V
"Sonny"
Ori 5h51'36"12d50'
Montgomery,Rita
Crt 10h55'20"-8d9'
Montgomery,Ruth Ann
Mon 7h51'5"-1d42'
Montgomery,Saundra Kay
Lac 22h2'24"38d56'
Montgomery,Sharon & Terry
Crb 16h0m52"31d45'
Montgomery,Sheila Huff
Sgr 19h14'34"-27d20'
Montgomery,Suzanne G
Uma 9h47'26"51d22'
Montgomery,Westley
Dra 18h36'43"70d9'
Montgomery,William Gardner
Uma 12h11'13"60d29'
Montgomery-MMM,Margie
Mon 8h7'8"-2d34'
Montgomery-MMM,Margie
Lyr 19h18'28"41d10'
Monthlie,Savigny-les-
Beaunes
Per 4h4'0"37d57'
Monti,Kara
Eri 3h11'18"-5d49'
Monti,Megan M
And 0h10'58"35d32'
Monti,Piero
Lac 22h33'43"38d30'
Monti,René
Aqr 20h55'30"0d13'
Montibeller,Katherine Melissa
Cas 0h52'22"63d57'
Montiel,Michael
Lac 22h14'0"54d32'
Montini,Silvia
Lyn 8h20'47"48d24'
Montion,Mauricio
Mon 6h53'36"-0d26'
Montiverdi,Forever Yours-
Michael J
Dra 16h4'47"68d10'
Montizon,Raq
Ser 15h14'44"1d9'
Montler's Star
Lmi 10h4'1"41d19'
Montoro,Deanna Marlene
Olague Perez
Mon 8h1'53"-9d56'
Montoro,Deanna Marlene
And 1h12'55"40d43'
Montoya,Brenda
Her 18h1'53"26d38'B
Montoya,David
Her 18h1'53"26d38'A
Montoya,Manny "Kito"
Cet 2h6'38"2d56'
Montoya,Monique Yvette
Lib 14h17'50"-8d9'
Montoya,Mr Michael D
Equ 21h20'1"11d39'
Montoya-La Luz De Mi
Alma,Angelica
Mon 6h54'32"0d17'
Montpetit,John Robert Florian
Ori 5h53'24"21d6'
Montreal,Jose Fernandez
Aur 6h3'43"45d55'
Montressa
Mon 7h3'25"-1d33'
Montri,Baby Greens- Robin
Crt 11h15'59"-10d16'
Montrone,Gina
Cas 1h54'41"58d58'
Montross,Barbara & Carter
Crb 15h50'49"27d55'
Montserrat,Dominique
Ori 6h1'40"0d20'
Montuori,Lauren Kay
Vul 20h5'16"28d30'
Monty
Cep 22h32'0"60d19'
Monty
Aql 18h57'1"-5d49'
Monty
Lmi 10h42'43"27d29'
Monty
Mon 8h2'36"-2d36'
Monty "My Love" Jordan
Aql 18h56'47"-1d12'
Montz,Robert Dale
Her 18h7'0"31d46'
Monumentum Ad Smittya
Cinnamoneque
Per 1h35'20"54d8'
Monya
Lmi 11h2'49"32d40'
Monyak,David J
Per 3h22'57"40d10'
Monė,Robert P
Her 16h38'58"34d50'
Moo
Cam 7h15'1"68d26'
Moo Moo & Bumpa I Love
You
Aql 19h24'30"13d37'
Moo Star
Dra 20h12'47"68d7'
Moody
Per 2h37'49"56d53'
Moody Blues:Justin Graheme
John Ray,The
Cmi 8h0'45"6d42'
Moody's Flame
Ori 5h23'22"-3d2'
Moody,Benjamin W
Cep 0h3'38"66d43'
Moody,Brad
Ori 5h46'45"18d33'
Moody,Edward Alistair
Psc 0h47'20"31d28'
Moody,Felicity Harriet
And 1h52'11"41d12'
Moody,Ian
Peg 22h24'16"20d42'
Moody,Jr,Dan Ray
Cep 0h14'17"66d23'
Moody,Kyal & Colleen
Crb 15h16'18"31d37'
Moody,Melanie Vanessa
And 0h42'60"30d9'
Moody,Philippa Louise
Lyn 8h30'47"47d42'
Moody,Sankey Alan
Gem 7h18'7"21d34'
Moody,Scott Runnells
Oph 17h54'18"12d22'
Moody,Sheri Sue
Cet 0h24'54"-11d58'
Moody,Zoe
And 1h56'50"47d19'
Moogalian,Pasqualena
Uma 9h0'30"70d40'
Mook,Sarah Gail
Tri 2h6'40"30d54'
Mookie
Aur 5h19'41"41d28'
Mookie's Star
Vir 11h47'11"4d14'
Moomaw,Erika Fern
Cas 0h46'12"71d13'
Moomin's 80th
Uma 10h20'22"57d52'
Moon
Cap 22h0'27"-16d25'
Moon Dog
Umi 15h47'0"71d19'
Moon L W
Peg 22h56'56"25d45'
Moon Over Moyer
Lib 14h42'0"-8d14'
Moon Stars Auntie Carole &
Uncle John
Eri 4h54'38"-5d8'
Moon Unit Zappa
Ori 6h15'42"7d51'
Moon Watcher 7-1-53
Cnv 13h41'50"45d26'
Moon,Carl
Boo 14h43'55"28d56'
Moon,Chris
Aur 6h14'39"50d8'
Moon,Dorothy Jean DeShon
Cyg 19h27'38"32d39'
Moon,Jason
Per 3h12'38"50d13'
Moon,Judi Ann Mandel
Del 20h14'25"9d38'
Moon,Judy Bond
Aqr 20h37'17"-11d4'
Moon,Julie
Del 20h22'42"10d32'
Moon,Mary
Com 12h21'1"20d6'
Moon,Michael
Boo 15h2'41"15d14'
Moon,Seoun & Sung
Uma 11h34'37"50d34'
Moon,Teresa
Lyn 8h1'40"40d41'
Moon,True Parent Hak Ja Han
Cyg 19h56'36"33d8'B
Moon,True Parent Sun Myung
Cyg 19h56'36"33d8'A
Moon-Anthony
Cet 3h12'32"1d20'
Moonbeam
Per 4h44'35"51d50'
Moonbeam
Uma 11h57'0"60d59'
Moonchild
Aur 4h59'1"51d43'
Moondance
Dra 19h42'56"61d4'
Moondoggie
Cam 6h9'19"60d51'
Moondoggy
Vel 10h5'21"52d51'
Mooney's "Wedding
Star",Samantha & Dan
Cyg 19h42'39"41d8'
Mooney,"Uncle Donnie"- D J
Oph 18h19'50"7d8'
Mooney,Andrea L
Uma 10h24'48"51d11'
Mooney,Bob
Hya 9h31'47"-5d55'
Mooney,Daniel P
Ori 5h25'25"-0d31'
Mooney,David Alan
Cmi 7h19'21"4d39'
Mooney,David James
Cmi 8h2'22"6d28'
Mooney,Elaine
Eri 4h32'0"-1d5'
Mooney,Elaine
Lyn 7h53'32"42d19'
Mooney,Francis
Uma 10h35'56"53d23'
Mooney,Jacqueline
And 0h45'27"39d54'
Mooney,James Gerard
Ori 5h59'35"21d26'
Mooney,Jimmy
Aur 5h35'51"50d8'
Mooney,Josh
Dra 14h22'0"63d39'
Mooney,Jr,Family Star, John
Lac 22h52'35"54d43'
Mooney,Laura Robbins
Cet 1h33'25"-12d23'
Mooney,MD,Robert M
Dra 20h21'46"62d4'
Mooney,Morag Mulholland
Stewart
Ori 5h24'30"-4d13'
Mooney,Natalie
Lyn 7h29'12"41d9'
Mooney,Richard Benjamin
Ori 5h6'32"1d39'
Mooney,Todd
Peg 0h6'50"13d37'
Mooney-Love Lives On,
Michaelene
Cyg 20h56'28"30d26'
Mooneyham,Edward
Her 18h16'14"14d57'
Moonfairy
Lyn 08h56'26"46d18'
Moonhee
Boo 14h12'0"32d22'
Moonier,Douglas Michael
Aur 6h55'48"37d0'
Moonier,Jacqueline Marie
And 0h50'60"34d11'
Moonlight
Lup 15h17'36"-42d42'
Moonlight Bandit
Her 16h2'54"50d23'
Moonlight Love
Cyg 21h32'48"36d0'
Moonlight Ride
Lyn 8h24'29"44d25'
Moonpupy Reeds
Ser 15h15'50"-2d58'
Moonshadow
Uma 9h34'11"55d12'
Moonstar
Cam 6h11'42"58d29'
Mooradian,Anee
Lyn 7h33'39"52d13'
Moore "Little One", Janice
Lynette
And 23h19'60"48d44'
Moore "Tesoro",Moore
Ori 5h51'50"15d18'
Moore & Michaud
Cep 21h51'13"55d47'
Moore & Moore
Peg 21h51'12"33d34'
Moore 1912,Clara
Lyr 19h16'16"37d35'
Moore Golden Jubilee, Sister
Helen
Ori 5h45'53"10d19'
Moore III,George Harold
Cmi 7h30'14"8d16'
Moore III,Richard R
Dra 17h5'1"68d24'
Moore In The Name Of "C
B",For Sheila
And 23h25'40"47d55'
Moore IV,Richard R
Boo 15h7'53"15d33'
Moore Star
Aql 18h58'42"-7d42'
Moore's-Unreachable Star,Dot
Lyr 19h9'10"38d6'
Moore,"Bry"Bryan Kristopher
Per 1h56'21"53d54'
Moore,"Moore's Dragon" Pat
& Deb
Dra 17h18'1"69d36'
Moore,Adeline Nichols
Lyr 18h54'0"30d20'
Moore,Alice
Peg 21h57'30"23d42'B
Moore,AMPM
Uma 14h3'23"53d42'
Moore,Anita Karen Walker
Aql 19h1'13"16d14'
Moore,Anita L
Lyr 18h56'53"30d18'

Moore,Austen Criley
Dra 16h10'0"64d39'
Moore,Barbara
Del 20h13'11"14d55'
Moore,Barbara
Equ 21h1'55"2d56'
Moore,Bevan Masterson
Aur 7h1'0"39d28'
Moore,Bil
Dra 16h12'18"62d23'
Moore,Bill
Uma 12h6'44"46d26'
Moore,Billy & Ginger
Boo 15h8'55"48d10'
Moore,Bob & Jean
Eri 2h55'1"-2d34'
Moore,Bobbie
Gem 6h56'46"13d34'
Moore,Brenda
Vir 13h31'1"12d16'
Moore,Bridget Lynn
Peg 22h57'11"27d13'
Moore,Carrie Ann
Lyn 7h53'22"43d8'
Moore,Carver Austin
Cet 2h10'1"2d12'
Moore,Catherine Elizabeth
Mon 6h54'30"-6d24'
Moore,Charles A
Umi 14h40'13"74d22'
Moore,Chris
Del 20h15'23"10d35'
Moore,Christian Greene
Aql 19h49'31"10d16'
Moore,Christian Peter
Sex 9h52'11"1d6'
Moore,Christopher D
Uma 9h2'12"52d50'
Moore,Claire
And 0h50'48"37d2'
Moore,Claire Elizabeth
Mon 7h52'45"-1d47'
Moore,Clayton
Vir 13h35'11"6d35'
Moore,Connor Dean
Cyg 20h54'44"30d14'
Moore,Craig M
Aur 4h59'20"40d9'
Moore,Dalen Thomas
Ser 15h55'55"4d10'
Moore,David Knight
Lmi 10h57'36"25d12'
Moore,David Teal
Hya 9h8'0"2d26'
Moore,Dayna
Cyg 20h57'46"48d52'A
Moore,Delores Everyone is
Less she's
Peg 23h1'59"8d13'
Moore,Derek
Per 2h29'1"50d58'
Moore,Diana E
Peg 21h55'46"28d43'
Moore,Dolores
Boo 13h37'54"15d3'
Moore,Douglas
Dra 17h34'54"63d52'
Moore,Douglas Wayne
Per 4h20'47"52d28'
Moore,Dr Thomas O
Oph 18h21'53"6d21'
Moore,Dylan- Christopher
Sct 18h58'19"-5d27'
Moore,Earl & Joyce
Ori 5h42'11"-0d43'
Moore,Eddie
Her 17h6'21"37d39'
Moore,Emily Charlotte
Cyg 21h29'42"33d26'A
Moore,Emma
Aql 19h48'25"12d56'
Moore,Enloe
Sex 9h45'1"3d48'
Moore,Eric Charles
Dra 16h30'20"63d45'

Moore,Essie Santerre
Crb 15h47'41"27d16'
Moore,Faith
Her 18h4'53"40d21'B
Moore,Florence D
Cyg 20h33'15"30d43'
Moore,Frances Marie
McConnell
Hya 9h19'1"-6d42'
Moore,Fred
Eri 3h47'29"-0d21'
Moore,Gary
Dra 17h3'30"62d34'
Moore,Gary
Oph 17h27'0"10d54'
Moore,Gary & Joan Donellan
Cyg 20h19'11"38d11'
Moore,Gary Wallace
Psc 0h58'16"11d60'
Moore,Gene
Dra 16h25'47"64d31'
Moore,George L
Uma 10h48'58"51d8'
Moore,Geraldine Elaine
Cas 0h59'21"64d7'
Moore,Geri Frances
Eri 2h48'29"-17d58'
Moore,Gordon Dudley
Cmi 8h8'35"2d48'
Moore,Greg & Kim Fluty
Mon 7h3'39"-5d48'
Moore,Gregory Wayne
Cet 0h27'0"-6d36'
Moore,Guss
Ori 6h0'52"0d32'
Moore,Hannah
Cyg 19h29'55"30d26'
Moore,Heather Ellene
Tri 2h45'11"31d55'
Moore,Heaven Star
Mon 8h5'24"-4d29'
Moore,Helen Bianchi
Oph 17h7'24"-23d35'
Moore,Hillary Ann
Lyn 9h12'18"44d9'
Moore,Irving J
Ori 5h53'32"11d43'
Moore,James
Tri 2h23'46"32d1'
Moore,James A
Lyr 18h47'59"30d3'
Moore,James Blake
Boo 15h6'44"10d26'
Moore,James Randall "Randy"
Ori 5h52'42"14d20'
Moore,Janice & Emily
Sonnessa
Cas 0h33'31"68d37'
Moore,Jean'ne
Com 12h50'35"19d52'A
Moore,Jell
Com 12h50'35"19d52'B
Moore,Jeff "Slick"
Dra 20h14'1"64d36'
Moore,Jenifer Leigh
Lyr 19h21'1"38d60'
Moore,Jennifer Christine
Uma 11h12'18"32d19'
Moore,Jennifer
Cas 2h2'1"73d57'
Moore,Jennifer Jill
Cyg 20h42'37"45d39'
Moore,Jessica Kay
Lyr 18h42'53"36d1'
Moore,Jewell
And 23h17'44"51d19'
Moore,Jim A
Oph 17h58'25"11d51'
Moore,Joan M
Cyg 19h56'43"40d59'
Moore,Joanne C
Lyr 19h16'34"26d48'
Moore,John G
Cmi 7h59'29"8d15'

Moore,Jolene & David
Del 20h18'25"16d6'
Moore,Jon Lawrence
Sct 18h54'22"-6d21'
Moore,Jr,Adrien D(Ace)
Aur 6h2'23"38d14'
Moore,Jr,John Alan
Lyr 18h20'15"38d17'
Moore,Jr,Richard R
Aur 6h29'0"31d10'
Moore,Jr,William J
Aur 5h57'37"37d32'
Moore,Juanita
Eri 4h37'15"-18d54'
Moore,Judith E
Cyg 19h28'45"32d53'
Moore,Justin Alexander
Aql 19h30'0"12d57'
Moore,Justin Thomas
Cam 3h26'26"60d35'
Moore,Kathryn B
Eri 3h55'35"-2d2'
Moore,Kathryn Louise
Aql 19h58'39"14d56'
Moore,Kay & Marshall
Sex 9h52'14"4d46'
Moore,Kellie L
Lyr 19h3'12"28d41'
Moore,Kelly
Aql 18h43'1"10d17'
Moore,Kenneth Leonard
Hya 9h10'21"1d22'
Moore,Kenneth Tyler
Her 16h48'44"41d8'
Moore,Kim
Peg 21h50'41"31d8'
Moore,Kim E
Cmi 7h53'51"0d38'
Moore,Laurie
Lyn 8h45'22"42d35'
Moore,Leslie Ann
Cam 8h52'38"77d48'
Moore,Leslie Marlene
Ori 6h8'46"4d11'
Moore,Lily
Ori 5h54'58"7d37'
Moore,Linda
Ari 3h12'36"28d9'
Moore,Lonnie
Aql 19h39'45"11d4'
Moore,Louise C
Cyg 20h30'24"40d28'
Moore,Louise C
Cas 0h6'27"59d0'
Moore,Lowell
Cet 1h1'0"-8d31'
Moore,Lucas T
Aur 5h1'56"31d16'
Moore,Lyndsay Nichole
Lyr 19h12'0"38d18'
Moore,Mac & Rob
Eri 4h47'39"-7d33'
Moore,Madison Gaines
Uma 9h45'0"42d54'
Moore,Magdalena Lynn
Cas 1h16'24"63d17'
Moore,Marcie C G
Cam 11h34'37"81d41'
Moore,Marian M
Cas 1h43'26"60d60'
Moore,Marilyn
Vul 19h58'29"22d34'
Moore,Marion Ann
Cyg 19h15'20"44d15'
Moore,Marjie & Henry
Cyg 21h5'38"30d29'
Moore,Marjorie A
And 2h30'50"49d4'
Moore,Marlene
Peg 22h7'41"26d13'
Moore,Mary
Peg 23h29'54"21d19'
Moore,Mary Margaret
Cas 1h4'40"58d20'

Moore,Matthew Badalamenti
Per 2h29'21"58d53'
Moore,Maurice
Her 18h4'53"40d21'A
Moore,Michael
Aql 19h48'18"11d32'
Moore,Michael Sheridan
Lmi 10h35'12"30d35'
Moore,Michelle Annette
Uma 11h51'49"50d34'
Moore,Mike
Hya 8h33'57"-01d49'
Moore,Morris W
Ser 18h17'22"-14d52'
Moore,Nathan
Her 18h4'50"48d57'
Moore,Neen-Neen-
Butterbean-Foot-Foot
Her 16h39'17"50d38'
Moore,Nellie Henry
Lyn 7h27'31"44d11'
Moore,Nichelle- Lorraine
Naomi
Vul 19h48'44"28d57'
Moore,Nicholas
Per 2h31'0"56d37'
Moore,Nicholas E
Oph 18h31'0"6d38'
Moore,Nicola Joanne
Cas 2h10'38"68d32'
Moore,Nicoli Blanche
Crt 11h8'3"-17d15'
Moore,Nita
Psc 0h20'53"0d55'
Moore,Olivia Lange
And 1h59'19"37d51'
Moore,Oran & Pat
Crb 16h20'34"28d28'
Moore,Pat
Del 20h19'30"10d12'
Moore,Patricia Ann
Peg 22h32'45"29d1'
Moore,Paul
Ori 5h56'1"14d19'
Moore,Phil
Eri 2h58'0"-5d5'
Moore,Philip
Ori 5h56'0"16d10'
Moore,Phyllis
Cnc 8h5'41"18d49'
Moore,Rachel Bianca
And 2h29'36"45d36'
Moore,Ralpholene
Cyg 21h10'31"35d43'
Moore,Rebecca Jane
Cyg 21h29'42"33d26'B
Moore,Reece Alexander
Cmi 7h0'0"-15d2'
Moore,Regina
And 23h2'0"51d0'
Moore,Robin Ora
Cas 0h43'36"64d24'
Moore,Rodney S R
Ori 4h55'51"1d27'
Moore,Rupert
Crt 10h57'58"-18d60'
Moore,Sabrina Ellen
And 23h10'29"40d12'
Moore,Sarah & David
Umi 15h51'57"76d57'
Moore,Sarah Donovan
And 23h17'30"44d56'
Moore,Sharon L
Cam 3h44'44"77d6'
Moore,Special Agent Jim
Lmi 10h20'43"28d46'
Moore,Sport
Her 17h32'54"23d17'
Moore,Sr,Michael R
Cet 2h32'60"-10d58'
Moore,Stephen
Cet 2h16'48"3d56'
Moore,Stephen P
Uma 11h58'35"43d26'

Moore,Stephen"Stevie"
Dra 9h51'1"74d13'
Moore,Steven
Lac 22h22'0"48d51'
Moore,Susan
And 0h20'14"35d55'
Moore,Suzanne
Eri 4h13'27"-17d21'
Moore,Sylvia
Lib 14h44'20"-20d10'
Moore,Tamarina J
Peg 23h40'44"8d3'
Moore,Tennica-Tennie
Com 12h24'14"22d6'
Moore,The Star of Bob &
Margaret
Lyr 19h21'59"41d18'
Moore,Thomas Joseph
Cnv 13h40'15"32d48'
Moore,Thomas Oran
Aql 20h8'13"1d43'
Moore,Tiffany A
Com 12h7'57"27d30'
Moore,Tim
Ori 6h9'24"4d4'
Moore,Timothy W
Dra 12h48'40"68d21'
Moore,Timothy William
Boo 15h12'49"30d9'
Moore,Tobias McFeeley
Sco 16h56'10"-40d39'
Moore,Todd Chapman
Cma 6h55'10"-18d40'
Moore,Tommy E
Cet 0h30'32"-10d58'
Moore,Tracy
Lyn 8h0'26"42d15'
Moore,Trinka Marian
Peg 23h47'37"17d58'
Moore,Vanessa Puaihilani
Lyr 19h0'39"30d30'
Moore,Vic
Peg 21h57'30"23d42'A
Moore,Victor E
Boo 14h28'54"17d56'
Moore,Victoria Tracey
Mon 6h55'37"-6d50'
Moore,Virginia(Ginny)
Cam 4h50'39"68d53'
Moore,W Craig
Cep 23h11'1"70d29'
Moore,Wendy Sue
Ori 4h41'44"12d56'
Moore,William & Melanie
Cyg 20h6'28"58d41'
Moore,Zachary Langston
Oph 17h53'37"13d44'
Moore-"Sweaty 1"Karen
And 0h50'20"38d45'
Moore-35,Jane & Carleton
Crb 16h16'46"27d36'
Moore-HPA,Sarah Marie
Cas 0h23'26"50d3'
Moorefield,Rhonda
Cas 0h32'12"66d49'
Moorhead,`Paddy'
Her 18h2'28"28d15'
Moorhouse,-Across The
View,Tim
Psc 1h1'1"20d16'
Moorhouse,Ian
Ori 6h7'34"1d17'
Moorhouse,Paul
Aur 5h0'57"51d15'
Moorman
Eri 4h14'25"-10d57'
Moorman,Nathan Charles
Cet 2h39'38"-8d54'
Moorman,Perry Trevor
Oph 17h59'25"12d15'
Moosang & Inni
Sge 19h58'28"20d20'
Moose
Cmi 7h5'1"1d28'

Moose Clark
Cmi 7h53'53"0d15'
Moose,Lauren Ann
Cas 0h29'38"61d49'
Moose,Walter L
Cyg 20h31'11"38d40'
Mooser,Gerhard
Aqr 22h50'52"-6d57'
Mootrey My Little Puffy-
Puss,Ron
And 22h58'20"40d1'
Mopetelly
Cyg 20h21'11"38d60'
Mopl,Carla
Cru 12h36'6"-62d15'
Mopsick,Kenneth
Cep 22h52'1"70d17'
Mopsik,Delores
Peg 22h31'12"8d15'
Moquay,Rotraut Uecker Klein
Eri 2h47'0"-6d28'
Mor,Bretnach Gregor
Sge 20h0'58"20d3'
Mora,Carla & Al
Mon 6h34'0"-0d59'
Mora,Jodee
Mon 7h6'27"-6d35'
Morabito,Donald A & Jane I
Cyg 19h18'58"44d39'
Morabito,Jr,Nicholas A
Cep 22h30'12"58d13'
Morabito,Teo
Aqr 21h32'50"-0d53'
Morabito-Wedding Star, John
& Sharon
Cyg 20h9'35"40d44'
Moradshahi,Michael
Aql 20h1'28"10d42'
Morag
Ari 1h58'50"17d44'
Moraga,Mary
Mon 6h4'47"-8d2'
Moragas,Mari-Pili Horta
Gem 8h4'42"33d19'
Morais,Mike
Sgr 19h39'20"-30d6'
Morales Happy Birthday
45,Vincent M
Tau 4h57'31"20d14'
Morales,Danny Dee
Uma 8h5'36"68d31'
Morales,Francisco de Padua
Her 17h57'42"14d32'
Morales,Francisco G.
Boo 14h34'48"50d3'
Morales,Gil
Cnv 12h23'15"38d13'
Morales,Gustavo
Cet 0h56'25"-5d21'
Morales,José Daniel
Per 2h51'50"45d21'
Morales,Judith S
Lac 22h36'31"56d48'
Morales,Tyler James
Lac 22h26'32"52d46'
Morali,Sarah
Aur 7h21'0"43d45'
Moram Star,Raymond
Augustus
Dra 20h15'57"70d39'
Moran Star,The
And 23h32'53"44d55'
Moran Wedding Star, Michael
& Robin
Cyg 19h42'51"42d2'
Moran's Heartlight, John
Boo 15h4'11"38d32'
Moran,Angela-Elizabeth &
Anthony Pengelly
Cyg 19h27'23"35d34'
Moran,Angie
Cyg 19h47'1"37d51'
Moran,Angie
Cyg 19h47'1"37d51'

Moran,Anne D
Equ 21h1'55"2d56'
Moran,Arthur Charles
Her 16h57'42"38d9'
Moran,Bartholomew
And 2h21'0"37d19'
Moran,Bill
Dra 14h39'35"58d10'B
Moran,Brian Scott
Ser 15h32'28"19d18'
Moran,Christopher Alan
Cap 21h26'0"-18d54'
Moran,Dr
Uma 11h50'43"33d20'
Moran,Edward John
Per 2h53'28"32d58'
Moran,Elizabeth
Mon 6h40'32"7d5'
Moran,Helen
Lyn 8h12'13"39d60'
Moran,James F
Aur 6h17'59"32d27'
Moran,Jason Edward
Aur 6h2'46"31d25'
Moran,Jean & Bill
Dra 14h39'35"58d10'AB
Moran,Joan
Cma 6h47'16"-17d15'
Moran,Joe & Nancy
Cyg 21h52'18"40d6'
Moran,John
Lib 15h8'15"-5d57'
Moran,John "Jackie" T
Cep 0h7'48"70d40'
Moran,John Patrick
Cnv 13h44'41"31d42'
Moran,Jr,(Jim),James Dana
Ser 15h56'28"4d56'
Moran,Kate Alexandra
Lyr 18h35'54"38d22'
Moran,Katylynn Hill
Peg 22h35'14"20d22'
Moran,Kelli
Peg 22h45'13"33d34'
Moran,Kimberly Ann
Ari 2h43'0"21d19'
Moran,Lee Edward
Aql 20h3'56"7d35'
Moran,Linda S
And 2h23'15"49d31'
Moran,Lois Burke
Equ 21h7'33"11d4'
Moran,Maria
Mon 7h24'29"-6d43'
Moran,Noah
Boo 14h37'22"36d34'
Moran,Patrick J
Aur 6h30'37"35d58'
Moran,Raewyn Charlotte
Uma 15h10'11"78d34'
Moran,Robert Frances
Aur 5h32'10"29d35'
Moran,Sarah J
Lyr 18h55'1"31d59'
Moran,Scott Meyers
Dra 19h46'50"68d30'
Moran,Siobhan Margaret
Cas 0h0'49"61d3'
Moran,Stephen
Aur 6h29'56"41d1'
Moran,Tasha Kaye
Mon 6h42'38"10d38'
Moran,Thomas J
Cas 1h44'19"60d35'
Moran,Thomas Joseph
Ser 15h43'58"21d16'
Moran,Tom
Aur 5h36'47"37d38'
Morana,Claudia
Cam 3h54'54"56d26'
Morand,Michele
Boo 14h59'11"12d21'
Morano,"Iron Mike" Michael
W
Her 16h46'37"34d43'

Morasch,Sarah
Mon 6h26'20"7d21'
Morash,Archie
Aur 6h32'0"38d44'
Morat,Bill & Jan
Lyr 18h42'11"42d59'
Morata-Bedoya
Oph 18h16'39"12d19'
Moratelli,Barbra Joyce
Cas 23h32'35"62d44'
Moratus
Aur 7h0'0"43d9'
Moravec,Raymond &
Germaine
Lyn 7h44'20"48d45'
Morawetz,Mag Bernhard
Ori 6h0'44"-1d50'
Morawitz,William
Boo 15h3'14"14d47'
Morawski,Michael (The
Mogul)
Cnv 12h28'15"33d37'
Morbidelli,Laura
Her 18h9'21"38d37'
Morchower,Karen Rose
Mon 6h57'48"-0d34'
Mordan,Phil
Per 2h9'56"57d14'
Mordaunt,Will
Ori 5h47'41"10d18'
Mordente,Lisa & Donnie Kehr
Cyg 20h30'59"48d58'
Mordini,Adolph "Ruby"
Her 18h2'0"48d24'
Mordini,Isabella Raine
And 0h19'12"30d45'
Mordue,Elizabeth Courtney
Cas 1h9'48"62d10'
More Infinity
Peg 23h5'59"21d57'
More Than This
Uma 10h4'14"47d45'
More Than Wisdom Or A
Drink
Cet 2h55'57"0d42'
More Than You Know
Aql 18h56'0"-0d50'
More,M Christine
Eri 2h58'19"-6d44'
More-Gordan,Charlotte
Cas 0h23'34"64d42'
Moreau's Millennium
Sge 19h10'40"17d6'
Moreau,Brother Justin
Vir 11h55'21"-4d23'
Moreau,Eileen
Cas 0h37'40"67d37'
Moreau,Jean Pierre
Cyg 21h6'35"30d50'
Moreau,Louis
Dra 18h36'42"80d24'
Moreau,Marc-Andre
And 1h11'54"47d20'
Moreau,Olivier
And 2h20'50"46d48'
Moreau,Philippe
Dra 15h4'35"63d40'
Moreau,René
Mon 7h51'58"-2d8'
Moreaux,Flavia
Cyg 21h19'23"28d42'
Moree,Patricia
Cas 1h4'12"64d2'
Moreen
Cnc 8h27'13"32d54'
Morefield,MD,Steven Quentin
Dra 17h50'34"68d37'
Morehead,Lucky
Ori 5h53'1"9d52'
Morehouse,Diane
Lac 22h51'25"38d11'
Morehouse,Don
Lac 22h2'39"38d38'
Morehouse,Duane
Cyg 21h5'0"32d46'

Morehouse,Frank
Aql 19h30'59"13d17'
Morehouse,Sheilah I
Lyn 7h25'57"44d3'
Morel,Claude
Cyg 20h42'6"46d4'
Morel,Denis
Cet 2h29'27"4d13'
Morel,Jean-Pierre
Cyg 19h45'20"31d11'
Morel,Jefte R
Ori 4h53'1"-2d1'
Morel,John George Alastair
Cyg 21h15'11"28d22'
Morel,Marie Pierre
Mon 7h54'27"-4d29'
Moreland My Eternal
Love,Chuck
Cep 23h1'52"70d56'
Moreland,Frank
Cep 21h5'48"67d49'
Moreland,Hal
Cet 2h55'34"4d20'
Moreland,Jr,Osbie
Her 16h54'11"50d46'
Moreland,Ramon George
Cep 22h2'1"60d53'
Moreland,Sheryl
Peg 22h33'22"8d1'
Moreland,Tom
Her 16h26'27"28d4'
Moreland,William O & Vera L
Lyr 18h43'37"32d54'
Morell,Lutz
Lyn 7h55'33"50d26'
Morell,Peter J
Lac 22h38'16"55d18'
Morelli
Cnv 12h16'43"47d36'
Morelli,Anthony
Cep 22h49'1"57d45'
Morelli,Enrico
Per 3h33'33"51d20'
Morelli,Giovanni
Lyn 7h26'1"40d8'
Morelli,Wendy Warren
And 23h1'56"51d34'
Morello,Charlie
Sco 16h44'23"-23d20'
Morello,Gino Francesco
Lyr 18h40'41"30d18'
Morello,Star Joe
Her 16h29'43"41d36'
Morelock,Virginia
Peg 23h35'0"27d40'
Morena
Cyg 21h22'39"38d1'
Morena
Lep 5h56'23"-24d40'
Morena,James J
Her 18h8'50"38d17'
Morena,M
Ori 5h7'30"1d32'
Morency,Mont
Lyr 19h15'51"41d52'
Moreno
Tau 4h38'33"15d14'
Moreno,Angelina
Cas 2h35'56"58d28'
Moreno,Antonio Lopez
Per 2h6'14"58d37'
Moreno,Carmen Maria
Barderas
Her 17h4'1"43d53'
Moreno,Diana
Peg 21h54'55"22d59'
Moreno,Holly Diane
Com 12h0'13"21d51'
Moreno,Michael
Lyr 18h44'27"45d5'
Moreno,RN,Anna M
Tau 5h51'28"23d3'
Moreno,Sonny
Eri 4h41'53"-21d11B'

Moreno,Yolanda Lyn 7h34'0"41d12'
Moreno-Valle R,Rafael Cnv 12h53'48"39d34'
Morentz,Grace Schrock Lyr 18h40'1"29d34'
Morentz,Paul Ernst Dra 14h21'28"63d50'
Moreo,Christopher Lac 22h36'25"53d36'
Moret,Dr Michael Oph 17h56'47"12d40'
Moretti,Jane Carolyn Lyr 18h58'57"38d15'
Moretti,Manuela Aql 19h1'0"10d28'
Moretti,Matteo Uma 11h41'0"51d60'
Moretti-Lanzano Uma 8h52'51"48d54'
Morey,Brad Dra 17h59'12"52d51'B
Morey,Christine Breisacher Dra 17h59'12"52d51'A
Morey,Ernest William Aur 6h0'55"35d51'
Morey,Joel Kevin Boo 14h39'24"32d30'
Moreé Boo 14h32'37"20d27'
Morf,Pascal-Alexandre Uma 8h45'13"50d9'
Morf-Steudler,Ruth Uma 10h41'59"52d25'
Morffi,Angela Uma 8h49'50"48d52'
Morford,Jr,Robert W Her 16h19'24"4d13'
Morgaine Peg 22h41'38"35d17'
Morgaine Cnc 8h32'30"31d2'
Morgan Cnc 8h33'15"31d3'
Morgan Lyr 18h31'1"42d53'
Morgan "Ricky",Eric Boo 15h2'11"15d18'
Morgan & Kathy Wedding Star Cyg 19h29'45"31d44'
Morgan 1-2-94,Heather Cet 2h20'16"-1d17'
Morgan 19AJP,Julie Dawn Uma 8h56'53"51d38'
Morgan Alexis Eri 2h50'39"-1d44'
Morgan Elizabeth Lyn 7h3'48"52d52'
Morgan Empowering Families 1991,Linda Mon 7h39'24"-0d56'
Morgan Fay Cep 22h54'16"65d58'
Morgan IV,Joseph David Boo 14h38'27"35d52'
Morgan Lee Cyg 21h55'16"52d41'
Morgan Lee Uma 11h35'54"61d4'
Morgan Leigh Cam 14h15'1"81d32'
Morgan Leigh's Wookie Star Aur 6h2'19"35d53'
Morgan Lynn And 1h28'30"50d9'
Morgan Star,John Cep 22h45'11"63d29'
Morgan Star,The Aql 18h57'47"10d52'
Morgan"Ma Mere Jolie", Rose Marie Peg 23h4'60"8d26'
Morgan's Diamond, Betty & Arthur Ori 6h2'16"8d51'
Morgan's Family Star, Tim & Tessa Cam 3h33'40"61d47'
Morgan's Guide Peg 22h32'42"29d41'
Morgan's Little Star Lyn 6h15'20"60d25'
Morgan's Spirit,James Oph 17h36'0"-16d6'
Morgan,"Bunny",Cynthia N Sct 18h47'28"-7d8'
Morgan,"Granny" Grace V Cas 22h57'31"56d32'
Morgan,"Molly Mouse" Vul 20h57'53"28d27'
Morgan,"Pops" Richard C Aur 6h1'37"30d21'
Morgan,Alan Cyg 19h32'58"37d26'
Morgan,Allie Elizabeth Aur 5h28'45"37d57'
Morgan,Amanda Sue Sco 17h27'51"-30d8'
Morgan,Amber Faye Mon 7h0'44"4d34'
Morgan,Ann And 0h58'26"37d2'
Morgan,Anthony Thomas Ori 6h7'36"4d22'
Morgan,Ashley A Cyg 20h53'0"30d32'
Morgan,Ashley Nicole Peg 22h57'58"20d11'
Morgan,Barbara Bradford & Taylor Mon 6h56'49"10d41'
Morgan,Barbra Vul 19h50'44"20d11'
Morgan,Bethany Nims Ori 5h52'17"20d25'
Morgan,Beverly Adel Schultz And 2h16'20"38d46'
Morgan,Brian & Cyndie Lyr 18h46'12"31d4'
Morgan,Charles F Dra 15h46'19"60d45'
Morgan,Christian Ori 5h51'38"18d40'
Morgan,Christine Lyn 7h46'21"38d59'
Morgan,Cindy J And 22h57'10"50d22'
Morgan,Connor Smith Her 17h10'21"48d20'
Morgan,Dan D Lac 22h28'1"38d58'
Morgan,David Garyth Uma 9h46'1"54d53'
Morgan,Dennis Uma 8h31'39"57d45'
Morgan,Doris Sct 18h41'39"-6d8'
Morgan,Duncan James Hya 8h19'1"-10d40'
Morgan,Elizabeth And 2h25'51"40d55'
Morgan,Elizabeth Anne Charlotte Crb 15h39'50"26d45'
Morgan,Erin Lynn Lyr 18h53'58"30d54'
Morgan,Evan Samuel Aur 6h27'1"30d53'
Morgan,Ewa Uma 10h44'58"58d49'
Morgan,Frieda Peg 23h16'59"31d15'
Morgan,Gregory Charles Uma 9h47'55"53d17'
Morgan,Heather Linda Del 20h30'16"10d4'
Morgan,James David Ori 5h7'30"1d32'
Morgan,James David Raymond Thomas Her 17h18'16"43d5'
Morgan,James Vaughan Dra 15h44'12"62d7'
Morgan,Jami Cyg 21h24'43"41d13'
Morgan,Jared Lac 22h3'56"46d24'
Morgan,Jason Ser 15h12'10"7d27'
Morgan,Jennifer Lee Cam 3h48'0"72d25'
Morgan,Jennifer Lynn And 23h18'46"47d48'
Morgan,John Alexander Vul 19h2'45"22d11'
Morgan,Josh A Aql 19h22'26"14d20'
Morgan,Jr,Irvin Dra 13h2'37"70d22'
Morgan,Jr,Wayne R Mon 7h18'36"-6d28'
Morgan,Kathrine' Kristine & William Sge 19h26'13"16d10'
Morgan,Kelly Umi 16h16'39"70d32'
Morgan,Kerri Ri Com 12h54'22"23d34'
Morgan,Kimberly Ann And 2h23'38"41d56'
Morgan,L L Cam 3h30'0"55d25'
Morgan,Lance Dra 19h16'10"67d49'
Morgan,Leslie Ser 17h32'16"-14d33'
Morgan,Linda Elizabeth Cas 0h25'16"74d13'
Morgan,Linda May Lyr 18h36'1"37d6'
Morgan,Makenzie Winnett Cyg 20h17'34"38d6'
Morgan,Marianne (Meme) Lyn 8h56'49"44d25'
Morgan,Mark Alan Hya 8h44'24"1d39'
Morgan,Martha Jackson Aql 19h29'16"7d42'
Morgan,Marybeth Ori 5h26'36"1d13'
Morgan,Matthew Tien Her 17h32'53"23d53'
Morgan,Michael John Ori 5h13'34"-1d51'
Morgan,Michael P Aur 6h10'1"33d21'
Morgan,Mickey Ori 5h54'47"6d52'
Morgan,Mr & Mrs Robert Mon 6h22'0"-5d56'
Morgan,My Sweet Mom Ruth Doyal Uma 11h21'1"71d46'
Morgan,Nancy A Uma 11h54'27"51d06'
Morgan,Pamela Kay Cet 3h13'45"3d10'
Morgan,Patty Equ 21h20'45"3d1'
Morgan,Paul Ser 15h11'1"8d3'
Morgan,Peg Aql 19h28'26"10d36'
Morgan,Randall James Lmi 10h40'30"31d4'
Morgan,Randy Sct 18h44'1"-6d25'
Morgan,Rayce Bradley Vul 19h35'48"27d29'
Morgan,Robin Cnv 12h27'12"48d21'
Morgan,Sabrina Lyr 19h10'54"40d48'
Morgan,Samantha And 2h2'30"40d25'
Morgan,Samantha Leda Cristina Rossi Mon 8h7'15"-0d48'
Morgan,Sarah Lise Cas 22h56'57"55d21'
Morgan,Scott & Alora Aql 18h51'13"11d59'
Morgan,Scott Andrew Aur 6h13'52"31d36'
Morgan,Sir David Cep 22h16'1"68d43'
Morgan,Sr,William H Oph 16h59'50"-25d6'
Morgan,Stacy Lynn Cas 1h53'19"68d53'
Morgan,Stacy M Her 16h57'0"40d16'
Morgan,Steth Dra 19h57'33"68d29'
Morgan,Teresa Ori 5h24'21"1d35'
Morgan,Terry James Umi 16h30'31"76d52'
Morgan,Tobin James Ser 15h57'33"9d44'
Morgan,Tyler Dra 16h44'49"61d37'
Morgan,Virginia Conway Lyn 7h44'0"45d49'
Morgan,William A & Cecilia E Cyg 20h24'22"40d3'
Morgan,William Edward Per 1h37'1"53d6'
Morgana Boo 14h50'37"30d20'
Morgane Adrian Cas 0h6'38"60d50'
Morgane G K Oph 18h42'12"6d44'
Morgane,Bomor Bonnaud Cnv 12h51'40"40d57'
Morganis Ori 5h42'22"8d55'
Morganna's Wish Cas 0h21'34"58d56'
Morganne Peg 23h1'1"30d44'
Morgans,Max Ori 6h3'60d0'36'
Morganstein,Sandi Vul 20h2'1"28d52'
Morgante,Cookie Lyn 8h36'48"45d40'
Morgen Star Tau 3h55'38"28d16'
Morgen,Daniel Benjamin Cnc 8h51'51"7d34'
Morgenroth,Herbert Leslie Hya 8h48'19"-0d19'
Morgenroth,Lawrence Vir 13h17'5"-5d14'
Morgenthaler,Lissa Lyn 9h0'41"41d35'
Morgera,Kathy Lyn Uma 10h18'41"53d21'
Mori,Lisa A Sge 20h3'40"16d47'
Mori,Takahiro Boo 15h36'41"48d26'
Moriah Hya 8h14'25"0d46'
Moriarty JP Ori 5h25'35"15d7'
Moriarty,Betsy Boo 15h5'15"14d34'
Moriarty,GO DEO- Patrick J Dra 16h8'41"61d13'
Moriarty,Kyle Thomas Aur 5h19'16"42d46'
Moriarty,Mary Lyn 7h13'1"58d21'
Moriarty,Nana Uma 0h16'40"44d36'
Moriates,Taran Nicholas Per 1h45'0"52d32'
Moriconi,Dr E Steven Oph 17h32'22"-23d43'
Morid,Mohammed Cyg 21h44'47"37d56'
Moriell,Caramante Schiano Cyg 20h3'30"31d37'
Morilla,Roberto Boo 14h20'0"13d13'
Morillo Family Star, The Mon 7h36'31"-5d41'
Morimoto,Allen Nagatoshi Cma 6h51'4"-18d23'
Morimoto,Faye Tsugie Sct 18h54'40"-7d38'
Morimoto,Ned Kenji Hya 8h36'49"1d31'
Morimoto,S52311 JQ3TAN Yuuki Psc 1h14'21"20d29'
Morin,Christian Ori 5h57'1"19d5'
Morin,Claude Lac 22h43'0"55d56'
Morin,Clémence Dra 15h43'54"57d55'
Morin,Hope Lyr 19h24'59"40d28'
Morin,Isabella Lyr 18h41'27"26d46'
Morin,Jean-Claude Psc 0h8'1"0d57'
Morin,Jeffrey Cep 21h22'32"67d51'
Morin,Jr,Robert W Cam 7h48'21"70d4'
Morin,Julie Cyg 21h15'0"38d20'
Morin,Kent Dra 17h31'25"75d32'
Morin,Lori And 1h37'48"39d5'
Morin,Maman Solange Pouliot Uma 9h4'10"48d25'
Morin,Melinda J Her 16h11'1"40d54'A
Morin,Pierre Uma 10h54'40"51d44'
Morin,René Uma 12h6'31"61d20'
Morin,Sarah Catherine Cas 0h25'46"68d54'
Morin,Shaela Rollande Sco 17h32'0"-38d53'
Morin,Sister Alfred De Marie Uma 12h24'26"53d27'
Morin,Tyler Pierce Cet 0h55'50"-10d37'
Morinello,Eric John Peg 21h51'34"34d54'
Morino,Richard Anthony Dra 14h12'21"63d21'
Morinville,Keith Maurice Lyn 7h10'12"51d34'
Morioka,Kenson K K Cet 1h59'21"-4d7'
Morison,Karen Cyg 20h17'0"38d0'
Morisseau,Catherine Cyg 20h23'38"39d38'
Morissette,Annie Lyr 18h47'59"36d18'
Morissette,Janet Cas 0h12'0"63d41'
Morita,Reiko Peg 23h25'1"16d56'
Moritz,A Star Called Ser 16h7'50"14d49'
Moritz,Christine Vir 11h50'34"-2d55'
Moritz,Craig P Boo 14h56'23"39d52'
Moritz,Heinrich Lyr 19h19'28"30d48'
Moritz,Karl & Maria Boo 14h57'1"31d4'
Moritz,Roger Tri 1h33'39"35d23'
Moriwaki,Stella of Shiho Ari 2h16'21"17d24'
Moriyama,Yasuhide Aql 19h26'31"-8d42'
Moriyasu Miwa 1967- 1007-0325 Lib 14h48'5"0d39'
Morizio,Anthony Aur 6h31'29"35d31'
Mork,Deborah Uma 11h1'45"47d32'
Morken,Sigurd Umi 14h6'33"70d16'
Morkovsky,Wayne Sct 18h44'14"-6d2'
Morlan,John Boo 14h11'37"32d6'
Morlet,Armel Cmi 7h29'26"5d59'
Morley,Ann Lyr 18h55'0"38d18'
Morley,April Aqr 20h58'12"-1d19'
Morley,James William Leonard Ind 20h49'28"-59d23'
Morley,Josephine Aur 7h0'53"40d18'
Morley,Jr,John Tri 2h21'41"28d46'
Morley,Melody Sco 17h54'17"-38d21'
Morley,Nicholas Tau 4h1'56"10d35'
Morley,Nigel Her 17h6'36"38d39'
Morley,Tisha Psc 23h6'39"6d57'
Morley,Veronica Gem 6h51'2"18d41'
Morlino,Joseph Robert Aur 4h57'26"33d52'
Mormor Aur 5h36'19"54d37'
Mornard,Michael Francis Cas 1h0'21"50d25'
Morneau,In Memory of Jonathan Dra 13h7'31"68d47'
Morning Star Leo 9h56'16"33d10'
Morning Star,Jennifer Ori 5h57'1"18d42'
Morning Star,Mate To Jennifer Ori 5h57'43"19d22'
Morning,Deborah Kaye Gatlin Her 16h23'36"27d47'
Morning,Heather Julie Louise Cas 23h24'51"60d45'
Morning,Matthew Thomas Cep 21h6'35"68d31'
Morning,Roy Akio Lmi 10h14'20"32d26'
Morninglight,Marcia Lyn 8h14'47"51d5'
Morningstar Constellation,The Aql 19h24'21"0d56'
Morningstar Boo 15h2'0"30d0'
Morningstar,Kyle Charles Uma 11h17'18"44d0'
Morningstar,Ronnie Cyg 21h24'1"40d44'
Moro,Federica Pic 5h6'47"-44d52'
Moroch,Sylvia Susan Smerek And 0h7'0"38d2'
Morolla,Antonio Col 6h25'4"-37d12'
Moroney,JoAnn Cam 7h52'17"61d37'
Moropito,Nicole Aql 20h6'39"0d57'
Morosetti,Anna Cnv 13h11'50"51d14'
Morovitz,Burton & Pauline Morovitz Cra 18h0'49"-37d6'
Moroz,Barbara L Vir 15h4'21"6d19'
Morphis,Jamie & Betsy Mon 6h39'1"10d44'
Morr,Deborah Mon 6h26'0"-6d40'
Morreale(mouse),Lori Cas 0h27'0"68d46'
Morreale,Alexander Jerome Her 15h53'1"42d27'
Morreale,Anna Leigh Cas 2h59'18"70d25'
Morrell 5-7-78,Lisa Marie Tau 5h47'34"23d14'
Morrell,Laurence David Com 12h41'12"30d40'
Morrella,Grace Boo 14h30'37"45d7'
Morrice,Marty Ori 5h57'14"14d45'
Morrie Umi 12h50'16"86d1'
Morrie & Rae Forever Vul 19h20'54"25d11'
Morrill,Cate Tri 1h47'25"27d53'
Morrill,Hanna Katharine And 1h4'44"47d54'
Morrill,My Star Forever,Michael Vir 13h57'23"-0d20'
Morris Birthday Star, Donna Lynn Cyg 21h21'27"38d44'
Morris Forever,Gary Ari 2h58'44"22d11'
Morris Jr Boo 15h5'27"13d28'
Morris Rising Hya 8h17'45"-0d49'
Morris,Alex Jayme Cet 0h53'57"-2d45'
Morris,Always Loved, Shana A Vul 19h47'22"28d31'
Morris,Andrew Patrick Her 16h34'47"32d25'
Morris,Ann Michele Vul 21h21'39"24d7'
Morris,Antoinette Vul 19h44'38"22d52'
Morris,Beverly Darlene Mon 6h51'57"11d40'
Morris,Brandon Cep 21h7'41"55d31'
Morris,Brenda Mon 6h1'30"-4d48'
Morris,Bruce & Nicki Lac 22h55'8"42d44'AB
Morris,Carol Lorraine Umi 14h58'43"70d54'
Morris,Carter Rocc Cam 5h4'22"67d48'
Morris,Christian Emil Alan Lac 22h20'23"37d57'
Morris,Christine Suzanne Mon 6h28'0"3d42'
Morris,Christine Kay Vir 13h28'2"-7d36'
Morris,Clare Ori 6h6'25"-0d43'
Morris,Clare And 23h20'1"45d31'
Morris,Courtney L Lyn 8h24'23"41d58'
Morris,Dale & Trudy Eri 3h42'33"-6d49'
Morris,David & Helen Cet 2h54'27"-0d30'
Morris,Deborah Lynn Cas 0h33'53"64d29'
Morris,Desmond Ori 5h34'29"-0d42'
Morris,Dorothy Lyr 18h59'11"30d45'
Morris,Edward Per 2h24'0"54d23'
Morris,Eleanor Eileen Cas 0h15'27"58d32'
Morris,Ethel Manning Cam 4h40'23"67d58'
Morris,Fielding Westbrook Per 2h7'44"57d36'
Morris,Fran Vul 19h18'29"27d13'
Morris,Freddie Mon 7h50'55"-1d51'
Morris,G R Peg 21h38'20"25d4'
Morris,Gayle Elizabeth Oph 17h39'37"-21d34'
Morris,George Edward Uma 12h15'1"55d55'
Morris,George Herman Dra 14h55'55"62d25'
Morris,Gerwyn Dra 9h56'1"73d42'
Morris,Glen Uma 9h41'22"47d52'
Morris,Graeme Oriel & Roma Janice Ind 20h49'57"-52d31'
Morris,Hannah Lyn 7h56'39"43d31'
Morris,In Loving Memory Elaine Starr Lyr 18h40'24"45d3'
Morris,J T Aur 6h1'15"30d13'
Morris,Jack & Jean Crt 10h58'48"-20d33'
Morris,Jacob Ross Peg 21h24'11"2d47'
Morris,Jacqueline Cyg 19h31'0"39d26'
Morris,Jan Cyg 20h38'1"45d45'
Morris,Jan Com 13h2'28"18d42'
Morris,Jane L Mon 7h40'53"-8d39'
Morris,Jaqui Eri 5h7'1"-4d26'
Morris,Jenna Mary Ori 5h56'0"14d23'
Morris,Jennifer Lyn Sex 9h53'10"3d50'
Morris,Jennifer Lynn Cyg 21h0'57"38d4'
Morris,Joan Cas 1h47'36"63d53'
Morris,John Aql 19h31'43"16d2'
Morris,John & Melissa Sge 19h41'15"16d20'
Morris,John C Her 16h25'45"28d51'
Morris,Jr,Dwight H Ser 18h20'30"-0d10'
Morris,Julie Ann Peg 21h30'40"4d20'B
Morris,Karen Hya 8h17'34"-6d57'
Morris,Karen And 1h40'30"38d10'
Morris,Karen E Mon 6h20'53"8d56'
Morris,Katherine Elizabeth Cyg 19h33'0"37d53'
Morris,Kathleen Uma 12h57'25"57d47'
Morris,Kathryn Cnc 8h50'28"30^56'
Morris,Kathryn Cas 0h50'20"74d2'
Morris,Kay Goenne Sco 16h57'37"-43d46'
Morris,Kira Mon 6h23'11"8d54'
Morris,Kris Cam 4h29'1"68d30'
Morris,Kyle Jayme Oph 17h2'37"10d5'
Morris,Lauren Christine Mon 6h38'38"7d59'
Morris,Leonie Catherine Sge 19h53'1"19d3'
Morris,Lily Margaret Cas 2h1'54"61d16'
Morris,Linda Lyr 18h29'32"32d52'
Morris,Lisa Anne Brabender Uma 11h35'27"31d22'
Morris,Liz Cas 22h58'50"55d18'
Morris,Loraine Deborah Tau 5h52'20"28d31'
Morris,Lynn And 1h33'18"41d1'
Morris,Marc Leonard Aql 18h43'49"-1d51'
Morris,Marlena And 0h49'45"35d4'
Morris,Mary Kate Peg 21h25'52"3d13'
Morris,Melanie L Tri 1h43'36"30d45'
Morris,Mervin G Oph 18h5'0"7d56'
Morris,Michael Per 3h40'4"38d55'
Morris,Michael G Dra 17h45'24"76d33'
Morris,Miss René Cnc 8h38'16"18d59'
Morris,Paschal Christopher Uma 9h19'56"54d30'
Morris,Philip Ser 16h1'27"-1d35'
Morris,R Craig Boo 15h12'28"38d18'
Morris,Richard Alan Peg 21h30'40"4d20'A
Morris,Robert Hya 9h13'0"5d20'
Morris,Robert E Dra 16h35'33"70d27'
Morris,Robert Kevin Boo 13h36'26"14d47'
Morris,Robert Randall Vir 13h27'46"-7d22'
Morris,Robert Reese Hya 8h50'33"-7d2'
Morris,Sarah Avery Equ 21h6'41"11d56'
Morris,Scot Cma 7h2'11"-16d43'
Morris,Sidney William Hya 10h22'22"-12d6'
Morris,Stephanie Lynn Cas 1h23'0"60d57'
Morris,Suesan Cyg 20h0'44"31d19'
Morris,Terry Cam 4h0'30"61d24'
Morris,Vernon Aur 4h51'44"40d42'
Morris,Vivian Hurt And 2h21'25"40d37'
Morris,Westley Aron Hya 8h59'25"-0d10'
Morris,William Bancroft Cep 22h31'41"56d52'
Morris,William Earl Oph 18h17'14"0d19'

Morris,Wyatt
Oph 17h33'55"14d5'
Morris-Duke,Eileen F
Vir 11h58'1"-3d45'
Morris-You're Loved!, Nicole E
Mon 8h5'42"-2d19'
Morris/"Lovey Puddin", Tony Gene
Vul 20h16'13"25d55'
Morrisey,Jeanne
Cam 4h18'40"59d30'A
Morrisey,John "Mike" & Beryl
Uma 10h38'26"48d39'
Morrison aka "Cutie", Michael Clark
Del 20h49'50"7d43'
Morrison III,White Hall
Aql 20h0'1"13d54'
Morrison Magic
Tau 4h46'54"20d28'
Morrison Southworth
Ori 5h20'1"1d16'
Morrison's Red Star, Jane Yankovic
Cra 18h27'58"-41d6'
Morrison's Star
Cet 3h16'14"1d33'
Morrison,Anastasia Victoria
Cam 6h6'40"70d51'
Morrison,April
Cas 0h45'28"70d57'
Morrison,Audrey
Lib 14h53'52"-5d52'
Morrison,Beloved Maj Nina Cappy
Peg 22h39'31"29d38'
Morrison,Bethany
Cet 3h0'37"0d23'
Morrison,Bret
Aur 6h32'41"31d7'
Morrison,Bryce Allan
Her 16h6'49"41d46'
Morrison,Carol Leuenberg
Com 13h8'0"28d13'
Morrison,Daniel J
Her 17h12'57"47d39'
Morrison,Darlene, Justin & Andrea
Mon 7h3'3"-6d31'
Morrison,Elizabeth Francis
Mon 6h59'37"-0d47'
Morrison,Erin Nicole
Peg 23h6'35"20d12'
Morrison,Garrett
Boo 13h42'39"22d37'
Morrison,Gladys
Tau 3h57'24"20d32'
Morrison,Heather Mary
Cyg 19h30'54"38d9'
Morrison,Helen & Willie
Uma 13h2'40"53d16'
Morrison,Helen Alycia
Lyn 7h2'52"44d32'
Morrison,Jacob Adam
Peg 23h6'35"21d56'
Morrison,James E
Cam 4h16'35"70d31'
Morrison,John R
Sex 9h52'13"-5d32'
Morrison,Jonathan
Her 17h20'0"47d51'
Morrison,Katie Ellissa
Cas 23h33'40"61d22'
Morrison,Kelly Leeann
Peg 22h39'26"26d43'
Morrison,Kevin Shawn
Cep 2h56'11"80d5'
Morrison,Merrill & Mark
Crb 16h2'45"32d26'
Morrison,Mike & Connie
Cet 2h47'12"5d3'
Morrison,Patsy
Cas 0h3'19"58d34'
Morrison,Paula
Ori 6h6'54"3d32'
Morrison,Rebecca Grey
Cra 10h50'36"-11d52'
Morrison,Richard Craig
Cep 22h53'38"58d45'
Morrison,Robert S
Sex 9h55'60"-6d17'
Morrison,Sam
Cyg 19h22'48"44d48'
Morrison,Sandra
Lyr 18h55'37"41d31'
Morrison,Sondra
And 0h16'1"38d7'
Morrison,Steven D
Crt 11h45'0"-17d48'
Morrison,Terry L
Hya 9h35'39"-17d28'
Morrison,Terry L
Sct 18h20'38"-5d30'
Morrison,Thomas Price
Hya 8h54'47"-6d28'
Morrison,Tommy
Equ 21h22'47"8d39'
Morrison,Tray
Per 3h0'0"41d41'
Morrison,Uarda Kay
Ser 15h39'1"21d14'
Morrison,Valerie
And 0h26'48"21d44'
Morrison,Van R
Aur 7h4'43"43d57'
Morrison-McQueen
Cmi 7h7'13"4d3'
Morrison;My Celestial Joy,Joyce W
Eri 3h36'26"-11d25'
Morriss,Ann
Her 18h15'58"14d27'
Morriss,Anthony
Ori 5h59'24"14d46'
Morriss,Christy
Mon 7h1'1"-6d60'
Morrissette,Claire
Cas 0h40'12"73d28'
Morrissey
Lac 22h48'10"54d20'
Morrissey Star,The Annie
Cma 6h58'50"-16d12'
Morrissey,Aasom-Aaron Scott
Boo 14h18'36"47d31'
Morrissey,Alison
Eri 2h43'28"-6d48'
Morrissey,Amore Mio
Aqr 21h36'28"-6d0'
Morrissey,Daniel P
Aur 7h3'45"36d49'
Morrissey,Michelle
And 2h29'30"41d44'
Morrissey,My Guiding Star,Richard F
Her 18h1'30"31d24'
Morrissey,Reynilda Roman
Cnc 8h31'0"7d45'
Morrissey,Robert Fitzpatrick Edward
Gem 6h32'23"14d33'
Morrissey,Thomas E
Her 16h21'14"26d33'
Morro's Promised Land, Peter R
Aur 6h1'50"38d23'
Morron,Omerle
Cyg 21h2'18"28d53'
Morrone,Christopher Michael
Lac 22h11'0"46d47'
Morrone,Maureen & Nicholas
Uma 11h1'13"42d9'
Morrow Avalon
Lyr 18h28'31"45d45'
Morrow Family Star,The KJ
Ori 5h51'36"12d4'
Morrow's Twinkle, Grandma
Lyr 18h52'1"31d17'
Morrow,Amber Noel
Tri 2h5'16"33d18'
Morrow,Bryce Edsel
Leo 9h51'51"31d56'
Morrow,Constance Kay Hanrahan
Aqr 22h49'26"0d26'
Morrow,Jason Lee
Lac 22h46'40"52d46'
Morrow,Kara Megan
And 23h45'1"45d44'
Morrow,Katelyn Marie
Sgr 18h51'18"-27d53'
Morrow,Margaret D
Lyr 18h51'13"40d13'
Morrow,Marjorie
Crb 16h3'17"32d25'
Morrow,Mark Jason
Boo 14h47'1"34d5'
Morrow,Martha
And 0h6'47"40d12'
Morrow,Murray
Dra 16h38'19"61d59'
Morrow,Richard
Hya 9h34'34"0d5'
Morrow,Robert Delano
Her 17h16'58"44d37'
Morrow,Ronald Lee
Aqr 22h49'53"-0d8'
Morrow,Ryan Joseph
Cnv 12h53'54"38d11'
Morrow,Ryan Lee
Aqr 22h49'50"-5d18'
Morrris,Bailey Nicole
Peg 23h19'30"13d24'
Mors,Jill
Aqr 23h11'58"-5d26'
Morse November 2,1916, Woodrow Wilson
Uma 10h36'34"51d3'
Morse,Barbara J
Cas 0h12'38"58d1'
Morse,Brittney M
And 23h23'22"47d22'
Morse,Corky
Boo 13h51'28"19d9'
Morse,Damon Richard
Dra 14h23'1"64d18'
Morse,David Brian
Dra 19h4'38"48d11'
Morse,Gregory Allen
Per 3h19'21"50d24'
Morse,In Loving Memory of Aaron
Peg 22h27'55"21d56'
Morse,John Casey
Dra 17h37'43"60d60'
Morse,Justin
Lyn 6h55'35"59d41'
Morse,Kelly
Mon 7h16'15"-6d38'
Morse,Kendall
Aql 19h57'37"13d37'
Morse,Kristen Michele
And 2h9'15"38d1'
Morse,Kristin Anne
Aur 5h27'46"31d42'
Morse,Lynn J
Cas 0h10'40"61d30'
Morse,Madeleine Dubios
Cas 1h43'26"76d30'
Morse,Mary
Tau 4h56'30"16d33'
Morse,Noah Prescott
Cet 3h3'0"4d14'
Morse,Sarah Nicole
Peg 22h41'1"33d18'
Morse,Susan C Rieur
And 1h5'0"37d36'
Morsellino,Salvatore
Dra 16h7'47"66d9'
Morsello,Casper
Lac 22h8'0"48d1'
Morshaed,Katherine
Umi 15h16'20"66d2'
Morson,Christopher Alan
Peg 21h47'42"30d50'
Morss,Kerynn Elizabeth
And 23h23'0"49d31'
Morstad,Ann Kennedy
Lyr 18h21'0"38d4'
Morstein,Sylvia
Lyn 9h15'57"34d7'
Morstest,The
Aql 19h1'11"4d54'
Mortagne,Patrick
Cam 13h10'32"82d14'
Mortelecque,Laureen
Mon 7h54'27"-4d33'
Mortellaro,Darlene
Mon 6h48'0"11d37'
Mortellite II,John Gerard
Lac 22h25'19"40d44'
Morten,Amy
And 0h0'1"40d7'
Morten,Terry Richard
Lyr 18h25'39"37d51'
Mortensen III, Theodore T
Ori 5h47'38"21d19'
Mortensen,Bebs
Sco 17h3'50"-33d59'
Mortensen,James Larry
Sct 18h35'12"-4d29'
Mortensen,Martha
Cas 1h6'44"61d5'
Mortensen,Walter
Boo 15h24'37"50d14'
Morter,Darren Van
Cep 20h56'47"55d25'
Mortham,Sandra Barringer
Eri 4h6'57"-19d51'
Morticia
Sgr 19h16'32"-22d2'
Mortier,Graham Ethan
Per 2h44'1"43d15'
Mortier,Logan Alexander
Per 2h44'31"41d13'
Mortillaro,Abraham
Lac 22h16'54"50d26'
Mortimer II,Peter Augustus Jay
Tau 5h22'48"16d49'
Mortimer's Marvellous Star,Ian
Ori 6h2'17"1d13'
Mortimer,Austin Wayne
Psc 0h47'36"32d22'
Mortimer,Clare
Cnc 9h9'0"33d6'
Mortimer,Grant & Bonnie
Lyr 18h37'12"29d21'
Mortimer,Jessica Zara
Eri 2h47'52"-5d29'
Mortimer,Madeline Rose
Com 12h45'27"21d21'
Mortimer,Michelle Elizabeth
Peg 22h38'0"20d47'
Mortimer,Nicki
Mon 6h6'41"-6d38'
Mortimer,Phyllis Dean
Sge 19h42'49"18d50'
Morton
Per 1h49'0"48d26'
Morton Star,Peter
Aql 19h18'1"15d25'
Morton Star,Tarlton
Aql 19h23'17"15d13'
Morton's Memory Ryan's Future
Cep 0h17'24"75d48'
Morton,Albert W D
Uma 8h13'53"70d50'
Morton,Andrew
Cep 22h34'47"56d53'
Morton,Brenda Joyce
Cyg 20h24'42"38d17'
Morton,Carol Ann
And 0h45'0"34d16'
Morton,Catherine
Cas 0h48'0"73d27'
Morton,Douglas Kevin
Ori 6h11'46"7d38'
Morton,Dr Peggy
Psc 1h43'56"20d21'
Morton,Emma Lee
Cas 23h14'38"60d12'
Morton,Eric
Oph 16h56'29"0d41'
Morton,Eva
Lyr 18h15'55"45d20'
Morton,Flora
Lyr 18h35'35"40d56'
Morton,Ginny
Lyn 7h39'57"51d59'
Morton,Ivy
Lyr 18h58'34"30d33'
Morton,Jack
Aur 6h13'0"32d8'
Morton,Jennifer Anne
Cas 0h4'1"61d2'
Morton,John Spencer
Aqr 22h30'24"-11d2'
Morton,Leonard P
Per 3h54'19"38d26'
Morton,Michelle B
Lyn 7h48'0"38d34'
Morton,Nicolas Bracchi
Lib 14h45'0"-20d14'
Morton,Nikki Marie
Tri 1h34'15"28d36'
Morton,Randy
Oph 18h8'10"13d40'
Morton,Susan Lea
Uma 9h48'19"48d35'
Morton,Tye & Casey Hembree
Aql 20h7'42"8d17'AB
Morton,William F
Her 17h25'21"21d55'
Morton-"The Dougstar", Douglas
Boo 14h59'43"18d35'
Morton-Cecestial Glory Allantial Glory
Uma 11h3'11"67d40'
Morton-Loved & Cherished,Stuart
Cep 22h8'20"60d54'
Mortrell,Talon Bryce
Aql 19h5'49"3d51'
Mortrude,Stuart J
Dra 18h14'60"70d38'
Mortson-Boyenko,Cheryl
Cas 0h31'59"62d18'
Mortstar
Uma 9h8'43"53d45'
Mortt,The Ray
Ari 2h44'49"28d53'
Morus,Steven
Aql 19h30'22"14d5'
Morvay,Heather M
Lyr 18h41'1"31d12'
Morvay,Zoe Rebecca
Tau 3h58'41"8d27'
Morven Lindsay
Peg 23h7'52"11d34'
Morway,Tacita Oriol
Uma 8h59'0"62d17'
Mory,Bertrand
Mon 6h27'37"7d50'
Mory,Frédérique
Uma 11h13'32"50d6'
Morzio
Dra 9h55'49"73d19'
Morzuch,Chuck
Lac 22h9'1"47d5'
Mosaquites,Danny
Cep 23h34'13"68d57'
Mosbah,Jessica Asensio
Uma 10h59'17"38d36'
Mosbrook,Sharon
Cas 23h15'48"60d19'
Moscarella,Ralph Vincent
Cam 12h45'55"80d25'
Mosch,Guenter
Cmi 7h28'42"7d34'
Moschella,Salvatore N
Her 18h8'1"38d22'
Moscheo,Angela Nicole
Cyg 21h6'30"40d20'
Moschetta,Jr,Anthony B
Per 4h4'51"51d5'
Moschettini,Luciana Torsello
Cma 6h24'0"-16d2'
Moschimaus I
Sgr 19h39'15"-42d6'
Moschovos,Elena "Kleine Prinzessin"
Vir 13h31'15"-13d42'
Moscowitch,Jane
And 23h23'0"37d40'
Mosea,Ingrid
Psc 0h16'41"11d50'
Moseback,Chelsie Kathleen
Sct 18h45'0"-7d4'
Mosele,Phyllis A
Cyg 19h45'41"38d16'
Mosele,Rocky
Lib 15h3'25"-10d6'
Moseley,Beth
Cyg 21h37'51"39d17'A
Moseley,Christian
Per 1h54'0"53d38'
Moseley,Fred & Oma
Uma 9h46'30"68d19'
Moseley,George & Carole
Peg 23h4'24"22d26'
Moseley,George William
Cmi 7h56'43"8d29'
Moseley,John G & Catherine S.
Aql 19h29'40"1d17'
Moseley,John R & Edward A
Ori 6h8'52"8d32'
Moseley,Mitchell
Peg 23h33'1"33d54'
Moseley,Sarah Marie "Sam"
Cas 0h26'36"63d12'
Moseley,Suzetta
Boo 14h6'1"52d44'
Moseley,Tucker
Cyg 21h37'51"39d17'B
Mosely,Kermit Jefferey
Cyg 20h2'56"31d49'
Mosely,The Star Of
Aur 6h29'59"34d21'
Moseman,Verna
And 0h58'30"38d15'
Moser,David James
Sct 18h43'28"-5d52'
Moser,Franz
Cep 22h0'1"63d53'
Moser,Jr,Peter T
Per 3h1'0"40d31'
Moser,Katie
Cas 1h19'33"66d57'
Moser,Margaret
Hya 9h10'49"1d49'
Moser,Scott Jon
Her 17h58'1"40d31'
Moses
Sco 17h34'27"-38d21'
Moses,Catherine
Lyn 9h1'25"44d33'
Moses,Jackie
Lyn 7h37'32"38d35'
Moses,Jason
Ori 5h51'49"15d16'
Moses,Jean
Sge 20h17'24"16d31'
Moses,Karya
Cas 0h35'28"62d59'
Moses,Katherine Anne
And 0h52'48"36d24'
Moses,Norman Glenn
Ori 6h1'56"10d1'
Moses,Robert John
Aur 5h4'0"37d37'
Moses,Sr,William D
Cyg 20h2'32"58d30'
Moses,Steven R
Dra 19h6'59"48d43'
Moshe Adam
Hya 8h54'16"0d51'
Mosher,Donald Edgar
Peg 22h2'56"7d35'
Mosher,Scott & Teresa Aleppo
Crb 15h56'38"30d31'
Mosher,Stanley
Cep 21h46'48"65d34'
Mosher,Todd Jon Jon
Aur 7h5'1"39d52'
Mosher,Virgina
Mon 6h54'1"-10d31'
Mosher,Vurden Edward
Uma 11h33'39"42d48'
Moshova,Elle
And 23h1'1"49d11'
Moshova,Elle
Lyn 8h19'59"47d49'
Moshé,Rachel Jessica & Joshua
Tri 1h36'1"31d48'
Mosier III,John R
Boo 14h22'26"32d11'
Mosing,Jeffrey Louis
Cep 1h5'43"85d15'
Mosinski,Danelle
Uma 9h27'16"55d18'
Moskal,Anna Marie
And 23h49'46"44d13'
Moskal,Eric Michael
Aur 5h6'1"40d60'
Moskal,Evan Thomas
Aur 7h22'43"40d32'
Moskowitz,Irwin & Elaine
Cyg 19h26'57"30d42'
Moskowitz,Rose
Cas 0h4'33"56d5'
Moskowitz,Thomas Barry
Aql 19h7'0"3d45'
Mosley,Capt T Alton
Leo 9h19'51"8d2'
Mosley,Elwood Alvin
Per 1h52'32"48d3'
Mosley,Jeffrey
Cet 1h50'50"-6d43'
Moss
Cnv 12h5'59"36d50'
Moss 65
Aur 4h54'46"38d50'
Moss Star Bob & Kathy
Mon 7h42'7"-1d10'
Moss The First of Leeds,Paul
Dra 16h12'29"61d11'
Moss,Douglas G
Gem 6h46'57"30d39'
Moss,Duncan
Ori 4h56'18"5d16'
Moss,Edwin
Gem 7h33'26"25d16'
Moss,Elizabeth
Crb 15h17'42"31d17'
Moss,Janet E
Com 13h1'38"15d51'
Moss,John Aquila
Aql 20h0'39"9d47'
Moss,Jr,Joe P
Ser 18h52'54"5d59'
Moss,Katherine Mary
And 2h32'34"41d10'
Moss,Kenneth Augustine Seward
Aql 20h0'25"4d52'
Moss,Lori A
Lib 14h19'21" 23d22'
Moss,Marven
Cep 22h14'44"60d15'
Moss,Michael H
Per 2h34'1"56d31'
Moss,Paul
Ori 5h50'59"15d10'
Moss,Phyllis
Com 13h17'20"28d40'
Moss,Ralph & Norma Jean
Cnv 12h15'25"43d14'
Moss,Raymond (Juan Spanish)
Lac 22h5'41"47d16'
Moss,Richard Charles
Dra 14h35'19"63d0'
Moss,Robert N
Lac 22h4'36"38d22'
Moss,Roz
Cyg 20h57'32"50d9'
Moss,Sabreena Marie
Cyg 21h24'14"31d8'
Moss,Shaun Mitchell
Crb 15h56'26"30d49'
Moss,Steve & Lori
Peg 22h23'40"30d14'
Moss,Susan Rosenberg
Crb 16h5'1"30d12'
Moss,Thomas
Cet 2h55'0"7d10'
Moss-Fletcher,MD, Rebekah
Lyr 18h55'57"37d7'
Moss-McKenna,Tamire
Cep 21h15'0"67d33'
Mossberg,Karyn Kay
Lyr 18h59'49"30d26'
Mossia,Petia
Uma 11h5'53"37d27'
Mosso,Scott
Per 1h26'48"54d20'
Mossy,Glenn
Lmi 10h19'12"31d55'
Most Beautiful Beth
And 1h55'12"46d54'
Most Beautiful Girl in the World,The
And 0h55'0"38d13'
Most,Michelle
Cyg 21h19'39"28d36'
Mostad,Maris Josephine
Uma 11h12'53"67d58'
Mostafa,Khaled
Dra 18h22'25"68d58'
Mostert,Harry Johan
Aur 6h33'44"40d32'
Mostest
Del 20h21'44"2d48'
Mostillo,Ralph
Boo 13h43'19"22d25'
Mostyn
Lac 22h3'0"51d36'
Moszer,Fabienne
Cep 22h25'51"57d48'
Mot-Nic
Uma 10h8'27"53d59'
Motes,Monica Louise
Lyr 18h39'23"27d4'
Mother Butterfly
Cas 0h32'56"61d42'
Mother Christina
Eri 3h14'14"-1d60'
Mother Fabec
And 23h39'35"46d22'
Mother Magoo
Uma 10h11'19"50d60'
Mother of the Fantastic Five-Jennie
Crb 15h50'46"31d54'
Mother's Star For Bobby
Cet 1h5'48"-6d42'
Mothershead,John Wesley
Cyg 19h22'1"28d29'
Motherway,Thomas Patrick
Cep 20h28'24"76d49'
Motin's Magic
Ori 5h57'30"16d4'
Motolko,Michael
Lac 22h1'25"51d12'
Motorola GmbH
Uma 11h55'0"61d48'
Motsinger,Arnie
Peg 22h36'18"24d6'
Mott's Star,David Tracy
William & Lamar
Oph 17h0'13"-26d9'
Mott,Charlotte Douglass
And 0h49'1"39d21'
Mott,Dior
Cam 12h7'38"81d34'
Mott,Orbin Darrell
Cas 0h34'24"61d50'
Mott,Peggy Joyce
Cas 0h33'55"60d50'
Mott,Robert L
Lac 22h3'36"46d27'
Mott,Ruth
And 1h49'48"37d59'
Motta,Derek Joseph
Sgr 19h36'51"-30d9'
Motta,Jacob Anthony
Hya 8h11'35"0d13'
Motta,Marina
For 2h6'24"-28d8'
Motte
Leo 9h18'0"12d10'
Motten,Betsy "Kip"
Cas 2h32'49"57d34'
Motter's Star,Dawn
Cyg 19h55'34"38d45'
Motter,Krista
Lyr 18h35'10"36d51'
Motter,Serenella Serrini
Boo 14h34'59"47d23'
Mottola,God Bless You Forever,Linda
Cas 23h16'44"61d3'
Mottoros,Theresa
Cep 21h55'55"80d27'
Mottram,A Lynn
Peg 23h5'49"22d5'
Mottram,Ron The Wolf
Ind 20h54'48"-55d59'
Motyka,Christopher & Jennifer
Cyg 19h26'1"40d18'
Motyka,Zachary Joseph
Boo 14h14'14"20d6'
Motz,Angelika
Com 13h3'50"28d11'
Motzer,Renate Manuela
Aur 5h6'55"37d14'A
Moubry,Karen
Eri 3h18'29"-2d48'
Mouche-Aimée
Cyg 20h24'31"39d18'
Mouffarrige,Oriana
Cas 23h4'55"58d58'
Mouffette
Cam 4h9'29"70d35'
Moukarze,Charles Meral
Boo 13h52'1"16d58'
Moulder,Brenda Sue
Boo 14h55'31"24d1'
Moulding,Eleanor Kathryn
Cas 1h6'28"53d51'
Moulds,Hayden Elizabeth
Cmi 7h46'59"7d59'
Moulin 24
Peg 23h23'27"10d35'
Moulsdale,Donna
Cas 0h47'34"60d50'
Moulton,Bronwyn Fall
Del 20h27'57"20d13'
Moulton,Jacob C
Cep 22h11'1"55d6'
Moulton,James
Aql 19h2'23"16d10'
Moulton,Teresa A
Gem 7h57'45"28d25'
Moulton,Veronica Plaza
Uma 11h22'21"38d9'
Moundy,Monique
Per 3h5'29"38d5'
Mount,Casey Lynn
Del 20h20'11"20d15'
Mount,Chad Alan
Peg 22h37'28"8d20'
Mount,Courtney Clay
Hya 8h52'52"3d52'
Mount,Emily Elizabeth
Cet 1h28'1"0d27'

Mount,Hildegard Anna Schlecht
Uma 8h50'40"49d47'
Mount,Jamie Lee
Aql 19h8'45"1d39'
Mount,Kimberly Ann
Oph 17h8'25"10d39'
Mount,Laura Lee
Tri 2h35'1"31d33'
Mount,Leslie Tsang
Mon 7h3'21"-5d39'
Mount,Lindsey Sarah
And 1h23'21"37d23'
Mount,Steven D
Her 17h4'50"40d50'
Mount-Actor,Thomas Edward Richard
Cap 21h44'0"-22d52'
Mount-Essex Sharilyn T
And 0h42'54"21d48'
Mountain Woman Star, The
Cas 0h52'16"62d5'
Mountain,Sandy
Uma 8h43'0"71d35'
Mountain,Simon Patrick
Ant 10h33'32"32d12'
Mountain-Climber
Vir 14h3'52"7d29'
MountainoftheHarvest Moon
Lac 22h9'36"49d51'
Mountcastle,Fiona Helen
Uma 8h19'39"61d36'
Mountford,Sarah Gail
Uma 12h45'54"54d42'
Mountjoy,Marian Jo
Cam 3h57'56"71d26'
Mounts,Jaime
Ser 18h37'57"4d18'
Mouopputos
Lyr 18h29'33"38d39'
Mouquet,Sophie Demouter
Cam 7h13'0"80d4'
Mouracade,Kimberly
Com 12h23'57"21d5'
Moure,Diana
Eri 3h49'32"-1d4'
Mourgue,Tony
Psc 1h38'52"28d16'
Mouritzen,Elsebeth
Cas 23h42'1"61d48'
Mousa,Sahab
Cam 4h44'46"68d5'
Mouse
Boo 15h23'0"48d1'
Mouse Face
Psa 22h30'15"-28d53'
Mouse Pratt
Dra 9h50'42"78d31'
Mouse Star,The
Cas 1h4'26"58d44'
Moussa,Farouk
Per 2h50'38"31d56'
Moussa,Oreett Lillian
Sco 17h1'59"-31d9'
Moussairoux,Bernard
Umi 13h52'1"77d43'
Moussou,Mara
Lyn 7h49'0"51d59'
Moustakas,Michael Stanley
Ser 15h41'13"19d45'
Moustgaard,Brita
Ori 5h59'33"14d39'
Moutchcc
Umi 13h52'54"76d10'
Mouton,Annette Marie
Equ 21h2'54"10d11'
Mouton,Manuel
Cam 4h31'59"58d25'
Mouton,Steve
Cet 2h9'29"-1d42'
Moutschka
Uma 9h1'29"50d14'
Moutte,Gilles
Aur 6h8'0"37d50'

Mouw,Harriet
Del 20h15'14"9d17'
Mouw,Russell
Aql 19h24'30"-1d13'
Mouwen II,Peter John
Oph 17h22'23"-20d4'
Mouwen,Herman C "Bud"
Ser 15h38'19"0d8'
Mouxaux,Jean
Crb 16h19'26"31d40'
Mouyiaris,Nicos
Uma 8h33'29"54d9'
Mouyiaris,Philip
Per 1h56'31"52d52'
Mouzakis,Lea Elise
Eri 3h54'0"-7d19'
Mouzouris,Anastasia (Tassia)
Cam 5h12'40"68d24'
Movo
Aql 19h25'29"15d32'
Movsesian,Fran & Moe
Eri 3h1'46"-10d42'
Mowad,Joseph James
Lac 22h30'38"41d14'
Mowad,Josephine M
Cas 0h18'40"60d7'
Mowad-Murphy
Cam 6h10'36"68d45'
Mowbray,Daniel Allen
Boo 15h14'57"26d35'
Mowbray,Lorene
And 23h23'17"43d31'
Mower,Garth C
Cet 2h32'18"8d54'
Mower,John David
Cet 1h32'37d-2d2'
Mowers,John
Boo 14h6'22"41d21'
Mowery Elementary School (Quincy,PA)
Uma 8h27'53"62d30'
Mowery Sr,Bruce A
Dra 18h38'0"67d33'
Mowery's Star,Papa
Umi 16h30'22"75d48'
Mowery,Diane
And 1h1'21"40d3'
Mowgli
Oph 17h2'1"11d13'
Mowry-EM-LR,Emily
Com 12h51'1"20d56'
Mox,Torsion
Aur 5h0'35"41d16'
Moxley,Inch Hop John Vaughn & Dixie
Cnv 12h42'51"33d47'
Moxley,Michele Renee
Peg 22h29'53"25d33'
Moxley,Robert Stewart
Tau 3h42'33"2d1'
Moxley,S H
Aql 20h10'56"4d47'
Moy,Ashley Sue
Cas 0h14'45"61d5'
Moy,Doreen Sue & Jeffrey
Cas 0h54'32"61d34'
Moy,Mary
Cyg 21h23'34"30d33'
Moy,Peter David
Per 2h37'12"56d30'
Moy,Prezleigh Teal Mikala
Vul 19h48'1"26d58'
Moy,Stanley
Cep 21h24'57"55d18'
Moy-In Loving Memory, John "Jack" B
Aql 10h19'29"15d44'
Moya K
And 23h42'0"41d35'
Moya,Sofia
Cam 5h6'45"68d6'
Moye,Karen Marie
Sco 17h53'53"-31d39'

Moyer "Bee Bee",Dr Ruth Anne
Cas 0h30'39"63d17'
Moyer's Alpha & Omega
Umi 15h18'53"71d14'
Moyer,Alexander Jason
Per 2h57'32"32d55'
Moyer,Benny Lyman
Uma 14h4'1"53d46'
Moyer,Beverly L
Com 12h38'54"20d11'
Moyer,Charlotte
Mon 6h28'30"11d40'
Moyer,Clayton Jay
Aur 5h17'1"52d47'
Moyer,Cody
Boo 14h44'51"23d10'
Moyer,Dan
Dra 20h6'0"68d57'
Moyer,Joe & Carol
Cyg 20h59'11"28d51'
Moyer,John T
Per 4h25'51"51d47'
Moyer,Keith A
Dra 17h5'36"62d42'
Moyer,Kimberly Joyce
Cap 21h45'1"-23d32'
Moyer,Mark
Aur 6h57'28"40d48'
Moyer,Mary Susan Marie
Lac 22h38'49"55d49'
Moyer,Nancy Ann
Mon 8h3'13"-1d8'
Moyer,Valerie Lynn
Gem 5h59'0"26d38'
Moyet,Philippe
Uma 10h32'39"61d3'
Moyett,Ashley Maria
Cam 6h1'0"68d52'
Moylan,David
Uma 11h34'28"37d41'
Moylan,James W
Boo 15h22'41"51d47'
Moyle,Madison Walker
Cnv 12h59'58"32d36'
Moyle,N Shepard
Cyg 21h27'14"30d57'
Moyle,Wendy Walker
Dra 18h29'0"70d17'
Moynahan,Brunilda
Uma 10h46'45"56d29'
Moynahan,Paul Daniel Vincent
Cep 21h0'49"60d56'
Moynihan
Her 16h57'59"30d5'
Moynihan,Daniel P
Her 17h59'24"40d0'
Mozambique
Lac 22h7'49"51d38'
Mozer Pascal Adrienne Caroline René
Per 4h7'14"39d57'
Mozer,William Donald
Her 18h18'52"12d17'
Mozie
Cyg 21h20'29"40d20'
Mozzillo,Michela
Gru 22h14'49"-56d1'
MPS Enterprise
Ori 6h4'20"4d7'
Mr & Mrs "Briz"
Cyg 21h10'36"37d13'
Mr B,The
Per 3h10'48"50d1'
Mr Baby Doll
Del 20h53'37"3d40'
Mr Bean
Ori 6h7'12"0d41'
Mr Bill
Cep 21h48'21"56d9'
Mr C
Her 18h2'27"31d14'
Mr C
Her 16h7'44"16d15'

Mr Chamler's Lady
Vul 20h15'31"23d3'
Mr D (Robert H Doolan)
Lmi 9h58'53"33d27'
Mr Dan
Hya 10h1'7"-17d6'
Mr E
Boo 13h47'12"14d41'
Mr Ed
Aur 4h56'15"40d59'
Mr G
Cep 1h24'0"87d21'
Mr G
Per 3h5'1"41d12'
Mr G
Peg 23h43'56d31d37'
Mr Hall Of Fame- "Mr Chad E O"
Uma 9h23'60"54d57'
Mr II,Christin & Matthew
Cyg 20h2'53"40d52'
Mr Jeff
Ser 18h16'34"-13d54'
Mr Joe & Rawk
Vul 19h22'17"26d36'
Mr Les
Aur 5h1'0"49d10'
Mr Lucky
Cnv 12h21'19"40d52'
Mr Mac
Per 2h51'21"50d10'
Mr Mark
Lyn 8h26'40"38d36'
Mr Marty
Umi 15h8'48"71d16'
Mr Mentals World Of Acid
Peg 23h41'42"16d16'
Mr Mikey
Aql 19h58'58"10d3'
Mr Mon
Per 3h1'43"47d57'
Mr Nite
Hya 9h16'3"-14d12'
Mr Perfect
Hya 8h13'30"0d53'
Mr RDT
Aql 19h52'18"14d57'
Mr Rooomantic
Her 16h48'32"30d5'
Mr Soup
Ori 4h43'23"0d51'
Mr Steve "You Make Pretty Too"
Aur 5h2'22"51d22'
Mr T's Wrinkle In Time
Uma 11h51'52"50d7'
Mr Tiff Peaches
Lyn 8h20'37"47d26'
Mr Walrus
Lyn 7h33'1"41d10'
Mr Weekend
Aql 18h53'11"-2d53'
Mr Wonderful
Ori 5h59'1"15d45'
Mr Wonderful
Uma 11h6'54"50d16'
Mr Wonderful CMFD
Aql 20h14'30"0d2'
Mr"D"
Aur 6h0'0"34d18'
Mrauk,Rosina Demarin
Lyr 18h50'42"42d31'
Mrdalj,Rita
Lyn 7h8'57"58d11'
Mrdjenovich,Don
Dra 11h0'39"78d38'
Mrion
Cet 1h41'16"-6d2'
Mroch,Starlene Clark
Hya 9h32'13"5d40'
Mroz,Darryl
Aql 19h55'15"14d25'
Mroz,Jeannette
Cep 3h10'47"77d52'

Mroz,Morgana Lynn
Cyg 21h4'45"38d53'
Mroz,Sue
Vul 19h44'0"28d54'
Mroz,Walter
Vir 13h29'34"-9d47'
Mrozek,Marcus
Cap 21h24'42"-20d15'
Mrozinski,Jeff Allen
Per 2h58'0"50d10'
Mrs Balken
Cas 23h30'50"61d58'
Mrs Supreme
And 0h56'1"34d58'
Mrs T
Mon 7h13'34"-5d6'
MRT * ADT Friends Forever
Ser 15h31'1"1d10'
Ms Amy
Cas 23h41'30"50d24'
Ms Carol
Peg 22h2'39"10d51'
Ms D
Peg 23h3'42"14d24'
Ms Holly
Uma 11h22'21"61d21'
Ms Inez C
Lyr 18h53'13"30d43'
Ms Janice
And 0h2'50"44d34'
Ms Mom's Star A K A Stellar Sheila
Crb 16h15'24"34d57'
Ms Q
Cas 0h28'0"50d28'
Ms Rosemary
And 1h20'0"38d59'
Ms Threasa
Cam 3h55'25"72d53'
Ms Viv
Uma 11h2'45"57d57'
MSB 7-24-92 BFAALFNBDLP
Uma 10h8'29"58d42'
MSNM Foley Star,The
Cep 5h6'0"80d15'
MSSO-143
Peg 0h1'0"22d5'
Mt Loeffler
Oph 17h40'27"-18d49'
MTM I
Vul 19h57'59"26d12'
MTM II
Cet 2h6'18"4d2'
Mu
Gem 6h51'34"12d38'
Mu
Dra 16h7'33"61d6'
Mucchetti,Michael R
Cep 23h3'43"64d11'
Mucchetti,Stephen G
Lyr 18h50'29"40d21'
Mucci III
Mon 7h4'10"-5d55'
Mucci,Alexa LaRayne
Vul 19h23'15"26d42'
Mucci,Eloisa
Umi 17h11'42"76d26'
Muccin,Valerie
Dra 12h6'0"70d22'
Muccino,Trooper John M
Cep 1h17'20"86d32'
Muccio,Homa C E
Cyg 21h56'53"53d26'
Muccio,Rose C
Aql 18h41'18"1d42'
Muchachita
Ara 17h57'19" 52d51'
Muchnicki,Jr,John Robert
Dra 16h57'21"73d33'
Muck's Star
Uma 10h28'18"41d26'
Muckerman,Alexandra Camille
Cam 3h35'44"70d58'
Muckjian,Deran
Aql 19h30'16"0d29"

Muckleroy,Richard Scott
Cma 6h56'50"-17d54'
Mudarra,Stella P M
Uma 11h25'39"44d29'
Mudbugs 8-31-95
Aql 19h7'52"1d49'
Mudd,Alexander Thomas
Aur 7h21'1"40d9'
Mudd,Elizabeth Rose
Cyg 21h38'42"38d48'
Mudd,Jim
Aur 6h26'1"38d6'
Mudd,Tony
Cep 22h30'36"70d47'
Mudd,Vicki Sue
Cam 8h33'29"73d38'
Mudgett,Benjamin John
Peg 22h39'59"20d37'
Mudgett,Carmen Rebecca
Cet 2h51'13"1d54'
Mudgett,Diane
Tau 3h44'43"25d53'
Mudick,Samuel Dugan
Vir 13h23'51"-10d49'
Mudrick Star 5-TTTKM
Per 1h47'48"50d8'
Mudry,Bnjamin
Per 2h53'55"46d15'
Mudry,Daniel
Her 16h43'40"28d22'
Mudry,Odette
Per 4h6'18"36d33'
Mudry,Paul
Cep 21h43'11"68d27'
Muecke
Ori 5h14'14"-8d36'
Muecke,Brandon H
And 23h17'57"40d21'
Muecke,Judge Carl
Ori 5h53'58"15d13'
Muecke,Larina
And 0h41'23"45d21'
Muecke,Natalie J
And 23h3'44"46d29'
Muecke,Ross C
And 23h17'46"40d47'
Muecke,Sammi
And 2h9'32"42d57'
Mueller aka Uschi, Nicole Ursula
Cyg 20h14'19"45d10'
Mueller Always & Forever,Gary W
Dra 19h7'52"65d0'
Mueller,Adelheid
Leo 10h2'1"10d56'
Mueller,Adolph & Erna
Del 20h26'53"10d13'
Mueller,Albert
Umi 13h49'10"71d4'
Mueller,Amanda
Del 20h27'59"11d0'
Mueller,Ashley
Lyn 7h52'20"43d42'
Mueller,Bert
Aur 7h5'48"41d8'
Mueller,Bill "Dude"
Lac 22h10'51"49d34'
Mueller,Bob
Aur 6h32'31"33d2'
Mueller,Brenda
Lyn 8h10'22"41d49'
Mueller,Brian Jeffrey
Aql 20h5'26"9d15'A
Mueller,Cassandra Wallace
Aur 5h59'54"30d16'
Mueller,Charles Wallace
Hya 8h44'22"3d55'
Mueller,Charlie
Cyg 21h5'28"37d27'
Mueller,Christiane
Gem 6h37'42"28d26'
Mueller,Diethelm
Peg 23h32'25"20d55'

Mueller,Donna & Lyle
Cam 5h42'20"73d1'
Mueller,Donny
Dra 19h21'53"56d3'
Mueller,Dr Edeltrud geborene Zimmer
Sgr 18h54'21"-27d7'
Mueller,Emily Elizabeth
Cam 3h34'51"60d2'
Mueller,Eric
Cep 23h5'41"70d47'
Mueller,Frank
Lac 22h50'56"37d47'
Mueller,Gertrude
Umi 16h15'38"70d26'
Mueller,Gregory C
Per 3h25'1"50d32'
Mueller,Guenter
Cmi 7h29'46"8d57'
Mueller,Hans-Joachim
Cnv 14h2'42"34d19'
Mueller,Jackie (Jax)
And 1h27'15"37d13'
Mueller,Jacob Reinhold
Aql 20h4'50"7d50'
Mueller,James C
Lac 22h24'1"38d7'
Mueller,Jeff & Lisa
Crb 16h6'24"26d27'
Mueller,Joela
Aql 19h47'15"14d17'
Mueller,John D
Aur 6h22'59"31d49'
Mueller,Kathleen
Tau 5h53'54"27d51'
Mueller,Kay & Herb
Oph 17h0'59"8d42'
Mueller,Lewis & Eileen
Aur 5h25'28"38d58'
Mueller,Lora
Lyn 9h14'11"42d7'
Mueller,Ludwig
Per 4h42'47"51d48'
Mueller,M Erik
Lac 22h21'1"52d31'
Mueller,Margaret E
And 0h53'1"40d28'
Mueller,Mark Wayne
Hya 8h58'4"-7d43'
Mueller,MD,Elmer J
Her 17h43'34"40d23'
Mueller,Meaghan Mitchell
Peg 21h49'0"33d44'
Mueller,Norbert
Cyg 21h54'39"38d23'
Mueller,Peter Christopher
Oph 17h54'32"12d25'
Mueller,Princess Brittany Carol
And 0h43'18"40d28'
Mueller,Ray
Per 2h31'29"57d12'
Mueller,Robert Martin
Boo 14h59'13"41d1'
Mueller,Sally P
Vul 19h44'0"20d19'
Mueller,Suzanne
Cyg 21h51'12"38d56'
Mueller,Suzanne Lee
Aql 20h5'26"9d15'B
Mueller,Terry & Joe
Aur 7h15'53"40d7'
Mueller,Theresa Ann
And 0h22'36"44d54'
Mueller,Tim
Ori 5h53'59"12d15'
Mueller,Tony
Aur 6h0'14"37d33'
Mueller,Tracey Lynn
Lyr 18h32'48"42d43'
Mueller,Ulrike
Gem 6h41'28"30d56'
Mueller,Werner
Equ 21h4'55"3d36'

Mueller,Wilhelm H.
Boo 15h29'12"40d8'
Mueller-Rech,Peter- Georg
Lib 15h15'4"-22d9'
Muenstermann,Daniel
Sco 16h30'48"-40d33'
Muenzer,Mayor Paul
Oph 18h6'39"0d25'
Mueth,Erin Rita
Mon 7h55'52"-2d54'
Muetterties
Boo 15h5'16"50d60'
Mufasa
Aql 19h44'60"14d3'
Muff
Cas 0h6'56"62d29'
Muffin
Cma 7h14'59"-13d47'
Muffin
Leo 10h54'37"-0d55'
Muffin
Lyn 7h54'51"50d6'
Muffin Bear
Uma 11h43'40"38d7'
Muffin Gold
Cam 13h11'0"81d47'
Muffin Meek
Cam 6h12'58"58d22'
Muffy
Cet 2h58'51"5d5'
Mugan,Patricia
And 2h31'47"47d23'
Mugar,David G
Tau 4h30'44"15d35'
Mugford
Cmi 7h57'12"5d18'
Mugg,Thomas Henry
Lac 22h8'0"40d44'
Mughal,Star of Mahira P
And 23h4'11"37d42'
Muglia,Rosalie
Uma h39'33"50d24'
Mugnaini,Donatella
Cma 6h10'1"-31d41'
Mugnier,Peer
Mon 6h42'41"10d55'
Mugridge,Kenneth Gordon
Cep 21h17'18"67d55'
Mugridge,Paul T
Hya 8h14'49"-1d21'
Muhich,John & Anna
Cyg 20h16'47"41d58'
Muhl,Helmut
Her 17h10'13"48d20'
Muhlenberg,Mattias
Boo 14h20'0"35d50'
Muhlenkamp,Ashley Janelle
Cet 0h30'67"-2d15'
Muhlenkamp,Gregory S & Cara M
Peg 22h44'0"5d3'
Muhler,Doug
Aur 6h20'49"31d58'
Muhlstein,Herman
Ser 16h6'33"5d13'
Muhr,Markus
Cyg 19h52'54"47d26'
Muhunney,Alpha
Aql 19h3'33"0d10'
Muir,Alexander
Aur 5h0'0"46d2'
Muir,Brad
Lac 22h34'45"37d50'
Muir,Ian
Ori 6h1'29"0d19'
Muir,Tla Elaina
Mon 7h50'32"-6d41'
Muirhead,William
Uma 11h33'0"60d52'
Muirneach Alexandra
Boo 14h6'32"32d3'
Muirneach Maitin
Lmi 10h4'1"37d56'
Mujtaba,Usman
Peg 23h29'29"32d9'

Mukai,Chikako
Eri 4h4'48"-22d7'A
Mukon,Chelsea Lynn
And 0h1'23"34d13'
Mukon,Christopher Mark
Her 16h43'43"30d56'
Mulanax"Grandparents", Bob & Anita
Aql 19h42'13"10d48'
Mularczyk,Eve
Cas 2h10'29"59d12'
Mularkey,Martin James
Her 16h51'27"50d47'
Mulatier,Emanuel
Aql 19h58'26"10d59'
Mulberry,Richard
Ser 16h7'59"14d17'
Mulcahy,Ashley Elizabeth Faverio
And 0h26'14"41d13'
Mulcahy,Kimberly
Lyr 18h41'39"46d14'
Mulcahy,Nancy Louise Fasci
Del 20d32'46"6d6'A
Mulcahy,Patrick
Tau 5h50'38"23d47'
Mulcahy,Robert & Margaret
Cyg 21h53'59"41d4'
Mulcahy,Stephen Paul
Her 16h11'0"40d2'
Mulder,Hollie
Sct 18h54'21"-8d57'
Muldoon,June & Lawrence
Cyg 19h56'10"41d14'
Muldoon,Patrick
Her 16h18'20"23d22'
Mulei,Evelyn
Cas 2h38'1"60d25'
Mulei,Jack
Lyn 7h38'48"36d50'
Mules,Caroline Patricia
Lup 15h4'26"-45d46'
Mules,Donk
Equ 21h4'13"10d42'
Mules,Kathy
Mon 6h54'49"-5d39'
Mulford's Wolf,Amy
Lup 14h20'26"-43d44'
Mulford,Nancy Johnston
Eri 5h0'28"-5d29'
Mulford,Tripp & Andrea
Peg 23h5'1"17d40'
Mulgrew,Kate
And 23h41'23"46d52'
Mulhall,Mandy
Lyn 9h0'50"33d44'
Mulhaupt,Dorothy Hazel
Lac 22h18'49"51d23'
Mulhaupt,Richard Carl
Lac 22h18'0"51d12'
Mulhaupt,Richard Verne
Lac 22h18'0"51d27'
Mulheisen,Alexander
Dra 20h1'0"62d47'
Mulhern,D
Dra 20h3'38"68d23'
Mulhern,Maureen Samara
Her 16h21'58"25d8'
Mulhern,PL
Crt 11h50'58"-10d1'
Mulhern,Todd
Boo 13h43'54"23d52'
Mulholland,Anthony Francis
Oph 18h19'51"7d2'
Mulholland,Bud
Hya 8h27'40"-8d47'
Mulholland,James
Dra 17h40'1"75d42'
Mulholland,Jason T
Dra 17h41'35"61d14'
Mulholland,Jonathan R
Cep 0h19'45"78d46'

Mulholland,Richard Francis
Cet 3h0'45"5d6'
Mulhollands Star,Janet & Tom
Cnv 14h2'26"40d29'
Mulichak,Nicholas Anthony
Her 16h48'37"34d20'
Mulkey,Dianne I H
Cas 0h40'29"64d55'
Mulkey,Gina Leann
Oph 17h13'38"-20d23'
Mulkey,Linda Dawn
Cyg 21h4'38"50d2'
Mulkey,Rebecca C
Cas 1h29'26"71d32'
Mulkey,Samuel A
Her 17h29'19"23d13'
Mullally,Ryan M-9 Ryan
Uma 10h48'11"61d14'
Mullally,Tara M-7 Tara
Vul 21h22'12"26d24'
Mullaly,Kevin Patrick
Cep 20h37'14"75d7'
Mullane,Elisabeth
Ari 1h55'17"23d7'
Mullane,Jill Catherine
Cep 21h4'25"61d57'A
Mullane,Thomas Kevin
Cep 21h4'25"61d57'B
Mullaney,Betty
And 2h22'28"38d47'
Mullaney,Carol
Lyr 19h23'59"35d26'
Mullaney,Jr,Gerald Russell
Hya 10h47'46"-18d23'
Mullaney,Margaret
Lyn 8h20'46"58d21'
Mullany,Mike
Hya 8h11'51"1d25'
Mullarkey,"Lotta"
Del 20h14'25"10d23'
Mullavey,Gregory
Ser 15h16'41"6d6'
Mullavey,Ryan
Her 17h1'1"21d4'
Mullavey,Tarryn
Lmi 10h49'0"25d8'
Mullen's Star of Love, James (Moon)
Her 16h7'26"48d42'
Mullen,Anne
Cas 22h58'57"55d59'
Mullen,Benjamin Alan
Crb 15h30'1"31d29'
Mullen,Bradley
Boo 14h43'57"25d31'
Mullen,Dennis
Per 3h13'1"55d36'
Mullen,Edwin Raymond
Lac 22h12'0"48d24'
Mullen,Gertrude W
Lyr 19h4'0"47d7'
Mullen,Hailey Taylor
Lyn 6h59'31"59d30'
Mullen,Jacob Peter
Aur 5h31'22"30d49'
Mullen,Joe
Aql 19h34'33"0d4'
Mullen,Joe & Rena
Aql 19h53'30"12d11'
Mullen,John P
Gem 7h54'27"30d29'
Mullen,John Richard
Oph 17h54'47"12d40'
Mullen,John William
Ori 5h52'18"17d7'
Mullen,Joy Kaufman
Ori 5h58'18"18d14'
Mullen,Jr,Danny Francis
Per 1h30'0"52d55'
Mullen,Kristen
Mon 7h39'43"-8d59'
Mullen,Len
Aql 18h58'20"-6d47'
Mullen,Marty G
Oph 18h17'33"11d53'

Mullen,Mary Ann
Peg 21h54'25"3d8'
Mullen,Molly
Cyg 20h35'57"45d17'
Mullen,Regis
Ori 5h27'33"-0d3'
Mullen,Robert Peter
Aql 20h11'0"0d44'
Mullen,Ryan Mathew
Lac 22h23'41"54d12'
Mullen,Vernon Andrew
Dra 17h14'32"63d50'
Mullen,Veronica Rain
Del 20h49'17"7d50'
Mullen,Viola Mary
Cas 1h40'24"60d34'
Mullen-Higgins,Kayla Ann
Lyn 8h10'33"41d18'
Mullenbach,Robert & Katherine
Crb 15h21'35"31d2'
Mullenix,Cherri
And 2h16'24"49d43'
Mullens,Kansas
Cet 3h0'0"8d6'
Mullens,William Lewis
Eri 4h47'16"-6d8'
Muller
Uma 11h23'11"30d27'
Muller,Adelyn
Eri 3h51'25"-4d53'
Muller,Christoph
Oph 16h59'53"10d40'
Muller,Claud B
Aur 4h57'20"50d19'
Muller,Claudia
Uma 11h7'32"51d21'
Muller,Dr Martin
Ori 5h53'42"1d20'
Muller,Gary Lee
Cet 2h18'58"9d57'
Muller,Jeane
Ori 5h9'58"-5d33'
Muller,John Thomas
Ser 16h16'16"1d43'
Muller,Linda
Cyg 20h32'0"42d58'
Muller,Lydia Ann
Aql 19h29'57"-6d5'
Muller,Margaret D
And 23h33'51"41d47'
Muller,Peter G
Hya 8h57'1"-6d51'
Muller,Ralf
Per 01h53'16"50d37'
Muller,Rev Donald
Boo 14h34'34"41d34'
Muller,Virginia "Mimi"
Cet 0h39'52"-7d1'
Muller,Ward
Umi 17h27'34"80d1'
Mullor Jones, Carol-Anne
Ari 3h23'51"28d20'
Mullery,Connor Thomas
Peg 21h32'23"9d31'B
Mullery,Joan
Vul 19h51'31"20d13'
Mullery,Joseph Michael
Peg 21h32'23"9d31'A
Mullet,Dean
Ori 5h53'46"18d17'
Mullett,Jon "Swist Fox"
Aur 6h26'43"35d53'
Mullibet
Eri 3h44'24"-2d41'
Mulligan,Abigail Elizabeth
And 23h21'0"38d18'
Mulligan,Darrah
Aur 6h27'53"33d35'
Mulligan,J
Mon 6h55'39"-8d57'
Mulligan,John
Dra 16h10'55"68d33'
Mulligan,MD,David Cobourn
Ori 5h3'0"10d44'

Mulligan,MD,Michael J
Aur 7h4'0"40d3'
Mulligan,Patricia
Lyr 19h26'12"40d46'
Mulligan,Patricia Ann
Aur 7h7'22"38d39'
Mulligan,Theresa
Uma 11h0'18"40d28'
Mulligan,Tom L
Tau 4h34'43"15d16'
Mulligan,Virginia
Del 20h14'59"15d3'
Mullin,Frances A
Eri 2h45'21"-5d44'
Mullin,John J
Sct 18h22'22"-10d6'
Mullin,Peter
Hya 8h42'52"3d38'
Mullineaux,Gary Michael
Uma 10h15'35"52d28'
Mullineaux,Shawn Morgan
Aql 19h53'22"15d43"
Mullins Star,Weldon & June
Aql 19h44'45"12d32'
Mullins,Antoinette T
Uma 10h25'18"51d45'
Mullins,Justin (7-12-93)
Aur 4h54'1"51d16'
Mullins,Kay
Del 20h52'58"2d15'
Mullins,Kerry
Cas 0h24'24"60d19'
Mullins,Mary-Kay
Cam 5h21'7"24d30'A
Mullins,Michael Aaron
Her 18h17'24"28d26'
Mullins,Randall
Aql 19h8'1"1d14'
Mullins,Randy Lynn
Lmi 10h5'51"30d10'
Mullins,Russell Keith
Cnv 12h25'49"37d35'
Mullins,Walter C
Her 16h11'57"11d1'
Mullis,Destiny Lynn
And 2h5'44"41d29'
Mullis,Gwendolyn Norman
Cyg 21h25'24"28d35'
Mullowney,Thomas Bartholomew
Per 2h20'0"54d30'
Mulloy,Mike
Per 1h46'22"53d59'
Mully
Mon 8h2'17"-0d16'
Mulnix,John Arthur
Aur 5h11'54"44d42'
Mulokozi,Claudia
Lyr 19h13'22"31d41'
Mulqueen May 23,1993, Greg & Alice
Cyg 20h50'0"40d31'
Mulrain,Donald I
Cet 1h17'25"-4d13'
Mulrain,Gary James
Cep 22h19'48"65d18'
Mulry,Chris
Per 2h29'41"57d41'
Mulsow,Juergen
Mon 7h52'2"-2d46'
Multi-Market Radio
Her 16h28'32"39d28'
Multz,Adam Scott
Vir 13h22'7"-7d28'
Multz,Jeffrey Scott
Aql 20h0'1"4d7'
Mulvaney,Abbie
Eri 4h3'55"-11d11'
Mulvaney,Carrie
Mon 8h3'24"-1d54'
Mulvaney,Millie
Mon 8h3'7"-1d57'
Mulvaney,Patricia
Aql 19h6'12"-0d30'

Mulvaney,Sean Gifford
Psc 1h30'22"27d39'
Mulvany,Wonderful Colm
Ori 6h5'60"4d55'
Mulvey,Kara Lee
Mon 6h36'22"-0d32'
Mulvey,Katherine G
Cyg 19h24'60"35d5'
Mulvey,Peter Vincent
Vir 13h35'11"2d38'A
Mulvihill,Brian James
Oph 17h32'44"-0d38'
Mulville,Karen Ann
Sco 16h36'42"-31d37'
Mulé,Lisa Nystrom
Cas 3h8'19"57d28'
Mum & Dad 25th Anniversary
Ind 20h54'51"-54d33'
Mumaw,Terri Lynn
Mon 6h31'42"3d38'
Mumford,Kelci Hanalei
Cet 2h37'25"1d37'
Mumford,Paul
Hya 8h59'0"0d2'
Mumma,Lois
Tau 5h45'52"16d42'
Mumma,Sr,Charles William
Cep 21h26'48"67d48'
Mumme,Courtney
Mon 6h20'46"0d49'
Mummert,Baby G
Cnv 13h40'22"46d9'
Mummert,James
Aur 5h56'47"50d34'
Mummert,Steven Ray
Cnv 13h46'12"45d24'
Mun,Devin
Uma 9h15'53"42d51'
Munch,Avery Kyle
Umi 15h16'53"69d48'
Munch,Geraldine
Cas 0h27'30"61d15'
Munch,Robert Allen
Sct 18h52'41"-6d42'
Munch-Rotolo,PhD, Thomas
Cam 3h51'35"61d26'
Muncy,Margaret J
And 0h14'0"37d46'
Muncy,Todd
Aur 7h15'46"39d35'
Mund,Ruth
Cyg 21h10'50"37d30'
Mund,Timothy Charles
Cap 21h5'36"-22d59'
Munday,Oliver Charles Johnson
Aql 20h0'45"4d26'
Mundell,Billy & Sandy MacBride
Cyg 21h38'50"41d42'
Mundell,Sue
Peg 21h40'50"21d49'
Munderville,Sarah Ann
Mon 6h56'19"-8d45'
Mundi,Lianda Faith
Peg 22h30'0"24d24'
Mundt,Ray
Boo 15h13'48"47d40'
Munduk
Dra 20h1'0"62d47'
Mundwiller,Lee A
Lac 22h25'54"40d1'
Mundy,Donna Margaret
And 0h16'55"38d27'
Mundy,Michael
Her 18h55'43"18d56'
Muneer Hammudeh
Boo 15h6'16"11d30'
Munera
Ori 5h52'34"14d4'
Munger,Rick
Aur 6h25'32"38d56'
Munger,Seth Harrison
Her 18h18'36"14d14'

Munguia,Robert Trujillo Pipo
Aql 19h47'14"14d1'
Muni
Oph 16h58'58"10d28'
Muni Muni Marco
Vel 10h12'32"-47d23'
Munkvold,Jessica Anne
And 23h8'28"40d15'
Munley,Dave P
Peg 21h55'50"2d54'
Munley,Ned
Hya 9h35'42"-8d50'
Munns Star,The Collins R
Cet 0h58'0"-0d45'
Munns,Ryan Thomas
Dra 17h24'14"61d46'
Munns,ST,CPB,S,BG,V, Ditto,Brenda M
Vul 20h39'24"20d14'
Munoz,Claude Christian
Aur 7h23'49"41d8'
Munoz,Irene Rios
Cmi 7h21'17"1d42'
Munoz,Katherine Anne
Cam 7h38'43"70d14'
Munoz,Ray
Aql 18h41'59"-2d34'
Munoz,Veronica Francisca
And 23h2'40"49d58'
Munoz,Wladimiro
Per 4h4'40"37d54'
Munro,Agnes
Cyg 19h30'32"34d52'
Munro,Christine
Com 12h35'22"14d15'
Munro,Greg
Oph 17h7'30"0d31'
Munro,Paul & Gloria
Crb 15h24'27"31d58'
Munro,Robert Jackson
Cep 23h1'28"62d18'
Munro,Roderick Christian
Dra 16h37'13"62d40'
Munroe,Chandler Grant
Leo 10h15'0"13d25'
Munroe,Glenn
Ser 18h12'1"-9d56'
Munroe,Matthew David
Ori 5h37'37"1d33'
Munroe,Norman Wilson
Aur 4h53'29"41d13'
Munroe,Sarah Anne
Cyg 20h16'17"30d54'
Munroe-Lang,Tamea Moana
Col 5h50'9"-42d30'
Munsch,Charlene- Barbara
Uma 9h37'30"49d4'
Munsch,Paul
Mon 7h1'12"4d24'
Munsell,Jane
Peg 22h40'42"22d52'
Munsell,William P
Sex 9h59'44"0d21'
Munsey,Cheryl Ann Purslow
And 23h26'0"47d3'
Munsey,Robert
Lyr 18h19'49"40d55'
Munson,Cammie Lou
Lac 21h57'21"36d56'
Munson,Erin Dianne
Lyr 18h49'23"40d34'
Munson,Georgia Lynn
Aql 20h11'42"6d26'A
Munson,Jack F
Her 15h50'49"42d3'
Munson,Joan Marie
Lyr 18h37'29"38d24'
Munson,Mildred Jean Hume
Cas 23h16'31"63d21'
Munson,Nadelle
And 0h17'19"38d43'
Munson,Patricia
Cas 1h32'31"61d27'
Munson,Ronald Steven
Aql 20h11'42"6d26'B

Munson,Scott Ramsey
Boo 14h18'40"12d59'
Munster,Mickey
Lyn 9h7'49"39d8'
Munter,Gregory Michael
Hya 9h10'53"0d11'
Munter,Leo
Cam 9h3'41"80d22'
Munyak,Nancy
And 0h39'54"40d38'
Munyan,Blair
Tri 2h19'11"30d56'
Munz,Jr,Charles
Cnv 12h27'34"37d35'
Munzer,Aaron James
Cmi 7h52'45"4d34'
Mura,Mary
Eri 5h3'44"-6d58'
Mura,Ron
Boo 15h18'51"38d60'
Muraca,Ralph John
Boo 15h1'53"14d46'
Murack,Kyle Thomas
Aur 5h17'59"40d23'
Murad,Sultan
Lac 22h27'1"53d46'
Murakami,Ikuo
Cam 5h55'59"70d58'
Murakamiito,Keiko
Dra 16h3'14"60d13'
Murakh,Elina
And 23h28'58"43d39'
Muramatsu,Ken
Ori 5h2'16"-0d54'
Murar,Arleen G
Lyr 18h43'1"32d18'
Murase,Riku
Uma 12h51'50"57d51'
Muratone,Sylvie
Mon 6h27'39"0d59'
Muratore's Star,Anne Louise
Peg 21h57'60"32d42'
Muratore,Jayne Lynn
Cnc 8h59'16"20d27'
Muratore,Sarah Gloria
Sco 16h52'21"-40d9'
Muratore,Thomas
Cep 0h15'28"75d54'
Murauer,Timon
Cas 0h13'0"60d22'
Murawski,Robert W
Aur 5h3'14"51d45'
Murch,Suzanne
Ori 5h58'35"11d48'
Murchison,Billy
Cam 14h12'18"80d28'
Murchison,Lucille, Gannon
Cas 0h40'58"73d34'
Murcic,Janis
Uma 10h29'53"68d56'
Murcott,Gail
Aql 20h11'57"1d52'
Murdach "Happy Valentines",Robert
Cep 20h22'33"76d13'
Murdick Forever Bright,James
Cep 21h4'28"55d8'
Murdoch,Anthony J
Boo 14h33'37"48d47'
Murdoch,Hannah M
Cyg 21h32'43"48d50'
Murdoch,Janet Kubala
And 2h3'35"38d3'
Murdoch,Joelle Seonaid
Dra 17h41'13"61d43'
Murdoch,Sandie
Umi 13h13'1"76d45'
Murdoch,Sharon
And 2h16'37"39d56'
Murdoch,T E
Peg 22h28'31"8d7'
Murdoch-Williams,Rev Sharon
Vul 20h4'20"28d44'

Murdock's Hi Fashion
Cma 7h5'39"-28d33'
Murdock's My Dear Watson
Cma 7h19'24"-16d4'
Murdock's Star,Matthew
Umi 15h32'17"68d2'
Murdock,Adrian Jon
Aqr 22h24'38"-17d58'
Murdock,Courtney
Cma 6h15'27"-18d51'
Murdock,Harold
Per 2h15'54"58d48'
Murdock,Nancy Jean
Mon 8h4'21"-9d54'
Murdock,Paul
Boo 15h1'10"30d14'
Murdock,Roy William
Dra 17h19'26"69d3'
Murdock,Sonia
Lyr 18h40'13"42d38'
Murdock,Stephanie Reichstein
Eri 3h6'1"-1d38'
Murdock,Zachary Jay
Aur 6h8'0"37d50'
Muren,Dennis
Hya 8h9'42"0d57'
Murez
Col 5h50'8"-40d9'
Murf's Star
Vul 19h40'22"20d30'
Murgatroyd
Ori 4h44'14"0d25'
Murgitroyde III,Thomas Paul
Boo 15h2'33"25d22'
Murie,Marty W
Uma 11h56'42"37d54'
Muriel & Denys
Cyg 20h27'1"31d42'
Muriel Alexandra
Cas 0h43'13"64d39'
Muriel TTC
Umi 15h48'3"73d10'
Murielle
Cyg 21h37'26"40d54'
Murin,Peter
Uma 10h36'37"50d30'
Murinchack,Jr,Robert James
Lmi 11h3'19"32d42'
Muriset,Jennifer
Peg 21h57'18"34d13'
Murkowski,Frank & Nancy
Cyg 19h31'20"31d21'
Murmelwurf's Moorea Love Star
Oph 17h20'1"10d40'
Murner,Duane
Oph 17h13'45"-18d49'
Murnion,John P
Aur 5h28'1"40d14'
Muro,Philip
Aur 6h29'43"38d42'
Muroff,Gerald & Sandra
Eri 4h4'6"-13d7'
Muroff,Hy
Ori 5h2'43"12d20'
Murph
Cam 12h18'52"82d15'
Murphey's Star, Rebecca
Sco 17h51'48"-40d34'
Murphey,"Papa Bear"
Uma 11h49'24"62d47'
Murphey,Debrah Yvonne
Oph 18h41'54"7d58'
Murphey,Forever Douglas & Giuseppina
Lyn 7h32'49"52d15'
Murphey,Gladys
Cam 7h7'53"83d11'
Murphey,Mary Margaret
Cas 0h59'48"50d28'
Murphey,Michael Andrew
Cep 22h28'1"58d17'
Murphy
Aur 5h8'33"38d21'

Murphy
Ori 5h19'33"-9d55'
Murphy
Her 17h17'59"28d13'
Murphy Family,Robert & Jean Carol
Aur 7h2'23"40d40'
Murphy October 1,1918, Doris Ann
Uma 11h14'33"47d12'
Murphy Of Wayland, Michael B
Cep 0h33'16"80d23'
Murphy PGM,Bess
Com 12h20'59"28d33'
Murphy Star
Aql 19h30'58"14d1'
Murphy Star,A V
Ori 6h17'24"8d37'
Murphy TM,Cassandra A
Eri 4h3'0"-10d4'
Murphy TM,Priscilla E
Peg 22h3'51"5d49'
Murphy's Star
Lyn 7h29'29"37d25'
Murphy's Star,Albert
Lac 22h23'28"37d58'
Murphy,"My Soulmate" Robert Michael
Uma 11h15'39"14d11'
Murphy,"Papa Duke"
Uma 9h34'1"55d54'
Murphy,"S A M 1" Stephanie A
Sex 9h54'34"1d12'
Murphy,"S A M II" Stacey A
Aql 18h56'18"-5d31'
Murphy,Aaron & Lauren
Boo 14h47'49"35d4'
Murphy,Alan P
Cma 7h1'30"-20d34'
Murphy,Andrea "Bunny"
And 23h32'58"40d12'
Murphy,Ann
Lyn 8h4'1"58d21'
Murphy,Aubree "Miss Aubree"
Her 17h21'55"44d52'
Murphy,Barbara Ann
Com 12h31'39"20d3'
Murphy,Brendan Raymond Adrian
Lyn 6h57'34"50d27'
Murphy,Brendan Raymond Adrian
Lac 21h58'16"41d59'
Murphy,Brian Joseph Christian
Psc 22h56'55"5d30'
Murphy,Caitlin
Cep 22h10'29"59d5'
Murphy,Captain Robert
Her 18h5'0"40d42'
Murphy,Carmen Marie
Mon 6h37'28"8d54'
Murphy,Carol Lynn
Vul 19h48'38"27d30'
Murphy,Carter James Henry
Boo 15h6'27"21d26'
Murphy,Catherine
Lyr 18h23'16"45d53'
Murphy,Chris
Crt 11h40'21"-10d41'
Murphy,Christopher A
Lac 22h20'56"51d45'
Murphy,Commander Robert E
Aql 20h6'51"4d37'
Murphy,Courtney
Mon 6h59'17"-10d27'
Murphy,Daniel J
Her 17h32'49"28d48'
Murphy,Daniel Joseph
Her 18h54'35"12d56'
Murphy,Daniel William
Cep 22h7'42"58d18'

Murphy,Darrel Wayne
Cet 2h0'41"1d31'
Murphy,Dave
Lmi 10h58'22"25d59'
Murphy,David Edmond Christopher
Cnc 9h1'36"18d47'
Murphy,Dawn Marie "Sunrise"
Cyg 21h28'30"31d48'
Murphy,Delores & Frank
Tri 2h12'33"33d38'
Murphy,Delphi Lynn
Sct 18h46'24"-7d19'
Murphy,Donna Marie
Lyr 19h3'20"38d54'
Murphy,Dorothy Elise Haldi
Cyg 21h28'32"40d57'
Murphy,Dr Christopher R A
Dra 20h7'30"62d35'
Murphy,Dr Gerard E
Aur 5h5'0"40d41'
Murphy,Eddie
Dra 16h57'11"69d53'
Murphy,Edward Leo
Her 16h32'25"38d29'
Murphy,Erin Olivia
Psc 1h19'32"20d4'
Murphy,Eternally Yours Louis
Per 2h25'53"57d25'
Murphy,Fr Thomas J
Aur 5h16'52"49d19'
Murphy,Frances & Paul
Lac 22h19'32"48d53'
Murphy,Frances Mary
And 0h41'21"40d32'
Murphy,Frank J
Sco 17h29'19"-31d1'
Murphy,Frederick F
Per 2h3'0"57d42'
Murphy,Gail J
And 0h11'57"47d2'
Murphy,Gary
Aur 5h18'43"45d44'
Murphy,Geoff
Ori 5h35'0"-6d17'
Murphy,Gerard
Boo 13h48'30"18d58'
Murphy,Guy & Joanne
Cyg 21h28'22"53d51'
Murphy,Gäel Ruth
Aqr 20h57'28"-8d18'
Murphy,Helena Elizabeth
Cas 22h56'17"56d24'
Murphy,Hugh
Aql 19h57'1"10d53'
Murphy,J Clay
Sct 18h44'46"-6d23'
Murphy,Jack Franklin
Boo 14h26'19"48d20'
Murphy,James Edmond
Boo 15h2'58"53d20'
Murphy,James V "Grandpa"
Her 17h5'18"40d33'
Murphy,Jane
Cyg 20h54'12"41d10'
Murphy,Janilyn
Aur 5h18'47"40d32'
Murphy,Jayne Marie
Tau 4h39'23"0d20'
Murphy,Jean
And 0h22'21"32d15'
Murphy,Jeanne B
Mon 6h54'42"-10d11'
Murphy,Jeffrey J
Uma 11h52'51"57d39'
Murphy,Jeri
Cas 0h34'22"60d19'
Murphy,Jim
Ori 5h55'47"15d50'
Murphy,Jim & Laura
Uma 13h31'38"57d47'
Murphy,Joel
Aur 6h51'34"37d9'
Murphy,John P
Aur 6h0'21"30d34'

Murphy,Joseph Per 1h54'55"48d14'
Murphy,Jr,Dennis A Uma 10h44'0"40d43'
Murphy,Jr,Frank E Ori 6h10'0"8d14'
Murphy,Jr,William D Cet 0h34'0"0d30'
Murphy,Karen & Bill Oph 17h12'29"11d31'
Murphy,Kassidy Danae Peg 0h4'0"27d55'
Murphy,Katherine Mon 6h11'22"-10d40'
Murphy,Katherine Michelle Peg 22h44'28"33d36'
Murphy,Kathleen And 23h31'20"44d6'
Murphy,Kathleen Elizabeth And 0h52'26"35d4'
Murphy,Kathleen Cas 23h14'32"61d1'
Murphy,Kathleen A And 0h48'60"36d3'
Murphy,Kathy Aql 19h56'0"8d54'
Murphy,Katlyn M Cam 12h50'54"78d5'
Murphy,Kayla Marie Lac 22h1'0"49d36'
Murphy,Kelly Lynn Cet 0h3'29"-18d11'
Murphy,Kenneth Mon 8h8'38"-9d18'
Murphy,Kyle Ser 15h11'34"12d4'
Murphy,Larry R Lac 22h31'20"50d22'
Murphy,Laura Elizabeth Vul 20h39'0"23d11'
Murphy,Lois A Com 13h16'32"20d21'
Murphy,Loretta Aqr 21h59'44"0d50'
Murphy,Lorraine Lyn 8h52'1"43d10'
Murphy,Lorraine "Mom" Cas 0h5'1"66d18'
Murphy,Maggie Jayne Lyn 7h54'12"39d45'
Murphy,Marcella Helena Shea Cas 1h25'43"50d16'
Murphy,Marilyn Boo 14h8'29"42d16'
Murphy,Marilyn Cet 2h31'1"-10d52'
Murphy,Mark Daniel Lib 15h4'1"-28d54'
Murphy,Mark Donavon Lyn 7h35'26"41d16'
Murphy,Mary Cas 0h45'25"67d38'
Murphy,Maxwell Jay Cam 6h15'1"70d59'
Murphy,Megan Cyg 20h41'1"30d33'
Murphy,Megan Cnv 13h22'33"30d31'
Murphy,Megan D Crt 11h8'23"-15d43'
Murphy,Megan Labourc Cam 5h48'60"70d41'
Murphy,Melva Uma 11h43'36"45d48'
Murphy,Michael & Judith Peg 21h55'31"5d42'AB
Murphy,Michael F Sex 10h29'26"4d42'
Murphy,Mildred Litchfield Lyr 18h47'0"42d5'
Murphy,Molly Danielle Cam 4h15'55"60d54'
Murphy,Nancy Ruth Lyn 7h33'55"38d58'
Murphy,Naomi H Lyn 8h19'23"57d40'
Murphy,Paddy Lyr 18h19'58"41d1'
Murphy,Patricia & Kirby Cam 3h16'12"61d2'
Murphy,Patricia Jean (Irish) Lyr 19h1'14"35d8'
Murphy,Patrick Gerow Ser 15h10'35"21d52'
Murphy,Patti L Vul 20h39'44"28d53'
Murphy,Paula (Mouse) Peg 23h41'27"12d53'
Murphy,Peter Francis Aur 5h33'42"40d43'
Murphy,Richard D Hya 8h41'57"-1d50'
Murphy,Robert Francis Patrick Boo 14h10'36"38d13'
Murphy,Robert John Hya 9h0'12"-8d36'
Murphy,Rosebeth And 2h25'52"45d32'
Murphy,S W Ori 5h59'0"-2d50'
Murphy,Sarah Elizabeth Cas 2h0'0"58d33'
Murphy,Sean Lac 22h16'44"49d29'
Murphy,Shane King Com 13h31'51"18d1'
Murphy,Sherry Tri 2h11'0"32d55'
Murphy,Sinead Elizabeth And 0h50'27"45d31'
Murphy,Sophie Mary Kolonko Cas 1h5'42"70d13'
Murphy,Sr,John A "Babe" Cep 2h4'27"78d56'
Murphy,Star Jessica Com 13h9'48"20d49'
Murphy,Stevey Hya 8h48'1"5d41'
Murphy,Sue Peg 22h41'19"21d23'
Murphy,Susan Celeste Darueau Cam 4h55'16"58d20'
Murphy,Terrance Patrick Oph 16h28'20"-5d52'
Murphy,Thomas Her 17h0'15"28d35'
Murphy,Thomas Francis Joseph Aqr 21h23'44"-10d42'
Murphy,Timothy J Aur 4h48'37"50d49'
Murphy,Tina Marie Sco 17h14'11"-37d12'
Murphy,Vanessa And 23h49'40"41d42'
Murphy,William Cep 22h15'38"61d33'
Murphy,William Joseph Cep 22h59'31"71d1'
Murphy,William Robert Aur 6h21'18"30d7'
Murphy/Franklin Boo 14h51'46"27d37'
Murphys'Star-50 Years to Eternity Cyg 19h22'25"44d41'
Murray Dra 17h8'25"63d9'
Murray Became An Angel 9/14/94,Mary Lyn 9h7'16"37d13'
Murray III,William D Cep 23h13'23"60d20'
Murray IV,James Francis Her 17h7'17"40d42'
Murray Star,The Jim & Judy Cyg 19h20'33"28d12'
Murray's Beulah Uma 8h51'1"68d14'
Murray's Dominus Vobiscum Ori 6h7'34"6d32'
Murray's Moon-Prince of Darkness Her 16h52'1"47d43'
Murray's Valentines Star '93 Her 16h54'29"37d31'
Murray,Alexandra Vul 19h16'15"21d52'
Murray,Allison Jean Aur 4h54'52"40d13'
Murray,Andrew Robert Cma 6h42'55"-15d45'
Murray,Annelle K Oph 16h33'30"-8d2'
Murray,Benjamin Lac 22h53'26"40d31'
Murray,Brian John Cmi 7h13'33"9d45'
Murray,Carl Cam 4h16'17"65d37'
Murray,Carol & Paul Aql 20h7'28"1d7'
Murray,Cathie Elizabeth Peg 23h20'17"32d25'
Murray,Charles Per 2h48'32"32d19'
Murray,Christopher Lincoln Uma 10h15'20"50d26'
Murray,Christopher William Aql 19h27'43"15d24'
Murray,Christy W Lyr 18h31'0"40d52'
Murray,Claire Lmi 10h56'17"30d13'
Murray,Coleman Tyler Peg 22h2'21"24d15'
Murray,Colin Peter Boo 15h8'12"28d27'
Murray,Cornelius E Aur 6h42'34"35d38'
Murray,Dale W Dra 17h40'41"60d28'
Murray,Daniel Wayne Boo 15h15'13"41d38'
Murray,Danielle Sge 19h32'38"16d25'
Murray,David Dra 17h6'30"63d34'
Murray,Debbie Aql 18h42'1"-1d18'
Murray,Dennis & Gail Cyg 21h5'22"30d48'
Murray,Dixie M Pcs 1h27'23"20d56'
Murray,Donald C Cam 12h39'16"80d54'
Murray,Donald D Uma 11h10'25"46d3'
Murray,Douglas S Uma 9h57'40"67d48'
Murray,Dr Francis J Oph 18h41'55"8d59'
Murray,Dylan Joseph Vir 12h1'43"-8d23'
Murray,Edward Lloyd Oph 18h0'39"11d33'
Murray,Elizabeth Mon 7h39'30"-5d52'
Murray,Ethelyn A Cas 0h31'34"60d31'
Murray,Francis Daniel Her 17h8'51"48d21'
Murray,Grayson Bradford Tau 3h40'55"20d23'
Murray,Gwen Aql 18h53'57"11d4'
Murray,Hazel E And 2h18'28"40d20'
Murray,Helen Lyr 18h14'23"41d12'
Murray,Iain Her 18h2'45"20d34'
Murray,Iain Patrick Joseph Boo 14h30'0"50d58'
Murray,Ian Trean Ori 4h59'19"5d1'
Murray,Iggy Lyr 18h59'12"26d3'
Murray,Jeanne Marie Lyn 8h14'35"51d44'
Murray,Jim Cyg 19h27'0"34d54'
Murray,John Her 17h56'22"40d43'
Murray,John George Allen Umi 14h54'1"70d36'
Murray,John Patrick Aur 6h30'53"31d57'
Murray,Jonathan D Boo 13h37'13"20d48'
Murray,Jr,James K Lyn 9h15'17"39d32'
Murray,Judith Lyn 8h8'14"47d14'
Murray,Judith Dorothy Lyr 19h35'45"43d41'
Murray,Julia Com 12h42'35"21d55'
Murray,Justin Aur 5h8'37"44d28'
Murray,Kathleen Lyn 7h45'22"50d28'
Murray,Keith Mon 7h13'9"-6d40'
Murray,Kelly Lynn And 2h18'1"40d59'
Murray,Kevin Her 16h28'43"36d33'
Murray,Krista Lynn Elizabeth Cas 2h33'24"57d55'
Murray,Kristen Peg 21h34'52"20d0'
Murray,Kristin Danistar And 0h52'58"39d15'
Murray,Lara Renee Leo 10h56'37"14d25'
Murray,Lauren Catherine Cas 0h43'1"62d11'
Murray,Louise Uma 11h55'0"42d7'
Murray,Maria Peg 22h38'35"26d50'
Murray,Martin John Aql 20h6'21"6d40'
Murray,MD,James P Ser 15h18'31"8d53'
Murray,Merlin Lee Leo 9h38'54"8d53'
Murray,Michele Cyg 19h18'1"44d53'
Murray,Patricia Jeanne Eri 4h4'0"-1d39'
Murray,Paul Lac 22h33'38"55d37'
Murray,PhD,Gregory Lee Umi 15h8'21"70d46'
Murray,Rick Cyg 20h25'45"38d46'
Murray,Rita D Uma 11h23'34"37d58'
Murray,Robert Aql 20h6'50"6d21'
Murray,Robert L Cep 0h27'39"76d44'A
Murray,Robert N Cep 20h49'16"65d6'
Murray,Rosemary Lac 22h30'58"41d5'
Murray,Susan Welch Lyr 18h22'0"45d49'
Murray,Sweet Eva Sco 17h23'31"-38d52'
Murray,The Star, Kathleen And 23h41'35"45d57'
Murray,Timothy S Aql 19h9'29"12d33'
Murray,Vera Margaret Cep 0h27'39"76d44'B
Murray,Viola And 1h28'39"39d35'
Murray,Walter Irving Dra 16h22'35"61d13'
Murray,William Boo 14h37'0"17d45'
Murray,William F Aur 5h44'53"50d19'
Murray,William Leonard Eri 3h40'28"-1d35'
Murray-Alissa's Heart, Alissa And 0h7'1"47d13'
Murray-Davis,Halie Emerson Cas 2h2'1"67d43'
Murray-Stoldt Umi 15h15'51"78d56'
Murrell,Cathy & Donnie Peg 23h4'36"8d57'
Murrell,Clara Eloise Crb 16h4'39"28d20'
Murrell,Heather Mon 7h59'43"-5d30'
Murrell,John R Hya 9h6'42"1d9'
Murrell,Tom Cmi 8h0'42"6d4'
Murrells,Connor James Ieuan Lac 22h15'41"38d47'
Murrells,Joshua William Francis Sct 18h41'46"-6d16'
Murrey,Lawrence Woodrow Umi 17h27'1"80d21'
Murrey,Melissa Eri 3h57'53"-0d46'
Murrey,Rosemary Uma 11h40'48"50d2'
Murrgailion Boo 15h5'34"20d27'
Murrie,Thomas Moncrieff Her 16h10'0"20d15'
Murrillo,Lupita And 2h35'32"41d3'
Murrin,Jack & Wendy Murdock Umi 15h5'44"68d38'
Murrow,Gerald Roderick Aur 6h54'35"37d58'
Murrow,James Ebbert Hya 9h9'55"-8d53'
Murry-Pittman,Brenda Crb 15h51'10"26d35'
Mursell,Mary Lyr 18h18'17"42d36'
Murtaugh,Jr,Bernard J Aql 20h10'47"11d59'
Murtha,Myra M Psc 1h24'29"23d2'
Murtha,Robert W Her 18h38'39"12d39'
Murty,Sean Joseph Aql 19h3'59"-0d58'
Murwaski,Mary Beth Cet 3h1'32"3d17'
Murzly-Stern Ser 16h6'39"9d15'
Musacchio,Ario Antonio Boo 14h54'39"30d42'
Musal,Helene Sgr 19h59'53"45d1'
Muscara,Maria Katrina Peg 22h24'0"21d46'
Muscarello,Kenn & Lynn Lyn 9h7'23"37d41'
Muscariello,Sabrina Aqr 22h5'12"0d30'
Muscat,Sharon Lynn- Jacques Roger Girard Umi 13h43'51"76d47'
Muscato,Paul Charles Uma 9h57'30"50d29'
Muschalik,Guenther Oph 18h24'28"8d16'
Muschalik,Reid Daniel Hya 10h46'50"-11d47'
Musche,Chester G Cet 2h41'14"-10d41'
Musci,Lindsay Ann Uma 10h5'1"50d48'
Muscolino,Kathleen Cas 0h36'35"75d11'
Muscolo,Angelo Aur 5h2'38"40d12'
Muscott's 21st,Lisa Cas 2h53'26"58d51'
Muse de Claudia Uma 8h53'0"50d57'
Muse Niec Tri 1h52'54"26d9'
Muse,John Mount Her 17h19'0"48d11'
Muse,Louis Alexander Dra 9h49'0"73d17'
Muse,Michael & Elizabeth Aur 6h31'19"38d41'
Muse,William & Deborah Vul 19h48'27"23d13'
Muser,Linda Pearl And 0h1'57"34d3'
Musette Eri 3h51'43"-5d3'
Musfeldt,Ashley Amanda Gem 6h43'29"18d39'
Musgrave,Jessica Anne Dra 17h27'18"72d7'
Musgrave,Megan Elizabeth Cas 23h22'41"61d31'
Musgrave,Ray L Per 2h24'21"55d16'
Mushat,William Cep 0h4'35"78d52'
Mushnick,Kristine Radgowski Cyg 19h59'60"30d8'
Mushpie & Penguin Cep 20h43'50"61d16'
Mushrush,Michael P And 2h30'19"40d9'
Musial,Deandra Lyn Cam 9h3'41"80d22'
Musial,Donald W Her 17h10'43"45d43'
Musial,Milysa Marie Cas 0h32'23"61d1'
Music By Jason Cet 0h48'1"-4d38'
Music-Pool Eri 4h51'0"-8d59'
Musical Aviator Lyr 18'50'40"38d50'
Musicco,Riccardo Per 2h7'30""56d24'
Musico,Paul Uma 11h18'19"52d6'
Musil,Hannes Cap 20h22'48"-26d42'
Musillami,Allison Eri 4h12'37"-19d15'
Musille,Viola Kay Aql 19h11'1"19d38'
Musillo,Matthew Her 16h21'27"26d26'
Musino And 23h3'0"51d51'
Musket,Chris Cas 0h26'48"60d53'
Musky Eri 4h9'52"-12d47'
Muslimat,Rina Pho 0h43'6"46d49'
Muss,Heather Eri 3h41'0"-18d19'
Mussachia,John James Ori 4h54'22"0d42'
Musselman,Jr,Robert H Hya 8h44'14"5d25'
Musselman,Rhonda And 1h10'40"40d53'
Musser,Amie Elizabeth Ari 2h57'13"28d40'
Musser,David L Ori 5h26'37"0d3'
Musser,Jason Hya 8h15'24"3d55'
Musser,Joanna Mon 8h4'1"-5d36'
Musset,Charles Boo 15h42'19"42d3'
Mussett,Deborah Sge 18h57'10"18d48'
Mussett,Roy Per 3h25'56"40d12'
Mussman,Brianna Kuhs Lyn 8h26'1"40d22'
Mussmann,Sylvia Tau 5h49'50"27d35'
Musso,Diana Peg 21h52'1"36d3'
Musso,Pauline Hallford Cyg 21h22'52"37d35'
Musson Alpha Mon 6h23'1"5d0'
Musson,Elsa Carter Cas 1h8'58"61d0'
Musson,Jr,William Lewis Cas 1h10'0"61d48'
Musson,Pamela And 23h3'36"47d36'
Musson,Woody Sge 19h53'38"18d47'
Musta,Shines for And 0h13'1"47d55'
Mustacchio,Tanya Cas 0h51'44"61d4'
Mustach,Matthew Boo 14h8'0"23d7'
Mustache Mike Cep 20h47'55"68d53'
Mustaka,Andrew & Rocky Cyg 20h36'30"60d10'
Mustang Mike Forever Peg 21h52'32"18d47'
Mustapha,Nadia Frances And 0h1'11"44d37'
Mustard,Katherine Lyr 18h15'37"47d7'
Muster,Wendy Ori 6h3'50"20d46'
Muston,Ma Ma & Pa Pa Cyg 21h3'60"39d18'
Mustroph,Axel Uma 9h17'1"48d52'
Musulin,George Edmund Cet 2h19'14"5d22'
Mutch,Thomas J Per 1h43'21"52d45'
Muth,Andy S Per 2h56'21"45d6'
Muth,Barbara Cnc 8h28'37"7d11'
Muth,David Aur 6h11'54"54d58'
Muth,Heidi Cyg 20h5'30"58d21'
Muth,Julia Ori 5h56'58"14d43'
Muth,Lynn G Mon 7h0'1"7d35'
Muth,Sr,Lawrence Tri 1h55'1"27d19'
Muthu,P Her 18h28'47"12d56'
Mutlak-Hunt Uma 9h39'59"53d31'
Mutschler Family Star, The Mon 6h36'38"-6d46'
Mutter,Fred C Aur 6h47'38"35d59'
Mutti & Zizi Aql 20h8'41"1d42'
Mutti Luv & Papa Bear Uma 11h4'0"62d2'
Mutti Martha's Del 20h16'47"9d29'
Mutz,Barb Uma 9h16'12"48d36'
Mutzy Aur 6h13'1"30d57'
Muxie,Mary Cam 4h14'31"68d20'
Muxworthy,Ian William Dra 18h1'40"70d45'
Muyshandt,Jane Cas 0h54'42"59d10'
Muzer "5th Grade Class Of 95",Fred Cep 21h44'0"61d48'
Muzi,Christopher William Per 3h10'0"47d56'
Muzikar,La Bellissima Stella di P F Ori 6h0'26"8d49'
Muzquiz,Jordan Taylor Cmi 7h32'37"0d41'
Muzquiz,Mary Ann Naylor Cyg 21h20'32"31d25'
Muzyk,Barbara Cam 4h24'28"65d10'
Muzyk,Patricia Regina And 0h13'33"41d8'
Muzyka,Lecina Sarah Cas 23h39'17"50d8'
Muzyka,Libertee Elisa Cyg 21h31'24"50d28'
Muzyka,Lindzay Dean Cam 5h41'57"67d54'
Muzzi 131293 Cep 1h45'32"78d42'
Muzzy,Adam Ori 5h53'16"21d29'
Mu Ori 5h53'16"21d29' Lyr 18h38'40"28d58'
Mu Lyr 18h38'40"28d58' Eri 5h5'17"-8d0'
Mu Eri 5h5'17"-8d0' And 0h4'23"37d44'
Mu And 0h4'23"37d44' Her 18h19'1"14d22'
Mu Her 18h19'1"14d22' Vir 13h1'38"-12d22'
Mu Vir 13h1'38"-12d22' Cyg 20h52'32"37d49'
Mu Cyg 20h52'32"37d49' Cnv 12h24'12"37d1'
Mu Cnv 12h24'12"37d1' Mon 7h5'11"0d54'
MWUH Lyn 7h49'19"45d4'
My "Ancy" Love Per 1h55'0"47d27'
My Amanda 12/5/84-4/30/96 Cas 1h9'47"61d22'
My Angel Del 20h53'31"3d53'
My Angel Eva Cas 2h8'32"68d6'
My Angel Trish Cas 3h1'59"61d9'
My Ann Mario Gru 22h28'10"-53d56'
My Aunt Barbie Tri 2h8'50"33d9'
My Baby Aql 20h1'25"7d19'
My Baby Aur 6h14'39"30d39'
My Baby Darling Honeysuckle Delight Her 17h13'49"20d39'
My Beatuiful Star, Vicki Peg 21h59'18"29d48'
My Beautiful Cristina Cyg 20h4'53"41d9'
My Beautiful Debi Cyg 21h51'45"53d42'
My Beautiful Jonathan Per 1h58'11"56d29'
My Beautiful Mother, Denise Carol Cas 1h15'0"64d30'
My Beautiful Priness And 2h22'21"40d56'
My Beautiful Rie Peg 23h3'16"8d52'
My Beauty Del 20h33'35"10d49'
My Bedroom Bulldog Cnv 12h20'35"48d46'
My Belle Her 16h35'38"37d43'
My Belle Michelle Lyr 18h26'33"40d30'
My Beloved Cyg 21h4'58"35d44'
My Beloved Her 17h11'55"27d6'
My Beloved Louis, Eternally Yours Aql 19h4'0"-1d24'
My Beloved Star,Bonnie Bonnie Mon 6h15'34"-6d30'
My Beloved Tommaso, Super Star Sco 17h54'33"-30d50'
My Best Friend Christopher's Star Vul 19h02'22"25d15'
My Best Friend Brad Oph 18h8'1"13d34'
My Best Friend Julie Elm Uma 10h57'14"48d29'
My Best Friend Rolf Sct 18h47'52"-8d48'
My Big Jim,Eternally- Hibiya Per 3h25'16"38d27'
My Birkenstar Uma 10h52'15"38d29'
My BJ Vir 13h27'49"10d44'
My Bobby Ray Ori 5h56'45"11d6'
My Bri Uma 12h56'17"53d12'
My Brother Dick Tau 5h51'1"28d49'
My Brother,The Big Skipper Uma 10h7'25"56d17'
My Bucket Mon 6h53'27"-1d32'
My Buddy Aql 19h10'33"15d26'
My Buddy Cheryl Equ 20h57'18"10d16'
My Buddy Inky (G B H) Aql 19h50'20"14d47'
My Carlitos As Always Forever With My Love Mon 7h0'37"8d46'
My Cathy Her 15h54'37"45d46'
My Cherub Ori 5h46'40"20d0'
My Cherub Jennifer Cas 1h24'0"73d36'
My Cheryl,Princess of The Universe Cet 2h59'46"6d41'
My Children-My Children Uma 11h51'52"43d42'
My Cousin Amy And 23h1'25"49d33'
My Cousin Tracey Cas 0h57'57"63d6'
My Cowboy Love Forever Your Cowgirl Uma 9h38'39"59d40'
My Cyn And 23h33'37"41d36'
My Dad Oph 17h52'59"7d45'

My Daddy
Cep 22h29'0"61d29'

My Daddy
Aur 6h24'38"32d24'

My Daddy & Me
Lmi 10h6'28"33d57'

My Daddy Star
Cmi 7h39'1"10d2'

My Daddy's Favorite Star
Uma 11h23'33"38d56'

My Dana G
Lmi 10h44'51"33d8'

My Darlin Julie
Cas 2h34'17"73d14'

My Darlin'Joe Beachboard
Lmi 10h31'36"32d5'

My Darling Bin
Ori 4h43'23"-0d3'

My Darling Christine
Sct 18h46'7"-8d15'

My Darling Cindy
Uma 10h39'28"47d46"

My Darling David
Sge 19h5'30"19d10'

My Darling Debbie
Her 15h58'24"40d18'

My Darling Denise (Brown) HB 1993
Com 12h54'48"30d55'

My Darling Diggy
Lyr 18h29'30"31d39'

My Darling Gayle
Cas 23h13'34"61d30'

My Darling Gill
Cyg 19h56'42"47d33'

My Darling Janet
Aql 19h25'25"12d40'

My Darling Karen- Heaven Is With You
Lyr 19h14'15"41d57'

My Darling Lynne
Cas 1h51'34"73d26'

My Darling Olenka Moyá Lapúshka Olenka
Cas 0h45'32"63d39'

My Darling Paula
Com 12h0'57"28d45'

My Darling Rick
Per 2h30'52"57d4'

My Darling Sandra
Cyg 19h25'1"34d7'

My Darling Vinny
Per 3h17'44"40d1'

My Darling,James
Hya 8h29'24"-0d1'

My David Chuck-A-Lump
Uma 13h13'34"62d18'

My David-My Honey
Aur 6h24'15"32d58'

My Dawn
And 23h1'58"51d4'

My Dear One
Ori 5h32'14"-1d57'

My Dear Stranger
Vul 19h23'40"26d33'

My dear Tomo
Leo 10h16'11"12d32'

My Dearest Bonnie
Lyn 7h34'22"40d15'

My Dearest Tim Love Forever,Sandi
Aql 19h4'19"2d59'

My Denny
Boo 15h18'45"52d29'

My Destiny
Mon 6h54'57"-8d59'

My Diamond Jacque
Lac 22h35'47"40d37'

My Diane
Cra 18h19'43"42d48'

My Diane
Mon 6h48'18"10d25'

My Dolly
Ori 6h5'0"8d54'

My Doodie
Ori 4h44'1"0d28'

My Endless Love Brandon
Hya 8h9'43"-9d27'

My Evan
Aql 18h54'54"-0d11'

My Eyes Looking At You,Reed
Vul 19h48'1"27d45'

My Fair Lady
Cas 3h31'35"68d21'

My Favorite Baby
Mon 7h0'0"5d27'

My Favorite Boog
Cep 21h11'20"68d5'

My First Angel Shiori
Leo 10h15'36"14d10'

My Forever Otto Star
Tau 4h6'51"20d17'

My Forever Shining Ohio
Oph 18h2'18"12d24'

My Four Leif Clover
Aql 19h56'38"15d27'

My Fragile Beauty Brenda B
Lac 22h24'57"50d15'

My Freckle
Per 4h0'53"51d48'

My Friend
Sgr 19h38'35"-41d34'

My Friend
Lib 15h8'1"-23d6'

My Friend Fred
Peg 22h2'18"10d16'

My Friend Susan
And 23h23'43"47d29'

My Gem
Uma 11h34'29"32d25'

My Ginger
Mon 8h1'22"-8d29'

My Girl Heather
Oph 17h16'12"-16d15'

My Girls
Mon 6h28'15"-0d54'

My Gloria-LT II
Cmi 7h41'51"4d44'

My Goofy Tim
Cep 22h44'0"70d15'

My Gorgeous Annie
Cyg 20h30'45"40d50'

My Gramma
Uma 10h52'39"59d6'

My Guiding Light
Lyr 19h21'0"42d39'

My Happiness Jo Ann
And 2h12'40"50d22'

My Heart Christine Ann
Peg 22h18'52"20d31'

My Hearts Desire Only For You
Aql 19h50'29"14d45'

My Heavenly Father, Love Michael
Cep 20h39'55"58d46'

My Hero
Her 17h3'20"48d33'

My Hero Jonny
Aql 19h31'35"10d11'

My Hon,Clarke
Tri 1h50'45"26d22'

My Honey
Her 17h15'51"45d12'

My Honey
Ori 6h4'13"9d27'

My Honey Bun
Her 17h20'48"38d12'

My Honey Christine
Del 20h24'1"8d7'

My Honey Jim
Lyr 18h51'33"35d21'

My Honey Star
Her 17h3'48"38d0'

My Honey's Birthday Star
Sgr 19h2'33"-24d20'

My Immaculate Mama
Mon 6h19'17"5d58'

My Impossible Dream
Lyr 18h51'39"40d59'

My Irene
Peg 22h57'26"33d42

My Jackson
Per 2h37'56"40^34'

My Jani
Leo 10h0'20"15d10'

My Jeffrey
Cep 23h10'59"60d48'

My Jenny Love
Mon 6h36'25"10d11'

My Jewell
Lac 22h8'55"50d49'

My Jim Your Gem
Sex 9h44'43"2d20'

My Jimmy
Per 2h51'36"31d19'

My Jon-Forever
Her 16h5'46"48d31'

My Jonny
Cet 2h6'10"6d34'

My Juan & Only
Ori 5h56'27"11d9'

My Judge
Ori 5h56'37"15d13'

My Jules Of The Night
Mon 6h26'36"-1d14'

My Kalen-Infinite Love
Sex 10h32'53"2d55'

My Kathy
And 1h7'28"38d37'

My King Of Hearts
Her 16h30'0"35d25'

My Knight In Shining Armor
Cmi 7h53'59"0d38'

My Knight In Shining Armour,Scott
Per 1h53'34"48d42'

My Lady & I
Sge 19h25'52"17d15'

My Lady Mary Lee
Cas 0h8'36"62d43'

My Len
Uma 9h52'1"43d31'

My Leonard
Aur 6h32'14"37d59'

My Linda
And 1h24'53"36d30'

My Little Angel Samantha Rae 143
And 2h14'23"38d41'

My Little H
And 22h58'18"51d9'

My Little Lindsey
Cyg 20h15'55"30d8'

My Little Marjorie 12/20/41
Cas 0h59'26"54d32'

My Little Mermaid- Marilyn
Cyg 19h55'0"38d56'

My Little Pony
Equ 21h21'51"10d13'

My Little Princess Lucia
And 0h57'1"45d5'

My Little Sister,Anne Marie
Cas 22h58'1"54d40'

My Love
Boo 14h39'25"50d25'

My Love
Boo 14h19'0"46d54'

My Love
Cyg 19h25'50"31d47'

My Love For Keith
Ori 4h51'1"4d34'

My Love For You
Boo 15h29'44"33d39'

My Love For You,(Deb & Tom Heed)
Cnv 12h55'11"50d30'

My Love is Eternal as The Star Luie R
Aur 6h31'57"37d24'

My Love Jerry
Her 15h49'1"42d17'

My Love of Kathleen
Cyg 19h28'23"35d8'

My Love Tom
Lmi 10h0'1"41d24'

My Love's 55
Cyg 21h10'28"37d57'

My Lovely
Uma 9h18'17"53d51'

My lovely Ursula
Cnc 8h34'32"30d30'

My Lovely, Lively Lilo
Cam 4h13'54"67d43'

My Lover Corey
Tau 4h12'24"24d10'

My Lover's Lane
Cyg 21h32'44"52d41'

My Loving Blaine
Hya 9h29'47"-6d9'

My Lucky Stars
Uma 11h18'45"37d59'

My M E O
Lac 22h34'60"53d20'

My Main Man Michael
Her 16h22'0"22d50'

My Mairead
Col 6h0'9"-34d30'

My Martin
Her 16h5'24"48d21'

My Mary Beth
Cap 20h5'0"-12d2'

My Matthew
Cmi 7h19'42"1d28'

My Meryl
Ori 5h54'26"11d23'

My Mickey Klang-Klang
Uma 8h25'12"71d38'

My Mitchell
Cep 20h10'19"76d20'

My Michelle
Pup 7h57'0"21d57'

My Mohamed
Per 1h47'0"48d13'

My Mom
Peg 23h35'23"28d21'

My Mom & My Partner
Uma 11h49'1"64d24'

My Mom's Star
Lyr 18h49'1"36d23'

My Moofin
Uma 9h55'18"58d8'

My Mother,Viola
And 23h16'50"48d31'

My Mum's Star
Cep 22h10'35"62d11'

My Music
Cam 5h7'32"68d14'

My Neil
Aur 6h32'15"32d18'

My Neilo Noodleheimer
Cnv 12h45'28"38d3'

My Neisha
Mon 7h5'12"-6d51'

My one & only "Ronnie"
Per 1h28'56"53d41'

My One & Only Contact Man
Cep 2h6'25"77d34'

My One And Only
Aqr 23h2'21"-4d57'

My One And Only Hero
Her 17h31'29"40d57'

My One And Only Tommy
Her 16h35'38"47d54'

My One J E M
Lyr 18h58'59"31d47'

My Own Star Chantal
Mon 7h59'56"-0d35'

My Pal
Cnv 13h59'34"32d13'

My Pal
Crb 16h13'24"38d25'

My Pal Charlie
Cet 3h11'46"-0d18'

My Papi Jerry
Boo 14h56'32"31d48'

My Perfect Lady- Heather C
Cyg 19h56'59"44d41'

My Pet
Cam 6h0'24"68d48'

My Peter
Per 2h57'1"50d9'

My Pooh Bear
Vul 19h38'38"27d45'

My Precious Amelia
Crt 11h14'56"-10d15'

My Precious Angel Natalie
Aql 19h41'47"10d41'

My Precious Babe Kelly
And 2h6'16"37d47'

My Precious Baby Kenneth
Aql 19h48'50"11d18'

My Precious FG Today & Always DL
Dra 16h36'45"62d30'

My Precious Mijo Nikko
Aql 19h46'1"10d54'

My Pretty Lady Fran
Peg 21h29'1"24d53'

My Princess
Cyg 19h32'25"30d51'

My Princess Colette
Ari 1h55'13"15d4'

My Princess Leslie Ann
Lyn 7h33'18"50d59'

My Princess Lucia
Cyg 21h36'0"42d6'

My Princess Miss Cosgrove MLBRB
Lyr 19h14'40"35d12'

My Prissy Girl
Peg 21h56'51"28d33'

My Puppy
Del 20h54'30"30d36'

My Queen Colleen
Aqr 21h30'14"-6d50'

My Rainman
Her 17h0'1"40d13'

My Romance
Her 16h6'0"42d6'

My Rudy
Her 15h50'0"40d38'

My Rudy
Boo 13h39'42"22d20'

My Sandra Lynn
And 1h25'1"48d59'

My Sandra's Star
And 23h32'11"44d48'

My Sheree Amour
Lyn 6h37'13"58d58'

My Sherri
Com 12h16'16"20d50'

My Shining Light
And 0h16'35"32d0'

My Shining Love Trish
Lib 15h16'52"-23d54'

My Shining Star Anthony
Dra 16h22'21"66d44'

My Shining Star "William"
Her 16h54'51"32d55'

My Shining Star "Kris"
Lyn 9h3'1"36d54'

My Shining Star Millie
Lyn 9h3'0"35d31'

My Shirley
Cas 23h3'50"54d6'

My Significant Other
Crb 15h58'28"26d58'

My Sister Carol
And 0h57'14"37d24'

My Skipper
Dra 19h2'59"65d32'

My Son Robert
Her 16h21'50"42d25'

My Soul Mate Kirk
Per 3h9'23"40d28'

My Soul Mate Lynne
Cas 23h29'15"60d44'

My Soul Mate-TGD
Aur 5h4'16"38d59'

My Spaceman Merlin
Peg 22h17'0"4d46'

My Special Martin
Per 3h6'12"31d49'

My Split-Apart David
Per 2h58'27"40d58'

My Star
Uma 10h31'40"70d17'

My Star
Cyg 21h37'1"41d17'

My Star Wood
Lac 22h50'41"53d15'

My Star,My Love,My Marianne
Ori 6h2'0"8d37'

My Stefan
Per 3h51'58"38d57'

My Stellar Epitaph
Aqr 22h0'53"-16d8'

My Stephanie's Superstar
Gem 7h58'11"30d40'

My Steve, My Inspiration
Dra 19h54'47"67d35'

My Stuff Forever
Lyr 18h22'1"45d50'

My Sunshine
Ori 5h44'1"-5d45'

My Sweet Andy
Her 16h58'21"40d17'

My Sweet Baby Boy
Vul 19h23'1"22d53'

My Sweet Baby Doll Grayce
Cyg 19h28'1"32d48'

My Sweet Jane
Cas 2h25'56"65d19'

My Sweet Lorraine
Gem 6h58'39"14d50'

My Sweet Louise
Lyr 19h16'25"40d57'

My Sweet Luis
Aql 19h6'32"-1d41'

My Sweet Michael's Star
Aql 19h3'33"2d12'

My Sweet Sally
Cyg 21h31'32"38d26'

My Sweet Star Helen
Umi 13h10'18"76d45'

My Sweet Valentine, Michael
Uma 9h49'17"71d35'

My Sweetheart Beverly's Star
Psc 1h3'38"18d10'

My Sweetheart, Assuntina
And 23h15'27"40d15'

My Sweetie,Jonathan
Cnv 13h51'1"32d28'

My Ten Little Heavenly Bodies
Uma 9h49'39"55d54'

My Texas Tom
Aql 19h2'53"-5d38'

My Texas Tornado
Cet 3h15'23"9d28'

My Tom, "Forever Cherished"
Per 1h39'24"53d10'

My Toots
And 0h30'60"38d26'

My True Campanion
Lac 22h31'22"48d53'

My True Companion Forever And A Day
And 2h24'33"44d51'

My True Love
Sco 17h52'39"-31d46'

My True Love
Uma 9h7'31"56d20'

My True Love
Her 16h40'20"51d5'

My True Love-143
Aur 6h31'1"38d48'

My Turtle's Heart
Umi 16h25'50"70d52'

My Twinkle M & M
Cma 6h53'53"-16d37'

My Two Grandparents
Boo 14h44'25"36d26'

My Unforgettable Harv
Cnc 8h38'53"18d17'

My Universal Flow of Life-CHC
Lmi 10h0'39"28d54'

My Valentine Baby Eileen
Mon 7h1'16"-6d49'

My Valentine Tammy
Mon 6h53'18"-1d17'

My Valentine, Jill-Marie
Peg 22h4'32"25d18'

My Very Special Cousin-GNL
Cam 6h9'14"80d8'

My Very Special Mother Helen
Peg 22h25'0"3d46'

My Vinny
Aur 5h30'13"31d2'

My Wedding Star
Lib 15h16'0"-28d14'

My Wonderful Dad-Ray
Per 2h20'32"54d25'

My Wonderful Kathleen
Cas 3h4'45"61d0'

My Zeta
Uma 9h26'39"68d59'

My-Angel-P S
Lyr 19h1'54"35d23'

My-e-gun Mis-kee-sick (Wolf's Eye)
Cep 22h54'27"59d16'

MY-STARR
Sge 19h54'18"16d0'

My-Von
Ori 6h2'23"0d21'

Myarg,Anja
Umi 14h5'37"71d11'

Myatt,Barbara
And 23h8'34"42d2'

Myaud
Aur 6h23'28"34d53'

Myer,Casey Todd
Aur 6h18'14"37d37'

Myer,Diann
Eri 3h29'31"-6d3'

Myer,Leona K
Lyr 19h16'55"40d32'

Myers Family,Joe & Chris
Mon 7h0'31"5d14'

Myers Pride
Aql 20h12'33"11d16'

Myers Star,The Sarah
Cyg 20h15'46"39d2'

Myers,Aaron & Jennifer
Cyg 21h0'23"30d38'

Myers,Adam Christian
Equ 20h58'13"10d44'

Myers,Alvin
Boo 14h41'40"20d19'

Myers,Anna
Cas 1h19'36"60d4'

Myers,Annie
Lyn 8h42'0"43d50'

Myers,Arleen
Aql 19h43'57"14d51'

Myers,Ashley Marie
Cas 1h42'59"60d28'

Myers,Beulah
Aur 5h8'0"41d8'

Myers,Brian E
Cep 0h50'11"78d52'

Myers,Carl
Hya 8h12'48"-0d50'

Myers,Charles Earl
Cep 22h10'11"61d24'

Myers,Ciro
Boo 14h54'51"52d30'

Myers,Clifford F
Aur 5h18'29"42d51'

Myers,Daniel H
Aur 6h27'44"38d9'

Myers,Dorothy
Peg 21h28'18"34d46'

Myers,Elle Kathryn- Forbes
Lyn 7h48'12"51d46'

Myers,Emmye Elizabeth
Lyn 9h3'44"39d45'

Myers,Grover H
Dra 19h26'58"71d3'

Myers,Helen Pauline "Tootsie" Shaver
Cyg 21h52'53"53d53'

Myers,Janet Yvonne
And 2h7'1"38d20'

Myers,Jared Edward
Dra 16h42'15"69d0'

Myers,Jason Collette
Her 17h33'0"28d16'

Myers,Jean Marie
Uma 12h40'51"61d46'

Myers,Jeff & Alison
Crb 15h58'50"34d25'

Myers,Jimmy
Cyg 21h35'26"42d46'

Myers,John Douglas
Cep 21h4'57"61d18'

Myers,John Trotter
Aql 18h56'12"17d28'

Myers,Joseph Alan
Vul 19h47'16"28d9'

Myers,Jr,Aubrey James
Aql 20h21'11"2d49'

Myers,Jr,Francis Arthur
Aur 7h12'28"40d15'

Myers,Jr,John Clark
Aur 6h28'0"38d24'

Myers,Julie Dunseth
Crb 16h22'15"31d54'

Myers,Julie Lynn
Gem 6h43'31"15d10'

Myers,Karen Marie
Lyn 8h32'26"40d56'

Myers,Kimberly MarieAnge
Gem 6h40'1"35d27'

Myers,LG "Buck"
Cep 22h19'37"56d3'

Myers,Martha Yowler
And 2h33'1"39d16'

Myers,Robert William
Umi 15h19'51"69d45'

Myers,Russell
Ser 18h4'0"-14d15'

Myers,Ruth Potter
And 0h47'35"35d2'

Myers,Samantha & Craig
Peg 23h36'1"10d12'

Myers,Sarah Bayla
Del 20h13'16"15d30'

Myers,Tiffany Anne
Mon 7h2'57"-6d40'

Myers,Tim & Barbara
Equ 21h6'37"11d32'

Myers,Van
Leo 11h47'44"22d54'

Myers,Violet
Lyn 7h48'19"51d30'

Myers,Wanda K
Cas 1h15'0"64d35'

Myers,William Robert
Cet 0h41'59"-2d1'

Myers-Gebhardt
Cyg 20h4'40"41d6'

Myginamy
Umi 13h29'25"70d14'

Myhill,Anne
Mon 6h54'56"-10d52'

Mykala
Eri 2h49'15"-2d1'

Mykkänen,Raimo
Umi 13h44'37"70d6'

Myklebust,Rita
Lyn 8h12'1"47d27'

Mylah
Dra 16h27'0"63d42'

Mylan,Steven Christopher
Ari 2h24'46"12d9'

Myler,Catrin
Lyn 8h27'10"42d14'

Myles,Bonnie E L
Mon 6h54'0"-10d44'

Mylinh
Lac 22h42'57"54d0'

Mylrea,Jane Marie
Cas 0h36'23"63d25'

Mylse
Cas 23h27'16"60d33'

Mymick
Crb 15h18'0"30d58'

Mynhier-Pieschel
Crb 16hh0'14"30d12'

Mynra
Uma 12h50'1"60d6'

Myo
Lyr 19h13'42"42d51'

Myone
Ori 4h59'30"7d47'

Myoung Sook Lim,My Love Forever
Cnv 12h31'56"41d8'

Myra & Elliott Forever
Ori 5h5'0"10d49'

Myra Daphne V Sta Iglesia
Per 3h12'18"40d35'

Myra Louise
Mon 6h21'50"7d32'

Myrenne,Michelle Christel
Ser 16h0'17"0d54'

Myriabelle
Cyg 20h48'30"38d9'

Myriam
Lyn 8h40'52"45d18'

Myriam LW 20-40
Boo 15h17'0"42d5'

Myriam und Florian
Cep 22h2'54"60d47'

Myriame Et Frederic
Cam 6h3'14"67d37'

Myrian
Lac 22h27'42"40d29'

Myrianne
Per 2h59'40"32d24'

Myrick,Benjamin Alfred
Aur 6h43'0"38d46'

Myrick,Jordan Lee
Peg 21h57'27"36d2'

Myrick,Matthew Dewayne
Ser 16h1'28"0d32'

Myrna Joy
Aql 19h57'59"11d31'

Myron My Love
Lac 22h31'59"40d28'

Myrone
Lib 15h42'43"-8d36'

Myrtiou G'earree
Uma 12h2'0"37d41'

Myrtle Estelle
Lyr 18h31'0"32d42'

Myrtle,Diane
Com 13h2'53"20d24'

Mysam
Ori 5h56'13"21d1'

Myslinski,Peggy
Peg 21h57'60"26d45'

Mysterious Mark
Boo 15h6'12"23d42'

Mystic
Uma 8h52'46"51d37'

Mystic J
Cet 3h13'11"0d56'

Mystic Moonbeam
Sct 18h48'27"-6d24'

Mystical Magic
Cyg 21h1'22"40d30'

Mystical Missie
Cas 1h53'21"61d25'

Mystérieuse
Uma 10h30'11"56d0'

Myszczak-#1 Teacher in the Universe,Mrs
Uma 11h4'13"57d46'

Myszkowiak,Mary Ann
Mon 6h21'37"3d49'

MYTJAMR
Lmi 9h55'1"37d49'

Mäkelä,Satu
Umi 13h12'11"71d59'
Mäkinen,Hannele
Cas 1h19'22"64d57'
Mäkiö,Eero
Dra 20h10'15"68d57'
Mäntymaa,Mavno
Ori 5h55'45"12d36'
Märkle,Leni
Oph 18h5'29"8d38'
Mäusekind
Sge 18h58'0"19d60'
Mäuserich
And 23h3'46"41d9'
MÜgÜrd,Lene
Cam 3h30'12"60d49'
Médard,Annie
Tri 1h58'52"33d50'
Méghane
Cas 0h1'56"61d32'
Mélanie
Umi 13h9'49"71d30'
Mélanie
Cyg 20h50'42"38d8'
Mélanie
Uma 11h59'17"40d31'
Ménard,Franáois
Uma 11h57'51"42d32'
Möhring,Ralph
Del 20h21'31"10d26'
Möller
Peg 23h27'55"10d17'
Möller,Carola & Thomas
Uma 10h54'45"67d50'
Möller,Hans M
Her 18h20'31"12d28'
Möller,Lutz
Aqr 22h44'37"-1d28'
Mönnig,Hans Ulrich
Dra 18h28'21"67d54'
Mönnig,Kornelia
Vir 12h4'52"-6d12'
Mörl,Jiona geborene Ensin
Ari 2h40'31"26d4'
Mössnang,Josef
Aql 18h56'46"17d59'
Möst,Thomas Mercator
Cep 23h18'37"65d10'
Möstel,Dietrich und Christine
Dra 17h42'42"85d2'
Mövenpick,Mutter
And 23h10'43"40d11'
Mü
Aur 6h23'25"33d12'
Mücke,Rainer
Leo 9h57'28"18d46'
Müffelmann,Karsten
Umi 14h7'12"68d34'
Mühape
Gem 6h38'41"31d6'
Mühlbauer,Josef
Cyg 20h25'1"30d4'
Müjde
Uma 10h5'57"57d54'
Müller,Alexander Dr
Sco 17h5'17"-30d18'
Müller,Anki
Sco 16h13'40"-22d2'
Müller,Christa
Cnv 12h16'0"35d30'
Müller,Christel und Heinz
And 1h42'45"41d58'
Müller,Christian
Cap 20h35'1"-10d15'
Müller,Elke
Dra 11h52'45"72d53'
Müller,Erich Emmerich Franz
Gem 6h39'17"34d12'
Müller,Erika Margarethe
Martina
Cap 21h41'39"-11d58'
Müller,Erika
Cam 5h43'38"70d11'
Müller,Frank-Mario
Lib 14h21'37"-22d43'
Müller,Franz X
Lyr 19h12'0"31d6'
Müller,Fritz-Konrad
Aqr 20h58'12"-10d34'
Müller,Georg
Lmi 10h45'0"25d47'
Müller,Gerd-Gustav
Uma 9h31'41"45d29'
Müller,Gerhard
Sco 16h58'55"-40d37'
Müller,Herr
Cmi 7h18'1"3d55'
Müller,Ilse
Tau 5h47'14"23d13'
Müller,Iris
Cas 1h49'36"73d27'
Müller,Jörg
Dra 10h5'21"80d5'
Müller,Laurens Melchior
Dra 14h35'11"62d24'
Müller,Lothar
Aqr 23h32'1"-8d56'
Müller,Lothar
Uma 10h30'33"48d30'
Müller,Lucca Jeremy Yves
Per 4h5'57"36d20'
Müller,Ludwig
Sgr 19h3'21"-22d40'
Müller,Manfred
Boo 14h14'14"17d45'
Müller,Maria geborene Berger
Cap 20h29'49"-26d59'
Müller,Olaf
Ser 15h40'52"4d11'
Müller,Scharnhorst
Tau 3h59'58"20d24'
Müller,Sonja
Cmi 7h20'1"4d17'
Müller,Sylke
Her 17h58'39"42d27'
Müller,Thomas
Peg 23h35'29"17d57'
Müller,Wolfgang
Uma 9h29'40"47d29'
Müller-Rüttinger,Jana
Gem 6h49'28"18d30'
Müllmaier,Silvia und Bernd
Ser 15h32'32"18d4'
Münch,Anneliese und Hans-Peter
Sco 16h39'1"-44d11'
Münch-Arke,Agnes
Gem 6h42'32"13d33'
Münstermann,Sabine 14 Juli 1995
Ori 5h59'1"19d56'
Münzberg,Erika
Tau 5h53'59"26d51'
Müschen,Otto
Cep 22h5'53"60d46'
Müssig,Ferdinand Jun
Sge 19h6'26"17d42'
Mütschard,Gisèle
Cam 5h57'53"58d13'
Mårez,Gloria
Aql 19h43'47"10d37'
MØller-Vetlov,Claus
Cas 1h31'37"64d45'
MØller-Vetlov,Jeanette
Cyg 19h21'36"54d29'
M'
Uma 10h1'22"59d11'
M' Anniversary
Ori 5h51'13"17d14'

N

N
Cmi 7h35'28"11d26'
N & D Star 11/12/94
Mon 6h49'1"11d26'
N & J Forever Shining
Cyg 19h37'47"37d53'
N C Acceleratus I
Ori 5h25'52"-5d42'
N J
Aql 20h13'24"3d55'
N L P
Uma 10h9'54"48d48'
N O L A
Crb 16h3'40"28d14'
N R B S
Hya 9h0'42"4d44'
N'Dar
Cet 2h55'55"7d56'
N-Nagy
Cep 1h22'1"80d27'
N.E.D.
Dra 18h59'0"65d7'
N7RH
Lac 22h25'53"56d35'
Na Hoku O'Kalani
Oph 17h1'19"-18d60'
Na's Hakuna
Eri 4h12'0"-13d52'
Nabarro,Ariane
Cas 23h42'23"61d41'
Nabatoff,Jacob Stewart
Ori 5h25'44"-2d11'
Nabisco-Toledo Flour Mill
Uma 9h57'42"42d24'
Nabors,Eric Southworth
Psc 23h6'1"5d50'
Nabrass
Ser 16h8'0"1d10'
NABUTCO
Per 3h3'28"50d20'
Naccarato,Sandra Francesca
And 1h0'53"36d30'
Nace,Suzanne
Cam 4h27'1"68d12'
Nacham,Bella & Abe
Mon 7h3'43"0d56'
Nachman,Joshua "Sky King"
Boo 14h37'23"39d24'
Nachtigal,Jane Francesca
Cyg 19h33'36"34d55'
Nachtigall,Gunther-The Brightest
Tau 5h45'58"21d25'
Nachtigall,Margaret
Hya 8h37'28"-0d42'
Nachtmann,Lothar
Hya 9h17'40"-9d41'
Nack,Horst-Walter
Umi 15h48'50"72d17'
Nack,Vivian
Cas 1h50'22"61d0'
Nackley,Elinor & George
Cam 6h10'45"70d34'
Nacson,Cynthia Ellyn
Hya 9h3'52"1d49'
Nada
Sgr 19h2'12"-21d9'
Nadal
Dra 15h10'1"62d41'
Nadalini,Jr,James Taubert
Peg 21h29'0"24d56'
Nadean
Aql 20h20'0"1d28'
Nadeau,Barbara J
Cas 0h30'1"61d33'
Nadeau,David G
Eri 4h2'11"-15d7'
Nadeau,David Michael
Oph 17h52'31"12d9'
Nadeau,Dorothea W
Cas 1h46'11"58d58'
Nadeau,George A
Oph 17h33'24"-22d37'
Nadeau,Ryan David
Aur 6h13'32"32d48'
Nadeau,Steven E
Per 2h9'1"56d40'
Nadel,Stanley
Boo 14h27'44"37d58'
Nadell,Mike
Lmi 10h23'58"33d44'
Nadelman,Sheldon & Phyllis
Eri 4h56'55"-10d24'
Nadene & Drew
Boo 13h38'25"16d16'
Nadette
Vel 9h20'12"-46d32'
Nadia
Cas 23h34'13"60d9'
Nadia
Boo 15h5'1"37d43'
Nadia
Boo 13h52'1"16d46'
Nadia
And 23h44'26"41d37'
Nadia
Vir 12h57'30"-20d51'
Nadia
Vul 19h45'1"20d36'
Nadia & Alan
Cyg 19h43'19"29d32'
Nadia & Alexander - Forever
Vir 13h9'55"-1d24'
Nadia & Sandro
Peg 23h42'37"11d22'
Nadia C
Uma 11h0'24"58d2'
Nadia Dürr
Ori 5h57'36"15d37'
Nadia Jane
Ori 5h57'25"21d4'
Nadia Joan
Ori 6h2'0"-2d43'
Nadia, Sweetie Darling
Mon 6h32'55"11d19'
Nadig,Barbara Marianne
Peg 23h38'23"30d34'
Nadine
Cas 2h33'53"65d7'
Nadine
Cas 23h2'47"58d10'
Nadine
Lyn 8h37'35"43d9'
Nadine
Cet 2h40'31"9d42'
Nadine
Dra 12h33'46"64d22'
Nadine A M
Aur 4h59'30"40d16'
Nadine's Star
Lyr 18h38'39"38d15'
Nadino 9-26-86,Anthony D
Cep 20h54'51"61d4'
Nadison,Danielle Tuite
And 0h58'42"37d18'
Nadja
Ari 2h43'45"28d29'
Nadler,Hannah
Cas 1h11'13"60d5'
Nadler,Jean
Lyr 19h25'11"38d15'
Nadolski,Spanky
Lac 22h25'39"52d36'
Nadon,Jacqueline & Jean Guy
Uma 10h7'11"51d27'
Nae Kim
Mon 7h4'0"4d41'
Naebyhtomit
Crt 11h19'7"-19d1'
Naegel,Jean-Jacques
Cyg 20h28'42"31d26'
Naert,Farah
Cyg 20h29'16"42d23'
Naftali R
Lib 14h57'0"-18d55'
Nagakawa,Eilo
Umi 16h4'26"71d35'
Naganalf
Cam 6h12'59"67d45'
Nagaoko,Nobuko
Aql 18h59'59"-2d33'
Nagasawa,Taki & Haru
Ori 5h37'34"11d52'
Nagatani,Joanie
And 2h23'46"42d40'
Nagel,Carl & Lillian
Aql 20h2'35"8d8'
Nagel,Frank
Per 4h23'22"50d2'
Nagel,Günter
Gem 6h41'39"33d51'
Nagel,Horst
Equ 20h58'36"10d39'
Nagel,Ingrid
Uma 10h54'40"61d27'
Nagel,John & Renee
Umi 15h9'36"68d4'
Nagel,Mary Jane
Cmi 7h28'40"7d32'B
Nagel,Randy
Sge 19h29'53"17d11'
Nagel,Ronald John
Cet 2h30'60"-0d24'
Nagel,Sharon
Peg 22h2'47"27d7'
Nagel,Stephen
Cmi 7h28'40"7d32'A
Nagel,Theodore Charles
Lac 22h2'17"38d27'
Nagel-Peterson
Cnc 8h32'60"32d28'
Nagel-Tornau,Philipp- Peter
Uma 10h39'1"41d48'
Nagelmann,Eric
Ari 1h58'18"11d25'
Nagengast,Joe
Cmi 8h4'40"1d32'
Nager,Jordan Serlig
Aur 6h14'58"31d20'
Naggle,Peggy R
Equ 21h22'1"10d54'
Naghmeh 143
Cas 2h1'55"60d27'
Nagi
Uma 10h29'28"51d16'
Nagi,Catherine Rose
And 1h24'58"40d22'
Nagle Love Je,Mary Catherine
Cas 22h57'1"55d35'
Nagle,Blondie
And 1h24'44"38d14'
Nagle,David J
Mon 6h20'56"5d28'
Nagle,James Robert
Aqr 23h1'19"-6d26'
Nagle,Joanne E
Aur 6h24'0"40d26'
Nagle,Leigh
Cas 0h42'50"66d5'
Nagle,Nicholas A
Hya 9h4'43"1d11'
Nagle,Patrick
Uma 11h22'0"38d55'
Nagle,Sallie M
Del 20h59'26"18d56'
Nagle,Steven Michael
Per 3h33'52"38d42'
Nagler,Thomas H
Her 18h19'38"18d47'
Nagorski,Diane T
Aql 19h29'58"-7d45'
Nagy Antalné-Urbán Mária
Peg 22h6'42"4d1'
Nagy,Diana M
Cas 3h9'27"58d43'
Nagy,Gabor
Lac 22h53'27"56d26'
Nagy,Randy Sir Hossalot Michael
Aur 5h1'0"40d39'
Nagy,Shirley Ann
And 2h10'56"50d19'
Nagy-My 100%,Lawrence Stephen
Per 2h4'49"57d53'
Nahani
Uma 11h35'57"46d32'
Nahin,Carol
And 0h49'54"33d44'
Nahla Abdel Azim- Weskamp
Ari 2h34'47"30d42'
Nahmias
Crb 16h8'51"27d30'
Nahmias,Dave & Molly
Aql 19h12'59"10d44'
Nahooikaika,Jaret-Levi Keliikinui
Cru 12h27'42d-61d32'
Nahoun,Jeanette Janice
Uma 9h21'48"51d12'
NAHRAJULIA
Cas 0h28'0"61d46'
Nahte
Aql 19h6'28"3d50'
Nahum,Maurice
Per 2h55'14"37d31'
Naidrich,Phil & Brenda
Crb 16h22'43"30d7'
Naigles,Elinor
Com 16h12'0"30d7'
Naigles,Nancy Liraz
Uma 11h41'21"53d20'
Naigles,Paris Ann
Cyg 21h31'32"30d59'
Nail,Angela
Lyr 18h44'54"42d19'
Nail,Jackie
Hya 8h57'25"4d2'
Naila
Ori 4h57'15"5d54'
Nailor,Donald L
Aur 6h24'50"30d20'
Nailor,Irene
Aqr 23h36'20"-5d37'
Nails,Odell
Cam 14h13'16"84d44'
Naiman,Kristen
And 0h13'25"31d28'
Nain,Michael & Jeanne
Cnv 13h47'1"31d36'
Nair,André N
Cep 21h51'16"60d23'
Naisbitt's,Patti & John
Uma 10h29'28"58d39'
Naito-Chan,Edna T
Cam 4h20'43"68d14'
Naitre
Sco 16h16'21"-8d58'
Najafabadi,Dr Ali
Oph 17h35'36"1d7'
Najarian,Allison Kathryn
Del 20h51'29"7d45'
Najarian,Andrew
Dra 18h12'11"58d43'
Najarian,Carolann & George
Cam 3h58'42"56d20'
Najarian,Christine Powers
Lmi 10h44'18"27d29'
Najarian,Dikran & Stavroula
Lac 22h17'19"35d34'
Najarian,Rachel Catherine
And 2h15'18"37d3'
Najdenova,Ljudmila Ljubova
Cnv 13h23'44"50d25'
Najera,Terry
Cyg 21h16'11"28d58'
Najette Et Frederic
Com 12h27'53"21d31'
Naji Bou-Hamdan XX=Love Lois
Aql 19h4'17"-1d17'
Najjar,Wadih El
Ori 6h5'51"0d13'
Najmi 458
Dra 18h31'44"58d53'
Najran,Vinty
Ori 5h57'43"16d37'
Najy,Marie Helene
And 1h6'10"41d11'
Nakagaki,Andrea
Lac 22h49'37"38d45'
Nakama,David
Her 17h23'16"49d1'
Nakamura,Masae
Aqr 21h58'11"0d50'
Nakamura,Shigeyuki
Lib 14h42'51"-8d22'
Nakanishi,Toshihiro
Ori 5h55'35"20d52'
Nakauchi,Mr T 4 August 1996
Leo 11h33'55"-0d2'
Nakazawa,Joshua
Gem 6h51'48"30d11'
Nakfoor,Madeline
Cyg 21h46'30"38d26'
Nakoa
Peg 22h8'55"13d50'
Nalan,Nothan
Aur 6h33'25"34d54'
Nalbant,Belgin & Java
Cyg 21h15'15"28d45'
Nalepa,"Steffie" Jasinski
Lib 14h26'0"-23d56'
Nalepka,Brenda
Cyg 20h57'31"30d42'
Nalepu,James & Katy
Cyg 20h53'54"31d52'
Nalinee Nippita
And 0h18'31"32d9'
Nalinikanta dasa - Ratnesvari dasi
Leo 9h32'0"7d46'
Nall,Andrea Lynn
Eri 2h50'43"-1d43'
Nall,Doc
Oph 18h28'14"7d59'
Nall,Gerald Scott
Hya 9h1'38"-6d43'
Nall,Lauriana Jane
And 2h28'28"50d7'
Nall,Nancy A
Peg 23h25'11"33d51'
Nall,Ruth
Lyr 19h25'51"38d41'
Nalle
Mon 7h1'32"-6d49'
Nalluri,Purna
Del 20h18'0"9d27'
Nally,Melinda
Her 17h36'28"14d52'
Nalpak,Hubert
Aur 6h18'28"38d36'
Nalven,Goldie
Cam 4h56'23"68d9'
Namaste
Aur 5h27'0"50d7'
Namasté
Peg 22h44'32"12d27'
Namasté
Her 17h3'13"42d55'
Namasté Angel
Cas 0h8'30"62d12'
Namasté,Mitch
Ser 16h1'55"14d49'
Namba,Hiroyuki
Uma 11h14'33"38d4'
Namer,Sarah
Peg 22h56'1"8d13'
Namie
Mon 6h28'35"10d13'
Namo
Lib 15h0'0"-20d8'
Nan
Uma 11h20'50"42d30'
Nan
Cas 2h50'38"65d20'
Nan & Bob
Cyg 21h42'1"28d26'
Nana
Aql 20h0'52"10d20'
Nana
Eri 3h33'2"-12d13'
Nana
Lyn 7h54'37"50d47'
Nana
Ori 6h17'34"-1d30'
Nana
Lyr 18h49'29"30d7'
Nana
Cnc 8h50'47"17d11'
Nana
Aql 18h46'24"11d55'
Nana & Gumpy
Mon 7h4'43"-6d26'
Nana J's Star
Lyr 18h30'17"41d5'
Nana Loretta
Equ 21h2'29"11d44'
Nana '72
Cas 1h37'46"63d55'
Nana's Star
Ori 6h0'46"-1d3'
Nanapuggy
Aur 5h29'0"50d6'
Nance Elizabeth
Mon 6h23'41"8d34'
Nance My Shining Star
Niece,Cherice Dawn
Eri 3h59'0"-16d26'
Nance's Star
Cas 23h14'48"61d32'
Nance,Christopher B
Cet 2h50'28"3d2'
Nance,Emily Elizabeth
Eri 4h1'28"-13d51'
Nance,June Ann
Sct 18h43'49"-6d10'
Nancito
Mon 6h20'35"-6d12'
Nancwa
Cmi 8h4'1"6d34'
Nancy
Cam 7h47'34"70d10'
Nancy
Lyr 19h15'0"41d18'
Nancy
And 0h48'59"37d55'
Nancy
Lyn 7h32'47"58d39'
Nancy
Cyg 19h50'42"40d26'
Nancy
Ori 6h1'20"10d15'
Nancy
Vel 9h19'13"-45d47'
Nancy
Sge 19h52'59"18d57'
Nancy
Mon 6h23'32"10d1'
Nancy "My Little Buddy"
Lmi 10h15'1"30d60'
Nancy & Bob
Sge 19h52'38"16d23'
Nancy & Joe Forever In Harmony
Lyn 7h37'40"37d14'
Nancy & Nizar
Cyg 21h1'58"37d52'
Nancy & Rob
Her 16h37'1"26d28'
Nancy & Rod Forever
Cyg 19h34'37"36d4'
Nancy Ann
Ori 5h16'0"10d54'
Nancy Ann
Crb 16h0'29"33d15'
Nancy Ann
Cas 23h25'15"53d18'
Nancy Ann
And 1h40'55"40d41'
Nancy Ann's "Lucky" Star
Mon 8h3'9"-10d9'
Nancy Anne
Lyr 19h5'1"26d45'
Nancy Beth
And 0h49'16"40d10'
Nancy Carol
Mon 8h8'60"-10d11'
Nancy Darlin'
Sco 17h33'36"-43d39'
Nancy Ellen
Ori 5h56'40"16d22'
Nancy Eloise Diana
Sgr 19h0'8"-20d47'
Nancy Grace
Tau 4h20'1"15d22'
Nancy J
Peg 22h25'21"31d23'
Nancy Janet
And 2h3'37"45d0'
Nancy Jay
Cas 2h8'40"59d33'
Nancy Jean
Uma 14h27'1"59d45'
Nancy Jean-The Mother Star
Mon 7h30'20"-5d46'
Nancy Lee
Peg 22h22'36"21d54'
Nancy Lucy
Cas 1h40'30"67d33'
Nancy Lynn
Vul 19h38'59"20d29'
Nancy Rose
Cnv 12h49'54"48d13'
Nancy Sue
Peg 22h39'59"32d18'
Nancy's Bill Through Eternity
Peg 23h31'40"18d6'
Nancy's Billystar
Eri 3h21'22"-6d45'
Nancy's Circle Of Light
Mon 6h20'1"5d56'
Nancy's Nocturne
Crb 15h37'57"26d14'
Nancy's Number
Cas 0h37'12"64d35'
Nancy's Number 1
Uma 12h37'0"56d14'
Nancy's Reverie
Uma 10h49'43"70d29'
Nancy's Spirit
Tau 4h16'21"24d7'
Nancy's Star
Cas 0h5'11"58d26'
Nancy's Star
Lib 15h17'42"-8d14'
Nancy,Anne
Eri 3h50'40"-4d52'
Nancy,G
Mon 6h54'16"-10d16'
Nancy-21
Cas 0h8'49"61d19'
Nancy-Forever My Love, David
Mon 7h4'25"4d55'
Nancy-Honey
Tri 2h36'44"31d17'
Nancy-Whom Ed Loves Forever
Cyg 20h11'51"37d39'
Nancylee
Tau 4h56'26"20d19'
Nanda Velue M
Aur 5h52'1"54d47'
Nando
Mon 6h19'43"7d50'
Nando
Dra 14h55'12"60d38'
Nando's Lapaloma Star
Lac 22h55'59"55d40'
Nandoo,Anderson
Ori 5h51'56"15d35'
Nandy 2005
Cam 7h58'59"61d41'
Nangle,Paul J
Umi 16h23'40"88d26'
Nangpa
Per 3h15'13"41d25'
Nani
Oph 17h21'7"-20d36'
Nani
Peg 21h44'43"28d13'

Nani I Magda
Cep 22h23'58"61d33'
Nani Makamae Tiffany
And 0h14'0"36d12'
Nani's Love
Tau 4h7'26"20d3'
Nania,Christina Virginia
And 2h18'36'46d3'
Nankervis,Mary Ann & Craig
Crb 16h3'43"32d30'
Nanlee
Uma 10h49'33"59d52'
Nanna
Nor 16h21'39"-50d35'
Nanna Elsie
Aql 19h42'0"10d32'
Nannery,Tom
Dra 16h12'1"64d20'
Nannie Grace
Cas 0h9'1"59d44'
Nanny
Cas 0h33'27"63d20'
Nanny
Cas 1h45'43"73d31'
Nanny Alice
Cas 1h4'60"56d8'
Nanny's Evan
Cas 3h5'0"61d46'
Nanny's Light
Eri 3h22'17"-2d22'
Nano
Cep 21h29'17"68d39'
Nano
Dra 17h0'1"60d49'
Nano 040568
Uma 11h3'29"61d35'
Nano 35
Boo 15h25'50"51d11'
Nano Spaziale- Spacedwarf H 1169
Scl 23h40'13"-28d56'
Nanook of the North
Dra 12h39'29"75d7'
Nanrus
Cyg 21h29'56"31d7'
Nante,Christian B
Dra 18h58'50"48d39'
Nantel,Linda
Cyg 19h27'32"33d48'
Nantucket
Vul 21h3'34"20d13'
Nany
Vel 9h23'35"-46d45'
Nanys,Peggie Eve
Leo 10h42'54"15d39'
Naoko
Aur 5h18'31"45d2'A
Naomi
Cep 23h12'25"70d14'
Naomi
Cas 0h29'51"61d54'
Naomi
Tri 1h45'1"30d8'
Naomi
Del 20h13'57"12d44'
Naomi & Andy
Mon 7h56'2"-2d53'
Naomi & Elvis
Cnv 13h45'19"46d46'
Naomi & Rain
Mon 7h5'43"-6d35'
Naomi Ann
And 1h16'32"37d19'
Naomi Beauty
Cnc 9h19'34"30d56'
Naomi E 4.5
Lyr 18h41'1"32d35'
Naomi Jean
Cnc 9h5'57"30d23'
Naomi Michelle
Cyg 19h25'17"31d54'
Naomi Nell
Mon 6h59'38"-1d29'
Naomi's Happiness
Aqr 22h2'35"0d49'
Naomi's Star
Umi 15h7'41"65d51'
Naomi,My Beloved Marine
Sct 18h51'16"-7d32'
Naoroz,David H
Ori 6h8'25"8d23'
Nap,Andrea Belinda Lucia
Aqr 21h58'27"0d16'
Nap,Elaine
Cnc 8h59'12"30d52'
Napau
Hor 3h11'55"-46d2'
Napier,Claire
Cas 0h29'30"60d24'
Napier,Dan "The Man"
Gem 6h58'12"13d14'
Napier,Daniel A
Per 1h47'27"48d39'
Napier,David Lee
Aql 20h5'37"8d36'
Napier,Holly Lane
Lyr 19h1'54"30d9'
Napier,Jason Douglas
Cnv 13h20'40"30d45'
Napier,Nicole Anne
Cas 0h57'25"62d58'
Napier,Victoria L
Peg 0h0'53"28d15'
Naples,Dion
Lac 22h42'30"37d35'
Napoleon,Luigi Rose
Aur 6h54'24"40d36'
Napoleon,Lynne M
Cas 0h38'13"58d8'
Napoleon,Tony
Per 4h9'58"38d60'
Napoli,Alyssa Marie
Cas 23h45'30"48d53'
Napoli,Danny
Aur 6h9'0"38d54'
Napoli,Michael & Jean
Uma 10h47'1"47d35'
Napoli,Rose Marie
Vir 11h39'0"9d29'
Napoli,Stephen & Judy
Cas 1h48'1"58d31'
Napoliello,Cameron Wayne
Cnc 9h9'27"32d12'
Napoliello,Theresa
And 22h58'21"40d7'
Napoliello,Wayne Anthony
Leo 10h56'41"22d20'
Napolitan,Michele
Lac 22h25'37"54d10'
Napolitano,Anthony Paul
Uma 16h41'48"78d42'
Napolitano,Anthony
Aur 6h3'54"33d9'
Napolitano,Carly Rose
Lyr 18h58'1"37d28'
Napolitano,Daniel Peter
Ori 5h56'47"15d6'
Napolitano,David Orazio
Boo 14h34'45"43d57'
Napolitano,John & Anne
Aur 4h49'0"38d37'
Napolitano,Justin
Uma 10h9'12"53d23'
Napolitano,Nick
Dra 14h3'56"65d13'
Napolitano,R J
Lac 22h14'35"51d33'
Napolitano,Reneé
And 23h1'21"51d36'
Napolitano,Ryan Christopher
Her 15h55'22"41d21'
Napolitano,Vanessa Maria
Cas 0h43'16"66d55'
Nappa,Rosaria Marfella
Mon 6h58'1"-6d34'
Nappe III,Anthony
Ori 5h23'51"-4d45'
Nappen,Audrey I
Cyg 19h56'14"38d42'
Nappez,Christophe
Cep 23h11'1"68d2'
Nappi Love Forever
Mom,Danny M
Cep 23h27'13"65d55'
Napua
Cet 2h34'38"5d35'
Nara
Cyg 19h59'31"47d42'
Nara
Tau 4h2'24"11d35'
Narai
And 23h31'51"44d8'
Narang,Gunda
And 1h4'49"40d31'
Narayan Star,The Jane
Cas 2h35'49"65d25'
Narayanan,Jack
Gem 7h17'23"21d51'
Narbaits,Jauregay Daniel
Boo 14h35'24"11d53'
Narbik-306
Del 20h22'26"2d53'
Narcisco,Ava
Cas 0h45'15"60d35'
Narcisi,Lara
Lyn 7h56'15"38d39'
Narciso,Michael
Sct 18h49'36"-7d8'
Narda
Ori 5h53'17"7d38'
Nardelli,Elizabeth Ann
Ori 5h33'51"-0d23'
Nardelli,Miss Mary
Vul 20h16'23"23d4'
Nardi,Jason Taylor
Dra 14h46'48"62d21'
Nardi,Joan
Uma 11h51'43"43d9'
Nardi,Luisella
Cep 23h34'0"64d11'
Nardick,Krystina Tamara
And 0h24'42"37d31'
Nardiello,Jason
Her 17h36'53"27d8'
Nardino
Sco 17h27'38"-30d24'
Nardo,Nickole Danielle
Cyg 19h41'56"28d51'
Nardone,Margaret
Cas 1h25'19"50d15'
Narducci,Angela
And 23h1'1"44d35'
Narducci,Diana Little Girl
And 22h58'0"50d4'
Narducci,Frankie
Uma 9h7'42"70d32'
Narducci,Julia
Uma 9h8'15"71d27'
Narducci,Pasquale
Uma 8h39'21"70d44'
Narduzzi,Mark A
Dra 15h58'24"68d31'
Nareau,Daniel Ernest
Boo 15h4'1"13d25'
Narendra III
Uma 11h6'33"43d5'
Narens Star,The
Dra 16h22'48"60d8'
Nares,Rafael Perez
Tau 5h46'36"27d44'
Nargaux Hubert
Oph 18h6'32"10d10'
Nargentino,Francesca
Cam 4h0'0"58d44'
Nargentino,Gabriella
And 0h16'41"33d7'
Nargentino,Isabella
And 0h13'0"39d56'
Nargisa Achmedowa und Ruslan
Sco 17h51'33"-30d12'
Narhisalo,Helena
Cas 0h22'1"58d59'
Narnia
Del 20h33'58"11d8'
Narodick,Philip H
Per 3h34'10"31d44'
Naroski,Krisha Ellen
Uma 13h47'0"52d20'
Naroski,Shayna Ellen
Boo 13h58'56"8d27'
Narrlangia
Tri 2h29'28"31d38'
Narron,Virginia
Cas 1h55'38"61d45'
Narson,Marilyn
Cam 3h56'25"60d44'
Narsus-Morrigan
Ser 15h13'29"4d52'
Naruo,Masami & Tomohiro
Cru 12h34'43"-57d48'
Narusis,Therese Marie
Cas 2h57'47"58d26'
Narvaez,Caroline R
Cam 7h41'18"60d25'
Narvell,Stephanie B
Cas 23h34'1"60d21'
Narveson,Barbara Lee
Ori 5h50'36"17d11'
Narwal,Jaswant Kaur
Ori 4h55'59"5d11'
Narzt,Karin
Cra 18h19'55"-42d20'
Nasa 1
Boo 14h4'10"26d26'
NASA Ames Research Center
Umi 14h45'34"75d36'
NASA Dryden Flight Research Center
Umi 14h44'21"76d13'
NASA Geo C Marshall Space Flight Center
Umi 14h7'49"75d52'
NASA Goddard Space Flight Center
Umi 14h44'14"77d7'
NASA Jet Propulsion Laboratory
Umi 14h39'49"77d18'
NASA John C Stennis Space Center
Umi 14h5'15"75d13'
NASA John F Kennedy Space Center
Umi 14h14'0"77d12'
NASA Langley Research Center
Umi 14h6'18"77d10'
NASA Lewis Research Center
Umi 14h9'22"76d57'
NASA Lyndon B Johnson Space Center
Umi 14h29'0"77d3'
Nasca,Mary
Com 13h16'17"28d56'
Nascentl,Domonic Paul
Her 18h14'1"47d35'
Nascimben,Maurizio
Ori 6h15'57"8d51'
Nascimento,David
Cnv 12h11'51"35d51'
Nascimento,Kimberli Suzanne
Ori 5h53'32"16d36'
Nash 2/16/65,Laurie Rae
Com 13h12'58"21d35'
Nash Elementary School
Hya 8h17'11"4d52'
Nash's Nova
Aql 19h49'0"14d11'
Nash,Barbara & Jack
Uma 10h26'20"61d14'
Nash,Bustin
Sco 16h25'23"-40d31'
Nash,Charles
Sco 16h35'32"-30d4'
Nash,Erin Jacquelyn
Eri 4h55'57"-8d27'
Nash,Haley Adams
Cet 2h55'19"2d5'
Nash,James
Aur 5h19'1"52d58'
Nash,James Edward
Cep 22h29'0"59d59'
Nash,Joseph Michael Sapitowicz
Aql 19h23'37"-0d1'
Nash,Jr,James Howard
Her 16h51'21"51d12'
Nash,Leona Knox
Oph 18h2'56"11d24'
Nash,Martin W
Tri 2h18'18"31d42'
Nash,Michael Bennett
Oph 17h28'1"0d46'
Nash,Nancy
Cyg 21h23'21"40d0'
Nash,Roberta A
Lyn 7h39'1"45d19'
Nash,Ronald K
Mon 7h4'11"-6d21'
Nash,Steve (WM7)
Cam 3h14'0"63d50'
Nash,Violet
Cas 1h58'1"63d55'
Nash,Zöe
And 1h5'26"38d52'
Nashick,Camella
And 23h42'55"42d23'
Nashick,Jerome
Aur 7h20'0"38d34'
Nashirah of Yayu & Miyu
Cap 20h57'36"-16d35'
Nashland,Maggie Elizabeth
And 1h22'36"48d50'
Nashonakee
Lac 22h25'1"37d55'
Nasilowski,Anette
Gem 7h31'53"20d45'
Naske,Martina
Aqr 20h47'52"-6d30'
Naslund,Lauren
Lyr 18h41'11"28d31'
Naso,Richard Frank Jude
Lac 22h18'45"55d7'
Naso,Stefania
Del 20h12'46"15d1'
Nason,Brett Alexander
Aur 7h4'37"35d44'
Nason,Karen Sue
Lyn 7h4'22"52d7'
Nasrin
Cam 4h13'35"68d23'
Nassaney,Mr & Mrs Richard & Melissa
Lac 22h10'45"47d43'
Nassau"L",Richard Joel
Aqr 21h56'25"-10d42'
Nasser Mohammed Nasser
Her 18h25'43"12d29'
Nasser,Jared Michael
Cam 3h56'18"74d26'
Nasser-Bailey
Umi 15h10'50"67d20'
Nassif,Michel R
Cas 23h29'44"61d30'
Nassoura,Kimberly A
Vul 19h45'34"20d28'
Nastir,Toby & Bernie
Cyg 20h6'26"40d40'
Nastre,Bibou
Cam 4h39'17"58d8'
Nastro,Anthony
Boo 14h36'44"51d12'
Nastro,John F
Dra 17h46'1"68d33'
Nasworthy,Karen
Cyg 20h19'34"31d15'
Nat
Per 4h23'15"50d10'
Nat - Stellie
Aur 5h7'55"54d23'
Nat'l Aeronautics & Space Administration
Umi 14h44'35"74d54'
Natal Midlands Centre
Cru 12h4'24"-59d24'
Natale,Anna M
Uma 8h42'17"49d41'
Natale,P J
Lmi 10h11'31"30d10'
Natale,Rocco
Lyr 18h41'47"33d18'
Natale,Sandy O'Toole
Ori 5h37'45"-0d21'
Natale,Susan
And 0h32'56"31d14'
Natale,Vanesa
Equ 21h2'23"8d0'
Natalee's Brook
Aql 18h43'47"8d57'
Natalia
Cep 1h37'0"77d54'
Natalia Beth
Cyg 19h25'17"44d29'
Natalie
Oph 17h50'15"-7d54'B
Natalie
Ari 2h35'46"30d10'
Natalie
Lyr 18h59'38"26d18'
Natalie
Cas 1h13'25"63d13'
Natalie
And 23h32'43"45d32'
Natalie
Cet 2h59'26"5d41'
Natalie
Ori 5h49'26"18d52'
Natalie
Peg 22h47'40"5d29'
Natalie
Uma 11h59'12"41d11'
Natalie & Guy
Uma 10h13'23"56d43'
Natalie & Randy
Uma 10h21'50"61d5'
Natalie & Rob Forever
Mon 6h23'0"5d10'
Natalie 12
Cnc 8h31'57"7d44'
Natalie Brianne
And 23h39'59"46d49'
Natalie Dawn
Sge 20h6'51"16d32'
Natalie Jeanne
Cmi 7h15'1"4d29'
Natalie Jeanne
Gem 7h2'59"30d5'
Natalie Julia Ann
Vul 19h42'29"20d4'
Natalie Lee
Cas 0h49'11"70d37'
Natalie Olivia
Cas 0h49'58"63d58'
Natalie Rose
Com 13h7'52"20d43'
Natalie To Pio Omorfo Asteri
And 1h48'22"37d41'
Natalie Virginia
Uma 12h11'50"61d21'
Natalie's Eahtatiene
Ari 2h54'1"30d21'
Natalie's Nova
Eri 3h21'28"-13d22'
Natalie's Star
Aql 20h34'48"-6d35'
Natalie's Star
Ori 5h56'53"13d60'
Natalie-Tony-Kaniesha
Lac 22h39'47"56d36'
Natalya
Cas 2h0'55"70d48'
NataMich
Cap 20h36'23"-10d14'
Natascha
Gem 6h57'13"14d43'
Natascia
Nor 16h19'32"-52d5'
Natascia
Pho 0h38'22"-49d51'
Natasha
Mon 6h29'55"-6d56'
Natasha
Cyg 20h33'30"48d50'
Natasha
Sge 20h5'54"16d20'
Natasha
Cet 0h39'50"-3d23'
Natasha Courtney
Ori 5h52'22"9d6'
Natasha Elizabeth
Cas 2h30'16"60d29'
Natasha Louise
Del 20h32'34"10d15'
Natashka
And 0h50'19"36d1'
Natasja Anna
Peg 21h39'55"20d5'
Nate
Leo 9h32'15"7d47'
Nate The Great
Per 4h3'1"35d32'
Nath,Dr Anil
Aql 20h2'12"10d16'
Nathalie
Ori 5h37'1"7d59'
Nathalie
Aql 20h0'20"14d18'
Nathalie
Lyn 7h57'26"41d1'
Nathalie
Cas 2h54'0"57d45'
Nathalie
Gem 8h1'45"30d26'
Nathalie
Umi 15h11'1"68d23'
Nathalie
Per 3h43'25"37d14'
Nathalie et Albert
And 1h3'57"41d7'
Nathalie et Benjamin
And 0h21'29"40d43'
Nathalie et Pierre-Luc
Uma 8h57'53"47d17'
Nathalie,Laurent
Cas 1h0'35"58d35'
Nathan
Aur 5h5'58"41d31'
Nathan
Ser 15h32'22"21d9'
Nathan
Dra 19h30'46"71d14'
Nathan
Her 17h29'38"23d17'
Nathan
Cet 3h3'15"2d54'
Nathan & Rio
Cyg 21h7'31"37d44'
Nathan Anthony
Aur 6h32'45"35d53'
Nathan Morris
Lmi 10h50'50"32d38'
Nathan's Gift
Aur 6h34'60"34d37'
Nathan's Nebula
Leo 10h16'29"13d1'
Nathan's Star
Per 3h8'17"41d24'
Nathan,Darrell
Ser 15h9'19"22d33'
Nathan,David James
Per 3h14'0"42d36'
Nathan,Jessica L
And 0h13'56"33d54'
Nathan,Miranda & Nicholas
Mon 6h37'38"10d43'
Nathan,Philip
Ori 4h41'57"5d52'
Nathan,RD
Her 15h52'48"48d8'
Nathan,Sheila Fern
Lyn 8h54'47"42d39'
Nathana
Aql 19h19'34"12d50'
Nathaniel Trevor
Vir 12h33'0"-8d28'
Nathanson,Allyson
Vul 20h3'43"25d48'
Nathanson,Dorothy
Eri 4h0'12"-11d17'
Nathanson,John David
Peg 23h0'47"28d22'
Nati
Cnc 9h0'56"8d40'
Natilia Cristina
Peg 21h41'50"28d22'
Nation II,Charles S
Per 2h37'19"37d45'
National Ed Center - Capitol Hill Campus
Mon 8h8'19"-1d13'
National Education Center-Fort Worth
Vul 19h42'53"26d8'
Nations,Steven R
Boo 15h17'27"38d14'
Nativi
Cam 7h37'30"60d3'
Natividad Martin "Mami"
Cam 4h11'40"67d45'
Natlage,Clemens
Oph 18h27'10"8d34'
Natleelis
Peg 22h0'41"34d40'
Natoli,Carmelo Joseph
Her 15h55'18"48d3'
Natoli,Cosimo
And 0h46'20"45d42'
Natoli,Nancy
Hya 8h53'23"5d14'
Natonio,Tony & Jennifer Waldman
Uma 12h4'33"37d6'
Natow,Harry
Aur 6h36'52"37d52'
Natterer,Regina
Mon 7h59'37"-9d14'
Natural Selection
Mon 6h54'41"-1d37'
Natvig,Renee Marion
Mon 8h8'0"-9d25'
Natwin,Sr,Kathleen
Lyr 18h31'0"36d44'
Natzya
Mon 8h2'8"-9d2'
Naudin's Forever Bright Star,Elizabeth
Cam 4h37'57"67d43'
Nauer,Bryan Francis
Per 1h59'36"52d35'
Nauer,Xavier F
Per 2h5'39"47d26'
Naughton,Joanne
And 2h25'43"45d1'
Naughton,Michael Edward
Boo 14h32'27"31d47'
Naughton,Paul
Per 3h0'1"46d8'
Naughty Sam
Ori 5h58'30"19d51'
Naugle,Eileen
Leo 10h57'50"20d12'
Naulin 20,J S
Peg 22h43'0"26d4'
Nault,Mildred T
Boo 13h59'24"10d40'
Naumann,Alfred
Uma 10h46'36"55d60'
Naumann,Helmut
Lyn 8h5'1"45d23'
Naumann,Johanna Augusta Lobe
Uma 9h53'0"56d12'
Naumann,Jürgen
Lyr 19h23'30"30d13'
Naumann,Professor Claus
Gem 7h23'37"35d15'
Naurel,Jean
Peg 21h51'0"20d29'
Naus,Charles Carson
Lyn 7h3'12"51d50'
Nause,Klaus
Cmi 7h29'37"8d14'
Nauseda,Alex
Dra 18h10'1"78d57'
Nauslar's Angel
Cam 5h59'49"60d17'
Nauta,Brad
Her 16h40'19"29d28'
Nautilus 88
Aql 19h55'13"-6d31'
Nava,Danielle Marie
Mon 6h22'20"4d9'
Nava,Yvonne
Peg 22h52'59"22d5'
Navagato,Angelo
Lyn 9h13'48"37d53'
Navalisa
Aur 7h25'12"40d10'
Navara,John Cipriano
Uma 8h33'59"58d13'
Navara,Kimberly Dawn
Uma 11h7'44"68d5'
Navarra,Silke
Per 4h3'23"36d19'
Navarre,Nicole
Cnc 7h58'30"10d22'
Navarro,Anna
Cyg 20h56'57"39d23'
Navarro,Gloria & Brian Rutledge
Cam 3h42'48"60d44'
Navarro,Gustavo Prieto
Her 18h4'26"40d16'
Navarro,Jose
Umi 15h17'14"83d39'
Navarro,M
Umi 13h51'59"78d43'
Navarro,Sr,David
Lyr 18h45'50"33d3'
Navarro,Susan
Mon 6h55'38"8d35'
Navarro-Anderson,Gina
Peg 23h42'0"31d48'
Navas,Dr Ricardo
Aql 20h4'12"12d32'A
Navas,Susan
Aql 20h4'12"12d32'B
Navasky,The Star Of Helen & Bernie
Cas 1h53'1"60d16'
Naveed
Dra 17h5'1"63d45'
Navi
Cae 4h51'46"-31d28'
Navickas,Elizabeth Marie
And 0h57'46"39d54'
Navickis,Ida
Uma 8h47'20"52d40'
Navila
Cam 3h18'51"67d32'
Navillus
Ori 5h49'30"11d39'
Naviwala,Humza
Uma 9h56'14"43d23'
Navlen,Charles
Leo 10h32'20"21d45'
Navzio Family Star, Roeckl
Crt 11h23'45"-10d33'
Nawal,Kristina
Cnv 12h27'25"32d14'
Nawcl
Ser 15h44'33"2d31'
Nawrocki,Lila Louise
Cam 11h3'1"82d8'
Nawrocki,Lynda Victoria
Cyg 20h16'40"30d49'
Nawroski,Alanna Skye
Lyn 7h44'18"48d45'

Nay,Billy
Her 16h54'17"23d31'
Nayar,Arun
Cyg 21h38'1"40d28'
Nayer
Cyg 20h2'22"30d46'
Nayi,Ricky
Her 17h38'59"27d11'
Naylor,Beth
Lyr 19h22'16"30d36'
Naylor,Charles Michael
Dra 20h24'26"62d11'
Naylor,Claire Glenora
Lyn 7h36'37"50d50'
Naylor,Deborah
Cas 0h19'13"62d13'
Naylor,Master Cody Michael
Lac 22h49'13"52d32'
Naylor,Mitchell Dmytryk
Cet 2h27'34"5d10'
Naylor-Legend,Philip
Per 1h47'58"56d52'
Nayrolles,Nicole
Boo 15h6'28"7d42'
Naysmith,Richard
Ori 6h0'22"20d51'
Nazan
Cnc 8h36'1"7d23'
Nazan K
Mon 6h32'1"3d28'
Nazari,Christine Tracy
Ori 6h17'18"-2d15'
Nazemi,Iman
Umi 14h34'33"65d58'
Nazim Star Shines On Me
Ser 17h53'39"-14d60'
Nazim,Nafisa
Leo 10h42'41"15d35'
Nazimek,Steven Joseph
Her 16h19'25"40d55'
Nazoa-Ruiz,Princess Annabella
And 23h47'1"41d21'
Nazzaro,Andrea L
And 1h49'19"47d22'
Nazzaro,Timothy J
Aur 6h23'15"33d16'
Naéj
Sgr 19h29'29"-39d50'
NBS Listen
Cma 6h56'49"-19d49'
Ndays,Michel Leconte
Uma 13h5'16"62d44'
NE Twin Lights
Cas 1h47'38"64d36'AB
Neads,Lotchie Corinne Phillips
Mon 7h2'1"5d1'
Neal Family Star,The
Aql 19h57'28"11d23'
Neal J
Cas 23h37'0"58d22'
Neal Nathan
Cep 22h44'57"68d59'
Neal Star,The
Uma 9h15'31"57d1'
Neal,Catherine Louise
Peg 23h0'11"20d16'
Neal,Christine Thomas
Ori 5h32'1"8d20'
Neal,Curtis R
Aur 7h22'0"40d10'
Neal,James Brian
Psc 0h20'30"18d4'
Neal,Jean Ann
Cap 21h55'22"-20d59'
Neal,Jeffrey I
Lac 22h0'50"37d37'
Neal,Jessica
Sge 19h55'0"18d50'
Neal,Jon
Cmi 7h44'1"1d58'B
Neal,Kyle Andrew
Ser 18h11'59"-5d12'
Neal,Louise Rothlisberger
Del 20h35'46"20d24'
Neal,Marion Pearson
Sgr 19h8'1"-27d7'
Neal,Matthew
Oph 17h53'38"11d31'
Neal,Megan
Cmi 7h42'24"0d53'
Neal,Natalie
Peg 22h59'44"27d31'
Neal,Nicolin
Equ 21h6'23"11d23'
Neal,Paula K
Crb 16h7'43"30d47'
Neal,Rebekah
Cmi 7h58'37"7d60'
Neal,Richard Evans
Cet 2h44'0"4d37'
Neal,Robert
Lac 21h58'29"38d48'
Neal,Stephen
Sex 10h43'41"-1d6'
Neal-Judd,Sarah
Cas 0h15'27"63d43'
Neale,Allen R
Hya 9h30'48"-0d22'
Neale,Anoushka Helena Theresa
Tau 3h34'1"30d22'
Neale,Michelle
Cas 1h50'48"70d34'
Neale,Shelby Ann Madison Humble
And 23h44'1"42d44'
Neale,Sir Charles Eric
Uma 11h0'52"55d58'
Nealeigh,Thomas T
Aur 5h25'13"38d7'
Nealon Star,The
Sge 19h42'54"16d29'
Nealon,Eileen B
Cyg 20h22'27"37d35'
Nealon,Mary
Lac 22h19'11"54d58'
Nealon,Matthew
Dra 9h51'23"74d14'
Nealon,Michele
And 0h29'51"44d49'
Nealsky
Cnv 12h36'0"51d19'
Nealy,Millicent T
Lyn 8h50'13"46d49'
Neaners
Cas 0h10'51"50d34'
Nearhood,Beth Ashley
Aql 18h58'19"-5d27'
Nearman,Cynthia Marie
Com 12h30'1"30d44'
Nearman,Jr,Richard Edward
Aql 20h1'42"6d51'
Nearman,Robin Eileen
Cas 0h57'43"63d17'
Neary,Kathleen
And 0h22'41"37d56'
Neary,Patrick W
Aur 5h28'55"40d54'
Neary,Ryan Kevin
Aur 5h6'26"41d15'
Neary,Sir Edward
Per 2h51'17"38d55'
Neary,Thomas J
Dra 16h7'49"68d38'
Neatherlin,Johnnie June True
Lmi 9h58'48"40d35'
Neathery,Ed
Lac 22h20'16"50d37'
Nebbia,Nathan J
Uma 12h58'28"58d42'
Nebediah
Ori 6h1'1"6d32'
Neben Star,The Peggy
Lyr 18h48'46"30d56'
Neberieza,Amy M
Cas 1h11'1"61d40'
Nebesnaya Natalia
Com 13h4'0"21d12'
Nebgen,Wyatt Henry
Cyg 20h49'46"37d47'
Nebouy,Gerard-Michel
Cyg 19h38'32"38d13'
Nebreda,Jose Luis Rodrigues
Cas 23h3'46"53d17'
Nebula Of Coty
Dra 20h1'0"63d23'
NEC Arkansas Col of Tech Camp-Little Rock
Oph 17h18'15"-21d7'
NEC Bauder College Campus-Fort Lauderdale
Aql 20h35'24"-6d32'
NEC Bryman Campus - Houston North
Oph 17h17'20"-21d39'
NEC Bryman Campus - Rosemead
Tri 2h24'43"28d38'
NEC Bryman Campus - Atlanta
Cnv 12h14'0"46d38'
NEC Bryman Campus - Wilshire
Aql 20h35'15"-6d33'
NEC Bryman Campus - Houston South
Ser 17h55'0"-13d46'
NEC Bryman Campus- Long Beach
Mon 8h1'34"-2d25'
NEC Kee Bus College Campus - Richmond
Ser 17h54'51"-14d22'
NEC Kee Bus College Campus - Norfolk
Cet 2h32'60"-10d58'
NEC Kee Bus College Campus-Newport News
Aql 20h35'37"-1d6'
NEC Kentucky Col of Tech Campus-Louisville
Hya 10h27'30"-18d49'
NEC Natl Inst of Tech Campus - Dallas
Cet 2h33'18"-11d20'
NEC San Antonio
Sex 9h58'19"-0d26'
NEC Sawyer Campus - Sacramento
Mon 8h8'21"-1d28'
NEC Sawyer Campus - Commerce
Sex 9h58'0"-6d27'
NEC Skadron Bus Coll Campus-San Bernardino
Cet 2h31'31"-11d1'
NEC Tampa Technical Institute Campus
Tri 2h1'0"28d13'
Necciai,Erik Thomas
Her 17h32'32"22d54'
Necco,Francesca Betteto
Peg 22h42'14"4d48'
Necessary,David
Aur 7h18'1"36d2'
Neckameyer,Bill & Pamela
Cyg 21h52'34"41d29'
Neco
Tau 5h30'55"29d15'
Necrason,Major General Conrad Francis
Uma 11h27'21"49d40'
Nectere
Aur 6h59'34"38d40'
Ned & Elizabeth
Peg 23h38'1"25d28'
Ned 40
Ori 5h8'16"-4d53'
Neda
Cyg 21h38'36"40d40'
Nedbalek,Mary Katherine
And 2h12'33"42d10'
Neddo,Debra
Cam 3h30'40"58d24'
Nedea,Monica-Lygia 30071994
Peg 23h16'25"33d49'
Nedelec,The Dave & Tamie
Aql 19h28'35"12d56'
Nederlander,Marjorie
Mon 6h42'59"6d8'
Nedjma
Cyg 20h21'29"39d42'
Nedoma,Anton
Cap 20h22'26"-20d18'
Nedoschil,Willy
Aqr 22h42'1"-5d23'
Nedra Jane
Aql 18h54'28"-0d6'
Nedra Jayne
And 23h47'30"47d2'
Nedsram
Ori 5h4'40"-1d50'
Nee,Sarah Brennan
Sge 19h2'35"19d28'
Neeb,Chip & Halle
Vul 19h48'18"28d58'
Neeck,Geraldine M
Cas 0h59'23"58d57'
Need,Astronomique Ronna
Boo 14h41'45"31d2'
Needels,Mara
Sgr 19h24'43"-40d33'
Needham,Chris
Cep 21h39'18"55d45'
Needham,David T
Aur 6h39'16"38d21'
Needham,Ginger L
Aur 7h26'22"39d33'
Needham,Kevan
Her 17h15'23"29d8'
Needham,Tobias H
Lib 14h57'34"-22d35'
Needle's Star
Aql 18h58'14"10d3'
Needle,Jack
Cyg 19h27'35"35d37'
Needle,Olive
Cyg 19h27'56"35d0'
Needleman,Doctor Stuart W
Oph 17h4'29"10d14'
Needles,Alison L
Aql 19h54'15"1d10'
Neefe,Cornelia
Psc 1h17'21"18d17'
Neek Banana
Cep 23h25'18"70d16'
Neel Family,The
Cmi 7h22'28"5d12'
Neel,Olivier
Mon 6h27'30"8d52'
Neel,Preston Hunter
Her 17h35'11"24d30'
Neel,Shawn & Tina
Com 13h16'46"21d50'
Neelands,Matthew James
Cep 20h25'47"60d21'
Neeley
Per 3h27'17"50d10'
Neeley,Jan Marie
Lyr 18h36'38"45d57'
Neeley,Mareena Katherine
And 2h5'55"38d34'
Neeley,Nicole G
Peg 22h19'52"20d28'
Neels,Jenny Lynn
Cas 0h32'23"63d8'
Neely,Debra
Lyr 18h46'1"31d16'
Neely,Greg
Sex 10h31'50"2d17'
Neely,Ives W
Per 4h33'40"34d11'
Neely,Mary
Dra 12h30'2"75d5'B
Neely,Wilford
Dra 12h30'2"75d5'A
Neena
Aql 19h6'47'15d39'
Neener
Uma 10h1'0"52d12'
Neenus Orbis
Oph 17h3'24d11d5'
Neer,Mark Thomas
Leo 10h50'15"-5d39'
Neerach
Uma 10h45'58"40d52'
Neese,Susan
Cas 0h49'22"73d40'
Neeson,G E
Lac 22h43'22"56d16'
Neff II,David Lee
Oph 17h59'15"12d23"
Neff,Barbie-Jo
Lyn 8h54'45"41d0'
Neff,Ernest Lindsay
Aql 19h5'40"-0d2'
Neff,Jason Joseph Girard
Ori 4h41'32"10d15'
Neff,Jerrad
Dra 16h51'53"63d38'
Neff,Stewart & Inge
Lyn 7h58'21"38d13'
Neff,Wallis Ann
Mon 6h43'0"7d34'
Neft,Beba-Katarina
Cet 2h54'59"0d10'
Neger,Nancy
Crt 11h13'48"-14d12'
Negev
Lac 22h21'58"52d42'
Neglia,Katherine
Peg 23h5'0"11d46'
Negovan,Pat
Lyn 8h23'45"43d55'
Negra,"The Dreamer" Talia
Lib 15h44'15"-22d40'
Negra,Perla
Vul 19h19'52"26d48'
Negrel,Jany & Pierre
Cnv 12h11'15"47d50'
Negrete,Yelitza Cacho
Lyn 7h1'21"58d36'
Negri,Giuseppe
Scl 23h10'40"-32d7'
Negri,Matthew Carmen
Cyg 21h31'43"38d50'
Negri,Tony Gene
Cma 6h56'19"-19d30'
Negrin,Carmen
Peg 23h44'50"27d21'
Negrini,Allison
Peg 21h43'59"23d15'
Negron's Star,Special Son Scott
Uma 10h31'26"54d6'
Nehal Joshi
Sge 20h14'56"17d47'
Nehamen,Gail (Gutzicle) Ann
Cyg 20h44'60"45d40'
Nehemiah,Marcia
Cas 1h52'0"58d39'
Nehez,James "Big Stick"
Lyr 18h21'40"38d25'
Nehez,Jim
Ori 6h0'26"8d49'
Nehez,Jr,James R
Aur 5h58'18"31d51'
Nehls Family Star,The Pete
Lac 22h44'35"54d38'
Nehls,Laura
Cas 23h31'22"60d47'
Nehmsmann,Michael
Dra 14h57'52"57d27'
Nehmzow,Detlef
Cap 20h40'49"-26d56'
Nehoda,Forever Liz To Steve R
Aqr 20h37'0"-1d3'
Nehoray,Ron & Michal
Ori 5h23'12"0d28'AB
Nehrbas,Brendan Edward
Oph 18h2'54"12d16'
Neider I Karlheinz
Cnc 8h58'0"11d25'
Neider,Only You Craig A
Uma 10h39'28"47d46'
Neidermire,Steven
Aql 19h30'10"13d49'
Neiffer,Dawn Marie
Peg 22h48'59"21d27'
Neige
Cep 21h29'34"67d47'
Neige,Blanche
Her 18h2'20"14d57'
Neigenfind,Ludmilla
Leo 11h49'43"21d28'
Neighbors,Doug (HB)
Ser 16h7'58"0d34'
Neigher,Geoffrey Mark
Hya 8h17'29"4d42'
Neiglick,Olof
Ori 5h32'54"0d60'
Neil
Cep 1h11'14"86d42'
Neil & Heather- Together For Ever!
Sge 20h4'37"18d51'
Neil & Karen's Stardust Embrace
Uma 11h10'17"71d25'
Neil & Laura
Cyg 19h33'0"30d18'
Neil & Michelle
Cyg 21h22'37"28d33'
Neil 6593
Ori 5h53'11"16d46'
Neil Louise Christopher
Dra 20h21'41"62d20'
Neil's Forever Shining Star
Cep 22h47'49"70d49'
Neil's Star
Lyn 8h16'30"41d26'
Neil-Salomon Maman Belecen
Per 3h26'53"51d35'
Neilbert
Sex 10h3'1"-5d30'
Neill,William R
Ser 16h6'27"0d55'
Neillie II,William A
Sex 10h11'44"-4d31'
Neilsen,Robert & Marie
Del 20h20'32"20d27'
Neilson,Bob
Eri 4h18'38"-19d27'B
Neilson,Randi & Rob
Ori 5h55'21"20d35'
Neilson,Sandy
Eri 4h18'38"-19d27'A
Neilson,Scott
Boo 15h10'14"51d36'
Neiman,Jeffrey Scott
Uma 9h46'18"70d27'
Neises,Aunt Tillie
Uma 11h22'51"45d6'
Neith The Star
Sex 9h50'0"-1d6'
Neithercutt,Marc
Oph 18h35'47"10d36'
Neitz,Bill
Crt 11h47'26"-18d56'
Neitzke
Cyg 20h21'0"39d54'
Neitzke,Michelle Lee
Aur 7h21'47"39d44'
Nekkrik Thirty Six
Aql 20h9'16"1d45'
Nekris,Evelyn
Cyg 21h3'1"48d54'
Nel's Lucky Star,Gay
Cam 3h16'0"61d46'
Nelem,Jennifer Francis
Aql 19h55'29"15d28'
Nelesena
Cyg 21h7'0"48d60'
Nelia
And 0h17'30"39d56'
Nelialex
Ori 6h15'51"-0d37'
Nelida
And 0h2'0"38d52'
Nelius,Kurt
Mon 7h53'54"-2d26'
Nell
Cyg 19h57'28"44d29'
Nell,Barbara
Vir 13h57'14"6d15'
Nell,Louise Clarissa
Lac 22h7'28"46d18'
Nell,Olinka
Sco 17h51'59"-31d19'
Nell,Patricia
Crt 10h49'35"-12d27'
Nella C
Cas 0h41'0"66d11'
Nella Nella
Cyg 21h37'17"40d19'
Nellene
Uma 8h56'41"70d48'
Nelles,Hunter Hanna Downs
Cnv 12h46'1"39d43'
Nellie
Uma 12h6'15"47d0'
Nellie
Lyn 7h53'42"33d22'
Nellie und Ewald
Cnv 14h1'1"39d36'
Nelligan IV,Luke A
Ser 17h33'1"-14d2'
Nelligan,Patrick
Cnv 13h43'43"32d42'
Nellis,Alice Mildred
Cas 0h32'42"75d31'
Nello
Boo 14h33'11"7d42'
Nelly
Peg 23h38'17"11d37'
Nelly Marguery
Boo 14h29'13"17d32'
Nelly Patricia
Cas 1h45'0"73d53'
Nelmes,Benji
Umi 13h33'42"72d13'
Nelms,Norma Lynn
Aur 5h55'10"31d38'
Nelms,Phillip
Boo 13h37'11"19d51'
Nelsen,Derek Spencer
Boo 14h7'1"32d33'
Nelsen,Donovan James
Her 17h38'43"22d59'
Nelsen,Ronald Keith
Her 17h16'11"41d7'
Nelson
Ori 6h16'59"-2d21'
Nelson
Boo 14h6'30"44d37'
Nelson #52 Forever Young,Danny
Aql 19h17'57"13d18'
Nelson & Denise Forever
Aur 5h59'19"23d50'
Nelson & JC Steiner, Love of HA
Crt 11h2'9"-18d40'
Nelson & Phyllis
Eri 3h32'37"-7d6'
Nelson Five
Cen 11h52'40"-43d43'
Nelson Gum
Boo 14h5'60"30d48'
Nelson Simmons Star, The
Cnv 13h41'34"38d48'
Nelson Star,The
Uma 12h7'55"56d29'
Nelson The Dutchmasters,James L
Dra 17h46'57"68d20'
Nelson,"Nano"
Cas 0h30'15"60d9'
Nelson,Alfred M
Boo 13h51'11"20d46'
Nelson,Amy
And 1h22'21"37d43'
Nelson,Andrew David
Leo 10h36'25"13d29'
Nelson,Ashley Danielle
Cmi 8h6'29"6d35'
Nelson,Aubree Genteel
Cyg 20h37'34"39d48'
Nelson,Becky
Lac 22h41'28"55d35'
Nelson,Benjamin E
Sco 16h56'0"-44d22'
Nelson,Bethany Erin
Lyn 7h54'51"52d11'
Nelson,Beverly
Lyn 8h50'35"44d52'
Nelson,Bradley David
Boo 14h41'11"38d56'
Nelson,Calvin
Dra 13h6'46"68d13'
Nelson,Carolyn & Jim
Dra 18h29'17"58d37'
Nelson,CC & Martha
Hya 9h4'37"1d3'
Nelson,Charles A
Cet 2h41'28"1d33'
Nelson,Charles A
Umi 15h33'17"70d36'
Nelson,Charlotte Amelia
Mon 6h37'12"10d19'
Nelson,Charlotte
Vul 21h2'28"20d3'
Nelson,Chris
Aql 20h9'12"6d48'
Nelson,Christine
Crb 15h27'23"30d26'
Nelson,Curtis W
Lyn 8h34'18"38d47'
Nelson,Cynthia E
Cam 4h2'30"60d57'
Nelson,D Brad
Her 16h26'56"27d23'
Nelson,Dale & Jewell
Her 16h0'48"41d20'
Nelson,Denise
Peg 22h46'49"30d54'
Nelson,Denny
Cet 2h6'43"7d1'
Nelson,Diane M
Cyg 21h35'0"40d30'
Nelson,Dianna
Del 20h35'11"20d12'
Nelson,Douglas
Aur 7h23'47"40d11'
Nelson,Edie M
Her 18h13'32"48d59'
Nelson,Edwin Albert
Eri 2h52'2-"-1d44'
Nelson,Elizabeth Le Bahn
Crb 15m17'1"30d55'
Nelson,Emelia Rae
Cas 0h55'55"75d12'
Nelson,Emma Marchelle
Eri 3h32'56"-6d4'
Nelson,Eric
Her 16h8'17"23d15'
Nelson,Eric Hilliard
Ser 18h16'47"-14d40'
Nelson,Fiona Jane
And 23h10'20"41d11'
Nelson,Frances Sedgwick
Cyg 19h53'59"38d39'
Nelson,Gary Sheldon
Uma 9h17'40"55d36'
Nelson,George Gale
Dra 14h59'60"64d0'
Nelson,Ginny Lee
Uma 12h57'14"53d53'
Nelson,Gordon L
Dra 16h4'22"65d9'
Nelson,Gregory Donald
Aql 18h56'44"16d38'
Nelson,Gregory Scott
Uma 10h23'59"40d48'
Nelson,Harold "J"
Cep 4h23'40"80d32'
Nelson,Hazel
Lyn 7h58'41"50d31'
Nelson,Helen
Cas 0h0'20"56d13'
Nelson,Helene Marie
And 23h37'51"46d12'
Nelson,Hilde DeVoin
Vul 19h48'40"23d43'
Nelson,Hilde Dorothea
Cas 0h23'26"50d3'
Nelson,Ian R
Her 17h15'0"44d36'
Nelson,Jacob Cyrus
Dra 20h1'1"62d19'
Nelson,Jacquelyn C
Cma 6h56'21"-20d29'
Nelson,James Daniel
Umi 16h46'24"75d45'
Nelson,James William Bradford
Cet 1h0'44"-1d10'
Nelson,Jayme Jo
Uma 9h16'52"61d33'B
Nelson,Jean Marie
Cam 7h51'18"61d8'
Nelson,Jeffrey M
Eri 3h51'11"-6d46'
Nelson,Jennifer Lynn
Vul 20h18'16"23d15'
Nelson,Jimma
And 2h26'1"40d41'
Nelson,Joanna
And 23h35'1"44d18'
Nelson,John Henry
Hya 10h14'21"-18d48'
Nelson,John L
Aql 20h6'47"0d55'
Nelson,Jonathon Scott
Uma 8h21'15"70d14'
Nelson,Joshua
Ser 15h46'18"20d53'
Nelson,Jr,Douglas J
Aql 19h2'28"-6d29'
Nelson,Jr,Kenneth F
Oph 16h59'15"-26d12'
Nelson,Jr,Robert E
Boo 14h19'53"23d33'
Nelson,Judy
Lyr 19h0'34"38d14'
Nelson,Karen Susanna
Crt 11h43'46"-11d14'
Nelson,Karl Edward
Ori 5h55'39"8d55'
Nelson,Katherine Marie
Cyg 20h21'39"31d49'
Nelson,Katherine Mary
And 23h31'37"49d34'
Nelson,Keith Gordon
Cet 3h0'16"8d36'
Nelson,Kelly M
Cas 1h19'37"68d31'
Nelson,Ken R
Hya 8h37'59"-1d41'
Nelson,Keri Ann
Peg 23h4'33"8d29'
Nelson,Krista Hope
And 23h29'58"44d22'
Nelson,Kristen J
Cyg 21h41'34"38d56'
Nelson,Kristin Marie
Eri 3h1'15"-13d26'
Nelson,Lance
Uma 11h18'34"52d56'
Nelson,Laura Elizabeth
Lmi 9h36'30"38d12'
Nelson,Laura Elizabeth
Peg 22h57'15"30d19'
Nelson,Laura Jean
Com 12h46'39"21d33'
Nelson,LTC Gene L
Dra 18h26'42"70d58'

Nelson,Lucy McKown
Peg 23h31'23"21d20'
Nelson,Lucy Pe23"21d20'
Ori 6h1'32"8d5'
Nelson,Margaret
Eri 2h54'1"-4d11'
Nelson,Marian (Micki)
Cmi 7h22'22"4d60'
Nelson,Marilee
Hya 9h32'1"1d13'
Nelson,Marilyn & Glen
Cyg 20h42'27"47d33'
Nelson,Martina Helene
Eri 2h55'1"-7d0'
Nelson,Mary Ann
Eri 3h41'48"-11d43'
Nelson,Mary C
Cyg 20h28'21"40d6'
Nelson,Marybeth
Dra 12h12'59"71d22'
Nelson,MD,James N
Sct 18h45'29"-7d16'
Nelson,Melanie
Ori 4h41'14"5d1'
Nelson,Michael D
Ori 4h54'33"-1d7'
Nelson,Nancy
Uma 11h54'25"43d3'
Nelson,Nancy
Aql 18h41'28"1d28'
Nelson,Ned "Wolfe"
Aur 5h9'51"40d12'
Nelson,Patricia Brook
Vir 13h33'37"-6d51'
Nelson,Patti
Peg 23h29'60"30d52'
Nelson,Peter
Cet 0h41'33"-2d23'
Nelson,Peter Christopher
Tri 1h54'28"28d46'
Nelson,Peter Dines
Cet 2h38'43"0d17'
Nelson,Polly & Korey
Cyg 20h22'30"41d6'
Nelson,Prince Roger
Lib 14h20'50"-20d4'
Nelson,Ray
Ari 2h7'13"18d14'
Nelson,Robert Alan
Boo 15h6'17"22d46'
Nelson,Robert F
Lac 22h47'19"55d13'
Nelson,Sandy
Lyn 7h25'44"44d41'
Nelson,Sarah Jean
Mon 6h27'55"11d53'
Nelson,Scott James
Cet 2h4'14"-6d33'
Nelson,Scott T
Aql 19h6'35"0d42'
Nelson,Scott William
Cet 2h35'12"-2d55'
Nelson,Shannon
And 0h47'13"45d22'
Nelson,Sheffield
Boo 15h9'0"30d42'
Nelson,Steve
Aqr 22h12'12"0d52'
Nelson,Steven A
Aql 18h54'46"-0d16'
Nelson,Steven Paul
Cep 21h4'55"61d13'
Nelson,Ted
Aur 6h0'37"30d33'
Nelson,Thomas
Aur 6h53'36"37d21'
Nelson,Thomas & Marianne
Cyg 19h18'14"45d33'
Nelson,Valerie
Cma 6h42'25"-15d23'
Nelson,Vera
Eri 2h59'1"-5d17'
Nelson,Walter John
Her 16h24'1"40d49'
Nelson,Walter S & Louise A
Lyn 9h9'1"44d49'
Nelson,Wayne H
Cep 21h56'25"55d36'
Nelson,Wendy Ann
And 0h39'44"40d18'
Nelson,Wendy Michelle
Umi 15h17'23"68d27'
Nelson,William Eric
Cam 3h23'0"60d12'
Nelson,William John
Dra 16h15'11"68d32'
Nelson,William Thomas
Uma 13h8'27"62d28'
Nelson,Zachery Mathew
Leo 11h32'45"-1d27'
Nelson-Gardell PhD, Debbie
Ori 6h3'18"8d58'
Nelson-Ross,Ozella Jo Anne
Cet 2h47'37"5d41'
Neltner,Dr Garry W
Oph 17h18'32"10d36'
Nelyda
Cas 0h36'13"64d20'
Nelzen,Ross L
Hya 8h58'1"4d36'
Nemecek,Alice & David
Crb 15h50'37"31d15'
Nemecek,Jacob John
Dra 16h53'18"67d17'
Nemecek,Joshua John
Aur 6h28'45"31d7'
Nemecek,Jr,Joseph John
Lac 22h31'50"52d38'
Nemenz,Otto
Oph 18h18'28"8d58'
Nemeroff (M A N), Martin A
Cnc 8h12'20"31d9'
Nemes,Sarah Elizabeth
Peg 22h0'37"27d44'
Nemeth,Frank B
Per 2h58'51"38d14'
Nemeth,Frank F & Ann
Cipolloni
Lyn 8h25'41"41d53'
Nemeth,Joan
Lmi 10h10'00'36d47'
Nemeth,Marika
Cet 1h43'52"0d53'
Nemeth,Mary Eileen
Uma 10h40'27"48d42'
Nemeth,Matthew Robert
Cep 21h21'25"65d15'
Nemeth,Michael A
Aur 6h25'46"30d14'
Nemeth,Peter Jason
Uma 10h58'0"47d14'
Nemeth,Theresa
Lyn 7h59'41"34d58'
Nemetz,"Laurie" Jean
Umi 15h44'51"76d0'
Nemiroff,Jarret
Aur 5h15'0"40d27'
Nemmers,Kathleen Bridget
Ori 5h50'38"15d5'
Nemo
And 23h3'43"50d39'
Nemoto,MD,Takuma
Her 18h1'1"45d21'
Nemura "Little
Princess",Courtney
And 23h15'0"49d35'
Nemy,John
Mon 6h23'56"-10d0'A
Nena
Lyn 8h13'22"58d37'
Nene
Cyg 21h3'29"34d49'
Nenkovski,Nenko
Peg 23h28'41"24d55'
Nenni,James
Aur 6h20'1"52d53'
Nenni,Kimberly
And 23h1'21"46d58'
Nentwich,James Christopher
Cep 22h13'42"61d59'
Nentwig,Marc
Sgr 19h45'56"-40d53'
Neoershai,Tiffany
Uma 12h2'33"36d50'
Neola,Tatiana
Gem 7h29'0"20d18'
Neophytou,Alexis
Peg 23h3'54"31d39'
Nepimach,Ing Anton
Aqr 23h32'23"-11d42'
Nepotismecca
Aql 19h59'41"15d59'
Nepper,Gary D
Aur 6h23'27"35d34'
Nepper,Jessica Rose
Ari 2h42'19"25d28'
Neppl,Mary & Steve
Lmi 9h55'40"34d18'
Nequell,Shantell
Peg 23h42'31"30d15'
Nerad,Pat
Umi 16h25'32"71d7'
Nerak-Ttai-Uh
Cam 4h5'31"80d1'
Nerbonne,Kendra
And 23h22'44"46d40'
Nerenberg,Patricia Mary
Eri 3h18'0"-13d52'
Neri,Anthony
Lac 22h32'1"50d26'
Neri,Claire
Sct 18h56'18"-5d31'
Neri,Geno
Umi 16h18'32"79d52'
Neri,Leeanne
Peg 23h1'1"20d2'
Neri,Marina
Cyg 20h34'54"48d47'
Neri,Philip G & Michele A
Cioffi
Cyg 20h30'25"40d40'
Nerina-Valli & James- Bian
Rosa
Dra 18h52'34"78d51'
Nering,Diana
Cmi 7h35'23"10d21'
Neroui,Gabriele
And 23h35'48"49d32'
Nerrault,Patrick
Cyg 20h42'44"47d3'
Nersinger,Carmen Veronica
Uma 9h24'30"73d7'
Neruda,Ian
Her 17h37'57"48d49'
Nervo,Paolo
Lmi 10h53'58"31d37'
Nerys
Cas 1h57'11"61d51'
Nesbeda,Nora
Ari 2h28'51"21d12'
Nesbit,Aquia Michelle
Vul 19h0'14"25d4'
Nesbitt,Donald Lee
Ori 5h48'0"10d31'
Nesbitt,Shane Robert
Alexander
Ori 4h49'39"0d58'
Nese,Joseph
Cnc 8h34'59"18d24'
Nese,Melissa
Cnc 8h35'0"18d30'
Nese,Patricia
Cnc 8h39'20"18d59'
Nesemann,Sam
Her 16h36'57"21d6'
Neser 54
Her 16h43'22"37d19'
Neshama of Ryan &
Bridget,The
Mon 6h43'55"8d54'
Nesher,Gisel Koll
Del 20h33'25"10d49'
Nesher,Jill & Robert
Lyn 7h52'59"42d53'
Neshika
Cnv 13h46'52"46d23'
Nesie 3147
Cam 5h45'25"68d12'
Nesius,Bradley Wayne
Uma 10h34'55"58d56'
Nesmith,Aileen
And 23h35'0"48d51'
Nesmith,L W
Aql 19h31'1"12d15'
Nespola,Jr,(Scruffy) Richard
Leo 9h42'37"18d49'
Ness 60,Maria
Cas 2h32'55"61d39'
Ness 80,Magne J
Eri 2h55'27"-4d38'
Ness,Barbara J
Vir 13h32'14"-3d19'
Ness,Jim
Uma 11h8'52"53d18'
Ness,John Milton
Lac 22h33'54"38d28'
Ness,K W
Sct 18h19'20"-13d54'
Ness,Ryan Anders
Aur 6h10'51"31d29'
Nessa
Psc 22h56'11"5d15'
Nesselmann,Frank
Her 17h17'56"42d57'
Nesselt,Jr,James John
Lyr 19h23'0"31d14'
Nesser,Elizabeth Williams
And 2h16'37"48d21'
Nessler,Jr,William E
Lyn 6h30'33"60d57'
Nessling,Michael
Per 3h1'32"41d36'
Nesslinger,Bettina
Peg 23h32'37"12d24'
Nessmith,Dave
Ser 18h3'47"-14d32'
Nest,André
Cyg 20h42'41"47d11'
Nest,Klaus
Vir 12h2'49"2d26'
Nester,Jr,Paul
Aur 5h8'59"41d10'
Nesti,Kaden & Gene
Eri 3h42'24"-11d50'
Nestler,James & Linda
Vul 20h29'30"28d46'
Nestman,Rachel Teresa
Cas 1h43'50"61d37'
Nestola,Claudia
Cas 1h28'12"58d44'
Nestor,Michelle
Cas 1h52'1"61d10'
Nestore,Chris & Louise
Leo 9h57'43"11d30'
Nesty
Aql 20h18'32"0d27'
Nesvold,Ronald Kevin
Dra 17h8'49"69d42'
Net,Juan Antonio
Ori 5h24'35"-6d53'
Neterval,Jeffrey Peter
Boo 15h16'56"48d58'
Nethercutt,William R
Ser 15h40'0"21d55'
Netherton,Rene
Del 20h19'36"10d39'
Netri,Patricia Ann
Cyg 21h20'54"37d48'
Netta's Star
Leo 9h45'49"7d13'
Nette's Star
Eri 2h47'37"-4d31'
Nettekoven,Heinz
Peg 23h28'20"10d13'
Netter,Aileen
And 0h17'29"37d57'
Netter,Jake
Cma 7h53'52"00d38'
Netter,Lani
Cma 6h31'15"-18d4'
Netter,Rick Jayson
Dra 19h31'15"65d7'
Netti,Matthias
Psc 0h22'55"8d49'
Nettie & Prindle
Aql 19h8'40"5d12'
Nettleton,Kristy & Steven
Crb 15h48'49"34d27'
Netty's Star
Cap 21h21'25"-20d2'
Netusil,Forever Young Robbie
Boo 14h10'1"52d33'
Neu,Jeanine
Cas 0h42'40"65d59'
Neu,Jr,Vincent J
Cep 20h47'32"68d51'
Neubecker,Ellen
Cas 0h4'31"58d25'
Neuberger,Catherine Ann
And 23h41'55"38d35'
Neubig,Hermann
Aqr 21h56'40"-8d5'
Neuborne,Ellen
And 1h14'55"40d58'
Neubrech,Karen
And 23h32'42"45d37'
Neubuck,Leigh
Cyg 19h21'33"28d39'
Neuburg,Hildegard
Sgr 18h53'6"-24d59'
Neudecker,August
Ori 5h26'48"1d51'
Neues Frank
Dra 16h48'28"65d22'
Neufeld,Dianne Lyn
Cnc 8h56'34"32d48'
Neufeld,MD,Robert James
Oph 17h4'17"11d31'
Neufeld,Randy Alexander
Cas 0h20'55"60d5'
Neugebauer,Joerg
Oph 18h28'48"8d57'
Neugebauer,Liane
Aqr 21h55'0"0d30'
Neuhalfen,Kyle
Uma 11h51'37"32d23'
Neuhamm,Professor Thomas
Tau 4h1'0"20d25'
Neuhaus,Diana "Sweety"
Uma 14h20'24"60d12'
Neuhaus,Friedrich Karl
Mon 7h1'36"-5d15'
Neuheisel,Richard G
Aur 6h25'36"30d7'
Neuhoff-Flanagan, Margaret
Lyr 19h14'1"37d31'
Neuhäusel,Jenny
Aqr 21h1'37"-6d29'
Neumaier,Janet
Umi 15h37'58"72d21'
Neuman,Alvin
Dra 20h6'34"63d36'
Neuman,Denise
Cam 12h58'58"81d40'
Neuman,Elizabeth C
Cas 0h35'45"60d39'
Neuman,Fredrick
Boo 14h40'0"50d26'
Neuman,Jim
Ser 15h56'44"1d14'
Neuman,Jr,Walter G
Boo 15h15'36"38d37'
Neuman,Marsha Wright
Cam 9h9'25"78d14'
Neuman,Michael William
Cep 23h34'1"63d27'
Neuman,Tanya
Cas 1h38'39"60d0'
Neumann,Barbara J
Lyn 8h51'44"34d9'
Neumann,Bonnie Ruhkala
Cam 3h46'58"56d7'
Neumann,Brunhilde
Psc 22h50'51"6d10'
Neumann,Friedrich Wilhelm
Lib 14h23'56"-23d39'
Neumann,Gisela & Alfred
Umi 14h50'11"68d40'
Neumann,Grit
Psc 23h23'26"1d60'
Neumann,Harald
Lyr 19h21'41"31d28'
Neumann,Jaime
Cas 0h43'40"61d44'
Neumann,Kurt Philipp
Tau 4h39'27"28d35'
Neumann,Lauren Simone
Cam 7h28'0"78d54'
Neumann,Lennart Petrus
Ari 1h48'31"24d29'
Neumann,Lothar
Cam 5h41'43"70d39'
Neumann,Michael James
Dra 15h2'1"65d21'
Neumann,Michael John
Her 16h9'20"40d41'
Neumann,Paul
Vul 19h45'48"29d5'
Neumann,Peter-Ernst
Gem 6h46'44"14d40'
Neumann,Wolfgang
Aqr 21h56'48"-17d31'
Neumayer,Burkhard
Cap 21h37'51"-23d29'
Neumayr,Bettina
Cnc 9h14'30"10d49'
Neumeister,Karl-Heinz
Cmi 7h19'40"8d39'
Neumeister,Marietta
Lyr 19h6'48"26d55'
Neumuller,Richard W
Aur 4h38'34"30d34'
Neumuller,Virginia
And 0h21'15"40d34'
Neuner,Allen G
Cet 2h16'1"4d56'
Neupert,Abigail Soojin
Lyn 6h15'46"58d23'
Neurohr,Margo
Mon 6h34'54"-6d12'
Neurohr,Michael
Dra 18h27'49"65d37'
Neuroth,Dewight A
Dra 15h52'39"57d36'
Neuroth,Gary R
Ori 6h4'40"0d57'
Neus
Dra 17h16'46"60d25'
Neus,Hans-Joachim
Aur 5h30'42"37d49'
Neusch,Natalie A
Mon 6h30'56"-6d57'
Neuschel,Barbara
Sco 17h27'31"-38d26'
Neustadter,Robert (Chick)
Her 16h25'11"35d55'
Neut Family,The
Per 3h57'1"31d19'
Neutag,Berndt
Tau 5h1'19"15d32'
Neuts,Amanda Leigh
And 23h38'59"42d25'
Neutzman,Leslie Marie
And 1h15'1"36d21'
Neuville,William Joseph
Lyr 18h54'40"34d46'
Neuwirth,Matthew
Her 17h37'20"42d35'
Neuwirth,Peter H
Aql 18h57'23"7d3'
Neuzil,Michael
Per 2h47'43"43d28'
Nevada,Elisa
And 1h23'29"38d36'
Nevara,Carolyn Rose
Eri 3h54'26"-2d15'
Nevarez,Alberto
Dra 18h15'47"65d2'
Nevarez,J Antonio
Sct 18h44'39"-6d19'
Nevarez,Rudy
Dra 17h32'52"60d8'
Neve,Alexander Joseph
Sct 18h54'0"-6d13'
Neve,Anthony Mario
Aql 19h31'30"-6d53'
Neve,Michael
Gem 6h39'46"30d28'
Nevel,Harper Layne
Uma 11h48'0"63d29'
Nevel,Sue
Lyn 8h24'52"41d47'
Neveling,Ursula
Lib 15h10'52"-21d40'
Nevenka
Cyg 19h33'53"33d8'
Nevenka
Mon 6h57'13"-1d7'
Never Be Lonely, Kelly C
Lyr 18h30'34"38d38'
Never Ending Love
Cyg 21h50'38"40d17'
Never Ending Story
Boo 15h3'33"23d19'
Never Far
Uma 11h23'43"38d10'
Never Nervous Pervis
Cyg 21h8'0"38d33'
Never Never Land
Eri 3h44'18"-3d59'
Never Too Far
Lyn 8h45'55"45d24'
Nevera,Sr,Joe
Uma 11h43'43"45d48'
Nevera,Susan Porto
Uma 11h40'1"47d18'
Neverisky,Joseph Donald
Cet 0h50'11"0d51'
Neverland
Uma 12h25'34"62d22'
Neverland
Oph 18h6'0"1d27'
Neves,Erin
Peg 22h39'22"11d17'
Nevett,Christopher Edward
Per 4h3'5"37d42'
Neveux,Pierre
Lyr 19h22'1"30d32'
Neveux,Regis
Cam 13h9'1"82d24'
Nevia
And 1h19'29"40d60'
Nevile,Christopher
Cyg 21h0'16"28d34'
Nevill-A Real Sweetheart,Cary
Sge 20h13'55"17d53'
Neville
Aql 18h44'11"8d58'
Neville's Nova
Mon 8h5'11"-2d59'
Neville,Kristin Marie
Ser 16h4'56"13d31'
Neville,Linda Joyce
Peg 22h29'40"18d48'
Neville,Micheal
Aur 4h57'33"48d56'
Neville,Whit
Cnv 12h12'57"37d43'
Neville,William Joseph
Her 17h21'36"46d57'
Nevin,Philip
Uma 10h17'23"41d16'
Nevin,Robert M
Ori 5h29'48"-1d59'
Nevins Star,The Pauline
Cyg 19h27'0"34d7'
Nevins,Caitlin
And 1h52'36"36d37'
Nevins,Michele
Mon 7h42'46"-2d22'
Nevins,Sheila
Mon 7h56'42"-1d17'
Nevirs,Thomas Ryan
Lac 22h8'35"49d49'
Nevo Nebula
Lmi 10h48'59"31d6'
New Athens
Her 16h50'15"48d9'
New Beginnings
Peg 23h32'31"21d29'
New Bern/Wilmington
Cougars
Aql 18h58'1"-5d18'
New Experience
Aur 5h15'1"40d37'
New Greenland
Cmi 8h5'0"0d4'
New Horizon
Lup 15h29'36"-45d50'
New Horizons Landscaping
Boo 14h51'11"32d56'
New Life Mills
Dra 17h35'42"64d45'
New Lyn
Lyr 19h20'35"30d53'
New Theatre
Aql 19h57'45"13d25'
New,Benjamin
Lyr 18h31'51"41d26'
New,Daniel Glen Brink
Her 18h0'40"28d30'
New,Happy 60th
Birthday,Ronald
Crt 11h31'56"-12d7'
New,Jo
Cyg 19h43'49"30d39'
Newark, New Jersey
Uma 11h30'46"41d1'
Newberry Star,The Marion
Ori 6h3'23"2d34'
Newberry,Arielle Natalie
Eri 5h4'36"-5d42'
Newberry,Charlotte & Ralph
Eri 3h0'55"-18d56'
Newberry,Donald N
Ser 15h11'18"21d54'
Newberry,Heather Llyynne
Lyr 18h29'55"34d37'
Newberry,Leslie
Peg 22h26'23"20d28'
Newberry,Virginia
Mon 6h28'1"8d37'
Newberry,William J
Her 16h22'23"20d36'
Newbery,David P
Lac 22h12'23"54d28'
Newbery,David R
Cam 8h31'45"77d40'
Newbery,Elizabeth Anna
Cyg 21h31'41"50d22'
Newbill,Danielle Nicole
Aqr 22h26'1"0d35'
Newbold,Amy Joan
Umi 17h21'39"75d14'
Newbold,Ashlee Katana
Mon 6h37'14"7d45'
Newburg,Roy Carson
Uma 11h59'16"62d43'
Newby
Lmi 10h34'37"30d8'
Newby,Anthony Claude
Per 3h26'11"40d11'
Newby,Betty Sue
Aur 6h31'42"37d6'A
Newby,EmyLee Lynn
Aur 6h31'42"37d6'B
Newby,Micah Rae
Lyr 18h31'14"46d59'
Newby,Sarah Marie
Cas 23h15'40"62d32'
Newcomb "Star Wars
Writer",Charlene
Vul 19h45'29"28d34'
Newcomb,Muffy
Lyn 9h13'58"40d51'
Newcomb,Sarah May
Lyr 18h45'50"35d5'
Newcombe & Elvia's Wish Star
Lyr 19h8'1"37d58'
Newcombe's Christmas
Star,Susan
Lyn 7h55'22"38d19'
Newcombe,Jemma Fiona
Aur 5h3'30"47d21'
Newcomer,Garrett
Cep 21h52'39"61d37'
Newcomer,Jason
Cep 21h52'23"60d60'
Newcomer,Warren Claire
Sex 10h3'33"-0d12'
Newell's Hope
Ori 5h21'25"1d8'
Newell,Amanda Alexandria
Aql 20h4'23"0d28'
Newell,Andrew
Aur 5h29'38"29d2'
Newell,Connie Soder
Aur 6h2'17"38d34'
Newell,Craig A
Uma 11h23'50"45d53'
Newell,Evelyn
Mon 6h19'20"4d15'
Newell,Jillian Ellen
Cam 7h30'27"68d23'
Newell,Lillian
Cyg 21h2'54"37d12'
Newell,Marshall D
Aql 18h58'0"3d50'
Newell,Michael Frank
Her 16h7'23"48d57'
Newell,Michael Patrick
Per 4h34'27"37d35'
Newell,Nancy
And 2h21'55"44d48'
Newell,Rick "Clover"
Peg 21h26'40"7d38'
Newell,Ryan Paul James
Umi 13h16'19"76d34'
Newell,Sr,Robert Carlyle
Cyg 21h11'46"37d43'
Newell,Vanessa Kirsten
Lyn 6h34'1"54d30'
Newell,Virginia Hadley
Cyg 19h22'0"30d5'
Newell,Zoe
Cyg 20h4'32"39d26'
Newell-Father Of No Sun,KJ
Sct 18h43'1"-6d51'
Newgent,Wilametta (Billie)
Francis
Cyg 21h53'36"38d39'
Newham,Jessica Ann
Cra 18h16'56"-40d17'
Newhard,Michael
Per 4h3'14"37d44'
Newick,Lorne
Lyr 19h25'26"41d33'
Newiger,Nicola- Alexandra
Cas 1h58'24"58d51'
Newirth-Chanin,Wendy G
Psc 1h37'41"27d36'
Newjack
Sct 18h47'35"-6d50'
Newlan,Debra A
Mon 7h3'32"-0d41'
Newland,Angela Shawn
Cam 4h54'29"61d23'
Newlands,Sam & Jean
Del 20h38'12"2d54'
Newley,Doris
Sgr 20h9'31"-36d29'
Newlin,Stephen R
Lmi 10h0'16"35d9'
Newlove,Tracey
Vir 13h15'35"-1d0'
Newman "Anchor Par
Excellence",Kevin
Her 18h1'38"14d53'

Newman's Star,Thomas
Her 17h9'16"20d35'
Newman,A Wink & A Smile,Beth Merion
And 0h50'0"36d52'
Newman,Andrew
Ori 5h55'34"19d45'
Newman,Arthur
Uma 9h8'0"49d14'
Newman,Austin David
Ser 15h34'12"18d10'
Newman,Barb
Ori 5h58'37"14d48'
Newman,Barry
Aur 6h54'25"38d39'
Newman,Becky
Cas 1h20'41"53d26'
Newman,Bob
Aql 20h9'51"0d28'
Newman,Cathy
Peg 22h10'27"4d16'
Newman,Cory Randal
Cet 2h14'35"-10d25'
Newman,David P
Cet 0h58'7"-1d55'A
Newman,Dessa Reneé
Cet 1h0'21"1d17'
Newman,Drew
Ser 15h53'26"18d26'
Newman,Erna-Bernard
Lmi 10h12'0"36d52'
Newman,Forever Gary
Cep 21h21'49"58d47'
Newman,Heather T
Cas 0h46'27"62d5'
Newman,Jeffrey
Lac 22h45'40"37d38'
Newman,Jeffrey Bryan
Her 17h52'30"14d50'
Newman,John Gordon
Boo 13h53'21"21d50'
Newman,John J
Her 17h6'12"49d47'
Newman,Josephine H
Cet 0h58'7"-1d55'B
Newman,Jr,T K
Cep 0h2'28"69d52'
Newman,Judith
And 2h26'22"47d17'
Newman,Kayla Paige
Oph 18h16'25"8d7'
Newman,Kelly
Del 20h17'33"10d15'
Newman,Kristin Denise
Crt 11h22'41"-19d54'
Newman,Lisa Diane
Equ 20h57'22"2d27'
Newman,Maren L
Mon 7h4'52"-7d8'
Newman,Mary Jane
And 2h18'45"38d6'
Newman,MD,Glenn Nathan
Aql 19h6'50"3d43'
Newman,Michael
Aur 4h59'23"38d19'
Newman,Nicholas Soderstrom
Her 16h51'1"40d23'
Newman,Pamela J
Lyn 8h5'53"50d18'
Newman,Peter H
Dra 12h38'1"76d31'
Newman,Peter Henry
Tau 5h2'31"16d36'
Newman,Randall Lee
Lac 22h6'21"49d24'
Newman,Richard
Cep 20h12'37"60d27'
Newman,Rick
Aur 6h5'36"32d7'
Newman,Ricky
Tri 1h55'51"28d46'
Newman,Roger
Lac 22h52'40"40d57'
Newman,Roger Allen
Cam 3h45'36"73d20'
Newman,Sarah Marie
Cet 1h33'0"-0d4'
Newman,Star Bruce
Cnv 12h29'26"32d51'
Newman;My Shining Star,David Paul
Crb 16h14'22"30d15'
Newmark,Matthew
Cam 8h3'39"60d50'
Newmyer,Crista Lee
Sex 10h33'33"-1d10'
Newner IV,John
Sco 16h26'55"-38d6'
Newnham,Maxwell Simon
Ori 5h27'24"-6d50'
Newport Heights P T A
Peg 23h35'25"23d4'
Newport Leasing
Peg 21h58'21"28d50'
Newport,Jay
Sex 10h39'56"2d30'
Newsham,David Lee
Cet 1h38'19"1d9'
Newsom,Dorothy
And 0h15'27"37d38'
Newsom,James Graham
Cet 2h30'14"8d45'
Newsome,Dana & Reggie
Mon 6h34'21"-6d40'
Newsome,Donald
Her 16h4'23"48d49'
Newsome,Linda
Vul 20h19'11"28d35'
Newson,Anthony
Leo 10h55'0"-0d48'
Newspaper for the Woman
Cas 1h1'48"61d17'
Newth,Jennifer E
Lyr 18h49'36"40d50'
Newton The Midnight Idol,Wayne
Her 16h12'1"41d24'
Newton,Alexis Marie
Aql 18h58'37"-4d51'
Newton,Allen Dean
Her 18h13'1"38d8'
Newton,Bert
Pho 0h51'32"-48d33'
Newton,Bobby Gene
Per 4h25'47"51d8'
Newton,Brenda Gail
Cyg 20h52'16"30d5'
Newton,Brian
Uma 11h47'56"60d10'
Newton,Burt
Cra 18h29'19"-41d58'
Newton,Carl
Aql 19h10'22"13d33'
Newton,Christopher
Oph 18h6'41"1d8'
Newton,Don & Liz
Crb 15h57'50"28d0'
Newton,Elizabeth Ella
Peg 0h2'42"30d34'
Newton,Emma
Cyg 21h6'34"48d53'
Newton,Eric James
Oph 17h58'14"11d9'
Newton,Eric Wayne
Lac 22h11'17"51d43'
Newton,Helmut
Her 17h34'41"40d56'
Newton,James DeWitt
Dra 16h23'13"68d43'
Newton,Kathleen & Wayne
Cyg 20h46'0"37d58'
Newton,Kenneth James
Cep 22h20'22"80d32'
Newton,Marge
Cas 0h58'1"63d53'
Newton,Pamela
Cyg 20h20'31"40d28'
Newton,Robert
Cep 21h43'29"70d17'
Newton,Rohan Murray Bannister
Cep 3h2'31"78d40'
Newton,Sarah Anne
And 0h21'29"38d37'
Newton,Stuart & Tracey
Lyr 18h31'44"31d49'
Newton,Wayne
Cmi 7h40'13"1d7'
Newton,Wayne & Kathleen
Cyg 20h59'49"31d27'
Newton-Smith, Christopher
Her 17h47'27"40d24'
Newville,Suzanne
Ori 5h57'1"18d4'
Next One,The
Cam 9h57'52"82d22'
Next One,The
Ori 5h45'1"12d9'
Nexus
Equ 21h1'22"8d35'
Nexus & Jacqueline's Te Tauaroha
Cet 3h15'17"2d26'
Nexus 6
Cyg 21h51'19"41d16'
Nexus 6-230779,Emma Jane
Cas 1h57'26"61d1'
Nexus,Shannon
Mon 6h20'35"8d39'
Ney,Linus
Ori 4h54'17"4d44'
Neybert,Dr Hilary F
Lyn 7h44'32"48d48'
Neye,Lucille Lorena
Cyg 20h9'20"39d26'
Neyedly,Grace Milly
Ari 3h0'56"25d43'
Neylin,Jo
Umi 14h39'20"81d53'
Neyman,Benny
Dra 17h41'55"63d56'
Nez & Artie's Aloha Kai
Mon 7h47'42"-7d49'
Ng
Aqr 22h4'1"-8d52'
Ng & Alan Shu,Angela
And 1h45'55"38d37'
Ng,Jennifer
Mon 6h42'26"11d52'
Ng,Karina
And 23h36'11"40d59'
Ng,Keith & Julianna
Lyr 19h14'50"42d59'
Ngai,James G
Cnv 13h53'33"37d37'
Nghia N Vo
Aur 5h23'14"40d4'
Ngo,Kelly
Mon 6h29'24"8d46'
Ngoâ Sao Tu-Huong Noël
Cas 3h2'59"58d38'
Nguyen January 25,1959 Lan Mai
Tri 2h9'11"32d27'
Nguyen Scientist,Van
Aql 19h28'43"12d4'
Nguyen Thi Em
Lyr 18h59'22"38d55'
Nguyen,Alan
Lyn 7h32'13"35d44'
Nguyen,Douglas
Ori 6h4'12"10d51'
Nguyen,Jacqueline
Tau 5h40'0"26d57'
Nguyen,Judy
Aql 20h18'0"7d32'
Nguyen,Lan Thingoc
Uma 11h41'30"38d9'
Nguyen,Paula
Mon 6h36'18"7d9'
Nguyen,Sidonie
Aur 5h53'46"50d34'
Nguyen,Thien
Aql 19h12'12"19d31'
Nhu-Ngoc
Lmi 11h4'1"31d20'
Ni Ste Star
Ant 10h31'25"-30d3'
Ni,Debbie
Peg 23h37'26"17d37'
Niagara Sails Forever
Lac 22h10'0"49d7'
Niall
Lac 22h47'54"54d56'
Niall
Vir 11h52'44"1d44'
Niall
Ari 1h58'40"20d14'
Nialy
Lmi 10h17'12"36d40'
Niamir,Kambiz
Her 17h17'21"44d38'
Nibbles
Lyr 19h16'0"28d12'
Niblock's Trigon
Lyn 9h4'59"42d48'
Niblock-Smith Art Star,Mark
Del 20h20'32"18d58'
Nic & Nina
Leo 11h14'44"-1d43'
Nicaise,MD,Robert J
Oph 18h16'30"1d27'
Nicastro,Laurie Ann
Cyg 20h23'1"40d14'
Nicci
Cep 22h23'13"63d49'
Nicci The Special Bunnygirl
Cyg 19h58'25"47d2'
Niccol,Andrew
Aql 19h1'1"4d55'
Niccoli,Patricia
And 0h11'29"47d42'
Niccolo & Josephine
Lyr 18h59'1"40d9'
Niccum,Elissa
Mon 6h29'42"-6d42'
Nice'n'Naughty Nancy
Cet 2h56'0"2d11'
Nice,Philip David
Cep 4h13'13"80d23'
Nice,Sally Anne
And 0h34'45"30d53'
NicEli
Peg 22h34'40"25d32'
Nicelli,Alessandra
Cyg 21h28'35"48d49'
Nichele,Briale
Del 20h18'53"10d60'
Nichi Loves Ryan
Lyr 19h0'28"26d53'
Nichik,Peter T
Aur 7h19'12"39d57'
Nichol,Margaret
Aql 20h6'55"0d20'
Nichola Teresa Paulo Christmasstar
And 2h26'16"42d0'
Nicholas
Umi 15h1'17"66d8'
Nicholas
Dra 16h30'40"70d25'
Nicholas zottirodilattanzio
Cae 4h46'38"-34d8'
Nicholas
Ori 6h4'18"-0d21'
Nicholas
Aur 5h31'29"37d46'
Nicholas
Aqr 20h54'52"0d51'
Nicholas
Lac 22h24'23"37d33'
Nicholas
Cep 22h47'31"58d32'
Nicholas
Uma 10h46'12"40d53'
Nicholas & Alexandra
Crb 16h11'15"37d56'
Nicholas & Firefly
Aql 19h55'17"13d34'
Nicholas & Matthew D
Peg 0h2'42"31d16'
Nicholas Christopher
Aur 5h5'45"40d52'
Nicholas Erik
Oph 17h52'30"-6d49'
Nicholas James
Per 3h32'20"31d41'
Nicholas Kent
Leo 10h16'1"13d2'
Nicholas Milo
Lac 22h21'48"50d21'
Nicholas Rae
Cnv 13h10'54"41d10'
Nicholas Raymond
Per 3h59'56"51d34'
Nicholas Ryan
Tau 4h12'26"20d53'
Nicholas Stephen
Her 17h25'50"18d46'
Nicholas' Night Light
Her 17h33'53"22d49'
Nicholas,George M
Uma 10h22'17"61d32'
Nicholas,Ivan
Tau 3h59'49"1d6'
Nicholas,Jane E
Cyg 20h54'1"30d43'
Nicholas,José Amy
Dra 16h38'52"67d9'
Nicholas,Michael Christopher
Lyn 9h4'33"37d45'
Nicholas,Novella
Lyr 19h21'13"38d3'
Nichole
And 0h47'34"40d32'
Nichole
And 1h33'29"37d37'
Nichole
Her 17h35'28"40d12'
Nichole Charmaine
Mon 6h35'26"-0d39'
Nichole Marie
Crb 16h12'47"32d19'
Nichole Victoria (Slamka)
Lyr 18h40'28"29d21'
Nichole's Halo
And 0h10'38"31d40'
Nichole,Brandy Nichole
Peg 23h28'14"24d43'
Nicholet Renae George
Uma 11h58'18"45d46'
Nicholi
Hya 8h12'33"4d30'
Nicholl,Trevor
Per 1h51'27"47d15'
Nicholls,Barrie
Cyg 19h27'18"37d5'
Nicholls,Charles Herbert
Boo 13h44'34"24d57'
Nicholls,Donna
Vul 20h14'49"25d48'
Nicholls,Meredith Lee
Cam 8h33'20"77d52'
Nicholls,Mr Dave
Ori 5h56'1"7d23'
Nicholls,Roy
Cyg 19h30'37"30d9'
Nicholls,Stephen C
Cam 8h0'18"68d60'
Nicholls,Steven Henry
Aql 20h6'35"7d27'
Nicholls,William Stephen
Per 2h59'21"43d35'
Nichols Star,The
Cnv 12h20'46"41d29'
Nichols,"Cowgirl" Lorie G Russell
Peg 22h42'44"10d9'
Nichols,Alexandra Elizabeth
Peg 22h47'1"32d53'
Nichols,Becky
Uma 9h6'25"59d15'
Nichols,Bobbie
Lac 22h11'51"48d46'
Nichols,Brice
Dra 11h37'0"73d19'
Nichols,Carol
Hya 8h14'38"6d21'
Nichols,Charles R "Chuckmeister"
Dra 16h52'0"51d37'
Nichols,Cheryl Ann
Cam 3h54'11"60d24'
Nichols,Cheryl L & Ward
Cyg 20h1'53"30d28'
Nichols,Claire R
And 0h46'27"34d46'
Nichols,Claudia & Gary
Crt 11h15'26"-10d19'
Nichols,Debbie
Umi 14h46'1"68d43'
Nichols,Diana
Peg 23h48'26"31d51'
Nichols,Don
Cet 2h40'40"-6d53'
Nichols,Doyal
Boo 14h24'21"16d38'B
Nichols,Gerald Keith
Per 2h25'48"56d55'
Nichols,Gloria
Cas 0h39'24"66d45'
Nichols,Gregory
Aur 6h25'0"38d51'
Nichols,Guy Lewis
Sgr 20h11'20"-28d47'
Nichols,Ian
Aur 6h3'0"35d50'
Nichols,Jeffrey Cole
Her 18h8'45"31d48'
Nichols,Jennifer
Lyr 19h18'37"38d56'
Nichols,Judy's Nick, Richard A
Per 3h11'51"56d39'
Nichols,Juli
Cyg 21h32'29"40d8'
Nichols,June
Cas 1h15'20"64d18'
Nichols,Karen Lee
Uma 8h39'1"68d25'
Nichols,Katherine Mary Matthews
Cas 23h33'38"61d5'
Nichols,Kathleen A
Cas 3h9'1"67d55'
Nichols,Kelly Maree
Cas 0h30'55"64d59'
Nichols,Malcolm Henry
Per 1h48'0"53d39'
Nichols,Martin VanArsdale
Aql 20h10'17"13d16'
Nichols,Martin R
Aur 7h3'51"38d4'
Nichols,Mary Beth
Eri 3h57'18"-11d6'
Nichols,Melissa Dawn
And 2h13'37"50d37'
Nichols,Norman Lee
Sgr 19h37'35"-32d19'
Nichols,Randy Ray
Aur 6h55'57"44d19'
Nichols,Richard Dean
Her 17h39'44"40d51'
Nichols,Star Of Paul
Dra 17h49'42"64d7'
Nichols,Vince & Sarah
Eri 3h39'16"-11d53'
Nichols,Wanda
Boo 14h24'21"16d38'A
Nichols-Klock,Karen
Peg 22h1'38"31d32'
Nichols-Warren,Austin
Dra 16h0'35"68d38'
Nicholson Star,In Loving Memory Betty
Uma 11h31'10"63d24'
Nicholson Stars
Per 1h30'1"53d29'
Nicholson,Barbara Irene
Cyg 19h30'29"35d39'
Nicholson,Brian
Ori 5h33'48"0d56'
Nicholson,Bruce Wiley
Cet 1h57'14"-3d16'
Nicholson,Dan & Cecilia
Cmi 7h7'21"9d45'
Nicholson,Doris B
Equ 21h3'0"2d26'
Nicholson,Elissa Alexandra
Cyg 20h29'0"37d44'
Nicholson,Elizabeth
Sgr 18h0'18"-28d60'
Nicholson,Emily Ann
Cyg 21h03'54"50d15'
Nicholson,Harriet Elizabeth
Mon 6h30'49"-1d2'
Nicholson,Jeri
Oph 18h34'48"10d37'
Nicholson,Joyce
Lyn 6h37'38"58d28'
Nicholson,Jr,James Leonard
Aur 6h17'26"46d37'
Nicholson,Kathleen
Cam 5h6'44"67d36'
Nicholson,Keith James
Uma 12h59'51"53d18'
Nicholson,Kristen Heather
Sco 17h34'44"-43d25'
Nicholson,Lady
Cas 1h45'0"74d14'
Nicholson,Maureen B
Cam 4h3'31"65d36'
Nicholson,Mike
Cet 2h11'40"4d2'
Nicholson,Norma M
And 23h29'25"43d1'
Nicholson,Scott,Cindy Reed & Revy
Uma 10h47'28"50d41'
Nicholson,Sian Marie
Aur 6h3'59"30d15'
Nicholson,Sr,Ray
Her 18h13'18"38d59'
Nicholson,Stella
Psc 1h27'48"30d20'
Nicholson,Stella Anne
Cas 0h26'1"61d53'
Nicholson,Tonijean
Cas 23h30'13"61d5'
Nicholson,William R
Cet 3h5'45"0d31'
Nicholson,William W
Aql 19h59'52"10d12'
Nick
Uma 11h7'41"43d49'
Nick "My Hero"
Her 18h3'54"28d22'
Nick & Becca
Sge 19h37'58"16d34'
Nick & Becca
Sge 19h38'53"18d53'
Nick & Cindy
Sge 18h59'15"19d10'
Nick & Tess
Sge 19h0'51"20d36'
Nick Always
Cyg 19h31'17"35d34'
Nick The Night
Lyn 8h12'0"38d36'
Nick's Force
Her 17h59'20"20d19'
Nick's Star
Cep 22h29'28"80d26'
Nick's Star
Dra 16h13'23"61d57'
Nick,Christine & Randy
Cnv 13h24'21"50d46'
Nick,CLU,ChFC,Paul M
Sct 18h30'47"-6d11'
Nick,Gary s
Dra 14h56'0"62d43'
Nick,Julian & Cheryl Ann Verde
Ser 17h54'52"-14d21'
Nick,Phyllis
Vul 19h58'52"22d58'
Nick-las
Lyr 18h59'15"30d57'
Nick-Rosa
Lac 22h6'28"38d43'
Nickel,A J
Uma 9h56'21"50d7'
Nickel,April Bliss
Lyn 6h15'1"60d49'
Nickel,Harald
Cnc 8h27'1"33d9'
Nickel,Lisa Tiffany
Cam 3h35'55"60d21'
Nickel,Lt Col Jack "Triple"
Aql 19h36'39"-0d4'
Nickell's Debra Ruth
Mon 7h41'1"-0d38'
Nickelly
Cyg 21h53'42"41d23'
Nickels
And 0h12'44"31d16'
Nickels,David
Her 17h26'32"40d38'
Nickelson,Anita JoAnn
Oph 17h56'20"10d37'
Nickens,Jr,Roy L "Imzadi"
Per 2h49'43"31d54'
Nickerbacher Star
Lib 15h3'1"-28d38'
Nickerson,Sylvia
Cas 1h17'34"64d49'
Nicki
Tau 3h54'47"30d30'
Nicki
Boo 13h43'25"11d7'
Nicki
Cet 2h21'42"-8d48'
Nicki & Greg Soulmates Forever
Cyg 21h4'42"32d20'
Nicki Annelle
Cet 1h47'54"0d39'
Nicki,Katie,Emily
Cep 22h35'41"58d12'
Nickle Hotdog
Cam 4h49'29"70d7'
Nickle,Andrea "Sweety"
Crb 16h3'54"28d13'
Nickles Star,The Jill
Lyn 6h33'17"60d30'
Nicko's Star
Ori 6h7'16"8d30'
Nickolas
Lmi 9h58'1"38d37'
Nickolas
Cyg 21h50'25"37d32'
Nickolas My Shining Star
Ser 15h25'41"1d26'
Nickolas,Katie Elizabeth
And 23h29'20"42d33'
Nickolas,The Brightest Star of All
Cep 23h15'54"64d28'
Nicks The Star of Rhiannon,Stevie
Uma 10h31'42"41d10'
Nicks,Matthew
Her 17h34'49"20d53'
Nickvena
Dra 16h10'33"65d12'
Nicky
Lmi 10h9'49"37d39'
Nicky & Mary "Always & Forever"
Cyg 19h32'35"38d54'
Nicky 29
Lyr 18h56'19"31d20'
Nicky Forty 25-12
Aur 5h26'14"50d16'
Nicky's Diamond in the Sky
Ori 5h52'54"14d31'
Nicky's Home Base
Eri 3h54'26"-10d28'
Nickyvon
Cyg 19h29'41"35d3'
Nico
Ori 6h3'1"8d28'
Nico
Her 17h54'53"14d43'
Nico
Tri 2h33'36"34d46'
Nico
And 1h8'25"47d34'
Nico Diaz,Mi Estrella Brillante
Aql 19h59'32"10d16'
Nico x Laura
Cyg 19h53'44"47d27'
Nico,M Cristina
Nor 16h22'36"-59d28'
Nicogossian,Elizabeth
Aql 20h1'55"13d49'
Nicol,Fred And Bonnie
Crb 16h23'0"3d2'
Nicol,John"Golden Boy"
Aur 6h27'48"38d42'
Nicol,Joseph
Ori 5h31'0"-1d30'
Nicol,Manfred
Hya 8h24'1"1d17'
Nicol,Scott (Nae Hassle) Murray
Uma 9h58'34"53d31'
Nicol,Stuart James
Aur 5h52'15"41d11'
Nicola
Vir 12h47'55"-0d44'
Nicola
Peg 21h51'53"30d33'
Nicola
Cyg 19h27'51"36d46'
Nicola
Ara 17h55'35"-52d24'
Nicola
Del 20h23'52"3d3'
Nicola
Per 2h10'12"58d25'
Nicola & Paul Forever
Cyg 20h1'31"31d33'
Nicola Jane
Lyr 19h11'0"41d12'
Nicola K
Cep 20h31'15"62d46'
Nicola Lannie
Per 4h9'52"40d48'
Nicola Lisa
Peg 23h42'31"30d15'
Nicola S
And 23h44'45"41d46'
Nicola Tracy Child Memorial Star
Cyg 20h11'29"41d28'
Nicola Wonderous Beauty
Lyr 18h34'13"28d8'
Nicola's Star
Umi 13h15'1"70d2'
Nicola,Barbara
Lyn 7h49'37"52d22'
Nicola,Cheril
Lyr 18h56'26"30d12'
Nicola,Gian
Cet 1h30'29"-3d9'
Nicola,Nassira
Peg 21h11'45"18d54'
Nicolacopoulas, Elizabeth "Liz"
And 0h8'36"34d59'
Nicolanna
Col 6h28'45"-36d9'
Nicolas
Dra 20h3'43"63d46'
Nicolas
Lac 22h39'31"38d60'
Nicolas
Lac 22h38'40"55d3'
Nicolas
Uma 12h1'1"34d33'
Nicolas Anthony
Lyn 7h50'56"44d40'

Nicolas C
Sgr 20h17'43"-37d33'
Nicolas,Jean-Michel
Dra 10h11'36"74d52'
Nicolas,L
Per 2h37'36"56d23'
Nicolas-Stern,Victoria
Cnc 9h1'38"18d48'
Nicolau,Teodor
Uma 10h19'32"47d39'
Nicolaus,Karl-Heinz
Cap 21h41'1"-22d31'
Nicole
And 2h28'19"50d3'
Nicole
Cnv 12h7'0"38d26'
Nicole
Cas 1h17'50"63d53'
Nicole
And 0h20'44"30d13'
Nicole
Eri 3h17'37"-1d49'
Nicole
Lyn 7h29'38"42d45'
Nicole
And 23h18'20"37d59'
Nicole
Lyn 8h43'48"37d43'
Nicole
Lyr 18h49'41"32d11'
Nicole
Vir 12h56'30"10d27'
Nicole & Colin 25
Cyg 19h47'43"30d13'
Nicole & Kevin,The Star of
Cyg 19h32'12"34d54'
Nicole & Michael
Aur 5h17'21"43d44'
Nicole & Stuart Forever In Time
Lac 22h32'39"35d36'
Nicole Ashley
Del 20h57'13"12d28'
Nicole Danielle
And 0h58'38"37d11'
Nicole Danielle
Cas 1h27'0"60d7'
Nicole DS-1
Lyn 7h48'48"44d37'
Nicole Elise
And 23h21'23"48d8'
Nicole F
Boo 14h28'16"8d49'
Nicole K
Cas 22h57'36"54d43'
Nicole Krystine
Vir 12h29'51"7d34'
Nicole Loves Brad
Lyn 7h52'37"33d22'
Nicole Lynn
Tri 2h19'57"33d19'
Nicole Lynne
Com 13h18'47"28d38'
Nicole Maree
Lyr 18h54'17"41d39'
Nicole Maria
And 0h3'57"40d3'
Nicole Marie
Cas 0h27'38"56d12'
Nicole Marie
Cyg 21h33'35"30d48'
Nicole Marie
Peg 23h6'1"10d28'
Nicole Marie
Hya 9h16'34"1d19'
Nicole Marie
Lac 22h35'27"56d31'
Nicole Marie
Cet 2h17'27"9d59'
Nicole Marie
Com 12h29'13"25d50'
Nicole Marie
Peg 22h13'42"20d47'
Nicole Michelle
And 0h21'45"35d52'

Nicole Rose
Vul 20h14'59"25d35'
Nicole Tina
Cas 0h17'59"61d40'
Nicole Wawra-Ich Liebe Dich-Erwin
Cnc 9h16'1"18d44'
Nicole's Echo
And 2h9'0"38d23'
Nicole's Fire
Aql 20h0'24"4d21'
Nicole's Golden Twinkle
Mon 6h54'30"-10d51'
Nicole's Guiding Light
Peg 22h0'40"31d12'
Nicole's One & Only
Psc 22h55'32"5d49'
Nicole's Star
Lyr 18h36'20"35d55'
Nicole's Star
Cas 1h30'0"60d34'
Nicole's Stern
Lib 15h2'45"-28d20'
Nicole's Wishing Star
And 1h19'53"38d25'
Nicole-Ryan
Ori 5h52'32"14d31'
Nicoles Star
Cnc 9h2'24"22d21'
Nicolet,Roland "60"
Per 3h57'31"50d45'
Nicoletta
Mon 7h58'52"-4d5'
Nicoletta
And 0h4'53"47d8'
Nicoletta
Pyx 8h46'9"-21d42'
Nicoletta
Vel 9h43'5"-45d55'
Nicoletta,Marianna
Cap 21h16'13"-26d51'
Nicolette
And 23h45'38"38d47'
Nicolette
Cyg 20h39'22"41d13'
Nicolette & Wayne
Eri 4h9'52"-13d44'
Nicolette's "Happy Sparkle"
Psc 1h32'12"20d39'
Nicoletto
Oph 17h15'27"10d5'
Nicolina
And 2h20'38"50d14'
Nicolle
Vel 9h22'3"-40d52'
Nicolo,Giuseppina
Cep 20h24'38"75d16'
Nicolopoulos,Georgina
Cas 1h10'45"62d50'
Nicolò
Boo 15h7'0"47d40'
Nicos
Peg 23h3'27"30d42'
Nicosia,Domenico Giuseppe
Cyg 21h51'13"42d25'
Nicotera,Vito Leonardo
Cmi 7h41'21"5d26'
Nicoud,Ray
Leo 11h22'25"-1d56'B
Nid Du Hibou
Lac 22h8'1"51d47'
Niday,Joshua Lawrence
Aur 6h7'23"33d32'
Niday,Michael Richard
Her 17h7'52"20d28'
Nidecker,Sean A
Dra 18h38'37"67d50'
Nidgren,Christopher Justin
Per 3h59'1"48d59'
Nidia
Pic 4h47'18"-49d15'
Nidia Esther
Equ 20h55'10"2d46'
Niebel's 50th ANR
Crb 16h14'39"34d56'

Niebergall,David Cooke
Ser 15h31'1"24d51'
Niebergall,Pearl G
Sco 17h52'15"-31d32'
Niebes,Jonathan David
Cap 20h26'48"-13d41'
Niebla,Alexander
Per 2h57'32"34d27'
Niebling,Michelle Lea
Lyn 8h19'50"34d60'
Niebling,William Vernon
Dra 18h35'24"58d33'
Niebor,Sevy
Cyg 20h58'59"50d15'
Niebruegge,Penelope Jean
Peg 22h24'60"25d41'
Niebs,Heather Lynn
Psc 23h1'46"5d10'
Nieburg,Deborah B
And 1h51'57"40d4'
Niebylski,Jessica
And 0h13'13"35d10'
Niedbalski,Donna
Ori 5h3'35"0d4'
Niederberger,Loïc
Peg 23h4'51"31d26'
Niederecker,Brenda
Cas 1h57'20"60d51'
Niederehe,Mary
Lyn 7h7'46"58d57'
Niedermann I,René
Vir 12h4'25"8d3'
Niedermayr,Hans
Her 17h39'24"46d57'
Niederprüm,Richard
Ari 2h25'15"21d38'
Niederstein,Annelie & Peter
Vir 13h6'57"-6d26'
Niedlich,Anne
And 2h34'24"50d0'
Niedzielski,Marek
Hor 3h34'4"-48d9'
Niedzielski,Susie & Peter
Cyg 20h44'53"38d6'
Niedzwiecki,Edith A (Haske)
And 23h24'12"47d11'
Niefer,Werner
Vir 14h6'20"-6d6'
Niel Hans Gylling
Per 2h53'52"34d54'
Nield 37-243,Vic & Thenetta
Cyg 19h30'49"56d59'
Nield,Jim
Aur 7h3'38"38d42'
Nieli,Dale
Per 4h2'27"51d32'
Niels,Billy
Ser 15h18'54"-1d59'
Nielsen,Adrian Lillegaard
Lib 15h0'13"-7d31'
Nielsen,Ali Richter
Her 17h53'16"14d39'
Nielsen,Ayan Biltoft
Uma 9h49'56"46d14'
Nielsen,Brian
Her 18h2'15"38d4'
Nielsen,Corky
Cnv 12h31'38"34d9'
Nielsen,Joseph Edwin
Her 17h21'44"44d16'
Nielsen,Karen S
Cam 3h29'57"60d35'
Nielsen,Katherine Diane
Cam 7h56'1"60d42'
Nielsen,Lisa
Mon 6h40'46"8d45'
Nielsen,Micayla Elyse
Mon 6h18'38"5d55'
Nielsen,Michael Wm
Gem 6h44'18"18d37'
Nielsen,Peyton W
Peg 22h48'37"27d18'
Nielsen,Ralph Odin
Aur 7h5'29"36d8'

Nielsen,Reagan L
Peg 22h53'27"27d41'
Nielsen,Simon Martin
Aur 5h36'42"38d24'
Nielsen,Sofie Marie
And 23h16'1"43d24'
Nielsen,Thomas Christian
Per 3h38'30"38d59'
Nielsen-Collins
Aql 20h5'57"1d7'
Nielsen-Mastick,Carol Ann
Aqr 20h45'26"-0d16'
Nielson,Gordon
Vir 14h0'27"-7d57'
Nielson,Julianne K
Ori 5h51'54"6d36'
Nielson,Kristine Lykke Ostergaard
Cas 0h33'32"50d15'
Niemann,Bonnie Ann
Cam 3h48'53"55d52'
Niemann,Gregory A
Dra 16h59'23"70d35'
Niemczyk,Joan Colleen
Lib 14h59'55"-10d42'
Niemeyer,Karl-Heinz Prof Dipl-Ing
Cnc 8h0'55"10d46'
Niemeyer,Lynn A
Peg 22h38'24"27d30'
Niemi,Christopher
Her 16h15'48"41d38'
Niemi,Kevin John
Aur 5h19'25"42d56'
Niemi,Loreen H
And 23h24'12"41d11'
Niemi,Matthew
Per 1h55'39"50d25'
Niemi,Nicholas
Cam 6h6'1"58d45'
Nienaber,Julie Ann
Aql 19h57'12"14d21'
Niendorf,Werner
Lyr 18h34'26"27d50'
Nienhuis,Clifford Charles
Boo 15h26'41"50d24'
Nienstadt,Katherine K
Lyr 18h27'44"38d39'
Nierenberg,Amelia Jane
And 23h36'28"48d29'
Nierman,Charla
Cas 0h26'47"58d20'
Nies,Becky
Cas 0h43'1"68d37'
Nies,Jennifer Elaine
Cet 0h27'26"1d60'
Niesloss,Kathy
Tri 2h5'0"30d58'
Nieszel,John
Her 16h40'0"34d54'
Nieters,Charles Joseph
Lib 15h29'16"-10d58'
Nieto,Aude
Lyr 19h22'21"31d18'
Nieto,Natalie
Cyg 19h51'51"38d5'
Nieto,Sally
Mon 7h49'45"-6d50'
Nietupski,Vera & Paul
And 0h13'52"46d48'
Nieudorp,Peter
Lac 22h23'46"40d47'
Nieuwenhuis,Kainoa
Cas 2h41'11"70d28'
Nieuwenhuizen,Eric Armand SH van
Cam 4h5'11"58d16'
Nieve Star,The
Lyn 9h14'1"36d10'
Nieves,Ann Marie Nicole
Umi 13h9'10"74d10'A
Nieves,Carmen
Lyr 19h21'1"41d33'
Nieves,Nora
Lyn 7h54'59"58d52'

Nieves-Santiago,Luis
And 0h18'52"35d50'
Niezabytowski,James T
Dra 17h53'1"76d35'
Nig
Cam 6h54'46"80d10'
Nigel
Ori 5h11'16"-5d0'
Nigel
Ind 21h20'55"46d10'
Nigel
Ind 21h21'1"46d49'
Nigel & Sarah
Cyg 19h22'41"44d19'
Nigel Rose
Per 3h33'0"40d31'
Nigeline
Cas 2h34'27"67d58'
Nigg,Nancy,Larry & Derek
Cep 20h12'56"60d44'
Niggli,David
Her 17h54'44"31d47'
Night Flower
Aql 20h31'51"-1d46'
Night Magic
Boo 15h1'29"18d42'
Night Sparkel
Cep 23h10'55"80d17'
Night View Of Mt Jun
Vir 14h25'36"-2d2'
Night,Jeannette Bunny
Cam 9h3'59"73d12'
Night Jewel
Vel 10h7'6"52d8'
Night,Nagel
Uma 11h24'25"63d5'
Night-Al's
Uma 14h24'1"58d41'
Nightengals
And 0h19'43"31d32'
Nightingale Accelerated School
Uma 10h58'57"37d47'
Nightingale Star,The
Mon 8h5'19"-5d51'
Nightingale,Eugenia M
Tri 1h54'26"26d45'
Nightingale,Matthew James
Per 1h47'38"56d44'
Nightingale,Sarah Louise
Cyg 21h30'32"53d57'
Nightly Passions
Uma 11h34'27"32d12'
Nigl,Otto William & Helen Dorothy
Umi 15h20'39"72d5'
Nign III,Daniel
Ser 17h30'53"-10d55'
Nigoom,Amar Khalid
Cam 4h33'16"68d43'
Nigro,Chloe Constance
Psc 0h53'0"7d55'
Nigro,Gregory David
Cap 21h54'40"-22d0'
Nigro,I Love You,Bruce
Hya 8h46'57"-8d1'
Nigro,Jr,Donald L
Sge 19h34'1"18d47'
Nigro,Kimberly Ann
Lyn 7h32'1"58d40'
Nihon-Ongaku-Daidouha
Aqr 21h56'38"0d10'
Nijankin,Robin Beth
And 23h23'39"41d14'
Nik
Uma 11h42'49"38d13'
Nik & Josh
Aql 20h11'33"12d43'
Nik & Lauren
Cyg 21h7'49"31d11'
Nik Est Viginti Unus
Cep 2h20'51"87d11'
Nik-Nak's Canonero Valentine
Oph 17h43'21"12d60'

Nik-Nik
Peg 23h4'48"30d37'
Nika Esther
Cyg 21h38'54"40d55'
Nike
Cep 20h25'35"60d17'
Nikhil
Crb 16h7'40"32d24'
Niki
Gem 6h43'49"31d49'
Niki (Nichola Cottenham)
Per 4h0'47"38d17'
Niki's Night Light
Lyr 18h43'14"41d57'
Niki's Star
And 23h44'16"41d30'
Nikia II
And 23h2'51"48d56'
Nikiforov,George
Cep 23h15'34"64d25'
Nikita
Cae 4h58'24"-33d27'
Nikita
Cas 1h55'23"71d4'
Nikka
Lyr 18h33'28"44d34'
Nikka Jatem
Vul 19h42'50"22d53'
Nikkel,Joelle Katrine
Hya 8h29'16"-9d9'
Nikki
Tri 2h30'45"31d46'
Nikki
Uma 10h59'49"50d45'
Nikki
Cnv 12h22'0"43d13'
Nikki
And 2h5'16"41d36'
Nikki
Aur 5h3'30"40d48'
Nikki
Cmi 7h54'58"8d19'
Nikki
Com 12h56'0"21d25'
Nikki
Cas 0h49'59"71d59'
Nikki "Our Little Shining Star"
Cma 6h50'15"-15d12'
Nikki & Johnny's Star
Uma 10h22'0"60d55'
Nikki 33
Lyr 18h43'34"32d25'
Nikki W
Uma 10h29'0"42d6'
Nikki's Dream
Uma 9h36'29"48d53'
Nikki's Light
Peg 22h30'19"10d55'
Nikko Jose
Her 17h38'0"41d1'
Nikko's Fire
Aql 19h46'44"14d53'
Nikky,Maro
And 0h23'0"40d19'
Niklas Bastian
Cnc 8h36'25"7d17'
Niklas Star
Uma 10h42'1"58d27'
Niklasch,Gustav Ulrich
Cep 22h38'0"61d16'
Niko
Lib 15h18'32"-8d23'
Niko & Karina 6-18-93
Cyg 19h41'24"38d23'
Nikola,Jean M
Cmi 7h27'13"8d4'
Nikolaidys,George + Peppy
Ori 5h57'37"21d50'
Nikolla,Anna
And 2h22'35"45d22'
Nikolla,Anna Chrisanthy
Lyn 8h24'36"44d42'
Nikopoulou,Wasiliki
Aqr 20h57'45"-10d32'

Nikos
Aqr 21h5'60"-1d49'
Nikos & Lisa
Aql 18h58'1"-7d35'
Nikula,Vickie Sue
And 0h24'0"28d17'
Nilan,Patty
Vir 11h45'1"9d33'
Niland,James W
Aur 4h36'0"30d8'
Nile,Anthony
Her 17h30'13"40d33'
Niles
Lib 15h8'0"-1d20"
Niles,Christopher
Her 17h7'28"43d11'
Niles,J Richard
Vul 19h40'48"26d4'
Niles,Jr,Frederick A
Cyg 20h57'51"30d11'
Niles,Michael Arthur
Cet 0h50'19"-2d27'
Niles,Paige
Lyn 8h41'51"46d33'
Niles,Sedona Jae
Vul 19h40'33"26d19'
Niloo
Crt 11h17'22"-13d8'
Nilsen,Cliff & Bennie
Hya 8h43'17"-6d4'
Nilsen,David & Pam
Cam 5h52'45"65d5'
Nilsen,Frh52'45"65d5'
Cam 5h45'44"61d13'
Nilsen,Gina B
Uma 9h52'58"56d4'
Nilsen,Ron
Her 16h48'15"50d50'
Nilsen-Schroeder, Elizabeth
Aql 20h7'29"1d13'
Nilsson
Aql 19h54'16"15d38'
Nilsson,Birgit
Tau 5h52'21"24d1'
Nilsson,Dan & Beth
Eri 3h39'48"-13d50'
Nilsson,Eric
Aql 19h23'1"14d3'
Nilsson,Ernst
Dra 20h29'22"67d54'
Nilsson,Mike
Cas 0h34'18"66d15'
Nilsson,Steven Carl
Per 4h18'54"50d31'
Nimac,Mario
Ori 4h47'27"5d43'
Nimarity (Nick Demos & Mary Yermal)
Cyg 20h7'45"38d15'
Nimatallah,Leila
Cas 3h6'25"58d8'
Nimble Nobby
Aql 20h8'11"4d37'
Nimbus,Cacoilai
Cyg 20h5'46"31d47'
Nimika
Peg 22h40'48"33d46'
Nimmich,Elizabeth Jean
And 0h14'16"38d10'
Nimmo,Caroline Elizabeth
Cas 0h25'31"72d54'
Nimmo,Nicholas Carothers
Per 3h10'32"47d12'
NIMMS Cardinal
Cam 3h43'31"70d22'
Nimr,Ramzi Amer
Cnv 13h41'39"46d51'
Nimr,Rashad Amer
Uma 9h23'42"55d50'
Nin,Provina V
Oph 17h14'48"-23d21'
Nina
And 0h28'58"30d40'
Nina
Eri 3h56'44"-0d18'

Nina
Com 13h25'47"26d31'
Nina
Umi 13h38'19"71d59'
Nina
Aur 5h0'54"45d38'
Nina
And 1h25'41"40d18'
Nina
Uma 11h52'57"60d21'
Nina & Robert
Lac 22h42'52"37d58'
Nina Belle
Ori 5h53'57"16d20'A
Nina Diane
And 23h28'1"42d39'
Nina Elaine
Cyg 19h24'49"35d3'
Nina Elizabeth
Lmi 10h10'36"36d10'
Nina Jean
Eri 3h38'43"-16d18'
Nina Jo
Cas 0h6'59"50d5'
Nina Lee
Mon 6h57'1"7d48'
Nina Ramona-Spirit Of Kwani
Cas 23h20'1"60d16'
Nina's Dreams
Vul 20h15'59"23d46'
Nina's Hope
Aql 18h58'43"-6d4'
Ninburg,Eliot Patrick
Leo 10h45'59"8d19'
Ninchen,Pete élskar
Hya 8h37'15"-11d9'
Nindu
Aur 4h56'39"40d16'
91221-Lisa
Aqr 23h7'20"-11d3'
937
Cet 3h8'52"1d41'
9A
Aql 19h30'56"10d27'
Nine Lives of Frans & Catherine,The
Cet 2h27'34"5d10'
Nine Mäus
Eri 3h32'55"-6d49'
Nine,Carol
And 1h55'10"40d14'
Niner Ben #16
Ori 5h17'37"0d19'
Niner,Elaine Cecelia
Lyn 6h17'45"54d13'
Nines,Greta M
Lyn 7h41'21"44d23'
Ninesling,Cathie
Mon 6h57'1"0d59'
Nineteen
Cnv 13h7'51"50d37'
19 Hermine 13
Cnv 13h48'43"37d44'
1980 Hannah's Light
Gem 6h37'52"13d30'
1993 CNS Northstars Field Marching Band
Peg 23h4'0"20d6'
1994
Tau 5h28'27"15d44'
1994 Advancement Of Excellence #98
Umi 14h17'1"68d24'
1994 Christmas Star, The
Lmi 10h59'28"31d15'
1995 District 18 Sutherin Stars
Lac 22h16'22"46d4'
1995 Frisco VIII Explorers
Ori 5h18'24"0d52'
1995 Great Bridge Piranhas
Aur 5h1'15"49d28'
1995 Rites Of Passage Nubian Star
Mon 8h5'4"-5d44'

Ninetta
Lyr 18h38'31"27d38'
Ninetti,Vittoria
Lyr 19h19'48"42d4'
Ninfa Poo
And 0h54'27"40d33'
Nini
Mon 7h6'41"1d10'
Nini
Ari 2h39'23"20d2'
Nini et Lolo
Uma 10h30'51"52d28'
Nini,Maria
Cap 20h7'1"-10d31'
Nini-Major
Uma 12h5'13"61d40'
Ninia,James Gerard
Boo 14h48'30"28d21'
Ninjy
Aql 19h5'0"4d54'
Ninna Nanna
Pic 4h42'19"-49d58'
Ninneman,Paul Alan
Peg 22h26'20"20d42'
Ninnie
Dra 16h56'49"61d6'
Nino e Raffaella
Pyx 8h31'10"-24d0'
Nino,David
Aql 19h27'41"1d6'
Nino,Marica Gabriel
Mon 7h4'17"-6d37'
Nino,Nora,Carla,Andrea e Paola
Cet 2h6'28"-5d7'
93 Christmas Present From Y to M
Gem 7h13'59"19d41'
Niny's Eyes
Cas 3h8'0"68d18'
Niola,Johnny
Boo 15h1'47"28d59'
Niper,Helen & Joel
Eri 4h13'56"-15d43'
Nippard,Deborah Anne
Ori 5h58'43"15d46'
Nippon Challenge
Cru 12h53'18"-59d37'
Nipsby
Dra 16h7'26"61d57'
Nique
Lyr 18h40'60"28d14'
Nirah 2-19-81
Aql 19h52'12"10d54'
Nire A
Umi 14h42'10"65d50'
Nirenberg,Ira
Cas 3h1'48"57d17'
Nirenberg,Juliet-Lewit
Uma 11h7'0"33d14'
Nirmala "Babe"
And 23h29'20"41d19'
Nirmala George
And 23h17'21"40d39'
Nirovka,Nikolai Asov
Cmi 7h59'46"0d48'
Nirovka,Nikolai Asov
Cma 7h13'16"-13d35'
Nirta,Emile
Ori 5h55'54"19d5'
Nirvana
Lyn 7h27'1"39d40'
Nis Domi
Peg 21h52'52"20d2'
Nisbet
Uma 9h14'31"52d9'
Nisbet,Andy
Cyg 19h30'1"35d25'
Nise
Her 15h49'1"50d13'
Nish,Harry T
Boo 15h1'1"12d23'
Nisha
Del 20h16'11"10d10'

Nisha's Wish
Peg 22h54'0"20d51'
Nishell,Chelsea
Aql 19h31'44"13d17'
Nishida,Mark & Todd Neuman
Tri 2h26'23"30d23'
Nishimura,Kazuko
Her 17h2'56"46d43'
Nishizuka,Star of Noriko Irara
Aqr 22h7'57"-8d30'
Nisida
Gru 22h13'8"-50d36'
Nisie
Mon 8h4'23"-6d47'
Nisley,Jenifer Lynn
Mon 6h53'45"-8d12'
Nissa
Mon 6h47'59"10d55'
Nissa Christine
Aql 19h31'39"13d30'
Nissen,Delos J
Lac 22h4'42"47d33'
Nissen,Kelly Lee
Aql 19h30'20"13d28'
Nissen,Wanda
Boo 14h5'55"13d14'
Nissenbaum,Princess Rebecca Leigh
Mon 7h0'36"8d25'
Nissenfeld Family
Peg 0h2'28"20d55'
Nissenson Star,The Harry & Sylvia
Per 2h37'0"45d50'
Nistoris Owner-DeKelaita,Homer
Lac 22h27'39"50d7'
Niswonger,Barbara L
And 23h15'35"48d32'
Niswonger,Denzil W
Aur 5h3'1"44d11'
Niswonger,Scott M
Aql 20h32'32"0d35'
Nita Bright
Vul 20h1'24"28d51'
Nita's Nova
Leo 11h54'21"26d18'
Nitahara,Keith & Teresa Miyagi
Lyr 18h22'0"42d5'
Nitch's Notch,The
Her 16h49'58"39d32'
Nitcher,Keith Noble
Cmi 7h41'49"4d41'
Niteflyer
Dra 15h49'32"60d25'
Niteknight's Neige
Cep 20h59'42"60d17'
Nithael
Cas 2h26'1"77d11'
Nitkin,Ruth T
Cnc 8h25'23"18d55'
Nito
Ari 2h34'35"21d53'
Nito
Col 6h37'45"-37d35'
Nitsche,Haio
Crb 16h1'24"26d29'
Nitschke,Diane & Werner
Tau 3h54'51"20d22'
Nitta,Mary Kikue
Lyr 18h45'0"39d1'
Nitti,Louis Frank
Her 17h42'49"40d30'
Nitz,Dawn Marie
And 23h37'34"45d5'
Nivard's Light
Uma 8h4'12"70d19'
Nivek,The
Aur 5h27'0"50d7'
Nivers,Skyler Jordan
Her 16h51'45"50d59'
Nives Prima
Col 6h29'16"-35d26'
Nivolo,Elisa
Hor 3h25'40"-47d8'
Niwergall,Alexandra
Sco 16h51'50"-40d11'
Nix is Fix-Stern Rainhard Fendrich
Psc 23h6'27"6d43'
Nix,Jonathan Lawrence
Dra 10h40'24"78d10'
Nix,Matthew James
Cnv 12h12'0"48d13'
Nix,Michele René
Aql 19h51'12"12d19'
Nix,Steven J
Oph 17h6'1"7d39'
Nix,Werner
Aur 5h14'33"42d37'
Nixel,Socorro
Cap 21h28'40"-22d56'
Nixo
Cet 1h18'19"-14d39'
Nixon"HEADS LIGHT", Anthony Toren
Ser 16h18'33"1d52'
Nixon,Carol Davis
Leo 11h0'23"-0d50'
Nixon,Clifford Eugene
Cam 6h10'16"60d49'
Nixon,Drew
Ser 15h11'58"9d2'
Nixon,George W
Boo 14h37'37"35d40'
Nixon,Grace A
Crt 10h57'31"-11d24'
Nixon,Lyn Zachariasen
And 0h12'20"46d18'
Nixon,Melanie & Larry
Sge 20h5'0"20d2'
Nixon,Pat
Tri 1h46'34"25d58'
Nixon,Penny
Aql 20h15'11"5d30'
Nixon,Teresa Dawn
Lyn 7h25'0"44d28'
Nixor,Lyndia Idem
Her 17h20'34"40d13'
Niyati
Col 6h36'30"-36d39'
Nizon,Marcel
Per 3h32'49"38d33'
Nizzo,Emilu Maria Liliana
Cas 3h8'10"60d39'
Njoku "Peaches", Juliana Cizona
Cas 1h43'21"70d19'
NJR
Cam 3h42'0"61d21'
Njus,O H
Lmi 9h50'40"33d31'
Nkomo-Buettgen,Siphiwe
Sgr 19h34'15"-34d22'
NLD's Child Life Star
Vul 19h48'37"28d13'
NMN
Umi 16h26'25"72d15'
No Limits World
Dra 16h15'0"60d15'
No Neck #50
Boo 13h59'54"14d14'
No No
Ser 16h18'1"2d38'
No-Name Star
Aur 5h31'39"31d36'
No.1 Star of Junichi
Gem 7h11'47"19d47'
Noa
Equ 21h22'0"11d53'
Noa,Charlotte
Aqr 20h53'17"0d16'
Noa,Kurt
Cap 20h39'30"-26d36'
Noack,Dietrich
Dra 18h24'53"68d30'
Noack,Shari L
Del 20h57'1"10d0'
Noah
Dra 17h5'15"50d58'
Noah
Lac 22h38'25"54d8'
Noah 25694
Cep 23h20'59"70d50'
Noah's Dream
Oph 17h54'14"12d0'
Noah,Judy
Peg 22h17'49"33d40'
Noakes,Ronald
Her 17h12'51"42d27'
Noakes,Thomas George Alan
Ori 5h40'17"-0d45'
Noar,Marie Anzalone
Lyn 8h44'15"46d53'
Nobbs,Jason
Cep 3h20'42"80d23'
Nobell
Aql 20h10'22"11d16'
Nobile,Donat
Per 4h41'47"51d34'
Nobile,Ryan Vincent
Per 1h46'43"54d3'
Nobilis,Stella
Cas 1h34'27"60d2'
Nobis,Gunter Steve
Cyg 19h56'15"38d31'
Noble Star,The
Her 18h25'41"28d41'
Noble's Delight #116
Uma 13h40'29"60d38'
Noble,Adrian John
Cyg 19h26'55"34d32'
Noble,Alan John
Ori 4h51'20"0d16'
Noble,Cara Elizabeth
Crb 15h58'11"26d37'
Noble,Christine
And 23h24'35"46d39'
Noble,Cynthia Camille
Mon 8h8'33"-01d42'
Noble,Drake Westin
Ser 15h50'27"20d34'
Noble,E M
Lac 21h56'32"40d23'
Noble,Heather
Cyg 20h9'42"41d9'
Noble,Jack
Lac 22h38'43"52d36'
Noble,Janice E
Del 20h20'1"18d60'
Noble,Jay
Hya 9h35'18"-10d19'
Noble,Kimberly
Lyr 19h26'1"37d36'
Noble,Linda
Cru 12h52'30"-60d22'
Noble,LTC Ronald K "Riverboat Ron"
Her 17h34'27"20d22'
Noble,Mica
Lyn 7h52'12"47d53'
Noble,Nick
Aql 20h6'23"-0d53'
Noble,Ona Marie
Cas 0h39'19"72d4'
Noble,Rachel Marie
Uma 13h27'34"62d20'
Noble,Sally
Mon 7h46'50"-2d13'
Noble,Sally Kilik
Com 12h19'54"19d40'
Noble,Siobhan Louise
Ari 2h33'0"20d31'
Noble,Stefanie
Lyn 7h56'20"58d57'
Noble,Todd
Cet 2h38'0"-7d48'
Noblecourt,Michel
Umi 15h39'1"82d43'
Nobles,Clayton
Cam 5h52'37"67d59'
Nobles,Clyde & Betty
Uma 11h34'27"46d56'
Nobles,Kelly & Debbie
Uma 11h27'58"49d12'
Nobles,Stephen Ronald
Lib 15h37'36"-21d4'
Nobles-Baker
Mon 6h19'30"3d24'
Noblin,Allyn & Sandy
Boo 14h23'33"50d33'
Nobo,Linda
Lyr 18h31'39"39d43'
Nobu
Ari 2h12'38"22d23'
Noccia
Psa 22h22'1"-26d15'
Noce,Pisello
Uma 10h11'21"52d57'
Nocentini,Silvia
Lac 22h27'15"50d15'
Nocera,MD,Lisa Marie
And 2h31'59"47d3'
Nocero,Joanne & Frank
Cyg 21h8'15"38d36'
Nocham
Mon 7h41'51"-2d34'
Noche,Nicholas Harold
Cet 2h48'50"5d46'
Nocho,Lobo
Cnv 12h49'59"36d26'
Nochta,Isabelle Ann
Cyg 20h50'38"40d21'
Nochton,Evelyn
Equ 20h58'13"2d26'
Nocita,Skylar Lane
Lac 22h37'0"37d43'
Nocito,Nancy
Cas 0h29'20"50d37'
Nocivelli,John & Marjorie
Eri 2h54'1"-5d12'
Nocker,Alexandra
Cep 22h0'24"67d57'
Noden,Jo-Anna
Lyr 18h28'58"40d35'
Nodianos,Victoria Little Sheep
And 0h45'33"28d12'
Nodie The Eternal Light
Oph 18h4'42"13d10'
Nodrag
Lyr 19h26'19"38d39'
Nodvin,Sondra Isonis
Mon 7h13'48"-5d8'
Noe,Christopher Andrew
Hya 8h52'34"-0d40'
Noe,W E(Nub)
Sex 10h6'29"5d12'
Noel
Cyg 20h31'56"38d19'
Noel
Mon 6h32'54"7d39'
Noel Anthony RAAF
Cru 12h35'0"-57d4'
Noel John
Dra 17h0'35"51d41'
Noel,Brian Paul
Ori 5h1'12"-2d37'C
Noel,David George
Per 4h25'51"51d28'
Noel,Heather
Mon 7h14'10"-5d8'
Noel,John J
Aur 4h51'18"51d4'
Noel,Karen
Cas 2h18'23"75d19'
Noel,Karen Jayne
And 1h20'1"34d45'
Noel,Michelle R
Lyn 9h14'14"39d11'
Noel,Sean Kelly
Ori 5h1'12"-2d37'B
Noel,Sean Robert
Dra 11h46'14"71d51'
Noel,Sharon
Sge 20h13'27"17d24'
Noel,Sheba
Cma 6h15'1"-15d53'
Noel,Stacey
Her 18h17'48"14d50'
Noel,Stacey Suzanne
Ori 5h1'12"-2d37'A
Noelani,Gift From Heaven-Mackenzie
Equ 21h5'53"11d40'
Noelle
Peg 22h14'44"34d47'
Noelle
Cas 0h35'32"74d46'
Noelle's Doorbell
Mon 7h14'18"-10d44'
Noelle's Eyes
And 1h23'39"33d49'
Noelle-Martin
Cyg 21h16'46"28d59'
Noelwindalawr
Per 1h58'33"56d38'
Noemi
Lmi 9h53'10"33d54'
Noerdlinger,Hal L
Sct 18h53'30"-6d51'
Noesges,Aaron Peter
Aur 7h10'40"36d44'
Noferi,Robert Edward
Aur 7h24'34"37d59'
Noff-Zoit
Mon 7h5'25"-0d59'
Nofsinger,Luke William
Cmi 7h42'42"7d35'
Nofsinger,Terrance Marshall
Per 1h38'33"54d2'
Nofsinger,Trevor Reed
Peg 22h57'13"28d30'
Nofzinger Truly Forever,Ron & Becky
Vul 19h42'55"25d22'
Noga,Heather
Ari 1h47'59"15d34'
Noggle,David H
Oph 17h11'13"11d53'
Noggle,Patricia J
Vul 20h14'48"25d55'
Nogle,Erin
Ser 18h53'0"2d25'
Nogo
Aur 7h3'40"40d53'
Nogosky,Sr,William Alexander
Aql 19h56'42"15d46'
Noguera,Luis Eduardo
Ori 5h32'27"1d47'
Nohad
Lyn 9h38'17"40d15'
Nohemie
Cas 0h59'30"60d50'
Noi
Cae 4h44'0"-31d43'
Noilhetas,Alexandre
Crb 16h18'40"32d3'
Noirot,Odile
Ori 6h7'11"0d27'
Noiseux,Janice Alice
Lyr 18h18'39"41d2'
Nokes,Gina
Del 20h32'33"18d50'
Nola Ann
Lyr 18h39'40"47d32'
Nolan Robert
Boo 15h12'31"40d35'
Nolan Star,The
Cet 2h15'57"4d23'
Nolan(Pip),Peter John
Per 2h52'0"32d60'
Nolan,Bradley Joseph
Dra 20h16'0"64d37'
Nolan,Chris
Ser 18h2'40"-14d44'
Nolan,Christi
Cet 0h51'14"-2d16'
Nolan,Claire
Col 6h14'51"-42d15'
Nolan,Erin
Boo 15h4'28"28d26
Nolan,Frank Arthur Gleason
Ori 6h3'42"7d51'
Nolan,Gloria
Equ 21h0'19"10d37'
Nolan,Grace
Sct 18h33'21"-4d58'
Nolan,James D
Per 3h11'24"44d30'
Nolan,Jelveh Jaferian
Ori 6h8'42"6d12'
Nolan,Jesse Francis
Ser 18h54'1"2d27'
Nolan,Joel Edward & Deborah Susan
Ori 6h16'38"8d58'
Nolan,Katie Mae
Cas 0h22'54"65d33'
Nolan,Kevin
Her 16h3'39"41d23'
Nolan,Penny
Peg 21h37'52"23d12'
Nolan,Phillip S
Eri 4h1'0"-14d48'
Nolan,Rebecca Angelle
Lib 14h36'45"-18d50'
Nolan,Stephen
Hya 8h44'30"-1d3'
Nolan,Terry
Dra 12h16'28"75d4'
Nolan,Terry
Uma 9h13'44"57d34'
Nolan,Tim
Cet 0h53'22"-3d39'
Nolan,Tom
Her 17h34'52"25d31'
Nolan,Tyson
Dra 16h30'49"69d41'
Nolan,William
Cep 21h36'1"70d18'
Nolan-Cook,Rosemary Patricia
And 2h30'31"45d23'
Noland,Derek Hunt
Cep 22h15h1"68d32'
Noland,Kimberly Sue
Peg 22h4'48"26d7'
Noland,Lauren Elizabeth
Lyn 7h46'60"44d39'
Noland,Marjorie Marena Marie Schubert
Cnv 13h21'32"41d52'
Nolet,Ethel
Cas 2h3'12"59d41'
Nolet,Gordon
Aur 7h8'13"40d57'
Nolet-Blais,Jérémie
Lmi 10h51'21"32d46'
Noley
Dra 14h58'59"62d11'
Nolfo,Paolo
Cnv 13h16'1"51d13'
Nolin,Franáois
Uma 10h56'38"47d27'
Noling,Harold Orville
Per 3h10'23"56d15'
Noling,Lawrence Jan
Her 16h57'35"30d9'
Noll,Cecilia
Cas 23h31'49"60d32'
Noll,Elizabeth
Cyg 20h22'17"39d31'
Noll,Erin Kathleen
And 0h0'22"47d44'
Noll,Gary
Lmi 9h39'43"38d16'
Noll,Johann-Heinrich
Leo 10h59'42"7d44'
Noll,Lisa
And 1h47'46"37d14'
Nollmann,Jennifer Michelle
Mon 6h44'0"3d39'
Nolry
Mon 6h56'59"-10d3'
Nolsheim,Robert Lewis
Hya 8h43'38"5d35'
Nolte,Benjamin Michael
Cyg 20h36'40"47d53'B
Nolte,Brad
Ori 5h3'30"-0d25'
Nolte,Cailin Mae
Psc 23h6'29"6d54'
Nolte,Christopher Robert
Uma 8h6'1"71d47'
Nolte,Omi Marie & Opi Hans
Aur 5h27'59"47d29'
Nolte,Susan Renee
Cyg 20h36'40"47d53'A
Nolte,Tai
Cet 2h53'19"-0d9'
Nolte,Ute
Gem 7h27'53"27d29'
Nolting,Kelly Denise
Lyr 18h21'1"42d40'
Nomamis
Gru 22h13'12"-54d28'
Nombel-Blanchard
Uma 11h43'32"51d28'
Nomey,Albert T
Per 4h5'60"48d35'
Nomi
Dra 20h17'33"64d25'
Nomis
Ori 5h28'19"1d21'
Non
Tau 5h28'23"27d42'
Non Plus Ultra
Lmi 9h52'40"34d33'
Nona
Ori 6h15'57"8d51'
Nona
Peg 21h46'60"34d31'
Nona's Nova (Blanchard - Lockhart)
Gem 7h22'44"34d53'
Nong,Diane P
Peg 21h59'21"31d9'
Noni
Cas 1h14'41"61d21'
Nonis,Rosalba
Cnv 13h21'25"28d52'
Nonna
Tri 2h6'48"32d48'
Nonna Anna
For 2h15'13"-27d27'
Nonna Maria Teresa
Vel 9h22'41"-48d26'
Nonnenmacher,Anja
Cas 0h35'54"62d18'
Nonnie
Cyg 20h0'15"40d17'
Nono G-One
Her 17h14'48"40d54'
Nononda Yolonda
Aql 20h2'14"12d46'
Nonte,David
Eri 4h13'18"-17d3'
Nooa's Ark
Cep 20h12'54"60d50'
Nooch
Lmi 10h4'36"31d57'
Nooch,The
Lyn 6h33'17"58d18'
Noodles
Boo 15h1'50"48d17'
Noogie
Peg 21h38'38"25d36'
Nook's Peavine
Hya 9h16'46"3d55'
Nookie Bluette
Cyg 19h20'0"28d18'
Noon
Aql 19h8'20"0d59'
Noon,Grandma & Grandpa
Cyg 21h52'42"38d44'
Noonan,Bill
Her 18h25'42"24d39'B
Noonan,Claire
Del 20h15'11"10d31'
Noonan,Emily
Cyg 19h15'0"49d59'
Noonan,Madeline & Samantha
Gem 6h49'1"30d52'
Noonan,Matthew Ryan
Her 16h36'56"24d22'
Noonan,Matthew"Poppop"
Cyg 21h7'29"30d22'
Noonan,Martha Maxine Griffin
Per 2h3'24"57d43'
Noone,James T
Lac 22h16'12"51d38'
Noone,Sinead
Lyr 18h15'42"45d6'
Nooneyville
Dra 12h4'34"68d19'
Noony
Aur 7h7'43"43d44'
Noor Qutub
And 0h22'14"35d44'
Noor Vl Ain
Sco 16h30'51"-30d30'
Noor,Khouloud
Lmi 10h17'55"34d36'
Noorda,Ray-CRPG
Per 1h45'49"54d6'
Nooren,Bettina
Cas 0h4'13"60d5'
Noot
Uma 9h50'25"43d38'
Nopar,Gail & Bob
Cet 1h51'50"-3d30'
Nor Nor
Ori 5h56'52"14d58'
Nora
Sgr 19h39'14"-43d11'
Nora
Cep 22h9'32"68d6'
Nora
Cas 0h47'35"75d57'
Nora Ann
Crb 16h22'13"32d2'
Nora Jean
Uma 8h58'58"49d13'
Nora Mary
Cas 23h28'57"53d39'
Noralouise
Lyr 18h29'43"46d52'
Norb & June
Peg 23h24'33"30d53'
Norback,Our "Golden" Star:David
Cam 5h30'0"61d7'
Norberg,Billie Brown
Oph 18h30'59"8d13'B
Norberg,Colleen (Frog)
Cas 2h45'31"61d12'
Norberg,Debra
Cnv 12h19'1"37d23'
Norberg,Jennifer A
And 1h14'55"34d13'
Norberg,Julie & Paul
Cyg 19h44'10"29d44'
Norberg,Ralph
Oph 18h30'59"8d13'A
Norbert
And 0h14'30"40d16'
Norbert
Lac 22h17'53"49d5'
Norbert
Lyn 8h19'12"49d29'
Norberts & Cordulas Glücksstern
Vir 13h5'0"-9d58'
Norbury's Nightlight
Cep 0h1'0"66d41'
Norby,Jason
Dra 15h51'32"61d36'
Norcross
Cep 20h23'1"60d31'
Norcross,Dean Scott
Aql 19h11'17"15d32'
Norcross,Gertrude
Eri 3h59'13"-18d21'
Norcup,Raymond
Ori 5h58'50"16d58'
Nord,Everett
Umi 14h52'11"71d40'
Nord,John Lamont
Oph 17h34'22"-22d30'
Nord,Joshua Robert
Aur 6h15'49"31d13'
Nord,Martha Maxine Griffin
Mon 6h53'48"10d17'
Nordbrock,Shelly A
Boo 56h4'15"18d9'
Nordby,Erin
Aql 18h59'29"6d42'
Nordell,Johanna Margarite Olsen
Cyg 19h23'53"54d53'
Nordhausen,Mitchell Bruce
Uma 13h55'30"51d5'
Nordick,Beth Ann
Vul 19h15'47"24d34'
Nordin,Caroline
And 23h38'37"45d58'
Nordling Family,Carl & Christine
Lac 22h26'19"54d54'
Nordling,Lieschen
Tau 3h27'43"28d29'
Nordlinger,Richard E
Boo 14h36'15"15d5'
Nordlof,John August
Lac 22h6'51"51d31'
Nordmann Louise DeVillars-Sur-Glane
Aur 7h22'19"40d57'
Nordmann,Bill
Oph 16h55'0"-28d42'
Nordmann,Georg
Cep 22h22'35"70d14'
Nordmeyer,Dr Henning
Ser 16h0'45"2d25'
Nordquist,Glen & Marie
Uma 12h59'47"58d5'
Nordstar
Lyn 9h8'15"44d33'
Nordstern
Her 17h50'43"42d40'
Nordstern
Cet 2h59'22"3d17'
Nordstern '93
Aqr 21h47'48"-6d39'
Nordstern,Kaddy 18/08
Cyg 20h52'1"37d38'
Nordstrom,Beck
Uma 9h14'28"51d37'
Nordstrom,Christopher
Her 16h16'17"48d42'
Nordstrom,Diane
Equ 20h55'58"4d36'
Nordstrom,Douglas Jay
Oph 18h15'24"11d4'
Nordstrom,Fredrik
Ori 5h23'36"0d20'
Nordstrom,Margaret Ann
Mon 6h40'15"7d58'
Nordstrom,Russell Oliver
Vir 11h47'49"4d59'
Nordyke,Mark
Tri 1h59'26"33d57'
Noreen
Cas 0h38'40"72d40'
Noreen
Hya 8h58'37"-7d17'
Noreen
Peg 23h29'22"20d15'
Noreen & Jim's Love Star Forever
Uma 9h54'31"61d52'
Noreen's Honor with Friendship & Love
Cyg 20h7'22"40d13'
Noreikairam
Dra 18h58'48"48d41'
Noret,Gina Reppucci
Peg 21h30'35"20d3'

Noretta
Leo 9h57'23"18d45'
Norfolk,Ira E
Vul 19h48'10"28d31'
Norfolk,Jr,Calvert William
Per 3h17'0"54d38'
Norgaard,Sr,John Francis
Her 17h26'54"21d32'
Norgard,Eugene Keith
Lmi 9h23'0"38d20'
Norgard,Keith
Ori 5h50'35"18d46'
Norgard,Matthew Gene
Equ 21h6'0"11d12'
Norgay,Tensing
Lac 22h8'20"40d27'
Norgrove,Dorothy
Umi 13h36'53"73d19'
Norick,Ron
Her 17h29'26"20d37'
Noriega,Henry
Sex 10h33'23"4d23'
Noriel,R Joseph
Peg 23h5'13"33d3'
Noriko
Uma 9h16'21"42d14'
Noriko's Star
Aqr 22h7'12"0d33'
Norimaki
Sgr 19h9'21"-16d22'
Norina
Psc 22h55'33"1d38'
Norina "Our Innocent Love"
Cas 1h45'57"58d47'
Norini-Johnson,Star
Uma 10h0'19"56d40'
Norkaitis Nebula,The
Lyr 18h56'56"40d48'
Norkey,Dennis D
Per 4h36'22"36d4'
Norkus,Tyler
Dra 16h36'44"73d7'
Norlander,John & Diana
Cyg 21h46'43"33d53'
Norlise
Cyg 20h22'19"38d2'
Norm
Cet 0h59'0"-6d58'
Norm & Carol Forever
Crb 16h16'0"37d37'
Norm & Sandy Eternal Solemates
Eri 3h51'31"-6d23'
Norm & Tracy
Eri 4h4'48"-16d51'
Norm's "Twilight"
Oph 18h4'51"12d35'
Norm's Lucky 39
Dra 16h39'36"62d35'
Norma
Aur 5h7'43"40d20'
Norma
And 2h17'23"38d4'
Norma
Lyr 18h46'43"42d38'
Norma
Cet 2h37'13"3d49'
Norma
Cnv 13h34'15"40d22'
Norma
Uma 10h59'19"43d59'B
Norma & Charles
Aql 19h1'50"1d21'
Norma & Jack
Eri 3h41'29"13d54'
Norma Jean
Cas 23h31'24"58d16'
Norma Jean
Lyn 7h32'56"40d4'
Norma Jean
Lyn 7h44'0"47d37'
Norma Jean's Night Light
Mon 7h38'59"-6d54'
Norma Lee
And 2h23'19"47d44'
Norma Marie
Peg 23h26'32"23d51'
Norma's
Eri 3h40'50"-4d54'
Norma's Star
Cyg 19h14'0"47d40'
Normajean's Nova Indiana
Com 12h10'34"32d33'
Norman
Boo 15h14'0"48d58'
Norman
Aur 5h0'0"29d45'
Norman
Com 13h0'16"28d41'
Norman & Kelly Forever & Ever 4/30/88
Mon 6h23'58"7d58'
Norman & Kendall Friendship Star,The
Peg 23h26'14"26d9'
Norman & Rita
Mon 8h2'42"-1d52'
Norman & Teresa's Anniversary Star
Cyg 29h21'12"30d1'
Norman Charles
Her 18h7'13"28d20'
Norman's Bottom Line VII
Cep 2h23'0"78d53'
Norman's Star
Cep 20h54'47"55d21'
Norman's Star
Peg 21h48'11"34d28'
Norman,"The Highest!" Loulie Jean
Uma 12h11'43"58d41'
Norman,Becky
Gem 7h24'56"25d6'
Norman,Cheryl Bug
Mon 7h5'0"0d27'
Norman,Darren Michael
Cet 2h55'43"1d29'
Norman,David L
Cep 22h56'45"67d40'
Norman,Debra M
Lyr 19h19'25"35d24'
Norman,Ethel Glenn
Com 12h18'50"20d9'
Norman,Gregory Michael
Cet 1h48'41"-0d22'
Norman,Ian
Cep 21h57'10"78d48'
Norman,Kenneth Elsworth
Aur 5h0'14"46d55'
Norman,Kim
Aql 19h56'32"-01d33'
Norman,Laura Kaye
Lyr 18h41'12'30d11'
Norman,Marc
Umi 14h41'59"77d33'
Norman,Matthew Michael
Dra 16h6'35"66d56'
Norman,Michael Ray
Cep 21h44'41"70d9'
Norman,Morgan-Leigh
Vul 19h15'39"22d10'
Norman,Neé "The Silly Girl"
Mon 6h42'16"11d48'
Norman,Noah N
Cet 1h31'53"-0d17'
Norman,Patricia
Mon 6h21'44"4d56'
Norman,Robert J & M Frances Norman
Umi 17h5'31"75d20'
Norman,Ronald A
Hya 8h16'59"5d33'
Norman,Roseanna Robinson
Boo 14h30'46"46d40'
Norman,Russell
Aur 6h25'39"38d59'
Norman,Shane Henry
Cep 22h6'40"68d16'
Norman,Tyson LeRay
Ori 5h7'0"-0d48'
Norman,Van
Cet 3h9'44"1d40'
Norman,Vera L
Peg 22h21'44"21d22'
Norman-Audenhove, Liselotte
Sco 17h0'45"-33d58'
Norman-Joan 40
Cyg 21h35'34"31d19'
Normand Jean
Cam 3h51'50"62d49'
Normand,Eugene(Gene)L
Dra 19h32'15"67d55'
Normand,Richard
Her 17h31'18"20d40'
Normandeau,Alison Claire
And 2h23'48"41d29'
Normandeau,Francyne
Umi 16h39'0"79d20'
Normandeau,Thomas Steven
Cet 2h42'1"-18d24'
Normandin,M C A
Uma 8h37'57"68d32'
Normandin,Michele
Lyn 6h56'43"54d32'
Normanjean
Lyn 7h52'19"44d28'
Normary
Lyr 18h52'19"30d9'
NORMAX
Lyr 18h33'56"33d57'
Normile,Joe & Kathleen
Peg 0h1'47"28d11'
Nornie
Ori 6h7'13"0d29'
Norquist,Melissa Renée
Lac 22h25'24"56d43'
Norrell,Jr,Cornelius S
Cep 20h29'19"75d14'
Norrey,Andrew
Uma 12h3'57"47d60'
Norrgard,Alycia Marie
Sco 16h38'22"-30d22'
Norrgard,Jr,Dwight Frederik "Derik"
Cap 21h54'19"-20d20'
Norrie,Betty B
Vul 19h47'28"27d45'
Norris I Love You,Geno
Cyg 21h50'15"37d25'
Norris Star,The Marguerite Lucille
And 2h31'40"44d4'
Norris,Aaron Gabriel
Mon 7h1'31"-5d7'
Norris,Alice L
Boo 14h45'57"52d20'
Norris,Billy Joe
Aql 19h22'37"15d58'
Norris,Chandler Nicole
Mon 8h4'44"-5d35'
Norris,Charles D
Boo 13h37'0"17d0'
Norris,David E
Her 16h55'57"32d35'
Norris,Eleanore T
Lyr 19h2'0"31d52'
Norris,Ella Cleo
Cas 2h1'0"60d56'
Norris,Fowler Barnes
Eri 2h48'40"-4d37'
Norris,Gail Warren
Ser 18h35'46"-3d14' B
Norris,Georgia Kathleen
And 0h28'50"22d24'
Norris,I Love You,Dia
And 2h13'52"42d4'
Norris,Imani
Dra 19h11'25"68d8'
Norris,Jane
Crb 15h27'40"31d60'
Norris,Jane
Peg 21h28'50"22d53'
Norris,Jim
Ser 15h56'12"1d13'
Norris,Joan Beardsley
Com 12h21'43"27d38'
Norris,John P
Per 2h56'54"43d23'
Norris,John Randy
Dra 17h34'0"76d46'
Norris,Joseph A
Dra 19h3'0"65d11'
Norris,Joseph N
Oph 17h49'24"14d3'
Norris,Kelly F
Vul 19h50'43"20d35'
Norris,Larry Ronald
Ser 18h35'46"-3d14' A
Norris,Louise
Umi 16h53'49"79d44'
Norris,Mandy
Cyg 21h16'47"28d18'
Norris,Moria Lynn
And 1h2'53"38d54'
Norris,Paul Michael
Per 3h43'57"50d12'
Norris,Rachel Erin
Eri 4h3'14"-10d39'
Norris,Rebecca Shiney
Mon 6h30'28"10d46'
Norris,Richard & Melanie
Sct 18h45'38"-7d6'
Norris,Robert Anthony
Oph 18h20'56"10d14'
Norris,Sandra Haynes
Cas 1h58'49"61d24'
Norris,Sharon Lee Trifaro
Lyn 7h48'44"51d20'
Norris,Sheila
Her 18h34'30"22d2'A
Norris,Silas George
Aur 5h19'0"41d29'
Norris,Spaulding A
Peg 23h49'1"11d47'
Norris,Theodore Joseph Essayian
Per 4h5'53"36d52'
Norris,Tiffany
And 23h42'53"47d34'
Norris,Timothy Earl
Cmi 7h7'24"3d53'
Norris,Vic
Her 17h13'25"28d4'
Norris-Cook,Cristina Mae
Lyn 7h53'58"34d3'
Norris-Ivy
Del 20h18'52"10d21'
Norristar
Hya 8h14'24"-1d20'
Norskog,Lyle N
Vir 11h51'14"9d52'
North
Dra 17h31'23"64d4'
North Amathe
Crb 15h16'11"31d25'
North Arlington Junior Woman's Club,The
Peg 23h31'31"22d44'
North Circular
Sge 20h1'47"16d7'
North Country Learning Center
Ori 4h57'26"4d16'
North Fulton All-Stars
Her 16h9'18"41d48'
North Green
Cnc 8h49'27"24d9'
North Grove Associates
Her 16h42'37"8d2'
North Town
Ser 17h33'35"-14d6'
North,Abigail Elizabeth
Equ 20h58'37"4d40'
North,Andy
Oph 17h32'43"-20d41'
North,Bob
Aql 19h56'37"12d8'
North,Drew H
Lac 22h6'34"51d53'
North,Hollie
Mon 6h53'10"0d33'
North,J Dylan
Per 3h59'51"50d51'
North,Jacqui et Sandrine
And 2h7'50"40d24'
North,Jan Arthur
Mon 7h59'14"-8d50'
North,Janson
Hya 9h7'40"5d58'
North,Jeffrey Pierce
Hya 8h16'23"5d23'
North,Jon Maynard
Aql 20h6'42"3d59'
North,Jon Robbins (JR)
Cet 2h51'55"2d32'
North,Kathy Archer
Eri 4h1'17"-14d3'
North,Kristopher
Aql 19h49'11"14d28'
North,Nikki
And 23h31'58"48d52'
North,Stacey Renae
Cmi 7h55'34"4d3'
North,T J
Uma 10h34'56"52d31'
North,William Frederick
Aql 19h9'36"3d38'
North-Boerth
Lac 22h3'1"48d48'
Northage,Mary
Cas 0h37'37"71d45'
Northcutt,Dan
Ori 5h27'1"-0d11'
Northcutt,Nelson
Per 3h7'1"47d49'
Northcutt,R Bruce
Hya 8h52'57"0d52'
Northcutt,Tracy Deeann
Mon 7h20'23"-8d8'
Northern Light Library Staff's Star
Dra 17h50'33"65d31'
Northouse,Brent Lyle
Aur 6h1'57"30d9'
Northridge,Emma
Cyg 19h43'59"29d30'
Northridge,Kate
Cyg 19h43'50"30d4'
Northridge,Richard
Cyg 19h46'19"30d18'
Northrop,Jason
Per 3h1'29"40d51'
Northrop,Michael Scott
Cnc 8h21'48"16d26'
Northrup,Daniell Ann
And 1h0'32"47d37'
Northrup,Erna Lynn
Cam 7h53'18"68d18'
Northrup,Joan
Cas 1h22'53"58d15'
Northup,Sara Haley Nevada
Mon 6h19'36"1d6'
Norton
Uma 10h2'10"58d10'
Norton,Alyssa
Lyn 7h53'0"51d32'
Norton,Audrey
Peg 22h27'38"20d9'
Norton,Caitlin
Cyg 19h44'22"30d39'
Norton,Cathy
Cyg 20h39'11"40d20'
Norton,David P
Cep 2h29'17"86d42'
Norton,Glenna Forrisdahl
Her 17h21'39"42d20'
Norton,Harriette "Holly"
Aur 5h16'55"47d54'B
Norton,Jane
Tri 1h57'34"33d54'
Norton,Jay L
Lmi 10h58'46'26d37'
Norton,Jeanette M
Lyr 19h24'31"38d46'
Norton,Jessica
Uma 9h56'44"58d24'
Norton,Kerry Lisa
Crb 16h20'1"27d9'
Norton,Mary Lou
Vul 19h47'1"23d32'
Norton,Michael
Aur 6h2'26"33d53'
Norton,Michelle Athena
Cyg 20h21'45"40d39'
Norton,Peter Charles
Uma 12h2'31"45d47'
Norton,Phil
Aql 19h48'1"12d39'
Norton,Wade Black
Aql 18h57'0"-5d32'
Norton,Walter
Ser 15h38'1"4d27'
Norvell,Owen & Elizabeth
Sge 18h59'20"18d38'
Norvelle,Everette & Ronda
Uma 11h34'28"46d20'
Norwood,"My Sweetie" Colleen Marie
And 1h25'12"38d11'
Norwood,Deanna Lynn
Lyr 18h35'44"40d9'
Norwood,Steven
Her 16h56'1"47d38'
Nosal,George
Her 16h24'24"30d46'
Nosal,John A
Dra 15h1'1"62d7'
Nose,Connie Frances
Peg 23h26'27"32d54'
Nosel,Keith Alan
Lac 22h31'14"53d14'
Noster Amatrix Pemes Altum,Jenny Lynn
Com 12h30'43"20d8'
Noster Pars De Caelums
Com 12h45'40"23d47'
Noster,Ursa
Mon 8h3'54"-0d14'
Nostin,Marlis L Mother of Dana Nostin
Scl 23h24'55"-26d47'
Nostollitym
Vir 13h5'11"-1d10'
Notarangelo,Lisa M
Peg 22h51'19"21d20'
Notarian,Evelene
Uma 12h17'1"60d60'
Notary,Kenneth E
Sct 18h55'52"-5d5'
Notes,Evan
Per 4h0'42"51d49'
Noteware,Christine
And 23h29'44"37d49'
Nothaft,Gisela
Hya 9h3'22"5d6'
Nothing In Particular For Karen
Cas 0h4'60"65d56'
Nothing New Under The Sun
Aur 7h11'16"35d53'
Noto,Jr,Anthony
Tau 5h20'54"28d49'
Noto,Madeline F
And 0h10'41"31d38'
Noto,Matthew Peter
Dra 10h10'43"80d19'
Noto,Melissa Teru
Uma 9h46'27"57d57'
Noto,Philip
Lmi 9h48'13"33d40'
Notre Amour
Lyr 19h22'55"30d18'
Notre Vie
Her 17h14'31"49d15'
Notta,Della
Eri 3h29'30"-2d27'
Nottaris,Nelly
Aql 19h59'13"-10d47'
Notter,James Francis
Cep 21h58'36"55d57'
Notti
Ori 5h57'1"15d47'
Nottidge,Rupert
Lyn 7h8'28"58d30'
Nottidge,Rupert
Aur 6h1'39"37d55'
Nottoc 50
Lac 22h0'47"51d3'
Notturno Italiano, Romantica Atmosfera
Ant 10h31'15"-30d40'
Notz,Christopher Edward
Cam 3h59'26"52d55'
Notz,Connie G
Sct 18h51'10"-6d22'
Notz,Edward W
Lac 22h14'60"48d58'
Nouhi,Alex
Aur 5h2'12"30d5'
Nounoute
Dra 17h55'38"60d30'
Nouri,Oulaïd
Cam 5h42'55"60d24'
Nourse,Kim
Peg 22h48'48"25d41'
Noury,Romain
Oph 18h1'36"10d53'
Nous Sommes Un
Ori 5h59'12"21d6'
Nousset,Jean Franáois
Equ 21h3'13"8d41'
Nova D DI
Uma 11h44'43"40d3'
Nova Kelly Anne 25,The
Cam 7h56'37"61d37'
Nova Peggy-Jayne
Gem 6h42'15"12d59'
Nova,Johnny
Cep 21h22'58"55d21'
Novacek,Hutch
Lac 22h20'14"40d13'
Novack,Paul & Linda
Lyn 7h57'20"40d5'
Novak's Star,Steve
Cmi 7h53'49"0d36'
Novak,Carolyn
Cam 4h10'38"67d36'
Novak,Charles W
Cep 20h52'26"63d8'
Novak,Chris
Ori 5h48'0"20d11'
Novak,Clayton R
Cnc 8h31'26"8d31'
Novak,Dan Mitchell
Aur 6h32'23"38d33'
Novak,Daniel Andrew
Cet 2h51'59"0d10'
Novak,Dennis
Aur 5h1'1"38d38'
Novak,Farrell F
Her 16h39'41"42d3'
Novak,James S
Oph 17h59'60"12d15'
Novak,Jeff & Maggie
Crb 16h23'43"37d40'
Novak,Joyce Ann
And 0h52'27"40d16'
Novak,Kay M
Vir 12h7'40"1d36'
Novak,Larry & Cindy
Lyr 18h16'20"38d43'
Novak,Matthew G
Lya 18h53'39"41d5'
Novak,Raye & Sol
Lyn 7h6'20"52d60'
Novak,Ronald John
Boo 15h16'22"48d45'
Novak,Sr,Matthew William
Her 18h8'48"30d34'
Novakovics,Walter
Cap 20h23'47"-26d39'
Novander,Alan
Lac 22h28'28"56d18'
Novantatre,Carola
Gru 22h13'38"-55d38'
Novara,Michael
Aur 6h13'0"38d55'
Novatney
Oph 18h17'32"1d27'
Noveck,Gregory (Master)
Aql 19h55'33"-8d23'
Novell,Joshua Adam
Lmi 10h0'39"36d44'
Novella
Nor 16h19'45"-52d55'
Novella
Cam 4h59'36"68d17'
Novella Jack
Ori 5h32'1"-8d37'
Novella,Kathy
Lyr 19h14'1"41d6'
Novelli,Anthony
Per 3h47'27"35d38'
November Joy
Del 20h18'16"10d21'
November Rain
Her 16h14'57"48d42'
Novesl,Ariel Victoria
Lyn 8h11'1"40d7'
Novesta,Ashlee
Cam 5h39'51"58d33'
Novia,Michael V
Her 18h2'49"30d28'
Novich,Alisha Nicole
Mon 6h19'50"5d20'
Novick Star,The M & I Albert
Umi 16h49'36"79d52'
Novick,Jason David
Lib 14h19'50"-20d24'
Novick,Marcie
Aur 5h4'45"38d12'
Novickis,Barbara
Cam 7h10'40"70d20'
Novicky,Rev Monsignor William N
Sgr 19h37'7"-34d4'
Novikoff,Christina Ann
Peg 22h24'40"27d33'
Novikova,Svetlana Nikolaevna
Umi 13h23'19"70d19'
Novinski,David L
Dra 20h8'53"74d33'
Novo,Esteban Ventura
Cap 21h38'59"-21d30'
Novosiltseff,Krystina- Nina
Aur 6h21'23"38d48'
Novotney,Grandpa John
Mon 7h14'14"-1d32'B
Novotny,Karen
Lep 15h28'58"42d1'
Novotny,Karen
Pho 23h43'40"48d44'
Novotny,Petra
Sco 17h5'53"-38d1'
Novotny,Ronald N
Sex 10h37'0"2d20'
Novus
Cnv 13h24'1"40d47'
Novus Angelus Patricea
Tau 4h4'4"21d28'
Novy,Brian B
Sex 10h3'38"5d14'
Novy,David B
Aql 19h29'26"-8d 6'
Now & Always
Peg 23h42'18"31d37'
Now & Forever Bill
Per 1h38'42"53d59'
Now Voyager
Peg 22h46'26"32d34'
Nowack's Nova
Ori 5h38'22"1d7'
Nowack,Daniel P
Boo 14h45'51"29d6'
Nowadly-1923,Stephen
Her 15h59'25"20d16'
Nowak
Dra 18h46'41"67d57'
Nowak,Adam Michael
Oph 17h29'27"8d45'
Nowak,Austin George
Cet 2h28'51"3d49'
Nowak,Cheryl Lynn
And 22h56'1"37d51'
Nowak,Darek
Per 1h46'56"52d45'
Nowak,Dr Wolfgang
Ori 5h1'0"13d28'
Nowak,Eddie
Sco 16h48'37"-28d16'
Nowak,Matthias
Umi 13h37'49"74d53'
Nowak,Mike
Her 17h53'58"42d58'
Nowak,Rita B
Cam 4h31'19"61d50'
Nowak,Rocket Rod
Hya 8h40'0"2d8'
Nowak,Sandra & Erin Carl Ivins
Cyg 21h50'50"42d60'
Nowak,Vera M
And 2h14'53"50d0'
Nowak,W Friederich
Umi 15h49'16"72d14'
Nowakowski,Anthony
Her 18h9'25"47d37'
Nowakowski,Luke Edward
Dra 14h35'23"63d17'
Nowell,Connor Darren
Cep 20h29'49"60d33'
Nowicki,Maria Fe E
And 0h45'44"31d39'
Nowicki,Paul J
Her 16h23'16"23d53'
Nowlan,Craig
Ser 16h4'35"13d34'
Nowley,Emma
And 0h2'1"43d25'
Nowlin,Jim (Daddy)
Cnv 13h5'38"51d43'
Nowrocki,Donald P
Per 3h40'10"35d33'
Noyes,Dwayne Aaron
Uma 9h1'21"49d51'
Noyes,Lois
Cam 6h42'60"67d60'
Noyes,Rachel Lee
Lyn 7h33'36"38d38'
Noyes,Sheldon Cedric
Uma 10h43'37"48d28'
Nozza,Napoleon
Lyn 7h7'10"53d11'
Noëlle
Lyr 18h20'59"42d39'
Nr 1/Ingrid Kuna Berlin
Cap 21h53'13"-19d12'
NSI 46
Cam 6h47'52"67d51'
NT 60
Dra 16h6'28"60d40'
Nu Nu
Umi 13h46'41"70d39'
Nu,William Omega
Peg 22h38'53"10d15'
Nuala
Hya 8h43'57"2d18'
Nuan
Lyn 7h54'1"41d6'
Nublat,Anthelme
Boo 14h39'41"44d2'
Nucci CN
Gem 6h47'13"17d36'
Nuccio
Per 2h6'48"58d57'
Nuccio Geraci
Aur 5h2'54"41d15'
Nuccio,Alessandro
Peg 22h38'58"21d58'

Nucifora's Flower Star 9-20,Camela
And 0h2'50"44d40'
Nuckelnatz 53
Lyn 8h15'29"42d37'
Nucker
Aql 19h8'53"15d25'
Nuckles,Robin Diane
And 1h38'55"38d29'
Nuckols,Mary Alice
Cas 0h32'1"63d4'
Nuclea
Oph 17h5'16"10d46'
Nuessen,David
Aur 4h54'31"40d35'
Nuessle,Marjorie Marie Egner
Cas 1h42'50"61d8'
Nuessle,Robert David
Cas 1h36'11"60d16'
Nuessle,Ruth Mary Lumley
Cas 1h54'28"68d55'
Nuestro Deceo
Cmi 7h6'17"3d50'
Nugent III,William H
Cmi 7h17'30"3d51'
Nugent,Barry Thomas
Cyg 21h52'20"52d53'
Nugent,James M
Sct 18h41'20"-6d3'
Nugent,Tara Kristen
Cyg 21h3'1"38d7'
Nughring,David Matthew
Eri 3h53'25"-5d50'
Nugnes,Christine M
Cas 0h18'56"60d7'
Nuguid,Clarita
Sco 17h3'15"-30d13'
Nuklegeus Gabriellus
Cen 11h53'33"-45d49'
Nuland,Anette "Shopping Star"
Umi 14h40'56"86d9'A
Nulf,Sandra Ziegler
Lmi 9h52'34"38d24'
Null,Gina & Anthony
Uma 12h58'50"60d41'
Number One Snoo
Aur 5h2'0"48d52'
Number One Toni
Aur 6h34'55"52d52'
Numerramar,The
Per 1h56'46"47d36'
Nummi,Anne
And 2h19'46"50d27'
Numnut,Ali Gato Fishead
Equ 20h54'15"3d36'
Nuna
Cyg 21h37'48"40d14'
Nunemann,David Brien
Aql 19h41'28"11d37'
Nunes,Alyssa Nichole
Peg 21h29'0"27d29'
Nunes,Brenda Jane
Cas 1h18'44"60d53'
Nunes,Desirae
Aql 20h6'38"0d20'
Nunes,Kathleen
Uma 9h29'13"48d15'
Nunes,Susan
Uma 11h54'27"40d1'
Nunes,Tony
Tau 4h56'21"16d17'
Nunez
Her 17h0'53"28d17'
Nunez A Single Point Of Love,Maripat
Cyg 21h0'57"48d57'
Nunez Star of Peace, Irene
Del 20h30'29"20d7'
Nunez,Danitzia
Mon 7h55'12"-6d48'
Nunez,Patricia
Cet 2h37'27"5d18'

Nunhuck,Chantal Marie-Claire
Peg 22h28'53"10d24'
Nuni
Aql 19h30'29"12d20'
Nunn,Daverica
Lyr 18h31'1"31d46'
Nunnemacher,George
Cet 2h4'39"1d55'
Nunnerley,Sandra
Cas 0h33'20"60d14'
Nunnrivargentonunziani nomarcopatty
Ind 21h9'54"-55d11'
Nuno,Marie
And 23h2'31"49d33'
Nunzia
Lep 6h4'14"-11d10'
Nunziata,Katie
Cas 0h52'58"61d54'
Nunziato,Judi
Cyg 20h15'47"40d36'
Nuova Compagnia delle Indie
Vel 9h22'6"-41d39'
Nupi
Lyn 7h28'33"40d12'
Nuptiale
Boo 15h28'54"38d52'
Nureyev,Rudolf
Oph 17h15'52"-22d56'
Nuria
Gem 6h48'37"19d23'
Nurmi,Stan
Cep 22h39'55"71d14'
Nurney,Samantha
And 23h47'13"41d26'
Nurre,Forever Young Tom & Kitty
Per 1h38'14"54d3'
Nurse Anne
Oph 18h40'40"6d39'
Nurys Iza
Vul 20h4'48"22d58'
Nusl,In Memory Of Edward John
Aql 18h43'20"11d39'
Nuslein,Ricky Lane
Eri 4h33'0"-19d4'
Nusloch,Jerry P
Dra 16h30'17"62d56'
Nuss,Alfred
Cet 2h33'39"-5d34'
Nuss,James I
Lac 22h36'30"38d45'
Nussbacher,Linda Jo
Cas 22h59'23"55d31'
Nusser,Tammy Sue
Cyg 21h19'43"37d48'
Nussy
Uma 9h42'0"48d57'
Nuszer,Princess Jennifer
And 22h57'51"50d27'
Nutcracker,The
Lyr 18h38'40"38d26'
Nutellamus
Ori 5h35'44"7d34'
Nutile,Bernadette
And 1h38'0"37d56'
Nutile,Danielle
And 2h26'27"38d33'
Nutley,Andrew John
Aur 6h31'60"34d25'
Nuts
Cyg 21h15'27"28d52'
Nutt,David John
Per 2h36'55"37d39'
Nutt,Jessica Louise
Ara 17h45'22"-50d26'
Nutt,Mike A & Dave E
Lac 22h27'35"53d35'
Nutt,Sandi
Peg 21h20'25"22d36'
Nuttall,Ann
Lyn 8h5'41"44d55'

Nuttall,Galen Clark
Dra 12h27'15"68d49'
Nuttall,Tracy Lynn
Peg 23h19'23"33d6'
Nutter,Bobbi
Boo 13h57'51"19d7'
Nutter,Chele
Cyg 20h25'11"42d4'
Nutter,Julia Michelle
Uma 10h13'0"42d22'
Nutting,Radley C
Per 3h5'10"47d18'
Nutz,"Immer Eins", Rudolph & Judy
Lyn 8h52'1"44d17'
Nuxoll,Elisabeth
Oph 18h3'54"8d52'
Nuyens,Laurie Kay
And 23h1'55"49d43'
Nuzzi,Mark
Lac 22h54'59"51d21'
Nuzzo,Amy
Peg 23h35'45"31d6'
Nuzzo,Amy Jean
And 0h10'49"47d56'
Nyara
Crb 15h44'50"26d35'
Nyars,Denise Marie
Lyn 8h19'32"48d36'
Nyback,Jane
Peg 21h58'26"28d5'
Nyberg,Ethel & Bo-Fjalar
Cnv 12h59'14"50d43'
Nyberg,Monica Ingegerd
Uma 12h6'35"47d34'
Nye,Aaron L
Her 17h23'38"46d41'
Nye,Curt & Linda
Her 15h52'27"40d29'
Nye,Karen
Cas 1h11'58"61d3'
Nye,Leigh Daly
Lyn 7h2'25"50d44'
Nye,Mary & George
Dra 17h47'39"60d2'
Nye,Robbie
Her 17h21'0"47d16'
Nye,Robert Scott
Boo 15h43'55"41d40'
Nye,Sandra Gayle
Crb 15h27'23"30d26'
Nyeholt,Collin H
Dra 16h1'24"67d27'
Nygaard,Anita
Cam 4h54'51"68d43'
Nygaard,Oscar E
Per 3h8'47"50d22'
Nygren w/daughter Angela,Terry S & Paula
Uma 10h56'12"48d48'
Nygren,Nancy L
Cas 1h20'39"68d28'
Nyhoff,Dorothy M
Vul 19h47'19"28d54'
Nyhuis,Claudia
Sgr 19h2'12"-29d2'
Nyisztor
Cep 21h5'1"80d27'
Nyklewicz,Christopher
Dra 17h34'0"61d2'
Nykor I,Jason Paul
Cam 5h39'57"68d31'
Nyl E Il Tutti
Lac 22h6'24"38d13'
Nyla
Her 17h54'32"39d26'A
Nylander,Erik
Aur 6h4'12"31d24'
Nyman,Sonja
Peg 23h29'52"30d15'
NYNEX Mobile Communications 100%
Her 17h48'21"40d60'
Nynex NIRC
Uma 11h10'12"41d30'

Nyquist Family Star, The
Eri 3h39'29"-10d17'
Nyquist,Carl-Erik
Cep 22h37'1"56d55'
NyrDyv
Lac 22h36'58"53d53'
Nysse,Douglas Philip- Julie Ann Wilson
Crb 15h52'42"27d1'
Nystrup,Stephen
Cep 20h52'1"63d49'
Nyström,Nils
Umi 15h52'10"80d3'
Nyvas Viking
Uma 11h7'20"40d54'
Nèmet,Petra
Vir 14h8'22"-2d13'
Nöel,Ellen W
Mon 6h21'43"8d33'
Nölle-Mahena
Lib 14h55'56"-23d8'
Nürnberg,Ulrich
Ori 5h49'37"21d7'
Nürnberger,Herbert
Ser 15h10'57"1d51'
NØrhave,Mogens
Aur 5h27'26"29d45'

O

O Cunningham,Anthony
Cep 23h8'40"61d38'
O D
Aur 6h22'23"41d49'
O J N Y A M
Cet 1h6'57"-5d21'
O N B 1-1-95
Boo 14h50'18"28d3'
O Pave A
Boo 15h17'36"37d58'
O T Ch Meadowpond Keepin in Stride
Cma 6h12'34"-11d5'
O'Alice,Beth
Peg 23h34'27"17d39'
O'Banion,Brittany
Aql 20h1'11"8d22'
O'Banion,Christina Carol
Vul 19h46'28"24d50'B
O'Banion,Donna
Aql 20h1'1"0d16'
O'Banion,Jr,William Jennings
Vul 19h46'28"24d50'A
O'Banion,Kenny
Aql 20h10'51"1d31'
O'Banion,Reneé
Aql 20h10'44"5d33'
O'Banion,Samantha
Aql 20h11'13"11d13'
O'Banion,Taryn
Aql 20h1'21"10d19'
O'Bannon,Kathleen Lee
Mon 6h20'54"2d41'
O'Bannon,Ronnie Tyshon
Uma 10h6'22"59d0'
O'Bayley,Byron Markus
Cma 7h12'57"-13d29'
O'Bie
Uma 9h53'13"58d7'
O'Boyle,Evelyn
Hya 8h43'1"5d49'
O'Brein,John Daniel
Cet 0h53'57"-2d45'
O'Brian,Zoe Olivia
Dra 18h18'0"68d29'
O'Brian-Hurtgen,June
Mon 8h5'21"-3d18'

O'Briant,Havilland Brooke
Mon 7h21'23"-8d2'
O'Brien
Ori 6h3'59"8d42'
O'Brien's Happy Land, Nigel & Carol
Eri 4h30'0"-6d36'
O'Brien,Alexandra
Mon 7h39'41"-8d14'
O'Brien,Anne
Uma 11h13'35"47d39'
O'Brien,Brett Russell
Tri 1h29'1"31d3'
O'Brien,Carol
Cyg 21h6'33"38d11'
O'Brien,Caroline Coohill
Cas 0h18'34"61d22'
O'Brien,Cathy
Mon 7h2'48"-0d22'
O'Brien,Chad
Uma 11h17'30"50d28'
O'Brien,Chelsea Kay
Mon 6h30'27"4d31'
O'Brien,Constance Berdella Espeland
Equ 21h10'1"10d4'
O'Brien,Daniel
Aur 6h32'22"40d31'
O'Brien,Daniel P
Her 17h21'50"46d28'
O'Brien,Deb
Lyn 6h13'17"54d9'
O'Brien,Declan
Cnv 13h30'7"37d4'A
O'Brien,Doreen Elizabeth
Cet 0h26'32"-8d29'
O'Brien,Dorothy Barnes
Cyg 20h28'1"40d60'
O'Brien,Emily
Lyr 19h25'33"38d0'
O'Brien,Father Donald
Per 1h45'49"54d6'
O'Brien,Genevieve
Del 20h14'1"14d30'
O'Brien,Geoff & Brenda
Eri 2h45'0"-5d57'
O'Brien,Gerald J
Per 3h1'0"48d50'
O'Brien,Glen Ronald
Cep 22h32'0"60d19'
O'Brien,Hannah Colleen
Lyr 19h20'13"38d46'
O'Brien,Jacinta
Cas 0h24'59"59d10'
O'Brien,Janice L
Umi 14h50'16"65d51'
O'Brien,Jen
Lyr 18h30'23"31d23'
O'Brien,Joan
Mon 7h5'28"-0d8'
O'Brien,Joan K
Cyg 20h37'1"53d43'
O'Brien,Joe
Dra 19h58'12"60d46'
O'Brien,John
Lyn 8h5'58"41d50'
O'Brien,John David Jamoc
Dra 18h26'1"70d16'
O'Brien,John E "Jack"
Aql 19h32'44"1d42'
O'Brien,John Fletcher
Her 16h21'23"50d40'
O'Brien,John Michael
Aur 5h33'45"38d24'
O'Brien,Joseph & Patricia
Aur 7h59'8"41d31'
O'Brien,Joseph P
Aur 7h4'43"43d30'
O'Brien,Jr,John & Susan
Cyg 21h18'0"28d56'
O'Brien,Jr,Michael J
Cep 0h2'27"68d36'
O'Brien,Kaitlin Bartley
Vul 20h19'15"23d34'

O'Brien,Kasey E
Uma 10h31'37"52d54'
O'Brien,Katherine (Katie) Ruth
Mon 7h51'30"-5d55'
O'Brien,Kelsey Anne
Lyn 6h28'16"59d26'
O'Brien,Kiana
Cas 1h7'42"60d57'
O'Brien,Larry
Aur 5h7'16"40d26'
O'Brien,Linda
Equ 21h22'32"12d38'
O'Brien,Margaret D
Lyn 8h3'54"33d42'
O'Brien,Marlene Moore
Cas 0h13'45"65d24'
O'Brien,Michael Gerard
Per 3h25'29"52d32'
O'Brien,Miss Amanda
Lyr 18h33'17"30d53'
O'Brien,Nicole
Cas 0h38'0"62d42'
O'Brien,OLC,Sister Pat -Hans Cox Award 1993-
Lyr 18h19'14"47d16'
O'Brien,OP,Fr Joseph
Aql 19h24'20"14d16'
O'Brien,Patricia Ann
Peg 23h29'21"33d23'
O'Brien,Paul
Cas 0h11'17"60d45'
O'Brien,Paul Stephen
Her 17h55'1"14d11'
O'Brien,Philip
Hya 8h52'59"3d49'
O'Brien,Reverend MSGR Joseph M
Aql 19h49'44"14d15'
O'Brien,Roger J
Hya 9h35'18"5d12'
O'Brien,Sandra
Lyr 18h45'27"37d59'
O'Brien,Shannon
Dra 17h14'34"60d34'
O'Brien,Shannon
Peg 22h29'41"28d14'
O'Brien,Shannon Elizabeth
Gem 7h21'59"30d2'
O'Brien,Shawn & Sue
Crb 15h50'0"26d55'
O'Brien,Staci
Aql 20h12'0"0d5'
O'Brien,Star Of Peggy
Cet 0h43'23"-6d6'
O'Brien,Susan Elizabeth
Cam 10h50'59"81d51'
O'Brien,Thomas Patrick
Boo 14h55'44"28d37'
O'Brien,William
Tri 2h0'35"32d44'
O'Brien,William Gunner
Boo 14h49'0"53d40'
O'Brien-Wickham,Marie
Tau 5h57'0"23d58'
O'Bryan,Sr,William
Lyn 6h53'35"56d10'
O'Bryant,Charles Alex
Cep 23h9'19"60d57'
O'Byrne,Colleen
Peg 21h55'19"24d24'
O'Byrne,Mary Ann
Cam 9h19'41"81d6'
O'Byrne,Patrick
Aur 6h29'20"37d31'
O'Byrne,Réamonn
Umi 15h31'43"75d33'
O'Byrne,Victoria Jane Elizabeth
Cas 0h45'36"71d7'
O'Coney's Irish Eyes
Aql 19h0'20"10d54'
O'Connell
Aql 20h13'10"7d39'B

O'Connell's Valentine Star,Richard
Aur 6h35'52"33d1'
O'Connell,Amy Catherine
And 2h11'34"41d12'
O'Connell,Brian
Boo 15h4'45"21d55'
O'Connell,Claudia Jane
Cyg 20h22'53"39d23'
O'Connell,Damian
Cyg 21h24'11"38d56'
O'Connell,Daniel Burgess
Cep 21h48'45"61d5'
O'Connell,Edward
Aur 5h28'0"48d51'
O'Connell,Gert
Aur 5h1'37"41d16'
O'Connell,J J
Cep 1h5'40"77d39'
O'Connell,John T
Her 18h5'0"31d16'
O'Connell,Jr,Rick
Sex 10h14'1"-5d39'
O'Connell,Kari Ann
And 23h16'38"45d1'
O'Connell,Kathleen Alyssa
Peg 0h9'16"13d37'
O'Connell,Katie O/ Katie
Cas 23h32'28"58d18'
O'Connell,Kevin J
Aur 7h22'28"39d24'
O'Connell,Margaret Delaney
Vul 21h16'21"20d25'
O'Connell,Megan Eileen
Lyr 18h22'37"42d44'
O'Connell,Michael
Cep 21h41'0"55d18'
O'Connell,Michelle
Ara 17h49'30"-50d52'
O'Connell,Patricia A
Cas 23h35'1"61d11'
O'Connell,Patrick James
Cnv 12h58'1"41d6'
O'Connell,Paul Michael
Aur 6h32'21"33d54'
O'Connell,Sarah Elizabeth
And 23h15'17"51d31'
O'Connell,Sheila
Lyn 8h1'51"45d17'
O'Connell,Toni
Ori 5h55'32"16d37'
O'Connell,William Gerard
Cnv 12h12'57"34d49'
O'Conner,Traci
And 0h52'38"39d50"
O'Connor
Cam 6h12'19"56d7'
O'Connor (Aloc),Amy Leigh
Uma 10h8'45"50d12'
O'Connor (Caritas), Tom
Per 3h8'48"38d59'
O'Connor The Rifleman, Chuck V
Aql 19h19'36"15d38'
O'Connor,Amy Kelly
And 0h17'0"45d54'
O'Connor,Barbara & Joe
Cyg 21h5'39"35d22'
O'Connor,Bryan Patrick
Cet 2h37'18"4d21'
O'Connor,Caitlin
Cam 7h47'14"80d20'
O'Connor,Casey
Lyn 7h51'36"35d25'
O'Connor,Christie J
And 2h19'19"48d1'
O'Connor,Christine Marie
Lmi 10h10'27"31d54'
O'Connor,CLU,Timothy J
Cep 3h31'15"78d55'
O'Connor,Cornelius
Crb 16h8'50"38d42'
O'Connor,Erin Marie
Aur 4h52'15"51d3'

O'Connor,Gene
Cep 21h57'50"68d17'
O'Connor,His Eminence John Cardinal
Cap 21h27'40"-23d50'
O'Connor,James Crecca
Per 3h9'10"40d37'
O'Connor,Jan
Peg 22h2'31"28d29'
O'Connor,Joe & Cynthia
Cyg 19h41'0"31d29'
O'Connor,John J
Cam 6h3'38"61d22'
O'Connor,Kathryn Conly
Peg 22h47'27"25d33'
O'Connor,Kenneth
Mic 20h32'46"-31d16'
O'Connor,Kristin
Eri 3h41'54"-17d43'
O'Connor,Larry
Aql 19h28'0"10d5'
O'Connor,Manuel
Boo 14h46'1"26d29'
O'Connor,Mark James
Pup 7h56'40"-25d35'
O'Connor,MD,Francis M
Lyn 8h52'45"41d37'
O'Connor,Michael J
Aur 5h32'54"30d30'
O'Connor,Mike & Kim Levine 4/9/94
Dra 17h12'56"68d27'
O'Connor,Monica
Aqr 21h7'28"-1d4'
O'Connor,Neil
Dra 16h53'43"62d14'
O'Connor,Rosemary Immaculata
Lyr 19h14'59"33d58'
O'Connor,Ryan Patrick
Cep 22h5'21"53d22'
O'Connor,Steve
Ser 18h6'20"-7d23'
O'Connor,Suzanne Carol
Cas 1h38'14"60d11'
O'Connor,Thomas M
Dra 16h54'30"63d59'
O'Connor,Tracey
Mon 6h55'44"-0d16'
O'Connor-Robbins
Cyg 19h33'19"30d2'
O'Connor-World's Greatest Dad,Mike
Cep 22h41'56"59d47'
O'Cozzolino
Cam 3h54'25"53d31'
O'Daniel,Rickey Douglas
Ser 18h15'26"-14d35'
O'Daniel,Ryan
Cet 1h36'44"-4d60'
O'Day,Matthew Brian
Cyg 21h35'12"42d49'
O'Dea,Jim
Aur 5h1'50"51d29'
O'Dell,Bernice
Crb 15h27'53"31d38'
O'Dell,Chloe Alice
Umi 14h39'49"69d60'
O'Dell,Dorsa Rattenbury
Aql 20h6'13"0d21'
O'Dell,Duff Johnston
Per 4h43'47"37d11'
O'Dell,Janet Fay
Eri 3h57'33"-13d28'
O'Dell,Marla Jo
Eri 3h56'35"-14d14'
O'Dell,Mary Jean
Lyn 6h35'57"58d39'
O'Dell,Megan
And 1h28'59"36d25'
O'Dell,Rick
Uma 9h16'54"51d13'
O'Dell,Robin Ann
Peg 22h30'40"8d30'

O'Dell,Terry
Peg 23h5'16"31d30'
O'Dell,William W & Grace L
Her 18h6'42"38d55'
O'Dell-Haines,Karen Lea
Eri 3h55'45"-16d12'
O'Donnell Special Daddy,Barry
Uma 11h31'57"62d3'
O'Donnell,Alsonya
Cas 2h33'22"60d29'
O'Donnell,Brian
Aur 6h22'37"31d26'
O'Donnell,Casey Lee
Uma 11h58'0"58d45'
O'Donnell,Daniel
Per 3h21'30"40d58'
O'Donnell,Daniel
Uma 9h40'27"44d20'
O'Donnell,Donna Marie
And 0h21'33"36d22'
O'Donnell,Edward James
Ser 18h4'51"-7d20'
O'Donnell,Erin JoAnn
Cas 0h8'16"54d38'
O'Donnell,Helen & Ben
Uma 10h53'50"37d31'
O'Donnell,Hester M
Lyr 19h22'45"31d37'
O'Donnell,John & Jeanne
Psc 1h6'32"27d39'
O'Donnell,John E
Cyg 19h45'43"30d33'
O'Donnell,Judith "Zimmy"
Uma 8h57'29"60d32'
O'Donnell,Justin Daniel
Her 18h19'30"12d11'
O'Donnell,Kathleen Ann
Aur 5h25'56"38d24'
O'Donnell,Kay
Vul 20h14'54"25d10'
O'Donnell,Kevin
Lyn 8h11'41"49d28'
O'Donnell,Kevin Joseph
Her 16h16'38"48d4'
O'Donnell,Laura M
O'Donnell,LauraM
Lyn 8h58'57"44d48'
O'Donnell,Lauren Dorothy
Lmi 11h1'54"30d10'
O'Donnell,Lee
Cnv 12h47'0"51d20'
O'Donnell,Luke McCauley
Peg 23h36'12"11d18'
O'Donnell,Marge
Lmi 9h58'0"40d18'
O'Donnell,Martha Guadalupe Martinez
Cet 2h41'58"-10d26'
O'Donnell,Matthew Joseph
Aur 5h12'21"42d21'
O'Donnell,Matthew
Dra 14h49'20"60d48'
O'Donnell,Moira Kaitlin
Lib 14h51'58"-1d11'
O'Donnell,Morgan Ryan
Cnv 13h35'16"46d54'
O'Donnell,Nancy A
Aur 6h1'12"54d42'
O'Donnell,Parker Jaren
Uma 9h8'56"49d28'
O'Donnell,Patricia Aine
Cas 1h10'24"63d55'
O'Donnell,Patrick
Her 16h43'45"10d12'
O'Donnell,Patrick
Oph 17h0'17"7d41'
O'Donnell,Paul Thomas
Dra 17h11'49"50d45'
O'Donnell,Roobert
Peg 0h1'54"14d24'
O'Donnell,Rosie
Cam 11h2'48"80d55'
O'Donnell,Shannon
And 23h15'30"43d3'

O'Donnell,Wayne Francis Oph 18h16'50"0d32'
O'Donnell,William Hya 8h59'53"2d34'
O'Donoghue's Star Cep 22h5'1"60d26'
O'Donoghue,Denise Umi 17h14'19"75d8'
O'Donohue,John & Agnes Cyg 21h8'19"31d51'
O'Donovan,Eoin Edward Daniel Ori 5h56'0"16d57'
O'Dowd,George Cep 21h8'21"70d58'
O'Dowd,Shannon Cyg 22h0'10"50d15'
O'Dowd,William Thomas Boo 14h56'57"26d15'
O'Driscoll,Richard Cep 22h14'36"62d55'
O'Driscoll,S F Rouviere Oph 18h16'46"11d54'
O'Dwyer "Surf It Jr", Stephen P Cep 22h42'36"70d5'
O'Dwyer,Cathy Marie Cyg 21h17'58"35d42'
O'Farrell Family,The Uma 11h34'49"31d5'
O'Farrell,Charles Evans Boo 14h32'17"46d10'
O'Farrell,Dave Lac 22h6'0"38d42'
O'Farrell,Winna Uma 9h13'24"48d19'
O'Feld Oph 17h13'2"-22d7'
O'Flynn,Glenda Kay Lup 15h26'3"40d26'
O'Flynn,Patrick B Sct 18h53'58"-4d39'
O'Gallagher,Malachi Ori 5h0'0"10d46'
O'Gara,Anna Cas 0h24'47"64d8'
O'Gorman,Tara Ann And 2h17'42"39d55'
O'Grady 10-9-60 10-21-94,Christopher Uma 12h3'53"58d7'
O'Grady,Caroline And 0h45'34"30d59'
O'Grady,Gregory Her 16h50'41"50d32'
O'Grady,Jack Aur 5h15'45"40d42'
O'Grady,Michael Boo 14h31'47"21d46'
O'Grady,Michael "Telellaa" Cnv 14h3'0"45d47'
O'Grady,Ryan Michael Gem 6h53'32"30d46'
O'Grady,Sean R Dra 16h4'0"66d40'
O'Grady,T K Cam 4h10'19"80d1'
O'Guin,Melissa Kay Lyn 8h1'42"34d6'
O'Gwynn,Mary Mon 6h30'58"0d50'
O'Hagan,Brian Joseph Cep 20h21'44"75d28'
O'Halloran,Kendra And 0h0'19"40d3'
O'Halloran,Patrick Her 16h32'1"36d46'
O'Halloran,Sally Uma 9h39'45"48d55'
O'Halloran,Thomas Aql 19h6'1"15d4'
O'Hanlon,"Quiet Joy" aka Katie Eri 3h48'19"-1d20'
O'Hara III,John Grady Her 17h5'46"41d1'
O'Hara,"Santa's Little Helper" By Michael Uma 11h0'54"60d17'
O'Hara,Betsy B Cyg 19h40'14"41d8'
O'Hara,Brian John Patrick Uma 11h47'53"52d12'
O'Hara,Colleen Cyg 20h41'26"42d12'
O'Hara,David Bigelow Aql 18h59'45"-5d10'
O'Hara,Eillen Mary Dorthy And 23h41'53"45d5'
O'Hara,Elizabeth Kathryn Vul 19h41'41"26d38'
O'Hara,Elizabeth Aql 20h33'18"-0d48'
O'Hara,Jacqueline Beth Peg 22h39'28"26d12'
O'Hara,Marty Del 20h16'25"9d53'
O'Hara,Meghan Campbell Cas 1h44'18"60d45'
O'Hara,Michael Scott Aql 18h41'40"1d52'
O'Hara,Nancy And 1h8'29"38d33'
O'Hara,Ruby Cyg 19h25'36"35d45'
O'Hara,Saren Uma 9h20'45"51d18'
O'Hare TM,Katherine Harvey Cmi 7h37'17"11d29'
O'Hare,Brian Lac 22h46'51"53d52'
O'Hare,Deborah Louise Uma 11h43'17"31d37'
O'Hare,Eileen Aql 19h31'16"11d24'
O'Hare,Karen Morris Cyg 19h54'25"38d22'
O'Harrow,Hailey Hudson Peg 22h53'57"21d2'
O'Hearn,Ann And 23h23'29"43d55'
O'Heeron,Connor Timothy Ori 5h51'23"18d50'
O'Heeron,Parker Kinney Hya 8h17'50"6d19'
O'Hern,Maureen Peg 22h50'57"27d47'
O'Joseph,Brightest Smiling Star Cep 21h20'1"71d12'
O'Kane,Charles V Vul 19h47'15"28d25'
O'Kane,Dorothy M Vul 20h16'14"22d42'
O'Kane,Michael J Vul 19h47'10"25d34'
O'Kane,Nancy Elaine Vul 19h45'10"23d52'
O'Kane,Pat Aql 19h56'25"8d50'
O'Kane,Reid James Cep 22h41'12"57d35'
O'Keefe,Bob's Love, Marie Cas 1h5'0"58d14'
O'Keefe,Cheryl Vul 19h44'21"27d11'
O'Keefe,Craig Stephen Per 3h20'58"40d30'
O'Keefe,Denis Cep 3h13'32"78d7'
O'Keefe,Frank Peg 0h6'50"21d37'
O'Keefe,Jodi Vul 19h3'53"25d1'
O'Keefe,Karen Cas 0h30'56"73d40'
O'Keefe,Kenneth J Per 3h36'1"38d22'
O'Keefe,Lynn & Robert Maffei Cyg 20h21'26"38d58'
O'Keefe,Tara Catherine And 2h26'21"44d56'
O'Keeffe Star,The Miles Oph 17h5'35"10d59'
O'Keeffe,Elise Lyr 18h35'0"27d49'
O'Konski,David Paul Cep 20h31'42"75d39'
O'Koon,Charles Aur 6h7'48"35d31'
O'Leary Star,The Stephanie Cas 0h32'34"66d27'
O'Leary,Christine Desmond Cas 0h47'0"74d44'
O'Leary,David Her 18h2'59"37d44'
O'Leary,John Michael Dra 18h57'22"50d3'
O'Leary,Juanita Dora Lyr 18h56'34"30d51'
O'Leary,Leanne Michele Lyr 18h59'25"28d57'
O'Leary,Mary E(Loftus) Cep 21h21'24"58d55'
O'Leary,Pamela Rose Lyr 18h54'16"30d53'
O'Leary,Patricia Vul 21h15'43"22d52'
O'Leary,Philip D Dra 16h54'13"67d31'
O'Leary,Richard Raymond Boo 13h58'37"19d17'
O'Leary,Scott Boo 15h13'39"51d10'
O'Leary,Simon Cep 21h8'46"70d42'
O'Lone,Jr,Paul Aur 5h17'46"43d48'
O'Loughlin,Elizabeth A And 23h2'58"50d11'
O'Loughlin,Frank Oph 16h50'36"11d36'
O'Loughlin,Monika Elizabeth And 23h40'37"38d24'
O'Mahony,Micheál Uma 10h3'54"55d11'
O'Malia's Hole In One Her 17h39'48"22d44'
O'Malley Star 69 Aqr 21h58'34"-8d19'
O'Malley's Diplomat Travel,The Crb 15h31'0"27d46'
O'Malley's Little Bit of Heaven Ori 5h12'54"-5d4'
O'Malley,Danielle Marie And 1h21'32"39d21'
O'Malley,Dennis J Lac 22h25'24"48d57'
O'Malley,Elsie Aur 5h59'21"40d26'
O'Malley,J J Her 17h17'45"41d52'
O'Malley,James Per 4h0'16"51d51'
O'Malley,Jamie Ann Lyr 18h48'25"34d41'
O'Malley,Lauren Peg 22h59'18"30d18'
O'Malley,Michael Sex 9h52'58"0d3'
O'Malley,Michael & Sandra Cyg 21h50'35"42d15'
O'Malley,Patrick Francis Ori 5h59'0"11d15'
O'Malley,Quinn Boo 14h45'32"32d15'
O'Malley,Ryan Michael Cap 21h26'14"-24d52'
O'Mara,Jay Aur 6h45'11"38d33'
O'Master Oph 18h40'36"7d47'
O'Meara's Smile,The Lori L Lyn 7h47'28"48d49'
O'Meara,Jackie Uma 13h43'57"48d40'
O'Meara,Sandra J And 23h30'48"42d16'
O'Mohundro,Candy And 23h19'0"46d47'
O'Mohundro,Larry Hya 9h8'1"0d50'
O'Neal XKYX,Brian M Boo 15h34'58"42d20'
O'Neal's Rainbow-1995, Rachael Erin Vul 20h14'51"25d40'
O'Neal,Beth M And 2h4'45"40d40'
O'Neal,Jr,Jesse Oph 18h19'50"6d34'
O'Neal,Larry Kent Sct 18h42'51"-5d28'
O'Neal,Mattie Mon 8h6'10"-1d44'
O'Neal,Tahira Mon 6h38'53"7d60'
O'Neil's Eyes,Maggie Peg 23h41'29"15d17'
O'Neil,Benji Dra 16h22'1"64d21'
O'Neil,Bill Cep 20h40'52"75d54'
O'Neil,Deborah Ann Cas 1h5'1"62d55'
O'Neil,Dreama Eri 3h23'19"-3d33'
O'Neil,Esmond Dra 19h3'25"58d14'
O'Neil,Gerald Emmett Her 17h28'0"30d34'
O'Neil,John J Lac 22h46'56"56d47'
O'Neil,Kelly And 0h25'18"45d23'
O'Neil,Lillian Cas 1h46'1"60d14'
O'Neil,Mattie Aql 20h2'39"-6d48'
O'Neil,Ross Charles Aur 5h4'24"40d10'
O'Neil,Tom Aur 7h25'44"43d44'
O'Neill "Tjantik Bintang",Dr Tom Crt 11h23'53"-8d40'
O'Neill,Amanda Cyg 21h19'40"28d20'
O'Neill,Anne Tau 4h0'14"30d29'
O'Neill,Arlene Uma 11h11'28"40d59'
O'Neill,Beverly Lewis Mon 7h52'20"-5d19'
O'Neill,Bobbi Cyg 20h41'34"38d57'
O'Neill,Brendan Patrick Crb 16h12'41"32d13'
O'Neill,Bridget & Catherine Cam 7h43'32"68d2'
O'Neill,Brittnay Cas 1h17'29"65d17'
O'Neill,Cassidy Shannon Cas 0h52'51"62d58'
O'Neill,Christopher P Dra 15h24'10"57d55'
O'Neill,Darren & Suzanne Sct 18h34'56"-6d55'
O'Neill,Debbie And 2h16'51"38d26'
O'Neill,Diana M Sgr 19h2'47"-20d59'
O'Neill,Frank Aql 19h32'50"0d10'
O'Neill,James Jasper Hya 10h13'12"-11d36'
O'Neill,Jr,Thomas P Aqr 21h22'13"-13d33'
O'Neill,Kathleen Genevieve Cas 0h22'33"60d11'
O'Neill,Katie Lmi 10h54'37"30d15'
O'Neill,M Tyler Umi 15h0'1"71d11'
O'Neill,MarthaLou Elizabeth Aql 19h3'59"16d21'
O'Neill,Mary Claire Mon 7h11'30"-5d18'
O'Neill,Mary Elizabeth Cas 0h25'31"61d27'
O'Neill,Meghan Gem 7h26'27"28d28'
O'Neill,Michael Anthony Dra 19h20'21"68d27'
O'Neill,Michael Alan Cep 23h26'35"64d19'
O'Neill,Nancy Peg 21h29'0"2d60'
O'Neill,Nicholas Ari 2h0'37"26d32'
O'Neill,Patricia Maureen Cam 4h16'31"78d53'
O'Neill,Petrona Crb 16h13'15"32d36'
O'Neill,Robin Noel Cas 0h50'40"68d26'
O'Neill,Scott Oph 17h53'16"8d21'
O'Neill,Tammy Ray And 2h18'58"45d12'
O'Neill,Terri Hya 8h15'46"-5d55'
O'Neill,The Star of Love,Alyson Lyn 7h17'55"51d18'
O'Neill,Tim Vir 13h33'55"1d51'
O'Neill,Tina Marie Cas 0h42'0"65d51'
O'Neill,Wayne Paul Peg 23h4'48"18d25'
O'Neill,William M Cep 22h14'13"68d54'
O'Oe Ka I Pane Mai Cam 5h57'31"61d45'
O'Pelt-Swanson, Isabelle Ida Cet 2h3'0"-0d19'
O'Quin,Tyler MacKenzie Oph 18h19'17"7d27'
O'Reilley,Anthony John Francis Ori 6h2'31"0d15'
O'Reilley,Mary Cam 3h26'17"61d49'
O'Reilly,Brendan Timothy Lyn 7h52'11"33d27'
O'Reilly,Chryssanthie Goulandris Cyg 19h33'21"39d23'
O'Reilly,Gwen Aql 18h58'41"-6d49'
O'Reilly,Hannah And 23h4'18"37d49'
O'Reilly,In Memory of Colin James Uma 10h30'30"57d53'
O'Reilly,John Eugene Shane Lac 22h14'27"51d23'
O'Reilly,Jr,Michael John Her 18h11'25"40d48'
O'Reilly,Shane Francis Ori 5h0'52"14d17'
O'Reilly,Tara Patrice And 2h30'42"50d4'
O'Reilly,William P Per 2h26'11"56d49'
O'Reily Judy Mon 6h30'24"-10d32'
O'Reily,Susan Cameron Cyg 19h27'1"36d31'
O'Rene,Kelly And 2h8'52"41d24'
O'Riley,Declan James Sean Umi 13h48'0"73d38'
O'Riley,Marcel D Cmi 7h53'31"4d3'
O'Riordan,Julia Rose And 0h19'40"31d13'
O'Riordan,Sean Her 18h24'21"28d35'
O'Roo,Vic Psc 0h56'11"28d38'
O'Rourke,Casey Boo 14h53'35"46d19'
O'Rourke,Desmond Aqr 22h2'1"-5d46'
O'Rourke,Dolores Dicks Aql 18h47'0"11d8'
O'Rourke,Emilienne Marie Psc 22h55'53"1d20'
O'Rourke,Jennifer L & Evans,Michael T Boo 14h36'17"21d30'
O'Rourke,Katelyn A Cas 1h54'32"61d52'
O'Rourke,Kevin Michael Aur 5h17'41"45d28'
O'Rourke,Lloyd Sex 10h31'51"3d30'
O'Rourke,Michael Ryan Per 2h50'21"46d42'
O'Rourke,Sawdey Sge 20h16'22"16d11'
O'Rourke,Theresa Boo 14h5'1"32d23'
O'Rourke,Thomas Robert Per 2h58'19"35d17'
O'Ryan Ori 6h12'57"8d42'
O'Saraus,Sharon & Vicki Peg 23h26'37"16d36'
O'Shaughnessy,Connie Tau 4h42'50"20d32'
O'Shaughnessy,Patrick & Maryalice Peg 22h13'59"2d29'
O'Shaugnessy,Norah Lac 21h57'41"41d55'
O'Shea Eri 3h36'34"-17d15'
O'Shea Fortieth Anniversary Aql 18h47'11"10d25'
O'Shea's Star Eri 5h8'10"-4d53'
O'Shea,Brandon Lac 22h9'22"49d33'
O'Shea,Collin Michael Cnv 13h55'36"40d22'
O'Shea,Deirdre Cam 3h25'12"53d25'
O'Shea,Elizabeth Cas 1h1'19"55d23'
O'Shea,Erin Denise Psc 1h1'1"27d55'
O'Shea,Liam Cep 22h25'1"70d54'
O'Shea,Mary Ann Uma 8h14'41"68d3'
O'Shea,Michael James Cep 21h1'13"80d7'
O'Shea,Mordecai Ali Van Allen Tri 2h5'50"32d59'
O'Shea,Terry Her 17h34'49"20d0'
O'Shia,Maureen And 23h20'42"38d25'
O'Shields,Rebecca Lynn Ori 5h33'38"-1d35'
O'Smith,Reece David Cnc 8h51'57"11d18'
O'Steen,Cindy Louise Eri 4h18'54"-19d25'
O'Steen,Steven David Aql 19h54'1"10d8'
O'Sullivan,Caroline Karin And 0h28'27"31d24'
O'Sullivan,Erin Anne Her 16h9'56"8d32'
O'Sullivan,Jeremiah Lib 15h35'0"-28d34'
O'Sullivan,Jessica A Cas 0h37'0"62d24'
O'Sullivan,Joanne Boo 14h10'19"51d5'
O'Sullivan,Kathleen Eloise Her 16h10'0"5d33'
O'Sullivan,Lorraine Cam 4h14'45"61d13'
O'Sullivan,Michele Boo 15h24'38"50d27'
O'Sullivan,Nancy Eloise Her 16h39'34"8d48'
O'Sullivan,Steve Uma 9h41'31"47d2'
O'Sullivan,Tim & Maureen And 2h22'1"48d58'
O'Sullivan,Zoe Diane Lyr 18h58'55"37d55'
O'Tain,Beth Ann Hynes Tau 5h57'34"28d50'
O'Toole,Bob & Doreen Cyg 21h24'43"41d13'
O'Toole,Catherine Peg 23h38'21"10d30'
O'Toole,Cynthia Lynn Eri 4h11'14"-17d15'
O'Toole,Danny Aql 20h35'29"0d27'
O'Toole,Declan Her 17h20'1"27d44'
O'Toole,Maureen Mon 5h58'13"-5d39'
O'Toole,Tommy Ori 6h3'34"8d56'
O,Connors' Irish Twinkler,Dick & June Dra 16h59'50"52d19'
Oag,Michael David Her 17h38'43"40d7'
Oak,Wayne L Boo 14h47'17"50d10'
Oake-50 Gem 6h51'34"18d19'
Oakes' Parrothead Pavilion,Mike Cep 23h13'36"62d19'
Oakes,Brian & Kari Aur 5h9'15"40d46'
Oakes,Denise And 2h8'51"37d41'
Oakes,George B & Patricia Aql 19h50'23"15d1'
Oakes,Kitty Uma 11h35'58"32d1'
Oakes,Stephen Scott Dra 19h34'0"65d4'
Oakfield School, Chesire Boo 15h24'36"48d45'
Oakley Peg 22h8'60"20d18'
Oakley The Boofy Girl, Christi Aql 19h9'47"15d31'
Oakley,Beverly Cyg 19h40'17"30d20'
Oakley,Edna Cas 2h16'46"73d25'
Oakley,Faith C Lyr 18h49'30"31d49'
Oakley,Forever Linda Lyr 18h31'37"37d32'
Oakley,Jack Ray Aql 20h12'1"11d12'
Oakley,Jacob James Cep 23h10'43"70d52'
Oakley,James Hya 10h1'39"-17d51'A
Oakley,James & Mavis Hya 10h1'39"-17d51'C
Oakley,Mavis Hya 10h1'39"-17d51'B
Oakman,Jack H Aur 6h30'25"52d56'
Oakman,Jean And 23h6'1"42d57'
Oaks,Dorothy Marie Cas 0h27'12"70d43'
Oaks,Margaret Mary Hayes And 23h40'20"40d55'
Oaks,Tim Cet 2h31'26"1d21'
Oas,Adrik Viktor Ser 15h44'1"-2d17'
Oasis Water Systems, Inc Cam 3h45'27"60d33'
Oates,Bill Hya 9h3'16"3d14'
Oates,Joan Cas 1h15'39"75d24'
Oates,Julie D Eri 3h59'34"-13d23'
Oates,Marvin L. (Buzz) Boo 14h7'58"40d9'
Oates,Travis W Aur 6h26'58"34d48'
Oatfield,Danielle Mon 6h19'57"8d51'
Oaties,Crystal Lynn Cma 7h22'14"9d30'
Oba,Brandon Masami Ser 15h56'12"4d56'
Obadia,Edward Lup 15h20'16"-43d38'
Obaretin,Ben Aur 6h31'1"33d21'
Obbie Cam 8h15'51"81d20'
Ober,Jutta Cnc 8h27'59"30d18'
Oberbürgermeister,Dr Werner Ludwig Lib 15h1'20"-28d25'
Oberg,Haley Dana Lac 22h42'11"53d40'
Oberg,Jesse Todd Dra 13h0'22"68d59'
Oberg,King C Dra 17h51'21"78d53'
Oberg,Lindsey Colleen Lyn 7h46'42"58d24'
Oberg,Lydia Cas 1h29'33"71d16'
Oberg,Shaylynne Nichole Cas 1h26'36"61d34'
Obergottsberger,Hugo And 23h9'22"43d27'
Oberhelman,Daniel Schwind Aql 20h11'34"14d45'
Oberhelman,Sara Schwind Cyg 19h57'25"37d54'
Oberhofer,Charles A Oph 16h48'0"11d44'
Oberhofer,George Paul Hya 8h19'49"2d9'
Oberhofer,Robert Lawrence Lmi 10h36'59"26d16'
Oberhuser,Klaus Cep 22h0'19"65d26'
Oberkfell,Larry Per 3h42'41"37d49'
Oberlag,Hendrik Lib 15h15'21"-20d0'
Oberlander,Charleen Eri 3h20'53"-1d48'
Oberle,Shirley Mae Lyr 18h33'39"35d17'
Obermaier,Donna Jean Peg 22h9'23"21d39'
Obermaier,Rita Vir 11h50'21"1d47'
Oberman,Jacob Johannes Cet 2h50'0"2d59'
Obermayer,Dr Gertrud Psc 23h6'59"2d17'
Obermayr,Johannes Hya 9h0'1"2d22'
Oberrieder,Hans Aql 18h57'14"17d59'
Oberste-Schemmann,Uta U Cam 6h10'25"68d33'
Obert,Stephane Cam 3h25'60"60d33'
Obertelli,Jamie Dra 17h57'1"64d44'
Obertin,Mike Ori 5h53'1"14d59'
Oberwinder,Jr,John F Uma 10h47'46"71d10'
Obie Dra 16h3'21"62d6'
Obie Ori 5h56'14"16d31'
Oblly,Leo A Her 16h27'27"41d32'
Oblow,Mark Richard Eri 4h7'23"8d28'
Oboczky,Michelle Cyg 20h2'44"41d2'
Obolensky,Prince Peg 22h31'11"24d8'
Oboler,Lillian Lynn Cyg 20h23'36"31d47'
Oborde,Nancy Orani Ori 5h56'27"21d56'
Obregon,Mattie Uma 11h34'1"44d1'
Obrigewitsch,John Paul Dra 16h46'39"67d38'
Obringer,Gerard Cyg 19h37'51"38d33'
Observer Life Magazine Ori 6h3'36"8d35'
OBSIDIAN Uma 11h20'37"71d44'
Obstoj,Jeanette T Cyg 19h29'32"36d20'
Obukauskaite,Dalya Stasevna And 1h24'35"39d42'
Ocala's Sister Star Cet 2h35'45"-11d23'
Ocana,Olivia Peg 22h31'0"33d48'
Ocasio,Luis Cep 22h12'46"68d29'
Occhiogrosso,Benis Boo 13h58'27"26d9'
Occhipinti,Chelline Uma 11h11'35"61d11'
Occhipinti,Mark & Arlene Sge 19h52'20"16d14'
Ocean Sunrise Per 3h2'48"45d34'A
Oceane,Florence Del 20h19'50"14d3'
Ocello,Tali Ann Claire Mon 6h56'13"-10d51'
Ochis,Robert Per 4h1'56"37d32'
Ochitwa,Rob Sgr 19h40'7"-37d5'
Ochman"Tiger",John S Her 17h55'14"18d49'
Ochman,Daniel Joseph Boo 14h57'1"28d18'
Ochoa,Ellen Mon 7h18'19"-8d5'
Ochoa,Irma Claudina Geronimo Cru 12h2'0"-59d22'
Ochoa,Joseph John Cet 0h54'11"-2d3'
Ochoa,Jr,Frank Joseph Her 18h7'34"40d33'
Ochs,Jill Aql 18h59'0"12d16'
Ochs,Joanne And 0h20'0"36d29'
Ochs,Rachel Cnc 8h35'20"18d7'

Ochs,Sharon Marie Diana
Lyr 19h25'53"42d32'
Ochsner,John D
Cet 1h15'42"-3d22'
Ockner,Lee
Ari 2h1'13"10d48'
Ocram Issor
Peg 22h42'54"3d46'
Octavia Jane
Mon 6h44'48"10d20'
Octavian,J S
Gem 6h47'56"14d57'
October 13th
Aur 5h13'27"42d16'
October 23rd Movement
Gem 6h32'15"17d10'
Oczepeck,John E
Lac 22h17'54"47d34'
Oczkowki,Theodore
Lyn 7h30'33"41d8'
Océane
And 2h31'20"44d54'
Oda & Klaus
Dra 10h44'1"73d37'
Odalys
Hya 8h42'52"-8d7'
Odam,Marie Houlette
Lyn 7h57'57"34d2'
Odd Couple,The
Lep 15h17'25"40d44'
Oddo,Dana Lauren
Lyr 18h42'26"45d44'
Oddo,David Howard
Peg 23h19'14"13d57'
Oddo,Joey
Lyr 18h37'38"38d2'
Oddo,Joseph "Joey" Peter
Her 17h30'51"20d49'
Oddo,Linda Nolen
Aqr 21h1'0"-10d56'
Oddo,Papa Ross
Lmi 10h46'31"31d36'
Oddy,Prince Of Love - Ben
Cep 3h51'28"80d35'
Odeeseus
Cyg 19h59'29"45d6'
Odegard,Charles Benedict
Boo 15h11'56"52d8'
Odegard,Ken Hegg
Cmi 7h57'17"0d14'
Odehnal"The Vic Star", Vicki
Peg 21h23'38"3d34'
Odel,Arthur
Ori 5h55'51"15d14'
Odell,John D & Merry Kathryn
Her 17h13'17"40d2'
Odell,Mary
Peg 23h16'50"33d3'
Oden
Dra 16h38'18"64d36'
Oden,Dick
Ser 18h5'24"-0d18'
Oden,James "Odie"
Aql 19h44'40"14d58'
Odenbach,John Matthew
Dra 16h59'1"61d6'
Odenbach,Mark John
Uma 11h24'18"32d33'
Odenthal,Anja K
Peg 22h2'23"30d25'
Odermatt,Sandi
Eri 3h37'1"-6d54'
Odessky,Debbie & Jeremy Erickson
Cyg 21h26'19"37d38'
Odgaard,Line Sander
Ori 5h40'57"10d14'
Odgen,Elizabeth Hunt
Cet 1h7'44"1d25'
Odierna,Steve
Uma 11h9'30"52d8'
Odile 5555
Tau 3h41'43"1d7'
Odile,Christine
Cep 22h5'12"67d44'
Odilia S
Cas 0h24'1"61d21'
Odin,Thomas
Dra 20h32'39"67d27'
Odinak,Abigail Marissa
And 22h57'17"51d8'
Odlath,MD,Dr Robert W
Oph 16h57'57"-25d41'
Odle,Jeffery
Lac 21h59'40"41d31'
Odom,Becky Logan
Lyn 8h8'12"37d30'
Odom,Brian Michael
Mon 8h4'41"-6d6'
Odom,Lisa
Cyg 20h22'57"38d36'
Odom,Marvelous Magical Mary
Com 13h1'52"27d20'
Odor,Erin
Lyn 7h31'20"58d43'
Odorfer,Charles F "Chuk"
Cet 1h30'12"-10d39'
Odos
Cam 5h52'37"67d59'
Odyssey House
Pho 0h38'37"-47d20'
Odén,Dr Robert R
Oph 17h57'17"11d33'
Odén,Nancy Newcomet Clow
Aql 18h56'59"16d19'
OeanasuS
Cam 7h57'14"60d36'
Oechsle,Ronald D
Lac 22h2'20"48d8'
Oehler's "Dragon Star",Robert
Her 16h14'24"4d58'
Oehler,Gisela
Her 17h52'30"14d50'
Oehrlein,Paul
Her 17h0'39"49d36'
OEI
Cnv 13h11'59"40d53'
Oelke,Hinrich
Lyn 8h7'58"42d12'
Oelkers,Sr Family Star,Robert F
Cet 3h16'45"6d45'
Oelschig,Melissa
Equ 21h18'16"3d3'
Oemiro
Tri 2h15'43"28d48'
Oeo,Raquel
And 23h8'25"40d29'
Oery,Christopher
Lac 22h9'41"40d10'
Oery,Michael
Lac 22h9'17"40d18'
Oeser,Jeanne
Com 12h58'44"17d34'
Oesterle,Shane
Dra 16h23'12"69d49'
Oestreich,Klaus
Aur 5h14'16"42d17'
Offerle,Leslee Frances
Sex 10h34'30"1d56'
Offerman,Charlotte
Lyn 6h21'46"60d21'
Offman,Cindy
Cas 1h50'1"61d48'
Offzanka,Detlef
Dra 18h18'13"68d11'
OFN Sunshine
Cas 1h9'49"61d60'
Ofner,Jenny
And 1h7'38"39d37'
Ofsthun,Sherman & Dawn
Cyg 21h9'34"50d23'
Ogan,H L
Cnv 13h24'0"50d50'
Ogar,Karen
Cam 6h12'19"70d2'
Ogata,Dr Roger & Agnes
Lyn 7h53'60"58d11'
Ogborn,Aaron D
Sex 9h50'39"2d12'
Ogburn,Ruth Micer
Mon 6h21'54"7d25'
Ogden's Rimworld
Uma 9h52'36"48d13'
Ogden,Alice
Lac 22h24'37"54d37'
Ogden,Allison Esther
Cyg 19h44'50"29d50'
Ogden,Derek
Oph 18h2'31"1d10'
Ogden,Donald W
Dra 19h49'20"61d4'
Ogden,Eric R
Her 17h24'33"30d37'
Ogden,Jennifer Lynn
Cyg 20h5'57"30d25'
Ogden,Kyle
Boo 14h20'1"39d3'
Oger,Eric
Peg 23h16'46"31d8'
Öger Travel Star I
Aqr 23h29'56"-18d36'
Ogg,H Lorraine
Lyr 19h19'34"38d56'
Oggie
Ori 6h6'34"1d27'
Oggioni,Paola
Cnc 9h1'40"30d34'
Ogier,Ronald
Mon 6h29'24"2d53'
Ogilvie,Elizabeth Anne
And 23h7'1"40d17'
Ogilvie,Steven Thomas
Oph 17h13'19"-20d54'
Ogisu,Akiko
Aql 19h33'35"1d47'
Ogle #1,Floyd Lelano
Boo 13h47'23"19d44'
Ogle,Jim
Aql 19h0'19"13d7'
Ogle,Serene
Tri 2h7'0"33d15'
Ogle,Shirley Ann
Peg 21h55'25"24d15'
Ogle,Thomas Peter "18"
Her 17h30'36"26d41'
Oglesbee,Amber Rene
And 1h55'55"47d2'
Oglesby,Garry
Per 1h50'56"54d5'
Oglesby,Jr,William Cooper
Ori 5h57'12"12d47'
Oglesby,Sydney Blue
Sge 19h57'28"20d35'
Oglesby,Tonja Paris
Aql 18h39'55"-2d43'
Ognar,Sarah Courtney
Leo 10h31'34"20d15'
Ognek,Ryan Elizabeth
Tri 1h36'47"35d14'
Ogo,Ramon & Maria
Peg 23h47'48"10d1'
Ogram-Lady of Science, Sandy
Lyr 19h23'1"30d42'
Ogren,Heather Ann
Del 20h13'47"15d23'
Ogunbanjo,Chief Chris
Dra 16h38'1"51d47'
Ogunmawo,Tylei Hakeem
Oph 17h22'34"-5d39'
Oh Baby Ron
Aur 6h50'46"37d19'
Oh Claire
Aql 18h58'27"16d45'
OH Maurey!
Cas 23h32'42"53d20'
Oh MY I I's
Oph 18h5'55"12d55'
Oh My Honey
Sgr 18h56'18"-27d48'
Oh Nathaly I Love You
Ori 5h59'0"17d59"
Oh Richard
Lmi 9h42'51"33d7'
Oh Stephen!
Per 1h53'15"52d47'
Oh! Cassy-My Universal Love
Cas 3h4'1"68d20'
Oh,Eugene
Cep 3h0'10"80d9'
Oh,My!
Cep 22h15'0"70d29'
Oh-Kaaayy Aww-Right
Ser 18h18'43"-2d54'AB
Ohanessian,Jean
Sex 10h14'45"-4d25'
Ohara,Deborah
Cam 12h42'20"77d14'
Ohara,Kaisho
Cep 0h36'25"77d51'
Ohara,Sandra "Lambchop"
Mon 7h31'19"-6d37'
Oheix,Daniel
Dra 10h11'0"78d26'
Ohhh Craig:Gen M M(New Moon)Thor
Aur 5h56'22"41d10'
Ohimesama,Ikuko
And 0h11'58"39d20'
Ohl,August
Her 17h56'41"42d31'
Ohl,Kathlene A
Uma 9h30'11"67d1'A
Ohleth,Kris
Uma 11h14'32"42d51'
Ohlfest,Donald Edward
Aql 18h44'30"-0d6'
Ohlhues Love Star, Dean & Rhonda
Cyg 21h6'24"31d43'
Ohliger,Lara
Boo 14h7'33"10d51'
Ohlmann,Dieter Erich
Uma 11h2'35"71d52'
Ohlmeyer,Kathleen F
Lyr 18h49'0"31d26'
Ohlsen I,Gunnar
Umi 16h27'15"79d27'
Ohlsen,Stefan Blizzard
Umi 15h14'19"70d53'
Ohlson,Mark R
Lac 22h19'53"49d43'
Ohlsson's "Wedding Star"
Crb 16h20'25"37d37'
Ohman,Rachael Lee
Com 12h38'49"21d43'
Ohme,Elyot W
Per 3h14'38"56d6'
Ohmen,Douglass John
Aql 19h15'31"13d11'
Ohmert,Dick
Per 2h35'34"57d9'
Ohnesorge,Nikki
And 0h11'11"32d53'
Ohnika's Light
Cas 1h11'1"64d38'
Ohnmacht,Tamara
Psc 22h56'49"0d50'
Ohno Family,Stella of
Psc 1h16'45"20d36'
Ohrel,Sierra
Cas 1h17'56"62d38'
Ohta,Tadayuki
Lyr 19h13'17"31d19'
Oibharriet
Aql 19h41'1"12d25'
Oien,Theodore Burton
Dra 18h46'58"70d32'
Oikawa,Stephen
Umi 15h18'24"68d6'
Oiknine,Michael
Boo 14h20'57"51d38'
Oink #50
Oph 18h9'59"12d9'
Oirad`94
Scl 0h51'16"-29d41'
Oiticica,Maria Christina Bastos
Cma 6h52'60"-16d46'
OJ N BJ
Uma 11h6'19"45d52'
Oja-Faraj,Beverly Elvira
Oph 17h59'43"-5d59'
Ojala,Eelis
Uma 10h50'15"44d36'
Ojala,Pirkko
Uma 11h35'58"32d1'
Ojalvo,Saara Victoria Guibert
And 1h32'20"41d2'
Ojapelto,Ari
Lyr 18h28'10"38d47'
Ojeda,William
Her 18h12'12"40d33'
Ojima,Professor Iwao
Gem 6h56'56"20d18'
Ojjeh,Lana
Sgr 19h48'0"42d36'
Ojjeh,Lia
Aqr 22h48'45"-1d38'
Ojjeh,Sara
Gem 6h41'51"31d19'
Okado,Kenji
Uma 10h17'32"59d13'
Okamoto,Manabu
Psc 1h14'17"20d34'
Okamoto-Kearney,Mauri
Aur 7h3'51"38d4'
Okamura,Hisao
Oph 17h53'36"13d43'
Okanaki,Karen
Cet 3h9'44"1d40'
Okano Yukiko
Ori 6h7'11"4d9'
Okasick,Ryan Michael
Aur 6h6'0"45d23'
Okene,Paul Bradford & Ketevan Ninua
Uma 12h10'24"53d6'
Okerstrom,Sylvia May
Cnv 12h13'33"51d43'
Okhotnikoff,Danielle
Lyr 18h39'35"45d6'
Okhotnikoff,Tatiana
Cam 5h56'11"70d44'
Oki-Doki
Her 16h21'2"10d42'B
Okker,Victor Charles
Equ 21h19'31"10d54'
Okonek,C C
Sgr 19h49'56"40d30'
Okoniewski,Chris
Dra 16h56'32"60d38'
Okonoski,Glen Thomas
Cep 21h23'19"68d15'
Okorn,Steven Frank
Aur 5h6'14"44d43'
Okostari
Lmi 9h25'12"38d29'
Okrasinski,Samantha Zoe
Lmi 9h34'14"37d33'
Okray,John
Uma 11h49'48"42d55'
Okray,Nancy Lee Oliver
Tau 3h39'59"24d35'
Okrent,Steven
Mon 6h21'59"8d54'
Oksnee,Dave
Dra 18h0'41"65d2'
Oksten,Ginny & Larry
Cyg 20h58'20"31d5'
Okubo,Christopher Taylor
Aur 7h9'25"43d24'
Okun,Linda Marion
Lac 22h48'32"56d11'
Okun,Mark Steven
Aqr 22h42'14"-2d10'
Okun,Scott Matthew
Cnc 8h13'59"30d24'
Okun,Stacy
Lac 22h35'54"37d52'
Okuniewski,Vincent Rocco
Her 17h35'1"26d8'
Okura,Hiroshi
Leo 9h29'47"7d6'
Okyle,Lisa
Lyn 7h39'57"38d23'
Ol' Blue
Cnc 7h56'38"18d51'
Ola & Anna Forever!
Cyg 21h53'0"53d22'
Olaf
Lac 22h5'53"40d7'
Olaiz,Timothy
Cep 23h7'18"62d39'
Olajuwon
Ari 3h1'45"22d53'
Olak,June Susan
Cas 0h37'48"60d54'
Olav,Anita og
And 0h6'52"41d10'
Olazar,Denise
Cas 0h40'29"68d46'
Olberg,Nancy
Cas 0h55'21"61d39'
Olbert,Boyd C
Aur 6h3'41"34d2'
Olbrich,Britta
Oph 18h27'36"8d34'
Olcott,Larry
Ori 5h56'21"5d50'
Old 80
Lac 22h36'22"50d35'
Old Boy 13
Dra 18h17'33"48d14'
Old Chapel House
Peg 23h36'1"31d12'
Old Honey
Ori 5h13'40"-5d52'
Old Man & Bounce
Vul 19h23'40"27d15'
Old Man Tom
Aur 6h14'48"45d54'
Old Mundo,The
Sge 19h36'15"16d41'
Old Princess
And 0h7'48"46d16'
Old Soul
Boo 15h1'38"18d55'
Old Watash
Aur 4h46'12"37d45'
Oldag,Anna
Lyn 8h8'52"42d21'
Oldemoppen,Casey E
Lac 22h27'15"40d20'
Olden,Leah
Lyr 18h14'14"52d2'
Oldenburg,Adam Joseph John
Her 17h22'30"38d14'
Oldenburg,Cheryl A
Lib 15h36'41"-21d35'
Oldendick,Elisabeth Leigh
Vul 19h48'45"28d14'
Oldendick,Natalie Christine
Uma 10h32'55"59d47'
Oldendorf,Margaret Lynn
Mon 7h23'40"-8d6'
Oldenkamp,Roger E
Boo 13h54'58"15d55'
Oldfield,Emma Jayne
Lyr 18h15'1"45d27'
Oldham,Alan
Boo 15h6'30"14d30'
Oldham,Brenda Sue
Uma 8h55'17"61d35'
Oldham,Connie Lee
Uma 8h36'1"60d49'
Oldham,Donna Delores
Uma 8h35'54"60d34'
Oldham,Eric
Cep 22h13'0"68d57'
Oldham,Jana LE
Cam 5h50'0"58d54'
Oldham,Johnathan
Cet 1h15'36"0d24'
Oldham,Joseph
Aur 5h58'37"30d36'
Oldham,Leslie Marie
Com 12h8'47"30d57'
Oldham,Linda Lue
Uma 8h36'35"71d58'
Oldham,Lori Jean Nicodemus
And 0h19'51"30d20'
Oldham,Patricia Kay
Uma 8h53'57"70d40'
Oldhoff,Henriëtte
Lyn 8h43'1"41d39'
Oldies Hinrichs
And 2h23'11"45d46'
Oldroyd,Kate & Steve
Sge 20h17'32"17d60'
Olds,Jeffery Roger
Dra 10h13'15"81d10'
Olds,Stephen Carl
Per 3h17'45"41d6'
Ole Bob S - Rg Snyder
Per 3h33'52"39d9'
Ole Faber
Cep 23h8'41"60d40'
Ole Miss
Mon 8h1'13"-8d20'
Ole Misty Eyes & My Little Orienta
Eri 4h35'40"-13d7'AB
Oleary,Jan
Lyr 18h36'11"40d13'
Oleas,Gina
Mon 8h5'48"-5d2'
Olefeldt,Palle
Cas 2h19'21"61d24'
Oleg Dal'
Uma 13h28'41"62d5'
Olejniczak,Tom
Aqr 23h49'22"-6d35'
Olena
Aur 4h53'27"40d18'
Olenick,Cheryl
Lyn 8h29'16"45d57'
Olenka-Ihor
Equ 20h58'49"6d0'
Oleo W
Cam 5h30'1"60d4'
Oler,Sr,Ralph Franklin
Cet 1h37'20"-4d55'
Oles,Ralph K
Dra 16h52'38"64d11'
Oleshansky,Marvin
Aur 6h34'57"32d26'
Olesko,Thisbe & Richard
Leo 10h39'15"13d6'
Oleson,Eric & Cathy
Cyg 19h40'10"40d42'
Oleson,Michael
Aql 19h29'0"8d35'
Oleszkowicz,Laura Ann
And 0h45'50"21d37'
Olewiler,Doris Rosemarie
Boo 13h45'25"11d45'
Olewiler,Krista Lynn
And 2h24'51"46d19'
Olga
Tel 20h20'25"-48d23'
Olga
Uma 13h2'37"61d50'
Olga
And 23h33'25"44d37'
Olga
Umi 15h39'21"68d40'
Olga
Cas 0h33'12"60d59'
Olga
Cas 1h45'48"61d15'
Olga Andreu
Mon 7h17'18"-2d3'
Olga Claudia
Sge 20h0'42"20d9'
Olga e Popi
Pic 4h32'53"-48d32'
Olga Orloff
Cnc 8h26'24"30d41'
Olga's Song
Cas 1h14'34"60d56'
Olidan,Tina Mater
Cas 0h16'60"47d36'
Olie,Dick & de Smet, Atie
Crb 15h27'48"30d29'
Olijnyk,Arsen & Natalie
Lmi 10h39'53"30d28'
Olila,Kari J
Vir 11h59'21"0d5'
Olimpieri,Giovanni
Ind 21h11'7"-59d31'
Olin's Deja Vu True
Mon 6h15'1"-10d16'
Olin,Cynthia
Cas 1h32'41"60d59'
Olin,Jacquelyn Smalling
Cam 3h56'12"57d5'
Olin,Jonathan F
Cet 2h41'35"-17d57'
Olindo,Lucchetti
Lyr 19h9'56"38d59'
Oliphant,Carol E
Oph 17h52'54"13d56'
Oliphant,Hugh
Cet 2h55'45"1d18'
Olis
Lmi 10h53'31"31d44'
Oliva,Alexandria Isabella
And 0h22'35"35d12'
Oliva,Cecilia Maria
Sgr 19h43'36"-40d4'
Oliva,Danielle
Lyn 8h48'1"36d47'
Oliva,Erik
Lac 22h4'48"47d32'
Oliva,Fran
And 23h41'48"47d4'
Oliva,III,Hilarion A
Peg 23h37'0"11d22'
Oliva,Lydia
Sgr 18h51'26"-23d7'
Oliva,Pepito
Sco 17h52'12"-31d17'
Oliva,Susan
Vir 13h28'11"-10d47'
Oliva,Teresa
Cam 4h53'32"60d25'
Olivares Family Star, The
Ori 6h0'28"5d56'
Olivari,Ann M
Lyn 8h0'34"37d36'
Olive
Ori 5h54'39"16d11'
Olive Ammelia
Cyg 20h3'42"30d27'
Olive's World,Frank
Per 3h4'40"47d24'
Olive,David & Karen
Cyg 21h57'15"53d47'
Olive,Keith Lewis
Cam 5h36'11"80d24'
Olive,Nicole
Com 12h45'1"21d43'
Oliveira,Daniel Mark
Hya 8h13'14"2d12'
Oliveira,Edgar & Zila Santoro Kramer
Lyn 7h53'32"42d6'
Oliveira,Guy Phillip
Aur 5h18'14"43d51'
Oliveira,MD,Michelle
Oph 18h32'40"10d30'
Oliver
Cnc 8h55'48"21d27'
Oliver
Ori 5h13'44"15d20'
Oliver
Cnv 12h9'60"37d51'
Oliver
Del 20h13'14"12d22'
Oliver
Cet 0h49'1"-3d27'
Oliver
Lyr 18h33'60"35d44'
Oliver
Aqr 21h7'22"-5d38'
Oliver
Cyg 21h39'33"41d44'
Oliver
Cep 21h44'1"55d46'
Oliver & Lisa!
Uma 9h14'15"48d5'
Oliver Eternal,Lynne & John
Cyg 19h27'48"38d50'
Oliver F
Umi 15h43'10"71d44'
Oliver Family Star,The
Ori 6h6'11"5d27'
Oliver Star
Cet 2h57'0"1d30'
Oliver"Nanu",Mary P
And 0h5'47"43d35'
Oliver's Nova
Vul 19h12'56"23d4'
Oliver,Aaron
Boo 13h41'30"23d8'
Oliver,Aly
Aqr 23h6'56"-6d53'
Oliver,Angela Michele
Lyr 19h10'37"40d32'
Oliver,Artelia
And 0h10'40"47d29'
Oliver,Christine Margaret
Umi 14h15'21"68d5'
Oliver,Cyrena
Aql 19h58'55"0d42'
Oliver,D'Bo-David Brian
Aql 19h27'57"-8d22'
Oliver,Daniel Conrad
Aql 19h44'23"14d44'
Oliver,David Nathaniel
Per 2h49'19"48d56'
Oliver,David R
Aql 19h33'35"1d47'
Oliver,Dominic
Boo 15h4'13"24d50'
Oliver,Donald McCreery
Boo 14h29'28"30d7'
Oliver,Edward
Uma 10h35'41"72d10'
Oliver,Fiona Ishbelle
And 23h31'22"48d57'
Oliver,Gary
Cmi 7h35'44"10d56'
Oliver,Hayley Goldsmith
Cyg 19h30'19"33d19'
Oliver,James A
Dra 9h44'1"74d30'
Oliver,Jeffrey J
Crt 11h13'1"-18d19'
Oliver,Joanne
Crb 16h19'44"27d11'
Oliver,Jr,David R
Peg 0h4'57"22d5'
Oliver,Karl
Dra 17h46'24"63d51'
Oliver,Katherine Marcia
Cnv 13h35'44"45d37'
Oliver,Kathleen Ann
Peg 23h38'21"30d55'
Oliver,Kathryn
Uma 10h35'0"72d25'
Oliver,Lisa R
Cet 2h50'0"6d16'B
Oliver,Marian Lee
Peg 22h44'0"20d20'
Oliver,Mark & Kim
Cyg 21h43'14"38d23'
Oliver,Milton D
Cet 2h50'0"6d16'A
Oliver,Nalini
Ori 4h45'12"0d56'
Oliver,Nicholas Todd
Cam 3h17'50"60d33'
Oliver,Patrica
Del 20h17'48"10d36'
Oliver,R J
Dra 20h9'44"63d29'

Oliver,Robert Lewis
Ser 15h36'27"18d34'
Oliver,Robert(Bert)
Per 3h10'11"41d56'
Oliver,Ronnie
Lyn 8h6'37"44d47'
Oliver,Samantha
Car 7h28'51"-59d29'
Oliver,Someone Special Carol
Cas 0h58'49"58d9'
Oliver,Sophia Tugela
Rosamund
Uma 10h23'52"71d13'
Oliver,Sr-Gramps, Raymond P
Uma 12h7'36"58d1'
Oliver,Stephen
Sge 20h1'53"16d28'
Oliver,Susan Marie
Crb 15h43'11"28d6'
Oliver,Taylor Leigh
Cyg 21h21'30"40d8'
Oliver,Wayland J
Cet 2h6'1"2d51'
Oliveri,Marco V
Aur 5h4'14"37d49'
Olivero's Setting Sun, Derek
Dra 16h53'0"66d6'
Olivia
Cyg 20h37'0"45d4'
Olivia
And 23h16'21"38d31'
Olivia
Cas 1h58'34"61d14'
Olivia
Cet 2h57'17"2d16'
Olivia
Psc 1h0'1"20d30'
Olivia
Umi 15h46'25"73d56'
Olivia Christy
Peg 21h55'20"33d9'
Olivia Fae
And 0h4'46"44d10'
Olivia Francesca Dianna
And 2h25'39"45d20'
Olivia Jane
Vul 19h21'52"26d48'
Olivia Mary
Cyg 19h30'13"31d25'
Olivia Rose
And 1h7'29"40d19'
Olivia Star,The
Lyr 18h43'46"32d25'
Olivia,Alice
Lyn 7h13'38"50d38'
Olivier Et Nelly
Boo 14h46'51"38d17'
Olivier,Ariane
Umi 13h37'24"75d13'
Olivier,Charles
Aur 5h51'28"31d1'
Olivier,Stephen Richard
Uma 12h29'21"60d18'
Olivieri,Antonella
Vir 13h7'17"-1d41'
Olivieri,Dominic Antonio
Crb 15h56'0"38d32'
Olivieri,Jennifer Lee
And 23h41'51"47d18'
Oliviero,Rose Marie
Cas 2h21'59"70d56'
Olivo,David Andrew
Dra 16h52'16"52d12'
Olken,Elizabeth & Michael
Abbott
Cyg 21h18'1"38d15'
Olkiewicz,Mark
Aql 19h24'49"8d41'
Olla,Dorothy C
Lac 22h26'31"50d25'
Olle's Refuge
Her 17h1'1"51d18'
Oller,Christie
Mon 6h53'40"10d49'

Ollie
Oph 18h17'51"11d38'
Ollie & Jim
Uma 8h29'58"62d10'
Ollie Allyne
Boo 15h6'40"13d34'
Ollie M
Her 17h30'34"20d57'
Ollifa
Cam 3h42'24"61d34'
Olliff,Jenni
Mon 6h23'18"5d59'
Ollom,Crystal Sue
Gem 7h23'42"34d33'
Ollomus,Chromos Domus
Uma 14h23'50"59d11'
Ollove,Phyllis Basson
Cet 1h24'37"-2d27'
Olly
Lyn 6h54'12"44d33'
Olly
Ori 6h0'50"1d5'
Olly
Uma 9h47'54"53d54'
Olly Bee
Cas 0h26'43"61d4'
Olmos,Martin
Her 16h38'54"48d18'
Olmstead III,Jack
Vul 19h48'29"28d11'
Olmsted,Paul A
Dra 17h46'1"76d26'
Olney,Alaria Evelyn
Cas 2h48'10"61d30'
Olney,Greig Robert
Cnv 12h40'20"38d8'
Olney,James
Per 4h58'27"38d29'
Olore,Stephen
Lmi 9h49'16"36d32'B
Olore,Tawny
Lmi 9h49'16"36d32'A
Olschewski,Carla
Gem 6h0'40"26d21'
Olsem,Stacy
Peg 0h3'26"31d26'
Olsen Family,The
Vul 21h25'17"27d11'
Olsen III-My Friend, My
Dad,Thomas T
Tri 2h18'57"33d47'
Olsen My Angel Star, Arlene
Marilyn
Cyg 19h47'56"29d34'
Olsen,Agnes
Lyr 18h36'0"40d26'
Olsen,Aleksander
Dra 13h19'24"63d55'
Olsen,Anika Jean
Aql 20h2'54"-1d8'
Olsen,Anna
Lyn 8h2'35"35d18'
Olsen,Bob & Silvia
Dra 12h43'46"75d11'
Olsen,Brooke
Aur 7h6'53"40d47'
Olsen,Bryant Tracy
Uma 9h27'0"48d49'
Olsen,Carolyn Elaine
Aqr 21h24'15"-5d42'
Olsen,Catherine A
Cam 7h25'51"80d9'
Olsen,Christopher John
Per 1h40'48"52d44'
Olsen,Dan
Ori 5h34'35"8d28'
Olsen,Daniel Thomas
Ori 6h16'52"-0d57'
Olsen,DDS PC,Steven K
Oph 17h19'20"11d23'
Olsen,Dr William L
Lac 22h48'48"37d44'
Olsen,Emma
Cas 0h25'30"62d20'

Olsen,Eric
Cep 21h5'23"60d50'
Olsen,Eugene Dale
Uma 11h0'60"59d32'
Olsen,Jack B
Vir 13h29'39"-4d11'
Olsen,James G
Uma 8h47'38"55d49'
Olsen,Jean,Tom & Caroline
Cyg 21h21'39"40d50'
Olsen,Jennifer
Psc 1h18'52"22d25'
Olsen,Jim & Tina
Sge 19h53'52"16d8'
Olsen,John R
Dra 16h49'34"73d0'
Olsen,Jon Hammer
Ori 6h0'0"10d32'
Olsen,Kimberly K
Peg 21h52'15"30d15'
Olsen,Mary Jean
Uma 8h36'22"53d51'
Olsen,MD,Amy Caroline
Oph 18h21'59"6d17'
Olsen,Pamela Katherine
Cas 23h6'1"58d13'
Olsen,Ralph
Eri 3h55'32"-12d22'
Olsen,Robert A
Her 15h59'50"42d34'
Olsen,Roxane
Mon 7h21'37"-5d50'
Olsen,Ryan Edward
Her 18h1'54"31d22'
Olsen,Shari
Aqr 22h4'29"-11d25'
Olsen,Sten
Umi 13h19'34"71d55'
Olsen,Timia Telsa
Cam 3h45'30"61d42'
Olsen,Victoria Catherine
Cyg 20h45'21"43d13'B
Olshan,Sylvia
And 23h9'44"37d42'
Olshefski,Jonathan
Aur 6h1'37"45d12'
Olson "CLO",Christi Leigh
And 23h18'11"49d32'
Olson "Teacher Par
Excellence",Gene
Dra 17h11'24"64d2'
Olson Anniversary Star Victor
& Ann
Crb 15h50'56"26d22'
Olson Re'alta-1,Bob
Her 16h36'23"28d43'
Olson Star,The
Aur 6h1'57"38d57'
Olson's Heart (Chuck &
Louise)
Ori 5h55'42"17d6'
Olson,"Ammo" Ann Marie
Manning
And 1h22'13"38d51'
Olson,"Peaches" Karlyn Marie
Lyr 18h57'1"30d54'
Olson,65,Bud
Aur 5h17'34"45d42'
Olson,Angela N
Umi 14h39'36"68d16'
Olson,Ardis Alvina
And 0h8'0"38d17'
Olson,Arnold Jack
Hya 8h19'40"3d16'
Olson,Ashley Mae
And 2h30'52"50d2'
Olson,Bartholomew J
Cnv 12h21'20"50d58'
Olson,Benjimin J
Aur 7h24'0"40d22'
Olson,Bill
Ser 15h10'1"-1d20'
Olson,Brock T
Cam 3h43'34"74d12'

Olson,Captain William Nickola
Sex 10h36'32"1d56'
Olson,Carl
Her 17h28'57"361d1'
Olson,Carla Kimberly
Mon 7h55'33"-3d23'
Olson,Cedrick
Del 20h19'47"9d23'
Olson,Christy
Peg 22h21'56"28d14'
Olson,Connor William
Aql 20h3'14"7d58'
Olson,Craig M & Nancy
Barnes
Lyn 8h45'48"36d55'
Olson,Crystal E
Cas 2h23'23"63d48'
Olson,David Jeffrey
Cep 22h29'10"59d58'
Olson,Dorothy
Peg 22h45'1"29d21'
Olson,Drew Wheeler
Cyg 19h55'44"58d54'
Olson,Dylan Stephen
Dra 14h54'51"64d20'
Olson,Ellen Marie Durand
Uma 9h28'23"48d47'
Olson,Eric
Lac 22h24'24"53d11'
Olson,Erika
Lac 22h29'33"38d41'
Olson,George
Boo 15h4'14"48d9'
Olson,Gerenee Anne
Aql 20h15'46"5d2'
Olson,Gregg Gilbert
Ori 4h44'27"-2d58'
Olson,H E
Aur 5h39'21"38d10'B
Olson,Heidi Lea
And 2h9'19"41d43'
Olson,Jean Archer
Cas 23h39'36"64d39'
Olson,Jennifer Marie
Sge 20h16'58"17d41'
Olson,Jo-Ann
And 23h0'24"50d7'
Olson,K E
Cep 0h3'22"68d27'
Olson,Karen Bradford
Cas 0h9'29"58d57'
Olson,Kaye Susan
Cyg 20h22'0"38d56'
Olson,Lynn Miley 143
Cet 1h49'53"-1d13'
Olson,Marianne T
Mon 7h0'39"-1d52'
Olson,Mark Paul
Cep 21h23'59"67d50'
Olson,Martha
And 23h31'0"45d52'
Olson,Matthew Alexander
Hya 8h45'0"-0d51'
Olson,Maynard & Emma
Cyg 21h52'1"37d33'
Olson,Miles T
Cep 23h2'20"70d40'
Olson,Nathan Alexander
Cet 2h33'1"6d32'
Olson,Nathaniel Benjamin
Butler
Peg 22h22'47"30d29'
Olson,Neale Edward
Eri 3h45'48"-0d53'
Olson,Oscar W
Cet 2h28'43"8d46'
Olson,PhD,Dr Wade Robert
Dra 19h45'59"67d59'
Olson,Piggy Bear
Per 2h1'17"57d55'
Olson,Quinten Iver
Del 20h57'43"10d11'
Olson,Roger
Cep 21h9'26"55d45'

Olson,Ronald G
Her 17h7'0"42d52'
Olson,Sandra Lee
Cas 1h19'29"60d37'
Olson,Scott
Cet 2h37'0"5d45'
Olson,Sharon Ann
Sgr 18h53'52"-24d25'
Olson,Tawnya D
Cas 0h9'0"62d41'
Olson,Theodora Colleen
Cet 2h37'19"1d11'
Olson,Tyler Shawn
Aql 20h4'60"4d17'
Olson,V D
Aur 5h39'21"38d10'A
Olson-The Grand Viking,Dan
Her 16h43'1"21d56'
Olsson
Uma 11h23'58"43d37'
Olsson's EL KADA
Tri 2h34'41"35d10'
Olsson,Jerry
Her 17h53'48"40d34'
Olsson,Rebecca
Cas 2h3'47"70d7'
Olsson,Vivian A
Lyr 18h27'1"31d36'
Olstad,Dave
Ori 6h6'58"9d33'
Olstein,Cortlandt Tracey
Cnc 8h58'45"17d25'
Olstowski,Franciszek
Ser 18h16'15"-14d53'
Olszewski,John Edward
Dra 16h22'13"64d39'
Oltarzewski,Evelyn Kurlinski
And 23h1'59"51d50'
Olten,Walti Egger
Dra 10h41'53"81d23'
Olthoff,Mark William
Dra 19h5'33"47d59'
Olton,Frank Thomas
Oph 17h54'56"11d11'
Olvera,Adriana E
Aur 4h56'58"41d3'
Olwyn Brothers
Per 3h6'42"41d4'
Olyha,Natalie S
Cas 0h35'31"75d32'
Olympia
And 1h15'49"37d1'
Olympia Fourtounis
Cap 21h16'58"-22d27'
Om
Cyg 20h19'33"41d35'
Oma
Uma 11h53'21"49d23'
Oma Anna
Lac 22h50'33"40d38'
Oma's Star
Dra 16h5'32"68d31'
Omaha
Boo 14h34'0"17d38'
Oman,Donald Alan
Lib 15h56'48"-18d54'
Oman,Jack Edward
Ari 2h25'33"11d50'
Oman,Noah Jeffery
Her 16h42'15"23d5'
Omans,Justin
Dra 20h5'32"64d52'
Omar
Eri 4h48'38"-6d8'
Omar
Eri 3h48'32"-2d4'
Omar
Aqr 22h27'44"0d3'
Omar,Said
Lyn 7h57'44"39d55'
Omar,Sarah
And 2h21'56"44d49'
Omar,Sharif
Aur 4h58'26"38d4'

Ombak
Mon 7h18'51"-6d51'
Omega Constellation
Tri 1h51'19"25d56'
Omega,Claire
Sge 20h1'52"17d13'
Omeis,Clifton Lloyd
Cep 22h4'38"58d48'
Omel,Sue
Cas 0h24'44"58d33'
Omelczuk,Heather
And 1h25'21"40d2'
OMI
Lac 22h54'0"35d27'
OMI-MAJA
Cnc 9h17'0"8d51'
Omlie,Jeanne Marie
Tau 4h1'50"28d37'
Omlie,Mary Fran
Gem 7h9'46"24d52'
Omlie,Randall Matthew
Vir 11h40'44"1d33'
Ommanney,Richard & Louisa
Peg 23h28'18"28d49'
Ommundson's 25th
Birthdaystar Kim
Gem 6h31'55"12d24'
Omni,Terra
Cas 1h53'0"75d38'
Omni-MAN & LKN,Amor
Vincit
Mon 6h20'42"8d3'
Omolara
Aql 19h0'40"-6d22'
Omran
Ori 6h1'10"1d39'
On the Wings of Your Love
Lyr 19h2'10"25d54'
On to Mars-Katydid
Uma 10h37'1"52d41'
Ona Ruth
Dra 19h29'53"80d15'
Onalee
Lyr 18h58'44"34d50'
Onalee S
Vul 19h13'29"21d41'
ONASLAB-48
Boo 14h12'27"40d18'
Onaslab-AI-13
Cyg 19h58'45"50d26'
Onassis,Blackie
Cyg 19h45'51"30d40'
1000 Wishes
Cyg 20h2'44"40d29'
123121491
Cam 3h57'19"60d24'
138 KDM
Ori 5h29'59"-0d16'
1920522511441411432519
Morning Love
Cas 0h48'0"60d52'
Once in a Lifetime
Crb 16h1'35"30d3'
Once Upon A Time
Ori 5h47'56"10d52'
Onda Rambla
Ori 5h54'0"1d3'
Onderbeke,For My Loving
Wife Deanna
And 23h39'28"42d6'
Onderbeke,My Beloved
Husband Richard
And 23h40'0"47d21'
Ondich,Wayne & Nancy
Cyg 20h32'42"50d2'
Ondish,Natalee Marie
Lyn 7h35'1"39d36'
Ondrejicka,John Anthony
Sex 10h15'11"-8d53'
Ondy-Busch
Cam 7h57'1"61d45'
One
Sct 18h53'19"-5d28'
1 4 3 Shannon
Cet 1h20'16"-14d6'

1 K Bailey
Cam 5h35'12"60d37'
1+1=1 Forever & Beyond
Lmi 10h3'1"40d44'
1-4-3 K W J
Del 20h14'44"10d1'
1-4-3 Lifetime
Boo 14h58'1"25d48'
143
Uma 10h3'36"58d33'
143 Babe
Cas 2h45'51"73d42'
143 Forever
Her 17h26'53"38d52'
143 Forever Jerry
Ser 16h12'54"2d48'
143 Forever Matter Hatter
Oph 17h58'52"12d51'
143 Forever Princess Charlie
Aql 20h15'47"8d2'
143 Gigi
Aur 5h23'12"38d48'
143 Infinite & More
Cep 20h29'28"76d14'
143 Jaime & Kelly 143
Eri 4h33'48"-11d50'
143-Always & Forever
Ori 4h49'34"0d15'
143Pb
And 0h49'49"22d21'
1Krum,Howard
Per 4h39'49"36d26'
103AVH
Lmi 10h2'18"32d30'
One & Only Mark,The
Ori 5h55'54"12d38'
One & Only Terri Lee, The
And 23h41'50"32d43'
One And Only
Sco 16h9'39"-28d0'
One Day At A Time (P J
32394)
Uma 8h30'27"52d23'
One For Lucas,The
Psc 1h0'59"21d38'
One Heart & Soul
Lyr 18h33'20"34d27'
One In A Million
Aur 6h19'0"35d36'
One In A Million
Her 16h31'46"24d37'
One of God's Stars The Donna
Jean
Cas 1h1'1"61d22'
One Old Shoe
Hya 8h54'38"2d54'
One Spirit
Ori 6h2'1"8d20'
One Wild-Wild-,The
Dra 15h56'53"66d22'
One Wish
Uma 9h19'28"52d16'
One,Joseph T
Her 17h21'28"44d44'
One,The
Uma 11h39'1"32d26'
One,The
Aur 6h17'9"31d31'
One-four-three-seven
Umi 16h13'33"76d44'
Oneal,Christopher Andrew
Aur 4h47'10"50d16'
Oneday
Cas 22h59'37"56d50'
Onee
Cam 14h2'30"82d0'
Onefourthree
Cyg 19h33'30"34d35'
Onegora
Cam 4h28'12"61d46'
Onestinghel,Patricia A
And 0h18'1"33d58'
Oney I
Uma 9h51'0"48d50'

Ong,A Great Realist &
Dreamer,Al
Sgr 18h48'15"-36d31'
Ong,Kimberly
And 1h58'55"37d34'
Ong,Suzanne M
Vul 20h3'12"25d33'
Ongaro,Tina Ann
Cas 0h26'33"64d6'
Onie,Shirley
Cas 0h9'15"50d10'
Onion Woman M M P MCM-
LYV III
Lyn 8h17'22"47d6'
Onjalique
Aur 6h11'50"30d34'
Onkelbach,Pia
Ser 15h12'1"-3d5'
Online Connecting Point
Ori 5h26'28"-0d7'
Only & Always Adrian
Cas 2h55'11"58d9'
Only Love
Mon 6h18'51"5d1'
Only One
Lib 14h30'51"-22d6'
Only One Barb!
Peg 22h0'40"29d42'
Only You
Psc 1h19'48"23d32'
Only You Marie
Cam 5h54'1"58d54'
Onnen's Outstanding ORB
Oph 18h40'13"8d28'
Onnen,Steven LeRoy
Oph 18h36'1"10d52'
Ono,Cheryl Michiko
Aql 19h26'58"0d8'
Ononiba,Vincent Ude
Cep 20h43'28"73d57'
Onorato,Daniel A
Lac 22h26'46"53d37'
Onorato,Jennifer
Lyn 7h41'45"45d14'
Onslager,Sheila Ann
Lyr 19h25'13"38d38'
Onslow,Richard
Per 2h53'36"46d10'
Onthank,Tari Lee
Cyg 19h27'1"32d17'
Ontje
Uma 10h58'49"40d44'
Onuma,Michael Tsuyoshi
Ori 6h17'29"7d31'
Onyx
Cmi 7h30'53"0d16'
Onyx
Cnv 13h8'15"38d23'
Oodie
Boo 14h20'1"31d38'
Oogie
Cep 22h19'25"78d60'
Oogie's Corral
Cma 6h56'50"-19d19'
Ookpik,JT
Cnv 12h28'17"37d59'
OOLS
Dra 17h33'53"68d56'
Oosterhouse,Roland
Tau 5h49'53"23d40'
Oosterhuis,Jim & Pamela
Oph 18h30'1"6d35'
Oosterlink,Rene
Ori 5h56'28"8d11'
Ooyman,Christa Mien
Ori 4h59'40"4d35'
Ooyman,Michael Fletcher
Ori 4h59'23"4d21'
Opa
Ori 5h55'20"11d25'
Opa Pauli
Cnv 12h34'46"39d33'
Opacity,Robyn
Lyr 18h20'37"44d15'

Opal S J
Com 12h57'29"31d17'
Opalack,Kelsey C
Lyr 18h35'36"27d2'
Opalka,Andrew
Cnv 13h52'10"30d9'A
Opalka,Lily
Cnv 13h52'10"30d9'B
Opatt,Christopher Mark M
Ori 4h43'20"8d8'
Opatt,Gabrielle Noel A
Com 13h4'35"30d6'
Opdyke,Gregory
Dra 10h32'20"81d16'
Opdyke,Patricia
Uma 10h12'0"50d19'
Opendo,Hannah
Cas 1h9'1"60d28'
Openshaw,Thomas
Oph 16h54'0"-28d55'
Openshaw,Trina Danielle
Peg 21h59'52"35d43'
Operacz,The
Ser 15h26'53"1d2'
Operation True Love Mission
Accomplished
Ori 6h2'44"8d52'
Opfell,Andrew
Oph 17h38'52"-16d24'
Opfermann,Annerose
Cmi 7h59'1"8d53'
Ophals,Elizabeth Granata
And 2h2'1"40d17'
Opipari,Elizabeth Ann
Lyr 18h46'10"32d59'
Opitz,Elizabeth
Mon 6h23'58"3d41'
Opitz,Susan Ione
Vul 20h21'48"25d49'
Oppedisano,Rocco
Boo 14h11'49"47d40'
Oppenheim,David Glen
Boo 14h33'44"47d37'
Oppenheim,Ray
Lib 15h3'56"-3d21'
Oppenheim,William
Sct 18h54'29"-5d36'
Oppenheimer,Frances Reese
Cam 6h10'53"61d28'
Oppenheimer,Laurie
Sge 20h16'1"16d58'
Opper,Irene E
Peg 22h49'0"8d22'
Opper,William
Aur 4h47'59"50d33'
Opperman(OPS),Frank
Dra 11h40'37"71d14'
Oppers,Karen Marie
Cnc 7h59'0"19d6'
Oppliger,Marcel
Cas 0h3'24"60d19'
Öppling,Claus
Ser 15h55'36"-2d15'
Oprah
Cyg 21h6'51"30d34'
Opresko,Andrew Gregory
Aql 19h52'60"12d57'
Opresko,Michael Gregory
Aur 5h0'17"46d60'
opta data hard-und software
Uma 13h36'27"48d32'
Optatum
Del 20h30'32"18d46'
Opénn
Aql 19h53'30"15d53'
Oquendo,David E
Cyg 21h20'31"40d42'
Oquendo,Nicole L
Uma 9h29'43"57d40'
Or Shel Chaim
Mon 6h18'41"5d11'
Ora
Vul 20h0'1"25d29'
Oracle "Sangwa Rikma" Lina
Krassa-Vafia
Uma 11h21'59"52d47'

Oram,Bernice "Bunny" Cyg 19h30'39"36d21'
Orange Blossom Special,The Aql 20h19'26"7d43'
Orange Dawn Cet 2h11'43"3d28'
Orange,Jason Umi 14h8'37"88d27'
Orantek,Fabian- Sebastian Lyr 18h26'1"47d6'
Orban Aur 4h54'53"51d1'
Orbe-Kirk,Milagros Boo 14h10'24"40d33'
Orbit Aql 19h53'49"-5d48'
Orbit Systems Integrators Ori 5h28'0"0d6'
Orchard Star Boo 14h30'21"11d55'
Orchard,Julia Cas 1h0'22"60d14'
Orchard,Mark & Sharon Uma 8h37'47"60d23'
Orchidaceae Cep 23h39'41"64d29'
Orchidée Cyg 20h49'58"38d19'
Orchidée Cep 20h58'15"62d14'
Orco Cyg 19h30'12"38d16'
Orcutt,Alice L Tri 2h34'1"34d53'
Orcutt,Jr,Charles E Aur 6h20'19"39d45'
Ord,Andrew Simon Her 16h39'9"19d54'
Ord-Wyness,Sheena Peg 22h29'30"28d20'
Order Of The Eastern Star,The Cas 1h33'34"60d43'
Ordille,Tiffany Lmi 10h32'0"33d41'
Ordino,Joan T And 2h22'11"39d4'
Ordonez,Mark Gabriel Leo 9h36'15"7d12'
Ordway,Jerry Bear Ori 5h52'24"14d29'
Ordway,Teri Sue Uma 10h55'46"62d24'
Ore,Matthew Rogers Cep 22h13'12"61d18'
Ore,Michelle D Aql 19h34'56"-6d40'
Orefice,Juno F Aur 4h55'45"51d12'
Orelac Cep 22h7'48"60d60'
Orella,Laura Marie Aql 20h8'47"4d53'
Oren Cnv 13h12'1"40d43'
Oren Star,The Her 18h7'1"28d22'
Orenstein,Maryanne & Ed Cep 21h12'40"64d11'AB
Orent,Jake Stephen Aur 6h32'46"31d3'
Oresman,T Ori 5h31'9"-0d35'
Oresta Vul 19h59'51"28d40'
Oreste Cam 4h54'50"71d1'
Oreste Psc 22h55'56"5d45'
Orfanedes,Dean & Kathleen Aql 20h12'0"4d13'
Orfas,Helen Uma 11h36'49"31d59'
Orford-Ashley,John Charles Mon 6h43'42"-1d53'
Organ Donor Spirit Star Her 16h56'36"35d45'
Orgelet,Stephanie Sgr 19h27'31"-36d9'
Orgeron,Laura Charlotte Mon 6h19'17"3d17'
Orgill,Cathy Peg 21h35'43"20d19'
Ori,Maryse Cas 0h19'45"63d51'
Oriana Pup 8h24'10"-22d44'
Oriana's Wishes Uma 9h36'30"67d31'
Orianne,Berthon Sgr 19h27'28"-33d51'
Orietta Aur 4h36'58"31d37'
Orig Anthropomorphic Personification Psc 0h47'17"31d21'
Originos,Dina Cas 0h33'52"66d17'
Origo,Lucis Ant 10h46'22"-35d36'
Origone,Austin A Aql 19h42'17"12d28'
Oril,Kenneth & Jean Sge 20h0'13"18d44'
Oriolo,Galactic Aqr 22h41'26"-1d47'
Orion Ori 5h12'30"-5d32'
Orion,Channa Lyr 19h13'50"41d7'
Orion,Nicklas Ori 5h42'30"-5d38'
Orisha Aql 19d3'55"2d1'
Orista "My Prince", Peter Aur 4h53'0"38d26'
Oristaglio,Jason Aur 6h22'50"33d5'
Oristaglio,Jett Michael Dra 16h26'0"63d18'
Oristaglio,Siena Michelle Cyg 19h34'26"32d26'
Oritz,Jennifer Marie Vul 19h18'51"25d31'
Orix Tri 1h58'42"28d55'
Orkin,Jocelyn Leigh Cas 0h35'21"63d10'
Orla-Bukowski,Adam Vul 19h48'39"23d33'
Orlandi Dance Center Aur 6h14'0"32d27'
Orlandi,John Alan Per 2h50'38"38d41'
Orlandi,José & Barclay Scott Crb 16h9'26"37d37'
Orlando Aur 6h57'23"37d58'
Orlando Star,Michael Dra 18h28'11"50d16'
Orlando's South Uma 8h37'1"59d51'
Orlando's Star Lyr 19h23'11"31d4'
Orlando,Carole A Cas 2h17'16"68d43'
Orlando,D Ori 5h59'11"16d25'
Orlando,David Christopher Tri 2h15'40"31d5'
Orlando,Katherine & Nicholas Cyg 20h15'1"41d55'
Orlando,Louis & Melissa Lang Sge 19h3'36"16d42'
Orlando,Paul F Aur 7h12'39"35d53'
Orlando,Stephen J Boo 15h26'58"32d57'
Orleman,William Uma 10h21'34"50d14'
Orlette Lyr 1h34'21"29d29'
Orlich,Michael Aur 6h0'58"30d9'
Orlick,Arnold Tri 2h15'0"31d9'
Orlik,Glücksstern für Natascha Umi 15h17'28"68d8'
Orlins,Edna Lewis Crb 15h58'47"32d7'
Orlob,Lillian May Lyn 8h2'15"48d2'
Orlosky,Jason Adam Dra 17h7'55"63d52'
Orlovic,Lidia Cru 12h32'5"-57d57'
Orlovic,Zdenka Ruza Ant 10h33'24"-39d19'
Orlovsky Edward J Aur 6h7'0"46d31'
Orluck,Steven G Cnv 12h28'0"32d12'
Orly & Milton Boo 14h17'38"35d47'
Ormaas,Roberta Gates Vul 19h41'21"23d52'
Orman,Shauna Cyg 19h24'30"31d39'
Orme,Lila Morton Ari 1h55'58"19d42'
Orme,Stan Aur 5h18'50"50d29'
Orme,William G Equ 21h23'35"2d45'
Ormes,Bruce & Laura Cyg 21h22'1"40d45'
Ormond,Chris & Juslaine Costanza Boo 14h36'13"41d41'
Ormond,Douglas Cep 21h55'0"55d5'
Ormond,Gavin Robert Boo 15h3'60"20d18'
Ormonde,Evelyn Aql 20h13'47"1d38'
Ormonde,Ronnie Cas 0h52'0"73d56'
Ornelas,Brandon Gem 7h26'39"30d39'
Ornelas,Carmen Lyn 8h56'47"44d34'
Ornella Cep 21h34'1"67d51'
Ornella,Vanoni Lyr 18h15'20"31d47'
Ornellas,Dan Aql 19h18'0"15d17'
ORO Dra 19h59'54"68d5'
Oroojico Cep 22h42'55"65d34'
Oros,Jason Carl Aur 6h8'1"46d36'
Oros,Lisa Nicole Vul 20h17'0"28d56'
Orosz Family,The Mon 7h58'56"-6d42'
Orosz,Travis Dylan Aur 6h3'23"50d30'
Orourke,Mr & Mrs Brian & Donna Cnv 13h52'11"40d3'
Orovic,Jaime Uma 11h1'51"45d26'
Orozco,Elizabeth Crb 15h22'0"30d60'
Orozco,Elvira B Leo 10h12'12"17d53'
Orozco,Hernando Her 10h7'20"38d48'
Orozco,Ivana Elisabeth Leo 10h3'25"18d23'
Orozco,Jessica Elvira Jesusa Leo 10h2'17"11d10;
Orozco,Jorge Ivan De Jesus Leo 10h4'12"17d45'
Orozco,Joseph P Hya 8h58'40"-11d8'
Orquidea,Zandra Lmi 10h54'56"26d36'
Orr & Family,Mr (Captain Kirk) Cep 0h14'27"68d55'
Orr III,Robert Joseph Aur 6h7'60"32d20'
Orr Star,Doris Newman Umi 14h39'0"81d10'
Orr,Adria Athena Cas 0h45'54"64d46'
Orr,Andrea Lac 22h55'58"53d49'
Orr,Colin Neville Cep 21h5'10"70d56'
Orr,David Buckner Aur 5h59'27"37d41'
Orr,Leslie Ann Lyn 7h48'37"58d52'
Orr,Margaret Cam 3h14'1"63d34'
Orr,Nicola Louise Cas 0h29'13"60d12'
Orr,Philip Ori 4h44'43"1d35'
Orr,Thomas Paul Dra 16h8'33"64d16'
Orr,Wesley "Tex" Tau 4h38'25"9d55'
Orr,William Hutchins Cyg 20h31'25"40d4'
Orrel,Wally Cep 21h22'59"58d14'
Orrence,Kimberly Anne Del 20h55'39"10d10'
Orrico,Michael Giancarlo Boo 15h4'36"26d14'
Orringer,Alta Lyr 18h56'58"31d14'
Orrs Elementary 1995, Robert V Nix Boo 13h46'20"14d13'
Ors 06-15-82,Lori Elena Mon 6h40'16"10d28'
Ors 11-03-79,Leslie Christina Peg 22h14'1"10d2'
Orscheln,Cara Hya 8h37'15"0h20'
Orsi,Arnaldo Cas 23h3'13"53d47'
Orsi,Janelle Vul 19h47'37"27d52'
Orsi,Jennifer Ann And 0h42'43"45d14'
Orsi,Jessica Ann And 0h21'39"30d43'
Orsi,Peter Pompeo Dra 14h58'1"64d18'
Orsini,Vincent P Del 20h22'1"10d3'
Orsulak,Stephen F Aur 6h55'40"37d17'
Ort Star——Joseph Michael Jarema,The Ori 5h53'39"18d21'
Ort,Lewis J Uma 11h7'39"59d34'
Ortalano,Mario "Billoy" Cas 22h57'32"56d34'
Ortau,Raymond Cet 2h37'1"6d33'
Ortega,Ariel Victoria Crb 16h16'39"31d20'
Ortega,Eddie Aql 20h7'49"0d4'
Ortega,Judy Peg 21h38'57"23d51'
Ortega,Julian George Lac 22h25'38"38d56'
Ortega,Lillian M Aql 18h54'36"6d52'
Ortega,Rocky & Linda Lacemann Boo 15h10'37"52d45'
Ortes,Domitille Sgr 20h18'6"-33d13'
Ortez,Alicia M And 23h15'0"44d27'
Orth Uma 11h59'37"48d53'
Orth's JDT Rainbow,KC Aql 18h43'26"11d18'
Orth,Franklin L Dra 17h52'1"67d54'
Orth,Jr,Lawrence Harold Uma 13h37'57"50d8'
Orth,Matthew Joseph Ser 15h20'38"5d42'
Orthaus,Alan Uma 11h0'53"58d51'
Orthaus,Steven Uma 11h7'23"59d21'
Orthmann,Karin Lac 22h2'1"50d48'
Orthopedic Stars Cnv 13h51'19"41d28'
Orthwein,Jason Thornley Ori 5h51'25"16d59'
Ortiz I,David Anthony Her 17h20'1"38d17'
Ortiz,"Martina's Star" Uma 8h42'57"51d36'
Ortiz,Alexander Godfrey Cnc 9h10'0"30d40'
Ortiz,Alexis Peg 22h38'1"26d32'
Ortiz,Ashley Nicole And 0h53'14"35d37'
Ortiz,Christen Cmi 7h14'50"9d19'
Ortiz,Christopher B Aur 7h17'0"39d47'
Ortiz,Diana Lind Aur 6h10'45"54d34'
Ortiz,Enrique Rangel Eri 4h13'45"-12d37'
Ortiz,Gloria Ori 4h55'1"-1d13'
Ortiz,Jaime Uma 8h40'0"53d15'
Ortiz,Jr,Carlos S Lyr 19h23'14"31d15'
Ortiz,Julianna And 23h0'1"41d1'
Ortiz,Lydia E And 1h34'0"39d28'
Ortiz,Marilyn Claire Mon 6h3'17"-5d7'
Ortiz,Miranda Jean Cas 22h57'48"55d43'
Ortiz,Monsieur Frederic Cam 7h41'15"68d27'
Ortiz,Nichole Charlene Dra 17h16'1"65d9'
Ortiz,Nilsa And 0h58'29"40d41'
Ortiz,Rafael Lanausse Mon 6h55'16"-10d29'
Ortiz,Ronald James Cep 22h47'28"59d43'
Ortiz,Uriel Aql 19h31'52"12d22'
Ortiz-Glass,Mimi Aur 5h26'27"38d29'
Ortiz-Spenst,Marilyn Sex 10h42'1"-5d33'
Ortland,Dan Boo 13h39'34"19d22'
Ortman,Daniel T Aql 19h2'0"11d14'
Ortner,Anne Lac 22h53'21"53d8'
Ortolani,Carla Cyg 21h26'15"48d55'
Orton,Cynthia Karen Mon 6h35'17"7d35'
Orton,Gladys Cam 5h47'40"73d32'
Orts,Nicole Christine Uma 8h37'37"48d31'
Ortuso by Max,Paola And 0h45'0"40d27
Ortuso,Lucia Dra 17h0'37"72d2'
Ortwein Cam 6h18'47"80d37'
Ortwein,Elisabeth Ari 2h55'0"21d6'
Oruch,Mark A Cnv 12h50'58"38d26'
Orum,Harold B Eri 3h24'43"-4d54'
OrvegUrd Indicium-Rolf 1994 Cam 14h11'57"82d2'
Orwig,Amy And 2h3'1"46d24'
Orwig,Michael K Dra 19h39'57"70d32'
Ory,Annie Cyg 21h1'22"40d30'
Ory,Margaret Katherine And 2h25'42"44d20'
Oryan Per 3h29'49"50d38'
Orzo,Allyson Rose Oph 17h5'38"-1d0'A
Orzo,Christopher Daniel Oph 17h5'38"-1d0'B
Osanna,Andrea Aql 20h0'35"4d29'
Osbeldiston,David Alan Cep 21h15'16"58d34'
Osbeldiston,Kenneth Per 3h10'45"41d0'
Osborn Chips Ori 5h48'53"11d4'
Osborn,Christopher Allen Aql 18h53'53"10d6'
Osborn,Colleen Claire Umi 20h30'14"88d40'
Osborn,Jan Tau 4h44'1"20d2'
Osborn,Jaydie P Peg 23h16'48"31d9'
Osborn,John & Darlene Sge 19h52'15"16d29'
Osborn,Joseph John Her 18h12'26"37d36'
Osborn,Kimberley Lafarge Lib 15h32'21"-28d51'
Osborne (Bob)Together Forever,Richard Cyg 20h53'57"38d18'
Osborne's Star,Ruth Cyg 20h21'12"38d30'
Osborne,Amy Sue Cyg 20h3'34"39d23'
Osborne,Annamarie And 23h28'18"41d29'
Osborne,Beverley Mary Joyce (Cope) And 1h43'11"40d31'
Osborne,Bonnie Lac 22h30'39"56d35'
Osborne,Bruce W Cyg 20h18'0"41d44'
Osborne,Craig Cas 0h33'60"62d8'
Osborne,Cupra AKA Dan Ori 5h50'21"20d59'
Osborne,David Richard Duke Aur 6h19'31"45d4'
Osborne,Davis Lee Sgr 19h30'34"-43d34'
Osborne,Debbie Mon 6h35'26"0d51'
Osborne,Dorothy Mae Lyr 18h40'28"45d19'
Osborne,Eugene John Hya 8h35'22"1d15'
Osborne,Gina Aql 19h54'18"13d35'
Osborne,Gorgeous Claire And 1h17'1"37d37'
Osborne,Howard W Aql 19h55'22"14d25'
Osborne,Jennifer Ann Cas 0h22'48"50d1'
Osborne,Judy And 0h4'52"38d16'
Osborne,Kate Dra 13h47'30"64d19'
Osborne,Lewis L (Craig) Cep 20h46'27"73d32'
Osborne,Os & Mick Vul 19h48'46"25d33'
Osborne,Regina Gail Peg 21h55'56"31d41'
Osborne,Robert Dean Ser 15h24'0"8d53'
Osborne,Shannon Marie Mon 7h58'30"-8d36'
Osborne,Sharon Peg 22h33'36"2d29'
Osborne,Todd Kyle Per 3h28'25"50d7'
Osborne,William Burton Cam 11h4'58"80d14'
OsborneGinter, Christine Cam 8h38'1"74d28'
Osbourn,Melinda Marie Cyg 21h9'33"48d46'
Osbourne,Amanda Sge 20h1'33"20d56'
Osbourne,David John Per 1h59'1"48d19'
Osbourne,Eileen Lyr 19h17'28"26d17'
Osbourne,Ozzy Uma 10h53'32"60d26'
Osburn,David Walden Hya 8h15'55"5d42'
Osburn,Katheleen Sharon And 0h6'13"43d22'
Osburn,Richard Glen Lyn 8h1'58"40d3'
Osburn,Sinda Woods Mon 6h22'0"-6d18'
Oscar Aur 6h1'58"40d26'
Oscar Per 1h58'57"56d21'
Oscar Oph 17h54'24"12d45'
Oscar Cae 4h51'45"-32d14'
Oscar Cet 3h18'37"8d11'
Oscar Aur 5h3'23"45d13'
Oscar & Maureen Lyr 18h50'47"42d44'
Oscar Annie-Claire Aqr 22h45'32"-2d2'
Oscar's Smile Hya 8h18'57"3d51'
Oscar,Bob Dra 19h4'36"48d2'
Oschin,Irving Ser 15h16'1"20d20'
Oscye,Gilles Aur 5h3'5"29d20'
Osel918 Colleen1213 Lyn 7h32'22"44d8'
Osgood,Sharon & Alfred Crb 16h3'56"37d45'
Osh-kosh Cas 22h57'44"55d43'
Osheroff,William J Cep 23h21'12"70d19'
Osherow,Caroline Alyssa Equ 21h2'12"8d35'
Osherson,Toby Aur 6h29'0"38d26'
Oshins,Bowl-A-Rama Aqr 23h12'29"-6d40'
Oshman,Dawn Michele Cas 3h35'45"74d32'
Oshman,Ken Oph 17h54'0"8d49'
Oshry "Doll Face", Barbara Cas 0h55'40"60d43'
Oshua Lac 22h9'46"49d46'
Oshun Boo 14h26'15"20d24'
Osin's Star,Chaim Aql 19h14'20"10d29'
Osinski,The Ari 1h55'26"11d16'
Osito Mon 8h2'54"-2d15'
Oskaloosa Aql 19h43'58"14d5'
Oskam,Frans Cas 1h58'32"61d40'
Oslac,Jordanna Elizabeth Cyg 20h52'12"40d21'
Osland,Raymond Daniel Peg 23h6'30"11d20'
Osley,Virginia & Donald Crb 15h50'22"26d38'
Osmak,Gregory J Aur 6h51'18"35d55'
Osman's Star,Ann Cyg 19h18'48"44d50'
Osman's Star,Dahlia And 0h53'57"40d9'
Osman's Star,Sherifa Uma 9h17'24"50d17'
Osmond,Christian Scott Cam 5h51'16"60d24'
Osmond,Eric Ed Cnv 13h52'46"40d36'
Osmulski,John E Cet 1h53'59"-0d45'
Osmun,Elizabeth Bruce Cyg 20h16'36"31d24'
Osnate Morais CD Cas 1h45'0"58d58'
Osofsky,Lisa Jill Tau 5h48'38"23d8'
Osorio,Alexis Marlene Lac 22h54'47"55d3'
Osowski,Sarah Lyn 8h47'38"41d35'
Osric Per 3h11'58"50d2'
Ossakow,Ralph & Geraldine Oph 17h32'24"-20d51'
Ossiboff,"Rachel" Victoria And 23h19'51"46d25'
Ossidiana Tel 19h6'6"-48d33'
Ossie & Dot Uma 11h38'42"47d54'
Ossip,Albert E Cyg 21h22'58"40d41'
Ossowski Star of Love, Patricia Tri 2h20'59"31d19'
Ostapenko,Gina & Vladimir Lac 22h15'27"37d36'
Ostberg,Kellie Elise Mon 6h52'47"-6d30'
Ostberg,Robert Cma 6h52'40"-16d27'
Ostendorff,Joan Cas 0h14'53"62d49'
Oster,Bob & Grace Lyn 9h30'1"41d18'
Oster,Carli Elizabeth Cyg 19h41'12"37d36'
Oster,Simone & Claus Ser 15h14'43"5d28'
Osterberg,"Nani" Margaret And 2h11'38"42d58'
Osterberg,Uncle Louis Aur 5h9'50"43d58'
Osterholt,Jane Marie Lyn 7h57'14"51d58'
Osterholt,Jen Sct 18h54'44"-7d36'
Osterhout,Amy Reneé Psc 1h33'1"21d33'
Osterloh,Carol Peg 23h22'40"15d59'
Osterman,Howard C Lac 22h54'11"54d29'
Osterman,James Peter Cet 0h33'34"-0d41'
Osterman,Janet E Per 3h16'10"40d47'
Osterman,Kiki Mon 6h34'43"-1d36'
Ostermann,Herbert Ori 5h57'49"1d30'
Ostermeyer,Edward Thomas Her 17h18'1"21d10'
Ostertag,Ruth H Eri 4h55'52"-7d37'
Osterwalder,Arnold Vir 12h25'35"-8d18'
Osthoff Peg 23h33'53"18d16'
Ostlund,Sonja R Cet 2h3'30"1d2'
Ostman,Jason Aur 6h0'29"36d52'
Ostojic,Ruth E Cas 0h25'1"63d46'
Ostrand,Gary G Vul 19h36'39"20d4'
Ostrand,George L Dra 16h3'27"66d45'
Ostrand,Helen M Lyn 7h58'21"39d51'
Ostrand,Karin A Cas 3h9'11"68d29'
Ostrand,William C Cyg 20h2'0"40d11'
Ostrander,Jordan Paige And 1h27'31"35d55'
Ostranders,Tracy Cyg 20h21'59"31d3'
Ostroff,"Big Red" aka Bruce F Aqr 23h38'25"-5d0'
Ostrom III,Lyle Ray "Skipper" Oph 17h58'1"12d15'
Ostromecki,Regina (Renee) P Uma 10h51'1"70d31'
Ostrov,Holly Cas 1h7'35"60d52'
Ostrow,Cookie Eri 5h2'26"-6d24'
Ostrow,David Seth Aur 6h26'11"38d44'
Ostrow,Ruth Sco 16h24'45"-38d11'
Ostrowska,Basia Uma 8h41'39"56d30'
Ostrowski Family Star, The Aur 4h46'27"51d20'
Ostrowski,Darlene Lyr 19h23'28"31d14'
Ostrowski,Lorrie Eri 3h41'29"-4d40'
Ostrum,Jane Dyas And 23h25'30"41d42'
Ostry,Larry Gerard Aur 4h46'21"38d18'
OstuO Cam 4h30'59"68d41'
Osuog,Lisa Dawn Cet 0h55'43"-2d6'
Osur,Jill Cet 0h24'43"-6d1'
Osvaldo Lyn 7h47'0"40d0'
Osvaldo,Daniela Peg 23h45'10"12d47'
Oswald 1-8-1968,Justin Patrick Cam 3h54'29"70d13'

Oswald,Anastasia Noelle
Sex 10h40'57"-1d56'
Oswald,Andrew James
Aql 19h55'35"7d43'
Oswald,Beverly Hubbard
Del 20h12'27"9d19
Oswald,Brianna Marie
And 23h27'42"48d44'
Oswald,Diane
Cyg 20h37'1"38d52'
Oswald,Herbert Ernst
Her 17h8'19"48d48'
Oswald,Jeffrey
Her 17h30'59"27d2'
Oswald,Renate und Martin
Lyr 19h16'58"30d38'
Oswald-September 13th, Ken
Her 16h41'44"28d57'
Oswalt,Douglas Dean
Hya 8h19'56"4d4'
Oswalt,J Harris
Aql 19h58'1"8d58'
Oswalt,Mark James
Hya 8h33'20"-10d12'
Oswill,Duncan Monroe
Cet 3h4'21"2d26'
Otaki,Masako
And 1h28'47"39d52'
Otaki,Setsuko
Aql 18h58'14"13d28'
Otaki,Sonia
Peg 21h56'49"31d42'
Otania Y Emery Amor Para Siempre
Aql 19h17'55"10d16'
Otchis,John
Cet 2h31'19"-4d30'
Otchis,Marsha Lee
Mon 7h44'57"-2d4'
Otepka,Gunthild und Christian
Uma 11h34'49"33d14'
Otermans,Geurt
Aur 7h25'19"35d29'
Otero,Duarte
Aql 19h7'0"3d39'
Otero,Edwin
Lmi 10h40'30"31d4'
Otero,Jose Antonio Garcia
Boo 14h48'49"22d49'
Oth,Robert G
Aur 4h59'48"31d9'
Othuse,James
Her 15h58'44"40d36'
Otilia
Cep 4h29'0"80d27'
Otis
Cep 21h59'28"56d14'
Otis
Per 2h9'18"57d28'
Otis 1995,Carr'e
Ori 6h5'26"6d5'
Otis's (Otie)
Cma 7h48'0"10d17'
Otis,Alexander Tyler
Umi 15h50'53"76d16'
Otis,Anne-Marie
Umi 13h11'52"71d6'
Otis,Don
Cet 1h15'0"-6d35'
Otis,Donnie
Aur 7h10'28"35d29'
Otis,Lisa-Marie
Her 16h59'16"15d1'A
Otis,Meryl
Tau 5h14'28"16d15'
Otjen Birthday Star, Rodney Joel
Dra 16h35'49"60d50'
Otness,Tricia
Cam 8h1'44"67d59'
Otoski,Christina
Aql 20h12'52"0d14'
Otranto,Adriana
Tau 4h6'60"20d23'
Otranto,Alexandra
Leo 9h51'0"30d7'
Otsason,Juri
Dra 17h46'24"76d40'
Otsubo
Dra 10h46'60"74d40'
Ott,Carol Diane
Lyn 7h58'18"36d43'
Ott,Dennis George
Per 3h6'34"47d9'
Ott,Gregory Kent
Per 3h13'18"56d22'
Ott,Guenther
Hya 8h46'28"-8d2'
Ott,Michael Eugene
Hya 9h13'12"0d54'
Ott,My Soulful Warrior - Stephen M
Ori 5h54'23"14d47'
Ott,Thomas
Lmi 9h57'41"33d22'
Ott,William
Sct 18h52'56"-5d43'
Ott,Wolfgang
Umi 14h8'21"67d37'
Ottaviano,Dominic Angelo
Ser 15h44'1"20d18'
Ottaviano,Gianni
Uma 14h8'20"60d47'
Ottavio
Ari 3h10'16"16d13'
Ottawa Elementary School
Dra 14h21'28"63d58'
Otte,Andrea
Her 17h8'37"50d23'
Otte,Katharina Cornelia
Ari 2h40'54"20d45'
Otte,Laura Patricia
Cnc 8h41'13"18d49'
Otteman,DeWayne G
Cet 1h31'32"-1d35'
Ottenbruch,Dieter
Per 4h6'7"49d49'
Ottensmeier,Cornelia
Peg 23h33'27"21d37'
Ottensmeyer,Gregory & Gabriela
Aql 20h1'36"10d23'
Otter,Jacob Robert Steinberg
Uma 11h14'20"46d57'
Otterbeck,Hans Georg
Mon 7h2'45"-5d6'
Otterlei,Stephen William
Dra 16h54'31"67d14'
Ottersky,Herbert
Vir 13h4'47"-17d55'
Otterson,Linda Lee
Eri 4h7'23"-13d40'
Ottewell,Meleta Marie
Her 16h49'0"50d46'
Ottie-20
Vul 20h58'14"20d19'
Ottilie
Cas 0h24'1"62d58'
Ottilie & Chris
Uma 11h25'13"44d50'AB
Ottino,John
Her 18h6'43"48d11'
Ottman,Dwight William
Her 17h23'0"47d13'
Ottman,Kimberly Marie
Peg 23h28'46"22d56'
Ottman,Nancy Jo Young
Cas 2h37'41"57d33'
Ottman,Ryan Jeffery
Sgr 19h24'16"-40d11'
Ottman,Zachary Steven
Boo 14h37'27"21d32'
Ottmann,Klaus-Dieter/ Glücksbote
Sco 17h51'0"-30d34'
Ottnad,Philippine
Ori 5h56'38"5d30'
Otto
Peg 23h1'50"31d46'
Otto & Christian
Cas 0h13'30"60d32'
Otto & Clark's Magnificence
Her 18h11'19"31d13'
Otto & Shirley
Peg 22h46'0"33d55'
OTTO's 50th
Ori 5h58'28"8d18'
Otto,Annmarie Katharina
And 2h33'14"45d36'
Otto,Brian Patrick
Aql 19h46'17"14d56'
Otto,Christian
Lyn 8h3'0"48d10'
Otto,Christian
Psc 22h51'39"5d32'
Otto,Cynthia Frances
And 1h23'31"35d58'
Otto,Dale
Dra 17h1'11"66d59'
Otto,Daniel
And 1h22'30"39d14'
Otto,Dr Michael
Uma 9h5'24"48d30'
Otto,Frank L
Boo 14h0'49"22d0'
Otto,Jörg
Vir 12h7'47"2d22'
Otto,Miriam
Aqr 21h21'21"-8d10'
Otto,Steven Byron
Cnv 13h49'40"32d4'
Otto,Vivien 30-11-1994
Lyr 19h17'32"31d29'
Otto-McEvoy,Jan
Uma 11h21'54"40d25'
Ottofeld,Günter
Cmi 7h17'34"5d6'
Ottomano,Susan
Vul 19h44'22"28d24'
Otway,Jamie
Mon 7h1'14"-8d35'
Otwell,Lynda
Aql 20h5'52"1d47'
OTZ
Cmi 7h54'16"8d43'
Ouaddaadaa,Nicole Nina
Cas 1h5'22"63d39'
Ouchi,Hideo
Ser 15h17'53"8d18'
Oudelhoven,Rene W M
Per 2h56'55"40d2'
Oudinet,Roland
Aql 19h7'56"14d54'
Ouellet,Jay Q
Uma 12h53'49"60d31'
Ouellette,Dena & Scott P Sevigny
Gem 6h50'55"30d4'
Ouellette,Gerald O
Aur 5h55'46"30d44'
Ouellette,Grace
Cas 0h50'23"67d48'
Ouida
Ori 5h34'15"-2d32'
Ouimet,Estelle
Aql 20h1'31"14d50'B
Oukhiar,Hassan
Vul 19h23'1"22d33'
Ouo (Mark Louis)
Aur 5h22'27"50d25'
Our Angel Alice
And 23h38'21"40d54'
Our Anniversary Star Of 20 Years
Sge 20h14'0"16d58'
Our April Forever
And 1h52'50"40d15'
Our Bell
Uma 8h31'0"62d18'
Our Beloved Peggy
Cyg 20h23'0"38d57'
Our Betty Honey
Peg 22h31'25"27d9'
Our Buddy Russ
Mon 6h40'24"0d56'
Our Castle In The Sky
Boo 13h40'0"21d24'
Our Children
Aur 7h22'1"40d29'
Our Connection
Uma 11h23'0"48d57'
Our Dad's Star
Her 16h55'43"50d53'
Our Dreams
Cyg 19h59'28"31d41'
Our Dreams Come True!
Leo 10h16'41"9d27'
Our Elysium
Uma 10h27'13"40d34'
Our Eros
Lyr 18h31'23"34d9'
Our Family Star-AMT
Cep 20h27'47"60d11'
Our First Kiss
Oph 16h18'48"-5d60'
Our Forever Wishing Star
Ori 5h56'30"16d55'
Our Future
Lyr 18h28'45"31d52'
Our Gift of Love & Friendship
Ori 5h4'31"8d58'
Our Gilly
Lyr 19h11'1"47d31'
Our Grandma Sallie's Loving Star
Eri 2h52'26"-6d31'
Our Grandmama's Star (Walli)
Lyr 18h15'50"38d26'
Our Guiding Light
Boo 14h23'51"31d39'
Our Guy Jerry
Dra 19h39'19"70d19'
Our Harp
Lyr 18h56'53"31d38'
Our Jack 1962-1985
Sco 17h27'29"-40d39'
Our Joanna
Cas 0h24'34"61d43'
Our Karl
Ori 5h56'0"20d43'
Our Lady Of Fatima
Vul 20h45'24"28d32'
Our Lady of the Rosary
Cas 0h47'32"68d23'
Our Larry
Aql 20h7'29"0d21'
Our Little Hide Out
Boo 14h18'0"19d34'
Our Little Man
Her 16h58'27"51d4'
Our Love
Cyg 20h38'0"40d5'
Our Love
Cam 12h47'0"81d8'
Our Love
Cyg 19h25'60"32d43'
Our Love Will Shine Beyond Eternity
Lyr 18h28'58"44d56'
Our Mama
Cas 1h43'38"60d30'
Our Montana Nitelight
Lmi 9h58'31"34d8'
Our Moon
Lyr 18h42'14"33d28'
Our Mother Marilyn
Peg 22h41'36"20d8'
Our Natural Lives
Aur 6h37'25"38d5'
Our Parents Are A Blessing "Tellez"
Psc 1h32'43"28d11'
Our Phoenix Star
Hya 8h13'30"0d54'
Our Place
Sge 19h57'38"16d9'
Our Place To Dream
Cyg 21h53'59"40d9"
Our Power Of Love
Aur 4h51'58"37d30'
Our Precious Christopher
Mon 7h59'16"-8d1'
Our Quiet Love
Uma 10h5'54"68d33'
Our Secret
Cep 21h4'50"55d19'
Our Shining Glories
JamieAngelaRhianStewar
Ori 5h52'22"9d2'
Our Shining Star"Dad"
Dra 19h41'0"68d12'
Our Son Doug
Aur 6h4'22"48d52'
Our Spot
Uma 11h51'28"49d32'
Our Star
Dra 16h18'46"60d23'
Our Star
Boo 14h28'59"25d25'
Our Star
Aql 20h9'47"6d59'
Our Star
Gem 6h58'46"14d44'
Our Star
Eri 2h59'26"-1d31'
Our Star (James & Kim)
Cyg 19h14'58"45d39'
Our Star Brian
Umi 16h38'42"79d58'
Our Star Donatiello
Lyn 7h3'0"50d20'
Our Star Elise
Ori 5h19'32"-9d44'
Our Star For David
Mon 7h3'53"-6d53'
Our Star Frank & Sue
Uma 9h42'37"70d2'
Our Star Meghan
Peg 22h36'50"20d9'
Our Star"The Babalu" Francis & Eva
Crb 15h26'14"30d14'
Our Starry Night
Peg 22h48'40"4d9'
Our Stellar Gabriella
And 1h34'26"38d20'
Our Steve
Uma 10h40'38"47d38'
Our True Diamond- Bridget
Lyr 18h53'23"31d58'
Our Twins Our Stars in the Heavens
Lib 14h19'14"-11d13'
Our Ultimate Ecstasy
Uma 8h31'17"50d3'
Our Union In Life Will Last For Eternity
Lyn 7h1'21"50d28'
Our Wedding Star
Crb 15h57'12"26d40'
Our's-The One That'll Last 4-Ever
Cyg 19h14'1"49d43'
Ourada,Danielle Mulvey
Sex 9h54'22"-1d48'
Ouret,Marie-France
Sge 18h58'15"19d29'
Ourique,Jaime
Uma 12h2'23"36d15'
Ourso,Robert Joseph
Sgr 18h48'30"-24d25'
Oury,Alexandre
Aql 20h0'29"1d11'
Ousley,Jr,Richard Lee
Cnv 12h22'54"40d41'
Outcalt,Dana Ellen
Ori 5h27'50"15d7'
Outcalt,Rick
Cep 22h48'28"67d51'
Outi Francke
Ari 2h6'33"18d22'
Outland,Donna
Cma 7h14'0"-13d13'
Outlaw West
Uma 9h44'27"70d14'
Outlaw,John William
Aur 6h55'55"36d38'
Outrageous J
Del 20h21'25"10d56'
Outten,Alan
Ori 4h48'57"0d15'
Outten,Jr,Edward F
Cma 6h58'42"-18d46'
Outwater,B J
And 1h26'53"37d60'
Outzen,Mildred & Robert
Uma 8h52'38"47d37'
Ouziel,The Star of David
Psc 1h43'23"21d58'
Ovchinnikoff,Paul
Uma 9h48'13"57d54'
Ovchinnikoss,Zina
Lyn 7h59'28"42d49'
Ovcjak,Mary
Cyg 19h55'19"40d42'
Ove Design Intérieurs
Per 2h38'27"43d57'
Ovens,Simon Niall
Her 17h2'17"20d34'
Overbaugh,Lee
Lyn 8h11'35"36d59'
Overbey,Lauren Elizabeth
Peg 22h36'0"22d23'
Overbey,Merrill Grace
Aql 19h6'54"15d33'
Overby,Michael S
Her 18h19'51"14d35'
Overcash,Marguerite Deloris
Ori 5h53'20"15d60'
Overheim,Alicia
Lyn 6h24'1"59d26'
Overholtzer,Carol
Vul 20h20'55"25d15'
Overhulser's Faith, Paul Gene
Ori 6h6'1"5d44'
Overjoyed
Boo 14h36'39"20d41'
Overly,Kelly Elizabeth
Mon 6h26'42"-1d2'
Overman,Jay Kirby
Tri 1h29'14"34d51'
Overman,Michelle Renée
Mon 6h35'56"-0d48'
Overmann,Megan
And 23h15'17"51d44'
Overmyer,Kayo
Crb 15h59'52"37d59'
Overs,Ronald R
Cep 21h7'0"58d15'
Oversohl,Verena
Lyn 8h12'32"42d15'
Overstreet,Justin Mark
Mon 7h37'1"-5d49'
Overstreet,Nora
Lyn 8h12'53"38d54'
Overstreet,Sean Steven
Cma 6h26'33"-20d23'
Overton,Darren Neal James
Lyn 9h8'12"38d50'
Overton,Gay
Aql 18h58'53"-1d54'
Overton,James Bowen
Hya 9h35'18"-10d19'
Overton,Kevin L
Tau 4h17'36"20d53'
Overton,Michelle Lynn
Cas 0h48'20"61d43'
Overton,Rick
Hya 8h11'14"0d28'
Ovidio-Because of You
Vul 19h21'1"26d25'
Ovington,Chris
Aur 6h0'18"38d50'
Ovnicek,Sandi L
Uma 9h53'32"42d29'
Ovsay,Juliette
Eri 3h59'24"-19d55'
Ovsay,Juliette
Lyr 18h44'1"34d27'B
Ovsay,Sam
Eri 3h22'16"-18d50'
Owana,Norma
Per 4h5'49"50d44'
Owczarzak "Flying Solo",Jill
Lyn 7h30'53"36d53'
Owen Ocean Falls BC, Jim
Her 17h3'29"43d43'
Owen Star,The
Peg 23h26'47"24d46'
Owen's Hole in One
Cep 22h11'18"62d28'
Owen's Star
Per 2h36'57"38d58'
Owen's Starlight
Cam 7h13'0"80d4'
Owen,Andrew
Aql 20h12'40"10d34'
Owen,Barry F
Cep 23h40'57"66d39'
Owen,Clare Gillian
Peg 22h2'35"10d36'
Owen,Clark Jackson
Lib 15h41'17"-23d16'
Owen,Darlene
And 23h36'1"47d14'
Owen,Davis Gerhardt
Lac 22h7'56"48d57'
Owen,Diana Jo
Cyg 20h7'0"38d25'
Owen,Eira M
Cas 0h38'31"70d36'
Owen,Gareth David
Cep 20h44'30"60d39'
Owen,Jacqueline
And 23h41'50"41d56'
Owen,Janet Brainard Spooner
Mon 6h18'0"0d4'
Owen,Joshua
Her 16h24'27"29d23'
Owen,Lance Falcon
Dra 18h56'13"68d46'
Owen,Louis
Peg 23h7'25"10d53'
Owen,Marie
And 2h22'18"41d40'
Owen,Martha Eskridge
Cam 4h14'38"60d6'
Owen,Mary Lyn Weber
Peg 22h4'56"8d48'
Owen,Matthew
Eri 3h12'30"-15d38'
Owen,Melissa Ann
And 23h22'10"45d39'
Owen,Pauline Jane
Cas 0h34'41"62d4'
Owen,Ray Shelton
Hya 8h44'54"1d20'
Owen,Susan Jane
Dra 19h1'1"50d6'
Owen,Taylor Irene
Sgr 18h56'49"-34d58'
Owen,Virginia T
And 2h23'0"48d19'
Owen,William Otway
Ori 6h7'23"9d2'
Owen-Everlasting, Christopher
And 2h28'29"40d37'
Owen-Meller
Aur 5h34'44"37d41'
Owen-Tanner 10-10-1941,Helena
Lib 15h38'40"-23d46'
Owenby,Carla
Cma 7h16'45"-15d42'
Owens S J,Bernie J
Cam 3h20'50"60d47'
Owens Shooting Star
Lmi 10h16'37"38d1'
Owens,Annie
Ori 5h58'18"16d39'
Owens,Avery Elizabeth
Peg 22h30'18"20d32'
Owens,Brian & Sheri
Cyg 20h53'35"30d14'
Owens,Brian J
Boo 13h38'30"14d5'
Owens,Charlotte M
Lyr 18h53'58"40d22'
Owens,Christy Denise
Cet 2h41'16"9d34'
Owens,Darla Jean
Cas 23h36'28"60d52'
Owens,David Neil
Dra 18h8'48"71d2'
Owens,Donna Jeanne
And 1h26'55"48d45'
Owens,Eugene Richard
Her 17h26'44"21d27'
Owens,G W
Cap 21h28'13"-23d58'
Owens,Glenda
Eri 3h40'0"-10d17'
Owens,J C J
Lyr 19h4'46"28d51'
Owens,Jack & Rosel
Cyg 19h29'57"33d30'
Owens,Jimmie L
Her 16h1'59"21d16'
Owens,Joan
Cas 0h53'45"54d54'
Owens,John & Janet
Mon 7h20'18"-5d38'
Owens,John Cecil
Cam 4h45'59"68d57'
Owens,Joseph H
Sgr 19h32'49"-44d17'
Owens,Jr,Robert J
Cep 22h31'45"59d34'
Owens,Jr,William Gordon
Hya 8h36'29"0d4'
Owens,Kelli
Lyr 19h2'40"40d10'
Owens,Marla
Leo 10h45'46"16d25'
Owens,Mayor Fred
Uma 10h28'38"41d2'
Owens,Michael
Lyn 8h15'50"50d29'
Owens,Michael Lee
Aql 19h24'39"-1d49'
Owens,Ray
Her 16h41'26"26d1'
Owens,Stephen William
Ori 5h58'47"9d22'
Owens,Steven
Cyg 20h18'26"39d56'
Owens,T-Ray
Aql 18h44'36"10d5'
Owens,Tamie
Cyg 20h18'22"30d35'
Owens,Tara
Mon 7h54'32"-5d35'
Owens,Tavy Jane
Aqu 23h30'55"-4d7'
Owens-Neads,Evon Ellen Veronica
Ori 6h5'56"3d50'
Owings,Brian
Lac 22h29'1"40d19'
Owl's Love Nest,The
Her 18h22'12"18d52'
Owl,Anne
And 23h22'48"41d6'
Owles,John Maxwell Harding
Ori 6h6'22"2d1'
Owsley,Noel
Oph 17h4'38"8d42'
Oxborrow,Joyce
Her 18h0'21"14d30'
Oxborrow,Mark
Tau 4h36'15"15d1'
Oxbow Creek
Lac 22h37'60"53d33'
Oxendine,Lee Ann Echo
Leo 10h34'36"22d17'
Oxenhandler
Vul 19h48'25"27d38'
Oxford,Carol Mary
Ori 5h54'43"17d24'
Oxford,Cheryl
Lyr 18h58'30"32d59'
Oxford,Cindy
Psc 22h54'27"6d30'
Oxford,Schnitzel
Aql 19h53'52"10d13'
Oxley Family,The Roy
Peg 21h47'55"31d35'
Oxley,Stephanie
Uma 10h35'0"53d1'
Oxley,Tracey Caneel High
And 23h43'12"38d36'
Oxley-Superstar,James Bryan
Cma 7h15'44"-16d4'
Oxman,Samuel & Edith
Cyg 21h3'0"38d4'
Oxman,Star Of Daniel
Uma 13h43'48"51d27'
Oxy Boy
Her 16h37'57"4d41'
Oyabu,Shawna
Cet 0h46'12"-5d14'
Oyler's Star To Wish On,Bill
Cet 0h33'51"0d31'
Oyler,Priscilla Lee
Cyg 19h59'36"37d56'
Oyola,Raul
Hya 8h41'1"-5d44'
Oyola,Sandra
Boo 15h5'29"38d32'
Oz,The
Cep 22h36'16"70d32'
Oza,Joe
Aur 6h35'16"37d48'
Oza,Nimish Virendra
Aql 20h0'30"4d48'
Oza,Sangita
Psc 23h25'1"5d54'
Ozaki,Nadine
Eri 3h2'58"-6d22'
Ozalis,Sheila
Vul 20h4'40"28d48'
Ozark Mountain Cosmic Nugget
Cet 2h49'0"7d14'
Ozaruk,Markian Bohdanovich
Cnv 12h57'30"50d20'
Ozaruk,Tatiana Zenaida
Lyr 19h11'41"40d24'
Ozcomert,Kemal Michael
Lyr 18h44'36"46d19'
Ozeki Yoshihiro Love
Aqr 22h4'0"-15d52'
Oziama,Ali
Cyg 21h34'35"41d43'
Oziama,Tyler
Lac 21h58'44"38d47'
Ozmond,James William Charles
Cep 0h9'37"67d14'
Ozols,Juris Viktors
Per 1h58'58"54d22'
Ozonnat,Jean Pierre
Sgr 19h33'47"-39d51'
Ozoux,Baba (Julie)
Cra 18h33'4"41d42'
Ozyck,P Christopher
And 1h56'41"37d35'
Ozzie
Sge 20h3'0"20d10'
Ozzie
Mon 8h3'58"-4d45'
Ozzie
Per 2h25'18"51d42'
Ozzie
Dra 16h28'55"62d1'

P

P & S
Eri 3h42'32"-2d15'
P A L
Cam 4h20'43"68d14'
P A T S
Per 2h38'1"35d26'
P B 3
Aql 19h6'15"4d56'
P B J,Jr
Her 16h26'33"39d24'
P B Max
Tau 3h50'37"22d10'
P Bear
Vul 19h18'41"26d29'
P C Angel (Gayle M Palmer)
Vir 13h33'24"5d4'
P C Bill
Cep 20h21'35"76d22'
P C's Night Light
Uma 9h9'0"68d10'
P D R Star,The
Tau 3h40'1"28d14'
P Double,The
Lmi 10h51'42"38d6'
P E R L
Crb 15h52'34"26d38'
P G 79/23/09 Pepin- Godbout
Uma 12h4'41"60d9'
P H P M S
Cet 0h28'1"-0d1'
P J
Uma 9h30'28"70d41'
P J & Amanda
Cyg 19h23'0"28d29'
P J M 50
Lyn 7h56'50"43d52'
P J S
Lac 22h7'0"47d52'
P J's Bright Star Twinkles
Lmi 9h46'37"40d6'
P J's Light
Per 2h49'44"50d6'
P J's Place
Boo 14h44'13"48d10'
P K
Dra 19h37'40"61d12'
P K H
Ori 6h22'14"10d32'A
P M P #138
Ori 5h15'16"15d58'
P M S
Uma 10h52'42"40d38'
P O G
Ser 15h58'1"0d48'
P R
Aur 6h50'43"38d22'
P S I
Her 16h18'48"48d13'
P S J K
Leo 9h55'45"29d43'
P Squared (P^2)
Vir 11h37'35"8d34'
P T True
Lac 22h31'21"53d52'
P'tit Bijou Wendling
Her 17h51'52"40d59'
P'tit Bijou Wendling
Uma 13h9'53"62d16'
P'tit Bouchon
Oph 18h42'1"7d34'
P'Tite Lune
Dra 12h6'50"63d31'
P-
Mon 6h32'0"11d6'
P2
Cam 4h59'0"61d23'
Pa kae eoia
Cnc 8h31'31"7d57'
Pa Te Je Pa Di Ka Ma Scalici
Aql 20h11'0"11d24'
Pa's Star
Uma 11h21'0"44d9'
Pa-Jeanne 1-2 & Junie Too
Ori 5h55'27"18d36'
Pa-Trish
Lyr 18h28'47"42d3'
Paananen,Marko Petri Pirkka
Dra 10h59'27"73d34'
Paananen,Tanja E
Lyn 7h30'50"44d12'
Paassen,Arno von
Tau 5h53'0"23d0'
Paavig,Johanna Sofia
Umi 14h57'1"80d50'
Pabillore,Evelyn
And 1h38'1"37d31'
Pablito
Lac 22h44'26"53d20'
Pablo
Aur 6h30'13"32d48'
Pabon,Jessenia Enid
Sge 18h58'45"24d25'
Pabst,Bob
Lmi 10h58'10"25d31'
Pabst,Jeremy William
Cet 2h13'52"6d26'
Pabst,Sr,Jean & Howard
Cet 3h17'37"2d43'
Pacak,John Patrick
Sgr 18h2'38"-28d26'
Pacasoni,Sara
Eri 2h44'44"-5d29'
Pacaud,Alain
Cep 2h47'45"80d16'
Pace
Peg 23h0'39"28d12'
Pace,Bernice
Peg 23h7'43"10d0'
Pace,Gregory DiAngelo
Peg 23h4'22"17d60'
Pace,Jason Douglas
Sco 17h54'26"-30d30'
Pace,Jesse
Hya 9h6'21"1d35'
Pace,Lesli
Cmi 7h22'14"5d16'
Pace,Rhonda K
And 1h8'1"40d59'
Pace,Tammy (Reno)
Boo 14h3'18"12d5'
Pace-The One & Only, John Joseph
Aur 5h58'32"37d46'
Pacelli,Carol V
And 0h44'14"37d30'
Pacelli,Nanceé Sherilynn
Cas 1h17'32"62d58'
Pacem
Crt 10h51'16"-12d25'
Pacera
Aur 4h45'18"34d22'
Pacey,Tommy
Her 17h35'28"27d6'
Pach
Aur 5h56'39"40d35'
Pacheco,Chandler Dean
Lmi 10h51'59"26d26'
Pacheco,Charles A
Ori 4h53'0"-2d53'
Pacheco,David Anthony
Uma 11h43'20"40d23'
Pacheco,Francisco
Hya 8h11'51"0d18'
Pacheco,Jodi
Uma 12h20'50"56d22'
Pacheco,Martha
Peg 22h35'54"10d44'
Pacheco,Randy
Boo 14h46'16"38d27'
Pacheco,Sally
Dra 18h56'13"68d46'
Pacheco,Sr,Dennis P
Aur 5h7'29"38d56'
Pacheo,Rickey
Mon 6h18'17"7d58'
Pachi
Cep 22h1'58"68d55'
Pachter,Jeffrey S
Dra 16h18'20"60d56'
Paci,Chrisy
Hya 9h31'17"2d18'
Paci,Michela
Vel 10h12'33"-43d22'
Paciello,Stephanie Marie
Lyn 7h34'22"52d24'
Paciello,Valentina
Cmi 8h4'27"5d55'
Pacificare's Star
Uma 12h13'1"58d3'
Pacini,Dixie Marie
Peg 22h15'40"31d19'
Pacini,Jr,Thomas J
Dra 11h31'0"70d44'
Pacini,Sergio
Per 1h55'23"56d47'
Paciulli,John R
Cet 2h0'21"-0d24'
Pack Scientist,Marci
Mon 6h32'54"7d43'
Pack(Le Lion),Nathan
Lyn 8h37'17"41d22'
Pack,Ernest
Mon 6h43'38"-3d58'
Pack,Jeffrey L
Her 16h24'25"48d45'
Pack,John Ronald
Peg 23h43'10"30d25'
Pack,Jr,Amo-Amamus Christo B
Lyn 6h59'19"60d45'
Packard,Andrew
Dra 19h47'50"68d50'
Packard,Chuck
Cma 7h14'1"-13d9'
Packard,Debbie
Cet 3h16'30"-0d54'
Packard,Jennifer
Lyr 8h16'40"35d25'
Packard,Justin
Lac 22h8'1"46d10'
Packard,Mark Andrew
Peg 21h58'24"28d30'
Packer,Kathryn Marie
Aur 4h47'12"50d45'
Packer,Lori
Cas 1h43'23"75d41'
Packer,Simon
Cep 20h35'44"76d31'
Packheiser,Maxine & Don
Cyg 20h50'55"40d15'
Packman,Elliot
Per 4h5'53"51d7'
Packy
Cnv 12h45'14"36d38'
Paco y Leo
Ser 17h34'23"-11d10'
Pacono,Anthony Michael
Boo 15h6'0"52d17'
Pacquer,Smiley
Lac 22h3'10"47d53'
Paczuski,Ernest
Lac 22h55'53"51d17'
Pada"Schatzi"152 & 155
Her 17h21'12"45d37'
Padalecki,Lucille
Lyr 18h45'54"33d34'
Padavan,Susan & Jeffrey
Cyg 19h29'15"31d52'
Padberg,Sr,Dr Frank Thomas
Uma 11h48'16"62d55'
Padden,Dana
Mon 7h44'55"-2d16'
Padden,Mr & Mrs Joseph S
Cnv 13h20'28"48d60'
Padderud,Allan B
Cyg 19h26'19"32d35'
Paddington,Barry
Ori 5h55'40"16d53'
Paddock,John Nicholas
Aur 6h11'1"45d33'
Paddock,Lavonne & Jim
Aql 19h8'18"15d55'
Paddy
Cma 6h27'52"-13d60'
Paddy Eternal
Peg 22h14'1"32d2'
Paddy Paul
Ori 4h59'0"0d37'
Paddy Vincent's Star
Cam 4h43'47"67d32'
Pade,Helen
Vul 20h58'54"28d42'
Padel,Giles & Marie
Lyn 7h6'1"50d16'
Paden,Christopher Scott
Boo 15h6'1"40d55'
Paden,Jodie
Leo 10h53'26"20d2'
Padgett "The One", Lauren Marie
Cas 2h33'41"57d26'
Padgett,Alberta F
Eri 3h46'0"-4d8'
Padgett,Charles Edward "Ched"
Her 17h11'1"48d57'
Padgett,John Willard
Aur 4h51'33"37d32'
Padian,Ian
Lyn 9h5'1"33d53'
Padilla III,Mauro T
Her 17h22'0"40d38'
Padilla,Alia Gabriela
Uma 9h53'22"42d8'
Padilla,Jason
Lac 22h21'49"53d35'
Padilla,Natalie & Matt
Cyg 20h44'16"38d22'
Padilla,Tony J
Oph 17h34'42"-22d32'
Padlo,Joseph
Her 17h56'41"14d19'
Padmore,Pairlene Eleanor Thomas
Gem 6h52'27"17d16'
Padrona,Jeanine
Cam 6h22'47"68d60'
Padrutt,Geliebte Rose Anita S
Cep 22h1'11"68d51'
Paduano,Robert
Dra 10h37'31"74d25'
Padubrin,Manfred
Per 4h5'43"47d56'
Padubrin,Roland
Per 4h5'43"49d1'
Paduck,Manfred
Cap 20h4'59"-10d26'
Padula,Michelle Lynne
And 2h16'58"41d50'
Padulo,Andrew
Her 16h44'27"37d26'
Paea,Nadezhda G L & Nikolas
Eri 3h15'23"-17d57'
Paeltaenre
Uma 10h7'1"56d13'
Paessun,Bruce
Aur 6h37'15"37d52'
Paetow,Uwe
Equ 20h58'1"8d11'
Paetowicz,Bettina
Hya 9h13'34"-7d14'
Paez,Ernie & Dee
Tri 1h58'27"26d5'
Paez,Margarita Rosa
Cmi 8h7'16"5d58'
PAF:Susan,Elizabeth, Susie
Uma 10h1'0"68d53'
Pagac,John T
Cep 23h21'1"70d5'
Pagac,Michelle
Cas 1h35'36"61d20'
Pagani,Paolo
Cam 14h9'1"84d41'
Paganin,Fabrice-Jean & Frederique
Pho 0h41'2"-47d46'
Paganini,Bart
Aur 6h24'42"30d8'
Pagano,Angelo
And 23h45'41"37d37'
Pagano,Caterina
Boo 14h26'43"50d32'
Pagano,Jim
Boo 15h4'17"19d37'
Pagano,Joseph A
Boo 14h53'52"23d33'
Pagano,Maria Alice Zamora
Uma 11h11'58"58d41'A
Pagano,Michael George
Uma 11h11'58"58d41'B
Pagano,Thomas Anthony
Aur 5h0'0"29d7'
Page
Lac 22h40'52"56d35'
Page "I Love You",Tom
Aur 6h0'1"32d28'
Page Of Swords,The
Ori 5h55'55"16d38'
Page's Mermaid Star, Sharon
Tau 4h2'49"20d16'
Page,Alice King
Del 20h13'34"11d28'
Page,Alison Peale
And 2h2'35"40d60'
Page,Andy David
Ser 16h16'54"1d18'
Page,Autumn Lee
Uma 12h23'31"54d38'
Page,Avril Moira
Tau 5h25'1"20d25'
Page,Cameron Ward
Mon 6h53'43"-1d23'
Page,Christopher John
Cet 2h1'40"2d11'
Page,Claire Jenise
Mon 6h53'1"-1d1'
Page,Fréderic
Per 3h3'14"37d48'
Page,Gary & Vicky
Aql 19h8'14"4d17'
Page,Hollie Ruth
Peg 22h5'15"25d27'
Page,John Edward
Boo 14h45'17"50d14'
Page,Joshua
Per 2h53'0"40d23'
Page,Joyce
Peg 22h50'27"29d15'
Page,Karen Lisa
Vul 20h42'1"28d10'
Page,Kenneth Aldon
Vul 20h15'34"26d5'
Page,Kymburley Lorrayne
Peg 22h40'19"27d49'
Page,Lauren
Cas 0h40'33"68d33'
Page,Linda Hartley
Peg 21h58'35"20d19'
Page,Lothar
Her 17h41'30"14d43'
Page,Marie Elana
Uma 12h50'40"53d49'
Page,Mr & Mrs Timothy
Umi 14h33'25"65d46'
Page,Paul Ewart
Aur 6h3'53"45d10'
Page,Peter Edward
Aql 18h55'1"10d47'
Page,Polly
Mon 5h55'16"-4d12'
Page,Richard A
Ori 5h54'39"15d40'
Page,Richard Kevin
Aur 7h12'41"39d28'
Page,Robert J
Crt 10h58'52"-8d51'
Page,Rosemary DeSarno
Cas 1h32'27"58d8'
Page,Sue
Cas 23h42'26"61d34'
Page,Thomas A
Aql 19h34'49"-6d33'
Page,Thomas Andrew
Dra 16h34'43"60d26'
Page,Thomas James
Aql 19h54'39"12d16'
Page,Travis William
Dra 14h52'55"60d54'
Page-20 05 1968, Stéphane
Cam 5h0'0"58d44'
Page-Pageanna One, Charles
Aur 4h53'27"51d26'
Pagel,Nancy R
Lac 22h24'48"38d8'
Pagel,Ruben
Lac 22h22'0"38d44'
Pagel,Simone
Leo 9h20'12"12d26'
Pageler's Shining Allstar
Lac 22h44'46"38d28'
Pages,Gérard
Dra 15h46'56"51d56'
Pagett,Roderic
Uma 11h37'32"30d54'
Pagley,Lianne M
Cas 2h10'30"60d1'
Paglia,Ruth Johansson
Lyn 7h28'0"42d6'
Pagliafora,Raffaele
Her 16h46'29"41d9'
Pagliari Star
Lac 22h45'46"55d1'
Pagliaro,"Eternal Star",Frank
Per 4h18'46"50d6'
Pagliuca,Al
Her 16h36'48"25d49'
Pagliuco,Ada
And 1h25'39"34d15'
Pagliughi,Tara Rene
Vir 13h14'31"-0d4'
Pagnozz,Dominic
Cma 6h26'54"-15d58'
Pagny,Jean Claude
Mon 7h41'40"-4d9'
Pagoota,Christopher Lee
Lyn 7h58'1"43d23'
Pagoria,Joyce
Mon 7h39'17"-1d28'
Pagstar,Paulie
Cas 0h50'20"63d44'
Pagteriafe
Lyn 9h1'42"38d26'
Paguro
Scl 23h7'43"-30d59'
Pagë Abé
Uma 9h59'22"50d42'
Pagé (fils de Jean- Paul),Jean
Lmi 10h14'58"39d27'
Pagé,Gino
Cam 6h58'37"65d7'
Pagán,Angelina Marie
Lyn 8h33'1"45d39'
Pahia,Elaine & Neal
Lyr 18h26'38"38d6'
PAHKA
Sct 18h43'59"-6d49'
Pahl,Markus
Gem 7h14'17"21d10'
Pahlavan,Jalil
Uma 11h5'23"32d21'
Pahlka,Clark Kerby
Uma 12h8'36"56d42'
Pahlow,Joseph Haffey
Cam 13h21'14"77d50'
Pahlow,Mary Lynn
Cet 1h29'48"-6d59'
Pahn,Peeter & Gail Armadas
Ara 17h54'2"-52d48'
Paholsky,Donna
Cas 3h10'1"67d41'
Pahrizia
Per 3h50'11"36d41'
Pahud,Yves
Ori 5h52'0"17d51'
Paice,Kathleen Ann
Cyg 19h30'23"33d13'
Paide
Lep 5h56'36"-20d6'
Paiement,Danielle
Per 3h7'25"41d12'
Paiewonsky,Judith Dwyer
Cam 6h10'39"58d14'
Paige
Peg 23h24'45"31d31'
Paige
Cet 0h53'30"1d14'
Paige
Gem 6h57'13"35d18'
Paige
And 23h47'58"46d38'
Paige & Doug
Lyr 18h43'54"32d23'
Paige & Melody
Eri 4h8'22"-11d48'
Paige Cheyenne
Cyg 21h8'18"31d14'
Paige Emma
Com 13h18'28"26d21'
Paige I Love You,Joy
Del 20h19'20"10d3'
Paige Marie
Umi 14h48'54"81d40'
Paige Meredith
And 0h5'14"47d32'
Paige's Sugar Beaches
And 0h38'33"40d22'
Paige,James Collington
Aur 4h47'45"36d41'
Paiges
Dra 17h15'1"73d59'
Paille,Ixilia
Ser 16h4'31"8d30'
Paille,Marie Pine
Dra 11h58'37"66d46'
Pailleron
Umi 16h25'42"79d49'
Paillez,Jacqueline
Mon 6h27'11"8d54'
Paillou,Nathalie
Cyg 21h20'29"40d20'
Pain,Alexandre
Oph 18h34'48"7d24'
Pain,Brian
Aur 4h39'29"33d53'
Painchaud,Joan
Per 2h56'44"34d46'
Paine,Jonathan Scott
Lyr 18h58'1"34d42'
Paine,Michelle Sarah
Lyr 18h30'16"32d20'
Painter,Jacob Michael
Aql 19h40'14"12d57'
Painter,Lana Rachelle
Mon 8h1'49"-5d56'
Painter,Veronica Louise
And 1h12'34"36d44'
Painter,Victoria Rose
And 1h13'17"39d55'
Painting,John Kyle
Sgr 19h8'3"-28d37'
Paisley Morgan Star
Sge 20h1'46"20d0'
Paisley Anthony
Her 17h33'0"24d37'
Paisley,Fred
Cep 22h10'37"59d39'
Paisner,Paul H
Lac 22h2'43"51d40'
Paitchel,Donna Lynne
Cas 0h51'32"76d56'
Paitsel,Carol Diane
Equ 20h54'49"5d46'
Pajennidan
Lmi 10h40'21"27d40'
Pajer,Daniel
Aur 4h50'22"51d33'
Pajot,Marc
Peg 21h45'48"26d41'
Pajot,Yves
Peg 21h47'40"20d27'
Pakala,John & Irene
Cam 3h35'51"62d2'
PakaLani's Dream
Ori 5h59'52"14d32'
Pakarinen,Ulla M
Mon 7h49'15"-1d31'
Pake,Wendy
Peg 22h20'27"3d51'
Pakex 92,Katrina
Boo 15h13'0"31d6'
Pakula,Brandon Alexander
Uma 8h39'18"61d38'
Pakula,Claudio
Cet 2h35'37"-11d56'
Pal
Tri 2h25'14"28d40'
Pal
Umi 15h36'33"68d7'
Pal Tommy
Cmi 7h33'43"1d0'
Palacios,Marco
Her 17h57'0"42d57'
Palacios,Pamela
Cmi 7h16'55"8d45'
Palacin,Cèlia
Cas 0h30'14"60d1'
Paladin Of Aurora
Oph 17h9'17"-24d5'
Paladino Star-Principal,Robert J
Lac 22h49'16"53d37'
Paladino,Carl P
Oph 17h26'60"1d19'
Paladino,John Anthony
Her 16h16'32"26d16'
Paladino,Robert
Sct 18h41'17"-4d21'
Palamand,Shashi
Sgr 19h31'23"-41d15'
Palamara,Joe & Vera
Uma 9h47'58"56d11'
Palamara,Rocky J
Hya 9h16'43"5d7'
Palamara,Tony J
Oph 17h7'31"-20d26'
Palangio Family,Frank & Vera
Lyr 18h59'1"28d4'
Palansky,Charles
Cet 2h28'55"0d22'
Palatnik,Leonid
Aur 5h19'52"43d58'
Palau,Elisabetta
Col 6h34'53"-36d12'
Palazuelos
Leo 9h54'25"12d5'
Palazzari,Gina Marie
Uma 8h41'49"51d32'
Palazzo,Joe T A A C
Lib 15h17'43"-22d19'
Palazzo,MD,Wm L
Per 1h52'52"47d38'
Palazzo,Paul
Per 2h55'1"57d1'
Palazzo,Paul Michael
Crt 11h13'3"-17d34'
Palazzola,Justina Marie-Grace
Uma 10h29'52"54d20'
Palazzone,Antonio Rocco
Her 17h27'38"18d49'
Palchik,George Raymond
Sex 10h33'52"5d32'
Palder,Saul "Bubbles"
Aur 6h29'44"32d39'
Paldino,Christopher
Aur 5h58'30"28d44'
Palekas
Psc 0h20'31"2d22'
Palello,Kelsey Burke
Sco 16h51'37"-37d44'
Palen,James F
Cep 21h31'45"55d5'
Palen,William George
Per 1h53'1"52d34'
Palentine
Lyn 7h20'50"44d42'
Palermino Of Rochetta, Leonardo S
Boo 14h15'27"35d51'
Palermo,Jeremy & Margot Whittemore
Uma 11h48'0"41d31'
Palermo,Mary Rosaly
Cyg 20h1'25"38d19'
Palermo,Philomena
Uma 10h19'53"60d30'
Palesky,Louris Iudici Lourie
Lyn 9h3'23"40d15'
Palestro,B
Cam 5h7'17"60d54'
Paleta
Hya 9h25'18"-5d46'
Paletta,George A
Aur 5h11'0"44d19'
Paley,Bert
Lyn 7h30'28"39d60'
Paley,Muriel
Peg 21h53'59"21d37'
Paley,Sybil Rosenbaum
Cas 2h33'40"57d40'
Paley-Meus Somnium Venio Verus,Lisa
Aql 19h30'18"-6d33'
Palfey,Susan Rose
Cyg 19h51'18"40d46'
Pali Star,The Bob
Boo 13h41'0"21d23'
Pali,Asok Sangha
Hya 8h17'42"-1d34'
Palillo,Floralyn
Sct 18h53'57"-4d38'
Palividas,Nicholas Stephen
Lyr 18h59'38"31d33'
Palivoda XXVII,Michael J
Per 2h24'12"55d31'
Palivos,Desiree
Boo 14h27'40"38d52'
Palko,Michael
Hya 8h59'1"1d10'
Palkovic,Sherrie A
Cet 2h56'12"1d46'
Pall,Hunter Scott
Her 17h23'56"43d45'
Palladino,Anthony
Dra 16h46'21"61d51'
Palladino,Celeste R
Cas 0h22'0"63d9'
Palladino,Joanie
Cas 23h29'26"60d31'
Palladino,Mary Lou
Cas 0h13'17"61d8'
Palladino,Michael A & Rosanne C
Aur 6h14'28"38d58'
Pallan Bicentennial Star,Richard
Dra 10h44'29"74d9'
Pallas-Lucky Star, Hope Christine
Mon 6h59'26"-10d29'
Pallasch,Gerhard
Gem 6h20'42"14d23'
Pallay,John James
Cam 3h57'56"69d56'
Palleschi,Baby Angel Anna
Lyr 18h57'41"30d11'
Pallett,Maggie
Com 13h15'1"27d25'
Palley,Myrna & Sheldon
Tri 2h21'47"30d20'
Pallidino,George J
Cam 3h37'14"73d43'
Pallister,Jennifer Joan
And 23h46'14"42d56'
Pallme,Daniel Carter
Aur 6h7'26"31d8'
Pallone,Thomas Matthew
Boo 14h10'50"34d26'
Pallot,Richard Allen
Aql 19h4'47"-1d45'

Palluck,Joyful Chuck
Uma 9h50'42"49d4'
Palm Desert (In Pegasus) A New Life
Peg 23h3'0"15d58'
Palm,Barbara J
Cas 0h2'52"58d42'
Palm,Chantal Antonia Louisa
Ori 5h31'45"7d39'
Palm,Janet E
Del 20h37'21"7d23'
Palm,Kristin D
Cmi 7h26'53"0d38'
Palm,Phyllis Mossberg
And 0h19'25"35d47'
Palma
Lyn 8h58'32"42d55'
Palma,David M.
Her 17h50'29"14d43'
Palma,Dr Russell L
Oph 17h37'24"-20d11'
Palma,My Dad Ralph
Lac 22h23'41"55d0'
Palmacci,Louis & Eileen
Cyg 20h59'39"39d54'
Palmaka
Cyg 20h23'55"38d29'
Palmar,Jesse N
Mon 6h34'1"-5d38'
Palmarini,John
Her 17h31'37"27d11'
Palmas,Alberta "tatòa"
Lac 22h29'38"50d30'
Palmasano,Eugene Mark
Ori 5h59'10"20d29'
Palmasano,Giselle Collazo
Umi 13h10'44"70d46'
Palmatier,Louise & Roger
Crb 15h52'53"28d59'
Palmay,Robert
Lac 22h53'20"37d38'
Palmer
Cyg 19h28'56"32d18'
Palmer Rose
Cyg 21h50'28"38d39'
Palmer's Dashnee,Ron
Boo 15h6'49"8d22'
Palmer(Palmerania)
Cet 1h31'1"-10d40'
Palmer,"The Cosmos Star",Frank
Cep 21h21'1"80d9'
Palmer,Alec & Bronia
Com 13h33'40"17d50'
Palmer,Ann
Mon 6h18'38"8d44'
Palmer,Ashley
Eri 4h8'34"-17d31'
Palmer,Brian
Her 16h39'18"21d39'
Palmer,Brian & Paula
Uma 11h53'56"31d12'
Palmer,Burt
Cam 13h58'42"82d22'
Palmer,C Edward
Cnv 12h10'37"37d33'
Palmer,Caitlin
Cas 0h49'58"71d45'
Palmer,Carla Marie
Cas 1h14'33"61d41'
Palmer,Charice Diane
And 0h15'43"30d34'
Palmer,Christopher Ludovic
Per 3h19'40"41d10'
Palmer,David Brent
Per 3h47'4"36d32'
Palmer,Dennis Wayne
Ser 15h49'32"19d55'
Palmer,Diane Elizabeth
Ori 6h4'28"2d4'
Palmer,Donald W
Cep 0h30'47"78d7'
Palmer,Dr Robert
Her 17h21'56"46d12'
Palmer,Florien A
Uma 11h0'32"70d0'
Palmer,Gabriella Nicole
Lyr 19h2'46"31d23'
Palmer,Heidi J
Cas 0h52'0"61d29'
Palmer,James Gilbert
Lib 15h38'20"-28d35'
Palmer,Janet Larsen
Mon 8h5'26"-0d58'
Palmer,Jean
Cet 0h26'34"-18d53'
Palmer,Jessica A
Crb 16h7'0"30d8'
Palmer,Jo & Bob
Lyr 18h28'28"32d6'
Palmer,John
Dra 18h42'14"70d18'
Palmer,Joshua
Her 17h53'52"14d13'
Palmer,Keenan Nicole
Mon 5h55'36"-8d57'
Palmer,Larry Richard
Ser 15h18'0"8d19'
Palmer,Lauren
Cas 0h50'1"71d56'
Palmer,Licieur
Eri 3h10'20"-5d46'
Palmer,Linda & Dan
Cyg 21h55'0"53d15'
Palmer,Marc L
Sgr 19h26'48"-42d26'
Palmer,Marian Roberts Blanchfield
Mon 6h54'0"-10d8'
Palmer,Michael Shane
Aur 5h2'28"37d39'
Palmer,Mitchell
Peg 22h13'27"30d45'
Palmer,Patrick
Per 2h39'35"37d18'
Palmer,Patron Saint of Brewers,Prof Geoff
Cep 20h51'28"70d43'
Palmer,Peter
Aql 20h2'12"10d32'
Palmer,Philip
Her 16h2'22"16d15'
Palmer,Rachel Elizabeth
Ari 3h13'34"28d36'
Palmer,Renate V
Peg 22h7'37"20d46'
Palmer,Robert From The Chgo Resch & Plan Grp
Aur 6h46'54"38d44'
Palmer,Robert M
Gem 6h41'44"30d6'
Palmer,Samuel David Francis
Lyn 9h14'1"33d34'
Palmer,Sarah
Mon 6h22'57"8d38'
Palmer,Scott & Athena
Cep 23h21'0"65d47'
Palmer,Scott Michael
Her 17h19'0"43d8'
Palmer,Sidney MacBeth
Cnv 13h22'47"38d27'
Palmer,Taryn J
Per 3h23'13"40d15'
Palmer,Travis Aaron
Aql 19h24'45"-10d11'
Palmer,Victoria Tori
Oph 18h37'52"6d48'
Palmer,Wendy M
Lmi 9h49'14"34d7'
Palmeri,Hannah T
Sgr 18h58'17"-22d11'
Palmeri,Lisa C
Gem 7h24'5"27d32'
Palmerstar,Laura
Mon 6h55'27"10d36'
Palmese Family Christmas Star,The
Aql 19h50'27"11d10'
Palmieri Family,The
Lac 22h22'40"50d21'
Palmieri,Amy Rose
And 1h47'0"36d33'
Palmieri,Gillian
Her 15h48'54"48d56'
Palmieri,Kristine M F
Lyn 8h8'28"50d27'
Palmieri,Sr,Peter Paul
Gem 7h26'25"21d13'
Palmiero,Robert
Her 18h17'56"31d26'
Palmino,Jerry
Her 16h56'53"23d33'
Palmisano,Kimberly
Vul 19h48'36"20d7'
Palmisciano,Nicola
And 22h56'38"48d55'
Palmisciano,Paul Dante
Uma 11h55'14"40d9'
Palmquist,Lisa Marie
Ori 6h6'49"7d6'
Palmroth,Tero
Aql 20h8'0"4d12'
Palo,Annikka
Cas 1h59'29"61d28'
Palomba,Freed & Seed
Per 3h15'51"40d41'
Palombi,Simona
Her 16h46'59"10d35'
Palomo,Eduardo e Carina
Psa 22h29'53"-27d50'
Palomo,Pina
Cet 0h58'36"-6d9'
Palonis,Bill
Cep 21h9'19"68d15'
Palote Paloma Gil Del Alamo
Dra 16h15'14"60d49'
Palovich,Shano
Sco 17h54'11"-30d20'
Palumbo,Cindy
And 2h3'50"40d7'
Palumbo,Jim & Andrea
Cnc 8h35'24"31d50'
Palumbo,Paul Stephen
Aur 4h56'26"40d40'
Palumbo,Sue & Vick
Lyr 19h0'0"26d8'
Paluszewski,Michael
Dra 9h46'32"80d7'
Paluszka,Roseanne "Snookums"
And 2h27'15"38d34'
Paluzzi,Paul
Her 17h16'59"42d26'
Palys,Nancy A
Cas 1h5'0"60d27'
Pam
Cas 3h9'58"68d30'
Pam
Cma 7h3'45"-15d52'
Pam
Lyr 18h36'57"29d48'
Pam
Lyn 8h53'0"34d32'
Pam
Dra 17h18'58"72d44'
Pam & Allen
Ori 4h43'21"-2d21'
Pam & Catherine
Hya 9h7'36"5d6'
Pam & Ken's Star Of Love
Cyg 20h0'13"38d31'
Pam 21
Ori 5h56'16"7d7'
Pam's "Paul Star"
Vul 19h48'0"25d52'
Pam's Dream
Cas 1h12'60"61d13'
Pam's Guardian Angel
Uma 11h38'58"46d8'
Pam's Place
Ori 5h17'30"10d7'
Pam's Sweet 16
And 1h3'27"37d47'
Pam's Wishing Star
Mon 6h21'50"0d22'
Pam,Jonathon & Jason We Love You
Aur 6h4'1"32d24'
Pam,R J
Lac 22h13'11"51d58'
Pamani,Kareena
And 23h45'52"40d46'
Pambianco A Star of Stars,Lou
Leo 10h54'48"-0d35'
Pamela
Mon 7h39'27"-2d17'
Pamela
Cam 8h35'0"73d39'
Pamela
And 2h8'24"50d32'
Pamela
Aql 19h31'38"7d58'
Pamela
Cam 4h58'25"58d25'
Pamela
Cas 0h54'49"62d25'
Pamela
Aql 18h58'42"-6d46'
Pamela
Crb 15h44'29d48"
Pamela
Cyg 19h26'50"31d4'
Pamela
Cas 0h39'17"71d33'
Pamela
And 23h11'0"37d28'
Pamela Ann
Cam 9h52'35"84d40'
Pamela Anne
Ori 4h59'26"13d7'
Pamela Daniele
Tri 1h45'0"28d2'
Pamela e Fabio 26/01/1992
Col 6h26'2"-35d29'
Pamela Gail
Com 13h1'27"18d36'
Pamela Jane
Peg 23h26'26"33d37'
Pamela Janna
Aqr 20h52'46"-6d2'
Pamela Jean
Tri 2h22'47"34d49'
Pamela Jean
Lyr 18h29'33"35d30'
Pamela Joy
Peg 23h44'1"15d56'
Pamela Joyce
Cyg 19h28'29"56d16'
Pamela June
Com 12h24'46"26d45'
Pamela Lynn
Mon 6h36'34"7d35'
Pamela Margaret & Victor Kochi Lin
Crb 15h53'57"30d25'
Pamela Miranda
Cas 1h56'59"71d2'
Pamela Pal
Cam 8h48'18"74d13'
Pamela Rose
And 0h6'40"38d4'
Pamela Roxanne-All Honey
Cas 0h2'21"63d8'
Pamela Sue
And 1h56'42"36d58'
Pamela Susan "My Bright Star"
Com 12h9'17"22d9'
Pamela Sweetheart
Del 20h18'47"14d14'
Pamela Terese
Sct 18h46'44"-7d21'
Pamela's Dream
And 0h23'51"37d33'
Pamela's Star
Lyr 19h21'1"41d48'
Pamela's Star
Lac 22h0'55"38d14'
Pamela's Wishes
Cas 0h28'33"63d44'
Pamela's `Lil Bit O'Heaven
Cyg 20h28'50"42d36'
Pamela,True Companion
Lyr 19h23'40"31d48'
Pamelot
Aur 4h58'1"36d29'
Pamfort
Lmi 9h52'54"34d5'
Pammel,James
Boo 14h55'43"41d16'
Pammy
Lyn 6h24'25"59d5'
Pammy
Uma 10h48'21"47d36'
Pammy
Aql 20h2'25"14d34'
Pampelune
Lac 22h48'29"53d18'
Pampinella,Dorothy
Cam 5h31'26"68d43'
Pampino,Carolyn & Nomee Dee Altschul
Aur 6h55'1"44d21'
Pamulastrohmyrus
Lyn 8h38'12"41d41'
Pamy
Cep 23h33'1"65d40'
Panades,Oscar
Ori 6h2'48"0d44'
Panagioths Ioannidhs
Ori 6h4'49"0d34'
Panagiotidou,Dimitra
And 23h24'51"44d27'
Panagiotis
Cyg 19h20'55"30d2'
Panalaëna
And 2h31'13"40d43'
Panama & Tami 12-13-93
Uma 10h38'0"68d13'
Panama,Nicholas
Cnv 12h10'53"51d55'
Panama,Zoe
Dra 16h4'0"66d4'
Panarella,Jr,Nicholas
Her 16h22'0"42d15'
Panaro,Claire Forchheimer
Cyg 21h24'53"31d6'
Panashida,William
Per 3h8'41"47d12'
Pancho
Lyn 8h46'34"34d59'
Pancirolli,Roberto
Eri 3h45'22"-1d48'
Pancoast,Christine Vought
Lyr 18h29'21"31d25'
Panczyk,Dirk
Ori 5h59'57"8d43'
Pandekew
Her 17h19'27"29d25'
Pando,Denise
Lyn 7h27'24"44d37'
Pandolfo,Frank Joseph
Boo 15h27'47"47d53'B
Pandolfo,Rose Lillian DeDonato
Boo 15h27'47"47d53'A
Pandora
Peg 23h40'33"10d34'
Pandora
Mon 7h40'38"-5d44'
Pandora
Uma 9h14'20"52d51'
Pandora-Aphrodite
Mon 7h21'10"-0d28'
Pane,Joseph M
Cep 22h11'0"60d11'
Panebianco,Barbara Jean
Sgr 18h47'55"-22d38'
Panel,Alain
Mon 7h50'36"-4d35'
Panella & Family,Guido & Anna
Cyg 20h29'1"42d35'
Panella,Ray
Aur 7h15'31"36d35'
Panepento,Patty "Bright Eyes"
Crb 15h37'1"26d59'
Panes Love Chris, Maureen
Lyn 9h14'0"45d54'
Panetta,Barbara
Lac 22h49'13"38d9'
Panetta,Tia
Cas 0h10'51"50d34'
Panetta,Toniann
Cyg 20h3'43"31d5'
Panfil,Fred
Aur 7h22'0"38d49'
Pang,Andrew Yin-Fung
Lyn 9h4'17"37d53'
Pangaea
Cet 1h27'24"-2d4'
Pangalos Feb 18,1934, Jo Ann Yvonne
Cet 0h52'1"1d50'
Pangaro,Antonella
Cnv 13h12'0"51d2'
Pangborn,David Blake
Dra 19h6'59"70d50'
Pangelina
Ser 15h12'40"1d56'
Pangua,Amaia Rodriguez
Peg 23h32'38"11d1'
Pani,Anthony
Oph 18h22'49"10d49'
Panian,Marlene
Dra 20h14'8"-7d3'
Panichi,Ann
And 0h2'36"46d28'
Panichi,Licia
Lyn 7h27'60"41d6'
Panico,Louis "Peanut"
Dra 20h13'19"74d4'
Panicucci,Melissa Ann
And 2h26'33"45d24'
Panicucci,Michele
Aur 6h51'1"37d57'
Paninos,"One In A Million",Greg Athan
Oph 16h21'1"2d52'
Panio,Laura & Joseph
Lyn 7h7'49"59d10'
Panitz,Arthur
Aur 5h9'25"38d57'
Paniz,Liora Tamar
Lyn 6h36'0"60d23'
Panjabi,Vinitha
Lyr 18h30'49"30d59'
Pankau,Roger
Aur 6h5'49"45d47'
Pankewicz,James
Aur 5h14'12"40d43'
Pankey,Light of my Life Richard
Her 16h43'24"35d38'
Pankey,Phil
Aql 19h39'49"14d54'
Pankey,Phillip
Peg 23h38'43"30d14'
Pankey,Robert "Bob"
Uma 11h58'1"56d44'
Pankhurst,Nikki Anne
Ori 6h1'49"6d28'
Pankoff,Peter
Uma 10h28'37"40d17'
Pankoff,Rolf
Peg 23h35'1"13d53'
Pankoninny
Lyn 8h42'25"33d30'
Pankow,Jennifer Louise
Gem 6h45'2"12d44'
Pankrath,Klaus Rüdiger Alfred Otto
Psc 23h9'43"6d16'
Pankratz
Dra 19h52'43"60d0'
Pankratz,Frank
Per 1h49'43"47d57'
Panky
Uma 11h42'31"49d33'
Pannabecker,Jr,Gerald K
Vul 19h48'1"28d12'
Pannell,Guru Ken
Boo 14h58'15"36d1'
Pannier
Ori 6h14'12"0d41'
Pannill,Reid
Vul 21h20'1"28d10'
Panno,Corrado
Dra 16h33'40"67d38'
Pannocchia
Leo 11h6'30"23d57'
Pannwitz,Michael
Sco 17h5'46"-30d49'
Panny,Jonathan Taylor Maxwell
Aur 6h27'15"37d0'
Panoke
Dra 16h48'13"60d4'
Panontin,Ugo
Umi 15h29'30"71d14'
Panormus
Scl 23h44'17"-27d36'
Panos
Umi 16h19'41"74d36'
Panos,Jim & Cee Dee
Cam 3h35'25"61d23'
Panos,Nick
Cnv 13h45'40"39d33'
Panosian,Gary
Dra 20h1'26"67d37'
Pansadoro,Vito
Lac 22h2'22"51d57'
Pansy's Panda Bear
Cyg 20h32'4"34d30'A
Pantalemon,L Hazel
Cyg 20h56'59"41d12'
Pantaleo,Bonnie
Uma 8h52'1"51d2'
Pantaleo,Doreen Catherine
And 1h28'44"34d7'
Pantano "The Shoe King",Kelly
Her 18h6'37"31d19'
Pantanus
Hya 9h34'0"-6d29'
Pantazi,Helen Regina
Cyg 19h46'14"38d28'
Pantella 01-30-68,R
Aqr 21h1'44"-13d19'
Panter,Ruth
Cas 3h0'0"58d38'
Pantera,Jr,Dr Matthew J
Dra 15h6'25"60d48'
Panther & Butterfly
Aql 19h43'41"12d29'
Pantlla,Courtney Aliesha
Cas 1h59'28"58d60'
Pantsar,Lissu
Ori 5h55'35"15d7'
Pantz,Evelyn Hyson
Lyr 18h58'34"41d31'
Panza,Albert Gene
Dra 17h5'25"70d24'
Panza,Janet Lee
Cas 0h18'45"60d53'
Panzariello,Lisa Antonia
And 1h3'26"38d34'
Panzarino,Gail Georgette
Cam 5h57'27"61d26'
Panzek,Alexander und Regina Donath
Umi 16h37'36"76d33'
Panzella's Pride,2003
Cet 2h27'20"3d45'
Panzer,Christian
Lac 22h9'37"38d25'
Panzer,David M & Patricia M
Psc 23h9'54"2d6'
Panzera,William R
Per 4h2'56"43d17'B
Panzeri,Monica Adèle
Lmi 10h58'23"33d33'
Panzica,Susan A
Cam 8h21'54"73d16'
Paodan
Cae 4h47'40"-31d18'
Paola
Pho 0h28'8"-47d8'
Paola
Pup 8h24'22"-20d34'
Paola
Eri 3h52'23"-3d21'
Paola
Aur 4h46'17"51d48'
Paola
Cep 23h33'34"64d5'
Paola
Oph 18h5'18"10d28'
Paola
Boo 14h25'1"48d36'
Paola E Paolo
Tau 3h59'56"30d44'
Paola T
Pho 0h36'23"-49d46'
Paola V C Car '40
Ori 5h54'20"-1d7'
Paola,Collesei
Umi 14h43'53"77d57'
Paolaromano
Lyr 18h39'11"29d58'
Paolastra
Cae 4h48'47"-28d23'
Paolella,Judy Whiteman
Cyg 20h30'40"41d9'
Paolera,Jr,Anthony R
Hya 8h54'1"1d43'
Paoletta,Reid
Uma 8h43'45"52d38'
Paoli,Renato
Peg 21h27'52"20d15'
Paolillo,Joseph
Her 16h36'41"42d18'
Paolin,Jerry & Laura
Peg 23h29'43"21d14'
Paolini & Family, Pasquale
Ant 10h46'21"-38d33'
Paolini,Michel
Aql 18h56'33"3d35'
Paolino
Mon 7h59'39"-4d7'
Paolino
Her 18h9'59"31d16'
Paolino
Hor 3h18'49"-48d5'
Paolino,Roe & Nicole
Sge 19h53'35"16d16'
Paolino,Taylor Nicole
Peg 22h0'0"34d57'
Paolisilla
Lac 22h15'49"38d25'
Paolo
Gru 22h31'14"-55d34'
Paolo
Umi 13h50'43"77d49'
Paolo
Gem 6h53'36"16d26'
Paolo
Lyn 7h28'0"40d11'
Paolo
Umi 13h19'1"70d33'
Paolo Dei Castellinaria
Gem 7h13'11"21d39'
Paolo e Giulia
Per 2h6'53"58d57'
Paolo's Star
Her 18h1'45"30d32'
Paolo,Jacopo
Eri 2h44'51"-5d44'
Paolucci,Elyse
And 23h38'31"38d46'
Paone,MTB RON 24, Dominick A
Per 3h26'0"41d10'
Pap Pap's Eternal Love
Vul 19h23'1"26d58'
Pap-Pap Mel
Lac 22h16'37"49d46'
PaPa
Ori 5h54'39"0d7'
PaPa
Eri 3h53'30"-5d39'
PaPa
Per 2h59'37"45d10'
Papa
Vir 14h28'48"2d15'
Papa
Cet 3h10'1"2d10'
Papa
Per 3h1'48"40d23'
Papa
Cmi 7h53'0"0d57'
Papa
Cep 22h37'36"65d8'
Papa & Chack
Cnc 8h48'24"20d30'
Papa Bear
Aql 20h6'0"3d57'
Papa Bear
Uma 11h26'0"66d33'
Papa Bear
Dra 17h18'49"61d47'
Papa Bear
Uma 10h5'41"56d32'
Papa Bear
Sct 18h43'31"-6d49'
Papa Butch
Boo 14h0'32"23d48'
Papa Charly
Her 17h25'24"42d58'
Papa Don
Dra 17h14'1"60d24'
Papa George
Ser 15h35'23"9d9'
Papa Impatiens
Cep 21h58'60"55d6'
Papa Jay
Ori 5h53'24"21d24'
Papa Richard
Uma 10h0'29"57d28'
Papa's Lake
Aqr 22h4'43"0d27'
Papa's Pride
Ori 5h36'57"8d53'
Papa's Star
Ori 5h54'28"1d31'
Papa's Star
Per 2h36'55"43d25'
Papa's Star
Her 16h28'27"50d34'
Papa's Star
Aur 5h16'11"54d35'
Papa,Jr,Eric Michael
Aur 5h3'11"48d13'
Papa,Maria Assunta
Cyg 19h51'15"40d54'
Papa,Pierre N
Cas 1h0'1"55d21'
Papa-Jean,The
Oph 18h8'45"12d13'
Papa-Mack
Ori 4h56'38"0d35'
Papa-son
Aql 18h57'18"-4d46'
Papadimitriou, Alexandra
Uma 14h20'55"61d48'
Papadimitriou,Kathy
Vul 20h0'49"23d18'
Papadopoli,Anna Catherine
Vul 20h4'45"28d27'
Papadopoulis,Dimitris
Umi 17h56'35"80d12'
Papadopoulos,Antonia
Cam 5h57'50"70d21'
Papadopulos,Christina
Mon 7h31'22"-1d51'
Papaionnou-Malakouli, Ioannis
Cyg 19h34'0"38d14'
PapaJohn
Cet 0h50'17"-02d27'
Papaleo,Nicholas Fantino
Her 16h45'45"20d37'

Papalimberis,Ted
Lac 22h30'32"53d59'
Papalos,Theodore J
Hya 8h19'34"2d21'
Papandrea,"The Metermaid",Rita
Cam 4h58'42"60d51'
Papandreas,George John
Oph 17h56'35"-6d49'
Papanek,Janet A
Uma 8h30'18"68d27'
Papania,Lee Ann
And 0h22'11"37d53'
Papanikolas,John E
Tri 1h48'11"28d28'
Papapietro-Redfield, Orion Zebediah
Ori 6h10'1"7d45'
Papapostolou,Efi
Aur 7h1'36"38d42'
Paparella,Steven
Her 15h57'21"48d51'
Papariello,Allison Kathryn
Peg 23h46'46"17d43'
Paparigian,Joan
Aqr 22h1'54"-5d34'
Papas,Elizabeth
Psc 23h1'53"2d7'
Papason
Ari 1h49'16"22d50'
Papastathopoulos- Pathos,Dr Spyridon
Aql 20h2'46"0d18'A
Papastathopoulos-Pathos,Katerine Anna
Aql 20h2'46"0d18'B
Papastathopoulos-Pathos,Spyro Pio Stavr
Aql 20h2'46"0d18'C
Papataros,Arthur
Aur 6h11'0"31d34'
Papazian,Ara
Vul 20h21'0"28d27'
Papciak,Walter William
Dra 11h42'59"71d40'
Pape,Gregory T
Aur 5h25'35"40d24'
Pape,Pauline Hutchinson
Cyg 21h3'35"37d51'
Papelino,Brianna Lynne
Eri 3h29'42"-2d43'
Papen,Andrew Collin
Ori 5h55'39"6d33'
Papen,Beatrix
Psc 0h8'52"-1d8'
Papendick,Marilynn
Lyr 18h58'13"30d40'
Paperno,Carolyn Beth
And 0h29'1"40d14'
Papero
Lyn 9h1'44"37d54'
Papert,Sammy
Cet 1h31'0"-0d56'
Papetti,Jr,Arthur Joseph
Aur 5h18'11"42d20'
Papetti,Marcus Stelvio
Uma 11h34'20"41d31'
Papiano,Neil
Aql 19h5'38"0d38'
Papich,Timothy James
Dra 19h1'53"48d21'
Papilion-Smith Philippe
Cep 23h11'1"68d16'
Papillon
Cyg 20h22'53"39d11'
Papillon de ma Vie
Cas 22h58'29"56d3'
Papillon I,Montana
Cep 23h9'12"64d35'
Papillon,Chow-Que
Cmi 8h9'26"1d32'
Papillon,Eleanor
And 23h47'21"45d46'
Papillon,Eve
Mon 7h47'46"-5d58'
Papillon,John Ray
Ori 6h3'0"7d30'
Papillon,Mathieu
Her 16h48'0"8d55'
Papillon,Sarah
Aql 18h57'44"10d40'
Papillon-Smith,Jessica
Cam 6h4'31"72d19'
Papke,Lyn C
Uma 8h42'21"50d35'
Paplanus
Aql 19h54'26"14d21'
Papou Niko & Yiayia Penny Lambert
Del 20h20'16"10d52'
Papou,Maxim
Uma 8h45'1"54d33'
Papoulias,Cathy
Tau 4h56'48"16d3'
Papovich,Melita
Cam 4h56'23"68d9'
Pappadis,Thomas James
Dra 17h52'1"58d21'
Pappagallo,Brian Anthony
Cam 6h11'27"68d25'
Pappageorge,George
Lac 22h34'27"53d4'
Pappageorge,Helen
Mon 8h8'52"-1d25'
Pappageorge,Jim
Cet 2h28'22"-10d47'
Pappalardo,Henry Pierre
Ser 18h0'23"-14d36'
Pappalardo,Yolande Centonze
Mon 6h58'1"0d23'
Pappas,Allison Rose
Cyg 19h21'13"28d38'
Pappas,Alysia
Dra 17h53'57"61d19'
Pappas,Constantine
Lya 18h59'39"27d25'
Pappas,Cornelia A
Del 20h18'0"16d5'
Pappas,Georgia
Mon 7h27'42"-6d53'
Pappas,James
Cma 6h56'9"-17d22'
Pappas,Leonidas
Tri 2h21'24"31d16'
Pappas,Marcella Milano
Cas 1h2'41"63d60'
Pappas,Nathan
Boo 15h1'32"12d40'
Pappas,Priscilla "Cookie"
Lyn 8h57'19"41d28'
Pappas,Robin
Lyn 6h42'39"50d17'
Pappas,Shirley
Mon 7h6'47"-7d11'
Pappelbaum,Wolfgang
Sge 19h5'37"17d53'
Papperitz,Siegfried Norbert
Uma 11h51'57"33d7'
Pappis,Nick
Peg 22h18'12"4d26'
Pappy
Lyr 18h56'36"30d28'
Pappy
Oph 17h6'21"11d20'
Pappy A Forever Ich Liebe Dich
Dra 19h49'17"60d48'
Pappy Doc
Aql 19h55'16"13d35'
Paproski,Hunter John
Ori 5h58'21"6d41'
Paproth,Horst
Vir 13h35'0"-11d2'
Papso OSF,Sr Melania
Uma 10h38'15"58d51'
Papson,Alexandra
And 23h45'12"45d36'
Papuga,Steven Thomas
Lac 22h9'40"51d54'
Papurkio Giorgio
Ori 5h31'40"8d30'
Papworth,Iris Elizabeth
Uma 11h11'1"31d40'
Papworth,Tracy
Lyr 18h16'22"41d13'
Papyuta
Tri 1h47'24"27d31'
Paquet,Elaine Rose
Cas 0h54'49"61d22'
Paquet,Shane Daniel
Cep 21h13'26"58d45'
Paquet,Virginia
Ori 5h57'10"18d11'
Paquette,Charles- Edouard
Tri 1h35'0"28d43'
Paquette,Félix
Tau 5h49'14"24d19'
Paquette,Julie Coucounie
Ori 5h59'26"15d15'
Paquette,Luc
Sgr 19h7'53"-21d17'
Paquette,Marielle Antoinette
Cam 5h54'1"61d35'
Paquette,Raymond E
Boo 14h27'39"17d59'
Paquette,Rolande Durand
Cep 23h9'13"67d46'
Paquette,Tracy
And 23h43'25"47d25'
Paquignon,Muriel
Crt 11h19'22"19d30'
Paquin,Brandon Rodrigue
Hya 8h39'49"-10d52'
Paquin,Joseph & Genevive
Lyr 19h17'50"28d18'
Paquin,Trudy
Ori 5h34'20-3d8'
Paquita
Ori 6h5'41"0d53'
Paquiva mon Amour
Uma 13h13'0"61d1'
Par Colleb Xnarth
Ori 5h58'40"21d7'
Para Siempre
Mon 6h21'38"6d1'
Para Siempre Mi Helen
Lyr 19h15'41"26d50'
Para Siempre Paul
Hya 9h25'33"-1d52'
Parab,Suresh & Shakuntala
Lyn 7h53'34"50d15'
Paracchini,Jean Guy
Ori 6h4'57"8d6'
Paradies,Dr Louis
Oph 16h53'58"-28d16'
Paradis Family,The
Vul 19h46'14"28d45'
Paradis,Fernand
Lyr 18h44'0"35d48'
Paradis,Francine
Ori 5h57'14"10d28'
Paradis,Joma
Cyg 20h41'30"45d51'
Paradis,Katherine Henley
Sge 19h30'35"16d40'
Paradis,Rachel Louise
And 2h29'58"48d56'
Paradis,Vanessa
Cma 6h57'28"-15d35'
Paradise
Uma 11h7'38"49d23'
Paradise
Ori 5h11'1"-5d41'
Paradise Cafe
Mon 7h2'56"-6d39'
Paradiso Island II
Cap 21h27'49"-23d38'
Paradiso,Julia
Oph 17h30'33"-6d5'
Paradiso,Margaret E
Eri 3h20'14"-14d50'
Paradiso,Mary E
Eri 3h34'34"-13d59'
Paradox,Peabody
Boo 13h59'32"26d2'
Paradysz,Alexandra E
Vir 12h28'15"7d44'
Paradysz,Daniella K
Ari 1h48'37"15d55'
Paragano,The Great Star of Nazario
Lyr 18h38'50"44d51'
Paraic The Walrick
Dra 11h45'31"71d1'
Paramount-40
Aql 19h3'11"0d55'
Parant,Anita R
Eri 3h13'38"-6d29'
Paratore,Robert Andrew
Cam 3h50'14"52d45'
Parcell,Nathan Harrison
Cet 0h34'48"-6d19'
Parcher,Lloyd M
Her 18h17'52"14d58'
Parda,Karin Veronika
Cap 21h42'46"-24d42'
Pardee IV,Calvin
Lac 22h5'44"49d22'
Pardee,John & Eileen
Eri 3h21'30"-5d36'
Pardee,Roger
Crb 15h47'13"28d22'
Pardina,Michael C
Aur 5h2'38"37d60'
Pardio,Patty
Sgr 19h39'46"-44d30'
Pardo's Little Twinkle Ron
Cet 2h47'49"3d57'
Pardon,Manfred
Cnc 9h2'21"7d18"
Parducci,Patrick Gerard
Aql 19h19'14"12d29'
Parducci,Patrick Gerard
Peg 22h40'31"27d44'
Pardue,Joe Leonard
Cep 21h19'16"68d12'
Paredes,Angel
And 0h9'28"35d35'
Paredes,Frank D
Leo 11h6'39"-2d22'
Paredes,Ivon Patrizia Lopez
Cma 6h15'49"-18d49'
Paredes,Tommy
Aql 19h9'31"3d2'
Pareja,Amalia
Uma 13h18'19"61d23'
Parekh (Pinky), Vaishali P
Sgr 20h19'0"-28d57'
Parekh,Sarita
Cas 0h18'53"61d24'
Parela & Philippe
Ori 5h46'29"19d34'
Parent,Alexandre
Aql 19h23'19"0d26'
Parent,Ann Miller
Mon 7h44'16"-5d25'
Parent,Bernard
Her 17h31'46"20d43'
Parent,Bernie(Bumpkin)
Cam 4h18'1"68d60'
Parent,Caleb Michael
Aur 6h26'50"35d0'
Parent,Ethan Thomas
Aur 5h45'40"54d23'
Parent,Frank
Ser 18h6'18"-14d7'
Parent,Lorraine
Cam 4h20'46"60d48'
Parent,Mélanie Lynn
Ari 2h35'46"21d59'
Parent,Philip
Hya 8h54'37"3d52'
Parent,Philip
Lac 22h50'51"53d39'
Parent,Richard
Lac 22h54'27"38d11'
Parente,Noel Paul
Boo 15h4'1"23d4'
Parente,Patrick David
Eri 4h0'57"-10d16'
Parente,Paula Teresa
Ori 5h27'29"15d21'
Parenteau,Frenchie
Aur 6h3'23"31d41'
Parenteau,Joseph & Elinor
Cyg 20h6'1"40d35'
Parenteau,Judith L
Umi 16h5'39"72d50'
Parenteau,Meaghan
Cyg 19h30'45"33d15'
Parenteau,Michelle
Peg 21h56'1"30d57'
Parenthia
Ori 5h54'0"19d45'
Pareschi,Antonio
And 0h46'41"40d54'
Pareti,Doug
Her 16h35'53"8d12'
Pareti,Vincent
Oph 18h2'1"8d41'
Parette,Laura & Bob
Lyn 7h49'39"36d29'
Parfait,Dominique
Ori 6h6'27"20d26'
Parfitt Star,The
Cyg 21h10'49"37d34'
Pargin,Preston Riley
Cet 1h35'0"-10d8'
Pargin,Raleigh S & Tiffany Sheelar
Oph 17h39'59"-28d54'
Pargmann,Klaus
Cmi 7h19'55"7d40'
Parham,Alyse Nicole
Cas 0h58'37"60d19'
Parham,Garrett Carson
Lac 22h22'0"48d51'
Pari
Uma 8h56'36"49d22'
Parial's Star,Lynda
Cas 0h51'24"65d25'
Pariota Kiranos
Lyn 7h25'49"44d32'
Paris
Mon 7h48'18"-1d60'
Paris
Dra 11h36'30"78d11'
Paris
Oph 18h39'40"7d36'
Paris
Cyg 21h56'45"53d45'
Paris & Janie
Uma 8h14'23"60d4'
Paris Autumn
Tri 1h55'30"26d11'
Paris Fabien
Dra 9h34'23"80d39'
Paris,Bella Stellar Ruth
Cyg 20h21'15"39d38'
Paris,Charles B
Ori 5h21'28"7d39'
Paris,Doug & Margie
Cnv 12h35'48"40d22'
Paris,Eugene & Bernadine
Tri 2h28'42"28d24'
Paris,G G
Vul 19h22'44"26d18'
Paris,Grégory
Cas 2h49'37"77d17'
Paris,Karine
Com 13h12'33"22d3'
Paris,Kelli
Leo 11h8'40"0d16'
Paris,Kevin Holiday
Per 2h59'1"40d3'
Paris,Shirley Ann McMillen
And 23h40'0"44d39'
Paris,Stuart A
Cep 1h11'1"78d49'
Paris,Tom
Per 2h3'20"58d45'
Paris,Wilbur L
Per 4h2'59"37d45'
Paris,Yvette
Lyn 9h5'35"40d25'
Parisa,Aubrey Rose
And 1h26'29"50d5'
Parisa,Sydney Marlena
And 1h19'40"39d4'
Parise,Robert
Per 2h56'41"40d8'
Parise,Vincent David
Uma 11h18'38"32d23'
Parish III,Freddie Leon
Cet 2h28'24"5d38'
Parish,Elise Robin
Oph 18h42'11"7d37'
Parish,Jerry D
Lmi 11h4'47"26d22'
Parish,Sandra L
Lmi 10h13'18"32d6'
Parisi,Carl Brian
Aur 7h17'56"39d41'
Parisi,Daniel A
Dra 11h59'36"66d60'
Parisi,Joseph & Nancy
Cyg 20h31'18"42d30'
Parisi,Richard J
Aql 19h40'28"14d21'
Parisi,Tracy
Cas 1h15'49"60d34'
Parisian,Ricky
Boo 14h57'21"52d3'
Parisot
Ori 5h25'49"1d23'
Parisot,Jean-Marie
Tri 1h58'42"31d43'
Parisot,Kurt Adam
Sge 19h56'53"20d9'
Parisot,Mousieur Thierry
Boo 14h38'23"48d35'
Paritsky,Shira & Yoni
Uma 10h8'0"60d13'
Park 9-143,Steven
Boo 14h22'26"50d21'
Park Family
Ori 6h3'54"6d2'
Park III,James W
Lmi 9h58'33"40d20'
Park,Arthur S
Per 4h20'0"50d39'
Park,Bobby Ray
Uma 10h32'50"70d38'
Park,Cecelia
Sgr 19h39'56"-30d42'
Park,Clyde Sei-Yong
Lib 15h19'52"-23d7'
Park,Donald Bryant
Aur 7h24'1"41d19'
Park,Dylan Gregory Lloyd
Cep 23h3'1"64d1'
Park,Eddie
Cep 22h10'27"62d2'
Park,Jamie Noelani
Crt 11h47'19"-10d46'
Park,Jason David
Cap 21h3'5"-22d22'
Park,Korby Kathleen
Cyg 21h28'51"37d57'
Park,Lewis E
Her 17h11'13"42d53'
Park,Robert W
Dra 20h17'37"67d56'
Park,Seeun
Tri 1h40'60"31d33'
Park,Susan Charlotte
Cas 1h40'0"60d50'
Park,Taylor Raye
Aur 6h35'17"32d57'
Park,William
Ori 4h46'25"0d2'
Parke,Jacqueline Joy Randall
Vir 13h59'0"2d12'
Parke,Nancy
Cet 3h6'30"3d44'
Parke,Nick
Cyg 19h39'58"30d46'
Parke,Pat & Dawn
Uma 11h12'41"31d48'
Parker
Cam 4h52'33"65d18'
Parker
Boo 15h14'51"26d28'
Parker B
Lyn 7h49'34"35d51'
Parker II,James E
Sct 18h55'54"-6d45'
Parker II,Robert R
Boo 14h46'33"31d29'
Parker O
Her 17h32'12"25d31'
Parker Sparkle Plenty, John
Cnv 13h51'60"48d2'
Parker's Star,Carolyn
Leo 9h53'1"10d9'
Parker,Adam
Boo 14h39'58"52d9'
Parker,Alexander Dirk Leroi
Peg 22h52'51"8d56'
Parker,Alexandra Elyse
Lac 22h14'39"54d55'
Parker,Andrew
Her 17h39'29"40d59'
Parker,Anne Elizabeth Gordon
Peg 21h39'30"27d37'
Parker,Annie
Sex 9h46'11"-5d2'
Parker,Ashley Morgan
Pup 7h28'45"22d17'
Parker,Belinda
Cyg 19h46'48"38d29'
Parker,Benjamin Lancaster
Uma 11h6'0"38d0'
Parker,Benjamin Michael Bruce
Cet 2h17'20"4d57'
Parker,Bernice Stimmann
Vul 19h18'36"26d47'
Parker,Bobbie Sue
Peg 22h1'0"31d43'
Parker,Brandi Lesley
Peg 21h29'11"27d32'
Parker,Bryce & Jade
Cet 3h16'25"1d29'
Parker,Catalina
Peg 23h24'34"10d12'
Parker,Catherine Margaret Ann
Cas 0h13'48"61d0"
Parker,Cathy
And 0h22'18"38d34'
Parker,Cathy
Mon 6h40'20"-10d7'
Parker,Chandra
Sct 18h45'44"-7d25'
Parker,Charles H
Sct 18h41'11"-6d55'
Parker,Charles Roy
Aur 4h49'17"50d2'
Parker,Charlotte & Alex
Crb 16h22'52"37d38'
Parker,Chris
Aql 18h59'46"13d13'
Parker,Clinton Mead
Cnv 13h7'1"51d44'
Parker,CLU,ChFC, William T
Cep 5h13'51"80d29'
Parker,Colonel Tom
Oph 17h12'51"-10d45'
Parker,David
Ser 18h1'49"-14d20'
Parker,David Charles
Ori 5h53'42"20d41'
Parker,David John
Lac 22h6'43"51d6'
Parker,David Keith
Her 16h31'1"39d24'
Parker,Deborah
Gem 7h25'32"28d22'
Parker,Dennis Henry
Peg 22h1'15"10d38'
Parker,Dewey J
Lac 22h15'35"50d23'
Parker,Dorothy
Cas 23h33'60"58d29'
Parker,Dorothy
Eri 3h46'24"-5d24'
Parker,Doyle Lindzy
Sex 10h33'51"-1d21'
Parker,Dr Beverly
Aql 18h54'12"-2d13'
Parker,Dr Robert Eugene
Per 4h21'32"50d22'
Parker,Farrell Amaria
Lyn 7h12'42"59d54'
Parker,Frank
Oph 18h42'48"8d12'
Parker,Fred A
Cet 3h0'37"0d23'
Parker,Geoffrey
Uma 10h11'56"71d36'
Parker,Ginger L
Cas 0h22'37"68d18'
Parker,Greg
Boo 14h1'13"23d3'
Parker,Harvey & Roberta
Ori 5h48'0"10d14'
Parker,Hedy
Del 20h27'12"11d8'
Parker,Herbert M
Lmi 9h59'48"33d5'
Parker,Isabel
Ori 5h39'31"-4d39'
Parker,Isaiah Gordon
Tau 5h34'43"28d36'
Parker,Jacqueline
Lyn 9h7'40"38d41'
Parker,James Henry
Aur 5h1'16"50d7'
Parker,James Lawrence
Uma 11h19'22"35d22'
Parker,Jeff Thomas Roache
Uma 12h57'26"57d47'
Parker,Jerry
Equ 21h21'58"2d44'
Parker,Joe
Her 16h5'50"21d20'
Parker,Joel
Dra 17h59'45"60d13'
Parker,John
Sex 10h19'13"2d29'
Parker,John
Her 18h4'40"40d55'
Parker,John Edward
Ser 17h55'14"-14d46'
Parker,John Edward
Dra 19h29'25"70d3'
Parker,Joshua
Dra 17h10'34"60d47'
Parker,Justin Jerome "Hawk"
Ori 6h15'51"-2d37'
Parker,Leslie
Peg 22h2'27"3d24'
Parker,Lewis K
Her 17h33'22"27d11'
Parker,Lillian B
Eri 4h4'30"-6d14'
Parker,Lisa K
Del 20h33'14"20d26'
Parker,Marie E
Eri 3h53'51" 1d10'
Parker,Megan
And 23h38'21"42d52'
Parker,Michael Lawrence
Dra 15h10'56"57d57'
Parker,Michele
Cam 13h9'1"82d24'
Parker,Mr James M
Boo 13h58'16"26d2'
Parker,Nancy Sharon Culbertson
Mon 7h55'50"-4d24'
Parker,Pam
Tri 1h47'54"27d26'
Parker,Patricia & George
Cyg 20h56'52"40d29'
Parker,Philip K
Her 16h15'11"21d18'
Parker,Phyllis Cutts
Lyn 7h56'45"39d56'
Parker,Richard & Rosie
Her 16h54'20"27d46'
Parker,Richard F
Cep 21h20'55"84d38'
Parker,Ronald C
Dra 9h46'21"74d2'
Parker,Rosanna Elizabeth
Cas 0h20'29"61d47'
Parker,Roy
Her 16h42'50"48d14'
Parker,Russell Roland
Hya 8h30'28"-8d22'
Parker,Samantha
Cas 0h10'0"61d15'
Parker,Senator Carl A
Cet 0h59'41"-8d12'
Parker,Shea Daniele
Lyn 8h21'23"50d45'
Parker,Shirley
Aql 20h31'16"-6d54'
Parker,Stephanie J
Hya 9h14'21"3d54'
Parker,Sue
Sge 19h41'32"16d43'
Parker,Susan
Boo 15h12'14"52d19'
Parker,Teresa
Eri 2h59'52"-5d39'
Parker,Timothy Patrick
Per 2h52'18"34d26'
Parker,Timothy Patrick
Vul 20h15'56"26d11'
Parker,Vera Jo
Cas 0h36'48"75d13'
Parker,Virginia Mayer
Oph 18h41'19"6d33'
Parker,Vivien
Cru 12h4'36"-62d51'
Parker,W F P, R E "The Kid"
Tri 2h3'48"30d2'
Parker,Wendy Susan
And 1h25'43"48d55'
Parker,Yvonne
Cyg 20h39'18"45d58'
Parker,Yvonne
And 0h11'41"39d0'
Parker,Yvonne Marie
Peg 23h26'27"17d21'
Parker-Watson,Cynthia E
Uma 11h28'17"60d51'
Parkerhorn 1
Boo 14h29'1"53d49'
Parkes Wedding Star, Chris & Angela
Cyg 21h2'52"38d12'
Parkes,Donna
Lyr 18h14'41"42d45'
Parkes,Michael
Ori 5h58'30"0d7'
Parkhouse,Geoffrey
Ori 5h2'59"0d38'
Parkhurst,Karen L
Cet 2h59'41"7d44'
Parkhurst,Pat
Dra 19h24'30"56d53'
Parkin,Grace
Lyn 9h3'53"38d11'
Parkin,Moira Kelly
And 23h46'58"41d47'
Parkinson,"Fat Boy"
Aur 6h26'59"37d9'
Parkinson,Adam
Peg 23h43'23"30d33'
Parkinson,Carol Lindh
Aur 5h26'21"31d49'
Parkinson,Cory Carrier
Peg 23h15'59"31d4'
Parkinson,Gemma
Lac 22h4'57"51d18'
Parkinson,Heidi
And 0h27'39"31d38'
Parkinson,Jim
Hya 8h52'50"3d0'

Parkinson,John C & Maria L Rivas Dra 17h18'1"68d4'
Parkinson,Julie Paige Sco 16h56'47"-4047'
Parkinson,Kraig T Dra 16h52'26"67d49'
Parkinson,Meta Rowe Mon 6h32'54"10d55'
Parkinson,Michael Cam 6h4'13"70d42'
Parkinson,Michelle Peg 22h1'49"21d33'
Parkinson,Nicola Crb 16h17'48"34d37'
Parkkola 40, Taistelutoverit Dra 17h13'0"67d56'
Parks,Arva Moore Del 20h26'44"11d11'
Parks,Bill Her 15h59'57"45d44'
Parks,Carol Cyg 19h15'39"44d52'
Parks,Carol Lynn Thornton Cyg 21h33'11"37d50'
Parks,David Michael Her 17h3'1"47d44'
Parks,James E Per 3h57'58"51d33'
Parks,Janice Vul 20h4'17"28d50'
Parks,Jeffrey A Dra 18h59'49"50d5'
Parks,Jordan Rebecca Cmi 7h44'15"1d31'
Parks,Joyce & Jim Cyg 21h53'13"41d42'
Parks,Jr,Malcolm Vul 19h22'40"25d10'
Parks,Katie Michelle Lyr 19h8'52"37d52'
Parks,Kelci Danae Del 20h16'1"14d26'
Parks,Laci Danielle Aql 19h30'26"1d23'
Parks,LeRoy Orien Lac 22h3'38"50d56'
Parks,Maggie Rose And 23h43'1"43d47'
Parks,Margaret Davis Cyg 21h32'0"40d5'
Parks,Nancy Jean Vul 20h42'10"28d52'
Parks,Paul Edwin Ser 15h37'53"3d9'
Parks,Sean M Dra 16h49'12"62d47'
Parks,Shelley L Sgr 19h16'31"-28d41'
Parks,Steven Leigh Ser 16h3'0"23d11'
Parks,Taryn Aql 20h14'45"3d47'
Parks,William Allen Lac 22h47'19"38d1'
Parky's Star Cyg 19h48'15"38d34'
Parlagreco Uma 9h59'15"42d6'
Parlante,Philip E Her 17h28'35"30d1'
Parlanti,Star Of Robert Ori 6h1'41"8d46'
Parlapiano,Darrel Her 17h50'46"40d22'
Parlee,Lachlan Oliver Ori 5h57'16"12d57'
Parlett,J Wisher Cnv 13h25'56"40d53'
Parlon,Jr,Michael Vul 20h0'49"23d40'
Parlon,Matthew Lmi 10h34'20"33d2'
Parlon,Melissa Crb 16h20'31"30d9'
Parlova,J & L Sge 19h35'46"16d31'
Parlow,Sabine Vir 14h41'59"-0d59'
Parma,Leon W Lac 22h17'28"50d11'
Parmegiani,Monica For 2h8'1"-24d12'
Parmeter's Glory Her 17h29'0"37d57'
Parmigiani,Giuseppe Pho 0h4'59"-47d45'
Parnaby,Philip Cep 22h21'37"56d40'
Parnassius Pic 4h59'22"-47d40'
Parnel,Chandler Rane Boo 15h14'0"37d58'
Parnell,Anthony Her 17h4'34"48d44'
Parnell,Brian K Oph 17h18'16"11d46'
Parnell,Denise Peg 22h11'29"3d22'
Parnell,Frances Cet 2h37'27"5d18'
Parnell,Joshua David Aql 20h12'0"0d18'
Parnell,Laura Joanne Cas 1h53'0"74d34'
Parnell,Robert Allen Cet 0h55'30"-7d6'
Parness,Leonard J Sgr 18h54'39"-22d20'
Parness,Mr P Equ 20h58'40"7d60'
Parness,Solomon & Rachel Schoenfeld Lyr 19h2'29"30d29'
Parnevik,Mia & Jesper Uma 11h9'51"56d46'
Parnham,Lynda And 0h15'21"38d5'
Paro,John J Her 17h14'15"26d32'
Paro,Joshua Roy Aur 6h43'20"37d16'
Parodi,Starr Ori 5h43'0"-0d44'
Paroisse,Agnès Heyberger And 23h0'57"38d37'
Parola Eternal Love, Irene Lyn 8h34'53"41d29'
Parola,Joseph & Lillian Sge 19h35'14"19d12'
Parolari,Jo Ella Mon 6h40'43"10d9'
Parolini's Light,Jack Psc 0h59'24"32d33'
Parolini,Martha Lyn Tau 4h10'2"20d2'
Parolise,Patricia Cep 22h26'0"58d55'
Parot-Losco,Daphne Cep 21h39'11"60d29'
Parowski,Al Cam 4h53'43"58d38'
Parr Star,DJ Aur 6h0'33"30d21'
Parr's Eastern Star,C Wayne Hya 8h15'43"0d28'
Parr,Hazel Jarvis Cas 0h12'16"60d2'
Parr,Janet Lea Mon 6h41'37"6d11'
Parr,Joseph Donald Her 17h15'21"42d43'
Parr,Kent C Her 15h52'1"41d35'
Parr,Shanna Kaye And 0h56'1"35d44'
Parr,Star Of Boo 14h57'39"41d50'
Parr-Handerson 50th Anniversary Star Cyg 21h10'30"39d33'
Parra,Carlos Fuentes Ser 16h6'25"5d33'
Parra,Lionel Her 17h57'18"41d6'
Parrack II,John Sex 10h40'48"5d11'
Parrack,Jennifred Ann Aur 5h31'18"29d1'
Parral,Jose Equ 21h6'33"10d20'
Parramore,Kristin Lynn Mon 6h46'49"11d36'
Parrella,Andrew Her 17h8'55"46d15'
Parrello,Jr,Pasquale Aur 5h54'1"40d12'
Parrett,Emily Katherine And 0h29'44"40d50'
Parrilla,Edward & Janet Lyr 18h17'36"47d3'
Parrinello,John Cep 20h50'41"60d51'
Parris,Alexandria Christian Peg 22h41'44"21d17'
Parris,Helen M Lyr 18h23'13"46d55'
Parris,Joe Wylie Sco 17h52'37"-40d48'
Parris,Kathleen E Aql 19h4'25"3d12'
Parris,Richard K Cet 0h41'46"0d10'
Parrish's Christmas Bride,Dot & CT Del 20h18'1"20d28'
Parrish,Alicia Barry Her 16h10'2"47d53'A
Parrish,Barby Lyn And 2h30'28"48d13'
Parrish,Brooke Elizabeth Peg 23h47'59"13d51'
Parrish,Cliff Aql 18h55'25"-0d8'
Parrish,Erin Elizabeth And 0h23'13"32d33'
Parrish,Jerry Michael Lib 15h34'47"-28d42"
Parrish,John Marshall Ori 5h53'0"0d47'
Parrish,Mrs Sandipa Michael Cas 2h38'36"63d46'
Parrish,Robert Earl Aur 6h58'54"38d50'
Parrish,Ron Joe Her 16h39'52"34d50'
Parrish,Scot Joseph Her 16h10'2"47d53B
Parron,Charles Per 3h32'39"51d26'
Parrondo,Laura J Tri 2h35'1"31d39'
Parrott,Danny Lyn 7h35'18"50d6'
Parrott,Dennis & Vivian Crb 15h47'15"38d18'
Parrott,Lane Aql 19h29'25"1d15'
Parrotta,Clea Cma 6h49'4"-18d39'
Parry,Angela N Cas 0h32'13"50d25'
Parry,Christopher Jeffrey Vir 11h45'46"9d29'
Parry,Clive Morgan Cep 0h6'23"69d48'
Parry,E Wesley & Ruth Cma 7h12'41"-16d31'
Parry,Katie Marie And 23h14'0"38d37'
Parry,Robert John Uma 8h13'38"61d52'
Parry,Sharon Maree And 1h12'19"37d58'
parsec Aur 6h53'36"41d8'
Parsegan,Edward L Uma 13h45'1"50d6'
Parsival Cyg 20h24'12"39d5'
Parsley,Caitlin Anne Aql 19h31'31"14d46'
Parsley,Michael Uma 11h44'34"50d39'
Parsley,Nancy Kay And 0h49'56"40d2'
Parsnip The Navigator, Jude Peg 23h30'17"10d54'
Parson's Providence Cep 22h17'50"60d7'
Parson,Ashlee Shannon Cas 1h35'27"61d3'
Parson,Joel W Aur 6h28'43"34d25'
Parson,Judy Mae Lyr 18h38'16"47d37'
Parsonage,Robert Rue Cep 23h6'1"64d58'
Parsons Christmas Star,Debbie Aql 19h56'11"15d21'
Parsons Elightenment, The Joseph Lyn 8h5'35"51d51'
Parsons,Carl Eugene Her 17h34'45"28d40'
Parsons,Courtney Marie Lyn 8h2'38"35d36'
Parsons,Dave Cet 2h22'18"-11d7'
Parsons,David Craig Cet 1h20'47"-11d1'
Parsons,Donna Marie Gem 6h49'54"18d14'
Parsons,Doris Ernest Cyg 19h27'0"38d24'
Parsons,Holly Lyn 7h30'56"52d11'
Parsons,John F Lac 22h19'47"46d34'
Parsons,Joshua Layne Peg 22h41'57"12d6'
Parsons,Lawren Charles Lyr 18h58'41"29d17'
Parsons,Matthew William Oph 18h6'49"8d17'
Parsons,Megan Kathryn And 1h28'0"39d45'
Parsons,Robert Charles Mon 7h53'17"-0d8'
Parsons,Theresa Crb 15h17'19"31d35'
Parsons,W Brooks Dra 16h59'26"67d51'
Parsons-Brown,O la Fois Uma 11h20'40"30d58'
Parsonson,Sally Lmi 10h53'26"30d26'
Parsser,Sven Ori 5h49'33"21d21'
Partain,John W Oph 17h53'52"-5d46'
Partenfelder,Werner Her 17h4'40"46d33'
Partie Family Star,The Eri 4h13'33"-14d14'
Partin III,Will Oph 18h35'16"10d12'
Partin,Thomas E Cet 2h59'51"2d46'
Partington,Tim Ori 6h4'59"10d44'
Partis,Louise Anne Ori 6h6'41"3d43'
Partl,Steven Michael Her 16h38'14"24d42'
Partner,Samantha Cyg 20h33'42"48d57'
Partners Uma 8h51'18"47d26'
Partners L L Mon 6h21'48"5d3'
Parton,Connie Carabine Mon 8h3'0"-5d45'
Parton,Julia And 2h14'46"42d18'
Parton,Marshall Wayne Aql 18h57'36"-6d9'
Parton,Sandra "Sweetyheart" Mon 7h4'23"0d7'
Partridge,Lee & Lindsey Cyg 20h24'20"38d18'
Partridge,Robert Charles Mon 8h6'17"-8d13'
Partridge,Sandrine Cas 1h57'41"60d37'
Partridge,Vicki Gaylene Mon 7h54'59"-9d0'
Partridge,William Kenneth Per 3h4'10"43d7'
Party "Artie" Uma 12h22'42"61d43'
Party Parade Aql 19h42'0"14d55'
Partyka,Ray Lac 22h23'24"50d29'
Partynski,Julie Lyn 8h1'45"40d41'
Paruisse,Agnes Heyberger Her 18h17'12"28d11'
Parv,Kayelar Paul Ara 17h55'59"-52d26'
Parveen I Cam 4h53'37"67d33'
Parvin,Eric Scott Aql 18h58'36"6d12'
Parvis,Joanie Crb 15h55'43"28d28'
Parys,Bernadette Ser 15h18'21"20d48'
Parz,Linzee Aql 19h24'54"10d8'
Parziale's Star 143 Aur 6h26'25"31d48'
Parzival Her 17h50'16"46d58'
Parzych,Theresa Com 12h11'53"20d24'
Paré,Charles A Her 17h15'1"45d0'
Paré,John Z Oph 17h2'45"10d58'
Paré,Michele Cas 2h15'0"68d56'
Pasalodos,Omar Joseph Ori 6h1'31"8d51'
Pascal Aql 10h59'52"-7d47'
Pascal Pyx 8h45'31"-21d42'
Pascal Et Myriam Mon 6h25'19"7d52'
Pascal,Adam Aur 6h23'51"32d35'
Pascal,Alexander Brett Psc 22h56'34"5d31'
Pascal,Bogaert Cam 3h47'11"60d28'
Pascal,David & Susan Umi 15h4'0"82d25'
Pascal,Irene J And 2h4'31"40d23'
Pascal,Melanie Ari 1h50'1"11d54'
Pascal,Nikky Aql 18h58'52"-6d57'
Pascal,Susan Ella Cyg 21h30'19"42d9'
Pascale Cep 0h57'56"80d16'
Pascale Cam 6h9'17"68d15'
Pascale,Annalora Gem 6h59'5"14d38'
Pascale,Dey Ser 18h31'20"1d41"
Pascale,Elaine Lyr 18h15'19"30d48'
Pascale,Madeleine Gem 7h4'26"29d57'
Pascale,Walter And 0h45'45"45d6'
Pascard,Jean Tri 1h58'44"34d34'
Pascarella,Alfred Cep 21h0'29"80d15'
Pascariu,Lelia & Beni Ser 15h31'24"18d54'
Pascee Sge 19h12'58"16d10'
Pascente,John E Cnv 13h31'48"41d19'
Paschall-Goodman,Linda And 2h4'42"38d35'
Paschke,Meranda Kay Mon 6h55'11"-8d59'
Pascia,Teresa Psc 1h21'23"31d36'
Pasciscia's Star,Sal Cyg 20h57'21"40d51'
Pasciucco,Kathy Karas Cas 0h36'1"54d57'
Pasco,III,Joseph A Cet 1h41'17"-5d12'
Pascrell,Joseph Ronald Dra 16h33'1"58d18'
Pascrell,Stephanie Rita Aur 4h51'50"51d26'
Pascual,Louis Lyr 18h53'26"39d56'
Pascval,Maria Girlie Uma 9h5'33"56d34'
Pasden,Mark Stanley Psc 23h34'14"2d18'
Pasden,Nicole Leanne And 2h17'49"40d11'
Pasek,Joseph Robert Aql 19h59'30"10d43'
Paseur,George Hya 8h14'15"1d37'
Pasha Sgr 18h58'40"-22d16'
Pasha Maria And 1h44'48"41d1'
Pashalis Australis Sgr 18h49'44"-20d18'
Pashby,Jason T Lmi 10h4'38"30d39'
Pasi,Stephanie Susana Lac 22h37'0"40d8'
Pasinato,Elisabetta Cnv 13h24'0"38d39'
Pasini,Bart R Leo 9h28'45"27d33'
Pasini,Carmelo & Mario-Sandro-Francesco Uma 11h45'57"50d12'
Pasini,Ralph Cep 21h15'34"61d1'
Paske,Chad Curtis Lac 22h31'43"52d58'
Paskel,Eric & Lisa Cyg 21h12'15"38d7'
Paskell,Tim Uma 11h9'18"51d57'
Paskevic,Mary Sct 18h43'14"-7d58'
Paskewitz,Robert Her 17h55'1"38d55'
Pasko,Augustus & Beverly Lyr 18h59'23"30d60'
Pasko,Kyle Michael James Aur 6h27'1"30d4'
Pasko,Loretta Eleanor Cooper And 23h35'50"48d36'
Pasolli,Scott A Oph 18h42'48"8d12'
Pasqarelli,Treva Marie Tau 5h45'33"18d49'
Pasqua,Giovanna Tel 20h9'7"-49d35'
Pasqual,Roberto Enrique Cet 1h25'56"0d39'
Pasquale III Aur 7h23'47"38d42'
Pasquale,Anne Marie And 2h33'16"41d15'
Pasquale,Mary Jo Aql 19h31'1"8d39'
Pasquale,Nick J Cam 7h47'28"60d24'
Pasquale,Steven Peter Hya 8h19'30"0d13'
Pasqualetti,Marina Ind 20h46'20"-52d31'
Pasquali,Guglielmo And 0h46'57"40d28"
Pasqualina Boo 14h33'44"18d8'
Pasqualine Cam 3h58'41"58d37'
Pasquarella,Artie Aur 6h27'25"41d12'
Pasquarello,Amelia Dra 15h52'19"67d1'
Pasquet,Marie Catherine Uma 11h8'0"40d49'
Pasquier,Corinne Aql 19h25'0"-10d19'
Pasquini,Ludovica Cam 6h26'12"65d10'
Pass,Michael & Maureen Cam 13h6'51"78d35'
Pass,Nancy Eri 2h53'31"-11d51'
Passalino,Jennifer Ann And 2h30'31"48d21'
Passamano,Allison Lyr 18h43'0"33d42'
Passamano,Christopher Her 17h19'29"21d30'
Passamonte,Chantal Cyg 21h6'28"30d47'
Passamonte,Jr,Dana Vincent Cnv 12h19'31"47d42'
Passannante,Dr Mario Per 3h24'21"41d1'
Passarella,Ann Lyn 7h6'1"58d30'
Passaro (AM),Anne Marie Tau 5h48'50"26d15'
Passaro,Douglas Anthony Lmi 10h4'1"36d41'
Passaro,Evan James Her 18h3'23"30d54'
Passelaigues,Mr D Peg 23h1'31"10d56'
Passemard,Jean-Marc Aur 5h2'14"50d16'
Passerelli,Jr,Moon John Dra 16h58'24"60d41'
Passet,Karl Cam 6h18'21"70d46'
Passey,Mary Ellis Lyn 8h25'33"43d54'
Passini,David Uma 12h2'57"40d48'
Passino,Celeste Lyr 18h58'1"34d26'
Passion Peach Peg 22h1'38"10d2'
Passion's Light Uma 9h25'18"50d29'
Passionate Pathologist,The Cet 3h3'52"0d20'
Passionate And 1h25'26"35d25'
Passler,Cassidy Elizabeth Lyr 19h20'1"40d20'
Passmann,Jenniffer Sco 15h54'56"-22d7'
Passmore Umi 15h22'54"71d34'
Passmore,Christopher Ryan Boo 14h56'25"28d50'
Passmore,Natahsa Louise Crb 16h22'37"37d1'
Passmore,Thomas W & Cristin B Cyg 21h0'48"30d34'
Passphroph Aql 20h7'20"1d21'
Past,Present & Promise Hya 8h24'54"-10d15'
Pastelak,Barbara And 23h2'42"51d51'
Pastelle Lucille Lyn 8h45'38"37d57'
Pasternack,Bruce Vul 19h44'34"28d26'
Pasternack,Florence Cas 0h7'42"60d6'
Pasternack,Karen Lea Peg 23h47'10"30d21'
Pasternack,Marcus Jay Ser 15h15'23"8d56'
Pasternak Knight Extraordinaire,S J Cep 21h36'45"80d16'
Pasternak,Marie Com 12h30'33"20d50'
Pasteur,Nina Flore Cas 0h55'34"68d29'
Pastl,Jeff Cet 3h2'40"0d43'
Paston,Gary S Boo 15h22'0"38d7'
Pastor's,The Aur 5h35'48"50d27'
Pastor,Norman Albert Oph 17h5'38"-10d9'
Pastore,Angela And 23h41'59"46d27'
Pastore,Coach Albert Cyg 20h1'34"50d19'
Pastorino-June 1993, Susan Cam 7h30'0"68d37'
Pastorius 6-9-92, Kevlyn Ferguson Lmi 9h49'35"40d57'
Pastrana,Louis Boo 14h14'54"52d10'
Pastre,Shellie Leigh Mon 6h19'30"-6d26'
Pastuch,Terriann Lyr 18h53'21"32d9'
Pastula,Richard Joseph Cep 21h4'40"55d41'
Paszkowski,Dawn Dra 17h2'40"67d51'
Paszli,Timothy J Per 2h25'40"58d39'
Pat Eri 5h0'38"-8d43'A
Pat Boo 14h57'35"26d8'
Pat & Amy Star Peg 21h54'56"28d29'
Pat & Andyville Uma 13h36'1"50d52'
Pat & Diane Cam 04h26'17"68^51'
Pat & Lynn-Kindred Spirits Uma 13h4'30"60d31'
Pat & Marge;Forever Cyg 20h32'32"42d0'
Pat & Ronnie's Marta Star Cyg 20h33'37"44d33'
Pat & Wendy Cyg 21h27'57"53d19'
Pat Maloy A Star Forever Joan Her 17h58'59"37d57'
Pat's "Shining Star Of 1994" Peg 22h46'45"4d59'
Pat's Birthday Star Cas 0h38'25"65d55'
Pat's Buddy Gem 7h18'35"21d49'
Pat's Little Piece Of Heaven Cas 0h20'36"61d31'
Pat's Paradise Cas 3h3'48"75d2'
Pat's Sparkle Dra 19h35'32"60d36'
Pat's Star Cam 11h36'0"80d5'
Pat's Star Cep 22h55'34"68d37'
Pat,Kim,Keri And 0h52'16"45d3'
Pat-Look up at the Sky,I'm there Del 20h17'50"9d53'
Patakos,Nick Aql 19h52'40"10d5'
Pataky,Rhiannon Maree Ind 20h32'48"-52d55'
Patalano,Suzanne Cas 0h56'0"70d40'
Patalex Lyr 18h30'26"32d59'
Patalski,Ellen Lynn Uma 10h52'30"52d24'
Patamonni,William Cep 2h44'32"80d37'
Patane,Christopher & Lauri Ori 4h59'55"0d32'
Patane,Frank Joseph Her 16h39'19"48d54'
Patanella,Joseph Her 18h18'35"12d47'
Patch Aur 5h14'21"44d55'
Patch, Grey Knight Cep 22h24'51"60d36'
Patch,Glenn & Barbara Ori 5h4'58"13d30'
Patch,Katie Elizabeth Lyn 6h15'52"59d24'
Patch,Robert John Aql 19h6'47"-1d43'
Patchen,K Bonnie Uma 13h35'0"50d6'
Patchen,Michael Aql 20h0'17"9d33'
Patchen,Robert s Lac 22h28'26"40d27'
Patchett,Linda Eri 2h53'1"16d40'
Patchin,Margy Cam 7h34'12"68d0'
PatDom the Celestial Connection Boo 13h35'20"22d4'
Pate,Boeman Lmi 9h42'33"37d59'
Pate,Christopher David Aql 20h3'0"3d47'
Pate,Doris Ellen Durham Eri 4h30'1"-8d3'
Pate,Keith Lac 22h32'42"55d12'
Pate,Nathalie Mon 7h3'19"5d37'
Pate,Richard Augustus Cet 0h55'36"-1d15'
Pateisat,Peter Aql 20h9'56"8d47'
Patek,Jack Oph 17h36'44"-24d27'
Patek,Rose B Peg 22h28'46"25d54'
Patel's Star Of Faith, Shaleen Lyr 18h40'14"36d15'
Patel,Alexsa Peg 23h39'1"31d18'
Patel,Dilip Boo 13h46'1"18d31'

Patel,Harshad
Aql 19h50'39"12d43'
Patel,Jasma
Cyg 21h6'48"30d50'
Patel,Joan Kaczynski
Cam 6h55'48"60d46'
Patel,Nicky
Eri 4h5'42"-17d53'
Patel,Priti "Tina"
Oph 17h55'21"13d29'
Patel,Rajiv Dilip
Oph 16h58'50"8d15'
Patel,Sanford & Claudia
Her 17h17'0"44d15'
Patel,Shefali
Sge 20h3'47"18d53'
Patel,Sima Siara
Aql 19h13'39"14d11'
Patel,Virel
Mon 6h33'52"2d50'
Pateo
Cet 3h6'55"6d3'
Pater-Lumen Soli Mutuum Das
Aql 19h0'0"4d9'
Paterno,Carlo Middaugh
Sge 19h55'20"16d9'
Paterson "Little Mickey",Michael A
Umi 16h35'55"77d25'
Paterson,Gina Malloy
Lyr 18h48'37"33d45'
Paterson,James
Uma 10h28'53"42d22'
Paterson,Jim
Her 17h24'32"38d13'
Paterson,Kimberley
Lyr 18h16'48"40d46'
Paterson,Mary Martha Schalk
Mon 7h29'44"-8d22'
Paterson,Mary Martha Schalk
Cas 1h28'22"60d15'
Paterson,Patricia
Aql 19h0'11"-0d45'
Paterson/Conley
Cyg 19h31'17"34d45'
Path,Jamy-Mark Kazanoff
Mon 6h27'47"8d6'
Pathak,Dr P D
Dra 15h52'53"66d25'
Pathfinder
Per 4h4'25"37d51'
Pati Y Javier
Lyn 8h11'49"48d45'
Pati,Anita A
Uma 10h21'1"61d15'
Patience
Lmi 9h32'1"38d26'
Patience
Ori 5h33'21"-6d25'
Patience
Cyg 20h24'48"40d25'
Patillas
Aur 6h6'21"38d51'
Patin de Normandie- Alsace
Aqr 22h59'46"-6d23'
Patin,Michael Scott
Boo 15h6'19"26d14'
Patino,Elizabeth Hurley
Dra 16h57'27"68d53'
Patino,George Matthew
Del 20h16'21"10d39'
Patino,Gerry
Umi 15h12'11"65d46'
Patisan
Mon 8h7'50"-1d28'
Patitucci,Simona
Ant 10h37'28"-36d16'
Patley,Dr Jason
Per 2h37'22"56d16'
Patmythes,Beth Ann
And 0h47'59"45d35'
Patock,Aimee Lynn
Uma 11h37'30"45d54'
Paton,Dorothy
And 23h31'27"39d32'
Paton,Edward
Cnc 8h31'22"32d13'
Paton,Elaine
Lyr 18h31'23"33d39'
Patou,Colombine
Mon 7h47'35"-4d34'
Patras,Michael
Dra 16h44'1"63d52'
Patreace
Lyr 18h46'28"34d24'
Patria,D Scott
Lib 15h2'41"-18d54'
Patriacca,Jr,Samuel Joseph
Aql 19h5'33"3d8'
Patrianni,Lascelles
Ori 5h57'31"15d2'
Patrias,Eilene Ann
Mon 6h38'48"6d3'
Patrice
Mon 6h39'47"6d11'
Patrice
Lyr 19h3'22"26d4'
Patrice
And 22h56'36"38d1'
Patrice
Peg 23h06'0"13d52'
Patrice Noel
Cep 22h14'60"55d28'
Patricia
Cyg 20h10'31"58d56'
Patricia
And 0h9'1"47d15'
Patricia
Cas 2h31'35"58d39'
Patricia
Cet 0h59'0"-18d48'
Patricia
Cyg 20h1'43"58d32'
Patricia
Gem 7h28'19"35d14'
Patricia
Psc 0h22'23"7d53'
Patricia
Cet 3h15'52"8d27'
Patricia
Cyg 20h35"34'31d20'
Patricia
Cas 0h21'43"60d2'
Patricia
Mon 7h20'55"-0d51'
Patricia
And 23h3'10"51d33'
Patricia
Eri 3h5'47"-1d41'
Patricia
Tri 1h39'42"31d16'
Patricia
Lyn 7h37'26"38d45'
Patricia
Cas 0h18'22"63d30'
Patricia
Umi 14h34'39"68d19'
Patricia
Sge 20h6'45"20d14'
Patricia
And 23h21'51"49d27'
Patricia
Lyn 8h2'0"35d37'
Patricia
Eri 3h21'22"-6d23'
Patricia & Christian
Her 17h48'47"40d20'
Patricia & Dennis' "Wishing Star"
Cyg 19h37'53"38d6'
Patricia & Peter, Forever
Cyg 21h50'50"42d6'
Patricia & Richard-25
Cyg 20h8'35"41d2'
Patricia - B
Aur 5h20'53"37d32'
Patricia 2260
And 2h19'51"49d32'
Patricia 39
Cru 12h37'37"-61d43'
Patricia A
Uma 8h40'38"51d45'
Patricia Ann
Cyg 20h29'22"40d46'
Patricia Ann
Cas 2h46'18"75d10'
Patricia Ann
Cas 2h43'32"70d19'
Patricia Ann
Mon 7h4'47"5d9'
Patricia Ann
And 2h12'30"38d24'
Patricia Ann
Peg 22h9'29"25d59'
Patricia Ann
And 2h33'1"45d40'
Patricia Ann My Number One Star
Cas 1h59'25"61d20'
Patricia Ann,The
Lyr 18h34'34"28d45'
Patricia Anne
And 1h26'37"36d25'
Patricia Anne's Star
Mon 6h58'16"7d45'
Patricia Avonne
And 23h43'58"43d11'
Patricia e Giuseppe
Lyr 18h39'19"26d1'
Patricia Forever
Cas 23h24'50"61d36'
Patricia Gail
Equ 21h6'45"11d11'
Patricia II
And 0h49'46"37d3'
Patricia Jean
And 1h41'26"39d16'
Patricia Jeanes
Lyr 19h6'23"40d41'
Patricia Jo
Lyr 18h48'44"40d45'
Patricia Kay
Eri 2h55'57"-6d58'
Patricia Kay
Peg 22h55'29"29d54'
Patricia Love
Cyg 21h8'0"30d16'
Patricia Lynne
Mon 7h7'0"-1d12'
Patricia Margaretta- "Peggy"
Cas 0h14'15"62d26'
Patricia Marie
Lyr 19h8'52"38d17'
Patricia O
Cam 5h27'30"65d3'
Patricia Rebecca
Eri 3h44'23"-12d50'
Patricia Who's Light Always Shines
And 1h56'41"41d9'
Patricia's Christmas Star
Cas 0h44'48"60d58'
Patricia's Dream
And 1h52'35"41d4'
Patricia's Guiding Light
Cas 3h3m23"60d21'
Patricia's Paradise
Ari 2h36'30"30d56'
Patricia's Star
Cyg 19h34'34"31d12'
Patricia,Heart Of The Desert
Dra 18h57'59"40d11'
Patrick
Boo 15h8'15"51d28'
Patrick
Sct 18h56'0"-6d53'
Patrick
Per 1h52'47"50d18'
Patrick
Boo 15h21'47"50d31'
Patrick
Psc 2h4'54"5d2'
Patrick
Aql 20h16'27"5d16'
Patrick & Amy's Star
Lyn 9h16'39"37d57'
Patrick & Emily: It's In The Stars
Crb 15h57'0"26d54'
Patrick & Gaye
Cyg 21h47'27"53d19'
Patrick & Hilldie 7-17-94
Umi 13h9'49"71d30'
Patrick & Jennifer
Cet 2h58'59"1d5'
Patrick & Mary Bright Star
Tri 2h20'11"32d24'
Patrick & Nancy's Bright Hope
Eri 4h32'0"-11d49'
Patrick & Patricia
Cyg 20h9'0"40d35'
Patrick (TW)
Sct 18h43'56"-5d30'
Patrick A T
Cmi 7h56'20"3d57'
Patrick David
Cyg 21h13'0"37d39'
Patrick Star,The David
Ori 5h34'32"1d14'
Patrick's Gift
Hya 8h24'13"-1d20'
Patrick's Link To Elvis
Uma 9h51'11"52d24'
Patrick's Playground
Dra 18h6'56"58d50'
Patrick's Polaris
Hya 8h44'16"2d30'
Patrick's Ponce
Her 17h42'38"40d27'
Patrick's Pub Costa Mesa,CA
Cet 2h31'45"7d24'
Patrick's Star,James
Sgr 19h7'57"-25d53'
Patrick's Sweet Dream, Larry
Boo 14h19'57"36d48'
Patrick,Alexa Laurel Clemmons
Aql 19h7'1"-5d53'
Patrick,Ann
Boo 13h36'55"15d40'
Patrick,Annie McDermid
Lyr 18h15'40"40d46'
Patrick,Brittany- Nicole
Mon 6h35'0"7d21'
Patrick,Dan & KSEV Listeners
Del 20h18'28"13d8'
Patrick,David W
Aur 6h12'35"50d20'
Patrick,David W
Per 2h53'44"48d49'
Patrick,Gloria E
Aql 18h58'46"12d21'
Patrick,Hillary Lee
Peg 21h51'39"34d28'
Patrick,James K
Tau 4h58'19"20d12'
Patrick,Kelly J
Cmi 7h42'50"0d48'
Patrick,Kristian Thomas Clemmons
Cnc 9h1'12"21d31'
Patrick,Lena Marie Reyes
Mon 7h52'5"-6d43'
Patrick,Maxie
And 1h13'47"35d14'
Patrick,Melanic Brett
Vul 19h47'38"28d16'
Patrick,Michael C
Boo 14h36'49"37d55'
Patrick,Monica Loraine Reyes
Eri 4h33'54"-18d40'
Patrick,My Love
Aql 19h55'33"10d9'
Patrick,Nicole Renee
Cyg 20h29'45"50d33'
Patrick,Picon Favre
Ser 15h16'17"7d52'
Patrick,Spencer Lanier
Aql 19h5'56"3d6'
Patrick,Taylor Lynn
Cam 4h12'23"70d15'
Patrick-Ramseyer Bryce Matthew
Uma 14h1'42"59d46'
Patrick-Shemaya Maman Belecen
Boo 15h21'30"38d57'
Patrickus,Lawrence George
Cep 21h55'1"55d20'
Patricolo,Lydia & Frank
Cas 0h16'47"61d54'
Patridge,J J
Lac 22h13'30"46d58'
Patridge,Michael
Dra 19h27'60"65d27'
Patrinely,Jr,James Randall
Ser 15h58'1"24d22'
Patriot,Leonard Peltier Ojibwa
Boo 13h41'54"13d48'
Patrizia
Pic 4h46'2"-46d41'
Patrizia
Pyx 8h40'49"-29d35'
Patrizia
Tel 19h6'3"-48d45'
Patrizia
Pup 8h24'18"-29d10'
Patrizia
Dra 16h51'57"62d15'
Patrizia
Pyx 8h46'8"-25d49'
Patrizia
Mon 7h55'35"-4d7'
Patrizia
Her 17h25'13"40d12'
Patrizia
Aqr 21h20'36"-13d17'
Patrizia 95
Pyx 8h46'14"-26d56'
Patrizio,Alexis S
And 2h18'50"42d28'
Patrizio,Joseph J
Cnv 12h22'19"46d50'
Patron,Stefanie
Cep 22h10'15"61d20'
Patronella Lake
Cyg 21h6'0"31d46'
Patronia
Peg 22h33'38"11d1'
Patruno,Laura
Del 20h19'36"16d22'
Patry et Cie
Per 2h55'10"47d29'A
Patry,Famille Gilles
Per 2h55'10"47d29'B
Patrycja
And 23h31'15"49d45'
Patrylo,Robert
Sct 18h21'1"-14d46'
Patschke,Wolfgang
Peg 0h12'1"18d2'
Patsy
Oph 17h56'49"13d11'
Patsy J
Uma 12h52'18"60d27'
Patsy Sue
Boo 14h30'55"50d20'
Patsy,The
Peg 23h32'1"22d24'
Pattcc,Mark
Sct 18h55'0"-4d45'
Pattee,Samuel Richard
Aur 6h12'49"33d35'
Patten's Star, Stephanie
Mon 6h54'25"11d22'
Patten,Cynthia
Crb 15h23'34"30d1'
Patten,Ian
Ori 5h54'36"18d11'
Patten,Larry G
Lac 22h34'40"56d49'
Patten,Marie Anna
Lyn 7h4'19"44d56'
Patten,Stephen F
Aur 5h34'45"41d10'
Patterson
Her 17h12'31"27d21'
Patterson II,Joseph Richard John
Ori 5h50'50"14d9'
Patterson III,Kaida M & William A
Cyg 19h34'15"36d11'
Patterson,Alycia M K
Aql 20h12'1"13d49'
Patterson,Anna Francesca
And 0h31'28"41d10'
Patterson,Bella
And 23h22'1"51d36'
Patterson,Bernie
Umi 15h13'20"70d36'
Patterson,Brian Thomas
Aqr 22h24'12"0d36'
Patterson,Carol
Lyr 18h50'51"41d11'
Patterson,Cathy
Aql 20h11'18"4d59'
Patterson,Charles Wesley
Equ 21h16'59"2d38'
Patterson,David Scott
Dra 14h24'45"64d40'
Patterson,Douglas Ryan
Vul 19h47'24"27d16'
Patterson,Dr James R
Cet 2h2'24"-1d7'
Patterson,Edgar
Her 18h0'27"38d58'
Patterson,Elizabeth Anne
Lac 21h56'35"37d23'
Patterson,Elizabeth P
And 0h18'43"33d52'
Patterson,Ella Virginia
Tau 5h48'1"23d53'
Patterson,Jack
Tau 5h19'11"28d37'
Patterson,Jamie
Lac 22h8'52"51d17'
Patterson,Jane
And 2h1'43"37d7'
Patterson,Janet
And 2h23'1"38d25'
Patterson,Jeffery P
Cam 14h11'44"80d28'
Patterson,Jennifer Gail
Aqr 22h25'0"0d10'
Patterson,John
Per 2h51'34"34d36'
Patterson,John D
Her 16h43'1"27d2'
Patterson,John Markley
Aur 6h6'11"48d58'
Patterson,Joyce
Cet 1h3'31"-0d23'
Patterson,Jr,Ralph William Bernard
Sct 10h42'11-7d37'
Patterson,Kecoclna
And 2h24'39"42d39'
Patterson,Laura Suzanne
Aqr 22h24'0"0d21'
Patterson,Lee
Lyr 19h14'60"42d53'
Patterson,Linda
And 23h35'58"48d17'
Patterson,Margaret Broschart
Cyg 20h14'44"38d34'
Patterson,Mark E
Cep 21h36'14"70d53'
Patterson,Mason
Cet 1h7'16"0d51'
Patterson,Megan Leigh
Tau 4h46'50"16d33'
Patterson,Nancy M
Boo 14h20'18"23d6'
Patterson,Paul C
Ori 5h56'29"15d15'
Patterson,Raymond L
Per 4h3'28"50d39'
Patterson,Richard
Lyr 18h56'34"45d57'
Patterson,Robert Brian
Her 17h37'1"40d46'
Patterson,Shawn T
Dra 17h32'13"60d36'
Patterson,Shirley
Cet 2h20'47"-5d30'
Patterson,Steven G
Ori 5h15'54"15d39'
Patterson,Sylvia G
Mon 6h55'43"8d33'
Patterson,T G "Pat"
Aql 20h18'55"7d50'
Patterson,The Rt Hon Percival James
Per 2h11'58"56d43'
Patterson,Timothy John
Ori 6h4'30"6d58'
Patterson,William David
Her 18h15'40"38d47'
Patterson-Turpin, Melanie Renae
Cap 20h26'40"-12d7'
Patterson-WGM 1993-1994,Louise H
Cyg 21h29'40"41d4'
Patteuw,Nicole
Lyn 8h43'45"44d40'
Patti
And 0h13'48"33d4'
Patti
Vul 19h48'14"25d45'
Patti & Arlene
Mon 7h42'55"-0d55'
Patti & John's "Wedding Star"
Crb 15h58'36"26d30'
Patti Ann
Equ 21h8'30"11d15'
Patti Ann
Cas 0h53'35"68d35'
Patti Cakes
Cas 0h13'15"62d11'
Patti Eileen
Dra 19h32'15"58d23'
Patti Lynn
Peg 22h33'16"25d39'
Patti My Skye
Cyg 19h28'49"30d17'
Patti P
Mon 6h54'57"-6d28'
Patti Purrfect
Lyr 18h41'20"33d20'
Patti Sue
And 0h37'41"41d3'
Patti Sue
Cas 0h57'41"68d46'
Patti W
Sge 19h31'54"17d56'
Patti's & Clint's Valentine Star
Uma 10h29'0"41d28'
Patti's Birthday Wish Star
Lyn 8h5'52"48d27'
Patti's Gem
Cas 2h30'11"60d23'
Patti's Place
Aur 5h36'0"40d12'
Patti's Star-143
Lyn 7h52'7"48d13'
Patti,Patricia
Cas 23h28'29"60d34'
Patti-Patti
Cet 2h15'25"7d6'
Pattimo
Oph 17h10'20"-20d32'
Pattison,Bonnie Lee
Com 13h4'1"15d25'
Pattison,Michael David
Her 18h3'52"28d55'
Pattison,Rod
Uma 11h31'32"50d38'
Patton
Aql 18h58'44"-5d24'
Patton Forever,Terri & Randi
Cam 5h31'0"80d8'
Patton II,Robert E
Gem 6h52'1"30d15'
Patton,Amy Beth Charleston
Mon 7h58'53"-0d52'
Patton,Constance
Cas 0h31'50"58d10'
Patton,Courtney Wayne
Aql 19h53'39"15d3'
Patton,Gary
Aql 19h7'47"5d37'
Patton,Gayle Ann
Cyg 21h20'19"40d31'
Patton,James & Sandra
Cyg 21h17'47"38d47'
Patton,Jean
Peg 0h0'26"18d52'
Patton,John Thomas
Gem 6h42'51"18d33'
Patton,Jonathan William
Ari 2h37'1"20d2'
Patton,Jonathan Butler
Per 2h56'28"31d7'
Patton,Jr,Jerry Max
Boo 14h14'1"47d37'
Patton,Peggy
And 23h32'19"41d15'
Patton,Richard R
Aql 18h56'1"-5d12'
Patton,Roberta Ann
Mon 6h43'35"11d59'
Patton,Stacey
Cam 6h8'42"60d45'
Patton,Thomas Allen
Uma 10h13'43"53d53'
Pattridge,Greg
Uma 8h57'56"48d33'
Patty
Com 12h17'52"20d47'
Patty
And 2h0'59"41d10'
Patty
Lyr 19h2'57"31d4'
Patty
Cas 0h13'1"56d1'
Patty
Lmi 10h1'0"41d3'
Patty & Freds Angel Star
Cet 2h25'0"5d45'
Patty & Tom's Star
Cyg 19h47'0"29d33'
Patty 72
Col 6h32'39"-33d3'
Patty Ann
Ori 4h49'47"5d41'
Patty Ann
Cam 4h59'56"70d6'
Patty B-LLLLL
And 23h31'19"49d58'
Patty Cakes
Lyr 19h14'50"42d10'
Patty G
Peg 22h4'25"18d56'
Patty Guardian To Earths Creatures
Ser 18h52'30"6d3'
Patty Jean
And 0h18'52"30d14'
Patty May II
Ori 6h6'20"20d57'
Patty's Light
Aur 4h34'0"31d10'
Patty's Place
Eri 3h29'38"-1d55'
Patty's Promise
Ari 1h58'14"15d25'
Patty's Star
Tau 5h3'51"16d15'
Patty,Michele
Eri 3h18'32"-17d55'
Patty-San & The Cypriot Prince
Lyr 19h19'13"41d33'
Patur,Sara Schipper
Lyn 8h2'1"46d21'
Paturzo,A Josephs
Aur 6h56'0"37d44'
Paty & Kevin's Star Café
Cnv 13h43'40"42d23'
Patz,Shirley Ann
Leo 9h45'45"7d19'
Patzelt,Natascha & Ralph
Cyg 20h16'54"41d49'
Patzer,Mary Frances
Cyg 21h8'56"31d22'
Patzer,Savannah Skye
Cnc 9h2'47"22d4'
Patzer,Scott Dean
Cep 0h18'28"76d29'
Patzl,Harald
Cep 22h10'12"68d11'
Patzschke,Roger
Boo 14h24'43"45d56'
Patzwaldt,RN,Shirley
Cas 1h13'36"62d44'
Paté-Al
Mon 7h6'24"-6d2'
Paudoie,Laurence
Aql 19h27'27"-10d50'
Paudrups,Anastasia K
Cas 1h19'1"61d50'
Pauer,Mary Etta
Lyn 6h58'0"60d15'
Paul
Lyn 8h20'47"45d6'A
Paul
Uma 8h46'1"50d52'
Paul
Aql 20h1'16"12d32'
Paul
Her 16h19'39"26d37'
Paul
Boo 14h47'20"22d49'
Paul
Lac 22h22'54"54d9'
Paul
Her 17h18'36"45d16'
Paul
Lac 22h17'37"46d38'
Paul
Cet 2h58'27"7d41'
Paul
Aql 18h53'55"10d48'
Paul & Amy
Cyg 21h3'17"33d42'
Paul & Amy
Cam 5h52'37"67d59'
Paul & Arlene
Sgr 18h59'35"-23d39'
Paul & Barbara's "Lovelight"
Cyg 21h34'15"31d2'
Paul & Cheryl
Sge 20h4'0"20d13'
Paul & Cheryl-Dream Lovers
Mon 6h53'1"-1d18'
Paul & Danotto
Cep 23h25'42"73d50'AB
Paul & Dee
Cyg 19h33'33"30d43'
Paul & Denise
Mon 7h44'38"-5d23'
Paul & Diane
Crb 15h16'13"30d17'
Paul & Gerry
Cet 1h20'0"-13d13'
Paul & Jean
Crb 15h18'49"32d25'
Paul & June "Bellofatto"
Cyg 21h51'11"38d2'
Paul & Kim Forever
Uma 11h43'40"38d7'
Paul & Lisa:May They Shine On Forever
Cyg 20h32'51"39d51'
Paul & Mary
Cyg 20h50'39"37d57'
Paul & Moose
Del 21h2'10"18d55'

Paul & Nanette
Cyg 19h19'1"44d58'
Paul & Nathalie
Cyg 21h45'0"31d31'
Paul & Pearl's Golden Memories
Sct 18h50'16"-7d0'
Paul & Racheal
Aql 19h13'38"13d1'AB
Paul & Robin's Star
Lyr 18h46'1"31d47'
Paul & Shirley's Silver Celebration
Aur 5h17'37"50d8'
Paul & Stacy -In Love Forever
Uma 12h2'0"35d26'
Paul & Steven
Cyg 21h6'60"30d38'
Paul & Tricia
Vul 19h47'36"29d2'
Paul & Wanda's Tomorrowland
Uma 10h4'13"57d8'
Paul Andrew
Her 17h22'0"45d34'
Paul Annie
Sge 20h17'15"20d7'
Paul B 40
Her 18h2'41"28d55'
Paul et Jeanette (Una in aeter-num)
Aql 19h5'16"1d21'
Paul G
Cap 20h33'0"-14d60'
Paul George D
Vul 19h57'38"25d39'
Paul II,Ron G
Sgr 18h54'56"-29d5'
Paul John
Aql 20h7'17"8d26'
Paul LaLena
Her 16h55'41"37d51'
Paul Michael "Peno"
Boo 15h18'57"42d5'
Paul Michael & Jeannette
Umi 15h18'33"66d42'
Paul My Precious Love
Aur 6h8'52"45d10'
Paul Noel
Dra 18h13'26"65d8'
Paul und Lucie
Uma 8h40'11"52d53'
Paul"You Are The Light Of My Life"
Aur 5h5'33"41d11'
Paul's 50th
Lac 22h4'27"49d41'
Paul's Celestial Embrace
Del 20h23'40"20d2'
Paul's Deb
Mon 6h23'11"8d8'
Paul's Diversion
Uma 9h27'1"49d44'
Paul's Dreams
Ori 5h38'50"15d18'
Paul's Dylan
Ari 2h32'31"20d51'
Paul's Finest Serve
Tau 5h45'24"27d40'
Paul's Fire
Dra 20h3'45"68d25'
Paul's Guiding Light
Ori 5h54'37"14d54'
Paul's Infinity
Mon 7h5'22"-5d24'
Paul's Laughter
Aur 5h1'0"50d46'
Paul's Logic
Per 2h0'0"56d33'
Paul's Peaceful Place
Ser 15h57'30"1d54'
Paul's Place
Her 17h38'46"21d46'
Paul's Point
Her 16h39'49"24d7'
Paul's Star
Sco 15h57'40"-25d54'
Paul's Star
Aql 18h45'26"10d51'
Paul's Star In The Universe
Cmi 7h44'1"4d55'
Paul's Star of Harley
Aur 5h18'29"49d3'
Paul's Stars Chanelle, Ugo, & Diandra
Uma 8h56'29"59d8'
Paul,Arthur James
Her 16h31'1"36d11'
Paul,Barbara
Peg 22h36'56"31d17'
Paul,Charles William
Oph 17h57'49"14d4'
Paul,Courtland "Scott"
Lyr 18h59'44"30d33'
Paul,David William
Uma 10h4'14"47d47'
Paul,Dean
Aur 5h24'1"40d21'
Paul,Deborah
Vir 13h31'44"11d32'
Paul,Diana
Mon 6h27'14"8d18'
Paul,Dr Jay
Cep 3h53'24"86d16'
Paul,Elisabeth
Crb 15h40'15"37d56'
Paul,Endless Love- Julie & Russell
Cyg 21h15'18"36d52'
Paul,Gaynor
Lyr 19h16'43"26d10'
Paul,Gregory
Cet 0h30'32"0d28'
Paul,Heinz & Melanie
Uma 10h7'46"52d12'
Paul,Herb
Mon 7h21'44"-8d11'
Paul,Hubert
And 23h45'46"33d3'
Paul,I Love You,Ninny
Her 18h2'34"31d49'
Paul,James S
Boo 13h47'52"18d37'
Paul,Jr,Mack Gibbs
Aql 19h6'19"4d1'
Paul,Judy E
Cyg 19h24'0"32d40'
Paul,Julie
Per 1h50'31"48d51'
Paul,Karen's Bit of Heaven
Lyr 10h58'50"38d50'
Paul,Katie M
Mon 8h7'22"-8d9'
Paul,Kimberly & Mark Villandrie
Aur 6h4'24"45d4'
Paul,Margaret Batchelor
Cas 1h3'0"60d6'
Paul,Martin
Uma 11h51'13"44d43'
Paul,Miel Marie Rachel
Uma 11h30'39"42d49'
Paul,Norbert
Hya 9h13'50"1d38'
Paul,Ralf & Christine
Uma 10h4'14"52d16'
Paul,Steve
Her 16h15'53d42d6'
Paul,Thomas J
Oph 18h18'41"10d36'
Paul,W Desmond
Lac 22h1'16"48d1'
Paul,Warren W
Lac 22h22'30"53d20'
Paul/Thor
Uma 10h29'38"56d26'
Paula
And 23h2'55"51d3'
Paula
Tau 4h13'52"21d5'
Paula
Lyn 9h0'30"41d21'
Paula
Lyn 8h3'16"35d54'
Paula
Crt 11h17'55"-11d58'
Paula
Sgr 18h57'37"-23d20'
Paula
Lyr 18h55'18"30d44'
Paula
Hya 9h19'55"1d1'
Paula
Uma 8h49'10"72d27'
Paula "Mami" R'o'ser
Peg 22h46'54"33d40'
Paula & Fiona's Lucky Star
Ind 20h50'48"-54d21'
Paula & Larry
Cet 2h40'1"-17d53'
Paula & Patrick
Cra 18h9'26"-39d27'
Paula & Wayne
Cet 0h52'56"-8d55'
Paula Ann
And 0h57'44"40d51'
Paula Ann's Star
Uma 10h51'15"61d43'
Paula B
Peg 23h4'37"20d54'
Paula C
Cyg 21h29'25"38d24'
Paula Elizabeth
Cas 0h5'1"62d25'
Paula Ellen Star,The
Cap 21h21'21"-22d2'
Paula Jayne
Crb 15h16'20"31d6'
Paula Lee
Lyr 19h11'26"41d9'
Paula Louise & Joseph 4-Ever Eternal
Uma 9h59'59"51d16'
Paula M
Hya 9h3'34"2d40'
Paula Maria,The
Cas 0h17'31"65d52'
Paula Renay
Tri 2h43'1"33d52'
Paula Super Star
Aur 4h39'43"34d31'
Paula Theresa
Cas 2h31'1"61d41'
Paula Therésé
Sgr 19h40'14"-38d59'
Paula's Eternal Opal
Uma 11h27'16"50d2'
Paula's Pearl
And 0h52'0"45d3'
Paula's Shining Star Forever
Cep 23h20'1"65d44'
Paula's Star
Mon 8h8'56"-7d7'
Paula,Bill,Don,Chris, Linda,Molly
Aur 6h59'15"38d7'
Paula,Peter Rudolf
Lib 15h31'18"-18d47'
Paula-Planet Absolute Unltd Love & Affection
Cyg 20h24'0"38d41'
Pauladoric
Cyg 21h45'23"38d1'
Paulaine
Mon 6h27'0"3d25'
Paularcia
Mon 7h1'16"4d39'
Paularis-The Weberling Star
Her 18h3'0"30d24'
Paulat,Raymond
Sge 20h1'46"20d0'
Pauldine,Jack
Uma 10h46'1"51d12'
Paule
Cyg 19h31'48"31d50'
Paulenka,Clifford Robert
Dra 16h27'20"65d6'
Paulenka,Clinton William
Aur 6h25'42"37d37'
Pauletich,Ed
Uma 10h6'31"48d25'
Paulette
And 0h53'52"41d13'
Paulette
Hya 9h35'33"-1d5'
Paulette
Vul 19h42'57"26d52'
Paulette 222
Cam 6h3'0"60d8'
Paulette Lynn
Tri 2h17'44"33d50'
Paulette's Bright Sparkling Star
And 0h50'22"40d17'
Paulette's Star
Lyn 7h54'33"40d0'
Paulette's Star
Com 12h54'18"27d52'
Paulette's Star
Aur 6h31'35"34d44'
Paulhan,Kathy
Cyg 21h37'30"40d40'
Pauli
Boo 14h50'16"32d13'
Pauli's Star
Leo 10h0'29"14d32'
Pauli,Daniel & Michele
Cam 3h49'28"58d47'
Pauli,Marianne
Sge 19h6'28"17d33'
Pauli,Russell
Per 2h22'55"54d55'
Paulicivic,Brigitte
Aql 19h2'1"-1d29'
Paulie Tee
Aur 6h5'43"31d31'
Pauligeoff,Twenty- Seven
Sge 20h17'31"20d31'
Paulin,Beverly
Cmi 7h37'16"1d51'
Paulin,Dominique
Peg 22h30'1"30d2'
Paulin,George Edward
Oph 18h2'52"10d51'
Paulin,Jean & Frank
Mon 7h43'20"-1d20'
Paulin,Michael Evans
Lac 22h20'1"40d10'
Paulina
Psa 22h55'13"-26d58'
Paulina Rae
Cam 8h4'30"70d21'
Pauline
Peg 22h17'58"7d57'
Pauline
Mon 7h56'52"-5d58'
Pauline
Aur 4h47'0"38d28'
Pauline
Del 10h53'1"9d9'
Pauline
And 23h20'0"44d49'
Pauline
Lyn 8h12'44"33d26'
Pauline
Pic 4h37'37"-47d49'
Pauline & John
Cyg 20h24'56"39d39'
Pauline *37*
Cnc 9h18'51"30d4'
Pauline 9-9-57
Hya 9h9'47"2d32'
Pauline Kay
Lyn 7h18'17"58d22'
Pauline Marie
Cam 3h26'41"63d52'
Pauline May (Pria)
Cam 4h29'50"65d14'
Pauline The Beautiful
Lyr 18h37'46"30d22'
Pauline the Tooth Fairy
Mon 6h52'43"-6d47'
Pauline,Will You Marry Me?
Ori 5h39'14"-0d51'
Paulino,Katrina
Lyn 7h37'0"51d35'
Paulino,Kay
Lyr 19h17'59"26d52'
Paulinus
Boo 15h16'0"34d28'
Paulis,Christopher
Lac 22h24'60"54d8'
Paulish,W Jeffrey
Cep 23h3'12"65d42'
Paulita
Cyg 21h6'19"31d25'
Pauliukonis,Pranciskus J
Cep 22h25'11"67d58'
Paull,D J
Eri 3h59'13"-19d7'
Paullen
Aql 19h55'54"10d13'
Paulleystar
Lac 22h53'19"50d0'
Paulolo
Ori 6h4'31"10d11'
Paulos Star,The
Aur 5h36'45"54d24'
Pauls,Lara
Leo 9h24'17"17d12'
Pauls,Marco
Vir 12h1'43"-8d23'
Pauls,Michael
Her 16h14'43"50d48'
Paulsen,A B C
Sge 19h29'11"16d10
Paulsen,Bruce & Lori
Umi 15h8'13"68d17'
Paulsen,Bryan W
Sgr 19h10'51"-22d35'
Paulsen,Joseph
Cep 21h57'18"55d26'
Paulsen,Kameron Axel
Cam 4h5'1"69d48'
Paulsen,Kollin Sean
Mon 7h47'38"-1d6'
Paulsen,Mary Melody
Ser 15h24'13"9d49'
Paulsen,Sarah Lynn
And 0h56'40"39d46'
Paulson (Kelly),Leslie Ann
Cyg 20h16'0"38d10'
Paulson,Ken
Cep 20h56'24"68d60'
Paulson,Kevin Joseph
Boo 14h55'18"41d58'
Paulson,Richard N
Aur 6h32'19"37d2'
Paulson,Skyler James
Umi 15h2'28"67d30'
Paulstar System,The
Aql 20h5'46"4d53'
Paulus
Per 2h27'55"56d50'
Paulus
Boo 14h20'0"48d43'
Paulus Leppardius
Lyn 8h23'27"48d29'
Paulus,Pauli
Ori 5h29'30"0d49'
Paulussen,Elsa
Aur 6h0'18"38d50'
Pauly
Umi 17h3'1"88d34'
Pauly V
Her 16h21'2"10d42'A
Pauly-Lane
Aql 19h53'17"15d25'
Paumgarten,Gerda
Psc 1h25'27"32d34'
Paumier,Jean-Pierre
Cyg 19h38'58"37d35'
Pausz,Gustav
Cep 23h28'39"70d45'
Pauta,Andrea
And 1h10'41"41d8'
Pauvert Christian
And 1h4'36"41d8'
Pauwels,Judith
And 23h35'59"46d2'
Pav's Eternal Hog
Dra 14h53'37"-61d25'
Pavain,Alexandra Paige
And 22h55'0"51d27'
Pavano,Sr,William F
Cep 22h17'48"65d24'
Pavarotti,Luciano
Ori 6h9'56"9d13'
Pavel
Lac 22h11'18"46d19'
Pavel
Cam 5h34'19"61d39'
Pavelites,Joseph John
Dra 10h54'32"78d43'
Pavelka,Katie Lynn
Umi 15h11'53"68d10'
Pavell,Bernard
Cep 22h15'25"61d24'
Paver,Jane E
Cmi 7h56'0"5d11'
Pavey
Lmi 10h27'25"33d20'
Pavia,Joanne Locurcio
Mon 8h4'53"-7d10'
Pavich,Christen Paige
Cas 1h8'47"61d27'
Pavich,Kari Marie
Lyn 8h45'59"35d16'
Pavini,Valeria
Cet 2h42'0"-1d34'
Pavlak,Vanessa
Eri 3h58'21"-17d49'
Pavlakovich,David Phillips
Lac 22h51'26"52d49'
Pavlat,Ann Margaret Rose
Mon 7h1'35"-1d19'
Pavlichek
Uma 10h22'57"51d19'
Pavlides,Chritopher
Ori 5h11'16"-5d39'
Pavlidis,Themis
Lyn 8h48'14"45d58'
Pavlis' Star,Chrissy & Steve
Uma 10h0'19"51d55'
Pavlo,Marusia
Equ 21h21'43"8d44'
Pavlovsky,Daniel Joseph
Lyn 7h43'20"44d12'
Pavlovsky,Katherine Rose
Lyn 8h53'18"46d34'
Pavlovsky,Michael William
Lyn 7h43'19"44d11'
Pavlovsky,Susan B
Peg 22h30'33"21d54'
Pavlu,Melissa
Per 2h59'0"31d57'
Pavolis,Daniel S
Tri 1h58'26"26d51'
Pavone,Joan
Cap 21h43'40"-20d17'
Pavone,Terri
Psc 1h21'0"31d47'
Pavé,Michael
Per 3h19'51"41d27'
Paw
Mon 8h5'21"-6d16'
Paw Paw
Lmi 9h54'11"34d56'
Paw's Star
Per 2h4'42"57d58'
Paw,Jean-Claude
Boo 15h26'26"37d53'
Pawelek,Wendy Blair
And 1h19'42"36d0'
Pawk,Maxwell Alexander
Cam 4h12'20"69d55'
Pawl,Joseph
Uma 10h58'0"56d7'
Pawlak,Little Mary
Mon 6h55'17"-6d12'
Pawlak,Peter John
Oph 17h38'57"11d29'
Pawlak,Robert D
Per 3h0'49"40d36'
Pawley Terrific Critic Martin
Lyn 9h7'43"38d41'
Pawlicki,Danielle Marie
Peg 22h39'55"20d44'
Pawlicki,George
Per 1h53'14"54d17'
Pawling's "InnerLight",Patricia
And 23h15'0"45d54'
Pawloski,Ann M
Tri 2h8'53"32d5'
Pawlowski,Joseph Steven
Ori 6h6'59"8d53'
Pawlowski,Mark Antoni
Lyn 6h35'19"59d7'
Pawlus,Monika Marie Mei
Cas 23h15'20"61d2'
Pawlyk,Rose
Cas 1h53'37"75d12'
Paws Lady Aisha of Jorss
Del 20h35'52"10d43'
Pawsey,Stuart
Ser 15h56'59"24d38'
Pawson,Janet
And 0h35'14"45d52'
Pax & Cheryl Star,The
Cyg 20h23'1"38d49'
Pax Times Two
Cep 22h11'39"55d39'
Paxman,Janie F
Peg 22h32'14"21d12'
Paxon,Susan Ruby
Cas 1h5'30"63d56'
Paxson,Robert & Margaret
Cyg 19h26'56"33d44'
Paxton's Star World, Robbie
Ori 6h6'24"10d25'
Paxton,Diana Lynn
Ari 2h0'47"25d22'
Paxton,Larry & Terry
Aql 19h54'25"12d18'
Payan,Christopher
Cnv 12h44'12"41d17'
Payer's Prayer
Aql 20h1'47"7d45'
Payet,André
Ori 5h57'0"20d42"
Payette,Gilbert A
Hya 9h2'24"4d19'
Payeur,Charles-Rafaël
Per 3h0'40"41d19'
Payeur,Geneviève
Per 3h6'0"40d0'
Payeur,Gilles
Per 3h6'57"41d33'
Payeur,Janine Samson
Per 3h6'30"40d48'
Payeur,Teddy
Per 3h0'1"40d5'
Payle,Sally Vann
Mon 7h59'37"-1d26'
Payne's Excalibur, Lance
Oph 16h59'1"10d16'
Payne's Light
Aur 4h52'0"40d39'
Payne's World
Aql 19h59'52"15d32'
Payne,Alfred
Uma 11h49'16"52d26'
Payne,Allison
Peg 22h2'20"25d21'
Payne,Andre Lamar
Lyn 8h16'50"36d28'
Payne,Barry Lee
Per 3h11'53"47d20'
Payne,Bennie
Boo 15h15'0"48d53'
Payne,Bryon
Uma 8h36'1"48d44'
Payne,Caroline Mills
Lyn 8h47'36"35d15'B
Payne,Cherra
Peg 21h25'0"22d32'
Payne,Christopher Clay
Oph 17h54'45"8d36'
Payne,Clara LaVerne
Oph 18h5'30"13d3'B
Payne,Darren Lawrence
Uma 11h40'0"53d5'
Payne,Debbie
And 23h26'29"46d12'
Payne,Doreen Mary
Cyg 19h31'60"37d45'
Payne,Gregory Robert
Aqr 22h1'0"-1d4'
Payne,Harold
Ser 15h13'40"23d51'
Payne,Irene Bianchi
Cas 2h7'1"59d41'
Payne,J R
Mon 6h58'18"-6d20'
Payne,James
Oph 17h4'0"11d6'
Payne,Jeffrey
Uma 10h42'14"70d38'
Payne,John Edward
Lyn 8h47'36"35d15'A
Payne,Jr,James Carter
Per 1h42'31"53d33'
Payne,Jr,Norman Emory
Oph 18h5'30"13d3'A
Payne,Kalebh
Aql 20h7'1"1d23'
Payne,Kathy Charlene
Peg 21h41'0"21d50'
Payne,Kevin Andrew Thomas
Per 1h52'0"56d36'
Payne,Lee
Cep 22h9'59"61d37'
Payne,Maria Louise Mary
Cyg 19h26'22"33d56'
Payne,Marion R
Cep 22h13'43"59d22'
Payne,Mark Gibson
Aur 5h2'44"44d41'
Payne,P P P-Paula
Pennington
Tau 5h0'39"28d54'
Payne,Pamela May
Cas 2h51'31"67d44'
Payne,Preston Ferris
Oph 18h39'22"7d2'
Payne,Robert Raymond
Psc 23h7'39"6d45'
Payne,Sarah Louise
Del 20h21'34"10d31'
Payne,Sumner Courtlynn
Oph 18h38'22"7d49'
Payne,Taylor Austin
Cet 0h49'54"0d24'
Payne,Tony (Jude)
Lyn 8h13'51"44d48'
Payne,Torrence Michael
Cep 2h0'30"77d24'
Payne,Tracy Jean
Mon 7h16'30"-1d13'
Payne,Trevor
Dra 20h18'25"62d55'
Payne,Vivian Antoinette
Mon 6h54'40"-0d57'
Payne,William Carlos
Ser 18h5'20"-13d56'
Paynter,Stephanie Suzette Bailey
Vul 20h2'22"23d12'
Payo,Barbara
Lib 15h40'12"-28d24'
Payret,LeTicia
Peg 21h41'14"23d45'
Payseur,John
Dra 16h9'44"67d13'
Payson,Carol Ann
Uma 11h59'0"32d12'
Payson,Michael Jeffrey
Sex 10h20'49"2d21'
Paysse,Eileene & René
Cyg 20h58'25"39d49'
Payton Michele
Equ 21h7'45"11d0'
Payton,Jerry
Mon 7h44'23"-4d26'
Payton,Jr,Edward
Per 1h49'54"54d19'
Payton,Matthew Alexander
Cet 2h35'26"3d42'
Payton,Mid Nance
Mon 7n10m37"-7d10'
Paytongavio
Eri 3h32'28"-6d58'
Pazder Star,The
Umi 15h17'45"70d19'
Pazzynski,Janet Ann
Cam 4h8'29"61d2'
Paälar 021994
Aql 19h4'0"0d9'
PBC Star,The
Her 17h11'18"47d12'
PBI Inc
Boo 13h38'12"23d12'
PBK 001
Boo 15h15'34"53d27'
pcp;startrek;280295 cigar end
Cosmic Bruce
Psc 1h36'24"28d6'
pcr"Monster Man"ber
Her 15h54'60"44d3'
PDP-4015695
Eri 3h2'34"-6d47'
Pe Wo Be
Tau 4h8'49"21d50'
Pea Pod Productions
Cyg 20h31'1"31d26'
Pea,Light & Love For Infinity
Dra 19h56'45"61d40'
Peabody,Barbara
Cas 2h4'0"70d1'
Peace
Peg 21h55'43"20d35'
Peace
Lup 14h38'20"53d3'
Peace
Cyg 19h28'10"38d54'
Peace & Love From Shenna
Peg 22h31'11"24d25'
Peace & Love for Carol
Cas 3h4'39"57d44'
Peace Child
Cyg 21h28'26"31d7'
Peace in the World
Cep 2h7'49"78d4'
Peace Star
Leo 10h17'3"13d42'
Peacc Star-93
Peg 23h3'21"31d16'
Peace To Achieve
Aql 19h55'57"-0d45'
Peace,Jacqueline L
Cas 1h33'26"60d34'
Peace,Jean
Cmi 7h40'13"4d35'
Peace,MD,Dr James
Oph 18h1'27"10d6'
Peacelove Star "Jayne",The
Eri 4h42'46"-6d30'
Peach
Ori 6h0'12"18d48'
Peach
Cas 0h56'1"54d44'
Peach I
Aur 5h42'40"50d20'
Peach,Jason Christopher
Dra 10h49'20"78d1'
Peach,John
Lac 22h31'34"55d13'
Peach,Margaret Ann
And 1h46'27"37d50'
Peach,Megan
Peg 22h39'45"21d26'

Peacher,Cynthia Anne
Mon 6h43'10"1d39'
Peacher-Ryan,Carla
Peg 23h27'20"32d52'
Peaches
Cyg 20h30'33"58d20'
Peaches
Cas 23h43'0"50d19'
Peaches
Ori 5h21'34"11d9'
Peaches
Ori 5h37'47"-0d34'
Peaches & Honey
Cet 2h31'1"-9d15'
Peaches for Eric
Lmi 10h9'14"32d1'
Peaches George
And 0h50'59"39d8'
Peachey,Trevor Mark
Ori 5h54'27"11d46'
Peacock,John James
Per 3h8'16"40d29'
Peacock,Joshua Ryan
Her 16h36'13"35d3'
Peacock,Keith
Per 3h47'11"39d50'
Peacock,Kyle
Per 3h54'0"38d15'
Peacock,Mark
Ori 5h52'45"16d14'
Peacock,Matthew Bryan
Cep 2h34'57"78d56'
Peacock,Philosopher, Thomas Irving
Cnv 12h19'47"38d23'
Peacock,Robin Leslie
Aur 6h3'14"37d36'
Peacock,Tara Nicole
Cyg 20h20'34"40d45'
Peake,Jr,Word Day
Cas 1h6'44"50d18'
Peake,Rose & Bob
Cyg 20h44'1"42d19'
Pealer,Cash
Sex 10h11'46"-6d55'
Peana,Jennifer Frances Kristine
And 23h38'60"46d55'
Peanne,Caroline
Sgr 19h27'42"-30d10'
Peanut
Dra 15h11'51"63d7'
Peanut de Silvia Alvarez
Peg 21h57'54"2d16'
Peanutty Girl
And 1h45'13"47d21'
Pear,Gidney
Aur 5h2'1"38d37'
Pearce D 250365
Uma 11h28'29"42d56'
Pearce,Bernadette K
Oph 17h36'47"3d34B
Pearce,Bob & Marie
Cyg 20h43'28"46d38'
Pearce,Christine Joy
And 23h20'35"40d0'
Pearce,Connie
Peg 23h21'38"17d18'
Pearce,Doris P
Aql 18h58'0"14d37'
Pearce,James A
Per 2h59'36"40d32'
Pearce,Janet Kay
Mon 7h53'25"-5d7'
Pearce,Joel
Dra 18h1'55"65d35'
Pearce,Michael Francis
Ori 6h5'51"1d23'
Pearce,Michael Wayne
Sgr 19h19'33"-43d18'
Pearce,Michele Anne
Com 13h6'12"18d11'
Pearce,Mrs Marie
Cam 7h54'31"60d57'
Pearce,Reginald Brown
Cam 4h11'29"70d37'
Pearce,Richard J
Oph 17h36'47"3d34A
Pearce,Sandra Kay
Sco 17h25'57"-38d12'
Pearce,Steven James
Tau 3h35'12"23d45'
Pearch,Big Hugh
Cet 2h57'43"5d8'
Pearcy,J L
Cet 2h41'48"2d49'
Peard,Carine
Ori 6h0'28"1d26'
Peardon,Jason
Cep 0h19'18"68d15'
Pearl
Tri 2h6'28"31d52'
Pearl
Cas 1h13'55"64d49'
Pearl of the Sky
Uma 10h39'12"48d17'
Pearl of Wisdom, Forever Gail
Tau 5h53'1"26d54'
Pearl Star
Cyg 19h33'26"30d41'
Pearl,Alyson J
Lyr 19h0'11"37d42'
Pearl,Battlestar Marvin
Dra 18h28'1"70d8'
Pearl,Ben
Uma 10h38'0"50d37'
Pearl,Dianna L
Lmi 9h53'37"38d40'
Pearl,Edward William
Per 3h23'44"40d4'
Pearl,Korry & Denise
Uma 11h21'45"37d41'
Pearl,Marlene A
And 0h25'34"22d3'
Pearl,Robert B
Per 4h6'36"51d35'
Pearl,The
Peg 22h51'38"29d31'
Pearl,Tony Lee
Cmi 7h31'0"8d9'
Pearlion
Cnv 13h13'37"40d49'
Pearlstein,Joel P
Aur 6h3'0"48d48'
Pearlstein,Judy
And 1h4'56"40d1'
Pearman,Christopher
Oph 17h7'45"-20d1'
Pearman,Robert W
Aql 19h25'36"15d10'
Pearman,Tia N
Umi 14h50'29"81d7'
Pearn,Bruce Starr
Del 20h19'47"10d59'
Pears,Alan William
Her 17h12'27"28d6'
Pears,Carols Price
Psc 1h2'45"20d3'
Pearsall,Leander Franklin
Cam 6h12'47"61d11'
Pearson (The Humbug), Harry James
Per 1h55'28"56d54'
Pearson Shining Star, D C
Lmi 10h10'42"33d14'
Pearson(Waif),Peg
Per 1h58'54"53d44'
Pearson,Arthur R
Aql 19h0'57"10d21'
Pearson,Bob
Oph 17h2'51"-20d48'
Pearson,Bobbie Evelyn
Uma 14h12'43"61d22'
Pearson,Buddy
Ori 5h1'17"7d32'
Pearson,C L
Cam 13h34'0"80d22'
Pearson,Cyprien Jane
Mon 8h2'41"-7d21'
Pearson,David Wesley
Cmi 7h42'51"4d56'
Pearson,Dawn Susan
Cyg 19h47'0"31d28'
Pearson,Frances Jane
And 0h42'0"45d41'
Pearson,Gail H
Cyg 19h42'25"38d32'
Pearson,Gary Malcolm
Per 2h39'46"37d28'
Pearson,James M
Uma 10h14'35"54d35'
Pearson,Jeffrey Eugene
Per 3h8'33"41d13'
Pearson,Joanne
Cas 23h28'39"61d2'
Pearson,Jr,John
Lac 22h38'1"53d53'
Pearson,Karen
Peg 22h12'20"4d53'
Pearson,Karen Joyce
Lyr 18h54'35"31d22'
Pearson,Krista
Lyn 7h52'45"42d32'
Pearson,Kristian A
Peg 23h29'42"18d9'
Pearson,Mary
Aql 19h24'26"-1d48'
Pearson,Michael John
Ser 15h42'50"0d12'
Pearson,Michelle Joy
And 23h26'43"48d10'
Pearson,Morgan Elizabeth
Cyg 21h6'54"30d23'
Pearson,Natalie
Cyg 21h30'26"37d2'
Pearson,Patricia Elizabeth
Cyg 19h27'59"35d51'
Pearson,Patricia
Cas 1h23'54"55d58'
Pearson,Robbie
Her 15h48'0"48d5'
Pearson,Roger
Cet 1h34'0"-1d29'
Pearson,Simon Kristian
Crb 16h9'23"33d38'
Pearson,Sr,Ray
Uma 11h11'11"70d30'
Pearson,Virginia Withington
Cyg 20h35'53"42d36'
Pearson-Bess
Dra 17h33'32"64d37'
Pearson-Stanford, Eileen
Umi 13h30'19"73d46'
Pearston,Caitlin Ann
Mon 7h3'24"-1d47'
Peart,Thomas
Aql 18h58'15"-8d32'
Peas & Carrots
Peg 23h38'23"30d34'
Pease,Christopher
Hya 9h17'1"1d5'
Pease,Courtney
Eri 4h9'27"-5d51'
Pease,Diane
Aql 19h2'0"-0d3'
Pease,Gregory Charles
Vir 14h9'43"-7d49'
Pease,Helen T
Cnv 12h56'53"32d48'
Pease,Henry
Boo 15h12'28"31d24'
Pease,Jean Campbell
Sge 18h58'0"20d33'
Pease,Jr,Herb
Ori 6h0'1"-2d44'
Pease,Kenneth
Cet 2h19'25"5d7'
Pease,Kimberly
Mon 6h57'60"-8d2'
Pease,Mary
Cmi 7h35'0"0d35'
Pease,Sr,William Henry
Cnv 12h59'33"32d36'
Peaslee School
Uma 8h32'18"70d20'
Peaslee-Our Sweet 16 Angel,Donna
Lib 15h38'41"-23d9'
Peasley,Theodore L
Peg 22h29'56"24d55'
Peau,Douce
Lmi 11h4'29"31d59'
Peavey,David
Her 17h2'0"43d59'
Peavler,Rhea
Tau 4h32'38"30d33'
Peavyhouse,Kenneth Charles
Cep 1h11'0"78d3'
Pebblechevy
Lmi 10h19'0"32d4'
Pebbles
Umi 13h48'20"76d45'
Pebbles
Cnv 12h24'40"39d9'
Peberdy,Amanda
And 23h16'57"51d18'
Pecaro,Kyle
Lac 22h33'32"56d16'
Peccini,Norman
Cep 21h53'27"58d55'
Pechalat,Florence
Per 3h29'25"39d35'
Pecher,Rosemary & Howard
Lyn 7h48'59"47d44'
Pecheux,Jacqueline
Aur 6h6'56"31d10'
Pechner,Shirley Tootsie Mom Bubie
Tri 1h59'0"30d59'
Pecho,Cheryl Lynne Thomas
Cyg 21h29'48"48d39'
Pecho,Mark Anthony
Cyg 21h22'28"38d14'
Pecho,Nicholas George
Cyg 19h27'38"32d25'
Pechota,Walter E
Lyr 19h1'21"26d23'
Pechter,Morton
Her 15h57'36"47d23'
Pechtol,Richard
Lac 22h33'0"40d49'
Peck III,George Charles
Tri 2h13'0"33d44'
Peck's The Kid,Bob
Psc 1h1'39"20d59'
Peck,Aubrey Noel
Lyr 18h44'14"35d28'
Peck,Carolyn Marie
Lyn 7h28'1"42d29'
Peck,Dotsi Powell
Mon 7h10'26"-6d28'
Peck,Griffin Tolles
Cnv 12h30'23"32d17'
Peck,Jack & Helen
Cam 5h48'38"71d15'
Peck,Jacqueline
Lyr 18h58'27"38d48'
Peck,Janele von Voigtlander
Aql 18h46'55"10d24'
Peck,Jim
Lib 14h59'60"-20d23'
Peck,Kathryn Mary
Cas 0h24'48"65d30'
Peck,Samantha Kay
Tau 4h6'1"21d59'
Peck,Sara Lou
Cas 2h33'40"57d40'
Peck,Stephanie Lynn
Lyn 8h4'1"35d33'
Peck,Stuart
Hya 8h57'0"-5d54'
Peck,The Suzanne C
Peg 22h17'11"34d36'
Peck,Thomas Henry
Per 3h18'42"41d37'
Peckham,Hillary Ann
Gem 6h48'54"12d10'
Peckham,Mimi
Cam 5h3'21"68d52'
Peckham,Nigel
Per 2h59'45"43d14'
Peckinpah,Garrett Denver
Vir 12h2'26"1d20'
Peckinpah,Garrett
Her 18h2'41"14d15'
Peckio,Deborah LePre
And 0h9'52"43d12'
Pecora,Irene
Cam 13h29'21"78d7'
Pecora,Robert
Aql 19h57'46"0d10'
Pecora,Veronica
Cnv 13h5'41"51d11'
Pecoraro,Dolores & Frank
Lyn 7h36'44"52d10'
Pecoraro,Edward Wolf
Hya 8h54'12"5d18'
Pecoraro,Patricia
Ari 2h58'0"21d23'
Pecoraro,Terry
Dra 18h57'56"48d51'
Pecukonis,Gretchen
Cas 1h13'20"63d32'
Peczynski,Kate Marie Elizabeth
Lyr 18h37'35"45d24'
Pedalino,Nicholas D
Her 16h44'17"21d37'
Pedalino,Victor
Cep 4h33'53"80d37'
Pedarzani,Jean Noël
Cas 0h1'14"61d27'
Peddicord,Carol
Ari 2h1'30"10d36'
Peddie,Colin
Aql 20h6'0"4d25'
Peddie,Harrison
Ori 6h14'20"9d40'
Peddle
Lyr 18h31'23"30d14'
Peddy,Debby
Com 12h24'22"20d17'
Pede,PhD,Jeffery H
Mon 6h54'0"1d30'
Pedersen,Berit Isabel
Peg 21h38'59"27d18'
Pedersen,Charlie P
Cnv 12h57'26"43d57'
Pedersen,Dawn Melanie
Mon 6h56'29"10d16'
Pedersen,Diane
Cas 1h0'51"58d50'
Pedersen,Errol G
Per 1h58'24"47d36'
Pedersen,George C (Chris)
Ori 5h50'16"16d35'
Pedersen,James Severin
Peg 22h37'1"27d48'
Pedersen,Jeanne Karina
Cas 1h40'20"75d11'
Pedersen,Julie Brynjolf
Cam 6h10'57"72d38'
Pedersen,Laura Helen
And 0h9'56"46d1'
Pedersen,Michael D
Her 16h50'47"39d58'
Pedersen,Ole
Tri 2h17'1"33d14'
Pedersen,Richard D 27/ 5/66-21/3/93
Ant 9h37'38"-37d29'
Pederson,Carmen Faye
And 2h25'58"50d10'
Pederson,Ed
Lac 22h27'27"55d16'
Pederson,Jack A
Hya 8h53'51"2d17'
Pederson,Katie
And 0h11'21"33d3'
Pederson,Pender
Uma 11h42'0"42d26'
Pedeville,Ashley Nicole
Lyr 19h15'41"42d8'
Pedezzi,Annalisa
Her 17h21'0"40d33'
Pedicin,Sheryl Rossman
And 23h0'41"51d43'
Pedicine,Marie Troncelliti
Lyn 8h1'37"52d7'
Pedicini,Amy
Uma 8h46'46"70d35'
Pedigo,Clayton Andre
Vir 14h23'12"-8d7'
Pedigo,Dr Francoise
Vir 14h15'48"-8d21'
Pedigo,Lise Walker
Peg 22h52'57"21d43'
Pedigo,Louise
Cas 0h41'0"60d9'
Pedigo,Patty L
Lyn 8h42'50"40d25'
Peditto,Vicki M
Aql 19h59'22"11d15'
Pedley,Martin John
Per 1h50'0"56d44'
Pedone,Rose "Nono"
Tri 2h0'57"32d17'
Pedone,Salvatore
Cnv 12h34'29"33d22'
Pedota,John & Christopher
Cyg 20h28'1"39d58'
Pedretti,Emanuela
Col 6h32'28"-38d18'
Pedretti,Principal, Robert
Lmi 10h58'40"31d38'
Pedro & Ciria
Cyg 21h25'47"37d32'
Pedro,John N
Her 18h1'49"38d42'
Pedron,Carlos Manuel Quintela
Cet 0h0'53"-17d39'
Pedrosa,MD,Victor Manuel
Umi 17h53'35"88d12'
Pedroso Scientist, Jasmine
Cet 2h28'34"4d49'
Pedroza's Very Own Star,Bob
Boo 13h38'50"20d15'
Pedroza,Hugo Alexandro
Gem 7h23'19"30d44'
Peduto,Mary
And 23h19'1"47d19'
Pee Wee
Oph 17h7'25"-20d8'
Peebles Love Star,The
Cyg 21h0'0"37d30'
Peebles,Doreen
Aur 6h13'34"36d16'
Peedin,Jonathan Ray
Boo 14h23'12"18d26'
Peedin,Tyler Vann
Lyr 18h25'16"45d26'
Peek III,Samuel E
Boo 14h19'46"37d37'
Peek,Angela Marie
Mon 7h29'38"-10d42'
Peek,Cheryl
Mon 7h7'22"-6d34'
Peek,George Francis
Ori 5h56'52"16d40'
Peek,Jackie Marion Buckalew
Cas 1h10'53"61d3'
Peek,Ronald Gene
Aur 6h28'53"32d57'
Peek,Stacey Liston
Peg 23h6'50"11d30'
Peekers
Lmi 9h35'37"37d41'
Peel,Christopher Holman
Per 2h39'52"40d60'
Peel,Don
Her 16h10'1"23d12'
Peele,Constance Nicola
And 0h37'0"40d14'
Peeler,Bradley A
Eri 4h1'41"-13d58'
Peeler,Kimberly Michelle
Ori 4h56'11"4d28'
Peeler,Phelps
Ori 6h5'15"7d54'
Peeley's Star
Dra 15h5'49"60d29'
Peeling
Hya 8h12'44"2d44'
Peell,Andrea M
Cet 1h9'45"1d14'
Peeman,Ohko O Jan, Kitty,Ilse & Hans
Uma 10h47'1"41d4'
Peemöller,Ursula *470429*
Tau 5h54'18"27d47'
Peep & Periwinkle
Sge 19h18'60"16d51'
Peeper,The
Cep 2h24'26"80d27'
PeeperBinks
Aql 18h59'49"-6d44'
Peeps,Astral
Cnv 12h45'0"39d26'
Peercy,Lt Gary
Oph 17h17'29"-20d9'
Peerman,Patricia J
Cyg 19h27'17"31d18'
Peers,Brian
Cep 22h4'46"60d35'
Pees Star,Beege
Uma 9h29'58"58d23'
Pees,Brandon Lee
Aur 6h10'22"30d39'
Peet,Edward C
Hya 8h40'28"6d34'
Peetsy
Mon 7h0'34"3d54'
Peffley,John Franklin
Her 18h0'0"48d34'
Peg
Peg 22h25'11"33d51'
Peg & 4 Plus 8
Crb 16h22'40"37d45'
Peg McN
Cas 2h35'18"57d32'
Peg's Star
Lyn 7h14'56"58d38'
Pegasus
Sco 17h51'41"-30d52'
Pegasus
Peg 23h16'12"31d22'
PEGEVEN
Ori 5h38'37"-0d36'
Peggie
Lyn 8h13'43"51d42'
Peggy
Cam 13h3'31"81d31'
Peggy
Cas 0h55'57"61d34'
Peggy & Hamlet Star, The
Cyg 21h54'18"53d4'
Peggy Ann
Eri 3h43'45"-15d9'
Peggy Jean
Cyg 20h7'27"40d17'
Peggy Jean
Peg 22h23'25"8d31'
Peggy Jo
Aql 20h30'28"0d32'
Peggy Jo
Vul 20h57'1"28d48'
Peggy Lynn
Uma 9h54'14"50d37'
Peggy Shining Irish Star
Dra 11h53'59"68d15'
Peggy Sue
Aqr 22h34'26"-6d49'
Peggy Sue
Del 20h15'0"10d47'
Peggy The Eagle Soars
Umi 14h46'31"78d22'
Peggy's (Perkiness)
Lyr 18h13'41"40d9'
Peggy's Abiding Love
Del 20h19'26"9d37'
Peggy's Star
Gem 7h57'0"32d9'
Peggy's Starlite
Cas 1h0'43"60d14'
Peggy-Kay
Cas 1h27'23"58d37'
Peggy-N-Jimmy-D
Mon 8h3'33"-5d60'
Peggylea
Equ 21h7'0"11d15'
Peggyray
Cas 0h31'32"70d59'
Peggytaria
Sgr 18h56'54"-28d46'
Pegidis,Yasilis
Sgr 19h1'39"-21d35'
Pegler,Rob
Lac 22h8'21"40d12'
Pegram,Larry
Ser 18h5'34"-14d29'
Pegsstar
Peg 23h29'56"10d18'
Peguin,Marpa
Aql 19h33'49"-0d25'
Peguy,Karinne
Sgr 20h17'37"-36d10'
PegVern
Cam 8h1'31"80d9'
Pehlivanov,Billy
Boo 15h3'52"32d0'
Pehr,Janet Mary Margaret
Sex 10h11'44"-3d45'
Peiffer,Gérard
Lyr 18h58'0"30d46'
Pein,Petra & Peter
Tau 5h29'29"28d52'
Peine,William John
Cet 2h43'11"3d47'
Peirce,Robert
Ori 5h25'0"-1d46'
Peise,Dr Peter
Peg 22h6'23"18d52'
Peisel,Veronica
Lmi 10h46'0"25d7'
Peitler,Elke
Sgr 19h18'56"-23d4'
Peitzmeier,Peter
Per 3h50'31"36d15'
Peitzmeyer,Friedrich
Cas 0h22'0"63d33'
Peixoto,Jose De Oliveira
Her 18h11'40"31d15'
Peixoto,Odile
Aur 7h6'47"38d58'
Peixotto,Jr,Roland
Aql 19h54'57"-0d41'
PEJI
Cet 0h29'0"-6d6'
Pejsa,T C
Ser 18h53'46"4d5'
Pekar,Becky & Regis
Dra 17h0'32"67d46'
Pekar,M De Fatima Guimaraes-Ferenc
Lyn 7h56'44"33d35'
Pekara,Jeanne
Sex 10h44'16"-11d15'
Pekkip
Cap 20h37'17"-14d30'
Pelayo Great Lord Eagle Of Fire,Ruben
Aur 6h12'27"38d47'
Pelayo,Julian Mikal
Cep 21h46'29"61d43'
Pelchat,Mario
Aqr 23h34'24"-11d39'
Pelda
Lmi 10h53'39"31d40'
Pele,Jean-Jacques
Lmi 10h44'1"28d21'
Pelegrini,"Beloved Mother"Roseanne
Lyn 6h55'0"54d60'
Pelesko Astronomer, John D
Per 1h44'1"53d45'
Peletz,Drew
Aur 6h53'1"37d35'
Peletz,Rachel
Sge 19h55'59"16d14'
Pelfini,Eddy
Aur 6h10'21"31d25'
Pelham,Colonel John
Boo 14h8'54"35d39'
Pelican,Looster Elizabeth Loren
Gem 6h49'18"30d49'
Pelikan,Bob
Uma 11h11'1"53d44'
Pelissie,Anne-Laure
Cnv 13h8'50"50d48'
Pelka,John Robert Zachary
Her 16h59'0"29d38'
Pell,Neta Beth
Cas 23h26'57"63d14'
Pella
Ori 5h6'40"1d51'
Pelland,Jeannine
Lyr 18h21'35"44d6'
Pelled,Irene
Cas 0h8'29"58d20'
Pellegatta,Nicolo
Lyr 19h16'26"41d19'
Pellegrin,Robert George
Lac 22h0'36"51d54'
Pellegrini,Christopher Ross
Her 17h36'47"27d7'
Pellegrini,Dina Erika
Mon 7h48'5"-5d42'
Pellegrini,Laura
Pyx 8h51'49"-24d42'
Pellegrini,Paola
Peg 23h36'0"22d10'
Pellegrini,William
And 23h17'0"50d13'
Pellegrino,Amina
Cet 2h31'12"-5d23'
Pellegrino,Drew T
Aur 5h6'59"42d2'
Pellegrino,Elaine Anne
Crb 16h18'46"28d11'
Pellegrino,Francis V
Ori 5h56'60"6d48'
Pellegrino,Jason
Cep 21h38'43"55d13'
Pellegrino,Maria I
Cyg 19h42'48"30d11'
Pellegrino,Shannon
Eri 3h20'51"-15d18'
Pellegrino,Sheryl Ann
Cas 23h21'23"60d45'
Pelleing,Marie
Cet 0h52'30"0d26'
Pellen,Sylvie
Dra 12h20'43"64d45'
Peller,Jacquelyn Marie
Uma 10h43'12"56d41'
Pellerin,Jean-Pierre
Aql 19h0'32"-8d39'
Pelletier Amour et Or, Laurent
Ari 1h46'14"18d42'
Pelletier Nicole
Uma 10h35'14"41d59'
Pelletier Ultimate Light,C Roger
Cep 22h10'31"56d2'
Pelletier's Star -A Special Venus,Ronnie
Lac 22h23'46"40d47'
Pelletier,Beulah Evelyn
Cam 12h7'38"81d34'
Pelletier,Dale
Cet 2h10'13"1d26'
Pelletier,Jean-Hugues & Pauline
Cyg 19h28'1"35d34'
Pelletier,John
Per 3h56'21"37d43'
Pelletier,Laurie J
Vul 20h0'0"28d41'
Pelletier,Lisa Daigler
Com 12h32'20"24d31'

Pelletier,Lise
Cyg 19h28'10"31d30'
Pelletier,Sylvia
Cyg 19h24'27"32d32'
Pelletier:60 Years In
Galaxy,René W
Uma 11h0'41"44d46'
Pelley-Witch III,Nan
Tau 5h46'14"28d32'
Pellicane,Avrey Gabriell
And 0h23'18"33d44'
Pellicano-Patcraft-
Designweave
Umi 13h46'10"70d17'
Pelliccia,Fran Rocca
Cas 2h32'14"57d46'
Pelligrino,John
Aql 20h10'47"0d59'
Pelling,Natasha
Cas 0h20'40"59d1'
Pellino,Michael
Uma 9h4'0"58d18'
Pellissier,Daniel Francois
Equ 21h6'24"10d36'
Pellitier,Maria
Peg 21h56'15"2d25'
Pellock,Randy
Dra 15h49'59"57d46'
Pelloni,Brian Anthony
Ari 2h37'35"30d42'
Pelloni,Gregg"With Two G's"
Ari 2h39'15"30d55'
Pelloni,Robert "Bob" "Bobby"
Cnc 9h5'48"31d44'
Pellow,Vernon E
Mon 8h7'59"-4d19'
Pellrine,Joseph & Linda
Cam 3h29'51"52d53'
Pelonero,Mae
And 2h30'55"49d51'
Pelose,John & Lillian
And 1h5'16"40d7'
Pelosi,Geraldine Virginia
Leo 10h1'57"15d5'
Pelosi,Jennifer
Cyg 21h23'57"28d36'
Pelosi,Lee J
Lac 22h34'0"56d31'
Pelosi,Lorenzo
Lib 15h1'51"-8d6'
Peloso,Anthony
Lac 22h8'24"47d48'
Pelotte,Lindsay Marie
Cet 1h7'39"-6d16'
Pelotte,Rick "Slick"
Per 2h57'32"32d22'
Pels,Maria
Aur 7h23'47"38d42'
Pelsue,Adrienne Dee
Cam 7h19'42"68d38'
Pelsy,Christine
Cra 18h32'15"-44d28'
Pelt II,John Larry
Lyr 18h13'33"40d29'
Pelt,Laurie Ann
Mon 7h0'31"8d58'
Peltier,Carl Vincent
Aur 6h17'0"32d26'
Peltier,Lydia May
Cas 1h4'50"61d6'
Peltier,Michel
Mon 6h23'31"-0d57'
Pelton,Jim
Ser 15h33'44"19d3'
Peltz,Daniel Learned
Per 2h20'40"58d24'
Peltzman,Robert Steven
Hya 9h52'52"-17d55'
Peluso 8,Mike
Sco 17h55'14"-40d56'
Peluso,Charles Peter Fredrick
Dra 16h5'59"68d46'
Peluso,Sal
Ori 6h14'0"1d42'
Peluso,Tony
Ori 4h43'1"4d9'
Pelve,Jacques
Per 3h56'58"37d15'
Pelz,Lou
Hya 8h42'42"2d25'
Pelzer,Daniel J
Per 3h24'60"40d58'
Pelzl,Karina Therese
Sco 16h38'34"-41d8'
Pember,Maryann Pohle
And 2h1'44"38d44'
Pemberton
Cam 7h55'25"60d22'
Pemberton,Gene"Dad"
Oph 17h29'55"-23d7'
Pemberton,John Kevin
Cep 22h1'55"54d52'
Pemberton,Roy
Sge 20h3'8"16d39'A
Pembroke,Michael Robert
Per 2h40'49"37d17'
Pement,John Michael
Cyg 20h55'19"40d21'
Pen-The Midnight Light
Cyg 19h34'0"37d58'
Pena,Angel
And 2h29'31"40d56'
Pena,Manuel G Silvia
Cet 3h0'60"5d36'
Penalver,Lucia Labaut
Cas 0h30'52"60d13'
Penatzer,Cassidy Jade
Cyg 19h27'42"37d50'
Pence,Sarah,Andy,& Michael
Tri 1h50'29"28d6'
Penczak,Bill
Ser 18h21'33"-2d2'
Pender,Christina Oswald
Cas 3h8'37"58d39'
Pender,Gordon G
Per 3h57'55"38d19'
Pender,Rachel Oswald
Cas 3h8'44"58d5'
Pender,Shane Patrick
Her 17h19'12"20d43'
Pendergast,Henry Thomas
Aur 6h10'23"45d21'
Pendergast,Jr,Daniel J
Lac 22h6'58"49d32'
Pendergast,Nicholas Bennett
Cep 1h18'14"77d39'
Pendergest I,John Hurley
Augustus
Cma 6h54'35"-18d21'
Pendergraft,Ted
Ser 15h21'55"5d0'
Pendergrass 831,Curtis
Del 21h2'54"12d29'
Pendergrass,Roger
Lmi 10h48'19"31d46'
Pendergrass,Victoria
Herrington
Cas 2h27'22"59d0'
Penders,Dennis
Ori 5h50'34"11d31'
Pendeville,Alexander
Aur 5h1'51"42d42'
Pendland,Virginia
Cet 2h13'24"2d7'
Pendleton III,Forrest C
Cet 2h28'18"3d51'
Pendleton,Brad
Hya 8h59'44"1d25'
Pendleton,Carole
Lyn 9h12'31"33d33'
Pendleton,David "Unit III"
Her 17h10'24"47d22'
Pendleton,Edwin Pyam
Vul 19h48'27"20d19'
Pendleton,James C
Lmi 9h54'32"40d32'
Pendleton,John E
Aur 4h56'0"40d32'
Pendleton,Kelly Hunter
Peg 22h43'41"20d21'
Pendleton,Larry Harlan
Dra 19h1'30"50d26'
Pendleton,Steve & Kathy
Cmi 7h57'43"3d53'
Pendo,ViTal
Cep 22h16'43"61d24'
Pendragon
Oph 17h29'23"8d25'
Pendred
Cyg 19h33'23"35d38'
Pendy,Agnes
Cas 0h37'1"68d55'
Pendzinski,Kimberly
And 23h36'10"47d23'
Penel,Carla Et Fred
Boo 15h39'49"42d27'
Penel,Sylvie
Cyg 19h47'35"37d31'
Penelope
Lyn 8h53'51"43d21'
Penelope Anne
Umi 14h44'20"67d28'
Penelta
Umi 15h39'32"68d57'
Penfield,Holly
And 0h59'60"37d51'
Penfield,Susan Jackman
Cyg 19h53'59"41d7'
Peng Family Star,W
Cam 3h52'22"57d19'
Peng,Hedy
Aql 19h44'0"11d39'
Peng,Yuan
Aur 4h41'56"31d20'
Penhallegon,Renee
Stoutenborough
And 23h47'22"46d6'
Penicaud,Bernard
Umi 16h27'23"71d42'
Penin
Cma 6h55'53"-15d6'
Penix,Jr,Dr Jack L
Oph 18h0'51"8d37'
Penk,Renate
Cap 20h17'42"-26d50'
Penkalski Family, Robert &
Clair
Cyg 20h58'29"50d30'
Penkoff,Magnolia Moon
Cam 6h59'1"80d0'
Penkoff,Wyatt Richard
Cam 5h45'30"67d48'
Penley
Her 17h6'6"31d16'C
Penman,Joseph William
Her 16h32'0"21d57'
Penman,Madelyn Damico
Cas 23h38'51"61d28'
Penn
Cam 3h28'1"60d33'
Penn,Julia Faxon
Peg 22h17'38"18d46'
Penn,Melanie Meador
Equ 21h5'22"3d42'
Penn,Steven
Cma 6h55'40"-18d36'
Penna,Elisabetta
Uma 12h0'32"31d46'
Penna,Gregorio Giannini Dalla
Scl 0h1'56"-28d8'
Pennacchi,Lorenzo Maria
Her 16h43'44"40d20'
Penndleton,Arthur
Aur 7h7'43"43d44'
Pennebaker,George Alvin
Hya 8h54'55"-6d48'
Pennebaker,Jr,Edward L
Cma 7h0'45"-15d0'
Pennell,Galen Abney
Peg 23h4'41"12d10'
Penner,Ethel Kentor
Peg 23h16'50"33d22'
Penner,John C
Aql 18h57'0"-5d52'
Penner,Les
Hya 8h15'36"0d45'
Pennes,Bo
Peg 23h5'58"10d16'
Pennetta,Summertime B
Lyr 18h30'34"38d38'
Pennewell,Jr,C W
Oph 16h51'33"-25d45'
Penney,Alexandra
Lyn 7h56'53"41d6'
Penney,Brendan B
Lyn 7h48'16"44d40'
Penney,Janine Rowsey
Pup 7h33'50"-36d42'
Penney,John Charles
Crb 15h53'2"34d30'B
Pennie & Andrzej 1963-
Forever
Lac 22h11'14"38d45'
Pennie's Star
Uma 9h2'56"51d24'
Pennington,Alan
Aur 5h23'1"50d4'
Pennington,Alice
Tri 2h35'59"31d51'
Pennington,Brian
Uma 11h34'33"40d51'
Pennington,Carolyn
Cyg 19h30'38"31d47'
Pennington,James Paul
Aql 20h2'34"1d2'
Pennington,Jane
And 1h46'0"40d33'
Pennington,John
Per 3h6'58"46d48'
Pennington,John Francis
Ori 5h55'25"12d37'
Pennington,K Kristina
Uma 13h59'32"58d56'
Pennington,Paula
Aql 19h4'34"-6d53'
Pennington,Susan Rabbeth
Cyg 19h31'1"34d37'
Pennington-Wells
Sge 19h19'50"18d50'
Pennink,Maarten Jacobus
Cep 22h27'1"59d15'
Pennjan
Peg 21h21'11"23d34'
Pennocchietti, Gabriella
Peg 23h31'22"20d4'
Pennock,Richard Freeland
Aur 5h3'18"38d8'
Penny
Cas 0h1'41"54d42'
Penny
Aur 6h40'46"38d35'
Penny
Cyg 19h29'28"33d28'
Penny
Del 20h34'21"20d21'
Penny & Bob
Dra 17h6'50'67d4'
Penny Angel Star,The
Cas 0h48'45"71d37'
Penny Ann
Umi 3h31'40"88d16'
Penny From Heaven
Del 20h23'37"11d7'
Penny From Heaven
Cam 8h6'49"80d3'
Penny Lyn
And 2h19'45"45d7'
Penny Lynn
Lyr 18h23'0"40d9'
Penny Lynn Star,The
Lyr 19h16'41"40d58'
Penny M
Aql 19h49'43"13d11'
Penny McLean Duell Team
And 2h2'49"40d43'
Penny McLean Duell Team
And 1h23'59"42d13'
Penny N Heaven
Peg 22h2'0"27d22'
Penny's Love Star
Cyg 20h59'43"31d22'
Penny's Reach
Cnc 8h54'32"30d29'
Penny,Bret Tillicum
Ori 5h5'38"-2d42'
Penny,Catherine Elizabeth
Sco 16h56'11"-41d2'
Penny,Harold A
Ori 4h55'27"-0d4'
Penny,Jason Travis
Hya 8h9'12"2d12'
Penny,John C & Deane B
Boo 15h25'37"38d28'
Penny,Patrick John
Her 16h30'15"20d14'
Penny,Violet
Ori 6h3'43"-1d3'
Pennygal
Ori 5h53'35"15d39'
Pennywell,Penny
Boo 14h0'48"18d32'
Penree,Jennifer
Oph 18h41'56"8d6'
Penrose,Craig & Ashley
Crb 16h7'0"33d49'
Penrose,Gilbert L
Her 16h31'15"41d4'
Penrose,Steven L
Per 3h29'33"40d0'
Pensées
Aur 5h56'29"50d2'
Penta
Cas 0h13'0"60d45'
Pentassuglia,Eve
Uma 9h12'46"62d11'
Pentassuglia,Nicolas
Uma 9h11'57"67d48'
Pentasuglio,David Michael
Lac 22h46'55'53d51'
Pentelei-Molnar, Nemelyn
Alcantara
Cet 2h55'24"0d1'
Pentelei-Molnar, Stephen E
Cet 2h55'38"5d47'
Pentelei-Molnar,Helen
Cet 2h54'50"3d27'
Pentelei-Molnar,Jr, John C
Cet 2h55'14"1d20'
Pentelei-Molnar,Sr,MD, John
C
Cet 2h55'37"2d12'
Penton,Nora Wright
Cas 23h25'27"53d17'
Penton,Patricia
Sge 19h54'28"16d10'
Pentz,Joseph James
Per 2h53'29"32d54'
Penuel,Douglas Tatman
Lac 22h40'59"54d9'
Penwone,Velella
Peg 23h30'58"21d2'
Peny Ann
Uma 9h58'15"48d34'
Penz,Helmut
Cnc 8h9'12"31d7'
Penzabene,Marsha Marie
Cas 0h1'15"65d52'
Penzo Maurizio
Uma 12h14'10"60d23'
Penzo Travel It DD 1994
Ind 21h43'45"-59d5'
Peoples,Abigail Jane
Vul 20h39'42"25d35'
Peoples,Brenda H
Peg 22h13'44"33d39'
Peoples,Cynthia L
Mon 8h1'10"-5d44'
Peoples,Dorothy M
Cas 2h0'18"61d11'
Peoples,Ian Douglas
Lac 22h4'47"40d15'
Peoples,Karen Michele
Gem 5h59'0"26d30'
Pep Figuls
Her 16h12'57"8d28'
Pepe Gil
Ari 2h59'10"30d36'
Pepe Willie
Per 3h17'27"41d24'
PePe's Star
Dra 19h35'13"67d44'
Pepe,Dena K
Umi 16h26'44"70d12'
Pepe,Susan H
And 0h19'55"32d49'
Peper,Vonda
Lac 22h35'19"53d1'
Peperone,Rich
Dra 17h12'42"61d5'
Pephyrs,Amber Star
Psc 1h40'40"27d51'
Pepiciello,Stefania
Her 16h12'12"11d4'
Pepin,Louis Ferdinand
Lyn 7h59'0"40d1'
Pepin,Tyler James
Cep 22h33'42"61d45'
Pepina,Joanna
And 1h20'32"41d13'
Pepitone,Blaise Tolon
Cas 0h56'18"63d21'
Pepitus
Dra 15h3'59"62d18'
Pepp,Jr,Edward J
Lac 22h51'1"56d3'
Pepper
Hya 8h41'55"6d28'
Pepper
Uma 10h9'42"56d37'
Pepper Light
Sgr 18h51'30"-22d38'
Pepper,Ann Harrison
Del 20h23'1"18d45'
Pepper,Bridget Ginger
And 2h21'0"39d5'
Pepper,Laura
Crt 11h46'30"-8d16'
Pepper,Nancy Hensley
Eri 3h42'21"-7d12'
Pepperall,Bonnie
And 23h16'17"42d26'
Peppermint Patty
Umi 13h8'48"70d45'
Peppers,Eva
Peg 22h7'51"3d49'
Peppin,Dr Bruce H
Ori 5h51'52"15d5'
Peppin,Karen
Cas 2h6'0"59d24'
Peppler,DO,Craig D
Cam 3h12'38"60d40'
Peppy
Aql 19h6'47"3d27'
Pepè
Pup 7h56'45"-22d49'
Pequep 7h56'45"-22d49'
Uma 14h24'49"61d44'
Per in Perpetua
Ori 5h56'14"16d39'
Per Mio B
Her 16h56'0"40d31'
Per Owe Gunnar Frödin
Cyg 19h58'40"29d28'
Per Sempre Claudio
Cas 3h5'43"60d58'
Per Sempre Harry
Per 2h27'46"57d9'
Per Sempre Jim
Sex 10h40'2"3d50'
Per Sempre Michael
Cep 2h41'1"78d22'
Per Sempre Tony Immordino
Lac 22h41'0"48d57'
Per sempre V E R A
Ind 21h9'16"-51d43'
Per Tutti La Mia Vita Janet
Ari 2h3'1"20d57'
Per Tutti La Mia Vita Jenifer
Cyg 21h33'17"30d58'
Per Tutti La Mia Vita Richie
Peg 0h8'21"20d29'
Perales,Richard R
Cyg 21h54'58"37d25'
Peralta,Holly
Aql 18h54'40"10d49'
Peralta,Scott
Aur 7h11'24"40d24'
Perau,Steven Douglas
Her 17h55'50"18d58'
Peraza,Fernando
Eri 3h31'27"-7d12'
Perazzelli,Robert
Ori 6h1'33"7d33'
Percan Till Eternity, Nicholas
Wayne
Aql 19h30'32"10d43'
Perces,Paul & Marjorie
Cnv 12h44'32"41d6'
Perch
Cep 6h13'0"85d37'
Perchevitch,Alexandria
Eri 4h53'21"-8d21'
Percival,95
Leo 9h19'33"18d55'
Percival,Aspen Rochelle
Uma 11h11'49"70d36'
Percival,Darren Lee
Per 2h12'35"57d5'
Percival,Maria
Cas 2h33'14"57d45'
Percy,Helen L & William F
Crb 16h4'54"32d26'
Percy,Lenny
Dra 17h8'25"63d52'
Perdelwitz,Marilyn Barrett
Eri 2h48'11"-2d13'
Perdomo,Lucy
Cas 1h10'35"75d55'
Perdue's Great White
Cas 0h4'42"61d30'
Perdue,Aggie
Aql 19h5'22"15d20'
Perdue,Howard
Aql 20h13'13"0d48'
Perdue,Randall
Boo 15h12'47"31d24'
Perduto,Bambino
Her 17h9'26"46d43'
Perea Family,The Mona
And 2h9'0"40d51'
Perea,Joaquin y Elisa
Mon 6h35'57"7d23'
Perea,Karen L
Peg 22h3'46"5d27'
Perea,Tracy Lynn
Cas 0h2'22"64d47'
Perego,Ivana
Tel 19h4'33" 47d6'
Peregretti Family,The
Aql 19h30'51"13d25'
Peregrina,Kerry Marie
Vul 20h16'20"22d48'
Pereira,Antonio Divino Alves
Her 18h4'23"47d49'
Pereira,Hannah Marie
Cas 0h29'17"63d49'
Pereira,Helder C
Lyn 7h8'21"44d17'
Pereira,John "Red"
Uma 10h27'0"50d31'
Pereira,Joseph
Aql 19h30'40"0d26'
Pereira,Mark E
Eri 3h27'3" 13d12'
Pereira,Michael Thomas
Uma 8h55'0"52d0'
Pereksta,David Michael
Ser 17h59'19"-11d10'
Perel,Daniel
Eri 4h31'10"-0d50'
Perel,Erica
Tri 1h58'1"30d36'
Perel,Julius
Oph 18h37'38"6d15'
Perelandra
Peg 23h5'15"30d18'
Perella Futurus
Ori 5h55'54"14d20'
Perella,Bob
Aur 6h28'34"40d17'
Perella,Kelly & Rob
Uma 10h37'36"48d51'
Perelman,Ronald Owen
Uma 11h24'1"62d38'
Perenchio,The Heavenly
Margie
Peg 22h18'0"5d9'
Perera
Aur 5h57'6"31d21'
Perera,L Wimal
Cnv 13h57'1"46d29'
Peres U Light Up My Life
Moi,Joseph
Aql 19h6'24"3d8'
Peresse,Jacques
Lac 22h4'1"37d50'
Peressini,Laura Ann
Lyr 18h50'12"31d30'
Peresson,Annick
Cet 0h57'59"0d35'
Perez
Boo 14h22'36"11d8'
Perez IV,Andrew
Aur 6h23'36"33d10'
Perez,Almendra Marisa
Peg 21h41'0"21d23'
Perez,Ana
Aql 20h10'54"11d7'
Perez,Ana
And 1h56'31"37d6'
Perez,April Marie
Ari 1h55'45"12d30'
Perez,Asuncion
Per 4h41'10"37d9'
Perez,Aurelio Y Aranzazu
Odriozola
Per 1h26'48"53d1'
Perez,Carlos & Gretchen
Lmi 10h41'18"27d38'
Perez,Carmen L
Boo 15h21'17"40d24'
Perez,David
Ser 15h11'0"10d11'
Perez,David
Per 3h37'56"38d25'
Perez,Elba
Eri 4h1'37"-6d25'
Perez,Elias C
Cas 23h31'35"61d33'
Perez,Enrique J
Mon 6h53'30"-1d15'
Perez,Eric Antonio
Cep 22h2'49"58d14'
Perez,Jandi
Aql 20h12'37"10d35'
Perez,Javier
Ori 6h1'11"5d45'
Perez,Jennifer
Mon 6h56'18"10d49'
Perez,Joe A
Cet 3h4'17"1d33'
Perez,John M
Her 18h7'49"30d3'
Perez,Jorge Nicolas
Peg 23h1'46"11d50'
Perez,Juan José
Ori 6h3'58"3d6'
Perez,Julio Enrique
Cma 7h15'30"-15d34'
Perez,Leilani Marie
Eri 3h8'28"-1d38'
Perez,Marianne
Sge 20h17'37"16d56'
Perez,Mariano Jrurre
Cas 2h27'40"68d26'
Perez,Maritza
Cam 11h37'58"81d21'
Perez,Mark Jose Benito
Uma 9h46'27"45d34'
Perez,Matthew Adam
Aur 5h7'45"43d17'
Perez,Mireille
Cnv 12h26'59"40d38'
Perez,Monica Marie
Cyg 21h41'56"28d48'
Perez,Natalie
And 0h10'39"30d29'
Perez,Neni
Uma 9h3'17"68d56'
Perez,Nicolas Jaime
Leo 9h20'32"14d45'
Perez,Orly
Cmi 7h27'0"8d38'
Perez,Patty
Ori 5h54'42"7d51'
Perez,Paul
Boo 15h3'24"20d13'
Perez,Piscean Partners Bob &
Claudia
Psc 1h6'24"28d37'
Perez,Salvador Led37'
Boo 15h41'0"42d27'
Perez,Sandra
Cas 2h55'44"57d43'
Perez,Selena Quintanilla
Cyg 20h44'49"45d33'
Perez,Shelby Rae
And 0h6'57"38d45'
Perez,USN,Capt Ramon Luis
Peg 23h39'31"11d9'
Perez,Val
Cnv 13h10'53"41d3'
Perez,Vernon
Ori 5h10'54"-6d1'
Perfect Peter
Per 3h1'31"47d21'
Perfect Song
Cyg 19h55'0"40d28'
Perfection
Lyn 7h23'51"58d46'
Perfetto,Lawrence "Chubby"
Lmi 9h58'33"33d44'
Perfetto,Ronde A
Crt 11h3'45"-14d19'
Pergament,Lizzy
Cam 3h21'24"61d17'
Pergantis,Steve
Her 15h53'1"41d17'
Pergere
Psa 22h23'49"-27d48'
Pergola,Brianne Marie
Ori 5h13'56"-8d55'
Pergola,Ilana Christina
Peg 22h36'49"20d18'
Pergola,Perry M
Crt 10h51'17"-12d24'
Perhach,Robert E
Per 2h41'24"40d53'
Peri,Karin
And 0h58'28"38d19'
Peri-Troy
Vul 19h44'1"22d58'
Pericaud,Marie Pierre
Cnv 13h17'33"32d9'
Pericle
Umi 15h37'56"68d35'
Perier,Lilian
Aur 5h8'59"38d43'
Perikleous Dazzler
Peg 22h16'48"34d6'
Perillo (68),Angela
Cep 23h30'32"63d41'
Perillo,Olga
Lyr 18h58'54"31d50'
Perin 20th Anniversary Star
Her 17h22'46"45d17'
Perine,Gloria
And 23h1'35"50d23'
Perinelli,Walter
Cmi 8h4'35"1d30'

Perini,Armondo J
Her 16h59'43"33d51'
Perkins III,Robert Delman
Peg 22h38'53"27d10'
Perkins Smiling Spirit James Grady
Cet 2h33'59"9d42'
Perkins,"Dear Aunt Toady",Ida Belle
Lyr 18h58'1"37d9'
Perkins,Antoinette
Mon 6h53'58"-10d35'
Perkins,Chris T
Boo 14h5'48"25d50'
Perkins,E Stuart
Aur 7h19'51"38d11'
Perkins,Gail Pierce
Peg 23h30'1"22d38'
Perkins,Garry Ring
Lac 22h30'11"50d16'
Perkins,George
Cnv 12h45'14"38d46'
Perkins,Goldie Jarvis
Cmi 7h58'21"8d57'
Perkins,Grandpa's Longest Drive W W
Aql 19h8'15"3d42'
Perkins,Jack
Boo 15h1'57"40d2'
Perkins,Jane
Lyr 19h16'17"33d57'
Perkins,Jeffery
Aur 6h19'20"46d35'
Perkins,Jeffrey David
Her 18h30'14"12d36'
Perkins,John
Boo 13h57'49"22d23'
Perkins,Karen Susan
Eri 3h55'17"-18d29'
Perkins,Mary Jane "Tinman"
And 1h13'29"35d17'
Perkins,Michelle Irene
Aql 20h0'27"10d24'
Perkins,Patricia A
Aql 18h56'48"14d28'
Perkins,Pauline Ann Wolfstone
Cas 0h17'25"60d30'
Perkins,Robbie Lee
Aur 6h20'51"35d49'
Perkins,Seana Elizabeth (Field)
And 0h20'25"32d23'
Perkins,Susan
Lyn 9h19'50"41d33'
Perkins,Thomas Johnathon
Cep 22h55'21"78d59'
Perkins-Hailey,Terry Ruth
And 0h50'0"37d16'
Perko,Mark Andrew
Aql 18h50'57"11d33'
Perkowski,Heather Marie
Cas 2h29'45"59d24'
Perkowski,Kelly Jean
And 0h12'33"47d14'
Perkowski,Megan Alexandra
Vul 19h18'38"23d18'
Perless,Cheryl Foy
Lyr 19h11'0"40d57'
Perlin,Allison Lindsey
And 0h9'39"44d43'
Perlin,Frances Lucky
Lyn 7h4'1"44d43'
Perlin,Marilee F
Mon 7h10'40"-0d10'
Perlin-Morales
Aql 18h41'47"0d51'
Perliskey,Tracy
And 0h58'1"39d24'
Perlman,Hillary Leigh
Cma 6h55'44"-17d56'
Perlman,Lynn
And 0h6'23"30d20'
Perlmuttek,MD,"The Neurologist" David
Ori 5h47'60"19d40'
Perminoff,Gregory
Aql 20h6'25"0d32'
Perna,Lori Beth
Cas 1h18'25"62d22'
Perna,Sr,George W
Boo 14h55'16"41d40'
Pernaselci-Locurcio
Mon 8h1'15"-10d0'
Pernell Penn Louise Kiki
Scl 23h30'36"-29d35'
Pernici,Destinée F
Uma 8h58'1"58d3'
Perniconi-Bicicchi, Mamie
Aqr 22h9'57"-8d15'A
Perochain,Florence
Oph 18h2'17"11d0'
Peron,Kevin Leslie
Boo 15h4'49"7d35'
Perone,Neil
Ori 5h3'0"0d55'
Peroni,Janice M
Lyr 18h30'58"40d7'
Perot,H Ross
Cnc 8h29'34"7d37'
Perotti,August
Her 17h10'17"22d26'
Perotti,Gianlucca
Cra 18h12'47"-39d43'
Perotti,John E
Boo 15h6'44"10d26'
Perova,Alla S
Eri 4h37'26"-6d18'
Perovanovic,Michael Bagge
Cnv 12h14'0"34d26'
Perozo,Laura & Rafael
Oph 16h57'30"-22d39'
Perpetua
Eri 2h53'53"-3d19'
Perpetual Essence
Lyn 8h25'55"45d47'
Perpetue Felecite Lehmann-Rosoanindrainy
Cru 12h27'14"-61d10'
Perpetuum
Uma 10h41'26"40d31'
Perpignano,Christina
Lyr 19h24'1"41d36'
Perr
Uma 11h9'1"43d25'
Perras,Alexandra Nicole
Cas 2h35'34"57d9'
Perras,Alicia Ann
Aur 6h56'38"44d23'
Perrault One
Cet 1h22'43"1d17'
Perrault,Suzanne & David Rago
Crb 16h19'31"38d27'
Perreault's 25th,Harry & Barbara
Cyg 19h20'29"28d44'
Perreault,Carl
Cep 23h18'21"64d10'
Perreault,Karen Jean
Cam 4h13'54"67d43'
Perreault,Lise
Lyn 9h0'10"34d44'
Perreault,Stephen L
Per 1h47'47"53d50'
Perrell (Pezz),Andrew
Ori 5h52'16"21d49'
Perrelli,Christina Marie
Lyn 8h58'43"46d41'
Perrelli,Maria(Sis)
Cyg 19h55'13"38d48'
Perrera
Sgr 20h17'49"-31d19'
Perrerall,Henry (Jack)
Dra 17h46'59"60d7'
Perret,Connor Gordon
Uma 10h23'58"47d56'
Perret,David
Crb 16h10'35"37d34'
Perrett,Norma
Psc 1h29'27"10d41'
Perretti,Jr,Frank Pasquale Ricca
Per 2h23'22"58d47'
Perri's Star
Cep 0h1'23"70d39'
Perri,Amelia
And 23h2'44"37d58'
Perrier Jouet
Ser 17h32'45"-14d6'
Perrier,Alfred et Jeannine
Umi 16h42'57"75d10'
Perrier,Virginie
Her 17h28'16"40d17'
Perrigo,Craig & Chérri
Cyg 21h38'53"42d4'
Perrin's Lucky Star, Doug
Boo 13h35'17"18d12'
Perrin,Kennon
Cep 22h21'56"63d25'
Perrin,Marie Franáoise
Per 3h56'48"37d22'
Perrin,Patricia Ann
Crt 10h53'32"-8d17'
Perrin,Sr,James
Lyn 8h13'47"38d60'
Perrine,Jerry & Diana
Crb 15h17'1"31d27'
Perrins,Ida V
Vul 20h41'0"20d14'
Perrins,My Perfect Anne
And 23h38'46"41d30'
Perris,Davis Harry
Her 17h18'52"44d2'
Perris,Enzo
Her 18h54'1"18d57'
Perrish,Kathy P
Aur 5h59'45"37d52'
Perro,Doctors Michelle & Richard Bodony
Crb 15h15'15"31d3'
Perrochet,B
Lmi 10h49'31"31d4'
Perrodo,Joel
Boo 15h23'56"37d53'
Perron,Chandler James
Dra 15h6'33"64d19'
Perron,Jedediah James
Ari 2h23'48"12d13'
Perron,Luc
Umi 15h58'40"71d19'
Perrona,Martina
Uma 11h59'16"31d32'
Perrone,Anthony Frances
Cmi 7h35'27"0d16'
Perrone,Debra
Lmi 10h36'20"26d8'
Perrone,Richie
Her 16h44'21"37d48'
Perrot,Gérard
Lyr 18h56'48"31d26'
Perrot,Sarah
And 2h25'46"40d10'
Perrotta,Christina Maria
Leo 10h26'60"11d8'
Perrotta,Joseph
Eri 4h10'55"-12d5'
Perrotti Galaxy
Crb 16h3'1"30d25'
Perrotti,Ann
Vul 19h40'39"26d50'
Perrotty,P Sue
Cyg 20h18'35"30d60'
Perrou,Caroline L
Lac 22h26'21"54d2'
Perrucci,John
Dra 19h3'48"48d37'
Perry
Lyn 7h42'1"45d34'
Perry
Dra 18h57'37"48d9'
Perry II,Craig Patric
Cet 0h52'46"-1d60'
Perry II,Lenny
Cep 22h30'45"60d29'
Perry II,Paul Michael
Aur 5h30'48"38d7'
Perry III,Charles Reid
Cnv 12h24'53"35d37'
Perry Star,The
Dra 9h38'12"78d0'
Perry Winkle
Ser 18h3'29"-14d27'
Perry's Patches
Leo 11h5'56"-5d50'
Perry's Star
Aql 20h2'1"11d53'
Perry's Star
Umi 15h18'60"71d46'
Perry,Ann
And 1h57'45"40d49'
Perry,Annaka
Ori 5h15'17"11d55'
Perry,Antoinette
Lyr 18h17'0"43d50'
Perry,Barney
Aur 5h1'46"48d40'
Perry,Billie
Aql 18h56'0"-7d59'
Perry,Blondie
And 0h16'0"37d45'
Perry,Brandie Janette
Cam 5h45'25"68d12'
Perry,Brennen Davis Leahy
Cet 3h12'27"5d27'
Perry,C Robert
Aur 5h25'17"30d22'
Perry,Child
Cmi 7h5'50"4d33'
Perry,Christopher D
Tri 1h58'28"26d2'
Perry,Dakota Levi
Per 23h0'0"30d38'
Perry,Dallas D
Aur 5h20'21"37d52'
Perry,Daniel S B
Her 18h0'26"37d45'
Perry,David
Ori 5h51'1"10d40'
Perry,David I
Aur 5h19'48"45d4'
Perry,Dennis Martine
Cep 22h57'49"63d18'
Perry,Devon Tyler
Lmi 9h48'19"34d7'
Perry,Don B
Aql 19h24'13"13d18'
Perry,Dora
Cam 7h59'11"61d27'
Perry,Dora Vachon
Lyr 19h20'33"37d60'
Perry,Faye
And 1h22'34"40d32'
Perry,Florence
Mon 6h24'1"5d53'
Perry,Frank A
Cet 2h5'0"4d55'
Perry,Garland Clyde
Aql 19h7'20"2d55'
Perry,Grandfather William Wade
Her 16h55'0"33d28'
Perry,Honeyguy
Her 17h8'24"46d56'
Perry,Illysa & Noam Izenberg
Cyg 21h30'30"40d41'
Perry,J G
Ori 6h15'1"-0d2'
Perry,James E
Her 17h49'1"40d37'
Perry,Jean Marie
Lac 22h5'45"46d54'
Perry,JoAnn Rose
Tri 2h8'51"31d40'
Perry,Joe
Aur 4h54'60"40d33'
Perry,Jonathan
Her 18h6'31"30d30'
Perry,Jordan Kaleikaumaka
For 3h29'32"-36d47'
Perry,Jr,William Ewing
Ori 5h16'37"-8d56'
Perry,Karen Marie
Tri 2h3'16"32d17'
Perry,Katherine "Katie" Susan
Com 12h31'1"22d16'
Perry,Kathy L Adkins
Cas 1h5'47"63d44'
Perry,Lynn
Cyg 19h47'52"30d16'
Perry,Matt
Per 2h41'28"34d26'
Perry,Matthew Allen
Aur 4h58'58"34d18'
Perry,Matthew L
Oph 17h18'26"-22d41'
Perry,Nicholas Robert
Psc 0h58'24"31d59'
Perry,Pam
Cyg 20h54'42"40d42'
Perry,Paul A
Her 16h33'1"41d23'
Perry,Peyton
Aql 20h19'30"1d33'
Perry,Rachel Caroline
Cas 0h32'58"63d18'
Perry,Ray & Sylvia
Cyg 19h47'57"58d59'
Perry,Robert E
Cnv 12h35'17"40d43'
Perry,Ronald (Papa)
Aql 19h31'31"8d53'
Perry,Ross
Ori 5h16'42"1d23'
Perry,Roxanne
Peg 21h30'0"20d33'
Perry,Seth Stanton
Aur 6h6'45"46d39'
Perry,Shawn Michael
Hya 8h18'44"-6d60'
Perry,Stan R
Aql 19h14'6"-11d4'A
Perry,Stephanie & Brittany
Mon 7h17'26"-5d15'
Perry,Teresa
Lyr 19h2'39"25d34'
Perry,The Michelle
Lyr 19h23'45"37d49'
Perry,TRP1-Thomas Richard
Her 16h21'32"23d6'
Perry,Zachary Daniel
Aur 6h6'57"45d5'
Perryman,Daniel
Tau 3h45'30"0d28'
Perryman,Laura Kiker
Aur 7h24'22"38d50'
Perryman,Russel
Sco 16h52'57"-41d4'
Perryment,Val
Cyg 21h45'43"38d0'
Persampiere,Albert Joseph
Tri 2h14'37"32d32'
Persampiere,Krysta Lee
And 0h53'57"40d9'
Persampiere,Victoria Lynn
Vir 13h27'15"-6d48'
Persaud,Andre A
Cnv 12h54'18"48d48'
Persaud,Rebecca
Cas 0h38'1"65d12'
Perschke,Caroline
And 23h41'33"46d10'
Perschke,Nicole Marie
Mon 8h0'12"-0d34'
Perschke,Susan
Sge 20h5'25"16d42'
Persdotter,Hanna Marie
Leo 10h15'54"14d34'
Perse"Bella",Linda Marie
Com 12h22'27"31d60'
Perselay,B & G
Mon 6h18'23"0d60'
Persello,Claude
Aur 5h1'25"38d15'
Persephone
Cyg 19h34'37"28d34'
Perseus 81193
SFTH/CN:SOSTAR
Per 2h7'25"56d34'
Pershin,Len
Cep 22h33'26"80d27'
Persico,Adrian Robert
Per 1h54'25"56d51'
Persist,Pleasantly For Peace
Boo 14h58'45"28d11'
Person III,Matthew M
Uma 12h51'48"59d49'
Person,Hans
Aql 20h8'25"6d18'
Person,Jessica Ann
Peg 22h44'45"24d20'
Person,Jr,Ray Kearney
Cma 7h52'1"11d55'
Person,Judith
Uma 11h34'40"47d50'
Person,Kristina Elizabeth
Mon 6h30'32"0d5'
Person,Ruby Marie Constant
Eri 4h43'21"-2d21'
Personally Dennis
Aur 4h55'17"40d20'
Personett,Rachel Noelani
Sgr 18h57'54"-25d 47'
Persson,Thérèse Maria
And 23h16'56"48d10'
Persyn,John
Peg 0h1'29"31d47'
Perszyk,Danielle Christine
Peg 22h8'23"3d16'
Perseg 22h8'23"3d16'
Ori 5h1'28"11d12'
Persri 5h1'28"11d12'
Oph 18h7'25"10d20'
Pert,Ashley Gerrard
Per 1h28'1"50d30'
Pert,Charlotte Baxter Mearns
Cyg 20h37'1"45d20'
Pertsas,Basile Ch
Boo 14h34'34"47d40'
Perugini'93,Faye
Cyg 19h25'20"31d27'
Perugini,Joseph Anthony
Cam 3h52'15"60d13'
Perugini,Matthew
Cep 21h12'0"67d60'
PERUNA
Cma 7h19'55"-15d1'
Peruzzini,William
Ori 6h8'19"4d27'
Perwaiz,MD,The Star Of Javaid A
Ori 6h4'14"9d23'
Perzik,David & Jane
Tri 1h46'48"28d54'
Perzik,Jodi
Hya 8h39'21"1d47'
Perzik,Jordan
Per 2h55'25"32d38'
Pesca
Mon 6h0'39" 4d24'
Pescatore,Robert T
Lac 22h50'27"53d15'
Pesce,Gaetano
Sco 17h32'0"-38d35'
Pesce,Patricia
Lyn 6h32'17"60d47'
Peschel,Michael J
Cyg 21h35'38"38d52'
Peschel,Samuel Brian
Aur 5h22'0"38d15'
Peschina,Susan und Mario
Umi 16h37'39"75d54'
Peschl,Hans-Joachim
Ori 5h42'49"8d52'
Pesci,Caterina
And 0h44'55"22d10'
Pesci,Tollo
Her 15h50'23"42d5'
Pesek,Mary Susan
Peg 22h47'18"4d29'
Pesenacker,Michael
Cas 0h32'50"63d31'
Pesenti,Rev Paul
Tri 2h35'38"35d38'
Peshca Bela
Sge 19h29'41"16d3'
Pesicek,April
Lyn 9h21'54"41d42'
Peska Love
Equ 21h19'36"8d46'
Peskett,Mollie Elizabeth
Lyr 18h59'53"28d23'
Peskey,Rose Marie Thurmond
Cyg 20h59'24"40d10'
Peskoff,Dorothy
Cas 1h38'31"75d53'
Peskoff,Laura Ann
And 2h23'28"45d27'
Pess,Dr Anne Rothstein
Oph 17h27'24"-24d5'
Pessereau,Patrice
Umi 16h43'47"76d5'
Pessin,Allan Harley
Lib 14h57'37"-23d32'
Pestana,Alan D
Dra 20h18'41"74d53'
Pesty
Ori 6h15'31"8d42'
PET
Uma 10h18'49"41d1'
Petalanie
Hya 9h3'9"-12d4'
Petardi,Lisa
Cas 0h37'49"62d48'
Petas,Sarah
Lyr 19h22'16"31d25'
Petasager
Lyr 18h28'40"47d27'
Petaura
Ori 5h56'48"14d57'
Pete
Cnv 12h26'41"41d2'
Pete
Per 2h38'55"40d45'
Pete
Cet 3h1'46"4d13'
Pete & Adrienne
Cyg 21h12'1"39d33'
Pete & Carol
Dra 17h49'30"60d20'
Pete & Cotton
Uma 11h42'19"57d3'
Pete & Dorothy,Sister Leah
Cam 4h23'30"70d48'
Pete & Maggie 1981
Uma 12h55'47"60d54'
Pete & Meredith's Wedding Star
Cyg 20h23'38"39d14'
Pete & Sally
Uma 10h52'10"52d8'
Pete & Tess In Heaven
Crb 15h54'13"38d10'
Pete Mommy Friend & Wife,Roberta E
Lyn 8h29'11"38d38'
Pete Ryan
Boo 15h15'47"40d46'
Pete's Eternally Magic Super Star
Ori 6h1'11"0d16'
Pete's Hot Dogs- Chicago
Aur 6h55'51"38d52'
Pete's Starr*Maui'93
Aur 7h5'30"40d8'
Pete's Wishing Star
Boo 14h17'60"52d47'
Pete-Star,The
Her 17h23'33"48d41'
Petedio
Cet 2h20'28"5d34'
Peteness
Psc 1h28'1"31d50'
Peter
Lac 22h17'51"47d34'
Peter
Cep 22h9'57"54d1'
Peter
Per 2h29'29"56d33'
Peter
Aur 6h3'34"37d54'
Peter
Cep 22h42'51"57d7'
Peter
Boo 14h22'0"34d4'
Peter "Spinner"
Sgr 19h8'44"-29d57'
Peter & Amy's Sunrise & Sunset
Uma 9h44'0"48d47'
Peter & Chandra's Star
Uma 11h48'22"45d10'
Peter & Ellie
Lyr 18h37'0"27d50'
Peter & Gerti
Cyg 20h48'32"37d37'
Peter & Jennifer 9-14-91
Mon 6h43'31"-2d52'
Peter & Lois 25 Years
Eri 4h33'25"-18d30'
Peter & Nicole's Luminous Wonder
Lyr 19h3'1"40d28'
Peter & Pauline's Star
Cyg 21h8'34"37d54'
Peter & Solange
Cyg 20h33'24"46d48'
Peter & Sonja's Wedding Star
Ori 5h38'4"-0d25'
Peter & Theresa
Lyr 18h51'36"41d5'
Peter & Trish's Wedding Star
Cra 18h14'44"-39d44'
Peter 10-17-94(Sweet Pea),Aimee N
Peg 22h4'49"3d40'
Peter 40-143
Psc 1h41'54"27d54'
Peter All-Star
Uma 10h24'45"48d6'
Peter Clarence
Ser 16h1'1"14d22'
Peter D Heavenly Body
Dra 18h18'18"64d6'
Peter der Grosse
Sgr 19h4'11"-24d36'
Peter der Große von Hamburg
Ser 15h27'43"2d45'
Peter DRD
Per 3h47'25"35d5'
Peter Emmanuel
Lac 22h11'39"54d25'
Peter Emmel
Her 17h47'43"41d10'
Peter Francis
Aql 20h2'30"8d10'
Peter John
Her 18h3'0"28d21'
Peter Joseph
Cet 2h9'32"4d4'
Peter Joseph
Dra 19h34'5"71d29'B
Peter Kooi
Lac 22h45'33"54d34'
Peter Marcus
Cep 22h8'31"60d29'
Peter Michael
Her 17h8'11"22d28'
Peter My Sweetie
Per 4h43'23"51d27'
Peter N Anju
Eri 4h43'23"-0d3'
Peter Pan
Eri 3h37'18"-6d14'
Peter Pan
Dra 19h37'47"78d54'
Peter's Hope
Lmi 10h17'17"33d4'
Peters Memory & Love Bruce L & Dorothy J
Cma 6h50'14"-16d36'
Peter's Muse
Aur 5h11'36"50d33'
Peter's Perfection
Per 1h31'1"53d60'
Peter's Point
Cep 21h19'0"56d15'
Peter's Princess Anne Elizabeth
And 23h36'30"47d30'
Peter's Providence
Lmi 10h37'31"28d18'
Peter's Star
Aur 6h25'42"33d51'
Peter's Star
Ind 21h18'26"49d0'
Peter's Terrain
Cep 21h52'0"70d2'
Peter's Watchful Eye
Aql 19h47'38"12d42'
Peter's Wishing Star
Sgr 18h49'43"-34d36'
Peter, Peter, Peter
Aql 20h1'41"12d44'
Peter,Christine
Aur 5h16'43"43d21'
Peter,Dan
Cam 3h20'0"60d21'
Peter,Daniel
Scl 23h21'43"-26d17'
Peter,Daniela
Lyn 7h33'40"41d18'
Peter,Ellen Clark
Lac 22h6'36"42d3'
Peter,Kathy
And 23h24'37"48d59'
Peter,Melitta & Franz
Boo 14h16'1"48d14'
Peter,Michael J
Eri 2h53'32"-1d51'
Peter,Regine
Cet 2h40'42"3d28'
Peter,Wolfgang
Aur 5h18'39"43d33'
Peter-Charles-Fehlbaum Stern
Sco 16h58'48"-41d13'
Peter-Schnuppi-Simon
Lib 15h17'20"-23d53'
Peter-Wagner-Stern
Aqr 22h41'35"0d39'
Petereit,Paul-Bertram
Peg 23h31'13"18d36'
Peterellen's Lovestar
Tau 4h14'54"1d44'
Peterhansel,Melanie
Cam 7h34'46"68d22'
PeterJoyAndie
Aql 18h42'0"0d38'
PeterKim Star,The
Peg 23h19'0"32d8'
Peterkin,Jr,Thomas J
Boo 13h34'1"26d5'
Peterkin,Lisa Nicole
Vul 19h53'0"20d15'
Peterman,Jeremy
Cep 20h39'24"76d20'
Peterman,Kimberly Marie
Cas 1h2'1"66d46'
Petermann,Cornelia
Leo 10h36'0"18d25'
Petermann,Hans-Ulrich
Lib 14h21'43"-22d54'
PeternalEternal
Aql 19h14'30"18d57'
Peters
Aur 5h14'32"40d47'
Peters
Cnv 12h16'25"35d42'

Peters & Family And 23h20'48"45d33'
Peters 1-14-91(Star Man),Tyler A Lac 22h21'29"53d6'
Peters 10-10-92 Shooting Star,Ryan J Uma 12h44'57"61d0'
Peters Dew Cnv 13h12'34"32d28'
Peters Forever My Beacon,Randy Lee Dra 19h26'0"61d37'
Peters"Captain Chaos", Jon J Cnc 7h58'49"10d17'
Peters,Alexander Spence Cmi 7h58'0"8d39'
Peters,Andreas Cap 21h38'57"-22d56'
Peters,Barbara Cet 2h37'32"3d34'
Peters,Barbara & Steve Cyg 20h6'0"40d19'
Peters,Brian J Hya 9h11'27"4d21'
Peters,Calvin Her 16h29'34"32d23'
Peters,Carl Cam 5h3'22"67d44'
Peters,CPA,Karen S Eri 2h59'26"-6d21'
Peters,Dale (Pete) Uma 11h9'39"59d20'
Peters,Daniel Del 20h14'21"14d39'
Peters,David A Hya 8h52'50"0d42'
Peters,Dean & Kim Mon 6h19'11"6d49'
Peters,Duane Aql 19h9'35"0d6'
Peters,Edwin L Cep 22h14'24"62d50'
Peters,Elle Kate And 1h30'1"39d2'
Peters,Eric Aql 20h1'1"8d29'
Peters,Ethel M And 2h31'17"44d14'
Peters,Frances Schengrund Cep 23h34'12"70d33'
Peters,Gerald Per 4h0'0"33d56'
Peters,Graham Layton Cet 2h59'29"0d57'
Peters,Ilene Mon 8h1'0"-0d33'
Peters,J Mark Cyg 19h29'29"31d4'
Peters,Jack D Tri 2h37'30"34d13'
Peters,Janice Equ 20h58'20"9d39'
Peters,Jessie (Gammy) Aur 5h15'16"41d49'
Peters,John "JP" Dra 18h6'45"65d28'
Peters,Joy Sge 19h43'18"16d31'
Peters,Jr,Clarence H Boo 14h9'1"31d22'
Peters,Jürg und Martina (Tegethoff) Sgr 19h24'22"-20d45'
Peters,Karen R Ori 6h5'33"7d48'
Peters,Kendall Harris Dra 15h14'48"61d29'
Peters,Kimberly Mon 7h2'13"-5d26'
Peters,Linda Del 20h19'19"20d30'
Peters,Maja Lyn 8h49'0"36d31'
Peters,Marie Mon 6h50'38"-1d6'
Peters,Melanie Lynn Mon 6h57'40"1d14'
Peters,Melinda Ori 5h26'40"0d45'
Peters,Morgan Mary Cam 3h18'56"61d17'
Peters,Nicole Vir 13h50'26"-5d29'
Peters,P J Cam 13h58'1"80d58'
Peters,Patricia Ann And 23h30'42"41d8'
Peters,Paula Mon 7h46'34"-1d8'
Peters,Paula Ari 1h46'0"15d20'
Peters,Perry Milo Lyn 9h9'30"39d43'
Peters,Ramona (Honey) Uma 11h2'48"59d35'
Peters,Robert R Aur 6h29'22"31d45'
Peters,Ronald A Aur 4h58'28"48d48'
Peters,Sari & Desirée Cas 0h29'20"73d18'
Peters,Trevor Cameron Cet 0h55'22"2d2'
Peters,Valerie Cyg 21h3'28"40d20'
Peters,William F Aql 19h30'43"8d50'
Peters-Weber,Elke 02/08/1958 Leo 10h7'14"18d51'
Petersen,Alexandia Lyn 7h36'12"42d44'
Petersen,Anna Mahoney Tau 5h45'24"23d30'
Petersen,Avalon Cet 2h22'26"-3d0'A
Petersen,C Retch Her 16h38'59"24d5'
Petersen,Carl Edward Cnc 8h10'51"31d18'
Petersen,Charlotta C Peg 23h37'50"18d28'
Petersen,Christine M Gem 7h16'56"28d48'
Petersen,Christy Cam 10h51'22"82'13'
Petersen,Connie & Doug Del 20h18'7"12d13'
Petersen,Delancy Cet 2h22'26" 3d0'B
Petersen,Earline Aql 18h56'48"-5d15'
Petersen,Emily Helen Peg 21h58'11"33d32'
Petersen,Harald Lyr 19h19'19"31d45'
Petersen,Jason Lac 21h58'17"41d42'
Petersen,Jill & Michael Cyg 21h27'27"40d45'
Petersen,Jon Q Her 17h47'1"14d39'
Petersen,Joseph Oph 18h19'1"7d47'
Petersen,Kathleen Marie And 2h30'38"47d42'
Petersen,Laurie Eri 4h4'38"-7d56'
Petersen,Lindsey Beth Peg 21h51'1"34d53'
Petersen,Lisa A And 23h33'19"43d20'
Petersen,Mackenzie Peg 22h4'38"10d4'
Petersen,Nancy Vul 20h14'17"23d36'
Petersen,Norman Dean Hya 9h0'39"2d46'
Petersen,Rachel Cet 0h30'27"-17d49'
Petersen,Samantha Kate Cyg 21h29'1"38d12'
Petersen,Sterling Cet 2h22'26"-3d0'C
Petershagen,Jana Leo 11h22'16"-1d13'
Peterson 10-19-93, Megan Lmi 9h51'46"38d13'
Peterson Forever, Steven & Veronica Cyg 21h3'54"38d26'
Peterson Star,The James A Her 16h20'54"23d49'
Peterson WM 1992, Frances H Mon 7h50'37"-3d35'
Peterson*Star Salesman,Conrad Boo 14h59'56"25d30'
Peterson,Alvina Marie Lyr 19h4'0"40d40'
Peterson,Barbara And 23h33'1"41d19'
Peterson,Beatrice V Umi 15h11'26"69d51'
Peterson,Benjamin Robert Silver Her 16h17'48"21d34'
Peterson,Bessie Mon 7h55'54"-6d49'
Peterson,Blake Allen Ori 6h16'11"7d45'
Peterson,Brenda E Cas 23h20'37"60d15'
Peterson,Carl Stanley Aql 20h10'54"0d5'
Peterson,Carroll V & L Maxine Mon 6h45'25"10d34'
Peterson,Cathy Sue Arundale Aql 19h28'18"-8d8'
Peterson,Charlene Marie And 1h13'0"35d49'
Peterson,Charles Aql 20h8'1"4d46'
Peterson,Cheree Cet 1h52'48"-3d35'
Peterson,Chesley Vul 19h19'30"25d6'
Peterson,Christine Com 12h30'22"20d54'
Peterson,Christopher Hadon Cma 6h57'51"-18d0'
Peterson,Clark Sct 18h46'7"-6d5'
Peterson,Clifton E Aur 6h55'39"35d38'
Peterson,Conny Ori 5h6'0"11d11'
Peterson,Cynthia Mon 7h19'1"-8d8'
Peterson,Dave Ser 15h20'1"7d12'
Peterson,Dave "Star Of My Life" Her 16h40'13"8d51'
Peterson,Deeth Lee Cam 4h56'59"58d10'
Peterson,Della Dra 18h46'12"49d22'A
Peterson,Dennis B Her 16h27'34"41d12'
Peterson,Dennis L Cep 21h11'25"58d55'
Peterson,Derek Roy Lac 22h25'40"55d3'
Peterson,Don Aql 19h35'19"1d52'
Peterson,Dr Richard G Oph 16h50'41"-28d49'
Peterson,Eddie Ori 6h1'60"-0d0'
Peterson,Edvin Hya 8h45'43"0d44'A
Peterson,Elena Brownwood Sge 19h4'18"19d28'
Peterson,Elwin L Uma 11h51'54"40d18'
Peterson,Francis Deen Vul 20h5'1"25d44'
Peterson,Fred Ori 5h14'15"-5d33'
Peterson,Gabriel Eden Aql 20h4'23"4d27'
Peterson,Gary & Jane Cam 6h10'16"58d18'
Peterson,Georgia Tri 2h17'38"32d27'
Peterson,Glen J Aur 7h3'23"40d44'
Peterson,Gunnar Ray Aur 6h1'1"37d31'
Peterson,In Memory of Dylan Michael Peg 23h1'15"11d55'
Peterson,James W Aur 6h16'11"37d30'
Peterson,Jayme LeeAnn And 0h5'57"35d48'
Peterson,Jeff Her 17h30'23"27d9'
Peterson,Joseph Cep 22h39'16"61d27'
Peterson,Joseph O Per 4h6'0"48d30'
Peterson,Joseph Ryan Peg 23h6'35"12d19'
Peterson,Joshua Andrew Dra 11h37'47"70d11'
Peterson,Jr,Reuben Walter Her 17h55'55"37d46'
Peterson,Jr,Robert Leroy Ori 5h56'51"15d46'
Peterson,Karan Ann Gibbons And 0h21'19"32d56'
Peterson,Karen Beth Peg 22h47'45"35d3'
Peterson,Kathleen Charlotte Lyr 18h41'19"45d38'
Peterson,Kevin Hya 9h6'56"3d10'
Peterson,Larry Charles Lac 22h6'41"37d48'
Peterson,Leonard Otto Oph 16h49'59"11d43'
Peterson,Louie Cet 1h12'58"-4d1'
Peterson,Marie Jane Cas 2h45'31"70d27'
Peterson,Mary "Maruch" Lyn 7h54'49"38d33'
Peterson,Mary Edith Lorena Hobert Cyg 19h46'54"30d52'
Peterson,Matthew G Dra 20h21'15"62d34'
Peterson,Mei-Hua Sha Scu 18h55'49"-10d49'
Peterson,Mildred Clarice And 0h57'37"36d59'
Peterson,Muffin Man, Derek Cnv 13h51'41"38d7'
Peterson,Nels Karl Aur 5h59'13"54d29'
Peterson,Patrick Oph 17h29'53"-22d20'
Peterson,Paul Arthur Cam 6h12'59"67d45'
Peterson,Paul F Umi 15h15'31"67d48'
Peterson,Peter C Ori 5h2'1"7d37'
Peterson,Rachel Marie And 2h34'41"42d43'
Peterson,Ramona Hya 8h45'43"0d44'B
Peterson,Raven Jade Cyg 21h13'1"28d55'
Peterson,Richard D Aur 5h24'16"37d39'
Peterson,Robert Dra 18h46'12"49d22'B
Peterson,Robert Tri 2h19'1"31d13'
Peterson,Robert & Jonni Mae Burns Cyg 21h6'41"38d56'
Peterson,Robert & Rosella Lac 22h17'43"37d42'
Peterson,Ronald David Aql 18h58'39"14d48'
Peterson,Sam Lyr 19h25'1"38d45'
Peterson,Samantha Cas 2h35'18"57d32'
Peterson,Samantha C Lyr 19h23'22"38d33'
Peterson,Sr,Jack Huston Dra 16h41'45"64d7'
Peterson,Stephen E Cmi 7h40'50"1d32'
Peterson,William R "Pete" Lyr 19h24'34"37d56'
Peterson,Zelma C Lyr 18h48'12"41d26'
Peterson-McCall 8-25-94,Suzanne Ori 5h25'12"-2d20'
Peterson-Peaceful Warrior,Chris Cap 21h23'0"-22d30'
Peterus Allanus Her 18h25'54"28d33'
Petervary,Dr Nicolette A Ori 4h58'40"-0d0'
Petetin,Eva Boo 13h54'17"16d33'
Pethers,Steven Boo 14h25'50"51d50'
Pethick,Ray & Helen Cyg 19h53'41"58d41'
Petillo My Guardian Angel, Mike Leo 10h50'53"-5d52'
Petillo,Danny Boo 15h4'34"14d24'
Petina-Dawn And 0h29'44"31d16'
Petiot,Guillaume Com 13h13'12"20d45'
Petit Bé Aur 6h20'36"31d28'
Petit Caribou Per 1h44'39"53d34'
Petit Peu Peg 23h30'54"15d48'
Petit Popo Sgr 18h55'58"-26d15'
Petit Prince Cet 2h58'0"5d38'
Petit,Cecil Mon 6h53'54"0d5'
Petit,Christiane Mon 7h44'16"-3d27'
Petit,Patrice Oph 18h0'38"10d34'
Petit,Reneé Lynn Eri 3h51'45"-7d7'
Petit,Sr,Stephen Michael Uma 10h51'39"67d35'
Petitclerc-Ivanov, Nicole Cas 3h0'30"61d24'
Petite Bunny Vul 21h21'1"24d3'
Petite douceur Ursidée Cyg 20h23'48"39d6'
Petite Renée Uma 13h46'15"61d4'
Petite Ruth Jane Mon 6h55'0"-6d48'
Petito,Barbara Vul 19h19'1"26d35'
Petiton,James Cet 2h33'23"-1d13'
Petitt Star,The Cet 2h31'58"1d15'
Petitt,Thomas Eugene Aql 20h1'29"12d11'
Petitte,Jan Lyn 8h11'13"46d22'
Petitti,Anthony Per 4h23'33"51d12'
Petitti,Jennifer And 2h24'1"39d30'
Petix,Michael J Cet 3h17'1"2d3'
Petko,Christopher Aur 6h21'1"32d29'
Petko,David & Colleen Lmi 10h0'11"38d32'
Petkovsek,Harry Hya 8h12'12"1d2'
Petley,Eileen Umi 15h30'58"78d4'
Petmeg Umi 15h44'29"70d53'
Peton,Anne-Claire Aur 5h59'53"54d52'
Petose,Bernie Boo 14h53'28"39d2'
Petr Per 2h42'18"38d40'
Petr,Mrnka Lmi 10h36'41"26d44'
Petra Ari 3h22'11"20d34'
Petra Umi 15h23'27"65d55'
Petra Cas 2h57'43"68d18'
Petra Pup 8h24'32"-26d47'
Petra Boo 14h1'56"12d3'
Petra "Trapefles 1" Vir 13h35'58"-14d3'
Petra & Marcel's Twinkle Sco 16h13'51"-21d44'
Petra & Peter 1995 Mon 7h1'52"-5d17'
Petra Gabriele Maria Aqr 22h54'43"-4d19'
Petra Maria 16081970 Leo 9h55'1"10d11'
Petra Träumerle Psc 0h17'41"17d35'
Petracco,Patrizia Boo 14h30'11"48d31'
Petraglia,Cindy Cas 1h7'38"58d21'
Petrakis,Tina Cyg 19h34'50"28d26'
Petralia,Sr,Nicholas A Per 3h0'53"40d21'
Petrarca,Paul David Dra 13h13'18"68d28'
Petrarota,Mauro Cas 23h20'47"53d39'
Petras,Ann Marie Uma 11h38'46"42d30'
Petras,Joseph J Psc 1h2'57"18d42'
Petrauskis,Donna Marie Cas 0h9'15"58d44'
Petre,Daniel J Vul 20h39'50"22d56'
Petre,In Memory of Timothy E Crb 15h45'12"27d56'
Petree,Stacy Com 12h7'50"22d24'
Petrella,Carol Lyn 7h36'43"43d49'
Petrella,Michael Lance & Christine Cam 3h34'50"61d50'
Petrella,Rayna Cnv 13h59'1"47d50'
Petrella,Sam Crt 11h21'1"-8d27'
Petrello,Jessica Peg 23h30'53"33d17'
Petri Del 20h52'51"6d12'
Petri,Colin Robert Per 3h29'0"52d11'
Petri,Ludwig Ser 15h11'27"1d9'
Petri,Madeline Irene Cmb 12h12'22"20d16'
Petriccione,John Michael Aql 19h27'55"13d36'
Petrie "Peachtree", Hubert Eugene Her 17h29'13"42d30'
Petrie "Pete",Matthieu Umi 16h40'50"75d2'
Petrie,Carolyn Maxwell Cyg 21h57'0"53d44'
Petrie,Jean Uma 12h54'23"60d26'
Petrie,Jim Cep 21h53'49"55d48'
Petrie,John Guy Aur 4h55'23"38d59'
Petrie,Kevin Uma 11h18'1"53d8'
Petrie,Rayne Scott Her 16h41'37"4d37'
Petrigliano's Soulmate George T Cep 9h58'23"80d5'
Petrigo,Colleen Cyg 20h10'21"37d54'
Petrik,Martin Lac 22h12'31"54d55'
Petrillo,Anthony J Aur 6h0'0"36d58'
Petrillo,David Michael Aur 5h2'54"42d56'
Petrillo,Dena Cam 3h18'50"55d20'
Petrillo,Joseph C Boo 14h38'39"45d5'
Petrillo,Maria Rosaria And 23h0'42"52d54'
Petrillo,Raymond Michael Cet 2h56'10"1d16'
Petrillo,Robbin Mae And 1h22'34"38d38'
Petrina Cru 12h36'42"-62d9'
Petrine Ori 5h58'51"14d36'
Petrini,Cathy And 1h52'59"36d45'
Petrini,John S Aur 6h6'31"38d32'
Petrini,Marie Umi 16h5'18"70d11'
Petris-Patel,Bradley Lac 22h20'24"50d9'
Petrisko,Paul Cam 4h56'1"70d60'
Petrizzi,Nicholas Dra 20h21'43"73d45'
Petro,Joseph Edward Boo 14h12'46"36d19'
Petro,Rose G Eri 2h54'29"-4d49'
Petro,Scott & Sue Cyg 20h51'1"30d24'
Petro,Stephen William Hya 9h24'1"-1d23'
Petrocci,William J Her 18h15'43"48d19'
Petrocelli,Barbara & Paul Uma 10h18'33"68d57'
Petroff,Paul David Boo 14h30'12"41d42'
Petrolo,Claudia Col 6h32'34"-35d10'
Petrone,Brian Lyr 18h31'40"41d32'
Petrone,C M Leo 9h57'52"22d6'
Petrone,Frank R Per 2h56'33"40d27'
Petrone,Loretta Cyg 21h32'14"38d10'
Petrone,Richard A Lac 22h35'48"48d51'
Petrone,Stephanie Cas 1h5'39"50d35'
Petronella Com 13h6'0"20d6'
Petronella,Ryan Per 2h28'26"58d56'
Petrongolo,Lisa Marie Lyn 7h35'13"44d49'
Petronus Cam 3h59'39"56d39'
Petronzio,Mark & Barbara Crb 16h1'14"30d34'
Petrosino,Anthony Per 3h23'33"40d20'
Petroski "A Star is Born",Zachary Aur 6h0'1"37d54'
Petroski's Starry 20 Seconds,Chris Peg 0h2'0"21d47'
Petroski,Julie E Cas 23h42'0"61d13'
Petrovsky,Dr Maurice E Her 17h38'0"26d34'
Petrucci,Frank Eugene Boo 14h28'38"27d57'
Petrucci,Kim And 0h10'35"33d37'
Petruccio,Joseph Edward Vul 19h18'41"20d36'
Petrucelli,MD, Salvatore R And 1h0'17"40d2'
Petruk,Jerome Dra 14h53'14"55d49'
Petrun,April Michelle Cet 2h34'1"1d6'
Petrunick,Ida Peg 21h59'57"36d4'
Petrunyak,Michael Vul 19h46'59"28d16'
Petrus,George Her 18h44'41"12d35'
Petrus,Helen M Cas 0h57'27"62d19'
Petruzzi,James Lac 22h6'0"49d35'
Petruzzo,Andrew Thomas Peg 22h4'35"10d27'
Petruzzo,Jeana Lynn Lyr 18h37'27"40d3'
Petry,Karen Ann Klopp Psc 1h26'14"20d17'
Petry,Kurt Cnc 8h0'21"11d3'
Petry,Oscar Lyr 19h21'0"30d23'
Petrylka,Alysyn J Tri 2h15'12"32d29'
Petschek,Nicholas Aur 5h1'53"48d51'
Petsikas,Nicholas Hya 8h59'58"-7d59'
Petska,Becky And 23h40'45"42d3'
Petta,Maurizio Umi 16h16'29"82d46'
Pettee Family,The Lyn 7h49'37"44d19'
Pettee,James Lombard Cep 22h6'1"54d12'
Petter,Eunice And 1h18'60"39d29'
Petter,Jane Cet 2h40'1"-12d17'
Pettersen,Richard Revie Cet 2h29'26"0d18'
Petti,Stephen William Dra 11h52'46"66d0'
Pettie,Penny Dee Cas 0h33'59"68d31'
Pettie,Steven Wayne Cep 23h28'34"70d57'
Pettifer,33",Honorable William Tri 1h58'36"33d47'
Pettifer,Joan Cyg 19h58'22"40d25'
Pettifer,William Ser 17h52'44"-13d39'
Pettinari,Benjamin Her 17h36'27"25d33'
Pettinato,Julia Joe Barbara JoAnn Vul 20h16'1"26d5'
Pettinato,Suzanne Lyr 18h29'0"30d41'
Pettinger,David Oph 17h7'0"10d12'
Pettingill,Emma & Andrew Cyg 19h43'0"29d48'
Pettis,Daniel Cet 2h18'29"6d25'
Pettis,David Richard Hya 9h30'40"-1d1'
Pettis,Susanna Irene Cam 12h37'1"80d22'
Pettit Y Jones,Fred Virgil Hya 8h59'0"-6d56'
Pettit,Athena Leo 11h46'54"22d18'
Pettit,Gilbert Dolan Lyr 19h22'40"38d52'
Pettit,Hailey Madison Uma 9h36'43"42d58'
Pettit,Karen & Tod Cyg 21h5'1"32d15'
Pettit,Levi Boo 14h10'0"41d47'
Pettit,Mark Lawrence Sex 10h33'1"1d14'
Pettit,Michael John Per 1h45'35"47d39'
Pettit,Rachel Bales And 1h28'27"34d59'
Pettit,Robert Calvin Per 3h12'16"48d52'
Pettit,Stacy Jo Aql 19h31'1"-8d19'
Pettit,Stephen R Cep 21h18'22"61d4'
Pettit,Steven D Aur 6h0'37"30d51'
Pettit,Steven Paul Cep 22h1'57"58d27'
Pettitdemaderaca, Juanita Jianoran- Oph 17h5'38"10d39'
Pettry,William "Bill" Aur 7h18'51"37d24'
Petts,Stephen John Uma 9h6'30"68d46'
Pettus Lmi 10h7'56"32d12'
Pettus,Al & Ruth Cyg 21h51'57"53d33'
Petty Love Star,John Neely Michele Sge 19h57'31"16d49'
Petty's Fireball,Lee Gem 7h52'33"31d14'
Petty,Alfred "Snooky"& Wilbert "Baby" Her 16h19'12"21d1'
Petty,Barbara Doler And 0h19'56"30d21'
Petty,Courtney And 23h14'31"40d22'
Petty,Don Uma 11h11'25"52d17'

Petty,G Shane
Boo 13h58'51"25d11'
Petty,Linda Ann
And 1h8'55"38d26'
Petty,Rebecca Jane
Lmi 9h58'34"30d17'
Petty,Rita
Eri 3h37'51"-3d25'
Petty,Sarah Kate
And 23h3'16"50d25'
Petty,Tom
Aur 5h1'44"48d35'
Pettyjohn,William J
Aur 5h28'54"38d38'
Petuch,Dr Edward James
Cmi 7h54'40"1d11'
Petza/Wolfgang
Ori 4h44'11"2d49'
Petzi
Ser 16h3'51"2d11'
Petzi Pens in a Pod
Boo 15h7'27"21d43'
Petzold,Nancy
And 2h25'32"42d23'
Petzoldt,Juergen
Uma 9h31'46"55d27'
Peueri,Simon Pietro
Aql 20h35'49"0d21'
Peugeot
Umi 14h53'44"65d52'
Peuster,Tilman
Gem 6h42'26"12d11'
Pewe Star,The
Lyn 8h19'0"41d36'
Pewett,Dee Anne
And 23h18'40"44d13'
Pewter,Nigel
Cep 23h8'38"61d25'
Peyer 1 03 1968, Laetitia
Per 4h5'7"36d49'
Peyravi,Maryse
Aur 6h2'59"31d30'
Peyrodes,Brices
Equ 21h0'30"8d33'
Peyron,Bruno
Peg 21h47'44"20d16'
Peyron,Christiane
Ser 15h54'52"18d49'
Peyron,Loïck
Peg 21h46'43"20d13'
Peyron,Oliver
Her 17h3'29"43d11'
Peyron,Stephane
Peg 21h46'1"28d18'
Peyronnet,Dr Gerard
Cma 6h56'37"-19d49'
Peyton
Cep 22h40'1"65d20'
Peyton (Bernie),B H
Aql 19h50'58"13d24'
Peyton Mountain Man, Big Gene
Gem 6h49'25"13d25'
Peyton,Diane
Aql 19h53'16"-1d28'
Peyton,Ellen
Lac 22h22'13"38d48'
Peyton,John
Hya 8h45'53"1d4'
Peyton,Trenten Anthony
Uma 10h20'53"72d8'
Pez-Fivel-Mar
Aql 20h14'44"0d17'
Pezhmon
Aql 19h31'0"12d32'
Pezone,Emanuele Serena Maria
Cet 2h42'1"-0d36'
Pezza,Joseph Gregory
Aur 6h8'54"46d22'
Pezzali,Marzia
Dra 15h19'36"58d37'
Pezzeca,Nicholas S
Cmi 7h32'1"10d50'
Pezzicola,Jr,Michael Louis
Boo 15h28'20"50d36'
Pezzullo,David
Her 15h52'51"46d10'
Pe Her 15h52'51"46d10'
And 1h0'24"36d32'
Pe And 1h0'24"36d32'
Del 20h12'1"12d6'
PF & Naomi's Bright Light Eternal
Cyg 19h22'60"54d59'
Pfaabstern
Ser 16h4'20"2d11'
Pfadenhauer,Patrick
Aql 19h57'36"15d38'
Pfaff,Donald
Lmi 10h15'53"31d58'
Pfaff,Juergen
Cnc 8h54'34"20d34'
Pfaff,William Wallace
Lac 22h38'33"55d54'
Pfankuch,Page Katelyn
Umi 13h46'33"71d57'
Pfannenschmidt, Christian
Her 16h40'49"28d2'
Pfannenstiel,Lisa
Lyr 19h5'21"37d37'
Pfannenstiel,Patrick & Eva
Mon 7h1'30"-6d26'
Pfanner,Waltraud & Anton
Uma 11h42'1"33d15'
Pfarrer Roland Herpich
Cmi 7h21'46"4d23'
Pfau,Jacob Michael
Lac 22h28'26"53d14'
Pfeffer,Andrew N
Aur 6h24'45"30d6'
Pfeffer,Edward Paul
Boo 15h0'59"11d38'
Pfeffer,Patrick
Per 2h20'58"58d32'
Pfeffer,Robert
Sgr 19h0'20"-26d2'
Pfeffer,Sr,Adolph A
Cma 6h48'38"-18d52'
Pfefferkorn,Karen Miller
Lyn 7h41'22"42d28'
Pfeffernusse,Darwin
Cma 7h12'31"-18d50'
Pfeifer McGrail Union
Cnv 12h10'28"45d12'
Pfeifer,Ella
Psc 23h7'49"1d20'
Pfeifer,Erika Lynn
Cas 0h58'46"63d60'
Pfeifer,Hans-Wolfgang
Tau 4h15'0"1d15'
Pfeifer,Janet
Lyr 18h39'48"46d33'
Pfeiffer,Audrey Josephine
Uma 11h0'54"48d5'
Pfeiffer,Ben
Crb 15h53'23"37d56'
Pfeiffer,Chuck
Sct 18h53'23"-5d34'
Pfeiffer,James August
Cep 20h48'0"68d4'
Pfeiffer,John Frank
Eri 2h43'42"-15d55'
Pfeiffer,Julia Elizabeth
And 2h11'32"38d33'
Pfeiffer,Karola
Sco 16h39'18"-44d19'
Pfeiffer,Lisa
Cyg 19h53'42"38d29'
Pfeifman
Ori 6h1'23"10d44'
Pfeil,Eberhard
Sgr 18h50'53"-20d1'
Pfeil,Emily Mae
And 0h3'43"46d53'
Pfennig,Sven
And 2h23'0"45d47'
Pfeuffer,Elizabeth
Lyr 18h52'37"30d7'
Pfile,Leroy
Aur 4h59'33"30d39'
Pfindel,Arlette Et Charles
Crb 15h29'1"30d26'
Pfister,Andreas
Lib 14h18'1"-22d54'
Pfister,Matthew C
Dra 19h29'0"60d41'
Pfister,Richard G
Aur 6h1'26"32d41'
Pfister,Robert William
Her 17h21'13"48d10'
Pfisterer,Thomas R
Per 1h40'20"53d55'
Pfitzenmeier
Lac 22h12'25"46d48'
Pfitzer,Haley Payton
Aqr 21h2'32"0d18'
Pfitzner,Birgit
Lyn 7h48'24"44d44'
Pflanz,Jr,Louis W
Aur 5h30'38"31d27'
Pflaumer,OD,Marvin L
Oph 17h18'13"12d13'
Pfleger,Danielle
Lac 22h7'33"46d53'
Pflieger,Julie & Robert
Cep 23h7'24"60d46'
Pflueger,John
Cet 1h8'40"-5d58'
Pfluger,Thomas
Mon 7h54'10"-1d34'
Pflugfelder's Star, Angela
Lac 22h54'31"55d41'
Pfoff,Dolores
Lyn 7h24'55"44d10'
Pfoh,Lawrence Joseph
Cmi 7h13'53"9d26'
Pfohl,Dmitri Wilson
Oph 18h20'16"8d58'
Pforr,Bruce Raymond
Her 17h11'11"44d49'
Pfrommer,Carl & Margaret
Peg 23h47'41"10d48'
Pfrommer,James Michael
Aur 4h59'1"50d26'
Pfrunder,Eric
Her 18h12'55"48d9'
Pfuetze,Gerhard
Cyg 20h53'46"58d26'
Pfundstein,Sr,Herbert E
Cep 21h46'42"60d6'
PG & RA
Uma 9h50'51"52d44'
PH2
Lyn 7h57'21"51d49'
PHAD-24
Sct 18h52'33"-9d24'
Phaedra Brasil
Ant 10h32'36"-34d49'
Phaenicia
Mon 8h8'26"-8d39'
PHAESS
Ori 5h7'18"0d51'
Phalen,David Lyle
Dra 12h58'42"71d37'
Phalen,Keller Frances
Cam 5h18'1"80d36'
Phalkin's Snow Angel
Lib 15h16'1"-23d15'
Phaneuf,Kim
And 2h33'36"49d10'
Phaneuf,Tyler Dominic
Sex 9h50'13"-6d50'
Phanie et Jean- Philippe
Lyr 18h34'0"39d53'
Phantom Spirit
Hya 10h12'25"-15d15'
Pharaon
Oph 18h3'15"11d52'
Phares,Sheri-Kelsi- Cory-Craig
Lyn 7h3'1"44d35'
Pharo,Stacey Marie
Cyg 19h29'48"31d40'
Pharoah,Anne
Cyg 20h49'34"48d54'
Pharos,Sam Jones
Ori 5h22'51"1d21'
Pharr,Daniel Lee
Cmi 7h30'47"8d53'
Phaysomphot Anousone
Cep 22h53'15"68d27'
Phee,James-Susan- Matthew & Gregory
Lyn 9h10'1"41d12'
Pheister
Sge 19h57'27"16d28'
Phelan,Angela M
Her 18h53'40"12d28'
Phelan,Daniel P
Cet 2h52'1"0d50'
Phelan,David W
Lac 22h19'58"46d58'
Phelan,Joanne
Com 12h7'31"28d40'
Phelan,John Patrick
Boo 14h48'60"50d4'
Phelan,Maureen Elizabeth Adrianna
Aqr 22h35'1"2d6'
Phelan,Monica L
Mon 8h6'22"-1d26'
Phelan,Tonia M
Ori 6h2'41"-0d28'
Phelps II,William L
Cma 7h16'32"-15d55'
Phelps,Anthony James
Ori 5h57'46"15d48'
Phelps,Audrey
And 0h22'26"40d44'
Phelps,Brightest Star of Amanda René
And 2h5'23"38d41'
Phelps,Charles R
Her 18h0'57"28d48'
Phelps,Corey Stephen
Ori 5h53'57"20d41'
Phelps,Debbie & Melissa Marshall
Lac 22h5'19"46d25'
Phelps,Dianne E M
Mon 6h35'43"11d19'
Phelps,Douglas W
Lac 22h8'51"50d16'
Phelps,Jane E M
Eri 3h43'58"-6d53'
Phelps,Jason M
Per 1h55'45"56d52'
Phelps,Jr,Bobby
Cyg 19h57'38"38d57'
Phelps,Julia Ann
Cet 3h0'24"-0d46'
Phelps,Kevin
Del 20h15'59"10d5'
Phelps,Larry & Charlotte
Lyn 9h8'0"45d11'
Phelps,Mont
Aur 5h49'56"50d13'
Phelps,Patricia
And 2h27'45"42d17'
Phelps,Robert Wayne
Lac 22h42'49"54d3'
Phelps,Scott D
Cet 1h25'15"-6d49'
Phelps,Tyler Benton
Dra 16h30'23"61d21'
Phelps,Zachary Scott
Aur 5h0'18"49d48'
Phene,LeRonica
And 23h25'34"47d10'
Phenomenal
Cam 3h30'1"52d57'
Phenomenal WCHS Class of 1944
Uma 12h15'24"56d6'
Phibbs,Christa Michelle
Aql 20h2'17"14d39'
Phil
Cep 21h20'4"80d8'B
Phil
Boo 15h15'1"40d41'
Phil & Hester's Star
Peg 22h55'53"25d25'
Phil & Irina
Crb 15h31'21"31d2'
Phil & Ken
Cyg 21h1'39"36d20'
Phil & Nicky's Love Star
Cyg 19h30'42"34d57'
Phil & Star
Dra 16h34'1"62d43'
Phil '93
Cet 1h40'52"-1d24'
Phil et Val
Per 1h29'13"53d33'
Phil's Brillance
Boo 14h25'15"39d16'
Phil's Dulcimer
Uma 8h32'55"71d34'
Phil's Paperclip Planet
Ori 5h50'43"19d8'
Phil's-Osopher
Lib 15h57'33"-8d14'
Phil,Our Star of Brightness & Strength
Her 17h24'46"30d46'
Phil-Arae
Mon 7h44'59"-6d32'
Philana
Leo 10h40'59"15d20'
Philando 40
Cyg 21h58'37"52d33'
Philbin,Alison Malia
Mon 7h4'30"1d10'
Philbin,Regis
Uma 13h48'0"51d52'
Philbin,Tara
And 1h28'32"39d32'
Philbin-Bowman, Jonathan
Mon 8h3'3"-2d2'
Philbrook,Kristen Ruth
And 23h49'57"44d1'
Philbrook,Matthew Bauer
Cam 3h56'0"78d25'
Philcon
Ori 5h54'38"13d52'
Phileo Eros Agape, Martin & Hauver
Ori 5h3'16"-1d41'
Philhower,Don
Ori 5h58'49"10d36'
Philia
Peg 23h27'11"17d6'
Philibert,Jean-Marie
Cam 6h40'1"67d60'
Philie-Leblanc,Thérèse
Per 3h16'0"41d44'
Philion,John
Ari 1h48'37"11d13'
Philip
Ori 5h4'57"8d50'
Philip
Dra 16h7'1"66d37'
Philip
Uma 9h59'33"60d10'
Philip
Ori 4h53'47"0d1'
Philip "The Light & Love Of My Life"
Per 4h22'50"50d21'
Philip & Susan
Cam 5h57'29"68d29'
Philip Aquarius
Hya 9h6'0"1d45'
Philip Eric
Aur 6h34'57"40d42'
Philip James
Lac 22h1'41"48d47'
Philip John
Ori 5h57'38"18d53'
Philip My Love
Her 16h50'55"33d22'
Philip Re
Per 3h14'24"50d9'
Philip's Isabella
Cyg 19h29'56"35d32'
Philip's Star
Cam 3h41'15"60d10'
Philip's Wish
Boo 15h9'44"38d25'
Philip,Bryan
Eri 4h50'50"-6d38'
Philip,Cariad
Uma 8h45'1"50d52'
Philipp I Love You Always 1994,Bob
Boo 15h14'40"41d8'
Philipp Simon
Tau 4h18'1"20d18'
Philipp,Carsten
Umi 13h55'39"74d0'
Philipp,Toshiyo
Cap 21h1'31"-27d9'
Philippa
Aur 4h49'57"40d30'
Philippe
Mon 7h56'42"-2d37'
Philippe
Aur 5h50'15"54d40'
Philippe
Del 20h42'31"12d7'A
Philippe & Irene Forever
Cyg 19h57'15"31d41'
Philippe Guy
Cnv 13h58'1"38d37'
Philippe Raphael
Ori 6h15'45"8d9'
Philippin,Michel
Cep 20h45'0"80d19'
Philipps,Caroline Ann
Uma 11h42'45"33d28'
Philipps,Robert
Cep 0h52'0"80d8'
Philips,Kristle Dawn
Peg 21h46'25"33d34'
Philips,Lia
Com 13h8'16"20d7'
Philips,Lil
Lac 21h55'36"42d8'
Philips,Loren
Aql 18h59'35"-2d47'
Philipson,Michele Cupolo
Uma 11h40'24"55d51'
Philivia
Her 16h17'31"23d51'
Phillingane,Kimberly Caudle
Aql 19h29'31"7d54'
Phillingane,Louis Ardell
Cet 0h51'18"-8d18'
Philliou,Marina Tsimenis
Cas 23h29'1"61d16'
Phillip
Cam 12h20'22"81d39'
Phillip C
Sex 9h55'16"2d9'
Phillip Lynn
Boo 14h56'59"44d1'
Phillipp,Peter
Ser 15h29'1"7d47'
Phillips Clan,The
Lyr 18h43'40"45d25'
Phillips Fish Tank For Kathy,The
Peg 22h21'1"25d5'
Phillips I
Hya 10h1'15"-16d41'
Phillips Of The Heavens,Princess Tania
Crb 16h17'26"37d32'
Phillips,"Phlips" Timothy F Frank
Her 16h34'30"33d55'A
Phillips,"Weeser" Louise Ellen
Her 16h34'30"33d55'B
Phillips,Alexandra Emily
Ori 5h50'53"12d11'
Phillips,Alma Osceola Beadle
Cet 2h35'0"-6d27'
Phillips,Amber
Lyn 7h17'0"50d2'
Phillips,Anita
Mon 6h19'0"4d31'
Phillips,Anita & Jim
Umi 15h5'57"67d43'
Phillips,Anne
Lyr 18h30'36"31d29'
Phillips,Annie M
Mon 6h26'36"-0d53'
Phillips,Barbara Ann
Cet 2h50'50"1d7'
Phillips,Becky Grace
Umi 13h7'38"72d58'
Phillips,Beverly Sue
Cas 23h41'60"65d13'
Phillips,Blakely Lauren
Cet 2h48'29"-1d17'
Phillips,Brian
Per 3h25'41"52d36'
Phillips,Brian Chatham
Lac 22h20'48"50d13'
Phillips,Brian Coye
Uma 12h22'43"61d28'
Phillips,Cheryl
Cas 0h53'30"73d42'
Phillips,Christopher Hallowell
Dra 12h1'34"68d54'
Phillips,Clifford Clayton
Her 17h38'44"21d30'
Phillips,Clint
Boo 14h8'49"40d14'
Phillips,Confrey
Ori 6h7'45"1d39'
Phillips,Daniel
Her 16h41'18"29d27'
Phillips,Daniel
Aur 4h46'53"34d42'
Phillips,Danny Eileen
Uma 10h21'19"70d43'
Phillips,Dave
Ori 5h56'1"14d20'
Phillips,David & Patti
Lac 22h30'33"56d38'
Phillips,Deb
Cyg 21h46'58"52d53'
Phillips,Deborah Lynn
Lyn 7h30'51"50d5'
Phillips,Dina
Boo 14h28'43"19d30'
Phillips,Donald B
Dra 14h2'0"68d11'
Phillips,Elizabeth Claire Fink
Peg 22h48'12"21d54'
Phillips,Elvis & Jean
Del 20h55'23"7d8'
Phillips,Emma
Lyn 8h20'58"41d3'
Phillips,Emma L
Peg 23h39'42"31d40'
Phillips,Eugene S
Aur 4h59'56"37d5'
Phillips,Eva
And 0h25'58"40d19'
Phillips,Francis
Lyn 7h4'59"50d56'
Phillips,Gary Lynn
Aur 6h2'33"32d57'
Phillips,Glen
Per 2h38'34"34d56'
Phillips,Glen L
Cet 2h52'32"4d33'
Phillips,Harvey Harold
Lac 22h1'56"37d34'
Phillips,Hellen & Bud 60th Anniversary Star
Crb 16h29'52"31d40'
Phillips,Howard B
Per 3h0'51"41d6'
Phillips,J Scott
Ari 2h36'57"20d7'
Phillips,Jacob Thomas Andrew
Aql 19h54'24"10d30'
Phillips,James E
Vul 20h56'29"20d28"
Phillips,James H
Ori 5h59'43"20d51'
Phillips,James H
Aur 6h5'11"31d49'
Phillips,James Noel
Ori 5h59'13"20d23'
Phillips,Janna
Psc 1h0'56"10d45'
Phillips,Jim
Dra 11h11'57"73d44'B
Phillips,Jo Anne Claire
Lyr 18h23'55"38d24'
Phillips,John & Muriel
Cyg 21h5'34"31d50'
Phillips,John F
Ori 5h6'0"1d45'
Phillips,Joseph
Cep 22h24'13"59d60'
Phillips,Joshua David
Aur 6h26'56"37d16'
Phillips,José LaCavalier
Uma 12h30'53"61d8'
Phillips,Justin Nicolas
Hya 8h46'30"-6d14'
Phillips,Kay Michelle
Eri 4h59'49"-4d42'
Phillips,Kendall
Aql 20h2'24"8d49'
Phillips,Kiki
Sct 18h52'36"-6d20'
Phillips,Lani & Jack
Sge 20h0'24"16d16'
Phillips,Laura
Cyg 21h32'35"38d13'
Phillips,Leona Anna Rasmussen
And 0h19'1"37d48'
Phillips,Liam
Dra 11h11'57"73d44'C
Phillips,Linda
Pic 5h0'54"-43d34'
Phillips,Lindsay F
Cyg 20h2'48"41d8'
Phillips,Lynnellen
Vul 20h15'41"23d29'
Phillips,Margaret Alexander
Lyr 18h34'1"35d53'
Phillips,Margaret Hansen
Lac 22h54'57"54d52'
Phillips,Margo Annette
Uma 11h23'17"38d18'
Phillips,Marshall
Aur 5h1'28"49d46'
Phillips,Martha
Dra 11h11'57"73d44'A
Phillips,MD,Clyde W
Per 2h41'35"34d23'
Phillips,Melissa
Cet 1h16'37"-11d46'
Phillips,Michael Alexander
Her 17h52'41"18d54'
Phillips,Morris & MaryEllen
Uma 11h22'0"32d11'
Phillips,Nicholas S
Her 16h49'0"41d11'
Phillips,Pat
Cet 1h14'45"-3d8'
Phillips,Pete & Doreen
Uma 9h15'29"43d6'
Phillips,Phyllis
Boo 14h38'11"11d25'
Phillips,Preston
Dra 12h3'47"71d31'
Phillips,Richard Aylmer
Her 17h34'51"24d59'
Phillips,Richard Henry
Per 2h58'38"50d23'
Phillips,Richard Hugh
Her 17h9'51"22d31'
Phillips,Robert
Aur 6h20'25"32d3'
Phillips,Roberta Stouch
Lyr 18h56'17"32d8'
Phillips,Roberta Serena
Lyr 18h17'44"42d29'
Phillips,Roxanne Marie
Mon 7h19'28"-6d26'
Phillips,Sawyer Anthony
Lac 22h12'36"48d6'
Phillips,Shawn
Aqr 23h37'41"-6d13'
Phillips,Shirley Beach
And 23h5'27"37d55'
Phillips,Sr Mary
Cas 0h26'54"54d36'
Phillips,Sr,Rev Lee Allen
Ser 16h2'0"-3d20'
Phillips,Sr,Thomas Charles
Ori 5h50'14"12d16'
Phillips,Stephanie
Tri 2h15'53"30d4'
Phillips,Stephen & Jamie
Eri 3h41'14"-5d50'
Phillips,Susanne Carol
And 23h32'28"41d6'
Phillips,Suzan R
Cas 0h30'33"58d19'
Phillips,Suzanne Hilary
Oph 18h4'55"11d39'
Phillips,Tamela Lyn
Cma 6h28'36"-15d40'
Phillips,The Man of Bosham-Roy
Ori 6h7'1"1d59'
Phillips,Thompson
Ser 15h16'24"24d58'
Phillips,Tim
Uma 11h11'32"52d15'
Phillips,Timothy Allen
Per 2h58'53"34d13'
Phillips,Tom Paxton
Her 18h18'51"28d41'
Phillips,Traci Lee
Lyr 18h53'45"40d52'
Phillips,Trae Marcus
Lac 22h40'53"54d18'
Phillips,Tyler
Mon 7h3'57"-1d1'
Phillips,Vince
Aur 4h58'39"34d39'
Phillips,Wendy
Cyg 19h34'22"33d37'
Phillips,William
Per 2h57'35"32d57'
Phillips-Grego,Gregory Todd
Oph 16h57'24"11d4'
Phillips-Hallinan, Debbie Gayle
Cet 1h7'34"-1d7'
Phillips:Kind & Gentle Spirit,Michael
And 23h39'56"42d21'
Phillipstar
Cet 0h37'16"-5d10'
Phillius
Sco 17h20'26"-38d3'
Phillpot,Lynne Margaret
And 0h2'0"47d25'
Philly Ann
Mon 7h17'25"-5d42'
Philly-Boy
Per 3h30'0"51d2'
Philo,Andrea
Lyn 8h16'58"39d21'
Philo,Nelson & Dorothy
Sge 19h55'59"16d27'
Philo,Patricia A
Leo 9h20'18"17d10'
Philo,Patricia Lee
Cyg 20h22'18"37d58'
Philomena,Jacquie
Cas 0h15'0"47d2'
Philopena,Amy Marie
And 2h34'34"50d23'
Philotess
Umi 16h6'39"72d23'
Philounat
Per 1h35'1"53d40'

Philp,Alice
Cas 0h23'26"71d8'
Philpot,Gifts of Love Hannah & Bradley
Lyr 18h41'41"38d12'
Philpot,Jacqui-Anne
Car 7h28'51"-58d38'
Philpot,Rebecca Lynn
Eri 3h19'0"-11d48'
Philpot,Sherby Renee
Ori 5h52'25"20d38'
Philpott(Faithfully), Joseph Stanley
Her 17h26'56"38d30'
Philtricia
Del 20h52'43"7d35'
Phinney,Donna Patricia
Oph 16h54'1"-28d9'
Phinorbil Blessings Of Family Love
Cep 0h35'19"78d22'
Phip's Star
Aql 20h6'1"1d30'
Phipps,MD,Henry K
Oph 18h27'25"8d17'
Phipps,Michelle L
Cma 6h54'3"-19d30'
Phjab La Stella Eterno
Cyg 21h41'39"31d43'
Phoebe
Lac 22h49'0"56d47'
Phoebe
Cas 1h58'44"58d20'
Phoebe
Cet 0h21'54"-8d13'
Phoenix
Ari 2h6'32"25d6'
Phoenix
Uma 9h39'20"58d8'
Phoenix Aimée
Cet 1h43'41"1d25'
Phoenix Computer
Pho 0h37'21"-41d55'
Phoenix Ice Star,The
Sct 18h41'34"-9d53'
Phoenix,Judy & David
Lyr 18h42'51"42d33'
Phoenix,River
Pho 2h15'18"-42d42'
Phoenix,The
Lyn 8h12'18"51d38'
Pholly
Cam 3h21'59"61d20'
Phommachit,Dala
Peg 22h24'20"29d49'
Phonekeo SaySongKham
Tri 2h5'0"30d1'
Phonsine
Cyg 21h50'43"53d12'
Phoo
Uma 9h50'27"58d7'
Phrosa A
Cyg 19h57'50"38d14'
Phrygia
Cnv 13h32'39"40d54'
Phuc,Paul & Patricia Ann Tran
Aql 19h43'10"11d8'
Phuong,Em
Cas 0h3'0"62d35'
PhuongHoang
Mon 7h40'1"-1d44'
Phylbert Elaine
Lyr 19h20'15"38d39'
Phyliss Shines Brightly
Lyn 8h53'18"45d13'
Phylissia
Cyg 20h29'51"37d52'
Phyllis
Lyn 8h22'17"43d31'
Phyllis
Cet 2h46'15"5d46'
Phyllis
Sex 9h52'0"5d12'
Phyllis
And 0h56'0"38d44'
Phyllis
Cyg 20h8'51"40d51'
Phyllis & Dan
Boo 14h30'0"19d3'
Phyllis & Giorgio
Lyr 18h48'23"35d31'
Phyllis & Leland
Aql 20h35'16"0d9'
Phyllis Anne
Cam 12h55'32"77d4'
Phyllis Douglas Star, The
Peg 21h23'33"22d59'
Phyllis Gail
Oph 18h29'35"6d44'A
Phyllis M B
And 23h44'0"45d3'
Phyllis!
Cas 1h0'32"63d44'
Phyllis(Mucheon's)
Lyr 18h40'1"42d26'
Phyllis-Nut #1
Cyg 21h50'24"37d52'
Physical Ed
Lac 22h2'58"46d16'
Pi
Col 6h35'2"-33d55'
Pi
Boo 13h57'0"17d47'
Pi,Chloe
Mon 6h58'46"10d31'
Pia
Peg 23h33'35"10d38'
Pia Alexandra
Cet 0h1'47"-10d30'
Pia Aura
Cap 20h58'40"-15d59'
Pia Caj 90
Dra 15h44'44"62d15'
Pia Maria Kathleen
Her 17h13'59"47d54'
Pia Stolz-One Dream
Tau 5h45'47"20d25'
Pia Urioste de Vidal
Cyg 19h55'1"58d25'
Pia's Piece Of Heaven
Lyr 18h21'25"47d24'
Pia,Doug
Ori 5h26'36"-0d15'
Piacentini "Ann's Star",Ann Marie
Lyn 7h54'11"43d42'
Piacentini,Gloria I
Cam 6h2'22"60d53'
Piana
Cas 23h23'59"58d12'
Piana,Salvatore e Giusi
Vel 9h43'49"-44d24'
Piana,Stefano
Cmi 8h4'35"1d30'
Piangelawi
Lib 15h2'1"-3d39'
Pianin,Sidney "Sid"
Dra 16h53'58"68d29'
Pianissimo
Gem 7h10'57"16d39'
Piantoni,Nadia
Cet 2h33'32"1d8'
Piaskowsky,Jill Meryl
Aql 19h53'0"12d47'
Piatek,Emilie Ilse
Lyr 19h23'22"38d27'
Piatkowski,Anna Christyna
Tau 4h39'23"1d51'
Piatt,Howard Scott
Aur 6h2'0"37d23'
Piazza,Barbara
Lyr 18h40'27"39d38'
Piazza,Bryan Alexander
Uma 9h58'24"61d18'
Piazza,Jennifer
Sge 18h58'47"19d11'
Piazza,Joseph & Annie
Ser 15h21'24"21d1"
Piazza,Judy
Cyg 19h30'19"32d24'
Piazza,Lynda-Lee
Peg 0h4'39"14d22'
Piazza,Margaret Monroe
Cyg 21h30'0"42d46'
Piazza,Marie Antonia
Umi 15h35'25"68d26'
Piazza,Michael A
Oph 17h17'21"10d4'
Piazza,Nicholas Erik
Uma 9h3'12"62d16'
Piazza,Nicole Ashley
And 23h40'56"38d30'
Piazzale,Karen
Cas 0h48'49"61d37'
Pica,Jacqueline
Lyr 18h55'0"41d40'
Pica,Jim & Muriel
Cyg 21h53'57"52d39'
Pica,Sara
Lyr 18h37'10"27d60'
Picard,Betjski
Cam 7h4'45"82d51'
Picard,Cody Jacob
Vul 19h21'59"27d45'
Picard,Lara Suzanne
Cas 1h2'56"55d44'
Picard,Patrick
Vul 19h21'59"27d11'
Picard,Paul
Psc 1h26'30"31d40'
Picard,Scott
Sex 10h44'35"4d19'
Picard,Stéphane
Ori 5h58'41"14d53'
Picard,Sylvie Marie Thérèse Contin
Oph 18h5'54"11d28'
Picardi,Michael J
Cyg 20h23'0"40d6'
Picarello,Frances & John
Crb 15h28'0"31d30'
Picarello,Melanie J
Sco 17h52'19"-30d3'
Picaro
Gem 6h51'17"18d51'
Picascia Family,The
Del 20h50'16"8d55'
Picayo,Lydia Gloria
Cyg 20h1'50"40d58'
Picayo,Mario
Cyg 20h2'30"40d18'
Piccinini,Anthony
Umi 10h54'42"40d40'
Piccinini,Irene "Pio"
Uma 10h43'49"40d11'
Picco,Elena
Uma 11h55'31"32d17'
Piccolina
Her 17h20'1"40d24'
Piccolo,Ambrogio
Lmi 10h2'16"32d10'
Piccolo,Andrea
Per 3h0'0"31d26'
Picconi,Harvey
Aur 6h56'11"44d24'
Picerno,Antonello
Pup 8h8'30"-28d27'
Pich,Melinda Katherine
Cyg 20h19'45"30d21'
Picha,Dean
Lmi 10h33'38"31d30'
Picha,Victoria Jean
Lyr 18h31'1"30d53'
Pichauant-Ruty,A
Lac 22h16'0"55d22'
Pichcuskie,Donna
Vul 20h16'58"23d34'
Piche,Corey A
Dra 16h29'53"69d27'
Pichette,Jacques- Bertrand
Umi 14h31'1"65d43'
Pichette,Jules
Dra 17h27'45"50d54'A
Pichierri,Melissa
Aql 19h13'49"13d38'
Pichler,Alexander
Gem 7h15'41"24d27'
Pichon,Bill
Dra 11h48'22"71d13'
Pichot,Bernard
Del 0h19'41"11d26'
Piché,Estelle Blais
Cas 0h53'32"74d51'
Piché,Martine
Tau 4h34'57"15d18'
Piché,Paul
Leo 9h57'30"19d36'
Piché-Larose,Maude
Umi 16h55'50"75d57'
Picini,Jordan Kyle
Peg 23h0'0"31d51'
Picinich,Matt
Aur 6h27'23"30d2'
Pick,Beatrice
Lyr 19h23'39"31d16'
Pick,Greg
Aql 19h8'20"3d47'
Pick,Jr,Colonel Lewis Andrew
Lac 22h26'25"55d13'
Pick,Mel
Her 16h53'59"28d0'
Pick,Sabine Nicole
Cru 12h37'56"-61d42'
Pickard,Barbara Branneky
Cyg 19h19'12"44d36'
Pickard,Gary William
Cep 21h31'22"58d29'
Pickard,Michael W
Cam 6h6'44"61d12'
Pickel Love of My Life Larry Wayne
Sex 10h32'34"4d26'
Pickel,Father Damian
Her 18h1'30"31d13'
Pickell,Sherry
Lac 22h39'32"54d16'
Pickelny,Ivan
Per 1h47'41"48d44'
Pickens,Carl M
Boo 14h7'20"33d60'
Pickens,Jr,Dr John E
Crb 16h17'37"28d49'
Pickens-Kriebel,Heide
Lyn 7h30'54"36d56'
Picker Melvin Susan Heidi Chris
Peg 22h35'39"8d11'
Picker,Allison H
Peg 23h1'54"33d45'
Pickering
Aql 19h56'0"14d2'
Pickering,John T
Her 17h16'37"21d46'
Pickering,Rosalinda
Cas 1h36'58"68d26'
Pickerings,The
Cyg 20h44'0"42d59'
Pickett,Charles M
Cet 1h7'1"-5d19'
Pickett,Hilton Pamela & Amanda
Peg 23h19'18"33d44'
Pickett,John Collins
Her 16h28'0"31d41'
Pickett,Julia Yohn
Cas 0h9'28"64d50'
Pickett,Keary Elison
And 1h4'1"40d45'
Pickett,Leland A
Mon 6h14'39"-10d36'
Pickford,Jack Aultman
Hya 10h12'15"-15d48'
Pickford,Michael Aloysius
Umi 16h57'31"79d24'
Pickford,Michael Robert
Cyg 19h26'55"34d44'
Pickford,Patricia Ann
Umi 16h23'24"70d31'
Pickford,Sacha
Lyr 18h14'1"32d43'
Pickford,Sean Thomas
Umi 16h31'58"78d31'
Pickle,Douglas J
Her 18h12'42"30d22'
Pickles & Rita
Mon 6h33'35"-6d46'
Pickrell,Frederick
Oph 16h50'53"10d7'
Picolla,James R
Cam 13h15'0"80d51'
Picone,Alfred J
Uma 11h49'39"40d36'
Picone,Raymond
Cep 0h17'35"73d20'
Picone,Rose
Cas 1h19'0"60d0'
Picotte,V H
Uma 9h22'52"62d6'
Picozzi,Elizabeth
And 1h0'1"37d9'
Picsou Forever
Uma 10h37'20"52d14'
Pidock,Kathleen
Equ 21h19'40"3d39'
Pie In The Sky
Cas 2h32'34"63d48'
Pie in the Sky
Cas 0h3'35"59d55'
Pie Thomas
Lac 22h38'45"55d23'
Pie's Star
Cyg 21h58'44"52d46'
Pie-Pie
Hya 8h59'36"3d52'
Piearcy,Eric Tyler
Per 3h12'13"48d41'
Piearcy,Katelyn
Tri 1h52'19"25d51'
Piechotta,Ludger
Mon 7h54'42"-1d36'
Piecko,Paul & Helen
Cyg 20h53'58"40d26'
Pied Piper Of Pandwyck
Uma 8h46'24"53d49'
Piedimonte,Kate Paduano
Gem 6h33'8"13d5'
Piednoël,Rolande
Cam 6h54'12"64d32'
Piehl,Lisa & Richard Armijo
Uma 12h35'40"56d12'
Piehler,J Evan Miles
Uma 9h45'3"47d17'
Piekielniak,Jason W
Boo 15h6'30"24d51'
Piel
Her 16h30'25"33d23'
Piel,Isaac
Aur 6h1'57"30d4'
Piel,Jürgen
Cam 5h37'26"70d44'
Piel-Brunner,Karin
Aqr 22h5'47"-7d43'
Pielmeier,Michelle A
Cas 1h45'18"73d28'
Piencipessa Condiotto
Dra 9h26'18"81d10'
Piening,Kimberly
Cyg 19h44'0"38d37'
Pienta,Aubrey Devonné
Sco 17h7'14"-38d58'
Pieper,Annegret
Peg 22h0'0"3d44'
Pieper,Eric James
Ser 17h53'45"-10d18'
Pieper,Jason C
Oph 17h8'41"-24d9'
Pieper,Meredith Reynolds
And 23h13'1"36d26'
Pieper,Rex
Her 18h5'1"31d29'
Pieper,Sandra
Lyr 18h24'56"47d25'
Piepgras,Don
Aql 19h30'37"10d34'
Piepkorn,Lara
Leo 11h23'19"-0d55'
Piepkorn,Sven
Sco 17h53'41"-30d31'
Piepul,Paty
Cnv 12h51'10"40d5'
Pier
Ori 5h6'18"1d6'
Pier Stella del Marinaio
Peg 22h28'17"30d49'
Piera
Cet 2h6'0"-2d27'
Piera L C Star
Gem 7h8'42"21d32'
Pieraccini,Luigi
Cep 23h29'23"64d58'
Pierangelo,Elisabetta E
Cet 1h32'1"1d40'
Pierannunzi,Kathy
Peg 22h57'41"30d58'
Pierard,Simone
Crb 15h56'18"35d46'
Pierard,Suzanne L
Cas 1h6'42"60d54'
Pierce & Family Lew Betty Bill,Gwen
Gem 7h12'57"24d47'
Pierce Grading & Pipe Line,Co,Inc
Hya 9h13'50"1d38'
Pierce Healer Of Body & Heart,J
Hya 8h12'13"3d11'
Pierce Star,The Jean
Cyg 20h10'55"38d59'
Pierce,Alice B
Cas 23h42'41"62d2'
Pierce,Andrew David
Aur 6h51'21"35d32'
Pierce,Andrew John
Vir 12h59'44"7d49'
Pierce,Berlyn
Ori 5h53'42"1d20'
Pierce,C Andrew
Cma 7h13'35"-13d8'
Pierce,Cindy
Aql 19h31'1"14d23'
Pierce,Damon S
Tri 1h45'38"25d42'
Pierce,Daniel Robert
Aql 19h50'11"14d23'
Pierce,Daniel W
Eri 3h30'42"-1d56'
Pierce,David Edward
Lib 14h59'38"-10d37'
Pierce,Geoff
Aqr 22h2'54"-8d25'
Pierce,Ginger
And 23h39'31"47d16'
Pierce,Hans G
Her 16h46'12"33d11'
Pierce,Harold E
Uma 9h16'29"50d59'
Pierce,Hunter M
Tau 4h1'42"11d25'
Pierce,Justin
Her 18h8'43"40d19'
Pierce,Kristina
Cyg 20h36'41"38d29'
Pierce,Lonnie
Uma 12h43'57"62d19'
Pierce,Matthew Ryan
Oph 18h16'58"8d16'
Pierce,Michael
Cma 6h45'22"-15d13'
Pierce,Nicholas
Aur 6h28'25"37d57'
Pierce,Ninette H
Ori 5h23'46"1d51'
Pierce,Pattie Lee
Boo 15h6'17"48d23'
Pierce,Randy Lee
Per 1h48'41"50d18'
Pierce,Rev Anne
Cas 0h16'15"63d34'
Pierce,Robert
Umi 15h38'32"77d34'
Pierce,Robert & Regan
Ori 5h57'12"15d24'
Pierce,Robert G
Tau 5h44'53"28d34'
Pierce,Robin Griffin
Eri 2h56'56"-16d39'
Pierce,Ron
Boo 15h39'28"41d56'
Pierce,Rory Michael
Cet 3h9'52"4d56'
Pierce,Sarah
Cyg 20h6'11"41d5'
Pierce,Sarah
Cas 3h1'59"57d28'
Pierce,Sarah Elizabeth
Vir 12h30'16"-8d24'
Pierce,William
Her 17h22'26"44d7'
Piercy,Hertha Anna Augusta Gleich
Lyr 18h46'20"42d33'
Piercy,In Memory Of Dorothy "Chubby"
Lyr 18h40'14"41d30'
Pierini,Dr Kenneth W
Lac 22h34'14"52d55'
Pierini,Maria
And 23h21'1"48d14'
Pierman,Jean
Uma 10h19'1"50d58'
Piermatti,John Michael
Per 2h59'43"38d26'
Piermatti,Laura Ann
And 23h15'52"49d17'
Piermatti,Valerie Jacqueline
Lyr 19h14'0"37d42'
Piero
Cam 6h43'53"67d36'
Piero,Laura E
Cyg 21h37'48"40d14'
Pieronek
Dra 15h14'54"65d13'
Pieroni,Julia
Lyn 8h26'36"42d43'
Pieroni,Michelle M
And 1h0'41"39d0'
Pierpoint's Angels
Cep 21h1'1"61d51'
Pierpoint,Heather
Uma 11h48'1"38d1'
Pierpont,John & Jinny
Cnv 13h56'1"47d33'
Pierra,Patrick
Ori 5h56'42"15d5'
Pierrard,Helen Ann Huitt
Lyn 8h4'55"35d10'
Pierre David
Vir 13h31'13"-21d25'
Pierre Denis
Dra 11h56'60"71d9'
Pierre et Armelle
Mon 7h49'19"-4d23'
Pierre et Christine
Vul 19h19'51"26d30'
Pierre et Marie Christine Gr
Her 16h10'21"48d15'
Pierre et Véronique
Uma 13h27'44"61d31'
Pierre G
Peg 0h4'36"20d39'
Pierre Nicolas
Peg 21h59'12"21d27'
Pierre,Cameron Isiah
Lyn 7h25'41"44d42'
Pierre-Auguste,Léopold
Lyr 18h38'39"34d8'
Pierre-Edouard Et Angelique
Lyn 8h1'49"50d43'
Pierret,Rita
Lib 15h8'10"-22d25'
Pierrette-Miche-Aile
Per 4h0'53"36d36'
Pierrick
Aql 20h4'18"1d49'
Pierrick
Lac 22h51'0"53d52'
Pierro,Jeannette
Lyn 9h16'57"41d3'
Pierron,Valérie
Cnv 13h19'54"38d54'
Piers
Her 16h5'21"40d53'
Piersall,Gladys "Granny"
Eri 5h7'28"-5d24'
Piersall-Kaiser, Jennifer
Cet 2h59'58"7d59'
Pierson's Passion
Gem 7h2'37"35d24'
Pierson,Andrew
Aur 7h18'0"41d17'
Pierson,Carol
Cet 0h5'33"-10d8'
Pierson,Dorothy
Oph 18h8'20"13d11'
Pierson,Géraldine
Cas 1h39'49"58d25'
Pierson,Jack & Robbie
Crb 15h55'58"27d34'
Pierson,Jessica
Equ 21h6'23"3d24'
Pierson,Jr,William P
Cnv 12h10'31"39d27'
Pierson,Phillip
Cmi 7h55'1"8d28'
Pieruccini,Deena R
And 2h30'52"40d20'
Pieruccini,Lee M
Aur 6h0'58"45d41'
Pieruccini,Lucas L
Per 2h37'23"43d48'
Pies My Only Love, Jennifer S
Sge 19h53'20"16d15'
Pietarila,Elizabeth
Cas 0h4'42"58d12'
Pietig,Ray & Mary Fran
Cyg 19h30'29"36d11'
Pietila,Kari P
Cas 2h9'53"67d40'
Pietrantoni,Natascia
Pup 7h56'28"-26d45'
Pietrasz,Edward Paul
Per 1h50'22"56d22'
Pietro
Lep 5h22'59"-20d33'
Pietro
Ori 5h52'45"1d13'
Pietrois-Chabassier, Thomas
Ser 15h16'46"18d19'
Pietrull-König,Helen & Dr H König
Dra 17h4'25"72d30'
Pietruszewski,Brian
Her 17h4'23"41d10'
Pietrzak,Connor Thomas
Her 16h20'12"24d45'
Pietrzak,Daniel J
Per 3h3'35"47d49'
Pietrzak,Eric A
Aur 6h27'0"35d5'
Pietrzak,Moochie-Face
Aql 20h4'20"1d19'
Pietrzyk,Rene Alexander
Aqr 22h5'54"2d16'
Pietsch,Ulrich
And 23h20'1"38d13'
Pietszak,Eleanor T
Cas 1h3'39"55d47'
Pietta,Nicolletta
Cas 2h14'18"50d57'
Piette,Gilbert
Cep 22h53'49"63d3'
Pietzie Igel (Jutta Freye)
Mon 7h35'39"-0d31'
Pievac,Samuel Dylan
Her 18h26'43"12d56'
Piezzemolo,Rosetta
Lyr 19h16'34"26d41'
Pif
Cam 7h13'16"68d13'
Pifer,Douglas
Sco 17h21'25"-38d51'
Pig
Oph 17h51'34"12d18'
Pig's Nose
Cmi 7h42'16"1d33'
Pigatti,Carla
Cyg 20h43'38"46d33'
Pigeon Star,Radcliffe J
Peg 22h36'57"10d11'
Pigeon,Jeffrey Scott
Boo 14h33'54"48d5'
Piggy
And 23h29'37"46d15'
Piglet
Mon 6h24'0"-10d8'
Pigmeat
Aur 7h24'54"38d35'
Pignataro,Augustus
Aql 19h0'31"13d1'
Pignatelli,Luca Alexander
Cam 9h9'54"78d28'
Pignault,Emmanuel
Peg 23h46'15"27d59'
Pignoulette
And 23h2'28"40d36'
Pigott,Alexandra Lila
Lyr 18h35'26"36d27'
Pigott,Mary Zita
Uma 11h4'35"58d21'
Pigott,Steve
Cam 5h23'35"68d42'
Pigott,Sue Strickland
Vul 20h56'48"28d21'
Pigott,Thomas Galvin
Per 2h53'1"40d3'
Pigozzi,Sadie
Cyg 19h42'11"41d52'
Pihko,Topi
Cam 6h14'0"80d2'
Pihlar,Oliver John
Per 1h50'16"48d15'
Pihlgren,Geoffrey
Oph 18h26'49"7d36'
Pihota,Richard "Ricky"
Aur 7h3'0"40d29'
Piippo,Ouli-Minna & Mannel Elgorriaga
Cyg 19h27'14"38d38'
Pij,Jashdhai Maganbhai Patel
Uma 8h54'12"50d23'
Pijper,Kinderkoor Jody
Vul 19h19'33"25d37'
Pik,Ying
Cyg 20h26'0"46d15'
Pike School-1926,The
Umi 16h29'37"78d40'
Pike,Blair E
Gem 6h0'53"26d44'
Pike,Elizabeth Watson
Mon 7h59'56" 2d47'
Pike,Frances "Grammie"
Peg 22h40'34"29d27'
Pike,Gavin
Her 16h54'38"38d15'
Pike,Georgette J F
Ori 5h54'20"17d18'
Pike,Glyn
Cyg 20h21'0"39d42'
Pike,Jesse Joseph
Aur 6h21'43"41d11'
Pike,Joan
Cnv 13h54'1"38d11'
Pike,Julia
Cas 0h26'52"61d2'
Pike,Sawyer Carson
Mon 8h7'6"-3d16'
Piken,Ablert William
Cet 1h2'25"-1d57'
Pikora,Joseph J
Dra 16h55'1"66d13'
Pikovsky,Adrien
Lib 15h36'19"-20d23'

Pikus,Nathanial Paul
Ori 6h5'14"8d25'
Pikuzinski,Rudy
Dra 18h55'59"70d26'
Pilant,Tony A
Her 17h49'12"18d51'
Pilarczyk,Kimberlee
Cas 1h34'56"77d24'
Pilarowski,Karen C
Cam 4h1'20"67d30'
Pilat,Henryk
Cep 20h30'35"61d46'
Pilato,Donna
Cas 23h13'42"62d9'
Pilcher,Ann Jackson
Cas 1h58'58"65d16'
Pilcher,Kathy
Ori 5h34'40"-1d57'
Pilcher,Ron
Ori 5h39'46"-0d56'
Pile,Frank
Lyn 7h48'56"35d58'
Pileggi,Kathryn Marie
And 1h35'52"38d37'
Pilet,Florence
Peg 23h5'52"30d12'
Pilet,Gerard
Her 18h0'36"28d51'
Pilgram-Rupprecht, Verena
Cep 22h43'14"59d45'
Pilgrim,Alexandria Elizabeth
Gray
Umi 15h30'37"68d27'
Pilgrim,Anna Campbell
And 1h56'1"38d32'
Pilgrim,Johanna
Cyg 19h34'17"35d25'
Pilgrim,Katherine
Cep 20h56'1"70d52'
Pilgrim,Mark T
Lac 22h0'18"48d54'
Piliero,Giuliano
Boo 14h27'25"27d4'
Piligra,Bryan Thomas
Boo 14h6'35"53d9'
Pilipchuk 4-8-94, William &
Susan
Cnv 13h14'0"50d34'
Pilipski
Cnv 13h47'29"35d27'
Pilipski,Ian
Per 1h49'35"50d18'
Pilipski,Mary
Aql 19h43'45"11d4'
Pilita Forever
Cyg 20h0'0"40d4'
Pilkington w/Joe,Ben &
Abigail Star,The Carol
Eri 5h1'49"-4d38'
Pilkington,Michael
Ori 4h56'28"15d26'
Pilkinton,Tina
Ori 5h38'57"-6d25'
Pilla III,Joseph John
Dra 17h12'40"69d36'
Pillard,Lois Louise
Woodworth
Cyg 19h26'1"31d24'
Pillay's Star,A J
Ant 10h33'24"-36d3'
Piller,Amanda "Emunah
Chaya"
And 23h22'57"41d48'
Piller,Jakob
Cmi 7h21'52"5d18'
Pillery,Jerry Lee
Hya 8h16'41"-6d56'
Pillor,Michelle
Peg 21h56'24"28d10'
Pillsbury,Chris & Ned Baker
Crb 15h23'60"30d29'
Pilltibaer I
Dra 10h25'53"80d21'
Pilo
And 0h45'32"40d59'
Pilo-Pilo
Mon 8h3'46"-8d30'
Pilon,Luc
Per 3h8'19"40d11'
Pilon,Stéphane
Lib 15h19'11"-24d26'
Piloni,Dario
Dra 11h28'43"78d11'
Pilot,Carl & Patricia
Cyg 19h24'36"48d50'
Pilot,Jacob Daniel
Crb 16h20'43"32d7'
Pilot,Sylvia
Aqr 22h7'11"-5d33'
Pilot,Zechariah Tyler
Aur 6h51'39"35d32'
Piloue
Com 13h32'0"22d25'
Pilsbury,Susan
Lyn 8h38'45"43d22'
Pilson Family
Cas 0h36'55"60d20'
Pilsucki's Star Of Dreams,Rob
Hya 8h13'57"1d2'
Pilt "The Star of
Happiness",Sandra
Cas 0h22'14"72d27'
Piltz,Veronika
Gem 6h42'33"14d12'
Pilz,Horst
Cmi 7h17'22"7d41'
Pim's Night Light, Jacei
Cyg 20h4'40"41d6'
Pim-85th Birthday Star Lorin
C
Com 13h17'33"22d13'
Pimblett's Little Star Ronnie
Umi 15h22'31"66d50'
Pimbley,Georgia Jade
Vir 14h55'13"7d25'
Pimenta,Tam & Goncalves
Cas 2h0'34"61d8'
Pimentel,Marcia
Lyr 19h13'11"41d16'
Pimienta-Bey,José Vittorio
Her 17h19'58"18d54'
Pimm,Bryan
Equ 21h2'38"3d21'
Pin
Aur 5h30'26"30d26'
Pina USMC,Joseph H
Aql 19h53'36"0d40'
Pina,Horace M
Cmi 7h56'50"4d50'
Pina-Daniela-Chiara- Franco-
Fasanaro
Pic 5h4'6"-47d10'
Pinafore & L K DuBarry's
Lac 22h0'41"51d52'
Pinalina
Cam 3h30'59"60d45'
Pinard "My Mom",Joan Mary
Aur 6h15'14"48d51'
Pinard,Brian Scott
Aql 20h4'54"0d3'
Pinaris,Stephen
Cmi 7h59'0"4d40'
Pinch
Dra 20h22'58"68d32'
Pinch We Love You C&B,
Robert McCall
Aql 19h54'33"10d59'
Pinch,Paul
Cep 22h54'16"56d60'
Pinchuk,Judy
Sct 18h44'5"-7d28'
Pinciotti,Monica M
Cnv 12h54'24"42d51'
Pincott,Charles & Sandra
Cyg 21h32'13"42d38'
Pincus
Cet 3h15'31"1d47'
Pincus 8
Peg 22h33'32"2d27'
Pincus Accounting Expert,A
Per 1h43'20"53d4'
Pincus,David
Aur 6h3'24"36d53'
Pincus,Suzanne
And 23h31'51"41d43'
Pinder,Edwin
Ori 5h53'31"18d9'
Pinder,Star of David
Uma 9h2'1"53d53'
Pindilli,Elizabeth Anne
And 23h24'34"42d30'
Pine,Sammi Jo Bubba
Aur 5h4'1"40d51'
Pine-50 Musical Years, May &
Irving
Lyr 18h31'27"30d32'
Pineau,Pierre
Uma 11h25'19"32d13'
Pineau-Valencienne, Didier
Leo 9h24'26"19d12'
Pineda,Liliana Maria Gonzalez
Nor 16h27'6"-44d42'
Pineda,Martha Bertha
Sarmiento
Peg 22h8'1"3d4'
Pinelle,Michelle Ann
Lyr 18h37'35"27d31'
Pinelli,Betty
Mon 7h48'27"-3d9'
Pinelope Marie
Lyn 7h31'41"44d21'
Pinero,Moises & Elsie
Lmi 10h4'0"30d50'
Pineschi,Paola
Cet 1h8'59"0d36'
Pinezic,Valda
Uma 10h56'57"53d30'
Ping
Aur 5h55'19"38d39'
Ping,Hui
Umi 15h27'36"77d38'
Pingree,Alyson Rose Mikaela
Lyr 18h59'11"26d1'
Pingrey,Carson Frederick
Dra 19h54'1"70d3'
Pinhas' 80th,The Maxi Star-
Max
Aqr 23h7'57"-6d37'
Pinhas,Victoria
Vul 20h19'22"25d48'
Pinheiro,John Christopher
Ser 16h1'31"0d28'
Pinho,Janet Kristie
Peg 22h25'60"30d3'
Piniella,Mary
Cas 23h33'15"61d43'
Pinion,Becky Sue
Mon 7h2'49"0d47'
Pink Chihuahua
Cma 6h55'32"-11d12'
Pink Courtney
And 0h11'16"37d56'
Pink Elephant
Lyr 19h4'46"40d41'
Pinker,Joel Lawrence
Dra 13h46'24"68d20'
Pinkerton,Deborah Noel
And 23h2'44"51d2'
Pinkerton,Marion Baldwin
Lyr 18h41'34"47d21'
Pinkerton,Phyllis H
And 0h50'23"35d27'
Pinkerton,Shannon Belle
Lyr 18h40'52"47d7'
Pinkerton,Trevor Glenn
Aur 7h20'53"38d40'
Pinkett,Dr Olivia
Lyn 7h54'30"42d0'
Pinkey Orio
Cnv 13h5'36"48d54'
Pinkham,Chase Alexander
Cet 1h7'11"-2d32'
Pinkham,F Jessica
Eri 3h39'0"-13d12'
Pinki
Sgr 19h2'2"-24d59'
Pinki Do Zobaczenia!
And 2h19'1"40d49'
Pinko,In Memory Of Emily
Lynn
Cas 0h31'23"60d41'
Pinkus,Michelle Lynne
Psc 23h27'1"1d50'
Pinkus,N
Sgr 19h17'41"-43d41'
Pinky
Sgr 20h3'50"-20d28'B
Pinky
Lac 22h8'55"51d33'
Pinky-Chérie
And 0h23'38"44d35'
Pinn Star
Uma 9h54'46"54d48'
Pinna-Panio,Louise Dorothy
Cas 23h0'41"58d57'
Pinnacle!,The
Cam 5h31'58"61d20'
Pinnacle,The
Uma 8h34'0"57d14'
Pinner,Sarah
And 0h50'16"40d43'
Pinney III,James S
Sgr 18h55'26"-30d51'
Pinney,Julia
And 0h30'58"45d30'
Pino,Giusy
Peg 21h50'55"33d49'
Pino,Orso
Tel 20h9'15"-46d14'
Pinocchio
Dra 16h3'27"67d53'
Pinola,Marjatta
Cas 0h21'46"58d20'
Pinonzek,Victor S
Her 15h57'52"46d11'
Pinotsis,Nicolaos
Cas 0h14'55"60d31'
Pinsent Aura,The
Cyg 19h22'30"28d28'
Pinsevaekkelsens Hoejskole
Mariager
Her 17h55'13"14d52'
Pinson,Barbara
Cas 22h56'35"54d16'
Pinta,Edmund
Per 4h2'0"38d25'
Pinter,Jason L
Umi 14h7'37"71d46'
Pinter,Mrs Gundula
Aql 19h32'30"-0d0'
Pintner,Allan B
Aur 5h30'27"31d49'
Pinto,Erika Chaves Loureiro
Eri 2h51'25"-17d3'
Pinto,Maria José Ferreira
Peg 21h54'1"30d0'
Pinto,Stefan J
Her 17h33'1"40d36'
Pinto,Stephen Michael
Per 2h50'56"37d40'
Pinzon (DMP 76),Dimas &
Maureen
Cyg 21h5'51"33d8'
Pinzon,Cecily Stone
Oph 18h1'15"10d39'
Pinzon,Dimas
Oph 17h43'12"12d26'
Pio De Corso
Cyg 21h37'14"41d35'
Pioch,David Andrew
Lyn 6h34'0"56d11'
Piodi,Fabien
Ori 6h7'13"0d29'
Piolanti,Yaël
Vul 19h21'39"23d19'
Piombino O K A Sis &
Catrina,Joann
Cyg 19h31'50"32d30'
Pioneer Woman
Aql 19h41'23"14d59'
Piontkowski,Brad Reid
Aur 5h4'41"40d5'
Piontkowski,Jennifer Lynn
Ari 2h33'36"21d59'
Pionton,Brian
Ori 5h19'18"10d8'
Pioppo,Mireille Luce
Mon 6h52'44"-1d34'
Piora,Lorna
Lac 22h28'43"52d33'
Piorier,Kelsie Alecsandra
And 23h29'45"42d55'
Piorun,Marissa Lee
Cyg 20h17'30"40d12'
Piotrow,Phyllis Tilson Cntr for
Comm Pgrms
Cyg 21h34'29"41d49'
Piotrowski,Becca & Kenn
Cyg 21h12'57"39d42'
Piotrowski,Laura
Crb 15h43'50"26d31"
Piotrowski,Robert Michael
Boo 14h6'10"10d38'
Piotrowski,Wayne
Ser 18h52'58"3d18'
Piotter,Sir James A
Per 2h53'56"37d53'
Piovane,John Rocco
Boo 15h6'34"12d59'
Pip
Ser 17h58'22"-13d52'
Pip
Lyr 19h23'15"31d32'
Pip's Paradise
Boo 14h6'50"52d22'
Pip's Planet
Umi 15h59'43"78d48'
Pip's Star
Lyr 19h13'52"40d45'
Piper
Cyg 20h19'1"40d40'
Piper 16
Cap 21h55'37"-18d3'
Piper IV,William Francis
Aur 6h54'0"38d42'
Piper Laurel
Aql 19h40'17"13d2'
Piper Vista Boza
Ori 5h25'31"-0d21'
Piper's Dream
Aur 6h30'21"32d32'
Piper,"Jake" Jacob
Her 18h3'0"31d2'
Piper,Alison & Alan
Cyg 21h0'1"35d43'
Piper,Gordon Ernest
Uma 10h21'25"54d18'
Piper,James David
Lmi 10h4'1"32d38'
Piper,John Leo
Boo 14h53'47"28d4'
Piper,Joyce E
Cas 23h20'14"60d16'
Piper,Kimberley Anne
Cyg 19h29'57"39d24'
Piper,L G
Sct 18h52'22"-7d47'
Piper,Robert Lee
Ser 15h28'36"8d35'
Piperno,Claudia
Lup 15h19'7"-45d32'
Pipes,Daniel
Lac 22h35'54"37d37'
Pipher,Dennis Lee
Ori 5h5'55"0d27'
Pipia,Jr,Richard John
Aur 7h24'14"37d5'
Pipkalejs,Mimi
Lyr 19h13'0"38d18'
Pipo-Capricornus-Star
Cap 20h17'25"-20d24'
Pipolo,Louis
Boo 15h16'23"52d10'
Pipomica,Wendy C
Peg 21h59'29"2d36'
Pipouche
Vul 20h15'43"22d40"
Pippa J
Cas 0h14'34"63d53'
Pippin,Joelle
Boo 14h37'27"46d59'
Pippin,Jr,Edward Bo
Boo 14h33'16"20d7'
Pipprich,Dr Bridgitte
Aqr 22h44'18"-0d31'
Pips
Cnv 12h10'10"45d8'
Piquemal,Gerard
Ser 15h18'52"11d35'
Piquet,Jacques
Peg 23h46'59"27d55'
Piquet,Robert
Aur 6h5'39"37d46'
Pirang,Alexander Jürgen
Ori 5h29'29"1d43'
Piraquive,Nancy
Cas 23h28'55"60d5'
Pirasteh Star,The
Ori 5h51'31"18d54'
Pirate Euge
Ori 5h58'22"15d19'
Pirate Prince
Oph 16h4'36"-5d44'
Pirbauer,Veronika 240371
Ari 2h1'46"26d33'
Pirdmore,Libby
Cam 7h46'0"68d54'
Pires,Sergio da Silva
Venancio
Oph 18h8'24"13d56'
Pirie,Edward & Doris
Lyr 18h59'19"46d5'
Pirie-"Nonie's" Star, Jessica
Mon 7h31'55"-6d45'
Pirilä,Lallah
Dra 15h11'46"60d50'
Piriou,Catherine
Ser 15h13'1"10d26'
Pirk,Judi
Cam 7h8'55"71d5'
Pirkey,Jennifer Suzanne
Aql 19h8'32"4d8'
Pirkey,Patrick Harding
Tau 5h50'1"23d37'
Pirman,Thomas R
Cep 21h6'1"58d29'
Piro,The Illuminating Karen
Mon 7h45'53"-6d41'
Pirocchi,John & Michelle
Lac 22h2'15"37d38'
Pirojnikoff Star,The Olga
Ori 6h2'25"8d46'
Piroli,The Jo
Cyg 20h53'10"38d1'
Pirollo,Mario
Aur 6h32'30"37d32'
Piron,Christian
Aql 19h7'0"13d53'
Piron,Cristiano Alexander
Cnc 9h0'22"31d44'
Piron,Isabelle
Per 3h8'28"40d28'
Piron,Julian Christopher
Cnc 9h0'16"31d4'
Pirone,Bernadette
Lyr 18h38'28"35d18'
Pirone,Marco
Pyx 8h46'21"-24d36'
Pirovano,Daniel
Cmi 7h43'35"1d5'
Pirro,Achille
Dra 19h2'48"48d50'
Pirtle's Ship of
Dreams,Matthew
Gem 8h4'22"28d48'
Pirtle,Jason "Face"
Lac 22h30'27"40d22'
Piru
Sge 20h16'20"16d41'
Pisani,Vincent Francis
Oph 18h4'52"10d50'
Pisanic,Janice
Sge 19h22'31"16d20'
Pisaniello,Jerry
Aur 7h12'34"36d4'
Pisaniello,Jordan
Aur 5h13'1"40d42'
Pisano,Guy
Dra 19h6'55"48d30'
Pisano,John Anthony
Cep 21h46'31"55d58'
Pisano,Luciano Silvestro
Vel 9h43'5"-45d11'
Pisano,Michael James
Dra 19h0'26"56d50'
Pisano,Ross
Lac 22h1'0"49d1'
Pisano,Sr,Joseph Edward
Lyn 7h56'51"47d60'
Pisapia,Cotu's Love- Edward
Lac 22h31'12"38d58'
Pisarowitz,Liselotte
Gem 6h34'20"20d19'
Pisaturo,Bruno
Pho 0h5'14"-45d47'
Pisaturo,Michael
Aur 6h53'1"38d50'
Pisauro,Jacqueline
Aur 4h58'27"38d37'
Piscean Dream Sisters Forever
And 23h5'51"44d41'
Pisces Peg
Psc 0h25'1"2d7'
Pisces' Pats
Psc 23h6'30"6d37'
Pisch,Coach
Lac 22h53'37"55d18'
Pischalko,Melanie Rose
Cam 13h30'47"78d28'
Pischel,Sylvine & Steven
Lyn 6h58'28"60d17'
Pischke,Haidi & Hans-
Juergen
Uma 10h0'36"51d29'
Pisciotta,Richard
Lac 22h5'17"49d36'
Pisciotto,Amanda "Mandy"
And 2h18'52"38d26'
Piscitello,Michael John
Her 16h48'21"30d32'
Piscopio,Karen
Cam 3h52'16"71d15'
Piscopo,Anthony Vincent
Her 16h55'0"40d40'
Piscopo,Diana
Dra 15h7'55"64d44'B
Piscopo,John
Dra 15h7'55"64d44'A
Piscopo,Sr,Stephen J
Her 16h2'39"47d57'
Pisculli,Billie June
Mon 7h2'53"-1d35'
Pisculli,Joseph
Aql 19h0'35"-1d35'
Pisel
Tri 1h58'36"33d47'
Pisell,Tracy
Eri 3h44'30" 5d16'
Pisello,Dawn
Lyr 18h49'50"37d20'
Piserchiac,Doreen "Super
Star"
Lyr 18h37'1"36d21'
Piskolti,Anne Rodak
Vir 11h40'43"8d18'
Piso-Kfoury,Ann
And 23h20'50"44d48'
Pistilli,John
Aql 19h55'12"0d13'
Pisto,August P
Boo 14h59'29"51d35'
Pistol
Sex 9h54'32"-1d17'
Pistol Patty
Peg 22h10'0"4d32'
Pistolis,Dena
Cyg 19h24'57"33d9'
Pistore,Mildred
Lyr 18h56'15"30d18'
Piszcz,Martha & Clemens
Uma 9h39'0"59d56'
Pitas
Lac 22h30'19"53d46'AB
Pitcher,Dr Beatrice
Tri 2h16'16"28d34'
Pitcher,Joanne
Lyr 18h15'37"30d42'
Pitcher,Kimberly A
Cyg 21h14'48"35d22'
Pitcher,Mary Ann
Oph 18h17'19"11d43'
Pitcher,Peter N
Aur 6h21'58"38d20'
Pitcher,William M
Cep 22h40'24"67d56'
Pitchoune
Uma 12h4'39"62d17'
Pitillo,Pamela Stefani
Lyn 18h37'1"46d10'
Pitipitumpa,Anna
Her 16h10'21"10d56'
Pitkaranta,Seija & Jarmo
Aur 5h15'23"47d1'
Pitman,John Kirby
Cet 2h58'13"-0d55'
Pitner,Kristy Gayle
Cas 0h1'15"64d19'
Pitner,Richard V
Cam 3h14'30"68d15'
Pitney,Stephen & Claire
Cyg 20h42'51"45d17'
Pitofsky,David
Her 16h46'32"21d58'
Pitolli,Vanessa
Cyg 19h48'27"29d33'
Pitra,Wanda
Ari 1h57'49"17d49'
Pitre,Jr,Robert Presley
Peg 23h22'36"11d4'
Pitre,Ralph
Peg 21h24'59"8d14'
Pitt,Brad
Ori 5h57'56"12d10'
Pitt,Ermil
Crb 15h18'0"30d33'
Pittack,Carmelina Agnes
Paolercio
Uma 14h54'11"78d22'A
Pittack,Claudie Clarence (Red)
Uma 14h54'11"78d22'B
Pittaras,George Dylan
Per 3h4'48"41d0'
Pittaras,Luca Alexander
Peg 21h48'18"33d12'
Pittaro-Shine Forever, Teresa
Ann
Lyr 19h16'17"42d39'
Pittenger,Donald
Cyg 20h23'1"40d19'
Pittenger-Star of
Transition,Mona
Eri 4h56'38"-10d5'
Pitter Pat
Cyg 21h51'0"41d35'
Pittman
Aur 4h48'41"40d23'
Pittman Honor
Boo 13h43'58"24d32'
Pittman II,Frank Eugene
Sex 10h19'7"-1d42'
Pittman III,Alfred Roland
Ser 17h33'47"-13d30'
Pittman,Andrew
Lmi 10h35'48"30d27'
Pittman,Best Friend, Euna
Cas 0h11'60"61d46'
Pittman,Dick
Aql 18h59'27"-2d40'
Pittman,Dod
Aql 19h51'60"10d20'
Pittman,Douglas L
Her 16h31'47"42d24'
Pittman,Douglas Mathew
Aur 5h16'36"48d31'
Pittman,Elizabeth Ruth Jones
Cas 1h52'24"58d16'
Pittman,Josh
Vir 12h36'21"2d19'
Pittman,Kim Annette
Aql 19h40'53"10d8'
Pittman,Mrs Linda S
Peg 21h55'39"20d27
Pittman,Rocio Del Mar
Ori 6h9'13"3d59'
Pittman,Rod & Dolly
Vul 19h40'30"26d35'
Pittman,Sarah Jeanette
Mon 6h44'1"9d3'
Pittman,Tiana Chelsea
And 0h0'50"43d38'
Pittman-Love Forever, Janet
Rose
Lyn 8h31'23"58d29'
Pittman-McDonald, Rachel
Elizabeth
Lyr 18h36'1"41d13'
Pittner,Dr Gerhard
Aql 18h53'0"10d53'
Pittock,Paul Derek
Dra 16h12'33"63d20'
Pittorelli,Mirella Mario
And 23h1'55"51d2'
Pitts"Happy 21st",Josh
Aur 4h52'57"40d17'
Pitts"Pitter",Dr Tom
Aur 4h44'32"34d30'
Pitts,Brandon J
Her 16h36'14"50d41'
Pitts,Christopher
Boo 14h57'39"22d2'
Pitts,David
Boo 15h7'55"12d31'
Pitts,Doris Searcy
Vir 11h52'19"8d13'
Pitts,Kathy
Cas 23h35'1"62d12'
Pitts,Laura Jeane
And 23h39'0"47d49'
Pitts,Mary Louise
Cep 0h16'26"68d3'
Pitts,Matthew Joseph
Per 2h59'33"41d1'
Pitts,Russell Anthony
Aur 4h47'32"36d31'
Pitts,Shannon Lynn
Lyr 19h16'47"25d55'
Pitts,Tom & Betty
Peg 23h3'23"20d54'
Pitts,William
Eri 2h55'25"-14d15'
Pitts-Bryan,Elizabeth Anne
Cyg 20h23'57"39d6'
Pittwood,Heather (Pandora)
And 0h47'0"22d6'
Pittz,Troy Lynn
Per 2h40'59"35d9'
Pitu
Dra 17h6'34"60d39'
Pitula,Peter & Phillipa
Cyg 21h13'21"37d41'
Pitulski,Carol
Aql 20h13'10"7d39'A
Pitusa
Cyg 20h53'25"40d21'
Pitz,Marie Elizabeth
Cas 2h33'34"57d33'
Piucci,Laura
Lac 22h11'28"51d31'
Piuspi
Ind 21h2'30"-52d34'

Piveronas,Barbara B
Mon 6h32'46"-5d52'
Pivinski,Robert Martin
Cep 20h36'25"76d3'
Pivowar,Linda C
Cyg 21h20'22"39d23'
Pix's Paradise 2
Peg 21h49'56"30d46'
Pixie Dust
Peg 21h53'44"33d56'
Pixie,Sue
Lyr 19h21'30"35d14'
Piza,Jorge A Rodriguez
Aql 20h1'52"10d43'
Pizzeck,Jr,John Amory
Aur 5h59'42"37d31'
Pizzeghello,Luigi
Ari 3h1'0"28d14'
Pizzino,Chris & Becky
Cyg 20h30'50"40d13'
Pizzio,Kelly Ann
Peg 22h45'41"21d34'
Pizzo,Ronald J
Cmi 7h5'44"5d23'
Pizzo,Stephen Peace
Hya 10h41'33"-17d32'
Pizzolato,Anthony
Aur 6h36'43"38d23'
Pizzuto,Angela Catherina
Cas 2h22'1"67d42'
Pizzuto,Michael
Boo 15h4'27"28d5'
Piáa
Lyr 18h53'1"40d60'
Pi Lyr 18h53'1"40d60'
Aur 5h8'41"38d6'
PJ
Per 2h54'13"40d14'
PJ
Ori 4h51'60"4d53'
PJ & Dave's First Date
Cyg 20h20'12"40d22'
PJ's
Per 1h55'0"56d58'
PJ's & DJ's Star (11788)
Dra 16h58'37"60d41'
PJ'S 20th 12-27-95
Aur 7h18'47"39d27'
Pjura,William J
Aur 7h24'1"37d46'
PK
Uma 12h5'48"47d37'
PK 1
Lyr 19h40'19"40d31'
Plab,D D
Aur 6h26'30"33d44'
Place In This World
Uma 12h3'23"55d7'
Place,Clarence S
Aur 6h17'53"35d19'
Placent,Sarah
Cas 3h0'14"58d34'
Placer,Iris
Aql 18h56'57"-1d7'
Plachetka,Emillie
Lyn 8h56'15"42d40'
Placios,Julian
Her 17h39'28"40d4'
Placko,Marilyn Sophie
Froelich
Lib 15h15'29"-17d36'
Placzek's Juwel,W
Boo 14h19'45"11d11'
Plagnol
Per 3h52'46"37d44'
Plagor
Ori 6h3'20"2d44'
Plain,Craig Dean
Dra 18h39'34"58d32'
Plainte,Fredrick J
Cyg 21h35'32"42d1'
Plaisance,Jr,Richard Robert
Her 17h8'19"48d48'
Plaisance,Mark Earl
Cet 2h8'53"-5d47'

Plaks,Nina
Mon 6h57'50"10d15'
Plamondon,Amy
Cyg 19h49'26"38d6'
Plamp,Alfred C
Hya 8h13'39"1d8'
Planas,Trena Simon
Ori 5h57'1"20d25'
Planche,Sacha- Stephanie
Per 3h36'33"50d6'
Planes,Philippe
Sge 19h2'42"20d21'
Planet Dave
Cnv 13h37'22"41d35'
Planet Dodds
Her 17h36'14"25d3'
Planet Dog
Cma 6h28'46"-28d40'
Planet Fraser
Cam 3h29'52"63d17'
Planet Hyack
Dra 16h5'13"67d48'
Planet Interactive
Tri 2h8'34"31d23'
Planet Janet
Cas 23h41'24"62d24'
Planet Janet,The
Cyg 20h27'0"40d3'
Planet Janet,The
Eri 3h38'0"-16d45'
Planet Jaz
Lyr 19h6'36"28d57'
Planet Joe
Aql 20h34'1"-1d41'
Planet Long
Hya 10h40'27"-17d34'
Planet O'Brian
Hya 8h18'56"2d57'
Planet of Love, Hidetaka &
Tomoko
Cap 20h53'9"-25d22'
Planet of the Guppies
Aql 19h57'1"13d45'
Planet Oz
Mon 7h5'38"-6d50'
Planet Parker
Uma 14h5'58"51d16'
Planet Pooh
Umi 17h50'56"80d23'
Planet Queen Mari
Aqr 22h6'24"-2d14'
Planet Rio
Dra 15h58'24"68d31'
Planet Roy
Aur 5h11'51"42d41'
Planet Witus
Gem 6h43'50"34d48'
Planetary Citizens
Cyg 20h21'13"38d28'
Plank "Laluna",Lindsey Ryan
Cas 0h59'0"60d22'
Plank,Alexander
Uma 9h59'52"50d47'
Plank,Gregory Newman
Aql 20h35'15"0d37'
Planner,Harry & Edythe
Tri 2h10'47"31d13'
Plano's Price
Cyg 1h27'17"40d36'
Planovsky,Mark J & Mary D
Peg 22h0'15"21d21'
Planovsky,Victoria "Tori" Shea
Crb 15h23'34"30d1'
Plant,Blair
Aur 5h2'48"45d13'
Plant,Elizabeth M
Lyr 19h6'23"28d10'
Plant,Geoffrey
Peg 23h30'43"12d7'
Plant,Pauline
Cyg 19h33'40"35d20'
Plant,Steve William
Hor 3h54'36"47d54'
Plante,Andre "Bear"
Aur 6h12'24"31d2'

Plante,André
Umi 14h42'0"69d9'
Plante,Brian
Cnc 8h6'21"32d15'
Plante,Chris & Melissa
Peg 21h59'25"30d0'
Plante,Jill
And 1h52'42"38d37'
Plante-Gagnon,Anaïs
Cas 1h15'45"61d6'
Plaskitt,Paige Nicole Mallabey
Cas 0h6'27"64d40'
Plaskoff,B & B
Ser 16h0'42"14d37'
Plassart,Gael
Sgr 19h33'27"-31d43'
Plassmann,Jacob
Aql 19h31'31"-8d10'
Plastus,Gillo
Her 16h59'0"26d40'
Plata,Emanuele
Col 6h25'27"-37d4'
Plate,Herbert
Cmi 7h18'17"1d31'
Plate,Herbert
Dra 18h32'26"67d46'
Plate,Kim
Peg 22h22'1"30d9'
Platina
Cam 13h11'44"78d36'
Platinum Bullets
Uma 10h31'48"58d15'
Platko,Michael Andrew
Per 1h55'42"53d12'
Platnick,Geoffrey Scott
Cep 22h5'21"70d36'
Plato,Kerri Louise
Crt 11h43'25"-21d32'
Platridis,Nicholas
Cep 22h10'47"62d21'
Platt,Arnold W
Per 2h38'32"35d21'
Platt,Campion Acheson
Cam 4h54'42"61d5'
Platt,Cortney Elaine
Peg 23h35'49"22d9'
Platt,David C
Cmi 7h54'14"4d45'
Platt,David Philip
Lac 22h7'22"40d18'
Platt,Erin & Chris Brown
Her 17h9'1"21d41'
Platt,Janet
And 23h35'43"47d41'
Platt,Jeremy Acheson Spear
Her 17h0'20"28d38'
Platt,John Richard
Uma 13h29'36"58d25'
Platt,Lewis
Aur 4h51'44"38d36'
Platt,Michael Jarod
Aur 4h56'49"51d42'
Platt,Thomas Heritage
Per 2h21'31"58d31'
Platt,Wendy
Ori 4h53'44"-1d20'
Platte,Holly
Ori 4h45'50"2d37'A
Platte,Sierra
Ori 4h45'50"2d37'B
Plattner,Claire
Cyg 20h51'16"31d20'
Platto
Uma 10h51'27"70d55'
Plattoian-23
Mon 6h54'39"-8d48'
Platz,Melanie Kristina
Lib 15h3'20"-20d22'
Platzer,Douglas
Dra 14h22'57"63d57'
Platzl
Lyn 8h4'1"43d33'
Plauché,Christopher Jude
Equ 21h21'1"8d35'

Plaumann,Stefanie
And 0h32'0"45d31'
Plaxin,Scott Michael
Aur 5h7'26"41d46'
Play,Marie Christine
Ser 16h7'38"0d2'
Playa Las Couchitas Kim-
Collins
Ori 5h4'12"8d1'
Player,Evelyn
Mon 6h46'55"10d20'
Playful Apolla, Unbounded
Luminary
Cet 2h6'1"3d30'
Playhouse
Aql 18h57'15"-6d30'
Playmaker
Cnv 13h0'1"38d55'
Plazanet
Lyn 8h28'0"40d31'
Plazza,Richard
Per 4h39'19"36d31'
PLB 49
Dra 14h24'0"63d51'
Pleasure
Aql 18h41'50"0d23'
Pleba,"Titti" Maria Cristina
Vel 9h43'0"-49d27'
Plebani,Cornel Dennis
Aur 5h1'17"50d10'
Pledger,Connie
Peg 22h0'0"25d45'
Pledger,Jr,Luther Louis
Her 18h2'57"28d50'
Pledger,Kyle & Madeline
Crb 16h2'15"33d51'
Pleinert,Kimie
Umi 14h31'20"67d37'
Pleinert,Sussi
Cas 2h56'49"60d48'
Pleitez,Ana
Mon 6h55'18"-6d27'
Plenker,Peter
Cap 21h58'57"-26d4'
Plentyn Sara
Lyr 18h28'58"37d32'
Plescher,George Edward
Uma 9h46'1"70d46'
Plesha,Plesh Fred
Aur 5h1'11"40d18'
Pleshette,Suzanne
Aqr 21h43'42"0d33'
Plesic,Sondra Marie
Cap 21h9'9"-15d9'B
Pleskus,Molly M
Lyn 7h4'48"52d25'
Pless,Wilhelm Junior
Gem 6h45'9"16d31'
Plessas,Patricia Coughlin
Eri 3h36'56"-2d39'
Plessner,Richard
Hya 8h14'20"2d2'
Plett,Karen Marie
Lyn 9h14'12"39d30'
Pletz Jr,Alfred
Per 3h7'57"50d30'
Pletzke,Steven R
Per 2h4'13"58d43'
Pleus,Anna
Lac 22h8'14"51d47'
Pleva,Lawrence T
Cyg 19h55'19"38d32'
Plew's "Wedding Star"
Cyg 21h4'58"48d55'
Plewa,Steven Michael Patsy
Her 17h36'19"24d12'
Plewe,Eleonore
Uma 12h35'20"62d38'
Pliertas,Robert
Cet 0h3'37"-17d32'
Plies,Levi Walker
Aql 20h5'53"8d25'
Pliley,Sandra K
And 23h7'39"41d25'

Plimption,Calvin Hastings
Lac 22h51'57"53d57'
Plimpton,Susan B
Cas 23h53'24"57d34'
Plinke,Kira Fabienne
Boo 14h19'14"10d1'
Plinner,Saar
Per 2h38'6"55d16'B
Pliske,Thomas
Aql 20h15'1"5d9'
Plisson,Dannielle
Cam 5h53'54"60d10'
Plitt Star,Irwin
Oph 18h 41'18"8d36'
Plitt,Laurie "Peaceman"
And 1h24'58"48d47'
Plock,Shelly Matusoff
And 1h15'43"38d52'
Ploetz,Olaf
Leo 10h28'59"10d12'
Plohn,Carol
Cam 4h11'40"67d45'
Plonchak,Patricia E
Vul 19h16'16"25d29'
Plonka,Irene Mary
Aur 6h3'45"35d13'A
Plonka,Linda S
Cyg 21h7'35"48d51'
Plonka,Patricia Ann
Aur 6h3'45"35d13'B
Plonski,Ken
Lyn 7h24'36"44d42'
Ploof,Adam Andrew
Boo 14h58'49"51d44'
Ploppa,Astrid
And 0h46'27"41d12'
Ploshay,Patti
Cyg 20h59'30"37d34'
Plosker Family Star, The
Oph 17h58'1"-8d52'
Plotkin,Andrew
Lac 22h10'22"50d7'
Plotkin,Rebecca Miriam
Mon 6h23'40"8d2'
Plotnik,Martin Alvin
Boo 14h37'21"39d51'
Plotrowski,Richard T
Dra 18h52'1"80d23'
Plott,Anne
Aql 20h6'15"8d33'
Plotts,Catherine
Cnv 12h46'45"38d6'
Plotzker,Jason Garrett
Leo 11h7'31"0d47'
Ploudre,Greg
Sct 18h34'1"-6d18'
Plourde,Myriam Laplante
Bernard
Per 2h52'15"37d45'
Plouvier,Jean
Vul 19h21'56"27d39'
Plowman,Bill
Uma 11h12'28"32d5'
Plucas,Bettina
Mon 7h53'35"-2d36'
Pluchino,Janet
Ori 4h53'56"-2d50'
Plucker's (Pop-Pop's)
Star,Arthur
Aur 7h24'41"40d40'
Plude,Henry J
Hya 9h35'13"0d1'
Plum,Sandra
Sco 16h49'1"-25d26'
Pluma Helguera
Hya 8h10'18"0d8'
Plumb,James
Cyg 21h35'12"42d49'
Plumer,Sandra S
Ser 18h20'38"-0d20'
Plumeri,Sr's 80th, Sam J
Per 4h8'33"37d39'
Plummer,Cameron C
Dra 17h12'30"61d8'

Plummer,Debbie
And 1h16'29"40d0'
Plummer,Eric Matthew
Lac 22h34'41"54d22'
Plummer,India Scarlet Sarah
Her 17h21'33"27d53'
Plummer,Kristen E
And 2h9'17"41d60'
Plummer,Lloyd J
Dra 19h54'0"61d36'
Plummer,Margot Ann
Cyg 19h33'37"30d24'
Plummer,Marty Lenora
Cam 6h4'52"58d47'
Plummer,Thomas
Cet 0h38'23"1d30'
Plummer,Victoria Ashlee
Cas 0h49'31"75d38'
Plummer,William F
Lyr 18h33'51"42d24'
Plumpy
Cyg 21h16'39"28d16'
Plungis,Charles William
Aur 6h31'0"37d57'
Plunkett,Angie
Cyg 20h17'35"39d15'
Plunkett,Peter
Cap 21h2'48"-16d23'
+R4th=1
Cam 3h22'54"60d44'
Plutis,Diana
Lyn 7h11'13"58d28'
Plöhn,Alfred
Cap 20h59'51"-14d54'
PM Buni Sylvia zu Bremen
Cap 20h25'5"-26d12'
PMA II
Umi 16h11'25"71d34'
PMC
Ori 5h55'21"21d23'
PMPL Naidraug Legna
And 0h36'48"37d43'
PMS Class Of 1996
Ori 5h50'12"17d60'
PMS-40
Uma 11h48'45"51d21'
Pniewski,Grace Nicole
Peg 22h44'11"3d8'
Pnina
Vir 11h47'45"3d55'
Pntoreau,Pascale
Cas 0h50'17"76d57'
Poandl,Father Robert F
Cyg 20h24'0"41d14'
Poarch-Fleming Penny Lynn
Cas 2h31'32"60d56'
Poate,Sr(Papa Bear), Robert G
Lac 22h29'0"50d37'
Poblet,Pau Pavella
Cyg 21h51'28"52d39'
Pobliner,Kathy
Cas 0h41'24"68d1'
Pobor,Jennifer
Cct 1h29'51"-10d55'
Pobre
Ser 15h21'36"4d39'
Pocacuttitta
Aur 5h26'0"31d49'
Pocalujka,Louis & Eugenia
Cyg 21h6'37"31d32'
Pocaterra,Letizia
Ind 21h8'36"-57d25'
Poce,JD
Cet 1h18'58"-5d52'
Pochard,Kimberly Nichole
Uma 10h31'29"55d53'
Pochert,Klaus
Cas 23h30'52"63d27'
Pochie
Uma 9h51'50"45d52'
Pocho
Aql 19h3'29"-6d15'
Pochop's Pride
Cep 20h39'52"58d12'

Pocius,Denise
Crt 10h51'36"-18d0'
Pockie
Peg 23h43'52"31d17'
Pocock,Ian David
Her 17h13'1"43d13'
Pocost,Michael M
Tau 4h28'26"0d58'
Podagrosi,Leonard "Toes"
Lyr 18h43'0"41d33'
Podawiltz,Dr Alan L
Uma 10h20'39"48d23'
Podawiltz,Jul Noël
Uma 10h58'57"48d48'
Podborny,Brandon Paul
Ori 5h32'51"-1d18'
Podborny,Raymond Paul
Ori 5h33'19"-0d50'
Podda,Tony
Gem 6h45'44"17d9'
Poddig,Scott & Patty
Sge 20h2'47"16d23'
Podell,Jeffrey Ian
Her 17h3'26"42d7'
Podemski's Angel, Kathleen
Hya 8h23'42"-6d55'
Poderys,Eric Kastytis
Dra 16h4'40"66d25'
Podesto,Janice
Cyg 21h10'60"35d45'
Podmore
Sge 20h17'0"20d51'
Podolsky,Scott Harris
Cnv 13h1'13"38d52'
Podracky,Dale Stephen
Aur 7h0'52"35d50'
Podracky,Ryan Jeffery
Her 17h34'21"27d2'
Pods,The
Cyg 21h58'1"53d55'
Podschwadek,Hans- Joachim
Dra 18h26'40"65d5'
Podulka,Amanda Jayne
Peg 22h30'18"12d9'
Podurgal,Sam
Aur 5h6'25"38d34'
Podzielinski,Lynn
Cas 23h33'21"61d47'
Podzun,Ulrich O
Dra 19h39'55"80d30'
Poe,E Carey
Cep 22h43'31"57d28'
Poe,Ninny
Ori 5h53'10"7d36'
Poehl,Geraldine "Gerri" D
Eri 4h5'43"-18d36'
Poehlmann,Christopher
Cas 1h34'47"62d5'B
Poehlmann,Thomas
Cmi 7h16'16"4d28'
Poehner,Ryan Michael
Cep 1h39'41"80d14'
Poeschke,Hans-Juergen
Lac 22h32'1"37d58'
Poeta,Joan Patricia
Cas 3h0'50"60d7'
Poeta,Joseph John
Aur 4h56'31"40d36'
Poeta,Patricia Ann
Lac 22h20'19"52d18'
Poete,Christian
Sge 19h3'0"16d20'
Poetter,Elisabeth
Per 3h51'29"38d10'
Pofahl,David
Cet 3h7'24"4d16'
Pofcher,Steven
Her 17h56'13"14d44'
Poffenbarger,Larry D
Lmi 9h38'16"39d10'A
Poffenbarger,Nancy J
Lmi 9h38'16"39d10'B
Pogey
Uma 9h20'41"56d56'

Poglitsch,Brian
Ori 5h46'1"10d18'
Pogodine,Baris Petrov
Uma 11h34'52"32d12'
Pogoler,Jason Matthew
Ori 5h56'49"19d29'
Pogoreutz,Ing Erwin
Ori 5h25'50"15d31'
Pogson,Jennifer Lee
Lyr 18h58'32"40d43'
Pogtis,Tiffany
Mon 6h53'17"0d44'
Pogue,Charles Edward
Dra 17h35'59"63d51'
Pohl,Joanna Meredith
Lyr 18h45'11"39d48'
Pohl,Norval F
Aur 6h38'16"40d39'
Pohl,Patty
Cam 7h15'46"68d48'
Pohlad,Carl & Eloise
Cam 3h18'41"60d53'
Pohland,Wolfgang
Cep 22h2'19"61d25'
Pohle,Helmut
Sco 16h38'51"-41d14'
Pohler,Kelsey Alexandra
Cas 0h35'45"54d39'
Pohlman,Andy
Cet 0h31'37"-3d25'
Pohlman,Darlene Marie
Peg 22h6'34"27d12'
Pohlman,Glenn "Bud" & Sue
Uma 8h41'24"50d22'
Pohlman,Jim & Jody (Robb)
Uma 8h38'25"50d24'
Pohlo,Ross & Hildur
Hya 9h1'55"1d34'
Pohmer,O Stanley
Dra 20h23'0"63d14'
Pohu,Brigitte
Aql 19h30'1"-8d46'
Poiet,Alain
Dra 14h57'17"60d44'
Poilbout,Michel
Cep 21h27'14"60d54'
Poilbout,Pascal
Cep 21h26'47"68d9'
Poindexter,Jr,MD, James M
Oph 17h52'56"1d13'
Poinsignon,Martine
Uma 12h58'0"59d29'
Point Boro Panther
Lmi 10h18'35"28d49'
Point Guard-Nick
Uma 13h41'13"50d10'
Point on a I,The
Ari 2h19'53"18d10'
Point Star
Dra 15h36'58"58d24'
Pointer,Adrian
Cep 20h25'41"66d7'
Pointer,James Russell
Ser 15h21'44"0d60'
Pointer,Stephen Xiao Hong
Uma 10h35'0"47d42'
Pointon,Craig
Sgr 18h50'0"-30d12'
Points,John
Cet 1h29'47"-3d27'
Pointyhead
Vul 19h15'38"21d48'
Poirier,Alain
Her 16h53'0"33d7'
Poirier,Anne Lear
Cas 1h18'1"65d18'
Poirier,Guy
Cep 23h38'1"64d18'
Poirier,Jessica A
Cas 1h52'53"70d19'
Poirier,Marie-Eve
Uma 11h35'52"31d49'
Poirier,Marion
Cas 0h48'16"76d8'

Poirier,Monique
Cep 21h27'47"68d7'
Poirier,Sonia
Cas 22h56'35"54d13'
Poirot,Marie-Pierre
Lac 22h4'57"38d31'
Poirrier,Claude
Cep 2h39'0"80d15'
Poiré,Elizabeth Ann "Napua"
Aql 18h47'0"10d16'
Poison
Cet 3h3'27"1d15'
Poissant,Liselotte E
Lyr 19h1'60"28d35'
Poisson,Robert
Her 17h22'20"20d44'
Poitras,J-Guy
Lac 22h3'53"47d59'B
Poitras,Sénateur, Jean-Marie
Uma 9h40'51"61d7'
Poitrenaud,Michel
Dra 9h22'54"73d34'
Poker,Mary Elizabeth
Cas 0h53'48"61d1'
Poklemba,Joyce
And 0h20'37"30d7'
Pokora,Bill & Bobby
Vir 14h2'0"-21d8'
Pokorny,Ann & Ernest
Aur 5h2'50"29d9'
Pokrivtsak,Steve
Ori 5h22'40"1d21'
Polacek,Desideriu
Tau 5h57'34"28d46'
Polacek,John W
Her 16h42'1"34d54'
Polachek,Julie & Gary Parr
Crb 15h56'43"32d25'
Poladian,Jacklin
Crt 10h56'40"-17d60'
Polakow,Jason KA1111
Ant 9h37'42"-32d41'
Polanco,Greg
Mon 7h0'47"7d50'
Poland,Carleen
And 2h30'31"38d43'
Poland,Fred
Aur 4h59'60"38d5'
Poland,Sarah Glen
Peg 22h3'17"20d7'
Polaris,God's Mark Paul
Umi 15h20'43"66d19'
Polastar,Barbara Ann
Vir 13h32'6"-6d36'
Polastar,Daniel John
Psc 0h8'26"-0d24'
Polatsek,Amy
Sge 20h1'52"17d28'
Polcari,Theresa Anne
Cas 3h9'1"58d54'
Polchert "With Love Paul",Pat
Cas 1h41'11"61d19'
Polda
Cam 5h35'13"67d51'
Polder,Eric Smulders Van Heer
Jans
Pyx 9h7'10"-29d57'
Pole,Stephen Charles
Cet 3h7'0"5d8'
Polese,Marcia Ann
And 23h45'57"44d15'
Polese,Nuno Thomas
Her 17h29'47"37d51'
Poleski,Walter Daniel
Her 17h35'1"20d47'
Poletti,Marguerite
Psc 0h20'58"1d12'
Poletti,Therese Anne
Ari 2h35'31"30d41'
Poletti-Frederick,Mary
Sgr 18h59'23"-25d35'
Poley,J S
Mon 8h3'19"-1d20'
Poley,Samuel G
Aur 6h4'49"45d33'

Polf The Doug-Star
Boo 15h7'56"22d24'
Polglaze,Daniel J
Cep 21h4'20"68d43'
Poli,Ibelina
Vul 19h23'20"26d27'
Poli,Professor Philip James
Uma 10h38'49"42d23'
Poli,Stefano
Lac 22h10'25"51d9'
Policano,Bianca Jenise
And 0h11'51"39d44'
Policarpo,Michael
Cnv 12h47'29"38d24'
Policastro,Austin Paul
Dra 16h0'45"63d23'
Polikarpou,Kiriakoula
Lyr 19h4'32"26d7'
Polikoff,Judy
Dra 19h58'22"61d11'
Polimeni,Louis
Her 17h15'21"22d27'
Polinchock,Vincent & Celine
Mon 6h19'53"5d43'
Poling,Georgie
Sex 9h51'11"-2d11'
Poling,Sabrina Michelle
Cas 23h24'35"60d23'
Polini's Paradise
Dra 17h53'47"68d34'
Polinsky,Eugene
Cep 23h1'49"61d42'
Polinsky,Lynn
Equ 21h4'35"10d18'
Polinsky,Mary
Cas 0h45'1"70d38'
Polish American
Princess,JTR,The
And 23h39'28"48d21'
Polish Prince,The
Per 2h27'52"57d45'
Polish Wildcat
Aur 5h7'38"38d52'
Polished William
Uma 11h34'27"43d27'
Polit,Anne and James
Cam 7h21'38"68d0'
Polite,Lealyn
Mon 7h4'34"-1d8'
Politis,George
Oph 16h58'48"11d31'
Politis,Peter
Aql 19h17'0"12d55'
Politis,Tina
Peg 22h20'14"27d39'
Politte,Craig D
Oph 17h52'34"-1d31'
Polivka,Craig N
Dra 20h22'1"64d44'
Polk,Austin James
Vul 19h21'52"26d48'
Polk,Jacqueline Danielle
Psc 23h6'0"1d49'
Polk,Jr,Gene Allen
Cet 2h32'1"5d9'
Polk,Madison Victoria
Psc 23h5'54"2d10'
Polk,Marina Janelle
Peg 23h39'33"10d12'
Polka,Darrell R
Per 2h40'45"40d16'
Poll,Jane
And 0h56'21"39d50'
Poll,Rachel E
Peg 21h52'14"33d22'
Pollack,Barbara & Mike
Ori 5h47'45"21d37'
Pollack,Brian Paul
Cep 21h50'25"55d60'
Pollack,Dave
Dra 13h5'20"67d52'
Pollack,Ellie Hannah
Cas 0h23'48"61d49'
Pollack,Joshua Andrew
Ori 5h48'1"21d3'
Pollack,Rebecca Mali
Lyn 8h12'21"46d25'
Pollak,Elissa
Uma 9h21'1"61d35'
Pollan,Homer
Cnv 12h47'1"39d55'
Pollard 031930,Newman
Lloyd
Cyg 20h1'13"40d18'
Pollard 041460,Michael Lee
Lyr 18h26'30"38d39'
Pollard 110328,Imogene
Lyn 8h1'60"41d9'
Pollard II,Oliver
Mon 6h35'21"1d9'
Pollard,Anne & Brett
Uma 11h23'50"42d2'
Pollard,Charles
Cet 3h3'53"4d59'
Pollard,Claire Brady
Eri 3h48'39"-10d31'
Pollard,Donald Lee
Oph 17h35'0"-6d26'
Pollard,Laura Landry
Cas 1h20'23"53d9'
Pollard,Mary Ann
Vul 20h37'24"20d5'
Pollard,Michael B
Lmi 10h0'27"38d44'
Pollard,Nancy
Ser 15h13'50"9d34'
Pollard,Samuel George
Cyg 21h15'44"28d58'
Pollard,Stephen J
Mon 8h1'40"-6d22'
Pollarine,C Frank
Tau 3h53'35"20d31'
Pollaro,Vincent Giovanni
Oph 18h2'33"11d11'
Pollart,Star of David- David
George
Dra 12h3'24"70d57'
Pollary,Dr Rodney A
Cep 23h35'37"70d59'
Polley,George Samuel
Boo 14h32'36"48d56'
Polley,Jarid Rickley
Lac 22h13'0"49d0'
Polley,Samuel S
Cyg 19h24'47"33d21'
Pollick,Joe
Boo 15h0'1"51d40'
Pollick,Mark
Uma 12h15'0"62d5'
Pollina,DeLynn Elizabeth
Boo 14h39'54"40d58'
Pollina,Ken
Peg 23h3'12"18d10'
Pollitt,Jonathan
Aql 19h5'40"1d18'
Pollmer,Johannes
Uma 8h46'31"51d55'
Pollnow,Nancy Bowling
Cma 6h57'25"-30d5'
Pollock,Alexander Charles
Her 17h35'40"25d16'
Pollock,David
Lyn 8h16'0"48d10'
Pollock,Emma
Vir 11h43'35"8d57'
Pollock,Franklin B
Aql 19h6'1"-0d17'
Pollock,Helen
Peg 21h59'1"31d48'
Pollock,John Wesley
Per 3h14'1"42d39'
Pollock,Kathryn Anne
And 0h14'0"37d12'
Pollock,Marge
Tau 4h10'41"21d43'
Pollock,Richard
Aql 18h59'29"16d58'
Pollock,Samuel Richard
Uma 11h57'22"41d1'
Pollock,Sophie
Lyn 8h29'38"48d31'
Pollock,Tom
Eri 2h44'53"-4d39'
Polly
Boo 14h6'19"52d1'
Polly
Cas 0h45'16"71d54'
Polly
Lyr 19h4'54"26d49'
Polly
And 22h59'11"38d51'
Polly Paige
Peg 22h52'15"27d5'
Polly's Planet
Cas 0h23'59"56d2'
Polly-Dale Heavenly Bliss
Vul 19h18'15"41d40'
Polly-Gone
Oph 17h30'24"-20d48'
Polmon,David A
Dra 19h9'28"67d56'
Polome',Beloved
Lioness,Sharon
Leo 11h44'18"27d20'
Polone,Polly
Aql 18h58'50"16d14'
Polonsky,Abraham & Sylvia
Sge 19h59'48"20d28'
Polonsky,Deborah
Tri 1h45'42"27d15'
Polony,Millie
And 0h46'42"39d49'
Polos,Mrs Laura
Cep 22h20'41"68d32'
Polovina"Dad",Michael N
Aur 6h25'0"32d1'
Polsinello,Julia Lacy
Vul 19h23'0"22d58'
Polsky,Andrew Jonathan
Leo 11h53'23"23d40'
Polsky,Elaine & Norman
Lyr 18h55'54"37d5'
Polso,Jukka
Aur 7h7'15"40d23'
Polson,Louise Dianda
Aql 19h12'36"14d6'
Poltorak,Kelly Ann
And 0h12'42"37d34'
Polutta,Christa
Aqr 21h19'18"-10d42'
Polverari,Adam Paul
Cep 22h51'43"58d32'
Polverari,Kendall Eve
Lyr 18h53'60"35d27'
Polvere,Michael Marshall
Peg 21h19'21"21d35'
Polverini,Andrea
And 23h21'1"52d40'
Poly,Murlin Patterson
Cyg 21h8'0"30d56'
Polzin,John Willard
Cep 21h24'26"61d0'
Polzin,Kathryn S
And 23h22'33"47d2'
Polzin,Tommy Andrew
Ori 4h44'19"0d11'
Poma,Margaret Myrlene
Maggie Doodle
Peg 22h42'0"29d1'
Pomahac-My Kind of Music
Star,Bruce
Lyr 18h40'26"45d11'
Pomarius,Nicholas
Aqr 21h23'15"-8d19'
Pomaro,Madeline Lee
Eri 3h52'47"-2d27'
Pombal
Cep 22h4'59"60d55'
Pomerantz,Andrea
Lyr 19h13'22"42d44'
Pomerantz,Howard
Boo 15h4'0"17d50'
Pomerantz,Joseph
Her 17h31'0"37d59'
Pomerantz,Kathrin Lynn
Lib 15h1'1"-10d7'
Pomerleau,Georgette Roy
Uma 14h4'37"52d56'
Pomeroy,Our Friend Kay
Cam 5h14'24"78d56'
Pomilia,Debby
Aql 19h2'31"16d41'
Pommerenke,George C S
Aql 20h1'29"14d36'
Pommersheim,William James
Cmi 7h26'0"1d22'
Pommier Jean
Lmi 10h48'14"33d4'
Pompa,Costanza
Ori 6h7'50"4d33'
Pompano,Eos
Vul 19h38'24"20d20'
Pompantinois,Theresa Childs
And 2h19'53"48d23'
Pomparau,Vasile
Cet 2h31'12"1d16'
Pompei,Gabriella
Del 20h17'0"11d9'
Pompeo,Edward
Uma 11h59'25"58d34'
Pompey,Jr,Russell
Aur 6h30'25"33d14'
Pompi
Sco 17h30'11"-31d43'
Pomponi,Angela
Lyr 19h4'47"38d18'
Pomposo,Antonia A
Sex 10h43'48"1d4'
Pompougnac,Fabienne
Aur 6h9'32"37d31'
Pompret,Mary
And 1h20'30"35d17'
Pomroy,Edward L S
Tau 3h36'47"15d52'
Pomroy,Terry
Ori 5h36'16"-1d31'
Poms,F B
Cam 13h48'1"81d57'
Ponak,Kyle Louis
Aql 18h42'16"-2d18'
Ponak,Maegan Jenna
Peg 22h19'31"10d36'
Ponce,Ana Georgina Sierra
Cyg 21h27'0"40d21'
Ponce,Danna Crescencia
Ser 18h2'49"-13d38'
Ponce,Jimmy Dabe
Aur 6h13'24"38d58'
Ponce,Mary Ida
Peg 23h49'46"31d50'
Ponce,Mikhail
Cam 5h48'25"60d6'
Ponce,Stephanie
Com 12h23'1"21d44'
Poncelet,René
Ori 6h5'11"8d37'
Poncet,Anne Marie
Cet 2h34'31"3d44'
Poncho's Equinox
Sct 18h55'33"-7d5'
Ponczocha,Scott Louis
Cam 4h20'38"70d25'
Poncé,Jr,José R
Her 18h7'16"14d19'
Pond,Beverly
Cyg 20h15'40"30d59'
Pond,James J
Sco 16h9'44"-24d41'
Pond,Lucy
Cas 2h49'0"68d10'
Pond,Nancy
Com 12h26'43"20d32'
Pond,Zachariah James Wesley
Her 18h6'0"30d11'
Ponder,Jim
Her 17h6'1"49d48'
Ponder,Margaret
Mon 7h29'51"-10d13'
Ponder,Michael S
Ori 5h2'15"14d39'
Ponder,Robert Lee
Aql 18h54'43"-0d4'
Pondt,David Robert
Hya 8h16'21"2d60'
Ponerantz,Arnold
Equ 21h1'1"8d38'
Pones,Steve & Julie
Cyg 21h6'44"30d59'
Poness,"Chris" Christine
Monique
Lyr 19h18'14"42d11'
Pongo,Chef Michael's
Boo 14h59'21"28d51'
Pongrac,Anthony Joseph
Per 1h33'51"52d46'
Ponko,Debbie
And 23h28'11"44d28'
Ponko,Trisha
Uma 10h15'21"52d47'
Pons,Juan
Boo 14h28'39"32d24'
Pons,Laure-Helene
Cam 7h37'55"80d36'
Pontarelli,Erika
Cas 2h32'35"57d32'
Pontbriand,André
Cam 12h58'58"81d40'
Ponte,Kay
Cas 3h26'11"73d12'
Ponte,Robert C
Her 16h37'54"36d4'
Pontegnier,Laurent
And 0h26'31"40d52'
Pontello,Arianna
Cam 4h15'55"60d54'
Pontello,Micaela
Lyn 6h18'28"58d15'
Ponti-Zins
Per 2h36'41"40d38'
Pontiac
Cam 8h14'0"73d23'
Ponticello,Roseann
Cnv 12h50'1"47d27'
Pontier II,Joseph
Oph 18h42'21"8d56'
Pontikis,Anastasia
Sge 20h0'30"20d1'
Pontin,Rick
Cap 20h5'1"-12d35'
Pontious,Samuel Preston
Aur 6h21'19"30d38'
Pontius,Marilyn
And 1h50'32"36d47'
Ponton,Marvin D
Aur 7h11'57"38d36'
Ponus,Cass
Dra 20h23'51"74d53'
Ponz,Casey Charles
Aql 19h24'55"14d60'
Ponzio,Alexandra
And 0h2'55"35d9'
Ponzoni,Cochi
Com 12h2'0"27d2'
Poo,Little Cindy Lou
Cet 1h30'56"-0d37'
Poochie
Ari 1h17'26"32d17'
Poodle,Margaret Linn
And 23h47'47"37d34'
Poodles
Cnv 12h47'0"36d29'
Poofins & Keener
Del 20h14'59"10d33'
Pooh
Cas 9h9'50"64d1'
Pooh
Uma 11h37'26"30d30'
Pooh
Uma 10h45'0"42d47'
Pooh
Cmi 7h40'56"7d50'
Pooh Bear
Lac 22h12'27"49d26'
Pooh Bear
Uma 10h17'14"42d18'
Pooh Bear
Uma 11h41'51"57d15'
Pooh Bear Ludy
Dra 14h45'19"60d9'
Pooh's Corner
Cep 23h13'19"71d11'
Pooh's Kiss
Per 3h35'25"38d58'
Pooh's Mumsy
Dra 17h39'43"60d12'
Pooh's Night Light
Per 4h30'12"38d42'
Pooh's Princess
And 2h23'0"42d26'
Pooh's Wish
Uma 9h54'53"42d7'
Poohbear
Cet 2h23'0"-10d41'
Poohie
Lyn 6h35'0"59d34'
Pook
Boo 14h31'22"44d13'
Pooka
Cas 1h39'36"63d54'
Pooka
Her 18h2'0"38d55'
Pookaleepoo
Aur 5h5'53"40d28'
Pookers
Lyn 8h21'31"58d49'
Pookey Bear's Star Home
Ara 17h56'23"-54d21'
Pookey's Mr Perfect
Hya 8h16'29"6d20'
Pookie
Uma 10h30'31"59d2'
Pookie
Aql 19h43'43"14d28'
Pookie
Peg 23h28'11"18d53'
Pookie
Cap 20h22'28"-26d4'
Pookie
Cet 1h39'26"-14d24'
Pookie
Cep 22h36'0"58d26'
Pookie
Her 18h4'37"37d47'
Pookie Bear
Uma 13h59'35"51d3'
Pookie DDS
Peg 23h37'54"17d43'
Pookie Michelle
Lac 22h23'50"37d42'
Pookie,M G
Oph 18h35'24"11d13'
Pooky
Equ 21h1'36"8d11'
Pooky
Umi 16h28'16"70d15'
Pooky
Lyn 7h54'38"40d14'
Pooky Terry-Lapine
Tri 1h49'48"28d27'
Pool,Kayla
Lyn 7h44'25"42d21'
Pool,Kristina Lynn
Cas 0h22'58"60d25'
Pool,Loren Richard
Aur 6h1'18"36d5'
Poole,Arthur
Her 17h18'56"44d13'
Poole,Barbara Ann
Crb 16h22'27"28d18'
Poole,Brandon Davis
Aql 20h0'35"7d53'
Poole,Jean Ann Ruffley
Uma 10h53'29"40d39'
Poole,Karen
And 23h35'56"44d35'
Poole,Lucy Anne
Cas 1h43'28"73d2'
Poole,Marcia
Aur 7h2'45"38d38'
Poole,Mark Allen
Her 17h36'52"18d46'
Poole,Sam
Lac 22h52'57"53d15'
Poole,The Light Of Judy
Peg 22h55'13"29d40'
Poole,William McKinley
Oph 18h15'37"0d34'
Pooler,Sally
Ori 5h52'37"11d46'
Pooley,Jerry
Her 18h12'57"38d37'
Pooley,Marcus James
Oph 18h18'57"8d53'
Poon,Andrea Young
Oph 18h42'48"7d3'
Poon,Chi-Lok
Per 2h40'1"40d20'
Poon,Rosita Yee
Oph 17h0'51"12d29'
Poonikas
Peg 22h17'54"35d23'
Poopie
Lmi 9h43'1"37d51'
Poopie Scoopie,The
Dra 19h1'44"48d47'
Poops
Lac 22h21'27"45d5'B
Poopsky
Cam 13h34'39"77d48'
Poopy Woopy
Uma 10h25'14"59d20'
Poore "The Great White
Hunter",David
Ori 5h54'59"20d52'
Poore,Jimmie Knox
Cnv 12h29'52"33d14'
Poore,Ryan P
Per 3h4'0"40d25'
Poorman's Star,Gerry Gessie
Per 2h55'36"40d31'
Poorman,Crystal Blue Leigh
Mark
Lyr 18h32'37"36d29'
Poorman,Forrest Lynn
Cnv 12h23'56"37d11'
Poos-Butcher,Kathryn Lynn
Equ 21h6'31"2d11'
Pootie
Aur 5h36'30"54d56'
Pootwaddle
Aql 19h53'22"12d54'
Poozas,Kevin
Oph 16h56'26"11d23'
Pop
Her 16h39'27"38d49'
Pop
Umi 16h4'19"70d7'
Pop's Star
Aql 20h10'48"10d7'
Pop's Star
Ori 6h0'1"-0d19'
Pop,Iggy
Aur 5h16'39"47d36'
Pop,Leonard
Cep 2h13'49"80d10'
Pop-pop
Crb 15h30'1"32d6'
Pop-Pop's Little Peace of
Heaven
Aur 6h26'10"37d26'
Pope Crook
Cet 1h44'45"-5d6'
Pope,Betty Ann (Lampi)
Vul 21h21'1"28d9'
Pope,Charlie
Aur 5h21'37"37d52'
Pope,Christine
Mon 8h0'25"-6d2'
Pope,Clarissa Lynn
Mon 7h15'53"-5d41'
Pope,Dominic M
Lyn 6h55'55"58d9'
Pope,Doris Pepe
Eidenback"Do Do"
Del 20h25'21"11d59'
Pope,Jennifer
Sge 19h56'56"19d9'
Pope,John
Eri 2h53'41"-12d23'
Pope,Jr,John
Per 3h27'47"51d21'
Pope,Jr,John H B
Hya 9h35'13"0d1"
Pope,Les
Aql 19h8'36"3d26'
Pope,Matthew Chandler
Boo 14h28'29"37d59'
Pope,Michael S
Lac 22h30'28"40d37'
Pope,Samantha
Lyr 18h23'19"47d36'
Pope,Sarah
Sct 18h54'52"-10d32'
Pope,Tamra
Lyn 8h26'52"58d19'
Pope,Terry Robert
Cet 1h52'41"-2d31'
Popeck,Stan & Vicki
Cyg 19h24'12"30d23'
Popejoy,Stephen C
Per 2h22'22"55d30'
Popelar,Rollin & Dorothy
Vul 19h48'49"28d59'
Popeye
Sgr 19h18'32"-28d51'
Popeye
Lyr 18h50'48"34d30'
Popeye Wagenblasts' Happy
Kitty
Sct 18h56'19"-6d30'
Popiel,Arline
Cmi 7h26'13"1d30'
Popiel,John & Stella
Lac 22h11'0"38d31'
Popielarski,Lois Marie
Uma 11h54'14"56d51'
Popineau-Derhy,Myriam
Aql 20h3'12"1d29'
Popis
Uma 10h59'53"50d16'
Popkin,Rebecca Louise
Cyg 21h54'54"52d44'
Popko,Noel C
Cep 21h6'52"60d14'
Poplar
Cnc 8h49'41"19d31'
Pople,Stephen
Cyg 19h28'33"35d10'
Poplin,Janice S
Cas 0h25'32"64d44'
Poplin,John & Suzanne
Cyg 21h30'10"50d5'
Popoff,Pete William
Dra 16h17'30"68d54'
Popour,L Todd
Dra 9h46'32"80d7'
Popov,Kristina & Edward S
Mikulas
Cam 14h25'34"81d35'
Popova,Sonia & Leon
Perigaud
Pic 4h47'10"-47d34'
Popovich (PoPo),Carol
Uma 10h19'49"72d50'
Popovich,Arlene Lynn
Cas 1h23'39"52d39'
Popovics,Marie & Anne
Cas 1h23'0"55d38'
Popovski,Anat
Cyg 21h8'18"31d25'
Popovsky,Yelena
Lyn 8h6'31"38d32'
Popow,Catherine
And 0h11'36"33d36'
Popow,Matthew R
Her 18h3'36"38d4'
Popp,Christopher A
Aql 19h49'49"14d53'
Popp,Christopher James
Her 16h17'40"25d24'
Popp,John
Uma 9h52'0"45d13'
Popp,Mark Lug
Lyn 7h52'38"43d36'
Popp,Robert Alery
Aql 20h11'52"11d15'
Popp,Sandra und Markus
Gem 7h1'45"24d38'
Poppa Gene
Aur 6h43'26"38d47'
Poppa Jim
Ori 6h11'36"0d5'
Poppa's Star
Her 16h26'26"40d34'
Poppa,Kim & Dick
Uma 10h55'52"51d57'
Poppe,Christa
Tau 5h0'55"16d20'
Poppe,Karl-Heinz
Lmi 10h42'48"25d8'
Poppe,Theodore Hans
Aur 4h51'10"50d18'
Poppele,June Patricia
Cas 23h38'21"61d31'
Poppert,Kathleen Gail
And 1h41'17"36d24'
Poppes,Jean
Vul 19h19'28"25d42'
Poppie Mel Star
Her 17h14'26"42d51'
Popplewell,Fay
Mon 8h3'31"-2d47'
Poppy
Cas 23h24'21"61d5'
Poppy
Ant 10h53'31"32d31'
Poppy
Lup 14h27'25"54d19'
Poppy
Sco 17h51'51"-33d53'
Poppy
Umi 15h48'6"74d55'
Poppy
Uma 9h58'1"47d46'
Poppy
Uma 9h14'0"48d5'
Poppy
Cam 8h2'27"67d49'
Poppy David
Lyn 7h50'22"39d46'
Poppy Gerry
Per 2h20'34"58d45'
Poppy Star (S3),The
Aur 5h2'24"30d24'
Poppy's Campfire
Per 2h39'43"38d2'
Poppy's Star
Hya 8h33'33"5d38'
Poppy-90
Uma 11h42'34"33d42'
Pops
Umi 13h17'14"70d6'
Pops
Cas 0h54'11"59d50'
Pops
Dra 17h2'0"60d7'
Pops-90th
Leo 10h51'41"15d13'
Popsicle Toes
Uma 10h43'0"40d31'
Popsicle Toes... Forever
Per 1h49'40"50d11'
Por El Amor De Marcos
Cet 0h35'55"1d52'
Por Siempre Martha's Estrella
Peg 23h1'22"17d41'
Por Siempre La Estrella De
Rose Marie
Del 20h19'49"10d20'
Por Siempre Tere
Mon 7h3'48"-6d37'

Por Tous Jours
Uma 8h57'52"51d56'
Porada,Jessica Erin
And 1h16'16"40d25'
Porada,Kaitlyn Veronica
Lyr 18h34'36"40d43'
Porcaro,Sarah
Cnv 12h52'15"40d51'
Porcel,Franck
Per 3h59'48"35d50'
Porcelli,Augustine J
Cmi 7h43'58"0d26'
Porcelli,Domus
Cep 22h14'0"61d32'
Porcelli,Ernestine
Peg 23h28'32"22d26'
Porcelli,Marcella
Cet 0h22'51"-11d42'
Porcelli,Vincent Joseph
Dra 20h31'21"70d22'
Porchet,Famille
Umi 16h13'46"73d31'
Porchetta,Sarah Nicole
And 0h24'26"32d13'
Porcillo,Amanda Michelle
Peg 22h30'28"8d39'
Porcillo,Christopher Anthony
Peg 22h34'14"8d59'
Porciuncula,Leslie
Ori 5h55'26"11d51'
Porco,Jr,Louis
Per 2h2'19"50d1'
Pordy,Stuart
Cep 0h19'53"70d51'
Pore,David
Uma 9h28'36"48d49'
Porfilio,Daryll & Stacey
Ser 16h1'27"8d48'
Porky
Tri 1h49'19"27d9'
Porky
Cas 1h40'58"57d17'A
Porky & Hobknobin
Ori 5h57'1"14d34'
Porrata-Doria Harding
And 0h58'48"34d10'
Porreca's Love Star
Boo 15h8'25"38d57'
Porreca,Joseph & Dianne
Mon 6h38'37"-10d30'
Porretta,Michael Patrick
Cep 21h34'15"60d32'
Porrino,Marilyn R
And 1h14'1"48d48'
Porry,Alain Henri
Uma 9h14'38"54d29'
Porsché Bob * The Lone Wolf
Cma 7h14'54"-13d39'
Porsia Maria
Cas 23h28'0"58d25'
Port Betty
Cas 3h9'56"75d17'
Port,Orin
Ori 5h55'20"16d4'
Port,Poppy Lou
Boo 14h18'15"12d54'
Portale,Carl
Boo 15h20'0"41d18'
Portalupi
Aur 5h8'29"44d17'
Portanova,Melissa
And 0h17'1"38d34'
Portela,Andrés Armas
Cyg 21h32'48"40d59'
Portenga,Jacob William
Cep 23h3'1"70d49'
Porter "Escape",Frank
Del 20h33'40"10d44'
Porter Boys Dan,Josh &
Eric,My
Ori 4h45'29"4d3'
Porter III,J Larry
Equ 21h7'16"11d29'
Porter Star,Colton Evert
Ori 6h6'25"20d4'
Porter's Star
Lyn 8h41'36"39d26'
Porter,Ashley Morgan
Hya 8h54'59"1d29'
Porter,Bob
Boo 14h59'1"42d41'
Porter,Bob
Ori 5h55'51"15d14'
Porter,Brian Law
Cnv 12h15'33"46d57'
Porter,Buzz
Gem 6h56'12"15d23'
Porter,Chuck & Barb
Uma 9h43'43"50d1'
Porter,Claire Louise
And 0h48'0"38d12'
Porter,Clifford George
Dra 19h5'13"70d45'
Porter,Conor Joynt
Aql 20h13'59"4d52'
Porter,Donald
Dra 20h8'23"64d26'
Porter,Donald J
Aql 18h56'35"13d12'
Porter,Elizabeth Anne
Her 17h21'43"41d11'
Porter,Franklin
Per 4h32'29"31d21'
Porter,Glen Scott
Boo 13h38'15"17d45'
Porter,Glyn
Ori 4h59'1"5d47'
Porter,Harry E
Per 2h25'35"55d34'
Porter,Henry Daniel
Cep 21h58'1"60d23'
Porter,Holley
Vul 20h3'36"23d9'
Porter,James E
Per 1h58'26"56d41'
Porter,Jennifer
Del 20h56'43"10d35'
Porter,John Edward
Hya 8h23'0"0d18'
Porter,John Stacy
And 2h27'13"45d25'
Porter,Kaley
Cet 3h0'21"0d15'
Porter,Karen
Crb 16h7'27"33d42'
Porter,Kendra Danielle
Lyr 18h54'53"30d21'
Porter,Kristen
Mon 6h22'42"3d25'
Porter,Lois Scott
Lyn 7h3'44"44d57'
Porter,Lyn Mitchell
And 23h1'28"47d55'
Porter,Margaret Lynn
Mon 6h23'42"8d42
Porter,Mark
Ori 4h41'34"0d55'
Porter,Mary Alice
Peg 21h51'1"31d20'
Porter,Maryanne
Cam 5h30'0"68d58'
Porter,Matthew James
Ser 18h40'33"4d59'
Porter,Melissa
Del 20h14'14"11d7'
Porter,Michael C
Cam 14h8'33"80d11'
Porter,Pat & Bill
Cyg 19h57'13"50d36'
Porter,Paul
Boo 14h25'0"52d33'
Porter,Sally
Cyg 21h44'19"28d28'
Porter,Sandra C
Del 20h25'0"20d13'
Porter,Scott Ian
Oph 18h15'26"0d44'
Porter,Tommie
Cet 1h3'34"-3d14'
Porter,William & Hilda
Uma 9h22'12"68d51'
Porter-Sir Knight, David
Her 17h36'42"41d12'
Portera,Annette
Peg 23h6'44"8d10'
Portera,Karen
Peg 23h29'39"31d28'
Portereiko,Michael
Ori 5h11'43"-6d0'
Porterfield,John Michael
Ori 5h33'42"-1d12'
Porterfield,Stephen Greg
Del 20h21'33"8d54'
Porth,Barbara Anne Schubert
Cyg 21h11'11"38d46'
Porth,Catherine Maria
Vul 20h16'17"28d54'
Portholio,The Great Valentine
Star
Uma 10h25'0"58d21'
Portier,Zaire
Com 12h22'1"32d15'
Portillo,Claudia Carolina
And 23h36'20"48d28'
Portland Actors Conservatory
1996
Cam 6h11'23"83d23'
Portman III,John C
Aql 18h54'33"-0d13'
Portman,Jodie S
Lyr 18h37'41"36d40'
Portmann,Jean Pierre
Ori 6h5'41"7d60'
Portmore,David
Dra 16h37'36"70d32'
Portnow 618J
Gem 7h22'57"20d11'
Portnoy,Morton L
Her 18h1'0"28d26'
Portnoy,Sandra Kalichstein
Cas 0h4'25"63d3'
Portnoy,Spodi Odi O
Mon 7h5'45"0d13'
Portnoys Marriage, Ellen & Hy
Dra 20h1'17"64d34'
Porto,Dave
Lac 22h15'47"49d31'
Portsmouth FC
Ori 5h45'0"11d51'
Poruban,Justin Thomas
Dra 16h1'58"67d21'
Posch,Tyler Gunnar
Her 17h7'1"22d26'
Poschenrieder,Petra
Cnc 9h0'11"7d41'
Poser,Alexander John
Aur 6h27'0"32d11'
Poser,Ashley June
Aur 6h27'30"32d45'
Posey,Garrett
Cyg 21h17'32"38d55'
Posey,Patrick L
Aql 20h0'58"10d0'
Posey,Samantha J
And 0h18'1"33d49'
Posey,Terry Wayne
Dra 19h2'26"56d29'
Posez,Christine
Boo 15h3'17"7d34'
Poshard,Jr,(Posh), William E
Cep 20h32'35"76d44'
Posillico,Dominic J
Her 16h28'45"20d5'
Posillico,Mario A
Oph 18h5'28"7d40'
Posin,Roger
Mon 7h1'1"-8d51'
Posipanko,Kimberly
Umi 16h0'57"72d19'
Positivity
Boo 14h56'12"52d18'
Poskitt,Peter
Ori 6h1'40"0d20'
Posner,Douglas Scott
Aur 5h2'0"40d41'
Posner,Virginia
And 0h11'17"33d46'
Poss II
Aur 6h3'1"34d33'
Poss,Marcus
Del 20h22'52"10d55'
Possert,Roe
Leo 10h57'1"20d5'
Possum
Equ 21h22'26"10d30'
Post Tenebras,Lux
Boo 14h49'0"33d8'
Post,Anna Othilia
Psc 1h18'1"20d35'
Post,Barbara Lynn
Aql 19h11'51"10d56'
Post,Carl E
Uma 12h2'24"42d20'
Post,Christine
Aur 4h48'34"40d31'
Post,Don
Her 16h55'20"35d50'
Post,J Williams
Aur 6h10'1"37d40'
Post,Jason Michael- Brian
Ser 15h12'1"7d22'
Post,Larana-David
Cyg 21h7'51"30d55'
Post,Sharon Ann
Lac 22h13'34"54d27'
Post,Shelly & Hershel
Uma 10h33'22"48d58'
Post,Suzette
Car 7h29'55"-58d20'
Post,Ted & Jo
Com 13h2'0"30d57'
Post,UWE Edgar
Cep 22h57'42sd70d16'
Post-Triplett
Equ 21h0'40"10d30'
Postel,Conrad Abraham
Aql 19h36'18"-1d16'
Postelle,Lynn
And 0h50'40"34d22'
Posten,Peter Matthew &
Robert John
Equ 20h58'30"7d1'
Poster,Thomas & Eleanor
Aur 5h21'19"40d58'
Postert,Hermann
Aur 5h14'47"42d47'
Postl,James Leonard
Cep 20h58'17"60d19'
Postle,Wendy B
Lyn 8h15'60"37d33'
Postlethwaite,Kathleen
Equ 21h23'57"2d34'
Postlewait,Gregory Tylor
Uma 8h46'0"68d54'
Postma,Rik
Aur 7h25'46"38d2'
Poston,Brittany Ashlyn
Mon 6h39'22"11d32'
Poston,Melanie Ann
And 0h4'13"47d37'
Poston,Spencer Maxwell
Cet 2h11'28"5d59'
Posvic,Holly Jill
Com 12h44'40"23d35'
Potamkin,Lexie & Robert
Lyr 18h41'1"38d30'
Potas,Elizabeth Ann
And 0h10'33"30d21'
Potash's Deerling Star,Mark
Uma 9h49'41"48d52'
Potash,Alan Richard
Sco 17h2'31"-31d33'
Potato
Cma 6h10'16"-16d8'
Potchoiba,Catherine
Cyg 20h19'25"38d31'
Pote,Andrew Ryan
Her 16h59'29"31d6'
Pote,Sr,Robert A
Cep 21h17'37"80d35'
Pote-Hunt,Alicia Jane
And 0h1'1"46d4'
Poteat,Leslie Cade
Tau 4h12'23"21d0'
Poteat,Mickey Franklin
Cet 3h5'34"5d12'
Poteet,Kiersten Victoria
Uma 11h13'57"49d53'
Potempa,Michael & Rhea
Cyg 19h31'48"36d23'
Potenza,Sam
And 2h20'54"42d14'
Poteshman,Michelle & Robert
Crb 16h0'35"31d60'
Potestio,Richard
Aur 6h18'0"38d49'
Poth,Daniel
Aur 6h37'0"37d49'
Poto,Great Grandma
Cas 0h20'26"63d28'
Potocki,Maria Nicole
Cas 0h22'59"61d48'
Potoli
Cyg 21h7'35"30d5'
Potorti,M & M
Aur 4h49'37"40d7'
Potso
Cyg 21h0'37"38d43'
Potter Extraordinaire, Don
Bradford
Per 1h45'46"53d57'
Potter,Carolyn S
Cyg 20h33'1"30d41'
Potter,Claire Ann
Cam 4h3'44"60d19'
Potter,Debbie
Cyg 21h33'38"30d12'
Potter,Delaney Ann
Mon 7h17'15"-6d56'
Potter,Devon Lee
Ser 18h1'54"-14d50'
Potter,Elaine H
Lyn 9h16'31"35d10'
Potter,George Randall
Ori 5h47'53"11d21'
Potter,Helen & Jack
Crb 15h59'0"37d36'
Potter,Jacqueline
Cas 0h44'19"63d22'
Potter,Jennifer A
Cas 1h54'0"63d50'
Potter,Jo Ann
And 23h13'48"37d37'
Potter,Joanna Rachel
Peg 23h32'1"11d51'
Potter,Lawrence
Cyg 19h31'46"36d30'
Potter,Leah
Cyg 20h4'38"40d2'
Potter,Linda Lee
Lyn 7h50'51"34d52'
Potter,Lindsay
Cas 0h56'58"50d28'
Potter,Marc
Dra 17h26'60"58d28'
Potter,Margaret
Crb 16h1'60"38d8'
Potter,Michael Elizabeth
Cep 22h16'23"61d50'
Potter,Michael Irving
Tri 1h46'0"25d42'
Potter,Michael Parker
Aql 19h4'56"-0d30'
Potter,Mike
Umi 16h13'13"73d39'
Potter,Mr Rory
And 0h22'57"40d55'
Potter,Mrs M
Lac 22h4'17"51d15'
Potter,Pamela Ann
Tri 1h34'43"30d25'
Potter,Randall L
Lmi 10h4'27"36d29'
Potter,Richard G
Cet 0h40'56"-3d44'
Potter,Scott
Her 17h16'23"27d51'
Potter,Suzanne "Doodee"
Cyg 21h30'32"53d1'
Potter,The Star Of My
Life,Deanna
Vul 20h17'1"28d49'
Potter,Todd
Per 3h16'18"41d28'
Potter,Walter F
Cet 3h9'16"5d35'
Potter,William Randall
Dra 16h32'21"61d55'
Potter-Efron,Pat & Ron
Lyr 18h31'42"32d39'
Potters,Hal K
Uma 9h34'42"50d12'
Potthoff,Katrin
Cas 0h48'54"60d41'
Potts III,"Bill's Star",William C
Her 17h36'1"21d31'
Potts Star,The
Peg 21h56'31"33d40'
Potts,"Best Friends" Barbara
Peg 23h5'54"21d7'
Potts,Alicia Marie
And 23h20'1"44d38'
Potts,Angelina
Vul 20h1'18"28d18'
Potts,Carole Leigh
Lyn 8h0'21"34d13'
Potts,Christopher James
Lac 22h11'36"46d49'
Potts,David
Aql 19h26'1"15d45'
Potts,David Oldham
Uma 10h28'59"53d18'
Potts,Douglas J
Hya 8h35'52"-6d29'A
Potts,Hollis Nicole
Cam 13h11'46"84d47'
Potts,Jordan F
Aur 6h31'28"32d39'
Potts,Lawrence Ray
Dra 15h58'13"68d27'
Potts,Lisa
And 23h16'21"40d0'
Potts,Lloyd L
Leo 9h22'33"17d35'
Potts,Louise Edith
Vul 21h22'44"27d14'
Potts,Mary Margaret
Del 20h22'26"7d31'
Potts,Melva
Peg 22h0'47"33d18'
Potts,Monica Rose
Cas 1h3'40"60d12'
Potts,Rebecca
Peg 22h23'43"25d7'
Potts,Rebecca M
Hya 8h35'52"-6d29'B
Potts,Robert Roy
Aql 19h25'45"-6d56'
Potts,Stefanie
Eri 3h24'0"-5d56'
Potts,Sylvia
And 0h11'37"38d0'
Pottstown Destiny 5-24-95
Aur 5h11'48"41d3'
Potuck Happy 50th,John W
Hya 8h41'37"2d40'
Potuck Happy 50th,Mary H
Eri 3h27'27"-5d27'
Potvin's Pied-A-Terre
Ori 6h8'50"5d32'
Potvin,Chantal
Cnc 8h31'28"30d2'
Pou "Ebbie Sue",Evelyn Susan
Mon 6h57'58"7d54'
Pou,Jr,"Bill Pou" William
Knox
Aql 20h2'12"1d26'
Pouchain,Fabienne
Dra 9h58'48"73d47'
Pouget,Alain
Crb 15h36'33"35d8'
Pouget,Thierry
Vul 19h21'37"23d57'
Poukha
Tri 2h14'24"31d17'
Poulenard,Laurent
Umi 15h44'51"73d16'
Poulet,Jessica Taylor
Mon 6h27'42"3d25'
Poulides,Nicholas G
Aur 5h54'51"30d19'
Poulin,Claude
Cep 22h29'44"63d31'
Poulin,Donald
Gem 7h0'38"11d32'
Poulin,Eveline et Philippe
Cyg 19h27'15"33d45'
Poulin,Laurie Ann
And 1h17'26"36d58'
Poulin,Vicki
And 0h11'38"47d32'
Pouliot,Martin
Dra 20h31'0"70d41'
Poulos,Tina
And 2h16'49"41d19'
Poulsen,Anker Fuglsang
Lyn 7h56'19"42d13'
Poulsen,Christopher
Her 16h38'20"26d2'
Poulsen,Timothy & Rebecca
Ori 5h28'51"1d30'
Poulson's Point
Aql 19h56'35"7d30'
Poultney,Chad F
Tau 4h3'3"21d7'
Poulton,Chip
Cet 2h5'19"1d35'
Pouncett,Mary & Jack
Cyg 21h38'14"30d9'
Pound,Anne-Marie
And 0h49'1"37d47'
Poupoune
Lac 22h38'49"40d47'
Pour l' Amour d'Alyson
Cas 23h26'33"61d3'
Pour L'Amour De Toi Theodore
Oph 18h39'15"7d12'
Pour Toujours
Vul 19h18'58"25d29'
Pour Toujours Ensemble
Cet 0h55'1"0d43'
Pour-Azar,Aaron
Cnv 13h1'50"50d50'
Pourcelet,Michel Groize
And 23h4'40"40d23'
Pourciau,"Forever", Lynn
And 22h59'10"41d6'
Poussin d'Amour et Nouno
Cnv 12h26'56"38d54'
Poutchnine,C
Lac 22h17'45"55d7'
Poutiatine,Marcia
Cas 0h21'24"58d17'
Pouxviel,Mousieur Pierre
Peg 23h48'26"31d2'
Povah,Sylvia
Ori 5h56'21"15d29'
Povah,Trevor
Cnv 13h6'48"41d14'
Povey,Joseph
Oph 18h2'27"12d25'
Povi
Uma 11h52'24"41d44'
Povich,Jennifer
Del 20h18'18"10d17'
Povich,Jeremy Steven
Boo 13h46'31"21d59'
Povich,Laura
Aql 19h7'12"1d29'
Povich,Michael
Cmi 7h35'50"10d37'
Povkovich,Larry Eugene
Oph 17h36'53"14d9'
Povman,Morton
Per 2h37'1"45d5'
POW
Aur 4h50'46"50d55'
Pow,Jill Ann
Del 20h23'0"10d35'
Powell
Hya 9h30'30"-5d24'
Powell Family's Star, The
George Macutchen
Cnv 13h41'50"35d54'
Powell III,Robert L
Cet 1h14'34"-6d44'
Powell III,Robert L
Ser 17h31'58"-11d3'
Powell,Albert W
Aql 18h53'47"8d55'
Powell,Andrea Elise
Peg 22h19'19"8d23'
Powell,Ashley Grace
Cnc 7h59'22"10d21'
Powell,Athena Brianne
Lac 22h14'46"49d1'
Powell,Barbara Ann
Cas 23h31'15"62d55'
Powell,Buster
Sct 18h52'45"-6d58'
Powell,Clara Margaret
Hya 9h24'38"6d27'A
Powell,Daniel Scott
Per 3h30'60"50d54'
Powell,Darlene A
And 23h4'16"40d54'
Powell,Darryl
Her 16h19'47"26d29'
Powell,David McConnell
Sge 20h7'12"20d27'
Powell,Debbra Jane
Cas 1h36'50"60d46'
Powell,Devin Taylor
Cnv 13h15'47"32d1'
Powell,Dr Robert
Dra 18h49'44"84d30'
Powell,Elsie
Uma 10h45'28"51d58'
Powell,Ervin & Jeanette
Lyr 19h20'19"38d12'
Powell,Evelyn L
Lyr 18h44'26"47d13'
Powell,Fred & Loretta
Cyg 21h58'27"50d34'
Powell,Gary Lewis
Dra 16h57'43"58d45'
Powell,Geoff K
Per 2h6'52"57d30'
Powell,Heather
Lyn 7h30'48"42d47'
Powell,James
Dra 20h38'29"68d49'
Powell,James Ian Adams
Aql 18h43'22"10d6'
Powell,Janet Lesley
Ori 5h41'59"1d43'
Powell,Janet Lou
Mon 7h5'22"-5d51'
Powell,Janice
Lyr 18h33'52"41d31'
Powell,Jennifer
Ori 4h53'1"-2d1'
Powell,Jennifer Jo Catherine
Aql 19h25'48"12d9'
Powell,Joe M
Cet 0h51'28"1d54'
Powell,John & Gail
Cyg 20h43'32"46d24'
Powell,Joseph F
Boo 15h2'40"41d44'
Powell,Jr,Frank J
Ori 5h55'59"14d27'
Powell,Julianne
Eri 4h4'52"-19d46'
Powell,Kellie Lovoe McDaniel
Eri 2h59'24"-4d43'
Powell,Kenneth
Boo 13h36'44"17d24'
Powell,Kerry Chivon
Oph 17h54'28"12d34'
Powell,Krishanna Renee
Gem 7h1'59"30d38'
Powell,Lauren Ashley
Eri 3h55'0"-2d30'
Powell,Lawrence "Cap"
Uma 11h40'40"62d18'
Powell,Linn McCarter Norris
Mon 6h19'25"8d44'
Powell,Mary Rachel
And 0h5'0"46d33'
Powell,Mavis Hollingsworth
Cet 1h26'27"1d7'
Powell,Natalie Ann
Oph 17h53'0"12d56'
Powell,Robert Alexander
Per 1h38'24"52d39'
Powell,Rocky
Sex 10h0'45"-10d23'
Powell,Roger Dale
Lac 22h33'27"50d3'
Powell,Sally J
And 0h5'32"46d55'
Powell,Samuel Andrew
Morgan
Hya 9h1'1"3d7'
Powell,Shannon
Sgr 18h4'39"-28d11'
Powell,Shannon Leigh
Mon 6h30'38"11d60'
Powell,Suzanne
Mon 8h2'27"-8d51'
Powell,Suzanne
And 23h30'49"42d40'
Powell,Tammy
Cas 1h54'50"75d25'
Powell,Tammy Marie Upman
Equ 21h9'17"11d31'
Powell,Thomas Harrison
Trostel
Hya 9h24'38"6d27'B
Powell,Todd
Cep 22h47'26"78d60'
Powell-Custer,Baby
Uma 8h56'23"71d31'
Powell-Stamm,Tristan
Darrowby
Aur 6h39'25"38d8'
Powelson,Jr,Richard Joseph
Aur 6h56'10"44d24'
Power Bulge,The
Dra 11h48'0"70d52'
Power IV,Thomas R
Her 17h28'46"28d11'
Power Of Friendship
Sgr 19h50'19"44d41'
Power of Love Forever Debra
Vul 19h23'1"25d8'
Power Of Love,The
Cyg 19h29'51"31d27'
Power of Polizzi,The
Oph 16h18'26"-6d60'
Power's Guiding Light
Mon 6h38'38"10d19'
Power,Brendan P
Per 1h57'36"50d18'
Power,Brian J
Her 18h12'56"30d57'
Power,Chris
Dra 10h36'0"77d52'
Power,David Joseph
Aur 5h55'29"31d13'
Power,Diane T
Com 13h17'46"26d22'
Power,Dr Helen
Oph 16h20'43"2d36'
Power,Janice
Mon 6h49'47"10d38'
Power,Jeffrey
Her 17h54'1"20d23'
Power,Matthew D
Boo 14h43'1"48d15'

Power,Michelle
And 0h6'12"47d13'
Power,Michelle-Paul- Charlie
Uma 8h35'29"59d38'
Power,Nicholas John
Ori 5h56'31"8d47'
Power,Sandy
Uma 8h47'16"50d48'
Powers Forever,Robert &
Eileen
Lyr 18h29'26"35d11'
Powers' Eternal Love, Jim &
Peggy
Uma 10h1'11"72d26'
Powers,Aeron Elizabeth
Oph 18h5'24"10d30'
Powers,Alice
Cam 4h14'25"60d16'
Powers,Anna L
Uma 10h33'45"42d9'
Powers,Barbara Ann
Cas 0h15'19"63d57'
Powers,Bernadette
And 1h17'56"38d4'
Powers,Betty & Richard
Aql 19h3'12"-5d44'
Powers,Bonnie
Eri 4h38'16"-8d56'
Powers,Bonnie-Annique
Katayama
Cas 2h7'0"70d18'
Powers,Carol L
Cas 1h21'12"50d36'
Powers,Catlin Ishihara
Cas 2h10'31"70d45'
Powers,Charles H
Lac 22h32'1"53d53'
Powers,Chilli Metz
Hya 8h55'1"2d31'
Powers,Dennis
Cnv 12h48'41"51d53'
Powers,Emily Katherine
Vir 13h33'3"-8d59'
Powers,Gaston Paul
Ser 15h38'15"3d40'
Powers,Harold & Violet
Uma 8h43'55"73d3'
Powers,Hilary Leigh
Cyg 19h23'19"44d23'
Powers,Jamie
And 0h18'43"31d31'
Powers,Janice L
Cyg 21h14'49"35d48'
Powers,Joan
Peg 22h0'1"22d31'
Powers,Jordan Jean
Mon 8h7'0"-8d38'
Powers,Jr,Robert
Her 17h33'59"21d11'
Powers,Kate
Mon 8h2'19"-10d9'
Powers,Laurene A
Aur 4h54'30"38d15'
Powers,Liam Seal
Cma 7h15'44"-15d29'
Powers,Mara
And 2h21'36"44d17'
Powers,Matthew
Dra 14h56'11"63d10'
Powers,Mildred C
Lyr 19h17'29"41d13'
Powers,Myron C
Uma 10h54'12"58d58'
Powers,Nancy M
Eri 3h55'43"-5d26'
Powers,Ragina
Mon 6h43'27"7d20'
Powers,Robert D
Cet 2h51'16"0d33'
Powers,Thomas Alan
Cnv 12h16'32"41d49'
Powers,William T
Dra 18h40'60"70d11'
Powerstar Bastian
Cas 0h10'0"60d46'
Powloski,James
Aur 7h7'0"38d31'
Pownall,Aaron
Dra 16h22'32"66d25'
Pownall,Amanda Kate
And 1h47'36"39d29'
Pownall,Stella Louise
Ori 6h7'47"2d44'
Powner,Stevoe
Pho 0h45'51"-45d9'
Poynter's Pinnacle
Umi 16h35'36"75d22'
Poynter,Denise Lyn
Mon 6h32'0"3d4'
Poyourow,Douglas
Ori 5h27'32"-6d11'
Pozarek,Steve & Pat
Cyg 19h33'34"27d53'
Pozo,Karla
Mon 6h23'17"7d58'
Pozorski,Diane
Cyg 19h34'1"28d29'
Pozza,Joseph
Her 17h17'22"43d13'
Pozzi,Cecile
Cmi 7h23'20"5d5'
Pozzoli,Kelly C
Peg 22h57'31"34d28'
PPG "Euro Vision",The
Oph 17h18'0"11d54'
Prabhakaran, Velupillai
Ori 5h56'57"20d52'
Prabhat Jain
Peg 21h24'10"10d19'
Prabis,Sylvain
And 23h2'32"50d56'
Prachar,Merrill Reece
Hya 9h3'15"1d50'
Pracht,David
Her 16h54'1"28d18'
Pracht,John C
Her 17h30'49"26d49'
Pracht,Larissa
Vul 19h22'1"25d21'
Pracht,Mechthild
Ser 16h16'52"0d50'
Pradel,Jean-Pierre
Umi 15h15'58"71d59'
Pradella,Barbara Ann Francis
Pho 23h41'10"-43d31'
Pradera,Caroline
Crb 16h18'1"38d50'
Pradhan,Tara
Cam 3h55'28"56d20'
Prado
Del 20h22'30"2d13'
Prado,Mary Dawn
Mon 6h21'19"5d26'
Pradon,Jerome
Aur 5h15'0"47d26'
Prady,Norman
Her 16h4'1"40d5'
Pradéra,Caroline
And 23h3'54"38d42'
Praeceptor H-J Waschke
Cnc 8h33'30"31d28'
Praeclara Rosa Flara
Ori 5h36'0"-6d21'
Praed,Michael
Vul 19h48'0"28d38'
Prael,Fredrick Traynor
Dra 16h51'32"67d12'
Prael,Katharine Elizabeth
Lyn 9h4'57"40d18'
Praetsch,Barbara
Gem 6h59'18"10d50'
Prager,Arthur
Per 3h22'41"40d14'
Prager,Michael William
Hya 8h42'0"5d48'
Prailes,Edward
Umi 15h33'53"66d13'
Prain,Marlene
Lac 22h11'10"51d19'
Prall,Nancy Elizabeth
Cas 23h27'55"54d13'
Prange,Andreas
Hya 9h17'6"-6d59'
Prange-"The Gangster",
Christopher
Dra 10h59'21"73d27'
Pranger,Pam
Her 17h29'57"30d43'
Prangnell,Lee Verna
Lyr 18h27'55"46d31'
Prantil,Lori
Cnc 8h20'14"6d48'
Prantner,Sabine und Gert
Uma 11h46'59"33d3'
Prasad,Rameshwar
Her 18h3'19"14d22'
Prasad,Urmila
Cam 13h26'18"80d59'
Prasad,William A
Dra 20h14'1"67d37'
Praschnik,Salomon
Ser 16h3'21"14d28'
Prasil,Jesse
Mon 6h31'55"10d1'
Prater 7-6-32 To 8-17-
95,Teena J
Lmi 10h47'0"30d8'
Prater,George
Per 4h21'18"50d30'
Prater,James Howard
Aur 6h27'1"37d38'
Prater,James Howard
Dra 9h41'23"81d15'
Prather 25JDM,Andrew J
Uma 8h52'56"51d31'
Prather II,David Ross
Vul 19h22'15"26d49'
Prather,Amanda
Cam 6h5'48"68d29'
Prather,Danielle Roxanne
And 0h21'35"43d38'
Prather,Edward W
Dra 15h42'54"66d1'
Prather,Georgia Cass
Leo 10h58'43"20d19"
Prather,Mary
Lib 15h35'57"8d37'B
Prather,Samantha
Cyg 19h45'37"30d21'
Pratik
Cam 3h37'40"70d45'
Prato,Gregory Joseph
Uma 10h9'12"59d19'
Prato,Gérard
Ori 6h21'33"10d10'
Pratt,Brendan Stewart
Lyn 7h42'1"50d46'
Pratt,Carrie Van Vleck
Cep 22h35'16"58d12'
Pratt,Christopher Thomas
Allen
Cet 1h55'41"-2d7'
Pratt,Christopher Gardner Lee
Aur 6h28'45"35d36'
Pratt,Col Robert H
Per 2h31'31"57d38'
Pratt,Colin Adam
Lac 21h59'52"41d2'
Pratt,Deborah
Gem 6h44'17"18d59'
Pratt,Forever & Always Chris
& Pam
Umi 16h25'53"71d22'
Pratt,George F
Lyr 19h4'36"28d30'
Pratt,Jamie Ellen
Peg 22h32'10"31d36'
Pratt,John
Aql 18h58'39"14d9'
Pratt,John David
Boo 15h4'30"30d29'
Pratt,Jonathan James
Ser 15h17'26"8d30'
Pratt,Lori
Del 20h16'48"14d37'
Pratt,Lorraine
Cet 1h2'49"1d54'
Pratt,Mary V
Peg 22h38'41"24d43'
Pratt,Penny
Sge 18h55'57"19d34'
Pratt,William W
Aql 20h1'43"10d44'
Pratt,Wilson Philip
Cas 2h3'58"60d6'
Prause,Melissa Marie
Lyr 19h20'58"41d8'
Pravato,Rocky & Lisa
Sge 19h35'29"16d52'
Pravosudovich,Mikhail
Ori 5h56'52"13d38'
Prax Point
Lyn 8h4'33"34d31'
Prax,Dominique
Aql 19h30'43"-8d48'
Pray (Dear Mom),Margie
Cas 0h27'48"61d22'
Pray,Jr,Natalie & Malcolm
Umi 15h32'47"67d27'
Pray,Randy Scott
Per 3h58'1"51d14'
Prayer,Nana & Papa "You are
the Best"
Dra 15h58'33"63d56'
Prchal,Sharon Rose
Aql 18h44'0"-2d8'
Preacher's Promise
Aur 6h33'0"35d55'
Preas,Maxine Christenberry
Psc 22h57'35"2d29'
Prebil,Susan M
Aql 19h54'58"10d28'
Preblud,Joseph Gregory
Her 16h40'24"35d24'
Preciosa Joanna
And 2h13'37"42d50'
Preciosa,Jane La Estrella
Cyg 21h7'37"39d39'
Precious
Sgr 19h40'5"-31d4'
Precious
Mon 6h53'37"-0d26'
Precious
Lyr 18h33'31"37d55'
Precious
Peg 23h41'0"17d25'
Precious
Mon 7h1'30"5d15'
Precious
Lyn 7h53'19"43d42'
Precious
And 2h2'0"39d24'
Precious
Aql 19h42'49"14d54'
Precious Alana Fay
Lyr 19h21'32"31d2'
Precious Amy
Mon 8h4'13"-1d2'
Precious Angel
Ori 5h55'48"17d16'
Precious Brittany
And 23h44'54"45d38'
Precious Connor James
Hya 8h35'46"-11d1'
Precious Deborah
Mon 8h6'56"-8d20'
Precious Elizabeth
And 23h46'50"47d0'
Precious Frintrup Friendship
Cyg 21h18'54"38d27'
Precious George
Boo 14h21'24"54d3'
Precious Hawk (Prezioso
Accipiter)
Cyg 20h37'26"41d13'
Precious Jeanine
Cas 0h53'16"61d19'
Precious Mary
Ori 5h53'0"14d35'
Precious Megan
And 1h36'20"37d18'
Precious Memories
Peg 22h40'45"32d22'
Precious Memories
Lyr 19h21'52"40d18'
Precious Mother Cara
"Granny"
Lyn 7h54'32"43d49'
Precious Nona
Peg 22h28'13"29d22'
Precious One Taylor Nicole
And 0h26'21"27d40'
Precious Patrick
Aur 5h10'15"40d11'
Precious Patty- Treasured
Friend
Peg 21h46'0"34d33'
Precious Peter
Boo 15h11'25"49d57'A
Precious Steven
Her 16h46'1"48d48'
Precious Susan
And 23h48'20"45d46'
Precious Suzanne & Patrick
Uma 11h1'29"46d57'
Precious Terese
Leo 10h39'34"20d10'
Precious Vaughn
Per 2h12'39"56d33'
Pred,Gabrielle Anne
Mon 7h48'40"-2d29'
Preddy,Donald Eric
Ori 5h27'35"-6d52'
Predergast,Patricia M
And 1h53'47"37d21'
Predovich,Dana Milan Clinton
Cas 23h39'42"61d10'
Preece,Emma
Lyr 18h23'12"46d56'
Preece,Joy
Lyr 19h20'0"38d25'
Preece,Roger c
Umi 17h28'25"85d54'
Preece,Shannon
Lyr 18h40'59"28d17'
Preeti va Manish
Cnv 13h30'46"47d47'
Prefontaine,Steve
Cnc 9h0'47"31d47'
Preg,Patricia A
Peg 23h6'10"11d35'
Pregent,Kristina Nicole
Mon 7h50'11"-3d22'
Pregent,Lauren Christen
Peg 21h49'60"33d56'
Pregent,Sara Anne
And 0h11'36"47d37'
Pregent,Timothy Michael
Peg 23h50'59"30d37'
Pregerson,David Brady
Aql 20h3'14"8d34'
Prehoda,Patricia K
Uma 11h54'44"40d44'
Preinfalk,Gabriele
Lib 14h23'34"-23d46'
Preisinger,Christianne
Aqr 22h51'38"-6d58'
Preisler,Sandy & Dick
Cnv 13h43'53"45d26'
Preismeyer,Tom
Aql 20h7'47"8d33'
Preite,Stephanie
Cyg 19h29'1"33d14'
Prelin
Cas 1h53'30"68d17'
Prelin
Cas 1h54'40"60d36'
Preller,Lt Col Robert H
Lyr 19h15'45"40d3'
Prem,Our Guiding Light
Lyr 18h32'1"42d28'
Prema Sai Baba
Dra 19h54'32"61d25'
Premazon,Michael Arthur
Cyg 19h26'32"50d8'
Premel,Thomas
Boo 15h7'0"10d28'
Prendergast,John Thomas
Peg 23h16'39"32d49'
Prendergast,Robert Lewis
Mon 6h15'20"-5d38'B
Prendergast,Sally Anne
Nicholas
Mon 6h15'20"-5d38'A
Prendergast,Sophie Lee
Peg 22h21'12"35d13'
Prentice,Amanda Mary
And 1h24'1"40d5'
Prentice,Barbara Susan
Crb 16h17'55"37d10'
Prentice,Nena
Cas 2h4'32"59d0'
Prentice,Phyllis
And 1h4'0"38d46'
Prentice,Stephanie Anne
Gem 7h2'19"24d28'
Prentiss,Robert Henry
Sex 10h13'48"-2d37'
Prepsky,Devoree D
Mon 6h7'19"-4d52'
Preschel,Milly
Cas 23h14'51"62d53'
Prescott,Fred H
Oph 16h21'0"-6d36'
Prescott,Heather
Vir 11h50'37"2d15'
Prescott,Irene
Cas 0h32'34"66d27'
Prescott,Jr,Curtis & Gwyn B.
Fordyce
Cep 23h0'15"60d27'
Prescott,Jr,Henry Emil
Dra 16h34'57"51d25'
Prescott,Patricia H
And 0h16'44"37d33'
Prescott,Toody
Lib 14h43'33"-20d16'
Prescott,Wendy R
Cam 3h57'39"56d37'
Presentati,Armand William
Sex 9h52'40'2d27'
Preshlock,Dennis
Her 16h33'32"39d43'
Presley,Jacob
Umi 15h16'48"66d57'
Presotto,Marion
Ser 15h15'53"17d40'
Press,Brian
Aur 5h10'42"40d59'
Press,Denise
Crt 11h11'45"-13d31'
Press,Kristen
Lyr 18h39'1"46d45'
Press,Mark & Nancy
Dra 16h15'26"66d2'
Press,MD,Robert
Oph 17h42'22"12d16'
Press,Neal
Boo 15h9'31"53d40'
Press-45
Lac 22h24'43"37d51'
Pressentin,Erin Nicole
And 1h3'36"47d9'
Presser,Deedee
Ori 5h42'38"1d14'
Presser,Garrett Andrew
Aur 5h33'55"38d43'
Presser,Matthew Aurthur
Boo 15h42'27"50d33'
Presser,Ross Joel
Her 16h25'24"22d48'
Pressey,Pamela Sue
Cyg 20h5'19"39d34'
Pressler,Mark
Her 16h27'34"23d40'
Pressley,Deanna B
Eri 3h4'42"-01d57'
Pressman,Gussie
Tri 2h5'54"32d17'
Pressney,Nicholas Escargot
Ori 5h58'35"11d29'
Prest,Marshall John
Uma 10h50'22"55d6'
Prestedge,Stanley
Uma 9h50'35"54d54'
Prestegaard
Cam 14h19'41"82d13'
Presti
Tau 5h33'0"20d10'
Presti,Vince
Aql 20h33'50"0d50'
Prestia,Jay
Hya 8h14'55"-6d23'
Prestige Accommodations
Peg 23h43'14"25d34'
Prestileo,Vincent
Dra 16h18'37"67d5'
Prestininzi,Susanne
And 0h30'10"28d0'A
Prestis,Andrew C
Hya 8h11'21"-6d40'
Preston B
Lac 22h20'47"53d19'
Preston,Barbara
Lyr 19h21'18"42d32'
Preston,Caroline
Gem 6h47'10"19d22'
Preston,Darryl
Cep 21h15'53"58d17'
Preston,Desiree
Vir 13h32'55"-1d10'
Preston,Don & Genie
Dra 19h13'56"71d14'
Preston,George R & Mildred
Cam 6h33'47"68d23'
Preston,Gia
Aqr 23h39'15"-5d57'
Preston,Gillian
And 0h33'49"40d28'
Preston,Harold Lloyd
Ori 4h56'8"14d28'A
Preston,Heather Lynn
Cas 2h19'19"73d28'
Preston,Ken
Dra 16h10'35"62d42'
Preston,Lorraine M
Ori 6h6'17"8d53'
Preston,Louisa Frances
Lyr 18h34'0"28d8'
Preston,Moira Ann
And 1h51'38"37d36'
Preston,Orvis J
Ori 6h6'30"7d59'
Preston,Pamela Lynn
Cet 0h50'40"-10d54'
Preston,Patrick Michael
Sex 9h54'13"1d47'
Preston,Peter & Sally
And 2h33'1"40d35'
Preston,Robert
Cet 2h51'0"0d55'
Preston,Robert & Dorothy
Cyg 19h30'36"35d54'
Preston,Robert Bear Creek
Cep 21h28'23"58d37'
Preston,Scott & Denise
Cyg 19h32'48"30d51'
Preston,The Jesse Alexander
Aur 5h58'11"37d35'
Prestwood,Roberta Jean
Mon 7h44'41"-6d56'
Preti,Annamaria
Vel 10h12'41"-44d37'
Pretiosus Illuminatus Sandy
Cmi 7h15'55"1d42'
Preto-Rodas,Jonathon
Christopher
Tau 5h39'53"16d14'
Preto-Rodas,Meghan Elizabeth
Cas 0h22'51"61d36'
Pretorius,Joseph William
Aur 7h2'1"41d22'
Pretsch,In Memory Of Paul
Boo 15h0'13"8d32'
Pretti Christi-1
Peg 22h53'0"25d38'
Pretty
Peg 23h3'32"10d4'
Pretty Girl
Cyg 20h24'49"38d52'
Pretty Girl
And 1h21'43"39d3'
Pretty Mary L
And 23h31'44"42d16'
Pretty One
Crb 16h0'32"31d43'
Preusker,Steven James
Aql 19h7'19"2d47'
Preuss,Helmut
Cmi 7h28'16"7d48'
Preuss,Jesse James
Boo 13h59'18"25d6'
Preussel-Duke
Aql 18h58'38"-6d45'
Previtali,Amy Lynn
Uma 11h20'38"45d17'
Previte
Cnv 13h0'18"40d3'
Prevost,Julien
Cep 22h0'23"53d32'
Prevost,Patrice
Vul 20h15'28"22d41'
Prevost,Paul
Tri 2h37'30""35d58'
Prewett,Judy K
Crt 11h22'59"-18d8'
Prewitt,Karen J
Vul 19h17'44"25d2'
Prewitt,Marlene
Cas 1h47'23"60d10'
Prezeau,Navia Veronique
And 23h31'53"49d31'
Preziosi,Federica
Cap 20h39'47"-18d45'
Prezioso Star,The Nicole &
Kimberly
Mon 7h12'39"-10d56'
Prezioso,Anthony F
Lac 22h19'11"46d9'
Prezioso,Vincenzo
Her 16h38'25"36d18'
Prezorski,James C
Aur 5h35'12"48d53'
Prezorski,Paul
Aur 6h31'32"30d22'
Pri,Patrick
Sct 18h44'22"-7d21'
Pribek,Frank
Boo 15h45'39"50d32'
Pribluda,Bagg
Ant 10h46'37"-35d40'
Price
Eri 3h21'58"18d49'
Price "Bo",Kristopher Lane
Lyr 19h2'46"28d39'
Price Numero Uno Tiger
Lawrence W
Tau 5h21'26"16d22'
Price Star,The Toni & Neville
Mon 6h34'53"0d13'
Price,Alexander
Pyx 8h45'22"-29d1'
Price,Alexis Ashton
Mon 7h15'18"-7d1'
Price,Andrew Bryan
Per 1h49'1"56d19'
Price,Barbara Jane
Cyg 21h4'12"40d26'
Price,Beverly Minten
Cas 3h10'23"58d28'
Price,Bob
Cyg 20h53'1"38d50'
Price,Brent
Dra 14h54'1"63d35'
Price,Brian GC
Ari 2h42'19"25d38'
Price,Cameron Alexander
Hya 9h4'1"5d22'
Price,Carla Beth
Sge 18h57'57"20d3'
Price,Carol
Ori 4h54'20"4d33'
Price,Cindy Dyan
Peg 22h14'36"8d58'
Price,Clark William
Oph 17h34'47"-0d51'
Price,Colin Russell
Cmi 7h56'24"4d36'
Price,Corine & Justin
Cyg 19h34'0"37d33'
Price,Daniel Michael
Dra 19h1'52"50d25'
Price,Dean Boyd
Lmi 10h50'51"31d41'
Price,DeAngelo
Aur 6h36'38"37d53'
Price,Dennis
Cyg 21h16'18"38d13'
Price,Dr Phillip D
Lac 22h3'55"40d20'
Price,Dwayne
Aur 6h32'30"37d32'
Price,Eric Christopher
Aur 5h53'1"40d34'
Price,Fred R
Ari 1h59'48"23d1'
Price,Frederick James
Dra 20h16'1"68d7'
Price,Greg
Her 17h36'15"26d13'
Price,Harry Douglas
Uma 9h33'0"48d49'
Price,Hayley
Pyx 8h41'35"-26d47'
Price,Herschel
Boo 13h59'0"12d29'
Price,Jay P
Per 1h47'0"48d4'
Price,Jeanne M
Lyr 18h44'17"42d35'
Price,Jennifer Kathleen
Peg 23h44'52"17d16'
Price,Jennifer Anne
Lyr 19h22'49"30d35'
Price,Jesse C
Her 18h4'15"31d33'
Price,Jessica Renee
Peg 21h50'32"34d51'
Price,Jill Kristin
Sgr 18h49'19"-25d2'
Price,Joe Allen
Aql 18h58'52"3d27'
Price,Jona Marie
Del 20h14'17"10d42'
Price,Josephiné
Umi 16h4'17"70d25'
Price,Jr,Mark A
Ori 5h53'14"15d52'
Price,Julie
And 2h22'37"44d25'
Price,Juno
Dra 16h38'1"70d12'
Price,Karen Jean Dunlap
Cyg 21h2'50"31d47'
Price,Kristi Robin
And 1h23'53"36d27'
Price,Larry
Hya 9h15'37"0d25'
Price,Lee & James Hayes
Lyr 18h30'11"38d4'
Price,Lesley
Aur 4h47'39"36d2'
Price,Lisa
Uma 12h16'39"53d23'
Price,Loraine
Mon 8h2'49"-6d47'
Price,Lorena Jo
Cet 0h47'17"-5d13'

Price,Marc David
Lac 22h32'30"55d36'
Price,Marian
Lyr 18h58'45"37d27'
Price,Mark Andrew
Hya 8h58'54"-7d18'
Price,Marsha Anne
Equ 21h6'0"11d16'
Price,Mary Elizabeth
Cas 0h31'41"61d13'
Price,Maureen Ann
Cas 0h36'10"60d46'
Price,Michael James
Cet 2h35'36"-12d17'
Price,Michelle
Peg 21h55'57"31d50'
Price,Michelle Heather
And 0h4'55"30d33'
Price,Moreton & Ruby
Crb 16h17'37"28d5'
Price,Norman Michael
Lac 22h12'22"38d20'
Price,Patrick
Aql 19h47'54"14d16'
Price,Paul
Hya 9h5'19"1d56'
Price,Peggy
Cyg 21h14'24"37d29'
Price,Penelope Jane
And 0h46'47"36d0'
Price,Peter Leon
Ari 2h38'0"30d16'
Price,Philip & Charles
Per 2h27'23"56d44'
Price,Ray
Aur 6h1'33"31d26'
Price,Rhonda
Lyr 18h27'12"38d55'
Price,Richard & Rhonda
Dra 16h7'29"62d29'
Price,Richard George
Lyr 18h32'0"33d10'
Price,Rita Ann
Oph 17h31'60"0d47'
Price,Robert
Dra 15h38'0"54d37'
Price,Samantha Jo
Eri 2h59'46"-6d4'
Price,Sandra S
Mon 6h57'49"7d35'
Price,Sarah
Equ 21h7'60"10d53'
Price,Sherylyn
Cmi 7h22'34"8d4'
Price,Sonia E
Sgr 18h50'38"-34d56'
Price,Spencer Jeffrey
Lyr 18h45'17"38d56'
Price,Susan
Ori 6h8'0"7d52'
Price,Terri J
Ari 2h49'18"30d21'
Price,Valerie
Uma 14h3'24"57d56'
Price,Veronica Anne
Cas 23h14'25"60d52'
Price-Bassett Star,The
Ori 5h4'46"1d48'
Prichard,Kathleen
Equ 21h7'1"11d35'
Prichard,Ken Scott
Aur 6h47'20"37d51'
Prichard,Vincent
Ser 17h55'14"-14d46'
Prickett,Melody Jo
Cma 6h59'32"-18d50'
Prickette,James G
Her 17h13'55"41d19'
Priddle,Bradford Harvey
Aqr 20h57'20"-0d26'
Priddy,Patricia F
Crb 16h5'0"30d36'
Pride of Starlord Bernie
Uma 14h2'20"71d56'
Pride,Dr James R
Oph 17h1'16"10d11'
Pride,James R
Hya 8h19'1"5d23'
Pride,Sr,Jeffrey Paul
Per 2h36'57"43d28'
Prideaux,Jack & Lynda
Ori 5h53'54"0d59'
Prideaux,John
Peg 22h10'42"4d39'
Prideaux-Brune, Nicholas
Per 3h29'11"40d24'
Pridgen,Julia Rusher
Peg 23h25'17"11d48'
Pridham,Mark
Boo 14h19'56"51d53'
Priebe,Malley
Lyn 8h52'13"33d43'
Priebe,Peter
Peg 23h34'29"17d23'
Priem,Ivan
Dra 10h40'42"80d23'
Priese,Irma
Cap 20h35'39"-20d7'
Priest,George & Anna
Aql 19h2'56"-0d21'
Priest,John C
Oph 17h29'30"-24d6'
Priest,William Stansel
Hya 8h56'37"-6d52'
Priest-Poppa Bear, Donald John
Uma 11h47'34"30d36'
Priester,Evan Hanson
And 0h24'1"45d43'
Priester,Thomas Morgan
Boo 14h56'11"28d13'
Priestley,Keith Ferrin
Per 3h4'43"42d11'
Priestley,Mae Irene
Cyg 21h13'22"39d24'
Priestley,Maxfield Von
Aql 20h2'0"9d55'
Priestley,Stephen C
Boo 14h26'0"51d5'
Priestley,The Andrew Mark
Ori 6h6'24"3d2'
Priestly,Duncan
Uma 9h48'0"54d55'
Priestman,Henry
Cyg 20h22'45"38d40'
Priestman,Paul
Lac 22h11'27"48d43'
Prieto Shines Forever
Tau 4h2'45"30d32'
Prieur-Cinquallbre
Peg 23h29'39"21d16'
Priff,Nancy & John Hubbard
Crb 16h6'36"26d32'
Prigge,Dr D
Uma 11h21'33"40d20'
Prigione,Elisabetta
Cyg 20h6'49"37d45'
Prignano,Bobby
Uma 10h35'31"52d8'
Prignano,Loretta
And 2h30'1"47d57'
Prijatel,Caroline Smith
Cet 1h43'40"-1d59'
Prijatel,Liza Carroll
Aql 19h3'48"-1d36'
Prikasky (Alex, Carol, & Daughters)
Leo 9h48'58"30d60'
Prilliman,Lewis A
Sgr 19h7'37"-28d53'
Prima Donna
Aql 20h1'38"10d17'
Prima,Terry
Pic 5h0'6"-44d54'
Primarius
Peg 22h44'38"3d44'
Primavera
Leo 10h50'16"-0d15'
Primavera
Hor 3h36'28"-46d14'
Primavera,Alissa Marie
Cas 0h56'44"63d3'
Primavera,Dr Louis H
Per 3h57'26"50d41'
Primavera,Gerald
Hya 8h22'55"5d49'
Primavera,Paul
Eri 3h17'40"-4d12'
Primdahl,Erin Kathryn
Cas 1h19'57"61d2'
Prime,Andie
Lyn 8h56'58"36d25'
Prime,James Miller
Aql 19h0'58"10d39'
Prime,Tricia Zachary
Oph 17h53'20"12d32'
Prime,Ursula Lipari
Cas 0h28'39"58d36'
Primeau,Bradley John
Hya 8h59'52"0d4'
Primeggia,Michael A
Aur 5h0'37"42d25'
Primenta,Costas D & Vasiliki Gigiakos
And 23h22'0"50d33'
Primis,Blair
Aur 5h59'36"40d48'
Primrose
Cas 1h28'22"58d43'
Primrose,Melissa Katherine
Lyn 7h42'44"40d14'
Primrose,Ryan Scott
Boo 15h38'0"47d34'
Prince
Her 16h47'15"48d49'
Prince
Uma 11h24'20"49d41'
Prince Albert
Ori 6h0'50"1d5'
Prince Angel
Peg 23h2'49"21d21'
Prince Brat
Lyn 8h13'0"47d33'
Prince Charming
Sge 20h6'59"16d6'
Prince Dawsa
Eri 3h10'0"-4d28'
Prince Dean
Lac 22h10'1"48d60'
Prince Eric
Per 3h12'45"48d49'
Prince Fumiya
Psc 1h21'38"4d11'
Prince James
Gem 7h31'58"31d4'D
Prince John's Fantasy
Ser 18h1'28"-14d30'
Prince Karaba
Aqr 20h53'59"-1d46'
Prince Lawrence
Per 3h5'39"50d3'
Prince Michael
Lac 22h1'44"51d18'
Prince Nickolas
Aql 20h15'20"5d33'
Prince Rick
Umi 16h17'21"71d57'
Prince Rudolf
Aqr 22h1'57"-11d9'
Prince Sean
Ori 5h54'1"-1d49'
Prince William
Aql 18h59'14"-6d46'
Prince,Alan
Her 16h26'13"28d16'
Prince,Amanda Faye
And 0h15'16"46d26'
Prince,Bernard
Sgr 19h33'43"-31d50'
Prince,Birgitte
Mon 6h41'54"7d18'
Prince,Charles L
Oph 18h7'1"11d43'
Prince,David Loren
Dra 14h42'1"62d33'
Prince,Frank & Esther
Cyg 21h8'19"48d59'
Prince,Jennifer R
Cet 1h57'51"0d44'
Prince,Kaitlyn Ann
And 23h45'17"46d42'
Prince,Kelsey Jane
And 23h26'25"48d19'
Prince,Ken
Mon 8h8'6"-8d18'
Prince,Megan Beth
Cam 3h49'36"58d42'
Prince,Peter
Cnc 9h9'1"30d1'
Prince,Rae Deane
Aql 19h27'10"10d47'
Prince,Tammy Louise
Cnv 12h46'28"40d15'
Prince,Thomas S
Aql 19h43'36"12d50'
Prince,Vanner
Cyg 21h2'55"39d45'
Prince-Schmidt,Marcia
And 23h20'29"42d25'
Princesa
And 1h7'38"40d46'
Princesa Edurne
Cru 12h26'8"-60d39'
Princess
And 0h30'44"31d45'
Princess
Cas 2h39'1"58d43'
Princess
Cas 3h23'1"75d16'
Princess
And 23h4'29"41d46'
Princess
And 2h5'20"41d12'
Princess
And 22h56'1"51d43'
Princess
Peg 22h10'43"25d5'
Princess
Del 20h35'52"20d12'
Princess
Peg 21h27'17"22d41'
Princess
And 0h8'16"34d54'
Princess & Frog
Aur 7h11'16"41d15'
Princess Alana
Com 13h1'34"21d46'
Princess Angela
And 0h17'1"46d14'
Princess Anne
Lyr 19h26'1"37d41'
Princess Aziza
Col 6h37'13"-34d14'
Princess Beverly
And 1h53'20"41d10'
Princess Bixby
And 23h17'23"47d33'
Princess Brenda
And 1h22'0"39d31'
Princess Bunny
Mon 7h42'44"-1d35'
Princess Buttercup
And 1h56'59"40d25'
Princess Carolyn
And 0h38'18"40d12'
Princess Catherine
Tau 3h35'25"24d4'
Princess Charlene
Lyn 7h38'21"37d32'B
Princess Cheryl
Cap 21h23'22"-14d50'
Princess Chiara
Cas 0h23'58"60d31'
Princess Dana
And 23h30'16"40d15'
Princess Darcy Michelle
Ori 6h15'36"8d32'
Princess Debbie
And 0h24'30"36d8'
Princess Diana,The
Sco 17h31'45"-30d39'
Princess Donna Louisa
Mon 7h0'54"8d42'
Princess Dumas
Lmi 9h29'16"38d31'
Princess Elisa
And 0h52'24"40d22'
Princess Elizabeth
Ori 6h7'38"20d8'
Princess Flo Prince Alex
And 2h31'23"40d48'
Princess Gerry
Del 20h17'12"9d15'
Princess Haifa
Tri 2h27'36"30d48'
Princess Hiver
Tau 4h59'32"28d16'
Princess Inna
Ari 2h38'13"30d39'
Princess Jan
And 23h20'57"38d5'
Princess Judy
Peg 22h41'47"25d23'
Princess Julie
And 1h29'51"39d49'
Princess Kathleen
Peg 23h49'30"30d55'
Princess Katrina
Com 12h52'19"20d32'
Princess Katrina
And 1h50'59"40d40'
Princess Kay
Gem 7h31'58"31d4'A
Princess Kay
And 23h30'60"49d52'
Princess Kelly Rae
And 1h8'35"38d26'
Princess Kimberly
Mon 6h32'37"-0d5'
Princess Leann
Aql 19h33'21"0d26'
Princess Leanne
Del 20h18'58"13d32'
Princess Lesley
And 0h29'35"28d27'
Princess Linda
Cas 1h38'53"58d37'
Princess Linda
And 0h59'12"45d31'
Princess Lisa
And 1h36'37"38d14'
Princess Lisa
Peg 23h25'37"10d10'
Princess Lisa Ann
And 23h17'58"49d12'
Princess Lori
And 22h56'22"51d9'
Princess Madeleine Fair-Star
Mon 6h24'23"10d36'
Princess Manto
And 0h12'15"39d59'
Princess Martie Jae
And 1h57'42"35d28'
Princess Mary Beth
And 23h34'25"41d10'
Princess Mary Lee
And 0h24'1"33d43'
Princess Megan Rose, My Love
Peg 21h37'48"23d46'
Princess Melanie
And 2h29'31"45d2'
Princess Muriel S
Per 3h31'29"48d59'
Princess Nicolette
And 23h36'40"47d21'
Princess of Snow
Gem 7h4'39"17d38'
Princess Pam
Peg 22h31'0"21d29'
Princess Patricia
Sco 17h39'29"-40d15'
Princess Patti
And 1h42'0"40d5'
Princess Paula
And 2h34'45"38d5'
Princess Peace
Eri 3h39'15"-6d8'
Princess Philina
Uma 14h25'32"60d42'
Princess Poco Kela
Cma 6h15'42"-28d18'
Princess Ronaye
And 23h24'56"44d22'
Princess Roseann
Psc 1h0'59"21d38'
Princess Sandy
Gem 6h16'46"25d2'A
Princess Sarah
Cyg 19h57'1"30d3'
Princess Sarah Ann of Foxmoor
Cas 1h14'51"64d32'
Princess Scarlet Of Hartsville
Her 16h40'1"35d50'
Princess Scooter
Eri 3h10'28"-3d25'
Princess Star Lisa
Lyr 19h2'59"35d14'
Princess Star-Yvette Luna,The
And 0h0'40"44d52'
Princess Starr
And 23h4'32"36d46'
Princess Stephanie
Del 20h26'1"20d22'
Princess Syndie
Cyg 19h57'52"38d2'
Princess T
And 2h17'0"49d2'
Princess Tasha
Cma 7h19'16"-15d0'
Princess Tawni,The
Cas 1h50'45"75d48' A
Princess TK Love
And 1h59'26"39d24'
Princess Tonya Lee
Leo 9h37'1"10d50'
Princess Tootsweenie
And 23h36'25"46d55'
Princess Tracie Haynes
Uma 9h15'34"67d55'
Princess Tracy
Lyr 19h17'29"28d59'
Princess,Sonia
Ori 5h54'47"19d58'
Princess-1
And 2h7'37"41d7'
Princessa AKA Ron
Eri 4h1'30"-10d8'
Princesse De La Nuit
Lac 22h19'29"51d18'
Princiotto,Marc Antonio
Ori 5h23'24"-4d35'
Principato,Lawrence
Her 16h23'23"32d21'
Principe
Boo 15h15'23"51d39'
Principe
Del 20h14'52"11d41'
Principe Dmia Amore Siempre Oophie
And 0h28'26"37d30'
Principe,Jeanne
Vir 11h41'46"3d21'
Principe,Julie
Mon 6h55'37"-1d37'
Principe-Gillespie, Susie
Cas 1h52'60"73d12'
Principessa
Uma 12h1'0"62d36'
Principessa Melania
Per 2h12'20"58d32'
Prindelley,Maristopher
Lac 22h17'33"49d17'
Pring,Bradley T
Dra 11h26'1"70d46'
Pringle,Alan K
Cet 3h12'44"0d31'
Pringle,Kirsty Emma
Lyr 18h23'44"46d40'
Prinos,Monique E
Cam 7h54'45"61d20'
Prins Joachim
Cyg 20h16'32"39d16'
Prinselaar,Wilhelmina Maria
Mon 6h35'3"-0d11'
Prinsesse Alexandra
Cyg 20h17'0"39d51'
Printz,Hilary
Psc 1h36'19"21d40'
Printz,John W
Per 3h18'31"41d54'
Prinz
Lib 14h39'33"-23d28'
Prinz Wolfgang II
Ser 15h58'19"2d3'
Prinz,Dr Matthias
Cmi 7h17'53"8d53'
Prinz,Josef
Cam 6h54'12"70d9'
Prinzessin Belinda
Cep 22h11'54"60d43'
Prinzo,Debbie
And 2h24'56"49d25'
Priola,Candy
Cas 1h58'31"73d39'
Prioleau's Trinary Distinguished André
Dra 18h37'42"52d17'A
Prioleau's Trinary Illustrious Danielle
Dra 18h37'42"52d17'B
Prioleau's Trinary Majestic Jean
Dra 18h37'42"52d17'C
Prioli,Toby A
Cet 2h5'51"2d8'
Priolo,The Great Star
Aql 20h2'37"1d21'
Prior,Paul
Boo 14h12'1"37d43'
Prior,Robert Andrew
Aur 6h3'40"30d34'
Priore,A J
Cep 22h10'22"61d33'
Pripfl,Christine
Dra 15h4'24"62d48'
Prisbrey,Robert Ellis
Aql 19h57'32"8d16'
Prisciandaro,Sr, Michael
Dra 18h19'31"71d2'
Priscilla
Cmi 7h59'26"0d29'
Priscilla
Ori 5h47'36"18d12'
Priscilla
Gru 22h14'30"-54d0'
Priscilla
And 1h54'33"38d53'
Priscilla
Equ 20h59'53"7d21'
Priscilla Lagas-Lisor
Cas 1h5'28"53d57'
Priscilla The Teacher
Lyn 8h43'50"44d39'
Priscilla's Lucy
Cmi 8h0'12"1d36'
Priscilla's Valentine Star, Love MJD
Cas 0h10'22"63d25'
Prisco,Sam & Rosemary
Umi 14h3'19"67d14'
Priscott,Marcus
Cyg 20h41'60"45d20'
Priselac,Albert & Mary
Cyg 21h48'0"53d52'
Priser,Alexandra Marin
Ori 5h2'23"10d60'
Priser,Jessica Elysse
Del 20h21'32"9d19'
Priser,Martha
Com 12h18'26"20d8'
Prita
Tau 3h43'50"11d4'
Pritchard's Star, Andy & Jason
Eri 4h16'4"-19d35'
Pritchard,Amy
And 23h38'10"48d5'
Pritchard,Andrew
Peg 22h58'37"8d37'
Pritchard,Benjamin T
Cnv 12h43'38"40d53'
Pritchard,Dr Donald William
Del 20h17'29"13d31'
Pritchard,Guy W
Ser 15h53'11"24d20'
Pritchard,Lauren Marie
Mon 6h18'41"8d51'
Pritchard,Mark
Cep 21h6'12"55d36'
Pritchard,Michelle A
Cam 13h5'32"81d12'
Pritchard,Noah Yong
Aur 5h3'17"40d8'
Pritchard,Ronald E
Her 17h57'47"18d56'
Pritchard,Susanne Lyn
And 0h3'43"46d53'
Pritchard,Vivien
Cam 5h21'17"67d47'
Pritchett,Adrian & Melissa
Aur 6h9'50"32d58'
Pritchett,Autry Lennon
Per 4h1'53"38d41'
Pritchett,Courtney
Mon 6h22'58"2d50'
Pritchett,David C
Cnv 13h40'40"38d48'
Pritchett,Don R
Lac 22h32'53"52d43'
Pritchett,Geraldine
And 23h29'56"49d7'
Pritchett,Grace Louise
Mon 6h25'1"10d55'
Pritchett,Michael
Aql 18h57'0"13d35'
Pritchett,Sheleah
Lac 22h33'32"56d16'
Priti
Sco 17h52'29"-30d32'
Priti
Mon 7h3'5"-5d17'
Pritsky,Jodi
Sco 17h31'1"-31d21'
Pritten
Cam 10h53'28"80d32'
Pritts,Jr,Harry
Dra 17h19'17"61d52'
Pritz,Alan L
Cep 21h33'23"63d45'
Pritzkat,Caitlin, Austin & Brandon
Cyg 20h12'27"37d60'
Privett,Erik B
Lac 22h30'47"53d22'
Privett,Jennifer L
Hya 8h14'59"1d42'
Privitera
Cnv 12h26'11"40d5'
Privitera,MD,Vincent John
Sct 18h43'59"-6d49'
Privitera,Therese Marie
Lyn 7h48'27"48d5'
Priya
Eri 4h7'33"-18d4'
Prize,Daniel Eliot Edward
Lac 22h6'22"49d47'
Prize,Dora
Crt 11h39'38"-22d44'
Prizer,Troy Alan
Lac 22h5'14"49d21'
Prizio,Jutta
Cas 2h4'0"61d13'
Priére Do Mami
Vir 14h26'24"-2d24'
PRL
Cam 14h0'18"81d6'
Pro Meus Viri Amor Aeternus
Lyn 9h17'52"39d8'
Pro Photo Of Indiana Board Of Dir 1994-1995
Uma 11h56'24"57d13'
Pro-Idee
Cap 20h5'50"-20d32'
Probert,Dream So Real Ted
Umi 15h0'29"71d1'
Probert,Stephanie H PPPHB-SSBGDDASPPF
Cas 0h29'33"60d21'
Probert,Susan J
And 0h57'16"38d36'
Probst,Anouk
Sex 10h12'13"-2d7'
Proby,P J
Uma 11h57'40"57d39'
Proc,Theodore et Irene
Com 13h10'1"26d54'
Prochazka,Karen-Jason Walker
Uma 9h50'34"58d40'
Prochniak,Dorothy C Kuhnwald
Uma 11h56'56"51d03'
Prochnow,Jurgen
Com 12h44'18"23d37'
Prochorena,Walter
Dra 14h49'35"60d56'
Prochoroff,MD,Nicholas N
Oph 18h18'41"7d52'
Prochuk,John
Dra 17h36'21"65d19'
Procter,Dorothy Helen
Gem 6h48'2"12d17'
Procter,Jane
Dra 16h3'30"62d54'
Procter,Simon
Ori 6h6'17"0d17'
Proctor,Andrew Jeffrey
Ori 6h6'35"0d15'
Proctor,Debbie
Cet 2h19'12"0d46'
Proctor,Gordon
Aql 19h1'11"13d23'
Proctor,Gregg
Cmi 7h31'45"7d3'
Proctor,Kathy Ann
Peg 21h38'38"26d10'
Proctor,Michelle S
And 2h16'22"40d50'
Proctor,Shaun
Her 16h30'50"39d53'
Proctor,Sian Hayley
Ori 6h17'24"8d37'
Proctor,Vera-Robinson Hanley
Lac 22h1'16"38d21'
Prodhomme,Serge
Umi 16h7'5"73d34'
Prodigee's Dream
Uma 10h22'20"40d43'
Prodigy
Cam 8h3'34"70d11'
Prodoehl,Alex John
Cnv 12h21'21"46d35'
Prodoehl,Matthew Kennedy
Her 16h19'26"51d13'
Prodromos
Dra 17h6'45"67d24'
Profer Gerhard
Ori 5h59'13"19d41'
Profeta,Judie
Aql 19h14'10"14d30'
Proff,Alfons
Peg 22h53'1"8d17'
Proffer,Spencer
Aql 20h2'0"11d36'
Proffitt,James Edward
Hya 9h1'11"1d55'
Proffitt,Margaret Taylor
Equ 20h58'14"6d40'

Proffitt,Max Arthur Her 17h14'1"45d55'
Proffitt,Ralph Dra 17h49'1"65d3'
Proffitt,Shannon Luke Vul 20h58'57"28d43'
Proffitt,Vince Aur 6h16'37"45d8'
Profile Coverage Corp Aur 6h6'48"45d5'
Proft,William J Dra 14h54'18"64d4'
Progar,Jerome Psc 1h2'57"30d4'
Prograf Dra 20h22'0"70d48'
Progress Ari 2h20'0"11d21'
Prohaska,Robert & Lauriann Lac 22h3'0"49d59'
Prohaska,William Aql 19h30'59"13d17'
Proia,Nicholas G Dra 17h54'46"61d6'
Proietti,Joseph Boo 14h13'0"41d0'
Proinseas Maire-An mna alainn Psc 1h25'1"11d35'
Prokell,Nellie Anastasia Cam 4h54'37"67d37'
Prokop,Alan S Lac 22h26'0"55d2'
Prokop,Keith Matthew Her 17h9'48"38d39'
Prokopis,Mick & Mary Crb 16h16'32"28d10'
Prokopp,Patrizia Lyr 19h20'33"30d12'
Proksch-Troope,Sandra Aql 19h59'0"8d51'
Promessa Veneziana Infinita Boo 14h23'54"48d13'
Prometheus Cam 5h43'56"60d38'
Promise Ori 5h52'27"8d57'
Promise Uma 9h55'12"59d22'
Promise Uma 10h29'1"41d60"
Promise For Paul Dra 16h35'40"62d25'
Promise To Love By Junko & Noriaki Tau 5h23'26"28d16'
Promise To Marry By Noriaki & Junko Tau 5h23'57"28d30'
Promises 7-30-94 Cyg 20h19'1"31d31'
Prommer,Friedbert Umi 13h45'34"74d25'
Promotion Partners Eri 5h6'39"-5d45'
Pronatti,Eliana Ind 21h2'1"-57d56'
Pronio,H A Equ 21h3'30"7d54'
Pronos,Charmaine A Vir 11h42'36"5d37'
Proper Star Boo 14h33'30"41d9'
Proper,Lynne Lyn 7h27'42"44d15'
Properous Pcte Lyn 8h23'24"40d24'
Propescu Cmi 7h17'59"4d6'
Propharm Ori 5h58'1"11d10'
Prophecy of Our Love, The Ori 5h23'30"1d21'
Propper,Ruth Anne Maness Oph 17h0'0"-18d55'
Propst,David Cep 20h42'27"63d58'
Propst,David Harold Cep 23h15'31"64d14'
Propst,Martha Mon 6h40'54"3d0'
Prosch,Bill Aql 19h44'15"10d33'
Prosdocimi,Maria Luigia Eri 3h47'41"-6d16'
Prose,Tommy & Kim Sge 19h40'29"16d10'
Prosek,Amy Lyr 18h50'0"40d31'
Proske,Joseph Oph 17h31'60"1d18'
Proske,Udo Lyn 8h0'28"40d14'
Proske,Udo Cyg 20h26'43"40d15'
Prosperi,Sandra Lyr 18h38'51"27d11'
Prosperi,Stephen Robert Cep 21h43'54"58d12'
Pross,Nicole Michelle Jacqueline Mon 6h55'20"-6d33'
Prosser,Abbie Milly And 0h20'0"38d36'
Prosser,Andrea Umi 14h3'46"77d47'
Prosser,Bob Lac 22h24'13"54d10'
Prosser,Dale V Oph 18h5'26"10d60'
Prossima Futura Prospera Nor 16h19'23"-57d17'
Prost,Anne-Marie Cyg 21h8'27"30d35'
Prost,Ruby Keith Peg 21h45'49"23d34'
Prostko III,Edward Richard Aur 7h13'31"38d25'
Protas,Nathan Cas 2h29'35"70d47'
Protch,Cyril Peg 23h39'18"15d37'
Protector,The And 1h4'38"46d32'A
Prothro IV,Lannes Cuthbert Del 20h33'49"20d20'
Prothro,Sara Kathleen Peg 21h59'44"21d36'
Protko,Joseph R Per 3h15'50"50d30'
Protsko,Amanda Noelle Dra 19h22'12"70d21'
Protz,Rüdiger Cnc 8h31'25"31d12'
Protzman Per 1h44'58"54d14'
Protzmann Aql 20h8'31"7d18'
Proud,Alastair Reed Her 16h53'52"39d46'
Proud,Tonii Sex 9h54'12"4d36'
Proulx,Clyde & Emma Uma 9h50'43"52d2'
Proulx,Joan Carole Cnc 8h24'1"7d26'
Proulx,Kellie Lee Lyr 19h3'45"40d25'
Proulx,Mark joseph Cam 3h55'46"60d38'
Prout,Alan Aur 5h58'32"37d54'
Prout,Jr,William J Her 16h46'39"30d33'
Prouty,Cindy S E L Com 12h12'35"22d13'
Prouty,Janis Renae Cas 1h1'41"60d8'
Provan,Donald Robert Cet 1h38'38"-4d58'
Provencher,Jason C Her 17h7'11"46d30'
Provencher,Jean-Marie Ori 6h0'36"7d28'
Provencher,Nathalie Cas 1h10'43"60d56'
Provencio,Al-Henry Hya 8h42'30"3d21'
Proville,Jonathan Her 17h54'23"14d46'
Provines,Deborah A Del 20h22'1"20d26'
Provins,Stewart Richard Peg 22h47'0"33d37'
Provolt,Kristy Mon 6h35'60"10d2'
Provost Celestial Viking,Emilie N Lyn 6h34'23"54d25'
Provost,Cliff Dra 16h51'0"62d44'
Provost,Star of Heather Cam 14h12'58"80d45'
Provost,Tracey J Lyr 18h30'1"43d26'
Prowse,Kev Dra 15h19'33"65d30'
Prowse,Lilian Cyg 21h53'51"52d39'
Prowse,Mary Bunniss Stevenson Lyr 18h49'0"35d40'
Prowse,Tyrus Raymond Lyr 18h49'17"35d1'
Prowten,David Wheaton Her 16h6'32"48d50'
Prozanski,Suzi Eri 3h8'29"-5d46'
Prozeller,Für Luisa Boo 13h35'14"10d8'
Prsoper 95 Umi 14h21'19"69d44'
Pru Lyr 19h16'57"25d52'
Pru Uma 9h21'53"70d17'
Prud'Homme,Pascal Peg 23h8'21"8d49'
Prud'Homme,Pierrette Peg 22h29'40"18d48'
Prudhomme,Jacques Aql 19h37'19"1d26'
Prudon,Delphine Lyr 19h3'16"28d39'
Pruett,Eugenia Louise Eri 2h52'10"-6d39'
Pruett,Jane Hardy Cyg 21h32'12"41d18'
Pruett,Jill Lynn Lyr 18h59'32"26d42'
Pruett,Julie Anne And 2h26'39"44d35'
Prufer,Nico Per 1h56'32"53d3'
Prugnetta Cae 4h44'48"-32d53'
Pruitt,Betty Peg 22h14'1"33d53'
Pruitt,Billy Her 17h34'10"27d29'
Pruitt,Chelsea Nicole Aqr 23h22'45"-4d21'
Pruitt,Cynthia Ann Mon 8h6'45"-8d31'
Pruitt,David & Kristin Uma 8h9'58"60d46'
Pruitt,Kenneth Daniel Sco 16h54'33"-43d52'
Pruitt,Peggy Sco 15h57'1"-25d16'
Prunckle,Jon Ori 6h1'0"-0d33'
Pruneaud,C Lac 22h20'52"55d18'
Prunes & Plums Cyg 19h19'41"49d46'
Prusakiewicz,Ursula Cas 0h23'54"61d20'
Prutky,Robert Tri 1h53'23"27d45'
Prutz,Robert Charles Aql 19h5'0"3d49'
Pryke (Harley),Emma And 23h6'19"38d12'
Prymula,David Ser 16h4'50"23d32'
Pryor February 9,1953, Jacquelene Eri 3h52'13"-6d21'
Pryor,Adam Edward Peg 21h54'1"34d57'
Pryor,Charles Davies Her 18h18'52"14d59'
Pryor,Douglas Ray Cmi 7h24'49"8d59'
Pryor,Ginseng Hya 8h42'42"3d19'
Pryor,JF Hya 8h26'44"5d39'
Pryor,John Donovan Dra 14h43'52"61d60'
Pryor,Raymond Francis Cet 3h6'17"1d44'
Pryor,Sarah Elizabeth And 0h57'1"35d37'
Prystajko,Michelle And 23h42'1"45d32'
Przedborski,Felix Lyr 19h20'29"30d19'
Przemek Sco 17h29'39"-30d6'
Przemyslaw Sapieha Cas 0h18'16"65d46'
Przybilla,Horst Lmi 9h24'21"38d54'
Przybyl,Donna Marie Vizzini Uma 11h29'47"43d40'
Przysucha,Hans-Joachim Lyr 19h23'55"30d39'
Prünster Duell-Team, Harald Uma 8h25'43"68d33'
Pr Uma 8h25'43"68d33'
Ori 5h52'38"14d40'
PS 236-The Class Of 1995 Lyr 19h0'46"31d7'
PS143 Cep 2h42'30"78d29'
Psaltis,Gianni Oph 16h54'48"-6d26'
Psaras,Diane Cyg 19h34'23"38d3'
Psaridis,Harald Ari 2h55'47"30d22'
Psomas-Sheridan,Haley Jeanné Mon 6h39'0"11d55'
Psomos,Penelope Vul 19h47'53"28d35'
PT's Hope Cam 4h1'0"69d33'
Ptacek,Denise Yvette And 23h57'49"50d37'
Ptachcinski,Eugene Eri 4h32'50"-9d50'B
Ptak,John Lac 22h24'50"37d44'
Ptaczyk,Thorsten Peg 23h29'56"12d23'
Ptaszynski,Joey P Dra 12h6'10"70d23'
PTO-60 Tau 4h47'36"16d3'
Ptohopoulos,Nyky Ori 4h44'40"0d36'
Publio E Maltida Cam 5h9'21"67d55'
Pucalka,Klaus Aur 5h13'51"43d21'
Pucci Cam 3h50'58"60d35'
Pucci,James Arthur Oph 17h55'1"12d3'
Pucci,Michael Lee Boo 14h57'0"30d51'
Pucciarelli,Michael Lac 22h7'1"51d5'
Puccio,Mark Vincent Boo 14h7'1"54d53'
Puce et Sweety Cnv 12h26'20"46d28'
Pucette Volatier Cep 20h11'39"60d39'
Puchalski,Donna Cas 1h42'25"58d27'
Puchalski,Laura And 23h18'20"42d38'
Pucheu,Bill Aql 18h44'56"11d52'
Puchungo Agosto Oph 18h42'10"8d55'
Pucine,Iris Mon 8h2'32"-0d23'
Puckett's Miracle- Austin Gem 8h0'57"33d10'A
Puckett's Miracle- Blake Gem 8h0'57"33d10'B
Puckett,Edna Nicholson Cmi 7h27'0"8d59'
Puckett,Emily Lauren Peg 22h22'36"21d16'
Puckett,Gary D Per 4h21'11"51d21'
Puckett,James E Aql 18h59'23"-1d19'
Puckett,The Marriage of Jane & Jim Sge 20h5'44"16d11'
Puckett,Wanda Rae Vul 19h6'32"24d5'A
Puddin' Cas 1h40'58"57d17'B
Puddin's Star And 1h8'37"39d39'
Puddles Cmi 7h36'30"10d36'
Puddles,Rick Love Dra 14h52'0"58d36'
Puddy,Avril Aqr 22h28'51"-17d45'
Pudela,Joe & Barb Cam 4h10'53"61d19'
Pudelski,Mark Boo 14h34'50"20d37'
Puderbach,Regine & Kurt Uma 10h54'50"71d35'
Pudles,Danielle Nicole And 23h39'45"46d32'
Pudney's Star in Heaven,Agnes Hansen Ser 15h33'42"21d24'
Pudoka's Star 1994, Maurice Cet 3h14'1"5d13'
Puech,Philippe Mon 7h44'32"-4d34'
Puen,Leni Sct 18h53'1"-6d18'
Puente,Sarah A Peg 22h43'31"27d25'
Puerari,Carmelinda Cas 0h40'45"65d11'
Pueraro,Margo And 1h14'47"39d49'
Puetz,Albert Peg 22h0'27"28d28'
Pufahl,Julie Lyn 8h1'27"33d25'
Puff's DeaconBlue Ari 2h34'1"30d41'
Puff,Luke Vincent Dra 20h19'25"64d4'
Puffer,Jr,Samuel H Boo 14h30'41"21d39'
Puffer,Lara Shey And 23h37'32"40d41'
Puffin Cyg 19h31'1"33d21'
Puffo Luciano Peg 23h0'46"20d10'
Puffy Boo Cet 0h36'59"-3d58'
Puffy Cheeks Cam 7h7'25"61d15'
Pugh IV,James Roy Cet 2h55'25"9d42'
Pugh,Anne Tipton Lac 22h6'58"51d30'
Pugh,Bliss Renae Peg 23h21'1"10d28'
Pugh,Dale H Cet 2h33'0"-6d4'
Pugh,David W Cet 2h28'1"0d23'
Pugh,Ellen T Aql 20h12'12"10d59'
Pugh,Linda Aur 5h57'47"37d48'
Pugh,Linda Sge 19h57'1"20d3'
Pugh,Linda Peg 0h1'50"27d39'
Pugh,Linda Lyr 18h59'16"35d47'
Pugh,Magan Elizabeth And 0h25'32"44d46'
Pugh,Rose Marie Del 20h34'45"10d32'
Pugin,Michèle Cas 1h53'13"61d8'
Pugliano,John S Her 17h6'0"43d29'
Pugliese,Antonella Psc 23h24'11"5d58'
Pugliese,Gary Dra 17h30'37"58d40'
Pugliese,Joseph & Rose Star,Kandi Her 16h39'23"29d31'
Pugliese,MD,Peter T Boo 15h4'20"23d56'
Pugliese,Sal Cam 8h34'34"74d12'
Puglise Forever Loved Star,Kandi Cyg 20h23'16"37d42'
Puglise's Star,Melissa Anne Lyn 7h46'26"41d56'
Puglisi,PhD,Rev Msgr Guy J Boo 14h58'57"32d16'
Pugnet,Christine Aur 4h53'53"40d44'
Pugnetti,Alix Aur 5h11'44"41d59'
Pugs Pugliese Cyg 19h43'57"30d60'
Pugsley,John & Elizabeth Mon 6h22'19"-6d13'
Puhala,Anita Louise Peg 22h36'1"27d38'
Puhala,Terri Ellen Peg 22h34'1"20d20'
Puhl,Kristin L Dra 17h24'16"70d37'
Pujalt,Mirella Lac 22h53'0"51d18'
Pujol,Marc Cam 7h46'0"67d48'
Puklin,David Steven Lac 22h26'19"55d12'
Pulaski,Anna Lyn 8h58'20"44d35'
Pulce Uma 12h2'54"40d48'
pulcherrima JODI Sco 17h28'26"-30d44'
Pulchrae,Papilia Vul 20h21'0"23d59'
Pulchrissima Ros Cyg 20h0'33"30d23'
Pulchritudinous, Sabrina Lyr 18h20'20"38d44'
Pulcina Cnv 13h37'40"32d10'
Pulcini,Marissa Stampf Cyg 21h27'18"30d5'
Puleio,Emily Cas 1h59'37"58d20'
Puleio,Margaret Cyg 21h27'18"30d5'
Puleio,Melissa Aur 4h59'25"51d13'
Puleio,Philip Cnv 12h59'34"31d52'
Puleo,Jamie Marie Cecelia And 0h11'34"35d37'
Puleo,RN,Joanne Hanna And 0h8'22"47d35'
Pulford,Fallon Francis Boo 14h57'0"37d60'
Pulford,Gurdon Cet 1h35'37"0d45'
Pulford,Sarah Jean And 2h17'35"46d6'
Pulfrey,Charles Walker Uma 10h45'38"40d31'
Pulfrey,Margaret Harrold Cam 7h7'1"60d41'
Pulichene,Toni Cas 0h43'29"67d42'
Pulido,Victoria Rae Peg 22h33'26"35d8'
Pulis,Pete Dra 14h0'28"68d55'
Pulito,Michael F Lac 22h55'16"51d43'
Pulitzer,Roslyn K Cyg 20h54'31"52d38'
Pullar,Douglas Cyg 20h22'41"41d19'
Pulle Eri 3h40'13"-5d9'
Pullen,Charles W Aql 19h53'54"12d2'
Pullen,Frank Christopher Cmi 7h55'28"0d15'
Pullen,Ken & Inez Aql 19h48'56"10d33'
Pullen,Rena Ardell Cam 3h28'17"61d33'
Puller,Lewis Lac 22h28'0"40d21'
Pulley,Carlyn Del 20h14'1"12d54'
Pulley,Diana Marie Com 12h53'42"21d2'
Pulley,Pat Cmi 7h26'25"7d33'
Pulley-King,Micah Thomas Her 17h38'41"38d14'
Pulliam,Hannah Marie Lyn 6h55'1"59d19'
Pulliam,Pat Lyn 8h58'30"38d11'
Pulliam,Preston Psc 1h0'59"21d38'
Pullin,Bob & Alice Lyr 19h4'58"25d47'
Pullinen,Joshua Alexander Pho 0h46'23"-47d23'
Pullman,Craig Lac 22h2'42"51d10'
Pulpito,Massimiliano Her 16h11'0"10d43'
Pulsatilla,Pauline Mon 6h47'39"10d9'
Pulsfords Beandog Uma 8h5'1"61d17'
Pulsifer,Randall F Cam 4h57'16"70d1'
Pulsinelli,Francesco Ant 10h44'41"-34d23'
Pulteney,Henrietta Laura Uma 12h14'22"55d50'
Pulver,Cynthia Ann Aur 7h5'42"40d48'
Pulver,Fred Lawrence Aur 7h4'48"38d37'
Pulver,Muriel Bregman Lyn 7h31'33"51d48'
Pulver,Paul Her 16h18'32"28d19'
Pulver,Seth Henry Aur 7h0'51"40d27'
Pulver,Stefanie Jo Aur 7h7'20"37d47'
Pulver,Valerie Aur 7h24'0"40d22'
Pulvermüller,Willhelm Georg Vir 12h28'32"-5d36'
Pum-Pum Star Ori 5h55'19"0d6'
Puma,M Lyn 8h23'1"40d8'
Pumky Monkey Jennifer Lyn 7h44'32"38d8'
Pummer's "Golden Star" Marge & Bill Ori 5h5'15"0d14'
Pumpernickel Bread Dra 20h3'50"75d45'
Pumpernig,Maximilian Aqr 23h40'0"-4d3'
Pumphrey,Anthony Joseph Cet 0h26'26"-5d49'
Pumpkin Mon 8h2'24"-6d58'
Pumpkin Lyn 7h53'0"42d5'
Pumpkin Leo 9h19'21"17d14'
Pumpkin Oph 18h39'47"8d44'
Pumpkin Cyg 21h21'1"28d36'
Pumpkin Cyg 20h50'48"37d38'
Pumpkin Uma 8h46'1"51d56'
Pumpkin Duster Leo 10h30'24"25d47'
Pumpkin Head & Sweetie Uma 9h35'30"43d48'
Pumpkin Michael & Allyson Forever Cas 0h58'14"62d8'
Pumpkin Star Umi 15h8'58"68d35'
Pumpkin Star,The Her 17h50'59"41d14'
Pumpkin's Pie Sct 18h39'12"-4d23'
Pumpkins Tri 2h37'37"34d29'
Pumpy Uma 9h4'10"58d31'
Punch Her 16h46'53"36d0'A
Punches,William Ori 4h42'1"10d58'
Punessen,Isabell Lib 14h22'32"-23d22'
Pungarscheg,Gabriela Dra 16h9'32"60d32'
Punkin Lyr 18h59'39"31d1'
Punkin Cmi 8h9'10"1d9'
Punkin Peg 23h4'58"30d23'
Punkin BL Kirwan Cnc 9h6'25"31d19'
Punky Cnv 12h50'40"41d29'
Punky Mon 7h26'50"-6d54'
Punky Aql 19h3'46"0d0'
Punnuse,Melvin Per 1h47'1"56d55'
Punongbayan Lyr 18h34'26"33d47'
Punster Nebulae,The Mon 7h0'12"7d49'
Puntman,Josepha Fransica Veronica Her 17h54'21"14d47'
Punto Rosa Del 20h17'31"12d9'
Punzo,Anthony Paul Boo 14h20'58"25d19'
Pup Cep 22h15'44"60d33'
Pup in Mood Cas 3h2'26"57d39'
Pup Star,The Cet 3h15'10"4d25'
Pup-A-Saurus-Rex Cam 7h57'21"61d22'
Pupach,Stephen & Margaretta Umi 15h13'56"67d6'
Puppy Love Cma 7h15'18"-11d13'
Puppy Love Star Lyn 8h5'50"50d41'
Pupsicle Ori 5h59'35"15d53'
Pupsta,Mayme Jewel Vul 19h43'16"27d22'
Pupster Cep 22h58'0"68d35'
Purbaugh,Alison Leigh Hya 8h55'1"-6d23'
Purbaugh,Elyse Uma 9h0'1"52d30'
Purburgh,Justine Ori 4h54'13"4d30'
Purcell's Terran Voice Oph 16h54'19"-5d52'
Purcell,Bob Cep 21h32'44"60d29'
Purcell,Dorothy Mon 7h42'1"-5d35'
Purcell,Dr Mary Oph 17h16'36"-24d26'
Purcell,Estelle Sgr 19h8'30"-24d31'
Purcell,Jennifer Peg 22h6'12"5d47'
Purcell,Jeremiah S Cet 2h20'48"5d11'
Purcell,Mary E Cet 2h47'13"3d31'
Purcell,Pamela Umi 14h59'57"69d43'
Purcell,The Dave & Lynn Mon 6h6'0"-4d15'
Purcell,Tony Cet 1h12'49" 0d17'
Purden,Sue And 1h7'39"37d52'
Purdon,Fiona And 23h19'0"40d23'
Purdy's "Nova Tar Heel",William R Aql 19h54'22"12d54'
Purdy,Christine Marie Lyr 19h22'14"35d27'
Purdy,Eileen Cas 0h45'53"63d5'
Purdy,Elia Mares Mon 6h44'27"10d12'
Purdy,Jennifer Ori 5h24'22"-6d54'
Purdy,Jessica-Nicolle Mon 7h47'29"-6d47'
Purdy,John Hya 8h54'0"2d13'
Purdy,Jr,Esq,Robert D Boo 13h34'54"14d26'

Purdy,K R
Lyn 08h41'15"46d34'
Purdy,Kevin Gerard
Lyr 19h5'53"28d30'
Purdy,Megan Blythe
Lyn 8h12'0"40d7'
Purdy,William Grant
Cet 2h34'38"5d35'
Pures,Ann Cynthia
And 0h16'37"35d28'
Purificato,Caprie Destinee
And 0h51'12"22d17'
Purinton,Zachary Watson
Cnv 13h32'33"48d33'
Purita,Daniela
Mon 6h57'41"-5d48'
Purkat,John & Myrtle
Her 16h8'1"24d48'
Purkiss,Claire
Peg 23h31'42"12d28'
Purkiss,Stuart
Peg 23h30'1"11d45'
Purnell,Angela Gail
Vul 19h59'1"28d56'
Purnell,Helen Georgina
Cas 23h32'24"58d49'
Purnell,John David
Dra 16h34'29"62d25'
Purnell,Marcus Gerard
Per 2h58'32"34d16'
Purnell,Willard Dale
Cep 2h40'38"87d7'
Purnell-Burson,Hattie
And 0h17'1"46d28'
Purnick,David James
Ori 6h7'30"20d30'
Purnick,Howard Bertram
Her 17h24'29"44d28'
Purol,Sherry, Michael, & Alexander
Uma 11h39'32"50d43'
Purple Haze
Eri 4h1'19"-16d4'
Purple Pixie
Cas 0h49'20"73d16'
Purple,Ah' Mah'Lei' Ahh's Jake B
Boo 14h55'1"37d9'
Purpora,Suzanne F
Uma 9h1'0"56d27'
Purrfect Grandmeow
Lyn 8h5'0"40d17'
Pursell,Erin Foote
And 23h27'35"47d34'
Pursell,Haley O'Dell
Lyn 8h51'54"35d49'
Purser,Coral Elise
Umi 15h4'53"67d18'
Purtell,Thomas P
Peg 22h29'17"8d7'
Purtilar,Chester
Aql 18h55'48"10d53'
Purtilar,Elyse
Peg 22h41'55"23d60'
Purton,Richard J
Per 3h28'42"40d45'
Purvez,Saher
Aql 19h5'23"15d57'
Purvis,Jennifer A
And 1h20'25"36d21'
Purvis,JW
Peg 23h5'16"30d49'
Purvis,Pamela Lynn
Cyg 21h26'42"28d34'
Purvis,Steven
Cyg 21h15'46"35d14'
Purzewski,Paul
Cep 23h9'36"60d14'
Pusateri,Jr,Joseph A
Cet 3h15'0"9d39'
Pusateri,Sam F
Her 17h38'41"23d53'
Pusateri,Tina
Uma 9h37'23"58d54'
Pusch,Dr Rolf
Sco 17h29'52"-30d17'
Puschelchen Gisela
Her 17h26'43"42d55'
Puschner,Thomas
Cep 2h14'46"80d4'
Puseratze
Gem 6h44'8"13d55'
Pusey Ruby Anniversary
Cyg 21h55'21"52d43'
Pushcar,Eleanor T
Del 20h18'51"14d5'
Pushkin
Cmi 7h58'26"7d50'
Pushkin,Sophia Ania
Lyr 19h5'41"28d50'
Pusic,Joseph T
Per 3h8'0"50d34'
Pusic,Teresa
Crb 15h51'1"38d21'
Puski,Gabor
Uma 11h15'48"70d1'
Pussy
Lyn 8h47'17"40d57'
Pussycat
Uma 13h24'23"62d58'
Pussycat's Tiger
Lmi 10h41'33"32d27'
Pust,Dieter
Boo 14h3'0"11d17'
Puszka,Romuald
Aql 19h0'33"-8d49'
Pusztai,Violetta
And 0h16'36"46d17'
Putchen,Elisabeth
And 0h55'28"33d53'
Putchsky I
Lac 22h42'10"54d28'
Putensen,David & Katherine
Uma 10h58'28"40d12'
Puthoff,Michael
Tri 2h1'41"31d26'
Putirka,Betty
Cyg 19h51'48"38d52'
Putis,Andrew & Carol
Sge 19h54'30"16d43'
Putman,Jewell
Peg 21h56'0"30d44'
Putman,Sr,Edward Paul
Per 2h24'43"54d26'
Putnam,Marylee
Cyg 20h37'19"31d30'
Putnam,Robert David
Dra 16h31'54"52d2'
Putnam,V R
Her 17h28'16"21d42'
Putney,Sharon Lee
Lyn 7h40'48"39d53'
Putnick's CORSAIR, Michael
Aur 6h1'26"31d13'
Putnoi,Dr Donald Wm
Lac 22h1'23"48d52'
Putterman,Catherine A
Peg 22h51'19"27d44'
Putterman,MD,Eric A
Her 16h7'19"21d12'
Putz,Margit Maria
Uma 11h57'44"32d18'
Putz,Viktor
Cnv 12h30'26"32d11'
Putze Anniversary Star
Cyg 19h27'31"33d9'
Putzi,Simone Gabriela
Lyn 9h13'1"45d44'
Putzker Star,The Ralph
Ori 5h57'0"15d35'
Puyau,Tammie R
Cet 0h51'0"-10d50'
Puybonnieux,Anne
Oph 17h4'27"10d27'
Puydebat,Jean
Vul 19h19'35"26d36'
Puyé
Aql 19h46'10"11d55'
Puzaitzer,Michelle Claire
And 23h42'48"43d5'
Puzo,Giovanni
Uma 11h20'39"48d32'
Puzycat Face
And 0h55'28"34d46'
Puzzitiello,Roger A
Cma 6h57'43"-31d12'
Puáanou
Cas 1h19'11"50d37'
PVM III
Uma 11h51'31"38d23'
PVTK Larkin Star,The
Dra 16h8'0"68d42'
Pwebs
Uma 9h53'20"48d47'
PY In The Sky
Uma 10h9'53"67d59'
Py-Leduc,Catherine
Peg 23h4'0"31d20'
Pyburn
Her 17h55'54"18d50'
Pyburn,Robert Armond
Her 17h5'0"30d50'
Pycho
Hya 8h54'53"-0d13'
Pye,Emily Rebecca
Mon 7h3'45"1d23'
Pyecroft,Jonathan Paul
Cep 22h21'38"61d35'
Pyes,Craig Harrison
Mon 6h33'56"0d7'
Pyett,Anthony Leonard
Cyg 21h0'19"28d53'
Pygmalion Galatea-Paul & Gayle
Cyg 19h55'19"37d52'
Pyke,David
Cam 3h35'23"61d24'
Pyke,Diana Medina
Oph 16h52'0"-28d26'
Pyland,Carolyn Lee Gillis
Cyg 21h22'25"38d12'
Pyle,Mary Keta
Cap 21h4'17"-22d54'
Pyle,Ray
Aql 19h21'10"15d0'
Pyles,Erich Clinton
Aur 6h51'18"37d55'
Pyles,Kristopher Aaron
Sex 10h34'30"1d56'
Pyles,Robert Harris
Her 17h39'39"22d45'
Pylling,Kay Marie
Vir 14h56'26"7d20'
Pyne,Millie
Lyr 18h50'0"40d15'
Pyott,Joanne
Vul 21h0'0"27d9'
Pyra,Kimmi Andersson
Cas 1h0'45"48d47'
Pyron,Margaret Barenthin
And 2h25'58"41d59'
Pyrtle,Kit Wai
Lyr 18h38'18"33d14'
Pyrz,Tim
Her 18h12'39"41d5'
Pyson,Elizabeth Tapool
And 1h45'55"46d56'
Pytte's Glowing Light, Ingjerd
Dra 12h1'30"71d34'
Pyykko,Seppo
Cam 4h22'19"70d50'
Pyzik,Thomas William
Her 15h55'21"40d19'
Pyzowski,Joe
Per 3h11'31"40d50'
Pàcheur,Daniel Plourde Martin
Peg 22h54'12"24d44'
Péloquin,Christopher J
Her 17h22'1"20d45'
Péloquin,Frederick L
Umi 13h51'46"75d1'
Pépel
Dra 20h11'27"80d11'
Pérez,César Augusto
Cnv 13h3'37"45d32'A
Pöhlmann,Gertrud Annemarie
Sco 16h49'26"-25d44'
Pölcz,Vera
Sgr 19h4'14"-22d48'
Pölzer Starwriter I, Robert
Cnc 8h57'24"8d22'
Pötters,Wanda
Cap 21h29'0"-23d35'
Püppi Angelika Klein
Cap 20h33'30"-26d37'
Püs,Eva-Maria
Vul 19h44'36"27d18'
Pütz,Günter
Cap 20h34'49"-10d30'
P˝
Vul 19h5'50"25d10'
P˝
Uma 10h1'52"56d10'

Q

Q
Cyg 21h20'50"31d13'
Q & V
Her 17h32'39"26d26'
Q Dog 2
Oph 18h25'0"8d2'
Q T
Her 17h56'32"28d49'
Q's World
Cma 7h15'25"-13d18'
Qajaq George
Ori 6h17'15"8d42'
Qelthan,Anthony
Aur 5h50'54"40d27'
Qian,Wendy Hui Min
Sco 17h25'26"-43d24'
Qiera
Ser 15h57'55"19d53'
Qo'NoS
Boo 13h40'37"23d40'
Qo'noS
Aql 19h0'52"12d4'
Qo'noS Hovtay' (Klingon Star System)
Uma 10h4'37"57d47'
Qosina Corp
Umi 16h15'58"70d41'
QSJZ 5
Com 12h57'24"27d30'
Quach,Lan
Lac 22h25'44"56d29'
Quackenbush,James
Her 18h19'41"20d36'
Quackenbush,Richard
Aur 6h18'47"32d9'
Quade,Mary
Tri 1h32'16"28d48'
Quaderer,Tamara Beth
Vul 21h21'24"26d30'
Quaderer,Ty Dane
Uma 11h21'21"71d8'
Quadrel,Jr,Nick
Cyg 21h18'29"38d8'
Quadrelli,Francesca Maria
Aql 19h25'45"-6d28'
Quadt,Ursula
Eri 4h54'44"-9d35'
Quae Academico Gradu Ornata Est
Tri 2h5'0"35d44'
Quagliariello, Pasquale J
Aur 4h59'33"37d25'
Quagliariello,Anthony
Per 2h22'49"55d23'
Quagliariello,Edward Thomas
Cnv 12h29'23"33d40'
Quagliariello,Felicia
Cas 22h56'46"54d17'
Quaglieri,Bertha
Lyn 7h54'22"39d45'
Quaid,Evi
Oph 18h5'0"7d48'
Quaile,Herb
Sgr 18h50'15"-29d58'
Quain,Beauty Briggs
Cyg 21h23'55"38d28'
Quain,James Michael
Cet 1h27'33"-4d11'
Quain,Leon
Aur 4h52'19"51d52'
Quake's Star,Richard
Her 17h22'17"46d7'
Quakkelsteyn 1995, Anita J
Mon 6h24'55"8d36'
Qualcuno per sempre
And 23h22'20"50d11'
Quality Nights
Dra 13h44'15"68d25'
Qualkenbush Family Star,Pat & Fuzz
Cep 20h12'11"60d21'
Quall,Phyllis
Uma 8h46'0"49d9'
Qualliotine,Erica
Uma 11h45'48"49d54'
Qualls,Chrytal
Peg 22h1'1"20d49'
Qualls,Dorothy Frances Thorne
Cas 23h32'23"60d40'
Qualls,Edith
Crv 12h14'15"-17d59'
Qualls,Jr,Grady Calvin
Cep 21h6'46"68d11'
Qualls,Misty Dawn
Peg 21h51'1"34d8'
Qualter,Donna Kim
And 0h1'52"46d13'
Quan,Ginny
And 0h32'17"45d40'
Quan,Ky Cao
Vul 19h21'34"26d51'
Quan,Linda
Mon 6h36'24"11d15'
Quane,David
Her 17h3'35"45d1'
Quanlock,Ann
Cam 5h57'25"72d57'
Quann,Lindsay Edward
Aqr 22h3'44"0d34'
Quantus Mechanicus
Eri 4h2'29"-17d42'
Quaranta,Claudio
Uma 13h0'17"53d59'
Quarenghi,Mariane Elisa
Mon 6h29'49"3d39'
Quarles,Richard
Leo 11h21'25"-5d55'
Quarmby Star,The
Cyg 19h40'52"41d7'
Quart,Christine
Del 20h24'22"11d13'
Quartarella's Baby
Ori 5h53'37"8d51'
Quass,Barbara
Peg 23h33'16"11d52'
Quast,Robert S
Aur 6h31'1"30d49'
Quatman,Fran
Boo 14h8'54"45d21'
Quatrella,Margaret
And 0h59'38"34d7'
Quattlebaum,Charles W
Hya 8h53'0"4d16'
Quattro,Memorial, Norine Marie
Lyr 18h58'33"30d23'
Quattrone,Mark
Her 18h9'56"38d40'
Quattrone,Vittorio
Col 6h30'30"-35d56'
Quay P L E A S E,Dawn
Mon 7h44'2"-5d27'
Quayle,Carolyn
Aur 5h12'58"40d49'
Quayle,Susan E
Cas 2h34'49"57d38'
Quayle,Voirrey Madelaine
Cas 1h57'28"75d36'
Que & Clarence
Lac 22h21'11"54d20'
Que,Dr Hsu Yau
Psc 23h7'41"-2d6'
Queen AmyLynn
Ori 5h54'53"14d26'
Queen Bliss (Andrea Syverson)
Aur 6h56'41"40d57'
Queen Janis Of The Bridge
Lyn 8h22'30"40d5'
Queen of Babelon
Mon 7h12'42"-5d25'
Queen Of Hearts
Eri 3h41'1"-2d12'
Queen Of Pentacles- Knight Of Rods
Sct 18h41'21"-7d25'
Queen of Stars Yuki
Sgr 19h13'17"-15d41'
Queen of the sea
Eri 4h57'18"-5d52'
Queen Rose
Cet 2h23'13"-11d14'
Queen Shannon
Cas 0h6'0"66d22'
Queen Victoria
Aql 19h55'30"11d12'
Queen,Bobbi
Com 13h15'19"20d24'
Queen,Elizabeth Cavender
Cas 0h32'38"68d48'
Queenan,Sarah McAfee
Umi 14h22'26"70d6'
Queenie's Light
Cmi 7h27'42"8d3'
Queheille,Denise
Cyg 21h3'1"40d4'
Quejarse
Com 12h27'50"27d34'
Quell,Norbert
Cyg 20h42'2"47d23'
Quelle,Jr,Joseph E
Boo 15h4'1"14d48'
Quenan,Kathryn
Vul 20h39'10"20d26'
Quenneville,Brenda A
Cyg 21h10'53"48d57'
Quenneville,Sylvie "vie"
Lyr 18h59'41"34d20'
Quentin l'homme
Ori 6h10'47"8d60'
Quentin's Star
Peg 23h3'39"18d16'
Quentin's Star
Cra 18h1'46"-39d29'
Quenton Kari
Ori 5h28'40"1d39'
Querarlt,Vicente Moliner
Cam 13h52'19"81d59'
Querceto,Mark Anthony
Uma 9h57'24"51d12'
Querer
Cap 20h58'2"-22d44'
Querfurt,Hans Werner
Ser 15h14'0"4d12'
Querida Gerburg Maria
Leo 11h53'0"20d47'
Querida Karen
Cas 0h14'0"63d53'
Querida Rosa
Umi 16h2'40"70d43'
Querido Richard El Amor Es Para
Ori 5h22'17"1d6'
Quernheim,Josef
Psc 1h4'30"2d23'
Quero,Nicole Yvonne
Cas 1h22'34"55d54'
Querrard,A Star Named Jim-Jim,Warren
Uma 11h22'1"40d38'
Quertier,Karen
Ori 6h15'59"8d47'
Query,Janis B
Uma 9h55'18"50d24'
Quesenberry,Denise Day
Tri 2h16'48"30d28'
Quesnelle,George
Sct 18h55'39"-4d47'
Quetton,Frances
Ori 5h18'12"-5d42'
Quezada,Carlos
Cet 1h28'45"-5d55'
Quibell,Cara Martine
Sgr 19h6'48"-24d15'
Quick,John R
Her 16h46'19"35d51'
Quick,Jr,Mr & Mrs Buren
Lac 22h17'43"51d34'
Quick,Lee Tanya
Aql 19h58'13"14d42'
Quick,Michael C
Hya 8h59'35"-7d38'
Quick,Owen Wesley
Tau 5h34'35"28d32'
Quick,Rachael Elizabeth
Lyn 8h29'0"40d4'
Quiedeville,Florence
Cas 1h19'10"50d8'
Quiero
Uma 10h7'1"58d29'
Quieted
Hya 8h53'48"0d41'
Quietsche-Entchen & Bugs Bunny PIEP
Ori 5h37'1"8d58'
Quigg,Jennie M & William M
Crb 16h1'54"33d4'
Quigley "Hanna's Star",Maureen R
Cas 1h24'58"61d39'
Quigley,Allison Margaret
Cas 1h11'37"77d7'
Quigley,Alpha Shepard
Del 20h15'0"14d27'
Quigley,Blaise
Mon 6h27'47"-6d20'
Quigley,Frances Foley
Mon 6h54'43"-6d4'
Quigley,Francesca Ann
Cas 1h1'41"68d53'
Quigley,Ian Alexander
Ori 5h56'59"16d50'
Quigley,India
Cam 10h52'17"82d19'
Quigley,Jacqueline
Lyr 18h29'13"47d21'
Quigley,Merissa
Lyr 18h47'54"31d54'
Quijano,Jessica Carmella
Ari 2h56'14"30d26'
Quijano,To a Skylark Rosie
Aqr 20h45'1"-6d59'
Quilisch,"Lady Di" Diana Regina
Aql 19h59'1"14d42'
Quillen,Congressman James H
Aql 19h55'19"10d40'
Quillen,Jr,Malcolm P
Ori 4h52'47"1d7'
Quillian,Elizabeth Sampson
Cam 6h59'22"70d43'
Quillinan,Damian Michael
Boo 14h37'40"20d5'
Quillinan,Diane Elizabeth
Cam 4h15'42"78d51'
Quillinan,Sean Patrick
Her 17h23'15"43d20'
Quillot,Didier
Sgr 19h27'1"-30d59'
Quimby
Uma 10h4'27"50d8'
Quimby,Donna
Uma 10h29'26"50d5'
Quincey
Cam 3h48'49"60d45'
Quincy School,Josiah
Hya 8h11'53"-0d28'
Quinell,Mary Anne
Cas 0h35'53"71d31'
Quingsene,Han Jingmiao-Li
Vir 11h35'45"4d19'
Quini & Isabel
Crb 15h54'56"30d33'
Quinlan,Barry
Aql 20h18'30"7d54'
Quinlan,Colin Vincent
Boo 15h25'44"42d11'
Quinlan,Ellie Danielle
And 0h49'17"34d1'
Quinlan,Ingrid Barbara
Com 13h19'13"26d58'
Quinlan,Margaret "Sissy"
Peg 22h39'43"32d23'
Quinlan,Patricia Ann
Tri 1h39'45"34d33'
Quinlan,Susan B
Eri 3h55'13"-13d12'
Quinlan-Ritchie,Jan Louise
Gem 06h6'27"20d26'
Quinlin,Thomas
Aql 19h30'38"10d41'
Quinlivan,Sarah & Sean
Cas 1h19'10"50d8'
Quinn
Eri 3h55'0"-2d30'
Quinn
Aur 6h2'16"31d23'
Quinn
Aql 19h45'24"14d49'
Quinn "Wedding Star", Tim & Chrissie
Aql 18h48'58"10d45'
Quinn III,Dr Charles Francis
Oph 17h59'43"-5d52'
Quinn Lauren-79
Leo 10h59'37"-6d11'
Quinn Star,The Karen "K"
Vir 12h31'36"-8d31'
Quinn,Alexandra Sharon
Umi 15h0'1"84d40'
Quinn,Anita Bernice
Psc 23h1'38"0d49'
Quinn,Aurelia Parma
Aql 20h14'30"1d32'
Quinn,Austin John
Tau 4h39'38"16d13'
Quinn,Brian James
Psc 23h1'1"5d45'
Quinn,Brigid Lee
Mon 6h33'38"4d49'
Quinn,C
Uma 8h43'22"61d49'
Quinn,Caleb Marcus
Vel 9h57'52"52d58'
Quinn,Catherine
And 0h56'26"35d20'
Quinn,Charles William
Cep 22h58'32"63d32'
Quinn,Christopher
Aql 19h0'24"13d27'
Quinn,Christopher Joseph
Sco 16h54'1"-43d46'
Quinn,Conor Andrew
Her 18h1'19"45d27'
Quinn,Dale Robert
Cam 6h0'41"60d27'
Quinn,Daniel J
Lac 22h37'1"41d11'
Quinn,David C
Boo 15h21'0"40d9'
Quinn,Dee Jayne
Lyr 18h42'54"41d2'
Quinn,Della
And 0h50'38"35d53'
Quinn,Donna
Ori 5h20'51"5d16'B
Quinn,Douglas Lee
Her 17h16'27"22d30'
Quinn,E R
Lyn 9h13'22"37d25'
Quinn,Eileen
Peg 22h8'59"25d58'
Quinn,Fran
Uma 10h19'1"52d31'
Quinn,Heather
Cam 6h50'53"70d26'
Quinn,James P
Dra 12h58'13"72d8'
Quinn,James T
Her 16h53'41"34d24'
Quinn,Jay
Aql 19h47'1"12d5'
Quinn,Jeffrey Mark
Per 3h9'48"47d25'
Quinn,Jeffrey Michael
Dra 11h42'41"72d49'
Quinn,Joseph John
Dra 10h28'51"78d17'
Quinn,Joseph Michael
Aql 18h43'25"10d55'
Quinn,Jr,Laurence T
Aur 5h57'19"50d21'
Quinn,Kaitlyn C
Vul 20h5'21"28d34'
Quinn,Katherine Ann
Lyn 8h18'42"40d46'
Quinn,Kathryn L
Tau 4h10'1"22d6'
Quinn,Kelly Therese
Dra 9h46'21"74d2'
Quinn,Kerry Sinead
Lyn 9h12'58"40d41'
Quinn,Kevin Francis
Dra 16h40'37"63d5'
Quinn,Kimberly Ann
Tri 2h32'32"31d49'
Quinn,Liam Edward
Per 3h10'42"50d32'
Quinn,Lindsey Elizabeth
Tri 2h19'25"32d2'
Quinn,Maggie A
Tau 4h7'5"23d14'
Quinn,Mandi
Mon 6h19'1"-1d44'
Quinn,Martin R
Per 3h13'54"40d7'
Quinn,Matthew Francis
Uma 10h15'51"59d51'
Quinn,Monsignor W Louis
Uma 10h10'47"60d34'
Quinn,Paul Patrick Kevin
Dra 16h43'40"66d28'
Quinn,Robert Patrick
Cep 23h11'27"64d3'
Quinn,Scott
Ori 5h20'51"5d16'A
Quinn,Shannon Lee
And 23h41'1"46d20'
Quinn,Shawn
Boo 14h21'13"18d6'
Quinn,Shawna Marie
Lyn 9h10'48"38d11'
Quinn,Suzanne
And 23h38'21"41d12'
Quinn,Terence
Cas 2h28'55"61d30'
Quinn,Thomas & Heidi
Uma 8h16'21"72d5'
Quinn,Thomas George
Cep 23h36'46"63d37'
Quinn,Thomas Russell
Cep 20h44'41"75d14'
Quinn,Thomas T
Her 17h32'0"27d22'
Quinn,Tom
Cnv 13h20'58"38d43'
Quinn,Tommy
Hya 8h11'15"4d44'
Quinn,Tommy
Del 20h53'46"3d39'

Quinn,will Cmi 7h41'21"1d31'
Quinn,Willie Dra 17h24'26"68d58'
Quinn,Zdena C Cas 0h33'49"61d31'
Quinn,Zizzy R Tau 4h1'47"23d5'
Quinn-Educo,Dr Tony Her 18h27'1"24d49'
Quinn-Guest,Carrissa Marie Sge 19h29'49"16d30'
Quinnan,Edward M Aql 20h5'35"1d30'
Quinnan,Michael Flanagan Per 2h52'1"50d7'
Quinnell,Bruce Lac 22h28'51"52d41'
Quinnett,Patrick D Aur 6h26'0"31d24'
Quinney,Brian & Sue Cyg 20h30'36"42d34'
Quinney,Edmond L & Catherine Uma 10h14'21"58d53'
Quinney,Lee & Marie Lyn 7h7'0"53d45'
Quinnstellation Uma 10h29'42"55d13'
Quinnton's Light Ant 10h33'56"31d5'
Quinones,Jr,Mario Aql 18h48'58"11d35'
Quinones,Wanda And 0h0'29"46d36'
Quinones-Esquilin Reinaldo J Mon 7h31'0"-8d42'
Quinquagenarius Per 2h53'11"46d43'
Quinquaginta,Rodney Webb Per 3h4'39"38d31'
Quinquis,Aubrey Lup 15h12'29"-44d56'
Quinsat,Balthazar Mon 7h53'27"-5d46'
Quint,Teresa & Thomas Aur 5h7'4"29d46'
Quintana,Dell Lenor Cas 23h15'41"62d22'
Quintana,Margie Sct 18h52'10"-6d10'
Quintana,Maria And 23h19'29"51d22'
Quintana,Patrick Lac 22h6'51"37d31'
Quintana,Peter A Lac 22h28'33"38d56'
Quintana/Meazle's Star Edith Lyn 7h59'28"51d44'
Quintanar,Anita And 2h6'32"42d16'
Quintanilla,Elida G Cma 7h0'19"-11d9'
Quintanilla,Marcela Villarreal Lib 14h50'58"-1d26'
Quintanilla,Priscilla Villarreal Vir 14h58'18"5d15'
Quintela 09-28-95, Gilbert & Anne Sco 17h52'28"-30d32'
Quintens,Herve Cam 6h5'44"70d6'
Quintessence Amor Cyg 21h12'39"35d32'
Quintessessential Quentin Cnc 8h58'15"32d6'
Quinti Consobrini- (M I L L E R) Uma 9h28'29"60d42'
Quintills Cas 1h56'16"60d57'
Quintin,Louise Cas 22h59'28"56d11'
Quinton Family Uma 10h23'1"56d39'
Quinton,Elyssa Robyn Del 20h53'15"6d43'
Quintrell,Michele Ann And 23h3'47"41d9'
Quinty,Laurie Cas 0h56'43"62d27'
Quiny Cnc 8h9'38"6d46'
Quique Altschul Delashmutt Lyr 19h7'44"40d3'
Quirico,Angelo P Per 3h16'58"54d40'
Quirk,Arlene And 2h22'39"44d43'
Quirk,Ian Mon 7h56'57"-2d15'
Quirk,James E,"TELL ME ABOUT IT STUD"Love Deb Aur 5h25'1"38d30'
Quirk,Peggy A Lmi 10h18'18"30d15'
Quirke,Tom Per 2h11'58"56d27'
Quiroga,Isabel And 0h15'35"38d46'
Quiroz,Debra Cas 0h40'0"72d7'
Quiroz,Larry Cmi 7h42'25"3d56'
Quirus,Nancy Weismuller Vul 20h15'19"23d22'
Quisalan Elevas Ori 4h49'36"4d24'
Quisbrock,Rolf Del 20h54'32"8d35'
Quistgaard,Mette Thousgaard Cam 6h1'28"72d13'
Quistgaard,Shirley L Vir 13h23'60"-3d44'
Quisumbing,Chad S H Ori 5h7'0"-2d32'
Quitadamo,Matthew P Her 17h21'24"45d2'
Quito Lyr 19h21'47"30d56'
Quituqua,Aida Linguete Oph 18h4'0"13d5'
Quix,Melissa Lynn Neubauer Mon 7h2'20"4d2'
Quiz,Jonny Aql 20h16'54"1d37'
Quiza Uma 14h5'18"57d37'
QuiUma 14h5'18"57d37' Aur 6h53'24"44d13'
Quincerot,Estelle Seyda Ser 15h14'29"9d31'
Quoc,Severine Bui Cnv 13h26'1"50d20'
Quraisli,Alexander Del 20h21'36"11d12'
Qureshi,Jamie Ann And 23h35'34"45d32'
Qureshi,Yasmeen Ori 5h56'1"21d20'
Qutob,Nahida Lourene Mon 6h57'36"-5d39'
Qutub,Laila Cam 3h59'37"55d51'
QVC Dra 16h50'1"51d34'

R

R Mon 7h47'0"-5d11'
R Mon 7h3'10"-7d11'
R & R Dibble Cam 8h2'35"82d7'
R A F's Star Fifty Aur 4h47'1"40d19'
R A M Tau 5h50'29"23d40'
R A M Star Lyn 8h49'0"37d5'
R A R E Peg 23h19'31"31d50'
R A W Jr Cep 21h57'51"55d18'
R Aaron 1 Boo 14h57'29"48d2'
R C Vul 19h35'48"27d26'
R C M & T R M-Mom Equ 21h0'26"11d18'
R C P-14 Her 16h45'36"30d29'
R C T - A G T Sct 18h33'57"-6d38'
R cor meum Boo 15h27'42"47d55'
R E A ~ R J M Cyg 21h5'25"32d24'
R E B Oph 16h4'12"-5d46'
R E C III Lmi 10h11'34"34d22'
R E M Pic 4h58'57"-48d28'
R E R Lac 22h35'58"53d33'
R F 26 Pup 8h8'48"-28d21'
R F Ochs Gem 7h13'51"24d42'
R G M Rutgers '56 Ari 2h30'0"24d38'
R H G M Y D A D Her 17h36'50"21d6'
R J C Aql 18h56'42"-1d55'
R J D The Best of The Best Per 1h45'51"53d59'
R J K Aur 5h18'34"49d49'
R J My Love Her 16h24'52"33d40'
R J S Lac 22h39'44"55d3'
R J S & T B I Cyg 20h58'46"37d51'
R J T Cam 12h36'1"84d44'
R L B Aur 5h40'18"50d24'
R L B III Cyg 19h50'31"38d51'
R L B-46 Aur 5h17'16"40d30'
R L D Ari 2h40'34"25d36'
R Linda-Sue Sge 20h7'23"20d13'
R Lloyd SI Her 17h6'50"44d36'
R M F 1 Gem 6h38'45"12d4'
R M T Enterprise Cma 6h54'4"-19d33'
R O C Star Lmi 10h55'59"30d25'
R O L-1903-92 Cyg 21h6'13"39d29'
R O M M E L Lmi 10h57'54"31d44'
R S (Russell's Star) Oph 17h54'16"12d2'
R S 1 Ori 5h12'38"-1d21'
R S R Star And 2h20'55"48d22'
R Star Crb 15h50'0"37d58'
R V G 2 Cam 12h9'53"80d31'
R&D 1716 Eri 3h43'20"-0d29'
R' Shooting Star Cet 0h25'0"-6d34'
R-Starr Sco 16h33'1"-40d54'
R2 Lyn 8h41'52"38d42'
R3 Pic 5h49'22"-52d46'B
Raab,Matthew Eldin Cep 23h13'28"60d22'
Raab,Ray Her 17h34'41"38d56'
Raabe,Christian Uma 9h27'52"47d31'
Raabe,Warren Hya 8h55'32"3d42'
Raadgever,Inge Cyg 20h43'1"45d26'
Raams,Pieter Oph 17h12'32"-10d48'
Raap,Monika Cep 22h23'48"61d40'
Raassina,Minna E Umi 17h38'33"80d6'
Rabach,Bernd Oph 17h23'26"10d45'
Rabalais,Lee Martin Aql 18h44'48"10d5'
Rabary,Jany Cyg 19h45'32"30d3'
Rabat,Valerie dit Per 3h21'21"41d14'
Rabbass Per 4h3'11"51d28'
Rabbi Marc Her 17h50'30"40d48'
Rabbitt,John Patrick Cyg 20h22'21"41d8'
Rabbitt,Joseph Dra 16h52'54"65d49'
Rabe,Brigitte Sco 17h27'36"-38d11'
Rabel,Cullen W Peg 22h48'47"25d56'
Raben,Larry Lyr 19h11'16"40d54'
Rabenhorst,Ellen Lyn 7h14'58"59d36'
Rabenou,Lori & Darren Aql 20h34'21"0d21'
Rabert,Curtis Michael Dra 19h24'16"58d15'
RABEVOD6 Uma 11h52'27"52d45'
Rabideau,Carol Peg 0h2'1"13d44'
Rabik,Patricia A Lyn 8h58'52"35d56'
Rabin,Beth Elise Cam 4h3'17"78d56'
Rabinovich,Polina Cyg 19h16'25"44d49'
Rabinovitz,Jack Cep 20h56'54"55d37'
Rabinovitz,Rubin Dra 16h7'59"62d48'
Rabinowitz,Craig Andrew Ori 4h52'56"1d32'
Rabinowitz,Lisa Cas 0h29'25"61d43'
Rabl,Birgit Umi 14h37'33"70d20'
Rabold,Matthew Edward Her 16h35'47"32d52'
Raboy,MD,Adley Per 4h5'0"51d5'
Raboy,Suzanne Lyr 18h30'48"30d48'
Raby,Ann Cas 0h7'56"61d46'
Raby,Hannè Leandèr Högberg Del 20h20'22"10d51'
Raby,Laura Janel Sge 19h40'25"18d52'
Raby,Linda Spruill Mon 6h35'1"-6d11'
Race's Wishing Star, Matthew Ser 15h33'54"21d56'
Racehorse Lil Cap 21h21'36"-22d38'
Racer,The Uma 10h39'42"42d12'
Racette,Kim L Cam 4h20'0"60d5'
Racette,Richard L Cnc 8h29'35"31d28'
Rachael Lyn 8h29'36"47d54'
Rachael & Joe Cyg 21h14'53"28d36'
Rachael Anne Sgr 18h51'49"-34d9'
Rachael Faye Lac 22h5'14"46d37'
Rachael Kristin Cas 0h59'59"64d1'
Rachael's And 0h37'11"45d21'
Rachel Ind 20h32'32"-52d51'
Rachel Cas 0h8'36"64d5'
Rachel Cas 2h57'0"65d35'
Rachel Leo 11h51'48"20d14'
Rachel Cae 4h57'15"-27d44'
Rachel Cet 2h17'53"9d29'
Rachel And 23h48'34"43d12'
Rachel Cam 3h29'35"58d53'
Rachel And 1h2'24"40d34'
Rachel Cas 3h0'23"60d21'
Rachel & Chip Mon 6h18'32"5d2'
Rachel & Jamie's Star Cyg 21h6'41"38d43'
Rachel & Randy Aql 19h54'1"13d28'
Rachel & Roland Cyg 19h32'50"37d8'
Rachel & Shannon Cyg 19h43'41"29d45'
Rachel 3 December Ori 4h59'45"12d9'
Rachel 69 Carol Aur 5h53'60"31d30'
Rachel Aimee Hya 8h12'38"0d10'
Rachel Alyson Com 13h32'18"20d19'
Rachel Amanda Lyn 8h30'15"40d24'
Rachel Ann Mon 6h8'41"-10d11'
Rachel Ann And 0h7'34"46d38'
Rachel Arising Peg 22h25'39"25d45'
Rachel C Uma 8h40'0"49d27'
Rachel Carmela Aql 20h5'54"1d27'
Rachel Elizabeth "Moon Basket" Mon 6h5'23"-10d38'
Rachel ET And 1h53'27"37d36'
Rachel et Philippe Dra 17h37'1"73d46'
Rachel Joy Of Caldwell And 2h31'29"50d7'
Rachel Lauren Lyn 8h13'37"58d48'
Rachel Lee Mon 7h12'35"-6d57'
Rachel Lee Cas 23h19'43"62d56'
Rachel Light Lyr 18h44'34"30d24'
Rachel Lyn Uma 10h35'17"47d32'
Rachel Margaret Cyg 19h34'28"37d4'
Rachel Marie Vul 19h20'35"26d26'
Rachel Sage Oph 18h3'23"11d59'B
Rachel Thy'la Del 20h54'18"9d6'
Rachel XL Cas 2h50'1"68d3'
Rachel's Aurora Sgr 18h51'38"-36d34'
Rachel's Brilliance Tau 4h41'15"28d34'
Rachel's Peace 1995 Peg 23h37'34"18d6'
Rachel's Resting Place And 0h57'18"36d47'
Rachel's Smile Cyg 21h9'53"40d18'
Rachel's Star Gem 6h44'2"18d16'
Rachel's Star Lyn 8h28'44"43d12'
Rachel's Starshine And 1h5'14"47d27'
Rachele "cintadi viole pura, riso di miele" Umi 15h33'18"68d26'
Rachele M And 0h42'11"40d24'
Rachele My Love Sge 20h2'17"20d32'
Rachelle Lyr 19h8'31"47d28'
Rachelle Lyr 18h47'26"36d40'
Rachelle,Amy Leo 10h57'1"23d22'
Rachels,Lee Anne Crb 15h25'24"30d18'
Rachels,Mae Cet 3h9'42"1d55'
Rachlin,David Oph 17h19'52"-23d59'
Rachlin,Sheila Marilyn Printz Cas 1h21'0"68d34'
Rachon,Anthony Dra 18h1'54"68d8'
Racine,Michel Uma 11h33'60"30d31'
Racine,Michel Aqr 21h41'30"0d54'
Racine,Pierre Oliver Ser 15h16'0"8d3'
Racine,Richard Ross Ari 3h1'57"28d25'
Racine,Sabbrina Michelle And 0h22'19"36d22'
Racioppi,Raquel Cep 3h29'37"78d39'
Rack,Dorothy Mon 6h18'42"8d19'
Rackson,Dr Chester Peg 23h18'31"33d33'
Racmas Lyn 8h0'54"51d34'
Raco,Thomas Joseph Cam 3h52'21"70d35'
Racquel-My World Cyg 21h33'21"41d33'
Racy,Elaine And 2h28'47"45d28'
Raczkowska,Janice And 1h55'2"41d9'A
Raczkowski,Richard And 1h55'2"41d9'B
Rad & Teresa Cet 1h1'14"-13d36'
Rad Dad Her 17h21'36"43d26'
Rad-Nik Cnc 7h56'58"10d0'
Radandt,Emil Cas 0h32'27"62d9'
Radbourne,Duncan J Lib 14h59'0"-8d17'
Radcliff,Byron William Ari 2h0'40"11d35'
Radcliffe,Cynthia Uma 11h53'0"40d7'
Radcliffe,Katharine Cas 1h17'45"60d53'
Radcliffe,Kelly Ann Cas 2h15'25"60d8'
Radcliffe,Norrie Ori 5h20'0"10d7'
Radcliffe,Richelle Katherine Her 18h5'19"19d39'A
Radda,George Boo 13h35'41"27d27'
Radde,Georg Dra 18h51'37"65d26'
Radding,Jennifer Ty & Douglas Benjmn Gardner Aur 5h51'2"30d29'AB
Raddish,John Ivan Cep 21h21'22"60d44'
Radecki,Nancy Cyg 20h15'42"30d10'
Radek,Johanna Blandine Vir 11h54'45"-5d58'
Radek,Katherine Lyr 19h17'48"37d42'
Radeker,Mrs Gwynn Eri 4h14'16"-13d34'
Rademacher,Mary Patricia And 23h30'46"42d10'
Rademaekers,Bill Aur 6h22'34"30d21'
Raden,Gary Aql 20h0'19"12d20'
Raden,Paolo Peg 0h12'31"13d50'
Radencich,Peter Anthony Dra 20h6'0"68d41'
Radenkovic,Michael Sreten Tau 5h47'42"27d15'
Radensky,Esther Mon 7h52'36"-4d25'
Rader's World,Dutch Dra 18h12'49"58d42'
Rader,Dale Aulton Ori 6h15'14"14d24'A
Rader,Genevieve A Ori 6h15'14"14d24'B
Rader,Katherine Alaina Aql 19h27'58"-10d23'
Rader,Linda Kay Sct 18h55'29"-11d8'
Rader,Lori Oph 17h38'15"-26d1'
Rader,Nicholas Allen Del 20h53'0"9d30'
Rader,Tori Evans Sgr 18h53'38"-33d11'
Radetzky,Valerie And 0h2'0"38d11'
Radford Cyg 20h52'0"38d44'
Radford,Hank Cyg 21h2'1"40d16'
Radford,Michael Victor Her 16h56'59"48d25'
Radha,Will You Marry Me,Please?! Umi 15h47'5"76d9'
Radhika Lib 15h40'0"-23d7'
Radi,Howriya Sayed Her 16h12'57"4d48'
Radi,Timothy A Cet 3h5'25"3d39"
Radiance Of Heather's Heart,The Cas 0h29'0"61d53'
Radiant Ryan Ori 6h6'20"10d15'
Radiant Susan, Jewel Of The Universe Cas 3h8'16"58d28'
Radiantly Rad Ryan Her 16h20'32"40d13'
Radice,Jr,Daniel F Dra 19h28'17"68d28'
Radice,Sharon & Johny Cep 20h18'21"78d56'
Radick,C M Uma 8h41'0"56d1'
Radieux,Berceau Peg 23h17'0"33d13'
Radigan,Abbey Elizabeth Lyn 7h19'11"59d15'
Radigan,Bridget And 23h37'51"39d44'
Radik,Frank Michael Cma 7h1'43"-20d23'
Radin,Lisa & Neil Cyg 19h59'53"40d20'
Radin,Lisa & Neil Aur 4h59'33"50d10'
Rading,Karin Peg 23h31'31"10d34'
Radio Lac 97,6 Per 4h2'21"37d16'
Radio Suisse Romande -La Première Per 4h5'50"36d29'
Radish,Christina Vel 9h19'18"-41d39'
Radka Cas 2h52'35"63d56'
Radko,Christopher Lac 22h17'27"38d2'
Radley,James Ori 4h45'41"15d6'
Radloff,Lawrence Michael Equ 21h7'45"11d0'
Radlsbeck,Iris Cas 0h31'50"63d24'
Radlsbeck,Werner Cmi 7h19'24"8d54'
Radnov,Rebecca Jane Sex 10h36'59"2d59'
Radosevic,Christopher And 23h0'56"46d19'
Radossich,John C Dra 16h9'43"67d13'
Radosta,Lena Lyr 18h31'49"47d9'
Radovic,Mrs Jacqueline Cas 0h46'1"74d13'
Radovich,Phyllis Gene Peg 23h40'14"25d41'
Radtke Love Lisa,Kraig Her 17h2'21"31d45'
Radtke,Corina Rose Cnc 8h21'44"19d2'
Radtke,Laura Taylor Uma 11h54'1"44d4'
Radtke,MD,TBOTB,Susan Anne Cas 23h40'1"60d41'
Raduege,Jr,General Harry Dale Aql 20h15'54"0d59'
Radvansky,Deborah Lee Eri 4h39'49"-6d54'
Radwanski-Class of 93, Brian Aur 5h29'15"40d53'
Radzinsky,Sr,Roy C Her 16h44'48"30d52'
Radziwon,Princess- Henry Cap 20h22'19"-20d12'
Rae Cyg 21h0'52"28d51'
Rae Ann's Light Cam 3h20'30"56d6'
Rae's Star Cnc 8h30'36"30d42'
Rae,Al, Rachel & Alexis Ari 2h1'39"17d59'
Rae,Cathy & Casey Cyg 19h43'50"31d38'
Rae,Frank Lester Lyr 19h1'15"33d50'
Rae,Jeanette Peg 23h2'20"17d50'
Rae,John Gordon Cep 5h6'36"80d28'
Rae,Shana Aql 19h58'17"15d55'
Rae,Val Cas 2h0'44"73d56'
Rae-Rae Cyg 21h8'44"30d13'
RaeAnn Lyr 18h50'52"38d43'
Raeber-Schneider,Elisa Cam 5h53'54"56d7'
Raeburn,Mickey Lyn 8h56'1"41d39'
Raemisch,Gerald G Per 3h19'39"40d32'
Raemisch,Thomas F Aur 7h11'59"38d48'
RAES 110694 Equ 21h1'35"2d38'
Raess,Sal Lac 22h38'0"38d0'
Raesz-My Friend Forever,Rick Aql 20h1'0"10d18'
Rafael Dra 14h57'1"62d10'
Rafael & Beatriz Cam 4h59'59"61d10'
Rafael Bruno Lmi 10h40'47"23d1'
Rafael(Sonhador Brasileiro) Ori 6h17'39"7d31'
Rafaela's Light Cnv 12h54'32"46d23'
Rafal,Alexandra Eve And 23h27'0"38d21'
Rafaldus Family Star, The Boo 14h45'33"50d24'
Rafalin Uma 11h21'44"37d50'
Rafano-ABCR Star,Ann Babb Crimmins And 23h0'56"46d19'
Rafaschieri,Dino MariaGrazia Cnv 13h1'21"51d52'
Raff,Valerie J Cas 3h3'25"58d55'
Raffa,Stephanie Nicole Lyr 18h57'51"37d8'
Raffael,Marchesa Lyn 9h13'53"41d32'
Raffaella Ant 10h45'55"-37d8'
Raffaella Ind 21h3'57"-55d56'
Raffalli,Yvette Del 20h23'53"16d5'
Raffanetti,Nikki Cam 3h19'17"57d52'
Raffaniello,Serena Cam 6h24'1"67d60'
Raffelt,Terry Farrand Lac 22h2'31"48d12'
Raffeniello,Serena Lyr 19h1'1"25d58'
Rafferty II,William J Aur 6h1'18"32d57'
Rafferty,Carole "Squirt" Peg 0h3'60"31d2'

Rafferty,Harrison James
Ori 5h35'11"-0d55'
Rafferty,Kathleen
Cas 23h19'47"62d59'
Rafferty,Maximilia Renée
Lyr 18h57'48"36d1'
Rafferty,Richard Kevin
Cnv 12h17'45"42d32'
Rafferty,Ronald & Carolle
Crb 16h15'12"34d38'
Raffler,Monique
Lac 22h49'0"38d5'
Raffles
Umi 14h42'12"66d19'
Raffo,Marlyn
Cas 0h20'29"61d25'
Raffol,Kimberly
Eri 4h13'53"-10d5'
Raffy & Cricri
Hor 3h26'36"-48d40'
Rafiki
Lyr 18h59'11"33d16'
Rafiq
Crb 16h6'37"32d22'
Raftery,Dr & Mrs Alan
Cam 4h13'10"68d49'
Raftery,Eugene
Cep 21h52'1"55d30'
Raftopoulos,Alexis
Georgeanne
Eri 3h43'26"0d2'
Raftopoulos,Christine &
Dionysios
Lac 22h6'11"38d29'
Rag
Dra 17h47'29"60d10'
Ragan(Haypooh),Hazel Louise
Vul 20h42'30"28d40'
Ragan,Christopher Patrick
Cmi 7h55'29"8d11'
Ragan-Pepper,Kalleen Gay
Cyg 19h29'37"33d31'
Ragantesi,David A
Hya 9h4'42"0d4'
Ragde,Kristina Elisabeth
Cyg 19h25'21"31d53'
Rager sen
Lyr 19h17'1"30d45'
Rager,Alan H
Oph 17h0'41"-28d17'
Rager,Inez T
Equ 21h1'56"10d26'
Raggio di Sole
Lyr 19h16'11"26d41'
Ragland,Dr Ray
Ser 15h21'16"7d41'
Ragland,Nicole
Cas 2h6'56"61d3'
Ragnar,Stella
Lyr 19h16'53"38d6'
Rago's Star,Renee & Lonnie
Lyn 9h5'53"33d48'
Rago,Louis P
Lac 22h8'41"51d12'
Rago,Sommer
Crt 11h2'29"-8d14'
Ragolia,Christine Jeanette
Cas 0h27'35"68d30'
Ragolia,Louis
Her 17h7'22"46d33'
Ragone,Edward Constantino
Ori 6h17'34"-1d2'
Ragone,Joseph
Aql 19h3'52"-1d16'
Ragoneau,Michel
Cyg 19h45'0"30d5'
Ragsdale,MD,Richard M
Lac 22h4'35"46d33'
Ragsdale,William M
Per 2h53'0"32d52'
Ragucci,Our Friend, Uncle
John J
Cas 0h37'33"75d54'
Ragucci,Renée
Aur 5h7'50"40d35'
Ragucci,Ronald Russell
Aur 5h7'26"41d56'
Ragusa,Christine Roberto
And 0h43'17"45d55'
Ragusa,Martin
Dra 19h4'38"58d48'
Raguseo,Nicholas Joseph
Uma 11h56'1"48d60'
Rahe,Kim
Cas 0h49'59"73d45'
Rahe,Petra und Teddy
Cnc 8h54'57"32d14'
Rahel,Kate Ono
And 0h24'21"37d53'
Rahffii & Kwijibo, Meredith &
Mark
Mon 6h44'14"10d28'
Rahidion
Lmi 10h19'52"38d8'
Rahill,Mary P
Cas 0h32'22"65d39'
Rahilly,Paul Michael
Dra 18h23'48"68d50'
Rahlmeyer,Silke
Aqr 22h6'31"2d14'
Rahman,Mohd Bazlur
Her 18h7'25"14d50"
Rahmeyer,Gerald Bruce
Hya 8h59'38"1d46'
Rahmutallah
Crb 16h7'1"37d51'
Rahn Family,The John
Cyg 21h37'19"38d27'
Rahn,Angelina
Cam 7h43'1"82d34'
Rahn,Dick
Aql 20h4'0"6d20'
Rahtz,Patty
Cas 23h13'46"60d22'
Rai,Cheyenne
Lyr 19h10'37"40d13'
Raia
Lyr 19h12'40"40d28'
Raia,Christopher Steven
Aur 6h8'56"32d2'
Raia,Ernest R
Cep 22h57'34"70d54'
Raible,Robert
Aur 6h2'17"38d34'
Raible,Robin Starr
And 23h25'45"43d28'
Raibman,Richard I
Lac 22h34'11"53d14'
Raic
Lmi 10h44'16"28d2'
Raiche,Romantique Guy
Ori 6h9'16"5d51'
Raichelson,Kimberly Elizabeth
And 23h18'20"42d3'
Raichle,Rebecca
Lyn 8h15'39"36d48'
Raiff,Alexander Clayton
Her 18h4'19"48d44'
Railey,Keith Lee
Cam 4h25'20"60d53'
Railsback,Linda
Cas 0h3'51"58d32'
Railton,Alissa L
Peg 22h2'13"10d38'
Railton,Meredith A
And 1h59'18"41d6'
railtour suisse
Uma 11h20'34"42d31'
Raimbault,Lucille
Peg 23h36'58"20d16'
Raimoili
Uma 10h22'26"47d37'
Raimondi,Pamela
Gem 7h6'9"21d33'
Raimondi,Ruggero
Lib 15h34'16"-28d26'
Raimondi,Susan
Gem 7h15'27"21d36'
Raimondo,Comm Carlo
Uma 8h15'1"62d22'
Raimondo,Victor Louis
Cet 1h59'49"-0d43'
Raina
Cyg 21h43'19"37d55'
Rainaldi,Louis F
Her 18h9'34"37d46'
Rainbolt,Heather Christina
And 2h23'58"42d35'
Rainbolt,Michael Paul
Cnv 12h42'33"36d55'
Rainbow Warrior,The
Aql 20h2'11"1d32'
Rainbow's End
Lyr 18h21'0"38d59'
Rainbow,Kirra Marie
Peg 21h55'45"3d33'
Rainbow-Emiko
Cam 4h6'58"60d40'
Rainbows & Roses Galleria
Crb 16h11'0"38d40'
Rainbows Forever
Crb 15h27'44"31d49'
Raine,Anne Marie
Cas 0h6'0"65d1'
Raine,John Peter
Ori 4h46'53"15d27'
Rainer
Psc 23h32'47"-2d42'
Rainer
Psc 0h44'13"20d42'
Rainer Rudy
Her 18h18'36"14d23'
Rainer,Fred
Cep 3h3'26"80d32'
Rainer,Marcel
Mon 6h53'54"-6d7'
Raines "Birthday Star" Kelly C
Gem 6h55'25"15d14'
Raines Super Star, Robert
Christian
Oph 17h31'30"8d2'
Raines,Charles Robert
Ori 5h23'28"-4d50'
Raines,Courtney
Cas 1h12'16"60d52'
Raines,Dorothy J
Peg 22h42'1"25d16'
Raines,Heath
Aql 19h56'31"-6d53'
Raines,Sarah Louise
Mon 8h5'20"-6d60'
Rainestar
Aql 18h58'38"17d59'
Rainey Family Star, The David
Charles
Ori 5h32'23"-6d43'
Rainey,Beatrice E
Lyn 8h0'52"38d25'
Rainey,Bill
Her 16h32'0"20d21'
Rainey,Cory
Aql 19h53'57"13d0'
Rainey,Edward Chrestman
Sct 18h54'48"-9d21'
Rainey,John
Per 3h13'47"42d57'
Rainey,John S
Cnv 14h4'38"30d31'
Rainey,Tamara Heidi
Cas 0h8'44"61d46'
Rainey,The Wayne
Ori 5h51'33"19d4'
Rainforth,John
Cep 22h42'36"63d39'
Rainier,Sterling
Eri 3h14'0"-5d3'
Rainone,Veronica
Cas 0h47'0"69d25'
Rains,Amitabha
Uma 10h11'1"71d29'
Rains,Jillian Lynn
Lyn 7h41'54"50d23'
Rains,Kimberly Annette
Crb 15h48'18"27d44'
Rainsborough,Thomas
Uma 13h59'13"48d59'
Rainsford,Richard
Boo 15h39'51"41d59'
Rainwater,Earl
Her 18h14'22"30d55'
Rainwater,Elizabeth
And 23h29'42"47d45'
Rainwater,Patricia
Mon 7h12'18"-5d22'
Rainy Night In Soho
Umi 13h51'50"76d12'
Raio,Lamar
Lmi 10h57'0"28d10'
Raisa Jaan
Cyg 19h59'31"41d22'
Raisch,Cullen Henry
Boo 14h18'0"18d20'
Raisch,Lucille Kathryn
Peg 21h56'47"32d43'
Raisch,Mitchell Godfred
Cet 2h16'18"4d6'
Raisler,John P
Dra 17h50'0"75d43'
Raissa-Emilie
Oph 18h17'16"10d47'
Raistlin X
Aql 18h57'37"16d3'
Raith,Vera Lea
Tri 2h41'16"31d15'
Raiß,Andreas
Cep 22h5'27"67d36'
Raj Cuteus Et Scrapytorius
Ramjotton
Ori 5h32'1"0d27'
Rajacic,Helen
Mon 7h1'44"-1d36'
Rajagopalachari,Shri
Parthasarathi
Mon 6h52'0"10d25'
Rajala,Fred
Cma 6h54'40"-19d5'
Rajala,Pekka
Ori 5h59'47"9d9'
Rajan
Eri 3h41'53"-5d56'
Rajanen,Petteri
Dra 19h42'0"68d20'
Rajczewski,Eddie
Ser 17h56'51"-14d58'
Rajeevan,P M
Aur 7h9'20"38d51'
Raji Boy
Cep 22h4'25"60d39'
Rajiv
Eri 3h35'11"-17d59'
Rajon "World's Greatest Hero"
Cmi 7h40'37"1d12'
Rajput,Kavita M
Lyr 19h16'13"41d18'
Raju,Suma
Oph 17h5'9"-23d50'
Rak,Brian Matthew
Per 2h52'0"37d3'
Rakar
Lup 15h39'55"46d57'
Rakastavaiset
Cam 3h43'0"70d13'
Rakel,Francine J
Cas 1h18'13"60d52'
Raker,Kate B
Cas 23h39'0"61d1'
Rakers,Aloysius G
Peg 0h3'40"22d12'
Rakers,Rita J
Peg 0h8'39"21d47'
Rakita,David
Her 16h40'19"29d37'
Rakowiecki,Karen Elizabeth
And 23h22'1"40d51'
Raley,Bob & Sue
Umi 13h11'10"71d16'
Raley,Marion Amy
Umi 14h47'0"66d16'
Raley,Michael
Dra 17h43'27"63d57'
Ralf & Jana
Uma 11h25'18"41d17'
Ralich,Sava
Cnv 12h5'21"35d30'
Ralin,Elea
Lyr 19h5'11"41'1'
Rallen R III
Cep 21h7'23"55d32'
Ralph
Ori 4h59'24"1d32'A
Ralph
Her 16h26'28"38d2'
Ralph
Lac 22h21'52"53d31'
Ralph & Grace
Uma 8h59'17"57d30'
Ralph & Janette's Dream
Crb 15h27'34"31d12'
Ralph & Kathy
Cyg 20h35'33"40d19'
Ralph & Terry
Peg 23h28'29"31d40'
Ralph None & Alice Marie
Forever
Cnv 13h4'39"50d38'
Ralph One & Two
Eri 2h57'13"-12d41'
Ralph's Rose of the Heavens
Eri 2h56'60"-11d36'
Ralph,Jacquie S
Cas 1h36'43"61d26'
Ralph,Zoanna
Cyg 21h5'33"30d14'
Ralph-Prince of Paradise
Tau 3h43'41"10d39'
Ralph-Semper Fi
Dra 16h20'44"63d12'
Ralphi d'Amour
Ori 5h58'0"21d8'
ralphus ursulinus rhein-
bachisiensis
Her 17h47'20"42d24'
Ralstin,Carolyn M
Mon 5h57'58"-6d1'
Ralston's Sky Diamond, Candy
Mon 7h54'38"-1d19'
Ralston,John Richard
Sex 9h53'39"0d14'
Ralston,Kay
Eri 3h58'20"-16d17'
Ralston,Madeline
Peg 22h14'14"5d48'
Ram
Peg 21h25'0"23d47'
Ram,Doreen
Cam 8h7'46"80d14'
Ram,Harvey
Her 15h59'47"21d18'
Ramadan,Maud
Cyg 19h31'17"36d3'
Raman & Vasanthi
Aql 19h7'13"0d50'
Ramanoski,Christopher Scott
Lac 22h13'13"50d12'
Ramasocky
Mon 8h8'32"-6d8'
Ramazzini,Judy
Cas 0h20'19"67d46'
Rambaud,Alain
Ori 6h1'21"0d44'
Ramberg,Pa & Betty
Uma 11h32'0"63d17'
Rambis,Natalie Paige
Cyg 20h9'24"40d58'
Rambo
Cet 3h5'23"1d27'
Rambour,Nicolas
Tri 2h23'55"33d40'
Rambow,Alan James "Buzz"
Oph 17h16'20"-20d41'
Ramer,Yolanda Lee
Vir 11h38'48"2d7'
Rametta,Matthew
Aur 5h1'13"41d9'
Ramey's Star Over Kentucky
Aur 5h23'44"37d35'
Ramey,Christopher Dale
Vul 19h48'48"28d22'
Ramey,Deann Rae
Vul 19h48'36"28d15'
Ramey,Helene
Sgr 18h48'22"-22d23'
Ramey,Jon Michael Aloysious
Aql 19h30'16"8d16'
Ramey,Michael Christopher
Vul 19h47'1"28d49'
Ramey,Morgan Kristine
Vul 19h47'0"28d13'
Ramich,James M
Ori 6h10'0"8d52'
Ramie
Cam 3h38'1"71d42'
Ramie's Nite Light
Ori 5h52'19"12d59'
Ramierez,Kimberly
Cas 23h20'29"58d42'
Ramillon,Helgard
Cam 5h5'36"61d27'
Ramirez Star,The
Mon 7h47'16"-1d48'
Ramirez,Alberto Javier
Hya 8h18'60"2d53'
Ramirez,Alejandro Burgos
Aur 6h3'31"30d36'
Ramirez,Amy Elizabeth
And 23h36'38"49d41'
Ramirez,Belina Brazil
Ser 15h12'29"7d50'
Ramirez,Captain Juan
Guillermo Baidal
Cet 1h33'54"-12d14'
Ramirez,Claudia Valenzuela
Cma 7h15'59"-15d19'
Ramirez,Elizabeth
Gem 7h12'23"28d50'
Ramirez,James Alfonso Merz
Eri 4h17'26"-18d27'
Ramirez,Kenneth
Cmi 7h22'41"8d50'
Ramirez,Laura
Ori 5h50'25"17d41'
Ramirez,Linda J
Aqr 20h37'1"-1d1'
Ramirez,Marco Mascara
Dra 19h4'17"48d50'
Ramirez,Mario
Aur 5h0'42"45d58'
Ramirez,Melissa Odette
Cap 21h37'59"-23d14'
Ramirez,Robert
Hya 8h26'15"0d6'
Ramirez,Rudolfo Alejandro
Cardenas
Aql 18h59'39"-4d48'
Ramirez,Salvador
Sex 9h44'52"4d16'
Ramirez,Sergio "Nokis"
Aql 19h26'44"-8d49'
Ramirez,Sergio Garcia
Cet 2h56'12"6d37'
Ramirez,Sylvia Elena Lizarraga
Mon 7h19'12"-8d42'
Ramirez-IFriends, Andres &
Cindy
Cyg 19h18'0"28d39'
Ramisch,Detlef
Mon 7h52'10"-1d3'
Ramkumar,Lalitha
Lyr 19h16'21"41d34'
Ramler,Hugh
Her 18h7'12"31d0'
Ramlo,Steven P
Aur 6h6'25"45d20'
Ramm,Elke
Cmi 7h20'60"0d2'
Ramm,Rick
Lac 22h5'22"49d51'
Ramm,Wilhelm
Eri 4h49'5"-8d52'
Rammage,Ashlee Nicole
Umi 15h23'27"67d7'
Rammage,Steve
Aur 7h12'44"40d43'
Rammel,Sharon Lea
Lyr 18h47'25"30d18'
Rammelt
Eri 4h50'27"-8d31'
Rammer,The
Her 16h50'56"40d32'
Rammie
Ori 5h59'25"14d6'
Ramneek
Cet 1h22'32"-11d54'
Ramnout,Guillaume
Ser 16h7'39"8d22'
Ramona
Cep 23h16'34"70d27'
Ramona
Cam 11h37'58"81d21'
Ramona
Peg 23h16'59"30d47'
Ramona
Pho 0h31'45"-49d22'
Ramona Jane
Peg 23h26'56"8d53'
Ramona's Glory
And 1h5'12"40d32'
Ramos & Alice,Abram &
Gabriela
Cyg 21h8'23"38d23'
Ramos Eagle
Aql 20h30'17"0d36'
Ramos,Augusto C
Boo 13h56'23"21d8'
Ramos,Carl A "Headbanger"
Sct 18h41'15"-6d7'
Ramos,Deborah
Com 12h21'17"32d59'
Ramos,Edward
Aql 19h0'1"-0d10'
Ramos,Jose(Tony)
Aql 18h59'60"10d58'
Ramos,Lorenzo
Sgr 18h49'37"-22d21'
Ramos,Maria Avelina
Cas 0h1'36"64d20'
Ramos,Michael
Per 1h52'42"56d45'
Ramos,Rudy
Vir 11h40'1"4d41'
Ramos,Steven Charles
Aur 6h15'55"31d2'
Ramoth,George
Cnv 14h0'60"31d8'
Rampillon,Marie Thérese
Her 17h53'41"40d38'
Rampini MRC,Rosangela
Lac 22h52'34"50d35'
Rampling,Coralie Ann
And 23h0'60"50d32'
Rampo,Massimiliano
Her 18h10'0"30d14'
Rampton,Ian N
Per 2h52'1"45d43'
Rampulla,Jack
Lyn 7h27'1"42d16'
Rampulla,Marguerite Ann
Eri 4h10'9"-19d23'
Ramrayka,Peter
Cep 5h37'23"85d27'
Rams
Mon 6h18'47"8d19'
Ramsaier,Alexandra Carley
Umi 15h44'35"77d31'
Ramsay,Frances Helen Star
Cas 1h4'46"58d3'
Ramsay,John Paul
Boo 15h14'40"26d41'
Ramsay,Kathy
Mon 7h4'45"0d41'
Ramsay,Kristine
Lyr 18h31'50"44d11'
Ramsberger,Jane
Cmi 7h26'1"1d15'
Ramsberger,Maggie
Cmi 7h26'0"8d40'
Ramsden,Katherine Emily
Eri 4h2'31"-12d28'
Ramsden,Peter
Her 17h20'45"29d18'
Ramsden,Suzanne M
Cas 23h29'25"60d58'
Ramsey
Her 17h8'17"42d20'
Ramsey
Uma 9h31'55"51d4'
Ramsey II,Donald R
Dra 17h12'1"72d16'
Ramsey,"Derrick-Time"
Uma 10h9'40"61d47'
Ramsey,"Dookey" Christopher
L
Aur 5h30'51"31d20'
Ramsey,Bruce
Her 17h15'44"44d11'
Ramsey,Clara Anita Paquette
Cas 0h48'50"75d32'
Ramsey,Daniel & Barbara
Aur 5h16'36"49d41'
Ramsey,Dannelle
Cep 22h24'60"70d19'
Ramsey,Diane K
Uma 8h37'20"60d37'
Ramsey,Janet Ruth
Peg 23h26'55"31d2'
Ramsey,Jennifer Drew
Cap 20h28'43"-12d1'
Ramsey,JoAnn
Lyr 19h25'29"38d21'
Ramsey,John
Ser 18h6'47"-5d41'
Ramsey,John Joseph
Lac 22h13'28"51d9'
Ramsey,John Keith
Sct 18h49'40"-7d8'
Ramsey,Jonathan David
Her 17h8'33"37d34'
Ramsey,Lance & Erin
Ori 5h54'0"14d30'
Ramsey,Megan P
Com 12h57'40"14d42'
Ramsey,Robert
Aql 19h58'0"14d54'
Ramsey,Sinbad
Aql 19h51'0"15d41'
Ramsey,Stacy(Jezebel)
Cap 21h21'29"-20d5'
Ramsey,William L
Boo 14h59'0"48d17'
Ramsey,William Thomas
Lyn 7h51'16"44d31'
Ramsey,Woodrow & Mabel
Boo 14h36'48"19d43'
Ramskjell,Nils
Dra 15h53'17"60d59'
Ramspacher,Charline Nathanël
Uma 12h24'35"62d30'
Ramspacher,Jean Claude
Uma 12h21'34"62d58'
Ramstad,Lewis J
Cep 22h16'1"65d1'
Ramtallie
Ori 4h42'26"4d47'
Ramundo,Lucille
Del 20h30'11"20d10'
Ramundt,Sarah Elizabeth
Mon 8h1'26"-6d47'
Ramzan,Mohammed Amin
Ori 5h55'22"19d7'
Ramírez-Siempre Amado,
Adan
Cam 5h35'0"60d32'
Rana,Anil
Lyn 7h8'29"53d2'
Rana,Bandhana R L
Tri 2h12'46"31d51'
Ranado,Stephen M
Cam 5h39'1"60d8'
Ranaglia,Lori Jean
And 23h42'1"38d36'
Ranalli,Joyce Ann
Uma 10h10'22"52d25'
Ranalli,R Linda
Lyn 6h45'31"60d32'
Ranally,Jr,John Andre
Per 3h25'54"40d25'
Ranata
Cas 0h27'46"61d36'
Ranauro,Steven J
Lac 22h24'34"53d6'
Ranch,Garrett Creek
Oph 17h53'52"14d6'
Rand,Paul
Aql 18h57'30"-6d30'
Rand,Roger R
Boo 13h55'1"17d57'
Randall
Cet 3h14'17"9d35'
Randall "50",Cam & Les
Cyg 21h24'41"40d56'
Randall 9-1-79,Donald James
Peg 23h0'42"32d46'
Randall,Alice Fracker
Aql 19h29'34"8d38'
Randall,Brian & Suzanne
Uma 10h51'46"58d58'
Randall,Carol
Gem 6h24'17"14d28'
Randall,Charlotte Catherine
Cas 0h51'50"61d11'
Randall,Chrissi & Michael
Uma 10h40'33"44d28'
Randall,Jeanette
Peg 22h21'0"3d18'
Randall,Jerri Lynn
Lyr 18h23'1"40d22'
Randall,Jon
Tri 2h7'45"33d57'
Randall,Jr,Maurice E "Randy"
Lac 22h21'27"55d35'
Randall,Kathy
Cyg 20h17'57"41d33'
Randall,Matthew
Boo 13h38'50"15d17'
Randall,Melvin T
Her 18h14'51"37d42'
Randall,Mitch Leroy
Boo 15h11'23"26d40'
Randall,Ross Paul
Dra 16h59'0"68d45'
Randall,Sally Belinda
Cyg 19h57'27"48d50'
Randall,Scott E
Per 2h45'10"43d47'
Randall,Sheena Marie
Lyr 18h38'50"41d54'
Randall-Star Of My Heart
Ori 4h48'55"4d14'
Randalls' Radiant
Remembrance
Cyg 21h55'42"40d12'
Randalls' Star
Uma 8h42'1"70d10'
Randar's Star
Aql 20h7'14"0d57'
Randazzo,Vincent Thomas
Her 18h50'25"18d58'
Randee & Gary
Sge 18h59'38"18d54'
Randell,Linda Knudsen
Peg 22h47'41"29d52'
Randeria,Aneal
Her 17h18'0"42d54'
Randhahn,Kurt
Aur 5h18'41"43d53'
Randhawa,Amber
Lyn 8h22'1"49d31'
Randi
Lyn 8h1'24"57d41'
Randi & Peter
Umi 14h14'0"70d3'

Randi Helen
Peg 23h33'49"10d46'
Randi Maylene
Dra 18h44'19"68d21'
Randi's Poppaw
Aql 18h59'24"-5d34'
Randle,Jimmy
Cet 3h19'13"4d30'
Randle,John "Newcastle Brown"
Uma 11h15'52"50d34'
Randle,Warren Gaylord
Hya 9h32'56"5d1'
Randlett-Forget Me Not Doug
Per 2h38'1"38d28'
Randol,Burton
Per 3h54'40"38d41'
Randol,Crackers
Lac 22h22'17"52d45'
Randol,Fritzl
Uma 8h42'44"67d33'
Randolph
Aur 6h32'55"37d1'
Randolph 7/4/81, Phillip & Karen
Dra 16h1'0"63d49'
Randolph's Eternal Glow,Ron
Aql 18h57'34"12d47'
Randolph,Bobbi Marie
Aql 20h1'1"10d22'
Randolph,Catherine Nelson
Lyr 19h3'30"47d40'
Randolph,Craig L
Aql 19h5'30"0d34'
Randolph,Gwen "Estrella Brillante"
Cyg 20h8'33"43d47'B
Randolph,Helen Garside
Cas 0h2'32"63d34'
Randolph,Herbert & Jeane
Crb 15h54'20"32d24'
Randolph,J Michael
Cet 2h21'25"2d3'
Randolph,Lauren Hadley
Aqr 22h30'52"0d42'
Randolph,Patrick Hart
Peg 21h49'46"28d6'
Randolph,Stewart Elliott
Per 4h18'56"50d23'
Randon,Jerome
Sge 19h58'30"20d33'
Randox Laboratories Limited
Her 16h47'49"35d19'
Randtopia
Eri 4h53'24"-4d48'
Randy
Aur 7h24'39"37d33'
Randy
Lac 22h26'29"53d53'
Randy
Hya 8h10'51"1d18'
Randy & Robin
Sex 10h5'0"-1d9'
Randy Reach For The Stars
Cep 0h54'16"77d59'
Randy Two
Boo 15h17'59"38d18'
Randy's 40th Wishing Star
Ori 5h59'27"20d37'
Randy's Lone Wolf
Sct 18h46'57"-7d54'
Randy's Sun
Aur 7h4'47"40d24'
Randy's Target
Cap 21h1'23"-27d16'
Randy,You're A Star
Cep 21h15'0"68d32'
Randy: Tonia's Lucky Star
Peg 22h2'10"31d53'
Randyland
Cet 2h9'1"-0d36'
Rane aka RaeRae, Evelyn Victoria
Peg 23h4'37"11d25'
Ranee
Umi 13h11'47"71d2'
Raneen
Lyr 18h59'47"40d45'
Ranellucci,Karen L
Cmi 8h7'47"5d41'
Raneo,Albert H
Aur 4h56'24"41d8'
Ranery,David Michael
Gem 7h23'44"35d14'
Raney's Star,Chad
Cmi 7h38'20"4d32'
Raney,Joel
Cyg 21h26'33"28d56'
Raney,Julia Adelle
Aql 19h25'41"-6d13'
Raney,Larry
Aql 19h56'41"12d26'
Raney,Liz
Eri 3h54'17"-4d17'
Range,Dale
Cma 6h55'26"-18d27'
Rangel,Linda
Aql 19h55'51"12d33'
Rangel,Miguel A
Cep 20h37'0"60d5'
Rangel,Rosalinda
Mon 7h56'0"-1d45'
Ranger Joe
Her 17h7'1"22d23'
Ranger,Rainne
And 2h17'26"38d38'
Ranger-Alburtus
Dra 19h2'23"50d22'
Rania
Per 2h29'28"58d14'A
Rania-Koulis
Per 3h36'0"38d57'
Rania-Maria
Cyg 21h31'26"36d8'
Raniere,Elizabeth Kathryn
Cam 3h24'38"58d55'
Ranieri,Paul
Cnv 13h28'18"41d37'
Ranjoué,Heinz
Per 3h25'30"50d6'
Rank,Sterling W
Hya 9h34'22"0d24'
Rankin's Radiance
Com 13h6'20"27d42'
Rankin,Amanda
Aqr 20h37'32"-10d10'
Rankin,Andy
Aql 18h57'41"16d31'
Rankin,Nyla Dallas Whitney
Lyr 18h24'51"42d59'
Rankin,Patricia S
Peg 21h19'26"20d2'
Rankin,Penny L
Lyr 18h25'29"44d33'
Rankin,Polly Gowans
Crt 11h16'39"-14d35'
Rankin,Scott
Oph 17h13'52"-20d57'
Rankin,Thomas Glenn
Aur 4h56'14"51d22'
Rann,Jason
Gem 7h3'32"28d39'
Rann,Reinhard
Mon 7h53'56"-1d42'
Rans"Arctic Explorer", Robert
Umi 16h24'1"80d6'
Ransom,Edward Anthony
Aur 6h24'46"30d59'
Ransom,Edye
Cas 0h52'11"62d37'
Ransome-Forever Young, Forever Missed,Danny
Cyg 20h26'15"46d14'
Ranson
Lyr 18h32'54"37d11'
Ranson,Delores
Aql 18h59'38"-1d54'
Ranstrom,Dorothy
Mon 7h45'10"-3d24'
Rantanen,Pertti
Uma 14h4'19"51d4'
Ranton II,Laroy Princeton
Lac 22h55'19"51d5'
Ranton,JoAnn
Peg 22h30'36"24d3'
Rantuccio,Matthew
Lac 22h11'13"46d29'
Rantuccio,Star of Nicholas
Lac 21h57'0"36d42'
Rantz Love Forever By Rusty,John W
Hya 8h40'58"-1d51'
Ranus,Cassie
Boo 13h36'43"14d55'
Ranwa
Cep 20h57'1"62d44'
Rao,Candace Marvel
Lyn 7h3'59"60d35'
Rao,Thara Bai & Rama
Vul 19h48'26"28d49'
Raouf,Samir
Lmi 10h49'50"33d8'
Rapaglia,Mario
Cyg 21h52'27"37d20'
Rapagnani,Richard G "Rich"
Aql 18h59'46"11d24'
Rapale,Ivette Lucille
Mon 6h19'28"3d17'
Rapaport,Neil Mitchell
Cmi 7h42'0"5d29'
Rapczynski,Lori Ann
Cas 0h3'32"61d12'
Raper,Dr Steven E
Aur 5h17'1"42d25'
Raphael
Cam 8h15'19"74d23'
Raphael
Her 17h1'18"40d19'
Raphael Family Star Susan,Richrd,Justn,Adm
Uma 11h30'19"51d6'
Raphael Rosa
Per 2h57'20"41d15'
Raphael,Ana
Lyn 7h45'1"43d9'
Raphael,Dylan Simon
Uma 11h26'35"50d23'
Raphael,James Matthew
Boo 13h39'34"19d30'
Raphael,Valerie
Lyn 8h5'0"50d18'
Raphaelle
Ori 5h2'0"1d40'
Raphaël
Cam 5h5'21"68d31'
Raphaël
Aur 5h52'15"54d55'
Raphaële
Peg 21h54'16"21d24'
Rapier,C
Cam 5h52'18"60d59'
Rapillo,Amy Johanna
Cyg 20h25'46"37d46'
Rapisarda II,Alonzo
Umi 16h47'28"76d46'
Rapisarda,Blanche C
Lyn 8h54'21"46d45'
Rapisarda,Blanche C
Gem 6h58'21"13d16'
Rapkin,Ruth & Yale
Crb 15h22'34"32d13'
Rapley II,William "Buck"
Aur 6h33'11"37d12'
Rapoport,Joseph J
Aql 18h56'0"-6d46'
Rapoza,Marion Woodman
Boo 14h28'1"23d12'
Rapp(Rainbows & Dolphins),Michael
Cnv 12h56'10"31d40'
Rapp,Britleigh Lynn
Lyn 8h37'1"40d21'
Rapp,David Adam
Cet 2h37'0"0d28'
Rapp,Diane Rebecca
Cam 4h14'12"60d11'
Rapp,Dillon Howard
Aur 6h57'21"36d53'
Rapp,Michael
Sco 16h56'1"-38d21'
Rapp,Robert Andrus
Cep 22h48'36"58d45'
Rapp,Stephanie Beatrice
And 2h32'17"49d51'
Rappa,Lisa M
Cet 1h32'40"-5d52'
Rappaport,Amanda
Lyn 7h30'44"36d7'
Rappaport,Gregory Alexander
Cma 6h56'15"-19d50'
Rappaport,Ileen Spivack
Sco 16h56'0"-40d43'
Rappaport,Jack
Per 3h0'31"50d29'
Rappaport,Jason Robert
Cmi 7h25'46"1d18'
Rappaport,Kenneth Dean
Crt 11h15'36"-8d59'
Rappaport,Marc Saul
Peg 21h33'51"20d14'
Rappaport,Peter
Boo 15h7'14"22d4'
Rappaport,Rob Davis
Cet 1h22'57"1d44'
Rappin,Jeff
Cyg 21h55'37"53d44'
Rappoport,Janet
Oph 18h0'50"10d33'
Rappoport,Paul S
Per 2h54'17"37d19'
Rapsys,Aldis
Aur 7h12'59"36d26'
Rapture D 14753
Dra 19h19'11"80d24'
Rapue,Carol Ann
And 0h42'48"38d4'
Raquel
Cam 4h6'1"67d58'
Raquel
Cam 6h4'11"61d16'
Raquepas,André Marcil Diane
Ori 5h40'49"12d13'
Rarey,Eli Nathaniel
Oph 17h28'42"0d24'
Rarey,Helen
Cyg 20h0'59"30d20'
Rarey,Jessica Ann
Mon 6h27'15"8d28'
Rarick,Olivia & John
Peg 0h1'24"30d50'
Rarig,Lloyd C
Vir 12h34'34"0d11'
RAS
Her 18h15'37"38d56'
RAS
Dra 19h2'31"48d14'
Rasch,Jo Ann Hansen
And 2h4'1"39d27'
Rasch,Suzanne Marie
Aql 18h59'28"-8d39'
Rasco-Hechim,Maxine Cletice
Mon 7h58'53d-5d16'
Rasely,Tim
Her 17h25'29"38d53'
Raser,Lois A
Mon 7h54'12"-9d1'
Raser,Roger
Sex 9h43'0"2d33'
Rash,Cheryl Ann
Equ 21h2'0"7d57'
Rash,Dorothy L
Lyn 7h45'43"41d59'
Rash,Janet Faye
Mon 80h7'50"-1d46'
Rasha,Esther
Lyn 6h26'17"59d39'
Rashan,Talyn
Cma 6h39'14"-11d9'
Rashanda
Aql 20h10'21"12d28'
Rasheed P M
Cam 4h40'50"68d46'
Rasheed,K J
Aql 18h58'0"11d23'
Rasheed,Omar Abdul Aziz Hassan
Per 1h47'1"52d57'
Rashid Dara
Eri 3h1'0"-1d47'
Rashid,Laura Melacina
Umi 16h48'37"85d56'
Rashid,Mark Moses
Cyg 21h35'26"42d46'
Rashida Ameer
Cyg 20h9'22"39d33'
Rashmi
Lyn 7h54'33"39d44'
Rasia Forever
Boo 14h30'60"51d15'
Rasizzi,Frank
Aur 5h31'50"50d4'
Rask,Eleanor
Mon 6h20'57"8d18'
Raskin,Phil
Aql 18h59'15"-2d46'
Raskin,Sheldon
Aur 5h30'14"38d58'
Raskin,Shelly
Cep 23h26'0"70d12'
Raskin,Susan
Cas 0h55'48"73d36'
Raskovsky,Ruth
Cas 0h23'0"70d50'
Rasmann,Dale N
And 1h34'18"37d46'
Rasmus,Kevin Michael
Her 17h4'0"48d8'
Rasmussen MA PhD,Dr Garry Allen
Her 18h2'39"27d6'A
Rasmussen's "Star Of RAZ",Rick
Dra 17h49'1"61d33'
Rasmussen,Amy Anne
And 1h25'0"34d2'
Rasmussen,Amy Leigh
Cas 0h24'47"66d16'
Rasmussen,Bertha
Her 17h9'51"40d9'
Rasmussen,Birgit
Com 12h7'20"32d53'
Rasmussen,David Curtis
Sex 10h45'0"-8d2'
Rasmussen,Desy
Cyg 20h49'46"50d23'
Rasmussen,Emma Maltha
Cap 21h22'17"-22d13'
Rasmussen,John Wayne
Her 18h4'28"38d30'
Rasmussen,Keith Brian
Ori 5h55'59"17d56'
Rasmussen,Kevin
Aur 5h11'46"40d32'
Rasmussen,Ret,Judge
Cet 0h1'18"-8d56'
Rasmussen,Willow Dean
Hya 8h31'50"0d6'
Raso,Patricia Alice
Com 13h16'53"21d50'
Raspa,Connan Anthony
Dra 16h12'22"68d11'
Raspa,Logan Manzo
Her 17h8'58"46d5'
Raspberry Freesia Gimlet
Uma 9h53'1"48d32'
Raspberry Joseph
Aql 20h7'0"0d57'
Raspen,Jay
Lac 22h36'0"53d48'
Rassbach,Hannes
Aqr 23h29'46"-11d57'
Rassmussen,Krista
Cyg 20h44'2"46d54'
Rassmussen,Mikkel
Umi 14h5'52"70d17'
Rast,Uwe
Lyn 8h14'57"47d54'
Rasta Princess
Mon 8h6'51"-9d37'
Rastar
Ori 5h51'48"17d47'
Rastoul,Manon
Peg 23h37'48"15d42'
Rastrelly,Chris
Leo 11h20'32"-1d42'
Rat Emmanuel De La Deule
Aql 20h15'42"1d38'
Ratajczyk,Craig R
Cep 22h14'20"60d26'
Ratcliff,John B
Cep 3h13'30"80d0'
Ratcliff,Steven Joe
Ser 18h6'16"-14d45'
Ratcliffe,Ann Currie
Psc 1h22'37"32d14'
Ratcliffe,Anthony John
Uma 10h9'18"48d38'
Ratcliffe,Carol Ann
Umi 15h35'43"68d26'
Ratcliffe,Darren Neil
Ori 6h7'37"10d1'
Ratcliffe,James F & Mary E
Cyg 21h14'12"34d38'
Ratcliffe,Paul
Ori 6h4'28"2d28'
Ratcliffe,Richard
Hya 8h19'1"0d32'
Rath Sisters Emily & Carolyn,The
Uma 8h40'13"56d7'
Rath,Beate
Vir 13h9'59"-5d46'
Rath,Bob & Melva
Lyn 8h0'59"38d31'
Rath,Connie
And 0h39'44"40d18'
Rath,David
Cam 5h21'7"24d30'B
Rath,Joel
Per 2h4'1"58d59'
Rath,William J
Lyr 18h59'28"43d4'
Rathbone,Robert Streeter
Cet 3h5'29"2d1'
Rathbun Loves Pete Levesque,Karen
Uma 13h8'16"75d6'
Rathbun,Robert
Her 17h56'16"40d36'
Rathert,Kaytee
Peg 22h58'1"32d3'
Rathey,Horst
Hya 9h35'1"5d19'
Rathka,Daniel C
Cep 22h27'23"61d49'
Rathke,Kimberly Ann
Lyr 18h35'1"45d50'
Rathod,Rahul
Mic 20h45'48"-29d7'
Ratica,Rhonda Weiland
Cas 2h33'1"75d2'
Ratingen,Christiane Müller Dürrst10
Boo 14h4'15"13d3'
Ratisbonne,Roger
Equ 21h1'58"8d2'
Ratley,Bobbi
Lyn 6h50'25"60d21'
Ratliff,Beccy
Mon 6h43'50"10d39'
Ratliff,Heather
And 0h12'33"47d14'
Ratliff,James Vernon
Leo 11h9'16"23d45'
Ratliff,Luke Robert
Aur 5h1'24"49d35'
Ratliff,Luke Robert
Uma 8h35'33"58d11'
Ratna-Manek
Cet 3h18'45"8d45'
Ratnam,Ms Anita
Lyn 8h6'1"37d58'
Ratner,A J
Cma 6h54'46"-18d1'
Ratner,Ilana Fay
Lyn 7h41'12"58d8'
Ratner,Jon
Sge 19h55'47"16d14'
Ratner,Julia Doree
Mon 6h19'22"4d58'
Ratner,Suzanne
Peg 22h16'53"7d59'
Ratte,Ruediger
Ari 2h52'25"20d2'
Ratterree,Katsy B
Mon 6h19'32"5d24'
Rattna
Cas 23h5'49"58d55'
Rattray,Daisy Alys Mererid
Lyr 19h20'30"42d35'
Rattray,Polly Jean Selina
Lyr 19h20'24"41d51'
Ratushinskaya,Irina
Eri 4h57'16"-5d44'
Ratze
Ari 2h23'29"26d24'
Ratzlaff,Karl-Heinz
Oph 17h5'51"1d3'
Ratzlaff,Lori & Mike
Vul 19h20'57"26d29'
Rau,Beate
Ari 2h53'35"21d36'
Rau,Dave
Oph 18h42'18"8d12'
Rau,Marion R
Aur 4h56'11"50d0'
Rau,Michael
Aql 20h9'23"4d54'
Rau,Mike
Lyn 8h27'25"41d33'
Raub,Wilbur & Dorothy
Cyg 21h7'0"31d8'
Rauch,"Little Smoky", William
Peg 23h43'46"31d3'
Rauch,Cindy L
And 1h21'58"34d31'
Rauch,Nicholas Andrew
Aur 5h58'15"30d32'
Rauchfuss,George
Boo 15h46'25"41d26'
Rauda,Eriks Guntis
Lac 22h14'15"51d29'
Raudes-Reyes,Luz
Lmi 9h55'38"38d20'
Raudes-Schoner,Ricardo J
Ori 5h3'1"-2d28'
Rauenet,Jean
Mon 6h22'56"-0d21'
Rauf,Amna
And 1h10'39"39d42'
Rauf,Natalie
Mon 7h52'34"-3d20'
Raugh,Anne
Cas 0h22'26"68d1'
Rauh,Evan Reed
Her 16h48'56"47d48'
Rauh,Karl
Peg 23h33'29"27d38'
Rauh,Rebecca DeAnn
Lyn 7h56'1"45d2'
Raukar,Logan Nicole
Cet 3h13'19"4d59'
Raul Julia
Dra 19h24'33"58d47'
Rault,Stefanie
Lac 22h19'27"50d11'
Raum,Andreas
Gem 6h40'1"31d42'
Rausch,Manfred
Cmi 7h19'34"8d44'
Rausch,Michael A
Lac 22h45'20"55d45'
Rauschen Bach
Uma 13h20'27"58d42'
Rauscher,Joseph
Aur 6h24'14"31d41'
Rautenberg,Iris
Vir 12h30'16"-8d24'
Rauva-My Heart
And 0h47'25"36d35'
Ravaioli Simone
Umi 15h58'43"70d41'
Ravassard,Tess-Heloïse
Crb 16h18'49"37d40'
Ravee's "Gottarun"
Aur 6h28'37"30d12'
Ravella,Lance
Oph 17h28'56"-22d27'
Ravelli,Joseph L
Boo 13h47'59"17d38'
Raven
Cep 20h21'25"76d35'
Raven,Lily
Cyg 20h24'18"39d21'
Ravenfire
Uma 11h53'27"42d8'
Ravenscraft,Megan Joan
Cyg 19h31'52"34d39'
Ravenscroft,"The Lowest!" Thurl
Umi 16h58'35"80d32'
Ravenscroft,Andrew William
Her 17h35'14"21d3'
Ravenscroft,Claire Elise
And 0h20'37"30d7'
Ravenstad,Grant Ronald
Vul 19h19'28"26d53'
Raver Star
Cep 21h56'0"61d2'
Raver,Don
Sct 18h44'18"-6d12'
Ravetta,Laura
Lyn 7h27'24"44d37'
Ravi,Alison & Angela's Star
Uma 13h2'50"52d55'
Ravikoff,Howie
Cam 11h7'18"81d60'
Ravin,Darlene
Mon 6h37'11"0d20'
Raviolo,Giampaolo
Boo 15h0'44"30d4'
Ravis,Herbert J
Ori 5h51'56"15d16'
Ravisha
Dra 16h47'42"63d30'
Ravishing Rhoda,The
Eri 3h5'41"-5d57'
Ravitz,Evonne
Mon 7h58'38"-5d53'
Raviv,Gideon
Lmi 10h8'56"31d47'
Ravo,Jude Victor
Uma 14h6'43"58d15'
Raw,Ben
Her 17h18'55"41d24'
Rawal,Sunil
Cyg 19h41'30"28d32'
Rawald,Bert
Oph 18h7'15"13d57'
RawButt
Aql 20h10'11"3d53'
Rawcliffe Star
Peg 22h40'0"34d29'
Rawdin,Our Tattooed Angel Jesse Daniel
Dra 11h25'13"70d43'
Rawley,Karen Jean
Cas 0h46'0"64d30'
Rawling,Darren
Per 1h52'0"50d23'
Rawlings,David W
Cep 22h0'37"62d32'
Rawlings,Leslie
Mon 7h3'32"-6d32'
Rawlings,Steven Kent
Her 17h38'1"40d47'
Rawlins Family Star
Cam 3h57'36"57d35'
Rawlinson,Caroline
Cas 0h47'53"60d42'
Rawlinson,Gertruda Johanna
Aqr 21h53'18"0d29'
Rawlinson,Joseph & Nancy
Per 2h36'10"46d20'
Rawls,Amy Lee
Tri 1h48'51"26d30'
Rawls,Gayle Catheryne
Mon 6h18'60"8d53'
Rawls,Patti Sue
And 23h41'42"46d22'
Rawls,William & Christine
Boo 14h53'23"26d32'
Rawlyk,Raymond Vincent
Dra 14h24'19"63d12'
Rawson,Archie
Cet 1h18'29"-12d42'
Ray
Aql 19h0'15"-0d7'
Ray
Lmi 9h42'13"33d56'
Ray
Uma 11h50'1"31d6'
Ray
Lyn 7h39'50"41d58'
Ray
Ori 6h3'11"4d15'
Ray & Amy
Dra 17h44'46"60d19'
Ray & Claire "Eternally"
Cyg 20h20'25"30d13'
Ray & Eileen
Vul 19h17'16"22d6'
Ray & Gwen
Hya 8h58'17"-7d53'
Ray & Karen Escape
Eri 3h46'22"-6d58'
Ray Adam
Aqr 23h18'59"-6d17'
Ray Lawrence
Aqr 23h8'38"-10d53'
Ray Lawrence
Lac 22h42'31"35d35'
Ray Shine
Cet 2h6m55"8d26'
Ray Star,The Rosemary
Sge 20h1'23"17d47'
Ray Teri
Cma 7h48'47"11d14'
Ray's "Child"
Boo 14h34'10"7d39'
Ray's Kahnstellation
Cmi 7h37'49"4d7'
Ray's Place
Dra 17h11'25"68d21'
Ray's Radix
Aqr 23h28'34"-19d30'
Ray's Star of Captured Dreams
Ori 5h52'22"20d15'
Ray's Star,Suzy
Peg 22h52'13"8d52'
Ray, Ann & Carla
Crb 16h16'45"34d28'
Ray,B J
Dra 17h50'52"64d18'
Ray,Betsy
Peg 21h38'30"23d16'
Ray,Dennis Lee
Equ 21h1'53"11d2'
Ray,Douglas W
Mon 6h26'57"-10d41'
Ray,Gary C & Joyce Williamson
Uma 11h54'13"40d8'
Ray,Giles
Uma 11h31'29"40d53'
Ray,Glenn-Kim-Kyle & Tyler
Aur 6h26'20"37d16'
Ray,Gregory
Cet 3h2'28"9d6'
Ray,Harold
Sge 20h1'36"19d5'

Ray,Isabelle Gloria
Lyr 19h10'38"47d12'
Ray,Jaime Dawn
Cyg 19h43'46"31d31'
Ray,James E
Cep 23h6'55"62d45'
Ray,Jamie L
Ori 5h44'36"1d28'
Ray,John Alfred
Boo 13h45'0"24d18'
Ray,John K
Leo 9h39'29"7d3'
Ray,Michael J
Cep 22h45'59"57d21'
Ray,Michael Steven
Per 2h51'0"43d31'
Ray,Peggy
And 2h19'37"45d24'
Ray,Rebecca Michelle
And 0h45'56"36d34'
Ray,Shanda
Lyn 7h52'32"35d18'
Ray,Susan & Pete
Cyg 20h20'19"41d34'
Ray,Tara & Brandon
Cet 2h32'1"5d9'
Ray,The Randy
Oph 17h30'35"-20d7'
Ray,Timmy & Claudine
Cnv 12h50'31"47d58'
Ray,Walter D
Her 16h29'26"38d30'
Ray,William Weldon
Ori 6h4'15"8d48'
Ray-Ray
Sex 9h59'1"-5d59'
Raya,Margarita
Cam 3h48'24"70d8'
Raybin,Haley
Boo 14h48'53"24d4'
Raybin,Shelby
Lyn 7h48'34"44d6'
Rayboudl,Simon
Aql 19h2'44"17d3'
Rayburn,Rita L
Lyn 7h4'18"44d45'
Rayburn,Thomas E
Her 17h55'0"30d34'
Rayden Star
Vul 1948'36"26d55'
Raye,Ashley
Cyg 21h48'52"37'13'
Raye,Leah
Uma 9h15'0"67d50'
Rayglo
Aql 20h10'16"10d36'
Rayher,Wolfgang
Ori 5h49'14"20d33'
Raylene
Cyg 21h5'58"31d30'
Raylon Nichole
Mon 7h59'32"-2d5'
Rayman,Mully
Eri 3h10'33"-2d11'
Rayment,Mary Christine
Vul 19h47'19"23d0'
Raymer,Alison Lisa
And 1h37'1"40d27'
Raymer,Irwin Michael
Lyn 6h56'50"54d59'
Raymond
Uma 10h34'1"61d50'
Raymond
Her 18h54'39"12d22'
Raymond
Per 3h13'29"42d40'
Raymond
Lib 14h58'59"2d57'B
Raymond
Cep 22h43'38"65d45'
Raymond "Stargate"
Per 2h50'52"37d35'
Raymond Alphonsus
Cep 23h10'23"61d35'

Raymond Jeff
Cep 3h0'1"77d29'
Raymond O
Vul 19h50'54"20d32'
Raymond Star I,Skyler Dean
Cas 2h19'51"68d5'
Raymond,Abby,Alec,& Sarah
Uma 10h41'18"72d31'
Raymond,Anne Billon
Aur 5h2'43"37d48'
Raymond,Christopher S
Cep 23h9'47"68d55'
Raymond,Dan
Aur 6h3'0"31d45'
Raymond,Dan
Hya 8h49'11"0d30'
Raymond,Derek-Austin &
Kendal
Aur 6h3'60"31d26'
Raymond,Elizabeth & Jeff
Cyg 21h34'15"42d1'
Raymond,John (Johnny Wave)
Ari 2h33'0"30d45'
Raymond,Juanita Susan
Eri 3h37'33"-6d31'
Raymond,Julia
And 23h33'39"44d57'
Raymond,Patrick
Aur 6h2'54"37d39'
Raymond,Robert Francis
Her 16h25'51"50d20'
Raymond,Rochelle Leigh
Tri 2h11'57"30d57'
Raymond,Rodrigue
Uma 10h28'1"41d19'
Raymond,Stéphane
Tri 1h44'19"30d22'
Raymond,Thomas
Peg 23h28'28"11d3'
Raymond,Timothy G
Cnv 12h50'37"40d18'
Raymond,Tracy L
Mon 6h54'37"-5d42'
Raymos,Goeroge
Hya 8h12'55"-6d21'
Raymundo,Adeelmo
Cma 7h0'35"-11d12'
Raymy
Umi 15h25'27"71d39'
Rayna
Eri 3h30'34"-7d25'
Rayne 1
Cas 0h56'1"75d36'
Rayne,E C
Sct 18h48'57"-9d50'
Rayne,Madge
Cyg 20h36'58"38d34'
Rayner Star,The
Cet 2h45'58"6d4'
Rayner,Des
Her 16h22'12"18d29'
Rayner,Katharine & William
Cyg 20h18'33"38d10'
Rayner,Lt Col (Ret) Jay C
Aql 19h58'20"-0d37'
Rayner,Marcus Newcomb
Her 15h52'45"42d46'
Rayner,Margaret Woods
Mon 7h13'46"-6d25'
Rayner,Sandra
And 23h2'39"51d27'
Raynne-Johnson,Rebecca
Elizabeth
Peg 22h21'32"18d45'
Raynor,Chuck
Aur 5h8'40"40d23'
Raynor,John
Cnv 12h15'60"37d19'
Raynor,Russell Jason
Aur 4h51'36"40d52'
Raynor-Jenkins,Kyle Patrick
Her 16h56'16"31d58'
Raynéé
Hya 8h21'35"0d12'

Rays
Aur 6h9'39"46d10'
Rayser,Kimberly
Vul 20h19'32"28d39'
Raysha
Eri 4h8'10"-7d33'
Raytown Flash,The
Mon 7h42'42"-5d60'
Rayven
Cnv 12h21'42"43d18'
Rayven My Star Bright
Umi 15h18'38"73d55'
Rayzannette
Ori 5h54'44"12d41'
Razi,Fariba
Psc 0h1'20"8d24'
Razi,Khalil
Crt 11h41'29"-10d59'
Razim,Aninka & Venda
Cyg 19h19'53"28d18'
Razina
Lyr 19h17'15"40d17'
Razny,Sascha
Her 17h48'54"46d54'
Razor
Hya 9h2'52"1d28'
Razorback Joe
Cet 1h3'49"-4d15'
Razvin-Bolin,Alexandr AMJ
Per 2h7'0"56d39'
Razzell,Terri
Lyr 18h36'52"32d13'
RBF's Glimmering Girl
Uma 11h26'57"48d3'
RB˜
Ser 15h37'34"19d9'
RCBH
Her 17h55'49"14d7'
Re Antonio
Cae 4h55'42"-27d48'
Re Re
Uma 10h37'31"59d56'
Re's Smiling Star,Tom
Her 17h7'27"38d49'
Re,Mary
And 2h26'1"38d26'
Rea
Aql 18h43'39"1d36'
Rea,Anthony Charles
Per 2h55'12"35d47'
Rea,John Michael
Cep 22h35'43"58d16'
Rea,Steven Michael
Cmi 7h34'49"11d38'
Reach For Your Star Tammy
Cep 20h31'49"66d57'
Read,Donald W
Per 2h49'17"32d28'
Read,George
Aur 6h8'54"30d33'
Read,James Richard
Sex 10h12'24"-1d41'
Read,John F
Oph 17h20'1"12d43'
Read,John Marion
Ori 5h41'24"1d10'
Read,Jr,H Milton
Aur 5h52'59"37d46'
Read,Kathleen Margaret
Aql 18h59'60"14d32'
Read,Lindsey Jane
Cas 0h4'12"58d44'
Read,Mair
Lyr 18h57'37"45d54'
Read,Meg
Cmi 8h6'24"1d2'
Read,Nancy Nelson
Lib 14h57'0"-5d49"
Read,Natalie Dawn
Lyr 18h36'44"27d53'
Read,Peter R
Umi 14h54'51"71d17'
Read,Robert B
Boo 15h4'53"48d44'

Read,Sr,Charles R
Aur 7h23'33"39d60"
Read,Suzanne
Cyg 19h50'59"37d59'
Read,Willa Elizabeth
Lyn 7h37'24"42d5'
Reader,Emily Louise
Cyg 19h26'37"34d1'
Reading,Jr,William Joseph
Lac 22h54'11"55d40'
Reading,Robert J
Oph 18h4'41"11d39'
Ready,Kathleen T
Hya 8h57'29"1d19'
Ready,Oliver
Ser 18h52'39"5d51'
Reagan,Bonnie
Aql 19h8'17"4d1'
Reagan,George Peter
Her 17h12'38"41d16'
Reagan,Heather Elizabeth
Cas 1h5'1"50d26'
Reagan,Ian Patrick
Cet 0h50'32"-14d27'
Reagan,Kevin James
Leo 10h0'54"13d52'
Reagan,Kyle
Lac 22h53'40"54d39'
Reagan,Larry Gay
Hya 8h16'22"4d16'
Reagan,Leo Robert
Boo 13h54'37"19d26'
Reagan,Missy Nicole
Boo 14h20'58"28d23'
Reagan,Ross
Sgr 18h56'47"-32d22'
Reager,Geraldine Margaret
Cet 3h19'23"0d3'
Reager,Leo Vernon
Aql 19h3'43"11d3'
Reagin,Marlene
And 23h35'0"32d36'
Real
Cam 5h49'55"67d55'
Real Humanity's Star
Tri 1h47'18"31d10'
Real Jennifer Rose
Com 12h18'13"32d21'
Real Neil,The
Aur 6h14'28"38d56'
Real,Dhany
Tri 1h58'54"34d11'
Real,Joshua Edward French
Ori 6h8'31"8d7'
Reale,Andrew
Boo 14h27'53"38d26'
Reale,Dan
Ori 6h5'60"7d39'
Reale,David
Her 16h26'31"35d36'
Reale,Joan
Lyn 6h42'44"60d54'
Reale,Joseph
Lmi 9h56'29"34d8'
Reall Forever & A Day, David
T
Cep 22h12'48"68d29'
Reall,Christopher Loren
Lac 22h1'31"47d44'
Realmuto,Michael
Ori 4h49'19"5d57'
Reals,James H
Boo 15h6'46"27d50'
Realschule Kamp-Lintfort
Del 20h19'13"10d4'
Realta of James
Cnv 13h55'17"31d27'
Ream,Ashlee
Lyr 19h19'1"37d45'
Ream,Christian David
Per 1h43'43"53d38'
Ream,Heavenly Adonis,
Michael Charles
Ori 5h59'27"18d14'

Ream,Michael Matthew
Ser 18h19'29"-2d9'
Ream,Steve Larkin
Aql 19h11'17"15d31'
Reamer,Frank Titus
Leo 9h35'55"7d40'
Reames,Charles Philip
Oph 17h10'15"-10d58'
Reaper,Monika
Cyg 19h40'42"38d48'
Reardon,Edward
Aql 19h3'30"-6d1'
Reardon,John Albert
Her 17h39'56"18d49'
Reardon,Joseph
Dra 16h25'52"63d42'
Reardon,Kristen Lea
Cas 0h53'58"54d47'
Reardon,Laura Lyn
Vul 19h48'22"25d36'
Reardon,Traci Ann
Cyg 21h39'17"41d42'
Rearick,Jr,John "JR" C
Aur 6h13'49"37d33'
Reason,Christopher Lee
Her 17h3'56"48d25'
Reason,Philip
Ori 6h9'14"3d9'
Reason,Richard Charles
Lac 22h19'46"51d26'
Reasoner,Carrie Michelle
Lmi 10h19'24"31d16'
Reasoner,Paul D
Her 16h55'52"31d2'
Reasons,William Steele
Tau 4h47'36"16d36'
Reasor,Harold D
Dra 19h29'28"58d51'
Reaves,Kathy
Eri 3h9'53"-6d54'
Reavill,David
Oph 18h5'30"8d58'
Reay,Jackie
Cyg 21h2'25"38d10'
Reay,Jennifer
Uma 10h43'41"57d8'
Reaza,Narcizo Ernie
Her 18h9'14"31d42'
Reback,Jodi
Hya 9h6'49"2d45'
Rebecca
Ari 2h1'27"18d7'
Rebecca
Oph 17h20'31"10d9'
Rebecca
Ori 5h31'53"0d1'
Rebecca
Peg 21h23'0"10d49'
Rebecca
Uma 8h24'11"68d9'
Rebecca
Boo 14h0'1"19d6'
Rebecca
Lib 15h1'52" 28d33'
Rebecca
Tri 2h5'16"31d5'
Rebecca
Uma 10h22'36"48d58'
Rebecca
And 2h30'47"41d42'
Rebecca
Lac 22h46'21"38d34'
Rebecca
Ori 5h39'34"10d36'
Rebecca
Cyg 20h39'41"30d4'
Rebecca
Lyr 18h33'44"40d32'
Rebecca
Boo 14h6'1"53d6'
Rebecca
And 23h25'14"44d38'
Rebecca & Benjamin
Cet 1h29'57"-13d5'

Rebecca ((Winzpuh I)
Lib 15h8'11"-21d6'
Rebecca 93
And 23h4'18"37d44'
Rebecca Ann
Boo 15h15'22"50d10'
Rebecca Ann
Cam 5h46'44"60d50'
Rebecca Ann
Lyr 18h47'53"30d19'
Rebecca Anne
Cyg 20h22'20"38d37'
Rebecca B
Sct 18h32'48"-4d33'
Rebecca Corrine
Cyg 19h28'1"38d35'
Rebecca Dawn
Vul 20h3'39"28d36'
Rebecca Elizabeth
And 0h37'13"31d25'
Rebecca Erin
Lyr 18h56'10"41d54'
Rebecca Irene
Aql 19h27'53"12d30'
Rebecca Jane
Peg 23h39'24"12d28'
Rebecca Jane"Boing"
And 23h28'14"40d50'
Rebecca Janeen
Peg 22h35'16"18d51'
Rebecca Jayne
And 23h17'30"35d26'
Rebecca Jeanne
Lyn 7h40'42"38d38'
Rebecca Jessica Barbara
And 1h16'56"39d1'
Rebecca Kay
And 23h46'1"40d1'
Rebecca Leigh
Cas 23h44'8"64d1'B
Rebecca Louise
Com 12h29'11"24d24'
Rebecca Lynn
Vul 20h14'13"25d3'
Rebecca Lynn
Cnv 12h34'1"40d21'
Rebecca Lynne
And 1h56'57"40d20'
Rebecca Mee-Love
Always,Stef
And 0h3'22"31d41'
Rebecca Rae
And 1h25'54"34d46'
Rebecca Sola & Guy
Peg 23h34'1"30d18'
Rebecca the Shining Star
Cyg 19h56'18"47d15'
Rebecca's 21st Dream Star
Cas 0h29'14"60d29'
Rebecca's Angel
Lyr 18h56'46"36d49'
Rebecca's Christmas Star
Cas 0h58'34"66d45'
Rebecca's Dream
Aur 5h4'33"40d19'
Rebecca's Fire
Sge 18h56'11"20d8'
Rebecca's Immortality
Vul 20h46'23"20d1'
Rebecca's Lavaliere
Mon 7h13'6"-5d13'
Rebecca's Reach
Peg 23h36'28"31d30'
Rebecca's Rising Star
Cas 1h20'12"56d3'
Rebecca's Star
And 0h23'49"40d15'
Rebecca's Star
Cyg 20h23'28"38d47'
Rebecca's Star
Mon 6h57'25"11d40'
Rebecca's Star/Eternal Love
For Seth
Cas 23h39'0"50d55'

Rebecca,Beloved Sister
Uma 9h26'1"51d39'
Rebecca-Derald Beta
Vul 20h38'22"21d44'B
Rebecca-Jane
Peg 21h58'24"30d48'
Rebecq,Jean & Jeanne
Lac 22h29'22"40d17'
Rebeiz-Nielsen, Alexander
Junker
Uma 14h4'30"48d20'
Rebeka
Sgr 18h49'57"-36d28'
Rebekah
And 1h15'41"37d16'
Rebekah
Mon 7h41'5"-1d37'
Rebekah Ellen
Mon 6h23'18"-6d35'
Rebekah Sidne Monique
Lac 22h30'50"41d10'
Rebekah-Lee
Mon 7h14'35"-10d55'
Rebekah-Mother
And 0h21'50"33d39'
Rebekkah Lynne
Aql 19h31'1"10d24'
Rebel
Dra 19h32'28"60d25'
Rebel Holiday
Uma 11h54'1"52d27'
Rebello,Ken
Lac 22h47'1"53d32'
Reber,Holly & Bob
Uma 12h4'54"38d54'
Rebeul,Stéphanie
And 0h32'56"40d29'
Rebholz,Janet M
Lyr 18h58'43"42d21'
Rebière,Denise
Cep 21h27'18"68d38'
Rebman,Matthew Thomas
Sgr 18h48'51"-21d35'
Rebmann,Ashley Elizabeth
Ori 5h51'16"19d17'
Rebolledo,Alvaro
Cep 23h4'31"65d19'
Rebottaro,Sage Maddox
Uma 11h0'12"60d3'
Rebouché,Russell Joseph
Cmi 7h7'1"4d2'
Reboul,Belinda
Peg 22h29'16"27d51'
Rebours,Morgane
Cam 3h28'16"61d52'
Rebustillo,Lisa & Tony
Cyg 20h5'11"41d1'
Reca,Bill & Sara
Cyg 19h41'0"38d52'
Recagni,Katherine
Mon 7h1'58"3d49'
Recca,Debora
Lmi 10h28'58"38d25'
Recca,Gino
Lmi 10h29'36"37d58'
Recca,Roberto
Lmi 10h1'22"32d19'
Recchia,Sheri
And 0h5'25"47d39'
Reccoppa,Lori Ann
Cas 2h2'45"59d43'
Receveur,Ellen Paige
Sge 20h17'1"17d50'
Rechin,Karen
And 2h9'20"42d44'
Rechis,Ruth Pendleton
Oph 17h56'16"12d28'
Rechmann,Herwi
Peg 22h54'19"8d41'
Rechner,Amber Marie
And 0h14'26"34d7'
Recht
Cyg 21h14'53"38d47'
Recht,Cornelia
Equ 20h56'0"8d30'

Recine,Christopher
Boo 14h43'36"20d2'
Recke,Dieter Baron von der
Lib 14h21'43"-22d54'
Recker,Gregory (Ishna) John
Ori 5h53'11"1d52'
Reckord,Keith
Her 15h49'48"50d29'
Reckord,Lee
Dra 16h41'30"62d51'
Recksieck,Antje
Cmi 7h3025"7d48'
Reclus,Madame
Cep 2h34'55"80d29'
Recor,Christopher
Cep 23h27'27"65d48'
Recor,Johnathan
Dra 10h49'1"74d16'
Record Star,The
Umi 41h51'42"65d31'
Record,Byron & Margaret
Sct 18h42'59"-7d25'
Record,William
Dra 11h47'26"72d48'
Recoulley,Jennifer Lynn
Peg 23h2'59"33d30'
Recovery
Cas 0h46'0"71d40'
Rector,Ethan Lyn
Boo 14h27'50"31d43'
Rector,Forever Marie M
Crb 16h6'60"31d3'
Rector,James Michael
Hya 8h13'26"6d17'
Rector,Jeanne
Lyr 19h1'42"25d34'
Rector,Joshua L
Per 2h37'1"45d50'
Rector,Priscilla T
Cas 0h45'57"69d43'
Recupero,Dolores
Lyr 18h31'32"46d32'
Recupero-Mettraux, Micheline
Uma 9h51'30"51d43'
Red
Dra 19h36'28"60d17'
Red Car,The
Per 4h2'14"37d10'
Red Dwarf's Data Base
Umi 14h35'21"70d38'
Red Hawk
Aur 6h24'50"32d40'
Red Letter Days
Cas 0h58'50"60d48'
Red Light
Ori 5h1'1"-2d30'
Red Rachael
Ori 5h55'15"7d17'
Red Ronnie
Lac 22h6'57"38d10'
Red Rooster
Her 18h6'39"31d21'
Red Roses
Del 20h38'0"18d50'
Red Rum
Peg 21h57'17"31d55'
Red Special,The
Cyg 19h44'22"30d35'
Red Star,The
Peg 21h54'1"31d13'
Red Storm,St John's
University
Lac 22h53'51"55d7'
Red Wolf Shadow Warrior
Ser 18h2'1"-11d6'
Red's Star
Sct 18h54'54"-8d14'
Reda's Star
Cyg 20h31'17"58d45'
Reda,Frankie R
Uma 11h53'57"32d53'
Reda,Jonathan Kurt
Per 3h12'47"48d2'
Reda,Matthew Arthur
Dra 17h5'30"66d51'

Reda,Tamara Linette
Cas 1h2'1"61d2'
Reda,Todd J
Dra 19h1'1"61d6'
Reda-Greaney, Jacqueline
Cnv 14h2'53"40d44'
Redante
Lyr 19h18'12"30d25'
Redasky,John
Aur 5h53'1"30d42'
Redasky,Sharon
Cyg 20h57'43"39d25'
Redavid,Patricia H
Mon 6h39'44"1d6'
Redcad 926
Cam 6h0'33"60d23'
Redd,Brian J
Hya 8h57'0"-17d35'
Redd,My Shining Star,
Herman Jennings
Aql 19h23'20"-0d16'
Reddan,Dr Stacy
Oph 16h50'33"11d14'
Reddell,Darynn Loraine
Peg 23h1'49"32d25'
Redden,Brandy Reneé
Uma 9h16'52"52d50'
Redder,Matthew
Lac 22h18'22"46d32'
Redderson,John Garvin
Aql 18h45'53"8d56'
Redding,Dave
Ori 5h13'58"-6d37'
Redding,James
Ori 5h59'47"0d11'
Redding,Jennifer
Cnc 8h55'32"32d38'
Redding,Martha K
Oph 18h0'21"12d14'
Reddington,Dan
Aur 6h33'15"32d22'
Reddington,Madeline
Lyr 18h59'34"38d59'
Reddington,Martha P
And 23h22'15"40d52'
Reddy,Barbara M U
Boo 14h21'39"11d28'A
Reddy,Jean Mary
Lyn 7h48'33"50d56'
Reddy,Reddygari Chinna
Hanumantha
Dra 14h58'32"63d16'
Reddy,Robert P
Boo 14h21'39"11d28'B
Reder,Barbara & John
Crt 11h17'26"-11d9'
Redfern Star Of My Life,Bob
Boo 14h42'22"38d33'
Redfern Star,The Judith
Cyg 20h15'14"39d17'
Redfern,Maureen Louisa
Cep 0h6'0"73d31'
Redfield
Peg 0h6'39"13d7'
Redford,"Katie" Katherine Jean
Lyr 19h18'17"40d14'
Redick,Kristin
Mon 6h54'20"-3d30'
Rediger,Andrea L
Crb 15h38'25"29d60'
Reding,Thomas Francis
Boo 15h1'60"38d51'
Redinger Star Man, Mark
Charles
Her 17h23'49"40d27'
Rediscovery
Lyn 8h21'21"41d6'
Redlin,Kirk B
Dra 19h1'31"70d20'
Redlingshafer,Teresa Ann
Lmi 10h24'40"33d15'
Redman,Beatrice
Com 12h12'49"32d12'
Redman,James Jefferson
Cnv 12h23'20"37d36'

Redman,Lloyd & Loretta Ori 5h56'23"10d20'
Redmann,Dieter Cap 21h28'57"-26d50'
Redmer,Andrew James Vul 19h21'18"26d52'
Redmer,Anthony Joseph Cnv 13h58'0"39d58'
Redmer,Ronald D Aur 6h8'0"40d11'
Redmon,John Howard Aql 20h11'33"13d22'
Redmon,Melissa Ann And 0h23'50"37d53'
Redmon,Richard & Anne Marie Lyr 18h46'0"37d50'
Redmond,9421 Jack Boo 14h20'19"27d5'
Redmond,Aaron Bradford Aql 20h7'43"3d57'
Redmond,Carla Lyr 18h39'38"45d16'
Redmond,Gregory & Sonja Peg 23h47'18"16d26'
Redmond,Joshua Alexander Hya 9h11'20"5d59'
Redmond,Lucy Morgan Eri 3h58'26"-2d12'
Redmond,Marguerite Tau 3h43'1"25d7'
Redmond,Orin Dra 16h21'45"58d5'
Redmond,Ralph Craig Sct 18h48'19"-6d16'
Redmond,Sr,John J Cam 7h34'12"68d0'
Redmond,Timothy M Aql 20h4'32"4d52'
Redmond,William E Dra 20h22'1"63d44'
Rednaxela Cnv 12h15'36"35d9'
Redon,Anahita Cam 5h45'30"67d48'
Redpath,Donald Eugene Her 16h18'0"23d4'
Reduce,Raymond & Dawn Marie Uma 11h14'57"50d25'
Reduto,Dr Lawrence A Lac 22h6'12"40d32'
Redwind,Jerry Lac 22h37'42"40d6'
Redwine,Ashley Elizabeth And 0h33'14"45d20'
Redwine,Robert John Ser 18h3'57"-13d43'
Redwood Sex 10h32'12"-1d56'
Redwood Falls MN Class of 2000 Uma 9h44'32"71d53'
Redwood Star,The William M Dra 20h15'21"80d26'
Ree,Annie Cyg 19h35'60"27d54'
Reeb Ori 5h38'11"-0d20'
Reece Heavenly Being, Aster Andrea Aql 19h53'28"15d26'
Reece,Amy Beth Peg 22h46'52"20d10'
Reece,Bridgie E And 1h42'42"37d1'
Reece,Brit-Brit E And 1h21'1"38d45'
Reece,Christopher Mon 7h54'44"-8d59'
Reece,Daryl Cep 1h54'25"78d24'
Reece,David W Per 1h45'1"53d21'
Reece,Infinitely Steve Oph 17h56'54"11d12'
Reece,Joan E Del 20h25'33"20d11'
Reece,Laura Elizabeth Lyr 18h57'1"31d24'
Reece,Sslm Uma 8h16'28"61d16'
Reece,Unconditionally Yours,John Adrian Psc 1h6'42"23d5'
Reed "Cherish" Love 25 to Eternity,Ed & Barb Uma 10h58'0"37d42'
Reed "Our Mama",Grace Phyllis Cas 3h28'16"70d26'
Reed "Our Papa",James Robert Cep 22h25'21"60d24'
Reed Star,The Anne Elizabeth Lyn 6h33'0"56d12'
Reed,Alvin & Adele Lyr 18h54'33"37d25'
Reed,Anna And 0h9'37"40d23'
Reed,Baby Uma 8h47'38"47d14'
Reed,Betty Lou Del 20h13'0"9d57'
Reed,Bob & Joyce Uma 9h29'45"57d12'
Reed,Bradley W Aur 5h18'32"46d2'
Reed,Brian Cep 21h3'59"58d30'
Reed,C A Ori 5h36'52"-0d48'
Reed,Chad Lewis Per 1h26'33"53d56'
Reed,Charlotte Mary Mon 7h39'29"-0d24'
Reed,Cherylle Rae Cam 4h8'27"60d37'
Reed,Chris & Jon Umi 16h29'0"78d26'
Reed,Christopher David Ser 15h57'30"20d52'
Reed,Clinton Ray Her 16h22'29"48d10'
Reed,Connor Ori 5h53'10"14d57'
Reed,Cory Her 16h18'58"8d28'
Reed,Cynthia Truver Cmi 7h46'26"8d38'
Reed,Daniel M Aql 18h58'29"8d39'
Reed,Darbee Kayla And 0h15'0"37d55'
Reed,Dawn Elizabeth Aql 19h4'24"0d7'
Reed,Debra Diane Sge 19h16'47"16d34'
Reed,Dexter C Aur 6h29'24"35d16'
Reed,Donald J Aur 5h33'20"37d36'
Reed,Donna Vul 21h15'31"20d15'
Reed,Edward M Equ 20h58'37"10d26'
Reed,Electra Aql 19h55'16"12d6'
Reed,Elizabeth M Peg 21h42'13"28d11'
Reed,Eric Uma 10h53'35"40d56'
Reed,Gary L Cet 3h11'36"3d35'
Reed,George Andrew Uma 9h16'1"58d59'
Reed,Harry L Cep 21h34'57"58d14'
Reed,Helen F Cyg 19h26'19"31d28'
Reed,Helen Yvonne Cas 0h28'38"60d21'
Reed,Holvion George Ori 4h50'24"0d13'
Reed,Ian Daniel Eri 3h20'23"-10d52'
Reed,Jacob Patrick Lmi 10h39'28"23d52'
Reed,Jaime Ellen Mon 5h57'28"-5d56'
Reed,James Maxwell Scott Per 2h21'14"58d52'
Reed,Jane Sunder Her 18h2'28"18d50'
Reed,Jessica C And 0h5'41"34d38'
Reed,Joshua Hamilton Dra 16h44'32"60d29'
Reed,Jr,Sherman W Aur 7h6'1"40d27'
Reed,Justin William Cyg 20h1'55"50d31'
Reed,Kimberly Katherine Cmi 7h39'21"11d29'
Reed,Lacey Ann Ori 6h0'57"20d41'
Reed,Landa Sge 19h54'35"18d45'
Reed,Lloyd A Aql 19h56'49"11d10'B
Reed,Mabel Twyman Eri 3h23'55"-13d28'
Reed,Marie & Clarence Sge 19h59'44"19d46'
Reed,Marjorie Louise Cas 23h20'57"60d45'
Reed,Mark Cet 1h19'28"-6d36'
Reed,Mary Beth Wofford And 23h0'1"35d8'
Reed,Mellissa Jane Lyr 18h39'1"41d13'
Reed,Michael Alan Cmi 7h5'26"5d11'
Reed,Michael Edward Cyg 20h4'38"40d2'
Reed,Michael T Cep 21h14'21"60d4'
Reed,Morgan Lac 22h20'32"40d24'
Reed,Nita Bertoia Lyr 18h36'31"47d20'
Reed,Old Man Her 17h57'0"14d52'
Reed,Patrick Scott Aql 20h7'15"4d7'
Reed,Rebecca & Joseph Her 16h23'58"28d26'
Reed,Richard M Ari 2h43'56"24d37'
Reed,Ricky Cep 23h26'19"65d22'
Reed,Sally Jo Crb 15h34'17"27d17'
Reed,Sarah Sex 10h38'35"0d50'
Reed,Scott Bradshaw Hya 8h57'38"2d31'
Reed,Shannon & Claudia Layne Com 13h16'0"21d12'
Reed,Skyler Dean William Oph 17h55'1"12d49'
Reed,Spencer Erwin Her 17h13'39"26d59'
Reed,Teresa S Aql 19h56'49"11d10'A
Reed,Theresa Gayle Com 13h15'36"20d8'
Reed,Todd Anthony Oph 16h59'12"-28d14'
Reed,Tony & Katherine Crb 16h16'14"28d23'
Reed,Troy Aur 4h52'1"51d22'
Reed,Warren A Lib 14h57'25"-23d43'
Reed,Wendy Cas 2h39'0"61d14'
Reed,William Shedd Aur 5h12'24"54d41'
Reed-Fitzgerald, Charlotte Cas 0h7'29"59d4'
Reed-Jarvis Aur 6h0'12"46d11'
Reed-Keil,Nancy Cnv 12h32'51"33d19'
Reed-Travers Shared Sky 10-4-93 Cra 18h22'37"-40d4'
Reedaling Del 20h28'1"18d49'
Reeder,Anne Cra 18h8'49"-39d46'
Reeder,Julie Cas 0h6'45"62d30'
Reeder,Kristin Ori 6h16'29"1d40'
Reeder,Loving Linda Del 20h17'36"9d40'
Reeder,Samantha Cas 1h58'45"60d55'
Reeder,Theresa Matsue Quigley Com 12h54'52"31d2'
Reedham Park School Star,The Peg 23h5'0"30d1'
Reedster,The Uma 9h15'49"43d13'
Reedville School District No 29 Cma 7h21'53"-16d50'
Reedy,Melinda Eri 3h27'58"-1d47'
Reedy,Tessa Lynn Cyg 19h55'1"37d39'
Reef,The Lyn 7h58'59"39d30'
Reeger,Manfred Lyr 19h17'60"30d13'
Reeling,Lisa D Peg 22h6'23"21d18'
Reeman,Adrian Ori 4h52'17"5d30'
Reena Cas 0h19'44"63d22'
Reenster's Destiny Aql 19h28'0"12d1'
Rees,A John Leo 9h57'28"18d33'
Rees,Angela Mon 8h0'32"-10d7'
Rees,Billie Dawn Tau 5h18'55"28d57'
Rees,Carol Anne Ori 6h8'30"6d58'
Rees,David S Her 18h6'31"31d8'
Rees,Graham Her 17h19'10"29d59'
Rees,James C Aur 5h35'11"50d29'
Rees,Kelly A Cas 23h20'43"54d3'
Rees,Rebecca Jo-El Peg 22h40'40"20d20'
Reese Ori 5h38'15"11d14'
Reese Elizabeth Dra 20h17'1"67d35'
Reese III,Michael Boo 14h54'48"52d24'
Reese's Piece Dra 19h31'60"60d28'
Reese,Brett Allen Ori 5h48'49"21d14'
Reese,Charlotte Turner Sge 19h56'0"20d23'
Reese,David A Lyn 7h58'20"33d39'
Reese,Erin Peg 23h18'18"33d58'
Reese,Eugenia Tau 4h8'45"20d47'
Reese,Gregory Allen Uma 10h52'54"41d1'
Reese,Jack Wheeling Lmi 10h32'54"30d51'
Reese,James Lac 22h16'44"46d59'
Reese,Jeff Aur 6h9'28"35d20'
Reese,Jennifer C Peg 23h17'42"33d30'
Reese,Joice Peg 22h29'19"25d59'
Reese,Karen Sex 10h1'58"-2d14'
Reese,Karen And 1h36'53"41d6'
Reese,Laura Cas 3h33'29"70d27'
Reese,Mamie Lyn 7h11'12"58d15'
Reese,Margaret Elizabeth Aur 6h24'51"37d46'
Reese,Robert R Her 18h8'23"30d0'
Reese,S Marly Ann Mon 8h4'2"-2d30'
Reese,Teri And 23h29'0"45d39'
Reese,Trevyn Richard Mon 6h30'55"-0d6'
Reet's Night Star Estonia Forever Equ 21h0'49"8d8'
Reev,The Cam 3h55'0"52d53'
Reeve,Barbro Com 12h25'0"23d32'
Reeve,Jayne Cas 1h2'27"60d29'
Reeve,Lauren Summer Umi 15h58'1"77d44'
Reevell,Amanda D And 22h55'16"40d17'
Reeves Star,The Edie & Ron Cyg 20h25'10"42d2'
Reeves,Alan Per 2h54'0"43d57'
Reeves,Anna Mae Del 21h4'45"12d30'
Reeves,Barbara Colleen Cam 3h31'33"60d37'
Reeves,Dean Thomas Her 17h12'57"47d39'
Reeves,Donald Sharp Dra 16h42'28"68d39'
Reeves,Dylan James Dra 10h18'28"80d50'
Reeves,James Marshall Aql 19h55'28"13d26'
Reeves,Jeff & Sharon Cyg 20h17'1"31d9'
Reeves,Kael Michael Kristian Cet 1h34'32"-6d4'
Reeves,Katherine Eri 3h12'33"-3d53'
Reeves,Kimberly Kay Lyn 7h43'1"48d30'
Reeves,Kimberly Lane Uma 12h0'44"58d34'
Reeves,Leslie Kaye Cet 1h25'40"-3d45'
Reeves,Linda Marie Iannuzzi Uma 12h56'49"53d43'
Reeves,Marc Christopher Dra 16h4'1"64d55'
Reeves,Matthew L Lac 22h6'40"49d20'
Reeves,Noah Nathaniel Dra 16h16'39"65d55'
Reeves,Sally-Meredith- Joann Cas 2h35'1"57d27'
Reeves,Soraya And 2h31'1"39d38'
Reeves,Star of Kimberly Susan Aqr 22h30'1"-18d24'
Reeves,Thomas Matthew Boo 15h24'28"50d19'
Ref,The Cep 23h37'56"64d18'
Refait,Denis Vul 19h19'46"26d32'
Reffett,Dorothy Lyr 19h25'27"38d22'
Reffner,Joel Dra 13h4'30"68d58'
Refield,Addison Lac 22h44'45"53d12'
Refle,Petra Augsburg Kulturstr 25B Sco 17h8'43"-38d27'
Reflection of Stepas Uma 9h46'0"44d19'
Reg Ori 4h48'58"0d29'
Reg & Tucker's Star Peg 21h54'33"22d49'
Reg Gair Her 16h29'58"40d38'
Rega Constellation Aur 5h28'34"31d32'
Regalia,Laura Ann Mon 7h47'0"-1d10'
Regalo De Mi Corazon Cam 4h4'0"68d32'
Regan Aql 20h2'27"12d55'
Regan,Andrew Douglas Aur 6h33'43"34d35'
Regan,Brett W Ori 5h26'1"-0d45'
Regan,Elizabeth H Peg 23h18'56"30d33'
Regan,Frank & Toni Aql 19h54'59"13d16'
Regan,Kevin David Per 3h1'23"47d36'
Regan,Marilyn A Mon 6h30'13"-10d46'
Regan,Michael W Peg 23h5'17"13d10'
Regan,Randy James Dra 16h7'19"68d34'
Regan,Richard Gerard Cmi 7h42'38"4d35'
Regan,Rita F Mon 6h18'42"1d51'
Regan,Rob Aur 6h8'35"32d41'
Regan,Robert Kevin Cnv 12h36'0"38d42'
Regan,Stephen John Her 16h11'34"24d37'
Regan,Thomas R Her 18h0'10"28d32'
Regan,Virginia And 1h27'24"40d4'
Regas,Angelique Lyn 8h52'27"41d2'
Regazzoni,Carlo Ant 10h44'48"-37d4'
Regazzoni,Marisa Cet 1h8'30"0d43'
Regel,Franziska Lib 14h56'27"-20d23'
Regen,Jay Vul 21h23'25"28d25'
Regenbogen-Brigitte Gem 6h38'53"28d13'
Regenbogenland Uma 9h18'25"43d30'
Regenstein,Lewis G Sex 10h35'43"2d23'
Regent,Patrick Cep 22h13'16"55d52'
Regent,Wilton Wade Vul 19h22'49"26d51'
Reges,Melissa Aur 6h0'0"30d38'
Regette,Michél Traherne Her 18h21'18"13d5'
Regette,Philippe Francois Cma 7h49'0"11d6'
Reggie Boo 14h51'20"30d13'
Reggie Sct 18h40'54"-6d6'
Reggio,Dylan James Aur 6h20'35"37d27'
Reggiori,Elisabetta Lyr 19h19'40"42d29'
Regie's Rapture Her 17h2'60"37d46'
Regier,Rob Eri 4h53'42"-7d44'
Regimbal,Denis Cas 0h9'37"66d16'
Regin,Peter Ori 5h55'27"1d50'
Regina Oph 17h56'53"10d50'
Regina Vir 12h56'13"-18d45'
Regina Cyg 21h22'55"38d33'
Regina Cas 3h21'29"71d4'
Regina Peg 22h51'1"27d38'
Regina A D O Gru 22h24'13"-50d19'
Regina Ann I Lyr 19h14'0"40d21'
Regina Ann I Peg 23h16'1"30d10'
Regina Cara Mea Cam 5h47'50"60d51'
Regina Dawn And 2h31'44"41d43'
Regina Elise Lyn 7h11'10"58d53'
Regina G Mon 6h58'18"10d28'
Regina I Peg 22h2'43"25d49'
Regina Ida Frida Gem 7h29'23"35d17'
Regina Margherita Del 20h17'1"12d15'
Regina Maria Cas 0h52'28"61d47'
Regina Michelle Lyr 18h26'18"38d28'
Regina,Marion Aql 20h4'47"-0d7'
Regina,The Lovely Cas 0h11'51"63d2'
Reginal Jose Her 16h45'48"33d18'
Reginald Mon 8h1'1"-1d21'
Reginald's Melody Cas 2h35'1"57d27'
Reginaldo i Regina Uma 10h52'44"40d4'
Reginato,Cameron Per 1h28'52"52d57'
Reginato,Ryan Her 16h46'22"47d43'
Reginia Christina And 0h15'58"33d45'
Regis Eri 3h3'1"-12d23'
Register,Gloria Ann And 1h35'0"41d15'
Register,Kenneth Mark Hya 8h30'18"-8d49'
Regitz,Carl David Her 17h55'15"20d33'
Regn,Susan Cas 0h55'25"65d8'
Regna,Forever #39 Peter J Cep 23h6'31"70d22'
Regnault,Frédéric Lmi 10h27'42"39d12'
Regner,Katherine Cas 0h40'51"69d31'
Regnier,Diane Cyg 19h45'14"31d13'
Regnier,Francois Lyr 18h37'0"41d18'
Regnier,Holly Erin Lyr 19h6'0"37d40'
Regno,George Del Uma 8h11'59"72d19'
Regnvall-McClure Dra 16h50'0"62d32'
Rego,Anthony A Her 16h17'38"41d26'
Rego,Louis D Per 3h15'44"50d29'
Rego,Sara Austin And 1h42'29"37d43'
Regor Cyg 21h20'1"40d56'
Regula,Pamela J Vul 20h15'41"25d15'
Regál Ori 6h4'48"-0d39'
Reha,Janice M Uma 8h43'23"56d58'
Rehac,John Per 3h17'57"40d7'
Rehak 80th Birthday Star,Margaret Equ 21h23'1"8d21'
Rehak,Arianna Deborah Cowan Tau 3h54'43"0d51'
Rehak,Arthur Her 17h1'37"20d16'
Rehan,Brittany Ann & Brandy Lee Dra 17h34'56"70d6'
Rehberg,Geraldine Cyg 20h13'27"37d33'
Rehbuck 3 Cam 6h14'57"70d36'
Reheil,Jack & Kristin Lyr 19h11'52"37d46'
Rehfeldt Star,The Ant 10h33'56"37d21'
Rehl Cam 7h26'30"60d44'
Rehm,Jeanne Cas 1h31'1"58d31'
Rehm,Kathryn Cas 1h31'25"54d18'
Rehm,Shirley Cas 1h31'1"58d24'
Rehman,Dr Abdul Dra 20h14'44"62d7'
Rehmer,Rena And 0h19'13"32d15'
Rehmus,Thomas Walter Lmi 10h56'37"26d35'
Rehrer,Hal Fisher Per 3h33'25"35d26'
Rei Hya 8h44'12"4d14'
Rei Marie Uma 9h47'1"51d4'
Rei und Moni's Stern Peg 23h16'49"33d56'
Reich,Beverly Lyr 19h18'0"41d21'
Reich,Bruno Oph 17h52'46"8d1'
Reich,Cordy Vir 13h27'44"-2d24'
Reich,Dr Melvin Oph 16h49'28"10d55'
Reich,Emil & June Eri 3h17'46"-5d36'
Reich,Gary Leo 10h19'1"12d12'
Reich,Heio Dra 15h55'41"65d31'
Reich,Jamie Herschenfeld Vir 11h40'31"6d15'
Reich,Jody Uma 10h49'0"70d41'
Reich,Michael Louis Boo 14h32'1"30d9'
Reich,Roger Psc 22h56'1"2d9'
Reich,Sheldon Uma 9h52'46"42d55'
Reichard,Austin Gregory Ser 18h55'47"2d55'
Reichardt,Kristina Ori 5h39'57"8d8'
Reichardt,Wally C Cmi 6h55'8"-16d51'
Reichart,Walter Per 3h20'48"40d15'
Reichbach,Cary Noah Tri 2h22'34"28d28'
Reichbauer,Alfred Cas 0h4'10"60d22'
Reiche,Alwin Ori 4h41'60"10d29'
Reichel Zone,The Her 17h26'32"18d59'
Reichel,Jennifer & William Cyg 20h32'1"40d36'
Reichelt,Dr Helga Psc 23h8'47"0d52'
Reichelt,Rene Boo 14h38'25"14d28'
Reichenbach,Rick Cap 21h20'11"-15d52'
Reichenbacher,Ilonka Per 4h29'21"50d55'
Reichenberg,Margaret Raye Mon 7h41'28"-3d37'
Reichert 1 Tau 3h43'29"11d11'
Reichert,Alice Wolfe Com 13h16'42"20d33'
Reichert,James Dra 16h51'12"70d10'
Reichert,William M Uma 8h56'31"49d12'
Reichhardt,John Cam 3h35'33"63d42'B
Reichley,William Edward Oph 17h53'0"13d9'
Reichling,David Tau 4h13'55"22d44'
Reichling,Joan Marie Crb 16h7'58"31d16'
Reichwald,Morgan Dana Mon 8h3'27"-10d10'
Reichwald,Phyllis And 2h26'14"38d21'
Reichwein,Bonnie Jean Cyg 20h22'0"38d45'
Reid III,Robert F Sex 10h11'56"-0d52'
Reid Number 1 Star,Tom Cap 20h44'28"-26d15'
Reid,A Aql 19h54'31"-5d46'
Reid,Addison William Lac 22h13'22"54d54'
Reid,Alex Cep 22h14'0"60d48'
Reid,Antonio "L A" Oph 17h57'1"13d18'
Reid,Calaya Cas 2h10'35"59d49'
Reid,Carlene M Peg 22h59'47"20d8'
Reid,Chris Uma 10h6'33"55d10'

Reid,Christopher Anthony
Boo 14h36'12"45d6'
Reid,Clark Davidson
Her 16h12'33"25d1'
Reid,Deborah Louise
Com 12h54'48"30d60'
Reid,Edward M
Ori 4h41'32"13d40'
Reid,Geoff
Aur 5h24'13"50d6'
Reid,Glenda
Eri 2h46'32"-7d21'
Reid,Gordon David
Aur 6h9'35"30d38'
Reid,Heather
And 2h23'1"44d58'
Reid,Helen Fuller
Peg 22h24'56"21d17'
Reid,Jennifer
Cas 2h33'32"68d18'
Reid,John
Dra 13h41'0"63d57'
Reid,Joy Theanne Hurlburt
Lib 15h17'0"-21d15'
Reid,Kailey Nicole
Cnc 8h38'33"18d6'
Reid,Lawrence Carter
Her 18h30'22"24d29'
Reid,Leslie
Aql 18h59'27"-2d5'
Reid,Linda
Cmi 7h35'53"0d25'
Reid,Linzi
And 23h1'40"51d31'
Reid,Maeghan Thole
Mon 7h35'20"-1d47'
Reid,Margaret Elizabeth
Cas 0h20'21"59d55'
Reid,Mary Hereford
Ori 5h53'14"21d55'
Reid,Mary Ruth
Com 13h32'24"25d55'
Reid,Melissa
Peg 22h46'32"25d5'
Reid,Michael
Cep 1h26'18"80d29'
Reid,Moira
Lyr 19h19'14"41d39'
Reid,P Carey
Uma 8h45'1"68d33'
Reid,Phillip Sean
Cyg 21h14'46"28d48'
Reid,Rhonda Mathisen
Peg 22h2'29"8d51'
Reid,Sarah Alexandra
Aql 19h2'0"16d14'
Reid,Scott
Aql 20h14'0"8d16'
Reid,Stewart A
Her 15h49'1"40d57'
Reid,Super Dad Clive
Aql 20h1'16"1d49'
Reid,Valerie Melville Chrystie
Uma 9h14'0"53d42'
Reid,Winston S
Cnv 13h41'44"37d39'
Reid-Pugliano
Lac 22h9'1"38d44'
Reid-The Adventurer, Jenny
Vir 11h39'1"7d16'
Reidenbach,Peggy Johns
Sct 18h5'13"-7d52'
Reider "Infinity", Matthew
Henry
Aql 20h2'59"8d32'
Reider-Valenta
Cyg 19h37'29"28d20'
Reidmiller,Robert Scott
Dra 19h18'56"80d18'
Reidsma,Barbara
Vul 20h57'46"28d35'
Reidy,Denis Michael
Lyn 8h2'31"50d29'
Reidy,Jennifer Dawn
And 2h12'30"43d0'
Reidy,Leslie
Cmi 7h24'16"0d38'
Reif,Doris
Cas 0h16'13"61d0'
Reif,Elaine
Aql 19h6'18"-5d47'
Reif,Taryn
Boo 15h14'41"47d30'
Reiff,50 Year Milestone
Dennis R
Per 3h9'55"46d28'
Reiff,Julie Ann
Cas 1h0'31"61d33'
Reiff,Mitchell Jay
Oph 17h50'18"12d9'
Reiff,Simon
Ori 5h38'44"11d46'
Reiff,Zoe
Her 16h34'24"21d53'
Reiffel,Nancy
Lyr 18h34'0"31d4'
Reifgerste,Markus & Reem
Salha
Mon 6h28'44"2d56'AB
Reifler,Colby Adam
Uma 10h49'30"72d24'
Reifsnyder,Stephanie Lynn
Lyr 19h2'55"38d10'
Reifsteck,Shannon Carroll
Leo 10h36'12"22d44'
Reig,Kathryn
Cas 0h4'10"54d50'
Reig,Michael
Cep 22h53'38"57d9'
Reignier,Barbara
Cyg 21h22'29"40d21'
Reignsborough,Jean
Cam 4h18'40"59d30'B
Reihl,Erich
Ori 5h12'31"15d26'
Reihl,Sara Catherine
Cyg 19h32'18"37d43'
Reiko,Selene
Cas 0h42'53"62d51'
Reikofski,Shelly Ann
Cas 0h48'22"69d34'
Reille,Christophe
Aql 20h4'11"0d24'
Reilley,Patrick Christopher
Cnv 12h23'28"44d27'
Reilly Star,The Malcolm &
Christine
Aqr 20h45'32"-5d53'
Reilly's Field Of Dreams
Sge 19h3'50"16d31'
Reilly,Alison Brady
Cam 7h11'51"70d14'
Reilly,Brian
Dra 15h55'34"67d55'
Reilly,Dennis Paul
Boo 14h10'11"31d4'
Reilly,Ellen
Cas 1h31'59"76d5'
Reilly,Endlessly Devoted To
Paul J
Peg 21h59'21"26d41'
Reilly,Hlandtk
Uma 10h53'52"38d4'
Reilly,Hobart L & Lydia E
Uma 9h1'59"47d31'
Reilly,Ian
Her 18h3'0"48d29'
Reilly,Jr,James Gordon
Sgr 18h56'26"-34d32'
Reilly,Judy
Lyr 19h3'0"38d0'
Reilly,Kelsey Ann
Peg 22h46'11"33d5'
Reilly,Kevin Christopher
Dra 15h54'1"61d14'
Reilly,Kevin George
Leo 10h43'45"15d29'
Reilly,Lacey
Boo 13h39'19"18d35'
Reilly,Margaret Pearson
Her 16h57'26"20d31'
Reilly,Marie
Cas 0h26'45"50d36'
Reilly,Maureen A
Uma 12h8'58"53d42'B
Reilly,Nancy A
And 23h9'29"37d53'
Reilly,Peggy
Peg 22h46'0"29d52'
Reilly,Philip A
Per 4h38'23"37d12'
Reilly,Phyllis
Lyn 7h51'49"36d23'
Reilly,Sarah Sargeant
And 0h50'0"38d59'
Reilly,Sherrie
Lyr 18h41'22"41d59'
Reilly,Sir John
Dra 17h28'33"73d10'
Reilly,Sophie
Cet 2h21'56"-10d8'
Reilly,The Janine dePeyer
Lyn 8h47'28"40d0'
Reilly,Thomas F
Lac 22h36'29"53d58'
Reily,David
Boo 14h59'0"51d33'
Reiman,Barry & Rozanne
Sge 19h37'53"16d32'
Reiman,Cynthia Good
Gem 7h29'1"34d35'
Reiman,Frank
Boo 14h57'24"32d23'
Reiman,Phillip J
Hya 9h5'19"-10d17'B
Reimann,Barb & Kell
Crb 16h18'35"33d1'
Reimar 1
Gem 7h25'24"27d32'
Reimbold,Joseph Maximillian
Aql 19h26'26"15d32'
Reimer,Joshua
Cma 7h51'58"11d30'
Reimers,Lorna Anderson
Peg 0h0'29"18d2'
Rein
Cyg 21h33'51"30d1'
Rein,Charlotte
Cas 0h6'41"64d12'
Rein,Sebastian Hart
Aur 6h3'13"37d31'
Reina
Cam 13h30'52"77d4'
Reina,Abeja
Per 3h28'16"52d8'
Reina,Amy
Lyn 7h40'45"42d45'
Reina,Jennifer Melissa
Cam 7h1'36"71d14'
Reinarz,Gilbert
Cmi 6h59'32"-18d53'
Reinarz,Lorraine
Ori 5h15'1"11d28'
Reincke,Richard A
Aur 6h16'32"45d22'
Reineck,John Paul
Cep 22h22'1"63d2'
Reinecke,Friedhelm
Ari 3h4'0"28d41'
Reinecke,Gisela
Cas 22h58'41"55d47'
Reiner's Rocket,John
Ori 6h6'1"1d44'
Reiner,Oliver
Cep 22h1'21"68d29'
Reinert,Eric William
Cam 6h4'41"58d17'
Reinert,SJ,Paul C
Oph 17h27'43"-6d15'
Reinertsen,Sandra
Gem 6h5'40"26d39'
Reinertson,Dr Jim & Bev
Uma 11h57'32"52d53'
Reinertz,Roy H
Lib 14h48'0"-7d51'
Reing,Ellen
Cas 23h14'11"60d23'
Reingold,Erna
Uma 8h39'30"56d56'
Reingold,Michael Louis
Cet 1h51'15"0d26'
Reinhard,Mathias
Her 17h40'51"42d30'
Reinhardt
Cam 3h47'52"56d15'
Reinhardt Star for Senter &
Rachel,The
Lyr 18h33'51"38d7'
Reinhardt,Barbara
Lyr 19h23'12"30d10'
Reinhardt,Cynthia M
Cyg 19h49'46"38d16'
Reinhardt,Laura Jean
Hya 8h53'12"-1d5'
Reinhardt,Marcus Alexander
Sgr 19h18'54"-21d44'
Reinhardt,Myranda Nicole
Cyg 21h52'0"42d58'
Reinhardt,Richard
Cyg 20h19'1"47d9'
Reinhart,Louis & Mary
Cyg 20h36'20"38d50'
Reinhart,Peter Thomas
Gem 6h24'36"14d6'
Reinhart,Wolfgang
Aur 5h5'1"43d16'
Reinheimer,Karen
Crt 10h56'53"-10d3'
Reinheimer,Rev Daniel
Uma 13h34'43"50d23'
Reinhold
Lyn 7h32'25"43d8'B
Reinhold Messner Duell Team
Uma 12h3'22"51d19'
Reinhold's
Traumhochzeitsstern
Crb 15h38'12"38d46'
Reinhold,Karl
Hya 9h33'2"-9d41'
Reinholz,Karl-Heinz
Gem 6h26'38"14d19'
Reininger,Nika Lianne
And 23h0'14"36d47'
Reininger,Scott Biber
Per 3h13'21"40d54'
Reininger,Troy Patricia
Boo 14h34'19"9d27'
Reininghaus,Sabine
Ori 5h39'11"8d59'
Reinke,M C
Cyg 19h58'55"37d48'
Reinke,Robert
Lac 22h49'12"56d30'
Reinke,Terry L
Mon 7h43'49"-8d14'
Reinking,Pamela Clark
And 1h54'25"20d37'
Reino #1,Monica
And 0h21'42"40d36'
Reins,Eiko
Lyr 19h21'34"31d46'
Reinsel,Hazel Milnes
Peg 23h27'35"8d58'
Reinstatler,Krissy
Mon 6h48'1"11d23'
Reinstein,Francine Paula
And 0h20'12"33d27'
Reinstein,Jack
Hya 9h33'41"5d43'
Reis,Helen
Cas 0h58'16"58d30'
Reis,Jane Hines
Cyg 19h27'51"56d55'
Reis,Laurie Elizabeth
Oph 17h40'51"10d23'
Reis,Nicole
And 22h55'40"50d31'
Reis,R C
Aur 5h8'20"40d29'
Reis,William George
Aql 19h14'46"14d35'
Reisberg,Erhard
Umi 15h22'0"68d24'
Reisboard,Beth
Cyg 19h42'58"28d21'
Reisch,Ernst Ludwig
Tau 5h32'45"26d15'
Reischauer,Rafael
And 23h10'1"42d31'
Reise,Mary Ann
Tau 5h2'0"20d37'
Reisel,Dr Johanna
Ari 2h5'1"22d26'
Reisener,Wayne
Aur 7h5'23"36d33'
Reiser Ascending
Umi 15h5'31"68d19'
Reiser,Ezra Samuel
Cmi 7h44'58"8d5'
Reiser,Paul
Aur 5h48'55"50d27'
Reisert,Lisa Maria
Cap 21h24'47"-26d59'
Reising,Anna und Peter
Cas 0h32'58"62d40'
Reisinger,Bonnie
Lyr 18h13'0"35d12'
Reisinger,Reinhold
Sco 17h29'29"-30d8'
Reisman,Neal Robert
Her 16h59'55"30d33'
Reisner,Jeanne E
Lyr 18h15'48"38d48'
Reiss,Guenter
Oph 16h50'58"1d43'
Reiss,Jeffrey Lee
Her 18h13'31"30d55'
Reiss,JoEllen
Peg 22h33'41"21d48'
Reiss,Joseph F
Hya 8h53'18"3d4'
Reiss,Julia
Gem 7h16'36"21d35'
Reiss,Roslyn
Mon 7h51'42"-6d48'
Reisschneider,Jason 2-17-73
Dra 17h33'40"64d6'
Reisser,Blaise Baron
Uma 9h38'26"56d4'
Reissig,Margaret Rose Jenne
And 2h15'54"37d18'
Reissnecker,Heinz
Gem 6h35'21"26d41'
Reissner,Elisa & Frank
And 2h18'46"39d55'
Reitano,Peggy & Paul
Aur 6h5'14"46d42'
Reitemeyer,Ernst- Ulrich
Gem 6h44'8"12d6'
Reiten,Richard
Cet 1h29'58"-5d44'
Reitenbaugh,Sr,Vernon J
Cam 3h46'20"70d2'
Reiter Star,David & Pam
Crt 11h36'20"-22d33'
Reiter,Aidan James
Vul 20h19'34"25d40'
Reiter,Christopher J
Cap 21h22'31"-27d11'
Reiter,Dr Pamela Benyas
Oph 17h37'59"-18d59'
Reiter,James Franklin
Lac 22h42'34"54d10'
Reiter,Lauren A
Cyg 21h47'39"52d35'
Reiter,Samara Elaine
Vul 19h47'58"28d14'
Reiter,Thelma & Bernard
Umi 14h11'17"70d8'
Reiter,Thelma & Bernie
Dra 12h48'58"72d7'
Reithmiller,Peggy
And 23h23'43"48d26'
Reitinger,Gert
Aqr 23h33'0"-10d15'
Reitman,Les & Sonia
Mon 6h53'48"-6d55'
Reitter,Rev Elizabeth
Cyg 20h45'12"38d19'
Reitterer,Michaela
Boo 14h7'22"13d18'
Reitz,Janet
Cas 2h3'43"70d34'
Reitz,Thomas Joseph
Lmi 9h48'42"37d51'
Reitz,Timothy Rudolph
Lyn 7h17'25"50d37'
Reizer,John L
Cap 21h21'30"-20d21'
Reißhauer,Thomas
Dra 18h15'27"65d1'
REJ Goddess Spirit
Lmi 9h50'26"40d44'
Rejane et Eric
Boo 14h49'22"48d51'
Rejdak,Edward A
Per 1h55'52"56d36'
Rejeter La Clef
Aql 19h59'1"-6d39'
Rejo
Cep 22h17'47"60d7'
RejCep 22h17'47"60d7'
Gem 7h19'53"31d49'
Rekhi,Ben
Aur 6h28'18"30d31'
Rekhi,Rag-Ann
Cyg 20h16'18"31d11'
Reklame,Dansk & Tekstil Tryk
Per 4h5'23"35d39'
Releford Luminary,The
Ori 6h6'37"-0d2'
Relentless Bliss
Aql 19h54'0"12d27'
Relentless Love
Cyg 21h59'38"53d17'
Relentless Rob
Her 16h4'26"40d31'
Reliable Tool & Machine
Cet 0h25'14"-1d50'
Relick,Jeffrey Richard
Ori 5h52'34"18d17'
Relihan,Courtney
And 0h23'54"34d13'
Relkin,Stanley T
Lac 22h21'46"55d1"
Rella J
Vul 19h43'28"20d14'
Relmuk Notae Nayrb
Cep 21h37'41"60d19'
REM-7
Lyn 7h29'0"38d54'
Remack,Gene A
Uma 9h35'12"51d19'
Remak,George Darren
Aur 6h18'36"37d28'
Remalc
Boo 15h3'59"37d38'
Remark Marketing Services
Cyg 19h48'10"29d53'
Remarkaboo
Her 16h4'0"20d31'
Rembar,Lilianna Rebecca
Cam 3h29'1"60d6'
Remberg,Susanne
Equ 21h6'25"3d39'
Rembold,Fred & Deborah
Dusty
Cyg 19h31'56"33d42'
Remember
Uma 10h54'49"37d39'
Remember Rolando
Uma 10h39'1"68d28'
Remember Rosemary
Cru 12h53'56"-60d9'
Remembering Bob
Aql 18h59'45"13d26'
Remembering Ray
Ori 5h47'1"11d48'
Remenar,Andrew Stephen
Aur 5h1'35"41d23'
Remeny,Janiece Louise
Eri 4h55'39"-5d25'
Remer,Hannah Ellen
Uma 12h19'21"60d25'
Remi Jacques F A
Ori 5h55'56"18d31'
Remich Family Star,The
Vul 20h40'52"20d20'
Remick,Brenda
Cas 23h4'1"53d9'
Remick,Marilyn
And 0h17'43"37d57'
Remick,Sean James Lee
Ori 4h59'42"4d4'
Remillard,Pierre
Uma 11h54'26"41d20'
Remillong,Maestro
Cep 6h18'53"85d16'
Remines,Joseph William
Aur 6h15'1"37d59'
Remington,Deb
Tri 1h49'29"26d11'
Remington,Russell L
Boo 14h56'32"51d49'
Remke,Paul
Ser 15h30'49"17d39'
Remkus,Rebecca Dawn
Sco 17h51'38"-40d55'
Remley,Daniel C
Per 3h9'46"40d16'
Remm-Juszczyk
Cnv 13h54'28"47d49'
Remmers,Mary Frances
Lyn 7h30'39"58d49'
Remmert,Katherine B
Lyn 8h16'32"48d37'
Remo
Aqr 22h43'51"-6d33'
Remona
Cyg 19h53'53"40d34'
Remondi,Michael Anthony
Boo 15h21'0"50d49'
Remow,I Love You Jennifer
Com 12h29'37"22d42'
Rempfer,Richard
Boo 14h20'54"32d16'
Remund,David R
Hya 8h12'12"0d56'
Remus,S E R Sarah Elizabeth
And 1h48'35"38d9'
Remy
Dra 17h54'25"73d33'
Remy of the Phantom Pair
Peg 23h47'18"27d24'B
Remy Why
Boo 13h55'39"16d17'
Remy,Jean
Uma 13h14'0"61d26'
Remy,Marie
Umi 16h10'36"70d53'
Ren 1
Tel 20h12'17"-48d33'
Rena "The Star Of Inspiration"
Eri 4h4'13"-12d8'
Rena B G
Peg 23h5'0"32d13'
Rena Lee
And 0h10'33"35d56'
Rena's Star
Aur 4h50'53"50d52'
Renae
Mon 8h5'49"-1d52'
Renae
Peg 21h28'45"22d53'
Renae Ann
Vel 9h57'42"50d12'
Renal Hemodialysis Unit 4700
Oph 17h17'14"11d27'
Renan,Jr
Per 3h8'23"40d20'
Renard,Erin Ann
Cas 0h59'20"62d19'
Renard,For Love Of Annabelle
Boo 15h24'22"37d54'
Renard,Jr YAWILY, Robert
Henry
Her 16h32'12"48d0'
Renard,Olivier
Cep 20h15'14"60d51'
Renardy,Paul
Sgr 19h36'1"41d18'
Renata
Peg 22h6'17"5d24'
Renata
Boo 15h0'41"32d21'
Renate
Peg 22h31'41"28d38'
Renate
Sge 20h0'30"20d1'
Renate
Lyn 8h10'13"42d26'
Renate Buchmueller 18051992
Tau 4h2'1"0d3'
Renate S 1995
Lib 15h38'59"-23d38'
Renate Schleif's
Orientierungsstern
Vir 12h27'59"-7d43'
Renate Zos
Tri 1h58'4"32d55'
Renate,Mathias
Sgr 20h3'27"-40d55'
Renate,Yvette
Lac 22h50'49"40d52'
Renato e Ornella
Pic 5h3'24"-44d30'
Renato Francesco
Peg 22h17'46"4d30'
Renato-Stelo
Ari 2h4'47"20d22'
Renaud,Anthony & Stephanie
Lup 15h20'27"-44d46'
Renaud,Dan Clyde
Cam 6h6'57"70d54'
Renaud,Franchini
Cam 3h22'38"60d0'
Renaud,Marc
Dra 20h6'29"68d17'
Renault,Bernard
Ser 16h2'1"0d3'
Renaworld
Vul 19h48'19"28d19'
Renay Beth
Tri 2h7'26"30d33'
Renda Light,Renda Bright
Cyg 19h42'20"30d34'
Renda,Elizabeth Carol
Eri 3h51'11"-06d46'
Rendel
Her 16h9'56"4d50'
Rendina,Derek Joseph
Dra 17h47'41"60d28'
Rendina,Judy
Uma 13h37'11"51d3'
Rendrick,Loretta
Eri 3h36'54"-13d32'
Rene
Cyg 20h25'47"37d36'
Rene
Mon 6h30'11"0d53'
Rene
Mon 7h45'58"-1d37'
Rene & Rick's Aaura of Light
Umi 15h40'15"73d48'
Rene Bucaram
Col 6h1'27"-32d20'
Rene,Shelly
And 1h22'15"38d54'
Reneau Family Star,The
Uma 11h46'49"50d46'
Renee
Lyn 9h10'0"45d28'
Renee
Cas 1h28'0"60d2'
Renee
Mon 6h26'46"01d19'
Renee
And 0h45'38"30d40'
Renee
Cyg 21h11'25"36d25'B
Renee
Ser 15h54'45"2d25'
Renee & Brian
Mon 6h56'11"1d19'
Renee & Michael
Crt 11h15'23"-15d57'
Renee & Walt's 60 Years of
Heaven
Sge 20h0'26"18d47'
Renee Angela
Sge 19h58'41"18d47'
Renee Bobbie
Cas 1h27'11"60d8'
Renee Dolores
Peg 23h27'42"11d55'
Renee Elizabeth
Peg 22h1'12"22d22'
Renee Eternal
Lyn 8h7'48"39d16'
Renee Jackie Dee
Peg 21h47'39"34d47'
Renee Marie's White Rose
Lyn 8h54'53"41d40'
Renee Rising
Lyr 18h43'1"30d27'
Renee S
And 1h28'17"34d5'
Renee Sara V
Cam 4h12'45"67d31'
Renee TC
Peg 21h55'57"34d32'
Renee's Shining Light
Tri 2h43'21"31d9'
Reneese
Lmi 9h37'41"38d24'
Renel,Maureen Martine Janice
And 1h18'53"36d57'
Renella,Nikki
Lyn 7h46'11"44d42'
Reneé
Vul 20h0'1"28d35'
Reneé
Lyn 8h41'26"36d57'
Reneé
Peg 23h16'24"31d29'
Reneé Suzanne
Crb 16h17'16"30d54'
Renfro,Darlene
Aql 19h41'25"14d57'
Renfro,Joseph Arthur
Cet 3h17'29"4d28'
Renfro,Wheeler E
Her 17h54'31"20d16'
Reni
Aur 5h58'18"31d28'
Renick,My Eternal Light,James
Tau 4h0'29"11d41'
Renier,Monica Anne
Cas 0h56'59"63d56'
Reninger,Lee Ann
Peg 22h21'50"4d35'
Reninger,Lives On,Judy
(Frack)
Psc 0h48'14"30d42'
Reninger,Ric
Boo 14h39'55"19d40'
Renk,Erwin K
Tau 3h49'48"0d20'
Renke,Skyler Clark
Dra 11h59'52"68d60'
Renkin-Hochzeitsstern, Karin
Ori 5h58'30"0d7'
Renna,Albert E
Aur 5h9'38"43d24'
Renne
Cas 0h38'36"61d8'
Renneke,Robert R
Sex 9h42'27"-5d23'

Renner,Barbara Sidlosky Cyg 19h29'21"31d16'
Renner,Brian Scott Mon 7h54'0"-5d53'
Renner,Candy Lorraine Cas 1h20'41"60d24'
Renner,Dalphine Lyn 6h28'19"59d20'
Renner,Elizabeth Marie Cam 5h58'50"61d28'
Renner,Frank Dra 16h57'12"71d58'
Renner,George Ori 6h9'51"8d7'
Renner,James M Cep 20h59'10"76d21'
Renner,Lee Ori 6h6'0"0d53'
Renner,Mario Cam 5h7'19"70d31'
Renner,Robert Bruce Oph 18h17'54"12d58'
Renner,Robert John Uma 15h38'54"66d17'
Rennet,Lauren Lesley Cep 0h12'44"66d41'
Rennie,Collin Carruthers Ori 5h55'55"16d32'
Rennie,Helen Cas 2h42'23"61d39'
Rennie,Jean S Cyg 19h48'26"37d3'
Rennie,Peter-Anastasia Helen-Elizabth&Christi Ori 6h1'49"7d16'
Rennie,Robert R Her 18h18'1"13d3'
Renno Cam 3h45'53"72d8'
Reno & Mari Aql 20h11'40"5d15'
Reno's "Moonbeam Rainbo" Peg 22h32'20"8d60'
Reno's Daddy Ori 5h6'29"0d37'
Reno,Dorothy Lyn 8h7'23"45d12'
Reno,Ginette Lyr 18h41'45"35d20'
Reno,Jim (JR) & Betty Hya 8h19'21"2d45'
Reno,Kelly Marie Anne Com 12h52'37"28d47'
Reno,Rick Uma 8h37'52"52d36'
Renock,Devon J Her 16h32'23"51d8'
Renoud,Christopher James Bacorn Vir 15h3'44"5d53'
Renowden,Edward Ser 18h29'54"1d44'
Renshaw,Karen Heslin Hya 9h9'1"6d6'
Renshaw,Leigh Umi 15h28'57"68d43'
Renski,Loretta V Lyn 8h40'37"34d49'
Renteria,Caitlin Lee Ori 5h49'12"10d20'
Rentero,Cristina Per 2h53'19"50d27'
Renth,Torsten Peg 23h32'17"18d23'
Renton,Malcom J Eri 4h9'44"-7d41'
Renton,Maryann Mon 7h24'41"-10d47'
Renucci,Lara Cas 0h29'57"72d20'
Renwick,Fay Cas 1h29'0"58d9'
Renwick,Rick Uma 11h19'27"60d58'
Renz,Klaus Per 4h5'44"49d57'
Renz,Nancy Eri 3h44'14"-6d19'
Renz,Rita Com 13h7'1"17d44'
Renzi Gem 7h23'51"35d29'
Renzi,Matthew Joseph Per 4h24'31"51d21'
Renzi,Nicholas John Ori 5h27'56"-3d48'
Renzo Ugo Negrini Cam 4h13'0"68d20'
Renzulli,Virgil Dra 19h7'52"50d30'
René Umi 15h26'55"68d33'
René Et Odette Aur 5h55'34"54d58'
René My Dancing Angel Lyn 8h24'31"34d35'
Renée Cyg 19h28'51"31d37'
Renée,Jill Cas 2h35'33"73d48'
Reolon,Martine Aur 5h2'31"38d38'
Repasky,Anthony Joseph Lac 22h21'0"40d55'
Repath,Aaron William Cet 0h55'15"-5d49'
Repenich,Marie Aql 20h2'33"10d38'
Repenshek,Victoria M Com 13h12'47"27d42'
Repetti,Christopher Per 2h50'47"40d15'
Repine,Thomas Cap 21h23'32"-15d53'
Repka,Joyce Jayne Cas 1h20'14"58d52'
Repka,Patricia Aqr 21h19'12"-13d37'
Repke,Martha Ari 3h0'59"28d16'
Replogle Ori 6h8'37"7d32'
Replogle,Dorothy Mae Psc 0h58'46"32d6'
Replogle,Holley Dawn Peg 22h46'56"33d52'
Repoll-Petruzelli,Amy Peg 22h52'14"29d15'
Repp,Daniel Cyg 21h31'30"42d47'
Repp,Sarah And 1h54'1"40d11'
Reppe,Christine Ser 16h6'27"0d55'
Reppert,Alan Dra 19h26'29"58d46'
Reppert,Chelsea N Peg 22h8'26"3d2'
Reppert,Kathy Lyn 7h49'39"36d58'
Reppert,Lena Marie Andrus And 23h42'38"38d25'
Reppi Cep 23h8'38"61d25'
Reppuhn,Lydia Cet 2h19'37"9d39'
Republic Of Junichi & Rieko Psc 1h21'45"9d21'
Repus Eri 3h54'58"-2d19'
Rera (Aunt Rera) Lyr 19h25'31"37d42'
Rere Cam 4h13'19"71d1'
Rerko,Frederick J Sct 18h47'32"-7d52'
Resca,Roget Per 2h36'1"50d10'
Reschke,Bruno Cmi 7h17'1"1d44'
Resenhoeft,Rolf Lac 22h30'0"38d44'
Reses,Abigail Louise Lyr 18h33'36"46d49'
Reses,Philip Andrew Her 18h11'42"31d52'
Reseska Cyg 20h21'25"41d18'
Resetar,Thomas S Cnv 12h54'49"51d25'
Resker,Jerome Per 3h27'53"51d7'
Resmer,Adele Stiles Cas 0h32'21"72d15'
Resnick,Elaine Bette Ari 2h26'1"10d54'
Resnick,Leslie Com 12h53'21"24d26'
Resnick,Louise Claire Cas 1h3'1"64d17'
Resnick,Lynda & Stewart Sct 18h52'9"-7d48'
Resnik,Judith Arlene Aql 19h4'0"-0d36'
Respect-95 Vul 20h16'18"23d39'
Resplendent Margaret And 2h34'17"41d21'
Resplendent,Jill Roxanne Murray And 1h9'35"41d8'
Response Inc Lyr 18h58'0"40d15'
Responsum Dallmeri Tau 4h2'41"7d46'
Resse,Rex Vernon Lep 5h33'21"-11d3'
Ressler II,K J Cnv 13h28'45"42d7'
Resso,Kyle John Dra 19h22'36"56d6'
Rest,Walter Garon Lac 22h19'58"50d6'
Restelli,Jude Del 20h33'27"20d24'
Restemeyer,Kelley & K R Peg 23h15'60"31d36'
Resti,Gina Ann Aur 5h22'0"41d14'
Restivo,Andy Aql 20h17'1"0d45'
Resto,Pete Her 17h9'35"38d38'
Restrepo,Rafael Hya 8h54'16"2d36'
Restua,Danilo Lyn 8h9'24"46d58'
Restucci,Jeff Lac 22h40'0"50d24'
Reszel,Marc Aur 6h13'12"35d19'
Reta And 23h37'41"45d40'
Retep Xam Ailenroc Uma 13h54'1"50d10'
Rethage,Glenn W Lac 22h34'58"53d43'
Rethore,Elisabeth Dra 19h5'23"80d32'
Retseck,James Wayne Her 18h51'52"51d3'
Rettagliata,John P Dra 17h11'19"68d50'
Retter,Joe "Pop" Aur 6h28'17"52d32'
Rettew,Patrick Charles Ori 5h59'39"9d57'
Rettig,Carol Lyr 18h55'48"31d34'
Rettig,H Wayne Aur 4h53'1"38d18'
Rettig,Jesse Aql 20h11'58"0d59'
Rettig,Lena Cas 3h26'24"73d22'
Rettinger,Stanza Aleta Boswell Cam 4h4'1"60d9'
Rettke,Eric Gerald Uma 15h38'39"80d59'
Rettke,Mari Jean Kubicki Uma 15h55'0"82d0'
Rettke,Nina B Bougartchev Pollard Uma 15h56'51"81d24'
Retzar,Mario & Katalin - Makabo Cap 21h15'35"-26d43'
Retzlaff,JeanLux Tom Cyg 21h3'50"50d29'
Retzler,Kurt & Rali Ori 5h50'1"20d55'
Reu,Michel Cyg 19h45'1"30d6'
Reube,Bambi Lynn Cmi 7h32'51"0d37'
Reuben Per 1h42'1"52d48'
Reuben,Neville Philip Per 2h27'46"56d58'
Reubenstein,Alexander King Boo 14h54'0"42d49'
Reubenstein,Benjamin Daniel Her 18h6'46"48d18'
Reubenstein,Robert Drake Aur 7h1'50"35d31'
Reubold,Elinor Cas 3h0'49"61d46'
Reuer,McKayla Marie Lyn 7h44'58"40d44'
Reuke,Irmgard und Kurt Ari 2h29'1"21d33'
Reunion Star,The Cmi 7h41'54"5d24'
Reusch,Eric Dra 12h28'46"64d42'
Reuter,Anne Cas 0h35'1"61d11'
Reuter,David Cep 22h5'1"53d2'
Reuter,Hans Boo 14h59'13"36d58'
Reuter,Juergen Eri 4h49'57"-6d11'
Reuter,Kurt Del 20h24'39"8d42'
Reuter-Showalter, Patricia Lyr 19h16'46"28d46'
Reuther,Devin Cam 4h59'17"68d34'
Reuther,Roger "Roddy" Aur 5h20'59"37d38'
Rev Ed's Rays Of Peace Joy & Love Cyg 19h30'45"36d34'
Reva Jane Vul 19h39'52"27d3'
Revak,Bailey Frances Per 1h57'48"50d20'
Reveilhac,Valerie Lib 15h39'18"-28d33'
Revel,Couville Lac 22h7'42"38d58'
Revelle,Helen Ori 5h31'1"1d16'
Revercomb,Ginger "Snow Bunny" Cep 23h13'51"71d3'
Reves,J R Cet 2h47'48"3d29'
Revest,Michel Cep 2h24'27"80d27'
Reville,James Aur 6h0'1"35d11'
Revis,Donna Uma 11h14'26"67d32'
Revson,Martin Per 3h54'0"40d21'
Rewwer,Jay Cep 21h16'1"65d3'
Rex Hya 9h7'1"4d58'
Rex & Linda Cet 3h8'48"0d47'
Rex Sylvae Tau 4h43'1"20d14'
Rex,David Cep 21h32'38"70d8'
Rex,John Cet 2h56'53"0d20'
Rex,John O Her 16h30'29"20d7'
Rex,Karin Uma 8h24'17"61d6'
Rex,Katie "Mississippi" Ori 6h7'20"1d37'
Rex,The Wonder Dog Cep 22h17'10"61d40'
Rex,Tim Her 17h33'35"20d49'
Rex-1977,The Cep 21h32'0"67d34'
Reximus Aleximus Boo 14h28'36"22d43'
Rexinia Del 20h25'14"20d10'
Rexrode,Jr,Richard Arthur Her 17h39'1"20d59'
Rexrode,Rob Dra 16h41'32"67d28'
Rexrodt,Günter Vir 13h24'58"-9d7'
Rey,Dr & Mrs & Herminia Farne Cyg 21h29'0"53d7'
Rey,Florian Ser 18h29'54"1d44'
Rey,Guy Cyg 20h4'49"39d54'
Rey,Johnny Mi Mon 6h3'37"-8d50'
Rey,Michel Del 20h19'44"10d41'
Rey,Robert Peg 22h27'29"33d46'
Reyes,Adele Cet 2h42'14"-18d60'
Reyes,Alberto Daniel Cnv 12h51'17"38d6'
Reyes,Ana Maria Ori 4h57'0"0d18'
Reyes,Angelica Com 12h27'50"23d34'
Reyes,Antonio Francisco Calvin Ser 18h18'36"-2d36'
Reyes,Christina A Ori 6h1'13"8d59'
Reyes,Christine T And 23h20'23"45d23'
Reyes,Elizabeth Lac 22h25'36"55d44'
Reyes,John David Dra 13h18'0"64d17'
Reyes,Ryan Alan Ser 17h55'48"-14d35'
Reyes,Samantha Taylor Del 20h58'20"18d53'
Reyes,Sarah Ori 5h1'39"10d13'
Reyes,Thomas J Cnv 13h8'50"41d12'
Reyes,Tina And 1h3'1"39d56'
Reymont,Joanne Lac 22h23'0"37d36'
Reyna,Ebelia Marie Cet 3h10'1"6d22'
Reyna,Father Cecilio Aur 6h56'32"36d26'
Reyna,Theresa Lynn Uma 11h33'53"47d47'
Reynertson,Mark N And 2h28'60"41d51'
Reynes,Brian Buchanan Uma 8h37'58"70d33'
Reynolds Her 17h11'1"44d57'
Reynolds BSMS,Craig Paul Aur 6h57'0"40d41'
Reynolds Family,The Lac 22h12'41"37d37'
Reynolds,Alexis Mon 6h8'11"-10d8'
Reynolds,Blair Elizabeth Creedle Sge 19h53'1"16d15'
Reynolds,Christopher S Her 16h51'20"38d23'
Reynolds,Dan Boo 14h9'0"51d42'
Reynolds,Dave & Dolly Ori 6h7'13"9d51'
Reynolds,Debbie Louise Per 3h5'37"35d16'B
Reynolds,Don & Marge Cyg 21h11'42"38d31'
Reynolds,Dr Whitman M Psc 1h29'0"32d4'
Reynolds,Ed G Lyn 7h3'32"44d56'
Reynolds,Edith F Peg 21h28'14"26d25'
Reynolds,Gabrielle Uma 11h18'30"61d25'
Reynolds,Gary & Paula Peg 22h46'15"34d49'
Reynolds,Gloria Ballengee Cet 1h0'54"-12d56'
Reynolds,Howard L Cet 2h20'55"-10d41'
Reynolds,Jacqueline Marie Lynch Ori 5h54'52"14d0'
Reynolds,James Lawrence Lac 22h54'58"51d34'
Reynolds,Janice Lyn 7h46'20"42d56'
Reynolds,Jason Ser 18h21'38"-2d2'
Reynolds,Jessica Jayne Lyr 18h19'1"38d54'
Reynolds,Joanne Elizabeth Cet 0h53'30"-8d46'
Reynolds,John A Dra 19h15'13"67d51'
Reynolds,John Andrew Dra 14h58'25"64d7'
Reynolds,Jr,Thomas W Her 17h56'14"28d27'
Reynolds,Kelsey Rae Eri 4h47'4"-8d13'
Reynolds,Kimber Michelle And 0h9'56"27d56'
Reynolds,Kristin Elizabeth Cnc 8h1'54"10d22'
Reynolds,Langston Octavius Waddy Aql 18h44'44"8d46'
Reynolds,Mark Oph 17h29'9"-20d47'
Reynolds,Mark A Ser 15h54'28"1d58'
Reynolds,Mark A Aur 5h55'1"30d58'
Reynolds,Mark Stephen Mon 7h59'30"-8d34'
Reynolds,Maureen And 23h1'57"50d45'
Reynolds,Morgan Elizabeth And 23h38'1"47d4'
Reynolds,Myrna Darlene Cas 23h15'51"62d15'
Reynolds,Neil Umi 14h39'42"67d51'
Reynolds,Norma Bruen Dra 10h31'23"80d20'
Reynolds,Patrick Kelly Uma 9h26'24"51d2'
Reynolds,Peter Ori 5h59'46"16d51'
Reynolds,Randal Carlin Cet 3h7'58"2d19'
Reynolds,Rebecca Lyr 18h40'41"40d43'
Reynolds,Rhonda Lyn 7h1'15"44d44'
Reynolds,Robert Her 17h7'51"42d27'
Reynolds,Robert Allen Her 16h21'57"42d1'
Reynolds,Robert Dale Vul 19h22'19"26d23'
Reynolds,Rosemary Lyr 18h18'43"42d39'
Reynolds,Sara Peg 23h25'0"32d7'
Reynolds,Seth Elijah Her 17h12'51"42d27'
Reynolds,Shane & Misty Oph 17h7'46"-23d38'
Reynolds,Stacey Gem 6h52'21"12d53'
Reynolds,Thomas Hill Per 2h57'14"34d23'
Reynolds,Toni Arnette Vul 19h18'0"26d52'
Reynolds,Victoria C Lmi 9h44'59"38d14'
Reynolds,W Jay Aur 7h16'46"36d38'
Reynolds,William Her 18h4'57"47d23'
Reynoldsburg Library Total Tomatoes Oph 18h42'25"7d9'
Reynoso,Rafael Oswaldo Her 17h20'45"50d9'
Reza,Jamshid Ori 5h18'15"15d55'
Reza,Shahram Hya 8h54'13"-5d42'
Rezac,Kayci Rae Aql 19h53'41"13d7'
Rezac,Kimberly Rae Eri 3h37'51"-14d53'
Rezac,Lindsey Mariah Peg 22h15'39"34d5'
Rezba,Jennifer Ann Uma 11h27'48"51d40'
Rezny,Michael N Tri 1h50'51"28d34'
Rezwana Aql 18h57'17"16d42'
Rezzonico,Connor R Dra 16h51'28"68d32'
Rezzonico,Justin J Her 17h57'32"14d36'
Rezzonico,Melody J And 23h18'50"43d37'
RF-7 Dra 17h29'57"75d41'
RFC 50 Uma 9h17'43"55d11'
RFM III Dra 19h3'27"71d5'
RG-1 Dra 16h46'0"51d24'
RG5 Uma 9h10'29"52d42'
Rhapsody Cet 2h44'58"7d19'
Rhea Cas 22h56'0"55d23'
Rhea Bethine Vul 19h47'55"28d50'
Rhea Cunnyngham Crb 16h7'23"30d4'
Rhea Dell Oph 17h7'48"-20d22'
Rhea,Pearl L Boo 14h14'28"53d10'
Rhealee And 0h24'38"41d4'
Rheanna Kathleen Lyr 18h24'14"38d4'
Rheanna's Star Umi 14h12'53"66d12'
Rheault,Guylaine Lac 22h43'41"55d57'
Rheaume,Roger P Cep 22h3'33"61d31'
Rheberg,Philip J Cep 21h21'0"61d18'
Rhee,Laila Elizabeth Cecilia Mon 7h4'45"5d8'
Rhein,Arwen Dra 16h41'28"61d21'
Rheinlaender,Randy Mon 7h4'16"-5d28'
Rheinländer,Matthias Mon 6h52'38"10d22'
Rhett & Leesa Cyg 21h27'35"53d45'
Rhi,Cati And 23h45'1"44d40'
Rhian Vir 14h37'23"7d17'
Rhiannon Ori 5h52'1"11d17'
Rhiannon Lyn 7h54'58"43d28'
Rhiannon Rain Lac 22h34'57"54d1'
RHILYFWYA Cma 6h45'5"-15d10'
Rhindress 13 Much, Nancy Louise Tri 1h48'37"26d14'
Rhine,Gregory S Oph 17h57'38"13d0'
Rhinehart "Forever", John & Frances Lyn 8h50'1"35d1'
Rhino Cep 23h4'27"64d31'
Rhoades,Caitlin Rebecca Mon 6h37'0"11d32'
Rhoades,Jason M Ori 6h15'44"-1d17'
Rhoades,Jeremy W Her 17h43'34"40d10'
Rhoads II,James Bradley Boo 14h35'33"40d21'A
Rhoads,Colleen Lyr 18h15'19"30d1'
Rhoads,David Lac 22h46'33"55d31'
Rhoads,Dustin Hya 8h44'22"4d4'
Rhoads,Julie Vul 19h43'48"25d35'
Rhoads,Michael J Per 3h7'49"46d51'
Rhoda Marie Lyn 8h28'14"45d9'
Rhoda's Gem Equ 21h18'53"3d25'
Rhode,Alma Cam 5h39'57"68d31'
Rhode,Jennifer Michelle And 1h16'43"39d40'
Rhode,Lori Lynn Cas 1h51'39"58d3'
Rhode,Marc Leo 9h59'16"17d43'
Rhode,Ronald & Margaret Cmi 7h31'17"0d6'
Rhodebeck,Virginia Byrd Cyg 20h29'22"40d46'
Rhoden,Donald Dee Ori 6h9'30"8d47'
Rhoden,Matthew Alexander Hya 8h24'54"5d41'
Rhoden,Virginia Lee Steiger Sct 18h46'18"-7d52'
Rhoderick,Mary Ellen Lyr 18h42'22"32d10'
Rhodes,Amy B Mon 7h20'51"-8d29'
Rhodes,Barbara Per 3h39'24"38d19'A
Rhodes,Catherine Lyr 18h58'52"26d18'
Rhodes,Charles Per 3h39'24"38d19'B
Rhodes,Christopher Aql 18h59'14"-6d46'
Rhodes,Emily Ann Cas 3h4'40"63d56'
Rhodes,Gary L Aql 20h2'26"1d46'
Rhodes,Joshua James Allen Aql 20h2'43"1d24'
Rhodes,Lee "Where's Waldo" Aur 6h26'54"35d34'
Rhodes,Mary Louise & Terry Connollly Boo 13h42'50"14d55'
Rhodes,Megan Cas 23h35'41"60d25'
Rhodes,Michael W Per 2h0'0"50d13'
Rhodes,Stacy L Crt 11h13'4"-19d29'
Rhodes,Travis Tate Her 16h1'0"40d16'
Rhodey,John & Kathleen Lyr 18h57'27"30d21'
Rhodora Uma 11h19'43"42d14'
Rhodri's Star,Owen Dra 17h53'53"68d45'
Rhody,Allassandra Fulkerson Cas 0h48'0"61d15'
Rhody,Kristofer Wolf Ray Boo 14h18'59"32d51'
Rholand Her 16h50'26"35d56'
Rhome,Joseph A Hya 8h56'49"6d0'
Rhome,Joseph Aloysius Michael Hya 8h52'48"2d55'
Rhon,Carlos Hank Sgr 19h40'12"-44d25'
Rhonda Cas 0h16'40"47d14"
Rhonda Sgr 19h23'37"-45d19'
Rhonda Eri 2h45'44"-17d21'
Rhonda Lyn 8h23'51"4d27'
Rhonda Mon 7h19'53"-0d57'
Rhonda & Bill Lyn 8h7'47"47d18'
Rhonda & Don Cyg 20h35'34"44d55'
Rhonda & Marty Uma 11h46'13"51d57'
Rhonda & Todd's Star Aql 18h59'0"-5d17'
Rhonda 1437 Cnv 12h14'12"39d52'A
Rhonda Gail Aql 19h1'56"16d48'
Rhonda Loves Bill Forever Ori 6h17'27"-2d55'
Rhonda Lynn's Inspiration And 23h41'20"47d17'
Rhonda M Her 17h14'31"48d19'
Rhonda Marguerite,The Lmi 10h2'0"28d25'
Rhonda Michelle Leo 10h58'38"12d43'

Rhonda Sue
Cas 2h39'0"58d36'
Rhonda's Flawless Beauty
Peg 23h1'50"21d35'
Rhonda's Guiding Light
Tau 5h53'29"23d13'
Rhonda's Realta
Lib 15h42'0"-20d35'
Rhonda-Boo
Eri 2h55'29"-5d4'
Rhonda-Jim
Aql 19h26'47"-10d13'
Rhondus KGEO
Equ 21h5'37"11d25'
Rhone,Andrew
Cyg 21h16'23"28d22'
Rhone,Mark
Oph 17h52'13"12d55'
RHONWYN
Sgr 19h7'46"-28d10'
Rhoton,Patricia
Cyg 19h31'49"34d18'
Rhuda-Gramma,Patricia A
Cas 22h57'1"53d39'
Rhue,Douglas Moore
Boo 14h39'28"34d28'
Rhyan,Edd
Lmi 10h45'27"24d7'
Rhyan,Jessica Ashley
Peg 21h52'58"30d30'
Rhyann Nicole
Cas 2h7'31"59d12'
Rhymann,Irfan
Ori 4h58'39"0d58'
Rhyne,Barbara
Mon 7h24'1"-8d40'
Rhys-Davies,Licidus
Aql 19h57'37"-6d37'
Rhyshek,Pamela
And 2h20'44"37d33'
Rhéaume,Ghislain
Lyn 8h24'52"43d12'
Rhü
Eri 3h53'14"-10d7'
Ria 96
Peg 22h15'16"21d1'
Ria Sal
Her 17h22'21"38d37'B
Ria,Constance
Lyn 8h4'27"37d47'
Riady,Nita
Mon 6h40'0"1d35'
Riall,Alan
Peg 22h33'0"28d58'
Rian
Del 20h16'25"15d37'
Rias & Babs A8-02031994
Ori 5h55'0"7d30'
Riazzi,Audrey Korn
And 23h15'46"40d21'
Riba,James
Uma 11h31'12"41d10'
Ribal,Lucille
Vul 19h2'27"21d32'
Ribal,Raymond R
Aql 19h6'39"3d11'
Ribandit,Deiesse
Aur 5h2'35"50d28'
Ribbans,Kristin
Cyg 20h56'13"39d29'
Ribbens,Nicole L
Del 20h58'39"12d38'
Ribeiro De Sá,Mark S
Dra 20h1'0"63d38'
Ribeiro Dec 24,1967, Maria José
Cap 20h17'1"-26d20'
Ribeiro,Richardo Fernando Inacio
Boo 14h56'41"27d30'
Ribé,Teresa Minguill30'
Boo 13h42'11"11d44'
Ric
Aql 19h13'10"13d2'
Ric & Jo-Star Of Love
Dra 17h0'32"67d18'
Ricard 607 Cat et Alex 143
Uma 8h58'58"47d38'
Ricard Harvey Sophie 143
Umi 16h5'1"79d45'
Ricard,Marc & Laura
Umi 15h41'16"78d47'
Ricard-Wolf,Angelika
Cam 14h24'33"80d9'
Ricarda,Carlotta
Lyr 19h4'48"25d51'
Ricardo
Lac 22h55'35"55d35'
Ricardo
Her 16h1'57"16d44'
Ricardo,Paloma S
Ser 18h3'41"-14d20'
Ricart,Fred
Per 2h53'40"38d15'
Ricart,Nadine
Lyn 9h4'47"33d27'
Ricasa,Erik Jacobsen
Cnv 12h26'0"48d20'
Riccagata
Eri 3h40'20"-6d3'
Riccardelli,Susan
Cam 4h58'1"67d47'
Riccardo
Peg 23h35'21"21d23'
Riccardo
Per 3h0'26"38d19'
Riccardo
Boo 14h16'40"48d49'
Riccelli,Nick D
Her 17h5'57"44d2'
Riccelli,Rita
And 23h40'44"42d31'
Riccello,Rich
Her 17h33'12"26d42'
Ricchiazzi,Lisa Gail
Cas 1h46'17"75d31'
Ricchione,Leo
And 23h3'11"50d3'
Ricci
Ori 5h58'57"16d50'
Ricci 1949,Sandra Ann
Lac 22h5'22"46d31'
Ricci,Adri
Uma 10h15'58"60d50'
Ricci,Alexander John
Boo 15h10'1"53d50'
Ricci,Leslie A
Cnv 13h55'46"45d58'
Ricciardi Will Shine Forever,Vera
Cyg 21h31'33"37d32'
Ricciardi,Amy Lippi
Cas 0h43'33"75d19'
Ricciardi,Laura
Cyg 19h26'0"30d46'
Ricciardi,Linda
Sco 17h29'0"-40d29'
Ricciardi,Robert Louis "Bobby"
Her 16h56'18"21d50'
Riccio Star,The
Ori 5h55'59"15d35'
Riccio,Biagio
Uma 10h39'45"68d51'
Riccio,Jenna
Cam 9h6'1"77d51'
Riccio,Lauren
And 2h2'1"41d39'
Riccio,Richard Paul
Lyr 19h27'23"40d58'
Riccio,Richard Steven
Per 2h59'60"45d9'
Riccitelli 81193,Logan Rhys
Cyg 21h17'0"36d17'
Riccitelli's A Salut, Melio
Boo 14h58'32"46d48'
Riccitelli,Tessa Maria
Cnc 7h58'60"11d13'
Ricciuti,Robert
Cep 23h12'0"64d4'
Ricco,Larry
Her 17h14'26"42d51'
Ricco,Ronald J
Boo 14h24'19"28d17'
Rice III,Joseph E
Ori 5h0'42"-0d41'
Rice Marry Me,Miss M L
Sge 19h29'34"18d40'
Rice Star,The Brooke Elise
Eri 4h17'53"-14d43'
Rice's Star,Jacquie
Aql 19h45'0"11d41'
Rice's Star,Jim
Aql 19h45'25"11d13'
Rice,Abigail Marilyn
Lyn 8h55'18"40d15'
Rice,Adam M
Cyg 20h54'36"38d11'
Rice,Andrew T
Her 16h6'0"23d4'
Rice,Andria Dawn
Uma 9h16'52"61d33'A
Rice,Ann Blakely
Aql 19h15'32"19d38'
Rice,Bernadette Hanley
Lyn 7h9'35"58d6'
Rice,Brian
Uma 13h28'29"62d59'
Rice,Brian
Ser 15h20'25"6d19'
Rice,Buffy
Uma 10h20'23"68d8'
Rice,Carol
Mon 8h7'40"-9d40'
Rice,Carter
Aur 6h6'56"31d10'
Rice,Cassandra
Cas 2h6'28"59d38'
Rice,Chaurice Briana
Vul 19h23'28"23d43'
Rice,Cindy Louise
Aql 19h25'48"-10d25'
Rice,Daryl Leo
Cep 2h43'19"78d49'
Rice,David Alan
Aql 19h55'35"13d8'
Rice,David Joseph
Aql 19h7'1"5d4'
Rice,Deborah Lee
Crb 15h25'0"31d60'
Rice,Debra M
Ori 5h50'42"14d10'
Rice,Ed & Gloria
Cyg 19h34'16"30d24'
Rice,Elizabeth Marie
Peg 21h48'37"31d53'
Rice,Elyn & Shane
Mon 8h1'33"-4d54'
Rice,Eva Helena Seibel
Vul 20h2'23"28d37'
Rice,Faye
Lyn 7h56'39"48d8'
Rice,Fr Patrick
Boo 15h4'24"30d44'
Rice,Garrett Senna
Lac 22h36'25"37d42'
Rice,Ginny June
Aur 5h53'16"38d56'
Rice,Harold
Aql 19h53'50"12d37'
Rice,Hollie
Cas 0h33'27"60d44'
Rice,In Memory of Jimmy
Crb 15h54'45"28d51'
Rice,Jane Loree
Vul 19h45'0"28d51'
Rice,Joseph B
Her 17h0'1"48d12'
Rice,Jr,David M
Sgr 20h0'46"-43d4'
Rice,Jr,George H
Eri 4h4'38"-7d37'
Rice,Jr,William J
Sge 19h54'22"16d6'
Rice,Katrina L
Cam 3h29'58"61d9'
Rice,Kenneth D
Crt 11h50'53"-7d43'
Rice,Kurt
Cet 0h1'37"-8d31'
Rice,L Gregory
Tau 4h9'41"23d52'
Rice,Lindsay Ann
Lyr 18h30'15"31d42'
Rice,Marina Martinez
Cap 20h30'13"-13d57'
Rice,Mark Andrew
Dra 15h58'13"68d27'
Rice,Megan
Lyn 7h40'23"38d48'
Rice,Melissa Leanne
Aql 20h2'29"10d51'
Rice,Misty Lynn
Del 20h19'21"10d27'
Rice,Patrick Malone
Equ 21h2'31"7d59'
Rice,Philip
Dra 13h26'39"68d18'
Rice,Sigrid M
Cet 3h6'1"5d18'
Rice,Stephanie Lynn
And 23h0'51"47d10'
Rice,Sue
Cet 1h22'1"-1d50'
Rice,Teen Queen,Angela Nicole
Lyr 19h17'1"40d15'
Rice,Valeta Mae
Peg 21h49'1"34d53'
Rice,Virginia Trapp
Sge 19h36'22"16d58'
Rice,Walter F
Per 3h9'0"50d30'
Rice-Mandigo,Linda
And 0h6'14"34d35'
Rich
Lac 22h1'55"51d1'
Rich & Beth
Ori 5h56'44"15d24'
Rich & Bev "Will Shine Forever"
Cyg 21h33'0"37d37'
Rich & Gerious Aniverious
Peg 0h5'24"13d7'
Rich & Kathy's Wishing Star
Her 16h50'46"34d22'
Rich & Margie's
Crb 16h19'27"38d30'
Rich & Melissa
Eri 3h14'53"-3d3'
Rich & Michelle's Precious Moments
Tri 2h0'1"31d18'
Rich & Nancy "Forever"
Mon 6h54'11"-1d43'
Rich & Pegg
Umi 14h21'37"67d39'
Rich (Susie),Susan Elizabeth
Cyg 21h53'1"40d45'
Rich Rx
Dra 19h12'44"70d47'
Rich S
Her 16h22'18"23d37'
Rich"10"
Cnv 13h53'0"40d19'
Rich's "Unforgettable" Star
Cep 22h14'19"80d0'
Rich,Alexander Vincent
Aql 20h2'0"10d6'
Rich,Andrew Michael
Hya 8h13'40"5d11'
Rich,Bridget B
Lyn 9h29'18"40d5'
Rich,De Anna
Lyr 18h35'52"45d4'
Rich,J P
Mon 6h24'1"8d2'
Rich,Jordan
Hya 8h37'17"5d39'
Rich,Michael & Christine Thomas
Cnv 13h44'39"41d22'
Rich,Michael J
Cep 20h56'0"55d31'
Rich,Nancy
Mon 7h56'53"-2d59'
Rich,Nick
Her 16h38'37"35d23'
Rich,Sharman
Aql 19h51'54"15d13'
Rich,Sheila Irisia
Vul 20h27'11"28d19'
Rich,Victoria
Umi 14h43'13"68d4'
Richan,Professor Willard Cooper
Tri 2h39'49"31d2'
Richand,Pierre
Peg 22h29'26"20d24'
Richard
Tri 2h42'12"32d55'
Richard
Sgr 18h53'19"-23d57'
Richard
Cam 5h8'48"69d46'A
Richard
Boo 14h23'16"33d44'
Richard
Peg 22h33'1"28d28'
Richard
Sge 19h55'0"16d22'
Richard
Cep 23h12'21"64d25'
Richard
Dra 14h55'42"63d7'
Richard & Annette
Aql 20h8'1"1d20'
Richard & Beth Wedding Star
Cyg 21h37'27"30d12'
Richard & Dan Star,The
Aur 5h10'58"40d33'
Richard & Deborah
Com 13h32'44"16d49'
Richard & Gillian 1994
Vul 19h44'11"20d19'
Richard & Gloria Forever
Tri 1h58'43"31d41'
Richard & Jaki
Lyr 18h57'32"30d31'
Richard & Julie Star, The
Uma 9h15'54"48d44'
Richard & June Forever
Aqr 21h53'54"-1d44'
Richard & Kathryn
Boo 15h12'33"47d51'
Richard & Keith, July 6th, 1985
Cap 21h36'41"-20d2'
Richard & Lourdes Forever In The Heavens
Lmi 10h12'57"32d25'
Richard & Lynne
Cap 21h10'15"-22d16'
Richard & Meredith
Lyn 7h50'52"42d39'
Richard & Michael's Star
Cet 2h1'0"0d55'
Richard & Michelle (aka Bucket)
Cyg 21h45'20"31d24'
Richard & Pearl
Dra 17h40'29"60d4'
Richard & Randi
Sco 17h28'28"-33d54'
Richard & Sue
Sge 20h7'25"16d19'
Richard & Victoria's Stellar Love
Ori 6h7'0"0d20'
Richard & Yvonne
Cyg 21h41'26"30d47'
Richard Alan
Dra 13h35'58"67d46'
Richard Alexander The Great
Leo 10h56'1"20d56'
Richard Earl
Her 17h3'11"20d15'
Richard Edward
Cep 21h43'53"55d34'
Richard et Christine
Cam 6h54'12"67d31'
Richard Gregory
Lac 22h24'15"40d47'
Richard Joseph
Boo 13h40'49"20d27'
Richard Kyle
Ori 5h55'50"-1d45'
Richard Lenard
Boo 14h54'0"45d60'
Richard Star & Love Of My Life
Boo 15h5'33"53d56'
Richard Thomas
Hya 10h43'39"-18d18'
Richard W & Marjorie A
Ser 18h53'40"2d40'
Richard's "Shining Star Of 1994"
Aql 20h3'52"1d43'
Richard's Angel
Eri 2h58'38"-6d33'
Richard's Brilliance
Aql 18h43'0"6d51'
Richard's Dream
Cet 2h43'23"3d6'
Richard's Equinox
Dra 11h37'18"72d24'
Richard's Golden Phoenix
Aql 19h55'49"0d2'
Richard's Gone Public
Per 3h18'56"40d3'
Richard's Haven
Mon 7h4'2"-5d22'
Richard's Hope
Cep 22h25'54"61d47'
Richard's Quest
Equ 21h22'42"11d24'
Richard's Star
Cep 21h50'22"60d23'
Richard's Star
Aql 18h49'34"9d58'
Richard's Star
Lac 21h58'34"40d40'
Richard's Star
Lac 22h20'48"55d18'
Richard's Star
Boo 15h0'1"27d41'
Richard's Think Of Me Star
Gem 6h52'40"18d59'
Richard,Jacques
Cam 3h26'1"61d1'
Richard,Jocelyne
Ari 2h40'10"20d57'
Richard,Kimberley
Uma 9h1'0"59d30'
Richard,Margot
Sct 18h45'28"-6d32'
Richard,MD,Norman B
Boo 13h55'13"22d2'
Richard,MD,Warren E
Aur 5h52'11"30d35'
Richard,Murielle
Mon 6h27'21"7d48'
Richard,Patrick
Aqr 23h24'14"-6d17'
Richard,Richard, Richard
Aur 6h13'25"46d20'
Richard,Rose
Aql 20h0'16"1d3'
Richard,Sophie
Crt 11h17'21"-19d42'
Richard,Thomas
Lup 15h12'35"-42d39'
Richard,Tina
Cyg 19h24'0"35d50'
Richard-Marie
Peg 22h13'38"20d55'
Richardean Advee
Crt 11h25'17"-18d4'
RichardElizabethRath Suter
Dra 20h17'26"64d36'
Richards Celestial Dream
Ori 4h42'29"10d39'
Richards Star,The Leah Katherine
Mon 7h46'48"-1d0'
Richards,Adam
Per 3h8'0"47d22'
Richards,Alexandra Claire
Cyg 19h33'12"39d28'
Richards,Alexandra Nicole
And 22h58'46"38d41'
Richards,Alice Lee
Lyn 7h33'34"45d60'
Richards,Alison Hope
Lyr 19h26'0"38d3'
Richards,Anne Louise
Cas 1h24'41"52d55'
Richards,Brian Daniel
Boo 14h45'1"52d10'
Richards,Camden Marie
Cas 22h56'12"54d20'
Richards,Carol
Cep 2h59'21"79d13'B
Richards,Christine
Peg 23h42'0"11d11'
Richards,Christopher Lee
Peg 21h42'13"28d12'
Richards,Chuck-Star
Aur 5h0'11"49d54'
Richards,David,Donna, Dustin & Drew
Peg 23h35'36"20d38'
Richards,Diana L Ruppel
Com 12h55'52"21d47'
Richards,Elizabeth Diane
Mon 6h23'44"1d26'
Richards,Elizabeth Rusling
Mon 6h51'0"10d5'
Richards,Emma & Gordon
Mon 6h39'44"10d19'
Richards,Gordon Edward
Gru 22h9'44"-51d57'
Richards,Helen
And 1h46'40"40d39'
Richards,Jacqueline Rosemary
Cas 1h58'1"61d16'
Richards,Jeff Frank
Hya 8h10'1"-0d45'
Richards,Jewell
Del 20h29'22"18d47'
Richards,Jim
Aur 5h46'49"50d12'
Richards,John
Hya 8h23'34"1d23'
Richards,John & Susan
Cyg 21h45'31"31d37'
Richards,John Y
Aur 6h26'16"35d11'
Richards,Joseph H
Dra 19h26'43"61d29'
Richards,Jr,Dean C
Cet 2h9'0"4d33'
Richards,Keith Charles
Aql 20h2'0"11d26'
Richards,Keith Michael
Lac 22h52'33"40d32'
Richards,Lane
Lyr 18h37'37"44d36'
Richards,Lesley
Eri 2h59'13"-5d52'
Richards,Linsay Elizabeth
Lyr 18h49'0"42d12'
Richards,Lisa
Lyr 18h41'1"31d10'
Richards,Lloyd & Jane
Crb 15h54'56"38d3'
Richards,M L
Cap 20h37'44"-26d23'
Richards,Olivia Lee
Aur 6h7'56"54d28'
Richards,Pamela
Cam 3h19'1"58d17'
Richards,Patric
Cep 2h59'21"79d13'A
Richards,Paulette M
Cas 1h9'32"63d34'
Richards,Robert
Lmi 9h27'24"38d13'
Richards,Scott Gerard
Lac 22h52'18"40d11'
Richards,Shannon
Uma 10h19'27"56d26'
Richards,Taylor
Cam 3h57'35"61d21'
Richards,Theodora Dupree
And 22h55'12"37d23'
Richards,Wanda
Peg 23h29'38"33d37'
Richards,William T
Aur 7h8'19"40d30'
Richards/Campbell
Her 16h50'0"48d45'
Richardson (Parijian), Carol J
Uma 11h29'52"50d17'
Richardson (Pig), Lisa Christina
Leo 10h59'27"21d22'
Richardson 1-25-1904, Margaret Brown
Sex 10h29'20"-1d53'
Richardson 7-29-51, Robert Howard
Cep 2h4'49"77d44'
Richardson's Starlight,Stuart
Cas 2h6'0"59d42'
Richardson,Alexandra Steele
And 23h19'1"40d20'
Richardson,Alicia
Cru 12h45'9"-62d45'
Richardson,Anita & Jeff
Cyg 20h15'32"30d58'
Richardson,Ashley
Lyn 7h9'1"52d55'
Richardson,Christine
Lyr 18h16'1"42d12'
Richardson,Darrel
Aur 4h37'36"30d20'
Richardson,Darren Wayne
Vel 9h59'0"51d27'
Richardson,David Michael
Cam 3h31'16"53d52'
Richardson,David
Dra 14h10'43"64d23'
Richardson,Doris
Lyn 8h27'50"45d41'
Richardson,Doris Elbertine
And 1h3'54"39d35'
Richardson,Fonita Angela
Lyr 18h44'1"39d9'
Richardson,Frederick Seibert
Aql 19h48'23"14d11'
Richardson,Geniffer Susan
Cam 7h26'1"60d35'
Richardson,Harold
Pyx 8h31'4"-22d17'
Richardson,Holly
Vul 20h16'11"23d26'
Richardson,J Donovan
Gem 7h0'14"11d0'
Richardson,Jaimé
Del 20h30'28"20d37'
Richardson,James & Maurine
Peg 22h8'46"4d40'
Richardson,Jana
Lyr 18h40'0"43d1'
Richardson,Janine
Uma 13h5'15"53d41'
Richardson,Jobe Connor
Hya 10h43'29"-17d60'
Richardson,Jr"Bubba", Glendon
Sex 9h50'1"-5d35'
Richardson,Julia Mills & Edward Earl Brighton
Uma 13h38'24"50d46'AB
Richardson,Julienne
Peg 23h42'56"31d52'
Richardson,Kellie
Peg 23h4'42"10d58'
Richardson,Larry Donald
Her 18h11'39"38d26'
Richardson,Laurel
Lyr 18h42'24"32d52'
Richardson,Laurie Espe
Crb 15h43'1"26d17'
Richardson,Malcolm
And 23h45'12"41d29'
Richardson,Margaret
And 23h36'42"40d41'
Richardson,Margaret Brown
Aql 19h58'51"0d51'
Richardson,Mark William
Per 1h48'0"53d59'
Richardson,Mary Lynn
Del 20h17'33"13d40'
Richardson,MaryAnn Alton
Peg 23h37'45"21d55'
Richardson,Mavis Irene (Driver)
Aur 7h5'0"38d10'
Richardson,Opal
Mon 6h51'31"10d8'
Richardson,Paul
Aql 20h9'0"7d31'
Richardson,Paula Ann
Peg 22h19'56"21d37'
Richardson,Ralph & Margaret
Lyr 18h45'51"41d22'
Richardson,Robert Allen
Lac 22h44'1"38d4'
Richardson,Rose
Del 20h20'21"9d32'
Richardson,Ruben Bru
Aql 19h3'46"-0d26'
Richardson,Sasha
Eri 3h27'35"-3d24'
Richardson,Scott K
Lac 22h9'58"50d9'
Richardson,Sheryl Lynn
And 1h59'14"37d40'
Richardson,Sophie Koken
Cas 23h1'28"58d11'
Richardson,Susan Bird
Lyr 18h41'22"28d20'
Richardson,Teresa Eden
Cas 2h2'0"58d21'
Richardson,Thomas Louis
Lyn 9h14'0"40d7'
Richardson,Tony
Aql 20h5'56"6d18'
Richardson,Victoria A
Lyr 18h46'42"39d9'
Richardson,Will
Oph 18h3'48"11d4'
Richardson,Zachary Duke
Boo 14h57'1"40d23'
Richcreek
Her 17h33'0"28d30'
Riche,Ryan
Aql 19h58'1"14d34'
Richei,Christine
Psc 22h49'1"5d21'
Richele
Ori 6h3'0"20d37'
Richelen Ridge
Leo 11h29'12"10d0'
Richelieu,Alysia*Linda *Jessica*Paul*
Uma 11h12'30"47d44'
Richelle
Lyn 7h24'14"44d46'
Richelle
Mon 6h53'34"0d20'
Richer,Linda Ramage Pierre
Peg 22h59'11"31d13'
Richert,Catharine Irene
Cet 3h4'16"2d40'

Richert,Hedyanne
Lyn 7h41'48"52d9'
Richert,Jane
Mon 8h0'53"-8d47'
Riches,James Alexander Embley
Equ 21h10'0"10d48'
Riches,Luke & Susan Rush 1989
Cyg 20h17'14"39d28'
Richet,Murielle
Eri 3h41'0"-6d30'
Richey
Her 17h13'0"28d38'
Richey's Wish
Peg 0h3'20"11d46'
Richey,Cherie Pride
Com 13h16'31"22d9'
Richey,Dylan Michael
Lac 22h15'38"51d40'
Richey,Elizabeth
Peg 22h26'22"31d0'
Richey,Kaleigh Alexa
And 0h41'30"41d3'
Richey,Suzanne
Peg 21h29'0"20d25'
Richey,Verona Gwendolyn Starley
Peg 21h26'44"23d48'
Richichi,Salvatore
Lyn 7h55'12"58d12'
Richie
Cyg 19h20'0"54d53'
Richie & Joey
Dra 16h52'48"61d59'
Richie 810
Lac 22h55'19"53d57'
Richie G
Aur 4h59'45"38d52'
Richie's Sailing Star
Del 20h29'60"20d0'
Richie's Star
Her 17h43'0"40d12'
Richie,Dawn Annette
Ori 5h25'30"-6d39'
Richie,Steven Linley
Ori 4h55'11"1d17'
Richison,Kathryn
Aur 4h57'1"30d58'
Richley,Maria Anne
Vul 19h35'53"27d28'
Richlu
Aql 19h31'50"11d23'
Richman,Barbara
Sct 18h43'14"-7d57'
Richman,Matthew Scott
Boo 15h25'28"47d49'
Richmand,Jesse Alexander
Aur 5h5'26"44d38'
Richmar
Ori 5h30'33"8d13'
Richmond,Alison Rachel
Sco 16h39'48"-37d45'
Richmond,Carol
Lyr 19h0'22"25d29'
Richmond,Charles R
Dra 19h4'44"50d23'
Richmond,Dr Harold Wayne
Ori 6h3'39"0d15'
Richmond,Elizabeth Ann
And 0h5'44"34d22'
Richmond,Jessica Lauren
Lyr 18h41'46"42d5'
Richmond,Matthew Frank
Sco 17h23'1"-38d5'
Richmond,Robert William
Peg 22h20'36"34d43'
Richmond,Samantha Marie
Cam 4h11'0"61d13'
Richo,Anthony & Catherine
Uma 11h39'42"31d55'
Richter
Lyr 19h17'26"28d49'
Richter,Dorothy Wille
And 2h18'58"48d18'

Richter,Howard
Dra 19h41'36"67d51'
Richter,Irmgard
Mon 6h17'32"-6d26'
Richter,Joyce & Robert
Cyg 21h13'42"38d30'
Richter,Justine Louise
Aur 5h8'40"38d56'
Richter,Karl-Hienz
Aql 20h9'27"0d7'
Richter,Klaus
Aur 6h30'0"38d59'
Richter,Käthe
Ser 15h40'48"4d34'
Richter,Louis
Aur 4h37'15"30d44'
Richter,Re'Han Cook
Lyn 7h54'46"58d12'
Richter,Todd Anthony
Aql 19h30'33"13d27'
Richter,Udo
Uma 9h15'14"45d9'
Richter,Yvonne Brigitte
Cap 20h27'28"-26d33'
Richter-1967,Rebecca Anne
And 0h2'1"47d25'
Richter-Kemper,Susanne
Vir 13h14'30"-6d3'
Richtol,Nancy A
Equ 21h21'35"12d34'
Richwine,Christopher Joseph
Cep 22h19'39"58d56'
Richwine,Jeffrey Todd
Cep 22h55'18"70d15'
Richwine,Jerry R
Cep 22h2'54"53d38'
Richwood,Steven James
Her 17h32'15"50d26'
RichYon94
And 2h23'18"43d4'
RICIGCHOOSK,The Pals- For-Life-Star
Crb 16h12'11"32d27'
Ricimba
Cnv 13h14'58"48d47'
Rick
Cnv 12h5'0"34d27'
Rick
Cyg 20h26'46"40d39'
Rick
Aur 7h25'46"38d2'
Rick
Ser 15h14'11"20d43'
Rick
Ori 6h2'47"0d21'
Rick & April
Crb 15h53'22"26d17'
Rick & Betty
Eri 4h0'0"-19d11'
Rick & Courtney
Tau 4h13'49"20d11'
Rick & Diane
Eri 2h56'1"-11d19'
Rick & Els
Cep 1h29'28"85d52'
Rick & Jamee
Crb 16h9'39"31d26'
Rick & Kathy 06-20-70
Aql 19h52'40"15d24'
Rick & Krista's Wedding Star
Sex 10h24'17"0d45'
Rick & Merritt
Lyr 18h51'49"31d16'
Rick & Pam
Uma 9h24'27"56d27'
Rick & Sandra
Lyn 7h46'39"48d52'
Rick Center's Kids of PL 94
Cam 11h57'60"78d57'
Rick Jason
Ori 6h1'54"10d56'

Rick Loves Audrey
Tri 2h0'1"32d19'
Rick My Love I Give You A Star
Cep 21h27'23"61d15'
Rick Parfitt *Ramp*
Lib 15h15'18"-28d46'
Rick Star,The
Gem 8h4'0"28d21'
Rick's Dream
Cnv 12h54'56"33d22'
Rick's Pick
Cnv 12h57'43"48d47'
Rick's Place
Ori 5h54'33"8d46'
Rick's Recovery
Her 17h36'51"27d57'
Rick's Right
Hya 9h25'12"-6d50'
Rick's World
Oph 18h41'0"8d60'
Rick,Brandon Logan
Aur 6h26'1"32d39'
Rick,Happy Anniversary Love,Tresh
Aur 6h18'35"38d11'
Rick,My Heavenly Body
Per 4h6'0"50d26'
Rick-A-Roo
Aql 19h2'47"5d38'
Rick-My Light,My Life!
Ori 6h20'15"16d32'A
Rickard"Wedding Star", Tony & Traci
Ori 5h7'32"-1d29'
Rickard's Sparkler
Ori 6h15'31"8d29'
Rickard,Billie
Peg 22h50'55"29d28'
Rickard,Mabel Ozella
Cnv 13h23'47"30d51'
Rickard,Ronald & Sandy
Dra 17h0'32"65d16'
Rickard,Sarah
And 0h27'55"30d1'
Rickauer,Franz
Sge 19h36'59"16d43'
Rickenberg,Kevin James
Uma 8h34'49"67d53'
Rickens,Jr,Albert J
Dra 15h8'45"55d13'
Ricker,Charlotte Andrea
Cma 6h59'45"-18d23'
Ricker,David Joseph
Aur 6h17'16"37d8'
Ricker,J William & Mary
Eri 3h3'35"-12d5'
Ricker,Michael Anthony
Lac 22h14'33"51d51'
Ricker,Nicole
And 2h12'14"50d9'
Ricker,Stephen F
Her 17h30'48"31d1'
Ricker,Wallace F
Her 17h32'27"30d9'
Rickerby,Matthew
Boo 15h24'0"52d36'
Rickerd,Maxine
Peg 23h0'15"33d17'
Rickert,Andre
Lyr 19h13'33"31d19'
Rickert,Manfred
Cas 2h59'47"60d30'
Rickert,Sandy
Del 20h13'25"11d35'
Ricketson,Tommy James
Cep 22h11'55"55d44'
Rickett,Barry
Boo 14h14'38"16d6'
Ricketts,David Charles
Dra 17h54'36"61d3'
Ricketts,David Martin
Cet 1h48'57"-0d44'

Ricketts,Edward William
Aql 18h58'43"17d1'
Ricketts,Erin Katrina
Lyn 7h27'60"41d6'
Rickie's Love To My Angels
Cnv 13h39'12"30d15'
Ricklin,Abigail Bailey
Cas 23h30'57"59d44'
Rickman,Chris
Per 4h2'17"31d44'
Rickman,Michael E
Her 17h6'14"48d52'
Rickmond,Michelle Josette
Peg 22h20'0"24d21'
Rickomoosius Michael
Uma 9h30'59"52d41'
Ricks,Alta Mere
Her 16h16'33"24d13'
Ricks,Miranda
Cas 2h49'57"71d1'
Ricks,Robert R
Aql 18h57'31"-5d40'
Rickson,Winifred Alice
Ori 6h6'16"0d23'
Rickster,The
Her 15h47'11"45d38'
Rickwa,Warren
Mon 7h2'11"4d48'
Ricky
Cep 0h2'14"68d15'
Ricky
Umi 14h54'18"68d27'
Ricky
Her 18h19'0"14d50'
Ricky's Heart
Cet 0h52'40"-6d58'
Ricky's Peace
Cet 3h10'36"5d42'
Ricky, My Heart is Yours Forever
Her 16h50'12"39d45'
Ricky,"Tougher Than The Rest"
Her 16h4'43"48d32'
Ricky-My Love
Ori 6h7'19"1d14'
Rickyanto,Frans
Cep 22h37'34"70d37"
Rico
Lyn 7h24'14"44d46'
Rico
Mon 6h18'52"8d57'
Ricpen Dawley
Uma 10h42'26"60d20'
Ricquart,Vincent
Peg 23h5'25"21d8'
Ridd,Ecco Promotions- Jan E
Cas 1h47'14"61d5'
Riddar,Margit
And 1h15'1"40d32'
Riddell,Anthony Edward
Cep 20h51'20"65d10'
Riddell,Carol
Vul 19h34'36"24d19'
Riddell,Catherine
Aur 6h30'52"36d45'
Riddell,Jimmy
Ser 15h34'33"14d7'
Riddering,Sherryl Ann
Cyg 19h45'52"30d16'
Riddle,Amber Mackenzie
Cas 0h38'40"73d31'
Riddle,David William
Per 4h43'11"50d11'
Riddle,Jr,JN,D/C, George A
Sex 10h34'1"-0d2'
Riddle,Katherine Keller
Cap 21h53'24"-18d34'
Riddle,Leah Nora Bunch
Mon 6h42'1"6d10'
Riddle,Lynn
Ant 10h39'54"-39d6'
Riddle,MD,John Marion
Boo 15h7'41"16d51'

Riddle,Scott
Cmi 7h22'50"8d29'
Riddle,Terria
Mon 6h36'39"-1d38'
Ridenour,Brenda W
Lyn 9h12'58"34d40'
Ridenour,Cynthia Kay
Lyn 7h38'14"50d8'
Ridenour,Jr,Robert "Robbie"
Aur 6h24'57"31d28'
Ridenour,Julie A
And 23h39'56"39d41'
Ridenour,Marjorie Mitchell
Peg 21h56'45"2d35'
Rider II,Charles
Aur 5h42'40"50d15'
Rider Star,The
Peg 23h4'58"30d23'
Rider,A J
Aur 4h45'1"31d38'
Rider,Josh
Ser 15h21'0"6d58'
Rider,Julia J
Cam 3h25'14"61d1'
Rider,Michael Barton
Her 18h53'25"18d56'
Rider,Michelle
Cas 1h37'58"61d18'
Rider,Miles Irving
Peg 23h2'56"32d12'
Rider,Ruth
Cas 1h30'27"65d10'
RIDGE RANCH STARS OF 1994
Cyg 20h0'38"30d26'
Ridge,Christine
Lyn 6h12'50"54d27'
Ridge,CLU,ChFC,Scott
Her 17h22'40"41d6'
Ridge,Monica
Cyg 19h40'28"37d55'
Ridge,Regina
Peg 23h25'57"30d8'
Ridge,Russell
Ori 5h53'49"18d44'
Ridgely,Lilyanne W
Mon 7h43'58"-1d50'
Ridgeway,Marsha L
Lyr 19h17'16"42d36'
Ridgeway,Nancy
Vir 11h46'56"9d7'
Ridgio,Joe
Dra 16h41'12"72d16'
Ridgley,Grant & Margie
Sge 20h16'42"17d53'
Ridgley,Robert L
Cet 1h29'31"-2d38'
Ridgway,Eric Stein
Cnv 13h35'20"45d54'
Ridgway,Kenneth
Uma 9h10'35"55d32'
Ridgway,Peter Terence
Cep 22h2'24"60d12'
Ridilla,Brianna Rose
Cas 1h8'24"61d6'
Riding Kazuyoshi & Yukihide & Hideyuki
Ari 2h19'23"27d16'
Riding,Alex
Cnv 12h25'58"33d47'
Riding,Simon Louis
Ori 5h43'51"10d14'
Ridings,Dory Erenwert
Cyg 20h23'0"41d42'
Ridings,Jeanne P Michitson
Cas 1h8'54"61d24'
Ridings,Robert Lewis
Aql 19h47'31"14d47'
Ridley's Star
Aqr 23h2'41"-11d14'
Ridley,Benny
Aur 6h8'27"30d6'
Ridley,Christine Maria Theresa
Cyg 21h48'52"52d39'

Ridley,Eric
Cyg 21h0'38"40d56'
Ridley,Nissa Delanie
Cas 2h53'37"60d7'
Ridley,Peter
Per 1h55'49"47d59'
Ridolfi,Kerry Anne
Crv 12h52'25"-12d26'
Ridout
Her 17h8'28"47d50'
Ridout,Angela & Tim
Lyn 7h30'55"48d42'
Rie & Noriko
Sgr 19h6'29"-15d38'
Riebe,Jimmie C
Aql 20h30'39"-0d3'
Riebling,Jeannie Dianne Druck
And 2h24'13"49d54'
Riebman,Avery Bennett
Dra 10h56'52"77d33'
Riebman,Avery Bennett
Aql 19h0'1"-0d19'
Riebman,Jerome Brian
Hya 9h2'43"2d25'
Riebock,Joan Amy
And 1h4'37"40d2'
Rieboldt,Susan Kathryn
And 1h31'42"36d26'
Rieck,Michele
Eri 4h2'0"-10d46'
Riecke,Travis McCoy
Eri 4h11'57"-14d8'
Rieckhof,Elke
Aqr 23h33'1"-19d0'
Ried,Ronald Francis
Psc 1h43'39"28d3'
Riedel,Clark Ward
Sex 10h49'26"4d11'
Riedel,Helga
Uma 8h31'19"48d54'
Riedel,Jim & Lynda
Uma 9h27'1"57d38'
Riedel,Melinda Marie
Cyg 21h35'32"40d33'
Rieder,Jay Charles Joseph
Gem 6h44'1"31d27'
Riedweg,Jacques
Ori 5h4'60"10d38'
Riefe,Karen Jeanette
Cas 2h59'1"70d24'
Riegel Family Star
Cyg 19h59'37"38d56'
Riegel,Charles J
Lac 22h41'30"54d32'
Riegel,Karl
Peg 23h34'41"16d32'
Riegel,Toni Rich & Krystal
Cyg 20h42'57"38d50'
Rieger,Alison & Gerald
Cyg 21h0'55"35d56'
Rieger,Wendy Loreen
Lyn 7h44'49"50d42'
Riegler,MD,Christopher FX
Oph 17h3'22"11d30'
Riehl,Jr,John
Sex 9h59'38"-1d35'
Riehle,Robert Allan
Cep 22h15'32"55d50'
Riekels 1
Oph 17h52'49"10d55'
Riekstins,David & Terri
Aur 4h58'0"34d23'
Riekstins,Valdis Janis
Cam 5h48'12"67d31'
Rieling,Edward
Ser 15h36'40"20d25'
Rielley,Jacqueline
Crb 16h18'22"37d42'
Rielly,Maureen F
Per 3h22'23"40d29'
Riemann,James Michael
Cet 3h8'48"0d47'
Riemenschneider,Klaus
Leo 9h19'20"17d59'

Riemer,Pauline
Hya 8h57'25"2d22'
Riemersma,Matthew Kirk
Aql 19h55'57"14d55'
Riener,John & Bernie
Ori 5h50'0"21d8'
Riensche,Marjorie E
And 23h1'34"37d24'
Riensche,Paul T
Dra 17h40'47"67d35'
Rienton,Laura
Lyr 18h17'0"40d18'
Riepe,Mervin M
Aql 19h55'26"13d15'
Riepen: Die beste Antwort auf Durst
Uma 13h46'1"50d57'
Riepl 12-14,John Andrew
Aur 6h29'16"37d54'
Riera,Gabriel Luis Otheguy
Eri 2h44'42"-1d51'
Riera,Nicolás Andrés Otheguy
Ser 17h34'27"-13d41'
Ries,Michael Charles
Cap 20h22'59"-13d22'
Riesbeck,Eva
Umi 14h46'31"78d31'
Riese,Horst
Aql 20h9'47"0d6'
Rieselman,Nicole
Cas 0h29'47"64d35'
Riesen,Dean & Bambi
Ori 5h50'1"21d52'
Rieser,David
Her 18h4'49"38d53'
Riess,Berg-Ing Helmut
Ori 5h26'49"1d49'
Rietschel,Baerbel
Oph 17h29'38"1d26'
Rietveld III,William
Lac 22h52'24"54d28'
Rietveld,Jr,William
Lac 21h59'0"36d36'
Rietveld,Kelly Suzanne
Sge 19h25'0"16d17'
Rietveld,Susan Ann
Lyn 8h42'1"42d49'
Rietz,David William
Her 17h49'24"18d50'
Rieu-Sicart,Mary Dorothy
Cyg 19h57'0"30d48'
Rieux,Olivier
Mon 7h57'27"-4d10'
Rifati,Benjamin Bloom
Equ 21h3'54"3d13'
Rife,Jacqueline
Lyn 7h14'1"59d48'
Riffe,James
Cep 22h50'48"70d53'
Riffendilly-C
Mon 6h34'54"0d27'
Riffle,Eric James
Cyg 21h5'51"36d9'
Riffle,Rogor
Uma 8h54'33"49d30'
Rifka Abraham
Ari 2h1'13"21d30'
Rifkah,Miriam
And 1h8'0"41d2'
Rifkin,Jennifer
Peg 23h5'55"10d21'
Rifkin,Judith
Lyr 19h21'49"37d43'
Rifkin,Sarah
Vul 20h22'14"28d13'
Rifkind,Ruth
Cnv 12h18'13"46d43'
Rigal,Michel
Dra 15h37'51"61d28'
Rigalega,Star of Richard
Aur 6h2'1"38d2'
Rigato,Kumudini
And 23h21'48"52d32'
Rigatti-Sciarrino, Rhea N
Lyr 18h51'17"30d59'

Rigazzi,Sibyl Emerson
Sct 18h41'56"-6d0'
Rigby,Aaron
Peg 23h18'37"32d18'
Rigby,Andrew David
Uma 10h0'16"58d34'
Rigby,Dr W K
Vul 19h41'48"23d33'
Rigby,Kelly Ann
Uma 11h41'23"32d6'
Rigby,Michael
Ori 5h50'44"19d20'
Rigden,Olivia
Lyn 9h3'24"33d51'
Rigdon My Rising Star, Larry W
Sgr 18h5'16"-28d10'
Rigdon,Chuck R & Susan L
Hya 9h52'36"-15d57'AB
Rigel
Tel 20h10'21"-46d42'
Rigg,Francis
Leo 10h52'38"-2d19'
Rigg,Kim DeHaven
Lmi 9h53'0"38d44'
Rigg,Tyler Linfert
Aur 6h34'45"35d34'
Rigger's Lantern
Aql 19h7'26"0d49'
Riggert,Debbie
Eri 4h42'18"-5d38'
Riggi,Margaret Cecere
And 0h15'36"33d46'
Riggie,Father Kloman Francis Xavier
Her 18h3'31"31d4'
Riggins,Tracy
And 0h50'26"34d48'
Riggio
Cet 0h40'45"-4d39'
Riggio,Kara Ann
Vir 13h7'28"-20d53'
Riggio,MD,David W
Aql 20h11'37"1d0'
Riggio,Richard
Her 17h56'27"28d20'
Riggle,Ellen Jean
Aql 19h4'26"-0d4'
Riggs,Amanda Jayne
Leo 11h8'1"23d26'
Riggs,Charles Rogers
Aur 5h56'51"30d7'
Riggs,Derrick Michael
Vul 20h0'16"25d7'
Riggs,Elaine Caroline Der Jones Und
Cam 9h8'22"81d7'
Riggs,Gladys
And 1h48'31"40d34'
Riggs,James C
Aql 19h53'37"10d56'
Riggs,Linda
And 23h36'13"48d58'
Riggs,Philip
Per 1h55'1"53d57'
Riggs,Ronald & Jacki
Ori 5h34'8"-3d36'
Riggs-Shining Star, Jennifer
Crb 15h27'11"30d56'
Riggsbee,Todd Allison
Her 16h45'45"21d39'
Riggsus Major (Constellation EmCare)
Uma 8h57'40"48d29'
Righetti,Roberto et Marina
Lmi 10h56'14"31d34'
Rightnour,Jr,George E
Cep 1h37'15"87d39'
Rightway Ad Specialties
Uma 11h50'15"32d24'
Rigly,Stephanie Renee
Gem 7h9'16"21d50'
Rigney,Clare Rose
Peg 22h24'42"30d50'

Rigney,Joseph Robert
Dra 11h33'57"71d42'
Rigney,Laurel
Vul 19h23'22"26d13'
Rigney,Ryan Judson
Lac 22h4'1"49d24'
Rigney,Stephanie Nicole
Mon 6h23'12"8d49'
Rignola,Janet
Cam 3h50'29"58d19'
Rigo D
Lmi 9h52'46"33d48'
Rigo-Porretta
Mon 7h12'1"-10d30'
Rigola,Cristina
Scl 23h22'40"-27d28'
Rigoni,Christopher F
Aql 18h56'27"15d49'
Rigori,Pelliccerie
Her 16h12'1"10d10'
Rigoulot,Kenneth G
Per 4h4'31"50d33'
Rigsby
Cam 13h14'0"80d16'
Rigsby,John & Janice
Boo 15h6'20"18d54'
Riha,Emily Rose
And 23h25'1"45d18'
Rihoy,Paul & Michael
Umi 13h56'11"78d37'
Riisness,Catherine F
Psc 1h28'35"20d54'
Rik
Cep 23h7'43"80d36'
Riker,Cassandra Ann
And 23h17'18"45d9'
Riker,Francesca Lauren
Tau 4h32'12"30d31'
Riker,Jessica Anne
Mon 6h19'57"8d51'
Riker,Jr,Jimmy
Boo 14h22'37"37d47'
Riker,Ruth Evelyn
Com 12h56'38"27d21'
Riker,William G
Her 16h37'21"28d29'
Rikki
Uma 9h55'1"49d42'
Riklis,Ira D
Aur 7h13'52"37d6'
Riklis,Meshulam
Cet 2h9'26"5d41'
Rilea,Bill
Del 20h17'31"14d44'
Rilea,Sonja
Del 20h17'50"14d50'
Riles,Mary E
Tri 1h46'51"28d32'
Riley
Cet 3h6'35"4d0'
Riley "Buck",T A
Lac 22h41'0"50d2'
Riley Celestial Mother,Hattie C
Cas 0h13'0"63d36'
Riley's Rising Star Of Joy
Vul 19h47'48"25d26'
Riley's Star
Dra 19h4'15"60d3'
Riley, My True Love, Dave
Ori 5h56'41"8d55'
Riley,Alice Mary
Peg 23h26'39"23d32'
Riley,Amanda Elisabeth
Cam 12h55'51"77d3'
Riley,Anthony
Peg 22h16'32"30d23'
Riley,Betty Darlene
Gru 22h59'10"-40d1'
Riley,Betty Jean
Vul 20h14'14"22d39'
Riley,Bob
Boo 15h7'54"52d17'
Riley,Britt Elizabeth
And 2h0'52"42d19'

Riley,Bruce Christopher
Aur 6h28'19"33d6'
Riley,Christopher James
Cnv 12h47'38"50d9'
Riley,Claudia
Cas 1h20'45"70d30'
Riley,Donna Lynn
Hya 10h17'48"-11d16'
Riley,Eleanor M
Lac 22h24'45"37d34'
Riley,Erica
Mon 6h22'46"6d49'
Riley,Gary"Joe Bob"
Mon 6h23'21"8d5'
Riley,Grant P
Boo 15h5'1"20d39'
Riley,J D
Uma 11h2'10"46d4'
Riley,Jim
Per 3h11'46"50d22'
Riley,Jr,Joseph J
Lac 22h2'30"38d7'
Riley,Jr,William L
Aur 5h4'15"41d36'
Riley,Judy Sharon
Aql 19h16'45"15d11'
Riley,Keep on Truckin' Walter
Cap 21h39'59"-22d42'
Riley,Kevin
Dra 16h33'1"68d19'
Riley,Kitty
Gem 7h0'38"31d35'
Riley,Maria Kimberly
Lyr 19h0'0"38d57'
Riley,Mark
Ser 15h51'16"18d51'
Riley,Megan
And 2h30'41"40d45'
Riley,Michael
Per 2h37'11"40d9'
Riley,Michael
Lac 22h41'55"53d15'
Riley,Open Door School,Sue
Spayth
Peg 22h17'23"2d21'
Riley,Parker
Cet 0h57'0"-0d53'
Riley,Patrick Bryan
Aql 19h23'32"15d25'
Riley,Patrick Joseph
Cnc 9h13'19"31d32'
Riley,Sandra Jean
And 23h22'57"48d49'
Riley,Savannah
Peg 22h32'0"21d51'
Riley,Shelby Rae
And 23h22'19"48d21'
Riley,Taylor Gabrielle
And 23h23'27"48d46'
Riley,Walter
Aur 4h49'27"40d19'
Riley,Walter A
Her 16h42'25"35d29'
Riley-Mazzo
Her 18h6'1"38d25'
Riley-Nuell,Cynthia
Gem 7h1'12"30d39'
Riley-Sorem,Alexandra Marie
Lyr 19h5'0"26d28'
Riley-White Light
Forever,Allen J
Her 17h1'18"13d7'
Riley:Carolyn's Epitomé Of
Love,RH
Hya 8h56'18"-11d14'
Rill,Ethan Manatis
Peg 21h27'1"3d38'
Rilling,Uwe
Umi 13h50'12"73d46'
Rilling-Gale
Dra 18h18'14"70d50'
Rilovich,Nicholas Antone
Peg 21h50'1"28d32'
Rilu
Vir 13h56'1"1d20'

Rim,The Star of John
Per 2h8'21"56d25'
Rim,The Star of Julie
And 0h3'36"43d34'
Rim,The Star of Mi
Cam 7h24'37"67d46'
Rima,Haley Amanda
Eri 3h10'1"-5d18'
Rimbach,Arnold "Bud"
Cnv 12h26'18"37d32'
Rimbey,Delwin H
Aur 4h56'38"40d31'
Rimbey,Glenn & Gladys
Aur 4h56'43"40d1'
Rimdzius,Stan & Kathy
Cnv 12h6'0"40d45'
Rimel
Cyg 20h24'37"39d15'
Rimes,Dennis John
Ori 6h2'1"1d5'
Rimgaudas
Oph 17h5'34"1d4'
Rimiheb
Cam 3h51'55"61d35'
Rimkus,Elizabeth Anne
Cyg 20h24'0"40d46'
Rimkus,Jeffrey Alston
Cet 2h29'27"4d13'
Rimler September 18
1995,Sofia Klein
Mon 6h43'51"10d16'
Rimmer,"Nonnie" Pat
Cam 6h4'37"80d30'
Rimmer,Jenny
Cyg 21h56'20"53d24'
Rimmer,Lynsey
And 23h39'48"32d48'
Rimmmer,David William
Aur 6h27'43"34d6'
Rimolt,Pap Pap & Grammy
Lac 22h27'33"53d48'
Rina
Lyr 18h57'0"45d39'
Rinaldi,Happy Birthday Peg
Harty
Lyn 8h12'52"37d54'
Rinaldi,Jr,Joseph Nicola
Ari 2h31'47"20d29'
Rinaldi,Matthew Edward
Cnv 12h47'18"51d13'
Rinaldi,Paul W
Per 2h55'49"32d5'
Rinaldi,Ruggieri Michele
Scl 0h8'43"-25d6'
Rinaldo Tranquillo
Ori 6h8'49"2d12'
Rinaolo,Bob & Mary Jo
Cyg 19h30'59"34d38'
Rinas,Marie-Kristin
Leo 10h59'22"13d0'
Rinck,Little Miss Lyn
Cas 3h6'26"61d29'
Rincoo
Uma 9h54'45"47d57'
Rinder,April
Tri 1h29'0"34d50'
Rinder,George & Shirley
Mon 6h18'53"8d11'
Rinderer,Jayne M
Mon 6h39'56"6d2'
Rinderknecht,Keith
Cmi 7h23'19"8d18'
Rinderknecht,Lester (PaPa)
Eri 4h14'50"-17d24'
Rindner,Daniel Mark
Cnv 12h35'0"33d18'
Rindner,Dorothy
Com 13h24'21"26d25'
Rindorff,August
Lib 14h30'33"-20d35'
Rinehart's Revelations
Peg 21h47'57"28d42'
Rinehart,Bethany Stallter
Peg 23h43'56"31d37'

Rinehart,Deputy Dan
Aql 18h59'18"-5d36'
Rinehart,Donna Jan Poulsen
Ori 6h6'1"10d48'
Rinehart,Karol
Crb 15h52'12"30d41'
Rinehart,Ralph Earl
Aur 4h49'12"50d34'
Rinehart,Raymond Henry
Del 20h16'41"10d46'
Rinehart,Susan
And 0h4'47"44d30'
Riner,Gwen Alaine
Cas 2h9'12"70d23'
Ring Of Stars
Ari 1h45'58"23d16'
Ring,Andy
Cet 1h18'54"-6d18'
Ring,Donna
Lyr 18h17'26"38d47'
Ring,Gertrude
Oph 17h55'29"12d6'
Ring,Paul J
Vul 19h4'0"24d54'
Ring,Rachel
Cas 1h1'47"53d43'
Ringe,Bob
Sex 10h35'46"2d9'
Ringel,Daniel Scott
Aur 4h48'50"40d17'
Ringen The Best, Trenton
Aql 18h53'41"10d26'
Ringer,Corey H
Cnv 13h18'1"28d6'
Ringer,Erika Valentine
Aql 19h53'31"1d46'
Ringer,Keith J
Vir 12h37'0"0d50'
Ringes,Brigitte und Hans-
Peter
Cap 20h58'54"-26d49'
Ringler,Ann
And 23h46'56"42d59'
Ringler,Betty
Aql 19h0'46"16d20'
Ringo
Cmi 8h3'28"0d3'
Ringo
Mon 7h3'0"3d47'
Ringo Starr,The
Her 16h54'53"26d8'
Ringo,Haley Rynn
Mon 7h47'22"-3d25'
Ringo,Justin Stacy
Her 18h6'43"47d26'
Ringquist,Neil A
Aur 6h26'30"31d47'
Ringston,Jo-Ann
Cas 0h45'47"67d33'
Ringö,Anna
Uma 10h13'0"52d18'
Rini,Vivian
Eri 3h42'13"-1d20'
Rinieri,Marie Claire
Aql 19h59'40"14d3'
Rinke,Annette
Gem 8h1'33"30d1'
Rinkel,Deborah Moonstar
Lyr 19h3'15"41d0'
Rinnus,Helga Alma
Tau 3h40'55"2d4'
Rinsma,Dann
Ori 5h55'44"14d45'
Rinzel,Mitchell Steven
Cet 3h18'18"7d55'
Rinösel,Edeltraut
Leo 10h54'27"20d10'
Rio
Psc 1h22'9"11d1'
Rio
Dra 20h18'31"64d23'
Rio
Lyr 18h32'59"42d29'
Rio
Peg 21h59'1"22d35'

Rio' Ney Moon
Lyr 19h21'20"30d51'
Riocreux,Audrey
Lmi 10h56'13"26d11'
Riordan,Cleo Owen
Uma 10h54'39"38d60'
Riordan,Debbie
Lyn 9h1'25"44d48'
Riordan,Dr Benjamin Raphael
Aur 5h9'29"40d39'
Riordan,Michael Sacco
Dra 14h9'17"63d56'
Riordan,Peggy
Cas 0h22'10"64d35'
Riordan,Richard Louis
Dra 19h20'52"56d50'
Rios,Alicia Hope
Oph 17h57'21"0d12'
Rios,Ascary
Lac 22h36'0"40d46'
Rios,Brittany & Ashley
Mon 6h22'22"1d51'
Rios,Carlos M
Ori 4h55'42"4d37'
Rios,Emilio
Aql 19h6'16"15d52'
Rios,Hilda Valentine
Mon 6h22'0"5d24'
Rios,Joe R
Boo 15h1'59"19d3'
Rios,Justin
Ori 5h46'0"20d24'
Rios,Marcos
Cet 1h39'0"-6d22'
Rios,Miranda Mariah Juanita
Lyr 18h39'34"27d31'
Rios,Monique M
Aql 19h56'0"13d44'
Rios,Nellie
Peg 22h0'1"30d41'
Rios,Reggie
Per 3h11'0"41d3'
Rios,Roberta
Cas 1h56'51"67d45'
Rios,Sarah Elizabeth
And 0h28'48"43d57'
Rios,Wendy Lisa
Lib 15h37'19"-5d41'
Rios,Zenia
Crb 15h26'16"31d1'
Riot,Lidia
Cyg 20h9'1"40d43'
Riou,Jacques
Aur 4h49'37"40d57'
Riou,Yvon
Tri 2h23'53"33d13'
Rioux,Christian
Aqr 21h41'51"0d39'
Rioux,Claude
Del 20h58'30"14d19'
Rioux,Philippe
Uma 9h10'49"60d36'
Rioux-Lévesque, Isabelle
Uma 9h11'58"67d59'
Rip & Barbie
Cam 4h4'51"58d30'
Rip Chords-Del Arnie Phil
Lac 22h32'1"56d17'
Ripa
Boo 14h40'10"39d11'
Ripandelli,Carol Janet
Del 20h27'54"11d4'
Ripley,Caitlin Clare
Cas 0h33'22"73d48'
Ripley,Geraldene
Ori 5h44'55"1d33'
Ripley,Julie
And 2h11'51"39d37'
Ripley,Linda
Eri 2h58'49"-5d6'
Ripley,Theodore James
Ori 5h43'37"1d46'
Ripley-Vigil,Mitchell Brian
Cep 22h18'0"62d22'

Rippel,Christiane
Sco 17h26'29"-31d13'
Rippel,Marianne Chr
Lyr 19h23'50"41d44'
Ripper,Dorris Byrd
Peg 22h8'0"24d20'
Ripper,The
Aql 20h31'58"0d30'
Ripperger,Ellen Joyce
Psc 1h28'40"10d27'
Rippich,Friedrich Dr
Tau 3h40'20"1d0'
Ripple,Jeffrey A
Cyg 21h33'33"38d46'
Ripple,Juli Michelle
Lyn 8h22'0"43d31'
Ripple,Kevin
Lac 22h43'22"35d35'
Ripple,Laura J
And 0h18'13"33d51'
Ripple,Tom
Cet 3h0'27"0d22'
Rique and Louise
Lac 22h3'23"49d38'
Ris,Amara
Cnv 13h12'0"38d7'
Ris,Bernard Henry
Lac 22h17'41"51d14'
Risa Kanemura
Vir 14h22'47"0d48'
Risbecker,Ann
Lyr 18h31'1"31d4'
Risch,Hans
Cmi 7h17'18"8d28'
Risch,Joshua
Boo 14h12'1"33d56'
Rischall,Leon J
Aur 5h11'59"42d35'
Rischall,Richard Yale
Lmi 9h20'0"37d56'
Risdale,Marlene
Cyg 19h23'21"50d25'
Risdale,Vincent
Cyg 19h22'32"50d2'
Risebrough,Robert
Cep 21h46'22"68d42'
Riser's Fire
Aqr 20h56'37"-1d9'
Rishel,Beverly
Uma 10h28'39"40d51'
Rishon
Eri 3h12'52"-3d57'
Rishty,Jacob M
Her 17h2'45"51d13'
Rising Star
Oph 17h17'42"-21d31'
Rising Star Jena Madelaine
And 2h22'53"42d58'
Rising,Curtis D,Alisa A 1995
Sct 18h54'1"-4d41'
Rising,Shanna G
Cyg 21h38'52"38d28'
Risinger,Jay "the Runt"
Crt 11h42'43"-22d47'
Risk,Queen of the Star Heidi
Peterson
Cas 0h21'13"70d2'
Riskin,Richie
Dra 11h51'0"70d29'
Riskowski,Kathleen
Cas 0h50'38"60d2'
Rislakki,Henrikki
Dra 9h46'34"73d13'
Risley,John G
Lac 22h28'18"56d33'
Risley,The Family
Cam 3h15'0"57d42'
Rispoli's Eternal
Interdependence
Cnv 13h7'25"40d18'
Rispoli,Chet
Aur 5h54'26"30d25'
Rispoli,Gabriel Thomas
Dra 16h5'27"62d23'

Rispoli,Joseph
Lac 22h21'24"53d5'
Riss
Tri 2h14'0"31d15'
Riss,Toni Pallett
Cet 1h47'44"-10d59'
Rissa
Del 20h13'46"14d31'
Rissanen,Tod Randall
Per 3h11'39"45d18'
Risse-Lessel,Birgit
Cnc 8h11'15"31d21'
Risser,Elvia "Cary"
Mon 6h54'38"-0d37'
Rissler,Laura
Lyn 8h0'21"51d44'
Rissman,Tom
Her 18h24'1"20d24'
Rist,Andrew Owen
Aur 6h11'39"31d59'
Rist,Harold
Lac 22h7'13"51d40'
Ristau,Joanne
Lyr 19h7'26"38d43'
Ristau,Peter
Mon 8h4'10"-4d12'
Risteard's Real
Aur 6h4'24"31d36'
Ristorcelli,Jr Brat of the
Cosmos,J Ray
Mon 7h21'48"-5d51'
Ristori,Yany
Uma 13h25'15"61d25'
Ristow,Ruth Secrist
And 1h0'54"34d3'
Rita
Cas 2h13'12"68d48'
Rita
Pyx 8h41'17"-20d55'
Rita
Ori 5h55'27"1d50'
Rita
Vul 20h15'30"25d8'
Rita
Lyr 18h48'12"32d5'
Rita
Psa 22h36'27"-25d35'
Rita
Boo 15h13'32"38d60'
Rita
And 23h39'54"40d24'
Rita
Cas 1h4'53"70d36'
Rita
And 23h30'1"41d14'
Rita & Phil "2 Wishes 2
Dreams"
Lib 14h5'1"-23d35'
Rita 2555
Uma 11h58'35"37d49'
Rita Ann
Lyr 18h43'42"40d45'
Rita Ann
Lyn 7h11'27"58d47'
Rita e Michele
Cas 1h59'43"58d33'
Rita Estrellita "Little Star Rita"
Mon 7h56'47"-3d45'
Rita Gene
Cas 0h28'17"75d47'
Rita Jean
Lmi 10h16'0"31d30'
Rita My Love
Ori 5h50'41"17d34'
Rita The Superstar Mum
Peg 23h0'37"13d10'
Rita Z
Cas 2h35'28"68d34'
Rita's Child
Umi 14h39'19"68d38'
Rita's Love
Mon 6h35'28"0d58'
Rita's Night Light
Oph 18h0'1"13d19'

Rita's Revelation
Del 20h16'29"9d38'
Rita,Adam & Jocelyne
Peg 23h3'20"11d36'
Rita,The
And 2h19'54"48d57'
Rita-Varouj
Lyn 8h55'28"42d52'
Ritacco,Joe
Cnc 9h7'24"30d22'
Ritacco,Papa-John Anthony
Aur 4h50'47"40d22'
Ritaine,Claire
Aur 5h51'18"31d41'
Ritchard,Brian Paul
Lac 22h22'39"50d33'
Ritchay,Susan Beauty
Crt 10h50'12"-11d41'
Ritchey,Katherine
Scl 23h13'31"-33d29'
Ritchie II,Joseph James
Aur 6h9'17"50d33'
Ritchie's Star
Per 1h51'56"53d39'
Ritchie,Allan
Per 3h24'1"54d30'
Ritchie,Charles
Tau 4h3'40"20d10'
Ritchie,Dan
Lac 22h28'39"48d49'
Ritchie,Helen
And 2h27'53"50d28'
Ritchie,James Alexander
Steele
Cyg 21h41'16"31d32'
Ritchie,Jane
Lac 22h3'16"49d13'
Ritchie,Jim
Aur 6h34'13"34d13'
Ritchie,Jim
Ser 15h51'53"21d20'
Ritchie,John G
Cep 22h45'1"70d58'
Ritchie,Josephine "Gran"
Lyr 19h14'26"41d57'
Ritchie,Karen Marie
Cyg 19h27'16"30d11'
Ritchie,Lois Rena
And 2h0'48"46d54'
Ritchie,Merle & Ed
Sge 19h36'4"16d52'
Ritchie,Michael
Aur 5h1'47"45d29'
Ritchie,Micheal Meredith
Hya 8h17'50"2d51'
Ritchie,Nicholas Albert
Sco 17h29'11"-31d40'
Ritchie,Omar
Dra 9h42'0"80d15'
Ritchie,Tiffany
Lyn 7h5'1"44d15'
Ritchie,William
Crb 16h19'48"37d46'
Rito,Kari
Umi 13h11'47"71d2'
Ritsuko
Lyn 7h32'13"51d26'37'
Ritt,Robert James
Ori 6h0'0"-0d12'
Rittenbacher,Karl
Sgr 19h56'49"-44d49'
Rittenhouse,Bernice
Cas 1h18'1"60d44'
Rittenhouse,Jessica Michelle
Peg 22h52'14"20d12'
Ritter III,In Memory of
Kenneth L
Lyr 18h18'43"47d2'
Ritter,Aaron Charles
Lac 22h29'50"53d25'
Ritter,Andrew Roy
Her 18h2'57"50d26'
Ritter,Barbara
Cas 0h28'28"67d7'

Ritter,Col James Alford
Eri 4h6'38"-19d15'
Ritter,Danielle Jeanette
And h58'50"40d51'
Ritter,Danielle Jeanette
Cyg 19h52'33"38d27'
Ritter,Emily
Lac 22h8'55"38d44'
Ritter,Florence
Cas 1h27'36"67d35'
Ritter,Gabriele Maria
Cnc 8h1'21"10d1'
Ritter,Harry C
Cep 3h20'20"80d36'
Ritter,James Damale
Vul 19h40'0"20d16'
Ritter,Jo Anne T
Lyr 18h34'31"36d26'
Ritter,Johannes
Per 3h56'28"39d35'
Ritter,Madge
Peg 21h25'51"22d49'
Ritter,Stefan
Hya 8h46'42"-8d15'
Ritter,Steven Patrick
Aur 6h1'17"35d24'
Ritter,Thomas Justin
Boo 14h38'37"15d37'
Ritter,Uwe
Lyr 19h17'44"30d44'
Ritter,Yvette
Aur 6h19'40"31d59'
Ritterhouse Star,Merle &
Mildred
Aql 20h10'19"4d35'
Rittersbacher,Daniel
Vir 11h53'44"-1d42'
Rittersbacher,Georg
Tau 5h0'53"28d53'
Rittersbacher,Gisela
Cnc 8h29'39"31d39'
Rittersbacher,Kati
Sgr 18h53'38"-27d18'
Rittmanic,Deborah Darling
Mark-Maya
Cam 3h17'1"60d3'
Rittschof,My Dear
Husband,Bill
Ori 5h56'1"9d33'
Rittweger,Jennifer
Uma 12h4'58"31d51'
Rittweger,MD,Edward
Oph 18h42'35"8d54'
Ritvo A Little Star Mom &
Dad,Max
Ser 15h29'34"22d19'
Ritvo Love Shines
Forever,Riva & Ed
Eri 3h3'12" 13d55'
Ritvo Our Star Love Mom &
Dad,Skye
Aql 19h11'57"15d35'
Ritvo Our Star Mom &
Dad,Victoria
Eri 3h17'33"-16d2'
Ritwo-Panzino
Cep 21h3'51"68d52'
Ritz,Captain Ross
Oph 17h31'38"-20d25'
Ritz,Helen K
Cas 1h14'42"60d34'
Ritz,John G
Cep 22h20'59"55d41'
Ritz,Mary Sue
Mon 6h25'24"10d32'
Ritz,Patricia
Equ 21h16'46"7d23'
Ritzman,Dann F
Aql 18h39'12"1d24'
Ritzmann,Paul
Per 3h34'57"38d18'
Ritzow,Gerald
Cep 21h58'53"55d8'
Ritzrow-Julian
Uma 11h32'0"30d44'

Rius,Juan et Rubia
Cam 7h38'17"68d42'
Riva "Tita",Nancy Susana
Ramona Martin
Mon 6h39'1"6d56'
Riva,Gerald
Aur 6h0'59"35d39'
Riva,Serena e Sandro
Lyr 18h39'20"28d12'
Rivard,Nathalie
Umi 16h4'31"79d4'
Rivas II,Ricardo Blas
Cet 3h4'27"2d23'
Rivas,Rossman
Aur 4h49'14"40d50'
Rivas,Victor
Boo 14h50'47"48d57'
Rivenbark,Dana
Mon 6h24'58"3d43'
Rivera,Carmen Dahlia
Uma 9h3'56"68d58'
Rivera,Chita
Com 12h51'51"26d15'
Rivera,Daniel Velez
Cnv 13h30'7"37d4'B
Rivera,Dick & Leslie
Del 20h35'10"3d23'
Rivera,Felix Maldonado
Per 2h51'1"38d47'
Rivera,Frankie A
Aur 6h27'22"35d57'
Rivera,George
Her 17h44'38"48d49'
Rivera,Janet & Nelson
Peg 23h41'17"30d30'
Rivera,Jr,José Antonio
Her 18h8'55"38d56'
Rivera,Jr,Julio Angel
Her 16h59'21"29d37'
Rivera,Julie Ann
And 0h10'55"27d55'
Rivera,Laura Rosa
And 1h47'34"40d44'
Rivera,Maximino
Sex 9h41'53"-11d14'
Rivera,Olaya
Uma 8h33'20"57d8'
Rivera,Olga
Aql 18h57'26"-2d19'
Rivera,Peter & Diana
Dra 18h36'0"70d24'
Rivera,Ralph
Her 17h27'28"26d54'
Rivera,Richard
Mon 6h21'34"-0d20'
Rivera,Rudy Ramone
Lyn 9h11'30"37d48'
Rivera,Sgt Israel
Per 2h39'27"35d19'
Rivera,Squeegie
Lmi 10h57'0"28d5'
Rivera,Tracy
Lac 22h35'42'56d41'
Rivera,Trish
Uma 11h6'48"48d49'
Rivera,Vivian
Lyr 19h12'57"40d23'
Rivera,Yolanda
Boo 15h8'0"19d32'
Rivero,Nestor & Diana
Lyn 9h1'1"46d12'
Rivero,Orlando
Ori 5h49'24"20d24'
Rivers
Oph 16h48'19"10d60'
Rivers Sigh
Cyg 19h55'48"45d7'
Rivers,Darryl Eugene
Cep 23h2'57"64d56'
Rivers,Emma & Julian
Cyg 20h22'54"39d40'
Rivers,Jennifer & Michael
Cet 1h21'9"-12d33'
Rivers,Joan
Uma 11h15'48"33d42'

Rivers,John Morrison Dra 18h33'35"58d56'
Rivers,Joseph Foy Mon 6h58'1"0d27'
Rivers,Melissa Umi 14h56'58"78d53'
Rivers,Pamela Susan Cas 1h40'50"60d60'
Rivers,Richard Equ 21h16'28"2d29'
Rivers,Richie Hya 8h20'19"-10d45'
Rivers,Tina R Peg 22h32'0"20d20'
Riverside Surgery Center Her 17h34'42"21d60'
Rives,Christian Per 3h5'15"38d55'
Rivet,Marcel Ori 4h48'15"5d4'
Rivet,Rachel Claire Lyn 7h27'36"38d33'
Rivett,Diana Dra 16h10'52"60d26'
Rivett,Monte Dra 20h3'19"63d30'
Rivette,Jacob Ryan Cnc 8h32'34"8d29'
Rivette,James Donald Aql 19h9'42"12d47'
Rivette,Scott Gray Cet 0h50'0"-3d34'
Riviello,Bianca Laren Uma 9h27'49"57d1'
Rivier,Patricia Mon 6h27'33"-5d38'
Riviere,Jeny & Laura Cma 6h57'2"-16d25'
Riviere,Pierre Pho 0h42'2"-45d15'
Rivkah Mina And 1h51'0"47d9'
Rivkelegraysonsofih Aiaooh Baca PhD Cnv 12h21'30"43d36'
Rivkin,Marni Uma 11h58'36"38d13'
Rivkin,Miriam Charlotte Cyg 20h10'51"38d59'
Rivkind,Perry Cep 2h32'52"80d21'
Rivoli,Giorgia Cam 14h14'12"80d45'
Rix,Jacqui Hya 8h14'28"-6d3'
RIXTAR Her 16h28'33"48d19'
Riz Ser 16h7'59"14d17'
Rizer Bill Ser 17h31'0"-14d20'
Rizk,Bryan Edward Uma 11h25'36"43d14'
Rizk,Randall Reid Cet 2h31'12"1d16"
Rizor,Bruce A Dra 17h51'22"75d3'
Rizpah Eri 4h34'13"-0d9'
Rizvi,Pashi Lac 22h23'45"50d29'
Rizwan Boo 14h5'57"38d48'
Rizza-Kellogg Boo 14h48'39"30d27'
Rizzi (Musettino), Mauro Leo 10h24'52"10d43'
Rizzi,Andrea Pic 4h50'10"-46d49'
Rizzi,Jr,Angelo C Lac 22h10'12"46d7'
Rizzi,Lia Uma 11h19'41"47d32'
Rizzi,Roberto Per 3h7'20"37d45'
Rizzio,M & J Cyg 20h24'1"38d56'
Rizzo The Wonder of You,Joseph Cnv 12h20'34"50d51'
Rizzo,Al Cet 0h53'19"-1d53'
Rizzo,Albert Dra 20h8'55"62d13'
Rizzo,Anna Aql 20h19'44"0d36'
Rizzo,Anna Sophie Leo 11h30'0"11d5'
Rizzo,Anthony Dra 20h13'48"63d25'
Rizzo,Bruno Cep 22h13'14"61d51'
Rizzo,Dawn Elizabeth And 0h46'0"39d12'
Rizzo,Fallon Cnv 13h50'17"38d39'
Rizzo,Frank P Aql 20h35'11"0d19'
Rizzo,Joe Ori 5h49'11"20d28'
Rizzo,Karina Christine Crb 15h58'38"32d22'
Rizzo,L B Ori 5h15'18"1d52'
Rizzo,Louis S Lac 22h51'42"55d11'
Rizzo,Maria A And 2h17'25"40d20'
Rizzo,Nicole And 0h12'25"47d21'
Rizzo,Paul Her 16h25'39"18d58'
Rizzo,Steven Wilhelm Lmi 9h31'40"38d30'
Rizzo-McClarty Boo 14h12'38"39d19'
Rizzotti,Carol Amalia And 23h38'35"47d28'
Rizzuto,Angelo Ori 6h0'0"8d53'
Rizzuto,Jr,Daniel Bernard Cep 21h0'39"56d7'
RJL #1 Mon 7h50'4"-2d51'
RJM 751 Cnc 8h55'45"22d25'
RJR & Family Boo 14h9'56"40d4'
RLD III Aur 6h36'23"30d13'
RMA 81879 Per 2h54'57"40d32'
Rmago Cas 0h40'35"61d11'
RMS Cam 7h50'23"70d39'
RMS-129 Per 2h57'32"46d18'
RNF5 Ori 5h48'55"11d56'
Ro's Rocket Lac 22h24'1"48d57'
Roach III,CDR John James Aur 6h18'0"33d9'
Roach,Chuck Cyg 20h56'30"39d38'
Roach,Corey Mitchell Cma 7h22'17"-15d56'
Roach,Doris Lee Uma 12h7'0"52d1'
Roach,Edward J Per 3h42'49"39d24'
Roach,Franklin Lyr 19h8'33"45d6'
Roach,Jason Oph 17h27'21"10d50'
Roach,Joseph Lorrigan Per 4h6'29"51d13'
Roach,Katie Lib 15h27'39"-8d40'
Roach,Kevin Fussell Cnc 9h7'55"30d36'
Roach,Larry Daniel Lac 22h3'56"47d46'
Roach,Leia Christine Peg 22h33'0"20d46'
Roach,Lisa Mon 6h59'59"1d32'
Roach,Mary-Jo And 23h34'19"49d38'
Roach,Regina Vul 20h45'12"20d7'
Roach,Star For Susan Mon 6h37'1"7d18'
Roach,William R Cam 4h0'54"61d6'
Roads,Mike Aur 6h16'44"33d15'
Roalf,Chelsea Walsh Aur 6h25'27"32d25'
Roan Star of Khan Lyr 18h56'23"34d39'
Roan,Conor J Gem 6h39'5"12d24'
Roan,Joseph Her 17h35'0"40d27'
Roark,Allison Ori 5h38'24"10d39'
Roark,Cortney Ann And 1h55'0"41d1'
Roark,Lucy Lyn 8h14'47"45d25'
Roark,Rollo Aur 5h11'37"42d34'
Roasa,CLU, Michael L Cep 1h1'1"77d29'
Rob Aql 18h58'36"15d1'
Rob Peg 22h33'51"29d21'
Rob & Amy Uma 9h46'14"48d6'
Rob & Denise - Forever Your Love Cyg 19h58'16"30d2'
Rob & Marie's Kooky Star Per 2h49'22"45d14'
Rob & Mark Star,The Boo 15h14'35"38d5'
Rob & Melissa Aur 6h31'15"48d49'
Rob & Stephanie Cep 0h11'33"66d46'
Rob & Sylvia Cyg 21h13'46"38d42'
Rob N Deb Per 1h47'58"56d52'
Rob's Romance Ori 6h9'44"1d28'
Rob's Shining 27th Birthday Star Her 17h38'40"27d0'
Rob's Star Her 15h58'43"48d38'
Rob's World Boo 15h25'14"38d30'
Rob'Star Per 1h41'41"53d2'
Rob,My "Lucky" Star I Love You Del 20h21'20"3d34'
Robandra Lmi 9h57'53"38d56'
Robards,Elizabeth Lyn 8h4'24"40d13'
Robb & Leighann 2 Very Special People Aql 20h7'57"1d40'
Robb III,Albert James Cmi 7h42'0"5d10'
Robb's Daisy Per 1h46'44"53d48'
Robb,Carla Ser 16h16'31"1d36'
Robb,James J Ser 15h26'12"18d13'
Robb,Jr,James A Lac 22h43'28"53d56'
Robb,Katharine And 23h22'40"41d0'
Robb,Maxine & Norman Peg 23h6'12"11d11'
Robb,Paul William Cnv 12h26'40"32d48'
Robba,Anthony Per 3h59'2"35d36'
Robbe,Alexia Ori 6h16'13"10d44'
Robbert,Emilie "Hatsy" Sco 16h56'18"-38d15'
Robbie Del 20h53'58"3d27'
Robbie Per 2h52'43"31d40'
Robbie & Sabine Sge 20h0'34"17d12'
Robbie Lou,The Ori 5h49'40"11d22'
Robbie's Star Del 20h37'46"3d15'A
Robbie's Star Companion Del 20h37'46"3d15'B
Robbie's Star Uma 11h22'22"42d55'
Robbie,Fred Per 3h26'12"40d52'
Robbie,Tim & Ann Mon 8h1'1"-1d32'
Robbiebobbieboogle marsh Boo 13h54'12"20d10'
Robbieland Uma 11h20'47"71d14'
Robbin's Favorite Sun Cnv 13h56'25"46d32'
Robbins Builders Star, Gene A Hya 9h3'34"4d15'
Robbins Family,The Cnv 12h51'0"40d15'
Robbins,Alexandra Elizabeth Leo 10h58'13"-2d18'
Robbins,Angel Valieda Cam 7h58'40"70d19'
Robbins,Annemarie Conway Aql 20h3'34"1d32'
Robbins,Bibiana Lyn 7h9'22"58d11'
Robbins,David Allen Her 18h14'0"47d56'
Robbins,Dewey Dra 11h1'14"73d43'
Robbins,Douglas L Her 16h47'23"33d59'
Robbins,George Cam 3h54'49"61d12'
Robbins,Ginger Louise Equ 21h6'32"11d24'
Robbins,Herman Oph 17h34'19"-22d38'
Robbins,Jeanne Marie Dowe And 0h16'26"40d43'
Robbins,Jeannie Cas 0h24'55"68d51'
Robbins,Jim Uma 9h7'14"58d5'
Robbins,Jim Per 2h41'37"35d49'
Robbins,Joe B Equ 21h15'43"7d8'
Robbins,Jr,David Lodge Dra 10h35'23"74d28'
Robbins,Justin Her 17h0'44"49d51'
Robbins,Linda Cas 0h25'33"60d39'
Robbins,Linda Lou Peg 23h27'16"33d36'
Robbins,Matthew Eli Dra 16h54'58"72d56'
Robbins,Nancy K Uma 9h57'39"57d15'
Robbins,Paul Forrest Cet 0h55'1"-2d34'
Robbins,Penny And 0h25'33"38d25'
Robbins,Robert Jay Cep 21h48'17"58d23'
Robbins,Rosalie Mon 6h28'53"11d19'
Robbins,Ruth Uma 11h6'33"33d7'
Robbins,Ryan L Dra 14h53'44"63d33'
Robbo Her 17h11'59"48d56'
Robbs,Jan Lyr 18h17'58"30d25'
Robbyn Arianne And 22h59'52"36d58'
Robbyn-N-Nicky Tri 2h17'59"30d4'
Robeck,Loren & Theresa And 23h18'24"38d47'
Robelatend Aql 20h1'21"4d15'
Robelski,Edwin Vir 13h30'1"12d33'
Robenault,James Michael Thomas Boo 15h1'47"8d31'
Robenault,Melissa Ann And 22h58'41"38d23'
Robenseifner,Jeffrey Scott Cep 21h38'45"58d36'
Robenseifner,Jennifer Lynn Cam 5h7'19"61d35'
Roberg,Bjorn Dra 19h29'46"65d5'
Roberson,Cosmic Ori 5h32'20"-1d24'
Roberson,Dr Lee Oph 17h37'37"8d11'
Roberson,Jackie E And 1h41'44"40d5'
Roberson,Jerome Scott Cep 22h15'0"61d6'
Roberson,John Barron Cet 2h34'1"-12d16'
Roberson,Joseph Baxter Oph 17h35'17"11d16'
Roberson,Kay Lyr 18h59'15"38d46'
Roberson,Kelli Ann (McNally) Peg 23h50'38"30d46'
Roberson,Lisa Kay Cas 1h4'32"53d43'
Roberson,Otho Charlie Hya 8h44'49"5d18'
Roberson,Sarah Elizabeth Cas 0h32'50"63d40'
Robert Aur 5h50'25"40d40'
Robert Cep 21h4'24"63d11'A
Robert Hya 8h19'26"5d8'
Robert Ori 6h6'16"0d23'
Robert Ori 4h51'38"0d51'
Robert Lyn 8h47'56"42d53'
Robert Her 16h40'18"48d12'
Robert Per 3h25'56"46d45'B
Robert Cep 23h2'0"70d49'
Robert Leo 11h21'43"-1d31'
Robert Ser 17h58'49"-14d31'
Robert "Sailing on Stardust" Mon 7h2'51"-6d21'
Robert & Aprile's "Dream Come True" Eri 3h37'16"-11d34'
Robert & Carole's Heavenly Ponderosa And 0h4'28"47d1'
Robert & Denise Forever Peg 22h56'31"20d48'
Robert & Elizabeth Forever Cyg 19h49'51"50d1'
Robert & Jaime Dec 14 1992 Aql 18h56'13"2d19'
Robert & Jennifer Cep 22h33'37"63d35'
Robert & Kay Boo 15h25'21"34d14'
Robert & Myrtle Cep 22h33'0"68d30'
Robert & Norma Star Eri 3h59'0"-5d52'
Robert & Suzan Forever Uma 10h28'41"51d32'
Robert & Tammy Cyg 20h6'49"40d60'
Robert Alan Aql 20h7'28"1d7'
Robert Allen Leo 11h7'58"2d23'
Robert Anthony Cep 23h3'52"60d6'
Robert Anthony Cam 3h40'12"61d19'
Robert Arthur Boo 15h21'1"50d44'
Robert Charles Aur 6h55'34"38d59'
Robert David Ori 5h44'13"11d59'
Robert E Cmi 7h57'31"1d48'
Robert Eddie Cet 2h45'51"4d33'
Robert Edwin Her 17h29'17"37d31'
Robert et Nancy Peg 22h21'34"15d1'AB
Robert Henry Cep 22h51'35"63d36'
Robert John 10 Boo 15h45'44"42d23'
Robert Keith Uma 9h37'60"70d33'
Robert R Rada Rex Her 17h27'21"20d58'
Robert Roy Peg 22h58'18"31d53'
Robert Steven Cet 1h31'45"-14d55'
Robert The Bright Lib 14h42'33"-23d21'
Robert The Great Her 17h15'53"20d16'
Robert Thomas' Star Her 16h59'41"29d52'
Robert Vance Oph 18h5'21"10d53'
Robert William Lyr 19h2'43"43d48'A
Robert"Almost Perfect" Dra 12h27'37"72d8'
Robert's Light Her 18h16'55"38d50'
Robert's Light AKA R A Trussell Ori 6h0'47"11d1'
Robert's Magic Lac 21h57'35"38d6'
Robert's Rising Star Oph 17h16'48"-16d31'
Robert's Royal Palm Ori 5h55'54"11d30'
Robert's Star,Michael Aql 20h33'39"0d21'
Robert's Vision Ori 5h32'60"-6d50'
Robert,Ginette Ori 6h12'27"0d54'
Robert,Jean-Claude Cyg 19h45'43"30d33'
Robert,Laura Cyg 20h24'1"30d41'
Robert,Mark A Dra 18h52'47"80d24'
Robert,Mary"Star" Cep 23h2'58"64d36'
Robert,Stephen Gem 7h17'25"30d59'
Robert,The Absolute Perfection Per 2h32'45"57d8'
Robert-Apollo-93 Ari 2h1'36"25d26'
Robert-Tissot,Alice India Cyg 19h27'55"37d11'
Roberta Pup 7h56'25"-26d2'
Roberta Cet 2h32'22"1d1'
Roberta Peg 21h51'0"31d33'
Roberta Lac 22h23'38"37d38'
Roberta Cas 2h50'35"60d43'
Roberta And 0h23'35"36d4'
Roberta Pyx 8h51'48"-26d19'
Roberta Cma 7h1'37"-28d9'
Roberta Lac 22h49'46"38d2'
Roberta & John Cyg 19h34'0"28d11'
Roberta Gayle Mon 6h33'40"7d29'
Roberta HYDU Cyg 19h21'19"28d21'
Roberta T Lac 22h23'45"37d48'
Roberta(Roby) Peg 23h29'22"22d29'
Roberti,Nicole Michelle Uma 9h48'26"42d30'
Robertico Per 1h59'32"56d19'
Robertine,Katie Wargo And 1h45'12"48d57'
Roberto Uma 12h12'22"55d8'
Roberto Dra 16h25'32"62d39'
Roberto Cyg 20h52'24"31d15'
Roberto Vel 9h43'24"-48d17'
Roberto Ori 6h0'41"8d28'
Roberto Hor 3h26'0"-47d58'
Roberto Hya 8h59'1"0d6'
Roberto Ori 5h56'54"16d1'
Roberto Del 20h13'14"15d1'
Roberto "Alessia", Antonella And 23h2'32"52d45'
Roberto's Light Aur 4h59'18"30d52'
Roberto,Diane Cas 2h0'37"73d16'
Roberto,Michele Cyg 21h35'27"41d47'
Roberto,Stephanie Cas 0h19'22"56d9'
Roberts & Amber Wilson ,Sarah Lac 22h46'33"54d21'
Roberts AKA "Nana", Stella Osa Mon 6h24'54"0d54'
Roberts Aura Ori 6h4'24"2d8'
Roberts IV,Lucien W Vir 15h0'0"6d30'
Roberts Sanfran Sydney Athens,Julie Peg 23h25'14"13d40'
Roberts Star,Brett & Carina Sct 18h56'56"-6d46'
Roberts"Twinkle Of Love",Scott Her 16h53'47"40d8'
Roberts,"Heart Of Gold" Debbie Uma 9h17'57"68d19'
Roberts,Alexander Uma 12h2'1"32d56'
Roberts,Alison & Keith Oph 17h55'1"12d58'
Roberts,Alison Schenk Oph 17h52'1"12d51'
Roberts,Amanda Lyr 19h25'0"40d52'
Roberts,Amanda And 1h28'0"48d49'
Roberts,Armond Albert Her 15h57'55"46d19'
Roberts,Avril Tau 4h11'8"23d0'
Roberts,Barbara Cyg 19h44'58"37d37'
Roberts,Beloved Parents,Russ & Julia Cnv 12h26'45"33d13'
Roberts,Bethany Ruth Beal Col 6h33'50"-33d10'
Roberts,Bill Her 16h24'41"36d47'
Roberts,Brenda Lee Cyg 21h32'1"34d59'
Roberts,Brian Her 18h11'22"30d53'
Roberts,C D Aql 19h43'26"11d19'
Roberts,C G Aql 20h1'60"-0d39'
Roberts,Carolyn Jean Umi 15h2'17"66d15'
Roberts,Caryn Crb 16h3'33"37d30'
Roberts,Chanelle Bird Per 2h46'40"41d10'
Roberts,Chris Cam 6h56'24"67d40'
Roberts,Christina Marie Com 12h39'17"23d32'
Roberts,Christine Michelle Lyr 18h33'34"46d47'
Roberts,Clark Ser 18h4'18"-14d35'
Roberts,Craig D Her 16h31'25"39d0'
Roberts,David A Ari 1h59'53"14d17'
Roberts,Dawn L Cam 5h30'11"60d26'
Roberts,Donza Mon 8h2'36"-8d54'
Roberts,Dorian Lyn 7h45'58"40d53'
Roberts,Douglas (Nino) Aur 5h19'22"43d41'
Roberts,Edna Mae Com 12h18'22"20d43'
Roberts,Elizabeth "Tina" Tinutti Nana Equ 21h3'29"2d58'
Roberts,Elizabeth Colette Viviano Lyr 19h25'48"42d54'
Roberts,Estelle & Daniel Lyr 19h19'17"40d28'
Roberts,Esther Pearl Sco 16h55'34"-44d27'
Roberts,Garth Alastair Aur 5h18'1"49d47'
Roberts,Gary M Per 2h7'0"47d55'
Roberts,George Cep 22h13'37"61d29'
Roberts,George Woodrow Ari 2h57'0"20d6'
Roberts,Georgia Crb 15h27'34"31d12'
Roberts,Graciela Ottilee Cet 0h39'10"0d27'
Roberts,Haley Henderson Lyr 19h12'47"40d42'
Roberts,Heidi Fox Cet 2h4'21"6d58'
Roberts,Helen Olive (Garrelts) Vir 13h0'44"-20d22'
Roberts,Jack Her 17h8'28"47d50'
Roberts,Jack Anthony Per 3h38'0"37d47'
Roberts,Jason Per 4h41'59"51d39'
Roberts,Jeanette And 2h17'17"41d29'
Roberts,Jeffrey Michael Aql 19h8'42"3d29'
Roberts,Jennifer Elaine And 1h17'49"38d35'
Roberts,Jennifer And 0h58'17"37d12'
Roberts,Jennifer Ellen Peg 22h14'47"33d8'
Roberts,Jessica Mary And 23h39'50"38d51'
Roberts,Joan Gladys Cyg 19h24'41"35d4'
Roberts,John Alexander Per 2h57'46"40d36'
Roberts,John Cypron Leo 9h52'55"8d25'
Roberts,Josh Aql 19h52'1"13d15'
Roberts,Joshua Cai Ori 5h52'25"6d21'
Roberts,Jr,Frank "Bub" Aql 19h51'45"11d17'
Roberts,Jr,James L Lac 22h13'59"50d30'
Roberts,Julia Mon 7h56'38"-1d24'
Roberts,Julian Lyn 8h9'16"41d52'
Roberts,Karen Ann Vul 19h19'0"26d39'
Roberts,Katie Elizabeth Lyr 18h15'39"40d20'
Roberts,Katie Louise Cas 0h55'1"74d13'
Roberts,Ken Cep 22h30'19"56d53'
Roberts,Ken & Mary Cyg 19h44'15"31d13'
Roberts,Kenneth Donald Eri 4h10'48"-17d51'
Roberts,Kim M Cet 1h18'48"-13d31'
Roberts,Laura Catherine Mon 6h20'46"8d52'
Roberts,Leslie Arlene And 23h18'1"51d27'
Roberts,Linda Lyn 8h4'45"51d49'
Roberts,Loren Chris Aql 18h44'44"11d2'
Roberts,Louise And 1h19'53"48d51'
Roberts,Luke Umi 13h57'58"75d58'
Roberts,Lynda A And 23h50'45"33d52'

Roberts,Maria
Crt 11h39'34"-22d60'
Roberts,Martin Dale
Tau 4h31'1"30d36'
Roberts,Mary
And 1h1'27"38d3'
Roberts,Meghan Kathleen
Cas 0h39'1"69d47'
Roberts,Michael Sebastian
Her 17h26'58"18d59'
Roberts,Michael Bradley
Aur 6h12'38"45d44'
Roberts,Michelle Nicole
Cyg 19h50'33"40d39'
Roberts,Monette Allison
Com 11h56'28"14d8'
Roberts,Monica Rochelle
Com 12h59'40"21d21'
Roberts,Natalie Ann
Equ 20h57'45"10d51'
Roberts,Neal Alan
Cma 6h57'43"-17d56'
Roberts,Norma
Mon 6h23'33"4d50'
Roberts,Patricia Ann
Com 12h22'34"27d36'
Roberts,Paul E
Aql 20h11'1"11d9'
Roberts,Phil,Karen, Nick & Kellie
Peg 23h46'57"18d12'
Roberts,Philip
Ori 5h50'16"10d34'
Roberts,Rob
Aur 7h20'21"40d19'
Roberts,Robert Willard
Dra 15h55'21"68d6'
Roberts,Robin Curtis
Vul 19h43'0"22d45'
Roberts,Roger
Aur 5h4'27"40d42'
Roberts,Rozilla
Aur 5h0'0"50d26'
Roberts,Silver Wedding Albert & Dianne
Cyg 21h38'33"30d27'
Roberts,Steve
Cam 8h1'31"80d9'
Roberts,Stewart Marvin
Lyr 18h37'47"40d16'
Roberts,Susie Lauren
Eri 3h15'31"-12d20'
Roberts,Susie Nicole
Uma 11h22'57"50d29'
Roberts,Suzi
Com 12h52'57"23d7'
Roberts,Tara
Eri 3h5'14"-1d56'
Roberts,Tasha Renee
Cam 4h56'48"68d29'
Roberts,The Mr John
Per 2h11'31"57d1'
Roberts,Thomas Brandon Phillip
Per 3h11'21"47d51'
Roberts,Timothy Adam
Lmi 11h4'38"28d38'
Roberts,Timothy C
Aur 5h4'25"40d17'
Roberts,Tina
Cas 0h26'56"62d47'
Roberts,Victoria Alice
And 0h47'0"34d60'
Roberts,Vigina Lee
And 2h30'57"48d51'
Roberts,Virginia Sykes
Hya 9h13'18"1d44'
Roberts-A Star Is Born,Clara Kate
Mon 7h29'51"-8d23'
Roberts-Bachelor of Arts,Janis
Cnc 8h26'41"31d15'
Roberts-R C D A,Kaysy
Aql 20h0'0"6d22'

Robertshaw,Tom S
Aql 20h0'57"12d49'
Robertson
Eri 3h57'11"-19d31'
Robertson II,Hugh C
Hya 8h53'35"-7d53'
Robertson III,Lake
Cot 2h30'25"8d48'
Robertson Jr,"Bapa", William C
Cnv 12h31'40"41d12'
Robertson's Heavenly Body,Dick
Ori 5h56'20"20d34'
Robertson's Popa I, Heather
Uma 11h20'43"62d52'
Robertson,A Haeworth
Uma 10h3'19"60d52'
Robertson,Bonnie Sue
Uma 8h37'24"56d42'
Robertson,Bradley Joseph
Oph 17h15'51"10d42'
Robertson,Brian Collins
Cam 3h26'57"61d33'
Robertson,Brooke Ruel
Peg 22h43'1"25d31'
Robertson,Charles John
Peg 22h14'43"32d52'
Robertson,Craig
Dra 18h20'37"48d10'
Robertson,Danica Dawn
Lyr 18h21'44"38d21'
Robertson,Darrell G
Tau 4h37'18"15d13'
Robertson,Dennis Michael
Dra 19h3'30"50d26'
Robertson,Ecce Sacerdos Father Gene
Cep 22h59'55"70d60'
Robertson,Frank J
Boo 14h11'44"51d36'
Robertson,Glen Edward
Ser 15h38'0"18d31'
Robertson,Gordon & Debbie
Sge 19h35'55"19d25'
Robertson,Grace Kathleen Ayres
Lyn 8h11'0"34d25'
Robertson,H Gene
Hya 8h54'42"0d36'
Robertson,Isobel Gall
Cet 1h2'20"-3d29'
Robertson,Jack D
Aur 7h9'39"40d55'
Robertson,James Earl
Ser 15h10'48"8d41'
Robertson,Janet Lee Robinson
And 0h11'35"37d20'
Robertson,Jennifer Michelle
Peg 22h21'55"21d58'
Robertson,Jr,Wa Robbie
Cet 1h58'52"-8d10'
Robertson,Juan Paul
Uma 10h47'11"52d3'
Robertson,Kai A K
Aur 5h53'48"38d50'
Robertson,Keity
Lyr 18h26'14"45d2'
Robertson,Kelvin
Aur 7h24'54"38d35'
Robertson,Lee
Sex 10h18'13"-8d18'
Robertson,Linda J
Vul 20h2'30"28d14'
Robertson,Linda Sue
Aql 20h18'28"5d32'
Robertson,Lindsay
Mon 8h6'4"-1d40'
Robertson,Lise
Psc 1h19'26"21d37'
Robertson,Michael Austin
Ori 5h57'21"15d43'
Robertson,Michael Ray
Ori 5h58'0"12d38'

Robertson,Michelle Fay
Leo 9h32'15"7d31'
Robertson,Muriel
Cas 0h3'0"61d37'
Robertson,Nathan, Bridget,Mark
Lyn 8h23'0"42d30'
Robertson,Otis C
Aur 5h19'39"48d5'
Robertson,Richard
Dra 17h6'33"61d54'
Robertson,Ryan
Oph 17h54'57"12d41'
Robertson,Samantha
Vul 19h47'18"23d1'
Robertson,Sarah Erwin
Cep 21h42'0"67d35'
Robertson,Scott
Cep 21h22'1"61d52'
Robertson,Scott
Dra 13h21'0"64d46'
Robertson,Terry & Lynn
Del 20h38'5"8d16'
Robertson,Timothy B
Uma 9h54'31"43d13'
Robertson,Wendy
Ori 6h7'46"5d47'
Robeta,Polpettina
Cnv 13h2'49"50d59'
Robeth
Cmi 7h5'42"1d26'
Robginia D
Hya 8h27'48"-6d0'
Robichaud,Stephen A
Lac 22h8'34"48d23'
Robideau,Ken
Cep 23h17'49"70d39'
Robillard,Hank
Cnc 8h48'22"31d44'
Robillard,John
Cnc 8h34'36"30d21'
Robillard,Simone
Peg 22h0'1"20d5'
Robillerness
Eri 3h42'44"-4d53'
Robin
And 23h8'0"41d60'
Robin
Ori 5h59'50"8d45'
Robin
Cyg 20h18'17"39d57'
Robin
Cyg 21h25'56"40d15'
Robin
Peg 21h23'19"8d8'
Robin
Mon 6h29'29"-8d10'
Robin & Andy
Cyg 21h58'13"53d45'
Robin & Olaf
Lib 15h12'49"-24d16'
Robin & Pooh, Christopher
Lyr 18h40'1"30d52'
Robin & Rick
Cyg 20h24'35"38d54'
Robin Ann
Cam 3h26'1"60d48'
Robin Loving
Sct 18h42'5"-9d51'
Robin Marie
Com 12h31'49"26d18'
Robin Nicklas' Stern
Cnc 8h28'19"7d20'
Robin Paul
Boo 15h29'1"50d57'
Robin Renee
Peg 22h46'36"34d58'
Robin Reneé
Lib 15h41'47"-24d19'
Robin's Beauty
Uma 9h6'52"56d20'
Robin's Dream
Cyg 20h50'38"40d42'
Robin's Emerald
Peg 23h3'44"21d4'

Robin's Forever Wishing Star
Cyg 20h24'42"38d17'
Robin's Light
Tri 2h5'0"31d36'
Robin's Love
Her 18h12'52"30d50'
Robin's Rising Star
And 1h23'55"41d0'
Robin's Star
And 23h37'36"48d51'
Robin's Star
Cyg 21h4'33"31d16'
Robin's Star
Cet 0h46'49"-6d32'
Robin,Arnaud
Tri 2h3'26"32d1'
Robin,Didier
Tri 2h3'27"32d43'
Robin,Marc
Per 2h53'52"37d6'
Robin,Noah Brett
Her 16h17'24"10d40'
Robin,Sam & Sylvia
Sge 19h55'38"16d4'
Robin,Sébastien
Tri 2h3'20"32d5'
Robin,The Christopher
Aur 5h59'51"29d40'
Robin-Wood
Umi 15h55'20"76d58'
Robina & Gary "Meant To Be"
Peg 22h9'13"2d29'
Robinea
Cas 0h57'30"61d43'
Robinett,Hillary Anne
Uma 11h29'55"50d34'
Robinette,David G
Ser 18h3'0"-13d22'
Robinette,Renée
Aql 19h58'60"15d26'
Robinette,Teana Ann
Lyn 6h53'55"60d4'
Robins "Wedding Star", Glenn & Doris
Sge 19h56'22"20d15'
Robins Lucky Star
Del 20h18'13"14d17'
Robins,Herbert Frank
Her 17h15'45"29d10'
Robins,Marcia
Lyr 18h58'1"32d26'
Robins,Oscar & Shirley
Oph 18h17'21"10d51'
Robins,Thomas
Boo 15h17'0"38d56'
Robins-Meara,Amanda
And 0h33'26"40d18'
Robinson "Lulu",Katy Leah
Del 20h22'46"18d56'
Robinson "Spunky", Matthew Kevin
Boo 13h54'19"21d59'
Robinson III,James D
Per 2h46'10"40d40'
Robinson III,T Leslie
Aur 6h2'21"31d13'
Robinson IV,John James (Jack)
Sex 9h41'38"2d31'
Robinson IV,William Edward
Boo 15h13'38"26d35'
Robinson III,Morley Ernest
Aur 4h56'31"50d33'
Robinson Our Angel, Eric
Boo 14h58'12"39d54'
Robinson Star,Lyle M
Aql 18h57'14"-4d44'
Robinson Star,The Ann
And 1h2'56"37d16'
Robinson V,James Edmund
Ori 5h55'57"12d28'
Robinson"Granny", Rosevenia Marie
Cyg 19h45'29"31d6'

Robinson's Centennial, Steve
Ori 5h58'39"17d11'
Robinson's Dragon,Sir
Her 17h36'32"37d39'
Robinson's Light
Cyg 20h3'58"31d36'
Robinson's Lucky Star
Lyr 18h51'1"31d54'
Robinson's Rose, Kathleen
Lyn 8h57'0"46d48'
Robinson,"The Lion in the Sky",Charles
Leo 10h23'52"10d33'
Robinson,Alice Newsome
Mon 6h53'48"-8d45'
Robinson,Allen
Per 1h54'24"50d0'
Robinson,Amanda Jolene
Equ 21h3'13"2d43'
Robinson,Amanda Marie
Gem 6h38'26"35d4'
Robinson,Ami
Crb 15h14'0"30d43'
Robinson,Annabelle
Cas 23h30'43"60d43'
Robinson,Anne C & Francis Victor Place
Cyg 20h20'55"38d29'
Robinson,Benita
Cyg 20h39'42"45d53'
Robinson,Betsy
Cas 0h1'35"65d2'
Robinson,Brad
Boo 14h56'19"44d31'
Robinson,Brian Keith
Aur 6h19'53"38d48'
Robinson,Cathy
Cas 0h1'39"50d30'
Robinson,Cecil Brooks
Peg 22h36'26"11d13'
Robinson,Christina Diane
Mon 8h6'46"-1d40'
Robinson,Christopher James
Cep 20h35'43"65d43'
Robinson,Christopher M
Aur 7h13'50"35d59'
Robinson,Cindi
Eri 4h24'54"-0d9'
Robinson,Courtney Alexander
Per 2h56'1"37d13'
Robinson,Crystal
And 2h16'26"40d18'
Robinson,Cynthia Lynn
Cas 1h33'49"63d49'
Robinson,Daniel Burwell
Boo 15h5'46"19d22'
Robinson,Danny & Carmen
Lyr 19h20'33"41d13'
Robinson,David Hawkins
Uma 10h11'15"51d46'
Robinson,David I
Hya 8h30'57"1d23'
Robinson,Donna M
Lyn 6h54'12"60d22'
Robinson,Doris "Moy" Minich
Mon 7h54'30"-3d22'
Robinson,Dorna Marie
Aql 19h31'26"8d27'
Robinson,Dr Gregory
Oph 17h2'17"10d53'
Robinson,Elaine
Lyn 8h29'57"47d13'
Robinson,Elizabeth Cameron
Cyg 19h28'59"38d14'
Robinson,Elizabeth Ann
Aql 20h12'12"12d5'
Robinson,Ellen
Lyn 6h37'0"54d58'
Robinson,Emilie
And 2h3'10"45d29'
Robinson,Emily Maria
Lyr 19h22'39"31d50'
Robinson,Georgina
And 1h22'19"41d14'

Robinson,Gwennie
Cas 2h55'27"61d3'
Robinson,Hannah Morgan
Sge 19h56'0"16d10'
Robinson,Heath
Cnv 13h30'22"40d57'
Robinson,Heather & Nathan
Cyg 21h30'53"42d7'
Robinson,Herbert
Her 16h57'0"51d10'
Robinson,Ida Jeanette
Cap 21h54'46"-18d33'
Robinson,Ida Sue
Lac 22h55'17"55d46'
Robinson,Irene
Lyr 18h15'0"44d34'
Robinson,Isaac G
Ori 5h28'35"-4d2'
Robinson,Isabel
And 23h15'52"51d43'
Robinson,Jan & Lee
Mon 6h53'38"-2d22'
Robinson,Jane Margaret
Com 12h32'39"25d53'
Robinson,Jarred Drew
Sgr 19h42'17"-41d52'
Robinson,Jason
Boo 15h5'60"22d4'
Robinson,Jason John
Cnc 8h48'54"22d44'
Robinson,Jason Richard
Her 16h39'43"8d29'
Robinson,Jean
Eri 3h55'12"-6d5'
Robinson,Jill
Cas 0h19'28"61d59'
Robinson,John
Cep 21h30'20"60d6'
Robinson,Jonathan
Umi 9h27'42"89d10'
Robinson,Joshanna N
Cas 1h18'0"63d26'
Robinson,Joyce
Cas 0h16'51"62d7'
Robinson,Jr,Carter
Aql 19h9'38"15d26'
Robinson,Jr,Charles Bernard
Aql 19h12'51"12d35'
Robinson,Juara Knight
Eri 3h38'35"-17d56'
Robinson,Julie-Anne
Cas 1h56'11"73d56'
Robinson,Kathleen
Com 13h6'15"16d2'
Robinson,Kathryn M
Lyr 19h4'12"37d33'
Robinson,Kelly Kuaihelani
Aql 19h58'29"8d10'
Robinson,Krista
Cyg 20h50'29"38d11'
Robinson,Kristan Joy
Ori 5h30'40"-1d18'
Robinson,Kyle Edward
Gem 7h9'59"21d30'
Robinson,Lauren McKenzie
Tau 5h20'53"28d59'
Robinson,Lee James
Cyg 19h32'29"28d19'
Robinson,Linda
Oph 18h6'40"12d13'
Robinson,Lois Fay
Cas 2h45'43"60d20'
Robinson,Lynn Bailes
Peg 22h1'22"30d53'
Robinson,Lynne
Umi 14h36'12"82d23'
Robinson,M D
Aur 6h1'21"37d55'
Robinson,Malcolm George
Ori 5h27'51"-4d40'
Robinson,Malcolm George
And 23h19'48"50d40'
Robinson,Mardi Ann
Mon 6h59'37"1d32'

Robinson,Mark
Cyg 21h24'41"39d30'
Robinson,Max Jordon
Cep 21h36'37"70d29'
Robinson,Megan Marie
Tri 1h38'52"30d25'
Robinson,Michael E
Her 18h6'23"14d54'
Robinson,Mitch A
Sct 18h55'21"-7d51'
Robinson,Mosher Marion Virginia
Aql 19h26'47"7d33'
Robinson,Mrs Barbara E
And 2h23'25"38d54'
Robinson,Nancy
Cas 0h14'20"61d9'
Robinson,Narielle Alyssa
And 1h46'26"39d23'
Robinson,Pamela
Lyr 19h20'23"40d10'
Robinson,Pamela J
Gem 7h30'48"33d56'
Robinson,Patricia
And 2h1'54"45d31'
Robinson,Peter Joseph
Per 1h34'24"54d16'
Robinson,PhD,Michael C
Aqr 22h31'13"-0d55'
Robinson,Phillip
Cet 0h3'54"-17d44'
Robinson,Phillip George
Uma 8h45'47"55d58'
Robinson,Raymond Joseph
Her 16h39'51"5d34'
Robinson,Rev Willie E
Her 15h59'57"47d32'
Robinson,Richard Louis
Cep 21h39'59"58d46'
Robinson,Richeal
Her 17h32'1"40d15'
Robinson,Robert John
Crb 15h18'42"30d36'
Robinson,Rodney
Aql 18h58'11"-2d52'
Robinson,Roger
Dra 17h29'14"61d34'
Robinson,Ryan
Umi 13h19'1"70d33'
Robinson,Ryan William
Dra 18h33'55"80d32'
Robinson,Sam Garrett
Tau 4h57'0"28d46'
Robinson,Shawn Z
Lac 22h12'37"46d39'
Robinson,Simon John
Ori 5h33'51"7d57'
Robinson,Stephanie
Lyr 18h17'13"40d33'
Robinson,Steve
Ser 15h39'18"24d37'
Robinson,Steve & Angie
Cam 3h51'39"58d13'
Robinson,Susan
Peg 22h0'0"31d29'
Robinson,Tammy Leah "Bowman"
Vul 20h15'1"23d58'
Robinson,Tara & Tom
Ori 5h50'13"12d21'
Robinson,Tim
Sge 20h0'1"16d1'
Robinson,Timothy Rupert
Umi 13h47'33"77d49'
Robinson,Tommy Jon
Boo 13h47'40"17d56'
Robinson,William & Amelia
Per 2h38'42"40d2'
Robinson,William L
Uma 12h30'29"56d35'
Robinson-Burnham
Ori 4h50'48"4d30'
Robinson-Telenko, Judith
Mon 6h51'47"11d33'

Robinson-`Eternity', David
Eri 4h1'35"-8d59'
Robisch,Herman
Her 17h58'38"38d46'
Robischon,John
Aur 6h19'55"38d4'
Robison,Cameron L
Her 16h6'23"40d23'
Robison,Charles Dimick
Per 2h51'39"37d24'
Robison,Deejah
Del 20h21'34"10d39'
Robison,GK
Crt 11h53'58"-8d24'
Robison,III,Joseph E
Dra 16h31'19"58d26'
Robison,IV,Joseph E
Lac 22h7'32"40d12'
Robison,Janet
Com 12h24'20"28d20'
Robison,Kathleen
Cyg 19h59'22"40d4'
Robison,Kim
Cmi 7h41'16"4d52'
Robison,Linda L
Cam 3h27'59"53d42'
Robison,Lisa D
Cyg 19h22'27"28d10'
Robison,Robert Drake
Lyn 9h16'35"39d8'
RobJam
Vir 15h1'30"5d45'
Robles,Javier E
Cam 13h18'0"77d4'
Robles,Monico
Cas 0h20'13"68d41'
Robles,Nora
Mon 6h36'25"0d15'
Robles,Wayne Joseph
Sct 18h52'36"-6d20'
Robles,Xavier Rafael
Cep 22h45'1"57d29'
Robles,Xelina Gabrielle
Cas 22h57'55"54d55'
Roblin,Caren Chase
Cas 0h35'59"71d21'
Robmeistar
Uma 9h47'24"68d21'
Robnett,"Mighty Mite Joe",Thomas M
Hya 8h24'31"1d46'
Robnett,Thomas Rory "Bugs"
Mon 7h41'26"-5d42'
RobRoy
Oph 17h0'23"11d13'
Robson,Bernice & James
Cam 7h7'1"61d41'
Robson,Connie Marie
Vir 11h38'22"8d5'
Robson,Dennis
Aql 19h37'12"0d17'
Robson,Jocelyn Corrine
Cas 0h50'40"61d21'
Robson,Lawrence John
Cyg 20h20'23"39d19'
Robson,Lynne
Uma 9h15'40"49d11'
Robson,Mark John
Ori 5h23'12"0d55'
Robson,Scott
Hya 9h5'41"-11d1'
Robson-Foster,Lesley
And 23h41'58"45d46'
Robustelli
Uma 10h32'54"41d44'
Robustelli,Jessica Lauren
Cyg 21h12'23"40d56'A
Robustelli,Marianne
Cas 23h15'50"61d14'
Robustelli,Robert
Cyg 21h12'23"40d56'B
Robustelli,Stacy
Dra 15h7'26"63d57'
Roby
Nor 16h22'11"-51d1'

Roby
Cyg 20h2'21"40d19'
Roby Baby
Uma 11h58'0"32d13'
Roby Vali
Cyg 20h22'39"39d36'
Roby,Christi Diane
Cyg 20h54'35"30d52'
Roby,Claude
Cnv 13h19'20"31d29'
Roby,Robert John
Per 3h8'56"40d13'
Robyn
Del 20h20'22"11d9'
Robyn
Mon 6h30'47"-1d0'
Robyn
Lyn 6h49'30"51d34'C
Robyn & Eddie
Peg 23h16'25"33d49'
Robyn Alexandra
Umi 15h44'33"81d54'
Robyn Emily Mary
Umi 15h5'0"66d42'
Robyn Eve
Umi 13h52'49"72d57'
Robyn Jane
Cyg 20h52'36"38d39'
Robyn Joy
Cam 3h57'21"61d48'
Robyn Light Of Kindness
Cnv 13h48'38"41d18'
Robyn's Dream
Cam 3h55'0"52d31'
Robyn's Radiance
Peg 23h29'54"17d32'
Roc'n Doc Young
Lac 22h16'1"50d19'
Roca,Tiffany Ann
Del 20h48'50"9d49'
Roca,Vincent Luigi
Uma 9h16'1"62d18'
Rocca Loving Heart Star, H. Della
Her 17h38'0"37d41'
Rocca,Elissa Della
Peg 22h18'12"32d44'
Roccabianca,Lara
Ser 15h44'29"2d25'
Roccaforte,Robert "Rocky"
Ari 2h56'59"28d54'
Rocchi,Cinzia Ferrarini
Lyn 7h29'18"41d1'
Rocchi,Michael Anthony
Cam 4h49'46"70d11'
Rocco
Aqr 21h5'20"-6d17'
Rocco
Her 16h41'21"48d57'
Rocco
Dra 12h3'45"71d9'
Rocco
Dra 11h49'57"74d42'
Rocco
Aur 6h5'1"38d54'
Rocco Too-42
Ori 5h29'50"-1d2'
Rocco's Angel
Per 2h50'28"40d43'
Rocco,Charlie
Her 18h55'1"12d37'
Rocco,Francesco
Del 20h53'38"4d29'
Rocco,Laura E
Lyr 19h16'16"25d47'
Rocco,Pat Della
Tau 5h52'43"28d46'
Rocco,Patricia Sandin Della
Cas 1h56'57"63d50'
Rocco,Richard
Her 16h35'27"13d4'
Rocco-1,Donna Marie
Cas 1h53'37"75d12'
Roccofordyce
Mon 6h33'59"-1d31'

Rocereta,Aaron W
Lac 22h17'1"51d39'
Rocereta,Sarah Ellen
Cas 0h57'35"62d38'
Roch
Ori 6h9'41"4d42'
Roch,Taryn Johanna
Lyn 8h31'35"40d54'
Rocha My Love Forever, Tony
Her 17h51'1"41d5'
Rocha,Elerd & Elke
Ori 5h56'40"17d59'
Rocha,John Joseph
Dra 20h23'13"63d18'
Rochad
Oph 18h24'42"8d53'
Rochambleu,Pierre Louis
And 0h32'0"40d8'
Rochard,Jocelyne
Aur 5h59'54"54d29'
Roche "Bugly
Rocheleua",Darrell
Leo 9h53'25"33d22'
Roche Family
Boo 14h56'13"17d45'
Roche The IV,James Francis
Eri 4h7'26"-8d39'
Roche,Brian William
Peg 21h53'54"21d31'
Roche,Dave & Mary
Vul 19h59'18"23d4'
Roche,Emily-Grace Kathleen
Per 3h8'30"41d29'
Roche,Michael J
Aql 19h46'47"14d12'
Roche,Myr Celestien Boswell
Ser 17h55'60"-10d23'
Roche,Sunriser,Martin W
Lac 22h18'45"37d41'
Roche,Susan Ann
Crb 16h11'0"36d16'
Roche,The Best Mom &
Friend Terry
Uma 11h19'48"40d30'
Roche,Thibault
Del 20h19'42"13d28'
Roche,Tim
Cnv 12h30'14"33d42'
Roche-Lilliott,Brianna Celeste
Cet 2h19'12"-10d39'
Rochek,Christy Ann
Peg 0h1'37"30d4'
Rocheleau,Bert
Cep 23h0'51"65d34'
Rochelle
Psc 23h2'44"0d16'
Rochelle
Tri 2h6'11"33d7'
Rochelle
Lyn 9h3'12"39d58'
Rochelle
And 0h20'23"30d48'
Rochelle
Lac 22h14'38"50d33'
Rochelle Forever
Cyg 19h31'39"31d40'
Rochelle,Susan
Peg 21h55'36"28d13'
Rochellory
And 2h20'31"39d44'
Rocher,Jean Pierre
Mon 7h45'49"-5d44'
Rochermorchlo
Cma 6h59'17"-15d56'
Rochet,Noelie
Cnv 12h18'33"38d9'
Rochette,Catherine
Mon 7h44'39"-4d28'
Rochette,Luc
Umi 15h34'49"67d58'
Rochford,Nicholas
Lac 22h27'56"56d43'
Rochin,Albert
Hya 9h8'1"0d8'

Rochlitz 70,Imre
Uma 11h17'56"55d39'
Rochlitzer,Claus
Sco 16h13'23"-24d57'
Rochman,Martin
Dra 14h37'11"60d37'
Rochman,Tali
Cam 4h27'1"68d12'
Rochon,Isabelle
Peg 22h39'41"21d10'
Rochow,Joel
Leo 10h58'41"-5d20'
Rochow,Marilyn
Cam 3h41'40"60d8'
Roché,Jacob
Aql 20h1'50"7d56'
Roché,Michael
Aql 20h1'44"6d3'
Roché,Ricky
Aql 19h58'22"8d59'
Roché,Tania (O'Banion)
Aql 20h10'1"3d52'
Rocio
Lyr 18h36'22"28d20'
Rocio,Baby Albert
Cmi 7h41'33"8d30'
Rock Star
Lyr 18h31'11"37d48'
Rock Star's Star
Uma 10h27'1"48d52'
Rock,Carolyn
Eri 3h27'10"-2d50'
Rock,David Spencer Percival
Ori 4h47'1"15d9'
Rock,Nicole Marie
Psc 23h3'12"1d31'
Rock,Philip Matthew
Per 4h43'22"40d44'
Rock,Rita
Cyg 19h53'22"38d8'
Rock-A-Rooney
Lac 22h33'0"50d31'
Rockafeller II,Harry Joseph
Her 16h36'17"29d15'
Rocke,Bonnie Gomez
Cas 0h6'26"58d24'
Rockee
Her 17h22'1"21d20'
Rockenbach,Gary R
Ori 4h49'12"5d20'
Rocket-Mary 1
Umi 15h3'43"72d12'
Rocketeer,The
Lac 22h22'56"40d47'
Rockets,Lucy
Cma 6h28'24"-20d6'
Rockey,John David
Aql 18h54'48"-2d29'
Rockhill,Agnes
Hya 8h29'44"-5d52'
Rockhill,I
Aql 20h11'0"11d40'
Rockhill,Thanè Vicky
Hya 8h29'37"-5d39'
Rockin B
Ser 15h13'12"1d5'
Rockin Roger
Ori 4h52'33"1d5'
Rockin' R
Aur 5h2'29"40d37'
Rocking Horse
Peg 22h57'12"20d30'
Rockingham,Vera
Lyr 18h30'16"32d51'
Rocklage,Carolyn Christian
Vul 19h40'37"25d22'
Rocklage,Eric Benjamin
Vul 20h3'39"25d26'
Rocklage,Gary Wayne
Vul 20h1'60"25d2'
Rocklage,Megan Noel
Vul 20h0'35"25d41'
Rockmore,Ethan
Cnv 13h40'56"39d59'

Rockowitz,Rachel Ann
And 23h2'16"51d3'
Rockwell IV,Garrett Willis
Aur 5h5'12"31d47'
Rockwell,Alexander James
Dra 12h33'38"68d40'
Rockwell,Deanna Marie
Peg 22h3'24"8d16'
Rockwell,Dominique Alexis
Com 13h21'12"26d46'
Rockwell,Jonathan N
Aur 6h25'55"38d8'
Rockwell,Marvin J
Oph 17h0'59"8d42'
Rockwell,Nathan Cuyle
Boo 15h7'1"24d13'
Rocky
Cma 7h14'58"-13d50'
Rocky
Aur 5h15'50"45d15'
Rocky
Her 16h49'52"37d46'
Rocky & Micky
Aql 19h18'52"12d17'
Rocky Regina
Lyn 7h41'0"42d16'
Rocky's
Her 16h14'21"25d8'
Rocky's Never Never Land
Uma 9h8'14"51d7'
Rocquet,Sandra
Umi 15h15'21"78d12'
Rod
Per 3h18'31"41d54'
Rod
Uma 8h53'32"61d13'
Rod's Magic
Hya 8h28'15"0d60'
Rod,Marianne
Uma 9h0'39"50d38'
Roda,Frederic
Her 18h30'13"12d38'
Rodabaugh,Thomas
Her 17h27'13"31d19'
Rodawald,Rod K
Uma 9h28'43"49d43'
Rodby,Roger Alan
Vir 13h33'10"-7d7'
Rodddy V IV Star,The
Ser 18h36'21"2d29'
Rodden,Esq,John
Ori 5h56'32"14d43'
Roddenberry,Gene
Cnv 13h51'15"42d25'A
Roddenberry,Gene
Cep 22h46'39"57d37'
Roddis,Adelle
And 0h11'30"46d56'
Roddy
Per 3h16'57"41d38'
Roddy,Elizabeth Duncan
Lyr 18h55'1"30d36'
Roddy,Pat
Ori 5h59'55"17d60'
Roddy,Sarah Elizabeth
Cas 2h48'11"61d12'
Rode,Astrid
Hya 8h10'32"2d49'
Rode,Carolyn Jean
And 0h7'37"43d41'
Rode,Constance Jeannette
Schweiger
Cas 2h33'0"58d49'
Rode,John David
Cep 0h5'37"80d11'
Rode,John Joseph
Lac 22h5'25"41d9'
Rode,Pamela Jane
Aur 7h3'17"38d35'
Rode-Nov 22,1969,John &
Jennifer
Cyg 21h5'0"32d51'
Rodeck,Donna Marie
Lyr 18h39'0"30d45'

Roden,Arthur Albert
Dra 18h18'24"47d45'
Rodenbaugh,Eric
Cep 2h11'0"80d14'
Rodenbaugh,Marcia
Eri 2h49'56"-2d33'
Rodenberg,Joan
Mon 6h45'1"10d47'
Rodenborn,Leo
Per 3h5'32"50d21'
Rodenhauser,Alexander
Douglas
Lac 22h22'59"38d40'
Rodenkirchen,Geordie Allen &
Kim
Lyr 18h42'59"38d6'
Rodenkirchen,Jutta
Ari 2h1'54"26d12'
Rodentia,Stella
Umi 15h42'39"78d18'
Rodeo Ben
Cep 21h52'0"70d12'
Rodeo Roger
Boo 14h36'26"51d42'
Roder,Angie Lee
Cas 1h16'0"62d42'
Roderich
Gem 7h23'1"35d14'
Roderick's Special Star,Nicole
Lyr 19h22'20"35d30'
Roderick,Chris
Hya 9h0'20"3d32'
Roderick,Crispian Elwyn
Ori 4h42'16"0d51'
Roderick,Jeffery Vernon
Per 3h12'35"45d8'
Roderick,Joseph
Aur 4h47'21"40d8'
Roderick,Oliver John
Per 3h1'23"37d47'
Roderick,Robert L
Aur 5h23'45"38d20'
Roderick,Sian
Lyr 18h17'58"40d16'
Roderick-Reynolds, Susan
Vul 20h45'13"28d21'
Rodericks,Paul
Aur 6h22'21"33d27'
Rodes,Dana Lesh
Oph 17h39'55"-23d39'
Rodewald,Amy
And 23h28'1"37d34'
Rodewald,Kathleen Elise
And 23h18'39"51d27'
Rodger's Star
Pup 7h59'19"-29d21'
Rodger,Nicola
Ori 6h6'46"2d48'
Rodgers "T",Terence James
Leo 10h50'53"-5d16'
Rodgers,Christine
Sge 19h37'9"16d27'B
Rodgers,Cyrena
Aql 18h40'31"-0d11'
Rodgers,Donna
Cet 2h33'18"-11d20'
Rodgers,Francis R
Cyg 21h20'0"40d2'
Rodgers,Gene
Peg 21h24'17"13d28'B
Rodgers,Jade
Lyr 18h14'1"47d35'
Rodgers,John E
Uma 10h44'25"58d57'
Rodgers,Jon
Aur 5h56'48"29d14'
Rodgers,Kristin Renee
Uma 9h8'54"52d37'
Rodgers,Lori Beth
And 0h2'31"38d27'
Rodgers,Paige
Cas 1h36'30"61d0'
Rodgers,Patrice & George
Cet 1h54'25"-0d49'

Rodgers,Patricia Kristin
Aur 5h7'51"40d7'
Rodgers,Richard Joseph
Aur 6h0'42"32d31'
Rodgers,Richard E
Sge 19h37'9"16d27'A
Rodgers,Ryan Allen
Cep 22h19'0"60d32'
Rodgers,Sally
Del 20h12'10"9d49'
Rodgers,Sandy
Lyn 6h24'29"60d53'
Rodgers,Suzanne
Cam 3h50'29"60d16'
Rodgers,Theresa Robin
Peg 23h41'37"28d59'
Rodgers,Vicki Ann
Vul 19h19'43"26d37'
Rodgers,Virgil & Virginia
Aql 19h29'43"12d54'
Rodgers,William
Cep 23h12'10"62d7'
Rodgers,Zachary Wade
Vul 19h18'1"22d34'
Rodier,Linda
Uma 11h50'48"47d22'
Rodifer,Timothy Cleve
Boo 14h52'41"30d44'
Rodiguez,Danny Master
Cep 20h19'36"75d48'
Rodil,Rodil Rivera
Umi 15h22'40"70d46'
Rodin,Nicole Lauren
Cas 0h26'19"60d39'
Rodin,Robert
Vul 19h35'58"27d17'
Rodino,Peter & Carol
Cyg 21h13'0"38d35'
Rodjinske,Emily
Cam 3h10'1"60d40'
Rodman,Coby Aaron
Cap 20h33'55"-26d53'
Rodman,Dennis-Alexis
Boo 14h59'39"20d10'
Rodman,Kelly
Lyn 8h57'14"35d35'
Rodman,Robert
Cet 2h56'1"2d23'
Rodman,Shirley
Cnc 8h26'11"30d11'
Rodman-A String of
Pearls,Judith
And 23h48'41"46d15'
Rodney
Cep 2h56'25"78d46'
Rodney Lee
Aql 20h6'1"8d48'
Rodney's Golden Dream Star
Ori 5h57'11"12d49'
Rodney's Star
Ori 5h55'34"15d40'
Rodney's Starlight Express
Lac 22h35'37"37d43'
Rodney,Dawn
Lyr 18h56'34"32d33'
Rodnite,Elizabeth Mary
Mon 6h47'45"10d49'
Rodnite,Katherine Rose
Fri 4h13'42"-12d52'
Rodolfo
Sct 18h54'32"-8d57'
Rodolfo ed Erminia
Cas 2h52'29"60d25'
Rodolfo,Vera
Lac 22h22'1"50d26'
Rodolphe Mon Amour
Tri 2h33'13"34d59'
Rodrigues 29 Stars, General
Dolores
Aql 18h59'0"-7d35'
Rodrigues,Ignatius H
Aur 4h55'34"40d19'
Rodrigues,Kathryn
Cnv 13h54'51"40d53'

Rodrigues,Vavø & Vav53'
Uma 8h52'33"55d5'
Rodriguex,Alexandra Reneé
Eri 3h5'44"-2d9'
Rodriguez,Adilersia
Uma 10h2'1"68d33'
Rodriguez,Alejandro
Eri 4h14'1"-17d22'
Rodriguez,Alex Edward
Aur 6h27'55"35d58'
Rodriguez,Basilisa
Cam 3h39'28"61d11'
Rodriguez,Bert
Hya 8h53'47"5d52'
Rodriguez,Betsy
Sge 20h17'35"20d31'
Rodriguez,Brianne Raquel
Lyr 18h40'34"31d44'
Rodriguez,Christian Edward
Ori 5h51'41"11d2'
Rodriguez,Christina Marie
Mon 6h43'53"10d20'
Rodriguez,Christina Rose
Aql 19h2'15"-5d46'
Rodriguez,Corey M
Cyg 21h50'56"37d52'
Rodriguez,Damien Michael
Cnc 8h28'11"31d17'
Rodriguez,Daniel
Lyn 8h1'52"40d42'
Rodriguez,Dario Joseph
Dra 16h48'21"67d10'
Rodriguez,David
Dra 19h58'29"70d54'
Rodriguez,Deanna
Mon 6h56'58"7d59'
Rodriguez,Diana
Peg 22h3'16"25d29'
Rodriguez,Dina Renae
Peg 22h49'1"27d45'
Rodriguez,Dulce M
Mon 7h18'36"-6d44'
Rodriguez,Eddie
Lmi 9h45'0"34d40'
Rodriguez,Eric William
Vir 13h38'42"-1d14'
Rodriguez,Fernando De la
Torre
Cep 20h18'56"60d26'
Rodriguez,Fred
Per 3h39'38"38d14'
Rodriguez,Geraldine
Com 12h54'10"27d53'
Rodriguez,Gerzon
Eri 4h27'48"-8d38'
Rodriguez,Granville
Mon 8h6'39"-4d1'
Rodriguez,J Emilo
Cnc 8h1'34"8d37'
Rodriguez,Jackeline Marie
Tau 3h41'25"12d20'
Rodriguez,JB,David
Dra 15h45'12"61d43'
Rodriguez,Jeff
Hya 8h35'41"-1d11'
Rodriguez,Jesica Galit
Cam 12h53'23"77d20'
Rodriguez,Jesus
Cam 11h3'1"82d8'
Rodriguez,Jesus
Ori 6h4'53"20d14'
Rodriguez,Joe S
Boo 14h37'0"40d49'
Rodriguez,Jose Luis
Her 18h3'18"31d52'
Rodriguez,Joseph Delores
Dra 18h56'54"58d60'
Rodriguez,Juan
Aql 19h58'52"-10d21'
Rodriguez,Julio A
Cep 1h45'30"78d53'
Rodriguez,Karl Eric
Lmi 10h11'20"33d47'
Rodriguez,Katherine
Sge 19h36'39"18d51'

Rodriguez,Les
Del 20h50'38"9d46'
Rodriguez,Lisa Charlene
And 1h52'22"39d30'
Rodriguez,Lisa
And 0h17'59"37d8'
Rodriguez,Lora Nicole
Del 20h19'28"9d26'
Rodriguez,Lucy
Cas 0h27'0"61d16'
Rodriguez,Luis & Magui
Crb 15h58'24"37d52'
Rodriguez,Mandy
Cet 0h30'38"-11d25'
Rodriguez,Maria
Cas 1h42'33"75d39'
Rodriguez,Maria Balbina
Sco 17h26'39"-40d38'
Rodriguez,Marie Ange
Aql 19h8'53"13d18'
Rodriguez,Marina
Hya 9h39'30"-11d12'
Rodriguez,Mary Jane
Cnc 7h57'30"10d0'
Rodriguez,Mary Louise
Cam 7h49'12"68d47'
Rodriguez,Oscar P
Leo 10h48'1"8d21'
Rodriguez,Philip Andrew
Psc 0h44'13"20d42'
Rodriguez,Providence
Aur 4h50'26"40d24'
Rodriguez,Ramon Rafael
Cep 21h56'1"70d6'
Rodriguez,Raul
Ser 18h35'13"3d27'
Rodriguez,Richard
Cep 22h6'32"60d14'
Rodriguez,Richard A
Hya 9h32'44"1d47'
Rodriguez,Rosa
Cet 2h16'0"-18d50'
Rodriguez,Saturnino &
Carmen
Cyg 19h26'12"31d34'
Rodriguez,Sophie
Mon 7h42'47"-4d11'
Rodriguez,Sue
Cam 9h35'17"84d47'
Rodriguez,Thomas
Uma 11h23'58"38d34'
Rodriguez,Vanessa Ann
And 0h59'0"39d41'
Rodriguez,Vincent Rafael
Gem 6h43'0"35d9'
Rodriguez,Wayne Charles
Mon 8h7'51"-3d23'
Rodriguez,Wilma "Billie"
Uma 11h26'1"57d38'
Rodriquez,Junita
Mon 7h56'55"-8d11'
Rodriquez,Miriam & Emilio
Aql 19h30'50"8d21'
Rodriquez,Neysa
Mon 8h0'44"-6d60'
Rodríguez,Danyssa Joandís
Rosa
Peg 23h46'14"15d24'
Rodríguez,Karyssa Joandís
Rosa
Peg 23h46'43"18d14'
Rodway,William
Dra 14h50'50"60d3'
Roe "Tulip" Star,The Mark
Cam 3h31'13"62d6'
Roe's Star
Per 2h31'11"57d31'
Roe,Andrae Mackenzie
Sgr 20h21'15"-28d9'
Roe,Brian Denis
Lib 15h3'12"-4d38'
Roe,Charles Barnett
Cyg 21h31'10"49d45'A
Roe,J B
Tri 1h59'52"30d26'

Roe,Jackson & Nona
Ori 5h56'55"15d3'
Roe,Jazmyne Mariah
Peg 23h34'25"12d28'
Roe,Josephine
And 1h18'1"38d56'
Roe,Marilyn Marie
Cyg 21h31'10"49d45'B
Roe,Nancy
Eri 3h17'49"-1d51'
Roe,Paula
Lyn 7h53'1"58d17'
Roe,Randy
Peg 23h1'10"30d11'
Roe,Sharon Lynn
Cyg 20h3'20"38d19'B
Roe,Sr,David E
Boo 15h7'1"14d27'
Roe,Sr,James W
Her 17h5'48"41d43'
Roe-"Matter Of Time", David
Ori 4h52'49"1d16'
Roebuck,Blaque Alexandria
Uma 14h7'50"60d19'
Roech,Robert B
Her 17h0'57"21d12'
Roecker,Steve
Dra 19h41'31"80d16'
Roedell,Rikki Kristina
And 23h42'54"42d53'
Roeder,Bud
Crb 15h56'47"37d58'
Roeder,David Fred
Vir 11h37'24"4d58'
Roeder,John Henry
Ori 4h59'26"-2d8'
Roeder,Luana
Uma 9h23'0"61d18'
Roeder,Meryl Jean
Cnv 13h20'1"41d50'
Roeder,Sonja K Rugh Dodd
Cas 1h46'0"61d17'
Roederer,Ian Ulysses
Leo 10h15'56"13d23'
Roeglin,Gordon
Lac 21h59'19"41d30'
Roehr,Elle Marie
Vir 13h37'51"-4d21'
Roehr,Orine & Ray
Crb 15h51'0"38d58'
Roehre,Mary Dorine
Mon 6h2'0"-4d22'
Roehrick,Greg D
Her 16h26'21"48d58'
Roelker Family Star
Eri 4h59'1"-4d17'
Roell,Lynn
Cyg 21h32'49"44d38'
Roeltgen,Marie
Ori 6h6'18"10d25'
Roemer
Umi 13h27'45"70d15'
Roemer,Barbara & Emil
Sammak
And 23h10'0"40d59'
Roensch,Eleanor "Jerry"
Cyg 19h25'53"33d43'
Roenspies,Cathy S
Lyr 18h34'0"35d38'
Roenspies,Christy A
Peg 21h55'18"34d38'
Roenspies,Kenneth
Her 18h2'25"40d22'
Roenspies,Lynne M
Lyn 8h37'0"43d58'
Roenspies,Marcia A
Cas 22h58'54"54d9'
Roeper,Ruth
Vul 19h15'57"24d43'
Roerden,Al
Dra 12h24'41"71d22'
Roesch,Ellen Eloise
Mon 7h24'22"-10d38'
Roesch,Mark William
Boo 13h43'37"14d32'

Roesch,William Mark
Sgr 19h38'18"-35d4'
Roese,Janis
Lyn 6h18'39"58d20'
Roese,Sheri
Ori 5h59'19"6d19'
Roesel,Meghan Lynn
Mon 6h17'25"-5d25'
Roeseler's Class '92- '93-
GIER,Mrs
Aur 5h7'43"40d32'
Roesener,Dittmar
Oph 17h5'40"1d43'
Roeser,Bonnie
Crt 11h22'54"-15d53'
Roesler,Kyle G
Aql 20h4'26"0d59'
Roessel,Jenna Hayley
Cas 0h13'58"50d50'
Roesslein,Forest Amanda
Cam 6h31'44"82d57'
Roesslein,Michael John
Lac 22h8'52"49d19'
Roessler,Daniela
Uma 9h9'14"72d29'
Roethig,Jason
Her 17h55'31"20d12'
Roethlein,Christopher John
Boo 14h27'30"20d4'
Roettger,Roxanne
Cas 1h3'58"63d29'
Roetto,Paul J
Ser 18h19'59"0d1'
Roff,Karol
Uma 9h3'28"52d37'
Roffe,Michael Pessin
Aur 5h1'9"29d1'
Roffman,Marvin B
Lac 22h54'40"55d58'
Rofhang,Monika
Sge 19h8'13"18d21'
Rofofsky,Matthew Stephen
Cam 12h32'43"77d9'
Rog,Matthew David
Ori 5h57'17"18d33'
Rogala,Hank
Her 18h6'27"40d13'
Rogale,Matthew
Her 16h38'38"28d6'
Rogalski,Jens
Lmi 10h44'17"26d24'
Rogalski,Jr,John William
Aur 6h33'1"32d59'
Rogalsky,GünterG
Cmi 7h18'1"0d16'
Rogan,Madeline
Psc 0h48'27"32d50'
Rogath,Leo Alexander
Aur 5h58'1"31d1'
Rogene,Lana
And 2h7'52"39d37'
Roger & Anastasia
Per 3h53'50"40d49'
Roger & Caroline
Uma 11h36'35"49d24'
Roger & Fiorella
Cyg 19h22'29"28d24'
Roger & Melissa Together in
the Heaven
Lmi 10h9'53"35d54'
Roger & Shiela
Sge 20h2'18"20d43'
Roger Allen
Ori 5h52'50"1d24'
Roger Being
Per 1h51'22"50d21'
Roger Trinary ABC,The John
Tau 3h31'28"24d17'ABC
Roger's Dream Girl
Lyr 19h8'47"47d5'
Roger's Hideaway in Heaven
Aql 19h52'27"13d15'
Roger's Joy
Cmi 7h33'41"0d7'

Roger's Spirit
Lac 22h48'0"38d14'
Roger's Star
Per 3h30'13"37d59'
Roger,Adrian
Vul 19h41'44"20d30'
Roger,Cecilia S R
Cas 1h33'29"58d28'
Roger,Darlina
Umi 16h54'32"77d17'
Roger,Jonathan
Umi 16h56'34"77d59'
Roger,Laurence
Aur 6h4'0"38d40'
Roger,Patricia Sue
Lyr 18h54'36"31d42'
Roger,Stellar
Aur 5h58'0"31d27'
Roger,Timothy
Her 17h1'33"30d2'46'
Rogeria Star Of Brazil
And 0h40'0"41d4'
Rogers Starship,Wayne
Leo 11h28'0"-5d23'
Rogers,Abigail
Peg 22h56'27"21d12'
Rogers,Alexandra Thurston
And 23h20'44"44d33'
Rogers,Alice
Cas 3h6'45"70d46'
Rogers,Amy & Wayne
Eri 2h50'43"-1d43'
Rogers,Andrew Peter
Lac 22h13'15"50d16'
Rogers,Autumn L
Aql 19h26'49"10d16'
Rogers,Aïda
Sge 19h52'15"16d33'
Rogers,Barbara A
Lyr 18h33'34"30d32'
Rogers,Beverley Jean
Lyr 19h14'21"42d21'
Rogers,Bill
Psc 1h6'56"23d31'A
Rogers,Carol
Eri 3h30'43"-6d7'
Rogers,Carol M
Cas 23h17'24"61d26'
Rogers,Carrie
Cnv 12h7'14"37d0'
Rogers,Charles Donald
Mon 8h5'29"-2d34'
Rogers,Charlie
Aql 20h19'12"7d39'
Rogers,Christopher Barry
Her 17h59'38"30d17'
Rogers,Chrsitopher Peter
Per 2h52'0"34d44'
Rogers,Cynthia Ann
Oph 18h6'45"13d5'
Rogers,Dave
Boo 15h9'31"50d43'
Rogers,David & Tara
Aql 19h25'43"10d43'
Rogers,David Frederick
Cep 1h56'17"77d35'
Rogers,Debi
Hya 8h28'49"-9d56'
Rogers,Don Eaglerock
Dra 16h35'18"51d54'
Rogers,Doug
Uma 11h16'43"30d46'
Rogers,Dr Harry
Ori 5h57'41"10d55'
Rogers,Elizabeth Rose
Peg 21h28'49"26d50'
Rogers,Estel Carrington
Cam 3h49'42"58d41'
Rogers,Estelle
Cnv 12h25'0"32d45'
Rogers,Forrest Wendell
Per 3h29'1"51d5'
Rogers,Gary L
Cet 2h13'12"4d30'

Rogers,Glenn B
Hya 8h25'37"1d17'
Rogers,Gwen J
Crb 16h11'47"37d51'
Rogers,Heather Colleen
Vul 20h58'0"20d9'
Rogers,James John
Aql 19h3'43"-0d3'
Rogers,James T
Uma 11h40'18"31d40'
Rogers,Jane A
Peg 22h28'57"29d44'
Rogers,Janelle Amelia
Cas 2h57'17"63d54'
Rogers,Jeff
Dra 17h36'52"60d31'
Rogers,Jessica Diane
Cas 2h56'0"63d54'
Rogers,John
Ori 6h4'49"18d59'
Rogers,John Wayne
Hya 8h14'28"5d15'
Rogers,Joseph "Big Daddy"
Aur 5h4'18"42d19'
Rogers,Jr,David H
Boo 15h2'1"22d27'
Rogers,Jr,Donald F
Oph 17h11'32"-22d53'
Rogers,Jr,Travis M
Aql 19h59'31"10d34'
Rogers,Julie "Fluffy"
Cas 0h28'29"61d43'
Rogers,Kathleen Elizabeth
Lyr 18h48'31"42d24'
Rogers,Kathryn R
Peg 22h46'12"35d26'
Rogers,Kenneth Bruce
Ser 17h32'27"-11d7'
Rogers,Kevin Walter
Boo 14h6'29"39d55'
Rogers,Kristina Marie
Cas 0h20'0"73d35'
Rogers,Larry E
Boo 14h54'47"50d38'
Rogers,Lisa Marie Queisser
Cas 2h10'30"61d7'
Rogers,Lisa W
Mon 7h1'29"3d51'
Rogers,Lissette G
Peg 23h28'46"17d13'
Rogers,Lois
Cam 13h23'11"81d59'
Rogers,Lucille
Psc 1h6'56"23d31'B
Rogers,Lynda Mae
Com 12h58'58"28d40'
Rogers,Margaret
Lyr 18h49'60"39d9'
Rogers,Marie J
Cma 7h1'26"-31d22'
Rogers,Mark Alexander
Aur 6h18'35"32d29'
Rogers,Mark Terence
Per 2h56'20"40d36'
Rogers,Mary
Peg 22h27'49"20d46'
Rogers,Matt
Ori 6h15'51"-0d37'
Rogers,Michael
Lac 22h24'1"55d52'
Rogers,Michael & Malissa
Cet 2h50'20"-0d14'
Rogers,Nicole Lynn
Ori 5h11'38"-5d3'
Rogers,Nina & Richie
Cyg 21h3'50"31d35'
Rogers,Patricia Carol
And 2h31'24"38d9'
Rogers,Rachel Atteberry
Mon 7h42'0"-5d5'
Rogers,Raymond A
Dra 18h35'59"58d32'
Rogers,Richard
Cep 0h27'1"78d29'

Rogers,Richard C
Ori 5h57'0"17d13'
Rogers,Ryan Michael
Aur 6h11'27"50d35'
Rogers,Shelly
Lyn 7h23'31"58d52'
Rogers,Simon & Clare
Medden
Cyg 19h57'1"48d55'
Rogers,Stephen A
Her 16h23'19"25d8'
Rogers,Summer Lea
Peg 22h17'22"20d28'
Rogers,Suzanne Darlyne Watts
And 2h15'27"47d50'
Rogers,Tiffany
Equ 21h6'16"11d51'
Rogers,True Lawson
Lyr 19h22'22"31d8'
Rogers,Warren
Aur 6h32'22"30d5'
Rogers,Whitney
Aql 19h58'33"-1d1'
Rogers,William Herman
Dra 17h5'18"66d18'
Rogers,William L
Lac 22h2'23"48d50'
Rogers-Wardle,Jeffrey Gray
Gem 7h54'45"30d56'
Rogerson,Alistair Robert
Cep 22h33'45"70d33'
Rogge,Bernhard
Sgr 19h18'38"-28d50'
Rogge,Siegfried
Ari 2h37'26"21d6'
Roggio,Tina M
Eri 3h12'53"-1d52'
Rogic,Theodora & Tony
Cyg 21h5'53"30d10'
Rogich,Erin Ann
And 23h37'30"49d46'
Rogier,Genevieve
Lyn 8h29'0"34d46'
Rogin,Gloria
Cas 23h40'33"60d52'
Roglen
Ser 16h7'10"5d39'
Rogman
Boo 14h54'26"40d45'
Rogner,Hermann
Psc 1h0'37"20d58'
Rognoni,Daniel
Del 20h19'39"10d24'
Rogotzke,Debbi
Eri 3h24'15"-1d52'
Rogovitz,Howard
Her 15h54'1"40d24'
Rogovsky,Elizabeth
Lyn 7h35'0"50d45'
Rogowiec,Elena Melissa
Scl 23h9'27"-32d11'
Rogozinski,Virginia
Mon 8h0'37"-8d4'
Rogy Baby
Lyn 8h29'48"57d41'
Rohal,Darlene
Aqr 23h36'1"-6d14'
Rohaly,Alyssa Jean
And 1h35'49"48d52'
Rohaly,Michael Andrew
Ori 5h20'12"-10d25'
Rohan,Jack Walter
Boo 14h34'45"15d23'
Rohan,Jimmy
Cyg 21h31'54"36d17'
Rohde,Connie
Lyn 8h30'42"52d12'
Rohdes,Dustin Mark
Her 17h55'25"37d39'
Rohe,Jr,Robert
Oph 17h12'60"-22d37'
Rohel,Dorothy
Agr 23h32'56"-5d40'
Rohini
Gru 22h17'13"-51d21'

Rohland,Angelic Elizabeth
Psc 1h27'58"20d34'
Rohlfs,Anne
Mon 7h7'1"0d51'
Rohling,Jackie
Lyn 8h40'54"45d26'
Rohloff,Carrie
Cas 2h9'28"59d54'
Rohovsky,Stephanie & Jeff
Sternberg
Gem 7h19'48"28d34'
Rohowits,John Thomas
Aql 20h33'51"0d52'
Rohr,Manfred
Leo 10h59'0"10d45'
Rohr,Matthew Levi
Dra 19h22'34"80d26'
Rohr,Maximilian Alexander
Lib 14h39'40"-6d35'
Rohrbach,Anna
Peg 22h4'45"22d15'
Rohrer,Benjamin Stephen
Cnv 12h54'34"42d53'
Rohrer,David
Boo 14h57'13"43d32'
Rohrer,Gene
Cnv 13h5'45"41d4'
Rohrer,Josefine
Sgr 19h35'17"43d9'
Rohrer,Natalie Elizabeth
Peg 22h4'31"20d17'
Rohrer,Tajah
Mon 7h6'17"0d53'
Rohwer,MiMi
Eri 2h54'23"-13d11'
Rohwer,Nancy
Cet 1h2'59"0d17'
Rohé,Gerd
Boo 14h7'26"11d42'
Roick
Lyr 19h20'26"31d20'
Roicki,Stacia
Cas 1h36'14"70d31'
Roinnel,Jules E
Uma 13h36'37"50d40'
Roissard,Georges
Equ 21h3'1"8d32'
Roiston,Lauren Diane
Lyr 18h17'46"33d57'
Roiz,Theresa
Cas 23h34'55"61d56'
Rojano,My Love,Marco
Cet 2h10'1"6d21'
Rojas,"Unforgetable" Dr
Gilberto
Oph 18h40'12"6d27'
Rojas,Sean
Her 17h21'15"22d3'
Rojas,Sigrid
Aur 5h27'5"29d41'
Rojas,Taylor Marie
And 23h0'44"41d6'
Rojerre
Boo 14h27'20"28d6'
Rojezal,Souche
Aur 5h51'31"30d38'
Rojoyce
Umi 13h29'25"70d14'
Rojs,Doris
Sgr 19h2'14"-22d23'
Roksund,Finn
Uma 9h49'1"50d45'
Roland
Cam 7h10'31"80d17'
Roland
Umi 17h5'31"78d53'
Roland
And 0h26'56"43d43'
Roland
Aur 7h0'35"41d8'
Roland
Ori 5h55'60"7d48'
Roland et Francoise
Aql 18h58'38"13d3'

Roland,Katie
And 2h20'20"45d0'
Roland-Lisa Marie- Mathieu
Per 4h5'39"36d3'
Rolande
Uma 10h33'36"48d5'
Rolande,Mohamed Fage
Per 3h53'1"39d59'
Rolander,Dan
Equ 21h22'41"10d1'
Rolando
Mon 8h0'1"-6d26'
Rolando,Ramon & Sandra
Alice Pulido
Dra 16h43'1"71d18'
Rolape My Angel, Kristie Ann
And 2h18'1"42d38'
Rolat,Sigmund A
Aur 5h3'35"38d21'
Rolcke 16-8-1945, Rudolf
Ori 5h54'44"1d19'
Roldan,Hope
Cas 0h7'45"62d21'
Roldania
Ind 21h10'1"-50d0'
Rolf
Ori 6h17'29"7d31'
Rolf & Jackie
Cyg 21h20'1"28d53'
Rolf and Leah's Happiness
Forever
Lac 22h32'43"41d6'
Rolfe's Seven,The Star of
Ori 6h3'31"9d5'
Rolfe,Gladys Evelyn
Lyn 8h41'13"34d25'
Rolfe,Robert Hoagland
Aql 19h58'28"14d38'
Rolfe,William George
Per 4h32'39"38d23'
Rolfsen,Rick "My Peep"
Equ 21h3'40"2d57'
Rolike
Cam 6h56'0"67d32'
Roll,Erik Michael
Per 2h0'1"50d19'
Roller,Nikki Lee
Ser 17h51'20"-11d19'A
Roller,Susan Jane Cathrine
Ori 4h41'16"11d48'
Rollercoaster John &
Stephanie
Aql 18h59'31"13d16'
Rolleston,Paula Haines
Com 13h5'47"27d25'
Rolley
Eri 4h6'32"-7d51'
Rollie & Marlene
Sge 19h59'38"16d17'
Rollings,"Puppy Eyes" Martin
Cnv 13h17'16"37d58'
Rollings,Katie
And 0h1'47"31d51'
Rollings-A Star Forever,Tim
Oph 16h52'44"0d49'
Rollins,Doug
Per 2h26'48"58d29'
Rollins,Frank
Aur 5h9'36"41d23'
Rollins,Kevin Joseph
Per 2h59'0"46d15'
Rollins,Lisa
Cyg 19h29'51"35d4'
Rollins,Lisa Ann Daniel
Mon 8h1'22"-8d29'
Rollins,Marie
Lyr 18h17'1"47d35'
Rollins,Roderick Marcus
Lmi 10h41'57"30d5'
Rollinsius,Jennius Janius
Sgr 18h47'49"-34d31'
Rollison,Vicki
Ori 4h57'14"5d24'
Rollman-Rice
Cep 23h18'39"64d26'

Rollo,Merle
Per 2h58'19"40d23'
Rollow,Larry
Cmi 7h58'32"4d46'
Rolls,Alison
Lyr 18h51'49"41d23'
Rollwage,E Jeanne
Cas 2h7'20"60d30'
Rolly
Cma 7h12'39"-13d55'
Rolnick,Alvin Mortimer
Cam 11h55'18"81d29'
Rolo Contendre
Sct 18h30'53"-6d52'
Roloff's Star,Lynn
Oph 17h9'10"-20d4'
Roly,Bianca
Cyg 19h18'1"28d51'
Roma
And 0h24'16"40d16'
Roma 40
Cyg 19h26'15"35d29'
Roma-Darlene
Cyg 21h1'20"30d12'
Romack's Star Light, Taylor
Browning
Mon 8h1'54"-3d51'
Romagnoli,Lucia
Cet 2h31'1"-6d20'
Romagnoli,Marie France
Cnv 12h53'44"38d24'
Romain,Charles C
Uma 11h6'30"71d17'
Romaine & Arlene
Her 16h48'0"50d43'
Romaine,Coles Connor
Boo 15h19'41"48d48'
Romaine,Helene Storie
Hya 9h16'12"-6d39'
Roman
Cet 2h39'1"-1d4'
Roman Gabriel
Oph 18h5'36"2d12'A
Roman,Anna
Mon 6h21'38"8d45'
Roman,Craig Michael
Per 2h52'41"40d59'
Roman,Franáois
Dra 20h1'27"80d27'
Roman,Jack & Daisy
Cyg 21h22'36"50d35'
Roman,Joseph James
Dra 12h21'47"64d0'
Roman,Joseph L
Per 3h30'45"38d55'
Roman,Juan Carlos Vidal
Uma 12h4'54"38d54'
Roman,Neil
Cma 6h55'27"-18d11'
Roman,Ruth Cohen
Ori 5h52'54"14d21'
Roman,Sarah M
And 22h58'42"50d2'
Roman,Stephan Anthony
Aur 6h52'59"38d8'
Romancier,Jack
Boo 14h54'0"42d10'
Romanelli,Michael Thomas
Cep 3h17'58"77d20'
Romanello,Patricia
Lmi 9h41'47"38d21'
Romani "K",Katherine Abrami
Mon 7h42'1"-0d13'
Romani,Karen
Vul 20h4'43"22d38'
Romanick,Robert
Cep 3h25'21"78d30'
Romano "Our Star",Joe &
Cathy
Aql 20h15'1"7d31'
Romano,Angelo Salvatore
Dra 17h22'1"68d22'
Romano,Bonnie
Vul 19h45'57"26d12'

Romano,Evelyn
Lyr 18h27'23"30d9'
Romano,Francesco
Her 18h9'36"37d36'
Romano,Geraldine
Dra 13h44'23"68d15'
Romano,Hazel Leslie
Eri 3h38'60"-14d58'
Romano,Jason Scott
Ori 5h56'51"15d52'
Romano,Joe & Jean
Cep 22h52'28"58d59'
Romano,John Michael
Aur 6h0'15"54d29'
Romano,Joseph J
Her 17h5'49"40d37'
Romano,Jr,Salvatore Dominic
Aur 6h29'25"31d43'
Romano,Kim Ellen
Cnc 9h1'0"11d8'
Romano,Leigh Alexander
And 23h23'40"50d31'
Romano,M E
Aur 5h50'40"50d26'
Romano,Maria
Cam 5h45'9"73d50'
Romano,Marilee
Eri 3h8'50"-3d49'
Romano,Meryl
Cam 6h57'16"63d50'
Romano,Michael Steven
Boo 13h39'45"19d56'
Romano,Michele Concetta
Lyn 8h10'1"35d28'
Romano,Nicole Vivian
Cas 0h32'20"65d43'
Romano,Nina
And 1h6'58"38d52'
Romano,Nina Alise
Uma 8h42'39"48d15'
Romano,Piero
Aur 6h8'14"37d39'
Romano,Richard M
Aur 7h23'35"38d9'
Romano,Rodney Guy
Oph 17h10'35"11d3'
Romano,Rose
Cas 2h10'59"61d22'
Romano,Ross
Cep 23h24'0"70d54'
Romano,Tony
Aur 6h7'54"33d52'
Romano,Vince
Aql 19h12'38"12d21'
Romano,Vincent Louis
Aql 19h30'59"13d11'
Romanoski,"Scrubadub" for
Joseph T
Her 17h15'32"26d17'
Romanoski,Mary
Cas 22h58'0"55d52'
Romanoski,Shawn
Cep 3h27'27"80d30'
Romanov,Helena
Peg 22h44'21"31d20'
Romanov-Stark,Lucia
Cas 2h38'30"67d46'
Romanow,Jr,Edward B
Dra 11h41'53"71d56'
Romans II v 33 fff
Aur 6h29'22"34d58'
Romans,Chelsea Danielle
And 0h39'28"40d38'
Romans,Christopher
Oph 18h16'0"11d57'
Romans,Joseph A
Mon 6h18'22"8d45'
Romantic Fantasy
Uma 10h51'11"40d2'
Romasco,Jay
Her 16h39'23"22d60'
Romashko,Mary
Lyn 6h13'16"54d14'
Romberg,Sevastian Nils
Tau 4h4'1"10d37'

Romberg,Stian Kristov
Psc 23h27'55"6d17'
Romberger,David L
Cep 23h33'51"68d12'
Rombey,Dona N
Cas 3h6m12"60d53'
Rome
Dra 20h2'59"60d45'
Rome,Alana Marie
And 1h18'55"40d26'
Rome,C J
Sgr 20h0'22"43d58'
Rome,Catherine Tinney
Vul 21h19'1"28d14'
Rome,Kathleen Darcy
Mon 6h11'15"-10d21'
Rome-Bouchard
Cyg 20h22'24"38d53'
Romeiko's Wedding Star
Nancy & David
Cyg 21h38'53"42d4'
Romeo
Her 16h24'1"26d44'
Romeo
Lac 22h40'34"51d22'
Romeo & Juliet
Uma 9h0'27"57d56'
Romeo & Juliette
Lac 22h5'1"46d41'
Romeo Papa
Ori 5h10'48"-8d33'
Romeo,Joseph M
Dra 19h7'0"61d39'
Romeo,Laura & Steve
Cyg 19h59'60"30d31'
Romeo,Laura Ann
And 2h30'24"44d18'
Romeo,Robin
Lyr 18h40'50"32d18'
Romer,Janet Lee Crawford
Lyn 7h57'0"50d4'
Romero's Angel,Carlos
Her 17h23'17"50d4'
Romero,Charles
Sex 9h42'1"4d50'
Romero,David
Cet 0h56'42"-6d9'
Romero,Diane M
Cas 1h46'0"58d20'
Romero,Francis
Oph 18h41'48"7d23'
Romero,Joseph
Tri 1h30'32"28d53'
Romero,Kate
Peg 21h55'0"3d42'
Romero,Lisa
Mon 7h56'42"-9d2'
Romero,Mary
Vir 11h48'39"6d9'
Romero,Melanie Annn
Aql 19h13'1"10d38'
Romero,Melissa Anne
Figueras
Aql 19h45'1"10d55'
Romero,Randall Lee
Lib 14h34'47"-10d18'
Romero,Roderic
Her 17h37'46"27d24'
Romero,Ronald Joseph
Peg 23h5'54"33d57'
Romero,Yhina
Lyn 8h13'39"58d58'
Romes(Patty,Murray &
Chipper),The
Her 17h39'1"18d49'
Romey,Mildred Carol
Peg 23h24'59"15d35'
Romich,Kyle Matthew
Per 2h27'21"50d55'
Romig,Melinda
Peg 21h49'31"32d16'
Romilly
Sge 20h0'46"16d31'
Romina
Sgr 19h24'28"-40d50'

Romine Irish Star
Umi 13h6'59"71d55'
Romine,Don & Mary Jane
Cyg 19h29'58"33d56'
Romine,Sarah Gayle
Uma 10h33'15"52d44'
Roming,Melissa Lynn
Dra 16h53'19"68d37'
Rominger,Yvonne,Mark, Alec
& Allison
Cep 20h17'53"60d49'
Rominski,Joseph
Aur 4h59'19"31d10'
Romita,Donna
Aql 19h34'18"0d25'
Romita,Joseph
Dra 19h50'0"68d58'
Romita,Roseann
And 1h4'48"38d17'
Romita,Susan
Cyg 21h3'42"30d38'
Romiti,Ginny
Cap 20h29'39"-10d1'B
Romiti,Maura
Lyn 7h4'32"60d10'
Romito,Nicole
Lyn 8h6'35"39d42'
Rommel,Gerlinde
Cnc 9h16'46"8d14'
Rommel,Greg
Her 16h10'59"5d28'
Rommel,Peggy
Lyn 7h58'40"43d23'
Romney,Alexandra Isabella
Cas 0h58'49"61d58'
Romo,Jayne
Aql 19h59'1"8d21'
Romona 7
Cas 0h1'32"58d59'
Romond,Cynthia
Boo 15h2'0"48d41'
Rompala,David John
Lac 22h51'0"37d36'
Romstadt,Nancy J
Mon 6h54'1"8d25'
Romvos,Judy
Umi 16h17'50"70d2'
Romy
Uma 14h25'49"60d59'
Romy Joe
Psc 1h1'49"23d25'
Romy Lauren
Lac 22h26'36"50d20'
Romy-Vaz,The
Sge 19h56'45"16d51'
Ron
Cmi 7h28'18"8d0'B
Ron
Sco 16h30'45"-41d15'
Ron
Dra 11h51'2"74d2'C
Ron
Cam 5h25'19"79d48'B
Ron
Aql 20h11'55"10d59'
Ron
Aur 6h23'31"33d34'
Ron & Helaina
Gem 7h14'44"28d49'
Ron & Kath
Cyg 20h34'31"58d27'
Ron & Peanut
Cep 1h23'43"77d35'
Ron & Peggy Starr,The
Eri 3h55'11"-10d9'
Ron & Robin
Uma 9h44'30"58d21'
Ron Roy
Dra 17h6'38"61d6'
Ron the Great
Boo 14h38'1"52d23'
Ron The Snow Plow
Boo 14h32'0"44d23'
Ron's Little Piece Of Time
Dra 17h6'34"64d7'

Ron's Rush
Aql 19h47'39"14d22'
Ron's Star
Cnv 12h22'25"43d54'
Ron's Star
Her 16h59'14"28d34'
Ron/Art Inc
Dra 17h13'42"70d44'
Rona
Peg 23h30'1"18d36'
Ronacher,Armin
Lib 15h24'21"-18d27'
Ronald & Lauren
Uma 9h17'60"56d49'
Ronald & Mary's Anniversary
Star
Lmi 10h35'0"24d46'
Ronald & Nora
Boo 15h2'56"16d29'
Ronald T's "Giftstar"
Cnv 14h2'26"31d39'
Ronald,Marilyn "Bunny"
Cyg 19h36'23"38d29'
Ronallo,Martin
Her 16h6'44"40d9'
Ronalo & Benno
Ari 2h55'46"30d28'
Ronan,Kayla
Cam 12h16'48"77d22'
Ronan,Nicholas Andrew
Cep 20h55'55"58d46'
Ronan,Terrence F
Sgr 19h37'2"-32d35'
Ronan,Whitney
Lyn 9h36'24"40d14'
Ronca III,William E
Vir 14h0'47"-1d47'
Ronca,Dinastia
Lyn 8h10'57"41d50'
Ronca,Guiseppe Luigi
Uma 11h25'0"32d29'
Roncari,Maria
Leo 9h52'40"10d6'
Ronchera,Alda Rodriguez
Cmi 7h17'53"8d35'
Ronchetta,Fabrizio
Her 16h44'44"48d20'
Ronchi's,"First Birthday
Star",Jake
Per 2h5'11"57d47'
Roncone,Jr,John L
Cam 3h32'40"60d34'
Ronda I
Equ 20h56'57"3d15'
Ronda Loraine
Oph 18h4'19"13d23'
Ronda Lynn
Aql 20h9'51"6d53'
Ronda-1993
Lyr 18h21'10"46d58'
Rondanice
Uma 9h33'19"57d6'
Rondeau,André
Ori 6h0'35"6d19'
Rondeau,Stephanie Gloria
And 23h36'0"42d50'
Rondeau,Stéphanie
Per 2h14'60"58d28'
Rondell,Mary
Del 20h17'34"10d30'
Rondell,Ronnie
Eri 3h30'42"-3d10'
Rondoletto,Joseph Michael
Aur 6h7'43"38d40'
Ronel
Boo 14h12'37"32d27'
Ronen,Daphna
Uma 8h58'54"57d48'
Roner,Jack Gordon
Cyg 19h29'46"32d37'
Rones,Clifford
Her 18h26'16"20d8'
Roney,Maya Moriarty
Cam 8h24'52"82d44'
Ronga,Matteo
Leo 9h21'50"27d34'
Rongey,Lydia M u Thomas E
Uma 9h36'19"45d23'
Rongo,Lucille
Dra 19h0'31"58d46'
Roni & Rene,Les Roi de
Figaro
Leo 10h3'59"18d6'
Roni's Hope
Aql 19h19'45"15d4'
Ronja
Cnv 14h1'37"38d41'
Ronja Andrina
Cyg 21h17'0"38d34'
Ronn,Orie Rechtman
Lyn 8h51'23"40d10'
Ronnel,E Lee
Her 17h34'17"21d42'
Ronni
Vul 20h4'33"25d21'
Ronni & Ed
Oph 17h5'44"-20d10'
Ronni's Heavenly Body
Cep 0h3'42"75d12'
Ronnie
Hya 8h18'34"1d45'
Ronnie
Aur 5h58'14"28d35'
Ronnie
Her 17h13'18"21d3'
Ronnie
Her 16h11'50"48d13'
Ronnie
Her 16h44'27"16d9'
Ronnie & Jayne
Cyg 19h30'39"38d48'
Ronnie & Jimmy
Cyg 21h12'18"39d46'
Ronnie & Sassy
Lmi 9h52'52"33d57'
Ronnie B
Cam 7h39'0"67d33'
Ronnie M
Her 16h22'59"26d5'
Ronnie Ray
Her 15h49'34"41d5'
Ronnie's Star
Per 2h2'26"56d35'
Ronnie's Star
Aur 7h25'54"40d6'
Ronnie's Star
Dra 16h29'15"67d46'
Ronnie's Star
Lac 22h52'59"54d45'
Ronnie's Star
Uma 10h14'58"53d2'
Ronny B
Sex 10h3'33"-1d33'
Ronny G
Cep 21h14'29"65d6'
Ronollo,Sr,Richard A
Ori 5h23'43"-3d22'
Ronquillo,Andrew
Cam 14h14'1"81d10'
Ronquillo,Karen Marie Marie
Uma 11h22'47"62d8'
Ronquillo,Sr,Allan Louis
Uma 10h57'0"61d22'
Ronsen,Steve
Boo 15h9'1"48d32'
Ronsivalle II,Nino- Joseph
Anthony
Uma 11h43'11"56d11'
Ronsstar Forever
Lac 22h15'1"54d58'
Ronster,The
Her 17h31'45"40d57'
RonSue
Cet 1h12'60"-0d15'
RonSun
Aql 20h1'50"11d33'
Ronulous
Ori 5h53'27"15d15'
Ronzo
Vul 19h22'59"27d9'
Ronzoni,Marco Emanuele
Dra 16h23'49"62d20'
Ronzoni,Mary Ann & Richard
Cyg 20h58'50"30d28'
Ron
Del 20h26'27"10d37'
Roo
Ser 15h27'1"0d48'
Roo Child
Lyn 9h3'17"42d34'
Roo's Star
Her 16h53'28"48d46'
Rooak
Sct 18h52'10"-6d29'
Roobik's Star
Peg 21h58'21"28d34'
Rood,Don D
Boo 14h29'41"21d59'
Rood,Michael Paul
Aql 19h13'28"19d13'
Rood,Stephen Andreas
Sex 10h23'12"-1d41'
Roof,Ryan L
Per 3h37'11"38d35'
Rook,David W
Lac 22h15'25"50d21'
Rook,Patrick
Hya 8h43'23"3d45'
Rooke,Jessica K You are not
Alone
Cru 12h35'26"-59d0'
Rooker,Amy Margaret
Peg 22h1'20"2d12'
Rooker,George L
Dra 18h18'58"65d4'
Rooker,Jane A
Hya 8h26'0"-8d41'
Rooker,Kelly
Aql 20h1'50"0d4'
Rooks,Jason R
Her 18h18'54"12d23'
Rooks,Jonathon Scot
Cap 20h34'35"-11d3'
Rooks,Sargent
Aql 19h27'45"8d31'
Rookwood,Teena Marie
And 23h41'11"40d36'
Room 17
Dra 20h30'53"68d15'
Room 8 Super Star
Peg 22h58'49"11d36'
Room I Fairmont 1994
Aql 19h57'43"14d28'
Rooney
Uma 8h31'52"52d1'
Rooney,Andy
Oph 17h32'21"-22d11'
Rooney,Avril
Cas 1h31'51"73d34'
Rooney,Brendon
Her 16h37'38"24d52'
Rooney,Gene
Aur 6h54'23"37d45'
Rooney,Jennifer-Jim
Eri 3h5'40"-1d59'
Rooney,Meg
Lyn 8h54'14"43d1'
Rooney,Mr Kevin Francis
Per 2h52'20"43d11'
Rooney,Priscilla D
And 2h34'0"39d37'
Rooney,Shaylyn
Peg 22h46'41"2d28'
Roons,Annet
Lac 22h21'30"40d13'
Roopchand,Susan
Lyr 18h22'0"42d42'
Roope
Umi 14h35'30"68d34'
Roophicks
Boo 13h54'19"17d43'
Roorda,Michael David
Uma 8h50'51"53d42'
Roos
Oph 18h33'50"11d12'A
Roos,Lloyd I
Her 16h13'53"47d36'
Roos,Maria
Lyr 18h51'45"42d55'
Roos,Michael
Mon 7h1'41"-6d21'
Roos,Ned
Cnv 12h48'47"32d8'
Roose,Sheryl & Robert
Sge 19h59'36"19d27'
Roosevelt,Katherine
Cas 0h18'10"64d60'
Roosma,Michael William
Dra 14h58'25"55d12'
Rooster's Gwiazda
Uma 10h26'26"62d13'
Root,Amanda Jane
Cas 0h51'1"74d53'
Root,Danny B
Her 16h10'53"41d30'
Root,David "Rooter"
Per 3h42'32"50d52'
Root,David Michael
Boo 15h6'0"14d24'
Root,Janet
Mon 6h23'26"7d35'
Root,Jerry & Eileen
Aur 5h47'19"50d4'
Root,Jr,David B
Cap 21h22'1"-15d43'
Root,Leon & Paula
Lmi 10h16'17"31d11'
Root,Mark Alan
Aur 6h2'10"32d49'
Root,Sarah Ann Bledsoe
Reeves
Crb 16h0'27"27d23'
Rop,Jennine Maria
Crb 16h14'48"32d41'
Ropa,Robert "Bob"
Aur 4h50'59"38d18'
Roper,Caitlin
And 0h46'38"38d20'
Roper,Jr,Kenneth H
Hya 8h11'36"6d8'
Roper,Lisa Jayne
Aql 19h12'12"15d26'
Roper,Nancy A
Cet 1h18'47"-12d37'
Roperto,Liza Jennine
Lyn 6h23'52"59d1'
Ropes
Mon 7h48'17"-5d20'
Ropiak,Robert F
Her 16h20'19"24d13'
ROQSI
Cnv 14h2'60"41d1'
Roques,Michele
Aur 6h1'11"38d36'
Rorer,Katherine Susan
And 2h33'49"50d5'
Rorer,Londeree
Sct 18h54'25"-6d43'
Rork,Billy
Cnc 8h30'41"8d44'
Rorke,Elizabeth Ann
And 2h27'40"49d42'
Rory
Umi 14h30'10"86d9'
Rory
Equ 21h1'45"3d9'
Rory
Per 2h55'58"50d33'
Rory's Guardian
Cru 12h26'12"-60d59'
ROS
Lac 21h59'55"40d35'
Rosa
Com 12h55'12"31d15'
Rosa
Cam 3h47'22"56d19'
Rosa
Lep 5h24'21"-20d36'
Rosa
Vul 20h15'54"23d49'
Rosa B
Aql 19h2'24"16d14'
Rosa Elena Estrella De Mi
Amor
Aql 19h8'19"1d49'
Rosa Geraldine
Umi 13h52'23"78d51'
Rosa,Beth
Lyr 18h28'37"31d44'
Rosa,Bob
Cam 7h48'51"60d39'
Rosa,Esten Watkins
Ori 5h56'31"8d47'
Rosa,Jesus R
Peg 0h4'1"18d16'
Rosa,Kellner
Cas 0h36'39"60d50'
Rosa,Laurie B Ackerman
Peg 21h38'55"26d28'
Rosa,Linda Sue
Cep 0h11'43"73d8'
Rosa,Nina
And 1h56'42"39d43'
Rosa,Pete
Boo 14h22'23"35d57'
Rosa,Salvator
Tri 1h29'45"34d39'
Rosa-Rolf & Sabine Forever
And 1h42'54"41d27'
Rosacker,Cory Thain
Boo 14h17'57"12d23'
Rosacker,Virginia Carol
Cyg 21h35'1"41d10'
Rosada,Silvia Gasparini
Dra 14h59'26"61d44'
Rosado,Jacueline
Mon 8h8'18"-9d47'
Rosado,Lizzie
Lyn 7h7'42"58d24'
Rosado,Sr,Peter D
Cep 23h0'51"65d43'
Rosalba
Cap 21h22'26"-23d59'
Rosalba
Peg 23h44'53"17d37'
Rosalba
Cas 1h46'22"75d39'
Rosales,Marisa Danielle
Aql 19h28'39"0d18'
Rosales,Porfirio Bruno
Dra 16h52'36"67d41'
Rosales,Rosa T
Mon 6h42'1"7d41'
Rosalie
Lib 14h19'1"-22d53'
Rosalie
Cap 20h5'0"-10d54'
Rosalind
Lyr 18h58'0"38d37'
Rosalind
Eri 3h58'27"-4d15'
Rosalind
Lyr 18h19'32"38d16'
Rosalinda
Eri 2h49'0"-2d3'
Rosaline
Mon 6h45'14"11d7'
Rosamilia,Frankie
Cyg 19h27'23"33d19'
Rosanna
Eri 2h43'24"-6d24'
Rosanna
Lyr 18h52'41"42d48'
Rosanna
Peg 22h30'60"20d29'
Rosanna
Cyg 20h50'17"39d36'
Rosanna
Sgr 19h28'30"-31d59'
Rosanna Dellapiana (Scottex)
Sco 17h50'23"-38d6'
Rosanne,Mark's Wish Come
True
Peg 21h39'16"25d49'
Rosano,Paul
Aqr 22h38'46"2d16'
Rosar,Edward
Aur 4h52'27"51d23'
Rosar,Frances
Cam 13h12'31"77d25'
Rosaria
Aur 4h46'1"51d1'
Rosaria
Boo 14h28'16"22d12'
Rosario
Gru 22h5'7"-53d36'
Rosario,Brenndan Robert
Lac 22h55'38"55d26'
Rosario,Catherine Luisa
Eri 4h4'27"-5d50'
Rosati,Lucilla
Uma 12h0'29"46d43'
Rosato,George & Patricia
Lac 21h55'31"41d53'
Rosback,Margery Kendall
Leo 9h22'59"18d57'
Rosch,Emily
Cam 3h50'41"57d11'
Roschen,Delphine Appleton
Cnv 13h30'43"40d22'
Roscher,Jean
Aur 5h16'1"43d9'
Roscigno,Diana M
Cas 23h17'30"60d30'
Rosciolo,James Michael
Aur 6h21'19"32d33'
Roscitt Family Star, The
Cyg 19h31'49"33d23'
Rosco My Love
Hya 8h33'31"-1d51'
Roscoe
Mon 7h3'1"1d49'
Roscoe Allen
Boo 14h55'55"53d32'
Roscoe,Chris S
Boo 14h48'1"36d56'
Roscoe,Clara Virginia
Christopher
Cap 21h8'56"-22d46'
Rose
Lyn 8h0'41"54d53'B
Rose
Uma 8h33'1"57d50'
Rose
Uma 10h47'22"68d5'
Rose
Crt 10h54'1"-22d35'
Rose
Peg 22h22'0"21d34'
Rose
Eri 3h55'1"-11d55'
Rose
And 1h6'48"38d41'
Rose
Aql 19h54'51"12d57'
Rose
And 0h46'14"37d25'
Rose
Cyg 20h15'39"38d33'
Rose & Bill Together In The
Stars
Lmi 10h9'0"35d49'
Rose & Fred
Boo 15h16'1"34d52'
Rose & Gary's Valentine Star
Sge 20h4'34"16d17'
Rose Ann
And 23h34'32"43d55'
Rose Anna
Cas 0h15'0"63d44'
Rose Annette
Cet 0h50'24"-1d6'
Rose Blanche
Uma 13h7'12"61d27'
Rose Bud
Mon 6h22'25"2d48'
Rose Catarina
Per 3h29'40"36d34'
Rose Gloria
Cas 0h59'28"70d39'
Rose Hwei Da Lin
Ori 6h3'60"8d52'
Rose K
Cyg 19h28'24"38d52'
Rose Marie
And 2h8'21"41d8'
Rose Marie
Cyg 19h50'0"37d51'
Rose Marie
Aqr 23h36'55"-8d37'
Rose Marie
Cas 0h43'0"62d31'
Rose Marie
Mon 7h20'28"-1d36'
Rose Marie
Mon 7h38'18"-6d58'
Rose Marie
Lyn 8h56'32"41d56'
Rose Mary
Cas 3h8'41"60d28'
Rose Of Shara
Lyn 7h52'24"58d58'
Rose of Sharon
Cas 0h58'1"60d33'
Rose Of Sharon,The
Peg 23h42'35"8d35'
Rose The Drummer, Morgan
Jay
Ser 15h28'60"22d58'
Rose's Schnook Of Damocies
Cam 6h5'35"58d34'
Rose's Star
And 2h31'49"50d33'
Rose(A J),Anthony John
Dra 19h0'36"65d24'
Rose,Alex Patrick
Ori 5h25'26"-3d38'
Rose,Alison
Cyg 20h38'58"30d38'
Rose,Allison
Peg 0h2'41"30d37'
Rose,Amanda & Michael
Lyn 7h54'29"41d4'
Rose,Amber & Elexus
Sge 19h36'54"16d30'
Rose,Andera
Hya 9h4'30"0d3'
Rose,Anthony
Sex 9h44'24"4d16'
Rose,ARCM,Michael John
Hya 9h3'45"2d22'
Rose,Betty L
Cyg 19h33'47"33d42'
Rose,Beverly Dawn
Lyr 19h16'1"25d54'
Rose,Bob
Aql 18h56'14"16d41'
Rose,Bruce Kenneth
Aur 6h17'45"30d14'
Rose,Bud
Cep 21h10'55"68d24'
Rose,Cager Constantino
Joseph
Tau 3h36'13"23d27'
Rose,Christopher
Uma 8h31'38"55d43'
Rose,Craig Allyn
Mon 6h36'21"-0d10'
Rose,David
Peg 23h15'29"33d38'
Rose,David Thoreau
Aql 20h22'42"1d35'
Rose,Deborah Jean
Sco 17h4'1"-31d17'
Rose,Diane
Cas 2h11'35"70d29'
Rose,Douglas C
Cyg 19h25'0"31d25'
Rose,Dr Dana Marie
Cas 0h34'43"65d24'
Rose,Faith
Mon 6h21'30"6d4'
Rose,Frank A
Lib 15h32'21"-10d5'
Rose,Greg
Boo 14h27'22"52d52'
Rose,Hanna Jo
Uma 14h18'26"60d6'
Rose,Haz,Dav,Mal,Su
Uma 9h13'35"55d57'
Rose,Holly Chase
Cyg 19h15'36"44d3'
Rose,Janet T
Cas 1h59'1"58d23'
Rose,Jason
Tri 1h50'0"26d29'
Rose,Johnathon Dean
Eri 2h58'18"-4d49'
Rose,Joseph D & Jacqueline
Cyg 19h41'46"38d20'
Rose,Judith
And 0h25'44"40d18'
Rose,Lily Louisa
Cyg 21h2'23"37d54'
Rose,Linda L
Cyg 20h12'19"37d45'
Rose,Louis F
Vir 13h23'47"-4d35'
Rose,Margaret Ware Zigenfus
Cam 8h36'21"78d36'
Rose,Martha Jo
Gem 6h47'3"19d51'
Rose,Maureen Elita
Ori 6h16'33"7d49'
Rose,MD,Robert "Bobby"
Mitchell
Psc 1h2'33"20d42'
Rose,Mia Olivia
Cru 12h3'35"-64d10'
Rose,Michael & Rona
Uma 8h54'52"49d49'
Rose,Paul
Cet 3h15'55"6d35'
Rose,Peggy
Sex 10h24'57"2d20'
Rose,Ronald E
Aur 6h28'33"38d10'
Rose,Sarah
Ori 6h2'38"20d50'
Rose,Sarah Langston
Com 13h26'32"26d19'
Rose,Shirley DeClue
Lyr 18h56'51"41d45'
Rose,Stacey & Andy
Ori 6h6'24"20d26'
Rose,Stanley
Her 17h13'14"28d20'
Rose,Steven John
Vir 11h51'41"9d28'
Rose,Susan Gail
Eri 2h43'44"-4d19'
Rose,The
Cyg 20h30'31"40d1'
Rose,The
Cas 23h34'19"61d52'
Rose,The
Dra 17h6'53"67d37'
Rose,Toni
Uma 11h2'15"48d27'
Rose,Vera Arnold
Sco 16h52'58"-40d18'
Rose,Wayne T
Her 17h4'35"48d50'
Rose,William Manning
Ori 5h2'0"0d6'
Roseabaum,Liz
Cyg 21h9'48"40d29'
Roseann
Crb 15h32'14"32d26'
Roseann
Lyn 6h49'30"51d34'B
Roseann Marie
Leo 11h8'40"-1d28'
Roseanna
Uma 10h15'1"51d24'
Roseanna Love
Lyn 7h44'11"43d7'
Roseanne
Cas 0h39'54"67d55'
Roseanne
Hya 9h14'40"0d22'
Roseanne Christine
Oph 18h42'18"7d58'
Roseapple
Uma 10h47'32"55d55'
Roseberry,Dory
Peg 21h53'52"33d3'
Rosebud
Ori 5h31'31"-8d15'
Rosebud
Peg 22h2'54"5d42'
Rosebud
Peg 22h50'39"20d44'
RoseBud Star
Cas 0h44'1"61d49'
Rosebud-Birdsong
Peg 0h3'1"31d34'
Rosebure-KJR,Kay Jones
Cet 2h49'40"1d55'
Roselinde
Mon 6h42'60"11d23'
Rosella
Cet 2h36'44"0d14'
Roselle,Joseph
Dra 16h45'46"68d13'
Roselli,Jude Anthony
Peg 22h54'21"8d19'
Roselli,Miss Jennifer Marie
Lyr 18h55'0"42d40'
Roseman's Star-69
And 2h12'0"41d39'
Roseman,Carolyn & Joel
Cyg 21h17'14"38d38'
Roseman,Jake Devin
Aur 6h32'15"38d56'
Roseman,Rachel Amy
Cam 9h6'0"80d27'
Rosemarie
Sge 19h55'51"18d55'
Rosemarie
Lyr 19h16'28"28d22'
Rosemarie
Cas 0h55'58"67d58'
Rosemarie
Cas 3h34'0"71d50'
Rosemarie "Light Of My Life"
And 2h33'33"50d26'
Rosemarie Luminous New
Star,The
Aur 5h53'48"40d8'
Rosemarie Zweydorf geb
Herwig
Dra 17h51'31"85d54'
Rosemarie's Hunner Star
Ori 6h6'1"-2d54'
Rosemary
Crb 15h43'31"28d15'
Rosemary
And 0h23'36"45d58'
Rosemary
And 23h35'58"49d40'
Rosemary
Cyg 19h27'23"32d11'
Rosemary
Lyr 19h22'60"38d57'
Rosemary
Cas 2h56'53"71d4'
Rosemary
Per 1h44'24"54d9'
Rosemary
Cas 2h53'1"61d15'
Rosemary
Cyg 21h35'46"38d3'
Rosemary
Peg 23h20'33"13d52'
Rosemary & John Together
Forever
Mon 7h38'54"-4d43'

Rosemary Ann Tri 2h5'0"30d53'
Rosemary I Vul 19h19'20"25d40'
Rosemary Marie Lyn 7h25'56"58d39'
Rosemary Paula Uma 11h7'16"57d30'
Rosemary S T And 2h24'42"46d54'
Rosemary Shining Sage Ari 1h55'51"17d2'
Rosemary's Evening Star Cnv 14h3'46"46d21'
Rosemary's Love Lyn 8h24'0"43d40'
Rosemary's Star Sge 19h20'37"16d35'
Rosemary-Forever In My Heart Eri 4h2'31"-14d28'
Rosen # 1 Teacher,Fran Cam 7h27'1"68d46'
Rosen Engineering,H Vir 11h51'44"2d57'
Rosen,Allan Eri 4h53'1"-6d7'
Rosen,Beverly & William Ori 5h44'41"11d17'
Rosen,Brett Justin Per 1h46'36"50d13'
Rosen,Brian Evan Per 1h52'33"48d19'
Rosen,Claire Cyg 20h44'58"37d44'
Rosen,Dale Thomas Ari 2h39'43"22d22'
Rosen,Dr Michael Boo 15h4'41"25d31'
Rosen,Elizabeth Sge 18h57'48"18d59'
Rosen,Eric Oph 18h1'18"7d53'
Rosen,Erica Alycia Uma 13h48'49"60d52'
Rosen,Fabienne Dra 14h59'51"61d19'
Rosen,Freda Cas 1h2'56"58d37'
Rosen,George Boo 14h53'32"30d22'
Rosen,Harry M Oph 17h18'35"10d28'
Rosen,Jaime Danielle And 0h3'50"46d19'
Rosen,Jared Per 2h10'44"56d24'
Rosen,Martin S Cep 22h54'12"70d38'
Rosen,MD,Allan S Oph 17h39'33"11d11'
Rosen,Michael & Pamela Mon 7h1'1"8d38'
Rosen,Mr Mint-Alan Cnv 13h46'0"37d34'
Rosen,Ned Cet 0h37'31"1d35'
Rosen,Norman S Sex 9h55'0"2d34'
Rosen,Phyllis M Sct 18h50'5"-7d50'
Rosen,Pola Aqr 21h24'38"-0d22'
Rosen,Rachel Adena Uma 10h31'53"71d60'
Rosen,Sara Lauren And 0h12'1"47d5'
Rosen,Temmy Uma 12h6'41"61d46'
Rosen,The Matriarch, Esther Uma 9h45'23"49d48'
Rosenbaum Star,The Lisa Umi 14h7'35"70d13'
Rosenbaum,Jacqueline Blair Cas 1h0'49"48d53'
Rosenbaum,Mickey "Desperado" Oph 18h40'56"8d36'
Rosenberg Bo-Bob,Dr Robert Cep 23h0'58"60d46'
Rosenberg,Bonnie Tri 2h10'44"32d26'
Rosenberg,David Gregg Her 17h1'12"28d48'
Rosenberg,DDS,Steven N Per 2h50'35"46d52'
Rosenberg,Dr Arlen Lac 22h20'1"40d5'
Rosenberg,Frank Aur 6h3'52"36d49'
Rosenberg,Happy Birthday!,Maurie Per 1h32'1"52d40'
Rosenberg,Joyce Com 12h18'0"22d10'
Rosenberg,Kim David Cyg 19h33'17"30d16'
Rosenberg,Lindsay J Dra 18h11'0"58d59'
Rosenberg,Michael Her 17h54'20"20d10'
Rosenberg,Michael Ross Cma 6h50'13"-18d35'
Rosenberg,Myrette M Peg 23h5'52"30d12'
Rosenberg,Phil & Jan Cyg 19h57'34"29d34'
Rosenberg,Rachel A Cas 1h57'1"60d3'
Rosenberg,Rebecca Serani Boo 14h58'53"52d19'
Rosenberg,Richard Issac Her 16h48'46"48d4'
Rosenberg,Rita Lehrer Cyg 19h47'24"29d33'
Rosenberg,Stanford L Aql 18h42'51"0d48'
Rosenberg,Sydney J Mon 6h28'1"8d50'
Rosenberger You Are A Star,Tiffany Cet 1h9'53"-5d27'
Rosenberger,Franz Crt 11h14'24"-10d17'
Rosenberger,Prof Dr Jürgen Sgr 18h49'20"-20d57'
Rosenberry,Wayne J Oph 18h3'55"12d42'
Rosenblatt,Stuart Her 15h50'33"44d58'
Rosenbloom,Nancy & Arthur Lyr 18h37'0"39d49'
Rosenblum "Blue Bell", Bette Bromberg Crb 15h30'48"31d37'
Rosenblum,Dr Ronald M Per 2h51'14"45d4'
Rosenblum,Ira Her 17h55'0"14d42'
Rosenblum,Jillian Brooke Cas 1h6'41"63d31'
Rosenblum,Sheila And 0h52'1"40d49'
Rosenborough,Tyllor Adrian Neil Her 16h36'58"37d57'
Rosenburg,Harold & Shaoli Cyg 21h3'39"38d41'
Rosenduft,Mona Leo 10h31'16"26d38'
Rosene,Jeremy Boo 14h30'38"48d38'
Rosenfeld,(PJ)Patti & Jim Overtoom Cnv 12h14'36"44d13'
Rosenfeld,Alexander Joshua Sco 17h55'43"-38d58'
Rosenfeld,Maury & Melinda Lawton Aql 20h1'36"0d9'
Rosenfeld,Mercer Joseph Aqr 21h56'59"-6d44'
Rosenfeld,Sidney Boo 15h16'53"52d46'
Rosenfield,Deborah & Burton Cyg 21h2'55"53d17'
Rosenfield,James R Eri 2h58'12"-2d42'
Rosengard,Mollie Cas 23h38'35"58d11'
Rosenhaus,Nicholas George Her 17h34'40"20d56'
Rosenkrantz,Annelise Cas 0h23'37"60d22'
Rosenkranz,John A Aur 6h17'37"31d37'
Rosenkranz,Yvonne Sco 17h26'26"-31d29'
Rosenow,Verl Elmer Aur 6h28'60"37d57'
Rosenthal,Andrew David Aur 7h22'43"40d32'
Rosenthal,Barry C & Noah V Cep 21h48'22"68d40'
Rosenthal,Bev Cyg 19h44'49"31d0'
Rosenthal,Debra K And 1h13'49"36d2'
Rosenthal,Dr Samuel Oph 17h12'45"-20d53'
Rosenthal,Eckhardt Hya 9h5'28"-6d31'
Rosenthal,Edna Peg 23h1'33"31d3'
Rosenthal,Eric Jon Lac 22h21'30"53d8'
Rosenthal,Gilbert Cep 21h52'37"58d60'
Rosenthal,Jenifer M Lyn 8h7'52"57d47'
Rosenthal,Jr,Richard A Lac 22h9'53"38d6'
Rosenthal,Lorelei And 23h46'18"43d36'
Rosenthal,Marvin Aur 6h15'33"31d46'
Rosenthal,Mitch Cep 22h39'37"61d30'
Rosenthal,Momma Cma 6h25'30"-13d25'
Rosenthal,Pearl & Paul Peg 0h2'12"18d51'
Rosenthal,Rose Eri 4h1'19"-13d20'
Rosenthal,Ryan Michael Per 4h42'26"41d9'
Rosenthal,Scott Matthew Lac 22h19'11"48d56'
Rosenthal,Thelma Floyce Aql 18h58'52"-7d31'
Rosenthal,Trixie Roxie Cma 7h12'0"-16d0'
Rosenzweig,Otto Dra 16h12'13"51d49'
Rosenärtner,Winfried Ser 15h54'0"1d47'
Roset,Eva Uma 11h23'25"38d1'
Roset,Jordi Lagares Aql 19h0'33"13d39'
Roseto,Julie Peg 21h42'54"27d38'
Roseto,Zachary Ryan Aql 18h57'49"12d32'
Rosetta Eri 3h59'39"-17d44'
Rosette,Joyce & Michael Aur 7h7'27"41d29'
Rosetti,Carlo Cep 21h21'18"55d10'
Rosevelt,Frederick Upton Ser 18h23'30"0d45'A
Rosevelt,Marjorie Dale Ser 18h23'30"0d45'B
Roseweir,Claire Cas 23h2'1"58d12'
Rosewell,Dudrey Lyr 18h27'55"46d36'
Rosewell,Edward J Aur 6h2'43"38d55'
Rosewell,Richard & Audrey Cyg 21h36'0"30d18'
Rosh,David John Crt 10h59'16"-10d57'
Roshani Uma 12h5'11"48d42'
Roshko,Alexandra Mari-Claude Lyr 18h38'55"36d25'
Roshni And 23h1'33"50d33'
Rosi Oph 17h5'51"-20d58'
Rosi Leo 10h3'53"17d38'
Rosi,Stephanie Ann And 1h35'28"48d54'
Rosica,Evan James Cnv 13h42'53"46d30'
Rosica,Karen Polan And 23h18'41"51d39'
Rosich,George Stephen Del 20h16'0"14d1'
Rosie Lyr 19h16'59"28d8'
Rosie And 23h49'38"40d53'
Rosie And 1h44'38"36d27'
Rosie Ori 5h45'12"7d57'B
Rosie and Pets Cas 1h58'35"68d36'
Rosie Bud Ori 6h9'1"8d50'
Rosie D Cas 0h1'38"62d23'
Rosie Maria Psc 1h3'17"23d50'
Rosie's Realm Lyr 18h32'50"33d22'
Rosimoto's Love Star 93 Aql 20h11'21"10d35'
Rosina-Susan Ori 6h6'38"1d9'
Rosinski,Emily Sarah Vul 19h22'26"23d39'
Rosinski,Robert Edward Vir 13h1'58"-8d13'
Rosinsky,Becky Melton Mon 7h24'1"-8d25'
Rosita Lyr 18h33'46"30d28'
Rosita Pup 8h24'12"-29d12'
Rosita & Carlos & Natalie Cnc 8h28'1"32d5'
Rosita Mora I Cas 0h4'0"58d59'
Roskam,Arjan Cnv 12h42'0"40d34'
Roskam,Carola Lyn 8h24'1"42d15'
Roskam,Celester Helena Crb 16h9'18"37d39'
Roskam,Eddy Ori 5h56'10"20d12'
Roskam,Kevin Her 18h11'15"48d19'
Roskam,Lou Lou Cheyenne Per 3h0'49"31d44'
Roskam,Marian Vul 19h19'32"20d31'
Roskie,Dariel Lyn 8h14'15"40d35'
Roslonek,Jason Aur 6h1'38"37d54'
Rosman,Aaron Joshua Del 20h39'15"10d57'
Rosman,Adam Johnathan Del 20h57'32"15d2'
Rosman,Ilan Chaim Ziegel Del 20h56'51"19d34'
Rosmarie Lep 5h22'36"-20d2'
Rosmarin,Celia Leo 10h57'0"-0d13'
Rosner,Gerhard MFN Uma 8h43'30"51d54'
Rosner,Howard Jay Ori 5h55'33"20d21'
Roso,Jr,Frank James Cma 6h55'19"-17d57'
Roson,Peter & Leslie Vul 19h48'21"28d20'
Rosowicz,Paul J Lac 22h40'22"53d38'
Ross Cep 21h4'0"65d21'
Ross Cam 7h47'37"68d2'
Ross Her 18h5'23"30d54'
Ross & Dale's "Rotterdam Shuffle" Boo 15h3'53"41d58'
Ross & Harriet's Anniversary Star Uma 11h31'46"32d35'
Ross Allen Oph 17h3'34"8d54'
Ross II MD,Stewart David Aur 6h12'53"37d38'
Ross II,James Gordon Crb 15h22'0"30d60'
Ross Inn,The Emily Cas 1h12'60"61d15'
Ross Lloyd Mon 6h31'1"4d22'
Ross Star,The Aql 20h9'39"1d32'
Ross with love Jak Sge 20h17'59"16d27'
Ross X 2,C Uma 8h59'11"54d7'
Ross' Never Never Land Philip & Sue Eri 4h3'48"-8d13'
Ross,Aaron Thomas Ori 5h56'16"14d45'
Ross,Adam Todd Lib 14h58'45"-0d29'
Ross,Agnes Murray And 0h46'42"22d4'
Ross,Alexandria Mary And 0h41'0"40d37'
Ross,Alice Oph 18h24'28"8d43'
Ross,Alison Cyg 19h39'56"30d27'
Ross,Angela Rita Tel 20h22'31"-47d43'
Ross,Anne Mon 6h57'16"-10d9'
Ross,Billie Jo Ellen Peg 21h38'56"25d34'
Ross,Bonnie Gail Cas 23h16'33"62d32'
Ross,Brian Boo 13h32'0"22d25'
Ross,Candace Peg 21h47'50"34d40'
Ross,Caroline K Uma 11h47'43"58d3'
Ross,Catherine Faye Eri 3h17'18"-11d56'
Ross,Cayla Marie Peg 21h35m30"18d51'
Ross,Charles Wayne Cet 2h20'19"5d17'
Ross,Cheryl D Boo 15h8'35"30d17'
Ross,Chuck & Kathy Cet 1h31'31"-12d39'
Ross,Clint Henry Aqr 21h58'52"-11d15'
Ross,Courtney Sale Mon 6h21'55"4d51'
Ross,Daniel Andrew Peg 22h33'33"31d32'
Ross,Daniel Stephen Cam 12h48'57"78d20'
Ross,David J Cet 2h14'1"5d5'
Ross,David M Sex 10h38'16"-1d49'
Ross,David Stewart Ori 5h34'12"-6d52'
Ross,David William Ori 5h39'36"-0d24'
Ross,Diana And 2h20'22"50d31'
Ross,Donald Wentworth Aql 19h54'42"12d55'
Ross,Eddie Boo 14h54'0"34d31'
Ross,Edsel Dale Mon 6h19'33"7d10'
Ross,Elaine Anne Cas 0h23'26"60d31'
Ross,Elizabeth And 0h38'33"38d17'
Ross,Elsie Fischbach Cas 0h33'44"58d23'
Ross,Emily-Danielle & Andrew Uma 9h19'21"52d17'
Ross,Frank Lincoln Tau 5h50'26"26d57'
Ross,Frederick Cep 23h15'55"64d43'
Ross,Friedrich Wilhelm Jannasch Cru 12h53'31"-60d24'
Ross,Gabe Her 17h53'22"28d22'
Ross,George G Eri 3h2'32"-14d33'
Ross,Gerald W Hya 8h54'0"3d19'
Ross,Holly Elizabeth Lyr 18h31'1"31d4'
Ross,Irwin Cas 1h3'1"61d49'
Ross,Jean And 23h47'38"33d15'
Ross,Jennie Aqr 20h55'23"0d48'
Ross,Joan Mon 7h43'56"-1d20'
Ross,Joe G Oph 16h50'21"10d15'
Ross,John Uma 10h48'28"62d28'
Ross,John Jerome Cyg 21h13'12"28d16'
Ross,Joseph Umi 15h31'58"68d1'
Ross,Joyce & Larry Cma 6h59'9"-18d23'
Ross,Jr,Steven Paul Lyn 7h3'33"58d38'
Ross,Jr,Thomas Lee Cet 2h51'39"3d7'
Ross,Judith & Graham Per 3h25'24"37d54'
Ross,Julie & Neil Del 20h19'16"10d49'
Ross,Julius Grayson Bright Lac 22h29'44"50d20'
Ross,Kathe Lyn Dra 15h59'46"64d23'
Ross,Kathryn V And 22h59'12"50d4'
Ross,Kelley Mon 8h5'21"-3d20'
Ross,Kellie Bryanna And 0h58'27"45d60'
Ross,Kevin M Ori 5h18'0"0d57'
Ross,Kimberly Anne Eri 4h13'1"-16d5'
Ross,Kris Aql 19h55'37"15d54'
Ross,Kymberly Cam 5h1'13"68d58'
Ross,Larry Aql 19h54'15"1d10'
Ross,Lauren Nicole Vul 19h22'37"23d37'
Ross,Linsey & Ward Peg 22h49'37"21d27'
Ross,Lorraine M And 23h17'39"49d24'
Ross,MA,Fern C Eri 4h53'23"-8d41'
Ross,Margaret Lou Oph 18h23'58"7d43'
Ross,Margaret MacBeth Cas 0h32'26"74d28'
Ross,Margaret Tarrant Cas 2h0'39"61d43'
Ross,Margie Molina Peg 22h36'25"24d36'
Ross,Marilyn Louise Zeigler Cyg 21h12'49"28d55'
Ross,Mark Elliott Per 2h58'27"38d10'
Ross,Michael Lyr 18h35'13"27d2"
Ross,Miss Jill Sgr 19h57'58"-42d59'
Ross,Mrs Super Wonderful Lesley And 1h56'11"37d45'
Ross,Natalie Kimberly Aur 5h6'49"38d54'
Ross,Patricia Lee Peg 0h1'10"14d14'
Ross,Paul-n-Etty Uma 11h48'0"37d55'
Ross,Paula Dane Stewart Mon 6h54'57"-6d28'
Ross,R Allan Tri 2h15'28"28d52'
Ross,Richard Sco 17h26'29"-31d13'
Ross,Robert A Dra 18h1'11"80d22'
Ross,Robert Weir Aql 18h53'47"8d34'
Ross,Russell Brunot Aur 5h13'0"40d40'
Ross,Saul H Dra 15h11'51"57d15'
Ross,Scott (CRPG) Chi Research & Plan Uma 9h58'14"52d40'
Ross,Shannon Kylene Cas 0h19'1"58d45'
Ross,Stephanie Aql 18h57'1"-2d5'
Ross,Tim Boo 13h44'32"13d18'
Ross,Tosha Colleen Uma 9h16'1"62d18'
Ross,Tresy Leigh Mills Ari 2h7'12"21d47'
Ross,Wesley Aur 5h45'0"50d18'
Ross,Wolfgang Aqr 22h40'25"0d59'
Ross,Zachary Paul Her 18h10'18"47d19'
Ross-"One in a million Mom",Marilyn Cas 1h9'45"66d51'
Ross-Bona Lac 22h30'1"53d17'
Ross-Loves-Stratton 1993 Uma 8h39'55"68d28'
Rossana V Lmi 10h51'49"32d54'
Rossano,Alexander Anthony Aql 19h47'17"10d26'
Rossano,Matthew FX Aur 6h19'30"37d47'
Rossano,Thomas Augustus Her 17h33'1"27d30'
Rossati,Chantal Dra 15h21'12"53d53'
Rossber And 2h8'10"40d14'
Rosse-Steinrotter, Diann And 1h29'45"48d48'
Rosseland,Marian And 0h58'11"36d50'
Rossella Cet 2h34'16"2d28'
Rossella Cas 1h58'26"58d11'
Rossella (Scarlett) Effe Lep 5h23'15"-25d45'
Rossella Righetti, Carlo e Bimba Umi 17h45'49"80d3'
Rosselli,Carmela Lyn 9h17'27"38d17'
Rosselli,Lucy Tri 2h17'48"31d24'
Rossello,Christian Lyn 9h6'20"34d48'
Rossen,Henry A Vul 19h15'21"25d15'
Rosser 1,A R Aql 20h1'0"4d24'
Rosser,Deborah Umi 16h12'40"76d36'
Rosser,Jr,Sherman A Her 17h3'20"42d50'
Rosser,Ronald Thomas Her 17h52'23"40d20'
Rosseter,Alexandra Jane Vul 19h21'50"25d32'
Rossetter,Kiley Anne And 23h14'14"44d36'
Rossetti's Rising Crb 15h16'13"30d17'
Rossetti,Federica For 2h42'59"-27d30'
Rossetti,Michael J Her 16h41'44"35d31'
Rossi & Roscoe Forever Del 20h17'17"14d37'
Rossi,Antonella Pho 0h29'26"-45d12'
Rossi,Aurora Gru 22h32'23"-50d17'
Rossi,Beth Tau 3h59'48"11d59'
Rossi,Beverly Cas 0h21'33"67d52'
Rossi,Camilla Cas 2h10'15"60d37'
Rossi,Camilla Pup 8h24'30"-29d17'
Rossi,Catherine G And 23h10'42"41d20'
Rossi,Christine And 0h49'41"40d14'
Rossi,Chyrisse Garrett Equ 21h3'0"2d33'
Rossi,Coach Ori 5h55'38"12d30'
Rossi,Don Gem 6h47'20"13d51'
Rossi,Donna Tri 2h6'54"32d52'
Rossi,Gianluca Carletto Equ 21h6'58"11d49'
Rossi,Giorgia Uma 12h9'47"62d1'
Rossi,Giovanna Lac 22h52'59"38d3'
Rossi,Grace Hya 8h56'59"3d4'
Rossi,Ida Lmi 10h7'1"38d1'
Rossi,J A N Peg 23h23'11"15d45'
Rossi,Jane And 0h11'49"36d48'
Rossi,Jerome Dra 20h6'14"62d23'
Rossi,Jerome Her 17h51'51"40d7'
Rossi,Judith Ann Marie Hatch Eustace Leo 11h10'0"-2d10'
Rossi,Kate Tau 3h58'23"11d28'
Rossi,Libby & Bill Eri 3h35'1"-13d39'
Rossi,Mark Oph 17h1'37"-1d49'
Rossi,Mark J Dra 18h35'60"58d45'
Rossi,Michael P Aur 6h25'23"38d31'
Rossi,Patrizia Cam 6h46'43"68d15'
Rossi,Paul Hya 9h36'42"-6d33'
Rossi,Raymond R Aur 6h26'1"38d49'
Rossi,Rosalba Pyx 8h46'3"-24d11'
Rossi,Thomas Cnc 9h1'28"30'5'
Rossi,Tina Cam 6h0'34"60d22'
Rossi,Tom Eri 3h39'55"-2d0'
Rossi,Valentina Stancio Ind 21h11'23"-50d9'
Rossi,Viviana Vanessi Cyg 20h52'27"31d18'
Rossi-Gauthier, Gabrielle Lyr 18h32'25"39d43'
Rossignol,Chantal Aur 6h5'45"31d2'
Rossini,Ralph J Lac 22h1'56"50d53'
Rossiter,Caroline Cyg 20h43'16"45d39'
Rossiter,Stephen John Her 16h16'12"48d48'
Rossler,Alexis Nicole Uma 9h22'42"47d14'
Rossman,"Christine" Cnc 8h51'14"30d10'
Rossman,"Silkie" Saraj Elizabeth Crb 15h28'41"32d51'
Rossmoore,Dr Harold W Boo 14h10'56"35d40'
Rossnegger,Elmer Cnc 9h14'59"30d48'
Rosso,Craig A Aur 4h53'19"38d46'
Rossol,Nichole Blake And 23h6'11"44d52'
Rossomando,Briana M Peg 21h50'0"34d6'
Rosson,Alfred M Ser 15h36'53"8d49'
Rosson,Shining Forever Edward H Lac 22h24'19"37d46'
Rossoni,Ralph Boo 14h36'45"37d6'
Rossovich,Roy Lmi 10h4'57"30d4'
Rossow,Karin Leo 10h52'52"0d9'
Rossow,Mary Boo 14h22'22"51d42'
Rossy The Clown Lac 22h11'58"48d59'
Rost,Peter Vul 20h45'48"20d11'
Rost,Timm Ori 4h41'33"15d10'

Rostal,Janice E
Peg 22h2'0"31d0'
Rostalski,Harald
Aqr 20h58'12"0d24'
Rostalski,Selma
Sco 17h27'49"-30d50'
Rostek,Deborah Lynn
And 1h14'36"41d11'
Rostin,Gerd
Sgr 18h52'25"-20d58'
Rosvall,Elizabeth Lynn
Eri 5h1'1"-6d43'
Roswell Cancer Ins, Michelle Woelfel
Lyn 8h9'12"35d47'
Roswell,Richard T
Cep 20h52'32"62d5'
Roswitha
Lyr 19h6'0"26d0'
Rosy
Mon 7h20'34"-5d52'
Rosy
Gem 6h55'44"17d21'
Rota,Brian Patrick
Cep 1h14'16"87d5'
Rota,James
Per 3h43'21"51d31'
Rota,Tommy
Aur 5h37'39"54d30'
Rotandi,Richard M
Cep 23h28'38"65d23'
Rotante,Robyn Joy
Cyg 19h26'17"30d58'
Rotbart,Dean Ira
Crt 11h8'27"-14d36'
Rotbert's Birthday Star,Stan
Cap 21h11'23"-22d14'
Rotchford,Debbie
Cam 6h1'41"67d48'
Rotella,Thomas
Hya 9h3'25"2d44'
Rotensztajn,Simon
Sex 10h43'12"5d51'
Rotensztajn,Simon
Her 18h19'0"12d34'
Rotgers,William
Hya 10h40'28"-18d19'
Roth 70th Anniversary, Mary & Harry
Cyg 21h14'0"38d13'
Roth 8th Dist Wis, The Honorable Toby
Lib 14h22'53"-10d49'
Roth Casino Frozen Foods,John
Ser 15h29'0"22d56'
Roth III,Paul N
Sct 18h52'56"-5d25'
Roth's Nifty 50,Patti
Aql 19h11'1"10d49'
Roth,Brian H
Aur 6h51'20"40d48'
Roth,Bärbel & Norbert
Ser 16h3'53"2d24'
Roth,Cameron
Cet 2h35'3"-2d9'
Roth,Casey M
Lac 22h9'12"40d10'
Roth,Deanna Lynn
Peg 22h35'0"33d46'
Roth,Dominica Thomas
Tri 1h59'18"34d1'
Roth,Dr Lane G
Eri 3h1'52"-6d18'
Roth,Elliott A
Per 1h53'59"50d20'
Roth,Ethan Michael
Uma 9h54'14"51d18'
Roth,Gary & Denise
Uma 8h32'41"52d5'
Roth,Goldi Martin
Cnc 9h11'42"8d19'
Roth,Graham
Mon 7h55'25"-6d39'
Roth,Harry
Aur 6h15'57"31d34'
Roth,Irene Mildred Lawrence
Lac 22h25'40"56d15'
Roth,Jamie
Cet 2h1'56"-6d4'
Roth,Jessica
Cas 0h53'1"66d57'
Roth,Jim
Her 17h10'27"40d60'
Roth,Josef SAD
Cap 20h29'35"-26d37'
Roth,Kimberly Cole
Peg 23h7'0"10d1'
Roth,Kyle Duncan
Lmi 10h56'1"30d20'
Roth,Larry
Per 3h18'46"41d9'
Roth,Linda Catherine
Cet 1h49'15"-1d31'
Roth,Marie
Peg 23h49'0"17d37'
Roth,Michael
Hya 8h14'1"-1d30'
Roth,Michael
Oph 17h4'1"11d30'
Roth,Mimi
Lyr 19h23'35"38d1'
Roth,Peter
Ori 5h56'0"15d0'
Roth,Rafael
Sex 9h57'36"2d29'
Roth,Richard D
Oph 18h17'36"10d7'
Roth,Robert Lee
Sex 9h40'31"-8d47'
Roth,Sandra Priscilla
And 23h43'56"40d50'
Roth,Scott A
Eri 5h0'38"-6d57'
Roth,Steven
Dra 15h1'44"60d54'
Roth,Tamara Lynn
Uma 10h34'45"41d3'
Roth,Thomas David
Cet 1h50'0"-4d8'
Roth,Thomas Vincent
Psc 1h23'38"23d2'
Roth,Udo
Vir 13h25'0"-6d39'
Roth,William E
Del 20h20'34"11d5'
Rothacker,Christine Robin
Ari 3h13'58"23d44'
Rothamer,Rocky & Princess Carol
Uma 8h29'55"61d44'
Rothaus,Finally-Judi & Perry
Cyg 19h52'44"48d48'
Rothbart,Yudda Lea Rosen
Aql 19h0'54"5d2'
Rothberg,Jacob
Ori 5h50'39"20d57'
Rothberg,L Patrick
Boo 15h18'0"48d20'
Rothe,Rainer
Dra 18h6'45"65d28'
Rotheiser "Cold Milk" Brenda Rénee
Boo 14h7'28"41d35'
Rothenberg,Adam L
Boo 14h39'18"52d9'
Rothenberg,Bradley
Ori 5h24'46"0d28'
Rothenberg,Daniel Alexander
Lac 22h39'11"54d17'
Rothenberg,Matthew David
Cep 20h24'49"75d25'
Rothenberg,Maxine
Com 12h18'16"18d51'
Rothenberger,Anni
Uma 11h56'13"60d40'
Rother,Karl A
Lac 22h25'1"50d21'
Rothermel,P J
Ori 5h54'41"5d38'
Rothermel,Rosi
Umi 14h14'47"68d5'
Rothery,"Glenbabey"
Cep 21h33'48"70d35'
Rothery,Lisa Marie Jessica
Lyr 8h34'0"31d47'
Rothfeld,Robert-Cindy- Cory
Lyr 18h56'0"31d2'
Rothgeb,Christopher Trent
Ser 15h58'27"11d6'
Rothman 60, Elliot & Irving
Cep 21h45'41"63d54'
Rothman,Frani
Cas 0h30'47"70d26'
Rothman,Rivca
Aql 19h57'40"14d22'
Rothmann-XL
Col 5h50'15"-42d1'
Rothring,Karen J
Mon 8h7'15"-0d48'
Rothschild,Carl
Aur 6h1'20"38d54'
Rothschild,Cody Lee
Her 16h24'24"37d34'
Rothschild,Elizabeth M
Lyr 18h49'1"39d33'
Rothschild,Lee
Lib 15h32'0"-28d28'
Rothstein,Brian Scott
Gem 7h12'30"21d32'
Rothstein,Eugene Francis
Cep 22h35'1"63d21'
Rothstein,Magna Cum Pedis,Star of
Aql 19h56'58"14d14'
Rothstein,Sandy
Lyn 6h56'50"50d24'
Rothstein,The
Aur 6h5'0"31d55'
Rothweiler,Karl A
Oph 18h17'48"10d35'
Rothwell,Jennifer
Peg 22h57'48"25d54'
Rothwell,Paul
Ori 4h41'26"0d29'
Rothwell,Stanley
Her 18h3'33"40d19'
Roti Roti,Donlad
Cam 3h55'47"79d30'
Roti,Christopher Allen
Uma 10h44'40"50d24'
Rotisimus Ratis Eternus
Dra 16h47'26"72d29'
Rotkvich,Debbie
Cet 1h26'51"-4d59'
Rotkvich,Debbie
Gem 6h58'17"35d3'
Rotkvich,Debbie
Cas 1h3'0"61d35'
Rotolo,Logan
Dra 16h32'43"61d52'
Rotolo,Mark William
Cep 21h10'15"65d34'
Rotondo,Christopher J
Per 3h36'1"38d30'
Rotondo,Madeline & Gary Johnson
Lac 22h48'34"52d53'
Rotramel,Casey W
Her 17h11'31"46d25'
Rotrou,Pierre
Cnv 13h20'27"31d19'
Rott,Betty L
And 23h13'0"40d48'
Rott,Carol & Sheldon
Cyg 21h39'56"42d58'
Rott,Chad
Aql 19h4'46"3d5'
Rott,Courtney "Corky"
Sco 16h21'43"-30d9'
Rott,Francis
Boo 14h39'21"51d17'
Rott,Hope
Aur 5h25'45"38d24'
Rott,Robert & Terry
Cyg 21h21'27"31d22'
Rotter,Bud
Her 15h57'0"48d10'
Rottger,Elaine
And 2h19'20"47d29'
Rottier,Amy Elizabeth
Lyn 6h24'27"60d23'
Rottkamp,Hans
Peg 23h30'22"11d60'
Rottmann,Gregory
Cam 5h43'29"61d29'
Rottmann,Marcel Michael
Dra 19h47'33"65d32'
Rottschafer-Warner
Boo 14h41'0"46d12'
Roubitchek,Adam
Ori 5h51'29"17d51'
Rouch II,Claud Albert
Aql 18h42'52"0d1'
Rouch,Bess & Fred
Cyg 21h4'23"38d43'
Rouda,Dylan Buchanan
Gem 7h26'46"34d49'
Roudabaugh,Teila
Dra 16h21'20"66d57'
Roudner,Dr Leonard A
Oph 17h53'16"10d7'
Roughgarden,Nancy
Mon 6h24'34"-0d38'
Roughton,Brandi Reay
Uma 12h3'52"39d24'
Rougier,Glenn
Cma 7h20'23"-15d5'
Rouhault,Bruno
Boo 15h0'41"10d58'
Roukéma,Jeannette
Peg 21h10'59"12d48'
Roula Liana
Vir 13h21'25"-9d11'
Rouleau,Claude
Cam 13h55'15"82d5'
Rouleau,Francine
Ori 5h1'10"11d3'
Rouleau,Guy
Ori 6h2'56"8d50'
Rouleau,Mary Ann
Cyg 19h29'42"33d31'
Rouleau,Philippe
Ser 15h44'11"2d37'
Roulet,Jeanie
And 1h11'0"48d51'
Roullier,S L
Vir 11h43'12"8d8'
Roult,Martha Miller
Lyr 19h17'36"26d12'
Roult,Sr,Charles Andrew
Dra 13h0'26"64d33'
Roumeihi,Luluwa Abdul Rahman
Lyn 8h9'31"51d37'
Roumieh,Samer
Aql 19h6'1"0d0'
Round Bellied Fuss Bottom
Cet 2h6'28"8d2'
Rounder,Jamie Blair
Cet 3h7'51"1d44'
Rounder,Peggy
And 1h2'34"40d39'
Rounds III,Robert
Cep 20h42'47"61d40'
Rounds,Andrew Paul
Peg 22h30'1"2d29'
Rounds,Cara Florio
Cas 0h52'15"75d10'
Rounds,Craig
Lac 22h18'29"51d44'
Rounds,Erik Thomas
Her 17h22'59"40d46'
Rounds,Paul
Dra 19h25'59"61d41'
Roundtree,John
Umi 14h43'36"78d16'
Roundy,Brandis & Scott
Cnv 14h2'54"40d57'
Rountree,Dory
Crb 15h31'0"31d30'
Rountree,Laura Marie
Com 13h2'54"15d29'
Roura,Jeane Nunn
Cas 0h8'35"50d29'
Rourke 1995,Mickey
Ori 6h7'58"7d58'
Rourke,Ernie & Irene
Cet 3h10'26"6d41'
Rourke,Kathleen
Cas 0h58'10"66d41'
Rourke,Michael L
Per 4h8'1"38d46'
Rouse,David Clinton
Boo 14h13'46"17d20'
Rouse,Elizabeth Meadows
And 0h59'27"45d32'
Rouse,Ina Sears
Ori 5h43'49"11d21'
Rouse,Jeffrey A
Lyn 8h53'1"34d50'
Roush's Heavenly Body
Her 18h12'1"40d19'
Roush,Barbara
Cam 12h20'16"77d3'
Rousseau Elementary School
Ori 5h56'24"15d44'
Rousseau,Anaïs Fiona
Per 3h29'27"36d57'
Rousseau,Geneviève
Umi 13h3'43"71d35'
Rousseau,Julie
Lyr 19h16'1"28d27'
Rousseau,Marie
Peg 23h16'14"31d29'
Rousseau,Patrice
Cyg 19h45'18"30d6'
Rousseau,Stéphane
Per 3h1'17"47d31'
Rousseau,Valentine
Cmi 7h7'34"2d59'
Rousseau-Dumarcet, Andre
Cas 22h57'15"54d24'
Roussel
Aql 19h54'21"0d42'
Roussel,Belinda Jayne
Lyr 18h43'47"42d50'
Roussel,Mallory Elizabeth
Lac 22h34'23"56d23'
Rousselle,Martial Andre
Her 17h0'36"40d26'
Rousset,Roger
Uma 13h26'33"62d11'
Roussot,Fabrice
Ori 6h7'19"18d49'
Roustai,Sophie Massoumeh
Cam 7h37'0"68d8'
Routenberg,Gary L
Lac 22h19'22"47d4'
Routh,Ed
Dra 15h39'43"58d58'
Routh,Robin
Cas 23h25'57"61d21'
Routhier,Alexandra Jordan
Cas 0h29'32"61d37'
Routhier,Pierre-Hugues
Cep 21h18'10"67d47'
Routhier,Zachary Cass
Aur 5h10'13"43d44'
Routledge,Michael
Ori 6h7'22"5d30'
Routt,Billy
Per 3h17'27"41d24'
Roux,Edith Beatrice
Cas 0h3'51"61d48'
Roux,Sandi
And 2h31'48"42d36'
Rouyer,Aurelie
Oph 18h16'38"10d39'
Rouzbeh & Laurie
Uma 9h10'45"48d12'
Rovano,Sarah Ruth Josephine
Com 13h16'58"27d7'
Roven,Talia
Lyr 18h15'37"38d56'
Roveneau,Mineille
Oph 18h5'34"7d60'
Roventine,Matthew
Aur 5h36'43"54d55'
Rover,Dieter
Mon 7h53'46"-1d47'
Rovera,Jr,Louis Frank William
Oph 17h33'38"-20d49'
Rovere
Scl 23h9'1"-31d35'
Rovinski,Connor Joseph
Aql 19h45'60"11d54'
Rovira,Alicia Ventura
Peg 22h17'27"5d3'
Rovner,David J
Cnc 8h39'42"18d30'
Rowaan
Her 18h27'24"12d40'
Rowan II,William Gregg
Hya 9h4'21"2d18'
Rowan,Aileen
Lyr 18h33'33"29d25'
Rowan,Lisa
Ori 5h25'19"-4d51'
Rowan,Mary's Sweet Annie
Ori 5h52'53"16d9'
Rowan,Molly
Sco 16h32'51"-43d53'
Rowbrey,Simon Peter
Sge 20h17'14"20d60'
Rowda Knoah
Vir 14h52'1"6d31'
Rowden,Katherine
Cas 0h36'35"63d48'
Rowdy
Lyn 9h6'24"45d19'
Rowdy Chester Jr
Ori 6h3'47"8d15'
Rowe
Aur 7h21'10"39d36'
Rowe 2',Becki
Cyg 20h37'48"44d9'
Rowe Legacy
Her 16h50'28"38d46'
Rowe,Ann Gerda
Lib 14h57'35"-22d37'
Rowe,Carol Ann
Cas 3h5'11"71d12'
Rowe,Charlotte Elizabeth
And 23h8'0"38d53'
Rowe,Falecia Nichol
And 23h2'0"46d56'
Rowe,Kathi
Oph 17h42'33"12d4'
Rowe,Kaylee Maeallii
Ori 5h15'45"-8d9'
Rowe,Kerry & Simon
Uma 10h5'36"58d52'
Rowe,Leslie Ann
Aql 19h0'24"10d25'
Rowe,Loving Rob
Ori 5h43'19"1d16'
Rowe,Marie Cheryl
Lyr 18h29'15"37d48'
Rowe,MD,Donald M
Per 2h4'51"48d45'
Rowe,Michael
Ori 6h3'0"1d11'
Rowe,Sally
Cas 1h41'1"72d16'
Rowe,Scott Chatman
Oph 17h6'17"8d6'
Rowe,Scott Michael Gregory
Lac 22h13'1"47d41'
Rowe,Terence B
Cas 0h14'18"59d41'
Rowe,The Muriel
Cyg 19h25'29"35d11'
Rowe-Touch The Future Star,Joan
Eri 4h49'41"-9d49'
Rowell (Orina), Terry-Sue
And 1h22'42"36d1'
Rowell,"Little Raeta" Norman
Hya 8h13'46"2d6'
Rowell,David E
Her 17h9'22"49d54'
Rowell,Melody
Oph 18h1'52"11d6'
Rowell,Sophie "Baby"
Umi 14h33'14"77d33'
Rowen,Barbara
Crb 15h32'21"30d17'
Rowings,James Raymond
Peg 22h51'42"8d52'
Rowland,Aileen
Aql 19h59'21"8d19'
Rowland,Cindy Lynn
And 1h52'47"40d54'
Rowland,Delores Alma
Aqr 21h51'59"-3d32'A
Rowland,Erin Marie
Mon 6h54'0"-1d24'
Rowland,Janet A
Sgr 19h36'6"-40d55'
Rowland,Kaylee Ann
Gem 6h47'57"14d26'
Rowland,Margaret Waterhouse
Cas 1h17'35"61d19'
Rowland,Patricia Louise
And 23h24'30"44d12'
Rowland,Paul Charles
Peg 22h58'46"10d18'
Rowland,Peter
Cep 21h59'0"60d49'
Rowland,Tyler Alexander
Ori 5h52'42"10d25'
Rowland,Wendy
Per 1h46'25"50d33'
Rowland-Hawkes
Equ 21h21'0"11d32'
Rowlands,Bryn Marston
Cyg 21h17'16"28d20'
Rowlands,Carmen
Cam 6h57'28"60d21'
Rowlands,Kevin
Her 16h27'33"50d45'
Rowles,Ann Christine
Lyr 18h29'0"44d49'
Rowley,Aviva
Vul 21h27'0"24d5'
Rowley,Carol Ann
Mon 6h36'0"0d11'
Rowley,David
Dra 14h40'19"61d4'
Rowley,Jennifer
Sge 20h1'55"16d28'
Rowley,John
Cnv 13h22'55"38d55'
Rowley,Michael Charles
Per 2h53'14"43d12'
Rowley,Paul
Cep 23h9'0"68d51'
Rowley,Penny Y
Mon 6h18'57"1d52'
Rowley,Sarah
Lyr 18h23'25"45d60'
Rowlingson,Hannah
Lyn 8h4'16"44d41'
Rownsend,MD My Brightest Star,Robert
Oph 18h7'11"8d25'
Roworth,Edwin Thomas
Aur 4h58'39"38d59'
Rowsby,Christopher
Aql 20h6'29"8d24'
Rowsell,Rev Bonnie
Mon 6h22'44"8d46'
Rowshnnigh,Maureen Key
Col 6h0'0"-38d9'
Rowson,Anne
Cas 2h3'19"61d7'
Rowswell,Kim
And 0h0'30"46d57'
Rowthorn,William James
Dra 18h2'21"70d37'
Rowton's Guiding Light Larry
Aql 18h40'1"-2d56'
Rox
Her 15h48'56"46d27'
Rox,Doque
And 23h32'55"44d55'
Roxan
Del 21h5'30"12d15'
Roxane
And 2h5'36"38d28'
Roxane
Lyn 9h11'16"41d28'
Roxane & Elkin
Lyr 19h19'13"37d50'
Roxanna
Com 12h18'21"32d5'
Roxanne
Cas 0h1'22"60d41'
Roxanne
Cyg 21h50'27"41d54'
Roxanne
Lyn 7h52'11"51d56'
Roxanne
Lmi 10h18'0"33d7'
Roxanne D B
Crb 15h57'50"38d12'
Roxas,Aurelia
Scl 23h10'1"-34d24'
Roxie's Mugsy
Cas 2h17'21"60d38'
Roxtar,Roxanne
Lyn 6h59'10"58d51'
Roxx,J C & Nicole
Equ 21h1'55"2d17'
Roxy
Boo 15h28'19"37d47'
Roxy
Lyr 19h8'27"38d24'
Roxy's Charm
Lyn 9h11'39"44d10'
Roxy's Star
Cyg 21h18'0"28d39'
Roy
Boo 14h30'44"38d56'
Roy
Her 17h21'27"22d22'
Roy
Cam 4h5'33"69d27'
Roy
Her 16h9'35"10d37"
Roy & Pam Star of Love
Eri 4h35'21"-8d16'
Roy & Penny
Cyg 19h56'27"58d55'
Roy '94,Peter & Jane
Lyn 7h49'1"44d44'
Roy (Morning Star), Danica
Cam 7h13'0"80d10'
Roy (Ted),Wayne Joseph
Dra 16h11'34"68d33'
Roy 201069,Patrick Archambault
Her 17h17'36"40d4'
Roy GBIV
Aur 4h50'46"48d57'
Roy Michelle
Uma 11h1'41"50d59'
Roy Re
Lmi 9h58'31"33d14'
Roy W
Hya 9h9'59"5d26'
Roy's Birthday Star
Uma 9h22'30"52d55'
Roy's Dream
Lib 15h37'43"-21d24'
Roy's Lucky Star
Boo 13h57'53"11d8'
Roy's Roost
Cnv 13h44'43"40d31'
Roy's Shiner"King of the Warthogs"
Per 3h5'14"56d37'
Roy's Star
Cet 3h17'0"0d17'
Roy,Alan
Aql 19h30'1"11d15'
Roy,Beth
And 23h20'36"49d34'
Roy,Bradley Joseph
Uma 13h34'1"51d51'
Roy,Carrie Lynn
Lyn 7h49'14"36d27'
Roy,Daniel Jason
Per 4h2'1"37d51'
Roy,David & Janice
Aql 19h2'26"4d23'
Roy,ingénieur,Clément
Dra 19h52'15"64d2'B
Roy,ingénieur,Denis
Dra 19h52'15"64d2'A
Roy,Jean-Aimé Olivier
Gem 6h46'23"31d47'
Roy,Joan L
Ori 4h59'33"14d45'
Roy,Laurie A
Mon 7h6'54"-6d34'
Roy,Meredith
Dra 17h39'44"61d37'
Roy,Michel
Aql 20h1'31"14d50'A
Roy,Michelle T
Uma 12h18'31"56d21'
Roy,Renée
Umi 15h21'0"65d60'
Roy,Robert
Per 3h2'26"40d28'
Roy,Robin Larocque
Uma 11h39'19"32d44'
Roy,Stephania
Cas 22h59'13"55d31'
Roy,Tommy Graeme
Cep 3h5'15"77d19'
Roy,Vivian F
Aql 20h16'13"1d52'
Roy-Rousseau,aieules, Yvonne-Léa
Uma 11h33'16"38d44'
Roya
Sgr 18h55'51"-27d43'
Royal Cadet
Eri 4h2'45"-19d53'
Royal Leamington Spa Rehabilitation Hosp
Uma 9h55'49"54d46'
Royal Majesty
Ori 5h53'0"15d2'
Royal,Deanna
Cas 1h19'13"61d29'
Royal,Emily
Cet 2h31'18"1d13'
Royandpauline
Lyr 18h30'17"39d55'
Roybal,Cynthia Lynn
Lyr 28h24'50"40d40'
Royce's Own Star
Oph 17h19'11"12d30'
Royce,Ashley
Cyg 21h23'40"28d57'
Royce,Christina Lee
Lyr 18h42'36"45d43'
Royce,Patricia Murphy
And 2h7'55"38d59'
Royce-Our "Star" Forever-oes,Jack
Cmi 7h55'15"0d13'
Royer,Jordan R
Aur 7h21'34"38d22'
Royer,Michel
Lmi 10h47'1"26d5'
Royerson,Kathryn Alix
Cas 0h7'56"63d20'
Royes,Robyn Alison
Peg 23h2'13"13d9'
Royle,Camille L
Cas 0h44'13"64d28'
Royo,Laura Obergon
Com 12h54'48"30d60'
Roys,Irazu Alvarez
Oph 17h53'27"13d7'

Royse 8-6-86,Joshua Douglas
Aur 4h56'0"40d32'
Royse,Jennifer Lee
Aql 19h11'0"19d43'
Royston's Universal Rhythms,Henry
Per 1h54'20"56d22'
Royston,Ryan Christopher
Hya 9h5'15"0d31'
Roythorne,Ann
Boo 13h40'37"15d56'
Roytman-Nxt Yr In Jerusalem,Mariana
Uma 11h9'56"44d29'
Roz
Ari 2h39'16"25d29'
Roz
Cas 0h56'1"66d18'
Roz
Lyn 7h21'13"58d24'
Roza,Leslie
Sct 18h43'1"-7d29'
Rozek,MD,Louis M
Per 3h5'19"47d17'
Rozella
Cam 4h19'29"60d54'
Rozelle Star,The
Uma 10h45'22"55d58'
Rozelle,Erica Anne
And 0h20'54"33d36'
Rozenblum,Dr Karen
Mon 6h36'16"10d6'
Rozendal,Andrew
Hya 8h12'0"1d20'
Rozendal,Damien
Dra 16h1'29"64d58'
Rozenia
Vul 20h37'40"20d20'
Rozentsvayg,Tanya
Lyr 18h26'40"38d32'
Rozenwald,Shelley
Peg 0h2'43"30d43'
Rozier's
Her 16h17'0"11d57'
Rozl
Aur 4h50'45"40d20'
Rozman IV,Matthew John
Boo 14h59'43"22d23'
Rozman,Eternal Love Alan & Arlene
Eri 3h34'55"1d58'
Rozman,Jeremiah
Ser 15h9'28"21d5'
Rozman,Natalie
Ori 4h55'58"5d32'
Rozner-My "Prince Charming",Tim
Her 16h21'1"50d45'
Rozsa,Joseph & Eva
Crb 15h24'15"30d45'
Rozsnyai,Imre
Cep 23h25'17"70d33'
Rozy
And 2h10'53"39d58'
Rozycka,Danuta
Uma 10h47'21"48d56'
Rozzi,Anita
And 0h14'1"32d29'
Ro^1
Lac 22h22'57"50d5'
RPM
Tri 1h59'29"26d49'
Rr Star
Peg 22h33'52"21d50'
RR/MR
Peg 22h12'59"8d47'
RRK Treon
Aql 19h42'0"12d20'
RSR 93
Uma 11h57'1"50d53'
RSW Kalic Leo Reed 2
Peg 21h26'26"8d26'
Ruadh,Domhnall
Cep 0h3'1"78d60'
Ruan,Adriana
Uma 9h48'13"51d51'
Rubado,Bob & Mary
Cyg 21h33'46"38d50'
RUBAL-50
Cyg 20h2'0"31d40'
Rubbelke,Renee M
Aql 18h54'22"-0d43'
Rube,The
Lac 22h21'38"40d55'
Rubel II,Carl McHenry
Ser 15h23'56"-0d26'
Rubel,Barbara
Peg 0h2'20"13d55'
Rubemeyer,Abigail Lynne
Cas 1h4'15"60d30'
Ruben
Aql 19h59'26"10d30'
Ruben
Aur 6h8'35"31d49'
Ruben & Johanna
Eri 4h5'1"-7d1'
Ruben Adame
Boo 14h54'47"35d7'
Ruben Elizalde
Boo 14h42'48"38d30'
Ruben Jr
Ori 4h41'41"10d32'
Ruben,Alexander John
Vul 20h18'32"26d3'
Ruben,Gregory
Hya 8h23'26"1d40'
Rubens,Emily-Caroline- Billy
Vul 19h57'41"28d19'
Rubenstein,Amy Kyle
Tau 5h56'1"28d44'
Rubenstein,Andrew
Lac 22h8'11"51d1'
Rubenstein,Bruce & Lori
Cyg 19h6'16"48d7'
Rubenstein,Hattie
Cep 1h24'1"77d56'
Rubenstein,Herb
Her 16h21'0"24d49'
Rubenstein,Irving D
Ori 5h11'0"-4d44'
Rubenstein,Rachel
Com 13h33'21"21d34'
Rubenstein,Randee Michelle
And 23h41'0"38d19'
Rubera,Debi
Aql 19h43'0"11d8'
Rubert,Christine
Cam 7h43'12"80d28'
Ruberti,Lori Ann
And 1h13'52"39d42'
Ruberti,Thomas J
Boo 14h34'60"39d43'
Rubiato,Marta Teresa
Ser 15h19'54"2d33'
Rubin Ed D,Esther Elka Revsin
Vir 13h33'54"2d13'
Rubin!,I Love You Marc S
Cet 2h53'22"5d17'
Rubin's Wish,Bruce Joel "My Life"
Per 2h26'56"56d30'
Rubin,Adam
Per 4h5'52"35d37'
Rubin,Alexander Leslie
Ori 6h4'19"1d25'
Rubin,Alijah Benjamin
Hya 8h12'29"5d37'
Rubin,Arline
Peg 23h46'14"30d59'
Rubin,Barbara Kaye
Uma 8h7'14"71d47'
Rubin,Daddy Stuart
Eri 4h17'53"-16d33'B
Rubin,Daniel & Dorothy
Cyg 20h56'51"40d11'
Rubin,David Benjamin
Ori 5h56'1"10d54'
Rubin,Earlene
Com 13h5'14"18d17'
Rubin,Elizabeth Anne
Aqr 21h43'23"0d30'
Rubin,Herman
Cam 4h9'59"67d43'
Rubin,Jaimee
Lyr 19h0'23"25d31'
Rubin,Jody Marsha
Oph 18h17'26"12d26'
Rubin,Jon & Beverly
Cam 12h51'34"77d15'
Rubin,Maria Carla
Uma 8h30'34"51d6'
Rubin,Mom & Dad
Cam 3h29'39"53d18'
Rubin,Murray (Mel)
Vir 12h49'44"-6d20'
Rubin,Naomi,Benjamin & Eleanor
Ori 5h58'56"1d23'
Rubin,Nathan
Sgr 19h28'51"-42d13'
Rubin,Robert
Per 3h14'1"41d40'
Rubin,Rochelle
Cas 1h32'47"65d32'
Rubin,Spencer Perry
Boo 14h48'45"34d9'
Rubin,Susan
Mon 6h30'35"8d46'
Rubina
Uma 9h16'28"58d38'
Rubini,Jessica Anne
And 2h25'55"40d45'
Rubini,Jr,Rudy
Aur 6h51'18"37d55'
Rubino,Jonathan
Psc 0h46'21"28d46'
Rubins,James
Aql 20h9'42"6d19'
Rubins,Nancy J
Peg 22h34'35"31d34'
Rubinsky,Edward R
Cep 23h28'0"70d51'
Rubinsohn,Tracey Beth
Psc 1h28'28"20d48'
Rubinstein,Josie-Laure
Aur 6h7'41"37d39'
Rubinstein,Rachel Ann
Lyn 9h14'50"34d50'
Rubio Krohne,Laura
Cru 11h58'0"-56d5'
Rubio,Alberto S
Ser 17h56'28"-14d52'
Rubio,Marco
Lmi 10h37'51"24d19'
Rubio,Pedro A
Ori 5h27'41"-6d47'
Ruble,Bob
Aql 20h0'48"11d0'
Ruble,Helen
Peg 0h11'0"14d22'
Ruble,James Mathew
Aur 7h5'51"37d47'
Ruble,Phillip L
Peg 23h7'1"14d24'
Rublino,Michael
Peg 23h20'34"8d6'
Rubner,Cagney Valentine
Equ 21h17'21"2d34'
Rubner,Shalin John Robert
Hya 8h58'0"5d31'
Rubner,Udo
Cas 0h32'53"63d44'
Rubnitz,Peter Block
Ser 15h42'42"10d54'
Rubscha,Rick & Joanne
Cas 0h15'51"65d37'
Rubush,Ellen Blaine
Cma 6h57'51"-16d37'
Rubush,John Lance
Lyn 8h0'1"52d27'
Ruby
Aqr 23h9'13"-5d53'
Ruby
Vul 19h48'17"23d3'
Ruby
Cep 22h22'13"56d8'
Ruby
Cyg 21h57'1"52d40'
Ruby
Cma 7h17'48"-15d42'
Ruby I,Kenith A
Boo 14h32'26"8d5'
Ruby Joan
And 0h59'1"37d13'
Ruby Lucille
Gem 7h23'15"34d39'
Ruby Lynne
Peg 22h39'40"26d39'
Ruby Mae
Boo 14h11'33"41d45'
Ruby Rena
Cyg 19h29'47"31d22'
Ruby Star,The
Vul 20h18'39"25d5'
Ruby's
Mon 6h43'34"-4d2'
Ruby's Diamante
Vul 20h39'38"23d13'
Ruby,Bradley David
Dra 15h51'1"62d18'
Ruby,Jack Edward
Aql 20h7'21"1d36'
Ruby,Jr,Harold L
Ori 5h12'54"-5d24'
Ruby-Bischer
Cas 0h13'0"62d36'
Rubykelly
Sex 10h2'53"-1d2'
Rucci Birthday Star, The AnnMarie
Psc 1h17'27"20d13'
Ruchay,Ulrich
Boo 15h20'1"40d33'
Rucinski,Haley Krajniak
Uma 11h59'30"55d12'
Rucinski,Patricia Ann
Cyg 20h29'49"37d46'
Rucker,Emily Caroline
Cam 3h52'1"60d19'
Ruckman,Danielle Nicole Samantha
Per 4h2'4"39d39'
Ruckman,Pamela S
Hya 8h34'10"-1d8'
Ruckteschler,AFC,Carol
Cas 0h58'52"63d11'
Rudaitis,Kaz
Mon 7h17'49"-7d6'
Rudd,"Billy"David & Michelle
Crb 16h7'0"33d5'
Rudd,Destini M
Aql 20h0'45"1d35'
Rudd,Howard
Lyr 19h21'28"42d60'
Rudden,Abigayle
Vul 19h36'54"20d9'
Rudderman,Randal,H
Mon 6h23'55"-1d30'
Ruddick,Mary Francis
And 0h23'23"33d45'
Ruddlesdin,John
Sge 19h58'57"20d18'
Ruddlesdin,Katharine
Cas 1h51'52"73d59'
Ruddock,Helen
Peg 23h1'23"30d9'
Ruddy,Jr,Jim "The Golfer"
Her 17h8'54"21d5'
Rude,Donal
Hya 9h2'11"0d26'
Rude,Jon R
Dra 17h48'46"75d40'
Rudel,Margaret Murchison
Del 20h18'37"16d28'
Rudel,Thomas
Cnv 13h1'35"31d8'
Ruden,Eric Michael
Lac 22h49'58"37d45'
Ruden,Stuart Michael
Vul 19h46'32"28d34'
Rudenko,Jessica
And 1h33'20"37d51'
Rudi
Cnv 13h36'15"32d12'
Rudi,Christer K
Cam 3h30'47"61d25'
Rudi-Andreas
Lyn 8h6'38"40d43'
Rudiak,John
Crt 10h50'36"-6d46'
Rudig,D A
Lac 22h32'0"53d27'
Rudiger's Guiding Light,Lance W
Ari 1h47'41"11d5'
Rudiger's Nautical Light,Heidi L
Leo 10h20'44"13d26'
Rudisch,Hofrat Dr Ansgar
Uma 8h53'27"51d34'
Rudisch-Kretschmar, Maria Magdalena
And 0h21'27"30d13'
Rudisill,Angela
Ori 5h11'0"-5d19'
Rudisill,Carol Ann
Cas 0h32'22"64d35'
Rudisill,Thomas
Per 3h26'41"51d43'
Rudl-Kaspar
Sgr 18h57'20"-25d0'
Rudman,Billy & Laney
Lyn 8h33'55"41d57'
Rudman,Emily H
Peg 23h1'0"20d12'
Rudman,Herbert Charles
Aur 5h29'51"38d27'
Rudman,Kate
Cyg 19h44'21"30d13'
Rudnick,Erica Lynn
Lac 22h20'34"40d28'
Rudnick,Jessica Lauren
Lac 22h20'1"40d5'
Rudnick,Michelle Linda
Cyg 20h22'1"31d2'
Rudnick,Wendy Susan
Eri 4h11'51"-13d48'
Rudnicka,Lucy Anna
Lyn 7h2'13"44d6'
Rudnicki,Barbara
And 22h58'23"40d14'
Rudnicki,Jeffrey Patrick
Aur 6h3'32"46d47'
Rudnicki,Jill Ellen
Cas 1h12'13"60d53'
Rudolf John
Aqr 22h40'44"2d20'
Rudolf,Tyler Allen
Cep 21h7'33"80d11'
Rudolff,Guy
Boo 14h30'1"8d57'
Rudolph's "Wedding Star",Brett & Amy
Cyg 21h28'35"48d49'
Rudolph's "Wish Upon A Star",Kathy
Peg 23h27'24"30d54'
Rudolph,Alison Lee
Com 13h11'31"21d12'
Rudolph,Baby J
Cam 4h6'25"60d4'
Rudolph,Cotzy
Lac 22h8'24"51d43'
Rudolph,Daniela
Leo 9h57'18"10d17'
Rudolph,Ellen Kean
Lyn 7h47'50"45d19'
Rudolph,Hans
Uma 9h21'48"47d34'
Rudolph,Joshua
Cam 3h20'1"60d20'
Rudolph,Lon
And 2h7'0"41d11'
Rudolph,Nicole Anne
Cyg 19h31'52"33d4'
Rudolph,Selma
Com 12h47'38"23d2'
Rudolph,Tyler Stephen
Lac 22h34'25"37d35'
Rudolph,Werner A
Aur 5h35'39"54d55'
Rudolphius,J
Dra 20h6'56"62d30'
Rudon-Ruth & Donald's Golden Star
Cnv 12h24'13"40d60'
Rudoy,Edward H
Per 1h58'1"56d55'
Rudy
Eri 5h0'38"-8d43'B
Rudy
Lac 21h59'20"37d43'
Rudy & Margie:Half a century love & laughtr
Cyg 21h1'30"33d45'
Rudy & Mary
Sge 19h56'15"20d6'
Rudy The Great Romantic Enthusiast
Cet 2h52'17"6d46'
Rudy's Guiding Light
Her 16h31'1"33d49'
Rudy-and-Brooke
Cyg 19h28'1"34d8'
Rudyhunt
Ori 5h3'46"14d57'
Rudziewicz,Joe
Aql 18h56'59"-2d41'
Rudzinski,Kevin
Cep 23h20'52"66d6'
Rudzinski,Samantha Lynne
And 2h28'0"39d33'
Rue,Charles A
Sex 10h43'53"-1d18'
Rue,Samantha
Lyr 18h16'23"47d5'
Rue,William
Sex 10h45'11"-10d4'
Ruecker,Alexander Danté
Cet 2h56'28"2d10'
Ruecker,Samantha
Equ 21h18'48"2d46'
Rueckert,Greg
Cam 6h57'10"65d16'
Rueckheim,Paul & Mary
Lac 22h20'25"52d6'
Rueda i Rufas,Antoni
Aur 5h16'41"43d43'
Rueda,Ana (Yjulia) Zamorano
Cyg 19h27'48"34d30'
Rueda,Ruben Ponce
Cma 7h12'16"-15d11'
Ruedi
Lib 14h57'46"-2d3'
Ruedisueli,Jon S
Oph 18h4'1"0d11'
Rueegg,Esther(Haelmli) 15 02 1967
Cas 3h4'49"57d52'
Rueff,Ann
Mon 7h1'0"7d58'
Rueff,Jr,Walter Thomas
Cet 2h43'47"3d18'
Rueff,Michael Andrew
Cma 7h15'23"-13d44'
Rueger,Joan
Mon 7h23'42"-1d51'
Ruegg,Jennifer Kayleigh
And 0h22'22"30d25'
Ruehling
Vul 20h21'40"25d38'
Ruel,Alexander Golden
Ori 4h54'22"0d42'
Ruel,Eileen Allyson
Ori 4h53'14"1d24'
Ruello II,Nicholas Joseph
Dra 15h11'0"61d24'
Ruepp-Verney,Rowena/ Claudio
Ser 16h2'29"2d40'
Ruesch,Richard & Margaret
Cep 22h36'22"60d33'
Ruetters,Heinz
Del 21h0'28"18d56'
Ruettgers 50
Boo 14h34'27"50d24'
Ruf,Keri
Cas 1h25'18"61d51'
Ruf,Melanie Rose
Lyn 8h45'21"34d47'
Rufca,Matthew Spencer
Aur 5h27'15"30d56'
Rufe,Magdelena Mae
Vul 19h13'25"22d13'
Ruff
Cma 7h12'10"-15d56'
Ruff,Eric Alexander
Ori 5h23'45"-4d60'
Ruff,Margee
Ori 5h56'22"8d51'
Ruff,Renate Anni
Vir 12h56'26"-18d20'
Ruffing,Rich
Hya 9h0'12"5d27'
Ruffino,Michael Anthony
Aql 19h29'52"12d1'
Ruffino,Noelle
Mon 7h27'35"-10d10'
Rufflebottom
Sct 18h42'35"-6d22'
Ruffner,Marion Candler "Boots"
Boo 14h12'27"30d44'
Ruffner,Van Natta
Cet 2h27'56"-10d28'
Ruffner,Zoe Morrison
Umi 15h27'36"68d20'
Rufka,Necha
Per 4h44'46"51d31'
Rufus
Cet 1h17'0"-13d34'
Rugala,Laura Jean
Cyg 21h56'35"53d34'
Rugby
Oph 17h53'31"13d19'
Ruge,Detlev
Leo 11h7'0"-0d28'
Ruggeri,Christina
Lyr 18h29'39"37d48'
Ruggeri,Enrico
Gem 6h47'7"17d29'
Ruggeri,Silvia
Per 3h53'5"37d52'
Ruggie,Mark Alexander
Lac 22h9'38"46d16'
Ruggieri,Sylvie
Cep 22h25'30"58d43'
Ruggiero,Jr,Louis
Cep 23h20'24"70d32'
Ruggiero,MD,Mary T
And 0h16'55"37d33'
Ruggiero,Patrick Spencer
Her 16h45'37"38d47'
Ruggiero-Forever Our Baby,Lucas
Her 17h30'0"27d15'
Ruggiero-Kersch, Frederick & Vivian
Peg 23h27'27"28d50'
Ruggles,Karla L
Lyn 8h1'57"39d9'
Rugiada Innocenti
Pup 7h56'58"-25d8'
Ruhe,Joseph
Tri 1h57'27"27d30'
Ruhf,Adrienne
Lyn 7h30'1"51d30'
Ruhland,Franziska
Leo 11h23'55"8d23'
Ruhlin,William Ritter
Cap 21h21'11"-26d23'
Ruhman,Ruth Anne
Cyg 20h4'54"40d26'
Ruhs,Captain J T
Boo 14h24'0"28d47'
Ruik,Laura
Cas 0h22'1"69d58'
Ruika Iruka
Cnc 8h50'0"8d54'
Ruiner,Ludwig
Mon 7h53'29"-1d49'
Ruini,Rossella
Her 16h41'40"18d46'
Ruiz,Cesar R
Per 1h25'0"52d48'
Ruiz,Hilda
Cyg 19h27'49"30d56'
Ruiz,James K
Per 2h40'1"43d9'
Ruiz,Lucy
Peg 22h25'55"30d44'
Ruiz,Margie
Ori 5h43'51"11d20'
Ruiz,Maria
Cas 0h36'0"61d24'
Ruiz,Marie-Solange
Cnv 12h20'50"40d0'
Ruiz,Michael Andrew
Cnv 13h25'13"38d40'
Ruiz,Paul
Hya 10h48'16"-18d15'
Ruiz,Rafael Rivero
Cep 3h3'30"80d21'
Ruiz,Sabina
Ori 5h39'15"12d24'
Ruiz,Thomas
Hya 8h44'30"4d10'
Ruiz-Conejo
Uma 11h19'38"57d12'
Rukavina,Cindy-Joe
Cyg 20h54'13"53d47'
Rule,Alverna G
Mon 7h15'50"-8d13'
Rule,Holly Christene
Aql 20h3'45"-0d37'
Rulein,Akiko S & Craig W
Lmi 10h6'1"31d31'
Ruler,Dominant
Ori 5h50'46"20d58'
Rulli,Sam
Dra 12h21'44"75d19'
Rulus
Lyr 18h57'25"31d30'
RUM
Aql 18h46'16"11d22'
Ruma,Delores
Lyr 18h56'15"32d45'
Rumana
Ori 6h3'44"9d14'
Rumble Fish
Cet 2h27'58"5d13'
Rumble,Kyra Lynne
Cas 0h57'30"60d30'
Rumbold,Margaret Evelyn & Alfred Eugene
Lyr 18h17'17"30d54'
Rumburg,Bryan Gary
Dra 19h51'56"70d42'
Rumdidit
Cnv 13h3'56"40d38'
Rumeczik,Herbert
Cep 22h3'45"60d14'
Rummel,Mae
Tau 5h48'43"26d52'
Rumney
Boo 15h7'55"16d25'
Rumoro,Salvatore & Karen
Boo 14h1'39"14d50'
Rumpca,Dave
Dra 16h53'34"63d53'
Rumsby,Frederick George
Cyg 19h37'37"28d40'
Rumsby,Joyce Bessie
Cyg 19h35'24"28d37'
Rumsey,Lyle Thompson
Hya 8h25'0"-6d41'
Rumsey,Sandra D
Ser 15h18'58"8d34'
Rumsey,Terry & Susan
Boo 15h5'49"50d20'
Runacres,H S
Dra 15h33'1"58d23'
Runcie,Patricia
Aql 20h14'42"4d52'
Rund,Jürgen
Aql 20h6'47"4d55'
Rundell,Charles G
Aql 20h21'25"7d58'
Rundgren,Todd
Uma 11h1'32"52d2'
Rundle,Clifford & Evelina
Sge 20h4'0"20d13'
Rundle,Jason & Dionne
Ser 15h20'35"2d7'
Rundle,Wright L
Sgr 18h2'26"-28d13'
Rundle,Zoë Ella
Umi 16h50'1"76d36'
Rundquist,Margaret & Larry
Eri 3h24'32"-6d57'
Rundt,Melissa Louise
Lyn 8h19'43"38d53'
Rune Brian
Hya 9h14'35"2d45'
Runeau,Pierre
Mon 6h25'41"7d38'
Rung,Al
Sct 18h52'48"-9d17'
Rung,Nicholas Christopher
Boo 14h55'47"31d57'
Runge,Glenn & Gloria
Cnv 13h53'52"40d45'
Runge,Marty & Rita
Cyg 20h51'14"48d49'
Runge,Wallace F
Lac 22h11'1"48d59'
Runion,Ora
Cmi 7h34'56"6d37'
Runkel Family
Cep 22h20'1"60d23'
Runkel,Edward Alan
Dra 16h59'17"62d29'
Runkel,Elyse Marie
Dra 16h7'1"66d37'
Runkle,Elizabeth Parker
And 1h20'19"37d22'
Runne,Stephanie
Lyr 18h51'1"31d8'
Runnells,Clive
Sct 18h52'49"-7d44'
Runnells,J Lovely
Cam 5h42'55"60d24'
Runnells,Nancy M
Peg 21h52'44"2d50'
Runner,Raye
Uma 10h47'29"68d50'
Running Bear
Uma 9h30'11"48d35'
Runowicz,Debra Ann
Sco 17h29'38"-33d59'
Runrig
Uma 10h1'39"55d26'
Runstein,Nina
Cas 23h35'41"63d16'
Runte,Dr Kennon
Uma 14h11'54"60d48'
Runte,Irmgard
Leo 9h21'24"8d38'
Runya
Uma 11h45'20"31d16'
Runyan,Elle
And 23h44'44"47d33'
Runyan,John Alexander
Lac 22h53'0"54d26'
Runyans,William Edward
Aql 19h26'54"14d43'
Runyon,Thom
Oph 18h40'1"10d33'
Ruocco
Aql 19h32'38"0d35'

Ruocco,Ambra
Cet 2h35'26"0d7'
Ruocco,Jane
Uma 8h30'0"50d49'
Ruona,Toddy
Aur 6h2'46"31d25'
Ruopp,Dr Richard R
Lmi 10h10'0"30d27'
Ruopp,Phillips B
Lmi 10h7'32"32d0'
Ruotolo III,George Charles
Aur 6h55'15"37d12'
Ruotsala,Tera
Cmi 7h6'55"1d31'
Rupar,Zackary
Dra 18h58'11"48d27'
Rupcic,Gregory
Boo 15h14'22"50d36'
Rupert
Uma 10h57'45"68d25'
Rupert
Uma 9h34'19"73d12'
Rupert & Jill
Eri 3h40'39"-5d38'
Rupert's Silver Star
Cnv 13h46'51"31d21'
Rupert,Katheryn E
Cnv 12h13'17"43d42'
Rupert,Scott W
Boo 14h40'18"52d3'
Rupert,Steve
Uma 11h34'49"41d2'
Rupert,Zachary
Ori 5h57'45"20d36'
Ruperts
Aql 18h57'21"-6d52'
Rupeter,B P S
Ser 15h37'0"9d28'
Rupff,Scott E
Aur 6h10'45"38d56'
Ruple's Star,Vance
Her 18h0'0"37d57'
Ruplenas,Andrius Kovas
Per 2h53'42"32d3'
Rupley,Nelle Milholland
And 1h13'11"34d39'
Rupner,Ilmar
Umi 15h29'58"78d3'
Rupp,Gordon(Sarg)
Cam 4h53'33"67d30'
Rupp,Gregory D
Her 17h3'31"38d37'
Rupp,James Michael
Cnv 13h11'24"40d53'
Rupp,Lee Roy
Aur 4h57'14"30d17'
Rupp,Ludwig
Cep 0h4'56"81d44'
Rupp,Verna Marie Adele Rath
Cas 1h32'32"65d25'
Ruppel,Henry Christian Jr
Uma 12h1'18"39d45'
Ruppel,William Reinhold
Dra 11h5'44"78d13'
Ruppenthal,Tom
Aur 6h5'37"45d23'
Ruppert,Jacqueline Susan
Com 12h16'37"32d54'
Ruppert,Thomas
And 23h46'1"38d44'
Ruppert,Tracy Lyn
And 2h9'30"39d41'
Rupprecht,Kay
Sgr 19h0'25"-24d10'
Rupprecht,Willibald
Del 20h23'45"3d8'
Ruprah,Simi
Cyg 21h24'53"28d32'
Ruprecht,Hilmer
Cmi 7h18'32"5d31'
Ruptus of Comcors
Lac 21h58'49"37d3'
Rurode III,William Ernest
Vul 19h23'18"23d16'

Rusch,Ken
Cma 7h23'21"-16d49'
Rusch,Silvretta 28670 Berndt
Ori 6h12'36"0d50'
Rusch,Sister JoAnne
Cas 1h19'53"60d23'
Rusch,Totzi 5672-Doris
Crb 16h10'51"37d49'
Rusch-My Christmas Star
Love Lynne,Frank
Aql 18h56'58"10d13'
Rusch-Stroud"Mary", Marcia
Ann
Cyg 21h52'44"38d24'
Ruscher,Mario
Lyn 8h8'21"43d10'
Ruschman,Donald James
Boo 15h34'24"41d26'
Rusciano,Jennifer
And 2h18'28"45d47'
Ruscigno,PJ & HG
Eri 3h52'54"-6d33'
Ruse,Erin Rachel
Cyg 19h32'0"31d16'
Rush,Amy
Cas 3h34'1"70d25'
Rush,Ashlee N
Vul 19h44'39"22d50'
Rush,Carol Lynn
And 1h58'41"36d10'
Rush,Dawn Cherie
Cam 6h11'38"58d41'
Rush,Eileen Maria
Psc 1h41'29"20d45'
Rush,Emily Elizabeth
Peg 22h48'16"27d58'
Rush,G L & Gladys
Lac 22h6'23"38d27'
Rush,Jr,Peter
Boo 14h53'49"26d15'
Rush,Katherine Maeve
Peg 22h28'1"20d23'
Rush,Kelly
Lyn 8h58'22"46d58'
Rush,Kelly A
Mon 7h1'30"-6d59'
Rush,Patricia Paynter
Com 13h2'52"21d49'
Rush,Rodney J
Cet 2h55'52"5d30'
Rush,Roy & Dorothy
Lyn 7h45'50"50d3'
Rush,Rue
Peg 22h1'1"5d36'
Rush,Sandra R
Aql 20h14'0"5d16'
Rushak,Gagoo
Peg 21h35'48"20d30'
Rushby,Simon Craig
Lyr 18h59'12"36d21'
Rushforth,Michael Allan
Crb 15h54'44"27d22'
Rushing,Dean Holden
Oph 18h39'46"8d21'
Rushing,Marda Kay Peters
Mon 7h28'0"-6d22'
Rushmer,Alan
Per 2h52'13"40d41'
Rushton III,James Edward
Cma 6h55'29"-18d38'
Rushton,Barbara
Aql 20h9'57"6d22'
Rushton,Vicki "What Ya
Doing"
Cas 0h23'16"60d4'
Rushton-(Jedi Knight) Phillip
Anthony
Ori 6h1'19"3d14'
Rushworth,Christine
Mon 6h54'0"-10d8'
Rusin,Kimberlee
Cyg 20h53'27"41d45'
Rusinskas,Wanda M
Ori 5h33'45"-0d45'

Rusk,Daniel
Aql 19h50'39"10d20'
Rusk,G L
Mon 8h8'38"-8d0'
Rusk,Jared Simon
Mon 7h16'15"-6d24'
Rusk,Melanie
Peg 22h58'50"33d22'
Rusk,Natalie Brooke
Oph 18h3'31"7d42'
Ruske,Tanja
And 2h3'32"40d52'
Ruskell,Mary Sara
Cas 3h1'1"57d24'
Ruski The Star
Sgr 18h54'37"-23d19'
Ruskin,Matthew Sean
Ori 5h12'35"-4d36'
Ruskin,Myrna
Sgr 18h56'15"-28d30'
Ruskowski,Kenneth M
Aur 6h9'14"46d26'
Rusky
Aur 5h15'26"43d26'
RusMar
Crt 11h25'37"-11d8'
Rusnica,Melissa Marie
Cas 0h59'27"63d36'
Rusomarov,Tracey Hooper
And 23h35'47"48d20'
Russ
Per 2h59'44"34d16'
Russ & Esther
Eri 3h53'10"-5d59'
Russ & Judy's 40th
Anniversary
Aql 19h51'50"12d18'
Russ F
Dra 18h19'39"67d42'
Russ, My Love
Her 17h21'19"49d1'
Russ,Autumn Kay
Cas 0h55'41"68d29'
Russ,Glenn Norman
Her 17h56'57"48d53'
Russ,Heike
Dra 16h54'38"51d60'
Russ,Joanne & Champ
Per 2h53'50"40d14'
Russ,Jr,Robert M
Mon 7h45'12"-5d53'
Russ,Lucian
Mon 6h46'42"11d44'
Russ,Maurine E
Aql 19h59'0"14d15'
Russ,Peter 20-04-1945
Lyr 19h16'18"31d34'
Russ,Princess Jennifer
And 0h46'59"39d46'
Russ,Stacy Lynn
Cyg 19h25'60"33d45'
Russ,Susan Bradley
Cas 0h37'17"61d40'
Russano,Domenic
Cnv 12h15'38"46d45'
Russeellara,Dora
Lyn 7h27'1"44d57'
Russel Tracy
Cyg 20h11'25"58d31'
Russel's Star
Peg 23h35'19"12d11'
Russel,MD,Hugh D
Oph 17h53'0"12d38'
Russell
Cyg 19h34'25"37d22'
Russell "A Guiding
Star",Jeanne M
Cas 1h11'12"61d6'
Russell & Stephanie
Lyn 9h13'55"35d33'
Russell Alexander & Tracey
Lynn
Boo 15h7'0"11d32'
Russell III,Hugh Mortimer
Ser 15h17'48"22d59'

Russell Star,The
Leo 10h32'0"10d30'
Russell V,Robert Lee
Her 17h1'26"20d51'
Russell Y
Aql 19h5'38"-0d24'
Russell's "Star Concerto"
Aur 5h25'21"30d22'
Russell's Light
Hya 8h30'27"-6d44'
Russell,Alice
Tri 2h38'32"32d33'
Russell,Alice Joyce
Eri 4h0'58"-15d32'
Russell,Alicia Ann
Mon 6h52'34"-6d48'
Russell,Amanda Ann
Cyg 20h27'50"38d31'
Russell,Aundrae
Cma 6h54'38"-16d46'
Russell,B Z
Cnc 7h58'0"10d33'
Russell,Barbara Elizabeth
And 23h17'26"37d49'
Russell,Betty Lou
Cas 0h18'56"58d36'
Russell,Brendan Corey
And 23h35'56"40d19'
Russell,Caroline J Jones
Oph 17h31'56"-20d27'
Russell,Casey Rae
Aur 6h13'57"30d53'
Russell,Christine Frances
Cyg 20h51'24"38d19'
Russell,Christopher Austin
Her 16h20'42"20d24'
Russell,Clifford Todd
Tri 1h45'44"27d18'
Russell,CLU,Joe C
Cep 1h39'17"77d56'
Russell,Cydney Eve
Lyn 7h53'28"33d48'
Russell,David
Peg 23h28'14"24d43'
Russell,Denise Kathleen
Lac 22h33'53"38d7'
Russell,Derek Edward
Ser 15h34'59"20d29'
Russell,Donald
Lac 22h53'1"51d11'
Russell,Douglas
Aur 5h28'51"38d13'
Russell,Duncan Weston
Uma 8h18'13"60d50'
Russell,Edward Randolph
Cep 9h45'58"80d37'
Russell,Elizabeth Carol
Cyg 20h55'30"30d33'
Russell,Emily A
Vir 13h5'29"-8d25'
Russell,Fran
Aql 19h54'59"7d52'
Russell,Gary Michael
Ori 5h57'33"16d26'
Russell,Gerry
Cap 20h19'10"-27d28'
Russell,Gregory Keith
Aql 19h54'14"11d8'
Russell,Homer B
Del 20h16'47"13d46'
Russell,John
Ori 5h55'13"15d03'
Russell,John
Per 3h33'26"39d44'
Russell,John M
Cep 21h25'18"58d44'
Russell,John Merritt
Hya 8h54'53"5d54'
Russell,Jonathan Cecil
Ori 6h15'11"8d37'
Russell,Jr,David Allen
Aur 4h50'47"37d44'
Russell,Jr,Kenneth Edward
Per 3h25'21"40d23'

Russell,Jr,William Alton
Cam 12h13'37"77d41'
Russell,Julie
Lyr 19h6'53"26d45'
Russell,June Dolores
Peg 22h8'28"3d57'
Russell,Kathryn
Cyg 19h30'11"33d9'
Russell,Katie Dean
Eri 3h0'56"-15d22'
Russell,Kristin Ashley
Vul 19h23'25"26d51'
Russell,Leo
Dra 19h1'38"58d59'
Russell,Lia
And 0h48'0"40d41'
Russell,Margaret
Lyr 19h14'34"42d36'
Russell,Maria Elena
Cyg 20h9'17"40d34'
Russell,Marise & Maurice
Uma 11h57'1"32d4'
Russell,Megan Melissa
Peg 22h56'0"20d54'
Russell,Melanie
Com 13h13'25"28d21'
Russell,Michelle Ann
Crt 11h9'41"-17d38'
Russell,Nancy Dunton
Lyr 19h18'39"38d15'
Russell,Nathan Alexander
Ari 2h40'14"22d7'
Russell,Neil & Elizabeth
Cyg 21h43'31"38d51'
Russell,Nell
Ori 5h29'45"-1d1'
Russell,Nicole Marie
Lyr 18h59'1"27d38'
Russell,Orville Kraig
Sge 20h17'42"17d43'
Russell,Patricia Ann
Uma 10h32'41"40d43'
Russell,Patrick
Hor 2h58'42"-49d30'
Russell,PhD,Sandra Lynn
Wilkinson
Ori 6h12'48"7d36'
Russell,Robert Ellsworth
Aql 19h46'53"12d30'
Russell,Robert D
Dra 15h38'32"60d46'
Russell,Robert E C
Per 2h59'50"38d51'
Russell,Ronald L
Aql 19h58'35"1d6'
Russell,Sandra
Eri 3h49'43"-6d17'
Russell,Sandra Froah
Cas 1h49'36"75d17'
Russell,Scott & Gail
Eri 3h39'1"-4d10'
Russell,Shannon
Sex 10h33'33"-6d10'
Russell,Sheila & Coco
Uma 8h53'29"50d38'
Russell,Sr,James H
Cep 0h56'1"78d38'
Russell,Stanley Martin
Her 16h36'11"10d60'
Russell,Tamora (Tammy)
Linise
Cas 23h32'1"61d38'
Russell,Tia H
Peg 22h19'34"26d31'
Russell,Trampus
Aql 20h10'47"10d39'
Russell,Wella
Peg 23h25'22"18d15'
Russell,Wendy Kathleen
Ori 6h5'43"5d27'
Russell,William Robert
Boo 15h16'11"52d23'
Russell,Winifred Y
Peg 21h37'46"26d38'

Russell,Xavier Vaughn-Ishi
Gem 7h27'25"31d10'
Russell-"Baby Love"
Umi 13h47'0"77d39'
Russell-Cross
Hya 8h12'51"-1d33'
Russell-Hiney,Shannon
Cyg 20h22'35"31d32'
Russenberger-Rosica, Phoebe
Cnv 13h19'14"38d38'
Russert,Franklin W
Dra 16h53'22"62d44'
Russey,Linda
Sct 18h55'59"-7d59'
Russey,Mary E
Eri 3h27'53"-6d1'
Russin,Roxanne
Lyn 7h6'0"50d31'
Russman
Dra 18h4'36"65d16'
Russo 651,Charlie
Aur 6h32'35"37d60'
Russo Happy Birthday, Blake
Boo 15h8'29"28d57'
Russo,"Tar Star" Teresa Ann
Cnc 8h27'39"11d10'
Russo,Albert
Aur 6h2'57"37d40'
Russo,Alexander Michael
Boo 14h51'12"31d43'
Russo,Andrew C
Dra 14h28'24"61d14'
Russo,Ann & Carl
Eri 3h23'31"-2d44'
Russo,Anthony James
Peg 22h36'47"31d29'
Russo,Betty
Lyr 19h3'32"38d1'
Russo,Bill
Aql 19h53'57"12d14'
Russo,Carrie Ann
Peg 22h17'44"20d3'
Russo,Charles
Cam 6h11'16"60d12'
Russo,David R
Her 18h1'60"40d55'
Russo,Delilah
Lyr 18h54'26"43d0'
Russo,Emily Clare
Gem 7h15'41"28d35'
Russo,Eunice (Nana)
Cas 2h26'40"63d60'
Russo,Frances Rose
Aql 19h47'44"13d4'
Russo,Frank M
Boo 15h6'31"26d50'
Russo,Gina Diana
Del 20h14'11"10d8'
Russo,Jackie
Cma 6h13'0"-13d13'
Russo,Jennie
Cas 0h9'46"61d20'
Russo,Jr,Matthew
Dra 16h3'36"66d16'
Russo,Judy Williams
Mon 6h29'60"8d41'
Russo,Lauren Marie
And 2h21'59"38d27'
Russo,Louis O
Aql 20h0'1"11d7'
Russo,Maria
Lyn 8h17'18"50d33'
Russo,Marie D
And 0h55'24"35d38'
Russo,Mark Joseph
Her 17h4'35"37d50'
Russo,Matthew Richard
Per 4h1'36"51d7'
Russo,Nelda Conger
Cnc 8h0'53"18d49'
Russo,Nick (Gabriel)
Ser 15h29'41"21d16'
Russo,Nora
Cet 0h53'1"0d10'

Russo,Patrick P
Lac 22h38'34"55d45'
Russo,Paul J
Cep 22h54'43"67d38'
Russo,Phil "Bear"
Cnv 12h48'45"37d13'
Russo,Phillip Blanchard
Dra 16h25'22"68d28'
Russo,Robert
Lac 22h55'49"51d37'
Russo,Ronald R
Her 16h29'23"41d42'
Russo,Ross
Cap 20h4'37"-10d22'
Russo,Sarah Jennifer
Com 12h19'22"31d53'
Russo,Shirley Ruth
And 2h18'27"48d11'
Russo,Stephanie
Cas 23h27'22"61d22'
Russo,Stephen
Gem 7h13'47"24d46'
Russo,Steven
Her 16h38'19"32d16'
Russo,Veronica
Del 20h18'59"10d7'
Russo,William Louis
Vir 12h58'21"-11d10'
Russo-Nankivell,Penni &
Dawn
Vul 19h23'25"26d51'
Russolello,Aubrae
Lyn 7h8'32"52d8'
Russoli,Ann B
Vul 19h18'60"22d48'
Russomano,Jackie
Cnv 12h27'21"43d22'
Russomano,Mark
Lac 21h56'48"42d20'
Russomano,Mickie
Aur 4h45'11"34d46'
Russos Odyssey
Aql 20h11'20"4d2'
Rust,Jack
Her 16h34'1"40d17'
Rust,Max
Aql 20h16'38"5d25'
Rust,Silke-Gertrud- Maria
Cap 20h44'10"-20d12'
Rustage,Adam Anthony
Cep 3h39'27"80d16'
Rustar
Aur 6h20'28"38d50'
Rusteberg,Frau
Boo 14h7'28"11d15'
Rustemeyer & Family, Richard
Franklin
Cam 8h50'41"78d52'
Rustic,Harry James
Dra 17h32'20"76d40'
Rustici,Claudio
Mon 6h15'18"-10d9'
Ruston,Betty
Cyg 19h28'32"39d26'
Ruston,Mindy Michelle
Ori 5h14'21"-10d12'
Rusty
Her 17h24'34"38d8'
Rusty
Nor 16h18'17"55d0'
Rusty
Cyg 21h29'40"41d4'
Rusty Lyn
Cmi 7h44'47"3d45'
Rusty Rule
Aur 6h37'25"38d5'
Rusty's Dreams
Cyg 19h26'52"30d39'
Ruszczyk,Walter S
Aur 5h9'22"40d15'
Rusznyak I Capricorn, Geza
Lac 22h20'22"52d1'
Ruta 15
Lyn 9h14'33"36d10'

Ruta,Sebastian
Boo 15h9'44"38d47'
Rute,Orrin K
Aur 7h13'36"41d35'
Rutenbeck,Mary Marjorie
And 0h10'10"31d56'
Rutgersen,Carolyn
Cas 0h16'45"59d3'
Ruth
Cam 3h20'0"55d21'
Ruth
Cap 20h33'0"-13d30'
Ruth
And 1h23'60"42d52'
Ruth & Greg
Cyg 20h26'46"40d39'
Ruth & Howard
Aql 19h1'46"16d38'
Ruth & Len
Sge 19h29'57"16d19'
Ruth & Philippe
Uma 10h19'14"58d53'
Ruth & Seymour
Aql 18h40'42"0d7'
Ruth & Skinny Together
Forever
Cet 0h26'58"-1d47'
Ruth & Todd Forever
Cyg 21h16'46"38d25'
Ruth Agnes
Aur 5h53'31"31d33'
Ruth Alpha Beta
Cas 0h22'35"50d19'
Ruth Angel
Lyn 7h29'58"44d21'
Ruth Ann
Cyg 20h50'16"38d35'
Ruth Ann
Psc 1h26'1"20d55'
Ruth Ann
Cyg 19h51'19"44d57'
Ruth Anne
Ori 4h59'33"10d8'
Ruth Bessie
Eri 3h41'0"-6d42'
Ruth L
Peg 22h17'39"8d6'
Ruth Marie
Lyn 6h28'49"58d38'
Ruth Marion
Leo 9h54'24"11d26'
Ruth Star,The David Bone
Ori 5h34'11"-6d31'
Ruth's Home
Uma 11h52'31"31d21'
Ruth's Light
Cas 23h39'41"50d49'
Ruth's Twinkle
Psc 23h6'11"5d35'
Ruth,David
Cet 0h51'0"-10d59'
Ruth,Dorothy,Henry & Marcy
Mon 6h56'32"-6d52'
Ruth,Grammy Marilyn
Cas 0h29'51"73d16'
Ruth,Joseph A
Per 2h2'45"50d36'
Ruth,Nancy Hewitt
Gem 6h45'1"18d58'
Ruth,Nelda
Mon 7h22'46"-8d51'
Ruthann
Mon 6h55'38"-4d32'
Ruthann
Cas 23h3'0"58d37'
Ruthas 4
Oph 18h1'39"10d19'
Ruthchild,Sunny
Umi 15h4'14"69d16'
Ruthelaine's Razzle Dazzle
Lyn 8h38'53"42d2'
Ruthellen
Equ 20h58'31"2d37'
Ruther,Joei
Tri 2h3'0"32d18'

Rutherford 62556 A Loving
Star,Gary
Lmi 9h56'26"34d7'
Rutherford's Anniversary Star
Cyg 21h33'0"50d18'
Rutherford,April
Cet 0h57'44"-7d17'
Rutherford,Brenda
Cas 0h22'13"72d24'
Rutherford,Cameron William
Her 17h12'47"40d34'
Rutherford,Clara Walls
Lyn 8h12'38"41d38'
Rutherford,Debbie
Cam 9h12'28"81d9'
Rutherford,Dorothy M
And 0h21'53"32d43'
Rutherford,E
Ori 4h57'30"-0d13'
Rutherford,Linn
Aql 19h56'58"8d13'
Rutherford,Linn
Sct 18h55'49"-5d29'
Rutherford,The Star of Georgia
Kate
And 0h26'45"40d25'
Rutherford,Trish
And 23h45'15"45d4'
Rutherford,William P
Aql 20h11'28"1d5"
Ruthi
Lyn 8h9'0"48d24'
Ruthie Baby
Cas 0h43'0"71d38'
Ruthie's Star
Lyr 19h23'1"30d21'
Ruthies Given Dream
Uma 10h33'24"52d13'
RuthiMac
Eri 4h11'29"-15d25'
Ruthless
Aql 19h47'59"11d54'
Ruths,Norbert
Dra 18h22'57"65d2'
Ruthven,Chris & Stephanie
Cam 4h31'0"68d34'
Ruthven-Murray,Ian
Lib 15h18'9"-19d33'
Ruti,John
Sco 16h36'0"-30d40'
Rutigliano,Frank & Chris
Cyg 20h24'50"38d19'
Rutilus,In de Betou William
Uma 11h50'17"32d6'
Rutkiewicz,Maxwell Thomas
Aur 6h28'13"35d23'
Rutkowski,Martina
Uma 10h21'59"50d28'
Rutkowski,Nolie
Aur 7h8'47"40d49'
Rutland,Alecia Juanita
And 2h17'53"42d24'
Rutland,Tania Louise
Cnc 8h29'1"32d39'
Rutledge,Adam Stuart
Ori 5h49'0"20d25'
Rutledge,David
Her 16h41'46"47d56'
Rutledge,Houston Cole
Aql 18h57'34"-5d57'
Rutledge,Jack
Mon 8h1'51"-6d3'
Rutledge,Michelle Petie
Lineberry
Cet 2h42'14"-18d60'
Rutledge,Ned
Aur 6h30'15"32d16'
Rutledge,Theresa Tessie
And 23h34'0"39d39'
Rutman,Doris
Cas 2h8'20"68d31'
Rutolo,Jr,David A
Aql 20h14'32"0d3'
Rutsch,Amparo
Peg 22h36'30"8d51'

Rutski,Michael
Her 17h17'36"45d32'
Ruttenberg,Loraine A
Cas 0h33'34"68d45'
Rutter,Eric John
Del 20h13'0"9d54'
Rutter,Jeffrey
Cyg 20h0'57"31d47'
Rutter,Jody
Cas 22h57'15"54d24'
Rutter,Steven James
Her 17h0'48"18d59'
Ruttledge,Carol
Aql 20h10'18"10d24'
Rutz-Okada,Etoile de Lisa
Per 4h5'2"36d55'
Ruvane,Eddie "Bugs"
Aur 6h0'29"35d36'
Ruvio,Pauline
Uma 13h4'25"61d44'
Ruwaldt,Rachael I.
And 0h33'12"45d56'
Ruwe,Jr,WP
Cet 1h28'10"-1d18'
Ruwe,Kurt
Her 17h41'55"14d16'
Ruwitch,Tucker Joseph Bates
Peg 0h6'29"14d51'
Ruxton,Big Jane
Crb 16h9'27"31d30'
Ruyam
Peg 0h1'49"21d28'
Ruzan,Jordan
Cep 22h10'52"59d28'
Ruzicka,Ludeane Fowler
Eri 3h43'48"-6d57'
RWK-93
Cyg 19h52'0"37d42'
Ry-Guy
Aur 6h0'48"46d2'
Ryall,Shannon Andrea
Cas 0h41'28"70d58'
Ryalls,Erin Sarah
Mon 6h24'27"10d1'
Ryals,C Blair
Hya 8h20'0"0d1'
Ryals,Howard Lewis
Oph 18h38'0"6d23'
Ryals,Randy Shane
Aql 19h15'29"19d46'
Ryan
Lac 22h41'36"53d14'
Ryan
Boo 14h13'28"15d51'
Ryan
Dra 14h2'44"68d2'
Ryan
Peg 22h49'2"26d7'A
Ryan
Aur 5h9m12d37"45'
Ryan & Amy
Cet 2h25'30"5d11'
Ryan & Karron
Lac 22h11'28"51d24'
Ryan & Misty
Lyr 19h12'43"37d38'
Ryan & Natalie's Star Forever
Crb 15h21'32"31d18'
Ryan & Patty - August 6,1994
Uma 9h20'46"51d41'
Ryan 50 Years Of Love, John & Claire
Cyg 19h29'41"30d59'
Ryan 83
Leo 11h1'11"-1d3'
Ryan Aaron
Aql 19h11'12"12d9'
Ryan Anthony
Her 17h37'28"14d9'
Ryan Christian
Her 17h38'18"21d39'
Ryan Christopher
Oph 17h52'43"11d39'
Ryan Christopher
Her 18h1'43"28d28'
Ryan Family Star, Patrick
Dra 16h7'42"62d50'
Ryan Family Star,The
Ori 5h32'12"-1d6'
Ryan II,Dennis Patrick
Cep 22h21'51"55d28'
Ryan III & IV,John J
Lmi 9h54'59"33d30'
Ryan Mary Ann
Aur 5h30'0"37d56'
Ryan Michael
Boo 14h43'1"20d4'
Ryan Patrick
Ori 5h56'1"12d53'
Ryan Patrick
Cyg 21h3'0"50d13'
Ryan Peter
Cep 22h17'38"62d25'
Ryan's Christmas Star
Cep 23h36'20"65d9'
Ryan's Destiny
Aql 19h51'0"12d47'
Ryan's Dream
Her 16h37'1"41d35'
Ryan's Hope
Her 17h16'50"49d35'
Ryan's Place,Maureen
Lyn 8h13'1"37d7'
Ryan's Reflection
Ori 6h17'35"0d52'
Ryan's Star
Her 17h4'34"48d44'
Ryan's Star of Wonder
Ori 5h55'51"9d55'
Ryan,Adrienne
Per 2h30'0"51d35'
Ryan,Agnes
Gem 6h43'22"15d27'
Ryan,Amy Elizabeth
Lyr 18h41'32"34d26'
Ryan,Arthur J
Aql 18h43'53"6d14'
Ryan,Beth
Aur 5h49'39"31d49'
Ryan,Bob
Dra 17h1'56"63d44'
Ryan,Carrie
Vul 19h18'51"26d50'
Ryan,Catherine
Peg 22h46'51"24d1'
Ryan,Christopher J
Her 16h24'16"51d3'
Ryan,Danielle
Cas '0h25'40"61d27'
Ryan,Darrelle Kelly
Cmi 7h40'19"5d29'
Ryan,David Kelly
Cep 22h3'0"61d55'
Ryan,Dennis J
Her 17h7'28"37d42'
Ryan,Dickie & May
Cyg 20h15'16"38d39'
Ryan,Dorothy
Mon 7h0'28"8d1'
Ryan,Edward
Dra 19h54'0"68d55'
Ryan,Eleanor O
Mon 6h57'51"-8d45'
Ryan,Elsie M
Lac 22h28'0"54d20'
Ryan,Emily
Peg 23h1'1"14d20'
Ryan,Florence E
Cyg 21h30'45"41d35'
Ryan,George David
Cap 21h16'11"-22d17'
Ryan,George H
Aur 6h4'13"31d37'
Ryan,George H
Del 20h14'11"10d22'
Ryan,George J
Uma 12h14'18"59d4'
Ryan,Harry's Therese aka Lorraine
And 0h12'58"47d9'
Ryan,Hazel B
Mon 6h30'36"-1d14'
Ryan,Ian Gallagher
Boo 14h1'47"21d13'
Ryan,J Paige
Vul 20h3'20"25d54'
Ryan,Jack
Aql 19h59'15"11d2'
Ryan,Jane F
Com 12h31'18"30d19'
Ryan,Janice Eileen
Uma 13h54'24"51d19'
Ryan,John Ashfield
Aur 7h24'26"35d32'
Ryan,Jr,Ronald Michael
Her 18h8'12"47d24'
Ryan,Jr,Stephen M
Lib 14h59'28"-23d17'
Ryan,Judith B
Cas 2h7'1"59d4'
Ryan,Kyle Anthony
Boo 14h54'1"25d1'
Ryan,Linnea
Lyn 9h0'38"41d45'
Ryan,Louis Stonefish
Ser 18h40'17"3d54'
Ryan,Luke
Her 16h27'30"25d32'
Ryan,Maria "Mia"
Hya 8h29'28"-5d45'
Ryan,Mary
Peg 22h58'0"32d21'
Ryan,Mary Katherine
Cyg 19h46'46"38d31'
Ryan,Mathew James
Uma 9h27'1"55d23'
Ryan,Matthew F
Cep 22h10'12"60d51'
Ryan,Melissa
Cas 0h29'38"63d36'
Ryan,Michael Duff
Aql 19h50'41"10d20'
Ryan,Michael Patrick
Her 16h58'30"27d53'
Ryan,Mr Michael
Cep 21h1'28"58d35'
Ryan,N Peters
Lyr 10h42'53"46d32'
Ryan,Neil
Aur 7h23'41"40d22'
Ryan,Patricia
Mon 6h22'49"1d24'
Ryan,Patrick
Her 16h31'19"20d3'
Ryan,Patrick J
Vul 21h20'59"27d36'
Ryan,Paul (G)
Cct 2h51'56"-0d50'
Ryan,PhD,Simeon P
Aur 6h4'48"34d40'
Ryan,Piszczek-327 34 5994
Del 21h1'60"12d16'
Ryan,Regina Maureen Siobhan
Cas 0h48'34"70d42'
Ryan,Rhonda Lynn
Com 12h28'1"21d19'
Ryan,Richard McCarthy
Her 16h58'50"26d8'
Ryan,Rita Elsroth
Mon 8h2'56"-0d56'
Ryan,Robert P
Cep 2h13'24"78d16'
Ryan,Rodger & Therese
Mon 7h45'33"-1d13'
Ryan,Rosary T
Cas 0h28'1"61d42'
Ryan,Samantha Lynn
Vul 21h13'39"20d0'
Ryan,Sara
Cet 2h17'26"8d56'
Ryan,Sean
Peg 21h55'20"18d46'
Ryan,Sean Andrew
Cnv 12h43'1"50d56'
Ryan,Sean Dixon
Sco 16h54'56"-40d54'
Ryan,Sean William Kelly
Lac 22h32'49"55d26'
Ryan,Sheila
Cas 0h58'19"70d33'
Ryan,Shelby Ann
Com 12h25'47"30d28'
Ryan,Shelby Dawn
Lyn 8h3'52"38d58'
Ryan,Sr,Martin L
Aur 5h1'36"38d8'
Ryan,Stacey Cooper
Uma 8h36'1"58d14'
Ryan,Stephen G
Per 2h57'44"55d6'
Ryan,Su Su,Sue
Cyg 21h32'30"40d29'
Ryan,Thomas Emerson
Ser 15h56'26"1d46'
Ryan,Thomas George
Cnv 12h43'36"50d21'
Ryan,Thomas Lindsay
Cyg 19h29'50"38d12'
Ryan,Tony
Mon 6h20'35"7d31'
Ryan,William F
Aur 6h26'50"34d0'
Ryan-Achenbach
Crb 15h27'16"31d29'
Ryan-Florence
Cyg 21h30'40"42d46'
Ryan-Roth
Dra 15h44'15"61d47'
Ryan-Turner,Patrice M
And 2h10'19"50d46'
Ryann
Uma 11h39'46"31d0'
Ryann's Star
Cet 3h8'55"4d16'
Ryari-Tinu & Kul
Aur 6h27'45"38d21'
Ryba,Connie L
Boo 15h0'59"8d53'
Ryback-Rye,Al
Lyr 19h10'48"38d19'
Rybarczyk Star; Josephus I
Oph 17h29'53"8d53'
Rybeck,Ted
Boo 14h11'59"52d17'
Rybicki,Carol A
Cas 0h1'47"64d35'
Rybicki,Kate Elizabeth
Cet 2h30'28"-8d59'
Rybicki,Lawrence E
Boo 15h0'54"30d27'
Rybicki,Mary Damien
Cyg 21h54'0"38d54'
Rybolt,Lisa Anne
Cra 18h18'0"-39d31'
Rychlik,Peter
Boo 15h39'14"41d24'
Rychnovsky AAF,Michael James
Lac 22h53'1"54d2'
Rychwa,Megan Nicole
Cam 5h44'50"61d18'
Rychwald,Wolfgang
Aqr 22h42'29"-5d2'
Rydbeck III,Edward Roy
Oph 18h37'18"11d26'
Ryder One
Ori 5h54'14"7d54'
Ryder's Star
Dra 14h22'39"63d32'
Ryder's Star,Barbara
Leo 9h48'21"7d10'
Ryder,Anthony John
Cnv 12h19'25"43d13'
Ryder,David
Uma 9h7'44"47d55'
Ryder,Fern
Ori 6h1'41"4d25'
Ryder,Gene Burnett
Aql 19h9'53"3d42'
Ryder,Jessica
Cas 0h18'35"62d48'
Ryder,Joseph
Sct 18h44'54"-7d41'
Ryder,Karen H
Mon 6h36'12"10d4'
Ryder,Nicholas Alexander
Aql 19h50'32"15d22'
Ryder,Patricia Evelyn
Lyr 18h31'0"32d24'
Ryder,Scott
Hya 10h18'33"-12d4'
Ryding 25
Cyg 19h59'50"42d45'
Rydman/Newton,D V
Aql 19h31'21"13d36'
Rydquist,Amy Raye
Ori 5h49'26"21d18'
Rydz,Mary"Zep"
Cas 0h29'57"60d39'
Rydzewski,Phillip "Our Star"
Oph 18h17'42"12d41'
Rydzowski,Randy & Roberta
Uma 10h58'0"34d23'
Rye,Courtney Lee
Vul 19h45'48"28d58'
Rye,Curtis A
Uma 8h52'10"58d60'
Rye-Walthall,Esther L
Ari 1h47'37"19d18'
Ryelle
Cam 5h3'42"54d41'
Ryerson,Julie Marian
And 0h10'0"39d52'
Ryerson,Lynn E
Lyn 8h37'38"37d40'
Ryfoss,Henning
Vir 13h30'19"11d20'
Ryglicki,David Ross
Cep 22h43'38"57d16'
Ryglicki,Jonathan
Per 1h46'0"54d18'
Ryjaianme
Cam 6h4'37"58d42'
Ryker,Jennifer Marie
Cnv 12h50'13"40d13'
Ryker,Michael Jay
Dra 17h1'12"67d46'
Ryland,Jacques Christopher
Ser 15h38'51"6d30'
Ryland,Lonnie Gene
Oph 18h19'58"8d48'
Ryland,Magdaline M Zienkiewicz
Oph 18h3'60"10d47'
Ryle,Kathyren
Cet 2h48'1"2d9'
Rylee Miranda
Peg 21h45'15"22d44'
Ryley,Roger
Cep 22h12'0"61d10'
Rym el Ghzala
And 0h7'57"40d22'
Rymal IV
Ori 6h8'24"6d10'
Ryman,Joel Robert
Cet 0h40'49"-4d55'
Rynearson,Raymond J
Per 3h2'51"40d35'
Rynes,Elfriede
Lib 14h58'48"-8d45'
Rynn,Melody Rose
Lyr 18h43'7"41d42'
Ryno 23
Sgr 19h1'31"-27d12'
Ryno,Brandon L
Aql 20h11'55"11d19'
RYO
Sgr 19h10'24"-17d12'
Ryokan College
Her 17h6'55"48d18'
Ryoko
Uma 9h24'44"51d10'
Ryolynn-The Star Of Love
Cep 22h33'0"65d26'
Ryoma(Stella of Your Name)
Cap 20h56'5"-19d6'
Ryon,Alan Vanness
Per 3h30'27"51d17'
Ryon,Jeane Blanz
And 1h45'1"40d19'
Ryosuke,Stella of Tsuji
Cap 20h59'12"-27d12'
Rypien,Andrew Robert
Cep 21h13'35"68d38'
Rys,James
Aur 6h28'1"30d17'
Rysedorph,Miss Karen
Cas 23h48'0"50d13'
Rysted
Lac 22h26'25"54d23'
Ryther,Christopher Thomas
Per 3h23'34"38d41'
Ryutaboshi
Ari 2h16'27"11d46'
Ryutaro
Leo 10h16'56"22d44'
Rywelski,Britta
Uma 9h52'30"58d20'
Rzehaczek,Angela
Crb 15h36'41"38d9'
Rzepkowski,Halina H
Lyn 7h30'33"58d36'
Rzucidlo,Patricia
And 23h34'27"47d48'
Răcci,Paola
Sco 16h39'24"-38d6'
Réal,Dhany
Tri 1h58'2"34d33'
Réalisation
Cmi 7h19'41"8d30'
Réant,Jean Louis
Lac 22h8'0"51d45'
Réger,Sylvia
Lyn 8h54'45"41d0'
Réjean dit Arthur
Her 16h55'40"25d30'
Rémi Fritz
Dra 9h21'27"74d35'
Rösa,Viktor
Cep 22h8'22"61d5'
Röcker's Glückersstern
Ari 2h37'33"21d47'
Rödemer,Günther
Aql 18h56'41"17d49'
Röder,Hans
Mon 7h36'47"-2d46'
Röder,Heinz
Cep 22h5'47"60d2'
Röger,Horst
Ser 15h14'34"4d19'
Röhricht,Eberhard
Uma 9h17'33"48d53'
Röner,"Susi"
Vir 14h6'52"-5d38'
Rössler,Hans-Willi
Vir 13h36'29"-8d44'
Röwer,Lothar
Cmi 7h17'1"0d31'
Rückes,Michael
Tau 4h15'56"23d53'
Rüdiger
Tau 3h55'51"11d39'
Rüdiger forever
Cap 20h56'54"-26d50'
Rüede,Hanspeter
Dra 12h2'31"70d25'
Rühlemann,Gerhard
Aur 5h15'1"43d37'
Rümmling,Michael
Lyr 19h13'14"31d34'
Rüter 1
Lib 14h56'29"-22d57'
RØnnov,Erik "Lucky Star"
Umi 14h40'56"86d9'B

S

(S B)James
Umi 15h41'45"70d2'
S "Latrobe"
Lyn 6h30'53"60d20'
S & C Star
Sge 20h0'59"16d36'
S & L
Umi 17h17'38"76d42'
S & S
Aqr 21h55'42"-1d7'
S & S Lucky Star
Cnv 12h37'50"39d39'
S 3+
Cnv 12h21'29"36d47'
S A M
Aql 18h57'57"-6d56'
S A Monico
Lyn 8h33'28"41d33'
S A Z
Hya 8h55'21"-7d12'
S D's Ojibwa
Lib 14h37'30"-10d53'
S Diane
Lyn 8h6'0"41d37'
S E A
Uma 11h12'52"41d13'
S E P-Steve's Star
Her 18h21'46"12d58'
S F O W G
Peg 22h24'1"31d6'
S Forever S
Cep 20h12'32"60d2'
S G
Sge 19h40'0"16d13'
S Genie
Psc 1h20'59"8d50'
S I L V I A
Del 20h53'11"2d49'
S J
Uma 9h53'14"45d15'
S J & Joey
Gem 6h48'14"12d19'
S L
Boo 14h52'37"47d54'
S L J
Leo 11h38'37"16d46'
S L T 1
Sgr 18h56'33"-25d29'
S M Angel Face
Cas 1h51'23"58d52'
S M R
Boo 15h19'50"48d41'
S M S '95
Uma 8h23'39"60d28'
S P
Per 4h5'53"51d1'
S R G
Cet 2h41'21"-0d6'
S R Kasper-"Hooper"
Uma 9h43'58"71d6'
S S Tar Clark
Cyg 20h28'38"50d23'
S T A A R of I U S B
Peg 21h55'24"36d11'
S'Agapo Eric
Aur 7h21'46"35d41'
S,Willi
Gem 7h33'1"30d1'
S-G '93
Dra 20h19'44"68d28'
S-K Atlanta Special Markets
Sct 18h55'25"-6d58'
S-K Baltimore Special Markets
Uma 14h39'36"81d34'
S-K Boston Special Markets
Lac 22h28'10"50d2'
S-K Chicago Special Markets
Cnv 12h27'32"40d20'
S-K Cincinnati Special Markets
Uma 11h38'14"44d12'
S-K Denver Special Markets
Aur 4h38'0"33d59'
S-K Detroit Special Markets
Cam 12h45'1"80d29'
S-K Houston Special Markets
Oph 18h38'45"7d29'
S-K Los Angeles Special Markets
Her 16h9'1"10d17'
S-K Minneapolis Special Markets
Lyn 6h26'54"54d54'
S-K New York Special Markets
Cam 12h45'55"80d25'
S-K Sacramento Special Markets
Vul 19h41'16"20d8'
S-K St Louis Special Markets
Tri 1h51'46"28d8'
S-K Tampa Special Markets
Cet 0h29'32"-11d30'
S-K Tulsa Special Markets
Mon 6h54'51"-10d12'
S;"Latrobe"
Cas 23h17'48"60d27'
Sa'D
Lmi 9h22'31"37d43'
Saad,Jr,Louis E
Dra 13h11'54"63d35'
Saad,Peter
Aur 4h50'26"40d14'
Saadey,Joe
Boo 14h30'35"20d7'
Saak,Dieter
Peg 23h30'55"17d37'
Saambra Li Kim Lian
Cyg 19h40'20"38d31'
Saari,Duane J & Nancy J
Gem 6h44'30"30d19'
Saari,Eric D
Psc 1h43'29"21d37'
Saarit,Emily
Lyn 8h1'29"38d48'
Saasto,Ernest Olavi
Per 1h45'0"52d42'
Saathoff,Marion
Oph 17h56'32"8d1'
Saathoff,Volker
Cmi 7h31'50"8d49'
Saavedra,Cesar
Her 18h42'0"12d26'
Saavedra,Charles & Evangeline
Cet 1h54'51"-4d32'
Saavedra,Lucia
Umi 14h40'13"69d44'
Saba
Lyr 18h38'1"40d11'
Saba
Per 3h52'41"36d13'
Saba,Anthony Charles
Lac 22h54'23"51d39'
Sabados,Robert & Michele
Aur 5h3'0"38d26'
Sabahattin Bilgutay
Per 2h12'0"56d16'
Sabaitis,Gene Bear
Dra 14h44'11"63d32'
Saban Family Star
Mon 6h57'14"-6d51'
Sabastian
Boo 14h55'54"45d46'
Sabatella,Matthew
Eri 3h28'53"-5d1'
Sabatello,Paul
Aql 19h58'39"12d7'
Sabater,Joan Sanifeliu I
Ori 5h57'0"-0d19'
Sabatini,Cristie
Cyg 20h0'1"30d11'
Sabato,Jr,Antonio
Aur 6h8'50"30d47'
Sabato,Samuel Louis
Peg 22h55'30"29d16'
Sabatou,Didier
Cep 2h23'1"80d28'
Sabatte,Maylani
Peg 22h13'58"20d20'
Sabau,Linda
Peg 22h1'45"4d42'
Sabb,David
Her 16h13'26"10d31'
Sabbagh,Florence
Peg 22h17'40"21d0'
Sabbagh,Joseph
Lyn 8h18'26"34d5'B
Sabbagh,Sweet Judy
Lyn 8h18'26"34d5'A
Sabbeth,Edgar
Psc 1h0'17"20d9'
SABBI
Cma 6h30'59"-24d54'
Sabelka,Paul C
Vir 15h1'46"5d38'
Sabella,Ann
Cet 0h40'50"-6d12'
Sabella,Joseph
Vul 19h48'19"20d18'
Sabella,Vince
Cep 22h55'0"70d4'
Sabelle
Ser 15h20'11"8d18'
Saben,Cameron Michael
Lac 22h22'36"38d3'
Sabet,Dr Joseph P
Oph 18h19'14"8d27'
Sabeti-Kolahi,Shahram
Cam 8h19'38"81d44'
Sabhlok,Mobim
Cet 3h5'52"2d7'
Sabin,Kenneth & Virginia
Crb 15h50'54"28d31'
Sabin,Nigel Paul
Cyg 21h11'31"34d27'
Sabin,Stanley Neville
Her 16h59'25"36d48'
Sabin,Val
Mon 6h34'46"-6d10'
Sabina
Cyg 20h1'15"40d54'
Sabina & Fabrizio
Pyx 8h45'30"-28d11'
Sabina '95
Vel 9h23'33"-48d3'
Sabina,Lou
Cep 21h07'47"61d51'
Sabine
Gem 6h38'34"30d40'
Sabine
Tau 5h29'41"28d33'
Sabine
Uma 13h51'18"50d20'
Sabine
Lib 15h11'14"-24d2'
Sabine
Sge 19h19'50"17d39'B
Sabine
Tau 4h19'11"20d23'
Sabine
Cnv 12h41'15"32d53'
Sabine & WolfgangOs Star
Boo 13h36'51"12d0'
Sabine La Rayonnante
Aur 4h53'55"40d51'
Sabine's Beautiful Boat
Aqr 23h14'16"-4d13'
Sabine-Maurice
Cas 2h35'50"77d11'
Sabine-Schimmel-Star- Of-Love
Vir 11h39'25"1d33'
Sabines & Didis Stern
Leo 10h54'59"22d2'
Sabino Michael,Our Precious Diamond
Cam 3h39'51"73d12'

Sabiron
Umi 13h50'30"72d18'
Sable,Mike
Cet 1h21'50"-0d22'
Sablik,Dave
Per 1h54'18"47d58'
Sablosky,Samuel Elliott
Aur 4h36'11"30d31'
Sabo Forever,The Ronald & Elizabeth
Cep 23h8'26"71d10'
Sabo,Cindy "Bugs"
Cas 0h48'1"61d53'
Sabo,Elizabeth "Liz" Giselle
Lyn 8h37'15"44d34'
Sabol,Bernard S
Her 18h25'1"24d38'
Sabol,Jeffrey R
Aur 4h54'1"41d8'
Sabol,Larry "Bud"
Lmi 10h1'32"32d27'
Sabol,Lisbeth G
And 2h30'47"44d48'
Sabolsky,Michael
Dra 11h33'41"70d55'
Sabor
And 23h5'48"52d49'
Saboundjiam,Vanouhy
Sco 16h10'58"-22d33'
Sabourin,Thérèse
Cas 0h51'55"61d48'
Sabra
And 2h24'15"39d54'
Sabra's Dream
Ori 5h48'54"10d23'
Sabre
Vul 20h4'34"22d33'
Sabre
Cyg 21h17'34"37d58'
Sabrednay
Cnv 12h47'0"40d27'
Sabrina
Cyg 20h38'58"40d43'
Sabrina
Scl 23h12'35"-31d15'
Sabrina
Ant 10h45'52"-38d22'
Sabrina
Per 3h36'42"51d13'
Sabrina
And 1h54'51"36d15'
Sabrina
Her 18h34'54"16d9'A
Sabrina
Umi 16h14'12"70d33'
Sabrina
Her 17h51'36"42d40'
Sabrina
Cet 2h28'10"-0d43'
Sabrina
Cep 0h43'30"78d41'
Sabrina Alexis
Gem 6h30'0"14d42'
Sabrina Bells "Our Special Star"
Peg 22h26'52"33d53'
Sabrina Christine
Cyg 21h17'18"37d57'
Sabrina Danyelle
Eri 4h34'1-19d50'
Sabrina Junior
Col 6h20'49"-35d13'
Sabrina Michelle
Com 13h7'34"22d23'
Sabrina Nicole
Peg 22h19'37"33d47'
Sabrina Ten Tricarico
Vel 9h19'12"-44d49'
Sabrina's Star
Lac 22h31'20"55d24'
Sabrina,25-05-1969
Gem 6h36'60"20d16'
Sabu
Cam 13h26'47"78d42'

Sabu
Aur 6h4'51"38d45'
Sabucco,Paolo
Lac 22h31'44"38d34'
Sabula,Thomas
Mon 7h46'20"-1d60'
Sacani,Annette
Aqr 22h40'0"-0d9'
Sacay,Aris
Cam 5h21'0"80d34'
Sacca,Richard Anthony
Lac 22h34'0"37d32'
Sacca,Teresa J
Lyn 8h15'0"38d12'
Sacchet,Giorgio
And 23h5'25"41d6'
Sacchetti,Paul Alfred
Dra 15h3'27"62d5'
Sacchini,Ida e Giosi
Cnv 12h11'15"51d39'
Sacci,Glenn
Cam 4h3'27"68d40'
Sacco,Giorgio Maldonado
Sct 18h46'17"-7d53'
Sacco,John
Ori 5h55'50"6d0'
Sacco,Joseph John
Aur 5h1'47"38d35'
Sacco,Jr,Ernest
Her 16h57'19"31d5'
Sacco,Marc Geoffrey
Aql 20h7'51"8d18'
Sacco,Philip
Aql 20h0'12"10d4'
Sacco,Thomas
Hya 8h33'43"-1d42'
Sacco,Todd
Ori 5h50'40"16d51'
Sacco,Ulderico E
Ori 5h38'55"15d29'
Saccoccio,Anna Marie
Cas 1h15'58"61d30'
Saccoccio,Charly Marie
Uma 9h21'47"56d55'
Saccoman,Spencer Jack
Hya 8h10'1"3d49'
Saccone,Donna M
And 2h28'30"45d59'
Saccone,Irene
Cep 1h30'1"78d30'
Sacha
Psc 0h1'42"-1d8'
Sacha Victoria
Gem 7h21'0"33d56'
Sacha's 16th Superstar
Vir 12h43'40"-7d37'
Sacharoff,Alice Welsh
Boo 14h0'24"8d45'
Sacharoff,Percival Welsh
Boo 15h6'18"12d15'
Sacharoff,Robert Welsh
Boo 14h1'36"8d18'
Sacharoff,Tupper Welsh
Boo 15h6'43"19d35'
Sachenbacher,Josef und Evi
Aql 18h58'42"17d38'
Sachika
Crt 11h21'11"-20d18'
Sachiko,Kuno
Lyr 18h29'51"31d48'
Sachman,Terrel
Vir 12h33'12"-8d11'
Sachowsky,Franz Josef
Gem 8h5'16"30d11'
Sachs,Andrew Elliott
Cep 20h55'43"55d35'
Sachs,Andy
Ori 6h1'20"8d35'
Sachs,Cathy
Lyn 8h31'0"40d11'
Sachs,Jeffrey M
Cep 21h48'0"61d10'
Sachs,Karen
Mon 7h10'47"-10d16'

Sachs,Mirja & Gunter
Dra 17h4'44"73d42'
Sachse,Elizabeth Justine
And 0h54'30"36d46'
Sachse,Lynne
Ser 18h13'2"-7d13'
Sachse,Vivian
Peg 23h35'1"13d29'
Sack,Cherie Anne
Cyg 21h35'17"44d15'
Sack,David Samuel
Aur 5h35'34"38d16'
Sack,Julie Renee
Cas 0h29'0"60d25'
Sack,Karen N
And 0h24'23"44d57'
Sack,Natan Yaacov
Lmi 10h10'41"30d32'
Sack,Robert F
Her 16h43'60"50d36'
Sacker,"Sacker The Walrus" Ira
Gem 6h58'20"15d28'
Sackett,Hugh & Claudette
Crb 16h18'0"38d43'
Sacketts Eternal Pride
Uma 9h53'54"56d26'
Sacki,Sandy Rose
Cas 23h5'0"54d8'
Sackler,Jenifer & Gregory
Vul 19h20'44"26d59'
Sackley,Stan Alan
Uma 10h50'24"42d11'
Sackman,Brian & Laura
Aql 19h51'31"11d2'
Sackman,Murray
Cnv 13h9'41"32d21'B
Sacks,Aaron V
Cep 22h39'27"65d9'
Sacks,Austin Eric
Aur 6h28'1"30d4'
Sacks,Marion & Larry
Umi 14h55'25"66d55'
Sacks,Michael J
Cep 22h22'27"70d24'
Sackstein,Joshua
Per 2h26'55"58d48'
Sackville's,The
Cyg 20h3'26"39d24'
Sackville,Gaylen Ray
Uma 11h46'1"41d55'
Sackville,Nina Marie
Uma 11h55'56"41d42'
Sada E
And 23h30'50"46d57'
Sadauskas Link,The
Vul 21h2'0"20d9'
Saddington's Super Star
Ori 6h9'1"4d7'
Sadie
Crb 15h24'27"31d58'
Sadik,Alsharifa Azza
Eri 2h53'53"-4d59'
Sadler,Cynthia Louise
Uma 13h15'1"63d4'
Sadler,David G
Psc 22h59'1"5d43'
Sadler,Dylan Matthew
Per 2h23'37"55d39'
Sadler,Inge
Boo 14h27'1"53d58'
Sadler,Karis Leigh
Cam 4h45'59"68d57'
Sadler,Kristina
Cyg 19h29'56"31d45'
Sadler,Richard
Lac 22h9'20"48d33'
Sadof,Jon
Lac 22h26'31"53d47'
Sadoski,Anthony John
Oph 17h4'14"11d49'
Sadoux 22-10-1937, Robert
Her 18h22'12"12d20'
Sadovskaya,Vlada
Vul 19h19'46"26d32'

Sadowske,Jacob Francis
Cep 21h15'28"65d37'
Sadowski,Elke
Her 17h49'20"41d14'
Sadowski,Kevin Michael
Sct 18h53'53"-6d4'
Sadowski,Patty A
Cam 3h13'0"58d56'
Sadowsky,Nina Ann
Crt 11h36'41"-20d32'
Sadowsky,Steven Judd
Cep 21h28'35"67d53'
Sadvary,Kathryn (Sis)
Ori 5h56'49"16d28'
Sadye Catherine
Vul 20h57'19"28d12'
Saeby,Hans
Aql 19h57'25"1d30'
Saeed,Hannah Irene
Cas 3h0'12"60d1'
Saeed,Saimah
Uma 9h35'20"54d16'
Saeger,Jacquie, Sean,& Treavor
Uma 10h2'16"57d40'
Saegesser,Scott D
Aur 6h7'12"45d27'
Saenz,George
Boo 13h57'42"22d27'
Saenz,Peter
Her 15h54'23"40d3'
Saez,Thierry
Aur 6h0'1"38d30'
Safan,Safaa
Peg 23h32'42"10d41'
Saffron,Matilda
Lyr 18h24'33"38d1'
Safia
Cyg 19h21'57"28d41'
Safinaz Bhai
Cas 1h39'25"74d18'
Safir,Erika Starr
Lmi 10h35'19"38d32'
Safiria
Ind 21h9'21"-51d51'
Safiya Courage
Uma 10h6'57"62d1'
Saft,Joan Viola
And 2h21'30"45d20'
Saft,Stuart
Cep 22h33'37"71d15'
Sagacious Sharon
Com 12h35'43"25d55'
Sagafjord,M S
Aql 18h43'45"-2d22'
Sagan,Harriette
Oph 18h0'41"11d57'
Saganice,Jennifer
Cet 1h39'35"-0d43'
SAGAPO
Uma 9h19'51"59d13'
Sagapo
Cae 4h49'23"-27d8'
Sagdhal,Nathalie
Umi 15h49'31"80d17'
Sage
Aql 19h53'1"13d28'
Sage
Cyg 19h43'37"30d39'
Sage,Pascal le
Crb 15h51'29"27d9'
Sage,Roger
Cas 22h59'37"54d47'
Sage,Stephanie
Lyn 8h22'35"46d35'
Sagehorn,Joanne Therese
Cam 8h36'45"78d22'
Sagemi
Cgy 19h55'54"47d27'
Sager 50
Cyg 20h38'29"53d48'
Sager,Joyce & Al
Aql 19h9'50"15d35'
Sager,Kim Lynn Damron
Mon 7h17'27"-1d9'

Sager,Maureen
Sge 19h53'54"18d47'
Sager,Patsy Jo
Aql 20h0'36"-0d12'
Sager,Sharon Denise
Mon 6h53'17"0d44'
Sager,Valerie & Scott Sager
Cyg 21h3'50"28d50'
Sagerman,Benjamin Joseph
Her 17h14'30"13d3'
Sagerman,Charles Louis
Aur 5h23'30"38d0'
Saggese,Ronald Joseph
Boo 14h56'19"21d34'
Sagi
Cma 6h51'13"-18d29'
Saginario,Carol
Cas 0h48'26"65d48'
Sagl,Christine
Boo 15h21'1"38d54'
Sagorsky,Anastasia
Lyn 7h45'50"40d41'
Sahagian,Lucy Sarkissian
Gem 7h29'36"20d12'
Sahajdak,Marie-Odile
Cep 22h15'28"55d29'
Sahjah
Cas 0h48'51"62d58'
Sahli,Joel
Per 2h55'18"35d28'
Sahli,Steve
Dra 17h8'1"62d39'
Saiainaru Haha Kiyoko Boshi
Ari 2h19'14"20d14'
Saibara,Kathryn Joyce
Mon 6h57'38"8d3'
Said,Geoffrey
Per 2h56'21"38d5'
Said,Karyn L/J Seth Sochacki III
Cam 7h7'54"60d6'
Said,Khaled
Aur 5h0'41"46d48'
Saie-Bordeaux,Nathalie
Uma 13h15'0"62d10'
Saiff,Jr,William J
Vir 11h58'0"-2d55'
Saifuidin,Ishaq
Her 18h26'58"12d21'
Saigger,Eleanor
And 0h59'14"45d47'
Saigger,Walter
Cnv 12h50'23"41d7'
Saija
Lyr 18h38'26"37d52'
Sailer,Tom & Pat
Lyr 18h50'20"34d49'
Saillant,Ray
Cep 0h2'13"80d18'
Sailler,Kerry Sunshine
Uma 12h9'29"51d56'
Sailor Sue
Sex 10h2'29"-8d4'
Sailors,Flossie Estelle
Oph 18h0'26"13d38'
Sain HD 27,Bill
Cet 1h23'11"1d31'
Sain,Ivan
Boo 14h19'34"35d6'
Saine,Tamera
And 22h59'41"37d34
Saint Christ
Peg 0h5'54"20d47'
Saint Dizier,Georges
Mon 6h26'34"-0d59'
Saint Ellen
Peg 22h28'10"27d37'
Saint Elmos
Cam 5h55'50"58d14'
Saint Georgie
Cep 0h46'52"78d55'
Saint Guilhen, Catherine
Per 2h37'46"56d58'
Saint Hill
Aur 5h50'1"40d39'

Saint John
Cma 6h30'13"-20d31'
Saint Marty
Cep 6h17'43"85d58'
Saint Matthew
Her 18h13'48"31d22'
Saint Michael The Archangel
Lmi 10h19'29"32d10'
Saint Mick
Aqr 23h2'58"-6d21'
Saint Roman Toledo
Boo 15h2'53"23d43'
Saint Thomas Bear
Boo 14h5'0"23d12'
Saint Vallier
Tri 1h58'47"32d39'
Saint's Starlight
Sex 10h28'48"-2d16'
Saint,Arthur Riblett
Dra 10h48'18"74d37'
Saint,Kenders
Umi 14h35'31"68d32'
Saint-Cyr,Jean-Thomas Francis
Sge 20h0'52"20d2'
Saint-Denis,Vivian
Eri 3h26'16"-12d45'
Saint-Elie,Regine
Uma 11h22'0"52d1'
Saint-Maurice,Simone
And 23h3'1"50d41'
Saint-Michael,Don
Boo 14h51'36"32d21'
Saint-Onge,Ginette
Cas 0h1'31"61d28'
Sainte Orlan
Del 20h19'54"11d7'
Sainthouse,Daniel
Per 3h30'16"52d11'
Saintonge,Jean Darribere
Aql 19h31'41"-10d2'
Sainvet,Sandra
Equ 21h20'30"10d18'
Sainz,Sandra
Lyr 19h6'41"26d10'
Saira,Kalevi
Cam 3h18'12"60d23'
Saira,Pauli
Uma 9h41'49"53d42'
Sairs,Haley
Cas 23h34'54"60d23'
Saito,Seiichi
Aql 18h56'18"-6d22'
SaJe
Cyg 20h7'53"40d39'
Sajewych,Ulana
Cam 11h59'53"81d40'
Sajid
Dra 18h4'0"70d38'
Sajjad,Asis
Uma 11h53'18"43d15'
Sak,Tony
Cas 0h31'25"75d23'
Sakai,Benjamin
Umi 15h29'0"69d6'
Sakai,Hisae
Lib 14h46'56"-18d43'
Sakai,Patricia
Umi 15h39'31"60d26'
Sakai,Reilly
Umi 15h33'38"69d3'
Sakai,Richard
Umi 15h27'18"69d5'
Sakai,Timothy
Umi 15h21'21"69d12'
Sakai,Yoshie Carrie
Cas 23h24'0"60d53'
Sakal,Holly
Cyg 20h15'36"38d7'
Sakalas,Laura Anne
Lyr 19h13'14"40d13'
Sakamoto-IMD,Joni
Cet 0h40'0"0d25'
Sakash,Jackie
Cyg 20h3'52"31d36'

Sakata,Frances
Mon 6h24'48"8d10'
Sakhel (Sakhels Star), Jamil
Cap 21h43'25"-19d35'
Sakhleh,Donna
And 0h17'41"34d19'
Sakija
Uma 9h24'1"53d7'
Sakina
Cas 0h25'1"59d26'
Sakina
Peg 23h5'52"33d56'
Sakis
Cnv 13h39'34"47d45'
Sakis K
Aur 5h56'34"40d28'
Saks,Pauline Greenberg
Cas 23h45'17"50d27'
Saks,Sol
Cma 7h15'14"-16d2'
Sakshaug,Eric N
Boo 15h8'1"27d36'
Sakura
Psc 1h22'51"12d36'
Sakurada,Star of Masashi
Sco 16h8'21"-9d21'
Sal & Vicki-Star Of Eternal Love
Lyn 8h12'47"51d2'
Sal My Everlasting Love
Ori 5h52'37"13d10'
Sal N Jen A Whole New World
Cyg 20h38'39"41d4'
Sal Ria
Her 17h22'21"38d37'A
Sal's Surprize
Ser 15h30'49"17d39'
Sal-49
Lyn 8h34'28"38d8'
Sala,Anna M Formiguera
Cas 0h16'30"60d7'
Sala,Catherine M
Cas 1h42'12"77d27'
Sala,Fabio
Cet 2h20'24"0d35'
Sala,Luciana
Lac 22h48'11"38d49'
Sala,Matthew R
Tau 5h57'42"24d6'
Sala,Rebecca Anne
Mon 6h37'57"7d57'
Sala,Xavier Armengou
Boo 14h9'26"51d35'
Saladak 20th Ann HVNB, Margaret
Cyg 21h50'49"42d32'
Saladini,Mary Jo
Lyr 19h1'44"38d40'
Saladini,Thomas Joseph
Her 17h35'28"40d12'
Saladino,Brenda
Mon 6h35'42"1d29'
Saladoff-TLPJ,Susan Vogel
Lac 22h33'0"53d11'
Salaga,Vonda
Cnv 12h13'54"44d27'
Salah et Sali
Lac 22h20'36"54d23'
Salah,Lili Ann
Dra 18h48'40"67d50'
Salak,Michael
Her 17h48'41"40d5'
Salak,Stefanie
Lyr 18h46'17"34d35'
Salama,Anne
Lac 22h1'48"48d54'
Salama,Audrey
Lac 22h6'32"40d22'
Salama,Sasha A
Sct 18h21'0"-13d23'
Salamacha,Walt
Cet 2h15'11"6d31'
Salamando,Zachary Alexander
Cyg 21h10'53"39d6'

Salamon,Johanna
Lyn 8h27'19"47d3'
Salamone,Sabina
And 23h0'32"47d32'
Salamy,Anoulla
Aql 18h59'36"-8d23'
Saland,Hermann
Lyr 19h17'17"30d34'
Salani,Christina Rose
Sge 20h17'43"17d25'
Salansky,Dana Marie
And 2h7'41"39d31'
Salapka,Sen Sei Ray
Vir 13h0'27"-20d23'
Salas,Lawrence J
Ori 6h14'5"18d58'
Salas,Richard Hirschfeld
Sex 9h46'26"-6d2'
Salasnek,Harold
Com 12h9'13"19d28'
Salata,Deena Christy
Aql 20h4'54"-1d13'
Salata,Donna Rose
Mon 7h5'7"-5d25'
Salati,Silvia
And 0h46'44"40d18'
Salatino,Joseph & Kerry
Crb 15h31'0"30d11'
Salatino,Roland Michael
Her 16h38'51"27d13'
Salatto,James A
Sct 18h46'3"-7d8'
Salazar I,Sir Bert "Sailor"
Sex 10h46'0"-8d3'
Salazar's Very Own Star,Delfino
Aur 4h48'1"40d45'
Salazar(Blaze),Jessie
Cam 5h52'47"56d9'
Salazar,Christina J
Crt 10h58'15"-8d2'
Salazar,Cory Michael
Cet 1h50'28"-5d47'
Salazar,Mary Alice
Eri 3h55'17"-16d59'
Salazar,Matthew
Cet 2h17'36"9d4'
Salazar,QRZ Gerardo
Ser 15h58'16"-1d36'
Salazar,Vanessa
And 2h31'41"40d41'
Salazar-Neuenschwander Steevy
Per 3h26'0"48d48'
SalBravo
Del 20h24'28"8d22'
Salcalbeam
Lac 22h41'29"54d53'
Salce,Nicholas P
Aql 19h30'51"7d53'
Salce,Nicholis P
Aql 18h58'60"-8d47'
Salcido,Nora H
Aql 19h49'51"14d45'
Salcito,Louis
Per 1h56'46"52d51'
Salcone,Katie Rose
Vul 21h19'21"28d13'
Salcumb,Tracy Kim
Crb 15h48'17"34d27'
Saldana,Melecio
Aur 5h5'15"42d18'
Saldivar,James Michael
Aur 5h41'29"50d31'
Saldivar,Thomas Robert
Aur 6h0'1"45d38'
Saldukas,Lois
Ori 5h54'46"14d18'
Saldutti,Emily Jo
Psc 23h9'1"5d7'
Salek,Jr,John L
Dra 16h28'1"66d6'
Salela,Scott Adam
Her 16h18'53"5d14'

Salem's Shining
Oph 17h53'44"11d22'
Salem,Daniel
Cep 22h11'47"61d30'
Salem,Jeffrey
Lyr 18h47'33"37d14'
Salem,Richard
Boo 14h7'24"33d14'
Salem,Richard Newman
Eri 2h50'20"-3d51'
Saleme,Mark
Dra 16h51'58"70d34'
Salemi,Charles A
Her 17h16'42"49d23'
Salemi,Michael
Cep 21h6'48"58d22'
Salen,Bradford Lear
Her 15h53'0"48d30'
Salenieks,Benjamin
Sgr 19h44'43"40d21'
Saler"Jimmy's Star", James Rodney
Peg 21h57'21"30d28'
Salerno Star,The
Lyn 9h7'12"42d33'
Salerno,C Lee
Cep 0h28'22"78d48'
Salerno,Jessica Victoria
And 23h2'0"48d40'
Salerno,Joseph
Umi 16h6'41"78d59'
Salerno,Stephen
Cep 22h15'0"62d33'
Salerno,Sylvester Ambrose
Lyn 7h31'57"36d8'
Salerno,The Sicilian Lion King Nick
Lmi 10h46'0"33d30'
Sales Amast
Sge 19h54'40"16d29'
Sales,Bryan Allen
Aur 6h53'1"38d50'
Sales,Elise & Raymond
Ori 5h56'45"10d21'
Salese,Stephen
Cet 1h15'39"-4d36'
Salesses III,William E
Aql 19h49'18"14d46'
Saletta,Meredith
And 0h16'55"45d47'
Salgado,Joao Luis
Lac 22h55'0"53d8'
Salgado,Pilar
Cam 4h31'48"58d10'
Saliba,Jordan Allison
Dra 14h7'52"62d50'
Saliba,Leon Charles Blythe
Hya 9h10'24"3d6'
Salima
Aur 6h11'0"30d50'
Salimbene,Jennifer R
And 0h56'0"45d15'
Salimbene,Michelle Juswinter
Cas 0h30'1"58d29'
Salinas,Claudia
Del 20h19'30"10d57'
Salinas,Noah Alexander
Cam 3h51'13"77d35'
Salinas,Zachary Michael
Lac 22h8'10"51d24'
Salinger,Lillie & Charles
Lyr 18h36'25"34d20'
Salinier,Guy
Peg 23h16'24"31d29'
Salinsky,Richard & Esther
Cam 5h57'20"67d33'
Salisbury & Crystal, Richard
Aur 4h56'53"40d49'
Salisbury,Brandy Nicole
Peg 21h55'28"28d19'
Salisbury,Carolyn J
And 0h55'29"21d41'
Salisbury,Colin
Cep 20h30'38"63d23'

Salisbury,Dorothy
And 23h4'24"41d6'
Salisbury,Estelle Marie
Lyn 8h12'29"48d2"
Salisbury,Helen
Cas 1h46'21"60d50'
Salisbury,Karina
Mon 6h39'17"-0d11'
Salisbury,Paul
Cep 21h10'1"68d15'
Salisbury-Hall,Jean Margaret
And 0h18'53"34d52'
Salize,Heins-Gerd
Hya 9h26'37"-18d49'
Saljon
Sex 10h0'1"1d51'
Salk,Evelyn
Lyr 19h8'54"38d29'
Salkeld Star,The Craig &
Annette
Dra 14h51'16"63d60'
Salkin,Rabbi Jeffrey K
Dra 17h34'55"60d49'
Salkowski,Robert Francis
Ari 2h6'32"18d56'
Sallabedra,Josephine
And 0h19'36"32d15'
Sallade,Melissa A
Lyr 18h41'59"37d13'
Sallai,Nathalie
Cyg 20h36'35"31d15'
Sallares,Joaquin
Lyr 19h24'24"41d19'
Sallee,Constance L
Cas 1h18'19"61d41'
Sallen,Leslie
Cmi 7h22'28"8d52'
Sallie & Noel Shining Forever
Lyn 7h58'50"42d51'
Sallie E=MC˝
Sge 19h53'22"18d50'
Sallie S-D
Cas 23h14'34"63d1'
Sallie's Star
Her 18h3'1"31d34'
Sallinen,Composer Aulis
Dra 16h37'19"61d45'
Sallinen,Pekka
Cep 23h8'30"70d51'
Sallis,Richard
Dra 11h36'20"70d55'
Sally
Gem 6h48'41"19d26'
Sally
Mon 6h54'39"-8d48'
Sally
Cas 2h53'52"65d6'
Sally & Bill's 35th
Sge 20h02'13"16d2'
Sally & Seth
Peg 21h19'39"21d29'
Sally Ann
Peg 23h43'20"11d31'
Sally Ann
Cas 0h43'45"71d2'
Sally Ann
Uma 11h53'15"32d23'
Sally C
And 23h1'20"49d32'
Sally Marie
Crb 15h18'0"31d12'
Sally P
And 0h6'15"31d36'
Sally Superstar
Mon 6h24'25"-6d12'
Sally Victoria
Umi 13h17'55"76d33'
Sally's Dream
Sge 19h57'1"18d60'
Sally's Fire
Cyg 19h30'21"35d16'
Sally's Shining Star
Uma 13h43'52"51d10'
Sally's Star
Tri 1h40'35"34d6'
Sally's Star
Peg 21h10'1"18d54'
Sally's Star
Peg 22h25'54"27d19'
Sally's Star
And 23h40'1"45d54'
Sally's Super Star
Cas 2h33'13"67d32'
Sally,Thanks For Being You!
Love,Pat
Cas 0h36'0"67d39'
Sally-Anne
Lyr 19h22'36"31d28'
Sally-Ray
Mon 6h30'21"8d37'
Sallés,Laia Valls
Aur 5h5'46"43d28'
Salm,Abraham A
Com 12h21'56"25d51'A
Salm,Edith Frankel
Com 12h21'56"25d51'B
Salm,Ralph
Lyn 7h43'0"50d43'
Salmagne,Patricia
Sge 19h57'59"20d27'
Salmann,Peter
Dra 18h7'59"67d56'
Salmen,Jay
Cas 0h37'39"56d11'
Salmento,Elaine Francine
Peg 21h40'13"20d14'
Salmon,Annajean
Aql 18h44'0"-2d23'
Salmon,Bennett Eugene
Boo 15h10'45"26d15'
Salmon,Christopher
Uma 10h37'11"40d45'
Salmon,Douglas Ralph
Aql 18h57'24"-0d44'
Salmon,Henry
Aur 7h24'47"40d52'
Salmon,Jayne Elizabeth
Umi 14h20'26"66d45'
Salmon,Jonathan Richard
Boo 14h48'29"33d52'
Salmon,Lorna
Lyr 18h16'39"30d27'
Salmon,Lucinda
Cam 3h52'1"60d15'
Salmon,Melissa Verletta
Lyr 19h3'48"28d14'
Salmon,Paul David
Umi 13h46'20"78d36'
Salmon,Philip Andrew
Umi 15h55'1"77d35'
Salmon,Sara Murphy
And 23h33'1"39d36'
Salmond,Ronald
Cet 2h26'39"7d31'
Salmonds Eternal Love Ross
Lambert,Lucy
Cyg 19h27'26"38d51'
Salmons,Linda Lee
Cam 3h13'0"60d20'
Salmons,Sherwood
Boo 14h21'1"16d30'
Salmonson,Jay David
Her 17h36'27"25d12'
Salna's Star,Erik
Oph 18h7'1"8d53'
Saloman,Avraham
Cep 22h31'15"78d60'
Salome & Simon
Cyg 21h54'18"53d4'
Salomon Family,The
Cam 3h24'30"55d1'
Salomon,Dean
Per 1h50'26"56d34'
Salomon,Jacob Eric
Lac 22h4'55"46d34'
Salomon,Jonas Edward
Ori 5h53'47"12d16'
Salomon,Julia Margarita
Martin
And 0h19'32"38d33'
Salomon,Lee
Her 16h34'1"32d29'
Salomon,Neomi
Cas 1h58'1"61d24'
Salomon,Siegfried
Lib 14h23'26"-23d47'
Salomon,Yaacov Valero
Ori 5h32'26"8d44'
Salomone,Giulia
Aur 5h0'51"41d31'
Salonen,Maarit
Uma 11h22'1"41d18'
Salonen,Pekka Ilmari
Cam 3h38'38"60d33'
Salonen,Wendy
Cas 23h20'59"62d21'
Salony,Stanley Fabian
Cet 0h32'12"-1d6'
Salotti,Lewis A
Aur 4h50'28"41d1'
Saloy,Father Tom
Cep 20h23'19"60d38'
Salsbery,Nicole Marie
Mon 8h7'36"-1d47'
Salsbury,Marvin & Marilyn
Cam 3h15'59"66d27'
Salt Of The Earth
Boo 14h42'41"47d23'
Salt,Jessica Amy
Peg 23h16'0"31d17'
Salt,Stephen Allenby
Cep 22h50'20"57d58'
Salter Anniversary Star
Crb 15h54'0"26d46'
Salter II,William K
Sex 10h39'39"1d38'
Salter,John Ralph
Equ 21h22'58"10d25'
Salter,Maria
And 2h27'58"47d24'
Salters-The Sapient Warrior
Aql 19h8'30"15d15'
Salthouse,Michael
Cyg 20h24'26"38d36'
Saltori,Rosa
Cyg 19h47'34"50d34'
Saltz,Florence & Jerome
Uma 9h28'11"50d13'
Saltzman,Daniel
Gem 6h57'53"15d35'
Saltzman,Danne Rafae
Her 17h24'0"47d34'
Saltzman,Mark Steven
Peg 23h18'21"25d6'
Saluga,Faith Bridget
Lmi 10h22'0"32d35'
Saluga,Joy Delia
Lyr 18h57'12"34d31'
Salva,Gary K
Psc 0h59'46"32d29'
Salva,Rose & Pasquale
Lac 22h53'39"51d17'
Salvadeo,Alicia Nicole
Cyg 21h6'23"30d29'
Salvadeo,Danielle Marie
Dra 17h39'58"63d41'A
Salvadeo,Lauren Michelle
Dra 17h39'58"63d41'B
Salvador,Richard
Aql 19h50'0"12d43'
Salvaggio,Jill Marie
And 23h30'40"39d29'
Salvagno,Joseph John
Lac 22h29'56"55d45'
Salvai,Serge
Dra 12h29'51"64d43'
Salvanos,Sandra Karis
Com 13h10'30"26d21'
Salvati,Giovanni
Lup 15h8'41"-38d15'
Salvati,Tiffany
And 23h46'18"42d19'
Salvatico,Maura
Dra 16h20'25"62d20'
Salvation,Captain James
Cep 22h10'19"56d13'
Salvato,Megan Nicole
Lyr 18h15'17"35d18'
Salvatore
Her 16h38'53"38d10'
Salvatore The Light In Our
Lives
Uma 11h31'0"46d46'
Salvatore's Dream
Sco 17h52'36"-30d6'
Salvatore's Star
Cmi 7h55'1"0d49'
Salvatore's Star
Her 17h30'42"14d37'
Salvatore,Alexandra Ann
Lyr 19h19'51"41d33'
Salvatore,Vincenza E
And 23h8'1"40d6'
Salvesen,Cara Diane
Lyn 9h1'13"42d15'
Salvesen,Steve
Gem 7h26'1"30d49'
Salveson,Lady Danielle
Jeannine
Mon 8h3'13"-1d38'
Salvesvold,Michele Laurren
Sgr 19h39'18"-31d29'
Salveter,Kennedy Morgan
Cep 22h35'56"61d0'
Salvignol,Laurent
Boo 15h20'57"37d59'
Salvin,THISILDO In Memory
of,Ted
Uma 10h35'45"60d33'
Salvitti,Reverend Jackie
Lyr 18h31'1"30d14'
Salvo,Clement
Aql 19h2'36"-5d42'
Salvo,Patricia
Ser 17h32'40"-13d33'
Salway,Toby
Boo 13h59'0"8d49'
Salwen,Harry
Per 1h59'36"53d3'
Salyer,Jacquelynn
Cas 23h0'0"53d31'
Salyers,Doug & Lyn
Aur 6h31'16"32d47'
Salz,Hannelore "Engel"
Cap 20h59'56"-15d24'
Salzburg's Star
Eri 4h33'55"-11d26'
Salzer,Kristofer Jon
Her 16h1'15"21d6'
Salzman,Betti Hellinger
Cas 0h58'35"61d7'
Salzman,Jerome
Lac 22h27'44"40d25'
Salzman,Richard H
Cep 22h45'37"70d27'
Salzmanensis,Stella Petri
Del 20h14'17"9d29'
Salzmann,Peter
Cep 22h23'0"61d32'
Salzsieder,Beulah
And 23h17'0"49d19'
Sam
Boo 14h30'0"27d24'
Sam
Hya 8h13'34"2d35'
Sam
Ori 5h58'52"17d55'
Sam
Cas 0h39'20"71d39'
Sam
Cep 3h46'28"80d20'
Sam
Cam 8h4'14"73d43'
Sam
Del 21h4'43"18d46'
Sam
Boo 14h17'0"54d23'
SAM
Sct 18h54'51"-6d26'
SAM
Ori 5h51'0"17d4'
Sam & Blinz
Cyg 19h47'1"30d2'
Sam & Dave's Star
Per 4h7'1"51d24'
Sam & Diana
Sge 20h1'55"19d8'
Sam & June
Mon 6h21'48"8d6'
Sam & Mary's Dream Catcher
Cyg 19h25'50"33d6'
Sam & Miriam
Umi 15h6'50"82d27'
Sam Bons'
Lyn 9h3'48"33d44'
Sam France
Del 20h58'1"10d5'
Sam Gor
Aur 6h34'25"32d27'
Sam Gregory
Cyg 20h41'36"45d38'
Sam I Am
Dra 16h58'21"60d11'
Sam Lloyd
Umi 14h55'32'67d29'
Sam Loves Rhett
Car 7h30'1"-57d53'
Sam Murry
Lyn 10'2"55'39d49'B
Sam W
Aql 19h57'51"14d34'
Sam's Star
Cep 22h50'14"57d10'
Sam's Keeper Of The Stars
Dra 16h46'44"68d53'
Sam's Secret Star
Aql 19h5'36"-5d41'
Sam's Star
Umi 14h15'1"67d28'
Sam's World
Uma 8h58'40"61d57'
Sam,Chen
Dra 16h15'42"62d41'
Sam,I love you always
Del 20h21'14"8d53'
Sam-n-Dave's Star
Cyg 20h38'24"45d23'
Sam-Till The End of Time-Star
Lyr 18h40'22"36d39'
Sama,Susanne
Leo 10h54'36"18d46'
Samad,M M Abdul
Her 16h12'40"8d27'
Samalin Annie
Cyg 19h31'26"35d44'
Samamtha
Lyn 9h2'36"36d14'
Samandarov Concord
Limousine,Rudolf
Leo 10h51'39"-5d9'
Samaniego,Joan
And 0h24'0"38d32'
Samaniego,Richard & Erinn
Equ 21h20'54"11d1'
Samanta,B K Katia
Del 20h18'1"10d49'
Samantha
And 0h23'10"38d27'
Samantha
Lac 22h19'12"51d13'
Samantha
Cas 23h41'10"62d38'
Samantha
Peg 23h32'29"31d18"
Samantha
Peg 23h43'20"30d33'
Samantha
Lyr 18h35'45"30d8'
Samantha
And 0h24'0"36d21'
Samantha
Gem 7h8'59"28d42'
Samantha
Sct 18h52'29"-9d21'
Samantha
Cyg 20h53'0"40d24'
Samantha
Mon 6h54'38"-0d37'
Samantha
Cas 1h0'14"63d52'
Samantha
Peg 23h33'60"33d55'
Samantha
Mon 6h26'36"8d19'
Samantha
Cas 23h16'12"60d27'
Samantha B
Aur 7h7'34"38d38'
Samantha Claire
Cyg 20h2'19"58d59'
Samantha Genevieve
Cyg 19h34'24"33d14'
Samantha Jade
Ori 5h55'51"15d53'
Samantha Jane
Cas 0h20'0"59d0'
Samantha Jean
And 2h20'0"42d33'
Samantha Jeanne
Cas 1h1'50"53d29'
Samantha Joy
Vul 20h57'23"28d34'
Samantha Julie
Ori 6h4'42"3d36'
Samantha Kate
Uma 10h50'0"48d0'
Samantha Lee
And 0h12'11"46d54'
Samantha Lee
Del 20h19'39"20d36'
Samantha Louise
Cyg 19h46'51"35d11'A
Samantha Margaret
And 0h7'57"31d25'
Samantha Marie
Cam 5h47'21"70d22'
Samantha Natalie
Peg 22h47'36"31d58'
Samantha Viktorya
Aql 19h27'38"1d21'
Samantha's Joy
Oph 18h3'11"11d52'
Samantha's Papa
Ari 1h47'1"15d56'
Samantha's Sonnet
And 0h23'0"36d21'
Samantha's Star
Mon 6h21'36"3d48'
Samantha's Star
Cam 6h59'22"63d20'
Samantha's Star
Crb 16h2'0"27d28'
Samantha's Sweet Smile
And 0h14'25"31d44'
Samantha-13 Years
And 23h9'49"41d10'
SamanthaBella
Vir 13h54'31"1d58'
Samantinka
Peg 23h48'15"28d8'
Samar
Lyn 7h14'50"50d37'
Samara,Konstaninos
Her 18h0'12"38d25'
Samardzich,Alexander
Cep 20h50'47"68d37'
Samarin
Cyg 19h15'53"44d37'
Samarin,Bill & Lorraine
Sge 20h1'44"19d15'
Samata,Evan Wilder
Sco 17h28'16"-30d15'
Samatan,Marine
Mon 7h43'26"-3d52'
Samayeu,Nathalie
Aur 4h55'33"52d10'
Samayoa,Carmen A
Boo 15h5'14"42d23'
Sambataro,Laura Jeanine
Hya 9h5'38"0d43'
Sambataro,Tina Marie
Mon 6h20'16"7d38'
Samberg's Nova,Cale AW
Cam 7h40'0"60d20'
Samberg,Benjamin Asher
Crb 16h23'54"37d37'
Samblas,Victor Marrero
Her 16h13'1"10d42'
Sambora
And 23h51'31"39d0'A
Sambora,Richie
Her 16h57'24"20d9'
Samchi
Lyn 8h59'1"36d56'
Same Time Next Year
Peg 23h31'39"10d21'
Same Time Next Year
Cap 20h55'45"-23d15'
Samel,Steve
Cnv 13h17'33"32d9'
Sameluk,Charles Edwin
Cnv 12h30'37"34d14'
Sameluk,Janis Signe
Peg 21h54'33"22d40'
Sameluk,Paul Charles
Aur 7h13'33"38d32'
Samen,Henry
Her 17h28'13"20d42'
Samet,Janie
And 23h15'30"50d34'
Sameva
Aqr 22h1'0"0d54'
Samford,Michael
Mon 7h12'1"-10d43'
Samgilger
Cas 2h58'18"77d8'
Samhat,Nayef H
Cet 2h11'48"2d43'
Sami's T˝
Ori 6h5'44"10d8'
Samia,Genevieve
Cas 0h54'30"61d22'
Samii,Madjid Dr med
Gem 6h49'41"13d35'
Samilenko,Oleh
Uma 9h51'29"42d1'
Samira
Aur 5h59'54"54d28'
Samis,Michael
Lyr 19h13'16"40d51'
Samitt,Bernice & Irving
Cyg 20h28'12"55d53'A&B
Samkakihonci
Cam 12h39'35"80d8'
Samler,Eli
Lyn 8h11'14"34d30'
Samma
Peg 23h43'38"13d48'
Sammataro,Nancy
Cas 1h43'53"58d25'
Sammel,Star Joyce
Lyr 18h38'31"39d42'
Sammler,Wayne
Dra 14h49'44"64d14'
Sammon,Christine B
And 2h7'0"38d4'
Sammon,John
Lac 22h3'39"40d36'
Sammons,Katie
Com 12h28'38"23d60'
Sammons,Melvin
Dra 17h13'56"68d2'
Sammut,Jean
Vul 19h21'48"27d24'
Sammut,Sadie Ann Queally
Cas 0h35'1"66d10'
Sammy
Cnv 12h37'0"33d43'
Sammy
Dra 20h21'47"68d37'
Sammy
Dra 14h49'54"59d8'
Sammy
And 2h17'56"42d14'
Sammy & Sylvia
Lyr 18h32'18"41d12'
Sammy Jo
Per 2h59'19"55d15'
Sammy Sir,Forever In Sophies
Heart
Boo 14h6'26"48d36'
Sammy's Star
Dra 11h10'0"77d41'
Sammy's Star
Ori 5h4'12"-1d54'
Sammy's Star
Peg 22h44'24"25d33'
Sammystar
Lyn 8h0'45"47d55'
Samnler,Wayne
Ori 5h14'51"-8d51'
Samodurov,Alexander
Lac 22h32'41"55d13'
Samos,Carl
Aur 5h59'0"31d7'
Samoyed,Basil
Cma 7h0'14"-31d46'
Sampaio,Jose Alfredo &
Graga
Col 5h45'46"-41d10'
Sample,Amber & Michael
Cmi 7h53'37"5d1'
Sample,Evan Michael
Lac 22h37'19"53d28'
Sample,Michael L
Uma 9h41'50"71d29'
Sample,Paul R
Oph 16h52'53"0d27'
Samples,Melvin Franklin
Mon 6h24'40"2d43'
Sampsel,Harry Eugene
Ori 5h57'57"21d12'
Sampson
Aur 5h1'27"40d43'
Sampson-Forever,Ian
Pup 7h28'50"24d33'
Sampson Sampson
Cyg 20h52'10"38d48'
Sampson,Charles
Lac 22h37'18"50d20'
Sampson,Christopher H
Ori 5h59'59"12d56'
Sampson,Gary E
Psc 0h57'51"31d54'
Sampson,Harold
Vir 14h58'18"7d19'
Sampson,Kim
Cas 0h52'39"68d55'
Sampson,Lindsey
Lib 15h35'1"-20d23'
Sampson,Phyllis Viola
Cas 0h59'0"64d24'
Sampson,Tracy
Eri 3h47'32"-2d12'
Sams,Dr James H
Her 16h59'54"20d23'
Sams-Barnard "New
Latitudes"
Eri 4h9'28"-16d9'
Samsara
Uma 9h44'18"43d16'
Samsing,James R
Her 17h57'46"20d16'
Samsky,Marissa Anne
Cas 23h15'34"62d45'
Samson
Cyg 19h22'15"28d46'
Samson 43,Paul
Cep 4h45'35"80d6'
Samson's Forté
Ori 4h58'14"5d59'
Samson,Chantale
Cas 0h21'0"72d23'
Samson,Kimberly Joy
And 23h40'27"47d46'
Samson,Lindsay Dawn
Mon 7h32'51"-1d40'
Samson,Marc "Emmy"
Andrew
Lyn 7h52'0"40d16'
Samson,Patrick
Per 3h2'1"40d17'
Samuel
Aur 6h8'0"38d24'
Samuel
Her 17h18'45"42d32'
Samuel Angela
Leo 9h52'43"33d1'
Samuel Christopher Oliver
Dewey Kott
Crb 15h50'1"38d34'
Samuel Owen Twinkle
Lyr 19h13'10"42d34'
Samuel R
Cam 8h15'0"73d35'
Samuel `94
Cam 3h56'0"62d57'
Samuel's Fire
Cnv 13h27'52"41d8'
Samuel's Star
Per 1h54'1"53d45'
Samuel's Star
Ori 6h1'40"1d34'
Samuel,Danny & Shannon
Aql 18h50'59"11d9'
Samuel,Frances Lee
Cet 2h29'1"-1d48'
Samuel,Kathryn
Crb 16h6'25"27d19'
Samuel,Michael A
Aql 19h44'37"10d47'
Samuela
Per 2h59'31"32d4'
Samuels,Alexander Harris
Equ 21h2'59"3d39'
Samuels,Ashlee Louise-Fae
Uma 10h8'28"50d22'
Samuels,Bryanne Elizabeth-
Fae
Uma 10h8'0"50d9'
Samuels,Grand Master Jim
Cep 21h48'21"55d25'
Samuels,Jack & Suellyn
Lyn 8h8'54"47d47'
Samuels,Jeffrey T
Boo 14h14'26"34d27'
Samuels,Molly
Peg 22h20'31"5d5'
Samuels,Myron S
Her 16h45'52"38d46'
Samuels,Nanny Evie Fyvolent
Del 20h21'29"10d23'
Samuels,Stephen Jay
Cet 2h33'31"-1d7'
Samuels,Vickie
Crb 16h20'39"27d12'
Samuels,Zachary David
Sge 19h17'0"16d37'
Samuels-Percussionist,
Lindsay
Tri 2h0'13"28d35'
Samuelsen,William
Boo 15h7'58"51d26'
Samuelson,Allegra Theo
Cas 23h32'36"58d16'
Samuelson,Jack
Vir 13h17'38"-8d30'
Samuelson,Julie
Cas 0h23'30"66d59'
Samuelson,Matthew Habas
Her 16h1'60"41d10'
Samuelson,Michael David
Sct 18h56'31"-4d18'
Samuelson,Tiffany Mary
Cas 23h33'41"58d33'
Samuelson,Toba
Lyn 6h22'36"60d16'
Samuelson,Virginia
Cas 1h20'15"75d13'
Samul,Margaret
Cas 1h2'25"58d15'

Samules,Derrick Anthony
Dra 17h31'21"64d13'
Samulis Star,Peter & Pearl
Ori 5h52'29"10d9'
Samwick,Evan
Cep 21h47'52"70d56'
Samy
Dra 20h0'40"70d51'
Samyr
Ori 4h53'14"1d24'
San Angelo,Marc Robert
Boo 14h58'49"48d58'
San Cristobal
Uma 11h34'51"32d11'
San Diego,Bernadette Reyes
Odet
Peg 21h50'34"34d41'
San Filippo,Hun, Honey
Cyg 20h15'45"31d46'
San Fillippo,Jr,Albert
Boo 13h42'34"14d59'
San Jose & Associates
Ori 5h22'51"-6d26'
San Jose,Martha B
Psc 22h53'50"6d40'
San Martin,Hugo E
Cet 0h33'30"-6d56'
San Roman,Eduardo
Cmi 7h54'1"0d33'
San,Pat
Cet 1h33'31"-3d31'
Sana
And 1h58'50"36d48'
Sanabia,Chelsea Ann
Mon 6h19'23"5d25'
Sanada
Peg 23h32'31"11d25'
Sanae
Eri 4h13'40"-18d53'
Sanae Star
Leo 9h20'10"10d42'
Sanakkayala,Bash
Lac 22h36'24"50d27'
Sanam
Per 2h37'58"38d30'
Sananes,Corinne
Cyg 19h25'44"33d9'
Sanasack,William Maxwell
Ori 6h16'36"-2d58'
Sanborn,Barbara G
Aql 20h2'1"4d4'
Sanborn,Cheryl Ann
And 0h6'59"40d8'
Sanborn,Deborah
Mon 7h0'25"8d39'
Sanborn,Richard & Gail
Lyr 18h57'48"31d59'
Sanches,Jr,Jeffery Lee
Aql 19h59'15"1d0'
Sanchez
Her 17h15'41"28d12'
Sanchez,Amador
Cmi 7h45'0"7d33'
Sanchez,Barbara A
Uma 11h52'28"51d37'
Sanchez,Brenda
Cas 0h2'51"63d58'
Sanchez,Christopher Allen
Cet 2h0'13"1d56'
Sanchez,George (Jorge)
Fernando
Mon 6h39'17"2d55'B
Sanchez,Gilda
Vul 21h0'49"27d4'
Sanchez,In Memory Of Steve
Aql 20h1'21"10d28'
Sanchez,Jacob Paul
Aql 18h40'1"1d5'
Sanchez,Janet
And 2h28'1"41d46'
Sanchez,Jr,John A
Mon 6h22'25"6d48'
Sanchez,Juanita
Cam 5h36'1"68d52'
Sanchez,Luis
Dra 17h38'23"61d17'
Sanchez,Maria Luisa
Uma 11h14'37"60d41'
Sanchez,Marie
Com 12h12'39"32d23'
Sanchez,Mario Perera
Cam 4h56'47"58d14'
Sanchez,Rachelle Lynn
Eri 4h9'1"-16d49'
Sanchez,Randy
Ser 17h32'45"-14d34'
Sanchez,Reuben A
Ser 15h20'55"1d24'
Sanchez,Roger René
Per 3h19'39"54d55'
Sanchez,Rudolph
Ori 6h5'56"7d57'
Sanchez,Ruth
Lyn 8h21'30"57d48'
Sanchez,Saudy Olinka
Mon 7h0'26"-1d41'
Sanchez,Steven Paul
Aql 19h29'29"-8d9'
Sanchez,Sunday & Tony
Crb 15h54'58"26d5'
Sanchez,Terry & Julie
Moseley
Mon 7h46'15"-5d6'
Sanchez-Pearce,Carol
Lyn 8h27'1"47d53'
Sancho,Jane
Cas 0h40'33"72d46'
Sancic,Adam
Oph 17h32'51"7d31'
Sancinito,Treacy
Aql 19h30'23"0d55'
Sanctuary
Ari 2h12'56"21d35'
Sand Castle Princess & Field
Of May
Cyg 19h19'32"28d33'
Sand III,Ralph Edward
Aqr 20h55'25"0d13'
Sand N' My Pants Alan
Sex 10h27'41"3d53'
Sand Our Shining Star, Brian
Anthony
Peg 21h28'1"2d43'
Sand,Deb
Cas 0h1'59"58d51'
Sand,Eric
Lyr 18h37'22"34d27'
Sand,Robyn Lucille
Del 20h14'53"14d11'
Sanda,Peter Joseph Charles
Cyg 21h29'31"40d15'
Sandahl,Andrew Curtis
Boo 14h56'0"51d47'
Sandahl,Jr,George
Cep 20h54'56"58d36'
Sandark
Her 17h24'30"27d4'
Sandbacka,Hakan
Lyn 7h0'35"50d35'
Sandbank,Luke Henry
Dra 16h39'24"67d26'
Sandbeck-My Piece Of
Heaven,Brad
Per 3h11'33"44d19'
Sandberg,Julian Caroline
Helland
Aql 19h24'17"-10d10'
Sandberg,Lynne H
Mon 6h37'38"7d18'
Sandborg,Paul Eric
Her 16h40'53"27d1'
Sandbrook
Her 17h16'52"28d13'
Sande,Glen
Per 3h44'26"51d17'
SanDee
Lyr 18h46'33"30d25'
Sandee's Star
Peg 21h55'19"24d23'
Sandefer
Uma 10h30'0"62d8'
Sandell,Eric Andreas
Cet 2h7'44"-2d19'
Sander,DDS,Michael H
Per 4h42'44"38d49'
Sander,Guy Henry
Per 3h11'58"42d15'
Sander,Gwendolyn Tara
Com 12h18'51"26d19'
Sander,Hans Jürgen
Boo 14h22'50"10d23'
Sander,Jeffrey P
Cnv 13h58'31"46d11'
Sander,Pia
Tau 5h47'53"27d34'
Sander,Rosel
Psc 0h3'19"-0d10'
Sandercock,Tucker Colin
Her 16h34'12"41d47'
Sanderlin,Kathy
And 2h20'37"46d53'
Sanders II,James Hampton
Tri 1h50'43"26d39'
Sanders III,Honorable Alfred
Foster
Boo 13h34'19"18d4'
Sanders,Alan
Per 2h53'52"40d29'
Sanders,Alden Hamilton
Her 18h41'28"13d5'
Sanders,Alison Crocker
Uma 9h38'43"44d13'
Sanders,Angela Christine
Miller
Aqr 22h49'54"-6d47'
Sanders,Angela Mary
Mon 7h3'27"-5d40'
Sanders,Brent Franklin
Cap 20h37'16"-20d2'
Sanders,Charles
Aql 20h1'13"0d14'
Sanders,Charles A
Boo 14h24'53"39d3'
Sanders,Chloe
Crt 11h21'34"-20d20'
Sanders,Christopher B
Her 17h59'30"14d15'
Sanders,Constance Weeks
Cam 5h36'1"68d52'
Sanders,Cynthia Lynn
And 23h43'29"44d4'
Sanders,Daniel P
Uma 12h3'55"58d52'
Sanders,Derek Scott
Lac 22h9'1"40d5'
Sanders,Donna Marie
Tri 1h55'33"25d46'
Sanders,E Jean
Aql 20h11'19"5d4'
Sanders,Edward Robert
Per 2h37'32"40d13'
Sanders,Fred
Aql 19h48'26"12d16'
Sanders,Gary Steven
Mon 7h1'28"-5d56'
Sanders,George Richard
Aur 6h55'42"38d11'
Sanders,Gregory Franklin
Cmi 7h6'33"5d33'
Sanders,Heidi
Lyn 7h5'58"52d49'
Sanders,Ina Meares
Uma 10h57'59"56d22'
Sanders,J B
Sct 18h54'26"-6d55'
Sanders,James J
Ori 5h58'25"10d41'
Sanders,Jane
Lyr 19h17'45"41d48'
Sanders,Jeff
Oph 17h5'56"0d56'
Sanders,Jeffery Douglas
Ser 15h48'37"24d18'
Sanders,Jennifer Lee
Sge 19h59'1"18d49'
Sanders,Jesse Lee
Del 20h29'21"10d32'
Sanders,Kaitlyn Victoria
Peg 23h39'42"30d54'
Sanders,Kara Nicole
Cmi 7h55'1"4d6'
Sanders,Karina L
Uma 11h53'33"51d42'
Sanders,Lisa
Cas 1h45'27"60d22'
Sanders,Margaret & Leon
Peg 0h0'52"27d38'
Sanders,Michael
Per 2h55'0"46d26'
Sanders,Molly Elizabeth
Cas 23h16'1"61d43'
Sanders,Nicholas Sean
Oph 17h38'51"-21d2'
Sanders,Nicole "Joie de Vivre"
Ari 2h44'45"28d39'
Sanders,Paula Louise
Cyg 21h48'1"53d26'
Sanders,R Scott
Her 16h45'46"20d37'
Sanders,Rebecca Jane
Vul 19h42'37"22d56'
Sanders,Richard Scott
Her 17h20'24"41d15'
Sanders,Robert
Ori 4h48'26"0d28'
Sanders,Robert
Lyn 7h52'27"48d37'
Sanders,Robert Frank
Cet 0h25'56"-6d4'
Sanders,Sandi
Crb 15h49'49"38d23'
Sanders,Sandra Maria
Peg 22h2'1"33d55'
Sanders,Scott E
Lac 22h19'18"37d37'
Sanders,Sendy
Peg 23h21'60"8d9'
Sanders,Steve
Ori 6h6'17"8d53'
Sanders,Teresa Dawn
Del 20h36'20"18d56'
Sanders,Teresa Kay
Peg 22h47'38"34d52'
Sanders,William Randolph
Boo 15h7'15"12d16'
Sanders-Curry,Jazmin Ashley
Lyr 18h37'0"27d11'
Sanders-Curry,Lori Anne
Vul 20h14'46"22d38'
Sanders-Curry,Madison Ida
Lyn 7h47'27"42d33'
Sanderson,Amy
Peg 23h40'31"18d45'
Sanderson,Anthony
Aur 6h35'37"30d26'
Sanderson,Barb
And 2h15'0"50d14'
Sanderson,David
Dra 19h8'51"70d16'
Sanderson,Eric
Cma 7h28'47"8d38'
Sanderson,Ida
Cyg 21h36'28"41d5'
Sanderson,Ivie Diann
Oph 18h5'54"13d1'
Sanderson,J R
Cet 3h9'29"3d1'
Sanderson,Joseph
Aur 6h39'30"37d48'
Sanderson,Kate
Peg 23h40'0"15d47'
Sanderson,Kevin
Cma 6h57'41"-18d2'
Sanderson,Leonard Keith
Per 3h18'12"40d42'
Sanderson,Marty
Mon 6h20'37"2d48'
Sanderson,McAfee
Eri 2h59'27"-10d32'
Sanderson,Peter
Sco 17h32'1"-30d28'
Sanderson,Ralph Durham
Oph 18h7'23"7d58'
Sanderson,Shelley Michelle
China
Lyn 8h1'50"38d25'
Sanderson,W C (Tim)
Boo 14h26'12"22d40'
Sanderson,Wayne
Boo 14h19'26"18d59'
Sandford,Grayson Douglas
Aql 18h59'32"10d27'
Sandford,Lauren Cherise
Peg 23h29'25"21d33'
Sandfoss,Debbie
Mon 6h44'1"11d35'
Sandham,Julie
Cas 1h47'53"70d25'
Sandher,Dildip Singh
Uma 9h28'16"49d14'
Sandi
Lac 22h50'10"53d43'
Sandi
Cnc 7h54'16"10d19'
Sandi
Gem 7h1'45"28d41'
Sandi
Uma 9h33'55"48d58'
Sandi
Mon 6h23'27"0d48'
Sandi & Steve
Cet 1h7'35"-1d9'
Sandi 92892
Peg 23h5'14"22d4'
Sandi Lynn & Suzi Lee
Mon 7h17'28"-4d29'AB
Sandi's Excel
Cas 23h27'1"61d53'
Sandi's Shining Star
Peg 22h45'10"12d24'
Sandi's Star
Aqr 23h39'24"-5d28'
Sandi's Tender Hearted Star
Peg 23h24'35"25d18'
Sandidge,Sandy
Aql 19h4'47"11d8'
Sandie
Cas 23h31'42"53d44'
Sandie
And 0h56'32"45d33'
Sandie
Cyg 19h39'15"31d33'
Sandifer,Frank H
Uma 11h43'16"55d16'
Sandilein
Uma 9h5'25"70d41'
Sandler,Fredric & Brook
Sge 19h29'46"19d22'
Sandlian,Donna Louise
Uma 11h4'16"40d0'
Sandman's Star
Cnv 13h29'24"41d38'
Sandman,Michael Jeffery
Uma 8h47'22"50d14'
Sandmore,Philip Sigaurd
Lmi 10h56'20"31d29'
Sandner,J Randolph
Dra 17h9'15"61d52'
Sandner,John F
Aur 6h28'48"33d6'
Sandok,MD,Burton A
Lac 22h3'58"38d14'
Sandomirsky,Noel
Vul 19h43'28"23d33'
Sandor
Umi 13h46'27"75d48'
Sandpiper
Cnc 8h35'24"7d26'
Sandra
Vir 13h30'4"-8d52'
Sandra
Mon 6h34'11"-0d52'
Sandra
Peg 23h19'34"33d38'
Sandra
Eri 3h52'15"-3d37'
Sandra
Cas 2h2'1"65d19'
Sandra
And 2h30'0"50d14'
Sandra
Tau 3h37'0"24d0'
Sandra
Lyr 18h48'33"34d16'
Sandra
Cap 21h52'0"-19d42'
Sandra
Del 20h12'17"11d21'
Sandra
Gem 6h49'44"16d21'
Sandra
Mon 7h39'10"-4d41'
Sandra
And 0h1'26"38d2'
Sandra
Oph 17h18'52"10d36'
Sandra
Cep 20h18'22"60d30'
Sandra "Little Twirp"
Cas 1h30'0"63d55'
Sandra & Brian
Cam 5h48'33"68d57'
Sandra & Christian
Cas 0h47'36"60d34'
Sandra & John Forever
Uma 8h43'40"68d1'
Sandra & Nadir
Ara 17h44'33"-56d52'
Sandra & Wolfgang
Cep 23h17'54"64d36'
Sandra Ann
Peg 23h44'0"31d22'
Sandra B
Cas 0h36'34"60d16'
Sandra B
Umi 16h21'21"72d26'
Sandra Dorothy
Peg 22h2'55"18d46'
Sandra E
And 0h50'21"41d8'
Sandra Ellen
Lyn 8h1'57"40d51'
Sandra et Jean-Paul T
Per 3h3'18"41d10'
Sandra G
Lac 22h45'1"38d42'
Sandra G
Aur 6h30'59"34d15'
Sandra Gail
Lyr 18h24'11"47d8'
Sandra H
Uma 11h35'58"33d9'
Sandra Irene
Aql 18h56'49"10d36'
Sandra Isabelle
Cas 23h1'56"58d22'
Sandra J
Sgr 18h49'49"-32d45'
Sandra Jayne
Uma 14h6'36"57d56'
Sandra Jean
Cas 1h2'24"60d25'
Sandra Jean-1/13/61
Com 13h33'37"17d58'
Sandra Kay
Com 12h29'39"14d8'
Sandra Lee
Com 12h0'18"18d47'
Sandra Lee
Aql 20h4'14"-6d1'
Sandra Lee My Love
Lyr 18h49'15"40d59'
Sandra M
Gem 6h29'22"12d54'
Sandra Martina
Lib 14h20'35"-20d26'
Sandra May
Cep 23h23'43"65d18'
Sandra Rose
Peg 22h24'53"30d36'
Sandra Valerie
And 0h29'18"31d51'
Sandra's Joy
Crb 16h10'59"34d31'
Sandra's Serenade
Cam 3h55'36"72d25'
Sandra's Song
Psc 23h2'30"0d24'
Sandra's Star
Vul 20h0'1"28d15'
Sandra's Very Own Star
Cyg 20h9'11"39d60'
Sandra,in aeternum
And 0h7'39"30d29'
Sandra,Simon
Sge 20h7'40"20d45'
Sandra,The
And 0h51'43"38d35'
Sandralyn
Aqr 20h47'19"-6d23'
Sandranthony
Mon 6h55'16"1d47'
Sandrew
Uma 13h35'16"54d12'
Sandridge,Jay & Betty
Cyg 21h19'38"30d41'
Sandridge,Spencer B
Cam 4h23'42"71d8'
Sandrine
Dra 11h33'17"68d48'
Sandrine Forever
And 23h27'13"40d30'
Sandrine J
Boo 14h4'56"12d28'
Sandrine Robert
Sgr 19h26'38"-39d25'
Sandrine,Wissenmeyer
Dra 20h38'37"80d13'
Sandro
Cam 13h30'0"81d52'
Sandro
Her 17h2'49"41d14'
Sandro Perfler
Cmi 7h18'7"1d9'
Sandruschka
Aql 19h46'24"10d18'
Sands,Ciara Elaine
Lyn 6h16'26"59d37'
Sands,Iain Christopher
Uma 9h2'27"58d2'
Sands,Jeremy
Dra 19h28'25"68d10'
Sands,Lillian Kaplan Albrecht
And 22h57'49"48d49'
Sands,Lynda
Cap 21h25'12"-22d38'
Sands,Mackensie Jo
Cyg 21h54'38"42d44'
Sands,Robert
Uma 9h3'1"53d38'
Sands-Forever in our
Hearts,Jamie
Cyg 19h33'11"38d45'
Sandschulte,Bus
Cmi 7h59'1"8d42'
Sandstorm
Cma 7h2'52"-31d34'
Sandstrom
Peg 23h17'31"33d57'
Sandstrom,Deborah Snapp
Lyr 18h49'1"39d0'
Sandt,Eric J
Dra 16h9'22"68d12'
Sandusky,Kate
Lyr 18h22'39"38d4'
Sandusky,T L
Leo 9h21'13"17d35'
Sandy
Eri 3h47'31"-1d52'
Sandy
Ori 5h20'19"1d13'
Sandy
Lyr 18h29'60"47d31'
Sandy
Peg 21h55'42"31d26'
Sandy
Sge 20h17'21"20d17'
Sandy
Crb 16h16'57"30d29'
Sandy
Peg 21h59'41"21d5'
Sandy
Peg 23h48'41"29d33'
Sandy
Lyn 7h5'13"50d52'
Sandy
Cet 2h59'0"1d31'
Sandy
Hya 8h16'30"-5d59'
Sandy
Cas 1h29'1"68d2'
Sandy
Vul 19h43'21"26d1'
Sandy
Cas 0h43'54"62d23'
Sandy & Arlene
Cam 5h43'58"61d13'
Sandy & Don
Aur 5h24'54"30d44'
Sandy & Elvis
Dra 17h16'22"60d55'
Sandy & Jim
Sct 18h41'34"-7d25'
Sandy & Steve
Crb 15h57'20"30d48'
Sandy G's Tai Sue
Cas 1h30'39"73d13'
Sandy Lindsay
Aur 5h19'57"49d23'
Sandy Love
And 1h21'59"37d7'
Sandy M
Cas 2h30'42"73d35'
Sandy Mac
Uma 11h43'36"30d13'
Sandy Rose
Cyg 20h9'58"40d33'
Sandy Seashell
And 1h45'1"38d28'
Sandy Star
Tri 1h40'51"31d35'
Sandy's Star Gazer
Cas 22h57'29"53d44'
Sandy's Way
Eri 4h9'58"-14d55'
Sandy's Wishing Star
Hya 8h9'19"4d46'
Sandy,Christian
Cyg 19h43'12"30d54'
Sandy,Gary Lee
Umi 15h14'12"68d28'
Sandy,Kathryn
Cyg 19h43'0"30d15'
Sandy,Trevor Christian
Ser 15h12'13"23d19'
Sandy,Tyler David
Aqr 22h4'44"0d17'
Sandy/Renee
Lyn 7h58'30"43d36'
SanFilippo,Alfred R
Aur 6h0'31"38d6'
Sanfilippo,Eugene & Rose
Ori 6h1'22"1d1'
Sanford
Cyg 19h26'54"34d7'
Sanford,Al
Aql 19h30'35"10d32'
Sanford,Bil
Her 17h33'0"21d29'
Sanford,Brian
Uma 11h21'11"38d1'
Sanford,Gena Michelle
Vul 20h18'15"22d31'
Sanford,Jeff
Cep 21h45'50"55d46'
Sanford,Joel
Del 21h4'54"18d55'
Sanford,Linda Kay
Lyn 9h15'57"34d7'
Sanford,Meagan Ashley
Umi 16h11'13"77d5'
Sanford,Tyler Kenneth
Sgr 19h36'58"-38d36'
Sang,Elsie O & Philip D
Cyg 21h29'19"40d36'
Sangalli,Giuseppe
Dra 14h59'44"60d0'
Sangbusch,Bill-Liz Billy-Katie
Uma 10h13'30"56d27'
Sange
Lyr 19h24'57"40d29'
Sanger,Rit & LaRaine
Ori 5h57'18"15d1'
Sangha,Devinderpal Singh
Sge 19h39'37"16d30'
Sangiacomo,Lurana
Ser 16h1'20"8d41'
Sangie
Cas 0h47'13"61d0'
Sangit,Sw Deva
Lyn 7h53'22"41d46'
Sangiuolo,Angelo
Uma 11h45'30"41d25'
Sangiuolo,Lisa
Cyg 19h39'12"28d21'
Sangster,Eline
Umi 13h41'36"77d45'
Sangster,Jamie
Uma 9h52'0"68d51'
Sanibel
Peg 22h38'10"12d18'
Sanjay
Cam 4h14'15"60d43'
Sanjay-My Star
Cyg 19h28'18"36d10'
Sankey Star,The Daniel A
Lac 22h9'51"50d6'
Sankey,Larry A
Aql 18h44'0"2d2'
Sankowski,Brian Ronald
Aur 4h57'46"41d2'
Sanmals,Amethan
Lyn 9h0'40"38d34'
Sann,Janice J
Crb 16h16'50"28d31'A
Sann,Thomas R
Crb 16h16'50"28d31'B
Sannebeck "Queen", Deborah
L
Cas 1h5'0"50d25'
Sannella,MD,Nick
Per 2h41'35"35d36'
Sanoski,Telemilo
Boo 13h41'10"17d42'
Sans,Brandle
Vul 20h29'23"28d20'
Sansale,Judy
And 23h25'26"41d10'
Sansaucie,Hazel Carmi Miller
Roth
Cyg 21h9'1"48d52'
Sanseverino,Jim
Cnv 12h15'12"43d36'
Sansing,Richard H
Sct 18h56'51"-6d9'
Sansola,Aleeyha Yankallis
Vul 19h51'39"20d17'
Sansom,M Geish
Vul 19h23'12"23d35'
Sansom,Ruqia
And 23h1'1"50d41'
Sansom,Samantha Elizabeth
Peg 22h40'15"25d54'
Sansone-Lynch,Gia
Boo 15h16'32"52d28'
Sansonetti,Lesia
Cas 23h1'43"53d38'
Sanstrom,William S
Ori 5h58'57"11d1'

Sansweet,Stephen Jay
Ori 5h47'31"10d46'
Santa Babylon
Peg 22h26'13"33d53'
Santa Lucia,Anthony
Cep 20h30'20"75d56'
Santa Maria
Per 1h50'47"50d6'
Santa Maria,Michael
Per 2h28'48"56d30'
Santa Maria-Lorton
Lyn 7h59'57"41d57'
Santa Rosa
Del 20h18'33"9d30'
Santa-Maria
Leo 11h49'41"22d27'
Santabarbara,Ruth Marie
Cas 2h32'10"58d13'
Santaguida,Karen & Rocco
Cyg 21h29'22"40d18'
Santaliz,Ricarda Altagracia
Cam 6h16'40"71d4'
Santalla,Annie
Ari 1h56'59"11d3'
Santambrogio,Luigi
Cas 2h59'35"60d54'
Santana,Angel
Mon 8h7'19"-9d24'
Santana,Melissa Marie
Mon 6h21'18"4d3'
Santana,Roman A
Uma 11h53'1"52d49'
Santana,Tito
Her 16h49'38"39d24'
Santanello,The Sparky
Her 17h31'56"21d52'
Santangelo
Per 1h52'27"56d55'
Santangelo & Family, George & Joanne
Lyr 18h41'48"32d21'
Santangelo,Charles A
Aur 5h35'0"50d21'
Santangelo,Diane Marie
Peg 23h36'11"31d32'
Santangelo,Donna Marie
Cas 1h30'1"70d23'
Santangelo,Louis
Boo 14h28'13"21d54'
Santano,Edward G
Aur 6h46'58"37d24'
Santarelli,Frank
Sex 9h51'54"3d33'
Santarsiero,Mary
Cas 23h27'37"60d41'
Santasieri,Stephen Ray
Lib 15h7'47"-0d57'
Santee,Christopher
Ser 15h39'0"3d26'
Santell,Luke Steven
Per 1h56'0"48d36'
Santelli,Patricia Ann
Sge 19h59'19"16d12'
Santerre,Josée LeFranáois Denis
Lmi 10h8'11"38d50'
Santha-Sue
Hya 8h31'25"-1d36'
Santhouse,Carl
Crb 15h55'33"27d27'
Santiago Cal
Uma 9h7'1"57d59'
Santiago Salmeron
Lyn 8h1'0"39d11'
Santiago,Adam
Aur 6h24'19"38d17'
Santiago,Alexis
Sex 9h56'44"4d55'
Santiago,Alice O
Cet 1h30'30"-4d34'
Santiago,Anahi
Lac 22h32'24"53d36'
Santiago,DC,DABSM,Dr Joann
Lyr 18h35'37"40d37'

Santiago,Felipe
Lac 22h8'23"47d5'
Santiago,Gabrielle
Peg 22h58'58"21d59'
Santiago,José
Boo 14h52'34"36d18'
Santiago,Peggy Jean
Peg 22h37'41"27d44'
Santillanes,Janelle Phyllis
Lib 15h3'48"-8d19'
Santilli,Lisa
And 1h15'39"36d45'
Santillo,David Gerard
Aql 19h27'59"12d42'
Santinello,Serena
Lyr 18h47'41"37d56'
Santinello,Shaina
Lyn 8h12'32"33d57'
Santino
Her 16h53'29"40d42'
Santodigiro
Umi 14h46'46"66d40'
Santodonato,Antoinette
Cas 1h31'18"58d48'
Santoni,Sue
Com 13h9'21"20d51'
Santopietro,Giuliana
Cae 4h49'25"-31d54'
Santora,Carmela
Cas 0h48'35"68d52'
Santora,Ciro & Lucy
Uma 10h47'1"59d58'
Santora,Jan
Del 20h22'47"20d37'
Santora,Jim
Uma 11h11'52"53d4'
Santorelli,Linda
Vul 19h18'57"26d48'
Santorelli,Nancy Ann
Peg 23h43'51"25d45'
Santori,Edward J
Dra 12h16'37"76d13'
Santoro II,Thomas Albert Vincenzo
Lac 22h36'22"55d59'
Santoro,Alexandra
Cas 0h30'27"61d18'
Santoro,Davide
Hor 3h26'54"-46d20'
Santoro,Emma Christine
Uma 9h41'17"68d55'
Santoro,Jill Meredith
Cas 1h7'60"60d48'
Santoro,Jr,Joseph Patrick
Uma 10h7'36"68d16'
Santoro,Lisa Marie
Lyn 6h39'56"58d43'
Santoro,Michaël
Uma 11h27'46"31d33'
Santoro,Nicholas F
Cep 21h56'0"55d43'
Santoro,Olivia C
And 0h55'55"35d7'
Santoro,Reese
Mon 8h8'18"-9d41'
Santoro,Susannah Lucy
Cas 0h26'51"62d39'
Santos,Beatriz L
Vul 21h1'16"23d47'B
Santos,Catarina
Lyr 19h2'26"38d21'
Santos,Dylan Rose
Peg 21h57'51"34d51'
Santos,Eileen A
Lyr 18h46'23"40d37'
Santos,Helder
Boo 15h27'51"52d3'
Santos,Jennifer
And 1h45'47"39d19'
Santos,Jonathan R
Aur 5h3'22"49d58'
Santos,Jorge & Jackie
Ori 6h5'32"7d47'
Santos,Lynn Ann
Lyn 7h56'36"35d51'

Santos,Maria
And 0h26'42"28d28'
Santos,MD,Clyde Dos
Aql 18h42'25"-0d19'
Santos,Star of
Crb 16h7'54"32d25'
Santoshisara
Hor 2h58'25"-49d34'
Santucci,Anthony Joseph
Boo 14h22'21"36d49'
Santucci,Lisa Marie
Cam 13h6'46"77d44'
Santucci,Michael R
Dra 9h49'15"73d25'
Sanville,Kevin & Laurel
Per 2h25'52"55d18'AB
Sanya
Cam 12h16'13"77d32'
Sanz,Angel Robador
Cnv 12h18'37"37d44'
Sanza,Robert
Lac 22h8'59"51d29'
Sanzanobi
Aur 4h59'11"40d9'
Sanzo,Michael Anthony
Per 3h16'10"40d47'
Saori K Rolan
Psc 1h16'9"29d36'
Saoud,Prince Mhamad Ben
Her 17h49'0"48d50'
Sapan,Claire Maxwell
And 23h29'42"43d28'
Sapan,Nathaniel Winston
Dra 16h22'17"60d43'
Saperstein,Dana Lynn
Aql 20h1'28"1d50'
Saperstein,Marian & Ben
Sge 18h56'45"19d43'
Saperstein,Nicole
Eri 3h3'55"-5d9'
Saphir
Cyg 21h6'59"31d28'
Sapiens Magister
Lac 22h48'41"54d6'
Sapienza
Uma 8h26'1"60d16'
Sapienza,Emily Ward
Cas 1h20'38"61d50'
Sapienza,Michael Ward
Peg 0h0'21"30d28'
Sapirstein-Duresky
Umi 15h42'25"76d2'
Sapo
Boo 15h10'1"47d35'
Saporito,Cristianna Sophia
And 0h45'16"36d49'
Saporito,Giavanna Marie
Peg 0h8'52"14d15'
Saporito,MD,Cynthia Eunice
Oph 17h57'42"0d16'
Sapourn,Helen K
Lyr 18h57'0"30d11'
Sapp,David Denise Colby
Aur 6h15'1"46d29'
Sapp,Douglas E
Cet 2h58'1"7d30'
Sappenfield
Cyg 19h26'31"30d42'
Sapphire
Cas 0h33'39"70d40'
Sapphire's Light
Mon 7h15'39"-6d29'
Sappho June
Mon 6h57'32"-8d7'
Sappho the Survivor
Ori 6h4'42"20d19'
Sappington,Glenn Franklin
Her 16h59'13"30d29'
Sappington,Richard Ray
Her 18h39'45"18d50'
Saputo,Tatiana
Lac 22h54'47"51d11'
Saputo,Tony
Lyr 18h46'0"38d13'

Saque,Marion
Aur 5h51'25"38d54'
Sar
Aql 19h28'38"16d8'
Sara
Cas 23h13'37"61d31'
Sara
Mon 6h28'23"-6d7'
Sara
Dra 18h13'48"56d34'C
Sara
Lyr 18h33'21"30d4'
Sara
Lyn 7h54'54"41d14'
Sara
Pho 0h30'0"-41d55'
Sara
And 0h17'39"33d29'
Sara
Tri 2h1'44"32d10'
Sara
Del 20h13'17"14d47'
Sara
Boo 15h8'36"38d5'
Sara
Mon 7h54'43"-5d44'
Sara
Cas 0h58'56"61d0'
Sara
Pup 7h57'7"26d57'
Sara
Cam 11h7'45"81d48'
Sara
Lmi 10h5'56"32d10'
Sara
Peg 23h5'25"14d11'
Sara
And 0h21'53"30d34'
Sara & Jamie
Uma 9h28'11"68d8'
Sara & Kerry
Cyg 20h23'1"37d52'
Sara & Will
Ori 5h53'45"8d7'
Sara B 11359
Lyn 7h5'34"50d24'
Sara Brittany
Cnv 12h55'1"45d29'
Sara Claire
Cyg 20h38'12"45d32'
Sara Dawn
Vul 21h16'56"20d10'
Sara e Michele
Hor 3h7'44"-49d9'
Sara Eliza
Cas 0h4'51"61d39'
Sara Elizabeth
Cas 0h6'14"61d25'
Sara Elizabeth & Mary Margaret
Lyr 19h22'49"42d58'
Sara G
Vul 20h23'28"28d39'
Sara Jo
Lyn 9h14'57"36d16'
Sara Katie
Umi 13h43'28"79d17'
Sara Lee
Lyr 19h9'38"37d39'
Sara Louise
And 0h5'59"47d28'
Sara Lynn
Aur 5h33'17"38d47'
Sara Marie
Mon 6h37'23"-0d26'
Sara Meighan
Lyn 6h58'0"59d50'
Sara Rose
Cyg 21h24'29"39d28'
Sara Ruthann
Lyn 8h47'39"37d11'
Sara Theresa
And 0h22'0"44d9'
Sara's Heart
Gem 7h55'37"20d29'

Sara's Star
Lyn 8h9'1"50d33'
Sara,Eternamente y Siempre Mia
Lyn 7h40'36"44d57'
Sarabella
Cet 0h56'17"1d26'
Saracco,Giuseppe
Cmi 8h4'36"5d48'
Saracino,Vincent
Per 1h47'11"50d21'
Saraf,Mazah Lahan Sasha
Sgr 19h5'2"-25d41'
Sarafina
Lib 15h15'53"-19d18'
Saragossa,Serge Franáois
Lmi 11h2'51"32d56'
Sarah
Aur 5h53'20"30d16'
Sarah
Aql 20h4'55"-0d33'
Sarah
And 1h33'47"40d6'
Sarah
Cyg 20h53'10"30d49'
Sarah
Aql 19h28'60"-10d19'
Sarah
Mon 6h35'53"-6d32'
Sarah
Aqr 21h55'50"-6d27'
Sarah
Cas 1h10'19"66d22'
Sarah
And 2h24'48"45d7'
Sarah
Cam 3h34'20"60d17'
Sarah
Vel 9h22'44"-40d28'
Sarah
Lyr 18h31'49"31d52'
Sarah & Brandon
Cyg 21h40'22"29d50'
Sarah 144
Sge 20h1'20"20d1'
Sarah Amelia
Peg 22h56'40"25d58'
Sarah Ann
Aur 5h4'45"40d3'
Sarah Ann
Uma 11h53'39"31d52'
Sarah Ann's Star
Cnv 13h56'58"45d58'
Sarah Bear's Forever There
Umi 15h57'8"74d5'
Sarah Catherine
Peg 21h56'29"28d29'
Sarah Dee Bat-Mitzvah Superstar
Mon 6h33'0"8d22'
Sarah Elizabeth
Cyg 20h19'58"40d46'
Sarah Elizabeth
Lyr 18h24'37"37d40'
Sarah Faye
Lyr 18h49'21"33d25'
Sarah G
Lyn 7h31'22"44d42'
Sarah G
Cas 0h36'28"60d18'
Sarah Grace
Cas 23h27'39"61d17'
Sarah Grace
Tri 1h57'1"28d17'
Sarah Isabel
Lyn 7h12'27"50d3'
Sarah J
Cap 21h0'19"-22d55'
Sarah J
Cet 0h36'11"1d12'
Sarah Jane
Cyg 19h47'15"31d44'
Sarah Jane
Lyr 18h20'12"38d49'

Sarah Jane
Lyr 18h31'44"33d18'
Sarah Jane
Cam 5h44'46"70d38'
Sarah Jane
Ser 18h37'53"-2d15'
Sarah Jeanne
Cyg 20h5'39"40d1'
Sarah K
Lyr 18h28'59"31d9'
Sarah Katelynne
Cas 23h14'33"62d51'
Sarah Leanne
Peg 23h17'53"30d35'
Sarah Lynn
Mon 7h57'14"-3d44'
Sarah Makayla
And 1h12'33"36d33'
Sarah Margaret G-22
Lyn 7h4'12"60d47'
Sarah Margrete
Tri 1h40'54"30d54'
Sarah Maria"Natividad"
Vul 19h3'30"21d31'
Sarah Marie
Lyn 6h25'10"59d12'
Sarah Marie
And 0h18'46"36d6'
Sarah Marie 1
Lyr 18h49'11"37d7'
Sarah Noelle D
And 23h12'0"37d17'
Sarah O
Umi 14h24'59"68d23'
Sarah Pearl Teresa
Del 20h22'1"3d18'
Sarah Renee
Lyr 19h6'0"40d12'
Sarah Smile
And 2h34'23"42d17'
Sarah Smiles
And 2h24'21"38d43'
Sarah Stella
Cas 0h52'57"71d2'
Sarah Suzanne
And 23h24'15"46d9'
Sarah Tracey
Lyr 18h55'53"30d11'
Sarah Victoria
Mon 7h27'41"-8d15'
Sarah's Beauty
Mon 8h4'50"-9d22'
Sarah's Celestial Sapphire
Sge 20h0'39"19d2'
Sarah's Rising Star
And 0h22'17"30d13'
Sarah's Star
Sco 16h56'12"-40d35'
Sarah's Star
Cas 1h39'36"58d11'
Sarah's Star
And 23h0'23"45d35'
Sarah's Wish
Cet 0h35'54"1d50'
Sarah, My Brown Eyed Girl
And 23h8'1"39d27'
Sarah-Mathilde
Cas 0h48'23"76d43'
Sarah/Michael
And 23h43'46"45d34'
Sarahbelle
Uma 8h31'46"49d10'
Sarahphil
Cyg 19h30'58"35d11'
Sarahs,Ronald Charles
Uma 9h0'59"53d38'
Sarajevo,Bambini
Aur 5h29'47"31d38'
Saralis,Lennon Sidney
Cep 22h51'12"56d52'
Saranpaa,Aarre
Umi 13h11'35"71d20'
Sararaebon
Per 1h28'48"52d41'

Sarasota
Sct 18h55'0"-6d44'
Sarath Jiana
Aql 19h58'13"12d39'
Saravia's Star,Robert J
Cet 2h26'37"7d50'
Sarazen,Alexandra Paige
Cas 2h45'1"61d37'
Sarbacher,Kristin & Kelly
Lyn 8h30'32"44d51'
Sarcia,Claude
Lmi 10h45'50"26d27'
Sarcone,Serina
Psc 1h30'38"27d48'
Sarda,Gisele
Per 3h59'37"36d43'
Sardella's Par Star, Bill
Ori 6h15'51"-2d37'
Sardella,Jimmy S
Ari 2h7'47"21d1'
Sardenane
Uma 10h13'13"47d35'
Sardeson
Per 3h33'13"38d10'
Sardi
Peg 21h52'54"20d2'
Sardi,Vincent
Dra 18h16'45"70d1'
Sardi,Vincent
Dra 19h20'28"84d36'
Sardini,Monica
Lib 14h47'30"-0d29'
Sareen Domaine of RamShazam
Peg 22h56'44"20d38'
Sarell,Sylvia Kokkinen
Umi 15h6'1"69d30'
Sarezky,Ethan
Boo 15h29'49"42d24'
Sarezky,Kevin
Cep 21h52'36"55d59'
Sarezky,Laurence
Her 18h2'11"40d27'
Sarezky,Ollie
Tri 2h43'0"31d4'
Sarf,Larry
Dra 18h5'23"68d20'
Sargeant,Jennifer Elizabeth
And 1h21'52"36d18'
Sargent's Love
Her 17h05'46"46d02'
Sargent,Brittni Rose
Lib 15h33'49"-8d53'
Sargent,Chelsie Christine
Aql 19h29'49"-10d29'
Sargent,Jennifer
Vul 19h41'23"22d43'
Sargent,John Edmondson
Her 15h51'32"47d23'
Sargent,Michael Daryl
Hya 8h14'28"-15d15'B
Sargent,Nanci Marie
Hya 8h14'28"-15d15'A
Sargent,Robert
Aql 20h1'29"14d5'
Sargentson
Cep 20h52'27"71d10'
Sargianis,James John
Aqr 22h19'21"0d15'
Sari
Cas 2h5'30"61d39'
Sari
Lyr 18h38'28"39d60'
Sariah Eileen
And 2h29'0"45d35'
Sariano,Jack
Dra 11h40'47"71d59'
Saric,Vesna
Cmi 7h33'49"0d1'
Sarica,Philippe
Vul 19h23'11"27d8'
Sarich,Safija
Cyg 20h57'21"39d45'
Sariego,Adam
Her 16h51'35"34d43'

Sariego,Ann
Sge 19h54'33"19d51'
Sarimar
Cra 11h7'3"-18d45'
Sarina
Mon 7h55'16"-4d6'
Sarina "World's Greatest Mom"
Lyn 6h56'38"60d32'
Sarion
Cep 22h34'1"60d59'
Sarita Keni
Cyg 21h6'14"31d33'
Sariyah
Ari 3h12'41"25d19'
Sarjeant,Joan Patricia
Eri 3h0'36"-1d53"
Sarkady,Marc David
Boo 14h19'55"36d13'
Sarker
Dra 14h54'40"62d45'
Sarkio,Kauko
Cam 13h30'40"80d40'
Sarkis,MD,Richard J
Oph 17h54'1"12d11'
Sarkis,Susan
Mon 7h28'22"-10d21'
Sarles,Marilyn Dawson
Cas 0h39'26"73d51'
Sarlos,Andy
Cep 2h57'0"80d25'
Sarmiento,Maria Teresa Bermudez
Cas 0h2'11"62d9'
Sarmite
Tri 1h47'59"31d38'
Sarmom
Cas 0h29'40"58d53'
Sarn,Audrey P
Eri 2h43'49"-3d48'
Sarna,Adolph J
Per 2h22'39"58d18'
Sarna,Anna K
Cas 0h24'38"61d12'
Sarna,John David
Per 2h54'14"31d28'
Sarna,Sajjan Singh
Umi 15h27'0"66d25'
Sarnecky,William J
Cep 21h38'0"55d16'
Sarner,Richard
Oph 16h38'56"1d12'
Sarney's Star,Harris M
Aur 7h11'20"37d7'
Sarnicki
Cyg 20h42'18"45d46'
Sarno,Anna Casella
Cnc 8h56'21"8d36'
Sarno,Mary
And 0h41'20"37d58'
Sarno,Richard S
Lmi 10h56'40"26d20'
Sarno,Stephen
Aur 6h13'0"32d50'
Sarnoff,Lili-Charlotte
And 23h45'38"46d25'
Saroj-Chandur
Cra 18h19'32"-37d22'
Sarosy,Ellen M
Lyn 7h36'45"45d9'
Sarote,Adam
Peg 22h7'0"21d44'
Sarote,Shari
Peg 22h9'23"25d12'
Sarozek,Ashley Marie
Peg 21h40'1"23d2'
Sarozek,Joshua Daniel
Aql 19h56'0"11d59'
Sarozek,Phillip Daniel
Aql 19h56'1"11d20'
Sarra
Ori 5h59'45"16d44'
Sarramia,Nathalie
Crb 15h54'52"30d57'

Sarraquigne,Francine
Oph 18h6'1"11d53'
Sarrat,Walter L
Aql 19h39'38"14d58'
Sarrels,Kelly Beth
And 1h31'52"48d51'
Sarrels,Megan Sarah
Peg 23h44'0"28d30'
Sarrels,Pamela Joy
Cyg 19h47'24"29d51'
Sarret,Christopher Michael
Per 2h21'15"55d12'
Sarret,Gordon James
Ori 5h50'48"14d1'
Sarret,JoAnn
Vul 20h1'0"28d10'
Sarret,Joseph Jules
Eri 2h59'34"-10d14'
Sarret,Jules
Cnv 12h48'17"32d9'
Sarret,Nicole Christine
Eri 3h43'36"-3d3'
Sarriau,Thierry
Per 3h28'54"51d4'
Sarro 7/16/93
Lyr 18h34'1"40d56'
Sarro,Francis & Adeline
Cam 5h4'20"54d58'
Sarrosy
Umi 16h22'55"79d23'
Sarsby,Dawn
Crb 15h15'34"30d48'
Sarsteiner,Norbert
Aur 5h0'1"38d10'
Sartell,Conner James
Boo 14h19'0"45d46'
Sartori Century,Nova
Umi 13h40'46"76d35'
Sarvadi,Lisa
Aql 19h33'21"0d26'
Sarvameenatmasukiransv ara
Boo 14h57'48"45d19'
Sarvela
Leo 10h56'0"-5d52'
Sarvello,Joseph Francis
Uma 12h11'1"55d39'
Sarver
Uma 10h20'23"51d15'
Sarwatka "Boo Bear", Owen Tyler
Lac 22h27'28"52d46'
Sarway
Uma 9h52'27"48d39'
Sas
Crb 16h6'59"30d8'
Sasa
Sge 20h1'26"16d42'
Sasaki,Kana
Lyn 7h51'33"48d60'
Sasaki,Miho
Lyn 7h51'1"40d32'
Sasaki,Nolan Shane
Peg 23h22'33"33d58'
Sasaki,Stella of Keiji
Vir 14h28'26"6d45'
Sasaki,Toshie
Aur 5h0'47"31d1'
Sascha Michelle
Lyn 7h25'19"44d29'
Sasco,Benoit
Boo 15h28'20"38d44'
Sasha
Cyg 21h7'38"37d40'
Sasha
Aqr 23h3'13"-11d4'
Sasha
Vir 11h41'21"1d36'
Sasha
And 2h23'52"47d38'
Sasha
Ori 6h6'35"0d15'
Sasha
Uma 8h39'49"51d23'
Sasha & Eglute
Uma 10h24'1"50d44'

Sasha Irene
Mon 7h42'5"-5d30'
Sasha's Star
And 2h23'42"39d37'
Sashai Dawn
Cmi 8h7'1"1d47'
Sashe
And 23h0'23"45d43'
Sashka
Lac 22h48'57"54d55'
Saska
Peg 22h33'12"11d36'
Saskia
Lyr 18h33'40"37d26'
Saskia Stern
Sco 17h34'23"-38d40'
Saskia/Ibolika
Vir 12h34'32"-8d24'
Sasmor,Dr James C
Ori 5h35'53"-1d25'
Sasparella,Daniella Arabella
Peg 23h6'0"11d19'
Sasquatch
Ser 15h13'11"21d6'
Sass,Stephen John
Lac 22h50'12"56d24'
Sassa,Taylor Alexandra
Cas 0h51'27"62d50'
Sasse,Tyler Thomas
Cet 2h44'15"5d3'
Sasser
Lyr 18h58'32"41d1'
Sasser,Donald
Ori 4h54'57"-2d16'
Sassi,Dante
Tau 3h44'41"25d47'
Sassie Kalaka
Ori 5h39'50"0d42'
Sasso,Dean Joseph
Per 4h22'0"52d2'
Sasso,Mary
Cyg 19h25'57"33d40'
Sasso,Raffaele
Cam 9h37'46"82d29'
Sassolini di cielo per sempre
Cam 8h21'48"80d60'
Sassoon,Deborah Joanne
And 23h20'0"49d15'
Sassy
Mon 6h57'44"10d52'
Sastaunik,Debbie & Blair
Cyg 21h14'17"38d13'
Sastravaha 1-26
Aql 19h51'11"14d7'
Sastry,Deepa
Lyn 8h13'27"37d30'
Satake,Keiske
Eri 4h5'11"-0d23'
Satchwell,Jake McClaine
Lac 22h48'37"54d16'
Satec
Boo 14h20'50"10d38'
Satellite Of Love Brian Sue
Cyg 20h44'59"38d54'
Sater,Daniel Christian
Peg 22h25'41"21d57'
Sater,Ryan David
Peg 22h1'19"31d6'
Sather,Joy
Peg 22h24'1"33d49'
Sather,Morton T
Mon 6h44'16"7d25'
Sathya,Joshua
Dra 16h38'1"68d53'
Satin
Crb 16h21'30"30d50'
Satinover,Julie Rachel Leff
Lyr 18h49'23"36d38'
Satkiewicz,Stephanie
Vul 19h58'26"28d23'
Satlov,Victor L
Aur 5h37'29"38d16'
Satmary,Vincent
Per 3h3'45"38d9'
Sato,Shane Kurata
Lmi 10h11'41"33d58'
Satoko
Lib 14h25'5"-13d34'
Satori
Lyn 7h56'51"51d28'
Satoru
Cnv 13h30'35"35d9'B
Satoru & Yuko
Tau 5h27'32"25d57'
Satoshi
Cam 3h44'16"61d33'
Satoshi Inherit Of Stella
Tau 5h25'2"21d21'
Sattazahn,Larry L
Aql 19h39'50"13d41'
Sattelkau
Vir 13h27'13"-2d19'
Satter,MD,Erby J
Oph 17h8'42"-22d36'
Satterdaze
Dra 20h1'15"74d39'
Satterthwaite, Clementine
Lyn 8h54'58"37d33'
Satterthwaite,Mary Elizabeth
And 0h20'45"33d19'
Satterwhite,Terri
Dra 17h2'21"65d36'
Sattler,Casey
Her 17h38'41"23d26'
Sattler,Erik Ryan
Her 16h58'43"26d50'
Sattler,Frank & Clara
Cas 1h0'0"66d32'
Sattler,Manfred
Ori 5h51'20"21d4'
Sattler,Sivia "Tubbsy"
Aqr 21h3'44"-14d8'
Satty
And 23h3'39"51d52'
Saturday,Tom,Linda & Kelly
Sct 18h56'51"-5d5'
Saturn Corporation
Dra 16h5'21"62d47'
Satya
And 23h43'45"41d27'
Saubade,Frederic
Cas 2h36'21"50d34'
Saucedo "The Code Man",Cody
Hya 8h41'14"2d31'
Saucier,Luc
Lyr 18h21'22"45d27'
Saucier,Lawrence K
Her 16h25'24"31d4'
Saud,Prince Nawaf Al
Peg 21h40'0"22d40'
Sauder,Gina Gurganus
Uma 10h30'42"53d8'
Sauder,Kati Lee
Uma 10h9'23"52d53'
Sauders,Pam
Cas 2h2'44"58d45'
Sauer Star-S
Aql 20h0'39"-1d39'
Sauer,Aud Itta 30-12-67
Oph 17h54'20"10d14'
Sauer,Dittmar Friedrich 02-01-34
Sge 18h59'22"19d33'
Sauer,Eleonor Theresa
Tau 5h12'52"16d52'
Sauer,Ernie
Aur 5h16'59"54d59'
Sauer,Gudrun Irina 23-05-65
Sge 19h3'0"20d23'
Sauer,Inge Emma Gertrud 27-11-34
Aql 18h44'12"7d19'
Sauer,Ralph D
Aur 5h8'15"41d7'
Sauer,Tristan
Cmi 7h20'1"1d1'
Sauerborn,Walter
Peg 22h53'46"8d22'
Sauermann,Mr & Mrs Frank
Ori 5h51'23"18d17'
Saugera,Jacques
Boo 14h26'50"40d14'
Sauget,Butler
Aur 5h2'45"52d3'
Sauk,Julie Jon
Boo 14h39'12"45d37'
Sauk,Nicole
And 2h22'1"42d35'
Saul,Harry
Ari 1h54'47"15d20'
Saul,Linet
Lac 22h22'50"55d52'
Saul,Scott
Dra 18h7'0"65d28'
Saulnier,Pierre
Ori 5h56'52"17d30'
Saulphyl
Crb 15h50'30"31d54'
Saulsbury,J J
Aql 18h58'24"-6d49'
Saum 3,Hugh Harris
Sex 9h40'44"2d29'
Saun
Aur 5h19'0"50d21'
Saunder,Robert
Sgr 18h52'24"-29d39'
Saunderlin,George Stephen
Lib 15h1'15"-28d17'
Saunders
Dra 12h36'40"71d36'
Saunders 2000,Michael
Aur 7h24'33"37d36'
Saunders III,Carleton Earl
Aql 19h39'42"10d22'
Saunders,Aimee & Jamie
Cyg 20h47'37"37d59'
Saunders,Amber Nicole
Cyg 20h42'54"38d50'
Saunders,Anne
And 23h0'41"51d43'
Saunders,Ashley Anne
Lyr 19h24'1"40d51'
Saunders,Bill & Lajuanna
Ori 5h55'23"15d26'
Saunders,Colin David
Lac 22h25'18"55d44'
Saunders,David Isidore
Her 15h51'18"40d3'
Saunders,Edwin M
Ari 1h47'49"17d28'
Saunders,Joan May
Boo 14h12'48"52d25'
Saunders,Jr,Gordon Davis
Ori 6h6'53"0d59'
Saunders,Jr,Stephen C
Peg 23h3'26"18d24'
Saunders,Kyle Thomas
Cyg 21h50'51"42d42'
Saunders,Mamie L
Lyr 19h14'0"40d57'
Saunders,Marc H
Cma 6h51'56"-17d2'
Saunders,Michael
Dra 16h3'34"64d38'
Saunders,Paula K
Vul 19h39'14"20d35'
Saunders,Rob
Ori 5h50'0"8d2'
Saunders,Ron
Lac 22h20'41"51d31'
Saunders,Rosy Alan
Her 16h39'13"41d40'
Saunders,Stephanie D
Lib 15h39'0"-20d20'
Saunders,Summer
Cyg 20h0'16"40d50'
Saunders,Thomas F
Her 18h3'52"30d4'
Saunders,Vivienne
Peg 21h58'0"24d21'
Saunders,Waheeda Jayne
Aqr 22h28'47"-12d14'
Saunders,Walter
Uma 12h6'23"48d39'
Saunier,Patrick S
Her 18h41'15"12d16'
Saurel,Hannah
Ser 15h52'45"-0d21'
Saurman Star,The Phillip W
Per 2h55'48"43d37'
Sauro,Bill
Ser 15h52'37"-0d36'
Sausen,Nicole & Daniel
Uma 8h13'20"68d40'
Sauvageau,Jean-René
Cas 0h22'57"72d48'
Sauvageot,Frim
Sgr 19h0'35"-20d12'
Sauvaggot,Louis
Sge 18h59'13"19d58'
Sauvé 210673,Erick
Ori 5h56'13"17d38'
Sava,Aaron Jacob
Sct 18h56'36"-6d1'
Savage,Anna (Chrastek)
Lyn 9h16'56"39d30'
Savage,Ben
Ori 6h0'35"8d9'
Savage,Catherine Elizabeth
Cam 5h47'51"68d5'
Savage,Christopher
Ori 5h18'20"1d4'
Savage,Clive Anthony
Aur 7h20'0"38d18'
Savage,Cynthia
And 23h37'0"46d5'
Savage,Doug & Jean
Cyg 20h23'14"39d33'
Savage,Elsie
Cas 0h5'55"62d58'
Savage,Erin
Aql 20h2'47"0d50'
Savage,Henry Salvatore
Aur 4h56'31"41d10'
Savage,James William
Lyn 9h18'40"39d35'
Savage,John Hill
Cnv 13h52'48"32d9'
Savage,Julie Alexandra
Lyr 18h33'59"27d3'
Savage,Lily
Her 16h10'1"47d41'
Savage,Marge
Uma 9h13'1"59d37'
Savage,Marshall Thomas
Lac 22h40'21"53d59'B
Savage,Mary Parmintor
Peg 22h47'39"29d49'
Savage,Michael Vaughan
Cap 20h30'32"-21d4'
Savage,P E
Cas 1h3'47"60d38'
Savage,Polly
Cas 1h26'44"58d57'
Savage,Ronald C
Cyg 20h39'55"52d53'
Savage,Rosiland
Boo 14h34'37"8d33'
Savage,Shannon
Ari 1h55'31"17d54'
Savage,Stefani
Mon 0h32'41"0d2'
Savage,Tami Alene
Lac 22h40'21"53d59'A
Savage,Thomas John
Her 18h6'10"40d15'
Savage,William Harold
Tau 3h54'41"30d57'
Savanna
Ori 5h56'14"7d14'
Savanna Marie
And 23h33'30"44d19'
Savannah
Cyg 20h22'41"40d46'
Savannah
Aql 20h32'16"0d38'
Savannah Elise
Mon 6h23'1"7d33'
Savannah Jane
Cet 3h19'16"3d6'
Savannah Jane
Mon 6h36'32"11d17'
Savannah Rose
And 1h12'25"39d18'
Savant
Sgr 19h28'38"-32d18'
Savard,Patricia "Sloopy"
And 2h26'19"42d56'
Savarese Associates
Uma 10h8'29"62d2'
Savarese,Jr,John W
Psc 1h21'0"30d35'
Savarin,Micheline & Wally Wilcox
Aqr 21h19'42"-8d47'
Savarino,Carleen
Lyn 8h25'45"41d12'
Savary,Isabelle
Uma 11h51'53"51d27'
Savary,Ollivier
Sgr 19h29'45"-36d6'
Savary,Paul
Ori 5h56'1"17d57'
Savas,Leila & Irakeli
Cyg 20h0'0"40d11'
Savastano,Patt
Hya 9h6'46"3d26'B
Savaya,Bonnie E
Cyg 21h23'25"30d0'
Savelesky,Kariann Michelle
And 2h34'32"37d39'
Saver,Margaret
Lyn 8h10'12"51d44'
Saveria
Pup 8h24'37"-24d23'
Saverien
Eri 4h14'58"-8d27'
Saverio,Angela
Pup 8h24'19"-21d23'
Saverwalt,Suzie
Ari 1h46'12"10d50'
Savery,Francine
Lyn 7h24'23"44d14'
Savery,Leif H
Her 18h6'36"38d50'
Savi
Lac 22h17'48"38d39'
Savia,Joanne
Leo 11h6'39"23d28'
Saviano,Doris & Joseph
And 0h45'0"37d14'
Savidge,Carolyn
Mon 8h7'42"-4d8'
Savidge,Kevin J
Cnv 14h0'20"38d55'
Saville,Buford
Boo 14h51'57"51d33'
Savin,Jr,Robert
Her 18h1'56"40d39'
Savine,John J
Dra 16h17'1"67d6'
Savini,Giulia
Aql 18h59'34"-6d50'
Savini,John
Cep 0h16'1"75d22'
Savini,Laura
Aql 18h59'20"-6d41'
Savino Star,The Martha Georgia
Cas 0h57'0"58d56'
Savino,Janet & Tom
Sge 19h34'59"17d5'
Savino,Jim-Always Right-Always Ready
Cep 22h32'24"60d14'
Savino,Joseph P
Vul 20h47'22"28d20'
Savino,Leslie & Barbara
Lyn 7h50'27"48d5'
Savino,Stephen V
Her 18h18'57"12d3'
Savino,Steven Joseph
Vir 13h13'31"-1d56'
Savino,Thomas James
Cnv 13h20'1"41d23'
Savino,Tom
Ori 5h6'23"-1d53'
Savino,Valerie Ruth
Cas 0h26'54"50d27'
Savio,Lisa C
Sco 16h22'17"-38d43'
Savoie (Goofy),Tenaj
Lyr 18h46'52"42d16'
Savoie,Kenneth Dillon
Lyr 18h57'29"45d27'
Savoie,Marie-France
Cep 22h9'1"60d56'
Savoie,Paul
Boo 15h36'40"43d45'B
Savoie,William Joseph
Del 20h37'44"9d25'
Savoy,Guy
Pho 0h41'0"-48d15'
Savoy,Norman
Aql 19h48'17"15d48'
Sawabini,Anna
Cas 1h1'19"58d55'
Saward,Ian Michael
Cam 3h43'44"71d28'
Saward,John
Boo 15h5'42"12d36'
Saward,Keith John
Aur 6h21'45"31d58'
Sawaya,Salah A
Del 20h17'34"10d13'
Sawchuk,Louise Marrone
Peg 21h41'54"21d45'
Sawdka,Maria
Oph 18h24'26"8d35'
Sawicki,Eloise
Cyg 21h36'1"38d18'
Sawicki,John W
Lmi 10h19'44"31d58'
Sawicky,Max
Boo 14h57'12"35d14'
Sawin,Evelyn Jennings
Cas 3h6'18"68d52'
Sawitz,Michael & Vicki "The Stars of AIM"
Cyg 19h24'1"30d16'
Sawtell,Margaret W
And 23h46'23"46d24'
Sawvell,Annika Ortega
Mon 7h52'21"-5d10'
Sawyer 50th Ann,Dot & George
Cyg 21h55'0"53d28'
Sawyer's Staron,Rhoda
Peg 0h1'1"18d16'
Sawyer,Andrew Davis
Cep 23h4'29"70d30'
Sawyer,Bridie
Peg 23h29'43"21d14'
Sawyer,Candace Jeanne
Cyg 21h41'11"28d38'
Sawyer,Doris Pinkerton
Uma 11h41'13"32d19'
Sawyer,Dr Greg
Oph 18h9'18"13d26'
Sawyer,Grant
Aql 19h5'14"15d30'
Sawyer,James Earl
Crt 11h36'54"-8d23'
Sawyer,Joan
And 1h6'54"39d1'
Sawyer,Jr,George
Her 17h6'17"42d1'
Sawyer,Julie Lynn
Cas 2h50'1"61d51'
Sawyer,La Pierre
Uma 10h1'18"53d26'
Sawyer,Nancy
Cas 1h30'42"58d41'
Sawyer,Nelson B
Sct 18h44'25"-5d33'
Sawyer,Paul
Dra 11h10'55"78d5'
Sawyer,Pauline
Tri 1h51'41"25d48'
Sawyer,Stephen Allan
Aur 6h56'40"38d29'
Sawyer,Stephen James
Cnv 13h56'57"37d37'
Sax Star,Alice Hurwit
Eri 2h57'51"-18d26'
Saxby,Amanda
And 1h11'25"46d58'
Saxe,Dorothy & George
Crb 16h9'37"32d48'
Saxe,John Rostlund
Cas 0h5'21"65d32'
Saxe,Mark A
Eri 4h14'0"-18d53'
Saxe,Virginia Laura
Cas 0h10'39"63d34'
Saxelby-Newall,Hannah
Cyg 19h41'30"40d9'
Saxena,Neha Saroj
Mon 6h20'29"4d59'
Saxena,Sameer Kanwar
Hya 9h4'36"4d37'
Saxena,Suchita Radha
Peg 21h52'29"21d1'
Saxman,Curtis W
Ser 15h58'50"-1d53'
Saxon
Cas 1h35'17"60d52'
Saxon,Aloah
And 23h32'1"48d59'
Saxon,James
Per 3h10'38"48d54'
Saxon,Karen
Aur 5h27'56"31d2'
Saxon,Maurice C
Dra 16h1'12"62d14'
Saxon,Spero James
Her 18h4'26"30d60'
Saxton,Andrew D
Ori 5h55'47"21d26'
Saxton,Brent Erik
Ser 15h21'39"-2d18'
Saxton,Farrell
Her 16h11'56"13d2'
Saxton,Jane Clair
Vul 19h48'20"23d32'
Saxton,Spencer
Her 17h24'24"45d34'
Say,Janet
Lyn 9h12'57"41d32'
Saya
Cas 0h15'37"62d56'
Sayag,Annie
Lyn 8h32'40"41d27'
Sayaka
Cap 20h59'26"-25d2'
Sayce,Paul Michael
Uma 10h58'14"71d6'
Sayegh,Famer
Psc 23h35'13"-1d46'
Sayegh,Victor
Per 2h30'1"52d3'
Sayell,Oliver Jordan (Oli)
Cnc 8h36'40"7d25'
Sayers,E J
Cet 2h4'36"0d50'
Sayers,Emily Margaret Kim
Del 20h20'54"10d7'
Sayers,Erin Marie
Eri 4h9'15"-11d60'
Sayers,Richard
Cyg 20h15'36"39d14'
Sayin,Neval
Lyn 8h22'1"40d31'
Saykal,Hasan
Ori 6h2'0"5d8'
Sayles,Anthony
Per 2h51'30"34d7'
Sayles,Corbin
Cep 22h5'39"68d34'
Sayles,Korey Ryan
Her 17h42'21"40d4'
Sayles,Zima
Lyr 18h18'50"42d10'
Saylor Family
Her 16h42'0"34d2'
Saylor,Jan "Mi Dulce"
And 0h59'54"33d43'
Saylor,Jo Ellen
Del 20h55'1"7d3'
Saylor,Samantha
Cas 1h23'41"58d29'
Sayne,Sheila Kaye
Mon 8h2'49"-6d47'
Sayre Elizabeth
Ori 5h55'14"8d27'
Sayre,Jennifer L
Peg 22h55'32"22d50'
Sayre,Jonathan
Cet 2h28'1"8d39'
Sayre,Stephanie Ann
Com 12h54'10"27d18'
Saysell,Colin Geoffrey
Cyg 21h17'2"39d32'A
Sayuko
Cyg 20h37'12"48d48'
Sazinsky,David
Boo 13h38'55"15d15'
Sbarboro,Jim
Her 15h49'43"43d45'
Sbarra,MD,Linda
Oph 17h50'38"12d29'
Sbecuzzi,Robertingno
Lyr 18h38'49"27d9'
Sberna
Aql 19h27'20"11d12'
Sbrana,T
Peg 22h55'40"29d26'
Scacci,Jackie
Uma 9h42'0"44d12'
Scacci,Jacqueline C
Cyg 20h24'0"38d56'
Scafella,Marcus Anthony
Aql 19h29'30"14d15'
Scaggs,"Silk Degrees" Boz
Vul 19h22'58"25d6'
Scaglione,MD,Peter R
Per 1h59'32"53d38'
Scaglione,Mitsie
Lyr 18h28'32"38d21'
Scagnetti,Tony
Dra 17h45'49"60d19'
Scags,Joe
Cam 4h3'1"67d59'
Scaife,Adrian James
Hya 8h40'32"-0d26'
Scala,Albert J
Dra 16h9'42"67d41'
Scalcione,Rose Cupo
Lyr 18h30'38"38d57'
Scale Skipper
Lac 22h53'24"56d26'
Scaler,Chantal
Uma 9h54'19"51d47'
Scalera,Dina & Sammy
Cyg 19h50'39"37d42'
Scales LivingisLoving You,Richard L
Lyr 18h59'0"45d32'
Scales,David John
Cru 12h5'32"-60d49'
Scales,Maggie
Aur 6h32'49"34d49'
Scales,My Dallas Cowboy David
Lyn 7h48'50"48d33'
Scalesciani,J B
Cep 22h48'12"58d49'
Scalfano,Alexander Anthony
Aql 19h0'55"1d4'
Scalfaro,Jr,Joseph Gordon
Cep 21h42'49"55d4'
Scalia,Jack
Cet 3h11'28"5d14'
Scalia,S A
Ori 5h32'55"-8d11'
Scalingi,Chantal
Boo 14h54'14"44d15'
Scalise,Anthony Mark
Per 2h57'28"32d11'
Scalise,Yvonne P
Com 13h17'33"28d7'
Scallywag
Lmi 9h40'32"33d42'
Scalzo,Anthony C
Ori 4h42'0"10d52'
Scalzo,Josephine
Boo 14h21'34"54d19'
Scandalito,Steven Paul
Vir 11h40'20"0d1'
Scandiffio,Michael Gabriel
Per 3h1'11"41d31'
Scandolera,Jean-Pierre
Ori 6h0'0"8d53'
Scandolera,Paul
Ori 6h0'1"1d32'
Scanlan,Darcy
Cam 4h31'26"60d55'
Scanlan,Mike
Ori 5h54'38"-2d30'
Scanlan,Roy Kevin
Aur 5h1'14"48d27'
Scanlan-90,Catherine "Mom"
And 2h21'36"45d6'
Scanlon,Charlie
Aur 6h29'46"31d13'
Scanlon,Cornelius J
Cep 22h12'60"55d47'
Scanlon,Jamie Michele
Her 17h16'1"40d39'
Scanlon,Joan Topar
Lyn 8h18'43"35d19'
Scanlon,MFR Cheryl A
Cyg 21h5'13"33d7'
Scanlon,Noel
Lyn 8h3'43"44d27'
Scanlon,Thomas D
Boo 15h0'44"21d29'
Scannaliato,Lynda Charmaine
Crb 16h9'13"37d9'
Scannell,Caroline Emily
Gem 6h37'34"31d21'
Scannell,Janet Bertrand Sharpe
Ori 5h23'17"-4d8'
Scannell,Shawn
Lyn 7h59'30"45d25'
Scantlebury,Joseph Lester
Lac 22h19'31"49d24'
Scarafile,Christopher Buck
Umi 14h54'57"66d1'
Scaramouche
Cam 5h4'34"68d14'
Scarano,Jessica Verna
Cas 1h19'18"63d13'
Scarano,Theresa H
And 0h18'35"32d53'
Scarantino,Jeanine Ann
And 23h2'0"40d12'
Scarboro,Matthew John
Vir 13h12'9"-5d32'
Scarborough,Heathere Dawn
Mon 6h56'0"10d42'
Scarborough,Patricia & Kismet
Peg 22h19'51"4d12'
Scarbrough,Abb Llewllyn
Mon 6h24'31"6d53'
Scarbrough,Bob
Uma 9h26'0"48d49'
Scarbrough,Ruth Werner
Peg 22h1'15"33d36'
Scarbrough-Liljedahl, Elizabeth
Oph 18h7'54"13d54'
Scarcella,Matthew Charles
Dra 19h10'28"68d35'
Scarcella,Prof Antonio M
Her 16h47'38"41d15'

Scarcliff,Daniel H
Her 17h38'20"14d51'
Scard,Victoria Frances
Peg 22h59'56"11d2'
Scarda,Frances
Cyg 20h36'42"42d5'
Scardino's Shield
Uma 11h31'17"32d49'
Scardino,Katherine Athena
And 23h1'15"50d4'
Scardolillo,Vanna
Boo 15h11'0"38d38'
Scarff,Brian K
Lac 22h14'21"49d15'
Scarff,Matthew David
Oph 17h29'45"7d49'
Scarfo's Oasis In The Sky,Joseph A
Aql 19h2'16"11d8'
Scarisbrick,Harry
Her 17h12'39"41d50'
Scarito,Bob
Uma 11h50'30"41d32'
Scarlata,John P
Sct 18h56'48"-5d15'
Scarlatella,Emily A
Boo 14h37'13"34d35'
Scarlet
Aql 19h46'49"10d5'
Scarlet,Pauline
Com 12h9'17"21d9'
Scarleth Mi Amor Por Siempre
Peg 22h6'59"4d24'
Scarlett
Cas 0h7'28"62d58'
Scarlett's Destiny
Cas 1h11'53"61d16'
Scarlin,Joanna Rachel
Cas 1h25'43"58d56'
Scarola,Stefanie
Cas 23h18'18"60d37'
Scarpa,Carmella
Peg 23h46'15"27d59'
Scarpa,Jr,Salvatore
Cep 23h5'0"64d16'
Scarpati,Karen
Lib 15h12'47"-28d55'
Scarpino,Joshua Riley
Aur 5h3'0"42d56'
Scarpino,William Michael
Aur 5h2'1"40d33'
Scarpuzza,Linda
Lyr 18h38'41"45d45'
Scarsella,Amanda Marie
Crb 16h1'25"33d19'
Scarter-Brown
Aur 5h9'50"40d33'
Scary Gary
Cet 2h38'31"0d57'
Scatena,Pierangelo
Cnc 9h17'0"8d50'
Scatter Lasker
Aur 6h39'37"38d9'
Scattergood,Sarah Jane
Lyn 8h44'0"39d43'
Scavelli,Edd
Cnv 13h41'0"33d54'
Scavo,Joe
Her 16h32'34"40d59'
Scawen,Peter R
Ori 5h55'59"0d46'
Scelba,Andrew Francis
Boo 13h44'32"26d34'
Sceno,Vincent
Boo 15h12'24"25d50'
Scerbo,Victoria
Vul 19h39'1"20d30'
Schaad,Andreas
Tau 5h55'23"24d14'
Schaad,Douglas
Lac 22h12'51"37d42'
Schaaf "Immer Eins", Peter & Renate
Cep 21h14'42"61d48'
Schaaf,Adele
Uma 13h45'38"50d39'
Schaaf,Joe
Uma 13h34'43"50d23'
Schaaf,Patricia & Charley
Cnv 12h51'37"40d56'
Schaal,Gregory L
Cyg 20h36'32"44d52'
Schaal,Rainer
Lac 22h1'24"51d49'
Schaap,David R
Aur 4h48'25"50d18'
Schaap,Helen Marie
Dra 14h52'0"58d36'
Schaardt Kulturbau GmbH
Lyr 19h20'39"31d16'
Schaarschmidt,Thorsten
Equ 20h56'16"8d38'
Schaarvogel,Jürgen
Tau 4h15'12"0d21'
Schaaza 95
Nor 16h22'58"-58d26'
Schabel Star
Cam 4h58'26"61d25'
Schabel,Julie Meyer
Cas 2h20'44"70d29'
Schaben,Mark
Aql 19h41'13"11d28'
Schaberg,Kaitlin Elizabeth
Cas 0h6'44"63d13'
Schabhueser,Werner
Del 21h1'13"12d45'
Schach,Charles A
Cnv 13h49'59"41d1'
Schachinger,Gisela
Cas 1h45'54"73d35'
Schachner,Hans Georg
Aur 5h14'45"43d38'
Schacht,Dale Joan
Cyg 20h3'19"39d39'
Schachte's Star
Ori 4h42'25"10d32'
Schachter,Samuel Evan
Leo 9h36'44"14d28'
Schachtman,Jack
Boo 15h28'0"33d45'
Schachtner,Leo
Hya 9h6'0"5d11'
Schachtsiek,Wilfried
Ori 5h56'14"18d55'
Schack,Leonard
Aur 7h8'48"40d7'
Schackert,Dieter
Hya 8h50'59"-8d7'
Schackman,Alan & Irene
Eri 2h49'33"-2d24'
Schad,Rawl
Aql 20h2'47"1d47'
Schadt,Jack S
Her 17h51'47"40d46'
Schady,Debbie
Cyg 19h26'58"30d50'
Schady,Joseph W
Aur 6h33'49"37d36'
Schady,Marianne
Uma 11h41'0"31d32'
Schaefer Family Star, The
Lmi 10h8'14"32d40'
Schaefer,Bradford Gerald
Ari 2h43'36"21d19'
Schaefer,Daniel Paul
Aql 19h27'54"1d28'
Schaefer,DC
Cam 3h18'48"55d40'
Schaefer,Dennis Aloysius
Ori 5h55'47"11d42'
Schaefer,Dick
Cet 1h56'1"-1d35'
Schaefer,Donna & Nicholas
Sge 19h36'54"16d13'
Schaefer,Dr John Paul Frederick
Oph 17h2'52"0d17'
Schaefer,Earl
Cep 21h46'38"68d14'
Schaefer,Egbert
Peg 23h32'1"20d32'
Schaefer,Eric Alan
Ser 16h13'55"1d23'
Schaefer,Friedrich
Hya 8h29'55"1d3'
Schaefer,Hans-Joachim
Lyn 9h0'33"38d11'
Schaefer,John Henry
Cep 23h0'51"65d34'
Schaefer,Les
Cyg 21h46'0"34d9'
Schaefer,Linda
Oph 17h12'51"-10d45'
Schaefer,Marins Steven Michael
Aql 20h8'1"4d59'
Schaefer,Mark David
Aur 6h59'44"43d57'
Schaefer,Mary Barber
Uma 13h43'55"61d17'
Schaefer,Reed Clinton
Cep 22h18'59"60d25'
Schaefer,Rick
Oph 17h5'53"-5d42'
Schaefer,Rita
Cyg 21h31'36"30d29'
Schaefer,Sr,Claude W
Cet 3h13'12"1d22'
Schaefer,Terron E
Cmi 8h7'25"6d10'
Schaefer,Thomas
Lyn 8h2'19"42d20'
Schaefers,Frederick Nicholas
Ser 18h40'46"4d51'
Schaefers,Udo
Del 21h0'31"12d13'
Schaefers,William Gregory
Sex 10h31'56"0d29'
Schaeffer,Edward
Cyg 21h8'35"48d56'
Schaeffer,Harley Gene
Lmi 9h48'45"34d47'
Schaeffer,Ina Christin
Ari 2h51'24"30d9'
Schaeffer,Joann Grace
Mon 6h54'12"-0d27'
Schaeffer,Kathy
Lyn 9h8'37"42d11'
Schaeffer,Kathy
Per 3h18'43"54d38'
Schaeffer,Mary Jean
Cas 0h1'45"60d42'
Schaeffer,Nancy Tameling
Mon 6h54'0"-10d27'
Schaeffer,Paul Sexton
Aql 19h1'52"-6d12
Schaeffer,Wayne K
Aur 5h8'29"38d43'
Schaeffer,Wayne K
Per 3h33'30"50d57'
Schaer,Peter
Umi 15h17'45"67d49'
Schafer,Catsy & Vernon
Eri 3h3'38"-10d44'
Schafer,David B
Her 16h15'1"26d41'
Schafer,Donna Joy
And 1h9'37"40d27'
Schafer,Mariann
Cas 0h17'21"56d2'
Schafer,Pamela
Vir 11h59'50"-1d53'
Schafer,Richard & Jennifer
Boo 15h2'18"28d15'
Schaff,David
Per 2h42'1"43d15'
Schaff,Kathleen Ann
And 2h21'49"45d23'
Schaff,Tom
Cap 21h45'44"-23d6'
Schaffer
Lyr 19h26'28"38d2'
Schaffer's "Wedding Star-1945"
Uma 12h11'40"56d38'
Schaffer,Abigail Jennifer
Vul 21h0'48"27d45'
Schaffer,Beth
Cas 0h39'21"60d19'
Schaffer,Denise Dondero
And 0h3'54"46d55'
Schaffer,Forever Dad- John
Per 2h52'59"38d43'
Schaffer,Gary Dennis
Leo 9h19'58"33d3'
Schaffer,James S
Aql 19h8'13"0d11'
Schaffer,Jeanne Susan
Aur 5h19'28"46d58'B
Schaffer,Jennifer Lee
Peg 23h35'49"11d45'
Schaffer,John N
Cnv 12h55'39"40d56'
Schaffer,Karl
Uma 12h0'31"35d46'
Schaffer,Lee B
Aqr 23h32'35"-6d39'
Schaffer,Lucky (Edward Alfred)
Hya 9h6'54"5d52'
Schaffer,Mark Alan
Peg 23h46'1"30d29'
Schaffer,Marvin
Aur 6h26'12"37d28'
Schaffer,Nicole Susanne
Aur 5h19'28"46d58'A
Schaffer,Peggy A
And 23h31'35"45d31'
Schaffer,Sigourney Catrice
Lyn 8h29'55"43d7'
Schaffer,Sir Lawrence C
Cep 22h9'16"61d34'
Schaffer,Sonja G E
Peg 21h56'54"32d1'
Schaffer,William James
Aur 4h54'56"38d18'
Schaffrina,Bruno
Hya 9h14'25"-6d53'
Schaffter,Steven S
Lac 22h7'19"38d28'
Schafler,Jonathan
Per 1h39'51"53d40'
Schafler,Michael
Dra 17h7'57"51d41'
Schafstein,Heinz Jürgen
Vir 13h28'28"-14d11'
Schaible,Richard Paul
Per 4h1'47"38d25'
Schaibly,Benjamin
Her 18h8'40"30d17'
Schajer,Abe & Maddy
Del 21h1'0"12d15'
Schall,Linda Ann
Cas 0h32'17"62d0'
Schall,Maurice Eugene
Cam 4h3'32"65d13'
Schaller,Donald Philip
Sct 18h55'25"-7d24'
Schaller,Ruth
Lyn 8h18'18"35d25'
Schaller,Sabine
Sco 17h32'35"-31d40'
Schaller,William & Alice
Peg 22h36'26"35d4'
Schalles,Harold P
Cet 2h44'49"0d30'
Schamalou
Peg 23h37'51"15d46'
Schamberger,Anton
Cyg 20h24'54"39d36"
Schamp,John Joseph
Lac 22h3'52"50d19'
Schaneman,Jill
Lyr 18h54'55"31d25'
Schank,Jeffery
Ori 5h58'26"8d23'
Schank,Paul Napoleon
Cep 1h10'41"78d46'
Schanker,Andrea
Crb 15h53'30"26d14'
Schantz,Stacey Meryl
Mon 6h29'17"10d59'
Schanz,Brian W J
Uma 9h1'49"51d16'
Schaper,Helena Petra
Leo 10h39'42"21d29'
Schaper,Maximilian Warren
Aur 6h22'1"52d35'
Schapiro,Burton
Aur 5h55'42"37d37'
Schapiro,Laurie Elizabeth
Sct 18h42'60"-6d18'
Schapka,Werner
Del 20h50'39"8d40'
Schapmire,Kate
And 1h43'19"37d50'
Schappell,Tricia L
Cma 6h54'53"-19d33'
Schappert,Marion
Cam 5h52'43"60d23'
Schar,Tom
Ser 15h34'51"21d41'
Scharf,Carolyn
And 1h56'41"38d16'
Scharf,Edna
Aql 18h45'57"11d24'
Scharf,Paul Joseph (Who-Man)
Her 16h11'0"48d29'
Scharfenberg,Fam Klaus und Christa
Sgr 19h24'7"-40d37'
Scharfenberg,Mark Anthony
Aql 19h55'42"10d41'
Scharff,Franáois
Oph 17h1'59"10d4'
Scharff,Mark
Vir 11h35'24"6d6'
Scharffenberger, Leonard J
Cyg 21h23'37"28d45'
Scharin,Raymond
Her 17h36'0"27d28'
Scharlin-Pettee,Hannah Abigail
And 23h38'28"48d14'
Scharlin-Pettee,Sophie Loraine
Aql 19h58'0"15d36'
Scharmann,Arthur
Uma 9h34'31"46d1'
Scharmer,Wolfgang & Meike
Uma 10h52'34"60d49'
Scharnberg,Stefan Enrico
Sco 15h46'7"-21d54'
Scharnhorst,Karen L B
Sct 18h40'47"-4d15'
Scharnweber,Kathrin
Per 1h34'17"53d56'
Scharpegge,Hans Joachim
Dra 16h56'53"72d14'
Scharrenberg,Matthew
Cet 2h37'32"3d34'
Scharringhausen,Hans
Cet 1h34'30"-3d19'
Scharschmidt,Sherry
Aql 18h44'17"-2d7'
Schartel,Michele Jean
Ser 15h53'18"19d32'
Schartel,Steven Frank
Ser 15h14'18"20d34'
Scharwey,Rita Justine Josepha
Cas 0h17'11"60d14'
Schatkun,Esther
Crb 15h17'1"31d18'
Schatkun,Marc
Cep 22h32'46"68d29'
Schatz
Boo 14h42'25"47d58'
Schatz,Mark W
Aur 5h59'12"31d42'
Schatz,Victoria Christine
Oph 18h1'32"11d46'
Schatzchen,Mein Sternchen Judy Fettig
Ari 1h59'25"11d17'
Schatzman,Craig "Buck"
Boo 14h21'42"39d10'
Schatzman,David Alexander
Dra 19h26'48"80d13'
Schatzman,Samantha Lynn
Lyr 18h30'25"40d56'
Schau,Mark
Lac 22h7'22"48d23'
Schauber,Jenny Thompson
Ori 4h42'0"0d57'
Schauberger,Sonja
Gem 7h33'11"33d52'
Schauberger,Steve
Aql 19h58'60"10d28'
Schauble,Joanne Margaret
Sct 18h45'46"-7d11'
Schauble,Robert Andrew
Oph 18h30'45"7d7'
Schauer,Herbert
And 23h15'10"42d2'
Schauer,Jerry
Her 17h59'35"20d29'
Schauerte,Achim
Peg 22h16'55"3d22'
Schauman,My Wish Come True,Skip
Sge 20h17'55"17d9'
Schaumann,Bernd Rainer
Ari 1h58'32"21d19'
Schaumberger,Irma
Cam 4h4'56"60d2'
Schaumberger,Trixi
Mon 7h11'46"-6d7'
Schaun,Liesesotte
Vir 14h3'59"7d23'
Schavier,Mary Elizabeth Pratt
Lyr 19h6'34"28d59'
Schavlan,Elizabeth Charlene
And 0h47'47"37d54'
Schawelson's,The
Umi 15h25'0"68d5'
Schawn,Loretta
Mon 6h56'19"11d55'
Schear Star,The Kim
Lyr 19h0'58"37d44'
Schear,Samuel David
Cyg 21h0'56"31d40'
Schearn's Star,June
Cas 1h24'39"58d43'
Schebo,Anne
Cas 0h8'20"56d12'
Schechner,Arthur
Dra 19h15'52"68d24'
Schechter,Cleopatra Susan
Lmi 9h48'1"40d22'
Schechter,Scott Jay
Aur 6h14'53"32d33'
Schechtman,Sheila
And 0h40'49"38d58'
Scheck,Doug
Cet 1h56'37"-4d58'
Scheck,Susanne
Cam 4h5'46"67d51'
Scheckelhoff,Ashlee Lauren
Mon 7h6'0"-1d33'
Schecter,Brandon David
Cnv 12h53'55"51d19"
Schecter,Darren James
Lmi 9h53'16"40d49'
Schecter,Joel D
Ori 5h7'42"-2d10'
Schecter,Jolie Marisa
Lac 22h12'56"54d27'
Schecter,Lloyd Vevrick
Lmi 10h27'14"32d11'
Schecterson,Alexandra M
And 2h29'24"49d7'
Schecterson,Ryan G
Her 16h53'0"32d58'
Schedlbauer,Marc A
Her 16h58'1"38d25'
Schedler,Daniel Richard
Lac 22h0'22"48d58'
Scheef,Sharon L
Peg 22h34'40"29d48'
Scheel,Florence
Uma 11h7'1"48d3'
Scheel,Linzey Ann
And 0h50'11"35d59'
Scheele 18,Shawna K
And 23h40'17"42d59'
Scheele,Alexa Michelle
Cas 1h15'28"60d3'
Scheele,Amy
And 1h12'43"39d35'
Scheele,Chloe Faith
Tau 4h36'59"18d58'
Scheele,Gregory
Boo 15h4'22"12d29'
Scheele,Horst
Ori 5h50'35"20d15'
Scheele,Luke
Her 16h52'29"32d2'
Scheele,Maddison Grace
Cas 1h52'20"60d58'
Scheeler,J B
Lac 22h4'56"49d41'
Scheeler,Jennifer
Cas 1h17'33"61d15'
Scheer Family Star,The Michael W
Lac 22h53'36"56d29'
Scheer,Sun of Nancy A
Cyg 19h22'56"28d37'
Scheerer,Volker
Boo 14h19'55"11d2'
Scheerschmidt,Michael
Boo 14h5'55"51d29'
Scheetz,Jeffery Michael
Ori 5h54'26"21d7'
Scheetz,Vincent
Cnv 13h48'41"30d17'
Schefer,Dorothy
Lac 22h33'33"38d40'
Schefers,Jan
Cep 22h3'16"68d0'
Scheffel,Cheri Ann
Cas 2h33'50"58d15'
Scheffey,Stacy
Cas 1h35'21"61d52'
Scheffler,Collin Joseph
Umi 3h48'1"88d47'
Scheffler,James Chester
Vul 19h18'28"27d46'
Scheffler,Lindsey Victoria
Tau 3h43'48"25d0'
Scheffler/Angel Plane, Larry
Cyg 21h7'57"31d52'
Scheherazade,Avalon
Aur 5h1'45"37d33'
Scheibe,Carsten "Typemania"
Ari 2h21'22"20d27'
Scheibe,Frank
Aqr 23h29'42"-11d54'
Scheibel,Karen
Lyr 18h40'13"26d51'
Scheid,Hannelore
Sgr 19h4'2"-21d30'
Scheid,Jürgen Hans & Jintana Scheid
Cap 21h24'40"-23d25'
Scheidegger,Shirley
Cas 0h33'1"63d19'
Scheideman,Karen Elaine
Lyn 7h55'16"58d33'
Scheidewig,Lindsey-Ann
Lyn 7h27'41"44d15'
Scheidges,Aisling
Uma 8h49'14"57d57'
Scheidges,Andrew
Her 17h36'0"27d13'
Scheidt,Jonny
Gem 6h34'13"26d60'
Scheinbach,Milton & Barbara
Cyg 20h16'45"30d12'
Scheiner,CHI-CHI, Michael
Ori 5h53'0"19d10'
Scheiner,Ivan "Duncan"
Cep 0h10'11"70d2'
Scheiner,Julie
Umi 14h11'26"69d7'
Scheiner,Patricia Ann
And 23h18'0"48d47'
Scheinholz,Nicole
And 0h5'0"47d3'
Scheinold
Cnc 8h33'0"7d31'
Scheip,Kathy
Crb 15h24'33"31d51'
Scheithauer,Glenn Robert
Cet 1h30'58"0d36'
Schekeryk,Peter
Hya 9h51'19"-19d44'
Schekirke,Roman D
Ori 5h58'0"12d4'
Schell,Amanda Voyles
Tri 2'25'41"28d7'
Schell,Georgina Hendry Anderson
Cas 23h0'0"58d10'
Schell,James Townsend
Per 4h44'33"40d8'
Schelnut,J Tim
Cep 21h59'58"67d39'
Schelp,Brian E
Lyr 18h50'46"40d1'
Scheltus,Sonja
And 23h3'18"42d35'
Schemmel,W Thomas
Cep 3h53'46"80d3'
Schempf,Cathy L
Aql 20h8'24"1d8'
Schena,Alison
And 0h18'43"38d45'
Schencker,Laura Annon
Boo 15h4'29"11d28'
Schender,Lisa Marie
Peg 22h43'52"34d20'
Schender,Michelle Lee
Lyr 18h42'44"31d31'
Schenk,Manfred Dipl Ing
Tau 5h29'51"28d56'
Schenk,Matilda Margaret
Cas 0h28'58"63d1'
Schenkel,Ingrid
Cas 2h7'29"63d47'
Schenkel,Kayla Deanne
Vul 20h39'51"23d11'
Schenker Star,The Andrew & Michael
Oph 18h1'29"11d13'
Schenker,Melissa
Vul 21h13'43"20d21'
Schenker,Richard
Lac 22h0'1"51d8'
Schepers,Sarah
Mon 6h21'53"7d53'
Schepi Und Stegi
Gem 8h2'20"33d12'
Scher,Alison Mollie
Cam 7h50'44"60d28'
Scher,Austin Jannuzzi
Per 4h40'46"36d52'
Scher,Gregory A
Cep 0h39'55"78d19'
Scher,Jacqueline Elizabeth
Lmi 10h35'30"28d49'
Scher,Rachel Marie
Peg 23h23'0"31d19'
Scher,Zachary
Per 4h3'12"51d60'
Scher-Star Videographer,Dayle
Lmi 9h45'58"40d48'
Scheraga,Carl A
Leo 10h40'48"15d1'
Scherer Heaven
Sge 19h11'52"16d44'
Scherer,Donna E
Cam 3h55'1"52d57'
Scherer,Harold
Cma 6h56'10"-19d20'
Scherer,John
Aur 5h12'17"40d46'
Scherer,Vera-Everett
Lyr 18h56'58"46d43'
Scherer,Violet
Mon 7h53'9"-3d26'
Scherfer,Ronni
Ori 4h55'34"1d23'
Scherhans-Ryan-Smith
Uma 8h12'52"71d49'
Scheribel,Victoria Anne
And 2h3'47"41d8'
Scherle,Sandra
Dra 17h52'58"65d36'
Scherling,Alice Hersh & Laura
Lyr 18h51'31"40d27'
Scherman,Rankin
Hya 8h48'0"0d33'
Schermer,Alexander M
Per 2h27'46"56d58'
Schermerhorn,Richard E
Lep 5h17'45"-11d18'
Scherr,Charles Adam
Per 2h50'37"38d39'
Scherr,Heidi Lynn
Lyr 18h54'31"42d57'
Scherr,William J
Dra 17h14'28"60d25'
Scherrer Notary of the Year 1995,R L
Ori 4h59'50"15d21'
Scherzinger,Patricia
Lyn 8h4'28"57d51'
Schettino,Susan M
And 1h48'45"36d7'
Schettler's Star
Del 20h13'46"15d52'
Schettler,John A
Eri 4h18'25"-17d2'
Scheuer,Helga
Gem 8h2'48"31d44'
Scheumbauer-Del Ray, Della E
Lyn 7h28'43"50d31'
Scheunemann,Werner
Boo 14h19'11"14d25'
Scheuplein,Edward A
Cam 12h54'0"82d7'
Scheuren,Fritz
Lac 22h12'19"38d52'
Scheurenbrand, Margarete
Ari 2h0'37"26d32'
Scheuring,Dawn
And 2h34'56"37d56'
Scheuring,Mike
Aql 19h53'36"15d21'
Scheurle,Ralf
Sgr 19h14'23"-20d48'
Schey,Sandy
Uma 8h45'36"73d9'
Schiano,Marina
Lmi 10h44'1"26d9'
Schiavi,Lynn D
Cas 23h32'38"62d19'
Schiavino,Jan Celeste
Lmi 10h35'15"33d21'
Schiavo,Anthony Paul
Her 17h13'40"20d30'
Schiavo,Melissa
Psc 1h0'37"20d58'
Schiavo,Paula
Tri 2h42'53"33d58'
Schiavone,Adriana
Lep 4h58'27"-11d27'
Schiavone,Alessandro
Aur 5h1'39"45d21'
Schiavone,Jade
Cam 3h57'55"61d36'
Schiavoni,Luigi J
Aur 5h52'23"54d49'
Schiavoni,Peter M
Her 17h14'53"20d13'

Schibig,Joseph
Cep 20h36'45"65d26'
Schibig,Rose
Boo 14h56'50"20d33'
Schibley,Ginger Elizabeth
And 0h12'17"35d35'
Schicht,Leslie
Sex 10h38'30"2d49'
Schichtel World
Uma 10h38'25"51d11'
Schick (w/l u v),Jim
Aql 19h51'27"13d12'
Schick van Allen, Ingrid
Cap 20h5'30"-10d46'
Schick,Fredrick A
Aur 6h35'0"35d1'
Schick,Kevin Gerard
Aur 5h19'0"54d56'
Schick,Paul Fredrick
Psc 23h8'47"5d2'
Schick,Robert J
Aur 5h17'23"48d42'
Schickling,William
Cep 21h9'56"67d56'
Schieb,Lindsay Anne
And 2h15'0"43d1'
Schiebener,Ulrike
Ari 1h45'47"20d57'
Schieber,Julie Rhodes
Cas 1h17'45"62d46'
Schieferstein,Udo
Sco 17h0'50"-31d9'
Schieffelin,Joy
Lyn 7h47'17"51d36'
Schieffer's Star,Mr
Per 3h18'47"41d46'
Schiefner,Franziska Sophia
Cas 0h15'48"60d51'
Schiegel,Sieglinde
Lib 14h42'0"-23d4'
Schiegg,Wolfgang
Cep 22h5'0"65d1'
Schield,Eileen M
Her 17h42'0"14d25'B
Schield,Wilbur L
Her 17h42'0"14d25'A
Schielke,Renate
Gem 6h25'19"14d26'
Schierbeek,Roxanne
And 22h57'1"38d47'
Schiessl,Fred Andrew
Per 3h21'13"40d57'
Schifano,Patrick
Aur 4h59'41"51d15'
Schiff,Bertie
Aql 20h3'37"1d40'
Schiff,Freddy
Aql 20h2'22"0d40'
Schiff,Max
Aql 20h6'39"1d4'
Schiff,Nelson
Sge 19h58'33"16d21'
Schiff,Steven Douglas
Her 17h24'32"48d29'
Schiff-Glenn,Bonnie Karin
Psc 1h33'12"20d38'
Schiff-Glenn,Emily Joy
Psc 1h25'42"21d52'
Schiffer,Beverly
Cas 23h24'60"51d9'
Schiffer,Fräulein Claudia
Vir 13h18'37"-0d5'
Schiffor,Imgriel
Cas 1h45'10"58d27'
Schiffman,Eric,Simon
Tri 2h37'58"31d41'
Schiffman,Jamie
Aur 7h0'55"38d25'
Schiffman,Zachary
Her 16h38'56"36d52'
Schiffmiller,Rebecca Lynn
And 23h20'57"38d16'
Schifty Schooter
Aur 5h54'41"48d55'
Schihl,Kirsten Ann
Peg 22h31'0"21d35'
Schikowski,Stephan
Lib 15h31'0"-10d3'
Schildcrout,Zachary Ross
Lac 22h4'20"49d6'
Schildwachter, Virginia Cecelia
Lac 22h9'24"51d33'
Schilke,Dieter
Sco 16h27'16"-40d45'
Schilke,P
Uma 10h12'55"52d24'
Schill II,Danny Joe, Kristin Lynn *
Crb 16h18'21"28d56'
Schill,Brigitte
Aql 19h2'27"16d42'
Schill,Jeffrey Robert
Lyn 7h50'51"42d27'
Schiller
Dra 17h19'24"70d55'
Schiller,Barbara & Howard
Lyr 18h56'36"35d11'
Schiller,Lindsey Rachel
Uma 10h23'33"61d27'
Schiller,Lisa
Cmi 7h22'30"5d33'
Schiller,Mary Beth
Cet 2h21'15"-10d56'
Schilling
Ori 5h54'34"16d7'
Schilling,Andreas
Cas 1h6'10"60d57'
Schilling,Angela Marie
Cas 0h36'38"69d39'
Schilling,Beth
Aur 6h2'1"35d7'
Schilling,Eberhard
Lyn 8h29'23"34d26'
Schilling,Frank
Vir 11h37'33"-3d38'
Schilling,Roger J
Cas 0h59'33"69d30'
Schilling,Sabine Rainer Lena Lisa
Cnc 8h27'16"8d45'
Schilling,The
Cyg 20h55'23"38d33'
Schilling,Tracy Renee
And 2h19'45"40d25'
Schilp,Jerry
Lac 22h54'49"54d29'
Schils Family Star,The
Uma 13h2'10"60d29'
Schiltz,Andy
Oph 17h27'27"-20d30'
Schilz,Denise
And 2h6'32"38d24'
Schilz,Helen S
Cyg 20h1'45"40d12'
Schimanski,Ulrich
Gem 8h2'40"31d50'
Schimberg,Jeffrey
Dra 14h17'25"63d35'
Schimek,Martin
Mon 8h4'14"-2d59'
Schimmel,Barry J
Mon 7h57'4"-3d15'
Schimmel,Eileen
Peg 22h2'60"2d19'
Schimmel,Eric
Aur 7h0'10"37d12'
Schimmel,Ian L
Lac 22h7'29"49d36'
Schimmel,Linda
Lyn 8h46'54"34d18'
Schimmel,Robert
Her 17h37'56"27d53'
Schimmelman,Mary Margaret
Cyg 19h51'21"37d58'
Schimpf,Sandy
Vul 19h38'35"20d23'
Schimpf,Sandy
Aur 4h58'22"50d39'
Schimpf,Sandy
Uma 10h25'23"62d3'
Schinasi,Olivier Ibram
Per 3h58'26"31d19'
Schincaglia,Franca Maria
Cas 0h40'16"67d20'
Schindele,Felix
Per 4h6'4"48d14'
Schindele,Peter
Peg 23h31'49"20d26'
Schindler's Starr
Vir 11h59'50"-4d2'
Schindler,Claudine
Aur 6h5'25"31d26'
Schindler,Dr David
Hya 8h58'56"6d32'
Schindler,Kelly
Lmi 10h5'26"38d46'
Schindler,Rose
Del 20h34'10"10d0'
Schinle,Karl
Per 4h5'32"47d7'
Schinzel,Erhard
Her 16h11'37"11d7'
Schioppo Star,The
Aur 6h22'27"31d59'
Schipani,Charles Benjamin
Cep 23h8'36"70d50'
Schiphorst,Bernd
Aur 6h30'0"40d26'
Schipilliti,Mark Adam
Boo 14h22'59"12d6'
Schipul-Always A Star, Cameron
Cet 2h13'34"7d30'
Schirmer,Samuel Louis
Vul 19h48'11"28d31'
Schiro,Thomas
Aur 7h5'57"40d42'
Schirra,Anja
Cep 22h2'32"61d46'
Schirru,Daniela
Aur 7h6'1"40d27'
Schisler,Danielle
Cas 23h40'1"58d13'
Schivelbein,Leo
Oph 18h42'16"7d39'
Schiveley,Michael Ross
Her 17h55'42"28d21'
Schivley,William W
Dra 17h47'1"75d3'
Schizophrenia,MJB
Cam 6h18'21"68d37'
Schlager,Katelyn Nichole
Peg 22h1'46"10d0'
Schlageter,Joseph Frances
Boo 14h42'60"37d38'
Schlageter,Jr,"Bobby"
Mon 6h58'1"-10d25'
Schlain's"1st Birthday Star",Gabby Star
Peg 22h1'1"31d42'
Schlais Star,The Lisa
Eri 3h41'10"-1d18'
Schlak
Cap 20h37'30"-14d30'
Schlam-Mimi's Star, Judith Ann
Cas 0h27'0"68d17'
Schlanger,Ruth
Peg 23h34'54"22d43'
Schlansy,Leonard & Adeline
Cyg 20h57'47"31d30'
Schlauer,Dipl Ing Dr Techn Rudolf
Per 2h53'55"50d23'
Schlechter-God's Surgeon,Dr Robert
Cet 2h3'42"-2d13'
Schlegel,Debbie
Lyn 8h38'0"40d57'
Schlegel,Glen Alden
Aur 6h29'58"30d16'
Schlegel,Volker
Mon 7h35'41"-3d58'
Schlehr,Matthew Paul
Sct 18h53'1"-6d21'
Schlehuber,Big Jim
Cnc 9h0'12"8d9'
Schleich,Werner
Ser 15h59'40"-1d48'
Schleif,Morgan Mekinzie
Cas 0h1'30"65d5'
Schleifer III, Theodore J
Her 17h11'21"48d23'
Schleigho
Lyn 7h48'44"35d15'
Schleitweiler,Clarence
Sct 18h43'13"-7d12'
Schlendorf,Peter C
Lac 22h30'35"55d4'
Schlenker,Andrea
Boo 15h37'0"41d4'
Schlervantes
Cam 6h53'1"71d11'
Schlesinger,Melanie
Mon 6h55'49"-10d19'
Schlesselmann
Cnc 8h9'1"30d12'
Schleyer,Iris
Cnv 12h40'53"33d10'
Schleyhahn,Bob
Aur 6h19'25"30d29'
Schlichter,Mark
Lac 22h49'45"54d47'
Schlichting,Georg
Boo 14h53'31"50d40'
Schlichting,Marianne & Jürgen
And 1h41'23"41d12'
Schlichting,Udo
Del 20h53'53"7d30'
Schlichtkrull,Kurt
Cap 20h45'34"-26d56'
Schlieman
Dra 16h53'19"62d2'
Schlienz,Shawn Douglas
Per 1h30'1"53d41'
Schlieper,Leni
Dra 17h56'57"52d13'B
Schlieper,Scott
Dra 17h56'57"52d13'A
Schlieper,Tim
Sex 9h51'1"-6d30'
Schlimbach,Hans 04-04-1922
Lyr 19h17'39"31d42'
Schlimbach,Mia 09-04-1923
Lyr 19h17'28"31d35'
Schlinke,Stephen
Lmi 10h45'56"24d53'
Schlitzer,Carmen
Sco 17h51'27"-31d9'
Schloberg,Barbi und Karsten
Gem 6h30'15"13d35'
Schlobohm,Kira Elizabeth Gui
Eql 21h9'49"11d6'
Schloesser,Suzanne Rhea Cohn
Cas 1h32'0"68d52'
Schlogl's Star,Prof Karl
Sgr 20h4'47"-31d19'
Schlomer,Melvin B
Lac 22h17'38"47d11'
Schlomer,Michael
Her 16h24'23"33d56'
Schlomski,Herbert
Boo 13h45'11"10d38'
Schlorff,Katja
Sgr 19h56'42"-41d43'
Schloss,Tyler Lee
Cnv 13h47'1"33d44'
Schlosser,Amber Lee
Peg 22h2'11"8d9'
Schlosser,Courtney Lynn
Cas 0h23'48"50d29'
Schlosser,Lucas Wayne
Del 20h16'1"10d22'
Schlothauer,Shirley Norton
Cas 0h42'48"65d6'
Schlotter,Theresia- Konrad-Anne-Gorh
Aqr 21h2'58"-14d36'
Schlough,Colleen
Ori 5h2'0"0d29'
Schlough,Samantha
Cyg 20h20'29"41d21'
Schluck,Thierry
Cep 22h24'51"57d30'
Schlueter,Helen & Arnold
Lmi 10h6'0"32d49'
Schlumi
Aqr 22h25'23"0d44'
Schlund,Thomas & Kerry
Cyg 19h11'45"48d10'
Schlupp Golden Anniversary Star
Sge 20h1'0"17d14'
Schlusselberg,Steven B
Cep 23h2'1"65d3'
Schluter,Marouska Natasha
Cas 1h48'12"68d55'
Schmader,Margaret Ann
Cas 23h21'23"53d40'
Schmahl,Juergen
Lac 22h1'46"51d49'
Schmalenberger
Dra 16h12'28"57d31'
Schmalstieg,MD,Walter
Oph 18h39'36"6d55'
Schmalz,Barbara
And 0h11'21"47d18'
Schmalz,Darin Blaine
Leo 9h59'17"19d41'
Schmalzried,Jodi
Uma 9h52'35"54d4'
Schmauss Star,Pete & Nancy,The
Sge 19h2'15"19d58'
Schmeckl,Herman
Equ 21h22'20"11d6'
Schmehil,Elizabeth Claire
Ori 5h39'25"12d31'
Schmeider,Linda
Peg 21h50'36"31d7'
Schmeil,Todd Carl
Uma 8h59'45"48d59'
Schmeing,Heinrich
Aqr 22h22'44"2d24'
Schmelig,Marnie
Ori 5h33'16"-8d6'
Schmeling,Frank H
Aql 19h27'32"0d25'
Schmelz,Anabel Rose
And 1h41'31"41d7'
Schmelzel,Jeffrey & Lisa Rich
Cyg 19h17'32"44d50'
Schmelzer,Edwina B
And 2h1'1"40d53'
Schmelzer,Jeffrey Scott
Cet 1h46'51"-0d49'
Schmer,Arthur
Lac 22h28'40"54d17'
Schmertz,Kennedy Brown
Uma 13h33'0"53d6'
Schmey,Torr
Boo 15h11'1"50d38'
Schmhil,Kristine-Kari- Robert III
Lmi 10h35'31"30d12'
Schmicker Fifty (50), Michael
Lac 22h38'14"41d11'
Schmickler,Gaby
Cas 0h41'51"60d39'
Schmid,Chuck & Judy
Ori 5h50'0"21d2'
Schmid,Daniela
Peg 23h32'37"11d58'
Schmid,David
Boo 13h35'0"18d35'
Schmid,Doris
Leo 9h58'54"19d22'
Schmid,Dr Heike-Dorit
Hya 8h54'38"-0d9'
Schmid,Eberhard
Aur 5h12'35"43d38'
Schmid,George J
Lac 22h29'57"55d2'
Schmid,Herbert
Leo 10h54'24"8d4'
Schmid,Martin
Vir 13h30'51"-2d23'
Schmid,Norbert
Umi 16h27'0"79d52'
Schmid,Oskar
Cyg 20h25'17"31d45'
Schmid,Patricia Jean
Cas 3h34'16"73d30'
Schmid,Siegfried Johannes
Ari 1h58'16"21d5'
Schmidgall,Alan
Oph 17h38'10"-24d32'
Schmidlin,Joseph Lucas
Dra 16h54'50"68d20'
Schmidt
Peg 23h29'13"11d40'
Schmidt (Cid),Cynthia
And 23h33'14"49d52'
Schmidt 07 März 1934, Lothar Hans
Ser 16h4'0"2d44'
Schmidt II,Peter A D
Aql 20h6'47"0d60'
Schmidt IV,Joseph John
Cam 3h48'42"67d30'
Schmidt's Piece of the Sky,George
Aur 6h26'16"30d43'
Schmidt's Star, Christine L
Peg 22h33'49"25d9'
Schmidt,Aaron
Aql 20h12'38"10d14'
Schmidt,Alfred Walter
Cep 3h9'15"77d24'
Schmidt,Allen Andrew
Aur 5h4'46"42d9'
Schmidt,Angela Leigh
Com 13h28'18"26d6'
Schmidt,Anne Frances
Uma 8h15'1"60d34'
Schmidt,Arlene & Leonard
Crb 15h27'40"31d60'
Schmidt,Brendan Riley
Eri 3h54'48"-2d52'
Schmidt,Brittany Gayle
Com 13h3'43"28d7'
Schmidt,Carsten
Boo 14h7'28"14d6'
Schmidt,Charles R
Dra 16h42'35"73d39'
Schmidt,Chris,Willi, Willi-D,Henry-C
Dra 14h11'35"63d32'
Schmidt,Daniel James
Boo 15h14'0"30d43'
Schmidt,Daniel Patrick
Hya 9h34'48"5d56'
Schmidt,Donald Gray
Cep 2h43'0"77d21'
Schmidt,Dorothy Virginia
Tau 4h11'43"20d13'
Schmidt,Edith
And 0h12'60"39d10'
Schmidt,Florence & Arthur
Hya 8h28'1"-8d13'
Schmidt,Gisela
Cap 21h14'50"-26d39'
Schmidt,Hans
Aqr 22h43'57"-1d39'
Schmidt,Horst
Sgr 19h3'26"-25d53'
Schmidt,Ilse-Marie & Walter
Leo 10h23'42"10d23'
Schmidt,Irvine Reed & Emily (Tiny) Lewellen
Cas 1h36'35"58d22'
Schmidt,James A
Aur 6h16'0"38d49'
Schmidt,Jennifer
Cyg 21h32'58"41d30'
Schmidt,Jill Ann
Cas 1h27'0"60d49'
Schmidt,John
Cep 21h41'58"65d22'
Schmidt,Jorg
Boo 14h5'52"50d59'
Schmidt,Jr,Arthur Trexler
Cet 0h58'1"-1d45'
Schmidt,Jr,Donald Richard
Cap 21h21'11"-23d7'
Schmidt,Jr,Edwin
Dra 16h59'19"69d49'
Schmidt,Jr,Gregory Phillip
Gem 7h23'9"20d22'
Schmidt,Kai
Ori 5h54'28"1d31'
Schmidt,Karen
Uma 11h45'40"53d31'
Schmidt,Kathy Farrell
Cyg 19h18'0"28d33'
Schmidt,Kathy Lynn
Cas 0h30'1"75d54'
Schmidt,Kaylee MaChelle
Lyn 7h37'55"41d16'
Schmidt,Kevin Christopher
Oph 18h2'41"8d12'
Schmidt,Kurt M
Mon 6h21'12"5d15'
Schmidt,Kyle A
Cet 2h19'59"5d36'
Schmidt,Larry
Oph 17h1'30"1d33'
Schmidt,Laurie
Boo 14h9'54"41d49'
Schmidt,Loriann
Gem 7h37'22"20d20'
Schmidt,Lovelight- Breath Of Life-Iris
Vul 21h1'56"22d37'
Schmidt,Manfred
Per 4h5'49"46d13'
Schmidt,Maria
Her 17h51'33"42d46'
Schmidt,Marion
Cnc 8h29'1"7d44'
Schmidt,Marion- Angelika
Cep 22h6'12"60d5'
Schmidt,Martin
Aur 6h22'43"30d43'
Schmidt,Megan Elizabeth
Crb 15h31'1"28d55'
Schmidt,Melissa Ann
Lyn 7h25'1"58d11'
Schmidt,Michelle
Del 20h13'32"15d21'
Schmidt,Norbert
Lyr 19h12'11"31d32'
Schmidt,Oma Maria und Opa Dieter
Uma 9h37'27"46d43'
Schmidt,Peter
Sge 19h7'39"18d51'
Schmidt,Peter Andrew
Ori 6h2'28"1d32'
Schmidt,Peter Anthony
Hya 8h42'37"-6d44'
Schmidt,Renate 05-03-1934
Lyr 19h17'21"31d11'
Schmidt,Richard Friedrich
Psc 1h24'16"22d46'
Schmidt,Richard A
Peg 22h38'0"7d34'
Schmidt,Robin Gayle
Cas 2h37'37"57d51'
Schmidt,Sara R
Cas 1h16'56"63d8'
Schmidt,Shane Edward
Dra 14h52'22"61d41'
Schmidt,Stefan
Cep 22h0'19"63d56'
Schmidt,Stephen
Her 17h50'57"38d45'
Schmidt,Susann
Tau 5h39'59"28d10'
Schmidt,Tammie Jean
Aql 19h29'44"10d36'
Schmidt,Thomas Craig
Gem 7h12'20"21d27'
Schmidt,Tina Marie
And 23h2'41"50d16'
Schmidt,Traude
Lyr 18h18'1"47d6'
Schmidt,USAF (Ret), Major J J
Aql 18h57'57"-6d56'
Schmidt,Victoria Ann
And 2h22'44"40d31'
Schmidt,Wilma Jean
Com 13h17'50"27d40'
Schmidt-Ackermann, Hartmut
Cep 22h5'46"65d19'
Schmidt-Ackermann, Miriam
Lyn 8h12'1"40d20'
Schmidt-Germeyer, Werner
Ari 2h55'35"21d38'
Schmidt-Hartmann,Inca + Achim
Gem 7h31'56"20d20'
Schmidt-Holtz,Rolf
Cep 21h24'11"60d52'
Schmidt-Purrmann, Horst
Aqr 21h55'24"0d21'
Schmidtke,Nonni
Ari 2h51'22"21d39'
Schmidy
Per 1h52'38"56d42'
Schmiedeknecht,Pam
Cyg 20h57'13"53d52'
Schmieder,Elizabeth Georgi
Vir 12h57'47"-7d47'
Schmieder,Herbert Albert
Aql 19h4'28"3d55'
Schmieder,Karl-Heinz
Oph 17h31'30"8d2'
Schmiegel,Sieglinde Edeltraud
And 0h8'46"38d39'
Schmieler's Star,A C
Ori 6h2'38"20d12'
Schmierer,Lynda
Mon 7h53'58"-0d22'
Schminke II,David T
Lmi 10h51'35"38d30'
Schmit,Dr Herman H
Aur 6h17'43"32d34'
Schmit,Rachel
Cas 3h10'56"57d48'
Schmitt,Alan B
Dra 15h9'51"57d31'
Schmitt,Alan Ray
Cet 2h22'40"-5d40'
Schmitt,Alice
Uma 9h33'31"50d9'
Schmitt,Catherine Emi
Peg 22h40'57"21d11'
Schmitt,Dieter
Lyr 19h17'1"30d53'
Schmitt,Harry
Tri 1h51'19"27d8'
Schmitt,Jeffrey Alan
Aur 6h42'36"38d22'
Schmitt,Joanna Lynn
Cas 0h29'30"08d55'
Schmitt,Kathleen
Cyg 21h31'57"41d6'
Schmitt,Mike
Her 16h50'1"50d34'
Schmitt,Randall K
Oph 17h3'20"-1d32'
Schmitt,Susanne
Vul 19h18'13"26d19'
Schmitt,Thomas Murphy
Oph 18h36'13"10d25'
Schmitt,Ulrich und Renate
Gem 6h53'16"17d24'
Schmitt,Victor
Per 2h36'43"56d42'
Schmitt-Caruana
Lyn 8h54'1"41d46'
Schmittel,Ralph Markus
Sgr 19h59'58"44d44'
Schmitter,Alain
Peg 23h45'17"30d55'
Schmitz,Bernhard Bruno
Sco 17h5'50"-31d21'
Schmitz,Catherine
Eri 3h31'28"-5d25'
Schmitz,John Charles
Lac 22h30'53"55d2'
Schmitz,Laura Ann
Uma 9h36'1"55d49'
Schmitz,Margery M
Cam 12h45'21"76d47'
Schmitz,Michael
Ari 2h30'41"28d42'
Schmitz,Norma & Glen
Oph 16h56'22"11d6'
Schmitz,Peter Josef Günter
Her 17h21'29"46d55'
Schmitz,Richard James
Her 18h1'56"14d39'
Schmitz,Robert J
Lac 22h18'53"47d41'
Schmitz,Stefan A
Cep 21h59'11"60d37'
Schmitz,Ute
Boo 14h21'29"11d14'
Schmitz-Knierim, Rosemarie
Sgr 19h18'46"-23d6'
Schmoller,Markus von
Leo 9h20'35"18d53'
Schmoo
Lyn 7h6'59"56d3'
Schmoogi
Aql 20h1'31"9d55'
Schmoop
And 0h53'50"39d55'
Schmoopie's Star
Lmi 10h2'12"36d48'
Schmoopy
Uma 10h4'34"70d25'
Schmoozen
Cet 2h17'37"9d58'
Schmoozer
Umi 15h2'49"72d29'
Schmuch,Wilma Bridgewater
Eri 2h47'27"-2d41'
Schmuck,Herbert
Hya 8h41'37"-8d9'
Schmucker
Equ 21h20'20"10d43"
Schmucker,Harvey D
Lyr 18h46'53"38d48'
Schmuckermaier,Robert
Her 17h19'31"42d10'
Schmuda,Sandi
Cmi 7h56'25"8d6'
Schmuelling,Rob
Her 16h43'0"21d26'
Schmöe,Jürgen
Uma 9h27'0"46d37'
Schmöger,Julian
Cap 21h39'58"-23d21'
Schnabel,Adam M
Lac 22h13'51"47d13'
Schnabel,Bettina
Tau 5h47'0"20d51'
Schnabel,Claudia
Vir 11h54'17"-6d56'
Schnabel,Cornelius
Gem 6h37'17"34d37'
Schnabel,Gerhard
Sgr 19h38'32"-41d37'
Schnabel,Hans-Jürgen
Umi 13h8'12"75d48'
Schnabel,John
Aur 6h25'22"38d38'
Schnackilein,mein Stern der Sterne
Psc 1h18'0"18d21'
Schnaidt,Jack E
Her 17h9'54"46d9'

Schnal,Jamie Ann
Vul 20h38'1"20d15'
Schnapka,Herbert
Cnc 8h6'55"31d11'
Schnapp,Monica Alexis
Eri 3h38'27"-6d2'
Schnarr Star,The Sue
Cas 0h3'20"63d13'
Schnatz,Commodore
Cnc 8h10'1"30d37'
Schneck
Cap 21h4'34"-20d41'
Schneck,Peggy
Aur 6h3'27"31d41'
Schneckchen
Cnv 12h41'0"39d34'
Schneder,Savannah Taylor
And 23h16'41"40d42'
Schnee,Raynald
Tau 5h53'53"26d58'
Schneebeck,Paul Otto
Aur 5h1'18"41d14'
Schneeberger II, Karoline
Peg 22h8'46"26d59'
Schneeflocke
Hya 8h29'38"-17d18'
Schneegurt,Dr Mark
Aur 5h57'0"29d33'
Schneemilch,Frank
Cnv 12h35'42"35d43'
Schneider "Angelle", Angela Beth
Ori 5h50'12"18d56'A
Schneider #2064,Mary Beth
Cma 6h54'22"-19d22'
Schneider 1/6/29-4/23/95,Anna Platt
Lyr 18h48'53"36d14'
Schneider Ma,Burgi
And 23h21'24"51d15'
Schneider,Adalbert
Aql 18h57'29"17d42'
Schneider,Alexander Charles
Aur 7h4'1"37d3'
Schneider,Allison Meg
Cas 23h37'25"63d19'
Schneider,Andrej
Sge 20h16'42"17d16'
Schneider,Astra Familiae
Her 18h1'1"31d32'
Schneider,Aurelie
Lyr 18h19'51"39d39'
Schneider,Barbara
Lyr 18h58'30"30d57'
Schneider,Barry Stevin
Boo 15h5'33"26d21'
Schneider,Bonita Miller
Ari 2h43'12"21d34'
Schneider,Britta
Cnc 8h9'1"30d6'
Schneider,Brook
Tri 1h58'37"30d24'
Schneider,Caroline B
Lyn 8h3'18"40d49'
Schneider,Charanna
Del 20h19'18"16d35'
Schneider,Charlotte
Lyr 18h24'0"44d34'
Schneider,Christy Renee
Mon 6h21'54"8d54'
Schneider,Dieter
Boo 14h6'59"11d26'
Schneider,Dirk
Dra 16h2'35"63d48'
Schneider,Dr Med Wolfgang Axel
Sco 17h29'50"-30d37'
Schneider,Faith Anne
Lyn 7h29'23"45d54'
Schneider,Franz Josef
Aqr 20h57'1"-1d12'
Schneider,Gabi
Peg 23h32'1"18d56'
Schneider,George
Per 2h55'21"43d17'
Schneider,Heinz
Dra 18h18'49"68d29'
Schneider,Heinz-Dieter
Cas 0h31'56"63d21'
Schneider,Helmhold
Cep 22h10'1"63d46'
Schneider,Hey Pretty Heidi
Eri 4h40'43"-1d1'
Schneider,Howard A
And 23h1'0"41d1'
Schneider,Ileene
Cyg 20h38'18"41d8'
Schneider,Irmgard & Rudolf
Cep 22h7'41"68d48'
Schneider,James H
Cep 22h11'1"58d32'
Schneider,James Robert
Boo 13h41'11"26d1'
Schneider,Jami B
Cyg 19h14'56"44d1'
Schneider,Jamie Ann
Eri 4h59'12"-8d12'
Schneider,Jeffrey
Her 17h11'60"43d35'
Schneider,Jeruel
Uma 10h55'28"68d40'
Schneider,Jessica Pauly
Lyn 7h37'42"50d10'
Schneider,Jessica & Eric
Mon 6h34'52"8d55'
Schneider,Jürgen
Cep 22h1'47"67d58'
Schneider,Jürgen
Her 17h42'53"42d33'
Schneider,Kathryn Moore
Cam 3h27'44"61d26'
Schneider,Kathryn
Vul 19h4'34"25d12'
Schneider,Leon
Cnc 8h1'0"8d58'
Schneider,Lloyd"Snitz"
Ori 5h39'14"-0d51'
Schneider,Lynette Lorraine
Ari 2h3'49"22d14'
Schneider,Marietta & Hans
Sge 19h55'57"16d34'
Schneider,Martin
Cyg 20h26'35"30d9'
Schneider,Maryanne O'Toole
Cyg 20h1'21"50d37'
Schneider,MD,Dr Peter A
Oph 16h55'0"-25d48'
Schneider,Michaela "Schlafmütze"
Cet 2h40'18"5d19'
Schneider,Mondo
Vul 19h46'40"28d25'
Schneider,Morton N
Cam 9h12'57"78d54'
Schneider,Orietta
And 0h24'0"44d53'
Schneider,Patricia
Cas 0h24'1"65d34'
Schneider,Ryan Stephen
Per 4h33'44"34d51'
Schneider,Sabine
Peg 22h3'1"18d52'
Schneider,Sandy
Cet 1h46'30"-0d19'
Schneider,Saul
Lyr 18h24'45"44d45'
Schneider,Schorling
Uma 11h9'42"46d29'
Schneider,Stefan
Cas 0h31'52"63d9'
Schneider,Stephen William David
Boo 14h5'42"10d48'
Schneider,Susan
Cas 1h25'35"65d29'
Schneider,Terena
Her 18h2'39"28d54'
Schneider,Tony
Aur 7h24'42"38d41'
Schneider,Ursula
Peg 23h32'10"17d27'
Schneider,Ute-R
Sgr 19h54'2"-42d10'
Schneider,Van Robert
Ser 17h21'29"-13d14'
Schneider,Veronica Anne
And 2h0'11"41d31'
Schneider,William Charles
Per 4h0'35"51d18'
Schneider,Zachary Allen
Vul 19h20'0"27d21'
Schneiderman,Bernard Joseph
Ori 5h53'1"13d50'
Schneiderman,David Crawford
Hya 8h54'46"5d46'
Schneiderman,Jodi & Jeff
Aql 18h51'10"11d32'
Schneiderman,Mark U
Mon 8h4'33"-8d42'
Schneitzer,David Robert
Lac 22h9'57"40d57'
Schnell,Claudette Y
And 2h19'24"47d31'
Schnell,Glen
Lac 22h13'51"50d20'
Schnell,John Arthur
Dra 12h23'39"70d35'
Schnell,Lee,Marji,Eric Luke & Lora
Her 16h29'18"40d23'
Schnell,Matthew George
Her 17h52'33"14d14'
Schnell,Sharon
Mon 6h24'36"1d2'
Schnell,Werner
And 1h23'34"42d3'
Schneller,Dieter
Eri 4h49'15"-5d26'
Schnellmann,Eric M
Ser 15h55'19"19d52'
Schnepf,Deryk
Lyn 7h27'41"39d54'
Schnepp,Ingrid
Vir 13h35'31"-8d36'
Schnepp,Kevin M
Her 16h29'57"33d18'
Schnepp,Kyle D
Lac 22h7'10"47d1'
Schneyer,Kathy Stafford
Ori 5h53'13"14d8'
Schnider,Nevada Julia
Cyg 20h5'0"40d23'
Schnieder,Lisa Ann
And 0h48'14"37d52'
Schnier,Ana Maria De Almeida
Mon 7h3'38"-6d54'
Schnitger,Victoria Louise
Lyr 18h59'27"29d45'
Schnitker,Gerda M
Sgr 18h54'6"-23d56'
Schnitker,Wilhelm
Peg 23h34'57"12d41'
Schnitkey,Junior Walter
Per 2h52'22"34d36'
Schnittkowski,Dietmar
Ori 5h51'39"21d36'
Schnitzer,Ashlee Brooke
Aur 5h9'56"41d6'
Schnitzer,Hilde
Sgr 19h35'34"41d14'
Schnitzer,Jordan
Dra 17h3'45"64d14'
Schnitzer,Zachary Adam
Aur 5h2'40"41d59'
Schnitzler
Lyn 7h53'1"45d3'
Schnobrich,Leon "Bumps"
Umi 16h33'52"75d41'
Schnookie's Star
Uma 8h56'49"58d24'
Schnoor,Jürgen
Mon 7h35'40"-2d39'
Schnubbel Helga Kentenich
Lib 14h40'56"-0d17'
Schnuck-Schatz
Ori 5h58'0"19d54'
Schnuckiputzis "Asc" Jork
Her 17h56'56"42d52'
Schnuerl,Karl
Sco 17h27'46"-30d27'
Schnupp,Corinna
Lib 14h44'1"-20d33'
Schnur,Anthony J
Aur 5h9'0"42d32'
Schnur,Austin Laura
And 23h18'43"41d13'
Schnur,Bettina
And 23h19'31"38d38'
Schnur,Robert
Aur 4h38'19"34d50'
Schnuriger,Heidi
Lyn 8h28'20"33d38'
Schnurmacher,Sylvia
And 1h3'0"37d46'
Schnurr,Vicki
Lyr 18h19'48"38d15'
Schnurstein"All Star", James Walter
Her 18h4'15"50d17'
Schnyder-Gauch,Chrigi und Werni
Lib 15h28'17"-22d3'
Schnyer,Dick & Margo
Uma 8h38'19"47d41'
Schobert,Reinhard
Lyr 19h23'57"30d15'
Schoch,Sandra Lee
Tau 4h9'55"22d53'
Schock,Corthay Moreau
Sct 18h56'37"-6d32'
Schock,Lunsford Daniel
Lyr 18h42'38"41d25'
Schock,Lynda
Aql 20h7'57"0d16'
Schodowski,Stacy
And 23h25'49"44d18'
Schodowski,Tricia
Peg 23h49'1"17d54'
Schodtler,Penny
Cas 23h31'52"53d35'
Schoeche,Ursula
Sco 17h29'43"-40d57'
Schoedel,Joseph Paul
Oph 17h38'59"11d44'
Schoedl,Jr,Heinrich A
Boo 15h5'15"38d10'
Schoeffler,Andrea Lynn
Uma 14h20'40"58d37'
Schoeffler,Ronald Evan
Her 16h58'20"30d44'
Schoeffler,Steven Todd
Per 3h46'35"35d33'
Schoen,Donna Lynn
Com 12h0'21"22d13'
Schoen,Douglas R
Cep 20h48'19"63d58'
Schoen,Mary Lou
Com 13h2'34"26d30'
Schoen-Detering,Birgit
Peg 22h58'0"33d53'
Schoenbaum,Mike
Aur 6h25'0"33d3'
Schoenbeck,Traves W
Ori 5h1'42"-1d34'
Schoenberg,Brian
Hya 8h23'21"-0d18'
Schoenberger,Ke-De
Uma 11h23'1"58d50'
Schoendorff,Jean Michel
Cep 22h16'1"55d51'
Schoeneman,Marcelle
Cyg 20h23'1"30d5'
Schoeneman,Norbert
Cep 22h29'23"67d49'
Schoenenberger,James Guy
Mon 7h59'0"-2d6'
Schoener,Nicholas John
Aur 6h35'50"32d40'
Schoenfeld,Sherry
Cas 1h43'32"60d59'
Schoenherr,Richard A
Her 17h30'37"21d59'
Schoenhut,John William
Tri 2h36'1"35d21'
Schoeni,Dorothy
Lyn 9h7'30"37d20'
Schoenknecht,Carol
And 0h45'28"39d36'
Schoenleber,Wayne C
Lyr 18h51'41"38d45'
Schoenman,Betsy Jill
Sex 10h11'51"-6d24'
Schoentag,MD,Diane J
Aur 5h56'59"30d52'
Schoepe,Jürgen
Peg 23h31'56"16d9'
Schoepf,Anna
Lyn 7h14'55"47d13'A
Schoepf,Richard
Lyn 7h14'55"47d13'B
Schoessling Star,The Ray & Jo
Eri 3h25'12"-12d46'
Schoessling,Jaimee
Ori 5h40'29"-1d24'
Schoff,Graham Kahler
Cam 3h24'13"61d15'
Schofield's Reach For A Star,Mike
Ori 4h43'53"1d49'
Schofield,Brenton V
Cep 22h16'55"70d46'
Schofield,Charlotte Alice
Umi 16h26'36"75d29'
Schofield,Daniel Joseph
Her 16h27'57"34d7'
Schofield,David
Uma 10h37'50"50d37'
Schofield,Jimmy V
Lac 22h27'45"41d7'
Schofield,Lauren
Mon 8h0'30"-1d17'
Schofield,Michael Alan
Sex 10h43'36"1d52'
Schofield,Paul
Cep 22h25'48"60d46'
Schofield,Ronald
Her 17h27'1"28d20'
Schoggen,Martha
Eri 2h54'25"-3d24'
Schoh,Edgar
Gem 7h9'7"21d10'
Scholars Always Sandra & Ryan
Cyg 19h45'35"31d10'
Scholastica Chittick Elementaricus
Lyn 7h48'32"47d49'
Scholer,Bobby
Her 16h53'23"50d42'
Scholl,Anne & Charlie
Dra 16h22'15"60d7'
Scholl,Eric
Her 16h22'1"24d14'
Schollmeier,Matthew Steven
Tri 1h50'1"27d2'
Scholte,Birgit
Lac 22h7'38"51d37'
Scholten,Frances M
Equ 20h59'53"4d31'
Scholten,Ludger
Lyn 8h14'59"42d40'
Scholtz IV,Whitten Walter
Aql 19h31'0"13d23'
Scholtz,Edward K
Cnc 8h55'1"31d3'
Scholz,Astrid
Peg 23h33'0"10d36'
Scholz,Daniel
Cnv 13h40'21"31d30'
Scholz,Erin C
Vul 20h19'58"25d26'
Scholz,Gerard Margaret
Hya 9h6'56"4d43'
Scholz,Heiner
Equ 20h57'21"8d37'
Scholz,Herbert
Del 21h5'54"12d25'
Scholz,Manfred
Lib 14h54'58"-5d39'
Scholz,Rainer
Aqr 21h53'56"0d27'
Scholz,Willi
Ori 6h7'60"5d10'
Scholze,Birgit
And 23h15'29"45d9'
Schomer,Howard
Her 16h27'56"42d8'
Schonbaum,Risa
Cas 1h17'50"63d53'
Schondelmaier,Heike
Lib 14h30'26"-22d59'
Schondorf,Reneé
Leo 11h10'24"-5d53'
Schonerstedt,Donna Renee
Mon 6h40'32"0d38'
Schonfeld,Bernard M
Ari 2h57'1"30d28'
Schonstal,Molly
Cas 0h45'33"60d46'
School,Yvonne Ramona
Lyn 7h55'52"44d33'
Schoolcraft, Christopher A
Ori 5h51'60"14d46'
Schoolcraft,Albert
Ori 5h51'16"14d23'
Schoolcraft,Carol J
Hya 9h12'51"1d10'
Schoolcraft,Caroline
Hya 9h9'54"3d28'
Schoolcraft,Lorri L
Hya 9h12'31"1d15'
Schoolcraft,Ruby Vaudine
Hya 9h10'0"1d8'
Schooley,Amanda & Graham
Aql 19h50'23"13d8'
Schooley,James Jimbo Ray
Aur 6h33'12"33d24'
Schooley,XOXO Annecy B
Aql 19h52'10"12d59'
Schoon,Paul Gerdes
Her 18h37'10"20d53'B
Schoon,Suzi Snezana Cekarmis
Her 18h37'10"20d53'A
Schoonmaker Forever, Remie
Uma 13h36'28"48d30'
Schoonover,Dean L
Sgr 19h40'40"-34d48'
Schoop's Star,Judy
Cnv 13h32'31"47d52'
Schoor,Samuel E
Aur 5h15'0"45d39'
Schopen,Kenneth
Oph 17h1'2"-21d51'
Schopper,Raymond Freemont
Lac 22h16'30"55d13'
Schoppers,Marcel
Oph 16h48'34"10d52'
Schorer,Marilyn
Ori 6h10'12"0d42'
Schork,Peter F
Lac 22h31'26"50d34'
Schorman,Carly Ann
Peg 0h1'52"10d26'
Schorman,David
Sct 18h54'55"-6d50'
Schorman,Matthew Allen
Cet 2h45'1"2d10'
Schornstein,Edward Price
Aur 6h55'42"37d23'
Schorr,Seth Alexander
Psc 0h57'22"32d34'
Schorsch
Aql 19h4'38"0d12'
Schortemeier,Harold
Peg 22h32'44"14d21'A
Schortemeier,Kristina
Peg 21h50'0"34d32'
Schortemeier,Lenore
Peg 22h32'44"14d21'B
Schortemeier,Linda
Ori 5h59'29"12d47'
Schortje,Frank
Peg 22h36'38"11d45'
Schoschkoff,Aleksander
Dra 16h28'26"60d28'
Schosheim,Samantha
Equ 21h20'53"12d17'
Schott,Andrew
Boo 15h29'26"41d8'
Schott,Doris E
Lyn 8h27'54"45d7'
Schott,Thais M
Cam 3h57'33"55d8'
Schottmiller
Aql 20h4'45"0d10'
Schottmüller,Karl- Heinz
Ser 16h1'23"2d48'
Schou,Dirck Theodor
Per 3h15'1"42d3'
Schouest,Carroll
Lyr 19h16'40"35d17'
Schouten,William
Boo 14h36'0"50d3'
Schoutteten,Gery
Aur 4h47'14"40d17'
Schrader,Austin Bernard
Cep 23h27'52"64d19'
Schrader,Bernd
Aqr 21h19'17"-14d14'
Schrader,Erika
Dra 18h23'48"68d50'
Schrader,Jr,Albert E
Ori 5h57'11"16d16'
Schrader,Viola M
Boo 14h24'55"51d24'
Schradle,Susan
Leo 10h49'20"-5d30'
Schraeder,Tamara Lee
And 0h21'25"38d44'
Schrag,Timothy
Cep 6h2'45"86d11'
Schrage,Liz
Aql 19h30'42"12d22'
Schram III,John Michael
Cet 3h4'17"2d42'
Schram,Audrey Bess
Mon 8h3'56"-6d14'
Schram,Janice Dorothy
Psc 22h52'24"2d14'
Schram,Kela Elaine
Aql 20h13'35"1d29'
Schram,Kristoffer
Uma 9h20'33"54d7'
Schram,Margaret Georgina Mae
Ori 4h44'19"0d11'
Schram,Ron
Aur 7h11'0"39d49'
Schrambling,William E
Per 1h53'0"56d59'
Schramm,Brigid Elizabeth
And 0h50'48"33d59'
Schramm,Dierk Christoph
Sgr 19h25'18"-42d13'
Schramm,Holly & Richard M
Uma 9h52'27"52d29'
Schramm,Kennith G
Cep 23h1'20"64d30'
Schramm,Marc Alexander
Gem 7h6'51"24d24'
Schrantz,Grant Edward
Boo 14h31'0"20d53'
Schraplau,Elaine Clark
Vul 19h23'1"23d45'
Schreck,Paul
Lmi 9h32'38"37d44'
Schreck,Paul Michael
Aur 5h7'51"44d34'
Schrecongost,Tucker Randall
Lac 22h16'48"46d5'
Schreer,Alex Matthew
Lyr 18h57'14"30d37'
Schreffler,James L & Connie L
Ori 5h52'47"16d4'
Schreiber Oreno Connors X-Mas 95-BW
Aql 19h58'28"11d52'
Schreiber,Allie
Cet 0h30'1"-11d12'
Schreiber,Kristin Alcantara
Mon 6h26'37"10d33'
Schreiber,Lynn
Vul 19h13'22"21d51'
Schreiber,Paul T
Mon 6h20'33"0d50'
Schreiber,Wendy & Jim
Crb 15h29'36"30d4'
Schreibman,Howard
Cep 22h21'30"63d28'
Schreibman,Marcy
Tau 4h26'59"28d28'
Schreibman,Maura Jean
Cas 0h59'22"66d9'
Schreier,Lorraine
Mon 6h53'39"0d2'
Schreinemakers, Margarethe
Leo 10h6'53"18d54'
Schreiner Family Star, The Steven
Cam 5h53'16"68d24'
Schreiner,Dave
Boo 14h29'37"31d16'
Schreiner,Elliot Baldwin Anderson
Uma 8h19'43"71d36'
Schreiner,Ernst
Per 4h6'1"47d56'
Schreiner,Mary
Dra 19h40'32"70d17'
Schreiner,Ricky
Cmi 7h6'12"1d22'
Schreiner,Riley James Banfield
Mon 6h40'36"8d8'
Schretter,Jason
Ser 15h30'27"18d60'
Schretzlmeir,Robert
Dra 13h4'53"63d45'
Schreurs Meu Amor Para Sempre,Marcia
For 3h29'1"-30d8'
Schrewe,Friedhelm
Peg 23h32'1"20d43'
Schreyer,Isabel
Lyr 18h52'57"37d44'
Schreyer,Jennifer Christine
And 0h6'28"40d46'
Schreyer,Konrad
Cnc 9h15'15"8d24'
Schrider,Theresa Francis of Assisi
Del 20h18'32"11d3'
Schrieber,Katherine
Cyg 20h18'1"30d44'
Schriener,Leslie
Peg 21h59'23"22d35'
Schrier,Mercedes Rottinghaus
Lyr 18h37'44"40d8'
Schrier,Robert Eugene
Lyr 18h38'52"41d38'
Schrimsher"Captain Snow",Frank Snow
Cet 2h59'1"1d31'
Schring
Sct 18h45'58"-7d9'
Schriver,Bob
Sex 10h39'34"2d59'
Schriver,Dr David B
Umi 16h48'31"84d30'
Schriver,Edward Michael
Cep 23h8'16"78d53'
Schrock,Eli Aram
Crb 15h41'18"27d23'
Schrock,Ezra Levi
Cep 21h7'1"60d26'
Schrock,Heather Ariane
And 23h1'31"37d58'
Schrock,Julian Merle
Com 13h14'51"27d16'
Schrock,Kevin Philip
Her 18h5'17"14d57'
Schrock,Mabel Frederick
Lyr 18h40'19"29d50'
Schrock,Marjalie Mae Swihart
Aur 6h2'18"45d13'
Schrock,Mark Yost
Cnv 13h21'25"38d32'
Schroder,Kate
Mon 7h4'0"0d8'
Schroder,Priscilla Jeanne
And 0h7'1"30d41'
Schroder,Timothy L
Cep 23h8'20"65d56'
Schroedel,Carol
Cep 21h37'58"60d54'
Schroedel,Guenter
Oph 17h31'45"8d35'
Schroeder "Coco Puff", Pamela K
Mon 6h56'12"-10d11'
Schroeder 100 yrs Young,Pauline (Gup)
Ori 5h57'13"21d46'
Schroeder's Noonday 03/28/94
Peg 23h34'1"20d31'
Schroeder,Al
Sgr 19h14'30"-29d47'
Schroeder,Barbara Radcliffe
Aur 6h28'45"35d16'
Schroeder,Bruce D
Tau 4h37'15"15d8'
Schroeder,Bryce Wilson
Her 17h16'25"47d31'
Schroeder,Carmella
Peg 21h54'28"20d12'
Schroeder,Clarence Anthony
Gem 7h24'42"31d14'
Schroeder,Douglass
Crt 10h50'52"-8d45'
Schroeder,Edward Andrew
Boo 14h23'1"48d54'
Schroeder,Emily Kaitlyn
Cam 7h2'28"67d31'
Schroeder,Eric Alan
Lac 22h6'0"51d43"
Schroeder,Eugene D & Marjorie L
Sge 19h36'42"16d32'
Schroeder,Gail
Psc 1h17'21"20d50'
Schroeder,Hermann
Oph 17h58'23"11d8'
Schroeder,Janice
Mon 6h5'0"-10d60'
Schroeder,Jeffrey D
Her 17h21'19"45d51'
Schroeder,John R
Aur 5h1'46"48d24'
Schroeder,Kenneth
Her 17h16'1"22d15'
Schroeder,Kevin
Aur 6h37'25"38d19'
Schroeder,Larry
Tau 4h31'0"30d27'
Schroeder,Mary D
Lyr 19h23'1"30d42'
Schroeder,Mary J
And 0h5'17"34d47'
Schroeder,Matt J
Cet 2h19'38"5d46'
Schroeder,Melissa Agnes
Cas 2h55'1"70d48'
Schroeder,Michael
Peg 23h32'10"21d52'
Schroeder,Miranda Alees
Peg 22h0'1"28d49'
Schroeder,Molly & Michael
Cyg 21h14'41"28d40'

Schroeder,Paul & Arlene
Cyg 21h1'38"40d13'
Schroeder,RE RE Maria Theresa
Uma 11h37'48"61d23'
Schroeder,Robert P
Dra 9h29'56"77d36'
Schroeder,Roslyn Noble
Lyn 7h55'11"40d18'
Schroeder,Shannon
And 2h21'46"46d29'
Schroeder,Shayna
Peg 22h17'36"7d43'
Schroeder,Tricia & Lance
Boo 13h53'1"20d7'
Schroeder,Wiebke
Ari 2h2'13"20d13'
Schroeder,William
Boo 15h20'18"42d2'
Schroeder,William R
Dra 19h10'54"68d50'
Schroepfer,William G
Aur 6h19'0"31d43'
Schroer,Jeffrey B
Aur 4h44'35"30d12'
Schroeter,Peter
Lac 22h4'43"51d32'
Schroetter,Angela von
Sco 16h13'42"-23d29'
Schroff,Casandra Laine "Casie"
Cyg 21h18'36"31d26'
Schroll,Gina M & Ken F
Her 18h7'0"31d7'
Schrom,Bernice
Cas 1h54'0"65d33'
Schron,Sandra Elaine
Cas 0h3'51"65d12'
SCHROSAL
Dra 19h28'39"60d45'
Schrotberger,Ethel
Sge 19h59'14"16d39'
Schrottmann,Ria E
Eri 3h0'52"-17d25'
Schrottmann,Ruth
Crt 11h38'27"-10d38'
Schroyer,Holly Beth
Cyg 20h17'13"38d50'
Schrubba Loves Grubba
Uma 8h50'27"67d57'
Schrudder,Kay
Vul 21h15'13"20d8'
Schrumpf,Christina Anne
Gem 7h7'18"21d50'
Schry,Andrew J
Her 16h46'1"20d8'
Schryver,Jack
Lac 22h20'21"53d51'
Schröder,Birthe
Lyn 8h27'13"42d35'
Schröder,Brigit
Vir 14h5'1"-6d5'
Schröder,Dagmar
Leo 10h54'59"18d39'
Schröder,Klaus M
Lyn 8h7'21"48d14'
Schröder,Matthias
Ari 2h29'32"30d31'
Schröder,Wolfgang
Umi 13h35'30"74d7'
Schrödter,Bernd
Ari 3h21'45"28d20'
Schröther,Eberhard
Cep 22h38'52"60d6'
Schu & Flyver
Lyr 18h20'40"42d55'
Schubart,Günter
Aql 20h9'39"1d30'
Schuber,William A
Her 17h19'45"41d12'
Schubert,Erich Bigstar
Tau 4h1'0"30d13'
Schubert,Saundra
Uma 10h19'33"58d12'
Schubert,Scott Owen
Aql 19h57'60"12d14'
Schubert,Shirley Marie
And 0h18'0"31d39'
Schubert,Stephanie Ann
Tri 1h56'29"28d3'
Schubpi,Klaus
Ori 5h52'1"8d3'
Schuch,Jon
Hya 9h30'40"-6d15'
Schuchert,Paul
Her 17h24'40"42d32'
Schuchman,Stephen M
Ori 5h16'0"15d31'
Schuck,Guy
Dra 14h58'44"60d26'
Schude,Jay F
Aur 6h33'38"40d47'
Schuder,Coming In Hot- Joe
Cep 22h58'36"59d33'
Schuder,Karl
Cas 0h31'0"63d39'
Schudroff,Kitty
Cas 0h30'54"63d13'
Schuebler,Hilde
Sco 16h13'23"-22d27'
Schuele,Else
Aqr 23h32'18"-18d34'
Schuermann,Natalie
Aql 19h55'42"10d43'
Schuessler,Barbra
Com 12h16'16"26d59'
Schuett,Beate
Sgr 19h19'49"-22d29'
Schuett-Meyer,Theresia
Gem 6h34'53"26d59'
Schuette,Kelli Lee
Uma 10h23'28"42d13'
Schuette,Thomas J
Dra 16h27'34"67d30'
Schuetter,Jolene & Phillip DeMatteo
Her 17h11'43"47d31'
Schuetz,Annemarie
Lib 15h7'46"-24d14'
Schuetz,James P
Cam 6h1'16"58d42'
Schuffert,Robert Lance
Dra 19h23'1"56d8'
Schufferts,The
Cam 6h5'15"58d51'
Schug,Cornelia
Peg 23h32'23"15d50'
Schug,Horst
Vir 13h59'23"-10d44'
Schug,Ryan
Mon 8h1'38"-4d37'
Schuh,Barbara
Aql 19h57'27"15d14'
Schuh,Dorothee
Her 17h7'1"48d22'
Schuh,Erich
Cyg 20h28'24"30d29'
Schuhknecht,Jennifer
Lyr 19h3'49"40d40'
Schuhmann,Berta
Oph 18h26'34"0d4'
Schuhmayer,Richard
Ori 6h17'1"0d54'
Schuite,Opa Hendrik & Oma Femmetje
Peg 23h3'0"10d35'
Schuitema,Todd Michael
Cep 21h42'58"78d58'
Schukat,Susan
Uma 12h21'34"62d18'
Schukoff
Ori 5h7'16"1d15'
Schul,Mary Louise
Cam 3h47'31"53d57'
Schulenburg,Carl Robert
Lib 15h20'29"-23d5'
Schuler,Deborah
And 0h11'37"30d53'
Schuler,Kristen "Thumper"
Vul 19h23'20"26d46'
Schuler,MD,Pamela M
And 1h2'0"41d10'
Schuler,Steven H
Aur 6h31'12"30d46'
Schulien Magic
Peg 23h0'56"18d23'
Schuller,Catharina
Boo 15h0'34"50d16'
Schuller,Hannah
And 23h57'27"38d34'B
Schuller,Lucia
And 23h57'27"38d34'A
Schulman "Mimi", Minnie L
Cam 4h19'44"61d48'
Schulman,Brian
Cyg 21h1'37"53d23'
Schulman,Dr David (Calvin)
Cyg 20h34'45"40d52'
Schulman,Irv & Marcie
Lyr 18h34'25"37d39'
Schulman,Laura
Cyg 21h0'32"30d21'
Schulman,Lori Ann
Cas 23h39'0"61d59'
Schulman,Shelley Cara
Lyn 8h6'32"38d57'
Schulmerich,Jason M
Ari 1h47'0"15d57'
Schult,Eckehardt
Cyg 20h26'46"30d19'
Schulte,Kilian
Lyn 8h28'37"42d8'
Schulte,Rudolf R
Hya 9h18'54"2d1'
Schulte,Stephen Robert
Cyg 20h45'21"42d13'A
Schulte,Tabea Alexa Thea Maria
Leo 10h35'42"12d28'
Schulte,Walt & Bev
Ori 6h17'0"8d48'
Schulte,Willi
Uma 9h29'32"46d18'
Schulte,Willi
Lyn 8h4'1"48d20'
Schulteland,Junior
Aur 6h29'28"37d58'
Schultz "Salesman Extraordinare", Jay
Peg 22h52'39"8d22'
Schultz Family
Hya 9h12'24"5d24'
Schultz Star,Robert Franklin
Aur 6h28'11"31d48'
Schultz,Aaron Kieth
Vir 13h3'1"-5d13'
Schultz,Allen Robert
Hya 8h31'44"-8d27'
Schultz,Calvin
Sct 18h44'15"-5d58'
Schultz,Captain Dad Sir-Robert F
Per 1h47'42"53d55'
Schultz,Christine Dawn
Lyr 18h45'39"33d53'
Schultz,Christofer Michael
Oph 18h37'56"6d41'
Schultz,Christopher John
Cyg 19h31'13"34d12'
Schultz,Connie & James A
Crb 15h51'41"27d51'
Schultz,Craig
Cep 21h42'0"68d34'
Schultz,Cristel
Cas 0h43'57"75d51'
Schultz,Culley Gracemarie
Cyg 21h6'57"31d6'
Schultz,David "Codyman"
Aql 20h1'60"8d10'
Schultz,David Nelson
Cep 22h44'0"70d5'
Schultz,Dennis
Aql 18h57'52"12d46'
Schultz,DO,Joseph J
Aql 19h27'21"10d21'
Schultz,Donna & Richard
Sge 20h1'52"17d9'
Schultz,Dorna J
Dra 17h53'39"76d2'
Schultz,Douglas Fifty
Cep 23h2'45"65d36'
Schultz,Dr William S
Oph 17h7'32"-22d56'
Schultz,Edward & Nora
Sge 20h15'43"17d50'
Schultz,F Michael
Aql 20h8'14"0d5'
Schultz,Frances Hurst
Uma 11h53'42"42d44'
Schultz,Harold A & Doris M
Eri 3h38'1"-6d8'
Schultz,Heidi Ann
Crb 15h26'48"30d7'
Schultz,Helen
Cas 0h5'33"58d6'
Schultz,Jack
Hya 9h12'33"2d29'
Schultz,James Edward
Aur 6h26'54"52d32'
Schultz,Jeffrey,Lynn
Sco 16h56'25"-40d33'
Schultz,Jeremy Francis
Her 18h6'55"38d4'
Schultz,John H
Uma 9h16'29"56d31'
Schultz,Jr,A Michael
Per 2h56'22"35d3'
Schultz,Kathy & Steve
Sge 19h54'58"20d33'
Schultz,Klaus
Her 17h55'0"14d28'
Schultz,Kristina DeAnn
Sco 16h34'13"-28d54'
Schultz,LaVonne Leona
Lyr 19h23'40"38d27'
Schultz,Lee Edward
Ser 15h54'51"19d30'
Schultz,Linda D
Aql 20h10'59"0d9'
Schultz,Lisa Noelle
Cet 1h53'13"-2d10'
Schultz,Mary "Bunners"
Cyg 21h21'1"40d35'
Schultz,Maryann
Cas 2h2'1"77d11'
Schultz,Matthias
Cmi 7h29'17"0d23'
Schultz,Michele Diane
Cma 6h50'41"-16d17'
Schultz,Mr & Mrs Mortimer Lincoln
Cyg 21h22'53"50d25'
Schultz,Pamela A
Lyr 19h16'26"25d58'
Schultz,Robert
Dra 19h4'18"50d32'
Schultz,Robin
Uma 10h23'1"58d2'
Schultz,Ronald P
Uma 12h20'1"60d57'
Schultz,Scotty
Dra 16h57'41"63d29'
Schultz,Steven Lee
Hya 8h26'44"5d60'
Schultz,Travis Kane
Ori 5h45'14"10d55'
Schultz,Trisha
Com 13h6'46"27d20'
Schultz,Wesley William
Lmi 9h58'27"37d46'
Schultz,William Robert
Ari 2h31'50"30d29'
Schultz,Wish Upon A Pam
And 2h31'34"47d32'
Schultze,Dan & Linda
Mon 6h28'24"8d3'
Schultze,Jerry
Per 2h0'24"56d18'
Schultze,Norbert
Equ 20h58'60"8d44'
Schultze,Prof Dr Arnold
Cep 20h54'53"55d41'
Schulz's Dog Star
Cep 20h25'57"60d40'
Schulz,Bradley
Sex 10h40'30"3d23'
Schulz,Christian
Lyr 19h21'36"30d12'
Schulz,Christina Hansen
Equ 20h57'28"3d12'
Schulz,Christine
Umi 14h52'54"65d41'
Schulz,David R
Cet 0h32'40"-4d50'
Schulz,Elisabeth
Lyn 9h0'0"34d48'
Schulz,Evelyn Charlin
Eri 3h34'38"-5d30'
Schulz,II,MD,Malcolm
Oph 17h40'0"-16d22'
Schulz,Jörg
Peg 23h16'33"31d50'
Schulz,Karlene Marie
And 23h36'16"38d42'
Schulz,Katja
Lyr 19h19'1"30d58'
Schulz,Kent C
Cyg 19h28'54"34d8'
Schulz,Kristine
Lac 22h2'1"51d31'
Schulz,Manuela
Vir 13h14'22"-8d21'
Schulz,MD,Valerie M
Oph 17h1'54"-24d24'
Schulz,Michael
Tau 4h17'0"0d50'
Schulz,Michael Joseph
Ori 5h50'31"17d53'
Schulz,Olaf
Gem 7h24'44"24d55'
Schulz,Ron
Per 3h7'0"47d60'
Schulz,Sara L
Eri 3h16'16"-5d53'
Schulz,Siegfried
Cas 0h31'0"63d25'
Schulz,Susanne 50
Boo 13h58'0"10d3'
Schulz-Dievneow
Ori 4h42'41"13d19'
Schulz-Klatt,Ingeborg
And 1h16'48"34d18'
Schulzc,Kristie Lynn
Lmi 9h53'24"38d41'
Schulze,Bill
Aql 19h48'20"14d36'
Schulze,Christopher
Hya 8h46'47"-0d3'
Schulze,Eric Nolan
Aql 18h55'44"8d11'
Schulze,Hermann
Peg 23h35'0"18d56'
Schulze,Kurt
Cep 21h36'1"65d17'
Schulze,Lois M
Del 20h18'5"10d30'
Schulze,Robert
Vir 12h56'55"-17d57'
Schulze,Shannon
Peg 22h3'47"3d6'
Schulze,Wolfgang
Ari 2h0'1"22d21'
Schulze-Schlüter,Ines
Aqr 22h54'11"-4d26'
Schum's "1995 Fireballs",Kim
Cyg 20h19'0"40d26'
Schumacher,APM Europe Hans-Georg
Ser 15h40'15"2d52'
Schumacher,Col David J
Vul 19h59'53"28d51'
Schumacher,Cristina
Cyg 20h20'1"40d34'
Schumacher,Erwin
Her 18h44'58"12d43'
Schumacher,Hans Josef
Dra 17h17'0"69d9'
Schumacher,Helen
Vul 20h22'12"22d39'
Schumacher,Jennifer Andrea
Vul 20h20'34"23d55'
Schumacher,Karl
Sge 19h53'31"16d40'
Schumacher,Kenny
Her 17h1'31"30d7'
Schumacher,Lance & Allen
Cyg 19h35'35"28d21'
Schumacher,Mark & Yolanda
Cet 3h15'55"1d22'
Schumacher,Martha Maurine
Cep 20h49'35"68d27'
Schumacher,Monashka by Heidi
Uma 8h38'17"56d1'
Schumacher,Papa Gail
Aql 19h53'32"15d27'
Schumacher,Phil
Umi 15h16'1"66d35'
Schumacher,Sara Jean
Cas 0h5'43"61d11'
Schumacher,Sheyenne Leigh
Lyn 8h48'48"41d14'
Schumaker,Melanie
Com 12h3'26"19d54'
Schuman,Dale A
Cnv 12h52'42"50d30'
Schumann,Annett
Gem 7h33'13"20d10'
Schumann,Anthony J
Per 2h20'0"55d36'
Schumann,Bill & Joanie
Aql 20h2'33"9d23'
Schumann,Ernst
Umi 16h37'47"78d60'
Schumann,Heide
And 1h9'39"40d10'
Schumann,Tammy
Equ 21h6'59"10d41'
Schumann,Tatum Kristine
Peg 22h24'0"5d49'
Schumann,Wolfgang
Peg 22h43'0"11d33'
Schumer,Dylan Charles
Hya 8h41'32"4d11'
Schumer,Maya Claire
Cap 21h13'50"-26d23'
Schumm,Richard
And 23h28'12"45d1'
Schuplee,Sylvia
Mon 7h19'22"-6d25'
Schuppener-Reh,Susanne
Lac 22h7'47"38d10'
Schuppisser,Cornelia
Umi 15h37'33"66d36'
Schupps,Papa Bear
Uma 14h19'0"59d25'
Schur,Jerry
Cnc 8h58'15"18d9'
Schure,Peter Brian
Uma 8h45'14"71d21'
Schurig,Andreas
Sco 16h38'40"-44d23'
Schurig,Claudia E F
Gem 7h7'50"24d26'
Schurle,Darrell
Dra 18h56'34"47d53'
Schurman,Florence Leach
Cas 0h15'52"61d16'
Schurmann,Lisa René
And 23h1'44"49d42'
Schursky,Brian & Dawn Cataldo
Cyg 19h25'55"31d39'
Schust,Jr,John W
Dra 16h1'58"67d21'
Schuster,Annemarie
Vul 20h46'43"20d30'
Schuster,Cecelia
Del 20h21'1"8d21'
Schuster,Christian Hanson
Lac 21h58'11"38d53'
Schuster,Gabriele
Her 17h59'32"42d51'
Schuster,Gilly
Peg 23h30'56"18d45'
Schuster,James & Yvonne
Cyg 19h46'56"38d55'
Schuster,Rolf-Dietmar
Ser 15h55'56"1d13'
Schuster,Ruth Jackie Noble
Cas 0h41'34"67d20'
Schuster,Steve
Aql 18h44'0"-2d23'
Schuster,Sylvia Katharina
Sco 16h59'32"-44d5'
Schuster,Werner
Del 20h19'11"11d25'
Schut,Lisa Marie
Cyg 20h35'14"42d0'
Schutt,Edward Philip
Aql 18h58'14"12d4'
Schutt,James B
Uma 10h37'14"47d48'
Schutt,Kevin Thomas
Boo 14h4'54"12d5'
Schutt,Michael C
Her 16h42'43"7d52'
Schutte,Elizabeth B
Ori 5h41'0"8d31'
Schutte,Francis X
Uma 10h54'24"48d16'
Schutte,J Claiborne
Uma 11h59'38"56d3'
Schutte,Lonnie Arthur
Boo 14h19'34"36d1'
Schutte,Nicole Anamarie
Cet 1h52'0"-1d48'
Schutte,R Rex
Her 16h3'1"20d18'
Schutte,Richard Earl
Her 17h57'1"18d53'
Schuttenberg,Lynn
Uma 11h37'24"40d45'
Schutz,Helen T
Lyn 6h59'12"59d43'
Schutz,Ted
Uma 10h30'31"71d14'
Schutz,Zachary Melech
Boo 14h22'1"38d17'
Schutzer,In loving Memory of Bobbie
Tau 5h53'57"28d24'
Schuyler
Vul 19h39'20"20d34'
Schuyler
Mon 6h43'13"10d26'
Schué,Konrad
Peg 0h5'51"18d14'
Schveetie
And 1h26'58"35d34'
Schwaab,Faby
Lyr 18h17'29"30d18'
Schwab,Dolores Margaret
Uma 9h22'42"52d15'
Schwab,Erin Kathleen
And 1h7'11"40d59'
Schwab,Hans
Cyg 20h26'1"30d36'
Schwab,Heribert
Cam 6h20'10"80d17'
Schwab,Kenneth Patrick
Cam 3h41'48"60d0'
Schwab,Mary
Tri 1h59'0"26d38'
Schwab,Nicholas A
Ori 4h53'19"-0d5'
Schwab,Rasso
Cyg 20h22'14"30d34'
Schwab,Vibeke C
Ori 6h5'56"-1d40'
Schwabenitzky,Reinhard & Elfi
Uma 13h38'18"48d17'
Schwabenland,Bernd
Dra 16h58'24"70d54'
Schwadler,Kristen Nicole
Del 20h14'20"10d47'
Schwaerzler,Hermann J
Dra 17h26'55"61d23'
Schwahl,Holger
Peg 23h32'10"10d25'
Schwalb,Ruby
Ori 4h42'1"1d4'
Schwalbach,Christel
Leo 11h5'27"26d47'
Schwalke,Geraldine (Gerry)O'Leary
Cmi 7h44'40"4d5'
Schwall Anniversary Star
Umi 14h17'12"72d3'
Schwall,Donna Lynne
And 2h7'44"40d3'
Schwam,Edna
Lyr 18h27'0"38d41'
Schwan,Aric R
Lac 22h1'23"51d44'
Schwan,Friedhelm
Lyr 19h19'33"30d24'
Schwan,Michaela
Her 17h56'20"42d39'
Schwanke,Marlene
Cyg 20h18'0"31d25'
Schwanstern
Peg 23h34'45"11d51'
Schwantz,The Kevin
Ori 5h59'19"19d18'
Schwarb,Willie
Tri 2h10'35"33d55'
Schwarting,Kathleen
Uma 10h55'0"71d27'
Schwarts,MD,Robert
Lac 22h29'26"48d47'
Schwartz "The Star", Gary
Her 17h4'36"21d19'
Schwartz,Alan
Dra 15h46'31"62d39'
Schwartz,Amy Lauren
And 0h10'1"47d17'
Schwartz,Andrea Lynn
Gem 6h51'6"18d23'
Schwartz,Andrew Lee Julian
Peg 22h0'32"31d7'
Schwartz,Arthur
Hya 8h53'57"6d27'
Schwartz,Barbara Sue
Cet 0h53'18"-10d60'
Schwartz,Basha
And 23h38'58"40d5'
Schwartz,Bernard S
Lac 22h44'0"37d59'
Schwartz,Betty
Lyr 18h25'19"40d6'
Schwartz,Brian Paul
Boo 15h11'34"38d1'
Schwartz,Charles B
Cet 2h18'29"-11d12'
Schwartz,Constance R
Cam 9h8'41"81d54'
Schwartz,Daniel Richard
Cet 0h58'1"0d40'
Schwartz,David
Per 2h21'47"54d50'
Schwartz,David Bryan
Lac 22h24'31"38d23'
Schwartz,Don
Cep 22h16'13"61d18'
Schwartz,Edward
Cep 22h14'0"60d40'
Schwartz,Eli Marc
Vir 11h46'1"8d31'
Schwartz,Emily
Lyn 7h35'35"58d58'
Schwartz,Fastboy Todd
Per 4h3'44"51d56'
Schwartz,Gary
Uma 11h17'38"30d15'
Schwartz,Ginny S
Aql 19h48'51"13d39'
Schwartz,Haley Morgan
Peg 0h2'34"31d42'
Schwartz,Harold & Linda
Cyg 20h56'54"30d59'
Schwartz,Ida
Lyr 19h21'1"41d23'
Schwartz,In Memory Of Mortimer Mayer
Uma 9h31'0"58d15'
Schwartz,Iris & Si
Tri 2h18'57"30d26'
Schwartz,Jane
Boo 15h7'1"18d57'
Schwartz,Janelle Marie
And 0h54'55"35d37'
Schwartz,John-Georges
Aql 20h8'32"8d39'
Schwartz,Julius Lambert
Aur 7h7'38"40d12'
Schwartz,Karen Dianne
Peg 22h47'29"5d10'
Schwartz,Kathy
And 1h23'35"40d37'
Schwartz,Kelly "Little Teapot"
Leo 10h33'44"15d3'
Schwartz,Law Offices of Joel H
Cep 22h34'15"63d54'
Schwartz,Lawrence Adam
Per 3h15'11"50d23'
Schwartz,Lisa A
Cmi 7h24'52"1d27'
Schwartz,LOML Marc J
Gem 7h14'56"21d33'
Schwartz,Louise Ellen
Com 12h6'28"31d51'
Schwartz,Mary Rose - Sydney
And 23h37'11"42d51'
Schwartz,Matthew Justin
Dra 17h31'45"64d60'
Schwartz,Michael
Cep 22h40'1"80d1'
Schwartz,Murray M
Dra 17h33'43"61d9'
Schwartz,Pamela
Ari 2h1'28"21d3'
Schwartz,Pepper
Dra 19h46'44"70d36'
Schwartz,Pvt Jason E
Aql 19h15'49"19d4'
Schwartz,Rali John
Uma 13h27'35"62d34'
Schwartz,Reuben
Per 1h46'56"52d55'
Schwartz,Sara Lu
Cas 0h45'40"69d37'
Schwartz,Scott
Boo 15h31'25"47d57'
Schwartz,Sherry
Cas 0h59'12"62d36'
Schwartz,Sierra
And 23h16'18"42d2'
Schwartz,Steven Daniel
Aur 6h26'12"31d48'
Schwartz,Tony & Stephanie
Sct 18h52'42"-4d2'
Schwartz,Vivian
Leo 10h33'14"26d21'
Schwartz,Zoe Perach
Mon 6h56'58"0d44'
Schwartz-Stein
Mon 7h2'13"-5d15'
Schwartzbard:True Stars,Marv & Carol
Crb 15h53'18"30d6'
Schwartzenberg,Ruth
Vir 13h1'45"-23d23'
Schwartzenberger, Tammy Jo & Mark
Dra 20h4'49"64d0'
Schwartzey,King
Aur 4h53'35"41d0'
Schwartzman,Jack
Her 17h31'28"40d32'
Schwartzman,Jeffrey C
Her 16h26'41"26d40'

Schwarz
Dra 18h23'28"67d39'
Schwarz Family, Berthold E
Eri 3h9'35"-4d31'
Schwarz Of The Firmament,Jason F
Dra 16h57'1"62d51'
Schwarz Semper Fidelis,Carla
Uma 10h33'10"41d55'
Schwarz,Bernadette Sarah
Ori 6h6'1"0d3'
Schwarz,C Blake
Cet 1h19'23"-3d40'
Schwarz,Elke
Ari 2h52'29"21d41'
Schwarz,Ewald
Per 4h5'39"49d24'
Schwarz,Guenter
Lyn 8h0'0"42d52'
Schwarz,Harry
Lyn 8h5'17"48d43'
Schwarz,Joachim
Aur 5h12'22"43d7'
Schwarz,June & Erich
Cyg 19h43'47"30d5'
Schwarz,Pettitt
Umi 15h17'50"77d45'
Schwarz,Roland
Ser 15h11'20"7d53'
Schwarz,Sister Denise
Lyr 18h49'0"36d15'
Schwarze
Uma 9h18'20"48d23'
Schwarzkopf,Freidmann-Eckhart
Leo 10h58'14"-5d16'
Schwarzmann,Robert und Michael
Ser 15h40'22"1d26'
Schwass,Sigrig & Joachim
Vir 13h6'1"-20d13'
Schwede,Marc
Dra 16h52'30"69d52'
Schweer,Kurt Wood
Boo 14h12'1"21d35'
Schwehm,Petra
Cnc 8h57'1"11d39'
Schwehm,Terri
Mon 6h36'27"10d15'
Schweig,Richard E
Cnv 14h1'1"40d5'
Schweiger,Wendi Rae
And 0h3'24"46d45'
Schweikert,Emil
Lyn 8h5'53"42d18'
Schweikher,Robert M
Hya 10h14'54"-16d15'
Schweinebaer
Lac 22h4'10"51d3'
Schweinfurth,Louise
Cas 1h30'59"61d17'
Schweinle,Albert
Cet 1h50'0"-1d54'
Schweinle,Elverda
Aql 19h25'57"10d48'
Schweinsmann,Bernd Michael
Sco 16h30'26"-40d23'
Schweinsteiger,Karin
Psc 23h0'1"0d47'
Schweiss,Karl-Heinz
Ori 6h17'16"0d28'
Schweitzer,Christina Maksudian
Cas 0h1'50"59d03'
Schweitzer,Greg
Aql 19h25'32"0d50'
Schweitzer,Martin
Cet 2h30'0"1d10'
Schweitzer,Peggy
And 1h29'15"39d30'
Schweitzer,Stacey
Aql 19h30'20"12d11'
Schweitzer,Steven Richard
Dra 18h0'41"65d2'
Schweizer,Fred
Dra 11h30'49"70d10'
Schweizer,Lisa Joy
Cyg 19h15'1"44d49'
Schweizer,Rolf
Per 4h5'40"49d27'
Schwelter,Frank
Ori 6h17'12"0d26'
Schwender,Brian
Her 17h56'1"38d55'
Schwender,Claudia
Cnc 9h15'16"10d43'
Schwendig,Priscilla
Uma 11h57'45"51d35'
Schweninger,Heather Paige
Peg 21h44'1"26d50'
Schwenk,Gerhard
Her 16h18'0"50d31'
Schwenk,Simon 65326 Aarbergen
Sgr 19h47'59"-42d13'
Schwenker,Carl
Oph 18h18'59"8d41'
Schwenn,Alexandra Christina
Uma 10h58'45"60d45'
Schwenn,Eva Maria
Uma 10h58'0"70d20'
Schwenn,Hartmut C
Uma 10h57'45"71d52'
Schwenn,Mario Felix
Uma 10h59'0"60d9'
Schwenn,Mary C
Mon 6h24'30"5d45'
Schwenn,Maximilian Cyrus
Uma 10h58'1"70d17'
Schwenn,Wm F
Cet 0h55'12"-5d58'
Schweppe Spotlight
Ori 5h53'42"13d19'
Schwer,Douglas
Boo 14h28'47"21d24'
Schwertner,Victor
Peg 21h42'11"26d20'
Schwesterchen
Lib 14h19'55"-22d41'
Schwetje,Ken E
Her 16h10'23"8d41'
Schwick,Bili-Ruth
Leo 9h58'56"7d54'
Schwiete,Carl
Cmi 7h19'55"8d59'
Schwimmer,Adam
Cet 0h43'35"-4d41'
Schwimmer,Carrie
Cas 23h37'40"60d6'
Schwind,Wolfgang Julius Ignaz
Lyn 8h3'51"48d35'
Schwindt,Paul Henry
Cnv 13h15'15"50d40'
Schwing,Tommy "T B"
Lyr 19h18'48"38d24'
Schwinn,Charles Lester
Her 18h3'19"28d26'
Schwinn,Ronald W
Aql 18h57'1"-8d43'
Schwinte,Gerard
Cas 22h59'43"54d24'
Schwitz,Rudolf
Hya 9h3'21"5d58'
Schwitzer,Dino DNA
Cep 22h18'44"61d46'
Schwitzke
Uma 9h13'0"48d22'
Schwolow,Albert R
Cet 2h22'32"-1d1'
Schwonke,Hans Juergen
Peg 23h36'49"30d46'
Schwormstede,Elsa
Tau 5h56'48"26d24'
Schwuchow's
Lyn 9h12'46"35d45'
Schyma,Fatima & Christian
Uma 13h51'45"48d21'
Schäfer
And 23h12'40"42d23'
Schäfer 08/08/1994, Marielle
Cet 2h58'51"4d24'
Schäfer,Franziska Alisia
Del 20h21'10"18d46'
Schäfer,Fred
Cep 22h5'34"60d56'
Schäfer,Hannelore
And 23h6'37"43d53'
Schäfer,Inge
Psc 0h53'0"8d54'
Schäfer,Jutta
Eri 4h50'17"-6d55'
Schäfer,Lothar
Uma 10h28'40"48d35'
Schäfer,Pascal
Psc 23h3'54"1d56'
Schäfer,Sandra
Cae 4h41'44"-32d35'
Schäfer,Werner "Baer"
Cas 0h40'52"60d5'
Schägger,Susanne
Sgr 18h50'46"-23d5'
Schäling,Heike
Psc 1h1'39"21d43'
Schätzel,Wolfram
Cnc 9h19'59"32d46'
Schöber,Wolfgang
Leo 11h14'22"-5d52'
Schöfmann,Hans-Jürgen
Sgr 18h53'27"-28d59'
Schögge,Susanne
Sgr 18h51'6"-22d38'
Schölen,Martina Maria Theresia
Psc 1h39'1"20d11'
Schönberg,Claude Michel
Mon 6h25'0"8d29'
Schönbrod,Karl
Lib 14h23'18"-23d48'
Schöndorfer,Hubert
Aql 19h1'10"17d43'
Schöne Aroma
Aqr 22h18'51"-12d55'
Schönenborn,Petra
Sgr 19h4'4"-26d24'
Schöner,Andreas
Vir 12h34'0"-7d56'
Schönherr
Leo 9h20'48"10d20'
Schönlechner,Annette
Lyn 8h0'36"47d56'
Schüller,Karl
Aql 19h57'33"10d55'
Schünemann,Brigitta
Lmi 10h41'42"27d37'
Schürmann,Fleming
Vir 12h6'21"2d16'
Schürmann,Lara Maria
Cnc 8h49'48"30d45'
Schüssler,Manfred
Ori 5h50'57"21d22'
Schüssler,Volker
Vir 14h0'32"7d13'
Schütte,Sarah
Lyn 8h1'36"47d10'
Schütze
Her 17h49'45"42d22'
Sciacca,Carl R
Lac 22h14'21"55d11'
Sciachitano,Sal
Sgr 19h58'3"-42d37'
Scialdo,Mary Ann
And 0h24'51"32d14'
Sciarretta,Anthony John
Cmi 7h45'41"8d21'
Sciarrillo,Victoria Rose
Lac 22h48'24"56d5'
Scibelli,Maria J
Lmi 10h4'59"34d27'
Scidmore,Don L
Cam 3h32'13"53d41'
Scieszka,Carl Dennis
Per 1h47'28"56d18'
Scime,Lisa M
Cyg 19h59'54"50d35'
Scimeca,Margaret Riley
Cyg 21h12'53"35d15'
Scintilating Sydney
Her 16h39'59"50d3'
Scintillans Mariana-Maureen's Sparkler
Cyg 19h34'17"35d33'
Scintillant Steve
Her 16h17'0"48d28'
Scintillating Dolly
Cas 1h28'1"74d6'
Sciortino,Gino Salvatore
Dra 11h28'48"71d37'
Scipione,Christina "Just Do It"
Com 13h4'12"22d25'
Sciranka,David
Cep 3h10'1"78d42'
Scirica,Sonja & Joseph
Aur 6h7'52"45d58'
Scirocco,Donna Marie Christina
Cas 0h2'20"64d22'
Scirocco,Mirello
Dra 9h27'1"73d33'
Scism,Debra L
Lyr 18h50'45"41d47'
Scism,Nellie Quinn
Cas 0h46'14"62d19'
Scisson,Brenda
And 23h7'0"41d49'
Sciulli,Luke
Dra 19h20'25"58d51'
Sciulli,Paul William Geoffry
Per 2h41'40"35d36'
Sciuto,Mary Margaret
Mon 7h16'1"-0d21'
Sciutto,Debra Lynn
Cas 1h21'44"55d12'
Sclafane 6/29/62-11/6/94,Anthony
Her 17h14'33"46d35'
Sclafani,Ivan
Vul 19h23'38"25d37'
Sclafani,Maureen
Lmi 10h39'40"31d33'
Sclove,Chad William
Equ 21h6'19"10d50'
ScoBar Family Star
Ori 5h56'39"5d11'
Scobey,Tricia
And 1h16'30"35d12'
Scoby,Nick
Hya 8h19'43"1d52'
ScoCiny
Aql 19h40'42"10d38'
Scodi
Lyr 19h9'18"41d6'
Scofero,Joseph James
Vul 19h46'31"28d28'
Scofield,David Michael
Boo 15h1'15"19d1'
Scoggins "Love Bug", Jamie Lynne
Lyn 7h5'38"58d22'
Scoggins,Cathy Ann
And 1h32'33"39d57'
Scoggins,Hanna
Peg 23h5'28"8d1'
Scoglio,Baby Boy
Ori 5h50'1"18d21'
Scohier,Jennifer
Mon 7h6'47"-01d13'
Scola,Fred Anthony
Lmi 10h57'50"33d26'
Scolamacchia,Marilyn
Cas 0h6'33"50d32'
Scolini,Michael F
Vul 20h4'59"22d37'
Scolio,A J
Her 17h27'45"20d40'
Scollay Starfire
Mon 6h4'31"-8d13'
Scollieri,Chivas
Cma 6h26'53"-16d18'
Sconyers IV,George Hanson
Oph 17h55'29"11d14'
Scooby Doo
Cam 5h36'42"65d11'
Scooby Duke
Umi 17h27'52"85d6'
Scooby-Pilot Star, Damien
Aur 6h11'40"36d8'
Scoot
Lyn 8h38'1"41d9'
Scooter
Hya 8h14'22"-5d47'
Scooter
Uma 11h33'0"63d7'
Scooter Pie in the Sky ,Jimmy J
Cap 20h4'40"-10d33'
Scooter's Midnight Angel
Cet 2h52'41"3d39'
Scopin,Gayle Allison
Lyn 7h12'0"58d29'
Score,Gina
Uma 10h2'56"50d1'
Scorer,Jennifer Mary
Cnc 8h55'38"18d24'
Scorpio Delphinus
Sco 16h58'20"-30d18'
Scorsone,Mary Virginia
Mon 6h54'1"7d33'
Scorup,Carl & Ginny
Boo 14h44'47"37d36'
Scostelau Diamond
Oph 18h27'60"8d30'
Scot,Rob Zaea
Aql 19h57'55"14d3'
Scotcher-Lane,Ben Charles
Cep 1h3'31"78d5'
Scott
Cep 21h1'0"61d51'
Scott
Ari 2h59'48"30d59'
Scott
Cmi 7h38'56"4d51'
Scott
Aur 6h17'56"30d31'
Scott
Ori 6h17'48"7d40'
Scott
Per 4h7'21"52d7'
Scott
Cep 22h46'22"59d58'
Scott
Ser 15h18'0"0d18'
Scott & Amber Eternal Love
Cyg 21h55'14"53d37'
Scott & Amie
Peg 22h36'45"27d32'
Scott & Chanda
Cyg 21h6'1"30d10'
Scott & Dawn To Infinity & Beyond
Uma 10h15'1"58d1'
Scott & Donna I Will Be Your Guide
Umi 12h57'34"88d57'
Scott & Donyelle,I Love You I + 1
Sge 19h3'14"16d45'
Scott & Joanna
Cet 1h58'10"-10d27'
Scott & Katrina's Star
Ori 5h24'24'0d53'
Scott & Kay's Gorgonzola Star
Uma 10h46'18"52d18'
Scott & Lori
Crb 16h1'38"32d51'
Scott & Lynn (Bro & Sis)
Cyg 19h54'34"45d13'
Scott & Renee's Star
Eri 4h23'25"-1d14'
Scott & Shanna
Cyg 20h2'40"37d55'
Scott & Shannon
Mon 7h1'52"-1d10'
Scott & Sharon Love Forever After
Sge 20h14'15"17d45'
Scott & Susan's First Anniversary
Ori 5h14'38"-4d49'
Scott & Trish
Crb 15h15'19"32d15'
Scott 2 7
Cep 22h54'1"63d33'
Scott 20th Anniversary Paula & Terry
Cyg 20h55'1"30d34'
Scott Dick
Per 3h5'27"40d12'
Scott Donald
Per 2h56'13"35d16'
Scott Forever
Uma 10h54'31"44d43'
Scott Gregory
Sge 18h59'1"20d9'
Scott Jared
Lac 21h58'0"40d44'
Scott Jay
Aur 6h33'55"31d42'
Scott Jr
Per 3h59'60"38d32'
Scott Michael
Hya 9h7'40"2d34'
Scott My Dream Come True
Leo 10h58'0"11d16'
Scott Spot
Uma 10h32'1"52d55'
Scott Star,Barbara & Will
And 23h45'42"45d35'
Scott's "Nightlight"
Vir 14h9'1"-8d17'
Scott's Claire
Lyr 19h21'10"31d21'
Scott's Destiny
Cnv 12h7'51"42d39'
Scott's For Never & Ever Star
Cam 3h30'14"59d34'B
Scott's II Eternity
Aur 5h15'51"45d7'
Scott's Lone Star
Sct 18h54'14"-4d31'
Scott's New Gold Dream
Her 16h21'47"47d44'
Scott's Scorpion-Eagle Penumbra
Sco 16h22'10"-40d40'
Scott's Star
Ari 1h59'38"17d34'
Scott's Star
Eri 5h4'47"-6d39'
Scott's Star
Her 16h26'23"28d56'
Scott's Star All My Love Kerri
Dra 12h59'49"75d20'
Scott's Star of Serendipity
Aqr 21h25'25"-0d12'
Scott's Synergy
Dra 16h5'48"60d46'
Scott's Valley
Lmi 10h34'1"24d5'
Scott,Aaron E
Sco 16h34'58"-44d39'
Scott,Alexandria Kathleen
And 0h30'23"40d58'
Scott,Alpha Joyce
Lmi 10h9'30"31d40'
Scott,Analda
Cam 7h37'19"71d14'
Scott,Andrea
Mon 6h36'26"0d29'
Scott,Andrew H
Sct 18h52'56"-6d23'
Scott,Angel Eyes of F Troy
Lyr 19h24'45"41d31'
Scott,Ann & Thomas
Tri 1h53'0"25d32'
Scott,Anne-Marie
Eri 4h3'0"-11d49'
Scott,Anthony James
Cep 23h33'12"68d5'
Scott,Austin
Aur 6h55'31"37d56'
Scott,Awesome-Twosome-Ronda & David
Cyg 19h31'26"33d51'
Scott,Benjamin
Boo 14h31'1"8d31'
Scott,Beverly Montgomery
And 0h14'19"39d19'
Scott,Bill
Her 17h46'41"48d49'
Scott,Boogie Face Roger N
Mon 5h56'0"-6d30"
Scott,Brad-The Oracle Team, CCH OMS Project
Her 17h20'10"14d50'
Scott,Bradley
Her 16h24'13"29d16'
Scott,Brenda Joyce
Lyn 7h27'17"38d2'
Scott,Brian
Ori 5h54'48"13d0'
Scott,Brian J
Boo 15h0'1"31d43'
Scott,Candice K
Cam 7h5'36"68d11'
Scott,Carey
Her 16h12'28"8d36'
Scott,Catherine Bowen
Lyr 19h15'28"25d51'
Scott,Cécile Labranche
Peg 22h50'34"16d34'A
Scott,D
Oph 17h1'52"-21d4'
Scott,Dane Geoffrey
Ser 15h13'43"6d36'
Scott,Darly
Sct 18h44'35"-6d35'
Scott,David Robert
Cet 2h9'1"-2d19'
Scott,Deborah Jean
Cas 0h32'49"50d4'
Scott,Dennis Richard
Ser 15h24'55"0d40'
Scott,Dennis Richard
Cep 22h52'30"70d28'
Scott,Denny & Marti
Aur 6h11'19"31d36'
Scott,Desirée Evelyn
Uma 10h10'19"52d12'
Scott,Donald Robert
Dra 17h23'39"58d34'
Scott,Donna
Mon 7h52'45"-7d18'
Scott,Duncan
Aql 19h57'57"15d0'
Scott,Duncan
Uma 9h2'39"53d53'
Scott,Edwin Sean
Cyg 19h20'36"44d13'
Scott,Elizabeth Anne
Aur 5h53'24"30d20'
Scott,Elizabeth Anne
Peg 22h33'1"30d33'
Scott,Emily Marie
Lyn 7h52'19"44d28'
Scott,Eric Wayne
Lmi 10h1'14"40d37'
Scott,Fiona
Lyr 18h36'0"28d40'
Scott,Franklin
Her 17h0'26"47d30'
Scott,Fred M
Dra 17h34'32"75d58'
Scott,Gary W
Cnc 8h29'34"7d55'
Scott,Great Grandma Mary
Cyg 20h56'34"40d19'
Scott,Gregory
Oph 16h56'56"-23d34'
Scott,Hayley Morgan
Peg 23h4'1"13d10'
Scott,Hazel Marie
Cam 7h3'28"71d6'
Scott,Ian Robert Hunter
Aur 6h23'1"34d44'
Scott,Jacob
Lib 15h2'57"-28d49'
Scott,James
Ori 4h50'33"0d25'
Scott,James & Janice
Umi 13h12'40"75d34'
Scott,James E
Her 16h47'1"34d11'
Scott,Jamison J
Boo 14h24'43"28d19'
Scott,Jane
Cas 1h27'10"61d43'
Scott,Jean
Lyr 18h37'21"34d5'
Scott,Jean Charlotte
And 23h20'42"38d55'
Scott,Jean-Robert
Peg 22h50'34"16d34'B
Scott,Jeana Sciarappa & Lar-Bear
Sct 18h51'8"-7d39'
Scott,Jeanne
Peg 22h13'15"4d40'
Scott,Jeanne
Cyg 20h30'1"42d51'
Scott,Jeffry
Vir 12h4'23"8d31'
Scott,Jennifer
Sge 19h57'12"16d6'
Scott,Jennifer Lynn
Vul 20h14'12"23d38'
Scott,Jill Mopsy Scorgie
Lyn 7h48'9"40d56'
Scott,Jim D
Aur 6h11'23"38d45'
Scott,Jock McCollum
Ori 5h53'26"17d20'
Scott,Joe
Ori 5h55'13"13d3'
Scott,John & Jennie
Cap 20h4'22"-10d17'
Scott,John Lana
Lyn 7h50'21"44d22'
Scott,John Robert
Aur 6h50'60"37d11'
Scott,Joline Blanche
Lyr 18h17'47"44d23'
Scott,Joshua Andrew
Aur 5h13'0"41d3'
Scott,Joyce
Lib 14h28'23"-20d9'
Scott,Jr,"Spooky" Joseph Hurlong
Per 2h56'52"31d21'
Scott,Jr,John Deal
Cam 3h56'1"74d2'
Scott,Jr,Walter L
Uma 12h48'35"53d22'
Scott,Judy Ellen
Cet 0h53'23"-0d47'
Scott,Julie Ann
And 2h5'49"42d48'
Scott,Kaji Travon
Her 17h53'17"14d52'
Scott,Karen
Peg 22h55'0"28d12'
Scott,Karen
Mon 8h4'34"-7d17'
Scott,Karla Denette
Mon 6h22'0"8d43'
Scott,Kathy
Aur 5h15'25"47d46'
Scott,Kelly
Cas 0h51'31"63d55'
Scott,Kenneth
Her 17h6'11"45d23'
Scott,Kenneth McKay
Cap 21h27'47"-23d1'
Scott,Kenneth W P
Dra 16h41'0"61d20'
Scott,Kenneth(Kenny)
Dra 16h56'15"69d55'
Scott,Kevin James
Uma 9h15'55"55d28'
Scott,Kirsten A P
And 2h29'14"42d35'
Scott,Kyle F
Cas 0h53'1"62d57'
Scott,Larry Marcus
Ori 5h52'42"15d8'
Scott,Lee Roy H
Lac 22h33'58"56d41'
Scott,Leigh Antonie
Dra 18h18'13"68d11'
Scott,Lindsay
Cyg 21h27'60"40d50'
Scott,Luke Frederick
Ori 5h34'35"8d13'
Scott,Lynda
Lyn 6h56'48"48d46'
Scott,Lynn Ann
And 23h37'60"46d37'
Scott,Malcolm
Cet 2h16'1"8d45'
Scott,Margaret
Lyn 8h24'46"47d43'
Scott,Margaret Helen
And 1h44'39"40d54'
Scott,Mark
Aur 5h55'10"38d48'
Scott,Mark "Gusman"
Lyn 7h53'22"51d46'
Scott,Marlys & Jeffrey
Lyr 19h17'17"40d26'
Scott,Mary
Cas 2h22'0"71d2'
Scott,Mike"Yipe"
Oph 16h59'14"11d11'
Scott,Miss Helen Louise
Aql 19h1'1"-10d11'
Scott,Nicole
Cyg 19h31'38"34d22'
Scott,Olivia Ann
Tri 2h19'37"30d2'
Scott,Oscar
Oph 17h17'36"10d36'
Scott,Pamela Elizabeth
Cam 8h22'46"78d51'
Scott,Paul
Cmi 7h17'0"3d13'
Scott,Rachel Leigh
Vul 19h18'0"26d58'
Scott,Ralph Robert
And 23h43'29"37d48'
Scott,Rena
Uma 10h20'59"68d12'
Scott,Rhoda
Mon 6h25'26"8d54'
Scott,Richard A
Cep 20h3'41"60d42'
Scott,Robert
Cnv 12h23'31"36d39'
Scott,Robert Avery
Cep 3h11'16"78d47'
Scott,Robert C-Bruce W Coon
Psc 23h9'51"2d12'
Scott,Ronald G
Hya 8h53'58"6d22'
Scott,Ronald Stafford
Ser 18h17'1"-0d1'
Scott,Rosanna Carolyn
Cas 1h46'1"65d25'
Scott,Sally R
Cas 23h14'31"61d41'
Scott,Samantha M
Cas 22h59'21"54d47'
Scott,Sandi
Cyg 19h33'1"38d6'
Scott,Sharon G
Peg 22h40'0"25d29'
Scott,Sheila
Cet 1h56'34"-3d32'

Scott,Sophia Marie
Equ 21h3'18"8d1'
Scott,Stacia
Uma 9h8'36"70d40'
Scott-Stanmore,Pamela Anne
Pho 23h48'45"40d45'
Scott,Susan
Eri 3h14'0"-7d14'
Scott,Susan & Frank
Ori 5h30'56"-3d12'
Scott,Suzan L
Lyn 8h7'28"38d2'
Scott,Syrona Renae Horton
Com 13h12'42"30d11'
Scott,Teri Ann
Uma 8h30'48"48d25'
Scott,Tipponey
Cnv 14h0'20"46d40'
Scott,Todd Richard
Mon 8h5'20"-4d54'
Scott,Tom Foolery
Cep 22h31'19"58d28'
Scott,Total Eclipse of the Heart
Her 17h30'34"26d59'
Scott,Tracy Lynn
Mon 6h20'40"8d46'
Scott,Trey Hamilton
Cet 2h54'57"1d39'
Scott,Troy Allen
Psc 0h53'10"24d7'
Scott,Valerie
Cyg 19h39'35"31d3'
Scott,Victoria Katherine
And 0h31'19"40d32'
Scott,Wanda McKnight
Uma 11h3'13"70d47'
Scott,Willard
Boo 15h3'17"21d48'
Scott,William Paul
Gem 7h58'0"30d9'
Scott,William Provan
Lac 22h36'29"53d7'
Scott-"50",Sam & Evelyn
Cyg 19h27'58"31d39'
Scott-Green,Amelia Elizabeth
Gem 6h51'47"31d14'
Scott-Johnson,Joan H
Vul 19h36'38"22d59'A
Scott-Moore,Peter B
Boo 14h5'23"36d53'
Scott-Rogers,Richard
Ori 5h27'25"-1d28'
Scott-Whittaker,Sherri Lanan
Equ 21h3'20"11d48'
Scott/Danielle-Forever
Uma 10h3'0"47d33'
Scottie
Per 2h3'30"48d57'
Scottie
Ori 4h51'39"5d40'
Scottie & Diny 4-Ever
Aur 4h58'33"37d57'
Scottie Bob
Dra 19h29'53"71d9'
Scottish Princess Sandra
Cas 23h38'33"61d7'
Scottles
Cma 6h12'56"-16d9'
Scotto,Gregory Matthew
Leo 10h56'10"23d41'
Scotto,Joseph David
Leo 10h34'23"23d37'
Scotto,Michael Jeffrey
Aur 5h4'1"50d53'
Scotto,Samantha Starr
Lyn 7h25'0"58d43'
Scotton,Mary & Dierdre
Shedlow
Lyn 8h54'54"37d13'
Scottonius
Sct 18h32'28"-6d17'
Scottstar
Cep 21h38'32"61d13'
Scotty Boy
Ori 5h54'19"12d40'
Scotty D
Cep 22h23'39"71d3'
Scotty S
Eri 4h53'35"-4d55'
Scotty Vincent
Cam 6h19'27"83d48'
Scotty's 24 Watts
Aqr 23h8'19"-8d3'
Scotty's Beam
Aqr 23h2'59"-3d49'
Scotty's Place For Kids
Aur 4h52'0"40d6'
Scotty-O
Aql 20h0'57"-0d6'
Scotty-Pooh
Cyg 20h42'49"37d42'
Scoubidou
Ari 1h55'52"15d53'
Scouler,Daniel Nicholas
Aur 6h28'0"52d49'
Scout
Boo 14h37'47"38d32'
Scouten,Dylan Douglas
Cep 22h7'37"60d52'
Scovell,Jane
Cap 21h40'59"-23d29'
Scovill,Scott
Cep 22h52'32"58d8'
Scoville,Sr,Dennis H
Peg 21h49'16"31d25'
Scragg,Michael L
Cep 22h1'22"60d48'
Scranton,Mary L
Lac 22h54'35"55d58'
Scranton,Paul & Barbara
Cyg 21h31'47"40d42'
Scribner,Dorothy Ann
Mon 7h59'15"-3d38'
Scribner,Jim
Cet 3h6'38"1d9'
Scribner,Karl & Judy
Sex 10h11'45"-8d34'
Scricciola,Beatrice
Umi 16h22'0"80d26'
Scrim,Catherine Ann
Uma 14h4'43"50d56'
Scrima "Vinnie", Vincent
Pasquale
Del 20h18'28"20d27'
Scrima,Edna & Fred
Cam 3h43'16"61d44'
Scrimali,Eric
Lyr 18h43'38"31d58'
Scrip & Little Joe 38
Cep 22h14'60"80d23'
Scripice
Psa 22h32'33"-27d36'
Scripps,Suzanne Michele
Del 20h18'30"10d59'
Scripter,Cheri L
Cas 23h23'24"54d17'
Scripter,Jeremy James
Her 17h5'33"38d0'
Scriva,Darlene
Cas 0h41'22"61d44'
Scrivano,Vanessa
Cas 0h24'44"68d51'
Scroggin,Garrett S
Dra 16h4'55"52d14'
Scrudato
Lmi 10h55'28"33d33'
Scruff
Her 18h7'25"14d57'
Scruggs,Janet Elaine
Cas 0h35'19"66d54'
Scruggs,Tom
Aql 18h44'51"-2d56'
Scrupulloyd
Umi 14h21'19"69d44'
Scudella,Robert
Lyn 7h7'26"58d29'
Scudiery,Patty
Lyr 19h25'56"38d37'
Scudillo,Louis J
Her 18h8'1"30d28'
Scuglia,Maria
Lyr 18h50'20"32d21'
Scull,Margaret Redman
Mon 5h59'12"-5d54'
Scull,Sam Wingo
Umi 13h9'15"75d42'
Scull,William
Dra 12h8,52"75d8'
Sculley Gods Gift To Kids,Viki
Cas 0h41'31"62d58'
Sculley,John
Lac 22h12'35"50d5'
Scullin,Naomi Elizabeth
Cas 0h49'28"63d39'
Scullion,Jonathan Alexander
Hya 9h12'24"0d26'
Scullion,Nicholas Terrence
Aql 19h25'45"8d25'
Scully F L C,Stephen Patrick
Oph 17h1'27"10d42'
Scully,Marie
Ori 6h3'49"0d2'
Scully,Thomas P
Sex 9h42'0"2d51'
Scully,Tony
Her 17h19'14"42d29'
Scully-CYR
Umi 13h52'0"70d48'
Sculrz,Monica
Scl 0h0'25"-27d2'
Scupin,Michael
Equ 20h57'24"8d13'
Scupper
Aql 19h47'32"10d48'
Sczechowicz,Jr,PhD,Ed
Cma 6h56'48"-19d41'
Sczepan,Johannes
Sco 17h51'58"-30d37'
Sczytko
Per 1h44'21"53d59'
Scé,Amanda E
And 0h16'0"31d14'
Sdad,Gary W
Ser 18h42'49"4d26'
SDC-l'étoile éternelle
Lyr 18h37'30"32d43'
SDR
Cet 0h28'56"1d26'
Sdrupa
Peg 22h0'60"28d8'
Se Enamoré de la Amiga
Scruffy
Cet 1h33'22"-0d4'
Sea Captain Sam
Cyg 20h32'4"34d30'B
Sea Fox,The
Uma 10h7'44"67d50'
Sea Goddess
Vir 11h59'11"-2d37'
Sea King
Cep 21h3'57"58d26'
Sea,Air,Space;You Me Always
Cyg 21h10'12"35d11'
Seaberg,Robert B
Cep 0h26'46"78d21'
Seabolt-Seguin
Uma 11h30'0"31d24'
Seabrands,Norma Viola
Lyn 7h57'41"44d55'
Seabright Millennium
Star,Bette
Com 12h29'40"23d15'
Seacrest,Jay Steven
Cep 22h4'58"53d56'
Seadcayt "Carisima", Carmen
Per 2h10'54"58d38'
Seador,Ernest A
Aur 6h35'1"32d0'
Seagal,Steven
Her 16h20'33"24d29'
Seagrave,Leslie B
Lyn 9h15'24"37d17'
Seagraves,John Kootenai
Lac 22h11'49"48d54'
Seagrief,Margaret
Cas 1h50'0"74d58'
Seal,Damian
Ori 5h55'10"19d51'
Seal,Tiffany Lynn
Sex 10h37'1"-5d35'
Seal,Tiffany Lynn
Peg 22h5'21"21d29'
Sealander,Leslie
Lyn 7h55'16"44d33'
Sealey for Ever More, Michael
Her 18h0'36"40d41'
Sealls,Gretchen Marie Kapaun
Aql 20h35'59"-7d42'
Seals,John Carleton
Cnv 13h17'32"40d47'
Seals,Patty
Equ 21h19'15"3d14'
Seals-Our Mom-Our
Star,Bessie Lee
Tri 2h18'10"31d35'
Sealy,Gladys"Glenn"
Umi 15h5'38"68d58'
Seaman Family
Del 21h5'14"12d53'
Seaman,Ellen Regina
Dra 17h4'23"63d7'
Seaman,Matt & Leigh
Crb 15h52'34"27d1'
Seaman,Thomas M
Lac 22h10'42"38d2'
Seaman,Todd Anthony
Boo 14h48'30"50d10'
Seaman,Tony
Uma 11h43'50"38d60'
Seamans,Brad
Dra 11h53'14"71d7'
Seamus
Cmi 7h55'37"8d13'
Sean & Brenda's Star Of Love
Vul 20h18'42"23d47'
Sean & Cory Star,The
Cyg 20h30'29"42d29'
Sean & I
Aql 19h9'23"0d53'
Sean & Julie
Aql 19h29'23"-8d12'
Sean & Tom's Guiding Light
Sge 19h53'38"19d25'
Sean & Valerie
Uma 9h5'34"56d47'
Sean 25
Lac 22h55'34"38d7'
Sean Forever & Always
Per 3h28'19"52d32'
Sean Gilbert
Lac 22h22'32"38d8'
Sean Lindsay
Cep 3h3'30"77d17'
Sean O' My Heart
Per 2h55'37"40d31'
Sean O'B-Star-70-94
Oph 16h0'60"-5d57'
Sean Star
Ser 15h55'0"0d26'
Sean World
Lac 22h5'17"46d26'
Sean's Ambition
Cet 1h34'1"-1d2'
Sean's Big Bright Star
Aql 19h8'58"3d9'
Sean's Christmas Star
Lmi 9h28'57"37d52'
Sean's Light
Aql 20h9'14"1d8'
Sean's Star
Sgr 19h19'17"-41d24'
Sean's Star
Dra 16h2'44"63d35'
Sean's Sun
Cep 22h51'14"70d11'
Sean's Wish
Cep 22h10'51"59d39'
Sean-E-Man
Cep 0h5'23"75d22'
Seanain,Caitlin Anna Ni
Peg 22h51'46"21d45'
Seanathair
Ori 5h4'33"1d6'
Seander,Ileana
And 23h49'49"40d26'
Seane & Richard "Till The
12th of Never"
Cnv 12h17'1"46d58'
Seane,Harold
Cep 22h17'1"67d52'
Seanez,Joe
Boo 13h44'16"27d47'
Seanmeister
Cnc 8h32'1"8d22'
Seany
Aur 7h6'14"40d40'
Sear,Stanley L
Ori 5h53'37"21d34'
Sear,Ward S
Aql 19h50'0"11d41'
Search,HRH Kerri Ann
Gem 6h45'2"26d58'A
Searching For Marilyn
Uma 9h30'0"52d43'
Searcy 6-25-87,William Cody
Cep 21h7'59"58d51'
Searcy,Our Star,Martin /Loue
Anne
Her 17h4'13"31d34'
Seariac,Bernard Edward
Leo 11h25'35"-0d46'
Seariac,William (Buzz)
Aur 5h54'49"40d45'
Searing,Bernard Anthony
Cnv 12h22'24"36d36'
Searing,Brianna Marie
Cas 23h26'44"62d55'
Searing,Laurann Carrol
Lyr 18h44'33"38d48'
Searing,Ryan Patrick
Del 20h13'0"10d33'
Searle,Gordon & Betty
Aur 5h1'40"29d59'
Searle,Laurie R
And 23h48'50"42d48'
Searle,Louise Cowles
Lyn 8h29'35"40d40'
Searles,Donald James
Lac 22h35'40"53d23'
Searles,Katherine Anne
Peg 23h24'23"32d53'
Searles,Lisa
And 23h2'30"40d5'
Searles,Natasha Poppy
And 23h22'15"50d9'
Sears,Gertrude Belle
Cyg 21h9'49"40d29'
Sears,Kenneth
Dra 16h37'55"60d2'
Sears,Maggie
Lyn 8h3'17"47d22'
Sears,Nathaniel James
Vul 19h48'18"26d15'
Sears,Samantha
Vul 19h18'54"27d18'
Sears,Shari
Peg 22h11'0"25d10'
Sears,T L
Uma 8h43'0"71d35'
Seastrom Star,The Gordon D
Uma 10h30'30"53d31'
Seastrom,Kristina
Lyr 18h31'1"30d28'
Seaton,Lauren Reed
Mon 6h43'45"2d59'
Seattle
Aur 7h6'40"40d59'
Seaver,Edward Cletis
Aur 6h8'45"40d1'
Seaver,Pearl
Lyn 7h43'32"41d34'
Seavey Anthony
Aql 20h15'1"8d29'
Seawall,Todd J
Her 16h53'17"37d25'
Seawell,Mrs
Lyr 18h59'33"26d47'
Seay Jr,Paul
Aur 5h57'21"30d28'
Seay Sr,Paul
Cam 3h53'26"60d18'
Seay,Bea
Cyg 21h3'40"38d13'
Seay,Gerald R
Per 3h4'53"41d5'
Seay,Harold
Cep 20h49'0"67d48'
Seay,Shelby Sarie
Dra 15h39'50"58d6'
Sebasti,Danielle
Lyn 8h3'32"39d1'
Sebastian
Peg 23h4'11"18d29'
Sebastian
Ori 4h54'26"4d58'
Sebastian Alexander
Hya 9h1'1"4d56'
Sebastian M
Umi 15h50'10"80d10'
Sebastian Skywalker
Ari 2h39'12"20d36'
Sebastian,Albert P
Her 16h22'25"48d15'
Sebastian,Jr,(Jojo), Hipolito D
Aql 18h59'28"10d26'
Sebastian,Julie Matyas
Cas 2h48'59"60d43'
Sebastian,Victoria (Vickie)
Ori 5h54'53"19d11'
Sebastien
Cam 7h40'55"68d53'
Sebastien
Tri 2h3'31"31d34'
Sebastien-Rodolphe- Jean
Cam 5h43'1"60d56'
Sebbo,Rosemary & David
Cyg 20h6'11"41d5'
Sebenius,Clare
Sco 16h50'22"-40d58'
Seberle,Patricia
Peg 21h57'0"20d14'
Sebok,Susan Marie
And 23h38'0"47d47'
Sebor,Ing Pavel G
Lac 22h39'57"56d48'
Sebree,Anita Marie
Cet 2h46'32"0d1'
Sebring,Terri Ann
And 23h27'19"45d28'
Sebyzia
Vel 9h43'8"-42d49'
Seca,Cabeza
Cmi 7h26'12"0d47'
Secade,Rolando Cubela
Cen 11h41'29"-59d43'
Sech,Megan Kelly
Eri 3h52'12"-2d17'
Sechan,Honey Bear
Cas 2h25'11"68d5'
Sechser,John & Linda
Cnv 12h18'34"38d18'A
Sechser,Sir Riley Spencer
Cnv 12h18'34"38d18'B
Seckelmann,Ingrid Gerda
Cas 1h3'37"55d15'
Secker,David J
Ori 6h0'53"0d56'
Secker,K Jayne
And 0h37'1"40d30'
Second Chance
Ori 6h5'54"0d36'
Secondwind Smile
Aql 20h7'23"0d36'
Secrest III,USMC, Major Lloyd
D
Aql 19h47'25"14d39'
Secrest,Teddy
Cap 21h21'50"-24d9'
Secret Love
Cyg 19h30'16"33d52'
Secret Squirrel
Cyg 21h28'47"39d54'
Secrist,William Lee
Vul 19h43'0"26d47'
Secundo,Andrew & Sally
Aur 6h4'1"46d31'
Securitas
Dra 16h50'15"62d50'
securitos
Aqr 21h25'1"-0d41'
Seda,Steven Matthew
Boo 14h10'0"31d37'
Sedaka,Neil
Boo 14h42'34"29d3'
Sedam,Tommy A
Ser 15h34'11"10d54'
Sedan,Antonella
Lac 22h17'1"38d28'
Seddon-Love of my Life,Roy
John
Aur 6h31'0"37d57'
Sedelmeier,Joachim
Dra 16h20'36"61d54'
Seden,Mr & Mrs P W F
Cyg 20h48'13"37d42'
Sedensky,Eric & Hiroko
Cyg 21h0'1"30d52'
Sederberg,James Arthur
Sex 10h22'0"5d55'
Sedereas,Peter C
Crb 15h51'44"30d12'
Sederstrom-Enzminger
Ori 4h59'30"11d3'
Sedgewick,Herbert
Lyr 18h38'0"39d11'
Sedgwick,Jr,Dale
Sct 18h55'28"-4d44'
Sedgwick,Nile Flemming
Aql 19h40'28"12d57'
Sedira,Yves
Ori 6h0'49"20d50'
Sedkey,Justine Patricia
And 0h18'1"33d50'
Sedlacek,Bridget Lea
And 23h16'19"40d13'
Sedlacek,James Donald
Her 16h57'40"26d3'
Sedlack,Diane Y
And 2h2'53"45d47'
Sedlack,Michael Thomas
Cam 4h4'20"61d46'
Sedlak,Debbi
And 1h0'40"47d56'
Sedlak,John Edward
Her 18h45'12"12d29'
Sedlik,Barry R
Cet 0h26'39"-5d36'
Sedlik,Mason Nathaniel
Oph 17h39'4"-23d56'
Sedláčková,Helena
Tau 4h0'50"0d8'
Sedon,Paul
Oph 17h17'24"-22d30'
Sedona Night Star
Umi 14h58'39"78d52'
Sedona Sky
Tau 4h39'58"20d5'
Sedor,Ann-Marie
Cas 23h3'37"58d47'
Sedway,R Bryan
Aql 19h46'37"14d19'
See,Arthur & Dorothy
Cam 8h20'0"78d49'
See,Jim & Jennifer
Per 3h42'41"36d58'
See,MacKenzie Fedora Lula
Cam 8h20'54"78d15'
See,Myles James Arthur West
Cam 9h8'0"78d5'
See,Raymond Bonavere
Her 15h55'21"40d22'
See,Stephanie M Podgurski
Lyr 19h15'57"28d35'
See,Victor
Oph 17h56'28"13d32'
See-Ra
Umi 15h29'42"68d9'
See-ra
Umi 14h5'32"68d28'
Seebacher,Axel
Lyn 9h16'11"37d47'
Seebald,Annemari
Cmi 7h20'50"8d58'
Seeberg,Karen
Sex 10h0'0"-5d36'
Seeberger,Bruno
Uma 9h56'1"59d57'
Seed,Sean S
Dra 17h1'10"66d19'
Seedlock,Maj Gen Robert
Hya 9h8'44"1d11'
Seeds,Eric John
Peg 23h7'41"14d16'
Seefried,Barbara A
Aql 19h57'58"10d14'
Seehagen,Eveline
Dra 20h24'10"68d39'
Seeiser,Milton D
Ori 5h53'50"16d53'
Seek,Amy
Eri 3h14'0"-13d49'
Seekely,Alexandra
Equ 21h21'1"10d18'
Seekely,Christopher
Hya 9h5'11"5d47'
Seeker-Go Irish!, George E
Aur 4h56'56"41d2'
Seekinglite
Boo 15h15'20"40d16'
Seel,Jamie R
Lyr 19h4'0"40d29'
Seel,Julie
And 23h21'42"43d24'
Seeley,John B
Cyg 19h32'30"32d37'
Seeley,Joseph L
Boo 15h3'54"13d6'
Seeley,Michael James
Cet 0h58'54"-6d40'
Seeley,Summer Raye
Mon 6h7'0"-10d36'
Seeley,Terri L
Cyg 19h25'1"30d12'
Seelhorst,Jason Glenn
Aur 6h19'21"46d9'
Seelig,Kathy,Jerry & Judy
Umi 15h46'19"78d32'
Seelmann-Eggebert, Dietrich
Peg 23h29'45"13d28'
Seelmann-Eggebert, Dietrich
Her 17h21'36"46d57'
Seely,Linda & Ed
Cyg 19h15'54"49d24'
Seely,Tony
Hya 8h52'57"0d52'
Seelye's Star,JoAnn & Doug
Lyr 18h35'0"37d53'
Seeman,Phomas
Cam 13h29'31"77d44'
Seemann,Parker John
Lac 22h38'30"56d44'
Seepe,Marita
And 23h13'58"40d54'
Seery,Brendan Thomas
Browne
Gem 7h12'26"24d52'
Seery,Dave
Dra 19h4'1"50d37'
Seesc,Peggy
Eri 2h57'1"-2d52'
Seeto,Sabrina
Lyn 8h8'40"33d34'
Seevers,Brenda
And 0h1'0"38d30'
Sefchick,Aaron G
Lac 22h40'23"50d13'
Seffert,Jörg Wolfgang
Vir 13h30'46"-8d46'
Seftar,Jim
Dra 20h7'1"75d14'
Sefton,Christine
And 23h15'46"46d8'
Segal Macy's 1995 Selling
Star,Rick
Ori 5h56'16"8d49'
Segal,Bernice
Cas 1h57'53"58d23'
Segal,Clara
Ser 17h27'6"-5d52'B
Segal,Debbie (G W)
Her 16h55'19"25d57'
Segal,Elysia Meghan
And 1h16'47"36d6'
Segal,Frederick Leslie
Eri 3h37'17"-9d5'
Segal,Fredrick
Cet 2h27'37"5d3'
Segal,Gordon E
Uma 11h59'58"64d54'
Segal,Harry
Ser 17h27'6"-5d52'A
Segal,Howard & Mindee
Lac 22h8'0"37d54'
Segal,Myra J
Aur 6h4'24"31d36'
Segal,Norman
Dra 10h29'38"78d37'
Segal,Paula
Cap 20h56'49"-16d42'
Segal,Robert J
Per 2h9'1"58d44'
Segal,Sabrina Meredith
Cas 0h54'51"72d5'
Segal,Sumner
Vul 19h47'34"27d58'
Segala,Erica
And 23h38'1"44d48'
Segalas,Daphne
Gem 7h9'18"21d32'
Segale,Emily Rae
Mon 7h5'21"-0d53'
Segall,Cory
And 23h45'35"45d12'
Segall,Jessica
Vul 20h14'43"22d55'
Segall,Michael Ronald
Lmi 9h42'30"40d14'
Segall,Morgan
Lyn 8h46'59"38d51'
Segalla,Anthony David
Uma 10h46'34"58d46'
Segasser,Madison Leigh
Umi 14h36'19"82d17'
Segatti,Anna Mary
Cas 0h9'39"58d55'
Segear's,The
Umi 15h4'45"77d2'
Segelke,Margaret
Lyr 18h42'31"33d51'
Segelstrom,Gary & Sharon
Lac 22h34'11"52d47'
Seger,Jennifer "Sassy"
Lyr 18h31'39"36d11'
Segerdahl,Kay
And 23h43'53"32d30'
Segers,Thomas
Boo 14h20'18"33d56'
Segesta,S Hunter
Tau 4h5'7"22d53'
Seghetti,Doris
Cas 2h22'46"63d46'
Segien,Peter
Aur 5h52'34"38d53'
Seglers,Maria Jose Gomis
Her 16h54'53"26d20'
Sego,Jeanette
And 1h39'48"37d54'
Sego,Robert F
Boo 15h0'11"15d56'
Segrave Star,Brian
Per 1h35'46"53'32'
Segro,Emma & Joseph
Lyn 8h51'0"41d25'

Segrè R D M
Ari 2h31'59"20d8'
Seguin True Love, E L Y A S Alban
Per 1h44'23"50d24'
Seguinot,Peter
Her 17h31'52"42d56'
Seguna,Loving Memory Of Katie & John
Umi 16h17'43"78d1'
Segura Star,The
Peg 23h32'42"33d36'
Segura,Christian
Peg 23h36'0"10d47'
Segura,Hector Leopoldo
Per 3h33'35"40d19'
Segura,Michele Ranee
Mon 7h10'43"-7d9'
Sehagal,Sangetta
Lyr 18h28'43"46d29'
Seherlis,Alexandra & Sotiris
Vul 20h16'35"23d31'
Sehi-Smith,Susan Sharon
Uma 14h14'49"59d14'
Sehloff,Joern
Ari 2h29'36"30d13'
Sehm,Bärbel
Cep 21h42'0"61d23'
Seholp,Katherine Louise
Eri 3h3'13"-8d16'
Sehs 94,Matt Morton
Ori 5h56'48"21d5'
Sei Una Ragazza Intel Piena Di Passione
Cyg 20h57'14"30d51'
Seib,Friedrich Georg
Cep 22h2'0"68d37'
Seibarth,Michael Charles
Cep 20h44'46"65d35'
Seibel,Oliver
Lyn 8h3'53"43d38'
Seibert,Catherine Caroline
Mon 7h38'57"-1d57'
Seibert,D C "Don"
Aur 6h23'1"31d53'
Seibert,Don
Cnc 8h33'39"30d2'
Seibert,Joseph Sloane
Cnc 8h40'36"17d56'
Seibert,Kyle Joan
Aur 5h39'30"38d8'
Seibert,Mary Ellen
Eri 5h2'41"-6d50'
Seibert,Max Manning
Lac 22h40'44"38d6'
Seibert,Nina
Psc 1h24'13"32d38'
Seibert,Nola Ann
And 2h27'25"48d35'
Seibert,Paul
Cep 22h35'52"70d8'
Seibert,Richard Manning
Aur 6h26'25"31d1'
Seibert,Russell & Isabelle
Uma 11h53'15"57d11'
Seibicke,Katja
Ari 2h38'53"21d33'
Seibold,Christopher
Cnc 8h9'45"30d35'
Seibold,Scott Ernest
Cnv 12h20'0"48d25'
Seide,Sylvia & Morris
Uma 8h45'37"71d4'
Seidel,Forever "Bud" Jeffery L
Dra 18h10'19"67d44'
Seidel,Robert Eugene
Aur 6h6'0"50d11'
Seidell,John Patrick
Aql 18h54'58"11d57'
Seiden,Amy
Cas 1h44'37"58d21'
Seidenberg,Ivan
Boo 14h9'16"31d35'
Seider,Robert R
Ser 15h55'42"-2d30'

Seidl Star
Mon 7h42'3"-5d7'
Seidl,Edward A
Her 26h56'47"21d36'
Seidl,Kristine Karen
Aql 19h9'55"13d12'
Seidl,Sepp und Marianne
Cap 21h34'53"-21d46'
Seidler,Lorraine"Bubi"
Ari 1h54'54"12d27'
Seidlin,Judy
Cet 1h54'48"-5d40'
Seidman's Diana,David
Per 2h59'27"32d58'
Seidman,Blair Morgan
Cam 3h54'25"74d36'
Seidman,DDS,Dr Alan
Oph 16h54'54"-6d50'
Seidman,Nellie
And 1h59'34"38d24'
Seifert,Anita Renée
Cas 3h3'35"65d11'
Seifert,Christie
And 0h13'12"37d14'
Seifert,Christoph
Cap 21h12'47"-20d33'
Seifert,Fabian
Uma 12h43'18"58d6'
Seifert,Gretchen Corie
And 0h9'1"47d9'
Seifert,Marie Elizabeth
Cas 0h49'49"64d53'
Seifert,My Star
Lyr 18h46'47"34d55'
Seifert,Phillip J
Her 16h8'1"42d14'
Seifert,Werner
Sco 16h52'35"-40d20'
Seiff,Matthew
Per 3h20'36"40d5'
Seiffert,Heather
Scl 23h22'40"-27d28'
Seifried,Jason Gary
Aql 20h12'48"1d20'
Seifried,Megan Joy
Del 20h17'58"10d32'
Seigal,Howard R
Ser 15h20'52"1d16'
Seigfried,Herbert
Mon 7h53'22"-2d1'
Seigle,Jacquelyn Lauren
Cas 0h31'20"75d24'
Seignadou
Cam 10h14'54"82d20'
Seigner,Andrew Jay
Ser 15h38'55"2d43'
Seignert,Bernerd
Cep 22h14'13"58d45'
Seigo
Cyg 19h57'20"31d32'
Seijo Lopez,Victor
Ori 6h4'57"8d30'
Seikel,Kenneth W
Aur 6h56'44"37d1'
Seiko's Dream
Sgr 19h14'54"-16d40'
Seiler & Associates, Inc
Ser 15h13'1"8d51'
Seiler,Chantal
Dra 11h57'11"70d59'
Seiler,Robert R
Mon 7h18'40"-6d39'
Seiller,Alexander Joseph William
Her 18h2'41"48d55'
Seim,Julia Lee
Del 20h19'32"10d21'
Seim,Sara
Cas 0h3'41"58d41'
Seim,Terri Lynn
Cam 7h55'0"67d45'
Seinen,André
Per 2h56'27"31d14'
Seinfeld,Jerry
Her 16h20'46"25d48'

Seinsoth,Robert Joseph
Dra 14h31'35"63d16'
Seipel,Eric J
Cet 0h39'19"0d54'
Seitel,Eugenie
Boo 14h51'25"38d52'
Seitel,Maximilien
Cnv 12h19'1"38d47'
Seiter,Donny P
Lac 22h27'24"52d31'
Seitfeld,Gertrud-Anna-
Cap 20h54'34"-15d9'
Seitler,David Roman
Dra 18h23'44"70d43'
Seitler,Shawn Dana
Her 18h6'24"41d7'
Seitler,Stefan Michael
Per 3h33'28"39d39'
Seitsive,Dr Lillian Paula
Oph 18h36'47"11d52'
Seitz Star I,Jarrett David
Boo 14h3'1"28d12'
Seitz Star I,Tamara Lee
Lyn 7h35'40"39d60'
Seitz,Alois
Gem 6h53'42"31d32'
Seitz,Diane
Hya 9h37'28"-6d31'
Seitz,Edward A
Ori 6h17'23"-2d24'
Seitz,Ingrid
Lib 15h33'39"-18d48'
Seitz,Irmgard
Vir 12h56'12"-17d41'
Seitz,Joyce Hooker
Mon 7h3'1"1d46'
Seitz,Reinhold
Ori 6h17'12"7d56'
Seitz,Walter
And 23h12'43"42d42'
Seiwell,Betsy
Tri 1h57'24"26d28'
Seiwert,Herbert
Crt 11h14'15"-18d31'
Seiwert,Ilse Dr
Aqr 21h1'42"-13d30'
Sejka,Michael
Cep 23h34'58"68d18'
Sekach,Michael James
Her 17h28'1"28d44'
Sekanina,Michael
Aqr 21h3'1"-14d16'
Sekellick,Sharon T
Aql 19h56'0"1d41'
Sekerak,Terese Michelle
And 23h36'53"48d51'
Seki,Bryce Tadahirio
Hya 8h53'11"1d20'
Seki,Hiromi Anne
Mon 7h0'27"-1d30'
Sekiya,Joseph N
Aql 20h5'31"4d27'
Sekiya,Kelly K
Eri 3h53'13"-4d42'
Sekiya,Stacy R
Del 20h29'22"20d19'
Seko's Star
Lac 22h29'37"50d24'
Selah
Peg 23h20'23"32d20'
Selberg,Kevin
Uma 11h38'15"44d22'
Selbig,Bronson DuBois
Boo 15h4'0"31d7'
Selby SW
Sct 18h46'3"-7d11'
Selby,Melinda
Aql 19h28'24"-10d46'
Selby,Pam
Lmi 10h18'54"38d12'
Selby-Lowndes,Greville
Aur 5h31'56"30d34'
Selby-Lowndes,Patricia
And 0h31'59"40d14'

Selby-Lowndes,Sally
And 0h41'0"40d22'
Selby-Lowndes,William
Aur 5h54'49"40d14'
Selcher,Craig Andrew
Dra 14h5'51"64d37'
Selden,Jeffrey
Per 3h4'16"48d53'
Seldin,Binah Kertzer
Aqr 22h31'0"0d25'
Seldin,Traci
Gem 6h44'55"18d57'
Seldom
Lac 22h52'49"56d17'
Seldon,Sheila June
Lyr 18h18'1"43d29'
Sele,Kyla Linnea
Cam 8h21'1"80d44'
Selecki,Daniel
Ori 6h5'0"0d53'
Seleire
Crb 15h15'33"31d29'
Selema,Luis E
Aql 19h48'1"12d56'
Selene
Cep 23h19'0"64d28'
Selene,Clossiana
Lyr 18h28'46"32d7'
Selengut-Good Soul, Rebecca
Cas 0h4'53"62d36'
Selenia
Cnv 13h1'20"33d17'
Selent,William Parker
Aur 5h15'19"46d32'
Selep's Star, The Dave
Boo 15h21'33"50d50'
Seles,Karol J
Eri 2h59'1"-15d37'
Seles,Zoltan
Mon 7h55'13"-0d26'
Seleski,Ryan Edward
Lac 22h24'1"40d44'
Seley,Robin Ray
And 0h12'1"46d60'
Self 6-14-83,Kristan Nichole
And 0h6'28"44d24'
Self,Dean Cameron
Per 2h54'24"43d12'
Self,Jill Marie
Cas 0h31'29"64d59'
Self,Lawrence Roy
Ori 5h42'19"11d17'
Self,Melvin
Her 16h35'59"34d40'
Self,Ronald
Aur 4h55'1"50d28'
Self,Tim
Ori 5h35'38"1d39'
Self,Willie
Aur 6h3'28"46d25'
Selgrath,Louis Francis
Cep 23h12'31"60d23'
Selic
Hya 9h5'15"5d58'
Selig,A Special Glow Peggy F
Com 13h7'28"30d40'
Selig,Jonathan Henry
Lyn 7h49'29"35d32'
Selig,Phillip
Boo 14h24'22"30d58'
Seligman,Barbara Bancrost
Lyn 8h43'47"46d41'
Seligman,Jesse Michael
Cma 6h47'36"-17d14'
Seligman,Peggy
And 23h20'59"41d6'
Seligman,Sally
Lyr 18h58'57"42d24'
Selina Anastasia
Cas 1h4'35"53d35'
Selinger,Al
Cmi 7h40'31"0d18'B
Selinger,Carol Sue
Mon 7h4'49"-0d41'

Selinger,Elizabeth
Cmi 7h40'31"0d18'A
Selinger,Ester Cynthia
Cet 1h32'21"-12d21'
Selinger,Gail M
Hya 9h2'1"0d48'
Selkie
Cet 2h52'16"-0d54'
Sell Star,The Joseph M
Oph 18h42'18"10d42'
Sell,Rodney -stands beside
Cen 11h33'24"-41d14'
Sellar,Polly
Lyr 19h23'57"40d53'
Sellares,Lady Mar
And 0h45'44"38d45'
Sellars,Eve
Ori 4h53'20"1d9'
Sellars,Mr Peter
Per 2h4'0"57d44'
Sellars,Robert Frederick
Ori 5h56'54"10d33'
Selle,Darlene
And 0h4'58"31d50'
Selle,Jill
Tri 2h17'1"33d24'
Selleana
Lyn 7h32'56"40d4'
Seller,Gary
Boo 14h48'35"33d37'
Sellerio,Sergio
Aur 5h3'16"30d34'
Sellers P 10,Linda
Gem 6h53'55"30d35'
Sellers,Christopher A
Vul 20h29'18"28d15'
Sellers,David Edward
Aur 6h2'1"40d44'
Sellers,Dean
Aql 18h58'40"17d9'
Sellers,Effie Mae "Gaga"
Cas 0h49'27"64d3'
Sellers,Jason Monroe
Sex 10h34'0"1d43'
Sellers,Julie Lynn
And 1h0'17"38d23'
Sellers,Kimberly Gayle
Aql 18h57'11"-7d48'
Sellers,Miss Dana
Cnc 8h38'1"18d13'
Sellers,Pupa Kishinchand
Uma 11h54'19"56d56'
Sellers,Thomas F
Sex 9h55'0"-0d51'
Sellinger Star,Joe Rock
Boo 13h55'29"30d5'
Sellinger,Melanie Kate
Cas 3h14'42"70d13'
Sellinger,Mr & Mrs John F
Lac 22h27'37"53d51'
Sellings,Peter Darren
Lyn 7h45'46"39d49'
Sellke,John Evard
Ori 5h55'1"11d17'
Sellmanstar-VI
Oph 18h15'15"0d7'
Sells,Callie
Cmi 7h54'37"0d24'
Sells,Dixie Lynne Daugherty
Mon 7h0'0"8d39'
Sells,Heather & Stephen
Ari 2h32'40"22d3'
Sells,Kathy
Lyr 18h31'0"30d53'
Sells,Russell P
Boo 13h57'24"26d15'
Selma
Cnv 13h26'48"40d8'
Selma & Sol
Aql 18h47'0"10d57'
Selma & Walter
Uma 8h59'25"71d55'
Selma Jane
Mon 6h26'10"-6d29'

Selman,Greg (Graham)
Uma 9h23'15"53d29'
Selman,John Christopher Hoke
Cmi 7h58'55"5d32'
Selover,Sharon Lea
Cas 0h56'49"69d50'
Selph,Betty Ann
Crb 16h11'12"30d20'
Selsek,Raymond
Ser 18h29'24"5d53'
Seltzer,Gary
Per 2h27'49"56d40'
Seltzer,Kim
Cas 23h39'23"50d0'
Selva,Graham Christian
Cet 2h25'1"3d30'
Selvaggia
Uma 9h57'1"47d47'
Selvaggia
Lyn 7h3'1"60d17'
Selvi 03/13/93,Olivia
Per 2h52'56"50d22'
Selvy
Sex 10h27'53"4d3'
Selway,Roy
Ori 5h44'18"10d55'
Selwyn
Uma 11h31'32"32d27'
Selzer,Gerhard Adalbert Königsberg
Cap 20h15'15"-26d43'
Selzer,Psychologist, Tehila L
Oph 17h24'58"1d8'
Sembrat,Eric
Cet 0h45'14"-3d50'
Sembrat,Kristen
Mon 6h46'34"11d49'
Semcal
Peg 23h15'60"31d36'
Semel,Ran
Ori 5h41'32"12d20'
Semenuk,John & Gerry
Lyn 9h5'59"35d23'
Semenzato 95,Giuliano
Aur 4h59'0"40d12'
Semerad,Jason J
Boo 13h43'15"24d4'
Semeraro,Troy-Jon Bruce
Cap 20h33'31"-26d47'
Semet,Pudge
Uma 10h23'39"58d7'
Semianczuk,Maria A
Cas 23h39'59"62d54'
Semiao,Andriano
Her 17h33'53"21d6'
Semidey,Luz
Tri 2h6'42"33d23'
Semigran,Lawrence B
Com 13h6'17"21d33'
Seminara,Carmen
Pho 0h3'41"-42d10'
Seminaris,Adele
Cmi 7h40'31"0d35'
Semkov,Peter
Umi 16h12'46"72d41'
Semkov,Vesselin
Umi 16h5'19"70d10'
Semkova,Metodka
Umi 16h48'56"78d52'
Semkova,Vesselina
Umi 16h10'28"74d33'
Semler,Amy Nicole
And 0h54'19"21d34'
Semley,Zoe
Peg 22h29'37"28d58'
Semlitz,My Shining Star the Edward
Aql 19h31'45"10d29'
Semmelink,"Big Brother" Dirk Jan
Boo 15h3'23"32d7'
Semmens,Cynthia C
Eri 3h26'22"-1d35'

Semmens,David
Dra 16h17'48"63d43'
Semones,Vicky M
Lyr 18h49'44"40d30'
Semonik,M D
Eri 3h13'32"-15d15'
Semoursan,Wendy Cheryl Toni
Dra 15h20'14"65d6'
Semper Amenus Ella Zuniga Ordono
Sge 19h41'1"16d12'
Semper Amicas
Lac 22h46'14"54d31'
Semper Amigos Christine,Gabriel/Fam
Mon 6h15'21"-6d26'
Semper Antonius
Mon 6h28'1"8d33'
Semper Eadem
Cam 4h32'48"68d0'
Semper EdandTia
Cyg 21h15'39"37d48'
Semper Fi
Uma 11h25'15"63d25'
Semper Fi
Per 2h3'42"50d8'
Semper Fidelis
Boo 14h12'55"40d18'
Semper Fidelis
Crt 11h15'15"-19d52'
Semper Fidelis
Lac 22h19'16"49d16'
Semper Fidelis To Kevin With Love
Lac 22h42'1"56d28'
Semper Fidelis,DRB
Oph 18h2'1"12d5'
Semper Joe-Anne
Dra 19h32'18"61d24'
Semper Memor
Uma 14h1'40"57d48'
Semper,Robertus
Her 16h42'25"47d38'
Sempiternity
Cyg 19h24'34"33d17'
Semple,Emma Alexandra
Uma 8h14'29"61d44'
Semple,Scott Alexander
Ori 6h0'50"10d6'
Sempre Fame
Lyn 7h28'27"58d57'
Semrad,Margaret Rose
Cyg 19h17'27"49d12'
Sena,Elizabeth
And 0h52'15"33d32'
Sena,Fabian
Ser 17h30'22"-14d37'
Sena,Michael A
Her 16h1'36"41d55'
Sena,Seliena Estelle
Psc 1h2'31"20d6'
Senak,Larry
Boo 14h30'33"43d27'
Senard,Luc
Del 20h17'0"14d33'
Sendek,Gary Michael
Her 17h0'19"27d44'
Sendel,Jane
Peg 22h15'27"34d37'
Sendra,Lionel Aurélien
Aur 5h52'28"38d26'
Sendzik,Charles E
Her 16h38'54"5d35'
Sendzimir,Dr Michael G
Sco 17h50'57"-38d27'
Sene,HLW
Cyg 21h0'28"37d60'
Senecal III,Robert Percy
Aql 20h2'41"-6d22'
Senecal,Bernard
Ori 6h0'32"0d1'
Senequier-Raffet, Dominique
Cep 2h22'29"80d18'

Senese,Robert J
Cep 23h39'19"65d41'
Senft,David V
Per 2h31'1"51d34'
Seng Family Star
Uma 12h51'59"54d33'
Seng,Christopher Edward
Aur 6h28'41"31d6'
Seng,Deborah K
Cas 0h38'47"68d22'
Senger,Jonathan T
Dra 16h26'23"69d20'
Senini,Eleonora
Vel 9h18'47"-42d50'
Senior (Ted),John Edward
Sex 10h38'35"2d36'
Senior,Kristyn T
Lyn 8h27'36"36d15'
Seniuk,Natalie
Cas 1h1'51"61d47'
Seniw,Alice Mae
Mon 6h37'56"10d22'
Senkbeil,Robert B
Ori 5h31'34"-3d1'
Senkovich Radiance, Barbara Jean
Mon 6h55'30"-0d50'
Senkowski,Ronald J
Hya 8h11'0"5d41'
Senn,Glen Wood
Boo 14h2'12"20d32'
Senn,Nathan Robert
Dra 19h21'1"58d40'
Senna
Dra 16h24'38"62d5'
Senna,Ayrton
Aur 6h53'55"37d56'
Senne,Teresa J
Cnc 9h9'31"30d42'
Sennett,ML GT
Cam 7h45'43"70d1'
Senney,Barbara
Cnv 12h26'58"41d34'
Senney,Steven
Cep 0h46'1"77d59'
Senor' Vic
Aqr 22h25'28"0d34'
Senorina
Ori 4h59'22"14d48'
Sensational Suzie
Equ 21h20'44"2d31'
Sensatus,Edmee
Cet 1h54'26"-5d11'
Senter,James P
Cet 2h2'29"-2d4'
Sentineri,Dott Giusy
Lyn 7h1'1"44d52'
Sentir,Jeryl Kay
And 1h32'34"36d42'
Senz,Opal Marie
Cyg 20h34'1"31d46'
Senés,Tony Richard
Per 4h27'49"50d30'
Seon,The
Cam 4h48'19"68d4'
Separated But Never Parted
Eri 4h53'1"-5d32'
Sepe,Claire
And 2h27'43"45d56'
Sepee & Paul Star
Uma 8h59'59"60d48'
Sephi
Peg 22h1'50"34d14'
Sepideh Hariri
Ser 16h17'29"1d13'
Sepotnick-Gottlieb
Uma 9h24'12"52d3'
Seppala,Bill & Ann
Tau 4h12'14"22d40'
Seppentino Educator of The Year-1996,Bob
Per 3h13'19"40d42'
Seppi,Sparky
Lyr 18h56'31"31d20'

Seppo Nuotio
Umi 14h42'0"69d38'
Sepriano,Angel
Vir 13h22'6"-7d55'
September 26th Star, The
Ori 5h48'24"21d39'
September Storm
Cyg 19h56'51"48d58'
Septernal G Fam
Lac 22h10'23"46d27'
Septka,Our Teacher Rod C
Mon 6h18'12"2d52'
Sepulveda,Annie
Com 13h10'1"28d40'
Sepulveda,Jr,Jesse Dean
Mon 7h41'34"-8d57'
Sepulveda,My Everlasting Love,Arman
Per 2h59'0"35d44'
Sepulveda,Raymond J
Boo 14h16'11"45d23'
Seqdgwi4ck
Cam 6h10'26"58d19'
Sera Since 1994
Sgr 19h6'17"-14d0
Serafin,John Paul
Aql 19h30'32"14d28'
Serafini,Charles P
Aql 19h8'0"0d55'
Serag,Carol
And 0h1'52"47d56'
Seragnoli,D ssa Daniela
Gru 22h32'22"-50d24'
Seragnoli,Isabella
Uma 16h10'15"82d41'
Serani Family
Aur 5h3'32"38d3'
Serani,Karen M
Vul 19h3'31"21d46'
Seraph,The J Dallas
Cnc 8h10'30"6d43'
Seraphin,Benoât
Aql 20h31'1"0d17'
Serba,Patrick
Aur 6h16'0"31d42'
Serber,Irwin Louis
Cet 2h14'22"7d20'
Serbi,Libby
Lyn 9h1'26"44d18'
Serbon's Star,Laura
Cet 1h52'42"-0d38'
Serbon,Laura
And 2h22'50"39d38'
Serbyn,Marika Brezden
Lyr 19h17'33"35d25'
Serda,Sergi
And 23h3'27"38d20'
Serdev,Aleksey
Boo 15h15'58"33d27'
Serdynski,Sandra Lee
Gem 7h1'33"21d28'
Serebro,Alona
Cyg 19h31'42"32d50'
Serefin,Octavio
Leo 11h0'14"11d29'
Serena
Ori 5h50'19"-6d53'
Serena
Peg 21h58'58"34d7'
Serena
Cnv 13h26'26"38d15'
Serena
For 2h42'52"-25d27'
Serena
Com 12h49'57"21d14'
Serena
Cyg 20h15'22"39d19'
Serena
And 1h5'28"41d6'
Serena & Darren
Sge 19h54'17"16d28'
Serena e Stefania
And 0h45'14"45d48'
Serena Nadia
Lyr 18h45'31"31d47'

Serena Rae
Peg 22h0'1"31d51'
Serena Rae S K
Her 16h54'34"37d48'
Serena-071184
Mon 7h33'58"-1d42'
Serendipity
Uma 8h34'52"47d2'
Serendipity
Uma 8h42'40"47d31'
Serenella
And 23h1'39"51d39'
Serenitas,Philipius
Per 3h8'37"46d28'
Serenity
Tri 1h39'28"30d39'
Serenity
Cyg 19h28'37"30d21'
Serenity
Aql 19h32'1"-0d54'
Serenson,Raymond Peter
Aur 5h6'40"38d42'
Serera,Olivier Segui
Ser 15h13'40"18d24'
Serex,JoEo Pedro
And 23h3'43"50d14'
Serge
Umi 13h6'51"75d12'
Serge
Cep 21h13'21"68d16'
Sergel,Gayle "Lizzie"
Lac 22h17'47"49d2'
Sergey III,John Michael
Umi 17h51'56"80d34'
Sergi,Antonio &
Powell,Deborah
Cas 23h28'51"62d59'
Serginski,Laure
Cet 2h37'49"8d17'
Sergio
Lyr 18h38'59"27d35'
Sergio
Aur 6h45'58"35d36'
Sergio
Her 16h54'24"33d10'
Sergio della Libera
Vir 14h37'0"7d1'
Sergio's Century Star
Ser 15h34'0"18d50'
Sergio,Nanni
Per 2h11'12"58d14'
Sergison,Zarah Dannetta
Cam 3h50'29"78d24'
Serido,Jr,Anthony Paul Joseph
Vul 21h21'50"26d26'
Serie,Christiane
Umi 17h0'43"76d17'
Serin,Bruno
Vul 19h33'50"27d26'
Serinese,JJ
Aql 20h1'42"12d50'
Serino,Bob
Aql 18h59'39"-0d2'
Serio,Joe & Betty
Ori 5h57'19"15d1'
Serio,Pablo A
Aur 6h15'15"31d7'
Seritour
Vel 9h23'25"-41d9'
Serjeant,Paul Howard
Her 18h28'56"24d49'
Serjooie,Mehrdad
Aur 5h23'51"40d28'
Serli,Cinudio
Cam 8h22'36"78d45'
Serlin,Camille Elyse
Lyn 7h6'17"50d59'
Sermersheim,Mike
Sct 18h50'25"-7d18'
Sermet,Christophe
Del 20h58'54"16d21'
Sermoneta,Nicholas L
Ori 5h46'12"18d4'
Serna,Arnulfo D
Tri 1h47'42"28d29'
Serna,Jr "Twinkle, -
Twinkle",John P
Lib 15h39'0"-20d20'
Serna,Russella
Hya 9h4'12"1d19'
Serniak,Barbara
Cas 23h21'36"60d19'
Serno,Angelika
Tau 4h5'13"22d7'
Sero II
Cet 2h50'37"-0d41'
Sero III
Vul 19h48'12"28d18'
Seroussi,Florian- Emmanuel
Jacques
Her 18h50'54"18d49'
Serpas,Dakota Lee
Cet 1h33'58"-0d36'
Serpelloni,Alessandro E
Francesca
Cyg 21h53'36"41d7'
Serpico,Alexander
Lyr 19h3'34"38d36'
Serra,Jacques
Aql 20h22'38"0d32'
Serra,Jesus Ma de Guisasola
Her 18h29'56"20d36'
Serra,Marcelino Pichay
Uma 9h41'15"61d16'
Serra,Marcella
Lac 22h40'56"38d4'
Serra,Sylvia Benet
And 1h56'27"47d9'
Serrah's Smile
Mon 6h36'16"6d21'
Serraino,Joseph
Cam 3h54'1"74d32'
Serrano,Amanda Beth
Mon 6h22'60"2d48'
Serrano,Ana
Lyn 7h54'1"40d13'
Serrano,Anna
Ari 2h40'21"25d7'
Serrano,Brenda Alicia
Peg 21h29'14"24d15'
Serrano,Cathy
Peg 22h42'1"11d29'
Serrano,Jose Bitrian
Cas 23h1'37"53d34'
Serrano,Jr,Ernie
Her 17h58'1"41d9'
Serrano,Nes Nicholas Espejo
Lyn 9h16'11"33d27'
Serrao,MaryLou
Sex 10h37'50"-0d13'
Serrapere,Frank
Cam 3h52'51"60d57'A
Serratelli,Margaret
Lac 22h12'1"54d52'
Serratelli,Mary Kate
Lac 22h11'56"50d6'
Serrault,Francoise
Sgr 19h33'38"-34d57'
Serre,Bernard C D
Uma 13h18'27"60d46'
Serrell III,Grant
Cep 22h30'20"58d42'
Serrette I,Creighton Carlyle
Ori 5h14'38"15d21'
Servantez,Joe & Frances
Hya 9h0'1"3d23'
Servedio,Barbara
Cyg 19h48'10"29d53'
Serventi,Jon Andrew
Umi 16h27'33"72d54'
Servien-Kenwood, Corrine
Patricia de
Cam 3h32'36"60d44'
Servitje,Tomas Tortosa
Umi 14h37'0"67d7'
Serwacki,Diana J
Cyg 21h12'37"35d54'
Seshens,David
Aql 19h28'58"14d52'
Seskin,Victoria Anne
Cam 4h27'0"78d51'
Sessa,Ernest
Dra 19h40'14"61d18'
Sessa,Raffaele
Cyg 20h22'36"40d16'
Sessa,Sam
Aur 6h17'27"31d12'
Sessions,Deanna Lynn
Cnc 8h55'48"21d27'
Sessions,Jacklyn Ann
Vir 13h53'26"1d56'
Sesti,Maria
And 23h30'41"45d23'
Sestini,Paolo
Lac 22h4'16"51d8'
Sesto,Maria
And 23h3'15"51d44'
Seten,Spenser Nathan
Her 16h50'0"30d34'
Setera,In Memory of Lois
Anne
Cyg 20h8'26"40d60'
Seth
Cep 20h15'38"76d1'
Seth 23
Hya 9h6'48"5d28'
Seth Allison MBS 0592
Cet 1h25'35"-10d59'
Seth G & Jenny N
Boo 15h8'58"26d56'
Seth Matthew
Lac 22h35'51"40d47'
Seth The Adventurer
Cet 0h49'31"1d16'
Seth's Star
Ser 17h57'50"-10d21'
Sethel-Quinn,Cynthia
Vul 19h48'47"27d31'
Sethi,Ravinder
Lac 22h15'1"49d49'
Sethman,Chad
Cet 2h14'33"6d8'
Seto,Robb & Lynn
Cyg 20h23'52"38d45'
Setos,George Andrew
Per 2h1'22"56d19'
Setser,Todd & Laura
Hya 8h45'36"12d45'AB
Sett 3000
Cet 2h18'51"5d14'
Setta,The Boss—-Joe
Cet 0h29'29"-0d1'
Settecasi,Elena
Eri 3h52'56"-5d57'
Setteducati,Michael L
Lac 22h19'57"54d34'
Settelmeyer,Ray
Aqr 22h1'21"0d19'
Setter,Michael Joseph
Her 17h6'57"42d39'
Setterfield,Lawrence W
Aqr 20h54'28"-0d0'
Settino,Ronald G
Her 17h8'16"43d5'
Settle,Austin Grafton
Vel 9h13'34"-44d42'
Settle,David
Aur 6h7'17"33d35'
Settle,Gary S
Her 18h41'1"12d32'
Settle,Harrison Houston
Vel 9h35'2"-41d7'
Settle,Judson Carr
Vel 9h3'50"-44d54'
Settle,Judy
Vul 19h37'43"20d8'
Settle,Kimberly Grant
Vel 9h16'42"-37d37'
Settle,Mark
Cet 2h0'10"1d16'
Settle,Roy Norman & Lisa
Jean
Aur 5h4'18"30d2'
Settlemier Star
Uma 12h38'21"59d19'
Setto,James Nicholas
Lac 22h51'58"56d20'
Setton,Helene
Cam 7h37'19"68d55'
Setzer,Gerhard
Gem 7h14'11"21d37'
Seuchan,Melissa
Mon 6h53'34"-0d50'
Seufert Integrity L O M L,Fred
Cep 21h55'45"55d25'
Seufert,Ludwig
Dra 17h47'11"68d58'
Seufert,Michael
Dra 13h23'56"64d15'
Seuffert,Michael Patrick
Dra 18h59'18"48d13'
Seumas
Lyr 18h40'25"35d57'
Seuvigné,Rachel Laïla
Per 3h32'0"41d7'
Sev
Aql 18h54'17"10d39'
Sevcik,Debra
Cas 1h50'10"73d39'
7¯ Cielo
Lep 5h52'1"-20d11'
Seven Springs
Aql 19h44'36"14d21'
1732
Boo 14h47'42"32d13'
Seventi,Anna
Cyg 20h16'33"38d45'
Seventytwo
Mon 6h35'0"-0d43'
Severance,Lucile Day
Lyn 7h36'35"52d28'
Severance,Patricia
Cas 23h31'56"61d25'
Severance-Nite Wind, Franklin
D
Aur 5h0'16"41d11'
Severen B
Uma 9h26'21"48d38'
Severine
Boo 13h43'30"16d14'
Severino,Linda Diane
Ori 6h3'21"20d31'
Severino,Steven Gregory
Per 3h46'22"37d7'
Severn,Joanne
And 0h50'0"38d49'
Severson,Hattie "Granny"
Cnv 12h21'13"4036'
Severt,Eliza
Vul 19h33'35"27d17'
Severt,Susan Delores
Cas 0h55'1"54d52'
Sevi,Francesca
For 2h46'42"-25d21'
Sevier,Richard Larry
Her 17h53'59"30d48'
Sevig,Lee
Uma 11h0'49"40d42'
Sevin,Earl
Cmi 7h6'17"3d47'
Sewall,Amanda
And 22h57'33"37d15'
Sewall,Joan & Tingey
Crb 15h30'39"30d12'
Sewall,Rita Spillane
And 0h45'55"28d23'
Sewall,Samuel Swanton
Her 16h8'29"42d15'
Seward #13,Curt W
Peg 22h24'0"5d0'
Seward,Bronwyn Dale
Boo 14h30'57"47d55'
Seward,Margaret "Meggie"
Theresa
Cas 0h0'1"58d24'
Sewdas
Her 17h38'20"14d22'
Sewell's Bit Of Heaven
Umi 15h18'33"71d15'
Sewell,Mary
Cas 0h46'43"75d53'
Sewell,Susan
Aur 4h36'30"31d36'
Sewell,Wanda
Del 20h14'51"10d9'
Sexagesimus
Per 2h37'22"40d42'
Sexgod
Ori 5h55'30"15d18'
Sexton's Star
Sex 9h54'27"1d56'
Sexton,Alex J
Ori 5h58'54"21d6'
Sexton,Andrew Christopher
Aur 6h8'41"50d36'
Sexton,Barbara
Cmi 7h56'0"8d40'
Sexton,Ben
Ser 16h1'20"10d19'
Sexton,Carolyn
Cas 0h32'33"60d35'
Sexton,Casey
Del 20h23'55"10d25'
Sexton,Chelsea Margaret
Mon 7h45'53"-1d59'
Sexton,Elaine
Lyr 18h51'35"38d8'
Sexton,Gabrielle Michelle
Sex 9h57'35"5d2'
Sexton,Joseph G
Cep 23h40'17"67d27'
Sexton,Katherine Agnes
Ori 5h26'16"-6d34'
Sexton,Kyle
Oph 16h20'0"-6d57'
Sexton,Laura Jean
Mon 8h3'40"-9d35'
Sexton,Lauren
Vul 19h48'22"26d24'
Sexton,Mitra N
Cam 5h44'1"60d28'
Sexton,Richard Brian
Umi 14h44'13"70d28'
Sexton,Tom
Lac 22h26'27"38d56'
Sexy Angie
Leo 10h1'33"10d53'
Seyda,Lucas Lluch
Sgr 19h0'54"-20d45'
Seydoux,Arnaud
Per 4h5'39"36d5'
Seydoux,Julien
Cam 5h3'14"61d50'
Seydoux,Monique
Uma 9h0'38"58d19'
Seyfert,Bryton
Cep 21h11'36"60d48'
Seyfert,Jennifer
Cas 23h16'46"62d47'
Seyffahrt,Dirk
Lac 22h3'31"51d53'
Seyffer,Steffen
Vir 13h32'16"-2d6'
Seyffert,Marion Brown
Uma 11h15'1"40d8'
Seyfried,Darlene
And 23h31'28"48d47'
Seyler,Marie-Charlotte
Uma 11h1'26"68d20'
Seymour's Heart, Where Lies
Dan
Cep 22h16'51"70d27'
Seymour,Allen & Carole
Lyn 7h47'1"42d51'
Seymour,Arsend
Lac 22h17'28"49d48'
Seymour,John
Dra 19h4'41"60d47'
Seymour,Lesley
Dra 16h3'11"62d55'
Seymour,Martin & Lou Amor
Cyg 21h4'38"30d12'
Seymour,Richard Michael
Umi 4h1'57"88d45'
Seymour,Sara
Sco 16h9'54"-31d14'
Seymous,Sister Cecelia
Cas 1h34'43"68d45'
Sezak,Jane
Cnc 8h45'31"20d52'
Sfetku,World's Best Parents
Nick & Diane
Aql 19h3'24"0d3'
Sfez,Fabien
Dra 11h57'32"72d12'
Sfingi,Beverly Hanson
Cet 2h22'22"-0d26'
Sforza,Bill
Vul 20h40'23"20d5'
Sforza,Cassandra
Com 12h9'1"21d32'
Sforza,Mari Anne
Peg 22h29'32"20d35'
Sgarbossa,Emanuela
Lyr 18h34'56"40d32'
Sgoplano
Aur 4h53'13"40d44'
Sgrecci,Brian Perry
Aur 5h18'26"40d39'
Sgrecci,Laura Louise
And 2h0'55"38d21'
Sgroi,Deborah
Sct 18h44'53"-7d34'
Squardi Star
Tel 20h11'58"-49d7'
SH-Star
Dra 13h35'45"70d12'
Sha Spa
Aql 19h6'43"0d18'
Sha-Mar
Mon 6hh58'44"11d46'
Shab,Grama
Aql 20h6'30"4d50'
Shabi Star of Stars, Ray
Uma 10h26'0"51d30'
Shablinskaya,Anya
And 1h59'40"38d46'
Shabot,Marilyn
Lyr 18h55'25"30d46'
Shabroligne
Peg 22h33'42"10d1'
Shack's Star
Uma 11h46'0"46d50'
Shackelford,Bill
Lac 22h6'38"38d10'
Shackelford,Cynthia
Mon 6h15'0"-6d46'
Shackelford,Steven
Aql 18h58'53"-1d54'
Shacklett,Antonia S
Mon 6h41'48"6d32'
Shacknai,Jonah
Cnv 13h29'59"40d30'
Shacknai,Kimberly & Jonah
Cnv 12h32'12"50d23'
Shacknow,Jaclyn Rachel
And 1h21'49"40d56'
Shadd,David E
Aur 6h23'39"40d13'
Shaddinger,Clay
Ori 5h0'39"-0d10'
Shaddix,Adam Tyler
Lib 15h43'21"-20d18'
Shaddy,Buhl F
Lyn 7h36'57"58d46'
Shade
Her 17h38'55"21d13'
Shade,Diddy & Bill
Boo 15h7'20"30d59'
Shade,The Star of Temmie
Lyr 18h41'19"31d43'
Shadeed,Suzanne
Uma 12h17'53"61d35'
Shader,Nora
And 23h22'55"45d37'
Shader,Patrick
Her 16h31'15"39d28'
Shadley,Kari Elizabeth
Lac 22h27'33"50d34'
Shadoe
Sco 16h25'15"-40d39'
Shadow
Cma 6h54'3"-17d56'
Shadow
Aur 6h9'15"0d29'
Shadow & Bob
Cyg 21h47'42"37d4'
Shadow Dancer
Lyn 6h54'49"59d16'
Shadow Seven
Dra 19h33'0"65d17'
Shadow Wolf
Aql 19h6'41"2d46'
Shadyside Church PNC 1996
Cam 3h17'42"60d18'
Shae
Pup 7h27'48"22d27'
Shae Francis
Mon 6h54'57"-4d35'
Shaeffer,Chandler Megan
Ari 3h12'51"23d48'
Shaeffer,Edward Thomas
Her 16h37'15"10d14'
Shafer Star,The
Aql 20h11'33"10d58'
Shafer,Craig & Nancy
Lyn 7h51'33"42d21'
Shafer,Daniel Adam
Vul 19h21'55"26d57'
Shafer,Gary Stuart
Leo 11h32'54"-5d17'
Shafer,James Ray
Gem 6h28'35"12d30'
Shafer,Jessica Fuchs
Del 20h13'46"10d22'
Shafer,Marie M
Mon 6h35'37"10d23'
Shafer,Mary Lou
Vul 19h47'23"28d18'
Shafer,Nathanial J
Cma 6h41'47"-13d23'
Shafer,Paquita Morton
Aql 19h41'34"10d16'
Shafer,Robert George
Her 17h39'36"20d53'
Shafer,Timothy Dale
Cnv 12h19'16"36d51'
Shafer,William
Lac 22h51'16"56d3'
Shaff,Richard & Marilyn
Lyn 7h33'51"44d10'
Shaffer (Pump Pump), Jane
Marie
Psc 0h2'14"-0d29'
Shaffer,Claudia
Cas 0h23'19"60d16'
Shaffer,Don & Berni
Cyg 19h32'18"38d23'
Shaffer,Elizabeth Allen
Aql 18h56'0"10d27'
Shaffer,Elizabeth
Uma 12h33'1"56d35'
Shaffer,Elon Daniel
Peg 22h25'13"21d8'
Shaffer,Eric
Uma 12h21'52"56d16'
Shaffer,Hannah Claire
Peg 22h42'56"22d57'
Shaffer,Harold Loyd
Her 16h58'1"38d59'
Shaffer,Jennifer Ellen
Vul 19h46'22"28d50'
Shaffer,Jessica Marie
Uma 9h57'60"50d38'
Shaffer,Jonathan Marshall
Boo 14h41'51"20d26'
Shaffer,Julie Kaye
Aql 18h56'57"-7d45'
Shaffer,Linda Page
Equ 21h3'1"3d24'
Shaffer,Peter L
Aur 7h18'0"41d11'
Shaffer,Randy Carl
Hya 8h13'43"0d51'
Shaffer,Robert Michael
Cep 22h18'59"58d16'
Shaffer,Suzanne "Bean"
And 2h8'37"38d8'
Shaffer,Tracy
Lyr 18h58'50"32d30'
Shaffer,Weldon Ezekiel
Aur 6h27'1"30d1'
Shaffer,William & Mary
Cet 3h6'1"6d55'
Shaffery,Alexa Louise
And 0h57'39"34d55'
Shaffner,Alex Michael
Aur 6h7'59"38d30'
Shaffner,Tyler Benjamin
Per 4h45'22"38d26'
Shaft,Ruth Nutting
Peg 23h3'11"33d55'
Shaggy
Boo 14h29'27"40d7'
Shagrin,Don
Cet 2h57'37"4d22'
Shah Singh,Pritam & Amrit
Eri 4h13'0"-15d45'
Shah,Ashmin
Per 2h5'57"58d60'
Shah,Chhotubhai B
Umi 13h46'16"78d19'
Shah,Kalpesh P
Oph 16h59'0"8d43'
Shah,Kinnari S
Uma 13h12'1"60d41'
Shah,Lillian A
Cyg 19h33'47"32d34'
Shah,Manoj
Cep 21h28'0"61d16'
Shah,Nyall
Cyg 19h30'11"38d41'
Shah,Prabhulal C
Her 17h12'31"28d48'
Shah,Trupti V
Peg 22h4'35"5d33'
Shahabi,Hamid
Ori 4h55'37"5d39'
Shahabi,Natasha
And 1h2'29"37d19'
Shaham,Ian Juri
Boo 14h41'40"46d36'
Shaheen,Michael C
Boo 15h2'0"28d41'
Shaheen,Susan
Cas 0h33'1"68d21'
Shahin
Gem 7h4m13"28d40'
Shahla
Cas 0h46'10"70d58'
Shahmoradi,Bijan
Ser 15h54'53"0d36'
Shahnaz
Lac 22h28'33"41d6'
Shai
Cam 3h40'34"61d16'
Shaid,Erika Ashley
Dra 15h59'58"61d11'
Shaid,Michael Dylan
Aur 4h37'12"30d25'
Shaikh Star,The Amir & Alia
Tau 4h1'35"20d36'
Shaina
And 2h28'15"39d52'
Shainaz J Karim & Oliver
Buggle
Gem 6h26'12"13d35'
Shainberg,Lawrence
Her 17h31'59"21d16'
Shairs,Savannah Raven
Aql 19h19'18"13d39'
Shaji Family Star,The Carl &
Janet
Aql 19h28'49"13d19'
Shaka,Celestina
Cnv 12h56'0"42d43'
Shakar,the Faber Star
Cnv 13h52'33"48d3'
Shake,Stuart,& Melanie Moss
Uma 11h12'32"51d53'
Shaker,H E Ghassan I
Lyr 19h2'45"30d48'
Shaker,Sherin
Ori 5h55'52"12d36'
Shaker,Tala
Ori 5h55'0"12d51'
Shakespeare,Ida Josephina
Cyg 19h30'31"39d47'
Shakey
Her 15h58'44"41d10'
Shakhasha
Peg 23h27'20"24d36'
Shakti-Sue
Cyg 20h16'24"38d3'
Shalam-Magnenat,Evelyn
Dra 10h59'27"73d34'
Shalamar
Cam 5h47'55"61d25'
Shale,Sheila M
Cas 23h39'41"64d29'
Shalisa & Ricky's Shooting
Star
Lib 15h44'1"-23d23'
Shalisa's Star
Uma 9h4'21"56d53'
Shalit,Malik
Cas 0h18'34"61d38'
Shall,Sharron
Peg 22h40'49"35d22'
Shallcross
Sex 10h42'26"-1d52'
Shallcross,Bob
Aql 18h44'1"-2d57'
Shallenberger,Joyce McDowell
Aur 5h24'47"38d24'
Shallman,John David
Sct 18h45'48"-7d15'
Shallow,Robert W
Hya 8h46'36"-0d29'
Shalmirane
Umi 14h4'39"70d59'
Shalom Children's Cntr
Kindergarten
Uma 9h15'40"56d53'
Shalom,Debra
Cas 0h56'21"58d22'
Shamah,Allan
Lac 22h14'12"48d21'
Shamar
Tri 2h19'32"30d26'
Shambala
Oph 17h57'27"10d15'
Shambaugh,Amy
Mon 7h27'54"-6d22'
Shamblin,Grant E
Per 2h40'0"34d33'
Shamblin,Jr,George L
Her 17h36'1"27d53'
Shamblin,Mildred
Per 2h57'25"32d45'
Shamblin,Ruby
Com 13h6'1"30d1'
Shameless Forever
Cam 4h57'17"68d24'
Shamenek,F Stephen
Aql 20h2'1"1d10'
Shamos,Francine Sharon
Peg 22h59'16"28d45'
Shames-Dawson,Ali
Crb 15h36'46"29d2'
Shames-Dawson,Sylvie
Crb 15h21'35"31d2'
Shamoo
Del 20h30'26"20d34'
Shamoon,Ramzi/Caryl- Sue
Silkie
Ari 1h45'1"24d53'
Shampa Banerji's Shining Star
Sco 17h3'59"-33d45'
Shampo's Star,Grace
Del 20h13'1"10d26'

Shamrock & Cricket
Peg 22h37'58"7d53'
Shamus Warren
Peg 23h38'20"25d14'
Shan
Peg 23h30'16"20d56'
Shana & Monica: My Eternal Friends
Peg 21h25'25"23d36'
Shana's Promise
Peg 23h7'15"11d53'
Shana's Star
Lyr 18h40'18"41d58'
Shana-Marim
Cyg 19h32'0"36d25'
Shanabarger,Virginia & Paul
Her 17h9'41"46d49'
Shanafelt,Phillipa
And 0h21'37"32d7'
Shanahan,Anne Marie
And 23h44'36"38d24'
Shanahan,Erin Kathleen
Gem 6h45'15"14d36'
Shanahan,Evelyn Boyle
Ori 6h8'59"6d15'
Shanahan,John M
Gem 6h2'23"26d52'
Shanahan,Kathleen Celia
Cas 0h10'17"61d20'
Shanahan,Peggy F
Peg 23h43'48"26d53'
Shanahan,Robert
Aur 7h17'1"40d60'
Shanaman,Michael
Per 2h43'21"40d39'
Shand,Dr Neil & Mrs Sheila Mary Farrar
Peg 21h40'14"27d12'
Shanda
Vul 20h0'13"28d51'
Shandles,DPM,Ira David
Ser 15h35'23"20d32'
Shandles,Martine Ellen
Cmi 7h6'53"5d32'
Shandles,Neil Aaron
Cet 0h50'26"-1d39'
Shandor,D & K
Cyg 20h15'1"38d49'
Shandra Anne
Vul 20h15'56"25d32'
Shandy
Leo 10h53'56"-6d16'
Shane
Cam 13h16'0"76d59'
Shane
Mon 7h19'0"-1d9'
Shane
Aur 6h29'1"35d41'
Shane's Kingston Trio, Bob
Aqr 23h14'58"-6d15'
Shane's Lucky Star
Her 16h46'53"21d20'
Shane's Oasis
Tau 4h11'25"1d43'
Shane,Andy
Her 18h7'0"40d6'
Shane,Christopher A
Dra 17h1'21"64d25'
Shane,Daniel
Dra 16h4'55"67d25'
Shane,David
Ser 17h55'22"-14d0'
Shane,Laurie
Uma 12h2'39"31d33'
Shane,Molly Elizabeth
Cas 0h33'30"60d36'
Shane-Our Little Slice of Heaven
Mon 6h40'24"8d52'
Shanel's Star
Cyg 21h31'20"44d54'
Shanes,A Great Husband & Dad Jeffrey
Her 15h49'1"43d33'
Shaney
Mon 6h19'37"8d38'
Shanfeld,Nancy Ellen
Vir 13h35'55"-7d40'
Shanghai Charity Foundation
Per 3h8'32"41d26'
Shango
Oph 18h16'19"10d3'
Shangold,Harlen Alec
Lac 22h23'1"54d45'
Shangri-La
Mon 7h4'34"-7d3'
Shangri-La
Aur 7h19'54"40d15'
Shangrila Star for M&M B&B,& C&C,The
Cyg 21h21'15"38d20'
Shank,Jonathan
Boo 14h50'59"34d47'
Shank,Stephen Anthony
Her 16h14'33"42d5'
Shank,Tyler Jeffry
Dra 20h5'55"68d52'
Shankar,Sri Sri Ravi
Cep 22h8'45"62d33'
Shanker,R H Sam
Her 17h33'59"28d12'
Shankle,Lisa
Tau 4h17'17"20d9'
Shankman A/K/A Mr Martin,Al
Vul 19h22'23"22d35'
Shanks,Christine
Cyg 19h32'1"38d24'
Shanks,Marjorie
And 23h35'33"41d16'
Shanley Born 18/3/61 Died 24/8/95,David
Cyg 21h17'1"35d2'
Shanley,Barry E
Her 16h21'26"22d40'
Shann
Tri 1h57'56"26d18'
Shann
Boo 14h57'38"43d37'
Shanna
And 2h27'57"44d37'
Shanna Christine
Aur 5h20'48"37d38'
Shanna's Star
Cas 23h4'51"58d59'
Shannen
Tri 2h5'1"30d19'
Shanner,Nicole Lynn Theresa
Cas 0h35'35"61d46'
Shanning,Robert E
Cet 1h11'47"-0d56'
Shannon
Cet 0h3'11" 7d43'
Shannon
Aql 20h2'10"11d1'
Shannon
Aql 19h4'53"1d45'
Shannon
Aql 19h29'48"8d6'
Shannon
Mon 6h54'45"-6d29'
Shannon
Eri 3h54'53"-5d23'
Shannon
Aql 20h12'55"12d10'
Shannon
And 2h28'0"44d40'
Shannon
Sge 19h14'36"18d52'B
Shannon
Uma 9h16'50"59d43'
Shannon
Aql 19h10'53"10d37'
Shannon & Niko
Sge 20h1'1"17d52'
Shannon & Ron
Lmi 9h34'32"38d14'
Shannon & Travis
Ori 5h57'52"12d3'
Shannon A Bridge Across Forever Paul
Cyg 19h28'26"37d41'
Shannon Elizabeth
Cyg 19h29'25"38d1'
Shannon Family,The
Ori 5h39'45"11d57'
Shannon I
Cas 0h28'48"68d5'
Shannon I,G P
Oph 18h42'53"7d20'
Shannon Joy
Peg 22h38'51"35d17'
Shannon Kelly
And 23h22'10"45d51'
Shannon Leigh
Vul 20h17'25"25d21'
Shannon Leigh
Aur 5h36'57"40d5'
Shannon Marie
Cam 5h46'55"67d45'
Shannon Marie
Cas 3h8'19"61d9'
Shannon Marie
Eri 4h6'0"-14d6'
Shannon Patricia
Boo 14h56'42"30d59'
Shannon Queen of the Stars
Cas 23h39'11"50d14'
Shannon's
Cam 11h59'50"81d24'
Shannon's "Golden Dreams",Harold R
Uma 10h46'11"40d37'
Shannon's "Golden Dreams",Evelyn M
Uma 10h38'26"40d11'
Shannon's "Wedding Dreams",Ron & Judy
Uma 10h46'26"40d2'
Shannon's Dreams & Wishes
Cyg 19h32'39"33d28'
Shannon's Love
Cas 3h4'32"57d45'
Shannon's Lucky Star
Crb 15h50'25"32d24'
Shannon's Shine
Lyn 7h58'10"58d57'
Shannon's Star
Uma 12h31'17"60d47'
Shannon's Star-Forever In Our Hearts
Cyg 21h6'0"30d13'
Shannon, My Daughter
Peg 21h39'32"27d54'
Shannon,Billy Wayne
Sgr 19h6'30"-22d27'
Shannon,Carol Ann
Lyn 6h56'24"54d48'
Shannon,David
Boo 14h28'54"28d21'
Shannon,Gerard Coldewey
Cep 20h42'42"65d35'
Shannon,Joe
Boo 15h33'1"48d14'
Shannon,Mercedes Andrea
Cet 2h19'25"2d3'
Shannon Myra
Ant 10h33'57"36d6'
Shannon,Riley Patrick
Peg 22h0'47"31d1'
Shannon,Susan Lynn
Mon 6h55'59"10d24'
Shannon,Teri A
Ori 5h3'0"0d42'
Shannon-Clarke,Dorothy A
Cas 0h34'0"61d47'
Shannon-My Beauty
Cas 0h5'15"58d1'
Shanny The Cloud
Tau 4h2'11"23d10'
Shanti
Cma 7h37'46"2d5'
Shanti
Mon 6h29'11"8d59'
Shanti's Inspiration
Eri 4h34'35"-6d33'
Shantra Starr
Mon 6h36'0"8d31'
Shao,Betty
Peg 22h16'58"34d23'
Shapack,Allan & Phyllis
Eri 4h14'1"-17d24'
Shapiro - The Star, Jeremy Alexander
Boo 14h29'46"21d5'
Shapiro Family Star, The
Boo 15h6'1"11d24'
Shapiro,Al
Dra 16h14'0"62d19'
Shapiro,Alfred
Lac 22h50'24"56d39'
Shapiro,Derryl L
Gem 6h51'11"30d43'
Shapiro,Evan James
Ser 15h14'56"21d38'
Shapiro,Gerald Bernard
Her 17h33'26"20d19'
Shapiro,Hannah Rochelle
Cyg 19h33'42"37d51'
Shapiro,Jack
Boo 15h46'1"41d30'
Shapiro,Jerome M
Aur 5h6'41"37d48'
Shapiro,Jill
Lmi 10h10'34"40d7'
Shapiro,Joseph Randall
Sgr 18h53'35"-30d5'
Shapiro,Lauren J
Lyn 8h17'32"46d11'
Shapiro,Lawrence Cyril
Sgr 18h1'18"-28d32'
Shapiro,Lee Tobey
Dra 11h48'23"72d8'
Shapiro,Linda Greene
Cas 23h28'1"60d59'
Shapiro,Michael Foster
Sct 18h54'0"-6d39'
Shapiro,Morris
Boo 15h1'1"20d51'
Shapiro,Nancy Saks
Cas 23h41'20"61d9'
Shapiro,Rachel Gilda Bohbot
Lyr 18h39'1"31d35'
Shapiro,Samantha
Cyg 21h28'1"53d28'
Shapiro,Steve Foss
Cet 1h30'50"-12d46'
Shapiro,Zachary
Boo 13h44'52"26d42'
Shapley,Dakota Jacob
Lib 15h42'24"-8d24'
Shappley,Jessica Elizabeth
Lyr 18h15'53"37d52'
Shappley,Linda Kim
Tri 2h6'40"30d38'
Shappy
Psc 0h23'26"21d35'
Shaq Star,The
Per 3h22'23"40d19'
Shar
Cyg 21h35'57"41d26'
Shar's Star
Per 2h53'11"40d22'
Shar-Ron
Cyg 20h25'40"38d0'
Shara Nicole
Lyn 7h5'18"50d36'
Sharad
Psc 1h3'52"20d6'
Sharaffa,Riana
Com 13h28'35"26d48'
Sharala
Mon 6h54'49"-10d13'
Sharanda Collette's Special Star
Uma 11h47'1"40d55'
Sharane
And 0h16'33"33d37'
Sharbono,Kati Marie
Mon 6h30'35"11d33'
Sharbono,Tayler Michelle
Cet 2h40'0"-2d4'
Share,Michael A
Uma 11h30'1"57d51'
Share,Sandra Blossom
Cyg 21h21'12"31d15'
Shared Star
Leo 10h49'59"-0d48'
Sharee
Cas 23h31'15"60d33'
Sharek,Danielle Elizabeth
And 22h57'48"38d44'
Sharena
Cas 0h1'33"63d33'
Sharentine MCMXCIII
Cep 20h27'0"66d59'
Sharett,Alan R
Cas 0h12'34"60d35'
Sharf,Erica
Cam 5h49'3"73d49'
Shari
And 2h4'27"37d58'
Shari
Cas 0h38'57"58d22'B
Shari Anne
Cet 2h41'36"2d35'
Shari D
Eri 3h12'42"-16d20'
Shari Lynn
Lyr 18h53'1"34d8'
Shari Rae Star,The
Ori 5h57'1"11d46'
Shari Ruth
Uma 10h34'1"70d24'
Shari Sword Earth
Ori 5h47'22"18d14'
Shari's Special Gift
And 22h56'13"50d15'
Shari's Star
Mon 8h4'1"-8d19'
Shari's Star
Umi 16h56'0"78d49'
Sharif,Karim Khalil Al
Lyr 18h57'16"30d49'
Sharis
Cyg 20h1'1"38d7'
Shark Bait-Ann
Sgr 18h56'35"-12d5'B
Shark-Chris,The
Sgr 18h56'35"-12d5'A
Sharka Birthday Star, The Walter J
Oph 16h43'16"8d39'
Sharkey,Bernard J
Aql 19h24'24"13d52'
Sharkey,Charles
Aur 6h23'17"35d23'
Sharkey,James
Boo 15h2'28"28d35'
Sharkey,John Kay
Oph 18h1'19"8d56'
Sharkey,June
Cyg 19h28'10"30d18'
Sharleen
And 2h21'32"41d55'
Sharlin,Norman
Tri 1h56'17"27d27'
Sharlykins
And 23h39'15"32d58'
Sharma's Star,Mira
Cap 20h21'0"-12d33'
Sharma,Bir Dutt
Ari 1h54'0"15d4'
Sharman,Sheri
Mon 8h0'0"-8d7'
Sharn
Tau 5h18'30"16d51'
Sharoff,Leighsa
And 23h35'38"49d40'
Sharon
And 23h1'35"45d1'
Sharon
Aur 5h19'38"40d19'
Sharon
Lyr 19h20'26"41d3'
Sharon
Lyr 18h59'13"28d28'
Sharon
Lyr 18h36'47"26d3'
Sharon
Boo 14h32'25"51d32'
Sharon
And 23h47'49"44d0'
Sharon
Sge 19h54'12"20d19'
Sharon
Mon 6h52'45"11d9'
Sharon
And 23h40'28"41d16'
Sharon
Cas 1h38'32"58d21'
Sharon
Cas 0h54'0"54d31'
Sharon
Del 20h55'1"8d59'
Sharon
Uma 11h12'34"48d25'
Sharon
Sco 17h52'27"-38d58'
Sharon & Bob's Star
Ori 4h55'1"1d32'
Sharon & Jennifer's Star
Mon 6h55'21"-4d27'
Sharon & Jon Stellar Marriage
Cyg 20h25'57"41d5'
Sharon & Keith
Aql 20h12'0"4d57'
Sharon & Mike
Cam 5h50'15"61d28'
Sharon & Ron
Mon 7h46'11"-5d11'
Sharon & Shayna's Graduation Star
Cam 6h36'17"80d3'
Sharon & Stuart
Cyg 20h23'43"39d23'
Sharon 07/03/1966
Ori 5h55'42"16d25'
Sharon 22-3-52
Lyr 18h35'1"33d9'
Sharon Alana
Psc 1h2'13"23d0'
Sharon Amanda
And 2h25'12"45d21'
Sharon Ann
Lyn 8h25'23"52d0'
Sharon Ann
Cet 2h32'39"-8d29'
Sharon Ann
Cet 0h53'53"-2d5'
Sharon Anne
Lyn 8h8'48"38d23'
Sharon Elizabeth
Crt 11h34'0"-22d17'
Sharon Eternal Love Stephen
Gem 7h36'8"25d1'
Sharon Gayle
Lyr 18h43'53"37d22'
Sharon K
Peg 21h26'15"23d43'
Sharon Kay
Lmi 10h34'45"33d26'
Sharon Leigh
Cas 1h20'16"53d3'
Sharon Lou
Cas 23h12'36"60d13'
Sharon Louise
And 0h42'52"30d30'
Sharon Lynn
Uma 10h41'47"44d8'
Sharon Marie
Lyn 8h32'24"45d34'
Sharon May
Cyg 20h14'18"38d47'
Sharon Patricia
Psc 23h43'25"2d36'
Sharon Rose
Peg 22h15'0"32d6'
Sharon's Celestial Star of Light
Aqr 21h39'34"0d26'
Sharon's Endless Sparkle
Cas 1h33'24"60d40'
Sharon's Estrella
Mon 7h4'57"-5d53'
Sharon's Eyes
Cas 23h42'0"60d28'
Sharon's Heart
Sct 18h52'50"-5d33'
Sharon's Light
And 22h58'55"40d56'
Sharon's Light 1
Lyn 8h28'57"40d22'
Sharon's November Ember
Cyg 20h19'26"38d37'
Sharon's Rae Of Luce Love
Peg 22h18'20"11d0'
Sharon's Smile
Mon 6h5'56"-8d49'
Sharon's Star
Lyr 18h41'23"41d9'
Sharon's Star
Mon 8h1'56"-9d42'
Sharon's Star
Lac 22h35'31"56d19'
Sharon's Star
Cas 0h39'51"64d4'
Sharon's Star
Peg 21h41'50"21d52'
Sharon's Star So Bright
And 22h58'21"51d31'
Sharon's Wishing Star
Lyr 18h59'32"28d40'
Sharon's Wishing Star
Com 12h30'24"27d18'
Sharon,Laura M
Boo 14h48'41"45d13'
Sharon,Sweetie
Cam 4h30'18"78d53'
Sharon-My Love
And 2h18'55"39d33'
Sharona
Com 12h20'50"32d39'
Sharona
Cyg 20h41'55"44d55'
Sharonius Companeilia
Cmi 7h44'38"-0d2'
Sharp "Our Love Forever",Bob & Denise
Sgr 18h58'36"-26d4'
Sharp,Ben Martin
Her 16h3'54"41d2'
Sharp,Bridget Sheila
Lmi 10h26'39"34d30'
Sharp,Cindy
Aur 4h41'1"30d38'
Sharp,Corrine Kay
Peg 22h32'15"26d1'
Sharp,Dennis Neal
Crb 15h56'11"31d37'
Sharp,Gina
Peg 22h20'27"33d41'
Sharp,Gregory Nesthus
Aur 6h8'54"33d40'
Sharp,Helen Laura
Cam 3h20'57"55d5'
Sharp,Jamie
Mon 6h44'1"11d34'
Sharp,Jeremy Jay
Eri 4h53'34"-5d7'
Sharp,Jr,Bobby
Hya 9h2'43"3d39'
Sharp,Jr,Francis H
Aur 5h4'23"38d37'
Sharp,Kerri Dyan
Cas 0h14'28"59d5'
Sharp,Lori Michelle
Eri 3h27'59"-2d45'
Sharp,Louis M
Cas 0h0'18"61d15'
Sharp,Martin
Ori 6h15'44"-1d17'
Sharp,Maxine Homeyer
Mon 6h33'3"-0d19'
Sharp,Michelle
Cyg 19h27'49"38d13'
Sharp,Mimi
Aql 19h24'25"15d11'
Sharp,Roger James
Sgr 19h40'3"-31d13'
Sharp,Sandra
Lyn 8h14'20"36d52'
Sharp,Stephen Douglas
Cep 22h5'17"62d59'
Sharp,Susan Ann
Lyr 18h17'19"46d58'
Sharp,Tim
Sge 20h1'13"16d11'
Sharp,Wes
Aur 6h29'51"38d55'
Sharp,Zachary Ryan
Aql 19h42'24"12d28'
Sharp-God's Oncologist Dr James
Hya 9h34'22"0d24'
Sharpe Star,The Robert M
Ari 2h43'0"30d39'
Sharpe,Granville & Barbara
Lyr 18h36'48"30d22'
Sharpe,Keith
Hya 9h25'58"-1d48'
Sharpe,Sr,E Ray
Cnc 8h50'1"32d55'
Sharpe,Stephen Thomas
Ser 15h39'50"6d53'
Sharpe,Vickie Mae
Mon 7h44'16"-1d16'
Sharpless,Anne Marie
Aql 20h1'31"14d2'
Sharpless,Walt
Cep 20h46'49"70d12'
Sharps,Jason S
Cmi 7h23'10"8d43'
Sharrar,Kent
Her 18h11'0"40d57'
Sharratt,Danny
Ori 6h3'33"3d14'
Sharrer,Stacie L
And 0h45'35"33d27'
Sharron
Vir 13h31'28"-4d20'
Sharrow,Gary
Cep 21h6'47"63d48'
Sharryn's Light
Ara 17h56'21"-56d21'
Shartel,John & Anne
Lyr 19h20'1"40d28'
Sharyn's Star "25th Anniversary"
Cas 1h49'12"65d15'
Shashy,Ranjes David
Boo 14h24'18"31d7'
Shasta's Star
Cet 2h38'25"-11d40'
Shastanalex
Peg 22h43'0"10d24'
Shastri,Neil Jay
Leo 10h56'1"18d44'
Shastri,Sanjay & Priti
Ser 16h0'0"9d40'
Shatoiya,GW
Aql 19h24'48"8d11'
Shatova,Luiza
Cas 0h55'48"58d9'
Shattuck,Janel Lynn
Lyr 18h28'29"30d18'
Shaturn
Aur 5h28'21"54d24'
Shatz & Shatzi Toll
Oph 17h31'45"8d55'
Shatz,Adrienne
Her 18h38'33"18d46'
Shatz,David
Cep 1h22'28"78d15'
Shatzi
Sct 18h50'51"-7d23'
Shaub,Susan Eileen
Cam 8h14'33"74d39'
Shaughaan
Lyn 8h51'38"35d13'
Shaughnessy,Jennifer Lee
And 23h42'15"42d27'
Shaughnessy,Mary Ellen
Cas 23h40'44"50d29'
Shaughnessy,Maureen
And 0h51'22"22d3'
Shaughnessy,Michael F
Ori 6h1'30"10d31'
Shaul,David
Her 17h19'30"42d10'
Shaul,Judith Ann
Cap 21h26'28"-22d51'
Shaulis,Alexis Anne
Lyn 8h4'43"33d33'
Shaulis,Glen
Per 4h7'21"50d22'
Shaun Richard Dicker, The
Cyg 20h37'11"45d51'
Shaun's Wild Flower
Boo 14h5'33"10d56'
Shauna
Cyg 19h20'11"28d60'
Shauna
Lyn 8h18'47"40d39'
Shauna Lynn
Cyg 19h28'37"38d8'
Shauna Marie
Lyn 8h45'43"38d13'
Shauna Michelle
Aql 20h1'20"13d11'
Shauna Rae
Vul 19h1'60"22d10'
Shauna's Star
Cet 2h25'37"0d55'
Shaver,Charles Michael
Uma 11h40'33"38d45'
Shaver,John C
Ori 6h17'36"8d16'
Shaver,Lord Christopher B
Ori 5h1'11"14d3'
Shaver,Robert O'Neal
Cet 3h1'1"0d9'
Shavitz,Gregory Joel
Ori 5h59'37"11d47'
Shavrick,Emma Lauren
Lyn 7h28'14"37d14'
Shaw III,Melvin Emil
Sge 20h14'26"17d29'
Shaw's Star
Uma 13h41'14"61d35'
Shaw's Tequila
Aql 18h52'39"10d55'
Shaw,"Pickle"-Sheri
Uma 10h21'1"72d27'
Shaw,Alex Hagan
Cet 0h44'41"-3d58'
Shaw,Andrea Jane Forrester
Cas 2h43'1"60d42'
Shaw,Anita
Lmi 10h40'15"33d20'
Shaw,Berenice
Boo 13h41'32"15d33'
Shaw,Beth J
And 1h21'39"39d53'
Shaw,Captain Colin
Her 17h39'12"40d18'
Shaw,Casey Allen
Cep 20h58'57"55d0'
Shaw,Christopher Buchanan
Aur 7h21'20"39d49'
Shaw,Cindy
Aql 19h53'12"15d52'
Shaw,Craig Garrison
And 22h57'36"51d46'
Shaw,Cynthia Tara
Peg 21h25'34"22d57'
Shaw,Daniel Slater
Boo 14h6'27"39d6'
Shaw,David Adrian
Aur 5h25'31"30d46'

Shaw,David Patrick Dra 16h59'30"67d17'
Shaw,Dorothy Alice Boo 14h30'15"21d42'
Shaw,Dyanna & Tom Eri 4h34'47"-11d56'
Shaw,Elba-Russell- Lorraine-Renee Aur 6h6'47"45d38'
Shaw,Eleanor Griffiths Aqr 22h35'58"-11d36'
Shaw,Elizabeth Danielle & Scott Mon 6h4'0"-8d12'
Shaw,Elizabeth Rauha And 1h17'21"38d45'
Shaw,Evelyn Cyg 22h0'43"50d24'
Shaw,Francia Lange Eri 3h55'17"-15d22'
Shaw,Francine Celeste Aql 19h1'25"16d50'
Shaw,Frederick J Cam 4h10'50"70d18'
Shaw,Geoffrey & Janet Cyg 21h27'30"49d11'
Shaw,Goddess Jane Cas 23h26'19"63d0'
Shaw,Graham Per 2h52'42"41d6'
Shaw,Heavenly Matrimony Mr & Mrs Cyg 20h18'46"39d25'
Shaw,James Her 17h1'20"43d19'
Shaw,James Robert Aql 19h29'17"13d1'
Shaw,Jenny Lee Cas 0h38'28"54d26'
Shaw,Joann Eri 3h46'44"-3d55'
Shaw,Joanne Sgr 18h59'19"-25d45'
Shaw,Joanne L Mon 6h6'27"-4d10'
Shaw,Jodi & Pat Eri 2h51'25"-5d13'
Shaw,John Aur 5h59'59"38d43'
Shaw,Joy-lim Aql 19h30'16"12d31'
Shaw,Jr,David Hya 9h38'13"1d7'
Shaw,Jr,Ronald Peter Cnv 12h33'15"38d45'
Shaw,Kevin & Natacha Cma 6h11'1"-16d36'
Shaw,Laura Judith Lyn 6h16'11"58d34'
Shaw,Lexy Vul 20h15'57"23d51'
Shaw,Linda Copenhaver And 23h37'43"43d54'
Shaw,Marlee Elizabeth Peg 21h37'56"24d26'
Shaw,Megan Elizabeth Uma 14h0'17"58d43'
Shaw,Michael Edward Dra 17h1'1"63d23'
Shaw,Michelle Lup 14h38'44"53d23'
Shaw,Patrick Thomas Geoghegan Uma 12h1'36"32d28'
Shaw,Richard Murray Cep 20h44'38"60d43'
Shaw,Rick Cep 23h2'31"60d4'
Shaw,Rick Cep 1h54'20"78d36'
Shaw,Rona Lyr 18h57'35"47d25'
Shaw,Rose Marie Lyr 19h14'44"35d9'
Shaw,Sarah Alison And 22h58'12"51d57'
Shaw,Stephen Bradley Per 3h18'13"40d13'
Shaw,Teresa Daly Leo 9h53'38"31d53'
Shaw,Thomas Per 2h57'19"43d16'
Shaw,Vera Nell King Cas 2h28'57"59d57'
Shaw,Veronica Peg 21h41'25"20d36'
Shaw,Victoria Louise Lyr 18h59'0"30d59'
Shaw,Violet Cyg 19h31'29"36d40'
Shaw,Wayne and Kelli Peg 23h28'37"31d55'
Shawcroft,Clerae Uma 11h28'39"31d20'
Shawen,Jeffrey Ori 6h4'42"7d57'
Shawl,William T Aur 6h17'0"52d55'
Shawn Her 17h4'26"41d60'
Shawn Aqr 21h58'49"-15d51'A
Shawn & Darren Ori 5h18'1"1d0'
Shawn & Mandi 1:Corinthians 13 Cyg 19h24'56"30d55'
Shawn & Michelle Tri 2h43'58"31d33'
Shawn Eric Sct 18h53'0"-6d43'
Shawn Michael Her 17h10'12"22d11'
Shawn Patrick Cam 4h24'40"68d15'
Shawn's Inspiration Aur 4h58'26"40d48'
Shawn-ing Star of Florida Mon 7h12'21"-6d40'
Shawna Peg 22h2'0"5d32'
Shawna Leigh Cyg 21h7'51"40d29'
Shawna Lynn Cas 2h36'31"70d57'
Shawna's Star Lyr 18h29'28"37d41'
Shawnee Sct 18h43'26"-6d2'
Shawnee Christy Lyr 19h3'47"38d17'
ShawnKim Dra 19h48'21"71d2'
Shay III,Paul Aur 5h29'55"38d55'
Shay's Falling Star Aql 19h29'58"-0d40'
Shay,Jeffrey Donald Cmi 7h44'38"8d47'
Shay,Jr,Andrew Colin Aql 19h0'41"13d27'
Shay-Shay Forever Peg 21h41'15"24d51'
Shaya,Suhail Theodore Uma 11h59'39"40d6'
Shayahulla Muntakoma Mon 6h43'46"7d3'
Shaye,Eternally Katie Sge 19h56'56"18d56'
Shayeb,John Victor Mon 8h2'28"-8d19'
Shayer,Kelsey Uma 11h59'39"40d6'
Shaykal-Linda Lee- Canus Alpha Tau 4h47'11"16d18'
Shayler Cas 0h34'20"62d18'
Shayna Peg 23h27'11"24d16'
Shayna's Twinkle Mon 6h29'31"-6d22'
Shayne Cyg 20h54'12"38d22'
Shayne Cyg 21h4'20"30d4'
Shayne & Casey Peg 23h37'25"12d44'
Shayne,Daniel Dra 17h47'32"64d3'
Shayne,Marianne Del 20h38'47"11d8'
Shayne,Scott And 0h7'54"44d38'
Shaz 381,The Cyg 21h57'53"53d27'
Shazia Mon 6h52'57"1d2'
She belongs to me And 1h37'19"38d21'
She-She Cas 0h45'37"61d53'
She-Wolf Lyr 19h20'1"37d52'
Shea Lyr 19h15'40"40d19'
Shea Aur 6h33'35"31d47'
Shea Nicole Mon 8h5'10"-5d38'
Shea's Love Star Cyg 21h7'53"30d60'
Shea's Shining Star, Kevin Aur 7h12'47"39d45'
Shea,Chris & Kevin Ori 6h0'0"8d53'
Shea,Christine And 0h5'13"38d5'
Shea,Denman MacKenzie Lmi 10h8'0"38d3'
Shea,George Michael Per 2h36'25"50d5'
Shea,George William Uma 8h45'58"61d29'
Shea,Jeffrey C Vul 19h48'28"28d50'
Shea,Jerry Lac 22h21'21"50d4'
Shea,John Michael Cet 0h34'0"-0d48'
Shea,Joseph J Aql 19h48'15"10d7'
Shea,Karan Peg 21h53'0"3d40'
Shea,Kathleen Mary Lyr 19h17'29"42d4'
Shea,Kathleen P Boo 14h30'53"47d28'
Shea,Kevin R Aql 19h5'40"2d48'
Shea,Marisa Noelani Peg 23h30'15"21d12'
Shea,Mary A And 1h23'36"40d26'
Shea,Mary Beth Cas 1h22'53"58d18'
Shea,Meghan Ann Uma 11h2'0"48d9'
Shea,Michelle Cas 23h15'0"60d25'
Shea,Molly Kathleen Boo 14h32'28"47d60'
Shea,Patrick J Boo 14h22'35"48d25'
Shea,Robert Per 1h52'49"56d53'
Shea,Spencer Michael Ari 1h48'20"17d22'
Sheaber,Maxine & Myron Cnv 12h37'0"41d6'
Sheafer,Matthew Joseph Lac 22h51'58"37d38'
Sheaffer,Robert Per 1h59'39"47d27'
Sheairs,Diane Lynn Cas 1h10'43"61d12'
Shealy,Ann Wilson Cet 2h38'21"2d27'
Shealy,Reggie Her 17h35'16"27d49'
Shean,Nancy Mon 6h36'29"0d17'
Shean,Randy C Cyg 19h53'16"38d28'
Sheandi,Thomas Cmi 7h30'42"7d30'
Shear,Torin Lyn 7h6'29"52d20'
Shearer,Charles John Dra 15h16'54"65d13'
Shearer,Forever, Marjorie & Jack Cyg 20h22'18"37d59'
Shearer,Ian Per 3h14'57"42d57'
Shearer,Karen Elizabeth Crt 10h57'10"-8d50'
Shearer,Louise Roberta Cyg 20h30'1"42d5'
Shearer,Lucy G Umi 14h29'44"67d51'
Shearer,Marlene Mark Com 12h47'29"21d10'
Shearer,Norma Jean Cas 23h39'12"61d7'
Shearer,Patti And 1h19'57"36d44'
Shearer,Ron M Lmi 20h1'52"40d48'
Shearer,Tim Cap 21h25'55"-22d60'
Shearer,Tyler Cma 7h3'17"-16d41'
Sheargold,Jason Ori 6h5'60"5d40'
Shearin,Dorothy And 2h35'13"40d25'
Shearin,Mary Elaine Mon 6h43'5"-3d29'
Shearon,Troy Dra 20h0'16"68d17'
Shearouse,Eric Gresham Leo 10h56'11"18d8'
Shearwood,Sophie Joe Lyn 8h3'31"46d44'
Sheba Lyn 9h11'13"39d0'
Sheba,Mark John Her 17h21'19"46d19'
Shebanek,Michael Brent Oph 17h57'50"12d49'
Shechtman,Arleah Uma 9h19'1"42d25'
Sheck,Carol Lyn 7h21'0"44d3'
Sheck,Larry G Aql 18h59'20"-5d24'
Shed Seven Cas 1h37'31"73d51'
Shedd,Hilary Suzanne Lyn 6h28'42"60d26'
Sheddon,Scott Uma 9h30'14"58d56'
Shedlock,"Ev's Star", Evelyn M Cas 2h29'15"70d36'
Shedlock,Ed Tau 4h16'11"20d50'
Sheeb Lyr 18h47'26"33d20'
Sheedy,Diana And 0h58'59"39d53'
Sheedy,Janet Cas 0h41'1"60d49'
Sheedy,Julie Peg 23h3'58"17d38'
Sheedy,Marty Aql 19h0'35"13d48'
Sheehan,Barbara Fee Lyn 9h13'58"38d10'
Sheehan,Bernadette Kathleen Maguire Ori 5h37'5"12d59'A
Sheehan,Carly Ann Cet 1h14'1"0d42'
Sheehan,Darlene "Toni-Mema" And 0h54'38"41d10'
Sheehan,Dolores Lyn 7h15'43"50d11'
Sheehan,Dot Lyn 8h53'41"42d40'
Sheehan,Forever Christian R. Uma 8h39'1"60d6'
Sheehan,Harold R Mon 6h13'1"-10d53'
Sheehan,Heather Ann Ori 5h37'5"12d59'B
Sheehan,J Lmi 10h39'22"26d8'
Sheehan,James J Tau 3h45'41"28d31'
Sheehan,Jesse Tanner Her 16h34'46"20d19'
Sheehan,Kellie "2Ks" Marie Del 20h13'42"12d27'
Sheehan,Margaret Ann Sge 19h39'1"16d22'
Sheehan,Patricia Evelyn And 0h19'46"35d25'
Sheehan,Philip James Dra 13h1'1"70d20'
Sheehan,Sandra Mon 7h18'44"-5d19'
Sheehan,Shirley Anne And 1h56'56"37d59'
Sheehan,Tim Cet 2h53'40"1d28'
Sheehan-Croll Cyg 21h29'43"40d55'
Sheehey,Madyson Cookie Peg 22h52'30"20d3'
Sheehy,Gail Marie Aql 20h9'13"7d53'
Sheehy,Miss Sharon Cas 1h43'49"73d2'
Sheehy,William B Lac 22h32'40"53d2'
Sheekey,Louise Cep 20h52'29"63d11'
Sheeks,Jason Scot Her 17h57'17"20d18'
Sheelah Cam 8h49'23"80d2'
Sheelar,Stacie D & David N Cherner Sct 18h53'30"-6d51'
Sheen,Charlie Ori 5h26'18"8d24'A
Sheen,Charlie Aur 4h34'55"31d51'
Sheena Del 20h13'1"14d34'
Sheena Cyg 19h25'48"34d39'
Sheena Cyg 19h27'57"36d1'
Sheena Lynn Del 20h56'45"13d9'
Sheenan Ori 5h56'43"15d21'
Sheenie Aur 6h35'44"40d6'
Sheer,Martin & Karen Dra 17h15'1"61d48'
Sheeran,James Cap 20h5'59"-20d46'
Sheeran,Thomas Dynamite Aql 19h28'28"14d10'
Sheesley,Linda Cnc 8h53'0"31d56'
Sheets,Helen B Lyr 18h44'31"39d10'
Sheetz,James D Per 2h48'52"46d2'
Sheetz,Katherine Marie Niles And 22h56'15"51d7'
Sheetz,Lloyd Norman Cet 1h38'47"1d4'
Sheffer,Bradley Cmi 7h16'54"4d47'
Sheffield,Diane Lmi 9h23'47"34d45'
Sheffield,Jill Stelynn Tau 5h21'26"16d13'
Sheffield,Linda B Ser 17h46'16"5d42' A
Sheffield,Ron Ser 17h46'16"5d42' B
Sheffielf,Simone Lac 22h20'37"50d13'
Shefrin Shining Forever,Jean & Joe Uma 12h4'32"51d4'
Sheftall Star,The Audrey Kelly Mon 6h35'48"-1d50'
Sheftman #80,Sylvia Aql 19h45'22"14d25'
Shefts,Nancy Uma 10h15'1"51d35'
Shefts,Stella Cyg 19h27'12"31d21'
Sheha,Michael Anthony Per 1h59'12"56d52'
Shehadi,Justin Paul Per 1h30'53"53d26'
Shehane,Penny Carol Aql 20h34'56"-0d54'
Shehorn,Rebekah Mon 7h26'16"-5d55'
Shei Uma 8h56'20"50d31'
Sheidy,Marian S Ori 6h11'19"8d46'
Sheika & Pepi Forever Cyg 20h17'1"38d58'
Sheikh Psc 22h51'43"0d35'
Sheikha Mahra Bint Tahnoon Bin Nahayyan Uma 11h12'41"57d36'
Sheil,Ashley Sweets And 23h20'0"43d55'
Sheil,Donald E Aur 5h7'43"40d20'
Sheil,Greg My Buddy Dra 16h49'59"70d8'
Sheila Peg 23h43'1"31d30'
Sheila Ori 6h14'57"11d3'A
Sheila Lyr 18h34'58"40d54'
Sheila Lib 15h17'1"-8d49'
Sheila Ann Cam 3h24'19"60d10'
Sheila Ann Cet 2h55'42"-0d47'
Sheila Anne Peg 23h31'1"18d43'
Sheila Beth Psc 1h27'1"30d36'
Sheila Dawn Lyn 8h16'24"58d39'
Sheila Diann Mon 7h1'39"4d36'
Sheila Jean Aql 18h57'1"17d52'
Sheila Marie Cas 1h0'47"55d29'
Sheila's "Dancing Daydream" Cyg 19h26'0"30d50'
Sheila's Rising Star Vul 21h26'15"26d54'
Sheila's Sparkling Delight Lyn 8h20'25"49d41'
Sheila's Star Vul 20h16'23"26d6'
Sheila's Star Com 12h23'21"30d5'
Sheilah My Sunshine Girl Sex 19h4'50"-6d43'
Sheilah W Cep 22h15'43"65d52'A
Sheilamarie Cas 0h52'0"73d53'
Sheild,Diane Hya 8h34'13"5d39'
Sheilds,David John Uma 9h56'12"45d34'
Sheils,James Sct 18h44'51"-7d51'
Sheiman,Evan Cnv 12h28'40"32d1'
Shein,Samantha Paige Lib 14h53'28"-1d27'
Sheinberg,Harrison Jay Ser 15h39'36"7d9'
Sheinberg,Pat Eri 4h57'54"-18d50'
Sheinberg,Sybil K Cam 9h12'28"81d9'
Sheiteres 1993, Charlotte Moore Cam 3h29'43"58d26'
Shekhtman,David Lewis Dra 17h3'30"62d34'
Shekinah Uma 10h20'47"70d25'
Shekinah Clothilde Uma 9h46'26"61d39'
Shekinah,Papaw Cep 22d27'48"60d59'
Shekut,Laura J Cas 3h10'1"68d58'
Shela Psc 23h2'44"5d15'
Shelain,Gavin Cyg 21h1'17"36d47'
Shelby Umi 16h2'37"71d28'
Shelby Mon 6h24'47"-6d58'
Shelby Alexis Peg 21h39'45"23d55'
Shelby Elizabeth Peg 23h30'49"11d14'
Shelby Leigh And 23h38'60"37d37'
Shelby Nicole Lyr 18h40'57"26d6'
Shelby Ruth Vul 20h58'32"20d24'
Shelby Ryan Sweet Love Her 17h2'33"38d52'
Shelby Township,Mr Per 2h12'13sd57d45'
Shelby's Star Gem 7h35'30"20d34'
Sheldan,Phoeby Cam 3h51'22"62d7'
Sheldon Uma 9h19'25"56d22'
Sheldon Ori 5h10'56"1d55'B
Sheldon "Angel",Lauren Ari 1h45'0"16d38'
Sheldon III,H Austin Aur 5h23'16"54d26'
Sheldon Star,The Cet 2h41'23"6d55'
Sheldon's Beautiful Universe Oph 17h46'55"6d12'
Sheldon's Mullett Man Star Cap 21h1'31"-15d21'
Sheldon,Alessia Cyg 19h24'36"34d18'
Sheldon,Allison Marie And 2h3'50"37d8'
Sheldon,C Franklin Cet 2h55'44"7d59'
Sheldon,Christopher Harrison Her 16h38'45"41d19'
Sheldon,Elizabeth D Ori 5h25'1"-6d33'
Sheldon,John Warren Lac 22h23'27"48d52'
Sheldon,Katie And 0h21'56"31d54'
Sheldon,Maureen-Curt Cyg 20h31'50"42d20'
Sheldon,Paul Peg 23h0'41"14d59'
Sheldon,Randall Wayne Cet 2h0'1"0d50'
Sheldon,Richard A Cet 2h32'52"-1d43'
Sheldon,William Kent Dra 14h24'11"63d43'
Sheldrick,George M & Blanche C Boo 15h42'47"50d25'
Shelene Forever Com 12h27'25"26d56'
Sheli's October Star Sex 10h6'1"5d1'
Shelia Ellen Cyg 20h16'12"39d19'
Shelia Marie Cam 3h28'54"61d49'
Shell Aur 5h59'36"40d48'
Shell Rae Uma 9h58'35"58d9'
Shell Star Cyg 20h19'17"31d31'
Shell's "Candlelight" (Benson-Blanchard) Psc 0h1'0"2d25'
Shell's Star With Love Cyg 19h32'57"35d16'
Shell,Holly Chilson Cam 7h50'40"82d48'
Shella's Stellar Star Sgr 19h57'18"-43d14'
Shellbar 11/67 Equ 21h22'14"11d18'
Shellemberger,Janet M Cyg 21h24'51"37d51'
Sheller,Julie Ann Lyr 18h43'40"38d30'
Shelley Tau 5h38'20"27d22'
Shelley Cas 0h59'17"50d34'
Shelley Lyr 18h55'52"30d57'
Shelley Lyn 7h31'48"50d31'
Shelley & Mike Crb 15h20'18"31d47'
Shelley Ann's Star Cmi 7h40'30"5d1'
Shelley Bear Peg 22h0'54"34d35'
Shelley D Aql 19h26'12"8d9'
Shelley's 21st Star Cas 0h46'30"63d33'
Shelley's Best Star Cam 4h40'50"68d46'
Shelley's Mistletoe Sgc 19h1'1"20d6'
Shelley's Shining Star Cyg 19h24'38"31d33'
Shelley's Star Cam 10h53'1"81d38'
Shelley,Catharine Regina Mon 6h35'42"6d21'
Shelley,Elizabeth Joyce And 0h21'42"36d43'
Shelley,Jamie Christine Aur 5h53'39"31d41'
Shelley,June Aur 4h38'26"31d29'
Shelley,Marjorie Caroline Peg 22h50'59"27d56'
Shelley,Matthew Frederick Her 17h1'22"37d54'
Shelleyon Lyn 7h48'0"41d39'
Shelli K Cyg 19h59'54"38d4'
Shellidene,Thomas Cam 8h0'26"60d11'
Shellie Lyn 7h50'21"44d22'
Shellie's Smile Mon 6h39'1"8d60'
Shellie's Valentine And 0h29'13"30d10'
Shelline Peg 21h59'32"11d12'
Shelly Cnv 12h56'49"37d42'
Shelly Cyg 19h55'55"38d26'
Shelly Cas 2h6'40"59d19'
Shelly Cas 23h15'41"62d20'
Shelly Cyg 19h44'24"30d8'
Shelly Mon 6h18'32"8d46'
Shelly & Brian Uma 9h37'1"51d37'
Shelly Dawn Sgr 19h42'21"-43d25'
Shelly Reneé Cyg 20h52'48"31d32'
Shelly's Cas 0h31'50"50d20'
Shelly's Eyes Cas 1h7'35"60d53'
Shelly's Star Peg 22h39'37"30d43'
Shelly's Star Sco 17h29'45"-31d9'
Shelly's Star Peg 23h30'34"22d30'
Shelly's Starfish Eri 4h30'27"-0d36'
Shelly,Kimberly Doughten Uma 9h16'58"62d4'
Shelly-Ayeshah Cru 12h24'47"-63d56'
Shelly-My Warmest Star Forever Kim Boo 15h17'0"53d19'
Shelov,Josh Cep 22h5'17"80d4'
Shelp,Russell Andrew Dra 19h26'35"56d40'
Shelpar Cyg 20h13'1"39d22'
Shelton Family Star, The Lyn 8h5'48"33d25'
Shelton Family Star, The Joseph W Cam 7h33'44"60d45'
Shelton,Anticipating Mike W Lac 21h59'20"42d11'
Shelton,Cale Broderick Ser 18h36'30"5d45'
Shelton,Charlotte Patricia Ori 6h1'39"7d45'
Shelton,Christopher Her 17h28'13"21d42'
Shelton,Daniel Bill Ser 15h29'1"18d56'
Shelton,Dawn Umi 14h28'48"67d42'
Shelton,Debbie Per 3h21'23"38d33'
Shelton,Dinah Aql 19h59'47"10d35'B

Shelton,Gary Frank
Tau 4h12'17"24d10'
Shelton,Jacob Adam
Cet 3h11'12"-0d59'
Shelton,Jesse
Umi 14h21'1"67d48'
Shelton,Joseph Orrey
Sex 10h31'36"2d18'
Shelton,Ken & Dena
Eri 4h10'56"-16d15'
Shelton,Linda Sandhop
Leo 10h53'54"20d15'
Shelton,Maureen
Uma 9h42'59"46d1'
Shelton,Phillip Sebastian
Her 17h13'41"41d16'
Shelton,Phoebe Rebecca
Cas 2h9'1"59d1'
Shelton,Rebecca Annie
And 23h39'42"41d44'
Shelton,Ricky Van
Her 18h19'51"12d21'
Shelton,Ricky Van
Psc 1h44'24"21d39'
Shelton,Ruth Cell
Del 20h18'0"20d34'
Shelton,Sheila
Vul 19h59'51"28d33'
Shelton,Stephen L
Her 17h39'37"21d5'
Shelton,Steven Edward
Cet 3h10'38"5d32'
Shelton,Taylor Marie
Lyn 7h10'31"50d38'
Shelton,Windell Lee
Ori 5h17'0"12d24'
Shelva,"St Peter's Assistant"
Uma 10h44'48"71d15'
Shemanski,Susan Marie
Aql 19h22'60"-0d55'
Shemaria,Victoria & Mary
Sweeney
Cas 23h27'36"58d20'
Shematek,Julia
Mon 6h54'12"-0d36'
Shemon,Rhonda
Ori 5h14'48"-5d23'
Shemonsky,Robert & Sandra
Cyg 20h30'13"40d21'
Shenefelt,Angela Jeanne
Aql 20h6'40"-6d57'
Shenefelt,Eric James
Ser 17h55'15"-14d36'
Shenfeld,David
Uma 10h43'20"59d23'
Shenfeld,Jessica
Aur 6h26'29"37d50'
Sheng,Suzanne
Cas 0h19'29"58d54'
Shenker,Flora Louise
Cyg 21h32'1"53d45'
Shenkman,Dasha
And 0h31'43"45d55'
Shenondilly's Beacon
Cas 2h37'44"65d8'
Shenrock,Msgr Joseph
Cep 20h57'53"58d38'
Shenson,Dr Ben
Crt 11h11'10"-14d50'
Shenton
Leo 11h16'10"3d24'
Shenton,Richard E
Aur 7h22'1"40d15'
Sheo
Aql 18h58'26"-6d50'
Shepard,Allen Willett
Her 18h6'58"28d9'
Shepard,Allen Willett
Dra 11h53'14"71d7'
Shepard,Beautiful Andrea
Cas 0h13'40"63d48'
Shepard,Douglas
Dra 19h23'18"58d43'
Shepard,Freddie J
Ori 5h37'26"8d48'
Shepard,Jack L
Dra 11h6'1"78d36'
Shepard,Jacqueline
Cas 0h38'32"60d11'
Shepard,Jim
Boo 14h26'29"17d58'
Shepard,Larry & Harriett
Cyg 19h59'43"50d14'
Shepard,Patricia M
And 2h22'34"48d34'
Shepard,Riley Chance
Ari 2h31'21"21d50'
Shepard,Roberta Michelle
Iannuzzi
Lyn 7h56'1"34d31'
Shepard,Rocky
Boo 15h20'23"41d29'
Shepard,Roddy
Eri 4h54'20"-6d46'
Shepard,Taylor Anne
And 23h21'30"44d48'
Sheperd,Dr Everett M
Cmi 7h39'32"5d6'
Sheperd,Shirley
Lyr 18h28'30"46d35'
Sheperd,Wendy Louise
And 0h10'19"37d54'
Shephard,Always My Love,R
Stephen
Oph 17h19'12"11d8'
Shephard,David
Cam 13h26'18"80d59'
Shephard,Jr,James "Jack"
Cep 21h36'47"67d41'
Shepheard,Nicola
Cyg 19h27'17"38d52'
Shepherd Sr's Fantasy, Robert
Dra 16h36'13"52d10'
Shepherd,Anne Marie
Peg 22h58'1"34d22'
Shepherd,Barnes
Leo 11h33'40"2d27'
Shepherd,Frances Lou
And 2h28'42"39d35'
Shepherd,Brenda
Uma 10h43'22"57d7'
Shepherd,Charles
Ori 5h14'44"-8d48'
Shepherd,Daniel William
Her 16h58'48"32d44'
Shepherd,Dawn
And 1h29'11"39d46'
Shepherd,Edward L
Per 3h24'35"41d8'
Shepherd,Elizabeth Christina
And 0h32'54"40d52'
Shepherd,Jacquelyn Jo
Del 20h14'22"12d6'
Shepherd,Janet
And 23h34'54"45d49'
Shepherd,Jim & Tracey
Lyr 19h23'45"31d52'
Shepherd,Julie Ann
Lyr 18h46'13"32d7'
Shepherd,Kara Lynn
And 0h29'23"41d13'
Shepherd,Lisa Marie
Lyr 19h1'49"31d32'
Shepherd,Lorraine
And 23h48'1"33d6'
Shepherd,Marilyn Louise
Ori 5h57'0"14d28'
Shepherd,Martin K
Aur 5h0'12"48d1'
Shepherd,Megan Marie
Peg 22h32'21"33d59'
Shepherd,Nancy
Tau 5h27'23"28d58'
Shepherd,Nola Rose
Aql 19h21'19"13d59'
Shepherd,Patricia
And 0h3'47"37d55'
Shepherd,Phyllis Louise
And 2h13'52"41d42'
Shepherd,Susan Gale
Aql 20h19'33"5d10'
Shepler,Jason Robert
Oph 16h57'17"-1d43'
Shepler,Joan
Lyn 8h22'31"46d52'
Sheposh,Anne
Leo 9h22'49"27d39'
Shepp,Carl
Aur 6h1'33"31d26'
Sheppard,Cameron
Cru 12h2'10"-55d8'
Sheppard,Carol Ann
And 2h31'33"37d47'
Sheppard,Gary
Her 17h12'0"48d0'
Sheppard,Grant N
Dra 20h5'51"73d52'
Sheppard,Jacqueline Ann
Cas 0h15'26"59d21'
Sheppard,Janice Ruth
Cas 0h51'20"67d16'
Sheppard,Nadine
Eri 4h42'18"-1d1'
Sheppard,Professor Edmund
Dra 18h57'43"50d37'
Sheppard,Samantha Antoinette
And 1h34'31"39d18'
Sheppard,Sandra Elaine
Del 20h37'0"10d23'
Sheppard,Sona
Uma 9h40'0"46d48'
Sheppard,Vicky H
Cyg 20h11'28"38d21'
Sheppard,William Melvin
Aur 5h32'21"37d51'
Shepperd,Joanna
Ser 16h9'24"0d16'
Shepperson
Aur 4h59'1"38d29'
Shepps Family,The
Cnv 12h51'32"40d21'
Shepps,Howard Zvi
Cnv 13h21'38"40d16'
Sher
Peg 21h47'1"31d26'
Sher/Grad Eternal
Aql 18h59'27"-5d18'
Shera A Bright Star
Uma 8h37'20"60d37'
Shera,Star Marge & Bill
Cyg 20h45'45"38d14'
Sheramar
Aql 18h43'44"10d52'
Sherard
Uma 9h6'39"59d12'
Sherbak,Eugene
Oph 17h10'52"-9d13B
Sherburne,David G
Oph 17h29'59"8d4'
Sherburne,Kelly
Cas 23h17'14"60d0'
Sherby,John Wm
Her 16h21'1"4d7'
Shere's Cosmic Star, Barbara
Mon 6h24'0"4d43'
Shere,Rodney Alan
Dra 9h29'45"74d13'
Sheree & Karen
Cyg 21h8'42"37d60'
Sheree Robin
Per 4h3'23"35d54'
Sheree's Fire
Cas 0h27'39"61d32'
Sheremeta,Mark James
Ari 2h23'16"11d37'
Sherer Dec 6 1918, Norbert
Richard
Aql 19h52'26"14d52'
Sherer,Mark Alan
Aur 6h59'15"38d7'
Sheri
Cas 0h57'55"62d31'
Sheri
Lyr 19h23'23"38d11'
Sheri
And 0h55'24"34d32'
Sheri M
Mon 6h22'52"5d8'
Sheri Theresa
Tri 1h56'41"28d40'
Sheri's Extra Point
Cet 1h25'56"-12d25'
Sherian One
Oph 16h41'0"2d47'
Sheridan
Aql 19h23'26"-1d11'
Sheridan
Cet 0h28'51"-17d35'
Sheridan
Boo 15h2'18"17d35'
Sheridan & Marion's
"Anniversary Star"
Cyg 19h17'32"44d30'
Sheridan Ann
Sge 19h31'0"19d12'
Sheridan DDS,Jack
Uma 8h58'20"52d26'
Sheridan III,Joseph Edward
Lac 22h36'25"38d33'
Sheridan '94 Superstar,The
Del 20h19'54"11d8'
Sheridan's Freedom
Aql 18h59'32"11d41'
Sheridan,"Mongo" Edward
James
Her 16h56'1"37d13'
Sheridan,Bev
Tau 4h3'33"21d36'
Sheridan,Beverley
And 2h32'31"38d35'
Sheridan,Debbe
Psc 0h57'48"27d58'
Sheridan,Desmond S
Her 16h28'23"36d56'
Sheridan,Elizabeth Lizzy
Her 17h50'21"14d58'
Sheridan,Frances Ruth
Cyg 20h5'13"30d50'
Sheridan,Kyle
Cet 2h46'30"4d56'
Sheridan,Lara Christine
Mon 7h0'24"-8d59'
Sheridan,Rita Theresa
Cas 1h35'36"65d15'
Sheridan,Sara
Cyg 21h55'42"53d48'
Sheridan,Sara Colleen
Aql 20h2'59"8d15'
Sheridan,Stephen
Cep 22h3'16"67d60'
Sheridan,Trish
Boo 14h9'19"44d4'
Sheridan,Trish
Lmi 10h37'55"26d24'
Sheridan,Trish
Lyr 19h0'0"38d44'
Sheridan,Trish
Cyg 19h22'37"44d15'
Sheriken
Peg 22h19'0"35d7'
Sherilin
Aur 5h28'20"37d40'
Sherin,Helen
Lyr 18h23'1"41d1'
Sherin,Thomas W
Boo 13h37'1"24d42'
Sherine,Alaa
Cyg 21h16'23"28d55'
Sherlee Win-Alpha
Cru 12h42'40"-60d42'A
Sherlee Win-Beta
Cru 12h42'40"-60d42'B
Sherlene
Mon 6h40'37"7d31'
Sherline,June A
Tri 1h49'23"27d36'
Sherlock,Russell & Wayne
Aql 20h0'32"10d56'
Shermak,Shawn Eric
Aql 19h8'60"3d47'
Sherman Forever,Denise &
Burton
Cyg 21h56'54"50d25'
Sherman's Star,Jason &
Grandpa
Mon 7h1'33"-7d3'
Sherman,"Stormy"
Mon 8h5'54"-5d37'
Sherman,Allan
Hya 8h9'1"-8d55'
Sherman,Andrew M
Lac 22h1'10"48d48'
Sherman,Arthur L
Cnv 13h49'46"38d58'
Sherman,Bernice D
Cyg 19h42'50"28d45'
Sherman,Brenda Gayle
Mon 7h16'26"-6d13'
Sherman,David Nathanial
Ser 15h32'55"-1d0'
Sherman,David
Cet 1h26'44"-2d57'
Sherman,Dennis
Ori 5h56'51"15d11'
Sherman,Diana
Lyr 18h37'19"40d21'
Sherman,Dr Robert E
Ori 5h30'12"-1d51'
Sherman,Edward M
Sco 17h35'23"-32d57'
Sherman,Emma E
Lyn 8h51'47"43d39'
Sherman,Eternally
Yours!,Edward J
Psc 0h46'0"31d58'
Sherman,Francine Spadone
Cas 3h6'53"75d20'
Sherman,Harriette S
Lib 15h30'33"-10d3'
Sherman,Howard Buske
Peg 21h32'41"20d34'
Sherman,Jacqueline
Ser 18h22'13"4d22'
Sherman,Janice Lynn
Uma 11h24'41"57d46'
Sherman,John H
Per 1h42'0"53d46'
Sherman,Jr,Jack M
Cet 0h27'29"-17d54'
Sherman,Larry L
Dra 17h53'28"70d9'
Sherman,Linda
Cas 23h14'53"60d8'
Sherman,Linda Laverne
Mon 6h10'35"-10d35'
Sherman,Lorelle M
And 0h21'1"34d45'
Sherman,Margaret Lowey
Aql 18h57'14"10d15'
Sherman,Mark
Umi 15h16'17"68d23'
Sherman,Michael Bradford
Cam 7h33'54"60d22'
Sherman,Michelle
Cet 1h0'48"-12d43'
Sherman,Nancy
Vir 13h23'58"-2d20'
Sherman,Nathan
Lyn 8h28'35"38d7'
Sherman,Samantha Elise
Peg 22h39'40"25d15'
Sherman,The
Uma 9h49'1"52d49'
Sherman,Thomas J
Aur 5h54'20"31d10'
Sherman,Troy & Shelly
Cyg 21h24'1"40d44'
Sherman,Veronica
Oph 18h34'26"10d36'
Sherman-McDonald,Missi
Peg 22h49'0"27d42'
Shermin
Lyn 7h13'0"56d59'
Sheroke,Mary
Cas 22h57'26"53d8'
Sheron's Serenity
Peg 22h32'39"7d39'
Sherr (ie) & Sherr (i) Alike
Cnc 8h51'40"31d1'
Sherrell Lynn
Gem 6h24'1"12d1'
Sherrell,David N
Aql 19h0'54"10d37'
Sherrell,Jake Albert
Aql 20h6'1"0d21'
Sherrerd-Smith,William
Windisch
Boo 15h37'0"40d45'
Sherri
Lyn 8h46'59"39d45'
Sherri
Mon 6h54'1"11d24'
Sherri
Peg 21h29'14"27d40'
Sherri & Kevin Forever
Cyg 20h58'46"31d7'
Sherri Jamie Jesse I Love You
Forever
Cma 6h51'26"-15d18'
Sherri Rene A Bright Star In
My Life
Eri 4h11'47"-17d13'
Sherri's Star
Tau 4h43'52"16d15'
Sherri-Wide Eyed Survivor
Cyg 19h32'59"38d40'
Sherrick,David
Eri 3h49'41"-1d13'
Sherrie
And 22h55'56"38d12'
Sherrie's Light
Ori 5h50'1"16d31'
Sherrie's Rowdie
Hya 9h14'1"5d17'
Sherriff,Briana Genevieve
Sutherland
Lyr 18h56'26"30d57'
Sherriff,Russell Lawrence
Lac 22h54'46"35d24'
Sherrill,Pierce McCammon
Lep 5h30'59"-11d3'
Sherrock,Janie Frances
Cas 0h38'30"62d18'
Sherrod,Elizabeth Llewellyn
Cas 3h35'0"74d45'
Sherron IV,William
Cyg 21h5'59"44d45'B
Sherry
And 23h22'41"50d23'
Sherry
Lmi 9h48'0"38d53'
Sherry
Del 20h13'40"10d23'
Sherry
Peg 22h18'1"4d27'
Sherry
Peg 23h31'34"33d41'
Sherry
Cep 21h50'22"65d31'B
Sherry
Cas 1h5'47"63d35'
Sherry
Peg 21h39'60"27d9'
Sherry
Cet 2h54'57"1d39'
Sherry & Brett's Wedding Star
Cyg 20h35'1"45d23'
Sherry 1-M92
Lyn 8h54'47"45d26'
Sherry 1958
Gem 6h51'53"14d17'
Sherry 70
Uma 11h51'38"45d23'
Sherry Ann
And 1h51'58"37d47'
Sherry Ann
Cet 1h49'57"-0d13'
Sherry Ann
Ori 5h53'42"16d18'
Sherry Baby
Sgr 18h51'50"-33d50'
Sherry Christina
Vul 20h46'32"20d31'
Sherry Dawn OMTA
Uma 12h17'51"62d26'
Sherry Gayle
Crb 15h57'38"28d17'
Sherry Jessee
Del 21h5'47"12d50'
Sherry Loves Ron-50 Times
More
Cet 1h28'21"0d57'
Sherry Lynn
And 0h11'58"39d25'
Sherry Lynne
Mon 6h53'50"8d4'
Sherry's College Graduation
Mon 7h58'56"-1d48'
Sherry's Star
And 0h25'28"44d7'
Sherry's Star
Sct 18h45'45"-6d39'
Sherry's Wish
Cam 3h40'49"60d53'
Sherry,Colleen
Peg 23h0'31"18d31'
Sherry,John
Ser 15h29'41"20d36'
Sherry,Marit
Mon 7h38'13"-5d38'
Sherry,Mik
Her 18h2'52"38d59'
Sherry,Ruth Ann
Cas 0h12'57"63d47'
Sherry-Marcus
Aqr 22h3'16"-11d29'
Sherston,Jack & Adele
Cyg 19h21'31"27d51'
Sherston,Nicola
Eri 4h2'33"-11d3'
Sherston,Peter
Eri 4h1'51"-12d8'
Sherston,Tom
Eri 4h1'56"-11d25'
Sherwin & Bobbe's Eternal
Love Star
Aql 19h4'22"0d10'
Sherwin Star,The Janet
Cas 0h55'47"74d56'
Sherwin,Sheila
Boo 13h41'56"15d57'
Sherwin,Teresa
And 23h26'58"49d53'
Sherwood Accelerated School
Eri 4h8'54"-7d34'
Sherwood,Alan & Michelle
Lyn 8h3'36"44d43'
Sherwood,Cynthia L
Cyg 21h39'25"38d2'
Sherwood,David William
Per 3h19'28"50d25'
Sherwood,Fredrika Anne
Cyg 20h32'48"58d19'
Sherwood,James
Tau 4h59'20"28d45'
Sherwood,Jessie Jean
Aql 19h42'52"11d10'
Sherwood,Judge William E
Per 1h45'1"48d33'
Sherwood,Larry E
Aur 6h30'17"32d37'
Sherwood,Lorinda Rita
Uma 8h35'24"51d28'
Sherwood,R C
Boo 14h29'27"40d7'
Sherwood,Ray
Sex 10h13'40"-1d38'
Sherwood,Rebecca M
Cas 1h18'1"60d24'
Sherwood,Roberta & Richard
Aql 18h43'24"11d9'
Sherwood,Robin
Lyn 8h55'25"38d18'
Sherwood,Sniggy
Umi 14h54'0"66d29'
Sherwood,Steve F
Sct 18h52'52"-6d52'
Sherwood-Barnard I
Cnv 12h27'25"44d22'
Sherwoods,The
Crb 15h26'16"31d1'
Shery
Cas 1h32'24"63d50'
Sheryl
Aql 20h18'60"5d17'
Sheryl & Carolyn's Friendship
Lyn 7h57'31"40d5'
Sheryl & Jack Star,The
Leo 9h34'1"10d16'
Sheryl 31992
Psc 22h59'30"2d21'
Sheryl Mae Grace
Mon 8h8'43"-1d49'
Sheryl Suzanne
Hya 9h27'24"-6d39'
Sheryl's Eternity
Lyr 18h29'35"41d9'
Sheryl's Outrageous Dazzling
Star
Cet 1h53'0"-2d34'
Sherylizgray
Lac 22h20'15"38d32'
Shesh
Peg 0h11'23"13d60'
Sheshin
Cas 0h21'15"56d0'
Shesol,Barry
Ser 16h7'29"14d23'
Shettlesworth,Jack Lee
Aql 20h15'36"5d19'
Shevach,Melissa
And 23h12'39"40d0'
Shevelew,Margaret &
Jonathon
Cyg 21h59'33"50d18'
Sheveron
Umi 13h5'0"70d54'
Shevlin,Darrell
Oph 17h54'24"-8d21'
Shew,David
Lmi 11h2'17"30d57'
Shewak,Linda H
Crt 10h52'41"-10d42'
Shewchuk,Nancy
Leo 9h19'38"17d5'
Shey,Dr Irving
Per 2h57'56"38d26'
Shi,Iwanuma
Cma 6h54'23"-18d50'
Shiabi,Mohd Al
Crb 16h11'21"38d33'
Shianne
Eri 3h40'55"-1d40'
Shianne Kellie
Cet 2h34'41"-5d4'
Shibel,Jordan Michael
Aur 6h57'28"36d47'
Shick,Erin Elizabeth
Cam 8h27'38"83d20'
Shick,Lara
Lyn 7h56'33"33d2'
Shideler,Mara Danielle
And 1h24'38"35d11'
Shideler,William
Dra 11h26'36"71d27'
Shideler,William & Elizabeth
Stock & Fam
Cyg 20h6'1"40d10'
Shidfar-Altobelli, Rosanna
Barnadetta
Lyr 18h18'54"42d28'
Shieh,Angela
And 2h22'0"41d58'
Shield,Michelle R
Crb 16h2'46"32d38'
Shields
Cmi 7h54'40"4d22"
Shields 082152,Susan Foster
Del 20h13'13"12d15'
Shields,A'Lynn
And 0h3'28"34d20'
Shields,Alexis Milan
Cas 1h51'22"58d46'
Shields,Angela M
Oph 17h3'57"-20d4'
Shields,Brooke
Lyr 19h4'39"41d20'B
Shields,Camille Brooke
Lac 22h55'12"38d53'
Shields,Caroline Loretta
Psc 1h43'36"20d17'
Shields,Chris
Lmi 9h42'33"34d43'
Shields,Clare
Mon 6h25'49"-1d43'
Shields,Dean-Melvyn- Hugh
Peg 22h0'40"34d48'
Shields,Dr James
Oph 18h42'34"6d41'
Shields,Gregory & LeeAnn
Cyg 21h8'52"30d57'
Shields,James
Aql 19h46'56"11d28'
Shields,Joe
Aur 6h16'42"31d48'
Shields,John J
Aur 5h18'52"43d41'
Shields,Jolynn Bagley
Lac 22h51'35"37d54'
Shields,Joyner
Del 20h37'55"18d54'
Shields,Kathryn
Sct 18h44'45"-7d21'
Shields,Lisa Hauk
And 0h13'13"35d27'
Shields,Lucas
Oph 18h2'49"10d53'
Shields,Michael
Her 17h39'26"38d55'
Shields,Michael P "Mike"
Per 3h38'32"38d59'
Shields,Sr,Gary Stephen
Cyg 20h1'28"31d23'
Shields,Stacy
Peg 22h47'13"33d24'
Shields,Stafanie S
Eri 3h30'60"-1d46'
Shields-Star Shine, Suzanne
Lynn
And 0h39'57"40d41'
Shier,Audree
And 0h25'42"43d38'
Shiff,Jonathan M
Sgr 18h49'46"-31d24'
SHIFFERMILLER MICKIE'S
GLOW,MICKIE
Del 20h18'53"11d13'
Shifflett,Crickett
Lmi 10h41'56"28d15'
Shifflett,Michael Jeffrey
Oph 18h0'35"12d3'
Shiffman,Corey Benjamin
Lmi 10h25'22"28d31'
Shifrin,Robin
Lyn 7h52'12"42d25'
Shifrin,Sarah Lin Fendrick
Cas 0h27'16"63d41'
Shifter,Helen
Peg 23h32'14"17d36'
Shige & Rika
Aqr 21h55'20"0d53'
Shige's Stella
Tau 5h12'18"21d24'
Shigemichi Take'- Shihan
Seagal
Aql 18h43'53"10d36'
Shigeyoshi,Yasuo
Equ 21h0'13"2d46'
Shigeyuki Tuneda
Lib 14h34'29"-20d54'

Shigo,Cathleen Uma 10h11'52"51d5'
Shih,David L Aur 5h56'21"31d15'
Shihata,Yasmine And 1h0'47"37d40'
Shihwarg,Alexander Aur 6h23'26"38d19'
Shiina Family, Stella of Cnc 8h48'56"16d21'
Shikha Cet 3h0'21"0d15'
Shikles,Earl Martin Her 16h59'48"38d37'
Shilanski,Rick Dra 14h33'58"60d6'
Shilen,Jacqueline Delaney Psc 1h35'34"27d31'
Shill,Wayne Cap 20h56'28"-15d1'
Shillander,Amy Sex 10h30'23"-0d58'
Shillet,Henry Ira Dra 16h56'14"69d29'
Shillinglaw,Claudine Aur 5h17'29"40d27'
Shillinglaw,Roger Dra 18h58'0"47d42'
Shilman,Perri "Sweet Pea" Lmi 10h48'0"32d43'
Shilvoch,Amy L Vir 12h58'0"-21d42'
Shilvock,Donald (Donnie) Aur 5h58'37"30d36'
Shima,Jeffrey Anthony Dra 16h28'1"63d29'
Shimansky,Andrew James Peg 0h11'0"14d26'
Shimansky,Kaitlen Ashley And 0h23'52"41d13'
Shimasaki,R Serenely Shimmering Eri 4h15'0"-17d18'
Shimazu Family Star, George & May Cet 2h50'53"2d18'
Shimer,Brianna Aql 20h6'16"0d43'
Shimer,John"Jack's" Thomas Cet 3h14'48"0d49'
Shimkus,Ronda And 23h22'47"43d43'
Shimmering Karen And 0h21'43"38d33'
Shimmie Uma 12h27'13"60d31'
Shimo Uma 12h23'31"53d23'
Shin's Star Uma 8h11'34"68d54'
Shin,Kwan Chang Nim Jae C Lyn 7h27'54"50d49'
Shinbo Aql 19h24'17"13d49'
Shine For The Love Of Charif Mon 6h39'41"0d19'
Shine On Adam Per 1h57'45"48d55'
Shine on Forever - Our Dakota Cyg 19h58'1"48d60'
Shine Simpson Per 1h55'51"56d55'
Shine Upon Us Michael Uma 11h50'27"40d17'
Shine,Martin Louis Mon 7h2'20"-6d30'
Shiner,David und Michaela Vir 11h50'49"-2d47'
Shines,Michael Lee Umi 13h18'25"73d46'
Shines,Mike Ari 1h54'31"11d25'
Shingler,John Spurgeon Her 17h30'36"41d10'
Shingler,Lisa-Jane Del 20h21'36"10d27'
Shingleton,Jeffrey Thorman Oph 17h36'13"8d31'
Shining for "Miles" Aur 5h52'0"31d11'
Shining For Jonathon K Del 20h31'0"10d26'
Shining Friend Forever Eri 3h1'20"-8d25'
Shining In The Center Of My Soul Cap 20h37'42"-15d14'
Shining Light of Paty & Rae,The Vir 13h21'45"1d39'B
Shining Pam Cyg 20h28'1"40d60'
Shining Peach,The Uma 10h41'0"55d11'
Shining Rick Uma 12h3'24"31d36'
Shining Skare Ori 5h56'12"16d51'
Shining Star "Christy" And 23h22'51"43d46'
Shining Star of Ed & Mary's Love,The Eri 2h50'26"-1d40'
Shining Star Sheila, The Cet 2h14'31"8d5'
Shining Susan Ann,The Cas 2h29'44"68d24'
Shining Tiger,The Lmi 10h1'44"32d20'
Shining Von Uma 10h54'1"54d53'
Shining Warrior, (Science Club) Aur 6h28'14"34d28'
Shinkle,Angela M Cas 0h54'29"66d55'
Shinko,Jeffrey Aur 5h54'13"29d38'
ShinLing 35 Aql 19h6'1"15d13'
Shinn Star,The Ambr M Aql 20h0'19"12d58'
Shinn,Cary Per 2h50'0"32d56'
Shinn-June 5, 1943, Betty & Roy Per 2h4'57"57d53'
Shinneman,Irene And 2h18'39"38d26'
Shinnick,Mark F Cmi 7h59'40"8d31'
Shino Mon 6h58'0"-0d18'
Shinohara,Etsuko Lac 22h46'41"52d60'
Shinohara,Tsubasa Dra 20h19'24"67d31'
Shinozuka,Shigeki Aql 19h31'13"12d26'
Shinsei Umi 14h14'38"66d56'
Shintani,Sandy Mon 6h24'11"11d38'
Shiobán Uma 10h48'18"70d38'
Shion Cnc 8h47'18"11d25'
Shiori Memory of Your Birth Aqr 22h17'18"-15d32'
Shipler,Thomas J & Betty D Per 3h2'50"48d55'
Shipley,Agatha Gail Lac 22h54'39"51d32'
Shipley,Donna Peg 22h10'18"24d43'
Shipley,Forrest A Uma 9h38'1"50d23'
Shipley,Leona Peg 21h59'37"23d37'
Shipley,Michael Her 16h46'1"51d17'
Shipley,Neal A Aql 18h43'23"7d52'
Shipley,Robert Aur 7h12'36"38d58'
Shipley,Ross Cet 1h14'49"-0d30'
Shipman Eternal, Mildred Janofsky Uma 12h45'11"63d3'A
Shipman,Joshua Hayden Hya 9h14'55"1d12'
Shipman,Love,Norman Uma 12h45'11"63d3'B
Shipman,Samantha Peg 22h36'0"29d36'
Shipp,Carol S Leo 9h42'16"16d18'
Shipp,Charlotte Peg 22h0'31"29d56'
Shipp,Nancy A Del 21h2'28"12d54'
Shippey III,William H Dra 17h58'0"76d0'
Shipton,Gordon Owen Cep 21h26'58"78d59'
Shipway,Eva Cyg 20h42'16"45d24'
Shipway,Raymond Lyn 8h13'51"41d22'
Shira Beth Cas 23h44'8"64d1'A
Shirar,Amy Lynn Cas 0h52'41"66d40'
Shirazi,Amir Mohammed Nejad Aql 19h42'16"10d50'
Shirazipour,Benjamin Her 17h13'34"20d40'
Shirby Lyn 7h8'45"58d57'
Shirdi Sai Baba Dra 19h55'35"61d16'
Shirdon,Diana & Jamie Cam 7h27'33"82d51'
Shire,Daniel Joshua Cnc 9h3'12"20d42'
Shire,Lydia Dra 19h34'35"65d24'
Shireen Oph 16h48'47"10d39'
Shirelle's Star Lyr 19h17'36"26d29'
Shirer,Peter J Cma 6h54'12"-19d6'
Shirer,Scott Dra 11h31'20"71d39'
Shires,Jack L Leo 9h46'21"18d48'
Shirey's East,The Ori 5h57'0"16d10'
Shirey,Harold Gene Cnv 13h0'56"38d16'
Shirey,Stewart Her 16h28'34"35d41'
Shirey-Col USAF,Clair L And 2h27'43"38d33'
Shirin,Look Up I Will Always Be There Aur 4h50'31"40d17'
Shiring,Bob Sct 18h37'47"-4d37'
Shirius Jr Psc 1h16'6"29d53'
Shirl J Del 20h14'1"15d14'
Shirley Aur 5h19'1"42d12'
Shirley Uma 10h46'31"55d12'
Shirley Her 17h42'57"31d9'A
Shirley Cas 0h3'39"63d57'
Shirley Lyr 19h18'21"42d5'
Shirley Tri 1h59'59"27d43'
Shirley Mon 7h1'1"-0d39'
Shirley Vul 20h39'57"20d20'
Shirley & Henry Eri 5h2'40"-4d37'
Shirley & Keith Still Shining Bright Cmi 7h7'14"4d17'
Shirley & Maher Peg 22h26'1"30d51'
Shirley & Mickey Aql 18h46'51"10d24'
Shirley A Uma 11h16'56"46d44'
Shirley Bay Cet 2h36'1"2d22'
Shirley C Ori 5h53'10"13d19'
Shirley Forever Denise's Valentine Lyr 18h25'1"46d46'B
Shirley Hazel Ann Cyg 19h40'55"38d27'
Shirley Inez Equ 21h6'29"2d40'
Shirley Jean Peg 22h22'55"21d47'
Shirley June Equ 21h7'31"10d32'
Shirley Kyle Peg 23h2'15"33d51'
Shirley Lee Ori 5h34'0"-0d1'
Shirley M Cas 22h58'1"55d25'
Shirley Maybell And 23h50'37"33d44'
Shirley Rosemary Crb 16h2'10"26d49'
Shirley Shine On Cyg 21h1'11"38d51'
Shirley Six "O" Cmi 7h23'32"8d4'
Shirley Theresa Lyn 8h45'25"41d40'
Shirley's Star And 2h24'36"41d41'
Shirley,Barbara Ann Lyn 9h13'1"38d32'
Shirley,Cailey Elizabeth Ann Peg 23h28'0"15d8'
Shirley,Chris & Elizabeth Reitzel Crb 16h1'24"31d35'
Shirley,Diedre D Aql 19h9'19"15d56'
Shirley,Jamie Lauren Cma 6h41'18"-15d38'
Shirley,Judith Marie Aql 18h57'50"-2d13'
Shirley,Melinda Ann Cet 2h45'0"0d22'
Shirley,Zachary Eri 2h56'47"-10d29'
Shirley-Ann Tri 2h21'1"31d3'
Shirleyann Cet 1h23'50"-6d51'
Shirlgirl (I) Cas 0h20'38"70d36'
Shirmel Aql 19h2'0"-6d46'
Shirra,Catherine Lyr 18h19'58"43d43'
Shirran,Gillian Ori 4h42'54"0d29'
Shirrell,Lucia Peg 22h26'0"26d1'
Shirvell of Shalden, Tricia Ori 6h1'12"1d3'
Shishman,Hedda Cet 1h8'12"-1d25'
Shively,Carol S Cas 0h12'55"58d13'
Shively,Marie Cet 0h0'60"-10d56'
Shivers,Charles Herbert Vul 19h19'1"26d35'
Shivers,Mitchell E Dra 17h41'57"68d13'
Shlapak,Beverly Mon 6h36'22"1d23'
Shlapak,Milton Sct 18h43'24"-6d49'
Shlaudeman,Kimberly Rae Lyr 18h38'23"30d19'
Shlenker,Marti Mon 6h43'20"3d57'
Shlomo R Cap 21h7'50"-22d26'
Shmalo,Nathan & Melanie Cyg 20h36'23"42d49'
Shmel Per 2h29'41"56d20'
Shmoltz Cet 2h31'10"-0d36'
Shmootz Aur 4h54'25"40d43'
Shnoook-Hum My Shnoook-Hum Cam 5h17'17"68d60'
Shoaf,Janelle Lyr 18h47'21"39d36'
Shoaf,Kimberly Kay Cas 0h59'34"69d8'
Shobita Aql 18h58'1"17d7'
Shocinski,Joseph- Evelyn Egan Lyr 18h34'10"37d15'
Shockley,Darby Maryline Lyn 8h7'15"58d36'
Shockley,Dorian Tyne Uma 11h44'1"48d58'
Shockley,James Mitchell Her 16h46'21"30d51'
Shockley,Joni Lynn Boo 14h19'54"54d19'
Shockley,Lee & Pannie Cam 12h53'58"77d3'
Shoemaker,Daniel Uma 11h12'13"38d7'
Shoemaker,Eric Boo 13h36'39"25d6'
Shoemaker,Marsha Cet 0h56'52"0d44'
Shoemaker,Mary Mon 8h1'50"-9d35'
Shoemaker,Peggy Lynn Ori 5h48'27"10d21'
Shoemaker,RLCJ Ori 5h59'12"12d16'
Shoemaker,Suzanne Massey Cyg 21h25'21"37d56'
Shoenfeld,Richard Super Stately Star Sex 10h26'49"1d10'
Shoffstall,Brenda Lou Hall Lyr 18h29'0"30d33'
Shogren A Star In Your Eyes,Everett Her 17h25'13"27d48'
Shohnt's Love Uma 11h53'55"64d60'
Shokai,Omakase Yamada Gem 6h56'33"25d58'
Shoko Tau 5h30'44"19d20'
Shoko (Japanese Crystal) And 23h4'23"41d19'
Shokoff,Dorothy And 0h21'11"30d30'
Shokriarv,Jila Cyg 21h7'47"30d21'
Sholes,Barbara & Roy Cas 0h24'38"61d20'
Sholl,Clint Aur 6h14'46"30d53'
Sholom Laura C F Cyg 19h54'29"44d47'
Sholom Nolan C F Cyg 20h31'16"37d46'
Sholty,Robert Edward Psc 1h23'51"17d43'
Shomaker,Edward Dra 18h27'24"58d44'
Shona Uma 13h56'45"58d24'
Shona Lac 21h56'54"41d9'
Shona B Cyg 19h31'52"33d42'
Shona-Bat Mitzvah, Emily B Peg 22h56'24"24d59'
Shonagh,Marry Me Cas 0h29'25"60d19'
Shondel,William John Aur 7h24'45"36d60'
Shondra Umi 10h29'17"89d5'
Shoog Cyg 20h53'29"38d7'
Shook,D J Gem 6h57'18"20d14'
Shook,Donovan C Cet 3h3'48"1d39'
Shook,Karen And 0h46'21"38d59'
Shook,Lester Eugene Aql 18h56'57"-1d7'
Shook,Mary Kathryn Campbell Mon 6h17'51"-6d41'
Shook,Megan Elizabeth And 0h0'43"37d46'
Shook,Ray Cet 2h38'56"0d19'A
Shook,Rebeka Ann And 1h3'40"40d52'
Shook,Ruby Cet 2h38'56"0d19'B
Shook-Merry Christmas, 1993,Scott D Ser 15h11'28"10d29'
Shooner Per 3h12'17"40d49'
Shoop-Kirkpatrick, Debron Vul 19h38'12"27d11'
Shoot For The Stars Woody Uma 11h53'10"53d46'
Shoot,Grandma Cas 0h8'20"60d17'
Shooter Dra 19h28'25"61d41'
Shooting BooBoo-Jam, The Lac 22h13'33"49d30'
Shooting Seymour Aur 6h26'59"31d38'
Shooting Stars Of Troop 423,The Uma 8h53'11"62d15'
Shope,Alexandre Michael Aur 5h0'55"50d26'
Shope,Melvin Leroy Ori 5h53'0"20d48'
Shorb,Terry Per 2h24'34"55d23'
Shore is Love Lyr 19h0'56"30d35'
Shore,Dustin Boo 15h34'51"48d20'
Shore,Elisheva S Peg 23h43'15"31d55'
Shore,Leonard & Aileen Lyn 8h54'21"41d7'
Shore,Michael J Dra 17h44'25"75d21'
Shore,Stacey Stern Tri 1h47'47"34d41'
Shore,Tiger Mon 7h12'48"-10d42'
Shore,William S Per 2h21'0"55d43'
Shorenstein,Walter Her 17h12'56"42d10'
Shores,Jenny Dawn Mon 7h53'1"-6d55'
Shores,Kimberly Ann Lyn 7h52'25"50d17'
Shorgrass,Flora Com 13h7'17"18d38'
Shorooghi,Homey Crt 11h43'31"-20d50'
Shorrock's Striker Boo 14h40'18"32d47'
Shors,Lady Stephanie And 1h25'35"40d14'
Shorsh Uma 10h5'34"48d47'
Short 30th Birthday, Russell S A Cep 21h59'46"71d13'
Short Star,John & Sue Cyg 21h9'51"38d23'
Short Stuff Leo 10h27'58"10d44'
Short World Of Hope, Jessica Dacey Lyn 7h58'35"45d7'
Short,Amy Peg 21h19'32"20d21'
Short,Aric Grayson Tau 5h47'0"26d22'
Short,Audrey Brooke Peg 23h39'36"31d30'
Short,Clifford & Mary Peg 22h36'0"31d17'
Short,David L & Tammy W Sge 19h58'20"16d21'
Short,John Doyle Cep 0h35'1"78d21'
Short,John L Aur 4h49'34"40d36'
Short,Justin Tyler Aql 19h27'36"13d30'
Short,Layla Umi 16h24'34"72d34'
Short,Lorraine M Cyg 20h48'34"37d33'
Short,Nigel C Cep 20h54'33"78d51'
Short,Peter Charles Ori 6h16'34"12d39'
Short,Sandy Lmi 9h41'53"40d42'
Short,Shirley Reed Aql 19h29'37"11d5'
Short-Sheridan, Kemberly Shiobhan Vir 13h29'36"-20d50'
Short-Spence,Norma Marie Lib 14h44'53"-0d43'
Shortell,Paul Per 2h45'44"40d21'
Shorter,Ana Maria Aql 18h50'16"11d21'
Shortnik Leo 11h32'42"11d11'
Shortt,Beth A Cet 0h29'39"-6d54'
Shortt,Jacqueline D Peg 22h36'21"25d55'
Shortt,James Ori 5h52'36"1d50'
Shortt,Jim Per 2h39'48"34d49'
Shorty Aur 6h24'54"32d21'
Shoshana Sgr 18h55'49"-22d32'
Shoshana Cas 1h34'45"70d13'
Shostack,John Dra 16h47'0"67d24'
Shotton III,Thomas E Psc 22h57'0"-2d49'
Shotwell Tri 1h57'44"30d33'
Shouel,Heather Lydia And 0h50'51"38d54'
Shoulders,Chad Cet 1h33'1"-0d41'
Shoulders,Jason Paul Boo 14h38'28"17d51'
Shoup,Elizabeth A Lyn 8h58'21"42d54'
Shoup,Linda Kay And 22h58'35"50d36'
Shoup,Mary Margaret Cas 1h33'22"60d39'
Shoup,Myna Harrod Boo 14h16'9"10d44'A
Shoupp,A James "Pete" Aur 6h30'25"37d18'
Shoupp,Casey Allan Dra 18h3'42"50d35'
Shova,Patricia Vul 19h12'37"21d15'
Shovelhead Cmi 7h30'47"0d27'
Show Aqr 22h19'35"-15d42'
Show Who Lib 14h46'6"0d57'
Show,Ormand C Lac 22h34'39"52d55'
Showalter,Lauren Paige Cyg 20h31'17"31d15'
Showalter,Nancy Patricia Gamble Cas 0h27'26"60d9'
Showalter-Bronson, Cherlyn And 0h10'54"33d33'
Showers Family Star Cam 9h20'24"73d13'
Showers,Ralph Vaus Dra 17h39'11"65d29'
Showler,Ethan Thomas Aur 5h52'24"30d30'
Showman,Hazel Irene Meier Lyr 18h28'0"45d48'
Shows,Janet Marie Theresa Borstner Mon 6h28'20"-8d18'
Shows,Susan Lyn 8h17'49"50d20'
Showtime Aur 6h7'30"36d37'
Shpakoff,Ginny Rose Lynn And 23h44'45"46d47'
Shpungin,Elaine And 23h2'45"40d39'
Shqipe Uma 10h17'24"53d39'
Shra Vul 20h21'34"25d30'
Shrader,Dennis J Aur 5h2'55"30d22'
Shreckengost,Ott Aur 6h36'53"37d21'
Shreeve,Terri Lynn Cas 2h4'35"59d18'
Shrekenhamer Wedding Star,The Eri 4h34'52"-0d35'
Shreter,"My Everything",Stephanie Lyn 0h55'40"41d7'
Shrewsbury,Kris Joseph Her 16h24'1"41d58'
Shrimpi & Neshi Cam 3h31'17"62d52'
Shriver,Daniel Aql 19h59'27"15d11'
Shriver,Debra K Lyr 18h35'46"26d15'
Shrock,Ronald L Lac 22h39'32"53d21'
Shrome,Bryan Arthur Cyg 19h28'45"32d21'
Shrout,Monica Cnv 13h22'47"41d35'
Shroyer One of a Kind, Terri Ann Mon 7h18'0"-6d58'
Shrubb,Sarahjayne Lyr 19h3'32"26d0'
Shrum,Bobbie Marie Uma 8h37'23"56d42'
Shrum,Connie Fay Minerich Uma 9h30'11"54d50'
Shrum,Courtney Anne Uma 12h7'36"46d21'
Shrum,Eva Marie Wishloff Uma 10h23'44"72d45'
Shrum,Jon E Peg 23h50'47"31d35'
Shrum,Tonya Marie Uma 12h7'39"47d46'
Shruti Ori 4h49'28"0d6'
Shtrang,Shipi Uma 8h40'19"62d10'
Shu-Shu Cmi 7h16'37"7d52'
Shubeck,Jr,John Robert Per 3h0'51"40d5'
Shubin,Genevieve Peg 23h21'56"10d20'
Shubin,Georgia Collins Sct 18h51'3"-6d50'
Shubunking Cam 5h47'50"60d51'
Shuell,Brad & Betty Tau 4h1'49"23d8'
Shuey,Michael Diven Ori 6h2'19"-2d9'
Shufeldt,Linda S Peg 21h27'1"22d48'
Shufelt,Susan Ann Uma 9h19'20"67d52'
Shuff,Mr & Mrs John Hya 9h10'12"1d30'
Shuford's Star Gem 6h48'15"14d43'
Shuford,Ben & Dede Sex 9h52'30"2d9'
Shuford,Laura Vul 19h46'56"22d59'
Shugart,Craig Per 2h20'49"54d59'
Shugdinis,Robert W Aur 5h11'34"54d52'
Shuji & Fusako Vir 14h7'54"-16d2'
Shuker,Mary Doreen And 23h3'33"50d12'
Shukla,Mita N Cyg 20h3'0"31d40'
Shukla-Mackie,Kenneth Per 1h46'8"47d38'A
Shukla-Mackie,Max Per 1h46'8"47d38'C
Shukla-Mackie,Neelam Per 1h46'8"47d38'B
Shula,Hannah Cas 0h59'58"51d31'A
Shuler,Beth Alisha Cyg 20h29'49"42d2'
Shulkin,Jeffrey Andrew Oph 17h24'57"-6d2'
Shull,Amy Lynne Uma 13h39'15"61d51'
Shull,Judy Cmi 7h55'12"7d48'
Shull,KD Dra 15h15'41"63d11'
Shull,Kelly Jean Uma 13h28'55"62d34'
Shull,Leo Aqr 21h32'0"-0d43'
Shull,Marcella H Peg 21h49'21"33d37'

Shull,Matthew Thomas
Uma 12h16'0"62d40'
Shullek,Toni
Lyn 7h22'1"44d48'
Shulman,Katie Uecker
And 23h27'29"44d17'
Shulman,Pam
Lyn 8h17'52"38d3'
Shulman,Shawn Alexander
Aqr 21h53'37"0d15'
Shulman,Terry
Hya 8h9'18"2d30'
Shults,Kim & Nancy
Lmi 10h35'18"28d6'
Shults,Lee A
Uma 9h45'23"57d58'
Shults,Marcy Rachelle
Cet 3h0'39"2d35'
Shults,Marcy Rachelle
Cyg 21h20'35"31d23'
Shultz,Andrea Allison
And 2h22'33"38d26'
Shultz,Christian
Oph 17h11'32"-24d50'
Shultz,Lee Allen
Aur 6h21'19"32d43'
Shultz,Lempi Maria
Sge 19h35'39"16d40'
Shultz,Nicole A
Lyr 18h31'16"34d57'
Shultz,Peter
Lmi 10h51'32"38d38'
Shum,Jenny Chow Ho
And 0h10'37"38d58'
Shumacher,Lance "Gnathostomata"
Psc 1h28'0"31d59'
Shumaker,Carol Ann
Cas 23h22'55"60d8'
Shumaker,Ethel Irene Warren
Cyg 21h3'36"31d48'
Shumaker,Julia Ann
Cas 2h40'19"67d52'
Shumaker,Michelle Lee
Mon 6h39'58"1d42'
Shuman,Elizabeth
Lyn 7h48'55"40d27'
Shuman,Lorna Elaine
Lyr 18h37'12"36d26'
Shuman,Yolanda Marie
Cas 23h28'25"61d33'
Shumate,Alice Dunn
Ori 5h9'1"-4d32'
Shumate,Christy
And 2h8'31"41d6'
Shumate,G Gray
Del 20h25'10"10d42'
Shumate,George J
Sex 10h42'24"0d49'
Shumskas,Edward Steven
Boo 15h36'27"50d7'
Shumway Luv U 4 Ever, Bruce & Michele
Cyg 20h18'44"38d40'
Shumy R
Cap 21h6'1"-22d5'
Shuop,William Walter
Boo 14h16'9"10d44'B
Shupak,Lizzie
Com 12h30'56"25d46'
Shupp,Bob
Cmi 7h25'25"9d37'
Shupper,Jason Beechey
Per 3h6'36"46d34'
Shupper,Rear Admiral Burton H
Ori 5h40'13"11d45'
Shur Good
Per 3h8'24"38d30'
Shur,Erica
Lyr 18h40'19"34d41'A
Shur,Rudy
Lyr 18h40'19"34d41'B
Shurery,Amy Ann
Lyr 18h14'16"47d20'
Shurgan,Elyce Lyn
Cas 0h44'43"70d4'
Shurie,John Edward
Uma 12h4'46"46d36'
Shurr,Russell
Cep 21h44'22"55d59'
Shurrolk,Christopher
Ori 6h3'15"0d32'
Shurte,Danielle Veotta Rose
Peg 22h45'14"26d0'
Shurtleff,Raymond Gregory
Cep 22h53'0"71d1'
Shurts,Margaret Mary
Cas 0h53'18"72d18'
Shusta,Allison Marie
Eri 4h27'35"-17d49'
Shuster,Bernie
Per 3h6'56"41d15'
Shuster,Kylie Rae
Cyg 20h34'57"40d29'
Shuster,Tracy L
Ori 4h57'17"-0d10'
Shute,June
Lyr 18h37'22"37d11'
Shutler,Tessa Marie Hall
Mon 7h21'20"-6d45'
Shutowich,Jaclyn Ann
Lib 14h46'37"-0d34'
Shutowich,Jaclyn Ann
Lyr 19h0'39"35d0'
Shuzo & Satomi
Ari 2h19'51"27d23'
Shvetankur
Cep 23h8'11"60d20'
Shwery,Lynne
Del 20h19'53"18d48'
Shy,Alison
Cnv 12h49'40"34d8'
Shy,James M
Cet 2h54'32"9d51'
Shylock
Dra 17h4'1"60d59'
Siamo Amici Per Sempre
Lyr 18h31'35"45d1'
Sian Anne Louise
Cru 12h8'44"-63d58'
Siani,Maria Christina Bimba Mommy
Del 20h22'27"20d6'
Siani,Papa Salvatore Farfallo Baby
Del 20h21'39"20d35'
Siano & Family,Tom & Natalie
Peg 21h20'49"22d36'
Siaura Leann
Ori 5h54'0"10d19'
Sibade,Mathilde
Boo 13h52'0"16d7'
Sibade,Thibault
Cnv 12h19'45"47d44'
Sibal,Cezar
Cam 3h56'47"61d28'
Sibbe,Rita
Lib 14h46'1"-23d3'
Sibel
Del 20h22'0"20d36
Sibel
Cam 6h8'33"83d20'
Sibel,Stacy Jo
Oph 18h1'59"13d44'
Sibelle,Jay
Cet 1h18'0"-4d53'
Siberia
Lup 15h0'46"-38d13'
Sibilia,Lynne Marie
And 1h51'35"37d31'
Sibley,Joan
Peg 22h21'46"29d18'
Sibley,Jr,H Robert
Crb 15h46'22"28d31'
Sibthorpe,Ellen C
Cas 2h16'1"68d10'
Sibylle
Psc 23h22'45"5d27'
Sica,Marc Vincent
Her 16h25'15"10d43'
Siccardi,Peter
Her 17h33'0"38d56'
Siceloff,Daniel Simeon
Aql 20h0'12"1d40'
Sicherman,Dara Kimberly
Lyn 8h13'19"42d13'
Sichky,Beverly
And 1h57'54"39d49'
Sichova,Lucie
Cas 0h25'39"61d10'
Sicia
Aur 5h57'50"54d35'
Sicilia,Dominique
And 1h5'35"40d57'
Siciliano,Avery Mae
Aqr 23h32'40"-5d10'
Siciliano,Klair Quincy
Tau 4h44'0"28d15'
Sick,Wolfgang
Cyg 19h52'41"44d7'
Sickel,Cody
Aur 5h16'27"46d33'
Sickler,Cookie
Aur 5h30'39"38d60'
Sickler,Kandace Bergh
Lyr 18h39'23"39d30'
Sickmen,Ronald M
Per 3h10'57"47d51'
Sicotte,Thomas Katchamess
Peg 22h44'1"31d35'
Sicurella,Joseph
Aql 20h7'47"0d25'
Sicurelli 1993 Papa's Star,Stefania A
Vul 20h5'22"28d54'
Sicurelli III 1993 Papa's Star,Robert J
Dra 12h23'40"75d44'
Sicurelli-Greber 1993 Pap's Star,R M
Aur 5h28'59"31d28'
Sicurezza,Lou
Cep 21h32'24"65d17'
Sid
Eri 3h44'54"-4d22'
Sid
Boo 13h55'29"17d20'
Sid & Antoinette
Peg 22h6'54"26d12'
Sid Louie
Aur 5h1'1"48d2'
Sid's Star
Ser 15h36'54"-2d42'
Siddall,David
Her 18h8'54"45d48'
Siddall,Joyce H
Peg 22h47'14"33d37'
Siddig El Fadil
Cet 2h41'56"0d32'
Siddiqui,Parveen
Lyn 9h7'59"40d55'
Siddle,John
Crb 15h17'30"31d21'
Siddle,John
Crb 15h17'1"30d55'
Side Out
Cmi 7h54'1"3d50'
Sider,Barbara
Uma 8h47'11"49d31'
Siderea
Aur 4h54'0"40d32'
Sideribus Inlustris Of Scott
Cam 4h9'34"61d52'
Sideris,Panayiotis
Cep 20h19'49"76d18'
Siders,Michael David
Oph 18h7'43"12d57'
Sides,Edwin G
Cam 4h3'0"68d59'
Sidhu,Baljit Kaur
Lyr 18h29'43"45d16'
Sidhu,Ganda Tara Singh
Cru 12h42'57"-60d16'
Sidhu,Nina
Peg 22h37'26"21d23'
Sidhu,Rory Eugene Singh
Cru 12h27'48"-61d7'
Sidler,Elizabeth Anne
And 2h1'60"45d6'
Sidlo,Andras K
Boo 15h10'40"40d14'
Sidlow,Mark Andrew
Leo 9h59'47"11d50'
Sidman,Joanne Elizabeth
Cap 21h22'52"-18d52'
Sidney Loves Maisie
Lyn 7h53'32"58d54'
Sidney Roger
Peg 22h15'53"7d39'
Sidney's Everlasting Light
Uma 8h33'57"68d23'
Sidney,Robert A
Sct 18h53'21"-9d13'
Sidney,Walter
Oph 16h56'15"-23d50'
Sidoli,Jr,Richard
Cep 22h10'0"59d27'
Sidor,Allen Joseph
Oph 18h2'1"0d9'
Sidoroff,Walter
Her 17h6'42"42d33'
Sidorowich,Catherine T W
Cam 7h55'25"70d30'
Sidorski,Carol & Ronnie
Cyg 21h27'32"40d32'
Sidorski,Mark Joseph
Dra 14h46'25"63d47'
Sidoti,Agnes F
And 2h1'44"39d34'
Sidoti,Orion Vincent
Ori 5h55'14"17d23'
Sidus Ioannae et Philippi
And 23h20'1"52d57'
Sidwell,Donna & Rick
Sge 19h32'1"16d31'
Sieban,Rob
Cet 0h37'40"-4d17'
Sieber,Heather Ann
Cas 0h30'50"71d8'
Siebers,Lynn Marie
Cet 1h18'27"-11d9'
Siebers,Susanne & Ingo
Ari 2h23'40"21d40'
Siebert,Amanda L
Com 12h31'35"30d38'
Siebert,Craig P
Aqr 22h28'32"-8d46'
Siebert,Hanspeter
Lyr 19h19'55"30d36'
Siebert,Jane Mara
Lyr 18h46'0"37d24'
Siebert,Jodelle
Sge 19h40'0"16d41'
Siebert,MD,John W
Aur 4h59'45"51d27'
Siebert,Rich & Sara
Cyg 20h42'19"42d33'
Siebert,Susan & Charles
Sge 19h3'53"16d13'
Siebertz,Günter
Sgr 18h57'35"-22d26'
Siebertz,Marlene
Vir 13h31'18"-3d36'
Siebler,Jackie
And 1h57'18"38d25'
Siebol's Stellar Estate
Uma 8h12'1"68d18'
Siebold,Dianne
Vul 21h23'32"26d57'
Siedenburg,Christopher
Her 18h10'15"40d56'
Siedlecki's Radiant Aura
Per 2h6'47"57d12'
Siefert,James Lee
Ari 2h58'49"28d14'
Sieg,Christopher Eric
Aur 6h8'19"32d56'
Siegel Kindness Hugs Safety,Dr Burton
Boo 15h0'14"11d16'
Siegel,Alexis
Del 20h20'44"10d16'
Siegel,Arnold Robert
Cet 2h14'1"3d51'
Siegel,Bettie I
Peg 22h10'1"29d42'
Siegel,Charles
Oph 18h35'39"10d19'
Siegel,David A
Aql 19h9'22"13d49'
Siegel,Fred & Maria
Uma 9h45'14"51d20'
Siegel,Gloria Petta
And 2h19'59"37d22'
Siegel,Lorenz
Cyg 20h26'42"30d36'
Siegel,Manfred
Umi 15h32'11"71d23'
Siegel,Marvin Selcer
Dra 16h35'30"68d39'
Siegel,Michael
Ser 16h14'34"2d14'
Siegel,Natasha Heather
Cas 2h58'59"61d7'
Siegel,Paulie
Uma 11h55'1"40d45'
Siegel,Phyllis
Cyg 20h6'49"41d14'
Siegel,Phyllis
Cas 2h13'0"59d49'
Siegel,Richard "Zaza"
Her 16h45'32"20d26'
Siegel,The Mark
Dra 16h35'49"60d50'
Siegel,Todd & Gigi
Eri 4h53'26"-7d56'
Siegel,Todd L
Per 3h11'13"41d32'
Siegel,Vicki
Crb 15h16'26"31d32'
Siegel,Wade
Cnv 13h47'12"38d23'
Siegelaub,Anne & Harold
Cyg 19h34'19"28d8'
Sieger,Robert R
Aur 4h58'47"36d1'
Siegert,Gram & Pop-Pop
Cyg 21h50'31"41d35'
Siegert,Manfred
Dra 18h18'30"65d18'
Siegfried
Ori 5h25'17"1d4'
Siegfried,Verne & Barbara
Her 16h54'33"38d40'
Siegl,Kathryn
Lyn 7h59'29"35d43'
Sieglaff,Joey
Aur 7h3'32"37d5'
Siegman,Alex
Cnv 12h59'47"42d56'
Siegman,Dennis & Nancy
Cyg 21h7'13"30d49'
Siegman,Wendy
Mon 6h53'29"-6d30'
Siegmund
Uma 9h58'37"42d22'
Siegmund,Hermann-K
Lyr 19h21'0"31d29'
Siegward"Siggi" Püttker
Siegwart,Debby A
Leo 10h37'43"15d54'
Sieja,Jr,Stanley
Uma 10h45'0"47d56'
Sieja,Tom
Her 16h57'18"24d42'
Siekmann,August- Wilhelm
Tau 3h41'15"10d39'
Sieler,Christina Teresa
Uma 11h52'0"62d8'
Siembida,Robert J
Lmi 10h35'44"24d2'
Siemens Superstar, Richard
Her 16h42'59"4d41'
Siemens,Blair
And 23h1'31"51d27'
Siemer,Rex Michael
Cet 0h49'40"-10d48'
Siemiatkaska,Veronica Zukowski
Cam 5h44'16"73d29'
Siemieniuk,Luba
Lmi 10h52'12"32d57'
Sieminski,Raymond Michael
Her 16h5'16"50d7'
Siempre
Lyn 7h5'1"50d34'
Siempre
Tri 1h51'22"27d59'
Siempre Manito
Vul 20h18'0"22d34'
Siempre Steve
Boo 14h38'28"43d5'
Siempre Tuya
Uma 8h33'54"62d8'
Siems,Angel Gary
Uma 10h25'32"40d28'
Siena-Felice
Del 20h56'36"16d21'
Sienkiewicz,Brian & Deanna
Cas 1h5'30"62d8'
Sienkiewicz,Henry
Cet 1h3'13"-2d1'
Sienknecht,Peter
Lyr 19h17'17"30d21'
Sienko,Troy Anthony
Aur 5h18'14"48d56'
Sienkowski,Randy & Jean
Aql 19h42'52"12d52'
Siepietowski, Christopher Michael
Dra 19h4'34"65d5'
Siepietowski,Jack
Her 16h54'27"32d55'
Siering,Raymond H
Per 2h39'37"35d33'
Siering,Raymond H
Cep 21h55'41"55d1'
Sierk,Ute
Cam 3h50'36"57d47'
Sierra De Nolla,Blanca
Ori 5h15'1"-8d54'
Sierra Vista Elementary School
Aql 18h53'57"10d59'
Sierra,Monte
Aql 18h55'26"11d32'
Siesta,Hondros
Ori 5h51'60"14d10'
Sievanen,Leo & Lillian
Cep 23h21'37"65d58'
Siever,Rick
Aur 5h59'0"30d1'
Sieveri,Kenneth
Dra 19h50'13"70d5'
Sievers Mr Wowo,Bob
Hya 9h18'1"2d15'
Sievers,Heidi Lynn (Reach)
Cas 1h9'13"63d36'
Sievers,Thomas Anthony
Cet 2h38'37"-12d27'
Sieverson,Pete
Cep 1h15'60"77d54'
Sievert 2000
Uma 11h53'44"58d20'
Sievert,Beverly Claire Watson
Com 12h25'33"27d6'
Sievertsen,Carsten
Uma 9h16'26"48d55'
Siew Forever
Per 4h44'52"40d20'
Siewert,Christa
Sgr 19h8'39"-26d9'
Siewert,Cynthia
Uma 9h6'0"48d8'
Siewruk,Maximilian Conrad
Mon 6h58'50"7d41'
Siff,Jessica Elizabeth
And 23h21'11"35d8'
Sifferlen,Albert & Dorothy
Crt 11h41'0"-11d48'
Sifnas,William J
Her 16h6'41"42d15'
Sifran
Ori 5h12'0"-0d58'
Sifuentes,Nilda
Cas 1h46'11"60d4'
Sigafoos,James Francis
Dra 19h25'1"56d53'
Sigal,Leah Elizabeth
Lyr 18h46'57"36d42'
Sigall,Dr Edward R
Oph 18h0'27"12d28'
Sigander
Uma 10h29'42"70d28'
Sigars,Vernon
Peg 23h7'54"10d58'
Sigel,Christine
Crb 15h28'12"31d53'
Sigelakis,Samantha Frances
Sco 17h29'22"-30d31'
Sigelman,Jonathan
Ser 18h37'24"0d10'
Siggy
Per 2h10'34"57d28'
Sighieri,Rossella
Del 20h18'32"14d32'
Sigler Star of Love & Friendship
Per 3h14'11"55d7'
Sigler,Stephanie
And 23h44'28"37d31'
Siglinde
Aqr 20h37'1"-0d16'
Siglore
Cet 1h35'50"-6d57'
Sigma Nova 57 Morris
Her 16h30'57"36d24'
Sigma Phi Epsilon
Ori 5h3'38"0d50'
Sigma,Denise
Ori 5h21'39"0d51'
Sigma,Paul Tau
Ori 5h21'1"0d28'
Sigmen,Carl-Heinz
Leo 9h36'30"14d9'
Sigmon,Kayla Jo
Cyg 21h39'59"37d48'
Sigmon,Michael
Boo 14h39'0"9d57'
Sigmund's Starry Starry Night,Rich
Her 18h5'46"31d33'
Sigmund,Laura Ann
Com 12h55'28"27d23'
Sign Of Strength
Mon 8h8'53"-9d21'
Signaigo Famiglia
Sex 10h25'57"-1d48'
Signer,Al & Marilyn
Ori 5h41'27"-6d49'
Signoracci,Comte Silverio
Per 3h53'6"36d59'
Signorelli,Francesca
And 23h21'1"51d0'
Signorelli,Jessica Marie
Vir 11h43'30"2d4'
Signorello Star,The John
Cep 2h15'58"80d10'
Signorini,Umberto
Boo 14h32'22"17d45'
Sigrid
Aqr 21h0'14"-14d10'
Sigrid
Cnc 8h35'17"17d41'
Sigrid
Cas 0h57'0"50d36'
Sigrid
Cet 2h40'38"2d24'
Sigrid,"Wiesel"
Gem 6h37'27"20d1'
Sigrist's Star,Adrian "Adi"
Lyr 18h45'22"31d49'
Sigrist,John Taylor
Her 16h28'1"39d32'
Sihaya (Amy's Heart)
Lyr 18h48'33"39d30'
Siik,Ida-Maria
Cam 5h12'29"67d32'
Siira,Julie C
And 23h24'0"45d6'
Siker
Dra 16h17'1"62d55'
Sikes,"My Belle" Elizabeth Sue
Vul 20h2'32"28d46'
Sikes,Evelyn
Eri 4h6'12"-18d45'
Sikes,John Alton
Ser 15h53'1"23d57'
Sikes,Michael S
Eri 4h32'10"-12d19'
Sikes,Thomas H
Hya 8h55'44"4d24'
Sikkel,Peter & Adriana
Cyg 19h40'40"28d25'
Sikkema,Art
Aur 6h6'59"31d33'
Sikkila,Shirley Ann
Cyg 19h25'11"30d48'
Sikkila,Sr,Russell Robert
Uma 10h3'1"56d18'
Sikora,Frances L
Uma 10h53'46"60d0'
Sikora,Kenneth M
Ori 5h51'16"7d46'
Sikorski,Adam & Jean
Cyg 19h23'17"28d0'
Sikstrom,Sara
Dra 16h2'1"61d16'
Sil,George Frederic White
Lac 22h53'1"54d34'
Silaste,Riitta
Cam 6h58'54"68d46'
Silavutiset,Pamela
Mon 6h42'13"1d15'
Silber Star,Jamie
Her 17h31'45"40d57'
Silber,Maia Rose
And 1h0'41"39d45'
Silber,Maurice & Ruth
Del 20h19'60"20d32'
Silber,Phil
Boo 14h25'19"38d15'
Silber,Rachel
Lyn 8h9'13"34d6'
Silberbordt,Birgit
Lac 22h2'16"51d34'
Silberhartz,Jamie Allison
Uma 11h45'14"55d11'
Silberhartz,MD,David Mark
Uma 11h41'56"55d11'
Silberhartz,Sheila Ann
Uma 11h4'48"55d33'
Silberman Family,The
Per 3h1'56"41d14'
Silberman,Frederic H
Aur 5h54'30"29d14'
Silberman,Leah Jane
Lyn 8h30'55"41d13'
Silberman,Scott Richard
Her 16h33'54"34d39'
Silberstein,Diane
Cas 23h21'30"60d11'
Silberstein,JoEllen
And 0h11'49"35d37'
Silcott,Jr,James W
Boo 14h5'11"38d31'
Silecchia,Denise Anne
Lyn 6h44'2"55d45'A
Sileikis,Darius Jon
Lac 22h21'18"55d37'
Sileikis,Justinas Adam
Cnv 13h48'0"32d7'
Sileo,Andrea M
Cas 3h1'17"70d38'
Sileo,Mary Ann
Cas 22h57'11"56d44'
Sileo,Michael & Ellen
Cyg 21h22'37"50d21'
Siler,Brooke
Cas 1h3'41"55d53'
Siler,Kourtney
Cam 4h11'26"67d50'
Siler,Linda Susan
Uma 13h39'48"61d2'
Siler,Mel
Boo 15h39'40"41d55'
Silfen,Shelley
And 0h43'59"38d36'
Silhany Star,Karen
Cam 4h15'10"68d1'
Silimperi,Annemarie
Lyr 18h59'11"31d12'
Siljestrom,William Scott
Del 20h14'41"15d25'
Silk,David Michael
Aur 6h37'0"40d48'
Silk,Stephanie Diane
Peg 22h2'0"28d42'
Silka,Theodore
And 0h18'0"38d36'
Silke
Psc 23h8'39"2d16'
silke
Lyr 18h59'54"34d3'
Silke die Feder
Cet 2h57'17"2d32'
Silke G
Cap 20h35'0"-10d13'
Silke Lovely - Can't Lose This One
Vir 13h31'38"-14d5'
Silkie
Lyr 19h25'1"38d45'
Silky
Lyr 19h16'17"40d53'
Silky Ray
Dra 17h58'57"58d39'
Sillerman,Mackinley Jo X
Aur 7h1'36"40d14'
Sillery
And 0h32'20"40d19'
Silliman,Jean
Lib 14h45'1"-8d13'
Sillman,Mathew Harris
Lac 22h51'34"37d33'
Sills,Gerry & Harold
Mon 7h52'36"-4d21'
Sills,Joan
Cam 3h48'0"68d41'
Sills,John Charles
Dra 18h0'12"67d31'
Silly P
Ori 5h34'22"0d41'
Silon's Legacy,Herb
Boo 14h23'0"31d26'
Silow,Adam Lucas
Ari 1h55'22"23d1'
Siltman,Melvin J
Her 17h0'42"20d48'
Silva
Cyg 21h17'34"28d21'
Silva Family
Eri 2h52'43"-5d20'
Silva For All Eternity Love Mike,Debra
Cyg 19h53'15"41d3'
Silva,Andrew
Her 16h10'57"8d8'
Silva,Barbara Louise Davey
Cas 23h19'32"60d31'
Silva,Barry
Sct 18h51'42"-6d4'
Silva,Carlos Daniel
Aur 5h27'1"54d53'
Silva,Chad Alexander
Aur 6h4'31"36d1'
Silva,Cheryl
Vir 11h51'18"8d8'

Silva,Christopher Brian
Lac 22h27'33"55d39'
Silva,Courtney
Tau 5h55'24"23d29'
Silva,David Allan
Her 16h35'42"33d7'
Silva,George M
Cmi 7h55'34"4d30'
Silva,Gregory
Pic 5h49'22"-52d46'A
Silva,Jon-Anthony
Sgr 18h4'1"-28d15'
Silva,Joshua
Ori 5h37'37"11d52'
Silva,Kristi
Mon 7h39'51"-1d43'
Silva,Marisa
Lyr 18h34'36"47d28'
Silva,Michele R
Cas 0h51'51"71f13'
Silva,Mike
Eri 4h55'57"-6d55'
Silva,Ralph
Aqr 20h37'0"0d57'
Silva,Rebeka G
And 23h23'1"46d49'
Silva,Renato & Diana
Uma 10h6'15"56d7'
Silva,Timothy,Scott
Her 17h7'32"49d5'
Silva,Tony
Cep 2h3'19"77d34'
Silva-Severin
Cyg 20h21'41"37d34'
Silvana
Uma 8h30'38"70d37'
Silvana
And 0h36'40"45d31'
Silvana
Lyr 18h59'47"36d14'
Silvana e Cipriano
Cas 2h4'18"60d6'
Silvana,Fratti
Sgr 19h30'7"-38d39'
Silvana,Maria
Peg 23h1'11"20d40'
Silvas,Abel
Cet 2h6'34"8d40'
Silveira,Andrea Salvo
And 0h2'54"41d4'
Silver
Per 3h9'24"38d3'
Silver
Ori 5h51'15"17d51'
Silver 'N Gold
Cyg 21h23'0"50d3'
Silver (Pawa),Patricia Walton
Lyr 18h55'42"33d24'
Silver Allew
Dra 20h8'28"84d42'
Silver Bear
Uma 11h2'33"40d49'
Silver Bullet
Boo 14h37'52"35d20'
Silver Dream
Cep 23h19'44"70d24'
Silver Fox
Uma 10h30'25"54d49'
Silver Fox
Vul 19h45'18"27d28'
Silver Lake Dental Arts
Eri 2h57'55"-16d50'
Silver Lining
Lib 14h40'32"-10d38'
Silver Mac
Peg 23h3'48"22d2'
Silver Pin
Cyg 21h31'48"42d51'
Silver Rain Irrigation Inc
Aql 19h46'35"12d6'
Silver Spoon
Uma 9h20'1"48d5'
Silver Star Harald & Ursula
Uma 8h49'18"62d7'

Silver Star of Mortman
Psc 1h4'1"27d49'
Silver Star,The
Tau 5h52'0"23d42'
Silver Streak
Uma 11h2'46"38d31'
Silver Wedding H & A
Cnc 8h49'21"11d26'
Silver XXV En "Sallakity"
Cyg 21h37'48"38d44'
Silver,Alexandra L
And 0h2'47"35d17'
Silver,Christian
Mon 7h50'10"-6d50'
Silver,Diane
And 2h3'12"37d15'
Silver,Gail
Lyn 8h48'55"37d20'
Silver,Ian
Dra 9h42'22"80d29'
Silver,James Leonard
Aur 6h2'38"32d7'
Silver,Lester Franklin
Dra 20h1'33"64d46'
Silver,Lets Fly My
Love,Georgia,Raymond F
Uma 10h53'0"38d14'
Silver,Lindy Michael
Sex 9h59'46"-5d37'
Silver,Mariden
Peg 22h59'1"8d20'
Silver,Matthew Ray
Hya 8h44'30"3d9'
Silver,Maurice
Lac 22h24'0"52d50'
Silver,Michael
Her 16h57'34"32d24'
Silver,Michele Claire
Oph 18h6'55"12d3'
Silver,Nicholas
Aql 19h29'49"-8d48'
Silver,Rachel Moriah
Peg 22h29'58"8d13'
Silver,Reverend
Per 1h57'19"48d23'
Silver,Rhonda
Lyr 18h56'56"30d30'
Silver,Samie-Lee
Sge 19h53'22"18d50'
Silver,Sol
Cam 6h57'34"63d46'
Silver,Stephen L
Per 4h25'1"50d24'
Silver,Steve
Cmi 7h57'15"1d20'
Silver,Steven Ross
Oph 16h42'42"2d40'
Silver-A True Warrior, Daniel
Ori 5h56'58"15d40'
Silverberg,Amy Rachel
Cas 23h46'39"50d5'
Silverberg,Nat & Evelyn
Uma 11h40'25"64d32'
Silveria,Alberta "Birdie"
Vir 11h44'19"9d21'
Silveria,Diane Louise
Sge 19h56'33"16d47'
Silverman's Silver
Her 18h2'39"37d55'
Silverman,Aaron J
Peg 0h2'12"11d19"
Silverman,Cori B
And 0h18'25"36d23'
Silverman,David Laurence
Dra 16h0'20"66d49'
Silverman,Deborah
Ori 5h49'43"11d58'
Silverman,Denis Gene
Boo 14h6'41"50d43'
Silverman,Dorothy
Cas 1h17'32"60d26'
Silverman,Dr Joseph A
Dra 20h8'22"62d28'

Silverman,Grandma Fay &
Grandpa Sam
Cam 7h54'35"61d12'
Silverman,Lauren & Dean
Sge 18h58'28"18d30'
Silverman,Linda
Lyr 18h49'48"40d40'
Silverman,Marlene Selzer
Uma 11h41'59"37d47'
Silverman,MD,Cary M
And 0h56'12"36d3'
Silverman,Mel
Ori 5h26'0"15d17'
Silverman,Mitch
Aql 19h17'1"14d23'
Silverman,Morris
Ser 16h1'28"0d32'
Silverman,Nicole
And 23h33'11"42d41'
Silverman,Roy
Dra 14h56'52"64d14'
Silverman,Shannon Stjarna
Lyr 19h22'43"42d4'
Silverman,Shauna-Matt
Cyg 19h50'13"40d53'
Silverman,Steven Paul
Cet 2h14'15"7d29'
Silverman,Tali
Uma 10h52'1"61d50'
Silvermoon
Sgr 19h17'20"-43d24'
Silvers Star
Boo 15h12'44"37d59'
Silvers,Cynthia Kay
And 23h28'0"48d20'
Silvers,David Ira
Psc 1h21'42"20d41'
Silvers,Shayna
And 0h55'16"45d24'
Silvers,Steven Hale
Hya 8h9'25"1d25'
Silverspoon's Hog Wild
"Fannie"
Cma 6h11'15"-11d8'
Silverstar
Boo 13h37'18"13d21'
Silverstein,Amalia Miranda
And 2h31'53"44d12'
Silverstein,ARI Daniel
Uma 10h15'35"60d36'
Silverstein,Bethlyn
Gem 7h5'59"-23d45'
Silverstein,Joseph E
Boo 14h38'24"39d19'
Silverstein,Roseanne
Lyn 8h4'23"51d52'
Silverstien,Nicholas
Cep 23h1'0"64d17'
Silverton's Sedulous Scholar Z
Lyr 19h1'59"31d41'
Silvester,David James
Ori 6h11'48"8d18'
Silvestri,Anacleto G
Oph 17h54'14"10d38'
Silvestri,Donato
Per 3h2'39"37d35'
Silvestri,Emanuela
Pho 0h33'18"-46d37'
Silvestri,Jo
Lac 22h38'1"52d51'
Silvestri,Joanne
Aql 19h31'53"12d3'
Silvestri,Michael Emilio
Her 17h48'1"40d18'
Silvestro,Ariana
And 22h56'37"51d3'
Silvestro,Dineen
Cyg 21h25'32"53d29'
Silvestro,Theresa Ann
And 2h8'28"42d15'
Silvey
Ser 16h7'11"10d19'
Silvey,K L
Peg 23h22'13"28d53'

Silvi,Michelle
Tri 1h40'34"30d1'
Silvia
Lyr 18h38'28"27d45'
Silvia
Scl 23h7'52"-30d24'
Silvia
Cas 2h9'59"60d47'
Silvia
Hor 3h11'37"-49d30'
Silvia
Pup 8h8'42"-23d9'
Silvia
Del 20h17'57"11d26'
Silvia
Ori 4h56'10"4d44'
Silvia
Lup 14h41'15"-42d7'
Silvia
Lyn 7h47'49"40d42'
Silvia
Cnc 7h56'12"10d35'
Silvia
Cnv 13h43'28"28d50'
Silvia
Cmi 8h4'23"5d41'
Silvia
Cam 13h52'58"82d16'
Silvia
Cyg 20h57'48"31d31'
Silvia
Cap 20h57'1"-11d9'
Silvia
Sco 17h28'23"-38d4'
Silvia
Cyg 21h18'0"30d32'
Silvia
Lac 22h13'37"55d14'
Silvia
Cas 23h20'0"53d45'
Silvia
Per 2h11'1"58d39'
Silvia & Carla
Lup 15h18'37"-43d53'
Silvia & Paco
Cyg 21h2'47"39d19'
Silvia & Uli
Gem 6h42'16"12d29'
Silvia et Beat
Dra 10h5'43"73d13'
Silvia In The Sky With
Diamonds
Cas 0h4'16"66d21'
Silvia Valério
Oph 17h5'43"10d40'
Silvia's Stern
Vir 13h28'36"-11d7'
Silvia,Terry
Lac 22h17'25"54d48'
Silviamei
Lac 22h18'46"51d42'
Silvio
Cep 1h6'41"80d20'
Silvio
Her 16h42'0"38d30'
Silvis,Gertrude A
Cyg 19h26'1"40d43'
Silvola,Alex Katherine
Gem 6h32'10"13d49'
Silvs
Her 16h37'39"38d15'
Silvy
Tau 4h7'56"1d19'
Sim's Unreachable Star,Ed
Ori 5h48'53"18d59'
Simala,Ranelle Jeanne
Kelleher
Lyr 19h22'42"30d34'
Simard,Chantale
Uma 10h3'57"53d26'
Simard,Denis
Uma 10h56'46"51d25'
Simard,Denis
Cep 22h33'44"63d20'

Simard,Jean Charles
Lac 22h18'20"51d28'
Simard-Lamirande, Berthe
Uma 9h35'16"62d4'
Simaron
Oph 17h51'1"12d3'
Simas,Jan & Gary
Cyg 21h51'12"42d16'
Simba-Yessa
Vul 20h20'15"26d14'
Simbro ORT,Susan
Lyr 28h49'27"38d21'
Simbu
Lyr 18h58'19"47d31'
Simcha Ab Im
Her 17h3'16"31d12'
Simcha Noah 80
Aur 6h57'1"36d51'
Simchick,Barbara
Cas 0h26'37"64d6'
SimDiego
Leo 10h57'29"12d19'
Simecek,Jr,Daniel Joseph
Aur 6h34'32"37d33'
Simelka,Andrea
Leo 9h19'41"10d34'
Simelka,Christel
Sco 17h51'31"-30d32'
Simelka,Holger-Walter
Aqr 22h56'0"-18d22'
Simels,Blair Matthew
Lac 22h16'52"50d20'
Simental,Ana Margarita
Equ 20h58'0"5d60'
Simental,Antonio S Javier
Sct 18h30'17"-4d24'
Simenton,Gladys
Cas 0h31'0"70d3'
Simenton,Lloyd
Cep 22h24'26"80d18'
Simeoli,Flavia
Hor 3h26'45"-48d17'
Simeon,Maria Linda
Lyn 7h47'22"42d23'
Simeone,Kewpie
Cep 20h38'34"63d51'
Simeone,Louis
Dra 11h22'1"74d25'
Simeone,Richard & Mimi
And 23h47'0"47d48'
Simeonova,Dr Tinka I
Oph 17h7'45"-2d15'
Simister 1908-1995, Gilbert
Cyg 19h27'31"33d6'
Simkins,Forever Richard
Cap 21h45'53"-24d12'
Simkiss,Jayne
Ori 6h3'43"5d2'
Simmel,Robert F
Aur 6h13'41"38d42'
Simmerman,Casey L
Eri 4h0'14"-7d46'
Simmerman,Ed & Janet
Her 16h20'23"40d36'
Simmers,Michael D
Aur 6d17'54"38d49'
Simminger,Greg
Uma 8h32'52"50d55'
Simmington,Graham
Cep 23h8'29"65d7'
Simmon,Michael Cantrell
Aql 20h1'13"10d5'
Simmonds
Aur 5h16'29"47d22'
Simmonds Star,The
Lyn 8h7'0"35d25'
Simmonds,Jon Royston
Ori 4h53'19"0d24'
Simmonds,Julian Anthony
Peg 23h40'25"10d43'
Simmonds,Maria & Alex
Peg 23h40'1"16d28'
Simmons 1944,Gregory Bruce
Ori 5h23'1"-4d56'

Simmons MKISA,R Alexander
Her 16h56'32"31d52'
Simmons,"Spifee Mom"
Norma
Cas 0h57'1"50d9'
Simmons,Amy Michelle
Mon 8h4'34"-8d37'
Simmons,Anais
Peg 23h0'46"28d56'
Simmons,Barbara J P
Cyg 21h46'43"36d55'
Simmons,Blanco
Peg 22h1'20"8d31'
Simmons,Brad
Per 3h5'33"57d12'
Simmons,Catherine Ann
Umi 14h31'27"65d50'
Simmons,Cathy M
Peg 21h57'56"26d44'
Simmons,Cynthia Ann Skols
Eri 4h14'54"-11d18'
Simmons,David K
Boo 14h3'11"11d30'
Simmons,Dennis R
Oph 17h32'46"-0d0'
Simmons,Dwayne L
Hya 8h12'11"5d41'
Simmons,Ella Rose
Cas 0h14'0"63d53'
Simmons,Hazel Maria, Shomo
Lyn 7h42'1"42d48'
Simmons,James Lamar
Sco 17h30'27"-30d55'
Simmons,Jamie Lynne
Vul 20h4'1"28d21'
Simmons,Jane Ellen
Cas 0h55'49"64d6'
Simmons,Jennie Lindsey
Cet 2h2'1"1d49'
Simmons,John Joseph
Peg 21h54'12"34d19'
Simmons,Karen Myslik
And 23h23'29"41d39'
Simmons,Katherine
Cyg 19h29'0"30d27'
Simmons,Kelvin
Tri 2h27'56"28d37'
Simmons,Lisa
Cyg 19h26'38"35d21'
Simmons,Lucas Marie
Uma 8h44'40"72d3'
Simmons,Ma Georgeana Fries
Cyg 20h0'20"40d60'
Simmons,Mabel Ruth
Mon 7h1'14"5d27'
Simmons,Madison
Aur 5h22'36"38d49'
Simmons,Marian Elizabeth
Phillips
Mon 6h39'35"10d46'
Simmons,Mary Ann
And 0h10'42"47d9'
Simmons,Megan Jeanette
And 1h11'55"41d11'
Simmons,Melissa
Uma 11h33'1"31d43'
Simmons,Melissa
Peg 22h53'55"29d49'
Simmons,Michael Ross
Oph 18h17'36"10d27'
Simmons,Mordecia Clifford
Vul 19h6'23"24d47'
Simmons,Nancy L
Cas 1h5'59"60d6'
Simmons,Patricia
Peg 22h32'53"25d21'
Simmons,Peggy Carol
Lmi 9h52'1"40d44'
Simmons,Rachel Robin
Peg 23h41'1"10d17'
Simmons,Richard
Boo 14h16'51"39d43'
Simmons,Richard
Uma 14h21'0"58d36'

Simmons,Richard G
Cam 3h42'55"74d6'
Simmons,Richard H
Her 17h2'57"48d48'
Simmons,Robert Thomas
Lib 14h42'50"-20d9'
Simmons,Rusty
Aql 19h55'32"8d56'
Simmons,Sr,Russell L
Dra 17h4'16"62d23'
Simmons,Tommy Jo
Dra 12h17'18"64d43'
Simmons,Tricky-Dicky
Ori 6h9'19"5d4'
Simmons,Truda Mai
Aur 5h31'27"31d2'
Simmons,Wendy
Lyr 18h40'20"47d11'
Simmons,William B
Dra 11h22'24"73d5'
Simmons,William Edward
Peg 22h16'28"10d18'
Simmons-Manson,Nancy
Cas 0h43'0"69d9'
Simmons/Guys & Dolls, J K
Per 3h13'14"55d42'
Simms "God's Light", Louise
Mon 8h5'59"-10d11'
Simms Guiding Star
Lyr 18h27'40"31d40'
Simms October 17,1937, Earl
Walter
Boo 13h40'59"21d37'
Simms,Carroll W & Dorothy J
Crb 15h55'33"27d27'
Simms,Kevin Michael
Cet 1h34'51"-2d6'
Simms,Madeline Marjorie
Uma 9h22'1"59d7'
Simms-Preston,Gower &
Caroline
Cyg 19h43'17"30d29'
Simo & Teo
Lep 5h23'53"-20d21'
Simon
Lmi 10h51'38"32d19'
Simon
Cep 23h13'59"80d7'
Simon
Ori 6h0'41"1d26'
Simon
Per 3h27'51"50d29'
Simon
Peg 23h2'55"18d7'
Simon & Russell Forever
Cyg 21h52'35"38d42'
Simon Rolla
Aur 5h50'42"38d13'
Simon's Star,Aaron Jacob
Mon 8h8'12"-8d15'
Simon,Abraham N
Aur 6h1'38"38d16'
Simon,Adrian Edward
Cep 20h33'57"75d43'
Simon,Alois
Cep 0h5'37"82d48'
Simon,Ami Opal
Cas 2h0'49"58d54'
Simon,Angela Marie
Lyn 7h42'42"48d36'
Simon,Benjamin J
Sct 18h47'8"-7d48'
Simon,Bruce R
Cep 22h36'42"58d8'
Simon,Christina Mari
Lyn 8h5'22"58d46'
Simon,Conrad
Aur 5h19'42"43d9'
Simon,D B
Sge 20h0'29"18d53'
Simon,Daniel Sol
Uma 10h7'0"57d1'
Simon,David
Aur 4h52'0"41d15'

Simon,David Herman &
Deborah Elizabeth Lyne
Her 17h29'40"31d35'
Simon,Douglas
Ori 5h35'58"-4d3'
Simon,Eric Marc
Her 17h20'45"49d13'
Simon,Fran
Cam 9h3'41"73d18'
Simon,G Richard
Oph 18h5'54"-8d54'
Simon,Jean-Christophe
Her 16h42'0"29d49'
Simon,Jillian Abigail
Cas 0h58'30"73d37'
Simon,Joel M
Boo 15h2'47"10d28'
Simon,Jonathon Stuart
Cam 3h53'45"52d57'
Simon,Jose Pastor
Cyg 21h56'37"50d3'
Simon,Judith
Vul 29h56'26"28d19'
Simon,Kyle D
Cep 21h1'59"61d47'
Simon,Lindsey Pam
Crb 16h1'49"31d47'
Simon,Liz
And 23h18'1"50d36'
Simon,Luis
Tau 4h18'11"24d7'
Simon,Lynn
Lyr 18h46'45"39d59'
Simon,Margaret
Lyn 6h15'38"60d38'
Simon,Margo
Lyr 18h58'45"42d52'
Simon,Marshall John
Cnv 13h22'1"41d50'
Simon,Melissa Kaye
Lyn 7h40'37"47d47'
Simon,Michael C
Aql 19h9'51"1d58'
Simon,Mr
Dra 14h48'47"61d38'
Simon,Nanette
Cnc 8h1'51"18d60'
Simon,Neil
Ori 5h57'24"20d2'
Simon,Oliver
Tau 5h34'51"26d6'
Simon,Pamela
Crt 11h8'10"-14d14'
Simon,Philippe
Del 20h56'15"14d11'
Simon,Robert J
Aur 5h2'28"40d46'
Simon,Samantha Linn
Lyn 7h34'23"41d30'
Simon,Sean Christopher
Uma 10h39'48"42d0'
Simon,Sherry
And 2h28'58"39d59'
Simon,Steven
Her 16h38'49"25d39'
Simon,Stéphane
Del 20h14'44"14d50'
Simon,Thomas John
Aql 19h30'56"10d27'
Simon,Ute
Gem 6h0'28"26d41'
Simon,William Lincoln
Boo 15h30'13"40d52'
Simona
Ari 1h59'50"18d54'
Simona
And 0h9'47"41d8'
Simona
Per 3h29'53"38d23'
Simona
Cep 20h14'0"60d49'
Simona
Aqr 21h23'51"-8d43'
Simona
Lup 15h2'0"-37d43'

Simona
Col 6h34'53"-33d31'
Simona
Ant 10h44'31"-32d15'
Simona
Cet 1h32'0"-2d59'
Simona
Mon 6h19'41"7d58'
Simona
Pho 0h38'25"-41d28'
Simona
And 0h56'26"37d4'
Simona Forever
And 23h12'36"40d25'
Simoncini,Sabrina
Peg 21h51'56"33d55'
Simonds,James Pearly
Aql 18h43'1"10d43'
Simonds,Virginia Barrett
And 23h27'39"46d23'
Simone
Ori 5h48'12"21d20'
Simone
Cma 6h28'29"-11d4'
Simone
Lep 4h59'35"-19d12'
Simone
Ari 2h41'1"20d28'
Simone
Per 3h32'38"52d3'
Simone
Uma 10h18'23"42d25'
Simone & Nicoletta
For 2h44'55"-25d21'
Simone Noel
Cas 1h0'36"53d30'
Simone Our Shining
Star,Matthew P
Her 16h54'0"32d42'
Simone's Bright Future
Cnc 8h0'44"11d1'
Simone,Anthony
Aur 6h45'12"37d7'
Simone,Clint Michael
Peg 22h42'1"11d41'
Simone,Gregory P
Umi 16h29'36"77d12'
Simone,Long & Lasting
Love,Geraldine
Cas 0h2'0"62d50'
Simone,Matthew Tyler
Dra 19h23'1"56d8'
Simone,Rick
Leo 10h58'48"14d50'
Simoneau,Joe
Ser 18h2'0"-14d36'
Simonelli Family Star, The
Mon 6h35'0"-0d21'
Simonelli,Robert
Aur 6h33'14"37d8'
Simones Traum
Psc 23h7'1"5d39'
Simonet,Joëlle
Ori 6h5'18"8d32'
Simonet,Marcel
Dra 14h57'30"59d28'
Simonetta
Lib 15h1'29"-6d8'
Simonetta
Cep 23h30'23"70d37'
Simonetta,Cara Rose
Sco 17h31'53"-30d31'
Simonides,Jonathon Thomas
Dra 19h55'20"80d30'
Simonova,Irina
And 2h29'40"49d7'
Simons 12/31/87, Katherine H
Vir 15h7'11"7d3'
Simons,Barbara J
And 0h11'40"39d8'
Simons,Charles
Boo 15h6'52"11d7'
Simons,Claire Frances
Vul 19h59'22"25d50'

Simons,Clayton William
Ori 4h41'48"8d35'
Simons,Dolph
Tri 2h5'0"33d41'
Simons,Dorothy
Cet 0h57'35"-5d0'
Simons,Leonard N
Peg 21h19'57"20d3'
Simons,Martin Daniel
Aur 5h2'25"42d10'
Simons,Rebecca L
Ori 6h4'54"-1d49'
Simons,Tayler C
Cep 22h7'37"60d45'
Simons,Thomas R
Cep 21h32'60"58d59'
Simons,Victoria Lynn
And 23h4'13"44d6'
Simonsen,Christina
Crb 15h47'1"38d53'
Simonton,Douglas
Lyn 7h10'31"51d52'
Simos Star,Mitchell G
Aur 5h4'59"42d9'
Simotas,Gerassimos "Kiddo"
Uma 9h40'52"68d35'
Simpkin,Samantha Louise
Cyg 20h44'38"44d59'
Simple Pleasure
Lyn 9h12'40"38d16'
Simplicio,Vito Frank
Hya 8h15'0"4d35'
Simply Jacy
Oph 18h6'10"12d39'
Simply Steve
Dra 17h28'59"60d44'
Simply Susie
Lac 22h5'57"50d3'
Simpson Family,The
Peg 23h0'48"32d31'
Simpson Star,The Janette Patricia
Cas 2h15'34"65d9'
Simpson's Class Rock Star '95,Mrs
Umi 13h57'24"70d30'
Simpson,Aileen & Elsie
Cmi 7h34'19"0d60'
Simpson,Alverda Grant
Umi 14h24'59"68d23'
Simpson,Anna
Lyn 7h45'0"48d46'
Simpson,Annie Mae
And 0h54'25"34d10'
Simpson,Arthur
Cet 2h45'1"5d59'
Simpson,Aunt Nancy L
Cet 2h38'12"-4d36'
Simpson,Bessie
Peg 23h29'31"33d25'
Simpson,Billy
Umi 15h54'40"83d46'A
Simpson,Chantelle Georgia
Lup 15h23'4"53d13'
Simpson,Colin
Nor 15h47'7"-43d44'
Simpson,Danielle
Mon 8h2'56"-6d47'
Simpson,Darlene
Cma 7h14'48"-15d15'
Simpson,Dawn
Eri 2h54'43"-5d45'
Simpson,Denis Edward
Ori 6h17'30"0d27'
Simpson,Elinore M
Sge 19h11'26"19d19'
Simpson,Eric A
Hya 8h57'16"4d51'
Simpson,Everett & Sally
Mon 7h2'42"-6d32'
Simpson,Freddie Joe
Cep 23h11'54"71d3'
Simpson,Gary
Dra 14h25'0"64d42'
Simpson,Gillian Rosemary
Umi 13h11'47"71d2'
Simpson,Gordon
Cep 21h27'25"70d28'
Simpson,Issy
Umi 15h54'40"83d46'B
Simpson,James F
Umi 13h3'1"74d36'
Simpson,Janet Y
Hya 9h35'46"-6d52'
Simpson,JoAnn
Lyr 18h57'41"34d32'
Simpson,Joseph D
Lac 22h0'15"51d57'
Simpson,Jr,John R
Per 2h59'58"50d7'
Simpson,Jr,Louis Reid
Cnc 8h5'38"7d3'
Simpson,Keith Edward
Cet 1h36'24"-10d1'
Simpson,Kirk Andrew
Per 3h0'1"31d30'
Simpson,Lewis
Ori 5h54'26"10d31'
Simpson,Lisa Gayle
Cas 1h19'30"67d56'
Simpson,Mark & Carol
And 23h33'49"46d42'
Simpson,Martin Van Buren
Eri 3h31'23"-5d8'
Simpson,Mary Melissa
Cyg 19h43'25"29d44'
Simpson,Michael J
Oph 17h58'45"8d14'
Simpson,Michele
Eri 3h56'1"-11d48'
Simpson,Mike & Wendy
Dra 16h10'49"51d46'
Simpson,Milford
Tri 1h31'41"30d43'
Simpson,Miranda
Lup 15h23'18"51d45'
Simpson,Nancy
Ori 5h48'43"20d4'
Simpson,Nancy
Cep 20h59'37"68d30'
Simpson,Pamela
Lyr 18h59'11"30d28'
Simpson,Peggy
Cas 0h23'37"63d51'
Simpson,Phyllis
Lyr 19h3'46"47d0'
Simpson,R Christian
Ori 5h39'1"0d23'
Simpson,Renee Marie
Aql 20h7'52"7d15'
Simpson,Rhonda
Cam 3h38'30"61d36'
Simpson,Rhonda W
And 0h52'37"39d31'
Simpson,Robin
Uma 10h59'25"56d45'
Simpson,Robina
Cyg 20h17'60"39d3'
Simpson,Roy P
Aql 19h6'60"3d51'
Simpson,Ryan & Jamie
Cyg 21h4'47"40d51'
Simpson,Scott W
Lmi 9h42'40"34d49'
Simpson,Shona Bianca
Gru 23h11'1"55d36'
Simpson,Thomas
Dra 16h25'26"64d12'
Simpson,Thomas "Dauden"
Cmi 7h56'36"3d54'
Simpson,Veronica Kay
Cas 0h14'55"56d8'
Simpson,Victoria Marie
Cas 0h10'57"64d56'
Simpson,William Glenn
Cet 2h8'1"1d35'
Simpsons,The
Cyg 19h19'37"27d51'
Sims,Arthur R
Ser 15h39'43"3d45'
Sims,Crystal Celeste
Peg 21h29'54"20d15'
Sims,Debra A
Mon 6h38'50"11d43'
Sims,Doyle
Cam 13h26'47"78d42'
Sims,Gary
Aur 6h56'21"37d47'
Sims,Hardy
Her 17h22'33"18d56'
Sims,Jack "Pinhead"
Peg 23h3'55"18d23'
Sims,Jacob Thomas
Tau 4h57'24"16d15'
Sims,Jeffrey T
Her 16h55'0"33d32'
Sims,Jenny
Lyn 7h10'21"50d13'
Sims,Lauren Teresa
Cas 0h9'36"60d28'
Sims,Lillian L Krotke
Cas 23h28'0"60d49'
Sims,LuAnn
Mon 7h9'27"-10d0'
Sims,Mike
Sex 10h32'54"-6d30'
Sims,Natalie Jean
Equ 20h58'58"3d21'
Sims,Patricia & Joe
Aql 19h50'46"15d46'
Sims,Patrick
Ori 5h58'21"16d36'
Sims,Penny
And 2h20'11"46d10"
Sims,Rufus T
Her 18h18'15"12d29'
Sims,Silvia
Ori 6h4'15"8d48'
Sims,Steven E
Cet 3h15'50"2d50'
Sims,Taylor Alexis
Uma 11h7'20"49d11'
Sims,Thelma
And 23h40'0"45d33'
Sims,Thomas G
Aur 7h8'32"40d56'
Sims,Twinkle
Aql 19h46'11"13d59'
Sims,William Patrick
Eri 4h14'40"-16d38'
Sims,Youncia Serene
Hya 8h54'30"2d7'
Sims,Zachary Daniel
Ser 17h19'11"-11d7'
Simsol
Sct 18h44'0"-7d58'
Simunek,Catherine
Uma 8h33'1"49d59'
Simunovic,Melina Nicole
Cyg 19h28'1"31d11'
Simuro,Marie Veronica
Cam 3h44'54"68d22'
Sin
Aur 5h59'11"29d59'
Sina & Da "Wedding Star"
Crb 15h55'43"28d28'
Sinacola,Richard S
Dra 20h1'1"74d39'
Sinacori,Paul J
Dra 20h25'53"63d42'
Sinagra,Ronald Christopher
Her 16h45'56"35d5'
Sinai
Cet 2h32'0"8d36'
Sinanian,Deacon Sage
Del 20h14'26"12d54'
Sinatra,Barbara A
And 0h58'37"33d47'
Sinatra,Hank
Boo 14h58'51"26d8'
Sinatra,Mary Ann
Lyr 18h56'47"40d30'
Since 1987 K & S
Lyn 7h35'1"51d46'
Sincerita'
Del 20h14'27"12d24'
Sincich,Virginia Marie
Mon 6h23'50"3d43'
Sinclain,Adam Charles
Her 17h26'29"30d13'
Sinclair
Cep 21h14'28"58d34'
Sinclair,Barbara
Cnv 12h22'44"44d51'
Sinclair,Bessie White
Cmi 7h28'1"0d52'
Sinclair,Brianne Elizabeth
Com 13h17'15"27d29'
Sinclair,Christie
And 23h22'1"50d54'
Sinclair,David
Aql 20h32'24"0d59'
Sinclair,Harold Paul
Per 3h43'1"36d57'
Sinclair,Howard
Ori 5h56'33"15d51'
Sinclair,James
Ori 5h59'14"8d24'
Sinclair,James Ralph
Aql 19h56'39"9d0'
Sinclair,Jan Adrienne
Equ 20h57'31"9d10'
Sinclair,Jr,Patrick John
Aur 6h55'34"38d59'
Sinclair,Linda Lee
Peg 22h53'15"29d43'
Sinclair,Nicki
Umi 16h7'35"71d8'
Sinclair-Pearson, Elliot
Cyg 21h32'24"42d41'
Sinclair-Pearson, Andrew
Her 16h9'0"47d45'
Sinco Cola
Ari 2h3'23"20d9'
Sindelar,Lenka Triticum
Uma 8h37'44"47d32'
Sinder,Rochelle Leah
Mon 7h0'12"-8d8'
Sinderby,Caroline Diana
Lyr 19h23'1"31d46'
Sindicic,Vera
Cnv 12h48'19"40d33'
Sinek,Michael J
Cam 7h6'39"61d38'
Sinex,Scott Alden
Per 1h41'39"53d6'
Sing Me To Heaven
Lyr 19h20'50"40d11'
Sing,Adam
Aur 5h58'0"30d16'
Sing,Chen
Per 1h41'38"53d47'
Sing,Fanny & Jackie
Crb 15h53'53"32d26'
Sing-Sing
Gem 7h55'1"20d18'
Singa,Krute
Cet 0h59'48"-0d52'
Singa,Timi
Aql 19h53'25"15d60'
Singahl,Anihl
Cnv 12h15'52"46d30'
Singapuri,Smeeta
Mon 6h22'25"2d48'
Singaracharlu,Sujatha
Cam 7h6'32"61d45'
Singdale,Chick
Lmi 10h12'0"31d58'
Singelton,Beckley/ Plane,Angel
Ori 5h25'23"-3^7'
Singer's Star,Joy
Sco 17h23'15"-38d21'
Singer's Star,Ray
Lib 14h59'13"-23d34'
Singer,Abygail & Joseph Singer
Sge 19h3'25"16d6'
Singer,Adam J
Per 4h42'54"38d52'
Singer,Adam Rayman
Cep 21h15'0"80d29'
Singer,Anne
Cas 3h7'38"71d22'B
Singer,Benjamin Stuart
Lmi 10h42'15"23d34'
Singer,C Randall
Boo 15h36'1"40d3'
Singer,Daniel Evan
Umi 13h42'48"77d42'
Singer,Gary
Her 17h21'37"50d11'
Singer,Harte F
Her 17h16'0"44d35'
Singer,Ilene
Cep 0h4'0"69d57'
Singer,In Beloved Memory,Stuart Adam
Her 18h4'32"31d41'
Singer,James Cornfed Allen
Aql 18h53'34"-2d11'
Singer,Joseph Anthony
Aur 6h31'49"35d35'
Singer,Kathleen
Cyg 19h18'0"28d19'
Singer,Laura Michelle
Cas 1h34'53"61d30'
Singer,Leonard N
Crt 11h3'1"-8d34'
Singer,Leslie B
Uma 11h47'0"58d47'
Singer,Lisa Jennifer
Peg 23h3'34"32d14'
Singer,Malody
Mon 6h44'24"1d37'
Singer,Marc
Aqr 22h31'25"-12d10'
Singer,Mark Ian
Cnc 8h6'11"6d57'
Singer,Michael
Aur 7h21'43"35d52'
Singer,Philippe
Per 3h52'55"36d46'
Singer,Stan
Hya 9h38'0"-5d6'
Singh Family,The Balwant
Cyg 19h30'1"58d2'
Singh,Angie
And 23h37'35"45d14'
Singh,Guru
Sex 10h6'25"5d51'
Singh,Radika
Hya 9h7'36"2d23'
Singh,Rodolfo
Aql 19h19'25"10d2'
Singh,Sanjay J R
Dra 18h11'37"70d14'
Singh,Shamara I D
Vul 19h16'44"25d29'
Singh,Sonny "The Houseboy"
Lyn 6h12'42"60d22'
Singh,Tara Jyote
Peg 21h57'37"23d33'
Singh,Terry
Cam 5h44'21"62d47'B
Singing Glory to God In The Highest
Ori 5h48'20"19d25'
Singletary,Alisa
Tau 5h52'0"27d4'
Singletary,Christopher Allen
Dra 20h5'31"64d45'
Singletary,Kasandra S
And 23h16'20"48d8'
Singletary,Khristie
Tri 2h14'12"32d17'
Singletary,Max Edward
Ser 16h13'33"2d37'
Singletary,William Clyde
Cep 23h4'55"60d15'
Singleton
Com 13h14'31"27d59'
Singleton's Light, Norma
Com 13h2'1"15d11'
Singleton,Adam Edgar
Ori 5h21'58"1d14'
Singleton,Allison Lynn
Eri 2h51'22"-4d9'
Singleton,Dawn E
Lyr 18h24'10"47d21'
Singleton,Gayle Yvette
Her 18h2'39"27d6'B
Singleton,Ian
Cap 21h53'48"-17d54'
Singleton,James & Ruth
Cyg 19h27'25"34d17'
Singleton,Jeffrey
Cep 23h11'26"70d47'
Singleton,Jr,John W
Boo 14h19'59"34d14'
Singleton,Katherine Mary
Oph 18h35'0"10d47'
Singleton,Lucky
Aur 6h14'13"45d30'
Singleton,Rosemary
Cas 0h33'30"70d52'
Singleton/Angel Plane, Beckley
Ori 5h25'23"-3d7'
Singley-Flynn "MUT", Rita Marie
Lyn 7h49'33"36d34'
Singmaster,Dara Winslow & Carlen&Ross
Per 3h26'0"41d5'
Sinha,Mona
Cas 22h56'24"55d6'
Sinha,Uttam
Lyr 18h39'16"27d11'
Sinibaldi,Luisa
Cae 4h43'57"-32d47'
Siniff,Paul Andrew
Boo 14h5'40"33d49'
Sinikan Tahti
Cam 9h42'27"82d6'
Sinikka,Raili
Aur 4h51'0"48d56'
Sinisgalli,Nick L
Cep 22h34'51"68d56'
Sinisi,Angelina & Michelle
Dra 18h57'10"48d48'
Sinivuori,Timi
Dra 17h17'31"71d12'
Sinivuori,Tomi
Cam 9h3'0"80d37'
Sink,Frank E
Cep 3h2'11"86d28'
Sink,Jordan William
Per 3h1'19"40d55'
Sink,Richard
Aur 6h45'36"37d21'
Sinkel,Amy
And 23h41'32"38d31'
Sinkinson,Carol
Cyg 21h32'53"44d18'
Sinkinson,John Richard
Per 2h2'15"57d53'
Sinn,Gordon H
Umi 14h57'39"78d42'
Sinn,Holly Harley
Cyg 21h24'56"38d12'
Sinn,Ida Laverne Z
Umi 14h38'1"78d49'
Sinn,Jack Henry
Boo 14h4'43"12d44'
Sinn,Melissa
Lyr 18h30'47"41d33'
Sinna,Anand
Cyg 20h39'12"30d16'
Sinned,M'Lady
Sct 18h52'35"-9d21'
Sinnott,Kari K
And 23h29'58"49d6'
Sinnreich,Michael Edward
Uma 11h19'32"46d42'
Sinoir Alexandre Laëtitia
Ori 6h16'54"1d41'
Sinon Star,The Jack
Cnv 13h20'37"40d23'
Sinopoli,Brittaney
Lyn 7h56'45"52d16'
Sinow,Jose
Cet 2h34'14"5d10'
Sinoway,Eric Charles
Boo 15h12'45"31d7'
Sinsigalli,Andrew "Andy"
Hya 8h10'34"4d19'
Sinsigalli,Christopher A
Aur 5h24'0"40d57'
Sinski,Shayna Ann
And 23h25'30"41d32'
Sinton-Wylde
Ser 16h7'49"13d41'
Sinéad
Peg 21h48'30"34d4'
Siobhan
Lyn 8h6'1"41d12'
Siobhan
Com 12h31'54"25d10'
Sioerring 8830
Dra 19h0'34"65d15'
Sioma,Elena Jurievna
Cam 5h50'1"70d27'
Sioniac,Michel
Uma 13h9'51"62d31'
Siow,James
Cnv 13h10'0"37d50'
Sipe,Helen A
And 23h5'0"45d47'B
Sipe,Sr,Chester J
And 23h5'0"45d47'A
Siperstein,Eleanor
Mon 6h36'31"2d49'
Siperstein,Marvin D
Sct 18h55'1"-5d47'
Sipes,Elizabeth Olga
Cyg 19h50'18"50d5'
Sipi
Boo 14h9'0"32d22'
Siple,Joanne JoJo
Mon 7h2'45"-5d1'
Sipler,Chad N
Cnv 13h46'43"40d16'
Sippel,Bobby
Aur 6h32'0"38d56'
Sipple,Elizabeth
And 1h57'56"36d12'
Siquier,Pierre
Sgr 19h26'34"-34d43'
Sir Anthony
Gem 6h16'46"25d2'B
Sir Bevis
Uma 14h17'1"61d36'
Sir Charles Isaac
Cep 23h15'26"64d9'
Sir Chuckles, Star Of Fun & Friends
Ilma 9h22'29"54d22'
Sir David's Star
Boo 15h0'52"11d17'
Sir Edric
Her 17h40'1"43d2'
Sir Gary
Cep 20h30'1"61d31'
Sir Gil-Lady Wanda
Cas 0h25'41"68d45'
Sir Guim
Her 17h31'16"31d52'
Sir Henry
Boo 15h4'51"21d49'
Sir Honorable James's Special Star
Aql 20h15'51"5d21'
Sir Ivar Nefarious
Cet 1h56'1"-10d58'
Sir James of East End
Cep 21h59'28"56d14'
Sir James of Nirvana
Her 15h51'34"48d5'
Sir Karl
Ari 2h39'0"20d3'
Sir Martin
Hya 9h30'44"0d19'
Sir Michael Prince of Love
Aql 19h50'11"14d47'
Sir Philip
Aur 5h34'57"38d35'
Sir Scott
Cet 0h9'28"-12d2'
Sir Scott
Cet 2h15'25"1d31'B
Sir Sherman of Niguel
Sct 18h41'39"-4d54'
Sir Wes
Aur 6h35'42"33d3'
Sir William
Aur 7h4'23"40d4'
Sir William Starlight-7
Cam 6h13'0"80d7'
Sir William Of Moose
Her 16h40'11"35d29'
Sira
Mon 6h32'50"3d34'
Siracusa,Frank
Lac 22h52'60"56d42'
Siracusa,Frank J
Per 3h18'49"40d53'
Siragusa,Paul
Sgr 18h3'14"-28d58'
Sire Renaud de soixante-quatre
Per 2h52'55"37d44'
Sire,Franáois
Dra 15h0'1"64d11'
Sireci,Philip A
Boo 14h4'30"28d6'
Sireka
Aql 19h27'1"0d26'
Siren 13 05 88
Ser 16h7'19"2d35'
Sirene,Ptite
Vul 19h23'15"26d13'
Siress,Zachary Thomas
Lac 22h53'42"55d41'
Sirgo,David
Boo 14h52'34"45d39'
Siri
Boo 14h29'14"48d18'
Sirianni
Uma 9h22'37"59d5'
Sirianni,Lisa
Cyg 21h31'42"42d8'
Sirica,Dolores Della Pietra
And 2h30'47"45d56'
Sirinek 9
Boo 14h50'11"33d11'
Sirinek,Kenneth
Eri 3h1'11"-17d13'
Sirinek,Margaret
And 23h30'17"42d6'
Sirinsky,Marc
Aur 5h18'21"45d26'
Sirio
Del 20h15'19"12d59'
Siripala
Cyg 20h1'23"30d25'
Sirisena,Mayura
Cas 1h2'41"63d60'
Sirka
Lac 22h41'31"56d8'
Sirko,Angela Marie
Lyr 18h59'11"40d24'
Sirko,Jennifer Lynn
Lyr 18h56'31"42d43'
Sirko,Loren D
Boo 13h45'22"21d35'
Sirmon,Lorraine
Del 20h13'20"15d11'
Siro
Dra 16h54'20"60d11'
Sirois,Guy
Cep 22h10'6"59d28'A
Sirois,Jr,Ronald
Cet 0h25'44"-0d45'
Sirokman,Caryn Eileen
Peg 22h40'28"32d50'
Sirolli,Jr,Louis
Lac 22h22'41"38d60'
Siroonian,Charles Bruce
Aql 20h2'12"10d49'
Sirota,Carol
Mon 6h44'42"1d36'
Sirota,Joyce Burn
Cas 1h9'48"58d36'
Sirota,Milton
Cmi 7h19'50"8d8'
Sirotin,Alexander
Cet 1h5'36"-0d10'
Sirotkin's Golden,Bev & Joe
Eri 5h6'1"-5d40'
Sirotkin,Paul & Celia
Per 2h51'19"43d25'
Sirotti,Laura
Dra 16h22'42"58d45'
Sirrom
Lac 21h56'18"40d4'
Sirtis,"Counselor Troi"-Marina
Lyn 7h45'0"42d54'
Sis's Star
Cas 0h44'14"60d27'
Sis,Scott Allen
Hya 9h23'1"-10d50'
Sisa Rie
Aql 19h31'42"10d15'
Sisbarro,Gail Diane
Leo 10h51'53"20d50'
Sisco
Dra 17h2'58"62d24'
Sisco,Catherine M
And 2h30'42"50d4'
Sisco,Dana
And 2h15'11"42d27'
Sisco,Donna
And 2h12'21"42d19'
Sisco,Mary Lombardo
And 1h54m20"38d24'
Sisco,Nina Alicia
Cyg 19h33'56"31d18'
Sisco,Rachele
Cmi 7h53'56"4d12'
Sisco,Steve
Her 16h40'0"38d12'
Sisel,Sandra & Wayne
Cyg 21h14'0"28d52'
Sisk,Adam Roger
Her 17h38'0"21d25'
Sisk,Jack
Per 3h1'56"41d14'
Sisk,Jennifer
Cyg 21h4'36"31d6'
Sisk,Maureen
Cyg 21h4'49"31d5'
Sisk,Patricia
Cnv 12h47'0"43d6'
Sisk,Randall J
Per 1h56'29"53d45'
Siskel,Gene
Oph 17h54'46"8d52'
Sisko's Star
Mon 6h35'25"10d55'
Siso
Lyr 19h8'1"38d16'
Sisoian,Paul A
Cma 6h56'46"-19d60'
Sissel,Matt
Cmi 7h35'32"11d6'
Sissi
Cet 2h33'0"0d54'
Sissi C
Cyg 20h2'53"31d3'
Sissine,Princess Of the Cosmos,Theresa M
Ori 5h46'0"18d33'
Sisson III,Lyle
Aur 5h56'21"30d23'
Sisson,Bridget C
And 1h28'39"37d3'

Sisson,Henry Winslow
Cnv 13h27'50"48d54'
Sisson,Jordan Stephanie
Cnv 12h31'40"50d20'
Sisson,Vance Frank
Uma 8h26'33"62d20'
Sissy
Lyn 9h7'47"39d54'
Sissy Winkler Vienna 50
Ari 2h27'0"26d56'
Sissy's Jewel
Peg 22h6'17"3d33'
Sissy,Löewenbaby
Vir 12h2'49"12d53'
Sistar
Lyr 18h14'34"30d13'
Sister Bea's Eastern Star
Ori 6h16'1"18d55'
Sister Debbie
And 0h13'14"36d5'
Sister Jane
Lyn 7h44'15"38d38'
Sister Katherin
Uma 12h9'23"61d59'
Sister Margaret
Ari 1h55'36"19d31'
Sister MarieAnn
Lyn 6h34'14"58d20'
Sister Mary
Uma 11h54'18"38d15'
Sister Mary
Uma 9h29'10"50d16'
Sister Norene
Cam 7h41'40"68d20'
Sister Rose Immaculata
Cas 23h25'25"60d39'
Sistermanns,Jochen
Dra 15h52'53"62d5'
Sisters Forever
Mon 6h53'31"0d4'
Sisters Forever- Cathy & Alyson
Peg 23h47'12"10d14'
Sisters Of The Heart
Cyg 21h31'1"50d23'
Sisters of the Heart
Umi 15h53'23"77d4'
Sisti,Scott
Aur 6h1'54"46d18'
Sisto,Jr,Richard & Frances
Dra 20h13'51"64d9'
Sisu,Rosa I
Del 20h53'1"7d57'
Sita & Lluis
Cyg 21h47'30"52d36'
Sitar,Bob
Per 3h9'12"40d34'
Sitar,Zachary Michael
Lac 22h23'34"50d14'
Sitarz,Laurie
Cas 3h10'43"63d52'
Site Search Star
Cyg 20h23'27"39d33'
Sites IV,Ome'r Jon
Dra 19h56'0"68d28'
Sitkin,James
Vul 21h3'36"27d30'
Sitko,Wanda & Jack
Cyg 21h23'31"38d50'
Sitlinger,Jill Louise
And 23h10'43"39d43'B
Sitongias L'etoile, Patricia Ann
Cam 12h43'0"78d13'
Sitorius,Taylor Virginia
And 0h33'11"45d23'
Sitruck,Guy
Mon 6h27'30"1d35'
Sitte,Ulrich
Cmi 7h19'49"8d54'
Sittig,Jodi Michele
Cmi 7h18'50"9d2'
Sitts,Jennifer H
Cas 0h21'58"60d30'
Sittu's Star
Tau 5h55'27"24d2'
Sitybell,Allan William
Dra 14h37'47"60d17'
Sitzler,Dee
Cas 0h31'43"60d41'
Sitzman,MD,Stuart
Hya 9h33'48"-9d36'
Sitzmann
Tri 2h15'57"32d21'
Sitzmann,Gary R
Aur 6h32'15"38d56'
Siu,Anthony
Cma 6h28'30"-24d41'
Siu,Lachlan Wilson
Cru 12h45'33"-55d46'
Siudmak,Dr Robert Conrad
Tau 4h11'30"20d52'
Siudym,Michael William
Cep 22h11'31"58d52'
Sivasegaran, Vijayamalar
Lac 22h8'48"49d20'
Siverhus In Memory of You Dad,Albin
Her 18h13'0"31d23'
Siversten,Marit Ingrid
Vir 13h38'36"-8d59'
Sivertsen,Shawna Starr
Mon 7h4'52"5d8'
Siverwright,Sandy
Cas 0h52'16"64d1'
Sivewright,Beryl Frances
Boo 14h21'0"53d36'
Sivill,Jodie Caroline
Cyg 21h0'59"28d56'
Sivils,Steven James
Uma 11h52'51"61d48'
Sivion
Cet 2h59'36"5d14'
Siwa
Peg 22h15'0"30d59'
Siwak,Carolyn
And 0h19'12"31d13'
Siwanowicz,Robert
Per 2h59'47"38d48'
Siway
Lup 15h28'8"-40d53'
Siwek,Caroline Leigh
Com 12h51'1"20d52'
605 Pearl
Aql 18h38'36"1d36'
Six of Hearts
Uma 12h16'59"60d30'
16 Penny Lux-The Sun Dad Never Had
Lyn 7h50'25"42d34'
Six Star Linie 6,The
Lib 15h12'54"-24d24'
Sixsmith,Sarah E
And 1h46'1"41d5'
Sixta,Adina
And 1h43'57"38d11'
65 Mark A Star As Bright As My Love
Vir 11h59'0"-2d38'
65th Eagle,The
Aql 19h7'42"14d59'
Sixty-Six
Lmi 9h59'21"28d21'
Sixx,Storm Brieann
Ari 1h55'58"22d40'
Sizemore,Brad
Per 3h9'25"47d52'
Sizemore,Jennifer Lynn
Lyr 19h14'38"42d27'
Sizemore,Martin
Lyn 9h29'38"41d4'
SiEn
Cyg 19h30'0"35d50'
SiEn
Uma 8h43'52"60d1'
SJF
Uma 8h38'18"51d45'
SJHS Graduating Students 1995
Uma 8h42'1"51d48'
SJMH 3-East Staff
Lmi 10h34'1"30d13'
SJMH CCU Staff
Uma 11h50'19"37d41'
Sk Chris
Tau 3h56'1"8d38'
Skaalerud Star,The
Uma 11h39'14"63d6'
Skadsberg,Jr,Willy
Cep 21h56'57"61d37'
Skaggs,Austin Taylor
Cet 0h46'1"-8d44'
Skaggs,James David
Eri 2h59'32"-8d37'
Skaggs,Traci
Cas 0h29'50"61d17'
Skaja,Eric J
Ser 16h0'0"14d21'
Skalabrin,Kathy Lynn
Peg 22h33'20"27d60'
Skaleski,Katita Chavez
And 1h36'30"38d41'
Skalitzky,Robert & Eva
Cyg 19h24'20"31d22'
Skalkos,Sackiess
Per 3h6'56"41d15'
Skalla,Jeanne McMahon
Mon 7h0'59"3d52'
Skalski,Amanda
Tri 2h17'49"30d42'
Skander
Eri 3h13'41"-5d48'
Skane,Patricia
Cas 0h58'35"62d54'
Skanke
Cru 12h5'27"-61d5'
Skapurovic,Vensan
Uma 8h44'0"50d13'
Skarda,Christine
Eri 2h50'30"-3d5'
Skarzynski,Dorothy
Cyg 21h7'35"48d51'
Skavlem,Sharon
Peg 21h58'36"28d3'
Skeels,Professor
Oph 17h0'59"10d17'
Skeen Star,The Benny
Her 17h35'40"26d23'
Skeens,Effie
Aql 19h47'32"11d54'
Skeens,Mary Katherine
Mon 6h18'42"1d9'
Skeeter
Cet 3h0'32"4d10'
Skeeter
Sct 18h49'0"-6d21'
Skeeter's Starshine
Aur 6h56'28"43d41'
Skeezicks,Puck Finlay
Aur 7h22'0"40d10'
Skeffinton,Don & Brigid
Cnv 12h10'55"48d33'
Skehan,Father John A
Eri 3h10'10"-5d16'
Skeim,In Memories of Bernice & Wally
Lyr 18h34'36"41d14'B
Skeith,Helen
Leo 11h1'29"8d12'
Skeith,Jack
Gem 6h1'45"26d26'
Skelley,Debbie Emory
Peg 22h29'43"26d38'
Skelley,Heather Nichole
Peg 23h37'21"12d19'
Skelly,Caralyn Patricia
Uma 11h56'14"45d51'
Skelly,Carolyn Mary
Boo 14h19'25"35d14'
Skelly,Chrystine Laurette Jondreau
Leo 10h32'55"16d0'
Skelly,Diane Colette Jondreau
Leo 10h46'19"16d59'
Skelly,Gregory A
Boo 14h32'16"38d53'
Skelly,Maureen
Tri 2h17'14"32d32'
Skelly,Meredith
Peg 23h1'15"25d55'
Skelly/Eternal Star, Timothy Cason
Aur 6h14'17"37d40'
Skelton III,George W
Sco 17h4'42"-31d23'
Skelton,Calvin E
Cyg 19h30'0"34d48'
Skelton,Jr,John
Aql 19h24'59"12d10'
Skelton,Pati Lu
Cyg 19h25'20"33d33'
Skene,Reja Jane
Sge 20h17'25"20d34'
Skentzos,Haroula
Vul 19h20'55"26d27'
Skerman,Glenn Lance
Cra 18h2'27"-37d26'
Skerrett,Jonathan Ryan
Cep 20h53'1"61d47'
Skerry Family,The
Ori 5h54'53"17d41'
Skerry,Philip J
Peg 22h18'21"22d48'B
Sketchley,Maurice John
Cep 22h17'1"68d4'
Sketl,Jr,Albert
Tri 2h11'45"33d37'
Ski
Cep 21h11'22"58d33'
Ski Inn,The
Cam 10h52'17"82d19'
Ski Star
Aql 20h16'32"0d31'
Skiba,Christina
Ari 2h50'42"22d19'
Skiba,Gloria
Per 3h50'32"39d41'
Skibba,Gisa
Ari 2h52'27"21d53'
Skibinski,John
Boo 14h35'36"37d44'
Skibsted,Marianne
Umi 15h11'43"68d46'
Skidmore,Chris M
And 1h57'17"41d9'
Skidmore,Jack
Aur 6h9'42"38d58'
Skidmore,Raquel
And 23h29'38"44d18'
Skiendziel,Agnieszka
Lmi 9h46'42"38d7'
Skiendziel,John Brian
Per 3h9'24"40d44'
Skies,Sanibel
Eri 2h56'52"-2d42'
Skiffington,Jacqueline
Per 2h19'58"58d26'
Skiles,Joan B
Cyg 20h18'44"41d9'
Skille,Jimmy
Her 16h29'52"35d13'
Skille,Sanna
And 2h18'1"48d21'
Skillestad,Gloria Ann
Cap 21h20'59"-27d41'
Skilling,Jeffrey K
Cet 1h18'49"-4d12'
Skillman's Star
Uma 9h21'44"60d33'
Skillman,Jeffrey B
Aur 5h39'51"37d45'
Skillman,Joe William
Cma 6h56'45"-18d56'
Skillman,Jr,Dean S
Lac 22h16'37"50d36'
Skillman,Susan Jane Taylor
Sge 19h58'29"18d52'
Skilton,Gilbert W & Isabel M
Lmi 10h5'0"30d18'
Skimblespats,Jocko
Cma 7h18'1"-18d59'
Skimhorn,Dalene Kay
Lyr 19h20'55"35d32'
Skimin-Barley,Benn Gary
Uma 8h14'31"70d33'
Skingsley,Caroline
Sge 20h1'29"20d27'
Skinner
Lac 22h44'24"38d43'
Skinner & Search's "La Barracuda"
Oph 17h2'0"11d58'
Skinner The Star of my Life,Elizabeth A
And 2h32'31"37d31'
Skinner's 40th,H Michael
Ser 15h53'44"21d43'
Skinner,Angela
Lyr 18h48'42"13'
Skinner,Barbara Claire Fornes Chambers
And 0h20'0"44d26'
Skinner,Bruce
Sgr 19h38'3"31d48'
Skinner,Donna
Del 20h35'23"10d56'
Skinner,Dorothy
Eri 2h52'18"-15d43'
Skinner,Faith & Roger
Dra 19h44'58"67d37'
Skinner,Kevin
Aqr 20h57'50"-0d43'
Skinner,Master Brandon Anthony "Snooty"
Cep 22h39'36"63d36'
Skinner,Papa George Herbert
Aql 20h11'45"10d51'
Skinner,Patrick J
Lac 22h32'25"56d20'
Skinner,Rebecca J
Cmi 7h22'48"1d39'
Skinner,Roland
Peg 23h2'49"21d34'
Skinner,Roxanne
Cyg 21h7'29"30d56'
Skinner,W B
Equ 21h3'49"2d43'
Skinner,Wayne
Oph 18h31'39"10d1'
Skinsley,Jacqueline Ann
Umi 17h45'32"84d35'
Skip
Aur 5h17'46"40d34'
Skip
Ori 5h55'26"15d45'
Skip To My Lou
Uma 8h37'15"54d26'
Skip's Glenda
Vul 20h18'59"22d35'
Skipper
Vel 9h43'23"-43d21'
Skipper
Cep 22h5'17"60d2'
Skipper's Spirit
Her 16h25'51"38d20'
Skipper-Dipper and The Suzie-Q,The
Dra 18h39'51"77d38'A
Skipper-Dipper and The Suzie-Q,The
Dra 18h39'51"77d38'B
Skippy
Ori 5h25'49"-0d13'
Skipsharmol
Cam 7h7'35"61d34'
Skittles
Gem 7h35'0"30d56'
Skiver,Devin Donald
Crt 11h43'28"-10d55'
Skjerven,Sean
Dra 14h39'33"61d21'
SKL Lucky Star
Aur 5h52'50"38d37'
Skladzien,Jr,Stephen A
Her 17h37'41"14d11'
Sklar Star,The
Cnv 12h53'45"41d51'
Sklaroff,Louise Gerber
Cas 3h8'24"58d19'
Sklarski,Dennis(Hawk)
Ari 1h57'43"11d32'
Sklena,Vincent John
Lac 22h26'38"50d31'
Sklenar,Christina Anna Marie
Cas 1h17'41"62d49'
Sklenar,Laura Leigh
Cas 1h13'59"63d29'
Sklenar,Rachel Alana
Cas 1h6'51"60d15'
Sklodowski,Krista Marie
And 0h21'28"32d51'
Skoff,Brian & Doris
Ori 5h0'54"-1d54'
Skoff,Sally
Aql 19h6'21"-6d21'
Skogen,Gordan R
Dra 17h15'49"69d35'
Skogen,Katelyn A
Lyn 8h21'32"41d31'
Skogen,Skip
Lac 22h47'1"52d44'
Skogh Clan Star,
Lmi 10h0'33"31d24'
Skok,Annette Suzanne
Del 20h53'0"7d10'
Skokic Zorica
Cas 1h47'25"60d37'
Skokna,Douglas
Aql 19h9'0"13d25'
Skold,Keith Allen
Dra 19h24'37"56d20'
Skole,Robert
Oph 17h4'38"11d53'
Skolik,Sonya & David
Umi 14h26'0"68d45'
Skolnick,Dr Judah L
Aur 5h59'46"37d43'
Skolnick,Marcia Lee
Vul 21h15'20"20d28'
Skolos,Rich
Cnc 7h54'24"10d44'
Skomal,Burton Shooting Star
Her 17h0'53"38d52'
Skominas,Walter "Butch"
Oph 17h0'10"0d39'
Skoneczna,Beata
And 1h28'29"37d44'
Skopek,Kathie
Cas 0h31'21"58d50'
Skopit,Stanley
Ori 5h56'20"15d18'
Skora,Jerzy
Sge 20h13'23"17d33'
Skora,John G
Her 16h44'38"51d7'
Skorina,Jane-Marie Grace
Cap 21h50'1"-14d43'
Skoronski,Jaime Heather
And 0h24'0"38d32'
Skorzewski,John B
Umi 15h48'34"78d7'
Skotnicki,Jerzy
Cam 3h17'10"61d50'
Skott,Jürgen
Psc 1h17'54"21d36'
Skovera,Ariel Linde
Boo 14h19'33"28d17'
Skovera,Joyce & John
Tri 2h10'58"31d21'
Skovermo
Ori 5h27'13"-2d52'
Skovron,Marilyn Elizabeth
Mon 6h19'1"5d52'
Skow II,Jon Buresh
Boo 14h2'53"26d29'
Skow,Carl Fredrik
Her 16h20'52"24d34'
Skowron,Robin
Del 20h24'0"8d2'
Skowronski,David B
Cep 21h45'59"61d49'
Skowsky,Ginger
Peg 21h29'48"20d3'
Skowsky,Walter W
Cmi 7h42'56"8d29'
Skraly,Frank A
Boo 14h49'22"31d14'
Skrbina,Robin Marie
Cas 1h32'41"61d3'
Skrebes,Kristine Anderson
Peg 22h35'27"8d55'
Skriba III
Vul 20h14'12"26d15'
Skriloff,Nathan
Aur 7h11'53"38d15'
Skrinde,Kelsey Marie
Vul 20h20'47"28d46'
Skrine,Rebecca Louise
Lyr 18h17'27"47d10'
Skrleta,Erich
Leo 10h45'1"8d8'
Skrobola,Stephen Frank
Ori 5h38'1"15d3'
Skrondo
Peg 21h52'46"20d2'
Skrovny,Lea
Cam 7h51'1"61d20'
Skrynski,Marion
Cyg 19h27'36"37d48'
SKS & SJB Forever In The Stars
Eri 2h49'37"-5d35'
Skubiak,Christina
And 23h24'55"48d57'
Skubisz,Steve
Sgr 18h25'19"-24d55'
Skully
Ori 6h15'18"-1d29'
Skvarna,Barbara & Mark
Sct 18h36'29"-6d43'
Skvasik,Joseph G
Boo 15h1'34"26d16'
Skwira,Alex
Her 18h17'55"38d33'
Sky Hinman
Uma 9h27'39"61d11'
Sky's Brightest Star Raija Elina
Dra 15h2'1"65d20'
Sky,Anna
Cas 2h16'30"68d33'
Skybolt
Her 16h3'54"41d22'
Skye,Allia
Cyg 19h57'0"37d45'
Skye,Lenny
Boo 14h17'35"36d8'
Skyfather,Carrier of Joy Miigwech
Tau 4h1'12"20d8'
Skyking
Boo 14h47'32"37d12'
Skylar
Boo 14h5'29"37d37'
Skyleah
Cam 8h1'54"83d58'
Skyler
Uma 9h59'17"56d2'
Skyler MacKenzic
Uma 10h17'41"62d14'
Skyline
Uma 11h56'28"31d58'
SkyWatch Information Systems,Inc
Umi 14h24'51"74d26'
SL Krol's Crazy River Star
Sct 18h52'21"-6d47'
Slack,Brian James
Cep 22h9'1"60d39'
Slack,Dale Michael & Johnnie Jill
Cam 4h24'29"68d24'
Slack,Jr,Harry Varley
Cmi 7h43'0"0d36'
Slack,Laura
Cyg 19h32'58"39d12'
Slack,Lina
Aql 20h2'52"0d46'
Slack,Mark
Cep 4h41'27"80d14'
Sladack,Nancy
Cnc 8h11'1"30d23'
Slade
Dra 14h40'52"62d48'
Slade,Clyde "Bikeman"
Aql 19h3'15"15d54'
Slade,Doreen
Cas 1h34'10"75d45'
Slade,I Love Laura
Sge 20h3'48"20d7'
Slade,John
Ori 4h46'39"0d16'
Slade,Linda Kay
Uma 14h9'20"59d48'
Slade,Lorna Carol
Cas 0h54'57"71d51'
Sladek,Adam Joseph
Hya 8h13'59"-5d51'
Sladek,Meredith Clarisse
Aql 18h59'0"12d55'
Sladek,Michael Jan
Uma 9h37'24"51d24'
Sladic Marco
Aur 5h1'33"37d39'
Slagle,Brian
Her 16h38'33"33d5'
Slagle,Holly Josephine
Cyg 21h54'12"42d28'
Slagle,Thomas Keith
Leo 11h7'40"-5d20'
Slam Dunken-Dave
Uma 11h25'0"55d33'
Slamon,Katherine
Cas 0h11'21"58d18'
Slan,Sandy & Hal
Vul 20h2'37"22d56'
Slaney,Kevin
Her 16h36'36"36d21'
Slang,Albert A
Aur 7h21'0"43d45'
Slanker,Wanda
Tri 1h59'52"35d5'
Slapanska,Tatana
Lyn 7h51'53"37d37'
Slapciu,George Alexandru Valentin
Cam 7h6'40"61d12'
Slapota,Daniele
Ari 2h3'18"21d7'
Slas,Sheri
Vul 21h16'1"20d19'
Slata,Jeanne M
Dra 20h22'0"61d50'
Slatcher,Peter & Erika
Peg 23h0'50"20d13'
Slate,John J & Catherine M
Dra 14h6'34"64d57'
Slaten,Arlene R
Oph 18h6'0"10d20'
Slater's Sillver Star
Cet 2h57'41"-0d28'
Slater,Aimee
Lmi 10h20'14"31d7'
Slater,Charlotte Oakley
Umi 14h29'54"66d31'
Slater,Christian
Aql 20h4'25"0d5'
Slater,Col Tip
Aql 19h41'20"14d41'
Slater,Gladys
Cas 0h17'45"61d56'
Slater,James
Her 17h18'20"12d59'
Slater,Jeffery Carroll
Oph 18h1'29"12d38'
Slater,John Paul & Lorraine
Nor 16h25'32"-45d25'
Slater,Justine
Dra 18h26'40"50d24'
Slater,Miles
Aqr 23h31'41"-4d25'
Slater,Stormie
Cam 3h52'10"55d37'
Slater,Terry
Her 16h19'28"41d38'
Slater,Tommy
Cmi 7h22'48"8d37'
Slater,W Stuart F
Sgr 18h0'47"-28d20'
Slatin,Arthur Joseph Kaltenborn
Boo 14h0'30"25d13'
Slatin,Arthur Joseph Kaltenborn
Ori 6h15'24"8d1'
Slatin,Kathryn Lee Kaltenborn
And 2h23'46"40d32'
Slatin,Patricia Katrina
Cas 1h55'23"61d18'
Slator,Nora
Lyr 18h21'16"42d53'
Slatterly,Thomas Lowry
Ori 5h56'54"13d8'
Slattery II,Joseph Austin
Her 16h39'1"7d40'
Slattery,Aaron Patrick
Per 1h48'45"50d34'
Slattery,Elizabeth J
Aql 19h1'23"-0d57'
Slattery,Kenneth M
Per 2h20'46"54d44'
Slattery,Robert Michael
Cnv 12h55'39"32d48'
Slatts
Uma 10h34'30"60d13'
Slaughter,Frederick D
Ori 5h32'50"-1d36'
Slaughter,Jr (CES2), Clarence E
Dra 16h3'47"66d12'
Slaughter,Ronald T
Dra 15h13'23"63d59'
Slaughter,Vicky
Aql 20h10'36"13d19'
Slavemaster Chuck
Boo 15h6'0"11d9'
Slaven,Ryan Jartman
Her 16h59'0"22d37'
Slavica
Boo 15h5'15"48d6'
Slavick,Shannon
Com 12h14'59"20d55'
Slavik Family Star,The
Uma 8h46'1"53d49'
Slavik,Frank R
Per 2h51'17"38d15'
Slavin Star,The Roberta
Lyr 18h56'19"41d0'
Slavin,Amanda Brianne
Lyr 18h40'42"29d18'
Slavinski,Jack
Her 17h1'55"40d41'
Slavyansky Bazar: Russia
Cnv 13h29'0"50d36'
Slavyansky Bazar: Ukraine
Cnv 13h3'44"50d59'
Slavyansky Bazar: Belarus
Cnv 12h37'36"49d6'
Slavyansky Bazar: Bulgaria
Cnv 12h25'0"46d5'
Slavyansky Bazar: Poland
Cnv 12h20'45"42d60'
Slavyansky Bazar: Slovakia
Cnv 13h9'29"37d42'
Slavyansky Bazar: Macedonia
Cnv 13h56'56"33d50'
Slavyansky Bazar: Serbia
Cnv 13h48'16"37d40'
Slavyansky Bazar: Croatia
Cnv 13h35'1"44d21'
Slavyansky Bazar: Slovenia
Cnv 13h15'40"44d38'

Slavyansky Bazar:Czech
Cnv 13h24'1"40d47'
Slawinski,Terra
Uma 9h19'39"57d50'
Slawnikowski,Florence
Gem 7h1'48"30d18'
Slawnikowski,Sylvester
"Pussycat"
Gem 7h0'0"30d11'
Slawsky,Lynn
Lac 22h5'54"46d58'
Slawsky,Stephen
Lac 22h4'31"48d16'
Slawter,Michael J
Lyn 7h49'46"44d7'
Slay,Richard Michael
Cep 21h21'14"80d1'
Slaybaugh,Craig William
Ori 6h8'44"17d23'A
Slaybaugh,Traci Jane
Ori 6h8'44"17d23'B
Slayer
Dra 14h30'1"60d25'
Slayer,Doris & Gerald/ Lila
Lyn 8h2'29"43d47'
Slazak,Heather
And 23h49'44"33d58'
Sleath,Damian Crisp
Cep 23h6'35"65d29'
Slebrch,Kathy"My Little
Noodle"
And 23h0'13"51d16'
Sleczkowski,Rebecca L
Aur 5h30'27"38d57'
Sled
Vul 19h22'45"26d57'
Sledesky,Anne & Tony
Uma 9h10'1"50d48'
Sledge,Mary Beth
Mon 7h1'1"8d38'
Slee,Michele
Cyg 21h16'60"28d35'
Sleeper,Chaucer "Shawn"
Peg 23h6'0"17d44'
Sleeper,Robert D
Ser 15h9'53"-2d37'
Sleepless In Seattle
Cam 3h30'30"60d9'
Sleepless...
Uma 10h38'43"51d30'
Sleepy
Aur 6h34'15"31d8'
Sleeth,Robert Glasford
Aur 5h7'42"43d47'
Sleigh,Kevin J
Boo 15h3'35"51d15'
Sleighter,Dick
Lac 22h26'40"38d42'
Sleiman,Janel Ashley
Cet 1h49'39"-3d44'
Sleiman,Sharif
Cet 1h6'37"1d53'
Sleitas,Eddie
Dra 12h50'44"75d3'
Slekis,Scott Alan
Dra 14h46'37"62d51'
Slemion,Todd R & Whitney W
Sgr 19h7'58"-25d45'
Slemmer,Tim
Cep 21h10'22"58d31'
Slepak,Paul A
Hya 8h41'55"0d12'
Slepetz,Jr,Stephen L
Aql 20h0'26"11d26'
Sless,Elliot J
Psc 1h2'45"20d3'
Sleury,Madeline M
Aqr 22h0'1"-1d4'
Slewett,Danielle Reneé
Psc 23h3'58"0d57'
Slicer,Monique C
Lmi 10h50'27"26d54'
Slichenmyer,Carol Beth
Cyg 19h59'58"37d50'

Slick
Per 3h34'41"38d55
Slick Rick
Her 17h53'49"38d51'
Slick,Brian Leland
Boo 15h1'49"8d59'
Slick,Ms
Lyr 18h46'0"35d6'
Slick,Ralph W
Tau 4h38'17"20d31'
Slicker,Laurie Anne
Com 12h23'1"20d12'
Slide Area THDRJABFJMH-
LKO
Eri 4h7'0"-7d43'
Slider,Jean Wells
Aur 6h39'22"38d46'
Slider,William W
Aur 6h16'52"38d2'
Slief,Laurance O
Per 2h40'19"35d17'
Sliger,Chloe
Tri 2h13'48"32d16'
Sligold,Jeremy
Boo 15h26'58"41d26'
Slingsby,Brian Taylor
Oph 17h33'0"-8d45'
Slingshot
Dra 17h7'34"67d30'
Slinker
Her 17h5'56"47d6'
Slinker,Alice Ann
Lyn 8h38'56"40d6'
Slip
Her 17h14'47"45d27'
Slipek,Karl
Cap 20h7'31"-11d12'
Slisz,Amanda Noel
Aur 7h6'44"40d16'
Sliva,Charlotte
And 0h57'22"35d55'
Sliva,Stan
Boo 14h27'21"38d6'
Sliverhead
Ori 5h55'0"13d20'
Slivka,John & Bertha
Cyg 21h0'1"28d23'
Sliwa,John
Ori 4h54'32"4d27'
SLM Forever Loved By AJC
And 0h0'0"37d58'
Sloan Family Piece of
Heaven,JCKM & G
Boo 15h9'50"52d4'
Sloan III,John Patrick
Peg 21h41'17"27d39'
Sloan,Carol & John
Aql 19h50'0"13d50'
Sloan,Claudia A
Uma 12h0'46"48d37'
Sloan,Erica Michelle
Cas 0h2'59"63d56'
Sloan,Geraldine Ellen
And 0h31'25"45d60'
Sloan,Gregory A
Cet 2h37'30"-1d57'
Sloan,Irene Gertrude
And 2h27'34"40d22'
Sloan,Larry Gene
Aql 19h6'37"4d31'
Sloan,MD,Susan
Oph 18h35'10"11d33'
Sloan,Nicholas Peter
Ori 6h18'58"18d56'
Sloan,Pat
Cnv 13h8'27"33d1'
Sloan,Rodney Toddy
Her 17h19'20"29d23'
Sloan,Scott Thomas
Boo 13h38'17"18d28'
Sloan,Susan
Com 13h4'20"15d16'
Sloan,Terry Lynne
Uma 11h23'33"71d10'

Sloan,Walter
Ori 4h57'29"0d19'
Sloane,Brianna Leigh
Aur 7h0'1"38d32'
Sloane,Sarah Elizabeth
And 0h12'1"39d33'
Sloane,Steve
Uma 10h34'49"40d56'
Sloat-Historian, Jerry A
Aur 5h58'43"37d48
Sloban,Tracy Elizabeth
And 0h36'39"40d38'
Slobin,Richard A
Cep 21h57'48"84d49'
Slocomb,John Alexander
Cep 1h47'59"77d42'
Slocomb,Richard Hawk
Sgr 20h1'13"-42d33'
Slocum,Barry
Aur 5h38'40"38d26'
Slocum,Jeffrey Scott
Uma 11h43'42"40d16'
Slocum,Richard Drake
Dra 14h57'39"64d35'
Sloly,Debra
And 0h19h0"38d19'
Slomka,Chester L
Sex 10h17'24"-4d15'
Slomka,Emily Lingna
Equ 20h57'16"8d41'
Slomowitz,Anna Michelle
Lyn 8h15'1"35d46'
Slomowitz,Rebecca Faith
Boo 14h57'11"42d55'
Slomski,Joshua Christoph
Per 1h53'40"54d8'
Slonaker,Kathy D
Cas 2h56'32"58d28'
Sloneski,Steve
Lac 22h4'16"40d57'
Sloopy
And 1h6'38"37d43'
Sloopy
Ori 6h3'0"7d38'
Sloss,Leon
Her 17h39'13"21d25'
Slosser,Sahar
Aql 19h34'35"0d36'
Slota,Frances (Fran)
Cas 0h21'1"64d49'
Slotemaker
Crb 16h18'0"28d25'
Slotkin,Alexander Mark
Dra 17h25'44"60d49'
Slovenski Family Star, Ruth &
Walter
Crb 15h59'43"26d60'
Slover,Lisa Marie
Cas 2h54'56"73d56'
Slovich,Audrey Ann
Boo 14h29'18"23d2'
Slovich,Laura Marie
Aur 5h9'15"38d15'
Slovich,Melissa Lynn
Umi 15h49'24"74d39'
Slovikosky,Linda
Lyr 19h9'39"38d58'
Slowik,Joseph
Aur 7h19'21"38d32'
Slubin's Star,Grandmom
Lillian
Lyr 18h17'1"46d56'
Slucter,Barbara
Lyn 7h55'50"44d29'
Slugger's Star
Aur 7h21'60"38d34'
Slumber
Tri 1h45'1"26d43'
Slupczynski,Madeline
Cas 0h43'50"69d16'
Slupinski,Detlef
Cas 0h31'46"63d15'
Slusarick,Marie Rose
Cas 0h56'10"60d49'

Slush
Aql 18h43'47"10d53'
Slusher,Fran & Michael
Aur 5h0'57"41d30'
Slusher,Jim
Ori 6h6'39"1d46'
Slusser,Patrick D
Boo 14h46'14"37d19'
Slutsky,Nathan
Cet 2h6'41"5d40'
Slutzky,Zoe
Lac 22h22'33"54d6'
Sluyter II,Ronald Lee
Mon 7h15'0"0d5'
Sluyter,Anne
Mon 7h7'0"0d57'
Sluyter,Dave & Kathy Fuller
Mon 7h3'29"1d10'
Sluyter,Jim & Marilyn
Mon 7h2'53"3d51'
Sluyter,Ronald Lee
Mon 7h3'1"1d14'
Sluyter,Steven Matthew
Mon 7h15'53"0d29'
Sly 930
Hya 8h13'29"1d26'
Sly Fox
Aql 20h10'28"14d7'
Sly's Star
Ser 15h14'56"23d36'
Sly,Richard Dwayne
Her 17h11'45"46d43'
Slykhuis,Andrew John
Ser 17h21'11"-11d7'
Slynn,Laura Jane
Cyg 20h1'34"38d52'
Slyper,Jane
Cet 2h28'55"0d43'
Slyvia Lucille-40
Cas 23h30'1"63d23'
Slywka-27,Andrij P
Gem 6h45'58"30d35'
Smaldino,Gregory Joseph
Dra 14h2'12"66d28'
Smaldon,Odette RK
Cmi 7h41'27"-0d0'
Smale,Laura Ann
Crt 11h21'3"-18d9'
Smalenberger,Ron
Aur 6h24'40"31d50'
Small Fry
Aql 19h31'1"13d38'
Small II,Richard B
Sex 9h57'28"-5d36'
Small III,James D
Oph 16h59'56"11d12'
Small"Smiley",Cindy A
And 1h40'34"36d21'
Small,Aaron JC
Dra 17h29'53"63d51'
Small,Christopher John
Per 3h14'0"42d45'
Small,Christopher John
Aqr 23h44'38"-4d23'
Small,Clayton G
Boo 14h6'30"33d47'
Small,Dylan Jacob
Aqr 22h15'0"0d51'
Small,Dylan Jacob
Her 18h55'1"12d7'
Small,Frances B
Lac 22h8'46"51d30'
Small,Griffin Oliver
Aqr 22h17'51"0d45'
Small,Griffin Oliver
Per 2h50'31"37d33'
Small,Henry Eugene
Dra 15h10'55"61d25'
Small,Hilary
Crb 15h34'22"26d57'
Small,Israel
Umi 15h33'39"68d45'
Small,Joan
And 2h35'16"40d18'

Small,John H
Aur 4h45'28"31d1'
Small,John W
Mon 7h26'50"-6d23'
Small,Jonathan William
Mon 7h28'1"-6d29'
Small,Lillian Marie
Mon 7h24'40"-6d21'
Small,Marvin
Hya 8h20'0"5d55'
Small,Olivia Paige
Peg 22h42'1"20d53'
Small,Rachel Michelle
Peg 22h54'54"21d38'
Small,Robert Stephen
Sct 18h47'33"-7d39'
Small,Vee
Cyg 19h55'26"48d6'
Small,William Gregory
Aql 19h58'28"12d43'
Small,William Hardin
Lac 22h25'50"55d21'
Small,William Rafferty
Her 16h21'50"42d29'
Smalley,Debbie & Johnny
Uma 11h19'18"48d55'
Smalley,Marc
Her 18h29'24"12d21'
Smalling III,HRH Raymond
George
Gem 6h45'2"26d58'B
Smalls,Jeanette Lynn
Lyr 18h35'50"37d40'
Smallwood 4-26-72, Jeffrey
Michael
Cep 21h0'43"61d48'
Smallwood Ross
Aql 19h20'21"15d53'
Smallwood,Agnes
Cyg 21h5'57"37d37'
Smallwood,Betty
Boo 15h5'24"23d4'
Smallwood,Jason Andrew
Her 17h54'15"30d2'
Smallwood,Kathryn
Cnc 8h27'39"11d10'
Smallwood,Nicole
And 2h35'1"40d17'
Smallwood,Susan
Lyr 18h56'56"37d25'
Smallwood,Thomas L
Boo 14h35'46"8d54'
Smalt,Homer & Loretta
Mon 7h42'44"-2d48'
Smalt,William Darwin
Hya 10h1'45"-17d6'
Smalt-Flynn,Bonnita Maureen
Mon 7h42'44"-2d48'
Smarkle,Kristina Lynne Allison
Cet 2h27'33"-10d10'
Smart As A Goose
Aur 6h49'12"38d51'
Smart Doc
Oph 16h57'0"-25d47'
Smart,Brian
Hya 8h56'43"0d36'
Smart,Cheryl
Cam 4h14'15"60d43'
Smart,Frank
Ser 15h56'55"19d48'
Smart,Gregory
Oph 16h59'60"10d2'
Smart,Jane Siobhan Furlong
Ori 5h24'32"1d4'
Smart,Marcia Jan
Eri 4h59'24"-10d57'
Smart,Maxwell B
Aur 5h15'54"41d27'
Smart,Mr William
Cet 0h30'45"0d53'
Smarz,Elizabeth Marie
Del 20h56'53"13d37'
Smathers,L'Etoile, Steven E
Cyg 21h54'45"53d41'

Smatusik,John
Dra 17h35'41"64d0'
SMC
Dra 15h22'45"53d50'
Smedira,Elsie B
Cas 23h32'34"61d35'
Smedley,Jackie M
Mon 6h43'10"7d13'
Smedley,Peter
Ori 4h50'27"5d34'
Smedley,Stuart
Cnv 13h34'0"42d14'
Smee,Robert A
Hya 9h1'18"2d24'
Smeed,James Edward
Mon 6h54'34"1d15'
Smeeton,Mike & Sue
Cra 18h42'42"-41d58'
Smeets,Graham Ross
Ori 5h2'13"8d44'
Smellwood,Peter William
Her 18h15'29"30d31'
Smeltz,Jeremy Lester
Dra 19h6'34"56d24'
Smentkowski,William
Cyg 19h56'58"37d53'
Smeraldo
Cep 2h13'49"78d15'
Smerdon,Scott William
Cmi 7h14'8"9d28'
Smereck,Geoffrey Arthur
David
Lyn 8h11'40"52d15'
Smerklo,Stacy
Lyn 8h0'29"48d39'
Smetana,Louise Rothman
Com 12h51'29"26d56'
Smewing,Arthur & Winnie
Cyg 19h31'28"36d23'
SMG CP Love Light
Sge 19h12'3"16d46'
Smicklas,Jordan Audrey
Ser 15h54'11"1d7'
Smiddy,Mark Alan
Her 18h3'24"14d54'
Smidt,Alexander Kenji
Her 16h33'1"32d50'
Smidt,Kelly Richard
Cam 13h36'57"78d3'
Smieja,Jennifer
And 2h14'0"42d56'
Smiell,Joe "Button Box"
Hya 8h13'47"1d2'
Smiell,Joseph E
Sct 18h27'52"-4d15'
Smietanski,Adam Michael
Aur 6h17'26"37d3'
Smigelski,Ms Marie
Cas 23h27'21"60d39'
Smiggen,Jimmy
Dra 16h47'59"67d3'
Smigiel,Daniel
Cnv 13h57'14"38d25'
Smigo,Robert
Aur 6h7'60"35d53'
Smilack,Susan
Mon 7h26'44"-10d18'
Smile
Vul 20h41'43"28d18'
SMILE HARUMI
Cap 20h51'53"-22d29'
Smile Na
Lyr 19h24'1"37d58'
Smile Of Misako
Tau 5h14'35"20d4'
Smile,Kyra
And 23h33'1"48d22'
Smiles,David Michael
Crb 15h27'29"31d32'
Smiley
Cra 11h8'58"-18d20'
Smiley's Star
Lyr 19h1'1"31d48'
Smiley,F David
Cnv 13h58'24"46d51'

Smiley,Merry Hope Trotter
Com 12h2'1"27d38'
Smiley,Michael
Sct 18h52'12"-7d22'
Smiley,Steven W
Hya 8h18'47"0d10'
Smilg,Ruth
Vul 19h33'56"21d28'
Smiling Angel
Cas 3h22'25"70d45'
Smiling Mighty Beavers
Cmi 7h43'15"7d40'
Smilnak,Sara
Lac 22h53'51"54d26'
Smilove,Robert Harley
Her 17h13'24"46d2'
Smirles,Gus
Aur 5h31'23"50d16'
Smisko,Paul
Ori 5h39'18"-6d39'
Smit,Anneke
Cam 12h23'42"77d24'
Smit,Edward Edison
Lac 22h28'26"41d13'
Smit,Nicky Henri
Dra 10h12'48"80d2'
Smith
Ser 17h54'32"-14d13'
Smith #22,Emmitt
Sct 18h42'0"-6d23'
Smith 57,T
Uma 10h20'37"56d25'
Smith Class Of 1960
Tau 3h39'1"28d23'
Smith Dad's Shining
Star,Amanda L
Cas 1h0'51"61d5'
Smith Family-Clay Kay Lance
Kyle,The
Del 20h13'0"15d32'
Smith I,William Patrick
Aur 5h52'23"30d17'
Smith II,Frank J
Cnv 12h33'0"40d7'
Smith II,Solomon Miller
Eri 3h20'1"-17d13'
Smith III #22,Emmitt J
Ori 5h42'16"1d25'
Smith III,Daniel Guy
Cep 2h14'37"80d0'
Smith III,George B
Cet 1h59'32"0d14'
Smith III,George W
Aql 20h12'56"12d22'
Smith III,Roger W
Ori 6h16'20"20d6'
Smith III,T J
Lmi 10h37'1"31d38'
Smith Kindergarten TCH Edna
Carman
And 2h15'14"41d39'
Smith Nurse Manager, Jean
Oph 17h14'6"-23d42'
Smith Star Of Hope,The Jeff
Cap 20h22'35"-13d60'
Smith Star,Laurence & Muriel
Cyg 19h46'26"50d14'
Smith Star,Shirley
Aur 7h13'45"38d7'
Smith Star,The Patricia &
Lewis
Cyg 20h42'0"42d36'
Smith Star,The Ivin & Jewell
Lac 22h17'44"50d32'
Smith Star,The Steven Robert
Sco 15h45'5"-21d17'
Smith Star,Tristan Laine
Peg 23h28'27"22d27'
Smith w/Steve Davis, Susan-
Shane & Shelby
Uma 10h50'0"44d46'
Smith's 50th Ann, Chuck &
Dolly
Cnv 13h17'11"50d13'

Smith's Bit Of Heaven, Wayne
Boo 15h5'1"20d35'
Smith's Family Star, Greg &
Linda
Sgr 18h48'49"-27d11'
Smith's Family Star, Glen E &
Doris A
Sgr 18h48'51"-20d55'
Smith's Gold Star, Dr J Kelly
Oph 17h40'0"-18d49'
Smith's Star,Michael & Carol
Boo 14h11'37"38d3'
Smith's,Not Here At
Work,Grant
Sgr 18h48'4"-23d6'
Smith,"Diggy Diggy", Jack L
Cet 1h25'36"-12d32'
Smith,"E King Special
Star",Gary D
Sgr 18h48'48"-26d46'
Smith,"Jon" P
Aur 6h41'30"39d46'
Smith,"Our Star",Dan J
Boo 15h4'13"53d27'
Smith,"Pennsylvania" Michael
Dra 17h45'1"63d60'
Smith,11-9/11-12, Rachael &
Arnold
Cam 3h50'57"61d54'
Smith,Aaron Michael
Ori 5h58'35"16d2'
Smith,Abram
Cap 19h9'4"-11d48'A
Smith,Adam
Ori 6h9'25"2d51'
Smith,Adam Paul
Sct 18h35'12"-6d24'
Smith,Aimee Lynn
Lmi 10h45'58"27d2'
Smith,Alden Dawson
Aur 6h4'0"38d44'
Smith,Alexander
Ori 5h50'37"20d34'
Smith,Alice Carlotha Howell
Cyg 20h7'40"40d32'
Smith,Alicia Marcelle
Vul 19h45'0"26d7'
Smith,Allan
Boo 14h17'17"18d43'
Smith,Allen Thomas
Ser 15h44'11"20d48'
Smith,Alvin Louis
Crt 11h32'24"-11d42'
Smith,Amanda Elizabeth
Lyr 18h47'57"41d35'
Smith,Amanda Jayne
Lyr 19h16'13"26d46'
Smith,Amy C
Uma 9h41'41"58d5'
Smith,Andrea Lee
Cas 2h57'0"61d13'
Smith,Andrea Lorraine
Dra 17h54'46"68d7'
Smith,Andrew Mason
Hya 10h26'1"-12d18'
Smith,Andrew Roger
Cyg 20h21'38"38d35'
Smith,Anita Parker Streeter
Tau 5h50'29"26d17'
Smith,Anne Josephine
Uma 12h58'11"62d42'
Smith,Anne Marie
And 23h37'37"48d60'
Smith,Antoinette
Mon 6h20'26"0d55'
Smith,Arthur
Ori 6h4'1"2d20'
Smith,Ashlee N
And 0h51'1"21d44'
Smith,Ashley Nicole
Lyn 7h45'0"41d29'
Smith,Aubrey O'Neal
Del 20h16'41"9d33'
Smith,Austin Cole
Equ 21h10'0"10d48'

Smith,Austin E
Ori 6h16'1"8d40'
Smith,Baby Noopy Bryan
Scott
Cet 2h57'18"5d21'
Smith,Bama Faye Strawn
Cyg 20h32'41"41d14'
Smith,Barbara
Mon 6h53'0"-0d9'
Smith,Barbara Ann
Cas 2h27'53"59d47'
Smith,Barbara Anne
Boo 15h5'7"9d24'B
Smith,Barbara E M
And 0h13'18"38d6'
Smith,Bea
Peg 22h16'11"7d36' A
Smith,Beautiful Dreamr Paula
Hamilton Denver
Ori 5h55'24"18d27'
Smith,Becky
Cas 2h40'33"73d34'
Smith,Benjamin
Cep 23h10'32"61d35'
Smith,Benjamin Audley
Umi 15h1'52"67d36'
Smith,Benjamin Todd
Oph 16h49'41"11d19'
Smith,Bessie
Sct 18h54'56"-6d32'
Smith,Bette Jean
Lac 22h45'24"56d41'
Smith,Beverly Pattison
Aur 5h47'31"50d16'
Smith,Bill
Ori 5h2'14"-2d50'
Smith,Blair Matthew
Cep 21h50'48"56d13'
Smith,Blue
Uma 10h8'34"61d30'
Smith,Bobbie
Dra 15h1'13"58d3'
Smith,Bobby L
Per 3h25'22"51d5'
Smith,Boo Bug a/k/a Charles
William
Her 16h37'1"37d17'
Smith,Braden Victor
Mon 6h55'59"8d46'
Smith,Brenda
Mon 6h35'16"7d34'
Smith,Brent C
And 0h0'30"37d53'
Smith,Brian
Dra 19h24'54"56d13'
Smith,Brian
Boo 14h54'31"28d5'
Smith,Brian Mark
Her 16h6'44"25d25'
Smith,Brian Norman
Per 1h26'46"52d56'
Smith,Brooke
Umi 15h33'12"71d34'
Smith,Brooke
Ori 6h5'56"0d29'
Smith,Bugs Barbara Grant
Com 13h24'52"26d12'
Smith,Byron Lee
Oph 17h33'0"7d54'
Smith,Capt R & Josephine W
Smith
Cyg 19h25'25"30d3'
Smith,Carl Allen
Aql 20h19'57"5d9'
Smith,Carla
Cyg 20h58'14"39d59'
Smith,Carlton Michael Ray
Hya 8h15'50"2d30'
Smith,Carol Ann
Cas 0h33'0"75d54'
Smith,Caroline Elizabeth
Mon 6h29'17"0d22'
Smith,Caroline
Peg 22h9'45"4d37'

Smith,Carolyn Marie Wood
Uma 9h36'37"57d54'
Smith,Cary Parker
Cep 22h30'18"67d36'
Smith,Catherine & John
Cyg 20h43'43"45d45'
Smith,Cecil O
Peg 22h45'49"29d53'
Smith,Chad Claybourne
Dra 17h42'17"63d49'
Smith,Charles Imbrie
Per 3h17'45"41d6'
Smith,Charles R
Her 17h49'32"14d59'
Smith,Chase
Uma 9h2'41"61d2'
Smith,Cheri Ann
Peg 22h50'56"27d31'
Smith,China Tew/Devin
Uma 11h54'0"52d33'
Smith,Christian Stanford
Her 18h3'19"28d26'
Smith,Christian L
Lac 22h12'31"51d30'
Smith,Christopher Roland Todd
Cet 0h33'31"-1d14'
Smith,Christopher Sa'id
Her 16h19'21"10d21'
Smith,Christopher J
Per 3h2'50"47d39'
Smith,Chuck
Per 2h56'35"56d33'
Smith,Chuckles R
Aql 20h19'52"5d6'
Smith,Cindy
Cas 0h48'0"66d2'
Smith,Clairbel
Crt 11h13'33"-13d39'
Smith,Claire Renee
Cma 7h2'40"-15d55'
Smith,Clark
Cam 4h19'0"67d48'
Smith,Clayton Holthoff
Ori 5h31'54"-0d50'
Smith,Clinton Frederick
Cet 0h50'40"-5d28'
Smith,Clive G
Boo 14h39'37"41d35'
Smith,CLU,G Scott
Her 17h23'38"40d40'
Smith,Colby James
Uma 11h39'52"61d1'
Smith,Colton
Uma 9h41'0"60d42'
Smith,Connie C
And 0h21'11"44d48'
Smith,Corinne Clark
Mon 6h30'33"1d24'
Smith,Cory Alen
Dra 16h52'21"63d31'
Smith,Courtney
And 2h4'21"41d23'
Smith,Courtney Lynne
Cas 2h38'13"61d26'
Smith,Crippled Creek- Dick & Joann
Lyr 18h34'36"41d14'A
Smith,Cynthia
Aur 5h55'49"40d11'
Smith,Cynthia & Christopher
Cyg 20h7'43"41d11'
Smith,D Donald
Hya 8h40'21"1d39'
Smith,Dale "Doc" R
Aur 5h59'40"31d23'
Smith,Dale Alan
Aur 6h59'31"40d21'
Smith,Dale Edward
Cep 20h54'42"66d48'
Smith,Dan & Mildred
Eri 3h58'53"-9d58'
Smith,Darlene Noel
Cam 11h47'37"78d41'

Smith,Darren & Missy Myers
Uma 13h56'50"51d30'
Smith,Darren & Roni
Uma 13h37'34"57d54'
Smith,Dave
Crt 10h49'11"-18d31'
Smith,David
Lmi 9h39'0"33d46'
Smith,David
Aur 6h41'15"38d15'
Smith,David Cullen
Cep 23h37'36"65d20'
Smith,David J
Ori 5h33'55"0d21'
Smith,David Jarod
Ori 5h30'10"-1d16'
Smith,David Kenneth
Oph 17h59'46"12d35'
Smith,David Louis
Boo 14h38'0"33d35'
Smith,Dawn Marie
Mon 7h51'53"-5d37'
Smith,Debbie Wharton
Mon 6h43'53"10d20'
Smith,Debra Ann
And 2h2'56"42d43'
Smith,Debra Kay
Cas 1h18'24"75d40'
Smith,Dennis
Lyn 9h16'0"37d50'
Smith,Derek
Her 17h12'0"46d10'
Smith,Derek & Netta
Uma 12h33'52"57d6'AB
Smith,Dominic Iassic
Boo 13h55'41"17d34'
Smith,Donald Russell
Dra 16h44'57"67d44'
Smith,Donald Wesley
Lac 22h48'52"52d34'
Smith,Donna
Aql 18h54'39"-0d5'
Smith,Donna & Joshua
Eri 3h19'39"-10d37'
Smith,Donna Sue
Uma 9h26'53"57d4'
Smith,Donya
Mon 6h39'54"11d22'
Smith,Dora
Uma 9h55'18"50d24'
Smith,Doris
And 0h30'10"31d50'
Smith,Dorothy Dewey & Richard J
Mon 6h44'27"8d49'
Smith,Dr Jeff
Uma 11h10'1"32d7'
Smith,Dr Len
Oph 17h16'11"-20d13'
Smith,Dr Mark Kennedy
Oph 18h19'20"8d35'
Smith,Dr Twyla & Holloway,Dr Jick
Lmi 10h13'46"34d46'
Smith,Dustin Peter
Dra 16h54'36"63d15'
Smith,Dustin S
Hya 8h10'34"0d2'
Smith,Eddie
Cep 22h16'31"62d2'
Smith,Edna Harmsworth
Eri 2h56'0"-12d52'
Smith,Edna M
And 23h37'52"49d15'
Smith,Edward D
Per 3h1'47"40d34'
Smith,Edward Earl
Dra 18h59'56"58d31'
Smith,Elayne Dianne
Mon 6h25'48"10d7'
Smith,Elizabeth Ann
Dra 19h26'27"80d15'
Smith,Ellen A
Cas 0h51'18"60d41'

Smith,Ellen Virginia (Gingy)
Aur 7h3'29"38d49'
Smith,Elspeth Miller
Lyr 18h16'37"30d7'
Smith,Emery & Cathy
Uma 10h7'22"61d25'
Smith,Emily Anne
Psc 1h35'14"21d16'
Smith,Emily Grace
Gem 5h59'12"27d3'
Smith,Emily Sue
Cam 9h1'47"80d19'
Smith,Emma Jane
Lyr 18h37'34"39d31'
Smith,Emma Lucy
Cyg 20h18'27"40d7'
Smith,Emma Seelbinder
Mon 7h59'21"-6d38'
Smith,Enid Barbara
Uma 11h17'59"50d9'
Smith,Eric
Cep 0h6'38"69d49'
Smith,Eric Scott
Boo 13h51'12"18d22'
Smith,Erin Lynn
Aql 19h48'42"14d12'
Smith,Ernest
Uma 9h39'52"54d14'
Smith,Esther & Charles
Aql 19h24'43"15d48'
Smith,Ethel
Cap 19h9'4"-11d48'B
Smith,Etta Mae Hargus
Mon 6h36'60"0d34'
Smith,F O
Oph 17h16'58"-18d48'
Smith,Father Nicholas W
Aql 20h5'0"0d34'
Smith,Flora Elizabeth
Lyr 19h0'21"28d40'
Smith,Florence
Uma 10h20'12"40d36'
Smith,Florence H
Cep 5h13'50"85d53'B
Smith,Floyd Monroe
Cam 7h4'8"65d16'B
Smith,Forrest L
Dra 17h1'42"65d39'
Smith,Frank & Margaret
Lyr 19h1'49"31d37'
Smith,Freda
Cyg 21h20'1"40d2'
Smith,Fuller Wayne
Ser 15h31'60"24d48'
Smith,G Michael
Cmi 7h56'50"7d42'
Smith,Garry Edward
Cnv 12h17'14"50d49'
Smith,Gary
Dra 17h5'1"60d39'
Smith,Gary L
Sex 9h55'38"3d50'
Smith,Gary Vernon
Uma 9h29'43"51d25'
Smith,Gavin Peter
Cyg 19h29'15"38d4'
Smith,Gayle M
Mon 7h6'23"-5d6'
Smith,Genevieve
And 2h25'1"47d29'
Smith,Geoff
Per 3h7'0"40d1'
Smith,Geoffrey
Vir 13h10'3"-1d55'
Smith,Gerald Bruce
Cyg 21h40'39"38d6'
Smith,Gideon E
Cet 1h0'13"1d43'
Smith,Gladys
Cam 3h20'26"60d10'
Smith,Glen Robert
Her 17h19'1"40d25'
Smith,Glenn W
Tau 3h41'36"22d7'

Smith,GM Christine Michele
Sgr 18h57'57"-22d53'
Smith,Gordon F
Per 2h0'57"57d1'
Smith,Grace Lyndsey
Cas 0h51'0"64d39'
Smith,Graham Dugdale
Equ 21h11'12"11d59'
Smith,Grandpa Alan
Boo 15h7'17"11d16'
Smith,Grandpa Fred
Aql 19h9'40"1d4'
Smith,Greg
Equ 21h7'1"11d53'
Smith,Greg
Aql 18h44'0"-2d8'
Smith,Gregory Alexander
Ori 5h38'34"-6d34'
Smith,Gregory Peter
Sex 9h48'56"-0d35'
Smith,Grover Cleveland
Cep 21h29'56"60d4'
Smith,Hank & Margaret
Per 2h57'28"50d34'
Smith,Harold Alfred
Cnv 12h42'58"38d37'
Smith,Harold David
Sct 18h43'33"-6d59'
Smith,Harold Edward
Aql 18h55'34"-0d41'
Smith,Harry
Ori 5h31'0"8d54'
Smith,Harry Allen
Uma 8h37'50"56d8'
Smith,Harry Lloyd
Aur 4h51'39"50d34'
Smith,Helen
Cnv 12h23'1"38d55'
Smith,Helen Claire
Uma 8h17'37"60d2'
Smith,Hideko
Dra 16h14'57"63d13'
Smith,Holly Alphonzo
Lac 22h54'23"55d3'
Smith,Holly L
And 23h22'1"47d15'
Smith,Howard
Cyg 21h20'44"40d41'
Smith,Ivan
Cam 3h14'56"63d53'
Smith,J Michael
Aql 18h46'0"11d38'
Smith,Jacob David
Ori 5h15'22"1d50'
Smith,Jacob Edward
Cep 23h9'11'64d35'
Smith,Jacob Lennart Cudole Bear
Ser 15h38'13"20d10'
Smith,Jacqueline
Cam 13h17'42"84d43'
Smith,Jake Tyler
Aqr 23h7'29"-10d11'
Smith,James Darwin
Cnv 12h35'0"38d58
Smith,James Douglas
Aur 6h18'21"37d56'
Smith,James Joseph
Aur 4h46'27"48d55'
Smith,James Lloyd
Cet 2h31'60"0d55'
Smith,James M & Mary F
Mon 7h12'15"-10d38'
Smith,James William
Aql 20h7'48"7d45'
Smith,Jan Berkes
Cyg 19h29'25"33d13'
Smith,Janet Sue AKA (Cindy)
Peg 0h9'33"18d15'
Smith,Janice
And 0h37'1"40d15'
Smith,Jasmine Ashley
Del 20h12'45"11d4'
Smith,Jason Keane
Uma 11h1'49"58d52'

Smith,Jeff & Linda
Aql 18h46'56"11d29'
Smith,Jeffrey J
Aur 5h32'21"38d39'
Smith,Jeffrey Jerome
Gem 7h55'0"20d16'
Smith,Jeffrey Ryan
Her 16h36'31"29d43'
Smith,Jeffrey Thomas
Cet 0h57'51"0d48'
Smith,Jenifer Ann
Peg 0h2'15"17d50'
Smith,Jennifer
Umi 14h59'15"68d24'
Smith,Jennifer Anne
Cmi 7h55'49"1d9'
Smith,Jennifer Lee
Cam 10h56'54"84d38'
Smith,Jeremy & Myrtle
Umi 16h18'46"73d58'
Smith,Jerry D
Aql 18h43'13"7d6'
Smith,Jerry Dan
Cet 0h38'0"1d9'
Smith,Jerry Lee
Her 16h50'12"39d5'
Smith,Jessica Elisabeth
Cas 0h54'14"56d13'
Smith,Jessica
And 23h13'12"37d38'
Smith,Jessica "Itz A Hit"
Mon 6h54'52"-0d23'
Smith,Jessica Barry
And 1h52'0"47d22'
Smith,Jill
Lyr 19h1'27"38d17'
Smith,Jill Farabee
Mon 6h22'27"7d46'
Smith,Jim
Cep 22h20'44"65d0'
Smith,Jim & Jana
Crb 16h8'21"31d50'
Smith,Jimmy
Dra 11h25'24"71d13'
Smith,Jimmy
Sgr 19h27'22"-44d51'
Smith,JM & NV
Dra 16h7'37"63d20'
Smith,Jo Lacy
Sge 19h57'47"20d26'
Smith,Joan
And 1h44'37"39d4'
Smith,JoAnne
Vir 13h55'1"6d15'
Smith,Joanne P
Uma 9h32'49"58d29'
Smith,Joelynn
Hya 9h14'40"0d22'
Smith,Joey Gwen
Boo 14h24'49"26d54'
Smith,John
Lac 22h36'1"55d37'
Smith,John & Nancy
Cap 21h21'40"-27d53'
Smith,John Derek
Boo 14h56'47"29d23'
Smith,John H
Per 4h27'0"50d7'
Smith,John Richard
Per 2h50'0"32d30'
Smith,Joseph Hughes
Mon 7h17'19"-5d38'
Smith,Joseph J
Cep 0h47'23"77d59'
Smith,Joseph John
Aur 6h56'23"43d54'
Smith,Joseph Michael
Uma 11h11'24"45d52'
Smith,Joshua Lee
Her 17h20'1"44d47'
Smith,Jr,"Jr",George D
Cet 2h35'11"-10d54'
Smith,Jr,Chester Taber
Dra 11h19'35"74d20'

Smith,Jr,Daniel Scott
Aql 19h59'58"10d11'
Smith,Jr,John Patrick
Dra 12h58'13"75d25'
Smith,Jr,Leland
Dra 15h53'36"60d50'
Smith,Jr,William R
Her 16h22'29"26d17'
Smith,Judy
Vul 21h20'46"24d2'
Smith,Judy Diane
Cas 0h59'41"64d9'
Smith,Julie K
Uma 9h39'1"58d42'
Smith,Justin
Ori 6h4'15"0d34'
Smith,Justin Richard
Eri 2h54'53"-15d39'
Smith,Justine Louise
Umi 15h3'28"66d49'
Smith,Kaitlin Annmarie
Aql 20h20'57"1d48'
Smith,Karen Stephanie
Tri 1h56'33"26d37'
Smith,Kate
And 23h43'32"43d20'
Smith,Kathleen
Umi 16h51'43"75d45'
Smith,Kathryn W
Peg 22h16'28"32d49'
Smith,Kathy
Mon 7h3'41"4d58'
Smith,Katie Lauren
Cam 4h4'25"78d57'
Smith,Keegan Gallagher
Dra 15h12'1"56d16'
Smith,Keith
Oph 18h6'26"12d18'
Smith,Keith Blair
Aql 20h10'22"10d56'
Smith,Kelli J
Cet 3h19'0"3d37'
Smith,Kellie
Cas 2h11'15"68d9'
Smith,Kelly
Lyr 18h14'59"45d60'
Smith,Kelsey Anne
Com 12h54'47"30d55'
Smith,Kenneth Gordon
Oph 17h7'35"-20d41'
Smith,Kenneth L
Aur 6h24'44"30d53'
Smith,Kevin
Ori 4h47'23"0d4'
Smith,Kevin Dawson
Her 16h14'33"7d33'
Smith,Kevin Thomas
Boo 13h53'1"18d10'
Smith,Kiesel
Vul 20h27'22"28d8'
Smith,Kim-Marie
Lyn 7h30'53"50d35'
Smith,Kirk Kelly
Lac 22h10'33"51d36'
Smith,Kit
Cyg 19h34'29"35d33'
Smith,Kraig L
Per 2h38'25"38d6'
Smith,Krista
Aqr 21h42'1"0d47'
Smith,Kristi
Cyg 20h21'44"40d13'
Smith,Kristian Lynn
And 0h19'56"31d13'
Smith,Kristin
Lyn 6h53'54"44d46'
Smith,Kristin Ashley
Lyn 7h34'14"45d24'
Smith,L'etoile de Dan
Aur 7h12'53"35d43'
Smith,Lane
Her 17h52'36"14d39'
Smith,Larry
Cep 21h57'42"84d33'

Smith,Laura Ann
Mon 7h39'0"-5d56'
Smith,Laura Gill
Eri 3h28'10"-2d9'
Smith,Lauren Elizabeth
Uma 11h34'0"43d52'
Smith,Laurie & Mary Witten
Peg 22h34'52"21d44'
Smith,Lawrence E
Lmi 10h1'1"36d4'
Smith,Lee W
Aur 6h16'59"31d15'
Smith,Leonard
Per 3h12'1"50d26'
Smith,Leonard Paul
Vul 19h35'33"27d8'
Smith,Leslie A
And 1h59'39"40d51'
Smith,Lester & Margaret
Cyg 19h17'52"45d14'
Smith,Linda
Cas 1h41'52"68d38'
Smith,Linda Ann
Cas 1h46'38"75d27'
Smith,Linda Meegan
Mon 8h3'45"-10d9'
Smith,Linda Vault
Peg 23h48'50"28d15'
Smith,Lindsey
Mon 7h45'22"-1d15'
Smith,Lisa
Cam 7h5'15"70d42'
Smith,Lisa Marie
Cas 0h17'52"63d56'
Smith,Lizzie
Leo 11h15'6"4d35'
Smith,Lowell & Rita
Eri 4h57'38"-5d1'
Smith,Lucas Emery
Aur 6h29'14"37d56'
Smith,Lucille McNaghten
Ori 5h23'12"0d55'
Smith,Luke
Boo 13h57'26"19d14'
Smith,Lyman Delke
Boo 15h5'7"9d24'A
Smith,Lynn
And 1h12'49"40d23'
Smith,Lynn Elizabeth
Aur 5h2'36"40d4'
Smith,Lynn Elizabeth
Cas 0h24'20"63d58'
Smith,M Veronica
Lyn 7h5'55"58d8'
Smith,Mackenzie Leigh
Mon 6h25'53"-0d13'
Smith,Madison Nicole
Cam 6h10'1"83d22'
Smith,Mandy Jane
Boo 14h15'32"50d30'
Smith,Margaret Louise
Aql 19h30'51"7d46'
Smith,Margaret Uhl
Eri 3h57'43"-2d19'
Smith,Mariah Allesandra Battiste
Cyg 19h43'50"29d38'
Smith,Marian "Marni"
Mon 7h38'55"-6d33'
Smith,Mario Bonduel
Boo 14h30'1"21d49'
Smith,Marissa Ann
And 1h58'39"38d21'
Smith,Marjorie & Jimmy
Sge 19h58'33"16d26'
Smith,Mark
Lyn 8h21'15"47d39'
Smith,Mark A
Per 3h47'25"38d50'
Smith,Marline T
Del 20h59'47"18d52'
Smith,Marlon D
Peg 22h32'38"25d44'
Smith,Martha & Eben
Cyg 21h14'0"39d18'

Smith,Marty
Mon 6h44'11"8d48'
Smith,Mary Anthony
Boo 14h30'60"8d37'
Smith,Mary
Aql 18h42'57"-2d56'
Smith,Mary
Cyg 20h25'22"40d14'
Smith,Mary Bernadette
Aql 19h11'36"10d59'
Smith,Mary Boyle
Cnc 8h27'0"10d11'
Smith,Mary Ellen & Robert Edward Smith
Ori 5h3'36"8d49'
Smith,Mary Eva
Cas 0h38'19"68d8'
Smith,Mary Margaret
And 2h8'42"39d46'
Smith,Mary Virginia
Mon 6h35'13"-0d13'
Smith,Matthew J
Cet 2h20'46"1d24'
Smith,Matthew J
Boo 15h7'25"20d24'
Smith,Matthew Jordan
Crb 15h54'57"27d32'
Smith,Matthew Philip
Umi 13h12'26"70d49'
Smith,Mavis
Sct 18h55'59"-7d56'
Smith,Maxwell Newton
Her 16h12'0"7d32'
Smith,MD,Donald S
Cep 5h13'50"85d53'A
Smith,MD,James P
Per 1h58'28"53d14'
Smith,Megan Elizabeth
Psc 0h59'53"32d6'
Smith,Melissa
Cet 0h36'0"1d49'
Smith,Melissa Orel Perry
Eri 3h51'11"-2d48'
Smith,Melody Ann
Cyg 19h52'51"40d36'
Smith,Michael
Her 17h27'46"37d37'
Smith,Michael
Lep 15h28'56"42d12'
Smith,Michael & Alex
Peg 23h0'38"17d32'
Smith,Michael Carson
Cep 22h39'14"65d1'
Smith,Michael David
Cmi 7h42'26"8d12'
Smith,Michael Lee
Per 3h40'5"38d34'
Smith,Michael Richard
Lac 22h16'52"46d60'
Smith,Michael W
Ari 2h1'41"25d14'
Smith,Michele Lynn
Lyn 8h14'1"36d52'
Smith,Michelle Margaret
Mon 7h32'48"-6d58'
Smith,Michelle & Kyle
Crt 11h40'49"-11d18'
Smith,Michelle R
Peg 21h55'0"24d18'
Smith,Mike J
Per 3h14'33"42d1'
Smith,Mildred
Cas 1h32'49"60d17'
Smith,Mitchell Kelly
Ser 18h3'0"-14d23'
Smith,Mitchell P
And 0h59'44"41d3'
Smith,Mom & Don Dad
Boo 15h10'41"48d31'
Smith,Morgan Chandler
Umi 15h35'52"78d27'
Smith,Morgan Elizabeth
Ori 5h39'39"15d9'
Smith,Mr Mike
Lac 22h19'12"50d2'

Smith,Mugsy Delaney
Cma 7h21'11"-18d54'
Smith,My Love,Michelle
Mon 6h20'16"7d38'
Smith,Myra
Eri 2h53'42"-5d51'
Smith,Myrtle Viola (Edwards)
Cas 0h46'35"69d22'
Smith,Nancy
Cas 0h34'37"62d5'
Smith,Nancy
Lyr 18h19'16"40d39'
Smith,Nancy F
Peg 21h39'53"26d57'
Smith,Nancy M
And 1h31'12"37d54'
Smith,Nathan
Cyg 21h17'24"35d26'
Smith,Nathan Daniel
Uma 11h24'41"30d38"
Smith,Nathaniel Stephen Devereux
Uma 9h4'21"50d7'
Smith,Naujokaitus, Joseph E
Her 16h25'1"18d54'
Smith,Nicholle Regina
Cma 7h0'40"-28d12'
Smith,Nicole Bethany
Cas 3h11'34"58d38'
Smith,Nikki Yvonne
Lyr 18h30'48"42d59'
Smith,Nikole Lee
Cyg 19h55'47"37d41'
Smith,Pam A
Lyn 8h20'33"50d2'
Smith,Pamela
Sgr 19h0'41"-29d1'
Smith,Pamela Faye
Peg 21h41'46"24d26'
Smith,Pamela Lynn
Mon 6h29'38"-10d57'
Smith,Pat
Cas 0h2'53"62d50'
Smith,Patricia B
And 23h47'45"47d47'
Smith,Patricia C
Cet 0h30'16"-3d59'
Smith,Patricia Danette
Uma 10h44'17"54d13'
Smith,Patricia Gail
Mon 6h36'22"6d21'
Smith,Patrick J
Aur 6h5'48"35d50'
Smith,Patrina
Umi 13h51'10"70d27'
Smith,Patti
Peg 22h17'0"33d44'
Smith,Patty Ann
Mon 6h54'34"1d19'
Smith,Paul Edward
Lac 22h15'31"49d47'
Smith,Pauline
Cyg 20h22'15"39d5'
Smith,Pauline
And 0h23'0"37d27'
Smith,Pearl Irene Anderson
Cas 0h30'27"76d3'
Smith,Peggy Brzeszkiewicz
Aur 6h3'50"31d10'
Smith,Perry
Sct 10h20'17"-4d4'
Smith,Peter
Ori 5h55'23"17d47'
Smith,Peter Charles
Lac 22h32'15"50d34'
Smith,Peter H
Sgr 19h41'33"-43d18'
Smith,Peter J G
Cep 21h27'38"61d23'
Smith,Phil
Tri 1h50'52"26d33'
Smith,Phillip Montaque
Hya 9h13'47"0d16'
Smith,Phyllis
Cyg 21h22'39"28d48'

Smith,Princess- Virginia Mae
And 23h34'44"40d23'
Smith,Quinton
Boo 15h11'39"28d33'
Smith,Rachael Anne
And 2h5'43"38d41'
Smith,Rachel
Cyg 21h37'54"42d12'
Smith,Rachel Ann
Lyr 19h21'1"31d52'
Smith,Rachel Katherine
Lyn 8h9'1"45d5'
Smith,RaeLee
Uma 12h46'40"54d32'
Smith,Randal Stephanie
Lyr 19h23'15"35d14'
Smith,Randy Boone
Vir 13h36'48"-7d43'
Smith,Randy Lee
Aql 16h26'54"-6d47'
Smith,Ray
Cmi 7h29'22"7d39'
Smith,Ray
Boo 15h26'10"38d43'
Smith,Raymond James
Cet 1h20'11"-1d37'
Smith,Rebecca Ann
Cas 2h48'0"70d12'
Smith,Rebecca Holland
Peg 23h2'17"31d49'
Smith,Reed V
Aql 18h57'25"4d1'
Smith,Rex
Hya 8h44'1"2d21'
Smith,Richalle Strong
Aql 18h58'32"8d9'
Smith,Richard
Umi 16h27'37"70d28'
Smith,Richard & Lori
Peg 22h40'0"27d47'
Smith,Richard Arthur
Dra 15h37'37"54d42'
Smith,Richard Earl
Lac 22h34'0"53d20'
Smith,Richard James
Umi 14h44'0"68d50'
Smith,Richard James
Umi 15h29'59"68d25'
Smith,Richard L
Aql 19h30'42"10d5'
Smith,Richard N
Cyg 19h27'13"34d37'
Smith,Richard Stall
Her 17h30'35"14d11'
Smith,Rick
Lac 22h36'12"38d25'
Smith,Rita J
Ori 5h52'38"8d20'
Smith,Robert "Bobby"
Lac 22h5'42"48d58'
Smith,Robert & Kate Goodwin
Peg 23h8'21"8d49'
Smith,Robert Andrew
Dra 16h29'45"64d5'
Smith,Robert Bandit
Her 17h58'1"14d20'
Smith,Robert C
Sex 10h11'0"-6d48'
Smith,Robert Dow
Eri 2h46'34"-4d1'
Smith,Robert E
Umi 13h12'37"71d37'
Smith,Robert E
Aql 19h9'58"3d8'
Smith,Robert William
Ori 6h14'50"0d46'
Smith,Roberta Kassig
Lyr 18h58'13"30d40'
Smith,Robin Michele
Ori 5h53'31"13d3'
Smith,Robyn
Uma 10h43'1"52d1'
Smith,Robyn Stoor
Cet 2h41'55"1d31'
Smith,Roger & Malgorzata
Lyr 18h48'1"30d32'
Smith,Roger F.
Cnv 12h54'46"41d5'
Smith,Roger M
Aql 19h57'57"1d48'
Smith,Rolland
Aur 5h2'0"48d49'
Smith,Ron
Hya 8h39'0"-11d10'
Smith,Rose
Cyg 21h4'0"38d11'
Smith,Ross H
Boo 14h14'39"16d46'
Smith,Rubye Marshall
Eri 4h5'58"-17d8'
Smith,Rudolph "Rhue"
Eri 2h56'42"-5d40'
Smith,Rupert W (Spike)
Aur 6h34'24"38d10'
Smith,Russell
Cep 22h10'32"62d47'
Smith,Ruth
Vir 13h10'19"-8d32'
Smith,Ruth & Jim
Cyg 19h28'53"31d50'
Smith,Ruth Loraine
Peg 23h3017"23d16'
Smith,Ryan Charles
Sex 10h29'29"-0d53'
Smith,Ryan Charles
Ori 5h58'32"16d13'
Smith,Ryan Mark
Cep 22h11'42"62d13'
Smith,Sadie May
Cas 1h46'1"61d27'
Smith,Sage
Mon 6h24'19"-0d17'
Smith,Sally Gordon
And 0h57'12"38d18'
Smith,Sally Jo
Com 12h39'42"21d35'
Smith,Samuel "Sam"
Lib 15h40'1"-28d55'
Smith,Samuel Drake Wise
Mon 8h0'0"-8d20'
Smith,Sandra Jean
Aur 4h53'35"41d1'
Smith,Sandra Jean
Cet 3h17'53"9d49'
Smith,Sandra Steele
Crt 11h16'5"-17d8'
Smith,Sandy
Lyr 18h17'59"31d52'
Smith,Sanford Daniel
Sex 10h32'27"5d51'
Smith,Sang-Sue/Susan
Ser 16h1'17"11d34'A
Smith,Sarah
Cas 0h52'1"68d49'
Smith,Sarah Kaley
Cas 0h26'1"61d47'
Smith,Sarah Laine
And 0h23'24"30d45'
Smith,Scott
Lac 22h25'33"54d3'
Smith,Scott A
Cam 3h55'44"67d38'
Smith,Scotty
Leo 9h54'18"10d58'
Smith,Seaborn Jackson Dean
And 2h24'53"41d58'
Smith,Sean
Cam 4h0'20"70d53'
Smith,Sean Thomas
Per 2h39'39"34d22'
Smith,Sempri
Uma 9h49'39"45d15'
Smith,Shane
Sex 9h54'12"-0d28'
Smith,Shane Matheson
Lmi 10h3'25"30d34'
Smith,Shannon Michelle
Hya 10h12'43"-17d9'
Smith,Shara Lee
Cas 0h32'33"63d35'
Smith,Sharon Dunlap
Lyr 18h28'51"30d30'
Smith,Shawn Altima
Lyn 8h18'1"58d25'
Smith,Shawnessey Lynn
And 2h28'20"44d34'
Smith,Sheila Ann
Ori 6h13'15"0d41'
Smith,Shi
Sge 19h28'10"16d22'
Smith,Shirley Ann
Peg 21h57'51"33d18'
Smith,Skye A
Tau 4h2'40"18d58'
Smith,Sr,Charles Michael
Cep 1h1'1"78d52'
Smith,Stacey Ann
Lyn 7h33'0"50d57"
Smith,Stacy Alan
Hya 8h25'1"-1d45'
Smith,Stanley
Peg 22h16'11"7d36' B
Smith,Star Rainbow
Her 16h52'57"13d41'B
Smith,Stephen Laura Doctor Tiberius
Aqr 22h14'43"-9d51'
Smith,Stephen,Taylor, & Molly
Oph 17h17'12"-23d57'
Smith,Steve & Jean
Her 18h8'1"38d44'
Smith,Steven Anthony Hudder
Mon 6h19'1"5d44'
Smith,Steven James
Aur 5h21'45"38d0'
Smith,Steven Scott
Cnv 13h48'54"47d52'
Smith,Stuart A
Cnv 12h18'51"51d1'
Smith,Stuart Allan
Lac 22h17'27"54d39'
Smith,Stuarts "Rummy & Dummy" Bruce
Her 17h22'48"2d14'
Smith,Sue & Lou
Peg 21h51'14"31d12'
Smith,Sunny
Equ 20h57'23"7d55'
Smith,Susan
And 22h57'10"36d7'
Smith,Susan Ann
Lyr 18h14'0"38d22'
Smith,Susan Claire
And 23h4'42"37d7'
Smith,Susan Elizabeth
Cas 0h3'0"61d28'
Smith,Susan Jane
Cas 1h2'41"71d13'
Smith,Susan Leigh
And 0h10'20"33d11'
Smith,Suzanne
Peg 21h10'28"18d47'
Smith,Suzie
Peg 23h43'1"31d18'
Smith,T D
Aur 4h54'31"40d2'
Smith,Tammy
Her 17h12'44"47d3'
Smith,Taylor Dee
Lyn 7h23'53"44d49'
Smith,Teresa
Oph 18h3'41"11d52'
Smith,Terri A
Crt 11h34'45"-8d56'
Smith,Terri Burdo
Aur 6h58'38"40d32'
Smith,Terri J
Cyg 19h56'28"38d12'
Smith,Terry
Uma 12h43'57"53d6'
Smith,Terry
Cyg 19h32'25"39d15'
Smith,The Forever Star Of James R
Ari 2h58'19"20d40'
Smith,The Jerry & Jennine
Cyg 21h31'35"40d40'
Smith,The Marilyn & Frank
Aql 19h30'42"7d51'
Smith,The Pal & Johnell
Vul 19h52'32"29d25'
Smith,The Rev Canon Rush WD
Cep 1h6'24"78d9'
Smith,Thomas
Lac 22h13'58"48d1'
Smith,Thomas & Tara
Sex 9h59'31"0d3'
Smith,Thomas A
Cma 7h11'44"-13d38'
Smith,Thomas E
Aur 6h34'26"30d51'
Smith,Thomas F
Crb 15h51'54"28d52'
Smith,Thomas R
Gem 6h42'46"31d30'
Smith,Timothy Victor
Cam 6h39'12"68d40'
Smith,Titian Glory, Emma Hamilton Denver
Ori 4h45'50"0d47'
Smith,Todd
Cma 7h3'29"-18d46'
Smith,Todd Hughston
Her 16h42'19"4d33'
Smith,Tommy E
Uma 10h40'0"52d8'
Smith,Tony
Dra 15h42'34"60d55'
Smith,Tony & Patsy
Cyg 19h44'23"29d22'
Smith,Tony Alonzo
Ser 15h21'30"22d36'
Smith,Tracie
Cas 2h11'13"60d4'
Smith,Travis W
Aql 19h9'48"0d59'
Smith,Trip
Cma 6h51'8"-18d36'
Smith,Trish
Tri 1h30'48"28d47'
Smith,Tyson C
Aur 5h2'1"51d29'
Smith,Valarie Kae
Mon 7h1'58"1d41'
Smith,Valerie
And 23h37'1"46d36'
Smith,Valerie Chan
Lac 22h29'59"48d53'B
Smith,Valinda Lee
And 0h13'22"47d24'
Smith,Vernon & Denise
Cam 12h33'0"80d33'
Smith,Virginia Georgette
Cyg 21h7'1"30d55'
Smith,Virginia
Sex 10h37'47"3d21'
Smith,Virginia "Ninny"
Ori 6h5'39"-1d59'
Smith,Von "AAbsolutely"
Hya 9h37'36"-10d36'
Smith,W Douglas
Ori 5h58'21"17d44'
Smith,Walter Lawrence
Cnv 13h30'15"48d12'
Smith,Wanda Lucille
Cam 7h4'8"65d16'A
Smith,Ward Alan
Oph 18h5'55"13d30'
Smith,Wayne R
Oph 17h9'43"10d27'
Smith,Wendall H
Dra 16h37'12"67d44'
Smith,Will
Cet 2h27'31"5d58'
Smith,William
Ori 4h50'51"0d6'
Smith,William & Shannan
Cyg 20h36'43"58d55'
Smith,William (Billy)
Dra 17h39'40"68d49'
Smith,William A
Cep 23h33'24"67d42'
Smith,William E
Ori 5h23'33"-2d59
Smith,William Eugene
Cet 0h26'0"-06d33'
Smith,William Felton
Her 17h37'11"37d39'
Smith-A True Star, Grandma Barbara
Mon 6h59'52"8d47'
Smith-Boulac,Donna Mae
Lyr 18h30'27"38d42'
Smith-Fagan,Eamon James
Vir 11h45'15"2d47'
Smith-Grimm,Ashley Lynne
Com 12h59'22"27d8'
Smith-Howell
Aql 20h20'17"7d30'
Smith-Kandoll, Jacquelin
Cyg 21h4'46"30d19'
Smith-Sipp,Diane Marie
And 23h18'0"48d59'
Smith-Star of Light, Ruth Hodges
Tri 2h39'21"31d21'
Smith-Stone
Aur 5h15'17"50d3'
Smith-Strover,The
Ind 20h49'20"-53d3'
Smith-Winnes
Aur 4h55'0"40d33'
Smithen,Kevin
Aur 6h2'1"41d10'
Smithen,Lindsay
And 1h9'0"40d5'
Smitheringale,Jordan Jazz
Peg 22h32'55"28d44'
Smitherman,Jake Alexander
Aql 18h58'1"-5d40'
Smitherman,Michael
Ori 6h14'10"0d2'
Smithers,William
Uma 12h9'19"46d45'
Smithey,Jerald W
Boo 14h59'57"52d49'
Smiths Treasure,Mark
Uma 11h47'30"32d21'
Smiths:Godson Steve-Debbie-Sten & Sky,The
Uma 10h49'35"47d28'
Smithson,Eddie
Her 18h5'0"31d3'
Smithwick,R Michael
Ori 5h30'35"-1d38'
Smithwick,Sally
Sct 18h53'32"-5d44'
Smithy
Uma 9h41'26"46d40'
Smithy,D J
Ori 5h49'35"18d3'
Smitley,Kimberly A
Hya 9h35'21"-11d9'
Smitley,Roselyn
Ori 5h33'26"-6d32'
Smits,Hans
Aur 7h1'1"38d31'
Smitty
Aur 5h9'35"44d49'
Smitty,Gramma & Grampa
Eri 3h23'33"-2d12'
SMMS-Sixth Grade Class Of 1994
Uma 11h12'0"33d15'
Smoak,Amanda Leigh
And 0h8'12"47d21'
Smoak,Rebecca Ann
Lyn 7h6'12"50d41'
Smock,Christopher Louis
Lac 22h12'32"47d10'
Smoke "Jell Bell-Kezar to Knox" Ily
Cyg 21h40'44"48d11'
Smoke,Lori
Lyr 19h15'45"41d52'
Smoke,Suzanne
Aql 18h47'0"10d60'
Smoke,Suzanne Foster
And 1h59'30"38d16'
Smoker,JoAnn Hern
Cen 13h33'40"-52d23'
Smokey
Per 2h59'59"50d6'
Smoky
Cam 6h58'1"70d30'
Smolen 2-14-95,Mi Amore Biscotti T W
Her 18h40'11"12d11'
Smolenske,Jack Ross
Ori 5h17'60"12d30'
Smolenski,Ben
Her 17h6'52"43d40'
Smoleroff,Mitchell
Aqr 21h0'17"1d10'A
Smoleroff,Molly
Aqr 21h0'17"1d10'B
Smolij,Aleksander William
Cnv 13h57'1"38d28'
Smolik,Marjean K
Ari 1h57'31"20d15'
Smolinski
Aur 5h14'0"44d27'
Smolka,Sydney
Del 20h20'52"20d27'
Smolkowski,HeavenLee
Psc 1h24'1"11d27'
Smollar,Jason
Her 18h15'51"14d56'
Smoller,Jeane Phillips
Cam 10h10'52"82d17'
Smoller,William
Her 17h32'23"40d34'
SMONK
Lyr 18h49'32"33d40'
Smoocher
Cyg 19h34'0"37d45'
Smooches
Sge 19h58'53"16d13'
Smoogie's Lover
Hya 8h53'23"-6d6'
SMOOT
Lib 15h0'43"-11d8'
Smoot,Benjamin Neeley
Cep 21h1'42"85d31'
Smoot,Brett
Cep 2h4'56"77d11'
Smoot,Diane Marie
And 2h12'0"42d8'
Smoot,Howard
Aql 19h51'44"11d30'
Smoot,John Michael
Her 17h32'46"40d15'
Smoron,Joan & Geoffrey
Crb 16h19'15"31d59'
Smothers,Dennis & Laura Lee Perdue
Sct 18h44'1"-6d14'
Smuels,Grand Master Jim
Her 18h4'40"45d29'
Smuksta,BS/Bob
Cam 4h49'36"70d24'
Smullin,Bernard
Sex 10h4'26"-0d55'
Smurf
Eri 3h56'32"-10d3'
Smutko,Frank
Cmi 7h23'31"1d21'
Smutny,Sr,George James
Per 2h57'56"46d44'
SMV 11793
Uma 11h47'1"31d50'
SMW I-IV-III
Sct 18h43'38"-7d20'
Smyer,Lee & Leisha
Lyr 19h6'0"25d35'
Smyke,Riley Patrick
Cep 22h15'24"61d14'
Smyrni,Sean Patrick
Ori 6h0'37"5d16'
Smyrnios,Ron Leroy & Arleigh
Boo 13h53'1"20d17'
Smyth, G27
Uma 9h37'15"48d49'
Smyth,Bill
Sco 17h5'0"-31d15'
Smyth,Lisabeth
Mon 6h42'0"7d36'
Smyth,Sara-Louise
Cas 1h54'36"73d9'
Smyth,Steven Edward
Cap 21h0'52"-15d58'
Smythe,Byron Meredith
Her 17h32'52"28d27'
Smythe,Jeannene K
And 0h26'52"31d46'
SmUgesjo,Sondre Romsaas
Cam 3h58'48"56d17'
Snable,Judi
Peg 22h53'0"25d35'
Snadles,Terence James
Her 16h51'0"47d48'
Snagglepus
Hya 10h13'36"-12d22'
Snake
Cet 3h14'29"9d3'
Snake et Babete
Uma 10h33'16"50d11'
Snake,Lesley Dianne
Lac 22h54'42"38d4'
Snapp,"The Snapper", Joe
Her 15h50'0"40d46'
Snapp,Albert B & Pauline R
Cyg 20h57'14"31d2'
Snapperhead
Her 17h19'38"14d43'
Snarks
Her 17h59'21"28d27'
Snarr's Starr
Del 20h25'56"11d12'
Snarski,Dolores L
Cyg 19h55'44"50d10'
Snax
Lac 22h54'53"50d36'
Snead,Carol Stephane
Vul 19h50'18"20d36'
Snead,Macy Lee
Cas 1h3'32"63d53'
Snead,Sarah Elizabeth
Lyn 7h53'38"43d59'
Snee,Patrick Allen
Aql 18h40'1"-2d58'
Sneed,Alexandria Nicole
Mon 8h6'49"-10d2'
Sneed,Barbara Gayle
Eri 4h30'53"-11d40'
Sneed,Sybil
Mon 7h48'54"-3d38'
Sneen,Edward Francis
Tau 3h59'44"30d12'
Snett,Eugene Matthew
Cet 3h0'50"7d7'
Sneider,Cailin Elizabeth
Lyn 7h53'56"58d15'
Snelgrove,James Thomas
Boo 14h14'42"16d34'
Snelgrove,Peter Keith
Lyn 9h12'51"39d51'
Snell,"Katt" (Kathie)
Cas 0h25'41"70d39'
Snell,Carol Yvonne
Cma 6h50'53"-17d33'
Snell,John
Per 3h44'17"50d23'
Snell,Loyal & Doris
Cyg 20h56'28"38d8'
Snell,Nellie Doodson FAGO
Tri 1h50'36"28d14'
Snell,Ricky
Cnv 12h49'59"47d8'
Snellen,Anne Marie
Lmi 10h13'27"30d41'
Snellen,Jerry Dee
Per 4h0'50"51d34'
Snellen,Josette Lee
Ori 5h5'51"12d57'
Snellgrove,Melody Marie
Cas 2h8'13"59d9'
Snelling,David E
Cep 20h55'1"67d55'
Snethen,Shawna Nichole
Peg 23h43'15"26d17'
Snickers-TK1,The
Cet 2h31'41"6d34'
Snider Star,The
Mon 6h25'28"10d7'
Snider,Barbie
Peg 21h44'0"23d21'
Snider,Cindy,Whitney, Jefrey & Jim
Uma 11h25'59"47d34'
Snider,Cole Ryan
Peg 22h2'39"8d25'
Snider,Erin
Uma 8h56'23"56d1'
Snider,J W & Geraldine
Sge 19h53'55"19d22'
Snider,Kerrick
Aql 19h52'34"10d55'
Snider,Zachary Brian
Ori 6h7'13"0d29'
Snider-Veaudoin
Her 17h21'0"44d39'
Sniffen,Katherine
Del 20h16'60"10d49'
Snigglefritz
Lyn 7h31'0"38d44'
Snijders,Franciscus
Eri 3h25'25"-6d9'
Snipstar
Lyr 18h36'51"28d13'
Snitkin,King Richard
Cma 7h0'48"-18d58'
Snitko,Big Tim
Lmi 10h56'24"27d5'
Snitzer,Brian J
Cet 1h49'0"-3d58'
Snitzer,Carolyn & Patricia
Oph 17h55'46"10d25'
Snitzler,Daniel B
Equ 21h19'35"12d39'
Snively,Carolyn Cay
Tau 4h5'9"20d47'
Snively,Molly Carlile
Lmi 10h21'0"31d30'
Snix
Lyn 6h33'55"54d26'
Snoddy,Dick
Her 16h25'32"48d49'
Snoddy,Kathleen Wood
Eri 3h30'39"-6d32'
Snodgrass,Anthony
Sex 10h36'0"0d49'
Snodgrass,Christopher Michael
Gem 6h45'31"12d21'
Snodgrass,Conrad Tillman
Aur 6h10'37"38d45'
Snodgrass,Dawn
Umi 15h5'26"67d16'
Snook,Fiona
Cmi 7h23'20"7d33'
Snook,Robert Gillen
Dra 18h9'16"58d40'
Snookie
Aur 5h23'37"37d40'
Snookie
Lyr 18h55'39"42d16'
Snookie
Cet 0h38'49"1d32'
Snookie's Diamond
Lyr 18h42'39"33d11'
Snookies Love
Sex 10h32'16"5d15'
Snooks
Uma 11h47'50"46d31'
Snooksie
Gru 22h41'53"56d8'
Snooth
Lac 22h20'19"53d6'
Snoppy
Dra 14h21'37"64d8'
Snorey Bear
Umi 16h8'31"80d10'
Snortfogger-Thudpucker
Aql 20h4'36"4d44'
Snot-Bad King
Lac 22h4'29"51d39'
Snow
Lib 14h25'54"-10d57'
Snow
Boo 14h5'57"23d38'
Snow Family Star,The
Tri 2h36'20"35d31'
Snow Star,The Bill & Zelma
Equ 21h2'50"3d31'
Snow White
Mon 6h42'41"10d55'
Snow White & Rudy
Boo 15h15'58"53d42'
Snow,Billy
Aur 5h6'48"44d55'
Snow,Charlie
Per 3h1'38"40d20'
Snow,Chelsea
Ori 6h6'53"20d20'
Snow,Crissie
Lyr 18h37'1"27d9'
Snow,Debra Pearl
And 23h46'23"42d9'
Snow,Denise
Uma 10h45'0"58d47'
Snow,James Lancaster
Cam 12h14'30"82d26'
Snow,Jhara
Ari 1h55'1"18d56'
Snow,Jim & Stephanie
Aur 4h52'56"38d3'
Snow,Joyice Annette
Peg 22h53'42"21d34'
Snow,Margaret
Lyr 18h35'60"40d52'
Snow,Maxwell
Lac 22h4'32"46d33'
Snow,Michael C
Oph 17h23'4"-20d23'
Snow,Sr,Alfred Lawrence
Aur 5h4'29"40d54'
Snow-Merritt,Leslie
Cas 1h1'37"58d12'
Snowball Star
Cnv 12h43'1"38d12'
Snowden,Christopher
Her 16h25'18"38d29'
Snowden,Jennifer Marie
And 1h19'49"33d35'
Snowden,Kara Marie
Cep 22h22'40"61d46'
Snowden,Paul Thomas
Ser 15h10'25"22d30'
Snowden,Stephanie
And 23h33'51"47d7'
Snowflake Orion (Elliot)
Ori 5h17'21"15d33'
Snowflame Diana (Beauty)
Ori 5h13'29"15d16'
Snowiss,Jeffrey
Ser 16h2'34"13d59'
Snufalupagus
Vul 19h22'26"25d30'
Snuff
Cnc 8h49'35"31d11'
Snuff
Cma 7h19'31"-15d8'
Snuffy
Lac 22h29'56"52d56'
Snuggle Bunny
Lac 22h53'0"53d19'

Snuggle Bunny Star,The
Uma 11h15'28"30d37'
Snugglebunnies
Vul 21h3'51"20d13'
Snugglepuss
Uma 11h36'29"42d36'
Snuggles
Eri 2h51'37"-5d31'
Snuggles
Uma 10h30'1"50d15'
Snuggles
Lyn 7h6'53"58d34'
Snuggly-Bunny,The
Lmi 10h3'40"30d28'
Snupkowski,Ellen J
Vul 21h0'51"27d23'
Snyder
Aur 5h30'40"38d30'
Snyder II,T J
Cep 21h48'41"58d59'
Snyder III,Victoria Z & Richard H
Crb 15h14'49"31d32'
Snyder IV,John J
Her 18h6'53"30d27'
Snyder,"Fred"
Tri 2h33'22"31d19'
Snyder,Alice McLoughlin
Peg 21h49'23"34d4'
Snyder,Amy Jennifer
Cet 1h0'35"-0d14'
Snyder,Anthony
Her 17h50'0"40d14'
Snyder,Charles
Per 3h44'41"51d57'
Snyder,Charles R
Lac 22h17'36"49d11'
Snyder,Cindy Ann
And 23h46'54"46d41'
Snyder,Corey Jean
Cas 0h59'49"66d7'
Snyder,Curvin
Mon 8h3'36"-6d30'
Snyder,Dave
Lac 22h13'55"47d19'
Snyder,David & Joanna
Eri 2h55'36"-14d5'
Snyder,Dawn P
And 1h42'11"39d17'
Snyder,Dr Steven B
Cep 22h29'36"65d18'
Snyder,Eric Jordan
Vul 19h42'54"22d43'
Snyder,Eugene A
Uma 11h12'29"62d10'
Snyder,George Morgan
Her 17h21'0"40d33'
Snyder,Georgine
Lyr 18h34'16"44d58'
Snyder,Hank
Aql 20h5'20"4d33'
Snyder,Harold Ray
Oph 17h59'51"8d11'
Snyder,Harry & Audrey
Sge 18h56'56"20d25'
Snyder,Howard
Aur 6h24'12"31d26'
Snyder,Jamie Lynn
Vir 11h43'20"2d15'
Snyder,Jeff & Josh
Cnv 12h28'12"32d42'
Snyder,Jeffrey Green
Cet 2h29'47"0d6'
Snyder,Joan
Lyn 7h52'34"34d45'
Snyder,John
Cep 0h3'38"66d28'
Snyder,John Alexander
Hya 8h34'52"-10d22'
Snyder,Joseph C
Lyn 7h4'0"58d32'
Snyder,Joy Renee
Mon 6h37'0"7d60'
Snyder,Jr,Arthur
Aur 5h2'26"38d39'
Snyder,Jr,David E
Cmi 7h16'0"4d29'
Snyder,Julia Ann
Lyn 7h29'0"58d33'
Snyder,Julie M
Aql 20h11'32"8d23'
Snyder,Katherine Ella
And 0h11'53"39d4'
Snyder,Ken
Ori 5h57'1"16d29'
Snyder,Ken & Mary
Uma 8h52'55"53d12'
Snyder,Laityn McKenna
Lyn 8h41'28"33d26'
Snyder,Larry D
Ori 5h5'53"10d6'
Snyder,Laura
And 1h6'29"40d29'
Snyder,Linda R
Lyr 19h2'39"38d35'
Snyder,Madison Leigh
Peg 22h41'18"29d45'
Snyder,Marge
And 23h46'44"47d57'
Snyder,Marilyn
Eri 4h18'12"-14d8'
Snyder,Marla
And 23h31'38"38d34'
Snyder,Martha
Cas 0h29'20"70d48'
Snyder,Mary Susan
Cas 0h43'39"64d7'
Snyder,Megan Brooke
Mon 7h44'57"-0d8'
Snyder,Michael Shane
Her 18h12'20"38d27'
Snyder,Mitchell Anthony
Lib 15h19'0"-23d49'
Snyder,Nicholas George
Ori 5h47'18"20d56'
Snyder,Nicole Marie
And 23h37'14"47d21'
Snyder,Nikki
And 23h32'49"38d53'
Snyder,Olivia Nichole
Mon 7h23'55"-0d19'
Snyder,Pinion 794 Chris
Uma 9h16'55"55d39'
Snyder,Reta
And 0h45'23"30d25'
Snyder,Robert "Dad"
Boo 15h40'46"41d3'
Snyder,Robert S & Albert L Patterson
Mon 6h39'0"11d25'
Snyder,Ronald
Her 17h55'49"40d29'
Snyder,Sara Ann
And 1h56'32"38d32'
Snyder,Sarah Mary
Peg 21h39'0"22d48'
Snyder,Shawn Michael
Her 17h1'36"44d21'
Snyder,Shirley J
Ori 5h19'15"11d12'
Snyder,Suzanne Marie
Com 13h8'19"20d45'
Snyder Thomas A
Boo 15h3'33"11d57'
Snyder,Thomas E
Oph 17h0'36"10d4'
Snydstrup,M Katherine Nelson
Aql 20h1'19"14d46'
So,Nitsaa
Aql 19h59'51"-5d57'
Soane,Joanne Mylrae
Aql 19h31'42"-6d59'
Soanes,Helen
Per 4h33'46"31d7'
Soapes,Christy
Aql 19h56'29"13d58'
Soares,Annie
Eri 2h58'40"-10d11'
Soares,Diane Hanley
Cas 1h40'0"58d14'
Soares,Sophie
Eri 3h58'17"-10d12'
Soares,Tony
Cnv 13h51'42"30d58'
Soave
Del 20h17'39"13d5'
Sobb,Camilla
Ind 20h49'32"-51d54'
Sobczak,Dorothy
Cas 1h2'11"60d11'
Sobczak,Peter
Lyn 8h2'18"42d46'
Sobczyk,Halina Helena
Peg 23h20'13"30d37'
Sobeck,Anja & Stefan
Leo 9h19'1"7d7'
Sobeit,Pukka Sahib
Cam 3h59'40"55d36'
Sobek,Drew
Cep 20h47'19"68d5'
Sobek,Stuart
Mon 6h55'18"-6d27'
Sobel,Aaron
Her 17h28'0"28d52'
Sobel,Babs
Cet 2h29'57"4d8'
Sobel,Ester
Cas 23h36'40"60d17'
Sobel,Garry Todd
Tri 1h53'40"28d57'
Sobel,Karen Ann
Tau 4h20'24"28d18'
Sobel,Lloyd
Lmi 10h10'51"38d15'
Sobel,Robin Goldstein
Lmi 10h1'22"38d49'
Sobey,Terry
Cma 6h56'38"-18d55'
SOBHY 'The Star of Shoots'
Per 3h3'29"41d51'
Sobieralski,Mary Clare
Sgr 19h34'28"-45d17'
Sobkowski,Thomas J
Dra 15h2'0"62d28'
Sobo
Cep 21h27'16"60d5'
Sobocinski,Jessica
And 0h14'0"32d57'
Soboleva,Olga Vladimirovna
Cas 0h35'1"58d40'
Sobolewski,Jolene
Cas 1h1'12"63d57'
Sobolowski,Tina Marie
Vul 19h44'58"28d31'
Sobon,Eugene
Aur 6h42'0"37d60'
Sobor,Candy
Cyg 20h33'1"39d30'
Sobor,Jennifer
Vul 21h16'25"20d25'
Sobotka,Aaron Paul
Ori 4h49'56"5d43'
Sobotta-Rostek,Ute
Sco 17h27'41"-31d8'
Soccorso,Nicole Lillian
Cas 0h1'22"60d36'
Sochaczewski,Peter
Uma 11h55'17"43d4'
Sochol,Ryan Daniel
Per 2h30'34"57d37'
Societe Rem
Cet 2h34'0"3d33'
Sockol,Laura Elizabeth
Sgr 19h4'47"-21d19'
Socol,Leonard
Lac 21h58'29"41d47'
Socorro,Maria
Lyn 7h59'1"35d27'
Socwell,Haley Marie
Del 20h52'59"4d52'
Soda
Boo 13h34'47"20d0'
Sodalis
Mon 7h6'55"-0d27'
sodalis+protector des Horst Modzel
Cap 21h28'0"-23d14'
Sodano,Guy
Ser 18h1'27"-14d50'
Sodbuster
Vul 19h20'27"26d27'
Sodd,Cecil & Minnie
Lyn 6h15'52"60d17'
Soden
Peg 21h57'1"2d26'
Soden,Christopher Adam
Boo 15h7'47"11d5'
Soderbeck,Michael J
Cyg 21h34'35"41d43'
Soderberg,Roberta Ann
Lyr 18h38'20"45d14'
Soderlund,Raymond
Her 16h41'43"50d52'
Soderman,Christine
Cas 0h12'58"66d5'
Soderman,Kenneth J
Ori 05h50'0"14d28'
Soderstrom,Christina Kathleen
Cas 1h40'57"76d10'
Sodikoff,Spencer
Boo 14h58'31"40d32'
Sodini,Delphine
Vul 19h3'20"22d17'
Sodini,Fr Pierre G
Ori 5h27'54"0d39'
Sodora's Crystal Cavern Star
Mon 7h55'49"-2d39'
Soeder,Lulu
Uma 9h43'57"43d19'
Soehner,Leigh Ann Shannon
Cyg 20h18'17"38d2'
Soekarno,Pudjadi
Lyr 18h38'11"33d15'
Soel,Otto
Hya 8h58'42"3d60'
Soesi Abhelakh
Cas 2h54'16"58d30'
Soester,Karen
Mon 6h25'48"-0d38'
Soeth,Cornelia
Del 20h23'14"18d56'
Soeur,Cerianne Lynn Mullins Ma Belle
Cet 2h49'1"7d26'
SOFA
Eri 4h3'54"-10d54'
Soffer,Justin
Cmi 7h7'30"7d25'
Sofferenza,Iana Hai
Mon 6h40'35"-10d29'
Soffiantini,Marta
Aql 19h11'30"13d13'
Sofge,Michelle
Cyg 19h44'22"30d39'
Sofia Rose
Peg 21h57'54"10d42'
Sofia Teresa 1-5-95
And 1h31'42"39d9'
Sofia-Loba
And 0h10'17"38d33'
Sofiah Jordanne
Umi 15h2'31"66d24'
Sofilion
Umi 16h7'13"73d38'
Sofjak
Peg 22h27'26"31d38'
Sofo,Philip
Cep 0h2'53"69d50'
Softic,Devon Peter
Cep 0h39'42"77d58'
Softy
Aql 19h50'19"14d32'
Sogamoso,Janet
Gem 6h48'33"19d23'
Sogemeier,Heinz-Jörg
Ari 2h55'12"22d7'
Sognare
Uma 12h7'0"56d47'
Soha,Louis S
Ori 6h4'40"8d29'
Soha-Amir
Umi 14h24'24"66d2'
Sohannita
Vul 20h15'28"22d37'
Sohel My Star
Ori 5h56'47"12d29'
Sohle,Mary Jane
Eri 4h1'27"-18d1'
Sohm,Baby Alycia Victoria Dowling
Cas 2h22'57"68d3'
Sohmer,Kenneth M
Her 16h10'30"41d8'
Sohn,Natalee
Cas 1h32'44"60d58'
Sohnrey,Angelika
Eri 4h48'51"-6d10'
Sohnsen,Ingolf
Cyg 20h42'50"46d40'
Sohren,Rachelle Sue
Peg 22h59'31"28d25'
Soich,A J
Aur 5h29'22"54d44'
Soika,Riso Star Richard
Cnv 12h35'35"39d26'
Soisson,Jenny
Cas 1h17'24"63d21'
Sojourner,Shirley
Aql 19h31'26"0d44'
Sokal,Michelle P
Cas 23h25'19"60d32'
Sokol
Uma 13h59'27"60d17'
Sokol's 90th Birthday, Michael
Hya 9h1'20'5d26"
Sokol's Star
Mon 6h28'45"-6d46'
Sokol,James
Her 17h2'14"40d33'
Sokol,Matthew Ryan
Her 16h42'60"39d5'
Sokol,Weatherly
Boo 14h33'35"21d43'
Sokoll,Rachel
And 0h12'47"31d31'
Sokoloski,Gary Matthew
Aur 7h23'0"41d18'
Sokoloski,Paul Bryan
Aur 7h21'27"41d14'
Sokolov,Lee
Lac 22h49'0"55d40'
Sokolowski,Antony Martin
Sex 10h31'12"5d18'
Sokolowski,David J
Her 16h59'1"29d37'
Sokolowski,Neta
Vul 19h44'59"22d40'
Sol
Per 1h25'1"53d26'
Sol Di Palo Bordenca
Cep 20h47'32"75d37'
Sol Genuine Superstar
Oph 17h16'44"12d7'
Sol Jack
Dra 19h26'55"70d34'
Sol,Jana's
Peg 23h30'0"30d17'
Sola,Anders E.
Uma 8h34'47"60d6'
Solak Star,The Barbara Ann Gunia
Cas 0h6'15"58d39'
Solana,Jose Luis
Aql 19h5'1"15d16'
Solange
And 2h32'45"44d31'
Solano,Charlotte Labouret
Cap 21h56'33"-20d2'
Solano,Daniel V
Sex 9h46'42"-6d56'
Solano,Deborah
Lac 22h46'10"56d20'
Solano,Romeo
Mon 7h32'1"-6d14'
Solar,Bramula
Aql 19h30'29"0d3'
Solari,Yia Yia Mary
Cyg 19h40'21"41d2'
Solarius,Mindy
Lyn 8h14'54"35d59'
Solarski,Matthew Ian Tosk
Tau 5h43'38"28d11'
Solarski,Maya Avery Tosk
Cas 0h13'1"64d29'
Solasz,Harry Press
Tau 4h7'52"20d13'
Solbach,Helmut
Lac 22h3'55"51d3'
Solbeck,Annette Kirstine
Her 18h54'1"12d6'
Solberg,Kaj & Grere
Cas 2h12'0"61d6'
Solch,Steven Lawrence
Dra 18h21'1"71d9'
Soldano,Dominick J
Cam 8h57'18"78d0'
Soldier Boy
Her 16h51'42"34d59'
Soldier Boy's Girl
Lyn 8h8'44"38d10'
Solecitto,Lillian
And 0h2'15"41d8'
Solecitto,Todd
Tri 2h14'19"31d57'
Solecki,Janette Vivian
Cam 5h56'52"56d1'
Soleil de Nuit
Peg 22h37'29"21d18'
Soleil X Ball Van Zee
Cam 3h26'47"60d18'
Solem,Anna Rose
Cyg 21h32'54"38d12'
Solemn
Tau 5h27'17"22d28'
Soler,Didier
Cyg 20h28'21"31d33'
Soler,Nora
Peg 22h39'1"24d19'
Solesbee,T David
Aql 18h53'58"6d54'
Soley,Paul
Ori 5h48'1"10d11'
Solich
Lyr 19h5'38"28d14'
Soline,Marie
Lyn 8h10'21"39d30'
Solinger,Jr,Fred R
Aur 4h50'32"40d42'
Solinski,In Memory of Margaret
Cyg 19h44'0"30d45'
Solis,Ana-Alicia
Peg 22h59'29"12d25'
Solis,Austin Bert
Mon 7h56'47"-2d17'
Solis,Carlos A
Cet 3h17'23"0d6'
Solis,Diogo Araya
Cma 6h25'49"-18d52'
Solis,Patrick Daniel
Peg 23h39'42"26d3'
Solis,Roberto Noel
Cma 7h20'17"-15d57'
Solis,Roxanne
Mon 6h43'0"7d22'
Solis,Rubia
Sct 18h22'55"-13d25'
Solis,Valeria Elizabeth
Peg 23h39'49"15d19'
Solis,Yolanda G
Sco 16h55'14"-40d32'
Solitaire
For 2h1'41"-26d30'
Solitaire
Nor 16h23'32"-54d0'
Solitude Aeternus
Cyg 20h41'0"30d54'
SOLITéR,Margit Christine Kerps
Umi 15h38'13"80d41'
Soliz,Jose Oscar
Aql 20h13'32"4d26'
Soljobson
Lmi 9h45'53"40d4'
Solk,Aldine
Uma 9h24'13"55d19'
Sollano,Robert
Cam 3h39'17"72d26'
Sollenberger,Diane
Tau 5h56'31"28d22'
Soller,Helene Elizabeth Mary
Cas 0h29'27"60d37'
Sollmon,Marc
Cet 2h44'22"3d55'
Solly
Uma 10h13'51"71d18'B
Solmor,Rachel Madison
And 0h21'26"32d47'
Solo,Holmegaard
Tau 3h56'24"28d40'
Solo,Honora
Peg 23h29'38"20d31'
Solock,Lee Stacey
Peg 21h55'0"23d28'
Solodyna,Alexander Anthony
Per 3h13'24"41d27'
Soloman,James Peter
Dra 17h24'48"60d5'
Solomita,Pauline
And 23h17'56"51d13'
Solomon
Oph 17h55'55"13d59'
Solomon T
Cnv 13h27'46"42d10'
Solomon's Starlight, C & R
Cyg 19h24'13"33d13'
Solomon,(Ist Cin)Cindy J
Cnc 8h53'51"32d50'
Solomon,Barry Ivan
Cma 6h56'41"-15d40'
Solomon,David & Lisa Siegel
And 0h43'45"30d40'AB
Solomon,Frances R
Dra 19h11'40"68d50'
Solomon,Gerson
Cyg 21h5'36"40d7'
Solomon,Harvey
Boo 13h38'21"18d53'
Solomon,Jennifer
Cep 20h54'34"68d55'
Solomon,Jerome "Our Brightest Star"
Oph 17h17'40"10d33'
Solomon,Jodi
And 23h48'14"42d28'
Solomon,Joseph Frederick
Boo 14h50'46"32d37'
Solomon,Maureen
Vul 19h45'32"22d36'
Solomon,Mikel
Dra 16h18'34"60d37'
Solomon,Mindy & Steven
Cnv 12h35'22"41d1'
Solomon,Nathan
Aql 19h29'53"13d13'
Solomon,Robert
Lac 21h56'42"36d55'
Solomon,Ronald Jay
Cet 2h39'33"-8d20'
Solomon,Saul
Aur 5h0'35"50d39'
Solomon,Stephen
Per 3h2'48"46d56'
Solomon,Suzette Joanne
Uma 8h31'20"48d9'
Solomon-The Big One, Forever Charles
Peg 23h6'28"10d38'
Solomonik,Christine
Lyr 18h40'0"26d36'
Solon
Cma 7h17'37"-15d8'
Solorzano,Benjamine Cathers
Aur 6h9'40"31d28'
Solosky,Anneliese
Aur 7h7'20"37d47'
Solosky,Archie H
Aur 7h9'37"36d54'
Solovey,Oleg
Ori 05h54'19"15d08'
Soloway,Edward Michael
Per 2h33'16"56d19'
Soloway,Jill K
Cam 4h33'0"60d31'
Solstad,James A
Cep 1h17'1"86d24'
Solstice
Ori 5h44'37"1d45'
Soltes'92,Gloria Frank
Peg 23h31'39"17d41'
Soltis Star,The Paul
Dra 17h42'28"64d45'
Soltis,Corrine
And 2h20'29"48d18'
Soltis,In Memory Of Earl
Aur 5h4'18"40d27'
Soltis,John
Lac 22h8'38"37d41'
Soltman,Jo
Com 13h5'28"27d54'
Soluk,Michelle
Lyn 7h0'29"58d49'
Solvason,Joseph
Aql 20h4'56"6d60'
Solvason,Stefan
Hya 9h10'50"2d26'
Solveig Aase
Cnv 13h2'60"51d60'
Solveira,Jorge M
Aur 6h29'49"38d44'
Soma,Anthony
Lyn 6h51'51"60d7'
Some Kind of Wonderful
Ori 5h56'46"12d39'
Somebody
Sge 19h37'40"16d30'
Somebody
Peg 21h55'0"34d12'
Someday
Oph 17h29'25"8d50'
Someday
Cyg 20h51'1"30d58'
Someday
Aql 20h5'56"6d8'
Someday
Vir 11h44'52"3d59'
Someday
Peg 22h57'31"33d53'
Someone Special
Uma 10h18'54"56d38'
Somero,Sandra Jean
Cnv 13h56'14"41d48'
Somers
Uma 14h4'0"51d29'
Somers,John A
Per 1h27'42"54d18'
Somerville,Anthony Neill
Aur 6h29'23"30d57'
Somerville,Ceara Robyn
Cyg 19h28'1"38d10'
Somerville,George Ross
Cet 0h56'55"0d5'
Somerville,Jamie Elizabeth
Uma 10h8'0"51d28'
Somerville,Lisa Marie
Cam 8h14'58"80d36'
Somes,Gavin William Anderson
Hya 8h17'1"0d33'
Something About You
Lyr 18h47'41"35d59'
Something Special
Del 20h14'37"13d13'
Something Spectacular
Cet 1h20'47"-5d29'
Something Very Special
Equ 20h58'52"8d3'
Somewhere In Time
Hya 9h14'12"2d41'
Somewhere In Time
Lac 22h34'56"53d8'
Somewhere In Time
Cet 1h24'54"-12d30'
Somewhere Out There
Lyn 7h6'14"50d13'
Somewhere Out There, Robert
Uma 10h36'53"58d35'
Somin,Elizabeth Stewart
Lyn 9h1'21"43d35'
Somin,Michael Mathew
Ser 15h41'0"24d4'
Somin,Violet Borkovich
And 0h49'1"39d28'
Somjee,Alisha
Cnc 8h11'1"6d59'
Sommagura,Aldo
Lyn 8h4'57"40d58'
Sommatino,John R
Ori 5h53'57"11d28'
Sommer Joi
Mon 5h54'47"-4d3'
Sommer,Franz Jun
Mon 7h53'31"-2d41'
Sommer,Gloria Lydia
Tri 1h58'29"26d58'
Sommer,Julien
Lyr 18h17'0"39d40'
Sommer,Kay
Cas 0h29'48"60d36'
Sommer,Kent
Aql 19h26'28"14d32'
Sommer,Lee Ann
Peg 22h13'25"33d35'
Sommer,Sara
Crb 15h34'49"28d39'
Sommer,Ute
Lac 22h2'15"50d41'
Sommer,Werner
Her 16h7'41"41d25'
Sommer,Wolfgang
Tau 4h16'51"0d4'
Sommerburg,Thomas
Aqr 20h57'35"-6d39'
Sommerfield,Sarah Ellis
And 0h59'43"35d9'
Sommerhalter,Benoit
Mon 7h58'24"-4d12'
Sommerquist
Cas 0h52'50"61d12'
Sommers,Becca
Cyg 20h21'0"31d30'
Sommers,Carina Leona
Oph 17h22'22"-20d24'
Sommers,Dave
Aur 7h6'36"37d48'
Sommers,Elizabeth Ann
Lyn 8h24'34"40d41'
Sommers,Jacqueline Marie
Lyn 7h33'54"50d2'
Sommers,Joey
Her 17h31'47"28d8'
Sommers,Kyle
Oph 17h7'19"10d20'
Somo,Joan M
Sge 19h57'27"16d30'
Somoano,Robert Bonner
Tri 1h57'57"27d49'
Somogyi,Dustin Hart
Per 1h43,59"52d54'
Somohano,Chelsey Nicole
Cmi 7h40'34"4d45'
Somrock,Margaret Seliga
Cas 1h55'42"61d48'
Son Ae
Uma 9h30'17"57d38'
Son Le
Lyn 8h49'49"38d15'
Son of Bud
Lmi 10h58'27"33d8'
Son-Ally's Cosmic Sunshines Can-Di
Aql 20h11'53"10d59'

Sonafrank,Lynann
Uma 12h25'57"54d29'
Sonaty,Patricia
Ori 6h9'53"7d42'
Sonberg,Eric
Mon 7h4'29"-6d22'
Soncire Ianthe
Cma 7h1'1"-13d48'
Sondergard,Brian Richard
Cep 23h11'46"60d21'
Sondergeld,David
Dra 11h48'0"67d60'
Sondermann,Udo, Germany
Tau 5h33'26"28d42'
Sondey,Edward
Aql 19h36'54"0d44'
Sondra
Cas 0h52'32"63d54'
Sondra Alyse
Uma 11h16'56"43d29'
Sondra's Smile
Ori 6h4'29"5d15'
Sondrini,Rudy D
Aur 7h3'26"36d25'
Song Of Deborah
Ari 1h47'32"22d46'
Song,Kathy
Mon 7h3'53"-0d3'
Songbird
Cyg 20h28'15"30d35'
Songbird
Peg 23h7'50"12d29'
Songbird
Aql 19h2'16"-1d49'
Songy,Larry
Ori 6h1'17"8d19'
Soni
Mon 8h3'17"-6d59'
Soni
Lyn 7h51'60"48d13'
Soni,Abe
Aur 6h28'59"31d43'
Soni,Jean-Noel
And 2h24'47"41d0'
Sonia
Sex 10h48'18"2d26'
Sonia
Cas 1h28'34"63d52'
Sonia
Dra 16h1'12"62d14'
Sonia
Peg 22h5'17"5d44'
Sonia
Mon 6h56'42"-10d7'
Sonia
Vel 10h11'58"-43d50'
Sonia
For 2h33'43"-28d50'
Sonia
Boo 13h56'0"16d37'
Sonia
Cyg 19h31'0"38d24'
Sonia K
Col 6h0'3"-31d49'
Sonia Michelle
Mon 6h35'0"3d2'
Sonia Miriam
Tau 4h38'35"15d31'
Sonia Natasha
Lmi 10h37'38"33d41'
Sonia per sempre
Col 6h37'18"-34d24'
Sonia's Love
And 23h4'30"42d55'
Sonia-Laetitia
Ser 15h42'22"2d58'
Sonigo,Alexandra
Ori 5h26'23"0d16'
Sonja
Gem 7h18'39"23d13'
Sonja
Cas 1h15'1"61d6'
Sonja & Randy
Cyg 19h58'32"30d34'
Sonja II
Sco 17h53'1"-30d40'
Sonja"Starlight"
Lib 14h58'59"-23d38'
Sonja-Cotton Candy
Com 12h5'16"27d7'
Sonjas Glücks-und
Gesundheitsstern
Ser 15h12'35"10d43'
Sonjetta
Oph 17h14'39"11d39'
Sonlight
Cas 0h5'23"61d10'
Sonne,Courtney B
Aql 19h31'49"10d19'
Sonneck,Rudolf
Sgr 19h50'53"40d56'
Sonnemann,Melissa Lyn
Cas 2h9'23"67d53'
Sonnenberg,Eva Barbara
Psc 1h0'17"20d9'
Sonnenberg,Gale
Sct 18h36'58"-4d48'
Sonnenberg,Patricia
Oph 18h35'47"10d6'
Sonnenberg,Paul
Cep 22h26'1"57d30'
Sonnenblumigste Mama
Lib 15h36'42"-22d12'
Sonnenburg,Michele Marie
And 0h10'19"27d32'
Sonnenfeld,Erik
Cep 22h22'54"61d40'
Sonnenschein,Emil August
Cap 20h34'31"-10d20'
Sonnenschein,Taylor Noelle
Cyg 19h28'1"33d48'
Sonnentag,Gerd
Lac 22h27'21"50d7'
Sonner,Jürgen
Ori 6h17'1"8d57'
Sonnleitner,Carl & Henrietta
Cyg 20h16'41"41d15'
Sonntag,Kevin W
Aur 6h3'12"31d4'
Sonny
Per 3h3'24"37d32'
Sonny
Her 16h37'51"41d4'
Sonny & Queenie
Ori 6h6'59"5d6'
Sonny Bob
Dra 19h45'25"61d39'
Sonny The Brightest Star in
My Life
Aql 19h30'34"12d27'
Sonny's Light
Dra 20h2'39"71d4'
Sonnye,Silver Fox
Aql 19h3'8"-0d59'
Sonoka
Lib 14h29'35"-13d58'
Sonomi
Psc 1h22'15"3d5'
Sonshine,Mildred & Joe
Uma 10h12'51"52d51'
Sonsky,Brian Justin
Aql 18h45'1"7d58'
SONSTAR
Sex 10h26'0"1d9'
Sontag,Mike
Her 18h9'0"47d9'
Sonya
Peg 22h7'58"25d49'
Sonya
Cyg 20h33'27"46d38'
Sonya & Stephen Whole Lotta
Love
Cyg 20h18'26"30d26'
Sonya Renee
And 2h19'36"45d15'
Sonya T
Per 2h56'13"38d6'
Sonya-3764
Lyr 18h58'1"30d4'
Soo Hill Husky Star
Uma 10h47'38"72d20'
Soo Hui Wen
Ori 6h9'31"0d33'
Soon
Sct 18h54'55"-9d14'
Sooner Milhous
Aql 19h7'13"3d25'
Soong,Chase C
Aql 19h29'48"-0d42'
Soorus,Alice Edwards
Cam 3h23'55"60d15'
Sooy Star,The Alma Hammell
Cas 1h2'24"60d49'
Sooy,Mary E Christy
Cyg 20h54'28"31d32'
Sopegacha ici et maintenant...
Per 2h53'12"45d39'
Soper,Mrs Nancy Kenealy
Mon 7h56'48"-8d4'
Soper,Neil
And 1h2'38"40d19'
Soper,Richard Graves
Sgr 18h59'41"-25d10'
Sopha-D'Onofrio,Shana Lynn
Uma 10h7'40"53d57'
Sopher,Farquharson Chieftain
Kenneth W
Her 17h15'50"47d44'
Sophia
And 0h52'0"22d29'
Sophia
And 2h8'26"42d4'
Sophia
Cma 7h23'27"-16d41'
Sophia
Lyn 7h53'32"36d47'
Sophia Charlotte
Lyr 19h17'1"28d42'
Sophia Katherine
Lmi 10h49'1"33d25'
Sophia Maria
Vul 19h31'47"24d58'
Sophia T
Cnv 13h16'32"38d1'
Sophia Theresa
Cas 23h15'24"61d13'
Sophia Violet
Mon 7h44'23"-1d31'
Sophia's Star
Aql 19h48'16"11d18'
Sophia,Teacher In Space
Aql 19h18'45"15d13'
Sophianos,Philippo
Eri 3h3'33"-5d38'
Sophie
Mon 7h1'54"-6d24'
Sophie
And 23h1'0"50d3'
Sophie
Cyg 21h0'53"38d13'
Sophie
Cas 0h56'12"58d13'
Sophie & Hellmut
Cmi 7h21'21"8d59'
Sophie Carol
Ori 5h56'11"8d59'
Sophie DK
Tau 4h10'38"22d2'
Sophie Elizabeth
And 23h22'11"51d24'
Sophie et Jean Marie
Lyn 8h27'44"38d7'
Sophie et Jerome
Lyn 8h33'0"41d53'
Sophie Kelly
Aql 19h3'0"0d2'
Sophie Louise
Cyg 19h33'1"33d14'
Sophie LWM
Per 3h0'60"37d43'
Sophie Marguerie
Vir 13h38'15"-4d19'
Sophie's Smile
Cet 2h5'21"3d43'
Sophie's Star
Lyr 18h47'29"38d32'
Sophie's Star
Cyg 19h56'14"38d31'
Sophie's World
Umi 14h44'26"66d40'
Sophie-The Beginning
Cyg 19h40'58"41d15'
Sopko,Wayne D
Dra 19h4'33"48d3'
Sopourn,Mary Teresa
Uma 9h52'1"57d30'
Sopp,Nigel A
Uma 11h55'16"45d21'
Soppet,Savannah
Com 12h46'22"21d44'
Soprych,James M
Boo 14h43'36"48d52'
Soprych,Karen S
Boo 14h49'52"48d33'
Sora
Lib 14h28'35"-13d28'
Sora
Psc 1h18'44"28d6'
Soragni,Maria Elena
Cam 6h41'56"68d29'
Soran,Mariana
Vir 12h56'56"-20d28'
Soranea,Giuna
Ant 10h45'45"-37d9'
Soranno,Patrick Carl
Aur 5h6'53"38d20'
Soraya
Uma 11h29'36"48d52'
Soraya
Cyg 21h28'55"30d3'
Soraya
Aql 19h9'52"12d54'
Sorbello,David C
Ser 16h2'52"22d39'
Sorbello,Luisa Pricipessa
And 23h33'31"52d33'
Sorbo,Kevin
Her 17h35'52"26d59'
Sorby,Peter
Vir 12h58'27"-9d22'
Soreca,Paolo
Ori 5h57'35"0d44'
Sorem,Mark J & Martha E
Riley
Her 18h18'49"27d30'AB
Sorensen I
Cam 6h13'15"68d6'
Sorensen's First Half
Century,Terje
Aql 19h17'14"10d5'
Sorensen,Bendicte Toft
Per 3h35'50"37d53'
Sorensen,Carol L
Vul 20h3'58"28d33'
Sorensen,Cookie
And 23h42'19"47d31'
Sorensen,Deborah
Crt 11h9'27"-16d48'
Sorensen,Diane Lee
Gem 6h41'43"35d12'
Sorensen,Evelyn Claudette
Cyg 21h52'34"41d29'
Sorensen,Inga Fonss Herskind
Aur 5h3'31"40d44'
Sorensen,Jeremy
Aur 6h56'31"37d35'
Sorensen,Joan
Lmi 9h47'53"33d17'
Sorensen,Jordan
Per 1h57'20"56d20'
Sorensen,Lee Edward
Mon 6h29'39"-10d7'
Sorensen,Mark Skibsted
Ori 5h35'30"-1d36'
Sorensen,Rick
Sex 10h18'47"-9d60'
Sorenson,Brian Joseph
Aur 6h4'1"46d28'
Sorenson,Doug
Cep 21h32'57"60d51'
Sorenson,Dr Kermit R
Dra 17h47'34"65d0'
Sorenson,Marion Todd
Ori 6h8'33"8d44'
Sorenson,Megan Lee
Cas 0h55'38"65d16'
Sorenson,Spencer Reed
Boo 14h48'44"45d40'
Sorese,Vince
Boo 15h26'45"38d26'
Sorg,Kay Lynnette
Cma 6h54'34"-17d51'
Sorgatz,Marion Marlies
Marina
Ari 2h55'0"21d31'
Sorge,Timothy C
Uma 10h10'0"42d12'
Sorgen,Papa & Grandma
Sge 19h2'17"20d26'
Sorgenfrei,Kennon
Her 15h48'15"44d17'
Sorgento,Jerry
Per 4h2'57"50d52'
Sorger,Dr Alfred
Her 18h27'55"24d46'
Soriano,Anthony Ray
Martelino
Lib 15h17'50"-8d47'
Soriano,Martha Louise
Lyr 18h37'0"41d2'
Soriano,Sr,Charles Alec
Cam 4h49'55"70d57'
Soriano,Stella
Cas 3h5'33"60d31'
Soric
Cyg 19h59'22"29d58'
Soricelli,Rhonda L
Sco 16h20'48"-38^40'
Sorige,Jan & Susanne
Cas 23h18'26"61d30'
Sorkin-West
Cet 0h41'53"1d15'
Sorlerer
Dra 20h11'60"62d46'
Sormilic,Carol Ann
Uma 14h18'27"61d34'
Sormus
Tau 5h27'23"18d42'
Soros,Paul
Gem 8h2'50"31d53'
Sorrell,Anita
Lmi 10h27'1"34d9'
Sorrell,Doris
Cas 0h53'1"72d14'
Sorrell,Helen Frances Walkley
Cmi 7h22'19"5d25'
Sorrell,Robert Hay
Vul 19h43'33"22d47'
Sorrell,Star Of
Crt 11h16'6"-14d21'
Sorrell,Victoria
Lyn 07h31'0"42d59'
Sorrells,Tanner John- Albert
Lyn 9h9'17"42d51'
Sorrels,Ryan & Nicole
Eri 3h42'25"-5d45'
Sorrenti,Brianna Page
Cas 22h56'51"54d26'
Sorrenti,Christopher
Per 2h20'0"58d23'
Sorrenti,James "Rockey"
Aur 6h15'18"38d54'
Sorrenti,Nicole (Princess)
And 23h0'43"50d51'
Sorrentino,Amanda Marie
Lyr 18h30'16"38d21'
Sorrentino,Anthony Franco
Dra 15h52'41"60d53'
Sorrentino,Cathy
And 0h18'43"36d54'
Sorrentino,Gina
And 1h22'60"38d24'
Sorrento,Solana
Cam 11h10'33"82d16'
Sorroche,Jack D
Her 16h29'44"34d50'
Sorroche,Patti & Jack
Cyg 21h5'24"38d53'
Sorsby,Our Mom,Jacie Sorrell
Carter
Cas 1h4'18"58d21'
Sorteni,Paola
Cnv 13h31'54"38d25'
Sortis,Chrystopus Melisus
Cep 1h12'15"77d44'
Sortore,Cheryl K
Cyg 20h53'24"38d5'
Sortore,L R Joan
Peg 22h16'10"32d5'
Sorum-Tarbox
Cam 13h12'19"82d9'
Sos Fenger
Ori 5h59'44"10d36'
Sosa,Louis
Cnv 12h24'23"37d53'
Sosa,Louis Edward
Mon 6h53'44"0d29'
Sosa,Petra Andrea
Dra 13h58'54"64d58'
Sosa,Victoria Brooke
Lib 15h57'27"-8d32'
Sosenko,Jennifer &
Christopher
Eri 3h33'26"-5d50'
Sosnovski,Gianni
Uma 10h1'57"68d18'
Soso
Cam 7h46'10"78d50'
Sossi,Victor J
Lac 22h40'47"50d13'
Sost,Timothy Michael
Psc 22h52'48"1d41'
Sotak,Joseph John
Oph 18h41'35"8d56'
Soter,Ernie
Boo 14h40'29"48d2'
Soter,William J
Per 2h3'21"48d55'
Soterakis,Gus
Cet 2h16'56"5d17'
Soteriou,Helen
Sgr 20h6'34"-41d59'
Soteris
Her 18h12'24"38d38'
Sotirios,Nick
Uma 9h49'25"52d2'
Soto,Bianca Jean
Del 20h14'34"13d14'
Soto,Carmen Elisa
Eri 3h47'0"-6d50'
Soto,G Hoffman
Oph 18h28'44"12d1'
Soto,Jaime Alexander
Lmi 10h2'27"31d6'
Soto,Linda Montgomery
Cas 0h28'54"61d37'
Soto,Madeline
Aql 19h24'58"15d24'
Soto,Richard Lewis
Dra 16h13'0"63d14'
Sotogrande Dos
Aql 18h57'28"16d2'
Sotomayor,Flora C
Cyg 19h33'52"36d2'
Sotomayor,Luis Anthony
Tri 1h58'23"27d1'
Sotomayor,Stephen Fancisco
Vul 20h30'0"28d20'
Sotomayor,Teresa Carroll
Aql 19h29'0"7d41'
Sottile,Erin Nicole
Peg 22h45'1"21d40'
Sottiurai,Romi
Ser 16h1'16"23d5'
Sottovia,Tiziana
Aur 5h26'54"31d41'
Sottung,Christopher Edward
Lac 22h7'1"50d9'
Souad
Cep 20h29'57"62d42'
Souba,Sharon J
Ari 1h46'0"22d15'
Soubeyroux,Laurence
Lyr 18h14'52"30d50'
Soucaret,Jean-Raphael
Cas 23h38'28"60d12'
Soucek,La Vaun
Lyn 8h12'29"39d25'
Soucie,Megan Courtney
Uma 10h54'0"37d43'
Soucy,Franáoise Gouin Michel
Lyr 18h49'0"34d6'
Soucy,G-Lady I
Uma 11h54'55"40d33'
Soucy,Tom & Jan
Lac 22h9'38"37d46'
Souczek,Monika
Her 17h24'40"42d32'
Soudan,Renda
Cep 20h44'45"68d24'
Souden,David Lee
Ser 15h9'15"8d20'
Souhrada,Scott
Cep 21h31'35"68d44'
Soukalova,Katerina
And 0h1'0"38d9'
Soukup,Sabine
Tau 5h53'12"23d7'
Soul
Del 20h17'44"15d56'
Soul Mate
Peg 21h46'33"34d55'
Soul Mate Hugh With Dreams
& Love
Cyg 20h56'44"40d11'
Soul Mates
Cyg 19h31'13"31d48'
Soul Mates Forever Eddie &
Marjie
Boo 13h44'26"15d18'
Soul Mates-David & Denise
Cam 13h28'0"84d44'
Soul Of Cheri
Eri 3h17'22"-13d13'
Soul of My Soul
Aql 19h30'16"10d30'
Soul of Taiyu
Vir 14h26'44"5d49'
Soul To Soul
Peg 22h5'32"3d45'
Soulard,Joanne "Wammy"
And 2h21'48"44d21'
Soule,Brian C
Lyn 7h54'28"40d35'
Soule,Geoffrey
Cet 2h11'32"8d15'
Soule,George F
Cyg 21h13'17"28d30'
Soule,Kristen
Uma 9h27'32"51d42'
Soule,MD,Bradley
Boo 14h1'21"21d53'
Souleyret,Elizabeth Monique
Com 13h5'29"30d11'
Soulier,Jean Pierre
Crb 16h3'33"39d9'
Souligny,Nanette
And 2h22'1"43d0'
Soulmate
Cnv 13h40'42"40d27'
Soulmate
Sct 18h41'46"-8d33'
Soulmate
Uma 11h44'29"30d35'
Soulmate,Wayne H Tyler
Dra 18h47'14"68d59'
Soulmates
Lyr 19h18'35"40d56'
Soulmates
Lyr 18h53'54"31d55'
Soulmates Eclipse
Boo 15h6'13"28d14'
Soulmates Warren & Linda
Cam 8h4'41"82d54'
Sourvanos,Dionyssios
Per 4h25'54"50d15'
Sousa
Mon 7h57'23"-3d28'
Sousa III,John Phillip
Tau 5h55'49"24d14'
Sousa,Amy
Cam 3h17'54"56d0'
Sousa,Dr Dean
Her 17h27'55"30d21'
Sousa,Jennifer
Cas 3h25'28"70d50'
Sousa,Wayne C (elestial)
Lac 22h51'0"40d48'
Soussa,Scott David
Cep 23h6'0"60d14'
South America Away
Anna,Pedro & Family
Cet 0h29'53"-12d29'
South Dakota Shorty
Lac 22h17'18"51d54'
South Euless Sparkler
Mon 6h30'57"-0d5'
South Wind-Creuza Bragunci
De Miranda
Ori 5h55'55"10d4'
South Winds Lady Kate
Cyg 19h31'16"33d18'
South,Ray "Bearrobe"
Aur 5h2'0"46d14'
Southall,Patricia
Mon 8h1'43"-8d58'
Southard's Star
Cnv 12h22'34"36d35'
Southard,Brandy
Mon 6h40'33"10d44'
Southby,Gary Lee
Cep 21h15'45"60d4'
Souther,Al
Aql 19h56'52"10d39'
Southerland,Donna P
Peg 23h48'21"12d56'
Southern Princess
Cyg 20h29'33"58d49'
Southern,Christopher Bryan
Per 2h11'0"57d14'
Southern,Dr Albert M
Oph 17h31'18"-0d37'
Southgate,Karen
Sge 20h5'56"16d3'
Southgates's Star,Sue
Cas 0h18'31"63d6'
Southwell,Timothy G
Boo 14h27'20"20d58'
Southwestco
Eri 4h2'48"-18d49'
Southwick,Dale Oliver
Her 18h17'28"18d47'
Southwold Star,Sarah's
Cas 1h29'56"75d2'
Southworth Dolly
Ori 6h3'13"2d28'
Southworth,Dean Michael
Boo 15h16'56"53d31'
Southworth,Paul
Cam 3h32'59"62d59'
Souvenir D'Amour
Ser 15h15'33"24d52'
Souvré,Christian
Ori 6h0'33"8d35'
Souza,Bryanna Leigh
Lyn 8h25'57"45d16'
Souza,Daniela
Aqr 20h37'14"-8d25'
Souza,Emily Perry
Del 20h34'47"10d28'
Souza,Kenneth Leon
Oph 17h16'17"-20d5'
Souza,Margaret Mary
And 1h46'35"36d11'
Souza,Nick
Cma 6h59'51"-16d35'
Souza,Noreen E
Cam 12h33'0"80d33'
Souza,Shirley A
Del 21h3'1"12d30'
Souza,Todd Michael
Lac 22h13'54"50d29'
Souza-Couch,Lorien Diedre
Cet 2h2'32"-11d5'
Sova,Mary Elizabeth
Mon 6h33'16"-0d34'
Sovany,Emily L
Eri 3h15'29"-11d46'
Sovereign
Cep 1h29'14"84d25'
Sowa,Ryan Joseph
Aqr 22h37'18"-0d7'
Sowarby,Harold "Pops"
Her 17h29'58"21d49'
Sowden,Ann
And 0h15'31"38d50'
Sowden,Benjamin Avery
Ari 2h36'46"22d10'
Sowinski,Dr John
Oph 17h2'52"-24d58'
Sowley,Alice Maude McCarty
Mon 7h4'44"3d55'
Soxie,John
Per 3h17'40"41d28'
Soya,William
Ori 5h32'11"8d59'
Soza,Sue
Ori 4h49'42"5d8'
Sozzani,Franca
Uma 11h13'33"50d8'
Space A Ice Sculptors, The
Ori 5h54'21"15d8'
Space Shuttle Challenger
Uma 10h54'39"38d26'
Spaceplay
Uma 9h17'41"57d42'
Spackman,Frances
Lyn 8h18'26"36d6'
Spada,Antonella
Cnv 13h5'50"32d53'
Spada,Joseph A
Lyn 7h54'11"41d16'
Spadaccini,Linda Jean
Cas 23h33'28"60d36'
Spadafore,Tom J
Cet 2h15'40"3d42'
Spadaro, "Unforgettable",Mary
Cam 8h38'51"73d33'
Spadaro,Francis Joseph
Aur 5h9'48"40d54'
Spadoni,Eileen
Aql 20h9'26"7d45'
Spady,Carolyn Helen
Sge 19h56'41"16d13'
Spaeder,Shirley Ann Wheeler
Cyg 21h32'55"53d34'
Spaehly III,Joseph Frederick
Boo 14h12'18"51d4'
Spaeth "Celestial
Splendor",Paul
Hya 9h19'34"-8d42'
Spaeth,Andreas
Vir 12h57'42"11d26'
Spaeth,Lisa Caron
Sgr 18h55'59"-33d33'
Spaeth,Peter
Per 4h6'1"45d9'
Spagna,Cecilia
Cyg 20h40'50"31d40'
Spagnolo,Corey & Becky
Peg 22h51'13"29d7'
Spagnolo,Vito
Vul 19h18'40"26d53'
Spagnuolo,Dominick R
Cep 1h7'22"78d10'
Spahn,Stephan
Ari 2h3'1"20d57'
Spahr,Austin Michael
Gem 7h14'56"21d50'

Spahr-Amator,William
Cnv 12h19'1"51d31'
Spaight,Billie Mudry
Dra 17h12'1"73d49'
Spain,Marriet De F
Cep 0h3'1"67d5'
Spain,Nettie Elizabeth
Cyg 21h38'31"40d58'
Spain,Teresa Ann
Aql 18h56'25"-0d45'
Spainhour,Harlan D
Lyn 9h5'39"34d29'
SpainStar
Lac 22h40'55"51d59'
Spakowski,Peter
Boo 13h53'10"21d8'
Spakowski,Richard
Mon 8h7'18"-5d21'
Spalding,Ann Talbot "Tobby"
Mon 7h0'20"-10d56'
Spalding,Anne Marie Rebecca
Cas 0h28'28"61d27'
Spalding,Jacob Michael
Her 16h36'1"47d45'
Spalding,Nyle Albert
Boo 14h41'33"36d57'
Spallino,Lucy
Cas 0h57'23"61d28'
Spallone ("Maxstar"), Marian R
Oph 18h23'35"7d37'
Spanagel,Eugen
Per 3h14'29"46d13'
Spanbauer,Judy Lynn
Peg 22h33'17"31d26'
Spaner,Howard
Per 3h34'1"37d32'
Spangenberg,Gary
Per 3h52'2"36d20'
Spangler Family,The
Aql 19h57'51"15d21'
Spangler's Diamond
Uma 10h51'26"48d12'
Spangler,Howard M
Cma 6h54'32"-18d5'
Spangler,John L
Ori 4h41'1"11d32'
Spangler,Jr,Richard E
Aql 18h58'27"13d1'
Spangler,Karen E
Cas 0h35'48"66d52'
Spangler,Kenneth M
Cep 22h19'11"56d3'
Spangler,Rebecca Kate
And 1h5'27"41d7'
Spangler,Robert
Her 18h49'32"18d46'
Spangsholm,Hans Ejner
Peg 21h39'57"26d50'
Spanier,Dieter
Ori 6h22'51"10d46'
Spanier,Karl
Cam 5h38'29"70d40'
Spanky
Her 16h35'0"21d3'
Spanky
Cet 2h0'1"0d53'
Spanky
Aur 5h8'20"40d2'
Spanky
Cam 4h14'31"68d20'
Spann,Diana M
And 23h30'47"40d22'
Spann,Doris
Leo 9h33'40"8d30'
Spann,Piper
Sge 20h16'19"20d16'
Spanner,Phillipa Kate
Cas 0h19'12"62d19'
Spano,Jennifer
Cyg 19h29'45"32d51'
Spano,Joseph
Aur 6h7'43"50d30'
Spano,Matthew Joseph
Mon 7h2'0"-6d15'
Spano,Michael
Boo 15h11'39"30d48'
Spano,Paola
Cas 2h2'50"60d35'
Spano,Timothy
Her 18h5'23"48d44'
Spano,Valter
Ori 5h35'0"8d58'
Spanos,Alex & Faye
Aql 19h0'28"-8d29'
Spanyer,J Carroll
Tri 1h57'57"33d55'
Spanziani,Frederick Francis
Aur 7h23'52"40d40'
Spanò,Gregorio Maria
Ant 10h30'58"-39d31'
Sparacino,Jr,Carlo Matteo
Cep 22h18'22"60d35'
Sparacino,Wayne
Lac 22h24'1"55d17'
Sparaco,Jacqueline MCginnis
Lyn 6h15'32"59d45'
Sparano,James Allen
Cet 2h24'29"-0d4'
Sparano,Willma Phal
Cas 2h50'42"61d23'
Sparber,Saul David
Lac 22h30'1"53d25'
Spargo,Elwyn E T Thomas
Cap 21h9'3"-22d47'
Spargo,Tony
Cyg 19h32'1"37d12'
Sparhawk
Cet 3h13'45"-0d14'
Sparkes,H Robert
Dra 12h3'46"71d35'
Sparkes,Ken - My Unique Star
Pho 23h57'47"-49d23'
Sparkes-KTS,Keith
Cas 0h15'53"59d16'
Sparkie's Place in Space
Cep 0h12'27"73d58'
Sparkle
Cap 21h7'59"-20d36'
Sparkle
Cyg 19h57'50"29d33'
Sparkle Farkle
Her 15h59'43"48d42'
Sparkle in Kathy's Eyes
Peg 23h20'0"18d33'
Sparkle Plenty
Ori 5h56'17"15d26'
Sparkle-93
Ori 5h4'48"8d30'
Sparkleboy
Peg 21h51'52"10d3'
Sparkler,Jeffrey
Hya 9h14'20"5d37'
Sparklie Barb
Mon 6h54'0"11d29'
Sparkling Heather
Psc 0h50'60"28d51'
Sparkling Iris
Psa 22h2'56"-26d52'
Sparkling JDM
Boo 15h7'35"16d51'
Sparkling Jon
Umi 14h54'28"71d44'
Sparkling Light Of Love From Above
Cam 13h20'27"77d39'
Sparkling Love
Boo 15h4'28"28d13'
Sparkling Sarah's Heaven
Com 12h33'1"25d26'
Sparkling Sheila
Uma 12h58'51"60d36'
Sparkling Stacey,The
Cyg 21h2'21"38d51'
Sparkling Susan
Peg 22h31'41"2d29'
Sparkling Suzette
And 0h50'0"38d59'
Sparkman,Briton Paul
Sct 18h55'1"-6d32'
Sparkosis
Del 20h38'45"18d55"
Sparks
Dra 12h43'59"72d23'
Sparks
Cep 21h42'37"58d57'
Sparks 50th Anniversry Eileen & Bob
Com 12h47'27"20d9'
Sparks,Alexander Nolan
Aql 18h41'41"0d59'
Sparks,Arthur Clifford
Del 20h20'1"9d26'
Sparks,Danny W
Aql 20h4'1"8d3'
Sparks,Denise Marie Tourville
Lyr 19h13'32"40d35'
Sparks,Destiny Lea
Uma 10h31'1"53d12'
Sparks,Francine
Crt 11h35'15"-11d44'
Sparks,Gary W
Aql 20h2'60"8d58'
Sparks,Jack David
Cep 21h55'36"60d40"
Sparks,Jay
Cma 7h1'0"-16d34'
Sparks,Josephine
Aur 6h23'13"35d12'
Sparks,Kenneth
Cet 2h59'21"1d44'
Sparks,Kevin
Ori 5h58'45"12d16'
Sparks,Kynan William
Mon 6h17'0"-4d59'
Sparks,Lillian
Umi 15h37'22"77d12'
Sparks,Matthew James
Lac 22h50'46"56d42'
Sparks,Paul Whitney
Aqr 21h58'1"-17d37'
Sparks,Stephen Michael
Per 2h52'23"43d42'
Sparks,Steven Carl
Eri 2h57'33'-1d56'
Sparks,Vida "Squeet"
Cma 6h53'6"-18d51'
Sparky
Dra 16h8'0"68d42'
Sparky
Her 16h13'1"48d33'
Sparky
Sex 10h32'54"3d2'
Sparky
And 2h28'14"40d2'
Sparky
Ori 6h4'21"7d4'
Sparky
Cyg 21h18'29"37d13'
Sparky Burns
Cam 8h24'59"78d34'
Sparky's Hideout
Lac 22h25'46"38d50'
Sparling,Emily O'Malley
Crt 11h25'19"-17d42'
Sparlow,Tira
Cyg 20h8'45"41d8'
SPARQ
Aur 5h32'13"38d21'
Sparrow,Robert Alex
Peg 23h3'16"11d29'
Sparshottll Luminii
Cnv 13h45'35"32d38'
Spartz,The Star of Thomas & Janet
Cyg 20h36'34"42d37'
Spataro,Betty Jean
Cyg 21h8'17"31d3'
Spatchie & Slimmy
Vul 19h20'54"26d43'
Spatz Carolin/CH-H
Aqr 20h56'1"0d30'
Spatz,Angela (Animal)
Lyn 8h24'31"57d53'
Spatz,Doris
Boo 13h38'44"10d8'
Spatz,Regina Ann
Sge 19h31'44"16d18'
Spatzele
Sco 16h13'52"-20d40'
Spatzl
Vir 12h4'50"10d50'
Spaulding,Brian G
Lyr 18h40'46"31d60'
Spaulding,Douglas Lee
Cma 7h12'43"-15d46'
Spaulding,Kimberly Sue
Cas 2h0'59"59d40'
Spaziale Galattica Universale
Her 16h27'18"18d46'
Spaziale,Calamaro
Cnv 13h5'33"33d18'
Speakman,Robii
Sco 16h37'57"-35d31'
Speaks,Jeffery
Her 18h2'1"14d24'
Speaks,Megan
Ori 5h48'43"21d29'
Spear "Babs",Barbara
Cet 2h23'13"-11d14'
Spear,Allison
And 2h26'39"44d39'
Spear,Amy
Cep 3h35'24"78d57'
Spear,Bennett Jeffrey
Hya 8h36'44"1d30'
Spear,Dennis
Dra 16h57'10"63d3'
Spear,Erica I
Lyn 07h31'54"42d37'
Spear,John M
Aql 20h0'35"8d59'
Spear,Lynn
Hya 8h18'13"2d44'
Spear,Mary
Peg 22h3'11"4d13'
Spear,Richard Kenneth
Her 17h13'0"20d34'
Speare,Our Smiling Star,Joel Carl
Dra 14h50'16"63d20'
Spearly,Ralph L
Dra 19h32'0"58d46'
Spearman,Yvette Denise
Peg 22h2'43"5d51'
Spears,Andrew
Aur 5h51'52"28d57'
Spears,Annette Kime
Vul 19h48'47"28d18'
Spears,Charles Nolan
Leo 9h21'47"18d28'
Spears,Cynthia A
Lyr 18h47'17"37d6'
Spears,Jerry
Aql 20h32'33"0d50'
Spears,Lucy & Paul
Cyg 20h55'19"30d44'
Spears,Todd Alan
Mon 6h56'1"8d20'
Spears,Wilfred William
Aur 6h27'15"30d9'
Speas,Robert L
Hya 8h14'33"0d1'
Speca,Corinne
Lyr 18h36'40"44d26'
Speca,Dominique Maria
Boo 13h47'39"20d10'
Speca,Genevra
And 1h2'59"38d25'
Specchio Star Of Mentors,The
Cet 0h46'12"0d59'
Specht,Angela Baby Fox
Peg 21h57'22"24d3'
Specht,Austen Taylor
Del 20h13'30"12d21'
Specht,Jr,James C
Peg 21h24'31"3d35'
Specht,Meredith
Lyn 8h13'29"48d0'
Specht,Sharon Lynne
And 0h23'23"30d33'
Specht,The Robin & Brian
Sge 19h12'14"21d6'
Special Angel
Cyg 19h22'0"48d50'
Special Ed
Cep 0h17'32"68d11'
Special J
Uma 9h49'33"58d13'
Special K
Hya 8h44'42"0d44'
Special K
Lac 22h37'21"41d9'
Special K
Cas 3h1'36"58d31'
Special Kay
Tri 2h12'54"32d15'
Special On Tap
Peg 22h42'59"21d40'
Special Sal
Ser 16h2'38"3d0'
Special-K
Del 20h37'1"10d53'
Speciale,Benjamin Barnes
Her 17h1'32"48d20'
Speck,Eva
Gem 6h39'39"31d19'
Speck,Jonathan M & Hillary F Stone
Aqr 19h31'13"7d46'
Speck,Peter
Cyg 20h26'41"31d29'
Speck,Vicki Sue
Cas 23h40'25"62d13'
Speckman,CLU,Trent V
Cep 0h4'59"80d17'
Spectacular Sandie
Lyr 18h49'33"41d55'
Spectacular Slout Star
Cet 0h25'48"1d49'
Spector,Elmor
Ori 6h14'32"0d35'
Spector,Hanakai
Uma 12h45'23"53d35'
Spector,Iris
Cam 4h4'33"60d28'
Spector,Judd
Aur 6h28'35"38d47'
Spector,Marc Adam
Ari 2h33'18"22d4'
Spectrum Diamond Star
Car 7h30'0"-56d44'
Speculation
Ant 10h53'43"36d15'
Spedicato 23,Léo
Dra 9h57'0"74d12'
Speed aka RKS Star, Ronald K
Cep 23h3'41"60d55'
Speed,Laurel Pierson
And 23h44'30"40d57'
Speedo Son of K-Bob
Cep 20h27'42"60d9'
Speedy
Aur 6h4'48"36d46'
Speegle,Christopher Palmer
Boo 14h53'46"22d53'
Speer,Brian & Tricia
Cet 1h27'23"0d29'
Speer,Harry
Cep 23h8'51"70d57'
Speer,Mark D
Her 16h41'45"40d55'
Speer,Michele Lee
Lyn 8h10'16"38d42'
Speer,Sharon Renee
Lyr 19h18'17"42d3'
Speer,Terrie Marie
Aql 20h7'20"1d26'
Speert,Elizabeth Klee
Cas 0h57'0"58d51'
Spei,Somnium
Cyg 21h56'22"53d24'
Speicher,Ken/Dolin Young
Ari 1h55'35"19d16'
Speidel,Birgit
Umi 14h40'57"77d51'
Speifer,Shawna Marie
Aql 19h56'0"15d13'
Speigel,Valerie Anne
Ari 2h25'54"21d58'
Speight,June & Randy
Aql 20h3'48"0d25'
Speights,Carolyn
Peg 23h44'32"28d37'
Speigner,C William
Gem 6h48'48"30d22'
Speldrich
Aql 19h56'36"8d11'
Spelic,Carey Anne
Lyn 8h25'52"42d46'
Spelic,Ronald
Cet 2h17'1"7d10'
Spelic,Shanon
Del 20h14'1"10d13'
Spelkoman,Kathleen Mulcahy
Cyg 20h22'53"38d56'
Spell,Audrey Diann
And 2h31'33"49d53'
Spell,George Robert
Cet 0h45'28"-4d42'
Spellbound '93
Com 12h19'32"32d40'
Spellbound In Shelley's Light
Cas 23h19'45"60d44'
Spellen,Jennifer
Peg 23h3'41"30d16'
Speller,Andrew J
Cep 3h29'59"80d26'
Speller,William T
Sco 16h50'21"-38d56'
Spellman II,James Earl "Dolch"
Lac 22h22'12"54d25'
Spellman,"Dick" Richard Deering
Ser 15h39'42"2d12'
Spellman,Joe
Lac 22h44'43"38d56'
Spellman,Kevin & Lisa
Crb 16h10'26"26d12'
Spells,Jaimi
Cet 2h41'22"-18d57'
Spelman,Bruce & Barbara
Cet 1h1'58"-1d56'
Spen's Star
Dra 16h24'46"60d27'
Spenadel,Cantor Irving N
Her 16h7'52"41d8'
Spence,Adam
Aql 19h30'13"12d57'
Spence,Angela Ann
And 0h17'28"34d49'
Spence,Barbara P
Lyr 19h18'16"42d4'
Spence,Carlee Mae
Cmi 7h53'30"5d8'
Spence,Jeffrey
Ori 6h6'55"9d58'
Spence,Judy
Hya 8h32'35"-5d39'
Spence,Lynda
Lyn 8h21'40"43d24'
Spence,M J
Cet 0h51'56"-1d21'
Spence,Maggie E
Cyg 20h6'50"40d29'
Spence,Michael
Oph 17h39'45"11d46'
Spence,Sally J
Cas 0h16'54"61d24'
Spence,Taylor "The Princess"
Peg 22h44'21"27d39'
Spence,Toran Marie
Cap 21h51'46"-14d25'
Spencer
Uma 10h51'49"48d13'
Spencer
Aur 4h54'26"41d13'
Spencer John
Boo 14h52'51"48d41'
Spencer's Star
Peg 22h23'28"31d29'
Spencer,Bill
Cet 1h17'39"-2d50'
Spencer,Brittany Bell
Peg 22h19'46"26d39'
Spencer,Carley J
Lac 22h16'23"54d31'
Spencer,Catharine S
Aql 18h58'32"-6d28'
Spencer,Charles
Lac 22h23'36"50d7'
Spencer,Clara Congdon
Aqr 23h14'13"-5d58'
Spencer,Curtis Dale
Lac 22h11'0"46d33'
Spencer,Dana Blair
And 22h57'33"37d52'
Spencer,Danielle M
Cnc 8h54'21"32d8'
Spencer,David
Sct 18h43'16"-5d9'
Spencer,David Cecil
Sct 18h54'0"-6d20'
Spencer,David Dwight
Her 17h30'43"14d55'
Spencer,Elizabeth
Lyr 19h4'51"28d33'
Spencer,Faye Alys
Umi 16h39'40"77d10'
Spencer,George
Cyg 19h57'1"40d28'
Spencer,Ian Joseph
Her 17h3'1"47d44'
Spencer,Jacquelin L
Sge 19h0'54"18d47'
Spencer,Jaime & Linda
Crb 15h25'44"32d25'
Spencer,Jean Elise
Peg 22h40'45"20d59'
Spencer,Jeff
Dra 15h41'1"62d2'
Spencer,John
Per 1h57'42"48d10'
Spencer,Jonathan Andrew
Her 16h47'14"19d48'
Spencer,Julia Marie Grace
Peg 21h41'42"20d30'
Spencer,June Alison
Cyg 19h29'26"34d41'
Spencer,Katherine Bernadette
Sct 18h51'16"-7d45'
Spencer,Kieran Jude
Vir 13h11'13"-8d25'
Spencer,Lauren
Lyr 18h15'25"45d59'
Spencer,Lisa-May
Mon 7h15'54"-10d5'
Spencer,M A
Lyn 8h21'31"58d54'
Spencer,Martin Roy
Aql 20h0'32"1d39'
Spencer,Mary Kaye
Cyg 20h39'27"40d2'
Spencer,Michael
Boo 14h58'43"37d48'
Spencer,Nelle May Stewart Johnsey
Aql 19h3'55"0d51'
Spencer,Nicholas William
Dra 16h3'27"61d19'
Spencer,Omer
Sct 18h53'57"-4d38'
Spencer,Pamela Faith
Lyn 8h2'23"37d44'
Spencer,Patrick Michael
Aql 19h57'23"10d11'
Spencer,Polly
Lyn 6h36'48"54d53'
Spencer,Randy Alan
Her 18h4'22"38d59'
Spencer,Robert
Cyg 19h50'0"40d47'
Spencer,Rodney & Carole
Cyg 19h27'48"38d5'
Spencer,Sammy
Boo 14h6'30"26d4'
Spencer,Samuel Carter
Vir 11h40'28"2d35'
Spencer,Stephanie Dawn
Peg 23h15'0"33d37'
Spencer,Stephanie
Peg 23h0'25"18d18'
Spencer,Stephen L
Her 16h46'48"22d14'
Spencer,Steven Lee
Aur 4h38'32"30d19'
Spencer,Sylvia Susie
And 2h25'43"41d51'
Spencer,The S Seward
Aql 19h30'0"13d28'
Spencer-Davis,Jane Elizabeth
Cyg 20h1'0"30d30'
Spencer-Heyman
Cam 9h30'0"84d44'
Spencer-Nathan
Vul 19h0'49"25d25'
Spencer-Reed,Maura
Lyn 8h21'44"43d27'
Spencley,Lee Anne
Cas 0h54'34"62d15'
Spengler,Dina
Leo 10h52'42"-0d24'
Spengler,Kenneth C
Cas 0h57'40"56d8'
Spengler,The Real Ghostbuster,Egon
Pho 1h58'51"-44d4'
Spenik,Lynn Marie
Psc 1h28'0"20d32'
Spenillo,Jocelyn Kerry Ann
Cyg 21h31'0"31d22'
Spenuzza,Connie
Mon 6h36'18"7d28'
Spera,Ilaria
Gem 6h55'50"13d47'
Speranza
Lup 15h18'32"-42d4'
Speranza
Pic 4h41'59"-48d55'
Speranza,Franáois Yves
Dra 15h43'18"65d29'
Sperber,David Josef
Boo 14h37'50"32d50'
Sperber,Jennifer
Cas 22h56'14"55d33'
Sperber,Joella
Cas 23h48'20"50d6'
Sperduto,Nicholas
Dra 16h9'13"68d18'
Sperfslage,Kay
Cas 0h26'0"62d10'
Sperling,Anthony A
Aur 6h27'36"38d7'
Sperling,Frank-Michael
Ori 5h49'1"21d12'
Sperling,Hannelore
Peg 22h17'54"7d40'
Sperling,Jean
Cas 1h49'45"58d40"
Sperling,Julia
Aur 5h1'44"40d59'
Sperling,Justin Fuller
Lac 22h12'1"38d30'
Sperling,Rena Ethel
Lyr 18h30'0"31d44'
Sperling,Roland Stewart Vandijk
Her 16h26'31"15d30'
Sperm Whale
Cet 0h10'11"-18d52'
Spero (Mr & Mrs P Stringer)
Ori 5h57'1"16d6'
Spero,Carmela
Cas 0h32'43"63d5'
Spero,Coty Joseph
Uma 12h2'1"57d2'
Spero,Joseph & Evelyn
Cep 22h15'23"59d11'
Sperring,Keith
Cnc 8h10'47"30d48'
Sperry,Richard A
Dra 16h43'0"67d15'
Sperry,Sharon (S)
Del 20h30'18"20d22'
Speth,August C
Uma 9h23'44"51d15'
Spethmann,Sven
Sgr 19h4'8"-28d43'
Spevak,Stacey
And 1h4'53"39d43'
Speyer,David Leon
Aur 6h1'1"30d38'
Speziale 101092
Cam 6h11'24"58d30'
Spezzano,Peter & Sherry
Cyg 21h33'21"41d33'
Sphunt
Cas 0h36'21"70d10'
Spialters' Sheine Sterendl
Dra 17h27'57"73d19'
Spicer Spark,The Phillip
Ori 6h8'28"9d37'
Spicer,Christine Louise
Mon 6h55'0"-10d56'
Spicer,Janis
Mon 7h43'4"-5d33'
Spicer,John
Per 1h42'56"53d40'
Spicer,Margaret Jane
Aql 19h5'51"-0d11'
Spicer,Max & Anne
Cet 2h37'44"-6d39'
Spicer,Meade
Ori 4h56'53"-2d43'
Spicer,Michael
Her 17h8'57"44d47'
Spickermann,Sabine
Sco 17h8'21"-38d8'
Spicola,Francis C
Del 20h14'51"10d29'
Spicola,Shannon
Eri 4h8'11"-12d47'
Spicy Jalapeno,The
Tri 1h50'43"26d39'
Spicy Jalapeno,The
Del 20h15'23"14d40'
Spidall,Denise Annette
Del 20h19'34"10d55'
Spider
Lac 22h5'13"51d50'
Spider & The Web,The
Cyg 21h39'29"41d33'
Spider Man Pete
Uma 11h50'30"31d59'
Spider's Way
Peg 21h56'23"24d13'
Spidey's Web
Eri 4h4'60"-18d51'
Spiegel AKA "The Silver Fox",Norman
Ari 2h55'48"30d54'
Spiegel,Gary N
Aur 6h15'46"38d59'
Spiegel,Hans-Willi
Cap 20h6'12"-10d15'
Spiegel,Lawrence Stewart
Aur 6h6'13"35d46'
Spiegel,Roger Von
Per 1h56'11"56d44'
Spiegelman,Cabot
Uma 9h21'15"68d17'
Spiegelman,Susan Lorraine
Peg 22h33'33"27d38'
Spiegl,Marie
Lyn 8h36'48"43d11'
Spiegler sen
And 23h10'58"42d48'
Spiegler,Henry
Aur 6h22'49"32d26'
Spiegler,Jacqueline
Cet 2h2'40"1d22'

Spiekermann,Raimund Peg 23h33'33"20d8'
Spiel,To My Star Warren F Cep 20h56'40"55d43'
Spieler,Dr Gary L Cep 22h57'36"57d53'
Spieler,Larry & Gean Crb 15h31'0"27d45'
Spieles,Dr Gisbert Umi 15h9'24"66d40'
Spielman,Jr,John Francis Peg 21h50'49"28d52'
Spielman,Lauren Melissa Cas 0h6'34"64d1'
Spielman,Richard Aur 6h7'37"35d54'
Spielman-Ewan,Carol Sge 20h17'47"17d58'
Spielmann,Anne Eri 3h34'41"-2d18'
Spies,Jack Carlson Aur 6h31'0"32d56'
Spies,Jr,Robert C Gem 7h1'15"11d55'
Spies,Kiersten Marie And 1h52'47"40d54'
Spies,Megan Renee Cyg 20h50'0"40d31'
Spieth,Anna-Katharina-Salome Per 2h58'45"55d32'
Spieth,Armin Per 2h11'57"58d27'
Spieth,Michael Tyson Lac 22h24'40"53d35'
Spiezio,Joseph Samuel Sco 16h49'52"28d9'
Spigot's Star Boo 14h30'45"20d32'
Spike Gem 6h37'48"34d4'
Spike & RooBoo Eri 2h54'0"-2d12'
Spike's Star Umi 15h1'37"89d34'
Spiker,Benjamin David Dra 18h49'1"70d37'
Spilka,Adam Boo 15h32'22"42d6'
Spilka,Bruce Martin Cet 3h17'47"0d26'
Spilker,Chad Aur 5h14'1"44d47'
Spilker,Jörg Her 17h8'51"48d21'
Spillane III,James F Cep 22h10'24"55d24'
Spillane,Anne Marie Uma 9h54'37"57d18'
Spillane,Anthony J Uma 8h42'19"53d39'
Spillane,Curtis John Cma 6h56'44"-19d38'
Spillane,Jessica Collin Lyr 18h59'17"30d41'
Spillane,Lucille Picciano Uma 10h44'0"59d47'
Spillane,Timothy J Uma 10h33'27"56d20'
Spillane,Timothy J Psc 1h27'28"11d19'
Spillane,Timothy P Uma 9h51'33"56d30'
Spillard,Sophie Charlotte Lyr 18h37'32"37d38'
Spiller Love Forever Dave,Winfred Umi 15h38'12"72d16'
Spiller,Rosemary Aql 19h40'0"11d52'
Spillers,Steve Lac 22h6'38"50d24'
Spilling,Jr,Captain Henning J Aur 6h17'14"33d19'
Spillman Star,The Uma 10h18'1"48d34'
Spillone III,Frank Vito Cnv 12h32'34"38d21'
Spilman,Thomas Aql 20h12'29"11d23'
Spilotro,Chloe Lyn 7h30'14"41d3'
Spina,Catherine Lyr 19h17'0"38d40'
Spina,Erma Joy Vel 9h57'26"56d4'
Spinazzola Foundation, The Anthony Cep 22h2'35"61d21'
Spindel,Mary L Cas 1h8'18"61d19'
Spindler,Volkmar, Martin,Helmut Lib 15h30'36"-10d52'
Spinelli,Frank Aur 7h15'17"36d56'
Spinelli,Marina Del 20h17'22"13d7'
Spinelli,Rose Marie Mon 7h20'57"-6d12'
Spinelli,Sam Dra 16h3'0"68d41'
Spiner,Brent Aqr 22h0'1"-5d35'
Spingler,Elizabeth Anne Mon 6h21'10"5d43'
Spink,Joseph H Per 2h8'37"57d40'
Spink,Peter Tri 1h48'60"28d24'
Spinnenhirn,Carlo Uma 9h55'1"51d19'
Spinner,Justin Alexander Del 20h13'31"12d15'
Spinner,Marion Aur 4h47'38"36d5'
Spinner,Marshal B Eri 3h11'1"-4d41'
Spinner,Patti Peg 22h0'14"4d13'
Spinner,William J Per 1h55'56"50d25'
Spino,Gary L Boo 14h18'15"12d14'
Spinogatti,PhD,Peter Joseph Aur 4h49'11"40d9'
Spinola,Andrew Aur 6h11'50"38d33'
Spinola,Carli Lee Taylor Cam 7h55'38"60d18'
Spinola,Lois Cas 0h58'41"61d7'
Spinosa,R Anthony Cep 22h8'25"70d44'
Spiny Norman Uma 11h1'44"48d40'
Spiralis Cyg 20h19'51"31d33'
Spires,Judith Peg 22h46'55"33d56'
Spirig Family Hya 9h54'16"-17d53'
Spirio 10-01-1925, Lawrence Lib 15h38'17"-28d21'
Spirit O'Brien Per 1h26'31"50d28'
Spirit of Amy Sue,The Lyr 19h17'30"33d45'
Spirit Of David Cep 23h5'13"67d33'
Spirit Of M & M,The Crt 11h9'35"-14d32'
Spirit of Maggie Gem 7h22'4"27d51'
Spirit of Nashville, The Aql 19h56'26"10d28'
Spirit Of Newport Cyg 20h30'15"40d30'
Spirit Of The White Wolf Hya 9h11'46"1d3'
Spirit Partners Don & Janel Lyn 7h39'0"43d7'
Spirit Warrior Ori 5h53'0"21d16'
Spirit Wings Cas 1h22'50"60d55'
Spiritia,Stella Ruggieri Cet 2h20'40"1d7'
Spirito,Denise Aql 19h48'28"13d7'
Spiritus Brandon Per 4h34'26"42d1'A
Spiritus Dea Juris Per 4h34'26"42d1'B
Spiritus Lida Per 4h34'26"42d1'C
Spiro,Hannah Michelle Umi 16h21'26"70d16'
Spiro,Jeffrey Uma 10h41'17"70d22'
Spiro,Norman Lac 22h40'54"51d0'
Spiroff,Karen Cet 3h0'1"8d1'
Spittle,Margaret Cas 0h35'1"73d37'
Spittler,Maria Gem 6h38'41"30d22'
Spittler,Mary Jane Com 12h53'52"25d38'
Spittles Star,The Cnc 7h56'52"18d57'
Spitz,Alison Jane And 23h37'0"47d31'
Spitz,Garrett Her 16h38'0"34d37'
Spitz,Lawrence S Boo 15h24'32"40d28'
Spitz,William Jerome Aql 20h0'1"10d8'
Spitz,Wolfgang Peg 22h56'41"8d21'
Spitzer,Erin Lyn 8h2'46"33d26'
Spitznagel,Erik Sex 10h21'1"-2d2'
Spivack,Diane & Evan Lyn 8h7'1"45d12'
Spivack,Joseph "Y" Aur 5h7'1"40d50'
Spivey,Jr,William David Ori 6h5'56"-2d42'
Spivey,Lee Cet 1h15'49"-1d60'
Spivey,Norma Lmi 10h52'36"30d39'
Spizer,Harold David Ori 4h59'1"10d51'
Splendens Stella in Perpetuum,Johnetta Lac 22h18'45"49d40'
Splendida Italia Umi 14h45'19"71d22'
Splendidus D Coombe, The Ori 5h26'40"0d45'
Splendor Magnificus Cet 2h59'10"4d44'
Splendora Opening Night Lya 18h34'0"33d52'
Splitapart Sge 19h55'44"19d30'
Splosion Cnv 13h49'48"40d11'
Spock Casey Hya 8h41'1"0d7'
Spoden,Thomas Lyn 8h21'14"49d36'
Spoelhof,Patricia Ann Lyr 18h23'29"38d12'
Spofforth,Penny Gem 6h52'9"12d26'
Spohn,Herbert E Per 1h45'34"53d54'
Spohn,Jerry R Lac 22h15'27"46d57'
Spohnholz,Jr,Robert G Sco 16h59'30"-30d34'
Spohrer,Robert F Lac 22h23'46"50d28'
Spokes,Gary Aur 5h33'52"40d45'
Spomer,Uncle Don Uma 11h56'10"40d16'
Sponar,Irene Cas 1h34'38"65d35'
Sponseller,Karen Cyg 20h15'43"30d51'
Spontak,Mark Her 18h4'17"40d26'
Spoo,Mark Lac 22h26'44"56d32'
Spookie's Super Star Cas 22h58'15"58d43'
Spoone,Sterling Jennings Aql 19h54'44"12d39'
Spoone,Tee Cmi 7h6'45"5d16'
Spooner,Craig Spooy in memory of Cru 12h48'36"-59d0'
Spooner,Daniel F Hya 8h11'59"-6d54'
Spooner,James Mitchell Psc 1h32'14"20d23'
Spooner,Jr,Tom Oph 17h21'44"-20d18'
Spooner,Richard Sherman Dra 14h46'0"63d46'
Spore,Thomas Allen Sex 10h29'1"-2d7'
Sporing,George "Dad" Boo 15h8'42"8d44'
Sport Cet 3h6'54"0d10'
Sportini,Kara H And 0h5'25"35d0'
Sposato,Bill Oph 18h2'44"12d16'
Sposi "My Little Ravioli" Sex 9h43'30"4d19'
Sposito,Kevin Cap 21h44'43"-24d5'
Spot Lyn 7h5'41"53d2'
Spot Lyr 19h22'44"40d34'
Spoth,Kristopher Keelan Lyr 19h23'15"40d35'
Spoto,Jr,Frank J Cam 5h50'50"60d18'
Spradling,E J Dra 18h56'41"50d29'
Sprafka,Sr,John J Cet 0h41'59"-2d1'
Spragale,Matthew Cleary Per 3h20'42"40d46'
Spragga Lyn 9h13'18"35d17'
Spragins,Robert F Aur 6h17'14"31d44'
Sprague's Star Uma 8h59'51"58d20'
Sprague,DDS,Richard A Cep 21h45'52"68d29'
Sprague,Eugene Peg 23h24'14"17d56'
Sprague,I Love John Ori 5h56'51"15d60'
Sprague,June Clare Equ 21h8'50"10d19'
Sprague,Leah G Crb 15h55'49"31d45'
Sprague,Michael Mon 7h58'20"-2d26'
Sprague,Rita & Jim Aql 19h55'47"14d3'
Sprague,Tyler Gregory Oph 18h19'33"7d1'
Sprague-Williams Jay-Jay,Jan Mon 6h51'59"11d10'
Spraitz,Dr Anton "Tony" Aur 4h41'28"31d25'
Spranger,Symantha And 0h9'34"41d5'
Spratt,Allison Ann Sge 19h40'48"16d21'
Spratt,Brenda Kay Cyg 20h23'27"30d33'
Spratt,Steve Sgr 20h22'25"-28d51'
Spreadbury,Sherri Marie Cas 23h34'0"61d53'
Spreng,Rita Leo 9h59'59"11d54'
Sprenger,Michael Boo 14h48'13"24d7'
Sprenger,Vera Cnc 8h0'31"10d20'
Sprengnether,Ronald J Her 15h48'57"42d30'
Sprigg,Steve Her 18h2'22"48d48'
Spriggs,Rachel Ann Cas 2h36'10"60d48'
Spriggs,Rachel Ann Cas 1h14'54"60d10'
Spring Cas 0h48'48"67d39'
Spring,Alice Lyr 19h18'36"42d48'
Spring,Anna Cas 1h31'1"70d42'
Spring,Elliot Blaikie Cam 9h57'52"82d22'
Spring,Matthew Lac 22h32'52"52d54'
Springate,Martin Ori 5h17'12"15d21'
Springen,Dan Aur 5h1'50"31d7'
Springer"Robin's Nest",Robin D Uma 11h6'51"47d3'
Springer,Becky Cas 2h38'53"73d56'
Springer,Bryan Aur 6h32'59"35d4'
Springer,Carol Peg 22h40'47"31d23'
Springer,Christopher Oph 17h18'39"10d6'
Springer,David A Boo 15h4'17"10d48'
Springer,Jack Per 1h30'29"52d53'
Springer,Linda Ann Wisler Mon 6h59'45"-8d22'
Springer,Maggy Ann Uma 9h27'15"55d26'
Springer,Nicki Christine Vul 19h23'1"25d5'
Springer,Sarah Jane Cas 0h27'1"60d0'
Springer-Schultz,Faith Peg 22h8'0"25d50'
Springfeldt,Christina Cas 0h14'26"63d60'
Springfield,Adam Kenneth Cet 2h56'1"4d7'
Springhorn,Mary L And 2h18'0"49d49'
Springman,Lorraine Mary Peg 23h38'1"20d3'
Sprinkel,Robert J Tri 2h33'0"32d42'
Sprinkle,Jim Per 3h12'40"47d47'
Sprite Ser 15h38'39"8d50'
Spritzer,Hortense May Peg 23h29'0"21d57'
Sproelich's 28 Star, Col Thomas (TK) Aql 19h54'57"14d9'
Sproge,Kim & Curtis Uma 10h14'58"50d35'
Sproger,Philip C Aur 6h7'33"31d35'
Spross,Sarah Lyr 19h21'58"33d46'
Sprotte,Daniel Joseph Psc 0h21'18"1d52'
Sproul's Star,Margaret And 23h47'24"44d58'
Sproul,Jennifer And 23h43'0"43d18'
Sprouse,Agnes Rose Lac 22h9'26"50d8'
Sprout,Carmela Anne Cyg 21h35'1"41d31'
Sprowls,Craig Cyg 19h25'31"32d1'
Sprowls,George F Her 18h8'34"40d14'
Spruce,Nigel Talbot Per 3h11'25"42d33'
Spruiell's Splendor Her 17h9'43"46d14'
Spruiell,Jr,Thomas B Cnv 12h20'34"40d37'
Spruill,Eddie Aql 20h14'34"0d17'
Spruill,Jr,Jimmie Dan Ori 5h59'53"15d26'
Spruill,Melton Gerard Cam 6h11'38"56d2'
Spruill,Robyn Leigh And 2h26'0"49d56'
Spruill,Walter J Boo 15h2'19"31d26'
Sprunger,Donna Cyg 20h19'18"37d48'
Sprungle,Jennifer R Lyn 7h3'47"50d16'
Sprunt,Jr,Kenneth M Mon 8h8'46"-0d33'
Spry,Larry James Her 16h51'56"38d44'
Spumoni Per 4h25'16"51d44'
Spunk Cap 20h28'39"-26d16'
Spunk,Deni Cam 7h10'31"80d17'
Spunky And 2h19'42"39d25'
Spunky Cyg 21h34'35"41d6'
Spunt's Guiding Light, Anne & Sy Peg 21h39'60"25d52'
Spur,Rebecca Lyr 18h29'39"45d41'
Spurge Cnv 12h48'32"51d3'
Spurgeon,Alex Matthew Boo 14h57'34"20d41'
Spurgeon,Benjamin Michael Peg 22h34'57"10d42'
Spurgeon,Blake Michael Aur 6h54'15"44d14'
Spurgeon,Casey Dwane Uma 9h24'41"48d15'
Spurgeon,Cecil Mon 7h0'37"8d46'
Spurgeon,Christopher Daun Vul 19h44'50"20d34'
Spurgeon,Jennatte Leigh Lyn 7h55'26"47d39'
Spurgeon,Jr,Denny Lee Cyg 20h50'18"40d43'
Spurgeon,Lucas Paul Lac 22h28'54"53d33'
Spurgeon,Micah Leigh Cyg 21h34'55"40d60'
Spurgeon,Natalie Jean Cas 23h22'40"61d48'
Spurgeon,Sara Leigh And 1h52'50"40d4'
Spurgeon,Sydney Olivia Aur 6h54'0"44d2'
Spurgin,Sophia H Cet 1h32'58"0d40'
Spurlin,Doyle & Marie Eri 2h54'59"-6d58'
Spurlin,Maggie Mon 6h19'48"4d35'
Spurling,Tara Ann Eri 3h1'35"-7d45'
Spurlock,Jay Wesley Eri 4h6'50"-11d46'
Spurlock,Kelli Anne Hood And 0h48'58"41d12'
Spurlock,Rosemary Lac 22h25'0"50d34'
Spurlock,Stephen A Her 16h49'13"41d7'
Spurr,Listiany Oph 17h8'0"11d44'
Spychala,André Ori 5h41'11"1d42'
Spychala,Corinne Ruth Per 3h47'50"35d25'
Spyguy Dra 19h20'53"58d24'
Spyrakos Tri 2h45'1"31d40'
Spyridakis,Evangelia-Michelle-George Boo 14h35'18"41d45'
Squadrito,Anthony Her 16h52'0"51d9'
Squadron,Daniel Louis Sco 16h52'42"-40d18'
Squalabeep,Vincenzo Cyg 20h11'17"37d31'
Squeak Loves Clarence Forever Dra 16h55'16"71d7'
Squeaky Cyg 19h33'1"36d56'
Squealer Sheedy Peg 21h51'40"30d35'
Squeek Ori 5h56'36"19d57'
Squeeky Sct 18h46'58"-7d23'
Squeo,John Her 17h56'55"41d3'
Squidaly Star,The William B McHugh Lac 22h19'17"50d28'
Squidney 296 (Kate's Place) Ori 5h56'59"12d40'
Squillante,Stefanee Lynn Del 20h14'34"11d46'
Squire,B B B Tri 1h59'16"26d56'
Squires III,St Arthur Lee Cet 0h39'51"-0d6'
Squires,Bradley David Robert Oph 18h7'1"13d53'
Squires,Fredric Barton Cam 7h58'13"60d48'
Squires,Jonathan Charles Per 2h2'31"50d10'
Squires,Keri Lee Equ 21h11'1"11d30'
Squires,Mike & Dianne Del 20h35'39"3d22'
Squirewell,Warren H Uma 9h39'17"58d17'
Squirrel's Mermaid Lib 15h39'30"-23d11'
Squirrelamour Aur 5h2'28"41d15'
Squirrell,Thomas Cam 3h54'50"62d27'
Squitieri,Connie Cyg 20h52'14"31d20'
Squits Oph 17h10'35"-20d56'
Squush-A-Muffin Cas 1h15'21"62d11'
Sr Bonnie Marie Sge 19h18'22"18d36'
Sr Canabal 201 Uma 11h2'17"25d7'
Sra Modoc Lyn 7h35'26"42d0'
SRA-The SMF Per 1h52'0"53d57'
SRCB-Volcun & Nimrod Lac 22h55'31"51d19'
Srebnick,Barry Boo 13h55'19"20d31'
Sridhar,Thirumalai Cyg 21h1'23"40d47'
Srivastava,Amit Aql 19h54'31"-5d46'
Srivatsa,Zena Tri 2h16'54"33d26'
Srocki,Walter & Laurie Uma 9h43'17"42d32'
Sroczynski (Ski), Ronald George Cyg 21h49'58"37d40'
Sronce,Alexander Benjamin Her 17h5'35"20d56'
Sronce,Angela Sue Cyg 20h29'55"50d14'
Sronce,Breanna Emma Lyr 18h40'31"35d2'
Srpan,Elizabeth & Joseph And 0h21'30"21d45'
Srpuill,Tommy Lee Ser 17h33'51"-14d32'
SSS Leebeaux Mon 6h53'53"-1d35'
SSSB Lmi 9h41'13"33d25'
St Alexamarcus Nor 16h19'50"-59d26'
St Amand,Marcel Cyg 21h5'26"32d44'
St Catherine of Siena Aql 19h25'54"-10d16'
St Clair Harvey Per 2h2'45"57d25'
St Clair,Berry Bear Uma 11h56'32"62d3'
St Clair,Elizabeth Anne Cas 1h4'42"52d52'
St Clair,Rickey Hya 8h31'18"-0d53'
St Clair,Stimpy Lmi 10h7'48"41d29'
St Croix,Wild Bill Hya 8h16'51"0d23'
St Cyr,John Boo 13h58'15"22d28'
St Cyr,Sue & Frederil Peg 23h23'0"30d40'
St Gelais,Roger Dra 15h46'1"61d52'
St George's Sapphire for Kevin Charles Leo 9h22'55"18d9'
St George,Dillion Aur 5h50'37"50d5'
St George,Kathy Equ 21h22'0"8d20'
St Germain,Genevieve Lyn 7h55'24"48d13'
St Germaine,Arthur Adelor Per 3h37'52"38d6'
St Germaine,Mary Laverty Uma 10h53'3"39d17'
St Goar,Walter Treat Lac 22h36'48"50d32'
St James,Monte Cet 2h36'1"-10d55'
St James,Venus Crb 16h1'0"28d8'
St Jean,Dorothy & Ernest Oph 17h23'0"10d19'
St Jean,Sr,& Family Ned Cyg 21h10'42"37d52'
St Joe Med Center's Bumblebee Brigade Cam 10h55'1"81d51'
St John's Studio II, Carol And 2h22'53"41d57'
St John,Ann Lyr 18h31'1"32d8'
St John,Beverly Mae Mann Com 13h24'1"26d29'
St John,David Cep 23h0'45"64d30'
St John,Don Aur 5h35'51"41d12'
St John,James Thomas Cet 2h33'46"9d37'
St John,Jennifer Lyr 18h55'30"33d0'
St John,Kylie Nicole Aql 19h29'31"8d31'
St John,Marlene J Umi 15h4'27"79d9'
St John,Melville Cet 1h58'16"-8d27'
St John,Nicholas Peg 21h8'0"13d6'
St John,Robert Ori 6h15'1"-0d48'
St John,Shauna Peg 22h45'50"29d39'
St John-Claire,Pamela & Alex Cyg 21h2'31"38d30'
St John-Sullivan's Paradise Sco 16h25'29"-40d38'
St Laurent,Jean Claude de Kerdual Her 16h25'12"23d27'
St Laurent,William Peg 21h24'40"10d54'
St Louis,Fernand J Cam 13h42'34"80d24'
St Martin,Carlisle Oph 16h50'31"11d19'
St Mary's Hospital Lyn 7h0'46"52d17'
St Mary's Rising Stars Umi 16h29'39"71d21'
St Matthew Roman Catholic Church Cyg 21h54'19"53d24'
St Michael Eri 4h45'11"-8d16'
St Michael Sge 20h0'32"20d19'
St Michael Per 3h10'13"46d35'
St Michael,Faith And 23h34'25"48d5'
St Nicks Sidekicks Sex 10h4'59"-5d40'
St Onge,Roger T Sgr 19h0'20"-23d45'
St Pat's Lovers Uma 10h8'43"54d55'
St Patrick,Daniel & Jamie Leigh Uma 10h47'41"48d53'
St Patrick,Laurel Melissa Lyr 19h19'52"41d41'
St Philip's Academy Dra 18h1'48"50d35'
St Pierre In Loving Memory,Jeanette Aql 19h46'55"10d4'
St Rita's Volunteers Cam 12h21'44"84d57'
St Sandy Ser 15h54'53"0d4'
St Schmidt Lyr 18h26'32"40d4'
St Stanislaus Kostka Uma 8h38'20"67d59'

St Thomass,Percell
Cep 21h39'42"55d11'
St Tom
Aql 19h23'29"-5d39'
St-Amant,Andrée
Lyr 18h41'38"35d5'
St-Arnaud,Simone
Lyr 18h21'49"44d27'
St-Cyr,Johanne
Uma 10h50'31"51d33'
St-Hilaire,Monique
Cas 22h56'52"54d14'
St-Pierre,Francine
Sco 17h51'19"-30d37'
St-Yves,Liette
Cyg 21h15'0"38d2'
St.Clair,Donald John
Aur 5h4'10"38d20'
Staar,Virginia
Uma 12h9'57"46d50'
Staats,Emily K
Cyg 19h42'37"30d20'
Staats,George Pomeroy
Ser 18h54'18"3d35'
Staats,Margaret & James
Crb 16h18'52"38d49'
Staben
Ari 2h41'0"28d46'
Stabenau,John
Per 2h7'44"58d33'
Stabernack,Wilhelm Georg Josef
Cnc 8h8'38"30d13'
Stabile SSEN,Elisa Marie
Cma 6h55'55"-16d2'
Stabile,Dana Leigh
Cyg 20h40'46"45d44'
Stabile,Mario Lawrence
Aur 5h19'31"41d16'
Stabile,Mary Elizabeth Belcastro
Cas 0h55'49"56d6'
Stabile,Mattia
Umi 16h1'19"72d25'
Stabile,Melissa Joy
Peg 21h53'25"28d41'
Stables,Shelley
Cam 5h38'35"73d57'A
Stacey
Lyn 8h12'25"43d4'
Stacey
Cyg 19h59'57"48d49'
Stacey
And 1h25'36"37d59'
Stacey
Aql 19h55'42"12d36'
Stacey
Lmi 9h22'0"37d39'
Stacey & Jon
Ori 5h57'40"8d50
Stacey & Mo's Forever Love
Aql 20h11'41"0d53'
Stacey & Robert Forever, 2-14-96
Uma 12h3'24"39d59'
Stacey Ann
Crt 11h36'11"-11d31'
Stacey Ann
Lyr 18h43'12"46d32'
Stacey Anne
Aqr 21h24'54"-0d56'
Stacey Catherine Emily
Cyg 19h49'16"38d41'
Stacey Dawn
Mon 6h29'46"8d2'
Stacey Elyse
Vul 19h14'51"21d26'
Stacey Jill's Rising Star
And 0h9'10"46d12'
Stacey Leigh
Cas 1h50'1"77d4'
Stacey M
Lyn 8h5'46"45d32'
Stacey Marie
Cep 23h5'54"63d57'B

Stacey Renee-J1
Umi 19h29'18"88d32'
Stacey's Star
Lyn 8h3'55"38d52'
Stacey's Stupendous Star
Com 13h17'29"20d24'
Stacey,Adam Jarod
Lac 22h4'30"38d53'
Stacey,Erik Logan
Cet 2h25'44"4d55'
Stacey,Jessica Gene
Lyn 7h33'1"41d48'
Stacey,Michael
Dra 16h32'49"68d48'
Stacey,Natalie N
Lyr 18h35'8"40d35'
Stacey,Virginia
Cas 0h32'37"60d41'
Stacey-Jayne,The Gypsy Star
Cas 1h42'41"60d24'
Stacey-Robert
Cyg 20h24'24"41d20'
StaceyJohn
Lyr 18h52'21"30d54'
Stach,Engelchen Elke
Leo 9h52'36"11d18'
Stachler's Dream
Sct 18h54'45"-6d19'
Stachnik,Helcia
Lyr 19h17'54"40d24'
Stachnik,Walter A
Aur 7h12'0"35d50'
Staci Lee T
Lyr 18h39'16"40d49'
Staci's Star
And 1h19'36"39d47'
Stacia
And 2h3'30"45d44'
Stacie
Cyg 19h14'47"45d52'
Stacie & Jess's Star
Mon 6h55'26"-10d46'
Stacie Lynn
Lyr 18h48'46"39d48'
Stacie Marie
And 1h41'38"50d8'
Stacie's Sweet Sixteen
Tau 3h36'40"24d24'
Stack,Danna C
Cet 1h43'20"-5d7'
Stack,Denver
Ori 5h31'3"-1d22'
Stack,Greg & Mary
Sge 20h0'43"18d28'
Stack,Jennifer
And 1h24'29"34d6'
Stack,Jim & Daisy
Lyr 18h41'27"47d47'
Stack,MBA,Robert P
Boo 14h47'35"29d42'
Stack,Susan Meehan
Ori 5h25'9"-2d58'
Stackhouse,Robert M
Cep 2h48'22"80d21'
Stackman,Bobbie Forgays
Cyg 21h5'1"35d51'
Stacy
Lyn 8h6'55"50d22'
Stacy
Peg 22h20'42"27d48'
Stacy
And 23h36'46"43d32'
Stacy
Peg 22h25'45"24d20'
Stacy
And 23h39'36"39d56'
Stacy
Vul 21h25'43"27d36'
Stacy
Lyn 6h17'43"59d53'
Stacy & Adam's Star
Cyg 21h58'39"44d51'
Stacy I
Aqr 20h46'32"-0d26'

Stacy Lynn Junebug
And 1h57'13"39d33'
Stacy My Love
Peg 23h6'40"11d55'
Stacy Starr
Lyr 18h36'22"40d39'
Stacy's Star
Tau 3h57'31"12d21'
Stacy's Star
Del 20h20'47"11d9'
Stacy,Jennifer
Sct 18h44'46"-7d53'
Stacy,Robert Edward
Cam 4h55'19"71d7'
Stadelmeier,Josiah
Her 18h10'14"30d22'
Stadler,Andrea Kristine
Lyn 6h34'31"60d49'
Stadler,Hans
Cyg 20h27'59"31d28'
Staedman,Gillian
Lyr 18h29'53"32d55'
Staeheli Star
Cam 4h13'40"68d15'
Staehle,Carrie Washburne
Aql 19h7'22"15d8'
Staehle,Gerhard
Peg 23h30'46"17d19'
Staenberg,Toby Ina
And 1h58'43"37d54'
Staff,Julia
And 2h1'45"47d14'
Staff,Matthew
Her 16h45'50"35d14'
Staffieri,Tony
Per 1h34'24"53d53'
Staffin,Lois & Gerry
Cyg 19h31'27"33d2'
Stafford III,Roland Thomas
Hya 8h12'1"0d56'
Stafford's Star,Jim
Per 3h7'25"40d30'
Stafford,Brett
Cap 21h37'31"-21d42'
Stafford,Evalyn & Twinky
Aqr 22h32'41"-0d46'
Stafford,Florence Fisher
Mon 6h37'1"6d25'
Stafford,Gillian
Peg 23h29'49"21d10'
Stafford,James L
Boo 14h12'0"40d10'
Stafford,Mark
Cyg 20h56'29"31d2'
Stafford,Mary Ellen
Ori 5h4'22"14d2'
Stafford,Pam
Equ 21h2'36"10d4'
Stafford,Thomas Leonard
Aql 18h57'34"-5d35'
Stafford,Wendy
Lyr 18h31'0"30d53'
Stafford-Owens
Eri 3h43'20"-10d33'
Stagg,Bryn Nichole
Ori 6h9'32"4d39'
Stagg,Christianne
Cet 2h8'12"0d40'
Stagg,Kristen Delaney
Ori 6h12'18"7d55'
Stagg,Melanie
Ori 5h30'34"0d11'
Stagg,Paul J & Sheila A
Cyg 21h4'38"31d4'
Stagge,Roberta "Cissy"
Lyn 7h58'48"36d28'
Staggerwing 17,Steve's
Dra 20h21'1"62d33'
Stagl,Werner
Sgr 19h47'39"-41d9'
Stagno,Nancy Lee
And 0h57'1"40d40'
Stags
Cet 2h0'1"-1d12'

Stahl,Bradford Tod
Lac 22h29'0"55d18'
Stahl,David & Andrea
Cyg 20h39'54"38d40'
Stahl,Ernest & Ruth
Cam 3h52'1"58d12'
Stahl,J D & Sarah
Lyn 7h40'55"40d26'
Stahl,Myron Richard
Her 16h47'0"50d26'
Stahl,Peter
Eri 4h49'12"-6d56'
Stahl,Richard & Alma
Her 15h55'34"50d39'
Stahl,Stephen Ray
Per 3h3'1"50d33'
Stahl,Susan
And 0h11'33"32d6'
Stahl,Terri Lucinda Church
Hya 9h22'56"-10d51'
Stahl,Werner
Lac 22h1'34"51d31'
Stahli
Umi 15h14'41"68d3'
Stahlnecker,Eugene
Lac 22h8'34"50d32'
Stahlschmidt Wishing Star
Equ 21h11'53"7d52'
Stahly Star,Vernon & Francis
Sge 19h54'34"18d45'
Stahly,Michael Evan
Cyg 21h40'30"38d41'
Stahnke,Paul K
Per 2h7'29"57d22'
Stahoski,Florence
Cas 23h39'46"64d21'
Stahoviak,Jessica Monique
Mon 6h38'38"1d40'
Stahr,Dallas
Aql 18h40'42"-2d42'
Staiano,Kayla Nicole
Aqr 23h28'58"-6d38'
Staiert For My Beloved Sweets - Amy
And 0h15'29"38d17'
Staimmer,Dieter
Umi 16h27'37"70d28'
Stainbrook,Ian James
Sex 9h53'35"-6d43'
Stainer,Glenn James & Kelly Anne
Cyg 20h15'42"31d4'
Staines,Ron
Per 2h13'1"57d7'
Staino Antonio
Vul 19h19'14"23d17'
Stairway To Heaven
Cyg 21h14'57"38d2'
Stajger,Jonathan B
Aur 5h32'43"30d57'
Stajkowski,Donna N
Oph 17h30'32"11d33'
Stalder,Robert E
Lmi 10h15'1"33d8'
Staley III,Robert Bertel
Cet 0h52'26"-6d38'
Staley,Dell Marie
And 0h48'23"40d48'
Staley,Jacqueline
Cas 0h39'54"67d55'
Staley,Jourdan Elizabeth
And 0h12'1"41d8'
Staley,Melanie Marie
Peg 22h45'47"32d24'
Staley,Phillis Maxine
Cas 1h0'51"64d16'
Staley,Steven G
Her 18h1'51"38d34'
Staley,Timothy J
Lac 22h1'0"38d14'
Stalinsky,Debbie
Peg 21h56'55"23d39'
Stalisia
Ind 21h18'47"46d42'

Stalker
Lac 22h13'43"46d45'
Stalker,Katharine
Lyr 18h47'51"31d24'
Stall,Constance A
Ari 1h56'16"11d38'
Stallard,Robert T (The Friendship Star)
Cas 1h0'58"50d8'
Stalli,Antony C
Cyg 20h30'37"30d29'
Stallings,Ana-Maria Patricia
Peg 21h44'28"24d2'
Stallings,Joe W
Cet 3h10'40"5d39'
Stallings,Mary
Sgr 19h37'4"-20d24'B
Stallings,Norman
Sgr 19h37'4"-20d24'A
Stallings,Robert Andrew
Boo 15h0'42"20d14'
Stallings,Sr,Joan & James R
Uma 10h38'13"52d32'
Stallone,Jr,Frank
Ori 5h57'51"8d47'
Stallone,Sylvester
Hya 8h30'41"-6d1'
Stallwood,Marion E
Uma 13h39'48"60d41'
Stalpers,Hubert
Cap 20h36'54"-21d12'
Stalter,Sonny Liane Cecelia
Lmi 10h4'29"35d29'A
Staman
Lib 14h17'13"-8d56'
Stamatina
Uma 9h3'24"48d52'
Stamatinos,Isidoros
Lyn 7h55'47"50d34'
Stamatiou,Paulina
Sco 17h4'1"-31d14'
Stamats,Richard I
Per 1h40'41"53d53'
Stambaugh II,Edward Eugene
Vul 19h39'18"20d12'
Stambro,Dorothy Elizabeth
And 23h15'27"49d10'
Stamer,Robert Samuel
Cet 2h16'22"3d47'
Stamie-Marie
Cas 23h14'0"61d35'
Stamm,Doris
Psc 23h9'51"1d14'
Stamm,Karl-Heinz
Cap 20h36'47"-10d36'
Stamm,Kornelia
Tau 3h48'17"2d3'
Stammberger,Wolfgang
Uma 9h34'0"43d4'
Stamp,Annabelen Acu'
Eri 2h51'41"-3d20'
Stampahar,Master Star Nancy Jean
Eri 3h19'45"-1d40'
Stampe,Maureen
Psc 0h19'39"18d52'
Stamper,Berniece Josephine
Del 20h14'14"9d38'
Stamper,Bill
Aur 5h47'0"50d34'
Stamper,Margee
And 2h21'38"39d36'
Stamper,Stephen M
Uma 12h10'38"59d58'
Stamps,Jay
Aur 4h55'0"52d17'
Stamps,Sue Craig
Mon 6h26'51"-10d53'
Stan
Boo 13h57'32"20d12'
Stan
Aur 4h21'9"34d12'A
Stan & Geraldine- Always & Forever
Cyg 21h57'33"52d45'

Stan & Rhea
Cyg 19h33'31"30d55'
Stan & Tiffany for Eternity
Lyn 7h56'0"42d3'
Stan 1437
Cnv 12h14'12"39d52'B
Stan Star
Ori 5h28'24"0d21'
Stan The Man
Lac 22h4'45"38d41'
Stan The Star
Cam 3h27'18"63d27'
Stan's Heart
Lyr 19h1'31"30d44'
Stan's Heart
Peg 23h2'0"17d56'
Stan's Heavenly Body
Oph 18h7'23"12d36'
Stan's Sparkler
Her 17h29'36"37d43'
Stan's Star
And 1h22'12"40d52'
Stan's Star
Vul 20h14'39"26d9'
Stan,Mary Ala
Cas 0h30'1"73d51'
Stanaway Family,The Brian
Dra 16h58'1"51d36'
Stanbalk,Timothy Permar
Hya 9h54'42"-16d27'
Stanbridge,Michael
Her 18h1'13"28d58'
Stanbury,Sebastian
Cep 21h34'22"61d2'
Stancati,Nicoletta
Umi 14h46'39"71d11'
Stancell,Woodrow Vernon
Boo 14h48'52"24d53'
Stanco III,Roland C
Dra 17h20'25"70d13'
Stancy,Mary C
And 0h21'25"37d6'
Stanczyk,Judy
Lyr 18h22'21"44d52'
Standart
And 0h45'34"40d7'
Stander,Gabrielle
Crb 15h29'1"30d26'
Standifer,Lela
Mon 7h39'45"-4d13'
Standiford,Sally
Cas 1h54'55"77d2'
Standing,Aaron Matthew
Aur 6h17'48"37d53'
Standing,Joan & J Howard
Aur 6h3'34"48d54'
Standish,Paul
Cam 3h38'56"74d21'
Standke,Rosi
Psc 0h9'13"-0d45'
Standlee,R D Dan
Cet 0h47'32"-5d2'
Standley,George Michael
Lac 22h23'52"53d33'
Standley,Mark Robert
Uma 9h20'42"48d27'
Standridge,Robyn
Eri 3h31'17"-5d50'
Staneart,Edward Volney
Boo 14h49'0"38d44'
Stanek,Aaron Michael
Per 4h45'19"38d51'
Stanek,Bashia
Uma 11h38'24"49d48'
Stanek,Gabriel L
Lyn 8h46'30"43d19'
Stanek,Roger G
Lmi 10h2'39"33d9'
Stanek,Spencer C
Her 17h7'56"48d24'
Stanett,Chuck
Sex 9h50'30"2d1'
Stanford,Aaron
Gem 6h52'45"18d41'

Stanford,Ann
Aql 19h54'21"11d45'
Stanford,Mark W
Vul 19h22'49"22d45'
Stanford,Miranda Eve
Com 12h1'49"19d54'
Stanford,Rosemary
Lyr 19h20'17"40d13'
Stanford,Samantha
Lyr 18h17'44"42d55'
Stanford,Tom G
Gem 7h51'56"30d27'
Stang,Dick
Boo 13h55'25"20d1'
Stang,Don
Uma 9h3'18"51d25'
Stange,Hans K
Cep 22h14'51"60d37'
Stange,Marguerite & Secor,Harry
Lyr 18h50'45"40d15'
Stange,Martha
Cmi 7h44'56"0d1'
Stange,Norbert
Umi 13h49'0"73d52'
Stangebye,Monica
Umi 13h24'23"70d48'
Stangeland,Don M
Lac 22h23'1"50d14'
Stanger,William
Gem 6h50'48"13d53'
Stango,Darlene
Lyn 6h33'56"54d59'
Stango,Jr,Robert G
Her 18h5'32"48d38'
Stangorra,Manuela
Cnc 8h28'46"7d39'
Stanhope's Star, Beverly Gay
Cas 1h14'42"65d1'
Stanhope's Star, Carolyn
And 1h48'1"40d42'
Stanhope's Star, William & Dorothy
Uma 8h10'47"60d24'
Staniaszek,Zbyszek
Uma 11h25'0"32d14'
Stanich,Joseph Daniel
Cep 2h4'39"80d33'
Staniewicz,Kelly Eileen
Lyr 18h59'20"34d23'
Staniford,Sean
Cep 20h25'26"63d37'
Staniforth III,William M
Tri 1h43'47"34d52'
Stanish,Stan
Dra 16h55'1"68d24'
Stanislas et Marine
Vul 19h21'15"25d1'
Stanislaus "Standing Glory"
Lyr 18h59'16"37d3'
Stanislaus,Ina Smith
Oph 18h1'0"12d56'
Stankevich Esq, George C
Ori 5h48'33"10d45'
Stankevich,Margaret Ardsley
Ori 5h58'59"20d28'
Stankey,H Brad
Lac 22h15'31"54d26'
Stankiewicz,Trisha
Aql 19h52'0"12d33'
Stanko,Lisa
Lyr 18h28'49"32d46'
Stanko,Rosemary
Hya 9h17'0"1d9'
Stankovici,Magdolna "Maggie"
Eri 2h59'12"-2d56'
Stankus,Rev Gregory A
Per 2h25'55"57d14'
Stankus,Tanya
Aql 20h6'37"3d46'
Stanley
Lac 22h7'33"47d32'
Stanley & Dorothy
Umi 3h39'40"88d12'

Stanley Star,The
Cep 22h26'0"63d28'
Stanley's Destiny
Crt 10h59'10"-21d38'
Stanley,Carlin James
Vul 20h43'48"20d31'
Stanley,Curtiss Gardiner
Cep 21h55'0"80d7'
Stanley,Delores
Uma 11h10'27"61d22'
Stanley,Don
Lac 22h34'38"50d24'
Stanley,Grace & Duane
Cyg 21h1'41"30d42'
Stanley,Hugh Lyman
Lac 22h11'58"51d57'
Stanley,Janice S
Cas 3h1'48"58d58'
Stanley,Joanne
Ori 4h52'59"0d24'
Stanley,Jonathan Murbach
Cep 21h9'15"67d54'
Stanley,Joshua Paul
Lyn 8h6'21"45d35'
Stanley,Kathleen
Lyr 19h14'31"40d4'
Stanley,Lil
Sgr 18h49'44"-36d14'
Stanley,Logan Hyatt
Ser 15h56'37"17d38'
Stanley,Margaret
Uma 12h11'22"59d6'
Stanley,Meagan Shea
Peg 23h1'37"13d10'
Stanley,Nadine M
Lyr 19h17'22"42d15'
Stanley,Paul
Cep 2h36'46"77d13'
Stanley,Robert Carl
Tau 5h53'18"23d30'
Stanley,Sandra Lee Crowford
Vul 19h48'31"28d10'
Stanley,Timothy & Michael
Uma 10h4'53"48d19'
Stanmore,Michael John
Gru 23h17'47"50d41'
Stanmore,Michael John
Gru 22h38'27"52d9'
Stann,Marilyn Marie
Cas 23h41'40"61d21'
Stannard,Chase Patrick
Boo 13h36'1"21d44'
Stannard,Marguerite Sheridan
Aql 19h47'1"10d16'
Stannard,Robert
Hya 8h55'11"1d16'
Stannett,Nicky
Sge 20h4'53"18d57'
Stannish Star,The Richard
Cep 20h31'37"62d13'
Stansberry,Brad
Lac 22h26'33"56d16'
Stansberry,Jean MacDowell
Cam 7h57'45"70d59'
Stansfield,Evelyn Ester
And 1h51'25"38d28'
Stansfield,Natalie M
Lyr 19h17'27"38d24'
Stansky,Joyce Marie
Cas 0h10'34"63d16'
Stant,Margaret
Cyg 19h34'54"33d34'
Stanton
Her 18h6'58"40d60'
Stanton,Arleen D - Arleen Delight
Del 20h23'11"18d55'
Stanton,Arthur
Cap 21h33'1"-26d31'
Stanton,Eoin Orion
Ori 5h43'24"10d0'
Stanton,Jennifer
And 2h6'35"40d43'
Stanton,Kenneth C
Ser 17h21'20"4d53'B

Stanton,Laura A
Cas 1h20'18"53d55'
Stanton,Mark Fowler
Uma 11h25'12"38d2'
Stanton,Maureen Bistline
And 23h22'56"51d38'
Stanton,Michael Sean
Lyn 9h14'40"33d23'
Stanton,Patrick Michael
Aur 5h28'27"30d17'
Stanton,Randell L
Cyg 20h21'18"40d13'
Stanton,Raymond C
Ser 17h21'20"4d53'A
Stanton,Rebecca F
Peg 21h54'32"29d29'
Stanton,Robert Douglas
Hya 9h1'50"0d15'
Stanton,Siobhan
Tri 2h2'0"28d23'
Stanton,Wayne
Uma 11h7'44"68d5'
Stanton,William Bennett
Lac 22h50'19"55d52'
Stanton,William Bennett
And 2h31'49"39d52'
Stanton,William Mackenzie
Cnc 8h40'11"17d39'
Stantz,The Real Ghostbuster,Raymond
Pho 1h30'34"-48d52'
Stanyard,Tina
And 2h35'24"38d42'
Stanyer 36-14,Claire Elizabeth
Umi 15h16'0"66d26'
Stanyer 36-14,Claire Elizabeth
And 2h35'25"40d12'
Stanzi
Cmi 7h30'59"8d51'
Stanziani,Lorraine
And 1h27'50"38d32'
Stanziano,Alexandra Marie
Com 11h56'14"14d33'
Stanziano,Kathryn Alison
Ori 5h28'48"-3d39'
Stanzione,Anna (Annabelle)
Lyn 7h0'50"60d17'
Stanzione,Denise
And 0h21'27"30d13'
Stanzione,James
Crt 10h50'58"-12d27'
Stanzione,Nancy
Mon 6h31'55"-6d26'
Stape,Keith & Janet
Aur 5h11'49"43d8'
Stapel,Klaus-Peter
Umi 13h23'16"74d13'
Stapelfeld,Gerd
Aur 5h18'58"43d17'
Stapelfeld,Werner
Lyr 19h17'52"30d50'
Stapenhurst,Captain Michael
Dra 12h13'53"68d24'
Stapinsky,Alan
Oph 17h0'35"7d57'
Staples,Claire
Mon 6h37'27" 6d0'
Staples,Rita
Lyr 18h16'57"45d50'
Stapleton
Uma 15h29'22"71d52'
Stapleton ILY,Caroline
Ori 6h2'18"2d25'
Stapleton,Edd
Cyg 21h19'55"30d23'
Stapleton,Ian John
Peg 22h39'22"33d10'
Stapleton,John & Maguerite
Crb 16h10'23"38d16'
Stapleton,Joseph
Aur 5h19'1"42d33'
Stapleton,Michael Martin
Aql 19h4'31"4d13'
Stapleton,Tracy
Lyr 18h31'1"30d14'

Stapley,Diane
Cyg 19h33'29"39d16'
Stapper,Ilona
Lac 22h20'15"40d8'
@@ Banyan
Leo 9h53'50"10d45'
*A Dor' U
Mon 6h37'37"10d31'
*IFFA*LAX*
Mon 7h55'1"-3d30'
*IFFA*SFO*
Peg 21h24'1"18d51'
Prince
Gem 7h7'39"24d54'
Star
Lac 22h45'53"38d14'
Star
Uma 9h56'46"50d15'
Star "Chaser" From His "Star",The
Her 17h27'12"40d5'
Star #13 Bonomo
Uma 12h5'44"56d0'
Star #2
Lyr 19h7'47"37d37'
Star 215
Mon 6h19'44"5d31'
Star 9
Per 3h52'49"39d56'
Star ABO
Cep 21h52'55"68d51'
Star Adriana
Hya 8h25'40"-9d45'
Star Alana,The
Lyr 18h37'42"40d18'
Star Alba Elizabeth
Cas 0h58'16"65d27'
Star Aline
Com 13h0'15"26d46'
Star At Pooh Corner, The
Cep 7h2'16"86d25'
Star B
Dra 17h22'36"73d12'
Star Babies
And 23h27'33"43d58'
Star Baker
Uma 10h38'36"40d45'
Star Bank
Lyr 18h22'43"37d58'
Star Bar
Lyr 19h6'36"40d9'
Star Bavonese
Sge 19h53'11"16d50'
Star Bell
And 23h0'12"50d25'
Star Betsy
Peg 21h56'42"34d51'
Star Betsy-Bo
Cas 23h29'0"61d9'
Star Blaz
Umi 15h43'23"70d26'
Star Bonnie
Eri 4h9'34"-15d8'
STAR Boss #1
Per 2h52'37"34d33'
Star Brian
Aql 20h2'12"12d36'
Star Buckaroo
Aur 7h0'27"35d52'
Star C E Q
Ser 15h37'34"7d40'
Star Called Heidi,A
Sgr 19h16'16d-23d39'
Star Called Kraftwerk, A
Peg 21h39'38"27d13'
Star Called Kreissl,A
Ser 16h7'27"10d13'
Star Called Michi,A
Cas 2h23'57"60d5'
Star Called Niki,A
Cyg 20h27'23"31d47'
Star Carmela,The
Eri 2h51'30"-2d5'
Star Child
Lyr 18h55'32"40d1'
Star Child Cara
Vul 21h3'11"27d39'
Star Christopher
Lac 22h18'15"50d22'
Star Club
Tri 1h31'24"31d21'
Star Colleen
Mon 8h4'50"-2d58'
Star Crossed Lovers Judy & Tom
And 2h24'1"45d48'
Star Daddy Love Britt
Dra 17h46'39"61d41'
Star Damon
Lac 22h37'20"53d57'
Star Diavanni
Cyg 21h5'14"30d53'
Star Domona (S C G)
Eri 3h17'32"-15d44'
Star Drop on a Hydrangea
Cnc 8h47'6"14d7'
Star Elizabeth
Lyr 18h32'60"38d39'
Star Elysia:Leela in the Sky with Diamonds
Cas 0h31'0"71d8'
Star Emile,No 2 Mom
Psc 0h47'1"31d37'
Star Eva
Aqr 21h2'40"-8d19'
Star For Sue & Brooks
Ori 5h29'36"-2d52'
Star Fredric
Cep 22h39'18"67d59'
Star From A "Sherry-Tale",The
And 23h28'1"48d43'
Star Gemma G
Peg 23h31'30"32d16'
Star Gimmee Q
Lyn 8h46'48"45d20'
Star Gray
Lmi 10h32'41"32d14'
Star Greger
Aur 5h27'37"30d25'
Star in the Sky- Regalia
Lyn 8h44'1"41d5'
Star INC
Uma 10h38'37"48d14'
Star IrvAnnette 44
Lac 22h29'44"49d46'
Star Jennifer
And 23h34'58"44d47'
Star JJ
Aur 5h57'54"30d37'
Star Karenda
Peg 22h41'47"20d2'
Star Kathryn Anne, Goddess Of Love
And 0h35'1"38d52'
Star Kendall
Boo 15h4'56"18d31'
Star Lawyer
Aur 6h43'47"38d53'
Star Lee Elizabeth
Cmi 7h26'0"0d34'
Star Light On A Nativity
Ori 6h13'36"0d30'
Star Lighty
Lyn 6h35'20"58d18'
Star Louey
Ori 5h23'30"1d47'
Star Lucy
Lyr 18h59'21"28d41'
Star M
Boo 15h3'0"20d11'
Star Manwar Shines Over Us!
Dra 10h1'11"80d42'
Star Mary Harry
And 23h36'57"46d38'
Star McInerney,The
Cyg 21h19'60"36d14'
Star Medalla
Boo 15h4'0"27d23'
Star Melody
And 23h30'49"43d33'
Star Melvin
Aql 19h58'22"15d38'
Star Meredith
Cap 21h39'11"-21d53'
Star Micronics
Lyn 9h11'56"38d13'
Star Morgado
Uma 9h34'31"68d41'
Star Muffin
Cet 3h8'1"0d18'
Star Muhlenkamp 25 yrs
Aur 6h42'1"38d41'
Star N Rockn Robin
Mon 6h38'60"1d43'
Star Named Shinichiro, The
Ari 2h10'35"25d8'
Star Neighbors
Aur 4h53'1"51d9'
Star Nova 95
Lyr 19h1'26"28d38'
Star of "Michael - Shari Love"
Peg 21h56'54"23d21'
Star of A Dad
Cep 22h23'20"70d21'
Star of Abigaile,The
Cyg 21h0'52"36d42'
Star Of Abraham
Ser 15h53'12"-1d56'
Star of Adam,The
Aur 6h24'27"34d26'
Star of Airam,The
Aql 19h55'46"14d27'
Star of Akifumi & Ayumi
Cap 20h52'21"-21d17'
Star Of Akihiro & Chiemi
Ari 2h42'6"25d59'
Star of Akira & Sayoko
Lib 14h32'26"-12d32'
Star Of Alan
Boo 15h33'30"41d26'
Star Of Amaya
Cnc 8h59'27"21d10'
Star of Amaya
Cnc 8h59'27"21d10'
Star of Amie Christine ,The
Lac 22h22'13"54d37'
Star Of Andrew & Jessica,The
Ori 5h6'38"-0d52'
Star of Andrew James, The
Cep 20h49'57"70d59'
Star of Angela
And 2h20'42"46d45'
Star Of Angeline
And 0h59'48"34d15'
Star Of Angelo
Her 16h55'35"37d24'
Star of Asako Taka
Leo 10h11'47"17d15'
Star Of Ashwani
Ori 5h27'20"1d6'
Star Of ATOBE
Leo 10h12'56"21d43'
Star of Atsushi & Hitomi
Aqr 22h6'38"-7d30'
Star of August & October
Tau 5h23'23"24d2'
Star of Autologue,The
Lmi 10h26'0"31d34'
Star of Ayako & Shunichi
Gem 7h20'11"22d56'
Star of Ayako Yamada
Leo 9h46'21"29d41'
Star of Babs,The
Cas 2h4'48"68d47'
Star Of Beatrice
Crb 15h58'45"30d22'
Star of Bermuda
Oph 17h9'13"-20d5'
Star of Bernadette,The
Vul 20h17'21"23d1'
Star Of Bethanyhem,The
Crb 15h17'1"31d18'
Star Of Bethleham,The
And 1h4'1"39d2'
Star of Betty
Com 12h20'1"20d14'
Star of Big Bamboo
Tau 5h32'23"16d28'
Star Of Bina
Cyg 20h23'34"38d52'
Star of Bird,The
Peg 22h38'37"31d32'
Star of Blue Creek
Aur 6h3'34"46d31'
Star of Boght Hills
Cam 7h34'1"61d14'
Star of Bozo,The
Cep 4h13'17"86d42'
Star of Brian
Lac 22h35'44"53d11'
Star Of Bruce,The
Her 16h40'24"35d27'
Star Of Bubba,The
Ori 5h48'50"11d10'
Star of Bullwinkle
Ori 6h17'49"-2d26'
Star of Campfield,The
Peg 23h2'0"20d30'
Star of Cappel
Aur 5h26'1"38d23'
Star of Carol
Cas 2h23'43"75d22'
Star of Carolyn
Eri 2h49'42"-4d13'
Star Of Celeste,The
Eri 4h9'56"-12d60'
Star of Chadwick
Uma 8h58'58"51d26'
Star of Char,The
Ser 16h2'32"22d33'
Star of Chuck
Per 3h2'29"48d58'
Star Of Claudia
Lib 14h50'29"-0d49'
Star of Colleen
Lyr 18h53'51"35d35'
Star of Cona,The
Cyg 19h40'15"38d58'
Star of D J,The
Dra 15h49'30"62d9'
Star Of Dad,Thee
Aql 20h7'35"1d42'
Star Of Daisuke
Aqr 22h7'2"-6d38'
Star Of Damian
Boo 14h37'53"40d10'
Star of Daniel,The
Cam 3h26'1"60d48'
Star of Daniel,The
Aur 4h47'1"37d14'
Star of Daniel,The
Cru 12h54'42"-60d37'
Star of Dave
Uma 10h49'47"50d8'
Star Of David
Umi 13h16'0"71d42'
Star Of David
Lac 22h1'15"48d12'
Star Of David
Uma 10h4'45"68d31'
Star Of David
Sex 9h54'0"-1d27'
Star of David
Cep 22h19'1"56d14'
Star of David
Cep 23h6'27"60d53'
Star Of David
Cyg 19h47'10"37d38'
Star of David
Psc 23h24'34"5d57'
Star of David (David W Pursley)
Her 16h51'18"50d2'
Star of David (The Stellar Zeque Version)
Aur 6h1'1"32d42'
Star of David Ray,The
Dra 17h52'1"60d9'
Star Of David,The
Boo 13h36'14"19d46'
Star Of David,The
Her 18h13'47"38d48'
Star of David,The
Her 15h47'28"40d44'
Star of David,The
Her 15h47'28"40d44'
Star Of Davis Terrince
Lac 22h53'0"50d35'
Star Of Davis,The
Uma 8h44'25"52d50'
Star of Dawn,Mother of Janine,The
Del 20h15'46"9d46'
Star Of Debby
Lyr 19h10'60"38d21'
Star of Della Marie
Cyg 19h25'1"32d20'
Star of Diane
And 1h36'46"39d40'
Star of Diane,The
Cam 3h26'45"60d46'
Star Of Donald
Lyr 18h39'0"37d45'
Star Of Dong
Lib 15h16'23"-28d14'
Star of Dunblane
Cyg 20h34'1"44d18'
Star Of Edward
Uma 11h12'0"46d29'
Star of Eigner,The
Her 18h4'43"28d21'
Star of Eileen
Cas 1h3'18"58d21'
Star of Elio
Boo 15h7'18"8d31'
Star of Emiko & Hidenori
Ari 2h23'9"22d57'
Star of Erma
Vir 13h32'1"-7d23'
Star Of Esther
Cyg 21h23'24"31d14'
Star of Esther,The
Hya 8h25'49"-9d43'
Star of Felicia & Lee, The
Crb 16h0'22"27d21'
Star of Fields (Jessie & Jonell)
Crb 15h57'25"30d22'
Star of FinLupe
Lyn 7h2'28"59d43'
Star Of First True Love,The
Peg 22h15'18"30d3'
Star Of Firsts
Uma 8h18'59"68d49'
Star Of Forbidden Fruit,The
Sge 19h39'59"16d48'
Star of Forts Ferry
Cas 0h51'41"70d58'
Star of Francesca
Lyr 19h1'47"40d4'
Star Of Fumiya & Akemi
Sco 16h11'32"-14d3'
Star of Gaylene
Cet 2h19'59"9d21'
Star of Gen's Light, The
Lyr 19h6'19"40d36'
Star of Genichiro & Minako
Lib 14h30'47"-10d26'
Star Of Gino Elia,The
Uma 9h35'23"51d16'
Star Of Glorianne,The
Cyg 19h18'41"45d33'
Star of Grace,The
Hya 9h6'11"-1d33'
Star of Grace,The
Cyg 19h56'1"38d43'
Star Of Hajime & Tomomi
Sco 16h9'45"-11d5'
Star of Hannah,The
Equ 21h1'47"10d27'
Star of Heather
Uma 12h8'54"51d5'
Star of Heather Belinda
Eri 3h57'32"-16d10'
Star Of Heekin
Lac 22h10'30"49d5'
Star Of Helen Elizabeth
Eri 3h57'56"-18d17'
Star Of Helene
And 23h40'0"43d32'
Star Of Hibiki
Sgr 19h23'42"-13d15'
Star of Hide & Nori
Sco 16h12'14"-14d32'
Star Of Hideaki & Setsuko
Aqr 22h5'42"-7d41'
Star Of Hideki
Vir 14h1'48"-8d45'
Star Of Hideki & Juri
Psc 1h27'9"23d25'
Star of Hikochan & Takachan
Aqr 22h12'3"-2d47'
Star of Hiro & Kumi
Lib 14h39'12"-8d59'
Star Of Hiro & Kumi
Lib 14h39'12"-8d59'
Star Of Hirokazu & Eriko
Sgr 19h8'59"-12d46'
Star Of Hiroki & Hiroko
Sco 16h15'27"-9d22'
Star of Hiroki & Akemi
Aqr 22h10'50"-8d29'
Star of Hiroko & Norihiko
Leo 10h29'29"22d15'
Star Of Hironii & Tomonee
Cap 20h30'12"-13d39'
Star of Hiroshi & Tamao
Gem 7h23'57"24d3'
Star Of Hiroshi & Mika
Tau 5h37'18"24d26'
Star of Hiroyasu & Yasuko
Psc 1h11'27"21d11'
Star Of Hiroyuki & Yuriko
Ari 2h24'14"24d11'
Star of Honey Bears
Cnc 8h46'24"22d39'
Star Of Hope
Mon 6h57'34"-10d44'
Star Of Hope,The
Cyg 20h23'0"40d54'
Star of Hossein
Aql 18h49'52"11d4'
Star Of Ichiro & Tokuko
Lib 14h38'14"-11d37'
Star of Ichiro & Tokuko
Lib 14h38'14"-11d37'
Star Of Inge
Uma 11h23'8"52d24'AB
Star of Isao & Mayumi
Lib 14h44'29"-1d31'
Star Of Isao & Miyoko
Gem 7h25'2"25d9'
Star of ISAO & TAMAMI
Lib 14h37'3"-11d55'
Star of Jack,The
Aql 19h4'1"-5d44'
Star Of Jasmine
Cas 0h28'47"67d3'
Star of Jeanne
And 0h22'19"44d42'
Star of Jeffrey
Cnv 12h37'32"40d31'
Star of Jeffrey
Aql 19h56'0"11d37'
Star Of Jennifer
Lyn 7h52'58"42d10'
Star Of Jessica
Uma 9h44'58"56d13'
Star of Joan
Cyg 19h25'0"34d54'
Star Of Joanne
And 0h48'14"37d52'
Star of Jodie
Peg 22h31'16"20d15'
Star of Jordan
Uma 11h57'38"37d32'
Star of Joseph,The
Ari 2h54'0"21d27'
Star Of Judith Ann
Cnv 13h0'17"41d7'
Star of Jun & Taiko
Lib 14h44'36"-5d53'
Star Of Kaire/Mangeris
Ori 5h57'0"10d37'
Star of Karin,The
Mon 8h4'46"-6d46'
Star of Kathleen
And 2h27'20"49d59'
Star of KATSU & YUKARI
Lib 14h43'47"-10d18'
Star of Katsuaki & Chiharu
Tau 5h39'29"18d36'
Star Of Katsuhiro & Fumiyo
Sco 16h12'20"-9d38'
Star of Katsuhiro & Fumie
Cnc 8h47'20"26d33'
Star of Katsuhiro & Fumiyo
Sco 16h12'20"-9d38'
Star of Katsumi & Yasuko
Ari 2h39'15"17d30'
Star of Katsutoshi & Yumi
Tau 5h23'42"25d12'
Star Of Katsuya & Tomomi
Ari 2h23'3"26d20'
Star Of Kenichi & Toshiko
Tau 5h27'5"23d40'
Star of Kenichi & Toshiko
Tau 5h27'5"23d40'
Star of Kenichi & Chie
Aqr 22h5'50"-4d44'
Star of Kenji & Tomok
Leo 10h11'33"26d10'
Star Of Kenji & Tomok
Leo 10h11'33"26d10'
Star Of Kentaro & Mayu
Lib 14h43'3"-11d14'
Star of Kia,The
Mon 7h2'8"-6d21'
Star Of Kieran
And 2h21'31"44d35'
Star of Kishiko
Leo 10h16'41"17d10'
Star Of Kiss & Memory
Ari 2h23'8"21d17'
Star of Kiss & Memory
Ari 2h23'8"21d17'
Star Of Kiyoko & Takashi
Leo 10h11'15"25d1'
Star of Kiyotaka & Mayumi
Psc 1h23'15"25d55'
Star of Kohei
Psc 1h16'8"23d20'
Star of Koichi & Katsumi
Sco 16h14'5"-12d26'
Star of Koichi & Takako
Vir 14h7'1"-8d41'
Star Of Koji & Rie
Sgr 19h10'39"-22d37'
Star of Koji & Tomoko
Tau 5h32'3"24d14'
Star Of Kozue & Hiroaki
Gem 7h27'26"25d35'
Star Of Krystle
Peg 22h18'15"4d40'
Star Of Kunihiro & Tetsuko
Psc 1h27'32"22d4'
Star of Kyo & Motoe
Psc 1h26'15"23d56'
Star of Lamar
Lyr 18h31'29"30d30'
Star of Latham Ridge
Dra 9h58'52"74d9'
Star Of Laurence
Per 2h56'1"48d49'
Star Of Leo
Ser 15h24'36"8d4'
Star Of Leo
Cam 3h47'28"61d45'
Star Of Le47'28"61d45'
Cmi 7h46'56"7d46'
Star Of Lidia
Vul 19h22'27"27d29'
Star Of Life
Del 20h20'23"18d47'
Star Of Linus
Hya 8h40'31"1d48'
Star of Lisa
And 1h17'1"37d46'
Star Of Lisa Ann
Cyg 20h17'43"38d37'
Star of Lisa Michelle
Mon 7h1'31"-1d7'
Star of Lisa Renee
And 2h9'12"38d53'
Star Of Lotem
Uma 11h28'41"43d46'
Star of Loudonville
Lyn 7h54'58"50d38'
Star Of Madeline
Eri 3h4'41"-2d26'
Star of Makoto & Tomie
Cap 20h29'5"-9d34'
Star Of Makoto & Miki
Aqr 22h10'50"-1d54'
Star Of Makoto & Tomie
Cap 20h29'5"-9d34'
Star Of Marc
Her 17h54'31"48d53'
Star of Margaret
Peg 21h24'44"20d27'
Star of Marie,The
And 2h23'21"43d0'
Star of Marjie,The
Ori 4h57'14"1d45'
Star Of Martin
Cyg 21h35'0"42d3'
Star of Mary
And 0h54'31"38d47'
Star of Marybeth,My Angel & Love
Eri 3h31'22"-5d41'
Star of Masaaki
Lib 14h32'51"-13d15'
Star Of Masahiko
Sco 16h9'41"-13d38'
Star Of Masahiro & Naoko
Lib 14h32'35"-8d53'
Star of Masahiro & Emiko
Psc 1h16'14"22d8'
Star Of Masao & Miyuki
Lib 14h32'2"-10d40'
Star of Masato & Yuri
Cnc 8h46'12"22d27'
Star of Megumi,The
Psc 1h19'39"3d15'
Star Of Meijyo,The
Ari 1h45'50"21d56'
Star of MiBrig
Aql 19h30'42"12d11'
Star Of Michael,The
Cep 21h45'1"63d51'
Star Of Michael,The
Hya 8h59'10"-1d35'
Star of Michelle,The
Mon 6h2'11"-6d13'
Star Of Michiaki & Yuka
Cnc 8h57'11"15d41'
Star of Miho 21 from Yoshiyuki
Sgr 19h13'29"-11d54'
Star Of Miles
Psc 23h6'30"5d26'
Star of Milton
Cep 22h47'10"67d60'
Star of Monty-Face,The
Cet 2h6'16"4d7'
Star of Morris-Happy 85t-With Love
Uma 12h53'45"54d19'
Star Of Mose
Gem 6h22'33"18d48'
Star Of Muneharu & Yuki
Sgr 19h28'18"-25d52'
Star Of Mutsuyuki & Kazumi
Sco 16h14'51"-21d49'
Star Of My Handsome Prince Henry
Per 4h30'33"37d35'
Star Of My Life, Bonnie,The
Cas 0h4'41"50d7'
Star Of Nagafumi & Akiko
Ari 2h38'17"25d21'
Star of Nami
Gem 7h11'35"23d53'
Star of Naoki & Michiko
Cap 20h31'33"-20d17'
Star of Nina
Peg 22h3'43"20d26'
Star of North Cliff School,The
Uma 11h45'40"50d31'
Star Of Only One
Psc 1h14'42"26d1'
Star of Only One
Psc 1h14'42"26d1'
Star of P J,The
Cet 2h46'32"2d37'
Star of Pastrick,The
Sgr 18h22'12"-23d3'
Star Of Patsy Rose
Peg 22h0'16"35d24'
Star of PaulLois
Uma 9h11'1"50d23'
Star of Pearl
Eri 3h42'0"-18d29'
Star Of Power
Dra 11h28'42"71d27'
Star of Prado
Cep 0h22'38"78d7'
Star of Princess Lisa Marie,The
Uma 10h59'42"35d21'
Star of Rachel Dorian
Cyg 20h19'0"38d10'
Star Of Ramel
Boo 13h41'11"26d25'
Star Of Raven,The
Cas 1h44'25"60d37'
Star of Ravizza,The
Mon 7h15'46"-1d45'
Star of Reiko
Leo 10h15'0"20d12'
Star Of Remy
Mon 6h24'28"11d36'
Star Of Rena
Psc 1h21'47"23d42'
Star of Rena
Psc 1h21'47"23d42'
Star Of Riley
Per 2h55'12"38d40'
Star of Ritsuko
Sco 16h4'9"-16d0'
Star of Robyn,The
Crb 16h18'43"30d22'
Star of Ron
Del 21h1'25"18d51'
Star of Rose
Eri 3h2'41"-4d1'
Star of Rosemary
Ori 6h2'59"-0d5'
Star of Rosenburg
Umi 16h41'19"76d2'
Star of Ruth
Cam 5h27'59"80d19'
Star Of S&M
Lib 14h34'35"-8d49'
Star of Saburo & Naoko Silver Wedding
Sgr 19h9'3"-14d41'
Star Of Sadao & Mami
Aqr 22h2'3"-21d17'
Star Of Saki
Psc 1h20'57"22d40'
Star of Saki
Psc 1h20'57"22d40'
Star of Samuel
Aur 6h32'22"38d13'
Star of Sandra
Leo 9h19'27"28d43'
Star of Satomi & Akiniko
Gem 7h13'33"24d58'

Star of Sawako Cnc 8h54'56"16d45'
Star Of Sayuri Del 20h21'26"10d18'
Star of Sedona Lmi 10h58'42"30d55'
Star Of Seiichi & Michiko Tau 5h34'53"25d31'
Star of Senf Uma 10h50'42"70d38'
Star Of Senji & Junko Gem 7h26'29"27d3'
Star Of Seth Lac 22h17'42"49d37'
Star of Shaker High Per 3h41'13"51d37'
Star of Shaker Junior High Cam 7h31'1"61d18'
Star Of Shannon Dra 16h56'51"64d31'
Star of Sharon Cyg 19h26'40"33d23'
Star of Sharon Lee,The Cam 4h52'51"69d55'
Star Of Sher Peg 23h39'13"31d18'
Star of Shigenobu & Miki Ari 2h22'39"24d36'
Star Of Shiho & Toshimi Sco 16h9'29"-10d39'
Star of Shin & Yasuko Sco 16h10'33"-17d58'
Star of Shiro & Tomoko Psc 1h20'39"27d23'
Star Of Shoji & Kazuyo Sgr 19h12'57"-24d36'
Star Of Shouma Leo 9h53'9"27d2'
Star of Show Gem 7h14'36"20d10'
Star of Show & Wakako Gem 7h10'57"20d18'
Star of Shuhei Tau 5h26'8"24d9'
Star Of Snaquistador Cam 6h14'1"70d14'
Star Of Sokol Uma 13h44'35"51d37'
Star of Sooty,The Cep 21h17'46"65d4'
Star Of South Africa, The Cru 12h45'15"-60d5'
Star of Southgate Per 3h41'1"50d44'
Star of Stefanie Hya 8h26'35"-8d49'
Star Of Steve & Peggy Peg 23h28'1"18d44'
Star of Steve,The Cet 2h33'19"8d11'
Star of Stevens Mon 8h8'34"-3d23'
Star of Storrs,The Peg 22h37'0"30d39'
Star of Stute Uma 10h24'30"56d47'
Star Of Suekar Aur 5h9'28"37d52'
Star of Susan GFG Ori 5h55'15"7d49'
Star of Suzanna & Roger,The Uma 9h55'33"51d28'
Star of Sweet Lorraine Cas 0h3'44"64d51'
Star of Tadashi & Atsuko Sco 16h18'15"-11d3'
Star Of Tadashi & Mika Leo 10h24'30"22d57'
Star of Tafdon Boo 15h10'50"31d20'
Star of Takahiro & Masae Cnc 8h49'20"13d36'
Star Of Takeshi Kazuyoshi,The Sco 16h5'20"-28d51'
Star Of Takeshi & Hideko Leo 10h26'21"19d8'
Star Of Takeshi & Kazuyo Ari 2h43'3"20d27'
Star of Takeshi & Noriko Gem 7h2'6"16d17'
Star of Tamara,The Gem 7h28'17"34d40'
Star Of Tammy Tau 5h48'16"23d42'
Star of Tammy Cyg 19h28'24"32d52'
Star of Tara Uma 9h1'33"47d57'
Star of Teresa Rae & Raechel And 1h2'55"40d45'
Star of Terrence Aur 6h2'30"54d36'
Star Of Tetsu & Miya Leo 10h14'21"22d43'
Star Of The Devine Vul 19h53'29"20d8'
Star of the Harbor of Grace Sco 16h14'35"-23d44'
Star of the Lamar Lover Ori 6h15'28"8d23'
Star of Timmy Aur 4h55'17"40d3'
Star Of Tohyama,The Tau 5h24'17"20d9'
Star Of Tom & Jerry Tau 5h31'36"26d32'
Star Of Tomoko & Sadao,The Sgr 18h54'19"-30d57'
Star of Toru & Akiko Lib 14h31'39"-14d36'
Star of Toshihiko Sgr 19h12'42"-14d48'
Star of Toshiko Ari 2h23'26"18d41'
Star Of Toya Tau 5h19'3"26d59'
Star of Toya Tau 5h19'3"26d59'
Star of Toyokazu & Toshimi Cap 20h52'11"-26d15'
Star of Trust Tau 5h24'2"22d53'
Star of Truster Sco 16h20'9"-28d54'
Star Of Tsuyoshi & Atsuko Lib 14h32'12"-14d21'
Star of Tsuyoshi & Noriko Lib 14h37'20"-12d22'
Star of Tyrone Ori 6h2'1"20d31'
Star of UMI,The Uma 10h53'0"70d14'
Star Of Victoria Lyr 19h17'33"33d60'
Star of Waheda Lyr 18h36'12"28d60'
Star Of Wanda Vul 19h18'45"27d30'
Star of Wellington Aql 19h5'33"0d2'
Star of Wendy Beth Lyn 6h41'32"50d6'
Star Of Wieser Cep 21h39'17"60d17'
Star of Wilber,The Umi 16h57'0"79d5'
Star of Yamada Cap 20h32'29"-13d47'
Star of Yasu & Sako Sco 16h7'47"-21d31'
Star Of Yasue Aqr 22h4'50"-4d21'
Star of Yasuharu & Naoko Lib 14h34'6"-11d27'
Star of Yasuhiro & Masami Psc 1h16'26"24d21'
Star Of Yoko Tau 5h25'26"22d39'
Star of Yoshiki & Kimiko Gem 7h10'45"28d2'
Star Of Yoshio & Reiko Ari 2h22'8"23d4'
Star of Youichi & Makiko Cap 20h31'5"-12d54'
Star of Yuhiko & Hideka Sco 16h0'47"-8d19'
Star of Yuichi & Yukiko Cnc 8h58'45"19d14'
Star Of Yuji & Osami Lib 14h36'20"-20d9'
Star of Yuji & Yuki Cnc 8h36'47"8d11'
Star Of Yuji & Yuri Sgr 19h12'30"-14d38'
Star Of Yuki Gem 7h28'53"17d12'
Star of Yuki Sgr 19h12'47"-22d30'
Star of Yumi & Toshihiro Gem 7h20'26"23d4'
Star of Zafarana Lmi 10h51'36"27d4'
Star of Zoe,The Aur 6h7'27"30d28'
Star Of Zoeb,The Ori 5h59'1"14d41'
Star Ofimelda Mon 6h55'41"0d33'
Star Onafets Blue Vel 9h23'26"-46d43'
Star Pegg 1 Ori 6h7'28"1d41'
Star Place Dra 17h22'0"60d39'
Star Pockey Cet 3h12'46"-0d20'
Star Rae Cam 8h20'50"83d46'
Star Renae Cet 1h24'14"1d4'
Star Robro Aql 19h48'57"14d20'
Star Ruyle'rs Gateway To Dreams Lac 22h41'16"53d40'
Star Sanderson Aur 5h54'24"50d21'
Star Sapphire Lib 14h40'32"-9d9'
Star Sebastian Peg 23h32'42"21d52'
Star Service Per 4h2'4"35d32'
Star Sherle L D And 23h30'15"43d40'
Star Smalley Oph 17h19'47"-23d52'
Star Smee Umi 16h53'44"76d5'
Star Soldier for Christ Per 2h4'36"58d31'
Star Sosic Uma 8h58'11"52d37'
Star Spangled Love Oph 18h3'43"12d13'
Star Strickland Tri 1h53'42"27d12'
Star Stuff John,Barb, Krista & Becky Cet 1h20'0" 13d4'
Star Susan Del 20h15'44"12d50'
Star Sweeper Ori 6h2'49"-0d17'
Star T J Ori 5h15'33"1d3'
Star Television Ind 20h53'30"-50d42'
Star Thrower '93 (Rob Pegg) Hya 9h3'23"0d42'
Star to Make Kaoru Happy,The Cnc 8h44'50"23d16'
Star Topper Umi 15h17'35"78d26'
Star Town Rx3 Dra 16h11'44"68d21'
Star Trak Cep 22h24'22"59d53'
Star Trek Cap'T Nick Per 1h47'31"54d12'
Star vor Linni B Tau 4h59'47"15d39'
Star Wendy Sunshine Daydream Her 16h52'57"13d41'A
Star Wright,Star Bright Her 16h35'35"36d49'
Star Zachary Jason Aql 19h30'1"10d28'
Star,David Aur 5h7'24"44d35'
Star,Patty & John Cep 2h16'1"78d39'
Star-Spangled-Mimi Eri 2h50'10"-2d41'
Star-Ted Per 1h57'36"53d10'
Starace Star,The Ori 5h33'54"0d46'
Staratzke,Antonia Uma 11h4'28"37d35'
Starbak Mary Lac 22h45'48"56d45'
Starbase Rodd 1701 Per 2h1'59"50d29'
Starbliss-Dina '76 Cae 4h50'23"-31d45'
Starboat Willy Cep 21h55'18"68d36'
Starbody,Frank A Sct 18h56'53"-6d35'
Starbooty Her 17h57'58"30d22'
Starborn Lmi 10h3'38"31d41'
StarBreit Uma 11h9'20"44d48'
Starbright Kevin Aur 6h40'0"37d29'
Starbuck Leo 10h32'21"21d22'
Starbuck Edwards Per 2h8'21"57d18'
Starbuck's Star Her 16h5'1"16d34'
Starbuck,Marven Russel Ori 5h11'40"-0d48'
Starbuck,Oliver Marren Cam 3h31'51"62d41'
Starbuster Aqu 19h57'1"8d12'
Starce Lib 15h57'27"-8d32'
Starcey Theresa Marie 9-16-1961 Lyr 18h53'13"30d40'
Starch,Patricia-Berta Boo 14h22'41"13d41'
Starchaser Erica Mae Cas 2h49'12"61d48'
Starck,Marcia Uma 8h47'24"67d38'
Starczewski,Paul Lac 22h51'42"53d42'
Stardust,Merry's And 1h43'30"39d25'
Staren,Ted Pho 0h30'51"-45d23'
Stares,Sheila Cas 1h53'60"60d48'
Starfinkle "50" Ori 6h0'48"8d36'
Starfire Dra 19h38'12"60d11'
Starfish Lac 22h34'20"48d49'
Starfish Ori 4h41'38"0d36'
Stargentina Uma 8h33'28"61d21'
StarGregan S Ori 5h34'0"-0d24'
Starin at Erin Cyg 19h23'42"28d17'
Stark Aur 6h27'0"40d41'
Stark Dra 19h3'39"58d32'
Stark,Austin John Lyr 18h59'17"28d43'
Stark,Barry Umi 13h8'48"70d45'
Stark,Darrell Hya 8h41'12"-6d49'
Stark,Ferris Michelle Hya 8h44'1"4d37'
Stark,Fred P Hya 8h43'0"2d58'
Stark,Irene Claire Vul 19h4'45"22d9'B
Stark,Michael Aqr 21h42'15"0d50'
Stark,Michelle Joy Uma 9h23'0"51d45'
Stark,Richard & Evelyn Oph 18h17'24"13d36'
Stark,Robert Wayne Vul 19h4'45"22d9'A
Stark,Roger "Bullwinkle" Aur 6h34'48"37d47'
Stark,Samantha Joelle Cnv 13h19'59"40d54'
Stark,The Star of Christopher Cmi 7h57'49"4d36'
Stark,Thomas Brandon Dra 11h8'51"78d33'
Stark,Vincent J Boo 14h35'12"18d50'
Stark-Coan,Darlene Uma 11h39'45"55d1'
Starke,Dieter Del 20h16'44"13d44'
Starke,Hortensia Lopez Laguda Hodel Com 13h4'44"17d2'
Starke,Lorine Cas 0h52'0"66d55'
Starkel,James M Her 16h29'46"39d14'
Starkenburg,Kimberly Kelly Cam 6h50'18"68d18'
Starkey & Barrow Ori 5h56'19"12d25'
Starkey LPB2JL,Richard Cnc 8h29'14"30d23'
Starkey,Jason Leo 10h53'29"21d42'
Starkie's Golden Star Cyg 21h0'0"38d16'
Starkman,Chuck Dra 16h30'25"73d47'
Starkman,Guy Oph 18h1'31"8d34'
Starkovich,Heather Lyn 8h5'44"38d51'
Starla Com 12h54'10"27d18'
Starla Cam 5h9'10"71d1'
Starla,Mother Of Tara & Derek Eri 4h29'38"-11d19'
Starlet Cet 1h32'39"-10d38'
Starley,Laura Lyn 6h54'0"54d44'
Starlight Express Lyr 18h37'32"40d43'
Starlight Kiss Ser 15h14'0"-2d27'
Starlight Renate Lib 15h14'39"-20d16'
Starlite Magic, Love Katie Cep 21h22'57"55d15'
Starlynx Lyn 7h57'42"45d8'
Starmaker Lyn 7h9'56"56d18'
Starman Per 4h6'19"50d48'
Starmates Soulmates 4Ever Sam & Bob Ori 5h56'30"15d3'
Starmax Prima Psa 22h1'1"-28d19'
Starmer,Michele And 23h7'29"44d30'
Starmer,Ron Crt 11h15'25"-8d49'
Starn,Michael & Melissa Cyg 19h43'57"29d39'
Starner,B G Her 16h26'22"50d47'
Starnes III,William Wellington Dra 13h01'51"63d54'
Starnes,Darryl F And 1h48'36"37d26'
Starnes,Judy Lee Peg 22h46'59"29d4'
Starobin,Daniel Cep 22h18'14"55d45'
Staron of Morris Vul 19h23'1"22d44'
Staron,Michael Rubinowitz Psc 1h20'20"20d15'
Staron,Sarka Cas 2h35'39"59d37'
StaronLight Of Ellen Cas 2h0'46"59d24'
Starovich II Cam 7h45'53"70d60'
Starovoitov,Aleksandr Nikolaevich Lac 22h19'11"37d34'
Starowicz,Steve Hya 8h14'52"-5d40'
Starpearl,Pia Leo 11h1'17"10d47'
Starphanna 60 Peg 23h38'44"30d46'
Starr Crb 16h14'28"30d32'
Starr Cet 1h31'33"-11d12'
Starr Vul 19h0'40"22d12'
Starr 13,Lucky Cyg 21h54'10"42d42'
Starr Loya Cnc 8h53'0"32d22'
Starr Renea Uma 8h29'39"68d16'
Starr Theresa Eri 4h9'24"-7d40'
Starr's Boo 13h39'53"24d11'
Starr's,Debbie's Star Of Mon 6h23'1"10d6'
Starr,Brenda Lee Vul 20h15'46"25d2'
Starr,Celeste Annemarie Uma 11h21'41"46d38'
Starr,David & Elizabeth Lmi 10h1'18"36d11'
Starr,Denise West Uma 11h48'0"32d17'
Starr,Dixie C Cyg 19h17'20"44d34'
Starr,Dora Aql 20h16'46"5d35'
Starr,Edwin Lone Cru 12h9'29"-59d9'
Starr,Eric Todd Aur 6h4'43"30d59'
Starr,Jack P Lac 22h27'43"41d12'
Starr,James Dra 10h16'55"80d48'
Starr,Jon Aql 19h5'11"-6d24'
Starr,Joshua Uma 12h43'18"61d15'
Starr,Jr,James J Her 17h59'59"38d35'
Starr,Kathleen Elizabeth Gem 6h54'9"16d16'
Starr,Kristy's Peg 21h28'51"21d16'
Starr,Lisa Kathleen Sex 10h2'10"1d52'
Starr,Lyn Barbara Her 18h4'1"40d1'
Starr,Mark Mon 6h25'45"11d20'
Starr,Mark & Mary Psc 0h3'51"0d47'
Starr,Mary Eri 4h36'40"-19d9'
Starr,Maureen Haynes Cas 23h45'51"58d54'
Starr,Melissa Mon 6h25'43"10d28'
Starr,Morgan Lindsey Sex 9h45'24"-5d13'
Starr,Ona Kay Aql 19h44'14"14d29'
Starr,Peter Harrison Mon 6h22'1"5d36'
Starr,Richard Alexander Her 17h54'17"14d26'
Starr,Robert Tredway Hya 8h9'42"0d57'
Starr,Sam Uma 10h35'39"42d28'
Starr,Samantha Ayala Peg 23h5'41"33d49'
Starr,Shelia And 2h31'37"41d35'
Starr,Susan Lyr 18h49'54"40d39'
Starr,Sydney Her 18h10'57"37d54'
Starr,Theodore & Mary Elizabeth Cyg 19h49'36"50d22'
Starr,Twyla Eri 3h34'45"-7d17'
Starr,William Cook Uma 9h45'1"43d26'
Starrvieux Hya 8h43'60"4d48'
Starry Aqr 23h37'27"-10d19'
Starry Eyed Lisa Cas 3h6'58"57d28'
Starry Eyes Lou Com 12h27'53"25d6'
Starry Night Tri 1h46'14"27d9'
Starry Nites And 0h49'47"38d35'
Starry,Carolyn Peg 22h21'55"31d28'
Starry,Kristin Cam 7h57'1"61d11'
Stars Forever Valerie & Steven Uma 11h25'19"32d13'
Stars R Us Aur 5h34'20"50d31'
Starseyn Sierra Porkey Cas 23h41'53"58d55'
Starsmore,James Ori 5h40'24"10d44'
Starstruck Scooby Uma 9h19'50"56d41'
Startema's Starlets- HCS-3A Dra 20h22'20"68d57'
Starvest 50,Gaston Bottero Per 2h48'17"41d12'
Starward,Patrick Henry Ames Ori 5h58'1"15d42'
Starwas,Share Selket Mon 7h49'44"-3d26'
Stary Kat,The Lyn 8h0'55"39d1'
staRYANgel Dra 17h21'1"58d24'
Starzee,Geralyn Cam 5h32'31"60d24'
Starzi,Mary Lyn 8h27'20"41d59'
Starzwest Cet 0h52'11"0d32'
Starzyk,"Starzyk" Robert J Aqr 21h36'27"-5d45'
Starzyk,Ruth Merrie Lyn 8h1'52"41d33'
Stasch,Lynn Peg 23h48'10"11d13'
Staser,Bill Aur 6h17'21"33d57'
Staser,David Reynolds Ser 18h35'59"3d21'
Stashower,Helaine And 0h9'0"34d40'
Stasi,Silvana And 0h3'16"38d49'
Stasia's Angel Del 20h34'24"20d31'
Stasiakowski,Olivier Mon 7h48'1"-3d37'
Stasio,Heather Lyr 19h18'44"41d41'
Stasko,Greg Hya 10h14'22"-11d32'
Stasko,Mary Cas 1h0'1"67d43'
Stassi "Stass",Timothy Oph 18h16'30"1d37'
Stassi,Renata Hya 9h0'22"5d44'
Stassins,James Lyr 18h37'1"29d40'
Stastny,Linda Sue Lyn 7h54'41"41d4'
Stasulli,Nikolas Michael Dra 18h37'48"68d49'
Stasulli,Timothy John Cep 22h42'1"65d9'
Stasyshyn,Penny & Tony Cyg 21h18'52"37d31'
Staszkow,Richard M & Kimberly S Lac 22h29'38"40d16'
Staten,Patrick M Aur 6h33'33"33d27'
Staten,Scott Michael Aql 19h40'23"14d32'
Statham,Hugh Dyer Cep 5h15'54"87d6'
Stather,Manfred Cyg 20h27'39"39d45'
Statie-Rodriquez, Rafael Benito Aql 19h47'29"10d52'
Statile,Louis Lyn 8h8'14"36d21'
Statile,S Jane Lyn 8h3'0"34d58'
Staton (Eagle Scout), Eric James Aur 4h54'58"38d51'
Staton,Colette Uma 10h18'34"51d23'
Staton,John Mark Aur 6h26'33"34d3'
Staton,Meirah Cam 4h57'10"67d30'
Statos John:112495 Sgr 20h4'17"-20d25'
Statt I,Louis Peter Hya 9h5'58"2d28'
Statton,Christopher Timothy Cnv 12h15'58"40d14'
Statton,David Charles Lmi 10h25'45"30d58'
Statton,Timothy David Her 18h1'56"38d57'
Statz,Charles E Cet 3h14'43"7d51'
Staub,Blake Her 16h59'50"35d9'
Staub,Georg Cnc 8h35'32"15d21'
Staub,Maria-Elisabeth Barbara Sco 16h37'11"-41d6'
Stauber Forever,Eric Charles Ori 5h9'41"-0d35'
Staud,Dr Waldemar Cnc 8h51'56"32d26'
Staudenmeier,Sarah Eri 3h8'0"-4d24'
Stauder,Annie Vul 19h48'24"25d18'
Stauffer,John William Aql 18h56'43"-6d10'
Stauner,Christa- Veronika Her 17h24'1"42d26'
Staup,Dicky Lee Aur 5h28'31"38d59'
Stauter,Josef Lac 22h7'44"51d59'
Stauter,Marilyn Jean Uma 10h1'20"54d50'
Stautner,Martin Cep 20h34'23"67d7'
Stautzenberger,James E Aql 20h12'51"10d43'
Stavashe Ser 15h40'17"20d23'
Stavrakis Star,The Jolene Peg 23h37'27"18d12'
Stavropouls,Pantelis And 23h1'33"51d39'
Stavros Aur 5h5'30"42d20'
Stawinski,Paul Psc 1h30'39"27d58'
Stay Star,The Eri 3h8'32"-5d60'
Staycee,Our "Amaranth" Peg 21h26'43"3d44'
Stayton,Ronnie Aql 19h5'59"15d7'
Stazi,Marina Cas 2h36'58"50d25'
Ste Marthe,Julien Cam 3h41'54"68d5'
Ste-Marie,Horace Cam 3h50'34"63d27'
Stead,Clark Uma 9h9'22"51d52'
Steadham,Joseph Delano Peg 22h3'36"29d21'
Steadham,Joshua Ray Peg 22h3'39"29d57'
Steadman,Bethany Lynn Cas 1h39'27"73d49'
Steadman,Clarence Bell Aql 20h2'22"9d23'
Stealth,The Umi 13h09'59"75d56'
Steamboat,Bonnie Her 17h35'0"40d25'
Steamboat,Jr,Ricky Her 17h39'28"40d4'
Steamboat,Ricky Her 17h57'55"40d37'
Stearn,Jess Tau 5h56'56"24d22'
Stearn,Madeleine Mon 7h41'16"-1d11'
Stearn,Michelle Leverenz And 0h22'57"40d55'
Stearney,Pat Lmi 10h6'1"30d19'
Stearns,Andrew Lac 22h31'38"53d22'
Stearns,Brenda Peg 22h32'52"27d35'

Stearns,Brooke Elizabeth
Ori 6h7'1"20d1'
Stearns,Evan
Vul 20h20'18"22d53'
Stearns,Kelly Ann
Lyr 18h51'30"41d50'
Stearns,Mattie Elizabeth
Lyr 19h21'28"31d17'
Steatham,Stephanie
Lyr 18h17'23"46d30'
Steb-Lar-Mar-"40"
Lyr 19h22'1"42d31'
Steblay,Liz
Lyn 7h57'57"44d48'
Stec,Kimberly Katherine
Tri 2h19'49"32d40'
Stec,Leocadia
Lac 22h26'27'50d3'
Stecher,Anne
Lyn 7h4'46"53d24'
Stecher,Arne
Leo 11h50'10"27d28'
Stechison II,MD, Michael
Thomas
Aur 5h1'0"40d44'
Steck,Austin James
Cam 3h32'17"67d34'
Steck,Gloria
Tau 4h4'58"1d48'
Steck,Lila
And 23h20'46"41d10'
Steckel,Beth A
Cas 1h11'40"62d30'
Steckel,Richard S
Lac 22h10'38"47d27'
Steckenbiller,Angelika
Cep 22h8'33"68d26'
Stecker,Sandra C
Lyn 9h7'16"40d28'
Steckle,Mary Margaret
And 2h5'30"38d58'
Steder,Audrey L
Cyg 19h56'41"41d41'
Steder,Gretchen L
Cas 23h42'12"64d14'
Stedman,Tanya
Com 12h20'47"30d5'
Stedon
Eri 2h56'19"-2d21'
Steed,DVM,Marvin
Cam 6h2'18"61d38'
Steed,Frank & Gail
Umi 15h6'59"65d60'
Steed,Lorelie Rose
Gem 6h33'8"13d52'
Steed,Susannah
Lyr 19h22'59"31d47'
Steed-Johnson
Uma 12h51'12"60d23'
Steede,Harry M
Sgr 20h4'29"-20d7'
Steede,Neil
Cep 23h0'24"80d30'B
Steedley,Beverly S
Peg 21h53'54"30d42'
Steegmann,Renate von
Sco 16h48'13"-25d19'
Steel "Star Of My Life",Susan
Lyr 18h49'35"41d42'
Steel Star
Ori 5h59'45"15d39'
Steel,Forever Martin & Vivien
Uma 9h3'1"53d40'
Steele
Aql 19h6'1"0d9'
Steele III,Felix G
Cet 1h22'12"-2d44'
Steele's Star
Cet 1h36'22"-3d18'
Steele,Aidan Richard
Cam 6h9'55"83d49'
Steele,Alice Soule
Uma 9h3'0"70d5'
Steele,Andy
Cet 0h42'59"1d50'

Steele,Ben
Lmi 10h41'0"28d3'
Steele,Bonnie J
Cas 1h49'53"65d34'
Steele,Carolyn & Larry
Tau 4h56'28"20d1'
Steele,Carroll Robert
Ori 5h16'59"15d39'
Steele,Christopher Allen
Uma 9h57'10"44d59'
Steele,David & Marie
Davidson
Peg 21h38'32"25d33'
Steele,Deborah M
Sge 19h41'45"16d36'
Steele,Derek
Her 17h50'1"41d6'
Steele,Drew
Ser 18h36'44"3d35'
Steele,Earl
Dra 19h26'44"56d45'
Steele,Evan Thomas Butler
Her 18h13'11"40d46'
Steele,G Rudy
Crb 16h23'0"37d55'
Steele,Gayle D
Oph 18h39'11"7d16'
Steele,Jackie
Cas 1h55'30"73d11'
Steele,Jeannie
Cas 0h20'34"58d12'
Steele,John P
Del 20h19'25"10d7'
Steele,Joie Lee
Lyr 18h37'19"42d29'
Steele,Jon Gary
Aur 5h33'42"50d29'
Steele,Jr,Marvin J
Boo 15h15'51"38d37'
Steele,Lacy Ann
Peg 22h2'34"8d7'
Steele,Lea
Eri 4h56'11"-4d20'
Steele,Lyndon B
Per 3h2'25"40d58'
Steele,Marie Agnes
Mon 6h58'23"10d21'
Steele,Mark
Lac 22h31'16"37d36'
Steele,Mark
Sex 9h51'53"2d28'
Steele,Micky
Per 4h7'1"50d1'
Steele,Nicholas (Nilo) Peter
Aql 20h0'38"12d54'
Steele,Robert
Aur 7h26'1"39d29'
Steele,Roman Wardlaw
Ori 5h8'54"-4d29'
Steele,Steven Charlton
Cmi 7h57'1"0d18'
Steele,The Adam Paul
Per 2h39'14"37d41'
Steeley,David
Cet 1h12'13"-3d0'
Steelman's Star
Aql 19h54'54"13d48'
Steelman,Jason & Monica
Uma 10h56'29"40d15'
Steelwright,Deborah & Lee
Cyg 21h3'1"33d16'
Steely,Officer Jeffrey M
Lac 22h30'15"50d34'
Steen,Elsa Carolyn
Del 20h39'14"18d56'
Steen,Maria
Sgr 20h3'0"-20d1'
Steen,Nils
Uma 9h13'23"55d58'
Steenwaerder,Carsten
Equ 20h56'12"8d50'
Steenwerth,Christa
Del 21h4'0"12d9'
Steer II,Steven A
Per 2h37'15"48d46'

Steers,Carly Anne
Cas 0h50'23"74d0'
Steers-Crist,Nancy
Lmi 9h49'35"33d11'
Steevensz,Michelle
Lyr 18h35'47"47d21'
Steever-Frisbie, Crystal
Mignon
Uma 12h55'56"53d19'
Stef
Cma 6h26'55"-20d5'
Stefal
Pup 8h8'33"-24d3'
Stefan
Dra 17h39'54"60d16'
Stefan
Lyr 18h31'1"33d4'
Stefan und Moni's Lumpistern
Her 17h19'38"42d38'
Stefan,Johnny "Be Good"
Hya 8h32'51"-8d1'
Stefan,Radim
Ser 15h38'25"18d39'
Stefana e Claudia
Col 6h34'45"-34d38'
Stefanatos,Sophia
Del 20h14'22"12d43'
Stefanazzi,Richard Hugo
Uma 11h13'21"32d57'
Stefango,Jim
Lac 22h54'53"54d7'
Stefani Alexa
Cnv 13h12'34"40d56'
Stefania
Lac 22h32'45"38d33'
Stefania
Lup 15h18'30"-35d12'
Stefania
Cyg 20h4'19"31"52'
Stefania
And 0h58'37"41d2'
Stefania
For 2h10'0"-26d42'
Stefania
Vel 9h23'28"-40d0'
Stefania e Angelo
Her 18h10'10"31d12'
Stefaniak,Elaine
Cas 0h28'42"66d6'
Stefaniamaria Romeo
Cep 1h42'45"77d53'
Stefanick,Gerard M
Ori 6h3'1"8d2'
Stefanie
Umi 16h19'12"79d26'
Stefanie
Lyn 7h10'16"50d60'
Stefanie & Ricci
Sgr 19h37'9"-41d21'
Stefanie Michele
Cas 22h56'1"53d17'
Stefanie's Star
Peg 22h0'19"8d17'
Stefanie's Wish
Sco 17h54'11"-38d53'
Stefanik,MD,David Francis
Oph 17h20'55"12d11'
Stefanik,Melanie
Cas 1h0'0"62d32'
Stefano
Pho 0h8'1"-42d14'
Stefano
Her 17h9'41"16d39'
Stefano
Del 20h16'52"15d28'
Stefano
Del 20h14'60"10d3'
Stefano 95
Cae 4h54'41"-28d57'
Stefano e Ely
Cas 0h1'43"63d58'
Stefano's Sweet Stuff
Her 15h59'43"48d59'
Stefano-Lorena
Ant 10h44'28"-37d22'

Stefano-Massimo
Lac 22h2'20"51d59'
Stefanos II
Cam 8h38'46"77d46'
Stefanov,Claudia
Cas 0h33'36"61d7'
Stefanová,Hana
Com 12h18'41"18d51'
Stefanski,Brittany June
Lyr 19h25'37"40d18'
Stefanski,Ignatius L
Her 17h25'40"20d51'
Stefanski,Irene
And 23h21'54"48d38'
Stefanski,Marliese Junkenitz
Lmi 10h50'58"26d56'
Stefanski,Melissa Anne
And 0h58'23"34d33'
Stefanski,Nicholas Steven
Ori 5h56'14"19d39'
Stefanski,Rebecca Jean
Uma 8h39'59"51d20'
Stefanus
Cet 2h58'0"0d11'
Steffaro,"Chedomirka" Carol
Jean
Aur 5h1'1"37d35'
Steffe,Sandi Marie
Vul 19h2'0"22d11'
Steffek,Wendy
Eri 2h45'41"-15d21'
Steffen,Alex
Lyr 19h19'59"30d36'
Steffen,Georg
Umi 13h39'0"74d23'
Steffen,Robert & Cathy
Mon 7h0'22"8d1'
Steffen,Werner 22-7-21
Peg 23h31'40"18d6'
Steffens,Christine
Lyr 18h53'29"33d55'
Steffens,Duane
Leo 9h52'41"10d42'
Steffens,Jacques
Cam 4h8'31"60d13'
Steffensen,Svanhild
Aur 5h11'42"29d59'
Steffes,CSA Jean
And 0h10'21"47d55'
Steffes,Mike
Aur 7h3'23"41d15'
Steffey,Lauren Lynn
Mon 6h55'23"-10d2'
Steffi
Sgr 19h54'58"-42d49'
Steffi
Tau 5h55'19"27d11'
Steffi's Star
Aqr 22h42'53"-5d5'
Steffies,Mike
Aur 6h1'15"30d13'
Steffl,LeAnn & Tony
Sge 19h56'27"20d4'
Steffler,Alexander M
Ser 15h34'43"24d46'
Steffman,Jeremy Andrew
Boo 15h0'53"11d49'
Steffy,Harriet L
And 23h8'58"40d51'
Stefko,Dawn Sue
Cas 22h55'44"55d23'
Stefl,Baili Elizabeth
Mon 7h54'58"-5d23'
Stefy
Vel 9h41'7"-43d2'
Stefy 23
Pyx 8h50'23"-25d15'
Stegall,Arthur John
Her 16h29'45"40d49'
Stegall,Jill
Equ 20h58'0"10d14'
Stegall,Ken
Oph 18h2'15"11d45'
Stegall,Sr,Roy DeWayne
Ser 16h13'44"-1d31'A

Stegama-Gaudio
Umi 15h5'19"70d8'
Stege,Julie Jenner
Cyg 19h24'59"33d15'
Stege,Jürgen
Cep 22h5'39"68d34'
Stegeman,Catherine Anne
Cas 0h6'26"58d24'
Stegeman,William Dirk
Aur 4h50'44"50d31'
Stegemann,Jutta
Leo 10h59'24"10d20'
Stegen,Elizabeth J
Cas 1h2'0"50d1'
Steger,Carlton
Cep 22h50'48"59d60'
Steger,Sarah Ann
Vul 20h5'31"23d10'
Steggler,Frances
Cnc 7h57'43"10d42'
Stegmann,Renate
Cap 20h31'19"-26d6'
Stegmier,George P
Sco 17h28'21"-43d20'
Steguweit,Bruno
Ori 5h4'0"10d55'
Stehl,Shaylyn Elizabeth
Cnc 8h56'0"31d30'
Stehno,Henry
Hya 8h11'49"5d59'
Stehr Family,The Leonard
Per 2h55'46"31d23'
Stehr Family,The Larry
Per 2h55'52"34d20'
Stehr,Horst
Lyr 19h19'22"30d13'
Steiberg,Daniel
Her 17h9'51"37d53'
Steich,Samantha Paige
And 1h30'13"38d33'
Steidley,Barbara Jo
Vir 11h50'50"9d24'
Steidlitz,Audrey Clare
And 23h31'45"41d45'
Steigelman,Anne V
Cas 1h2'0"58d49'
Steigelman,Edward T
Lac 22h22'52"37d30'
Steiger,Birgit
Lyn 8h24'42"42d5'
Steiger,Charles Mark
Sex 10h31'43"-0d50'
Steiger,Michael George
Cep 0h45'34"78d37'
Steigerwald,Jessica
Peg 22h6'0"29d50'
Steimer,Isabelle
Lyn 8h25'46"42d8'
Steimetz,Joan
Aql 19h39'60"11d30'
Stein III,Alan W
Her 18h4'19"38d23'
Stein's Brilliant Star,Brett
Uma 10h57'1"34d34'
Stein's Shining Glory
Dra 10h3'0"80d14'
Stein,Alan J
Her 16h57'46"38d17'
Stein,Andrew C
Aur 6h11'55"37d32'
Stein,Andy
Aur 7h23'11"35d28'
Stein,Anna Theresia Vom
Oberen
And 0h3'22"44d6'
Stein,Austin Joseph
Dra 16h15'1"62d33'
Stein,Carmen Camille
Eri 4h12'1"-12d7'
Stein,Carol & Jay
Com 12h4'15"23d34'
Stein,Casey Marie
Cas 0h35'17"62d36'
Stein,Dawn M
And 23h9'10"40d44'

Stein,Diana
Cas 0h50'46"69d53'
Stein,Doctor,Craig M
Oph 18h38'55"8d35'
Stein,Donna Z
And 2h2'37"46d1'
Stein,Ellen C
Lyn 8h55'57"38d41'
Stein,Fran & Al
Ori 5h47'0"10d40'
Stein,Frederick Eaton
Dra 20h11'60"62d46'
Stein,Frieda
And 0h2'20"40d7'
Stein,Gary Earl
Aql 19h16'12"13d19'
Stein,Gregory
Her 16h18'1"10d2'
Stein,Ira
Her 18h53'41"12d41'
Stein,Jane G
Mon 6h20'0"5d50'
Stein,Judy
Mon 6h26'24"10d19'
Stein,Kenneth
Aql 19h51'35"12d1'
Stein,Kyle Noah
Del 20h49'52"9d54'
Stein,Lauren Jill
Cas 0h57'1"61d4'
Stein,MD,Harry Leo
Lac 22h5'44"38d15'
Stein,Mikah Samuel Feldman
Per 2h53'34"55d34'
Stein,Miranda Rose
Lib 14h31'36"-20d14'
Stein,Molly Ann
Peg 21h57'28"33d47'
Stein,Natalie Jeanne
Lyn 7h59'1"41d18'
Stein,Norris
Hya 8h34'47"0d41'
Stein,Reid Jordan
Her 17h34'1"24d13'
Stein,Walter,Virginia & Jack
Cnv 12h50'19"50d50'
Stein,Wilbur S
Ori 5h59'13"15d52'
Stein,William Richard
Lac 22h8'46"40d28'
Stein,Zoe
Boo 14h19'20"38d7'
Steinau,Bill
Boo 15h40'52"41d50'
Steinaway,Richard
Cet 2h41'52"7d32'
Steinbach,Derek Michael
Ori 5h7'19"0d43'
Steinbach,Sally
Lyn 8h1'60"45d55'
Steinback's Fat Free Star,Jyl
Peg 22h52'1"27d47'
Steinbauer,Franz Xavier
Cyg 20h24'58"39d28'
Steinbaum,Jerry
Oph 18h17'31"13d28'
Steinbeck,Kevin
Dra 19h3'1"60d12'
Steinbeck,Maria
Ari 2h53'1"22d27'
Steinberg,Barry
Dra 15h8'37"61d13'
Steinberg,Bruce Harris
Her 17h19'12"44d33'
Steinberg,Daniel Jeremy
Cep 22h0'44"61d42'
Steinberg,Dianne
Cas 0h6'56"62d36'
Steinberg,Edythe
Oph 16h54'7"-19d27'A
Steinberg,Jennifer Crystal
Mon 7h12'14"-6d55'
Steinberg,Joan E
Peg 21h47'1"28d22'

Steinberg,Mason
Ser 15h49'12"24d32'
Steinberg,Mitchell I
Her 16h42'60"11d43'
Steinberg,Murray
Oph 16h54'7"-19d27'B
Steinberg,Samuel
Aql 19h58'31"12d2'
Steinberg,Sandra & Zane
Cyg 19h58'39"30d23'
Steinberger,Melitta
Aqr 22h47'30"-2d21'
Steinberger,Richard Anthony
Her 16h44'34"25d45'
Steinberger,Susanne
Cnc 8h6'32"30d41'
Steinbergers' Love Star,Carrie
& David
Uma 11h0'21"58d31'
Steinbruch
Aql 20h9'18"1d28'
Steinbrugger,Chuck & Theresa
Eri 3h52'33"-1d26'
Steinbüschel,Klaus Willibald
Sgr 19h1'33"-21d32'
Steindl I,Peter Friedrich Walter
Lyn 8h11'0"49d58'
Steindl,Anna & Thomas
Uma 11h11'15"62d3'
Steiner
Cas 3h3'33"58d13'
Steiner & Co,Jeffrey
Sgr 18h55'11"-27d5'
Steiner,Cindy "Sugar Bear"
Cyg 20h7'25"38d14'
Steiner,Clara Carroll
Cyg 21h13'51"28d43'
Steiner,Jacquelyn Yetta
Cam 4h51'52"60d0'
Steiner,Patrick Russell
Dra 16h51'28"68d32'
Steiner,Robert L
Cma 6h56'19"-17d1'
Steiner,Sarah,Robert
Rebecca&DavidWeintraub
Umi 16h4'46"78d46'
Steiner,Whitney Leigh
Cyg 19h21'24"28d28'
Steiner-Pichler, Annemarie
Sco 17h29'1"-31d1'
Steinerman-svi-hesh, Harvey
Peg 22h3'14"2d28'
Steinert,Jeff H
Her 17h38'51"26d35'
Steinert,Sandra
Gem 7h21'26"27d30'
Steinert,Wolfgang
Per 3h50'31"38d50'
Steinfels,Martha
Dra 15h59'27"62d21'
Steinhart,William
Her 18h4'0"31d8'
Steinhauer
Lyr 19h17'52"30d9'
Steinhauer,Aaron J B
Cam 3h40'36"73d18'
Steinhauer,Alisa J
Eri 3h57'27"-19d13'
Steinhauer,Danna
Tri 2h14'0"32d55'
Steinhebel,Martha
Cyg 20h8'53"40d24'
Steinheimer,Elizabeth Victoria
Aql 20h31'12"-6d42'
Steinhoff,Jessica Stuart
And 0h20'15"43d12'
Steinhoff,Matthew Carson
Her 16h25'34"41d28'
Steinhoff,Sarah Anne
Gem 7h4'32"24d47'
Steinhoff,Seth Corey
Gem 7h18'34"24d50'
Steininger,Albert
Ser 15h32'43"18d37'

Steininger,Birgit
Sco 17h4'42"-31d23'
Steininger,Kenneth C
Cep 21h58'24"67d51'
Steininger,Kenneth C
Peg 23h39'38"26d37'
Steininger,Patrick J
Dra 19h35'31"68d32'
Steinitz,Hunter Elizbeth
Lyr 18h37'30"43d2'
Steinkamp,James Robert
Lac 22h38'50"52d44'
Steinkamp,Jena Marie
Cyg 19h32'1"32d1'
Steinkamp,Jessica Lee
Uma 10h36'52'55d57'
Steinkamp,Renè Victoria
Lyn 8h4'58"38d25'
Steinke,Charles
Crb 15h50'11"30d36'
Steinke,Gabriele
Cap 21h57'38"-26d43'
Steinker,Dr Gerald E
Oph 17h13'51"-24d14'
Steinkopf,R K
Tri 2h19'44"31d35'
Steinle,Anne
Leo 10h43'58"7d36'
Steinman,Barry
Ser 18h0'55"-13d21'
Steinman,Joan & Lewis
Cam 3h52'60"55d20'
Steinman,Thomas G
Per 3h11'46"50d30'
Steinmann,Charles F
Her 17h11'58"27d10'
Steinmetz,Charles Paul
Cyg 19h12'53"49d40'B
Steinmetzer,Helmut
Cnc 8h6'56"31d6'
Steinmüller,Detlef
Ari 2h29'22"26d44'
Steinn Logi Björnsson
Uma 11h1'34"43d58'
Steinshine LoveRhon, Arory
Aur 6h5'1"30d46'
Steinwurtzel,Jennifer Lynn
Cma 6h17'42"-15d29'
Steise's Stellar Caswell
Lmi 9h21'46"37d46'
Steivang's Norske Star
Cyg 20h26'32"38d55'
Steivang,Heather & Laura
Her 17h39'43"15d58'B
Stejoho-40
Crb 16h4'43"28d6'
STELALZA
Aur 7h1'47"41d32'
Stell,Joe
Per 2h55'0"38d50'
Stell,Julie Dory
Com 12h17'51"31d44'
Stella
Del 20h14'35"12d10'
Stella
Cyg 21h3'22"40d23'
Stella
Lyr 18h53'28"38d43'
Stella
Lyr 18h57'13"41d15'
Stella
Cas 0h53'17"61d32'
Stella
And 1h22'38"40d55'
Stella
Cam 3h55'1"57d3'
Stella
Sex 9h53'52"-1d42'
Stella
Boo 14h55'22"38d18'
Stella "Emmanuel" con due
emme
And 23h9'14"43d3'
Stella & Iian
Ori 5h24'10"1d24'

Stella (Asteroid B 613)
Vir 13h33'18"5d57'
Stella 03/09
Cnv 12h38'13"39d20'
Stella aimer Kazuo
Leo 10h17'33"8d56'
Stella Andrea Et Peter En Ciel
7
Leo 11h18'59"-2d20'
Stella Angela Felice
Cep 23h32'45"65d49'
Stella Appartenere Michelle
Uma 10h12'35"57d4'
Stella Atsuhiro & Mizue
Aqr 22h4'38"-9d27'
Stella Birgit Libra
Lib 14h40'44"-6d47'
Stella Bova
Cyg 21h23'21"40d1'
Stella Cathrina
Cam 3h31'0"62d48'
Stella Christopheri Sullivani
Ori 5h17'14"1d34'
Stella dei desideri impossibili
Pyx 8h41'19"-24d7'
Stella Dell' Amore;The Light of
my Life
Lyr 18h47'18"39d54'
Stella Di Fortunata
Cyg 21h3'19"36d48'
Stella di Nozze
And 0h14'26"33d16'
Stella di Stagno
Ant 10h43'43"-37d2'
Stella Don Briando De Escocia
Boo 15h5'52"31d38'
Stella Fatum De Margene y
Hugh
Cyg 19h31'31"33d1'
Stella Felicitatis alfred katjae
Lac 22h9'46"38d37'
Stella Fortunae Philippi
Vir 12h25'17"-5d10'
Stella Fortunae
Crb 15h32'44"37d2'
Stella Grimoldi
Per 3h32'12"50d45'
Stella Helene
Cas 0h32'45"50d5'
Stella Joanne Marie
Lib 14h24'14"-10d3'
Stella Leondina
Cep 23h12'39"70d32'
Stella Lore
Com 12h56'15"28d6'
Stella Madonna
Ari 1h45'0"16d31'
Stella Manoidem
Lac 22h7'38"38d17'
Stella Marci
Lyr 18h15'46"38d57'
Stella Mary
Pup 7h59'41"-20d21'
Stella Michaeli
Ori 4h52'30"1d41'
Stella Monica cum armae
And 0h12'34"30d34'
Stella Monocerox Matrae
Cas 0h43'37"64d3'
Stella Of "Atsuhiko Yamazaki
Is No 1"
Gem 7h12'36"22d15'
Stella of "For You"
Ari 2h26'18"21d20'
Stella of "Katsu-G"
Ari 2h18'33"19d24'
Stella of 15th Stardust Revue
Gem 7h11'38"28d32'
Stella of 777 Tetsuya & Eri
Cnc 8h48'42"18d26'
Stella Of Access
Sco 16h4'57"-26d13'
Stella of Access
Sgr 19h18'15"-27d59'

Stella of Ai
Cap 20h49'53"-23d43'
Stella Of Aiko
Leo 10h9'45"22d27'
Stella of Aiko
Leo 10h9'45"22d27'
Stella Of Aiko & Noriko
Ari 2h23'15"-8d43'
Stella of Aimer W Yoshiko
Sco 16h4'56"-15d30'
Stella of Airi
Vir 14h24'33"1d11'
Stella of Akari
Cap 20h51'8"-19d10'
Stella of Akaru
Lib 14h48'45"-22d29'
Stella of Akavi
Sco 16h24'15"-29d10'
Stella Of Akemi
Psc 1h19'39"28d56'
Stella of Akemi
Leo 10h4'27"18d49'
Stella of Akemi
Sco 16h7'47"-25d24'
Stella of Akemi= Takehito
Tau 5h25'9"22d51'
Stella Of Aki
Aqr 22h18'12"-15d11'
Stella of Aki
Cap 20h50'50"-17d4'
Stella of Aki
Ari 2h12'3"20d46'
Stella of Akihiko
Sco 16h3'50"-14d9'
Stella of Akihiko & Midori
Cap 20h52'11"-24d25'
Stella of Akihio's Life
Sgr 19h8'5"-16d34'
Stella of Akihiro
Sco 16h5'0"-13d23'
Stella of Akihiro & Kikue
Aqr 22h4'6"-23d52'
Stella of Akihiro & Michiko
Leo 10h16'8"6d54'
Stella of Akihito & Junko
Sgr 19h15'0"-16d26'
Stella of Akihito & Rie
Aqr 22h16'3"-9d31'
Stella of Akika
Lib 14h24'53"-11d6'
Stella Of Akiko
Gem 7h12'5"14d5'
Stella Of Akiko
Tau 5h21'3"24d53'
Stella Of Akiko Kajiwara
Cap 20h52'18"-22d17'
Stella of Akiko & Miwako
Leo 10h18'6"19d45'
Stella of Akinori
Psc 1h18'47"18d2'
Stella of Akio
Aqr 22h1'45"-23d26'
Stella of Akio
Ari 2h16'44"21d55'
Stella of Akio & Yukie
Sco 16h4'23"-13d56'
Stella Of Akira & Atuko
Vir 14h50'33"2d29'
Stella of Akira & Naoko
Cap 20h59'56"-16d35'
Stella of Akira & Sachiko
Tau 5h28'6"20d6'
Stella of Akira & Shigemi Family
Leo 1h4'23"14d23'
Stella of Akira & Masami
Aqr 22h0'14"-11d41'
Stella of Akira & Yoshiko Okada
Ari 2h15'33"21d47'
Stella of Akira & Naoko
Gem 7h14'29"28d31'
Stella of Akira & Chie
Psc 1h11'59"27d49'
Stella Of Akira & Yoko
Vir 14h27'1"5d36'
Stella of Akitoshi
Lib 14h49'27"-19d14'
Stella of Ako
Vir 14h1'38"-9d16'
Stella of Ako
Sco 16h24'57"-26d21'
Stella Of Ako
Vir 14h1'38"-9d16'
Stella of AKO
Leo 10h16'51"6d51'
Stella Of Akogare Horiyama
Tau 5h16'56"28d6'
Stella of Amantes del Mar
Sgr 19h11'15"-29d26'
Stella Of Amour
Gem 7h13'53"16d14'
Stella Of Aoi
Aqr 22h19'17"-15d8'
Stella Of Aoi
Lib 14h44'35"-11d11'
Stella of Aoi
Psc 1h22'12"18d14'
Stella of Aoki Hiroko
Tau 5h10'18"18d27'
Stella Of Arata & Chiemi
Gem 7h13'15"26d2'
Stella of Arisa
Leo 10h9'38"17d30'
Stella of Asuka
Gem 7h20'11"22d26'
Stella Of Asuna
Vir 14h34'19"5d52'
Stella of Atsuko
Lib 14h28'17"-13d16'
Stella of Atsuko
Sgr 19h14'11"-14d39'
Stella of Atsuko
Lib 14h41'0"-12d43'
Stella of Atsuo & Tomoko
Ari 2h16'33"14d41'
Stella of Atsura
Leo 10h9'56"20d42'
Stella of ATSUSHI
Sgr 19h10'50"-15d36'
Stella of Atsushi & Yuka
Aqr 22h9'12"-5d14'
Stella Of Ava
Vir 14h2'0"-9d20'
Stella Of Aya
Lib 14h27'39"-14d24'
Stella of Aya
Sco 16h19'47"-24d29'
Stella of Aya
Sgr 19h8'29"-14d6'
Stella of Ayachan
Sgr 19h8'21"-12d9'
Stella of Ayaka
Ari 2h23'38"12d41'
Stella of Ayako
Sgr 19h12'17"-29d56'
Stella Of Ayako
Tau 5h32'33"16d24'
Stella Of Ayako
Vir 12h3'22"2d1'
Stella Of Ayano
Tau 5h34'0"15d33'
Stella of Ayano
Vir 14h32'57"2d18'
Stella Of Ayumu
Tau 5h32'32"16d12'
Stella of Ayumu & Junko
Sgr 19h14'38"-16d11'
Stella Of Baba-P
Tau 5h31'41"17d4'
Stella of Bakuchist Miyazaki
Psc 1h22'24"2d49'
Stella of Bauer 434
Cap 20h52'8"-25d22'
Stella of Betty Sugano
Aqr 22h1'24"-13d16'
Stella of Birth
Psc 1h23'30"6d3'
Stella of Birthday
Ari 2h29'21"21d46'
Stella of Bok Soon
Cap 20h52'0"-23d2'
Stella of Bros
Aqr 22h1'35"-24d12'
Stella of Bunpei & Miharu
Sco 16h20'54"-25d16'
Stella of Chala
Ari 2h15'45"13d43'
Stella of Champion
Cap 20h50'29"-17d31'
Stella of Charchan & Chichan
Cap 20h54'56"-23d8'
Stella of Chi-bo
Psc 1h15'18"18d32'
Stella of Chiaki Y
Psc 1h14'23"27d42'
Stella Of Chie
Cnc 8h45'11"15d25'
Stella Of Chie
Ari 2h14'3"23d3'
Stella Of Chieko & Toshio
Ari 2h19'33"26d2'
Stella of Chiemi
Gem 7h1'45"16d14'
Stella of Chiharu
Cnc 8h45'57"10d37'
Stella of Chiharu
Sco 16h2'0"-14d45'
Stella of Chiharu & Sakiko
Lib 14h47'39"-24d48'
Stella of Chiharu & Masae
Sgr 19h7'33"-16d22'
Stella of Chika
Cnc 8h42'8"22d2'
Stella of Chikako
Ari 2h17'8"16d0'
Stella of Chikako
Lib 14h32'29"-18d36'
Stella Of Chitose
Cnc 8h47'45"11d37'
Stella of Chiyomi
Sgr 19h14'38d-27d40'
Stella Of Chiyori
Cnc 8h45'20"6d50'
Stella Of Chiyuri
Ari 2h39'12"10d20'
Stella of Chrrear
Lib 14h49'45"-23d55'
Stella Of Count 35
Ari 2h19'33"25d59'
Stella of Courage & Luck
Psc 1h18'18"21d20'
Stella of Couragi & Heart
Cnc 8h45'53"12d52'
Stella of Daichi
Lib 14h25'8"-14d10'
Stella of Daiki & Suzuka
Cnc 8h47'45"17d46'
Stella Of Daima Kouji
Gem 7h12'15"16d46'
Stella of Daisaku Ikeda
Cap 20h58'36"-24d0'
Stella of Daisaku & Kaneko Ikeda
Cap 20h58'54"-14d57'
Stella of Daisuke
Sco 16h18'48"-24d22'
Stella of Daisuke
Sco 16h16'59"-22d23'
Stella of Daisuke
Sco 16h24'39"-27d9'
Stella of Darling
Cnc 8h40'41"24d10'
Stella Of Darling
Cnc 8h40'41"24d10'
Stella of Deen
Sco 16h10'17"-13d6'
Stella of Doi & Masumi
Cap 20h52'29"-16d13'
Stella Of Doug & Hiromi
Sco 16h26'30"-29d5'
Stella Of Doux Sourire
Gem 7h13'18"22d45'
Stella of Dr Jo
Cap 20h54'33"-24d40'
Stella of Dr PaPa Toshio
Cap 20h54'32"-20d14'
Stella of E & K
Ari 2h17'15"26d17'
Stella of E amuri noa'tu
Cnc 8h47'53"17d53'
Stella Of Eiji & Youko
Sgr 19h10'36"-14d32'
Stella Of Eiji Miho
Psc 1h15'8"27d11'
Stella Of Eiko
Aqr 22h1'38"-14d18'
Stella of Eisei Noguchi
Leo 10h8'29"20d55'
Stella of Eisei & Fukuko Noguchi
Leo 9h58'39"21d11'
Stella Of Eisei & Fukuko Noguchi
Leo 9h58'39"21d11'
Stella of EKA
Gem 7h13'36"23d58'
Stella of Emi
Cnc 8h45'27"15d17'
Stella of Emi
Psc 1h16'6"19d1'
Stella of Emiko
Lib 14h33'2"-18d45'
Stella of Emiko
Psc 1h21'48"6d54'
Stella Of Emiko
Psc 1h21'48"6d54'
Stella Of Emiko & Tomoko
Leo 10h4'50"11d11'
Stella Of Endless Love
Tau 5h15'44"28d1'
Stella Of Erina
Gem 7h12'47"26d12'
Stella of Eternal Love
Gem 7h14'51"24d36'
Stella Of Eternally Shine Mari
Lib 14h26'51"-10d22'
Stella Of Etsuko
Lib 14h37'32"-15d2'
Stella of Etsuko
Lib 14h37'32"-15d2'
Stella of Etsuko & Junichi & Hiroyoshi
Ari 2h21'57"11d8'
Stella of Etsuko A
Gem 7h22'17"22d40'
Stella Of Etsusi & Mari
Lib 14h40'39"-10d24'
Stella of First Bell
Ari 2h19'53"18d49'
Stella Of Flower Festival
Ari 2h19'20"24d12'
Stella Of Forever
Gem 7h11'6"14d5'
Stella of Forever
Tau 5h26'5"20d22'
Stella of Forever
Sgr 19h13'48"-16d22'
Stella of Forever Aiko Kentaro
Cnc 8h49'8"12d5'
Stella of Fortunate Sisters
Sco 16h28'39"-25d53'
Stella of Fortune for K & Y forever
Leo 10h17'14"11d22'
Stella Of Fuchikawa Family
Ari 2h14'18"21d19'
Stella Of Fujio
Ari 2h8'41"25d51'
Stella of Fukushima Family
Sco 16h23'59"-29d41'
Stella of Fukuzo & Sachiko
Gem 7h10'33"15d17'
Stella of Fumiaki & Yuiko
Ari 2h26'57"19d40'
Stella Of Fumiaki & Yuiko
Ari 2h26'57"19d40'
Stella of Fumiaki papa
Sco 16h26'57"-26d26'
Stella of Fumiari & Eriko
Vir 14h21'18"5d37'
Stella of Fumie Kusachi
Sgr 19h11'11"-28d48'
Stella Of Fumihico
Aqr 22h3'5"-18d21'
Stella of Fumihiko & Hiroe
Gem 7h13'6"24d44'
Stella of Fumihiko & Seiko
Sco 16h2'35"-14d57'
Stella Of Fumiko
Psc 1h22'41"9d27'
Stella Of Fumiko T Fujiwara
Sco 16h25'39"-26d5'
Stella of Fumio & Fumiko
Gem 7h11'35"24d7'
Stella Of Fumiya
Cnc 8h43'20"25d29'
Stella of Fumiyo
Aqr 22h1'14"-22d1'
Stella of Furukawa Family & Chiko
Vir 14h26'8"-5d2'
Stella Of Furuya Family
Aqr 22h0'32"-21d57'
Stella Of Fusao & Rimi
Lib 14h30'42"-23d40'
Stella of Fuyuko
Cap 20h55'11"-22d40'
Stella of Gakuto
Sgr 19h14'50"-26d47'
Stella Of Ganbaru Hiromi
Psc 1h19'59"11d24'
Stella of GC
Sgr 19h10'26"-12d12'
Stella of Gen
Vir 14h8'17"0d22'
Stella of Gensai & Chie
Sgr 19h6'45"-14d39'
Stella Of Go
Ari 2h12'15"27d12'
Stella Of GON
Lib 14h43'6"-11d8'
Stella Of H & K
Ari 2h12'41"27d7'
Stella of H M
Tau 5h34'35"19d41'
Stella Of H Takusagawa Family
Ari 2h16'48"25d15'
Stella of H Tokunaga's Team:Junko
Ari 2h29'27"21d11'
Stella Of Hadzuki
Leo 9h58'33"22d12'
Stella Of Hadzuki
Leo 9h58'33"22d12'
Stella Of Hajime
Tau 5h16'54"27d0'
Stella of Hajime & Chiyoko
Sgr 19h6'51"-15d58'
Stella of Hajime & Izumi
Sgr 19h11'44"-27d41'
Stella of Hajime & Miki
Lib 14h52'44"-24d4'
Stella of Hakubun & Hiroe
Sgr 19h12'27"-11d58'
Stella of Hanaru
Cap 20h49'47"-24d23'
Stella of Happiness
Aqr 22h0'53"-24d54'
Stella of Happy Takeshiho
Sgr 19h5'15"-13d18'
Stella of Happy Yuji
Psc 1h18'39"25d24'
Stella of Harry
Cnc 8h46'56"9d2'
Stella of Haru S
Lib 14h50'56"-23d46'
Stella Of Haruhisa
Cnc 8h50'6"22d12'
Stella Of Haruka
Aqr 22h4'48"-21d0'
Stella of Haruka
Tau 5h27'20"20d28'
Stella of Haruko
Gem 7h13'6"15d25'
Stella Of Haruko
Gem 7h13'6"15d25'
Stella Of Haruko Namba
Gem 7h13'42"26d5'
Stella Of Harumi
Aqr 22h18'24"-15d15'
Stella Of Harumi & Rika
Cap 20h55'39"-20d20'
Stella of Harumi S
Sgr 19h12'51"-28d4'
Stella Of Harumichi
Lib 14h33'24"-22d31'
Stella Of Haruna
Psc 1h23'8"9d6'
Stella Of Harunobu & Hiroko
Cap 20h59'45"-25d31'
Stella of Haruo & Yuki
Cap 20h54'36"-17d56'
Stella Of Haruyo
Psc 1h24'14"16d8'
Stella of Hashomonde Mina
Gem 7h12'9"25d23'
Stella of Hatsuko
Vir 14h24'43"2d23'
Stella of Hayato
Gem 7h11'41"25d26'
Stella of Hayato Narushima
Aqr 22h4'36"-10d40'
Stella of Heian-Kyou
Aqr 22h2'56"-23d44'
Stella of Hekiru
Psc 1h22'35"5d12'
Stella Of Hero-B
Tau 5h11'0"17d24
Stella of Hide
Gem 7h2'36"15d26'
Stella of Hideaki Tokunaga
Psc 1h24'54"4d11'
Stella of Hideaki & Yukiko
Sco 16h23'33"-29d9'
Stella of Hideaki & Tomoko
Cap 20h59'0"-24d26'
Stella of Hidehiro & Masako
Tau 5h7'3"27d35'
Stella Of Hidekazu
Tau 5h13'44"16d57'
Stella Of Hidekazu & Hiroko
Cap 20h59'3"-21d19'
Stella Of Hidekazu & Terumi
Ari 2h10'53"25d21'
Stella Of Hideki
Vir 14h29'35"0d16'
Stella of Hideki
Sco 16h3'24"-13d45'
Stella of Hideki & Naomi
Vir 14h23'32"0d30'
Stella of Hideki & Harumi
Psc 1h18'11"19d20'
Stella of Hideki & Naomi
Cap 20h54'48"-16d13'
Stella of Hideki & Keiko
Cnc 8h47'51"23d17'
Stella Of Hideki & Kahori
Tau 5h28'54"18d9'
Stella Of Hideki I
Aqr 22h0'51"-17d30'
Stella Of Hideo & Ayako
Psc 1h23'6"9d18'
Stella of Hideo & Michiyo
Ari 2h17'20"21d15'
Stella of Hideshi & Chiaki
Psc 1h20'6"6d10'
Stella of Hideshi & Chiaki
Psc 1h23'32"6d23'
Stella Of Hidetaka
Aqr 22h1'23"-9d41'
Stella of Hidetami & Reiko
Sgr 19h7'21"-15d44'
Stella of Hideto Tsuchiya
Cnc 8h54'8"17d40'
Stella of Hideyuki & Sachi
Sgr 19h8'41"-16d21'
Stella of Hideyuki & Chiharu
Cap 20h54'12"-24d1'
Stella Of Hideyuki & Yoshimi
Lib 14h30'29"-22d42'
Stella Of Hideyuki 25
Leo 9h56'11"24d36'
Stella of Higeaki from Realize
Psc 1h24'42"6d42'
Stella of Hijiri
Cap 20h59'21"-24d48'
Stella of Hikaru
Gem 7h20'23"25d38'
Stella of Hikaru & Mama
Tau 5h12'53"19d36'
Stella Of Hikaru,The
Aqr 22h4'35"-19d49'
Stella Of Hikarugenji & Hiroki
Cnc 8h48'35"20d16'
Stella Of Hime
Ari 2h10'8"19d33'
Stella of Hino Family
Ari 2h16'9"11d59'
Stella of Hirano Family
Lib 14h27'21"-14d5'
Stella of Hiro
Gem 7h3'23"17d48'
Stella of Hiro
Gem 7h11'38"20d36'
Stella Of Hiro
Tau 5h16'44"25d6'
Stella Of Hiro
Cnc 8h48'20"8d43'
Stella Of Hiro & Noriko
Ari 2h13'29"26d46'
Stella Of Hiroaki
Sgr 19h11'44"-14d33'
Stella Of Hiroaki
Sco 16h9'27"-28d11'
Stella of Hiroaki
Gem 7h4'39"18d23'
Stella Of Hiroaki & Youko
Ari 2h13'42"22d54'
Stella of Hirochika
Psc 1h19'42"24d36'
Stella of Hirofumi
Aqr 22h0'8"-7d41'
Stella Of Hirofumi Okamoto
Aqr 22h17'20"-15d3'
Stella Of Hirofumi & Satomi
Sco 16h19'42"-23d1'
Stella of Hirofumi & Kazumi
Gem 7h3'2"19d48'
Stella of Hirohiko 7 Mieko
Sco 16h4'26"-17d48'
Stella of Hiroki & Yuki
Sco 16h0'0"-17d17'
Stella Of Hiroki & Miyoko
Gem 7h13'3"20d29'
Stella Of Hiroki & Mayumi
Sgr 19h10'42"-13d52'
Stella of Hiroki F & Setsuko Y
Lib 14h49'3"0d59'
Stella of Hiroko
Cnc 8h47'26"12d8'
Stella Of Hiroko Takano
Aqr 22h1'18"-18d2'
Stella of Hiroko & Tsuneo With Love
Ari 2h28'8"21d57'
Stella of Hiroko & Toshio
Aqr 22h18'44"-14d44'
Stella of Hiroko & Yuichiro
Leo 10h0'35"8d27'
Stella of Hiromi
Vir 12h57'44"8d33'
Stella of Hiromi & Athuno
Sco 16h28'54"-29d47'
Stella of Hiromi & Masaru
Tau 5h33'11"19d31'
Stella of Hiromi & Mogu
Aqr 22h16'26"-8d5'
Stella Of Hiromi & Rumi
Tau 5h28'27"18d57'
Stella Of Hiromi & Nakano Family
Cnc 8h48'41"21d58'
Stella Of Hiromichi
Ari 2h11'24"27d15'
Stella of Hiromichi & Akiko
Lib 14h42'9"-10d40'
Stella of Hironobu & Yoko
Cap 20h55'9"-15d21'
Stella Of Hiroshi He-kun
Leo 10h1'21"7d24'
Stella Of Hiroshi & Yasue & Ai
Vir 14h55'1"2d2'
Stella Of Hiroshi & Harumi
Vir 14h5'34"6d10'
Stella Of Hiroshi & Takako
Psc 1h23'42"16d51'
Stella Of Hiroshi & Yukiko
Ari 2h28'35"20d24'
Stella of HIROSHI & MIYUKI
Vir 14h27'21"-4d4'
Stella of Hiroshi & Masae
Vir 14h24'30"-1d37'
Stella of Hiroshi & Chiaki
Aqr 22h18'36"-12d27'
Stella of Hiroshi & Hiromi
Ari 2h29'27"20d53'
Stella of Hiroshi & Yukiko
Ari 2h28'35"20d24'
Stella of Hiroshi & Naomi
Tau 5h24'48"26d5'
Stella of Hiroshi & Riri
Cap 20h58'54"-22d45'
Stella of Hiroshi & Chiaki
Cnc 8h49'17"13d22'
Stella of Hiroshi & Masami
Sco 16h20'40"-25d40'
Stella of Hiroshi & Hisae
Tau 5h30'2"25d50'
Stella of Hiroshi & Kyoko
Sco 16h20'9"-26d14'
Stella of Hiroshi & Naoko
Cap 20h52'48"-23d27'
Stella of Hiroto & Miyuki
Gem 7h12'57"14d53'
Stella of Hiroyasu Utsunomiya
Psc 1h19'18"24d48'
Stella of Hiroyasu & Mayumi
Cap 20h53'14"-24d7'
Stella of Hiroyuki
Lib 14h33'53"-19d53'
Stella of Hiroyuki
Sgr 19h11'9"-16d8'
Stella of Hiroyuki
Cap 20h55'32"-17d13'
Stella of Hiroyuki
Sco 16h3'9"-18d51'
Stella of Hiroyuki
Sco 16h0'26"-18d8'
Stella Of Hiroyuki Namba
Lib 14h29'56"-9d33'
Stella Of Hiroyuki & Atsuko
Vir 14h0'32"6d24'
Stella Of Hiroyuki
Aqr 22h2'18"-19d6'
Stella Of Hiroyuki & Toshiko
Cap 20h55'33"-26d11'
Stella Of Hiroyuki & Hiromi
Ari 2h14'56"22d25'
Stella Of Hiroyuki & Keiko
Aqr 22h15'14"-14d46'
Stella of Hiroyuki & Mari
Sgr 19h12'57"-16d23'
Stella of Hiroyuki & Keiko
Aqr 22h15'14"-14d46'
Stella of Hiroyuki & Ikuko
Psc 1h22'0"2d44'
Stella of Hisa
Lib 14h49'20"-20d37'
Stella of Hisa
Lib 14h45'2"-20d39'
Stella of Hisae
Ari 2h16'15"19d27'
Stella of Hisakatsu
Lib 14h44'0"-1d15'
Stella Of Hisaki
Vir 12h1'38"2d3'
Stella of Hisakuni & Haruko
Cnc 8h43'47"21d48'
Stella of HISASHI & SETSUKO
Cnc 8h48'39"18d16'
Stella of Hisatada & Yumi
Sgr 19h11'15"-14d53'
Stella Of Hitomi
Sco 16h26'15"-29d57'
Stella Of Hitonori
Tau 5h22'57"25d41'
Stella Of Hitoshi & Naoko
Ari 2h13'29"19d28'
Stella Of Hitoshi & Hiromi
Tau 5h28'14"18d12'
Stella of Hitoshi S
Sgr 19h12'39"-28d5'
Stella of Hizuki
Sco 16h20'36"-29d38'
Stella Of Honzawa Family
Tau 5h17'11"24d56'
Stella of I Love You Darling
Cnc 8h48'39"8d32'
Stella of Ibuki
Gem 7h12'24"25d23'
Stella of Igarashi Family
Lib 14h53'59"-24d10'
Stella Of Iida Family
Lib 14h32'47"-22d57'
Stella of Ikaika
Sco 16h23'44"-29d0'
Stella of Ikue
Sgr 19h7'21"-13d28'
Stella of Ikuko
Lib 14h34'17"-18d7'
Stella Of Ikuko M
Tau 5h39'54"28d16'
Stella Of Ikuya & Fumiko
Tau 5h14'41"21d36'
Stella of Ikuya & Yumi
Gem 7h3'53"15d40'
Stella of Imami
Vir 11h39'21"6d10'
Stella of Iroha
Aqr 22h3'45"-16d54'
Stella of Isamu & Kumiko
Sgr 19h16'0"-28d12'
Stella of Isao
Tau 5h26'21"21d15'
Stella of Isao
Psc 1h22'17"5d59'
Stella of Isao & Akiko
Sgr 19h5'3"-13d22'
Stella Of Isao & Eiko
Tau 5h29'26"19d4'
Stella Of Isao & Keiko
Psc 1h23'51"13d43'
Stella of Ishikawa Family
Sco 16h8'39"-29d45'
Stella of Issei
Aqr 22h2'42"-7d51'
Stella Of ITO Family
Lib 14h32'27"-24d9'
Stella of Itsuki
Ari 2h19'38"21d54'
Stella of Itsuko
Cap 20h52'53"-17d18'
Stella Of Iwahira Family
Cnc 8h46'6"6d46'
Stella Of Izumi
Ari 2h29'39"18d41'
Stella of Izumi
Sco 16h3'36"-18d30'
Stella of Izumi Family
Sgr 19h11'23"-12d11'
Stella Of IZUMI-chan bear in mind
Psc 1h16'17"28d42'
Stella Of Jigger's Son
Gem 7h10'27"16d15'
Stella Of Jiro
Cnc 8h47'33"19d27'
Stella of Jun
Sco 16h4'14"-29d52'

Stella of Jun
Gem 7h4'48"17d6'
Stella of Jun
Cap 20h54'39"-24d43'
Stella of Jun & Michiko
Gem 7h10'57"25d6'
Stella of Jun & Yoshiko
Lib 14h30'47"-16d11'
Stella of Jun & Izzy
Gem 7h10'11"24d44'
Stella of Jun & Masako
Leo 10h5'30"18d59'
Stella Of Jun & Tomomi
Leo 10h0'39"10d59'
Stella of Jun Asaka
Sgr 19h13'15"-17d2'
Stella of Junichi
Cap 20h59'3"-27d16'
Stella of Junichi
Sgr 19h12'14"-28d31'
Stella of Junichi & Masami
Sco 16h25'59"-27d15'
Stella of Junichi & Kisako
Aqr 22h17'53"-12d32'
Stella of Junichi & Natsuki
Aqr 22h3'20"-15d7'
Stella Of Junichi & Machiko
Sco 16h8'18"-26d36'
Stella Of Junichiro & Hazuki
Vir 14h51'23"2d3'
Stella Of Junko
Leo 9h56'35"23d30'
Stella Of Junko
Cnc 8h43'14"25d32'
Stella Of Junko
Cnc 8h46'50"20d42'
Stella of Junko
Tau 5h9'54"18d6'
Stella of Junko
Psc 1h11'23"19d57'
Stella of Junko
Cnc 8h49'12"11d19'
Stella of Junko
Leo 10h2'45"12d45'
Stella of Junko
Lib 14h47'42"-20d33'
Stella of Junko
Lib 14h33'54"-18d37'
Stella Of Junko & Masaichi
Vir 14h6'24"2d14'
Stella Of Junko-27
Lib 14h46'41"-19d18'
Stella Of Junsei
Gem 7h12'57"22d6'
Stella of Juri
Sgr 19h10'44"-25d55'
Stella of K & K
Cap 20h57'57"-27d35'
Stella of K & T Love
Aqr 22h16'41"-8d47'
Stella of K T Pupu
Vir 14h26'57"-2d4'
Stella Of K&Y
Sgr 19h13'45"-12d55'
Stella of Kacchan
Tau 5h24'8"26d32'
Stella of Kaede
Gem 7h10'18"14d37'
Stella of Kaede
Leo 10h19'35"17d43'
Stella of Kagari
Sco 16h15'38"-22d46'
Stella of Kaito Narushima
Cap 20h56'38"-19d12'
Stella Of Kaito Narushima
Cap 20h56'38"-19d12'
Stella of Kamiyama Family
Sgr 19h19'14"-29d3'
Stella Of Kana
Sgr 19h17'48"-28d36'
Stella Of Kana
Aqr 22h3'3"-19d10'
Stella Of Kanako
Sgr 19h18'21"-28d18'

Stella of Kanako
Sco 16h25'48"-28d56'
Stella of Kaname Kawasaki
Aqr 22h4'47"-24d45'
Stella Of Kaname
Gem 7h14'29"21d6'
Stella Of Kaneaki & Hiroko
Cap 20h56'54"-21d45'
Stella of Kanedo Miki
Cnc 8h45'45"23d36'
Stella of Kanehito & Nobuko
Sgr 19h16'15"-29d7'
Stella Of Kanichi & Kimiya
Ari 2h28'50"19d3'
Stella of Kanko
Cnc 8h47'38"18d22'
Stella of Kanon
Sco 16h4'30"-29d37'
Stella of Kaori
Psc 1h17'3"18d18'
Stella of Kaori
Psc 1h15'35"26d38'
Stella Of Kaori
Gem 7h11'35"22d50'
Stella of Kaori & Shingo
Sgr 19h18'47"-27d12'
Stella of Kaori & Hideyasu
Sgr 19h14'0"-28d54'
Stella Of Kaoru
Cnc 8h40'45"29d26'
Stella Of Kaoru
Sco 16h9'14"-27d3'
Stella of Kasumi
Ari 2h22'6"13d26'
Stella of Katashi & Tsuya
Sco 16h25'26"-29d27'
Stella Of Katsu & Haru
Psc 1h21'6"7d56'
Stella of Katsufumi
Gem 7h0'27"15d17'
Stella of Katsuhiko & Miyoko
Sco 16h28'38"-26d28'
Stella of Katsuhiko & Fumiko
Ari 2h25'5"21d51'
Stella of Katsuhiro & Yumiko
Ari 2h24'59"20d15'
Stella of Katsuhiro & Junko
Cap 20h52'48"-18d46'
Stella Of Katsuhiro & Yumiko
Ari 2h24'59"20d15'
Stella of Katsuhisa & Kazuko
Sgr 19h6'8"-16d17'
Stella of Katsuhito
Psc 1h24'53"5d13'
Stella of Katsumi
Vir 14h4'13"5d29'
Stella of Katsumi & Mariko
Psc 1h18'20"24d53'
Stella Of Katsumi & Naomi
Ari 2h29'17"22d33'
Stella Of Katsumi & Midori
Ari 2h7'8"25d57'
Stella Of Katsumi & Takae
Psc 1h20'20"8d55'
Stella of Katsuo & Michiko
Vir 14h26'33"0d17'
Stella of Katsushisa & Maki
Gem 7h3'15"20d9'
Stella of Katsutoshi &
Mutsuko
Lib 14h52'12"-19d48'
Stella Of Katsutoshi & Kaori
Lib 14h44'36"-9d44'
Stella Of Katsuya & Fusayo
Ari 2h10'11"22d4'
Stella of Katsuyuki
Cap 20h59'38"-15d37'
Stella Of Katsuyuki & Sanami
Ari 2h29'12"18d8'
Stella of Katuhiko & Kaori
Toba
Cap 20h57'3"-27d29'
Stella of Kayo
Sgr 19h8'33"-15d18'

Stella Of Kayo
Vir 12h24'52"2d9'
Stella of Kaz & Mee
Aqr 21h56'44"0d3'
Stella Of Kazu & Aki
Sgr 19h15'38"-28d44'
Stella of Kazu & Chi
Sco 16h27'24"-29d22'
Stella of Kazu & Tomo
Cnc 8h48'26"16d11'
Stella Of Kazu & Tomo
Sgr 19h14'59"-13d9'
Stella Of Kazu & Yuki
Cap 20h59'18"-20d32'
Stella of Kazu & Yumi
Vir 14h21'1"5d40'
Stella of Kazuaki & Haruna
Cap 20h59'11"-21d38'
Stella Of Kazuaki & Yumi
Ari 2h12'0"22d21'
Stella Of KAZUE
Vir 12h2'1"2d20'
Stella of Kazuhide & Junko
Lib 14h48'18"-23d39'
Stella of Kazuhiko
Leo 10h0'59"18d26'
Stella of Kazuhiko & Minako
Aqr 22h17'53"-9d20'
Stella of Kazuhiko & Sanae
Tau 5h14'44"20d21'
Stella Of Kazuhiko & Noriko
Ari 2h25'5"17d50'
Stella Of Kazuhiko & Sanae
Tau 5h14'44"20d21'
Stella Of Kazuhiko & Kiyoe
Vir 14h39'33"6d29'
Stella Of Kazuhiro & Seiko
Tau 5h26'8"19d57'
Stella Of Kazuhiro & Mariko
Tau 5h30'36"18d43'
Stella of Kazuhiro & Fumiko
Sco 16h7'6"-25d18'
Stella of Kazuhiro & Sumie
Gem 7h4'30"20d33'
Stella of Kazuhiro & Miwa
Sgr 19h13'38"-28d46'
Stella of Kazuhisa
Vir 14h5'2"-18d58'
Stella Of Kazuhisa & Rika
Tau 5h17'33"23d14'
Stella of Kazuki
Aqr 22h1'14"-8d9'
Stella Of Kazuki Papa
Sco 16h21'15"-24d54'
Stella Of Kazumasa & Yasuko
Gem 7h11'53"16d41'
Stella Of Kazumi
Lib 14h38'36"-13d45'
Stella of Kazumi
Cnc 8h45'26"9d49'
Stella of Kazumi
Vir 14h36'32"5d53'
Stella of Kazunari & Kazumi
Cnc 8h49'15"20d0'
Stella Of Kazunori
Cap 20h57'42"-25d43'
Stella of Kazunori & Tuneko
Cap 20h55'54"-16d36'
Stella of Kazuo & Youko
Leo 10h10'18"18d23'
Stella of Kazuo & Sachiko
Leo 9h55'18"21d11'
Stella Of Kazuo & Chikako
Cnc 8h48'15"18d12'
Stella Of Kazuo & Sachiko
Leo 9h55'18"21d11'
Stella Of Kazuo & Miho
Ari 2h13'11"22d15'
Stella of Kazuo & Yuki
Cnc 8h44'5"21d49'
Stella Of Kazushi & Mayumi
Sco 16h7'8"-26d9'
Stella of Kazutoshi
Sgr 19h11'9"-16d47'

Stella of Kazutoshi
Ari 2h17'26"14d4'
Stella Of Kazutoshi & Tamami
Sco 16h8'32"-27d24'
Stella of Kazutoyo & Miwa
Sco 16h24'5"-29d15'
Stella Of Kazuya
Leo 10h2'9"13d5'
Stella of Kazuyo
Leo 10h19'36"25d19'
Stella Of Kazuyoshi & Midori
Lib 14h33'3"-21d59'
Stella of Kazuyoshi & Harumi
Sgr 19h16'3"-27d54'
Stella of Kazuyuki
Psc 1h19'32"18d24'
Stella of Kazuyuki
Sco 16h29'27"-29d9'
Stella of Kazuyuki & Sachiko
Lib 14h48'54"-24d5'
Stella of Kei
Aqr 22h4'3"-7d12'
Stella of Keiichi
Ari 2h19'12"12d3'
Stella of Keiichi & Amor
Gloria
Cap 20h58'24"-24d29'
Stella of Keiji & Akiko
Psc 1h19'47"21d48'
Stella of Keiko
Cnc 8h52'39"16d27'
Stella Of Keiko
Gem 7h12'15"23d20'
Stella Of Keiko
Tau 5h30'14"15d53'
Stella Of Keiko
Vir 14h1'37"-11d8'
Stella of Keiko & Tachuo
Psc 1h17'35"21d5'
Stella of Ken & Aki
Cnc 8h45'24"9d9'
Stella Of Ken & Tomo
Leo 9h59'45"22d29'
Stella of Ken & Yumi
Sgr 19h13'24"-16d25'
Stella Of Kenichi
Ari 2h12'33"18d36'
Stella Of Kenichi & Miwako
Ari 2h9'23"25d10'
Stella of Kenichiroh & Naoko
Cnc 8h45'56"12d43'
Stella of Kenji
Gem 7h2'30"16d26'
Stella of Kenji & Yuriko
Sgr 19h6'17"-12d56'
Stella of Kenji & Tamaki
Cnc 8h45'18"18d46'
Stella of Kenji & Naomi
Cap 20h59'41"-15d40'
Stella of Kenji & Fumiko
Aqr 22h17'?"-14d41'
Stella of Kenji's spit-fire
Ari 2h22'45"11d46'
Stella Of Kensho
Sco 16h17'6"-23d41'
Stella Of Kento
Leo 10h17'30"8d51'
Stella of Kenzi & Toshiko
Lib 14H43'3"-12d11'
Stella of Kiku
Gem 7h22'59"22d12'
Stella of Kimi & Megu
Lib 14h47'35"-23d23'
Stella Of Kimiko & Tameichi
Ari 2h28'42"18d33'
Stella of Kimio & Emi
Cap 20h51'20"-22d15'
Stella of Kimitaka & Michiko
Cap 20h51'35"-26d9'
Stella of Kimiyo
Sgr 19h13'56"-16d40'
Stella Of Kinichi & Mieko
Psc 1h23'11"16d41'
Stella of Kinji & Yumi
Leo 10h0'9"14d21'

Stella Of Kinu & Masako
Leo 10h17'14"8d13'
Stella Of Kira
Gem 7h27'59"29d18'
Stella of Kiyo
Psc 1h24'23"3d55'
Stella of Kiyofumi & Kyoko
Lib 14h48'56"-2d18'
Stella of Kiyoko
Lib 14h50'15"-23d15'
Stella of Kiyomi Seike Kiyomi
Tau 5h13'56"19d14'
Stella Of Kiyono
Gem 7h10'48"19d18'
Stella Of Kiyoshi
Gem 7h3'45"20d39'
Stella Of Kiyoshi
Tau 5h28'45"18d57'
Stella of Kiyoshi Maeda
Ari 2h11'8"20d34'
Stella of Kiyoshi
Cnc 8h53'23"15d54'
Stella of Kiyoshi & Miyuki
Ari 2h19'44"16d38'
Stella of Kiyoshi & Reiko
Vir 14h27'48"-1d49'
Stella of Kiyoshi & Hiroko
Leo 9h56'26"20d38'
Stella of Kiyoshi & Ryoko
Psc 1h22'0"6d45'
Stella Of Kiyoshi & Hiroko
Leo 9h56'26"20d38'
Stella Of Kiyoshi & Ryoko
Psc 1h22'0"6d45'
Stella Of Kiyoshi & Hisako
Tau 5h17'14"24d26'
Stella Of Kiyoshi,
Naoko,Yusuke & Shoko
Tau 5h32'14"15d33'
Stella Of Kiyoto & Yuko
Ari 2h25'29"17d8'
Stella of Knetaro & Yuko
Gem 7h10'51"15d28'
Stella of Koh & Natsu
Sco 16h24'0"-27d4'
Stella of Koharu
Psc 1h23'53"17d57'
Stella of Kohei & Taka
Leo 10h5'2"16d29'
Stella of Koichi & Yoko'
Tau 5h12'45"18d7'
Stella Of Koichi & Noriko
Cnc 8h47'12"22d57'
Stella Of Koichiro
Psc 1h15'50"28d24'
Stella of Koizumi Kazuhiko &
Samtomi
Ari 2h19'3"15d54'
Stella of Koji
Sco 16h25'12"-26d22'
Stella of Koji
Cnc 8h46'33"8d11'
Stella of KOJI & NORIKO
Cnc 8h44'30"21d43'
Stella of Koji & Kumiko
Cnc 8h45'29"19d51'
Stella Of Koji & Tomoko
Vir 14h29'1"1d27'
Stella Of Koji & Yasue
Sgr 19h12'11"-13d14'
Stella of Koji & Kyoko
Cap 20h50'57"-24d26'
Stella of Koji & Yoko
Tau 5h13'0"20d3'
Stella of Koji Hatano
Gem 7h4'27"22d47'
Stella Of Komatsu Family
Sgr 19h11'53"-14d30'
Stella of Koni
Aqr 22h4'53"-14d44'
Stella of Koro
Cnc 8h47'14"15d0'
Stella Of KOSUKE
Ari 2h18'53"25d30'

Stella of Kota & Kenji
Sgr 19h8'54"-13d51'
Stella Of Kotaro
Sgr 19h11'14"-14d7'
Stella Of Kotaro
Lib 14h42'9"-9d31'
Stella of Kotaro & Atsuko
Psc 1h18'57"25d20'
Stella of Koto
Gem 7h13'6"14d43'
Stella of Kotoyo
Aqr 22h2'51"-7d57'
Stella Of Kouhei
Gem 7h14'47"15d51'
Stella Of Kouichi Yamadera
Gem 7h10'14"18d59'
Stella Of Kouichi & Toyoko
Cnc 8h44'23"29d19'
Stella Of Kouichi & Toshie
Ari 2h28'39"17d18'
Stella of Kouichi & Kazuko
Cap 20h59'6"-21d32'
Stella of Kouji & Chika
Psc 1h15'26"23d26'
Stella Of Kouji & Junko
Ari 2h17'35"10d25'
Stella Of Koushin & Mayumi
Sco 16h15'26"-23d7'
Stella of Kousuke Terouchi
Psc 1h21'14"11d7'
Stella Of Kouzou & Takako
Cap 20h55'15"-21d32'
Stella Of Kumada Family
Psc 1h23'24"12d57'
Stella Of Kumagaya Family
Sco 16h7'6"-25d14'
Stella of Kumi
Cnc 8h48'33"18d4'
Stella of Kumiko
Vir 14h4'40"-8d45'
Stella Of Kumiko
Vir 14h36'20"0d22'
Stella Of Kuni & Naomi
Lib 14h30'18"-22d8'
Stella Of Kunihiko & Mayumi
Vir 14h5'52"5d31'
Stella of Kuniko
Leo 10h3'56"18d51'
Stella of Kuniko
Aqr 22h5'14"-2d36'
Stella of Kunio
Lib 14h48'11"-19d15'
Stella of Kunji e
Aqr 22h2'15"-13d15'
Stella of Kurumi
Leo 10h4'0"11d24'
Stella of Kyohei & Megumi
Cap 20h54'42"-18d53'
Stella of Kyohhei
Lib 14h47'23"-21d19'
Stella of Kyoko
Lib 14h34'24"-10d12'
Stella Of Kyoko-T
Gem 7h13'59"15d32'
Stella of Kyon & I
Lib 14h44'59"-12d46'
Stella Of Lemon
Gem 7h14'53"26d24'
Stella of Leny
Cap 20h55'51"-24d15'
Stella Of Lica
Aqr 22h25'9"-6d12'
Stella Of Lien
Vir 14h35'0"0d41'
Stella of Lilyka
Sgr 19h5'15"-13d8'
Stella Of Lina
Cnc 8h46'21"22d5'
Stella Of Love & Peace Akiko
Aqr 22h4'11"-9d8'
Stella of Love & Peace Zone
Ari 2h18'14"18d20'
Stella of Love & Truth
Vir 12h59'17"2d1'

Stella of Love Miwa
Lib 14h47'33"-23d2'
Stella Of Love Miwa
Lib 14h47'33"-23d2'
Stella Of Luckey
Psc 1h23'57"7d10'
Stella Of Léonie & Angela
Cnc 8h46'6"22d23'
Stella of M & Y
Aqr 22h18'12"-9d2'
Stella of M Fujio
Cap 20h54'24"-18d2'
Stella Of M T M C
Psc 1h16'47"27d41'
Stella Of Madoka
Cap 20h57'3"-20d10'
Stella of Madoka
Vir 14h25'54"6d59'
Stella of Mai Shimanaka
Ari 2h18'30"23d10'
Stella of Mai
Tau 5h34'8"20d17'
Stella Of Mai
Psc 1h21'6"17d32'
Stella of Maki
Cap 20h51'39"-19d1'
Stella of Maki
Psc 1h11'48"27d7'
Stella Of Maki and the people
loved by Maki
Cnc 8h45'12"21d38'
Stella of Makosana
Sco 16h27'59"-29d50'
Stella of Makoto Hinokida
Tau 5h33'11"19d27'
Stella of Makoto
Psc 1h20'36"10d23'
Stella of Makoto & Tamayo
Sgr 19h14'57"-17d1'
Stella of Makoto & Hisayo
Kojima
Tau 5h29'26"26d54'
Stella of Mami
Cnc 8h49'26"23d55'
Stella of Mami
Sco 16h8'51"-25d17'
Stella of Mami Kitajima
Takarazuka
Ari 2h13'14"20d27'
Stella Of MAMI & Everyone
Tau 5h30'35"15d56'
Stella Of Mamiko
Tau 5h34'33"15d50'
Stella of Mamoru & Mizuho
Aqr 22h18'23"-12d44'
Stella of Mamoru & Fumie
Lib 14h53'32"-24d29'
Stella of Mamoru & Chikage
Lib 16h19'32"-24d15'
Stella of Mamoru & Kotoe
Sgr 19h14'21"-28d36'
Stella of Manabu Ariga
Aqr 22h0'59"-8d3'
Stella Of Manami
Psc 1h22'21"11d6'
Stella of Manavu
Ari 2h18'38"13d42'
Stella Of Marble Cats
Lib 14h41'24"-20d27'
Stella of Mari & Nobuhiro
Cnc 8h51'30"17d23'
Stella of Maria
Sgr 19h6'41"-13d33'
Stella Of Marie
Lib 14h38'47"-14d16'
Stella of Marika
Gem 7h1'45"15d19'
Stella of Mariko
Cnc 8h49'9"17d23'
Stella Of Mariko
Gem 7h12'26"17d0'
Stella Of Marin Kamada
Cnc 8h46'44"21d24'
Stella of Marina
Cap 20h50'20"-22d8'

Stella of Mariya
Lib 14h48'26"0d23'
Stella of Marquis
Sgr 19h10'47"-15d31'
Stella of Masaaki
Ari 2h24'0"10d20'
Stella of Masaaki
Gem 7h13'27"29d19'
Stella of Masaaki & Hiroko
Sgr 19h14'51"-12d18'
Stella of Masae
Gem 7h12'51"19d55'
Stella of Masafumi
Leo 10h7'50"21d33'
Stella Of Masafumi & Harumi
Sgr 19h13'0"-13d8'
Stella Of Masaharu & Misako
Tau 5h24'35"15d58'
Stella of Masaharu & Keiko
Tau 5h12'54"18d34'
Stella of Masaharu & Yuko
Cap 20h57'35"-15d55'
Stella of Masaharu & Akiko
Cnc 8h47'38"12d3'
Stella of Masaharu & Yuko
Cyg 21h10'1"36d51'
Stella Of Masahide
Aqr 22h0'30"-9d21'
Stella Of Masahiko & Miyuki
Lib 14h43'50"-21d59'
Stella of Masahiko & Tomoyo
Ari 2h16'17"21d41'
Stella of Masahiro
Cap 20h57'41"-27d13'
Stella Of Masahiro Shimazaki
Gem 7h10'51"19d1'
Stella Of Masahiro
Cnc 8h49'39"7d6'
Stella Of Masahiro & Fumiko
Aqr 22h15'41"-15d16'
Stella Of Masahiro & Megumi
& Naomi
Aqr 22h1'51"-8d59'
Stella of Masahiro & Naomi
Ari 2h15'2"14d2'
Stella of Masahiro & Miyuki
Sco 16h19'14"-22d30'
Stella of Masahiro & Mika
Tau 5h34'39"15d39'
Stella of Masahiro & Kaori
Psc 1h15'41"26d28'
Stella of Masahiro & Mitue
Sgr 19h12'41"-29d39'
Stella of Masahito
Psc 1h21'14"5d3'
Stella of Masahito & Takako
Sgr 19h9'30"-16d22'
Stella of Masakazu & Kumiko
Lib 14h32'56"-19d33'
Stella of Masaki
Leo 10h2'2"11d15'
Stella of Masaki & Chieko
Aqr 22h5'27"-8d13'
Stella of Masaki & Maki
Sgr 19h11'30"-29d21'
Stella Of Masaki & Setsuko
Tau 5h15'35"21d42'
Stella Of Masaki Seino
Aqr 22h3'6"-10d32'
Stella of Masako
Cap 20h51'48"-18d59'
Stella of Masako
Leo 10h15'44"13d0'
Stella of Masako & Megumi
Lib 14h48'27"-18d43'
Stella of Masami
Cap 20h51'44"-22d15'
Stella Of Masami & Miyako
Vir 14h5'11"6d6'
Stella Of Masami & Yuri
Aqr 22h1'54"-17d48'
Stella Of Masami & Harumi
Lib 14h42'11"-14d11'
Stella of Masami & Kouzi
Psc 1h17'41"17d52'

Stella Of Masami's Song
Sco 16h9'9"-27d48'
Stella of Masamichi
Aqr 22h0'32"-24d6'
Stella Of Masamitsu Onishi
Cnc 8h42'53"24d58'
Stella Of Masamitsu & Satomi
Ari 2h18'30"11d30'
Stella Of Masani
Lib 14h32'51"-23d0'
Stella of Masanobu & Kimie
Lib 14h47'26"-24d14'
Stella of Masanobu & Mayumi
Aqr 22h2'51"-24d51'
Stella of Masanobu & Tomoyo
Sgr 19h14'24"-28d44'
Stella of Masanori & Junko
Cnc 8h48'0"18d11'
Stella Of Masanori & Hiroko
Ari 2h25'14"19d46'
Stella of Masao
Cnc 8h47'27"8d27'
Stella Of Masao & Chizuyo
Lib 14h34'32"-22d47'
Stella Of Masaru Shishido
Leo 10h17'38"11d6'
Stella of Masaru
Lib 14h47'23"-19d41'
Stella of Masaru & Akemi
Cnc 8h46'2"12d47'
Stella Of Masashi
Lib 14h41'12"-21d45'
Stella of Masashi & Hitomi
Sgr 19h8'17"-15d59'
Stella of Masashi & Akemi
Psc 1h20'39"9d33'
Stella of Masashiro &
Mutsuko
Cap 20h53'57"-24d22'
Stella of Masataka & Masami
Vir 14h5'33"-11d29'
Stella Of Masataka & Naruyo
Psc 1h23'12"15d32'
Stella of Masataka N
Cnc 8h54'33"16d29'
Stella of Masato
Cap 20h58'35"-19d6'
Stella Of Masato
Cap 20h58'35"-19d6'
Stella Of Masato & Yukie
Sco 16h9'38"-26d14'
Stella of Masato & Kaori
Aqr 22h17'38"-16d50'
Stella of Masato & Hal F A 69
Cnc 8h48'20"9d50'
Stella of Masato & Chika
Vir 14h6'0"-11d35'
Stella of Masato S
Cap 20h50'2"-17d12'
Stella Of Masatoshi
Ari 22h0'21"-8d45'
Stella Of MASAYO
Tau 5h33'9"16d4'
Stella Of Masayuki
Ari 2h13'48"22d14'
Stella of Masayuki
Cnc 8h45'32"11d33'
Stella of Masayuki
Sgr 19h11'41"27d38'
Stella of Masayuki Watanabe
Vir 14h4'47"2d28'
Stella of Masayuki & Midori
Leo 10h3'0"11d8'
Stella of Masayuki & Naomi
Lib 14h32'27"-19d26'
Stella Of Masayuki & Naoko
Vir 14h9'53"0d28'
Stella Of Masayuki & Yoshiko
Sgr 19h12'53"-29d45'
Stella Of Masayuki & Hiromi
Gem 7h13'18"21d11'
Stella Of Masayuki & Sachiko
Leo 10h2'30"8d53'
Stella Of Masazumi & Kimiko
Psc 1h21'9"9d6'

Stella of Masumi
Vir 12h51'46"2d12'
Stella of Masuya Shiga
Ari 2h18'2"18d38'
Stella of Matsushima
Gem 7h2'18"20d5'
Stella Of Maya
Cnc 8h48'8"8d4'
Stella Of Mayo
Psc 1h24'35"7d57'
Stella Of Mayu
Gem 7h11'59"23d12'
Stella of Mayumi & Norio
Cap 20h55'8"-23d44'
Stella of Megumi
Gem 7h12'11"24d45'
Stella of Megumi
Cap 20h50'47"-18d34'
Stella Of Megumi
Lib 14h37'50"-13d51'
Stella Of Memories
Gem 7h13'21"14d30'
Stella of Meoto
Aqr 22h2'29"-6d33'
Stella of Mermaid
Psc 1h23'48"5d22'
Stella Of Mibuki
Cnc 8h48'15"7d1'
Stella of Michiaki
Vir 14h3'11"0d45'
Stella of Michifumi & Sumiko
Tau 5h28'6"23d7'
Stella of Michiko
Psc 1h19'47"29d33'
Stella of Michiko
Gem 7h13'56"29d38'
Stella of Michio & Naomi
Cap 20h57'17"-23d25'
Stella Of Michiro
Ari 2h14'30"19d48'
Stella Of Michu-san
Ari 2h27'59"19d54'
Stella of Michu-san
Ari 2h27'59"19d54'
Stella Of Mick
Vir 14h4'23"-11d30'
Stella of Midori
Sgr 19h10'5"-29d56'
Stella of Mieko
Vir 14h6'31"-10d6'
Stella of Mieko
Psc 1h11'32"28d29'
Stella Of Mieko
Vir 14h9'22"1d12'
Stella of Mieko & Naoki
Sgr 19h9'0"-14d6'
Stella Of Mieko Agnes Ozaki
Psc 1h24'24"11d7'
Stella Of Miho
Gem 7h12'39"15d20'
Stella of Miho
Gem 7h12'39"15d20'
Stella of Mika
Cnc 8h49'39"24d8'
Stella of Mika
Sgr 19h13'15"-25d24'
Stella Of Mika
Lib 14h33'45"-24d15'
Stella of Miki
Cap 20h55'9"-16d38'
Stella Of Miki & Taka
Psc 1h21'12"11d43'
Stella Of Mikiko
Cap 20h54'56"-20d16'
Stella Of Mikiko
Vir 14h20'54"5d24'
Stella Of Mikio & Yuko
Psc 1h23'47"16d43'
Stella of Mikio & Yuko
Sgr 19h8'30"-15d24'
Stella of Mikiwa
Tau 5h28'33"22d6'
Stella of Miku
Lib 14h49'29"-21d31'
Stella Of Miku
Ari 2h19'33"25d54'
Stella of Million Loves
Cap 20h57'11"-22d32'
Stella of Mimi & Miwa
Sco 16h15'14"-22d59'
Stella of Mimo
Psc 1h21'3"3d3'
Stella of Minako
Cap 20h51'5"19d57'
Stella of Minami
Lib 14h26'42"-10d34'
Stella of Minami
Sco 16h5'54"-29d38'
Stella Of Minami
Sco 16h5'54"-29d38'
Stella of Mine & Kumi
Lib 14h26'2"-12d36'
Stella of Mineo
Sco 16h2'32"-18d31'
Stella Of Minoru & Shinobu
Cnc 8h42'41"29d27'
Stella Of Minoru & Tomoko
Cap 20h55'3"-25d0'
Stella Of Minoru & Toshie
Aqr 22h1'56"-19d19'
Stella of mio caro
Gem 7h1'47"17d5'
Stella of Miri
Leo 10h15'57"7d31'
Stella of Miri
Leo 10h18'54"12d1'
Stella Of Miru & Mana
Vir 14h32'23"6d2'
Stella Of Misa
Sco 16h4'54"-27d35'
Stella of Misao & Hero
Tau 5h32'6"20d11'
Stella Of Mitsu
Psc 1h24'15"3d16'
Stella of Mitsuaki & Hiroko
Vir 14h36'0"1d51'
Stella of Mitsugu
Sco 16h8'2"-25d16'
Stella Of Mitsuharu & Kumi
Ari 2h17'56"10d23'
Stella Of Mitsuharu & Chiemi
Vir 14h53'17"2d25'
Stella Of Mitsuhiro Maki
Tau 5h25'54"16d21'
Stella Of Mitsuhiro & Hideko
Vir 14h30'40"0d39'
Stella of Mitsuhiro & Yumi
Psc 1h21'38"6d12'
Stella of Mitsuho
Sco 16h1'59"-16d45'
Stella Of Mitsunori & Manami
Tau 5h18'42"27d21'
Stella Of Mitsuo & Sachie
Leo 9h56'35"25d42'
Stella Of Mitsuo & Sachiko
Sgr 19h14'30"-29d49'
Stella of Mitsuo & Kazumi
Ari 2h30'3"23d33'
Stella of Mitsuo & Yoshiko
Ari 2h7'5"26d14'
Stella of Mitsuo & Chiharu
Cnc 8h40'27"9d36'
Stella of Mitsuru
Cap 20h51'20"-25d8'
Stella Of Mitsuru
Psc 1h22'15"12d18'
Stella of Mitsuru & Saori
Sco 16h20'56"-26d0'
Stella of Mituharu, Emi & Narumi
Gem 7h1'51"15d52'
Stella Of Mituko
Tau 5h11'53"20d48'
Stella of MIW
Vir 14h31'1"6d9'
Stella Of Miwa
Vir 14h52'58"0d44'
Stella of Miyabi
Gem 7h10'51"28d38'
Stella Of Miyako
Aqr 22h2'6"-17d55'
Stella of Miyashita Tomoko
Ari 2h28'5"21d15'
Stella of MIYO & HIROSHI
Leo 10h6'33"21d59'
Stella of Miyoko
Ari 2h25'17"20d54'
Stella of Miyuki
Gem 7h1'32"18d53'
Stella of Miyuki
Cnc 8h49'44"16d40'
Stella Of Miyuki & Shoichiro
Lib 14h41'59"-12d59'
Stella Of Miyuki & Masayoshi
Cap 20h59'36"-25d26'
Stella of Mizue
Psc 1h24'5"8d20'
Stella Of Mizuho & Takeshi
Psc 1h22'50"13d8'
Stella Of Mizuki
Vir 14h6'44"1d55'
Stella Of Mizuki
Gem 6h33'15"14d1'
Stella of Mizuki & Marina
Aqr 22h4'20"-22d20'
Stella Of Mizuna
Ari 2h19'21"12d57'
Stella Of Momo
Leo 9h56'53"26d23'
Stella of Momotaro
Cnc 8h48'12"16d5'
Stella of Morimasa & Yumie
Tau 5h29'26"22d51'
Stella Of Morimichi Fujii
Cap 20h56'0"-19d21'
Stella Of Morio & Junko
Lib 14h42'44"-9d2'
Stella of Moritaka
Aqr 22h2'48"-23d15'
Stella Of Motoi & My Family
Cnc 8h49'53"22d36'
Stella Of Mrs Fujita
Psc 1h20'47"6d29'
Stella of Mrs Fujita
Psc 1h20'47"6d29'
Stella of Mutsumi
Psc 1h16'3"29d33'
Stella of My Baby 9 July 1994
Gem 7h1'26"18d7'
Stella of My Child
Lib 14h31'18"-19d19'
Stella of My Dear Hiroyuki
Sco 16h21'21"-29d34'
Stella Of My Dear Mother,Yoshimi
Cnc 8h44'0"25d39'
Stella of My Lover Mitch
Sgr 19h11'44"-28d20'
Stella of Nagase M & K
Psc 1h19'11"26d19'
Stella of Nagi
Cnc 8h48'48"12d20'
Stella Of Nakai Family
Lib 14h33'5"-24d38'
Stella of Namikawa
Vir 14h4'51"0d36'
Stella of NaNa
Cnc 8h49'24"19d4'
Stella Of Nana
Tau 5h25'30"16d32'
Stella Of Nanami
Cnc 8h49'20"22d12'
Stella of Nanami
Cap 20h57'6"-22d55'
Stella of Nao
Psc 1h12'5"19d58'
Stella of Nao
Sco 16h4'26"-17d44'
Stella Of Naoaki
Vir 14h2'25"-11d43'
Stella Of Naohide & Emiko
Vir 14h39'59"6d39'
Stella Of Naoki & Michiyo
Vir 14h50'43"2d25'
Stella of Naoki & Kimiko
Sgr 19h11'9"-16d47'
Stella of Naoki & Yoko
Leo 10h1'29"14d13'
Stella of Naoko
Cap 20h51'17"-25d6'
Stella of Naoko
Sgr 19h11'21"-27d26'
Stella Of Naoko
Leo 10h17'12"8d31'
Stella of Naoko & Yoshio
Aqr 22h0'35"-15d11'
Stella of Naomi
Tau 5h13'41"18d23'
Stella of Naomi
Sco 16h22'24"-26d2'
Stella of Naomi
Leo 10h17'11"7d52'
Stella of Naomi
Cnc 8h47'38"20d33'
Stella Of Naoshi & Yuki
Psc 1h22'38"15d59'
Stella of Naoto & Keiko
Lib 14h53'3"-24d3'
Stella Of Naoya & Keiko
Tau 5h25'14"16d23'
Stella Of Naoyuki & Tomomi
Lib 14h41'38"-11d25'
Stella Of Naruhisa
Vir 14h6'42"2d21'
Stella of Narukazu
Aqr 22h4'39"-24d10'
Stella of Naruyoshi & Naoko
Gem 7h10'45"28d34'
Stella of Natsuko Yagi
Cnc 8h47'42"17d13'
Stella of Natsumi
Sco 16h27'2"-24d59'
Stella Of Natsuyo & Ryuichiro
Gem 7h10'3"7d51'
Stella Of Natu & Lover
Tau 5h34'44"16d30'
Stella of Nekoyashiki No Jyunin
Tau 5h13'20"19d12'
Stella of Neo Sanagedai
Sco 16h23'59"-26d8'
Stella Of Night Hawk
Cnc 8h41'33"24d58'
Stella of Nishimura
Leo 10h3'20"20d47'
Stella Of Nishizawa Family
Leo 10h16'44"10d11'
Stella of Nobayashi Keito
Leo 10h3'14"18d19'
Stella of Noboru & Yuka
Lib 14h52'29"-24d33'
Stella Of Nobu & Toshi
Leo 10h19'9"10d56'
Stella Of Nobuaki & Yayoi
Lib 14h42'3"-13d39'
Stella of Nobue
Cap 20h57'33"-15d20'
Stella of Nobuhiro & Yoko
Sgr 19h11'12"-15d59'
Stella Of Nobuhiro & Haruo
Tau 5h28'12"18d30'
Stella Of Nobuhiro & Yasuko
Lib 14h40'47"-8d57'
Stella of Nobukatsu & Shukuko
Sgr 19h18'50"-29d37'
Stella Of Nobukazu
Lib 14h44'30"-11d21'
Stella of Nobuko
Psc 1h10'50"29d27'
Stella of Nobuo
Leo 10h0'29"16d31'
Stella of Nobuo & Masayo
Tau 5h22'38"27d53'
Stella of Nobuyoshi
Leo 10h4'14"18d29'
Stella of Nobuyuki & Hitomi
Leo 10h17'54"6d43'
Stella of Nobuyuki
Cap 20h58'41"-22d26'
Stella Of Nobuyuki
Aqr 22h2'48"-14d2'
Stella of Noda Family
Sgr 19h10'6"-17d5'
Stella of Non & Len
Sgr 19h14'26"-27d20'
Stella Of Nonoka
Tau 5h28'45"19d35'
Stella Of Noriaki & Naomi
Ari 2h13'27"22d58'
Stella of Noriaki & Hiromi
Aqr 22h1'59"-22d28'
Stella of Noriaki S
Tau 5h29'11"20d59'
Stella of Noriaki-T
Leo 10h19'2"6d53'
Stella of Norihito
Vir 14h4'52"6d18'
Stella of Noriko
Tau 5h26'8"20d6'
Stella of Noriko
Cap 20h55'29"-27d29'
Stella of Noriko Kawakami
Vir 14h22'19"6d54'
Stella of Noriko
Lib 14h29'32"-10d48'
Stella Of Noriko & Takemitsu
Sco 16h26'51"-18d35'
Stella of Noriko U Gaea
Sgr 19h14'56"-16d3'
Stella of Noriko uk
Lib 16h5'23"-25d31'
Stella of Norio Kitajima
Sco 16h28'23"-29d46'
Stella Of Norio & Yumiko
Psc 1h20'27"16d33'
Stella of Noriyoshi
Tau 5h28'59"20d40'
Stella Of Noriyuki
Tau 5h25'23"21d21'
Stella of Noriyuki & Jon
Aqr 22h1'51"-23d57'
Stella of Noviaki & Masae
Psc 1h13'15"20d12'
Stella of Now & Forever Yoko "Snaffy"
Leo 10h16'5"6d57'
Stella of Nozomi
Leo 10h14'59"11d31'
Stella of Nozomi
Cap 20h50'17"-26d32'
Stella Of Okidate
Cnc 8h48'27"22d24'
Stella of Okusan Naka
Aqr 22h0'8"-10d54'
Stella of Osamu & Noriko
Cap 20h50'45"-26d28'
Stella of Osamu & Kayo
Ari 2h17'6"21d14'
Stella of Osamu Kaneko
Ari 2h21'39"11d17'
Stella Of Oscar
Psc 1h16'45"27d27'
Stella of Ouji
Sgr 19h5'35"-13d53'
Stella Of Papa Koichi
Leo 10h17'45"9d11'
Stella of Point Shop Best Friends
Gem 7h1'36"15d36'
Stella Of Polepole
Sgr 19h10'2"-14d0'
Stella Of Ponta
Leo 10h0'47"13d42'
Stella of Preades & Rasalhague
Cap 20h52'3"-15d28'
Stella of Princess Yoshiko
Tau 5h25'47"21d37'
Stella of Princess Tomoko
Lib 14h25'54"-10d51'
Stella Of Prof Mario
Gem 7h14'45"21d23'
Stella Of Promised Night
Leo 9h55'9"23d59'
Stella of QUI
Sgr 19h13'42"-16d29'
Stella Of Realize Our Hopes
Vir 14h1'0"11d19'
Stella Of Rei & Koukei
Ari 2h14'48"25d37'
Stella of Rei Miyamoto
Gem 7h11'44"14d39'
Stella of Rei Tanaka
Lib 14h30'48"-20d51'
Stella of Reiko
Cnc 8h45'54"20d31'
Stella of Reiko
Cnc 8h51'11"17d39'
Stella Of Reiko
Cnc 8h42'9"28d49'
Stella Of Reiko Harmy
Sco 16h9'3"-28d54'
Stella of Remiko
Aqr 22h8'15"-2d21'
Stella Of Rena
Cnc 8h47'0"21d8'
Stella Of Rena
Gem 7h13'59"26d58'
Stella of Reo
Vir 14h36'32"6d19'
Stella of Rie
Sco 16h27'2"-27d48'
Stella of Rie
Tau 2h11'18"19d50'
Stella Of Rie
Sco 16h8'24"-26d57'
Stella Of Riho
Sco 16h7'11"-29d9'
Stella Of Rikako
Lib 14h32'24"-22d26'
Stella of Riki & Kyoko
Ari 2h28'8"29d50'
Stella Of Riko
Leo 10h11'32"20d39'
Stella Of Rina
Vir 14h26'43"2d26'
Stella Of Ritsuko
Leo 10h0'29"13d5'
Stella Of Ritsuko & Toshimitsu
Cnc 8h43'3"29d12'
Stella of Rui
Lib 14h34'51"-19d53'
Stella of Rumi
Sco 16h20'2"-26d41'
Stella of Rumi
Cnc 8h48'42"24d17'
Stella of Rumiko
Ari 2h26'3"20d57'
Stella of Rumiko
Lib 14h53'33"-24d3'
Stella Of Ryo
Cnc 8h49'54"11d27'
Stella of Ryo
Tau 5h26'44"26d51'
Stella Of Ryo
Psc 1h20'14"7d9'
Stella Of Ryo & Kciko
Leo 10h0'14"7d21'
Stella of Ryohei
Sgr 19h10'50"-15d52'
Stella Of Ryohsuke
Psc 1h39'41"27d56'
Stella Of Ryoichi
Lib 14h26'50"-9d56'
Stella of Ryoko
Tau 5h23'51"25d42'
Stella of Ryosuke
Lib 14h49'14"-22d49'
Stella of Ryosuke
Vir 14h1'55"5d59'
Stella Of Ryosuke
Leo 9h56'42"21d33'
Stella Of Ryotaro
Gem 7h11'47"27d0'
Stella Of Ryotaro
Tau 5h17'17"25d5'
Stella of Ryouko
Sco 16h1'48"-14d11'
Stella of Ryuhei
Leo 10h15'29"7d46'
Stella of Ryuichi
Psc 1h20'9"8d29'
Stella of Ryuichi & Tomoko
Sgr 19h8'3"-12d34'
Stella of Ryunosuke
Sco 16h21'27"-26d42'
Stella Of Ryusei
Ari 2h16'30"25d31'
Stella Of Ryuya & Yoko
Ari 2h14'45"22d0'
Stella of S 50413 R O
Ari 2h19'50"15d58'
Stella of S To S
Ari 2h15'3"19d27'
Stella of S Tsuyoshi
Aqr 22h14'48"-12d30'
Stella Of S52 Keiko & Ring
Tau 5h27'5"16d6'
Stella Of Saburo & Junko
Leo 9h55'45"22d31'
Stella Of Sachi
Vir 14h39'17"5d10'
Stella Of Sachi
Tau 5h24'33"24d40'
Stella of Sachie
Aqr 22h2'56"-15d7'
Stella of Sachika
Sgr 19h7'8"-12d56'
Stella of Sachiko
Cap 20h53'26"-20d47'
Stella of Sachiko
Sco 16h4'42"-14d36'
Stella Of Sachiko
Tau 5h10'41"17d13'
Stella Of Sachiko
Sco 16h20'20"-21d11'
Stella of Sachiko & Mayumi
Cap 20h49'50"-23d48'
Stella of Sadanobu
Tau 5h29'20"16d5'
Stella of Sadao
Aqr 22h3'50"-22d19'
Stella of Sadao & Takako
Lib 14h49'56"-24d6'
Stella Of Sadayoshi & Hiroko
Sgr 19h12'18"-29d35'
Stella Of Sae
Cap 20h55'11"-26d30'
Stella of Sae
Psc 1h15'38"29d18'
Stella of Sakae Wada
Aqr 22h15'8"-10d0'
Stella Of Sakai Family
Cap 20h51'15"-22d33'
Stella Of Sakura
Vir 12h55'59"2d30'
Stella of Sanpan & Jenny
Psc 1h24'38"3d33'
Stella of Saori
Sgr 19h9'32"-15d14'
Stella of Saori Kunihara
Cap 20h56'6"-15d48'
Stella Of Saori
Vir 14h21'24"5d15'
Stella of Sasaki Family
Cap 20h55'59"-15d54'
Stella Of Sato Family
Tau 5h31'36"15d44'
Stella of Satoko
Cnc 8h45'27"15d54'
Stella of Satoko
Tau 5h29'41"26d14'
Stella of Satomi
Leo 10h1'35"9d2'
Stella of Satomi
Aqr 22h1'20"-16d23'
Stella Of SATOMI
Lib 14h30'45"-24d9'
Stella of Satomi & Mariko
Gem 7h10'11"20d38'
Stella of Satoru
Leo 10h5'6"18d43'
Stella of Satoru & Mituyo
Sco 16h28'26"-29d32'
Stella of Satoru & Fumi
Aqr 22h4'45"-6d38'
Stella Of Satoru & Noriko
Tau 5h16'32"23d21'
Stella Of Satoru & Noriko
Psc 1h22'47"9d3'
Stella Of Satoshi
Vir 14h7'51"1d26'
Stella of Satoshi
Sgr 19h10'26"-12d9'
Stella of Satoshi & Kyoko
Cap 20h58'54"-18d21'
Stella of Satoshi & Tomoko
Lib 14h49'24"-23d45'
Stella Of Satoshi & Yuka
Ari 2h10'50"25d7'
Stella of Sawako & Yukiko
Lib 14h29'0"-10d53'
Stella Of Sayuri
Aqr 22h2'56"-10d9'
Stella of Seagal
Gem 7h1'47"18d14'
Stella of Sei & Oz
Sgr 19h9'30"-13d32'
Stella of Seiichi & Miyako
Leo 10h19'24"11d33'
Stella of Seiichi & Junko
Tau 5h12'35"19d23'
Stella of Seiji
Gem 7h13'9"24d33'
Stella of Seiji & Sachiko
Psc 1h15'2"23d14'
Stella of Seiji & Suemi
Tau 5h25'53"23d53'
Stella of Seiji & Yoko
Tau 5h24'2"27d33'
Stella Of Seiji & Sumiko
Tau 5h19'38"22d6'
Stella Of Seiji & Mika
Leo 10h4'27"9d1'
Stella of Seiki Mori
Gem 7h11'47"19d57'
Stella Of Seiko
Sco 16h8'18"-26d20'
Stella Of Seisa
Gem 7h12'41"22d38'
Stella Of Seiya
Cap 20h56'12"-27d11'
Stella of Serika
Lib 14h46'41"-1d31'
Stella Of Setsuko
Sco 16h27'23"-25d21'
Stella of Sharness
Sgr 19h13'6"-16d43'
Stella of Shiawase- boshi
Gem 7h13'30"29d57'
Stella Of Shigenobu & Kumi
Vir 14h27'39"6d42'
Stella of Shigenori
Sgr 19h13'54"-29d27'
Stella of Shigeo Kobayashi
Cap 20h54'18"-18d58'
Stella of Shigeo & Kumiko
Sco 16h19'48"-8d38'
Stella of Shigeri
Gem 7h1'17"17d48'
Stella of Shigeru
Sco 16h25'21"-26d33'
Stella of Shigeru & Moriko 25th
Lib 14h47'39"-19d0'
Stella Of Shigeru & Yoshie
Sco 16h6'38"-27d29'
Stella of Shigeyuki & Satomi
Sgr 29h12'36"-16d11'
Stella Of Shiho
Sco 16h15'20"-23d9'
Stella of Shin & Jill
Leo 10h16'33"12d13'
Stella of Shin-ichiro
Cap 20h56'33"-23d51'
Stella of Shingo
Vir 14h3'16"6d42'
Stella of Shinichi Akemitsu
Psc 1h21'26"4d58'
Stella Of Shinichi
Leo 10h1'26"8d6'
Stella Of Shinichi & Hiromi
Tau 5h23'47"16d11'
Stella Of Shinichi & Cora
Ari 2h12'14"22d23'
Stella of Shinichi & Akemi
Tau 5h14'56"19d53'
Stella of Shinichi & Akemi
Aqr 22h17'26"-12d10'
Stella of Shinichi & Kumi
Lib 14h42'38"-12d47'
Stella of Shinji
Sco 16h20'38"-27d13'
Stella of Shinji
Cap 20h50'12"-25d13'
Stella of Shinji & Tamaki
Leo 10h2'17"15d21'
Stella of Shinji & Hitomi
Sco 16h28'44"-29d57'
Stella Of Shinji & Mika
Sco 16h8'24"-26d50'
Stella Of Shinnichi & Keiko
Cnc 8h55'5"20d39'
Stella of Shinnosuke Miyanobu
Cnc 8h49'53"8d25'
Stella Of Shino Shiris
Cap 20h54'29"-22d6'
Stella Of Shinobu & Yuko
Leo 10h4'15"6d40'
Stella of Shinobu & Yoko
Aqr 22h4'12"-6d58'
Stella of Shinpei
Cap 20h53'17"-17d26'
Stella Of Shinya Masaki
Gem 7h14'17"26d27'
Stella Of Shinya & Yukiko
Gem 7h11'26"16d38'
Stella of Shinya & Chinatsu
Tau 5h29'0"21d57'
Stella of Shinya & Reiko
Cap 20h58'24"-23d16'
Stella of Shinzo & Mayumi
Aqr 22h16'44"-12d11'
Stella of Shinzo & Reiko
Gem 7h14'53"28d34'
Stella Of Shirakawa Daigo
Ari 2h17'45"25d13'
Stella of Shirasaka Michiya
Leo 10h19'5"23d24'
Stella of Shiro & Kazumi
Lib 14h46'18"-23d30'
Stella of Shizuko Ohtake
Aqr 22h9'18"-5d12'
Stella Of Sho
Psc 1h21'33"13d39'
Stella Of Sho
Vir 14h32'28"0d18'
Stella of Shogo & Tomoko
Gem 7h0'44"19d17'
Stella of Shoji
Cnc 8h40'59"22d53'
Stella Of Shoji & Ako
Tau 5h11'21"16d37'
Stella Of Shoko
Sgr 19h18'3"-28d35'
Stella Of Shouichi & Yukie
Tau 5h16'15"21d36'
Stella of Shouji & Kumi
Ari 2h11'51"21d6'
Stella of Shu & Kei
Aqr 22h7'35"-5d13'
Stella of Shuichi
Psc 1h16'44"24d9'
Stella of Shuichi
Aqr 21h58'20"0d58'
Stella of Shuichi & Chihami
Psc 1h21'8"6d12'
Stella of Shuichi & Katsue
Psc 1h20'14"11d6'

Stella Of Shuichi & Hiromi
Gem 7h14'23"20d50'
Stella of Shuichiro & Katsuko
Aqr 22h1'32"-24d57'
Stella Of Shuji & Etsuko
Aqr 22h0'5"-17d41'
Stella Of Shuji & Saeko
Leo 9h58'8"22d3'
Stella Of Shuji & Elica
Sco 16h18'12"-23d35'
Stella of Shun & Kei
Sco 16h29'24"-26d27'
Stella Of Shunichi
Leo 9h59'12"24d12'
Stella of Shunichi & Rie
Ari 2h19'45"21d3'
Stella of Shunji
Cnc 8h45'36"17d50'
Stella Of Shunji
Psc 1h19'30"24d23'
Stella of Shunji
Cap 20h59'30"-22d30'
Stella of Shunsuke Shimanaka
Gem 7h14'20"24d17'
Stella of Shunsuke & Mayumi
Tau 5h29'41"23d20'
Stella of Shutaro
Cnc 8h46'42"18d8'
Stella of Shuya
Tau 5h29'12"24d54'
Stella of Sigeko Twenty
Psc 1h20'2"10d8'
Stella of Silver
Sgr 19h16'39"-29d55'
Stella of Sincerity
Ari 2h18'11"27d6'
Stella Of Sincerity
Sgr 19h16'20"-28d38'
Stella of Sinji
Sgr 19h18'51"-29d44'
Stella of Sinvino
Cnc 8h44'57"24d13'
Stella Of Sinvino
Cnc 8h44'57"24d13'
Stella Of SIUGE
Lib 14h33'20"-24d11'
Stella of Snow
Psc 1h16'42"22d46'
Stella of Snow White
Ari 2h19'59"23d51'
Stella Of Souichirou
Lib 14h32'8"-24d44'
Stella Of Splendid Yumi
Ari 2h25'53"19d6'
Stella Of Stardust Revue
Gem 7h14'17"16d23'
Stella Of Story
Aqr 22h17'42"-14d49'
Stella of Story
Aqr 22h17'42"-14d49'
Stella Of Sugao
Vir 14h52'37"0d1'
Stella of Sugiko
Aqr 21h59'59"-12d27'
Stella of Sumiko
Leo 10h17'38"7d57'
Stella of Susumu & Mayumi
Sgr 19h6'9"-15d21'
Stella of Suzu
Cap 20h59'32"-18d9'
Stella Of Sylphide
Gem 7h14'17"13d57'
Stella of Syoichi & Hiroko
Leo 10h6'30"21d29'
Stella Of Syokazyu
Gem 7h12'39"22d3'
Stella of Syuji & Meriko
Cap 20h52'24"-16d45'
Stella of T & Miyuki
Cnc 8h49'8"10d35'
Stella of T & R
Ari 2h25'23"23d4'
Stella Of T & Y
Sgr 19h15'15"-28d26'
Stella of T & Y & Mam
Vir 14h6'54"-10d54'
Stella of T M Eternal
Lib 14h34'2"-21d56'
Stella Of T-Myho
Tau 5h18'2"23d57'
Stella of TachanEchan
Cap 20h52'30"-16d52'
Stella of Tachi
Aqr 22h1'56"-22d32'
Stella Of Tadahira & Tomomi
Lib 14h32'35"-22d52'
Stella of Tadao
Leo 10h3'50"13d1'
Stella of Tadao & Kyoko
Psc 1h22'15"18d49'
Stella Of Tadashi
Sgr 19h10'9"-13d54'
Stella of Tadashi & Hidemi
Psc 1h23'51"9d16'
Stella of Tadashi & Naoko
Lib 14h45'26"-20d18'
Stella of Tadashi & Kazue
Psc 1h15'0"22d44'
Stella of Tadayuki & Toshie
Gem 7h3'47"19d10'
Stella of Taichi
Psc 1h15'33"29d32'
Stella of Taisei & Mayumi
Aqr 22h4'47"-24d54'
Stella of Taisho & Hazuki
Sco 16h21'14"-25d35'
Stella Of Taka & Kayo
Ari 2h12'11"22d0'
Stella Of Takae
Aqr 22h1'33"-20d38'
Stella of Takahide & Kyoko
Cnc 8h48'9"17d41'
Stella of Takahiro
Gem 7h3'36"18d56'
Stella of Takahiro
Ari 2h26'24"20d36'
Stella Of Takahiro
Cnc 8h49'59"9d14'
Stella of Takahiro
Ari 2h26'24"20d36'
Stella Of Takahiro
Tau 5h14'32"17d36'
Stella Of Takahiro
Cap 20h56'45"-20d39'
Stella Of Takahiro & Shihoko
Sgr 19h11'51"-12d54'
Stella Of Takahiro & Yuki
Aqr 22h3'45"-14d21'
Stella of Takahito
Gem 7h2'0"18d14'
Stella of Takako
Leo 10h17'41"11d35'
Stella of Takako
Aqr 22h7'50"-5d27'
Stella Of Takako
Tau 5h17'53"26d56'
Stella Of Takako & Sotaro
Ari 2h18'53"26d14'
Stella of Takamasa
Aqr 22h2'23"-23d41'
Stella of Takamatsu Family
Tau 5h27'29"25d16'
Stella of Takami & Yoko
Sgr 19h17'23"-28d5'
Stella of Takamitsu "Starion" Masuda
Cnc 8h49'27"15d28'
Stella of Takamune & Yumiko
Sco 16h4'17"-16d49'
Stella of Takano Family
Cap 20h50'11"-23d44'
Stella of Takanobu Masuda
Psc 1h17'3"20d59'
Stella of Takanori
Psc 1h23'41"10d16'
Stella of Takanori & Miwako
Cap 20h56'6"-18d3'
Stella Of Takanori & Kiyomi
Gem 7h11'41"24d57'
Stella of Takao
Leo 10h19'23"6d46'
Stella of Takaoki & Sakiko
Lib 14h48'56"-24d44'
Stella Of Takapyon & Mitchan
Leo 10h0'12"7d57'
Stella of Takashi Sakamoto
Cnc 8h46'18"24d8'
Stella of Takashi
Sgr 19h13'48"-16d32'
Stella of Takashi
Sgr 19h10'11"-12d41'
Stella of Takashi & Masami
Sco 16h1'30"-19d17'
Stella of Takashi & Akiko
Sco 16h2'9"-18d24'
Stella of Takashi & Kei
Ari 2h27'50"29d48'
Stella of Takashi & Rumiko
Vir 14h9'49"2d27'
Stella of Takashi & Miho
Lib 14h48'54"-20d24'
Stella of Takashi & Yoko
Gem 7h1'45"16d48'
Stella of Takashi & Kumiko
Gem 7h12'47"24d30'
Stella of Takashi & Mikako
Vir 12h5'21"2d11'
Stella Of Takashi & Yuko
Sco 16h8'24"-27d2'
Stella Of Takashi & Rumi
Ari 2h12'50"26d23'
Stella Of Takashi & Mareko
Ari 2h14'21"26d29'
Stella of Takashi & Shoko
Tau 5h12'39"21d39'
Stella Of Takashi & Chikage
Sco 16h5'27"-29d1'
Stella of Takatsu Shin go
Sgr 19h5'39"-13d3'
Stella of Takayuki
Sgr 19h14'15"-29d34'
Stella Of Takayuki
Lib 14h29'50"-24d12'
Stella Of Takayuki & Sayuri
Tau 5h17'6"21d4'
Stella of Takayuki & Yasuko
Lib 14h47'3"-24d1'
Stella of Takayuki & Kaori
Aqr 22h4'17"-22d27'
Stella of Take & Chiharu
Sgr 19h6'17"-14d18'
Stella Of Take & Yumi
Cnc 8h45'0"25d43'
Stella Of Takehiko & Eri
Vir 14h21'1"5d24'
Stella of Takehiko & Kiyo
Cnc 8h49'33"10d4'
Stella of Takehiko & Kikuka
Lib 14h44'38'0d38'
Stella Of Takehiro & Sayoko
Gem 7h11'23"14d4'
Stella of Takeko
Psc 1h19'30"26d12'
Stella of Takeo & Michiko
Ari 2h18'53'23d23'
Stella of Takeshi Kouda
Cap 20h57'12"-23d48'
Stella of Takeshi
Psc 1h22'45"10d42'
Stella Of Takeshi
Ari 2h10'21"23d0'
Stella Of Takeshi & Sonoe
Lib 14h37'38"-13d54'
Stella Of Takeshi & Mariko
Lib 14h42'56"-13d51'
Stella of Takeshi & Hisako
Cnc 8h45'24"15d15'
Stella of Takeshi & Kiyoko
Tau 5h29'53"26d41'
Stella of Taketo
Sgr 19h6'32"-15d54'
Stella Of Takeyuki & Kyouko
Vir 14h5'55"2d4'
Stella of Taku
Lib 14h47'26"-20d27'
Stella of Taku
Gem 7h2'24"16d47'
Stella Of Takuma
Psc 1h20'2"16d2'
Stella Of Takumi
Psc 1h15'47"28d39'
Stella of Takumi & Naomi
Sco 16h3'54"-16d24'
Stella Of Takuto's Father (Teruaki)
Vir 14h51'19"5d41'
Stella Of Takuya
Tau 5h32'18"18d45'
Stella of Takuya
Sco 16h0'45"-16d15'
Stella of Takuya
Sco 16h11'27"-16d25'
Stella Of Takuya 11.13
Sco 16h6'27"-28d51'
Stella of Takuzo Oda
Lib 14h41'12"-12d56'
Stella of Tamaki & Mickey
Aqr 21h59'36"-1d19'
Stella of Tamaki
Gem 7h10'30"24d52'
Stella Of Tamao
Gem 6h22'1"18d55'
Stella of Tamayo, My Shining Star
Cap 20h58'12"-23d1'
Stella of Tamie
Cap 20h50'9"-20d51'
Stella of Tamiko
Psc 1h15'27"26d10'
Stella of Tamiko & Toshihiro
Sco 16h0'9"-13d41'
Stella of Tamio & Hitomi
Aqr 22h19'30"-11d8'
Stella of Tamotsu & Naomi
Gem 7h3'20"19d5'
Stella of Tamotsu & Maki
Aqr 21h57'17'-1d19'
Stella Of Tamotsu & Fukiko
Lib 14h41'39"-22d2'
Stella of Tania
Sco 16h29'14"-26d1'
Stella of Taniuchi
Leo 10h5'36"20d33'
Stella of Taro & Yoshiko
Aqr 22h4'39"-24d38'
Stella of Tatsuku
Sco 16h22'2"-27d33'
Stella of Tatekabe Family
Gem 7h13'12"19d40'
Stella of Tatsumi
Cnc 8h47'26"18d5'
Stella Of Tatsumi
Gem 7h4'15"18d29'
Stella of Tatsumi & Ayumi
Aqr 22h17'9"-21d34'
Stella Of Tatsuo & Kaoru
Sgr 19h14'38"-12d56'
Stella Of Tatsuya
Vir 14h30'42"5d32'
Stella Of Tatsuya
Leo 10h15'35"8d56'
Stella of Tatsuya & Mitsuko
Gem 7h3'5"20d33'
Stella of Tatsuya & Atsuko
Sco 16h22'18"-29d17'
Stella of Tatsuya & Chie
Cap 20h50'3"-17d44'
Stella Of Tatsuya K
Ari 2h12'50"22d57'
Stella of Tatuki
Aqr 22h4'59"-5d33'
Stella of Tatuya
Vir 14h2'37"6d50'
Stella Of Teacher Kubo
Vir 14h26'35"2d8'
Stella of Tenma Hashiba
Sgr 19h12'57"-16d21'
Stella of Terresia Kayo
Ari 2h19'53"15d47'
Stella Of Teruko
Psc 1h23'12"10d9'
Stella Of Teruo & Chiemi
Lib 14h42'30"-8d45'
Stella of Tetsu
Cap 20h56'53"-22d53'
Stella Of Tetsuo & Mayumi
Ari 2h6'32"25d43'
Stella Of Tetsuo & Michiko
Sgr 19h11'12"-12d51'
Stella Of Tetsuya
Sgr 19h10'57"-14d0'
Stella Of Tetsuya
Cnc 8h48'0"8d48'
Stella of Tetsuya
Sgr 19h6'50"-15d20'
Stella of Tetsuya & Hiroko
Vir 14h36'37"5d60'
Stella Of Tetsuya & Takako
Lib 14h49'45"-3d21'
Stella of Tetsuya & Shizuka
Cnc 8h43'30"24d3'
Stella Of Tetsuya & Shizuka
Cnc 8h43'30"24d3'
Stella Of Tetsuya & Michiyo
Vir 14h6'25"24"5d6'
Stella Of Tetsuya & Kiyoko
Leo 10h3'57"7d10'
Stella of Tetuo & Kumiko
Psc 1h12'45"20d8'
Stella Of The Ambitious
Psc 1h18'26"26d33'
Stella of The Beatles
Gem 7h10'39"25d3'
Stella of the M & F
Aqr 22h18'8"-13d51'
Stella of Tochitachie
Sgr 19h9'21"-15d59'
Stella of Tohru & Tomoko
Ari 2h17'47"13d27'
Stella Of Tohru & Mitsuko
Ari 2h7'33"25d9'
Stella of Tohru & Aki
Ari 2h10'29"20d21'
Stella of Tokaiteio
Sco 16h20'35"-26d8'
Stella of Tokomiya
Cap 20h55'23"-16d52'
Stella of Tokuko
Aqr 22h6'33"-5d53'
Stella of Tomiko & Hidekazu
Sco 16h17'41"-8d54'
Stella Of Tomiko & Kiyoshi
Sco 16h5'47"-28d3'
Stella of Tomiya & Akiko
Cnc 8h51'8"17d44'
Stella Of Tomo
Psc 1h20'36"12d24'
Stella Of Tomo & Yoko
Lib 14h34'29"-23d37'
Stella of Tomoaki & Yoko
Cap 20h51'36"-22d0'
Stella Of Tomoaki
Leo 10h9'17"19d9'
Stella of Tomoaki K
Sgr 19h26'36"-26d58'
Stella Of Tomoharu & Kimiko
Aqr 22h0'21"-20d22'
Stella Of Tomohide
Sgr 19h19'20"-28d40'
Stella of Tomohiko
Sgr 19h11'26"-29d57'
Stella Of Tomohiko & Hiroko
Aqr 22h16'6"-16d37'
Stella Of Tomohiko & Yoko Uno
Sco 16h26'42"-25d3'
Stella Of Tomohiro
Lib 14h33'44"-23d9'
Stella of Tomohiro
Lib 14h48'44"-23d36'
Stella of Tomohiro & Minako
Sco 16h27'21"-26d56'
Stella Of Tomohiro & Shima
Ari 2h11'15"23d28'
Stella Of Tomohisa
Lib 14h35'29"-14d44'
Stella Of Tomoki
Psc 1h20'3"16d56'
Stella Of Tomoko
Leo 9h57'51"23d44'
Stella of Tomoko
Leo 10h15'2"7d46'
Stella of Tomoko Matsumoto
Ari 2h19'20"21d32'
Stella of Tomoko
Lib 14h46'9"-22d17'
Stella of Tomoko & Preiades
Cap 20h56'36"-15d0'
Stella of Tomoko & Hitoshi
Gem 7h3'57"16d43'
Stella of Tomoko & Eisuke
Sco 16h28'51"-28d25'
Stella of Tomoyuki & Shoko
Aqr 22h16'59"-7d58'
Stella of Toraki & Kazuki & Satomi
Gem 7h0'9"17d56'
Stella of Toru
Sgr 19h8'41"-16d7'
Stella Of Toru
Sco 16h26'48"-24d51'
Stella of Toru & Ayako
Lib 14h50'36"-24d17'
Stella of Toshi & Akko
Sco 16h5'38"-29d47'
Stella Of Toshi & Nao
Sco 16h6'30"-27d46'
Stella of Toshiaki & Kunie
Cap 20h56'35"-24d2'
Stella of Toshiharu & Risa
Cap 20h50'53"-17d25'
Stella Of Toshihide & Fumie
Tau 5h28'54"19d29'
Stella Of Toshihiko & Satomi
Psc 1h20'21"12d37'
Stella Of Toshihiro
Leo 10h2'11"10d51'
Stella of Toshihiro
Vir 17h23'9"-1d0'
Stella of Toshihiro & Tokuko
Leo 10h8'56"18d31'
Stella of Toshikatsu & Miho
Sco 16h23'59"-25d59'
Stella of Toshikatu
Lib 14h41'47"-9d18'
Stella of Toshikazu & Sayaka
Cap 20h52'3"-23d53'
Stella of Toshikazu & Emi
Aqr 22h15'50"-14d54'
Stella Of Toshikazu & Emi
Aqr 22h15'50"-14d54'
Stella of Toshiko
Lib 14h52'2"-24d23'
Stella of Toshiko
Sgr 19h8'12"-12d14'
Stella Of Toshio
Sco 16h9'35"-26d45'
Stella of Toshiya
Sgr 19h14'2"-16d42'
Stella of Toshiya & Tomoko
Lib 16h8'42"-25d45'
Stella Of Toshiyasu & Yukiko
Tau 5h17'41"24d2'
Stella of Toshiyuki
Tau 5h28'59"21d37'
Stella of Toshiyuki & Hiroko
Sco 16h22'11"-27d36'
Stella Of Toshiyuki & Ryoko
Aqr 22h1'57"-20d56'
Stella Of Totchi & Ami
Leo 9h59'35"25d30'
Stella Of Totto
Lib 14h32'53"-17d49'
Stella of True Love
Psc 1h15'14"29d14'
Stella of Tsubara
Tau 5h34'23"18d56'
Stella of Tsubasa
Leo 10h4'44"14d9'
Stella of Tsubasa 1993
Aqr 22h15'42"-12d22'
Stella of Tsuchiya-shi
Sco 16h28'32"-26d17'
Stella of Tsugio
Ari 2h11'42"20d52'
Stella Of Tsugio & Yumi
Gem 7h13'3"16d50'
Stella of Tsukasa
Vir 14h8'50"-8d13'
Stella of Tsukasa & Shizuko
Vir 14h48'35"0d7'
Stella Of Tsukasa & Yukie
Sgr 19h16'26"-28d42'
Stella of Tsuneaki
Sco 16h20'32"-29d0'
Stella of Tsuneo & Yukari
Sgr 19h5'24"-12d49'
Stella Of Tsuneyo & Keiko
Lib 14h43'23"-13d35'
Stella of Tsutomu & Megumi
Leo 10h2'53"11d7'
Stella of Tsuyoshi
Sco 16h19'12"-24d15'
Stella of Tsuyoshi
Ari 2h16'12"27d8'
Stella Of Twin Star
Gem 7h10'18"17d0'
Stella Of Ueda Junichiro
Sco 16h27'56"-29d53'
Stella of Ueda Mai
Aqr 22h16'51"-10d3'
Stella Of Uichi
Vir 14h5'25"5d15'
Stella Of Uka
Psc 1h16'33"28d56'
Stella of Umeda Family
Gem 7h13'0"28d36'
Stella Of Unozawa
Gem 6h51'27"21d13'
Stella Of Utako
Tau 5h27'32"19d48'
Stella Of Uwe & Yasuko
Vir 14h27'18"6d42'
Stella of Victory
Sco 16h26'17"-29d44'
Stella of WAGAMAMA
Cap 20h54'33"-19d6'
Stella of Wakako
Gem 7h2'21"19d52'
Stella of Wakasa
Leo 10h2'26"19d39'
Stella of Wakko & Yoshimitsu
Ari 2h18'57"15d17'
Stella of Ware-Ware
Cnc 8h46'14"23d9'
Stella of Wataru
Tau 5h33'12"19d6'
Stella of Wataru Tsukinishi
Aqr 22h18'54"-12d49'
Stella of Wataru & Kana
Gem 7h12'36"28d57'
Stella of Win
Aqr 22h19'23"-9d2'
Stella Of With You
Gem 7h14'32"22d17'
Stella Of Y&M
Gem 7h14'5"17d26'
Stella of Yamamoto
Leo 10h7'44"21d8'
Stella of Yamamoto Mitugu
Vir 14h41'53"-6d13'
Stella of Yamamoto
Sco 16h5'11"-25d28'
Stella of Yasuaki & Chitose
Vir 14h29'51"-8d17'
Stella Of Yasuda Family
Gem 7h11'47"26d44'
Stella of Yasuhide
Lib 14h47'18"-22d51'
Stella Of Yasuhiko & Junko
Tau 5h16'30"24d35'
Stella of Yasuhiro Yamane
Leo 10h17'11"7d43'
Stella of Yasuhiro
Sgr 19h10'26"-27d29'
Stella of Yasuhiro
Cap 20h50'35"-19d18'
Stella of Yasuhiro & Tomomi
Cap 20h50'47"-23d41'
Stella Of Yasuhiro & Tamaki
Sgr 19h12'27"-14d32'
Stella Of Yasuhiro I
Leo 10h19'17"8d46'
Stella of Yasuhito & Miyako
Sco 16h24'20"-25d2'
Stella of Yasuhito & Takako
Aqr 22h7'26"-4d25'
Stella Of Yasuji Nakamatsu
Leo 9h58'50"24d3'
Stella Of Yasuko
Lib 14h36'35"-14d28'
Stella of Yasumitsu
Lib 14h59'32"-8d11'
Stella of Yasunori
Cnc 8h48'23"12d50'
Stella of Yasunori
Cap 20h52'47"-19d12'
Stella of Yasunori
Cnc 8h50'27"17d35'
Stella Of Yasuo & Mie
Lib 14h40'59"-11d24'
Stella Of Yasushi
Cnc 8h40'57"24d58'
Stella of Yasushi & Maya
Gem 7h4'36"19d16'
Stella of Yasushi & Hisayo
Lib 14h26'45"-13d37'
Stella of Yasushi & Masayo
Lib 14h46'59"-1d50'
Stella Of Yasuto
Leo 10h3'38"8d11'
Stella of Yasuyuki
Lib 14h47'50"-19d20'
Stella Of Yasuyuki & Kazuko
Sgr 19h10'15"-14d11'
Stella of Yo & E
Cap 20h57'0"-18d0'
Stella of Yohei Onozawa
Gem 7h15'0"29d27'
Stella of Yoiyo
Psc 1h19'38"29d38'
Stella of Yoji & Mari
Lib 14h49'0"-24d23'
Stella of Yoko
Leo 10h1'9"16d26'
Stella of Yoko
Aqr 22h2'9"-24d26'
Stella of Yoko
Cap 20h51'41"24d1'
Stella Of Yoko
Cap 20h50'33" 22d36'
Stella Of Yoko
Cap 20h53'50"-22d11'
Stella Of Yoko & Tatsuya
Psc 1h20'29"8d56'
Stella of Yoko & Holmes
Leo 10h15'56"13d2'
Stella of Yoko Family
Leo 10h7'24"18d23'
Stella of Yoneko
Psc 1h29'17"16d41'
Stella Of Yonetoshi & Katsue
Ari 2h27'41"18d25'
Stella of Yoriko
Sgr 19h12'24"-26d1'
Stella of Yoshi
Leo 10h19'14"12d41'
Stella of Yoshiaki & Mie
Tau 5h32'2"19d23'
Stella of Yoshiharu
Leo 10h0'59"12d20'
Stella of Yoshihide
Leo 10h4'56"15d21'
Stella of Yoshihiro
Vir 14h9'45"1d15'
Stella of Yoshihiro
Sco 16h26'44"-26d28'
Stella of Yoshihiro & Yasuko
Cap 20h49'59"-16d21'
Stella of Yoshihiro & Sachiko
Sgr 19h16'51"-29d20'
Stella Of Yoshihiro & Akemi
Vir 14h27'39"6d31'
Stella of Yoshihisa & Yoshie
Gem 7h12'30"23d57'
Stella of Yoshihito & Emiko
Lib 14h34'44"-21d51'
Stella of Yoshiki & Mari
Cap 20h58'5"-22d29'
Stella of Yoshiko Koumi
Lib 14h51'27"-24d38'
Stella of Yoshiko
Tau 5h33'47"18d51'
Stella Of Yoshiko
Gem 7h12'9"25d57'
Stella of Yoshimasa & Sanae
Sgr 19h12'56"-14d56'
Stella of Yoshimi
Cap 20h59'14"-22d56'
Stella of Yoshimi & Rie
Aqr 22h16'51"-14d35'
Stella of Yoshimi & Yumiko
Cap 20h51'44"-18d20'
Stella Of Yoshimi & Kumiko
Ari 2h29'47"18d15'
Stella of Yoshimoto & Kayoko
Gem 7h3'5"15d14'
Stella of Yoshinori & Chie
Lib 14h48'18"-24d39'
Stella of Yoshio
Gem 7h11'12"19d35'
Stella of Yoshio & Hiromi
Vir 14h32'40"6d8'
Stella of Yoshio & Junko
Lib 14h31'17"-19d0'
Stella of Yoshio Okano
Ari 2h18'29"27d8'
Stella Of Yoshitaka & Kazue
Tau 5h21'17"16d30'
Stella of Yoshito & Masayo
Sco 16h0'51"-19d27'
Stella of Yoshiuori & Kiyomi
Sco 16h26'5"-29d21'
Stella of Yoshiyuki & Chieko
Ari 2h18'9"21d6'
Stella of Yoshiyuki & Sawako
Leo 10h9'18"20d19'
Stella of Yoshiyuki & Junko
Sco 16h4'57"-16d22'
Stella Of Yoshiyuki & Fumiko
Lib 14h36'57"-13d53'
Stella of Youhei
Psc 1h16'41"17d53'
Stella of Youichi & Kayo
Leo 10h9'24"21d30'
Stella of Youko
Cap 20h57'24"-16d6'
Stella Of Youko & Yuseki
Lib 14h44'30"-9d34'
Stella Of Yu
Gem 7h10'57"26d33'
Stella of YU
Psc 1h19'8"22d16'
Stella Of Yu-Itchi
Tau 5h26'2"19d59'
Stella Of Yudai
Sco 16h27'44"-25d33'
Stella of Yuhei
Cnc 8h45'33"11d47'
Stella of Yui
Vir 14h32'20"5d28'
Stella of Yui
Aqr 22h0'53"-7d42'
Stella of Yuichi & Akiko
Sgr 19h8'24"-15d38'
Stella of Yuichi & Midori
Sgr 19h14'38"-14d35'
Stella Of Yuichi & Midori
Sgr 19h14'38"-14d35'

Stella of Yuika
Ari 2h16'3"15d14'
Stella Of Yuji & Akiko
Sco 16h7'45"-29d18'
Stella of Yuji & Ayaka
Psc 1h18'45"18d44'
Stella Of Yuji & Yoko
Gem 7h1'48"18d46'
Stella Of Yuji Kazumi Kobi
Ari 2h27'50"22d12'
Stella Of Yuka
Sgr 19h13'48"-30d0'
Stella of Yuka
Cnc 8h45'24"17d34'
Stella of Yuka Kaneko
Gem 7h11'56"29d56'
Stella of Yukako
Leo 10h1'0"20d38'
Stella of Yukari
Ari 2h17'54"22d11'
Stella of Yukari
Lib 14h30'51"-18d59'
Stella of Yukari
Sgr 19h14'23"-16d9'
Stella of Yukari Kurose
Leo 10h2'24"8d14"
Stella of Yukari
Lib 16h9'18"-24d57'
Stella of Yuki
Sco 16h2'53"-8d15'
Stella of Yuki
Cap 20h57'18"-15d5'
Stella of Yuki
Sco 16h9'11"-29d53'
Stella of Yuki
Aqr 22h16'57"-14d26'
Stella of Yuki
Sco 16h23'42"-27d0'
Stella of Yuki
Cnc 8h48'51"17d56'
Stella of Yuki
Cap 20h54'35"-25d3'
Stella of Yuki
Cap 20h28'42"-25d8'
Stella of Yuki
Sco 16h12'29"13d7'
Stella Of Yuki
Lib 14h44'39"-8d52'
Stella Of Yuki
Leo 10h15'51"10d23'
Stella Of Yuki
Cnc 8h45'42"20d52'
Stella of Yuki & Masahiro
Psc 1h20'41"10d35'
Stella of Yukie
Leo 10h6'54"22d1'
Stella of Yukie
Ari 2h23'23"10d57'
Stella Of Yukie
Cap 20h53'9"-22d15'
Stella Of Yukihiko Otagiri
Lib 14h36'57"-13d48'
Stella Of Yukihiro & Kumi
Sco 16h7'2"-26d47'
Stella of Yukiko
Cap 20h56'57"-26d44'
Stella of Yukiko
Psc 1h19'36"20d25'
Stella of Yukiko
Ari 2h5'59"27d20'
Stella of Yukiko
Aqr 22h7'6"-5d34'
Stella of Yukiko & Ayako
Psc 1h18'36"22d6'
Stella of Yukiko I
Psc 1h11'23"20d18'
Stella of Yukinori
Lib 14h29'2"-12d59'
Stella of Yukio & Asako
Lib 14h33'21"-19d59'
Stella of Yukio & Junko
Psc 1h21'56"6d41'
Stella Of Yukio & Junko
Psc 1h21'56"6d41'
Stella Of Yukiya & Atsuko
Psc 1h21'21"15d3'
Stella of Yukiyasu & Tomomi
Ari 2h29'33"22d5'
Stella of Yuko
Aqr 22h17'26"-12d29'
Stella of Yuko
Sgr 19h7'26"-13d54'
Stella of Yuko
Gem 7h1'2"16d13'
Stella of Yuko
Aqr 22h0'51"-24d8'
Stella of Yuko
Sgr 19h13'42"-27d13'
Stella of Yuko
Sgr 19h12'30"-27d58'
Stella of Yuko
Vir 14h1'39"-8d7'
Stella of Yuko
Aqr 22h9'21"-1d31'
Stella of Yuko & Daisuke & Ayako
Sco 16h15'6"-22d34'
Stella Of Yuko,The
Cap 20h55'59"-25d33'
Stella Of Yumi
Gem 7h14'5"21d50'
Stella Of Yumi
Sco 16h25'9"-25d21'
Stella of Yumi
Leo 10h24'45"7d28'
Stella of Yumi
Vir 14h29'24"-1d38'
Stella of Yumiko
Psc 1h19'51"22d3'
Stella of Yumiko Yamamoto
Psc 1h18'53"24d34'
Stella Of YUMIKO
Cnc 8h45'59"21d44'
Stella Of Yumiko
Cnc 8h48'51"20d54'
Stella Of Yumiko
Cnc 8h48'0"7d56'
Stella of Yumiko & Toshiyasu
Cap 20h52'41"-14d49'
Stella of Yurika
Gem 7h1'30"15d17'
Stella of Yuriko
Gem 7h4'24"15d52'
Stella of Yuriko
Sgr 19h11'24"-27d30'
Stella of Yuriko & Hajime
Tau 5h14'47"20d29'
Stella Of Yuriko & Hajime
Tau 5h14'47"20d29'
Stella of Yuriko-Tago
Tau 5h28'56"26d37'
Stella of Yushi II
Aqr 21h57'2"-1d9'
Stella of Yusuke Kurachi
Ari 2h17'45"19d53'
Stella of Yuta
Lib 14h44'9"-1d18'
Stella of Yuta
Psc 1h18'53"21d6'
Stella Of Yuta
Leo 10h19'44"10d4'
Stella of Yutaka
Aqr 22h7'9"-2d16'
Stella of Yuu
Lyr 18h53'51"35d6'
Stella Of Yuuki
Sco 16h6'38"-26d20'
Stella Of Yuya
Cap 20h58'14"-21d38'
Stella Of Yùya
Tau 5h20'48"16d26'
Stella Of Zyuntyan, Kaotan
Lib 14h25'51"-9d59'
Stella Patricius (Patty's Star)
And 23h22'39"45d48'
Stella Philaura
Boo 14h10'43"39d5'
Stella pp Beentje
Ori 6h2'32"5d48'
Stella Puchrae Birgitae
Cnc 9h16'47"12d10'
Stella Puffino
Vel 9h22'2"-46d50'
Stella Rosa
Sco 16h59'52"-44d7'
Stella Star
Ari 2h0'15"18d17'
Stella Susanne Kehl
Cap 21h37'42"-22d38'
Stella Valentina
Psc 22h55'52"1d21'
Stella Waltraut
Cam 7h37'23"61d9'
Stella's Little Star
Peg 22h30'28"24d25'
Stella,Kimberly
Lmi 10h9'17"33d8'
Stella,My Starlight
Cas 23h30'59"60d48'
Stella-Alessia- Chrystal
Cae 4h56'59"-27d60'
Stellae Jacobae
Aql 20h11'18"11d58'
Stellaliisa
And 2h4'53"39d41'
Stellar Akers
Mon 8h2'46"-6d29'
Stellar Carrie Lynn
Cas 23h22'49"53d7'
Stellar Helen
Aql 20h0'21"10d10'
Stellar January Two
Uma 10h49'22"70d50'
Stellar Jim
Cyg 19h25'0"31d52'
Stellar Jim
Lac 22h17'58"47d23'
Stellar Massie
And 0h41'0"45d31'
Stellar Michael
Boo 15h21'29"52d27'
Stellar Parents
Tri 1h30'59"30d40'
Stellar Phillip
Boo 13h33'20"20d15'
Stellar Ruth Ann
Cas 23h22'23"53d18'
Stellar Scholars of St Bedes,The
Ori 5h32'52"-0d7'
Stellar Schuller
Gem 7h59'56"20d26'
Stellar Sisters Anne & Helen
Uma 11h57'49"44d37'
Stellar Stephanie,The
And 0h15'53"30d28'
Stellar Team 9196-FPD Operations
Ser 15h57'24"1d54'
Stellato,Alan
Ori 6h2'50"4d17'
Stellavarious
Lyr 19h5'59"26d15'
Stellex I
Uma 8h58'11"50d49'
Stelley,Patricia
Aql 20h19'34"7d39'
Stellina
Boo 15h8'45"26d38'
Stellina
Aur 5h52'48"31d4'
Stelling,Dennis Michael
Cet 2h58'57"2d17'
Stelling,Ida M
Eri 2h59'0"-17d40'
Stellman,Bill
Tau 5h1'1"16d25'
Stellmaszek,Robert
Cas 0h4'29"60d10'
Stellum Sandyius
Oph 17h56'13"13d31'
Stelluti,Anna Emanuele
Aqr 22h5'15"-0d42'
Stelluto,Maria & Angelo
Crb 15h22'25"31d44'
Stelly,Kathryn
Ari 2h56'56"28d46'
Stelly,Salvador Timothy
Dra 17h34'10"61d7'
Stelma,Henry John
Aql 19h6'51"-0d33'
Stelmack,Christine Marie
Cas 2h47'15"61d33'
Steltman,Karen
Cas 1h10'30"65d12'
Stelz,Hannelore
Ari 3h15'54"20d27'
Stelzer,C D
Uma 12h36'44"62d59'
Stelzer,Dr Herbert
Uma 11h45'16"33d7'
Stem
Cam 5h27'38"68d53'
Stember
Aur 6h1'1"50d34'
Stemen,Michael
Cnv 12h49'1"38d48'
Stemhagen,Kristin & Kurt
Lyn 6h26'54"54d54'
Stemler,Ronald
Cnv 12h32'31"32d44'
Stemmie & Bob
Cam 4h32'30"68d23'
Stemmler,Hannchen und Erich
Dra 15h4'1"62d53'
Stempniak,Matthew Joseph
Sct 18h43'12"-9d47'
Stempuzis,Matthew R
Aur 6h32'36"37d54'
Sten Hopper Nr Vedby Denmark
Ori 5h26'0'0d14'
Stenberg,Edwin E
Tau 5h52'57"23d45'
Stenehjem,Dalainy Leary
Aql 19h44'0"11d59'
Stenella Plagiodon
Del 20h18'31"14d24'
Stenfors,Sr,Robert James
Cep 20h25'37"60d43'
Stengel,Albert
Aql 19h53'0"1d40'
Stengell,James
Uma 11h4'19"40d7'
Stenger,Aileen
Cyg 19h28'19'30d41'
Stenger,Martine
Per 3h29'56"37d3'
Stenhouse,Janette
Lyn 8h23'15"44d47'
Stenius,Gertrud
Cas 0h20'14"61d0'
Stenseth,Richard "Bubby"
Ori 6h7'15"10d41'
Stenstrom,Darla J
Lyn 8h9'50"39d56'
Stentz,Carrie Lee
Cas 0h2'51"64d39'
Stentzel,Emmanuelle
Aql 19h33'26"-6d37'
Stenwick,Eric
Her 17h52'16"28d37'
Stenwick,Justin
Aur 5h2'47"40d20'
Stenzel,Eva Anastasia
And 2h17'39"46d26'
Stenzel,Michael
Cap 20h6'35"-11d1'
Stenzel,Susanne
Ari 2h20'38"20d7'
Stenzel,Ursula
Tau 3h41'50"7d58'
Stenzhorn,Greg
Cep 22h24'22"68d43'
Steorra Bai
Dra 17h37'60"67d31'
Stepahin,Sharon L
Lyr 18h20'47"38d40'
Stepakoff,Jeffrey Howard
Aur 5h5'44"38d16'
Stepancic,Stefan
Peg 23h48'0"16d36'
Stepanian,Gretta
Lyr 19h22'42"38d54'
Stepanie
Lyr 19h15'12"40d17'
Stepe
Aql 18h58'58"7d55'
Stepe,Juris
Dra 18h57'51"58d43'
Stepek,Christopher James
Uma 11h22'33"57d56'
Stepek,Jason Michael
Uma 11h19'36"56d59'
Steph (J M 25167LC)
Uma 11h11'13"43d36'
Steph L
Boo 14h31'26"50d11'
Steph N' Ax
Ser 16h3'44"10d21'
Steph's Star
Mon 8h8'60"-8d26'
Stephahy's Wish
Peg 22h40'36"25d33'
Stephan
Dra 18h28'20"70d1'
Stephan
Lib 14h44'48"-22d58'
Stephan & Daina
Sco 15h45'30"-28d30'
Stephan's Goal
Her 16h37'34"25d20'
Stephan,Heather Rae
Gem 8h5'29"28d38'
Stephan,Klaus
Ser 15h54'12"17d41'
Stephan,Paul Mercier
Aur 6h4'0"38d44'
Stephan,Sophie J
And 0h52'41"34d9'
Stephane
Uma 11h26'59"32d35'
Stephane Et Sandrine
Cam 7h28'30"67d35'
Stephane,Alexandra
Aql 19h3'37"16d7'
Stephanek,Indra
Sco 17h24'1"-30d34'
Stephanelle
Lyn 6h36'34"59d36'
Stephani Lynette
Aur 5h35'46"50d17'
Stephania
Mon 7h27'41"-10d41'
Stephania Behania
Cyg 20h2'38"38d47'
Stephanie
Cas 1h3'31"53d57'
Stephanie
Lyr 18h21'51"46d54'
Stephanie
Oph 17h57'12"12d1'
Stephanie
Cet 2h0'16"-1d39'
Stephanie
Com 12h12'1"21d30'
Stephanie
Cap 21h16'1"-26d50'
Stephanie
Lib 15h7'53"-23d55'
Stephanie
Lyr 19h6'51"38d46'
Stephanie
Cyg 20h36'38"41d0'
Stephanie
Peg 23h28'56"30d41'
Stephanie
Oph 17h11'53"-23d25'
Stephanie
Cas 0h5'1"64d1'
Stephanie
Cas 2h53'53"68d48'
Stephanie
Cyg 20h1'0"31d51'
Stephanie
Peg 22h42'16"34d58'
Stephanie
Aql 19h54'0"11d5'
Stephanie
And 1h18'1"39d24'
Stephanie
And 2h19'22"49d8'
Stephanie & Craig
Cas 1h58'1"70d3'
Stephanie & Dean
Eri 3h56'40"-4d6'
Stephanie & Hussein 143 ACCP USA
Cyg 19h25'37"48d48'
Stephanie & Mike
Lyr 18h17'32"38d51'
Stephanie & Richard
Cam 6h5'52"72d49'
Stephanie & Tony 2 Hearts-1 Dream
Aql 18h45'38"11d14'
Stephanie Ann
Ori 5h55'47"19d11'
Stephanie Anne
And 0h51'26"38d11'
Stephanie Beyond My Dreams
Cas 3h9'54"61d47'
Stephanie Elizabeth
Cyg 21h56'1"52d42'
Stephanie Et Fabienne
Ser 16h7'44"6d20'
Stephanie Faye & Randolph Earl
Cyg 19h47'17"29d55'
Stephanie Holly
Lyn 6h57'35"56d14'
Stephanie Jean
Lyr 19h14'43"40d24'
Stephanie L
Dra 19h27'49"58d4'
Stephanie L
Lyn 7h54'0"36d54'
Stephanie Louise
Lyn 9h9'52"42d19'
Stephanie Marie
Aqr 23h4'33"-18d15'
Stephanie Marie
And 0h20'30"34d44'
Stephanie Marie
Aur 7h20'19"43d52'
Stephanie Marie
And 1h50'51"41d8'
Stephanie Morgan
Lyr 18h47'41"38d5'
Stephanie Nichole
Peg 23h3'15"32d3'
Stephanie Noelle
Lyn 6h35'53"54d13'
Stephanie Rhea
Ori 5h53'10"13d7'
Stephanie Rose
And 1h9'36"39d40'
Stephanie Ruth
And 23h37'45"46d3'
Stephanie Suzanne
Cet 2h41'24"-11d35'
Stephanie The Love of My Life
Cet 0h50'1"-5d48'
Stephanie VV
Cas 0h29'54"50d11'
Stephanie's "Brite Lite"
And 2h12'30"40d47'
Stephanie's Dawn
Aql 19h31'41"10d8'
Stephanie's New Beginning
Aql 19h32'44"-6d49'
Stephanie's Night Blossom
Vul 21h0'46"20d27'
Stephanie's Rose
Uma 10h22'17"50d37'
Stephanie's Star
Lyr 19h15'36"41d4'
Stephanie's Star
Cmi 7h29'15"8d3'
Stephanie's Star
Aql 20h0'1"6d38'
Stephanie's Star
Peg 21h59'38"23d42'
Stephanie:Princess Du Soleil
Peg 22h41'47"20d47'
Stephanos
Hya 9h4'35"5d52'
Stephans Mr & Mrs
Cep 0h12'0"66d36'
Stephanus
Lib 15h8'15"-21d5'
Stephany,Lynley
And 2h20'32"37d34'
Stephaotis,Steven Michael Gregor
Boo 15h6'20"40d3'
Stephee 21
Her 17h56'16"40d36'
Stephelen
Crb 15h34'52"31d57'
Stephen
Peg 0h4'47"26d10'A
Stephen
Aur 6h35'36"31d27'
Stephen
Dra 12h4'56"63d22'
Stephen
Boo 14h27'24"10d50'
Stephen
Cep 22h20'12"63d36'
Stephen
Peg 21h59'39"30d1'
Stephen
Cas 23h21'50"57d15'A
Stephen & Elizabeth
Uma 10h1'0"50d1'
Stephen & Mieko
Her 17h36'10"27d35'
Stephen & Vikki's Stairway To Heaven
Umi 15h3'53"66d51'
Stephen Andrew
Cep 23h5'54"63d57'A
Stephen Arthur
Tri 2h10'50"33d46'
Stephen Cellestial 50
Cap 20h35'16"-12d27'
Stephen Christopher
Aqr 21h44'20"0d8'
Stephen E
Lac 22h21'15"55d22'
Stephen Eugene
Ser 15h14'43"-1d32'
Stephen Forever
Cep 23h12'42"64d14'
Stephen Jacques
Ori 6h7'60"5d44'
Stephen James
Her 17h19'14"40d50'
Stephen Kyle
Her 16h34'26"21d32'
Stephen L R
Aur 5h39'48"33d50'A
Stephen Malin,The
Dra 11h27'28"72d23'
Stephen Roy
Ori 5h18'0"15d59'
Stephen Scott
Ser 16h8'55"2d49'
Stephen Timothy
Cep 23h15'17"65d1'
Stephen's "Midnight" Star
Leo 10h57'56"8d10'
Stephen's Anniversary Star
Her 17h51'59"38d19'
Stephen's Dark Star
Per 3h46'50"38d18'
Stephen's Destiny
Per 3h50'1"38d48'
Stephen's Family Star, The
Aur 5h1'15"47d37'
Stephen's Flame
Ori 6h4'1"8d5'
Stephen's Light
Her 17h35'50"26d29'
Stephen's Princess
Uma 11h35'37"40d43'
Stephen's Sino 1
Aur 6h29'33"38d57'
Stephen's Star
Aqr 22h34'23"-0d3'
Stephen's Star
Ori 5h0'55"12d58'
Stephen's Sun
Cep 20h53'1"70d52'
Stephen,Douglas
Aur 4h54'11"40d47'
Stephen,Eileen
Eri 3h16'1"-13d28'
Stephen,Guppy
Equ 20h59'44"6d14'
Stephen,Jessie Lee
Mon 7h45'10"-7d49'
Stephen,Lynn & Drew's "Doorway"
Lyr 18h30'45"37d9'
Stephen,Robert Joseph
Aur 6h28'56"37d5'
Stephens Star,The John
Cep 22h3'29"60d36'
Stephens,Adelaida Kerr
And 23h41'54"42d26'
Stephens,Barbara
Cas 0h47'0"64d21'
Stephens,Basil James
Aur 6h13'13"36d32'
Stephens,Betty J
Mon 6h0'0"-10d33'
Stephens,Big Bri
Cep 22h20'1"55d36'
Stephens,Chad
Ari 1h55'0"18d3'
Stephens,David Robert
Cnv 13h23'13"40d26'
Stephens,Diane J
Cet 2h38'53"-8d7'
Stephens,Gregora Russell
Umi 14h42'39"78d19'
Stephens,Hunter Warren
Her 17h32'47"26d50'
Stephens,J B
Cma 6h53'1"-19d49'
Stephens,James A
Leo 11h0'38"-2d0'
Stephens,James Carl
Per 1h30'49"53d34'
Stephens,James M
Sct 18h44'15"-6d27'
Stephens,Jane Monroe
Crb 15h30'39"32d10'
Stephens,Janice Hart
Cyg 21h50'30"40d13'
Stephens,Jeri Lynn
Com 12h17'27"32d3'
Stephens,Joanna
Lyr 18h57'17"40d22'
Stephens,John M
Aur 6h18'50"48d50'
Stephens,Julie Caroline
Peg 22h58'49"12d7'
Stephens,Ken
Peg 21h9'34"12d48'
Stephens,Kyle Christopher
Cap 20h58'17"-20d32'
Stephens,Mikaylla Noel
Lyr 18h38'16"36d17'
Stephens,Roger DeWitt
Cyg 21h27'60"40d50'
Stephens,Roy & Rachel
Hya 9h16'29"5d53'
Stephens,Russell A
Cep 21h45'43"55d58'
Stephens,S A S Sheila Ann
Uma 9h10'56"53d13'
Stephens,Sandi J
Lyr 19h16'48"28d18'
Stephens,Shane A & Laura A
Lac 22h29'17"54d12'
Stephens,Stacey Anne
Cyg 21h54'18"42d39'
Stephens,Stacia
And 1h13'41"48d53'
Stephens,Stevie
Ori 5h58'49"16d13'
Stephens,Susan
Crb 16h20'50"31d9'
Stephens,Susan
Tri 1h41'29"33d52'
Stephens,Thomas & Regina Mallory
Cyg 20h17'55"31d3'
Stephens,Thomas M
Peg 23h48'37"8d18'
Stephens,Tracey Lynn
Uma 9h31'34"53d22'
Stephens,Valerie Hunter
Mon 6h32'20"0d45'
Stephens,Victoria LeBlanc
Crb 16h4'23"26d34'
Stephens-Ofner,Marcus
Aql 20h8'41"1d11'
Stephens-Wandell,Truda May
Del 20h14'40"9d46'
Stephenson,Amy
And 1h9'0"39d58'
Stephenson,John Michael
Vir 13h56'29"-0d12'
Stephenson,Joshua Charles
Ori 5h52'58"20d46'
Stephenson,Jr,A Emmet
Aql 19h53'12"15d27'
Stephenson,Katelin May
Mon 6h52'14"11d2'
Stephenson,Linda
And 23h21'49"40d33'
Stephenson,Lynn K
Peg 23h43'10"30d25'
Stephenson,Margot
Cas 0h58'25"75d59'
Stephenson,Nicole Kepler
Peg 23h19'39"10d16'
Stephenson,Philip Charles
Mon 8h6'4"-6d21'
Stephenson,Revis Lindsay
Peg 22h40'25"24d11'
Stephenson,Roger Alan
Sgr 19h0'23"-29d41'
Stephenson,Sparkling
Her 17h17'0"41d42'
Stephenson,Stephen
Ari 2h59'41"21d51'
Stephenson,Thomas Williams
Per 2h53'18"38d13'
Stephey,Laura Lee
Cyg 19h53'33"38d59'
Stephney,T L
And 1h36'28"39d42'
Steplitus,Ann Marie
Ori 6h14'51"0d46'
Stepoulos
Boo 15h2'35"42d28'
Stepp (1850-1955),John Myra
Uma 11h55'11"32d22'
Stepp,Hazel E
Her 18h55'46"18d60'
Stepp,Sommer Anne
Peg 22h33'1"25d11'
Steppan,Victor
Crb 15h28'12"31d53'
Stepparau,Maria Grazia
And 0h31'16"40d49'
Steppe,Dennis L
Hya 9h1'15"5d0'
Steppe,Jennifer
Lyr 19h17'55"40d47'
Steppe,Jordan Taylor
Mon 7h42'6"-8d58'
Stepps,Donald Joseph
Vul 19h4'48"24d27'
Stera of Amaranth
Sco 16h4'2"-13d47'

Stera Of Beer Tau 5h12'53"17d14'
Stera of Kasumisou Sco 16h0'30"-17d2'
Stera of Laksmi Psc 1h14'39"20d39'
Steransak,Marietta Lyn 9h27'29"40d48'
Sterba My Great Dad, Carl Aql 19h5'54"5d19'
Sterba,Ludvik-Marlene Hya 8h50'42"-6d1'
Sterbens,T J Leo 10h57'47"-1d43'
Stergios,Robert G Ser 17h31'28"-10d8'
Stergiou,Nikos Ser 15h21'51"6d10'
Sterlacchini,Rita Umi 16h22'51"72d18'
Sterlin Cet 3h15'21"3d17'
Sterling Aql 20h7'27"-5d39'
Sterling Lib 15h3'25"-10d6'
Sterling & Barbara Star Eri 3h34'55"-1d58'
Sterling Family,The Jeff Aql 20h11'51"10d23'
Sterling,Ashley J And 23h23'46"51d24'
Sterling,David Matthew Hya 9h32'53"5d48'
Sterling,Elizabeth Brooke And 23h18'0"51d20'
Sterling,Jessica Lauren And 2h25'16"41d30'
Sterling,Morry Gem 6h53'55"30d48'
Sterling,Roaslie Vul 19h45'19"28d59'
Sterling,Sheri Diane Del 20h13'43"13d56'
Sterling,Tao Del 20h34'1"20d5'
Sterling,William Grant Per 1h39'24"53d24'
Stern unserer Liebe und Hoffnung Psc 1h14'56"18d12'
Stern von Monika & Ingolf Lac 22h25'35"54d39'
Stern,Align...Intuit- Dr Larry Her 18h12'27"30d13'
Stern,Allan Herbert Tau 4h43'0"20d4'
Stern,Carly Erin And 1h31'31"36d55'
Stern,Carol H Mon 6h39'1"1d20'
Stern,Constance J Lyn 6h15'39"60d13'
Stern,Daniel R Dra 17h45'40"60d57'
Stern,Ellery Alan Psc 22h53'42"5d45'
Stern,Freündlich Tri 1h59'32"25d38'
Stern,Gabriele Holzer Cap 21h21'50"-20d34'
Stern,Harry & Deborah Boo 14h59'36"32d48'
Stern,Heimat Peg 22h19'51"28d56'
Stern,Helmut Aqr 23h29'57"-12d29'
Stern,Henry Aql 20h13'21"4d14'
Stern,Jack T Gem 6h44'6"16d1'
Stern,Jacquie Cas 23h3'53"53d46'
Stern,Jay "Bubby" Her 17h46'18"40d55'
Stern,Johnnie Dra 16h51'55"67d14'
Stern,Jonathan Per 2h41'28"34d26'
Stern,Jonathan Dwight Aur 6h32'36"32d49'
Stern,Katy Cam 5h48'40"65d31'
Stern,Leo Lyr 18h56'21"40d12'
Stern,Marilyn Cas 0h59'1"67d4'
Stern,Marla Garil Cyg 20h31'49"31d6'
Stern,Maurice Peg 23h41'11"15d46'
Stern,Norbert Lac 22h19'0"48d32'
Stern,Rachael S And 0h24'41"30d25'
Stern,Unserer Ew'gese Cam 5h34'51"60d6'
Stern,Wolfgang Horsthemke Vir 11h35'38"-1d35'
Sternberg,Fred Uma 11h19'1"56d5'
Sternberg,Marc Benson Cet 3h15'44"4d51'
Sternberg,Mervyn Howard Aur 6h5'10"46d21'
Sternchen Vir 13h24'33"-9d52'
Sternchen Ingeborg Her 17h40'30"43d0'
Sternchen of Bruce Cep 22h35'50"71d11'
Sternchen,Iris Meine Peg 23h5'59"17d56'
Sterne-Mond-Sonne- Party Rosemary-Bogman Psc 1h19'58"18d41'
Sternen-Pünktchen- Doris Lib 14h21'27"-23d20'
Sternenkind Tau 4h18'23"1d24'
Sterner,Ginny Lee Uma 12h55'15"53d21'
Sternkucker Peter Joachim Velte Gem 8h4'27"30d31'
Sternlicht,David N Lmi 9h22'39"33d21'
Sternschnuppe Tabaluga Uma 10h35'54"41d9'
Steronko,Robert Jon Cnv 13h59'56"28d16'
Sterrett,Reverend William A Boo 14h28'40"40d58'
Stetekluh,Paul Robert Her 16h28'0"50d27'
Stetler,Scott Benjamin Aur 6h22'58"37d54'
Stetson,Ann Mon 6h31'37"8d26'
Stetson,Gena Cyg 21h48'60"38d9'
Stetson,Julia A Crb 15h58'57"37d51'
Stetson,Margaret Alice Umi 16h15'40"78d32'
Stetson,Roger T Dra 16h34'0"69d31'
Stetta,Dotty Kupfer Crt 11h41'0"-21d23'
Stettner,Michelle Aql 19h23'1"-1d49'
Steuart,Elizabeth J & Snowden Cet 0h7'40"-7d55'
Steve Tau 5h55'53"24d14'
Steve Ori 6h4'36"20d44'
Steve Aur 6h37'54"37d0'
Steve Aql 19h9'1"13d43'
Steve & Amanda-1 Year Aql 20h8'1"0d3'
Steve & Andrea Cyg 21h14'44"28d60'
Steve & Betty Uma 12h27'32"61d38'
Steve & Cathy Cas 3h7'13"63d53'
Steve & Denise Crb 16h7'14"31d30'
Steve & Desiree's Dream Star Lyr 19h21'1"40d1'
Steve & Gretchen Cyg 19h31'31"33d26'
Steve & Irene Cet 0h29'25"-5d40'
Steve & Jo's Sacred Day Cyg 20h23'34"38d27'
Steve & Julie,The Crb 16h15'32"37d10'
Steve & Michele Cyg 21h33'54"40d2'
Steve & Morgan: Together Forever Cyg 20h41'56"42d53'
Steve & Sandra Cyg 19h28'28"30d12'
Steve & Tracee's Love Star Uma 10h26'18"56d37'
Steve & Tracy's Twilight Ori 6h2'48"5d25'
Steve & Vicki Forever Per 2h53'0"40d23'
Steve B (The Big One) Uma 8h57'30"56d47'
Steve Loves Angel Cet 1h25'53"-12d28'
Steve `N' Steven Gem 6h52'52"14d1'
Steve's 101 Boo 14h54'0"52d18'
Steve's Argosy Cnv 12h34'26"32d11'
Steve's Dream Sex 9h56'47"4d3'
Steve's Eternal Shepherd Equ 21h8'40"11d45'
Steve's Lucky Star Uma 10h41'16"40d41'
Steve's Rappture Aqr 21h26'1"-0d10'
Steve's Robert Per 3h26'50"48d48'
Steve's Song Her 16h30'21"41d18'
Steve's Star Cep 20h16'27"78d52'
Steve's Star Uma 9h51'29"54d26'
Steve,With Love Kim & Zach Lac 22h17'1"47d17'
Steve-Lee 111294 Peg 23h48'57"15d9'
Steve-My Friend,My Angel In Disguise Dra 17h7'44"64d6'
Steven Ori 5h53'24"10d21'
Steven Oph 17h53'57"0d38'
Steven & Kathy-A New Star In The Sky Peg 21h48'23"34d49'
Steven & Malinda 4-5-92 Crb 15h17'16"31d42'
Steven & Terry Cyg 19h21'0"28d59'
Steven & Tim Leo 11h0'1"22d46'
Steven & Tracy Lmi 10h39'15"32d31'
Steven Ba Doo Dra 18h39'21"58d17'
Steven Carl Lac 22h5'43"47d38'
Steven Chibuzor Chukumba Tau 3h50'49"1d1'
Steven Daniel Ori 6h3'1"-1d14'A
Steven Deane Cep 22h6'44"60d12'
Steven Francis Aql 19h35'43"-6d47'
Steven John Her 16h17'25"25d34'
Steven Joseph Ser 15h16'18"10d36'A
Steven My Love Always Star Oph 17h8'2"-22d19'
Steven Tee Aur 5h51'1"50d8'
Steven Wayne Aur 5h29'39"30d35'
Steven's Dream Per 2h50'44"38d34'
Steven's Grand Star Dra 17h35'0"76d9'
Steven's Heavenly Music Maker Cet 0h51'47"-1d23'
Steven's Light Her 16h59'24"24d29'
Steven's Star Cma 6h59'27"-18d59'
Steven's Time to Shine Aur 6h28'0"33d51'
Steven,Gary Cep 0h6'1"70d2'
Steven,Heather Lorraine Lyn 8h36'22"38d6'
Steven,Maureen,Jason, Jaime,Ryan,Matthew Cyg 19h59'27"31d47'
Steven,Tubman Dra 16h35'60"60d27'
Steven-A Star For A Star Cap 20h26'1"-20d28'
Steven-Bruce Cyg 20h16'53"41d25'
Stevenastra Zoppigesternte Ori 6h2'26"20d13'
Stevenish II,aka Bondfire,Robert John Cam 4h48'47"70d36'
StevenJon Her 16h50'49"40d34'
Stevens 30,Amanda Aql 20h5'0"-6d1'
Stevens Light Uma 11h20'39"62d25'
Stevens Star Uma 9h14'19"54d29'
Stevens' Anniversary Star Crb 16h10'11"30d40'
Stevens,"FIAT" Thomas Per 1h58'52"47d49'
Stevens,Aaron Aql 20h2'11"6d15'
Stevens,Allison Cas 0h7'16"54d33'
Stevens,Andy Cep 0h7'26"69d49'
Stevens,Angela Cas 2h4'42"59d45'
Stevens,Betty Jo Lyr 18h41'50"27d55'
Stevens,Bill Aql 20h2'57"4d33'
Stevens,Billie Amber Umi 14h33'30"68d4'
Stevens,Bruce N Cnv 13h21'35"51d46'
Stevens,Carol Cyg 20h12'1"43d13'A
Stevens,Colleen Peg 23h3'10"10d48'
Stevens,Craig & Barbara Mon 6h23'52"6d18'
Stevens,Dana D Sex 9h53'40"-1d34'
Stevens,Daniel Lee Lac 22h36'0"38d5'
Stevens,Danielle L Cam 11h8'0"80d1'
Stevens,Douglas George Per 3h6'51"46d21'
Stevens,Edward Dra 15h54'29"51d22'
Stevens,Elizabeth (Bette) Umi 16h56'42"76d21'
Stevens,Frank Albert Per 4h43'56"38d48'
Stevens,Gary Sct 18h52'35"-5d26'
Stevens,George Aql 20h35'40"-1d9'
Stevens,Greg Cep 1h50'33"78d60'
Stevens,Homer Darling Ser 15h52'11"20d20'
Stevens,Howard Her 18h6'18"31d11'
Stevens,Hélène M Uma 9h23'0"49d54'
Stevens,Irene Cyg 21h0'56"35d41'
Stevens,Ivy Saxe Cyg 20h25'53"42d6'
Stevens,Janet Peg 23h20'12"18d11'
Stevens,Jeffery Lane Per 2h55'27"50d20'
Stevens,Jennifer Jean And 1h44'39"39d38'
Stevens,Jerry Wayne Aql 19h0'34"13d23'
Stevens,Jim Hya 8h47'25"1d31'
Stevens,Josephus Cma 7h3'0"-20d8'
Stevens,Jr,Milton Lewis Sco 16h20'24"-38d36'
Stevens,Julie Mon 6h55'0"-6d53'
Stevens,Julie And 1h13'23"37d33'
Stevens,Leah Marie Uma 12h11'0"60d23'
Stevens,Lee Her 17h2'15"43d28'
Stevens,Lee J Aur 5h16'31"46d17'
Stevens,Mark Peg 0h1'16"20d27'
Stevens,Mark Elliott Lac 22h24'1"55d23'
Stevens,Martha Peg 22h0'1"5d37'
Stevens,Mary L Cyg 21h20'32"31d3'
Stevens,Mary McCarty And 0h11'32"34d34'
Stevens,Melissa And 2h22'47"42d43'
Stevens,Merle Cas 0h44'13"63d48'
Stevens,Michael S Dra 20h31'15"73d46'
Stevens,Michelle Jayne Eri 5h1'1"-4d28'
Stevens,Mikell Vul 20h3'29"25d29'
Stevens,Mildred N Sex 10h6'24"5d30'
Stevens,Nicholas Aql 20h6'56"8d18'
Stevens,Nicholas James Lyn 7h19'27"50d6'
Stevens,Nicole Ashleigh And 0h18'1"38d29'
Stevens,Pauline & Trevor Aur 5h18'14"46d10'
Stevens,Peter Oph 18h1'0"1d34'
Stevens,Phoebe Eliza Uma 11h20'55"37d39'
Stevens,Robert Michael Aur 6h0'20"35d9'
Stevens,Roberta And 1h25'59"36d34'
Stevens,Ron & Marty Sge 20h0'20"16d27'
Stevens,Ryan Anthony Hale Her 16h5'20"21d11'
Stevens,Sean Francis Mitchell Her 18h6'46"40d44'
Stevens,SF-1 Meredyth M And 23h42'46"37d44'
Stevens,Teri Dra 17h47'15"64d7'
Stevens,Terry L Aql 18h53'57"8d51'
Stevens,Tracey Margaret Cas 0h47'0"71d51'
Stevens,Tyler Joseph Cam 12h20'37"78d47'
Stevens,Vivian And 2h22'25"42d42'
Stevens,Walker Lee Aur 7h24'38"40d45'
Stevens,Wally Eri 2h59'24"-18d40'
Stevens,Warren L Cam 3h51'46"79d18'
Stevens-Sauers,Lynette Lyn 8h20'43"50d39'
Stevenson 1,George Lac 22h51'35"56d11'
Stevenson Robert Louis Sex 10h41'1"-0d7'
Stevenson,Alfred J R Ori 5h22'55"0d0'
Stevenson,Amy Aql 19h31'0"8d22'
Stevenson,Anita Ant 10h46'30"-37d47'
Stevenson,Ava Cnv 12h23'57"51d38'
Stevenson,Benjamin Haynes Per 2h55'24"40d35'
Stevenson,Bobby Cap 20h36'40"-18d40'B
Stevenson,Darla Cyg 19h42'54"30d56'
Stevenson,Deborah Cam 10h11'59"82d5'
Stevenson,Elaine Cas 2h40'42"68d41'
Stevenson,Elizabeth Lyr 18h47'13"32d15'
Stevenson,Jill Peg 22h20'37"28d36'
Stevenson,Jocelyn Lyn 8h28'38"57d49'
Stevenson,Karen A Cyg 20h34'19"41d0'
Stevenson,Lindsey Mon 7h1'56"-6d2'
Stevenson,Michelle Cyg 19h24'51"31d23'
Stevenson,Patricia Lynn Creal Lyn 7h15'0"58d21'
Stevenson,Patrick Per 3h0'56"38d29'
Stevenson,Rachel Crb 16h8'11"28d11'
Stevenson,Robert Warren Aql 20h4'47"1d51'
Stevenson,Sharon Crb 15h56'47"38d22'
Stevenson,Stella Umi 16h0'13"70d17'
Stevenson,Steve & Adelaide Crb 15h29'25"31d53'
Stevenson,Stewart Cep 20h50'0"68d21'
Stevenson-Michener, Debi Mon 5h55'55"-8d10'
Steveo Dra 16h54'39"71d31'
Steves,Chris Uma 10h15'14"56d19'
Steves,Ryan Philip Cep 21h50'44"58d55'
Steves,Toby Umi 14h14'0"70d3'
Stevie Lyn 7h54'23"42d4'
Stevie Aur 5h0'0"50d27'
Stevie B Ori 4h55'16"5d32'
Stevie Doris' Guiding Light Cet 3h10'32"2d33'
Stevie Reh Lac 22h14'24"54d29'
Stevie Star Ori 6h7'0"4d21'
Stevie's Star Dra 9h45'0"80d8'
Steward II,The Spirit of Charles Cyg 20h23'1"40d19'
Steward,Michael Vul 19h23'1"26d39'
Steward,Shaun Mon 8h1'59"-6d45'
Steward,William Turner Cet 2h8'45"1d59'
Stewart (Dolly),Alice And 0h35'48"30d10'
Stewart 12/29/88,Diane & Jeffery Ori 5h26'0"-0d21'
Stewart III,James Lawrence Aur 7h1'1"36d40'
Stewart The Shark Aql 20h10'28"0d28'
Stewart's 50th,Howard & Evelyn Cyg 20h38'0"31d7'
Stewart's Gluecksstern Umi 15h1'45"68d9'
Stewart's,40th Birthday Star,John Cnv 13h11'41"37d39'
Stewart,"Woogie Wadgie" John F Uma 11h44'54"54d56'
Stewart,Abbie Louise Cru 12h24'51'-62d11'
Stewart,Alison Cas 1h17'19"63d33'
Stewart,Alison Jane Cas 2h29'38"60d34'
Stewart,Allison McKinley Mon 7h12'0"-10d53'
Stewart,Amanda Morgan Crb 15h27'23"30d26'
Stewart,Amy Francis Cas 2h13'14"59d40'
Stewart,Ann Armour Ori 4h42'55"0d37'
Stewart,Anna K E And 23h21'36"35d1'
Stewart,Beinetta Lyr 18h49'50"35d8'
Stewart,Bernard Per 1h59'1"50d8'
Stewart,Brian Lac 22h19'25"48d53'
Stewart,Carol & Richard Cyg 19h47'18"30d9'
Stewart,Catherine-Anne Cas 1h8'1"76d8'
Stewart,Charles Hurd Cnv 13h52'40"30d30'
Stewart,Christopher Evert Her 18h45'1"12d57'
Stewart,Corrine Lyr 18h31'56"44d36'
Stewart,D K Tau 4h27'1"28d32'
Stewart,David Christopher Cma 7h15'35"-16d46'
Stewart,David Murdo Oph 18h5'13"12d0'
Stewart,Dillon Patrick Peg 23h19'1"25d39'
Stewart,Donald Robert Aur 6h24'12"38d41'
Stewart,Edna Aldridge Vul 19h46'10"25d16'
Stewart,Eleanor H Cas 2h17'0"65d0'
Stewart,Elsa Gould And 23h3'57"42d16'
Stewart,Emily Dickson And 2h20'44"45d43'
Stewart,Enid Llort Psc 1h0'59"10d39'
Stewart,Ernest L Lmi 10h36'22"23d38'
Stewart,Francis Douglas Ori 5h4'46"1d26'
Stewart,Frank Per 3h43'52"50d40
Stewart,Gail D Oph 17h54'30"1d12'
Stewart,Gary E Cmi 7h17'5"1d25'
Stewart,Gloria Cas 0h1'59"58d51'
Stewart,Heath Wayne Her 16h18'58"11d38'
Stewart,Ian Ori 5h58'0"19d54'
Stewart,Izabela M Del 21h5'48"12d34'
Stewart,J E Wayne Aur 5h24'39"37d47'
Stewart,Jacquelyn Cma 7h1'19"-1623'
Stewart,James Dean Aql 20h0'55"10d25'
Stewart,Jennifer Lyn 8h56'45"45d10'
Stewart,Jennifer Cyg 20h23'29"38d20'
Stewart,John J Aql 20h5'14"6d40'
Stewart,Joseph Lee Hya 8h59'22"4d1'
Stewart,Joseph Scot Cep 23h1'57"71d12'
Stewart,Joyce & Gus Umi 14h56'1"69d8'
Stewart,Jr,Edwin Uma 8h39'1"60d6'
Stewart,Karen And 0h41'1"45d15'
Stewart,Karrie Elizabeth Ori 5h54'39"14d44'
Stewart,Katherine Alexandra Cas 23h16'12"60d32'
Stewart,Keith Maxwell Ori 5h50'1"10d10'
Stewart,Kelly Eugene Aur 5h23'0"38d26'
Stewart,Keri Umi 14h22'37"67d23'
Stewart,Kevin Per 2h52'30"43d55'
Stewart,Kimberly And 1h3'59"38d27'
Stewart,Larry Eldon Per 4h9'29"37d51'
Stewart,Laura R Mon 6h58'57"-10d20'
Stewart,Lauren Crb 15h17'59"30d41'
Stewart,Linda Ori 6h9'36"1d33'
Stewart,Linda Lou And 0h20'12"32d30'
Stewart,Lisa And 1h41'57"41d14'
Stewart,Lisa Dawn Cas 0h7'0"60d6'
Stewart,Lucille & Jerald Cyg 19h30'1"31d18'
Stewart,Lydia Athalie Mon 7h8'8"-5d28'
Stewart,Margaret And 2h20'54"45d24'
Stewart,Mark Cnv 13h53'0"37d3'
Stewart,Mark L Boo 14h34'51"30d22'
Stewart,Matthew W & Debra H Cnc 8h9'21"31d11'
Stewart,Mildred,H Peg 23h0'11"30d5'
Stewart,Mrs Margaret Lyr 18h29'54"30d56'
Stewart,Nolita Peg 21h57'32"29d58'
Stewart,Norma ULM Schalk Cas 23h30'48"61d28'
Stewart,Patrick Cnv 12h23'57"44d38'
Stewart,Paul Ruth P D Doug Cet 2h12'54"4d38'
Stewart,Paul W Cet 2h16'47"0d1'
Stewart,Paula Peg 21h42'59"27d24'
Stewart,Prudence Kathryn And 2h23'37"44d60'
Stewart,Ralph "Stew" Psc 23h32'18"2d21'
Stewart,Rita Gray Mon 6h19'21"8d47'
Stewart,Robert I Aur 7h16'31"39d56'
Stewart,Robert Paul Oph 17h6'21"-23d19'
Stewart,Ronald Steve-Alan Tri 1h41'1"33d53'
Stewart,Ronald R Oph 18h3'31"10d45'
Stewart,Ruth And 0h18'34"35d7'
Stewart,Susan S Aur 6h7'30"38d46'
Stewart,Ted Renfro Cmi 7h30'59"6d48'
Stewart,Thomas J Per 2h55'0"38d58'
Stewart,Valerie Cmi 8h8'32"3d48'
Stewart-"Byube",Robert F Cet 3h16'54"8d37'
Stewart-1st Year, Jennifer & Scott Crb 16h14'51"37d45'
Stewart-Beevor,Fiona Lac 22h43'34"38d38'
Stewball Her 16h32'26"38d21'
Stewie Sco 16h50'39"-44d31'
Stewli's Star Cam 3h36'35"60d0'
Stewwed Pruner Cet 1h45'19"-1d37'
Stewy Mon 6h26'46"2d52'
Stibel,Gary & Elaine Sge 20h1'17"18d49'
Stibrik III,John Paul Her 16h50'21"30d4'
Stich,Gerlinde Sco 17h4'16"-31d24'
Stichauner,Karen & Scott DeCarolis And 0h20'59"38d42'
Stick Her 17h41'1"48d59'

Stickel,Helen & Ralph Cyg 20h52'0"40d9'
Stickland,Fredrick John Cma 7h12'29"-15d33'
Stickland,Liesl Cyg 19h31'40"35d26'
Stickle III,John R Dra 15h3'45"61d11'
Stickle,Elizabeth Dra 17h5'0"58d39'
Stickle,Jim Aur 4h58'54"41d3'
Stickler,Edwin & Gladys Ori 6h4'46"9d43'
Stickler,Lawrence W Aur 6h8'27"30d6'
Stickles,Howard Mon 8h7'22"-2d31'
Stickney,Genevieve Anna Cyg 19h49'59"37d46'
Stickney,Jephreaux Ray Lac 22h21'50"38d19'
Stickney,Joseph B Ori 5h38'40"-0d32'
Stickney,Nelda Ray Lac 22h31'57"38d1'
Stickney,Roland & Phyllis Uma 9h39'16"49d4'
Stiddard,Sarah Jo And 23h25'0"40d56'
Stidham,Gwendolyn Mae Com 12h22'32"20d5'
Stidhem,Alton Marice Boo 15h46'11"41d24'
Stidolph,Donna C Mon 6h21'10"3d50'
Stidum,Ryan Taylor Hya 9h13'0"6d11'
Stiebel,Joy Danielle Mon 6h42'39"7d43'
Stief,Thomas Frederick Vir 11h37'55"9d49'
Stiefel,Henny & Walter Cyg 21h13'1"39d47'
Stiefel,Leopold And 1h48'29"40d49'
Stiefel,Marco-Carl Sco 17h29'0"-31d36'
Stiefvater,Erich Matthew Ser 17h30'56"-14d35'
Stiefvater,Keith Paul Per 2h24'27"54d26'
Stiefvater,Marie Louise Com 13h4'57"15d13'
Stiegeler,Albert Cap 20h42'20"-23d45'
Stiegelmaier,Kevin Aur 6h28'1"37d39'
Stiegemeyer,Jodi Cas 0h21'59"67d59'
Stieghorst,Emilie & Georg Cas 3h0'53"58d13'
Stieglitz,Christian Ori 4h41'31"10d51'
Stiehl,Jeanne E Lyn 8h51'52"37d19'
Stiehl,Jennifer Nicole Lyn 9h6'24"37d5'
Stieler,Dietmar Cas 2h57'58"68d33'
Stiemmerling,Andreas And 23h3'22"43d22'
Stier,Georges et Paulette Cyg 20h22'33"40d10'
Stiff 1914,John M Aql 18h57'18"-6d30'
Stiffey,Stuart Alexander Her 16h16'24"4d35'
Stigen,Veronica Huaal Uma 9h52'12"46d47'
Stigers,Natalie Maurine Vir 13h34'6"-7d34'
Stiles Dra 17h24'0"61d8'
Stiles Family,The Mon 6h31'29"10d32'
Stiles,Jade Adam Oph 17h16'41"-20d38'
Stiles,James Edward Cma 6h56'5"-16d54'
Stiles,Jr,Hubert M Cep 4h43'43"80d23'
Stiles,Ms Jack Cet 3h1'15"9d10'
Stiles,Sue And 2h33'23"39d37'
Still Marie Emilie Aur 5h1'19"38d37'
Still,David Harold Cet 3h19'56"9d38'
Still,Odell Paul Boo 15h6'0"10d5'
Still,Ronald W Cep 21h20'34"85d30'
Stillerman,Saul Tau 4h20'25"28d52'
Stilley,Josh Boo 14h56'33"28d16'
Stillings,Jo Lyr 18h39'16"40d27'
Stillman,Janice K And 23h28'52"43d10'
Stillson,Alyce Irene Willis Cam 4h21'32"70d52'
Stillwell,Dan "The Reds Sweep LA!" Aql 18h51'37"11d42'
Stillwell,Jack R Cep 21h47'58"66d33'B
Stillwell,Jan M Cep 21h47'58"66d33'A
Stillwell,Jeni "McDougal" And 23h38'37"39d46'
Stiloski,Daryl Her 17h30'39"20d7'
Stilson,Donald David Per 3h38'31"38d20'
Stiltner,J W Dra 16h10'14"67d32'
Stilwell Ser 17h29'52"-10d7'
Stilwell,Christina Garza Lyn 8h52'57"37d9'
Stilwell,Denzil & Ellen Cas 0h29'32"61d51'
Stima,Michelle & Jonathan Lyr 18h30'20"30d6'
Stimac Star 9 Boo 14h1'0"16d47'
Stimac,Jr,Anthony John Her 16h36'46"22d58'
Stimac,Mary Beth And 2h29'25"44d40'
Stime Boo 14h16'1"31d19'
Stimming Cet 3h10'42"0d10'
Stimpert,Warren Per 1h52'14"52d58'
Stimson-Johnson Her 17h17'42"28d34'
Stimus,Linda Ann Peg 21h56'0"28d46'
Stina Sge 19h31'40"18d36'
Stincer,Cami Mon 6h19'54"5d46'
Stine's Star,Tina Louise Mon 6h37'0"10d1'
Stine,Alta Per 2h41'30"37d6'
Stine,Charles Erik Lac 22h36'27"55d57'
Stine,David Her 16h22'0"28d46'
Stiner,Sandy And 23h38'50"38d23'
Stines,Connie Cyg 20h18'36"41d34'
Sting Her 17h29'33"41d9'
Stinga,Eva y Vicente Tau 4h16'18"11d13'
Stinger,Gabriella Y Lmi 10h36'0"32d49'
Stinger,Keith W Aur 4h54'55"48d56'
Stingers,The Cam 5h30'42"61d34'
Stingl,Nora Aqr 22h34'40"2d6'
Stingo,Anne Sco 17h26'40"-40d47'
Stinklepisus Ori 5h18'25"-6d47'
Stinky Mika Cmi 7h59'18"1d42'
Stinky Nick Aur 6h3'24"33d24'
Stinnert,Pamella Mon 6h53'41"-10d35'
Stinnett,Michelle Elizabeth Crt 10h56'18"-8d17'
Stinski,Cheryl And 0h45'54"33d31'
Stinson,Drake Her 17h30'32"42d43'
Stinson,George Aql 19h54'59"13d9'
Stipicevich,Paul Lac 22h34'60"48d49'
Stipkala,Edward J Cep 22h29'0"60d47'
Stipkala,Frances Ann Cas 1h19'1"68d33'
Stipkala,Gregory Edward Boo 15h23'55"33d44'
Stirken,Peter Cam 5h38'56"70d10'
Stirlen,Ned William Aur 6h29'35"31d8'
Stirling Lmi 10h9'27"40d21'
Stirling Rita Vul 20h57'47"28d34'
Stirling,Margaret Ann Cas 3h0'41"67d33'
Stirling,Robert A Dra 16h44'55"52d23'
Stirm,Greetish Alice Com 13h6'26"21d52'
Stirman,Robert Her 18h27'15"20d32'
Stirner,Alois Her 17h8'17"42d20'
Stirrett,Zara And 23h34'43"41d59'
Stirton,Carol Ann Cyg 20h50'26"40d24'
Stirwalt,Harry Allan Oph 17h16'22"-20d14'
Stiskal,Lydia- Christiane Aqr 21h58'0"-18d59'
Stitt,Britt Sex 9h50'42"3d42'
Stitt,Peggy Boo 14h23'58"52d40'
Stival,Ed (Fast Eddie) Cep 1h5'0"80d33'
Stival,Gina Cep 0h56'5"81d25'
Stivers,Jerrold L Ori 6h0'35"0d36'
Stiwitz,Heidi und Gerd Peg 22h16'57"3d21'
Stjärna,Kerstins Cas 2h0'0"67d53'
Stjärna,Toves Umi 16h24'47"77d52'
Stluka,Julie And 1h58'0"39d40'
Stobart,John Aur 5h11'11"54d27'
Stobaugh,Scot Ser 15h57'19"-2d16'
Stobo,Robert Douglas Cep 23h19'0"65d10'
Stoby "Bellofatto", Tanya Ori 5h56'1"16d5'
Stocchetti (financier) Jean-Aymé Ori 6h2'47"7d55'
Stock,Aribert Lyn 8h2'1"42d22'
Stock,Jeanne L Lyr 18h58'23"42d7'
Stock,John L & Bonnie Cam 12h55'57"78d58'
Stockbridge,Holly Cyg 19h49'19"38d55'
Stockbridge,Valerie And 1h15'30"37d24'
Stockdale,Gretchen And 0h47'27"39d23'
Stockdale,Shannon Lyn 8h17'0"38d24'
Stocker Star Cnv 13h57'1"41d51'
Stocker Star,The Uma 9h38'35"52d13'
Stocker Star,The Kevin Lac 22h39'41"38d58'
Stocker,Deborah Ann Peg 23h34'54"17d55'
Stocker,Janice Hall Cyg 19h14'1"45d59'
Stocker,Mark Steven Anthony Per 2h57'35"41d13'
Stocker,Sassy-Bear Cam 3h21'56"55d17'
Stockerl-Goldstein's Star,Chuggs Tri 2h21'52"31d38'
Stockey,Bill Cep 22h27'17"60d4'
Stockford,Shawn Aur 5h6'41"38d36'
Stockhausen,S J,Gerry Ori 6h2'19"-2d9'
Stocking,Elaine D Cas 23h16'10"60d10'
Stockinger,Moritz Phillipp Vir 11h36'0"-4d8'
Stockland Star,The Ruhama Psc 0h20'0"2d0'
Stockland Star,The Becky Psc 0h20'29"0d7'
Stockly,William H Aql 19h31'59"-0d3'
Stockman,Kathleen Uma 10h27'5"51d15'A
Stockman,Larry Lac 21h59'16"38d4'
Stocks,Cody Davis Lyr 18h19'1"38d54'
Stockton,A Star of Dreams J C Peg 21h56'16"28d3'
Stockton,Candace Anne Mon 6h34'11"-6d21'
Stockton,Donald Boo 15h3'0"28d12'
Stockton,Elizabeth Marie Peg 23h39'44"18d10'
Stockton,Frances Ann Aqr 23h30'20"-4d51'
Stockton,John E Cet 0h54'11"-5d50'
Stockton,Julia Aql 19h10'55"15d46'
Stockton,Sydney Chapman Dra 14h26'28"64d49'
Stockwell Star,The Ori 5h23'33"1d35'
Stockwell,Robert Dra 16h24'33"60d55'
Stockwell,Tom & Laura Cyg 20h51'49"38d45'
Stocus,Sophie Margaret Cyg 21h46'59"34d42'
Stoddard III,William H Cep 20h54'28"66d41'
Stoddard,Grandma Cyg 20h33'47"38d55'
Stoddard,Grant & KayLyn Cyg 20h6'26"40d40'
Stoddard,James B Lyr 18h23'12"37d32'
Stoddard,Jessica, Lindsey & Danielle Cyg 20h6'20"40d9'
Stoddard,Rachel Elizabeth Boo 13h38'0"21d16'
Stoddard,Richard & Josephine Dra 18h16'48"70d11'
Stodgell,Judy Cra 18h18'46"-42d57'
Stoeber,Michael Ryan Mon 7'17'0"-8d2'
Stoeckl,Wilhelm & Gertraud Uma 10h11'28"50d33'
Stoeckle,Thomas J Cnv 14h0'11"46d39'
Stoecklin,Timothy Oph 18h0'34"13d8'
Stoeke,Shannon Jack Lac 22h7'57"37d31'
Stoelting,Anna Fredricka Louisa Leo 9h53'45"31d44'
Stoertz,Spencer Lewis Peg 23h19'45"33d4'
Stoessel,Shlomit & Yoav Finkelstein Lac 22h5'44"42d42'AB
Stoewe,Michelle Aur 4h48'55"51d37'
Stofac's Tryda-Findme, Dr,Robert L Aur 5h34'24"29d37'
Stofer,Deborah Kay White Tande Cam 10h42'14"82d18'
Stofey,Gayle Ann Com 13h17'24"30d18'
Stoffel,Fred Aql 19h35'55"-6d54'
Stoffel,Herr Cmi 7h17'37"1d6'
Stoffel,Paul Umi 15h47'52"89d13'
Stoffer,Jeff & Mary Plaia Lyn 8h48'41"37d26'
Stoffregen,Werner Aur 5h11'59"43d51'
Stofsky,Lianne Lyn 6h59'57"60d40'
Stogdill,Joyce Aoi Her 17h46'3"34d17'B
Stohr,John P Aql 18h57'42"-4d19'
Stojan,Daniel J Dra 17h38'33"78d58'
Stojinski,Robert Her 17h53'59"18d57'
Stojkovic,Caitlin Joanna And 0h13'1"30d53'
Stoker Laser Acoustics Ron & Stacy Her 16h7'47"16d14'
Stoker,Brandon Orion Joseph Crb 16h6'23"30d6'
Stoker,Christopher Joseph Crb 16h10'19"30d48'
Stoker,Jaimee Hollowbush Crb 16h20'41"30d2'
Stoker,Jr,George R Her 18h15'21"37d42'
Stoker,Marline Koucheck Crb 16h21'58"30d12'
Stoker,Roderick Joseph Crb 16h11'59"30d3'
Stokes,"81"Greg Per 3h3'0"47d60'
Stokes,Alicia Lyr 18h18'38"42d3'
Stokes,Gregory Holmes Uma 10h35'25"50d56'
Stokes,James E Del 20h18'46"10d19'
Stokes,Jeremy G Sex 9h50'26"-5d5'
Stokes,Kathy Cas 0h6'39"61d36'
Stokes,Nancy E Vul 20h45'36"20d30'
Stokes,Nathan Michael Cep 22h8'23"60d41'
Stokes,Paul Ori 4h50'0"0d54'
Stokes,Paul Mark Ori 6h3'42"7d51'
Stokes,Raylon Lee Aql 19h55'32"13d23'
Stokes,Rebecca Shankle Peg 23h0'20"33d42'
Stokes,Robert Ori 6h8'1"0d43"
Stokes,Robert Eric Aur 6h7'16"46d0'
Stokes,Sally Cas 23h29'15"61d21'
Stokes,Shawn Marie Cmi 7h31'51"0d18'
Stokes,Taylor Nicole Peg 22h13'0"10d41'
Stokes,Zachary George Lac 22h13'60"51d43'
Stokes-Wall,Kenneth Richard Cyg 21h15'39"39d32'B
Stokes-Wall,Valerie Ann Cyg 21h15'39"39d32'A
Stokka,Roger Aur 6h17'36"30d60'
Stoklas,Debbie L Mon 6h36'1"6d26'
Stoklossa,Cornelia "Conny" Vir 13h28'43"-12d23'
Stokovich,Nicholas Jacob Dra 20h4'14"62d28'
Stolen Moments Vul 19h48'1"28d11'
Stoler,Megan Cam 11h3'33"81d10'
Stoleson Boo 14h20'20"31d48'
Stolfi,Dee Cas 2h18'54"73d12'
Stolfi,Francesco Vincenzo Cep 21h53'29"63d54'
Stolin,Margaret "Gram" Cas 0h2'1"63d31'
Stoll,Amy E And 2h21'44"37d9'
Stoll,Barbara Gem 6h39'18"31d37'
Stoll,Russell Joseph Dra 15h9'56"58d22'
Stoll-Fein,Tutti Aqr 20h58'50"-14d2'
Stolle,Christine Lib 14h31'25"-20d27'
Stollenmaier,Viola Dra 16h59'0"66d40'
Stollenwerek,Richard Lac 22h10'11"54d28'
Stoller,Herbert Umi 13h5'58"71d34'
Stolley,Constance Velde Skillman Cas 2h26'15"60d57'
Stolley,Richard Brockway Per 2h59'32"48d57'
Stolman,Hy Eri 4h23'34"-1d31'
Stolowitz,Alan Lac 22h30'56"37d59'
Stoltmann,Kathryn "Katy" Elizabeth Mon 6h38'29"1d42'
Stoltz,Carol Jeanne Cyg 19h46'58"56d20'
Stoltz,Michael P Lac 22h18'10"55d1'
Stoltzfus,Heather Marie Cma 6h50'15"-19d50'
Stoltzfus,Heather M Lyn 9h11'16"38d51'
Stolynski,Sylvia Mon 7h44'39"-1d34'
Stolz,Dr Thomas Aur 4h56'19"40d46'
Stolz,Ev Aur 5h56'52"48d55'
Stolz,John H Boo 14h39'37"21d47'
Stolz,Katie Grace Cas 1h42'0"63d52'
Stolz,Scott Boo 15h3'58"50d0'
Stolzenberg,Harald Cnv 12h34'51"51d5'
Stolzenberg,Laura Cas 0h58'51"67d7'
Stolzenberg,Ursula Cam 12h47'52"77d1'
Stolzenburg,Dieter Sco 15h56'22"-21d55'
Stolzer,Catherine Burdette Cas 0h57'0"63d16'
Stom,Monty Cet 3h8'39"1d25'
Ston Promise Lac 22h29'29"38d29'
Stone 9-23-80,Shauna Leigh Cas 0h31'55"54d41'
Stone ML,Sarah Tri 2h2'50"31d34'
Stone Star,The Jeffrey Ross Cet 3h6'21"5d53'
Stone Star,The Terry Ann Peg 21h51'59"34d55'
Stone,Adam Cnv 13h4'21"51d39'
Stone,Aileen Her 17h13'56"17d53'
Stone,Brent Joseph Her 16h43'0"47d35'
Stone,Brian N Cam 3h36'47"73d43'
Stone,Brittany Danielle Eri 4h3'1"-18d19'
Stone,Carole & Tim Cyg 21h33'15"41d34'
Stone,Christopher Russell Sex 10h9'26"-0d10'
Stone,Col Robert Cnc 8h9'45"30d38'
Stone,Connie Sue "Sissy" Uma 8h43'27"47d31'
Stone,Dan Boo 15h47'58"43d20'
Stone,David Cep 22h10'0"56d6'
Stone,David Gregory Aur 6h15'14"50d0'
Stone,Diana Ari Cas 0h1'43"65d16'
Stone,Dolores Lyr 19h29'1"40d4'
Stone,Donna MacGibeny Cam 8h50'43"78d45'
Stone,Eileen Cyg 20h30'24"40d3'
Stone,Gary William Cmi 7h35'43"0d39'
Stone,George Cep 22h43'40"57d18'
Stone,Gerry Mae Lac 22h11'55"37d57'
Stone,Gregory Alan Hya 9h34'1"0d43'
Stone,Harrison Clay Ori 5h50'47"18d29'
Stone,Helen & Irving Boo 14h40'1"31d56'
Stone,Ian William Psc 1h27'33"20d59'
Stone,James Ori 5h27'58"0d51'
Stone,James M & Anna F Per 2h36'18"41d9'
Stone,Jason Max Vul 19h45'1"28d43'
Stone,Jeff Aql 20h11'24"5d36'
Stone,Jeff & Amy Cyg 21h15'54"37d48'
Stone,Jennifer Blake Cyg 21h8'19"48d59'
Stone,Jennifier Lyr 18h50'17"42d24'
Stone,John Boaz Cyg 20h15'24"39d10'
Stone,Josh Tropauer Per 2h54'33"50d24'
Stone,Jr,Howard Dra 15h56'48"65d53'
Stone,Julie Cas 2h4'33"65d15'
Stone,Julie Cyg 19h32'20"36d49'
Stone,Julie Shira Cyg 19h19'0"45d21'
Stone,Justin Lee Hya 9h0'40"-11d9'
Stone,Katherine M Vul 19h42'0"20d17'
Stone,Keith Randall Cyg 21h48'55"53d12'
Stone,Kim & Rick Cet 1h54'22"-0d38'
Stone,Krisan Vul 20h1'43"28d59'
Stone,Laura Jill Peg 21h44'30"28d9'
Stone,Lauren Peg 23h44'10"15d32'
Stone,Leona Cyg 19h30'1"37d42'
Stone,Loren Richard Uma 14h15'0"60d58'
Stone,Lylia Jean Umi 15h9'32"78d24'
Stone,Madeline & Elliot Cyg 20h21'24"39d27'
Stone,Martin Her 18h26'27"24d59'
Stone,Martin & Anna Cyg 19h28'18"35d14'
Stone,Martin J Cep 22h16'40"71d11'
Stone,Michael Kelly Ser 15h39'42"3d2'
Stone,Millard Maxwell Uma 8h46'43"48d60'
Stone,Myron & Rosalind Lyr 18h18'43"38d24'
Stone,Nanci M Cam 7h56'0"60d41'
Stone,Nicholas Alexander Aur 6h28'40"31d15'
Stone,Norris Byard Boo 14h43'51"23d45'
Stone,Paul Aql 19h26'59"13d21'
Stone,Raymond George Cyg 20h39'58"30d10'
Stone,Rebecca And 2h0'22"38d3'
Stone,Richard Paul Her 18h12'36"40d34'
Stone,Richard Stuart Her 16h38'26"21d35'
Stone,Robert Aur 7h1'48"36d29'
Stone,Robert Andrew Ori 4h41'38"0d9'
Stone,Roger Tropauer Cep 22h55'39"58d49'
Stone,Ryan & Drew Hya 9h54'34"-18d59'
Stone,Stacey Eri 3h40'44"-15d54'
Stone,Sylvester H Uma 11h34'53"41d23'
Stone,Tadami Hya 9h2'29"-1d31'
Stone,Tanner Carter Her 16h43'1"20d2'
Stone,Toni Lyn 6h20'39"61d3'
Stone,Willie Nolan Del 20h14'45"11d14'
Stonebraker,April Marie Ari 2h38'23"22d13'
Stonebraker,Carl Aur 6h32'39"52d48'
Stonebrook,Pamela Cas 1h25'58"52d34'
Stonecipher-Harmon, Lori Peg 22h58'38"27d23'
Stonehocker,Anna Uma 8h38'28"60d47'
Stonehouse,Olivia Cas 1h6'23"63d49'
Stoneman,Dick Cmi 7h15'23"4d38'
Stonemetz,Benjamin Otis Uma 10h25'21"53d52'
Stoner 41342 Lac 22h14'35"47d18'
Stoner Mon Per 2h55'46"38d10'
Stoner's Dance Cet 2h42'18"4d58'
Stoner,Adam,Aubrey, Alex-Starbase Peg 21h50'44"33d49'
Stoner,Jeffrey Owen Lmi 10h18'52"31d33'
Stones,John Aur 5h50'52"54d52'
Stoneson,Cathina Eri 4h24'24"-2d16'B
Stoneson,DES Eri 4h24'24"-2d16'A
Stoney Cyg 21h36'18"42d3'
Stoney & Moet Aqr 20h58'44"-10d47'
Stonier,Charity Ann Cas 3h11'0"66d6'
Stoodard,Grandpa Her 16h25'27"28d7'
Stoonie,Eugene & Anna Marie Weems Cyg 19h24'0"32d26'
Stoops,Denice A Aql 19h53'0"15d60'
Stoops-McClain Sebastian Aaron Her 17h27'10"30d45'
Stoots,Charlotte Rhaye Cas 0h34'18"66d27'
Stoots,Sammy Lee Her 17h10'37"49d1'
Stopar,Gustav Lac 22h2'59"50d10'
Stoppiello,Carmen Michael Uma 9h47'46"68d19'
Storako,Sara Jane Cyg 19h26'1"30d54'
Storch,Anne Lyr 18h58'52"38d4'
Storch,Ken W Aql 20h6'56"7d4'
Storch,Patricia Aql 20h34'21"0d21'
Storch,Zoe Rachel Boo 15h16'0"47d45'

Storck Star,Timothy Ryan
Per 2h57'37"40d1'
Storck,Günter
Per 3h0'1"50d11'
Stordahl,Kristi Kay
Ari 2h44'49"30d33'
Storek,Catharine Munds
Vul 20h21'13"23d52'
Storer,Holly
Cet 1h56'48"-3d44'
Storer,Scott M M
Ori 6h7'0"10d59'
Storer,Stan
Cet 1h48'24"-5d21'
Storey,Alan
Aur 5h6'28"42d21'
Storey,Alexander Peter Shade
Uma 11h10'0"58d19'
Storey,Derek
Aql 20h6'58"8d5'
Storey,Hannah Rachel
And 23h47'59"38d27'
Storey,Janet
Cas 0h36'33"60d22'
Storey,Janice
Ori 5h36'23"1d1'
Storey,John & Barbara
Cyg 21h1'1"36d49'
Storey,Margaret Mary
Cyg 19h58'1"29d31'
Storey,Mary Kathleen
Lyr 18h54'17"42d45'
Storey,Megan Erin
Ori 5h30'46"-8d11'
Storey,My Star Home- Beverly A
Com 12h28'1"30d20'
Storey,Neil
Per 2h37'16"40d31'
Storey,Peter
Ori 6h6'14"20d19'
Storey,Peter
Per 4h43'23"40d2'
Storey,Sarah
Del 21h4'41"13d58'
Storey,Steve
Cep 20h51'58"70d52'
Storey,To Our Star- Jamison Everett
Her 17h44'1"14d46'
Storey,Tom
Dra 14h2'38"64d33'
Storin,Rose Lane
Cyg 20h23'45"38d53'
Storino,Vincent J
Lac 22h17'60"51d50'
Stork,Great Grandpa Fred L
Vul 19h45'44"28d28'
Stork,Krystine
And 1h28'53"35d14'
Storlee,Dana Michelle
Eri 3h59'11"-7d20'
Storlee,Dawn Marie
Mon 6h44'38"11d44'
Storlee,Douglas Manly
Cet 0h40'25"-3d30'
Storlee,Virginia Anita
Peg 22h10'43"29d8'
Storlle,Daniel Douglas
Aql 19h44'32"14d13'
Storm
Ori 5h7'19"0d43'
Storm & Chevvy
Ori 5h34'51"0d39'
Storm Of Reinhardt
Del 20h36'1"20d5'
Storm Rider
Her 18h41'54"13d2'
Storm,Adrienne
Cas 1h7'30"50d14'
Storm,Angie
Peg 22h8'1"26d53'
Storm,Bonnie
Lyn 7h56'35"35d30'
Storm,Charlotte
Cyg 20h25'0"42d20'
Storm,Connie
Uma 14h20'49"61d15'
Storm,Crystal
Cet 0h56'11"-5d22'
Storm,Denise Nolan
Cas 1h20'59"50d18'
Storm,Dorothy Cave
And 22h59'24"38d24'
Storm,Francis R
Cyg 20h7'55"39d49'
Storm,Gabriel
Lyr 18h14'19"33d52'
Storm,Hannah
Cyg 19h47'56"30d43'
Storm,Jesse
Dra 18h58'27"84d31'
Storm,Kayleen
Cas 1h30'36"76d14'
Storm,Leslie Andrea
Cas 1h20'48"50d17'
Storm,Morgan Rosanne Burgo
Peg 22h6'46"5d41'
Storm,Soren
Uma 10h25'36"58d19'
Storm,Stephen G
Per 4h33'37"31d13'
Storm,Thoughts of Killian Rudolph
Ser 18h44'1"3d13'
Storm-"Your Very Own Star",Porsche Lelani
Mon 6h53'24"0d14'
Storman,Dr Philip
Aur 6h1'21"37d55'
Storment,Malinda Dawn
Peg 22h5'1"8d27'
Storms,Keelyn
Tri 2h4'0"31d54'
Stormy Weather
Lyr 18h32'51"41d57'
Stornetta,Jr,Ron
Aur 6h7'43"32d12'
Storozuk,Don "Mickey"
Cet 2h32'58"-5d57'
Storr,Kimberly G
Lyn 8h42'23"34d47'
Storrar,Lisa Marie
Peg 22h21'36"26d10'
Storseth,Barbara
Cmi 7h30'4"7d15'
Storseth,Marjorie
Mon 7h49'44"-3d39'
Storves,Roger "Starry Eyes"
Lac 22h37'59"55d28'
Story,Eleanore
Mon 6h39'0"6d52'
Story,Jr,James L
Sex 10h34'36" 6d41'
Story,Keith William
Ori 5h35'19"0d23'
Story,Lisa Marie
Cet 2h1'30"-10d23'
Story,Mildred L
Tri 1h55'18"28d46'
Story,Susan Marie Simmons
Lyn 7h53'43"38d32'
Storz,Mika
Boo 15h19'50"50d4'
Stosh
Oph 18h4'13"10d45'
Stossel,Alexandra Marie
Lyn 7h28'51"43d23'
Stostrom,Margaret Ann (Croke)
Peg 23h4'1"21d15'
Stosz,Sandra Leigh
Sge 20h2'42"20d10'
Stotelmyer,Callie
Peg 22h43'42"26d14'
Stotler,Cathy
Peg 22h43'29"33d23'
Stott 94,J & H
Col 6h1'38"-33d4'
Stott,Christopher
Aql 20h8'1"6d55'
Stott,Kate
Cas 0h50'54"63d9'
Stott,Richard
Hya 8h34'26"-5d53'
Stott,Robert K
Dra 9h35'49"78d58'
Stottlemyer,Pat
Aql 18h55'45"-0d1'
Stotz,Anna
And 2h5'48"40d2'
Stotz,Ken
Peg 22h4'50"34d16'A
Stotz,Marge
Peg 22h4'50"34d16'B
Stouchbury,Jacqueline
And 0h12'34"38d12'
Stoudt,Theodore Charles
Aur 5h59'14"28d36'
Stough 52nd Annivrsary Gene & Willie
Her 17h31'10"28d30'
Stough III,Morrow Franklin
Ser 15h14'1"20d34'
Stough,F R Bob
Her 16h59'1"30d3'
Stough,Patty Ann
Cam 3h47'1"61d19'
Stoughton,Charlotte Julia
Cam 4h23'38"69d43'
Stout's Excuse
Ori 5h59'42"20d16'
Stout,"Sigmund"
Cep 21h27'29"60d26'
Stout,Anthony Kyle
Per 2h56'18"37d19'
Stout,Barbara E
Cyg 21h17'26"37d12'
Stout,Bernice & Hurley
Lac 21h58'17"42d16'
Stout,Dillon K
Uma 11h31'1"56d48'
Stout,James E
Boo 14h17'54"52d2'
Stout,Kelly C
Peg 23h35'19"30d4'
Stout,Kelly Lee
Uma 9h8'41"49d3'
Stout,Mary Louise Ashley
Eri 3h52'28"-1d58'
Stout,Nina Lorraine
Uma 8h47'51"55d41'
Stout,Rochelle
And 23h7'43"44d29'
Stout,Thomas E
Ori 5h58'57"12d31'
Stout,Veronica
Lyn 9h8'33"39d52'
Stoutenburg,Dale
Eri 3h37'1"-15d7'
Stovall,Charles
Oph 17h19'20"-20d10'
Stovall,Deborah L
Eri 3h44'0"-0d48'
Stovall,Happy 1st Lisa & Scott
Crb 16h7'46"31d48'
Stovall,Minta Mae
And 1h41'54"48d57'
Stovall,Rusty
Aql 18h53'52"7d56'
Stovall,Saba Wiley
Cet 2h2'38"0d42'
Stoveken,Eric
Boo 14h10'60"43d8'
Stovell,Helen Penrose
Aur 6h1'32"31d20'
Stover
Lac 22h43'24"37d56'
Stover Bunch,The
Cet 1h26'16"1d15'
Stover's Star Is Born, Martin Alan
Her 16h7'53"21d38'
Stover,"Flag-A-Lucky!" Linda Belle
And 0h21'14"33d41'
Stover,Blake Patrick
Hya 8h59'0"6d11'
Stover,Catherine
Peg 22h42'1"20d54'
Stover,Jim
Aur 5h3'51"42d15'
Stover,Jimmy Louis "Smokey"
Hya 8h17'59"3d45'
Stover,Ty
Lac 22h35'49"52d49'
Stow,Alvin Dale
Aql 19h30'27"10d16'
Stow,Jack Edward Conway
Ori 5h24'19"0d18'
Stowe Ailanjian The Empereur's Star
Aql 18h58'23"13d56'
Stowe,Dorothy Parker
Mon 6h43'15"7d4'
Stowe,John C
Ser 16h18'30"0d31'
Stowe,Ronald Floyd
Her 18h10'19"38d17'
Stowell,Marilyn
Cas 23h37'55"60d49'
Stowell,Ronnie
Cam 4h55'45"67d43"
Stoy's fairy-tale
Her 17h28'22"42d34'
Stoychoff-Inman,Karen
Cyg 19h44'17"30d22'
Stoycos,William
Lac 22h28'51"56d19'
Stoyle,Sarah
And 23h41'24"41d15'
Straatsma,Ryan Teniss
Sex 9h39'38"-8d33'
Straavaldson,Bill
Aur 6h37'0"35d17'
Strable,Leilani Lamar
Lyn 7h4'1"58d56'
Stracchi,Chiara
Uma 10h23'51"71d50'
Strachan,Doris
Del 20h39'43"18d58'
Strachan,J Allan
Boo 14h23'41"15d59'
Strachan,Jr,Thomas Alexander
Boo 14h43'20"33d8'
Strachan,Nancy
Mon 6h47'18"11d49'
Strachan,Sharon Renia
Mon 7h2'39"-7d6'
Strachota,Mark Andrew
Sco 16h20'35"-40d56'
Strachota,Michael Roger Sylvester
Lac 22h22'50"49d4'
Strack,Doc
Aur 7h21'46"40d23'
Stracke,Angelika
Mon 7h48'4"-3d41'
Strackhaar,Ernst
Cnc 8h6'17"30d29'
Straczynski,Michelle
Mon 7h1'23"5d22'
Strada,Catherine
Ori 6h2'0"-2d37'
Strada,Mia Bella Stella
Peg 23h26'28"11d47'
Strada,Rossella
Gru 22h5'14"-59d27'
Strader,Ellen Lois
Del 20h20'49"10d36'
Strader-Monaghan,Kier Suzanne
Mon 6h28'12"5d19'
Straffi,La Stella De Gianluca
Umi 14h59'15"68d24'
Strahan,Shelda
Ori 5h13'1"-9d13'
Strahl,Konrad Paul Ludwig
Sco 15h57'28"-22d13'
Strahl,Leslie Robyn
Cyg 21h1'16"30d6'
Strahlende,Rodolfo der
Her 17h55'34"14d51'
Strahlendorf,Traci Michele
Lyr 18h48'45"39d59'
Strahman,Cora Ann
Cas 0h1'14"61d19'
Straight,Melanie
Sge 10h0'37"20d36'
Straile,Friedrich
Lyr 19h16'38"30d58'
Strain,Ethan Thomas
Her 16h36'0"21d54'
Strain,Florence G
Cas 1h41'39"70d5'
Strain,PhD,George
Cet 0h25'40"-17d49'
Strait,George
Tau 3h57'1"0d30'
Strait,Jr,John Howard
Aur 4h52'41"38d1'
Strait,Matthew Tyler
Lib 15h38'13"-23d26'
Strait,Richard Evan
Ari 1h58'27"22d19'
Straiton,Ian
Crb 16h11'24"33d19'
Strakeljahn,Gerd
Peg 23h33'23"15d42'
Straley,Kandi Sue
Lmi 10h18'0"30d35'
Strand,Collins Weir
Peg 23h55'12"11d11'B
Strand,Dick
Cma 6h53'41"-16d50'
Strand,Michael Wallace
Lac 22h10'24"38d27'
Strand,Pattie & Bert
Mon 6h23'55"3d41'
Strand,Pauline Holmes
Peg 23h55'12"11d11'A
Strandberg,Madeleine Maria Stenmann
Cas 23h1'35"58d42'
Strandness,Lillie Borg
Vul 20h1'19"23d4'
Strane,Kate Murphy
And 0h21'15"33d53'
Strang,D P
Cet 0h4'37"-10d3'
Strang,Sascha Ingo
Her 17h7'0"42d46'
Strange,Ashley Nicole
Ari 1h50'59"23d25'
Strange,Josephine
Crb 15h43'13"28d31'
Strange,Laura-Anne Yvonne
Aur 5h33'23"38d19'
Strange-Preferred Star ,Mark
Dra 17h0'1"62d59'
Stranix,Bennie
Peg 23h23'1"33d49'
Strano,Rosario
Per 3h7'1"46d10'
Stransky,Eric D
Vul 21h1'16"23d47'A
Strapko,John Mark
Dra 19h4'46"48d22'
Strasburg,Katherine
Cas 0h34'11"60d59'
Strassburger,Kapitän
Lyn 8h11'30"48d38'
Strassel,Julius
Cyg 21h15'0"34d42'
Strassel,Marie Louise Schwartz
Eri 3h52'37"-6d4'
Strasser,Michaela
Peg 23h19'13"13d56'
Strassman
Vir 11h45'25"6d50'
Strate,Valerie J
Cas 3h7'37"58d37'
Strategic Systems
Lac 22h6'11"51d31'
Strategic Systems Southeast
Her 17h33'20"33d45'
Stratennio
Scl 23h42'57"-25d18'
Strater,Elizabeth Ann
Lyn 8h33'11"41d49'
Stratevest Star
Umi 16h58'22"78d60'
Stratford Cheerleaders 96 & Mrs A
Peg 22h10'0"20d50'
Strathmann,Bernhard
Mon 7h51'19"-9d10'
Strathmann,Winnie
Uma 8h32'35"51d40'
Strathmore
Cam 4h32'18"61d14'
Strati Two Dogs,Robert
Ant 9h38'19"-30d30'
Stratigos,Anthea
Sgr 19h8'19"-25d35'
Stratman's Soulmate, Timothy M
Cet 0h25'48"-2d17'
Stratman,George A & Nina M
Oph 17h2'0"-0d44'
Stratmann,Jörg
Lyn 7h28'42"40d8'
Stratter,Samuel Harrison
Her 17h34'20"40d47'
Strattman,Steven Paul
Aqr 21h53'32"-1d52'
Stratton IV,Edward
Per 3h8'44"46d5'
Stratton,Iva Muriel
Uma 11h22'17"47d5'
Stratton,Jr,Paul Herman
Cep 9h47'0"81d25'
Stratton,Michael Joseph
Aur 6h9'41"37d12'
Stratton,NC,USN,Rear Admiral Mariann
Cas 1h2'44"61d52'
Stratton,Nelson M
Lac 22h26'10"50d25'
Stratton,William Grant
Psc 23h1'23"5d37'
Stratton-Kennedy, Victoria
Cas 0h32'33"63d4'
Straub,Betty Waters
Del 20h16'51"12d28'
Straub,Davis Paul
Dra 16h56'34"67d40'
Straub,Diane Reyes
Lyr 18h51'52"32d58'
Straub,Donald B
Lib 15h31'12"-18d52'
Straub,Dorothy & Gene
Peg 23h5'24"8d40'
Straub,Josef
Per 3h18'29"47d44'
Straub,Thomas
Sco 17h6'1"-38d50'
Straube,Dany
Cam 6h52'12"68d50'
Strauch,Barb
And 1h9'13"37d45'
Straus,Judy Ann
Cas 23h42'20"62d21'
Straus,Stanley R
Vul 20h22'15"22d44'
Strause,Portia
Cnv 12h33'42"40d6'
Strauser,Nik
Ori 5h9'25"-6d3'
Strauser,Sara R
Lyn 6h33'13"59d8'
Strauss Clan,The
Boo 14h2'0"14d4'
Strauss,Dorris Baxter
Aqr 22h42'0"2d21'
Strauss,Fran
And 0h23'52"34d50'
Strauss,Gloria A
Vir 13h1'12"-5d19'
Strauss,Jonathan
Her 17h59'21"28d27'
Strauss,Keith
Ser 15h17'20"11d42'
Strauss,Kimberly Lauren
Aql 19h53'13"15d27'
Strauss,Mary
Cas 0h35'18"67d25'
Strauss,Samantha Leigh
Lyr 18h48'22"31d4'
Strauss,Savannah K
Cyg 20h18'0"38d30'
Strauss,Tawny
Mon 6h54'24"-1d38'
Straut,Laurie
Cam 4h14'38"60d6'
Strautmann,Ursula
Peg 23h32'0"17d37'
Strauβ,Karl-Heinz
Lyn 8h17'20"43d17'
Stravinski,Lynn
Lyn 8h10'20"37d27'
Straw Star,The Mary Ann & Gordon
Crb 15h52'20"26d42'
Straw,Brandow J
Ari 2h1'34"21d32'
Straw,Wesley
Her 17h56'0"14d32'
Strawberry
And 23h27'56"42d1'
Strawberry
Boo 15h4'69"29d23'
Strawberry
Lyr 18h20'22"47d47'
Strawberry & Max
Cyg 20h17'42"31d12'
Strawberry Kisses, Please Midori
Psc 1h23'27"5d39'
Strawberry White
Lyr 18h37'52"40d14'
Strawberry Yoghurt Kisses
Cyg 19h26'20"35d16'
Strawn,Keith
Sct 18h53'35"-7d11'
Strawn,Phara Ila
Cnc 9h4'47"7d27'
Strawser,Dean Alan
Aur 7h0'31"39d58'
Strawser,Miriam Ruth
Aur 7h9'45"40d19'
Strawser,Oliver Dean
Aur 7h20'23"38d59'
Straza,Robert Michael
Hya 8h27'1"1d33'
Streak,Nishi
Ori 5h24'14"-4d1'
Streaker,Shirleen
Lyn 9h13'26"38d10'
Streaking Blonde
Sex 9h45'1"2d2d44'
Stream,Gloria
And 2h7'60"40d18'
Streamer,Joy
Sct 18h46'33"-7d14'
Streams,Barbara Jayne
Cas 22h57'41"55d15'
Streat Star,The Elizabeth Ann
Peg 23h22'22"15d41'
Streb,Bill
Aql 20h9'0"1d32'
Strebert,Wolfgang
And 23h13'51"40d2'
Strecher,Julia Claire
Peg 23h23'60"17d33'
Strecher,Rachael Halley
Ori 5h38'31"-10d24'
Streckbein,Ingrid
Lyn 8h6'60"41d45'
Stredler,Nathan H
Uma 11h43'37"48d22'
Street,Jackie Sue
Mon 7h38'19"-0d37'
Street,Janice Kergan
And 0h9'20"39d1'
Street,John"Wings"
Boo 13h39'22"20d21'
Street,Margarete Primrose
Lyr 18h36'22"28d36'
Street,Michael J
Aur 5h36'15"48d52'
Street,Picabo
Uma 11h14'55"71d5'
Street,Sherry Lynn Breese
And 23h3'0"48d59'
Street,Theodore James James
Ori 6h9'50"5d43'
Streeter,Jonathan
Per 3h4'0"40d25'
Streeter,Lord Myron
Lmi 9h57'32"40d24'
Streets,Lisa Lynn
Mon 6h24'21"8d30'
Strega
Leo 9h55'42"11d30'
Strega Popolina
Ori 6h4'51"1d22'
Strehle,Christopher & Sandra
Peg 23h18'42"33d16'
Strehlow,Alana Mary
Cmi 7h31'57"10d16'
Strehlow,Joann "Jo"
Cam 13h25'37"81d31'
Streich,Erwin L
Uma 11h56'42"48d20'
Streicher,Jean Marc
Per 3h30'19"36d20'
Streicher,Mary-Jo
And 2h26'0"38d27'
Streifel,George John
Aur 5h21'32"40d49'
Streiner,Katrina René
Cas 22h59'16"56d38'
Streitwieser,Katherine Rose
Uma 8h58'29"48d54'
Strekai,Jack J
Dra 13h30'56"64d50'
Strelec Love Chris, Justin
Hya 8h56'39"5d29'
Strelitzia 92
Mon 7h48'58"-4d23'
Streller,Jenny Hunnicutt
Uma 10h5'0"51d2'
Streller,Tim
Uma 10h20'28"53d58'
Streller,Timmy Hunnicutt
Uma 9h46'0"49d52'
Strelo,William Arthur
Ori 5h25'16"0d57'
Strem April 1,1936, Michael E
Her 17h35'38"21d2'
Strenfel,Robert "Bobby" J
Her 18h13'56"31d32'
Streng,Jacques
Gem 6h57'41"18d58'
Strength,Dedication & Cooperation
Ori 5h18'39"0d30'
Strenk III,Joseph John
Lyr 18h44'40"33d45'
Strenz,Fred
Per 3h39'1"40d59'
Strese,Brigitte Astrid
Sco 17h52'12"-30d38'
Stretch Darlin'
Sgr 18h49'45"-22d27'
Stretchberry,Donald
Sex 10h41'53"0d33'
Strey,Fred & Estelle Strey
Aql 20h24'51"-2d16'B
Strey,Murray
Cet 2h51'1"0d52'
Striano,Maria
Lyr 19h16'29"25d34'
Striar,Ruth
Cas 1h6'56"68d30'
Stribley,Bill
Sgr 19h48'0"-45d18'
Stribling,PhD,Dr Elizabeth
Oph 17h59'30"1d34'
Stricker,Augie E
Dra 17h38'57"75d37'
Stricker,Sacha
Peg 22h6'54"5d2'
Strickfaden,Gerry
Hya 8h38'39"-10d34'
Strickland,Breanna Lynée
Mon 8h2'60"-4d33'
Strickland,Eleanor
Aql 19h23'52"10d55'
Strickland,Ginger
Lac 22h32'48"38d44'
Strickland,Howard Dee
Lac 22h50'33"55d52'
Strickland,Jennifer Lynn
Cet 1h43'33"-6d43'
Strickland,Lori Ann
Ori 5h56'44"10d56
Strickland,Mark B
Per 2h53'1"40d14'
Strickland,Sandria Vonesia
And 0h34'31"40d48'
Strickland,Susan Christine
Vir 11h59'13"-0d31'
Strickland,Warren Davis
Aql 20h0'29"7d18'
Strickland,Zoë Elizabeth
Ori 6h5'0"3d4'
Strickler,Annie
And 23h39'59"36d54'
Strickler,Art
Gem 6h30'55"13d43'
Strickler,David
Aql 20h7'19"-8d31'
Strickler,Grant Richard
Aur 5h18'0"43d10'
Strickler,Kelsey Louise
Cas 0h51'1"67d7'
Strickler,Mabel
Lyr 18h56'17"31d23'
Strickler,Stacy Lynn
Cam 6h0'60"70d34'
Stricklett,Lanny
Vul 19h22'23"26d42'
Stricklin,Mary Lou
Lyn 8h24'35"45d9'
Striegnitz,Peter
Aur 5h14'46"43d27'
Striese,Dr Richard
Uma 13h41'1"48d31'
Strietzel Star,The Donald H
Cma 7h19'25"-18d51'
Strife,Stuart C
Mon 8h7'20"-8d56'
Striglia,Saverio Hugho
Her 18h3'1"28d8'
Strike
Cyg 21h34'42"40d40'
Strikos,Elizabeth D
Hya 8h24'0"5d55'
Strimel,Florine
Eri 4h32'57"-8d1'
Stringari,Lawrence J
Lac 22h2'11"38d46'
Stringer,Collin Dean
Ser 17h56'34"-13d49'
Stringer,In Mem Of Our Dearest Friend Dr Paul
Uma 12h0'47"64d6'
Stringer,James Mark
Lac 22h53'16"50d9'
Stringer,Jennifer
Boo 14h7'12"28d0'
Stringer,Margarett & David
Cyg 21h36'1"30d37'
Stringfellow,Victoria
Lyn 8h4'1"38d37'
Stringham's Way,Guy
Cnv 12h21'11"42d52'

Stringham,Robin
Aql 19h30'43"12d56'
Stringle,Julian Marc
Her 16h59'40"50d39'
Strining,Kathy
Lyr 18h59'30"47d13'
Strite,The Star Of David J
Uma 9h35'34"51d39'
Strittmatter,Jeff & Melissa
Peg 22h8'0"4d45'
Stritzelberger,Rolf
Mon 7h53'29"-3d60'
Strobel,Austen Allan
Cyg 20h23'18"40d45'
Strobel,Barbeangel
And 23h15'29"48d36'
Strobel,Dieter
Hya 9h7'27"5d43'
Strobel,Enrico
And 23h1'28"42d53'
Strobel,Roland
Psc 23h6'25"1d25'
Strobell,Peter H
Sex 9h53'13"-2d10'
Strobl,Kelly David
Sco 16h35'47"-37d45'
Strodtmann,Keith Alan
Cep 21h19'28"65d12'
Stroger
Eri 3h33'26"-6d12'
Strohl,Asardiman- Richard J
Aur 6h3'39"45d43'
Strohl,William Wayne
Vul 19h47'46"28d12'
Strohm,Rudi
Vul 19h47'40"29d2'
Strohm,Tiffany Giselle
Ori 5h54'14"10d8'
Strohmayer,Jake & Mary
Mon 6h55'22"-3d11'
Strohmeier,Anne Marie
Ari 2h42'52"22d9'
Strohmeier,Michael
Per 3h0'59"50d16'
Strohmeyer,Gregory Mark
Vir 12h2'36"1d37'
Stroinski,Dan & Jan
Cyg 21h50'28"37d60'
Strole,Archie
Her 17h36'1"27d31'
Strolz,Lila Marie
Aqr 20h58'38"0d28'
Strom,Angela Lucia
Cyg 21h4'57"40d9'
Strom,Cheryl
Cas 0h52'16"64d46'
Strom,Deborah Bubba
Aql 18h58'1"13d15'
Strom,John
Her 17h54'57"40d2'
Strom,John A
Cet 1h21'57"-1d17'
Strom,Jozlyn Light
Peg 22h1'31"8d28'
Strom,Mabel M
Cas 1h14'44"61d37'
Strom,Mechthild
Aqr 20h56'12"-14d6'
Strom,Rachel
Ori 5h49'41"21d18'
Strom,Shaina
Uma 8h59'49"57d59'
Stromaier,Horst
Cmi 7h20'40"4d12'
Stroman,Guy
Her 17h0'56"48d7'
Stromberg,Susan L
Cas 23h26'1"60d43'
Strombom,Betty & Paula
Trusty
Peg 22h41'0"22d38'
Stromeyer,Cecilia Francis
And 2h33'11"50d19'
Strommen,Jay Michael
Boo 14h45'46"36d12'
Strommer,Adam Lidell
And 23h36'4"34d45'A
Strommer,Trina "Houghton"
And 23h36'4"34d45'B
Strompen,Ulrich
Cmi 7h5'25"1d49'
Strondak,Larry
Boo 14h45'43"30d52'
Strong Alice & Toby
Cyg 21h1'46"40d32'
Strong,Adam Michael
Her 18h9'1"31d2'
Strong,Alexandr Ora
Cam 6h11'1"71d5'
Strong,Christopher Sanderson
Cmi 7h23'33"0d43'
Strong,Diane C
Vul 20h2'1"25d55'
Strong,Donald Roy
Mon 6h30'23"-10d40'
Strong,Earline
Com 12h2'51"26d45'
Strong,Edith Rose
And 23h3'0"37d56'
Strong,Elizabeth
Cas 1h30'54"75d11'
Strong,Elizabeth Ann
Mon 7h1'44"0d16'
Strong,Gary Michael
Aql 18h43'17"8d60'
Strong,Greg & Lisa
Sge 20h0'49"19d20'
Strong,Hanne M
Lac 22h36'23"40d57'
Strong,John
Her 16h33'47"38d31'
Strong,John Marshall
Uma 8h49'28"53d35'
Strong,Kay & Nate
Aur 5h3'59"37d56'
Strong,Keith Allan
Aur 5h27'21"40d56'
Strong,Marie
Lyn 8h26'42"50d60'
Strong,MD,Dr Eric
Cep 20h53'46"58d26'
Strong,Rachel & Terry
Cyg 21h50'35"40d3'
Strong,Robert H
Hya 8h13'30"1d54'
Strong,Robert L
Hya 8h29'1"-8d32'
Strong,Russell
Cru 12h32'39"-57d40'
Strong,Shiloh
Cas 3h3'1"58d38'
Strong,Steven Reed
Per 1h59'1"56d32'
Strong,Tim
Hya 8h10'23"2d42'
Strong,Waldo
Tri 2h35'1"34d53'
Strong,Wayne Peale
Cet 2h44'25"4d0'
Strongheart,Tedrick
Tri 2h15'20"28d48'
Stroobandt,Adam Wade
Aql 19h28'14"-8d48'
Stroot Is 40 Yrs Old, Hey
World Molly
Lmi 10h53'14"26d28'
Strosnider,Vicki Lynn
Com 12h22'32"20d37'
Strother,Glen
Gem 6h15'10"25d56'B
Strother,Melodie Lynn
Sex 10h29'35"-5d43'
Strothman,Linda Jean
Gem 7h17'39"21d52'
Stroub,Rebecca
Cmb 12h34'42"14d22'
Stroud Our Sun,Aaron
Michael
Lac 22h55'48"53d11'
Stroud,Charles D
Aql 18h44'17"8d31'
Stroud,Dacia
Dra 17h59'38"58d36'
Stroud,Travis
Lmi 10h0'49"35d26'
Stroud,William
Her 16h42'44"48d55'
Stroud,William Blake
Crt 11h38'30"-8d59'
Stroud,William M
Her 17h55'44"48d53'
Strough,Kathy
Lyr 19h1'57"41d15'
Stroup,Billy & Alicia
Cet 1h50'29"-3d1'
Strouss,Daniel Joseph
Dra 15h56'26"66d55'
Strout,Tyler John
Uma 11h40'24"42d22'
Strozdas,Blake Alexander
Umi 15h17'0"66d2'
Strozdas,Jay
Uma 11h38'33"50d4'
Strubbe,Betsy R
Lyn 7h57'53"35d12'
Struck,Frank
Her 17h32'21"30d4'
Strudwick,John
Cep 20h37'32"75d45'
Struletz,Adam
Aql 20h7'36"8d12'
Strulouici,Christian
Peg 23h45'15"31d5'
Strumor,Matthew A
Lmi 9h54'37"40d22'
Strumor,Shirley L
Ori 5h7'55"-2d55'
Strumpf,Lynn & Lawrence
Cyg 21h30'37"38d38'
Strunk,Harold "Hal"
Cam 8h0'35"70d39'
Strunk,Manfred
Aqr 22h58'38"-5d54'
Struss,Mary Christina
Psc 0h56'26"27d52'
Struthers Stellarnaut, Wm
Armour
Lac 22h21'47"38d50'
Struthers,Troy Michael
Oph 18h15'36"10d28'
Struthmann,Winfried
And 23h26'59"45d25'
Strutt,Kirsty Anne
Ari 2h26'1"21d5'
Strutz,Al
Cma 6h26'28"-15d31'
Struve,Laura
Gem 7h2'41"28d47'
Struzziero,Ralph E
Per 3h0'32"40d54'
Stryker,Ron
Uma 12h0'57"31d8'
Stryker,Scott S
Cet 2h32'1"-5d46'
Stu & Lou
Boo 14h11'50"30d16'
Stu & Lulu
Ori 5h14'22"-5d43'
Stu's Way
Lac 22h13'15"40d23'
Stu-24
Oph 17h17'38"11d8'
Stuard,Joe
Per 4h0'40"50d14'
Stuart
Cam 6h22'32"70d33'B
Stuart "CAFS",Hazel M
Lmi 10h48'43"27d15'
Stuart & Jan
Sge 20h4'39"16d39'
Stuart & Karin's Wedding Star
Ori 5h43'53"12d18'
Stuart & Shirley
Com 13h1'55"20d30'
Stuart 40
Ant 10h30'53"-33d16'
Stuart Alexander
And 2h18'24"44d22'A
Stuart Cape Light,The
Cep 0h6'0"71d11'
Stuart IV,James Leo
Aur 5h1'24"49d7'
Stuart IV,Robert Edward Lee
Hya 9h34'43"2d21'
Stuart Randall
Equ 21h2'1"8d10'
Stuart's Star
Ori 5h56'50"18d21'
Stuart,Aisannette
Per 1h57'56"50d31'
Stuart,Bunny Lee Martin
Lyn 8h19'19"37d14'
Stuart,Carly Diane
And 0h23'15"37d24'
Stuart,Charles Iain
Sge 20h2'12"20d53'
Stuart,Dr Wayne J
Oph 17h25'10"-23d0'
Stuart,Frances & Harold
Mon 7h2'57"3d49'
Stuart,George Robert
Aql 20h9'19"7d39'
Stuart,Jaimie
Lmi 10h54'28"27d52'
Stuart,Jason
Uma 8h22'60"62d6'
Stuart,John
Aql 20h11'36"12d27'
Stuart,John
Cep 0h3'1"78d58'
Stuart,Marty
Crb 15h45'40"27d57'
Stuart,Marty
Boo 14h13'40"32d51'
Stuart,Mary T
Mon 6h2'23"-8d45'
Stuart,MD,Carlos A
Oph 18h42'0"10d32'
Stuart,Michael Patrick
Her 17h39'54"22d53'
Stuart,Randall
Her 17h9'58"48d5'
Stuart,Seewee
Per 1h55'45"56d55'
Stuart,The Hillbilly Star,Marty
Tri 1h55'30"26d56'
Stuart,Thomas Alan
Ori 6h0'57"2d6'
Stuart,Valerie J
Lyn 8h0'37"58d36'
Stuart-Good Night & Forget
Me Not
Cyg 21h26'58"38d46'
Stuart-Lilley,Laura
Vir 11h59'48"5d28'
Stubbington,Georg
Sgr 18h53'19"-29d38'
Stubblebine,John G
Umi 10h43'35"72d54'
Stubbs,Angie
And 23h21'28"45d1'
Stubbs,Christopher Clark
Lyn 6h55'25"52d32'
Stubbs,Christy L
Aql 19h25'12"12d8'
Stubbs,Collin Forrest
Cet 0h55'33"-0d44'
Stubbs,Debbie
Ori 6h4'19"2d45'
Stubbs,George Walter
Per 2h27'55"56d50'
Stubbs,Kathleen May Davis
Boo 15h20'14"37d45'
Stubbs,Laura Massey
Peg 21h43'41"28d10'
Stubbs,Matthew Howard
Hya 9h13'26"1d10'
Stubbs,Stephen
Lac 22h54'58"50d29'
Stubbs,Yolande
Tri 2h4'48"31d19'
Stubby's Star
Umi 16h19'15"70d48'
Stubenvoll,Marianne
Mon 7h1'43"-7d8'
Stuber,Steven Phillip
Cep 21h8'36"68d30'
Stubitsch,Jessica Mary
And 0h55'29"40d25'
Stubler,William
Her 16h27'16"48d42'
Stucchio,Diane
Cas 2h59'24"58d2'
Stuchin,Blake
Gem 6h51'7"12d52'
Stuchkus,John
Her 18h19'39"12d26'
Stuchl,Vit
Gem 8h2'15"31d42'
Stuckey,James E
Dra 11h49'22"67d46'
Stuckey,Mike
Aql 19h30'30"14d39'
Stuckey,Opal H
Uma 11h43'0"49d32'
Stuckey,Robert M
Aur 7h8'37"40d35'
Stuckey,Wes
Tri 2h40'1"34d1'
Stuckey,Wes
Ori 4h41'36"10d22'
Stud Muffin
Dra 15h54'52"62d29'
Stud Muffin
Cep 21h4'50"58d41'
Stud Muffin
Ori 5h46'20"10d13'
Studdard,James A "Star"
Cnv 13h59'21"30d42'
Student,Madeleine Duchesne
Com 12h28'43"21d44'
Studer,Andrea D
Uma 11h37'25"42d19'
Studi,Daniel
Tau 4h1'12"20d30'
Studi,Khelan
Tau 4h1'56"22d29'
Studi,Leah
Tau 4h3'53"23d4'
Studio Disco I
Her 17h24'44"21d20'
Studt,Laura
Peg 22h35'23"27d16'
Stueley,Fran & Bill
Cyg 21h37'1"42d3'
Stuetz,Dora
Ari 2h43'39"25d40'
Stuey
Uma 11h33'44"32d38'
Stugard,Leif Varin Markle
Mon 7h7'48"-7d7'
Stuhlmiller,Cori
Lmi 10h31'22"32d35'
Stulich,Peter
Boo 14h53'0"31d0'
Stull,Wendy
Del 21h1'49"18d59'
Stuller,Jr,Matthew Gordy
Her 18h19'43"14d24'
Stullich,Dorothy
Mon 6h56'1"-5d45'
Stullick Family Star, The
Lyr 18h50'26"42d23'
Stultz,Barbara
Cyg 21h32'1"44d45'
Stultz,Bradley William
Oph 17h3'43"10d43'
Stumbo,Brittani Alayne
Mon 7h43'32"-2d15'
Stumm,Frederick
Aur 6h9'1"46d26'
Stumme,Sarah
Cas 2h29'34"60d41'
Stump,Bo Dickson
Aql 19h50'16"14d11'
Stump,Lindsay Lee
Ori 6h6'53"7d4'
Stump,Marion J
And 2h30'18"39d3'
Stumpf,Alexis Victoria
Crampton
Sge 20h16'22"16d39'
Stumpf,Kelly
Lac 22h34'35"55d13'
Stumpf,Manfred
Cas 0h31'0"63d56'
Stumpf,Michael Horst
Ser 15h50'54"19d51'
Stunar
Aur 4h55'24"50d35'
Stunard,Joan
Lyn 8h27'57"41d2'
Stunner,Ickle
Aur 5h0'1"46d22'
Stunner,Stableford
Ori 5h57'20"16d51'
Stunson,Clyde Barry
Boo 15h13'15"50d25'
Stupak,Bob
Boo 14h30'47"21d45'
Stupakis,George
Aur 6h51'50"37d12'
Stupecky,Bohmir
And 23h2'24"43d16'
Stupies-25,The
Peg 22h20'10"32d57'
Stupka,Elsie C
Cas 0h43'45"67d12'
Stupp,Jack
Her 17h26'58"28d25'
Sturdevant,Molly Ann
Peg 22h28'13"29d49'
Sturdevant,Shelley
Cas 0h22'1"67d32'
Sturdivant,Jocelyn
And 23h0'58"51d38'
Sturdy,Rebecca
Peg 22h16'0"30d50'
Sturgeon,Doris Ann
And 1h52'46"38d31'
Sturgeon,Lisa Marie
Cas 23h22'59"54d1'
Sturgeon,Philip
Ori 5h58'25"0d5'
Sturgeon,Wesley
Her 17h18'0"20d33'
Sturges,Gabriel Clowe
Boo 13h39'47"22d9'
Sturgill's Star,Traci
And 0h46'38"33d37'
Sturgill,Michele Renee
And 23h39'1"40d7'
Sturgis,Harvey
Aur 5h17'52"48d40'
Sturgis,Ralph-Rita Davidson
Gorgias
Ori 4h59'26"-2d8'
Sturguess,Kyle Thomas
Lyr 19h22'0"30d39'
Sturkey,Roxanne
Cyg 20h22'22"30d55'
Sturley,Walter H
Aur 6h28'14"40d26'
Sturm,Astrid
Sgr 19h19'47"-20d25'
Sturm,Fletch
Cet 1h15'1"-3d43'
Sturm,Johann
Sge 19h11'54"18d58'
Sturm,Josef
Boo 14h20'11"14d5'
Sturm,Justin Rodney
Per 2h50'15"43d58'
Sturm,Michael
Ser 15h15'41"2d19'
Sturm,Rosemarie
Cas 2h18'17"67d44'
Sturms,Faith
Cyg 20h14'23"38d40'
Stursberg,Craig Robert
Oph 18h15'47"10d44'
Sturtevant,Jim
Aur 7h11'13"41d5'
Sturtevant,Kathryn
Lac 22h20'32"40d24'
Sturtevant,Renee A
Cet 1h51'15"0d26'
Sturtevant,Suzzette
Cas 1h49'31"58d35'
Sturtz,Bob
Boo 14h58'11"37d34'
Sturtz,Jess Roger
Per 2h56'46"55d44'
Sturtzer,Michel Olivier
And 0h21'15"40d34'
Stutes,Alexander Lachlan
Greene
Cnv 12h6'15"34d38'
Stutler,Michael Lee
Ser 16h16'44"-3d23'
Stutsman,Sharon Marie
Henslin
Peg 21h38'55"25d1'
Stuttard,Natasha Naomi
Ori 5h54'29"7d45'
Stutts,Cynthia
And 23h36'17"39d59'
Stutts,Salvador Jaime
Antonetti
Oph 17h37"13d-28"52'
Stutz,Mike
Her 16h23'19"21d20'
Stutzke,Anja
Leo 10h51'12"0d40'
Stutzki,Thomas
Lyr 18h26'53"47d51'
Stutzman,Altagracia
Cmi 7h7'14"4d18'
Styburska,Zoe
Cas 0h18'37"62d51'
Styburski,Agnes
Sgr 18h48'40"-36d29'
Styczynski,Nancy Sherlock
Cas 1h32'42"65d29'
Style By Franco
Cyg 20h2'32"31d42'
Styles,Samantha
And 2h18'26"40d25'
Styner,David Marc
Cnc 8h3'25"6d57'
Stypa,Paul
Aur 6h57'0"35d30'
Styrt,Sailing in San
Diego...Stacy
Del 20h15'26"14d33'
Stéfane et Natacha
Lyn 8h29'22"39d36'
Stéfano
Cyg 20h58'10"31d7'
Stéphane
Dra 17h43'52"70d32'
Stéphane Daniel
Uma 9h5'0"62d17'
Stéphanie
Umi 13h38'26"71d22'
Stéphanie
And 1h2'47"41d8'
Stéphanie
Cam 5h52'38"58d37'
Stéphano et Corinne
Per 4h4'22"50d39'
Stöckl,Elisabeth
Cmi 7h16'16"1d33'
Stöcklin,Mischa
Cam 5h57'1"61d35'
Stöhr-Meix,Dorothea
Cap 20h34'24"-10d2'
Stölck,Lita
Tau 3h56'27"10d1'
Stülpnagel,Calla V
Gem 7h23'1"35d4'
Stünkel,Saskia Tamara
Sgr 19h1'13"-28d56'
Stürmlinger's Star
Vul 19h23'0"26d53'
Su,Dennis
Cep 2h53'24"77d28'
Suardini,John
Ser 15h57'1"20d31'
Suarez,Angela
Cas 0h13'22"47d24'
Suarez,Bernadine
Lyr 18h50'51"40d47'
Suarez,Dayana
Lyn 8h2'12"33d55'
Suarez,Hiram Anthony
Cep 21h38'21"55d55'
Suarez,John William
Her 17h3'30"43d51'
Suarez,Ramiro
Hya 10h45'55"-17d57'
Suarez,Teresa
Peg 0h6'1"14d45'
Suave,Rico
Cet 3h3'27"5d16'
Subalusky,Amanda Lee
Michelle
Mon 6h29'0"-8d53'
Subbiondo,Ralph
Per 1h58'47"53d2'
Subhash Kapur
Cam 12h29'44"80d27'
Subi
Dra 20h26'14"67d41'
Subia,Joanne
Lyr 18h19'19"38d31'
Sublett,Tara Rhnea
Cyg 19h33'42"31d48'
SUBLIME
Uma 13h54'18"51d1'
Sublime KD
Ori 5h4'29"13d19'
Sublime Sandra
Cas 0h31'47"61d5'
Subokow,Edward
Per 3h22'13"50d4'
Subotich,Gina Christine
And 2h18'46"41d12'
Subotich,Michael M
Cam 6h11'40"58d38'
Subotich,Patricia Penman
Ori 5h38'50"11d18'
Sucato,George & Pauline
Peg 21h59'32"30d59'
Sucato,Peter Augustine
Cep 20h21'33"60d34'
Success
Leo 10h1'23"8d59'
Suchao
Aql 18h56'0"-5d49'
Sucharski,Frank J
Dra 17h48'23"61d15'
Sucher,Kathleen Ann
Vul 20h1'42"28d19'
Suchie
Peg 22h14'0"31d8'
Suchko,Daniel Joseph
Lac 22h1'31"51d10'
Suchowij,Cathcrine M
Cas 2h12'34"73d30'
Suckle,Benjamin
Equ 21h18'54"2d38'
Suckman,Barry & Judy
Lmi 10h57'1"27d6'
Suda,Mari
Dra 16h43'1"69d30'
Suda,Yoshie
Uma 11h43'54"43d1'
Sudarius Fortitude
Uma 9h39'29"71d17'
Sudduth,John Christopher
Dra 18h56'35"68d26'
Suderman,John Randolph
Hya 8h12'22"3d37'
Suders"Suds",Donald E
Her 16h36'29"49d56'
Suders,Rod
Sct 18h42'14"-6d17'
Sudesh
Cas 1h52'36"73d19'
Sudi
Lyn 8h12'59"43d59'
Sudie
Eri 2h46'41"-4d58'
Sudjian,Debra S
Mon 7h50'59"-2d29'
Sudol,Richard
Cet 1h5'59"0d36'
Sudol,Ryszard
Per 3h53'12"39d11'
Sudre,Guy
Cam 13h5'32"81d12'
Sudy
Aqr 22h28'50"-18d43'
Sue
Lyn 7h57'47"41d36'
Sue
Lyn 7h28'19"51d47'
Sue & Andrew;Voor Altigd
Cyg 20h33'23"46d1'
Sue & Bert's Corner of the Sky
Uma 11h10'1"40d32'
Sue & Greg
Per 3h11'44"46d53'
Sue & Kit
Peg 22h34'30"2d25'
Sue & Kris's "So Much"
Psc 0h47'47"32d39'
Sue & Loc's Small Squair
World
Lyr 18h41'14"41d17'
Sue Anne
Tri 1h30'33"31d21'
Sue Bal
Cyg 21h54'47"53d1'
Sue du Ciel
Lyr 18h36'18"37d38'
Sue Lee
And 0h38'34"40d32'
Sue Love
Per 2h30'34"57d37'
Sue Mai
Cma 6h16'0"-26d4'
Sue Star
Equ 21h7'1"11d53'
Sue's Christmas Carrell
Lyr 18h44'39"45d50'
Sue's Star
Aql 18h58'1"15d49'
Sue's Star
Aql 19h53'18"14d34'
Sue's Star
Mon 7h6'43"-7d9'
Sue's Star
And 1h43'22"40d14'
Sue's Star In The Sky
Dra 17h33'55"68d55'
Sue's Star, Barnaby's Bauble
& Tom's Toy
Ori 6h1'27"2d40'
Sue's Valentine
Cas 23h42'1"61d48'
Sue's Wantakiss
Cas 23h28'59"61d32'
Sue's Wish
Cas 3h0'46"60d7'
Sue-Kyong
And 2h21'41"42d18'
Sue-n-Tom
Eri 3h12'30"-15d35'
Sue-Q
And 0h19'31"34d10'
Sueanndy
Peg 23h20'32"32d29'
Suehr,50th Anniversary Dick
& Ruth
Crb 16h16'55"37d55'
Suel III,James L
Lac 22h0'46"51d1'
Suelan
Lyn 9h21'37"41d25'

Suella-Jo
Com 12h41'22"21d13'
Suelly e Alan
Aur 5h18'24"45d23'
Suepreme
Peg 21h59'40"11d14'
Suerila's Star
Cas 1h2'0"58d6'
SueRyan
Uma 11h35'24"61d5'
Sues,Ann Lignell
Hya 9h36'22"2d12'
Sues,Barbara Laura Clark
Eri 4h13'34"-16d47'
Sues,Phillip Michael
Her 18h44'46"12d53'
Sues,Robert Phillip
Hya 8h15'51"4d30'
Suetholz,Sandy
Del 20h13'56"11d15'
Suez
Lyr 18h30'1"35d20'
Suffern,Rebecca Lynn
Del 20h18'44"20d28'
Suffi,Elio
Dra 16h33'19"62d39'
Suganuma,Perry K
Cet 2h41'12"6d28'
Sugar
Uma 10h26'0"61d45'
Sugar
Her 16h30'51"40d13B'
Sugar Bear
Boo 13h57'37"25d29'
Sugar Bear
Cyg 19h33'10"30d59'
Sugar Britches
Com 12h11'53"20d29'
Sugar Cube
Cep 3h22'1"78d5'
Sugar D
Peg 23h6'26"20d3'
Sugar Magnolia
Ari 3h0'11"22d56'
Sugar Sparkle
Eri 2h51'35"-15d56'
Sugar,David Michael
Oph 18h17'35"10d39'
Sugarfoot
Lyr 19h22'57"38d1'
Sugarman,Dr Sigmund
Oph 18h6'32"8d45'
Sugarman,Joe
Per 3h15'41"40d5'
Sugden,Sadie
Lyr 18h31'35"30d54'
Sugg,Gary R.
Dra 12h30'10"75d27'
Suggs,"Captain" Garland
Aql 20h10'57"10d35'
Suggs,Renae Elizabeth
Com 12h33'21"20d56'
Sugiyama,Hiroshi
Ori 6h16'24"20d30'
Sugizo
Sgr 19h15'36"-28d18'
Sugleris
Uma 11h31'18"30d8'
Suglia,Family
Uma 10h36'50"58d55'
Sugnaux,Véronique
Cam 5h5'1"60d6'
Sugumele,Dennis
Aql 19h30'14"14d23'
Suh Mei
Cma 7h0'0"-15d50'
Suhle,Hans O
Gem 7h11'7"27d42'
Suhr,Geraldine
Cas 23h39'15"48d50'
Suhr,Juergen
Cap 20h33'34"-23d54'
Suhr,Rich
And 1h11'42"40d12'
Sui Generis "Peggy" LX
Cas 0h3'35"61d45'
Suiluli
Del 20h54'35"6d20'
Suitcase
Sct 18h45'21"-7d15'
Suitts,Allison & Hayley
Ori 6h12'44"7d47'
Sujéamour
Lyr 18h13'20"31d36'
Suk,Leonard Victor
Cnv 12h59'39"47d6'
Suk,Ruth Mary
Cyg 20h58'29"30d46'
Sukalac,Jeanette
Uma 8h30'39"58d11'
Sukenick,Jerry
Aql 18h42'38"-2d8'
Suki
Aql 16h26'36"-1d52'
Suki
Sct 18h53'57"-10d54'
Suki 18
Lyr 18h28'46"40d35'
Sukitsch,Mary E
Cyg 19h32'56"33d21'
Sul
Uma 11h44'36"32d37'
Sulaiman,P A
Her 18h25'30"13d1'
Sulasteri
Cas 0h34'18"66d15'
Sulecki,Maria
Com 13h4'19"28d52'
Sulecki,Maria
Eri 4h30'38"-1d3'
Sules,John T
Dra 16h54'56"63d58'
Sules,Louise M
Sge 18h58'49"18d48'
Sulich,Shirley
Cas 1h41'55"65d24'
Sulich,Sr,Michael
Her 17h48'11"40d13'
Sulis,Michael & Kristin
Dra 17h14'53"67d42'
Sulista,Mme Renee
Cas 1h12'20"66d19'
Sullentrup,Garry
Aur 6h12'53"48d49'
Sulli's Peace
Ori 5h11'49"-4d17'
Sullinger,Jean Marie
Vul 20h19'33"25d18'
Sullins,Alfred Lee
Dra 16h50'59"67d46'
Sullivan II,Wilfrid Thomas
Cep 22h45'43"59d13'
Sullivan III,"Jamie" James
Daniel
Tau 5h50'16"28d31'
Sullivan III,Maynard West
Lyn 8h42'29"33d46'
Sullivan Star,Star Katie-
Kathleen
Peg 23h38'40"27d53'
Sullivan Star,The Laurence E
Aur 6h21'1"37d48'
Sullivan Star,The
Mon 6h18'1"7d36'
Sullivan's Eye On Georgia
Her 16h40'33"47d33'
Sullivan's Shamrock Oasis,J
Stacey
Mon 6h34'28"4d43'
Sullivan,"Bob's Lite" Robert E
Per 1h34'46"54d5'
Sullivan,Amanda C
Lyn 6h56'17"59d48'
Sullivan,Barry Thomas
Lyr 18h58'56"26d1'
Sullivan,Barry Thomas
Per 1h53'25"48d0'
Sullivan,Benjamin Heath
Her 16h3'18"40d11'
Sullivan,Bryan Christopher
Her 17h16'28"45d38'
Sullivan,Bryce A
Cet 0h59'27"-5d39'
Sullivan,Cara
And 2h17'0"42d5'
Sullivan,Carolyn Jeanette
Mon 8h1'5"-5d51'
Sullivan,Christi
Cas 0h26'0"58d31'
Sullivan,Christine
Peg 23h41'15"31d3'
Sullivan,Colleen
Uma 11h7'32"32d49'
Sullivan,Daniel Patrick
Cet 2h18'1"7d11'
Sullivan,Debra Ann
Cyg 20h23'45"38d53'
Sullivan,Devin John
Lac 22h40'29"51d2'
Sullivan,Dianna Lynn & Kevin
A
Eri 3h19'51"-17d28'
Sullivan,Dizzy Teenager
Cep 21h44'49"68d37'
Sullivan,Eileen Ann
Cyg 20h18'48"31d11'
Sullivan,Erin
Ori 6h0'47"-2d4'
Sullivan,Erin P J
Aql 19h12'25"15d45'
Sullivan,Forrest Hillyer Quaid
Lac 22h38'44"40d51'
Sullivan,Haleigh
Dra 16h59'22"67d11'
Sullivan,Jean
And 0h0'10"46d47'
Sullivan,Jeanne Rose
Lib 15h33'52"-8d33'
Sullivan,Jeremiah Patrick
Aur 6h3'0"46d35'
Sullivan,John
Per 2h50'48"45d12'
Sullivan,John
Aql 19h51'1"14d51'
Sullivan,John B
Boo 14h20'21"46d11'
Sullivan,Joseph
Oph 18h6'50"11d27'
Sullivan,Joseph P
Her 18h3'1"47d11'
Sullivan,Jr,John Martin (Jack)
Hya 9h2'49"2d52'
Sullivan,Jr,T W
Aql 19h52'20"12d9'
Sullivan,Katy
Cma 7h14'57"-13d42'
Sullivan,Kelly Anne
And 1h24'0"40d0'
Sullivan,Kelly J
Uma 11h51'40"50d2'
Sullivan,Kerri Lynn
And 2h29'13"47d28'
Sullivan,Kevin Joseph
Boo 13h56'10"15d7'
Sullivan,Kimberly A
Lyr 18h57'48"31d53'
Sullivan,Laura Erin
Cas 3h1'48"60d36'
Sullivan,Lauren Lisa
Cet 1h4'22"-4d36'
Sullivan,Lawrence E
Cmi 7h57'23"8d45'
Sullivan,Linda
Oph 17h31'6"-24d37'
Sullivan,Lisa
Cas 23h22'44"54d21'
Sullivan,Lucky
Boo 14h29'53"19d47'
Sullivan,Marge Nichols
Eri 3h32'47"-6d2'
Sullivan,Marie
Cas 0h59'59"68d34'
Sullivan,Mary
Cas 0h57'33"67d10'
Sullivan,Mary R
And 0h43'12"45d27'
Sullivan,Matthew James
Boo 14h55'52"48d58'
Sullivan,Meaghan Elizabeth
Gem 6h41'60"35d18'
Sullivan,Melissa Anne
Cas 0h32'46"62d23'
Sullivan,Michelle Marie
Lyr 18h34'56"40d32'
Sullivan,ML,Joseph J
Her 16h23'1"25d54'
Sullivan,M"
Uma 9h17'59"52d49'
Sullivan,Namasté Patrick &
Susan
Mon 6h19'52"8d41'
Sullivan,Patricia "Patch"
Joseph
Tri 2h1'54"31d40'
Sullivan,Regan O'Keefe
Lyn 7h28'39"43d10'
Sullivan,Richard Albert
Boo 14h19'41"35d44'
Sullivan,Robert Kevin
Cet 1h54'11"-8d5'
Sullivan,Ryan Joseph
Oph 18h19'57"8d11'
Sullivan,Samuel & Kathleen
Cyg 21h16'18"36d7'
Sullivan,Sandra
Aql 20h31'49"-6d28'
Sullivan,Sean Patrick
Lac 22h20'24"50d29'
Sullivan,Sr,John M
Per 4h19'47"51d18'
Sullivan,Stephanie
Hya 9h4'22"5d47'
Sullivan,Stephen
Hya 9h39'27"-0d11'
Sullivan,Summer Ann
Peg 23h28'32"33d3'
Sullivan,Susan
Lyr 19h6'13"38d5'
Sullivan,The Josie
Hya 8h59'36"4d6'
Sullivan,The Vic
Ori 5h56'0"11d16'
Sullivan,Thomas Phillip
Oph 18h35'58"11d8'
Sullivan,Thomas A
Aql 19h1'0"-1d7'
Sullivan,Tim & Sue
Cep 20h34'0"63d28'
Sullivan,Timothe
Her 18h13'45"40d52'
Sullivan,Timothy R
Uma 8h47'16"51d24'
Sullivan,Tom-Dad- Grampy-
Papa
Uma 11h12'54"62d1'
Sullivan,W Price
Cep 22h16'14"60d6'
Sullivan,William
Eri 3h27'44"-6d8'
Sullivan-Horvath, Judith Ann
Lac 22h25'0"37d53'
Sullivan-My Shining
Star,Timothy Patrick
Leo 9h58'1"10d22'
Sullivani,Stella Briani Dundori
Ori 5h28'53"1d27'
Sully
Lyr 19h13'36"41d11'
Sully & Irma
Ori 5h27'48"-1d9'
Sully Forever,Chris
Per 2h37'34"40d41'
Sully-Morgan,Julia
And 0h30'46"45d19'
Sulmanis,Uwe Heinz
Dra 19h13'19"84d60'
Sultan,Hisham
Lac 22h13'43"47d54'
Sultan,Lucienne
Aur 6h5'36"31d51'
Sultan,Prince Khled Bin
Cep 21h8'1"70d36'
Sulzer,Jesse A
Cam 3h27'20"61d20'
Sumansky,Taylor
Boo 15h32'0"47d38'
Sumard,Franáoise
Uma 9h33'0"48d18'
Sumarlin,J B
Sgr 19h40'16"-32d38'
Sumaylo,Dioscora L
Cas 1h20'44"60d45'
Sumbeam Of My Life
Aur 5h16'27"43d13'
Sumdayus
Peg 23h33'10"31d43'
Sumi
Ori 6h19'59"10d22'
Sumida,Kyle
Aqr 23h17'0"-5d33'
Sumit & Chhavi
Aql 19h30'55"12d59'
Sumkatz
Peg 23h28'30"21d27'
Summer
Lyr 18h41'44"29d37'
Summer "Ad" Ventures I 1994
And 0h3'37"37d47'
Summer "Ad" Ventures I 1995
Peg 0h2'29"28d40'
Summer Adventures I 1993
Uma 9h24'54"48d42'
Summer Faith
Del 20h18'23"20d32'
Summer Joy
Cam 9h4'50"82d23'
Summer Joy
Hya 8h20'34"0d47'
Summer Leigh
Lmi 9h27'37"38d24'
Summer Michelle
Eri 3h41'12"-1d51'
Summer Moon
Lyr 18h30'12"32d18'
Summer Pipes
Hya 8h15'20"5d44'
Summer Rose
Cas 1h47'0"61d12'
Summer Star of Friendship &
Love
Aur 5h28'32"37d46'
Summer Wind,The
Uma 11h2'20"45d58'
Summer `93
Eri 4h56'58"-8d1'
Summer,Lester J
Aur 6h42'22"38d10'
Summerall,Charles
Aql 19h50'0"13d18'
Summerall,Jody
Sct 18h44'46"-6d27'
Summerfield,Alec
Mon 6h56'42"8d35'
Summerfield,Melanie
Cas 0h12'1"61d15'
Summerhayes,Kathleen &
Lillian
Uma 10h15'18"56d13'
Summerhays,Grannie Annie
And 23h3'1"48d39'
Summerlin,Buddy
Ser 16h3'58"10d39'
Summerlin,Valiska Renegar
Mon 6h58'1"8d44'
Summers Star
Ori 5h59'35"10d15'
Summers' Eve
Cnv 12h13'28"44d50'
Summers,Anne Marie
Lyn 7h24'13"44d39'
Summers,Arvel & Bertha
Eri 4h0'22"-15d30'
Summers,Brooke Alexandra
Jordan
Cnv 12h11'59"43d43'
Summers,Cody Lee
Cnv 12h16'31"43d35'
Summers,Donna Lee
Peg 22h1'18"10d49'
Summers,Donna Marie
Lyn 9h0'1"37d30'
Summers,Giselle Dulcie
Com 12h20'1"21d11'
Summers,I Marlene
Cas 3h19'1"75d5'
Summers,Jeffrey Allen
Lmi 9h50'51"33d52'
Summers,Lance R
Dra 15h55'39"61d56'
Summers,Pamela
Cas 2h48'15"65d25'
Summers,Shawnna
Peg 23h21'0"32d54'
Summers,Sr,Joseph L
Dra 16h55'10"68d1'
Summersell,James Bradford
Her 18h17'42"28d19'
Summerville,Berl & Shirley
Peg 23h19'21"30d31'
Summerville,Mark K
Oph 18h39'41"7d5'
Sumner
Lac 22h41'50"56d34'
Sumner,Blair A
Mon 7h44'36"-6d40'
Sumner,Carl Phillip
Oph 17h14'40"10d40'
Sumner,Dorothy Louise
Cyg 21h28'17"31d26'
Sumner,Giacomo Luke
Per 3h8'32"41d26'
Sumner,Jamie
Boo 15h38'59"48d5'
Sumner,Lona
Peg 22h43'51"11d38'
Sumner,Norman E
Aql 19h36'0"1d38'
Sumner,Paul "Pepper"
Mon 8h1'21"-9d1'
Sumner,Stephen
Ori 5h37'20"10d45'
Sumner,William Clay
Ser 15h38'32"20d7'
Sumnler,Timmy Mark
Aur 6h54'54"37d46'
Sumpter,H R Butch
Aql 19h30'22"10d25'
Sumrall,Benjamin Phillip
Aur 6h13'45"32d14'
Sumski,Joan Ann
Sco 17h54'14"-30d20'
Sun City Tucson
Ant 10h39'50"-33d26'
Sun Data
Vul 19h38'34"20d32'
Sun Of Rudiak
Lyr 18h28'55"45d25'
Sun,Christine Hyun & Dong
Uk Kim
Lac 22h14'1"46d23'
Sunahara,Amy
Cmi 7h25'49"1d42'
Sunaina
Uma 13h59'58"53d57'
Sunanda
Cyg 20h11'50"37d46'
SunAngel
And 23h41'36"46d21'
Sunapoee
Peg 23h3'21"20d8'
Sund,Herbert
Lyr 19h18'0"31d36'
Sundance
Cyg 21h14'41"37d34'
Sundance Primary
Teachers,The
Cam 9h34'49"82d2'
Sunday,John
Sco 16h1'58"-28d31'
Sundberg,Hannu
And 2h24'53"48d34'
Sundberg,Mary E
Mon 6h55'22"10d5'
Sundblom,Christine L
Cap 21h22'15"-16d24'
Sundby,Blake Arden
Sco 16h53'57"-40d25'
Sundby,Jordan Russel
Cap 21h37'58"-21d38'
Sundby,Jr,Harold C
Aur 6h18'50"45d25'
Sunden,Nancy
And 0h11'57"47d27'
Sunderland,Rita & Jack
Peg 21h41'54"21d34'
Sunderwirth,Keenan Alexandra
Olivia
Cnv 13h22'13"31d14'
Sundman,David M
Uma 9h41'60"55d`14'
Sundman,Maynard
Dra 19h55'21"60d4'
Sundmark,Astarte
Uma 10h12'39"71d35'
Sundmark,Johann
Cnv 12h39'1"51d44'
Sundowner
Cep 22h24'12"68d55'
Sundquist,Melanie M
And 1h50'51"36d46'
Sundqvist-20000, Sven-Ivan
Dra 17h23'56"71d33'
Sundria,Jean
Lyr 19h16'14"35d29'
Sundsmo,Herbert Alan
Aur 7h16'41"36d40'
Sundstrom,Kim
Cam 5h44'54"67d34'
Sundén,Bertel
Cep 21h59'16"68d59'
SunElaine
Cam 3h58'21"57d26'
Sunflower
Tri 2h6'59"33d25'
Sungy,Mora
Com 12h23'37"24d35'
Suniadane
And 0h53'52"41d13'
Sunier,Marlon
Lyn 7h5'46"60d31'
Sunil
Ori 5h55'18"18d0'
Sunio,Bituing Ryan Oliver
Fqu 21h1'52"2d57'
Sunita
And 1h2'34"36d10'
Sunkel IV,Phillip C
Cnv 12h41'32"41d9'
Sunkel,Robert L
Dra 18h35'52"70d15'
Sunkin,Lori Yves
Sge 20h6'24"20d17'
Sunnervik,Claes
Umi 13h43'51"76d48'
Sunni
Sgr 19h40'35"35d22'
Sunny
Cyg 21h39'57"28d50'
Sunny
Vir 14h3'52"7d29'
Sunny
Lyr 18h49'34"31d59'
Sunny "T"
Umi 17h25'54"75d45'
Sunny Moon
Cnc 8h49'41"9d16'
Sunny's Acre II
Ori 5h1'31"15d36'
Sunnyboy GM
Cnc 8h29'1"8d38'
Sunnye Dawn
Ori 5h34'0"-2d11'
Sunrise
Leo 10h32'59"15d17'
Sunset Memory With My Dear
Sgr 19h16'29"-28d56'
Sunset/Dawn-Amanda's
Valentine
Peg 22h20'1"27d51'
Sunshine
Lyr 18h28'23"31d22'
Sunshine
And 23h18'23"46d22'
Sunshine
Cas 23h40'0"62d17'
Sunshine
And 2h31'57"39d37'
Sunshine
Lyr 18h17'46"42d53'
Sunshine
And 23h32'36"45d32'
Sunshine
Lyn 9h1'23"40d46'
Sunshine
Lyn 7h6'42"58d46'
Sunshine
Mon 7h13'55"-7d0'
Sunshine
Cas 2h31'0"60d25'
Sunshine
Mon 7h37'43"-3d28'
Sunshine
Boo 15h0'11"30d19'
Sunshine
Cyg 19h42'39"30d59'
Sunshine
Sge 19h59'34"16d39'
Sunshine & Rainbows
Mon 6h13'37"-10d43'
Sunshine Always Bill & Jen
Cyg 19h32'21"33d18'
Sunshine I
Uma 8h31'11"53d18'
Sunshine Loves Tony Forever
Aql 18h44'43"11d41'
Sunshine Lynn K
And 0h9'0"34d44'
Sunshine Princess Dolly
(CTL)
Boo 15h2'19"50d20'
Sunshine Sheri D
Del 20h25'1"20d27'
Sunshine Star,The
Cet 1h37'42"-12d35'
Sunshine's Destiny
Mon 7h7'55"-6d27'
Sunshine's Starshine
Lyn 7h33'29"38d59'
Sunshine's Treasure- Mr
Happy
Uma 12h2'0"36d40'
Sunshine,Dana C
Lyr 18h17'1"35d28'
Sunspirits Esprit De Corps
Uma 10h56'13"47d54'
Sunus
Aql 19h56'10"11d58'
Suny
Aql 18h57'50"10d36'
Suojala,Ari Antero
Peg 22h59'50"30d50'
Suozzimoto
Ari 3h7'58"16d44'
Super Colette
And 1h1'25"47d54'
Super Duper Daddy Dupe
Ori 5h58'37"17d39'
Super Duper JHC
Tri 1h29'45"30d17'
Super Enos
Lyn 8h27'20"47d23'
Super H
Cam 3h23'1"61d15'
Super Noëlle
Lyn 8h23'55"43d7'
Super Petard
Her 18h20'59"18d49'
Super Rob
Umi 14h22'16"68d29'
Super Roland
Tri 2h19'21"31d11'
Super Sandy
Ori 6h1'17"-2d4'
Super Scott
Ari 3h0'23"22d16'
Super Snake
Oph 17h58'25"0d54'
Super Staff
Cyg 21h6'1"30d36'
Super Star
Cet 3h1'20"4d12'
Super Sue
Dra 17h43'34"70d4'
Super Sue
Cyg 19h55'52"38d50'
Super Trace
And 23h46'1"46d18'
Supera,John
Cep 22h27'57"59d45'
Superache
Vul 19h22'55"22d31'
Superbubu
Eri 3h0'58"-6d10'
Supergrandad
Peg 22h59'18"32d35'
SuperJohn
Cet 3h18'34"9d34'
Superman-Doig
Her 18h12'0"40d42'
Supermazzu
Cyg 20h23'52"40d5'
Supershow Nevada
Aur 6h5'1"45d1'
Supersiggi-the Ultra Mom
Cma 6h55'15"-16d4'
Superstar "Elfi"
Ser 15h53'43"1d17'
Superstar Dad
Her 17h37'43"41d15'
Superstar Micou
Peg 22h36'15"28d21'
Supertonno,Annalisa
Cas 2h10'37"60d48'
Superus Artifex Kwi Nan
Ori 5h50'39"16d52'
Suplee,Janet
Cas 1h54'24"58d42'
Suppe,Carol
Cas 1h54'50"60d45'
Suppies,Zachary
Aur 6h6'21"38d50'
Supple,Bart Patrick
Uma 11h22'39"37d46'
Supple,Dr Edward W
Oph 18h31'1"10d21'
Supple,Marcia Brown
And 23h22'0"44d56'
Supple,Matthew Edward
William
Her 16h25'45"39d56'
Supplee,Dorothy F
Lib 14h29'44"-23d8'
Supreme Star Chambers aka
BobStar
Cam 12h29'41"84d49'
Sur les rives d'un rève
Tri 2h38'11"34d15'
Surace,Ian Daniel
Aqr 21h3'43"0d18'
Suranie,#1 Father, Mitchell
Anthony
Her 16h9'55"10d10'
Suranyi,Sandor Matthias
Ari 2h39'12"21d40'
Surbrook,Ceil Scott
Boo 14h8'42"54d39'
Surely Valentines of Denise's
Heart
Lyr 18h25'1"46d46'A
Surerus,Marianne
Cam 13h11'41"80d46'

Surette,Mary Lou "Chickie" Cas 0h44'0"73d16'
Surette,Robert F Cnv 12h43'18"51d39'
Surgot,Noel Sgr 19h28'5"-31d19'
Suri's Destiny Hya 8h9'18"1d42'
Suriani,Valentina Lyn 8h13'14"57d40'
Suriano,Spencer Per 2h51'59"35d30'
Suring,Derek S Cnv 12h44'36"48d48'
Suris,Adam Aur 6h1'31"35d43'
Surma,Stan Cep 20h45'0"61d14'
Surmont,Jacques Crb 16h14'50"38d31'
Surniak,Forever Dazzling Laura Lac 22h6'53"51d48'
Surowiec,Sharon J Aur 20h39'37"-13d43'
Surprises Lmi 10h5'11"38d37'
Surratt,Lorraine Cyg 20h2'30"40d18'
Surtees,Jean Lyr 18h17'0"46d58'
Surufka,Papa Leo Leo 10h42'35"14d16'
Survivor Hya 8h20'23"-6d24'
Surwillo,Chester Raymond Her 16h42'20"32d29'
Susa Nicjeza Lavjlla Ser 15h14'51"0d53'
Susan Cyg 21h0'53"38d44'
Susan Umi 14h51'53"70d32'
Susan Uma 11h59'58"50d1'
Susan Gem 6h55'35"18d27'
Susan And 0h52'43"35d24'
Susan Peg 22h40'55"27d43'
Susan And 1h52'1"36d30'
Susan Lyn 7h53'0"44d39'
Susan Lyn 9h39'56"40d42'
Susan Psc 1h0'32"20d24'
Susan Sge 19h35'26"16d34'
Susan Uma 12h3'1"30d5'
Susan Uma 9h32'27"47d29'
Susan Mon 6h57'1"8d34'
Susan & Bailey-Antigua 95 Ori 5h4'28"-1d22'
Susan & Dick Car 8h0'11"-54d23'
Susan & Fancy Fri 4h10'1"-8d4'
Susan & Gordon Sex 10h38'13"-5d28'
Susan & Igloo Forever Sge 20h1'43"16d8'
Susan & Jon Peg 22h37'0"8d28'
Susan & Kitty 30th Anniversary Star Cas 1h19'1"62d41'
Susan & Loren (Madonna & Her Biker) Lyr 18h30'28"45d18'
Susan & Shaun Lyr 18h38'53"34d38'
Susan 1 Mon 7h4'21"3d60'
Susan Ann Lyr 19h4'47"25d53'
Susan Ann And 0h17'28"36d41'
Susan Ann Cas 0h31'27"74d43'
Susan Ann Vul 19h47'30"20d1'
Susan Ann Cas 1h18'1"60d44'
Susan Anne Cas 0h7'10"60d19'
Susan Anne Lmi 10h9'42"38d25'
Susan B Cyg 19h57'20"48d3'
Susan Beth And 23h35'56"47d17'
Susan Claire Katherine Cas 1h28'0"71d15'
Susan E,The Peg 23h38'44"30d46'
Susan Elaine Sgr 19h16'17"-29d6'
Susan Eve Cas 0h9'24"64d6'
Susan Forever Cas 2h56'21"67d41'
Susan Forever And 0h53'13"45d50'
Susan Jane Com 12h53'23"28d50'
Susan Jane Aur 6h0'56"36d40'
Susan Jean Mon 7h2'0"5d6'
Susan Jean Umi 16h13'24"80d34'
Susan Lee Mon 7h47'16"-1d41'
Susan LeRae Uma 8h34'59"60d27'
Susan Lindsey And 0h13'35"36d0'
Susan Madeline's Star Sct 18h50'31"-7d51'
Susan Marie And 2h18'24"44d22'B
Susan Marie Peg 22h20'0"27d31'
Susan Marie And 0h10'35"36d18'
Susan N Cyg 20h50'16"38d13'
Susan N Peter Lyn 8h2'1"58d46'
Susan Renee And 0h53'57"40d31'
Susan Sara Rose And 1h21'52"38d40'
Susan Sweetpea Cyg 20h31'34"42d1'
Susan Swing Mon 8h7'5"-1d57'
Susan's Eli Aql 19h30'40"1d6'
Susan's Freesia Lyr 18h29'51"31d8'
Susan's Guiding Light And 0h57'0"39d34'
Susan's Guiding Light Cyg 21h17'34"38d37'
Susan's Light Vul 19h18'28"26d43'
Susan's Light Lyn 7h36'39"41d48'
Susan's Love And 2h28'23"49d2'
Susan's Precious Moment Mon 7h38'23"-5d49'
Susan's Rising Star Cyg 19h22'1"54d59'
Susan's Rising Star Peg 22h45'16"35d3'
Susan's Shining Star And 0h48'58"22d10'
Susan's Shining Star Del 20h36'27"10d28'
Susan's Smile And 2h31'32"40d13'
Susan's Spirit in the Sky Cyg 21h35'1"34d36'
Susan's Star Cas 0h24'16"50d18'
Susan's Star Cet 1h7'53"-0d60'
Susan's Star Lyn 8h4'18"46d14'
Susan's Star Sgr 18h48'11"-26d47'
Susan's Star Aql 19h2'51"16d41'
Susan's Star Oph 17h51'53"11d8'A
Susan's Star Barkers Cas 2h22'0"63d56'
Susan's Success Cas 0h52'56"61d36'
Susan's Super Star Vir 15h5'37"6d23'
Susan's Sweet Smile Peg 23h21'50"25d54'
Susan, My Layla Cas 2h3'29"59d41'
Susan,Susan,Susan Cas 0h0'41"54d37'
Susan-Blair Cmi 7h25'53"1d45'
Susan/Dale 04-12-92 Love Eternal Lyr 19h16'47"42d42'
Susan/J K Ori 5h43'41"11d53'
Susana Espinosa De Los Monteros Rosanes Cas 0h35'43"60d22'
Susanbenny Peg 23h31'59"22d54'
Susanel Aur 5h3'41"47d15'
Susanna's SteppingStar To Angel Land Lyn 8h53'56"43d54'
Susannah's Twenty One Cap 20h59'37"-23d42'
Susannah-Luke Lyn 7h55'27"58d51'
Susanne Hya 8h10'1"-0d38'
Susanne Vir 13h50'58"5d11'
Susanne Ari 2h43'47"25d23'
Susanne Lib 15h0'34"-28d22'
Susanne Leo 11h22'33"-6d5'
Susanne Cnc 9h15'13"7d9'
Susanne Lac 22h40'34"56d4'
Susanne 16 Ari 2h42'58"26d12'
Susanne und Martin Sco 16h48'24"-26d6'
Susanne-Stern Cnv 13h46'34"35d27'
Susannestar Ori 5h35'5"-0d23'
Susaphil Lac 22h14'46"38d34'
Susema Uma 10h59'36"57d46'
Susemichel,Diana Lyn 8h8'14"45d4'
Susen,Mrs Cas 1h18'23"75d53'
Susha Dra 13h53'38"69d47'A
Sushen Ori 4h58'57"5d58'
Sushilla (Blackman) Tau 3h42'35"25d60'
Sushinette Uma 13h45'29"60d49'
Sushko,Amy Lyn 8h46'0'42d52'
Susi Cap 20h44'0"-26d53'
Susi Boo 13h59'57"10d36'
Susi Cyg 21h59'46"50d35'
Susi & Rainer Uma 10h3'34"58d37'
Susi,Christine Theresa Lyn 8h52'56"43d17'
Susia's Star,Allan Cep 21h2'42"55d59'
Susie Cam 8h2'56"70d28'
Susie Peg 23h38'47"32d17'A
Susie And 1h11'46"39d20'
Susie DC Lyn 8h25'59"40d21'
Susie Lee Mon 8h3'56"-6d14'
Susie M - Beautiful Lady in Red Vir 12h58'36"-20d32'
Susie O Cas 0h50'35"68d27'
Susie Q Peg 22h44'0"18d47'
Susie Rachel Sge 20h2'28"16d14'
Susie Rain Cyg 21h8'55"48d47'
Susie Sparkle Cet 1h44'53"0d22'
Susie Woogee Ori 5h16'47"-8d40'
Susie's Cutie Peg 22h15'15"30d45'
Susie's Song Oph 17h52'56"1d13'
Susie's Star Crb 15h48'23"30d53'
Susie-Q Cmi 7h30'57"0d20'
Susie-Rickie Aur 4h50'43"51d2'
Susino-Forever Friend/ Fella,Sal Uma 8h54'31"47d32'
Suski,Jackson Oliver Ser 15h30'32"10d57'
Susnow,Arthur Sgr 18h55'43"-22d52'
Suspeedtia Dan Aur 4h59'32"50d10'
Susser,Dr Murray Oph 17h0'47"-0d13'
Susser,London Cam 12h43'39"82d14'
Susskind,Andrew B Oph 16h30'1"-6d39'
Sussman's,Brian Thermoluminescence Cmi 8h4'41"6d47'
Sussman,Alicia And 1h50'1"38d43'
Sussman,Howard & Sherry Cyg 21h5'41"38d9'
Sussman,Jacob Ried Aur 6h5'17"31d1'
Sussman,Kiwini Gann Ori 5h56'23"15d28'
Sussman,Rachel Andrea Cas 0h52'59"61d39'
SuSu Uma 9h25'48"49d27'
Susu Cas 0h5'45"60d50'
Susu Dra 15h11'51"63d7'
Susuki,Marian S Cyg 19h41'30"31d31'
Susumu & Tomoko,So Sweet Sgr 19h18'44"-28d2'
Suszka,Thomas John Cep 2h50'56"77d58'
Suta,Rickey Boo 14h7'58"43d39'
Sutch,Kerry Louise Cas 23h28'43"61d41'
Sutcliffe Ori 5h9'34"-5d4'
Sutcliffe,Jackie Lyr 18h16'52"42d8'
Sutcliffe,Jackie Umi 13h30'39"72d7'
Sutcliffe,John `Snookes' Cyg 19h59'37"41d12'
Sutcliffe,Jr,William J Her 17h17'58"44d53'
Suter,John,Patti Spinner Peg 23h19'44"31d31'
Suter,Lynn M Cet 2h48'13"5d24'
Suter,Michael William Sex 9h50'58"-2d2'
Suter-My Teddy Bear Star,Bbabette Jo Uma 9h13'36"60d3'
Suterwalla,Anis Ori 5h21'1"15d6'
Suther,June P Vul 21h13'23"28d19'
Sutherland Family,The Cet 0h30'45"-17d43'
Sutherland,Amanda Josephine Peg 22h43'30"35d0'
Sutherland,Andrea Estella Castro Lyn 6h29'55"60d20'
Sutherland,Benjamin Uma 10h4'37"55d40'
Sutherland,Gerri Ori 5h33'17"-6d25'
Sutherland,Ian Peg 22h1'11"32d40'
Sutherland,Joan Eri 4h14'23"-16d54'
Sutherland,John James Aur 5h51'27"40d9'
Sutherland,Kathleen W Vir 13h8'43"-2d8'
Sutherland,Lisa Sct 18h54'49"-8d13'
Sutherland,Mason Adam Dra 16h38'43"63d12'
Sutherland,Miles Uma 9h22'28"48d56'
Sutherland,Paul Michael Aur 6h54'35"37d58'
Sutherland,Ray Hya 9h0'15"4d6'
Sutherland,Robert Boo 14h52'44"25d49'
Sutherland,Robert Uma 10h3'20"55d56'
Sutherland,Ross Iain Lac 22h41'11"56d27'
Sutherlen & Wanda Johnson,Eric & John Sge 19h57'18"20d33'
Sutherlin,Barbara Cyg 19h59'58"30d32'
Sutherlin,Nancy Ingrid Leo 10h52'38"-2d19'
Sutila,Analise Rebecca Crb 15h28'1"31d36'
Sutin,Gerald Uma 8h6'44"60d30'
Sutliff,Nancy Lyn 6h49'35"60d14'
Sutliffe,Thomas A Her 18h40'38"18d47'
Sutlive,MD,William Greene Her 15h52'38"44d37'
Sutor,Julie Ori 5h39'20'-0d46'
Sutphin,Sr,ThD,John Boo 13h54'22"17d55'
Sutt's Star Uma 10h0'20"70d59'
Sutte,Jan Peg 22h2'40"2d15'
Sutter,Eleanor Koehler And 23h37'0"42d25'
Sutter,Irene And 0h6'48"43d54'
Sutter,Joshua Frank Lac 22h52'30"50d12'
Sutter,Karl Per 3h50'34"37d37'
Sutter,The Silver Lord David Scott Aql 19h41'53"13d2'
Sutter,Xuan Nguyen Vul 19h22'26"26d36'
Sutterley Star,Paul & Edward Her 16h2'22"18d57'
Suttin,Benjamin Samuels Mon 6h40'0"7d40'
Suttles,Scott Andrew Cet 3h2'54"2d31'
Sutton II,Lawrence T Aur 6h18'0"35d32'
Sutton,A Richard Lac 22h46'17"54d32'
Sutton,C J Ori 5h25'20"-6d47'
Sutton,Cole Oliver Her 16h50'0"50d56'
Sutton,David & Toni Aur 5h53'1"30d59'
Sutton,Dennis Grant Dra 16h45'57"61d38'
Sutton,Elizabeth Oph 16h50'0"-28d38'
Sutton,Grant G Vul 19h58'57"23d38'
Sutton,Haley Elizabeth And 0h21'38"37d1'
Sutton,Jack Pennington Cep 20h43'0"63d57'
Sutton,Jacqueline Starrie Peg 23h27'56"8d22'
Sutton,James Sex 9h39'27"-0d11'
Sutton,James Thomas Ser 16h3'24"10d25'
Sutton,John Michael Cmi 7h58'41"7d50'
Sutton,John Torres Hya 8h47'10"5d42'
Sutton,Laura Peg 21h58'1"29d27'
Sutton,Loyal Davis Aur 4h56'60"40d45'
Sutton,Martyn Leese Her 18h4'32"14d60'
Sutton,MD,Jonathan M Her 18h55'13"12d11'
Sutton,Peter Wallace Crb 16h7'50"36d2'
Sutton,Ray Aql 20h1'23"11d6'
Sutton,Rebecca Lyr 18h45'16"34d58'
Sutton,Rick (My Precious D) Lac 22h4'20"46d40'
Sutton,The Miss Meg Cam 3h12'28"60d52'
Sutton,Zach Aql 20h1'28"12d9'
Suwalkowski,Renee Suzette And 0h59'14"45d47'
Suwalkowski,Ryan James Aur 6h24'43"35d51'
Suyer I Love You Forever,Edward Dra 18h7'0"65d28'
Suyo,Katherine Elizabeth Lyn 7h47'0"42d49'
Suyo,Kevin James Cep 21h11'37"60d8'
Suz's Sparkle Sco 17h50'49"-38d24'
Suzanna Cas 0h14'11"66d8'
Suzanna Cas 0h39'1"75d34'
Suzanna-Jewel Of The Sky Sge 19h15'40"18d55'
Suzanne Eri 3h35'22"-4d17'
Suzanne Lyr 19h2'43"37d45'
Suzanne Umi 14h40'23"80d16'
Suzanne Cyg 19h34'22"30d58'
Suzanne Tri 2h16'19"30d29'
Suzanne Lyn 8h6'11"33d50'
Suzanne Peg 22h59'14"21d48'
Suzanne Com 12h31'38"20d47'
Suzanne Gem 7h10'24"24d56'
Suzanne Cas 0h36'26"58d17'
Suzanne Cyg 20h34'0"50d6'
Suzanne Vul 19h17'58"21d45'
Suzanne Hya 8h46'5"1d7'A
Suzanne Aur 6h0'47"46d3'
Suzanne & Jeffrey "Our Star" Peg 23h22'26"17d33'
Suzanne & Richard "Wedding Star" Cyg 20h5'11"41d1'
Suzanne 12-18-69 And 23h33'35"42d2'
Suzanne C Eri 2h55'26"-17d5'
Suzanne De Grandpré And 0h4'58"31d50'
Suzanne Elizabeth Cyg 19h41'55"40d14'
Suzanne Eve The Brightest One Oph 18h23'60"7d36'
Suzanne Forever Ori 5h4'31"13d51'
Suzanne M Lyn 7h52'31"34d42'
Suzanne Marie Eri 3h35'31"-13d21'
Suzanne Michelle Marie Vul 19h53'13"20d17'
Suzanne Ophelia And 23h22'42"42d14'
Suzanne Victoria Tau 4h39'11"15d19'
Suzanne's "Big Four-O" Peg 22h3'0"10d29'
Suzanne's Sweetie Mon 6h57'29"10d36'
Suzanne,Godess Of The Cosmos Sga 19h55'34"19d9'
Suzanne,Gwendolyn And 23h37'46"47d34'
Suzanne,Kalyn Cet 2h46'12"1d32'
Suzannes Star Cas 0h19'0"62d58'
Suzannia Jepsonicus Equ 21h6'0"11d12'
Suzee-D And 1h3'0"40d29'
Suzette Sgr 19h2'57"-27d5'
Suzette Mon 8h4'22"-4d9'
Suzette Ori 5h31'26"-0d59'
Suzette Cas 0h21'53"75d43'
Suzette Peg 22h10'21"26d29'
Suzette H A Uma 9h45'44"50d14'
Suzi Cas 2h19'21"61d37'
Suzi in the Sky With Diamonds Cas 1h46'20"72d47'
Suzi Jack Cas 2h38'30"75d48'
Suzi's Starbright And 23h39'47"47d51'
Suzie Sge 20h1'35"20d3'
Suzie Cas 3h8'12"61d22'
Suzie B Cyg 19h20'55"44d44'
Suzie et Marc Peg 22h46'18"31d46'
Suzie P 1-8-55 Aql 19h53'49"12d47'
Suzie Sparkle Cas 1h9'0"70d8'
Suzie-CB1 Uma 10h45'14"60d41'
Suzie-Q Love Krissi- Lizzi Uma 9h43'1"68d23'
Suzie-The Star Of My Heart And 2h20'34"39d53'
Suziura,Nagiko Aql 20h11'51"7d44'
Suzo Com 13h18'38"30d49'
Suzuki,Anne Cyg 21h9'45"41d1'
Suzuki,Chizuko (Chiko) Mon 7h0'36"8d15'
Suzuki,Kentaro Peg 23h0'48"24d40'
Suzuki,Lisa Miki Mon 6h25'0"10d51'
Suzuki,Misasha Lyr 18h51'19"40d17'
Suzuki,Mr Yuta Aur 6h3'21"30d44'
Suzy Vir 13h22'26"-2d25'
Suzy Peg 22h30'42"27d45'
Suzy Aql 19h50'1"15d34'
Suzy M And 23h21'34"50d25'
Suzy Parfum d'été Umi 13h42'16"71d30'
Suzy Q Dra 14h40'33"64d45'
Suzy Star Lyr 18h59'29"33d7'
Suzy's Angel Cas 23h22'42"63d23'
Suzy's Light Gallery Com 12h53'32"30d56'
Suzy's Star And 0h45'20"37d32'
Suzy-Q Lyr 18h42'0"30d59'
Suárez,Nancy Rosemary Lyn 7h54'54"51d2'
Svad Abdi Ali Cam 11h51'40"78d59'
Svanoni,Gaia Psa 22h33'13"-26d14'
Svartback,Henrik Cam 5h58'32"61d46'
Sveaas,Christen Cam 5h4'39"58d16'
Svec,Charles H Lyr 19h21'1"40d41'
Svec,Jordan Thomas Peg 0h3'1"20d4'
Svec,Laura Cas 0h35'18"65d34'
Sveen,Barbro Edla Cam 13h15'25"78d31'
Sveen,Michelle Marie Peg 22h9'1"26d37'
Svejkovsky,William T Hya 8h55'55"1d23'
Sven's "Nachtlicht" Per 2h54'31"37d12'
Svendsen,Kristen Cas 0h29'0"61d35'
Svendsen,Sarah Lac 22h0'1"50d4'
Svendsen,Sarah Umi 13h21'1"71d25'
Svenja And 23h23'43"52d58'
Svenja Lyr 19h14'28"31d0'
Svennevik,Kjell A Cyg 20h17'4"45d27'
Svenska Lac 22h44'36"53d36'
Svenson,Lise Vir 13h38'20"-4d12'
Svensson,Lola Cyg 21h5'56"48d50'
Svensson,Paul Cep 23h14'27"64d51'
Sverrisson,Siggeir Ori 5h39'23"8d8'
Sveta Ori 4h53'44"-1d20'
Svetal Aur 7h15'17"36d33'
Svetlana Lyn 7h40'0"44d12'
Svetlana Lya 19h21'19"31d35'
Svetlana And 0h21'39"37d59'
Svetlana Cnv 12h46'21"40d15'
Svetlana-Bio-122368 Uma 10h27'44"55d21'
Sveva,Barbara Cep 22h25'56"58d14'
Svoboda's Dream Vir 13h59'60"-1d36'
Svoboda,Jennifer Lynn Mon 6h23'46"10d18'
Svoboda,Scott William Her 18h10'33"38d2'
Svobodu Cyg 21h27'0"30d2'
Svoger Uma 9h59'57"67d45'
Swabon,Bob Aql 18h42'28"-2d36'
Swadling,Bonnie J Peg 22h27'21"33d49'
Swahn,Diana Peg 22h2'21"32d41'B
Swaidner Son,Scott A Boo 14h20'51"12d33'
Swailes,Linda "Moo" Lyn 7h51'15"50d23'
Swailes,Lisa "Toe" Cas 0h25'41"70d60'

Swails,Amelia
Vul 20h18'29"22d50'
Swails,Kalie
Mon 6h44'12"0d44'
Swaim,Christine Robalin
Boo 14h29'20"8d40'
Swaim,Tricia
Peg 21h57'46"24d53'
Swain,Barry E
Boo 14h57'41"41d38'
Swain,Catherine Leigh
And 0h14'14"35d42'
Swain,Lea
Com 13h14'17"22d20'
Swain,Samantha Michelle
Uma 11h57'33"41d26'
Swain,Sean P
Per 3h58'1"31d21'
Swain,Shawn Kenneth
Lyn 7h39'56"51d49'
Swain,Shea & Scott
Vul 20h18'50"23d19'
Swainson,Eliot (E/OT)
Aql 19h7'41"3d24'
Swainston,Scott Christopher
Hya 8h44'27"6d14'
Swale,Lord
Ori 5h32'1"0d8'
Swalley
Cam 13h10'32"82d14'
Swallow,Betty & John
Cyg 19h33'57"39d18'
Swallow,Tina Louise
Umi 14h57'50"71d28'
Swallows,Martin A
Cet 2h46'44"4d15'
Swan Star
Lib 15h30'25"-10d15'
Swan Star,The
Cyg 19h27'32"35d44'
Swan,Anna May
Cas 0h51'37"61d46'
Swan,Bryan C
Cnv 12h31'19"40d12'
Swan,C Alexander
Ori 5h50'48"17d40'
Swan,Dennis F
Hya 9h16'50"5d2'
Swan,Gabriella Ayres
Peg 23h32'33"20d20'
Swan,Helen
Cyg 21h56'36"53d39'
Swan,Jeanie Marie
Cas 0h33'1"67d40'
Swan,Jeanne Ellen
Cyg 20h16'19"41d17'
Swan,Jordan Hulpiau
Cyg 21h33'1"38d9'
Swan,Katie & Sarah
Uma 10h43'10"52d58'
Swan,Mary Bunce
Lyr 18h41'0"36d49'
Swan,Max Nathan Jr & Fran
Oph 17h12'24"11d9'
Swan,Peter
Lyn 8h19'30"41d48'
Swan,Philip Gibson
Aur 5h1'34"38d44'
Swan,Ryan
Per 2h57'32"32d55'
Swan,Tamara Sue
Cas 2h34'15"61d25'
Swan-Knaust,Donna J
Crb 15h53'50"28d48'
Swan;Sing'Unfold' Shimr' in new lite
Cyg 21h35'0"42d3'
Swanberg,Carl Richard
Per 4h10'0"51d59'
Swaney,Irene
Ori 5h56'1"15d35'
Swanger's Star,Frank Lee
Boo 15h3'30"12d14'
Swank,Connor Ryan
Aur 6h33'28"35d40'
Swank,Grandpa & Grandma
Uma 10h34'47"70d49'
Swank,Hanna Elisabeth
Mon 7h30'25"-8d31'
Swank,Michael & Janeen
Sex 10h10'52"-4d26'
Swank,Thomas F
Sct 18h44'1"-5d46'
Swank,Wesley Staton
Dra 14h58'37"64d7'
Swann,Margaret Gaines
Crb 15h54'54"30d23'
Swann,Margaret Gaines
Cas 2h38'19"57d23'
Swannell,Barnum Harry
Cyg 21h9'1"37d32'
Swannell,Kira Mae
Cas 1h54'60"71d8'
Swannie,Eric
Mon 6h54'39"-6d55'
Swanson Anniversary Star
Cyg 19h21'26"54d39'
Swanson Gods Little Angel,Jack S
Peg 22h40'15"27d47'
Swanson's S'Wonderful Silver
Cyg 21h54'1"42d21'
Swanson,Betty
And 0h9'38"35d19'
Swanson,Blakely
Cep 22h55'39"70d54'
Swanson,Bradley David
Cep 21h21'52"65d30'
Swanson,Brenda Jane
And 1h24'39"40d1'
Swanson,Chris
Aur 6h8'46"41d6'
Swanson,Dorothy M
Cas 0h49'21"66d41'
Swanson,Douglas Clark
Hya 8h51'16"2d0'A
Swanson,Edward
Cmi 6h59'51"-16d19'
Swanson,Floyd Elsworth
Ori 6h12'5"14d42'A
Swanson,Hazel Irene
Ori 6h12'5"14d42'B
Swanson,Jack Stanley
Aur 6h32'46"32d15'
Swanson,Jack Stanley
Lyr 18h58'28"37d10'
Swanson,James Michael
Aql 20h8'1"4d31'
Swanson,Jo
Cam 14h18'12"82d12'
Swanson,Judith Ann
Lyr 19h23'52"35d36'
Swanson,Kenneth & Rosie
Mon 6h26'25"11d54'
Swanson,Kristin Heather
Vir 13h35'11"2d38'B
Swanson,Lynn & Ron
Lyr 18h44'17"35d10'
Swanson,Marilyn Ann
Aur 4h56'22"40d29'
Swanson,Matthew Robert
Aur 6h8'56"35d29'
Swanson,P C
Aql 19h56'12"14d44'
Swanson,Pam & Tyler
Sge 19h37'7"16d43'
Swanson,Sandra L
Cam 10h25'1"82d7'
Swanson,Sarah
Sex 10h11'10"-1d44'
Swanson,Susan Diane
Hya 8h51'16"2d0'B
Swanson,Tamara Lynn
Leo 11h55'27"21d42'
Swanson,Virginia J
And 2h2'36"40d50'
Swanton,Keith
Cet 2h31'39"1d19'
Swanton,Ken & Veda
Uma 9h35'2"47d34'
Swapp,Jonathan
Ser 15h14'53"11d29'
Swarbrick,Robert
Per 3h4'43"41d23'
Swarez,Betty
Mon 6h24'12"-1d5'
Swart,Ann & Greg
Peg 22h24'40"35d25'
Swart,G William
Peg 23h33'14"31d48'
Swartout,Bernard J
Hya 9h16'14"2d15'
Swartout,Kelly Rae
Vul 19h21'58"26d57'
Swarts,Janine
Del 20h53'50"7d18'
Swartz,Denise R
Aur 5h51'52"28d59'
Swartz,Inez
Eri 3h43'41"-5d51'
Swartz,Joshua Carl
Ori 5h35'21"0d59'
Swartz,Matthew A
Her 17h29'26"38d10'
Swartz,Patricia Grace T
Cyg 19h30'48"38d1'
Swartz,Reba M
Cam 5h49'48"58d32'
Swartz,Sr,Alex Elias
Aql 19h6'1"0d0'
Swartz,Tracey L
Dra 18h26'15"80d14'
Swartzie
Peg 23h39'36"17d1'
Swasand,Henry
Cnv 12h35'38"33d8'
Swasey,Steve
Lac 22h26'32"54d42'
Swatek,Edward J
Her 16h41'20"20d10'
Swati
Sge 19h57'47"16d30'
Swattridge 1996,Jeremy
Aql 18h56'1"16d6'
Swauger,Catherine "Babs"
Cas 1h22'21"53d24'
Swaymar
Umi 13h25'1"73d48'
Swazey,John M
Boo 14h20'25"22d48'
Sweales,Roger
Peg 23h19'31"31d50'
Swearingen,Jeff
Lac 22h20'18"37d38'
Sweat,Ludy
Dra 13h41'34"68d10'
Sweazey,Christy
Lyn 8h26'50"34d51'
Sweda,Aunt Rennae
Mon 6h39'44"7d41'
Sweda,Stan
Cet 2h43'16"-1d18'
Swedish Song
Lyn 7h45'56"42d8'
Swedish Zeppelin In My Skye,The
Mon 6h36'58"-6d13'
Swedlow,Dave & Ruth
Aql 19h15'54"14d27'
Swedlund,Nils Eric
Cnc 8h56'1"22d9'
Sweeden,Nikki & André Bollaert
Lyn 7h53'1"51d34'
Sweeley,James C
Per 2h55'54"37d11'
Sweeney "A Stellar Teacher",Mary
Lyn 7h33'17"38d1'
Sweeney,Aaron Charles
Ari 2h36'22"21d42'
Sweeney,Connie
Her 16h54'49"14d45'A
Sweeney,Dennis W
Dra 16h43'27"67d4'
Sweeney,Edward Thomas
Her 18h38'40"12d50'
Sweeney,James
Aql 19h25'1"14d47'
Sweeney,Jeffrey Scott
Dra 15h12'29"58d21'
Sweeney,Jenna Marie
And 0h22'37"21d40'
Sweeney,Jennifer
Lyn 8h13'40"47d35'
Sweeney,John Thomas
Her 16h42'53"33d37'
Sweeney,John-Paul
Lyn 7h38'32"52d0'
Sweeney,Jr,Anthony C
Cam 8h39'21"78d6'
Sweeney,Kenneth Lee
Oph 18h8'51"12d28'
Sweeney,Margaret A
Cam 5h52'0"68d41'
Sweeney,Marie Ellen
Cam 5h53'27"68d14'
Sweeney,Mark Edward Rosendahl
Aql 19h6'26"2d45'
Sweeney,Mary
Cyg 21h6'23"30d31'
Sweeney,Mary Therese Boyle
Com 12h34'32"20d2'
Sweeney,Michael
Lac 22h35'41"38d31'
Sweeney,Mike
Ori 5h31'44"-0d17'
Sweeney,Parissa Lynne
Cam 5h50'52"61d42'
Sweeney,Patricia Ann
Mon 7h56'27"-1d44'
Sweeney,Rose-Ellen
Her 16h54'49"14d45'B
Sweeney,Sharon Lee
Cas 1h3'19"58d36'
Sweeney,Shereé Ann Stockton
Mon 7h55'32"-1d28'
Sweeney,Sinead Mary
Aur 5h0'20"48d55'
Sweeney,Terrance Michael
Per 3h7'25"41d12'
Sweeney,The
Cet 2h29'46"0d25'
Sweeney,Will 14
Her 18h2'51"30d43'
Sweeney,Will Atman
Cnc 8h9'57"30d17'
Sweeney,Wilma
Mon 6h39'20"6d31'
Sweeney,Zachary Taylor
Cep 3h12'1"78d3'
Sweep
Ori 6h4'22"20d13'
Sweep
Aur 5h26'49"38d59'
Sweet Amy
Lyr 19h26'28"38d2'
Sweet Andrea Beth
Vul 20h39'24"28d48'
Sweet Angel Laurie
Mon 8h2'13"-1d44'
Sweet Angel Mom
Cas 23h13'53"62d53'
Sweet Anne
Per 4h44'26"51d13'
Sweet Baboo
And 23h26'13"47d22'
Sweet Baby
Cyg 21h22'52"50d35'
Sweet Baby
Dra 12h34'57"68d8'
Sweet Baby James
Cam 4h51'33"68d6'
Sweet Baby Jean
Cyg 19h18'37"45d40'
Sweet Baby Snicky
Lyn 7h55'29"58d27'
Sweet Billy-Carole & Kick-Ass Gene
Cyg 21h34'22"30d32'
Sweet Bonnie
Cas 0h26'0"50d34'
Sweet Carole Ann
Cyg 20h31'45"42d26'
Sweet Caroline
Sct 18h46'47"-7d35'
Sweet Caroline
Hya 8h30'0"-0d10'
Sweet Carolyn
Leo 9h25'37"27d35'
Sweet Christy's Shining Star
And 0h54'0"38d26'
Sweet Danny
Aur 4h57'25"31d43'
Sweet Debbie
Umi 15h17'57"70d23'
Sweet Dove Bobby
Lac 22h23'54"38d54'
Sweet Dreams
Uma 11h47'45"51d18'
Sweet Dreams
And 1h26'36"39d27'
Sweet Dreams Danny
Dra 19h4'35"56d51'
Sweet Dreams Dennis Tanjeloff
Dra 12h33'30"64d6'
Sweet Dreams Jason
Dra 16h31'32"51d56'
Sweet Eating-Kate
Cyg 20h12'38"39d36'
Sweet Eternity
Cas 0h43'48"61d41'
Sweet Family Star
Gem 7h9'59"22d32'
Sweet Feet Johnny's
Boo 15h0'0"48d32'
Sweet Fenton
Del 20h50'51"7d44'
Sweet Genevieve
Lyr 18h15'18"30d40'
Sweet Guisy
Sgr 19h26'38"-32d0'
Sweet Inez
Mon 6h24'26"0d35'
Sweet J Heart
Sgr 19h5'39"-15d57'
Sweet Kristy
Cas 0h1'55"62d33'
Sweet Lady Marie
Del 20h18'26"13d17'
Sweet Leo
Cyg 19h28'26"38d57'
Sweet Lew
Hya 8h24'58"-0d40'
Sweet Lynnie B,The
Cas 23h20'50"62d7'
Sweet Marie
Uma 10h31'34"41d34'
Sweet Melissa
Uma 11h7'51"45d2'
Sweet Melissa
Peg 0h2'0"30d57'
Sweet Melody
Cyg 19h33'45"28d45'
Sweet Michael
Cep 0h9'43"70d7'
Sweet Moe
Ori 5h19'15"15d30'
Sweet Moriah
Uma 11h40'0"31d39'
Sweet Nick
Peg 23h30'1"10d57'
Sweet Olivia
Ari 3h23'50"28d20'
Sweet One-Six
Crb 16h11'47"31d37'
Sweet P
Ori 5h35'59"-0d19'
Sweet P-mer
Cam 3h47'27"52d33'
Sweet Pamela
Cas 0h30'59"75d42'
Sweet Pea
Aur 6h12'0"31d46'
Sweet Pea
Dra 15h56'30"66d56'
Sweet Pea
Peg 21h39'47"23d3'
Sweet Pea
Cas 0h38'0"71d28'
Sweet Pea
Lmi 9h22'34"37d50'
Sweet Pea
Lyn 7h55'1"40d18'
Sweet Pea
Aur 5h4'40"41d54'
Sweet Pea Carol
Cas 23h14'0"60d57'
Sweet Pea Miche'le
Sgr 18h59'23"-28d42'
Sweet Pea's Smooty- Tart
Cnv 13h48'15"46d48'
Sweet Prince
Per 1h49'14"54d22'
Sweet Princess Darlene
Peg 22h43'49"20d12'
Sweet Rebecca
Mon 7h17'43"-0d49'
Sweet Sandy
Peg 23h30'13"11d47'
Sweet Sara
And 0h22'60"40d6'
Sweet Shannon
Cyg 21h6'51"31d33'
Sweet Sherin
Dra 16h57'20"69d55'
Sweet Shining Sue
Cam 7h7'0"60d32'
Sweet Susan
Ori 4h49'0"13d34'B
Sweet Suzy Blue Eyes
Aql 19h6'27"15d22'
Sweet Tara
Cep 21h47'45"55d21'
Sweet Tara 6269
Cas 1h15'40"65d10'
Sweet Thing
Lyr 18h45'48"31d0'
Sweet Tod
Per 1h54'0"56d59'
Sweet Wally
Uma 11h34'1"38d24'
Sweet William
Cep 22h21'52"60d17'
Sweet William
Boo 15h7'21"10d54'
Sweet William
Umi 15h56'51"81d24'
Sweet Yvonne
Mon 7h41'55"-8d59'
Sweet,Alaena Brooke
And 1h18'0"33d44'
Sweet,Alison Marie
Sge 20h1'55"20d14'
Sweet,Bruce Russell
Peg 21h52'0"34d44'
Sweet,Daniel Eugene
Cet 3h10'30"2d28'
Sweet,Elizabeth Sofya
Cas 0h31'46"65d29'
Sweet,Emily Ann
Eri 3h58'1"-10d15'
Sweet,Jean Carson
And 2h24'13"49d46'
Sweet,Joseph
Boo 15h6'20"18d54'
Sweet,Lawrence Collins
Lac 22h38'1"52d56'
Sweet,Russell
Dra 19h30'51"71d3'
Sweet,William Robert
Lac 22h26'31"55d33'
Sweet-Pea
Aql 19h31'46"13d1'
Sweetbaum,Jodi
Crb 15h27'14"31d57'
Sweetcakes
Cyg 19h32'22"30d20'
Sweetep
Uma 11h52'36"43d40'
Sweetest Man,The
Her 17h3'1"48d34'
Sweetheart (5150)
Cet 2h23'29"10d3'
Sweetheart Bumpus
Vul 20h43'1"20d32'
Sweetheart Jinder
Cyg 21h57'53"52d53'
Sweetheart,Julie
Ori 5h30'41"-0d2'
Sweetie
Peg 0h11'1"14d2'
Sweetie
Ori 6h16'21"-2d25'
Sweetie
Mon 6h27'41"3d40'
Sweetie Helen Debbie HDA
Cyg 19h28'18"38d11'
Sweetie Muffin
Boo 15h8'44"48d17'
Sweetie Pie
Lyr 19h14'52"41d16'
Sweetie Pie In The Sky
And 23h20'59"45d10'
Sweetie Pig
Boo 14h9'44"51d54'
Sweetie Precious
Mon 7h39'6"-4d11'
Sweetie Truffle
Crb 15h58'36"26d30'
Sweetie-I Love You 3/17/93
Per 3h29'59"52d41'
Sweetiepuss
Cam 7h6'32"61d16'
Sweeting,Ralph Davidson
Cep 20h32'36"60d16'
Sweetland,Lauren Elizabeth
And 0h11'55"39d22'
Sweetman,William John
Umi 15h28'46"75d1'
Sweetness
Uma 11h5'34"35d25'
Sweetness & Phoo Forever
Cnv 13h47'14"39d59'
Sweetpea
Lyn 8h31'23"45d49'
Sweetpea
Peg 0h7'43"21d40'
Sweetroot
Dra 19h4'38"48d11'
Sweets
Lyn 8h1'49"50d43'
Sweets
And 2h6'36"38d38'
Sweetser,Mary
And 23h24'52"45d30'
Sweetwater Lady
Del 20h17'40"9d57'
Sweety
Tau 5h27'53"25d45'
Sweety W
Ser 15h59'0"4d49'
Sweillan,Manar
Cnv 12h39'0"50d9'
Sweisford,Tyler Jon
Cet 2h35'20"-1d35'
Sweitzer,Anna Aura
And 23h23'34"41d11'
Sweitzer,Sherry Lynn
Peg 21h25'17"8d25'
Swelstad,Matthew Reed
Dra 16h25'1"68d26'
Swender,Katy
Cas 1h27'36"61d51'
Swendsen,Francis J
And 0h26'11"30d8'
Swenie,Robert E
Per 2h25'1"56d52'
Swensen "The Moose", David F
Boo 14h36'0"42d23'
Swensen,Charlie & Joan
Lyn 7h50'14"42d54'
Swensen,Donald F
Cet 1h4'31"-0d55'
Swensen,Grace & Dick
Crb 15h51'47"34d56'
Swenson,Clinton
Cet 0h26'45"-12d19'
Swenson,Dr Rich
Vul 20h58'51"28d53'
Swenson,Grace & Harry
Aql 18h59'31"-1d15'
Swenson,Jr,Eric- Elaine-Sarah,Roland
Uma 13h29'13"61d44'
Swenson,Merri Jayne Stickle
Cyg 20h19'42"38d37'
Swenson,Paul R
Boo 15h12'60"52d55'
Swenson,Sr,Chester Albert
Aur 4h51'46"41d10'
Swensson,Norbert
Ori 5h46'18"20d42'
Swerdlow,MD,Charles D
Dra 17h55'25"64d46'
Swerdlow,Sherry Lee
Eri 3h54'40"-16d0'
Swersky,Dr Stanley
Uma 11h41'43"61d3'
Swertfager,Kelly & Jack
Sge 20h7'26"20d17'
Swesnik,Sylvia & Dick
Mon 6h36'28"1d1'
Sweson,Elsa
Uma 11h12'42"61d35'
Swetland,Rosemary
Peg 23h19'0"32d23'
Swetland,Russell
Cet 3h15'12"8d58'
Swett,Lynn E
And 2h21'33"42d53'
Swiat,Alexia Dauchy
Ori 5h52'1"19d21'
Swiatek,Katherine
Peg 22h12'1"20d52'
Swick Angelique
And 23h3'24"50d54'
Swiden,Elmer & Isabel
Cyg 21h2'43"28d50'
Swiderski
Lyn 9h20'22"41d45'
Swiderski,Jr,Tom
Per 3h7'45"46d57'
Swidersky,Paul
Boo 14h44'42"35d42'
Swieczak,Ania
Lyr 18h30'44"31d47'
Swieczkowski,R H
Cnv 13h40'40"46d8'
Swiencki,Rosemary T
Equ 20h56'11"2d30'
Swienton,Helen Charlotte
And 2h24'47"42d30'
Swierk,Elizabeth
Equ 20h55'0"2d47'
Swiers,Dean Malcolm
Uma 10h47'22"40d45'
Swies,Patrick J
And 2h11'44"42d31'
Swieter,John & Agnes
Cyg 20h37'24"53d51'
Swift,Carolyn
Lyr 19h3'30"26d46'
Swift,Christopher Garrett
Sex 10h12'43"-0d52'
Swift,Daniel R
Dra 15h3'37"63d23'
Swift,Elizabeth A
Aql 18h56'0"17d29'
Swift,Genora Katherine
Uma 10h43'0"53d34'
Swift,Jessica Leigh
And 0h23'1"34d50'
Swift,John Joseph
Lac 22h21'13"54d5'
Swift,Laura & P J Martin
Mon 7h48'16"-2d6'
Swift,Nicholas Paul
Cep 22h17'31"63d49'
Swift,Samuel George
Umi 16h36'25"77d3'
Swift,Taylor Mackenzie
Her 17h38'49"21d34'
Swig,Kim
Del 21h4'26"12d24'
Swig,Martin
Boo 13h37'45"21d39'
Swig,Richard Henry
Per 2h53'25"40d21'
Swig,Samantha
And 23h23'43"41d55'
Swig,Scout
Cet 2h18'19"3d33'
Swig,Steven
Equ 21h17'49"3d34'
Swig,Vicky
Cas 0h35'0"61d23'
Swiger,"Mamow"-Lilly Ann Welch
Vul 20h15'20"26d8'
Swiger,Tami
Crb 15h51'43"30d30'
Swiggard,Jack
Tau 5h33'24"20d20'
Swihart,George Ralph
Boo 15h2'48"30d58'
Swihart,Ruth Lucile Heestand
Cyg 21h7'52"30d16'
Swiklinski,Teresa
And 1h3'25"38d19'
Swilling,Keagan James
Dra 12h30'19"68d10'
Swimin Princess-Kim
Tau 3h39'57"23d54'
Swimm,Sigrun
Cyg 21h34'33"38d32'
Swimmer,Casey Elizebeth
Peg 22h28'22"27d35'
Swimmer,Natalie Taylor
Peg 22h18'33"3d41'
Swinburne,Annette
And 23h4'29"42d32'
Swindle,Gena Lynn & Oscar
Cyg 21h57'18"53d55'
Swinehart,Ann
And 0h10'44"28d34'
Swiney,Marilyn
Mon 6h43'44"8d47'
Swinford,James R
Aur 4h35'25"30d22'
Swinford,Jeff
Cet 1h35'30"-13d57'
Swing Dancing on a Star
Mon 7h12'5"-6d37'
Swing,Maggi
Aql 20h3'1"-0d9'
Swingle,Jane
Cas 2h27'49"67d58'
Swingle,Joseph
Aur 5h53'0"30d33'
Swingle,Kari
Lyn 7h49'45"51d58'
Swingle,Kathrine
Cas 2h26'46"67d43'
Swink,Estelle
Lyn 8h53'19"44d60'
Swinney,Leslie Ann Wife & Mother
Cyg 20h32'48"58d27'
Swinney,Scott Michael
Ser 15h23'43"18d7'
Swinny,Susan
Cma 7h0'50"-13d46'
Swinter,Michelle Marie
Umi 17h22'57"76d48'

Swirsky's 30th
Oph 17h14'28"-22d46'
Swiryna Our Life As One,Bo & Miki
Uma 11h31'36"43d6'
Swisher,Benjamin Mohler
Ori 5h16'51"0d44'
Swisher,Lesley Ellen
And 0h38'21"40d21'
Swisher,Patrick
Ori 5h39'0"11d43'
Swislow,Robert M
Cep 22h26'20"70d19'
Swiss
Boo 15h3'23"32d7'
Swissair-Curriculum 1
Per 4h3'18"35d48'
Swistak,Angela "Cookie"
And 1h53'28"39d43'
Swistak,Sandy
Peg 22h0'56"27d47'
Swit,Michael A
Cep 0h2'40"69d60'
Switek,Star
Aur 5h3'31"40d3'
Switlik,Richard
Cam 7h51'60"60d23'
Switzer's Eastern Star Glady
Mon 7h9'16"-6d38'
Switzer,Floyd L
Aur 6h50'0"37d54'
Switzer,Josephine Halladay
And 23h20'35"51d41'
Switzer,Stephen Lawrence
Peg 23h33'50"20d48'
Switzler,Jonathan Paul
Ori 5h53'30"-10d25'
Swoboda,Claudia
Ari 2h20'50"21d14'
Swofford,Lauren Elizabeth
Sge 19h52'0"16d34'
Swofford,Leslie Ann
Lyr 18h21'38"47d38'
Swomley,Art & Joyce
Aur 4h53'29"40d9'
Swope,Erica Jean
Lyr 19h18'1"41d4'
Swope,Opie K
Cet 1h35'0"-10d55'
Sword,Barbara Ellen
Sgr 19h4'50"-26d25'
Sword,Carl Richard
Aur 6h26'39"38d8'
Swordwolf's Star
Per 2h28'51"56d21'
Swyers,Austen Terry
Lmi 10h46'55"32d39'
SXJ-1 Saphire Blue
Uma 9h46'34"49d40'
Sy of Sighs
Her 16h52'0"48d49'
Sy Olson
Cep 22h59'39"57d45'
Sy,Cinzia
Mon 8h1'56"-8d25'
Syal-Lionheart,Hateesh David
Sco 16h34'45"-30d1'
Syawla Evol
Aql 18h54'49"11d58'
Sybille Et Bertrand
Lyn 8h53'43"40d30'
Syd & Bob
Uma 10h31'46"47d60'
Sydlar,Jessica Renee Marie
And 1h10'47"40d24'
Sydnam,Ben Allan
Cnv 12h58'44"50d29'
Sydnee Elizabeth (Tweedles)
Lyr 18h29'36"42d30'
Sydney Anne
Umi 13h19'16"75d39'
Sydney Elizabeth
Vul 19h46'0"25d19'
Sydney Jaye
Cas 1h16'54"62d8'
Sydney Luv
Aql 19h58'11"15d4'
Sydney Marie My 7243278
Cyg 21h35'56"38d48'
Sydney Plus Three
Cas 0h47'44"67d55'
Sydney,Colar
Boo 15h3'0"38d24'
SydneyMead
Lyn 9h10'42"34d55'
Sydnor,Fabian Royster
Tau 4h0'0"11d23'
Sydnor,John Mark
Cet 0h41'57"1d29'
Sydny
Cet 2h59'0"2d5'
Sygitowicz,Stan & Carlene
Tri 1h49'30"26d14'
Sykes "Be Love" 1931-1994,Rozzell
Cap 21h23'10"-24d52'
Sykes III,Frank Joseph
Uma 12h23'20"63d19'
Sykes Star 1
Lyn 7h29'52"35d31'
Sykes,Charles E
Lac 22h11'57"50d33'
Sykes,Christopher David
Hya 8h59'58"-8d38'
Sykes,David J
Cet 2h27'58"5d13'
Sykes,Dottie
Her 17h51'16"14d56'
Sykes,Frank J
Her 17h21'23"37d38'
Sykes,James Arthur
Ori 6h1'10"1d39'
Sykes,Jeannie Marie
Ori 5h55'45"11d18'
Sykes,Jessica Jordon
Cet 2h14'11"2d37'
Sykes,Jimmy
Cnv 12h51'42"38d1'
Sykes,John
Cep 3h52'1"80d28'
Sykes,Jr,John Leo
Aur 6h28'51"35d26'
Sykes,Kelly Louise
Umi 16h24'53"71d34'
Sykes,Leana Ray
Aql 19h22'47"-0d50'
Sykes,Lula Messick
And 0h9'48"27d52'
Sykes,Margaret Rayna
Aql 20h6'43"0d17'
Sykes,Maureen & Philip
Cyg 20h36'59"45d56'
Sykes,Michael Zachary
Oph 18h27'31"7d50'
Sykes,Robert Peter
Lac 22h30'47"40d15'
Sykes,Rory Mackay
Peg 23h7'15"11d53'
Sykes,Tamy Jo
Ori 6h6'43"1d44'
Sykes,Tom O
Hya 8h44'39"2d23'
Sykora,Suzanno Allioon
Cnc 8h6'57"7d20'
Syliangco,Marilyn G
Oph 17h19'41"12d4'
Sylid
Lyn 7h8'37"58d43'
Sylliaasen,Daniel K
Ori 6h1'41"8d13'
Sylphide,Mathilde
Sgr 18h58'51d-22d56'
Sylt,Juergen
Lac 22h5'42"51d33'
Sylvain Louis
Cyg 20h49'46"38d34'
Sylvain,Bareille
Ori 6h5'19"20d57'
Sylvalee
And 23h29'17"47d12'
Sylvan,Theodore
Tau 4h59'0"16d3'
Sylvana
Cas 3h2'28"58d24'
Sylvester
And 0h5'2"39d52'A
Sylvester
Aql 19h54'57"-0d41'
Sylvester "Quincy", Andrew & Louise
Aur 7h3'10"38d21'
Sylvester,Kirsten
Cyg 20h23'1"31d16'
Sylvester,Lee
Sco 16h58'13"-40d15'
Sylvester,Lois & Walt
Uma 9h53'58"46d1'
Sylvesters",Fred & Jane,"The
Sge 18h56'44"19d8'
Sylvestre
Cam 7h58'17"67d56'
Sylvestre,Annie
Cyg 19h28'49"35d19'
Sylvestri,Leo G
Aur 4h58'1"38d28'
Sylvia
Boo 14h51'19"39d20'
Sylvia
Cyg 20h59'20"53d16'
Sylvia
Mon 6h51'46"10d4'
Sylvia
Umi 14h14'34"77d54'
Sylvia
Aql 19h57'18"8d14'
Sylvia
Psc 23h23'15"6d45'
Sylvia & Billy
Vul 20h1'16"28d47'
Sylvia & Gary
Aqr 22h1'20"0d5'
Sylvia & Les
Lyr 18h58'24"38d20'
Sylvia & Tony
Cyg 21h0'39"38d29'
Sylvia Ann I
Her 18h0'25"28d35'
Sylvia Grace
And 0h59'38"36d58'
Sylvia H
Eri 2h53'41"-14d20'
Sylvia Rose
And 2h27'56"42d20'
Sylvia U
And 23h41'19"44d14'
Sylvia's Shining Star Von Noma #60
Cas 0h54'52"63d29'
Sylvia,Linda M
Lyr 18h43'0"30d25'
Sylvia,Monteil
Cnv 13h21'51"50d4'
Sylviane,Gerbert
Lac 22h28'55"40d5'
Sylvie
Dra 11h28'60"66d16'
Sylvie
Lyn 6h39'12"59d0'
Sylvie
Mon 6h52'57'-6d39'
Sylvie
Aur 6h19'21"31d59'
Sylvie & Dan
Cyg 19h29'49"31d51'
Sylvie Anne
Oph 18h6'58"11d47'
Sylvie De Barnier
Boo 15h18'33"37d44'
Sylvie Et Thomas
Boo 13h56'55"16d59'
Sylvie Pagé
Cam 10h57'26"84d53'
Sylvie's
Sco 16h15'53"-21d5'
Sym,Tadeusz
Cnv 13h54'1"40d4'
Symanski,Vincent
Per 3h47'8"36d57'
SYMAR Star
Umi 14h39'46"81d52'
Symbol Of Our Love
Cet 2h3'10"-18d52'
Syme,Robert Henry
Cap 21h4'1"-23d41'
Symeonidis,Benedikt
Dra 19h2'42"68d43'
Symes,Perry
Per 2h52'34"35d14'
Symes,Rebecca Mary
And 23h20'52"42d16'
Symkowick,Anthony
Aql 19h53'0"10d53'
Symm,Jeffrey
Lyn 8h19'32"48d36'
Symmonds,Jeff
Ser 18h17'21"-5d26'
Symms-JAMY,Amy
Cyg 21h30'51"52d41'
Symo
Uma 10h9'1"48d9'
Symonanis,Michael Alan
Aql 20h11'49"0d35'
Symonds,Pamela Joyce
Boo 13h36'14"15d16'
Symons,Glenda
And 0h45'27"31d41'
Symor
Uma 9h23'14"48d10'
Syms,Liz
Cyg 21h51'42"40d17'
Syms,Stevie
Ant 9h39'37"-33d13'
Synchronicity
Umi 14h24'39"66d4'
Synchronicity
And 1h13'54"39d55'
Synder,Jeffrey R
Ori 5h55'29"17d3'
Synk,Holly
Sge 19h16'24"16d17'
Synthonia
Mon 6h54'27"-2d17'
Sypek,Susan Marie
Peg 23h5'29"20d26'
Syran,Virginia Ellen
Equ 21h17'46"3d38'
Syrett,Gillian
Lyn 7h47'52"50d53'
Syrett,James P
Lac 22h35'1"38d7'
Syrop,Sheila W
Hya 9h16'21"-6d12'
SYS
Ori 6h3'39"6d19'
Sysack,Lawrence X
Cma 7h16'0"-13d24'
Syson Family,Donald J
Sge 19h12'22"16d36'
Syssau,Alain
Aur 5h2'34"30d40'
Systeme U
Vul 19h21'1"25d6'
Systeneu
Dra 16h48'17"62d47'
Syuzva
Cyg 19h25'34"32d39'
Syverson,Dean
Aur 7h0'38"38d20'
Syzygy CK
Mon 6h30'26"8d55'
Szabo,"The Pirate" Michael John
Cep 21h34'49"70d56'
Szabo,Igor
Umi 14h38'1"65d46'
Szabo,Istvan (Steve)
Per 2h43'36"43d13'
Szabo,Mark
Scl 23h19'40"-28d44'
Szabu,Maria
Aql 19h31'13"-0d56'
Szafrajda,Kerry
Mon 8h1'50"-10d15'
Szajowski,Casimir Jan
Boo 14h59'22"52d25'
Szakály,Carlo
Cnv 12h35'22"37d45'
Szala,Joseph
Per 1h43'16"53d38'
Szalak,Magdalena
Cas 3h0'11"57d27'
Szalay,Steven C
Ori 6h1'0"20d47'
Szalontai
Lyr 19h16'10"38d18'
Szanto,Sara
Cas 2h41'51"67d36'
Szaraniec,Barbara Ann
Lyn 8h4'59"34d12'
Szarlan,Alexandra
Mon 7h11'41"-3d48'A
Szarlan,R Carl
Mon 7h11'41"-3d48'B
Szary,Paul
Aql 18h42'20"-2d3'
Szatanski,Kristie Elaine
Uma 10h16'26"48d29'
Szatanski,Melanie Ann
Uma 10h17'32"48d10'
Szatkowski,Dr Jerry
Uma 9h13'38"48d24'
Szczepkowski,Brian Andrew
Lac 22h5'34"49d36'
Szczerba,Patricia
And 0h4'33"46d46'
Szczesny,Patricia Anne
And 0h0'48"34d35'
Szczotka,Troy
Dra 15h49'12"61d58'
Szeglin,John
Cep 2h40"12"77d44'
Szekely
Lyn 9h13'23"43d1'
Szenkiel,Elzbieta
And 1h3'16"37d45'
Szepanski,John P
Dra 15h58'45"68d26'
Szepseg,Juliette
Peg 23h37'13"30d53'
Szerelmem,Szabolcs
Aur 5h25'10"54d31'
Szerelmem,Yvette
Aur 5h18'57"52d36'
Szetela,Anthony Blaine
Aqr 22h39'60"-1d52'
Szewczul,David P
Per 3h12'0"46d16'
Szigethy,Katya McAuley
Mon 6h56'15"-1d23'
Szilagyi,Mike
Hya 8h42'28"4d12'
Szmanda,Sr,Robert L
Dra 17h57'21"61d2'
Szmidt,Christopher Roman
Ori 4h47'29"15d20'
Szmuk,Helga Daisy
Cru 11h58'16"-57d47'
Szodruch,Fred
Cnv 13h46'46"30d12'
Szoke,Arthur P
Mon 6h9'15"4d39'B
Szoke,Helen C
Mon 6h9'15"4d39'A
Szokoli,Matthew Sean
Cmi 7h19'38"8d28'
Szot,Dorothy
And 0h42'1"31d3'
Szotak,Amber Melina
Crt 11h13'3"-11d38'
Szpyrka,Scott
Cep 22h14'45"68d5'
Sztenderowicz,Joseph
Cet 1h22'11"-1d54'
Szuch,Starna Lee
Peg 21h34'0"20d4'
Szucsits,Justin Joseph
Lac 22h5'20"49d58'
Szudor,Ryan Robert
Vul 19h22'29"25d31'
Szulinski,Henry & Jean
Boo 14h17'39"30d36'
Szumilo,Angela Marie
Psc 23h20'16"0d10'
Szumilo,Sr,Frank
Lib 15h19'26"-21d47'
Szwarc,Leslie
Cam 3h19'47"60d34'
Szwarc,Mike E
Cet 0h50'33"2d1'
Szy In The Sky
Uma 10h14'0"59d27'
Szydlowski,Merla A
Vir 11h59'54"-2d27'
Szymaniak,Katherine Lynn
Com 12h31'57"20d40'
Szymanski,Diane Elizabeth
Uma 9h4'52"70d45'
Szymanski,Eleanore K
Cas 0h58'60"56d5'
Szymanski,Sophie
Lyr 18h25'33"46d15'
Szymanski,Stella
Lyr 18h12'16"40d59'
Szymber,Laura M
And 0h22'0"37d48'
Szymczyk,Peter
Her 16h37'59"38d34'
Szymkowiak,Carrie Ann
Lib 15h36'1"-20d26'
Szyp,Bernice D
And 2h22'14"43d0'
Szyslo,Manfred
Sco 15h50'59"-22d11'
Säan Bethany
Ori 5h58'24"11d1'
Sänger,Herr
Cmi 7h17'10"1d35'
Sängersen,Herr
Cmi 7h17'57"0d42'
Sébastien
Umi 13h49'10"71d4'
Séné,Fabienne
Crb 15h58'30"35d9'
Séverine Désirée
Cas 1h56'0"67d55'
Sézille,Mathilde
Mon 7h42'15"-4d30'
Söderberg,Anette Maria
Umi 13h55'15"71d14'
Sönnichsen,Karl-Heinz
Mon 7h36'46"-1d26'
Sülün
Cyg 20h21'55"39d5'
Sünje
Psc 22h51'36"1d27'
Süss Stephanie
Umi 14h49'22"66d43'
Süperchen
Cep 22h10'41"61d35'
Sánchez,Montserrat Porcel
Cep 22h9'13"68d37'
Sárka
Uma 11h15'52"41d35'
SØrensen,Brigit
Cas 1h26'43"64d40'
SØrenson,Jacob
Sge 19h5'46"17d30'
SØrensen,Tenna
Cnv 13h30'29"38d39'
SØrheim,Arna V
Cyg 19h26'17"57d55'
S"=Love
Cnv 12h23'46"44d16'

T

T
Eri 4h14'49"-10d4'
T
Aur 6h28'29"38d48'
T
Aql 18h59'27"13d32'
T & A
Cyg 20h51'0"30d29'
T & G's Destiny
Sge 19h56'14"20d34'
T A N S
Mon 6h55'16"-10d9'
T A O S
Pho 0h28'36"-43d34'
T B
Ori 5h56'0"8d55'
T C
Lyn 8h2'36"35d50'
T C
Tri 1h57'17"26d18'
T H E Star
Lyn 7h39'22"51d1'
T H N 50-Daddy's Star
Cep 21h33'21"70d4'
T J
Aur 5h3'15"38d48'
T J
Aur 6h46'39"38d16'
T J
Her 15h51'29"50d30'
T J
Cyg 21h17'28"35d18'
T J 1
Cam 5h50'50"60d18'
T J C
Cep 23h2'23"71d5'
T J F
Lyn 8h12'1"41d7'
T J N
Uma 11h59'49"48d12'
T K
Boo 15h14'11"48d7'
T L O M L
Lyn 8h13'48"50d38'
T Lisa
Cas 23h27'18"61d54'
T M B
Leo 10h34'0"20d35'
T N T Dynamite
Boo 14h27'47"54d6'
T O Y's Academy & Square
Lmi 10h49'21"24d7'
T R
Peg 21h50'28"28d37'
T R
Eri 3h25'57"-3d59'
T R B D M K Always
Vul 19h2'18"24d31'
T R R
Aql 20h6'48"0d46'
T S W of Mitsuyuki & Motoko
Lib 14h47'53"-23d11'
T Star
Uma 8h45'25"53d19'
T T
Del 20h26'50"11d6'
T T I T L to travel is to live
Ori 6h7'35"4d29'
T V Rentals
Uma 13h46'28""61d17'
T W 1
Aql 19h1'59"-1d30'
T W W J
Del 20h18'0"9d27'
T W's Star
Lyn 6h56'0"44d40'
T'Anne
Cas 2h34'20"58d56'
T's Lucky Star
Cyg 21h40'46"37d34'
T-13 North Dallas,TX
Peg 21h20'56"23d55'
T-246 Simi Valley,CA
Aur 6h31'44"38d58'
T-604 Kalamazoo,MI
And 1h24'59"38d42'
T-662
Uma 11h20'58"44d34'
T-664
Cnv 13h22'54"37d49'
T-677
Lmi 10h12'46"38d14'
T-694
Ori 6h7'14"8d8'
T-877 Hollywood,FL
Her 17h18'28"20d41'
T-901 Kalamazoo West, MI
Lyn 9h0'0"46d43'
T-Bone & Lex
Cet 0h54'1"0d31'
Taaffe,Mary Grace
Aql 18h59'27"-5d18'
Taalab,Aziz Fahmy
Cam 3h25'17"67d51'
Taamjunyer
Cnv 12h30'35"38d2'
Tabacchi M J
Sgr 19h31'42"-41d32'
Tabacek,Tracy Lynn
Vir 13h32'3"-7d20'
Tabachnick,Alison
Mon 7h0'38"-10d51'
Taback,MD (PS I Love You),Steven M
Oph 17h5'18"11d45'
Tabakotani Family, Stella of
Leo 10h3'18"19d40'
Tabaluga
Umi 15h50'19"82d18'
Tabanelli,Lorena
Eri 2h43'57"-4d10'
Tabares,Staci M
Aur 7h2'49"38d48'
Tabarez,Joe
Ori 6h15'20"-2d45'
Tabarez,Lottie(Bora)
Cyg 19h13'48"48d40'
Tabasco Cat
Cet 1h59'28"-1d59'
Tabatha
Lyr 19h22'1"30d18'
Tabatzky,MD,Alfred J
Aql 19h30'26"0d53'
Tabb,Neva Duncan
Oph 18h40'33"8d54'
Tabb,Sr,James Bowden
Lmi 9h21'0"33d48'
Tabenske,Curt & Patty
Cyg 21h19'1"35d29'
Taber,Carol Anderson
Lyr 18h27'0"34d4'
Taber,Clinton I
Sct 18h41'23"-7d7'
Taber,Richard Burt
Aur 7h24'54"38d35'
Taber,Timothy Paul
Her 15h55'43"47d52'
Taber-O'Keefe,Dale
Dra 18h31'1"58d16'
Tabernik,Anna Maria Fortunato
Vul 19h41'13"25d29'
Tabernik,Janice
Lyr 18h43'43"32d31'
Taberski,Carol
Vul 20h1'24"28d44'
Tabitha Babette
Umi 16h33'1"76d31'
Tabitha Christine
Lac 22h8'24"46d38'
Tabitha II
Lyn 8h3'52"35d8'
Tabitha Lea's Sunflower
Mon 6h39'0"6d48'
Tabken,Joerg
Oph 18h27'15"7d59'
Taboada,Pablo Fernandez
Cmi 7h17'31"8d54'
Tabone,Charles V
Per 2h38'41"40d7'
Tabor,Alexandra Lee
Peg 23h6'13"8d55'
Tabor,Donna Jean
Cyg 19h28'29"30d51'
Tabor,John Paul
Aur 5h1'23"50d26'
Tabor,Sean Andrew
Dra 19h3'33"56d38'
Taborn,Geri & Richard
Crb 15h30'38"30d29'
Tabron,Krystal Michelle
Peg 22h23'43"5d53'
Tabs 66
Pup 7h28'23"28d44'
Tabu
Cma 6h59'15"-31d31'
Tabuchi,Christina
Hya 8h44'13"4d15'
Tabuchi,Dorothy
Peg 23h35'12"30d29'
Tabuchi,Shoji
Aql 20h12'20"1d41'
Taccone,Danielle
Com 13h0'12"15d59'
Tachera,Steve
Per 3h18'49"50d9'
Tachibana,Yaeko
Mon 6h42'13"7d21'
Tachina
Lyn 7h59'53"51d19'
Tachiyama,Glenn
Del 20h19'12"9d42'
Tachney,D Min,Rev John Patrick
Aur 6h55'44"38d16'
Tackett,Amy Lucinda
Peg 22h47'15"5d40'
Tackett,C F "Buddy"
Lmi 10h39'24"24d48'
Tackett,Lucinda
Cyg 19h59'40"30d26'
Taconet,Ashley Viviane
Lyn 7h32'60"42d51'
Tacy's Star
Tau 3h54'0"2d29'
Tad James
Cet 0h5'45"-18d10'
TADAA!
Lmi 10h50'29"32d30'
TADAM
Cet 3h2'23"-0d56'
Taddchelle's Love Eternal
Cnv 13h40'1"35d36'
Taddei & Family, Domenick
Her 18h36'27"18d53'
Taddei,Camilla
Cmi 8h4'40"0d57'
Taddei,Robyn
Cas 23h15'45"60d36'
Tadden,George
Lyr 18h38'50"33d21'
Taddi,Jacopo
Peg 22h48'0"3d31'
Taddoni,Chris
Uma 0h57'33"70d17'
Taddy Lynn
Mon 8h1'60"-4d51'
Tadel,Ronald A
Her 16h57'16"32d50'
Tadley J
Gem 7h27'46"24d30'
Tady
Her 16h27'31"20d3'
Taerud,Mary
And 23h26'40"46d56'
Taff,Tracey
Com 12h20'49"32d50'
Taffinder,Lawrence
Per 4h6'23"40d8'
Taffy & Max
Crb 15h21'58"30d60'
Taflin,Heidi Lynn
Ori 5h56'0"16d36'
Tafreshi,Majid Karimi
Oph 17h18'34'-20d41'

Tafro's Terrific Triumph
Boo 15h11'34"27d25'
Taft
Peg 23h15'0"33d33'
Taft & Ruth
Sge 19h1'60"20d2'
Taft,Diana
Lyr 18h56'48"30d45'
Taft,Kara Elizabeth
Cam 8h35'13"78d0'
Taft,Ned
Ori 5h45'56"11d8'
Taft,Sarah K & Whitney R Jones
Cyg 21h31'58"50d23'
Tagestad,Kirsten
Uma 11h18'30"32d27'
Tagg III,Charles William
Per 2h58'21"32d30'
Taggart
Tri 1h48'27"28d30'
Taggart,Elaine & Donald
Uma 10h2'44"60d15'
Taggart,Helen Currie
Cyg 21h18'45"39d40'
Taggart,John Edgar
Dra 12h53'58"68d25'
Taggart,Meredith
Cas 0h48'31"73d55'
Taggerty,Kaitlyn Scarlett
Lyn 8h19'56"50d53'
Tagliabue,Eleonora
Vir 13h37'47"5d47'
Tagliaferri,Bill
Lyn 9h1'55"39d9'
Tagliaferri,Tina
Com 12h33'1"27d16'
Tagliaferro,Jean
Lyr 18h39'24"38d22'
Taglieri,Ronald A
Sgr 20h1'6"-43d42'
Tague,Skip & Anna
Crb 16h7'1"32d17'
Tahan,Patti
Mon 8h5'26"-0d58'
Taheri,Kevin David
Her 16h55'49"26d32'
Tahil,Nena
Lyn 8h12'23"33d48'
Tahoe 94
Aql 20h8'0"7d57'
Tai Chi Lovers
Hya 9h1'32"-8d18'
Tai,Karla Antoinette
And 23h41'41"37d47'
Taiani
Cyg 19h34'51"39d23'
Taiani,Kancho
Cam 5h39'51"68d33'
Taibbi,Genevieve DeMaio
Mon 7h1'1"-10d10'
Taico,Rosa A
Cyg 19h35'16"38d14'
Taid
Cnv 12h50'0"38d51'
Tailgunners
Her 17h32'1"27d19'
Taillefer,Jean-Paul
Lyr 18h53'37"34d45'
Tails-a-Waggin
Dra 20h21'1"71d8'
Taimanov,Sima
Ori 5h56'36"18d25'
Tainsh,Jacqueline Mary
And 2h0'46"43d0'
Taishoff,Lawrence B
Aur 6h3'1"36d19'
Tait,Sara Jessica
Lyn 8h54'33"40d32'
Taita
Vel 9h18'51"-41d8'
Taitano,Henry F
Aql 20h20'0"5d27'
Taittinger,Virgine
Her 18h23'1"13d1'

Taiwan
Dra 11h29'0"68d31'
Taj De Lisboa
Umi 14h46'31"66d43'
Tajzler,Emilia & Frantisek
Cyg 21h4'58"37d49'
Taka & Aki 1993
Gem 6h48'38"19d20'
Taka-Kazu-Mao-Chiko-Yu Mato-Nanu-Sei
Cnc 8h43'41"22d32'
Takaaki
Sgr 19h12'53"-26d44'
Takach,Alison
And 0h9'45"37d35'
Takach,Cathy
Lyr 18h17'19"47d20'
Takacs,Frank Alex
Cet 2h6'53"5d41'
Takacs,Loriana
Cas 1h34'17"60d18'
Takahama,Kazuhide
And 0h55'26"34d27'
Takahashi,Lukas Moto
Peg 23h42'1"31d44'
Takahashi,Nobuko C & Hideki
Cyg 21h11'19"38d13'
Takahashi,Nobuyuki
Umi 13h53'32"73d21'
Takahashi,Sae
Sco 16h58'25"-38d52'
Takahashi,Shuichi
Uma 8h46'32"47d26'
Takahashi,Tadeusz Olo Kauzo
Sge 20h5'1"20d7'
Takakawa,N S
Peg 23h22'60"30d28'
Takako
Cam 6h6'38"60d30'
Takanori
Psc 1h20'54"16d9'
Takara,Betty
Mon 6h58'1"10d34'
Takashi A No15
Leo 10h1'41"13d48'
Takashima,Alanna Hanako
Per 2h17'16"58d39'
Takata,Gary
Lac 22h41'15"54d8'
Takata,Sydney
Her 17h37'46"27d37'
Takayuki & Ritsuko
Cap 20h53'11"-19d13'
Take That
Sge 20h2'17"20d9'
Takemoto,Dale Toshiro
Aur 6h1'1"34d51'
Iakeshi & Hltomi
Psc 1h18'9"17d56'
Takeshita,Asti Piaget
Ori 5h59'1"8d34'
Takeshita,Dori
Cas 23h38'0"60d54'
Takeshita,Summerwynn Chivas
Peg 21h57'32"24d1'
Takiyah
Vul 19h23'0"22d30'
Taku
Vir 11h38'1"2d41'
Takuma to Yuka
Tau 5h31'15"20d25'
Talaat,Mohammed
Ori 5h56'17"21d5'
Talamas,Montse
Ori 5h39'38"1d38'
Talamo,Bernadette
Peg 21h25'55"20d28'
Talamo,Raymond
Her 17h37'23"50d1'
Talan,Jack
Cet 2h57'1"4d5'
Taland,Albert T
Ori 5h14'34"-9d12'

Talano,Rosina Maria Bertucci
And 23h25'58"46d33'
Talanya,Madison
Gru 22h39'31"56d37'
Talarico's,Dan "White Light"
Cep 20h42'59"75d14"
Talarico,Lorne Stephenson
Dra 16h42'31"62d28'
Talasek,Brandy Leah
Equ 21h6'59"10d10'
Talasek,Christopher Michael
Aql 19h52'49"13d44'
Talasek,Tara Elizabeth
Boo 14h36'55"19d55'
Talbert
Ori 5h57'22"16d28'
Talbert,JM
Sct 18h52'45"-4d43'
Talbert,John & Sarah Marie
Equ 21h0'0"8d44'
Talbot,John Christie
Aur 6h0'0"37d57'
Talbot,Joseph P
Uma 8h56'24"68d20'
Talbot,Lane
Cyg 20h16'0"31d45'
Talbot,Lumpy
Lac 22h9'12"40d10'
Talbot,Sam-Michele- Lindsay-Teddy
Vir 13h26'49"-2d36'
Talbot,Sophie
Umi 15h6'55"69d2'
Talbott,Joanne
Cap 19h46'50"-16d32'A
Talbott,John Lewis
Lac 22h9'34"38d15'
Talbott,Jr,Doug
Dra 16h32'44"67d30'
Talbott,Loubug
Cnv 12h31'48"32d46'
Talbott,Our Mawee
Ori 5h19'58"7d41'
Taldone Forever, Gaetano & Lisa
Lmi 10h8'29"40d20'
Taleb,Jasmin
Cnc 9h15'31"31d48'
Talen,Charles Raymond
Hya 9h2'56"1d17'
Taley,André Léon
Uma 11h7'55"51d41'
Talia
Psa 22h36'45"-25d53'
Talia-Anuschka-Lavater
Her 17h39'38"42d52'
Taliaferro,Kenneth Lynn
Aql 20h6'27"1d18'
Taliento,James & Christopher
Dra 19h33'40"65d24'
Talierco,Alain
Sex 10h19'38"-7d53'
Taliesyn & Devon
Cam 5h47'8"70d19'
Talin
Cyg 20h34'0"50d25'
Talise 1
Mon 7h16'39"-0d6'
Talkington,Iona McMillan
Peg 22h35'56"20d54'
Talkington,Monique Ionne
Peg 22h52'35"22d59'
Talkington,Stephanie
Peg 22h6'35"21d58'
Tallant,Ben Patrick
Aur 4h49'46"40d17'
Tallas,Alexandra Renata
Mon 6h21'36"8d17'
Tallent,Anna Faye
Aqr 23h39'15"-4d32'
Taller,Dr Stephen Lee
Her 18h18'20"12d54'
Tallerico,Pauline
Uma 10h15'22"52d4'

Tallerino,Tammi Jean
Del 20h51'49"9d20
Talley (Our Angel), Jordan Edward
Peg 21h52'37"20d12'
Talley Our Angel, Nicole Colleen
And 2h33'37"38d55'
Talley Star,I W W W D W B & B
Cyg 19h27'41"33d23'
Talley,Andre-Leon
Dra 16h0'18"63d35'
Talley,Dale & Patricia
Lyn 7h27'26"42d45'
Talley,David
Eri 3h29'29'-7d8'
Talley,Jamie Lynn
Cam 7h53'1"61d52'
Talley,Joel Mark
Hya 8h44'36"3d49'
Talley,Jr,Albert J
Oph 17h4'51"7d52'
Talley,June & Bill
Peg 22h1'42"24d8'
Talley,Linda Lee
Lyr 18h58'22"34d44'
Tallia,Alicia Rae
Cnc 9h13'0"31d13'
Tallman,Allen R
Boo 15h17'41"38d1'
Tallman,Cordelia Kirkendall
Cam 7h3'59"78d59'
Tallman/Winkler
Cam 9h41'55"82d19'
Tallon,Beatriz
Ori 5h58'35"11d26'
Tallulah
Peg 23h4'39"30d2'
Tallufo,Paul
Cep 0h50'48"78d32'
Tally
Eri 3h51'31"-6d23'
Talmage,Jay & Zona Lucile Brooks
Sge 19h55'19"18d46'
Talon
Her 18h10'12"31d50'
Taluntis,Povilas
Aur 7h20'0"38d45'
Taly
Uma 10h59'33"71d58'
Tam,Eunice
And 1h43'19"38d35'
Tam,Kitty
Aur 6h8'35"30d24'
Tama
Vir 11h54'55"-6d39'
Tamai,Marcel
Peg 23h35'23"15d11'
Tamaian
Cnc 8h30'31"8d51'
Tamala Lyn
Mon 6h24'48"0d5'
Tamalpais
Cnv 13h2'54"38d50'
Tamane,Carolyn Linh
Peg 21h39'45"24d32'
Tamanora
Aql 18h58'16"-5d35'
Tamaqua
Lyr 18h50'35"42d7'
Tamar The Palm Tree
Peg 21h23'35"20d30'
Tamar,Betsy N
Vul 19h45'35"20d06'
Tamara
Ori 5h57'13"15d24'
Tamara
Lyn 8h55'30"40d17'
Tamara
And 2h24'1"44d55'
Tamara
Mon 6h24'0"5d57'

Tamara Dawn
Lyn 8h22'26"47d33'
Tamara K
Vul 20h21'19"28d15'
Tamara Meus Aeternus Amicus Et Amor
Ori 5h57'42"7d48'
Tamara V
Cmi 7h53'58"0d39'
Tamara's Wishing Star
Aql 19h50'23"12d5'
Tamarald
Lup 15h17'47"-42d4'
Tamayo,Carlos Alberto Flor
Her 16h42'53"21d24'
Tamayo,Ignacio
Ori 4h56'0"-1d28'
Tamayo-PapaBear,John Patrick
Uma 9h55'40"61d28'
Tambellini,Juli Anna
And 0h28'12"45d19'
Tambo
Uma 9h35'27"67d38'
Tamborelli,Keith
Per 3h29'47"50d1'
Tamburello,Charlie
Dra 19h53'42"68d42'
Tamburin,Barry
Per 2h55'53"56d40'
Tamburrino,Bill
Dra 18h33'26"50d17'
Tamburro,James A
Her 17h38'58"23d28'
Tamedica,Bernadette
Aur 5h29'25"29d36'
Tameling,Joan M
Mon 6h34'46"-1d7'
Tameling,Stephen
Her 15h56'45"40d51'
Tameling,Stephen
Per 3h28'48"50d45'
Tami
Cep 21h27'46"58d21'
Tami
Mon 7h6'1"0d48'
Tami
Cas 0h52'41"68d3'
Tami T
Cas 2h40'49"60d25'
Tami's Bright Light
And 0h38'47"41d5'
Tami's Stella Amor
Cam 3h54'59"56d49'
Tami-Jo
And 0h3'21"40d47'
Tamiann
Lyr 18h49'36"31d12'
Tamiati,Paola
Pic 5h2'29"-46d22'
Tamietoad
Vul 19h34'13"27d48'
Tamiko's Light
Peg 23h19'40"30d16'
Tamilia,Laura
Uma 11h54'58"30d54'
Tamin,Caroline Jean
Peg 23h28'26"17d26'
Tamm,Brenda Ann
Mon 7h2'19"3d52'
Tamm,Jayanti
Boo 14h17'43"31d54'
Tammen,Paul E
Aur 6h30'39"31d26'
Tammera Lynn
Uma 11h58'17"32d17'
Tammi Lynn
Ser 17h33'17"1d1'A
Tammi Michelle
Mon 7h0'0"7d56'
Tamminga,George
Dra 16h41'12"72d27'
Tammy
Cas 1h21'51"52d41'
Tammy
Mon 7h2'38"-6d52'

Tammy
Sge 19h37'1"16d32'
Tammy
And 23h34'01"49d42'
Tammy "O"
Aql 19h44'1"14d28'
Tammy D
Mon 6h39'11"10d10'
Tammy Jean
And 0h50'0"39d6'
Tammy Jo
Crt 10h54'1"-10d46'
Tammy Lyn
Del 20h49'38"8d51'
Tammy Lynn
Aur 6h0'0"38d58'
Tammy Lynn's Shining Star Of Love
Cas 23h16'51"63d20'
Tammy Lynn(11-3-65)
Vul 20h14'1"23d59'
Tammy Lynn-(T J)
Ori 5h57'1"8d19'
Tammy Mae
Lyn 8h47'35"39d46'
Tammy Renee
Peg 21h31'54"20d5'
Tammy Shine on (TRLW)
Uma 9h11'43"60d7'
Tammy's Forever Love
Sex 10h12'56"-2d1'
Tammy's Light
Cmi 8h0'22"6d3'
Tammy's Lighthouse
Boo 14h36'0"45d9'
Tammy's Star
Psc 1h0'58"32d59'
Tammy's Star
Umi 14h55'58"66d31'
Tammy's Star Of Inspiration
Lyn 7h37'0"40d23'
Tammykins
Eri 3h14'17"-13d21'
Tammylyn
Lyn 7h53'48"58d40'
Tammys Lil Twinkler
And 2h27'38"41d48'
Tamona
Cmi 8h7'34"1d6'
Tamra Jade Dawn
And 23h22'56"49d1'
Tamrya
And 1h36'1"36d2'
Tamsin Samantha & Jacqueline Antoinette
Cyg 21h2'15"38d42'
Tamstar
Pup 7h58'5"23d52'
Tamtaouré
Cep 21h24'47"68d22'
Tamulonis,John
Ori 5h59'55"11d31'
Tamura,Spencer
Oph 18h17'50"10d30'
Tamzin Angelus Cygnus
Cyg 19h59'22"41d5'
Tan,Greg
Oph 17h2'58"11d30'
Tan,Joy Whei-Mien
Cas 2h23'0"61d9'
Tan,Kee-Meng
Umi 14h57'39"78d30'
Tan,Kuya José
Ori 5h35'45"7d43'
Tana Cristine
Mon 8h6'1"-9d45'
Tanabata Kiss
Boo 14h38'52"51d9'
Tanabe,Andrew Tsuneo
Aur 4h50'22"40d57'
Tanaiewski,Rita "Chickie"
Cas 0h8'23"64d12'
Tanasoiu,Oana
Peg 21h58'25"21d24'

Tancrede
Her 18h19'39"12d26'
Tandy,Ramsy Christyn
Tri 2h9'49"33d45'
Tandy,Sam
Aql 20h7'0"6d46'
Tane,Bunny S
Lyn 6h18'31"58d15'
Tanem
Cyg 21h2'18"28d53'
Tanen,Reni & Nat
Cyg 21h13'59"35d31'
Tanetschek,Franz
Ser 15h12'17"10d24'
Taney,Charles & Amy
Cnv 12h59'18"32d50'
Taney,Kate
And 0h54'11"33d38'
Tang,Kei Y
Ori 5h57'38"21d50'
Tangen,Janice
Cas 23h28'10"54d2'
Tanger,Stanley K
Uma 11h24'0"42d14'
Tangerine
Lac 22h3'0"37d56'
Tanghare,Timothy A
Dra 19h33'17"61d11'
Tangi & Ingrid
Crb 16h14'31"38d10'
Tango 76
Lmi 9h41'0"37d54'
Tangren,Sara
Cas 0h26'46"67d54'
Tanguay,André
Uma 10h4'19"47d54'
Tanguay,Geneviève
Lyr 18h31'51"38d43'
Tanha,Amir
Cep 20h51'32"60d42'
Tani
Tel 20h15'8"-48d44'
Tani
Ari 1h58'1"23d37'
Tania
Cam 5h3'22"67d44'
Tania Joy 30/94
Lyr 19h20'51"40d20'
Tania's Diamond In The Sky 1994
Cas 1h51'46"73d21'
Tanica Chantel
Lyn 7h57'19"43d53'
Tanith Lee
Cas 0h53'23"59d12'
Tanith-Lei Lily
And 0h1'22"47d38'
Tanja
Cas 0h42'41"60d54'
Ianja
Cet 2h58'13"1d11'
Tanja 22/10/1973
Cnv 14h1'59"38d29'
Tanja Steffi Sanna Behrendt Germany
Dra 15h57'43"62d18'
Tanja,Sandra Elisabeth
Leo 9h54'59"11d51'
Tankai,Jun Lucas
Ori 6h8'36"5d39'
Tankersley,Allen Wilson
Her 17h1'51"43d47'
Tanksley,Wendy K
Lyr 18h37'42"40d18'
Tannebaum,Papa Lou
Per 4h4'56"51d41'
Tannehill,Kenny
Per 2h44'58"36d36'
Tannenbaum,Maurice
Aur 4h49'43"40d49'
Tannenbaum,Michael Thomas
Boo 14h7'35"35d23'
Tannenbaum,Milton
Cnv 12h24'12"43d49'

Tannenbaum,Philip David
Cnv 12h43'47"51d36'
Tanner
Oph 16h49'23"10d43'
Tanner Brightly Shining
Her 17h38'11"22d5'
Tanner III,Donald B
Per 2h51'24"34d45'
Tanner,Bob
Her 17h54'21"14d51'
Tanner,Charles E
Aur 6h27'20"38d25'
Tanner,Cindy
Lmi 10h56'22"31d48'
Tanner,David Lee
Cep 22h37'15"58d11'
Tanner,Elizabeth Anne
And 2h24'31"42d17'
Tanner,Jr,Donald B
Peg 23h29'37"21d16'
Tanner,Linda Abbott
Ser 15h24'16"9d55'
Tanner,Rachel Ann
Cas 0h24'29"74d0'
Tanner,Tristan
Umi 16h26'40"72d11'
Tanner,Triston Rock
Vir 11h38'48"9d5'
Tannock Family,The
Uma 11h0'24"44d21'
Tannor,Erika Madison Angelica
And 2h24'31"38d15'
Tannous,Chad Elias
Her 18h4'48"47d41'
Tanny,Sandra Faye
Ori 5h40'1"-10d14'
Tanons
Dra 17h2'53"60d22'
Tanski,Janet
And 0h29'0"44d35'
Tanski,Sarah Elizabeth
Lyn 7h42'39"48d59'
Tanstaafl
Ori 6h1'54"8d53'
Tanstypides
Uma 13h50'26"61d52'
Tanswell Superlunary I
And 0h33'12"40d2'
Tantasia
Sco 16h17'29"-24d2'
Tanti Star,The
Lyr 19h22'0"31d36'
Tantleff,Alan
Per 2h25'36"58d18'
Tantleff,Rhoda & Ivan
Cyg 20h43'54"46d16'
Tanton,Philip Lawrence
Per 4h45'1"50d5'
Tanya
Umi 14h6'33"71d22'
Tanya
And 2h23'13"42d24'
Tanya
And 23h37'14"41d23'
Tanya & Dima
Lac 22h14'42"49d57'
Tanya Cutie
And 1h48'59"40d32'
Tanya Forever
Del 20h21'14"10d3'
Tanya Jade
Cru 12h36'23"-63d5'
Tanya Louise-Peter Andrew
Pho 23h43'12"-46d10'
Tanya Reneé
Peg 22h18'41"20d13'
Tanya Sue
Umi 15h28'1"78d16'
Tanya's Star
Uma 11h22'39"43d29'
Tanya's Wish
Cas 1h0'12"61d44'
Tanyaneel
Cyg 20h23'52"30d51'

Tanès
Boo 13h57'29"17d42'
Tao
Boo 15h35'13"40d19'
Taormina,Kelli
Lyn 6h30'50"54d58'
Taormino,Anthony Joseph
Cet 2h31'1"5d3'
Tapager,Eric C
Hya 9h1'34"0d24'
Taperman,Dennis M
Sgr 19h12'55"-28d2'
Tapia,MD,Eulogio Horacio
Oph 18h1'42"1d33'
Tapiero,Daniel M
Cas 0h11'25"64d12'
Tapio's Wish
Boo 15h14'22"50d20'
Taplett,Patricia Kingston
Leo 11h9'55"-5d2'
Taplett,Robert D
Boo 15h15'30"52d12'
Tapley,Edna Dinklage
Mon 6h59'1"1d12'
Taplick,Marlies
Sgr 19h14'16"-24d35'
Taplin,Dawn
Peg 22h1'33"31d22'
Taplin,James Richard Dominic
Aql 20h0'1"3d47'
Taplin,Toban David Blackshield
Uma 9h45'42"47d48'
Tappan
Aql 20h1'1"7d36'
Tappen,Paige Skye
Com 13h28'48"26d38'
Tapper,Gary & Cindy
Uma 10h45'1"57d54'
Tappy's Star
Peg 21h59'0"22d31'
Tapscott's Little Putz 1955,Bruce
Hya 8h49'23"-6d42'
Tapscott,Don CRPG
Tri 2h20'43"28d29'
Tapscott,Jenny
Cas 1h51'0"63d59'
Tapyrik,Lauren Elizabeth
Cam 13h4'40"78d41'
Tapyrik,Lauren
Lyn 7h36'13"45d34'
TAR
Lac 22h52'59"53d30'
Tara
Eri 4h55'24"-8d1'
Tara
Sct 18h46'9"-7d3'
Tara
Lyn 6h58'54"59d0'
Tara
Peg 0h1'37"30d4'
Tara
Crb 16h19'54"27d49'
Tara
Cas 0h31'1"50d33'
Tara
Lyr 18h46'35"42d8'
Tara
Cas 0h39'0"60d40'
Tara
Umi 15h17'45"70d19'
Tara
And 23h41'31"45d53'
Tara Ann
Cyg 20h40'1"30d53'
Tara Anne
And 2h27'0"39d38'
Tara Beth
Peg 21h44'24"24d44'
Tara Bridget
Lyn 7h37'0"37d7'
Tara Catherine
Vul 19h23'24"22d35'

Tara Ellen Buttercup
Dra 16h34'1"72d33'
Tara Felicia
And 0h21'49"33d15'
Tara I Will Love You Always... John
Lmi 10h1'26"32d39'
Tara Jean
Mon 7h19'15"-6d17'
Tara Kaley
Gem 6h30'23"12d52'
Tara Leah's Forever Shining Star
Lyn 8h1'29"38d54'
Tara Lee
Gem 6h52'37"14d8'
Tara Lee's Star
Cma 6h57'25"-20d27'
Tara Louise
Lyr 19h6'1"25d48'
Tara Lynn
Cas 0h37'35"60d44'
Tara Lynne
Cam 4h24'29"68d24'
Tara Of Many Colors
Peg 23h44'33"8d44'
Tara Rose
Mon 6h57'39"11d19'
Tara Serenita
Gem 7h23'27"30d48'
Tara Su
Peg 21h42'33"24d21'
Tara TEN
Aql 20h11'21"5d26'
Tara Theresa Marie
Cas 1h14'42"63d4'
Tara V
Cet 1h20'11"-10d18'
Tara Vara
Lyn 8h31'40"40d11'
Tara X
Oph 17h20'51"10d28'
Tara's Louis 2-14-96
Oph 18h9'12"13d28'
Tara's Wishing Star
And 23h39'54"46d27'
Tara-Jane
And 2h31'51"50d52'
Tarabini,Bruno
Ori 5h53'0"-5d60'
TaraGene's Wagon-Post
Cyg 21h5'32"38d58'
Tarah & Steve
Cyg 19h47'59"30d38'
Taramatt
Ori 6h8'22"9d9'
Taranaki
Aql 19h3'17"0d55'
Tarangelo,Thomas J
Aql 19h56'58"1d4'
Tarantino,Laura A
And 2h23'28"39d57'
Tarantino,Michael James
Aur 5h32'1"31d21'
Taras
Peg 23h5'49"17d34'
Taras' Grandma
Com 12h46'22"23d4'
Taras-Diamond
Lyn 8h45'42"36d39'
Taraskiewicz,Susan
And 2h12'28"38d52'
Taravella,James Robert
Per 1h28'56"53d21'
Tarbell,Elizabeth Weatherbee
Cas 23h21'38"60d40'
Tarcisious
Uma 11h43'37"53d44'
Tarde,Betty Ann
Uma 12h6'33"52d15'
Tarde,Mickey
Cam 6h5'1"67d39'
Tardibuono,Theresa M
Cam 7h32'19"67d49'
Tardif,Margot
Cyg 20h23'57"39d6'
Tardif,Robert
Ori 5h52'34"8d41'
Tardioli,Francesca Romana Mombelli
Cam 7h1'20"61d34'
Tardo For Pandora,Lisa Marie
Cas 1h42'11"61d43'
Tardo,James
Her 18h17'39"14d50'
Tardy,Dakota
Aql 19h28'24"7d38'
Tardy,Jean Pierre M
Crb 15h48'36"35d15'
Tardy,Justin Lee
Cam 5h25'34"80d19'
Tarek
Per 4h6'40"50d31'
Tarena
Aur 7h1'14"41d8'
Targosz,Benjamin Lee
Cam 5h57'20"67d33'
Targé,Anne-Pascale
Cam 3h59'29"61d46'
Tarif,Veronique France
Cas 0h8'0"61d25'
Tarin,Katelyn Renae
Cas 2h36'35"57d37'
Tariq's Wish
Her 17h21'35"27d4'
Tarke,David Leighton
Her 17h12'37"44d56'
Tarkington,Sally M
Lyr 18h55'1"42d31'
Tarko,"Gaborel"
And 0h9'28"47d10'
Tarlalyn
Eri 3h39'0"-3d57'
Tarlochan
Aur 6h7'47"36d42'
Tarmahomed,Adbul Gaffar
Uma 10h29'20"47d46'
Tarmy,Henry
Dra 16h11'44"68d21'
Tarn,Fiona Jane
Del 20h21'12"10d54'
Tarnastar II
Hya 8h17'1"2d38'
Tarni Star
Cyg 19h26'43"34d28'
Tarnoczy,Dana Lynn
And 23h33'44"43d54'
Tarnovskaja,Natalia
Peg 22h5'47"4d7'
Tarnow,Cathy G
Aql 20h7'1"1d1'
Tarnowski,Gary Edward
Aql 19h6'0"-0d32'
Taro
Lac 22h43'51"56d26'
Tarot
Cyg 21h20'1"53d44'
Tarpinian,Bruce Milton
Uma 12h5'55"63d15'
Tarpinian,Madeline
Uma 12h3'8"63d12'B
Tarpinian,Milton
Uma 12h3'8"63d12'A
Tarquinio,Sira
And 23h25'0"40d12'
Tarr's Star,Mom
Uma 11h26'27"47d2'
Tarr,Michael
Cep 23h33'12"67d56'
Tarrani,Melissa Lynn
Lyn 8h38'14"40d53'
Tarrant,Gina Clarice
Mon 7h26'47"-6d9'
Tarrant,Ken,Leann, Kevin & Elise
Cet 2h26'58"7d19'
Tarrant,Patrick Allen
Aql 19h4'47"4d46'
Tarrant,Victoria Ann
Cas 0h26'0"50d37'
Tarrat,Cécile Laure
Uma 13h15'1"62d57'
Tarrice 2-13-56, Michael Rhodes
Cep 22h43'33"82d0'B
Tarro,Ronald John
Hya 8h18'42"0d59'
Tarrytown 93
Uma 10h55'11"37d49'
Tarsi,Maray Lynn
Cas 23h41'28"61d4'
Tarsi,Peter & Josephine
Crb 16h20'58"32d32'
Tartabini,Sr,Joseph
Dra 19h7'56"70d15'
Tartaglia
Gem 6h54'12"35d2'
Tartaglia,Mary G
Cas 0h26'24"61d10'
Tartaglino,Lynda
Cet 0h27'11"1d20'
Tartan
Cma 7h22'22"-15d36'
Tartt,Pamela Melvin
Peg 23h35'44"21d25'
Tarulli,Dean
Per 1h36'59"53d4'
Tarverdian,Eileen
Oph 17h54'15"13d2'
Taryn's Wonder
Eri 3h1'0"-8d22'
Taryn,Marie Zeigler
Lac 22h7'52"49d46'
Tarzan
Aql 19h40'7"4d31'
Tarzian,Harry S
Per 3h55'53"40d28'
Tas 1
Cnv 12h23'1"45d28'
Tas Kardias TG8412.29
Ori 4h41'33"11d29'
Tasber,Annette
Cas 0h56'30"68d55'
Tasca,Scott Francis
Boo 15h10'26"31d25'
Taschen,Benedikt
Aqr 21h54'1"0d49'
Taschenberger,B R "Doc"
Oph 17h18'22"-20d8'
Tasha
Leo 9h27'43"28d11'
Tasha
Lyr 18h15'54"40d47'
Tasha
Mon 6h19'49"8d49'
Tasha
Cas 0h23'24"70d40'
Tasha
Aur 5h58'24"38d57'
Tasha & Bruno
Lac 22h52'1"56d50'
Tasha & Jerome Forever
Her 16h39'37"50d47'
Tasha & Ken's Cosmic Sparkler
Cyg 19h11'43"47d45'
Tasha Forever
Lac 22h11'47"51d10'
Tasha's Twilight
Cma 6h46'0"-11d13'A
Tashiro,Nobora
Cnv 12h32'43"33d46'
Tashjian,Hrair
Tri 2h12'19"31d7'
Tashman,Melanie
Peg 23h2'20"18d46'
Tashman,Rick
Her 18h8'32"31d32'
Tashman,Stephen
Her 17h23'44"40d19'
Tasker,Cassandra Lyn
Uma 8h32'41"47d57'
Tasker,Matthew Anthony
Dra 18h4'43"70d10'
Tasker,Spencer Christian
Ori 5h40'12"1d46'
Taskila,Anja K
Cam 3h53'38"55d3'
Tasmanian Devil
Sge 20h0'29"20d1'
Tasmin Alicia
Cas 0h27'57"64d6'
Tasnadi,Princess Christina Elizabeth
Aql 20h32'42"0d43'
Tasoula & Kostas
Sco 17h20'26"-31d0'
Tasoula & Kostas
Sco 17h4'31"-31d5'
Tassa,Aida
Lyn 7h34'0"41d15'
Tassi,Jeffrey
Her 16h56'23"39d41'
Tassie Star,The
Aql 18h40'48"1d36'
Tassie's Star
Del 21h5'1"12d58'
Tassie,Gerard
Sex 10h43'0"5d11'
Tassigny,Herve
Cyg 19h47'14"31d33'
Tassin,Trish
Cet 0h23'30"-11d9'
Tassitano,Susan
Hya 8h23'47"0d4'
Tassler,Zachary David
Lmi 10h39'58"24d5'
Tasso,Albert L
Her 17h32'58"21d3'
Tassé,Jérémie
Her 16h53'0"33d11'
Tasto,Harriet Elizabeth
Aur 4h56'32"40d8'
Tasto,Leo
Aur 4h37'10"34d2'
TAT x 2
Cyg 20h19'33"41d55'
Tata
Cae 4h56'31"-32d34'
Tata-Madre Veselo Bozic,Amare Vi
Dra 16h59'19"62d39'
Tatai,George
Ori 5h51'10"14d38'
Tatarano,"Jonathan" Marco
Per 2h8'49"58d48'
Tate 11-2-84,Adrienne Bianca
Hya 8h12'0"5d42'
Tate 7-14-82,Lauren Mercedes
Cas 0h39'1"64d12'
Tate's Dream
Her 18h8'20"40d44'
Tate,Athena Marie
Del 20h18'1"9d27'
Tate,Dylan Alexander
Per 1h44'38"50d21'
Tate,Fran E
Aql 20h9'50"4d15'
Tate,Haney V & Dorothy E
Mon 7h0'22"0d4'
Tate,Joshua
Aql 20h19'17"5d8'
Tate,Kylie & Susan Padilla
Tau 3h57'29"28d55'
Tate,Lynda Melanie
Cam 12h26'31"80d41'
Tate,Mary
Aqr 23h32'47"-5d36'
Tate,Robin
Ori 5h35'47"-6d42'
Tate,Shirley
Lyn 7h45'25"51d53'
Tate,Thomas
Cet 2h16'27"1d1'
Tate,William M
Ori 4h55'48"4d50'
Tate-Haag
Aur 6h26'40"33d3'
Tate-Welsh,Stephanie
Cyg 21h7'20"39d58'
Tatem,Margaret Calderhead
Lyr 18h56'53"31d38'
Tateossian,Robert
Ori 5h36'25"8d59'
Tates
Cas 0h57'46'66d44'
Tatiana
Cas 2h3'31"59d1'
Tatiana
Lyn 8h7'22"58d56'
Tatiana
Lyn 7h53'44"36d23'
Tatiana
Cyg 20h59'46"30d47'
Tatiana
Cyg 19h27'38"37d46'
Tatjana
Lmi 10h24'13"30d5'
Tatjana Buby
Cet 2h3'45"-5d47'
Tatman,David Robert
Cyg 20h0'40"31d37'
Tato
Aur 4h59'31"40d31'
Tato
Tau 4h11'27"0d47'
Tatoul Star,The Warren P
Leo 9h57'56"28d22'
Tattam,Susie
Umi 14h45'39"69d13'
Tattar,Marc James
Dra 16h54'19"67d54'
Tattersall,Kerri
Uma 12h2'24"31d17'
Tattini
Cyg 21h32'26"38d28'
Tatto,Daniela Rose
Cmi 8h8'23"1d43'
Tattoo
Dra 9h47'29"74d32'
Tattoo
Mon 7h59'7"-4d20'
Tatulli,Lucille
Cap 21h51'32"-14d44'
Tatum,Christopher Hunter Luis
Tau 4h43'27"16d41'
Tatum,Ida J
Aql 20h18'1"1d5'
Tatum,Jennifer
Ari 2h43'54"21'29d
Tatum,John
Her 16h44'21"21d48'
Tatum,Jr,George W
Her 18h13'24"41d1'
Tatum,Mary Ellen Barnes Stewart
Crt 10h53'56"-11d12'
Tatum,Willa Mae
Eri 3h59'41"-16d12'
Tatyana
Lib 14h26'59"-22d59'
Tatyana
And 2h11'18"41d3'
Tau,Stellina
Pyx 8h45'27"-24d12'
Tauaese,Misly
Psc 23h23'31"1d6'
Tauano,Nathalie
Crb 16h11'0"38d40'
Taub,Dennis
Peg 22h6'52"28d56'
Taub,Dr John S
Her 17h22'25"41d3'
Taube,Andreas von
Aur 5h13'1"43d40'
Taube,John
Ser 15h32'49"-0d28'
Tauber,Selma Bernice
Lyr 19h23'33"30d53'
Taubman,Andrew B
Her 15h49'57"42d12'
Taubman,Robert
Ori 5h52'56"17d44'
Taubman,Wendy
Cas 1h5'0"58d39'
Tauff,Betty
Eri 3h33'50"-7d18'
Taul,Marjorie Jean Quirk
Mon 6h35'0"-6d26'
Tauno
Aql 19h17'39"14d55'
Taupio für Immer
Cma 7h18'1"-15d46'
TAUREAN
Cyg 21h45'26"37d57'
Taurean Star:Queen Debra,The
Cas 1h39'27"67d56'
Tauriello,Joe
Per 2h50'27"45d26'
Taurus
Tau 5h21'39"20d22'
Tauscher,Ellen
Del 20h52'0"9d31'
Tauscher,John
Lyr 18h41'46"40d10'
Tausz,Gregory M
Aur 6h28'36"38d39'
Tauzier,Jr,Herbert J
Aql 18h57'33"10d21'
Tauzin,Angie M
Crb 16h7'1"30d58'
Tavano,Nathalie
Dra 19h4'60"80d18'
Tavares,Alex
Boo 14h58'30"30d36'
Tavares,Ernest H
Boo 14h19'0"38d53'
Tavares,Joseph
Ser 15h31'30"18d48'
Tavares,Lisa A
Cyg 20h22'0"38d7'
Tavares,Maygen S
Cas 23h30'59"60d49'
Tavares,Thomas
Mon 7h12'49"-10d50'
Tave,Franne & Richard
Umi 13h53'44"77d58'
Tavernellt,Tomaltn
Hya 8h58'51"6d18'
Tavrovskaya,Marina
Cyg 21h26'43"30d51'
Tavrow,Joshua
Aur 5h33'19"48d50'
Tavrow,Sara
And 0h2'1"46d58'
Tavs,Janette
Mon 6h57'20"-6d58'
Tawnee Carol
Cyg 21h19'37"38d12'
Tawney,Donna
Cma 7h0'43"-15d45'
Tawnya Renee
Peg 0h1'32"18d2'
Taxman,Myra
Cmi 7h18'54"9d50'B
Taxman,Royal
Cmi 7h18'54"9d50'A
Tay,Alyssa Maryk
Cas 0h43'46"64d11'
Tay,Jamie
Aql 18h59'41"11d42'
Taya,Andres Espinos
Aql 19h1'0"1d13'
Tayabas,Dani
Eri 4h31'55"-0d55'
Tayag,Jr,Romeo M
Cma 7h13'1"-16d30'
Tayansil,Nancy
Lib 15h3'43"-5d48'
Taybro
Uma 10h54'1"56d38'
Tayla Lou
Cyg 20h1'21"38d4'
Tayler
Aqr 21h56'58"-6d44'
Tayler,Agnus William Holroyd
Uma 8h34'21"62d18'
Taylor
Mon 7h0'29"-8d16'
Taylor
Aur 6h26'1"37d54'
Taylor
Hya 10h32'27"-12d3'
Taylor
Per 1h44'53"53d27'
Taylor "Forever Your Daddy's Angel"
Mon 6h28'53"8d44'
Taylor & Dallas
Sge 19h30'54"16d32'
Taylor & Madison
Aql 19h8'1"3d37'
Taylor (Bizzie Lizzie) Mrs Elizabeth Jane
Lyr 18h36'32"31d49'
Taylor 090693
Lac 22h54'11"38d44'
Taylor 1,Brian W
Leo 10h52'1"22d21'
Taylor A Proud & Brave Man,Ron
Ori 6h5'1"2d50'
Taylor All My Love TTA&F,Elizabeth A
And 1h58'52"40d20'
Taylor Ann
Crb 16h11'41"30d15'
Taylor Dawn
Lyr 18h23'18"42d15'
Taylor Diane
Cas 0h14'29"58d16'
Taylor Dyan
Cep 5h15'38"80d17'
Taylor Jae
Cnv 12h16'1"37d47'
Taylor James
Lac 22h14'13"51d29'
Taylor Jaye
Her 17h47'0"48d51'
Taylor Leigh
Lyn 7h10'19"58d15'
Taylor No Greater Love,Judy
And 1h23'32"40d17'
Taylor Selena
Eri 3h15'0"-15d34'
Taylor With Love Forever,Susan Jane
Lyr 18h59'35"28d29'
Taylor"Loverly"Trent N
Eri 2h59'53"-10d32'
Taylor"The Groovy One" Karen P
Lyn 7h9'38"59d19'
Taylor's Expression
Mon 7h19'55"-5d17'
Taylor, Vivian
Vir 13h33'42"-6d46'
Taylor,"A Special Grandson" Brian Davis
Ori 5h50'58"15d19'
Taylor,"CC"
Hya 9h8'1"4d33'
Taylor,1st Lt Stephen L C
Cep 21h27'56"58d17'
Taylor,Alan & Barbara
Ori 5h56'36"17d39'
Taylor,Alisha Rose
Her 17h34'40"14d18'
Taylor,Alison Leigh
Peg 21h20'33"22d37'
Taylor,Allen
Per 2h30'53"38d59'
Taylor,Amanda Michelle
Cet 2h34'37"-8d47'
Taylor,Amethyst Louise
Eri 4h3'39"-14d39'
Taylor,Amy
Cas 0h33'1"62d11'
Taylor,Andrew Scott
Cep 23h6'27"60d30'
Taylor,Ann E
Aql 19h57'1"10d33'
Taylor,Anne Marie
Cas 0h13'29"64d9'
Taylor,Arthur Joseph
Her 16h53'44"48d33'
Taylor,Barrett Stevens
Aql 20h0'10"10d1'
Taylor,Belinda J
Hya 10h29'0"-12d16'
Taylor,Beth
Lyn 7h55'59"34d41'
Taylor,Billy Joe
Aql 19h31'24"12d40'
Taylor,Blaize Nicole
Cas 23h19'0"60d24'
Taylor,Brenda Herron
Per 2h2'21"50d15'
Taylor,Brian
Eri 4h7'38"-11d8'
Taylor,Bruce
Hya 8h30'0"-8d6'
Taylor,Bruce R
Dra 15h10'30"63d50'
Taylor,Burdette
And 2h26'1"47d36'
Taylor,Carl & Merlene
Cyg 21h8'15"38d1'
Taylor,Caroline Ruth
Lyn 8h5'53"50d18'
Taylor,Catherine
Aql 19h2'22"16d15'
Taylor,Catherine Rycroft
Tau 3h36'1"24d54'
Taylor,Cathy
Cas 23h40'45"61d37'
Taylor,Charles Allen
Per 1h57'29"50d16'
Taylor,Charles Ray
Aql 19h58'45"12d51'
Taylor,Cheryl
Del 20h14'52"12d29'
Taylor,Christopher Anthony
Sge 20h3'1"20d3'
Taylor,Clark
Her 18h24'0"28d38'
Taylor,Coleen A
Cas 0h30'42"69d23'
Taylor,Craig
Lyn 8h11'26"44d42'
Taylor,Curt
Cep 4h48'49"87d14'
Taylor,Danielle Nicola
Lyn 7h15'47"50d11'
Taylor,Darcy Charlene
Mon 6h50'27"11d27'
Taylor,David "Dave" H
Tau 4h38'16"18d58'
Taylor,David Allen
Dra 16h35'44"52d25'
Taylor,David Andrew
Cep 22h17'20"70d57'
Taylor,David Carl
Boo 14h35'24"9d41'
Taylor,David Malcolm
Boo 14h19'31"52d17'
Taylor,Debra
Aql 19h22'0"11d58'
Taylor,Debra
Tri 2h8'47"33d37'
Taylor,Denise A
Cas 1h43'17"60d24'
Taylor,Denise J
Del 20h22'50"18d59'
Taylor,Denise Susan
Peg 22h22'34"5d55'
Taylor,Derek M
Lac 22h5'1"46d31'
Taylor,Diane & Gary
Lyr 18h55'15"30d53'
Taylor,Dolly & Rodney
Cnv 12h15'0"45d49'
Taylor,Don
Lac 22h14'1"50d12'
Taylor,Donald Franklin
Hya 9h18'40"-6d3'
Taylor,Donald Owen
Ori 5h31'1"7d53'
Taylor,Doug
Lac 22h2'1"46d52'
Taylor,Drew
Vul 20h15'51"25d19'
Taylor,Ernestine Marguerite
Uma 9h53'50"42d48'
Taylor,Florence & Cyril
Cyg 21h37'56"40d29'
Taylor,Foster
Uma 8h50'12"56d57'
Taylor,Freddie
Boo 14h19'59"26d33'
Taylor,Gary & Sandy
Cyg 21h46'0"37d48'
Taylor,George C
Aur 5h55'16"31d41'
Taylor,Glory May
Lyn 8h1'29"38d48'
Taylor,Glynn
Aur 6h36'0"37d57'
Taylor,Graham Scott of Tayco
Cep 21h16'19"55d0'
Taylor,Gwendolyn May
Cet 1h43'1"0d39'
Taylor,Gwendolyn R
Cet 3h1'19"4d60'
Taylor,Halfred E
Cet 0h53'56"-4d12'
Taylor,Herbert John
Ari 2h33'39"30d24'
Taylor,Ian R
Her 17h29'10"42d30'
Taylor,Jack & Bonnie
Umi 14h50'44"75d25'
Taylor,James
Her 15h54'11"42d10'
Taylor,James A
Aur 6h33'24"33d42'
Taylor,James K
Per 2h21'32"58d49'
Taylor,Jane
Vir 12h4'1"1d51'
Taylor,Jane S
Ori 6h11'1"8d32'
Taylor,Janeth Rask
Umi 16h21'18"70d56'
Taylor,Jason Lee
Cet 0h49'1"0d52'
Taylor,Jason Scott
Her 17h24'44"21d58'
Taylor,Jeffrey H
Ser 15h14'34"24d4'
Taylor,Jerry L
Mon 7h24'1"-8d34'
Taylor,Jessie
Aql 18h56'29"-10d23'
Taylor,Jessie Lucy
Cyg 19h47'33"50d18'
Taylor,Jill
Aql 18h56'44"7d18'
Taylor,Jill
Crb 16h6'42"27d56'
Taylor,Jim
Per 2h50'47"43d43'
Taylor,Jim
Dra 16h54'45"67d44'
Taylor,Joan M & Michael L Zahm
Cyg 19h26'11"33d40'
Taylor,Joanne
Aql 20h14'21"5d15'
Taylor,Joc
Her 17h0'36"41d42'
Taylor,John
Cep 22h7'1"62d6'
Taylor,John Alan
Ori 5h51'11"18d37'
Taylor,John Barry
Ori 5h56'18"13d14'
Taylor,John C
Sex 10h14'26"-5d9'

Taylor,John Hollon
Hya 8h47'54"-4d0'B
Taylor,John McKowen
Aql 19h51'12"12d36'
Taylor,John Scott
Cet 1h43'43"-4d0'
Taylor,Joseph
Per 3h59'33"40d9'
Taylor,Joseph
Cep 20h13'24"60d4'
Taylor,Joshua Allen
Hya 9h12'48"-14d38'
Taylor,Jr,Allen Benson
Leo 9h26'29"28d40'
Taylor,Jr,Lorimer R
Ori 5h13'27"-4d4'
Taylor,Jr,Robert Allen
Ori 5h4'33"-2d22'
Taylor,Jr,Robert W
Dra 16h3'0"67d35'
Taylor,Kai Gregor
Cam 9h8'22"81d7'
Taylor,Katharine
Crb 16h10'32"31d29'
Taylor,Katherine Preston
Lyn 7h14'22"50d36'
Taylor,Kathleen M
And 1h51'41"40d1'
Taylor,Kathryn Ann
Peg 23h5'54"32d51'
Taylor,Kathryn Lee
Cnc 7h54'59"10d28'
Taylor,Kathy-Rosemary
And 23h0'60"37d10'
Taylor,Kevin Flannery's
Her 18h11'15"47d18'
Taylor,Kimberly Faith
Lyn 7h46'57"42d23'
Taylor,Kitana Ruth
Cas 1h23'47"55d52'
Taylor,Krissy
Lyr 18h59'22"30d21'
Taylor,Kristen Elizabeth
Aur 7h22'54"37d32'
Taylor,Lana Jean
Cyg 21h4'0"39d59'
Taylor,Laura Irene
Peg 23h27'20"24d36'
Taylor,Lauren Ashley
Mon 7h41'53"-1d6'
Taylor,Laurie Ann
And 23h34'1"45d3'
Taylor,Leah Elizabeth
Cas 2h59'11"60d39'
Taylor,Lee
Uma 9h55'0"70d38'
Taylor,Leigh
Sgr 20h21'19"-29d7'
Taylor,Leo Edward
Dra 15h10'52"63d35'
Taylor,Leonard B
Dra 18h52'31"67d47'
Taylor,Linda Bea
Cas 0h0'38"62d21'
Taylor,Linda D
Aql 19h9'12"12d40'
Taylor,Liz
Cas 0h19'35"63d51'
Taylor,Louis Ervin
Cep 22h25'45"56d55'
Taylor,Louis Wilson
Aqr 20h55'22"-1d8'
Taylor,Luther Craig
Sct 18h42'33"-6d43'
Taylor,Margaret Graham
Vul 19h23'35"26d11'
Taylor,Marie Werner
Ori 5h12'45"-4d31'
Taylor,Marike
Cam 5h55'47"80d13'
Taylor,Mary English
Peg 22h58'0"31d46'
Taylor,Mary Kristin
Peg 23h23'0"31d19'

Taylor,Master Rich
Oph 16h41'40"1d28'
Taylor,Matthew Adam
Lac 22h21'0"52d35'
Taylor,Matthew Gorman
Oph 17h7'20"11d1'
Taylor,Mel
Cet 1h31'0"-1d59'
Taylor,Melanie Dawn
Her 17h33'36"14d45'
Taylor,Melba
And 0h53'34"35d11'
Taylor,Michael A
Oph 18h42'52"10d54'
Taylor,Michael Biscoe
Her 17h52'37"40d16'
Taylor,Michael Sutton
Dra 17h20'1"61d45'
Taylor,Mikki Garth
Umi 17h14'14"76d42'
Taylor,Mildred
Cyg 21h52'35"44d41'
Taylor,Mimi
Cam 7h54'17"67d43'
Taylor,Mollie Mei- Ouida
And 23h21'37"41d37'
Taylor,Monique Pauline
And 23h9'57"37d41'
Taylor,Mr Dana
Her 18h55'12"12d52'
Taylor,Mr Mark
Dra 14h17'33"64d36'
Taylor,Mr Ruél Aldophus
Per 3h12'31"42d43'
Taylor,Natalie J
Peg 22h46'40"35d19'
Taylor,Nora
Cyg 19h49'0"38d11'
Taylor,Peter & Kirsty
Cyg 20h52'57"38d52'
Taylor,Philip Lothian
Her 16h29'0"40d16'
Taylor,Ray Thomas
Ori 5h17'21"15d40'
Taylor,Richard
Lac 22h44'0"53d43'
Taylor,Richard Alan
Her 16h34'44"41d13'
Taylor,Richard D
Cet 1h25'35"-10d59'
Taylor,Richard L
Per 3h4'1"46d45'
Taylor,Riley Skip
Cyg 21h14'17"38d5'
Taylor,Robert Christopher
Aql 18h56'28"13d57'
Taylor,Robert & June
Aql 18h47'37"11d45'
Taylor,Robin
Per 3h4'42"42d28'
Taylor,Robin "Blue Eyes"
Aql 19h9'0"15d47'
Taylor,Rodney
Per 2h53'19"35d45'
Taylor,Roxanne
Lyr 18h23'34"46d51'
Taylor,Ruth
And 23h31'59"40d57'
Taylor,Sam & Philip Strout
Peg 21h21'1"23d5'
Taylor,Samantha Crelier
Eri 2h52'1"-6d15'
Taylor,Scott
Aql 19h9'27"12d48'
Taylor,Second Lt Stephen
Her 17h56'0"18d46'
Taylor,Sharon
Peg 22h14'18"4d42'
Taylor,Shelly
Ori 6h3'33"1d31'
Taylor,Sibyl
Cyg 21h9'37"38d15'
Taylor,Sidney
Cyg 21h1'40"36d55'

Taylor,Sofi
And 23h18'32"42d2'
Taylor,Stephan & Lisa
Cyg 21h59'0"52d48'
Taylor,Stephanie Teresa
And 0h37'43"40d7'
Taylor,Stephen"Honey"
Lac 22h19'47"51d9'
Taylor,Steve
Uma 10h22'34"72d23'
Taylor,Steve
Tri 1h57'12"28d32'
Taylor,Steven L
Lac 22h6'37"51d27'
Taylor,Steven Robert
Aur 6h4'39"32d13'
Taylor,Sue
Cam 9h16'50"81d43'
Taylor,Susan
And 23h21'45"51d4'
Taylor,Susan & David
Cyg 21h3'1"37d58'
Taylor,Sylvia Marie
Dra 17h16'21"72d34'
Taylor,Tamani Caroline Kellar
Ori 6h2'16"8d51'
Taylor,Tamara
Equ 21h18'55"10d7'B
Taylor,Tammy M
Crt 10h59'22"-12d29'
Taylor,Tanya Tanita
Cas 1h5'45"68d11'
Taylor,The June & John
Macleod Star
Cyg 19h20'29"28d21'
Taylor,Thomas Stanley
Boo 14h22'41"27d58'
Taylor,Thomas William
Ser 17h59'0"-13d47'
Taylor,Tom
Oph 17h6'1"8d28'
Taylor,Tom
Equ 21h18'55"10d7'A
Taylor,Tori
Aql 19h6'38"15d48'
Taylor,Tracy A
And 23h18'27"44d10'
Taylor,Trisha
Cmi 7h52'39"1d22'
Taylor,Ty
Dra 19h9'51"78d53'
Taylor,Unforgettable II,Betty
Jane
Lep 5h23'49"-19d44'B
Taylor,Unforgettable, Arthur
George
Lep 5h23'49"-19d44'A
Taylor,Venita Rose
Cas 1h44'26"63d56'
Taylor,Vicky Lawrence
Peg 22h45'33"25d57'
Taylor,Victoria Jessica
Cas 1h46'0"65d9'
Taylor,Warren
Cep 0h45'22"78d32'
Taylor,Wendy & Jeffrey
Mon 8h4'13"-7d14'
Taylor,William
Her 17h17'46"41d7'
Taylor,William F Cody
Lac 22h2'0"46d40'
Taylor-Davis,PhD, Stephanie
Anne
Tri 2h20'10"30d51'
Taylor-Love Always, Mike &
Mary Ann
Tri 2h45'48"32d12'
Taylor-Retired USN,Lt William
K
Equ 21h21'13"2d51'
Taylor/Angel Plane, Zack
Ori 5h0'54"-2d25'
Taylor:The Brightest
Star,Kathleen A
Uma 9h8'34"67d55'

Taylord Computer Marketing
Corp
Cep 21h2'54"70d34'
Taz
Oph 17h53'24"8d26'
Taz
Uma 9h51'0"68d27'
Taz Tkg
Boo 15h20'37"41d56'
Tazmania
Cyg 19h49'49"58d10'
Tazmanian Devil
Cam 3h49'44"60d36'
Tazzmania Rick-N-Michelle
Cyg 21h8'30"40d16'
TB's Twinkle KCL 945 SHM
Lac 22h3'1"46d5'
TC
Uma 12h8'0"48d58'
TCA-#1
Aur 7h4'33"38d9'
Tcatchenko,Julie
Cet 2h34'19"5d12'
TCB 1
Ori 5h42'58"8d54'
Tchakerian,Isabelle Afarian
Cet 2h29'31"2d11'
Tchakirides,Nelson
Her 18h4'30"48d42'
Te amo Kate
Cyg 21h58'47"52d33'
Te Amo Siempre Lou
Her 16h32'59"50d44'
Te Amo,Miguel
Cep 21h38'38"84d51'
Te Amoschetti
Crt 11h23'12"-16d39'
Te Quiero Mucho Michele
Mon 6h46'52"10d10'
Te Quiero Rita Maria
Cas 0h35'0"58d22'
Te-Urich
Ori 6h13'22"0d27'
Tea
Del 20h12'29"10d21'
Teacher Jan
Equ 21h2'50"2d51'
Teacher's Twinkle
Uma 10h56'14"38d31'
Teacher,The
Cam 7h8'28"61d16'
Teachout,Sarah Jean
Cnc 8h9'1"30d12'
Teagle,Sarah
Peg 23h43'31"31d12'
Teague IV,Samuel Wesley
Aur 6h14'7"31d33'
Teague,Abigayle "Abby"
Cyg 19h34'11"38d25'
Teague,Barry
Per 2h56'32"40d4'
Teague,Carrie
Eri 3h58'45"-8d55'
Teague,Christopher-Jo
Aur 6h3'34"37d54'
Teague,Don Henry
Hya 8h14'26"5d56'
Teal
Oph 17h12'25"11d32'
Teale,Sharon Kay
Peg 23h6'34"20d8'
Teall,Jacquelyn H & James E
Godette
Tri 1h52'24"26d47'
Team 925
Her 16h36'1"21d58'
Team Bob
Cet 1h29'19"0d54'
Team Fernco
Tri 1h50'30"27d20'
Teana Aliane
Cnv 13h33'20"42d17'
Teaney,Shawn Randolph
Costello
Uma 9h42'1"68d33'

Tear,Jessica
And 22h58'59"50d31'
Tears in Heaven
Eri 4h4'18"-19d42'
Tearte,Curtis H
Boo 14h30'20"22d42'
Teasdale,Diane Elizabeth
Cyg 19h57'49"45d15'
Teasdale,Mr Anthony William
Aql 19h5'60"0d52'
Teasley,Hunter Allison
Uma 11h22'57"58d59'
Teasley,Tim
Ser 18h5'49"-10d2'
Teaster,Matthew Clay
Boo 14h22'22"31d23'
Teater,Diana
Peg 21h42'29"23d1'
Tebbe,James Maxwell
Sex 10h38'12"0d34'
tebbs,patsy sue
And 1h20'1"38d1'
Tebeck,June
Lyr 18h42'19"34d22'
Tebel-Nagy
Cap 20h55'55"-23d17'
Tebel-Nidetsky
Sco 17h27'46"-30d27'
Tebinka,Ashley
Cam 6h37'1"80d27'
Tebrocke,Ralf
Lyn 8h0'46"41d27'
Tecam
Hya 9h6'49"-0d19'
Tech Computer,Inc
Aur 5h3'13"50d20'
Techant,Ruediger
Equ 20h58'50"8d54'
Techi
Ori 6h8'17"4d36'
Technovin
Cam 4h11'56"61d10'
Techow,Hans Eike
Cnc 8h28'0"8d52'
Tecihila
Del 20h19'26"10d43'
Teckman,Michael
Peg 23h44'1"30d56'
Tecla
Lep 4h59'23"-11d40'
Tecnec
Lyn 7h50'31"42d19'
Tecta,Alberti Stella
Ori 5h47'54"10d35'
Teczar,Nicole R
Psc 1h22'24"32d30'
Ted
Lyn 7h40'45"45d29'A
Ted & Elaine's Happiness
Cyg 19h25'47"48d58'
Ted & Kathy's Star
Uma 9h34'43"56d46'
Ted & Linda's Star
Cyg 19h28'23"41d6'
Ted & Muriel
Uma 11h12'28"48d8'
Ted's Delight
And 23h29'40"41d25'
Ted's Millennium Lifestar
Uma 8h39'0"70d37'
Ted's Place
Aql 19h26'1"-8d44'
Tedder,Fred
Cmi 7h35'48"10d32'
Teddy
Aur 4h49'49"40d51'
Teddy
Ori 5h39'52"15d5'
Teddy
Cet 0h47'1"-2d44'
Teddy 42
Cep 21h57'51"55d18'
Teddy Bear
Umi 14h46'0"80d8'

Teddy Bear
Boo 14h38'30"19d40'
Teddy For Ever
Sge 19h36'11"16d31'
Teddy's Star
Hya 8h11'25"0d3'
Teddys Twinkle
Boo 14h19'48"50d15'
Tedeschi Always &
Forever,Suz & Gian
Cyg 20h36'29"30d6'
Tedeschi,Katia
Lyr 18h37'1"43d57'
Tedeschi,Matthew Michael
Cam 3h54'36"70d14'
Tedeschi,Roberto
Per 3h29'10"50d45'
Tedesco,Anthony
Per 2h43'55"43d40'
Tedesco,Janine
Cyg 21h7'24"30d20'
Tedi "Princess Bear"
Umi 15h39'53"69d27'
Tedinel
Uma 8h32'60"55d30'
Tedly
Ori 5h31'38"1d40'
Tedone,David Matthew
Cet 2h17'15"5d35'
Tedone,Melissa Virginia Marie
Mon 6h28'40"8d43'
Tedone,Vincent
Her 17h35'37"28d52'
TEDSTAR
Aqr 20h55'15"-6d24'
Tee Ridder
Aql 18h42'23"0d53'
Teebken,Meike
Sco 16h51'46"-43d13'
Teebs
Aur 6h32'38"34d24'
Teed,Michael David
Sex 9h40'21"1d40'
Teeias,Mona
Cyg 21h30'43"37d55'
Teel,Doyle
Sge 20h3'0"16d51'
Teel,James Earl
Lac 22h32'46"55d10'
Teel,Kurtis
Uma 9h38'57"45d16'
Teel,Mildred Rita
Oph 18h38'36"6d53'
Teel,Richard H
Aur 5h29'30"36d17'A
Teel,Virginia J
Aur 5h29'30"36d17'B
Teele,Gerald A
Sct 18h44'0"-6d12'
Teeple,Brandon
Dra 16h3'13"63d51'
Teeple,Mary Christine
Lyr 19h21'59"37d53'
Teesdale,John Robert
Aur 5h51'57"31d15'
Teesta
Cyg 21h21'51"40d58'
Teeter,Emily
Cam 3h31'46"58d28'
Teeter,Jenny L
Cyg 19h33'1"38d18'
Teeter,Sylvia
Com 12h28'53"20d60'
Teeters,Jesse Coleman Russell
Sco 17h1'51"-37d48'
Teeters,Joseph Carter
Eri 4h46'34"-7d60'
Teeters,Star-Margaret "Mer"
Cas 23h33'25"61d6'
Teets,Jessica Anne
Lyn 7h42'0"43d4'
Teets,Jonathan A
Cep 23h13'50"68d29'
Teets,Rebecca Ellen
Cas 23h47'36"50d18'

Teetzen's Anniversary Star
Dra 20h16'34"67d44'
Tefertiller,Jim
Lac 22h47'1"53d27'
Teff,Iris & Terry
Cyg 20h34'15"31d28'
Tegan Paige
Mon 6h25'45"2d42'
Tegen,Charlotte Bocher
Her 16h49'26"30d18'
Tegen,Tyler W
Per 1h58'48"56d35'
Tegenborg,Alexandra French
Cyg 19h34'52"32d59'
Tegenborg,Kristen Ann
Lyn 8h50'51"41d11'
Tegti,Grace
Cyg 21h5'20"31d9'
Tegtmeyer,Wilma & Ernest
Cyg 21h1'23"40d47'
Tegwani
Dra 9h55'48"73d19'
Tehane
Cyg 20h21'32"39d50'
Tehrand,Edith
Cyg 20h23'21"40d44'
Tehrani,Parvin Dokht Sajadi
Uma 11h14'52"48d26'
Teich,Dr Clifford
Cep 22h57'14"57d60'
Teicher,Michael
Per 2h0'43"50d17'
Teicher,Rachel
Eri 4h4'24"-5d20'
Teichert
Dra 18h22'25"68d58'
Teichgraf Jochen Peter
Kentenich
Ari 2h1'0"21d37'
Teichman,Jacqueline Lee
And 2h28'49"47d16'
Teinturier,Iure
Her 18h7'33"38d51'
Teisen,Christy Lynn
And 2h31'12"42d3'
Teitel,Rebecca
Peg 21h54'50"28d50'
Teitelbaum,Herbert
Gem 7h29'35"30d38'
Teitjie,Mary Farrar
Sco 16h38'13"-33d57'
Teitsort,Arthur Joseff
Dra 12h49'38"70d36'
Teitsort,Sonja Talyse
Cam 4h11'26"61d17'
Teixeira,Amy May
Mon 6h44'35"10d44'
Teixeira,Charles August
Psc 1h35'11"27d53'
Teixeira,Joan M
Cyg 19h58'50"40d32'
Teixeira,Ronald J
Scr 15h17'0"20d9'
Teixeuo,Mélanie
Com 12h31'13"31d18'
Teja,Carole
Eri 3h46'46"-0d9'
Tejada,MD,Carlos T
Uma 11h48'1"41d29'
Tejani,Salini
Uma 9h37'26"48d32'
Tejano,Dan
Ser 16h0'28"23d33'
Tejera,R Jay
Peg 23h25'26"18d33'
Teklits,Kathleen
And 0h27'14"38d59'
Tekstar
Uma 9h53'27"57d21'
Tektronix Golden Galaxy,The
Her 17h33'35"24d57'
Tektronix Golden Galaxy,The
Her 17h37'29"21d22'
Tektronix Golden Galaxy,The
Her 17h30'1"25d17'

Tektronix Golden Galaxy,The
Her 17h33'45"22d35'
Tektronix Golden Galaxy,The
Her 17h34'54"23d43'
Tektronix Golden Galaxy,The
Her 17h38'46"20d47'
Tektronix Golden Galaxy,The
Her 17h31'45"24d2'
Tektronix Golden Galaxy,The
Her 17h39'57"25d1'
Tektronix Golden Galaxy,The
Her 17h36'53"20d0'
Tektronix Golden Galaxy,The
Her 17h34'21"22d11'
Tektronix Golden Galaxy,The
Her 17h35'36"22d56'
Tektronix Golden Galaxy,The
Her 17h37'34"22d30'
Tektronix Golden Galaxy,The
Her 17h26'52"25d51'
Tektronix Golden Galaxy,The
Her 17h0'55"25d40'
Tektronix Golden Galaxy,The
Her 17h11'56"25d33'
Tektronix Golden Galaxy,The
Her 17h13'0"25d19'
Tektronix Golden Galaxy,The
Her 17h17'50"27d46'
Tektronix Golden Galaxy,The
Her 17h18'14"25d58'
Tektronix Golden Galaxy,The
Her 17h18'53"27d43'
Tektronix Golden Galaxy,The
Her 17h16'56"27d2'
Tektronix Golden Galaxy,The
Her 17h20'14"29d5'
Tektronix Golden Galaxy,The
Her 17h21'54"29d23'
Tektronix Golden Galaxy,The
Her 17h22'15"28d24'
Tektronix Golden Galaxy,The
Her 17h23'37"28d57'
Tektronix Golden Galaxy,The
Her 17h25'18"24d35'
Tektronix Golden Galaxy,The
Her 17h26'56"29d32'
Tektronix Golden Galaxy,The
Her 17h33'18"25d14'
Tektronix Golden Galaxy,The
Her 16h16'1"50d14'
Tektronix Golden Galaxy,The
Her 16h44'48"36d30'
Tektronix Golden Galaxy,The
Her 16h50'45"36d10'
Tektronix Golden Galaxy,The
Her 16h28'33"30d22'
Tektronix Golden Galaxy,The
Her 16h47'31"37d48'
Tektronix Golden Galaxy,The
Her 16h29'15"30d2'
Tektronix Golden Galaxy,The
Her 15h59'0"50d49'
Tektronix Golden Galaxy,The
Her 16h1'39"50d17'
Tektronix Golden Galaxy,The
Her 16h2'26"50d58'
Tektronix Golden Galaxy,The
Her 16h8'1"50d38'
Tektronix Golden Galaxy,The
Her 16h11'37"50d8'
Tektronix Golden Galaxy,The
Her 16h9'35"50d27'
Tektronix Golden Galaxy,The
Her 16h19'39"50d60'
Tektronix Golden Galaxy,The
Her 16h21'23"50d40'
Tektronix Golden Galaxy,The
Her 16h23'27"50d27'
Tektronix Golden Galaxy,The
Her 16h48'1"37d10'
Tektronix Golden Galaxy,The
Her 16h33'36"30d50'
Tektronix Golden Galaxy,The
Her 16h36'11"30d23'

Tektronix Golden Galaxy,The
Her 16h29'11"30d5'
Tektronix Golden Galaxy,The
Her 16h38'49"30d9'
Tektronix Golden Galaxy,The
Her 16h39'41"30d44'
Tektronix Golden Galaxy,The
Her 16h41'20"30d58'
Tektronix Golden Galaxy,The
Her 16h44'0"37d56'
Tektronix Golden Galaxy,The
Her 16h46'42"37d1'
Teleciel Thierry Nicol
Ori 5h55'1"19d59'
Telencio,Gloria
Lyn 8h56'45"45d10'
Teles,Joana Neiva Olivia
Her 16h26'1"23d59'
Telesis
Aql 19h30'42"13d6'
Telford,Kathy
Lyn 9h19'22"36d59'
Telford,Lynn Ann
And 0h10'11"35d32'
Telisman,George G
Her 18h55'24"12d26'
Teller,Corinne
Cas 0h42'40"67d24'
Teller,Inge Stern der Liebe
Vir 12h25'35"-8d18'
Teller,Jr,James David
Boo 13h48'29"17d59'
Tellervo Stluka,Irma
Leo 11h7'18"2d2'
Telli "Elusive",Robert
Uma 11h54'35"57d40'
Tellian,Linda & Frank
Cyg 19h32'45"56d23'
Tellier,Amanda Surwilo
Cap 21h20'29"-18d56'
Tellier,Michel
Tri 1h58'44"30d54'
Tellini
Uma 8h53'21"53d33'
Telljohann,Richard Ellis
Oph 17h57'59"12d21'
Telloli,Fiorella
Cnc 9h1'57"18d38'
Tells,Virginia Ann
Mon 7h14'6"-5d8'
Telotha
Sge 18h57'0"20d16'
Telshaw-"Sonny", Clarence
Aql 19h45'1"14d26'
Telson
Her 18h2'54"28d37'
Telthorst,Margery
And 1h33'18"40d11'
Teman,Nora Jean #1
Aqr 23h21'9"-8d44'B
Tembra
Mon 6h58'38"11d58'
Temes
Vul 21h26'0"24d22'
Temesan,Elaine
And 1h33'20"39d57'
Temkin,Charles
Dra 14h15'24"63d5'
Temkin,Ronald Norman
Ser 17h34'56"-10d41'
Temkin,Z & The Big I
Crb 15h16'11"27d1'AB
Tempalski,Evelyn Vajda
Cas 1h54'50"75d25'
Temperten,Emma
Sge 20h7'40"20d3'
Tempest
Her 18h4'0"28d26'
Temple III,Joseph (Buck)
Uma 11h12'52"49d4'
Temple Star,The Victoria
Vul 20h3'18"25d9'
Temple,Charles Bennett
Del 20h16'0"11d3'

Temple,Douglas Clarkson
Ori 5h56'28"8d11'
Temple,James Mahaffey
Peg 23h30'11"32d53'
Temple,Poomer McKinley
Mon 7h25'52"-1d39'
Temple,Robert G
Her 16h28'48"35d13'
Temple,Russell Douglas
Lac 22h38'47"52d39'
Temple,Suezie
Equ 21h7m27"11d6'
Temple,Susanne Sophia
Uma 11h11'50"49d21'
Temple,The
Cyg 19h37'52"33d51'B
Templet,Janie Duncan
Cyg 21h59'31"52d51'
Templeton,Alisha
Cam 3h52'39"71d34'
Templeton,Farlie Robison
Peg 22h9'51"4d53'
Templeton,Heather Ann
Cas 2h6'52"59d60'
Templeton,Jim
Aur 7h24'0"43d22'
Templeton,Stephen L
Hya 9h17'25"6d10'
Templeton,Sven Eric
Oph 17h53'44"11d58'
Templeton,Travis Perry
Cep 23h10'1"64d46'
Templin,Michele
Cas 23h19'42"60d11'
Templin,Patricia Anne
Mon 7h1'59"-6d28'
10-23-1956 Hungarian Freedomfight
Ori 5h54'17"21d53'
Ten Enterprises/de Moetew
Cam 5h3'0"65d22'
Ten Times More
Uma 12h1'12"48d21'
Tena
And 2h32'12"38d45'
Tena'nche,Star Of Joshi
Peg 22h45'19"34d14'
Tenacious Kim's Comet
Lmi 10h25'42"30d1'
Tenaglia,Audrey Atkin
And 23h21'0"49d29'
Tenaglia,Ralph P
Sex 10h35'24"5d51'
Tenaj Apork
Hya 8h50'11"-5d40'
Tenanty-Nason,Patricia
Vir 12h3'43"2d7'
TenBarge,Anthony Ramond
Aqr 20h45'32"-0d58'
Tencer,Beloved Samuel
Equ 21h20'43"7d37'
Tench,Cathy
Gem 7h16'49"21d28'
Tencton
Dra 17h59'10"58d43'
Tencza
Dra 16h47'0"67d24'
Tenda,"Little Pilgrim" Alicia M
Peg 22h0'42"30d8'
Tenderfoot's Lady in Red
Cma 6h57'14"-30d30'
Tendil,Claude
Leo 10h32'19"10d21'
Tendler,Barry F
Per 1h32'56"53d48'
Tendler,Phillip
Cep 23h4'42"64d4'
Tendler,Rav M D
Lyr 18h43'50"33d16'
Tendress et Aladin
Ori 5h27'41"0d25'
Tenerife
Ori 5h39'39"1d7'
Tenesse,George & Betty Ellen
Sge 20h2'26"20d31'
Tenhagen,Herr
Boo 14h27'25"10d6'
Tenhunen,Tiia Tuulevi
Cam 5h33'27"60d20'
Tenille Joy
Vel 9h45'25"52d27'
Tenk,Ricati
Aur 4h35'45"30d31'
Tenk,Richard A
Aur 4h54'22"51d9'
Tenley & Victor
Sge 19h18'38"18d9'
Tenley-Frels,Bonnie S
Cas 0h56'52"60d31'
Tennant,Barbara A
Crb 15h23'40"30d50'
Tennant,Glennie
Lyn 9h1'44"40d36'
Tennant,Jeffery T
Oph 17h38'20"11d31'
Tennant,Steven P
Per 1h27'33"50d14'
Tenner,Carol
Cas 23h41'36"61d34'
Tenney
Lyr 18h43'24"42d37'
Tenney,Danielle
Mon 7h23'47"-5d56'
Tenney,Del & Margo
Cyg 19h24'51"31d24'
Tenney,Michelle Elizabeth
Peg 23h23'17"33d29'
Tennis,Wanda Lee
Lmi 10h18'12"31d31'
Tennison,Callie Lupton
Leo 10h55'41"15d44'
Tennison,Courtney Lee
Ari 3h10'59"15d29'
Tennison,Gloria Lupton
Mon 6h37'0"7d26'
Tennison,Harry Lee
Leo 10h52'0"16d13'
Tennison,In Memory Of Linda Marie
Peg 23h38'45"18d28'
Tennison,Lee
Hya 8h9'32"3d56'
Tennison,Margo Marie
Sct 18h34'1"-5d0'
Tennison,Patty Cartwright
Gem 6h39'30"30d57'
Tennyson-McFatter, Shirley Jean
Equ 21h0'56"7d51'
Tenore,Nicole R
Cet 2h14'49"1d40'
Tenorio "My Lovely Wife",Bonnie A
Sge 19h16'28"16d40'
Tentschert,Jeremy
And 23h38'34"40d6'
Tenzer,Christoph
Cep 22h3'10"61d33'
Tenzer,DMD,Jonathan A
Oph 17h4'48"1d32'
Tenzler Star,The
Cam 4h49'13"70d25'
Teodoro Asis de los Santos
Cap 21h39'18"-20d18'
Teodosio,Valdemar
Boo 14h36'0"31d37'
Teoh,Gary James Chern-Wei
Oph 17h3'53"-24d11'
Teoh,Melissa Sky Jin-Li
Sex 10h37'48"-6d23'
Teolis,Diane
Cyg 19h27'30"35d38'
TePastte,Jr,Albert Louis
Ari 2h16'8"22d56'A
TePastte,Lori Ann
Ari 2h16'8"22d56'B
Tepe,Amanda C
Cyg 21h17'11"38d47'
Tepe,Riley Ann
Peg 23h28'31"27d47'
Tepfer,Avi
Lyn 7h2'21"53d37'
Tepper,Celia
Cas 1h2'39"61d47'
Tepper,David V
Hya 8h26'36"-17d14'
Tepper,Jürgen
Tau 3h56'21"10d48'
Tepper,Susan L
Cas 23h42'42"61d20'
Tepper,William
Per 2h9'31"57d10'
Teppler,Gisela
Psc 1h20'0"21d53'
Teppler,Helmut
Tau 3h41'0"7d37'
Teppo,Tyrmi
Cas 0h40'13"61d42'
10-23-1956 Hungarian Freedomfight
Ori 5h54'17"21d53'
Tequila
Aql 19h31'55"12d9'
Tera
Peg 23h28'19"23d43'
Tera
And 0h48'30"39d31'
Tera Lynn
Ori 5h52'22"17d18'
Tera Toro Redstem 9
Ori 5h58'1"17d41'
Teracer,Carol
Vir 13h51'48"-5d40'
Terada,Stella of Daiki
Sco 16h22'38"-27d42'
Teragram Deer
Cam 6h8'35"60d27'
Teramo,Anna Claudia
Umi 16h25'0"78d54'
Teraoka,Traci Lynn
Lyr 18h29'15"30d23'
Tercho,Leona
Cas 1h5'23"60d25'
Terd Star,The
Uma 9h17'1"42d12'
Tere
Aur 6h59'41"40d17'
Tere Ros
Cep 22h23'53"60d4'
Terebey,Stacie A
Del 20h18'50"11d55'
Terek,Carly
Cyg 20h20'38"31d50'
Terence
Uma 9h36'21"53d19'
Terence
Dra 16h10'47"61d4'
Terence James
Cyg 19h43'59"34d53'B
Terenia
Lac 22h45'50"53d45'
Terenyi,Georg Dr
Cnc 8h25'30"30d34'
Terenzi,Lorraine
Cas 0h39'19"58d23'
Tereo,Michael Barret
Her 16h4'15"20d4'
Teresa
Oph 18h24'27"8d25'
Teresa
Vul 20h38'34"20d23'
Teresa
Cmi 7h42'42"5d35'
Teresa
Mon 6h21'10"4d13'
Teresa
Cet 1h34'36"0d51'
Teresa
And 1h16'0"35d45'
Teresa
Dra 11h52'1"73d5'
Teresa
Peg 23h39'36"31d10'
Teresa
Uma 10h31'14"48d46'
Teresa
Ari 2h59'22"30d56'
Teresa
Mon 6h24'12"-8d52'
Teresa
Cyg 20h16'25"30d27'
Teresa
Uma 11h51'1"32d7'
Teresa & John
Cyg 20h24'0"38d56'
Teresa Ann
Lyn 9h14'46"40d15'
Teresa Ann Elizabeth
Lyr 18h46'0"33d48'
Teresa Costa i Sala
Lyn 8h19'22"48d54'
Teresa Delores
Lyr 18h21'0"38d48'
Teresa Diane
Peg 23h45'22"27d11'
Teresa Eternal
Lyr 19h16'17"40d21'
Teresa Jane
Lyr 18h29'40"34d11'
Teresa Light Of My Life
Aql 19h54'0"14d41'
Teresa Lynne
And 1h2'55"47d9'
Teresa Marie
Cas 3h6'57"67d44'
Teresa Marie
Lyr 18h18'24"46d53'
Teresa Nicole
And 1h46'13"47d26'
Teresa Shirley
And 23h49'51"41d7'
Teresa Sue
Vul 19h46'53"22d40'
Teresa's Eternal Verse
Lyr 18h35'26"30d8'
Teresa's Samuel-Emily
And 23h32'40"45d28'
Teresa's Star
Com 12h36'35"24d27'
Teresa's Star
Lmi 9h53'54"38d46'
Teresa's Twinkle Star
Mon 6h30'36"7d49'
Terese
Cyg 19h31'1"30d15'
Terese The "E"
Uma 10h45'10"53d11'
Teressa,Debra
Lyr 18h34'1"40d56'
Tereza
Dra 15h58'54"61d28'
Teri
Peg 22h41'0"34d47'
Teri
Cyg 20h15'16"38d39'
Teri Lynn's Twinkle
Sge 20h17'47"17d57'
Teri's La Luz De Mi Corazon
Cet 2h42'60"6d39'
Teri's Twinkle
Sct 18h55'29"-7d14'
Teri,Kevin N
Boo 14h10'11"31d4'
Tericky
Lac 22h1'57"40d30'
Terjen,Henry Foster
Aqr 20h59'44"0d48'
Terlecki,Glenn
Her 17h11'1"45d37'
Terlep,Daniel Joseph
Her 16h33'30"37d39'
Terlingo,Andrew Edward
Aql 19h6'27"3d46'
Terlingo,Samuel Anthony Allport
Peg 0h11'29"13d45'
Terlizzi
Cep 22h59'1"65d24'
Terlizzi's Twilight, Anthony
Aur 4h59'60"38d40'
TerMaat,Rodney
Her 17h3'55"47d58'
Terme Continental
Pup 7h56'32"-28d45'
Termine,Florence
Cas 2h38'57"57d28'
Termine,Reverend Vincent J
Dra 19h25'46"56d26'
Termini,Deborah J
Peg 22h44'36"34d36'
TerminusEldorado
TheKillerDickOfGold
Her 16h42'1"23d14'
Ternes,John Frederick
Boo 14h26'19"28d37'
Ternes,Thomas Jeffery
Lac 22h52'15"37d34'
Ternisien,Didier
Cyg 19h47'1"38d23'
Ternyik 2-28-36,Joan Estel [Brady]
And 2h33'25"42d57'
Terote
Peg 21h55'36"20d14'
Terpstra,Bernadette
Mon 7h4'42"-6d35'
Terpstra,David M
Ser 15h58'25"1d35'
Terpstra,Liorah Pauline
Vir 11h38'38"5d0'
Terpstra,Samuel Damon
Sco 17h52'28"-33d48'
Terpstra,Theodore Rosevelt
Cep 22h7'31"60d38'
Terra
Peg 23h45'43"12d52'
Terra Byerly's Star
Del 20h51'21"2d13'
Terra Information Mngt Systems Co
Mon 6h35'59"1d17'
Terra Lee
Equ 21h20'35"10d6'
Terra Lynn
Lyn 9h8'27"46d37'
Terradista,Lauren
Lyn 7h55'32"43d27'
Terradista,Sarah
And 23h20'23"45d16'
Terradista,Steven
Her 16h49'54"38d38'
Terragni,Maddalena Milani
And 23h6'33"52d48'
Terrahe,Christoph B
Vul 19h44'30"27d11'
Terraine Landing
Aql 19h54'29"10d56'
Terranova,Christina Michelle
Cyg 19h32'38"34d45'
Terranova,Victoria
Mon 7h46'32"-5d34'
Terranova,Vincent
Cet 1h33'47"-14d57'
Terrant,Elizabeth
Cmi 7h55'1"7d43'
Terrapin
Cam 10h51'34"81d19'
Terras,Marie
Cas 1h40'1"60d59'
Terras,Robert (Scotty)
Lac 22h2'1"40d36'
Terrassin,Charlotte
Cyg 20h29'14"31d40'
Terrazas,Henrietta
Cam 4h4'44"61d6'
Terrell(Kiki),Kristine
Uma 10h57'0"57d49'
Terrell,Paul
Oph 17h56'1"12d40'
Terrell,William
Sgr 18h55'41d-22d2'
Terreri,Madelyn Nigara
Cas 23h28'27"62d17'
Terressa
Uma 9h51'1"49d41'
Terrey,Avril Margaret
Cas 0h54'25"63d3'
Terri
Cam 5h49'41"58d29'
Terri
Vul 19h23'29"22d31'
Terri
Vul 19h47'0"20d8'
Terri
Ori 6h14'35"1d39'
Terri Anne
Eri 4h3'30"-1d56'
Terri Donna
Uma 10h13'0"55d54'
Terri G
Peg 22h43'51"12d21'
Terri Jo
Cet 1h45'50"-5d17'
Terri Kaye
Vul 19h41'52"27d0'B
Terri Lea
Mon 7h43'7"-2d15'
Terri Lee
Lyn 8h7'34"40d60'
Terri Michelle
Pho 23h42'8"-45d31'
Terri's Birthday Star
Leo 11h5'29"-1d34'
Terri's Dream
Sex 10h46'9"-7d42'
Terri's Star
Cyg 21h59'46"36d18'
Terri-Lynn
Del 20h16'23"10d50'
Terriault,Olivier
Her 17h31'50"20d53'
Terrie Helene
Gem 7h21'2"20d21'
Terrien,Benjamin A
Her 17h7'16"20d6'
Terrien,Thierry
Per 3h57'18"36d13'
Terrific Tommy
Ori 4h49'0"13d34'A
Terrill,Bernard L
Aur 6h26'16"35d24'
Terrill,Kenny
Dra 19h2'30"68d59'
Terrill,Mary Pfister
And 23h18'40"41d29'
Terrison
Sct 18h54'40"-6d40'
Territoire Redskins
Her 18h4'10"14d47'
Terrovona,J Rick
Umi 16h10'22"80d35'
Terry
Lyr 19h26'50"37d54'
Terry "Mattie", Matthew Abel
Ori 5h56'48"14d17'
Terry & Bill
Cep 3h3'47"81d16'AB
Terry & Duane "Unforgettable"
Sge 19h0'32"20d27'
Terry & Fabienne's Silver Star
Crt 11h38'51"-22d28'
Terry (Nephtac),Steve
Ori 5h56'48"10d35'
Terry Alexander
Ori 6h21'49"18d49'
Terry Daniel
Lyn 8h3'47"42d9'
Terry Elaine
Lyr 19h6'0"38d40'
Terry Jo
Lib 14h35'22"-18d50'
Terry Lynn 8-03-85
Uma 11h14'17"55d22'
Terry Star Bucking Horse Kate
Aql 18h58'49"-5d19'
Terry Terrestial,The
Peg 22h34'12"21d41'
Terry Tori's Star
Mon 7h39'38"-4d36'
Terry's Carwash
Peg 22h0'1"30d6'
Terry's Heart
Cas 1h2'0"50d1'
Terry's Heart
Del 20h15'49"13d57'
Terry's Teddy Bear Star
Cas 23h40'11"64d53'
Terry,5'6' 1903,Herbert L
Ori 5h52'34"11d30'
Terry,Adrian & Margaret
Cyg 19h24'44"44d31'
Terry,Alex Mitchell
Her 18h18'47"28d14'
Terry,Amy
Cas 1h44'13"58d60'
Terry,Betty & Duane
Cam 3h48'1"61d38'
Terry,Don Walton
Sgr 18h57'56"-28d47'
Terry,Garrett James
Peg 23h37'21"18d16'
Terry,Gillian Lesley
Crb 16h10'42"34d55'
Terry,Helen & Frank
Boo 13h39'48"14d21'
Terry,Helen Mericle
Eri 2h59'1"-18d10'
Terry,Holly Elizabeth
Com 12h12'32"19d18'
Terry,Jack T
Per 2h21'55"58d58'
Terry,Joan
Cyg 20h33'1"38d22'
Terry,Kaitlin Marie
Peg 22h38'24"26d40'
Terry,Kathy & Duane
Cyg 19h36'31"28d40'
Terry,Larry Wayne
Cmi 7h52'56"3d55'
Terry,Mark Nicholas
Her 16h59'11"38d4'
Terry,Melissa
Cnv 13h32'55"40d8'
Terry,Michele D
Cyg 21h5'45"30d34'
Terry,Sydnel Beth
Com 12h58'43"26d26'
Terry,W M
Aql 19h30'20"14d40'
Terry,Warren Harding
Dra 19h19'11"68d32'
Terryle
Dra 17h22'46"61d48'
Tersaga,David A
Aur 5h7'60"38d56'
Terse,Christine A
Lyr 19h19'49"40d5'
Tersigni,Claire Elizabeth
Lyn 7h50'48"38d33'
Tersillo,Antonia Petrina
Cas 2h4'48"59d52'
Tersillo,Petrina Tantillo
Cam 7h55'50"60d2'
Tersteph
Aql 18h43'36"10d16'
Tersti
Lac 22h9'17"38d14'
Tersy,Raul Q
Aur 5h8'55"43d44'
Tertel,Vonnie
Crb 16h14'33"34d6'
Tertius,George Chapman
Lac 22h42'45"54d10'
Tervit,Jacqueline
Cyg 20h43'56"45d20'
Tervort,Dale D
Her 18h8'14"47d18'
Tervort,David G
Her 18h7'45"47d58'
Tervort,Marylne K
Her 18h8'28"48d19'
Terwilliger,Barbara
And 0h12'31"33d22'
Terwilliger,Daniel
Boo 13h52'11"19d50'
Terwilliger,Mary Lucille Fanatilo
Lyn 7h53'45"43d17'
Terwilliger,Rex D & Susan Kasperovich
Crb 15h52'15"32d29'
Terwilliger,Thomas
Per 4h6'44"48d56'
Teryl One
Aql 18h58'0"11d56'
Terzakis,Theodore G
Oph 18h2'53"11d52'
Terzi,A Jesse
Cep 22h23'49"59d33'
Terzi,Robin M
Lyn 7h3'18"58d8'
Terzian,Martin Love You Forever Linda
Tau 5h45'33"23d60'
Terzuola,Daniel
Per 1h48'45"50d37'
Terzuola,Matthew
Cep 21h2'57"58d51'
Terzuoli,Robert J
Per 2h20'14"55d9'
Tesauro,Joanne Trabold
And 1h5'0"38d12'
Tesch,Randy S
Aur 6h55'20"44d12'
Tesh,Prima Sellecchia
And 2h24'25"46d43'
Tesh,William H
Equ 21h1'28"10d57'
Tesi,Janeane
Peg 22h33'25"25d28'
Teske's Magnificent, Mister
Cet 0h58'46"-1d21'
Tesler,Jeny
Aql 19h27'59"-10d30'
Tesler,Marc Stanley
Peg 23h9'14"62d45'
Teslik,Maria
Cas 2h28'1"59d29'
Teslik,Nicholas
Eri 3h52'12"-2d17'
Tesluk,Harriet Mom
Lyn 8h3'0"50d19'
Tesmer,John & Alice
Boo 13h39'40"20d36'
Tesoriero,Andrea Rose
And 22h57'30"38d7'
Tesoriero,Bradley Drucker
Aur 6h32'55"38d59'
Tesoriero,Dr Richard John
Dra 19h2'46"50d37'
Tesoriero,Joseph L
Ari 2h1'29"22d7'
Tesoriero,Laura Rae
Cam 7h51'31"60d1'
Tesoriero,Terry
Lmi 10h9'1"38d4'
Tesoro Gina
Uma 9h55'39"61d23'
Tesoro Richard
Cnv 12h25'26"41d34'
Tesoro Thomas's Rionnag
Lyn 9h3'60"38d5'
Tesoro Timothy
Cyg 21h4'29"50d3'
Tess
Peg 22h13'0"4d15'
Tess
Tau 4h57'52"20d13'
Tess
Lyn 7h53'41"51d54'
Tess' Heart
Cet 0h39'37"0d37'
Tess,Teresa Terri
Psc 23h32'59"0d28'
Tessa
Equ 21h19'20"10d8'
Tessa
And 1h38'20"37d54'
Tessa
Cyg 21h38'22"42d33'
Tessa & Olli
Aur 5h0'14"47d38'
Tessa the Beautiful
Cas 0h32'54"50d26'
Tessari,Claus
Umi 16h46'38"75d18'
Tessarolo,Katy
Peg 21h29'15"28d2'
Tesse,James Briann
Aur 6h41'38"38d18'
Tesse,Jessica Katherine
Lmi 9h23'26"38d25'
Tesse,William Sean
Cnv 12h26'1"43d43'
Tessereau,Steve
Cmi 7h22'11"5d16'
Tessie
And 1h51'28"39d9'
Tessie's Twinkler
Uma 12h1'59"57d37'
Tessier,Franáois
Mon 6h24'13"8d53'
Tessier,Mario
Cyg 21h36'40"38d17'
Tessier,Peter J
Lup 15h12'2"-42d23'
Tessler,Jonathon
Boo 14h37'23"38d5'
Tessman,Robert Charles
Cep 20h56'16"62d35'
Tesson,Jean Michel
Umi 16h11'15"73d52'
Tesson,Prosper Anthony
Uma 10h1'51"71d22'
Testa Forever
Uma 10h51'18"40d24'
Testa III,Louis F
Cep 20h59'36"61d42'
Testa,Hank
Cam 3h58'55"78d44'
Testa,Hon Frank J
Lac 22h24'36"38d14'
Testa,Laura Sartori
Cas 2h18'59"60d17'
Testa,Nicholas T
Cam 6h13'55"60d45'
Testaert,Sabrina
Cam 7h37'39"80d25'
Testani,Fausto
Cyg 21h2'20"31d10'
Testardi,Doris
Cnc 9h5'36sd32d28'
Tester,Grannie
Peg 23h32'36"10d1'
Testolin,Reno & Beverley
Cet 2h14'10"1d33'
Teta,Alysha Ann
And 0h13'58"35d3'
Teter,Regina Ann
Mon 6h32'1"3d1'
Teti,Hilary
Cas 2h43'22"61d30'
Teti,Tonino
Cnv 13h28'33"37d46'
Tetler,Alison Marie
Lyr 19h16'1"40d44'
Tetley
Uma 11h26'25"51d11'
Tetlow,Catherine A
And 1h51'47"37d41'
Tetrault,Claire
Cas 1h8'32"60d47'
Tetreault's Pure Heart Michael
Lyr 19h13'0"35d14'
Tetreault,Aaron Mark
Equ 21h14'35"5d14'A
Tetreault,Lauren Leilani
Equ 21h14'35"5d14'B
Tetreault,Marilyn Lee
And 2h20'52"39d25'
Tetro,Rocco S
Her 16h20'48"40d5'

Tettaton,Nicholas Scott
Lmi 10h16'52"39d35'
Tettemer,Jessie Sara
Cas 1h40'50"58d33'
Tettonis,Konstantinos
Sge 19h18'47"19d7'A
Tettonis,Vasilios
Sge 19h18'47"19d7'B
Tetu,Mary Jo
Lyn 6h39'12"56d10'
Tetzlaff,Joel David
Cep 23h1'1"61d3'
Teuch
Her 17h20'52"41d9'
Teuchner,Anton
Dra 15h9'1"62d57'
Teufel,Eugen
Per 4h6'6"49d44'
Teuscher,Robert George
Aql 19h45'39"14d3'
Teutonico,Leonardo Cami
Uma 10h33'36"54d31'
Teutsch,Will Louise
Her 16h35'1"47d46'
Tevah,Koukla
Peg 22h0'38"11d3'
Tevenini,Sr,Louis Anthony
Ori 5h32'53"1d22'
Teverzczuk,Robert Leon
Dra 16h53'50"69d47'
Teverzczuk,Tracey Lyn
Cas 1h20'16"75d13'
TevRe Star,The
Dra 16h37'24"60d20'
Tewes,James
Her 18h19'16"18d58'
Tews,Maggie
Lyn 7h47'15"50d16'
Tews,Robin Ann
Ari 3h1'22"26d4'
Tews,Tami Jane
Aqr 22h2'34"0d38'
TEX-SCOT
Cet 1h4'1"-1d44'
Texaco
Vul 20h14'20"25d50'
Texas Monthly's 20th Anniversary
Aql 19h35'0"0d11'
Texas Scottish Rite Hospital
Aur 4h50'19"37d45'
Texier,Barbara
Umi 16h8'19"71d60'
Texier,Nathalie
Aql 19h30'14"-8d22'
Texsun-6
Aur 5h4'10"38d17'
Teyssedre,Birba
Cam 3h47'0"68d35'
Teyssier,Jean-Jacques
Per 3h42'20"39d21'
TFS3;SALIER
Aur 6h31'13"32d6'
TG-JG
Vul 19h48'41"23d33'
Thack,Gered
Lmi 10h7'21"32d5'
Thacker,Carol Ann
Cas 0h27'0"61d51'
Thacker,Jared
Lyr 18h41'0"30d51'
Thacker,Sharon Ann
And 1h9'51"40d42'
Thad
Lyr 19h26'26"40d54'
Thaddeus
Per 2h29'1"57d33'
Thaddeus-Fasano
Aql 18h59'33"11d44'
Thaeler,Charles S
Aur 7h8'0"38d46'
Thahir,Al
Her 18h3'45"14d28'
Thain
Uma 10h8'56"48d3'

Thain,Ian J
Per 3h10'31"41d0'
Thain,Mabel
Cas 0h30'26"61d51'
Thakkar,Chuck
Per 3h7'22"46d29'
Thakkar,Saroj Ambalal& Rajni Karsandas Somaia
Cyg 20h16'22"38d25'
Thakur,Randhir Paul Singh
Uma 10h40'47"61d59'
Thaler,Manley & Doriseve
Hya 8h12'0"-6d3'
Thaler,Nancy Lee
Aqr 23h11'51"-4d49'
Thalin
Cmi 7h54'1"0d33'
Thalken,Sally S
Lyn 7h35'54"37d5'
Thallas,Kassie Nicole
Peg 22h36'39"29d0'
Thalman,Dillon James
Ser 18h16'41"-13d25'
Thalova,Hana
Cyg 19h33'11"34d29'
Tham,Elaine
Lyr 18h28'50"32d58'
Thaman,Mark
Cnv 12h29'39"51d17'
Thamer,Faisal Ebrahim Abdul Rahim
Cam 4h40'23"67d58'
Thamer,Stephen Alexander
Leo 9h57'23"7d33'
Thamert,Andrew Ryan
Del 20h18'37"20d18'
Thames,Joan I
Hya 8h52'34"-0d40'
Thames,William James
Her 16h23'50"38d16'
Than,Dr Rolf Walter
Cnc 9h1'0"30d22'
Thanestar
Cyg 21h18'43"31d14'
Thank You...
Lac 22h43'38"56d17'
Thank-you Little Star
Aur 7h22'29"38d39'
Thankachan,Suji
Peg 23h3'41"32d15'
Thanner,Arthur
Peg 23h28'48"10d20'
Thanos,Christos Vasilios
Aur 6h11'48"32d12'
Tharet,Elodie
Uma 9h39'33"48d28'
Tharp Birthday Star, Duane E
Tau 5h21'19"28d23'
Tharp Christmas Star
Uma 10h42'37"52d2'
Tharp Family B A S B A,The
Cyg 20h11'32"38d5'
Tharp,Buz & Pat
Her 17h55'25"28d41'
Tharp,Chip
Oph 18h18'21"7d56'
Tharp,Steven & Diane
Cas 1h5'28"53d7'
Tharpe,Helen O
Com 13h17'51"20d47'
Tharr,Catherine Stowe
Mon 7h41'7"-2d38'
Tharrats,Dr Jesùs
Mon 8h3'7"-1d33'
Tharrington,"Miss Vicki"
Cas 2h12'43"59d42'
Thasos's Star
Dra 16h9'44"68d12'
That One J A
Lyn 8h22'1"43d35'
That Paper Place
Uma 11h33'44"32d38'
That's Not Funny Dad- Jacklyn Ann
Uma 11h31'47"52d2'

That's Us
Cyg 20h47'12"51d43'B
Thatcher,Betty Jo
Uma 10h46'50"57d29'
Thatcher,J Todd
Aur 5h17'23"42d25'
Thatcher,Jeffrey P
Boo 15h21'57"42d20'
Thatcher,Sr,Paul Rexford
Her 16h59'51"31d7'
Thatcher,Taylor Nicole
Ori 5h47'23"18d52'
Thaut,Lyle
Aql 19h4'37"2d52'
Thaxton,Barbara Claire
And 0h49'42"45d7'
Thaxton,Carolyn Doris
And 0h49'19"45d54'
Thaxton-Bearg,Myrna
Tau 5h19'32"20d29'
Thayer,Betsy May
Sgr 18h49'49"-28d12'
Thayer,Jeffrey R
Cet 2h2'39"1d17'
Thayer,Laura Lee
Cas 0h30'21"60d40'
Thayer,M J
Cyg 19h59'21"41d2'
Thayer,Richard R
Per 2h39'25"37d7'
Thayer,Valaree Rose
Peg 22h43'11"33d9'
The "L" Factor
Crb 13h32'15"18d48'
The 1/2 Now & 1/2 Later Jack Star
Her 17h10'1"21d12'
The Anniversary Of Yutaka & Yoko
Leo 10h15'39"11d4'
The Bear Key -A M L T
Uma 8h55'35"54d0'
The Commitment With Love
Sge 19h55'1"16d27'
The Eternal Soul of Hiromi
Gem 7h11'29"25d59'
The Greatest Star; Rick's Star
Aql 19h24'38"14d31'
The Hands On! Museum
Eri 4h43'54"-1d17'
The iTV Corporation
Umi 14h19'60"74d21'
The James
Gem 6h39'45"13d54'
The Sheep of OCTOBER
Leo 10h52'0"8d51'
The Star of Wedding Anniversary
Psc 1h17'29"25d46'
The Teacher Brazil's Gift,Lydia
Lyr 18h57'39"37d18'
Thea
Cyg 19h30'57"31d42'
Thea
Sgr 19h45'4"-41d5'
Thea & Klaus Star,The
Uma 9h51'45"53d59'
Theaker,Ernst
Sge 19h11'45"18d27'
Thebaud,Cynnie
Sgr 19h33'11"-30d40'
Theby Family,The Steve
Lyr 19h22'40"35d24'
Theda
Lyr 18h39'59"41d48'
Thee,Virginia D
Gem 6h50'23"17d43'
Theer,Jordi Alexander
Psc 1h3'17"22d15'
Theilen,Steve
Aql 18h42'15"-2d20'
Theiner,Daniela
Lib 14h39'51"-23d8'
Theira,Beatrice
Aql 20h12'40"12d55'

Theiry,Vaughn D
Dra 17h54'55"58d32'
Theis,Anthony & Sherry Mink
Lac 22h7'0"48d7'
Theis,Heiko
Lyr 18h34'19"27d2'
Theis,Leon J
Dra 16h3'29"64d52'
Theis,Lindsey Dianne
Cas 0h21'0"68d57'
Theise,Jay K
Cep 23h27'1"68d42'
Theisen,Garrett Anthony Joseph
Boo 14h41'27"32d55'
Theisen,Jolie
Com 13h2'20"20d3'
Theisen,Nicholas David
Her 17h7'17"20d1'
Theiss,Robert Lawrence
Cep 22h45'40"71d3'
Theissen,Stefan
Boo 14h6'1"11d17'
Thelander,Erin Maureen
Lyn 7h54'27"39d56'
Thelen,Susanne
Sco 16h29'1"-40d31'
Thelma
Tau 3h40'7"25d31'A
Thelma
Com 12h28'20"21d2'
Thelma & Arthur
Uma 11h20'45"30d13'
Thelma & Louise
Peg 22h47'35"25d27'
Thelma Amelia
Cyg 21h14'21"35d0'
Thelma Margaret
Cas 0h16'45"61d41'
Thelma's Star
Cas 2h1'0"58d19'
Thelvis
Ori 4h56'0"-1d28'
Themeli 1993,Maria
Lyn 8h34'1"42d19'
Themi
Tau 3h50'1"1d11'
Themistokles S
Uma 12h6'55"56d54'
Theo
Cep 22h31'0"58d21'
Theo
Hya 8h35'14"1d28'
Theo
Cma 6h41'49"-13d41'
Theo 191043 W
Lib 15h1'30"-28d10'
Theo's Light
Peg 22h44'32"24d18'
Theobald,Ronald C
Uma 10h51'27"70d55'
Theocharidis,Sarandos
Aql 18h57'39"-6d40'
Theodora
Cmi 7h19'9"2d46'
Theodore Jerome
Boo 15h2'36"20d50'
Theodoridis,Elefterios G
Lyr 19h18'14"30d28'
Theodosia
Crb 15h27'40"32d26'
Theodosiou,Constantine E
And 2h27'23"38d15'
Theola
Lyn 8h20'47"45d6'B
Theon,Christy
Tri 1h57'29"27d57'
Theonnes,Ken
Dra 17h28'23"72d12'
Thera
Per 2h59'53"45d58'
Thera 9-11
Cas 2h11'29"60d11'
Therault,Kristen Page
Peg 21h21'51"22d59'

There Is A Place For Us
Lyn 8h45'27"34d33'
There's Only One Robby Mütti was Right
Aql 20h10'52"10d40'
Theresa
Uma 8h59'0"62d17'
Theresa "21"
Cyg 20h57'0"37d41'
Theresa & Andy's Star
Cyg 21h11'28"34d28'
Theresa Ann
Vul 21h15'44"20d30'
Theresa Antonia
Cap 20h41'58"-26d31'
Theresa Leigh
Cas 0h5'27"65d33'
Theresa Marie
Cas 0h33'51"64d24'
Theresa Marie & Theodore James
Crb 15h28'0"31d30'
Theresa Mary & Beloved Son Joseph Orion
Ori 5h33'1"0d56'
Theresa Patsy
Peg 22h19'1"7d53'
Theresa's Christmas Star
Equ 21h10'59"10d21'
Theresa's Light
Del 20h54'26"8d37'
Theresa's Star
Lyn 8h41'40"44d28'
Theresa's Star
Ori 6h0'1"3d57'
Theresalinda
And 0h50'22"36d47'
Therese
Cas 3h1'22"58d22'
Therese
Com 12h34'46"23d52'
Therese
Cmi 7h58'39"1d19'
Therese
Cas 1h39'22"60d38'
Therese's Star
And 1h58'17"37d38'
Theret,Veronique
Umi 16h24'16"70d50'
Theriault,Gina Maria
And 0h22'11"37d44'
Theriault,Jerry I
Boo 14h55'16"41d32'
Theriault,Scott John
Aql 20h12'18"1d34'
Theriault-Mars 20, 1929,Grace
Uma 9h23'27"51d14'
Thernes,Becky & Mark
Tri 2h17'1"33d56'
Therrien,Jean-Yves
Cep 23h27'51"65d17'
Therrien,Louise
Cas 0h22'22"72d39'
Theryoung,Richard J
Uma 8h9'16"71d53'
Thesbias,Tobias
Cep 23h12'29"60d15'
Thespis
Gem 6h37'29"13d45'
Thetford-A Peaceful Warrior,Jeff
Aur 6h6'21"45d3'
Thevie
Equ 21h1'55"2d17'
Thevmasia,Thomasia Hughes
Peg 23h30'34"11d22'
Thew,Dan
Cam 3h54'0"70d26'
Thia
Aql 18h59'20"-6d58'
Thias,Scarlett
And 23h2'27"37d47'
Thibadeau,Joseph H & Rita H
Del 20h33'22"11d9'

Thibaud
Aur 4h53'20"40d21'
Thibaud,Dedieu
Mon 7h44'57"-4d13'
Thibaud,Nicolas
Cas 2h13'46"50d57'
Thibaudeau,Bruno
Dra 11h56'21"73d31'
Thibaudeau,Eric
Tri 2h4'10"32d37'
Thibault,Benoit
Uma 8h46'38"50d18'
Thibault,Erin
Mon 7h39'0"-0d31'
Thibault,Megan
Lyr 18h18'17"38d48'
Thibault,Meredith Shelby
Uma 9h19'24"62d3'
Thibault,Ryan
Ori 6h7'26"-1d40'
Thibaut,Starry Janie
Mon 8h7'31"-10d8'
Thibert,Julien
Psc 0h3'58"2d23'
Thibodeau,Kelly
Psc 23h31'41"1d44'
Thibodeau,Mark Andrew
Dra 17h30'0"64d21'
Thibodeau,Steven Charles
Dra 19h25'28"68d8'
Thibodeaux 2
Ori 4h41'55"14d45'
Thibodeaux,Charles Richard,Donna & Tina
Lac 22h54'11"40d14'
Thibodeaux,Mary
Her 16h47'15"13d20'B
Thibodeaux,Michelle Lea
Mon 8h1'29"-5d18'
Thibodo,Russ & Margie
Cyg 19h29'48"35d3'
Thie
Aql 19h2'30"16d52'
Thiebault,Joshua Edward
Cep 21h42'27"68d56'
Thiede,Cynthia Russeler
Lyn 7h31'27"41d1'
Thiede,Michael
Lyr 19h19'52"31d41'
Thiedemann,Horst
Ari 2h44'12"25d37'
Thiel Star,The
And 23h30'45"48d24'
Thiel,Doris
Lyn 8h14'59"48d9'
Thiel,Joel A
Dra 17h3'0"62d4'
Thiel,Roland
Gem 8h1'45"30d59'
Thiel,Tiffany
Eri 4h12'54"-18d14'
Thiele,Barbara
Mon 7h53'39"-1d22'
Thiele,Jon & Romy
Cyg 20h21'55"41d56'
Thiele,Mary Lou
Uma 11h3'22"51d3'
Thielecke,Marion
Ori 5h51'44"21d26'
Thielecke,Winfried
Dra 20h6'24"62d22'
Thielen,Sr,Dr Albert Ernst
Lac 22h21'50"53d60'
Thielet,Raymillia
Cep 22h6'11"67d32'
Thielkings Star,Kym
Tau 3h44'42"28d23'
Thiellet,Franáois
Dra 10h5'31"78d38'
Thielmann 25
Sex 10h4'1"1d42'
Thielmann,Florian
Cep 22h7'49"60d7'
Thielmann,Walter
And 23h2'1"43d46'

Thielsen,Esben
Her 15h9'18"18d7'
Thieme,Carrie
Eri 3h17'58"-17d28'
Thieme,Gisela
Cmi 7h20'42"4d49'
Thieme,Wolfram
Ser 15h11'43"10d54'
Thienel,Manfred
Oph 18h26'53"7d54'
Thier,Robert & Yvonne
Lyn 7h6'50"44d26'
Thiercelin,Guillemette
Lyr 18h14'37"30d49'
Thierry
Cyg 19h51'1"44d47'
Thierry
Aur 5h33'18"29d23'
Thierry Cavé "Lena"
Per 4h0'46"35d58'
Thierry,Berguerand
Ori 5h13'44"15d14'
Thierry,Reine
Cet 0h56'27"1d44'
Thierry,Robin Reneé
Peg 22h46'58"25d9'
Thiers,Wolfgang
Oph 17h59'0"0d14'
Thiery,Cara Besh
Uma 8h53'55"71d53'
Thies,Gregory Ralph
Peg 21h20'51"22d44'
Thiesing,Kurt A
Tau 3h57'11"10d10'
Thiessen,Carl
Lac 22h48'1"55d28'
Thiessen,Karla
Cnc 8h6'13"32d59'
Thiffault,Dorothée
Cas 0h39'29"64d50'
Thiffault,Jean-Roch
Her 17h22'27"20d45'
Thigpen,Allison Paige
Mon 7h16'54"-6d24'
Thilenius,Rob
Cet 2h33'19"6d25'
Thillet,Carlos
Dra 16h34'0"73d51'
Thillet,Valérie
Cyg 19h26'59"33d37'
Thimm,Angela
Vir 13h2'49"-0d48'
Thimmesch-Carmody, Debra Ann
Cyg 19h43'49"33d39'
Think of Me Kindly
Aur 5h10'39"44d6'
3rd Stone
Aur 6h6'60"45d31'
Third,Brenda Karen
Cra 18h2'1"-37d26'
Thiriet,Andre
Sge 20h0'54"20d11'
Thirion,Aude
And 23h19'40"50d21'
Thirkill,Dawn
Her 17h27'41"34d59'B
Thirkill,Jason
Her 17h27'41"34d59'A
Thirston
Cep 1h27'50"78d7'
Thirtle,Andrew Richard
Cep 20h31'19"60d9'
30 12 70 JB
Tri 2h0'49"30d59'
333 Chuck & Mur
Uma 9h30'12"68d15'
3711
Cam 3h44'57"74d29'
Thirty Three Story
Gem 7h12'18"25d8'
Thirty-Seven Blue (LS)
Lyn 7h57'10"50d20'
Thiry,Jared Scott
Cep 0h15'23"66d59'

Thiry,Pascale
Mon 6h25'1"-5d38'
This Much
Cet 1h20'1"1d46'
This Star Is As My Love,Forever
Per 2h57'14"38d21'
Thisong,Isabelle
Com 13h13'15"21d47'
Thistle Farm Star
Lac 22h8'26"38d33'
Thistle,Carol L
And 2h33'28"44d16'
Thivierge,Luc
Per 3h13'59"43d20'
Thivierge,Maryse
Cas 22h57'56"54d32'
Thobe,Tina Marie
Vul 19h48'36"28d46'
Thoburn,Nancy Ellen Van Buskirk
Uma 9h3'1"55d18'
Thoca 1
Aql 20h11'32"12d53'
Thocking
Peg 21h54'54"28d19'
Thode,Monique J
And 22h57'22"51d55'
Thoelecke,Amanda Eileen
And 0h20'33"36d51'
Thoelecke,Danielle Justine
Cas 0h39'48"62d49'
Thoeming,John Voll
Aur 6h2'51"34d22'
Thoenes,Edward G
Tau 4h44'57"20d30'
Thoeni,Kenneth Christian
Ser 16h16'1"-2d5'
Tholey,James M
Aur 6h55'16"43d52'
Tholt,Janie
Ser 17h32'35"-14d31'
Thom's High Star
Tri 1h46'31"28d20'
Thom,Amy Elizabeth
And 1h1'27"40d41'
Thom,Frank Andrew
Per 3h13'13"50d29'
Thom,Tara Jill
Uma 11h8'42"60d22'
Thom-Bertsch
Aql 20h36'31"0d42'
Thom-Hansen,Sten Tyler
Cnv 12h21'25"34d3'
Thoma,Alma Reetz
Lyn 7h13'23"58d22'
Thoma,Hermann
Cap 21h27'44"-23d40'
Thoma,Penny
Sge 19h37'23"16d33'
Thoma-Star
Uma 8h47'42"51d49'
Thoman,Danielle Ivette
Vul 19h48'33"27d53'
Thomas
Dra 19h14'43"68d30'
Thomas
Boo 15h20'31"52d22'
Thomas
Cep 21h29'30"58d44'
Thomas
Cmi 8h4'14"1d8'
Thomas
Com 13h8'38"28d46'
Thomas
Her 17h26'1"40d42'
Thomas
Aur 4h48'34"40d39'
Thomas
Cep 20h26'36"60d33'
Thomas
Cep 21h4'24"63d11'B
Thomas "Dada",Marvin
Ori 6h6'43"9d23'

Thomas "Forever My Bright Spot"
Cnv 13h46'30"41d60'
Thomas & Joy Forever In Love
Cyg 19h28'27"34d39'
Thomas & Kristen
Cas 3h7'23"58d54'
Thomas (Grosspuh)
Tau 5h27'36"26d17'
Thomas 1 2 3,Beverly K
Cam 12h4'30"77d26'
Thomas 1946
Cap 21h24'0"-14d56'
Thomas 1955-1995, Angela
Cyg 19h27'23"35d37'
Thomas C
Cnv 12h48'34"48d19'
Thomas Clyde
Oph 17h6'51"10d9'
Thomas Eternity,Andrew Craig
Cnv 13h17'31"38d9'
Thomas Eugene The Boy King
Aur 4h51'57"50d38'
Thomas F & Mary J
Cam 5h28'39"68d43'
Thomas F-Dynamite Dad
Her 15h57'35"40d20'
Thomas Family Star,The
Uma 11h4'45"40d56'
Thomas I Love You, Doreen D
And 23h18'35"51d43'
Thomas IV
Aql 20h16'16"5d10'
Thomas J
Aql 20h18'45"0d48'
Thomas James
Cam 3h29'48"60d15'
Thomas Jeffrey
Dra 17h46'39"64d24'
Thomas Leisse & Papa Whiskey
Cas 0h7'43"60d40'
Thomas Michael & Stephanie Helene
Her 16h58'0"40d35'
Thomas Neil
Lac 22h54'0"55d2'
Thomas O Light
Her 16h50'23"40d42'
Thomas Our Angel, Jeffrey
Umi 15h10'10"71d46'
Thomas P,The
Cep 21h53'27"58d29'
Thomas Ryan
Cam 4h48'44"67d24'A
Thomas S
Cet 2h40'15"3d46'
Thomas W & Katherine M
Del 20h56'56"12d41'
Thomas Wayne
Crt 11h12'39"-14d57'
Thomas' Fire
Sex 9h58'30"4d53'
Thomas,"Castus"-Kathy D Kinneman
And 23h1'56"48d4'
Thomas,"Phoenix"- Jennifer Susan
Cas 0h10'19"61d28'
Thomas,Alan S
Her 18h5'1"31d8'
Thomas,Alexander Louis Kostrinsky
Gem 6h45'21"14d21'
Thomas,Allison Marie
Equ 21h0'30"8d53'
Thomas,Alpha Frank Maxwell
Lmi 10h55'47"30d18'
Thomas,Amy E
Lyn 7h52'49"38d38'
Thomas,Andrea
Peg 22h33'35"26d55'
Thomas,Andrea Marie
Cyg 21h2'32"28d43'

Thomas,Anita Antoinette Lyr 18h52'1"33d33'
Thomas,Anthine Lyr 18h58'26"30d8'
Thomas,Anton Peter Sco 16h58'49"-30d7'
Thomas,Aschley Dorothy Vul 19h3'0"21d58'
Thomas,Ashley Mary Peg 23h43'33"13d22'
Thomas,Barbara Joanne Blose Lac 22h7'43"46d5'
Thomas,Benjamin Cap 21h54'11"-18d7'
Thomas,Benjamin Neale Cnv 12h22'44"37d5'
Thomas,Bernard L Aur 6h3'40"31d37'
Thomas,Bernhard Lyr 19h17'18"42d54'
Thomas,Bernie James Cep 22h4'58"58d9'
Thomas,Bethany Megan Sgr 19h8'36"-25d4'
Thomas,Brenda & Steven Cet 1h11'20"-6d60'
Thomas,Brian J Dra 14h28'1"63d16'
Thomas,Buck Leo 11h51'59"22d47'
Thomas,Cameron Per 3h55'47"41d6'
Thomas,Casey Aur 7h11'55"39d26'
Thomas,Celeste Aql 19h53'43"13d27'
Thomas,Chelsea Ann And 0h0'35"47d18'
Thomas,Cheryl M Sct 18h41'60"-6d29'
Thomas,Christopher M Gem 7h38'26"21d7'
Thomas,Claire Cyg 20h59'37"28d30'
Thomas,Colton Eri 4h14'55"-12d47'
Thomas,Cyndee Gem 6h49'20"19d56'
Thomas,Cynthia Crb 16h22'33"27d9'
Thomas,Cynthia M And 2h3'24"45d19'
Thomas,D C Aql 20h13'27"5d10'
Thomas,Dana Lynn Cas 23h14'53"60d39'
Thomas,Daniel Evans Equ 21h2'1"3d40'
Thomas,Danny Wason Equ 20h57'52"3d49'
Thomas,Dave Per 3h0'23"56d43'
Thomas,David Cyg 21h19'0"28d23'
Thomas,David Gareth Lyn 9h8'49"38d20'
Thomas,Deborah Cas 23h41'23"61d16'
Thomas,Debrah S And 2h22'54"39d42'
Thomas,Donald William Cep 20h37'35"76d38'
Thomas,Donna Sue Equ 20h56'30"2d33'
Thomas,Dr George Aqr 22h0'31"0d15'
Thomas,Dustin Aql 20h12'49"12d14'
Thomas,Ebby-lan Uma 8h38'22"51d41'
Thomas,Eleanor Breitwieser Uma 11h4'59"40d54'
Thomas,Emily Ann Mon 5h58'53"-6d48'
Thomas,Emily Frances Mon 7h1'59"-6d33'
Thomas,Emma L And 0h31'50"40d2'
Thomas,Ethel Mae Tau 5h38'21"26d9'
Thomas,Eugene Oliver Dra 18h23'53"50d34'
Thomas,Frank Cep 22h4'37"60d59'
Thomas,Franáois Dra 15h43'18"58d35'
Thomas,Freda Kathleen Cas 0h9'49"60d9'
Thomas,Gail Ann Mon 8h8'46"-4d12'
Thomas,Garret Lamar Her 17h58'0"37d35'
Thomas,George Joseph Aql 18h56'53"-6d35'
Thomas,Gertrude H Mon 7h20'15"-0d15'
Thomas,Gilly Cyg 19h40'60"40d56'
Thomas,Gina Umi 16h29'28"71d22'
Thomas,Harold Her 17h38'0"27d37'
Thomas,Heide Lyn 8h31'11"58d8'
Thomas,Helen Marie Cyg 19h27'1"32d9'
Thomas,Herman Cet 2h59'0"1d44'
Thomas,J.A. Uma 10h23'1"40d36'
Thomas,Jack & Susan Lynne Weres Cyg 19h50'21"44d55'
Thomas,Jack Allen Oph 17h3'18"10d48'
Thomas,Jack E Aql 19h34'29"1d22'
Thomas,Jack Peter Gru 22h9'29"-56d27'
Thomas,Jack William Sex 10h39'25"5d14'
Thomas,James Henry Cmi 7h25'45"7d48'
Thomas,James L Aur 6h16'19"30d32'
Thomas,Jane B Lyr 18h35'18"30d25'
Thomas,Janet & Mary Cam 6h15'34"67d47'
Thomas,Janet M Cas 2h7'51"60d8'
Thomas,Jason Martin Oph 17h38'55"-21d53'
Thomas,Jerry Uma 10h21'16"52d6'
Thomas,Jesse Burke Aur 5h35'0"54d54'
Thomas,Jessika A And 23h16'53"49d55'
Thomas,Jim Uma 8h13'53"70d50'
Thomas,Jo Ellen Cyg 20h8'20"40d57'
Thomas,Joanne Cas 23h0'49"58d10'
Thomas,Joel Hya 8h19'29"1d13'
Thomas,John D Mon 8h3'36"-0d4'
Thomas,Jonathan & Claudine Ori 4h55'29"4d51'
Thomas,Joy Carol Aql 19h10'26"19d46'
Thomas,Joy F And 23h31'34"44d21'
Thomas,Jr,Kenneth Ser 15h21'30"3d21'
Thomas,K Diane Com 12h54'13"24d55'
Thomas,Kathy L And 2h30'1"44d9'
Thomas,Keith Henry Aur 5h30'0"40d16'
Thomas,Kellie & Daniel Vul 19h48'11"20d33'
Thomas,Kelly Kristine Cmi 7h53'24"5d24'
Thomas,Kevina Kay Cas 0h12'58"56d5'
Thomas,Kit Cmi 8h0'22"5d42'
Thomas,Larry Cmi 7h44'1"7d31'
Thomas,Lata-LoriAnne Cyg 20h20'1"41d3'
Thomas,Laura Ann Aql 18h45'49"11d8'
Thomas,Lauren And 23h37'25"43d40'
Thomas,Lauren Grace Cyg 19h28'23"35d57'
Thomas,Letty S Tri 2h16'46"33d0'
Thomas,Lisa Allison Aql 19h18'11"10d46'
Thomas,Logan Michael Dra 18h1'55"65d35'
Thomas,Lori Renee Gem 6h56'45"16d29'
Thomas,Lynda Eri 2h45'18"-17d43'
Thomas,Lynn Michele Per 3h5'37"35d16'A
Thomas,Mandy Gale And 1h51'56"47d2'
Thomas,Marc Gordon Cnv 12h8'31"39d16'
Thomas,Marcia Lyn 7h15'51"59d52'
Thomas,Marie Bellrichard Ori 5h3'59"0d38'
Thomas,Marjorie B Ori 5h32'25"-2d8'
Thomas,Mark Anthony Cep 0h1'23"68d22'
Thomas,Mark Kevin Umi 14h32'51"69d47'
Thomas,Marley Rae Ori 6h5'60"-0d3'
Thomas,Mary Jo Cep 22h59'60"78d45'
Thomas,Mary Lourdes Mon 6h38'13"7d30'
Thomas,Mathew Andrew Equ 21h6'15"10d40'
Thomas,Maureen & Todd Cyg 19h25'0"33d57'
Thomas,Max & Marty Lyn 6h56'55"60d8'
Thomas,Megan Ashley Lyn 7h35'19"58d40'
Thomas,Megan J Com 12h1'24"25d36'
Thomas,Megan Patricia Cam 7h55'23"61d27'
Thomas,Melanie Sct 18h54'48"-7d7'
Thomas,Melvin A Cma 6h56'51"-19d46'
Thomas,Michael Ori 5h56'50"16d29'
Thomas,Michael Cep 21h15'49"61d3'
Thomas,Michael Tilson Dra 17h56'28"61d9'
Thomas,Michele Vir 13h31'25"-8d3'
Thomas,Michelle Vul 19h22'45"26d43'
Thomas,Michelle Lynn And 0h12'25"47d21'
Thomas,Nicholas B Boo 14h48'0"33d3'
Thomas,Nick Ori 5h58'26"8d23'
Thomas,Nigel Tansley Sgr 20h4'54"-34d54'
Thomas,Niniane Sioned Lyn 8h20'0"48d14'
Thomas,Norma Cet 2h50'53"2d18'
Thomas,Olivia Lindsey Com 12h9'41"20d3'
Thomas,Our Pa...Jay K Per 1h47'59"53d18'
Thomas,Out Of This World Garry Dra 16h18'40"66d3'
Thomas,Patricia D Uma 10h19'55"60d55'
Thomas,Paul B Boo 14h59'22"25d56'
Thomas,Peter A Peg 22h14'12"34d42'
Thomas,Peter Anthony Oph 17h57'0"12d3'
Thomas,Peter Philip Uma 11h49'41"30d11'
Thomas,Phyllis T Peg 23h24'25"8d32'
Thomas,Rachelle Cyg 21h52'21"53d0'
Thomas,Ray Lyn 8h2'42"41d34'
Thomas,Rececca Lyn 8h17'14"34d22'
Thomas,Richard Dra 16h27'1"61d21'
Thomas,Richard Dale Cyg 21h22'34"38d34'
Thomas,Richard J Hya 8h57'19"1d31'
Thomas,Richard Scott Cyg 21h50'50"42d60'
Thomas,Robert & Kathleen Uma 12h50'26"61d4'
Thomas,Robert James Cma 7h14'27"-16d11'
Thomas,Ron Uma 12h11'25"56d47'
Thomas,Roy S Per 1h59'15"54d21'
Thomas,Russell "Beebock" Per 2h31'59"56d42'
Thomas,Russell C Del 20h24'41"20d20'
Thomas,Ruth Marie And 0h8'47"34d16'
Thomas,Sam & Jess Peg 22h48'46"25d47'
Thomas,Sandra Weathersby Peg 22h59'55"25d1'
Thomas,Scott Alan Cyg 21h37'0"41d53'
Thomas,Sharilyn And 1h25'14"37d18'
Thomas,Sharon Marie Cyg 19h56'0"45d57'
Thomas,Shirley Marie Cyg 21h27'14"40d3'
Thomas,Stephanie Nicole Lyr 19h8'25"37d54'
Thomas,Stephanie Paige Lyn 9h12'1"40d11'
Thomas,Stephen J Aur 6h3'40"30d34'
Thomas,Steven Lee Aql 19h48'50"14d28'
Thomas,Terri Eri 3h37'46"-14d13'
Thomas,Terry Sex 10h15'21"-0d9'
Thomas,Tina Jean Cas 0h30'29"75d47'
Thomas,Tracy Peg 21h51'26"33d28'
Thomas,Tudor & Dorothy Ori 5h50'19"15d50'
Thomas,Victor C Cru 12h49'21"-57d22'
Thomas,Vivian Cas 1h33'41"63d46'
Thomas,W Curtis Aur 6h23'0"33d39'
Thomas,Wanda K Mon 7h4'30"-1d44'
Thomas,Wendy Samantha-Lou Vul 19h46'13"20d10'
Thomas,William Harvey Uma 11h6'0"37d31'
Thomas-"New Wings", Carl Vincent Cas 2h58'0"68d57'
Thomas-Buckel,Mary Jo Peg 23h3'0"13d60'
Thomas-Favreaux,Jill Marie Cyg 20h51'1"38d17'
Thomaschautzki,Claus Lmi 10h45'42"25d52'
Thomason 30th Anniversary Star Crb 15h27'29"31d32'
Thomason,James O Her 17h20'16"43d46'
Thomason,Jr,John S Aql 20h7'0"1d11'
Thomason,Sharlene M And 1h12'39"39d33'
Thomassen,Dr John Paul Oph 17h31'7"-20d19'
Thomassin,Samantha Per 3h29'28"38d44'
Thomasson III,James Nelson Hya 8h55'14"-0d17'
Thomasson,Margaret Leora Peg 22h3'1"10d7'
Thomassy,Richard David PaPa's Star Ori 5h53'10"10d23'
Thomes,Heinrich Oph 18h24'48"8d25'
Thommen's Little Star, Brittney Lynn Uma 9h47'43"42d52'
Thommen's Lucky Star, Valerie Giunta And 1h30'1"38d4'
Thompkins,Doris Aql 20h30'43"0d44'
Thompkins,Michelle Cnv 12h15'54"44d11'
Thompsett,Garry Lyn 9h38'31"41d39'
Thompson Cyg 21h12'1"38d36'
Thompson "Belly Bo", Christopher Dra 20h2'14"76d37'
Thompson "Granny", Alice K Lyr 19h5'1"38d39'
Thompson "Tomos Star", Stephen Per 2h59'17"43d41'
Thompson Family Star, The Eri 3h48'25"-6d38'
Thompson Never Forgotten,K T,Richard Cyg 21h36'22"41d14'
Thompson Star,The James Peg 22h42'1"31d23'
Thompson's "Smile", Barbara Lyr 19h14'34"41d29'
Thompson's Twinkle Dra 19h22'16"56d45'
Thompson,Aaron Paula Umi 12h0'4'34"62d40'
Thompson,Albert Grant Cnv 13h21'58"41d43'
Thompson,Alexandria Elizabeth Lyn 8h45'38"39d16'
Thompson,Alexis & David Murphy Crb 16h7'36"33d28'
Thompson,Alfred Charles Cet 0h27'23"-0d47'
Thompson,Alice Laughead Cyg 20h52'44"39d25'
Thompson,Always Kenneth Albert Lac 22h7'40"51d12'
Thompson,Amy Cyg 19h44'49"37d55'
Thompson,Andrew Cmi 6h55'53"-15d49'
Thompson,Andrew James Boo 14h0'29"25d29'
Thompson,Angela Ray Cyg 21h14'34"38d44'
Thompson,Ann Marie And 22h56'10"50d44'
Thompson,April Lahoma Com 12h51'54"26d47'
Thompson,Arthur Lavon Uma 9h58'53"49d24'
Thompson,Arthur Q Ori 6h16'1"20d14'
Thompson,Audrey"Mom" Cas 0h2'1"54d44'
Thompson,Ava Peg 22h23'53"5d42'
Thompson,Becca Lynne And 0h49'32"33d25'
Thompson,Blaine Dra 16h1'14"65d18'
Thompson,Brayden David Her 18h35'1"18d55'
Thompson,Bryan Keith Boo 15h1'39"25d41'
Thompson,Caitlin Peg 0h1'25"28d18'
Thompson,Cara Diane And 2h25'24"42d31'
Thompson,Carolyn Arlene Aql 19h56'26"12d28'
Thompson,Charles Adam Her 18h3'35"38d57'
Thompson,Charles John Hya 8h11'21"0d49'
Thompson,Cherene And 1h11'0"40d39'
Thompson,Christina Elisabeth Lyn 7h38'17"37d13'
Thompson,Christine Cyg 20h51'21"39d44"
Thompson,Christopher Ahn Her 16h18'21"8d40'
Thompson,Chuck Cam 3h32'15"61d40'
Thompson,Collin James Lac 22h27'28"53d44'
Thompson,Dananna Mon 7h0'1"-10d1'
Thompson,Daniel Andrew Umi 14h46'51"66d28'
Thompson,Dawn Ellen Peg 22h38'14"24d49'
Thompson,Dean J Aur 5h9'46"38d46'
Thompson,Deane Cep 22h2'51"61d40'
Thompson,Debi Cyg 20h38'53"31d0'
Thompson,Dennis Sct 18h43'19"-6d29'
Thompson,Desiree M Eri 3h49'33"-0d18'
Thompson,Don Boo 14h56'22"50d35'
Thompson,Donna M
Thompson,DonnaM Lyn 8h56'44"44d6'
Thompson,Doris Smith Cyg 19h32'29"36d26'
Thompson,Dottie Cam 3h33'58"61d5'
Thompson,Elise Nicole Boo 14h2'32"21d25'
Thompson,Eric Henry Ori 5h59'23"21d5'
Thompson,Ethel And 23h42'15"45d54'
Thompson,Garry & Cynthia Cam 5h57'44"61d23'
Thompson,Gary Lac 22h14'48"51d58'
Thompson,Gayla D Sex 9h51'15"-1d51'
Thompson,George L Her 16h39'57"24d57'
Thompson,Glenn Jeffry Per 4h21'51"51d8'
Thompson,Greg Aql 19h55'1"14d30'
Thompson,Gregory Warren Her 17h23'55"23d23'
Thompson,Harry & Nellie Uma 13h29'60"60d22'
Thompson,Herb & Sharon Cam 5h53'29"70d58'
Thompson,Isabelle Rosemary Marguerite Leo 10h35'39"13d16'
Thompson,Jack Aur 6h27'16"38d39'
Thompson,James Ryan Dra 12h7'33"70d6'
Thompson,Jason S Her 16h54'50"32d53'
Thompson,Jeanie George Uma 11h47'1"44d5'
Thompson,John Marshall Aql 19h52'38"14d50'
Thompson,Jr,Charles David Oph 17h24'52"-1d34'
Thompson,Jr,John A Her 16h53'58"25d44'
Thompson,Jr,Robert Lewis Peg 21h26'26"10d51'A
Thompson,Judy Cyg 21h55'0"53d27'
Thompson,Karen Peg 23h22'1"16d32'
Thompson,Karen A Cas 0h57'29"69d46'
Thompson,Karin Crb 15h29'21"30d47'
Thompson,Keith Aur 4h51'16"40d56'
Thompson,Kelsie Frances Grace And 0h4'22"34d31'
Thompson,Kenneth Lawrence Dra 15h42'1"60d49'
Thompson,Kevin Michael Lac 22h9'0"40d48'
Thompson,Kimberly Kae Mon 6h29'56"-10d55'
Thompson,Kippy Lyn 8h27'36"58d42
Thompson,L Millard "Chin" Ori 5h54'22"1d20'
Thompson,Laura Natalie And 2h20'47"38d55'
Thompson,Leslie Ackiss Vir 12h34'34"0d6'
Thompson,Lisa A Lyn 7h28'45"39d45'
Thompson,Logan Ryan Ori 5h55'32"15d11'
Thompson,Lois Lee Mon 6h40'40"11d48'
Thompson,Marc A Per 4h6'16"50d52'
Thompson,Margaret Lee Cet 2h33'0"-10d21'
Thompson,Marie Louise Vul 20h15'22"28d9'
Thompson,Marjorie Paige Cyg 19h54'0"37d59'
Thompson,Mark Oph 17h3'45"-23d29'
Thompson,Mark Alan Lyr 19h26'23"38d13'
Thompson,Martha A Uma 10h25'0"41d58'
Thompson,Mary Cyg 19h30'40"32d34'
Thompson,Mary Peg 22h40'41"29d22'
Thompson,Maryssa Renae Gum Mon 6h28'44"-10d40'
Thompson,Maureen Lynn Mon 6h57'30"8d46'
Thompson,MD,Jay Aur 6h11'25"37d42'
Thompson,Melissa Peg 22h46'58"32d15'
Thompson,Michael Cap 21h15'14"-23d6'
Thompson,Mike & Susan Aur 5h47'22"54d49'
Thompson,Millard Lewin Aur 5h3'54"37d40'
Thompson,Murray Arnold Boo 14h38'26"38d44'
Thompson,Neal Per 4h34'19"38d5'
Thompson,Norbert Boo 15h4'35"15d49'
Thompson,Oval Uma 12h17'49"62d38'
Thompson,Pamela Cnc 8h26'27"7d49'
Thompson,Patrick M Aql 19h53'28"12d41'
Thompson,Patti & Mark Ori 5h32'23"-6d43'
Thompson,Paul D Hya 8h54'21"2d9'
Thompson,Paul W Aql 19h30'53"7d35'
Thompson,Peggy Sue Aql 19h40'18"10d1'
Thompson,Ralph Cushlamochree Aur 5h19'31"41d29'
Thompson,Randall Creed Cnv 13h10'13"33d43'
Thompson,Richard & B'Linda Lyn 8h4'53"57d40'
Thompson,Richard R Per 3h15'59"50d24'
Thompson,Rion Sky Ori 5h22'13"-4d19'
Thompson,Robert Hya 8h34'51"-6d23'
Thompson,Robert V Cep 23h0'41"60d36'
Thompson,Roberta A Lyn 8h16'39"38d31'
Thompson,Robyn Elise Peg 22h36'16"29d5'
Thompson,Rosemarie Cas 23h34'44"00d27'
Thompson,Ross Richard Aur 6h22'0"30d56'
Thompson,S Lynn Peg 22h1'55"4d20'
Thompson,S Michael Per 2h55'14"43d39'
Thompson,Sara-Jane Cyg 19h34'51"34d5'
Thompson,Sarah Joy Sge 19h53'47"18d47'
Thompson,Sarah K And 1h11'0"39d53'
Thompson,Sharlene Cyg 19h26'17"32d48'
Thompson,Shelley Ara 17h56'31"-55d2'
Thompson,Soyina Nicole Mon 8h2'40"-3d1'
Thompson,Sr,Michael R Boo 14h14'19"45d47'
Thompson,Stacy Lynne Com 13h31'44"21d36'
Thompson,Stella Lee Cyg 21h52'1"27d59'
Thompson,Susan Cording And 0h22'44"38d51'
Thompson,Susan Jane Umi 15h12'43"65d57'
Thompson,Susanne And 2h33'26"44d30'
Thompson,Suzette Cas 0h50'40"64d29'
Thompson,Thrasher Gary Ori 5h38'25"15d9'
Thompson,Timothy George Cep 21h57'29"60d2'
Thompson,Tygh Cam 5h17'0"67d44'
Thompson,Van Aur 7h1'23"43d50'
Thompson,Vickie And 0h4'27"46d25'
Thompson,Victoria "Tori" Mon 7h6'50"-6d52'
Thompson,Victoria Dee andress Lmi 10h59'28"31d15'
Thompson,Warren Tri 1h39'21"28d60'
Thompson,William Kathy Her 16h56'39"38d23'
Thompson,Xander Sean Leo 9h27'37"27d60'
Thompson,Yayo Aur 6h6'11"48d46'
Thompson,Zoe Brainwood Cas 0h19'0"62d37'
Thompson/Beebe Marriage Star 7-27-94 Ser 18h0'1"-14d1'
Thomsen,Gary Mon 6h42'47"10d14'
Thomsen,Kalle und Uschi Lyn 8h5'1"42d59'
Thomsen,Sir Danish Dra 17h56'21"63d56'
Thomson"RGT's Candle", Reginald Her 16h18'59"48d58'
Thomson,Alastair Donald Cyg 20h25'28"40d21'
Thomson,Alice M Cyg 19h25'41"34d33'
Thomson,D Lawrence Eri 3h57'43"-2d19'
Thomson,Dave & Celeste Dra 16h41'43"73d5'
Thomson,Debbie Eri 2h49'31"-5d15'
Thomson,Deborah Anne Lyr 18h15'24"38d46'
Thomson,Dorothy Mon 8h8'12"-8d30'
Thomson,Franklin Greg Aur 6h23'38"38d53'
Thomson,George Uma 10h23'35"70d33'
Thomson,George & Janet Cyg 21h45'21"38d39'
Thomson,Lynne Uma 10h37'29"40d34'
Thomson,Marion Cyg 19h31'17"35d50'
Thomson,Morag Carol Cap 21h24'53"-14d35'
Thomson,Nan & Elder Umi 17h12'47"85d17'
Thomson,Natalie Ann Lyr 18h14'1"30d40'
Thomson,Paula Vul 20h1'16"22d33'

Thomson,s Star,Jamie
Aur 5h3'20"45d0'
Thomson,Susan
Lyr 18h33'56"31d17'
Thomson,William & Catherine
Cyg 19h26'34"30d40'
Thomä,Uta
Vir 14h7'27"-1d3'
Thoney,Sean Patrick
Her 17h37'27"26d56'
Thonnerieux, Tony & Faith
Ori 6h7'32"4d27'
Thor & Dita
Cet 0h26'1"0d55'
Thorborg
Her 17h53'38"14d16'
Thorburn,Graham John
Uma 9h53'49"54d37'
Thorin
Ori 4h47'36"4d37'
Thorley,Norman
Aur 5h0'11"48d44'
Thorman,Brent
Sex 10h30'43"-1d50'
Thormborow,Alfred
Boo 14h33'25"30d20'
Thorn "Mein Liebling", Dale
Boo 14h49'25"31d57'
Thorn,Andrew Stanton
Lac 22h22'31"50d17'
Thorn,Donald Paul
Boo 14h54'26"29d55'
Thorn,Raymond
Cam 3h45'23"68d40'
Thorn,Ruby
Aql 18h43'33"7d9'
Thorn,Sharon K
Aql 19h3'44"-6d57'
Thornberg,Amanda
Vir 11h50'43"4d7'
Thornberry,David
Per 3h16'35"41d30'
Thornbrue,Mary Margaret
Uma 11h20'43"43d10'
Thorne,Fredrick W
Dra 16h9'1"68d10'
Thorne,James Robert
Per 3h12'0"45d9'
Thorne,Mary Beth
Lyr 18h14'51"37d42'
Thorne,Michele Christina
And 2h28'1"44d23'
Thorne,Sonya Grace
And 0h50'1"45d40'
Thorne,Tom
Her 16h59'15"27d46'
Thorneloe,Emma
Lyr 19h15'49"25d37'
Thorner,Eddie & Oliver
Cyg 19h31'0"30d23'
Thorneycroft,Christina
Michelle
Com 13h4'38"18d58'
Thornhill,Cory Nathan
Aql 19h52'59"10d47'
Thornhill,Mitze
Cyg 21h19'13"34d11'
Thornley,Patricia Jenkins
Cas 1h38'54"61d17'
Thornsberry,Sandra
Lyr 18h35'44"42d6'
Thornton,Beverly Harris
Stamps
Lyn 9h29'1"40d46'
Thornton,Brenda Wynelle
Mon 6h22'51"7d57'
Thornton,Chloe Rebecca
Cam 8h33'20"77d52'
Thornton,Cory
Gem 6h45'14"12d23'
Thornton,Dave
Per 3h12'15"49d9'
Thornton,Dr Dan
Cyg 20h55'21"31d36'

Thornton,Eleanor M
Lmi 9h43'55"34d55'
Thornton,Elizabeth
Lyn 8h59'0"46d57'
Thornton,Erin
Cnv 12h12'48"50d8'
Thornton,GC
Aql 19h1'51"16d18'
Thornton,Gerald Clive
Her 17h32'13"30d50'
Thornton,Ivan
Aur 5h14'42"44d16'
Thornton,James Michael
Aql 19h5'20"2d21'
Thornton,Jenny
And 2h22'1"48d16'
Thornton,John Collier
Ser 16h12'21"0d35'
Thornton,Justine
And 23h35'33"49d44'
Thornton,Michael
Lac 22h9'48"49d56'
Thornton,Theresa
Lyn 7h34'14"45d43'
Thornton,Thomas
Aur 5h9'0"42d52'
Thornton,Verne F
Vul 20h4'48"25d40'
Thornton,William Langston
Cnc 8h2'1"10d8'
Thornton-"Sparkle", Leslie
Renee
Leo 10h31'45"20d22'
Thorogood,George
Her 17h38'33"14d50'
Thorogood,Gillian Catherine
Del 20h31'22"10d10'
Thorogood-Fanciulli, Rosanna
Ori 5h56'29"12d2'
Thoroughman,Michelle
And 22h59'13"50d29'
Thoroyan,Valerie Jean
Cas 0h4'42"62d32'
Thorp's Piece of
Heaven,Nicole
Lyr 18h52'12"31d13'
Thorp,Kassidy Noelle
And 0h10'1"34d54'
Thorp,Kelsee Marie
And 23h17'28"41d3'
Thorp,Scott
Dra 20h9'0"63d21'
Thorpe,Ann Boyett
Mon 6h58'1"10d20'
Thorpe,Brett Christian
Ser 15h37'52"2d36"
Thorpe,Doarese Cason
Oph 18h16'41"14d4'
Thorpe,Jill Valerie
Peg 23h2'53"30d19'
Thorpe,Madeline Rachel
Cam 3h51'40"58d25'
Thorpe,Ruby Pauline
And 0h42'16"38d18'
Thorsen,Bonnie Jean
And 23h42'30"43d46'
Thorsen,KelliJean
Cas 0h43'24"60d17'
Thorsen,Shannon Alayne
Cas 2h47'23"61d25'
Thorson,Flint
Hya 8h9'0"-8d24'
Thorson,Gertrude
Hya 9h14'24"5d46'
Thorstad,Ronald C
Aql 19h41'52"14d39'
Thorsten Dein Glücksbringer

Thorsten Kaye
Aql 19h7'42"4d13'
Thorton's Star,Shaun Michael
Aur 5h1'33"42d55'
Thorud,Karel Helen Law
Oph 17h0'22"7d59'

Thorumeko
Tau 4h3'34"23d60'
Thorup,Helle Eiberg
Mon 7h57'6"-5d56'
Thorup,Oscar & Barby
Cyg 21h19'30"30d13'
Thorvald
Tri 2h35'1"35d24'
Thoulouse,Michel
Cep 22h23'55"58d15'
Thoumire of Edinburgh,
Elizabeth
Umi 14h49'34"67d38'
Thoutte,Greg
Her 18h9'16"40d37'
Thrailkill,Michael & Wendy
Mon 6h33'26"-0d26'
Thrash,T Flynn
Aql 19h6'19"15d56'
Thrasher,Tonya Marie &
Joseph Robert
Aql 19h10'13"13d37'
Threatte-Sheaffer,Jill
Del 20h57'29"10d46'
3
Vul 20h14'27"26d4'
3 2 1 Gnitset Gnitset
Dra 16h31'37"70d56'
3 4 7 1 6
Cas 0h41'15"70d9'
3 B S Shining Star Bob
Cep 21h17'33"55d15'
3 L-Daddy-Star
Dra 12h0'42"70d25'
3-C Sunshine
Dra 14h26'1"61d57'
3/4 Time
Boo 14h51'0"22d59'
381 Hilary Jayne
And 0h40'44"45d55'
381 Tree
Cyg 21h19'18"28d23'
Three Bear's,The
Per 2h47'14"43d14'
Three Generations
Tri 2h4'48"32d18'
Three Hearts
Uma 11h51'13"38d9'
Three Wells
Boo 14h48'39"32d22'
Threets,Israel
Hya 8h30'21"0d44'
Threin,Bernd
Lyr 19h21'47"30d27'
Threlkel,Tyler Owen
Ser 15h22'0"2d6'
Threshold of a Dream
Cnv 13h47'31"38d22'
Thrift,Colonel Robert J
Cet 2h32'53"-0d42'
Thro
Ori 6h0'46"5d24'
Throckmorton,Robyn Kristin
And 0h46'22"35d51'
Throckmorton-Johnson
III,Jack Lee
Ori 5h52'16"15d2'
Throckmorton-Johnson,
Kathryn Nicole
Peg 21h58'22"2d54'
Throne,Nancy & Michael
Crb 15h53'22"26d18'
Throneberry,Coy Benton
Cep 21h32'42"60d7'
Throop,Kristen
Mon 6h52'39"10d60'
Through The Years
Uma 11h51'39"31d34'
Thrower,Quinton Albert
Dra 20h11'1"70d43'
Thrush,Chester
Lyn 9h6'36"39d43'
Thrush,Paula"Sweet P"
Com 13h9'15"21d26'

Thrushman,Carl Allen
Oph 16h2'16"-6d48'
Thu Nga: Brightest Blossom
Of Heavens
Vir 11h42'12"0d20'
Thu's Star
Oph 17h53'41"10d26'
Thuan Minh Thi Nguyen
Aql 19h30'46"13d13'
Thucydides
Pic 4h36'42"-49d54'
Thuerbach,Jan
Lac 22h1'1"48d55'
Thues,R Thomas
Hya 9h35'31"5d54'
Thul,Gregory Delano
Peg 23h1'33"33d51'
Thuli,Ted
Lac 22h8'39"46d45'
Thulin,Diane
Aql 19h54'54"12d19'
Thulin,Francis Martin
Cnv 13h56'1"41d3'
Thullier,Alain
Dra 15h43'21"63d47'
Thum,Wilhelm
Cmi 7h17'0"1d38'
Thuma,Nora K
Lyn 8h5'38"35d38'
Thumberger,Jacob Andrew
Dra 19h1'41"65d13'
Thumbs Up
And 0h1'1"47d31'
Thume,Sr,Jon Lockwood
Cet 1h5'44"0d56'
Thumper
Vul 19h46'15"25d22'
Thumper
Ori 5h31'14"1d23'
Thumper Junette
Cmi 7h58'19"8d34'
Thunby,Johnathan & Linda
Cyg 21h9'11"37d51'
Thunder
Leo 10h56'10"12d8'
Thunder Mountain
Cet 2h17'18"6d1'
Thunderfoot
Sex 10h19'8"-2d50'
Thurber,Davis Christopher
Boo 14h58'53"27d56'
Thurber,Katie Anne
Del 20h16'33"14d47'
Thurber,Michael
Uma 9h3'21"61d55'
Thurin,Ann Marie
Peg 0h2'29"27d30'
Thurman,John-Catrina
Dunagun
Lyr 19h7'1"38d19'
Thurman,Sam
Dra 16h55'29"67d11'
Thurman,Sharon
Oph 16h53'50"-28d50'
Thurman,Uma
Peg 22h32'21"21d35'
Thurman,Uma Karuna
Cet 2h25'21"7d20'
Thurman-Kristof, Suzanne
Com 12h19'57"24d5"
Thurmond,Kaye Collins
Peg 23h21'0"8d41'
Thurmond,Ken
Her 16h55'11"35d43'
Thurner,T Jeff
Cet 2h17'39"1d7'
Thurnherr,Colleen
Uma 10h57'37"47d53'
Thurnherr,Silvia
Dra 10h54'50"73d9'
Thurow I,Christopher Criag
Hya 8h43'1"4d45'
Thursby,Betty & Roy
Eri 3h43'12"-6d14'

Thursby,James Douglas
Aql 20h3'0"8d22'
Thurston,Jeff
Cep 21h45'1"58d57'
Thurston,Jeffrey Brian
Her 17h12'11"18d55'
Thurston,John
Cep 20h53'48"58d10'
Thvedt,Joshua
Aur 7h5'1"36d28'
Thwaites Star,The Mrs Sharon
Cyg 20h33'0"41d8'
Thwaites,Adam
Umi 16h13'57"76d32'
Thwaites,Alice
Cyg 19h33'60"33d55'
Thweatt,Jr,Herbert
Uma 9h25'1"71d13'
Théo
Uma 13h28'53"61d12'
Thériault,Lili
Lyn 9h37'40"41d23'
Thérèse
Cas 0h11'23"60d13'
Thérése
Gem 7h16'16"31d44'
Thérése Nicole Papa
And 1h14'17"39d30'
Thíebaut,Jacques
Lyr 18h58'1"30d30'
Ti Amo
Her 18h20'15"12d20'
Ti amo Richard
Aql 19h27'17"-10d18'
Ti Amo Sempre
Ari 2h4'13"21d35'
Ti Lin
And 0h58'52"45d48'
Ti-Luc
Per 2h58'28"34d30'
Tia
Cnv 12h47'0"32d50'
Tia
Peg 21h58'42"22d32'
Tia
Lyn 8h2'0"33d43'
Tia
Dra 13h56'53"68d37'
Tia's Reach
Cam 5h3'1"67d45'
Tiagonce,Adam Calixto
Boo 14h28'43"23d52'
Tiara Rose
Leo 11h9'54"-0d45'
Tibak,Joanne Piazza
Peg 22h33'0"20d26'
Tibbets,Billy Joe
Vir 11h45'42"7d59'
Tibbets,Phil
Aql 19h57'0"-0d33'
Tibbett,Douglas G
Ori 5h59'12"9d7'
Tibbetts,Don & Marge
Cyg 20h21'17"38d10'
Tibbetts,Jami Anne
Boo 14h2'0"29d18'
Tibbetts,Molly Catherine
And 2h0'12"38d41'
Tibbetts,Susan Marie
Lac 22h3'43"46d37'
Tibbs,Debra Ann
Eri 2h44'53"-5d28'
Tiberi,Louis A & Michele
Horne
. Psc 0h56'1"28d16'
Tiburcio,Rogelio De Jesus
Mon 8h1'47"-3d33'
Tiburzi Our Star, Zachery
Michael
Lac 21h56'45"37d21'
Ticaric,Patricia M
Cas 1h12'51"61d48'
Tice,Janie
Lyn 6h30'57"58d45'

Tice,Jason Anthony
Cep 22h36'44"60d10'
Tice,Jeffrey Alan
Per 3h3'1"47d40'
Tice,Senior Chief Clifford
Aql 18h50'0"11d57'
Tice,Thomas
Dra 9h47'0"74d49'
Tich #11
Dra 16h0'58"58d26'
Tichacek,Hannah Margaret
Peg 22h28'24"11d5'
Tichatschke,Anja
Leo 11h15'50"-5d48'
Tichenor,Neal
Cet 2h9'11"-1d58'
Tichi,Achim
Lyr 19h23'0"42d1'
Tichy,Fred & Debbie
Cyg 19h32'44"30d46'
Tick,Michael
Per 1h43'19"53d35'
Tickie
Lac 22h27'52"50d35'
Tickie
Lyr 18h44'1"35d43'
Tickle,Bogie
Cma 6h43'43"-15d39'
Tickle,Donald R
Aql 18h57'34"3d14'
Tiddels
Aur 6h8'35"30d24'
Tidigk,Petra & Rainer
Sco 17h54'33"-30d50'
Tidwell,Carolyn Fillingim
Eri 3h3'33"-5d30'
Tidwell,Sr,Calvin McMahan
Aql 19h57'27"15d54'
Tidwell,Tracy
Lyr 19h20'13"42d26'
Tie
Cnv 13h16'36"32d0'
Tie & Teach
Ori 5h43'34"-4d17'AB
Tiebe,Ric
Cam 5h50'1"56d6'
Tieche,Albert David
Aql 19h54'42"14d33'
Tieche,Nancy Monroe
Tri 2h17'0"30d26'
Tieche,Tiffany Annemarie
Cas 1h8'56"61d11'
Tiedtke,Gerhard
Cam 8h31'2"74d53'A
Tiedtke,Harro
Ori 5h57'55"19d57'
Tiefel
Cep 21h53'11"67d38'
Tiefenbach,Horst
Lib 14h57'58"-23d11'
Tiefenbacher,Annemarie
Ari 2h39'1"28d49'
Tiofenbrunn,MD,Larry J
Cep 2h26'1"86d22'
Tiefenthaler,Sigrid
Aqr 21h53'54"0d47'
Tieger,Harry & Dolores
Aql 19h48'44'13d38'
Tieman,Erin Michelle
And 23h1'1"50d41'
Tieman,Julie Ann
And 23h26'57"45d15'
Tiemann,Amy Lynn
Cyg 20h41'30"30d51'
Tien Thi
Cyg 21h4'0"50d7'
Tier Star,The Linda C
Cas 1h31'39"60d31'
Tier Wishing Star, Jennifer
Linn
Ari 2h5'16"22d20'
Tieramasue
Uma 9h18'41"57d41'
Tierce Basile
Boo 14h37'23"11d2'

Tieri,Jonathan C
Aur 6h20'19"30d21'
Tierney,Bill & Margaret
Lyn 7h44'28"50d31'
Tierney,Brian P
Her 18h1'54"38d52'
Tierney,John Norbert
Aql 19h52'45"-5d54'
Tierney,Patrick
And 1h56'54"47d2'
Tierney,Rebecca Daniella
Lyr 18h16'44"45d28'
Tierney-Mayer
Cam 9h8'41"81d54'
Tierra
Cnv 13h31'54"37d58'
Tierza
Hya 10h12'37"-18d49'
Tiesler,Donnie
Her 16h19'57"20d29'
Tietje,Klaus
Lyr 19h23'15"30d41'
Tietjen,Donald George
Aur 7h25'1"40d2'
Tietjen,Rudolf
Psc 22h57'35"0d34'
Tietsort,Jeremy William
Aur 4h58'31"31d2'
Tietz,Harold
Dra 15h11'1"64d37'
Tietz,Harry
Cmi 7h17'25"0d42'
Tietz,Patsy
Umi 15h7'0"79d36'
Tietzen,Lorraine A
Lyr 18h59'48"47d34'
Tiff
And 23h28'55"48d8'
Tiff 34
Lyn 7h53'52"43d5'
Tiff,Mickey
Uma 9h13'0"57d2'
Tiffanie
Uma 13h19'33"61d8'
Tiffany
Vul 20h13'58"22d57'
Tiffany
Del 20h19'15"11d41'
Tiffany
Peg 23h0'11"21d53'
Tiffany
Mon 6h16'52"-5d53'
Tiffany
Aql 19h27'35"-10d13'
Tiffany Leigh
Lyr 18h30'40"40d17'
Tiffany Marie
Cyg 19h59'53"40d47'
Tiffany Novell
Mon 7h55'30"-6d46'
Tiffany Rene
Sge 18h58'21"20d31'
Tiffany Sapphire
Lyn 9h16'0"36d33'
Tiffany Star
Ser 15h35'55"9d20'
Tiffany's
Lib 15h41'25"-28d51'
Tiffany's 21st
Ori 5h28'18"1d10'
Tiffany's Ichabod
And 0h59'31"35d8'
Tiffany's Light
Cet 3h1'1"0d9'
Tiffany,Chuck
Per 2h24'38"55d29'
Tiffany,Frank J
Aql 19h43'0"10d25'
Tiffeni
Cas 0h58'47"50d7'
Tig & Mo
Cyg 19h30'26"36d59'
Tig'r
Sgr 18h38'29"-27d59'

Tiger
Mon 7h48'11"-5d5'
Tiger
Lyn 6h56'55"59d29'
Tiger
Lyn 7h27'14"38d45'
Tiger Lily
Peg 21h54'1"35d54'
Tiger Man
Lyn 8h46'25"40d52'
Tiger Star,The
Cap 20h42'29"-18d13'
Tiger Team,The
Lyn 6h53'23"61d23'
Tiger und Teffy
Sgr 18h57'10"-26d15'
Tiger W
Lyr 18h54'28"30d15'
Tigg-Sbi
Cet 3h14'36"0d44'
Tigger
Ori 5h38'47"11d51'
Tigger
Lyn 7h44'42"39d38'
Tigger
Lib 15h2'29"-20d22'
Tigger
Dra 17h4'45"51d30'
Tigger
Lyn 7h0'11"50d17'
Tigger
Aur 7h24'16"43d29'
Tigger Blue
Boo 15h7'13"11d10'
Tigger Mears
Vul 19h44'22"23d58'
Tigger's Girl
Aur 6h14'28"30d50'
Tiggy
Uma 14h14'46"58d59'
Tighe Forever
Her 18h12'26"38d51'
Tighe,Ian & Dawn
Aql 18h55'53"10d40'
Tighe,Melissa Regina
Cas 1h6'28"53d13'
Tignor,Barbara Renee
And 2h14'51"38d58'
Tigram an Geerd - Lebe
Deinen Traum
Cap 20h41'37"-26d60'
Tigran
Lac 22h39'1"52d33'
Tigre
Cma 6h28'14"-28d15'
Tigress Tess
Uma 9h55'48"43d27'
Tigress,The
Uma 12h1'11"48d52'
Tigresse
Cnv 13h5h1"50d2'
Tigrett,Augusta King
Peg 23h26'51"24d11'
Tigrett,Isaac Burton
Sgr 18h58'48"-27d17'
Tigrett,Maureen Starkey
Leo 10h59'37"21d30'
Tigris Anas
Sco 16h58'55"-40d15'
Tijan
Cnv 13h34'1"41d12'
Tijerina,Arthur L
Cep 4h15'50"80d32'
Tijon
Cet 2h58'33"3d54'
Tiki
Del 20h32'47"18d46'
Tikki Tikki
Mon 6h32'53"-6d34'
Tilas,Stephen P
Her 17h55'49"14d35'
Tilberis,Elizabeth
Vul 19h42'30"26d30'
Tilburg,Tylor Scott
Cmi 7h39'51"4d54'

Tilden,Sally
Com 12h57'58"28d54'
Tildesley,William (Bill)
Cyg 20h15'32"38d9'
Tiley,Caroline
And 0h20'41"45d38'
Tilgner,Hannelore
Ari 2h57'58"21d19'
Tilko,Drew Edward
Vul 21h26'39"27d57'
Till Eternity
Lyn 7h47'0"40d0'
Till,Shar
Cyg 21h5'54"37d25'
Tillander's Shining
Star,Melanie
Sgr 18h51'13"-32d29'
Tillander,Susan Bailey
Peg 22h29'15"27d46'
Tiller,Evelyn
Lyn 8h12'17"45d17'
Tiller,Joan
Cyg 19h46'23"29d46'
Tiller,K.C.
Cet 2h19'34"-10d46'
Tiller,Linda J
Boo 14h3'17"8d6'
Tillery,Jordan Johanna
Lyr 18h24'23"46d25'
Tilley,Christine
Cas 0h45'58"72d37'
Tilley,Jennifer R
Vul 20h27'26"28d23'
Tilley,John
Cet 2h3'51"1d45"
Tilley,John McIver
Oph 18h19'55"7d27'
Tilley,Nadine
Uma 12h40'28"58d39'
Tilley,Roberta Jean
Sct 18h54'17"-10d44'
Tilley,Sr,Rice Matthew
Cet 0h51'23"-1d23'
Tillie & Michele's Walkwalk
Cam 4h57'17"68d24'
Tillinghast,Elaine
Cas 2h32'51"68d52'
Tillinghast,Matthew Cole
Eri 4h6'1"-15d47'
Tillis':Joe,Jessie, Terri & Tracy
Uma 12h24'42"61d18'
Tillis,James
Boo 14h7'56"39d56'
Tillis,Pam
Cas 0h35'25"61d30'
Tillman Star
Uma 10h8'0"61d24'
Tillman,Glen E
Cep 21h58'18"55d17'
Tillman,Joseph Martin Leven
Aur 6h53'30'44d20'
Tillman,Mary Elaine Titus
Cas 0h23'58"69d34'
Tillman,Michael Brian
Sor 15h21'36"5d17'
Tillman,Thomas James
Dra 17h49'11"64d59'
Tillotson,Romance - William
Her 18h2'0"50d5'
Tillson,Donald
Lac 22h24'51"37d38'
Tilly
Aql 18h44'0"8d53'
Tilly & Jack
Crb 16h21'49"34d34'
Tilly und Karl, 6-5-1996
Ori 5h25'14"-2d11 A&B
Tillyer,Melissa & Joe
Lyn 8h7'0"50d50'
Tilotta,Sandra
Cas 1h0'10"62d16'
Tilp,Benjamin Scott Maxwell
Aur 5h50'20"40d2'
Tilston,Marc Gary
Ori 5h54'58"21d23'

Tilton,Lawrence & Eva Cyg 20h18'42"41d1'
Tilton,Star of David & Jessica Cyg 19h47'1"38d55'
Tilvio Vel 9h18'50"-49d41'
Tim Per 2h52'45"32d51'
Tim Ori 5h56'50"6d59'
Tim Ori 6h1'43"3d27'
Tim Aql 19h2'27"0d57'
Tim Dra 12h9'51"71d33'
Tim & Angie Cyg 21h32'0"41d45'
Tim & April Forever Mon 6h7'54"-10d16'
Tim & Cheryl Forever Sge 18h57'49"18d46'
Tim & Dawn's-Eternal Love Mon 7h18'49"-6d57'
Tim & Deaun Star,The Vul 19h49'1"20d29'
Tim & Debbie Umi 13h9'23"70d51'
Tim & Elizabeth Lyr 18h49'27"41d5'
Tim & Francine's Eternal Bond Uma 9h56'27"44d10'
Tim & Karen Sge 20h7'41"20d36'
Tim & Kim Cyg 20h47'12"51d43'A
Tim & Wendy Mon 7h2'10"1d19'
Tim 2323 Mon 8h6'28"-7d11'
Tim 38 Boo 14h29'1"22d7'
Tim Loves Coco Her 18h16'18"20d0'
Tim's Guiding Light to Lyndi Uma 10h56'40"68d48'
Tim's Star Ser 15h37'55"-1d49'
Tim's Star 5/14/59 Boo 13h54'42"21d51'
Tim's Star,Our Guiding Light Her 16h36'31"33d47'
Tim,Andy,Lela & Rob Tau 4h33'25"15d59'
Tim-Tim Her 17h32'32"40d8'
Timakos Cmi 7h51'33"10d6'
Timaul,Ray Boo 14h44'22"21d58'
Timber Sco 17h4'35"-38d59'
Timberlake,Alex Cnc 8h28'0"7d40'
Timberlake,Kelsey Sgr 18h0'40"-28d11'
Timberlake,Peter D Dra 16h4'0"65d58'
Timberlake,Reese Elizabeth Vul 19h20'55"25d50'
Timbo's Star Dra 16h33'29"67d46'
Timbrell Star of Julie Ann Lyr 19h0'28"28d51'
Timbrook,Gerald E Cnv 12h53'51"40d49'
Timby,Justin Michael Lac 22h26'29"52d47'
Time Lac 22h14'47"50d32'
Time & Different Circumstances Ori 4h54'29"4d27'
Time Bandit Hya 8h54'1"2d7'
Timeless Troy Her 17h13'51"29d47'
Timewell,Charles George & Allalee Cyg 20h52'37"31d34'
TiMichael Uma 9h35'13"56d47'
Timko,John & Rosalie Lac 22h36'11"53d46'
Timm (Bratty),Marion E Aql 19h1'1"5d36'
Timm,Anke Her 17h26'59"42d56'
Timm,Jana Lynn Ser 15h33'31"-0d9'
Timm,Mirjam-Katrin Ser 15h32'52"1d16'
Timm,Nancy Thelma And 2h18'44"48d37'
Timma,Anne Elizabeth Boo 13h57'57"7d58'
Timmer,Jan Cmi 7h21'57"8d50'
Timmermann,Bryton Robert Her 18h15'1"47d46'
Timmie Cep 2h13'48"77d32'
Timmins,Bob Equ 21h7'16"10d18'
Timmins,Craig Her 17h19'49"40d12'
Timmins,Sean & Mary Alice Mon 6h19'17"3d17'
Timmons <<Gothic>>, Christopher R Lac 22h48'28"53d52'
Timmons,Beloved Healer-David Oph 17h7'5"-23d55'
Timmons,Buster Aql 18h45'34"10d16'
Timmons,James Edward Lac 22h20'1"51d8'
Timmons,James M Cet 2h49'19"2d45'
Timmons,JNR Aur 6h4'53"38d51'
Timmons,Nancy Marie Olas Cas 1h7'1"60d39'
Timmons,William Howard Boo 13h46'19"20d15'
Timms,Perry Mathew Her 16h56'27"41d8'
Timmy Uma 9h13'59"61d9'
Timmy Per 1h44'41"50d12'
Timmy & Cheri Love Always Uma 9h24'46"57d50
Timmy & Nancy Lyr 18h57'25"31d33'
Timmy's Star Aur 6h17'42"37d16'
Timon Cep 22h34'47"60d13'
Timoteus,Astra Lyn 7h11'1"59d29'
Timothy Sex 10h45'14"-2d22'
Timothy Per 4h38'0"37d10'
Timothy Hya 9h1'55"2d13'
Timothy Cep 21h31'0"58d26'
Timothy & Dawn's "Wedding Star" Cyg 21h54'58"53d34'
Timothy Brian Ori 5h50'10"-0d4'
Timothy Daniel Psc 23h27'34"6d35'
Timothy Edwin Ari 2h4'23"17d45'
Timothy John Boo 14h51'59"30d6'
Timothy Ryan Her 16h16'21"5d33'
Timothy's Enchanted Star Ori 5h59'11"-0d6'
Timothy's Paige Peg 21h39'1"23d37'
Timothy's Star Aur 6h26'52"31d53'
Timpe,Mark Her 15h51'22"43d25'
Timpson,Aleta May Lyr 18h29'16"34d11'
Timpson,Ashley Nicole And 23h19'1"45d4'
Timpson,Frederick H Dra 17h48'35"64d13'
Timrod Per 2h36'46"38d4'
Tina Mon 6h24'11"-6d34'
Tina Vir 13h18'34"-10d38'
Tina Del 20h20'21"20d18'
Tina Cyg 19h33'15"31d21'
Tina Ori 5h55'33"12d58'
Tina Cyg 20h24'48"40d31'
Tina Ari 2h41'44"20d33'
Tina Scl 23h13'20"-34d8'
Tina & Cole Sge 18h58'15"19d46'
Tina & Franz Uma 9h57'30"50d29'
Tina & Irene "Always & Forever" Lyn 8h24'1"43d33'
Tina & John Forever Cyg 21h23'48"53d56'
Tina & Willie Cyg 20h39'0"44d33'
Tina Ann Per 2h31'1"56d31'
Tina Lynn Com 12h32'14"27d29'
Tina Maria Cyg 21h6'0"40d1'
Tina Marie And 0h48'47"36d11'
Tina Marie Lyr 18h45'11"39d57'
Tina Marie & Maremar Aur 6h21'50"40d45'
Tina Marie Stella,The And 0h8'0"41d10'
Tina's 50th And 0h20'44"35d8'
Tina's Amorous Cet 0h56'11"-5d22'
Tina-Timmy Yes! Cnv 13h47'38"42d1'
TinaOs Schatz Cnv 12h41'42"39d17'
Tincher,Mary Aql 19h43'31"14d43'
Tindale,Grant Beal Lyn 8h26'36"48d10'
Tindale,Monaghan Aql 18h59'28"10d8'
Tindall Effect,The Mon 6h57'58"-10d5'
Tindall,Audrey And 23h15'0"47d42'
Tindall,Eric Boo 14h18'45"52d32'
Tindall,Nigel & Laura Cyg 21h18'25"38d41'
Tindall,Sara Elizabeth Cet 2h31'25"4d48'
Tindell,Mary Suzanne Hoorman Lyr 18h42'40"35d16'
Tineo,Mary And 0h1'20"35d47'
Tineo,Nick Dra 15h1'49"60d40'
Tiner,Jr,John Joel Sge 19h35'21"19d14'
TinesOs Teddy Vir 11h39'59"1d31'
Tinez & Melody Forever Uma 9h40'46"70d28'
Ting,'S M' Uma 9h23'14"72d7'
Tinger,Jeffrey Paul Her 17h16'0"47d1'
Tingle II,Damon Lee Boo 15h21'36"50d38'
Tingle,Heidi Eri 2h48'39"-4d27'
Tingler,Joanne Kelley Cap 21h2'1"-22d23'
Tingley's Tailgate Uma 12h9'21"60d50'
Tinglum,Trina And 2h24'40"47d19'
Tinikka's Star Ori 5h55'25"20d4'
Tinkel,Janet Lynn Lyn 8h59'52"36d19'
Tinken,Chad Cet 0h31'49"0d59'
Tinker Bell Vir 14h23'57"5d10'
Tinker,Emma Jane Umi 14h57'22"67d23'
Tinkerbell & Her Teddy-bear Cyg 21h54'1"42d21'
Tinkerbell's Hideaway Lyn 8h17'26"40d6'
Tinkerbelle And 0h23'22"30d24'
Tinkerbelle (Tina) Lyr 18h16'0"42d13'
Tinkle Forever,Dorothy Lac 22h40'44"56d7'
Tinkle Forever,Sally Ann And 23h17'33"51d41'
Tinney,Ralph T & Elvin M Hya 8h43'12"2d0'
Tinny Boo 13h55'1"17d57'
Tinny And 2h29'15"41d9'
Tino,Alyssa Marie And 23h18'33"44d56'
Tinouche And 23h47'53"41d6'
Tinsley 143 Eternally, Jan Tau 4h4'59"22d11'
Tinsley,Chelsea Lillian Mary Lyn 7h48'47"51d17'
Tinsley,Claire Tucker Com 12h0'10"21d28'
Tinsley,Elizabeth A Mon 6h59'0"7d35'
Tinsley,Jason & Lacy, LaSonya Eri 4h13'23"-10d51'
Tinsley,Walter Rhodie Per 3h35'35"40d37'
Tinsley-Giles,Clarice Aql 18h58'0"-2d24'
Tint,Star of Andrew Franklyn Boo 15h13'13"51d54'
Tintes,Mary C Cas 23h17'21"63d27'
Tintin Peg 22h38'16"21d24'
Tintle,Ann Margaret Uma 12h14'28"56d33'
Tintle,Jane Ellen Uma 12h9'29"55d29'
Tintle,John Bernard Uma 12h13'0"59d58'
Tintle,Mary Elizabeth Uma 12h12'48"59d38'
Tintoretto Crb 15h53'27"30d18'
Tiny Dancer And 2h6'52"38d58'
Tiny Dancer Maria Lyr 18h49'42"36d22'
Tiny Dancer-Janny Mac Vul 21h25'20"24d19'
Tiny Mana Cap 20h59'27"-25d20'
Tiny Princess Cam 3h21'44"60d7'
Tiny Tears Aur 5h1'59"49d58'
Tipellie Umi 14h41'1"65d54'
Tiphangne,Roger Oph 18h17'50"10d56'
Tipp,Jerold Eri 2h52'37"-11d58'
Tippi Boo 14h7'37"8d29'
Tippin,Stacey Lynn Mon 6h22'55"4d2'
Tipping,Joanne Cas 0h27'23"50d6'
Tippit,Remi Chaisson Miller Boo 13h44'31"16d21'
Tippit,Susan Gay Miller Cyg 21h26'0"28d56'
Tippner,Kristie Lynn Lyn 7h1'52"52d53'
Tippy "The Little Warrior" Dra 20h21'24"67d15'
Tippy's Star Mon 7h4'17"-6d8'
Tips,Scott Cameron Sct 18h42'9"-7d48'
Tipton,David Carl Milan Per 2h51'42"35d29'
Tipton,Harriet Peg 23h29'56"22d26'
Tipton,Jr,Joseph Reuben Ori 5h3'34"15d7'
Tipton,Laura Lynn Cas 2h35'53"58d30'
Tipton,Patricia Grace Tau 4h57'42"18d56'
Tipton,Sr,Joseph Reuben Boo 13h38'33"26d32'
Tirado,Michael Joseph Lac 22h5'17"46d26'
Tirado,Nilda Lyr 18h46'27"36d38'
Tirado,Sara Cas 2h6'28"59d17'
Tiralosi,Matthew Richard Cet 2h35'1"-6d45'
Tirebois,Xavier Aql 19h56'1"14d24'
Tirith,Minas Mon 7h19'42"-6d35'
Tiro Cam 3h27'23"53d16'
Tirone,Christopher & Julie Lyn 6h57'0"60d50'
Tirone,Mark Dra 9h28'42"80d22'
Tirone,Peter & Susan Cam 8h27'22"82d16'
Tirosh,Zoe And 0h41'10"33d2'A
Tirrell,Jaime D Dra 18h18'17"68d9'
Tirschwell,Barbara Weiss Cas 1h19'13"61d33'
Tirtei,Marina Lmi 10h32'40"32d4'
Tirtei,Marina Uma 12h29'0"60d12'
Tirza Ant 10h32'22"-33d30'
Tis I,The Big Guy Her 17h53'36"38d58'
Tisa And 2h29'0"38d16'
Tisch,Benjamin Ari 2h49'23"30d26'
Tisch,Charlotte Frances And 23h1'26"50d13'
Tisch,Gary Nelson Cyg 19h19'33"28d59'
Tischler,Mae & Saul Aur 4h47'50"48d50'
Tischler,Mae T Cyg 20h26'37"37d35'
Tischler,Nicholas Her 16h3'34"48d31'
Tischuk,Pat Ori 5h56'0"15d55'
Tisdale,David William Aur 6h54'25"38d39'
Tisdall,Vera Lib 14h59'22"-7d32'
Tish's Star And 2h22'37"42d5'
Tish's Wishes Mon 7h54'26"-0d11'
Tishamingo Princess And 1h1'18"33d41'
Tisherman,Flora & Sandy Cet 2h29'26"4d26'
Tishman,John L Uma 12h14'1"54d50'
Tismo,Juan & Dolores Cyg 20h15'27"30d51'
Tismo,Patrio Aur 6h31'20"31d45'
Tison,Claude Lyn 9h7'1"33d25'
Tisserant,Nicole Dra 11h56'32"70d1'
Tissie Bean Cap 21h23'5"-22d4'
Tissot,Jackie Crb 16h18'59"32d16'
Tissy Lyr 19h23'42"31d34"
Tita And 23h16'32"51d44'
Tita Aql 19h2'40"-6d60'
Titanda Psa 22h36'45"-25d53'
Titanica Hya 9h33'40"-5d11'
Titches,De De Oph 17h30'43"-1d37'
Titcomb,Megan Quinn Per 4h5'58"45d45'
Titensor,Steven Mon 7h56'46"-2d56'
Titheradge,Peter Richard Henry Aur 6h26'45"34d5'
Titley,Allison Elizabeth Cas 0h5'37"58d8'
Titley,John Edward Cep 23h5'20"80d12'
Titlow,Aimee Lynn Peg 23h35'0"21d30'
Tito Oph 17h56'59"7d44'
Tito Joe Ori 5h56'39"15d4'
Titou Cyg 20h5'35"30d59'
Titsworth,Kayla Marie Uma 9h39'24"60d16'
Tittle,Gary R Lac 22h52'52"56d27'
Tittle,Olivia Lyr 18h57'50"38d1'
Titty And 0h12'16"38d47'
Titus,Alexandria Nicole And 0h11'21"31d4'
Titus,Hannah Elizabeth Cas 2h41'29"60d43'
Titus,Jonathan Douglas Aur 6h6'31"33d6'
Titus,Keith (Courage) Per 3h14'42"40d37'
Titus,Matthew William Per 2h54'49"32d52'
Titus,Olivia Marian Peg 23h43'16"15d45'
Titus,Paul C Aur 6h27'58"33d33'
Titus,Robert Daniel Lmi 9h25'18"38d36'
Titus,Todd Lac 22h21'12"37d40'
Titusmvjmpcgjmkkccemmr ljsvcakjdmj92 Lyn 6h25'38"54d3'
Titza,John Cet 2h56'25"0d20'
Tivnan,Sheila McCarthy Cyg 19h45'59"47d46'B
Tivnan,Thomas C Cyg 19h45'59"47d46'A
Tiwul,Astra Ori 5h29'23"1d52'
Tix,Mattias And 23h2'1"42d52'
Tixi,Captain Giacomo Ori 5h35'29"7d38'
Tizer of Biston Hill (I Love You) Dra 16h26'17"60d12'
Tiziana Col 6h31'59"-39d36'
Tiziana Cyg 21h53'51"52d57'
Tiziana Ind 21h10'29"-58d43'
Tiziana Vir 13h34'20"5d16'
Tiziana Boo 14h33'15"12d30'
TJ Keel Uma 9h16'0"61d41'
TJ's Magical Mystical Starship Crb 15h33'51"26d23'
Tjaben-Stevens,Wilhelm Umi 13h9'47"75d17'
Tjaden,Helen Trottere Cas 2h34'31"60d10'
Tjej Special Sex 9h58'20"2d22'
TJF-8 Cam 6h2'41"60d18'
Tjiputra,Dewi Peg 22h7'43"26d53'
TJO Boo 15h24'1"40d7'
TJV 52790 Cam 6h9'20"68d28'
TK Gem 7h54'27"30d21'
TK Per 1h26'22"53d40'
TK's Star Gem 6h46'1"30d29'
Tkac,Paula Picha And 23h20'10"51d26'
Tkach MD,Alpha Stephen Lyr 18h41'18"46d42'
Tkalcic Vul 19h58'20"29d9'
Tkalec,Misha Mon 6h59'34"-5d51'
TKK Aur 5h25'15"30d40'
TKO Lyn 7h44'32"40d56'
TLC Aur 5h2'0"40d48'
TLT Chérie St-Tite Cyg 19h25'23"30d12'
TMN Tau 5h27'38"24d44'
TMP-My Love-LJC Hya 8h18'1"3d12'
Tnook Dra 16h25'32"69d42'
TNT For Eternity Aql 19h20'42"10d28'
To Bella,Your one & Only Lorenzo Umi 15h16'20"78d42'
To Buffy My Shining Star-Love Mike Ori 5h5'11"13d5'
To Celebrate Our "MaRich"#5 Crb 15h55'24"30d59'
To Dana—Miracle Hugs- - Love Tim And 0h53'1"40d42'
To Gary My King, Love Ellen Uma 9h57'32"62d27'
To Jams With Love Kathy Mon 7h9'37"-6d37'
To K C With Love Star, The Uma 9h12'48"59d35'
To Manny,With Love From Michelle Oph 17h59'46"12d15'
To My Dear Friend, Alice Lac 22h54'26"55d40'
To My Dearest Wife Nancy-Love Charlie Cas 1h31'15"75d3'
To My Eternal Love, Christine Peg 23h3'60"32d10'
To My Randall Patrick Oph 17h31'57"8d24'
To My Son Richard Aql 19h17'17"10d21'
To My Special Sis B & Alfie Hya 9h0'18"2d37'
To My Warm Fuzzy Jan Peg 23h1'43"17d44'
To Scott I Love You! Deborah Boo 15h14'59"26d8'
To Super Mom, Love Joe Cas 2h35'43"60d46'
To The One Forever Annette Mon 6h51'38"10d36'
To The Staff of Two South-CINH Uma 9h57'57"51d20'
To Tod From The Yank Ind 20h32'53"-51d35'
To Tony,Forever Yours, Love Kerry Per 1h27'37"54d22'
To Venus Thru Lissa Cyg 19h50'30"40d17'
To-Mas H Lac 22h5'14"47d59'
Toad's Place Dra 16h5'15"63d38'
Toad's Star Ser 15h41'4"17d31'
Toah et Moah et Nozamimr Uma 14h8'38"60d12'
Toastette Lyn 8h13'1"38d55'
Tobac,Zane Thomas Lac 22h54'46"56d38'
Tobaccorella Aql 18h58'0"17d17'
Tobasko,Monica Leo 10h35'47"14d1'
Tobe Cyg 19h20'31"28d39'
Toben,Marilyn Bridget And 2h20'13"45d19'
Tober,Barbara Lac 22h34'48"38d2'
Tober,Jan Lyr 18h28'40"32d36'A
Tober,Ronnie Dra 17h25'15"58d35'
Tobey Loves Bobby Lyr 18h44'29"41d41'
Tobey,Patricia Ann Cnv 13h20'12"40d9'
Tobey,Thomas Henry Her 17h50'26"50d11'
Tobiah & Amber In Jehovah Forever Equ 21h2'34"3d37'
Tobias Boo 15h12'1"52d19'
Tobias Cep 1h29'45"86d9'
Tobias Jr,Frank Per 2h40'46"43d39'
Tobias,Arthur L Dra 19h22'0"56d16'
Tobias,Baby Lyr 18h15'34"31d34'
Tobias,Cindy Ori 5h7'22"9d54'A
Tobias,Claire Cas 0h18'37"62d13'
Tobias,Daniel Morgan Ori 5h56'1"16d12'
Tobias,Misti A Mon 6h48'0"10d5'
Tobias,Terry Ori 5h7'22"9d54'B
Tobie & Bernie's Wedding Star Cyg 20h51'48"46d3'
Tobie Star,The Cep 22h36'11"60d28'
Tobin,Dennis Lac 22h53'20"55d23'
Tobin,James "Madame Zebu" Aur 4h56'56"41d2'
Tobin,James John Aql 19h41'43"14d6'
Tobin,Julie Ori 4h55'1"-2d43'
Tobin,Leta N Lyr 19h23'47"31d21'
Tobin,Linda Jeanne Aql 19h45'0"14d3'
Tobin,Mary Jane Tau 4h2'39"30d46'
Tobin,Mary Therese Cas 3h2'22"60d0'
Tobin,Michael Henry Ori 4h48'0"0d53'
Tobin,Patty "Kid" Lyr 19h24'58"38d27'
Tobin,Robin Kimberly Gem 5h59'1"27d28'
Tobin,Rosalyn B Tau 5h20'13"20d33'
Tobin,Toby Ori 5h56'23"15d52'
Tobin,Walter Cnv 12h48'1"51d49'
Tobisch,Lotte Peg 23h32'34"18d34'
Tobler,Douglas & Maxine Lyr 19h2'39"25d49'
Toblor,Jürgen Cam 5h43'15"70d32'
Tobman,Eva & George Oph 17h53'51"10d9'
Tobnorenity Cet 3h17'15"3d16'
TOBO Aur 7h3'23"41d15'
Toby Ori 5h21'31"15d46'
Toby Per 2h10'37"57d13'
Toby Aur 7h2'52"40d41'
Toby Lmi 10h37'55"33d38'
Toby Sex 10h17'30"-3d34'

Toby
Peg 21h6'30"12d11'
Toby
Her 17h47'20"14d54'
Toby E
Per 2h53'0"46d6'
Toby's Star
Umi 14h40'29"68d36'
Toby's Star
Pup 8h0'2"-25d55'
Toby's Wish
Aql 19h47'53"14d12'
Tobyn
Uma 11h20'21"33d16'
Toce,Anthony J
Tau 5h47'38"28d28'
Toche Stella d'Amore e Fortuna
Cnc 8h49'34"31d18'
Toczek OFM,Father Melchior
Ori 5h48'29"20d60'
TODA
Lyr 19h1'51"40d53'
Todavich,John
Lac 22h31'44"53d9'
Todd
Cyg 21h11'25"36d25'A
Todd
Cma 6h56'41"-19d40'
Todd & Gwen
Cyg 19h29'35"40d36'
Todd & Kris 5 Years & Forever More
Umi 14h13'58"69d23'
Todd & Lauren
Her 16h51'57"34d54'
Todd & Sarah Forever
Ori 6h7'14"11d4'
Todd & Sheri
Cet 0h59'55"-10d38'
Todd Dominique
Aur 6h22'17"38d59'
Todd Irvin
Cmi 7h15'37"1d42'
Todd Loves Wanda 7-29-93
Cyg 20h17'23"30d27'
Todd's Birthday Candle,Mary
Gem 6h51'8"18d39'
Todd's Guiding Light 12-19-1973
Sgr 20h23'40"-28d32'
Todd's Hope
Her 17h36'17"27d42'
Todd's Star
Cma 6h31'11"-17d9'
Todd's Strength
Aql 19h58'12"10d25'
Todd's Twinkler
Vul 19h18'10"27d37'
Todd,Abbie-Stuart
Peg 23h5'1"10d43'
Todd,Adam Ross
Aql 19h14'13"19d49'
Todd,Alison
Ori 6h3'45"5d33'
Todd,Anna
Tri 2h4'48"30d7'
Todd,Darby
Del 20h53'57"4d13'
Todd,Debra
And 0h1'1"43d48'
Todd,Devon Rebecca
Cet 1h35'25"-13d24'
Todd,Elise
And 23h15'57"48d2'
Todd,F Stuart De P
Dra 14h18'42"64d9'
Todd,G D W
Aql 18h43'57"7d16'
Todd,Gail Lee
Eri 4h10'1"-18d50'
Todd,I'll Love You Forever,Julie
Cep 20h37'26"75d29'
Todd,J Harrison
Dra 16h13'53"68d24'
Todd,Jennifer Elaine
Mon 6h30'23"-0d55'
Todd,Jennifer Lane
Vul 21h14'44"20d34'
Todd,John & Katherine
Sge 20h16'27"17d27'
Todd,Joshua
Cyg 21h52'27"37d2'
Todd,Karen Elizabeth
Peg 21h58'32"34d3'
Todd,Kelley Kathleen
Ari 2h1'42"20d14'
Todd,Kelly
Vul 20h14'48"23d18'
Todd,Londa
Cas 3h6'21"57d37'
Todd,Louise Alexandra
Cas 1h15'0"75d57'
Todd,Melaine Ann
Sgr 18h47'50"-32d56'
Todd,Patricia Ann
And 23h35'51"33d33'
Todd,Rebecca Kathryn
Ori 5h1'1"10d52'
Todd,Richard Anthony
Aur 6h16'46"38d7'
Todd,Robert Edward
Her 16h36'29"11d22'
Todd,Sr,Jerry
Cet 2h29'18"8d44'
Todd,Sr,Stephen Thomas
Dra 17h45'1"64d8'
Todd,Stephen G
Oph 17h14'1"10d40'
Todd,Suzanne Elizabeth
Cet 2h46'44"3d36'
Todd-Brown,Katherine E O
Cas 0h22'57"75d55'
Toddley
Ori 6h7'32"9d1'
Toddy
Dra 16h54'31"68d1'
Toder,Jonathan
Lac 22h17'16"55d12'
Toder,Paige
Lyr 18h38'47"41d9'
Toder,Spencer
Cep 20h25'48"60d19'
Todes,Myrtle
Lac 22h55'29"51d12'
Todino,Catherine Kindya
Sct 18h52'40"-9d15'
Todora,Michael & Helen
Aql 20h6'48"4d6'
Tody
Sco 17h51'36"-31d18'
Toedtman,Ryan
Her 17h33'29"21d9'
Toeppe,Frederick William Max
Cep 22h38'51"58d28'
Toeppen,R Paul
Hya 8h30'45"-8d11'
Tofallos Agape Mou, Carole & Harry
And 2h7'1"39d38'
Tofanelli,Sharon
Vul 19h3'26"24d31'
Tofani,Janice Marie
Lyr 18h20'47"38d54'
Tofel,Jr,Walter John
Uma 9h2'28"56d35'
Toff,Francis "Buttercup"
Mon 6h28'26"-10d34'
Toffolo,Luciano
Sex 10h14'17"-4d53'
Together
Aql 19h57'35"15d45'
Together Forever,I Love You Sandi
Boo 14h59'34"21d58'
Together, Forever
Cyg 20h42'30"45d37'
Together-Gina & John
Sge 19h40'28"16d49'
Togetherness-Lesley, Les & Ella
Uma 9h3'38"53d4'
Togher,Jr,James & Jean
Cyg 20h1'19"30d52'
Tognazzini,Daniel P
And 0h58'0"41d2'
Tognazzini,Terry D
Per 1h50'53"48d60'
Tognetti,Michael "Mick"
Eri 3h44'29"2d2'
Tognoni,Elisabetta
Cae 4h57'15"-31d53'
Tognucci,Ernest Louis
Lyr 19h1'10"28d32'
Togo,Yukiyasu
Equ 21h20'38"8d5'
Tohill,Tamara
Aql 18h56'29"7d20'
Toigo,James
Cam 3h26'0"60d59'
Toivanen,Terho
Uma 10h23'50"48d34'
Toivon Tahti
Dra 19h24'56"80d9'
Toivonen,Leena
Lyn 7h8'35"44d18'
Tokaleno
Cam 8h9'11"80d58'
Tokamo
Uma 9h19'46"59d43'
Tokarz,Tricia
Lmi 9h59'1"38d57'
Toker,Lorne
Dra 14h27'40"61d34'
Tokotch,Diane Lynn
Sge 19h19'48"16d18'
Tokunaga,Stella of Team Captain,H
Psc 1h12'32"19d33'
Tolan,Ahne
Lib 15h3'25"-1d36'
Tolan,Mike & Amanda
Cyg 21h51'38"53d39'
Toland,Brian
And 0h30'10"28d0'B
Tolbert,Gina F
Equ 21h9'50"11d5'
Tolcher,G B
Uma 8h56'55"50d22'
Toledano,Ralph
Uma 11h6'39"55d26'
Toledano,Roberto
Psc 22h56'43"6d37'
Toledo,Bon Scott
Aur 5h5'56"44d17'
Toler,Randy
Aql 18h43'49"10d50'
Toler,Stephanie
And 1h2'36"38d4'
Toles,Gregory Anthony
Her 17h14'15"43d41'
Toles,Marc A
Boo 14h33'31"47d26'
Tolga
Ori 5h29'27"-6d8'
Tolhis,Court
Cep 22h22'40"61d46'
Tolin,Thomas F
Boo 15h13'38"48d13'
Tolino,Michael
Per 4h31'28"34d58'
Toll's Star,Danny & PK
Vul 20h4'0"23d42'
Toll,Connie
And 1h52'23"39d30'
Toll,Deborah Marie
Cnc 8h0'54"8d54'
Toll,John
Ori 6h0'49"0d40'
Toll,Mary & Douglas
Lyr 18h30'21"37d35'
Toll,Sue Anderson
And 2h2'14"46d27'
Toll,Susan
Cas 1h36'17"60d7'
Toller III,James Andrew
Aur 6h39'10"37d48'
Toller,Amy
Cas 1h5'41"68d25'
Tolles,Joann Frances
Cyg 19h42'1"40d28'
Tolles,William Charles
Per 3h37'46"38d5'
Tolleson Family,The
Cyg 19h54'49"41d9'
Tolley,David Allan
Per 2h9'31"57d30'
Tolli,R F (====) "Troll"
Aql 18h55'42"-0d10'
Tollin,Georgia Faye
Leo 11h8'1"2d1'
Tollin,Georgia Faye
Mon 8h6'28"-9d39'
Tolliver,Cheryl
Cyg 20h9'46"40d34'
Tolliver,Christopher David
Aql 19h6'50"3d28'
Tolliver,James R
Aql 20h8'40"0d18'
Tolman,Donna Jo
Lyn 8h21'44"41d16'
Tolman,George Earle
Lmi 9h48'1"38d32'
Tolman,Kristie Marie
Lyr 18h58'43"26d43'
Tolman,Melissa Kay
Cas 1h53'33"68d41'
Tolpegin,William Doyle
Uma 10h40'27"48d42'
Tolson II,George Jeffries
Uma 9h54'30"42d20'
Tolson,Emma-Sophie
Umi 15h30'50"70d29'
Tolson,Hubie
Per 3h57'40"51d50'
Tolstoy Livingston Star,The
Sge 20h5'18"20d31'
Tolstoy,Anne Marie
Cas 0h30'1"61d53'
Toltezman,Anne Marie
Cas 2h8'21"60d25'
Tom
Cam 8h44'54"80d26'
Tom
Her 16h16'41"24d7'
Tom & Delayne's Star
Cyg 21h36'24"30d15'
Tom & Denise
Aql 18h57'37"10d53'
Tom & Edith
Crb 16h5'41"37d53'
Tom & Eileen
Del 20h14'38"10d57'
Tom & Eliz - Always
Uma 14h1'0"48d41'
Tom & Georgie
Cep 21h33'21"60d20'
Tom & Jill-Astro Lovers
Lep 5h15'19"-15d16'A
Tom & Jill-Cosmic Kids
Lep 5h15'19"-15d16'B
Tom & Judith's SnowFire Star
Aql 19h59'1"8d32'
Tom & Karen
Sge 19h16'21"16d10'
Tom & Kristal
Cyg 20h39'0"53d28'
Tom & Lenora
Aqr 22h22'25"-1d60'
Tom & Linda's Wish of Happiness
Cyg 20h31'52"48d51'
Tom & Lisa
Uma 11h5'27"31d57'
Tom & Mary's 12th Of Never
Peg 23h47'16"11d19'
Tom & Melinda El
Dra 17h7'1"68d37'
Tom & Michele
Peg 23h2'54"11d52'
Tom & Nella A Child Is Born
Aur 7h19'60"40d17'
Tom & Non
Uma 9h51'15"48d40'
Tom & OK
Cnv 13h39'58"46d18'
Tom & Patty For A Million Years
Peg 23h37'51"21d36'
Tom & Sheila-1994
Crb 15h27'43"30d14'
Tom & Teresa
Cyg 20h39'37"31d30'
Tom Allen
Eri 3h18'0"-17d44'
Tom Forever
Sco 16h25'25"-40d27'
Tom Forever My Love
Cnc 8h0'55"10d46'
Tom Francis
Aql 18h54'48"8d46'
Tom J. "Pelar" Pawlowski
Aur 7h16'60"37d27'
Tom Keery Super Star
Per 2h36'34"41d12'
Tom Loves Christine
Boo 14h58'34"20d8'
Tom Out of the Dust Brothers
Dra 17h55'50"68d54'
Tom R My Love Forever & Always
Per 3h22'44"40d42'
Tom Terrific
Lac 22h23'19"50d34'
Tom's "Little Piece of Heaven"
Aur 6h28'54"37d39'
Tom's Birthday Star
Ori 6h3'15"8d33'
Tom's Christmas Magic
Per 3h3'0"40d21'
Tom's Dream
Her 17h5'31"31d31'
Tom's Funkler
Tau 3h46'48"0d8'
Tom's Genesis
Ser 15h10'30"-2d10'
Tom's Goal
Mon 7h57'34"-5d39'
Tom's Guardian Angel
Crb 16h21'43"38d28'
Tom's Homerun
Dra 18h7'59"67d56'
Tom's Honor
Lac 22h39'15"37d43'
Tom's Joy
Cnv 13h3'45"40d53'
Tom's Lil Tital 077
Aur 6h56'56"40d11'
Tom, King of My Heart II
Cep 0h13'44"73d11'
Tom,Cendie & Whitney
Uma 9h6'55"56d48'
Tom,The "Twinkle" in my Eyes & Heart!
Cam 12h45'1"80d29'
Tom-Love Of My Life LLL
Sge 19h39'1"16d45'
Tom-N-Judy
Sge 19h53'39"16d14'
Tom44
Sco 17h27'15"-40d53'
Toma,Caroline Monir
Cyg 19h22'0"30d1'
Tomaculum
Ori 5h54'0"18d23'
Tomafisher
Cas 0h36'24"60d44'
Tomah
Aur 4h48'22"50d13'
Toman,Dick
Cas 1h45'0"63d53'
Tomancik,Gogan's Star & Grill,William J
Uma 12h5'1"56d3'
Tomanini,Craig
Uma 11h53'53"30d15'
Tomann,Kimberly Betanzos
Cyg 20h18'0"41d58'
Tomao,Christopher James
Peg 23h7'13"8d46'
Tomao,MD,Frank A
Per 3h11'26"45d30'
Tomar Industries
Cet 2h19'0"6d31'
Tomari,Jr,Alexander J
Dra 17h27'18"70d18'
Tomaro's Star,Jennifer
Lyr 18h35'0"41d12'
Tomas 25-12-86
Umi 14h10'47"68d51'
Tomas,Leo Luis
Cet 2h30'0"-10d41'
Tomasa,Abuelita
Lac 22h36'1"37d36'
Tomasco,Richard
Dra 16h30'13"61d38'
Tomasiak,Jim
Aur 6h35'33"30d19'
Tomasiewicz,Thomas Martin
Aur 5h6'48"42d57'
Tomasik,Brian
Cyg 19h59'32"58d24'
Tomasik,Eric Douglas
Her 17h13'41"26d46'
Tomasina
Uma 11h27'34"56d37'
Tomasini,The Ross W Star
Car 7h29'27"-52d47'
Tomasino,Laurin
And 23h36'0"49d46'
Tomasino,Whitt
Aur 6h29'52"33d6'
Tomaskovic,Sarah Marie
Cam 4h19'59"60d55'
Tomassetti,Victoria M
Cyg 19h55'1"38d3'
Tomassi,Lilli
Cyg 21h58'39"53d45'
Tomassini,Alice
Cas 23h4'34"53d1'
Tomasso,Robert P
Peg 21h49'45"31d40'
Tomasson,Kimberly Dawn
Lyn 7h24'51"44d45'
Tomaszewski,Elyse Catherine
Cyg 20h21'39"31d29'
Tomaszkiewicz,Peter
Psc 1h31'42"20d26'
Tomav
Mon 7h17'15"-1d47'
Tomazewski-Dunsmuir, Henrietta
Lyn 8h20'45"46d12'
Tomazi PhD,Keith G
Uma 10h49'49"44d38'
Tomb,Michael
Cma 7h0'40"-13d39'
Tombasco,Pat
Crt 11h12'26"-17d39'
Tombazzi,Hillary
And 2h30'1"48d13'
Tomblinson IV,Ben
Aql 19h54'21"-0d42'
Tomboys
Peg 22h26'15"31d8'
Tomcat
Lmi 9h22'49"33d36'
Tomchik,Joseph John
Aur 7h17'27"41d12'
Tomchin,Elaine
Cet 1h24'20"-1d45'
Tomcko,Steve & Tillie
Sge 19h55'20"20d26'
Tomczak,Kristina
Eri 3h4'32"-3d36'
Tomczyk,Mary Jo
And 23h12'39"37d5'
Tome,John Jack
Per 3h46'58"38d48'
Tome,Rudolph Harry
Aur 5h21'43"40d41'
Tomek,Lilo Gabriel
Per 2h52'14"50d15'
Tomes,Homer
Per 4h1'16"51d54'
Tomes,Michael Anthony
Dra 19h32'30"61d17'
Tomezsko,Edward Stephen
Sex 9h58'30"5d12'
TomGini
Uma 10h46'0"55d16'
Tomi Jo
Ori 5h55'58"16d58'
Tomi Kaaria
Cep 5h19'12"86d15'
Tomi,Lovely Ms
Aql 20h13'30"4d15'
Tomikazu & Miyoko
Sco 16h29'14"-28d57'
Tomiyasu Ikeda
Gem 7h14'32"16d36'
Tomko,Jr,crouch-Robert J
Eri 2h46'15"-2d36'
Tomko,Jr,Paul B
Aur 4h53'10"38d25'
Tomlin,Joyce
Aur 6h35'27"34d52'
Tomlin,Marcus Lynn
Cet 3h1'25"1d22'
Tomlin,My Hero,Stephen Craig
Ori 5h28'51"1d20'
Tomlin,Phillip
Cet 2h47'28"7d0'
Tomlin-40
Leo 10h59'33"23d16'
Tomlins,Mackenzie
Cam 6h13'21"58d9'
Tomlinson Sister Star, The
Mon 6h58'27"-5d46'
Tomlinson Star,The
Ori 4h54'33"-1d7'
Tomlinson,Corey L
Cnv 12h46'50"32d27'
Tomlinson,Doug E
Cet 2h9'48"-0d58'
Tomlinson,Jack E
Uma 8h36'1"51d47'
Tomlinson,James A
Boo 14h54'11"38d41'
Tomlinson,Jessica A
And 2h19'40"45d29'
Tomlinson,Johanna K
Lac 22h24'1"56d38'
Tomlinson,Jr,Harvey Gillingham
Her 18h12'56"38d48'
Tomlinson,Karen
Crb 15h16'35"30d49'
Tomlinson,Katherine
Lyn 7h28'30"38d16'
Tomlinson,Martha A
Uma 8h39'35"52d32'
Tomlinson,Ross E
Cet 2h32'54"-1d47'
Tomlinson,Ulrike Gerlinde
Peg 22h5'0"29d23'
Tomlonson,Tom
Ori 5h4'15"1d49'
Tommael,Soena
Cnv 13h0'13"32d23'
Tommasina
Ind 21h12'48"-55d3'
Tommasino "The One", Chris
Sct 18h42'33"-6d4'
Tommaso
Aur 6h2'1"30d6'
Tommaso,Monica E
Peg 21h53'17"21d42'
Tommie
Ori 5h55'31"17d4'
Tommie Jean
Per 1h30'23"53d36'
Tommie T
Oph 16h33'12"-6d46'
Tommijo
Cet 0h46'0"0d45'
Tommin Ja Majon Oma Maailma
Uma 11h57'11"30d5'
Tommis
Aur 6h6'56"37d39'
Tommy
Boo 14h29'24"26d52'
Tommy
Per 3h1'1"38d8'
Tommy
Boo 15h7'40"40d54'
Tommy
Ori 4h54'35"5d16'
Tommy
Tri 2h46'22"33d43'A
Tommy
Lac 22h17'55"54d48'
Tommy
Per 3h18'24"50d35'
Tommy & Elliott's Diamond In The Sky
Lmi 10h18'14"35d30'
Tommy & Honey
Vul 20h4'56"22d30'
Tommy & Lisa's Wedding Star
Cyg 20h20'14"41d7'
Tommy & Mare Star,The
Cyg 20h22'56"38d29'
Tommy D
Boo 14h46'37"29d55'
Tommy D's Shining Star
Dra 16h52'12"66d40'
Tommy D,The Guiding Light
Lmi 9h22'33"33d27'
Tommy G
Her 15h52'33"40d42'
Tommy Z
Her 17h32'44"28d55'
Tommy's Dad
Crb 16h19'49"27d52"
Tommy's Pig Star
Her 18h4'28"38d30'
Tommy's Sky
Her 18h11'35"40d13'
Tommy's Star
Her 17h15'48"41d5'
Tommye's Star 75
Sgr 18h48'19"-30d56'
Tomohiko Morishita
Vir 14h5'43"-10d10'
Tomohiro to Tomomi
Sgr 19h11'54"-26d12'
Tomohiro,Nakao
Gem 7h11'29"14d53'
Tomoka
Aql 19h12'25"10d49'
TOMOKAZU & MAMI
Cnc 8h47'20"18d2'
Tomoko 222
Eri 4h13'50"-12d13'
Tomori,Gregory Peter
Her 18h19'12"12d21'
Tomorra
Oph 17h4'52"-20d11'
Tompkin,Roy
Her 18h4'18"14d17'
Tompkins III,Bill
Aur 5h55'24"29d3'
Tompkins,Andrew Charles
Aur 5h55'29"29d51'
Tompkins,Anthony Mark
Hya 9h1'24"2d48'
Tompkins,Barry Marc
Sgr 19h36'50"-38d14'
Tompkins,Brian John
Her 16h11'1"25d13'
Tompkins,Casey
Dra 17h40'55"68d44'
Tompkins,Daniel Frank
Aur 5h56'34"29d48'
Tompkins,Darrius & Treisa
Aur 5h58'54"31d3'
Tompkins,Darrius & Treisa
Lmi 9h49'0"34d5'
Tompkins,James
Cma 7h14'19"-16d18'
Tompkins,Jennifer
Cas 0h20'50"67d59'
Tompkins,Judge Harold
Sge 19h5'57"17d43'
Tompkins,Mr Mark Antony
Uma 11h25'11"30d48'
Tompor,Karen
Aur 6h37'30"30d40'
Tomporowski,Jason "Catchus Troutus"
Aur 6h15'0"46d33'
Tomschaefer
Uma 9h15'11"50d49'
Tomy e Geo
Cas 2h3'1"60d58'
Tomé John
Aur 5h8'29"40d12'
Ton N' Peg
Uma 9h16'32"50d31'
Ton,Hans-Peter
Cep 21h40'0"61d1'
Tona Maria
Hya 9h34'37"2d22'
Tonatzin
Cet 0h54'18"1d5'
Tondelli,Franca Podestani
Tel 18h7'14"-48d16'
Tondolayo
Lac 22h24'0"52d54'
Tone Stone & Sky High
Umi 16h19'33"76d58'
Tone,Stephen
Cet 0h3'56"-10d4'
Tonella-Mauci,Giulia
Pic 4h36'47"-48d40'
Tonellier,Markus
Sgr 19h57'20"-44d7'
Toner,Ed
Boo 15h5'57"14d16'
Toner,Jim
Her 18h13'44"38d6'
Toney,Jack
Aur 5h30'53"38d49'
Toney,Mimi & Sheldon
Dra 18h19'44"58d51'
Tong,Esther
Lyn 8h4'59"38d7'
Tonganoxie Distinguished Readers
Lyn 8h12'26"45d53'
Tonge,Dawn
Lyr 18h44'34"41d3'
Tongen,Richard W
Cnv 12h15'18"35d7'
Toni
Ori 5h50'11"21d37'
Toni
Pup 8h24'13"-26d39'
Toni
Cet 0h52'18"1d2'
Toni
Dra 15h1'39"62d18'
Toni & Patricia
Cyg 20h51'49"50d11'
Toni & Richard
Cen 11h42'15"-41d31'
Toni Ann
Lyn 7h8'0"59d31'
Toni Ann & Jerry Star, The
Ori 5h43'0"11d6'
Toni B
Lyr 19h1'13"31d37'
Toni Gail
Cyg 20h2'14"38d32'

Toni Jane
Eri 4h12'19"-9d15'

Toni Jayne
Cas 0h46'52"71d31'

Toni Kaye
Vul 19h41'52"27d0'A

Toni My Lover,Wife And Bright Star
Sge 19h57'0"18d55'

Toni T
Pic 5h1'47"-47d48'

Toni's Star
Umi 15h39'55"77d4'

Toni-Anne & Robert
Lyn 7h54'0"45d49'

Toniamax
Uma 10h44'41"55d7'

Tonica Chenelle H
Cas 23h0'34"58d39'

Tonidandel,Henry & Patricia
Uma 8h37'38"56d16'

Tonight Live
Mon 8h6'37"-1d31'

Tonini,Larry
Lac 22h30'0"53d31'

Tonja
Equ 21h21'39"11d3'

Tonjeunesse
Uma 9h46'44"71d59'

Tonks,Fil
Uma 11h39'15"30d28'

Tonks,Paul
Aur 6h2'57"45d45'

Tonks,Terry & Kerry
Per 4h8'53"38d46'

Tonn,Jens
Hya 9h14'57"-9d33'

Tonnac
Sgr 19h30'36"-39d53'

Tonnancour,Robert C
Ser 18h1'57"-13d49'

Tonner,Gail Ann
Lac 22h35'30"56d48'

Tonner,Marcia D
Uma 9h1'43"57d39'

Tonniges,Parker James
Per 2h21'34"54d59'

Tonnisen,Julie
And 23h56'35"41d24'

Tonnley,Richard Maxwell
Ori 5h55'16"19d55'

Tono
Leo 9h21'37"8d17'

Tonsar
Aur 4h54'58"51d19'

Tonsgard,Hal & Muriel
Umi 14h58'43"71d13'

Tonsing,Elias Hale
Cyg 21h27'45"28d59'

Tont
Dra 16h13'57"61d1'

Tonti,Christina Margaret
And 0h3'50"46d31'

Tonton
Lyn 7h52'53"58d17'

Tonton,Barbu et Zézé, Noël 1992
Per 3h52'55"37d52'

Tony
Oph 16h20'17"1d51'

Tony
Per 2h2'36"50d1'

Tony
Hya 9h4'0"5d27'

Tony
Boo 14h6'28"39d23'

Tony
Per 2h12'1"57d29'

Tony
Her 18h7'37"28d32'

Tony
Aur 6h15'27"37d19'

Tony (Giovane Per Sempre)
Sge 18h57'51"18d56'

Tony & Charline
Aql 19h31'29"11d8'

Tony & Lori-One Love, One Life Forever
Uma 9h47'36"61d4'

Tony & Missy
Uma 9h47'12"47d3'

Tony & Monica Forever
Cyg 19h58'1"50d21'

Tony & Shannon's Star
Cyg 21h31'48"42d51'

Tony & Viki's Island In The Sky
Lyr 19h4'0"28d50'

Tony (Daddy's Star)
Uma 13h28'37"61d9'

Tony et Denise
Lib 15h39'56"-23d14'

Tony Hedley's Star
Cnv 13h30'48"48d47'

Tony Janné
Tau 4h5'60"20d25'

Tony Lee
Peg 22h44'0"33d29'

Tony P
Aur 4h36'1"30d56'

Tony The Star Of My Life
Lyn 7h29'18"40d29'

Tony's Angel
Lyr 18h54'0"30d35'

Tony's Dream
Cnv 13h40'26"36d18'

Tony's MA
Lac 22h15'49"52d24'

Tony's Star
Aql 18h56'24"17d32'

Tony's Torch
Her 16h34'35"36d8'

Tony's Triumph
Aur 5h4'24"40d25'

Tony,Meredith & Joshua "Forever"
Peg 23h29'46"17d47'

Tony-my 'major' friend
Uma 8h49'12"56d34'

Tonya
Ori 5h39'34"10d4'

Tonya
Mon 6h56'39"10d38'

Tonya & Connie Best Friends
Peg 22h13'58"32d57'

Tonya Dee
Dra 10h0'11"74d33'

Tonya Forever
Sct 18h41'39"-7d39'

Tonya Michelle
Eri 4h57'17"-4d13'

Tonya's Dream
And 2h25'1"39d59'

Tonyan,Joey Lee
Aur 6h8'29"32d24'

Too Good To Be True
Dra 17h43'33"60d41'

Too Good To Be True- Marge & Tony
Cyg 21h7'18"31d11'

Too Jazzy Hens
Uma 15h24'31"70d44'

Too You From Me
Cam 12h12'24"78d11'

Toodie
Aql 19h30'21"10d44'

Toogood,Chris & Ginne
Eri 3h40'20"-6d3'

Tooher,Ryan Patrick
Her 16h59'35"37d8'

Tooke,Pamela
Uma 10h30'56"40d15'

Tooker,Ralph Paul
Aur 6h1'30"46d44'

Toole,Charles Broome
Dra 20h8'0"64d26'

Toole,Diane Lyn (Silver)
Hya 9h14'0"5d52'

Toolen,Andrea
Cas 0h13'1"60d53'

Tooley,Richard M
Aur 5h0'47"45d17'

Tooley,Ryan
Cmi 8h8'0"5d31'

Toolie
Tri 2h35'0"34d26'

Toomey,Charles
Aur 7h18'16"41d0'

Toomey,Timothy F
Ori 5h59'56"5d59'

Toon,Lynn Georgina
Cyg 19h57'18"38d49'

Toone,F LaVern & Mary M
Cam 5h3'47"61d50'

Toops,Michael Jay
Boo 14h15'11"38d14'

Tootell,Neil & Donna
Cyg 21h43'0"30d28'

Tootell-Edwards,Sarah Angela
Lyr 18h57'19"37d56'

Tooth,June & Graham
Cyg 20h51'58"38d0'

Toothaker,Ulrike Uli Monika Münch
Uma 9h54'0"57d29'

Toothe,Gary William
Per 2h57'38"38d55'

Toothill,Ruby
Cyg 20h1'15"37d51'

Tootie
Cet 1h10'38"-4d21'

Toots & Cider
Cma 6h56'34"-11d12'

Toots & Scott's 1st
Crb 15h52'25"38d30'

Tootsie
Leo 10h51'41"20d10'

Tootsielou
Aql 20h1'59"1d31'

Top Dog Obedience School
Cmi 7h58'43"0d40'

Top Gun Tour
Ant 10h46'11"-36d50'

Topa,Alexandra Leigh
Cam 5h1'19"61d48'

Topa,Andrea Lynn
Lyn 7h40'24"40d31'

Topalian,Richard Peter
Aql 19h20'16"13d25'

Topaz
Cam 6h37'1"80d27'

Topaz Aurora
Uma 11h12'28"62d16'

Topeka,Boo C
Ser 16h4'54"11d9'

Topeka,Kira Lane
Lyn 8h6'19"52d21'

Topf,Frieda
Lyn 6h54'58"44d50'

Topf,Manny
Cep 22h5'0"61d9'

Toph
Psc 0h46'60"28d34'

Topher
Ori 5h53'0"12d31'

Tophoff,Alfons
Cam 5h47'23"70d1'

Topi
Cam 3h48'30"57d30'

Topolos
Hya 8h49'39"-7d56'

Topolski,Teresa M
Peg 22h19'41"26d49'

Toporowski,Joey
Cnv 12h14'31"50d33'

Topp My Precious Tall Person,Dave
Uma 10h22'24"42d5'

Topp,Grant Darren
Sge 20h7'40"20d38'

Topp,Martin Andrew
Uma 13h0'17"53d59'

Topp,Selina
Cyg 21h53'35"53d37'

Topper,Kelsey L
Lmi 9h41'53"38d47'

Toppett,Howard J
Sct 18h49'50"-6d22'

Toppin,Alan Matthew
Cnv 12h39'32"33d53'

Topping,Cuthbert
Cyg 19h36'52"28d51'

Toppings Pride
Dra 20h11'0"71d13'

Topple Superstar,The Craig David
Per 2h3'38"50d14'

Tops Club,Inc
Mon 6h52'59"-5d49'

Topy,Joshua Andrew
Cam 5h59'51"61d43'

Tora No Yumi
Peg 0h7'12"20d3'

Toral My Love,Babe, Barbara Louise
Gem 6h38'50"12d51'

Toran,Nicky
Cnv 12h45'0"38d41'

Torbet,Fiona
Cas 1h16'52"60d44'

Torbiner,Maxwell Phillip
Her 16h43'0"29d41'

Torch,Paula Antonella
And 0h52'1"35d48'

Torch,Robin & Nate
Lyr 18h59'26"47d16'

Torcivia,Catherine & Paul
Lmi 9h24'39"38d20'

Torcivia,Special Friend Chet
Tau 5h52'57"23d20'

Tordis Aune Austad
Aql 20h1'59"1d31'

Tordjman,William & Stacey
Cyg 21h7'45"31d44'

Tore
Umi 14h37'41"67d45'

Toreador,Bobbie
Aur 6h31'1"37d26'

Torell,Mark James
Mon 8h4'34"-3d55'

Torello,Jacqueline Marie
Lyr 19h15'35"26d54'

Torgersen,Blythe Kirsten
Umi 16h29'33"70d52'

Torgersen,Donald
Per 2h22'55"58d36'

Torgerson,Jane Marie
Cas 3h8'45"68d7'

Tori
Mon 8h3'19"-6d54'

Tori
Peg 22h16'10"33d57'

Tori
Del 20h49'11"9d37'

Tori
Umi 13h15'12"75d18'

Tori & Eric
Cyg 19h26'15"33d5'

Tori Don
Dra 19h27'51"61d37'

Tori's Diamond
Sct 18h38'59"-6d24'

Torick II,Stephen Paul
Peg 23h5'17"18d47'

Toriello,Adam Paul
Tau 4h43'43"16d46'

Torin & Kristin A Touch of Heaven
Cyg 20h15'12"38d32'

Toring,Sheila
Cas 0h10'33"62d38'

Torino-A Heavenly Body,Auralie M
Uma 8h32'29"52d46'

Torka,Andrzej
Peg 21h59'20"31d52'

Torkelson,Dean
Dra 10h48'42"77d52'

Torlone,Kelly Noel
And 0h24'15"44d16'

Torlone,Lauren Joanne
Cas 0h13'16"47d42'

Tormey,Diane
Aur 7h7'44"40d54'

Tormey,James Michael
Boo 14h53'14"25d35'

Tormey,Justin
Cmi 7h2'28"-18d58'

Tornberg,Larry
Oph 17h55'26"-6d42'

Tornberg,Rita
Tri 1h56'1"28d57'

Torneby,Samuel C
Her 18h11'56"30d57'

Tornek,Lawrence David
Aqr 21h52'47"-5d43'

Tornensis,Allison Lynnette
Uma 11h48'17"38d8'

Tornes,Maria J "Cuqui"
Mon 6h52'31"-1d24'

Tornillo,Mark & Rosita
Cam 3h13'18"60d43'

Toro,"Goddess" Evelyn Nelson
Ori 5h56'58"17d18'

Toroni,Jr,Albert E
Aur 5h41'0"54d31'

Torony,Tim
Her 16h35'15"50d41'

Torosian,Sheri Lynne
Cam 3h49'0"60d45'

Torosian,Sussan R
Mon 7h19'47"-8d19'

Torp,Cynthia K
Cet 3h4'22"0d44'

Torpedo Nick
Her 18h19'49"12d29'

Torpey, Linda & John
Aql 19h17'27"15d14'

Torpey,Paul J
Lac 22h28'36"55d14'

Torpey,Rev Michael J
Boo 14h51'53"52d50'

Torpey,Robin Shaer
Cas 0h30'54"60d24'

Torpinus
Ser 15h40'1"8d54'

Torralva,Ben
Psc 0h44'36"20d53'

Torrance,Ewan James
Cep 22h30'57"70d42'

Torrance,Lajauna
Cas 0h29'13"75d4'

Torre,Brian
Ori 5h54'12"12d47'

Torre,Gertrude Mary
Mon 6h29'57"-10d49'

Torre,Lou & Rita
Cyg 20h58'12"53d25'

Torre,Rocco Cono
Sex 10h22'1"5d16'

Torre,Sgt Scott
Aur 7h23'18"38d15'

Torrealday,Saioa
Oph 18h35'59"11d1'

Torredemert,Claudio
Ori 5h33'46"-2d4'

Torrence,Reverend, Steven M
Oph 17h38'7"-21d1'

Torres
Her 17h19'0"28d55'

Torres 40th Birthday, Miguel
Aql 19h51'14"13d32'

Torres,Alexandra Justine
And 2h31'22"40d38'

Torres,Cesario R
Mon 7h55'2"-1d40'

Torres,Craig A
Her 17h18'48"20d26'

Torres,Dennis & Averi
Crt 11h23'4"-18d31'

Torres,Emily
Umi 14h35'25"81d32'

Torres,George Charles
Cma 6h30'19"-18d15'

Torres,Iveliz
Lyn 7h22'0"58d47'

Torres,Jacob Anthony
Her 17h27'49"28d26'

Torres,Jay Christopher
Cnv 13h6'33"33d30'

Torres,Jerry
Lyn 07h40'18"50d24'

Torres,Jordi y Marisa Güimil
Psc 1h0'59"21d38'

Torres,Jose Antonio Martinez
Her 16h9'37"10d40'

Torres,Jose E
Aur 6h24'50"52d35'

Torres,Jose Paul
Lac 22h3'1"48d48'

Torres,Kimberly Jennifer
And 23h39'26"40d60'

Torres,Lee
Mon 6h26'0"1d43'

Torres,Linda B
Cet 2h47'31"0d21'

Torres,Livia Ann
Lyr 18h15'42"30d51'

Torres,Mayra
Lmi 10h1'39"39d50'

Torres,MD,DC,Gary S
Oph 18h4'43"7d51'

Torres,Michael
Ori 6h4'58"10d14'

Torres,Miguel
Sct 18h56'26"-6d36'

Torres,Prisciliano (Pete)
Per 3h57'47"34d54'

Torres,Reinaldo E
Cmi 7h10'21"7d57'

Torres,Robert M
Ser 15h21'31"2d35'

Torres,Sabine & Eduardo
Aql 19h27'0"10d52'

Torres,Traci Lee
Cas 0h54'13"61d55'

Torres-"Shortstuff", Maria Teresa
Peg 22h2'46"21d6'

Torres-Yap,Fortunato
Boo 15h19'18"50d6'

Torreson,Carol Louise
Del 21h5'0"12d30'

Torretti,Dawn Marie
Cas 1h22'45"53d20'

Torretti,Kim Rose
Cam 3h50'0"57d18'

Torretti,Nicholas Anthony
Boo 14h52'51"48d47'

Torrey S
Mon 8h6'0"-0d34'

Torrez,Carlos Adrian
Dra 17h25'34"64d44'

Torrez,Elisabeth Ann
Aqr 22h37'54"-6d4'

Torri Lynn
Aql 19h25'42"15d29'

Torrico,Alvaro
Boo 15h0'18"22d27'

Torrieri,Loretta Veronica
Lyr 18h43'0"31d50'

Torrisi,Amanda
Lyn 7h50'20"45d36'

Torro y Duby
Cet 1h28'12"-1d23'

Torrone,Daniel Joseph "Doctor D"
Oph 17h33'4"-23d32'

Torsani,Roberto
Cep 1h42'1"78d12'

Torsti
Cam 6h0'29"65d33'

Tortissier,Gilles
Dra 17h48'58"71d1'

Tortoise,Robbie
Uma 8h46'50"60d58'

Tortora,Kristi-Jo & John
Lib 15h44'25"-28d24'

Tortora,Mia
And 0h41'52"37d59'

Tortora,Noel
Lac 22h39'1"53d4'

Tortora,Pasquale
Dra 16h48'52"61d52'

Tortora,Ralph
Cam 6h6'1"60d46'

Tortorella,Lauren
Uma 8h51'14"49d3'

Torupkins,Jon & Danielle
Cyg 20h58'45"30d55'

Torvend,Alice Kjesbu
Peg 23h32'59"17d59'

Torviaggi
Hor 2h46'18"-49d52'

Tory
Aur 4h54'18"41d1'

Toryk,Trevor Nathaniel
Vul 19h0'27"24d59'

Torykian,Eric Aram
Cep 2h10'52"78d13'

Torykian,James Andrew
Cep 2h12'0"78d57'

Torykian,Jr,Richard Paul
Cep 2h7'43"78d17'

Torykian,Mary Lou Elizabeth
Cep 2h2'30"78d30'

Torykian,Sr,Richard Paul
Cep 3h10'18"77d19'

Tosca
Cma 7h12'42"-13d33'

Toscano,John M
Per 2h58'33"35d25'

Toscano,Jr,John Frank
Aur 7h24'54"35d33'

Toscano,Olivia
And 2h30'0"38d33'

Toscano,Olivia Rae
Ari 1h54'38"15d53'

Tosches,Michael Paul
Her 16h20'31"25d14'

Tosha
Cnv 13h51'60"38d44'

Tosha
Cyg 20h22'40"38d53'

Toshiaki & Megumi
Ari 2h15'45"21d38'

Toshifumi & Ayako
Ari 2h18'27"21d21'

Toshiki 1994-10-30, Shirane
Sco 16h22'15"-29d43'

Tosia
Uma 9h50'56"58d14'

Tosini,Luca
Pup 8h8'25"-21d42'

Toso,Steven Paul
Dra 16h38'33"63d43'

Tosoni,Zeno
Dra 16h28'47"58d33'

Toss,Ramon
Cap 20h36'52"-26d30'

Tossy Noosebond
Peg 22h59'46"11d18'

Tostanoski,Eugene
Boo 14h38'13"46d39'

Tostaraed
Ori 5h32'38"-5d60'

Tosto,Marilena e Michele Losito
Lep 5h22'0"-25d43'

Toszegi,Cindy
Mon 8h0'0"-0d15'

Total Bliss
Lyn 8h4'1"58d39'

Totaro,Rose & Al
Cyg 19h58'16"30d2'

Totdeauna Randy
Cep 22h20'13"70d53'

Totdeauna Rick & Gayl
Uma 10h10'0"48d30'

Toth II,John E
Aql 19h6'37"3d46'

Toth,1 Christopher Michael
Aql 18h59'44"-1d55'

Toth,Carey Marie
Cyg 21h11'25"38d40'

Toth,Chris
Boo 15h0'46"53d13'

Toth,Jan
And 0h14'23"37d32'

Toth,Ken
Aur 6h5'54"31d2'

Toth,Lorili
Cyg 19h45'18"30d6'

Toth,Paul
Hya 8h44'0"5d27'

Toth,William Justin
Leo 11h55'48"23d14'

Tothpal,Peter
Sex 10h2'1"1d28'

Toti,Bettie
Lyn 7h50'18"50d18'

Totin,Michael Richard
Per 2h49'32"32d3'

Toto
Her 16h33'18"47d32'

Toto
Cnv 14h2'47"31d51'

Toto,Jonathan Stephen
Lac 21h55'39"40d52'

Toto,Jordan Marie
Cam 3h26'23"53d28'

Totten,Kaity Leigh
Tau 4h33'31"30d6'

Totten,Rich A
Aql 19h26'23"14d29'

Totten,William L
Aql 19h19'28"12d0'

Toub,Jacqueline Michelle
Cet 3h19'58"1d22'

Touber,Liesbeth
Cas 1h40'43"67d45'

Touboul,Maurice
Per 3h57'6"39d13'

Touboul,Régine
Cnv 13h22'27"38d28'

Touch of Class Catalog
Peg 22h50'41"8d15'

Touchatt,Sandy
Peg 22h42'0"32d15'

Touchston,"Dr T"Dr W Joseph
Per 2h55'23"35d49'

Tougas,Marie-Soleil
Uma 10h10'31"47d41'

Toujours
Aql 19h51'27"11d15'

Toujours
Peg 23h2'47"18d27'

Toujours Amies, Line et 2
Cnv 13h18'60"40d40'

Toujours,Sean Et Kathy
Oph 18h0'53"8d20'

Toulouse
Oph 17h12'31"-18d46'

Touma,Jamil C
Cma 6h31'15"-16d35'

Touponse,Marcel Jule
Del 20h51'1"9d54'

Tourangeau,Jacques
Ori 4h50'14"5d28'

Tourdjman,Jean Michel
Lyn 7h58'11"47d46'

Tourneur,Olivia
Lmi 10h56'1"28d11'

Tournier,Helene
Dra 17h4'11"60d4'

Tournier,Mathilde
Lac 22h39'57"40d14'

Tournié,Justine
Crt 11h19'17"-18d57'

Tourso,Michael
Aur 5h1'19"41d6'

Tousignant,Betty Anne
Lyn 8h9'49"43d3'

Tousignant,Tammy
Uma 11h53'58"40d59'

Toussaint the 3rd, Anthony Cyrus
Per 2h28'54"56d39'

Toussieng,Robin Ann
Peg 21h42'0"23d38'

Tout,Donald Patrick
Her 18h0'51"37d31'

Touval,Robin
Aql 19h59'42"0d11'

Touw,Jason
Ori 6h2'50"3d10'

Touw,Sascha
Cas 0h48'1"71d3'

Tovah's Nova,Alix
Lyn 7h43'43"43d56'

Tovar,David J
Lac 22h22'40"53d3'

Tovar,J
Mon 6h39'58"10d6'

Tovar,Paul
Her 18h5'43"28d24'

Tovee,Paul
Lyn 9h9'19"37d39'

Tovello,Pamela H
Ori 6h3'1"6d35'

Toveri
Cma 7h18'0"-15d38'

Toves,Bertha Aurea
Aql 19h3'14"-0d29'

Towa
Psc 1h23'27"4d11'

Towar,Benjamin Charles
Aql 19h1'53"1d14'

Towar,Melissa Corine
Eri 3h6'38"-1d39'

Toward The One
Cyn 7h57'14"40d17'

Towe,Arthur
Per 3h26'58"52d6'

Towe,Carol
Crb 16h3'57"28d38'

Tower
Cas 2h15'55"65d24'

Tower,Hunter NBL
Per 1h49'44"53d4'

Towers,Christopher Carl
Boo 14h5'54"10d2'

Towers,Douglas
Per 2h50'45"38d26'

Towers,Elizabeth A
Mon 6h43'25"8d49'

Towers,Johnathan Edward
Dra 14h53'59"59d13'

Towers,Robert
Uma 12h4'22"42d57'

Toweson,Kaitlyn Ann
Mon 7h14'14"-1d32'A

Towey,Margaret
And 0h14'60"36d34'

Towle,Barnaby Ledyard
And 0h51'22"39d47'

Towle,Brodie Patrick
Aur 6h22'0"33d5'

Towle,Muriel Ayres
And 0h53'50"40d32'

Towler,Dorothy Joan (Koscielniax)
Aur 5h57'53"54d40'

Towler,Sr,Robert Joseph
Boo 13h46'25"21d20'

Towles,Kathleen Mary
Cas 0h28'25"58d14'

Town,Jeremy Michael Allen
Aur 6h19'19"38d51'

Towne,Ellen Allecea
Mon 6h50'23"10d10'

Towne,Iona
Tri 1h29'47"30d28'

Towner,Christopher James
Cep 21h36'53"78d53'

Townes,Anthony H
Uma 9h2'59"49d44'

Towning,Patricia
Lyr 18h14'33"34d21'
Towns,James Walter
Cep 21h25'27"70d40'
Townsell,Tab
Aql 20h15'32"0d30'
Townsend 9:18AM, Madyson Kaleigh
Equ 21h17'37"3d37'
Townsend Family, Angelica
Cet 2h4'0"3d35'
Townsend,Captain Michael J
Tau 4h11'48"22d32'
Townsend,Charles
Per 2h29'24"58d17'
Townsend,Don & Brenda
Peg 23h45'15"31d5'
Townsend,Floyd "Bud"
Aur 6h19'28"37d40'
Townsend,Irene
Cas 0h25'0"66d8'
Townsend,J David
Equ 21h22'28"10d21'
Townsend,James Arthur
Cep 21h35'12"55d25'
Townsend,Jennifer
And 0h25'16"44d20'
Townsend,Jr,Happy 1st Terry R
Peg 21h55'43"20d35'
Townsend,Jr,Michael G
Aur 6h11'11"31d5'
Townsend,Michael L
Cep 22h37'19"68d10'
Townsend,Nicola Emma Camilla
Uma 11h54'30"40d44'
Townsend,Peter E
Dra 17h0'41"71d12'
Townsend,Phoebe
Cas 1h20'40"50d6'
Townsend,Princess Kaiulani Judith
Oph 17h6'21"11d20'
Townsend,Tyler Toby
Aql 19h30'49"10d45'
Townsend,William
Hya 8h14'54"-0d1'
Townshed,Kevin A
Boo 14h44'15"27d27'
Townshend,Camille Lamarre
Per 2h51'54"32d29'
Townshend,Raphaël John Lamarre
Ori 6h2'23"7d36'
Townson,Edith Alice
Aql 18h58'38"17d59'
Townson,Howard
Aur 5h45'10"50d6'
TOY
Oph 18h16'0"11d16'
Toy Boy Star,The
Aql 19h13'38"13d49'
Toyoda,Dr Shoichiro
Uma 11h58'20"58d36'
Toyoda,Eiji
Uma 14h8'1"56d4'
Toyoda-Boshi Y H Y T
Tau 5h11'8"21d9'
Toys
Mon 6h20'33"7d35'
Tozer,MD,Randall K
Oph 17h3'57"11d19'
Tozer-Anniversary Star Phil & Bette
Mon 8h2'34"-4d46'
Tozzi,Giulia
Hor 3h19'45"-46d35'
Tozzoli,Guy F
And 2h16'51"48d23'
Tozzoli,Shirley
Vul 20h5'0"25d23'
TPF III
Cam 9h24'10"81d3'
TPO III
Cnv 13h11'21"48d48'
Traband,Georges
Sex 10h14'59"-8d14'
Trabattoni,Leo Robert
Her 17h10'59"43d13'
Trabeau,Brooke Lynn
Lyr 18h40'59"40d56'
Trabucco,David Joseph
Her 17h54'49"20d2'
Trabucco,Kristin Beverly
Lyr 18h43'0"47d2'
Trace,Laura
Cas 3h10'45"65d33'
Tracer,Ray
Dra 16h10'24"64d21'
Tracey
Vul 21h13'53"20d14'
Tracey
And 2h23'1"47d52'
Tracey
Cyg 19h31'33"36d3'
Tracey & Gerard
Cyg 21h6'41"30d43'
Tracey & Ian
Cyg 20h53'18"37d37'
Tracey + Mark TLA Star
Cet 2h17'20"7d15'
Tracey Ann
Cyg 20h3'58"41d6'
Tracey H
Cas 0h59'33"59d55'
Tracey I,Admiral Patricia
Cas 0h50'3"60d51'A
Tracey II,Admiral Patricia
Cas 0h50'3"60d51'B
Tracey III,Admiral Patricia
Cas 0h50'3"60d51'C
Tracey Marie
And 0h29'60"43d58'
Tracey R
Cyg 21h26'28"31d30'
Tracey's Place
Dra 20h22'26"62d24'
Trach,Juliana Elizabeth
Cas 1h43'16"71d53'
Trachta,Jeff
Dra 16h1'22"63d34'
Trachtenberg,Keri
Lyn 8h48'13"37d40'
Trachtenberg,Stephen Joel
Cep 22h23'45"59d59'
Trachtman,William M
Boo 15h14'49"31d32'
Traci
Aql 18h44'58"10d56'
Traci & Andre
Cam 6h58'52"65d24'
Traci's "Nightlight"
Aur 5h55'30"40d43'
Traci-Craig
Psc 1h3'13"20d6'
Tracie & Rob
Aql 19h3'10"15d31'
Tracie Lynn
And 1h49'0"38d41'
Tracie Rae
Mon 8h5'58"-1d54'
Tract,Laurence Todd
Cep 22h55'59"59d47'
Tractenberg,Stan
Her 18h11'13"40d1'
Tracton,Forever Red & Carolyn
Peg 22h53'0"27d23'
Tracy
Mon 6h43'51"7d18'
Tracy
Oph 18h23'41"7d40'
Tracy
Aql 20h11'12"13d23'
Tracy
Mon 6h55'0"-1d2'
Tracy
Hya 8h46'5"1d7'B
Tracy & George
Ori 4h51'33"7d17'AB
Tracy & Rob
Eri 4h43'21"-1d4'
Tracy 143
And 1h29'0"40d11'
Tracy Angelina
Com 12h3'0"24d56'
Tracy Ann
Cas 0h52'54"74d45'
Tracy Anne
Cas 1h56'15"75d60'
Tracy Arlyn
Peg 21h57'33"36d10'
Tracy Christine
Cyg 19h27'0"40d51'
Tracy K
And 2h25'31"44d37'
Tracy Lyn's Star
Cyg 21h32'40"41d3'
Tracy Lynn
Lyr 19h23'47"38d53'
Tracy Lynn's Love
Vul 20h19'22"23d16'
Tracy Margaret Snugglebunny
Cas 22h58'0"55d8'
Tracy Marie
And 23h4'1"37d56'
Tracy Marie
Uma 11h7'59"47d38'
Tracy&Eric Eternal 3-4
Teahoney Star
Sge 19h43'35"16d48'
Tracy's Comet
Uma 8h41'16"53d18'
Tracy's Heavenly Body
And 2h32'32"38d58'
Tracy's Star
Gem 7h22'34"34d12'
Tracy's Starlight Delight
Oph 18h39'1"10d53'
Tracy's Treasure
Ori 5h51'21"16d31'
Tracy,Arthur James
Aql 18h55'25"11d18'
Tracy,Bart G
Dra 11h34'20"72d17'
Tracy,Bryan Patrick
Oph 18h32'25"10d14'
Tracy,Dick
Aur 6h16'54"30d21'
Tracy,Forever My Angel
Eyes,Love TK
Hya 8h31'35"-6d31'
Tracy,I Love You! Christian
Cyg 21h22'31"53d42'
Tracy,Kelly John
Cnv 13h1'38"32d9'
Tracy,Lisa Mary
Lmi 10h8'16"38d32'
Tracy,Max Michael
Aur 7h1'46"43d20'
Tracy,Michelle Ann
Mon 6h33'59"-0d14'
Tracy,My Love Forever, Mike
Eri 2h45'12"-17d20'
Tracy,Roger Daniel
Aur 6h19'0"32d19'
Tracy-Marie
Peg 22h29'16"10d8'
Tracy/Kristin
Hya 8h12'52"0d56'
Tracys Stink Star
And 23h35'45"49d49'
Trader Horne
Lac 22h45'24"56d22'
Trader,George & Nell
Eri 4h56'20"-11d48'
Tradewind Airlines
Dra 16h3'17"64d2'
Trae
Cam 4h21'12"71d6'
Traecy,Lorcan Edward Michael
Cet 1h59'50"-8d35'
Traeger,Jörg
Aqr 22h49'57"-4d54'
Traficante,Thomas Michael
Per 2h56'1"43d52'
Traficanti,JoAnn,Susan Robert
Lmi 9h58'39"31d16'
Trafton,Joey & Kiyah Duffey
Dra 19h20'27"58d30'
Trafton,Lawrence
Ori 5h47'23"11d58'
Trageser,Jr,Timothy "Tink"
Equ 20h57'34"3d27'
Tragos,Alexa Marie
And 0h5'10"38d35'
Trahan,Miia
Aql 19h57'12"14d48'
Trahanovsky
Umi 15h19'24"69d1'
Trahas
Umi 15h55'41"83d15'
Traianou Rockstar, Helen
Psc 0h9'27"7d57'
Trail Blazer
Cyg 21h18'21"37d32'
Trail,Mervin Lee
Sgr 18h51'28"-23d2'
Traille,Justin John
Ori 5h54'40"17d40'
Train,Marybeth Margaret
Vul 19h59'36"28d13'
Traina,Edward & Noreen
Peg 23h36'0"31d39'
Trainman,Jodie & George & The Butterfly
Eri 2h50'26"-1d40'
Trainor,Katherine
Boo 15h13'2"38d29'B
Trainor,Kelli Ann
And 0h49'41"39d45'
Trainor,Norma
Cas 0h25'1"58d32'
Traister
Mon 7h41'29"-2d45'
Traister,Elena
Vul 19h49'0"20d34'
Traitel's Tribute,Dr Richard B
Tau 5h49'53"23d28'
Traitz,Barbara A
And 2h10'27"42d15'
Traji
Aql 19h19'45"11d53'
Trakalo,Kim
Peg 22h36'41"21d45'
Tralongo,Cayleigh Riaghna
And 2h17'47"46d36'
Tramacera,Michael D
Lac 22h20'17"40d23'
Trammell,Tanaka
Peg 22h44'1"29d34'
Tramontana
Crb 16h10'42"33d2'
Tramontano,Robert
Dra 18h29'25"50d19'
Tramp Loves Pldge
Cyg 19h47'43"50d5'
Tramuto,Anthony
Lac 22h27'17"53d12'
Tran Quoc Si
Vul 19h42'52"20d11'
Tran,Antony Andrew
Del 20h13'42"11d45'
Tran,Brynn Anh
Aql 19h41'34"11d46'
Tran,Jordan
Aql 19h5'1"3d10'
Tran,Tinh Minh
Mon 6h54'57"-8d59'
Tranchina Star,The
Cam 7h46'1"61d22'
Tranchina,Joseph
Her 18h17'28"18d47'
Trang Vu Dinh
Cet 1h31'27"-10d37'
Tranik,Betty Ann
Cyg 21h47'37"38d33'
Tranquilla,Tony & Clara
Sge 19h58'38"16d52'
Transeth,Jr,Norman S
Hya 9h26'29"-1d2'A
Transki,Donald E
Dra 19h6'1"48d16'
Transportation Unlimited
Cnv 12h47'1"39d37'
Transseth,Kathleen Cheng
Hya 9d26'29"-1d2'B
Tranter,Alexander
Umi 17h22'17"75d0'
Tranter,Jane (Gromit)
Uma 9h44'19"48d59'
Trantham,Larry D
Her 16h36'27"29d27'
Trantham,Sherry Lynne
Cet 0h55'12"-5d58"
Trantolo,Joseph J
Lac 22h20'52"53d45'
Trantolo,Jr,Joseph J
Per 2h55'52"41d8'
Trantolo,Mark D
Aql 20h11'14"0d17'
Trapane,Vincent Joseph
Cam 14h11'44"81d23'
Trapani,Rose
Aql 19h5'18"15d1'
Trapasso,Eileen T
Cyg 21h34'45"44d32'
Traphagan,Johnny
Ser 15h12'46"8d48'
Traphagen Water From the Moon
Dra 18h14'26"68d38'
Trapp L T 1935, H-Dietrich
Psc 0h46'42"20d56'
Trapp,Paul & Dotty
Cyg 21h28'0"28d51'
Trapp,Richard Edward
Boo 13h40'1"21d24'
Traseus
Cas 23h38'0"50d36'
Trash
Eri 4h42'28"-6d2'
Trasko,Charlotte I
Cnv 13h55'37"41d37'
Traskus,Thomas
Sct 18h21'22"-13d35'
Trasvina,Diane
Mon 8h2'45"-8d11'
Tratacus,Dominicus
Per 3h52'50"39d38'
Tratar,Larry A
Aql 19h9'36"10d57'
Tratar,Stephen John
Aur 5h21'21"38d54'
Tratner,Marilyn Patricia
Peg 0h0'36"13d40'
Trattles,Jane
Lyr 18h15'21"44d0'
Traub,Gordon
Oph 17h5'24"11d50'
Traub,Julie & Glenn Capra
Aql 19h56'41"15d14'
Traub,Manya Roxane
Cap 21h27'0"-24d34'
Traub,Tomi Paige
And 2h30'47"42d39'
Trauger,A Star Still Shining,Josie
Cas 0h54'56"72d51'
Traugott,Annie & Tommy
Aql 20h9'15"8d30'
Traugott,Lise P
Lyn 7h58'27"35d20'
Traum,Dennis
Boo 13h41'0"17d59'
Trauner,Alison Rose
Cma 6h53'20"-18d28'
Traurig,Harold
Aql 19h52'20"11d42'
Trauscht,Donald C
Per 3h58'1"48d54'
Traut,Frederick Grandon
Uma 9h53'28"68d59'
Traut,Glenna
Ori 6h13'38"8d35'
Traut,Virginia Warner
Uma 9h41'0"67d33'
Trauth,Zoë Lorenz
Ori 6h2'26"20d22'
Trautlein,Ramona
Peg 21h59'0"34d1'
Trautman,Gena Anne
Eri 2h52'31"-1d34'
Trautmann,Dieter
Aqr 22h47'38"2d2'
Trautmann,Tizian Gabriel
Boo 13h58'0"13d20'
Trautvetter,Rod
Her 17h1'23"21d26'
Travaglini,Domenic & Nancy
Cap 21h28'0"-23d35'
Travaglino,Enza
Cmi 7h16'38"4d36'
Travalino (The Trav Star),The
Dra 19h44'51"67d34'
Traven,John
Cep 21h47'50"55d7'
Traver,Janna Lee
Vir 14h5'1"-2d9'
Traver,Karen Jane
Psc 0h47'32"31d8'
Travers,Joe
Ori 4h51'49"5d4'
Travers,Robert
Cyg 21h0'0"31d46'
Travers,Sidonie
Cas 2h0'18"68d21'
Traverso,David L
Cep 21h7'32"61d40'
Traverso,Pete
Per 2h12'1"57d33'
Traversone,Paola
Cnv 13h5'22"33d40'
Travis
Cma 7h1'39"-19d58'
Travis
Uma 8h35'22"70d3'
Travis III,Daniel Franklin
Cam 4h58'1"65d24'
Travis,Cpt Kirk
Oph 18h17'0"11d23'
Travis,Cynthia Brooks
Lyn 7h26'22"40d9'
Travis,Douglas Jackson
Crt 11h15'7"-11d7'
Travis,Geri Kelton
Lyr 18h56'41"30d51'
Travis,Kelly Anne
Eri 4h2'0"-17d1'
Travis,Nancy
Mon 7h43'58"-8d25'
Travis,Robert James
Cap 10h23'19"-23d54'
Travis,Stuart
Cep 22h0'1"61d13'
Travous,Odell H
Aur 5h5'41"50d49'
Trawczynski,David
Her 17h4'59"37d36'
Traweek,Joan Elizabeth
Ori 5h59'0"17d46'
Trawick Legal Star, Leon A
Lac 22h10'32"46d24'
Traxler,John A
Her 16h59'32"40d48'
Tray
And 23h47'27"47d3'
Traybar
Lyr 19h1'45"28d26'
Traylor,Ruth Elizabeth
Cas 3h9'20"58d13'
Traylor,Sandra Lee Hopkins
Cap 21h54'0"-14d26'
Traylor,Steven E
Aur 6h48'0"35d28'
Traynham,Linda
Cyg 19h27'55"32d50'
Traywick,Karen Wyndham
Cet 3h16'29"7d54'
Trazzera,Michael A
Her 16h42'39"32d30'
TRC
Umi 15h36'25"69d7'
TRC-EEC,S Wish Come True
Cam 6h8'39"60d16'
Treacle
Per 3h6'13"41d51'
Treacle's Star
Eri 5h3'47"-4d17'
Treadway The Great Great
Dra 19h58'11"65d31'
Treadwell,Sr,James L
Per 3h0'47"40d23'
Treanor,John
Boo 15h0'56"12d7'
Treasured Thomas
Oph 18h0'36"12d27'
Treasurer,Bill
Aql 18h42'23"1d10'
Treat,Charles Joseph
Aql 19h6'0"15d37'
Trebbau,Helga
Cap 21h39'12"-23d57'
Trebino,Debbie
Psc 1h23'56"20d41'
Trebor
Psc 0h48'36"27d56'
Trebor's Wish Upon His Star
Sex 9h44'11"-2d1'
Trebor,Sivad Navi
Hya 8h12'0"2d28'
Treccia e Umbi
Uma 12h1'18"47d22'
Tredegar Star,The
Umi 14h20'50"68d2'
Tredwell,Connie
Peg 23h41'58"31d4'
Treece,Scott A
Ser 15h21'14"-2d22'
Treesh,Courtney Renee
Mon 7h51'3"-1d4'
Trefiletti,Samuel Paul
Boo 13h54'49"18d24'
Trefry,Sean
Dra 17h19'1"60d55'
Trefz,Margaret Mary Ponzini
Eri 3h59'41"-13d1'
Tregellas,Marjorie C
Sgr 20h0'46"-26d58'
Tregellas,S Staley
Sgr 20h1'56"-26d46'
Tregidgo,Sandra
And 0h22'47"45d24'
Treglia,Antonio Mike
Boo 13h42'19"13d0'
Treharne,John
Dra 12h10'18"72d27'
Trehubenko,Eric
Aqr 21h29'57"-6d44'
Treichel,Rudolf
Peg 23h29'46"12d21'
Treiman,Oscar Bard
Cep 20h25'53"62d21'
Trejo "Minja",Krystal Lee
Cyg 20h7'29"38d16'
Trellys
Cet 0h52'18"1d1'
Trem,Candy
Uma 11h48'19"57d6'
Trem-T
Cam 3h32'13"62d58'
Trembath Major
Cmi 8h8'13"6d22'
Tremblay 70,Eric
Her 17h4'42"22d23'
Tremblay,Adam
Del 20h14'0"11d55'
Tremblay,Benjamin Bruelle
Umi 15h19'42"68d10'
Tremblay,Danielle et Carol
Cas 0h48'17"73d26'
Tremblay,Gaétan
Her 16h52'44"33d15'
Tremblay,Ginette
Peg 22h8'22"2d23'
Tremblay,Jocelyne
Cyg 19h24'1"30d29'
Tremblay,Luke Thomas
Cet 0h39'1"0d44'
Tremblay,Normand J
Per 2h59'22"32d54'
Tremblay,Richard S
Her 18h13'36"38d29'
Tremblay,Samuel
Del 20h13'46"14d48'
Tremblay,Sasha Cøté
Del 20h14'59"15d13'
Tremblay,Thomas Bruelle
Umi 15h19'22"68d34'
Tremine-Hamon,Liliane
Oph 18h17'10"10d20'
Treml,Paul R
Aur 6h10'59"33d25'
Tremlin,Bonnie
Cas 0h24'24"66d11'
Tremoglie,MD PhD,David E
Aur 5h0'54"49d42'
Tremolet
Cam 7h29'16"80d21'
Tremp,Nelly & Martin
Boo 14h40'16"47d56'
Trenary
Uma 10h34'47"58d50'
Trendall,Catherine
Uma 9h29'33"47d48'
Trenery,Tamara Lee
Vir 11h45'60"7d58'
Trenk,Gregory James
Cet 2h56'1"6d54'
Trenkle,Wei & John
Aql 18h44'41"10d4'
Trenkler,Jennifer
And 0h37'52"40d15'
Trenn,Bob
Her 16h53'21"38d27'
Trent
Hya 8h43'46"6d31'
Trent Monster's Light
Dra 19h4'41"54d56'
Trent,Jr,John
Per 2h24'1"55d56'
Trenta
Per 2h51'52"50d4'
Trento,Andrew Fred
Cnc 8h29'56"8d5'
Trento,Eric Marc
Sgr 18h49'3"-20d26'
Trento,Matthew Stephen
Sco 17h54'29"-31d46'
Trentowski,Stanley Joseph
Her 17h36'37"38d17'
Trepton,Jessica Shirley
Cas 2h33'20"57d55'
Treptow,Earl J
Boo 14h37'1"20d26'
Tresa's Tiara (From Wiley)
Crt 11h2'11"-18d58'
Tresaloni,John P
Hya 8h15'19"0d36'
Tresanini,Gregory
Equ 21h3'32"8d36'
Trescott,Leon L S
Tau 5h55'0"28d30'
Tresemer,Frank
Cap 21h3'1"-20d14'
Tresenriter,Debbie G
Cyg 20h51'57"50d31'
Treshansky,Stephen
Hya 8h46'10"-7d35'
Treshock,Melissa
And 1h59'40"37d28'
Tresie's Light
Tri 2h7'1"31d6'
Treskow,Peter
Ori 5h26'42"1d42'
Tresnak,Ashley
And 2h17'43"41d27'
Tresnak,Katelyn
Cyg 21h2'43"38d35'
Tresnak,Ty
Tri 2h20'27"34d25'
Trespassers William
Cam 7h7'33"61d5'
Tress,Beanana
And 0h7'17"46d49'
Tress,William E
Per 2h56'18"31d53'
Tressier,André
Lyn 8h0'17"51d16'
Tressilian,Connie
Cyg 20h21'28"38d34'
Tresslar,Nini
Cet 3h13'21"2d35'
Tressler,Judd D & Angela J
Cyg 20h23'12"39d18'
Trester,Horst
Boo 14h20'41"14d4'
Trettacconi,Miria
And 0h45'22"45d7'
TRETTANY
Lyr 18h34'21"37d54'
Trettau,Ronald Ernest
Cep 22h23'49"58d34'
Tretter,Bertram
Vul 20h44'41"20d32'
Tretter,Stephen
Aur 4h51'47"41d8'
Treubig,The Future Mr & Mrs Ernest
Uma 11h7'39"46d49'
Treude,Brett Steven
Her 17h14'35"20d37'
Treude,Conner Joseph
Her 17h55'59"20d6'
Treude,Pam & Vaughn
Peg 22h19'47"34d13'
Treusch,Brian
Hya 9h15'39"3d30'
Treusch,Kristine
Dra 17h24'0"60d42'
Treut,Jocelyn Foskett
Ori 6h8'42"9d45'
Treutelaar,Little Nancy Lee
Aqr 20h56'54"-5d59'
Treutler,Mr O M
Uma 9h0'22"57d2'
Treva's Light
Oph 18h6'16"8d0'
Trevaskis,Jack Elliot
Cep 22h15'21"61d47'
Trevener,Valerie Jeanette
And 1h56'33"42d5'
Trevino,Tony
Uma 10h42'27"70d27'
Trevisanut,Jessica
Peg 22h42'31"11d33'
Trevor
Her 16h43'50"32d57'
Trevor B
Lac 22h6'24"50d28'
Trevor Jon
Aql 19h7'44"0d7'
Trevorrow,Jean Young
Lyn 8h16'48"38d21'
Trew,Anne Marie
Crb 15h56'25"26d21'
Trew,Jason
Lac 21h59'11"42d33'
Trewyn,Mary Ellen
Cmi 7h11'18"7d13'
Trexler,Patricia Ann Cromwell
Peg 22h17'16"33d50'
Trey
Ori 5h27'0"-1d50'
Treylee
Sge 20h17'10"20d13'
Treylon
Tri 2h46'19"31d33'
Trezza,Laura
Lyr 19h12'38"34d28'A

TRG IV
Her 16h21'36"40d4'
Tri Delta Resources
Lyn 8h31'18"41d21'
Tri Star Destiny
Uma 13h1'58"54d33'
Triad Tiger
Vul 19h35'0"27d9'
Tribbey,Jan
Lyn 7h55'35"41d52'
Tribbey,Jay Allen
Dra 19h41'41"68d4'
Tribble,Mary
Sex 9h51'43"1d27'
Tribble,Nita
Peg 22h36'45"25d28'
Tribe
Lyn 9h14'1"34d10'
Tribuna,Tara
Lyr 18h29'0"34d12'
Trice,Kimberly Renee
Del 20h34'22"10d32'
Trice,Rachel Alexandra
Lyr 18h47'46"39d0'
Trice,Sandra Kay
Eri 3h40'43"-6d25'
Trichter,Pat
Eri 3h20'0"-13d60'
Tricia
Lyr 18h54'51"41d12'
Tricia
Mon 7h48'36"-2d34'
Tricia & Steve
Del 20h30'50"20d5'
Tricia Bernadette
Ori 4h51'55"5d23'
Tricia Ellen
And 0h56'57"39d49'
Tricia M V
And 2h13'19"39d39'
Tricia My Love
Boo 14h37'18"10d51'
Tricia's Star
And 2h19'35"45d8'
Tricia's Twinkle
Umi 15h16'47"77d52'
Tricia-WB
Cyg 19h59'0"31d11'
Trick,Annabel
Peg 22h39'24"31d46'
Tricker,Andrea D "Rea"
Mon 7h24'12"-8d33'
Trickey
Lyn 6h55'31"60d1'
Tridekka
Ori 4h57'17"-0d10'
Tridico,Marjorie Seitz
Lyn 8h0'36"38d2'
Trieger,Terri
Uma 10h37'35"60d8'
Trieloff,Christiane
Gem 6h43'25"31d6'
Trienski,Tom
Per 2h57'20"55d18'
Trierweiler,Cindy
And 2h21'10"48d11'
Trietsch,Shelly
Mon 6h45'17"10d12'
Triffin,Emily Star
Vul 19h21'39"25d6'
Trigg,Jack A
Her 17h57'12"37d58'
Trigg,Margie
Lyn 8h11'1"48d17'
Triggiani,Jr,Frank Joseph
Uma 11h21'18"49d33'
Triggiani,Mr & Mrs Joseph A & Rosemary
Uma 11h40'45"45d10'
Triggs,Michael
Per 2h56'18"40d49'
Trillia,Scarlett
Tau 4h10'3"20d9'
Trilling,Neil
Per 1h54'27"53d22'
Trim,Gregory
Oph 17h13'22"-20d39'
Trim,Lindsay
Cma 7h16'20"-13d35'
Trimarkie,Marianne (Ann)
Lib 15h17'42"-8d39'
Trimbach,Dan & Sandra
Aur 6h1'21"50d37'
Trimble,Samuel James
Ori 5h33'45"-1d40'
Trimble,Stephen Joseph
Umi 14h57'0"66d49'
Trimborn,Horst
Ari 3h14'0"23d33'
Trimbur,Forever Barrie Lynn
Lmi 11h2'48"25d1'
Trimm,Robin
And 1h34'41"37d5'
Trimmer,Kenneth Malcolm
Her 16h56'19"27d26'
Trina
Umi 16h8'45"70d38'
Trina Denine
And 23h31'34"47d32'
Trinagal
Mon 7h24'0"-8d53'
Trinagel,Michael
Sge 20h7'20"18d51'
Trincado,Herman Anthony
Her 17h31'47"26d22'
Trincado,Sierra Josie
Vul 20h0'17"23d54'
Trinder,David
Uma 9h10'31"55d15'
Tringali,Dominic Louis
Lyn 8h17'23"43d45'
Tringali,Rita Cecile
Lyn 8h17'26"43d7'
Tringy
Aql 19h21'8"-0d13'
Trinidad,Wanda I
Cas 0h4'31"58d49'
Trinie
Peg 22h17'60"33d46'
Trink L
Uma 10h23'51"67d47'
Trinka
Sgr 18h35'33"-20d17'
Trinky-Dink '93
Mon 7h17'52"-7d11'
Trinler,Nicole
Sgr 18h55'46"-20d43'
Trinnis
Aql 19h49'41"13d14'
Triola,Marc
Cnv 12h15'0"43d10'
Triola,Scott
Lac 22h21'59"55d18'
Triolier Audrey
Cet 1h27'42"0d59'
Triolo,Gasper R
Aur 6h1'44"30d41'
Trip
Peg 22h53'0"26d23'
Triphon,E George
Ser 15h16'1"7d27'
Triple Crown
Sex 9h57'38"1d38'
Triplett
Dra 2h4'32"71d43'
Tripodi,Amie
Uma 11h13'51"56d26'
Tripodi,Louie
Dra 19h0'41"48d36'
Tripodi,Mary
Lyn 7h7'54"44d48'
Tripp'n Tracy
Lyn 7h33'11"42d42'
Tripp,Amy Christine
Eri 4h33'41"-0d27'
Tripp,D Edward
Boo 15h6'0"15d13'
Tripp,Joanne P
Mon 6h52'49"1d5'
Tripp,Louise
Peg 22h31'37"24d54'
Tripp,Paul Richard
Uma 10h45'32"47d49'
Tripp,Shari
And 1h40'11"38d31'
Tripp,Tobin Elliott
Her 16h5'38"47d39'
Trippel,Albert W
Hya 8h15'40"4d41'
Tripper(Far Out Man!), Doctor
Cam 3h25'52"63d55'
Trippett,Ann
And 0h21'32"36d45'
Trippett,Ken
Cep 0h16'51"69d29'
Trischa & Josh
Lmi 10h47'33"31d57'
Trischitta,Frank & Lisa
Tau 4h41'46"1d11'
Trischitta,Michael James
Dra 17h30'47"64d31'
Triscos Family
Eri 3h36'22"-1d34'
Trish
And 1h52'52"40d32'
Trish
Del 20h20'57"10d55'
Trish
And 23h36'25"47d56'
Trish
Eri 3h35'50"-15d51'
Trish
Cnc 8h40'48"18d54'
Trish & Ferdinand
Peg 22h1'41"28d18'
Trish 95
Aql 19h31'15"12d10'
Trish,The
Lyr 19h0'11"26d19'
Trisha
And 2h22'22"48d12'
Trisha
Crb 16h15'28"34d25'
Trisha
Cam 3h21'19"55d9'
Trisha
Cas 0h37'27"58d34'
Trisha 062373
Cam 12h40'40"82d22'
Trisha Lee
Cyg 19h41'0"37d39'
Trisha Marie
Cas 23h21'54"60d28'
Trisha's Star
Per 3h59'27"40d4'
Trished
Cru 11h57'59"-59d40'
Triska,Isaac J
Lac 22h4'29"46d27'
TristAmber
Lyn 7h20'40"51d29'
Tristan
Peg 23h38'18"16d38'
Tristan
Ori 5h41'18"11d25'
Tristan
Cmi 7h27'1"0d31'
Tristana
Boo 15h4'56"37d50'
Tristesse Aimee
Lyn 8h49'26"39d57'
Tritton,Alex
Uma 12h1'32"30d9'
Tritton,Matthew
Uma 12h1'25"31d36'
Trittonius Maximus 45
Aql 19h2'1"-0d49'
Triva
Aur 6h14'17"30d1'
Trivellone,Luisa
Peg 22h40'49"3d9'
Trivelpiece,Eric S
Aur 5h53'17"38d46'
Trivers,Julian Myers
Ori 4h52'33"1d40'
Trixi
Cap 21h21'19"-24d31'
Trixie
Sgr 19h29'19"-31d25'
Trignør'poc Maigh'eo
Ori 4h42'0"8d31'
TRL & MRC
Uma 12h4'53"39d57'
Trober,Jacqueline Hilary
Cas 0h19'36"62d18'
Trochu,Daniele
Boo 14h29'39"44d5'
Troderman,Dylan Katz
Cap 20h58'26"-26d41'
Trogden,Denah Kaye
Lyn 8h41'25"44d57'
Trogdon,John Edwin
Boo 14h8'0"45d54'
Troggy
Dra 20h21'11"62d28'
Troiano,Jessica
Umi 16h59'49"77d4'
Trois,Charles & Rebecca
Eri 2h45'0"-2d26'
Troise,Jared
Aur 7h11'51"39d35'
Troise,Marissa
And 23h33'30"43d0'
Trojacek,Jeffrey Wayne
Ser 15h44'53"-2d32'
Trojan
Peg 21h51'31"33d16'
Trojanowski,Carolyn (Shortstuff)
Eri 3h7'12"-5d55'
Trojanus
Per 2h9'27"56d20'
Trolio,Albert & Sandy
Tri 1h41'1"30d53'
Trombadore,James
Boo 13h46'0"16d50'
Trombini,Virginia
Leo 11h49'51"22d8'
Trombino,Pam
Com 12h17'27"32d3'
Trombley,Sr,Arnold E
Vir 13h26'44"-7d12'
Trombone Sam & Jen
Cyg 20h19'54"38d46'
Trompeter,Patricia
Hya 9h9'23"2d58'
Trondyke
Sct 18h46'7"-7d40'
Trongone,MD,Richard J
Per 2h43'29"40d45'
Trono,Luca
Lyr 19h22'29"31d16'
Trono,Peri Olivia
And 23h42'49"37d47'
Tronser,Konrad Berthold
Cnc 8h0'31"10d20'
Tronzo,Marc T
Aql 19h30'0"10d60'
Troop,Telstarr Stuart
Umi 14h38'32"69d1'
Trooper 1970
Uma 9h7'0"70d27'
Trop,Helen/Jerrold Zell
Dra 20h6'50"68d4'
Trop,Rhonda
Cas 2h41'30"73d21'
Tropiano,Maria
Cas 0h38'20"68d1'
Tropik Star,Sissi
Per 1h44'36"53d48'
Troppmann,Amy
Sex 9h50'27"-6d41'
Trosini,Milena
And 1h19'46"39d7'
Trosper,Willis "Sam"
Cnc 9h0'11"30d4'
Trost 11-18-48, Barbara S
Gem 6h44'12"16d34'
Trost,Christian Kenny
Cam 4h2'1"58d26'
Trost,James Nelson
Her 16h57'35"24d60'
Trost,Scott A
Dra 17h57'16"61d52'
Trotman II,Tsgt Peter J
Cnv 12h53'17"51d9'
Trotman,Peter J
Per 2h25'55"55d53'
Trott,Arthur J
Boo 14h32'16"22d13'
Trott,Jr,Alan Dana
Cep 20h53'49"55d48'
Trott,Ruth
And 23h36'0"44d43'
Trotta,Lynda
Com 12h37'51"23d21'
Trotter,Elizabeth Roy
Aql 20h31'45"-6d59'
Trotter,Frank C
Cet 1h0'47"-11d49'
Trotter,Mr & Mrs Thomas N
Mon 7h58'29"-6d59'
Trotter,Russell James
Boo 14h18'0"30d54'
Trotter,Sr,Donald W
Sex 9h55'0"-5d55'
Trottier,Ginette
Cas 0h6'0"64d12'
Trottier,Miss Katie
And 22h55'59"51d38'
Trottier,The Big Bright Robert
Per 2h26'43"56d24'
Trottman,Anita
Aur 4h49'35"48d55'
Troup,Isabella Galloway
Cyg 19h28'21"36d22'
Troup,John
Boo 15h3'21"23d54'
Troup,PhD,Jan M
Crt 11h17'45"-16d38'
Troup,Robin Jean
Leo 11h12'40"27d50'B
Trout,Catherine
Lyn 6h21'18"60d53'
Trout,Tracy
Sgr 19h7'35"-21d45'
Troutman,Aeko Y
Dra 15h12'26"57d26'
Troutman,Jr,Robert S
Aur 7h2'40"35d56'
Troutman,Leslie Noel
Peg 23h41'38"28d0'
Troutman,Robert
Lac 22h17'37"54d32'
Trovato,Linda Ann
Cam 05h58'50"61d28'
Trowbridge,Intercalary Mallory Ann
Lyr 18h48'14"39d12'
Trowbridge,James Everette
Per 2h41'57"37d9'
Trowbridge,Sheena "Enhydra Lutris"
And 0h15'14"30d57'
Trower,Spencer Eli
Boo 14h38'48"46d29'
Trowsdale
Cas 23h5'34"58d23'
Troxell,Beanye
Lyn 7h30'53"52d17'
Troxell,William T
Cnv 13h47'41"36d49'
Troxler,Barbara
And 23h18'51"45d31'
Troy
Del 20h50'46"9d22'
Troy & MacKenzie's Star In Heaven
Ori 5h30'35"-2d5'
Troy & Suzanne's Star of Love
Cyg 21h8'55"48d47'
Troy Daniel
Her 18h19'53"20d11'
Troy's Old Souls
Cep 21h37'20"56d6'
Troy's Star
Dra 19h27'1"68d51'
Troy's Treasure
Mon 8h5'17"-2d11'
Troy,Edward John
Lac 22h49'57"54d57'
Troy,Joseph W
Per 2h27'57"58d48'
Troy,Kenneth Michael
Dra 17h39'14"65d37'
Troy,Stephen Keith
Oph 17h15'56"-22d53'
Troy,Steven M
Lac 22h0'34"38d2'
Troy,Tylor Dina
Lib 15h2'1"-10d33'
Troyano,Pat (Sassie)
Tau 5h12'43"16d35'
Troyer,Scott Ryan
Oph 17h6'0"10d29'
Troyner,Michael Scott
Lac 22h8'33"47d26'
Tru
And 0h2'56"44d35'
Truax,Ed
Dra 14h27'55"62d33'
Truax,Michael Merritt
Dra 17h18'55"60d45'
Truba,Rita M
Gem 6h47'57"14d13'
Truba,Ron
Aur 5h30'27"38d54'
Truba,Stuart
Cep 21h54'51"68d5'
Trubak,Daniel
Lac 22h35'24"37d56'
Trubble
Umi 14h51'54"65d39'
Trubman,Kevin
Oph 17h15'29"10d35'
Trubshaw,Mark
Per 3h4'38"42d59'
Truby,Roy
Aur 5h59'0"30d1'
Trucchio,Bliss
Cas 0h56'0"58d27'
Trucchio,Nancy
Cam 3h52'24"55d32'
Truchan,John
Cet 2h0'16"-1d34'
Truchan,Veronica
Cet 2h0'20"-1d45'
Truchelut,Ryan
Cet 0h8'54"-10d2'
Truchot,Laure-Anne
Aur 5h9'56"29d28'
Trucksess,Donald J
Boo 14h0'51"27d20'
Trucvy
Cyg 21h51'0"41d35'
Trude,Bernard M
Oph 17h55'32"7d57'
Trudeau,David Anthony
Cet 1h51'27"-8d30'
Trudeau,J J Richard
Per 3h12'0"40d52'
Trudeau,Rory Scott- Brent E Bowes
Cyg 21h9'45"41d1'
Trudeau,Susan Kay
Tau 5h44'27"16d52'
Trudel,Claude & Diane Dubé
Uma 8h54'55"54d37'
Trudel,Laurent
Uma 8h58'47"55d25'
Trudel,Maurice
Dra 13h47'26"67d60'
Trudel,Paolo
Cep 21h29'36"61d51'
Trudell,Cal
Cmi 7h35'26"11d19'
Trudell,Carolyn Ann
And 23h0'11"48d57'
Trudell,Clifton Jurome
Her 18h1'1"38d53'
Trudell,Heather Lynn
Eri 3h12'53"-1d52'
Trudy
Cas 2h51'42"70d46'
Trudy
And 0h7'18"30d23'
Trudy J
Lac 22h19'12"51d40'
Trudy's Resplendence
Cyg 21h3'1"40d4'
Trudy's Star
Cas 0h34'51"60d50'
Trudy's Star
And 1h46'1"40d11'
True Companion
Ori 6h5'12"8d33'
True Companion
Lyr 18h31'26"30d47'
True Companion
Per 2h36'32"45d48'
True Companion II
Cyg 21h43'44"38d16'
True Companions
Ori 5h50'46"9d44'
True Companions Frank & Kathy
Cyg 21h17'1"38d17'
True Cooper
Equ 21h20'39"12d54'
True Dreams-The JRG Express
Boo 15h12'54"48d36'
True Energy
Ori 5h55'36"10d14'
True Heart & Mayumi
Gem 7h2'39"15d24'
True Hearts Kiss
Cyg 20h17'22"39d27'
True Love
And 23h22'47"46d51'
True Love
Eri 3h42'0"-17d31'
True Love
Sge 19h31'51"17d1'
True Love
Aql 19h26'20"-0d50'
True Love
Mon 6h39'49"10d9'
True Love
Uma 11h21'58"37d51'
True Love
Cnc 9h7'36"30d45'
True Love
Per 3h44'35"50d15'
True Love Of Kuniyoshi & Teruko
Ari 2h13'42"23d36'
True Love Terry
Aql 19h57'55"13d36'
True Love,Yoshi & Megu
Gem 7h22'44"25d6'
True Love-J & S 95
Cas 0h46'26"75d9'
True To You
Cam 13h19'18"76d48'
True,S D
Lyr 18h47'39"33d18'
Trueb,Guenter
Cam 3h18'48"57d16'
Truedell,Nathaniel
Lmi 9h46'37"40d21'
Truelove,Michele
Per 2h10'1"57d7'
Trueman
Her 17h14'16"29d27'
Truesdale,Dorenda Gregg
Lyn 7h4'16"52d37'
Truex,Brooke Ellen
Mon 6h30'41"-6d38'
Truex,Robert Ross
Per 1h28'57"53d23'
Truhan,Rosemary Egan
Cas 2h10'33"59d36'
Truilhé,Pierre
Lmi 10h9'13"32d22'
Truitt,Joseph Christopher
Cnv 12h24'22"38d25'
Trujillo of Evanston, WYO,Sharon
Vul 18h55'56"24d45'
Trujillo,Kendra
Cet 0h53'30"1d54'
Trujillo,Lisa
Tri 2h17'4330d57'
Trujillo,Marc E
Oph 17h24'40"-23d11'
Trujillo,Solomon
Sco 16h55'45"-40d1'
Trujillo,Susan J
Vul 19h5'50"25d14'
Truky
Pho 0h3'41"-44d17'
Trulsson,Nicholas Jon
Sge 19h56'37"16d15'
Truly
Cnv 13h3'53"48d50'
Truly Shining Star
Vir 14h31'0"1d16'
Truman,Barry John
Hya 9h2'31"5d33'
Truman,Brian
Boo 15h22'21"34d32'
Truman,Dan
Her 17h13'39"48d42'
Truman,Zane Michael
Dra 16h45'19"61d57'
Trumbauer,Sarah Lyn
And 0h22'36"32d26'
Trumbore,Anna E
Cas 2h15'27"65d6'
Trump,Daniel Patrick
Hya 9h15'21"0d44'
Trump,Donald J
Dra 16h5'34"66d29'
Trump,Tiffany Ariana
Umi 16h2'27"76d53'
Trumpet,B E P
Cep 0h30'25"86d59'
Trunk,Kevin My Bear & Starry Knight
Uma 9h14'14"68d12'
Trunzo,Lee
Lmi 9h58'55"33d29'
Trup
Sct 18h41'49"-7d35'
Trupiano,Angela Marie
Eri 4h2'1"-8d40'
Trupkovich,Tommie
Her 16h37'52"38d28'
Truscelli Beamer Guiding Ships Home
Per 3h19'18"40d17'
Truscott's Make A Wish Star
Cyg 20h2'26"37d37'
Trush's Everlasting Light
Cyg 19h26'30"30d49'
Trushke,Ken
Her 16h10'28"40d39'
Truska,Edward
Her 17h10'38"45d17'
Truskolaski,Franklin William Joseph
Aur 7h20'59"36d43'
Trusky-Lewis,Diani
Lyn 8h54'21"42d24'
Trusler,Wanda Elizabeth
Hya 8h14'37"-5d46'
Trussel,Garland
Her 17h24'14"40d26'
Trussell,Lois Blakley
Com 13h3'41"27d7'
Trusselle,Miss Tracey
Cas 0h20'24"59d42'
Trusty,Phyllis
And 1h27'10"50d8'
Truth Unlimited (Daro)
Lac 22h43'41"54d21'
Trutzenbach,Darren
Per 3h1'29"50d3'
Trutzenbach,David
Cep 21h30'10"55d38'
Trutzenbach,Kevin David
Peg 23h30'13"21d29'
Truxes,Mildred Ruth
Del 20h13'1"10d22'
Truxes,William Walter
Aql 19h52'1"12d20'
Truxton
Eri 3h2'26"-8d3'
Truxton,Elizabeth Ann
Tau 4h41'22"16d21'
Try,Travis
Cam 3h50'36"61d53'
Tryba,Andrew K
Crb 15h17'59"30d41'
Trybom,Victor B
Lyn 7h27'43"36d2'
Trybus "Wedding Star", David & Kathy
Sge 20h1'21"20d34'
Trygstad,Andrew Stevens
Dra 16h26'1"64d12'
Trygstad,Christian John
Ori 6h8'21"5d1'
Trygstad,Jr,John Gordon
Uma 9h0'54"55d58'
Trynka S
Mon 8h7'27"-0d20'
Tryon,Thomas Emery
Her 17h44'1"40d21'
Trythall,James
Lac 22h5'11"46d33'
Tryviane
Cep 22h25'27"58d17'
Trzaska,George & Clara
Umi 15h18'44"77d45'
Trzaska,Sunny
Mon 6h41'48"10d22'
Trzcinski,Gisèle
Ori 6h7'1"8d44'
Trzebunia,Alina
And 23h36'21"47d49'
Trzesinski,Michael R
Lmi 10h0'51"30d35'
Trzetziak,Joachim `Joey'
Aur 6h7'1"36d29'
Trznadel-Jablon, Bertha
Sgr 19h24'37"-40d56'
Trány
Ser 15h38'41"6d43'
Tré & Sue "Forever Together"
Gem 6h41'53"12d53'
Trébor
And 0h54'19"39d42'
Trésor
Cyg 20h22'56"38d36'
Trésor d'étincelles
Cyg 19h28'40"33d43'
Trösser,York-Enzo
Psc 1h0'22"32d6'
Tröstl,Friedrich
Aqr 22h53'12"-5d20'
Trübestein,Gustav
Ori 5h57'27"1d16'
Tsantis,George & Annette
Crb 15h34'57"27d57'
Tsao,Andrew Huei
Ser 15h14'18"8d56'
Tsarnas,Gina
Lyn 7h56'56"58d54'
Tsautakis,George
Per 1h59'0"56d44'
TSC I
Cep 0h5'57"69d43'
Tschachtli,Jane Richardson
And 23h27'22"43d11'
Tschalie,KE Herrbruck-Twobridges
Lib 14h58'50"-20d35'
Tscharner de Richard
Cam 4h58'42"60d51'

Tschauder,Gerd
Ari 2h58'44"21d57'
Tschentscher-Reimann, Dagmar
And 1h42'45"41d20'
Tschetter,Martin Benjamin
Cet 0h28'20"1d14'
Tschudi,Felix
Peg 22h27'58"30d45'
Tschudi,Janice
Com 13h31'50"21d0'
Tschudy,Virginia
Mon 7h3'25"-0d41'
Tserotas,Vivian
And 0h50'12"39d42'
Tsetsakis,Joann
Lyn 7h43'50"58d20'
Tsikis,Andrea
Vul 19h18'41"26d38'
Tsiknas,Gus
Aur 7h19'37"35d49'
Tsiknas,Michael
Her 17h37'48"27d60'
Tsiotsias,Alexander
Per 3h4'44"43d16'
Tsolis,James Anthony
Her 17h14'46"22d14'
Tsoulos,Theodore
Cnv 12h6'15"37d10'
Tsouris,Ariana Aleni
And 0h22'21"43d58'
Tsuchiya,Kyouko
Psc 1h2'29"22d15'
Tsuchiya,Shinichi
Lib 14h45'17"-24d25'
Tsuchiya,Suzanne
Sct 18h44'30"-7d17'
Tsuda,Robert Yukio
Aql 19h52'49"15d5'
Tsui,Dr Dan
Cma 6h57'41"-20d29'
TSURUCHAN BOSHI
Aqr 22h0'6"-9d8'
Tsuruta,Eddie Hisao
Aql 19h58'44"-6d54'
Tszo 1993
Lib 15h0'36"-20d1'
TT
Her 18h9'18"31d15'
Ttamarrac
Cam 8h10'27"83d41'
TTL
Tau 4h58'1"16d52'
Tu,Helen
Mon 6h44'13"11d41'
Tua
Tri 1h44'18"28d34'
Tuami,Charlene
Ori 5h55'21"13d15'
Tuba
Dra 15h44'16"58d55'
Tubb,Fred-Susan-Brian-Rachel
Cet 2h2'19"-18d54'
Tubbs "The Avatar", Clifford
Boo 14h39'32"50d14'
Tubbs,Paul Marvin
Cet 2h57'25"5d36'
Tubolino,Tina Lee
Lyr 19h0'53"30d26'
Tubster's Twin
Aql 19h4'44"-6d47'
Tuccelli,Sal
Cep 22h14'56"67d32'
Tucci,Louis Joseph
Oph 18h2'48"13d40'
Tuccillo,Anthony
Her 16h37'52"28d32'
Tuccio,Angela & Peter
Lyr 18h46'0"30d40'
Tuccio,May
And 1h35'29"39d54'
Tuccitto,Kevin Richard
Boo 14h45'57"31d59'
Tuccitto,Megan Richelle
Cas 1h7'21"60d34'
Tuccitto,Ryan Patrick
Boo 14h45'41"32d14'
Tucholski,Andrew Michael
Lac 22h20'55"38d44'
Tuck,James Richard
Hya 8h18'13"-0d50'
Tuck-My Loving Mother, Peggy M
Peg 22h34'51"26d14'
Tucker
Boo 14h43'0"27d54'
Tucker
Cnv 12h21'56"51d30'
Tucker Family,The
Peg 23h1'54"17d43'
Tucker III,James
Sgr 18h48'54"-35d12'
Tucker Star,The Peggy
Cnv 13h49'34"41d3'
Tucker Wedding Star, Neil & Shawn
Lac 22h2'0"38d53'
Tucker's Aunt
Aur 7h6'19"37d59'
Tucker's Light
Peg 23h2'56"18d38'
Tucker's Mom
Uma 10h2'46"50d6'
Tucker's Star
Cet 3h12'58"5d50'
Tucker's Star,J C & Shirley
Mon 7h4'0"3d59'
Tucker,Andrea
Cas 23h26'12"58d17'
Tucker,Anna Beth & Robert Lee Whittington
Peg 23h43'39"30d58'
Tucker,Benjamin Charles
Uma 9h0'26"58d14'
Tucker,Betty Nancy
Cyg 20h16'12"39d3'
Tucker,Caitlin Marie
Lac 22h22'57"38d58'
Tucker,Carol Jean Screeton
And 23h23'23"45d52'
Tucker,Christ
Del 20h15'1"10d58'
Tucker,Clyde Emerson
Her 17h46'49"40d0'
Tucker,Colleen Elizabeth
Lib 14h52'23"-1d4'
Tucker,Corey Allen
Tri 2h5'51"31d31'
Tucker,Courtney Rae
Cyg 19h28'55"31d9'
Tucker,David R
Umi 15h37'12"66d36'
Tucker,David Scott
Aur 6h31'15"33d19'
Tucker,Deborah A
Aqr 22h42'42"2d14'
Tucker,Desiree
Cas 0h58'16"62d37'
Tucker,Donald Marvin
Her 18h0'40"14d60'
Tucker,Dr Jesse
Oph 17h31'60"1d18'
Tucker,Dr Joseph C
Cnc 8h52'60"31d45'
Tucker,Harry Melvin
Dra 16h47'11"68d4'
Tucker,Howard C
Boo 15h17'27"38d14'
Tucker,Hoyt A
Ori 5h0'48"-2d23'
Tucker,Isador & Anna
Cyg 19h47'58"30d18'
Tucker,James Preston
Her 17h33'43"26d48'
Tucker,Jamie Duane
Sco 17h51'25"-30d26'
Tucker,Jane
Cyg 21h45'39"34d15'
Tucker,Joan
Peg 22h40'1"34d44'
Tucker,John
Uma 10h58'1"48d13'
Tucker,John James
Cet 0h42'0"1d18'
Tucker,John Paul
Aql 20h13'29"7d50'
Tucker,John Percy
Cyg 20h15'54"39d50'
Tucker,Joyce
Eri 3h37'29"-6d13'
Tucker,Joyce Lynn
Sgr 20h11'16"-29d22'
Tucker,Jr-World's Best Dad,Wilmark
Per 3h52'0"35d4'
Tucker,Julian J
Cas 1h5'1"62d1'
Tucker,Kalvin Anthony
Cam 3h20'36"61d17'
Tucker,Kenneth Spunky
Aur 4h35'12"31d18'
Tucker,Kirk Andrew
Lac 22h31'41"38d58'
Tucker,Lyndsay Martine
Leo 11h7'14"22d41'
Tucker,Marguerite
Cyg 21h34'32"41d43'
Tucker,Marian E
And 2h15'1"40d44'
Tucker,Marie A
Eri 3h23'12"-10d17'
Tucker,Martha Carolina
Lyr 18h29'24"30d55'
Tucker,Mary Elizabeth
Tri 1h41'0"30d56'
Tucker,Melanie Dawn
And 0h20'55"43d16'
Tucker,Michael D
Lac 22h23'40"41d7'
Tucker,Mildred Chittenden
Sgr 18h51'48"-29d45'
Tucker,Norma Jean
And 0h51'16"36d14'
Tucker,Patricia
Peg 22h21'0"29d55'
Tucker,Piper Ian Thomas
Gem 6h47'37"31d36'
Tucker,Randy Dale
Oph 18h4'44"11d32'
Tucker,Reginald
Lyn 8h44'1"39d47'
Tucker,Rick
Per 3h25'0"50d30'
Tucker,Robert & Aubrey Jo
Per 2h41'55"36d31'
Tucker,Ronald & Anne
Crb 15h27'34"31d12'
Tucker,Ronald Everett
Dra 15h36'43"53d29'
Tucker,Scott Jacob
Dra 16h39'37"52d30'
Tucker,Sherri Lynn
Lmi 9h50'1"40d40'
Tucker,Steaven
Cnv 13h48'16"30d45'
Tucker,Stella G R
Uma 11h23'59"44d12'
Tucker,Stephen M
Boo 14h53'38"40d4'
Tucker,Sydney Leigh
Per 4h5'1"48d50'
Tucker,Taylor Alexis
Uma 9h54'58"58d22'
Tucker,The Tommy
Tri 1h47'31"28d12'
Tucker,Travis Richard
Oph 17h8'33"10d0'
Tuckett,Iris Olive
Cyg 20h1'52"30d19'
Tuckey,Tierna Lynn
And 0h21'44"40d49'
Tuckley,Katie
Cyg 21h37'59"42d55'
Tuckwiller,Ellen Louise
Tri 2h45'28"32d42'
Tudela,Maria
Aql 19h6'14"-0d6'
Tuders,April Dawn
Peg 22h42'31"34d49'
Tudisco,Ines
Gru 22h21'31"-52d0'
Tudor,Jr,Ted Graham
Tau 3h54'19"2d22'
Tudor,Sanya
Peg 22h31'35"11d44'
Tuech,Henry "Pete"
Aql 20h1'0"11d6'
Tuell,Irene M
Cas 23h33'31"60d30'
Tuerack,Beth
And 1h52'35"36d30'
Tufexis,Mary K "Jenny"
Uma 11h12'40"40d31'
Tuffentsammer,Hartmut
And 23h1'23"42d49'
Tuffy
Cmi 7h54'17"5d34'
Tuffy,Gypsy,Ripple
Peg 23h22'22"11d22'
Tufts,Allen Phillip
Ori 5h14'51"-8d51'
Tufts,Floyd "Brother"
Her 16h57'17"35d47'
Tufts,Jennifer A
Lyr 18h41'0"36d11'
Tug's Star
Her 16h47'10"50d4'
Tugar
Uma 9h51'32"52d4'
Tugendreich,Stuart Michael
Aur 4h48'35"50d53'
Tuggle,Brett
Boo 14h13'1"41d53'
Tuggle,Jack H
Aql 19h31'15"10d5'
Tugnoli,Francesca
Aur 5h25'21"31d16'
Tuitasi,Devin Fiatau
Vul 19h53'21"20d31'
Tuite's Star
Per 2h52'31"40d41'
Tuite,Tracy
Dra 17h6'20"61d59'
Tukan,MD,PC,Dr Elizabeth Victoria
Uma 8h59'25"57d14'
Tuke,Sarah Jane
Lyr 18h29'42"31d18'
Tulacro,Brett
Psc 1h2'48"21d9'
Tulino,Ruth B
And 23h49'59"32d11'
Tulipan,Doctor
Cep 21h50'15"61d44'
Tulk,Glen Andrew
Cet 3h2'0"7d59'
Tull,J James
Uma 8h12'0"72d25'
Tullai,Mary
Cnv 12h51'14"42d18'
Tuller,Jeffrey Wayne
Lac 22h41'53"56d34'
Tulley,B
Lmi 10h58'18"27d16'
Tullier,Lisa Michelle
Uma 8h59'39"47d44'
Tulloch,Lee
Lyn 7h54'39"48d53'
Tulloch,Matt & Scott
Sge 19h43'60"16d16'
Tulloch,Michael Trey
Ori 5h51'1"7d33'
Tulloch,Theresa
Cas 0h56'53"50d3'
Tulloch,Tiffany Victoria
Peg 23h35'52"23d59'
Tulloh,Anthony John
Ori 4h58'33"5d23'
Tullus-Bergeron,Kay
Cet 1h50'0"-1d54'
Tully #1,R
Her 17h34'46"26d8'
Tully,Bernice
Mon 6h54'46"-8d50'
Tully,Carol June Summerfield
Cas 0h35'31"70d36'
Tully,Catherine
Cam 3h54'21"62d11'
Tully,James Michael
Cet 1h32'14"-13d7'
Tully,Patricia
Lyr 18h23'56"38d56'
Tulowitzky,Jeffrey
Sct 18h44'30"-7d20'
Tulpan,Fern & Marvin
Sge 20h0'11"16d0'
Tuma,Catherine V
Cas 0h9'40"63d57'
Tumarkin,Gertrude Plaine Friedman
Mon 6h57'28"8d38'
Tumbidge Star,The
Uma 8h57'43"50d34'
Tumble
Cyg 20h20'13"39d41'
Tumble Bug
Oph 16h56'11"-28d43'
Tumble Tots
Del 20h54'60"9d25'
Tumbleweed Louise
And 0h19'50"30d3'
Tumilowicz,Julius Julian
Aur 5h2'56"44d37'
Tumilowicz,Karen
Gem 7h36'23"34d37'
Tumilty,James Patrick
Gem 6h58'39"15d53'
Tumlin,Denny
Her 15h50'49"42d3'
Tummala,Neel
Cep 21h57'57"55d57'
Tummalapalli,Chandra Mouly
Mon 6h19'41"7d42'
Tummillo,Peter
Dra 20h13'59"62d43'
Tumminello,Samantha Jo
Lyr 18h54'36"31d15'
Tumulty,Charles James
Dra 16h32'59"67d21'
Tumulty,Charles James
Aur 6h25'45"30d57'
Tune
Lyr 19h12'1"37d39'
Tune,"Star Dancer" AKA Tommy
Boo 14h19'41"15d50'
Tune,Jeff & Denice
Lyr 18h36'27"31d47'
Tunesi,Jean
Del 20h20'24"16d10'
Tuney & Bruce True Love Forever
Lyr 18h17'58"34d56'
Tungseth,Allan Erling
Hya 9h0'22"1d9'
Tunis
Cep 21h2'0"63d48'
Tunis III,Edward Curry
Aur 6h57'46"38d35'
Tunis,Alexandra
Uma 9h46'18"59d58'
Tunis,Carol
Oph 17h2'55"-24d24'
Tunis,Christine Marie
Cas 1h13'1"62d45'
Tunis,Rebecca
Uma 9h49'56"42d31'
Tunkel,Volker
Hya 9h38'37"5d29'
Tunket
Cet 1h26'25"-1d38'
Tunks,Scott
Aur 6h23'1"32d55'
Tunney's Twinkle
Cam 3h55'27"57d14'
Tunney,Thomas Mark
Cep 21h21'30"84d51'
Tunon,Conrado Garcia
Sct 18h22'36"-14d55'
Tunstall,Always - Kimberly & Thomas
Cyg 19h31'38"34d8'
Tunzi,Michael Anthony
Uma 9h56'0"51d25'
Tuohy,Jenn
Aql 19h28'52"8d44'
Tuomi,Jacob
Dra 16h27'1"60d18'
Tupes,Constance
Lyn 7h49'39"44d18'
Tuppatsch,Raymond
Boo 14h59'21"47d51'
Tuppence
Ind 20h25'31"-58d7'
Tupper,Stephen C
Per 1h39'0"53d34'
Tura,Will
Lyr 19h22'11"30d17'
Turano,Danny
Aur 5h7'32"40d51'
Turansky,Jack
Aur 5h26'25"41d4'
Turay,Rick & Mary Lou
Cas 0h38'1"58d46'
Turbeville,Lisa E
Aqr 21h35'52"-6d27'
Turbi
Cas 1h49'47"71d37'
Turbon,Klaus
Sco 16h13'7"-24d34'
Turbow,Barbara Hope
Cas 0h47'50"66d31'
Turbowicz,Shira
Lyr 18h39'36"46d44'
Turbyne,Cathy
And 0h21'58"36d48'
Turcan,Ashley Rose
Ari 1h46'22"22d18'
Turchan,Sarah Nicole
Crt 11h29'37"18d17'
Turchet,Montana Brielle
Her 18h7'40"38d52'
Turchi,Jessica Dixon
Cnv 13h22'35"38d14'
Turcillo,Susan Rose
Aur 5h55'19"30d24'
Turck,Andrew Joseph Ciquyaq
Lac 22h13'0"46d19'
Turck,Daniel Frank
Dra 19h42'35"65d26'
Turco,Michael Alfred
Per 2h29'1"57d36'
Turcot,Alan P
Aur 6h57'45"40d9'
Turcot,Delphine Gervais
Ori 5h43'22"8d60'
Turcotte,Albert Henry
Cet 2h7'42"4d52'
Turcotte,Carol
Cas 23h18'56"62d1'
Turcotte,Cynthia Jean
Com 13h29'53"25d39'
Turcotte,Frédéric
Cep 21h19'31"85d6'
Turcotte,Réal
Cyg 19h27'39"33d5'
Turcotte,Sandra
Lac 22h49'56"53d16'
Turcotte,Serge
Her 16h57'58"27d51'
Turcsanyi,Jana
Aur 6h51'53"35d34'
Turczak,Chris
Aur 4h59'26"36d32'
Turek,Gina
Cyg 21h7'10"31d35'
Turek,Kim
Cnc 7h57'20"10d46'
Turek,Norman
Uma 11h14'46"56d15'
Turen Family
Cas 0h9'40"50d31'
Turenne,Noel-In Honor of Thomas Noel
Lac 22h24'15"38d31'
Turetsky,Stephen Joel
Dra 13h59'29"64d20'
Turgeon,Reneé
And 1h53'26"37d54'
Turi & Family,Joseph & Margaret
Umi 15h8'12"71d3'
Turini,William L
Dra 20h33'54"68d0'
Turinia,Tom
Cma 6h29'55"-18d58'
Turja,Tom A
Ori 5h25'0"1d10'
Turk
Her 15h49'29"42d22'
Turk,Theresa
Lac 22h21'31"55d52'
Turkel,Dylan Jake
Leo 9h59'43"18d0'
Turkel,Mark David
Lac 22h21'40"55d0'
Turker,Ugur
Eri 4h25'39"-0d2'
Turkie,Tina
Cma 6h15'12"-16d31'
Turkington,Michael
Aql 19h12'49"14d57'
Turknett,Lyn
Vul 20h14'30"22d48'
Turley,John P
Boo 13h48'1"15d56'
Turley,Jr,Thomas Ballard
Hya 8h18'41"3d11'
Turley,Michel J
Boo 14h13'29"51d12'
Turley,Robert E
Cyg 21h3'0"33d5'
Turmalina
Sge 19h54'54"18d49'
Turman,Philip Darknell
Her 18h5'58"30d0'
Turmel,Joan Dareth
Aql 19h31'21"8d29'
Turn It Into Love
Psc 1h23'57"7d1'
Turn Pike Joe
Boo 14h58'48"22d59'
Turn,Sr,Adam Charles
Cam 3h46'59"61d25'
Turnage,Jill Rebecca
Cyg 19h59'32"29d53'
Turnage,Lynn
Cyg 20h18'26"41d36'
Turnage,Melanie Jeanne
Cas 22h58'43"53d10'
Turnage,Ruth F
Cas 1h49'48"60d42'
Turnage,Scott Daniel
Aql 19h3'15"1d31'
Turnbaugh,Larry
Ori 6h4'41"8d12'
Turnbull,Ashley Kaitlyn
Uma 14h2'20"52d41'
Turnbull,Bobby
Uma 9h10'1"48d47'
Turnbull,Dennis
Cet 2h52'22"3d47'
Turnbull,Devin Michael
Uma 14h4'42"50d2'
Turnbull,Fallon Katherine
Uma 14h3'12"53d22'
Turnbull,Jr,William
Ari 1h47'33"25d21'
Turnbull,Kyle Chase
Her 16h38'54"26d23'
Turnbull,Rodney
Per 2h8'1"57d35'
Turnbull,Shannon
And 23h1'13"47d3'
Turnbull,Zachary Adam
Vul 20h15'13"23d49'
Turnell,Kenneth Lee
Her 17h27'46"31d47'
Turner 1981,Andrea Clayton
Cas 0h46'27"67d18'
Turner 5-13-84, Nicholas
Boo 13h34'32"17d27'
Turner Star Mother, Myra Holt
Aql 18h58'15"7d8'
Turner Star,Clarence Gertrude
Cyg 21h51'17"40d5'
Turner The Seer-Our Star,Judi
Leo 10h32'16"15d54'
Turner,Adam Joseph
Boo 14h1'1"26d22'
Turner,Alexader James Earl
Per 3h3'37"40d29'
Turner,Andrew Hancock
Aur 6h11'57"33d46'
Turner,Andrew Robert
Ari 1h54'32"23d10'
Turner,April
Cam 6h10'1"68d27'
Turner,Arcia O
Aur 5h57'13"30d36'
Turner,Barbara
Aql 20h4'37"4d25'
Turner,Carey B
Cap 21h36'0"-20d2'
Turner,Carol
Mon 6h29'54"1d23'
Turner,Clint
Her 16h59'38"31d10'
Turner,Dee
Peg 23h39'13"17d59'
Turner,Denece
Mon 6h25'52"5d31'
Turner,Dennis
Boo 14h16'53"15d14'
Turner,Derri Lynn
Cyg 19h42'21"41d22'
Turner,Diadem
Lyn 8h0'45"52d17'
Turner,Dianne Susan
Equ 21h19'20"10d49'
Turner,Donald James
Hya 8h54'30"2d48'
Turner,Doris Marion
Cas 0h29'19"60d25'
Turner,Eldridge John
Aur 6h15'43"46d17'
Turner,Emery C
Cnv 12h43'28"37d58'
Turner,Eric Heinz Goldfarb
Her 18h2'51"31d31'
Turner,Father J Murphy Ed
Cep 21h7'1"60d49'
Turner,Forrest Elliott
Boo 15h10'27"27d54'
Turner,Fred
Sex 9h54'13"1d47'
Turner,Fred W,Memory B/G Eugene H Beebe
Aql 18h57'12"7d25'
Turner,Gerald H
Lac 22h21'0"55d43
Turner,Gillian Sharon Brighteyes
Crb 16h4'27"34d39'
Turner,Gloria
Lyn 9h13'1"38d43'
Turner,Gregg
Her 16h18'43"41d19'
Turner,J C
Aql 19h55'1"-6d5'
Turner,Jack
Cet 1h28'21"-1d16'
Turner,James
Dra 17h1'0"68d41'
Turner,James & Stella M Wright
Ori 5h53'53"16d51'
Turner,James Ryan
Boo 15h10'10"27d42'
Turner,Jeff Scott
Aql 20h35'13"0d28'
Turner,Jennifer
And 23h25'19"43d24'
Turner,Jennifer Nicole
Cet 1h52'31"-2d17'
Turner,Jeremy
Boo 13h45'0"25d5'
Turner,Jim
Cet 0h33'18"-0d33'
Turner,Joanne Dayle
Lyr 19h24'57"40d35'
Turner,Joel Dugan
Cmi 7h58'37"0d50'
Turner,John
Ser 18h3'38"-14d11'
Turner,Joy
Uma 10h23'37"51d40'
Turner,Jr,Christopher James
Lac 22h19'13"51d37'
Turner,Jr,James E
Boo 14h1'18"16d2'
Turner,Jr,John M
Ori 5h58'26"14d5'
Turner,Jr,R Pelham
Cep 22h8'12"61d42'
Turner,Kali Alexa
Cyg 19h42'11"42d7'
Turner,Katherine Margaret
Cyg 19h41'48"30d16'
Turner,Katie Elizabeth
Umi 16h26'35"75d59'
Turner,Kelley Marie
Vul 20h58'42"28d12'
Turner,Kenneth
Ser 15h37'50"19d1'
Turner,Kenneth H
Aql 20h14'22"3d49'
Turner,Kiele Gabrielle
Her 16h19'0"25d28'
Turner,Letisha Ann
And 22h57'11"50d26'
Turner,Lisa
Mon 7h42'47"-1d41'
Turner,Lori
Cam 3h32'26"60d57'
Turner,Mary-Kate
Uma 9h11'1"59d54'
Turner,Matthew "Snuggle Bear"
Lac 22h36'30"37d52'
Turner,Monika
And 0h0'0"47d56'
Turner,Monique
Peg 22h13'12"8d29'
Turner,Nancy B
Lyn 7h35'59"58d54'
Turner,Nancy E
Mon 6h20'43"1d7'
Turner,Paul Alexander
Oph 17h10'13"11d55'
Turner,Paul R
Ori 5h57'36"20d45'
Turner,Paul T
Oph 17h54'42"11d3'
Turner,Perry Williams
Sco 17h25'45"-33d49'
Turner,Philip
Per 2h1'1"57d7'
Turner,Polli
Cyg 19h31'49"33d7'
Turner,Rachel Eve
Cmi 7h23'1"8d39"
Turner,Rachel Lynn
Mon 6h58'46"8d43'
Turner,Rhea Irene
And 23h18'0"51d14'
Turner,Robert A
Cep 21h20'49"55d46'
Turner,Ronald S & Patty B Fischer
Mon 6h35'17"1d24'

Turner,S Denise
Com 12h59'36"21d1'
Turner,Sean Colm MacKenzie
Uma 10h18'38"57d9'
Turner,Sheleen M
Her 18h3'1"14d49'
Turner,Sheri Beth
Lyr 18h40'41"33d41'
Turner,Stephen F
Her 17h33'33"23d25'
Turner,Susan Elizabeth
Lyn 8h37'29"40d28'
Turner,Thomas Mickey
Oph 17h58'52"12d46'
Turner,Valerie A
Cas 23h40'31"60d12'
Turner,William Joseph
Her 16h55'28"27d11'
Turner-Thompson,Johnny & Valerie
Per 3h36'40"39d33'
Turneri,Stella
Ori 5h0'30"13d3'
Turney,LLCS-6-94-95
Cmi 7h55'58"8d53'
Turneypops,Leigh
Crb 16h9'58"37d48'
Turning Point
Umi 14h55'0"67d3'
Turnini,Rosa
Aqr 20h55'19"-5d46'
Turnipseed,Beatrice
Cep 23h37'50"65d18'
Turnipseed,Darryl
Uma 11h17'11"43d22'
Turnipseed,Eugene
Lmi 9h33'40"38d22'
Turns,John E
Lac 22h26'28"55d34'
Turo,Betty-James
Aql 19h54'59"12d44'
Turoczy,Darleen Joy Ann
And 2h20'0"50d8'
Turone,Dad "The Star"
Uma 10h31'29"57d3'
Turowski,Joseph
Vir 13h0'1"-10d10'
Turpin,Jonathan Christian
Ori 5h44'0"10d4'
Turpin,Stéphane
Ori 6h5'33"10d58'
Turrell,Marjorie & Douglas
Cyg 21h2'1"30d46'
Turrentine,Christopher Aaron
Uma 9h16'52"61d33'C
Turrini,Renata
Pho 0h29'30"-47d30'
Turro,Gerard
Aur 5h7'1"44d11'
Turs,Kyla Nikol
Lyn 6h22'30"60d51'
Turs,Kyla Nikol
Dra 17h48'45"64d36'
Turtel,MD,Andy H
Cep 4h42'52"87d30'
Turter,Ellen Marie
Cas 0h21'58"68d19'
Turtle
Ari 1h57'27"22d9'
Turtle Creek Elementary
Lyr 19h10'27"38d46'
Turtle,The
Ori 5h53'0"9d35'
Turtle-Dove
Eri 4h31'42"-0d41'
Turtleman
Cmi 7h31'24"7d42'
Tusa,Helmi E
Peg 22h50'55"29d19'
Tuscaloosa
Uma 10h2'21"56d5'
Tuscan,Joseph John
Per 2h3'59"58d21'
Tuscarora Elementary School(Martinsburg,WV)
Uma 8h34'21"62d18'
Tuscher,Walter
Boo 14h3'47"11d59'
Tuschhoff,Dr-Ing Hilmar
Umi 14h11'48"68d30'
Tusek,Vlado
Cmi 7h19'21"5d23'
Tush,Bill
Boo 15h14'53"47d41'
Tushcayinja
Dra 12h3'34"72d27'
Tusindfryd,Lotte
Cas 1h3'36"55d28'
Tusindryd,Louise
Cam 3h44'1"60d16'
Tuski,Kim
Lyr 18h49'45"40d32'
Tuson,Keith W
Cep 20h57'14"71d12'
Tustin,Ella E
And 23h27'40"43d58'
Tutan's Star
Ori 5h57'34"15d3'
Tutcher,Margaret Edna Fawl
Mon 6h39'46"6d49'
Tuter
Lyn 8h2'50"36d20'
Tuteseins
Lib 15h11'8"-21d47'
Tuthill,Erin Nicole
And 2h25'1"40d12'
Tutin
Uma 9h28'32"48d5'
Tutman,Madeline Spring
Tau 4h9'10"20d2'
Tutman,Paula Lorraine
Tau 4h18'2"20d4'
Tuttiett,John Henry & Valmai
Peg 22h30'25"28d59'
Tuttle's Eastern Star, Lyle
Eri 4h0'13"-15d27'
Tuttle,Christopher
Aur 7h8'38"36d58'
Tuttle,Delbert E
Sex 10h20'46"5d58'
Tuttle,Glenn L
Dra 11h33'31"68d33'
Tuttle,Heather Nicole
Cas 1h20'24"55d18'
Tuttle,Tony
Lyn 7h20'53"44d51'
Tuttman,Andrea Lisa
Cas 2h41'0"60d5'
Tuumamo
Uma 8h55'59"68d52'
Tuuri,Anna Rodica
And 2h20'51"38d16'
Tuvonen,Oskar
Dra 16h57'49"69d24'
Tuxedo
Cam 3h44'0"71d59'
Tuxhorn,Caroline Anne
Eri 2h54'15"-17d16'
Tuxhorn,Henry Tyler
Sct 18h45'33"-7d35'
Tuyen's Little Allstar Philly
Peg 23h42'28"27d56'
Tuzman,Esther & Arnold
Lyr 18h41'15"38d47'
Tuzzeo,Joseph & Margaret
Cyg 19h43'0"30d36'
Tuzzeo-Pfeiffer, Kathleen J
And 2h27'32"41d48'
Tvedt,Rick "Howard"
Dra 15h7'27"63d8'
TVY INC
Boo 15h2'46"40d36'
TW's Restaurant
Uma 10h0'10"73d2'
Twaddle,Dale S
Cep 22h38'13"59d57'
TWD Smokey
Oph 18h41'0"10d20'
Twede,Mark Billiou
Aql 19h52'60"12d1'
Tweeddale,Cora
Cyg 19h42'13"38d5'
Tweedle,David
Her 18h16'40"14d56'
Tweeneema
Lyr 18h30'40"36d21'
Tweet,Rosemary Buckingham
Cas 1h7'0"62d31'
Tweetie In The Sky
Sco 16h24'41"-40d34'
Tweetle
Cnv 13h21'1"40d51'
Tweety
Lib 15h15'4"-22d18'
Twells
Her 17h24'58"49d52'
12-14-89
Aql 18h58'49"8d1'
1205 EJC 77-95
Aql 19h16'0"15d29'
20
Lib 15h9'9"-16d29'
20th Century Launder
Peg 22h12'58"30d17'
21K28JLOFCSRBRMOPIGRWD 46AG9114N94
Peg 23h19'53"14d45'
21st US Marine Regiment
Her 16h59'0"28d26'
24 x 7 WOBU
Umi 13h15'49"70d18'
25 Years of Charlotte & Philip!
Cyg 21h2'55"38d17'
25th July 1994
Umi 17h12'0"76d21'
26 Michael,143 Marchesa Raffael
Del 21h5'32"12d6'
28th Star of Biltmore, The
Lmi 9h59'56"28d51'
Twentey,Patricia Halley
Cam 5h41'19"61d18'
Twentyman's 50th Birthday,Peter
Peg 22h46'13"33d24'
Twidge
And 0h54'55"37d38'
Twigg My Imzadi,Tori
Lyr 18h47'28"37d22'
Twigg,Earle's Girl Taylor Delmonica
Peg 23h1'47"17d45'
Twigg,Justice
Cet 0h50'57"-4d53'
Twigg,Sylvia Joyce
Boo 14h59'53"33d58'
Twilight Tyres
Lyr 18h47'56"41d12'
Twilley,Byron Bill
Her 17h12'54"21d18'
Twink
Dra 14h7'30"64d2'
Twink
Cmi 7h43'11"5d20'
Twinkle
Cmi 7h55'1"8d28'
Twinkle
Uma 10h16'39"67d44'
Twinkle
Per 3h17'56"40d35'
Twinkle
Uma 9h50'38"55d7'
Twinkle
Uma 8h35'30"55d11'
Twinkle
Cyg 20h10'40"38d60'
Twinkle
Cep 21h8'29"60d40'
Twinkle Daddy
Cep 23h0'12"64d8'
Twinkle in Charlotte's Eyes,The
Mon 6h37'11"6d1'
Twinkle In Dina's Eyes,The
Cam 5h31'26"68d43'
Twinkle of Anna's Eye, The
Pho 0h33'27"-45d17'
Twinkle of Katie's Eyes,The
Lyn 7h7'25"59d44'
Twinkle Rudberg
Per 2h56'60"46d21'
Twinkle Toes
Boo 13h42'58"14d16'
Twinkle Toes Tidings
Uma 12h12'29"59d1'
Twinkle Tracey
Cyg 20h39'37"45d45'
Twinkle Twinkle Masayoshi & Kazuko
Tau 5h40'20"20d0'
Twinkle Twinkle
Cas 1h52'13"73d10'
Twinkle Twinkle Richard Pa
Cep 23h10'23"61d0'
Twinkle Twinkle "Carolyn's" Star
Cyg 21h25'11"39d45'
Twinkle Twinkle Little Mark
Ori 5h58'0"15d6'
Twinkle Twinkle MNS
Uma 9h15'15"61d53'
Twinkle's Destiny
Ori 6h6'38"1d9'
Twinkle,Twinkle Little "Bill"
Cam 13h2'16"78d42'
Twinkle,Twinkle Little Bo-Bo
Cnv 12h9'56"39d28'
Twinkle,Twinkle Little Dar
Uma 8h55'60"53d46'
Twinkle,Twinkle,Baby Bo
Hya 10h32'12"-11d34'
Twinkle-Twinkle Tyler's Star
Lyn 7h56'13"43d51'
Twinkler,Sweet Irene & Witty Stanley
Sge 19h43'48"16d20'
Twinkles of 3A,The
And 0h9'1"34d58'
Twinkling Bobo Wobo
Ser 15h43'53"19d57'
Twinkling Cardinal
Cam 3h47'37"55d52'
Twinkling Dads-MDY & HWF
Cnv 13h56'42"38d53'
Twinkling Emma Louise
Lyr 18h31'47"37d48'
Twinkling Helen
Peg 22h9'52"29d8'
Twinkling Hubert
Cep 21h47'38"80d0'
Twinkling Jenn
Cma 6h53'43"-18d19'
Twinkling Timothy
Ori 4h58'28"-2d54'
Twins Forever
Lyr 19h5'38"37d44'
Twintreess
Cnc 8h21'31"15d9'
Twitter-Duder
Cep 22h21'17"55d44'
2 B or Not to Be
Vul 19h49'38"20d33'
2 Pieces of a Puzzle
Cyg 20h2'58"40d5'
2B
Cam 5h0'38"65d22'
2E Norris
Hya 10h1'12"-18d53'
2Gether 4ever Mike & Kim "Soulmates"
Sge 20h17'43"16d28'
Two Bobs,The
Aql 20h10'10"-3d8'AB
Two Boston Bongs
Ori 5h50'11"16d46'
Two Doves Tara & Stephen Star
Peg 21h59'18"35d55'
Two Ed's R Better Than One
Her 18h8'0"28d58'
Two Elk
Aur 6h34'0"38d21'
Two Glyns,The
Uma 11h27'17"32d3'
Two Good To Be True
And 0h9'53"28d42'
Two Hearts
Cam 11h36'50"84d55'
Two Jewells-Chris
Sge 20h8'58"16d42'B
Two Jewells-Jan
Sge 20h8'58"16d42'A
Two Seas For My Cassandra/Christina
Cas 1h3'0"61d2'
Two-Step Weave
Peg 23h42'0"30d30'
Twogether Forever- Terry & Joy
Crb 15h27'53"31d38'
Twohig,Francis Jeremiah
Aql 20h6'1"7d10'
Twombly,Sr,Timothy Wayne
Dra 16h40'48"52d21'
Twomey
Her 16h32'51"38d16'
Twomey,Doris Marie
Cyg 20h58'57"30d15'
Twomey,Edward J
Dra 12h58'42"71d37'
Twomey,Helena Beth
And 2h29'12"48d41'
Twydell,Alison Sarah
Cyg 19h28'26"37d6'
Twyford III,William H
Boo 14h14'39"16d46'
Ty Troy
Lac 22h4'50"46d9'
Tyborczyk,Doris
Ari 2h2'15"21d59'
Tyche
Gru 22h7'8"-54d15'
Tye,Lindsey Paige
Cyg 21h32'55"53d3'
Tye,Terence Leif
Oph 17h6'11"8d30'
Tyer,Ashley
Lyn 7h50'1"44d27'
Tyer,Bubba
Boo 14h58'43"20d35'
Tygerr "The Star Of Best Friends"
Umi 14h23'46"68d22'
Tyler
Aql 19h6'1"15d13'
Tyler
Aur 5h23'15"54d60'
Tyler & Lori
Tri 1h58'52"33d50'
Tyler 2995
Dra 16h39'51"62d34'
Tyler 4 May 1922, Francis Robert (Bob)
Cyg 19h34'28"31d2'
Tyler Andrew Always Shining For Us
Dra 16h44'21"72d37'
Tyler C
Oph 16h52'24"-20d1'
Tyler Gregory
Lac 22h27'0"55d12'
Tyler M
Her 17h0'14"28d10'
Tyler Michele
Ori 6h4'35"7d1'
Tyler's Shot Star
Uma 10h52'21"40d24'
Tyler's Star
Hya 9h35'56"-10d47'
Tyler's Star
Her 17h39'31"18d52'
Tyler's Star,DeAnn
Peg 21h23'35"23d45'
Tyler's Touch
Peg 0h5'33"22d16'
Tyler,"Red Delight"/ SSG Sweetie Ann
Lyn 6h26'20"60d36'
Tyler,Adele
And 1h12'0"37d9'
Tyler,Brian A
Sct 18h54'0"-6d4'
Tyler,Brian Daniel Semler
Cyg 21h41'18"37d32'
Tyler,Daniel Thomas
Cyg 21h39'38"42d14'
Tyler,Evelyn & Jim
Cet 3h18'59"3d53'
Tyler,Gordon
Aql 20h13'51"0d6'
Tyler,Jason
Leo 9h33'0"7d6'
Tyler,Jeremy Dustin
Cet 0h56'54"1d1'
Tyler,Lorraine Elizabeth
Ori 6h3'1"2d59'
Tyler,Marjorie "Nana"
And 0h20'21"36d17'
Tyler,Mildred
Lmi 9h20'1"34d15'
Tyler,My Favorite Star
Psc 1h21'48"43d44'
Tyler,Nikki
Peg 21h38'40"24d35'
Tyler,Patricia
Cam 5h2'18"60d9'
Tyler,Patricia Alice
Cam 5h17'0"68d59'
Tyler,Robin
Mon 7h25'35"-10d25'
Tyler,Steven
Her 18h2'52"41d9'
Tyler,Tenaya & Art
Sge 18h58'1"18d37'
Tyler,Teresa
Cam 5h53'60"73d8'
Tyler,Three Martin
Cnc 9h6'16"28d46'
Tyler,Tim "Timmy T"
Cam 6h16'48"83d17'
Tyles,Christine Nicol
Lyr 18h59'31"30d10'
Tyll,Edward J
Cyg 21h38'40"41d48'
Tylman,Anthony
Ori 5h55'30"11d15'
TYM
Tau 5h55'0"23d31'
Tymko,Erin L
And 2h0'35"47d6'
Tymn,Christar For Chris
Tri 2h32'50"31d46'
Tynan,Frances Elizabeth
Cas 1h45'34"77d21'
Tynan,Julie
Mon 6h23'36"5d22'
Tyndall,Russell Lee
Ori 5h41'17"12d24'
Tyne,Kenly
Cam 6h48'22"68d53'
Tyner,Cassandra
Cas 0h32'29"62d23'
Tyner,Jr,William Henry
Ori 4h58'28"-2d54'
Tyner,The Star of Bob
Per 1h49'14"48d54'
Tynes,James & Doris
Uma 11h36'49"31d59'
Tynes,Veronica
Peg 0h3'30"30d21'
Tyng,Anna S & William W
Sge 19h28'38"19d21'
Tyree,George Wishar
Aqr 23h31'18"-5d4'
Tyrka,Ray
Lib 14h56'14"-4d47'B
Tyrone's Star
Cnc 9h8'17"31d37'
Tyrone-The Sunshine Star
Car 7h34'24"-54d0'
Tyrpak,Kenneth Andrew
Per 2h57'29"37d58'
Tyrrell,"Angel" Sydney Chyenne
And 1h46'54"40d52'
Tyrrell,Jessica
Aql 19h56'15"8d15'
Tyrrell,John Paul
Aur 5h27'25"29d13'
Tyrrell,Mr Derek & Mrs Helen
Ant 10h46'34"-32d49'
Tyrrell,Nick
Aur 5h5'51"41d22'
Tyrrell,Wesley P
Aql 19h17'0"19d43'
Tyrrell-Kenyon, Alexander
Dra 18h59'0"58d29'
Tyson
Dra 15h17'27"55d58'
Tyson,Alexis
Peg 23h43'45"18d26'
Tyson,Bruce & Julie
Cyg 19h28'35"32d23'
Tyson,Carla
Cet 1h25'16"-11d9'A
Tyson,Cassandra
Uma 10h52'0"38d17'
Tyson,Dick
Her 16h52'40"14d6'
Tyson,John Howard
Boo 14h2'56"8d19'
Tyson,Julie K
And 2h24'46"50d18'
Tyson,Matt
Per 3h12'0"46d13'
Tyson,Teri A
Vul 20h45'60"28d28'
Tystar
Uma 8h59'39"58d31'
Tytler,Hilary & Robert
Cyg 20h53'33"37d52'
Tyus,Jim
Sct 18h33'24"-6d3'
Tzenman,Paul David
Uma 9h17'14"67d41'
Tzevnick
Uma 11h49'0"30d50'
Tzortzi,Evagelia Christodoulidou
Lyr 18h18'32"40d51'
Tzva Alta Kakas
Vul 19h49'1"20d10'
Tähti,Mikan & Sirpan Onnen
Cam 5h49'32"80d5'
Töllen,Ellen
Sgr 18h47'8"-25d10'
Töllner,Dirk
Aql 20h9'14"1d8'
Tünkler,Gerhard
And 1h41'53"41d43'
Türk,Udo
Ori 5h58'25"1d12'
Türpitz,Wolfgang
Aql 18h58'31"17d37'
T" Lifestar
Uma 8h36'21"53d38'

U

U P C E
Per 3h43'36"37d55'
U R Z
Boo 14h12'34"36d8'
U S S Shadow Hawk
Dra 16h25'41"68d27'
U S Technologies
Lac 22h29'22"48d55'
U2 Sunshine
Per 2h28'18"56d49'
U4EREH
Cnv 13h1'50"50d50'
Ubben,Bradley W
Uma 9h7'1"48d25'
UBE
Ori 6h3'60"0d36'
Ubeda,Marta
Boo 14h16'39"19d16'
Uchida,Christy L Fulbright'
Cas 0h55'20"58d44'
Uckermark,Carsten Alexander
Sco 17h6'12"-38d20'
Udell,Jerry
Her 17h16'52"42d45'
Udesky,Robert
Oph 17h55'43"14d2'
Udich,Betty Louise
Com 12h32'25"23d26'
Udo's Glücksstern
Ari 2h52'16"30d27'
Udoni,Jenn Jennifer Beth
Uma 11h37'24"61d46'
Uebele,Volker
Lmi 9h20'12"37d56'
Uebelhoer,Aaron Michael
Cnv 12h14'31"48d19'
Uebelhoer,Adam Thomas
Aur 4h58'60"51d44'
Uejo,Kiahna Lauren
Oph 17h26'33"-23d17'
Uelk,Laura
Peg 23h42'39"30d10'
Uelk,Tim
Ori 5h59'15"15d57'
Uemura,Akiko
Vir 13h23'27"-3d39'
Uestal,Kathleen Lepore
Aur 6h27'18"40d46'
Uetz,Kurt Marcial
Dra 17h28'0"58d35'
Ufelman,Jeff
Aur 4h48m18"48d51'
Ufen,George R
Cyg 19h41'14"40d36'A
Ufen,Sandra A
Cyg 19h41'14"40d36'B
Ufer,Amy Diane
Aql 19h28'34"10d45'
Uffen
Cam 6h37'18"68d47'
Uffens,Alexis Morgan
Aql 19h30'16"0d29'
Uflorian
Peg 21h29'1"20d11'
UFO-TNT-Reuer
Aql 20h22'30"0d46'
Ugarow,Wincenty
Peg 23h7'50"12d19'
Ugarte,Paulina
Gem 7h29'19"30d36'
Uggi
Cyg 21h1'20"38d23'
Uggie
Cnv 12h26'43"46d7'
Ugh
Lmi 10h27'0"31d42'
Ughi Sonia
Cyg 20h40'1"31d25'
Ugielie,Adnan
Cyg 19h27'36"30d31'
Ugly
Cet 1h30'26"-11d6'
Ugly Moose
Dra 16h54'45"67d44'
Ugly Mus-tard
Boo 14h23'19"16d7'
Ugnon
Cae 4h51'19"-31d59'
Ugo
Lep 4h56'30"-19d52'
Ugo,Catherine
Ori 6h21'1"10d22'
Ugolini,Luigi
Hya 9h10'6"-14d47
Ugolini,Mario
Aur 5h54'1"30d38'
Ugrin,Ingrid
And 0h46'1"40d46'
Uguccioni,Giovanna
Ori 5h49'52"10d30'
Uhed 5
Ori 6h0'58"7d44'
Uhl,Doris J
And 23h29'0"45d12'
Uhl,Günther
Her 17h46'0"42d12'
Uhl,Margaret
Mon 6h25'0"7d27'
Uhle,D & A
Ori 5h57'41"15d55'
Uhler,Dale
Cep 2h47'50"59d6'
Uhlhaas,Doris
Gem 6h46'47"16d21'
Uhlig,Crystal L
Lyr 18h58'19"46d9'
Uhlig,Karen E
Cyg 21h17'32"38d29'
Uhmann,Hartmut
Cnv 12h16'33"37d22'
Uhran,John T
Her 17h54'60"14d47'
Uhrig,Josephine Garcia
Lyr 18h44'13"32d8'
Uhrskov,Elizabeth & Tage
Ori 5h32'46"0d28'
Uhter,Hans-Theodor
Lyn 8h3'37"47d13'
Uhuru II
Cyg 19h33'49"28d13'
Uinta Business System
Uma 9h36'32"48d21'
Ukena,Helmuth
Peg 23h33'55"18d1'
Ukita,Stella of Katsumi & Kazue
Aqr 22h16'26"-11d3'
Ulan,Angel
Del 20h13'10"15d41'
Ulan,Melissa Jill
And 0h19'59"36d4'
Ulaner,Gary Jay
Hya 8h35'46"0d49'
Ulatowski,Valerie
Cas 1h56'11"58d9'
Ulbricht,Bob & Betty
Cnv 13h18'20"28d21'
Ulbrick,Shannon & Fred
Cyg 21h54'10"42d42'
Ulderico
Cae 4h59'25"-32d58'
Ule
Leo 10h7'55"7d13'
Uli
Uma 9h40'59"48d11'
Ulibarri,Ernest
Mon 6h11'0"-10d22'
Ulibarri,Kenneth L
Her 17h37'41"43d0'
Ulibarri,Kenneth L (JR) De
Her 17h12'15"42d0'
Ulickas III,Walter James
Sct 18h20'43"-13d34'
Ulinski,Christina
And 2h34'53"50d16'
Ulisa - Fletcher
Cyg 21h57'1"50d27'
Ulises Star
Per 4h3'21"48d55'

Ulki
Eri 5h2'0"-4d30'
Ulla
Umi 15h24'0"69d12'
Ulla
Ari 2h5'26"23d38'
Ullah,Abraham
Per 3h0'1"56d42'
Ullas Traumstern
Cnc 9h17'32"31d4'
Uller,Sandy
Cas 23h27'48"61d45'
Ullersberger,Ma-mar & Pop
Uma 9h44'1"56d19'
Ulli
Dra 20h26'58"62d20'
Ulli u Karl 1971
Cep 22h3'47"60d57'
Ulli und Gert
Uma 11h10'28"60d3'
Ullman,Mike
Del 20h26'10"20d25'
Ullmo,Phyllis June
Uma 8h43'25"68d38'
Ullo,Doris May
Cas 0h35'0"63d8'
Ulloa,Ronnie
Boo 14h51'38"25d51'
Ullom,Kelsey Erin
Mon 7h46'8"-3d56'
Ullom,Matthew T
Her 16h54'44"28d27'
Ullrich,Brigitte
Cam 5h55'0"68d23'
Ullrich,Drew
Uma 9h56'55"57d23'
Ullrich,Frederick S
Aur 6h10'55"31d19'
Ullrich,Jürgen
Cap 20h5'21"-10d40'
Ullrich,Uwe
And 23h2'51"42d38'
Ullswater,The
Uma 9h18'43"57d30'
Ulman,A Jay
Aur 6h58'20"43d27'
Ulman,Leslie A
And 0h26'0"38d6'
Ulmer,Alexander "X"
Dra 17h34'47"63d60'
Ulmer,Dorithy J
Mon 7h46'16"-2d8'
Ulmer,Kelly & Jade Totman
Cet 2h25'23"4d37'
Ulmer,Sarah Marie
Peg 22h33'0"29d47'
Ulmicom
Cet 2h57'42"6d1'
Ulrey,Ivy Olivia
Uma 11h43'25"48d47'
Ulrey,Mark L
Peg 22h4'47"8d56'
Ulrey,Zoey Isabella
Uma 11h43'19"38d56'
Ulrich
Peg 23h31'24"12d17'
Ulrich
Ari 2h23'24"10d44'
Ulrich,25 Janre Christel und
Hans
Cep 22h5'17"68d59'
Ulrich,Debbie
Aql 20h16'54"7d31'
Ulrich,Helen & Herman
Cyg 19h58'43"48d57'
Ulrich,Jennifer
Lyr 19h3'33"28d41'
Ulrich,Jr,John M
Boo 15h2'27"20d13'
Ulrich,Kristi Lynn
Lyn 7h8'38"53d58'
Ulrich,Nina Morstad
Sge 19h5'56"19d23'
Ulrich,Steven
Her 18h13'17"31d51'
Ulrichsen,Lana
Lyr 18h40'41"35d35'
Ulshoefer,Gerhardt Hilde
Cep 22h9'27"67d56'
Ultimate D B E,The
Ori 4h53'56"-2d58'
Ultraria
Boo 14h58'36"30d31'
Ulusoy,Aly-Joy
Dra 14h30'0"64d29'
Ulvestad,Cato
Per 2h41'7"36d31'
Uly Gooch & Worth Kenyon
Sge 19h16'35"20d5'
Umanoff,Nancy
Mon 6h12'2"-9d1'A
Umaro
Cnv 12h19'0"47d43'
Umberger,Mark Alan
Ori 5h31'25"-1d47'
Umbert,Dolly Villalba
And 23h2'12"50d1'
Umbinetti,Kiersten Lee
Lyn 7h10'0"58d23'
Umbinetti,Lindsay Alyse
Lyn 6h54'1"59d20'
Umbra,Maria Marta
Ari 2h34'0"20d48'
Umebayaski,Mitsunori
And 2h2'21"38d13'
Umeboshi
Ari 2h24'57"15d53'
Umeboshi
Aqr 22h3'9"-16d14'
Umfante
Uma 10h39'20"58d54'
Umhau,MD,J B
Per 3h24'30"50d7'
Umla,Wolfgang
Umi 13h35'15"74d45'
Ummes
Tri 2h28'17"31d43'
Umphlett,Evelyn Marceron
And 23h2'32"51d48'
Una 1
Peg 21h56'52"31d39'
Una Amica In Lea
Ori 6h16'12"18d59'
Una Bella
Cas 23h38'51"61d36'
Una stella navigante in un
mare d'emozioni
Psa 22h3'45"-27d48'
Una stella è un amore infinto
Vania ed Ale
Cam 6h39'42"68d18'
Unalipas
Cet 2h40'1"7d47'
Unbehaun,Ulaf Siegmar Detlef
Cmi 7h5'51"1d38'
Unbridled Passion: Stephen &
Kathleen
Uma 8h40'1"70d38'
Unc
Her 16h41'0"32d28'
Unc & Auntie M
Cyg 20h59'42"31d40'
Unc O'Brien
Aql 18h59'15"-2d46'
Unchained Melody
Cyg 19h32'18"33d32'
Uncle 80's
Aur 6h35'32"32d42'
Uncle Alan's 50th Birthday
Star
Ser 15h43'58"0d46'
Uncle Arthur
Aur 5h12'38"41d15'
Uncle Barry
Her 16h48'41"48d23'
Uncle Bill
Cep 22h36'43"60d57'
Uncle Bob
Lac 22h15'1"51d1'
Uncle Bob's Star
Lac 22h38'0"56d16'
Uncle Bud's B&B
Aur 5h7'1"38d12'
Uncle Charlie
Boo 14h21'1"30d55'
Uncle Daddy Paul
Ori 5h33'0"-8d24'
Uncle Dickie
Oph 17h6'4"-23d26'
Uncle Doc
Per 2h58'21"37d15'
Uncle Don
Per 1h56'0"47d38'
Uncle Franken
Aur 5h14'14"40d8'
Uncle Freddie
Uma 10h55'30"70d11'
Uncle Gary
Boo 13h45'20"20d46'
Uncle Gene
Aur 5h5'0"50d52'
Uncle Harold
Aur 5h52'43"38d50'
Uncle Harry
Uma 8h43'48"50d19'
Uncle Jay
Ser 15h16'1"12d25'
Uncle Jimmy
Per 2h56'33"37d42'
Uncle Max
Cep 22h24'53"63d30'
Uncle Mikey
Lac 22h36'37"50d36'
Uncle Pat
Aur 5h7'55"40d23'
Uncle Pat
Uma 11h51'17"33d36'
Uncle Pete
Lyr 19h18'0"42d22'
Uncle Phil
Boo 13h58'0"20d5'
Uncle R C
Cet 2h43'16"1d22'
Uncle Ray
Cep 22h3'14"53d34'
Uncle Red
Ori 6h8'11"1d28'
Uncle Rick
Eri 4h14'7"-12d29'
Uncle Rom
Uma 9h48'26"50d38'
Uncle Rudy
Cet 3h19'17"7d21'
Uncle Sammy Star,The
Oph 17h4'30"10d10'
Uncle Scott Guardian Angel
Star,The
Ori 5h47'38"11d5'
Uncle Sethie
Per 2h25'47"51d60'
Uncle Si
Ori 6h7'43"0d24'
Uncle Tim
Ari 3h14'33"20d5'
Uncle Tony
Ori 6h12'1"8d2'
Uncle Wayne
Vul 19h48'36"27d42'
Uncle Yummy
Cnv 12h58'17"51d16'
"Un-Cola" Constellation
Uma 11h44'59"48d51'
"Un-Cola" Constellation
Uma 11h44'49"41d12'
"Un-Cola" Constellation
Uma 11h43'1"40d31'
"Un-Cola" Constellation
Uma 11h38'29"40d4'
"Un-Cola" Constellation
Uma 11h46'1"44d13'
"Un-Cola" Constellation
Uma 11h44'30"41d7'
"Un-Cola" Constellation
Uma 11h39'46"40d27'
"Un-Cola" Constellation
Uma 11h26'21"40d5'
"Un-Cola" Constellation
Uma 11h24'50"40d8'
"Un-Cola" Constellation
Uma 11h18'0"41d1'
"Un-Cola" Constellation
Uma 11h15'58"44d52'
"Un-Cola" Constellation
Uma 10h52'56"40d46'
"Un-Cola" Constellation
Uma 11h18'25"40d58'
"Un-Cola" Constellation
Uma 11h16'18"43d48'
"Un-Cola" Constellation
Uma 11h18'30"48d59'
"Un-Cola" Constellation
Uma 10h52'44"40d4'
"Un-Cola" Constellation
Uma 10h52'49"44d55'
"Un-Cola" Constellation
Uma 10h54'24"48d16'
"Un-Cola" Constellation
Uma 10h36'30"45d40'
"Un-Cola" Constellation
Uma 10h15'37"47d40'
"Un-Cola" Constellation
Uma 10h52'37"47d40'
"Un-Cola" Constellation
Uma 10h46'46"47d4'
"Un-Cola" Constellation
Uma 10h27'27"42d12'
"Un-Cola" Constellation
Uma 10h15'38"40d41'
"Un-Cola" Constellation
Uma 10h15'12"41d39'
"Un-Cola" Constellation
Uma 10h15'56"48d17'
"Un-Cola" Constellation
Uma 10h15'17"42d7'
"Un-Cola" Constellation
Uma 10h15'12"41d39'
Unconditional
Oph 17h54'41"12d17'
Unconditional Friendship
Uma 10h44'0"53d24'
Unconditional Love
Cnv 12h54'28"40d45'
Unconditionally Zoe
Lmi 9h32'49"38d51'
Uncow
Boo 14h20'13"15d54'
Uncy
Aql 18h58'20"-6d51'
Undercoffer,Orvis J
Sge 19h9'1"19d9'
Underhill,Byron D
Aqr 20h58'12"-13d9'
Underhill,Mike
Ori 6h6'23" 0d25'
Underwood,Blair
Aql 19h5'0"2d54'
Underwood,Denzil & Virginia
Cyg 19h34'53"38d25'
Underwood,Dorothy Helene
Mon 6h21'55"2d60'
Underwood,Désirée & Blair
Mon 6h59'43"8d10'
Underwood,Eileen
Cas 2h14'0"67d47'
Underwood,Eric
Dra 16h56'60"68d35'
Underwood,Faye Marie
Cyg 21h32'35"53d16'
Underwood,Franklin
Tau 3h55'43"30d5'
Underwood,Jason Michael
Uma 12h5'39"52d19'
Underwood,Marge (Betty
Boop)
Peg 22h2'1"31d48'
Underwood,Mary Margaret
Aql 20h32'33"0d50'
Underwood,Molly Rose Muriel
Cas 0h16'11"47d8'
Underwood,Nancy
Her 16h21'37"11d11'
Underwood,Roberts F
Aql 19h31'18"1d4'
Underwood,Robin
Aql 19h58'46"10d2'
Undying Love
Aql 20h7'14"0d57'
UndAql 20h7'14"0d57'
Ori 5h19'26"15d17'
Une Etoile Lise
Cyg 19h27'32"38d57'
Unforgetable
Cyg 21h29'47"30d14'
Unforgetable
Psc 1h14'47"22d0'
Unforgetable
Gem 6h49'32"19d11'
Unforgettable
Ori 5h53'54"13d16'
Unforgettable
Cam 5h51'18"56d6'
Unforgettable
Uma 11h52'27"62d11'
Unforgettable
Dra 14h32'26"62d11'
Unforgettable
Crb 16h22'11"30d39'
Unforgettable Carol
Cyg 19h29'20"30d24'
Unforgettable Lori
Cam 13h24'58"76d57'
Unforgettable Q
Uma 9h47'1"70d1'
Unforgettable You
Hya 8h39'19"-0d13'
Ungar,William Stephen
Cma 7h14'31"-11d14'
Ungari,Robyn Bernadette
Cet 2h30'0"-10d41'
Ungarino,Stephanie Ann
Sgr 19h9'23"-24d23'
Ungawa Julie
Lyn 7h25'0"44d43'
Unger'Fink,Allison Ann
Mon 6h36'23"1d14'
Unger,Barbara
And 0h46'15"37d57'
Unger,Bill
Her 17h8'19"42d50'
Unger,Edith C
Vul 21h18'1"28d14'
Unger,Erika
Aqr 21h55'5"-3d4'
Unger,Gabriele
Lib 14h39'56"-23d32'
Unger,Honoria C
Lyr 18h29'56"42d28'
Unger,Tammy
Lyn 8h30'10"41d11'
Unger,Walter-Zee Silver Fox
Gru 22h10'30"-52d30'
Ungestüme,Dr Ferdinand
Podkowicz d
Aql 18h40'42"0d37'
Uni
Lac 22h8'41"51d47'
Uniat,John & Leslie
Cyg 20h2'58"40d5'
Unica: Reina de las Amazones
de Amor
Peg 23h1'46"33d40'
Unicorn I
Cap 21h0'29"-20d36'
Unicorn Rachel
Mon 8h1'1"-1d21"
Union Station Lounge
Cam 4h53'33"67d30'
Union,Taylor Morris
Aur 6h6'1"46d31'
Unique
Dra 11h31'45"67d32'
Uniquely Tony
Lac 22h42'14"53d17'
Unity
Crb 16h15'17"27d2'
Universal Distribution
Services
Dra 16h51'1"71d36'
Universes
Ori 5h56'31"15d12'
Unk,Dan & Lois
Cyg 20h7'12"41d13'
Unka
Oph 18h42'43"10d38'
Unmistakably Sheela
Cyg 19h24'17"33d6'
Unpremeditated Passion
Sge 19h57'19"18d46'
Unruh & Wysuph
Cyg 21h13'21"35d59'
Unruh,Monica
Com 12h18m33'22d12'
Unruh,Telsa
Cmi 7h35'32"1d42'
Unser Paradies für Heinz
Gem 6h49'17"18d14'
Unsheim,Günther
Lyr 19h23'25"30d27'
Unsworth,Audrey
Cas 0h7'45"61d28'
Unterberg,Walter A
Lac 22h41'50"54d32'
Unteregger,Berthold Maria
Cap 20h37'31"-14d38'
Untervald,Jacques
Sex 10h12'3"-2d20'
Until The End Of The World
Mon 7h3'58"-1d9'
Until The Last Moment
Her 17h30'22"21d16'
Untinen,Kaitlyn Lenore
Karlena
Cam 6h5'24"67d55'
Untinen,Kirsten Geordie
Uma 12h14'31"55d59'
unu s
Uma 9h46'58"48d23'
Unus Infinitus
Uma 10h16'59"58d46'
Unverzagt,Alexander
Cap 20h38'33"-18d56'
Unwin,Clover
Cyg 19h28'21"35d60'
Unwin,Teresa A
Lib 14h53'35"-5d50'
Upah,Jake Brian
Peg 23h17'31"33d57'
Upchurch,Doris M
Peg 21h42'38"24d21'
Upchurch,Mollye Diana
Cam 6h36'19"83d5'
Updegrove,Sylvia & John
Boo 14h59'35"20d9'
Updike,Susan Hobbie
Cam 7h54'42"68d23'
Upham
Ori 6h7'44"0d1'
Upham,Lesley Leigh
Cas 1h1'37"62d8'
Uphues,Christopher
Her 17h35'44"26d38'
Uplegger,Hans
Lyr 19h22'42"30d34'
Uplifting Symbiosis
Mon 7h32'15"-1d37'
Upp,Ronald
Oph 16h22'19"-22d38'
Uppal,Bhim Sian
Aql 20h10'43"1d2'
Upper,Elizabeth
Lyn 8h27'51"51d43'
Uppey Alyse
Com 12h2'1"27d12'
Uppington,Shelia
Uma 9h42'14"47d53'
Upshall,Roland
Lyn 9h18'26"38d52'
Upshaw,Amelia
Crb 16h12'34"34d34'
Upson,Stacey Marie
Mon 7h6'29"-1d27'
Upson,Thomas John
Per 1h53'31"48d12'
Upston,Tommy Lee
Aur 6h3'42"45d41'
Upton,Allen Quinn
Lac 22h22'11"48d58'
Upton,Connor Reed
Aur 6h16'34"46d19'
Upton,Janet Hickey
Eri 3h21'56"-3d6'
Upton,Judy
Mon 7h58'9"-3d21'
Upton,Maria
Lac 22h35'57"53d53'
Upton,Mark William
Aqr 22h30'21"0d46'
Upton,Martin
Ori 6h7'40"2d48'
Upton,Robben Renè
Dra 10h18'45"73d52'
Ur Kat
Eri 4h8'20"-18d37'
Ura Stella di Makoto e Mariko
Gem 7h14'45"24d5'
Uralli,Patricia Paduano
Findaro
Sge 20h0'35"17d27'
Uram,Sidney
Aur 5h57'23"29d43'
Uram,Sue
Lyr 19h13'37"37d55'
Urania Evdoxia
Cam 4h55'1"60d39'
Uranut,Gottfried
Dra 20h5'1"75d31'
Uras,Stefano
Umi 16h0'41"86d5'
Uras,Stefano
Her 16h4'35"16d37'
Urasaki,Hiromi
Her 17h46'3"34d17'A
Uraue,Jean-Pierre
Aur 6h13'9"31d21'
Urbain,Bernardo
Ori 6h4'57"8d51'
Urban II,William Joseph
Aql 18h57'57"-4d56'
Urban's Light
Lyr 18h28'29"34d36'
Urban,Albert
Per 3h59'46"51d39'
Urban,Caitlin Starr
Cas 0h35'48"72d53'
Urban,Detlef
Lyr 19h20'22"31d7'
Urban,Hans W
Hya 8h34'1"-10d39'
Urban,Jack
Aql 19h57'13"10d23'
Urban,Jonathon
Aql 18h58'0"14d7'
Urban-Harvey's Rolls-
Royce,Ralph J
Hya 8h17'42"2d32'
Urbanik,MaryAnne
Lyn 7h27'30"42d31'
Urbano
Boo 14h6'43"42d9'
Urbano,Anthony
Dra 19h0'12"65d36'
Urbanovsky,Tony
Dra 16h5'22"63d7'
Urbanowski,Kate
Umi 15h14'17"84d52'
Urbanska,Malgosia
Del 20h16'46"10d30'
Urbanski,Dr David / Dawn
Doros
Dra 17h6'0"61d21'
Urbanski,Jeannie
Com 12h49'44"23d29'
Urben,John Albert
Her 16h9'35"50d27'
Urbina,Ricardo M
Per 2h7'51"47d39'
Urbinati,Maria Cristina
Pic 4h48'34"-48d50'
Urcan,Jr,Eugene-Rapid A/C
Lac 22h3'57"41d3'
Urchison,Karen Sue
Del 20h13'56"11d2'
Ure,The Fiona
Cyg 19h27'49"38d13'
Urge,Alexandra
Lib 15h37'0"-20d10'
Urgo,Mary & John
Cyg 20h53'50"30d23'
URGONOVA
Aur 6h18'37"37d32'
Urguhart,Jenny
Com 12h15'1"21d60'
Urguhart,Pamela Jean
Crb 16h18'11"28d53'
URI,Grace Marama & James
Germain
Cam 4h48'14"68d40'
Urias,Margaret Kristine
Lyr 18h50'16"42d36'
Uribe,Alexander Jon Ethan
Peg 23h33'12"17d34'
Uribe,Caroline LK
Cas 0h59'33"50d35'
Uriel
Aur 6h17'34"32d48'
Uriwal,Edmond Cashmere
Hya 8h36'0"-0d28'
Urk's Love
Aur 5h3'39"51d29'
Urmilajh Roma Disney
Cep 0h0'32"70d21'
Urmson,Vauna
Cyg 20h32'0"40d38'
Urmy-Chiapa,Jacqueline
Marie
Mon 7h6'0"0d16'
Urner CLU,James A
Cep 3h41'58"78d1'
Urosevich,Todd
Cet 2h51'56"6d33'
Urquhart,Kris Riley
Aql 19h3'25"5d28'
Urquhart,Patti-a Real Star
Gem 6h6'0"26d21'
Urquhart,Thomas Whitmel
Her 18h18'12"18d46'
Urquhart,William Bernard
Per 2h52'1"38d34'
URS
Umi 13h56'0"77d42'
Urs-Bert
Uma 8h42'37"68d34'
Ursa Derek - Derek "The Bear"
Uma 12h34'0"61d24'
Ursel-1
Psc 23h2'16"1d14'
Urso,Bob
Lac 22h10'14"51d29'
Urso,Lucia
Lyn 7h28'0"40d53'
Urso,Mary
Lyn 7h45'38"44d28'
Urso,Paul J
Ser 15h38'16"0d15'
Ursu,Katelyn MacKenzie
Lyr 19h24'13"40d42'
Ursula
Cyg 20h26'1"40d38'
Ursula
And 23h4'53"41d6'
Ursula & Jeffrey
Cyg 21h8'1"40d39'
Ursula Gabriel
Vul 19h18'19"26d31'
Ursula-Christl
Sco 17h54'30'-30d49'
Urth,Dietmar
Leo 9h59'0"11d59'
Urwin,Colin William
Aur 5h31'11"38d52'
Ury,Travis
Cyg 21h11'1"39d18'
Ury,Tyler
Cyg 21h4'1"39d35'
Us
Lyn 7h57'45"41d34'
US
Equ 21h10'40"11d38'
US(Neni & J C)
Her 16h30'0"35d26'
Uschi
Per 2h58'16"55d43'
Uschi
Sco 17h5'18"-30d31'
Uschi und Manfred 22-02-
1994
Psc 1h22'1"10d55'
Usen,Abigail Catherine
Lyr 19h14'15"41d19'
Usha
Lyr 19h18'14"38d51'
Usher III,Edward Stuart
Lmi 11h4'17"27d40'
Usher,Arthur & Nina
Com 12h5'17"22d10'
Usher,Jennifer
Cyg 19h51'11"44d9'
Usher,Steven
Uma 8h40'39"57d31'
Usher,Vaughn Paul
Cet 3h7'59"4d26'
Usher,Verlee
Mon 6h53'32"1d25'
Usher,William F
Ori 6h0'60"10d10'
Ushirikiano,Tanzania Green
Star
Cyg 21h32'12"38d5'
Ushman,Michael Paul
Boo 14h1'57"26d41'
Usquin,Bernard
Cep 2h18'41"80d3'
Usrah
Gem 7h10'54"24d43'
USS D I P Ship
Del 21h4'50"18d59'
USS Dennis
Her 16h48'1"40d47'
USS England DE 635,The
Uma 12h5'55"62d33'
Ustimenko,Galochka
Dra 20h20'1"70d44'
Uta
Vir 13h29'21"-8d47'
Uta
Cap 20h38'37"-20d7'
Ute
Tau 3h56'24"2d28'
Ute
Sco 17h30'29"-30d14'
Ute
Dra 18h19'1"70d6'
Ute
Dra 15h13'53"63d48'
Uteak,Charles Joseph
Per 4h2'27"51d39'
UTEOT
Peg 23h33'24"11d21'
Utermoehlen,Franklin & Janet
Boo 14h30'7"26d53AB
Utermoehlen,James & Jolene
Aur 6h0'51"45d35'AB
UTGARD-LOKI,King of
Jotunheim '92
Boo 15h2'48"29d43'
Uth December 1992, Beverley
Ann
Peg 21h31'51"20d24'
Utley,Greg
Cet 0h38'28"0d7'

Utopia
Cep 22h21'1"63d49'
Utopia
Cyg 20h2'50"37d58'
Utsman,Tinah Rhee
Aql 19h9'17"1d21'
Utta & Bill
Del 20h22'53"8d24'
Utter Star,Gary
Dra 16h47'34"68d23'
Utterback,Mitch
Cep 23h13'49"60d24'
Uttke,Adreanna Marie
And 23h40'25"47d59'
Uttley,David
Ori 4h59'25"0d51'
Uveges,Melissa Lee
Ari 1h48'13"23d39'
Uwe
Gem 7h31'31"27d29'
Uwe & Amy
Aql 19h0'25"16d29'
Uwe,Nicholas
Her 16h26'20"30d58'
Uyehara,Richard Joseph
Psc 0h21'23"8d2'
Uyemura,Edward
Cma 6h59'45"-19d48'
Uyemura,Tama
Cma 6h59'60"-19d54'
Uyen & Young
Umi 16h5'12"73d5'
Uzee,Mary Ellen
Aql 20h14'39"4d47'
Uzler,Herrmann
Per 3h50'39"36d36'
Uzma
Lyr 18h46'55"42d57'
Uzoziri,Ambrose
Ori 5h25'14"15d16'

V

V I P Ayers Rock
Cru 12h44'43"-58d8'
V K K B H d T
Aqr 22h23'1"2d8'
V M F Loves M A M
Ori 5h34'0"1d16'
V Shirl
Gem 7h26'38"20d19'
V-Anne
Cet 1h51'39"-0d29'
V-Day 1996 I Love You
Charlie,Casey
Aql 19h4'40"0d15'
Vaca,Timothy Paul
Cnv 12h34'13"38d22'
Vacani,Wendy Kathleen
Uma 9h13'24"68d53'
Vacanzeria
Boo 15h5'38"48d2'
Vacaro,Ann C "Bibsy"
Cas 0h45'14"67d24'
Vacca Star,The One & Only
Julie Anne
Leo 10h39'13"15d26'
Vacca,John & Kathy
Aur 4h51'34"38d16'
Vacca,Rocco
Aql 19h18'59"15d34'
Vaccarella,Dylan Joseph
Aur 6h11'36"31d47'
Vaccari,Marco Eugenio
Hor 3h25'33"-49d50'
Vaccaro,Amy Kathleen
Uma 9h50'20"56d30'
Vaccaro,Grace Ann
Cam 4h1'19"58d42'
Vaccaro,Gregory James
Aur 6h9'60"33d25'
Vaccaro,Jennifer
And 23h27'54"38d37'
Vaccaro,John
Boo 15h8'28"48d37'
Vaccaro,John A
Dra 17h3'23"64d58'
Vaccaro,Joseph
Aur 6h6'38"31d47'
Vaccaro,Lia Genevieve
And 2h23'22"47d14'
Vaccaro,Luke James
Dra 17h3'33"68d34'
Vaccaro,Natalee
Cnv 14h2'59"42d6'
Vaccaro,Patricia J
Equ 20h57'19"5d32'
Vacchelli,Giovanni
Mon 7h35'17"-0d55'
Vacchi,Patricia
Aql 19h26'18"13d20'
Vacek III,Reuben Paul
Cma 7h19'58"-15d15'
Vacek,Ruza
Mon 6h59'18"-8d24'
Vach,Gregory M
Aql 19h31'13"10d16'
Vachon,Franáois
Cyg 20h23'36"39d23'
Vachon,Jules
Per 2h58'41"48d57'
Vachon,Marraine Nathalie
Uma 11h33'16"30d3'
Vachon-Boutouil, Maryline
Uma 11h33'16"33d36'
Vachonfrance,G et P
Vul 19h21'45"23d55'
Vachy,Forever Grandpa
Per 1h47'59"56d23'
Vadala Te Adoro Dodi, Jeanne
Cas 0h24'0"61d47'
Vadala's Starlight, Donna
Lyn 7h54'15"40d40'
Vadala,Frank & Helen
Cmi 7h43'60"4d15'
Vadala,Marie J
Cam 5h5'44"60d54'
Vaden,Dr Reggie
Oph 16h50'47"11d53'
Vadgama,Mukesh
Uma 9h44'1"45d32'
Vaega & Skydancerr
Per 3h5'0"41d43'
Vaerst,Bodil
Ori 6h2'53"8d3'
Vaessen,Peter
Lyn 8h7'45"47d33'
Vafias,Alkis
Uma 11h23'17"38d18'
Vagabond
Lep 4h57'25"-19d40'
Vagja
Crb 16h22'0"28d46'
Vago
Eri 2h50'31"-5d5'
Vagts,Drew Donald
Aur 6h32'13"31d29'
Vahlbrauk,Karl Heinz
Aqr 22h40'60"-5d53'
Vahle,Janet
Aur 5h11'1"44d33'
Vahni
Aql 18h47'22"11d7'
Val,Nikki S
Cyg 21h14'28"38d38'
Vaid,Karen Kelly
Com 12h23'25"31d23'
Vail III,John J
Uma 9h52'57"51d4'
Vail,Alexandria Katarina
Cas 0h27'55"62d17'
Vail,Richard James
Aql 19h26'0"15d14'
Vail/Baker
Uma 9h19'15"42d34'
Vaillancourt,Josée
Her 17h54'25"18d34'A
Vaillancourt,Pierre
Aur 6h21'22"38d57'
Vaillant,Jocelyn
Cas 0h6'0"58d50'
Vaillant,Silvy
Cyg 19h53'55"47d55'
Vainio,Ari
Uma 11h27'46"44d46'
Vainio,Nina
Uma 10h47'1"40d17'
Vairin,Marshall Ben
Cap 21h27'23"-23d51'
Vairin,Taylor Kenneth
Aqr 22h6'49"0d47'
Vais,Charlotte Ann
Mon 7h46'49"-1d11'
Vaisnis,Patricia
Eri 3h14'15"-1d53'
Vaizey,Thomas Peter John
Umi 16h4'37"70d22'
VAJA
Uma 8h36'24"58d24'
Vajda,Alex
Cnv 12h49'15"38d18'
Vajda,Grace Stefan
Aur 6h56'11"44d24'
Vajda,Patricia
Ori 5h0'22"8d26'
VAJRAYOGINI
Tau 4h7'55"20d25'
Vakili,Akram
Umi 16h58'33"77d9'
Val
Cas 0h10'20"59d36'
Val
Cas 0h31'38"64d42'
Val
Lyr 18h31'16"30d10'
Val
Cam 4h2'16"70d55'
Val & Jimmy
Eri 4h2'38"-10d13'
Val D'Isère
Lyn 9h4'1"38d16'
Val Louise
Dra 17h2'24"62d11'
Val Loves Bob
Peg 21h52'55"28d21'
Val's Gal's Emma,
Punkin,Nellie
Uma 9h47'52"71d12'
Val's Star
Cas 0h10'1"59d30'
Val's Victory
And 23h25'40"48d49'
Val,Stefano Urbani
Cep 0h16'21"68d38'
Val-Eric
Boo 14h58'28"17d34'
Val-Lee
Aql 19h25'21"-8d28'
Valaperta,Angelo
Eri 3h50'31"-6d55'
Valarie
Cyg 19h40'24"30d43'
Valas,Eleni Abastasia
Eri 3h35'55"-2d52'
Valat,Etienne
Uma 11h27'27"52d25'
Valat,Sandrine
Ori 4h52'33"1d40'
Valcarcel III,Julio
Aur 6h4'48"45d1'
Valderas,Harold Michael
Sex 10h44'12"-8d38'
Valdes,Blasa
Cnc 8h37'53"18d49'
Valdes,Esperanza
Hya 9h5'43"-0d50'
Valdes,Victoria Esther Marie
Cma 6h53'26"-17d24'
Valdez "Dream" Star, Norma
Oph 18h3'20"10d27'
Valdez,Kimberly
Cyg 20h1'21"30d6'
Valdez,Stephen C
Cmi 7h52'51"1d52'
Valdez,Sylvia
Vir 13h29'31"-4d34'
Valdivetro
Gru 22h22'40"-50d4'
Valdivia "30",George
Ser 15h10'53"17d51'
Valdivia,Jesus
Dra 14h55'25"65d23'
Valdivieso,DDGM,OES Sr
Myrtle
Crb 15h27'34"31d12'
ValDona
Peg 22h45'37"5d20'
Vale
Cas 23h3'43"53d51'
Vale,Nina
And 0h48'25"38d34'
Valencia,Lisa Christine
Lyr 18h56'45"31d45'
Valencia,Luis O
Vir 13h35'28"0d38'
Valencia,Naomi
Lyn 8h23'0"50d59'
Valencic,Michelle Ninette
Christine
Peg 21h46'0"33d24'
Valencic,Nancy Neubacher
Cyg 19h29'32"30d58'
Valente,Annamaria
Cnc 9h14'22"8d4'
Valente,Paul Joseph
Cyg 19h28'22"36d29'
Valentella
And 0h11'12"38d6'
Valenti's Land,Giulia
Cyg 20h4'22"30d1'
Valenti,Alexandria Frances
Cas 1h3'1"61d30'
Valenti,Emilio T
Vul 20h58'1"28d59'
Valenti,Jerry
Her 17h37'14"18d46'
Valenti,Joseph,John
Per 3h58'27"50d43'
Valentin Calamel
Aur 6h11'37"31d28'
Valentin S
Uma 9h40'0"51d60'
Valentin,Christy
Crb 16h21'24"30d19'
Valentin,Jean Michel
Cet 0h56'15"1d13'
Valentin,Richard, Maggie &
Shane
Per 3h47'30"39d44'
Valentina
Mon 7h48'45"-1d29'
Valentina
Lyr 18h31'54"40d1'
Valentina
Sgr 18h55'26"-22d2'
Valentina
Cam 6h50'39"68d4'
Valentina
And 23h0'20"52d54'
Valentina
Umi 17h45'49"83d16'
Valentina
Cyg 19h25'1"30d54'
Valentina
Cam 6h56'0"64d41'
Valentina
Cnv 13h31'47"37d57'
Valentina
Lep 4h57'27"-19d6'
Valentina e Gianluca
Lup 15h17'23"-41d21'
Valentina L I
Equ 21h20'26"10d50'
Valentina,Ave
Lyn 7h35'16"45d9'
Valentine
Hya 8h28'42"-5d56'
Valentine
Eri 4h21'14"-1d49'
Valentine & Stephanie
Boo 14h43'36"42d35'AB
Valentine Emma Maria
Cyg 20h21'15"39d48'
Valentine Memory
Lyn 7h22'12"58d42'
Valentine's Love
Cyg 21h14'25"28d52'
Valentine's Star, Johnny
Her 18h9'47"48d2'
Valentine,Barbara
Cas 1h4'50"63d3'
Valentine,Bob
Ser 18h25'48"1d21'
Valentine,Caitlin A
Lyr 19h23'43"31d21'
Valentine,Diane M
Cas 2h5'37"73d48'
Valentine,Elaine
Sgr 18h52'13"-28d50'
Valentine,Gina
Com 12h29'27"20d58'
Valentine,Ina
Cam 8h17'37"74d1'
Valentine,Joan Louise
Vul 19h48'30"28d36'
Valentine,Laura
Lyn 8h3'59"40d56'
Valentine,Leo & Lenore
Crb 16h7'24"26d39'
Valentine,Maria Dolores
Cas 22h58'0"55d8'
Valentine,Star
Lyr 18h42'22"32d46'
Valentines' Star: RD & NA=ÏI
Love You
Oph 17h8'37"10d45'
Valentini,Felice
Pic 4h40'15"-48d34'
Valentini,Giorgio
Pho 2h15'7"-47d46'
Valentini,Luisa Frangi
Pic 4h40'53"-49d28'
Valentino
Lyr 18h38'45"27d17'
Valentino Star, MDLPLDJ-
CLLVTNJ'S
Per 3h12'33"40d36'
Valentino",MKV-1"Mary Kate
Tri 2h29'36"30d13'
Valentino,Alfred
Lac 22h55'44"51d37'
Valentino,Anthony Hop
Lac 22h10'50"54d59'
Valentino,Cole Michael
Lac 22h14'57"47d27'
Valentino,Denise
Cas 1h43'30"65d8'
Valentino,J Robert "Bob"
Cet 0h46'1"-4d58'
Valentino,Jerry
Boo 14h41'44"37d11'
Valentino,Rudolph Guglielmi
Tau 4h21'52"15d11'
Valentino,Tina
Cyg 19h29'17"31d18'
Valentovich,Alexis Michelle
Boo 15h47'28"40d41'
Valentín,Roland
Ori 5h39'36"15d16'
Valenza,Giovanna
Lyn 7h27'0"40d44'
Valenzuela,Cristina
Peg 22h36'32"27d54'
Valenzuela,Franz & Grethel
Monzon
Cyg 19h22'21"54d51'
Valenzuela,Joseph Daniel
Cma 6h51'40"-16d47'
Valenzuela,Karen Lee
Peg 22h25'49"24d45'
Valenzuela,Rachel
Cap 21h38'30"-22d58'
Valenzuela,Rick & Lori
Tri 1h42'30"28d47'
Valeran
Crt 11h14'39"-19d2'
Valerani-Knoblich, Matthew
Wyatt
Uma 10h18'0"53d29'
Valerani-Knoblich, Richard
Marcus
Uma 10h8'41"54d2'
Valerani-Knoblich, Taylor Gail
Uma 10h9'55"52d18'
Valeri,Anthea M
Cyg 21h6'1"31d26'
Valeria
Boo 14h1'56"17d51'
Valeria
Cet 1h32'0"1d21'
Valeria
Com 12h27'30"20d39'
Valeria
Boo 15h10'45"47d31'
Valeria
Uma 11h54'1"30d35'
Valeria
Cet 2h34'1"1d21'
Valeria
Ant 10h43'47"-36d36'
Valeria
Tel 20h11'40"-46d12'
Valeria
Del 20h12'42"12d12'
Valeria 4/6/65
Cas 0h2'36"59d25'
Valeria S
Cas 1h0'21"55d22'
Valerie
Uma 12h19'32"62d32'
Valerie
Mon 6h20'46"5d39'
Valerie
Cnv 13h38'55"45d18'
Valerie
Cas 0h10'48"58d14'
Valerie
And 0h9'0"40d11'
Valerie
Cam 4h26'52"68d50'
Valerie
Cas 1h24'50"60d37'
Valerie
Peg 22h24'22"20d22'
Valerie & Bryan
Aur 5h2'28"40d41'
Valerie & Eric
Cyg 20h19'26"38d37'
Valerie B
Ser 15h56'14"2d23'
Valerie Belle Enfant
Cas 1h0'0"55d54'
Valerie Ellen
Hya 9h14'46"2d20'
Valerie et Philippe
Peg 21h59'19"10d59'
Valerie Helen
Cas 0h14'0"60d13'
Valerie Ingrid
Sgr 20h5'29"-40d50'
Valerie Jane
And 23h44'42"41d24'
Valerie June
Lyr 18h36'21"46d28'
Valerie Lee
Lyn 9h0'0"36d4'
Valerie Lynn
Boo 15h47'15"47d52'
Valerie Michéa C
Uma 9h34'17"49d3'
Valerie Sue
And 0h12'36"33d40'
Valerie's First Star
And 0h14'30"37d43'
Valerie's For Never & Ever Star
Cam 3h30'14"59d34'A
Valerie's Starr
And 0h57'24"45d34'
Valerie,Ma Poulette
Per 2h37'16"50d27'
Valerio,My Gladiator, David
Lac 22h10'33"49d6'
Valerioti,Geri
Tri 2h6'30"31d35'
Valetta,Geoff
Ori 5h50'23"9d23'
Valianti,Vincent
Cas 0h28'25"69d14'
Valiga,Robert B
Boo 15h5'33"31d50'
Valigosky,Jamie
Ori 5h56'41"17d1'
Valinakis,Panos
Vul 19h21'56"25d11'
Valinoti,Joey & Janine
Lac 22h14'1"38d27'
Valitski,William Fredrick
Tau 5h45'31"23d18'
Valji,Matthew Tarcher
Dra 11h21'23"74d25'
Valk,Anne duPont
Boo 14h10'11"45d34'
Valko,George"The Bum"
Cep 22h27'23"61d49'
Vallade,Jean-Charles
Peg 23h5'46"31d26'
Valladres,Miguel
Aql 20h5'18"4d16'
Vallance,Evelyn
Lyr 18h39'49"38d53'
Vallance,Pearl
Eri 4h27'0"-11d19'
Vallario,Milvia
Pup 8h24'28"-25d47'
Vallario,Vincent D
Cnv 12h29'26"32d51'
Vallas,Peter S
Her 16h10'12"41d28'
Vallat,Etienne
Uma 11h27'27"52d25'
Vallat,Sandrine
Ori 4h52'33"1d40'
Vallauri,Natalia
Lmi 10h12'31"37d56'
Valle,Fabiana Della
Lep 6h10'59"-20d13'
Valle,Richard C
Per 4h0'18"51d25'
Valleala,Merja
Cam 5h40'53"80d23'
Vallee,Sherry Gayle
Crt 10h56'27"-8d37'
Valleix,Agathe
Per 1h44'1"53d33'
Vallejo,Boris & Julie
Cyg 21h28'29"38d5'
Vallejo,Sergio Hugo
Aur 5h59'24"30d3'
Vallely,Laughlin Hutton
Ori 4h53'31"1d43'
Vallenari,Brooke Nicole
Ori 6h5'33"-2d24'
Vallera,MD,Dino U
Lac 22h24'46"50d24'
Vallerle
Peg 0h2'47"18d47'
Vallery,Roger Marvin
Cet 1h12'32"-4d9'
Valles-Clifford,Nora
Mon 6h54'47"7d51'
Valles-Gary Wishee, Ellen
Dra 20h 21'12"68d51'
Valley,Kelsey
Eri 2h45'55"-17d6'
Valley-"Santa",Bertil E
Aur 6h16'0"38d49'
Vallez,Shannon
Umi 14h49'18"69d59'
Valli
Cnv 13h5'30"37d34'
Valli W
Aur 6h36'43"38d23'
Valliant,Holly White
Mon 6h35'23"10d3'
Valliant,Joshua Andrew
Mon 6h56'33"0d19'
Valliant,Wayne W
Cmi 7h40'47"8d31'
Valliere,Gloria I
Peg 22h25'51"29d28'
Valliere,Rene & Ernie
Cyg 21h29'29"39d38'
Vallin,Aimee Rebecca
Cas 0h18'0"62d35'
Vallis Star,The
Peg 21h53'36"31d23'
Vallivana
Ser 16h4'55"2d14'
Vallon,Marjorie
Mon 6h26'43"-8d30'
Vallone,Lori A
Peg 21h33'11"24d13'B
Vallone,Lucille
Per 4h44'50"51d12'
Vallone,Peter F
Lac 22h1'41"48d47'
Vallone,Trina Lynn
Uma 10h54'53"54d58'
Vallorani,Licinio
Sgr 19h29'0"-30d51'
Vallotton,Daniel
And 0h3'21"38d17'
Vallozzi,Anthony James
Tau 4h37'20"15d20'
Valls,Ugo
Her 18h22'40"12d22'
Vallée,Louise
Lyr 18h39'24"34d8'
Vallée,Ségolène
Umi 16h38'50"76d5'
Valo,Aamuni
Uma 11h42'24"50d3'
Valorie
Peg 23h39'39"30d24'
Valorz,David Christian
Ser 18h15'30"-14d53'
Valrie
Aql 19h40'33"11d6'
Valrus II
Lyr 18h38'12"34d22'
Valsasina,Valeria
Cep 23h29'26"64d38'
Valter
Cet 2h34'19"2d16'
Valter Vola
Pup 7h56'52"-26d9'
Valtiner,Gery
Dra 15h17'57"61d7'
Valtisiaris,Diane
Cas 1h55'51"75d16'
Valtrex
Boo 14h2'44"12d22'
Valvan
Tri 2h13'56"31d40'
Valvason,Roberto
Cnv 13h6'18"51d43'
Valvoda,Lindsey Marie
Lyn 9h31'58"41d44'
Valy's Star
Pic 5h3'3"-45d0'
Valzano in Pasquino, Vittoria
And 0h12'0"46d54'
Valéria
Mon 7h43'30"-4d6'
Valérie et Eric
And 1h7'52"47d20'
Vampier Luis,The
Mon 6h58'30"8d41'
Van Ackern,Karin gennant-
Stropp-
Cnc 9h15'22"10d26'
Van Aerden,Frédéric
Oph 17h21'41"11d5'
Van Aerden,Raphaël
Oph 18h5'41"10d10'
Van Aken,Monique
Tau 4h19'32"21d7'
Van Allen II
Equ 21h4'18"2d41'
Van Allen,Christopher
Cnv 12h10'60"51d51'
Van Allen,Leah
Lyn 8h35'59"41d18'
Van Alstin's of Lawler Ave
Uma 12h15'1"55d56'
Van Alstine,Joan
And 0h49'54"41d14'
Van Amber
Cam 3h27'1"61d25'
Van Amberg,Kirstie
And 23h25'25"45d47'
Van Antwerp,Joel
Peg 22h51'27"8d41'
Van Arragon"Angel", Elizabeth
Jane
Lyr 19h13'39"40d11'
van Arsdale,Lola
And 23h1'1"48d16'
Van Arsdale,Margette
Lyr 18h15'38"37d54'
Van Arsdall,Lisa
And 2h27'53"50d28'
Van Arsdol USN(Ret), Captain
Robert A
Cep 23h30'1"66d29'
Van Bebber,C A
Cnc 8h6'14"30d1'
van Beers,Chris E W
Cam 4h12'25"58d27'
Van Benthem,Marian
Vul 19h19'32"20d31'
Van Betten,Lana & Tom
Cyg 21h53'13"41d6'
Van Bibber,Virginia
Mon 8h7'30"-8d12'
van Biene,Kim
And 0h20'34"30d52'
Van Blarcom,Amy Louise
Moreland
Cam 3h20'14"58d17'
Van Blarcom,R David
Oph 16h57'23"-25d48'
Van Bodegom,John
Dra 17h19'16"64d52'
Van Bodegom,John S
Dra 17h40'1"61d1'
Van Boxel Frederik
Pup 8h8'15"-26d23'
Van Breeman,Patricia Ann
Cas 0h55'60"73d26'
Van Brunt,Daniel Wayne
Cet 2h30'36"-2d13'
Van Brunt,David William
Cet 2h30'0"-0d23'
Van Brunt,Dennis Lee
Cet 2h38'15"2d30'
Van Brunt,Jerry Wayne
Cet 1h39'0"-1d6'
Van Brunt,Sharon Marie
Cet 2h38'43"-6d58'
Van Buren,Alexandra
Cyg 20h23'11"40d54'
Van Buren,Donald
Oph 17h55'6"11d2'B
Van Buren,Louise
Oph 17h55'6"11d2'A
Van Buskirk,Joyce Ann
And 23h22'49"44d27'
Van Camp,Teresa Haugen
Uma 12h42'19"62d4'
Van Casteren,Anton
Boo 14h37'33"37d10'

Van Cleave,Austin Cole
Hya 9h6'59"6d11'
Van Cleave,Deborah Suzanne
Eri 3h10'1"-1d46'
Van Cleve,Ryan
Lac 22h8'59"46d10'
van Coevering,Bright As My Luv-Claude
Cnv 12h27'24"37d36'
Van Dahlen,Lisa Jean
Aqr 21h35'31"-1d13'
Van De Giesen,Nick
Her 18h2'33"47d31'
Van De Laar,Ria
Cnv 13h44'48"47d42'
van de Muts,Sophia Bernardina
Cyg 21h50'58"52d43'
Van De Pol, Branden Lane
Uma 11h25'12"38d2'
Van de Sype,Ferdinand
Ori 6h7'0"20d42'
Van De Voorde,Hilde
Lyr 19h21'27"30d16'
Van De Wiele,Bobby
Mon 7h9'21"-8d13'
Van Dehey,Joseph Andrew
Cep 2h4'3"81d14' A
van den Boogert,Emma Johanna (Moksi)
Vul 19h20'36"25d28'
van den Boom,William
Her 18h4'40"45d29'
Van Den Bosch,Frans CC
Ori 5h5'58"0d10'
Van Den Broeck,Sandra
Cnv 12h54'54"40d47'
van der Graaf,Wim
Cet 1h0'1"-11d5'
van der Leek,Paul
Boo 15h7'35"51d41'
van der Vecht,Carla
Aur 7h23'49"38d47'
Van Der Velde,Anne
Cas 2h1'58"59d0'
van der Velden,Boiten
Cep 21h0'29"80d15'
Van Der Zee,Julius Tremayern
Uma 9h34'53"51d24'
Van Derslice,William & Carol
Sge 20h1'19"17d48'
Van Doorne,Donna
And 0h54'36"41d4'
Van Dore,Quentin R
Cep 21h53'46"56d8'
Van Druten,Yvon
Cas 1h47'20"68d7'
van Duijn,Ms Josje
And 0h40'49"45d45'
Van Dusen,Michael Eaton
Oph 17h17'57"-20d9'
van Duynhoven,Joseph Mark
Sco 17h52'12"-31d17'
Van Dyck,Linda
Ari 3h1'17"25d5'
Van Dyk,Carol
Cas 1h41'19"70d54'
Van Dyke's,The
Cam 6h44'36"80d17'
Van Dyke,Chloe Carolyn
Pho 23h54'39"-42d22'
Van Dyke,Florence Sebilla
Pic 5h52'20"-45d42'
Van Dyke,Hugo Benjamin
Ind 21h4'38"-46d41'
Van Dyke,Jacquelyn Stacey
Mon 6h19'53"4d52'
Van Dyke,Jr,David Summerfield
Ser 15h25'22"8d53'
Van Dyke,Karen
Cyg 20h24'25"41d11'
Van Dyke,Paul
Aur 7h14'57"38d54'
Van Dyke,Richard "Gerry" & Kathleen M
Peg 21h30'39"20d29'AB
Van Dyke,Sarah Charlene
Eri 3h21'57"-16d35'
Van Dyken's Star, Charles
Her 16h40'1"34d25'
van Erp,Jr,Peter
Lac 22h30'44"54d5'
van Ess,Leon
Ser 15h53'45"-0d19'
Van Estenbridge,Donell
Sco 16h33'0"-28d14'
Van Etten IV,John W
Per 2h52'43"41d12'
Van Etten,Ellin
Crb 15h50'58"32d1'
Van Etten,John Edelen
Dra 17h39'1"64d42'
Van Etten,Kathleen A
Lyr 18h57'20"33d1'
Van Etten,Mark T
Per 2h51'35"34d6'
Van Fleet,George
Oph 17h59'33"12d29'
Van Gijlswijk
Uma 13h19'28"62d0'
Van Gilder,Derek Robert
Peg 22h59'23"18d43'
Van Gilder,Noah Jasen
Aqr 21h52'57"-1d5'
Van Ginhoven,Neil H
Boo 14h57'0"50d3'
Van Gogh,Michael
Per 1h57'1"54d3'
Van Gorden,Denice
Cet 0h25'50"-5d21'
Van Haeften,Fleur Learn
And 22h55'20"40d5'
Van Hagey,Rama R
Lmi 10h24'38"28d38'
Van Hecke,Chuckers
Dra 20h23'1"61d50'
Van Hee,Jacques
Lyr 18h57'41"47d2'
Van Heusen,James Kirk
Aql 19h45'47"14d9'
van Holstein,Peter
Cnv 13h36'1"47d55'
Van Hook,Daniel Adam
Tri 1h34'13"28d56'
Van Hook,Donald L
Hya 10h12'23"-16d15'
Van Hoose,Eric Richard
Boo 15h9'19"27d36'
Van Horn
Dra 12h20'29"75d54'
Van Horn,Jane
Crb 15h31'57"31d35'
Van Horn,Star of Evadna
Hya 8h27'26"0d59'
Van Houten,Fannie
Ori 6h4'19"8d13'
Van Houtte,David
Cma 6h56'24"-17d6'
Van Hove,Jody Alison
Lyn 7h55'0"44d58'
Van Huben,Michael Kenneth
Lib 15h4'1"-28d45'
van Kaathoven,Jordie
Vul 19h20'31"23d19'
Van Kampen,Kevin
Ser 16h6'11"10d24'
Van Kimmenade,Sally
Cas 0h5'51"64d26'
Van Laer,Jerome
Crb 16h13'58"38d27'
Van Landingham,Azura by Sandra
Uma 12h39'1"59d49'
Van Leer-Tang, Schuyler George
Cma 7h51'22"10d44'
Van Leeuwen,Cassandra Helena
Cas 0h6'1"63d53'
Van Lenten,David
Aql 19h1'27"0d31'
van Linde,Margreet
Cam 4h30'49"58d13'
Van Marche,R/B
Cet 3h13'12"1d22'
van Mechelen,Harold
Dra 19h37'38"68d16'
van Meeden,Rolina Smit-Kaman ster
Cnv 13h34'52"47d44'
Van Meter,Ian Wesley
Her 17h39'37"21d60'
Van Meter,Richard D
Per 1h55'39"52d51'
Van Natta,Raymond B
Cep 22h33'42"61d45'
Van Natter,Nicholas Amos
Cet 2h51'42"4d14'
Van Neck,Todd Gregory
Aur 6h21'0"37d8'
Van Ness,Susan J
And 23h43'17"40d7'
Van Nest,Jessica
Ori 5h36'0"-6d21'
Van Norde,Peter J
Per 1h43'52"53d45'
van Oidenborgh, Johannes
Cas 1h26'0"60d45'
Van Oostrom's Lucky Star,Gerald
Oph 17h54'20"14d9'
Van Orman,Jason
Her 18h6'50"14d55'
Van Otten,Auna Ca'reen
Uma 8h51'39"55d50'
Van Otten,Michael James
Uma 8h47'25"55d31'
Van Overloop,Dustin
Aql 18h58'44"13d4'
van Overstraeten,Herr
Boo 14h25'54"10d3'
Van Pamel,Kristen
Tri 2h7'19"33d22'
Van Patten,Bill & Marge
Eri 3h28'12"-5d12'
Van Patten,Daniel Ray
Lac 22h2'1"49d1'
Van Patten,Jr,Robert F
Leo 11h19'35"-5d24'
Van Patter,M Berthe (Forest)
Cap 21h14'19"-23d15'
Van Pay,Gary
Lyn 7h30'1"40d23'
Van Pragg,Sarah
Lyn 9h14'37"34d41'
Van Putten,Peter
Hya 9h47'2"-15d37'
van Remmen,Carin & Manfred
Lyn 7h27'16"40d53'
Van Rheeden,Mark & Sally
Cam 3h54'59"52d48'
Van Rheenen,Bea,Leen, Mandy,Manon
Cas 1h49'28"67d32'
Van Rheenen,Mandy
Cas 1h57'25"61d46'
Van Rheenen,Manon Kelly
Cas 1h56'45"60d49'
Van Riper,Sue Ellen & Edward
Crb 15h54'48"34d23'
Van Roekel Mark
Boo 14h27'47"28d26'
Van Ruissen,Erik
Her 16h36'31"34d32'
Van Ryn,Geoff & Karen Johnston
Ind 20h49'46"-59d53'
Van Sant,Muriel Helen
Lmi 10h52'45"28d33'
Van Schaick,Joseph P
Uma 13h27'0"62d46'
Van Schaik,Carolyn
Cas 0h35'13"63d22'
Van Schaik,Douglas Lance
Aql 20h4'15"1d12'
Van Scyoc,Chad Michael
Aur 7h2'58"38d10'
Van Scyoc,Christopher Colby
Cep 23h7'19"61d3'
Van Sicklen,Andrew
Ori 5h54'51"10d55'
Van Sicklen,Charles
Cep 22h29'0"61d22'
Van Skarda,Larry
Aql 20h35'49"0d48'
Van Skiver,Jr,Jon P
Dra 17h48'1"58d16'
Van Speybroeck, Laëtitia
Uma 11h19'27"37d53'
Van Straten,Gary
Peg 23h19'44"13d16'
Van Tassel,Gabrielle
Lyn 19h16'54"38d58'
Van Tassel,H E
Ori 5h53'58"11d15'
Van Tassel,Marisa
Ori 5h59'48"20d12'
Van Tine,Richard J
Her 16h41'29"51d13'
van Valderen,Iris
Cas 3h3'29"58d16'
Van Valkenburg
Cet 2h24'14"-0d36'
Van Valzah,My Love, Sarah Yeager
Peg 22h20'29"25d2'
Van Vaulkenburg,Sara
Lyn 7h16'29"58d39'
Van Vechten,William
Lac 22h1'57"40d36'
Van Vleet,Deborah
Peg 21h55'0"30d21'
Van Vleet,Willie & Barbara
Del 20h17'17"13d36'
Van Vliet-Johnson,Lady Kathryn
Peg 21h25'21"3d15'
Van Vlyten,Stefani
Uma 11h17'23"32d40'
Van Volkenburgh,David Scott
Cep 0h6'40"69d59'
Van Vollenhoven,Frits Willem
Ser 16h0'42"11d6'
van Vollenhoven,Renée
Vul 19h20'36"25d11'
Van Wagnen,Andrew Brian
Tri 2h33'15"32d5'
Van Wie,Christine
Cas 0hd26'27"65d12'
Van Will,Rien
Cas 1h49'48"60d48'
Van Winkle,PhD,Lon J
Her 16h48'42"30d52'
Van Woerkom II,Linn Dwayne
Aql 18h56'0"13d21'
Van Woerkom,Gwen Marie
Peg 22h16'11"34d2'
Van Woerkom,Todd Alan
Oph 17h53'17"13d13'
Van Wyck,Jr,Frederic Bronson
Lac 22h40'20"51d12'
Van Zandt Star,Townes
Uma 9h8'24"50d26'
Van Zile,Gretchen
Mon 6h24'47"-5d51'
Van Zyl,Margaretha Afina
And 1h33'1"40d20'
Van's Star-where dreams come true
Del 20h27'34"18d53'
Van-Main
Lib 14h23'39"-18d59'
Van-Maëlle,Anaïs
Her 18h7'42"31d48'
Vanacore,Linda M
Cyg 19h50'37"38d11'
Vanario,Anthony B
Per 3h16'25"50d36'
VanArsdale,Sharon Louise
Aql 19h44'11"11d36'
Vanasse,Constance Davis
Com 12h17'0"24d56'
Vanasse,Pierre
Sco 16h36'13"-25d45'
Vanatta,Rose
Vul 20h5'30"23d49'
VanBlarcom,Jr,Richard Peter
Vir 12h34'32"-8d24'
VanBlerkom,Elaine
Cyg 21h21'0"28d58'
VanBoxel,Bernard M
Cyg 21h33'1"50d32'
Vanburg,Jr,Dana
Dra 17h8'30"64d12'
VanBuskirk,John R
Eri 4h14'26"-10d0'
VanCampen,Michael
Cma 6h54'3"-19d16'
Vance III,James L
Peg 23h3'22"11d56"
Vance Zachary
Ori 5h42'41"8d39'
Vance's Star
Equ 21h2'25"2d43'
Vance,Chris
Hya 8h13'31"5d52'
Vance,Christy
Mon 7h2'31"3d57'
Vance,Garry
Boo 14h33'29"47d27'
Vance,George & Virginia Otis
Cyg 21h3'17"38d38'
Vance,James H
Aql 20h7'20"8d1'
Vance,Jane Clarke William Wintersole
Lyn 8h14'32"35d20'
Vance,John & Oneta
Mon 6h2'1"-5d6'
Vance,Jr,(Pumpkin) Joseph A
Cnv 12h5'51"36d22'
Vance,Karen Alicia Caldwell
Vul 19h19'26"26d30'
Vance,Michael
Hya 8h44'17"-6d54'
Vance,Phebe
Cyg 20h30'39"42d30'
Vance,Robert (Mr Smiley)
Cep 20h47'14"70d20'
Vance,Sara Haden
Lyn 8h20'16"52d20'
Vance,Scott Coleman
Vul 19h39'27"20d19'
Vance,Stephen
Cep 22h53'25"71d3'
Vanchieri,Adam Daniel
Cnv 13h52'57"31d23'
Vancil,Mary Ann Rice
Eri 4h27'37"-11d52'
Vancise,John
Lyr 18h56'46"45d41'
VanCreveld-World's Best Mom,Marcia
Leo 10h55'32"14d13'
Vancura,Charles
Aur 6h6'43"31d34'
Vanda
Lep 4h56'31"-19d30'
Vanda Io
Lac 22h33'27"38d57'
Vandaele,Gabriel
Aur 5h51'12"30d49'
Vandal,Jacinthe
Ori 5h59'46"11d38'
Vandale,John Fredrick
Aur 6h0'21"46d25'
VanDall,Donna Lee
Cam 3h34'26"61d35'
VandeBoom,Leon
Cep 21h31'37"61d30'
Vandecar,Richard L
Aql 19h2'40"15d2'
Vandegrift,Cathy Ann
Lib 15h26'14"-6d28'
Vandegrift,Emily Margaret
And 23h37'56"49d5'
Vandella
And 0h3'17"47d25'
VandeLogt,Mary Kay
Cas 3h3'16"61d29'
VandenBerg,Chad Dayle
Her 17h17'1"45d39'
VanDenberg,Elizabeth Hart
Leo 10h56'0"2d27'
Vandenberg,Greg
Aur 4h50'24"51d21'
Vandenberg,Terry
Cet 2h4'38"5d43'
Vandenberge,William C
Cnv 12h17'37"51d16'
Vandenberghe,Cyrus Alexander
Cet 2h1'51"-10d28'
Vandenburg,Donna
Cet 1h11'38"-3d21'
Vandenburg,Erin
And 23h18'27"42d49'
Vandenbussche,Hervé
Cam 3h47'11"67d33'
Vandenhende,Loic
Dra 18h29'19"50d35'
Vander Burgh Jr, Warren M
Dra 20h28'24"70d49'
Vander Heyden,Dr A Renee
Lib 14h27'1"-22d43'
Vander Vennet,Donald
Aur 6h2'36"32d50'
VanDera,Dick
Aql 20h14'16"04d52'
Vanderbilt Planetarium
Boo 14h46'13"46d58'
Vanderbosch,Kathy
Cam 6h1'24"60d45'
VanderBossche Pour L' Eterní,A Eric
Aql 20h22'44"1d42'
Vanderbrouck,Vianney
Aur 5h8'29"29d52'
Vanderbrug,Lucas Raymond Martin
Aur 6h19'1"45d11'
Vanderford,Mary E.
Eri 3h39'46"-15d8'
Vandergrift,David
Cma 6h53'50"-19d34'
Vanderheiden,Marion Allor
Dra 17h0'43"68d26'
Vanderheyden H75B, Louise Mom
And 0h19'41"31d45'
VanDerhoef's Wishing Star,Billy
Cyg 20h57'36"40d22'
Vanderlee,Pamela S
And 0h2'16"46d2'
VanderLoo,Nana & Pop Pop
Cyg 20h26'42"38d52'
Vanderpol,Netty & Ries
Lyr 18h29'56"32d17'
Vanderpool's Star, Julie
Peg 22h35'14"25d47'
Vanderpool,Duane
And 2h26'1"39d28'
Vandersteel,Mariel
And 23h28'47"45d33'
Vandervalk,Astrid
Cas 23h0'0"58d20'
Vanderveen,John & Frances
Uma 11h50'58"44d17'
Vanderveen,William
Cep 22h11'42"60d27'
VanderVen,William H
Aur 4h35'26"30d44'
Vandervort,Jashia Shane
Aur 5h12'59"41d34'
VanderWerf,Tayler S
Per 4h6'43"47d40'
Vandette,Sylvie
Lmi 10h55'17"27d25'
Vandeven,Karen "NQ"
Uma 12h12'6"57d39'
VanDevender,Jim
Vul 20h44'0"28d57'
Vandever,Ernest & Ann
Oph 17h6'37"-23d19'
VandeVisse,Richard Alan
Dra 11h29'32"71d7'
Vandewalker,Sara Elizabeth
Cyg 20h5'1"30d4'
VanDiest,Lady Renetta
And 23h49'60"44d4'
Vandis,Jasper Hall
Mon 7h15'49"-10d46'
Vandone,Lucia
And 23h8'25"41d11'
Vandor,Michael W
Aql 18h56'1"7d16'
Vandr
Uma 9h27'1"49d2'
VanDreese,Rhonda J
Cas 0h9'51"61d31'
VanDusen,Bob
Tau 3h53'21"20d1'
VanDusen,Dewana
Cas 0h58'27"71d21'
Vandusen,Dewana
Cyg 19h17'13"49d57'
Vandusen,Dewana
Aur 5h58'1"37d45'
Vandusen,Dewana
Cet 1h26'45"-3d30'
Vandusen,Mark Kenneth
Cam 8h26'56"83d45'
Vanduyne,Chelsea René
Mon 6h57'23"11d35'
VanDuyne,Lindsey Danielle
And 1h26'30"50d9'
VanDyne,Matthew Duane
Ori 4h44'18"0d35'
Vanecek Family,Edward & Marla
Cyg 21h4'56"36d53'
VanEerd,Laura Lynne
Uma 9h19'1"48d22'
VanEgmond & Daniel J Hoffman,Tilly
Cyg 19h16'12"47d26'
Vanek,David
Per 3h18'47"40d21
VanEman,Bruce
Her 18h55'25"18d53'
Vanepps,Michelle Naomi
Cyg 19h28'23"37d52'
Vanessa
Lyr 19h23'60"31d12'
Vanessa
And 2h29'53"41d52'
Vanessa
Cnv 13h22'21"50d23'
Vanessa
Cae 4h45'13"-32d2'
Vanessa
Cas 2h28'24"61d48'
Vanessa
Peg 22h0'28"30d11'
Vanessa
Cam 4h6'1"58d59'
Vanessa Darlin
And 1h22'10"35d30'
Vanessa Elizabeth
And 1h4'20"38d23'A
Vanessa Kathryn
Cap 20h33'16"-10d8'
Vanessa Love Forever
Lib 15h39'0"-28d41'
Vanessa Lynn
Mon 6h21'18"7d28'
Vanessa N
Cyg 19h43'1"37d35'
Vanessa Noel
Lib 14h41'39"-23d59'
Vanessa Rene-34
Cam 7h42'57"70d44'
Vanessa's (Charm)
Lyn 8h2'20"41d29'
Vanessa's Chronicle Of Maddog
Cyg 21h36'18"37d46'
Vanessa's Glückstern
Sco 17h29'1"-30d54'
Vanessa's Star
Peg 22h12'39"34d13'
Vanessa's Twinkle
Peg 23h1'42"14d5'
Vanetti,MD,Charles E
Oph 18h0'11"11d45'
Vang,Nancy J
Lyr 18h16'51"30d60'
Vangelof,Julie Joy
Lyn 8h18'24"35d33'
VanGorder,Lillian
Uma 8h32'55"51d56'
VanHagen,Sarah Harrison
And 1h47'0"40d26'
VanHeel,Kimberly
Peg 22h42'27"27d8'
Vanhevsen,Rose Jackie
Lyr 19h23'22"37d50'
VanHoosier
Cep 20h58'11"60d57'
VanHorn IV,William Henry
Ant 9h41'50"-29d6'
Vanhorn,Robert Wayne
Cmi 7h53'38"0d9'
Vania's Estrela
Peg 23h47'48"13d22'
Vanin,Lidia e Ivan Venturini
Tel 19h5'16"-48d42'
Vanitsky,Arseny N
Aur 6h3'30"31d46'
VanJuud,Morgan Sonnenkind
Cyg 19h26'58"30d50'
VanKampen,Dennis
Eri 4h2'27"-19d41'
VanKeulen,Kelly M
Mon 6h42'40"1d38'
VanKeuren,Donna
Cam 4h11'23"70d15'
VanKuren,Denise Marie
Sgr 18h17'1"-28d33'
VanLaningham,Craig
Her 16h29'20"41d57'
Vanlo,Juhani
Uma 9h35'31"54d3'
Vanluchem,Donald James
Ori 5h1'38"1d34'
VanMatre,Emily Renee
Cyg 19h20'37"28d37'
VanMeter-B
Tri 2h36'47"35d18'
VanMeter-S
Lac 22h42'39"53d25'
Vann,Colin L
Dra 19h13'43"70d7'
Vann,Mike
Per 2h53'29"56d34'
Vannais,Kevin
Her 16h36'24"50d20'
VanNane,Joanie K
Sex 10h37'59"-0d49'
Vannatter,Philip L
Hya 8h17'26"2d35'
Vannatter,Rita
Peg 22h40'37"25d44'
Vannatter-Dietz,Ruth
Cas 0h3'0"50d25'
Vannatter-Trempe,Anne
Mon 6h30'33"-10d2'
Vanner,Donna
Cas 0h31'43"61d54'
Vanness,Nancy
Cyg 19h29'1"38d41'
Vanni,Carla
Cam 4h9'14"65d15'
VanNice,Andrew William
Her 18h3'17"14d36'
VanNice,Faith Louise
And 1h45'58"40d23'
VanNice,Katherine
Cyg 20h1'18"40d29'
Vannieuwenhuyze,Hilde
Com 12h2'51"26d45'
Vannlandingham,Wm R
Eri 4h18'59"-18d10'
Vano,George
Per 4h34'40"38d7'
Vanoni,Elizabeth
Tau 4h17'39"23d4'
Vanoni,Otto
Per 2h41'16"37d12'
VanOrden,Abigail
And 2h25'53"37d40'
VanOsdol,William Woodson
Aql 20h3'12"1d50'
VanOver,Suzanne & Rick
Aql 18h58'49"11d38'
VanOver,William George
Cnv 12h49'10"47d16'
VanPatten,Guy
Cet 2h29'40"0d3'
VanRaaphorst,Nettie Riedel
Lyn 8h53'46"45d41'
Vanrees,Brent
Ori 5h6'16"11d8'
vanReken,Calvin & Suzanne
Uma 9h53'59"50d21'
Vanreys,Sandra
Dra 10h49'0"73d20'
VanRoekel,Stephen Ned
Crb 15h17'19"31d35'
VanRoekel,Susan Jean
And 2h2'1"40d12'
VanRoy,Amanda Lynne
Cet 3h11'14"3d1'
Vansaders,Courtney
Lyr 18h40'35"26d8'
Vansant,Jane
Lyn 8h25'55"50d7'
Vanschoelandt,Amy
Peg 22h2'1"22d45'
VanSchuyver
Uma 12h24'40"62d9'
Vanslyke,Phyllis Young
Cas 1h41'41"58d42'
Vansovics,Lydia
Cas 0h15'56"58d13'
VanStory,Terry P
Leo 10h1'42"8d13'
VanTassell,Carole
Lyn 8h9'0"50d40'
Vantu,Quynh Hoang
Cep 22h57'13"67d33'
VanVliet,The Jill
Cyg 21h18'26"37d49'
VanVolkenburgh, Jonathan Michael
Aql 20h1'1"10d29'
VanVorhis,Jeff
Aur 5h18'14"45d21'
VanZandt,Billy
Cmi 7h46'52"3d20'A
vanZandt,Pamela
Cas 23h25'37"53d6'
Vanzee,Katie Jo
Uma 10h12'0"57d9'
Vanzo,John
Lmi 9h51'24"37d57'
Vap,Daniel Joseph
Cep 21h42'34"55d15'
Vappi,Tara
Lmi 9h53'39"33d38'
Vara,Jordan Marc
Lyr 19h2'0"30d57'
Vara,Lisa
Mon 7h15'42"-10d42'
Varacchi,Cynthia Ann
Lyr 18h46'48"35d21'

Varacchi,Dawn Marie
Cyg 20h41'23"42d44'
Varacchi,John Rocco
Per 3h8'51"47d7'
Varacchi,Julianne Tamara
Uma 10h14'40"56d44'
Varacchi,Laura Ann
Lyn 8h18'1"33d60'
Varagnolo,Alessandro
Umi 16h8'45"85d4'
Varaljai-Laban,Georg
Cas 0h42'25"60d34'
Varallo,Jennifer M
And 0h9'45"37d35'
Varallo,Jr,John J
Cas 1h3'42"60d2'
Varano,Michael
Dra 17h15'54"67d19'
Vardeman,Grant Colton
Ser 15h12'28"0d18'
Vardinoyannis,Joanna N
Cam 3h49'16"58d31'
Vardis,Arietta Anastasia Kai Yorgos
Cam 3h24'29"50d46'AB
Vardo,Jean Claude
Sex 9h59'51"-4d32'
Vareil,Annie
Sex 10h13'33"-8d46'
Varela,Frank
Cep 22h25'17"70d53'
Varela,Santiago
Ser 15h55'27"24d49'
Varela,Selena
Mon 6h43'23"6d40'
Varella,Hazel
Uma 11h4'0"44d16'
Varey "M Canada", Matthew Bryon
Tau 4h31'1"30d5'
Varga,Layne
Aql 19h9'12"3d28'
Vargas III,Felix A
Cas 2h11'37"61d5'
Vargas,Carmen I
Crb 16h11'51"28d56'
Vargas,David Michael
Sco 16h31'60"-30d53'
Vargas,Ella Alscher
Cyg 20h14'53"46d46'
Vargas,Javier C
Oph 18h32'32"10d10'
Vargas,Mark & Courtney
Sex 9h57'0"2d13'
Vargas,Norma
Lyr 18h35'17"27d6'
Vargas,Otto W
Boo 15h20'14"38d10'
Vargas,Pablo
Aur 5h56'15"31d5'
Varghese,George
Crb 15h55'24"30d59'
Vargo,Ada O Freiberg
Peg 23h1'23"30d9'
Vargo,Danika
Lyn 7h38'21"52d11'
Varin,Sr,Paul A
Per 2h49'15"40d41'
Varini,Adriano
Cyg 21h24'0"48d56'
Varis,Agnes
Cas 1h10'0"60d10'
Varis,Agnes
Lyr 19h16'21"40d40'
Varlamov,Vladimir Ivanovich
Tri 1h39'16"30d53'
Varlamova,Marina Leonidovna
Cas 0h32'48"61d27'
Varlet,Aloin et Josiane
Lac 22h26'45"40d43'
Varley,Clayton Hazlett
Hya 8h23'33"-0d45'
Varley,Marie
Tri 2h15'44"32d21'
Varley,Maura Beth
Sge 18h58'31"18d60'
Varnadore,Heather Nicole
And 23h35'32"41d15'
Varner,Kathy Lynn
Aql 19h50'32"13d57'
Varner,Michael Austin
Her 17h25'37"38d21'
Varner,Toby & Brenda Steinacher
Lac 22h37'54"56d23'
Varnes,Richard
Aqr 23h2'43"-6d51'
Varney,Mr & Mrs David S
Cam 3h45'55"77d16'
Varnum,Keith
Per 2h28'55"52d19'
Varnum,W S
Ori 5h57'24"16d48'
Varriale,Anna Lisa
Ind 21h1'42"-50d41'
Varro "65",Alexander Paul
Cnc 8h32'18"32d48'
Vars,Dr Gordon F
Aqr 22h41'34"0d21'
Varsam,Janice Michelle
And 0h5'31"44d56'
Varsolona,Peter Gary
Lyn 8h29'19"50d23'
Vartanian Star,The Brent
Lyr 18h29'20"30d59'
Varvel,Pop
Uma 11h56'0"60d7'
Vary,Louise Marion
Peg 22h18'51"7d51'
Varzos,Anastasios Nicholas
Aur 4h52'59"40d26'
Vasaio,Fiorenzo E
Cas 0h16'55"59d9'
Vasarab,Joe
Lmi 10h22'20"33d50'
Vasconi,Richard Anthony
Her 15h56'55"46d18'
Vasek,Lisa
Cra 18h19'9"-43d28'
Vash,In Loving Memory Of George
Aql 19h0'22"4d17'
Vashel,Vincent J
Lac 22h3'10"41d6'
Vasi,DC,Sal Joseph
Ori 5h55'34"-0d39'
Vasick,David B
Her 18h9'29"40d4'
Vasilache,Stefan
Gem 6h39'43"30d18'
Vasilash,Jr,Daniel G
Vul 19h22'55"26d50'
Vasili,Vessica Maguerite
And 1h45'40"38d14'
Vasiliauskas 45, Chester
Aur 5h3'11"41d47'
Vasilika,Gula
Ser 15h54'20"17d47'
Vasko's Vision
Uma 8h59'34"70d39'
Vasko,Alison Justine
And 2h21'0"47d45'
Vasko,Nick Joice
Dra 19h22'24"56d49'
Vasmel
Cam 11h8'0"80d1'
Vaso,Sophia
And 2h0'0"47d9'
Vasquez,Angela Marie
And 23h36'52"48d55'
Vasquez,Chris & Cindy
Aql 19h2'0"15d1'
Vasquez,Desiree La Rae
Cep 22h28'16"59d26'
Vasquez,Gene
Her 18h5'47"14d14'
Vasquez,Hershey
Cas 0h7'1"62d18'
Vasquez,Jasmine Rose
And 23h31'19"49d36'
Vasquez,Jason
Aql 18h54'15"6d55'
Vasquez,Joshua Adam
Her 16h58'27"24d49'
Vasquez,Kira Nicole
Cam 3h15'51"60d16'
Vasquez,Lupe
Cnv 13h10'23"37d46'
Vasquez,Natividad L
Del 20h23'1"10d26'
Vasquez,Victoria Elizabeth
Gem 7h26'1"28d18'
Vasquez,Yvonne Ortiz
Tri 2h6'32"33d51'
Vass Star,The
Cyg 19h57'46"30d35'
Vassall,R
Hya 9h8'1"4d7'
Vassanella's Unmapped Territory
Leo 10h57'58"11d21'
Vassard,Océone
Dra 9h20'24"73d13'
Vassberg,Alan H
Hya 8h48'58"-7d42'
Vasseler,William Salvatore
Her 17h54'58"14d53'
Vasser,Leslie Renee
Aql 20h30'29"-6d30'
Vasseur,Marie
Ori 5h57'0"8d48'
Vassilieff,Svetlana
Sge 19h39'36"16d51'
Vassiliki Pizza
Cas 23h41'0"61d19'
Vast Nicolas
Boo 14h51'11"50d38'
Vasta,Sal
Cep 21h6'40"55d56'
Vastar
Ori 6h10'35"8d46'
Vasterling,Terry
Aql 19h55'59"10d51'
Vastez,Dominique
Del 20h18'59"10d10'
Vater,Mary Rose
Cas 23h17'29"60d16'
Vatinel,Sandrine
Cam 3h26'16"61d50'
Vatter,Horst
Peg 22h29'58"8d3'
Vaudene Rose
Cam 6h1'1"60d16'
Vaudescal,Patricia
Cam 7h35'13"68d45'
Vaudo,"Yogi" Anthony J
Her 16h41'28"23d37'
Vaudo,Rosemarie
And 2h21'32"44d51'
Vaugan,Billie Jo
Uma 9h43'34"50d5'
Vaugh,Emily Faith
And 0h19'51"30d20'
Vaugh,Jennifer Gay
Com 12h0'55"20d40'
Vaughan,Alta
Cas 0h1'27"63d15'
Vaughan,Anthony William
Aur 6h18'1"52d48'
Vaughan,April C
Aql 19h25'45"-1d50'
Vaughan,Barbara
And 23h15'24"44d21'
Vaughan,Bill
Aur 5h26'55"30d34'
Vaughan,Col Norman
Umi 13h2'41"70d17'
Vaughan,Craig D
Mon 8h2'0"-1d48'
Vaughan,Douglas
Uma 10h10'26"62d12'
Vaughan,Everett-Eugene
Leo 10h55'54"22d44'
Vaughan,Gail Helen
Cas 1h0'52"64d52'
Vaughan,Jimmy D"Busta"
Cet 1h15'6"-15d54'
Vaughan,Joseph Clayton
Boo 15h34'1"40d3'
Vaughan,Lee
Sge 20h4'43"20d38'
Vaughan,Lois
And 23h9'1"42d2'
Vaughan,Louise Faye
Sge 20h16'32"20d52'
Vaughan,Marty
Cet 1h54'14"-18d47'
Vaughan,Monica
Cas 1h12'39"60d1'
Vaughan,Patricia & Stephen
Cyg 19h30'0"39d45'
Vaughan,Peter & Alison
Crb 16h0'10"26d12'
Vaughan,Piers Allfrey
Cep 0h7'22"69d54'
Vaughan,Sela Claire
Cas 1h0'31"66d49'
Vaughan,Sr,David J
Uma 10h56'51"37d56'
Vaughan,Sr,Ron
Sex 10h14'19"-4d50'
Vaughan,Stevie Ray
Her 18h12'44"30d20'
Vaughan,Thomas J
Dra 16h42'31"68d59'
Vaughn My Lucky Star, Charles J
Uma 12h33'39"62d21'
Vaughn's Star
Peg 23h6'32"22d8'
Vaughn,Barbara Kaiser
Peg 21h47'51"33d7'
Vaughn,Benjamin Inman
Aql 19h4'1"10d8'
Vaughn,Bridget Elizabeth
Mon 7h38'52"-2d21'
Vaughn,Christy
Mon 6h23'48"1d11'
Vaughn,Claire
Peg 23h1'43"8d2'
Vaughn,Deborah F
Cas 1h1'19"63d41'
Vaughn,E E & Venoy
Aql 19h9'15"3d40'
Vaughn,Evitta
Sct 18h38'11"-5d19'
Vaughn,George William
Aql 20h13'38"5d26'
Vaughn,Joanne & Charlie
Sge 19h53'49"18d51'
Vaughn,John A
Dra 20h28'12"73d33'
Vaughn,Jr,Earl Franklin
Aur 5h6'22"40d17'
Vaughn,Katelyn Jane
Cas 0h9'25"62d47'
Vaughn,Kathryn
And 23h36'48"47d30'
Vaughn,Linda Carol
Mon 6h20'19"0d39'
Vaughn,MaryLou
Cas 23h22'24"61d1'
Vaughn,Michael
Boo 14h42'1"24d54'
Vaughn,Michael David
Sct 18h43'16"-7d10'
Vaughn,Patrick Reiss
Boo 14h42'1"26d38'
Vaughn,Royce Harris
Ori 5h16'0"1d24'
Vaughn,Star of David
Cep 23h0'38"68d16'
Vaughn,Susan
Eri 2h58'41"-17d3'
Vaughn,Viola
Lyn 8h12'48"58d26'
Vaulotan
And 23h16'1"50d18'
Vaumoron,Pascal
Cmi 7h18'30"1d59'
Vause,James E
Her 18h2'59"28d23'
Vauthier,Marie-France
Lac 22h20'18"50d24'
Vautrain
Equ 21h6'0"2d17'
Vautrin,James (The Dude)
Aql 20h5'59"7d16'
Vavaroutsos,Eva
Com 12h54'0"21d40'
Vaverchak,Cyril & Katherine
Sge 19h34'1"16d48'
Vaverka,Joseph Robert
Psc 0h47'17"27d41'
Vaxon,Robert Endicott
Sge 19h53'31"16d40'
Vayle
Aur 4h47'24"37d26'
Vaz,Rev,Francis
Lyr 19h19'33"42d11'
Vazquez Star of Love
Lyr 18h56'0"37d19'
Vazquez,Edo
Uma 10h48'27"48d60'
Vazquez,Kimalona
Aur 6h11'58"35d24'
Vazquez,Steven G
Cnv 12h12'48"50d8'
Vazzano,Agnes
And 23h1'50"51d15'
Vazzano,Deanna
Cas 1h43'26"68d36'
Vazzano,Douglas John
Boo 14h27'1"43d7'
Vazzano,Francis Jerome
Her 18h4'47"30d43'
VB One
Aur 7h2'55"43d48'
Veach,Christine
Del 20h23'25"10d14'
Veach,Kathryn L Jones
Mon 7h55'11"-2d36'
Veal,Paul Everett
Aur 4h53'0"51d44'
Veal,Richard Charles
Her 16h56'59"50d47'
Veale,Jennifer Christine
Ari 2h25'44"12d30'
VeAlletto,Carolyn Shaw
Lyr 19h22'58"42d30'
Veath,M Lois
Lyn 8h8'13"46d46'
Veats,Jacqueline L
Peg 23h29'49"18d50'
Veats,Stephen A
Ser 15h18'44"8d28'
Veazey,Darron Ross
Cep 2h31'52"78d30'
Veazey,Lawrence & Elizabeth
Aql 19h55'12"10d54'
Vebber,Nancy Simonis
Mon 8h5'17"-0d23'
Vecchi,Jr,E Joseph
Her 17h31'19"28d38'
Vecchi,Monica
Crb 15h29'49"31d4'
Vecchiarelli,Jim
Aur 6h22'59"38d44'
Vecchio,James Ralph
Cep 21h4'58"58d56'
Vecchio,Lisa Emily
Cyg 21h59'38"52d35'
Vecchio,Lori Marion
And 2h29'12"50d36'
Vecchio,Peter J
Dra 12h50'21"67d59'
Vecchiolla,Joseph Dennis
Per 2h25'53"57d12'
Vecchione,Jaclyn
Lmi 10h48'0"27d13'
Vecchione,Jill Rachel
And 2h25'51"41d22'
Vecchione,Matthew Robert
Boo 13h35'36"21d39'
Vecellio,Lisa
Lyr 18h35'55"41d46'
Veck,Elizabeth Anne
Eri 4h8'12"-11d29'
Vecvanags,Jade Priscilla
Aqr 23h13'50"-6d37'
Veda
Eri 4h8'55"-10d37'
Vedder,"I Believe" Herb
Cas 0h49'20"64d12'
Vedder,Anne
Lyn 7h55'56"58d37'
Vedel,Valérie
Cyg 21h3'38"48d46'
Vedrana
Boo 14h30'23"50d55'
Vedreivive,Eric
Boo 14h36'29"44d1'
Veech
Umi 14h21'41"71d50'
Veeder,Douglas & Stephanie
Cyg 21h0'26"39d46'
Veenar,J L T
Uma 9h57'43"54d10'
Veenman,Marianne
Cam 4h59'1"58d27'
Veerkamp,Alexander Butzer
Cam 12h47'41"80d51'
Veerman,Neeltje
Dra 12h4'32"71d43'
Veety,Ryan G
Per 1h40'53"53d12'
Vega
Lac 22h45'0"38d41'
Vega 3/29/69 KK,Daisy
And 23h36'53"40d47'
Vega Too
Lyr 18h56'60"31d32'
Vega,Bernardino R
Aur 6h17'25"30d58'
Vega,Clara Aurora
Lyr 18h35'33"37d43'
Vega,Fernando Alonzo
Aur 7h13'55"41d29'
Vega,Frank
Eri 3h3'19"-3d21'
Vega,Gil
Eri 4h43'26"-8d31'
Vega,José Antonio Sosa
Lyn 8h31'52"43d3'
Vega,Jr,Tony
Per 3h10'44"50d21'
Vega,Maria
Peg 22h7'36"4d5'
Vega,Mercedes Ortega
Lyr 19h13'11"31d14'
Vega,Osvaldo
Oph 16h55'0"-22d31'
Vegas
Aur 7h9'50"40d53'
Vegh,Alice
Cas 2h20'29"68d34'
Vegh,Sandor
Lyn 7h35'16"39d39'
Veguez Trinary ABC,The Rahma
Tau 3h57'58"23d3'ADC
Vehslage,Mark Robert
Lac 22h53'50"37d36'
Veicht,Barbara
Cyg 20h26'17"30d47'
Veikko Ylitalo
Boo 14h29'1"45d17'
Veile,Rebekah Diane
And 1h36'11"37d4'
Veillette,Eric
Lyn 8h40'20"42d14'
Veilleux,Corry
Ori 5h6'12"10d25'
Veilleux,Happy 21- Nicole Danielle
Lyn 7h25'20"50d37'
Veimeris,Andrea Janae
Hya 9h9'0"3d27'
Veirheller,Elizabeth Edna
Vul 20h23'0"28d37'
Veisvul,Or
Sex 10h6'31"2d11'
Veit III,Richard Lawrence
Boo 15h0'51"12d32'
Veit,Reinhardt
Umi 13h28'50"74d35'
Veit,Sue
Cyg 20h22'28"40d36'
Veit,Sue
Mon 8h6'22"-0d7'
Veitch,Andrea
Cas 0h49'11"64d11'
Veitch,Daisy Helen
And 23h29'52"46d57'
Veitch,Mary
Aql 18h43'45"6d17'
Veith,Charles Douglas
Dra 17h49'1"64d41'
Vejar,Lindy
Mon 7h49'42"-6d42'
Vejar/Happy Birthday, Lindy
Mon 7h57'30"-7d50'
Vejnoska,Mark Andrew
Tau 4h56'50"16d29'
Vejtzik,Myriam
And 22h59'41"38d49'
Vekas,Tonia Marie
Com 12h18'15"27d31'
Vel,Daisy
Aql 19h53'0"14d22'
Vela "Solomon Beckett" ,Damian
Aqr 22h24'33"0d2'
Velarde,Enrique
Hya 8h27'30"5d40'
Velasco,Gracia Parejo
Lyn 8h13'50"49d31'
Velasco,Keil E
Hya 8h16'20"4d22'
Velasco,Pilar
Ori 5h47'24"10d15'
Velasquez,Daniel A
Aur 5h27'0"30d1'
Velasquez,Esmeralda
Cma 7h3'1"-18d46'
Velasquez,Walter A
Cmi 7h46'10"7d35'
Velazquez,Elizabeth
Aur 5h35'10"41d48'B
Velesta
Vul 20h4'58"28d17'
Velez,Hortencia
Lyn 8h2'17"48d50'
Velez,Jasmine Marie
And 0h51'38"39d26'
Velez,Jon Bertrum
Lac 21h57'1"36d36'
Velez,Jr,Ernie
Eri 3h0'31"-11d25'
Velez,Mr Federico
Lac 22h55'36"37d38'
Velez,Ronald Luis
Lyr 19h25'1"37d45'
Velicer,Anthony D
Hya 8h55'0"2d24'
Velilla,Pilar
Uma 12h2'55"31d14'
Velimaxstefanell
Tel 20h20'58"-48d28'
Veliz,Julie E
Com 12h22'44"30d19'
Vella,Gabriel Alexander
Crb 15h56'47"28d42'
Vella,Nadia
Cas 2h23'41"61d13'
Vellano,Holly
Vul 19h16'27"21d28'
Vellios,Chris
Pho 23h50'22"-46d47'
Vellucci,Nina
And 23h23'14"50d39'
Velma
Uma 9h46'17"52d12'
Velma
Sco 17h27'38"-30d24'
Velocci,Ginina Grace
Cas 0h12'0"61d2'
Velocci,Jake
Dra 20h23'29"71d7'
Velosa,Kristine
Cas 23h31'11"61d11'
Veloso,Abraham
Per 4h3'38"35d1'
Velotto,Claire
Vul 20h19'14"25d21'
Veloz,Max
Lyn 7h49'32"43d23'
Velte,Werner
Dra 20h23'36"68d41'
Velten,George P
Aql 19h6'0"-0d58'
Veltman/Milner Family, The
Mon 7h24'18"-10d52'
Veltri,Dan
Cep 22h16'14"70d12'
Veltri,John
Per 1h43,1"52d42'
Veltri,Michael J
Her 18h3'17"18d49'
Vemp
Lac 22h2'43"51d40'
Vemuri,Rajendra
Lyr 18h28'53"35d4'
Ven Roy,Rozalind
Uma 10h22'33"40d18'
Venable,Dolly
Uma 8h46'46"49d51'
Venable,Patrick Ryan
Lac 22h20'0"54d46'
Venables,Elaine
Crb 15h35'55"31d6'
Venafra,Samantha Christine
Cam 12h52'48"77d57'
Venaleck,David H
Dra 17h13'18"60d37'
Venaleck,James J
Dra 11h45'0"73d48'
Venamore-Cox,Aaron Jordan
Cep 20h31'34"61d7'
Vendemia,Joseph Christopher
Her 17h54'12"18d56'
Vendler,Helen Hennessey
And 0h8'35"47d5'
Vene
Ori 5h54'0"15d34'
Venech Star,George's Tricia
Her 16h21'51"4d22'
Venegas,Thomas
Vul 20h19'30"26d14'
Venegas,Tico & Marlene
Cet 1h32'32"-4d22'
Veneranda Libera Romantica
Her 16h41'40"18d46'
Venet,Glo
Com 12h20'16"20d20'
Venetia-Monique
And 23h21'28"40d27'
Venette,Adam
Aql 19h33'30"0d31'
Venezia,Bernie
Ser 15h15'33"11d40'
Venezia,Donna
Sge 19h32'18"16d17'
Venezia,Sheila
Eri 4h1'52"-12d14'
Vengelis,Jason
Tau 5h54'1"24d7'
Venice Athene
Sgr 18h5'16"-28d11'
Venier,Alessandra
Pho 0h40'7"-44d48'
Venier,Florence (Mar-Mar)
Cas 23h0'0"58d47'
Venis' Star
Ser 17h54'30"-14d9'
Venis,David Keith
Dra 11h41'0"71d57'
Venisha
Uma 8h31'1"56d46'
Venita
Lyr 18h41'27"32d37'
Venkman,The Real Ghostbusters Peter
Pho 0h34'44"-42d2'
Vennamo,Pekka
Uma 12h1'18"31d10'
Venner,Joey
Lib 15h39'0"-28d33'
Vennes,Noel L
Her 15h54'54"48d1'
Vennetilli Star,The
Lyr 18h55'0"37d13'
Venteciola,Larry Joseph
Aql 20h14'33"4d35'
Ventimiglio,Sontino Santellano
Aql 19h31'51"12d57'
Ventolini,Claude
Aur 6h52'25"44d19'
Ventoso,Sandra
Lyr 19h10'45"38d23'
Ventre,Cesha
Cyg 21h6'35"39d36'
Ventre,Joe (Curly)
Cam 6h19'50"71d8'
Ventura
Cyg 20h51'0"30d41'
Ventura's Dream
Aur 7h21'1"37d56'
Ventura,Austin Michael
Hya 10h47'11"-18d11'
Ventura,Citadel G
Peg 22h19'49"8d23'
Ventura,F Paul
Dra 17h36'31"68d58'
Ventura,Jr,Gustavo Julian
Cet 0h58'1"-7d11'
Ventura,Kathy Marie
Lyn 7h35'1"50d57'
Ventura,Michael R
Aql 18h44'0"8d25'
Ventura,Sue & Jim
Sge 20h0'11"16d35'
Ventura-Jackson,Rena
Mon 6h23'26"8d55'
Venturi,Sabrina
Pic 5h1'38"-45d36'
Venturini,Carolyn
Cam 3h47'53"58d42'
Venturino,Deborah
Peg 22h12'35"31d3'
Venuleth,Dr Christian
Sgr 19h0'31"-21d4'
Venus
Lmi 9h59'22"34d48'
Venus
Cam 5h59'55"67d39'
Venus de Lisa e Marco
And 23h10'27"42d17'
Venus Paula,The
Vir 14h58'18"7d19'
Venutoli,Cristina
Tau 4h9'44"0d46'
Venzon,Avelina
Del 20h14'0"10d24'
Veprin,Anna
Eri 4h55'27"-6d29'
Vera
Cyg 19h43'50"30d4'
Vera
Peg 21h45'48"20d33'
Vera
Lyn 7h31'46"40d38'
Vera
Mon 6h19'60"4d35'
Vera
Cyg 20h35'46"42d41'
Vera
Cyg 20h15'30"38d31'

Vera
Ari 1h16'32"32d27'
Vera 55
Sco 17h31'46"-38d42'
Vera Jean
Gem 7h7'29"21d7'
Vera Lee
Gem 7h29'12"30d14'
Vera Loves You
Cyg 19h26'12"33d32'
Vera Rose
Cas 0h15'44"63d22'
Vera,Rhea Luisa P
Lyn 7h34'0"41d15'
Vera-Kacena
Uma 11h21'33"31d4'
Verania,Jacob Stanley Kaleikaumaka
Peg 23h0'0"31d40'
Verbarg,Dieter
Cep 22h22'54"61d40'
Verbeek,Thomas Geert
Ori 5h58'14"8d35'
Verbeke,J J
Cam 7h15'1"82d50'
Verbeke,Julius A
Boo 13h54'17"15d59'
Verberne,Pieter
Cam 8h29'35"81d26'
Verbert,Grandpa Roger
Dra 17h8'21"62d59'
Verbizier,Jacques
Lyr 18h56'48"31d25'
Verbryck's Sweetheart Star,The
Sge 19h43'1"16d52'
Verce Star,The Frank
Uma 12h5'59"57d49'
Vercel,Vanessa
Del 21h0'18"12d43'
Verceletto,Christine
Boo 14h33'11"11d29'
Vercellini,Marc Anthony
Dra 15h59'28"67d56'
Vercher,Michael
Aql 19h52'55"-6d36'
Verdani,Durgesh
Her 18h7'1"14d49'
Verde II,Joe
Hya 8h56'31"0d33'
Verderosa,Lauren Elizabeth
Her 17h24'24"44d56'
Verdesco,Massimo
Com 13h11'39"26d59'
Verdi Family Star,The Sybil & John
Cam 8h36'59"78d32'
Verdiamoremio 1/12/76
Cas 2h58'45"60d25'
Verdick,Always Leo
Her 17h44'20"18d58'
Verdier,Deann
Uma 10h27'29"51d14'
Verdier,George
Her 17h37'0"25d25'
Verdon
Mon 6h59'0"8d33'
Verdon,Stuart & Debbie
Eri 2h43'36"-3d37'
Verdugo,Frank
Crt 11h13'12"-19d14'
Verdun,Dan
Dra 18h58'15"80d3'
Vereeke,Rick
Cep 2h28'29"78d13'
Veremis,Brandon
Lib 15h0'35"-28d28'
Verena E
Cet 2h57'32"9d23'
Verena Regina Vera Mei Cordis
Umi 14h50'42"66d53'
Verenazi,Allyson
Aqr 23h5'59"-18d39'
Vereé
Umi 15h26'21"80d35'
Vergara Star,The Angelo
Mon 7h53'21"-6d27'
Vergara,Ana Louisa
Vul 19h53'0"20d3'
Vergara,Jimena
Vir 14h22'36"-7d42'
Vergelas,Patrick
Oph 18h16'16"11d42'
Vergilio's Magic Star, Ron
Sex 10h41'32"-0d17'
Vergissmeinnicht,Hanna
Aqr 23h38'34"-4d2'
Vergnaud,Celine
Lac 22h5'50"38d30'
Verguet,Agathe
Ori 5h58'35"18d13'
Verhaeghe,Kelly Jean Blumke
Uma 11h41'49"58d50'
Verhalen,Pearl
Uma 8h51'59"71d8'
Verheecke,Philippe
Sex 10h47'50"5d15'
Verhey,Robert S
Ori 5h56'48"-1d0'
Verhoeven,Paul
Cma 6h56'17"-19d59'
Verhoeven,Terrill B
Lmi 9h23'13"34d54'
Verhoff,Jeannie
Cyg 21h37'43"38d7'
Verhoff,Jeannie
Lmi 10h37'22"26d22'
Verhoff,Jeannie
And 2h9'23"42d24'
Verhoog,Norman
Hya 9h15'16"1d24'
VerHulst,John J
Hya 9h2'19"0d13'
Verilla,Tiffany Marie
Mon 6h19'46"4d7'
Vering,Bob
Dra 19h50'38"70d54'
Verini,Joseph
Boo 15h12'39"26d20'
Verite,Alain
Peg 21h47'34"34d50'
Verite,Christian
Peg 21h47'59"33d46'
Verity
Cam 7h14'21"82d37'
Verity Fox
And 1h2'55"39d39'
Verity,Blake
Cep 21h50'53"56d0'
Verity,Lynne
Ari 1h58'24"25d4'
Verive III,Peter Joseph
Aur 5h18'46"45d46'
Verkaik,Bart
Per 2h31'25"51d38'
Verkuil,Paul
Cam 3h53'37"69d15'
Verkuyl,Giles
Aur 6h21'58"38d8'
Verlander,Morris West
Oph 17h2'42"-23d13'
Verlie C
Cyg 20h22'31"37d54'
Verlin,Steven
Lac 22h34'20"38d55'
Verlinsky,Yury
Dra 15h7'1"65d12'
Verluca,Constance
Aur 5h38'26"50d26'
Verlynn
Umi 16h19'34"71d6'
Verma,Abhishek
Lyn 8h53'55"34d18'
Verma,Sangita
Cam 5h39'1"60d16'
Vermeulen,George
Cam 6h4'1"58d10'
Vermie
Ari 1h50'30"23d45'
Vermiglio,Lisa
And 1h44'1"38d52'
Vermoch,Marc M
Per 1h55'50"52d58'
Vermorel,Evelyne
Lyr 19h16'51"42d51'
Vern-N-Shannon
Eri 4h32'12"-11d6'
Verna Gayle
Cas 1h12'52"61d33'
Verna's "Lasting Light"
Sge 18h55'28"19d34'
Verna,Jr,Louis R
Leo 9h54'30"8d12'
Vernacchia,Louis & Sue
Cyg 21h56'29"50d23'
Vernassiere,Henri
Aur 5h8'29"29d37'
Verner,Doug & Diane
Lmi 10h18'17"32d26'
Verner,Harry
Ori 5h53'47"13d3'
Verner,Niels
Ori 5h51'22"18d4'
Vernie
Peg 23h40'19"26d38'
Vernon & Bobbi
Oph 18h37'26"7d18'AB
Vernon & Lisa Together Forever
Ori 5h17'13"-8d42'
Vernon,August
Her 17h30'52"26d27'
Vernon,Charles & Lorraine
Cyg 19h32'18"37d19'
Vernon,Christy L
Cmi 8h2'1"0d2'
Vernon,Estelle Renée
Cyg 21h26'38"38d15'
Vernon,Robert Fox
Oph 17h31'29"-1d5'
Vero
Umi 13h46'36"77d37'
Vero Fleury 25000
Cyg 21h30'26"50d15'
Vero,Mommy "Meat-Hooks"
Lyn 8h12'37"37d42'
Vero-Louise
Lyn 8h27'53"38d7'
Veron
Cam 6h20'10"80d17'
Verona,Chantal
Cyg 19h45'32"38d31'
Verona,MD,Steven
Uma 11h56'24"30d38'
Veronica
And 1h47'24"40d56'
Veronica
Mon 6h23'27"-10d46'
Veronica
Cas 1h48'0"58d14'
Veronica
Peg 22h44'1"5d5'
Veronica
Lyr 18h57'15"42d36'
Veronica
Crt 10h52'42"-17d47'
Veronica
Cam 5h58'1"78d52'
Veronica
Lmi 10h5'33"38d49'
Veronica
Mon 6h24'56"10d43'
Veronica
Ser 15h20'1"8d31'
Veronica
Cas 1h48'32"74d58'
Veronica
Crb 16h16'25"34d33'
Veronica Moonbeam
Uma 13h37'33"61d46'
Veronica Suzette
Cyg 20h0'29"40d58'
Veronica V V
Ari 2h51'17"30d13'
Veronica's Big Birthday
Ori 6h17'35"0d52'
Veronique
Peg 21h29'24"20d25'
Veronique
And 1h27'17"35d20'
Veronique B
Lac 22h27'44"40d25'
Veronique et Christian
Umi 16h14'18"71d28'
Veronique H
Aql 20h1'26"10d29'
Veronique Mon Amour
Lyn 8h10'49"39d58'
Veronique P
Dra 9h57'0"73d16'
Verplank,Theresa
Cet 1h17'39"-13d11'
Verrastro,Lucas Dollard
Lac 22h41'1"56d21'
Verrastro,Paul
Gem 7h4'55"28d44'
Verratti,Rob & Anne
Boo 15h8'51"50d5'
Verrechio,Mr Gerard
Lyn 7h41'43"45d52'
Verrelli,Gloria Julia
Peg 22h42'1"21d44'
Verrett,R J
Aur 5h8'0"42d39'
Verrette,Paul
Sco 17h51'21"-40d3'
Verrico,Bonnie Goodwin
And 2h33'57"44d10'
Verrico,Donald
Dra 17h46'31"63d49'
Verrips,Lael
Lmi 9h46'20"38d9'
Verro,Dick & Jane
Sct 18h39'24"-6d43'
Verry,Amy Nicole
Lyn 6h34'35"59d19'
Versaggi,Charles
Psc 1h0'57"27d56'
Versaggi,Lisa Koenig
Cas 0h51'0"64d1'
Versal,Duka Luna
Hya 8h58'49"4d31'
Versatile D
Del 20h32'21"10d42'
Versavage,Bill
Lac 22h40'45"51d46'
Verseck,Silke
Ari 2h54'0"21d31'
Versfelt,Divit & Joyce
Cyg 21h14'30"39d55'
Verspieren,Hubert
Del 20h15'38"10d41'
Versteegh,Hugh Robert
Ser 15h12'29"4d51'
Verstegen,Doris (Grandma)
Cyg 20h34'39"42d6'
Vertefeuille,Joan M "Best Buddy"
Aql 19h19'48"13d56'
Vertolli,Susan Price
Aql 19h55'12"13d25'
VerTreese
Mon 6h27'46"-6d10'
Vertullo,Anthony
Lac 22h20'1"51d2'
Verusch
Lyr 18h34'37"40d7'
Vervalin,Stephen & Jane
Lac 22h38'49"53d14'
Verver,Hans Henricus Johannes
Cru 11h58'0"-56d38'
Verville,Michel
Ori 6h8'38"5d8'
Vervoort III,Frank James
Ori 5h45'0"11d18'
Verwyst,Mr Douglas Marinus
Her 16h4'29"40d38'
Very Loving Joan
Cyg 19h26'36"30d59'
Verzi,Dennis
Aur 4h48'42"40d11'
Vesco,Daniel William
Aql 20h1'22"7d12'
Vescovo,Edward R
Hya 9h17'17"5d37'
Vesel,David
Cma 6h27'55"-13d20'
Vesey,Patrick
Mon 8h6'3"-4d35'
Vesio,William
Lyr 18h58'19"30d22'
Vesneski,Peggy Jane
Aql 19h30'57"10d6'
Vespa,Carl
Lac 22h43'53"53d0'
Vespa-Ehlen,N Cinda
Lib 15h36'35"-20d35'
Vespasiano Cum Caelum Undique,Tony
Per 2h51'0"43d41'
Vespucci,Gabrielle Michelle DaVita
Cas 0h7'11"61d59'
Vessels,Jennifer Karen
Cyg 19h41'20"28d54'
Vessicchio,Anthony
Ori 5h34'55"-1d39'
Vest,Jamie Marie
Cas 0h8'37"60d46'
Vesterinen,Dimitrij Karierich
Umi 17h11'43"80d5'
Vestice,Gene
Her 17h3'35"20d28'
Vestrand,Christine
Cam 6h3'31"60d41'
Vestri,Silvia
Vel 9h43'24"-48d8'
Veta V C H
Peg 22h4'10"5d59'
Veteri,Gabrielle Anna
Vul 19h3'49"21d57'
Veteri,Samantha Jo
Lmi 9h55'47"38d6'
Vetrano,Isadore "Phil"
Dra 11h40'1"70d53'
Vetrano,Michael Anthony
Boo 14h22'57"52d33'
Vetro,Ernestine Russo
And 0h4'30"38d17'
Vetrone,Mary
And 23h30'32"48d43'
Vetter's Delight
Aql 19h3'17"15d26'
Vetter,Francis Theodore
Boo 15h5'19"22d27'
Vetter,Frank
Ser 15h14'12"4d8'
Vetter,Stevens Lane
Ori 5h59'17"7d43'
Vetter,The Hob & Nancy
Vul 19h17'14"25d7'
Veuve,Hermance
Cnv 12h22'33"47d23'
Veva Shining Star
Cyg 20h37'0"53d4'
Veva-The Star of My Life
Cet 2h22'45"-8d39'
Vexler,Adrian
Cas 1h13'44"67d59'
Vezin,Didier
Sco 16h26'1"-40d36'
Vezina,Kathy
Lyr 18h44'29"47d28'
Vezzali,Jr,Russell James
Uma 9h13'32"70d26'
VGDS Kindergarten Pandas 1992/1993
Uma 10h7'0"42d3'
Vi Vi
Lyr 18h48'0"35d4'
Vi-Star
Cet 3h14'18"1d16'
Via,Kitty
Lyr 18h59'1"33d26'
Viaggi,Baino
Lep 5h54'1"-20d1'
Viaggi,Catia
Lmi 10h7'21"32d6'
Viaggi,Masolino
Hor 2h49'0"-49d4'
Viaggi,Svago
Cam 6h22'38"68d22'
Viakowski,Ralf und Sabine
Dra 17h2'55"67d20'
Vial,Béatrice
Crb 15h38'25"35d23'
Viala,Marie
Lmi 11h3'21"33d25'
Viana,Marisol
Vul 19h45'30"28d21'
Viands,Bill
Per 3h59'0"37d50'
Viands,Bill & Betty
Gem 7h11'24"24d53'
Viard,Daniel
Umi 16h24'31"70d4'
Viard,Virginie
Cnv 13h12'53"38d54'
Viau,Miguel
Cam 3h59'55"53d42'
Viavant's Vis Vitae
Ori 6h8'15"4d23'
Vibbert,Aubrey G
Eri 4h33'31"-3d42'A
Vibert,Mousieur Jean- Claude
And 0h27'47"44d26'
Vibert,Raymond (AW) Joseph
Lac 22h21'27"54d17'
Viblein,Jutta
Cap 20h32'58"-19d31'
Vic
Dra 17h39'45"64d39'
Vic & Ady
Ant 10h31'49"-38d9'
Vicari,Clement
Per 1h50'55"53d48'
Viccars,Sally
Oph 18h7'36"12d10'
Vicci
Lmi 9h45'23"34d24'
Vicens Forever
Uma 9h41'59"61d14'
Vicik,Dawn & Sonny
Lyr 18h43'0"30d7'
Vicinelli,Davide
Dra 16h21'21"62d49'
Vicious Vixen Mission Stardust-E B
Lmi 10h57'50"30d34'
Vick,Amanda Joy
Vul 20h45'59"20d9'
Vick,Austin Lafayette
Dra 18h30'25"50d15'
Vick,Devlyn Lane
Peg 22h6'26"26d52'
Vick,Stephen Lyon
Ori 6h4'0"6d34'
Vick,Traci Michelle
Peg 23h23'55"30d7'
Vick-A-Rue
Lyr 18h29'16"30d27'
Vickers,Archie D
Her 16h5'15"20d30'
Vickers,Bamboo
Her 16h50'19"13d6'
Vickers,Daniel F
Ser 15h52'52"-0d53'
Vickers,Julian Nicholas Ward
Uma 8h15'37"60d44'
Vickers,Marie
Com 12h54'54"24d16'
Vickers,Sweet Elizabeth
Sex 9h52'46"2d20'
Vickery,Ardennes Drew
Uma 9h40'36"48d40'
Vickery,Blanche
Mon 6h20'25"8d33'
Vickery,Brian
Aql 20h16'44"5d28'
Vickery,Colleen
Lyr 18h57'1"47d30'
Vickery,Jr,John David
Lac 22h53'23"55d53'
Vickery,Lynn Marie
Cas 0h50'42"60d49'
Vickery,Olivia
Crb 15h55'15"37d37'
Vicki
Mon 7h58'32"-2d2'
Vicki
Uma 14h0'41"61d15'
Vicki
Peg 23h6'0"17d50'
Vicki
Her 16h5'58"16d12'
Vicki
Hya 9h30'31"-13d17'B
Vicki
Mon 6h26'39"11d30'
Vicki
Lyn 9h0'10"39d42'
Vicki "Fantasy" & Diann "Passion"
Uma 9h6'35"48d20'
Vicki & Jerry's Wedding Jewel
Cyg 20h19'25"41d44'
Vicki & Randy's Love Star
Lmi 10h38'45"38d27'
Vicki & Ron
Aql 20h12'43"0d12'
Vicki & Walt "Star Parents"
Uma 10h4'33"68d11'
Vicki 22
Umi 14h55'15"67d46'
Vicki Ann
Cas 0h24'17"50d12'
Vicki Bear
Peg 22h34'17"25d37'
Vicki Dawn
Boo 15h46'45"40d35'
Vicki Lynn
Uma 10h30'41"70d37'
Vicki Lynn
Peg 23h38'56"27d21'
Vicki Star
Vul 19h33'21"25d12'
Vicki's Passion
Lyr 18h37'35"40d32'
Vicki's Star
Com 12h24'49"21d38'
Vicki's Star "Vitara"
Ori 6h6'50"-2d40'
Vicki's Toshi-Keeta
Cmi 7h55'40"5d33'
Vicki's Victory
Eri 2h58'58"-6d45'
Vicki's Wink
Crb 15h55'27"26d53'
Vicki-Leanne
Mon 6h28'58"3d26'
Vickie
Per 4h0'46"50d55'
Vickie
Sgr 19h28'31"-42d55'
Vickie L
Uma 11h9'50"43d59'
Vickie Lynn
Cyg 20h0'57"40d28'
Vickie's Place
Cet 1h34'28"-12d31'
Vickienbonotarius
And 22h59'54"51d19'
Vickner,Patrick A
Sco 17h50'30"-40d51'
Vickrey,Tanya Dee
Cyg 19h26'48"30d35'
Vicky
Cyg 19h57'0"31d40'
Vicky
Lyr 18h56'1"37d33'
Vicky
Ori 4h43'19"0d22'
Vicky & Jason
Cyg 20h23'12"39d18'
Vicky 21
Lyr 18h59'31"36d16'
Vicky C
Lyr 18h59'20"40d53'
Vicky D
Eri 4h55'4"-8d15'
Vicky Jenny Julie
Com 12h30'34"22d12'
Vicky Sue
Sge 19h2'21"21d3'
Vicky,My Angel
Lyr 18h17'44"38d52'
Vico
Cnv 12h5'1"41d34'
Vicom Computer Services
Cep 20h56'48"55d54'
Vicster
Lac 22h23'30"40d19'
Victoire Et Georges
Her 17h21'43"41d11'
Victor
Aur 5h1'1"47d19'
Victor
Mon 7h37'21"-6d7'A
Victor
Her 18h14'35"48d19'
Victor & Ana Maria Forever
Crb 16h4'55"32d29'
Victor John
Tau 5h1'20"28d28'
Victor Kirk
Cet 2h6'40"6d24'
Victor Paul-Robin's Forever Love
Cet 1h29'49"0d35'
Victor's Courage
Peg 23h8'21"10d14'
Victor's Fire
Boo 14h10'24"40d33'
Victor's Primeline
Peg 22h10'11"4d46'
Victor's Star
Dra 18h29'43"58d31'
Victor,Anthony James
Per 3h56'41"31d52'
Victor,Mary Ann
Aql 19h55'1"12d51'
Victoria
Leo 11h25'12"-2d25'
Victoria
Cyg 20h34'0"45d20'
Victoria
Lyr 18h30'1"30d59'
Victoria
Peg 22h59'56"25d4'
Victoria
And 1h1'40"35d4'
Victoria
And 2h28'1"47d57'
Victoria
Eri 4h4'27"-18d18'
Victoria
Mon 7h44'12"-1d59'
Victoria
Tau 5h1'12"20d12'
Victoria
Lib 15h3'46"-28d43'
Victoria
Lyr 18h40'14"27d42'
Victoria
Uma 12h7'16"46d22'
Victoria
Cas 2h56'16"73d27'
Victoria & Athanasios
Uma 11h32'0"31d6'
Victoria & Griff's Love 11/20/1992
Ori 4h44'27"-2d58'
Victoria & Stuart Star
Cyg 21h45'38"38d44'
Victoria & Tony
Crb 16h23'23"32d21'
Victoria (Mushy) 96'
Cas 1h16'1"60d8'
Victoria Anne (Tori)
Cyg 20h23'50"30d54'
Victoria Ayesha
Leo 9h33'46"7d22'
Victoria Dawn
Mon 7h21'14"-5d42'
Victoria Evelyn
Umi 13h45'39"77d13'
Victoria Gay 20-8-93
Leo 10h52'29"20d28'
Victoria Glorious
And 1h32'46"39d5'
Victoria Helen
Cas 0h32'0"58d51'
Victoria Jane
Mon 6h24'22"-10d22'
Victoria Jane
And 0h3'54"47d41'
Victoria Jane
And 1h15'0"37d37'
Victoria June
And 1h4'20"38d23'B
Victoria Linn
Mon 7h52'31"-3d5'
Victoria Louise
Cas 0h16'35"61d46'
Victoria Lynne
Peg 22h8'12"24d48'
Victoria M
And 0h4'29"46d52'
Victoria Magdalyn
Del 20h54'27"9d25'
Victoria Margaret
Com 12h15'39"21d48'
Victoria Marie
Lyr 18h16'1"30d14'
Victoria P
And 0h19'34"32d52'
Victoria Paige
Peg 22h13'56"33d29'
Victoria Rose
Cas 0h11'47"61d34'
Victoria Rose
Lyn 7h3'20"44d40'
Victoria Twinkle
Uma 13h40'39"61d57'
Victoria Wedding Chapel
Lyr 18h44'0"39d39'
Victoria's Fantasy Dad
Cep 2h44'0"77d41'
Victoria's Happiness
Dra 11h54'52"66d56'
Victoria's Star
Cyg 20h31'21"38d30'
Victoria-Melissa
Ori 6h1'33"10d8'
Victoria-My Guiding Light
Aql 20h11'20"0d18'
Victoria-Nicole
Dra 13h27'43"70d39'
Victorian
Umi 15h21'40"67d45'
Victorious Spirit
Boo 13h54'11"21d27'
Victory In The Sky
Gem 6h44'1"30d21'
Viculis,James S
Uma 10h9'20"47d39'
Vida
Per 1h39'43"53d18'
Vida-"I Am Life"
Cas 2h31'32"60d27'
Vidakovic,Valdan
Leo 10h18'22"13d9'
Vidal,Carlo Enrique
Per 2h10'45"57d26'
Vidal,Charlene
Lyr 18h39'53"30d60'
Vidal,Daniela Dulcinea Leona
Aql 20h6'0"0d10'
Vidal,David Emeric
Her 17h6'57"41d11'

Vidal,Delphine And 0h28'27"44d10'
Vidal,Desiree Anne Cmi 8h9'18"1d42'
Vidal,Gore Oph 18h17'27"11d43'
Vidal,Janie Cyg 21h38'46"42d49'
Vidal,Michel Aql 19h24'46"-10d55'
Vidal,Ramon E Her 16h27'1"42d22'
Vidal,Ryan James Lac 22h37'37"55d53'
Vide,Theresa Lyr 18h58'38"37d10'
Videan,Nap T Aql 19h30'54"12d17'
Viders,Barbara Del 20h50'46"8d46'
Vidhya,My Only Star (Fish) Uma 12h14'39"58d14'
Vidiella,Bernat Cobera Ari 2h44'20"21d39'
Vidiella,Martin Corbera Cnc 9h16'53"12d26'
Vidlund,Steven Charles Aur 7h17'39"36d41'
Vidmar,Karyn And 23h42'32"45d58'
Vidmer,Abigail Kate Peg 23h4'32"20d4'
Vidmer,Kurt Stephen Per 3h27'38"52d19'
Vidoni,Fabio e Gabriella Dra 14h56'0"60d16'
Vidou,Mr André Oph 18h3'0"10d30'
Vidrequin,Renee Boo 14h39'35"44d30'
Vidya & David's Star Cyg 21h12'53"38d30'
Viebach,Wolf Cnc 8h0'0"8d49'
Vieira,Edwin-Bonnie Lyr 18h56'34"46d57'
Vieira,Joseph Her 16h34'57"21d39'
Vieira,M L Lyr 19h19'56"41d47'
Vieiro,Mary & Remigio Cyg 21h18'54"28d27'
Viel,Andre Sex 10h47'11"5d54'
Viel,Andre Her 18h19'38"18d50'
Viel-Tolic,Mary Lyr 19h0'1"38d6'
Vielhaber,Michael & Lisa Her 15h55'26"50d26'
Vielwerth,Victoria Valentina Umi 14h32'51"69d47'
Vienna Lyn 6h58'58"54d13'
Vienna S F Johnny Cep 23h38'28"64d35'
Vierig,Max O Her 18h11'30"48d53'
Vierling,Jennifer Sgr 19h26'15"-40d46'
Vierling,Mark L Aur 6h19'0"32d4'
Vierra,Caroline Lee Sge 19h57'0"16d50'
Viertel,Walter Dra 18h31'50"67d55'
Viesca Jr's Anniversary Star,Sal Aql 20h17'58"1d40'
Vieser,Milford A Cyg 20h0'20"31d9'
Vieser,Ruth Crb 15h57'38"29d32'
Vieser,Walter Nelson Boo 14h12'0"32d51'
Viess Family Guiding Star of Harmony Uma 13h17'37"62d39'
Vietcha,Yves O'Reilly Tri 2h20'11"30d8'
Vieten,Cassandra "Spooky" Eri 3h46'45"-7d21'
Vieth,Franz Josef Peg 22h0'21"28d25'
Vieth,Georg Boo 14h18'36"13d17'
Vietor,Cornelia Christina Uma 9h10'32"50d26'
Vietor,Sr 12-5-1912, Hendrik Engelbertus Uma 9h32'20"58d46'
Vietzke-Oppermann, Gisela Lib 14h46'0"-8d49'
Vieyard's Vision Eri 3h31'1"-04d31'
Vigario,Michelle Kristen And 0h10'0"30d49'
Viger,Joseph Thomas Dra 15h3'53"65d6'
Viggiano,Christine McBride Com 12h55'35"25d50'
Viggiano,William Per 4h6'47"47d21'
Vigh-Czinkota,Ilona Hya 10h1'25"-19d52'
Vigil,Bob & Lynn Cyg 19h25'32"30d55'
Vigil,J Patrick Lac 22h13'32"49d58'
Vigil,Valerie Ann Uma 11h35'60"37d35'
Vigilante,Carla Peg 22h14'13"30d22'
Vigilante,John Aur 4h51'12"50d36'
Vigliotti,Terris Lee Del 20h16'57"13d34'
Viglotti,Laura Allison Lyr 18h27'33"30d26'
Vigna,Jane Elizabeth Peg 22h13'39"30d20'
Vigna,Kim Marie Cyg 20h25'57"41d5'
Vignaux,Jeffrey Cep 22h13'45"55d34'
Vigneault's Haven Uma 8h48'59"50d14'
Vigneault,Kim Lyr 18h47'19"34d8'
Vigneault,Patricia Ori 4h46'17"5d13'
Vignocchi,Joan Marie Lyn 8h47'18"45d28'
Vignola,Stevie Ray Aur 4h54'57"40d46'
Vignon,Jean-Paul Peg 22h55'36"27d10'
Vignona,Ralph Cmi 7h39'4"3d25'
Vigott,Raquel Rojas Cam 7h43'15"68d1'
Vigoureux,Philippe And 0h27'21"40d17'
Vigus,Aaron Cam 3h18'1"56d21'
Vigus,Roy Cam 3h18'45"56d6'
Viherlaakso,Heikki Cam 5h50'15"61d28'
ViJeff Dra 16h37'19"61d9'
Viki's Lonely Planet Dra 9h54'25"74d54'
Viking 9 S-3 Tau 5h0'56"28d30'
Viking Son of Love Del 20h55'52"9d55'
Vikki's Vesper Oph 18h39'13"7d5'
Vikler,Coby Z Ori 5h42'32"10d39'
Viktor & Marie-Louise Lyn 8h28'57"58d35'
Viktor,Guenter Aur 7h20'21"36d46'
Viktor,Kenneth B Aql 20h1'48"1d13'
Vilamosa,Helios Del 20h36'27"6d21'
Vileisis,Birute Mon 6h25'49"11d22'
Vileisis,Vita Eri 3h40'40"-6d23'
Viljakainen,Matti Cam 4h58'0"68d31'
Villa,Adriano Oph 16h52'54"10d26'
Villa,Alessandro Boo 15h4'37"42d12'
Villa,Anthony Steven Dra 18h38'37"68d15'
Villa,Augustine Fernando Lyn 6h54'57"59d16'
Villa,Cristina Cas 2h1'1"60d11'
Villa,Declan Ser 15h18'51"11d21'
Villa,Franca Col 6h34'55"-36d8'
Villa,Graziella Lmi 10h2'18"32d30'
Villa,Irene Cas 0h4'52"59d34'
Villa,Jean M Cyg 19h51'0"40d51'
Villa,Marianne Com 12h30'12"26d32'
Villa,Marisa Lmi 10h2'55"32d14'
Villa,Sam Her 17h39'54"18d52'
Villa,Shawna Dee Cet 0h5'40"-7d54'7'
Villacura Uma 12h50'56"61d32'
Villadsen,Rebecca King Lyr 19h5'12"26d53'
Villagra,Ivan Marcelo Oph 17h54'32"11d10'
Villain,Jacqueline Rene Cyg 20h51'22"50d16'
Villalba,Geraldine Vul 19h21'11"26d50'
Villalba,Lynne Cyg 20h0'16"40d44'
Villalba,Russell C Cep 21h58'38"55d27'
Villalobos Star,The Meece Cet 0h29'1"-0d3'
Villalobos,Juan Aql 19h45'17"11d22'
Villalobos,Juan Lib 15h40'32"-22d0'
Villalon,Patrick Peg 23h47'21"30d5'
Villaneuva,Daniel Cet 3h18'1"3d7'
Villani "Medicine Magic",Dr Michael Cyg 21h49'55"52d37'
Villani,Robert Martin Cam 8h2'17"70d6'
Villano II,Vannesa & Michael Umi 16h12'14"71d50'
Villano,Catherine Aur 4h56'43"40d1'
Villano,Marie A Uma 14h0'37"48d3'
Villanueva,Armando Fernando Hya 9h5'1"4d29'
Villanueva,José Carlos Per 3h42'38"51d53'
Villar,Diego Moreno Aql 19h31'0"12d51'
Villar,R L Uma 11h23'43"33d5'
Villard,Berkley J Cas 1h15'23"64d35'
Villard,Kris Del 20h12'58"14d47'
Villareale,Cynthia F Cam 4h3'13"61d24'
Villarin Her 16h37'31"36d51'
Villarreal,Alexandra Glander Aqr 22h21'19"-6d8'
Villarreal,Elisa Pellico Ari 2h57'56"22d4'
Villarreal,Lisa Uma 13h22'45"58d26'
Villarreal,Lydia L Pellico Ari 2h51'21"21d17'
Villarreal,Reginald Per 3h1'11"41d31'
Villarreal,Tere Mon 6h59'32"-6d4'
Villarreal,Veronica Sofia Leal Cnc 8h59'27"21d24'
Villarreal,Victor Vidal Her 17h57'27"40d23'
Villasoto,Helen V Lyr 19h16'55"41d30'
Villaume,Lena Cmi 8h7'44"5d45'
Villavicencio,Gina Carmela Uma 9h22'58"47d45'
Villavicencio,Princess C Alexandra And 1h13'24"37d54'
Villega,Candice Ser 15h19'0"8d54'
Villegas,Jeanne And 1h0'43"38d9'
Villegas,Michael Per 3h58'49"31d22'
Villegas,Tony & Kerin Cyg 21h52'33"41d14'
Villers,Emma Gay Vul 19h18'23"23d20'
Villers,Joanne Cyg 19h33'42"35d50'
Villers,John Marshall Aql 20h22'24"2d1'
Villez,Donna Cyg 21h33'48"40d27'
Villicana-55(FV55), Frederick Ori 5h58'58"10d1'
Villman,Christine Peg 22h35'27"8d1'
Villon,Prof Dott André Franáois Ori 5h31'0"0d19'
Villone,Concetta V And 2h24'42"45d49'
Villont,Denise E Dra 11h52'36"72d27'
Villotti,Andrea Lyn 6h12'46"54d55'
Villoutreix,Michele Cas 2h27'51"50d52'
Vilma Sgr 19h30'27"-34d38'
Vilma Sco 17h37'23"-40d22'
Vilnrotter,Felicia M Sgr 18h55'14"-21d39'
Vilord,Matthew R Her 17h21'27"44d57'
Vimala Cma 6h28'27"-24d35'
Vimal Cyg 19h46'56"38d13'
Vimkova,Lenka Aur 4h47'50"50d31'
Vinas,Alicia Lynn Aql 18h43'51"10d16'
Vinas,Faith Tatiana Cyg 20h24'0"41d1'
Vincar Dra 16h18'54"60d11'
Vince Her 16h37'37"29d10'
Vince Lyr 19h20'48"30d47'
Vince & Lynn Aqr 20h55'16"-6d30'
Vince & Phyllis Forever Lyn 7h31'39"58d27'
Vince & Rosemary 1995 Eri 3h3'1"-2d15'
Vince Family Star,The Peg 21h41'32"28d11'
Vince Neal Del 20h25'42"18d52'
Vince's Star Cep 1h27'34"80d32'
Vince's Vision Aur 7h7'34"38d38'
Vince,Carrie's Prince Per 4h41'0"38d10'
Vince,Kent James Her 18h4'58"28d27'
Vincelette,Sir Knight Roderick D Ori 5h33'34"-1d13'
Vincent Boo 15h7'29"16d14'
Vincent Cyg 19h33'1"37d8'
Vincent Ori 6h5'23"1d10'
Vincent Lac 22h7'14"49d1'
Vincent Per 2h3'43"56d40'
Vincent Uma 12h19'17"60d3'
Vincent & Dena Mon 6h18'1"5d47'
Vincent & Dina's Star Cam 8h52'41"78d49'
Vincent (Winzpuh II) Aqr 21h55'43"0d22'
Vincent Anthony Ori 5h47'14"18d60'
Vincent Et Yoann Lyn 7h52'39"51d51'
Vincent George Aur 5h2'27"46d9'
Vincent Lewis "Luminaire" Per 2h47'16"41d12'
Vincent Mon Amour Lyr 18h41'29"42d26'
Vincent Robert Cyg 20h50'59"38d10'
Vincent Théry Uma 13h22'13"62d53'
Vincent's Brillance Sct 18h46'35"-6d12'
Vincent's Flight Uma 9h23'45"52d35'
Vincent's Star Eri 5h1'51"-6d5'B
Vincent,Barbara Jo Lyr 18h37'16"42d24'
Vincent,Brian F Ori 6h17'38"7d43'
Vincent,Curt & Tricia Lyr 18h32'59"36d59'
Vincent,Franáois Dra 10h6'43"73d23'
Vincent,Jim Sge 20h5'0"16d18'
Vincent,Jim Aur 7h6'59"38d38'
Vincent,Jr,Francis Fitzgerald Ori 6h1'1"20d59'
Vincent,Jr,John J Cet 0h40'35"-2d12'
Vincent,Jr,PFC Paul W Per 2h22'27"54d50'
Vincent,Michael Perrin Lac 22h24'1"55d7'
Vincent,Morgan Talia Aur 4h58'45"50d57'
Vincent,Rett Cet 1h54'29"0d30'
Vincenti,Maria Pia Scarcilia Ant 10h37'52"-30d34'
Vincentini,Michael Uma 10h50'0"56d1'
Vincenza Uma 10h9'25"47d31'
Vincenza And 0h46'46"40d34'
Vincenzi,Paola Cas 1h17'0"61d10'
Vincenzo Cep 21h49'1"85d52'
Vincenzo Lyn 7h34'29"44d28'
Vincenzo Psc 22h52'0"1d13'
Vincenzo's Pork Store Lmi 10h1'19"30d18'
Vincenzo,Deborah Darlene Lyn 7h46'30"51d51'
Vinchon,Elisabeth Aql 19h3'0"16d30'
Vincicus "The Grateful Dead" Mon 7h39'54"-3d18'
Vincienne,Regis Per 3h28'0"50d32'
Vinciguerra,Daisy Cep 4h19'28"80d19'
Vinciguerra,Danielle Ari 1h59'47"19d20'
Vinciguerra,DeAnna Sgr 18h4'34"-28d18'
Vinciguerra,Debbie Psc 1h21'23"31d35'
Vinciguerra,Patricia Peg 21h42'54"21d52'
Vinciguerra,Philip Vir 11h51'32"8d4'
Vinck,Ria Aqr 20h56'58"-1d4'
Vincze,Nelsie Anna Lac 22h53'54"52d59'
Vinding,Jørgen Uma 9h28'59"53d9'
Vine,Howard Oph 17h57'0"14d10'
Vine,Stephen Glenn Per 2h54'58"37d15'
Viner,Christian Scott Uma 12h3'1"30d5'
Viner,Shari Cas 0h18'46"59d24'
Vines,Joe B Hya 8h12'43"0d13'
Vineyard,Angela D And 1h48'1"38d38'
Vineyard,Bill & Nancy Aql 20h12'24"4d44'
Vineyard,Christine Mon 7h40'24"-6d12'
Vinichris Cas 0h31'23"68d25'
Vinie Dra 17h1'30"65d31'
Vinnie Cep 22h59'35"70d54'
Vinnie's Special Shining Star Hya 9h2'0"5d6'
Vinnie's Star Aur 5h18'48"43d14'
Vinny Dra 14h55'39"59d37'
Vinny Aur 6h27'30"38d0'
Vinny Lyr 18h49'36"33d49'
Vinny & Jennifer Cam 8h2'14"78d18'
Vinny 143 Cep 20h53'57"76d8'
Vinny B Her 18h38'33"18d56'
Vinot,Christelle et Didier Her 17h2'48"50d29'
Vinson,Eugene Aql 19h31'38"13d10'
Vinster,The Cet 3h12'17"6d47'
Vinti,Rachel Oph 17h33'25"-23d51'
Vinton,Billy B Cap 20h28'39"-12d43'
Vinton,Elenore F Sco 17h28'33"-43d14'
Vinton,Elizabeth T Vir 11h59'60"-2d38'
Vinton,Joe Allen Lib 14h24'15"-20d13'
Vinton,Mary Stewart Sgr 18h49'28"-35d6'
Vinzani,Albert & Ann Uma 8h39'46"53d24'
Vinzani,Antoinette Lyn 8h8'12"52d7'
Vinzani,Dina Lee And 0h45'39"35d37'
Vinzani,Earl Her 16h57'30"21d46'
Vinzia,Frederic Cep 22h24'18"58d39'
Vinzons,Fernando Aql 18h41'59"-2d31'
Viola Cmi 8h4'21"0d58'
Viola Cyg 21h11'44"38d37'
Viola And 1h0'24"39d9'
Viola Louise Mon 7h56'24"-3d48'
Viola Mary Lyr 19h16'23"42d54'
Viola,Andrew S Dra 15h45'59"62d8'
Viola,Dianne "DeeDee" Marie Uma 10h56'26"50d33'
Viola,Jessica Lynn Cyg 20h39'28"37d58'
Viola,Richard Boo 14h28'29"52d22'
Viola,Ronald "Sugar Bear" Cap 20h30'0"-26d22'
Violaine Cam 7h26'44"67d35'
Violet And 2h18'0"40d39'
Violet Peg 23h30'26"12d14'
Violet Lyn 6h49'30"51d34'A
Violet & Ben Cam 7h25'22"78d50'
Violet & Bernard Cyg 20h41'54"45d52'
Violet John Cyg 19h34'48"31d31'
Violet's Shining Star Aur 6h6'43"48d48'
Violetta Leo 10h34'1"18d17'
Violetta Uma 11h56'29"30d58'
Violette LLB,Diane A Aql 19h6'50"-1d35'
Violette,Kip Uma 12h0'58"64d21'
Vion's Special Star, Joanne Tau 4h2'0"18d57'
Vipond,John Ori 6h7'48"0d11'
Virag,Adam Ori 5h31'19"-0d46'
Virag,Marine Cyg 20h28'16"30d24'
Virden,Maddison Aileen Cas 22h56'52"56d32'
Virelli,Dominic Aur 6h4'37"31d25'
Virgen-The Floater, Malaga Equ 21h4'1"10d38'
Virgil & Fern Cyg 19h57'37"30d48'
Virgil & Marguerite Tri 2h22'29"35d22'
Virgilee Aur 6h9'43"35d23'
Virgilio,Frances Boo 15h5'17"38d16'
Virginia Uma 10h30'52"50d2'
Virginia Allen And 23h17'28"51d38'
Virginia Anne Cyg 21h39'0"38d42'
Virginia Elizabeth Mon 7h49'57"-3d26'
Virginia Iva Com 13h15'1"21d48'
Virginia Mae Cam 14h12'58"80d45'
Virginia Mae Uma 11h33'32"47d43'
Virginia Mae Ari 1h59'21"22d13'
Virginia Mary Cyg 19h28'25"35d29'
Virginia Pearl Ori 5h45'26"19d59'
Virginia Rose Peg 22h33'47"26d25'
Virginia's Opus Cyg 19h26'21"30d21'
Virginia's World Cas 22h57'11"56d44'
Virginia-Mine Came True With You Cyg 19h27'14"38d38'
Virginiae,Stella Lib 15h13'10"-20d33'
Virginiaroz Lyr 19h20'46"40d36'
Virginie Tri 1h46'1"34d44'
Virginie Cam 4h35'34"58d38'
Virginie Vul 20h16'14"28d30'
Virginie & Bruno For 3h30'12"-31d44'
Virginie & Michel Cyg 19h33'40"30d26'
Virgo Star of Daniela,The Vir 13h22'45"-2d13'
Virgo Sue Until Forever From 02-16-96 And 1h48'19"39d31'
Virgona,Susan Psc 1h1'45"22d56'
viribus unitis Franco u Nicole Her 17h53'50"41d5'
Virideau,Laure Cam 7h28'52"67d35'
Virin,Josie Pic 4h44'21"-48d32'
Virkler,Marvin L Her 16h11'0"20d17'
Virostek,Hope Lyn 7h5'29"44d12'
Virrazzi,Anthony Per 2h6'31"57d49'
Virtanen,Rauli Per 2h1'0"56d47'
Virtel,Leo Her 18h13'33"30d5'
Virzi,Al Boo 14h30'34"24d55'
Vis,May Peg 22h45'50"32d14'
Vis,Paul Her 16h11'33"48d57'
Visca,Curtis Cet 2h9'54"6d54'
Visca,Ellen A Ori 5h52'47"15d9'
Viscardi,Uncle Gabe Aur 5h18'0"41d31'
Viscomi,Jessica VanLeer Mon 7h44'43"-1d55'
Visconti,Evangelina Lac 22h37'54"52d49'
Visconti,Jr,Charles Michael Dra 17h57'36"61d39'
Visconti,Martin I Cep 22h8'21"61d8'
Visconti,Ronald Boo 15h4'42"30d37'
Visconti,Sheila I Sge 19h30'44"16d20'
Visconti-Balsamo,Gia M Aur 7h2'40"35d30'
Visconto,Philip & Rose Tri 2h21'42"30d46'
Viscusi,Karen Jean Cam 6h4'22"61d45'
Visegrad D'Helene Lac 22h28'58"40d6'
Viselle,Carole Cam 5h40'59"65d35'
Visendi,Morning Star- Peter Ori 6h1'40"1d34'
Visger,"JEV" Judith E Cas 23h32'49"60d27'
Vishal Per 1h49'18"54d13'
Visich,Jr,Frank J Cmi 7h59'42"0d3'
Vision Dreamer Mon 7h2'46"3d54'
Viskocil,Daniel Dra 19h21'13"56d57'
Visners,Marite Her 16h55'14"40d20'
Visocchi,A M M A Cyg 1h50'15"37d2'
Visonti,Nicholas Dra 12h29'36"64d30'
Visoskas,Kyle Robert Aur 5h4'23"40d30'
Visotschnig,Beate Lib 15h1'48"-28d27'
Visquesmel,Celine Ori 5h57'31"18d41'
Visquesnel,Pierre Boo 15h21'40"42d21'
Visser,Jan Per 2h56'18"31d18'
Visser,John Per 3h19'47"42d50'
Vissicchio,Jack Joseph Boo 14h12'33"30d49'
Vita (hel) Muthatur Crb 16h13'48"38d17'
Vita I I Y Sge 19h57'22"20d12'
Vitacco,Kenneth L Per 2h56'0"50d33'
Vitacco,Michael John Dra 16h3'27"66d45'
Vitae,Carmen Meae Com 13h8'16"15d42'
Vitae,Nova Cyg 20h2'24"40d25'
Vitae,Stella Mae Com 12h27'25"26d56'
Vitagliano,Debra Anne And 0h22'48"34d33'
Vitagliano,Douglas Aur 4h39'12"34d1'
Vitagliano,Joseph Cnv 14h3'0"37d55'
Vitagliano,Melissa Jayne Cyg 19h29'29"31d52'

Vitakis,Maria Alexandra
Uma 9h21'18"47d45'
Vital,Elan
Cas 0h1'43"50d15'
Vitale Italian Princess,Lorraine
Del 20h22'34"20d26'
Vitale,Ashley
Lyn 7h7'42"59d8'
Vitale,Brienna
Cnc 8h36'15"7d20'
Vitale,Charles & Helen
Lib 14h28'49"-20d37'
Vitale,Joan Louise Fasanella
Sct 18h44'47"-6d53'
Vitale,John
Her 18h5'45"30d9'
Vitale,John E
Her 16h59'41"30d3'
Vitale,John P
Umi 16h15'1"79d12'
Vitale,Lisa Ann
Peg 21h37'57"24d18'
Vitale,Megan S
Mon 6h54'0"-10d54'
Vitale,Melissa
Lyr 18h40'19"47d14'
Vitale,Nicholas Mark
Per 2h22'35"55d36'
Vitale,Norman
Her 18h42'19"12d43'
Vitale,Paul John
Cyg 21h54'23"53d21'
Vitali
Sex 10h46'53"-5d3'
Vitali,Flavio
Cam 6h11'13"61d2'
Vitalis Hübel & Koch GBR
Tau 4h23'36"20d6'
Vitalo,James
Her 16h35'54"8d33'
Vitamin X
Aqr 22h3'23"-23d21'
Vitek,Henry
Uma 9h38'0"55d12'
Vitek...Simply The Best!,"Miss Dee"
Ori 5h53'59"14d6'
Vitelli,Carol Ann
Tri 2h22'28"35d49'
Vitelli,Marjorie
And 2h16'31"47d53'
Vitense,Russell R
Aur 6h14'59"31d57'
Vithana
Cas 0h48'1"70d48'
Viti,Mr & Mrs Brian & Kammie
Cnv 13h3'40"40d48'
Viticristar
Mon 7h50'7"-6d41'
Vitiello,Anthony
Her 16h26'27"33d38'
Vitiello,Thomas
Dra 15h50'46"51d32'
Vititis
Psa 22h28'0"-28d15'
Vititoe,Karin & Jim
Mon 6h22'43"8d28'
Vitolo,Albert (Buddy)
Lmi 10h10'53"31d24'
Viton,John E
Crt 11h14'54"-19d27'
Vitone,Catherine
Cas 1h7'51"60d12'
Vitry,Francois
Cet 0h58'1"0d40'
Vitt,Steven Lee
Her 18h7'12"40d13'
Vitti,Janna Marie Allen
Lyr 18h18'1"37d44'
Vittoria
Lmi 10h6'0"31d37'
Vittoria
Aql 19h51'56"12d9'
Vittoria
Psa 22h22'51"-27d35'
Vittoria,B Tracy
Cas 2h58'20"68d42'
Vittorina Paolo Elisa
Cae 4h59'29"-33d35'
Vittorini Family Star, The Joe
Her 17h8'24"46d1'
Vittorio
Cas 2h4'33"60d34'
Vittorio
Her 16h44'30"47d33'
Vittorio,Anthony
Oph 18h4'59"11d57'
Vitvcci,Val
Mon 6h31'1"-1d22'
Viv's Journey
Mon 7h9'40"-6d55'
Viva
Ori 4h54'28"0d15'
Viva Vivi
Mon 7h8'46"-5d24'
Vivan
Peg 22h33'37"29d31'
Viventi,Michele
Lyr 19h2'41"35d23'
Vivi
Com 13h12'33"27d42'
Vivian
Ari 2h38'11"22d12'
Vivian Marie
Eri 3h18'27"-6d27'
Vivian V
Aqr 21h4'45"-8d26'B
Viviana
Dra 17h59'15"58d18'
Viviana
Tel 19h5'50"-46d50'
Viviano,Alicyn
Cam 7h42'52"60d43'
Viviano,Megan
Ori 5h56'18"11d7'
Vividus
Lmi 10h58'29"31d39'
Vivien
Hya 8h32'44"-1d23'
Vivien
Ori 6h5'56"0d29'
Vivien Ann-"Stars & Numbers"
Ari 2h0'20"26d36'
Vivien Anne
And 23h2'23"51d26'
Vivien Rose
Com 12h35'28"21d27'
Vivien's Wish
Uma 10h46'56"55d21'
Vivienne
Cyg 21h18'22"39d50'
Vivienne
Lyr 18h31'35"33d27'
Vivienne's Star
And 23h34'17"40d42'
Vivki J
Mon 7h19'40"-6d11'
Vivo Rive Amor
Dra 20h34'41"73d40'
Vix,Christian
Cyg 20h28'41"39d37'
Vizard,John"Jack" T
Uma 11h44'48"43d57'
Vizor,Joanne-Marie
Cas 0h26'12"50d35'
Vizoso,Vera G
Vul 19h20'56"25d25'
Vizoyan,H Jeffrey
Oph 18h24'36"8d5'
Vizurraga,Alexander Luie
Her 18h44'16"12d28'
Vizzini,Jesse David
Cep 22h7'19"60d7'
VJM Pop
Cet 2h50'0"4d26'
Vladimira
Her 16h59'18"38d12'
Vladimira e Franco
Pup 8h24'26"-29d56'
Vlahek,Anton
Cep 22h15'18"67d60'
Vlasak,Kurt Franz
Ari 3h20'1"30d6'
Vlasta
Tri 2h5'58"30d55'
Vlastas,Anthony
Cap 20h28'48"-12d7'
Vlazny,Lois
Mon 6h44'29"10d30'
Vlcek,Robert F
Lac 22h25'17"50d25'
Vlies,Sarah Michelle
Umi 17h25'27"75d46'
Vlouhos,John Michael
Per 4h28'25"50d58'
VMQ Jr
Her 18h17'58"38d20'
Vnenchak,Joseph W
Per 2h41'18"34d47'
Voack
Vir 13h39'5"-2d15'
Vochekium
Ori 6h3'36"8d35'
Vocke,Manfred
Lyr 19h20'0"31d21'
Vodermayer,Martin
Sgr 19h39'10"-41d50'
Vodicka,Mary Eileen
Cas 2h38'43"57d29'
Voegeli,J & N
Cyg 21h19'43"30d20'
Voehl,Larry
Cam 7h58'1"60d21'
Voelker,Frank J
Per 3h11'52"41d5'
Voelker,Kay
Cyg 21h33'0"41d27'
Voelker,Raymond
Lac 22h51'22"52d36'
Voelz,Jack
Boo 14h57'0"24d2'
Voest-Alpine MCE
Uma 9h32'38"46d13'
Voge,Daniel R
Cet 3h11'47"-0d24'
Vogel Anniversary Star,Earl & Ginny
Cyg 20h52'46"30d22'
Vogel,Audra
Vul 20h20'20"28d19'
Vogel,Bernard Joseph
Lib 14h57'1"-10d15'
Vogel,Cody Allen
Per 3h12'12"41d29'
Vogel,Dale
Her 16h49'0"39d12'
Vogel,Dirk
Boo 14h18'19"14d39'
Vogel,Eugene Edward
Aur 6h18'38"32d43'
Vogel,Frederick
Aur 6h28'0"38d25'
Vogel,Gregory Gene
Cam 5h47'51"68d5'
Vogel,Hubert
And 23h0'43"42d36'
Vogel,John Douglas
Her 17h39'54"21d45'
Vogel,Karla D
Cam 9h20'0"74d13'
Vogel,Lis & Herluf
Cyg 21h32'24"42d41'
Vogel,Lyn K
Mon 7h2'46"-1d5'
Vogel,Marianne
Aql 19h17'0"15d45'
Vogel,Peter Nicholas
Dra 16h37'31"72d6'
Vogel,Samuel
Oph 17h54'0"11d54'
Vogel,Thelma & Philip
Cyg 20h54'32"37d45'
Vogel,Virginia
Lyn 8h54'0"42d51'
Vogelsang,Hans-Thilo
Uma 10h30'20"48d25'
Vogelsang,Peter
Boo 14h10'56"32d22'
Vogelvang-Splinter, Joop en Alie
Her 17h58'48"14d48'
Voges,Brandon & Cori Brown
Aql 19h5'40"15d13'
Vogg,Karl
Cyg 20h25'0"31d30'
Vogl Prarie Star,The
Lac 22h55'45"53d39'
Vogl,Douglas Keith
Dra 16h17'48"66d16'
Vogl,Frank & Emily
Cam 3h54'10"74d35'
Vogl,George B
Aql 20h12'46"5d27'
Vogler,Teresa Marie
Lyn 8h12'38"51d29'
Vogler,William Stephen
Her 18h15'1"47d46'
Vognild,Larry L Dorothy L
Her 16h59'16"38d56'
Vogt's "Dream",Joseph
Boo 15h4'54"15d49'
Vogt's Star,Johnny
Her 17h2'41"31d23'
Vogt,Benjamin Christopher
Her 18h5'60"30d36'
Vogt,Carl
Del 20h13'33"10d0'
Vogt,Heather Lynn
Peg 21h56'18"33d49'
Vogt,Jürgen
Lyr 19h21'1"31d27'
Vogt,Thomas
Mon 8h6'25"-1d4'
Vogt,Thomas Alexander
Aur 7h2'60"41d12'
Vogt,Tom
Lac 22h35'22"54d10'
Vogt-Nan,Great Nan, Gertrude Louise Bell
Cam 5h52'0"68d41'
Vohland,Karsten
Gem 8h2'45"31d3'
Vohwinkel,Robert Earl
Boo 15h42'36"40d22'
Voice Of Amy
Lyr 18h40'1"38d39'
Voigt,Bill
Her 16h57'48"47d35'
Voigt,Don & Jean
Cam 7h41'55"70d3'
Voigt,Elizabeth Williams
Cas 0h32'0"61d19'
Voigt,Ellen
Aqr 20h54'41"0d46'
Voigt,George
Crt 11h17'0"-14d54'
Voigt,Nicole Marie
Cas 23h30'0"60d14'
Voigt,Ray Lawrence
Uma 9h30'40"48d55'
Voigt,Roy William
Dra 20h34'49"67d42'
Voigtmann,Siegfried
Leo 11h17'52"-5d20'
Voirin,Christopher
Aql 19h17'22"15d26'
Voirin,Rebecca
Del 20h13'42"14d21'
Voiro,Salvatore "Sammy"
Boo 14h1'52"11d26'
Voisard,G
Lac 22h19'34"55d16'
Voisine,Jacques
Uma 10h13'1"54d29'
Voisse,Christian
Ori 6h1'1"20d14'
Voit,Angel
Ori 5h55'53"17d9'
Vojak III,Joseph Michael
Aur 6h8'0"50d32'
Vokac,Christina Nicole
And 1h52'1"40d42'
Volante,Susan
Lyr 18h36'30"39d39'
Volber,Klaus-Peter
Peg 23h29'49"12d53'
Volberg,Louise E
Uma 9h13'36"58d44'
Volbrecht,Jackielee
Cyg 19h28'12"38d16'
Volbrecht,Jason Shawn Michael
Boo 15h4'18"14d28'
Volckhausen,Walter & Eva
Hya 8h49'25"4d39'A&B
Voldal,Emily Christine
Cyg 20h32'18"37d45'
Volfovskaya,Innulya
Cas 1h32'55"75d43'
Volia,Johny
Crt 11h17'39"-19d13'
Volk,Ann Renee
Cyg 20h24'35"40d22'
Volk,Carol
Mon 6h27'13"-1d13'
Volk,Cynthia S
And 23h37'13"46d4'
Volk,Heidi J
Cyg 20h17'11"38d32'
Volk,Jerry Pierds
Aql 20h1'55"6d20'
Volker
Ser 15h41'39"0d56'
Volkert,Kristin
Peg 21h57'0"30d8'
Volkert,Robert Bruce
Cet 1h36'24"-12d57'
Volkland,Dietrich
Cep 20h43'16"76d36'
Volkmar,K David
Aur 4h48'0"40d35'
Volkmar,Sherry Lorraine
Cyg 19h20'38"54d56'
Volkmar,Sherry M
Ori 4h42'39"10d26'
Volkmar,Vicki S
And 22h55'0"36d46'
Volkoff,Sharon & Bill
Eri 4h13'23"-17d30'
Voll Armin
Aql 18h57'26"17d13'
Volla,Streya
Crb 16h23'60"38d28'
Vollan,Maureen "Sweetpotata"
Vul 19h59'0"23d5'
Vollborn,Sharon K
Lyr 18h32'0"35d22'
Vollbrecht,Haley Sherrene
Crb 15h35'55"28d7'
Voller I,Julie
Cyg 19h28'16"38d51'
Voller,Rich
Dra 18h23'30"80d6'
Vollero,Darcy Anne
Com 12h23'60"28d47'
Vollersten-Mueller, Brigitte
Tau 5h1'11"28d13'
Vollien,Floyd & Elise
Eri 4h18'9"-14d21'
Vollmer,Betty J
Lac 21h59'0"39d0'B
Vollmer,Birgit
Ori 5h54'29"16d35'
Vollmer,Evelyn
Lyr 19h10'18"37d42'
Vollmer,Franz-Josef
Her 17h25'0"42d5'
Vollmer,John
Cnv 13h26'16"42d29'
Vollmer,Jr,Charles G
Lac 21h59'0"39d0'A
Vollmer,Katharina- Ivana
Mon 7h18'59"-6d21'
Vollmer,Marietta
Psc 23h0'0"2d5'
Vollmer,Robert Stewart
Oph 17h31'55"-20d23'
Vollrath,Peter E
Cep 22h16'19"60d46'
Vollucci,Barbara A
Cas 0h30'24"60d10'
Vollucci,Eugene E
Lac 22h2'0"50d21'
Volmar,Margaret A Hagen
Lyn 8h22'39"42d41'
Volmer,Peik
Cas 0h14'18"60d8'
Volodya & Heidi
Cep 22h7'21"67d36'
Volos,Ms Martha Tanas
Cas 1h47'0"61d21'
Voloshin,Grandma Anca Makusi
Uma 11h0'1"71d48'
Volpatti,Tammy
Cas 1h4'1"53d39'
Volpe,Deborah A
Gem 6h39'58"13d1'B
Volpe,Donna M
Gem 6h39'58"13d1'A
Volpe,Georgann
Cep 1h23'14"77d47'
Volpe,Joanne
Cas 1h27'36"73d37'
Volpentesta R,Giordano
Cam 6h36'1"67d52'
Volpi,Roberto Maria
Hor 3h22'55"-47d1'
Volquardsen,Jerry
And 2h24'14"49d5'
Volta,Alessandro
Col 6h25'23"-33d14'
Voltura,Kathleen Ruth
Com 12h17'35"31d8'
Volz,Ellen J
Cas 0h48'34"70d37'
Volz,Travis
Tau 5h44'15"16d46'
vom Bären,Heiner und Mary
Gem 6h46'41"16d38'
Vomlehn,Loretta
Lyn 6h32'16"58d39'
Von
Vul 20h4'0"23d10'
Von Allmen,Linda
Cam 8h27'23"74d45'
Von Arnim,Piero & Szilvia
Pho 0h31'38"-44d21'
Von Arx,Anoki
Lyn 8h53'29"40d45'
von Ballmoos,Katja
Dra 15h51'1"62d18'
Von Bartheld,Barbara Frances Ahrens
Vir 12h34'20"1d27'
Von Bartheld,George Herman
Vir 12h36'19"1d60'
Von Bartheld,George William
Vir 12h37'28"1d40'
Von Bartheld,Olive Laura Glaspy
Vir 12h31'11"1d5'
von Bechtolsheim, Clemens
Uma 9h8'41"47d44'
Von Bette Jean,Jeanne Ray
Cas 0h33'1"68d21'
von Boecklin,Linda
Cas 2h4'42"70d22'
Von Brecht,Forrest
Cep 21h26'17"55d47'
von Buchholtz,Baron Wolf
Gem 6h25'3"14d5'
Von Bunca,Sheba
Cma 7h13'1"-13d36'
von Daacke,Ulrike
Ori 5h57'0"19d36'
Von Der Damerau, Christa
Cas 0h43'44"64d11'
von der Goltz,Erika
Cet 3h17'59"2d30'
von der Lieth,James Penders
Her 17h13'22"26d32'
Von der Schmetterlinge Ann
Cam 5h48'33"68d57'
Von Donze
Uma 12h46'27"60d2'
Von Droste,Kerwin
Ori 4h45'48"4d52'
Von Erdmannsdorff,Mary Kay
Mon 6h5'1"-8d16'
von Furstenbert,Diane
Lmi 9h42'39"40d51'
Von Gontard,Paul Theodore
Dra 14h4'51"64d26'
Von Habsburg-Lothingen Karl Philipp
Lib 15h36'50"-21d11'
Von Hilde
Boo 14h24'33"28d26'
Von Hoffmann,Kristina
And 1h18'12"38d58'
Von Holdt,Eric
Cnv 12h44'56"38d40'
Von Holstein,Christina Staël
Cas 1h2'1"60d38'
von Huth,Katrine
Umi 15h34'12"77d45'
Von Kanel,Beverly
Cas 0h39'58"62d40'
von Kasylan,Charmo
Ori 5h58'40"21d7'
Von Kessler,Jill
Eri 3h38'52"-5d24'
von Liebig,Suzanne & Bill
Eri 3h37'33"-3d57'
Von Lintig,Dr Richard D
Uma 11h45'60"30d14'
Von Maydell,Olaf
Boo 15h27'0"47d60'
Von Mehren,Susan & Robert
Cyg 21h21'39"40d50'
Von Miller,Phyllis
And 23h37'1"48d41'
Von Oswald,Alexander
Cep 23h3'37"71d12'
VON PLOENNIES
Lmi 9h42'1"33d46'
Von Preussenthal, Umberto Menzinger
Uma 10h11'57"42d23'
Von Roos,William N
Cas 1h12'34"61d5'
Von Rüden,Robert
Cas 23h38'12"63d3'
Von Scherer's Star
Lac 22h47'13"54d13'
Von Schmittou,Pauline Farr
Uma 10h56'1"70d24'
Von Schmitz,Ewige Liebe
Aql 19h53'58"12d7'
von Selchow,Viktor
Boo 15h19'49"38d23'
Von Sichlan,Tina
Cas 2h35'43"50d38'
Von Summer
Vul 19h24'5"26d29'
von Thun-Hohenstein,CD Natango
Vir 14h20'47"-7d37'
von Thun-Hohenstein,SH Charlotte
Sgr 19h59'15"41d33'
von Thusis,Samuel Gartmann
Uma 10h57'1"40d26'
Von Whaley,Russell
Aur 6h32'39"37d7'
Von Zandt,Cindy
Peg 23h28'26"33d46'
Von Zetto,Trixella
Vul 20h16'17"23d22'
Von's Sun
Lmi 10h6'40"35d54'
Von,Eric
Aql 19h0'56"-0d9'
Von,Kris
Cep 21h1'18"55d57'
Vonasek,Paula Proteau
And 23h38'1"46d48'
Vonbehren
Uma 9h54'60"68d36'
Vonda,The Beautiful One
Com 13h6'0"20d6'
Vonderheide "Dipper", Dolores
Uma 9h4'37"70d44'
Vondran,W Earl
Ser 15h37'34"2d42'
Vondunn,Debbi Cartesio
And 0h8'28"27d55'
vonEssen,Leah Rachel
Ori 5h55'41"20d5'
Vongsady,Eric
Del 20h16'56"10d29'
VonHünersdorff Owens, Gloria
Peg 21h49'58"21d2'A
VonHünersdorff Owens, Christina
Peg 21h49'58"21d2'B
VonHünersdorff Owens, Olivia
Peg 21h49'58"21d2'C
Vonie's Bright
Mon 7h3'0"4d16'
Vonna
Crt 11h21'29"-16d48'
Voogt,Gerrit
Vul 20h5'24"22d42'
Voorbraak Familie Fter
Cam 4h8'19"69d24'
Voorhis,Thomas
Per 3h43'53"37d8'
Voosen,Friedhelm
Mon 7h52'45"-2d27'
Voosen,Walter
Mon 7h52'48"-2d47'
Voran,Jim
Uma 9h46'21"51d24'
Vorbeck,Gisela
Psc 22h49'29"6d37'
Voreis,Chloe R
Boo 13h55'41"17d34'
Vorena
Lyn 8h22'52"44d38'
Vorfeld,Egon
Her 16h32'0"41d3'
Vorick,Micheal V
Per 2h52'20"32d33'
Vorisek,Jake
Umi 16h50'25"75d60'
Vormstein,Cornelia
Cnc 8h0'38"10d33'
Vorndamme sen
Lyr 19h4'21"28d12'
Vorperian,Kevork A
Dra 19h24'16"68d24'
Vorster,Randon M
Aqr 20h55'1"-0d11'
Vos Star,The Johann F
Her 15h58'18"40d19'
Vos,Damon William
Ant 10h39'55"-33d26'
Voshall,William Dale
Cet 0h54'37"-16d25'
Vosk,My World's Greatest Hubby Stu
Per 1h55'1"56d42'
Vosmik,P
Cet 1h22'23"-0d18'
Voss,Abbygail Fortune
Peg 22h15'49"8d48'
Voss,Connor Dean
Boo 15h0'19"53d16'
Voss,Delaney Shea
Lyn 8h52'22"41d59'
Voss,Diana
Cas 0h35'0"62d38'
Voss,Frauke
Cet 2h29'59"6d51'
Voss,James
Aql 19h30'35"10d38'
Voss,John Matthew Hunter
Her 16h0'27"21d31'
Voss,Jr,Jeff
Cep 22h36'41"61d10'
Voss,Jr,Joseph Millsaps Fitzhugh
Lac 22h31'39"50d2'
Voss,Lorel
Lyr 18h46'35"40d39'
Voss,Michael Josef
Mon 6h20'0"7d33'
Voss,Rudi
Umi 15h9'40"71d10'
Voss,Warren
Dra 19h20'39"56d43'
Voss,Wendell G
Ori 5h40'24"-0d46'
Voss,Werner
Sge 19h8'29"17d40'
Vosseler,Jana Marie
And 1h26'35"34d5'
Vossen,Robert
Boo 15h40'51"48d55'
Vossos,Vasilios D
Her 17h34'24"20d50'
Votava,Olga
Mon 6h21'1"0d51'
Voth,MD,Gayle V
Oph 18h40'15"10d57'
Voth,Ryan
Aur 6h22'49"30d52'
Vouga (Pessy),Elaine
Lyr 18h38'58"40d15'
Voulanger,Patrick M
Uma 15h18'0"70d25'
Voultsides,George E
Oph 18h39'36"8d44'
Vournas,Emily
Cam 8h37'0"80d37'
Vous et Nul Autre Hans & Sandra
Cnv 13h28'53"40d40'
Vous et Nul Autre
Lyr 18h52'30"41d7'
Vous et Nul Autre- Hans & Sandra
Peg 21h37'49"20d22'
Vous et nul ultra You and No Other
Cyg 19h31'57"31d10'
Vowels,Marshall L
Peg 21h50'17"34d2'
Vowels,William Connor
Cet 2h19'20"8d43'
Vox Dilecti Mei
Cyg 20h31'0"40d6'
Voyack,Jr,Michael John
Cap 20h21'31"-12d31'
Voyack,Nicholas Charles
Cnv 13h55'0"40d44'
Voye,Tay C
Cnv 12h10'36"43d46'
Voynovich,John & Spillers,Beth
Cyg 21h2'41"38d24'
Voyta,Villa
Cyg 21h39'13"37d34'
Voyton,Mark Francis
Aur 5h23'40"37d42'
Vozeh,Christian Matthew
Vul 19h57'48"26d2'
Vozeh,Colin James
Lac 21h58'50"37d28'
Vozeh,Mary Theresa
Cyg 19h26'0"40d46'
Vozzella,Angelo Anthony
Cnv 13h39'1"45d42'
Voß,Richard
Uma 9h16'16"45d18'
VPT I
Aqr 23h4'13"-12d10'

Vraciu
Cyg 21h8'20"30d40'
Vradenburgh,Clara
And 2h28'60"38d51'
Vrai Amie Avec Un Belle Coeur
Peg 23h33'42"21d30'
Vramish,Margaret Ann
Lyn 7h51'0"41d15'
Vreftos,John S
Her 16h57'50"36d14'
Vricella,Marco H
Ser 15h19'34"20d9'
Vries,Renate De
Vir 12h28'38"-8d17'
Vroni
Cnc 8h27'44"7d33'
Vrooman,Krystie Ruth
Mon 7h44'8"-2d44'
Vrtala "Finrod", Alexander
Ori 5h44'19"8d39'
Vruenhoek-Zant, Mariojke
Dra 16h8'1"60d11'
Vselman,Mark P
Cnv 13h2'56"32d12'
Vtelensky,Kerstin
Ari 1h54'25"25d15'
Vucci,(Dreams Of Oz), Julie & Suzanne
Eri 4h36'16"-18d19'
Vucichevich,Mr Rad
Tri 1h33'43"28d41'
Vuckovic,Ched
Peg 0h2'13"30d17'
Vuckovich, Mileko "Diamond"
Aql 19h33'22"7d26'
Vugrin,Jr,Joseph G
Boo 14h24'28"27d50'
Vuiches,Michael
Per 3h7'12"40d9'
Vuillaume,Thierry Katia
Del 20h19'49"10d25'
Vuillemot,Robert Lee
Dra 19h18'13"70d7'
Vuillien,Monique
Ori 5h57'28"18d11'
Vujko,Natalija
Cyg 20h41'34"42d55'
Vukelic,Gospava
Aur 4h41'55"31d30'
Vukmanic,Thomas
Boo 15h13'2"38d29'A
Vulcan,Carl H
Per 4h25'55"51d51'
Vullo,Henry
Aur 6h48'0"48d57'
Vulpi II,Marco
Tau 4h31'29"28d51'AB
Vulpi,Pat
Tau 4h33'1"30d43'
Vuono,Carleigh Ann
Cyg 19h34'20"28d50'
Vuono,Carleigh Ann
Vul 19h46'1"28d47'
Vuori,Olli
Ori 5h59'1"14d36'
Vuoso,Marisa Ann
Peg 23h36'45"12d7'
Vurchio,Julia
Cas 1h43'51"61d51'
Vus,Kuddl
Aqr 21h0'1"-8d6'
Vyborny,Dominique Teresa
Oph 17h23'37"1d50'
Vyborny,Stephanie Julia
Oph 17h22'47"1d22'
Vydra
Lyn 8h14'51"34d4'
Vye,Sage
Boo 13h45'0"16d9'
Vyhnanek,Marie & Irv
And 23h43'0"45d51'
Vyzas,Ray
Dra 20h20'49"67d36'
Vähäkuopus,Toivo
Peg 22h27'33"31d34'
Véga
Uma 9h15'1"44d30'
Vérena
Tri 2h2'27"35d2'
Véronique et Patricia
Boo 14h52'42"51d7'
Vézina,Charlotte
Cas 1h17'0"60d13'
Vézina,Daniel
Lmi 10h53'11"32d9'
Vézina,Jean-Maurice
Uma 11h34'39"42d48'
Vöhringer,Margarita
Cas 0h9'28"62d16'
Völkel,Markus
Her 17h57'52"42d53'
Völkel,Thomas
Aqr 21h57'1"-10d37'
Völkl,Helmut
Sco 17h20'11"-38d36'

W

W 10 Y LILO
Ari 1h45'36"11d43'
W A M L
Aql 18h43'55"0d2'
W A Y, Jr
Boo 15h46'41"44d41'
W D B
Uma 11h23'26"70d2'
W D H Victory
Ser 15h15'36"20d51'
W E M/E C R
Cet 2h36'52"1d13'
W E's "Peace"
Cet 2h59'1"0d26'
W G 1
Cnc 9h13'0"8d22'
W K W II
Leo 11h6'57"-5d36'
W P J The Stars,The Moon & Forever
Peg 21h27'57"3d18'
W P K III
Crt 10h57'39"-12d29'
W P K IV
Oph 17h2'57"-20d1'
W P X R Power 98.9
Uma 15h12'0"68d27'
W-2S-K-3B-N-4MC-P SHS GD7 96 R
Cyg 21h5'0"39d11'
W.A.G.
Lac 22h9'25"51d57'
W7MB/W7RVM
Tri 2h40'1"31d44'
Wa
Cet 3h17'18"9d34'
Waack,Gus
Dra 16h32'56"73d50'
Waage,John Arlos
Ori 5h51'1"11d1'
Waare,Patrick O
Ser 18h43'28"3d35'
Wabbersen,Tara
Cet 2h21'17" 10d57'
Waber,Charmagne
Mon 6h39'39"10d10'
Waccary,Cynthia
Equ 20h55'57"5d26'
Wach,Robert Errol
Cyg 19h26'57"33d43'
Wachs,Erica Faye
And 23h44'1"46d55'
Wachsberger
Aql 19h25'41"15d30'
Wachsberger,Rose & Sidney
Crb 16h2'10"30d48'
Wachsman,Flauri
Lyr 18h47'0"33d38'
Wachsmuth,Nina
Ori 5h38'15"8d11'
Wachtel,Richard Stephen
Aql 19h35'40"-6d9'
Wachtel,Suzanne Marie
Sct 18h49'26"-7d47'
Wachter III,Burt Carlton
Per 1h26'12"52d44'
Wachter's Wannit
Her 16h28'1"51d1'
Wachter,Catherine
And 1h4'27"38d9'
Wachter,Konrad
Ser 15h34'31"18d35'
Wachter,Margaret S
Lyr 19h16'1"41d3'
Wachtler,Karlheinz
Boo 14h7'28"11d15'
Wacinski,Antoni
Per 2h0'17"56d34'
Wack,Otto
Cap 20h19'22"-26d41'
Wacker,Klaus
Sco 17h28'49"-31d37'
Wacker,Steve
Com 12h26'38"21d1'
Wacker,Suzan
And 23h3'44"50d54'
Wackerly,Geneva Mae
Crb 16h18'0"37d58'
Wackerman,Matthew Joseph
Cnv 13h29'0"48d59'
Wackermann Star,The
Boo 14h50'1"24d13'
Wackler's "Sweet Love Remembered"
Cyg 20h16'11"31d20'
Wadas,Jr Shining Eternally,Frank C
Cet 1h32'42"-1d39'
Wadas,Jr,DDS,John J
Her 17h54'58"50d14'
Waddell,Bonhomme Richard E
Aur 5h31'19"38d31'
Waddell,Ellen Maxine
Mon 8h2'37"-9d32'
Waddell,Rachel Kimberlee
Crb 13h9'23"31d6'
Waddingham I,George Hamlet
Oph 18h31'39"10d31'
Waddington,Geoffrey
Uma 10h1'22"55d23'
Waddington,Nicole Paige
Crb 16h3'26"34d41'
Waddle,Dirk
Eri 3h40'11"-18d57'
Waddle,Heather
Boo 14h18'0"53d40'
Wade
Dra 19h29'57"58d27'
Wade
Ori 5h56'12"15d2'
Wade
Ori 5h56'37"14d7'
Wade & Lincoln's Light
Sct 18h42'0"-5d25'
Wade III,James Arthur Joseph
Aur 6h0'44"35d42'
Wade,Anne-Marie
Sgr 19h40'22"-45d12'
Wade,Brenda C
Cam 3h43'0"73d41'
Wade,Christopher Donohue
Crb 16h16'28"37d43'
Wade,James B
Cep 1h12'11"80d20'
Wade,James Crockett
Her 17h58'44"28d41'
Wade,Jessica Lynn
Leo 9h26'57"28d55'
Wade,John Michael David
Per 2h20'22"55d8'B
Wade,Lynn Fields
Mon 6h31'15"-6d41'
Wade,Meredith
Cam 7h58'1"70d11'
Wade,Michael
Aur 6h32'37"37d5'
Wade,Michael & Sharon
Aur 6h25'40"31d33'
Wade,Michael Musgrave
Ori 6h3'27"0d30'
Wade,Natalie
Mic 20h39'51"-44d15'
Wade,Nathalie Sarah
Vul 20h3'0"25d17'
Wade,Patricia Anne O'Neill
Per 2h20'22"55d8'A
Wade,Rick
Tri 1h42'41"31d27'
Wade,T Rogers
Lac 22h4'43"49d52'
Wade,Tammy
Peg 22h38'36"25d56'
Wade,Walter
Cet 0h44'53"-6d46'
Wade,William Clifford
Her 17h28'16"31d25'
Wadestar
Cyg 21h54'0"37d39'
Wadhwani,Anjali
Cap 20h33'37"-10d41'
Wadino,Michelle Maureen
Vul 21h24'30"24d1'
Wadis,Bill
Dra 19h47'50"60d57'
Wadley,Michael
Her 17h3'60"46d6'
Wadowski,Matthew
Cnv 13h0'27"38d55'
Wadsak,Walter
Cap 20h32'23"-26d29'
Wadsella
Cam 7h43'12"80d28'
Wadsworth,Katie Ann
Lyn 8h26'40"41d31'
Wadsworth,Logan
Gem 7h2'6"21d30'
Wadzeck,Ray
Hya 8h54'53"1d21'
Waelchli,Peggy Rae
And 0h3'30"46d8'
Waen,Hannah Rose
Lyr 18h37'11"37d7'
Waen,Jeremy
Lac 22h53'47"54d8'
Waever,Jr(Bud),Louis L
Dra 18h59'56"70d9'
Waffenschmidt
And 23h1'1"43d28'
Wagemaker,Letty & Scott
Sct 18h52'21"-7d51'
Wagenblast,Scott Michael
Aqr 22h49'17"-2d5'
Wagener,August Joseph
Ori 6h4'1"0d8'
Wagenheim,Adrienne
Peg 22h19'1"8d20'
Wagenhofer,Monika und Max
And 1h42'60"41d55'
Wager,David
Her 17h0'38"38d11'
Wager,Len
Her 16h55'43"26d29'
Wager,Reverend Win
Per 3h11'26"49d30'
Waggin Tail & Gentle Earwig
Cnv 12h24'19"36d53'
Waggoner,Jack B
Aur 6h24'42"37d40'
Waggoner,William J
Aql 19h7'18"0d15'
Waggonner,Joe D
Ori 6h6'50"8d60'
Waggott,Jessie
Cmi 8h8'56"2d7'
Wagle,Ayesha
Ori 5h48'15"18d40'
Wagle,Vivek
Ori 5h44'38"10d57'
Wagman,Bernard J
Oph 18h35'0"10d43'
Wagman,Scott K
Aql 18h43'26"11d51'
Wagman,Vivian C
Peg 22h50'47"21d35'
Wagner,Alfred
Cyg 20h51'30"37d53'
Wagner,Amanda
Cam 8h48'54"77d31'
Wagner,Audra Lee
Lyr 19h4'0"40d15'
Wagner,Barb
Cas 0h37'14"72d21'
Wagner,Bonnie
Lyn 8h27'14"50d43'
Wagner,Chanel Adriana
Cyg 19h26'53"33d5'
Wagner,Chris
Uma 9h26'15"52d13'
Wagner,Christian A
Cma 6h26'33"-13d54'
Wagner,Christine Schuller
And 23h2'42"51d19'
Wagner,Cindy
Cas 0h26'26"58d9'
Wagner,Cindy
Lyr 18h37'1"45d11'
Wagner,Clarence & Charlotte
Hya 9h8'40"5d16'
Wagner,Claus & Lisa
Uma 9h57'40"68d32'
Wagner,David
Cep 0h21'51"77d16'
Wagner,David W
Boo 14h35'34"40d25'
Wagner,Deanna
Cyg 21h54'43"52d36'
Wagner,Diane Rose
And 2h27'10"48d6'
Wagner,Dina Bilic
Del 20h21'1"10d47'
Wagner,Edythe J
Com 13h2'0"26d44'
Wagner,Eleanor C
Sge 19h56'1"16d33'
Wagner,Elizabeth R
And 1h13'14"36d34'
Wagner,Emily Christine
Lyn 7h55'50"58d17'
Wagner,Eugene N
Aur 4h52'33"38d46'
Wagner,Frank
Oph 17h2'32"1d44'
Wagner,Frederick Leroy
Uma 10h17'18"47d31'
Wagner,Friedhelm
Del 20h22'1"3d18'
Wagner,Friedrich Kurt
Dra 15h3'10"62d17'
Wagner,Heinrich
Mon 7h53'20"-1d8'
Wagner,Helmut L
Aqr 21h59'45"-21d43'
Wagner,Horst
Peg 22h50'1"8d22'
Wagner,Jack Jeffrey
Cam 3h54'22"77d30'
Wagner,James David
Gem 7h2'15"21d29'
Wagner,Jeffrey Reed Russell
Her 18h50'27"18d47'
Wagner,Jocelyn
Mon 6h58'40"11d46'
Wagner,John
Her 16h45'20"33d42'
Wagner,John & Patricia
Cep 22h39'39"70d35'
Wagner,Jonathan
Aur 6h3'31"30d36'
Wagner,Jr,Eugene B
Cnv 12h27'39"37d57'
Wagner,Jr,James Esro
Aql 19h26'0"-10d18'
Wagner,Karl
Her 16h45'26"30d2'
Wagner,Kathleen Ann
Sge 19h16'20"16d51'
Wagner,Kathleen Juan
Leo 9h52'50"31d47'
Wagner,Kathrin
Cmi 7h20'1"8d35'
Wagner,Kelly
Aql 19h5'30"15d0'
Wagner,Kenneth C
Ori 5h32'41"8d34'
Wagner,Laura
Equ 20h56'26"5d56'
Wagner,Lindsay
And 1h31'47"40d44'
Wagner,Lisa-Marie
Lyr 19h6'49"38d12'
Wagner,Mark Alan
Lac 22h25'49"50d0'
Wagner,Matthias
Dra 19h46'50"58d45'
Wagner,Michael MacKenzie
Aql 18h54'16"7d24'
Wagner,Michael David
Cyg 19h29'22"36d26'
Wagner,Nathanial P
Lmi 9h23'0"38d20'
Wagner,Nicole Marie
Peg 22h3'36"20d14'
Wagner,Patrick
Hya 9h5'50"4d32'
Wagner,Polly Wren Shepherd
Sex 10h1'59"-1d14'
Wagner,Rae Elaine
Mon 6h42'0"10d7'
Wagner,Raymond J
Tau 3h41'47"25d41'
Wagner,Rebecca M
Cas 0h39'29"64d30'
Wagner,Reinhard
Cas 0h47'54"60d32'
Wagner,Renee Ann
Cas 1h29'45"67d37'
Wagner,Richard "The Angel"
Crt 11h42'26"-21d53'
Wagner,Richard A
Cet 3h17'25"0d57'
Wagner,Richard Alvin
Ori 5h3'20"15d54'
Wagner,Robert Lynn
Cep 22h52'58"57d46'
Wagner,Samantha Lyenn
Uma 10h35'1"59d24'
Wagner,Sharon
Her 16h48'44"32d44'
Wagner,Sheila Ann
Lyn 8h5'58"38d3'
Wagner,Silvia
Psc 1h19'0"18d38'
Wagner,Sr,Richard F
Uma 9h59'44"50d3'
Wagner,Sr,William J
Cep 20h42'1"65d23'
Wagner,Steve
Aql 18h41'59"-2d24'
Wagner,Susan "Susie"
Peg 23h26'25"33d25'
Wagner,Ulf
Dra 18h16'45"68d55'
Wagner,Ursula
Sco 17h31'32"-38d43'
Wagner,Vernon Carson
Boo 14h29'54"22d19'
Wagner,Virginia M
Cam 3h13'53"60d13'
Wagner,William
Lyn 6h28'50"60d46'
Wagner,William Griffin
Aqr 22h22'20"0d10'
Wagner,Wolfgang
Vir 13h23'20"-9d46'
Wagner-Kasten
Her 18h28'31"12d25'
Wagnon,Manford K
Ser 18h15'29"-14d43'
Wagoner Family Star, The
Peg 21h53'1"31d19'
Wagoner,Billy D
Boo 14h49'50"38d27'
Wagoner,Bob
Ser 16h14'0"2d32'
Wagoner,Howard
Per 1h56'18"53d57'
Wagoner,Richard B
Per 3h4'56"41d12'
Wagstaff II,Joseph Robert
Vir 11h56'0"-0d48'
Wagstaff,Darren
Ori 6h6'20"4d36'
Wagstaff,Doug
Cep 22h4'32"53d35'
Wagstaff,Mark
Her 18h9'1"40d4'
Waguette,Christophe
Ori 4h43'12"9d48'
Wah Ai Ni
Sct 18h30'1"-6d4'
Wahab,Maya
Peg 22h39'20"22d5'
Wahe,Sherry
Umi 9h36'53"50d35'
Wahl QC,Mr Jack
Boo 14h13'1"34d55'
Wahl,Bill
Per 2h41'33"35d60'
Wahl,Carolyn
Leo 9h56'19"12d9'
Wahl,David Ronald
Aur 6h24'54"30d57'
Wahl,Harald
Per 4h6'0"49d25'
Wahl,James F
Her 18h43'48"18d51'
Wahl,Jochen
Per 4h6'57"48d19'
Wahl,Karl-Ernst
And 23h0'56"42d30'
Wahl,Stephen William
Hya 10h40'45"-18d44'
Wahlberg,Donnie
Per 3h5'41"50d6'
Wahle,Elsie & Ludwig
Lyr 19h5'32"28d14'
Wahler "Bill",Martin
Cnv 12h55'1"51d29'
Wahler,Stellar Maria Carmina Casale
Lyn 7h32'32"38d27'
Wahlers,Inge
And 0h6'1"30d7'
Wahlers,Star of Connor
Boo 14h52'56"37d10'
Wahlgren-June 13,1942, Kay
Gem 6h51'14"14d13'
Wahoski,Joyce Fedewa
Cas 1h47'22"63d55'
Wahoski,Stephen Michael
Cam 5h59'29"80d1'
Wahrenberger,"Max" Maximilian
Umi 15h49'45"80d16'
Wahrenberger Zumstein, Marthe v
Umi 15h45'49"80d12'
Wahrlich,Horst
Eri 4h50'37"-9d24'
Wai,Remona Lee Sui
And 23h16'31"37d53'
Wai-Ling
Ori 5h56'29"21d11'
Waidbacher,Christoph
Lyn 8h12'22"48d53'
Waightavia,Donnula
Mon 7h10'24"-10d18'
Waiglein,Helene
Vir 11h39'45"-2d59'
Waimo
Psc 0h23'1"-5d53'
Wain "I Love You Truly",H & W
Lyr 19h13'46"41d6'
Wain,Frances Yvonne
Cyg 19h32'36"35d13'
Wainscott's "Glance of an Angel",Chanda
Lyr 19h7'59"40d7'
Wainstead,Richard H
Tau 4h18'58"23d30'
Wainwright,Andrew S
Sge 20h0'0"20d1'
Wainwright,Carroll & Catherine
Sex 10h13'47"-4d5'
Wainwright,Daniel Bryan
Her 16h25'24"50d31'
Wainwright,Hazel Margaret
Ori 5h49'44"18d43'
Wainwright,Joseph
Cep 20h51'23"68d57'
Wainwright,Ron
Ori 5h48'16"21d6'
Waissman,"Tutu" In Memory Of Susan
Lyr 18h37'58"44d44'
Waistimmons,Ashley Rose
Hya 8h43'45"3d39'
Waite,Andrew
Aur 6h50'36"41d8'
Waite,Harlow E
Cam 11h59'36"77d50'
Waite,Jaime Leigh
Cyg 21h2'51"30d14'
Waite,James Frederick Weldon
Her 17h24'55"27d6'
Waite,Jr,Herbert
Aql 20h7'34"0d2'
Waite,Kathy
Cyg 20h23'17"40d48'
Waite,Lisa & Randy
Sge 19h34'25"19d5'
Waite,Lori Louise
Cas 3h10'28"65d30'
Waite,Stanley
Cyg 20h18'25"38d42'
Waites,Dr Larry
Cnc 9h6'28"31d27'
Waits,Laura D
Cap 21h0'57"-22d58'
Waits,Susan
Peg 22h38'12"22d5'
Waits,Tracy
Peg 22h23'58"4d41'
Waju
Boo 15h7'0"22d23'
Wakao Risa
Vir 14h36'48"6d39'
Wakatsuki,Taiki
Aql 19h3'50"15d17'
Wake,Richard L
Aur 6h21'55"37d42'
Wake,Steve
Per 2h53'1"45d11'
Wakefield Vons Company Jim
Ori 6h2'25"4d42'
Wakefield,Alan
Her 16h59'11"25d13'
Wakefield,Amy
Lyn 8h6'15"44d5'
Wakefield,James Walter
Peg 0h11'13"18d55'
Wakefield,Lloyd & Flo
Cyg 21h18'15"36d42'
Wakefield,Nicola
Lyr 18h29'31"31d14'
Wakefield,Wilma Boerhave
Eri 4h13'37"-11d50'
Wakeford,Stanley
Cep 21h6'17"70d47'
Wakeling,John Llewellyn
Vir 14h43'20"-6d5'
Wakeman,Jill Marie
And 1h5'1"38d59'
Wakin,Malham M
Ori 5h57'0"8d48'
Wakinyan
Sge 19h23'22"18d52'
Wakley,Lisa Maria
And 0h45'44"38d45'
Wakohyohshi
Sgr 18h54'15"-32d40'
Wakuzawa,Wynn
Mon 6h26'18"-0d14'
Wal
Dra 17h49'0"64d15'
Walas,Walter
Dra 14h29'29"60d2'
Walat,Catherine Guilfoy
Cas 0h1'13"50d23'
Walbert,Carol Joy Epperson
Cas 23h16'42"62d59'
Walbridge,Susan Cecelia
Ori 5h50'24"18d55'
Walbridge,Terry #2
Aqr 23h21'9"-8d44'A
Walby,Judith K
Cas 0h49'58"61d6'
Walcutt,Clifford A
Uma 9h35'51"59d7'
Wald
Cyg 20h14'55"43d58'
Wald,Barney
Per 1h53'1"56d24'
Wald,Darla Jean
Cas 1h0'52"53d19'
Wald,Jr,Mark David
Aql 19h58'39"14d59'
Wald,Monika
Psc 1h39'12"21d30'
Waldbaum,Phyllis
For 3h41'53"-34d52'
Waldeck,Wendy Kay
Sge 20h7'27"20d0'
Walden,Bob
Boo 14h1'0"11d39'
Walden,Charles
Ser 15h29'0"21d18'
Walden,Dave
Ori 5h17'1"15d13'
Walden,Dorothy Jackson Dillow
Cas 23h30'49"54d20'
Walden,Karen
Cam 5h4'11"67d35'
Walden,Kay
Cas 0h38'12"63d60'
Walden,Matt
Cet 3h2'11"1d32'
Walden,Morgana & Brad
Eri 3h30'27"-5d14'
Walden,Narada Michael
Uma 10h53'54"54d49'
Walden,Ronald L
Lmi 10h3'53"31d19'
Walden,Ryan Matthew
Aql 19h1'57"-0d13'
Waldenberg
Sex 9h51'38"2d24'
Waldenmaier,Shelley
Lyr 18h59'25"27d32'
Waldens,The
Ori 5h28'53"-6d34'
Waldherr,Robert E
Dra 17h2'55"60d35'
Waldi 16-10-43
Lyn 7h51'57"52d14'
Waldinger,Ira William
Per 2h21'1"58d26'
Waldl,Liane Und Alexander
Ori 5h14'46"-9d2'

Waldling,Rafael
Ori 5h0'56"8d51'
Waldman Radiant Light of
Love,Faye
Crb 16h9'35"27d52'
Waldman,Amy
Hya 8h16'41"-6d51'
Waldman,Heidi
Uma 10h3'32"58d44'
Waldman,Janet
Psc 1h26'48"33d21'
Waldman,Robert Harris
Aur 6h20'42"37d19'
Waldman,Summer
Cyg 21h25'52"37d54'
Waldman,Susan
Mon 6h19'26"8d23'
Waldmania
Aur 6h52'43"40d1'
Waldmann,Aline
Cas 0h54'53"69d6'
Waldmann,André
Lyn 8h39'54"42d11'
Waldner,Beth Marie
Aql 19h13'1"13d10'
Waldner,Charles Patrick
Peg 22h1'0"33d54'
Waldner,Leah Roselyn
Peg 22h16'60"31d59'
Waldo
Cam 3h20'42"61d12'
Waldo
Uma 11h27'18"41d12'
Waldorf,Marina Anna
Cas 0h30'34"67d12'
Waldraff,Herma
Ari 2h4'13"21d35'
Waldren,Joseph Andrew
Cma 7h2'27"-30d22'
Waldrep,Randall Craig
Her 16h20'17"11d43'
Waldrip,Irene Lahr
Ori 5h53'54"10d4'
Waldron,(LUPPIE)
Cet 2h5'59"6d32'
Waldron,Barry Alvin
Sgr 19h14'35"-25d39'
Waldron,Elizabeth Hunt
Uma 8h33'1"53d42'
Waldron,Faith
Aql 20h0'47"10d12'
Waldron,Harriette Owens
Cyg 19h28'48"30d12'
Waldron,Lisa
And 0h8'18"38d40'
Waldron,Mary Kay
Aur 5h18'1"40d30'
Waldron,Nigel
Aql 18h57'43"-5d59'
Waldron,Sarah
Sco 17h54'1"-40d32'
Waldroop,Enos Carroll
Mon 7h3'13"-6d21'
Waldrop,Bryan
Lac 22h29'0"38d52'
Waldrop,Forrest & Phyllis
Peg 22h15'18"5d44'
Waldrop,Ruth E
Mon 6h53'1"-1d1'
Waldshan,Poppy-Stanley
Uma 9h33'42"48d19'
Waldstein,Mark
Lac 22h18'31"55d19'
Walen,Norris Sanford
Per 2h21'37"54d23'
Walendy,Rita
Boo 13h37'49"12d39'
Walentiny,Georgianna
Cas 0h31'39"76d28'
Walerski,Lois Lynn
And 23h35'58"48d27'
Wales,Anita
Peg 22h36'19"30d7'
Wales,Reece Robert in
Memory
Sco 16h31'54"-31d41'
Waletzki,David Alan
Her 16h42'37"11d57'
Walfish,Steven Ira
Hya 9h35'17"0d31'
Walford,Joshua Richard
Umi 17h26'13"86d3'
Walhiem,Thomas
Lyr 18h59'52"28d56'
Walicek,Jimmy
Uma 9h56'25"43d6'
Walk on Faith Trust In Love
Lyr 18h42'46"34d8'
Walk,Alan J
Aql 19h8'37"1d57'
Walken,Christopher Amazing
Cyg 20h59'18"31d36'
Walker
Peg 23h0'0"30d39'
Walker
Aql 19h12'56"13d2'
Walker (Dumbarton) Star,Billy
Uma 9h45'1"46d50'
Walker 29,Julie
Com 12h27'58"21d58'
Walker Anniversary
Star,Ginger & Joseph
Cnv 13h49'1"40d8'
Walker Project Team, The
Lac 22h50'1"55d44'
Walker Star,Sharon E
Aql 19h1'26"0d19'
Walker's Place In Heaven,JD &
ML
Dra 17h40'10"68d1'
Walker's Red Velvet
Aur 6h6'18"48d50'
Walker's Star,Brenda
Eri 3h53'14"-3d32'
Walker's Star,Loraine
Ari 2h27'59"21d6'
Walker's Wish
Ori 5h11'37"-1d51'
Walker's World,Adam
Dra 11h39'1"71d53'
Walker,A D 66 Chevy
Aur 6h44'52"38d31'
Walker,Alissa Marie
Vul 20h14'13"23d8'
Walker,Alistar J
Ori 6h5'23"5d32'
Walker,Amy Susan
And 0h29'41"40d45'
Walker,Andrew & Jenny
Cyg 20h42'43"45d3'
Walker,Angela Kay
And 0h5'51"40d24'
Walker,Antony George Tutham
Dra 14h30'36"64d44'
Walker,Aron Patrick
Oph 18h38'56"10d49'
Walker,Beth Louise
Lmi 10h17'34"30d52'
Walker,Big Daddy
Ori 6h17'17"-2d42'
Walker,Bill
Ori 6h6'13"-0d0'
Walker,Bill
Ser 18h35'53"2d55'
Walker,Billy L
Cet 2h41'20"4d12'
Walker,Carina Katyarina
Alexandra
And 2h34'37"37d45'
Walker,Carol
Aur 5h19'53"40d20'
Walker,Cathy L
Vul 19h46'0"28d40'
Walker,Celestia Catherine
Wilcox
Lyr 18h40'37"28d18'
Walker,Charles Hoyt
Gem 7h17'2"21d10'
Walker,Christel
Cnc 9h2'1"32d18'
Walker,Cleon Ann
Mon 6h53'56"1d35'
Walker,Clifford James
Aql 18h53'0"-0d5'
Walker,Coco
Uma 11h4'23"51d20'
Walker,Cody,Annie,Kate &
Taylor
Aql 18h54'47"-1d16'
Walker,Craig
Ori 6h0'1"1d29'
Walker,Damien Jordan
Sco 17h20'55"-39d3'
Walker,Dana Marie
And 1h48'27"47d6'
Walker,Dawn Loree Schrock
Cyg 21h7'45"30d19'
Walker,Dawne
Eri 4h1'58"-18d5'
Walker,Deanna L
Umi 16h27'31"70d18'
Walker,Debra
And 0h9'0"27d44'
Walker,Dennis Elwood
Her 16h51'18"35d16'
Walker,Diane
Lyn 8h3'47"33d40'
Walker,Diane Alison
Cas 0h6'39"60d53'
Walker,E B (Buck)
Her 16h13'12"40d30'
Walker,Edward James Philip
Aur 6h2'43"46d17'
Walker,Elaine L
Gem 7h5'38"24d53'
Walker,Elizabeth Haley Francis
Psc 1h34'59"27d42'
Walker,Emily Jordan
And 1h53'12"41d2'
Walker,Emma Claire
And 23h3'18"50d55'
Walker,Emma Louise
Cyg 19h30'43"38d55'
Walker,Eric R
Lyn 7h36'32"42d43'
Walker,Gary
Lac 22h14'14"51d24'
Walker,Gayle
Peg 23h5'12"32d42'
Walker,Gersha
Cas 0h43'51"61d50'
Walker,Glenn
Hya 8h12'30"3d41'
Walker,Greg
Her 16h37'52"38d46'
Walker,Gregory
Lac 22h13'19"54d38'
Walker,Gregory Bruce
Her 17h11'52"27d31'
Walker,Heather
Lyr 18h32'12"33d21'
Walker,Heidi
Lyr 18h27'1"35d16'
Walker,Helen Premo
Uma 9h27'1"49d13'
Walker,Hunter
Cam 10h55'1"81d51'
Walker,In Loving Memory of
Dale Lee
Ser 15h28'40"18d1'
Walker,Ivan
Ser 17h55'42"-14d51'
Walker,J Bruce
Ser 18h1'44"-13d42'
Walker,Jacquelyne
Lyn 8h2'42"39d13'
Walker,James Ian Francis
Vir 13h16'23"-1d55'
Walker,James John
Cep 2h15'44"77d49'
Walker,Jameson Aleksander
Tau 4h22'0"16d15'
Walker,Jan
Cas 0h53'19"70d27'
Walker,Jay
Per 3h4'29"40d57'
Walker,Jean
Lmi 9h32'1"38d26'
Walker,Jeni
Cas 1h21'1"58d9'
Walker,Jenny
Peg 23h26'49"12d30'
Walker,Jerry Bar Shart
Michelle E
Cyg 21h5'0"40d6'
Walker,Jessie May Gallagher
And 0h1'13"38d36'
Walker,Jillian
Cyg 20h25'37"40d11'
Walker,Jim Marie & Cody
Aql 19h31'31"13d6'
Walker,Jimmy
Ori 5h30'55"-8d34'
Walker,John
Ori 6h7'31"3d35'
Walker,John & Elizabeth
Uma 11h56'34"57d13'
Walker,John & Jane
Cmi 7h35'55"6d56'
Walker,John & Joyce Harris
Walker
Sge 19h19'34"16d11'
Walker,Johnny
Cep 1h13'1"80d6'
Walker,Jonathan Ian
Her 17h19'39"41d9'
Walker,Jonie Denise
Mon 6h44'48"10d20'
Walker,Jordan
Ori 5h46'50"18d59'
Walker,Jordan Renee
Mon 6h58'39"8d35'
Walker,Joseph
Ori 6h5'39"-0d9'
Walker,Joshua Robert
Ori 5h25'0"-0d42'
Walker,Jr,Jack Arthur
Her 18h0'51"37d31'
Walker,Jr,Robert W
Sco 16h38'11"-30d24'
Walker,Karen Star
Mon 6h35'1"1d4'
Walker,Kevin Charles
Ori 5h11'33'-1d10'
Walker,Kimberly Ilene
Cyg 19h31'55"32d46'
Walker,Kristine Marie
And 23h21'42"48d8'
Walker,Laura Elizabeth
And 2h4'0"41d42'
Walker,Lauri "Ditty"
Cas 23h29'18"61d21'
Walker,Lisa
Cas 0h34'38"56d2'
Walker,Loren C & Gloria M
(Thatcher)
Cyg 19h44'24"30d8'
Walker,Love Always
Tammy,Joshua A
Dra 14h27'11"62d48'
Walker,Luke Sky
Ori 5h2'43"15d2'
Walker,M & M
Cyg 21h23'26"30d45'
Walker,Margaret F
Umi 14h53'21"65d34'
Walker,Mark
Aql 20h11'40"13d28'
Walker,Mark Everett
Lac 22h36'36"52d59'
Walker,Mark Heath
Ori 4h44'20"1d1'
Walker,Mark Richard
Lyn 7h56'55"41d31'
Walker,Marquie Layne
Vul 19h41'20"20d4'
Walker,Martha & Bryant E
Peg 23h29'30"21d52'
Walker,Matthew Alexander
Per 3h12'17"42d19'
Walker,Matthew C
Aur 5h42'57"50d27'
Walker,Matthew Michael David
Her 17h51'24"18d58'
Walker,Melanie Jayne
Sge 20h16'30"18d58'
Walker,Michael Allen
Aur 7h19'45"38d6'
Walker,Michael John
Uma 10h10'0"59d53'
Walker,Michelle Elizabeth
Lyn 7h36'47"40d46'
Walker,Milo D
Lac 22h1'36"38d16'
Walker,Missy
Cep 22h53'55"70d58'
Walker,Mort
Lmi 9h32'41"38d20'
Walker,Murray H B
Dra 15h44'1"60d36'
Walker,Nancy J
Cam 6h31'37"68d6'
Walker,Nancy M
Umi 15h17'58"71d58'
Walker,Natalie
Lyr 18h29'0"32d54'
Walker,Nathan Brock
Ori 5h56'1"6d18'
Walker,Nellie M
Mon 6h44'11"11d55'
Walker,Nina Ammidon
Gem 7h10'21"21d9'
Walker,Oliver Dewayne
Cnv 12h57'30"40d0'
Walker,Patricia Ann
Aql 20h11'1"10d48'
Walker,Patricia Ann
Crt 11h14'1"-17d16'
Walker,Paul Franklin
Dra 14h25'1"63d30'
Walker,Pauline
Cas 23h21'41"60d5'
Walker,Philip Marshall
Sct 18h52'33"-7d4'
Walker,Rebecca Kaye
Cyg 20h23'47"40d53'
Walker,Reed Harrison
Aur 5h29'35"48d51'
Walker,Robert Montgomery
Hya 8h44'36"4d55'
Walker,Robert M
Aur 6h12'57"37d11'
Walker,Robin Lynn
Lyn 6h28'26"59d1'
Walker,Ros
Cam 3h58'16"62d13'
Walker,Russell & Lani
Uma 12h34'40"56d19'
Walker,Samuel Ray
Aql 19h32'43"1d52'
Walker,Sandra
Cas 1h54'28"60d16'
Walker,Santino
Mon 6h48'28"10d30'
Walker,Sara Ella
Uma 8h25'33"72d5'
Walker,Sophie Bronyn
Cas 0h55'59"74d29'
Walker,Sr,James Thomas
Per 3h29'25"48d58'
Walker,Stan
Dra 17h2'48"69d47'
Walker,Steve
Her 16h3'58"50d26'
Walker,Thomas
Ori 5h53'43"-10d24'
Walker,Tommy
Boo 14h14'39"31d2'
Walker,Trent Thomas
Ser 18h0'21"-14d10'
Walker,Veronica Anne
Aql 18h58'31"-3d7'
Walker,Victor
Vul 20h43'39"20d16'
Walker,Victoria "Shelly"
Peg 22h27'13"20d12'
Walker,William J
Vir 11h43'14"7d13'
Walker,Willsey
Cam 3h33'34"61d21'
Walker,Woods
Dra 11h58'56"68d30'
Walker-Hutsell, Nicholas Kyle
Dra 17h20'19"61d19'
Walker-McMullen,Lynn
Cas 0h13'0"61d33'
Walker-Wedell, "Terwalk",Terry
Her 18h3'47"28d48'
Walkerwicz,Captain Ronald
Oph 18h32'48"10d42'
Walko,Lee & Kim
Lmi 10h38'0"25d54'
Walko,Leo S
Uma 10h54'1"38d51'
Walkup,Robert E
Eri 3h55'1"-2d0'
Wall 50th Birthday Star,The
Randy
Ori 5h58'57"-2d58'
Wall,Barbara
Leo 10h18'0"13d7'
Wall,Bernard John
Cep 22h31'0"58d29'
Wall,Bob
Cep 2h49'56"77d32'
Wall,Brian Michael
Cep 21h19'32"85d6'
Wall,Christine
Umi 15h2'19"69d45'
Wall,Christopher
Uma 10h13'1"71d17'
Wall,Connor Douglas
Aur 4h59'12"48d52'
Wall,Connor Douglas
Boo 14h37'1"20d12'
Wall,Darlene
Uma 12h5'51"58d22'
Wall,Donna Francis
(Litchfield)
Aql 19h0'26"13d32'
Wall,Hannah
Cyg 20h41'30"42d28'
Wall,Kelley Marie
Dra 11h41'13"70d45'
Wall,Michael David
Aur 6h22'51"32d26'
Wall,Regina Carlett
And 0h30'55"28d56'
Wall,Tony
Peg 22h20'23"31d0'
Walla,Manfred
Lyr 19h23'0"30d29'
Walla,Mark D
Ori 6h0'28"8d14'
Wallace 21,Heidi
Lyn 7h55'51"38d32'
Wallace Dean
Her 16h42'58"35d53'
Wallace I Love You...
Kat,Darryl W
Uma 10h40'23"40d35'
Wallace My Bubba, Cameron
Uma 8h33'60"48d12'
Wallace Ray
Uma 12h31'44"61d4'
Wallace Serve,The
Uma 8h57'27"68d58'
Wallace's Star
Vul 19h59'45"27d28'
Wallace,A
Cnv 13h14'43"32d11'
Wallace,Aaron
Lyn 8h4'22"46d56'
Wallace,Allan Steve
Aql 18h57'29"14d25'
Wallace,Allen
Oph 16h48'27"11d20'
Wallace,Autumn Dianne
Mon 7h53'57"-6d53'
Wallace,Bobbi
Cet 2h32'52"-0d25'
Wallace,C
Lac 22h8'45"51d51'
Wallace,D
Lyn 7h31'46"40d38'
Wallace,Debra & Matthew
Uma 11h2'34"52d50'
Wallace,Don Clinton Y
Pyskat,Døna René Y
Her 17h21'22"40d5'
Wallace,Dorothy
Umi 15h17'28"68d8'
Wallace,Dusty Scott
Aur 6h7'28"37d46'
Wallace,Elizabeth A
Peg 23h19'17"30d18'
Wallace,Father "Wally"
Tau 4h32'26"15d33'
Wallace,Forrest Robert
Her 15h51'59"46d14'
Wallace,Fred Edsel
Cap 20h9'37"-10d47'
Wallace,Gary Keith
Cet 2h7'42"6d1'
Wallace,Gregg & Debbie
Cyg 19h42'28"31d6'
Wallace,Jane
Peg 22h4'56"3d36'
Wallace,Janet
Tau 4h14'57"20d2'
Wallace,Jennifer R
Peg 22h49'32"21d43'
Wallace,Joe
Cyg 21h21'38"40d11'
Wallace,Johnny
Aur 6h17'0"38d47'
Wallace,Jordyn
And 1h27'25"34d22'
Wallace,Julia Rebecca
Vir 11h39'12"7d49'
Wallace,Kacy
And 1h21'36"38d46'
Wallace,Karen
Ori 4h54'38"1d7'
Wallace,Kelly Cohen
And 0h10'35"35d9'
Wallace,L
Cam 6h13'43"60d46'
Wallace,Leah Havens
Aql 20h11'24"1d6'
Wallace,Lee
Cep 22h11'34"68d19'
Wallace,Leigh Ann
And 2h3'17"37d52'
Wallace,Lydia
Lyr 18h27'23"38d7'
Wallace,Marissa Lynne
Peg 23h35'35"11d1'
Wallace,Mary E
Vul 19h48'19"23d17'
Wallace,Mary Lee
Del 20h53'0"8d32'
Wallace,Mary Lou
Cas 1h0'1"63d46'
Wallace,Matt
Aur 5h8'29"43d59'
Wallace,Maureen J
Uma 11h57'46"30d15'
Wallace,Michael James
Her 18h17'0"20d18'
Wallace,Michael Lee
Cet 3h1'13"0d44'
Wallace,Mike
Aur 6h4'45"38d8'
Wallace,Mort
Ori 5h52'53"16d18'
Wallace,Pamela
Com 13h8'24"20d14'
Wallace,Rachel J
And 2h18'14"46d44'
Wallace,Rickey Brandon
Ori 5h55'52"11d32'
Wallace,Robert Charles
Dra 15h58'11"65d17'
Wallace,S
Aur 6h15'41"45d23'
Wallace,Sharon
Cyg 19h32'33"35d52'
Wallace,Summer F
And 0h52'41"39d26'
Wallace,Sylvia Kathleen
Lyn 7h30'42"42d54'
Wallace,Thomas
Cep 22h50'48"59d59'
Wallace,Tiana L
Cam 3h53'55"52d56'
Wallace,Virginia
Eri 4h30'14"-10d10'
Wallace,W Clinton
Aql 20h13'21"1d18'
Wallace,Wally
Dra 17h39'36"68d12'
Wallace,William
Cep 4h1'33"80d33'B
Wallace,William Ashley
Her 16h29'18"40d32'
Wallace,William R
Aur 4h53'1"51d9'
Wallace,Willie
Cet 2h11'29"4d45'
Wallace-Schemo,Dalen Albert
Aql 19h24'36"-8d39'
Wallach,Alexis Elise
Cnc 8h29'23"10d1'
Wallach,Joseph
Aur 7h9'51"40d4'
Wallach,Larry
Dra 12h12'15"71d42'
Wallach,Michael Samuel
Leo 9h37'55"7d27'
Wallach,Virginia
Peg 23h40'48"18d22'
Wallander,Larry
Dra 16h39'13"61d60'
Wallast,Gerty
Cas 1h54'32"75d21'
Wallau,Taylor Sheridan
Crt 11h14'57"-10d13'
Wallen,Barbara
And 2h29'27"37d51'
Wallen,J Ray
Aql 19h31'1"0d1'
Wallen,Jeffery
Aql 19h9'0"3d25'
Wallen,Kelli
And 2h24'52"41d28'
Wallen-94
Cep 22h40'12"70d55'
Wallenberg,Sharon
Peg 0h5'17"14d17'
Wallenfels,Helmut
Lyn 6h57'54"59d46'
Waller Star,Gladys
Cas 0h33'50"73d57'
Waller With Love, Laura II,To
Jason
Her 18h38'46"12d44'
Waller's Star,Anne
Cet 1h51'1"0d35'
Waller,Arthur
Aur 7h9'23"36d57'
Waller,Charlie Sr.
Peg 21h20'1"18d49'
Waller,Cynthia Louise
Eri 3h15'46"-13d58'
Waller,Jennifer
Cas 1h1'53"62d24'
Waller,Nicholas James
Ori 4h57'50"5d4'
Waller,Steve
Oph 17h36'51"-23d26'
Waller,Vance F
Crt 11h52'37"-8d16'
Waller,Victoria Ritson
Peg 23h19'1"33d9'
Waller,W E W,Robert Daniel
Dra 19h6'29"60d37'
Wallers,Felicia
Peg 22h35'12"30d57'
Wallerstein,Herb
Aur 5h0'59"44d31'
Wallien,Gayle
Lyr 18h43'0"38d32'
Wallier,Ghislain
Lmi 10h35'57"32d5'
Wallin,"Big Charlie"
Her 16h40'48"10d13'
Wallin,Alexandra
Lyr 19h6'55"30d13'B
Wallin,Hannah
Lyr 19h6'55"30d13'A
Wallin,Lisa S
Peg 22h17'44"35d28'
Walling,Dick "The King"
Her 17h1'20"30d3'
Walling,Peyton Leeann
Cyg 21h50'39"37d59'
Walling,Richard C
Cnv 13h35'35"40d51'
Wallinger,Arthur John "AJ" &
Cindy R
Dra 16h4'47"62d29'
Wallingford,Jr,David Thomas
Dra 19h25'28"68d8'
Wallis,Benjamin Andrew
Lac 22h12'0"48d48'
Wallis,Dennis
Per 1h28'20"53d37'
Wallis,Erin Joy
Cas 3h10'42"58d11'
Wallis,James L
Lac 22h16'53"55d27'
Wallis,Jim & Pat
Cyg 19h44'50"30d29'
Wallitzek,Gisela
Lib 15h8'7"-21d19'
Wallman,Amy
And 1h4'36"39d8'
Wallner Star,The Frank A
Boo 15h7'58"28d11'
Wallner,Alfons
Peg 23h28'51"11d55'
Wallnig,Josef
Aql 19h8'23"4d31'
Wallower,Carl S
Uma 11h39'18"38d1'
Walls Always in Our
Hearts,Freddy
Sct 18h53'22"-5d7'
Walls V,John
Uma 12h6'1"48d3'
Walls,Elizabeth Mary
Cas 0h23'52"60d21'
Walls,Gareth
Ori 5h23'33"-0d4'
Walls,James Richard
Ori 6h6'38"1d7'
Walls,Jeanie Hughes
Sco 17h25'49"-33d52'
Walls,Jimmy Wayne
Her 17h0'38"20d2'
Walls,Kevin Lyn
Eri 4h10'34"-12d28'
Walls,Leah
Sex 10h6'36"-2d22'
Walls,Mike
Aql 20h3'1"8d56'
Walls,Wesley
Cet 2h5'40"1d18'
Wallster,Dale & Sue
Dra 18h17'12"48d29'
Wallworth,Elaine
And 23h16'19"42d29'
Wally's Guiding Light of Love
& Joy
Aql 18h58'0"8d54'
Wally's Star
Uma 10h9'13"58d12'

Wally's Stargate To The Heavens
Her 16h57'1"27d39'
Wally's World
Cep 22h30'31"68d3'
Wally,Margo
Del 20h23'54"20d14'
Walor,Cacey Nicole
Vir 11h42'25"2d42'
Walpole,Benjamin
Ori 5h56'29"17d41'
Walpole,Sandy
And 1h9'21"37d58'
Walrad,John Francis
Cep 21h33'45"55d24'
Walsh III,John Robert
Aur 7h10'25"40d34'
Walsh True Love Weddng Star,Maxine&Tim
Cyg 20h36'0"40d8'
Walsh's Wish,C M S
Sex 10h32'1"-1d53'
Walsh(A Little Bit Of Heaven)
Cnv 12h25'1"42d45'
Walsh,Alden Fidler
Ori 5h55'55"6d50'
Walsh,Alexander James
Cap 21h27'56"-20d26'
Walsh,Annette
Lmi 10h54'1"32d22'
Walsh,Barbara Jean
And 2h34'23"42d54'
Walsh,Betty
Lyr 18h19'22"42d25'
Walsh,Betty & Tom
Del 20h36'59"10d30'
Walsh,Brady
Sex 9h41'0"2d11'
Walsh,Brody Thomas
Her 16h30'27"41d22'
Walsh,Carol
And 23h48'29"46d21'
Walsh,Chloé Nouvel
Ori 5h53'36"6d41'
Walsh,Christopher Francis
Dra 20h8'55"62d13'
Walsh,Cynthia & Michael J
Uma 9h23'43"58d15'
Walsh,Darlene Frances
Leo 11h55'0"23d19'
Walsh,Donald David
Uma 9h25'18"51d40'
Walsh,Donald F
Lac 22h18'31"51d46'
Walsh,Dr Christopher Sean
Oph 17h54'28"8d33'
Walsh,Dr Michael Dennis
Oph 17h33'0"7d54'
Walsh,Elizabeth
Cnv 12h25'1"32d15'
Walsh,Ellen
Crb 15h28'12"31d53'
Walsh,Emily
Tau 4h17'15"22d27'
Walsh,Gary R
Cnv 12h38'34"40d52'
Walsh,Gillian Ann
Lyr 18h19'39"42d15'
Walsh,Gwen
Cyg 19h57'1"38d2'
Walsh,Isabel
Tri 2h7'59"32d58'
Walsh,Jack
Cep 2h22'38"80d9'
Walsh,James
Cep 22h3'37"61d33'
Walsh,James & Mary
Eri 3h48'39"-10d31'
Walsh,Jennifer
Peg 21h59'49"34d5'
Walsh,John "Spider"
Dra 17h4'1"63d21'
Walsh,John Edward
Crt 11h12'37"-10d59'
Walsh,John Gordon
Dra 17h7'20"64d43'
Walsh,Joyce
Cam 7h31'1"60d52'
Walsh,Jr,Ben B
Cet 2h56'29"0d39'
Walsh,Jr,Clune J
Cep 23h44'8"84d46'
Walsh,Jule
And 23h45'29"40d30'
Walsh,Julie Ann
And 0h8'58"46d55'
Walsh,Kathleen
Cas 0h51'56"65d14'
Walsh,Katlin Brie
Vul 20h2'39"23d40'
Walsh,Kevin & Nancy
Crb 16h5'12"30d8'
Walsh,Kevin P
Oph 17h54'13"8d58'
Walsh,Lori M
Psc 23h20'21"6d26'
Walsh,Lou
Hya 8h53'31"2d24'
Walsh,Marguerite
Ori 5h54'31"-8d16'
Walsh,Mary "Itsy"
And 23h23'53"43d24'
Walsh,Matthew S
Aur 4h48'19"50d23'
Walsh,Maureen
Cam 9h15'50"78d7'
Walsh,Molly Louise
And 0h13'41"46d42'
Walsh,Nicole
Leo 10h53'11"-1d3'
Walsh,Pam (Bindi)
Oph 17h57'49"12d45'
Walsh,Patrick Joseph
Gem 5h59'20"27d10'
Walsh,Paul
Ori 4h57'12"5d31'
Walsh,Ray
Dra 14h28'14"62d24'
Walsh,Rev Msgr Richard J
Cep 0h55'38"80d25'
Walsh,Robert & Ann
Cyg 20h9'37"39d25'
Walsh,Robert John Patrick
Aql 19h4'42"-0d59'
Walsh,Sandra L
Uma 12h4'39"62d17'
Walsh,Sara
Aql 20h1'54"11d25'
Walsh,Sarah Mae
Uma 9h38'54"48d58'
Walsh,Shane Kevin
Aur 7h13'29"41d57'
Walsh,Shannon
Mon 7h44'1"-6d18'
Walsh,Sherry
Vir 11h56'41"-2d47'
Walsh,Sidney G
Cam 4h52'20"65d23'
Walsh,Stephen John
Per 2h52'22"45d34'
Walsh,Steve
Aur 5h39'1"50d35'
Walsh,Susan Elizabeth
Peg 21h55'36"36d6'
Walsh,Suzy, Brandon, & Jamie
Lyr 18h51'0"42d45'
Walsh,Terrence Charles
Hya 8h19'43"3d4'
Walsh,Terry
Cnv 12h5'24"43d29'
Walsh,Thomas John
Lyn 7h20'52"44d26'
Walsh,Tracy Michelle
Mon 6h20'59"8d51'
Walsh-Moore 12-30-95
Cyg 21h5'50"33d26'
Walski,Jonathan David
Lmi 11h2'22"31d2'
Walston,Natisha Noel
Sct 18h56'41"-6d55'
Walston,Paul Kenneth
Cet 2h28'10"3d21'
Walston,Sherry
Lyr 18h21'45"42d3'
Walsworth,Allan
Cep 3h17'34"78d31'
Walt
Cnv 12h18'0"42d34'
Walt & Anita
Lyn 9h11'16"38d25'
Walt Marti
Boo 14h10'35"43d7'
Walta,Audrey J
And 23h29'12"49d50'
Waltemate,Laura-Danika Nicole-Amy
Uma 10h10'0"51d42'
Waltenburg,Melissa Ellen
Lmi 9h47'23"38d57'
Walter
Cet 3h0'33"0d20'
Walter
Cas 2h32'1"70d53'
Walter
Aur 5h59'49"50d23'
Walter
Vul 21h20'34"24d8'
Walter
Uma 8h55'43"60d16'
Walter & Cynthia Forever
Crb 15h53'59"28d30'
Walter & Valarie "Light Of Love"
Cyg 21h18'49"31d0'
Walter Alexander
Lac 22h23'35"40d41'
Walter II,Bruce Chadwick
Dra 13h12'23"68d31'
Walter IV,Harry Christian "Kit"
Boo 15h8'27"51d38'
Walter Marie
Cyg 21h38'0"40d50'
Walter Max
Ori 5h12'23"-5d52'
Walter's Dream
Cet 2h41'59"-8d6'
Walter,Alice M
Cas 0h35'15"72d30'
Walter,Andre
Peg 23h32'36"10d1'
Walter,Art & Fran
Dra 16h19'19"60d0'
Walter,Carmel
Dra 16h54'30"70d6'
Walter,Chanda
Cam 7h51'0"61d37'
Walter,Cynthia Ann Hanzak
Lyr 19h16'50"41d28'
Walter,David Earle
Dra 17h51'1"65d23'
Walter,David Paul
Boo 13h40'0"24d14'
Walter,Hans-Georg
Ori 6h6'47"10d49'
Walter,Janet M
Peg 22h7'11"21d43'
Walter,Josef
Ser 15h13'35"-1d40'
Walter,Lanette
Peg 22h1'1"10d3'
Walter,Martin
Cyg 19h53'51"47d3'
Walter,Mary Lou
Sge 19h30'48"16d38'
Walter,Monica
Gem 6h55'12"30d58'
Walter,Paul
Dra 20h6'53"63d12'
Walter,Saskia
Leo 10h39'53"14d6'
Walter-Kirk,Steven
Aql 19h55'24"13d3'
Walter-Ludwig-Stern
Leo 9h55'42"10d42'
Walters "Wart", Katherine
Vul 20h14'39"25d18'
Walters Star,The Jacqueline
And 1h13'0"37d36'
Walters,Ben
Ori 5h34'23"-0d43'
Walters,Bianca Jade
Cep 20h28'36"62d9'
Walters,Carl W
Uma 10h13'17"42d21'
Walters,Carol M
Cet 1h22'53"-5d55'
Walters,Charles Ray
Ari 2h6'18"22d18'
Walters,Danny Wayne
Sct 18h53'18"-6d54'
Walters,Don & Colleen
Mon 6h23'32"7d58'
Walters,Donna
Ori 5h28'48"-1d1'
Walters,Dr Shirley
Cyg 19h29'26"37d44'
Walters,Elden L
Aql 19h15'56"12d57'
Walters,Jeffrey Kyle
Cnc 9h0'51"10d50'
Walters,Joanne M
And 0h7'54"31d46'
Walters,Jr,John Andrew
Her 18h3'57"31d37'
Walters,Julie
And 23h27'26"47d21'
Walters,Juliet
Uma 13h52'11"51d12'
Walters,Katherine V
Cet 2h5'1"1d2'
Walters,Kathryn
Uma 11h42'1"38d46'
Walters,Kenneth Jay
Her 17h10'29"40d55'
Walters,Kevin Michael
Per 2h53'13"37d45'
Walters,Lauren Nicole
Oph 17h21'24"-22d13'
Walters,Mallory
Peg 23h40'56"31d11'
Walters,Margaret Jean
Peg 22h13'1"4d34'
Walters,Michael
Aur 6h0'35"38d7'
Walters,Nancie
Cyg 20h39'0"42d37'
Walters,Sage Lindsay Rose
Cyg 20h23'34"40d45'
Walters,Wesley Wade
Cet 1h26'20"-12d44'
Walters-Hathcock,Dawn M
Peg 23h15'7"18d1'A
Walth,Star Michael Owen
Lac 22h52'50"51d59'
Walthall-Tau Zeta Advisor,Mary
Mon 6h57'9"-1d10'
Walther C -100474
Per 4h3'49"35d47'
Walther,Christiane
Psc 23h1'52"6d57'
Walther,John Cain
Uma 9h2'14"57d54'
Walther,Laura
Cyg 21h33'42"44d38'
Walther,Ludwig
Aql 20h0'24"10d18'
Walther,Regina
Lib 14h20'24"-23d50'
Walther,Ron
Hya 8h21'43"1d11'
Walthers,Betty Howell
Cyg 19h44'43"30d0'
Walti,Heinz
Ori 5h36'14"-6d39'
Waltke "The World's Greatest Dad,Robert
Cep 22h16'16"61d33'
Walton Leasing Star, Ray
Dra 17h1'19"69d37'
Walton,Ada
Lyr 19h20'0"41d49'
Walton,Alia
Ori 4h51'59"0d1'
Walton,Allan R
Hya 9h5'24"4d50'
Walton,Betty
Cam 7h9'21"68d5'
Walton,Brian Todd
Her 16h38'16"34d18'
Walton,Carol "Tootie"
Cet 2h52'11"1d47'
Walton,Christy M
Ori 5h23'52"-1d23'
Walton,David Patrick
Dra 19h36'32"61d47'
Walton,Ethel Marie
Cas 1h12'53"75d14'
Walton,Hal
Aql 19h30'1"10d30'
Walton,Jack
Her 17h17'53"44d15'
Walton,Jaryd Niles
Aql 18h56'17"17d40'
Walton,Jason John
Her 16h53'21"29d9'
Walton,Jeffrey C
Aql 20h27'0"5d34'
Walton,Jennie
Cet 0h58'57"1d12'
Walton,Jess
Peg 22h15'42"30d0'
Walton,Jill R
Lyn 8h57'39"34d36'
Walton,Jo
Ori 5h46'1"19d59'
Walton,John Bernard
Cas 0h39'36"74d3'
Walton,Joseph Alan John
Aql 19h6'40"1d7'
Walton,Jr,Silvester
Uma 9h44'31"53d56'
Walton,Juliana Kathryn
Mon 7h53'16"-9d2'
Walton,Kimberly
Lyn 9h5'54"42d1'
Walton,L Dean
Lmi 10h30'19"38d18'
Walton,Marcelyn Ellen
Del 20h26'13"20d31'
Walton,Michael E
Cet 2h29'54"-11d7'
Walton,Michael E & Jeffrey C Walton
Her 18h6'31"18d55'
Walton,Mildred M
Cas 1h48'41"73d12'
Walton,Nicholas
Aur 5h0'21"45d33'
Walton,Pennie
Cyg 20h18'12"30d1'
Walton,Raymond
Boo 14h44'22"51d40'
Walton,Raymond H
Aql 19h8'45"1d25'
Walton,Sandra Elaine
Lyr 18h17'23"30d12'
Walton,Terry
Cru 12h44'14"-60d57'
Walton,Zachary J
Cnc 8h36'34"11d34'
Waltons Wonder
Boo 15h15'17"34d50'
Waltz,James
Cam 3h37'1"76d58'
Waltz,Marie Bertha Weber
Cas 0h38'17"64d53'
Walvoord,Erica
Del 20h37'35"10d60'
Walz,Andrea
Lyn 8h12'56"47d44'
Walz,Carrie Ann
Oph 18h17'50"10d56'
Wambach-Family-Star, The
Ser 16h6'43"8d44'
Wamper,Waldi
Dra 16h47'0"67d25'
Wamsteker,Kees
Mon 7h53'0"-1d31'
Wamsteker,Yvonne Marianne Louise
Eri 4h14'22"-17d6'
Wan Shan Chu
Cyg 19h52'53"38d2'
Wanada Sue
Cas 23h15'34"62d45'
Wanamaker,Duncan Seth
Cmi 7h44'46"4d41'
Wance,Donna Jean
Aur 7h1'36"40d14'
Wand,Milton
Per 2h7'0"57d18'
Wand,The
Crb 16h1'1"38d36'
Wand-Spurs Medal Old Cud
Lyn 7h53'36"51d30'
Wanda
Aur 5h17'38"41d40'
Wanda
Sex 10h12'52"-5d5'
Wanda Jean
Uma 8h39'37"67d59'
Wanda Maria
Cas 1h7'44"73d30'
Wanda's World
Ori 4h42'13"1d31'
WandaPeacemaker ListenerFriendFowler
Crb 16h2'23"33d17'
Wandas
Cet 2h33'0"-10d21'
Wandel,Karl & Ruth Drs med D7443
Uma 10h54'51"71d13'
Wandell,Lester E
Aql 19h6'34"-0d36'
Wander
Ori 5h59'58"20d28'
Wanderer,Bonnie
Ori 5h52'42"14d55'
Wandering Dingo
Ari 3h24'41"20d12'
Wanderlust-Scot,Mark, Rob,Jim
Lac 22h50'37"56d20'
Wandersee,Wolfgang
Aur 5h39'58"49d32'
Wandmaker Family Star, The
Uma 11h2'36"52d7'
Wandmaker,Helmut
Ori 5h51'46"21d50'
Wands,Richard Paul
Aur 5h18'33"45d54'
Wane,Samuel Oliver
Ori 6h5'23"6d41'
Wanessa
Cma 6h29'41"-20d22'
Wang Lian
Cas 0h7'29"60d35'
Wang,Charles B
Ori 5h25'15"-4d31'
Wang,J P
Uma 9h40'23"55d3'
Wang,Jordan
Lyn 6h39'25"59d23'
Wang,Lin Poo
Lyn 7h58'35"34d2'
Wang,Marco
Her 16h14'0"10d28'
Wang,Vanessa
Hya 9h23'3"-2d13'
Wang,Xiao Feng
Mon 6h24'47"8d43'
Wangart,Christian
Cyg 20h53'16"50d11'
Wangler,Black Bart, Brett,James
Aur 4h59'0"40d20'
Wanjsztok,Joseph
Del 20h15'0"10d54'
Wankier,Brent & Amron
Mon 6h15'18"-10d9'
Wankowski,Heather Marie
Lyr 19h20'24"38d10'
Wanlass,Leah Louise
Equ 20h57'45"5d40'
Wanless,Donna Louise
Umi 15h34'34"72d13'
Wanless,Frank A
Cyg 19h30'31"33d18'
Wannemacher,Stefan S
Boo 15h31'0"50d24'
WANNER Family Star,The
Uma 11h20'59"62d22'
Wannstedt Family
Aur 5h3'0"42d25'
Wantland,John & Barbara
Eri 4h14'3"-12d36'
Wappler,Patrick Carl
Cnc 8h59'16"20d33'
Wapskineh,Casey Joe
Uma 10h12'31"48d53'
Waqar & Lisa
Lyr 18h34'53"40d54'
War & Peace Forever- Armand & Louise
Uma 8h42'0"53d32'
Waraksa,Lynne
Cas 0h54'29"56d3'
Waran,Drs Sandy & Shantha
And 23h38'24"42d30'
Warbelow,Caitlin
Lyn 8h29'1"34d19'
Warburton,Jane Elizabeth
Lyr 18h16'47"32d44'
Warburton,Peter
Cep 21h14'36"70d34'
Warchal,Gladys
Lyr 18h25'25"37d56'
Warchol,Diane
Cmi 7h24'17"1d40'
Warchol,William Lee
Her 17h28'25"40d48'
Warchola,Jean
Ori 5h38'22"-1d22'
Warchola,Marty
Aur 6h8'17"31d42'
Warcholak,Sue
And 0h5'0"31d12'
Warcup,Gerald Wayne
Ser 18h17'44"-14d53'
Ward "Captain Magic", James
Hya 8h42'21"2d20'
Ward Family,The
Uma 9h48'24"44d19'
Ward II,Randy Joe
Cnc 8h51'39"31d43'
Ward Memorial Star,The Richard
Cyg 20h21'38"38d30'
Ward The Edge,Pete
Umi 14h17'53"69d19'
Ward's Gate
Lac 22h13'25"49d49'
Ward's Way Home
Vul 19h42'30"22d48'
Ward,"Bobbie Lynne" Hosfeldt
Aql 19h28'38"-6d55'
Ward,Ann & Mike
Boo 14h33'49"47d8'
Ward,Anna
And 23h43'13"43d17'
Ward,Arabella Fleur Helena
Gem 7h59'25"28d57'
Ward,Aubrey O
Boo 15h0'59"20d13'
Ward,Bob
Uma 8h55'18"56d27'
Ward,Brad Eugene
Ser 15h54'44"3d32'A
Ward,Bradford Jon
Cep 21h35'45"58d51'
Ward,Cherami Chantel
Leo 11h18'14"-5d11'
Ward,Christopher Bryan
Sgr 19h2'38"-26d43'
Ward,Connie Ann
Lac 22h22'11"38d50'
Ward,Daniel
Aur 6h25'58"38d11'
Ward,Darby Ashley
Her 17h12'0"28d31'
Ward,David Edward
Per 1h48'34"56d27'
Ward,Dawn Arnel
And 0h2'11"44d23'
Ward,Diane
Lyn 7h7'14"44d36'
Ward,Douglas James
Aur 7h1'34"36d36'
Ward,Edward
Her 17h31'59"20d11'
Ward,Elizabeth Jane
Cas 2h27'48"61d28'
Ward,Eric Lee
Ori 5h51'31"16d55'
Ward,Erika Ilda
Cam 11h54'55"77d31'
Ward,Frank
Her 16h50'45"38d32'
Ward,Frank A
Aql 18h59'36"11d46'
Ward,Frank Richard
Aql 20h3'29"6d39'
Ward,Gail
Lyr 18h56'0"40d45'
Ward,Gary
Leo 11h50'42"20d1'
Ward,Gerald Clark
Aur 5h24'51"30d41'
Ward,Gillian Louise
Dra 20h6'22"64d8'
Ward,Glenn R
Dra 20h12'23"71d10'
Ward,Gregory Artemas
Aql 18h56'1"16d13'
Ward,Griffin Berlin
Ser 15h54'44"3d32'B
Ward,Harry James
Ori 5h53'11"1d52'
Ward,Holly Mermaid
Aqr 23h8'53"-6d48'
Ward,Huston R
Mon 6h19'48"8d35'
Ward,Irma Herrmann
Cas 0h57'10"62d15'
Ward,Jack Kinsey
Cet 2h7'25"1d8'
Ward,Jacqueline
Ori 6h0'0"3d3'
Ward,James Timothy
Car 7h38'53"-56d37'
Ward,Jamie
Peg 23h31'40"18d28'
Ward,Janet
Dra 14h37'26"60d21'
Ward,Jarrett James
Sco 15h57'24"-21d6'
Ward,Jeff & Dina
Equ 21h21'48"12d27'
Ward,Jeffery
Crt 11h16'30"-8d8'
Ward,Jeremy Kalon
Aqr 23h37'49"-19d37'
Ward,Jerome
Her 16h48'32"32d57'
Ward,Jesse Beryl
Hya 8h57'17"2d51'
Ward,Jimmy Dale
Aur 6h17'29"37d8'
Ward,Joe/Jen Brower 1
Boo 14h23'55"28d14'
Ward,Joy S
Mon 6h19'26"8d45'
Ward,Jr,Paul Milton Albert
Boo 14h25'0"28d17'
Ward,Judy
Aql 18h58'12"17d41'
Ward,Kenny
Aur 6h26'16"31d46'
Ward,Kenny A
Crt 10h50'58"-12d27'
Ward,Laura Dorothy
Lyr 18h18'12"42d10'
Ward,Marcia
Aql 19h6'1"-05d45'
Ward,Maria Therese
And 0h51'25"35d49'
Ward,Marian Walker Gallahan
Lyr 18h29'53"47d4'
Ward,Marie Pamela
Cyg 20h18'32"39d54'
Ward,Mary
Cas 0h33'39"70d36'
Ward,Maureen & Peter
Cmi 7h34'21"1d5'
Ward,Michael R
Cma 6h51'10"-16d32'
Ward,Millie
Del 20h33'48"10d59'
Ward,Miss America 1982 Elizabeth
Lyn 6h59'27"58d24'
Ward,Natalie
Cyg 20h35'50"45d20'
Ward,OMI,Father Arthur F
Lyr 19h18'26"42d28'
Ward,Patricia
And 23h1'33"47d18'
Ward,Patricia Helen
Eri 3h34'37"-2d24'
Ward,Patricia Roberts
And 2h24'40"37d42'
Ward,Paul E.
Ori 6h0'0"-0d12'
Ward,Paw-Paw
Crb 16h14'60"28d54'
Ward,Peter
Per 2h55'36"35d13'
Ward,Rachel Colleen
Peg 21h9'11"18d45'
Ward,Randolph Allen
Ser 15h56'39"19d22'
Ward,Randy Joe
Tau 4h2'31"23d5'
Ward,Renee
Cas 0h56'49"60d59'
Ward,Robert Clellan
Aql 19h53'32"14d46'
Ward,Robert Wilson
Lyn 7h11'24"58d31'
Ward,Ronald Kent
Per 3h18'1"41d6'
Ward,S Churchill
Oph 17h24'11"-22d30'
Ward,Sally
Crb 16h10'34"38d26'
Ward,Sandy
Lyn 8h17'39"50d50'
Ward,Sara J
Lyr 18h28'42"38d38'
Ward,Shelly L
Cyg 20h27'0"42d54'
Ward,Steve
Cma 6h13'38"-13d34'
Ward,The William & Mary
Her 16h55'47"41d4'
Ward,Thomas Dean
Her 17h39'1"21d15'
Ward,Tinker's Crown Jewel Leslie
Peg 21h20'14"23d57'
Ward,Tom
Per 1h47'34"54d2'
Ward,Valerie
And 0h11'36"32d57'

Ward,William Lee Sex 9h50'51"5d57'
Ward-Byers Cyg 19h28'12"34d52'
Ward-Caldwell,Andre Peg 23h25'50"33d20'
Ward-My Quintessence Of Love Oph 17h6'40"-5d55'
Ward-Sanchez,Sandra L Cyg 19h59'35"30d37'
Warden,Christopher S Ser 15h10'59"24d28'
Warden,Gloria Jean Peg 21h28'60"22d46'
Warden,Robert Martin Sex 10h15'16"-1d23'
Warden,Stevie Dra 13h34'22"63d25'
Warden,Séan Michael Gem 7h28'32"33d55'
Warden,Zoe Lyr 18h34'27"29d54'
Wardet il Saharaa Psc 22h52'30"1d35'
Wardle 831,Jason Cep 21h17'13"63d51'
Wardle,Barbara Ann Aql 19h58'54"11d37'
Wardle,Chole Janine Umi 1624'55"72d38'
Wardle,Karen L Peg 21h19'43"20d23'
Wardle,Kym And 1h13'16"34d35'
Wardle,William David Per 1h54'10"52d57'
Wardleworth,Neil Cep 21h28'1"68d45'
Wardlow,Kathleen & Steven Lyr 18h34'1"44d50'
Wardrup,Elizabeth Mon 6h7'16"-4d33'
Wards,Craig E Oph 17h55'32"10d39'
Ware Family Star Uma 10h37'1"47d59'
Ware My Starry Eyes, William Aur 5h13'14"44d17'
Ware Star,Samuel Lac 22h28'0"50d17'
Ware's Star,Ruth Eri 4h8'0"-10d9'
Ware,Candyce Starre Hya 9h35'5"1d55'B
Ware,Chelsea Nichole Mon 6h41'11"1d14'
Ware,Clayton David Hya 9h35'5"1d55'A
Ware,Deborah Jayne Peg 23h0'57"10d40'
Ware,Denise Vir 11h50'44"3d53'
Ware,Kim Lyr 18h54'59"36d8'
Ware,Linda Dianne Peg 23h36'59"31d32'
Ware,Teresa Lynn Peg 21h50'48"34d57'
Ware,Terry(Queenie) Cas 23h17'39"60d56'
Ware,Tommye Sct 18h53'33"-7d4'
Warfel,John Bruce Sct 18h41'48"-8d48'
Warfield,Charlotte M And 0h21'52"30d13'
Warfield,Edward Aur 4h58'21"31d19'
Warfield,Rosalind Mon 6h38'41"7d48'
Warford,Garry Per 4h42'1"37d19'
Wargo Nature Center, Joseph E Aur 6h28'19"34d52'
Wargo,Ugene Michael Lac 22h47'14"54d44'
Warin,Daniel Uma 9h39'41"54d29'
Warin-hari Gem 7h23'36"35d17'
Waring,Barbara May Lyr 18h38'0"33d35'
Waring,Justin Kyle Sex 10h47'36"-5d20'
Waring,Randall M Eri 3h46'41"-0d10'
Waring,Ray Cam 3h55'22"62d26'
Warkentin,Victor Frank Per 2h1'34"50d29'
Warland,Marie- Antoinette Dra 14h44'0"64d28'
Warlick's Star Bright, Catherine Equ 21h8'41"10d53'
Warlick's Star Light, Caroline Peg 22h59'0"30d6'
Warlock-TWT Cnv 13h0'0"50d32'
Warlow,Phoebe Rose Ara 17h55'9"-52d34'
Warm Fuzzy Uma 11h10'0"31d30'
Warm Thing Cet 2h5'56"7d1'
Warman,Heather Cas 0h29'49"61d59'
Warmbier,Guntram Equ 20h56'35"7d50'
Warmbold Family,The Uma 10h54'50"61d9'
Warmington's Destiny, Drew P Ori 6h19'42"18d53'
Warmuth Ann E Lyn 7h38'1"44d40'
Warmuth,Daniel B Cma 6h25'14"-11d15'
Warmuth,Donald B Lyn 7h44'1"45d7'
Warmuth,Jennifer E Ari 1h48'54"14d22'
Warmuth,Tracey A Cmi 7h40'49"8d30'
Warne,Elizabeth A Mon 6h52'32"-5d53'
Warne,William E Her 17h1'1"46d30'
Warnecke,Chrisant T Uma 11h59'16"51d58'
Warnell,George Francis Her 16h51'37'40d58'
Warnell,Sally Lyr 18h36'40"26d8'
Warner's Star,Ryan Cep 22h16'0"55d50'
Warner,Allen Mark Cep 0h13'0"75d47'
Warner,Ann & Dudley Cyg 20h31'43"41d7'
Warner,Bethany Brooke Eri 2h56'44"-6d47'
Warner,Brittany Klynn And 2h25'41"42d38'
Warner,Cadence Margot Kaur And 1h39'45"37d10'
Warner,Carl & Jean Mon 6h23'1"-1d52'
Warner,Charlotte Cas 0h43'50"60d24'
Warner,Dennis Paul Aur 5h10'47"42d40'
Warner,Hans-Michael Ari 2h41'41"21d22'
Warner,In Honor Of Harriet & Jordan Sge 19h32'58"16d49'
Warner,Janet And 1h10'47"38d4'
Warner,Jr,"An Illusion",F James Her 17h56'32"28d12'
Warner,Jr,Benjamin Evans Her 16h58'43"26d38'
Warner,Julie E Equ 21h21'39"12d20'
Warner,Keegan Arthur Her 16h38'14"41d44'
Warner,Kipling Conrad Singh Crt 11h13'29"-14d12'
Warner,Kristine Ori 6h7'38"6d31'
Warner,Kyle Matthew Dra 17h49'34"60d58'
Warner,Lewis Cep 21h3'58"58d55'
Warner,Lex & Debbie Cyg 19h34'30"35d14'
Warner,Margaret Meroney Peg 0h2'42"30d34'
Warner,MD,Charles Aur 5h28'18"37d36'
Warner,Pamela Ann Uma 9h10'58"52d54'B
Warner,Peggy Jean And 0h19'51"31d51'
Warner,PhD,Barbara Mon 8h0'20"-8d4'
Warner,Philip Victor Aur 6h7'29"36d43'
Warner,Phillip Owen Uma 9h10'58"52d54'A
Warner,Ralph Dra 17h58'18"70d35'
Warner,Richard John Lyn 8h11'1"44d23'
Warner,Shane Michael Lac 22h27'30"54d22'
Warner,Susan Teresia Lmi 9h58'44"34d20'
Warner,Taylor Carolyn Sex 10h13'34"-1d18'
Warner,Taylor Marie Peg 21h58'42"23d59'
Warner,Thomas & Mary Kay Hocking Uma 9h54'12"52d20'
Warner,Thomas E Uma 9h57'12"55d25'
Warner,Tod Hatton Dra 15h10'38"63d32'
Warner,Torben Caroc Cep 22h43'1"65d60'
Warner,Victor Aur 5h49'25'50d0'
Warner,Zachary Mason Her 16h53'36"34d56'
Warnes,Robert Cep 21h20'37"71d13'
Warnet,Holly And 0h21'1"40d38'
Warnham,Jim Aql 19h6'43"1d51'
Warnick,Katarina Josefa And 23h2'58"51d38'
Warning,Glen Lac 22h38'54"56d42'
Warnke Aql 20h8'1"7d17'
Warnke,Vera Vul 19h20'58"26d51'
Warnken,Lorene Oph 18h25'48"8d39'
Warnock,Darlene Stewart And 0h50'1"21d37'
Warnock,Davia Uma 12h3'47"40d24'
Warns,Jr,Hugo John Dra 14h36'38"62d25'
Warny,Linda M Equ 21h21'0"2d17'
Waroquiez,Isabelle Dra 12h1'50"70d2'
Warr,Vincent Charles Lac 22h9'18"49d44'
Warren Aql 19h25'21"15d54'
Warren "Ange Gardien", Gary Lac 22h1'16"48d50'
Warren "Nip it" USA, Buzz Timboy Aql 19h29'48"10d54'
Warren & Debbie's Wedding Star Mon 6h18'25"5d3'
Warren Family Star, The Michael & Laura Cas 0h32'46"74d53'
Warren Man with Smiling Eyes Hya 9h3'37'6d26'
Warren Star Per 2h41'21"35d29'
Warren Star,The Carla Boo 14h17'27"35d37'
Warren's Wish Cam 4h16'32"69d54'
Warren's Wonder Aur 7h18'0"35d39'
Warren,Andrew James Per 2h59'40"43d37'
Warren,Angela Lyn 6h38'15"58d56'
Warren,Ashley Corwin Cet 3h19'0"0d35'
Warren,Benjamin Judson Cam 3h55'11"77d26'
Warren,Bill & Katie Sge 19h14'48"16d12'
Warren,Charles R Sct 18h44'11"-6d29'
Warren,Christopher Dowling Boo 14h29'40"38d21'
Warren,David Cep 21h46'20"70d1'
Warren,David Lee Cet 2h29'27"3d15'
Warren,Doris Miyoko Mon 6h55'13"-8d55'
Warren,Elton William Per 4h23'49"50d35'
Warren,Emilie Vaughn Cam 4h52'34"68d52'
Warren,Estelle Louise Zink Uma 8h54'12"49d58'
Warren,Floyd E Her 17h10'54"45d51'
Warren,Harry L. Lyr 18h25'53"37d58'
Warren,Hilda T Cyg 21h10'16"35d35'
Warren,Ian David Cyg 19h22'12"30d4'
Warren,Ileen Mon 7h45'29"-8d4'
Warren,Janet J Vul 19h22'60"25d57'
Warren,Jayme Dakota Aql 19h48'31"14d53'
Warren,Joseph Tremayne Boo 14h12'27"45d41'
Warren,Jr,Joseph John Cep 21h51'28"58d47'
Warren,Kayce Raye Ori 5h54'1"5d8'
Warren,Kyle C Aql 18h41'0"1d47'
Warren,Linda Anne And 2h32'25"45d47'
Warren,Linda Shelton Cas 1h6'31"64d33'
Warren,Lou B Aql 19h7'11"3d29'
Warren,Melissa And 2h13'53"39d48'
Warren,Mitch Her 17h30'54"21d39'
Warren,Mitchell R & Molly J Panko Cyg 21h25'56"40d15'
Warren,Paige Freeman Eri 4h6'13"-17d57'
Warren,Paul R Sge 19h59'22"16d35'
Warren,Pearl Cas 1h10'57"67d31'
Warren,Richard Michael Peter Aql 19h2'30"0d32'
Warren,Ron Her 18h18'54"12d23'
Warren,Samantha Jade Mon 6h44'52"10d50'
Warren,Spencer Charles Her 16h17'32"8h22'
Warren,Sr,Frederick Arthur Boo 14h54'13"48d41'
Warren,Sr,Kenneth Earl Cep 23h21'58"70d19'
Warren,Starla Lee Peg 22h31'20"29d52'
Warren,Steadman Cet 3h1'58"0d44'
Warren,Stewart Aql 19h28'49"13d21'
Warren,Susan Lucy Nathalie Aql 19h1'1"0d25'
Warren,Taylor Alexandra Barton Del 20h17'36"11d10'
Warren,Travis Colton Ser 18h5'26"-8d0'
Warren,Virginia Mon 6h37'26"-0d54'
Warren,William Zachary Aql 19h51'1"11d51'
Warren-Jones,Monica Aur 6h8'35"31d49'
Warren-Pace,Barbara Mon 7h18'7"-7d7'
Warrick,Emily Ann Aql 18h57'39"16d37'
Warrick,Florence A & Theodore F And 23h35'56"41d1'
Warrick,Michelle Lup 14h33'46"54d40'
Warring,Rees Boo 15h27'0"41d23'
Warrington,Chuck Cet 1h25'17"-6d29'
Warrior,Stanley Nathan Oph 16h21'0"2d59'
Warrvella Peg 22h25'59"33d45'
Warshavsky,Jordan Aur 6h52'16"35d30'
Warshaw,Aaron Nathan Ari 2h36'50"21d9'
Warshaw,Amelia Beatrice Peg 23h26'43"17d39'
Warshawer,Jenny B Cet 3h9'25"2d3'
Warszawski,Michael Sex 10h36'28"-1d44'
Warszawski-Petit Caramel,Eva Cas 0h33'1"75d15'
Wartel,Anne-Christian Alers (comédiens) Per 3h15'21"41d21'
Warter,Catherine Dra 18h38'53"70d14'
Warter,Rebecca Cnv 12h39'17"38d1'
Warther,Nicholas Lee Her 16h52'12"38d33'
Warwick 12-12-80,Gary & Cynthia Cyg 20h8'0"39d56'
Warwick Perry Vul 19h22'17"26d36'
Warwick,Elsa Lee And 0h43'52"30d20'
Warwick,Shawn Kevin Sex 10h8'29"5d36'
Warwick,Susan M Lyr 18h53'48"41d58'
Warwick,Tony Her 16h57'52"25d6'
Warwick-Ching,Fern Elizabeth Crb 16h22'59"34d55'
Waryn,Courynn Kole Cas 2h49'17"61d35'
Warzybok,Kyle Anthony Uma 11h8'0"40d7'
Wasco,August & Mary Uma 9h40'58"51d24'
Wash,Brian Palmore Cet 2h29'21"8d50'
Washack,Amy Ari 1h59'26"18d54'
Washam,Worth Tri 1h47'20"28d50'
Washburn,Sarah Elizabeth And 1h53'48"41d4'
Washburne,Matthew Courtenay Her 17h32'56"21d22'
Washington,Anna Marie And 0h49'36"35d19'
Washington,Dianne M Cet 0h56'15"-6d45'
Washington,Garrett E Aur 6h57'55"43d43'
Washington,Leonard J Cyg 21h54'49"37d26'
Washington,Marilyn & Donald Cyg 19h47'53"30d23'
Washington,Melissa Jean Peg 22h26'42"27d38'
Washington,Monica Aql 19h56'34"10d38'
Washington,Pam Eri 3h54'52"-6d50'
Washington,Tahvia Lyr 18h27'18"43d50'
Washkaviak,Paul Boo 14h29'0"38d1'
Washo,Tracy Anne Ori 6h5'22"8d34'
Wasilesky,Michael Henry Boo 14h28'44"47d31'
Wasilewski,Leona Cas 0h14'38"62d18'
Wasilewski,Wojciech Aql 20h8'14"8d49'
Wasinski,Marie H And 0h51'13"36d27'
Waska,Erika, Irattenbach 172 Sgr 19h18'50"-24d31'
Waskiewicz,Katherine Mon 7h42'1"-1d51'
Waskiewicz,Ryan James Dra 17h4'1"67d12'
Wasko,Michael Edward Peg 21h56'27"18d58'
Waslin,David J Aur 5h55'30"30d47'
Wasmus,Gail L Umi 16h18'0"71d59'
Wass Boo 15h0'55"41d19'
Wass,Lois Aql 19h55'1"7d32'
Wasser,Stephen John Her 17h33'55"40d21'
Wasserman 1/10/86, Shira Rachel Mon 6h23'51"4d19'
Wasserman,Amy Robin Aur 5h26'51"38d7'
Wasserman,Laura Jean Tau 5h27'0"18d53'
Wasserman,Leonard S Sgr 18h57'49"-23d21'
Wasserman,Sherri Scl 23h24'13"-25d52'
Wasserman,Timothy D Cet 1h27'26"-2d30'
Wassermann,Mary P Uma 10h31'41"47d34'
Wasserstein,Rebecca Danielle Com 12h43'14"20d41'
Wassi Cmi 7h39'58"5d10'
Wassmann,Frank Aur 5h1'14"29d22'
Wassmer,Janet Cam 3h53'34"71d11'
Wasson In The Sky With Diamonds,Jim Per 4h26'0"50d16'
Wasson,Bradley K Cnc 8h9'57"30d17'
Wasson,Chris Peg 23h24'39"12d28'
Wasson,Dianna Crb 15h49'56"27d56'
Wasson,Elizabeth "Biggie" And 1h52'47"41d10'
Wasson,Jr,Bradley Keith Hya 8h59'55"3d46'
Wasson,William John Aql 19h48'22"11d10'
Wassong,Roy Leo 11h7'44"23d56'
Wassweiler,Vera & Gary Ori 5h44'54"10d57'
Wasta,Stacey Ann Lyr 19h6'53"38d30'
Wastman,Serena Cas 1h23'0"72d18'
Waston (He Flies The Big Ones),Doc Aql 19h24'33"8d39'
Watabe Family, Stella of Lib 14h29'44"-12d47'
Wataker,Heidi & Odd Crb 16h7'18"31d45'
Watanabe,Bennett Carlton Ori 6h17'35"18d55'
Watanabe,Nancy "Shooting Star" Sgr 18h52'42"-33d19'
Watari,David Barrett Boo 14h23'39"18d23'
Watashi No Chiisai Tenshi Aur 6h34'48"37d47'
Watcher's Rise Umi 14h17'1"65d39'
Watching Rachel Boo 14h44'0"50d9'
Watchmaker,Robert James Ori 6h6'38"-2d32'
Watchman,Jr,Leo C Aur 5h22'1"40d14'
Water Aur 5h4'20"42d57'
Water Child X21 Boo 15h8'44"38d27'
Waterland,Robert Arthur Dra 20d2'16"60d44'
Waterloo Elementary School Uma 11h56'17"42d13'
Waterman (Jovian at Heart),William Lac 22h55'34"53d14'
Waterman,Anne & Charles Uma 14h2'17"60d57'
Waterman,Charles Alexander Per 4h35'1"37d36'
Waterman,David Ari 2h57'25"30d53'
Waterman,Olivia Marion Elyzabeth And 23h17'32"51d46'
Waterman,Patricia Peg 0h0'0"18d44'
Waterman,Roxanne Cet 1h46'25"-2d24'
Waters Family Umi 15h40'19"76d7'
Waters,Bobby Her 17h39'26"14d58'
Waters,Brian R Her 18h3'49"38d58'
Waters,Buck Boo 13h53'24"17d17'
Waters,Christopher Michael Cnc 8h54'0"31d33'
Waters,David Podbros Dra 16h28'35"61d28'
Waters,Deborah Uma 10h15'40"61d49'
Waters,Doris Blanche Com 12h24'34"30d37'
Waters,Emily And 0h15'22"37d30'
Waters,Esq,William L Her 17h28'50"28d15'
Waters,Gavin Nicholas Umi 14h35'1"65d44'
Waters,Graham & Carr Her 17h13'24"48d28'
Waters,Heater And 23h27'14"44d57'
Waters,Jack & Doris Cyg 20h15'0"41d38'
Waters,James Merlin Per 1h42'36"53d49'
Waters,Lee A Sex 9h52'18"2d24'
Waters,Michelle Cas 0h25'43"61d27'
Waters,Nancy Ossink Aql 20h4'24"0d37'
Waters,Rob Lac 22h11'38"37d41'
Waters,Ronya Crt 11h7'40"-14d19'
Waters,Shane David Uma 11h6'1"32d7'
Waters,Thomas Patrick Cam 6h32'37"80d14'
Waterworth,Vicki Sct 18h44'44"-5d27'
Watha,Hia Nancy Lyn 8h22'12"58d12'
Wathen,James Leo Dra 18h21'10"48d46'
Wathen,Joseph Page Aur 6h31'28"31d30'
Wathen,Jr,Charles Anderson Cet 3h9'51"0d50'
Wathen,Leo Leo 9h23'40"17d48'
Wathen,Michael J Lac 22h10'0"47d42'
Wathue,Helen Cas 1h35'2"64d36'
Watkin,Suzanne Frances Tau 4h4'54"16d39'
Watkin,Suzanne Frances Gem 7h35'27"27d47'
Watkins II,Mark Joseph Uma 11h53'0"31d34'
Watkins Our Star Is Born,Kelsey Reed And 23h16'40"40d33'
Watkins,Aron Taylor Cet 2h46'36"-0d2'
Watkins,Darren "D-Dog" Ser 16h1'55"9d46'
Watkins,George H Per 1h53'50"54d14'
Watkins,Gregory Curtis Cap 21h1'54"-26d31'
Watkins,Jessica Lynne Uma 11h29'53"49d25'
Watkins,Jim Sex 9h57'1"-0d30'
Watkins,John Sex 9h51'1"1d46'
Watkins,Joseph D Uma 11h51'16"40d37'
Watkins,Julia Storm Cyg 21h6'0"30d11'
Watkins,Karen Aql 20h10'43"10d34'
Watkins,Katie Harriet And 2h22'39"40d37'
Watkins,Kay Cam 11h28'42"81d8'
Watkins,Kelly James Her 16h49'48"50d39'
Watkins,Kelsee Lynn Cam 5h21'1"68d23'
Watkins,Kingsley Marie Uma 11h53'12"38d37'
Watkins,Linda Eri 3h54'49"-5d36'
Watkins,Meghan Nicole Cyg 21h35'21"41d30'
Watkins,Mitchell Jay Cet 2h33'22"-6d58'
Watkins,Olivia Moira Vul 19h44'54"22d46'
Watkins,Patricia Naomi Lac 22h46'23"56d20'
Watkins,Rebecca Lyn 7h22'42"44d5'
Watkins,Rhita M Oph 17h12'21"23d21'
Watkins,Richard Her 16h53'41"38d40'
Watkins,Stephen Boo 15h2'46"24d5'
Watkins,Todd Lac 22h54'55"55d59'
Watkins,Tom Cep 21h48'18"68d32'
Watling,Sr,William John Oph 16h59'34"8d31'A
Watlock,Helen & Steve Cyg 19h24'49"33d19'
Watring,Vera Peg 21h50'13"28d58'
Watrous,Bridget Lyn Per 3h53'52"40d25'
Watrous,Jill Aql 19h17'13"15d38'
Watschke,Nelson Lac 22h10'45"51d21'
Watson II,William A Aur 5h3'35"38d13'
Watson III,Raymond Alexander Ser 15h24'22"22d16'
Watson IV,Kenneth E Dra 16h25'34"69d25'
Watson Outshines Them All,Thomas J Equ 21h7'0"11d53'
Watson September 24 1994,Todd & Jill Lyn 7h0'50"60d2'
Watson Star,The Robert Lac 22h14'39"46d47'
Watson's European Lyr 18h18'41"42d15'
Watson,Anne Marie Cas 0h6'1"61d5'
Watson,Anthony Richard Ori 5h57'12"14d57'
Watson,Bradley Craig Oph 17h17'45"-22d18'
Watson,Bruce E Her 18h6'31"37d31'
Watson,Carol Aql 19h20'26"10d31'
Watson,Christopher James Lib 15h43'54"-23d8'
Watson,Dave Dra 20h9'56"76d26'
Watson,David Kutler Dra 11h46'54"72d21'
Watson,David Lin Vir 11h41'1"2d2'

Watson,Dorothea Kelly "Hully"
Cet 1h26'1"-13d15'
Watson,Elery Winston
Ori 5h52'49"14d44'
Watson,Garvey Ashhurst
Cep 21h34'30"55d17'
Watson,Gene & Sharon
Del 20h23'51"18d52'
Watson,George E
Cet 2h19'59"4d57'
Watson,Gordon Nicholas
Her 17h17'20"44d18'
Watson,Grant James
Ori 5h54'30"18d55'
Watson,Herbert L
Hya 8h59'50"6d20'
Watson,Jack Stuart
Umi 15h32'36"75d28'
Watson,James "Jimmy" D
Uma 8h40'0"71d19'
Watson,James W
Cep 23h24'44"68d33'
Watson,Jen Yarroll
Cyg 19h24'49"35d3'
Watson,Joan Sutton
Cas 0h28'60"62d3'
Watson,Joann & Bill
Aur 5h2'1"41d10'
Watson,John
Boo 15h39'1"41d52'
Watson,John Madison
Cet 1h5'45"-5d51'
Watson,John P
Uma 9h25'30"47d30'
Watson,Jr,Al (Babe)
Oph 17h22'34"-20d35'
Watson,Kathleen
Uma 10h27'5"51d15'B
Watson,Kelsey Taylor
And 2h33'1"48d59'
Watson,Kendall Van
Oph 16h53'37"-26d14'
Watson,Kyle Orris
Cet 2h57'37"0d22'
Watson,Lesley Trudy
Lyr 18h59'45"36d3'
Watson,Linda
Mon 6h58'52"10d24'
Watson,Louise
Cas 0h3'48"63d53'
Watson,Louise Ann
Cyg 19h57'23"58d19'
Watson,Madison René
Mon 6h47'30"10d44'
Watson,Mark
Lac 22h23'50"38d41'
Watson,Martha Elsey
Aql 19h11'25"10d7'
Watson,Martyn David Taylor
Cep 0h9'25"66d42'
Watson,Maureen & Roy
Cyg 21h15'16"37d49'
Watson,Michael Doc
Oph 18h38'21"6d52'
Watson,Michael Wayne
Hya 8h14'33"0d44'
Watson,Myrtle Millor
Uma 9h36'54"52d28'
Watson,Nicholas
Lyr 18h32'0"31d53'
Watson,Patricia & Mark
McDonald
Cyg 19h40'27"41d35'
Watson,Rebecca Michelle
Peg 21h55'50"28d32'
Watson,Richard A
Cet 0h32'1"0d28'
Watson,Robin Danielle
Sco 16h25'39"-40d6'
Watson,Sag to Sag, KK
Sgr 19h41'5"-41d14'
Watson,Sarah Lee
Boo 14h8'27"53d4'
Watson,Shannon Elizabeth
Hall
Aql 19h33'14"-0d26'
Watson,Shayna Lynn
Mon 6h37'34"6d7'
Watson,Sonny
Aql 20h17'12"5d1'
Watson,Stanley J
Gem 7h24'26"31d41'
Watson,Steve D
Boo 15h2'28"28d35'
Watson,Susan Patricia
Cas 1h51'22"75d36'
Watson,Tom
Her 16h43'0"34d39'
Watson,Tony
Cyg 21h18'26"39d57'
Watson,Tracey
Lyr 18h31'14"31d26'
Watson,Tracy Martin
Cas 0h32'47"54d39'
Watson,William Keith
Sct 18h38'50"-6d26'
Watt OBE,Brigadier Reddy
Cnv 13h35'29"46d34'
Watt,Alan
Aur 5h15'0"46d38'
Watt,Ann
Cyg 21h30'19"42d9'
Watt,Goddess-Dianne H
Cas 1h6'36"70d53'
Watt,Jason
Cep 20h36'26"75d21'
Watt,Kirsty M H
And 0h21'13"45d51'
Watt,Mimi/Tom Skilling
Cyg 21h17'20"28d45'
Watt,Rachel Frances
Cas 0h5'0"62d10'
Watt,Shawn & Joey Lynn
Cyg 21h53'29"38d31'
Watt,Shirley
Hya 9h8'12"3d13'
Wattenburg,Kyle Hudson
Ser 15h53'20"0d51'
Watter'Star Aries Four
Ari 2h1'14"12d13'
Watters,Edwin C
Vul 19h44'1"25d46'
Watters,Harmon & Irene
Lyn 7h57'25"45d21'
Watters,Michelle
Aql 18h59'17"15d38'
Watters,Theresa
Uma 11h8'55"32d29'
Watters,Vanessa
Tau 4h56'32"16d6'
Watterson,Kent
Ser 18h0'21"-14d58'
Watton,Curtis Lee Christopher
Boo 14h58'0"17d49'
Watton,Serena Julianna
Vul 19h45'0"28d15'
Watts,André
Gem 6h49'16"13d34'
Watts,Ashley Marie
Mon 6dh37'26"-0d4'
Watts,Brook
Sco 17h4'1"-31d14'
Watts,Darlene J
Aql 19h31'0"10d48'
Watts,David Charles
Per 2h57'0"31d41'
Watts,Deborah L
And 0h9'48"27d52'
Watts,Gruffudd William
Cep 23h7'15"60d50'
Watts,Jan Maree
Cen 11h50'23"46d17'
Watts,Jeffrey Arthur
Per 1h41'1"52d47'
Watts,Jesse Curtis Roy
Lyn 8h7'0"52d8'
Watts,John
Lyn 8h4'46"51d3'
Watts,Kathy M D
Lyn 8h9'46"36d17'
Watts,Marilou Senen
Lyr 18h27'25'46d20'
Watts,Melissa
Cam 6h58'60"67d57'
Watts,Michael J
Her 18h27'56"20d1'
Watts,Michelle
Del 20h52'57"4d18'
Watts,Michelle Elizabeth
Lyn 8h10'31"36d22'
Watts,Nancy Elizabeth
Sex 9h50'0"3d19'
Watts,Nathan Richard Resnick
Peg 21h25'21"2d57'
Watts,Nicole
Eri 3h31'0"-2d7'
Watts,Richard
Aur 6h4'15"50d36'
Watts,Rick & Julia
Lyn 7h4'0"58d57'
Watts,Robert Michael
Dra 15h54'41"52d20'
Watts,Robert Michael
Cam 4h19'26"68d26'
Watts,Rolonda
Cas 1h1'54"55d26'
Watts,Terry & Patsy
Uma 11h22'1"68d31'
Watts,Tyler Ryan
Ser 15h52'30"0d7'
Watzl,Mother & Guiding
Light,Fern
Com 12h24'35"30d28'
Watzlowik,Karl
Cam 6h52'56"70d50'
Waudby,Keith
Mon 6h47'20"10d32'
Waugh,Donna C
Boo 15h3'50"32d12'
Waullauer,William
Her 16h51'22"50d56'
Wauri,Sybille
Tau 3h56'54"11d20'
Wave,The
Per 2h2'21"50d15'
Waverly
Mon 6h26'52"8d36'
Wawrousek,In Memory Of
Joanne
Peg 21h59'46"20d34'
Wawryk,James David
Per 2h52'35"46d15'
Wawrzyniak,Michelle
And 0h58'42"39d11'
Wawzin,Ilse
Peg 23h30'35"11d2'
Wax,Roslyn
Uma 9h40'35"56d52'
Waxenberg,Alan
Dra 17h27'45"68d58'
Waxler,Phyllis N
Mon 6h58'33"-8d8'
Waxman,Marvin & Barbara
Cyg 20h45'30"38d25'
Way,Amanda Lucien
Cas 1h49'54"58d33'
Way,Ervin
Umi 10h26'15"41d36'
Way,Helen Ann
Cet 1h48'47"-1d26'
Way,Kevin Frank
Her 16h53'1"39d22'
Way,Madison Robin
Vir 13h29'44"-4d52'
Way,Rebecca
Cas 0h20'36"62d6'
Way,Robert
Aur 5h2'55"31d36'
Waybill,John"Fee"
Aql 19h6'52"3d13'
Wayble Star, The Reneè
Vul 19h43'34"22d34'
Waycott
Cmi 7h34'1"1d12'
Waycott,Amber & Brian Davis
And 23h35'54"42d44'
Wayland,James Robert
Cep 23h7'37"64d26'
Waylands Wonder
Aql 20h0'44"11d24'
Waymire,Heather
Mon 7h54'20"-7d19'
Waymire,Scott
Lac 21h58'14"42d43
Waymost,Matthew David
Her 17h38'39"22d44'
Wayne
Boo 14h27'1"41d5'
Wayne & Carole-25th
Uma 11h57'1"32d10'
Wayne & Cris
Mon 7h2'14"1d20'
Wayne & Pamela's Star
Crb 15h20'28"32d1'
Wayne & Phyllis
Cyg 21h52'40"40d6'
Wayne & Susan-Infinity &
Beyond
Sge 19h55'22"16d23'
Wayne & Tracie
Sge 20h5'31"20d14'
Wayne Andre
Lac 22h32'39"37d36'
Wayne's Runt,Barbara Seago
Ori 5h56'51"41d26'
Wayne's Star
Oph 18h18'15"8d55'
Wayne's Stern 143
Del 20h30'59"10d29'
Wayne's Wish
Aql 18h44'29"6d36'
Wayne's World
Cnc 9h3'3"15d43'
Wayne's World
Ori 6h2'0"1d5'
Wayne,Brandon"Guardian
Angel"
Her 15h52'0"46d12'
Wayne,C David
Dra 14h51'1"62d44'
Wayne,Christina
Cyg 21h8'47"31d18'
Wayne,Dede
Cep 21h41'59"58d15'
Wayne,Eric
Dra 13h58'45"63d10'
Wayne,Jackie
Cnc 8h52'1"30d48'
Wayne,Jerry
Tri 1h56'50"26d47'
Wayne,Judith
Cas 0h8'30"58d20'
Wayne,Nicholas
Ser 15h13'0"9d47'
Wayne,Palmer & Gary Jaffe
Tri 2h23'25"30d24'
Wayne,Peter
Aql 20h0'45"4d48'
Wayne,Tyler Scott
Gem 6h48'57"12d7'
Wayne,Yvonne,& Flea
Ori 5h31'31"-8d15'
Wayne-007-Rooke,1955,
Canada 40
Lmi 10h50'27"31d18'
Wayne-Garrrr
Cam 6h32'31"78d55'
Wayner,Scott E
Lac 22h24'13"40d30'
Wayson,Doris Phipps
Peg 22h43'37"31d17'
Waz
Mon 6h35'13"1d15'
Wazek,Andreas
Cmi 7h20'49"8d52'
WB 0115
Mon 6h21'28"-5d49'
WD 40
Lmi 10h27'26"32d1'
We Are
Uma 9h33'31"51d59'
We Love Grandmerm
Mon 7h28'31"-1d11'
We Share
Ori 6h9'36"9d6'
We Three
Cam 5h10'17"68d26'
We're
Cap 20h52'18"-16d19'
We're A Match Made In
Heaven
Lyr 19h15'29"41d11'
Weagant
Sge 19h35'16"16d33'
Weakley,Fred H
Cep 22h47'0"67d43'
Wean,David
Sct 18h41'9"-6d39'
Wear,Chad
Aur 6h34'17"33d33'
Wearn,Marion & Ernie
Lyr 19h21'14"40d51'
Weatherbee,Lawrence
Aur 6h24'1"31d3'
Weatherby,Nicki Lee-Anne
Lyr 18h19'14"42d46'
Weatherford,Don & Sally
Hya 8h59'52"3d10'
Weatherhead's Illusion
Boo 15h2'1"48d1'
Weatherley,Jason
Ori 6h5'38"5d8'
Weatherred,Duane
Ori 5h57'51"18d33'
Weathers,Angela
Cet 3h19'46"0d14'
Weathers,Cody Jevon
Aur 5h16'40"40d30'
Weathers,Deborah G
Cam 8h8'37"81d14'
Weathers,Jeffery Alan
Oph 17h53'14"13d46'
Weatherspoon,Scott Braden
Gem 6h58'10"18d6'
Weaver
Cep 22h45'1"59d16'
Weaver
Uma 10h59'19"43d59'A
Weaver Family
Lac 22h0'14"40d37'
Weaver's Star,JR (BOB)
Boo 13h54'51"21d11'
Weaver,Allison Christine
Peg 21h58'35"26d49'
Weaver,Alyssa Lane
Lyn 8h7'32"52d10'
Weaver,Annabella
Gem 6h50'34"30d57'
Weaver,Art
Lac 22h35'36"38d21'
Weaver,Bob MB
Her 16h19'14"10d51'
Weaver,Bradley Justin
Lac 22h48'1"56d49'
Weaver,Brian Richard
Boo 15h8'23"41d32'
Weaver,Charles E
Oph 18h7'27"13d13'
Weaver,Charles P.
Her 18h9'14"40d30'
Weaver,Colton James
Cep 5h57'39"85d37'
Weaver,Dana
Dra 17h58'1"60d33'
Weaver,David
Boo 14h30'34"48d28'
Weaver,Diana P
Cam 5h40'45"61d12'
Weaver,Emily Christine
Cas 2h27'0"59d34'
Weaver,Ethel
Lyr 18h37'47"41d57'
Weaver,Forever Carl & Kelly
Cet 2h15'27"4d7'
Weaver,Gary
Aur 6h8'60"45d48'
Weaver,James Thomas
Boo 13h51'40"17d39'
Weaver,Jr,MD Harry
Lac 22h28'18"50d28'
Weaver,Justin Scott
Dra 17h46'41"63d54'
Weaver,Katharine Cory
Sco 17h3'52"-33d49'
Weaver,Kelly
Mon 6h22'30"4d19'
Weaver,Kelly Lynne
And 0h5'58"31d20'
Weaver,Kenneth & Marian
Uma 12h5'23"50d21'
Weaver,Kevin
Ori 5h19'49"1d19'
Weaver,Kevin Robert
Dra 16h4'44"65d0'
Weaver,Kristen
Cas 1h40'12"58d13'
Weaver,Kyle R
Hya 8h36'49"1d41'
Weaver,Laura M
Cyg 19h29'47"31d54'
Weaver,Lori
Lyr 19h16'44"40d23'
Weaver,Loy Trudy
Lac 22h55'42"56d55'
Weaver,Lynn & Art
Gem 6h48'34"30d50'
Weaver,Matthew
Boo 15h2'52"23d50'
Weaver,Mitchell J
Ser 16h3'12"-0d33'
Weaver,Nancy
Crb 15h49'14"30d4'
Weaver,Paul
Ori 5h43'50"10d27'
Weaver,Phillip
Her 18h44'18"12d50'
Weaver,Richard Mellinger
Per 2h26'0"54d37'
Weaver,Ryan Owen
Cep 0h17'38"70d38'
Weaver,Shula Twinkle
Lyr 18h29'46"31d1'
Weaver,Thomas
Cep 22h2'0"60d55'
Weaver,Timothy Paul
Peg 22h23'14"30d51'
Weaver,William & Susan
Ori 5h51'54"16d26'
Weavil,Nyppsi
Per 2h54'46"31d28'
Weavil,Roman
Per 2h53'31"35d56'
Webb "Aroo",Paul Maurice
Aql 19h29'53"12d14'
Webb Biard
Cet 3h14'53"0d24'
Webb DC,CCSP,Khelly Susan
Jeanne
Cyg 21h48'0"36d34'
Webb"The Bear",Harold
Eugene
Aur 6h1'39"37d55'
Webb's Star,Julia
Cyg 20h59'13"28d46'
Webb's Star,V
Aql 20h1'46"14d60'
Webb,Alexandra Leigh
And 0h3'23"47d35"
Webb,Austin Edwards
Lyr 19h12'48"38d56'
Webb,Barrie A
Psc 1h36'44"21d47'
Webb,Betty Shepherd
Uma 11h24'49"56d55'
Webb,Bob
Sgr 19h38'10"32d19'
Webb,Carlyle S (Chic) &
Doris A (Pat)
Crb 15h20'46"30d58'
Webb,Clair Louise
And 23h1'34'51d18'
Webb,Cleve
Del 20h27'26"20d32'
Webb,Colin
Her 17h55'26"14d19'
Webb,Colton Joshua
Ari 1h55'0"16d49'
Webb,Craig Lee & Camile
Ori 5h48'54"19d6'
Webb,Deryck Nicholas Hamer
Dra 9h50'13"81d15'
Webb,E J
Hya 9h12'31"1d15'
Webb,Eavalyn Anne
Sgr 19h33'44"-44d23'
Webb,Frank Junior
Ori 5h54'33"8d46'
Webb,Grant Liam
Gem 7h4'45"21d3'
Webb,Helen
Equ 21h3'39"3d35'
Webb,Ian
Cep 21h41'22"78d45'
Webb,James
Aql 20h16'53"5d17'
Webb,Jason Gabriel
Aql 20h32'25"0d34'
Webb,Jean
And 1h36'57"40d43'
Webb,Jeff
Cet 2h10'22"1d58'
Webb,Joan L
Lyn 8h40'17"36d28'
Webb,John Michael
Cep 0h25'1"77d49'
Webb,Jordan Alexandra
Ori 5h48'27"21d45'
Webb,Jr,Dayton D
Cep 21h43'20"58d57'
Webb,Ken
Eri 4h41'53"-21d11A'
Webb,Lessie M
Cam 3h14'40"61d34'
Webb,Marcia
And 0h7'35"40d27'
Webb,Maude Lee
Com 12h16'59"28d5'
Webb,Michael "Scooby"
Her 17h28'0"31d52'
Webb,Michael C
Aql 19h53'23"15d20'
Webb,Michael John
Her 17h52'31"28d28'
Webb,Mike
Del 20h15'46"10d13'
Webb,Myrtice
Peg 22h16'40"20d17'
Webb,Neil Andrew
Her 17h55'31'14d33'
Webb,Noah Wolf
Aur 6h14'27"46d49'
Webb,Reid & Karla
Crt 10h53'38"-11d8'
Webb,Ricky
Aql 19h20'0"13d43'
Webb,Robert Allan
Aur 5h15'1"54d51'
Webb,Roger John
Uma 12h9'24"47d31'
Webb,Ruth W
Del 20h16'44"10d51'
Webb,Sarah Kristine
Peg 22h41'42"21d35'
Webb,Shelly & Terry
Crb 15h15'0"31d30'
Webb,Tatrene
Cas 0h14'1'63d9'
Webb,The Andrea
Cma 7h15'0"-16d22'
Webb-Carstensen,K B
Sge 20h7'0"20d5'
Webb-Peploe,Hamilton
Peg 23h6'27"13d16'
Webbe,Robert Bader
Dra 19h3'0"70d47'
Webber,Alexandra Lauren
Eri 2h56'14"-11d51'
Webber,Charlene (mama)
Cas 2h56'1"58d17'
Webber,Courtney Elizabeth
Sct 18h54'54"-7d4'
Webber,Dan
Oph 17h3'57"-23d25'
Webber,Devon
Sct 18h41'40"-7d8'
Webber,Jean M
Com 12h56'41"26d22'
Webber,Jeanette
Gem 7h21'55"20d9'
Webber,Jerrel Brookes
Aql 19h7'31"0d56'
Webber,Kennith Scott
Lac 22h22'46"38d52'
Webber,Randall Phillip
Dra 16h19'47"67d40'
Webbstar
Her 18h42'1"13d6'
Weber "CowboyStar", Daniel
Lee
Per 3h5'33"40d45'
Weber Birth Star,Lucas
William Gunnar
Her 17h29'21"21d29'
Weber Family,The
Uma 13h45'31"60d59'
Weber Insley
Cam 3h41'43"61d50'
Weber My Best Friend
Forever,Thomas Claus
Hya 9h14'56"3d23'
Weber Star,The Russ & Lois
Vul 20h16'13"23d6'
Weber,Alfred
Aur 6h15'50"46d32'
Weber,Amanda Victoria
And 2h0'58"41d21'
Weber,Anita
Sge 20h2'55"16d17'
Weber,Bernd
Peg 23h41'16"18d59'
Weber,Bonnie Jean
Mon 6h25'14"7d40'
Weber,Brian Keith
Aur 6h34'26"52d49'
Weber,Bruce
Aql 18h43'49"6d36'
Weber,Carol Susan
Lac 22h26'19"38d5'
Weber,Cathy
Eri 4h13'55"-10d56'
Weber,Chelsea
Lyr 18h58'0"30d46'
Weber,Christian
Cyg 20h21'52"39d32'
Weber,Christiane
Uma 10h20'21"60d20'
Weber,Courtney L
And 23h30'1"42d58'
Weber,Cynthia Kathleen "Tara"
Eri 2h47'23"-18d39'
Weber,Dale
Per 2h2'45"50d36'
Weber,David
Boo 15h32'36"42d15'
Weber,David Edmund
Her 16h43'39"22d38'
Weber,Edgar Edelweiss
Ari 1h45'33"11d12'
Weber,Elsbeth und Klaus
Her 17h6'46"43d0'
Weber,Eric A
Dra 16h39'39"69d51'
Weber,Gabriele
Boo 14h38'47"13d4'
Weber,Garry
Sct 18h33'0"-6d36'
Weber,Garry
Ori 5h58'31"16d23'
Weber,George
Gem 7h52'41"32d18'
Weber,Hans-Joachim
Oph 18h29'29"7d57'
Weber,Hermann
And 23h0'27"42d10'
Weber,Jackson Charles
Dra 16h56'34"63d34'
Weber,Jacob William
Cep 20h53'1"61d2'
Weber,Janet M
Lyr 19h19'13"35d21'
Weber,Jean Marie Reppert
And 23h43'25"38d28'
Weber,John Arthur
Cep 22h25'42"58d53'
Weber,John III & Muriel Mary
Nevins
Crb 15h57'51"26d16'
Weber,John P
Cmi 7h17'22"8d8'
Weber,Jonny
Mon 7h53'41"-2d56'
Weber,Jr,Ben R
Hya 8h55'1"0d8'
Weber,Jürgen 29/4/1960
Tau 5h31'28"20d3'
Weber,Kathleen
And 23h36'28"48d18'
Weber,Laetitia
Mon 7h54'37"-4d29'
Weber,Lewis John
Uma 12h15'39"56d41'
Weber,Lisa M
Psc 1h43'25"22d11'
Weber,Lynch "Butler"
Aql 19h55'51"12d12'
Weber,Maggie
Peg 0h12'20"22d0'A
Weber,Major
Cma 6h51'52"-15d26'
Weber,Maryann
Mon 7h55'35"-6d46'
Weber,Matthew Lawrence
Aur 6h6'11"54d51'
Weber,Megan Amanda
Sge 19h52'41"18d47'
Weber,Meghan Ross
And 22h56'50"51d6'
Weber,Michael
Cet 0h57'13"-11d47'
Weber,Michael David
Mon 6h52'40"0d32'
Weber,Michael J
Ori 5h3'45"15d17'
Weber,Michael John
Oph 17h51'53"13d55'
Weber,Mildred E
Del 20h22'0"10d39'
Weber,Morris
Boo 15h4'54"16d27'
Weber,Paul D
Lmi 9h57'1"34d46'
Weber,Peter
Lyn 9h1'1"40d4'
Weber,Ray
Aur 6h35'17"31d48'
Weber,Richard J
Lac 22h20'18"54d49'
Weber,Richard P
Boo 15h7'17"11d16'
Weber,Ronald
Cma 6h55'30"-19d45'
Weber,Ryan
Her 15h50'34"41d17'
Weber,Scott & Denise
Mon 8h6'53"-3d34'
Weber,Sharon
Cam 9h10'16"80d43'
Weber,Skip
Peg 0h12'20"22d0'B
Weber,Sr,John
Lac 22h25'0"55d5'

Weber,Stacey
Vul 19h44'46"28d37'
Weber,Stephan,Flüschen
Peg 22h16'1"35d21'
Weber,Steve
Lac 22h33'41"55d10'
Weber,Thomas E
Sex 10h22'20"-5d55'
Weber,Tim
Cep 23h9'14"70d33'
Weber,Viola I
Lyn 7h42'31"43d40'
Weber,William
Uma 11h47'26"49d33'
Weber,Wolfgang
Aql 20h6'28"1d32'
Weber,Wolfgang
Sge 20h13'14"17d43'
Webers Wedding
Leo 10h27'42"11d38'
Webley,Michael Ray
Sco 16h24'21"-29d10'B
Webley,Michelle
Vul 19h39'29"20d3'
Webostad
Ser 17h33'19"-10d25'
Webster (Nijak), Jacqueline Marie
Ori 5h30'24"0d36'
Webster's Star,Eric
Lac 2h40'42"53d43'
Webster,Braden
Umi 13h10'44"70d46'
Webster,Brian
Per 2h11'29"57d33'
Webster,Bryn
Boo 14h26'43"50d23'
Webster,Candice Marie
Cas 1h19'29"61d58'
Webster,Carol E
Boo 14h54'52"33d36'
Webster,Chadd Alan
Sgr 19h36'31"-31d52'
Webster,Christopher N
Per 2h10'1"57d37'
Webster,Curt K
Aur 6h3'17"46d50'
Webster,Daniel G
Hya 8h12'35"-0d14'
Webster,David Howard
Aur 5h42'1"50d26'
Webster,Derek
Cep 20h9'1"60d20'
Webster,Gordon
Hya 8h12'1"5d44'
Webster,Ian Austin
Cep 20h39'0"58d10'
Webster,Janette
Lyr 18h30'0"31d6'
Webster,Jennifer "Clyde"
Cas 0h14'44"63d6'
Webster,Jessica Marie
Ari 2h1'40"20d58'
Webster,John
Aql 19h37'30"1d16'
Webster,K C
Dra 17h6'15"69d37'
Webster,Kathryn Baney
And 2h28'0"45d39'
Webster,Kristin Taylor
Cas 0h3'54"50d15'
Webster,Mary Lou
Cep 20h38'58"78d48'
Webster,Neil
Ori 5h25'26"15d19'
Webster,Pamela K
Vul 19h22'27"25d8'
Webster,Patrick Charles
Cam 3h59'26"62d35'
Webster,Rebecca M
Lyn 9h6'15"44d20'
Webster,Susan Alice
Mon 6h31'50"-0d9'
Webster,The Star Stuart
Her 16h7'45"48d59'

Webster,V G
Cet 2h38'58"-9d57'
Webster,Vicky Anne
And 23h0'1"51d46'
Wechsler,Leon & Lilly
Crb 16h10'17"31d59'
Wechsler,Sunny
Boo 15h22'17"48d1'
Wecker,Al & Genie
Lyn 7h45'56"58d26'
Wecker,David S & Joann E
Her 17h33'11"28d34'
Weckerly,Janet Eleanor Nielsen
Sgr 18h51'14"-22d45'
Weckerly,Marcella K
Cas 0h40'31"61d35'
Weckert,Sarah
Sgr 19h18'51"-24d31'
Weckop,Roberto
Cep 22h16'19"59d11'
Weckström,Björn
Lyr 18h34'58"29d34'
Wecwar-Bruner
Cet 0h44'41"-3d58'
Weczerek,Dieter
Gem 6h37'32"30d57'
Wedder II,Mark J
Aql 19h30'45"8d44'
Wedding & Home's Superstar
Cyg 19h46'13"30d24'
Wedding Memories in Heaven
Sge 20h5'16"18d52'
Wedding on November 21,1993
Sco 16h23'21"-27d34'
Wedding Star of Gene & Heidi
Cyg 20h23'0"30d57'
Wedding Stella of Masashi & Nahomi
Cnc 8h46'21"9d17'
Wedding'92:Kim & Gary
Com 12h7'13"32d12'
Wedding,Angela
Lyr 18h35'26"34d34'
Weddington,Eda Rosella Hoida
And 23h29'1"42d2'
Weddington,Jackie Lynna
And 2h27'31"44d14'
Weddington,Sarah Ruth
Del 20h38'15"10d45'
Wedeking,Ann
Cas 2h34'30"67d59'
Wedell,"Mr Satchmo"
Cap 20h19'38"-20d13'
Wedor,Kjell
Uma 9h34'60"57d54'
Wedertz,Justin Keith
Mon 6h36'25"0d29'
Wedl,Kendra Louise
Vul 21h23'58"26d37'
Wedlake's Kenpo Karate,Lee
Aql 19h57'54"13d51'
Wedlake,Brian Richard
Ori 4h56'53"5d13'
Wedlake,Wayne J
Psc 0h46'21"30d21'
Wednesday,Jarvis The 25th
Aql 20h11'16"10d39'
Wednesday,Veronica
Peg 22h0'1"25d13'
Wee "C"
Umi 14h33'30"68d4'
Wee Sensation,The
Umi 13h37'1"73d48'
Wee Wee
Del 20h15'0"13d34'
Weeble
Cnv 12h24'12"37d31'
Weed
Mon 7h6'20"-0d52'
Weed,Joan P
Del 20h49'38"8d43'

Weed,Mary Hill
Ori 5h30'48"-1d49'
Weeden,Hildur Bartlett
Crb 16h16'35"28d36'
Weeder,Grace
Lac 22h34'20"52d38'
Weedermann,Verena
Sco 17h31'45"-30d39'
Weedon,Jason Matthew
Cmi 7h42'17"4d29'
Weeg,The
Cas 0h20'53"63d20'
Week-End (Mauritius)
Pho 0h42'36"-41d31'
Weekes,Charles "Bucky" Ravenel
Aur 5h9'25"42d35'
Weekes,Connor
Ori 5h55'14"20d3'
Weekly,Ral
Boo 15h4'48"18d3'
Weeks,Calvin Daniel
Ser 18h17'30"-13d37'
Weeks,Catherine Paige
Cas 23h35'49"63d9'
Weeks,Christopher A
Vul 19h48'46"22d34'
Weeks,Joseph Crane
Cep 23h31'28"66d52'
Weeks,Max
Ori 5h55'26"18d27'
Weeks,Paige Robyn
Lyr 18h32'45"35d32'
Weeks,Pam
Mon 6h53'35"-6d4'
Weeks,Robyn
Cam 5h50'0"58d31'
Weeks,Spencer Thomas
Cep 20h29'41"61d2'
Weeks,Wally
Aql 20h10'50"11d32'
Weema
And 23h22'56"44d45'
Weems,Thomas
Cep 2h30'17"78d33'
Weener,Arthur B
Dra 14h56'54"62d25'
Weenig,Joshua C
Cma 6h56'57"-19d37'
Weenig,Rachel K
Eri 3h54'14"-0d8'
Weeno
Sco 17h28'20"-40d1'
Weeraratne,Asoka
Dra 17h29'59"60d52'
Weers,Gerd und Ellen
Cap 20h6'14"-10d9'
Weese,Don & Marie
Lac 22h52'30"55d12'
Weeter,Greg & Betsy
Peg 21h48'12"30d9'
Weezer
Lyn 9h4'37"33d36'
Weezer,Kimberly Michele Bailey
Eri 4h46'27"-9d56'
Weg 144
Cyg 19h33'49"35d2'
Wegener's Star,Ken
Tau 4h20'31"15d21'
Wegener,Teri Dawn
Peg 21h59'1"23d46'
Weger,John C
Ori 6h2'44"8d52'
Wegert,Klaus H
Gem 6h41'42"30d25'
Wegesin,Jr,Watcher of The Stars,Bob
Ari 1h59'42"14d46'
Wegge,Bradley S
Uma 10h59'43"44d24'
Wegge,Liz
Peg 22h1'47"5d59'
Weging,Harrison Clay
Sco 16h34'43"-43d28'

Weglarz,B M
Aql 19h48'32"13d3'
Weglarz,Caroline
Peg 22h45'15"33d55'
Weglehner,Berta
Umi 13h34'18"71d9'
Weglewski,Casey
Dra 9h44'49"73d40'
Wegley,Kimberly Ann
Cas 1h48'21"58d2'
Wegman,John Wayne
Ori 5h3'57"10d16'
Wegner,McKenna Lee
Lyr 18h49'0"31d10'
Wegner,Ryan Michael
Tri 1h42'1"30d20'
Wegner,Sara Elizabeth
Cas 0h10'22"62d50'
Wegscheider,Michaela
Cap 20h24'1"-20d2'
Wehenkel,Detlef
Lyn 8h2'14"42d46'
Wehling Star,Andrew Loren
Peg 21h42'15"21d52'
Wehling,Kevin
Cnc 7h55'23"11d22'
Wehmann,Lonie
Vir 13h25'34"-9d58'
Wehmer,Heather
Lmi 9h51'20"33d48'
Wehmeyer,Jeanette
Lyn 9h14'42"44d4'
Wehner,Elizabeth
Cet 1h54'15"-2d30'
Wehner,Patrick
Vul 20h1'36"28d39'
Wehres,Hugo
Lac 22h1'0"51d44'
Wehrfritz,Adolf
And 23h28'33"45d25'
Wehrley,Katie
Cas 1h1'1"65d40'
Wehrley,Robert
Her 16h55'29"32d9'
Wehrli,Ronald J
Per 1h49'56"48d45'
Wei,Fang Chi
Cep 23h9'1"60d22'
Wei-Li
Umi 14h39'22"68d21'
Weiand,Kristine Lynn
Lac 22h40'34"54d38'
Weibel,Kristina Michelle
And 23h20'57"48d8'
WeiBenborn,Janine
Leo 10h57'17"10d25'
Weibert Star
Her 16h39'51"48d13'
Weibhauser,Bernhard
Sgr 19h53'57"-41d8'
Weible,Melissa Belle
Cas 0h33'27"63d11'
Weichert,James M
Boo 14h29'11"40d9'
Weideman,Lisabeth
Peg 21h11'7"15d46'B
Weideman,Michael
Peg 21h11'7"15d46'A
Weidemann,Celia Jean
Vul 19h58'42"25d17'
Weidenbaum,Adam
Lyr 19h23'25"38d9'
Weidenbaum,Amy
Cas 0h17'25"60d58'
Weidenbaum,Hannah
And 0h2'1"47d11'
Weidenbaum,Sarah
Cyg 21h56'33"53d22'
Weidenburner,Mara McKenna Marion
And 23h37'39"47d21'
Weidenfeld,Dr Irwin
Boo 15h3'40"40d53'
Weidenfeller,Ronald William
Dra 12h1'51"70d2'

Weidenhamer,Bradley Carlton
Boo 14h34'19"52d29'
Weidenhamer,Henry
Aur 5h9'42"44d22'
Weidenschlager,Günter
Ori 5h58'56"19d55'
Weidhorn,Drew
Ari 2h7'34"21d46'
Weidler,Suzanne Lipp
Cas 23h45'48"50d37'
Weidler,WF
Crt 11h46'45"-17d52'
Weidman,Amanda
Com 12h42'54"20d59'
Weidman,Kathy Lynn
And 0h21'60"37d22'
Weidman,Loleta Mae
Vul 19h17'28"21d24'
Weidmann,Patrik
Aur 5h6'11"43d32'
Weidmann,Raimund
Lmi 9h25'25"38d15"
Weidner,Ben
Lyr 19h16'59"31d26'
Weidner,Clinton Odell "Del"
Ser 16h3'19"14d43'
Weidner,Doris Dail
Uma 11h13'55"40d20'
Weidner,Emily Elizabeth
And 2h26'14"44d7'
Weidner,Hannah Lindsey
And 2h17'1"40d33'
Weidner,John Pershing
Uma 11h52'38"38d35'
Weidner,Samuel Albert
Vul 19h44'42"28d55'
Weidner,Virgil & Minerva
Cyg 19h26'11"30d31'
Weiers,Vic
Ser 16h2'15"2d47'
Weigand
Eri 3h46'59"-5d49'
Weigand,Emily Violette
Cas 0h32'50"64d4'
Weigand,Rudi
Ser 16h7'51"1d17'
Weigandt,Nicholas James
Aur 5h49'33"34d26'A
Weigandt,Ryan Richard
Aur 5h49'33"34d26'B
Weigel,Dennis R
Lac 22h27'52"50d35'
Weigel,Thomas
Lib 14h44'40"-23d38'
Weigelt,Katja
Cmi 7h28'58"8d52'
Weight,Scott Jacob
Cep 22h42'14"57d60'
Weightman,Roland
Her 17h4'57"22d5'
Weihs' Wonder,G J
Aur 4h37'28"30d29'
Weik,Margaret Rose
Peg 22h31'17"27d59'
Weil,Bernd A Dr
Sgr 20h3'25"-44d26'
Weil,Bonnie
Cas 0h38'39"58d13'
Weil,Christopher
Aur 5h7'13"38d9'
Weil,Gary J
Cap 21h56'39"-21d2'
Weil,Joseph Paul
Dra 16h43'24"69d22'
Weil,Leigh Ann
Cas 1h14'23"60d39'
Weil,Nicholas
Aur 6h4'13"31d34'
Weil,Otto
Eri 4h48'26"-9d33'
Weil,Rita
Ori 5h59'36"16d59'
Weil,Sugar Plum Laura
Cam 7h55'34"68d36'

Weil,TonyBeth
Cam 7h34'42"60d31'
Weil,William L
Cep 21h19'19"68d52'
Weiland Explorer, Travis Jason
Lac 22h14'0"51d8'
Weiland,John Markus Bruno
Sco 15h53'50"-22d3'
Weiland,Pia Veronika
Dra 15h13'45"63d36'
Weilbaecher,Doctor Donald G
Oph 18h19'48"7d52'
Weilbaecher,Marcelle G
Cet 0h2'12"-8d59'
Weilbaecher,Patti
Del 20h53'44"6d5'
Weiler,Darren
Per 2h53'36"40d13'
Weiler,Mathias
Cmi 7h18'38"1d10'
Weiler,Matthias
Uma 10h21'11"48d36'
Weiler,Richard E
Cep 20h45'22"65d18'
Weill's "Birthday Star",Stacy
Peg 23h22'22"18d4'
Weill,Michael & Margo
Sge 19h37'23"16d15'
Weill,Philippe P
Leo 9h25'15"19d54'
Weill,Sandy
Lyr 19h20'53"38d56'
Weiller,Paul-Annik
Per 4h5'45"37d57'
Weimer,"Dream Finder" Harry
Her 17h53'24"40d50'
Weimer,Jim
Per 2h59'0"48d57'
Weimer,Martha
And 1h25'28"40d14'
Wein,Irving L
Lac 22h39'1"40d20'
Wein,Rose Josephine
And 2h1'32"38d43'
Weinand,Mariah Rose
Vir 11h58'1"-1d59'
Weinbeck,Winny
Uma 9h18'16"44d46'
Weinberg Star,The Michael & Kimberly
Cas 1h16'1"62d42'
Weinberg,Alison Jennifer
Leo 10h30'35"26d51'
Weinberg,Audrey
Aql 18h56'45"-2d16'
Weinberg,Bob
Vul 21h24'43"27d34'
Weinberg,Bonnie\Tisha
Uma 9h52'42"58d47'
Weinberg,Jan & Robert
Del 20h13'33"9d26'
Weinberg,Jesse,Geri & Harold
Cyg 21h40'33"38d3'
Weinberg,Norma Lee
Lyr 19h1'1"41d12'
Weinberg,Philip I
Lac 22h6'47"37d43'
Weinberg,Richard Marvin
Oph 17h57'0"12d12'
Weinberg,Seymour
Oph 17h6'31"-24d8'
Weinberger,Janet
Lyr 18h46'15"33d5'
Weinberger,Perla & Joseph
Cyg 21h29'15"40d28'
Weinberger,Robert
Per 2h38'43"41d9'
Weinbrand,Lillian
Peg 22h46'39"5d40'
Weiner,Alan L & Carrie Berse
Cep 22h8'0"70d27'
Weiner,Alyssa Joy
Cam 4h3'44"61d7'
Weiner,Ari Michael
Her 18h5'31"40d58'

Weiner,Arthur
Her 17h32'14"26d57'
Weiner,Derek & Drew
Ser 16h0'39"7d18'
Weiner,Dr David & Dr Abby Phillipson
Oph 17h52'31"-3d22'
Weiner,Dr Gilbert R
Oph 17h39'0"11d44'
Weiner,Dr Harry I
Aur 5h9'14"41d7'
Weiner,Jack B
Ori 5h56'0"21d6'
Weiner,Jane Michelson
And 23h23'22"44d33'
Weiner,Jeffery & Cheryl H
Dra 14h36'1"65d30'
Weiner,Lindsay Faith
Cas 0h58'0"60d31'
Weiner,Mark S
Cnv 13h42'0"40d43'
Weiner,MD,Melvyn L
Sgr 19h2'21"-23d39'
Weiner,President
Her 16h0'11"24d29'
Weiner,Rosalyn
Lyn 9h26'15"41d31'
Weiner,Stacy L
Cam 5h47'44"68d18'
Weiner,Verdeen,Sidney & Andreana
Eri 3h26'23"-11d32'
Weinert,Hans-Ulrich
Peg 23h27'21"11d18'
Weinert,Jenna Nicole
Leo 10h42'52"14d53'
Weinfeld,David
Ori 5h41'37"11d35'
Weinger,MD,Mark
Oph 17h57'35"7d51'
Weingold,David
Ori 5h38'19"12d22'
Weingärtner,Sibylle
Lyn 8h7'0"45d0'
Weinhandl,Leslie D
And 0h54'38"40d34'
Weinhart,Jennifer
Ori 5h3'34"-2d25'
Weinhaus,Ronald L
Aql 20h14'43"4d32'
Weinhold,Col Aaron Z
Aur 5h0'47"29d30'
Weining,Deborah Ann Martha
And 0h45'11"33d34'
Weininger,Mickey
And 23h37'56"49d2'
Weinkauf,Franklin C
Aur 4h53'30"50d16'
Weinland,Daniel Timothy
Her 18h41'38"13d0'
Weinland,Dylan Stephen
Her 18h45'18"13d5'
Weinman,Beverly & Hy
Dra 19h48'44"67d40'
Weinman,David Nelson
Uma 11h23'0"40d39'
Weinmann,Dr Manfred
Cnc 9h14'13"8d19'
Weinmann,Pat & Alyce
Peg 22h45'36"27d55'
Weinreich,Cameron
Lyn 8h19'27"39d49'
Weinreiss,Malgorzata
Gem 6h45'17"16d20'
Weinroth,Alvin A
Aql 20h2'12"1d26'
Weinroth,Jake
Ori 5h52'47"15d12'
Weinschenk,Carvel
Aur 5h8'54"40d36'
Weinschenk,Jane E
Cyg 19h26'57"33d43'
Weinstein,Bernice & Eli
Ori 6h1'36"0d47'

Weinstein,Carly Diane
Lyn 7h52'47"58d32'
Weinstein,David "Radboy"
Aur 4h59'30"40d13'
Weinstein,Francis
Cyg 21h26'15"48d55'
Weinstein,Frederic
Aql 20h9'45"6d4'
Weinstein,Jeffrey Aaron
Her 17h50'31"40d56'
Weinstein,Jeffrey P
Aur 5h6'13"40d35'
Weinstein,Jennifer & Lloyd
Lyn 7h3'0"53d28'
Weinstein,Joshua David
Lac 22h5'42"48d58'
Weinstein,Lacey
Cma 6h11'31"-24d53'
Weinstein,Lillian
Cas 1h49'48"58d55'
Weinstein,Marc Adam
Cep 2h31'52"78d30'
Weinstein,Muriel
Mon 8h4'49"-10d2'
Weinstein,Nancy F
Lib 15h34'22"-8d54'
Weinstein,Naomi Drosdoff
And 0h21'34"30d25'
Weinstock,Amber L
Mon 6h26'44"10d8'
Weinstock,Camen Browning
Peg 23h34'0"18d14'
Weinstock,Michael J
Eri 3h41'46"-6d59'
Weinstock,Sam & Mary
Tri 1h57'23"30d39'
Weinstock,Sarah E
Cet 0h48'55"-2d44'
Weinstock,Steven
Cam 4h9'0"68d48'
Weintraub,Myron
Hya 8h39'51"-18d49'
Weintraub,Rosalind
Aql 20h3'29"1d25'
Weippert,Daren
Per 2h59'54"37d57'
Weir Hear
And 1h4'1"40d39'
Weir,David
Sct 18h42'40"-6d48'
Weir,Mama Marie Kern
Cas 23h1'32"53d59'
Weir,Mr James
Dra 18h58'45"48d57'
Weir,Robyn Smith
Lyn 9h30'22"40d34'
Weir,Roger M
Uma 11h46'29"55d57'
Weir,Sammy
Sct 18h50'54"-7d42'
Weis,Amy
Vir 11h44'29"3d58'
Weis,Angelika
Leo 9h24'29"19d4'
Weis,Charla Faye
Vul 19h37'40"27d16'
Weis,Diane & Bob
Crb 15h27'40"31d60'
Weis,Jr,Eric Fredrich
Ser 15h49'44"21d45'
Weis,Katherine
Cas 3h11'1"68d2'
Weis,Kristina Louise
Peg 22h1'12"3d10'
Weis,Mike/Virginia Bigelow
Her 16h6'27"20d10'
Weis,Pamela Sue
Mon 6h25'0"5d3'
Weis,Phyllis
Uma 9h45'37"51d29'
Weis,Theo
Uma 9h37'47"43d48'
Weisberg,Irving H
Cep 4h35'43"80d17'

Weisberg-Lippitz,Brad
Leo 9h58'1"19d21'
Weisberger,Matthew
Ori 6h6'47"7d49'
Weisbrod,Les a.k.a.The Pit Bull
Cma 6h26'10"-28d41'
Weisburger Forever, Andy & Susan
Cet 0h36'1"-6d16'
Weise,Hartmut
Vul 20h45'38"20d6'
Weise,Horst
Cam 5h5'37"70d30'
Weise,Sean
Cam 3h35'33"63d42'A
Weise-A Real Sweetheart,Sharon
Sge 20h17'18"17d16'
Weisenberg,Jane
Peg 22h35'36"25d34'
Weisensee,Michael Gerard
Dra 20h17'19"67d39'
Weisensee,Natalie Gabrielle
Cam 4h49'59"68d27'
Weiser Family Star,The
Dra 16h9'1"62d41'
Weiser's World
Lmi 10h31'0"32d5'
Weiser,Brian & Scott Wieser
Tri 1h52'15"25d58'
Weiser,Mindy
Ori 5h52'41"11d12'
Weiser,Paul
And 23h45'21"38d24'
Weiser,Sandra
Cas 3h6'28"67d53'
Weisert,Elise Michelle
Peg 21h50'50"31d35'
Weisfeld,Bill
Aql 19h59'35"10d29'
Weishaar,Leslie
Uma 10h0'40"52d40'
Weisheit,Randy D
Dra 16h32'0"61d57'
Weisman Star,Harold & Lois
Peg 22h27'55"35d26'
Weisman,Brett
Her 17h0'35"20d25'
Weisman,Dr Max
Cep 21h21'42"68d11'
Weisman,Gilbert
Per 1h44'22"52d31'
Weisman,Jeffry Howard
Peg 23h34'34"21d52'
Weisman,Neal Eliot
Ori 5h53'52"6d16'
Weisman,Nicole Brooke
Ari 1h54'54"12d27'
Weisman,Pauline Ashley
Cyg 22h0'19"52d56'
Weisman,Tova Miriam
Lyn 8h59'33"35d32'
Weisman,Yale Allan
Uma 12h2'18"57d11'
Weisner,Alexa
Aqr 22h44'24"-0d10'
Weisner,John T
Her 15h49'22"41d44'
Weiss 3-10-1948,Barry Alan
Psc 0h0'47"2d23'
Weiss Way
Uma 11h1'54"35d50'
Weiss"We Love You,Dad" Sumner B
Her 16h23'44"11d12'
Weiss,"A Star is Born" Harris
Aur 6h22'35"38d22'
Weiss,A J
Lyn 7h52'37"44d6'
Weiss,Andreas
Sco 17h31'1"-30d45'
Weiss,Ann & John
Aur 6h22'0"40d37'

Weiss,Arlene
Lyr 18h42'34"37d8'
Weiss,Barbara & Gilman,Allan
Cyg 21h7'46"38d50'
Weiss,Clara
Lyn 8h12'41"51d44'
Weiss,David
Hya 8h15'55"1d14'
Weiss,Debbie
Cyg 19h26'0"31d4'
Weiss,Debbie & Jay
Sgr 19h37'37"-33d31'
Weiss,Elaine
And 23h23'20"44d6'
Weiss,Erwin
Ser 15h11'55"8d26'
Weiss,Fred Toby
Oph 17h16'0"10d30'
Weiss,Gertrude
Cas 1h29'0"60d46'
Weiss,Howard
Lyr 19h20'57"40d0'
Weiss,Jamie
Lyn 7h43'41"51d47'
Weiss,Jonathan Winchester
Cep 21h15'56"68d49'
Weiss,Joseph Daniel
Boo 14h36'51"16d52'
Weiss,Judith Carol
Mon 6h34'37"1d44'
Weiss,Julia
And 1h33'15"39d31'
Weiss,Lydia
Del 20h21'0"10d22'
Weiss,Margaret Jane Hurst
Peg 22h16'35"34d13'
Weiss,Maria und Volker
Psc 0h17'1"17d36'
Weiss,Michael
Umi 14h50'31"65d58'
Weiss,Morris D
Aur 5h11'41"41d11'
Weiss,Mrs Mary J
Lyn 8h13'18"33d46'
Weiss,Murray Jules
Her 17h13'50"26d56'
Weiss,Neil
Boo 15h13'59"55d6'
Weiss,Pat
Sgr 19h6'33"-23d16'
Weiss,Robert K
Dra 16h59'59"52d9'
Weiss,Silke
Lib 15h8'37"-24d23'
Weiss,Susan L
Ori 5h27'38"1d20'
Weiss,Theodore "Teddy"
Peg 21h58'28"21d21'
Weiss,Ute
Sco 17h31'36"-31d20'
Weissbein,Sarah
Cas 1h42'20"61d3'
Weissberg,Marjorie J
And 23h0'0"51d47'
Weisse,Alva & Frank
Dra 14h18'20"63d24'
Weisse,Katherine Elena
And 0h11'29"31d44'
Weissenberg,Fred
Aur 6h44'39"37d17'
Weissenborn,Michael
Boo 13h35'56"10d1'
Weissenburger,Isabella & John
Aur 6h33'41"32d43'
Weissenfels,E
Uma 9h13'28"28d54'
Weisser,Kathleen Ann
Peg 22h0'38"3d55'
Weisser,Wolfgang
Dra 9h48'23"73d31'
Weisser-West
Del 20h13'13"14d53'
Weissgerber,Dr Reinhard
Aql 19h1'1"13d12'
Weissgerber,Frank
Mon 7h52'2"-2d59'
Weisshaupt,Karl-Heinz
Lib 14h46'24"-23d9'
Weisskopf,Victor
Aur 7h5'27"38d54'
Weissman,Arthur
Cep 22h47'13"70d44'
Weissman,Dr David E
Boo 15h1'44"26d41'
Weissman,Gracia
Cas 0h48'1"61d27'
Weissman,Joel Stephen
Dra 19h21'49"58d51'
Weissman,Linda
Aql 19h57'16"14d51'
Weissman,Lisa Jan
Mon 7h49'8"-1d38'
Weissman,Rose & Andy
Cyg 21h28'36"38d50'
Weistling,Michael
Ser 15h38'41"6d43'
Weitman,Aaron
Cet 2h2'16"0d5'
Weitner,Dr Lutwinus
And 23h3'15"50d26'
Weitsman,Our Star In The Sky Ruth
Sge 19h14'14"16d38'
Weitz,Dominique
Sct 18h45'36"-6d15'
Weitz,Michele
Mon 7h55'32"-1d11'
Weitz,Rudolf
Lyr 19h18'11"30d58'
Weitzel,Amanda
And 0h13'11"37d27'
Weitzel,Amanda Michelle
Cam 4h19'27"60d4'
Weitzen,Bobbie Jo
Ori 5h6'1"0d33'
Weitzman,Darlene Manes
Weitzman,Steven
Lyr 18h52'49"42d45'
Weitzman,James Paul
Cep 22h17'31"55d52'
Weitzman,Mathew Laurence
Lac 22h4'55"40d24'
Weizenhöfer,Lorette
Cap 20h31'39"-20d25'
Weiß,Horst
Uma 9h37'39"45d34'
Weiß,Ute
Peg 23h32'23"17d0'
Weksel,Our Starring Dad-William
Mon 7h58'41"-2d42'
Welbes,Sarah Christine
Mon 6h47'49"11d28'
Welbourne,Star of Mickey Jean
Del 20h49'19"8d1'
Welburn,Ross & Sheryn
Uma 10h7'11"50d27'
Welburn-Bluedorn,Linda
Uma 8h40'29"51d20'
Welby
Vul 19h49'21"20d32'
Welch III,MD,Robert M
Ari 1h47'1"14d39'
Welch's Eternal Love Star,Rick & Barbara
Cyg 19h29'0"56d53'
Welch,Ada Jane
Oph 17h30'7"-22d53'
Welch,Betty Lou Bonham
Cyg 19h30'16"31d21'
Welch,Bill & Debbie
Cyg 19h25'46"30d40'
Welch,Brian Matthew
Boo 14h29'37"21d36'
Welch,Candice C
Cap 20h20'27"-12d46'
Welch,Catherine C
Sco 17h34'45"-43d11'
Welch,Chauncey
Cet 1h55'11"-1d19'
Welch,Claude David
Cyg 19h24'0"30d6'
Welch,Cynthia Sue
And 0h4'10"31d17'
Welch,David
Dra 16h7'17"66d12'
Welch,Deborah Ann
And 0h17'38"36d38'
Welch,Del
Dra 10h5'20"80d5'
Welch,Doreen
Peg 21h59'1"28d52'
Welch,Dustin
Mon 6h31'39"-0d37'
Welch,Elizabeth Rose
Cam 3h59'1"60d27'
Welch,Elmer Alexander
Her 16h41'40"50d40'
Welch,Geralyn
And 0h53'41"34d50'
Welch,Gloria R
And 2h3'17"45d4'
Welch,Helen & Ernie
Cmi 7h6'40"1d33'
Welch,II,Grady D
Lac 22h13'47"51d6'
Welch,Irene F
Cyg 20h19'55"38d31'
Welch,Jane
Lyr 18h59'52"30d58'
Welch,Jean Malison
Cam 13h15'25"78d31'
Welch,Jeff
Aql 19h55'12"7d52'
Welch,Jeffrey Craig
Hya 9h7'28"2d20'
Welch,John
Aql 19h58'45"10d50'
Welch,Joseph
Cyg 21h32'26"30d58'
Welch,June M
And 23h2'31"51d49'
Welch,JV
Dra 16h31'23"64d54'
Welch,Laing Elizabeth
Peg 22h37'58"20d30'
Welch,Madison Nicole
And 0h20'0"32d17'
Welch,Mary Kathleen Laughlin/Kate
Lyr 18h35'37"38d15'
Welch,May
Cas 23h25'0"61d52'
Welch,Michael & Diane & Ziffy
Oph 18h37'56"6d50'
Welch,Patricia Gerard
Lyn 7h37'32"38d12'
Welch,Raven & Walter
Her 16h58'60"21d15'
Welch,Savannah Rose
Aql 19h0'1"16d34'
Welch,Timothy Robert
Aur 7h15'43"40d45'
Welch,Tonnye
Cet 1h41'49"-0d10'
Welchel,David Thomas
Cet 2h5'34"-5d60'
Welcher's Wonder,Judy & Harold
Lyr 18h35'58"37d16'
Welcher,Dorothy Starr
And 1h35'51"38d23'A
Welcher,Gus C
And 1h35'51"38d23'B
Welcome Baby Brian
Boo 14h12'25"46d8'
Welcome,Agnes
Aql 20h6'30"3d49'
Welcome,George
Ori 5h54'54"12d48'
Weldele,Cathy S
Eri 3h32'20"-1d42'
Welden,Julia Marie Wilcox
Peg 23h40'15"26d2'
Weldon,Clu,C Raymond
Cep 2h54'1"80d29'
Weldon,Craig Chapler
Dra 18h56'32"48d40'
Weldon,Danielle Jenny
Umi 13h55'15"71d14'
Weldon,George
Ser 17h52'8"-11d37'A
Weldon,Grover Cleveland
Gem 6h59'55"12d38'
Weldon,Jeannie Ann
Uma 13h40'32"50d55'
Weldon,Jr,William Coburn
Aur 7h23'42"38d7'
Weldon,Judy
Ser 17h52'8"-11d37'B
Weldon,Mike R
Leo 10h34'54"14d49'
Welford,Gail I
Cyg 20h15'11"38d46'
Welgs,Anthony John
Aur 6h27'42"33d9'
Welgs,Frances Marie
And 0h55'52"33d38'
Welgs,Theresa Marie
Cas 1h18'34"62d29'
Welhouse I
Ori 6h15'36"-2d55'
Welhouse II
Ori 5h56'51"14d56'
Welker,Geneva
Com 12h52'10"24d15'
Welks,David J
Her 18h3'25"30d47'
Welky,Robert & Mildred
Hya 8h20'55"-10d4'
Well,Lorraine
Lyr 18h55'13"38d58'
Well,Mary Margaret
Crb 15h31'40"27d27'
Wellace,Jr,Ellis H
Cas 0h45'21"61d5'
Welland Wisher,The
Per 1h51'0"53d39'
Welland,Albert
Aql 19h22'36"14d8'
Welland,Mei Chen
Aql 20h9'50"7d40'
Wellauer,Bob
Oph 17h19'46"-20d28'
Welleck's Star
Peg 21h53'50"20d2'
Wellen,David George
Peg 23h19'22"28d54'
Wellen,Nicole Danielle
Psc 1h17'0"21d52'
Wellenberger,Carol Ann
Cas 0h41'13"64d21'
Wellenraiter,William
Aur 6h29'33"35d7'
Weller,Alexander
Dra 15h8'34"62d1'
Weller,Burkhard
And 23h19'35"41d50'
Weller,David John
Per 2h6'34"57d47'
Weller,Erwin
Lmi 9h23'0"33d39'
Weller,Lee
Del 20h39'23"12d20'A
Weller,Norty
Del 20h39'23"12d20'B
Weller,Richard Cole
Hya 9h3'31"2d25'
Weller,Richard Ray
Aur 6h24'55"35d55'
Welles,Ralph Sidney
Dra 17h18'24"61d40'
Welling,Bea
Lyr 18h59'54"40d54'
Welling,Grant
Aql 19h5'53"3d0'
Wellinghoff,Gene
Aur 5h43'47"50d29'
Wellinghoff,Laura
And 23h0'13"51d16'
Wellman & Tom Taris, Eternity Lydia
Cas 2h59'34"58d27'
Wellman,"Lady Jean"
Sge 19h53'49"18d51'
Wellman,Abbey Lynn
Cas 1h31'41"73d16'
Wellman,Henry & Fromma
Cyg 19h18'1"28d18'
Wellman,M.Christine
Lyr 19h16'0"26d54'
Wellman,Martin Eric
Peg 22h15'1"10d31'
Wellman,Richard,Lynda & Franklin
Hya 8h34'56"-6d27'
Wellnitz,Barbara
Cas 23h12'35"60d20'
Wellnitz,Liane
Tau 4h58'48"15d43'
Wells
Cam 4h13'0"71d8'
Wells 119,Carol Lee
Mon 8h5'15"-1d37'
Wells Family Legacy
Aql 18h54'43"-0d4'
Wells III,Leroy Sebastian
Cet 2h9'34"7d2'
Wells,Agnes
And 23h19'22"44d44'
Wells,Allen Fredrick
Psc 23h5'28d1'7"
Wells,Anna Mae
Umi 15h16'28"68d19'
Wells,Benjamin Holland
Her 18h6'35"30d47'
Wells,Bob
Aur 4h49'48"38d7'
Wells,Brian Thomas Alfred
Aur 7h15'48"39d23'
Wells,Callie-Ann
Cyg 19h39'16"31d27'
Wells,Calvin B
Sct 18h44'1"-5d11'
Wells,Christopher
Ori 5h32'0"-8d3'
Wells,Cynthia Valentine
Eri 3h55'13"-7d23'
Wells,Daniel Myral
Uma 9h40'34"58d52'
Wells,Doylette
Ori 5h48'0"10d51'
Wells,Galynn
Hya 8h47'52"0d7'
Wells,George Edward
Cet 0h32'31"0d43'
Wells,Gizelle & Walters,Kurt
Cep 22h0'6"82d37'AB
Wells,Hal & Joyce
Mon 7h32'52"-0d27'
Wells,I Love You, Elizabeth,Len
Aur 6h59'0"43d10'
Wells,Jacob Samner
Umi 17h20'46"85d12'
Wells,Jamie Paul
Cyg 20h52'51"38d21'
Wells,Jamie Rose
Com 13h10'52"28d59'
Wells,Jennifer
Lyr 18h57'13"45d31'
Wells,Jr,David F
Aql 19h51'25"13d38'
Wells,Judy
Tri 2h26'43"28d14'
Wells,Karlene K
And 0h1'1"38d31'
Wells,Ken
Aql 19h10'38"12d15'
Wells,Krista (Krusty)
Lyr 18h54'38"32d49'
Wells,Laura
Del 20h38'2"7d18'
Wells,Lavella
Lyn 6h23'17"59d21'
Wells,Linda
Vul 19h46'0"27d20'
Wells,Lt Col Perry P
Aql 18h45'1"10d47'
Wells,M Diane
Vir 12h59'54"-21d47'
Wells,Mary
Hya 8h13'60"1d48'
Wells,Megan
Tau 3h28'34"30d46'
Wells,Michael A
Eri 3h52'1"-1d23'
Wells,Michael Adam
Cep 23h14'53"64d54'
Wells,Monica Maria
Cyg 21h33'41"50d20'
Wells,My Michel- (Gross-Pa Pa')
Crb 15h29'36"30d4'
Wells,Parks Tigrett
Aur 5h18'1"42d7'
Wells,Patricia
And 23h1'39"51d39'
Wells,Patricia Ann
And 1h25'41"35d19'
Wells,Prince Charming AKA Jerry Lee
Her 16h10'43"5d23'
Wells,Rachael
And 1h43'40"37d55'
Wells,Raeford M
Hya 8h55'1"3d4'
Wells,Roger
Dra 17h34'28"64d58'
Wells,Samuel Ryan
Crb 16h21'47"28d27'
Wells,Scott R
Aur 7h0'39"37d45'
Wells,Scott R
Ser 17h32'19"-13d8'
Wells,Sharren J
Lyn 7h34'44"50d13'
Wells,Shirley Alice
Lac 22h28'45"40d33'
Wells,Steven & Kimberly
Aur 5h2'14"40d59'
Wells,Tina
Peg 21h27'55"23d37'
Wells,Tom & Mary
Eri 4h4'18"-10d16'
Wells,Wes & Ruth
Dra 20h28'22"67d56'
Wels,Theo
Lyr 19h20'13"31d33'
Welsch,Gretchen Paulette
Lyr 19h4'0"28d59'
Welsch,Karl-Heinz
Ser 17h35'11"-13d30'
Welsch,Kenneth R
Cnv 12h59'32"31d44'
Welsh III,William
Lib 15h0'56"10d50'
Welsh,Adam Jonathan
Cas 2h37'55"57d39'
Welsh,Anne
Lyr 19h16'41"25d56'
Welsh,Gary
Boo 14h3'0"14d55'
Welsh,George Michael
Lac 22h7'0"49d55'
Welsh,James P
Her 16h18'40"26d28'
Welsh,Janice Mabel
Per 3h6'60"56d40'
Welsh,Jr,Harry J
Cnc 8h58'17"18d32'
Welsh,Jr,Kenneth Gordon
Ser 15h13'34"11d37'
Welsh,Kevin
Lib 14h42'46"-20d13'
Welsh,Linda Susan
Gem 6h54'15"31d20'
Welsh,Mark
Aur 6h21'19"35d55'
Welsh,Michelle
And 23h49'37"41d9'
Welsh,Mommie Janice
Boo 14h55'16"40d52'
Welsh,Neal
Aur 6h31'18"38d26'
Welsh,T Austin
Aur 6h7'32"45d54'
Welsing,Kerstin
Per 3h51'15"35d53'
Welsome,Thomas Garvey
Mon 7h4'49"-5d28'
Welte,Caitlin Ashe
Cmi 7h31'53"8d46'
Welte,Helen Wedekind
Tri 1h48'16"28d20'
Welte,Sumner Vaughn
Mon 7h39'15"-0d34'
Welter,Frank
Sgr 19h53'31"-43d44'
Welter,Heinz Peter
Equ 21h2'1"3d28'
Welton,Danny
Her 18h3'26"48d10'
Welton,Ellie
Cet 3h15'12"1d34'
Welton,Jennifer Merrit
Com 12h19'0"31d18'
Welton,Kenneth Earl
Lac 22h45'26"54d20'
Welton,Michael David
Aql 18h58'49"12d18'
Welton,Rachel Mair
Cyg 20h13'43"40d5'
Welton,Sherryl
Lyr 18h14'58"46d19'
Weltonia
Lac 22h45'58"53d19'
Welty,Paula Diane
Del 20h53'1"8d20'
Welzen,Colleen A
Cyg 19h26'53"40d2'
Welzmiller,Barbara
And 23h7'32"43d0'
Wemmert,Christian John
Aur 6h13'37"38d53'
Wemmert,Jill Michelle
And 1h22'58"37d16'
Wempe Star,The
Sex 10h41'0"-0d27'
Wempe,Hellmut
Cam 6h8'1"61d49'
Wempe,Kim
Uma 12h22'1"53d37'
Wempe,Kim-Eva
Lib 15h12'53"-20d13'
Wen's Star
Boo 14h47'38"36d22'
Wen,Chen Bei
Crt 11h15'17"-10d36'
Wenar,Corrine C
Peg 22h55'25"28d51'
Wenbenburg,Carl Albert
Hya 8h10'25"1d30'
Wenck,Allen & Maryanne
Oph 18h17'50"12d9'
Wend,Denise Marie
Cyg 20h24'50"38d19'
Wend,Saskia
Cmi 7h16'19"1d35'
Wendarella,Contessa
Tau 3h56'1"20d25'
Wendel,Iris
Lyn 8h35'1"42d36'
Wendel,Jenna Bricker
Cam 7h37'1"78d51'
Wendel,Karen R
Cas 3h21'51"75d26'
Wendelborn D
Uma 11h32'55"30d3'
Wendelin
Cap 21h1'53"-22d35'
Wendelken,Thomas R
Leo 9h55'28"11d24'
Wendell Keith
Eri 3h13'37"-6d28'
Wendell,Heidi Anna
Tri 2h43'31"31d3'
Wendell,Jonny Lee
Hya 9h5'11"5d50'
Wendell,Virginia Lee
Cyg 21h51'11"42d22'
Wendellyn
Com 12h57'46"27d2'
Wenden,David
Dra 16h1'22"61d53'
Wendenburg,Pamela Elizabeth
Cmi 7h28'46"8d31'
Wender,Brian Joshua
Mon 7h59'30"-2d2'
Wendi
Mon 8h3'51"-5d12'
Wendi
Peg 22h1'47"34d31'
Wendi Anne
Cyg 20h36'21"42d31'
Wendi Sue
Lyn 8h11'42"51d24'
Wendi,Keith & Michael
Lyr 18h42'38"33d55'
Wendl,Marcus Willibald
Vir 12h3'36"2d12'
Wendland,Barbara
Lib 15h20'39"-22d45'
Wendland,Patricia Ann
Uma 8h48'44"55d2'
Wendland,Rico
Cmi 7h20'0"0d1'
Wendland,Wendo
Peg 23h32'43"16d38'
Wendling,Karen A
And 0h3'54"35d27'
Wendolyn
Cyg 19h27'14"31d56'
Wendorf Family
Oph 17h34'27"14d1'
Wendt,Botho
Vir 14h40'39"-5d53'
Wendt,Clair & Pat
Lyn 7h54'59"52d26'
Wendt,Henry
Dra 18h49'49"84d52'
Wendt,Lani
Ori 6h6'50"20d26'
Wendt,Michael
Sgr 19h34'11"-32d6'
Wendt,Peter Carl & Barbara Jean
Cyg 20h37'37"52d42'
Wendt,Skyler
Uma 12h9'1"58d9'
Wendt,Wendie Lee
And 0h16'28"45d49'
Wendte,Peter
Uma 9h30'55"45d30'
Wendy
Cas 23h1'1"58d27'
Wendy
Sge 20h2'1"20d24'
Wendy
Peg 22h0'16"27d44'
Wendy
Cas 0h24'51"71d59'
Wendy
Ori 5h14'48"-8d59'
Wendy
And 23h6'37"42d5'
Wendy
And 0h14'27"39d46'
Wendy
Cnv 13h21'14"41d48'
Wendy
Peg 23h6'18"20d10'
Wendy
Lmi 9h25'0"38d24'
Wendy
Cyg 19h46'52"31d22'
Wendy
Lyn 7h57'40"34d10'
Wendy "5683"
Lac 22h7'11"46d34'
Wendy & Jeff's Star
Lyr 19h19'38"33d57'
Wendy & Peter
Crb 15h28'19"31d14'
Wendy & Peter
Cam 9h21'29"80d54'
Wendy & Reuven
Cam 3h35'1"71d57'
Wendy & Victor's Heaven
Cyg 20h58'48"30d11'
Wendy Dawn
Tau 4h22'14"15d10'
Wendy Diane
And 0h24'48"45d55'
Wendy Elisabeth
Lyn 7h50'16"50d45'
Wendy Elizabeth
Cyg 19h20'58"44d47'
Wendy Elizabeth
Lyr 18h36'0"33d50'
Wendy Gail
Cam 5h52'49"70d40'
Wendy Glow
Lyn 9h0'11"40d17'
Wendy I Love You With All My Heart & Soul
Ori 5h9'12"-5d11'
Wendy Jay
And 23h41'44"45d37'
Wendy Jean
Cyg 20h57'14"30d30'
Wendy Jo #25
Sge 19h30'3"16d20'
Wendy Lee
Aql 19h31'15"14d41'
Wendy Michelle
Mon 7h12'14"-6d26'
Wendy O
Mon 7h11'57"-7d9'
Wendy Rene
Peg 22h40'1"27d42'
Wendy Shirley
Lyr 18h39'0"34d56'
Wendy Sue
Mon 6h27'0"10d6'
Wendy"The Sweetie Star"
Cyg 19h51'51"37d58'
Wendy's Hope
Mon 7h4'53"5d19'
Wendy's Joy
Lyn 6h24'47"59d24'
Wendy's Light
Crb 15h30'43"31d59'
Wendy's Star
Cam 4h57'5"61d0'A
Wendy's Star
And 23h24'1"42d2'
Wendy's Star
Aql 19h27'35"13d13'
Wendy's Window
Psc 23h9'49"0d43'
Wendy's Wish
Cas 0h1'0"61d15'
Wendy,A Look Is Worth 10,000 Words
Eri 2h46'57"-6d55'
Wendy,I Love You,Steve
Peg 23h18'48"28d49'
WendyHeath's "Wedding Star"
Aql 19h6'31"4d7'
Wendykins-Great Admirer of Her Mütti
Cyg 19h41'51"30d0'
Wendys
Mon 6h40'1"7d20'
Wendys Star Always S L W
Peg 0h9'1"13d44'
Wener,Mark
Cep 20h55'15"58d49'

Wenger,Clayton
Cam 3h19'43"61d36'
Wenger,Lera
And 23h27'1"45d33'
Wengerd,Timothy Mather
Aql 18h57'13"-5d31'
Wengrover,Bernard
Aur 5h14'1"40d35'
Wengrowski,Mary
Lyr 18h54'26"33d10'
Wengyn,Sylvia
Eri 2h52'28"-17d43'
Wenham,Ryan Alan
Cyg 21h18'32"35d35'
Weninger,Mariella
Tau 5h46'0"20d13'
Wenke,Andrew Peter
Her 16h45'24"50d44'
Wenke,Robert Andrew
Per 1h54'37"48d41'
Wenko,Margot Marie
Vul 19h46'32"20d31'
Wenness I Love You Always Godfrey Shan
Pho 23h42'15"-47d37'
Wenngatz,Dr Harry H
Gem 6h54'20"14d41'
Wensdofer,John P
Dra 19h4'45"70d6'
Wensel,Chadwick D
Dra 17h0'26"50d23'
Wensell,Kyle Peter
Her 18h39'24"12d19'
Wensky,Joanne & Arthur
Vul 20h20'32"28d8'
Wenslow,Jarrick Austin
Lmi 10h17'21"36d11'
Wenslow,Skylar Jeffrey
Ori 5h57'28"10d60'
Wenstrom,Paul D
Dra 16h30'55"63d18'
Went,Robert
Ori 4h56'0"5d52'
Wenta,Edwin
Per 2h32'0"56d41'
Wente,Der Glänzende Stern Des Larry
Per 1h39'11"52d34'
Wente,Michael Lewis
Boo 13h38'15"15d22'
Wentink,Wallace
Psc 23h3'1"5d1'
Wentling 13,Tristen
Peg 22h38'1"26d15'
Wentworth,Eric Charles
Sco 17h29'25"-30d37'
Wentworth,James A
Sex 10h11'3"-1d27'
Wentworth,Tom
Cet 3h1'27"7d52'
Wentworth,Virginia "Ginny"
Aql 18h59'59"13d55'
Wentz,Philip & Elizabeth
Tau 5h48'48"23d16'
Wentz,Richard Carrington
Tau 4h10'0"1d29'
Wentzel,Christine Lynn
Cyg 21h6'25"38d47'
Wentzel,Jennifer Lynn
Lib 15h42'22"-24d20'
Wenuwishupona
Lyn 8h48'13"34d30'
Wenz,Carol Lynn Desiree
Cas 0h56'32"54d24'
Wenz,Fred J & Mildred M
Dra 19h41'32"70d21'
Wenzel
Com 13h17'49"18d1'B
Wenzel
Cap 20h6'40"-10d39'
Wenzel,Baerbel
Oph 18h24'13"8d59'
Wenzel,Carlie
Cyg 19h44'50"31d46'

Wenzel,Christian David
Aur 7h8'56"35d38'
Wenzel,George M
Aql 20h8'56"1d26'
Wenzel,Isi
Ari 2h52'0"20d3'
Wenzel,Jan
Ori 5h29'41"-8d41'
Wenzel,Kathi Lee
And 23h35'50"48d44'
Wenzel,Mark
Dra 16h10'11"65d26'
Wenzel,Richard Alan
Dra 17h50'27"64d5'
Wenzel,Richard C
Boo 15h3'53"20d31'
Wenzel,Rolf
Lyr 19h21'52"31d34'
Wenzel,Susanne
Leo 11h3'46"11d46'
Wenzky,Jürgen
Lmi 9h30'1"37d33'
Wenzl,Sarah
Ser 15h29'0"19d52'
Wenzler,John
Cep 21h0'49"60d56'
Wenzler,Margit
Leo 9h23'27"17d30'
Werbe,Mary French Jackson
Cam 6h36'17"80d3'
Werbitt,Rand L
Umi 15h6'56"71d18'
Werblow,Louis
Aql 19h56'53"1d48'
Werchovsky,Ignatius Boris
Ori 6h8'44"17d23'A
Werden's Eternal Love, Jeff & Rosa
Eri 4h34'13"-8d7'
Werden,Eugene T
Cep 20h55'0"65d43'
Werger,Arthur
Aql 18h38'40"0d57'
Werhane,U B
Cam 3h37'27"60d33'
Wering,Glenn
Cam 4h9'0"68d23'
Werker,Juergen
Cyg 20h19'42"47d8'
Werksman,Adam William
Lac 22h35'37"53d53'
Werle,Erhard
Aur 5h57'51"48d18'
Werley,Mariah Ann
Peg 22h24'43"30d14'
Werling,Maria S
Aql 19h30'38"10d38'
Werling,Susan Lynn
Cyg 20h36'1"50d7'
Werling,Volker
Mon 7h51'20"-2d45'
Werman,Jane C
Com 12h1'23"25d47'
Werneke,Edwin M
Dra 9h22'54"73d34'
Werner
Leo 11h44'38"22d8'
Werner 50
Lac 22h6'24"50d9'
Werner en Lianne veel geluk,Mama
Cam 5h1'42"54d24'
Werner Hermann Otto
Aqr 21h20'1"-8d57'
Werner,Armin
Aql 19h6'57"10d20'
Werner,Bianca
Cas 0h41'0"60d11'
Werner,Colleen
Aql 20h35'21"-8d16'
Werner,Cortney Gene
Cmi 7h23'20"7d40'
Werner,David Joseph
Oph 17h56'14"7d31'

Werner,Evan Forrest
Ari 3h13'17"20d29'
Werner,Hilbert
Lyr 19h18'50"31d39'
Werner,Jana
Cep 22h0'0"60d45'
Werner,John Thomas
Cnc 9h12'57"30d4'
Werner,Jordan Paige
Cas 23h17'48"62d41'
Werner,Julia
Cmi 07h53'14"00d26'
Werner,Lauren Kathryn
Oph 17h19'13"-22d42'
Werner,Leslye
Per 4h5'19"51d47'
Werner,Megan Alexandra
Aql 18h57'23"14d47'
Werner,Michael J
Lac 22h26'45"40d43'
Werner,Mone
Ari 2h29'33"21d10'
Werner,Paul Louis
Per 2h56'36"50d26'
Werner,Peter C
Oph 18h17'48"10d51'
Werner,R Jay
Her 16h37'48"7d35'
Werner,Stephen
Per 1h46'51"48d49'
Werner,Todd David
Her 18h23'1"28d25'
Werner,Ute "Frauchen"
Sgr 18h54'37"-29d46'
Werner,Veronica L
Per 3h9'25"41d5'
Werner,Walter Thomas
Cnv 12h34'14"33d22'
Werner-092290, Starkweather
Aql 19h50'30"13d52'
Werner-Meiers,Else
Tau 4h6'37"21d24'
Werners Planet,Lisa Erin Ryan Dan
Ori 6h2'41"-0d28'
Wernet,Joseph A
Oph 17h3'2"-23d32'
Wernette,Jerri
And 23h1'29"49d37'
Wernick,Evelyn D
Cyg 21h53'59"41d4'
Wernick,Justin
Per 3h7'19"41d5'
Wernicke,Claus
Aur 5h15'17"43d37'
Werns,Elaine
And 0h30'18"45d47'
Werrell,Jason Todd
Per 1h29'0"53d43'
Werrenrath,Eric Fleming
Oph 17h54'0"13d3'
Wersching,Patricia R
Hya 8h54'57"-0d13'
Werstler,John Edward
Aql 19h55'26"8d9'
Wert,Haley Christian
Aql 19h7'14"15d31'
Werth,Gayle Dianne
Dra 11h52'41"70d31'
Wertheim,Helga
Cap 20h47'21"-26d24'
Wertheimer,Jill
Vul 21h17'13"20d25'
Wertheimer,Naomi & Bruce
Lyr 18h40'0"39d33'
Werthmann,Dean Donald
Dra 16h4'1"66d40'
Werthmann,Kayla Anne
And 1h59'31"41d13'
Wertman,Bradley G
Boo 13h59'32"15d51'
Wertz,Catherine Spencer
Eri 4h7'55"-19d7'
Wertz,Jane L
Mon 6h20'1"4d23'

Wertz,John & Margaret Schlangen
Cyg 21h50'38"40d29'
Wertz,Scott Michael
Cet 0h55'36"2d0'
Wes
Her 17h25'53"40d5'
Wes & Alina
Eri 4h36'0"-1d25'
Wes' Star
Cnv 12h58'59"51d11'
Wes,Susie
Peg 22h54'1"28d54'
Wesala,Jacob Paul
Aur 6h9'36"45d46'
Wesch,Derek J
Cmi 7h59'36"0d17'
Wesche,Hans-Joachim
Lyr 19h19'43"31d32'
Wescon
Mon 8h1'18"-3d60'
Wescott,Lori Jean
Ant 11h1'53"-36d34'
Wescott,Susan
Cam 8h6'32"84d20'
Wescott,William
Boo 14h24'15"26d39'
Wesemann,Darren
Boo 13h52'1"19d16'
Wesemann,Hermann
Lmi 10h42'11"25d59'
Wesemann,Joseph
Aur 7h23'16"39d38'
Wesemann,Stacey
Lyr 18h41'46"34d17'
Wesley
Cet 2h54'44"1d19'
Wesley & Lucille
Her 16h50'0"35d43'
Wesley 50
Cyg 20h28'21"40d6'
Wesley Dennis
Cmi 7h44'23"8d10'
Wesley's "Margaritaville"
Dra 17h0'56"51d33'
Wesley's Ring
Her 17h57'1"40d28'
Wesley,Kenneth Charles
Cmi 7h59'37"0d13'
Wesley,Sharon Agnes
Tau 4h57'38"20d2'
Weslis' Passion
Eri 4h7'34"-17d23'
Wesner,Craig
Lac 22h40'23"54d24'
Wesoloski,Steve
Aql 19h8'17"3d17'
Wesselink,Ellard
Boo 14h51'20"47d32'
Wessell,Chuck & Beth
Eri 3h50'45"-4d12'
Wessels,Daniel J
Per 4h20'39"51d49'
Wessels,Hubert
Dra 18h59'0"65d7'
Wessels,Jeffrey L
Dra 20h9'31"68d45'
Wessels,Kathyrn
And 2h27'1"39d53'
Wessely,Gudrun
Mon 7h53'3"-2d54'
Wessely,Uli Stephan Sebastian
Peg 23h34'16"12d50'
Wessenberg,Renee Anne
Peg 23h7'24"12d11'
Wessendorf,Kathleen "Kat"
And 23h26'31"49d54'
Wessing,Timmy Jon
Aur 5h35'52"50d12'
Wesslen,Ellen/Arne
Lyn 8h56'25"46d1'
Wessler,Fred & Betty
Cyg 20h20'14"41d7'
Wessler,Michael
Psc 1h1'36"22d9'

Wessling,Nathan Thomas
Dra 16h24'33"64d16'
Wessman,Troy
Umi 16h23'54"71d1'
Wessoleck,Anikke
Mon 7h19'53"-6d57'
Wesson,Howard
Tau 4h31'56"30d36'
Wesson,Parks
Cet 1h21'17"-0d33'
West Canada Rock
Uma 11h26'32"62d14'
West III,John Raymond
Aur 6h25'33"37d60'
West Lake Middle School
Ori 5h38'0"7d35'
West Love Lite,Wade Through Bridges
Ori 4h43'14"0d12'
West Michigan Sprinkling, Inc
And 1h41'19"39d12'
West Star
Lib 15h35'56"-20d3'
West(Jody's Treasure), Helen Joan
Cam 5h16'20"68d58'
West,Adam Brinkworth
Cap 21h52'21"-22d23'
West,Alice M
Mon 6h24'39"1d15'
West,Amber D
Lyr 18h48'60"34d38'
West,Amy M
And 1h52'35"41d4'
West,Bob
Dra 15h54'27"52d11'
West,Bonnie
Cam 3h57'35"69d44'
West,Charles C
Per 2h50'1"36d30'
West,Charles Henry
Cep 21h44'43"60d36'
West,Cody Allen
Cet 2h17'45"4d44'
West,D Ardella
Lib 14h58'34"-0d51'
West,Dominique
Mon 7h56'55"-8d11'
West,George Benjamin
Per 2h54'0"35d21'
West,Gordon Robert
Her 17h39'0"21d42'
West,Grant
Boo 14h31'34"39d58'
West,Janet Lee
Cyg 20h19'55"38d29'
West,Jean Chin
Ser 15h55'0"-2d13'
West,Jessica
And 23h40'44"45d38'
West,John
Her 17h55'49"40d29'
West,Jon Justin
Aql 19h43'0"12d15'
West,Julia
Cam 4h1'36"61d17'
West,Kayla
And 23h26'58"46d33'
West,Kelly Lynn
Umi 15h11'54"68d59'
West,Kirby Ann
Psc 0h46'33"31d22'
West,Lauretta
Lyn 8h46'1"38d18'
West,Lucas John
Per 2h40'25"35d56'
West,Lucille Eileen Best Tilley
Cas 23h36'58"61d47'
West,Matthew Thomas
Per 2h13'1"58d24'
West,Meredith Nicole
Cnc 8h53'39"21d16'
West,Michelle
Mon 7h5'46"-6d5'

West,Myron & Alice
Eri 4h53'25"-5d33'
West,Patricia Ann & Daniel James Gaustad
Aql 19h6'35"15d10'
West,Patrick
Per 2h47'11"43d26'
West,Prince Charming- Jim
Aql 20h2'33"11d50'
West,Richard & Pacita
Mon 6h57'49"7d35'
West,Rick
Lep 5h34'1"-11d0'
West,Ricky Dale
Oph 18h41'55"10d5'
West,Rosemary
Uma 10h8'58"51d23'
West,Ruby D
Aql 18h58'24"11d12'
West,Samantha Lauren
And 23h1'21"45d44'
West,Sandra
Cet 2h4'0"-18d55'
West,Stanley
Cep 0h14'15"75d13'
West,Stephanie Erin
Cas 1h14'16"60d50'
West,Timothy James
Cyg 19h28'36"37d20'
West,Together Forever John & Lillian
Aqr 23h19'25"-6d13'
West,Tom
Her 17h19'15"14d35'
West,Tracey Jill
Lyr 19h24'37"39d39'
West,Zoe Michelle
And 0h37'43"40d7'
Westall,Christopher
Hya 8h19'34"1d45'
Westall,Harold
Ari 2h1'24"26d53'
Westall,Lynne
Oph 18h42'24"10d15'
Westberry ØB‰,John E
Aql 19h3'15"5d44'
Westberry,Richard Elliot
Leo 9h40'14"16d20'
Westbrook Star,The Robert T
Dra 15h11'1"61d51'
Westbrook,Dawn A & Ronald D Barnett
Cyg 19h29'21"36d39'
Westbrook,Dr Christopher
Oph 18h3'1"8d4'
Westbrook,Haley Christine
And 0h21'16"31d2'
Westbrook,Mackie Gilbert
Cam 4h12'0"80d3'
Westbrook,René Allene
Cnc 8h11'22"30d41'
Westbrook,Simon & Gillian
Lyn 7h31'37"40d1'
Westbury,Doodle
Cam 3h47'1"53d4'
Westbury,Dupree
Cam 3h49'20"53d49'
Westbury,Kenneth O
Ori 4h52'34"-2d57'
Westbury,Maggie
Lyn 6h21'28"60d15'
Westenbarger,Joe
Aql 19h58'45"15d31'
Westendorf,Kevin J
Aur 6h24'19"37d20'
Westendorf,Tanja E
Vir 13h29'39"-3d36'
Westenhoefer,Jr, George M
Lyn 7h55'56"33d47'
Westerbeck,Jeffrey S
Oph 17h31'37"11d48'
Westerberg,Michele
Sct 18h19'11"-13d57'
Westerbur,Leann
Cas 1h47'24"65d2'

Westerdale,John
Uma 10h59'57"48d47'
Westerfield,Dr Keith James
Dra 16h52'11"69d48'
Westergren,Joseph Carl
Aqr 23h22'11"-5d42'
Westerhoff,Jürgen & Uschi
Uma 11h12'25"41d38'
Westerhold,Wilfried
Uma 9h30'1"46d45'
Westerholt,Mildred
Cap 21h55'23"-22d18'
Westerink
Aql 19h59'51"7d35'
Westermann,Detlef
Uma 10h26'0"48d9'
Western Star,The
Cam 4h40'12"67d35'
Western,Giles
Per 2h57'57"50d9'
Western,Richard James
Cmi 7h42'13"-0d3'
Westervelt,David Selby
Cru 12h52'39"-63d20'
Westervelt,Melanie Sue
Cas 3h6'39"58d16'
Westervelt,Warren
Aur 6h10'17"38d8'
Westfall,Barbara L
And 1h55'15"47d17'
Westfall,Eugene J
Her 17h0'46"47d34'
Westfall,Jim & Pam
Dra 14h2'36"64d20'
Westfall,Margie
Cyg 19h19'12"44d57'
Westfall,Margo
Cas 23h40'22"61d14'
Westgate,Jr,Eternal Love,Thomas S
Sge 19h55'57"20d15'
Westgate,Terry
Per 2h56'29"31d49'
Westgate,Wendy Carol
Cas 1h39'46"60d14'
Westhoff,Jennifer L
Cas 23h3'36"58d48'
Westhoff,Matthew J
Dra 17h39'38"78d55'
Westin,Dr Glenn
Oph 17h54'1"10d41'
Westland,Casey Ryan
Cet 3h3'15"1d3'
Westland,Sabrina May
Mon 6h53'41"-1d40'
Westland,Tarah Marie
Vul 20h38'27"20d4'
Westland,Tiffanee
Hya 8h18'0"5d16'
Westler,Bill & Jenevee
Boo 14h38'48"19d33'
Westley,Randon A
Aur 5h5'0"40d21'
Westman,Virginia
Cas 3h34'17"69d44'
Westmoreland,Brooke Colleen
Cam 4h28'1"60d54'
Westmoreland,Sara Kathleen
Cam 4h51'58"68d50'
Westnaufer,Barbara
Cyg 19h54'0"45d24'
Weston
Tau 5h0'46"16d11'
Weston
Dra 17h30'1"63d54'
Weston
Her 17h20'54"21d33'
Weston Star,The
Equ 21h7'38"11d49'
Weston's First
Cet 2h39'46"-4d2'
Weston,Cotu's Heart- Charles
Lac 22h28'32"38d25'
Weston,Daniel
Cyg 21h35'0"37d2'

Weston,Deanne Virginia
Lyn 8h3'36"39d54'
Weston,Diana
Ori 5h32'40"7d55'
Weston,Dorothy & Ed
Aql 19h52'59"-0d22'
Weston,Glen
Cep 22h24'27"65d25'
Weston,Gregory E
Dra 19h1'35"48d4'
Weston,Jason & Susan
Per 1h56'21"56d20'
Weston,Julie (Sexpot)
And 0h2'52"47d55'
Weston,Louise
Cyg 21h30'17"37d30'
Weston,Ross
Her 17h0'30"30d49'
Weston,Ryan Hunter
Gem 6h44'31"30d30'
Weston,Sacha
Cyg 21h39'50"37d40'
Westover,Quinn Louis
Cep 21h48'28"55d7'
Westphal,Andrew & Jim
Oph 17h28'0"-1d29'
Westphal,Brian D
Per 3h58'45"52d23'
Westphal,Brian Elden
Ori 6h0'52"10d40'
Westphal,Guy
Cet 0h58'0"1d10'
Westphal,Jutta
Equ 20h55'55"3d27'
Westphal,Victoria Anne
Peg 22h2'31"2d3'
Westphal,Virginia L
Cap 20h46'58"-26d40'
Westphal,William D
Tri 2h19'35"30d23'
Westrick,Mary
Uma 10h36'10"48d56'
Westrup,Helen Muriel
Cap 21h3'55"-18d47'
Westwell,Jeanette Willena
Lmi 10h14'35"40d53'
Westwood,Elizabeth
And 0h30'44"40d59'
Westwood,Janet Sue
Aur 6h57'29"43d9'
Wetfox
Cyg 20h24'12"38d45'
Wetherall,Peter
Cep 21h15'19"60d33'
Wetherbee,James Daniel
Cep 22h52'46"57d54'
Wetherby,Anthony T
Aur 5h42'30"50d13'
Wetherille,Kevin John
Dra 14h18'1"64d3'
Wethy, The Star In My Life,Cindy
Lyr 18h55'43"41d15'
Wettach,Michael
Lac 22h1'45"40d32'
Wette,Krings
Mon 6h21'35"-6d52'
Wetter,Kerry Ann
Lyr 18h58'25"34d35'
Wettstein,Brian
Cet 2h37'1"2d5'
Wetzel,David
Dra 16h15'49"68d33'
Wetzel,Eric
Her 17h8'14"44d18'
Wetzel,Oliver & Margaret
Eri 4h18'28"-13d20'
Wetzel,Tom & Lou
Cet 2h39'7"-4d2'
Wetzel-Wiersma,Marion
Aqr 21h1'44"-14d34'
Wexler,Anne R
Peg 22h43'0"20d6'
Wexler,David Craig
Per 4h7'18"48d46'

Wexler,Jonathan
Cnv 13h47'28"31d12'
Wexler,Wendy L
Lyn 7h46'16"40d36'
Wexner,Hannah
Mon 6h44'16"10d28'
Wey,Dr Jur Renate
Leo 11h24'54"-0d22'
Wey,Gregory
Cyg 20h28'16"31d18'
Weyel,Ivo
Lyn 7h54'54"41d14'
Weyenberg,Charlotte
Cma 6h55'34"-11d15'
Weyer,Kimberly
Mon 6h32'1"8d5'
Weyer,Maximilienne
Cmi 7h23'5"5d43'
Weyerbrock,Herbert
Mon 7h51'37"-1d24'
Weygandt,Steven L
Aur 6h57'46"38d35'
Weyler,Klaus F
Lyn 8h11'54"43d40'
Weylon S
Mon 8h6'58"-0d26'
Weymouth,Jeffery
Cet 1h21'49"-0d28'
Weymouth,Wells Luke
Sgr 20h0'58"-45d5'
Weyna-Weir,Mark & Nancy
Ari 2h56'41"30d22'
Weynen,Heinz-Gerd
Cep 22h3'47"60d54'
Weyrich,Fred
Cas 0h16'41"60d56'
WGF 28
Uma 12h5'18"60d37'
Whait,Martyn
Uma 9h56'36"54d39'
Whalea
Uma 11h43'16"47d7'
Whalen "California Dude",Harvey E
Cep 23h3'39"70d22'
Whalen,Corey
Aqr 21h44'0"0d47'
Whalen,Dan
Dra 17h1'13"68d29'
Whalen,Dan
Cet 2h9'38"5d33'
Whalen,Elizabeth Kelty
And 2h18'56"38d9'
Whalen,Jason
Aur 5h18'17"42d16'
Whalen,Jenny
Cas 0h52'10"61d29'
Whalen,Joshua Matthew
Her 16h10'29"50d34'
Whalen,Kenneth Eric
Aur 5h32'55"40d4'
Whalen,Kristen Marie
Lyn 7h33'41"39d45'
Whalen,Mary Alice
Cas 1h45'42"68d38'
Whalen,Natalie Nicole
Uma 10h57'17"40d13'
Whalen,Philip Michael
Aur 5h5'42"30d45'
Whalen,Sheila-Mike Mullin
Peg 23h37'20"31d12'
Whaley,Emma Jane
Crb 15h54'17"27d51'
Whaley,Russ & Mary
Uma 11h4'28"46d56'
Whaley,Sherry
Lyr 18h48'36"30d29'
Whalin,John T
Dra 11h57'35"70d12'
Whall,Ken
Ori 6h5'22"1d12'
Whalley,Elaine
Lyr 19h16'49"28d10'
Whalley,Sheree J
Cas 0h35'42"56d13'

Whardo,Dennis Mitchell
Aur 6h44'40"38d21'
Whartenby,Mary F
Cap 21h8'32"-22d14'
Wharton,Betty Sue Wyatt
Lyn 8h24'54"34d10'
Wharton,Delicia
And 23h23'59"50d25'
Wharton,Donna
Uma 8h55'33"51d23'
Wharton,Evelyn
Her 16h26'48"26d22'
Wharton,James H
Dra 12h9'49"76d5'
Wharton,Kurt Lee & Sheryl Wharton
Cyg 19h29'43"58d8'
Wharton,Michael
Boo 14h11'59"40d14'
Wharton,Nan Marie
Mon 6h0'47"-6d51'
What Was,What Will Be, Mark Lynn
Cas 1h37'1"60d24'
What You Are Doing?
Uma 10h53'1"37d4'
What-Vanhoose
Mon 5h54'13"-4d19'
Whatley,Tonia
Gem 7h23'53"28d50'
Whatley,Yvette
Eri 2h49'58"-17d54'
Whatmore,Bill & Stella
Cyg 21h45'1"38d4'
Wheat,Cameron Eilene
Lib 14h21'19"-11d13'
Wheatley,Alua
Oph 17h15'47"-20d56'
Wheatley,David
Uma 8h45'0"55d25'
Wheatley,Debi Jo
Peg 22h24'30"35d27'
Wheatley,Donald & Anna Ruth
Peg 23h26'40"11d51'
Wheatley,Jacqueline
Uma 9h8'39"55d28'
Wheatley,Sarah Dallas
Del 20h17'28"13d23'
Wheatley,William A
Aur 6h56'32"40d60'
Wheaton's Special,John
Hya 9h6'1"2d30'
Wheaton,Christa
Uma 11h17'31"43d37'
Wheaton,Mark Jacob
Dra 9h32'14"77d55'
Wheaton,William Chase
Del 20h25'0"20d7'
Wheele,Michael Paul
Per 4h8'49"38d41'
Wheeler My O,Mandy Berg
Vul 21h13'1"20d7'
Wheeler Star,The Betty
Cmi 7h24'31"0d42'
Wheeler's Dream
Pho 23h43'22"-40d1'
Wheeler's Scoop,Tracy
Lyr 19h22'51"42d10'
Wheeler,Amy Elizabeth
Cyg 20h27'13"40d28'
Wheeler,Andrew
Cep 22h25'58"70d25'
Wheeler,Andrew Ramon
Umi 13h5'1"72d1'
Wheeler,Annalynn
Lyn 8h7'40"38d34'
Wheeler,Ashlee Ann
Leo 9h36'60"18d38'
Wheeler,Audra Sue
Cam 12h13'49"81d28'
Wheeler,Beloved DH
Boo 15h44'1"40d11'
Wheeler,Catherine
Peg 23h29'30"21d52'
Wheeler,Dennis
Dra 20h23'1"68d12'
Wheeler,Donald Douglas
Aql 19h11'0"12d17'
Wheeler,Gary W
Cet 3h7'43"-0d9'
Wheeler,George & Violet
Cyg 20h37'42"40d55'
Wheeler,Harris
Ser 18h16'29"-14d27'
Wheeler,J J
Crt 10h54'40"-21d34'
Wheeler,Josephus Peeler
Mon 8h6'45"-2d38'
Wheeler,Kenneth Hale
Aql 19h53'1"12d9'
Wheeler,Kimberly
And 23h38'38"45d56'
Wheeler,Linda Norris
Cru 12h46'29"-60d25'
Wheeler,Lloyd & Jean
Mon 6h26'48"8d51'
Wheeler,Madeline Gray
Mon 7h6'16"-7d8'
Wheeler,Magill
Uma 9h19'30"48d13'
Wheeler,Margaret Mary
Cas 2h43'1"61d36'
Wheeler,Martin John
Per 2h40'37"37d12'
Wheeler,Mary Alma Mithchell
Mon 6h56'0"-10d56'
Wheeler,Matthew,Thomas
Cet 2h37'0"5d44'
Wheeler,Maxwell P
Boo 14h6'45"50d49'
Wheeler,Michael Cooper
Per 2h39'33"43d43'
Wheeler,Michael Mortimer
Ori 5h24'24"1d42'
Wheeler,Mr & Mrs Gordon L
Umi 14h50'1"81d38'
Wheeler,Nicholas
Her 15h56'57"50d54'
Wheeler,Papa
Oph 18h20'1"8d45'
Wheeler,Peggy
Lyn 7h44'57"47d43'
Wheeler,Philip
Aur 5h0'0"45d27'
Wheeler,Robert Leo
Cep 20h59'1"55d19'
Wheeler,Ronald L
Aur 6h7'13"38d0'
Wheeler,Sandra
Peg 23h26'32"23d51'
Wheeler,Sara
Aql 18h58'54"-6d5'
Wheeler,Scott Martin
Dra 11h48'48"66d58'
Wheeler,Stephen Douglas
Aur 7h24'42"38d27'
Wheeler,Steve
Her 18h3'52"28d23'
Wheeler,Suzanne Helen
Cas 2h55'21"73d27'
Wheeler,Thomas
Cep 21h49'17"68d20'
Wheeler,Wendy Alvarez
Crt 11h42'21"-11d42'
Wheeler-Holes,Annette Jane
Cas 0h0'22"61d17'
Wheeler-Holes,Maisie Ellen
Cas 0h0'30"61d56'
Wheeleronia
Cet 3h17'19"4d16'
Wheelock,Jr,Eleazar Louis Russell
Cet 0h29'0"-12d3'
Wheelock,Kiki
Vir 13h3'19"11d33'
Wheelock,Laura
Cas 1h12'40"61d23'
Wheels 60,Helen
Sge 20h14'47"17d39'
Wheelton's Ruby
Peg 22h56'41"17d49'
Wheiler,Max
Her 17h27'45"20d57'
Whelan,Casey D
Dra 14h30'54"58d9'
Whelan,Clint
Psc 23h27'21"5d42'
Whelan,David Arthur
Tri 2h36'0"35d52'
Whelan,Jesse James
Uma 9h47'1"70d1'
Whelan,Joanne
Cyg 20h38'52"53d53'
Whelan,Marva Jean
Cmi 7h59'13"4d13'
Whelan,Matthew J
Vul 19h58'14"23d2'
Whelan,Thimothy Peter
Gem 7h10'37"24d43'
Whelpley (The Duck) Star,Chris
Cet 3h3'51"1d1'
Whelton,William B
Per 3h15'54"50d34'
When D (PLSYB)
And 0h59'40"39d53'
When I Found You There Was Meaning
Cyg 21h30'30"31d15'
When Jack & Maria Met
Ori 6h2'59"0d5'
When You Wish...T" Forever...
Aql 19h31'10"8d40'
Whenday,Edna Mary
Sge 20h0'1"20d5'
Where Rock-N-Roll Belongs
Lyr 18h45'22"41d33'
Wherry
Uma 12h32'20"62d50'
Wherry,Marisa Michelle
Lyr 19h3'49"38d60'
Whetnall,Sara
Leo 10h55'0"11d49'
Whetstone,Sally Sue Sadler
Hya 8h30'1"-6d57'
Whidbey
Uma 9h30'40"61d6'
Whiddon Super Star, Steve
Cnc 8h23'27"7d21'
Whiffle,Susan Joy
Cam 5h30'0"68d58'
Whildin,Lorraine Kindred Peters
Hya 8h56'44"3d5'
Whiles,Sarah Beth
Lyr 18h30'37"43d36'
Whipkey,Cameron Lee
Her 16h12'34"8d51'
Whipple,Mary Agnes
Hya 8h30'14"-6d18'
Whippstar `94
Uma 11h12'19"31d16'
Whirlow "God's Helping Hand",Jim
Vir 11h47'13"5d16'
Whiseanmor Gordons
Cnv 12h52'58"40d17'
Whisenant(9-26-37), Kate Skeen
Cyg 21h11'31"34d25'
Whisenant,Chris
Hya 10h46'53"-18d28'
Whiskers,Mr
Lyn 7h38'31"44d12'
Whiskey
Uma 9h7'30"58d3'
Whisky
Per 3h4'46"31d52'
Whisper Alley
Sct 18h56'22"-4d49'
Whistler
Vir 12h58'22"7d54'
Whit
Cet 3h17'52"9d23'
Whitacre,Aaron Patrick
Per 1h45'11"53d41'
Whitacre,Charles Kenneth
Boo 15h4'0"27d39'
Whitacre,David B
Vir 14h12'32"-8d13'
Whitacre,Russ
Dra 20h15'11"63d4'
Whitaker & Associates Inc,W
Aur 6h15'56"33d34'
Whitaker A#1 Dad, Robert
Per 1h56'0"56d50'
Whitaker's Hope,The
Sgr 18h49'8"-25d24'
Whitaker,Anita
Leo 10h25'57"12d14'
Whitaker,August Niccole
Boo 14h10'44"50d42'
Whitaker,Chuck
Eri 2h44'24"-6d60'
Whitaker,Debra
Lyr 18h30'47"43d16'
Whitaker,Derek
Per 3h24'59"50d15'
Whitaker,Louis Cody
Dra 17h45'41"65d24'
Whitaker,Margie
Cas 0h52'51"73d42'
Whitaker,Mitchell
Sex 9h39'51"-0d13'
Whitaker,Quinn Matthew
Mon 8h5'16"-4d39'
Whitaker,Robert
Ori 5h22'42"15d50'
Whitaker,Seth Neil
Boo 14h51'1"34d4'
Whitaker,Star King 44 Joseph David
Ser 18h1'25"-13d38'
Whitaker,Traci Nicole
Peg 23h6'51"13d20'
Whitbread,Jessica
And 23h19'41"51d48'
Whitby Whimsy
Mon 6h54'1"-0d7'
Whitby,Mary Josephine
Leo 9h27'49"28d50'
Whitcomb,Brad & Cheryl
Uma 10h56'33"58d35'
Whitcomb,Cindy L
And 23h22'37"43d2'
Whitcomb,Deborah Lynn Homeier
Boo 15h3'25"51d32'
Whitcomb,Erin Claire
Ori 5h51'53"20d26'
Whitcomb,John Joseph
Oph 17h39'20"10d49'
Whitcomb,Lee R
Dra 14h1'59"63d17'
Whitcomb,Taylor Kevin
Boo 15h29'32"37d52'
Whitcomb,Thomas Anthony
Lac 22h24'27"38d19'
Whitcomb,Trevor Wade
Lep 5h57'36"-18d52'
Whitcomb,Uncle Bud & Aunt Marion
Cyg 21h20'0"40d2'
White "Fat Boy",J Lynn
Cet 0h34'56"1d21'
White 1992,Hollis
Cas 23h25'1"60d29'
White Family Star,The
Dra 19h44'45"68d59'
White Feather's Fiddle Faddle
Lyr 18h14'44"37d38'
White Hart
Peg 21h53'1"3d6'
White Heaven Sent, Carol Griffith
Ori 5h38'34"10d31'
White III,Frank Xavier
Cep 21h5'10"55d55'
White III,William C
Ori 4h44'56"0d37'
White Lily
Mon 7h0'39"8d33'
White Pine Bowe
Uma 11h21'57"63d11'
White Roses
Sgr 19h18'23"-29d42'
White Star
Ori 6h7'47"0d20'
White Star,The Fred & Bev
Cyg 19h46'39"30d42'
White's Finest Glow
Mon 6h23'59"-1d38'
White,Ada & Bill
Boo 15h4'1"25d58'
White,Alexis Nicole
Lyn 8h36'59"40d32'
White,Alyssa Ann
Gem 6h42'22"16d58'
White,Amanda
Peg 21h25'13"22d51'
White,Amanda Kay
Lmi 10h18'0"32d15'
White,Anita
Lyr 19h25'1"38d20'
White,Anne Marie Dinardo
Aur 5h47'42"54d39'
White,Anne-Marie
Cas 0h16'32"61d33'
White,April D
Lyr 19h22'19"31d17'
White,Ariel N
And 2h15'40"45d59'
White,Austin
Aql 19h49'40"14d3'
White,Bam
Vir 13h30'41"-4d9'
White,Betty
Cas 2h33'17"60d4'
White,Bill
Aql 20h11'37"1d0'
White,Bonne
Sct 18h44'52"-6d45'
White,Bryan
Her 17h21'29"46d55'
White,Bryan
Hya 9h5'57"3d20'
White,Bryan Mathew
Her 16h48'41"48d7'
White,Carl & Rita
Lac 22h11'22"49d53'
White,Carole Anne Evans
Lib 15h1'15"-28d17'
White,Carolyne & Phil
Eri 4h28'25"-10d57'
White,Catherine Allen
Cas 0h51'41"62d51'
White,Catherine Robin
And 23h21'1"44d49'
White,Cece & Lee
Lyr 19h18'48"41d0'
White,Chanelle
Peg 22h2'0"20d48'
White,Cherri Ellen
And 23h39'34"36d60'
White,Cindy Jane
Lyn 7h42'35"42d6'
White,Claire
Lyr 19h15'43"40d3'
White,Clark
Lac 22h33'42"56d43'
White,Clyde
Per 3h25'15"41d14'
White,Coral
Uma 12h3'41"45d49'
White,Cory Richard
Cmi 7h54'28"1d41'
White,Danny
Hya 8h32'1"0d54'
White,Darlissa
Mon 8h4'43"-9d24'
White,David Michael
Her 16h9'14"40d42'
White,Dean
Peg 23h28'30"21d27'
White,Deborah A
Cas 0h57'12"59d31'
White,Dennis E
Aql 19h59'43"0d52'
White,Dennis Ralph
Uma 11h36'18"31d43'
White,Diane L
Mon 6h51'0"11d29'
White,Don & Lee
Cam 6h17'1"68d58'
White,Doris Dalby
Mon 6h19'50"9d2'
White,Edith
And 23h16'43"47d8'
White,Edward "Pal" Eugene
Ori 5h38'17"0d6'
White,Eileen Francis
Cas 0h16'32"64d25'
White,Elizabeth
Oph 16h30'10"10d27'B
White,Elizabeth M
Cas 1h16'0"75d45'
White,Emily Meagan
Lyr 18h39'49"41d11'
White,Eric Allen
Boo 15h7'15"18d28'
White,Floyd
Cep 21h48'58"80d34'
White,General James A
Her 17h54'59"14d55'
White,George
Oph 16h30'10"10d27'A
White,Gina
Psc 1h27'0"31d36'
White,Harry & Brenda
Peg 22h31'21"21d15'
White,Hazel Mae Tarbert
Peg 22h3'12"20d34'
White,Heather "Snow"
Cma 6h53'57"-17d42'
White,Heather & Jerry
Lac 22h34'20"50d15'
White,Helen Jane
Cyg 21h26'30"38d21'
White,Henry Fitzgerald
Aql 20h1'52"10d35'
White,Hensley
Aur 7h12'25"40d2'
White,Howard Victor
Aql 19h51'53"10d5'
White,Irene
Boo 13h41'1"15d32'
White,Isabel Tommie
Peg 23h18'55"18d6'
White,Jack
Uma 9h48'1"43d44'
White,Jackson Dail
Cma 6h56'19"-17d36'
White,James Lee
Aql 19h57'1"8d58'
White,James T
Cep 22h8'23"70d46'
White,Jamie
Lyn 7h56'0"50d17'
White,Jana B
Aur 6h55'20"44d12'
White,Janet Molly
Gem 6h29'42"13d6'
White,Jason Richard
Aur 6h30'0"32d17'
White,Jean
Uma 11h38'1"41d5'
White,Jenna
Peg 22h3'27"29d32'
White,Jennifer
Lyr 19h24'28"38d13'
White,Jennifer Louise
Com 13h32'52"20d27'
White,Jeremy Matthew
Hya 8h10'11"1d24'
White,Jeresa
Aql 19h17'22"15d8'
White,Jessica Andrea
Pup 6h40'33"-38d9'
White,Joan
Peg 22h1'47"32d43'
White,Jody Marie
Com 12h35'15"22d18'
White,John R
Aur 6h2'0"30d16'
White,Jon Ann
Cet 3h3'1"4d44'
White,Jordan Steele
Hya 8h51'15"-1d41'
White,Joseph Daniel
Her 18h8'20"28d34'
White,Joshua
Boo 14h49'45"39d19'
White,Jr,Ben
Hya 8h44'56"-1d38'
White,Jr,Donald Edwin
Aql 19h52'41"15d45'
White,Jr,James Coleman
Per 2h59'14"40d56'
White,Jr,Robert
Per 4h29'16"50d0'
White,Jr,Steady
Dra 15h12'10"64d24'
White,Kate
Lyr 19h16'47"42d56'
White,Kathleen
Cas 1h1'12"58d33'
White,Katrina Ruth
Cyg 21h6'58"31d17'
White,Keith Lloyd
Per 3h7'37"50d22'
White,Kenneth
Sct 18h42'58"-4d13'
White,Kenneth Jill Owens
Peg 23h23'37"15d16'
White,Kerri Ann
Cam 3h25'31"53d57'
White,Kim T
Lac 22h2'59"51d40'
White,Laura
Eri 4h1'29"-11d19'
White,Laurie D
Mon 6h24'53"10d17'
White,Lawrence Dennis
Cet 1h23'37"1d54'
White,Lena Ruth
Cmi 7h24'45'7d36'
White,Liana
Cyg 21h3'28"31d38'
White,Linda
Lyn 7h25'37"44d42'
White,Linda
And 2h33'47"40d17'
White,Linda
Lyn 8h6'57"58d50'
White,Lindsay Briana
Cet 1h53'24"-2d19'
White,Lisa
Lmi 10h52'57"30d48'
White,Lisa Ann
And 2h3'34"47d16'
White,Lorinne
Lyr 18h53'1"41d22'
White,Manami
Cas 3h10'13"63d15'
White,Margaret Louise Heller
Lyn 8h51'0"34d1'
White,Margrette Rose
Lyn 7h50'18"37d9'
White,Maria Schiefer
Sge 20h5'47"20d15'
White,Martin R
Del 20h56'12"16d14'A
White,Mary Catherine
And 0h52'59"40d57'
White,Mary Faith
And 23h46'47"44d20'
White,Maryann
Cyg 19h43'0"31d9'
White,Matt
Aql 20h0'51"13d40'
White,Matthew
Her 17h16'38"29d20'
White,Matthew R
Lac 22h48'30"56d20'
White,Maxine Lesley
Cyg 19h34'44"33d33'
White,Megan
Del 20h37'0"10d42'
White,Melissa
Eri 3h16'59"-4d31'
White,Michael
Dra 15h2'34"61d52'
White,Michael Dennis
Per 2h47'53"40d54'
White,Michael Ray
Oph 16h49'1"11d53'
White,Michelle R
Del 20h13'21"15d14'
White,Morgan Charlton
Lmi 10h0'16"28d40'
White,Morgan Herbert
Sct 18h44'14"-7d53'
White,Nancy N
Aql 19h54'44"11d57'
White,Nicola
Aql 18h58'13"16d17'
White,Oliver & Sara
Lmi 10h4'54"32d38'
White,Pamala Ann
Lib 15h15'58"-18d1'
White,Pamela
Mon 6h22'39"2d43'
White,Patricia Ann
Aqr 21h56'17"1d44'
White,Paul
Cma 6h54'22"-19d51'
White,Paul
Oph 17h58'21"11d22'
White,Paul E
Cnv 13h18'47"28d38'
White,Philip E
Uma 9h43'0"42d37'
White,Richard
Ori 5h47'55"20d32'
White,Rickey
Mon 7h40'24"-0d18'
White,Robert
Dra 10h30'11"78d50'
White,Ron
Cet 1h12'57"-3d22'
White,Ronald & Grace
Del 20h53'30"7d20'
White,Ronald E
Lac 22h26'25"38d34'
White,Sarah Beth
Lyn 7h48'11"38d2'
White,Sarah Hague
Peg 22h23'30"27d2'
White,Sr,Richard S
Cep 2h2'41"77d45'
White,Stephanie
Cyg 20h36'12"33d10'B
White,Stephanie Leigh
Cnv 13h41'33"33d46'
White,Stephanie Louise
Sco 17h53'53"-38d27'
White,Stephen & Karen
Umi 15h40'41"74d44'
White,Steven E
Her 16h25'54"26d48'
White,Summer
Gem 6h59'12"16d3'
White,Tammy L
Lyn 6h14'40"58d23'
White,The Graduate- Betty C
Cas 0h33'0"63d36'
White,Thomas
Cep 22h20'0"55d50'
White,Thomas James
Per 1h52'26"53d59'
White,Tom
Aur 7h3'34"37d20'
White,Tom & Pam
Cas 1h7'60"61d34'
White,Trevor Alexander
Aur 6h13'29"45d51'
White,Trevor Alexander
Cep 22h35'59"61d46'
White,Tyler Gavin
Ori 5h59'1"12d4'
White,Valerie Sue
Mon 6h25'1"11d14'
White,Vic & Rita
Uma 8h12'0"62d2'
White,Walter Finch
Lac 22h18'45"46d53'
White,Walter H
Peg 23h3'12"31d21'
White,Walter Miller
Cet 3h0'20"6d56'
White,Wendy
Lyn 8h24'37"47d3'
White,William F
Per 3h2'17"37d45'
White,William H
Cmi 7h59'48"8d26'
White,William Robert
Ori 5h3'1'8d34'
White,William Velma
Her 18h6'39"14d26'
White,Wyatt
Aur 5h12'0"40d0'
White-Pennel,Brittney
Eri 2h43'46"-16d53'
White-Salmon, Marguerite
Uma 10h57'19"59d45'
White-Wells,Leah
Psc 1h0'54"32d7'
White-Will You Be Mine,Michael Philip
Ori 5h55'23"13d59'
Whiteass,Lance Corporal
Ori 6h5'56"-1d40'
Whitebay,Evelyn
Lyn 6h38'23"59d14'
Whitebread,Benjamin Karl
Her 16h53'44"27d48'
Whited,Lauren Ashley
Peg 21h21'43"22d59'
Whitefield,Gordon
Dra 19h41'34"61d31'
Whiteford,Kimberly Elaine
Peg 22h42'45"33d13'
Whitehair,John
Lyn 7h50'45"36d5'
Whitehead Because I Love You,James
Lac 22h0'1"38d54'
Whitehead,33¯,III J William
Cet 2h37'58"1d36'
Whitehead,Fritzi-Beth
Cyg 21h20'29"40d20'
Whitehead,Jack
Per 3h3'31"41d24'
Whitehead,Jim
Per 2h48'45"45d48'
Whitehead,Joanna
Cyg 20h36'52'45d3'
Whitehead,Laura Anne
And 23h26'30"42d14'
Whitehead,Marie Carolina
And 0h42'1"31d16'
Whitehead,Marylee
Cyg 20h29'34"42d2'
Whitehouse,Brian John
Per 3h4'0"40d18'
Whitehurst,Mack G
Aur 6h51'19"35d56'
Whitehurst,Tara Lesley
Crb 15h25'0"31d60'
Whiteley,Canny
And 2h19'15"41d38'
Whiteley,Helen
Cnv 13h23'45"31d34'
Whiteley,Nicholas
Cep 20h57'59"59d1'
Whitelock-VanOrden, Sydney
Uma 8h42'59"48d57'

Whitely,Pat & George Shofner
Ori 6h6'51"7d58'
Whiteman,Athena Pauline
Vul 20h41'52"28d16'
Whiteman,Diane Marie
Psc 22h57'16"-1d25'
Whiteman,Margaret Bassett
Uma 8h45'44"50d8'
Whitemiller,Carol & Ken
Aur 5h3'0"50d5'
Whitener
Lyn 9h13'38"34d43'
Whitener,Jacob Cole
Aql 18h43'39"8d42'
Whitener,Stacia Lynn
Eri 4h8'30"-17d5'
Whiteside,Emily
Cyg 21h23'22"37d59'
Whiteside,Jimmy
Her 16h16'38"8d25'
Whiteside,Rose Kral
Peg 22h58'1"33d59'
Whiteway,Sachiko Doryunlee Takahary
Cyg 19h49'24"37d36'
Whitey
Peg 22h29'59"7d27'
Whitey-8 Ball
Uma 9h38'0"52d28'
Whitfield II,Jerry
Hya 8h10'17"5d49'
Whitfield,Diane Loves Scott James
Uma 9h25'12"56d59'
Whitfield,Jeff Lee
Dra 18h25'24"80d33'
Whitfield,Mary Elizabeth
Hya 8h12'23"-0d27'
Whitfield,Shyrl A
Cyg 21h5'45"38d52'
Whitfield,Susan
Peg 23h39'12"31d44'
Whitford,Brad
Boo 14h34'0"32d6'
Whitford,Karen
Cas 23h29'60"54d21'
Whitford,Peter L
Boo 13h43'1"17d46'
Whiting,Erin Katherine
Umi 14h39'47"81d34'
Whiting,Jackie L
Tri 2h5'0"33d9'
Whiting,James
Cep 21h41'14"70d43'
Whiting,Jason Paul
Boo 14h50'1"46d54'
Whiting,Marcus Ben
Peg 22h19'21"34d50'
Whiting,Michael Stephen
Cet 2h32'19"5d12'
Whiting,Peter D
Cnv 13h58'27"37d49'
Whitlam,Roger Lee
Ori 5h27'29"-5d4'
Whitley,Christopher C
Cam 6h2'21"60d31'
Whitley,Dr "Chet"
Oph 18h18'55"8d59'
Whitley,Georgia Faye
Eri 3h1'40"-11d1'
Whitley,Joyce C
Cam 8h17'38"82d55'
Whitley,Rita G
Cas 2h16'46"75d23'
Whitley,Sadie Jo
Cap 21h0'29"-23d36'
Whitley,Sandy
Cyg 20h22'38"31d29'
Whitlock,David"Shatzi"
Ori 6h5'28"-1d46'
Whitlock,James O'C
Per 3h24'26"40d4'
Whitlock,Jr,"Star", Hunter B-Eloise P-
Cet 3h18'53"9d26'
Whitlock,Julian Kirk
Peg 0h1'36"13d7'
Whitlock,Kate
Aql 19h58'45"11d3'
Whitlock,Peter Michael
Cyg 20h21'43"40d9'
Whitlow,AFIFA Saba
Aql 19h4'19"-1d15'
Whitlow,Laura Mae Fitzpatrick
Eri 2h53'38"-1d41'
Whitlow,Suzanne
Peg 22h25'44"25d12'
Whitman,Elizabeth
Aqr 23h33'44"-11d59'
Whitman,Helaine
Aql 18h56'38"-3d7'
Whitman,Russell
Aur 6h11'53"33d1'
Whitman,Woodrow Wilson
Her 16h39'51"27d50'
Whitmer "My Heart,Your Star",Ronald N
Hya 8h22'56"0d24'
Whitmer,Christopher A
Her 16h11'1"23d57'
Whitmire,Joni
Mon 7h8'22"-5d15'
Whitmont,Reed Ezekiel
Tau 4h0'0"28d49'
Whitmore Angel Of Our Dreams,Alex
Her 18h54'43"12d16'
Whitmore,Anita
Peg 21h59'25"20d26'
Whitmore,Drew
Aql 19h51'23"11d56'
Whitner,Drew R
Uma 11h52'36"38d35'
Whitney
Cep 4h58'0"87d17'
Whitney
Tau 4h21'35"15d8'
Whitney
Cas 23h38'18"61d34'
Whitney
Aur 6h9'52"33d24'
Whitney
Crb 16h18'59"30d31'
Whitney M
Aql 20h10'47"14d26'
Whitney's "Wishing Star"
Mon 7h0'46"8d56'
Whitney's Wish
Ori 5h58'43"17d33'
Whitney,Andrew Fisher
Cet 1h54'25"-0d49'
Whitney,Anne Elizabeth
Cyg 19h29'15"31d42'
Whitney,Cameron
Cas 22h58'1"55d30'
Whitney,Cindy
Crb 15h52'15"28d14'
Whitney,Eric
Aur 7h21'46"39d58'
Whitney,Errol D
Lac 22h20'58"38d38'
Whitney,John
Leo 9h59'27"10d6'
Whitney,Jr,Theodore R
Lac 22h50'33"54d36'
Whitney,Kim
Psc 1h23'32"31d46'
Whitney,Lisa
Sge 19h59'31"20d34'
Whitney,Nancy A
And 0h9'19"40d30'
Whitney,Nicholas
Aqr 20h37'14"0d45'
Whitney,Raymond Ernest
Cep 21h41'31"85d55'
Whitney,Rose M
Cam 6h7'1"68d37'
Whitney,Ruth
Lyn 7h56'0"48d17'
Whitney,Samantha Renee
Lyn 8h18'20"57d39'
Whitney,Samantha
Per 2h38'24"34d25'
Whitney,Sarah
Mon 6h19'40"8d23'
Whitney,William James
Cep 22h8'23"61d9'
Whitsett,Patricia L
Cyg 19h59'38"38d56'
Whitsitt,Antonette Marie
Cam 3h19'29"60d37'
Whitson,Bradlee Stutzman
Boo 14h4'31"26d18'
Whitson,Harold
Boo 15h7'0"50d13'
Whitson,Michael K
Aql 19h0'59"-6d10'
Whitson,Ryan Lyal
Boo 14h38'37"21d0'
Whitt,Bon
Per 1h54'0"53d44'
Whitt,Illia Marie
Lib 15h29'29"-8d55'
Whitt,Sandra Lynn
Eri 3h30'42"-4d58'
Whittaker
Per 2h56'18"43d59'
Whittaker,Allan & Noelle
Cyg 19h24'1"30d5'
Whittaker,Derek George
Ari 2h33'14"30d11'
Whittaker,Edmund John
Peg 23h19'35"30d32'
Whittaker,Edmund John
Crb 16h16'44"33d57'
Whittaker,Gabrielle Sarah Yildiz
Mon 6h35'1"-6d41'
Whittaker,Martin
Her 17h17'0"43d56'
Whittaker,Richard & Nikki
Cyg 20h23'36"39d23'
Whittaker,Robert A
Aur 7h15'54"41d37'
Whittall,Elizabeth Ann
And 23h39'41"37d37'
Whittam,Bruce L
Cmi 8h8'25"6d36'
Whittam,Jennifer
Cas 0h36'60"74d45'
Whittemore,Joe
Her 17h38'13"26d48'
Whittemore,Leslie Harvey
Eri 2h49'38"-3d46'
Whitten,Leah D
Cas 23h13'1"61d12'
Whitten,Madison Elizabeth
Aql 19h55'33"14d36'
Whitten,Melissa
Dra 17h52'0"76d18'
Whittenbarger,Jennifer Jill
Peg 22h59'27"31d48'
Whittepate
Uma 9h24'23"58d12'
Whittingham,Ellen
Cam 9h2'36"74d40'
Whittingham,Guy
Ori 6h1'18"7d14'
Whittington,Bethany
Cas 0h55'29"61d31'
Whittington,Dr John
Lyn 8h22'1"44d43'
Whittington,MD,H G
Ser 15h17'48"12d26'
Whittington,Shorty
Per 3h17'45"40d25'
Whittle,Beryl Virginia Weeks
Dra 16h40'32"62d24'A
Whittle,Gerald F
Dra 16h40'32"62d24'B
Whittner,Lee Edward
Cep 20h41'40"61d28'
Whitton,Katy
Cam 7h56'14"60d39'
Whitwell,Florian
Lmi 9h50'26"38d39'
Whitwell,Stefan
Oph 17h13'40"-24d20'
Who's Who
Lyn 7h58'35"47d47'
Who's Your Buddy-Who's Your Pal
Per 1h25'12"53d42'
Whodie
Lyr 19h20'52"38d26'
Wholihan,Mickey
Ori 5h31'15"-3d35'
Wholley,Charise Mae
Vul 19h18'29"25d56'
Wholley,Jay & Marie -Noelle
Vul 20h0'55"28d23'
Whoopie
Cyg 21h7'39"30d9'
Whriternour,Michael James
Her 17h13'20"20d52'
Whybrow (LJ's Star), Lisa Jane
And 23h6'14"42d46'
Whye,Reese
Per 4h0'12"34d18'
Whyley,Stephanie
Per 3h2'25"40d52'
Whynot,Austin Chase
Aql 20h1'1"10d42'
Whyte #9,Sean
Aql 20h0'56"9d58'
Whyte (The Poet), Hamish
Lyr 18h17'30"41d3'
Whyte Fearless Leader, Jean
Lac 22h47'38"52d33'
Whyte,Alan
Peg 23h28'18"21d23'
Whyte,Anny
Peg 23h27'1"15d45'
Whyte,Derek
Per 2h39'23"37d28'
Whyte,Evan Michael
Ori 5h54'57"14d55'
Whyte,Jacqueline
Boo 14h22'45"50d47'
Whyte,Marion L
Eri 2h51'59"-1d47'
Whyton,Chelsey Leigh
And 22h59'41"41d7'
Wi Wi
Cyg 20h26'13"46d19'
Wiant,Helen P
Lyr 18h57'46"41d43'
Wiard,Nancy B
Cet 0h53'30"1d12'
Wiberg,Bengt"Joe"
Boo 14h45'20"50d4'
Wibirt,Craig
Crb 15h57'12"32d14'
Wibke*2431963-LTE
Cmi 7h20'59"5d19'
Wible,Matthew Thomas
Tau 3h54'31"18d58'
Wicall,Barbara Jean
Cma 6h37'38"-19d48'
Wiccan Jim
Per 3h46'50"37d47'
Wichert,Su
And 23h42'33"47d44'
Wichert,Tye R
Sge 19h29'34"19d12'
Wichita Lineman
Her 18h8'24"30d40'
Wichran,Aneta Aguta
Uma 9h59'18"45d32'
Wicht,Walter
Aqr 22h22'11"1d38'
Wick,David Cameron MacIver
Hya 8h43'48"-18d59'
Wick,David Rolland
Aql 19h10'0"12d7'
Wick,Elijah David Kirpal
Cet 0h51'22"-2d20'
Wick,Jessica
Del 20h31'22"10d35'
Wick-1994,Georg
Peg 22h28'12"31d35'
Wicka,Helen
Lyn 8h57'54"36d42'
Wicka,Marianne "Snookums"
Cas 0h19'49"66d9'
Wickberg,Gus
Cas 22h59'46"54d48'
Wicke,Ms Rebecca
And 0h58'39"36d53'
Wickel,Richard John
Aur 4h49'0"51d19'
Wicken,Helen
And 2h10'49"40d31'
Wickenden/Sales,Janine
Cas 0h43'59"63d55'
Wicker,Dan
Per 4h2'24"31d15'
Wicker,Dudley Byers
Ori 6h11'23"0d7'
Wicker,Ernest
Dra 19h26'45"84d35'
Wickersham,Cameron Patrick
Peg 0h7'34"13d6'
Wickersham,Elizabeth A
Tau 5h57'30"24d9'
Wickersham,Marshall E
Psc 23h3'48"1d25'
Wickes,Jason Todd
Cep 22h0'54"53d4'
Wickes,Robert Jeffrey
Aur 6h1'53"30d22'
Wickham,Kathy
Uma 11h1'58"48d54'
Wickham,Tami
Peg 23h41'21"8d58'
Wickizer,Kelley L
Lyn 7h59'58"40d31'
Wickliffe,Alice E
Mon 8h2'9"-1d18'
Wickline Heavenly Husband Star,The
Ser 15h13'27"22d31'
Wicklow,Leona Braun
Cam 4h6'23"60d4'
Wickman,Anne
And 23h20'29"47d6'
Wickoren,Andy
Ari 1h46'59"13d11'
Wicks,Ben
Her 18h2'59"28d23'
Wicks,Christopher
Sex 10h26'43"-8d52'
Wicks,Kathleen Marlow
Cam 4h50'1"68d23'
Wicks,Louise
Cas 0h20'23"62d3'
Wicks,Sam
Ori 5h56'50"13d12'
Wicks,Scott J
Aur 7h3'0"36d2'
Wickstrom,Melissa Jean
And 1h53'49"46d57'
Wickward,Robert Shawn
Cyg 20h23'34"38d27'
Wickwire,Emily Melissa
And 23h20'46"40d10'
Widbin,Elmer
Ori 5h56'13"17d2'
Widder
Ari 2h54'32"22d15'
Widdis "Eternal Love Star",Bernie & Bobbie
Del 20h34'0"18d57'
wide awake jake
Her 18h20'20"12d50'
Wide,Adam George
Aur 5h0'32"48d14'
Widejko,Chris Ann
Ari 1h52'29"10d44'
Widener,James Forrest
Oph 16h4'47"-6d39'
Widerman,Tracie B
Peg 23h35'12"30d29'
Wides,Kristi Keller
Cet 3h9'41"2d9'
Widger,Richard M
Aur 4h46'56"40d29'
Widick,Matthew Paul
Her 18h3'32"30d30'
Widlock,Alan K
Per 3h19'15"41d37'
Widman,Cari Anne
Gem 6h21'19"18d51'
Widman,Nell
Del 20h20'32"8d40'
Widmann,Bryan Lawrence
Boo 14h27'53"20d40'
Widmann,Michael Jay
Dra 19h45'1"84d55'
Widmann,Randall Marc
Boo 14h34'1"42d2'
Widmark,Carlye Hailand
Cnv 12h48'1"39d40'
Widmark,Carlye Hailand
Uma 11h56'0"61d6'
Widmer's Fishing Star, Grandma
Uma 11h46'33"57d10'
Widmyer,William N
Ser 18h43'28"3d35'
Widomski,Christine M
Cam 5h3'14"61d50'
Widro,Scott Z
Vul 19h22'19"23d53'
Wiebalck,Kristin
Lyn 9h12'46"37d20'
Wiebe,Jörg
Uma 9h27'16"47d58'
Wiebe,Robin
And 23h16'48"40d56'
Wiebe,Viola
Eri 3h31'22"-6d7'
Wieber,Thomas M
Aur 5h2'1"40d20'
Wiebke
Lyr 18h57'15"31d40'
Wieboldt,Christina A
Vir 11h43'37"8d7'
Wiebouw,Benjamin
Peg 23h37'20"22d1'
Wiechert,Norbert
Leo 9h20'55"7d57'
Wieck,Reinhold
Peg 23h29'54"13d30'
Wieczorek,Heather Anne
Lac 22h3'55"49d48'
Wieczorek,Michaeleen
Lyr 18h58'38"41d57'
Wieczorek,Monika
Leo 9h55'1"10d11'
Wieczorek,Patrice T
Sgr 19h48'52"-42d45'
Wieczynski,Thomas W
Cet 2h4'1"2d17'
Wiedecker,Linda E
Lyn 8h42'31"42d52'
Wiedeman,Debrah Kay
Eri 4h32'50"-6d2'
Wiedeman,Ronald
Aur 6h7'49"32d59'
Wieder,Howard A
Aur 5h29'38"31d41'
Wieder,Karen L
Cas 1h40'55"75d1'
Wieder,William
Cnc 8h35'19"32d22'
Wiederhorn,Peter William
Boo 14h23'43"26d55'
Wiederman,Billy
Aur 5h32'29"48d47'
Wiederrecht,Linnea McDonald
And 1h35'11"40d26'
Wiederrecht,Thomas Paul
Cep 21h43'36"55d55'
Wiederrecht,Thomas McDonald
Cep 21h43'31"58d14'
Wiedling,Gertrude Elizabeth Korte
Uma 12h7'0"57d11'
Wiedman,Edith
Ori 5h14'55"-5d47'
Wiedman,Edward
Ori 5h8'54"-4d35'
Wiedmeyer,Warren
Aql 20h1'29"1d25'
Wiefel,Bruno
Sgr 19h1'45"-24d23'
Wiegand,Karen
Ari 1h47'35"16d11'
Wiegand,Kenneth F
Aql 20h1'49"10d4'
Wiegard,Bernard J
Ori 6h2'1"5d39'
Wiegers,Christiane und Andreas
Cep 22h5'20"67d51'
Wiegmann,Thomas Bernhard
Uma 10h32'13"40d21'
Wiegmann,Wilma
Uma 10h49'15"47d3'
Wieher,Samantha Marie
Cam 8h1'29"68d46'
Wiekamp,Johanna
Equ 20h57'15"7d33'
Wieland,"M&M's"Sherry
Boo 14h22'24"23d46'
Wieland,Brigitte Everdine
Cnv 12h34'42"35d51'
Wieland,Craig David
Boo 13h56'22"17d36'
Wieland,Drew
Vul 19h34'43"27d46'
Wieland,Gladys H
Cyg 19h25'46"30d40'
Wieland,John
Cep 1h40'11"80d34'
Wieland-S,Christa
Leo 9h19'17"12d2'
Wieler,Michael Stephen
Dra 12h49'38"70d36'
Wielgo,Sigrid A & Zachary A
Cet 2h59'54"0d28'
Wielitsch,Martin
Sge 19h8'19"17d45'
Wiemann,Hans-Udo
Dra 15h10'46"62d26'
Wiemer,Charles & Tiernie
Sge 19h40'58"16d34'
Wien,Simone Shire
Cnv 13h0'43"50d11'
Wien,Stein Tore
Cep 21h34'45"60d17'
Wienckoski,Tom
Hya 8h43'53"4d44'
Wiener,Andrea J
And 1h18'22"39d29'
Wienhusen,Fredrick
Cet 2h42'48"3d5'
Wiening,John & Nancy
Crb 15h55'59"27d33'
Wienken,Annette
Cnv 12h18'1"39d25'
Wienman,Pete
Aql 20h3'0"8d33'
Wienman,Theresa
Aql 20h4'18"4d46'
Wiens,David A
Oph 17h33'41"-20d35'
Wiens,Patrick Byron
Aur 5h25'1"30d0'
Wienski II,Edward E
Cep 20h25'17"76d15'
Wier,Austin
Her 16h37'40"25d27'
Wierdo
Sge 20h17'45"16d10'
Wierenga,Debra Ann
Aql 19h7'18"-5d42'
Wiersbitzky,Horst
Cep 21h49'18"58d10'
Wiersch,Hans-Ulrich
Tau 4h13'33"1d56'
Wiertel,Bill
Boo 14h26'0"26d19'
Wierzbicki,Barbara I
Lyn 8h43'1"45d31'
Wierzbicki,Matthew
Dra 16h5'27"58d57'
Wiese,Laurin Norman
Uma 10h38'1"68d13'
Wiese,Louis
Aur 6h27'1"30d1'
Wiesemann,Aljoscha Renatus Franzis
Lyn 8h0'34"42d42'
Wiesen,Edy
Aql 19h53'44"12d54'
Wiesenbauer,Ursula
Umi 16h25'39"74d5'
Wiesenburger,Michael B
Her 18h2'46"28d46'
Wiesenhuetter,Ursula
Gem 8h2'52"31d14'
Wieser,Elisabeth
Gem 7h23'1"35d4'
Wieser,Ernst
And 23h1'51"49d36'
Wieser,Herbert
Cam 7h59'6"73d56'
Wieser,Ludwig
Ari 2h39'1"21d44'
Wieser,Scotty
Cma 7h21'28"-15d2'
Wieske,Elizabeth Malia
And 0h54'23"35d8'
Wiesmann,Klaus W
Boo 14h19'47"10d47'
Wiesner,Automaten
Gem 6h54'45"14d19'
Wiesner,Juergen
Ori 5h54'20"1d47'
Wiesner,Pete
Dra 15h55'21"52d28'
Wiesner-Buckley I
Cnv 12h23'54"44d51'
Wiest,Donna M
Lyr 18h28'0"42d44'
Wiest,Jessica Jeanne
Tri 2h0'32"31d40'
Wigand Star,The Rob & Nikki
Lyn 6h15'1"61d2'
Wigand,Jr,William James
Cep 23h21'30"70d28'
Wigda,Weronika
Ori 4h42'1"14d26'
Wigert,Shea Marie
Mon 7h46'1"-1d43'
Wiggans,Lari L
Hya 9h36'12"0d8'
Wiggenhorn,John
Aql 19h6'48"0d12'
Wiggerman,Austin James
Leo 9h19'54"17d29'
Wiggin,Marley Sigrid
Umi 16h4'26"75d19'
Wiggin,Melba & Tom
Lyn 8h59'23"40d60'
Wiggins Evening Star, Mark Alan
Cma 6h15'57"-20d35'
Wiggins' Star
Gem 6h50'24"18d28'
Wiggins,Christine Ann
Cas 1h15'18"60d6'
Wiggins,James Livingston
Cet 1h41'53"-4d4'
Wiggins,Jessica
Cas 0h16'17"66d0'
Wiggins,Maurita J
Cas 1h4'1"55d19'
Wiggins,Princess Susie
And 23h29'1"49d30'
Wiggins,Shanna
And 2h29'49"45d23'
Wiggins,Stephen L
Cep 23h11'0"61d23'
Wiggins-North,Judith
Mon 6h18'21"5d26'
Wiggles & Tim
Aur 5h0'40"45d49'
Wiggleton,Buddy
Sex 10h33'59"1d20'
Wiggs,Whitney Helen
Eri 3h0'54"-11d10'
Wiggy II,I M
Sge 19h0'41"18d47'
Wighardt,Elaine
Peg 23h6'35"20d12'
Wight's Twilight X,Joe
Aur 6h28'1"38d37'
Wight,Michelle
And 23h37'1"43d28'
Wight,Robin
Uma 9h57'26"51d55'
Wight,Sydney Erin
And 0h53'15"39d40'
Wightman,Barbara
Cas 2h30'18"73d33'
Wightman,David
Aql 20h6'33"3d48'
Wightman,Elliott
Oph 17h52'38"8d23'
Wightman,Steven L
Cma 6h36'1"-16d40'
Wigmore,John Joseph
Cep 21h56'21"67d35'
Wigodsky,Elena
Aqr 21h24'36"-10d57'
Wigy
Ori 5h59'36"11d0'
Wihbey,William
Peg 21h24'7"19d9'B
Wiik,Randi Lynn
Lac 22h31'52"41d1'
Wijtse
Ori 5h13'23"-0d42'
Wik,Daniel Ryan
Aur 5h50'31"38d20'
Wika,Terry R
Aur 5h17'0"52d39'
Wikane,Karl P
Sco 16h8'28"-31d7'
Wikehart,James M
Dra 19h44'52"68d11'
Wikgren,Kai Erik
Umi 13h9'15"75d42'
Wikman,Rebecca Jane
Col 6h0'24"-30d0'
Wiktil,Thomas
Aql 19h58'24"14d36'
Wil 04
Vel 10h6'7"56d20'
Wil & Renee All My Love
Mon 6h58'37"-6d22'
Wil,Marco
Aur 7h24'39"38d18'
Wilber's Wright
Her 16h13'28"48d40'
Wilberforce,Zana Anne
Umi 15h12'29"66d13'
Wilberg,Leeann
Vul 19h21'0"25d2'
Wilberg,Nicolette Marie
Aur 7h1'48"41d6'
Wilbers,Jessica
Cyg 21h24'59"39d24'
Wilbert,Annegret
Ori 6h0'1"-0d19'
Wilbert,Frederick Leigh
Dra 16h17'1"60d1'
Wilbert,Gregory P
Lac 22h5'55"38d56'
Wilborn,Thomas Brian
Cep 5h47'50"85d51'

Wilbur
Boo 15h26'11"41d26'
Wilbur,Marie Chanel
Mon 6h37'46"6d59'
Wilbur,Robert
Aur 5h12'23"41d34'
Wilburn,Bryan Karl
Oph 18h16'0"10d23'
Wilburn,Debra
Com 13h25'1"26d48'
Wilburn,Richard
Her 16h12'51"41d47'
Wilcken,Rhonda
And 23h40'11"42d29'
Wilco,Alan
Peg 22h39'57"2d18'
Wilcock,Anthony C
Per 2h11'39"57d11'
Wilcockson,Vicki
Vul 19h34'50"26d27'
Wilcove,Neil
Per 1h36'0"52d50'
Wilcox,Allen Russell
Oph 18h7'16"12d7'
Wilcox,Beth Conner
Mon 6h59'17"8d42'
Wilcox,Betty Lou Barringer
Cas 0h54'58"69d53'
Wilcox,Bradley Joe
Per 2h5'48"56d23'
Wilcox,Daniel Carle
Psc 0h3'42"-1d33'
Wilcox,David
Her 16h56'59"50d47'
Wilcox,Douglas
Aur 6h57'1"44d13'
Wilcox,Douglas E
Hya 8h34'47"0d17'
Wilcox,Lauren Nicole
Eri 3h55'1"-2d0'
Wilcox,Leslie Allison
Vul 19h57'25"27d21'
Wilcox,Monnie
Cyg 20h43'40"46d10'B
Wilcox,Robert Willis
Cep 23h19'51"65d20'
Wilcox,Ryoan Reneé
Ori 5h55'24"5d58'
Wilcox,Samuel Andrew
Her 16h15'1"48d47'
Wilcox,Sheryl
Crb 16h20'31"27d21'
Wilcox,Steven Anthony
Cep 21h41'52"56d2'
Wilcox,Stud Muffins
Her 18h0'1"38d32'
Wilcox,Sylvia
Her 16h58'47"50d42'
Wilcox,Thomas Jefferson
Cet 3h2'37"4d51'
Wilcox,Tigue
Cyg 20h43'40"46d10'A
Wilcox,Tony
Aql 20h0'17"9d31'
Wilcox-Sherman,Grace -Born 26/1/20
Aqr 22h23'33"-0d11'
Wilcoxon,Jason Paul
Her 17h33'26"20d40'
Wilcoxson,Kory Thomas
Peg 22h45'47"4d56'
Wilczynski,Stephen Walter
Dra 16h47'50"61d52'
Wild 21,Helen
Umi 13h59'1"72d34'
Wild Bill
Ser 15h22'0"-2d35'
Wild Bill & The Wench
Lac 22h28'51"38d45'
Wild One,The
Cnv 14h4'57"37d37'
Wild Rose International
Cyg 20h24'12"40d38'
Wild Thing
Lac 22h55'36"55d38'
Wild William
Aur 5h26'53"30d34'
Wild Wonderful William
Oph 18h4'59"-8d44'
Wild,Aggie Englum
Cet 1h37'17"-13d30'
Wild,Douglas
Peg 21h47'20"34d15'
Wild,Werner
Aql 20h9'55"4d59'
Wilda,David E
Boo 14h54'46"25d38'
Wilday,Mark "Wizzard"
Lyn 9h4'33"37d33'
Wilde II,George
Per 1h57'42"47d51'
Wilde III,George
Per 1h55'41"50d6'
Wilde III,William Hamilton
Lyn 7h46'18"44d24'
Wilde IV,George
Per 1h56'1"47d30'
Wilde,Andrea Sylvia
Psc 0h19'38"10d37'
Wilde,Eben Peter Wilson
Ori 5h0'19"15d16'
Wilde,Kim
Cas 2h13'40"60d60'
Wilde,Skyler McGovern
Per 2h55'34"40d51'
Wilde,Tucker Windsor
Aur 5h32'35"54d23'
Wilde,Wayne N
Ser 15h36'58"20d41'
Wildeboer,Ralph
Cmi 7h53'20"1d2'
Wildeman,Albert
Tri 1h40'1"30d8'
Wilder,Claire
Peg 21h58'52"33d21'
Wilder,Cynthia
And 1h49'14"40d49'
Wilder,Douglas Brent
Ori 6h16'37"-2d30'
Wilder,Stephanie Cree
Com 13h24'25"26d52'
Wilder,Timothy Bryan
Aur 7h1'16"43d39'
Wildes,Sr,Richard C
Per 2h51'30"40d9'
Wildey,Eric
Boo 14h37'15"31d50'
Wildfeuer,Alex Ryan
Uma 11h14'19"45d9'
Wildflower
Crb 16h16'23"28d36'
Wilding,Ellen
Cep 21h59'0"56d1'
Wildman,Leo
Leo 11h32'25"10d27'
Wildman,Tarik Charles
Ori 4h51'58"0d11'
Wildmann,Petra Lisa
Ari 3h20'49"30d35'
Wildner,Gerold
Cyg 20h25'1"40d36'
Wildner,Heidrun
Psc 23h4'57"1d22'
Wilds,Dick
Aur 5h10'45"41d17'
Wilds,Marlene S
Cnv 12h21'41"37d1'
Wilds,William "Kuhuna"
Cma 7h14'14"-16d41'
Wildsmith,Charles DeHart
Aur 6h8'31"38d44'
Wildsmith,Ira
Aql 19h8'0"1d12'
Wildstein,Kate Rostek
Sco 16h56'25"-41d1'
Wile,Dylan P
Lac 22h21'14"38d18'
Wilemon,Margie B
Cam 4h3'16"58d28'
Wilems,Jon
Crt 11h46'52"-17d33'
Wiler,Gloria
Equ 21h20'58"2d20'
Wiles II,Mark
Her 16h55'1"38d2'
Wiles,Alice L
And 0h6'41"31d28'
Wiles,Dale
Aql 19h56'39"13d54'
Wiles,Kenneth "Putsy"
Aur 4h53'60"41d10'
Wiles,Nancy Lynn
And 0h20'1"36d12'
Wiles,Sue
Cas 1h5'21"73d31'
Wiles,Suzanne
Aql 18h56'13"-10d2'
Wiles,Wanda Jean
Uma 12h5'35"56d44'
Wiley 02-14-86
Sex 10h30'23"-0d58'
Wiley,Brandi Monique
Eri 4h8'11"-15d60'
Wiley,Brooke
Cet 1h24'13"-10d36'
Wiley,Catherine
And 1h45'1"40d22'
Wiley,Glenda Eloise
Peg 22h8'37"27d45'
Wiley,Gregory Stephen
Cet 3h11'52"2d55'
Wiley,Lyle
Per 2h59'53"38d2'
Wiley,Lysle Tobin
Aql 19h25'1"14d43'
Wiley,Madeline Linnea
Lyn 8h36'1"41d2'
Wiley,Ralph "Mick"
Equ 21h20'59"8d49'
Wiley,Robert G
Lac 22h36'1"53d35'
Wiley,Therese A Field
Ori 6h6'38"6d58'
Wiley,Tracey Renee
Lyn 8h4'36"33d59'
Wilfling Super One
Aql 19h0'14"-5d41'
Wilfling,Scott Robert
Gem 6h33'22"13d17'
Wilfong,Alba Tart
Cyg 19h22'23"28d15'
Wilfong,H Naomi
Peg 22h0'1"21d54'
Wilford
Boo 15h4'49"10d14'
Wilford Binary Star, The
Cyg 20h37'32"30d20'AB
Wilheim,Julian Robert
Her 17h8'30"42d32'
Wilhelm
Dra 9h39'0"80d53'
Wilhelm Herold
Cep 22h23'53"60d4'
Wilhelm Ruling The Host
Cep 1h58'18"77d14'
Wilhelm,Anita L
Uma 11h55'38"38d57'
Wilhelm,Bernd
Vul 19h44'58"29d9'
Wilhelm,Bernd
Peg 23h35'35"11d25'
Wilhelm,Hans-Georg
And 23h21'24"45d1'
Wilhelm,Helga
Vir 11h42'1"7d50'
Wilhelm,Leo Benedikt
Sgr 18h50'30"-25d39'
Wilhelm,Li
Cap 21h3'22"-23d16'
Wilhelm,Roy Carlos
Vir 11h35'56"2d45'
Wilhelmi,Douglas
Her 17h11'40"48d38'
Wilhelmson,Clifford G
Uma 8h58'16"50d4'
Wilhelmus-Fortis Custos
Uma 10h42'27"55d48'
Wilhelmy Family,George
Cam 11h22'47"80d17'
Wilhite,Ian "Buddy"
Aql 20h21'38"3d43'
Wilhoit,Zrteest Lynn
Cam 10h14'54"82d20'
Wiliam,Gerald
Boo 14h3'19"10d18'
Wiliamson,Tricia Jo
Mon 6h26'12"-6d25'
Wilinski,Kilika
Peg 22h47'55"31d44'
Wilinski,Marcellus Paul
Cep 21h5'54"60d57'
Wilk,Daniel Mark
Uma 8h35'55"70d35'
Wilk,David Mayer
Uma 8h54'0"62d17'
Wilk,Franciszek
Per 3h1'34"46d6'
Wilk,Heather Gayle
Uma 8h36'50"71d53'
Wilk,Larry
Lac 22h28'53"54d9'
Wilk,Wendy Ann
Uma 8h52'23"71d14'
Wilke B
Cep 20h56'37"70d50'
Wilke,Ursula
Cnc 8h31'40"30d55'
Wilken,Christine Marie
Crb 15h46'57"37d57'
Wilken,Heino
Oph 18h0'1"7d43'
Wilken,Tara Michelle
Crb 16h19'15"38d17'
Wilken,William
Lac 22h1'14"50d37'
Wilkens,Jonathan William
Lmi 9h28'1"38d23'
Wilkens,Mark
Aur 5h25'20"54d33'
Wilkerson II,Regina & Joseph
Cep 23h10'54"71d11'
Wilkerson,Cheri
Mon 7h43'25"-2d21'
Wilkerson,Joan
Cyg 21h7'0"30d43'
Wilkerson,Richard P
Hya 8h41'19"3d36'
Wilkes,Christopher
Aur 7h10'17"41d33'
Wilkes,Henry Alan
Ari 2h59'1"30d23'
Wilkes,Jennifer Jane
Peg 22h0'57"30d56'
Wilkes,Linda
Cyg 21h3'35"30d59'
Wilkes,Phil
Ori 6h1'42"3d47'
Wilkes,Randall W
Equ 21h3'1"8d5'
Wilkes,Richard Benjamin
Sct 18h43'3"-6d52'
Wilkes,Shernie Rae
Cam 5h11'51"68d58'
Wilkes,Teresa
Peg 23h36'40"31d34'
Wilkeson,Gloria K
And 0h21'42"37d46'
Wilkeson,Robert L
Aur 6h9'42"35d31'
Wilkie
Aur 5h17'53"47d51'
Wilkie,Alison J
Cam 4h52'1"70d43'
Wilkie,George John
Aql 20h12'52"10d59'
Wilkie,Hiawatha & S F
Cyg 21h54'23"36d15'
Wilkie,Leighton
Cnv 13h39'46"40d26'
Wilkie,Olivier
Umi 14h42'53"69d34'
Wilkie,William
Cep 23h20'24"70d22'
Wilkimson,Robert B
Cnv 13h51'35"38d51'
Wilkin,Bradley E
Uma 8h44'20"68d1'
Wilkin,Fred
Boo 14h26'27"27d14'
Wilkins Family,Mark J
Lyn 7h2'49"51d17'
Wilkins' Eyes,The Sparkle In Laurie
Cas 0h8'1"66d7'
Wilkins,Bill & Margie
Crt 11h36'46"-20d47'
Wilkins,Darlene
And 1h58'22"40d26'
Wilkins,Dennis
Per 1h51'1"47d38'
Wilkins,Fred Allen
Cet 2h32'56"6d27'
Wilkins,Holly Jane
Uma 9h21'29"53d31'
Wilkins,Mabel
Cyg 20h5'17"31d38'
Wilkins,Marlyn Susan Gardiner
Ori 6h0'18"3d36'
Wilkins,Marsha
Lac 22h14'28"50d10'
Wilkins,My Little Morgan Brooke
Mon 6h33'42"-0d18'
Wilkins,Paul Philip
Tau 4h19'9"20d8'
Wilkins,Ricardo
Mic 20h37'58"-29d38'
Wilkins,Richard L
Dra 20h13'22"63d31'
Wilkins,Wendy
Uma 10h27'24"51d27'
Wilkinson,Ann
Cas 0h17'52"62d44'
Wilkinson,Austin Wyatt
Cnv 12h57'52"51d38'
Wilkinson,Bethany Yvonne Michaela
And 23h22'42"51d60'
Wilkinson,Bruce Griffith
Ori 6h2'34"4d57'
Wilkinson,Carl
Per 2h8'59"57d48'
Wilkinson,Chad Michael Robert
Aql 19h15'57"10d24'
Wilkinson,Claire
Lyr 18h18'36"46d45'
Wilkinson,Clare
Lyr 18h29'15"44d49'
Wilkinson,Collin Russell
Peg 22h19'0"7d45'
Wilkinson,Dorothy
Peg 21h50'18"33d26'
Wilkinson,Dorothy Anne
Aql 19h59'0"15d40'
Wilkinson,Eric Todd
Oph 17h21'45"22d44'
Wilkinson,Hannah
And 0h10'56"33d48'
Wilkinson,Howard C
Her 17h38'26"14d43'
Wilkinson,J Anne
Mon 6h58'28"10d36'
Wilkinson,James
Cyg 20h0'46"50d29'
Wilkinson,John Griffith
Lmi 10h39'42"31d0'
Wilkinson,Katelyn Nicole
Lyr 18h34'41"36d43'
Wilkinson,Matt
Aql 19h31'39"10d59'
Wilkinson,Michael
Aql 20h10'0"1d43'
Wilkinson,Nicholas Scott
Aur 6h25'44"38d45'
Wilkinson,Patricia Ann
Mon 6h33'56"-0d56'
Wilkinson,Pauline Murphey
Cas 0h3'1"61d28'
Wilkinson,Robert & Mary
Cyg 19h29'27"30d7'
Wilkinson,Ronald John
Aql 19h57'39"15d16'
Wilkinson,Sarah
Lyr 18h55'17"40d46'
Wilkinson,Steven & Kristi Ann
Peg 23h31'55"21d22'
Wilkinson,Tracy
Dra 14h3'18"63d37'
Wilkisson III,Frank W
Her 18h38'1"18d57'
Wilkisson,Faith Elizabeth
Her 18h37'45"18d53'
Wilkisson,Jr,Frank W
Lyn 7h29'23"41d38'
Wilkisson,Timothy Edward
Ori 5h42'45"-10d25'
Wilkowski
Cet 3h18'57"5d15'
Wilks,Jane & Michael
Cyg 20h39'14"42d41'
Will
Cma 6h54'11"-17d27'
Will
Cnv 13h13'28"40d9'
Will Son Septaurus
Ari 2h0'47"26d16'
Will You
And 0h11'10"37d12'
Will You Marry Me Gia?
Lyn 8h56'51"38d10'
Will's Analytical Engines In The Sky
Cep 0h9'1"80d16'
Will's Electric Dream
Cmi 7h55'0"7d43'
Will's Kenya Star
Ser 15h27'23"10d10'
Will's Star
Lyr 19h0'0"26d4'
Will's Wonder
Aql 19h54'43"13d58'
Will,Alice J
And 23h15'39"51d0'
Will,Ileana Michelle
Peg 22h42'20"21d38'
Will,Justin Michael
Dra 11h4'46"74d7'
Will,Kathryn
Lyn 7h27'31"50d19'
Will,Keith P
Boo 14h32'26"45d40'
Will,Lane
Umi 14h35'0"67d12'
Will,Melanie
Cas 0h40'1"60d11'
Will,Michel
Ori 6h1'30"8d43'
Will,Roswitha
Cap 20h8'50"-10d28'
Will-Andress,Kerstin
Leo 9h19'0"17d56'
Will-Dor
Cep 22h14'51"61d32'
Willa
Uma 11h13'45"48d47'
Willa de Paul et Clélia
Uma 8h31'0"50d30'
Willa Sue Love
Equ 21h19'26"10d36'
Willa's Star
Umi 16h11'0"70d21'
Willadsen,Lisa
Vul 19h18'58"26d50'
Willahan,Susan
Vul 19h17'37"22d16'
Willaman,"US" Lisa & Gregg
Lyn 8h2'1"47d46'
Willard
Hya 8h9'46"2d26'
Willard,Aimee
Cas 1h7'0"63d44'
Willard,Bruce N
Hya 8h31'19"-6d8'
Willard,Charlotte D
Dra 19h24'60"68d56'
Willard,David Gene
Cyg 21h5'53"39d41'
Willard,Loralee S
Uma 9h59'37"44d55'
Willard,Robert
Hya 9h0'12"2d22'
Willard,Sarah Elizabeth
Cam 4h21'41"61d19'
Willard,Stephen Douglas
Her 17h53'57"14d37'
Willbanks,Jeffrey Noel
Gem 7h6'10"21d40'
Willbanks,Michael Shane
Oph 17h52'30"-6d49'
Willcox,Barbara
Uma 9h38'53"50d47'
Willcox,Ray & Connie
Cyg 21h51'53"42d13'
Willden,David Bret
Ori 5h45'37"12d16'
Wille,Howard Albert
Uma 10h25'37"40d6'
Wille,Judy
Sge 19h52'59"16d18'
Wille,Judy
Peg 0h0'49"30d20'
Wille,Leonard Ernst
Cep 22h32'1"63d12'
Wille,Petra
Mon 7h53'44"-1d41'
Wille,Wilbur
Cnv 12h12'30"37d46'
Willee
Ori 5h54'35"5d2'
Willem,Nancy Lynn
Mon 7h43'48"-4d23'
Willems,Julie
Cyg 21h31'1"40d48'
Willener,Beatrice Abbott
Cas 0h22'26"70d57'
Willenson,Elaine K
Peg 23h45'28"31d44'
Willer,Agnes
Cas 2h31'50"70d13'
Willer,Dory
Peg 22h23'23"21d1'
Willerup,Hope Elizabeth
And 23h33'12"44d24'
Willett,Alma
Mon 6h19'24"8d37'
Willett,Claire
Cyg 20h16'52"40d32'
Willett,Melissa
Peg 23h19'55"30d6'
Willett,Susan Elisabeth
Ori 6h14'27"3d53'
Willett,Walker Jennings
Per 3h7'41"46d7'
Willette,Darcy Ann
Crb 16h2'0"28d19'
Willette,Jr,Scott Edward Stanley
Per 2h42'23"40d2'
Willever,Cory Allen
Aql 19h24'36"-5d47'
Willey,Cheryl
Peg 22h30'57"21d42'
Willey,James
Ori 5h55'23"7d39'
Willey,Karen
Peg 22h56'15"17d41'
Willey,Marianne
Lyr 18h51'26"30d0'
Willey,Michael E
Aur 5h12'45"40d38'
Willey,Paul & Alice
Mon 7h42'14"-2d15'
Willheim,Johannes- Patrick
Dra 12h3'2"69d4'AB
Willhide,Paul C
Ari 3h24'44"28d16'
Willi
Lyn 8h11'38"43d30'
Willi & Trish
Cas 0h29'31"68d15'
William
Cyg 21h31'1"50d6'
William
Hya 8h18'27"-10d25'
William & Aliana
Eri 3h16'1"-12d1'
William & Carolyn
Cyg 21h23'28"31d51'
William & Carolyn's Golden Cynosure
Umi 13h50'46"75d59'
William & Gillian
Cyg 21h4'59"30d38'
William & Sondra
Her 16h8'46"25d23'
William & Sonja
Umi 15h6'22"70d34'
William Eros Aeternus, Tatiana
Vul 20h59'60"20d4'
William Eternal- Forever BV
Her 18h0'16"48d27'
William F-#1 Super Dad
Aur 5h2'19"38d9'
William III
Lac 22h30'16"55d30'
William IV,Zarian, Gregory
Cet 2h6'26"2d14'
William Ivy"Bill Lee"
Cma 6h53'16"-17d8'
William Martin
Cet 1h33'12"-10d50'
William Saul
Cmi 7h27'1"7d45'
William Stanley Sr
Ori 5h50'31"11d58'
William V:Ira
Dra 12h19'1"70d35'
William Wayne
Her 16h25'51"38d27'
William Zachary
Ori 5h49'19"11d37'
William's Delight
Oph 18h17'16"12d6'
William's Legacy
Cet 3h0'57"2d30'
William's Star Of The Abiding Presence
Ori 5h40'13"8d41'
William's Way
Sct 18h47'21"-6d30'
William's Wish Land
Dra 17h5'41"61d40'
William,Elizabeth Joy
And 23h38'18"40d52'
William,Henry
Ori 4h55'24"-2d13'
William,Jake
Ori 6h4'0"5d11'
Williams (Big Daddy) #55,Jayson
Lac 22h37'56"55d49"
Williams Escape,Eve
And 0h50'54"41d10'
Williams Family Star, The
Aql 20h21'18"7d38'
Williams Family Star, James S
Cet 2h20'29"-1d34'
Williams Family,The
Boo 14h58'15"38d33'
Williams III,David
Lac 22h36'50"38d37'
Williams IV,Henry Phillips
Leo 9h19'0"18d11'
Williams Love,Tom & Judy
Cas 0h37'43"70d27'
Williams Lux Vitae,G J
Per 3h10'31"41d0'
Williams Star Mom, Shannon Janet
Com 12h23'12"26d32'
Williams Star,Sheerin
Per 2h2'57"50d41'
Williams' Lucky Star, R Marc
Aur 6h4'57"45d35'
Williams' Night Light, Dwayne
Cet 1h24'1"-12d43'
Williams,Adriana Taylor
Aql 19h54'38"13d3'
Williams,Aimee
And 1h58'23"46d60'
Williams,Alan
Aqr 20h36'43'-1d3'
Williams,Alexander Francis
Uma 10h10'17"58d59'
Williams,Alfred Frank
Uma 12h20'48"54d29'
Williams,Alison Jane
Lac 22h31'57"56d24'
Williams,Allison Leigh
Sgr 19h30'31"-44d34'
Williams,Alyse
Cas 23h20'1"53d41'
Williams,Amanda
Cep 0h5'46"68d14'
Williams,Amanda
Mon 6h38'55"6d9'
Williams,Amelia B
Equ 21h19'39"10d19'
Williams,Amy C & John M
Cyg 21h18'11"38d16'
Williams,Andrew Keith
Aql 18h43'36"8d28'
Williams,Anna
Lyr 18h54'32"35d0'
Williams,Anna Corliss Maggio
Mon 6h44'1"7d60'
Williams,Art & Angela
Equ 21h19'53"11d9'
Williams,Ashley Marie
And 0h20'26"30d14'
Williams,Barbara
Cyg 21h6'18"31d17'
Williams,Barbara A
Lyr 19h18'57"42d13'
Williams,Beau Ethan
Umi 13h15'46"76d31'
Williams,Betty Ruth
Mon 6h53'48"-10d45'
Williams,Billy
Cyg 20h22'1"38d17'
Williams,Bob
Cet 1h29'51"-5d7'
Williams,Bonnie
Eri 3h58'49"-4d34'
Williams,Bonnie Toxby
Cam 3h59'48"70d19'
Williams,Brian Dean
Cnv 13h43'1"33d39'
Williams,Brian T
Lac 22h7'16"51d52'
Williams,Brody Michael
Cam 4h21'1"67d50'
Williams,Bruce I
Her 16h1'34"48d30'
Williams,Bruce Todd
Her 16h41'0"20d26'
Williams,Candice Rose
Com 12h53'16"24d28'
Williams,Candy
Cyg 20h37'42"31d12'
Williams,Carley Marie
Uma 12h27'56"62d17'
Williams,Caroline
Cas 0h59'25"50d21'
Williams,Carrie Marie
Lyn 7h55'41"35d37'
Williams,Catherine
And 0h38'54"40d35'
Williams,Catherine Ruth
Aql 19h30'30"12d15'

Williams,Cedric Charles
Per 3h12'17"40d49'
Williams,Charity
Peg 23h34'1"31d49'
Williams,Charles Richard
Boo 15h5'25"30d50'
Williams,Charles George
Ori 5h25'41"15d19'
Williams,Charles Dewey
Lac 22h28'1"53d40'
Williams,Charles Scott
Aql 20h5'16"0d41'
Williams,Cheney Hastings
Cep 23h5'32"65d9'
Williams,Cheryl
Leo 10h35'41"22d11'
Williams,Christina
Uma 10h39'29"47d10'
Williams,Christine
Pho 23h43'12"-41d21'
Williams,Christopher Thomas
Vul 19h43'18"23d40'
Williams,Christopher
Dra 20h1'1"62d19'
Williams,Cindy
Tri 2h2'23"31d5'
Williams,Claire
Ori 6h0'58"1d11'
Williams,Curtis
Uma 11h20'0"32d14'
Williams,Darlene
Mon 6h38'47"-0d8'
Williams,Darwin Lee
Vir 13h38'56"-9d47'
Williams,David James
Lac 22h10'0"50d15'
Williams,David John
Her 17h32'20"37d44'
Williams,Dawneda F
Peg 23h23'55"30d7'
Williams,DC,CCSP, Michael A
Aur 5h9'0"42d42'
Williams,Deidre
And 0h6'29"30d41'
Williams,Denen
Her 17h19'38"42d38'
Williams,Dennis L
Sct 18h53'0"-6d19'
Williams,Don D
Her 16h59'0"40d46'
Williams,Donald
Aur 6h22'51"33d28'
Williams,Donald C
Leo 10h26'31"11d18'
Williams,Donivan R
Cet 2h17'42"0d33'
Williams,Doris
Cas 0h32'17"60d27'
Williams,Dorothy
Cas 23h43'31"50d0'
Williams,Dot
Cas 0h48'55"73d28'
Williams,Dr Sheila Quinlan
Oph 18h16'47"1d9'
Williams,Edward F
Aur 6h6'39"31d58'
Williams,Edwin "Ashanti"
Oph 17h16'23"-23d57'
Williams,Eileen
Lyr 18h54'60"37d28'
Williams,Eileen A
Vul 20h39'18"28d35'
Williams,Elaine Louise
Cyg 21h25'22"38d43'
Williams,Eleanor
Cyg 19h37'57"38d36'
Williams,Elizabeth Rachel
Peg 22h48'42"27d53'
Williams,Elizabeth Ann
Mon 6h19'24"8d37'
Williams,Eloise
And 2h30'1"49d44'
Williams,Emily Elizabeth
Lyr 18h43'43"40d9'
Williams,Emily Kathryn
Cas 2h40'25"70d26'
Williams,Ernestine Towns
Ori 4h57'54"3d32'B
Williams,Evan T
Cmi 7h59'58"0d30'
Williams,Fanny
Dra 19h6'52"48d9'
Williams,Fernando H
Vir 11h55'0"-0d45'
Williams,Finty
And 0h1'14"46d57'
Williams,Fran "Muzzy"
Aql 18h43'17"11d2'
Williams,Frances
Ori 5h15'54"15d40'
Williams,Gabriella Martay
Peg 23h4'19"8d3'
Williams,Gail & Claudia
Lyn 8h6'1"50d43'
Williams,Gena L
Lyn 9h7'11"38d21'
Williams,Geoffrey D
Equ 21h2'27"2d38"
Williams,George Franklin
Sct 18h32'52"-6d31'
Williams,George J & Mary L
Cyg 21h36'13"38d19'
Williams,Gregory
Vul 21h3'31"26d54'
Williams,Gretchen
Crt 11h41'1"-10d56'
Williams,Gwen
Cas 0h56'15"67d1'
Williams,Heather Michelle
Eri 3h55'16"-10d8'
Williams,Hello Mother!
Connie
And 23h13'44"35d56'
Williams,Herbert Clinton
Ori 4h51'53"7d17'A
Williams,Hettie Olene
Cas 0h2'27"63d39'
Williams,Holly Marie
Sex 10h10'44"-0d41'
Williams,Imari Kimya
Mon 7h57'14"-2d3'
Williams,In Memory Of Brad
Cmi 7h22'28"4d58'
Williams,Iwan
Umi 15h22'28"70d24'
Williams,J Morgan
Dra 16h52'0"63d51'
Williams,James & Jennifer
Cep 20h24'1"75d13'
Williams,James Craig
Hya 8h12'16"-0d51'
Williams,James Howard
Leo 9h58'57"16d9'
Williams,James S
Cet 1h21'27"-0d9'
Williams,Janet
And 0h20'60"45d43'
Williams,Janice
Mon 7h22'31"-8d6'
Williams,Janice A
Lyr 19h21'14"38d15'
Williams,Jaunta Gui
Ori 4h51'53"7d17'B
Williams,Jay Wright
Aur 6h29'30"39d49'A
Williams,Jean-Pierre
Hya 8h57'39"0d30'
Williams,Jeff
Aur 6h1'24"30d38'
Williams,Jeffrey P
Her 18h31'36"20d31'
Williams,Jena
Vul 20h15'37"23d33'
Williams,Jennifer Lea
Lyr 18h31'34"33d0'
Williams,Jennifer Lynn
And 23h3'1"51d4'
Williams,Jessica
Peg 22h59'11"31d33'
Williams,Jill
Ori 5h19'53"0d2'
Williams,Jim
Aur 5h59'26"37d45'
Williams,Joann
Aql 18h58'38"13d29'
Williams,Joey
Aql 20h33'30"-6d59'
Williams,John
Aql 19h57'38"14d35'
Williams,John C
Per 1h47'1"56d39'
Williams,John Mills
Oph 17h31'26"-20d35'
Williams,John-Margaret
Com 13h4'37"15d1'
Williams,Jonathon Hunter
Aql 20h20'0"1d9'
Williams,Joseph"Sean"
Dra 15h59'27"68d36'
Williams,Joshua Adam
Mon 8h1'20"-6d52'
Williams,Joshua C
Boo 14h14'43"15d8'
Williams,Joshua David
Cep 21h29'29"55d52'
Williams,Joy Dean
Peg 23h1'24"32d57'
Williams,Jr,Douglas J
Uma 8h35'33"60d47'
Williams,Jr,Frank H
Cep 0h49'0"86d28'
Williams,Jr,Joahn A
Her 16h52'1"39d24'
Williams,Jr,Manning Augustus
Lyn 9h16'30"39d11'
Williams,Judy
Cam 5h29'57"68d2'
Williams,Judy
Cas 23h33'23"60d31'
Williams,Julian
Lyn 8h7'59"44d42'
Williams,Julian & Karen
Peg 23h17'13"33d55'
Williams,Juliann J
Vul 20h15'57"26d7'
Williams,Julie Ann
Cas 23h31'59"61d29'
Williams,Karen Kaye
Vul 19h0'21"25d37'
Williams,Karen Lee
Cyg 21h25'59"38d45'
Williams,Karen S
Peg 21h44'54"27d34'
Williams,Kathleen B Schindel
Aqr 20h53'1"-0d30'
Williams,Kelsey Elizabeth
And 0h4'31"28d28'
Williams,Ken
Cyg 19h25'58"35d46'
Williams,Kenneth
Cep 23h17'55'64d33'
Williams,Kenneth J
Dra 17h4'55"64d1'
Williams,Kevin D
Her 16h38'49"47d47'
Williams,Kevin Homan
Aur 4h58'59"52d19'
Williams,Kevin W
Vul 19h48'16"28d12'
Williams,Kevin Wayne
Dra 19h51'15"68d47'
Williams,Kimberly Michelle
Umi 15h44'34"78d23'
Williams,Kimberly S
Cas 0h57'30"50d14'
Williams,Kris
Peg 22h39'13"29d59'
Williams,Kristin Mourna
Com 12h7'39"27d7'
Williams,Lauren Alise
Cet 2h19'47"4d53'
Williams,Len
Her 17h52'0"38d20'
Williams,Leslie & Phillip
Del 20h23'55"10d18'
Williams,Linda
Oph 18h3'31"11d32'
Williams,Linda
Aql 19h56'52"14d6'
Williams,Linda M Bennett
Mon 8h3'1"-1d39'
Williams,Lisa
And 0h20'42"35d3'
Williams,Louise
Vul 19h48'28"20d21'
Williams,Louise Marie
And 2h18'46"37d6'
Williams,Lynn Caroline
Sge 20h0'24"16d31'
Williams,Lynsey Joanne
Crb 16h13'42"32d13'
Williams,Margaret Elaine
Cas 2h33'34"75d57'
Williams,Margaret Mennen
Peg 21h55'32"33d19'
Williams,Marie
Cas 0h31'1"58d8'
Williams,Marie Louise
Sgr 19h37'37"-34d2'
Williams,Mark
Cam 3h55'33"74d21'
Williams,Martyn Jon
Cep 21h19'47"55d17'
Williams,Mary Ellen
And 1h37'57"39d20
Williams,Mary Jo
Mon 6h54'1"-0d8'
Williams,Mary Rodney
Equ 21h19'54"2d19'
Williams,Mason Paul
Lac 22h10'1"47d56'
Williams,Matthew
Uma 8h34'26"67d54'
Williams,Maureen Ann
Cyg 21h27'14"48d45'
Williams,Max
Ser 15h20'13"20d55'
Williams,MD,Sonja
Peg 23h15'29"30d12'
Williams,Megan H
Peg 22h39'42"11d45'
Williams,Megan Spess
And 0h14'46"39d47'
Williams,Melanie
Peg 23h3'54"31d39'
Williams,Melissa
Peg 0h8'55"20d28'
Williams,Melissa Ann
Cyg 21h50'36"38d38'
Williams,Melissa Ann
Cyg 21h13'55"39d43'
Williams,Michael
And 23h20'51"35d17'
Williams,Michael Anthony
Per 3h14'38"41d48'
Williams,Michael
Ser 15h38'58"-1d16'
Williams,Michael
Cam 4h0'11"61d46'
Williams,Michael Emory
Peg 23h27'11"24d16'
Williams,Michelle Diane
Cas 23h40'12"48d48'
Williams,Mr & Mrs Mitch
Lmi 10h50'46"32d55'
Williams,Neal Earl
Hya 9h33'51"0d27'
Williams,Nerys
Lyr 18h26'49"45d23'
Williams,Nicolas Luke
Aur 5h0'11"46d11'
Williams,Norma Jean
Aql 19h48'52"14d6'
Williams,Norman David
Sgr 19h37'37"-38d1'
Williams,Olivia Danielle
Psc 0h25'12"8d5'
Williams,Owen Oscar David
Her 16h38'49"4d25'
Williams,Pamela
Aql 19h31'41"-0d5'
Williams,Patricia Tilson
Cet 1h33'0"-11d38'
Williams,Patricia L
And 0h19'39"38d26'
Williams,Patricia S
And 0h16'53"30d2'
Williams,Patrick
Uma 10h43'21"47d29'
Williams,Patrick & Diana
Lyr 18h34'42"37d1'
Williams,Paul
Lmi 9h49'48"34d57'
Williams,Paul
Aur 4h56'45"40d2'
Williams,Paul N
Her 17h19'52"29d47'
Williams,Penny
Cma 6h59'31"-19d32'
Williams,Peter Lavis
Her 18h5'59"45d11'
Williams,Phyllis A
And 2h2'0"45d46'
Williams,Rachel Elizabeth
Cyg 21h6'1"30d44'
Williams,Rachel A
Cyg 20h20'42"40d31'
Williams,Raquel
Ser 17h33'57"-10d2'
Williams,Ray J
Ser 17h31'0"-13d9'
Williams,Raymond
Cet 1h21'1"1d5'
Williams,Raymond Paul
Hya 10h18'42"-16d25'
Williams,Richard Coleman
Lac 22h3'0"51d25'
Williams,Richard & Carla
Cet 1h34'20"-6d7'
Williams,Richard F
Boo 14h43'1"21d46'
Williams,Rita Kranson
Aql 20h9'14"7d15'
Williams,RN,Kim
Cyg 21h5'38"35d21'
Williams,Robert
Cep 23h5'35"64d22'
Williams,Rodney
Oph 16h50'11"10d23'
Williams,Rodney Lee
Sex 10h34'21"-6d16'
Williams,Roger
Ori 4h57'54"3d32'A
Williams,Roger Earl
Her 16h21'14"23d38'
Williams,Rose
Cam 7h20'43"73d10'C
Williams,Rozan Reed
Aql 18h43'46'6d59'
Williams,Ruel Melbourne
Eri 3h44'11"-3d58'
Williams,Russel Todd
Gem 6h43'19"15d21'
Williams,Ryan
Her 16h35'16"21d33'
Williams,Sandy
Lyr 19h1'0"46d57'
Williams,Sara
Umi 13h46'57'77d45'
Williams,Sarah
And 0h26'23"38d22'
Williams,Sarah Louise
And 0h39'15"31d40'
Williams,Sarita
Lyn 9h5'1"41d57'
Williams,Scott Edward
Cyg 21h0'1"31d35'
Williams,Sharon L
Cet 2h34'58"-10d47'
Williams,Sherry
Uma 13h49'29"61d15'
Williams,SiÉn Louise
Aur 6h35'46"30d44'
Williams,Spencer
Umi 3h48'1"88d47'
Williams,Stancle Edward
Cet 1h32'49"-12d29'
Williams,Stanley
Cnv 12h43'19"40d15'
Williams,Stefan "Gowumpkey"
Boo 14h6'22"47d48'
Williams,Stell & Lew
Lyr 18h50'51"41d3'
Williams,Stephanie
Lyn 8h26'51"41d51'
Williams,Sterling Scott
Per 2h52'54"40d58'
Williams,Steven & Janet
Lac 22h2'1"37d43'
Williams,Stuart James (Kimo)
Oph 18h3'45"12d13'
Williams,Susan Elizabeth
Lyr 18h41'12"26d19'
Williams,Susanne Bliss
Lyn 9h17'30"37d21'
Williams,Suzy
Cyg 19h30'12"36d56'
Williams,Tamara Lynn
Uma 11h25'43"32d25'
Williams,Terry
Per 1h52'34"56d26'
Williams,The One & Only
Nancy
Mon 6h54'57"8d45'
Williams,Thomas George
Per 2h50'0"46d45'
Williams,Thomas Karl
Umi 15h18'1"72d3'
Williams,Tim-Kibby
Oph 18h2'59"11d7'
Williams,Todd
Her 17h6'29"38d6'
Williams,Ty Reed
Sex 10h9'49"2d20'
Williams,Varalyn Denise
Del 20h24'58"10d57'
Williams,Varina"Sugar"
Crt 11h16'15"-17d23'
Williams,Vicki A
And 1h30'29"39d35'
Williams,Victoria
Cas 0h58'34"58d57'
Williams,Wallace Audley
Cma 6h54'1"-19d13'
Williams,Wendell
Eri 3h49'14"-4d25'
Williams,Wendi Raechel
Cyg 20h55'38"30d13'
Williams,Wendy
Cyg 21h54'12"42d28'
Williams,Wilhelmine M
Cas 1h12'0"61d50'
Williams,Zarra Ann
Uma 8h41'1"56d58'
Williams,Zeek
Cap 21h2'0"-22d59'
Williams,Zoe
Lyn 8h12'0"44d33'
Williams-WGP 1993-1994
William L
Per 1h43'17"53d34'
Williamsen,C Douglas
Aur 5h54'33"40d4'
Williamson III,Jerry Bowen
Cma 7h5'46"-28d17'
Williamson,Christopher
Her 17h28'45"20d8'
Williamson,Don
Dra 16h16'11"67d4'
Williamson,Donald Lee
Hya 8h10'0"5d21'
Williamson,Donna Margaret
Lyn 7h46'45"51d5'
Williamson,Erick
Per 4h1'1"50d34'
Williamson,Erin
Dra 16h47'39"68d11'
Williamson,Erwin Blackstone
Cnv 13h57'58"31d32'
Williamson,Haley Benjamin
Peg 23h39'60"31d58'
Williamson,James Douglas
Lac 22h9'34"38d49'
Williamson,Jeffrey
Dra 15h44'17"58d48'
Williamson,Jesse Speight
Aql 18h59'37"10d58'
Williamson,John & Sharon
Lmi 10h11'1"33d38'
Williamson,Julie Ann
Cnc 7h55'52"10d42'
Williamson,Kristy Ann
And 0h6'38"44d27'
Williamson,Lavell
Peg 23h1'13"13d52'
Williamson,Louise Amanda
Cam 3h56'0"58d4'
Williamson,Luke Joseph
Ser 15h49'12"22d1'
Williamson,Marjorie Cherup
Peg 22h8'11"25d25'
Williamson,MD,David A
Oph 17h5'54"7d40'
Williamson,Nancy R
Lyr 18h32'0"46d17'
Williamson,Nina Louise
Cas 0h18'14"58d38'
Williamson,Patricia R
Cas 0h38'60"60d8'
Williamson,Patti
And 0h6'0"46d40'
Williamson,Rebecca Elizabeth
Mon 7h57'56"-0d22'
Williamson,Robert A & Helen
M
Sge 20h13'23"17d32'
Williamson,Robert Paul
Dra 19h41'39"67d30'
Williamson,Stephen Neil
Dra 20h9'51"62d42'
Williamson,Sunday Elizabeth
Cyg 19h40'46"38d59'
Williamson,Talon L
Aur 6h47'24"37d39'
Williamson,Trevor James
Her 16h50'23"40d43'
Williamson,Tyce James
Oph 17h32'38"-0d26'
Williamson,Victor
Aur 7h12'0"40d23'
Williamson,William
Aql 18h53'40"10d38'
Willibey,Tom D
Per 3h11'25"47d33'
Willick,Kelly Elizabeth
Ari 2h0'23"18d40'
Willie
Cet 3h12'23"0d37'
Willie
Aur 6h56'41"43d24'
Willie
Aur 5h1'37"42d59'
Willie
Del 20h28'41"20d11'
Willie
Her 16h58'1"32d23'
Willie
Boo 15h4'21"14d18'
Willie O
Dra 11h9'45"74d50'
Willie One
Aur 7h17'41"41d14'
Willie's Genesis
Mon 7h52'38"-4d20'
Willie's Guiding Light
Per 3h3'37"48d48'
Willie's Way
Ori 5h56'57"11d30'
Willie's World
Aur 5h20'54"38d8'
Williesan-Monkster
Cyg 21h14'22"34d37'
Willig,Christopher Robert
Boo 14h56'0"21d40'
Willimas,Jon
Per 3h11'0"45d58'
Willin "My Hero", Michael
Uma 13h22'60"54d16'
Willinger,Christina
Cas 1h47'1"58d2'
Willinger,Jan & Harding
Dra 17h51'43"57d39'
Willingham - Pacoima JHS
Music,Irene
Lmi 10h55'48"25d9'
Willingham,Dakota Ellis
Ori 5h2'1"1d28'
Willingham,Dean
Ser 16h15'0"2d57'
Willingham,John
Per 4h7'1"51d9'
Willingham,John Leonard
Oph 16h20'16"0d13'
Willingham,Judy
Peg 22h3'59"2d7'
Williquette,Barney
Cep 21h46'54"60d36'
Willis II,William Granville
Her 18h2'31"28d19'
Willis,Bobby Dean
Hya 9h9'0"5d11'
Willis,Bradley Joseph
Tri 1h57'51"30d40'
Willis,Cynthia
And 0h8'44"28d2'
Willis,Cynthia B
Peg 22h56'1"20d34'
Willis,David
Cnc 7h55'0"10d45'
Willis,Devyn Taylor
Equ 20h59'18"6d28'
Willis,Donald G
Aur 5h2'14"49d36'
Willis,Elizabeth
Eri 3h29'15"-3d13'
Willis,Frank Thomas
Cnv 12h49'19"39d35'
Willis,Ginny Turner
And 23h16'47"40d9'
Willis,Helen C
Equ 21h12'0"4d28'A
Willis,James Dean
Ori 5h29'33"-0d1'
Willis,Jean Pinder
Lib 15h1'12"-5d39'
Willis,Jennifer
Cam 6h8'49"60d43'
Willis,John Alexander
Sex 10h13'19"-4d0'
Willis,Kristen Marie
Aur 5h15'1"40d37'
Willis,Lawrence W
Hya 8h59'18"1d48'
Willis,Lee
Cet 2h17'19"3d42'
Willis,Lyndsay Gray
Uma 11h30'42"52d50'
Willis,Markus
Uma 12h58'14"54d29'
Willis,Matt M & Hoffman,Kelly
J
Crb 16h19'43"27d16'
Willis,MD,Amos Johns
Oph 17h29'34"7d47'
Willis,Patricia G
Aqr 20h52'35"-5d43'
Willis,Paul Michael
Cma 7h55'0"1d2'
Willis,Philip C
Equ 21h12'0"4d28'B
Willis,Richard S
Per 2h57'19"34d20'
Willis,Richard S
Tri 2h26'0"28d24'
Willis,Robin
And 23h36'34"39d29'
Willis,Roy Clifford
Her 17h10'34"46d40'
Willis,Ryder Blue
Dra 15h37'21"54d18'
Willis,Stephen Randell
Sge 19h40'42"16d24'
Willis,Steven Gray
Uma 11h27'29"50d30'
Willis,Teresa Thompson
Uma 11h48'20"53d32'
Willis,Timothy F
Boo 15h18'11"51d37'
Willis,William
Peg 23h41'47"15d17'
Willis,William Brinson
Ser 16h4'1"10d27'
Willison,Daniel Martyn
Her 18h3'40"20d14'
Willison,Gary & Nicole
Macklin '96
Uma 9h34'33"55d30'
Willison,Jr,William J
Boo 14h23'20"26d32'
Willison,Mason Cade
Cet 2h17'31"3d54'
Willits,Bill
Aql 20h11'0"4d2'
Willits,Neil
Cap 20h39'26"-18d47'
Willkomm
Cmi 7h40'0"0d39'
Willliams,Joanne
Uma 10h45'32"47d49'
Willman,Benjamin Charles
Her 18h10'1"40d38'
Willmann,Laura Mae
Cam 6h44'55"68d19'
Willmann,Monika Maria
Gem 7h4'22"24d41'
Willmott VP9 MU,Paul
Aql 19h56'1"10d41'
Willner,Jay
Aur 5h31'35"50d17'
Willock,Eternally Dan
Per 1h26'58"53d11'
Willoughby,Juanita & Hugh
Cyg 19h29'43"33d43'
Willoughby,Karen
Lyr 18h58'35"33d33'
Willow
Crb 16h14'1"37d47'
Willow Star
Cnv 13h28'17"41d8'
Wills (O&C),C Kevin
Aql 20h14'19"1d23'
Wills,Denise
Ori 4h48'13"0d8'
Wills,Diana Leigh Thomas
Peg 22h40'0"20d4'
Wills,Dr Klaus
Aql 19h5'42"-1d6'
Wills,Jeffrey Matthew
Crb 15h57'42"38d4'
Wills,Jospop
Ori 5h57'16"16d12'
Wills,Jr,Gregory Lee
Cep 22h34'21"60d51'
Wills,Patsy Jane Cecil
Cnv 13h22'59"42d10'
Wills,Paula
Peg 22h1'27"30d7'
Wills,T Stewart
Oph 17h15'30"10d43'
Wills,Tara Louise
Cas 23h5'57"58d22'
Wills,William H
Aql 20h36'35"-8d48'
Willsey,Alan
Cep 22h12'13"68d16'
Willsey,Catherine
And 2h9'0"40d50'
Willshaw,Alexander George
Cep 22h53'58"63d33'

Willson,Amy
Tri 1h51'60"26d51'
Willy
Aqr 22h49'53"-4d21'
Willy's Wonder
Ori 5h37'27"1d13'
Willyard,Albert H
Her 17h2'36"48d42'
Willyard,Jack
Her 17h13'55"48d43'
Wilma "Schnuffi"
Leo 10h54'14"2d22'
Wilma Faye The Wonderful
Cam 10h57'0"80d1'
Wilma Jean,The
And 0h6'23"38d30'
Wilma-Wish Upon A Star
Cas 0h58'56"59d19'
Wilmer,Ag
Gem 6h44'30"18d34'
Wilmer,Robert D
Vul 19h58'1"22d44'
Wilmer,Rudy
Gem 6h44'52"18d56'
Wilmink,Ronnie J H
Tau 3h43'52"25d16'
Wilmore,Winston
Ori 5h55'16"14d56'
Wilmot,Tomas
Cep 23h8'11"60d47'
Wilmoth,Nancy
Leo 10h42'36"15d5'
Wilms,Bernd
Lyn 8h7'58"41d29'
Wilmsen,Heidrun
Hya 8h13'60"1d48'
Wilmsmeier,Günter
Uma 9h31'41"45d56'
Wilner,Julia Leigh
Cam 3h24'0"61d32'
Wilner,Larry
Aur 6h35'30"37d30'
Wilner,Madison Lauren Elizabeth
Mon 6h34'19"0d41'
Wiloe
Cen 11h50'22"47d59'
Wilp,Charles -WXLP-
And 23h22'44"45d30'
Wilps,Ralph Felix
Aur 5h25'0"30d41'
Wilsack,Lisa Simone
Ori 5h55'58"14d56'
Wilsch Forever,Robert M
Aur 6h11'38"30d45'
Wilsch,William John Ambrose
Boo 15h44'51"48d3'
Wilsdon 75,Rene
Cas 2h55'26"63d60'
Wilshusen,Ginia
Peg 22h38'49"30d0'
Wilsim
Lup 15h17'57"-40d51'
Wilson
Aur 7h4'54"36d60'
Wilson
Vul 19h16'10"24d37'
Wilson & Betty Abbott's Ember
Boo 14h42'0"26d5'
Wilson & David,The Kenneth
Sco 17h27'19"-30d26'
Wilson & Missy,Sharon K
Cam 5h51'43"58d38'
Wilson (Cutie Pie), Teresa M
Cyg 20h26'53"37d57'
Wilson (Douglas Smith),William A
Cep 23h9'33"70d15'
Wilson 1994,Dawn
Cyg 20h37'0"38d18'
Wilson Family Star,The
Mon 6h21'44"2d44'
Wilson I,John Karol
Dra 16h7'24"66d45'
Wilson II,Colin M
Cep 21h28'22"66d49'A
Wilson III "Vince", John V
Ori 5h57'18"5d29'
Wilson III,John Claude
Dra 12h43'55"75d6'
Wilson III,Thomas Hampton
Oph 17h34'44"-1d25'
Wilson IV,Robert C
Ori 5h39'0"1d33'
Wilson MA,MFCC,Phyliss
Equ 21h20'44"10d20'
Wilson MD,Phillip O
Oph 18h6'53"8d28'
Wilson Star of my Heart,Jimmy
Her 17h38'24"23d31'
Wilson Star,The Frank & Alice
Cyg 19h28'20"30d50'
Wilson V,George Field
Aur 6h11'31"31d19'
Wilson's Bright Wonder Lloyd
Del 20h14'28"10d34'
Wilson's Rising Star 2000
Aur 7h9'44"37d47'
Wilson's Special Star, Shirley
Peg 21h53'15"33d4'
Wilson's Stars Of 1995
Lac 22h26'36"54d0'
Wilson,Adam John William
Boo 15h32'14"48d13'
Wilson,Albert H
Lac 22h23'40"41d7'
Wilson,Alex
Uma 2h12'27"56d4'
Wilson,Alexandra Marie
Tri 2h4'38"31d46'
Wilson,Allison Elizabeth
Cas 23h38'1"50d23'
Wilson,Anna
Eri 3h14'59"-11d11'
Wilson,Anne
Cet 2h44'14"-1d15'
Wilson,April Jodene
Lyn 7h9'54"44d45'
Wilson,Archibald Potts
Eri 4h3'60"-11d36'
Wilson,Arnold
Cet 2h27'37"5d3'
Wilson,Barbara R
Cas 0h28'0"50d9'
Wilson,Beau
Her 17h3'35"21d32'
Wilson,Berkeley
Ser 15h14'11"24d42'
Wilson,Bernice
Lyr 18h45'38"37d16'
Wilson,Bertha
Mon 7h53'17"-5d58'
Wilson,Betty
Cyg 20h42'58"38d49'
Wilson,Beulah
Com 12h29'18"30d20'
Wilson,Billie Jo
Mon 6h56'52"-10d11'
Wilson,Billy
Lib 15h43'56"-20d19'
Wilson,Bradley John
Gem 6h49'50"14d51'
Wilson,Brenda
Del 20h14'54"15d20'
Wilson,Brian Lyle
Sex 10h36'32"-6d18'
Wilson,Britney Lynn
Cet 3h5'14"3d50'
Wilson,Brittany
Cam 3h54'38"71d28'
Wilson,Bryan Michael
Tau 5h51'19"28d32'
Wilson,Bud & Freda
Vul 19h34'27"27d21'
Wilson,C T Fantasy- Claude Tyler
Uma 9h2'15"56d21'
Wilson,Candace Lee
Aqr 20h59'54"-10d55'
Wilson,Carla Valentine
Cyg 21h21'59"38d26'
Wilson,Carnie
Hya 9h3'23"2d53'
Wilson,Carol Renee
Com 13h24'52"26d12'
Wilson,Carolyn Marie
Peg 22h3'60"25d46'
Wilson,Cdr Joseph R
Hya 9h4'44"2d29'
Wilson,Chad Joseph
Cet 2h8'1"-2d3'
Wilson,Charles
Ori 6h2'44"8d49'
Wilson,Cheryl Jo
Lyr 19h13'15"41d44'
Wilson,Chris
Ser 17h33'11"-14d34'
Wilson,Christopher Ryan
Crt 11h18'45"-10d57'
Wilson,Christy Lynne
Cam 3h27'42"60d14'
Wilson,Cindy
Cas 3h4'22"60d41'
Wilson,Colin H
Her 16h56'45"30d5'
Wilson,Curtis
Del 20h53'51"2d24'
Wilson,Dan C & Helen M
Sge 19h54'39"16d12'
Wilson,Dana
Aql 20h1'11"4d49'
Wilson,Danica Reneé
Cas 1h58'41"73d36'B
Wilson,Daniel S
Mon 6h2'35"-8d34'
Wilson,Danny Ray
Aql 18h58'31"-5d35'
Wilson,Darrel Dean
Ori 5h21'0"12d59'
Wilson,Darrell K
Cam 3h57'31"68d18'
Wilson,David
Aur 4h53'44"51d16'
Wilson,David Bruce
Her 16h53'37"28d12'
Wilson,Del
Her 16h34'41"39d19'
Wilson,Dennis Alexander
Ori 6h2'19"2d19'
Wilson,Diana Marie
Peg 22h4'0"25d38'
Wilson,Don
Aql 20h2'21"1d31'
Wilson,Donald Edward
Ser 18h55'0"3d7'
Wilson,Donna R
Aur 7h3'17"38d35'
Wilson,Doris
And 2h34'19"44d40'
Wilson,Douglas
Per 4h28'14"50d17'
Wilson,Douglas Usher
Per 2h3'25"50d17'
Wilson,Dr Timothy R
Oph 18h19'21"10d34'
Wilson,Drue
Aql 19h53'25"13d52'
Wilson,Elijah Robert
Aur 5h32'39"30d37'
Wilson,Elizabeth
Cyg 19h43'49"35d58'A
Wilson,Elizabeth Ann
Leo 11h8'32"-1d29'
Wilson,Elizabeth Anne
Lyr 18h59'60"41d1'
Wilson,Enos Robert
Cet 2h10'54"5d38'
Wilson,Erin
Mon 6h22'25"5d43'
Wilson,Ewen Keith
Per 3h7'0"31d16'
Wilson,Floyd Leonard
Aur 5h11'33"41d55'
Wilson,Frank Jahile
Cap 20h32'44"-13d11'
Wilson,Gary Lee
Oph 16h58'58"11d8'
Wilson,Gary Wayne
Sex 10h13'48"-4d53'
Wilson,George & Maribel
Crb 16h0'39"33d56'
Wilson,George Paul
Cma 6h55'10"-19d15'
Wilson,Georgia L
Vul 19h3'31"25d22'
Wilson,Gordon
Aql 20h6'24"6d54'
Wilson,Guy
Per 3h4'51"41d50'
Wilson,Hamline & Dorothy
Cyg 20h28'51"40d2'
Wilson,Harley Joseph
Boo 14h39'54"15d53'
Wilson,Heidi
Del 20h22'1"2d31'
Wilson,Helen C
And 0h34'28"23d44'B
Wilson,Irene
Umi 15h21'0"68d52'
Wilson,Irene Arlene
Lyr 18h58'22"46d27'
Wilson,Irma R
Mon 6h12'31"-10d21'
Wilson,Jackie
Aur 6h57'1"43d57'
Wilson,Jacob Bard
Cmi 7h53'20"1d2'
Wilson,James
Lac 22h26'37"50d18'
Wilson,James Harold
Dra 14h17'18"64d33'
Wilson,Jane
Aql 19h54'1"10d15'
Wilson,Janet Maria
Equ 21h4'0"3d14'
Wilson,Janice Rose
Cas 1h25'14"68d1'
Wilson,Jean Louise
Lyr 19h25'38"38d34'
Wilson,Jeffrey Thomas
Ori 5h9'14"-9d16'
Wilson,Jennifer Rae
Eri 2h53'0"-3d28'
Wilson,Jess Edward
Sex 10h43'22"0d33'
Wilson,Jessica Andrea
Mon 7h49'56"-3d40'
Wilson,Jill Lenise
Cra 18h16'34"-43d27'
Wilson,Joanne S
Ori 5h52'11"12d12'
Wilson,John
Cma 7h2'57"-28d15'
Wilson,John A
Cep 21h39'22"55d41'
Wilson,John B
Equ 21h20'1"10d38'
Wilson,John B
Ser 17h59'14"-14d38'
Wilson,John Calvin
Lmi 9h57'35"37d33'
Wilson,John M
Boo 14h18'31"54d51'
Wilson,Jon Dale
Aur 5h25'51"50d8'
Wilson,Joseph G
Cmi 7h45'34"7d37'
Wilson,Josie Jayne Holdcroft
Lyn 8h48'11"37d31'
Wilson,Jr,George A
Oph 17h0'17"10d43'
Wilson,Jr,James Allen
Aql 19h50'52"13d59'
Wilson,Jr,Jim
Eri 3h49'53"-2d25'
Wilson,Jr,Steven John
Dra 16h50'20"66d54'
Wilson,Judith Gay
Peg 0h9'1"18d4'
Wilson,Judith S
Vir 13h2'46"-8d59'
Wilson,Judy
And 2h35'55"41d10'
Wilson,Kalsey Marie
And 2h22'11"38d53'
Wilson,Karen Annette
Peg 22h16'59"21d36'
Wilson,Kari-Emma Valentine Star
Lib 15h18'10"-21d39'
Wilson,Kathleen Ann
Uma 10h24'57"59d46'
Wilson,Kathy Lynn
Mon 6h54'46"-10d39'
Wilson,Kelly Nicole
Cmi 8h1'43"1d10'
Wilson,Kelsey
Vul 20h22'17"25d25'
Wilson,Kelsie Danielle
Peg 0h8'1"18d30'
Wilson,Kiel C
Oph 17h54'38"10d53'
Wilson,Kim
Dra 18h9'1"80d9'
Wilson,Kimberly Lynne
And 23h37'1"46d57'
Wilson,Krista Karin
Cas 0h32'59"61d3'
Wilson,Kristine
Peg 22h43'33"25d46'
Wilson,Larry
Aur 5h0'22"38d41'
Wilson,Lauren
Cyg 21h6'40"31d2'
Wilson,Lawrence Artimus
Oph 17h21'5"-20d4'
Wilson,Lee Edward
Peg 23h24'47"18d26'
Wilson,Lee-Anne
Cyg 19h30'1"30d45'
Wilson,Leon C
Uma 11h39'37"45d27'
Wilson,Leslie N
And 0h34'28"23d44'A
Wilson,Lillian
Eri 2h49'49"-6d46'
Wilson,Linda
Sct 18h55'57"-7d2'
Wilson,Lois Mae
Cyg 20h3'39"40d50'
Wilson,Lori Ann
Sge 19h53'22"16d5'
Wilson,Lucas Robert
Cep 22h40'51"57d40'
Wilson,Mabel
Ori 6h6'58"0d55'
Wilson,Marianne
Mon 6h39'2"-0d3'
Wilson,Marie
Sge 20h14'49"17d35'
Wilson,Marsha A
Mon 7h2'23"5d3'
Wilson,Martha Jane
Peg 21h25'51"23d15'
Wilson,Mary
Cas 0h7'41"63d5'
Wilson,Mary Echols
Cyg 20h34'1"37d60'
Wilson,Mary Louise
Mon 7h16'11"-6d48'
Wilson,Mary Lucille
Equ 21h6'58"11d49'
Wilson,MaryGayle
And 1h20'0"37d25'
Wilson,Matthew C
Sex 9h51'47"-0d24'
Wilson,Megan T
Peg 23h49'19"31d14'
Wilson,Michael Clifford
Her 17h20'1"44d29'
Wilson,Michael Jon
Lac 22h38'1"53d21'
Wilson,Michael Scott
Leo 10h59'23"2d8'
Wilson,Michael Shaun
Aur 6h4'59"36d56'
Wilson,Michael Vincent
Tri 1h41'0"33d55'
Wilson,Michelle Franklin
Uma 12h36'1"60d4'
Wilson,Mr Mark A
Aur 6h2'1"45d6'
Wilson,My Beebers, Laura
Sct 18h44'3"-7d1'
Wilson,Nancy
Cet 0h23'50"-18d18'
Wilson,Neil
Cet 2h4'37"4d31'
Wilson,Nicole Katherine
And 0h12'54"37d17'
Wilson,Norm John
Her 16h39'0"48d17'
Wilson,Norma K
Aql 20h14'17"1d22'
Wilson,P & H Cafe Wanda
Uma 12h38'47"60d59'
Wilson,Patrick M
Aur 5h5'12"31d5'
Wilson,Paul E & Irene L
Lac 22h8'29"38d33'
Wilson,Paul F
Cep 22h30'42"63d59'
Wilson,Paula
Peg 22h59'50"22d15'
Wilson,Peter
Cyg 19h43'49"35d58'B
Wilson,PhD,Jeanne
Dra 14h3'30"63d41'
Wilson,Phillip Craig
Peg 0h9'33"18d27'
Wilson,Phillip McEwen
Peg 0h9'0"13d23'
Wilson,Prima
Eri 4h28'42"-1d22'
Wilson,Rabecca M
Lib 15h12'46"-8d50'
Wilson,Rachel I
Cyg 19h31'12"33d21'
Wilson,Rachel L
Cyg 19h31'28"35d10'
Wilson,Rex Daniel
Dra 17h6'32"51d28'
Wilson,Rex Duane
Aur 5h58'27"29d53'
Wilson,Ricardo Aparicio
Sct 18h55'60"-6d46'
Wilson,Richard
Aql 19h49'34"12d12'
Wilson,Richard
Her 16h40'1"50d32'
Wilson,Richard & Ruth
Cam 5h44'18"60d53'
Wilson,Richard C
Boo 14h42'33"38d43'
Wilson,Robert Christopher
Boo 14h59'24"10d20'
Wilson,Robert Allen
Ori 6h8'30"5d53'
Wilson,Robin Durae
Cet 3h4'1"6d59'
Wilson,Robin Lee
Lac 22h20'0"51d29'
Wilson,Robyn Paige
Cmi 7h35'21"11d54'
Wilson,Rod & Jenny
Cam 7h42'28"72d25'
Wilson,Rulfus T
Lac 22h13'19"46d42'
Wilson,Samantha Ann
Lmi 10h51'43"26d1'
Wilson,Samuel Chester
Per 2h41'24"43d29'
Wilson,Sandra
Aur 5h17'0"40d40'
Wilson,Scott Adam
Her 17h21'55"21d37'
Wilson,Scott Cole
Ori 5h38'56"-10d54'
Wilson,Sean
Oph 16h59'42"-25d40'
Wilson,Shaina
And 1h53'60"38d19'
Wilson,Shelby Clara
Lyr 18h31'0"38d26'
Wilson,Shelby Louise
Lyn 6h20'30"60d46'
Wilson,Sr,Henry Woodrow
Lyn 07h40'01"52d08'
Wilson,Stan A
Aur 5h52'49"30d43'
Wilson,Starr
Her 16h45'1"35d4'
Wilson,Stella
Cas 0h6'33"64d8'
Wilson,Sterling Forrest
Equ 21h21'0"8d22'
Wilson,Steve
Lac 22h13'51"51d24'
Wilson,Steven Adam
Hya 8h55'31"3d46'
Wilson,Steven Wayne
Tri 2h43'46"31d46'
Wilson,Stuart W
Cep 23h3'43"80d4'
Wilson,Susan
And 0h51'17"39d26'
Wilson,Sylvester L
Dra 17h28'45"50d35'
Wilson,Sylvia Caryl Posch
Sex 10h7'55"-2d17'
Wilson,T
Del 20h52'53"9d60'
Wilson,Tedd T
Vul 19h45'54"28d48'
Wilson,Terry Alan
Lac 22h33'48"50d19'
Wilson,The Italian Pinto Roxana
Cet 2h44'34"-1d27'
Wilson,Thomas Buck
Her 16h39'30"50d34'
Wilson,Timothy S
Aql 19h26'40"7d55'
Wilson,Tom & Leo Medish
Dra 20h20'40"64d2'
Wilson,Trent Nicholas
Her 17h20'40"50d8'
Wilson,Trisha Lynn
Uma 11h44'0"47d0'
Wilson,Tug
Her 18h28'60"12d22'
Wilson,Unruley Julie
And 0h16'38"30d42'
Wilson,Wendell
Uma 13h21'42"58d46'
Wilson,Wendy
Cet 3h1'40"2d34'
Wilson,Wes W
Her 16h0'40"42d23'
Wilson-A Shining Star Forever,Ralph
Lyn 7h52'54"50d45'
Wilson-Angelo,Nadine M
Cam 3h13'7"65d28'A
Wilson-Best Bud,Bob
Dra 17h13'47"64d41'
Wilson-Littlejohn, Julie Ann
Del 20h36'45"18d46'
Wilson-Shine Forever, Johnadean
Crb 16h10'32"26d45'
Wilson-Underwood,Mary
Mon 6h35'13"11d1'
Wilsons
Cyg 20h5'0"40d23'
Wilt,Dr Edward F & Marie T
Cyg 21h32'0"53d1'
Wiltanger,Holly M
And 2h13'1"41d55'
Wilton,Betty D
Hya 9h0'26"4d0'
Wilton,Serena Elizabeth
Uma 10h59'48"59d34'
Wilton,William Jeremiah Teresi
Tau 4h1'18"21d15'
Wiltshire Happy 70th 8/10/25,Eric
Cep 21h22'46"67d55'
Wiltshire,Nancy J
Com 13h0'31"28d28'
Wiltshire,Peter
Ori 6h5'42"5d23'
Wiltshire,Richard
Dra 16h43'1"60d10'
Wiltz,Brady Lee
Ser 17h34'32"-13d54'
Wiltzius,David Paul
Her 17h32'20"26d4'
Wilusz,Amy
Her 16h32'15"42d9'
Wilusz,Thomas John
Ori 5h56'0"14d46'
Wim de Haas
Psc 1h0'51"31d13'
Wimberly,Jerry
Per 2h52'28"32d53'
Wimble,Irene V
Boo 14h58'43"20d31'
Wimer,David Blair
Dra 15h49'0"65d43'
Wimer,Eric Steven
Per 4h40'17"38d49'
Wimer,Kathleen
Vul 19h16'0"24d44'
Wimmer,John & Elizabeth
Cyg 20h18'49"31d2'
Wimmers,Ursula
Cas 1h45'29"61d18'
Wimpenny,David A
Oph 18h38'56"10d35'
Wimpy,Danielle Nicole
And 2h24'15"42d24'
Win's Star
Cas 2h44'31"67d37'
Winand,Bruno
Peg 22h30'51"8d25'
Winand,Shirley
Cam 6h55'1"68d39'
Winant,Asa Mercury
Ari 2h5'20"21d24'
Winar
Dra 16h48'17"61d59'
Winberg,Jordan Faith
Cyg 21h11'0"37d54'
Winch,Lisa & Shawn
Mon 6h0'53"-5d56'
Winch,PhD,Gil S
Lac 22h9'42"46d16'
Winch,PhD,Guy
Lac 22h30'0"50d13'
Winchell,Bob
Cam 5h5'14"70d9'
Winchester III,Eugene
Cnv 12h54'13"51d20'
Winchester,Charlie R
Eri 3h2'45"-1d52'
Winchester,Dennis
Aur 7h20'15"40d20'
Winchester,Dorothy & Ottis
Cyg 21h53'13"42d3'
Winchester,Matthew
Her 16h37'15"36d57'
Winchester-Sells
Cet 3h16'51"5d34'
Winckler,Brigitte
Aur 4h52'52"50d22'
Wind Beneath My Wings MCGD
Cyg 19h42'13"28d44'
Wind Dagger
Dra 19h59'56"70d59'
Wind Dancer G A H
Aur 5h18'33"45d38'
Wind On The Giburarutal,The
Aqr 22h3'36"-21d17'
Wind,Brian
Cet 0h25'1"0d56'
Wind,Jr,Daniel C
Oph 17h54'17"11d19'
Windau,Alexander Christian
Aur 6h3'31"33d38'
Windau,Amanda Marie
And 0h1'26"34d23'
Windau,Christopher Shane
Lyn 7h4'0"59d53'
Windel,Dennis & DeDe
Equ 21h22'45"3d39'
Windemuth,Donna Sue
And 22h59'0"50d41'
Winder's Christmas Star,Geri
Sct 18h30'54"-4d49'
Windette,William
Lac 22h24'51"50d18'
Windham-Klepper,Lawson
Aur 6h7'54"43d48'B
Windisch,Marcia
Lac 22h3'1"49d48'
Windischman
Lmi 9h30'46"38d18'
Windle,David J
Cru 12h40'14"-62d29'
Windle,Jerry & Georgetta
Cyg 20h17'18"30d50'
Windle,Samantha M
Cyg 20h20'1"39d21'
Windmiller,Elizabeth J
Cas 0h23'54"60d29'
Windmoser,Eva-Maria Ernestine Eva
Ari 1h52'23"11d21'
Windover,Donald Arthur
Ser 18h15'60"-0d11'
Windrider
Ori 5h56'38"14d42'
Windrum,Catherine L
Cas 0h36'29"67d58'
Windsome Farms Star
Aql 20h10'35"5d35'
Windsor,Evan Joseph
Her 17h16'34"48d53'
Windsor,Matt Finnish
Dra 16h15'10"63d43'
Windstein-Garrigan, Stella Mary
Lyr 19h15'44"42d14'
Windt,Daniel L & Susan Schwab
Cyg 21h37'47"28d35'
Windy
Crt 11h18'2"-18d21'
Windy
Uma 11h54'44"43d29'
Wine,Michelle
Mon 8h1'16"-8d12'
Wine,Roger Mark
Leo 10h34'23"21d29'
Wineberg,Robert & Kathy
Cyg 20h54'30"40d8'
Winegarden,Esther May
Eri 3h28'18"-5d43'
Winegarden,Norena Kathleen Ana Linn
And 1h14'0"34d45'
Winer,Bette Marcia
Crb 16h10'41"34d17'
Winer,Julius Max
Her 16h16'48"11d34'
Winer,Ronald
Dra 12h32'37"71d34'
Wines,Lawrence E
Oph 18h17'12"8d41'
Wines,Sandy's Man Bruce
Per 4h5'16"51d40'
Winfield,Christine
And 0h43'20"41d7'
Winfield- Rockabrand, Andrew
Cmi 7h40'31"1d34'

Winfred
Her 17h48'1"41d12'
Winfred
Lmi 10h8'56"38d45'
Winfrey,David L
Ori 5h1'24"0d55'
Winfrey,Mark
Her 17h30'12"27d60'
Wing & Bonneth
And 1h47'42"38d25'
Wing Family Of America
Oph 17h39'59"-20d46'
Wing,Alicia Marie
And 23h44'26"40d27'
Wing,Barbara L
Gem 8h0'57"33d10'A
Wing,Deborah Susan
Peg 23h38'29"26d54'
Wing,Earl
Aur 5h9'42"42d17'
Wing,Gary L
Ser 15h16'15"10d11'
Wing,Helen
Aql 18h43'12"8d41'
Wing,Jack Richard George
Aql 19h53'45"-1d22'
Wing,Jack Richard George
Del 20h53'12"4d33'
Wing,Simon
Peg 0h9'37"17d31'
Wing,Tammy L
Eri 3h56'58"-3d5'
Wing,William Hinshaw
Aql 19h55'32"10d3'
Wing-Waller,Penny E
Cnc 8h55'0"31d9'
Wingard,Grace Cornett
Umi 18h42'15"88d18'
Wingate Belle Ame, Michelle
And 0h20'31"31d4'
Wingate's "Wishing Star",Linda Lee
Ori 5h54'50"14d26'
Wingate,Phyllis
Cas 0h51'0"74d15'
Wingenroth,Ruth
Lyn 8h32'22"43d50'
Wingert,Melissa Lynn
Cas 0h51'43"61d25'
Wingerter,Carola
Oph 17h56'41"10d29'
Wingfield,Jeffrey L
Boo 14h24'13"47d34'
Wingfield,Jr,John W
Aur 5h19'37"48d4'
Wingfield,Rhonda
Peg 21h55'51"20d9'
Wingnut
Cep 0h42'14"78d58'
Wingo's Star
Boo 14h57'0"21d13'
Wingo,Herbie
Oph 17h36'35"-8d48'
Wingo,Michael P
Cet 2h26'25"-5d40'
Wingren,Adriane Renee
And 2h9'16"41d55'
Wings
Cep 22h59'26"70d9'
Wings Of Dreams
Her 18h3'49"28d34'
Winhall,Kelly
Umi 15h5'29"67d6'
Winiarski,Claude M
Her 17h21'41"42d54'
Winifred's Star
And 1h16'18"34d60'
Winings,Jesse Alan
Aur 4h48'44"38d36'
Winings,Mitchell Ryan
Lac 22h8'18"51d53'
Winistörfer,Bryn
Umi 14h11'15"72d26'
Wink's Guiding Light
Hya 8h44'22"5d32'
Winkelbauer,Lori
Crb 16h13'34"30d38'
Winkelman,Jaimie L
Cet 0h6'37"-11d1'
Winkle Winkle Bill's Little Star
Lac 22h6'36"51d19'
Winklepleck,G R
Cyg 21h30'30"40d41'
Winklepleck,Sue
Aur 6h55'42"38d11'
Winkler's Star,Lori
Del 20h50'30"9d12'
Winkler's Star,Tristan
Boo 14h30'1"20d16'
Winkler,Alicia Marie "The Doc"
Crt 10h54'43"-22d22'
Winkler,Diana
Aqr 21h29'33"-0d18'
Winkler,Dolores M
And 0h10'57"46d10'
Winkler,Frederick Louis
Cep 22h32'26"58d24'
Winkler,Gabe
Sex 10h27'50"-6d28'
Winkler,Genevieve & Walter
Ori 5h16'33"0d21'
Winkler,Henry
Boo 13h45'24"24d46'
Winkler,Jamie Brooke
Vir 13h33'57"5d9'
Winkler,Louis O
Uma 8h44'29"70d49'
Winkler,Richard A
Dra 19h7'0"65d14'
Winkler,Susan Helen
Lyn 6h55'58"59d15'
Winkler,Susan Therese
Hya 8h58'12"-6d12'
Winkler,Thomas Frank
Aur 6h30'58d37'52"
Winkler,Ute
Cnc 9h0'47"31d47'
Winkler-Smutny,Imelda Theresa
And 2h5'57"39d44'
Winkley,Morgan Elizabeth
Lyr 19h19'1"42d51'
Winklier,Megan Nicole
Peg 21h50'44"34d6'
Winkmann,Hermann (Günter)
Lyn 8h15'45"45d42'
Winkolak,Tom & Kathy
Cyg 20h43'31"46d55'
Winky
Her 18h5'52"30d22'
Winland Family Star, The
Eri 2h44'57"-4d2'
Winland,Steve Allen
Oph 17h14'21"20d13'
Winn's Guiding Light, Dana Irene
Cyg 21h10'47"38d14'
Winn,Bobby
Cep 20h37'41"60d13'
Winn,Dennis & Carla
Cyg 21h21'12"41d14'
Winn,Kathryn
Com 12h55'34"23d22'
Winn,Maria
Cas 0h35'33"58d18'
Winn,Thomas H
Cep 21h31'36"65d37'
Winnaman Family Star, The Steven R
Aql 20h10'26"12d28'
Winner,Christian F
Aur 4h54'58"40d58'
Winner,Cynthia
Ori 5h53'13"11d2'
Winner,Karen Lee
Uma 9h13'60"52d33'
Winner,Lauren Ashley
Oph 17h4'28"-24d47'
Winners At Work
Aql 18h58'40"-6d4'
Winners,Drew Patrick
Cet 0h46'57"-10d36'
Winnett,Baby Jordan
Umi 14h42'23"81d53'
Winnick,David
Aur 4h48'12"40d24'
Winnie
Cas 23h1'48"53d18'
Winnie
Aur 5h58'45"31d33'
Winnie & Bonnie
Lyn 7h0'1"51d40'
Winnie D Dog
Cyg 20h33'1"50d10'
Winnie D,The
Oph 18h40'41"10d0'
Winnie Star,The
Aql 19h45'19"13d38'
Winnie's Star
Pho 0h45'50"-44d2'
Winnie's Star
Cas 0h23'34"64d43'
Winning Couple, Michael & Ida,The
Cyg 19h57'10"40d11'
Winograd,Mel
Cnv 12h16'0"48d43'
Winokur,Joseph A
Umi 15h4'13"81d19'
Winona David
Mon 7h43'58"-7d17'
Winona Grace
And 2h30'58"47d30'
Winship,Fayette
Cas 1h56'36"70d7'
Winship,James Martin
Leo 9h55'19"11d14'
Winslow
Boo 14h9'37"38d41'
Winslow,Amy
Lyr 19h22'45"41d48'
Winslow,Darren Jay
Vul 19h46'52"28d44'
Winslow,Dillon James
Hya 9h15'25"5d55'
Winslow,Elda C
Cas 0h8'17"63d47'
Winslow,Ernie
Aur 6h37'15"37d33'
Winslow,Joe & Susan
Crb 15h27'48"30d29'
Winslow,Jr,George A
Hya 8h16'26"4d2'
Winslow,LaTonya Sheri
And 23h35'36"48d59'
Winslow,Oman
Del 20h37'20"10d7'
Winsman,Melissa
And 23h41'49"38d10'
Winsor,Derek
Ori 5h42'56"11d16'
Winstain,Leland S
Cma 6h59'59"-19d36'
Winstanley,Janette
Del 20h53'0"2d48'
Winstead,Anita Lynn
Aql 20h2'47"6d6'
Winstead,Samuel N
Oph 16h20'23"1d20'
Winsted,Shannon Ellen
And 2h30'1"49d29'
Winstin,Laura
Cam 3h17'27"55d58'
Winston
Cep 21h23'29"60d36'
Winston Graham
Per 2h56'27"37d12'
Winston,Arthur
Aur 5h40'58"50d19'
Winston,Buff
Lac 22h6'47"51d6'
Winston,Denise
Cnv 12h50'15"45d7'
Winston,Donald G
Per 2h57'52"40d59'
Winston,Leonard
Sgr 18h49'45"-28d37'
Winston,Patti Haley
Mon 6h54'56"-10d57'
Winter "Valley Flower",Stacy Lynn
Cas 0h59'21"62d49'
Winter MS,MFCC,Brenda
Aql 19h52'59"-6d33'
Winter Skize
Eri 3h33'24"-4d60'
Winter Star,The Sara & Rob
Lyn 7h56'55"34d9'
Winter's Rose
And 23h49'34"32d8'
Winter,Alissa Josephin & Taylor Brown
Sct 18h56'56"-6d5'
Winter,Christopher Mathew
Ori 5h53'57"15d13'
Winter,Debra Jean
Lyr 18h58'37"30d47'
Winter,Friedrich-Karl
Sgr 20h4'47"-40d31'
Winter,Ginger
Cnv 12h10'43"50d25'
Winter,Herbert
Sco 16h12'30"-24d5'
Winter,Jacob Thomas
Aur 6h24'51"31d49'
Winter,Karel
Cas 3h3'15"57d44'
Winter,Laura
Peg 22h5'53"5d45'
Winter,Nicholas
Per 1h52'36"48d41'
Winter,Paul Martin
Dra 16h7'16"61d21'
Winter,Ross A Mercury
Ara 17h55'2"-50d50'
Winter,Shauna Leigh
Tri 2h23'18"30d41'
Winter,Toni
Sco 17h28'47"-31d7'
Winter-A Winter Star Lights Up Life,Mark
Ori 5h38'1"0d30'
Winterbone,Jonathan
Boo 14h51'1"48d14'
Winterbottom,Bruce
Aur 7h20'19"43d52'
Winterburn,Winn
Aur 6h28'1"38d3'
Winterer,Roland
Her 17h43'1"14d45'
Winterhalter,Dorothy Theresa
And 23h30'33"45d43'
Wintcrhollor,Michaela
Aqr 22h43'10"0d21'
Winterich,Jack
Per 1h50'31"50d12'
Winterleitner,Sandy Sonja
Gem 6h38'31"33d59'
Winterly,Karyn Rose
Peg 22h22'38"21d47'
Winternheimer,Brian
Hya 8h28'53"0d10'
Winters Night
Per 1h28'43"53d36'
Winters,Alan F
Aur 7h25'20"37d44'
Winters,Alma
Cet 2h46'18"3d5'
Winters,Branndon James
Her 16h47'49"40d28'
Winters,Brett
Boo 13h41'49"23d46"
Winters,Donald P
Cep 2h47'30"80d15'
Winters,Elizabeth Desiree (Beth)
Lyr 18h38'59"34d16'
Winters,Elsie C
Mon 7h48'14"-5d25'
Winters,Geneva Amber
Lyr 18h37'43"37d55'
Winters,Jacqueline
And 1h12'57"38d57'
Winters,Janice
Lyr 18h33'29"42d34'
Winters,Jesse
Aql 19h8'1"0d24'
Winters,Julie
And 23h41'18"46d51'
Winters,Lois Mae
Lyn 7h29'16"38d21'
Winters,Louis
Her 18h2'25"47d25'
Winters,Miximilian J
Crt 11h20'55-18d44'
Winters,Nicole
And 1h46'14"36d22'
Winters,Priscilla A
Lyr 18h56'41"30d3'
Winters,Ralph E
Cet 3h7'30"2d13'
Winters,Sylvia
And 0h25'41"44d39'
Winters,Teena Ping
Cyg 20h6'1"40d27'
Winters,Thomas
Lac 22h22'1"53d7'
Winters,Thomas Patrick
Aql 19h31'1"13d16'
Winterton,Donald C
Tri 2h28'45"35d57'
Winthrop,Charlotte Brock
Per 3h32'52"37d20'
Winthuis,Marlene
Peg 23h33'54"20d59'
Wintjen,Henry
Aur 5h8'53"41d44'
Wintle,Stephen Edward
Cep 23h3'60"70d57'
Winton,Charles
Ori 4h43'10"12d18'
Winton,Kevin Michael Patrick
Cep 1h6'22"77d41'
Winton,Suzanne
Cam 5h47'44"68d18'
Wintour,Anna
Cas 23h1'26"53d42'
Wintour,Anna
Aur 4h52'20"50d4'
Wintrode,Dawn
Cnc 8h52'57"30d6'
Winyard,Eliza
Cas 23h28'59"61d58'
Winzenrieth,Jacqueline
Per 3h58'49"35d43'
Winzinger aka The Winz BG Robert J
Aur 5h4'48"40d55'
Wipper,Tom
Tri 2h15'13"30d0'
Wipperman,Beth
Lyn 8h0'40"34d16'
Wipperman,Greti,Werner & Dirk
Ser 16h4'22"2d18'
Wipple
Uma 10h54'41"68d53'
Wirag,"Poppy" Don
Her 15h55'0"44d38'
Wirdnam,Eve
Crb 16h18'34"37d19'
Wirsz,Francis P
Uma 9h31'33"58d37'
Wirt,Daniel Jordan
Lac 22h31'12"52d44'
Wirtanen,Erika
Lyn 8h12'23"38d40'
Wirth,Anthony Paul
Aur 6h2'31"45d49'
Wirth,Barbara Maria A Star is born
Hya 8h52'58"-0d12'
Wirth,Betty B
Peg 23h5'54"18d48'
Wirth,Daniel
Dra 16h15'20"67d26'
Wirth,Lisa
Lyr 18h20'52"38d49'
Wirth,Mary
Peg 21h57'21"34d46'
Wirth,Patrick
Uma 9h21'0"58d31'
Wirth,Ruthann
Cyg 19h31'51"32d31'
Wirth,Sonia
And 23h6'0"42d51'
Wirtz,Hans-Peter
Lac 22h5'16"50d36'
Wirtz,Karin
Uma 9h33'0"46d48'
Wirtzer,Alex
Aql 19h50'1"14d20'
Wisbaum,Wayne D
Gem 6h53'19"30d57'
Wischmeier,Elmer
Cnv 13h50'40"37d41'
Wischnewski,Mr Body & Soul Wilhelm
Ari 2h24'1"26d34'
Wischum,John T & Anne M
Dra 19h21'52"84d49'
Wisdom
Uma 11h28'0"48d60'
Wisdom
Tri 1h50'33"25d56'
Wisdom 11-15-74,Mary
And 1h19'10"40d52'
Wisdom Family
Cnv 12h16'30"42d13'
Wisdom,Grace Lucille
Lac 22h36'55"40d10'
Wise "Duke",Richard Dean
Cnv 12h24'18"37d59'
Wise "Superstars", Bente & Terry
Aql 20h2'17"6d27'
Wise II,Richard Allen
Cep 0h13'0"68d9'
Wise Wishes
Cnv 12h42'60"50d36'
Wise's 50th Anniversary Star,The
Crb 15h45'0"29d7'
Wise's Shirley,Jay
Lyn 8h55'10"38d50'
Wise,Caroline E
And 23h43'16"42d13'
Wise,Colleen Meredith (Puddin')
Lyn 8h14'30"46d59'
Wise,David K
Oph 16h58'29"-22d33'
Wise,Douglas Ray
Sge 19h36'31"16d44'
Wise,Enigma V For Clifford C
Cet 3h19'41"8d1'
Wise,James Murray
Per 2h1'20"57d54'
Wise,James Patrick
Equ 20h55'56"2d52'
Wise,Joe Lloyd
Ori 5h54'16"16d31'
Wise,Joyce
Sct 18h47'46"-7d56'
Wise,Kathleen Victoria
Sgr 18h54'47"-23d7'
Wise,Larry
Cep 22h14'56"68d33'
Wise,Matthew Ryan
Lac 22h13'34"51d21'
Wise,Robert Parker
Ori 5h57'20"15d20'
Wise,Shirley
Cyg 21h48'26"37d35'
Wise,Stephanie Leigh
Peg 21h38'37"26d23'
Wise,Sybil Z
Cas 1h5'31"63d58'
Wise,Taylor Reéne
Peg 21h42'0"24d46'
Wise,Truman
Ser 18h53'42"2d2'
Wiseman's Way
Aur 7h24'54"41d7'
Wiseman,Connie & Joel Wall
Eri 4h54'57"-6d28'
Wiseman,Joseph Nicholas
Her 18h20'0"13d2'
Wiseman,Luke Anthony
Peg 21h53'13"34d16'
Wisenbaker,Big John
Ori 5h30'16"-6d20'
Wisenbaker,MB
Ori 5h52'40"15d59'
Wish Upon The Eckelkamp's 17th
Crb 15h16'57"30d53'
Wish you were here
Tri 1h57'32"28d50'
Wishes Do Come True - Peter & Nicole
Crb 15h24'35"30d26'
Wishes For Ryan
Aur 5h54'55"54d46'
Wishing 2
And 22h56'0"38d43'
Wishing Star
Ori 5h2'23"0d39'
Wishnoff,Stanley
Ori 5h4'15"1d49'
Wiskirchen,Dr Franz Josef
Aqr 21h4'45"-1d50'
Wiskow,The Star Of Richard & Ruth
Cyg 20h30'24"40d3'
Wislar,Elizabeth Jane
Mon 7h1438"-5d16'
Wisler,Sharon S
Cyg 20h4'46"30d13'
Wismar 25,Cissy & Greg
Cas 0h15'0"63d12'
Wismer,Raymond A J
Vul 19h23'0"23d42'
Wismer,Seth Alan
Uma 10h40'22"47d19'
Wisne,Maxwell Mariea
Cam 3h18'34"61d52'
Wisner-5,John & Sophie
Cyg 19h26'46"32d43'
Wisnieski,Melvin G
Lac 22h43'17"53d26'
Wisniewiski,Rose S
Lyr 19h2'59"28d41'
Wisniewska,Renata
Uma 11h13'31"48d53'
Wisniewski,Donna J
Cap 20h23'25"-13d34'
Wisniewski,Marcella
Cmi 7h41'41"0d11'
Wisniewski,Sydney McKenna
Cam 3h17'27"60d31'
Wisniewski,USMC2353948 S S
Ori 5h44'39"1d43'
Wisowaty,Dorothy M
And 0h31'27"27d43'
Wissel,Delores & Wilbur
Uma 11h5'16"35d57'
Wissman,Ella Frances
And 0h52'1"37d5'
Wissman,Ginny & Wayne
Lyr 18h50'51"41d9'
Wissman,Herbert A
Aur 6h21'50"32d43'
Wissman,Thomas Joseph
Dra 12h11'23"71d57'
Wisspeintner,Thomas
Aqr 22h43'57"-1d39'
Wist 1993 Papa's Star, Milton C
Cep 22h17'25"59d16'
Wist 1993 Papa's Star, Ariana C
Cas 0h9'45"58d60'
Wist,George & Vivian
Umi 15h34'17"68d52'
Wistner,Marti
Ori 5h52'46"15d26'
Wiswall,Helen Lacobee
Peg 22h18'45"31d3'B
Wiswall,Willard Edwin
Peg 22h18'45"31d3'A
Witchdance
Lmi 10h51'20"26d56'
Witcher,J Byron
Her 16h18'0"26d29'
Witcomb's Refatinia, Terence Michael
Cas 2h13'15"61d51'
Witcover,Carolyn
Mon 6h29'28"-8d39'
Witenko,PhD,Barbara
Uma 10h41'1"40d47'
With
Sgr 19h13'27"-16d38'
With Love From A to Z
Crb 16h1'50"38d39'
Witham West Point Class Of 95,Jamie
Lyr 19h20'30"33d56'
Witherow,Jaime Jean
Uma 9h50'40"49d55'
Withers,Dana Ruth
Cnv 12h53'51"40d49'
Withers,Florence Ruth
Lib 15h17'0"-25d51'
Withers,Holly Elmer
Lib 15h38'0"-21d19'
Withers,John Joseph
Lyn 8h49'29"35d4'
Withers,Jr,W Russell
Dra 17h59'29"68d20'
Withers,Robin
Leo 9h47'38"33d19'
Withers,Stella
Sge 20h2'24"20d37'
Witherspoon Star,The George B
Per 4h23'39"50d55'
Withington,Joanne & Terry
Cyg 19h25'1"44d8'
Withrow,Gerald Wayne
Mon 6h43'12"-3d17'
Witke,Renee
Com 12h7'12"25d50'
Witkop,Daveena Ann
Cam 3h29'28"61d19'
Witkowski,Marie
Cnv 13h49'22"34d26'
Witman,Sarah E
Lyn 8h46'37"46d46"
Witmer,Michael John
Uma 8h39'31"70d7'
Witous,A Star is Born-Michael James
Ori 6h1'51"1d37'
Wits,Wessel
Vul 19h6'41"25d23'
Witt
Her 17h23'33"28d6'
Witt,Ekkehard
Mon 7h52'27"-1d56'
Witt,Hans-Juergen
Equ 20h58'55"8d32'
Witt,Helfried
Uma 10h39'44"48d13'
Witt,Jerry R
Aur 5h9'32"41d51'
Witt,Katarina
Sgr 20h5'31"-45d0'
Witt,Ken & Georgie
Cyg 19h43'0"38d38'
Witt,Kyle Thomas
Her 15h57'34"46d2'
Witt,Linda Sue
And 23h2'59"41d19'
Witt,Meg
Umi 15h37'29"68d8'
Witt,Nancy Geane
Cyg 21h31'27"42d10'
Witt,Patricia A
Cas 0h14'12"58d33'
Witt,Sonya
Ori 5h22'58"-0d35'A
Witt,Tanya
Ori 5h22'58"-0d35'B
Witt,Walter C
Ori 5h51'30"15d38'
Wittbrodt,Leona Marie
Cas 3h2'32"61d41'
Witte,Annegret
Mon 7h59'25"-5d54'
Witte,Linda
Peg 22h30'34"26d13'
Witte,Rebecca
Oph 18h4'19"12d55'
Witte-My Handsome Prince,Wayne
Peg 21h8'56"18d58'
Wittels,Howard Bernard
Per 16h9'54"42d1'
Wittels,Stephen Parker
Aql 20h4'54"0d38'
Witten,Theresa J
Peg 22h16'28"21d42'
Wittenberg,Udo
Uma 9h31'50"46d41'
Wittenberg,William Edward
Uma 10h51'51"47d36'
Wittenbreder
And 23h1'19"43d55'
Wittenheim,Cora
Cam 6h59'28"64d27'
Witter,Marc George
Cyg 20h29'52"44d58'
Witter,Matthias
Aql 18h57'22"17d39'
Witter,Scott B
Ser 16h4'33"10d37'
Wittgen,Kate
Lyn 8h54'51"38d13'
Witthauer,Jack
Her 16h13'0"41d3'
Wittich Esto Perpetua, Bobbie J
Aql 19h55'24"13d35'
Wittick,Debbie
Aql 20h1'12"13d8'
Wittig,MaryBeth
Lyr 18h24'40"45d20'
Wittmann,Barbel Monika
Lac 22h22'40"53d4'
Wittmer,Georg
Lmi 9h29'21"38d16'
Wittner,Theo
Per 3h18'47"46d27'
Wittrop,Lene
Equ 21h3'48"10d19'
Wittschier,Otto
Cnc 8h7'22"31d13'
Witty
Lac 22h2'0"51d53'
Witucki,Edward "GI3'
Aur 6h26'13"35d20'
Witz,Arthur I
Lac 22h15'35"48d17'
Witz,Janet
Eri 4h34'54"-6d54'
Witzany,Christian
Lyn 8h2'14"47d38'
Witzel,Jay & Sandy
Aur 5h1'53"29d49'
Witzig,Janus
Cep 22h7'23"67d43'
Witzke,Laverne Lucille Schmidt
Boo 14h36'0"8d10'
Witzke,Pamela Janice Schmidt
Cam 4h57'60"68d19'
Witzlsteiner,Katie & Bill
Eri 4h34'47"-1d19'

Wiwigac,Barry M
Per 3h7'23"54d58'
Wixforth,Thorsten
Cam 4h11'26"67d50'
Wixie
Pho 0h48'19"-44d24'
Wixom,Rhonda Salyer
Peg 22h50'1"27d50'
Wixted,Susan Mary
Peg 23h30'30"30d58'
Wizard's Nic Nac Nova
Uma 12h37'11"61d25'
Wizard,The
Boo 13h39'55"21d10'
Wizenberg,Dr Morris J
Cet 3h5'0"1d55'
Wizig,Andrew Scott
Ori 5h57'26"11d52'
Wizzle
Uma 9h31'11"48d19'
WJC Jr, My Knight & Shining Star
Her 18h24'1"28d42'
WJE & BSEV Eternal Kingdom
Aql 19h7'53"5d23'
Wlaker,William Lucas
Her 18h17'33"14d58'
Wlasiuk Twinkle In My Heart,Doug
Ori 5h54'17"16d40'
Wlassak's Star
Eri 2h53'15"-5d34'
Wlazlowski-My Polish Prince,Ed
Aur 7h25'0"37d57'
Wleklinski,Alexandra Rose
Mon 6h43'1"11d26'
Wlodarczak,Jeannine
Sgr 19h29'58"-30d47'
Wlodarczyk,John
Cet 1h3'35"-2d30'
Wlodarczyk,Tekla
Cas 23h29'1"62d30'
Wlodarski,Arthur
Dra 16h17'54"66d54'
Wlodawski,Brian Matthew
Boo 15h7'0"12d29'
Wlodyka,Carolyn
Aql 20h6'11"4d14'
Wmistar
Cam 5h48'12"67d31'
WMS 222
Per 2h29'45"57d35'
WMS/BPS/10-7-88
Sco 17'51"23"-30d52'
WNBC
Umi 15h54'3"74d26'
Wnek,Christina
And 2h29'0"42d50'
Wnorowski,Amy & Mark
Cyg 21h0'19"31d17'
Wnuk,Michel
Mon 7h55'48"-3d32'
Wobbe,Heather Ashley
Peg 22h10'41"28d5'
Wobbe,Winfried Aloys
Mon 6h22'0"8d15'
Wochner Star,Cori
Mon 7h47'10"-1d34'
Wochner Star,Lindsay
Mon 7h47'3"-1d25'
Wockenfuss,James
Aur 5h55'31"31d47'
Wocl,Edward A
Cam 12h18'52"82d15'
Wodicor
Vir 13h31'22"-4d46'
Wodinski,Karin
Cnc 8h6'32"30d41'
Woehle,Taylor Ingelore Lisette
Ari 2h7'19"17d54'
Woelky,Carina
Tau 4h9'38"23d34'
Woeppel,Stacy Lin
Uma 8h42'24"71d25'
Woepple,Stentzee Beth
Uma 12h0'1"62d34'
Woerly,Paula
Cas 1h9'1"75d56'
Woerner,Ed & Vivian
Lac 22h8'59"51d14'
Woerth's"Gardian Star"
Ori 6h12'57"7d34'
Woertink,Patrick
Cma 7h20'55"-15d1'
Woertz,Linda
Peg 22h46'36"33d7'
Wofford,Karen R
Del 20h25'30"8d7'
Wogahn,Tiffany
Cas 23h28'0"63d11'
Wogan,Thomas A
Hya 9h54'48"-17d34'
Woghin,Larissa Kathleen
And 23h32'49"43d54'
Wohl,Andrew Glenn
Per 1h50'56"48d2'
Wohl,Charles
Her 18h4'22"30d39'
Wohlbedacht,Wilhelmina
Cam 5h51'43"61d30'
Wohlfarth,David W
Dra 14h30'31"58d33'
Wohlfeil,Joachim
And 23h25'30"45d1'
Wohlford,John Charles
Oph 18h38'47"10d29'
Wohlgamuth,Roger Allen
Her 16h55'23"33d30'
Wohlmacher,William G
Aur 7h23'1"43d10'
Wohlschlaeger, Frederick & Mary
Dra 9h46'26"80d39'
Wohlschlaeger,George & Andrea
Aur 6h3'0"30d54'
Wohlstein,Kerry
Lyn 7h27'1"58d43'
Wois,Elaine & Alfie
Lyn 7h55'38"43d33'
Woisin,Sandy
Cas 0h56'23"66d23'
Wojciak,Ken
Aur 7h21'19"41d23'
Wojciechowicz,Amanda Lee
And 0h43'1"45d21'
Wojhan,Nahami & Thomas Nagelschmitz
Sgr 18h54'18"-24d7'
Wojick,Jim
Aql 19h30'1"14d4'
Wojnowski,Greg Alan
Crb 16h13'39"30d51'
Wojo
Aur 6h25'1"33d16'
Wojtowicz,Marcia
Cyg 19h28'42"38d17'
Wojtusik,Ted
Aur 7h23'56"35d56'
Wojtysiak,Alice
Cas 3h19'56"75d25'
Wokaty,Holly
Eri 4h32'0"-12d17'
Woko
Ori 6h3'53"2d52'
Wolak,Christa
And 23h0'50"46d13'
Wolanski,Kaz
Leo 10h56'0"15d18'
Wolchik,Honorable Judge Joseph J
Cep 21h47'12"60d20'
Wolcott's Star
Aql 18h57'24"2d51'
Wolcott,Bill
Aur 5h2'17"40d27'
Wolcott,John Lester
Sgr 19h7'47"-24d52'
Wolcott,Julia M
Crb 15h32'31"28d10"
Wolcott,Robert Ayers
Ori 5h2'51"1d14'
Wold,Dr & Mrs Kieth
Oph 17h19'34"-20d12'
Wold,Elizabeth Ann
Crt 11h21'25"-22d31'
Wold,Ronald Michael
Per 1h44'38"52d44'
Wolen,Casey Matthew
Cet 2h9'31"-11d15'
Wolen,Skylar Jarred
Cet 2h32'60"-10d58'
Woleske,Sioux Star No 31 "LJ"
Cam 6h20'0"68d49'
Wolf
Per 3h18'47"40d21'
Wolf I (Lupo)
Cnc 8h4'47"20d34'
Wolf Star
Dra 10h16'23"78d35'
Wolf's Destiny
Cmi 7h7'28"4d27'
Wolf's Love Star,Scott and Dianne
Uma 8h7'13"61d6'
Wolf's Star,P J
Cet 3h16'18"5d58'
Wolf,Alla
Cas 0h2'1"63d46'
Wolf,Andre
And 23h2'26"43d48'
Wolf,Andrea/Berlin 29/07/1967
Leo 10h25'16"11d6'
Wolf,Anne
Vul 20h19'32"23d37'
Wolf,Arnold Walter
Aur 6h26'45"31d6'
Wolf,Barbara Louise
Ori 5h53'60"19d44'
Wolf,Bradley Steven
Her 17h34'52"26d50'
Wolf,Cerstin
Lyn 8h6'49"35d34'
Wolf,Christian
Sco 17h28'56"-30d4'
Wolf,David Starbuck
Her 17h32'41"40d33'
Wolf,Devon
Ori 5h34'47"8d56'
Wolf,Dolly & Lloyd
Mon 7h0'28"3d50'
Wolf,Edith und Paul
Cmi 7h16'27"3d56'
Wolf,Edward "Pete"
Cnv 12h57'36"31d35'
Wolf,Elissa Victoria
Cmi 8h8'13"1d26'
Wolf,Elliot
Per 2h4'49"47d9'
Wolf,Harald
Ori 4h52'56"1d32'
Wolf,Inge Verena
Peg 22h44'48"20d39'
Wolf,Jack
Cep 22h4'1"60d9'
Wolf,Jeffrey Todd
Her 18h19'15"14d48'
Wolf,Jennifer
Lyn 9h4'15"40d8'
Wolf,Jessica Jeanne
And 0h35'44"38d7'
Wolf,Jodi
Uma 10h0'1"58d3'
Wolf,Joe & Shirley
Lyr 19h14'18"42d55'
Wolf,John
Aql 18h59'33"-3d19'
Wolf,Jonathan
Per 3h9'1"47d25'
Wolf,Jonathan Pomow
Mon 7h0'52"8d13'
Wolf,Jordan T
Aql 20h12'11"4d17'
Wolf,Jr,Richard A
Ori 5h48'50"10d36'
Wolf,Judith & Brian Goodwin 50th
Ser 15h58'34"13d24'AB
Wolf,Justin T
Mon 7h40'27"-0d49'
Wolf,Kelly Danielle
Sco 16h56'0"-38d27'
Wolf,Kurt
Mon 7h53'9"-1d6'
Wolf,Lee
Aur 6h58'1"43d40'
Wolf,Marty
Dra 16h9'42"67d41'
Wolf,Matthew
Aur 5h55'19"31d25'
Wolf,Matthew Ryan
Aur 5h26'0"41d7'
Wolf,Maya
Oph 17h16'16"-20d43'
Wolf,Michael H
Cnv 13h49'31"30d3'
Wolf,Michel J
Ori 5h59'14"8d24'
Wolf,Oliver
Cap 20h46'23"-20d19'
Wolf,Opal J
Mon 7h57'24"-1d17'
Wolf,Reginald
Sco 16h30'56"-43d28'
Wolf,Reverend Jivan
Her 16h59'1"32d46'
Wolf,Rudi
Cmi 7h19'43"4d57'
Wolf,Scott
Cet 0h35'41"-6d43'
Wolf,Susan Celestial
Cas 3h1'0"70d9'
Wolf,Thomas
Dra 18h1'34"67d31'
Wolf,Tiffany
Lyn 6h22'1"59d46'
Wolf,Wendy A
Equ 21h23'0"3d29'
Wolfberg,Jessica A
Cas 0h11'12"61d37'
Wolfe "R & R "Superstar",Gregory
Leo 10h19'20"13d23'
Wolfe & Sons Inc, Charles
Per 3h8'18"46d51'
Wolfe Forever,William & Pamela
Uma 9h14'44"56d21'
Wolfe,Bonnie Jane
Peg 23h23'26"25d44'
Wolfe,Craig & Pat
Dra 18h38'0"71d12'
Wolfe,Daniel
Lac 22h29'49"52d60'
Wolfe,Daniel Jacob
Her 16h29'1"40d29'
Wolfe,David
Her 17h31'52"38d7'
Wolfe,Deborah Fern
Cam 7h17'51"83d13'
Wolfe,Donna
Mon 6h26'1"11d44'
Wolfe,Douglas
Uma 10h34'56"54d2'
Wolfe,Edna & Rowland
Mon 7h59'26"-1d56'
Wolfe,Elliott
Ser 15h55'48"18d28'
Wolfe,Jeffery
Hya 8h43'1"5d49'
Wolfe,Jennifer
Lyn 7h59'43"37d9'
Wolfe,JoAnn
And 1h44'51"39d30'
Wolfe,Jolyn
Peg 23h6'45"26d38'B
Wolfe,Kayla
Aql 19h57'58"12d43'
Wolfe,Lisa
Lyr 18h37'57"47d16'
Wolfe,Maestro Duain
Dra 18h44'31"70d17'
Wolfe,Mark
Peg 23h6'45"26d38'A
Wolfe,Mary Nelle Miller
Peg 22h34'51"29d27'
Wolfe,Paul
Lac 22h53'11"38d13'
Wolfe,Rowie
Hya 9h30'34"-6d46'
Wolfe,Shawn Marie
Cet 3h11'44"0d22'
Wolfe,The Rev Douglas E
Aur 6h51'0"37d42'
Wolfe,Thomas G
Boo 14h17'32"33d15'
Wolfe,Tiffany
Cma 7h0'0"-13d9'
Wolfe,Wendell E
Aql 20h1'1"13d17'
Wolfe,William Robert
Her 16h58'31"32d52'
Wolfe-Crow Star,The
Equ 21h16'57"3d37'
Wolfendale,Clive
Cma 6h28'31"-24d49'
Wolfersberger,Charles & Olive
Boo 15h12'43"48d35'
Wolfert,Elizabeth Morgan
Vul 19h42'25"23d13'
Wolfert,Phillip Leslie
Equ 21h21'0"11d33'
Wolff "Mimi", Marilyn Ella
Ori 5h56'22"8d51'
Wolff,Andreas & Catherine
Lyr 18h58'13"38d58'
Wolff,Betty J
Lyr 18h40'40"36d17'
WOLFF,Bobstar
Uma 8h37'0"56d59'
Wolff,Carrie
Mon 7h11'54"-10d48'
Wolff,Eléonore
Per 1h44'21"53d59'
Wolff,Jan & Bob Martz
Cet 2h12'1"3d40'
Wolff,Janet Carey
Lyr 18h55'34"40d9'
Wolff,Johannes
Aqr 21h54'11"0d17'
Wolff,Karin
Dra 16h57'57"66d24'
Wolff,Olga
Cas 0h32'56"67d37'
Wolff,Patricia A
Uma 12h13'45"62d55'
Wolff,Raven
Aql 19h6'35"15d46'
Wolff,Sonnenscheinchen Andreas
Dra 16h50'11"65d28'
Wolff,Uwe
Cam 3h45'14"67d32'
Wolfgang
And 23h2'54"50d8'
Wolfgang
Gem 6h35'54"12d7'
Wolfgang & Tamara
Ori 5h43'19"8d17'
Wolfgang und Barbara
Cnc 8h58'0"12d6'
Wolfie's Star
Aur 7h21'35"40d21'
Wolfinbarger,Mel & Dell
Mon 7h42'29"-2d51'
Wolfinger,Frank K
Psc 1h21'32"32d39'
Wolfinger,Harvey"Red"
Aur 5h0'19"37d57'
Wolfinger,Violet
Cam 5h46'6"73d22'
Wolfington,Isola (Zolie)Jessica
And 2h14'1"42d1'
Wolfje En Boefje Semper Idem
Uma 11h16'25"70d9'
Wolford Star,The Captain Ron
Boo 13h34'55"17d49'
Wolfscomb
Cet 3h17'39"7d41'
Wolfslau
Her 17h24'58"28d27'
Wolfson is a Heavenly Teacher!,Mrs
And 0h56'31"39d25'
Wolfson,Amy B
Lmi 9h59'41"38d12'
Wolfson,Jeff
Boo 14h58'1"20d53'
Wolfson,Jeffrey
Aqr 21h23'15"-1d1'
Wolfson,Jeffrey Alan
Aqr 21h26'35"0d33'
Wolfson,Scott Matthew
Per 1h38'29"53d16'
Wolfson,Stephanie Danielle
Cyg 20h30'0"39d45'
Wolfy
Del 20h54'31"3d12'
Wolfy
Cmi 7h15'45"5d37'
Wolgast,Robert Douglas
Her 18h0'20"38d42'
Wolin,Luke Thomas
Uma 10h52'42"35d12'
Wolinsky,Todd Patrick
Boo 14h47'52"32d47'
Wolk,Edward J
Dra 14h52'0"61d9'
Wolk,Esq,Arthur A
Per 1h39'23"54d6'
Wolk,Marion Beth
Sco 16h52'1"-37d58'
Wolke,Frank G B
Lyr 18h34'30"27d34'
Wolke,Sherry Ann
Sgr 19h26'46"-27d5'A
Wolke,Stephen John
Sgr 19h26'46"-27d5'B
Wolkoff,Mindi K
Lyr 18h13'14"35d23'
Wolkowsky,Edna
Cet 2h14'28"7d29'
Woll,Alan J
Aur 6h5'21"32d5'
Wollbaum,Birgit
Lib 15h41'16"-20d13'
Wollen,Lloyd & Ethel
Cyg 21h3'12"33d17'
Wollensak,Jenny A
Cas 1h44'32"60d57'
Wollenschlager,Amanda Anne
And 2h31'23"50d19'
Wollersheim,Alex Logan
Per 2h37'56"37d31'
Wollersheim,Megan Jane
Cas 2h7'31"63d48'
Wollman,Jeffery M
Cmi 7h45'51"7d52'
Wollman,Michael
Cet 2h14'33"1d56'
Wollman,Our Angel Brandon Thomas
Aur 5h7'1"44d30'
Wollschläger,Horst
Cmi 7h19'23"5d33'
Wolman,Philip
Uma 10h27'0"58d34'
Wolmuth,Jessica Louise
Peg 22h0'49"20d60'
Wolney,Sydnie
Lyn 8h12'14"37d2'
Wolney,Taylor
Her 17h35'0"28d43'
Wolniak,Andre
Cas 22h56'57"58d45'
Wolnick,Michael
Boo 15h40'16"48d1'
Wolnitzek,Frank
Mon 7h46'1"-3d15'
Wolnlammipi
Dra 17h7'26"68d52'
Wolny,Dennis J
Per 4h44'54"50d4'
Wolosoff,Juliet Garrett
Cas 1h27'45"67d51'
Wolpert,Christopher Edward
Mon 6h56'19"3d40'B
Wolpert,Grayson R
Aql 18h57'59"-6d15'
Wolschlager & Family, Dave & Kelly
Aur 7h0'33"40d56'
Wolschlager,Wally & Donna
Dra 17h57'1"58d60'
Wolsey,Brandon Lamont
Vul 19h48'22"28d24'
Wolstencroft Star,The Paul
Cep 0h12'55"76d52'
Wolstenhome,Rachel
And 1h44'46"40d7'
Woltas,Hans Georg
Dra 18h56'54"65d12'
Wolter,Günther & Annelis
Boo 15h7'16"52d33'
Wolter,Jenny
Lyn 8h14'23"42d8'
Wolter,Joe
Aql 18h56'44"17d36'
Wolter,Marcus
Lyn 8h3'12"43d40'
Wolter,Petra
Gem 7h3'13"24d39'
Wolter,William J
Dra 12h55'1"76d8'
Wolters,F H
Cet 1h37'51"-14d10'
Wolters,Pascal
Dra 15h47'54"62d23'
Wolthausen Anniversary Star,The
Hya 8h24'44"-0d15'
Wolverine Lawn Sprinkling Inc
Lac 22h29'45"55d51'
Wolverton
Cnv 12h49'33"43d20'
Wolverton,Benjamin
Cep 0h11'49"69d58'
Wolyn,George "Popeye"
Boo 15h25'26"41d37'
Wolzinger,Mel
Eri 4h10'1"-15d33'
Womack,Gregory E
Cyg 19h31'26"33d32'
Womack,Richard
Cmi 7h56'31"8d27'
Wombacher,Alfred
Hya 9h17'11"-9d36'
Womble,Isabelle
Equ 20h57'40"10d39'
Womble,Janet & Carey
Sct 18h54'0"-5d36'
Womble,Russell Dean
Cet 0h55'15"-5d49'
Wombmate
Cet 0h10'11"-18d52'
Womelsdorf,Meryl
Del 20h37'39"6d58'
Women of Hot Wire,The
Cas 23h0'50"53d33'
Womer,Eleanor
Peg 22h17'1"21d8'
Wonder
Vul 19h47'0"28d48'
Wonder,Stevie
Ori 4h58'46"15d9'
Wonderful Ann
Cas 0h29'51"60d18'
Wonderful Fuad
Boo 15h1'37"26d38'
Wonderful Mary Ann,The
Uma 11h45'38"54d56'
Wonderful Tonight
Her 18h19'23"28d29'
Wonderful Wendy's Wish
Ori 5h50'42"21d9'
Wonderful Willy
Cep 21h15'33"68d60'
Wonderful World of Edelweiss
Vul 19h34'25"26d53'
Wonderful"E"
Her 18h30'13"24d43'
Wonderwoman
Ind 20h58'14"49d4'
Wondolkowski
Cnv 13h13'17"38d59'
Wondolowski,William
Hya 8h41'52"6d29'
Wone,Anita
Cas 0h14'28"46d44'
Wong Family's Shining Star,The
Ser 15h18'1"4d5'
Wong Fung Yin,Clara
And 1h44'15"38d40'
Wong Wing-On
Lyn 6h58'37"59d0'
Wong's Star,Lisa
Mon 6h58'1"11d7'
Wong's White Light, Pauline
Cas 0h46'32"73d6'
Wong,Alice
Aur 6h1'1"37d31'
Wong,Charles
Umi 14h30'41"68d20'
Wong,Connie Wye Yee
Aql 20h1'23"14d37'
Wong,David & Kelly Ann
Eri 3h57'1"-6d28'
Wong,Iris
Ori 5h27'44"1d23'
Wong,Kela Malia
Peg 23h43'52"31d17'
Wong,Kevin B B
Cep 0h10'27"73d37'
Wong,Lei & Winfred
Umi 15h19'0"70d19'
Wong,Marianne Fei Leam
Aur 5h54'57"31d23'
Wong,Odete The Brightest Star
Ari 2h36'22"21d42'
Wong,Pamela
Cam 6h7'43"58d12'
Wong,Patricia
Cam 6h8'20"58d27'
Wong,Ricky
Ori 5h33'37"-6d23'
Wong,Ronald T
Lyr 19h1'11"35d7'
Wong,Sara
Uma 11h59'0"49d27'
Wong,Stephen & Jacqueline
Cru 12h10'9"-61d37'
Wong,Steven
Aur 6h26'53"33d33'
Wong,Wesley
Cet 2h33'21"7d18'
Wong,Wynne
Mon 7h14'24"-10d16'
Wong,Yvonne Mayping
Cet 2h20'55"-10d41'
Wong-Ng,Winnie
Boo 14h12'56"31d30'
Wongsuwan,Samantha Dongchantr
Eri 4h0'21"-14d1'
Wonsek,Paul Fredrick
Aur 4h48'50"41d13'
Wonser,Matthew R
Vul 20h39'30"28d9'
Wonser,Melissa M
Vul 19h57'41"27d20'
Wonser,Vicki L
Vul 20h16'11"23d30'
Wonsetler,Thomas & Lynn
Boo 15h4'11"28d23'
Woo Woo
Her 17h50'31"40d56'
Woo,David Brian
Her 16h46'0"21d53'
Woo,Jim & Andrea
Uma 11h48'38"64d49'
Woo,John May Chris Kevin Sharyn
Cma 6h55'9"-18d33'
Woo,Kinglsey
Uma 12h5'24"58d59'
Woo,Tiffany
And 1h49'16"38d50'
Woo-Woo
Cep 2h41'28"78d50'
Wood III,Jefferson K
Sex 9h59'1"4d50'
Wood"Snoopy",Laurie
Cas 23h35'10"61d24'
Wood,Adam
Her 17h25'1"21d41'
Wood,Addison Elliot
Uma 12h2'10"39d60'
Wood,Agnes E
Eri 3h37'33"-6d59'
Wood,Alan L
Aur 5h8'0"40d16'
Wood,Albert
Her 18h17'29"28d27'
Wood,Alexander James
Her 18h11'49"31d30'
Wood,Alexandra Grace
Aqr 20h36'27"-0d35'
Wood,Alexandra Rachel
Vul 19h20'38"26d48'
Wood,Andrea Kathleen
Aur 5h55'49"40d37'
Wood,Anthony
Aql 19h42'15"14d24'
Wood,Benjamin
Ori 6h8'28"2d24'
Wood,Benny D
Aur 6h2'16"31d23'
Wood,Bradley Michael
And 1h25'35"36d9'
Wood,Brian
Lyn 7h52'40"40d31'
Wood,Bryant Stephan
Dra 19h18'24"68d23'
Wood,Cara Ashley
Ari 2h29'42"21d52'
Wood,Carol
Ori 6h16'1"8d1'
Wood,Carol Ann
Ori 4h56'28"5d3'
Wood,Charles & Tamara
Aql 19h0'20"16d28'
Wood,Charles Raymond
Lac 22h15'51"51d40'
Wood,Chelsea Lee
Del 20h37'0"10d51'
Wood,Christina R
Her 18h9'19"41d14'
Wood,Clare Margaret
Lyr 18h55'55"41d27'
Wood,Cody Joe
Gem 6h32'36"14d19'
Wood,CSM US Army (Ret),Clyde H
Aql 18h57'39"15d49'
Wood,Dana
Lac 22h26'17"55d4'
Wood,Daniel A.
Boo 15h14'41"32d16'
Wood,Danni Jo
Cep 2h5'26"78d11'

Wood,Danny Larry
Her 18h12'24"48d29'
Wood,Darcy Mary
Lyn 7h35'16"36d38'
Wood,Dave
Her 15h49'25"46d8'
Wood,David
Uma 11h34'1"32d34'
Wood,Dee
Del 20h15'1"10d18'
Wood,Derek Guy
Ser 15h13'34"11d4'
Wood,Diane Campbell
Cep 22h10'13"56d6'
Wood,Diane G
Ori 6h4'41"8d53'
Wood,Dr Don H
Oph 18h35'54"10d42'
Wood,Drew Ethan
Ori 5h57'39"-2d4'
Wood,Edward
Lyn 8h3'0"40d51'
Wood,Elaine Day
Vul 20h18'11"23d13'
Wood,Elise
Lyr 18h27'26"34d38'
Wood,Frank Boardman
Uma 10h31'1"50d41'
Wood,Gary
Crt 11h1'24"-11d59'
Wood,Gary
Uma 10h41'16"40d41'
Wood,Gene "Rocketman"
Aql 18h54'0"8d44'
Wood,George & Elizabeth
Cyg 19h25'30"31d47'
Wood,Gerald & Nanci
Aql 19h53'13"10d6'
Wood,Gerald Lewis
Her 17h37'15"42d22'
Wood,Grace I & John G
Aql 19h55'21"15d34'
Wood,Gregory Ross
Per 1h47'15"50d6'
Wood,Harold
Ser 15h11'57"1d44'
Wood,Heather Pulsar
Lyr 19h25'33"41d59'
Wood,HMCM "SS" Brent J
Her 18h4'57"38d46'
Wood,Jacquelyn Christine
Lyr 18h40'54"35d13'
Wood,Jaime Lynn
Cam 7h13'0"80d10'
Wood,James C
Aql 20h1'40"12d8'
Wood,Jeffrey Scott
Cnc 9h16'32"7d39'
Wood,Jessica Lynn
Aqr 21h19'40"-8d15'
Wood,Jill May
Cas 2h59'24"58d1'
Wood,Joe Marcus
Her 16h28'15"20d37'
Wood,John R.
Aql 20h2'25"8d24'
Wood,Josh
Her 16h58'39"23d4'
Wood,Jr,Forrest Glenn
Cep 22h29'1"58d60'
Wood,Jr,Parlia Brown
Ori 5h55'31"8d60'
Wood,Jr,Theodore
Boo 15h3'48"20d16'
Wood,Jr,Thomas Richard
Vul 19h5'39"24d23'
Wood,Judith
Mon 7h44'35"-2d2'
Wood,Karen M
And 23h19'39"43d23'
Wood,Katie
Cyg 19h34'1"39d5'
Wood,Kenneth E
Per 2h58'44"40d45'

Wood,Kenneth M
Dra 16h53'29"67d15'
Wood,Kristen Marie
Lyn 9h3'56"40d24'
Wood,Leah B
Mon 7h54'20"-5d54'
Wood,Leone Darlene
Cyg 21h4'36"38d46'
Wood,Linda
And 1h8'26"40d35'
Wood,Lyle
Cet 1h34'40"-11d19'
Wood,Mabel Margaret
Umi 16h20'37"79d55'
Wood,Madeline
Uma 12h26'52"62d55'
Wood,Madeline
And 1h22'38"33d34'
Wood,Maree
Lyn 7h53'1"42d7'
Wood,Maree
Ori 5h0'34"1d51'
Wood,Mariah
Cep 21h1'23"60d37'
Wood,Martin
Lib 15h20'31"-23d51'
Wood,Mary Elizabeth
Cas 1h5'51"50d31'
Wood,MaryHelen Cook
Cet 3h16'28"7d47'
Wood,Missy
Cyg 21h13'21"28d32'
Wood,Nancy
Uma 11h28'54"45d54'
Wood,Norma
Lyr 19h22'29"31d16'
Wood,Pamela Joy
Lyn 6h31'16"61d12'
Wood,Philip Burnett
Cep 21h20'49"58d58'
Wood,Renate F
Hya 9h6'28"1d14'
Wood,Richard Phillps
Dra 13h23'1"68d18'
Wood,Robert Eric
Cma 6h55'38"-18d56'
Wood,Ron
Per 2h54'35"54d49'
Wood,Roy
Cma 6h52'54"-16d39'
Wood,Scott A
Lac 22h1'16"48d50'
Wood,Sheri Lynn
Lyn 8h3'11"46d48'
Wood,Stacey
And 2h1'13"42d49'
Wood,Stephanie
And 0h6'57"43d34'
Wood,Sylvia
And 0h25'45"40d22'
Wood,Thomas B
Cmi 7h55'1"8d49'
Wood,Thomas Richard
Uma 11h0'39"46d44'
Wood,Tracy Lynn
And 1h17'0"40d59'
Wood,Wesley Charles
Umi 16h32'22"78d47'
Wood,William John
Her 16h16'44"48d35'
Wood,William Richard
Umi 16h41'56"78d1'
Wood,Win
Umi 14h40'56"67d33'
Wood,Winfield McGregor
Dra 20h37'14"70d11'
Wood-100!,Gladys Kieffer
Lyr 19h2'43"25d40'
Woodage,Robert
Her 16h59'30"30d37'
Woodahl,Jessica & Emily
Lyr 18h31'36"33d4'
Woodall,Jake G
Hya 8h18'22"1d0'

Woodall,Maryanne Noriko
Peg 23h7'42"8d18'
Woodard III,Otis Jack
Ori 5h51'47"7d9'
Woodard,Charles & Anne
Brown
Cmi 7h59'38"0d23'
Woodard,Elizabeth Louise
Lyr 18h38'1"35d44'
Woodard,Heather
Aql 19h23'17"-6d35'
Woodard,Hoyt & Stephanie
Sge 20h0'32"20d19'
Woodard,Jeff & Sarah
Uma 10h14'18"58d7'
Woodard,Kenneth B
Oph 18h18'24"8d4'
Woodard,Mary Elizabeth
Lawrence
Psc 23h27'0"5d7'
Woodard,Phyllis
Crt 11h14'25"-19d4'
Woodard,William
Hya 8h59'23"-7d3'
Woodburn,Richard
Lyn 8h4'42"38d43'
Woodcock,Christopher Bailey
Cep 22h12'44"56d9'
Woodcock,Dominic C
Her 17h19'53"47d30'
Woodcock,Patrick Ryan
Dra 9h26'42"73d8'
Woodcock,Robert Howard
Crb 16h14'11"37d57'
Woodcock,Stacey
Lyr 18h38'45"41d2'
Wooden,Diana Sue
And 2h27'36"50d14'
Woodfin,William L
Cep 20h11'25"60d43'
Woodhall (Senior), William
Per 3h27'1"40d29'
Woodhead,Ann
Lac 22h36'22"53d17'
Woodhead,Karen
Lyr 18h17'31"42d56'
Woodhouse,Aunt Julie Soper
Peg 23h32'58"31d18'
Woodhouse,Gillian
Cas 0h7'45"61d48'
Woodhouse,Michelle
Cas 1h42'38"54d15'
Woodie's Star
Her 17h38'59"26d39'
Woodin,Sheila Starr
Eri 3h50'59"-0d18'
Woodings,Courtney Lynn
Mon 7h1'54"5d20'
Woodle,Claire
Uma 11h44'23"64d57'
Woodlen,Anthony Phillip
Crb 15h56'1"28d34'
Woodley,Linda
Per 2h29'52"52d5'A
Woodley,Richard Lee
Per 4h23'43"51d15'
Woodley,Sr,Garry A
Her 17h12'14"22d51'
Woodley,Tom
Per 2h29'52"52d5'B
Woodliff,Scott Law
Her 16h58'43"35d39'
Woodman,Lindsey
Lyr 19h21'16"41d12'
Woodmansee,CLU,ChFC,
Ronald I
Her 17h23'49"40d27'
Woodroofe,Jeff
Per 3h51'43"40d36'
Woodruff,Chris R
Lac 22h24'20"54d14'
Woodruff,Delos W
Uma 10h45'11"54d46'
Woodruff,Gerry
Mon 8h8'31"-2d46'

Woodruff,Jr,Charles Alexander
Oph 17h11'15"-10d6'
Woodruff,Linda & Harry
Peg 22h37'43"30d50'
Woodruff,Melissa Ann
Cam 3h26'0"55d36'
Woodruff,Rus
Dra 14h38'31"60d58
Woodruff,Sally Hopkins
Mon 7h57'57"-8d26'
Woodruff,Terry Keith
Mon 7h42'19"-6d49'
Woodruff,Valerie
Tri 2h4'26"31d50'
Woodruff,Yvonne K
Cyg 19h28'0"32d22'
Woodrum,John
Her 17h7'29"38d36'
Woods & Bud,Princess Lori A
Mon 7h1'16"-6d52'
Woods Forever Special Love
Kylie,Anthony
Ind 20h50'37"-56d49'
Woods,Alyssa Helen
Lyr 18h56'58"31d48'
Woods,Angel
Peg 21h54'47"23d36'
Woods,Anne
Cas 3h7'0"65d31'
Woods,Bobby
Sct 18h52'44"-7d22'
Woods,Carolyn Hornsby
Cas 0h55'45"72d22'
Woods,Christopher Michael
Aur 6h17'17"46d43'
Woods,Dinah A
And 0h20'27"30d27'
Woods,Dorothy Grace
Cam 3h56'1"58d45'
Woods,Douglas Richard
Sex 9h58'38"-6d58'
Woods,Edwin
Peg 22h39'1"27d33'
Woods,Gavin
Uma 8h30'0"51d1'
Woods,Gerald Carl & Marilyn
Lee
Eri 4h14'0"-11d9'
Woods,Jeffrey Scott
Mon 6h51'36"-4d25'
Woods,Jennifer
Cas 0h40'19"69d20'
Woods,Joseph
Uma 8h43'1"56d38'
Woods,Jr,Dennis Edward
Hya 8h31'15"-6d11'
Woods,Karen
Lyr 18h17'27"40d30'
Woods,Kim Ellen
Com 13h16'0"28d24'
Woods,Marielle Elena
Cas 0h51'1"64d19'
Woods,Mary Ann
Mon 7h43'21"-1d55'
Woods,Mary Grace
Vul 19h15'54"25d11'
Woods,Michael J
Lac 22h33'1"40d31'
Woods,Michael Stephen
Cmi 7h57'48"8d26'
Woods,Mrs Jill Elizabeth
Cas 22h57'0"55d47'
Woods,Pamela Rita
Dra 13h17'31"64d7'
Woods,Patricia
Uma 8h41'13"55d43'
Woods,Rebecca
Mon 6h23'49"5d49'
Woods,Richard Henderson
Her 18h18'12"18d59'
Woods,Robert Jon
Vir 14h2'15"-5d20'
Woods,Robert Wayne
Aql 19h37'15"0d16'

Woods,Roger Lee
Ser 15h46'57"21d21'
Woods,Stephanie Cara
Cas 0h18'1"63d51'
Woods,Stephen C & Michelle
L Nelson
Lyn 7h31'0"58d13'
Woods,Sue
Cyg 19h35'1"27d55'
Woods,Tina Michelle
Cyg 21h36'29'40d52'
Woods,Vanessa-Ann
Cyg 20h2'53"40d52'
Woods,Virgle Lee
Aur 6h6'25"35d0'
Woods,Walter & Joyce
Sex 10h18'14"-0d29'
Woods,Willie E
Per 3h57'48"51d45'
Woods-21496,W K
Dra 14h57'43"61d2'
Woods-Baker,Catherine
Cam 7h7'23"68d39'
Woodson,Catherine Walton
Sge 19h25'58"18d47'
Woodstock,Nina
Lyn 8h0'43"44d37'
Woodward's Wishing Star,Amy
Lac 21h57'0"38d38'
Woodward,Brendan James
Her 18h30'50"24d25'
Woodward,Gill
Lyr 18h29'12"30d35'
Woodward,Helen
Peg 23h0'11"21d53'
Woodward,James S
Umi 17h10'1"85d55'
Woodward,John W
Ari 3h17'28"25d8'
Woodward,Katie
Equ 21h9'17"11d54'
Woodward,Lisa
Lyr 18h59'25"40d13'
Woodward,Marvin Seward
Her 17h27'47"40d33'
Woodward,Monte
Cam 8h26'13"80d27'
Woodward,Myrtle
Eri 3h5'0"-6d30'
Woodward,Peter David
Per 4h5'53"39d24'
Woodward,Queen Alice
Com 12h48'58"12'
Woodward,Ryan David
Sex 9h43'11"0d31'
Woodward,Ryan R
Dra 16h51'13"73d1'
Woodward,Serena Mary-
Elizabeth
Mon 6h34'50"-0d12'
Woodward,Steven M
Hya 8h53'16"0d33'
Woodward,Tracey
Cyg 19h26'18"31d15'
Woodward,Woodie
Cep 20h22'40"76d23'
Woodwind
Cnv 12h30'47"40d33'
Woodworth,Dolores "Dee"
Cas 0h14'56"62d58'
Woodworth,Ethel A
And 2h23'1"39d28'
Woody
Leo 10h19'29"9d13'
Woody
Sct 18h44'22" 6d8'
Woody
Cep 22h24'24"70d15'
Woody & Molly
Aql 20h2'36"7d27'
Woody & Noreen
Equ 21h10'26"10d40'
Woody,Anya
Lyn 9h14'22'37d39'

Woody,Jeanne
Sco 17h22'43"-38d23'
Woody,Larry R
Cet 1h43'13"-7d1'A
Woody,Virginia H
Cet 1h43'13"-7d1'B
Woodyard,Paul
Her 15h55'38"41d41'
Woodyard,Phillip G
Sct 18h55'45"-8d9'
Woodyard,Traci Michelle
Peg 22h8'17"26d56'
Woodzick,John H "Woody"
Cep 22h35'1"68d32'
Woog,John Paul
Boo 13h45'26"17d18'
Wooher,Lisa & David
Cyg 21h50'18"52d38'
Wookie
Boo 15h11'1"52d51'
Wookie
Lyn 8h4'1"50d8'
Wookie's Brown Eyes
Crb 15h49'49"27d44'
Woolard Star,Christie & Tony
Lmi 10h2'0"32d40'
Woolard,Arthur
Dra 18h22'14"48d37'
Woolard,Emily Anne
Ser 15h40'50"13d49'A
Woolcock,Bob
Per 2h52'53"46d7'
Wooldridge,Catherine
Cnc 7h55'47"10d8'
Wooldridge,CB
Cyg 21h34'29"37d32'
Wooldridge,Collette
Ari 1h59'53"15d21'
Wooldridge,Darlene
Gem 8h5'59"30d15'
Wooldridge,David
Leo 9h36'44"11d53'
Wooldridge,Donald
Leo 10h38'44"12d38'
Wooldridge,Elizabeth
Her 17h24'50"30d44'
Wooldridge,Jack
Tau 5h56'13"23d20'
Wooldridge,Joshua
Tau 5h56'0"23d10'
Wooldridge,Julie Kristene
Cyg 19h33'0"38d18'
Wooldridge,Kenneth
Lib 15h44'21"-22d35'
Wooldridge,Kerry
Psc 1h21'13"31d44'
Wooldridge,Kevin
Leo 9h55'35"10d48'
Wooldridge,Kirk
Gem 7h12'35"31d50'
Wooldridge,Philip Anthony
Aql 19h22'41"12d2'
Wooldridge,Ralph
Psc 1h22'54"31d2'
Wooldridge,Sandra
Ari 1h59'1"17d10'
Wooldridge,Susan
Cap 21h44'48"-24d21'
Wooldridge,Wendy
Sgr 18h4'41"-28d59'
Woolery,James
Mon 7h39'42"-1d57'
Woolery,Jan
And 2h31'55"45d34'
Woolery,Jerry T
Boo 15h6'28"50d38'
Woolever,Lorraine
Aql 20h35'22"-7d34'
Woolever,Paula Kay
Eri 3h36'11"-18d11'
Woolf,Laronna & Jim
Cyg 19h54'18"48d57'
Woolger,Laura
Cas 2h34'22"60d36'

Woolheater,John
Per 1h52'42"50d28'
Woollett,Christobel Elizabeth
Aql 18h58'29"17d50'
Woolley,Carol
And 23h1'43"51d22'
Woolley,Catherine Marczak
Lib 15h56'23"-18d55'
Woolley,Devon Allison
Lyr 18h43'34"36d7'
Woolley,Lauren Nicole
Sco 16h50'18"-40d46'
Woolley,Steve
Her 15h54'45"48d38'
Woollie Clinkstar
Aql 20h2'52"1d31'
Woolly
Uma 11h52'41"47d9'
Woolly Star,The
Uma 9h52'0"46d30'
Woolman,Bethany Lynn
Crb 15h32'49"26d31'
Woolman,William M
Mon 7h21'1"-1d49'
Woolston,Traci Lynn
And 1h53'30"39d22'
Woolwine,Ashley Michelle
Lyn 7h45'38"44d46'
Woolworth,Marcia D
Tau 5h2'17"16d47'
Woomer,David Charles
Her 17h3'24"30d23'
Woonch Bagoonch,The
Lac 22h23'30"50d6'
Woosey,Dominic
Ori 5h56'43"16d52'
Woosey,Laura Holly
Lyn 8h22'1"43d7'
Woosley-031360,Sandra Duff
Aql 18h59'58"15d18'
Wooten,Don W
Oph 17h53'49"-0d28'
Wooten,Joel Andrew
Ser 15h34'52"8d0'
Wooten,Jr,George R
Her 17h33'23"50d5'
Wooten,Jr,William G
Peg 23h18'38"32d52'
Wooten,Robin Nathaniel
Cmi 8h5'42"0d44'
Wooten,Steve
Aql 20h4'55"0d9'
Wooten,Virginia
Lyr 18h49'45"42d36'
Wooton,Christopher
Uma 12h4'5"31d21'
Wor Sue
And 23h20'56"40d5'
Worden,Lee
Del 20h15'32"10d45'
Worden,Mitchell
Her 17h10'1"20d58'
Worhack,Thomas "T R"
Her 17h26'15"40d47'
Woringen,Jr,James & Diana
Ori 5h57'0"14d50'
Work,Anna Elizabeth
Cam 6h40'46"68d41'
Work,Mary Ann
Cam 3h56'22"52d44'
Workman,Curtis
Her 18h25'42"24d39'A
Workman,Dads Star, Cindy
Roseann
Aql 19h29'27"10d30'
Workman,Kathleen M
Cas 0h34'1"67d22'
Workman,Nichole June
And 2h32'32"40d36'
Workman,W Kay
Cnv 13h41'30"30d32'
World Bonsai Friendship '93
Uma 11h42'48"38d37'
World Class Incentives
Cam 5h43'21"68d60'

World Mogul Team
Dra 18h13'26"65d8'
World Of Jay,The
Aur 6h6'0"40d54'
World,Huws
Mon 8h2'53"-8d3'
World,Tony
Ori 6h8'56"0d27'
Worley's Star Jocelyn
Uma 9h55'1"57d48'
Worley's Star,Chad
Per 2h56'20"35d23'
Worley's Star,Melanie
Cyg 19h59'25"37d40'
Worley,Chris
Lmi 10h5'0"37d32'
Worley,Daniel
Cep 20h17'57"75d34'
Worley,Martha Jean Blake
Crb 16h4'17"32d40'
Worley,Michaela
Cyg 19h32'52"35d30'
Worley,Thomas Benton
Dra 20h4'51"68d5'
Worley,Thomas Benton
Cet 2h51'19"3d8'
Worleymeister
Ser 18h42'47"4d17'
Worm,Thomas
Psc 1h33'1"27d37'
Worman,Ronald William
Cep 22h1'54"61d7'
Worman,Tara Ann
And 23h29'58"49d11'
Wormer-Lang Hilsbach
Vir 13h5'1"-8d10'
Wormley,Evelyn W
Eri 3h57'19"-1d35'
Wormstadt,Cindy Kay
Uma 8h51'49"61d2'
Wormuth,Jr,Howard
Cam 7h50'35"82d37'
Worrell,Clairetta F
Peg 22h18'34"21d22'
Worrell,Jane & Trix
Cyg 21h4'19"30d53'
Worrell,Jeff Allen
Cap 20h7'38"-10d45'
Worrell,Kaitlyn Danielle
Peg 23h3'12"17d32'
Worrell,Michael B
Her 16h9'53"23d32'
Worrell,Troy Patten
Boo 15h16'25"38d47'
Worrell,William George &
Yvonne Jean
Boo 14h0'40"27d19'
Worsnick,Gregory
Sex 9h42'12"2d55'
Worster,Catherine E
Cas 0h52'27"64d28'
Worswick,Ronald J
Sex 9h59'51"-6d22'
Worth,Harvey
Ori 6h12'25"8d50'
Worth,Matthew Lawrence
Sco 16h37'59"-31d38'
Worth,Michael
Cep 22h19'47"55d24'
Worthen,Barbara S
Eri 3h53'20"-2d3'
Worthington,Jonathan Davis
Ser 18h42'26"4d38'
Worthington,Jr,MD, Francis
Xavier
Oph 16h20'36"2d51'
Worthington,Patricia Koenig
And 23h40'57"45d27'
Worthington,R G
And 2h4'49"40d34'
Worthington,Rachael Elizabeth
Eri 3h22'1"-13d20'
Worthington,Rachel Faith Joy
Com 12h1'21"27d19'

Worthington,Richard
Pho 23h44'57"-45d36'
Worthy's Star,Jeb
Cet 2h43'31"5d35'
Wortman,James
Tau 4h4'30"24d12'
Wortmann,Rachel Sela
Lyn 7h49'49"43d12'
Woskey,Kathy
Lyr 19h2'0"25d51'
Wotten,Christopher Chadwick
"Chad"
Aql 18h58'52"-5d11'
Wotton,Diane
Lyn 8h24'60"49d48'
Woudstra,Martijn Aiko Jelle
Aur 7h23'33"38d57'
Woughter,Helen
Cas 1h2'16"68d38'
Wow
Aur 5h54'18"30d56'
Wowaglake,Micante Etan
Dra 18h46'47"80d12'
Wowina
Cyg 21h5'57"31d1'
Wowra,Jürgen
Aqr 22h8'58"-0d60'
Woycie
Cam 3h58'1"52d55'
Woyender,Neil Peter
Ori 5h0'1"15d23'
Woynar,Paul
Cmi 8h4'1"6d31'
Wozar,Janis Carlin
Eri 4h14'25"-8d51'
Wozney,Samantha Neal
Cas 0h33'40"63d10'
Wozniak Ma
Aur 5h2'29"38d24'
Wozniak,Christine
And 1h52'37"36d24'
Wozniak,Elizabeth
Lyn 8h53'29"40d45'
Wozniak,Gregory
Matthew,Beloved SonPvt
Her 17h53'30"14d26'
Wozniak,Lynette Victoria
Cam 6h8'38"60d46'
Wozniak,Michael J
Lmi 9h57'23"40d41'
Wozniak,Ross
Cep 22h19'0"60d9'
Wozniak,Roy Jason
Cmi 7h16'31"5d34'
Wozniak,Sandy
Vul 19h57'42"28d8'
Wozniak,Theresa June
Beloved First Dtr
Cyg 20h24'18"30d57'
Woznica,Joshua Gideon
Her 16h21'48"26d43'
Wozny Star "Woz",Ted
Vul 19h18'5"26d33'A
WPS
Cnc 8h54'35"32d14'
WR95
Tri 2h6'1"31d17'
Wrabel,Dorothe & John
Dra 12h2'41"70d37'
Wrack,Ami-Rose
Com 12h18'59"22d7'
Wragg,Yvonne Elizabeth
Umi 15h21'15"68d21'
Wraight,Cheryl
Cas 0h46'20"61d59'
Wraight,Derek
Uma 12h3'29"31d58'
Wraith
Her 17h31'56"20d12'
Wraith,Kristen Jade
Umi 15h13'31"66d5'
Wrana,Sherry & Ralf
Ori 5h15'50"10d43'
Wratten,Kevin
Aqr 23h2'16"-12d24'

Wray's Wish
Hya 8h59'58"-10d9'
Wray,Debra Ann
Lyr 19h4'45"40d26'
Wray,France Monique Tracy
Lyr 18h45'27"31d43'
Wray,Gary A
Boo 13h57'53"20d42'
Wray,Helen
Mon 7h24'14"-8d50'
Wray,Jennifer
Lyr 18h41'28"46d43'
Wray,Jonathanya
Cet 2h41'42"-5d56'
Wray,Jr,Col Bill
Her 17h59'57"20d2'
Wray,Lon S
Aql 19h31'25"11d5'
Wray,Marlena
And 23h21'47"42d14'
Wray,Richard & Janice
Cyg 20h3'43"40d31'
WRC II
Cet 2h27'20"3d45'
Wren
Peg 0h1'42"21d54'
Wren Day
Cyg 20h21'10"30d46'
Wren,Kimberly G
Cet 2h50'0"1d38'
Wren,Michael J
Aur 7h10'51"40d44'
Wren,Summer Rae
Lyn 6h17'32"59d10'
Wrenn's,The
Crt 10h59'44"-23d11'
Wrenshall
Cet 0h22'34"-8d29'
Wrenstar
Cas 0h38'23"60d38'
Wrentmore,Cynthia Y
And 2h25'33"46d25'
Wressell,Thomas George
Her 17h14'11"27d11'
Wride,Howard Tenison
Cep 21h56'0"80d24'
Wright #1 Dad & Mom, Gene & Joan
Uma 9h26'23"55d37'
Wright (Love of my Life),Susan Hallam
Cas 1h39'1"73d54'
Wright 5-23-22,David
Oph 17h25'52"1d50'
Wright A Rising Star! Joby
Dra 14h17'18"63d60'
Wright Dacre
Uma 8h14'39"68d39'
Wright Star,Larry Edward
Ori 5h4'38"10d32'
Wright Star,Susan
Com 12h30'0"21d15'
Wright Star,The
Cyg 21h37'34"40d46'
Wright Star,The
Uma 9h7'28"48d46'
Wright Star,The Brian
Ori 5h57'1"16d48'
Wright Star,The David E
Dra 19h28'44"68d48'
Wright"Nana",Christine
And 0h20'27"35d7'
Wright's Guardian Angel,Lonnie
Per 3h53'15"41d13'
Wright's Light
Uma 14h23'36"62d9'
Wright's Piece Of Heaven,Libby
Cyg 21h21'13"39d37'
Wright's Star, J
Cam 12h15'0"80d20'
Wright's Union
Cyg 20h18'59"40d38'
Wright,Alice
And 0h10'49"38d57'
Wright,Allison
Mon 6h23'20"10d11'
Wright,Alvin
Oph 16h20'12"2d23'
Wright,Alvonia M
Her 18h2'23"30d39'
Wright,Amy Beth
Mon 6h20'24"5d37'
Wright,Ann Litkenhaus
Ori 6h2'30"-2d7'
Wright,Ann Renie
Peg 22h2'58"33d59'
Wright,Anne P
Peg 22h1'35"4d2'
Wright,Anthony Edwin
Dra 16h2'35"63d48'
Wright,Ashley
Uma 11h5'1"44d14'
Wright,Bailey Marie
Dra 20h7'57"74d53'
Wright,Barbara
Cma 6h32'11"-11d11'B
Wright,Betty
Peg 22h6'55"5d46'
Wright,Bill
Uma 12h4'12"30d1'
Wright,Brenda Cejda
Eri 3h22'52"-11d49'
Wright,Bruce
Sge 19h55'46"18d51'
Wright,Carole Beth
Lyn 7h37'22"48d1'
Wright,Casey
Cyg 21h50'18"41d53'
Wright,Catherine Litkenhaus
Ori 6h2'33"-1d30'
Wright,Caz
And 23h16'1"51d51'
Wright,Christy
Cas 0h26'1"63d52'
Wright,Colby
Cep 23h35'43"64d25'
Wright,Commander Sarah Asher
Lyn 7h5'18"52d14'
Wright,Curtis & Juanita
Boo 13h53'58"21d25'
Wright,Darlia Dawne
Cyg 20h39'44"45d31'
Wright,Daughtry Taylor
Eri 4h32'59"-12d1'
Wright,David Leland
Her 16h2'51"48d29'
Wright,Deann Marie
Cas 1h14'54"60d6'
Wright,Deborah Louise
Del 20h15'43"8d55'
Wright,Diane Marie
Aql 19h54'57"11d37'
Wright,Eddie
Her 18h0'1"14d44'
Wright,Edwin Paul
Boo 14h7'50"28d33'
Wright,Elizabeth "Bubbles"
Lyr 19h1'18"35d37'
Wright,Elvis
Oph 17h31'7"-22d24'
Wright,Emily
Mon 6h23'1"11d40'
Wright,Eric
Cam 12h34'16"77d59'
Wright,Eric Lloyd
Ser 18h2'1"-14d27'
Wright,Erin Marie
And 0h17'43"33d44'
Wright,Ernest P
Dra 11h28'0"68d30'
Wright,Francis Edward
Cep 21h46'21"68d11'
Wright,Gene
Cma 6h32'11"-11d11'A
Wright,Gene Alan
Lyr 18h43'58"34d36'
Wright,George
Aur 5h52'26"40d60'
Wright,Glen
Ser 15h44'50"10d57'
Wright,Greg
Aur 7h3'31"38d50'
Wright,Gregory "Prince Noir"
Cep 20h19'34"60d20'
Wright,Harold
Dra 18h31'33"58d40'
Wright,Harry Franklin
Cet 2h2'1"1d49'
Wright,Herbert & Virginia
Vul 21h2'26"24d22'
Wright,Jacob Tyler
Lac 22h48'55"53d59'
Wright,James
Lyr 19h15'52"40d41'
Wright,James W
Hya 9h35'58"5d52'
Wright,Janet
Aql 20h6'41"0d19'
Wright,Janice
And 23h38'41"45d16'
Wright,Jared Anthony
Cnv 12h25'53"44d38'
Wright,Jay
Aur 4h51'28"38d57'
Wright,Jean Alexander
Cma 6h59'8"-19d48'
Wright,Jennifer
And 0h37'0"37d48'
Wright,Jennifer Nicole
And 23h24'22"45d44'
Wright,Jim
Sex 10h30'56"-1d36'
Wright,Jim R
Mon 6h25'8"-0d37'
Wright,Joan Carol
Peg 22h2'47"20d23'
Wright,John H
Boo 14h18'48"54d38'
Wright,Jr,Howard Wesley
Aql 18h58'1"-8d26'
Wright,Jr,Reached With Love,Edward F
Cyg 21h4'48"36d40'
Wright,Julius
Cma 6h58'18"-20d14'
Wright,Keith Harrison
Oph 18h2'16"12d57'
Wright,Kenneth Howard
Per 3h18'50"42d9'
Wright,Kevin Anthony
Per 3h28'1"31d11'
Wright,Lauren Emily
And 1h35'24"35d34'
Wright,Lawrence
Cep 23h36'31"63d27'
Wright,Linda Ann
Lyr 18h41'50"28d44'
Wright,Lindsey Mitchell
Cyg 20h19'57"38d22'
Wright,Lorraine
And 23h16'13"41d8'
Wright,Louise
Cyg 20h6'50"40d29'
Wright,Louise K
Eri 2h48'34"-5d45'
Wright,Loyd & Alix "Loyal"
Boo 14h0'44"12d11'
Wright,Lucy
Lyr 19h26'56"38d44'
Wright,Margaret Litkenhaus
Umi 16h13'60"70d31'
Wright,Marie
Cyg 20h43'0"45d19'
Wright,Mark
Her 16h37'43"40d2'
Wright,Martin Thomas
Cep 22h48'54"56d53'
Wright,Mary Margaret
Sex 10h7'41"-0d57'
Wright,Maryann
Cam 13h16'42"77d15'
Wright,Matthew Litkenhaus
Mon 8h1'24"-0d10'
Wright,Michael James
Her 17h24'26"42d42'
Wright,Michael T
Aql 18h56'25"-5d52'
Wright,Michelle
Mon 6h44'27"11d46'
Wright,Nancy
Del 20h13'34"11d52'
Wright,Odie
Aql 19h47'48"11d9'
Wright,Patty Karen
Cas 0h4'27"60d30'
Wright,Paul,Alwyn
Cep 20h35'52"78d57'
Wright,Peggy Eileen
Tri 2h5'48"30d15'
Wright,Peyton James Lee
Aql 19h59'18"14d23'
Wright,Philip H
Cyg 19h29'39"35d55'
Wright,Princess Caroline's Diane
Cnc 8h28'59"8d24'
Wright,R & D
Cet 1h32'1"-02d30'
Wright,Rachel Dawn
Mon 6h54'39"-10d10'
Wright,Rev Donald A & Dorothy
Crb 15h49'46"28d39'
Wright,Richard E
Dra 18h29'12"68d51'
Wright,Ringo
Eri 2h50'1"-6d16'
Wright,Robert A
Uma 11h50'49"61d25'
Wright,Robert E
Dra 10h24'59"78d58'
Wright,Robin Simonian
Cet 2h4'0"6d31'
Wright,Rusty
Hya 8h13'29"1d26'
Wright,Sally Jo
Uma 8h50'29"50d39'
Wright,Samuel Lawrence
Cet 1h34'58"-11d13'
Wright,Sandi
Cnv 12h54'24"50d16'
Wright,Sandra N
And 2h24'37"40d23'
Wright,Sarah Genevieve
Lib 15h33'29"-8d4'
Wright,Sheri
Cas 0h18'16"64d27'
Wright,Shirley Bruton
Aql 19h40'50"10d21'
Wright,Stephen
Per 2h4'1"57d21'
Wright,Steve & Jodi
Cyg 21h18'59"37d45'
Wright,Steven Roger
Sex 10h15'6"-3d42'
Wright,Susan
Mon 7h56'5"-5d28'
Wright,Susan McLaughlin
Lyr 18h29'32"37d42'
Wright,Teona Sari
Uma 9h21'20"51d47'B
Wright,Terry George
Cep 23h10'59"64d57'
Wright,Tom C L
Cet 1h23'16"-4d56'
Wright,Warren Kenneth
Her 17h25'19"27d19'
Wright,Webster
Per 1h58'40"48d1'
Wright,Wendy
Peg 21h19'38"20d5'
Wright,William
Dra 17h59'22"64d40'
Wright-Hackett, Florence E
Peg 22h34'26"20d17'
Wrightsman,Bill
Cet 0h39'29"-5d33'
Wrightstar
Ser 15h58'50"1d52'
Wrigley,Donald Kenneth
Dra 17h36'50"61d33'
Wrigley,Michael Joseph
Gem 6h33'26"14d15'
Wrinkle,Stephen F
Hya 8h12'53"1d34'
Writer's Best
Boo 15h20'0"40d33'
WRL
Uma 10h37'43"48d50'
Wrobel,Rainer
Peg 23h30'17"10d32'
Wroblewski,Bob(Bobby)
Cep 22h4'22"53d53'
Wroblewski,Isabelle
And 23h35'57"49d41'
Wroblowa,Halina
Uma 12h10'24"54d55'
Wrona Star,Stella
Tri 2h6'29"33d31'
Wrona,Eleanor
Peg 21h49'18"19d4'B
Wrona,Jill M
Peg 22h25'50"5d26'
Wrona,Peter
Peg 21h49'18"19d4'A
Wrona,W
Cam 5h56'46"60d26'
Wronski,Matt G
Aur 6h26'23"32d51'
WS "Jay"
Aql 20h4'48"3d53'
Wseeks
Oph 18h7'45"12d40'
Wszolek,Pam
Lyr 18h39'27"37d48'
Wu I Nee
Aur 6h7'36"38d6'
Wu Yuting
Peg 22h3'10"3d53'
Wu,Annie Nai-I
Cas 23h36'57"58d31'
Wu,William
Cet 3h6'27"2d10'
Wubba
Uma 9h22'56"51d37'
Wubba,Richie
Per 1h48'59"53d9'
Wubble-U 290593
Cas 0h34'0"61d47'
Wubbolding,Christine F
Eri 3h39'0"-17d52'
Wuckert,Rosemary Kreml
Cas 0h53'1"66d10'
Wuczynski,Pamela Marie
And 0h9'50"37d31'
Wuerfel,Heino Dieter
Ori 5h31'21"-6d10'
Wuerflein Edson
Mon 7h59'19"-8d27'B
Wuerflein Hollands
Mon 7h59'19"-8d27'A
Wuerker,Beppo Steffen Otto Michel
Cet 2h14'40"5d39'
Wuermohon
Cnv 14h2'36"33d16'
Wuerth,Catherine Leigh
Cas 0h39'18"66d13'
Wuethrich,Cindy
Lyr 18h46'36"32d57'
Wuggazer,Duane T
Aur 5h15'49"44d58'
Wujick,Fawn Elizabeth
Cyg 21h17'22"37d17'
Wulf,Cheryl Louise
Her 16h52'0"47d51'
Wulff,Ann
Crt 11h35'42"-20d52'
Wulff,Gustav
Hya 9h13'60"-6d45'
Wulff,Kirsty
And 0h43'50"38d14'
Wulff,Ursel
Sco 17h31'52"-38d39'
Wulfsohn,Janine Gabrielle
Cam 12h53'35"80d60'
Wulkan,Gale
Hya 8h54'43"4d48'
Wulssohn,Joseph
Peg 23h3'39"17d49'
Wunder,Jordan Scott
Cep 22h21'18"61d28'
Wunderlich,Fritz
Cyg 20h24'36"39d37'
Wunjo
Lyn 7h9'49"59d17'
Wunsch IV,Albert Henry
Her 17h9'1"21d4'
Wunsch,Wolfgang
Oph 17h58'45"8d14'
Wunschel,Carl C & Eleanor C
Lac 22h1'24"49d5'
Wurlitzer,Lee
Dra 19h30'53"61d27'
Wurstbauer,Claudia
Ari 2h1'30"25d16'
Wursteisen,Alexandra Louise
Lyr 18h50'51"34d37'
Wursteisen,Jacob Michael
Aql 18h48'48"11d54'
Wursteisen,Payton Joseph
Aur 6h26'32"30d20'
Wurster,Debbie & Mike
Lyn 7h0'53"59d26'
Wurtz,40th Anniversary ,Jim & Nancy
Umi 15h46'40"76d50'
Wurtzinger,Richard
Dra 19h3'39"48d38'
Wurzburg,Matil
Mon 6h32'30"-0d16'
Wurzel von der Quickborner-Heide
Cmi 7h19'40"4d46'
Wuschu
Cam 5h47'48"58d28'
Wussow,Joel Richard
Her 17h34'36"27d30'
Wustin,Donald C
Boo 14h2'11"25d9'
Wutever
Peg 22h35'20"26d6'
Wuttke,Alexandra
Cep 2h10'48"80d16'
Wuzzle
Lmi 10h58'31"26d43'
Wvening Flower #94 OES Pawlet,VT
Cam 13h24'36"77d21'
WW
Boo 15h25'38"50d21'
Wyand,Gage Michael
Aur 5h21'10"40d55'
Wyandt,Handsome Herbie
Equ 21h22'50"11d39'
Wyant,Jason Mark
Cmi 7h23'34"0d12'
Wyatt Michael
Cep 21h28'0"55d58'
Wyatt Paul
Her 17h20'30"49d11'
Wyatt's Wierd & Wonderful
Lmi 9h25'25"38d15'
Wyatt,Beatrice Dorthea
Cyg 19h36'0"28d2'
Wyatt,Cheryl Louise
Cyg 21h17'50"35d46'
Wyatt,Dreamer of Dreams,Kendra
Lyn 7h32'16"41d33'
Wyatt,James Bradley
Cet 2h28'45"-10d31'
Wyatt,James J
Lac 22h19'25"35d37'
Wyatt,Jane Alison
Lyr 18h14'27"30d12'
Wyatt,Jennifer Marie
And 2h20'11"50d16'
Wyatt,Jimmy & Melody
Mon 6h19'19"0d4'
Wyatt,Kathryn Lynn
Peg 22h28'36"35d11'
Wyatt,Lee Ann
Cas 0h25'14"63d32'
Wyatt,Lucy Ann
Cyg 19h14'13"48d26'
Wyatt,Mark Elton
Per 3h18'22"41d28'
Wyatt,Robert Alexander
Aur 6h2'15"30d36'
Wyatt,Robert Jason
Cmi 7h6'49"4d23'
Wyatt,Scott Joseph
Her 16h35'60"35d34'
Wyatt,Thomasina J
Com 12h14'49"22d23'
Wyckoff,Gary T
Mon 7h12'25"-10d3'
Wyckoff,Rachael
Peg 22h11'54"16d56'B
Wyckoff,Steven
Peg 22h11'54"16d56'A
Wyco,Bridget Ann
Umi 16h41'26"78d1'
Wycoff-Daley,Sandra Jean
Vul 19h19'10"20d27'
Wydler,Susan Hart
Aql 18h46'40"1d34'
Wygle,Matthew David
Uma 8h44'31"52d46'
Wyka,Jeffrey A
Dra 15h19'54"52d52'
Wykes,Julie Ann
Cyg 19h27'46"33d54'
Wykes,Keith
Lyn 9h15'44"39d18'
WyldAngelsWhoRunWith Egrets
Cet 2h17'55"4d17'
Wylde Charlie
Aql 19h1'50"16d54'
Wylder,Duane
Cep 2h56'13"78d41'
Wylder,Phyllis
Uma 11h20'23"62d27'
Wylder-J
Lac 22h55'1"54d47'
Wylen,Joseph
Per 3h57'46"50d47'
Wyler,Stéphanie
Tri 1h59'19"31d44'
Wylie
Lyr 19h8'18"37d46'
Wylie Coyote
Lac 22h12'1"47d28'
Wylie Coyote A K A Skip
Cas 0h18'23"65d15'
Wylie,Jeanne & Ruth
Boo 15h6'23"10d27'
Wylie,Jr,James A
Her 16h20'14"24d22'
Wylie,Robert David
Sct 18h42'1"-6d42'
Wylie,Sharon
Lyr 18h58'17"46d52'
Wylie,Skylar Rae
Lac 22h12'31"48d59'
Wylie-Hoskins,Doris K
Lyr 19h16'15"35d20'
Wyllie,Donald
Ser 18h11'29"-7d25'
Wyllie,Lauren
Peg 22h45'43"21d35'
Wyllie,Uncle Charlie
Per 3h0'0"41d41'
Wylly,M Dasher
Sct 18h53'38"-6d38'
Wyllys,C
Umi 14h51'50"70d52'
Wyman,Barbara L
Mon 8h8'6"-8d42'
Wyman,Donel D
Cep 20h39'35"58d38'
Wyman,Jean
Cas 0h32'12"58d10'
Wyman,Linda Lee Nuzum
Lyr 18h48'34"40d34'
Wymer,Jack
Cep 21h59'18"61d7'
Wymer,Patti
Peg 0h0'55"30d4'
Wymer,Patti
Cet 1h26'14"-6d6'
Wymer,William Lee
Pcs 1h39'19"22d5'
Wymore,Lee
Uma 11h1'0"67d56'
Wyn-40
Lmi 10h8'30"39d6'
Wynde,Matthew Thomas
Mon 6h51'53"11d13'
Wyndee
Peg 22h0'1"31d19'
Wyndham,Victoria
Mon 6h21'38"1d26'
Wyne,Glenn P
Aur 5h46'13"50d10'
Wyngarde,Peter
Aur 5h16'20"41d33'
Wynkenbrite
Peg 23h6'18"13d1'
Wynkoop,Christopher J
Cep 22h10'49"55d44'
Wynkoop,Zachary Mills
Ari 2h40'36"21d40'
Wynn,Bryan
Lyn 8h6'39"41d48'
Wynn,Dr George Allan (Bud)
Oph 17h18'22"-20d13'
Wynn,Elizabeth
Lyr 18h28'45"30d38'
Wynn,James E
Dra 19h1'16"70d29'
Wynn,Mark
Dra 16h0'29"63d17'
Wynn,Red
Oph 17h28'13"-22d6'
Wynn,Ron
Boo 15h5'41"10d28'
Wynn,SM,John Addison
Her 17h53'0"14d45'
Wynn,Tristan Dean
Uma 11h21'31"37d55'
Wynne III,James Riley
Cmi 8h2'20"5d41'
Wynne"The Kid", Shannon Shelmire
Sgr 18h55'36"-29d34'
Wynne,Donnie
Cet 2h30'46"8d11'
Wynne,Elizabeth
Cas 1h52'13"58d34'
Wynne,Helen Louise Lefevers
Lyn 7h0'48"50d18'
Wynne,Hilary
Cyg 19h55'27"38d32'
Wynne,James E
Dra 16h51'56"65d59'
Wynne,Tricia
And 1h44'16"39d36'
Wyoming Writer,The
Aur 4h48'29"50d53'
Wyon,Christophe
Boo 14h20'40"51d48'
Wyrick,Elizabeth Ann
Cas 2h5'24"60d31'
Wyrick,Gordon Franklin
Per 2h36'1"43d39'
Wyrick,Sr,Paul L
Ser 15h25'0"24d7'
Wyrobnik,Simon
Umi 13h58'58"74d4'
Wyrwa,Thomas
Her 17h24'22"31d9'
Wysiwyg
Cnv 12h21'24"48d55'
Wyskocil,Steve
Ser 15h54'22"1d37'
Wyslick,William R
Aur 5h29'0"50d6'
Wysocki,Jessie
Cyg 19h25'17"44d31'
Wysocki,Megan C
And 0h19'39"31d40'
Wysocki,Thomas Joseph
Ori 5h35'56"-1d10'
Wysocky Family,The
Aql 19h6'39"-6d57'
Wysoski,Jody
Lyr 19h9'11"38d24'
Wyss,Amy
Lyr 19h20'51"35d34'
Wystyrk,"Spatz" Edeltraut
Leo 11h24'51"-5d21'
Wytrwa's,Mainstream America,Dr
Tri 2h2'53"30d26'
Wytrykus,Sandie
Hya 8h57'36"-6d46'
Wyvonne
And 2h17'41"42d10'
Wyzkoski Summer Place, The
Cet 2h26'0"3d19'
Wächter,Johannes
Aql 18h56'15"17d5'
Wächter,Raimar
Dra 15h13'35"62d52'
Wächter,Wolfgang
Lyr 19h21'30"30d22'
Wélise
Cep 21h28'49"61d30'
Wöbcke Family,The Compliments La Prairie
Dra 17h1'12"67d46'
Wölfer,Uwe Compliments La Prairie
Cam 13h37'33"81d0'
Wörmann,Tobias
Leo 11h21'53"2d10'
Wörner,Lothar
Ser 15h28'53"8d44'
Wöstefeld,Kerstin
Lib 14h18'35"-22d41'
Wötzel,Silke
Hya 9h5'35"5d18'
Wübbena
Sco 16h31'0"-40d16'
Würth,Wolfgang
Ari 3h22'25"28d21'
Würzburger,Frank
Sgr 19h37'11"-41d54'

X

X Cybil
Lyn 8h2'12"52d15'
X Zer (Stephen Tenzer)
Cnv 13h22'26"40d23'
X-COMICS
Cep 22h2'0"60d21'
X-C'94
Cam 6h11'41"68d6'
X-tine
Uma 10h57'31"48d58'
Xan
Del 20h13'21"13d46'
Xanadew
Uma 9h54'1"45d30'
Xanadu
Aql 20h6'1"4d34'
Xanadu
Eri 4h52'59"-4d30'
Xander
Aur 6h30'54"32d43'

Xavier XVII
Lac 22h30'47"50d30'
Xavier,Tessa Ann
Cas 23h24'32"61d36'
Xejanala
Cet 2h40'18"7d33'
Xenia
Uma 11h14'0"57d30'
Xenia
Vir 15h7'34"0d13'
Xenia
Lyn 6h58'25"54d40'
Xenos
Cnc 7h55'47"10d8'
Xenos
Cam 5h4'14"61d13'
Xethalis,MD,John L
Lac 22h37'55"55d42'
Xexe
Lyn 7h35'49"52d10'
Xian
Lyn 8h27'56"48d37'
Xiao-Yun,Zhu
Pup 7h59'24"-27d35'
Xiaojin,Zhang
And 22h59'27"40d4'
Xibor
Dra 15h44'44"53d50'
XIO
Lyn 8h1'39"41d32'
Xiomarys
Lyn 8h42'26"46d38'
XOFOX Industries,Inc
Lyr 19h20'51"38d49'
XOXO
Psc 1h43'40"28d3'
XTI
Aql 19h6'20"-0d25'
XTY
Cam 5h5'0"68d34'
Xuan
Per 2h56'25"55d36'
Xueref,Stephanie Therese
Lyn 6h24'15"60d57'

Y

Y & R's Idletyme Star
Crb 16h19'55"28d59'
Y O C III
Boo 14h20'38"25d31'
Y Seren Fy Nhad
Cnc 9h17'0"10d9'
Y's
Tau 5h13'54"20d3'
Y's 3103
Cnv 13h48'32"34d47'
Ya Ya
Mon 7h39'45"-4d37'
Ya-Lei "Jonathan"
Aql 19h50'54"14d50'
Ya-Noor Kabar Britt
Dra 17h0'23"52d9'
Yaari,Liat
Umi 16h18'46"78d49'
Yabe Star
Cet 2h36'45"1d34'
Yablon,Gilbert Joseph
Per 4h1'54"51d4'
Yablonsky,Jane
Cas 2h7'59"67d35'
Yachnin,Ila
Cas 0h30'56"67d59'
Yachnin,Mara
Aur 5h54'23"40d28'
Yackell,Karen
Cas 0h28'27"56d3'
Yacko,The Star Of Wendy S
Com 12h34'28"20d17'
Yacomelli,Dorothy & Alfred
Cyg 19h44'17"31d42'
Yacos,Mathew
Per 3h58'0"37d41'
Yadegar,Daniel
Aur 4h47'45"38d48'
Yadira 1
And 1h5'22"39d46'
Yadron,Melissa
Cas 2h39'18"61d35'
Yaede,Billie Minton
Aur 7h22'44"38d35'
Yaeger,Big Al
Uma 12h5'27"59d21'
Yaeger,Christopher G
Lac 22h32'29"40d17'
Yaeger,Kurt Edward
Dra 9h37'34"80d0'
Yael
And 1h41'36"39d11'
Yaffe,Joe
Ori 5h17'53"10d53'
Yaffe,Valerie
Peg 23h29'42"20d3'
Yaffee,Nancy & Eric
Cma 6h57'2"-16d35'
Yagel,Terry M
Cmi 7h11'21"7d8'
Yager Star,The Murray "Big Guy"
Cep 23h0'58"60d7'
Yager,Chris & Ali
Ori 6h9'28"9d54'
Yager,Jennifer Christine
And 2h26'37"48d17'
Yager,Joan
Cas 2h6'51"59d39'
Yager,RN,Darlene
Cam 8h5'15"73d38'
Yager,Shawna Krysti
Ori 6h0'36"8d43'
Yager,Sheila & Bob
Cyg 21h1'11"30d9'
Yai Yai
Uma 10h51'59"70d30'
Yajima,Miki
Cet 2h32'19"5d12'
Yakabosky,John
Aur 6h24'43"31d59'
Yakes,Derk
Her 16h7'0"48d57'
Yakima,Susan
Lyn 8h34'14"45d38'
Yakuboff,Joan Susan
Lyn 7h28'46"45d42'
Yalango,Richard
Per 3h25'1"40d34'
Yalouris,The Lucky Star of Marc P
Uma 9h48'1"50d47'
Yamada Family, Stella of
Sco 16h22'33"-26d26'
Yamada,Stella of Tatsuya
Psc 1h10'12"20d12'
Yamaguchi,Ikuko
Cyg 21h12'19"37d33'
Yamaguchi,Kogo
Ori 6h10'47"1d33'
Yamaguchi,Kristi
And 23h47'44"46d34'
Yamamoto 7-26-1995, Laura Yuuko
Equ 21h20'60"11d37'
Yamamoto,Jeannie F
Cet 0h30'11"0d50'
Yamamoto,Stella Of Kengo
Gem 7h1'17"18d43'
Yamane,Traci
Mon 6h47'46"-0d41'
Yamashita,Kaoru
Mon 7h43'26"-5d55'
Yamauchi,Kent
Cet 2h4'58"0d29'
Yamauchi-Lynda's Friend,Sharon
Mon 6h30'15"-10d11'
Yamazaki,Stella of Ryo
Lib 14h28'38"-13d24'
Yamazaki,Teizo
Lmi 10h34'12"32d9'
Yambasky
Aur 4h55'34"40d19'
Yamini,George O
Uma 13h2'11"58d38'
Yamini,Sally M
Uma 13h0'44"58d20'
Yamini,Sara W
Uma 13h25'54"54d11'
Yanagihara,Yoichiro
Dra 19h26'0"56d59'
Yancey,Barbara
Cyg 21h14'51"35d46'
Yancey,Gaynor
Lac 22h42'25"56d23'
Yancey,Kathy & Jeff
Crb 15h56'18"26d30'
Yancey,Myrtle Viola Robinson
Vul 21h14'23"20d29'
Yanci,Lori
Uma 10h5'42"47d60'
Yandell,Charlie
Aur 5h1'48"48d47'
Yandle,Margaretta
Peg 22h11'0"5d26'
Yandle,Steven Chase
Aql 18h48'58"11d40'
Yandow,Fran "Ma"
Uma 10h24'29"48d42'
Yaney
Com 12h43'56"30d43'
Yang
Ser 17h49'28"-13d38'B
Yang,Blong & Shannon
Cyg 21h12'55"38d15'
Yang,Christopher Y
Her 17h34'54"25d44'
Yang,Danny
Mic 21h15'24"-32d6'
Yang,Seeka
Cnv 12h29'52"33d14'
Yangua,Rafiki
Uma 11h7'13"48d51'
Yani
Mon 7h45'19"-2d18'
Yank,Joann Sean Jennifer Sean Mathew
Uma 9h5'16"50d21'
Yanke,Barbara
Cas 1h46'19"58d23'
Yankee Clipper
Lac 22h50'17"37d36'
Yankee Man
Her 18h8'0"37d43'
Yankel
Cyg 20h4'12"39d54'
Yankey,Blanche B
Com 13h30'54"16d32'
Yankitis,Jeremy A
Cep 23h9'19"71d6'
Yanko,Alice V
Lyr 19h2'44"38d16'
Yankopolus,Alexis Katherine
And 0h51'55"45d26'
Yankopoulos,Judith Ann
And 1h42'1"40d40'
Yann's Sparkle
Ori 4h54'49"5d11'
Yann,Claude Pierre
Peg 22h28'52"20d7'
Yanna
Vul 19h20'33"26d26'
Yannaco,Ryan Joseph
Aur 6h10'1"48d50'
Yanne,Paumet
Ori 6h15'54"7d30'
Yannick Geuther "Yanic"
Per 4h5'14"35d41'
Yannick,Joseph George
Aur 6h43'43"37d43'
Yannuzzi,Marla E
Cyg 21h54'24"52d33'
Yannuzzi,Robert P
Cep 0h42'0"78d24'
Yanohira,Eric
Hya 8h41'39"0d40'
Yanovich,Jake T
Tri 1h32'33"35d8'
Yanuck,Carley & Alyssa
Peg 23h42'13"30d50'
Yao,Adolfo
Her 16h56'52"26d45'
Yao,Earl Kevin Beck
Boo 14h58'45"11d58'
Yao,Frank
Cep 23h23'19"65d27'
Yapor,Eneida
Cas 0h15'13"61d10'
Yapor,Joy Ann
Cas 23h4'14"54d3'
Yapor,Lance Nathaniel
Per 2h58'0"32d18'
Yapor,Myles Prescott
Boo 15h6'26"41d25'
Yarabeck,Ann Elizabeth
Lyr 19h5'51"37d46'
Yarasheski 3/26/41, Gail Anderson
Mon 7h1'14"5d34'
Yarawsky,Michael Thomas
Cmi 7h40'12"8d6'
Yarborough,Glenn Curtis
Cep 23h3'15"71d10'
Yarborough,Zackery Paul
Aql 19h27'52"-10d5'
Yarbrough,Blaine & Thomas
Cam 12h42'55"76d32'
Yarbrough,Douglas Leon
Aur 4h55'30"38d38'
Yarbrough,Justin Scott
Aur 7h19'19"35d51'
Yarbrough,Mary Elizabeth
Mon 6h41'0"6d9'
Yarbrough,Veronica
Peg 22h57'12"29d27'
Yarbrough,Vince
Oph 17h29'46"-0d38'
Yard,Robert L
Ori 6h16'41"7d41'
Yardena
Cnv 12h49'1"38d51'
Yardinoyannis,Pavlos N
Crb 16h3'32"32d25'
Yardley,William
Cet 1h10'56"-0d60'
Yardy
Sex 9h59'44"5d11'
Yarhi,Michel
Del 20h15'45"10d56'
Yarmark,Alan
Ser 15h39'49"4d16'
Yarmark,Anita Northwood
Peg 23h40'15"31d21'
Yarmark,Claire
Cmi 6h42'52"-13d23'
Yarmark,Harold
Oph 17h29'19"-22d14'
Yarmark,Justin
Aql 20h33'13"0d21'
Yarmel,Catherine L
Aur 4h56'32"40d8'
Yarmowich
Uma 10h21'18"40d35'
Yarmuth,Jennifer Rebecca
Ori 5h27'35"-0d7'
Yaroszeufski,Sr,James
Her 18h13'33"38d11'
Yarter,Harry J
Lac 22h22'10"55d24'
Yarwood,Roger Alexander
Boo 14h44'23"38d23'
Yarworth,David
Uma 8h30'44"55d53'
Yashue 25
Vul 20h22'1"23d15'
Yask,Albin S
Hya 8h14'39"1d42'
Yaskal,Randy Beth-Eve
Lyn 8h10'60"52d22'
Yaskal,Todd Harrison
Her 18h3'30"31d2'
Yasky,Precious Betty Jane
Crt 11h7'57"-12d42'
Yasmeen Sana
Umi 13h18'1"71d59'
Yasmin
Del 20h30'1"10d16'
Yasmin
Crb 15h53'12"26d56'
Yasmin Elliot
Cyg 20h26'54"39d25'
Yasmin,Kieran
Peg 23h30'0"18d48'
Yasmin-Happy 21st
And 22h57'50"40d2'
Yasmina
Peg 22h4'35"7d57'
Yasmina
And 0h23'22"43d230'
Yasmine
Lyn 7h7'54"58d59'
Yasminn,Dr Nina
Oph 17h54'45"10d11'
Yasparro,Rosemary
Uma 12h57'50"60d6'
YASSC-1995
Cam 3h46'1"68d34'
Yasu
Uma 12h2'50"61d56'
Yasuda-Terrones,Carol
Peg 21h19'40"21d25'
Yasuhito
Sgr 19h45'24"-44d21'
Yasui,Rikuji
Oph 17h56'45"11d39'
Yasuo & Hiroko Anniversary 26/6/96
Cnc 8h6'46"20d26'
Yasuoka Forever,Yukino
Gem 7h0'38"18d25'
Yasuto & Fumiyo
Tau 5h11'23"19d12'
Yatar,Peter S
Lyn 7h29'18"40d38'
Yatcilla,Stephen Gregory
Aur 5h8'13"40d29'
Yater,Sr,Robert Russell
Dra 16h54'37"70d0'
Yates,Amanda Kirk
Cyg 21h31'23"53d56'
Yates,Andrew Keith
Her 16h15'23"23d52'
Yates,Anthony J
Com 13h2'13"20d15'
Yates,Candace
Lyn 8h28'0"45d17'
Yates,Celestrial Cecilia Ann
Peg 22h17'16"31d44'
Yates,Christopher
Uma 8h46'32"60d51'
Yates,David Stanley
Her 18h4'50"28d20'
Yates,Delaine
Com 12h19'59"30d15'
Yates,Georgia
Dra 12h30'15"67d39'
Yates,Hilary Roberta
Uma 9h20'40"57d38'
Yates,Jennifer
And 23h0'18"37d59'
Yates,Jerry
Vul 21h25'0"27d7'
Yates,Jr,Joseph Edgar
Oph 17h34'0"11d3'
Yates,Judy
Cet 0h37'32"-2d16'
Yates,Karen & Garry
Umi 14h15'1"66d19'
Yates,Karen E
Vul 20h1'54"23d35'
Yates,Kevin E
Boo 14h57'38"24d11'
Yates,Leo
Lac 22h23'57"50d18'
Yates,Martyn
Dra 14h45'55"64d45'
Yates,Mary
Cyg 20h22'0"37d43'
Yates,Matthew
Her 17h4'55"48d1'
Yates,Melinda Christine
Tri 1h47'53"26d37'
Yates,Michael
Cma 6h55'16"-17d13'
Yates,Rickie
Lyr 18h29'0"32d39'
Yates,Sally
Aql 20h14'53"4d19'
Yates,Susan
Equ 21h5'1"2d32'
Yates,Thanen J A C
Aur 5h3'36"48d7'
Yates,Thomas Christopher
Aql 19h23'22"13d28'
Yates,Wilfred Alan
Cyg 19h31'36"37d37'
Yatesy
Lyn 6h37'0"54d10'
Yatishalaka,Apurva
Uma 11h47'56"46d51'
Yatman,Esther
Cas 0h15'1"65d33'
Yatman,Lon
Lyn 7h54'44"44d22'
Yatvin,Doris
Cas 0h59'22"58d33'
Yatzek
Car 7h34'18"-57d28'
Yauch,Steven Seymour
Aur 5h4'17"44d34'
Yauck,Phoebe Anne
Lyr 18h58'16"34d50'
Yaudes Star,The Lynn
Lyr 18h17'16"37d52'
Yaunches,Tamara
Peg 21h45'19"24d2'
Yaure,Philip Christopher
Cnv 14h1'15"42d2'
Yavanna
Uma 11h33'43"40d21'
Yax,Colleen
Lyn 9h27'45"40d49'
Yaxley,Andrew John William
Her 18h6'0"20d34'
Yaya
Eri 3h54'32"-4d5'
Yazdani,Farshad
Lyr 19h23'53"38d54'
Yazmin Louise
Ser 16h6'10"13d17'
Ybarra,Gregory Scott
Cet 0h50'48"-0d52'
Ybarra,Ricardo "Jessy"
Oph 16h56'1"-23d41'
Yco
Her 17h26'1"28d35'
Ycre,Jr,Louis R
Aur 5h13'55"44d22'
Yde,Marianne
Lib 15h30'11"-11d14'
ye ye
Lyr 18h34'11"43d3'
Yeager,Jacob Thomas
Dra 16h43'37"51d29'
Yeager,Julie
Mon 6h56'49"-0d11'
Yeager,Kimberly J
Peg 22h31'59"21d36'
Yeager,Marianne E
Cyg 20h6'58"40d12'
Yeager,Michael & Jill
Uma 9h47'56"51d59'
Yeager,Nicole Elizabeth
Cyg 21h30'50"44d9'
Yeager,Raeanne Dawne
And 0h39'16"40d31'
Yeager,Tammy Lynn
Cas 0h30'37"61d34'
Yeager,Thomas Roy
Lac 22h2'57"51d55'
Yeakle,Myrna
Cas 0h34'1"60d36'
Yeaman,Kristin
And 23h1'14"51d35'
Yeanette
Lyn 8h56'33"46d42'
Year of the Dot,The
Mon 6h32'1"11d51'
Yeardley,Kane
Leo 9h28'27"28d50'
Yeargin,Leonard
Dra 16h1'36"67d53'
Yearicks,Megan E
Cam 4h50'41"78d55'
Yearley,Norman
Ori 6h4'38"2d31'
Yearly,Colin
Uma 9h29'17"50d20'
Yearout's Christmas Star 94,Jan
Lyr 18h33'1"45d23'
Yeary III,James (Mack)
Per 3h53'25"37d50'
Yeater Family,J R L S
Cyg 21h19'0"38d27'
Yeater,Deanna Christine(Gibbs)
Com 12h32'32"30d31'
Yeatman,Dorothy Ann
Lac 22h38'53"37d41'
Yeatman,Trudi
And 0h57'17"37d2'
Yebisu
Leo 10h15'0"13d8'
Yedlin,Schuyler Elisabeth
Mon 7h17'55"-8d1'
Yee Jr,Min
Uma 8h58'45"48d48'
Yee,Alice
Mon 7h2'1"5d8'
Yee,Carsan
Uma 9h54'49"57d39'
Yee,David & Anna
Peg 23h5'36"32d59'
Yee,Jin
Uma 11h37'12"60d32'
Yee,Kit
Uma 10h46'27"60d6'
Yee,Min & Dianne
Uma 11h42'38"62d7'
Yee,Susan
Aql 19h24'39"15d52'
Yee,Tai
Uma 8h58'23"48d58'
Yee,Tyler R
Boo 15h3'36"24d11'
Yegishe's Dream
Cet 2h46'52"1d59'
Yeh,Lisa I-Ching
Cas 1h6'36"61d52'
Yehl,Michaela Christine
Cas 2h2'0"58d12'
Yekaterina "Katja" Star
And 2h26'41"42d30'
Yelcich,The
Peg 23h0'27"17d57'
Yelenik,Stephan
Aur 7h21'0"37d2'
Yelle,Gerard
Psc 0h57'20"22d49'
Yellen,Rosaria & Joseph
Per 2h52'55"35d13'
Yellin,David Jordan
Dra 19h22'52"56d16'
Yellin,Jared Ian
Lac 22h8'42"50d11'
Yellin,Jerry
Lac 22h30'10"50d19'
Yelling,Martin Rhys
Cep 0h17'33"68d56'
Yellor Sun-60
Aqr 20h54'51"-6d14'
Yellow M
Umi 17h38'44"86d11'
Yellow-Claudia-Jager- Stripe-mh-1995
Cmi 7h21'17"1d3'
Yellowman,Linda
And 2h30'0"50d41'
Yelm,Phillip Kyle
Cep 22h17'21"61d19'
Yelnats,Nipec
Aql 18h59'58"-6d3'
Yeltsin,Boris
Vul 19h6'42"22d53'
Yelvington,Browing
Lac 22h3'48"47d33'
Yelvington,K Lynn Hoffman
Aql 19h57'30"11d22'
Yencha,Joseph A
Cep 20h36'53"60d34'
Yenerall,David
Aur 5h19'0"46d55'
Yenney,Big Al
Per 1h44'16"54d21'
Yenzer,George R
Cep 22h45'57"68d21'
Yenzer,Gertrude C
And 2h2'12"45d18'
Yeoman
Cmi 7h59'15"8d45'
Yeoman,Karen Adel
Aur 5h50'55"41d4'
Yeoman,Paul Stanley Pressick
Ori 6h1'41"1d36'
Yeoman-Dona
Peg 21h28'32"21d44'
Yeomans,Donna Faye
Vul 20h17'48"25d43'
Yeomans,Zoe
Lyn 8h3'13"44d47'
Yeow-Stinson
Ori 6h8'38"4d23'
Yerbic,Bernice Elizabeth
Lib 15h27'48"-8d34'
Yerg,Harry
Per 2h50'51"45d20'
Yerike
Aql 19h30'39"-1d12'
Yerina,Patricia
Ori 5h57'11"14d42'
Yerka,Gero
Aur 6h36'16"37d19'
Yerkes IV,Norman J
Peg 0h11'33"17d41'
Yerkes,Piera
Cam 9h21'30"80d58'
Yerkes,Robert Stephen
Ser 15h10'0"-2d26'
Yerkes,Tazetta
Umi 13h12'41"75d34'
Yerkes,Tristan
Dra 17h2'17"73d45'
Yes Forever
Mon 8h8'49"-2d44'
Yes U Can
Dra 20h21'1"68d19'
Yes,Fig,Forever
Ori 6h0'34"-0d4'
Yesh,Angela Lynn
Lyn 7h12'0"59d26'
Yesmin
Cep 21h57'31"55d10'
Yester,Joseph Marion
Vir 12h33'12"-8d11'
Yester,Katherine Michele
Gem 7h52'58"32d1'
Yesterday,Today & Forever
Ari 1h53'12"11d24'
YesYesYes Len
Boo 15h4'46"21d25'
Yetman,Rick
Oph 18h6'19"1d33'
Yetta's Lighthouse
Uma 12h39'47"62d56'
Yetter,Kristen Marie
Lyr 18h42'19"39d13'
Yeva,Tami
Mon 8h2'47"-1d30'
Yevzlin,Larisa
Peg 21h59'37"28d46'
Yew-Ling,Suzanne Chin
And 1h28'56"39d49'
Yewman,Zachary Erik
Cep 22h17'12"68d50'
Yeznack,Gloria Buck
Mon 6h25'0"-8d32'
Yi,Kwan
Uma 9h2'58"68d27'
Yia Panda Effie
Cas 1h5'55"50d14'
Yianitsas,John & Debbie
Tau 5h51'27"26d36'
Yiatras,Harry & Alexandra
Aql 20h2'30"0d31'
Yidam,Wolf Lupo Farkas
Her 16h34'50"39d24'
Yih,Stephen & Olivia
Boo 14h59'57"40d44'
Yim,Bradley Weng Wai
Cep 22h15'52"68d27'
Yin
Ser 17h49'28"-13d38'A
Yin,Eric Shun-Tzen
Hya 9h6'0"5d59'
Yin,Mei Mei King
Uma 10h49'50"55d2'
Yin,Victor;Adele; Bili & Zoe
Peg 22h19'35"4d56'
Ying,Lau Siu
Umi 14h35'21"65d44'
Ying,Richard
Cep 23h37'1"64d43'
Yingling,Craig
Cet 2h58'1"7d34'
Yioshioka,Yasuko
Uma 11h13'38"51d51'
Yirmiahu
Oph 17h7'60"-22d6'
Ylvisaker,Trina & Jeffrey
Umi 14h26'0"68d45'
YMAEVOLI
Eri 2h58'1"-18d12'
YMCA Camp Ernst 1996
Ori 5h28'29"-0d31'
Yo Jo
Aur 4h37'31"30d46'
Yoakam,Dwight
Equ 21h20'51"3d13'
Yoanides,Jeffrey
Her 16h57'1"32d32'
Yoann
Dra 17h55'26"64d36'
Yob
Uma 11h22'38"63d57'
YOB and the KK
Cap 21h3'39"-24d50'
Yobo
Uma 9h52'50"46d52'
Yochabel
Uma 11h3'37"48d44'
Yochim,Sandra Jeanne
Cet 3h2'1"0d15'
Yocom,Karen
Sge 20h17'44"20d20'
Yocom,Luke Logan
Eri 4h13'18"-13d3'
Yocom,Meredith Grace
Peg 23h41'21"31d33'
Yoda Kaori
Vir 14h57'30"5d10'
Yoder IV,Glenn
Hya 8h14'38"5d57'
Yoder,Chriss
Vul 19h0'31"25d29'

Yoder,Dick
Per 2h58'26"43d17'
Yoder,Floyd R
Cam 12h17'47"80d44'
Yoder,Kathryn
Peg 22h25'23"30d20'
Yoder,Michael
Peg 22h25'26"30d25'
Yoder,Ralph
Boo 13h46'13"22d18'
Yoder,Rona Lee
Uma 9h7'0"59d31'
Yoder,Sarah Victoria
Cas 0h39'38"61d10'
Yoder,Savannah K
Peg 23h40'50"28d14'
Yoder,Ted
Peg 22h25'23"30d22'
Yoder-The Love Of My Life,Bryan Allen
Ori 5h56'0"17d26'
Yoel
Cnv 12h41'60"40d52'
Yogi
Uma 9h52'0"42d13'
Yogi Bhajan
Vir 12h28'25"8d22'
Yogi(Max Rossoff Nichols)
Cep 21h5'44"63d46'
Yoh,Mary P & Harold N
Tau 3h51'51"2d8'
Yohan,Poulichet
Ser 16h4'17"0d25'
Yoho,Frank
Aur 7h2'20"43d19'
Yoichi 1014
Lib 14h33'2"-24d23'
Yoji & Reina
Cap 20h50'9"-16d2'
Yoka
Cet 1h39'45"-0d56'
Yoklavich,1-13-48,John
Uma 11h14'1"43d35'
Yoko's Angelo
Uma 11h1'39"34d29'
Yoko's Dream
Peg 23h36'52"10d1'
Yokoi,Art
Lac 22h13'43"43d39'B
Yokoi,Sharon
Lac 22h13'43"43d39'A
Yokoyama Forever, Goichi & Miki
Lib 14h27'59"-13d23'
Yoku
Leo 9h57'17"24d46'
Yokum
Mon 7h37'53"-6d51'
Yolanda
And 2h20'60"45d40'
Yolanda
Peg 23h2'51"17d44'
Yolande
Cyg 20h24'15"39d21'
Yoli
Ori 5h53'1"12d41'
Yollie's Shining Star
Oph 17h52'28"11d22'
Yolonda
And 1h24'36"34d21'
Yolton,Susan A
Sge 19h57'1"19d1'
Yonan,Amanda Nicole
Cas 1h19'34"61d55'
Yonchak,Erin Allison
Dra 16h58'45"68d14'
Yonemura,Tracey Yumiko
And 22h56'12"51d21'
Yong,"SL"-Jacky K W
Uma 10h31'1"70d36'
Yong,Charmaine
Cra 18h27'49"-43d23'
Yoni's Star
Uma 9h59'19"54d14'
Yonke,Ingrid
Peg 22h56'13"29d32'
Yonke,Kristina
Cyg 19h19'21"45d37'
Yonker,Ben
Boo 14h4'56"26d25'
Yonker,Cathy
Cas 3h18'53"73d40'
Yonker,Jeremy
Aur 4h51'32"40d22'
Yonkers,Lance E
Aur 6h26'49"32d5'
Yonkers,Laura
Mon 7h0'53"8d45'
Yonkes,Loretta P
Lyr 18h29'15"37d53'
Yonroe
Her 18h6'41"14d43'
Yontosh
Cam 6h6'17"60d20'
Yontz,Brian
Her 18h13'18"38d59'
Yonushonis,John & Ann
Ori 4h58'56"4d57'
Yoo,Jane
Ori 6h7'15"-0d47'
Yoos,Stuart
Her 18h5'59"48d41'
Yopie-Girl
Cam 3h14'13"61d16'
Yorch,Debbie
Peg 23h6'26"20d3'
Yordy
Ori 5h53'40"16d28'
Yorgen,Gregory Scott
Vul 19h23'1"26d54'
Yorgen,Katherine Lynn
And 23h1'33"50d59'
Yorgey,Melissa
Cas 0h29'42"63d13'
Yori,Anthony Foley Hoshino Kazu
Uma 8h45'24"60d1'
Yorick
And 2h32'1"44d32'
Yorio,Cordes
Cam 4h15'16"61d43'
Yorio,Helen Todisco
Cyg 20h32'21"31d0'
York,Angelina Karma
Mon 6h38'44"-10d47'
York,Cecelia L
Dra 18h57'22"58d47'
York,Cheryl
Del 20h54'11"9d10'
York,Clinton Robert Fraser
Lyr 18h15'1"47d46'
York,Dan & Cindy
Del 21h0'30"12d11'
York,David
Cma 6h30'55"-28d15'
York,Dr David A
Ori 6h17'13"8d32'
York,Elizabeth Marie
Mon 7h0'1"-10d1'
York,Jane & Dennis
Ori 6h9'1"8d6'
York,Janet Brewster
Cma 6h24'58"-13d42'
York,Jeffrey
Boo 14h51'36"45d46'
York,Joshua Gene
Uma 8h33'25"50d56'
York,Justin Brigham
Cma 6h14'33"-18d51'
York,Keith James
Cam 4h42'22"67d58'
York,Lily Joy Louise
Umi 13h7'1"75d40'
York,Lily Joy Louise
Uma 9h56'18"60d1'
York,Michael B
Peg 21h52'44"18d60'
York,Monica
Uma 11h9'53"70d56'
York,Nancy Singleton
Cyg 20h4'55"38d17'
York,Shelly Ann
Cet 1h56'20"-4d57'
York,Steven & Nancy
Cyg 21h52'0"42d58'
York,Stuart
Lyr 18h59'37"47d26'
York,Timothy Thomas
Cmi 7h56'1"0d2'
Yorke,Jacqueline Vreeland
Cyg 20h16'10"31d0'
Yorke,Peter
Aql 19h31'1"10d24'
Yorker
Boo 15h11'54"40d48'
Yorkminster `95-96
Ori 4h48'0"4d28'
Yoroizuka,Miki
And 1h0'26"39d46'
Yort
Cam 6h19'16"65d16'
Yoshi's Star
Eri 4h27'27"-18d55'
Yoshiharu Fukuhara
Lmi 10h11'13"38d6'
Yoshikai,Alyce Wada
Cas 2h15'14"63d59'
Yoshikawa,Terrance
Cam 7h51'57"60d30'
Yoshiko By The Silent Sea
Sco 16h28'51"-27d56'
Yoshimura,Mari
Boo 14h50'35"37d0'
Yoshioka,Maximilian NCY
Cap 20h18'17"-26d56'
Yoshiyuki & Tomoko
Leo 10h4'51"17d56'
Yoshizoh
Gem 7h4'36"26d15'
Yosowitz,Brenda S
Eri 2h58'17"-10d29'
Yost's 50th Anniversary,Pearl&Roy
Peg 22h34'28"2d29'
Yost,Clementine Maureen
Sgr 20h2'36"-40d7'
Yost,John Phillip
Ser 15h31'39"18d15'
Yost,Kyle Joseph
Per 4h0'38"51d10'
Yost,Marion M
Psc 1h23'58"30d37'
Yost,Megan Lynn
Cyg 21h4'43"31d30'
Yoste,Tolley Olivia
Mon 6h43'17"7d27'
Yotuckder
Cyg 21h53'23"40d34'
You
Lyn 7h34'32"37d6'
You
Aql 18h48'48"11d51'
You & Me
Lyr 18h59'50"30d35'
You & Me
Cnv 12h13'39"41d22'
You & Me-Paul & Teri Paul & Teri
Uma 10h2'17"48d40'
You & Me...Everlasting as the Stars
Ori 5h39'0"11d19'
You & Yours
Ori 5h57'0"9d10'
You Are As Brilliant As Any Star
Cam 5h36'29"80d7'
You Are In My Heart Forever
And 23h32'47"49d48'
You Can't Hurry Love
Aqr 20h46'34"-1d22'
You Know
Hya 9h39'39"2d27'
You My Love Friend & Life Forever U
Sge 19h53'1"16d25'
You Too
Aur 5h54'59"48d52'
You'll Forever Be My Dearest,Johnny Baby
Cep 20h32'35"75d17'
You've Got A Friend
Aur 4h37'13"30d45'
Youdale,Diane
Cyg 19h39'14"31d18'
Youino,Jr,Vincent
Cep 3h43'15"87d26'
Youji
Psc 1h21'45"7d6'
Youki
Vir 11h41'19"9d51'
Young
Aql 19h53'0"1d40'
Young 12-28-85/ 12-28-95,Dennis R
Dra 16h58'50"68d46'
Young II,James William
Hya 8h10'29"1d41'
Young Lochinvar
Uma 11h41'18"32d40'
Young Love Forever,Hi Honey Joan
Umi 15h35'36"77d1'
Young Quan Gui
Cyg 20h58'43"31d14'
Young Star,The Tom & Mary
Cma 6h55'7"-19d19'
Young XTC 143,John Heath
Hya 8h56'47"5d28'
Young,"Epiphany" For Yvette
Cyg 19h59'1"31d33'
Young,"I Love You" Dwain
Peg 22h59'33"30d31'
Young,Aaron
Her 17h34'28"28d45'
Young,Adrea
Cmi 7h43'1"8d43'
Young,Alan Russell
Cma 7h1'55"-30d38'
Young,Alexander Martin
Aur 5h0'0"45d8'
Young,Alexis Eve
Ori 4h48'1"4d0'
Young,Alice May B
And 23h2'13"45d54'
Young,Amanda
Peg 22h43'50"35d28'
Young,Amanda Caroline
And 23h8'18"42d10'
Young,Andrew
Boo 14h22'48"48d3'B
Young,Angeline & Brannon
Uma 8h41'46"53d0'
Young,Arlene
And 0h55'48"36d24'
Young,Ashley Kristin
Cam 5h37'37"80d3'
Young,Audrey Rose
Peg 22h13'58"5d1'
Young,Barbara R
Cas 0h37'32"67d5'
Young,Bea & Morry
Aur 5h30'1"31d40'
Young,Betty
Peg 21h52'40"32d56'
Young,Bill
Aql 19h56'32"0d35'
Young,Bill
Cet 1h18'41"-4d8'
Young,Bill & Erin
Uma 13h24'56"58d2'
Young,Bob
Ori 6h16'59"-2d21'
Young,Bradley Robert
Her 15h55'52"50d26'
Young,Brian
Uma 10h21'34"50d14'
Young,Cari
Aql 20h11'47"1d38'
Young,Charlie Jordon
Cyg 20h21'26"38d39'
Young,Christopher
Dra 20h21'1"62d33'
Young,Christopher
Ori 4h52'55"1d20'
Young,Clifford Brent
Hya 8h44'60"3d5'
Young,Cookie/Marilyn
Peg 22h43'59"35d30'
Young,Cortney Eve
Ori 5h55'50"17d40'
Young,Curtis
Lac 22h42'23"38d30'
Young,Cynthia Jade JLP
Sge 20h16'58"16d20'
Young,Cynthia M
Com 12h55'41"27d12'
Young,Daniel E
Her 17h37'24"14d43'
Young,Daniel Russell
Dra 14h28'50"61d56'
Young,Darryl (Daaru) Ananias
Boo 14h12'13"52d59'
Young,Daryl Len
Eri 3h41'1"-5d22'
Young,David
Lib 14h59'58"-10d25'
Young,David
Her 17h19'25"41d13'
Young,David
Cyg 19h30'11"35d23'
Young,David W
Lyn 8h20'37"40d14'
Young,Deborah Dawn
And 2h30'27"48d1'
Young,Deborah L
Aqr 23h4'38"-5d30'
Young,Delia
And 1h6'48"38d8'
Young,Dennis Eugene
Her 17h3'13"42d55'
Young,Dolores A "Dee"
Lyr 18h58'17"30d16'
Young,Donald
Aql 19h53'48"13d7'
Young,Donald Wayne
Hya 8h10'17"5d49'
Young,Donna
Psc 22h56'43"6d37'
Young,Dr Sarah
Sex 10h38'55"2d53'
Young,Drew
Hya 9h17'36"0d10'
Young,Eddie
Lmi 9h57'1"38d51'
Young,Edward"Teddy"
Peg 22h25'27"30d40'
Young,Elizabeth
Cas 1h22'34"50d34'
Young,Eric Paul
Mon 6h55'0"8d14'
Young,Ernest George
Per 3h28'52"40d13'
Young,Etta Marie
Lyr 19h19'0"38d51'
Young,Frank
Cep 21h51'50"65d3'
Young,Frank M
Aur 4h53'27"40d42'
Young,George Eugene
Mon 8h8'4"-7d53'
Young,Graeme
Boo 15h8'42"20d34'
Young,Grant E
Hya 8h17'36"3d20'
Young,Gretchen Elaine
Com 12h59'38"31d1'
Young,Harold B
Boo 14h8'43"47d49'
Young,Herman
Cep 23h6'55"60d33'
Young,Ho
Lac 22h4'45"50d55'
Young,Howard
Aur 6h31'12"31d36'
Young,II,James Bernard
Sex 10h30'56"2d13'
Young,In Loving Memory of Vera Lee
Eri 2h55'0"-18d23'
Young,J J
Her 17h16'0"21d18'
Young,James
Lyr 19h3'48"47d11'
Young,James Christian
Ori 5h35'29"1d7'
Young,James Harvey
Cnv 13h40'17"34d9'
Young,James Robert
Cnv 12h48'27"40d26'
Young,James V
Cep 23h24'58"68d3'
Young,Jane Pauline
Lyn 7h17'36"58d24'
Young,Jason Arthur
Dra 16h27'57"63d58'
Young,Jayme Christopher
Hya 9h10'6"-14d27'
Young,Jean
And 23h20'1"51d53'
Young,Jeanne D
Vul 19h50'18"20d2'
Young,Jennifer Blakeney
Ori 5h36'27"-0d48'
Young,Jeremy
Per 1h27'23"50d25'
Young,Joanna
Per 2h47'26"40d19'
Young,John
Ori 6h7'44"0d1'
Young,John B
Peg 21h24'17"13d28'A
Young,John Daniel
Her 16h28'59"38d28'
Young,Jordan
Aur 6h29'29"35d46'
Young,Joseph Clifford
Del 20h16'0"10d3'
Young,Jr,Claude D
Aql 19h52'48"12d20'
Young,Julius W (Ned)
Hya 8h38'21"-10d47'
Young,Keely Elizabeth
Sgr 19h10'37"-22d27'
Young,Kevin
Boo 14h23'15"29d39'
Young,Kristine Kay
And 22h57'54"50d34'
Young,Laurel Nicole
And 2h23'31"41d54'
Young,Lisa Marie
Lyr 18h18'31"40d47'
Young,Lorraine
Lyr 18h16'24"47d13'
Young,Louise
Ori 5h55'19"16d42'
Young,Luke Amtsen
Her 18h2'58"37d44'
Young,Lynda Kay
Ari 1h58'19"21d56'
Young,Malcom
Uma 8h17'11"70d49'
Young,Margaret
Oph 18h1'0"13d9'
Young,Matthew Lawrence
Aur 5h17'40"41d50'
Young,Matthew Rohe
Her 17h53'58"14d22'
Young,Melanie Rose
Peg 23h21'55"30d14'
Young,Michael
Dra 20h16'39"64d51'
Young,Michael A
Aur 5h14'51"52d57'
Young,Ms Penny
Lyr 19h22'12"31d50'
Young,Nicholas P
Cnv 13h48'1"30d44'
Young,Nicola Fay
Aqr 22h20'10"0d51'
Young,Nicole Taylor
Cyg 20h58'23"38d15'
Young,Pamela
Peg 22h50'56"29d44'
Young,Pamela
Uma 9h10'56"72d4'
Young,Parrish Champion
Tau 5h50'32"23d55'
Young,Patricia Ann
Cas 23h33'37"61d50'
Young,Patrick Kelly
Per 2h32'1"56d40'
Young,Payton H
Dra 16h39'46"52d7'
Young,Rachael Kathryn
Peg 22h37'1"27d25'
Young,Reginald John
Cyg 19h25'34"35d11'
Young,Rev Adm James R M
Cnv 12h58'20"37d34'
Young,Rex
Boo 15h6'12"21d1'
Young,Richard F
Oph 17h17'6"-20d16'
Young,Richard W
Boo 14h22'48"48d3'A
Young,Robert
Umi 17h35'33"85d17'
Young,Roger
Boo 14h24'0"20d36'
Young,Roger
Her 17h15'54"27d3'
Young,Ryna
Dra 18h35'52"80d24'
Young,Samuel
Aur 5h12'34"43d41'
Young,Sandy
Cyg 19h23'10"28d38'
Young,Sara
Com 12h9'26"24d42'
Young,Sarah Andrea Sophia
Sco 17h51'41"-30d42'
Young,Shane
Ori 5h55'18"19d48'
Young,Sheila
Cyg 21h8'1"40d39'
Young,Skip Lee
Cet 2h59'40"0d45'
Young,Steve
Dra 13h35'0"70d24'
Young,Steve P
Peg 23h32'50"28d42'
Young,Susan Louise
Ari 1h56'0"11d24'
Young,Susanne Ashton
Ori 5h34'48"1d44'
Young,Thomas
Lmi 9h59'15"30d15'
Young,Timothy Lee
Lyr 18h43'31"38d55'
Young Travis
Cet 3h16'29"9d41'
Young,Victoria & Dick Harrington
Vir 13h14'30"-9d1'
Young,Vincent DePaul
Gem 7h18'21"24d54'
Young,Vivian
Aql 20h12'56"0d16'
Young,Wade E
Hya 8h34'39"1d19'
Young,William
Equ 20h56'1"5d30'
Young,William Davis
Cnv 12h22'45"37d19'
Young,William Gordon
Lac 22h47'25"55d58'
Young,Winifred Teresa
Cyg 21h28'38"40d18'
Young,Winona
Leo 10h17'11"12d53'
Young,Wurzel
Uma 14h8'36"57d49'
Young-Coyote Point, Angela J
Uma 9h58'31"48d29'
Young-Olson,Gage Nicole Christine
Equ 21h21'0"11d21'
Youngberg,Dr John
Oph 18h17'34"10d9'
Youngberg,Kristin Elizabeth
Aqr 23h22'57"-4d0'
Youngberg,Laura Ellen
Mon 6h27'12"7d22'
Youngblood,Karen Elaine
And 2h32'58"49d5'
Younger Friendship Star,The R & R
Cnv 13h50'13"40d19'
Younger,Ayala
Lyr 18h59'1"26d42'
Youngman,Betty
Per 2h44'4"53d44'A
Youngman,Judith
And 0h29'17"40d30'
Youngman,Laurel Sue
Lyn 8h8'49"48d40'
Youngman,Lisa Lee
Com 13h29'0"26d57'
Youngman,Lyle
Per 2h44'4"53d44'B
Youngs,Chase W
Cam 3h52'46"79d38'
Younker,Jr,Scott W
Cet 0h24'46"-1d33'
Yount,Andy
Lac 22h47'21"53d18'
Your Beloved Sam
Eri 4h4'57"-13d47'
Your Diamond in the Sky
Lyr 18h40'14"38d46'
Your Fur Stinks
Boo 14h56'34"21d12'
Your Goddess
Boo 14h27'1"41d29'
Your Magic Ladder
Dra 12h27'40"72d3'
Your Nation
Dra 19h32'1"61d12'
Your Peach
Dra 19h2'24"50d22'
Your Sam,Forever With You I Love You
Cyg 21h46'1"37d21'
Your Seven Dwarfs
Cnv 14h4'35"41d51'
Your Star
Mon 8h6'30"-5d44'
Your Twisting Kaleidoscope
Uma 9h31'0"57d18'
Your Wishing Star
Cas 1h33'35"61d19'
Yourgo Ziras,Je T'Aime
Umi 16h56'32"78d16'
Yours Eternal
Eri 3h40'39"-5d38'
Yours Forever
Sge 19h30'17"18d50'
Yours Forever & Ever Poot
Cep 21h50'28"71d14'
Yourtz,Joshers - Joshua Barricks
Ori 5h46'25"-0d23'
Yoush
Cas 2h3'0"61d16'
Youster,Karen Elaine
Cyg 21h35'39"42d30'
Youvan,Douglas Charles
Sex 10h30'24"3d38'
Yow,Michael Vernon
Boo 14h15'31"53d52'
Yoyo
Her 17h15'58"50d12'
Yrdnek,Yhtomit
Cam 8h48'30"78d56'
Yrieix
Per 3h32'51"39d47'
Ys Packaging
Uma 11h57'13"56d9'
Ysaye Marc
Lyr 19h21'52"30d10'
Yska
Ori 5h12'34"-6d14'
Yslas,Jr,Armando
Dra 16h22'11"67d42'
Yslas,Savannah Marie
Cet 2h19'1"0d23'
Ystraad,Jennifer Ann
Aqr 23h44'0"-4d11'
Yu,Bin & Weijie Yang
Lyr 19h9'25"38d54'
Yu-Chung Chou,Stanley
Hya 8h32'42"5d55'
Yubby
Cet 0h24'54"-2d41'
Yucius,Jennifer
Lyr 19h19'50"38d35'
Yudai No Hoshi
Gem 7h11'26"15d51'
Yudichak,Jr,Robert W
Cap 20h21'1"-26d12'
Yuhas Clan
Cnv 12h8'1"39d33'
Yuhl,Christopher Paul
Ser 16h3'27"14d34'
YUICHIE
Tau 5h27'21"27d39'
Yuizu
Dra 16h58'42"67d37'
Yuji
Boo 14h38'0"42d24'
Yuka
Sge 20h6'23"16d11'
Yuka & Hide
Leo 9h57'21"23d14'
Yukari-Yoiko 31-05- 1992
Gem 6h5'57"26d37'
Yukel is a Star
Cet 3h1'1"8d51'
Yukiko
Aqr 22h7'14"0d26'
Yukiko
Uma 10h54'39"38d26'
Yukimi
Tau 5h28'59"24d54'
Yukl,Tony
Ori 5h27'41"-0d16'
YUKO & RYOICHI
Sco 16h23'51"-19d2'
Yuler,Grinn
Com 12h26'10"22d22'
Yuliya
Cam 5h53'27"68d14'
Yum-Yum
Cma 7h21'42"-15d13'
Yuma
Lep 5h55'18"-24d38'
Yumet,Eileen (Farrell)
Tau 5h51'36"23d3'
Yumi
Uma 10h13'33"42d11'
Yumi Nakamoto
Leo 10h15'27"8d21'
Yumie
Gem 7h10'30"14d53'
Yumiko
Lib 14h54'23"-23d31'
Yumiko,The Star of His Hope
Leo 10h2'27"17d47'
Yuncza,Patricia Mary
Lyn 8h12'57"36d56'
Yuni & Missy
Umi 15h8'51"72d51'
Yunis,Lori
Umi 13h27'0"71d37'
Yunker,Colleen E
Cyg 21h17'36"38d57'
Yunker,Rich & Kim
Cyg 20h8'50"41d8'

Yup Yup
Cam 6h0'11"68d2'
Yurchisin,Mark & Kathy
Hya 8h10'16"4d22'
Yuri & Hidekazu Forever
Lib 14h42'17"-11d1'
Yuri Temirkanov & St Petersburg PO
Sgr 19h11'42"-12d41'
Yuriath
Gem 7h12'26"18d53'
Yuriko,Carol & Bruce Charles Stark
Peg 23h29'53"17d7'AB
Yurko,Reverend Paul E
Ori 6h2'52"8d42'
Yurkovic,Mary L
Boo 14h24'28"21d46'
Yurkowski,Vincent J
Cet 2h7'36"3d51'
Yusuf,Samina
Lyr 18h33'1"30d38'
Yuta 1
Cam 6h37'16"80d32'
Yuuga
Cap 20h54'23"-15d54'
Yuwana,Frances
Cas 0h38'22"66d18'
Yuy Pheng
Lac 22h25'13"40d5'
Yuzuru & Yumi
Tau 5h29'38"26d45'
Yve
Cas 2h28'24"61d16'
Yves
Boo 15h0'28"23d59'
Yves et Nicole
Per 3h32'0"41d7'
Yves Georgette
Sgr 19h33'58"-34d24'
Yves Laffont de Castelginest
Cet 1h0'1"1d8'
Yvette
Peg 21h54'1"30d19'
Yvette
Umi 15h19'39"68d41'
Yvette
Aql 19h53'55"12d36'
Yvette 4 Ever
Lyn 7h51'54"33d53'
Yvette Princesse Des E'toile
Uma 11h49'1"51d8'
Yvette's Destiny
Com 12h19'21"21d9'
Yvette,Tim P
Lyn 8h32'51"43d11'
Yvin,Franáoise
Dra 10h10'0"78d20'
Yvo
Ori 5h52'38"8d20'
Yvone
Lyr 18h48'19"30d53'
Yvonne
Uma 13h44'36"48d19'
Yvonne
Cnc 8h54'53"18d7'
Yvonne
Psc 0h16'58"20d39'
Yvonne and Woody #40
Cyg 21h1'20"28d40'
Yvonne of Red Cloud
Cru 12h8'53"-57d58'
Yvonne,Andrea
Aql 18h55'56"-2d49'
Yvonne-C
Peg 22h40'56"32d16'
Yznaga,Alicia V
Peg 22h56'26"22d38'
Yzquierdo,Ashlee & Brett
Boo 14h51'0"24d29'

Z

Z
Sgr 18h52'18"-32d5'
Z
Boo 14h37'21"40d14'
Z
Ori 6h7'27"7d56'
Z & K Wishes
Cam 6h0'11"68d2'
Z Bridget
Lyn 9h13'60"33d26'
Z Star
Eri 2h49'59"-12d10'
Z's JLZ CAZ SCZ,The
Ori 4h59'10"-0d21'
Z's Xanadu
Cyg 21h14'51"37d11'
Z-Force 70
Cyg 20h22'56"38d36'
ZAB
Cam 6h7'31"83d30'
Zabalza,Jared Phillip
Ori 6h16'0"-1d35'
Zabel,Nicole
And 0h34'24"37d42'
Zabel,Robert G
Her 17h5'1"48d7'
Zabelou
Dra 11h26'49"67d9'
Zabierek,Gabrielle Helen Carolina
Sgr 19h1'31"-28d59'
Zabierek,Matthew Aaron
Sgr 19h4'25"-21d41'
Zabinski,Susan M
Lyr 18h45'13"42d56'
Zabit,Bill
Aql 20h7'30"1d16'
Zabitka,Susan
And 0h34'23"31d18'
Zabka,Günther
And 23h14'53"40d46'
Zabka,Valerie
And 0h23'36"33d13'
Zabou
Vul 19h21'13"22d54'
Zabounette
Uma 10h31'15"62d13'
Zabretsky,Jack
Lmi 10h42'23"27d21'
Zabriskie,Laura Thompson
Cas 23h35'15"60d0'
Zabukovec,Salome
Ser 16h6'42"8d23'
Zac et Rata
Cyg 21h39'1"40d23'
Zac et Rato
Lac 22h27'18"40d42'
Zac's Star
Cap 20h58'1"-26d18'
Zaccaria,Albert Charles
Oph 18h25'20"8d27'
Zaccaria,Jr,AKA Felix, Anthony
Her 15h48'38"50d42'
Zaccaro,Joyce
And 22h56'33"38d22'
Zaccheo,Herbert Ambrose
Aur 6h11'39"50d30'
Zaccher,Thomas James
Oph 17h8'47"11d48'
Zacchio,Anthony
Cep 22h59'46"68d49'
Zach
Lac 22h45'56"53d35'
Zach's Star
Lac 22h3'45"49d4'
Zach's Star Forever
Her 16h27'50"41d16'
Zach,Heinz
Sgr 19h4'18"-24d15'
Zacharias,Carol Ann
Uma 11h34'48"63d56'
Zacharias,James Ralph
Her 15h58'15"46d11'
Zacharias,Margaret King
And 1h25'25"33d43'
Zacharias,Nancy Ann
Eri 2h48'29"-17d34'
Zacharias,Richard Dean
Her 15h55'11"45d34'
Zachariasse,Mrs Toni D.R.
Eri 4h21'49"-1d34'
Zachariasse-V Dijk,Mrs Nel
Eri 4h9'36"-1d3'
Zachariasse-V Toorn, Mrs Hanni
Cet 1h52'0"-3d51'
Zachariou,Peter
Per 3h11'39"42d16'
Zacharopoulos,Yiannis & Elena
Lyr 18h49'0"33d48'
Zachary
Her 16h20'14"23d48'
Zachary
Her 16h43'36"23d16'
Zachary
Aql 19h47'19"11d25'
Zachary
Lyr 18h45'28"31d12'
Zachary
Lyn 8h23'27"40d43'
Zachary Aaron
Aur 5h14'30"41d36'
Zachary K
Ori 5h51'14"15d14'
Zachary Lewis
Per 3h42'52"38d19'
Zachary Maximilian Jeffery
Oph 17h38'12"-20d44'
Zachary Ryan
Aur 6h40'27"35d59'
Zachary Star
Cep 22h33'14"59d1'
Zachary,Dana Susan
Cas 0h17'24"59d1'
Zachary,Sarah Marie
Cas 1h9'1"61d38'
ZacharyP89
Her 17h9'31"21d10'
Zachery,Annette
Eri 2h55'35"-18d53'
Zachery,Thomas David
Sex 10h24'1"-10d59'
Zachez
Peg 22h1'21"28d8'
Zachmeyer,Christine
Lyr 18h32'38"47d34'
Zachow,Arnold August
Dra 17h21'12"73d29'
Zachrich,Mr Craig
Lac 22h19'41"51d31'
Zack #1
Dra 19h43'54"70d25'
Zack Attack
Per 3h11'16"45d39'
Zack's Star
Peg 0h7'22"13d10'
Zack,Marian M
Cyg 21h12'38"38d24'
Zacny,Collin Robert
Hya 8h12'53"4d8'
Zacny,Parker Michael
Her 18h19'12"28d49'
Zacur,Tyler Aaron
Ori 6h4'25"10d49'
Zaczek,Cynthia Ann
Lyr 18h44'60"34d33'
Zadell,Helen
Lyr 18h25'23"38d59'
Zadernak,Bonnie Lynne
And 0h29'1"44d53'
Zadernak,Lauren Kathleen
Lyr 19h17'28"41d22'
Zadie
Lyn 7h57'24"35d48'
Zadikow,Dara Shrensel
Cyg 20h21'1"38d36'
Zadikow,Lauren Shrensel
Lyr 18h37'57"44d45'
Zadra,Katy Jocelyn
Lib 15h18'43"-25d59'
Zadravetz
Lyn 8h45'46"45d5'
Zaenglein,Eric
Ser 18h35'44"2d39'
Zaepfel,Pete & Cathy
Eri 2h55'55"-18d45'
Zafar,Sultana
Cam 6h16'39"68d25'
Zaffino,Jr,Danny
Cet 2h21'56"-10d8'
Zaffke,Greg
Sge 19h17'42"16d37'
Zafra,Vilma
Cnc 8h54'1"17d35'
ZAG
Aur 5h2'18"40d27'
Zagaglia,Sabrina
Cas 23h6'1"53d33'
Zagallo,Filippo
Pic 5h0'8"-49d38'
Zagari
Cam 7h54'47"70d21'
Zager,Franklin Walter
Peg 23h24'33"26d10'
Zager,Kathleen Jane
Peg 22h42'57"27d29'
Zagnut
Uma 9h8'36"56d25'
Zagorski Rex,Ricardus
Per 1h42'28"52d40'
Zagorski,Brian Michael
Aql 19h2'38"1d2'
Zagorski,Dr Joseph B
Oph 17h53'47"10d23'
Zagorski,Eric Laurence
Per 1h47'55"54d9'
Zahabian,Khanbaba, Parvin & John
Cyg 20h24'36"40d0'
Zaharchuk,Nicholas Paul
Her 15h55'43"42d24'
Zaharek,Susan
Cas 3h11'24"70d14'
Zahn The Coolest Star,Aaron
Uma 9h43'1"43d38'
Zahn,Douglas Jude Thaddaeus John
Cnv 12h18'14"50d13'
Zahn,Elliot
Lac 22h18'0"49d16'
Zahn,Friedrich
Cnc 9h15'57"7d48'
Zahn,Peter
Lmi 9h28'57"38d4'
Zahnow,Robert "Bob"
Del 20h21'46"11d1'
Zahylkiewicz,Irene
And 2h6'12"38d51'
Zai,Ruedi
Uma 10h1'30"62d7'
Zaiba Nanji
Dra 20h15'38"62d41'
Zaida
Lac 22h40'37"52d30'
Zaino,Audrey B
And 23h3'25"49d54'
Zaino,Buzz
Aql 18h53'45"-1d16'
Zaino,Tara
And 1h29'11"50d9'
Zaiontz,Jeanne M
Sgr 19h25'1"-40d24'
Zajac,Alexis Dominique
Lyn 8h17'34"40d17'
Zajac,Amber Marie
Gem 6h39'12"31d5'
Zajac,Amelia Regina
And 23h41'52"42d20'
Zajac,Jamie
Dra 16h28'0"64d16'
Zajac,Jerry
And 23h36'20"45d40'
Zajac,Sheri Marie
Peg 22h3'14"25d12'
Zajac-1960,Robert Henry
Aur 5h8'57"38d51'
Zajicek,Christopher Michael
Her 18h4'43"30d48'
Zak
Aqr 20h56'42"-1d19'
Zak Vonni
Lac 22h50'21"55d43'
Zak,Christina
Lyn 9h13'58"38d10'
Zak,Eileen
Cas 0h32'43"66d34'
Zak,Jim & Sis
Ori 6h8'53"0d4'
Zakarian,Paula
Lyr 18h37'41"40d16'
Zakarija,Daniel Mathew
Per 3h28'34"40d9'
Zakas,Jo
Tau 3h56'50"22d47'
Zakhary,Susan
Cyg 19h20'51"28d37'
Zaklina
Mon 6h55'39"0d55'
Zakroczymski,Laura
Lyn 8h37'38"44d24'
Zakrzewski Love Mom-Mom,Nicholas Mark
Dra 20h30'19"68d4'
Zalackas,Brandon
Lyn 8h1'55"51d55'
Zalackas,Erin Kathleen
Vul 19h58'10"28d42'
Zalanka,Rick
Cet 1h5'19"1d53'
Zalar
Aur 5h6'50"42d9'
Zalejski,Elzbieta
And 1h28'15"34d47'
Zaleski,Lena
Tau 4h21'40"16d39'
Zaleski,Sophie
Cam 8h38'34"78d39'
Zaleski-Delilah, Beverly Rae
Umi 16h19'18"72d24'
Zaleski-Samson,James Vincent
Cep 23h21'53"63d41'
Zalewski,Bridget Ellen
Mon 5h56'25"-4d22'
Zalkovsky,Gerry
Ser 15h55'19"20d20'
Zallar,Dick
Aur 4h54'0"51d16'
Zallen-Sevett,Samuel & Noah
Dra 9h37'59"77d45'
Zaller,Heidi
Lyn 9h18'24"37d15'
Zallmann,Manfred
Aql 19h54'54"10d27'
Zaloga,Alice V
Cas 0h9'45"63d58'
Zalovick,James A
Her 18h15'38"38d20'
Zalovick,Kyle M
Her 18h12'46"30d50'
Zalud,Sarah "Joseph"
Ori 6h6'56"7d40'
Zalupski-Provost,CL, Dr Vilma
Oph 16h52'0"-28d9'
Zamagni Alias "Mago", Marina
Cam 14h10'40"81d2'
Zaman,Natalie
Peg 22h4'45"4d14'
Zamarron,Joanna Arminjo
Cet 2h26'1"2d17'
Zambaiti,Lucia
Peg 0h6'50"13d37'
Zambaras,Stephanie
Cas 0h12'34"63d18'
Zambelli,Elena
Umi 16h23'59"80d26'
Zamboni,Claudia
Peg 22h31'20"28d48'
Zamborsky,Anna Catherine
Mon 6h57'1"10d47'
Zamborsky,David
Aqr 22h2'31"2d3'
Zamborsky,Rebecca Lynn
Mon 6h57'0"11d14'
Zambrano
Per 3h5'30"40d26'
Zambrano,Bob
Ori 6h17'23"0d37'
Zamek
Dra 14h48'41"60d49'
Zametzer,Philipp
Aqr 22h42'17"5d44'
Zamira,Frank & Kate O'Hara
Sex 9h54'19"-0d15'
Zamis,Abigail Marie
Cas 3h6'20"73d31'
Zammit,Edward Michael
Dra 18h56'20"70d15'
Zammit,Netty
Her 16h52'0"40d25'
Zammitto,Carmelo
Cam 7h9'0"67d48'
Zammuto,Nicholas
Ser 16h1'46"1d5'
Zamora,Damien M
Cmi 7h15'12"1d50'
Zamora,Greg
Aql 19h10'42"13d2'
Zamora,Guadalupe
Mon 6h40'1"10d5'
Zamora,Jaime Paz
Cru 11h58'39"-57d27'
Zamora,Jesus Espino
Lmi 10h7'40"38d13'
Zamora,Pedro
Sex 10h23'58"-1d33'
Zamora,Sandy
Mon 7h3'52"5d14'
Zamorano,Victor Alfonso (Vitoco)
Cet 2h42'42"-6d2'
Zamost Star
Oph 16h49'38"11d4'
Zampini,Claudia et Paolo
Oph 18h6'39"10d53'
Zamporlini,Pier Luigi
Cep 23h33'20"65d2'
Zamsky,Janet C
Lyn 8h13'51"58d47'
Zamzow,Faye W
Cyg 19h28'35"30d42'
Zana,Wendy
Cas 23h23'14"60d33'
Zanaglio,Leslie A
Cep 0h18'20"87d21'
Zanardi,Gil
Cet 2h32'39"0d37'
Zanca,Joseph Charleton
Aur 6h14'1"35d27'
Zanca,Kimberlee Jo
And 23h35'39"38d6'
Zanca,Nicole
And 23h0'0"43d53'
Zande's Dreams
Cnv 13h4'42"50d52'
Zander,Bob
Per 1h34'22"53d55'
Zander,Jennifer Marie Macha
Tau 5h18'54"16d15'
Zander-Lu 93
Cep 3h5'11"77d30'
Zanders,Debbie V
Mon 7h55'56"-0d53'
Zandman,Ruta & Felix
Lyr 18h41'58"38d40'
Zandnia,Vida
Eri 4h13'35"-8d23'
Zandra
Aql 19h5'1"15d41'
Zandy,David
Dra 16h43'10"63d29'
Zane
Dra 16h7'44"62d20'
Zane,Christopher
Aql 19h26'24"0d4'
Zane,Victoria Lee
Cnc 8h35'1"30d38'
Zanellato,Lorella
Per 3h35'54"51d15'
Zanetich Will Shine Forever,Anthony
Del 20h15'47"13d2'
Zanette,Toni
Cnc 8h35'11"16d10'
Zanetti,Anna
Cae 4h48'25"-32d52'
Zang
Oph 17h12'28"-20d52'
Zang,William Franklin
Dra 16h42'46"66d50'
Zangara,Thomas Michael
Lac 22h5'37"46d45'
Zangemeister,Andrea
Cap 21h19'26"-23d37'
Zanger,Bettina Maria
Cnv 12h36'54"38d35'
Zangheri "AM",Ennio
Cnv 13h7'1"51d44'
Zangriles,Alexander Gust
Per 2h52'23"32d26'
Zani,Jeffrey Frederick
Her 17h59'38"48d55'
Zanish,Michael
Dra 13h31'45"63d57'
Zanker,Celine Murschel
Hya 8h54'17"-6d60'
Zanko,George Michael
Uma 10h24'1"57d44'
ZANNECHAEL
Cyg 19h23'23"48d52'
Zanni,Jennifer Angela
Uma 9h19'25"68d36'
Zannini
Lib 15h3'22"-5d51'
Zannini,Beatrice
Per 1h37'16"53d17'
Zannini,Marie
Per 3h36'26"51d28'
Zannis,Irene
Aur 7h24'22"38d50'
Zannotti,Paolina
Cep 21h21'34"68d13'
Zanoni,Peter
Mon 8h5'50"-1d40'
Zanovich,Gary
Her 17h22'44"41d27'
Zanstar
Mon 7h59'2"-6d39'
Zantedeschi,Annalisa
Tau 5h50'42"23d17'
Zanutto,Cleo J
Mon 6h59'15"-1d19'
Zany,Bob
Aur 4h47'14"40d43'
Zanzarella-Bosco, Louisa E
Cyg 20h16'1"30d17'
Zaobiedny,Adam
Aur 6h14'32"30d47'
Zap GAW
Aur 6h46'44"37d48'
Zapalac,Kate Elizabeth
Aql 18h59'59"15d3'
Zapata,Luz Mery
Ori 5h54'52"14d13'
Zapata,Marie Rose Concepcion Gayanilo
Sgr 18h49'35"-28d47'
Zapata,Oscar Mario
Cet 1h25'35"-14d19'
Zapf,Andrew
Dra 10h53'51"73d20'
Zapf,Barbara (Murmel)
Psc 1h36'58"20d13'
Zapf,Edith
Lyn 8h10'18"43d50'
Zapiler,Aaron
Lac 22h38'36"54d4'
Zapisocki,Forever,Mr & Mrs Ron
Cam 7h3'26"67d41'
Zaplatilek,Yoan
Per 3h30'5"37d17'
Zapotochna,Wesley J
Boo 14h5'14"38d57'
Zappa,Frank
Hya 9h39'20"-18d54'
Zappa,Frank
Eri 2h56'1"-11d7'
Zappin,Stan
Her 16h32'58"32d17'
Zappulla,Bart & Sherry
Cyg 20h51'1"30d59'
Zar
Lyn 6h47'10"60d11'
Zara Mina
And 0h3'36"46d58'
Zara,Katie C
Mon 6h22'43"4d41'
Zaradich,Matthew Ryan
Cep 23h12'47"68d17'
Zaragoza,J Michael
Her 16h48'0"32d22'
Zaragoza,Véronique
Aql 19h27'60"7d37'
Zaramella,Jean-Marc
Ori 6h0'13"10d23'
Zarandona,Juan M
Ori 6h17'51"-1d20'
Zarate,Juan Carlos
Sex 10h26'22"-8d44'
Zarbock,Carolyn Preston
Peg 23h4'56"12d19'
Zarbock,Eric
Gem 6h52'34"18d39'
Zard,Nazih
Cam 4h57'16"70d1'
Zard,Vina
And 1h4'38"38d23'
Zardo,Lena
Cas 0h30'34"67d12'
Zards
Peg 23h29'49"21d10'
Zarem,Bobby
Aur 6h6'1"46d47'
Zaremba,Jack Louis
Ori 5h53'1"12d55'
Zaremba,Jeffrey Brockton
Her 18h53'38"12d48'
Zaremba,Jonathan Michael
Ori 5h54'12"12d50'
Zaremba,Julie
Peg 22h59'25"20d39'
Zaremba,Laura Allison
Ori 5h53'25"12d49'
Zaremba,Susan A
Cas 0h0'30"54d57'
Zaren,Anne Marie
Cyg 20h56'24"40d23'
Zaren,Christopher David
Per 2h51'30"43d56'
Zareski Sultan's Sparks,Sharon
Mon 8h4'57"-1d9'
Zareski,Carmela & Ed
Ori 5h47'23"20d53'
Zaretsky's Big 4-0 Super Star,Alan
Per 1h28'51"52d54'
Zaretsky,Marty
Per 1h50'58"53d30'
Zargon
Sge 18h55'37"20d23'
Zarifes
Per 3h7'24"48d48'
Zariffa,Nevine
Sct 18h53'15"-4d46'
Zarin
Sgr 20h6'54"-41d17'
Zarina
Peg 23h1'58"20d59"
Zarkowski,Joseph
Cnv 13h2'35"37d37'
Zarlenga,Mary G
Peg 21h57'40"24d57'
Zarneski,Joe
Aur 6h3'0"31d56'
Zarnoch,Helen & Joseph
Cet 2h7'18"2d21'
Zaroff,Jacob H
Per 3h11'43"49d39'
Zarou,Moneer
Vir 13h8'11"-19d46'
Zarrella,Thomas J
Aql 19h51'30"10d4'
Zarroli,Mauro
Hor 3h29'27"-46d46'
Zartner Star,The
Cam 9h22'56"81d45'
Zaruba,Josephine
Lyr 18h51'23"42d53'
Zarudyanskaya,Tamila
Aur 5h39'24"37d53'
Zary,Lyle
Lac 22h25'55"43d51'AB
Zarzycki,Dee L
Per 2h51'34"31d2'
Zarzycki,Wolfgang
Vul 20h42'12"20d21'
Zasowski 163-01-8232, Katherine Yaramin
Aql 18h59'0"2d14'
Zasowski 185-05-2713, Henry John
Her 16h37'26"7d59'
Zasowski,Joey
Dra 18h11'46"65d28'
Zasowski,Joseph
Aur 6h11'21"38d35'
Zasowski,Jr, AF13834823,AB Henry J
Aql 19h55'24"-6d14'
Zasowski,Kevin
Per 2h54'25"43d14'
Zasowski,USAF (Ret), TSgt Henry J
Aql 20h1'53"6d58'
Zass
Vul 19h45'46"29d3'
Zastrow,Mark Allen
Cep 22h56'52"65d26'
Zastrow,Stephanie Michele
And 2h15'14"45d14'
Zatawashyana
For 2h47'19"-27d20'
Zatkalik,Micah
Aur 5h15'1"41d53'
Zatorski,Ted
Lyn 8h10'46"58d22'
Zauberman,Nathan
Per 3h26'17"50d16'
Zauberstern Für Karlheinz Zervas
Aqr 22h45'1"-0d20'
Zaucha,Matthew Edward
Cam 8h3'44"70d16'
Zaug,Margaret
Lyn 7h50'17"48d28'
Zaumseil,Jane B
Cyg 19h16'10"44d51'
Zaumseil,Walter B
Cyg 19h17'42"44d24'
Zavala-Courage & Faith,JP
Cas 23h7'58"57d10'B
Zavala-Hope & Love,JP
Cas 23h7'58"57d10'A
Zavaleta,Mia Torres
Lup 15h4'33"-44d37'
Zavaroni,Laurent
Her 18h19'30"12d24'
Zavaroni,Marie-Ange
Her 18h19'33"12d53'
Zavatsky,Marybeth
Uma 11h40'45"43d12'
Zavilon,Angela
Lyr 18h51'39"40d49'

Zawacki,Kathleen=Diane
Boo 14h11'46"30d32'
Zawacki,Pop
Per 3h18'47"41d48'
Zawadzki,Dennis Sylvester
Her 16h45'27"34d27'
Zawadzki,Dorothy Pepe
And 2h28'37"47d51'
Zawelensky,Pete John
Tau 5h45'1"16d27'
Zawislak,Amy
Lyr 19h4'0"37d42'
Zawislak,Joel David
Her 17h32'60"31d37'
Zaworski,Leo
Per 2h38'36"35d10'
Zaya
Ori 5h52'20"15d57'
Zayas,Michael Gabriel
Sct 18h52'18"-7d8'
Zaza
Cnc 8h34'39"31d18'
Zaza
Cas 0h29'44"68d27'
Zazzaron,Jean Luc
Cet 0h56'21"0d24'
Zazzera,Elisa
Cas 2h37'45"73d47'
Zbarsky,Sandy
Boo 14h50'1"46d40'
Zborowski,Father Richard
Aqr 20h37'45"-8d59"
Zborowsky,Peter Alexander
Per 3h25'28"40d47'
Zboyan,Douglas Benjamin
Cnv 12h20'24"42d32'
Zdenek Teply
Cnv 13h54'0"32d3'
Zdeni
Cyg 19h53'50"37d38'
Zdrahal,Jennifer
Cam 7h39'56"70d45'
Zdravko Minck-The Lite I'll
Always Remember
Ori 5h56'42"15d40'
Ze Count
Cam 3h48'0"52d40'
Zeallor,Heather
Lyn 8h7'18"50d28'
Zebrowski,Albert Alexander
Oph 17h56'32"8d1'
Zec,David Francis Theobald
Hya 9h16'1"1d37'
Zecchillo,Anna
Aur 5h29'52"38d14'
Zecchin,Anna
Lyr 18h31'60"36d3'
Zeck,Andrew
Lac 22h54'35"53'18'
Zeddemore,The Real
Ghostbusters,Winston
Pho 0h28'41"-48d55'
Zedrick
Cyg 19h43'49"30d37'
Zeeh,Axel
Uma 9h28'0"47d18'
Zeender,Annelise Julia
Leo 10h19'39"13d3'
Zeender,Brigitta Margrit
Leo 10h39'31"13d5'
Zeender,Katarina Heidi
Peg 22h4'25"8d36'
Zeesman,Allen
Her 18h3'46"48d1'
Zeff,Jamie Edward
Peg 23h1'18"20d39'
Zeff,Jennifer Ann
Cas 0h26'52"56d0'
Zeff,Maurice L
Ori 5h50'24"16d26'
Zegarac-Martin,Kay
Crb 15h48'55"38d44'
Zegger,Raymond J
Her 16h28'44"41d17'

Zeglin,Andrew
Boo 15h15'28"32d55'
Zeglinski,Lothar
Lyr 19h19'22"30d12'
Zehethofer,"Harley" Johannes
Cep 22h50'13"58d47'
Zehetner,Martin Ludwig
Cas 0h9'0"60d48'
Zehetner,Nicole Jade
Cas 0h9'58"60d50'
Zehnder
Boo 15h0'15"38d41'
Zehnder,Daniela
Del 20h16'32"12d4'
Zehnder,Jane Marie
Mon 8h8'44"-9d13'
Zehner,Kelsey Orion
Ori 5h42'16"-5d51'
Zehngraff,Terri
Aql 19h13'25"10d46'
Zeiders Star
Cet 2h3'32'0d20'
Zeidler,Doris
Aur 5h13'15"43d19'
Zeier,David
Cyg 19h29'39"50d26'
Zeigler,Bessie "The Body"
Mon 6h32'49"1d3'
Zeigler,Jerry L
Dra 20h3'59"64d11'
Zeigman,Brandon Steven
Lyr 18h40'49"38d31'
Zeiher,Melissa Dawn
Lib 15h43'1"22d36'
Zeiler,Harold F
Aql 19h57'48"1d6'
Zeiner,Shirley M
And 0h13'16"47d29'
Zeira,Leona
Cnv 13h0'43"50d11'
Zeis,Father Gabriel
Aql 18h58'56"-6d60'
Zeiser,Bernd
Umi 13h22'53"75d1'
Zeiser,Donald
Her 17h17'51"44d5'
Zeismann,Gustav
Aql 18h58'53"17d18'
Zeiss (May),Mary A
Cyg 21h21'56"40d11'
Zeiss,Madison Emily
Cas 1h18'0"75d48'
Zeitlan,Barrett Mandel
Lac 22h31'25"37d38'
Zeitler,Albert
Her 16h41'19"23d51'
Zeitler,Corinne
Lyn 8h3'45"34d59'
Zeitler,Doreen
Cet 3h7'1"1d8'
Zeitler,Erin Lynne
Peg 22h3'15"20d30'
Zeitler,June E
Cas 1h38'0"65d4'
Zeits,PhD,Carol R
And 23h45'1"43d25'
Zekaria,Albert
Her 16h56'14"40d18'
Zeke
Uma 9h19'21"62d7'
Zeke 23
Aql 19h31'23"7d52'
Zel,Mohammed Khamis
Lyn 8h24'43"51d38'
Zelakowski,Cathy
Peg 22h2'53"3d57'
Zelakowski,Cathy
Cet 2h45'17"3d55'
Zelakowski,Henry
Cma 7h14'41"-16d4'
Zelano,Marcella"Popiz"
Scl 23h25'12"-27d43'
Zelasko,Daniel M
Cam 4h17'32"71d0'

Zelasko,Rick
Aur 5h2'52"41d16'
Zeld,Christine T
Umi 16h24'42"70d12'
Zelda
Ser 18h54'48"2d16'
Zelda
Mon 6h55'39"-8d57'
Zelda
Lyr 18h55'25"40d12'
Zelda's Enlightened Spirit
Aur 5h27'16"38d2'
Zeldes,Rich
Tri 1h50'0"27d2'
Zeldli,Jr,Joseph Angelo
Cyg 21h21'30"40d8'
Zelenko,Ronald Walter
Sgr 19h38'52"-30d26'
Zelenski,Alex
Uma 9h58'28"61d11'
Zelenski,Jane D
Uma 9h57'40"52d52'
Zelenski,Nicole
Uma 9h51'16"56d10'
Zelent,Boguslaw
Cma 6h11'46"-13d10'
Zelenyuk,Dimitri
Uma 15h14'48"80d6'
Zeliff,Peter
Dra 16h4'0"64d55'
Zelinka,Frank E
Her 16h10'19"10d34'
Zelinski,Jason Eric
Aur 6h26'47"31d27'
Zelinsky,Annette
Uma 9h10'15"68d3'
Zelissen,Bernard
Lac 22h5'19"51d14'
Zell,Bari
Dra 17h12'44"65d31'
Zelle
Uma 11h47'30"44d14'
Zeller Star
Lyr 18h38'47"41d9'
Zeller,Gail
And 0h53'58"36d20'
Zeller,John Sebastian
Her 17h38'34"40d40'
Zeller,Jürgen
Dra 18h24'46"68d26'
Zeller,Marjorie A
And 23h23'24"44d44'
Zeller,Poppy-A Star for Nathan
T
Lac 22h10'1"54d50'
Zeller,Princess Jennifer
And 0h16'45"41d4'
Zeller,Richard
Cep 20h51'59"62d11'
Zeller,Susan Cooper
Peg 23h24'44"10d12'
Zeller,W Catherine
Cyg 21h4'12"40d5'
Zellers,Emily Kathleen
Tri 2h15'26"32d42'
Zellers,Susan Jill
Cyg 19h41'47"31d21'
Zellin,Jay
Aur 6h12'43"41d9'
Zellman,Margaret Louise
Cmi 7h41'22"3d46'
Zellner,David "Clem"
Lac 22h24'18"54d48'
Zellner,Lynn
Cam 5h0'33"68d26'
Zellner,Nancy Jean
And 23h45'1"43d1'
Zelnik Dynasty,The
Cep 23h34'27"65d24'
Zelof,Shannon & Maher
Uma 12h3'15"35d0'
Zelten,Ebony
And 1h19'12"35d25'
Zelvis,Noah
Aur 7h3'14"40d3'

Zemaitis,Kathy Lynn
And 23h47'40"45d38'
Zeman,Carl & Mary
Cyg 20h47'11"37d32'
Zeman,Rachel Leah
Lyn 7h35'20"38d5'
Zeman,Richard Alan
Lac 22h16'39"37d57'
Zemanovich,Gary J
Her 17h0'15"20d40'
Zembaty,Ilse
Cas 2h0'32"58d23'
Zemeckis,Alex
Hya 9h6'1"4d45'
Zemel,Mark Leo
Dra 17h20'55"65d23'
Zemke,Angel Bertha
And 0h35'57"40d9'
Zemke,E Joseph
Aur 4h58'37"41d3'
Zemla,Alexander F
Boo 14h17'27"31d16'
Zemla,Drew A
And 23h17'0"48d8'
Zemont In Flight
Sgr 18h51'10"-20d59'
Zemont,Dale
Leo 10h57'36"22d44'
Zemp
Boo 14h25'40"51d7'
Zena Superterrestrial I
Cas 1h26'35"58d42'
Zenato,Alberto
Scl 23h15'34"-26d46'
Zendmera,Noelle
Boo 14h18'28"48d24'
Zenes,Mary
Lyr 19h18'35"41d53'
Zengeni
And 0h1'44"47d35'
Zeni,John & Megan
Aql 19h51'38"15d25'
Zeni,Victor
Ari 1h55'17"10d54'
Zenith,Alda Louise Gonzalez's
Leo 11h9'25"-0d6'
Zenith,Peter
Uma 9h28'12"53d13'
Zenith-1
Lac 22h39'37"55d38'
Zenke,Helga
Sco 16h14'33"-23d11'
Zenner,Peggy
Vul 19h6'23"24d46'
Zenner,Robert Walter
Cet 2h45'18"2d30'
Zenner,Robert William
Cet 2h45'38"2d6'
Zenon
Her 16h2'12"40d48'
Zenon
Ori 4h59'10"5d24'
Zenor
Per 4h18'49"50d26'
Zenor,Ryan Joseph
Com 13h6'31"20d54'
Zentner,Diana
Cmi 7h26'38"0d15'
Zeober,Andrew Michael
Aur 6h9'1"38d17'
Zeorlin,Alex
Her 16h49'37"30d48'
Zepernick,Christopher
Per 2h41'25"35d36'
Zephyrus/JA 173
Cma 7h16'11"-16d13'
Zepp Mazzeo
Vir 11h58'47"-1d2'
Zerbe,Eric Zane
Aql 20h6'1"1d2'
Zerbes,Happy Birthday,
William
Tri 2h1'39"28d46'
Zerblis,John
Oph 17h0'0"-18d55'

Zerbone,Giana
Sgr 19h30'28"-30d14'
Zerbone,Vanda
Sgr 19h30'2"-37d44'
Zeremes,Nicole
Lyn 8h30'1"41d38'
Zerfoss,Robyn Lee
Aql 19h25'50"-8d6'
Zeris Family,The
Uma 12h0'42"60d47'
Zerman,Robert James
Aur 5h5'1"40d53'
Zero Celeste Lx Wyattsphere
Sct 18h55'51"-6d44'
Zeron Ura
Umi 13h46'14"70d32'
Zerwetz,Dominique
Sex 10h47'1"5d1'
Zester,Opal
Eri 4h25'57"-11d49'
Zeszutek,Laura
Uma 10h55'37"48d55'
Zeta Tau Alpha Theta Delta
Del 20h15'26"15d15'
Zeterberg,Dorothy & Michael
Cyg 20h33'1"31d45'
Zethia
Cet 1h37'33"-2d23'
Zettel,Charles A
Cep 22h11'21"55d57'
Zetterman,Bonnie
Gem 6h24'3"12d42'
Zettl,Klaus
Cyg 20h27'43"39d32'
Zettler's Infinity,Al
Boo 14h54'31"36d20'
Zettos
Oph 18h17'16"11d11'
Zeuch,Kim
And 0h59'1"34d0'
Zeuch,Sandra
Cas 0h56'1"63d37'
Zeus
Lac 22h18'15"49d49'
Zeus Nozza
Lyn 8h1'9"48d22'
Zevan,Crystal Lauran
Cet 0h0'1"-8d12'
Zevit,Zehava & Eli Alshech
Peg 21h59'59"34d37'
Zewicki,James & Jill
Per 2h29'15"52d17'
Zewillis
Ori 5h36'38"-6d25'
Zeyack,Fr John
Tau 5h54'25"24d11'
Zeyen,Diane Saenz
Boo 14h56'1"44d25'
Zeysing,Joachim
Gem 7h16'40"24d42'
Zgonc Star
Lyn 8h38'1"40d43'
Zhaba
Cyg 21h15'26"28d53'
Zhanna
Uma 11h10'46"47d4'
Zhao,Li Ming
Tau 4h1'0"1d59'
Zhavez,Sandy & Alvino
Uma 13h35'50"48d55'
Zhenru
Umi 14h20'1"66d44'
Zhenya
Peg 21h20'47"23d7'
Zhidkova,Elena
Peg 21h55'44"32d49'
Zhuang,Christie
Uma 10h19'0"59d5'
Zia Lidio e Zio Elio
Cep 1h41'0"78d33'
Zia Marisa
Cnv 13h32'42"51d19'
Zia Nina
Dra 17h49'35"64d3'

Ziad
Ori 5h56'55"16d0'
Ziaja,Jimmy J
Ari 2h43'49"28d13'
Ziamandanis,John Peter
Lac 22h36'37"50d20'
Ziants,David Scott
Ori 5h20'0"-10d27'
Ziarko,Leslie Ann
Cas 0h1'26"61d38'
Zibluk,John
Lac 22h21'40"55d18'
Zicari,Audrey Jean
And 23h38'56"40d35'
Zicaro,Michelle Nicole
And 23h39'1"47d53'
Ziccardi,Marina
Sco 17h28'57"-30d13'
Zichittella,Jack
Tri 2h15'19"35d52'
Zick,Betty & Leonard
Vul 19h45'11"28d50'
Zickefoose,Tina Marie
Aql 19h26'1"0d40'
Zickert,Dallas A
Cep 22h18'50"60d37'
Zide,Bruce
Aql 19h1'50"3d47'
Zidell,Ted & Ruth
Sge 19h56'36"20d17'
Zides,Shirley G
Mon 6h58'30"8d21'
Zidiramm
Lac 22h9'12"51d52'
Zieba,Ella S
And 2h33'38"38d10'
Zieba,Judge Joseph C
Dra 16h58'30"61d8'
Ziebart,Marion
Cep 22h10'26"60d6'
Zieber,Stephanie
Cas 1h17'16"63d24'
Ziegelbauer,Brian Scott
Her 17h37'33"48d47'
Ziegenfelder,Noelle
Lyr 18h30'32"40d19'
Zieger,"Smokey",Alan Mark
Sgr 20h2'32"-45d7'
Zieger,Martin
Lac 22h17'26"51d55'
Ziegert,Svenja- Christiana
Cap 20h37'32"-10d10'
Ziegler,Arthur
Aql 19h53'56"13d24'
Ziegler,Brian Christopher
Peg 23h0'19"32d17'
Ziegler,Dale Allen Christopher
Boo 13h55'41"19d56'
Ziegler,Finn
Ori 5h3'19"8d35"
Ziegler,Jack
Ser 15h20'51"7d3'
Ziegler,Jaime Danielle
And 2h25'40"38d49'
Ziegler,Jessica Elizabeth
Peg 23h28'27"22d27'
Ziegler,Paula Julia
Oph 16h59'34"8d31'B
Ziegler,Preston
Lib 15h28'0"-8d32'
Ziegler,Ronald Edward
Vul 20h21'47"22d38'
Ziegler,Ruth Geraldine
Cyg 19h45'46"38d9'
Ziegler,The Star Of
Mon 7h0'17"-8d49'
Ziegner,Martin
Uma 9h39'41"44d54'
Ziehmayer,Daniel
Sgr 19h47'13"-41d59'
Ziel,Mary Ellen
Mon 7h55'3"-2d9'
Zielaznicki,Raymond A
Lac 22h13'50"54d51'

Zielinski,Andrew Shaw
Ari 2h24'34"10d54'
Zielinski,Christiane
Tau 5h47'22"27d59'
Zielinski,Dawn Michele
Cyg 20h36'0"39d37'
Zielinski,Eric R
Umi 16h1'25"70d12'
Zielinski,Gary Richard
Sct 18h53'15"-7d6'
Zielinski,Jr,Richard Walter
Aur 5h6'28"44d5'
Zielinski,Louis Stephen
Umi 14h10'43"69d47'
Zielinski,Louis
Lac 22h14'36"51d11'
Zielinski,Rita
Gem 7h24'40"35d22'
Zielinski,Sr,Mark
Per 2h22'39"58d9'
Zielke,Felix Igor
Sco 17h32'56"-31d50'
Zielke,James & Betty
Peg 21h59'34"34d9'
Zielonko,Casimir
Aql 19h54'11"10d59'
Zielonkowski,Barbara
Lib 15h12'16"-21d48'
Zieman,Thomas
Oph 17h10'51"-24d23'
Ziemann,Niklas
Sco 16h51'1"-40d7'
Ziemba,Peter
Boo 14h4'51"14d11'
Ziemer,Pam
Lyn 8h12'59"41d12'
Ziemienski,Mike
Peg 22h36'31"31d17'
Ziemke Star of my
Heart,Gabriele
Dra 17h54'11"83d49'
Ziemkiewicz,Jill Ann
And 23h8'25"39d30'
Zier,Cathy Diane
Aql 19h31'17"12d19'
Zier,Dennis Norman
Lmi 10h7'13"32d47'
Zierenberg-Werner, Lilian
Monika
Aqr 22h58'48"-18d37'
Zierer,Daniel Andreas
Per 1h56'1"48d29'
Zierer,Rebecca Laura
Cas 0h15'37"63d3'
Ziering,Madeleine Noelle
Mon 7h2'52"5d23'
Zies
Ori 5h34'10"9d27'
Ziesch,Eberhard
Cas 0h47'17"60d8'
Ziesel,Joshua Paul
Dra 17h3'35"50d48'
Zieseniss,Marie
Aql 18h45'20"10d46'
Ziewer,Jutta
Psc 1h18'42"21d24'
Ziff,Aaron
Aur 4h49'12"41d9'
Ziffels,Wolfgang
Lyn 8h7'0"48d40'
Ziffers Love Star
Com 15h58'49"27d28'
Ziggy
Her 16h44'0"48d7'
Ziggy
Lyn 7h54'46"42d59'
Ziggy
Cma 7h23'51"-16d36'
Zigler,Janet Simpson
Cas 0h55'45"75d36'
Ziglioli,Enrico
Pho 0h3'57"-43d11'
Zigman,Daniel
Ori 5h26'27"0d50'

Zigman,Emily Logan
Aql 19h31'46"11d46'
Zigman,Kerri Elizabeth
And 23h30'10"45d30'
Zigmond,Shara J
Aql 19h23'22"10d2'
Zigmund Pa'lffy
Umi 15h17'26"69d22'
Ziilch,David Nicholas
Cep 3h2'51"80d33'
Zijlstra,Ingrid
Cas 1h46'22"60d21'
Zikman,Tammy
Com 12h54'41"28d56'
Zilas Aeis
Crb 16h3'37"33d2'
Zilhaver,Laura
Cyg 20h4'39"40d51'
Zilhaver,Sara
Peg 23h4'0"20d6'
Ziliak,Patrick
Lyn 9h10'18"36d59'
Ziliani,Yaneth Quinones
Peg 23h48'20"11d14'
Zilker,Sandra P
Aqr 20h37'27"-6d32'
Zilli,Therese
Cyg 19h26'28"33d42'
Zilm,Nicholas Ryan
Dra 13h25'46"67d45'
Zilma
Mon 7h2'51"-1d52'
Zilmienski,Bruce V
Hya 9h17'58"-7d25'
Zilo,Philip
Oph 18h38'38"10d18'
Zima,Julia & Daniel
Cyg 21h7'0"37d35'
Zima,Kimberly & Mark
Eri 3h34'16"-6d45'
Ziman,Arthur L
Cet 0h58'27"-17d19'
Zimbardo,Alexandra Rose
Lyn 8h56'1"43d53'
Zimecki Star,Arthur
Dra 17h0'34"66d45'
Zimenna,John
Hya 9h1'36"4d12'
Zimick,Louis
Boo 14h19'0"51d34'
Zimkas,Daniel Charles
Cet 2h30'0"1d10'
Zimkas,Katelynn Carol
Mon 6h37'40"3d6'
Zimkas,Rebecca Ann
Mon 6h38'22"8d49'
Zimkas,Sean Michael
Her 18h4'52"31d2'
Zimmer "Eli",Elisa T
Aur 7h11'12"37d12'
Zimmer,Alexandre
Uma 11h23'43"38d10'
Zimmer,Andrea *02081963*
Leo 11h2'36"-2d15'
Zimmer,Erich & Esther
Lac 22h11'0"50d1'
Zimmer,Janina Swenja
Mon 8h2'25"-6d6'
Zimmer,Jill's Shining Star
Lyn 9h6'54"46d40'
Zimmer,John J
Aur 6h13'60"32d1'
Zimmer,Katrina Elizabeth
Cnc 8h21'0"15d10'
Zimmerli,Stephanie Lynn
Sgr 18h53'39"-24d41'
Zimmerly's Star, Richard
Boo 14h18'36"19d38'
Zimmerm'Anne '92
Per 4h3'29"39d12'
Zimmerman Star,The Sandy
Cyg 20h6'27"40d45'
Zimmerman,Carl Bernard
Boo 13h38'33"20d41'

Zimmerman,Catherine
Cas 22h56'24"55d17'
Zimmerman,Christine Clark
And 23h22'42"44d55'
Zimmerman,Craig & Nancy
Cyg 20h18'13"40d45'
Zimmerman,Dale E
Cep 21h8'1"60d15'
Zimmerman,Delbert D
Cas 3h18'14"70d15'
Zimmerman,Edward
Ser 15h56'39"18d46'
Zimmerman,Ellen
Mon 7h3'1"-6d25'
Zimmerman,Fred
Oph 17h55'1"12d58'
Zimmerman,Hartzel H
Lmi 10h56'26"33d14'
Zimmerman,Jennifer
Vul 21h0'29"27d59'
Zimmerman,John
Aur 6h2'29"31d12'
Zimmerman,Jr,Roy A
Cmi 7h30'32"0d23'
Zimmerman,Kori Gene
Aql 18h43'1"11d8'
Zimmerman,Kristin B
And 1h48'29"37d48'
Zimmerman,Lief
Per 3h42'20"50d5'
Zimmerman,Mariette
Dra 20h41'13"80d27'
Zimmerman,Mildred Faye
Mon 7h54'19"-6d51'
Zimmerman,Peter
Lyr 19h18'59"31d14'
Zimmerman,Richard & Connie
Cyg 21h52'1"41d15'
Zimmerman,Robert J
Aur 4h49'0"40d43'
Zimmerman,Sharon L
And 22h57'53"37d46'
Zimmerman,Tillie
Lyn 8h20'11"43d10'
Zimmerman-A True Love,
Sharon CE
Lyn 7h55'12"58d20'
Zimmermann,Bernhard
Oph 17h34'41"8d28'
Zimmermann,Blue Eyes
Lac 22h52'59"50d20'
Zimmermann,Dennis Robert
Vir 13h3'40"-20d26'
Zimmermann,Dieter
Cam 5h27'0"68d57'
Zimmermann,Edith
Leo 10h2'12"10d49'
Zimmermann,Eric
Lac 22h39'35"55d12'
Zimmermann,Katie
Leo 10h28'56"10d21'
Zimmermann,Michael
Dra 18h23'20"68d41'
Zimmermann,Peter
Leo 9h19'49"7d21'
Zimmermann,Peter
Cmi 7h6'23"5d36'
Zimmy
Aql 18h43'0"9d5'
Zinaida's Sparkle
Lyr 19h23'1"31d46'
Zinda,Allen
Aur 5h53'10"30d50'
Zindars,Sonya Louise
Cam 7h51'21"70d41'
Zingara
Pyx 8h41'26"-25d7'
Zingarelli III, Thomas P
Her 18h11'22"47d4'
Zinger,Pianoman-Craig L
Peg 21h54'23"30d43'
Zingerman,Sarah Lynne
Peg 22h19'37"31d4'
Zingg,David
Cep 23h7'39"70d29'

Zingg,Jay D
Sgr 19h36'28"-43d35'
Zingone,Lisa Marie
Lyr 19h0'0"31d1'
Zini,Gabriella
Cam 10h20'39"81d51'
Zini,Luca
Lyn 7h27'23"50d11'
Zini,Marina
Oph 17h53'30"11d29'
Ziniti,Katherine
Sex 10h2'14"2d24'
Zink & Son Jerry,Trudy
Cma 6h57'27"-11d1'
Zink,Gertrude "Trudy"
Cma 6h15'38"-18d46'
Zink,Jean-Bernard
Mon 6h21'0"-1d31'
Zink,Lance Leon
Hya 9h18'1"-14d20'
Zink,Norbert G
Cma 6h54'59"-11d7'
Zink,Paul Julius
Uma 8h33'17"50d28'
Zinke
Her 17h27'19"28d37'
Zinman,Betty
And 0h13'12"36d32'
Zinn III,Richard Stevenson
Cep 1h27'52"80d24'
Zinn,Gavin James
Lac 22h29'38"40d16'
Zinn,Sara M
Oph 18h21'11"7d33'
Zinnato,Stephen Jo
Eri 2h57'0"-2d52'
Zinner,Nancy Davis
Vul 19h0'0"22d26'
Zino,Vic
Per 2h33'0"57d1'
Zinzi,James
Boo 14h51'50"48d19'
Zinzi,Lawrence J
Mon 7h44'26"-3d13'
Ziobro,Michon
Dra 17h28'42"64d16'
Ziolek,Paul A
Aur 6h5'0"34d30'
Ziols,Gregory "Phantom"
Her 16h42'42"34d6'
Zion
Cam 5h0'1"65d2'
Zion,Arnaud
Crb 16h18'59"39d43'
Zip
Cyg 19h29'13"38d50'
Zip #1
Sct 18h40'58"-5d41'
Zipp,Gerald Paul
Dra 19h28'29"60d3'
Zipp,Michael
Ser 18h41'25"2d54'A
Zippeldeppel Noldi Deuber
Uma 9h2'41"50d11'
Zipperman,Sidney Anthony
Aqr 20h59'32"-5d57'
Zippy
Lyn 8h8'21"50d15'
Zippy
Lyr 19h23'53"35d12'
Zippy
Ari 1h46'11"18d58'
Zipursky,Jesse
Eri 4h43'16"-8d30'
Zirbs,Emilie Savigny
Cam 5h46'0"60d24'
Zirelli,Maryann
Peg 23h33'33"30d28'
Zirkel,Linda
Cas 23h31'46"53d55'
Zirkel,Walter
Aql 19h8'1"10d32'
Zirkle,Eric A
Boo 14h44'59"35d9'
Zirnheld,Albert
Cyg 20h28'32"40d40'
Zirotti,Franck
Cyg 20h55'43"50d10'
Zirpoli,Grandma
Cyg 19h59'1"30d59'
Zirpoli,Grandpa
Per 2h59'51"35d7'
Zirve
Cam 3h54'1"62d13'
Zis,Gayle
Gem 6h57'26"20d36'
Ziska,Russell
Oph 17h5'47"11d31'
Ziska,Zev
Cas 0h59'58"51d31'B
Ziske-Your Dreamstar, Sandra
Lyr 18h21'26"46d11'
Ziskin,Carol Anne & Randolph Eric
Peg 21h57'25"34d48'
Zisselman,Jeffrey
Cam 8h15'54"77d48'
Zissi-Memmos,Liana
Aql 20h1'30"11d50'
Zissman,Carole
Tri 2h6'38"31d0'
Zissman,Lorin
Tri 2h5'13"33d18'
Zitaglio,Victoria Anne
Lac 22h36'31"53d18'
Zitek,Ken
Boo 14h52'36"26d23'
Zitin Star,The Sam
Cmi 7h57'17"0d56'
Zitkus,Zukie
Lyn 6h57'22"58d22'
Zitney,Joseph Kyle
Cet 0h34'0"-0d32'
Zito
Cmi 7h23'48"8d3'
Zito,Anthony & Joseph
Psc 1h5'55"27d60'
Zito,Charlotte
Crt 11h15'31"-21d23'
Zito,Chuck
Psc 23h22'55"1d58'
Zito,Sherri
Mon 8h0'36"-0d2'
Zito,Taylor Marie
Aur 6h29'44"38d53'
Zito,William
Lac 22h6'13"40d38'
Ziva
Aur 5h2'44"41d14'
Zive,Carl & Trish
Cyg 19h29'47"32d6'
Zivin,Casey
Dra 12h4'18"75d39'
Zivin,Jeffrey Robert
Ori 5h57'1"14d38'
Zivin,Sparky
Vul 19h47'26"22d48'
Ziwes,Ernst
Lyr 19h21'52"31d2'
Zizine-Volta,Myriam
Boo 14h30'60"40d16'
Zizzo,David
Per 3h26'54"50d20'
Zizzo,Grace
Lyr 19h23'41"38d57'
Zjaba,Michelle Ann Laura
Sge 19h14'59"20d24'
Zlahtic,Anthony A
Dra 17h47'30"76d33'
Zlatnik,Frank J
Cnv 12h21'0"40d60'
Zlochenko,Arie
Cap 21h28'18"-22d43'
Zlotescu,Dodi
Aur 7h14'47"38d36'
Zlotkin,Joshua
Dra 18h15'27"65d1'
Zlowzower,Zak Taylor
Mon 6h41'28"7d17'
Zmaczynski,Nils
Per 4h6'32"51d34'
Zmich,Dez-Donald Edward
Peg 23h1'10"25d27'
Zmuda
Boo 15h21'1"32d54'
Zmuda,Louise A
Ori 5h33'1"-0d44'
Zmuda,Ronald A
Ori 5h31'54"-0d50'
Znoj,Alex John
Aql 19h57'47"13d39'
Zobel,Jerome Fremont
Sge 20h7'30"20d33'
Zobel,Louise Purwin
Sge 20h17'24"17d51'
Zobro,Melissa
Cam 3h26'36"61d11'
Zobèyda
Cmi 7h26'51"8d36'
Zocca,Marina
Gru 22h31'21"-55d43'
Zoccola,My Fa Fa Love Always Ton Ton,John
Her 17h6'13"46d14'
Zoch,David
Her 16h40'58"20d2'
Zoche,Kimberly Susanne
Mon 6h31'30"-0d19'
Zoe
Uma 11h50'1"43d49'
Zoe
Aql 20h10'55"14d38'
Zoe
Cas 0h9'18"61d57'
Zoe
Uma 8h38'40"53d43'
ZOE
Leo 10h59'30"11d4'
Zoe 458
Cra 11h1'55"-7d40'
Zoe 837
Lac 22h2'44"50d54'
Zoe Elise
Peg 22h18'35"34d51'
Zoe Jennifer
Cas 0h59'44"68d17'
Zoe's Star
Cas 23h24'41"61d36'
Zoeller,Caroline
Lyr 19h25'44"42d45'
Zoeller,Joan Lisa Thomas Leigh
Ori 5h44'39"12d1'
Zoepf,Katherine
Cyg 19h24'0"33d7'
Zoevalenstein
Uma 8h32'49"50d51'
Zoffel,Andreas
Lib 15h11'32"-20d37'
Zofia-Minnie
Lyr 18h35'1"44d37'
Zog 1
Lmi 10h11'51"31d34'
Zoh Tsekoura
Uma 10h37'41"50d35'
Zohab,Jessica Ann
Cas 1h45'54"73d35'
Zohar,Matan Yitzhak
Oph 17h5'1"11d30'
Zoia
Cnv 13h22'59"41d17'
Zoila
And 0h23'30"38d49'
Zoja,Gabriella Renata
And 0h29'30"44d37'
Zola
Cas 0h1'45"54d44'
Zola,Joseph Pasquale
Aur 4h54'1"50d32'
Zola,Michael Sher
Sgr 19h7'37"-21d27'
Zoladz,Jim
Her 16h15'0"22d48'
Zolbe,William Thomas
Uma 9h10'1"71d29'
Zoldock III,John A
Her 16h43'0"30d7'
Zolian,Edward
Ori 5h19'17"1d42'
Zoll,Brian Michael
Mon 7h44'50"-8d29'
Zoller,Lowell Kay
Aql 19h0'1"12d17'
Zolot,Benjamin M
Per 3h7'14"40d2'
Zolton,Cynthia Ann
Cas 0h33'25"61d24'
Zombie Woof
Dra 19h30'18"71d1'
ZombieDust
Lac 22h11'58"47d36'
Zomer
Lac 22h16'30"37d55'
Zommers,Irena
Del 20h22'19"7d38'
Zona's Ray
Dra 16h25'46"60d5'
Zona,Mark Christopher
Boo 15h2'31"25d48'
Zonca,Caroline
And 2h31'1"40d16'
Zonda's Light
And 2h21'43"44d20'
Zone,Nadene
Aur 5h1'52"47d44'
Zone,Steve
Sex 9h51'20"1d25'
Zonta,Mark Daniel
Boo 14h29'11"32d28'
Zook,Dean
Aur 6h0'0"50d27'
Zoom
Aqr 22h21'21"-0d12'
Zootis,Carol
Cet 3h17'0"8d4'
Zoppis,Christian Ludovico Gabriele
Cam 14h3'45"82d12'
Zoppé,Brian & Sandra
Oph 17h2'37"10d36'
Zora
Lac 22h50'32"40d52'
Zoraida
Aur 6h6'44"40d9'
Zoran's Continual Light
Aql 20h7'33"7d58'
Zoref,Anita Samuels
Dra 12h9'46"71d49'
Zorehkey,Alaysha
Aql 19h6'19"3d16'
Zorehkey,David
Aql 19h42'52"12d3'
Zorehkey,Michael
Oph 18h9'59"13d5'
Zorehkey,Rick
Aql 19h4'17"15d14'
Zorehkey,Skyler
Ser 15h12'17"9d26'
Zoretich,Josie
Mon 6h16'52"-6d56'
Zorianna
Boo 15h4'22"53d4'
Zorich,Toni Maria Sarah
Mon 6h35'40"8d40'
Zorich-Named By Stephanie,Chris
Hya 8h14'60"-6d11'
Zorika,Ursula
And 23h42'28"43d50'
Zorman,Brendon Shane
Per 3h46'13"39d34'
Zorn,Mr John
Aur 4h56'37"52d19'
Zornow,Karin
Mon 8h8'12"-8d30'
Zorr,William A
Aql 19h30'44"13d40'
Zoschke,Magda Lynne
Cas 0h24'42"50d24'
Zosh
Boo 14h46'1"35d24'
Zottoli,Debbie E
Cas 0h37'17"64d12'
Zoubek,JoAnn J
Uma 11h3'1"55d59'
Zouzounis,Nicholas
Sex 10h46'25"-0d22'
Zowie
Uma 9h26'51"51d49'
Zoya Naumov Corbera
Boo 14h0'55"12d54'
Zoë Ann
Uma 11h38'1"52d59'
Zoë-Jade
Cas 1h2'19"61d2'
Zoé
And 22h59'44"37d53'
Zsak,Mary Anne
Cas 1h35'1"60d1'
Zsamac
Ori 5h6'21"13d49'
Zschage,Myriam
Leo 11h23'23"0d9'
Zschoche,Jr,Robert Curtis
Cas 0h59'39"60d32'
Zschomler,Kristen Marie
Aql 19h13'19"12d39'
Zsifkovits,Wilhelm
Cap 21h44'0"-24d31'
Zsinko,Matt
Leo 9h32'55"14d37'
Zsinko,Michael
Leo 10h18'41"12d55'
Zuba,Sean Brennan
Her 17h10'57"22d45'
Zubchevich,Lara Whitney
Crb 16h22'44"32d12'
Zubeck,Elaine Maxine
Cas 0h27'11"54d35'
Zubel,Jr,Lawrence John
Uma 9h27'0"49d2'
Zubel,Lawrence John
Peg 0h11'40"18d20'
Zuber
Cyg 21h18'11"38d42'
Zuber,David
Cet 1h13'26"-5d58'
Zuber,David
Sct 18h44'14"-5d8'
Zuber,Violet M
Cas 0h46'29"69d26'
Zublena,Raymond "Poppy"
Per 3h16'1"41d16'
Zubrinsky,Bernice
Lyr 19h19'40"42d24'
Zuccarelli,John & Lillian
Cyg 20h8'45"41d8'
Zuccarello,Paul
Ori 5h11'21"-5d20'
Zucchero,Stephanie
Lyr 19h22'29"31d18'
Zucchetti,Enrico
Cmi 8h6'24"1d2'
Zucco,Joe & Alexandra
Lmi 10h52'36"27d36'
Zucher,Adam David
Cep 21h24'35"70d53'
Zuck,Sr,James Eldred
Aur 5h4'14"37d49'
Zuckarelli,Toni
Ori 6h17'56"10d47'
Zucker,Jenna Lauren
Mon 6h39'43"10d34'
Zuckerbrod,Adam Held
Cma 6h26'24"-15d14'
Zuckerman,Jonathan Eric
Mon 6h39'35"7d46'
Zuckerman,Murray
Boo 14h33'1"47d59'
Zuckerman,Paul
Dra 19h38'47"61d16'
Zuckmann,Jenna Ceil Stubbs
Cap 21h22'15"-27d37'
Zucoski,Joseph Richard
Boo 15h1'0"12d7'
Zuczek,Jennifer Nicole
Cas 2h26'48"70d31'
Zuendel,Amy & Mike
Lyr 19h17'26"42d49'
Zuercher,Fay
Lyn 8h8'58"49d3'
Zuffante 1207,Kim
Cam 6h56'26"71d8'
Zuggi,Wendy Marie
Lyr 18h50'60"33d49'
Zughaft,Miriam
Aql 18h55'39"11d1'
Zuhlke's 50th Anniv, Joyce & Arthur
Cyg 21h2'34"33d37'
Zuidema,Hendrik Jan Willem
Cet 1h20'41"-6d14'
Zuidma,Anne Lynn
Mon 6h55'13"10d11'
Zuka
Cet 0h25'15"-1d7'
Zukas,Gene & Martha
Ori 6h2'48"0d44'
Zukor,Mark
Uma 11h12'57"55d27'
Zukowski,Jonathan Vincent
Dra 18h11'46"65d28'
Zukowski,Rosemary Frances
Leo 9h22'16"18d42'
Zukowski,Tristan Philip
Uma 9h40'41"51d10'
Zulch,Cynthia Rachel
Cyg 20h18'51"38d10'
Zulfiqar,Mohsin
Aur 6h35'12"31d23'
Zuliani,Norma
Eri 3h5'18"-5d50'
Zulich,Michael
Cyg 21h6'11"31d12'
Zulim,Daisy Mae
Mon 7h14'7"-5d56'
Zulli,Alice Parsons
Boo 14h57'37"28d53'
Zulli,David Joseph
Boo 14h59'35"22d9'
Zullo
Lmi 10h41'26"25d45'
Zullo,Dr Edward A
Oph 17h25'53"-20d26'
Zulma
Lyn 7h49'54"35d54'
Zuly Ann I
Crt 11h11'14"-19d12'
zumBrunnen,Michael & Diane
Cyg 21h51'27"40d49'
Zumpella,Joanne Johnson
Aql 20h6'55"0d34'
Zumrick,Jonathon Richard
Her 18h55'30"18d54'
Zumsteg,Elsa
Cam 7h56'19"60d55'
Zumwalt,Harrison Gerald
Aur 6h47'41"35d46'
Zumwinkle
Per 3h8'30"38d56'
Zuniga Jnr,Fidel
Per 2h32'19"56d21'
Zuniga,Eric
Dra 18h50'24"68d59'
Zuniga,Luis Leonardo
Ser 15h36'54"18d20'
Zuniga,Osvaldo Daniel
Cam 6h9'11"67d58'
Zupanzick,Rudolf
Lyr 19h16'53"30d46'
Zuraw,Ded Moxie
Crb 16h11'0"32d49'
Zurbriggen,Jr,John Paul
Aql 18h50'1"11d16'
Zurbrugg,Sarah Margaret
Lyn 7h56'43"51d51'
Zurcher,Mary Nicole
Ori 5h55'0"10d2'
Zurek,Else
Ari 2h39'50"22d19'
Zurich,Zachary
Aql 19h8'1"15d17'
Zurlo,Francesca
Cas 1h58'37"58d23'
Zurmühlen,Frank
Lmi 9h32'41"38d35'
Zurosky,Stanley
Mon 6h20'36"-6d21'
Zusel,Yvonne Faith
Psc 1h23'19"10d26'
Zutler,Samantha Wilson
Com 12h59'53"20d26'
Zuzek,Adam
Aql 19h57'11"14d51'
Zuzek,Geno
Aql 19h57'10"10d40'
Zvaic,Saar Foox
Dra 10h0'1"73d28'
Zvara's Star
Aql 18h58'12"3d14'
Zvezda
Cam 4h27'28"68d21'
Zvezda Antona
Cyg 21h6'30"40d20'
Zvezda,Moya
Aql 19h18'12"10d20'
Zvolensky,Nicole Marie
And 0h10'16"28d34'
Zwaal,Jan Pieter Frank
Cas 1h46'1"61d12'
Zwack,Viktoria
Psc 23h22'48"5d24'
Zwackel
Cam 5h58'32"61d46'
Zwald,Charlie
Her 16h50'32"32d40'
Zweback "50",Stan- Dianne
Crb 15h55'18"32d26'
Zweiback,Nicole Isabella
Ori 6h5'60"-0d3'
Zweizig,Elizabeth Marie
Uma 11h28'12"43d13'
Zweizig,Kristen Shea
Uma 11h28'38"43d4'
Zweizig,Zachary
Lac 22h34'32"55d26'
Zwelling,Adira Gabrielle
Aql 19h31'52"10d30'
Zwick,Jochen Rudolf
Sgr 19h55'25"-41d17'
Zwick,Roman
Lac 22h18'12"55d5'
Zwiebach,Judy
And 0h12'11"32d48'
Zwiebel,Matthew Benjamin
Peg 22h27'18"31d51'
Zwiebel,Michael Ryan
Uma 10h55'1"57d46'
Zwieg,Andrew Lyle
Oph 17h57'13"12d57'
Zwillenberg,Gerhard
Ori 5h31'48"-6d28'
Zwiller,Jerome Joseph
Boo 15h17'34"50d46'
Zwilling
Her 17h27'27"42d36'
Zwinck,M & R
Per 2h43'21"40d46'
Zwisler,Evan & Taylor Rand
Cyg 21h2'15"38d57'
Zwygart-Leiser,Helen
Cas 23h15'50"61d14'
Zych,Eva
Cet 1h34'54"-6d48'
Zychek,Patricia A
Cet 1h16'19"-4d27'
Zydek,Dominique
Boo 15h7'11"7d3'
Zydor,Doreen
And 0h24'50"37d34'
Zydowsky,Lauren Kathryn
And 23h21'44"47d13'
Zygas,Jonas K P
Boo 13h36'11"21d7'
Zykina,Ludmila
Cnv 13h50'52"32d2'
Zylberberg,Charles
Aur 5h15'15"29d32'
Zylberman,Penina Ita
Cra 18h19'52"-42d56'
Zymodis
Lyn 8h29'14"55d31'A
Zysko,Ana
Tri 1h55'29"26d36'
Zyskowski,Lesley
Cas 3h1'10"57d48'
Zywiec,Cynthia
Com 12h57'16"23d43'
Zäsar
Sco 17h51'40"-31d3'
Zöe (Everlasting)
Dra 16h14'54"68d51'
Zöe (Meaning Everlasting)
Her 17h4'16"48d2'
Zörb,Gerd
Aqr 23h33'13"-12d24'

The Astronomer

Dr. James J. Rickard received his M.S. and Ph.D. degrees from the University of Maryland. He was a member of a team of radioastronomers who produced the first high resolution maps of neutral hydrogen in the spiral arms of our own Milky Way galaxy. The 300' dish antenna at the National Radio Astronomy Observatory in Green Bank, West Virginia was used to gather the data for his research in spiral structure. His interest in the interstellar gas led him to Palomar Observatory and then to the European Southern Observatory in Chile to use optical telescopes to continue his research. The years in Chile involved research, development of electronic cameras and the introduction of computers for telescope control. He is a member of the American Astronomical Society and the International Astronomical Union.

Dr. Rickard returned to the USA and the small desert community of Borrego Springs to work again as a radioastronomer at the Clark Lake Radio Observatory, a remote research station for the universities of Iowa,Maryland, and California. The research projects included studies of the sun, solar wind and scintillations of compact objects. One of the peculiar compact objects was later discovered to be the first *millisecond pulsar,* a neutron star spinning at over 38,500 RPM.

With the closure of Clark Lake Radio Observatory, due to funding cutbacks and consolidation in federally supported research, Dr. Rickard has turned his attention to education projects. He established popular monthly star parties for visitors to enjoy the dark night sky in the Anza-Borrego State Park, is a science writer for the local newspaper, and teaches at San Diego State University. He is an avid solar eclipse and comet chaser; and has led expeditions to Mexico, the Philippines, Brazil and Chile. He volunteers his time and technical talents to aid the Astronomical Foundation and environmental projects in Southern California.

The Author

James E. Magee is a native Chicagoan with an interesting array of diverse experiences, Magee flew in the U.S. Air Force during the Korean Conflict, was a boxer, a sculptor, a commercial artist, a seaman and a writer. In his tenure as a salt water sailor, Magee was principal salvage diver on the Lucayan treasure wreck, a Spanish galleon sunk in 1623 off the Bahamas.

Magee's writing credits include sales and documentary films for General Motors, Westinghouse, Union Carbide and many other Fortune 500 Companies. He is the author of dramatic episodes of G.E. Theater, Playhouse 90, the Zane Gray Theater and the DuPont Show of the month. Magee has also written comedy for the Jerry Lester Show, The Garry Moore Show, The Jackie Gleason Show, The Tonight Show with Johnny Carson, The Bob Newhart Show and many others.

His Scientific works include: *Oxygenation and Burn Trauma, The Magic Basophil Cell* and *Origin of the Cosmos.*

An accomplished Philanthropist and Astronomer, Mr. Magee is currently residing in San Diego where he is hard at work on his autobiography "Confessions of a Comedy Writer."

Magee's Interest in the heavens began as a navigator on the 50,000 mile oddessy of the "Harry W. Adams." The "Adams" was the largest privately owned sailing vessel of the 1970's. Just as early sailors navigated by the stars, Magee's interest blossomed as a navigator on the "H.W. Adams" and he has since become an accomplished astronomer. Mr. Magee is a member of the Writer's Guild of America West, and the Institute for Advanced Studies as well as a member of the Von Braun Astronomical Society.

The Calligrapher

Jane Johnson Nelson has chosen to use a modified rendition of Italic in this publication. Jane is also proficient at Old English, Celtic, Spencerian, and Italic. She has worked as a professional calligrapher for fifteen years throughout the Chicago Metropolitan area and has personally penned over four hundred thousand certificates for the International Star Registry.

Illustrations

Taken from Sheglov's reproduction of the 1690 Star Atlas by Johannes Hevelius, these lithographs appear to be drawn backwards because Hevelius drew his star atlas as though he was looking down on the celestial sphere. Johannes Hevelius lived from 1611 to 1687. He was born in Danzig, Poland (now known as Gdansk). His studies of the surface and the movements of the Moon laid the foundation for modern lunar topography. Hevelius was one of the first astronomers and although his star atlas was published after his death he is responsible for designating many of the constellations which are still recognized today.

The Photographer

Bob Ritt is a member of the staff of The Eisenhower Observatory. He is a dedicated astronomer, and works with local community groups on astronomy related tasks. He is past president of the Northwest Suburban Astronomers and is still an active club member. Ritt's main interest is astro-photography. He uses a variety of lenses, from regular camera lenses to telescopes with sixteen inch diameter lens.